D0742863
BAINBR R

CHILTON'S

IMPORT CAR MANUAL 1988-1992

Publisher	Kerry A. Freeman, S.A.E.
Editor-In-Chief	Dean F. Morgantini, S.A.E.
Managing Editor	David H. Lee, A.S.E., S.A.E.
Senior Editors	Richard J. Rivele, S.A.E.
	Nick D'Andrea
	Ron Webb
Project Managers	Peter M. Conti, Jr., Ken Grabowski, A.S.E.
	Martin J. Gunther, Richard T. Smith
Editorial Staff	Robert E. Doughten, Jeff H. Fisher, A.S.E.
	Jacques Gordon, Michael L. Grady
	Ben Greisler, S.A.E., Jeffrey M. Hoffman
	Steve Horner, Neil Leonard, A.S.E.
	James R. Marotta, Robert McAnally
	Michael W. Parks, John H. Rutter
	Don Schnell, James B. Steele
	Larry E. Stiles, Jim Taylor
	Anthony Tortorici, A.S.E., S.A.E.
Manager of Manufacturing	John J. Cantwell
Production Manager	W. Calvin Settle, Jr.
Assistant Production Manager	Andrea M. Steiger
Mechanical Artists	Lisa Gressen, Marsha Park Herman
	Lorraine Martinelli, Kim Tansey
Special Projects	Peter Kaprielyan
OFFICERS	
President	Gary R. Ingersoll
Sr. Vice President	Ronald A. Hoxter

Cover designed by Donna Monturo for DM Design

CHILTON BOOK COMPANY

*ONE OF THE **DIVERSIFIED PUBLISHING COMPANIES,***
*A PART OF **CAPITAL CITIES/ABC, INC.***

Manufactured in USA
© 1991 Chilton Book Company
Chilton Way Radnor, Pa. 19089
ISBN 0-8019-7907-2
ISSN 0271-3608

1 2 3 4 5 6 7 8 9 0   0 9 8 7 6 5 4 3 2 1

ACKNOWLEDGEMENTS

AB Volvo, Göteborg, Sweden
American Honda Motor Company, Moorestown, New Jersey
American Isuzu Motors, Inc., Whittier, California
BMW of North America, Inc., Montvale, New Jersey
Branick Industries, Fargo, North Dakota
Chicago Rawhide Mfg. Company (Fuel/Water Separators), Elgin, Illinois
Chrysler Motors Corporation, Detroit, Michigan
Hyundai Motor America, Garden Grove, California
Mazda Motors of America, Inc., Compton, California
Mercedes-Benz of North America, Inc., Montvale, New Jersey
Mitsubishi Motor Sales, Inc., Fountain Valley, California
Nissan Motor Corporation of USA, Carson, California
Porsche Cars North America, Reno, Nevada
Robert Bosch Corporation, Long Island City, New York
SAAB-Scania, New Haven, Connecticut
Subaru of America, Inc., Cherry Hill, New Jersey
Tokico America, Inc., Torrance, California
Toyo Kogyo, Ltd., Hiroshima, Japan
Toyota Motor Sales, USA, Inc., Torrance, California
Volkswagen of America, Inc., Troy, Michigan
Volvo, Inc., Rockleigh, New Jersey

TABLE OF CONTENTS

Car Sections

Unit Repair Sections

Kitsap Regional Library

HOW TO USE THIS MANUAL

This manual is arranged in two sections:

Car Section

Car sections are grouped by manufacturer and arranged in alphabetical order. The text and illustrations that comprise the service procedures in each Car Section are arranged in the following order of systems and components: Engine Mechanical, Engine Lubrication, Engine Cooling, Engine Electrical, Emission Controls, Fuel System, Drive Axle, Manual Transmission/Transaxle, Clutch, Automatic Transmission/Transaxle, Front Suspension, Rear Suspension, Steering, Brakes, Chassis Electrical.

Specification charts are always located at the front of each section. All illustrations are located as close as possible to the pertinent text. Procedures are for all models in the particular section unless specifically noted otherwise.

Unit Repair Section

The Unit Repair Section contains troubleshooting and overhaul procedures for the major components and systems of your car. This portion of the book is intended to be used in conjunction with the Car Sections.

Every major Unit Repair Section contains an Identification or Application chart to correlate the information contained in that section. The sections are usually arranged by brands, manufacturers or types of components rather than models of cars. All overhaul procedures in the Unit Repair Section begin with the component removed from the car. The reason for this division of material is an economic one. The steps involved in overhauling an engine are virtually the same for all engines. However, the operation of removing the engine from the car varies greatly from model to model. By combining where possible, and separating where necessary, we are able to publish the maximum amount of information.

Locating Information

The Table of Contents, at the front of the book, lists the beginning of each Car and Unit Repair Section in the manual.

To find where a particular Car Section is located in the book, you need only look in the Table of Contents. Once you have found the proper section, you may wish to find where specific procedures are located in that section. Turn to the Index at the front of the section. At the upper left-hand side is a listing of the main topics within the section and the page number they will be found on. Following the main topics is an alphabetical listing of all the procedures within the section and their page numbers.

Safety Notice

Proper service and repair procedures are vital to the safe, reliable operation of all motor vehicles, as well as the personal safety of those performing repairs. This manual outlines procedures for servicing and repairing vehicles using safe effective methods. The procedures contain many NOTES and CAUTIONS which should be followed along with standard safety procedures to eliminate the possibility of personal injury or improper service which could damage the vehicle or compromise its safety.

It is important to note that repair procedures and techniques, tools and parts for servicing motor vehicles, as well as the skill and experience of the individual performing the work vary widely. It is not possible to anticipate all of the conceivable ways or conditions under which vehicles may be serviced, or to provide cautions as to all of the possible hazards that may result. Standard and accepted safety precautions and equipment should be used when handling toxic or flammable fluids, and safety goggles or other protection should be used during cutting, grinding, chiseling, prying, or any other process that can cause material removal or projectiles.

Some procedures require the use of tools specially designed for a specific purpose. Before substituting another tool or procedure, you must be completely satisfied that neither your personal safety, nor the performance of the vehicle will be endangered.

Part Numbers

Part numbers listed in this book are not recommendations by Chilton for any product by brand name. They are references that can be used with interchange manuals and aftermarket supplier catalogs to locate each brand supplier's discrete part number.

Although information in this manual is based on industry sources and is as complete as possible at the time of publication, the possibility exists that some car manufacturers made later changes which could not be included here. Information on very late models may not be available in some circumstances. While striving for total accuracy, Chilton Book Company cannot assume responsibility for any errors, changes, or omissions that may occur in the compilation of this data.

Copyright Notice

No part of this publication may be reproduced, transmitted or stored in any form or by any means, electronic or mechanical, including photocopy, recording or by information storage or retrieval system without prior written permission from the publisher.

Acura / Sterling

Integra, Legend/825, NSX — All Models

1

SERIAL NUMBER IDENTIFICATION

Vehicle Identification Number

Acura and Sterling vehicle identification numbers are mounted on the left top edge of the instrument panel and are visible from the outside. There is also a vehicle identification number stamped on the center of the firewall in the engine compartment.

Engine Number

On Integra, the engine number is stamped into the block towards the front of the vehicle near the timing belt cover. On Legend and Sterling, when viewed from the flywheel, the engine number is stamped into the block in front of the left bank cylinder head. On NSX, the engine number is on the flywheel end of the block, behind the right bank cylinder head. The first 5 digits indicate the engine model identification. The remaining numbers refer to the production sequence.

Vehicle Identification Label

A complete vehicle identification label with specific model information and a paint code is on the driver's side door pillar.

Transaxle Number

On Integra, the transaxle number is on the top of the gear box or on the bell housing. On 1988–90 Legend and all Sterlings, the transaxle number is on the top near the back end of the case. On 1991–92 Legend, the number is on the bell housing on the right side of the vehicle.

SPECIFICATIONS

ENGINE IDENTIFICATION

Year	Model	Engine Displacement cu. in. (cc/liter)	Engine Series Identification	No. of Cylinders	Engine Type
1988	Integra	97 (1590/1.6)	D16A1	4	DOHC
	Legend	163 (2675/2.7)	C27A1	6	OHC
	Legend Coupe	163 (2675/2.7)	C27A1	6	OHC
	825	152 (2494/2.5)	C25A1	6	OHC
1989	Integra	97 (1590/1.6)	D16A1	4	DOHC
	Legend	163 (2675/2.7)	C27A1	6	OHC
	Legend Coupe	163 (2675/2.7)	C27A1	6	OHC
	827	163 (2675/2.7)	C27A1	6	OHC
1990	Integra	112 (1834/1.8)	B18A1	4	DOHC
	Legend	163 (2675/2.7)	C27A1	6	OHC
	Legend Coupe	163 (2675/2.7)	C27A1	6	OHC
	827	163 (2675/2.7)	C27A1	6	OHC
1991-92	Integra	112 (1834/1.8)	B18A1	4	DOHC
	Legend	196 (3206/3.2)	C32A1	6	OHC
	Legend Coupe	196 (3206/3.2)	C32A1	6	OHC
	827	163 (2675/2.7)	C27A1	6	OHC
	NSX	181 (2977/3.0)	C30A1	6	DOHC

DOHC—Double Overhead Camshaft
OHC—Overhead Camshaft

GENERAL ENGINE SPECIFICATIONS

Year	Model	Engine Displacement cu. in. (cc)	Fuel System Type	Net Horsepower @ rpm	Net Torque @ rpm (ft. lbs.)	Bore × Stroke (in.)	Com-pression Ratio	Oil Pressure @ 3000 rpm
1988	Integra	97 (1590)	PGM-FI	113 @ 6250	99 @ 5500	2.95 × 3.54	9.5:1	60–78
	Legend	163 (2675)	PGM-FI	161 @ 5900	162 @ 4500	3.43 × 2.95	9.0:1	71–82
	Legend Coupe	163 (2675)	PGM-FI	161 @ 5900	162 @ 4500	3.43 × 2.95	9.0:1	71–82
	825	152 (2494)	PGM-FI	151 @ 5800	154 @ 4500	3.31 × 2.95	9.0:1	71–82
1989	Integra	97 (1590)	PGM-FI	118 @ 6500	103 @ 5500	2.95 × 3.54	9.5:1	60–78
	Legend	163 (2675)	PGM-FI	161 @ 5900	162 @ 4500	3.43 × 2.95	9.0:1	71–82
	Legend Coupe	163 (2675)	PGM-FI	161 @ 5900	162 @ 4500	3.43 × 2.95	9.0:1	71–82
	827	163 (2675)	PGM-FI	161 @ 5900	162 @ 4500	3.43 × 2.95	9.0:1	71–82
1990	Integra	112 (1834)	PGM-FI	130 @ 6000	121 @ 5000	3.19 × 3.50	9.2:1	69–79
	Legend	163 (2675)	PGM-FI	161 @ 5900	162 @ 4500	3.43 × 2.95	9.0:1	71–82
	Legend Coupe	163 (2675)	PGM-FI	161 @ 5900	162 @ 4500	3.43 × 2.95	9.0:1	71–82
	827	163 (2675)	PGM-FI	161 @ 5900	162 @ 4500	3.43 × 2.95	9.0:1	71–82
1991-92	Integra	112 (2675)	PGM-FI	130 @ 6000	121 @ 5000	3.19 × 3.50	9.2:1	69–79
	Legend	196 (3206)	PGM-FI	200 @ 5500	210 @ 4500	3.54 × 3.31	9.6:1	71–82
	Legend Coupe	196 (3206)	PGM-FI	200 @ 5500	210 @ 4500	3.54 × 3.31	9.6:1	71–82
	827	163 (2675)	PGM-FI	161 @ 5900	162 @ 4500	3.43 × 2.95	9.0:1	71–82
	NSX	181 (2977)	PGM-FI	270 @ 7100 ①	210 @ 5300	3.54 × 3.07	10.2:1	50

PGM-FI—Electronic Fuel Injection
① With manual transmission; 252 @ 6600 with automatic transmission.

ENGINE TUNE-UP SPECIFICATIONS

Year	Model	Engine Displacement cu. in. (cc)	Spark Plugs Type	Spark Plugs Gap (in.)	Ignition Timing (deg.) MT	Ignition Timing (deg.) AT	Com-pression Pressure (psi)	Fuel Pump (psi)	Idle Speed (rpm) MT	Idle Speed (rpm) AT	Valve Clearance In.	Valve Clearance Ex.
1988	Integra	97 (1590)	①	0.039–0.043	12B ⑤	12B ⑤	135–192	36	700–800	650–750	0.005–0.007	0.006–0.008
	Legend	163 (2675)	②	0.039–0.043	15B ⑤	15B ⑤	142–171	36–41	630–730	630–730	Hyd.	Hyd.
	Legend Coupe	163 (2675)	②	0.039–0.043	15B ③	15B ③	142–171	36–41	630–730	630–730	Hyd.	Hyd.
	825	152 (2494)	②	0.039–0.043	3B ④	3B ④	135–178	36–41	670–770	670–770	Hyd.	Hyd.
1989	Integra	97 (1590)	①	0.039–0.043	12B ⑤	12B ⑤	135–192	36	700–800	650–750	0.005–0.007	0.006–0.008
	Legend	163 (2675)	②	0.039–0.043	15B ⑤	15B ⑤	142–171	36–41	630–730	630–730	Hyd.	Hyd.
	Legend Coupe	163 (2675)	②	0.039–0.043	15B ⑤	15B ⑤	142–171	36–41	630–730	630–730	Hyd.	Hyd.
	827	163 (2675)	②	0.039–0.043	15B ⑤	15B ⑤	142–171	36–41	630–730	630–730	Hyd.	Hyd.

ENGINE TUNE-UP SPECIFICATIONS

Year	Model	Engine Displacement cu. in. (cc)	Spark Plugs Type	Spark Plugs Gap (in.)	Ignition Timing (deg.) MT	Ignition Timing (deg.) AT	Compression Pressure (psi)	Fuel Pump (psi)	Idle Speed (rpm) MT	Idle Speed (rpm) AT	Valve Clearance In.	Valve Clearance Ex.
1990	Integra	112 (1834)	②	0.039–0.043	16B ⑤	16B ⑤	135–196	35–41	750 ± 50	750 ± 50	0.006–0.007	0.006–0.008
	Legend	163 (2675)	②	0.039–0.043	15B ⑤	15B ⑤	142–171	36–41	680 ± 50	680 ± 50	Hyd.	Hyd.
	Legend Coupe	163 (2675)	②	0.039–0.043	15B ⑤	15B ⑤	142–171	36–41	680 ± 50	680 ± 50	Hyd.	Hyd.
	827	163 (2675)	②	0.039–0.043	15B ⑤	15B ⑤	142–171	36–41	680 ± 50	680 ± 50	Hyd.	Hyd.
1991	Integra	112 (1834)	②	0.039–0.043	16B ⑤	16B ⑤	135–196	35–41	750 ± 50	750 ± 50	0.006–0.008	0.006–0.008
	Legend	196 (3206)	②	0.039–0.043	15B ⑤	15B ⑤	142–192	31–37	650 ± 50	700 ± 50	Hyd.	Hyd.
	Legend Coupe	196 (3206)	②	0.039–0.043	15B ⑤	15B ⑤	142–192	31–37	650 ± 50	700 ± 50	Hyd.	Hyd.
	827	163 (2675)	②	0.039–0.043	15B ⑤	15B ⑤	142–171	36–41	680 ± 50	680 ± 50	Hyd.	Hyd.
	NSX	181 (2977)	②	0.039–0.043	15B ⑤	15B ⑤	142–199	36–44	800 ± 50	700 ± 50 ⑥	0.006–0.007	0.007–0.008
1992	SEE UNDERHOOD SPECIFICATIONS STICKER											

NOTE: The Underhood Specifications sticker often reflects tune-up changes made in production. Sticker figures must be used if they disagree with those in this chart.

MT—Manual transmission
AT—Automatic transmission
Hyd.—Hydraulic
B—Before Top Dead Center
Hyd.—Hydraulic valve lash adjusters

① BCPR6EY-11
BCPR6EY-N11
Q20PR-U11
② BCPR6E-11
BCPR6EY-N11
Q20PR-U11

③ Vacuum advance hoses disconnected. White mark on crankshaft pulley
④ Vacuum advance hoses disconnected. Yellow mark on crankshaft pulley
⑤ Red mark on crankshaft pulley
⑥ Check idle speed in gear

FIRING ORDER

NOTE: To avoid confusion, al ways replace spark plug wires one at a time.

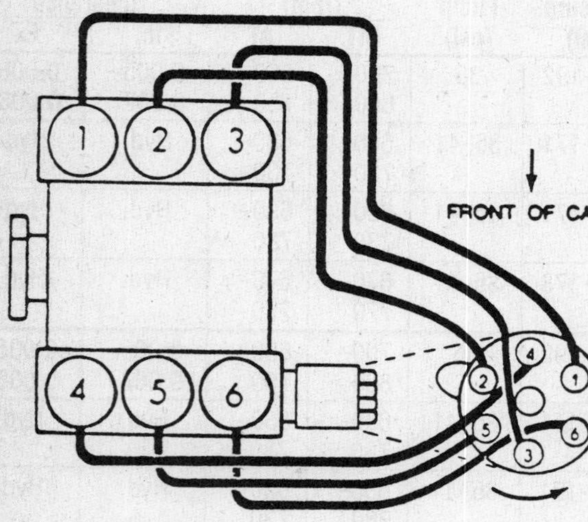

2.5L, 2.7L, 3.0L and 3.2L Engines
Engine Firing Order: 1–4–2–5–3–6
Distributor Rotation‹ Counterclockwise

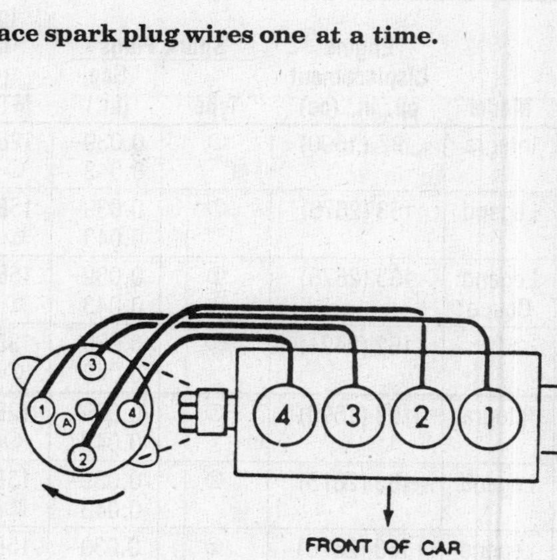

1.6L and 1.8L Engines
Engine Firing Order: 1–3–4–2
Distributor Rotation: Clockwise

CAPACITIES

Year	Model	Engine Displacement cu. in. (cc)	Engine Crankcase (qts.) with Filter	Engine Crankcase (qts.) without Filter	Transmission (pts.) 4-Spd	Transmission (pts.) 5-Spd	Transmission (pts.) Auto.	Drive Axle (pts.)	Fuel Tank (gal.)	Cooling System (qts.)
1988	Integra	97 (1590)	4.5	4.0	—	4.8	5.0 ①	—	13.2	6.0
	Legend	163 (2675)	4.8	4.2	—	4.6	6.8 ①	—	18.0	9.2
	Legend Coupe	163 (2675)	4.8	4.2	—	4.6	6.8 ③	—	18.0	9.2
	825	152 (2494)	4.8	4.2	—	4.6	6.8 ①	—	17.0	9.8
1989	Integra	97 (1590)	4.5	4.0	—	4.8	5.0 ①	—	13.2	6.0
	Legend	163 (2675)	4.8	4.2	—	4.6	6.8 ①	—	18.0	9.2
	Legend Coupe	163 (2675)	4.8	4.2	—	4.6	6.8 ③	—	18.0	9.2
	827	163 (2675)	4.8	4.2	—	4.6	6.8 ①	—	17.0	9.8
1990	Integra	112 (1834)	4.6	4.0	—	4.0	6.0 ①	—	13.2	6.9
	Legend	163 (2675)	4.8	4.2	—	4.6	6.8 ①	—	18.0	9.2
	Legend Coupe	163 (2675)	4.8	4.2	—	4.6	6.8 ①	—	18.0	9.2
	827	163 (2675)	4.8	4.2	—	4.6	6.8 ①	—	17.0	9.8
1991-92	Integra	112 (1834)	4.6	4.0	—	4.0	6.0 ①	—	13.2	6.9
	Legend	196 (3206)	5.0	4.5	—	5.2 ②	7.0 ①	2.2	18.0	8.0
	Legend Coupe	196 (3206)	5.0	4.5	—	5.2 ②	7.0 ①	2.2	18.0	8.0
	827	163 (2675)	4.8	4.2	—	4.6	6.8 ①	—	17.0	9.8
	NSX	181 (2977)	5.3	—	—	5.8	6.2 ①	—	18.5	③

① Oil change capacity.
② Including cooler, 4.8 without cooler
③ Manual transmission, 17.0 for overhaul; automatic transmission, 17.4 for overhaul; both, 12.7 for change.

CAMSHAFT SPECIFICATIONS

All measurements given in inches.

Year	Engine Displacement cu. in. (cc)	Journal Diameter 1	2	3	4	5	Lobe Lift In.	Lobe Lift Ex.	Bearing Clearance	Camshaft End Play
1988	97 (1590)	—	—	—	—	—	1.2822	1.2735	0.0020–0.0040	0.0020–0.0060
	152 (2494)	—	—	—	—	—	1.5578	1.5561	0.0018–0.0032	0.0020–0.0060
	163 (2675)	—	—	—	—	—	1.5535	1.5515	0.0018–0.0032	0.0020–0.0060
1989	97 (1590)	—	—	—	—	—	1.2822	1.2735	0.0020–0.0040	0.0020–0.0060
	163 (2675)	—	—	—	—	—	1.5535	1.5515	0.0018–0.0032	0.0020–0.0060
1990	112 (1834)	—	—	—	—	—	1.3160	1.3080	0.0020–0.0040	0.0020–0.0060
	163 (2675)	—	—	—	—	—	1.5535	1.5515	0.0018–0.0032	0.0020–0.0060

CAMSHAFT SPECIFICATIONS

All measurements given in inches.

Year	Engine Displacement cu. in. (cc)	Journal Diameter 1	2	3	4	5	Lobe Lift In.	Ex.	Bearing Clearance	Camshaft End Play
1991-92	112 (1834)	—	—	—	—	—	1.3160	1.3080	0.0020–0.0040	0.0020–0.0060
	163 (2675)	—	—	—	—	—	1.5530	1.5510	0.0018–0.0032	0.0020–0.0060
	196 (3206) Manual	—	—	—	—	—	1.5750	1.4870	—	0.0020–0.0060
	Automatic	—	—	—	—	—	1.5750	1.4870	—	0.0020–0.0060
	181 (2977) Manual Primary	—	—	—	—	—	1.4601	1.4393	0.0020–0.0040	0.0020–0.0060
	Mid	—	—	—	—	—	1.4975	1.4725	—	—
	Secondary	—	—	—	—	—	1.4695	1.4465		—
	Automatic Primary	—	—	—	—	—	1.4672	1.4393	0.0020–0.0040	0.0020–0.0060
	Mid	—	—	—	—	—	1.4825	1.4724	—	—
	Secondary	—	—	—	—	—	1.4765	1.4465	—	—

CRANKSHAFT AND CONNECTING ROD SPECIFICATIONS

All measurements are given in inches.

Year	Engine Displacement cu. in. (cc)	Crankshaft Main Brg. Journal Dia.	Main Brg. Oil Clearance	Shaft End-play	Thrust on No.	Connecting Rod Journal Diameter	Oil Clearance	Side Clearance
1988	97 (1590)	2.1644–2.1654	0.0009–0.0017 ①	0.0040–0.0140	3	1.7707–1.7717	0.0008–0.0015	0.0060–0.0120
	152 (2494)	2.5187–2.5197	0.0009–0.0019	0.0040–0.0140	3	2.0463–2.0472	0.0010–0.0020	0.0060–0.0120
	163 (2675)	2.5187–2.5197	0.0009–0.0019	0.0040–0.0140	3	2.0463–2.0472	0.0010–0.0020	0.0060–0.0120
1989	97 (1590)	2.1644–2.1654	0.0009–0.0017 ①	0.0040–0.0140	3	1.7707–1.7717	0.0008–0.0015	0.0060–0.0120
	163 (2675)	2.5187–2.5197	0.0009–0.0019	0.0040–0.0140	3	2.0463–2.0472	0.0010–0.0020	0.0060–0.0120
1990	112 (1834)	2.1644–2.1651 ②	0.0009–0.0017 ③	0.0040–0.0140	3	1.7707–1.7717	0.0008–0.0015	0.0060–0.0120
	163 (2675)	2.5187–2.5197	0.0009–0.0019	0.0040–0.0140	3	2.0463–2.0472	0.0010–0.0020	0.0060–0.0120
1991-92	112 (1834)	2.1644–2.1651 ②	0.0009–0.0017 ③	0.0040–0.0140	3	1.7707–1.7717	0.0008–0.0015	0.0060–0.0120
	163 (2675)	2.5187–2.5197	0.0009–0.0019	0.0040–0.0140	3	2.0463–2.0472	0.0010–0.0020	0.0060–0.0120
	196 (3206)	2.6762–2.6772	0.0008–0.0017	0.0040–0.0110	4	2.1250–2.0866	0.0017–0.0018	0.0060–0.0120
	181 (2977)	2.5187–2.5197	0.0009–0.0019	0.0040–0.0140	3	NA	NA	0.0060–0.0120

NA—Not Available
① No. 1 oil clearance—0.0012–0.0190
② No. 3—2.1642–2.1651
③ No. 3—0.0012–0.0190

VALVE SPECIFICATIONS

Year	Engine Displacement cu. in. (cc)	Seat Angle (deg.)	Face Angle (deg.)	Spring Test Pressure (lbs.)	Spring Installed Height (in.)	Stem-to-Guide Clearance (in.)		Stem Diameter (in.)	
						Intake	Exhaust	Intake	Exhaust
1988	97 (1590)	45	45	—	—	0.0010–0.0020	0.0020–0.0030	0.2591–0.2594	0.2579–0.2583
	152 (2494)	45	45	—	—	0.0010–0.0020	0.0020–0.0030	0.2591–0.2594	0.2579–0.2583
	163 (2675)	45	45	—	—	0.0010–0.0020	0.0020–0.0030	0.2591–0.2594	0.2579–0.2583
1989	97 (1590)	45	45	—	—	0.0010–0.0020	0.0020–0.0030	0.2591–0.2594	0.2579–0.2583
	163 (2675)	45	45	—	—	0.0010–0.0020	0.0020–0.0030	0.2591–0.2594	0.2579–0.2583
1990	112 (1834)	45	45	—	—	0.0010–0.0020	0.0020–0.0030	0.2591–0.2594	0.2579–0.2583
	163 (2675)	45	45	—	—	0.0010–0.0020	0.0020–0.0030	0.2591–0.2594	0.2579–0.2583
1991-92	112 (1834)	45	45	—	—	0.0010–0.0020	0.0020–0.0030	0.2591–0.2594	0.2579–0.2583
	163 (2675)	45	45	—	—	0.0010–0.0020	0.0020–0.0030	0.2591–0.2594	0.2579–0.2583
	196 (3206)	45	45	—	—	0.0010–0.0020	0.0020–0.0030	0.2157–0.2161	0.2146–0.2159
	181 (2977)	45	45	—	—	0.0010–0.0020	0.0020–0.0030	0.2156–0.2159	0.2146–0.2159

PISTON AND RING SPECIFICATIONS

All measurements are given in inches.

Year	Engine Displacement cu. in. (cc)	Piston Clearance	Ring Gap			Ring Side Clearance		
			Top Compression	Bottom Compression	Oil Control	Top Compression	Bottom Compression	Oil Control
1988	97 (1590)	0.0004–0.0024	0.0060–0.0140	0.0120–0.0180	0.0080–0.0280	0.0012–0.0022	0.0012–0.0022	—
	152 (2494)	0.0002–0.0013	0.0080–0.0140	0.0080–0.0140	0.0080–0.0280	0.0008–0.0018	0.0008–0.0018	—
	163 (2675)	0.0006–0.0015	0.0080–0.0140	0.0140–0.0190	0.0080–0.0280	0.0006–0.0018	0.0006–0.0018	—
1989	97 (1590)	0.0004–0.0024	0.0060–0.0140	0.0120–0.0180	0.0080–0.0280	0.0012–0.0022	0.0012–0.0022	—
	163 (2675)	0.0006–0.0015	0.0080–0.0140	0.0140–0.0190	0.0080–0.0280	0.0006–0.0018	0.0006–0.0018	—
1990	112 (1834)	0.0004–0.0016	0.0080–0.0140	0.0160–0.0220	0.0080–0.0280	0.0018–0.0028	0.0018–0.0028	—
	163 (2675)	0.0006–0.0015	0.0080–0.0140	0.0140–0.0190	0.0080–0.0280	0.0006–0.0018	0.0006–0.0018	—

PISTON AND RING SPECIFICATIONS

All measurements are given in inches.

Year	Engine Displacement cu. in. (cc)	Piston Clearance	Ring Gap			Ring Side Clearance		
			Top Compression	Bottom Compression	Oil Control	Top Compression	Bottom Compression	Oil Control
1991-92	112 (1834)	0.0004–0.0016	0.0080–0.0140	0.0160–0.0220	0.0080–0.0280	0.0018–0.0028	0.0018–0.0028	—
	163 (2675)	0.0006–0.0015	0.0080–0.0140	0.0140–0.0190	0.0080–0.0280	0.0006–0.0018	0.0006–0.0018	—
	196 (3206)	—	0.0980–0.0160	0.0160–0.0220	0.0080–0.0280	0.0014–0.0024	0.0012–0.0021	—
	181 (2977)	—	0.0100–0.0160	0.0140–0.0200	0.0080–0.0280	0.0010–0.0020	0.0010–0.0020	—

TORQUE SPECIFICATIONS

All readings in ft. lbs.

Year	Engine Displacement cu. in. (cc)	Cylinder Head Bolts	Main Bearing Bolts	Rod Bearing Bolts	Crankshaft Pulley Bolts	Flywheel Bolts	Manifold		Spark Plugs
							Intake	Exhaust	
1988	97 (1590)	①	46	23	83	②	16	23	13
	152 (2494)	③	④	32	83	⑤	16	⑥	13
	163 (2675)	③	④	33	83	⑤	16	⑥	16
1989	97 (1590)	①	46	23	119	②	16	23	13
	163 (2675)	③	④	33	83	⑤	16	⑥	16
1990	112 (1834)	⑦	56	30	87	⑤	17	23	13
	163 (2675)	③	④	33	123	⑤	16	⑥	16
1991-92	112 (1834)	⑦	56	30	87	⑤	17	23	13
	163 (2675)	③	④	33	123	⑤	16	⑥	16
	196 (3206)	③	⑪	33	174	⑤	16	25	16
	181 (2977)	56	⑧	⑨	⑩	⑤	16	23	16

① Nuts—1st step—22 ft. lbs.
2nd step—48 ft. lbs.
② Manual transaxle—87 ft. lbs.
Automatic transaxle—54 ft. lbs.
③ 1st step—29 ft. lbs.
2nd step—56 ft. lbs.
④ Cap bolt (9mm)—29 ft. lbs.
Cap bridge bolt (11mm)—49 ft. lbs.
Side bolt (10mm)—36 ft. lbs.
⑤ Manual transaxle—76 ft. lbs.
Automatic transaxle—54 ft. lbs.
⑥ 8mm nuts—22 ft. lbs.
10mm nuts—40 ft. lbs.
⑦ 1st step—22 ft. lbs.
2nd step—61 ft lbs.
⑧ Inner—48 ft. lbs.
Outer—29 ft. lbs.
⑨ Torque to 14 ft. lbs. plus 95 degrees
⑩ Torque to 203 ft. lbs., loosen, then torque to 181 ft. lbs.
⑪ Inner—57 ft. lbs.
Outer—29 ft. lbs.

BRAKE SPECIFICATIONS

All measurements in inches unless noted.

Year	Model	Lug Nut Torque (ft. lbs.)	Master Cylinder Bore	Brake Disc		Standard Brake Drum Diameter	Minimum Lining Thickness	
				Minimum Thickness	Maximum Runout		Front	Rear
1988	Integra	80	—	①	②	—	0.12	0.06
	Legend	80	—	③	④	—	0.06	0.06
	Legend Coupe	80	—	③	④	—	0.06	0.06
	825	80	—	③	④	—	0.06	0.06

BRAKE SPECIFICATIONS

All measurements in inches unless noted.

Year	Model	Lug Nut Torque (ft. lbs.)	Master Cylinder Bore	Brake Disc Minimum Thickness	Brake Disc Maximum Runout	Standard Brake Drum Diameter	Minimum Lining Thickness Front	Minimum Lining Thickness Rear
1989	Integra	80	—	①	②	—	0.12	0.06
	Legend	80	—	③	④	—	0.06	0.06
	Legend Coupe	80	—	③	④	—	0.06	0.06
	827	80	—	③	④	—	0.06	0.06
1990	Integra	80	—	③	②	—	0.06	0.06
	Legend	80	—	③	④	—	0.06	0.06
	Legend Coupe	80	—	③	④	—	0.06	0.06
	827	80	—	③	④	—	0.06	0.06
1991-92	Integra	80	—	③	②	—	0.06	0.06
	Legend	80	—	⑤	④	—	0.06	0.06
	Legend Coupe	80	—	⑤	④	—	0.06	0.06
	827	80	—	③	④	—	0.06	0.06
	NSX	80	—	⑥	④	—	0.06	0.06

① Front—0.67
Rear—0.31
② Front—0.004
Rear—0.006
③ Front—0.75
Rear—0.31
④ Front—0.004
Rear—0.004
⑤ Front—0.83
Rear—0.30
⑥ Front—1.02
Rear—0.75

WHEEL ALIGNMENT

Year	Model		Caster Range (deg.)	Caster Preferred Setting (deg.)	Camber Range (deg.)	Camber Preferred Setting (deg.)	Toe-in (in.)	Steering Axis Inclination (deg.)
1988	Integra	Front	1 3/16P–3 3/16P	2 3/16P	1 1/2N–1/2P	1/2N	1/32N	13
		Rear	—	—	1N–1/2N	3/4N	1/16P	—
	Legend	Front	11/16P–2 1/16P	1 11/16P	1N–1P	0	0	NA
		Rear	—	—	1N–1P	0	0	—
	Legend Coupe	Front	11/16P–2 1/16P	1 11/16P	1N–1P	0	0	NA
		Rear	—	—	1N–1P	0	0	—
	825	Front	11/16P–2 1/16P	1 11/16P	1N–1P	0	0	NA
		Rear	—	—	1N–1P	0	0	—
1989	Integra	Front	1 3/16P–3 3/16P	2 3/16P	1 1/2N–1/2P	1/2N	1/32N	13
		Rear	—	—	1N–1/2N	3/4N	1/16P	—
	Legend	Front	11/16P–2 1/16P	1 11/16P	1N–1P	0	0	NA
		Rear	—	—	1N–1P	0	0	—
	Legend Coupe	Front	11/16P–2 1/16P	1 11/16P	1N–1P	0	0	NA
		Rear	—	—	1N–1P	0	0	—
	827	Front	11/16P–2 1/16P	1 11/16P	1N–1P	0	0	NA
		Rear	—	—	1N–1P	0	0	—

WHEEL ALIGNMENT

Year	Model	Caster			Camber			Steering Axis Inclination (deg.)
		Range (deg.)	Preferred Setting (deg.)	Range (deg.)	Range (deg.)	Preferred Setting (deg.)	Toe-in (in.)	
1990	Integra	Front	1 1/8P–3 1/8P	2 1/8P	1 1/2N–1/2P	1/2N	1/32N	13
		Rear	—	—	1N–1/2N	3/4N	3/16P	—
	Legend	Front	1 1/16P–2 1/16P	1 11/16P	1N–1P	0	0	NA
		Rear	—	—	1N–1P	0	0	—
	Legend Coupe	Front	1 1/16P–2 1/16P	1 11/16P	1N–1P	0	0	NA
		Rear	—	—	1N–1P	0	0	—
	827	Front	1 1/16P–2 1/16P	1 11/16P	1N–1P	0	0	NA
		Rear	—	—	1N–1P	0	0	—
1991–92	Integra	Front	1 1/8P–3 1/8P	2 1/8P	1 1/2N–1/2P	1/2N	1/32N	13
		Rear	—	—	1N–1/2N	3/4N	3/16P	—
	Legend	Front	1 1/16P–2 1/16P	1 11/16P	1N–1P	0	0	—
		Rear	—	—	7/16N	1	—	—
	Legend Coupe	Front	1 1/16P–2 1/16P	1N–1P	0	1	1/32P	—
		Rear	—	—	7/16N	1	—	—
	827	Front	1 1/16P–2 1/16P	1 11/16P	1N–1P	0	0	NA
		Rear	—	—	1N–1P	0	0	—
	NSX	Front	—	3/4N–3/4P	1/2N–1/2P	7/16P	1/8N	—
		Rear	—	—	1/2N–1/2P	1 1/2N	1/4P	—

NA—Not Available
N—Negative
P—Positive

ENGINE MECHANICAL

NOTE: Disconnecting the negative battery cable on some vehicles may interfere with the functions of the on board computer systems and may require the computer to undergo a relearning process, once the negative battery cable is reconnected.

Engine Assembly

REMOVAL & INSTALLATION

1988–89 Integra

On these vehicles, the engine and transaxle are removed as a unit. The lower ball joints and the inner CV-joints will be separated but not removed.

1. Disconnect the battery cables from the battery; negative cable first. Remove the battery and the battery tray from the engine compartment, if necessary.
2. Using a scratch awl, scribe a line where the hood brackets meet the inside of the hood; this will help realign the hood during the installation. Disconnect and remove the washer fluid tube(s). Remove the hood-to-hinge bolts and the hood.
3. Raise and safely support the vehicle. Remove the engine and wheel well splash shields.
4. Drain the oil from the engine, the coolant from the radiator and the fluid from the transaxle.

NOTE: Removal of the filler plug or radiator cap will speed the draining process.

5. Remove the following items:
   a. The front and engine air intake ducts.
   b. The throttle control cable and/or clutch cable.
   c. The coil wire and spark plug wires.
   d. The cruise control cable.
6. Relieve the fuel pressure by slowly loosening the service bolt on the top of the fuel filter about a turn.

CAUTION

The fuel system may be under pressure and fuel will be sprayed. Be sure to take the appropriate safety and fire precautions.

7. Remove the banjo bolt and disconnect the fuel hose from the filter.
8. Label and disconnect the engine compartment sub-harness wiring connectors.
9. Most of the vacuum hoses lead to an emissions control equipment box or panel on the firewall. Do not disconnect the hoses, remove the box or panel and lay it on the engine. All other vacuum hoses can be labeled and disconnected.
10. Loosen the throttle cable locknut and adjusting nut, then, slip the cable end out of the throttle bracket and remove it.
11. Remove the power steering pump-to-bracket bolts and V-belt, then, without disconnecting the hose, move the pump aside.
12. If equipped with an automatic transaxle, working inside the vehicle, remove the center console. Move the shift lever to the **R** position and remove the lock pin to separate the shift cable.
13. If equipped with a manual transaxle, at the firewall, slide the spring pin retainer towards the front of the vehicle. Use an 8mm pin punch to drive the spring pin out to separate the shift rod.
14. Label the heater hoses and disconnect the heater and radiator hoses from the engine.
15. On vehicles equipped with automatic transaxle, disconnect the oil

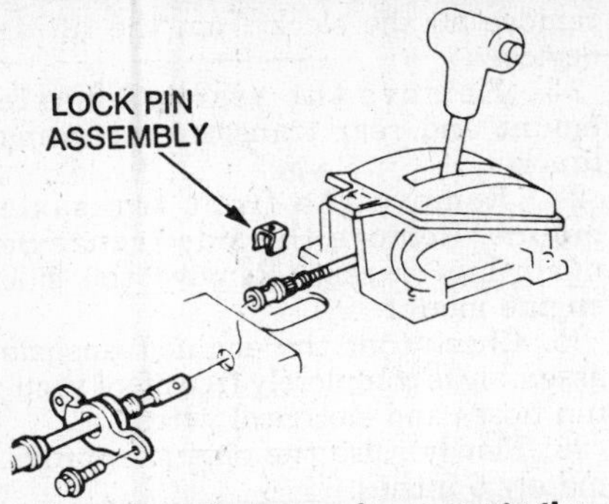

On automatic transaxle, separate the shift cable

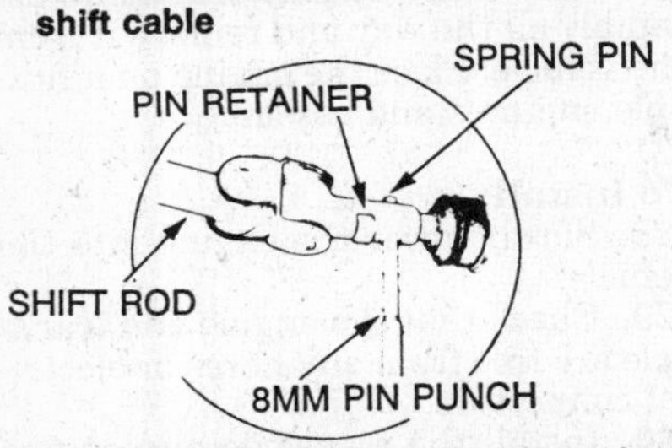

On manual transaxle, move the retainer to drive out the pin

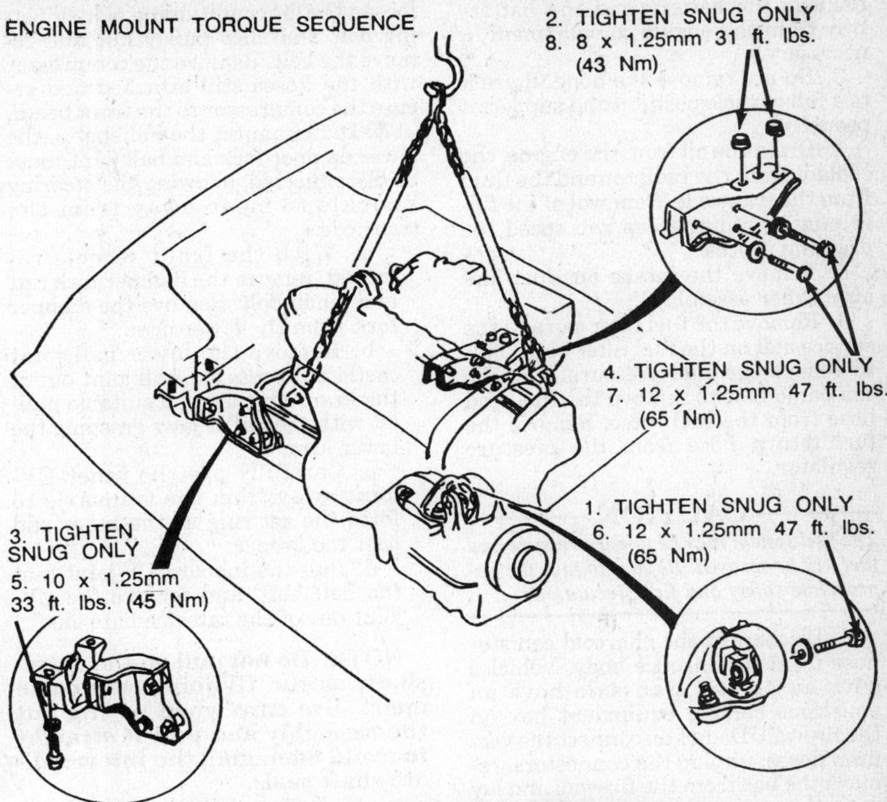

Engine mount torque sequence on 1988–89 Integra

cooler hoses from the transaxle and allow the fluid to drain from the hoses. Secure the hoses to the body near the radiator.

16. Remove the speedometer cable clip and pull the cable from the holder.

NOTE: Do not remove the holder from the transaxle as it may cause the speedometer gear to fall into the transaxle.

17. If equipped with air conditioning, loosen the drive belt adjusting bolts and remove the belt. Remove the compressor-to-bracket bolts and secure the compressor onto the front beam. Remove the lower compressor mounting bracket.

NOTE: Do not disconnect the air conditioning freon lines. The compressor can be moved out of the way without discharging the system.

18. Disconnect the alternator wiring harness connector. Remove the adjusting bolt and the alternator belt. Remove the mount bolt, bracket and the alternator.

19. Using a ball joint puller, separate the ball joint from the front hub.

20. Slowly, lower the floor jack to lower the control arm. Using a small prybar, pry out the inboard CV-joint approximately ½ in. (13mm) in order to release the spring clip from the groove in the differential. Pull the steering knuckle assembly outward. Pull the halfshaft from the transaxle case. Secure the halfshafts out of the way. Do not let them hang or the outer CV-joint will be damaged.

21. Attach a lifting sling to the engine block and raise the engine slightly to remove the slack from the chain.

22. Remove the rear transaxle mount and the bolts from the front transaxle mount and the engine side mount.

23. Check that the engine and transaxle are free from any hoses or electrical connectors.

24. Slowly raise the engine up and out of the vehicle.

To install:

25. Slowly lower the engine into the vehicle.

26. Check that the engine and transaxle are free from any hoses or electrical connectors.

27. Loosely install the rear transaxle mount and the bolts for the front transaxle mount and the engine side mount. Tighten the bolts in 2 steps in the correct sequence. This is important to help minimumize vibration.

28. Install the halfshaft assemblies. Use new spring clips and be sure the CV-joints click into place.

29. Reinstall the alternator and the air conditioning compressor.

30. Install the drive belts and adjust the belt tension.

31. Install the speedometer cable.

32. Reinstall the radiator assembly, coolant hoses, heater hoses and oil cooler lines.

33. Reinstall the shift linkage and adjust the clutch cable.

34. Reinstall the power steering pump and drive belts. Install the throttle cable bracket and throttle cable.

35. Reinstall the control box or panel to the firewall. Reconnect all disconnected vacuum lines and electrical connectors.

36. Reinstall the battery box, battery and battery cables. Reinstall the hood, be sure to line the hood up with the scribe marks made during the removal procedure.

37. Install the wheel well and engine splash shields.

38. Refill the coolant system, engine oil and transaxle fluid. Start the engine, bleed the cooling system and make all necessary adjustments.

1990–92 Integra

On these vehicles, the engine and transaxle are removed as a unit. The lower ball joints and the inner CV-joints will be separated but not removed.

1. Apply the parking brake and place blocks behind the rear wheels. Raise and safely support the vehicle. Remove the engine and wheel well splash shields.

2. Disconnect the battery cables from the battery; negative cable first.

Remove the battery and the battery tray from the engine compartment, if necessary.

3. Do not remove the hood. Raise it to a full verticle position and support it properly.

4. Drain the oil from the engine, the coolant from the radiator and the fluid from the transaxle. Removal of the filler plug or radiator cap will speed the draining process.

5. Remove the intake air duct and air cleaner assembly.

6. Remove the fuel filler cap and the service bolt on the fuel filter banjo bolt to relieve the fuel pressure. Remove the banjo bolt to remove the fuel feed hose from the fuel filter. Remove the fuel return hose from the pressure regulator.

CAUTION

The fuel system may be under pressure and fuel will be sprayed. Be sure to take the appropriate safety and fire precautions.

7. Disconnect the charcoal canister hose from the throttle body. Vehicles with automatic transaxle have an emissions control equipment box on the firewall. Do not disconnect the vacuum hoses; unplug the connectors, remove the box from the firewall and lay it on the engine.

8. Remove the ground cable from the transaxle. Disconnect the 2 distributor electrical connectors and remove the distributor. Be sure to make alignment marks on the distributor to help in the installation procedure.

9. Remove the throttle cable by loosening the locknut, then slip the cable end out of the throttle bracket and accelerator linkage. Take care not to bend the cable when removing it. Do not use pliers to remove the cable from the linkage. Always replaced a kinked cable with a new one.

10. Remove the mounting bolts and the V-belt for the power steering pump, then without disconnecting the hoses, secure the pump out of the way.

11. Near the brake booster, disconnect the engine wiring harness connectors and remove the wire from the clamp.

12. Remove the brake booster vacuum hose from the intake manifold. Disconnect the upper and lower radiator hoses and heater hoses.

13. Without disconnecting the oil hoses or cable, remove the bolt to remove the speed sensor as an assembly.

14. Disconnect the transaxle cooling hoses, if equipped, and unplug the radiator fan connectors. Remove the radiator and fans as an assembly.

15. If not already removed, remove the driver's side splash shield at the wheel well.

16. If equipped with air conditioning, loosen the air conditioning belt adjusting bolt and idler pulley nut and remove the belt. Remove the compressor with the hoses still attached and secure the compressor to the front beam.

17. To disconnect the halfshafts, the lower damper fork and ball joint must be disconnected, allowing the steering knuckle to move away from the transaxle.

a. With the front wheels removed, remove the damper fork nut and pinch bolt. Remove the damper fork from the lower arm.

b. Remove the lower ball joint castle nut press the ball joint out of the lower arm using a suitable puller, with the puller jaws grasping the lower arm.

c. Carefully pry the inner CV-joint away from the transaxle to force the set ring at the inner end past the groove.

d. Pull the inboard CV-joint, not the halfshaft, and remove the CV-joint out of the intermediate shaft.

NOTE: Do not pull on the driveshaft, as the CV-joint may come apart. Use care when prying out the assembly and pull it straight to avoid damaging the intermediate shaft seals.

e. Support the halfshafts or hang them from the body with wire. Do not let them hang from the outer CV-joint or it will be damaged.

18. Disconnect the alternator wiring harness connector. Remove the adjusting bolt and the alternator belt. Remove the mount bolt, bracket and the alternator.

19. Remove the front exhaust pipe assembly.

20. If equipped with a manual transaxle, perform the following procedures:

a. Remove the clutch cable. Take care not to bend the cable when removing it. Do not use pliers to remove the cable from the linkage. Always replaced a kinked cable with a new one.

b. To disconnect the shift linkage rod, push back the boot and drive out the pin securing the shift rod universal joint to the transaxle. Use a new pin when reassembling.

c. Remove the shift lever torque rod.

21. If equipped with an automatic transaxle, remove the torque converter cover. Remove the cable holder, cotter pin and control pin, then remove the shift control cable. Take care not to bend the cable when removing it. Do not use pliers to remove the cable from the linkage. Always replaced a kinked cable with a new one.

22. Attach a suitable lifting device to the engine. Raise the engine slightly to remove all the slack from the lifting device.

23. Remove the rear transaxle mount and rear transaxle mounting bracket.

24. Remove the front transaxle mount. Remove the side transaxle mounting bracket. Remove the side engine mount.

25. Check that the engine/transaxle assembly is completely free of all vacuum hoses and electrical wires.

26. Slowly raise the engine approximately 6 inches.

27. Raise the engine/transaxle assembly all the way and remove it from the vehicle. Place the engine on a suitable engine stand assembly.

To install:

28. Slowly lower the engine into the vehicle.

29. Check that the engine and transaxle are free from any hoses or electrical connectors.

30. Install the transaxle and engine mounts and bolts. Some of the transaxle mount bolts are designed to be torqued only one time and must be replaced whenever the engine is removed. Be sure to torque all bolts in 2 steps in the correct sequence. This is important to help minimize engine vibrations.

31. If equipped with an automatic transaxle, reinstall the control pin, cotter pin, cable holder and cable. Install the torque converter cover.

32. If equipped with an manual transaxle, install the shift rod, shift torque rod lever and clutch cable. Check the clutch adjustment.

33. Install the front exhaust pipe.

34. Reinstall the alternator assembly.

35. Install the halfshaft assemblies. Be sure to use new set rings.

36. Reinstall the air conditioning compressor and drive belts.

37. Install the speedometer drive.

38. Reinstall the radiator assembly, coolant hoses, heater hoses and oil cooler lines.

39. Reinstall the power steering pump and drive belts. Install the throttle cable bracket and throttle cable.

40. Reinstall the distributor assembly and ground cable to the transaxle.

41. Reinstall the control box brackets and control box connectors. Reconnect all disconnected vacuum lines and electrical connectors.

42. Reinstall the battery box, battery and battery cables.

43. Reinstall the wheel well and engine splash shields.

44. Refill the coolant system, engine oil and transaxle fluid. Start the engine, bleed the cooling system and make all necessary adjustments.

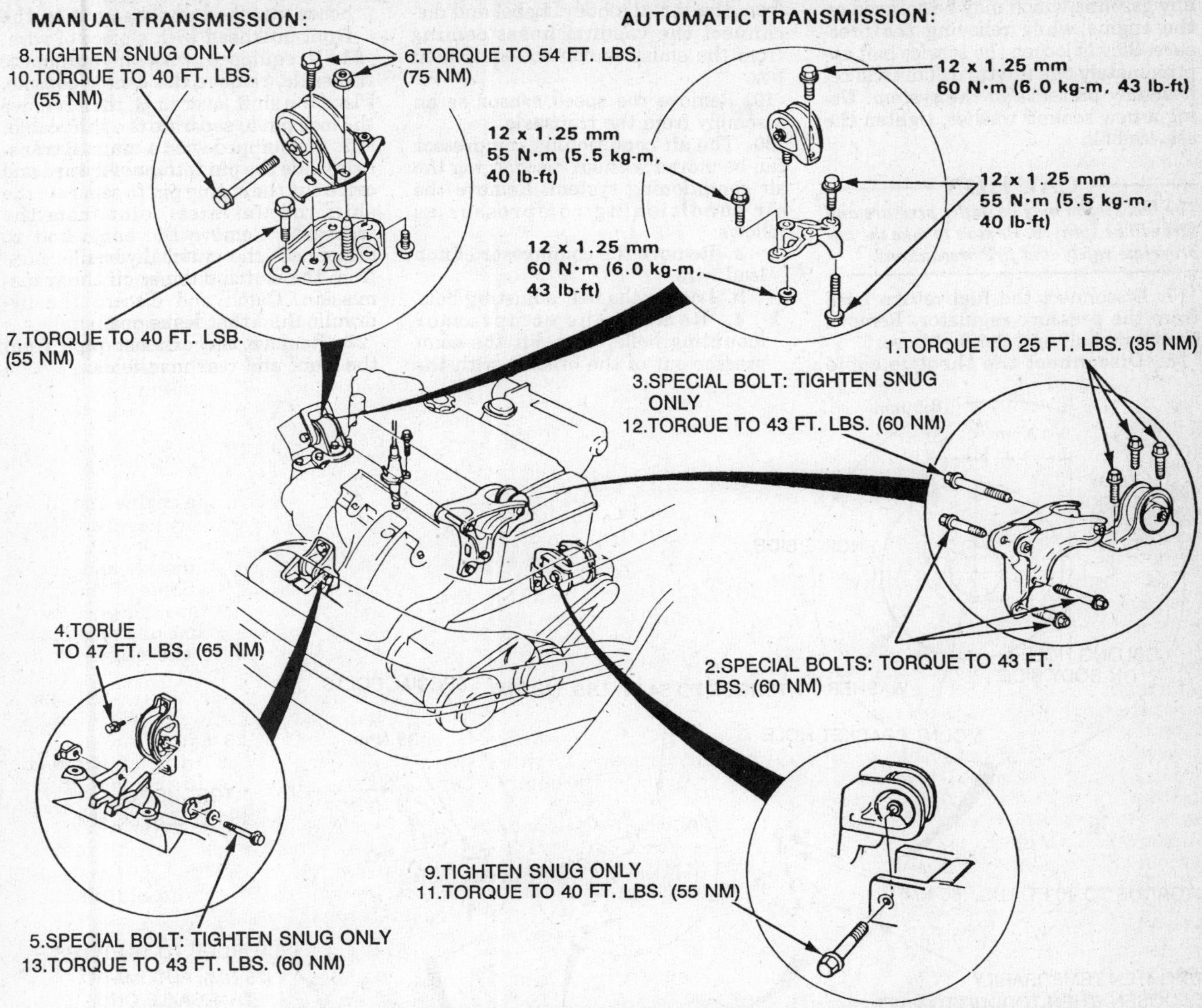

Engine mount torque sequence on 1990–92 Integra: the special bolts in steps 2, 3 and 5 must not be re-used, replace with new ones

Sterling and 1988–90 Legend

On these vehicles, the engine and transaxle are removed as a unit. The lower ball joints and the inner CV-joints will be separated but not removed.

1. Apply the parking brake and place blocks behind the rear wheels. Raise and safely support the vehicle.
2. Disconnect both battery cables from the battery. Remove the battery and the battery tray from the engine compartment.
3. Position the hood in the vertical position by removing the open stay mounting bolt on the hood side and fitting it to the mounting hole near the hinge.
4. Remove the air intake tube, air cleaner and resonator tube as an assembly.
5. Remove the splash guard from under the engine.
6. Remove the oil filler cap and drain plug to drain the engine oil.
7. Remove the radiator cap, then open the radiator drain petcock and drain the coolant from the radiator. There are also block drain bolts on each side of the engine.
8. Remove the transaxle filler plug, then remove the drain plug and drain the transaxle.
9. Disconnect the pressure switch wire from the oil filter case.
10. Disconnect both water hoses from the engine oil cooler.
11. Remove the drain bolt from the oil filter case to drain the oil. Remove the oil filter case from the engine block.
12. Disconnect the upper and lower radiator hoses from the radiator.
13. If equipped with an automatic transaxle, disconnect cooler hose from the bottom of the radiator.
14. Disconnect the following engine sub-harness connectors from the body side:
    a. Four right side connectors and clamp
    b. Both left side main fuse connectors
    c. Coil wire, primary lead connectors and the condenser connector.
    d. Both ground cables from the cylinder head and the transaxle.
15. Disconnect the connector from the power steering pump and both hoses. Disconnect the hose from the cruise control actuator and the hose from the power brake booster.
16. Using the following procedures relieve the fuel system pressure. Place a shop rag over the fuel filter to absorb

any gasoline which may be sprayed on the engine while relieving the pressure. Slowly loosen the service bolt approximately one full turn. This will relieve any pressure in the system. Using a new sealing washer, tighten the service bolt.

CAUTION

The fuel system may be under pressure and fuel will be sprayed. Be sure to take the appropriate safety and fire precautions.

17. Disconnect the fuel return hose from the pressure regulator. Remove the banjo bolt and the fuel hose.

18. Disconnect the throttle cable from the throttle body. Label and disconnect the vacuum hoses coming from the emission control equipment box.

19. Remove the speed sensor as an assembly from the transaxle.

20. The air conditioning compressor can be moved without discharging the air conditioning system. Remove the air conditioning compressor as follows:

a. Remove the compressor clutch lead wire.

b. Loosen the belt adjusting bolt.

c. Remove the compressor mounting bolts, then lift the compressor out of the bracket with the hoses attached and hang it to the front bulkhead with a piece of wire.

21. If equipped with an automatic transaxle, remove the center console. Place the shift lever in **R**, then remove the lock pin to separate the shift cable.

22. If equipped with a manual transaxle, slide the pin retainer forward and drive out the spring pin to separate the shift rod universal joint from the transaxle. Remove the banjo bolt to disconnect the clutch hydraulic hose from the clutch damper on the transmission. Catch and discard the hydraulic fluid that leaks out.

23. Remove the exhaust pipe from the front and rear manifolds.

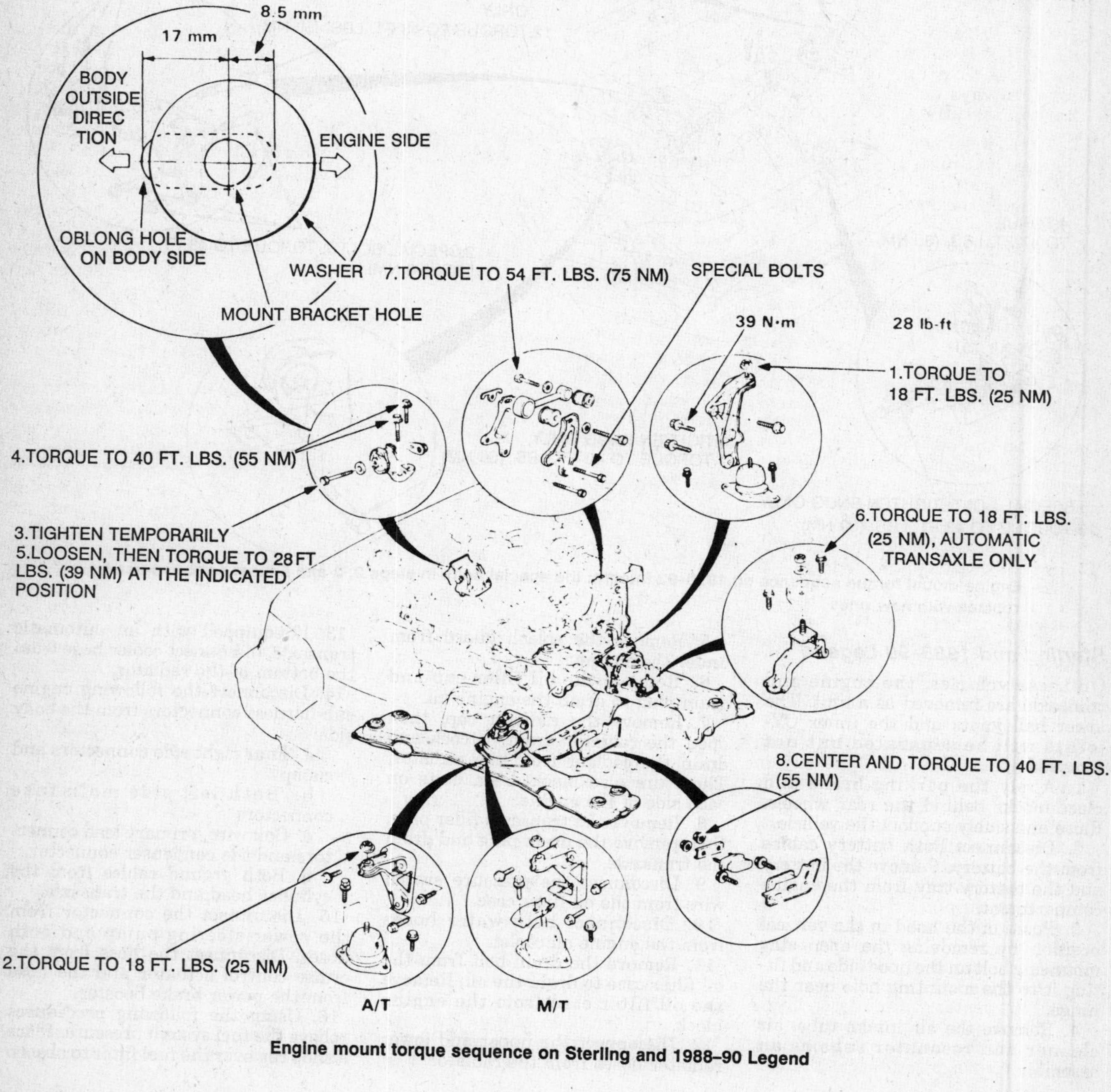

Engine mount torque sequence on Sterling and 1988–90 Legend

24. Remove the halfshaft as follows:
a. Raise and safely support the vehicle and remove the front wheels.
b. Remove the ball joint bolt and separate the ball joint from lower control arm.
c. Using a small prybar, pry out the inboard CV-joint approximately 13mm in order to release the spring clip from the differential, then pull the halfshaft from the transaxle case.
d. Cover the inner CV-joints with a plastic bag and support or hand the halfshaft from the body with wire. Do not let the halfshafts hang by the outer CV-joint or it will be damaged.

NOTE: When installing the halfshaft, insert the shaft until the spring clip clicks into the groove. Always use a new spring clip when installing the halfshaft.

25. Attach a suitable lifting chain to the engine and raise it enough to remove the slack.
26. Remove the engine side mount bracket bolts.
27. Remove the front engine mount nut, then remove the rear engine mount nut.
28. Loosen and remove the alternator belt. Disconnect the alternator wire harness and remove the alternator.
29. Remove the bolt from the rear torque rod at the engine, then loosen the bolt in the frame mount and swing the rod up and out of the way.
30. Tilt the engine about 30 degrees and raise the engine carefully from the vehicle checking that all wires and hoses have been removed from the engine/transaxle. Raise the engine all the way up and remove it from the vehicle.

To install:
31. Slowly lower the engine into the vehicle.
32. Check that the engine and transaxle are free from any hoses or electrical connectors.
33. Install the transaxle and engine mounts and bolts. Some of the bolts in upper rear mount are designed to be torqued only one time and must be replaced whenever they are removed. Be sure to torque all bolts in 2 steps in the correct sequence. This is important to help minimize engine vibrations.
34. If equipped with automatic transaxle, reinstall the control pin, cotter pin, cable holder and cable. Intsall the torque converter cover.
35. Reconnect the shift cable and reinstall the center console assembly.
36. If equipped with manual transaxle, connect the clutch hydraulic line, gear shift linkage and torque rod.
37. Install the front and rear exhaust pipes.
38. Reinstall the alternator assembly.
39. Install the halfshaft assemblies. Use new clips on the CV-joints and make sure they click firmly into place. Torque the lower ball joint nut to 72 ft. lbs. (100 Nm) and install a new cotter pin.
40. Reinstall the air conditioning compressor and drive belts.
41. Install the speedometer drive assembly.
42. Reinstall the radiator assembly, coolant hoses, heater hoses and oil cooler lines.
43. Reconnect the power steering hoses and refill the reservoir to the upper mark. Install the throttle cable bracket and throttle cable.
44. Install the distributor assembly and ground cable to the transaxle.
45. Install the control box brackets and control box connectors. Reconnect all disconnected vacuum lines and electrical connectors.
46. Reinstall the battery box, battery and battery cables. Install the hood back to its proper position.
47. Install the wheel well and engine splash shields.
48. Refill the coolant system, engine oil and transaxle fluid. Start the engine and bleed the steering system by turning the steering wheel lock–to–lock several times. Bleed the cooling system by opening the bleeder screw near the thermostat housing. Make all other necessary adjustments.

1991–92 Legend

Portions of the front sub-frame are made of aluminum alloy. Using normal steel bolts on aluminum will cause an electrolytic reaction: the aluminum around the fastener will corrode and the bolt will loosen. When replacing fasteners, be sure to use bolts that have a Dacro® coating specifically designed for such applications. Dacro® bolts can be identified by a dull grey finish, sometimes with a dull green finish on the threads for more accurate torque wrench readings. These parts should be available at the dealer.

On this vehicle, the engine and transaxle are lifted out as a unit. The battery will also be removed. Since the vehicle is equipped with a theft protected radio, it is important to have the security code before disconnecting the battery. After reconnecting power to the radio, the 5 digit code must be entered to restore operation.

1. Do not remove the hood. Disconnect the hood stay strut and reconnect it to hold the hood in a verticle position. Remove the battery and the battery box.
2. Remove the strut bar that runs across the engine compartment and the bracket. Working underneath the vehicle, remove the splash shield and drain the engine coolant and oil and the transaxle fluid.
3. Label and disconnect the starter and battery wiring from the main fuse/relay box. Remove the ground cable from the engine block and label and disconnect the main engine wiring harness.
4. Remove the throttle cable cover. Without turning the adjusting nut, loosen the locknut, which is closer to the throttle, and disconnect the throttle cable from the throttle and bracket.
5. Remove the air cleaner assembly and ducting.
6. On the right shock tower, disconnect the igniter unit and remove the wiring harness clamp. Disconnect the engine ground cable.
7. On the firewall behind the right cylinder head is a control box containing emission control equipment. Without disconnecting any vacuum hoses, unplug the electrical connectors and remove the control box from the firewall.
8. Label and disconnect the main engine wiring harness connectors and remove the bracket.
9. On top of the fuel filter, relieve the fuel system pressure by slowly loosening the service bolt 1 turn.

CAUTION

The fuel system may be under pressure and fuel will be sprayed. Be sure to take the appropriate safety and fire precautions.

10. Remove the fuel supply hose and disconnect the return hose from the pressure regulator.
11. Disconnect the vacuum hose to the brake booster at the check valve.
12. At the left rear of the engine compartment, disconnect the transaxle wiring harness and remove the clamp.
13. Disconnect and plug the transaxle cooling hoses at the radiator. Disconnect the upper and lower hoses and the fan and sensor wiring. Remove the radiator and fans as an assembly.
14. Near the power steering fluid reservoir, remove the solenoid valve assembly, vacuum pipes and air tank.
15. Without disconnecting any hoses, remove the power steering pump and secure it out of the way.
16. Raise and safely support the vehicle and remove the front wheels.
17. Remove the lower damper forks and remove the nut from the lower ball joint. Use a suitable ball joint removal tool to disconnect the ball joint from the lower control arm.
18. Carefully pry the inner CV-joints from the transaxle and hang the halfshafts from the suspension. Do not

let the halfshaft hang by the outer CV-joint or the joint will be damaged. Cover the inner joints with plastic bags.
19. Remove the lower plate from the rear beam.
20. Without disconnecting any hoses, remove the power steering speed sensor from the differential housing.
21. Disconnect the exhaust pipe from the catalyst and remove it from the manifolds.
22. Without disconnecting any hoses, remove the air conditioner compressor and hang it from the body with wire.
23. On vehicles with a manual transaxle, remove the clutch slave cylinder without disconnecting the hydraulic line. Disconnect the shift lever torque rod and disconnect the shift linkage by driving out the 8mm roll pin.
24. Remove the engine mid-mount nuts and bolts and the transaxle rear mount.
25. Working from above, remove one of the EGR valve passage bolts and install a lifting hook. Attach a chain hoist and take up the slack.
26. Remove all engine and transaxle mounting nuts and bolts. Raise the engine/transaxle slightly and check to make sure all hoses and wires have been disconnected. As the unit is raised, allow it to tilt up in front to provide proper clearance past the rear beam.

To install:
27. When installing the engine/transaxle, rotate the unit up in front to avoid hitting the rear beam with the transaxle. Check carefully to make sure the rubber mounts are not twisted or offset. Start all the nuts and bolts and then torque them in the correct order. This is important to help minimumize engine vibrations.
28. After the mounts have been properly torqued, install the mounting bolts on the rear transaxle mount and torque to 28 ft. lbs. (39 Nm). Install the engine mid-mounts and torque the nuts and bolts to 28 ft. lbs. (39 Nm). Remove the lifting hook and install the EGR valve passage bolt.
29. Reconnect the shift linkage. On vehicles with a manual transaxle, install the clutch slave cylinder.
30. Install the air conditioner compressor and torque the mounting bolt to 16 ft. lbs. (22 Nm). Adjust the belt tension for about ¼ in. of deflection with 22 lbs. (10 kg) force and torque the idler pulley nut to 33 ft. lbs. 45 Nm).
31. When installing the exhaust pipe, use new gaskets and self locking nuts. Torque the manifold nuts to 40 ft. lbs. (55 Nm) and the catalyst flange nuts to 16 ft. lbs. (22 Nm).
32. Install the power steering speed

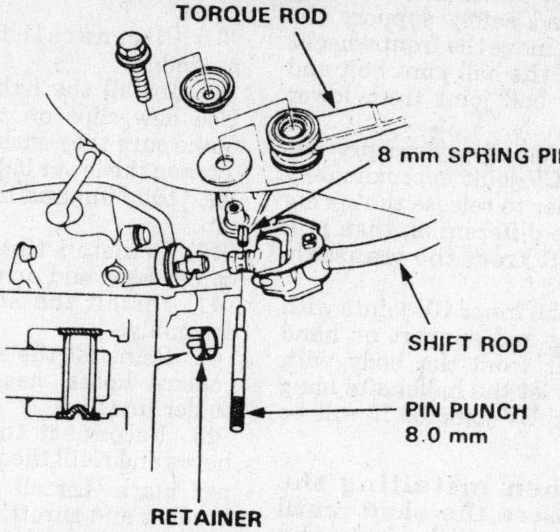

Manual shift linkage connection on 1991–92 Legend

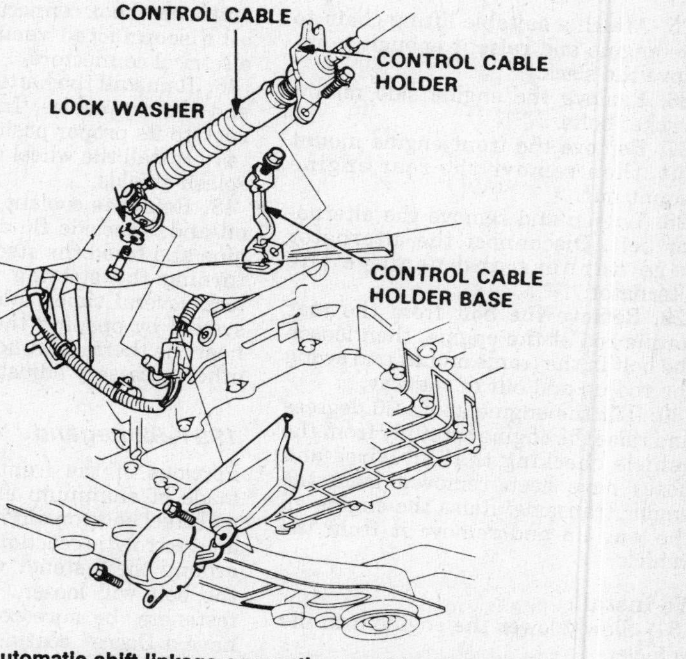

Automatic shift linkage connection on 1991–92 Legend

sensor and connect the wiring. Torque the bolts to 9 ft. lbs. (12 Nm).
33. Install the lower plate using the special corrosion resistant bolts. Torque to 29 ft. lbs. (39 Nm).
34. Install new clips to the end of the inner CV-joints and press the joints into the transaxle. Make sure joint clicks solidly into place. When assembling the suspension, torque the lower ball joint nut to 51–58 ft. lbs. (70–80 Nm). Torque the lower damper bolt to 51 ft. lbs. (70 Nm) and the upper damper pinch bolt to 37 ft. lbs. (51 Nm).
35. Install the power steering pump and adjust the belt tension for about ½ inch deflection with 22 lbs. (10 kg) force. Torque the mounting bolt to 33 ft. lbs. (45 Nm) and the nut to 16 ft. lbs. (22 Nm).
36. Install the radiator and connect the cooling hoses. When filling the system, open the bleeder where the upper hose connects to the engine.
37. Install the emission control equipment bracket and connect wiring and vacuum hoses.
38. Install the fuel supply pipe, using new gaskets. Connect the return hose and the brake booster vacuum hose. Finish connecting all the remaining wiring and vacuum hoses, except for the battery.
39. Install the throttle cable. If the adjuster nut (farther from the boot)

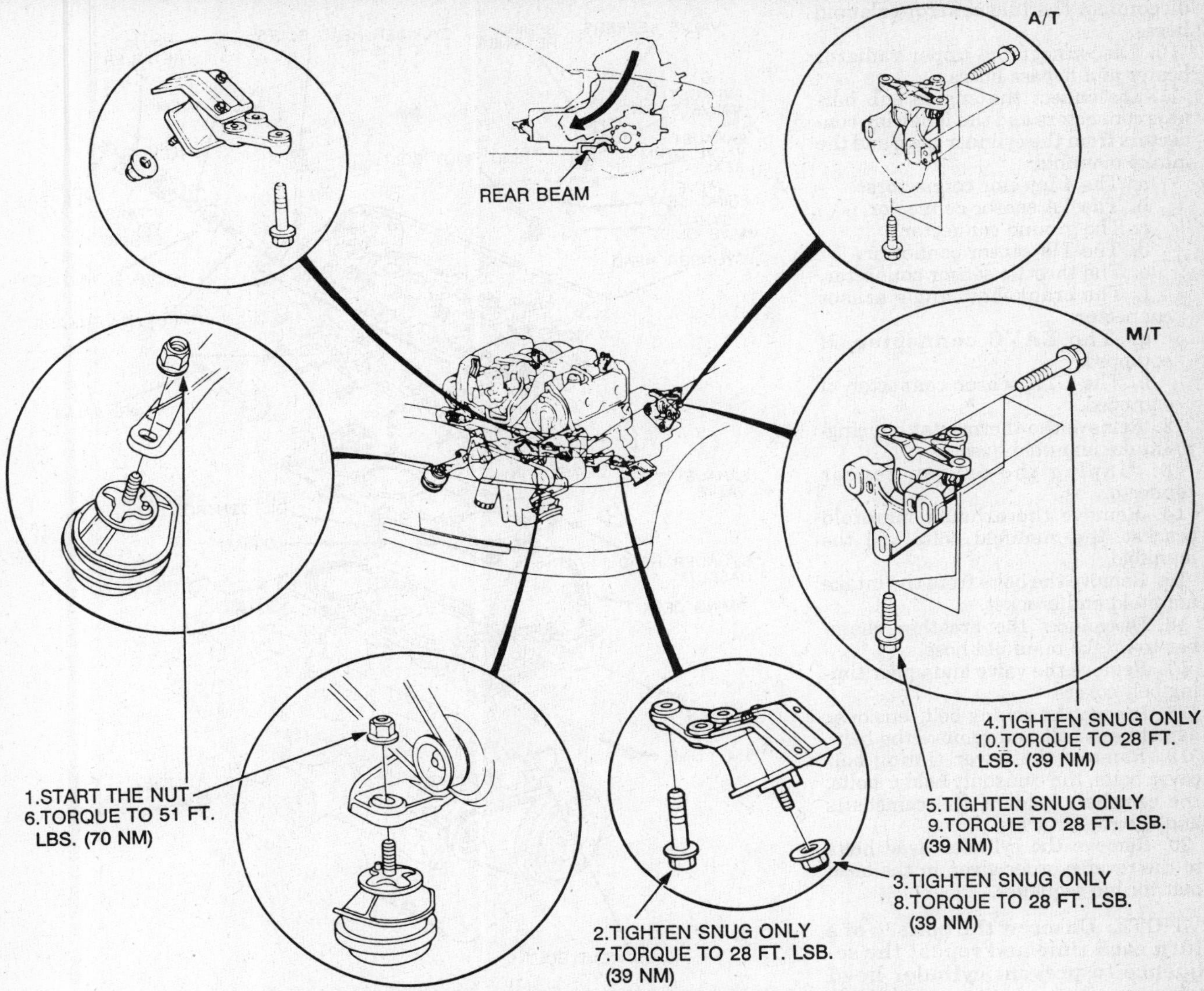

On 1991–92 Legend, torque the engine mounts in the sequence shown

has been moved, temporarily tighten the lock nut. The engine must be fully warmed up to adjust the cable.

40. Install the air cleaner and ducting, strut bar and splash shield.

41. Refill the engine oil and transaxle fluid and install the battery. When the battery is connected, turn the ignition switch **ON** and **OFF** a number of times to pressurize the fuel system and check for leaks.

42. After checking carefully and making sure everything is properly connected, start the engine and bleed the cooling system. When the engine is at operating temparature, stop the engine and adjust the throttle cable.

43. To adjust the throttle cable, loosen both nuts and take up the slack in the cable. Back the adjusting nut away from the bracket so there is 0.120 in. (3.0mm) gap between the nut and bracket. Make sure the throttle opens and closes fully with pedal movement.

Cylinder Head

REMOVAL & INSTALLATION

1988–89 Integra

NOTE: Cylinder head temperature must be below 100°F.

Before removing the cylinder head check the following:

Inspect the timing belt.

Turn the flywheel so the No. 1 cylinder is at TDC.

Mark all connectors and vacuum hoses before disconnecting them.

1. Disconnect the negative battery cable.
2. Drain the cooling system.
3. Remove the air cleaner:
   a. Remove the air cleaner cover and filter.
   b. Disconnect the hot/cold air intake ducts and remove the air chamber hose.
   c. Remove the air cleaner.
4. Slowly loosen the service bolt on the top of the fuel filter about a turn to relieve the fuel system pressure. Disconnect the fuel return hose from the pressure regulator.

CAUTION

The fuel system may be under pressure and fuel will be sprayed. Be sure to take the appropriate safety and fire precautions.

5. Remove the brake booster vacuum tube from the intake manifold.
6. Remove the engine ground wire from the valve cover. Disconnect the throttle cable from the throttle body.
7. Disconnect the spark plug wires from the spark plugs and remove the distributor assembly.
8. Disconnect the hoses from the charcoal canister and from the No. 1 control box at the tubing manifold.
9. If equipped with air conditioning,

disconnect the idle control solenoid hoses.

10. Disconnect the upper radiator heater and bypass hoses.

11. Disconnect the engine sub harness connectors and the following connectors from the cylinder head and the intake manifold:

a. The 4 injector connectors.
b. The TA sensor connector.
c. The ground connector.
d. The TW sensor connector.
e. The throttle sensor connector.
f. The crankshaft angle sensor connector.
g. The EAVC connector, if equipped.
h. The CYL sensor connector, if equipped.

12. Remove the thermostat housing-to-intake manifold hose.

13. Unplug the oxygen sensor connector.

14. Remove the exhaust manifold bracket, the manifold bolts and the manifold.

15. Remove the bolts from the intake manifold and bracket.

16. Disconnect the breather chamber-to-intake manifold hose.

17. Remove the valve and upper timing belt covers.

18. Loosen the timing belt tensioner adjustment bolt and remove the belt.

19. Remove the lower timing belt cover bolts, the camshaft holder bolts, the camshaft holders, the camshafts and the rocker arms.

20. Remove the cylinder head bolts in the reverse order given in the head bolt torque sequence.

NOTE: Unscrew the bolts ⅓ of a turn each time and repeat the sequence to prevent cylinder head warpage.

21. Carefully, remove the intake manifold from the cylinder head and the cylinder head from the engine.

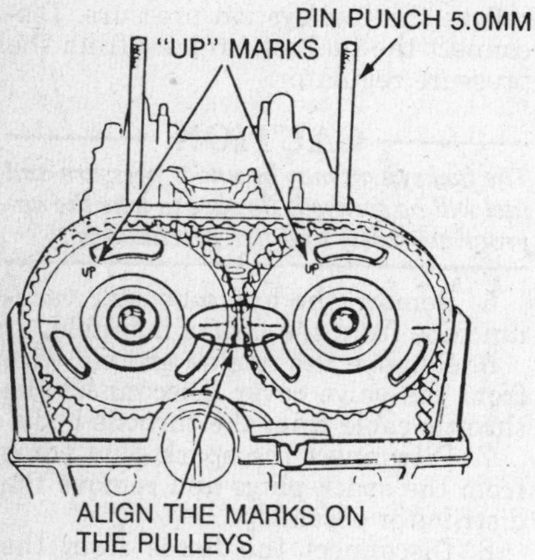

Alignemnt of camshaft marks on 1988–92 Integra

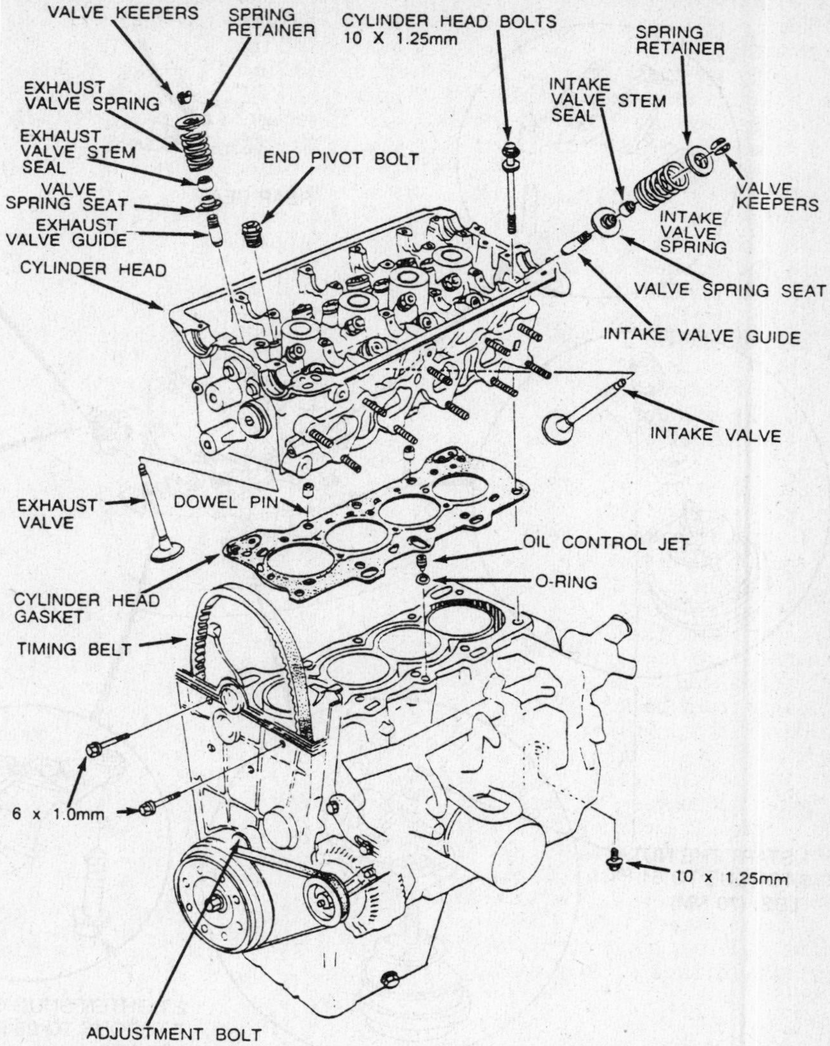

Install the oil control jet when installing the head on 1988–92 Integra

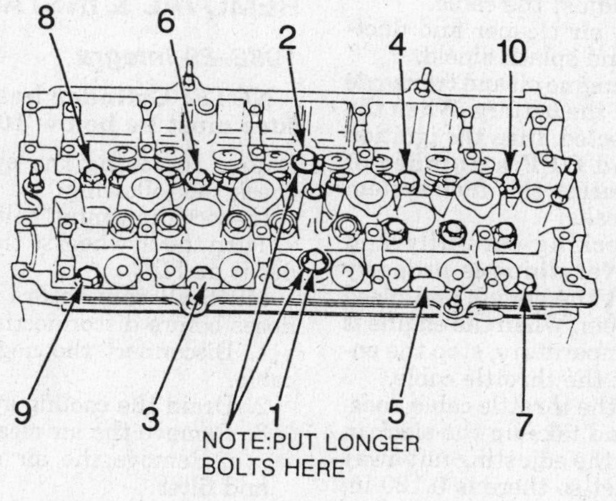

Head bolt torque sequence on Integra

22. Clean the gasket mounting surfaces.

To install:

23. Using new gaskets, install the intake manifold onto the cylinder head and torque the nuts to 23 ft. lbs. (32 Nm) in a criss-cross pattern in 2–3 steps.

24. Install the cylinder head onto the engine block, after making sure the mating surface was cleaned and a new gasket was installed. Be sure to pay attention to the following points:

a. Be sure the No. 1 cylinder is at TDC and the camshaft pulleys UP mark is on the top before positioning the head in place.

b. The cylinder head dowel pins and oil control jet must be aligned.

c. Torque the cylinder head bolts in 2 progressive steps and in the proper sequence. First to 22 ft. lbs. (30 Nm), in sequence, then to 48 ft. lbs. (67 Nm), in the same sequence.

d. Use the longer bolt in the No. 9 position.

25. Make sure the keyways on the camshafts are facing up. The valve locknuts should be loosened and the adjusting screw backed off before installation. Replace the rocker arms in their original position.

26. Place the rocker arms on the pivot bolts and the valve stems.

27. Install the camshafts and the camshaft seals with the open side facing in.

28. Be sure to note the I and E marks that are stamped on the camshaft holders. Do not apply oil to the seal mating surface of the camshaft holders.

29. Apply a liquid gasket to the head of the mating surfaces of the No. 1 and No. 6 camshaft holders then install them, along with No. 2, 3, 4 and 5. Tighten each bolt 1 turn at a time to insure that the rockers do not bind on the valves.

30. Torque the camshaft holder bolts and make sure the rocker arms are properly positioned on the valve stems. Start at the center holders and work out towards the ends, torque the bolts to 9 ft. lbs. (12 Nm).

31. Press in the camshaft seal securely with a suitable seal driver.

32. Install the keys into their groves in the camshafts. To set the No. 1 piston at TDC, align the holes on the camshaft with the holes in the No. 1 camshaft holders and drive 5.0mm pin punches into the holes.

33. Push the camshaft pulleys onto the camshafts, then torque the retaining bolts to 27 ft. lbs. (38 Nm).

34. Install the timing belt and adjust the timing belt tension. Install the lower timing belt cover and bolts.

35. Install the upper timing belt covers and valve covers.

36. Reconnect the breather chamber-to-intake manifold hose.

37. Install the exhaust manifold, the manifold bolts and the manifold bracket.

38. Reconnect all disconnected electrical connections and vacuum lines.

39. Reconnect the upper radiator hose and bypass hoses.

40. Reinstall the distributor assembly and the spark plug wires to the spark plugs.

41. Install the engine ground wire from the valve cover. Reconnect the throttle cable from the throttle body.

42. Reinstall the air cleaner assembly and duct work that goes along with it. Reconnect the negative battery cable.

43. After installation, check to see that all hoses and wires are installed correctly.

44. Refill the coolant system.

45. Adjust the valve clearance. Make all other necessary adjustments

NOTE: It is recommended that after completeing the cylinder head removal and installation the engine oil should be changed.

1990–92 Integra

NOTE: Cylinder head temperature must be below 100°F.

Before removing the cylinder head check the following:

Inspect the timing belt.

Turn the flywheel so the No. 1 cylinder is at TDC.

Mark all emission hoses before disconnecting them.

1. Disconnect the negative battery cable.

2. Drain the cooling system.

3. Remove the air cleaner:

a. Remove the air cleaner cover and filter.

b. Disconnect the hot/cold air intake ducts and remove the air chamber hose.

c. Remove the air cleaner.

4. Relieve the fuel pressure by slowly loosening the service bolt on the top of the fuel filter about a turn.

CAUTION

The fuel system may be under pressure and fuel will be sprayed. Be sure to take the appropriate safety and fire precautions.

5. Disconnect the fuel feed line. Remove the vacuum hose, breather hose and air intake hose.

6. Remove the water bypass hose from cylinder head. Remove the charcoal canister hose from the throttle body.

7. Remove the brake booster vacuum hose from the intake manifold. Remove the fuel return hose. Remove the PCV hose.

8. Remove the throttle cable from the throttle body. Take care not to bend the cable when removing it. Do not use pliers to remove the cable from the linkage. Always replaced a kinked cable with a new one.

9. Disconnect the ignition coil connector, TDC and crankshaft/cylinder sensor connector from the distributor.

10. Remove and tag the spark plug wirers.

11. Remove the emission control equipment bracket but do not disconnect the emission hoses.

12. Disconnect the 3 engine harness connectors on the left side of the engine compartment.

13. Disconnect the engine sub-harness connectors from the cylinder head and intake manifold. The connectors are as follows:

a. Four injector connectors.

b. TA sensor connector.

c. Throttle angle sensor connector.

d. EGR valve lift sensor; automatic transaxle only.

e. Ground cable terminal.

f. TW sensor ground.

g. Coolant temperature gauge sender terminal.

h. Oxygen sensor terminal.

i. EACV connector.

14. Remove the upper radiator hose and heater inlet hose from the cylinder head.

15. Remove the power steering belt and power steering pump. Do not disconnect the hoses from the pump.

16. Raise and safely support the vehicle.

17. Remove the left front wheel and then remove the left splash shield.

18. Remove the intake manifold bracket bolts and remove the exhaust manifold upper shroud.

19. Remove the exhaust manifold bracket. Remove front exhaust pipe and remove the exhaust manifold.

20. Remove the valve cover and engine ground cable.

21. Remove the timing belt middle cover. Loosen but do not remove the adjusting bolt and release the timing belt tension. Tighten the bolt to hold the tensioner in the released position.

22. Remove the timing belt from the driven pulleys. Be sure not to crimp or bend the timing belt more then 90 degrees in less then 1 in. (25mm) in diameter.

23. Remove the camshaft driven pulleys. Loosen all the camshaft holder bolts 1 full turn at a time to release valve spring pressure evenly. Remove the camshaft holder bolts, then remove the camshaft holders, camshafts and rocker arms.

24. Remove the cylinder head bolts and remove the cylinder head. To prevent warpage, unscrew all the cylinder

head bolts in sequence ⅓ turn at a time, repeat this sequence until all the bolts are loosened.

25. Remove the intake manifold from the cylinder head.

To install:

26. Use new gaskets and install the intake manifold onto the cylinder head and tighten the nuts in a criss-cross pattern in 2–3 steps, beginning in the middle. Torque the nuts to 17 ft. lbs. (23 Nm).

27. Install the cylinder head onto the engine block, after making sure the mating surface was cleaned and a new gasket was installed. Be sure to pay attention to the following points:

a. Be sure the No. 1 cylinder is at TDC and the camshaft pulley UP mark is on the top before positioning the head in place.

b. The cylinder head dowel pins and oil control jet must be aligned.

c. Torque the cylinder head bolts, in 2 progressive steps: First to 22 ft. lbs. (30 Nm), in sequence, then to 61 ft. lbs. (85 Nm), in the same sequence. This sequence is the same as on earlier Integra engines.

d. Apply engine oil to the cylinder head bolts and washers. Use the longer bolt in the No. 1 and No. 2 positions.

28. Make sure the keyways on the camshafts are facing up. The valve locknuts should be loosened and the adjusting screw backed off before installation. Replace the rocker arms in their original position.

29. Place the rocker arms on the pivot bolts and the valve stems.

30. Install the camshafts and the camshaft seals with the open spring side facing in.

31. Be sure to note the I and E marks that are stamped on the camshaft holders. Do not apply oil to the seal mating surface of the camshaft holders.

32. Apply a liquid gasket to the head of the mating surfaces of the No. 1 and No. 6 camshaft holders then install them, along with No. 2, 3, 4 and 5. Tighten each bolt 1 turn at a time to insure that the rockers do not bind on the valves.

31. Torque the camshaft holder bolts and make sure the rocker arms are properly positioned on the valve stems. Start at the center holders and work out torards the ends, torque the bolts to 9 ft. lbs. (12 Nm).

32. Press in the camshaft seal securely with a suitable seal driver.

33. Install the keys into their groves in the camshafts. To set the No. 1 piston at TDC, align the holes on the camshaft with the holes in the No. 1 camshaft holders and drive 5.0mm pin punches into the holes.

34. Push the camshaft pulleys onto the camshafts, then torque the retaining bolts to 27 ft. lbs. (38 Nm).

35. Install the timing belt and adjust the tension. Install the lower and middle timing belt covers and bolts.

36. Install the valve cover and engine ground cable.

37. Install the exhaust manifold and torque the nuts to 23 ft. lbs. (32 Nm). Install the bracket and upper shroud and attach the exhaust pipe.

38. Install the left front wheel splash shield and the left front wheel. Lower the vehicle.

39. Install the power steering pump and drive belt.

40. Reconnect all disconnected electrical connections and vacuum lines.

41. Reconnect the upper radiator hose and heater inlet hose.

42. Reinstall the spark plug wires to the spark plugs.

43. Install the engine ground wire to the valve cover. Reconnect the throttle cable from the throttle body.

44. Reinstall the air cleaner assembly and duct work that goes along with it. Reconnect the negative battery cable.

45. After installation, check to see that all hoses and wires are installed correctly.

46. Refill the coolant system.

47. Adjust the valve clearance. Make all other necessary adjustments

NOTE: It is recommended that after completing the cylinder head removal and installation the engine oil should be changed.

Sterling and 1988–90 Legend

NOTE: The cylinder head temperature must be below 100°F.

Before removing the cylinder head check the following:

Inspect the timing belt.

Turn the flywheel so the No. 1 cylinder is at TDC.

Mark all wiring and emission hoses before disconnecting them.

1. Disconnect the battery ground cable.

2. Drain the cooling system.

3. Remove the vacuum hose from the brake booster.

4. Remove the secondary ground cable from the cylinder head and the transaxle housing.

5. Disconnect the radio noise condensor connector, ignition coil wire and the ignition primary connector.

6. Remove the air cleaner cover.

7. Relieve the fuel pressure by loosening the service bolt on the top of the fuel filter about a turn. Disconnect the fuel return hose from the pressure regulator. Remove the special nut and the fuel hose.

CAUTION

The fuel system may be under pressure and fuel will be sprayed. Be sure to take the appropriate safety and fire precautions.

8. Disconnect the throttle cable from the throttle valve.

9. Disconnect the charcoal canister hose from the throttle valve.

10. Disconnect the engine sub harness connectors from the cylinder head and the intake manifold:

a. The 6 injector connectors.

b. The TA sensor connector.

c. The temperature unit connector.

d. The ground connector from the fuel pipe.

e. The TW sensor connector.

f. The throttle sensor connector.

g. The crankshaft angle sensor connector.

h. EGR valve connector.

i. The 4 wire harness clamps.

11. Disconnect the oxygen sensor coupler.

12. Disconnect the cooling system hoses from the cylinder head. Remove the hose between the water passage and the intake manifold. Disconnect the connecting pipe to the valve body hose and bypass outlet hose.

13. Disconnect the spark plug wires from the spark plugs and remove the distributor assembly.

14. Remove the intake manifold cover from the intake manifold.

15. Remove the wire harness cover.

16. Remove the alternator pulley cover.

17. Remove the alternator and belt.

18. Remove the power steering pump and disconnect the pump hoses. Also, remove the hose clamp bolt on the body.

19. Disconnect the idle boost solenoid hoses.

20. Remove the cruise control actuator.

21. Remove the exhaust header pipe and pull it clear of the exhaust manifold.

22. Remove the air cleaner base mount bolts and disconnect the hose from the intake manifold to the breather chamber.

23. Remove the air cleaner base from the intake manifold.

24. Remove the EGR tube nuts from the cylinder head.

25. Remove the exhaust manifold cover nuts.

26. Remove the air suction tube nuts from the exhaust manifold and air suction valve.

27. Remove the intake manifold assembly from the cylinder head.

28. Remove the water passage assembly from the front and rear of the cylinder head.

29. Remove the timing belt upper covers.

30. Loosen the tensioner adjustment bolt and remove the timing belt.

NOTE: Advance the crankshaft by 15 degrees before removing the timing belt to prevent interference between the piston and the valve.

31. Remove the front and rear camshaft pulleys using the following procedure:

a. Before removing the rear pulley, adjust the cam position so no valve is fully open.

b. Remove the pulley mounting bolts with a universal holder and a double-end wrench. For the rear pulley, first, remove the top 2 bolts and then the remaining bolt.

32. Remove the upper cover back plates.

33. Remove the valve covers and the head side covers.

34. Remove the bearing cap oil pipes, the bearing caps and the camshaft.

35. Remove the intake and exhaust inside rocker arms and pushrods.

NOTE: Label all valve train components to ensure installation in their proper locations.

36. Remove the cylinder head bolts and remove the head.

NOTE: Unscrew the cylinder head bolts ⅓ of a turn in the reverse order of the torque sequence until loose to prevent warpage to the cylinder head.

37. Clean the gasket mounting surfaces.

To install:

38. Use new gaskets and install the exhaust manifold onto the cylinder head and tighten the bolts in a criss-cross pattern in 2–3 steps.

39. Install the cylinder head onto the engine block, after making sure the mating surface was cleaned and a new gasket was installed. Be sure to pay attention to the following points:

a. Be sure the No. 1 cylinder is at TDC and the camshaft pulleys UP mark is on the top before positioning the head in place.

b. The cylinder head dowel pins and oil control jet must be aligned.

c. Torque the cylinder head bolts, in 2 progressive steps: First to 29 ft. lbs. (40 Nm), in sequence, then to 56 ft. lbs. (78 Nm), in the same sequence.

40. Pour engine oil into the cylinder head hydraulic tappet mounting hole, up to the level of the oil path.

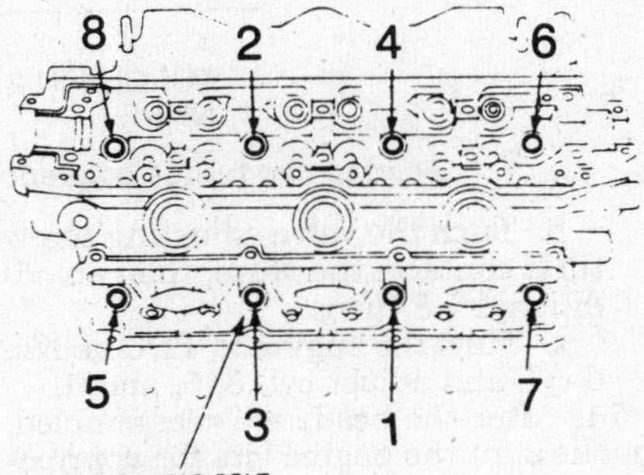

Cylinder head bolt torque sequence—1989–90 Sterling and Legend

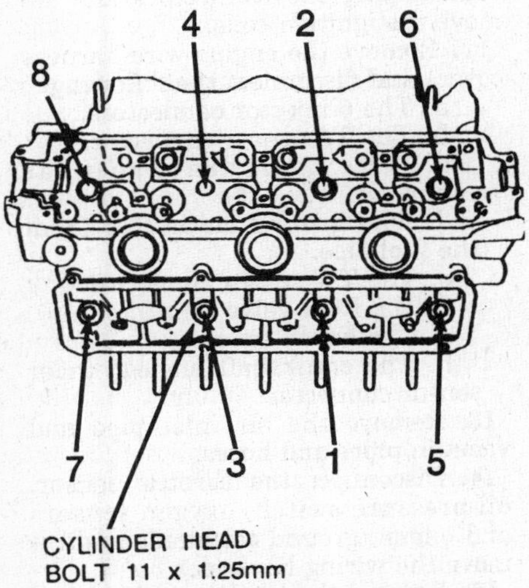

Cylinder head bolt torque sequence—1988 Sterling and Legend

41. Install the hydraulic tappet into the cylinder head. Do not rotate the hydraulic tappet while inserting it into the head.

42. Pour engine oil into the oil fillers on the cylinder head.

43. Install the pushrods and rocker arms. Be sure to install each part in its original position. Loosen the rocker arm adjusting screws and locknuts before installation.

44. Install the camshafts and camshaft oil seals. Be sure to take note of the locations of the camshafts; the front camshaft has a groove for driving the distributor.

a. Make sure the camshaft is mounted parallel with the rocker arm slipper surface.

b. Advance the crankshaft by 15 degrees from the No. 1 cylinder TDC of compression stroke to prevent interference between the piston and valve.

c. Place the rear camshaft on the cylinder head at the position where the cam is not pushing the valve.

d. Preset the oil seal, with its spring side facing inward.

e. Install the rear camshaft sealing rubber. Do not apply oil to the cam holder side of the oil seal.

45. Apply liquid gasket to the camshaft oil seal mounting surface and on the head contact surface. Finger tighten the bearing caps.

46. Carefully, fit the camshaft oil seal until it contacts the bearing cap.

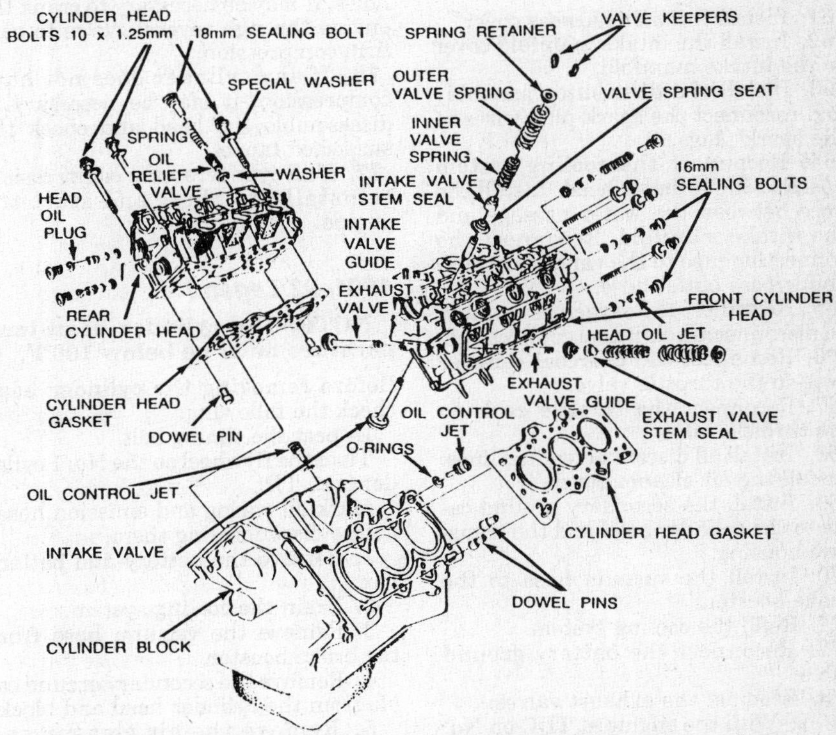

Cylinder heads and related components—1988–90 Sterling and Legend

Torque the bearing cap bolts in 2 steps to 20 ft. lbs. (28 Nm) in the sequence shown. Make sure the oil seal is properly positioned and torque those bolts last to 9 ft. lbs. (12 Nm).

47. Install the upper timing cover plate. Install the camshaft pulley and torque the bolts to 23 ft. lbs. (32 Nm). Install the timing belt.

48. Install the water passage assembly from the front and rear of the cylinder head.

49. Install the intake manifold assembly and torque the bolts to 16 ft. lbs. (22 Nm).

50. Install the air suction tube nuts to the exhaust manifold and air suction valve.

51. Install the exhaust manifold cover nuts.

52. Install the EGR tube nuts to the cylinder head.

53. Install the air cleaner base to the intake manifold.

54. Install the air cleaner base mount bolts and reconnect the hose from the intake manifold to the breather chamber.

55. Install the exhaust header pipe to the exhaust manifold.

56. Install the cruise control actuator.

57. Reconnect the idle boost solenoid hoses.

58. Install the power steering pump and reconnect the pump hoses. Also, install the hose clamp bolt on the body.

59. Install the alternator and belt.

60. Install the alternator pulley cover.

61. Install the wire harness cover.

62. Install the intake manifold cover to the intake manifold.

63. Install the distributor assembly and reconnect the spark plug wires to the spark plugs.

64. Reconnect the cooling system hoses to the cylinder head. Install the hose between the water passage and the intake manifold. Reconnect the connecting pipe to the valve body hose and bypass outlet hose.

65. Reconnect the oxygen sensor and all disconnected electrical connections.

66. Reconnect the charcoal canister hose to the throttle valve.

67. Reconnect the throttle cable to the throttle valve.

68. Install all disconnected fuel lines. Install the air cleaner cover.

69. Install the secondary ground cable to the cylinder head and the transaxle housing.

70. Install the vacuum hose to the brake booster.

71. Refill the cooling system.

72. Reconnect the battery ground cable.

73. Readjust the exhaust valves:

a. With the engine at TDC on No. 1 cyl., adjust cyl. 1, 2, and 4.

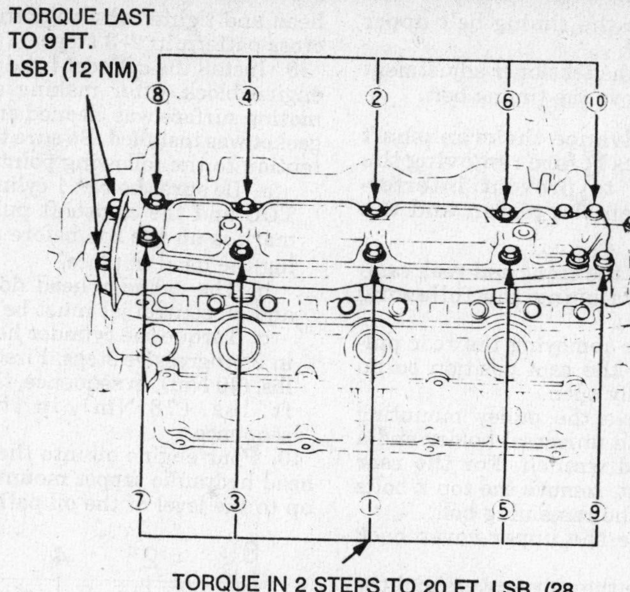

Bearing cap torque sequence—1988–90 Sterling and Legend

b. Turn the valve adjusting screw till it contacts the valve, then an additional 1.5 turns.

c. Turn the engine to TDC on No. 5 cyl. and adjust cyl. 3, 5, and 6.

74. After the heads are reassembled, make sure the engine sits for approximately 5 minutes to allow the hydraulic tappets to reach the proper oil level.

75. Remove the spark plugs and crank the engine, feel for compression at each cylinder at the spark plug holes. It may be necessary to crank the engine through several cycles to confirm compression.

76. If any cylinder does not have compression, it may be necessary to disassemble the head and check the suspected tappet.

77. If all cylinders have compression, reinstall the plugs and start the engine.

1991–92 Legend

NOTE: The cylinder head temperature must be below 100°F.

Before removing the cylinder head check the following:

Inspect the timing belt.

Turn the flywheel so the No. 1 cylinder is at TDC.

Mark all wiring and emission hoses before disconnecting them.

1. Remove the battery and battery box.

2. Drain the cooling system.

3. Remove the vacuum hose from the brake booster.

4. Remove the secondary ground cable from the cylinder head and block.

5. Remove the air cleaner and ducting.

6. Relieve the fuel pressure by loosening the service bolt on the top of the fuel filter about a turn. Disconnect the fuel return hose from the pressure regulator. Remove the special nut and the fuel hose.

CAUTION

The fuel system may be under pressure and fuel will be sprayed. Be sure to take the appropriate safety and fire precautions.

7. Disconnect the throttle cable from the throttle valve.

8. Disconnect the charcoal canister hose from the throttle valve.

9. Disconnect the wiring and remove the main fuse box.

10. Remove the injector resistor and the connector.

11. Unplug the connectors and remove the ignition coils.

12. Remove the engine wire harness covers and disconnect the following:

a. The 6 injector connectors.

b. The TA sensor connector.

c. The temperature unit connector.

d. The ground connector from the fuel pipe.

e. The TW sensor connector.

f. The EGR valve connector.

g. The knock sensor.

h. The crankshaft angle/cylinder sensor connector.

13. Remove the air inlet pipe and vacuum pipes and hoses.

14. Disconnect the throttle sensor, oil pressure switch, oxygen sensors and engine ground terminals and remove the wiring harness.

15. Remove the intake manifold.

16. Remove the upper timing belt covers.

17. Do not remove the timing belt adjuster bolt. Loosen it ½ turn, relieve the belt tension and tighten the bolt.
18. Remove the cam timing belt and the cam pulleys.
19. Remove the timing belt cover plates from the heads and remove the crank angle/cylinder sensor from the left head.
20. Remove the cylinder head covers.
21. Remove the bolts from the alternator and power steering pump brackets as required.
22. Remove the self locking nuts from the exhaust manifolds and slip the manifolds off. If necessary, remove the camshafts.
23. Loosen each head bolt about ⅓ turn in the correct sequence. This is important to prevent warping the heads. Repeat until all bolts are loose and the head can be removed.

To install:

24. It is easier to install the exhaust manifolds and their covers onto the heads before installing the heads to the engine. Use new gaskets and self-locking nuts and torque to 25 ft. lbs. (34 Nm).
25. Install the heads with new gaskets and O-rings, making sure the dowel pins and control orifices are properly positioned. Oil the threads and washers on the head bolts and torque in 3 steps in the sequence shown to 56 ft. lbs. (78 Nm).
26. Apply liquid gasket to the corners of the camshaft holders and install the cylinder head covers.
27. Install the crankshaft/cylinder sensor to the left cylinder head, then install both timing belt cover plates.
28. Install the camshaft pulleys and torque the bolts to 23 ft. lbs. (32 Nm). There is a left and right pulley; the left one goes with the crankshaft/cylinder sensor.
29. Align the timing marks on the crankshaft and camshaft pulleys, install the timing belt and adjust the belt tension. Install the timing belt covers.
30. Install the intake manifold with new gaskets. Torque the 6mm bolts to 9 ft. lbs. (12 Nm) and the 8mm bolts to 16 ft. lbs. (22 Nm). Connect the cooling system hoses.
31. Install the air suction, EGR, vacuum and air inlet pipes. Reconnect the vacuum and fuel system hoses, using new gaskets on the fuel supply hose.
32. Connect the sensor wiring and install the wiring harness for the injection and ignition systems.
33. Install the fuse box and connect the wiring.
34. Connect the throttle cable and refill the cooling system.
35. Make sure all other wires and hoses are properly connected and install the battery. Before starting the engine, turn the ignition switch **ON** and **OFF** a number of times to pressurize the fuel system. Check for leaks.
36. After starting the engine, bleed the cooling system.

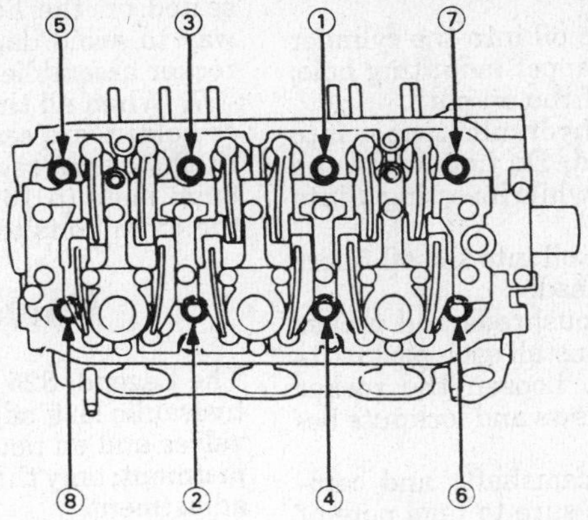

Cylinder head torque sequence—1991–92 Legend

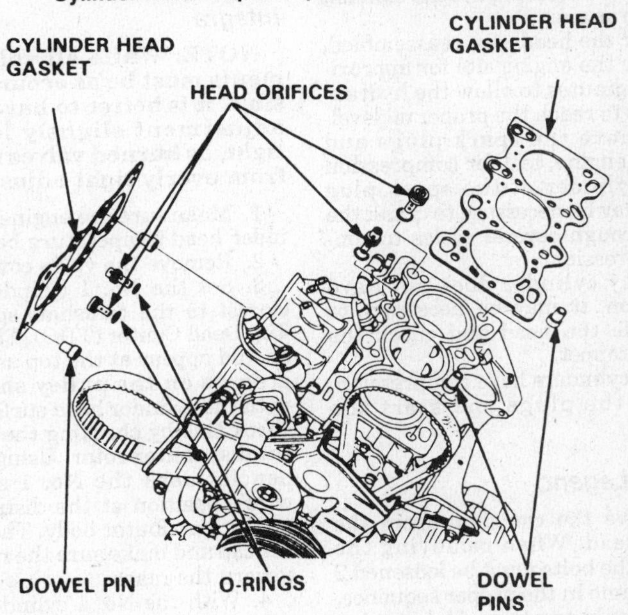

Control orifices and dowel pins—1991–92 Legend

Valve Lifters

REMOVAL & INSTALLATION

Sterling and 1988–90 Legend

1. Remove the camshafts from the cylinder heads.
2. With the camshafts removed, remove the rocker arms and the pushrods.
3. Use a suitable valve lifter removal tool and remove the hydraulic tappet (lifter) from the cylinder head hydraulic tappet mounting hole.
4. Use the following steps to inspect the hydraulic tappet (lifter).

a. Inspect the hydraulic tappet for wear or damage or for a clogged oil hole.

b. Measure the free length of each hydraulic tappet by attaching the hydraulic tappet bleeder to the tappet. Then push and release the bleeder slowly while in a container filled with 10W–30W engine oil. Be sure to keep the hydraulic tappet upright and below the surface of the oil while pushing and release the bleeder.

c. Continue operating the bleeder until there are no air bubbles left in the hydraulic tappet.

d. Remove the hydraulic tappet and try to compress it quickly by hand. Measure the compression stroke with a dial indicator on a surface plate. The standard compression stroke measurement should be 0.0004–0.003 in. (0.01–0.08mm).

To install:

5. Pour engine oil into the cylinder head hydraulic tappet mounting hole, up to the level of the oil path.
6. Install the hydraulic tappet into the cylinder head. Do not rotate the hydraulic tappet while inserting it into the head.
7. Pour engine oil into the oil fillers on the cylinder head.
8. Install the pushrods and rocker arms. Be sure to install each part in its original position. Loosen the rocker arm adjusting screws and locknuts before installation.
9. Install the camshafts and camshaft oil seals. Be sure to take note of the locations of the camshafts; the front camshaft has a groove for driving the distributor. Adjust the exhaust valve clearanace.
10. After the heads are reassembled, make sure the engine sits for approximately 5 minutes to allow the hydraulic tappets to reach the proper oil level.
11. Remove the spark plugs and crank the engine, feel for compression at each cylinder at the spark plug holes. It may be necessary to crank the engine through several cycles to confirm compression.
12. If any cylinder does not have compression, it may be necessary to disassemble the head and check the suspected tappet.
13. If all cylinders have compression, reinstall the plugs and start the engine.

1991–92 Legend

1. Remove the camshaft from the cylinder head. When removing the camshaft, the bolts must be loosened 2 turns at a time in the proper sequence. This is the only way to avoid damaging the valves or rocker assemblies.
2. Each rocker arm has a letter **A** or **B** stamped into the side. Before disassembling the rocker arms, note the position of each letter so they can be reassembled the same way.
3. Do not remove the hydraulic tappets unless they are to be replaced, they cannot be repaired or tested. Handle the rocker arms carefully so the oil does not drain out of the tappets. If replacing the tappets, also replace the O-ring on the tappet.

To install:

4. Place a new camshaft seal on the end of the camshaft, lubricate the journals and set the camshaft in place on the head.
5. Apply liquid gasket to the mating surfaces of the end camshaft holders.
6. Set the rocker arm assemblies in place and start all the bolts. Make sure the rocker arms are properly positioned and turn each bolt in sequence 2 turns at a time till the holders are seated on the head. This is the only way to avoid damaging the valves or rocker assemblies.
7. When all the camshaft and rocker holders are seated, torque the bolts in the same sequence. Torque the 8mm bolts to 16 ft. lbs. (22 Nm) and the 6mm bolts to 9 ft. lbs. (12 Nm).

Valve Lash

The Legend, 825 and 827 engines use hydraulic lash adjusters on the intake valves and do not require periodic adjustment; only the exhaust valves need adjustment.

ADJUSTMENT

Integra

NOTE: While all valve adjustments must be as accurate as possible, it is better to have the valve adjustment slightly loose than tight, as burned valves may result from overly tight adjustments.

1. Make sure the engine is cold; cylinder head temperature below 100°F.
2. Remove the valve cover.
3. Set the No. 1 cylinder, cylinder closest to the camshaft sprockets, to Top Dead Center (TDC). The word UP should appear at the top and the TDC grooves on the pulley should align with the cylinder head surface. Double check this by checking the position of the distributor rotor. Using chalk or a pencil, mark the No. 1 spark plug wire's position at the distributor cap on the distributor body. Then, remove the cap and make sure the rotor points toward the mark.
4. With the No. 1 cylinder at TDC, adjust the valves of the No. 1 cylinder by performing the following procedures:
   a. On the 1988–89 model, clearance should be 0.0051–0.0067 in. (0.13–0.17mm) on the intake valves and 0.0059–0.0075 in. (0.15–0.19mm) on the exhaust valves.
   b. On the 1990–92 model, clearance should be 0.006–0.007 in. (0.15–0.19mm) on the intake valves and 0.007–0.008 in. (0.17–0.21mm) on the exhaust valves.
   c. Place the feeler gauge between the rocker arm and the camshaft lobe. There should be a slight drag on the feeler gauge.
   d. If there is no drag or if the gauge cannot be inserted, loosen the valve adjusting the screw locknut.
   e. Turn the adjusting screw with a suitable tool to obtain the proper clearance.
   f. Hold the adjusting screw and torque the locknut(s) to 18 ft. lbs.
   g. Recheck the clearance.
5. Turn the crankshaft 180 degrees counterclockwise, the cam pulley will turn 90 degrees. With the No. 3 cylinder at TDC, the distributor rotor should be pointing to the No. 3 plug wire and the UP marks should be at the exhaust side, adjust the valves on the No. 3 cylinder.
6. Turn the crankshaft 180 degrees counterclockwise, the cam pulley will turn 90 degrees. With the No. 4 cylinder at TDC, both UP marks should be at the bottom and the distributor rotor pointing to the No. 4 plug wire, adjust the valves on the No. 4 cylinder.
7. Turn the crankshaft 180 degrees counterclockwise. The No. 2 cylinder will now be on TDC, this can be confirmed by the distributor rotor pointing to the No. 2 plug wire and the UP marks should be at the intake side. The valves on the No. 2 cylinder may now be adjusted.

Sterling and 1988–90 Legend

This procedure is used to adjust the exhaust valves; the intake valves require no adjustment.

1. Make sure the engine is cold, cylinder head temperature below 100°F.
2. Remove the valve and side head covers.
3. Rotate the crankshaft to position the No. 1 piston on the TDC of it's

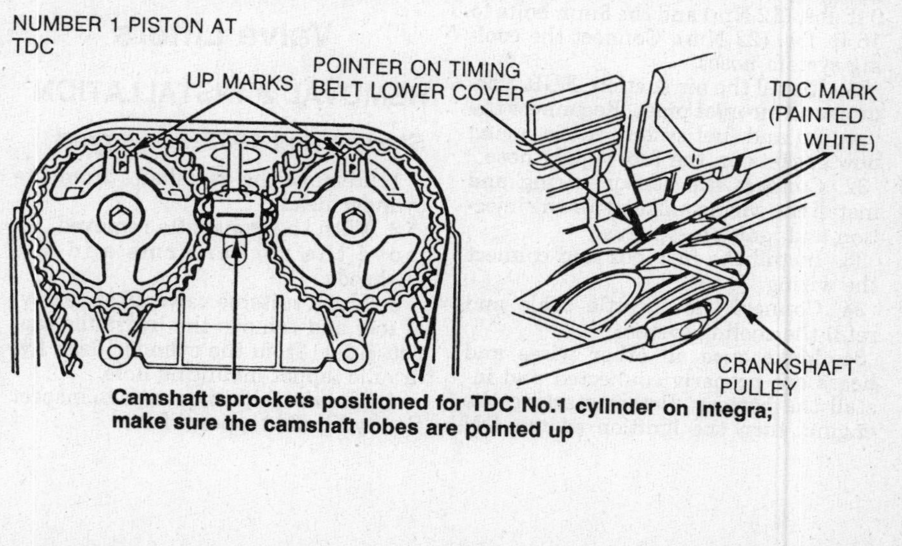

Camshaft sprockets positioned for TDC No.1 cylinder on Integra; make sure the camshaft lobes are pointed up

compression stroke; the camshaft sprockets should be in the upward position, aligned with the timing mark and the crankshaft pulley V-notch should be aligned with the timing pointer on the timing cover.

NOTE: Double check this by checking the position of the distributor rotor. Using chalk or a pencil, mark the No. 1 spark plug wire's position at the distributor cap on the distributor body. Then, remove the cap and make sure the rotor points toward the mark.

4. To adjust the exhaust valves, perform the following procedures:
   a. Loosen the exhaust valve locknuts on all of the cylinders.
   b. Tighten the adjusting screw of the No. 1 cylinder, until it contacts the valve and tighten it 1½ turns. Tighten the locknut firmly.
   c. Perform the same procedure for the exhaust valves No. 2 and 4.
   d. Rotate the crankshaft 180 degrees and align the crankshaft pulley's V-notch with the timing pointer on the timing cover; the No. 5 piston is at TDC of it's compression stroke.
   e. Tighten the adjusting screw of the No. 5 cylinder, until it contacts the valve and tighten it 1½ turns. Tighten the locknut firmly.
   f. Perform the same procedure for the exhaust valves No. 3 and 6.
5. After adjustment, use new gaskets and install the valve and side head covers.

Rocker Arms/Shafts

REMOVAL & INSTALLATION

The rocker arms ride directly against the camshafts and are serviced with the camshaft. The valves are designed to be adjusted after the camshafts are installed.

Intake Manifold

REMOVAL & INSTALLATION

Integra

1. Disconnect the negative battery cable. Drain the cooling system.
2. Remove the air duct from the throttle body.
3. Remove the intake manifold bracket and the EVAC.
4. Relieve the fuel system pressure by loosening the service bolt on the fuel filter about 1 turn, then disconnect the fuel supply and return lines from the manifold.

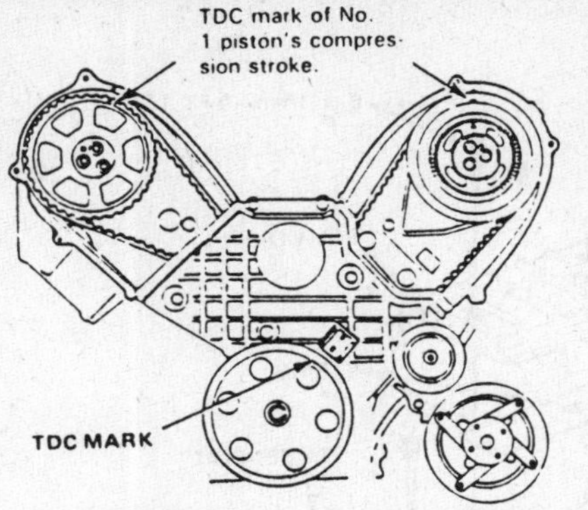

Camshaft sprockets and crankshaft pulley positioned for TDC No.1 – Sterling and 1988–90 Legend

CAUTION

The fuel system may be under pressure and fuel will be sprayed. Be sure to take the appropriate safety and fire precautions.

5. Label and remove any electrical connectors running to the intake manifold.
6. Remove the intake manifold-to-cylinder head nuts, in a criss-cross pattern, beginning from the center and moving out to both ends. Remove the manifold and the gasket.
7. Clean the gasket mounting surfaces. Inspect the manifold for cracks, flatness and/or damage; replace the parts, if necessary.

NOTE: If the intake manifold is to be replaced, transfer all the necessary components to the new manifold.

To install:

8. Use new gaskets and reverse the removal procedures. Torque the nuts/bolts, in a criss-cross pattern, in 2–3 steps, starting with the inner nuts, to 16–17 ft. lbs.
8. Start the engine, allow it to reach normal operating temperatures and check for leaks and engine operation.

Sterling and Legend

1. Disconnect the negative battery cable. Drain the cooling system.
2. Remove the air duct from the throttle body.
3. Remove the EVAC, the air suction valve and the EGR tube.
4. Label and remove any wires running to the intake manifold.
5. Relieve the fuel system pressure by loosening the service bolt on the fuel filter about 1 turn, then disconnect the fuel supply and return lines from the manifold.

CAUTION

The fuel system may be under pressure and fuel will be sprayed. Be sure to take the appropriate safety and fire precautions.

6. Remove the intake manifold attaching nut in a criss-cross pattern, beginning from the center and moving out to both ends and the manifold.
7. Clean the gasket mounting surfaces. Inspect the manifold for cracks, flatness and/or damage; replace the parts, if necessary.

NOTE. If the intake manifold is to be replaced, transfer all the necessary components to the new manifold.

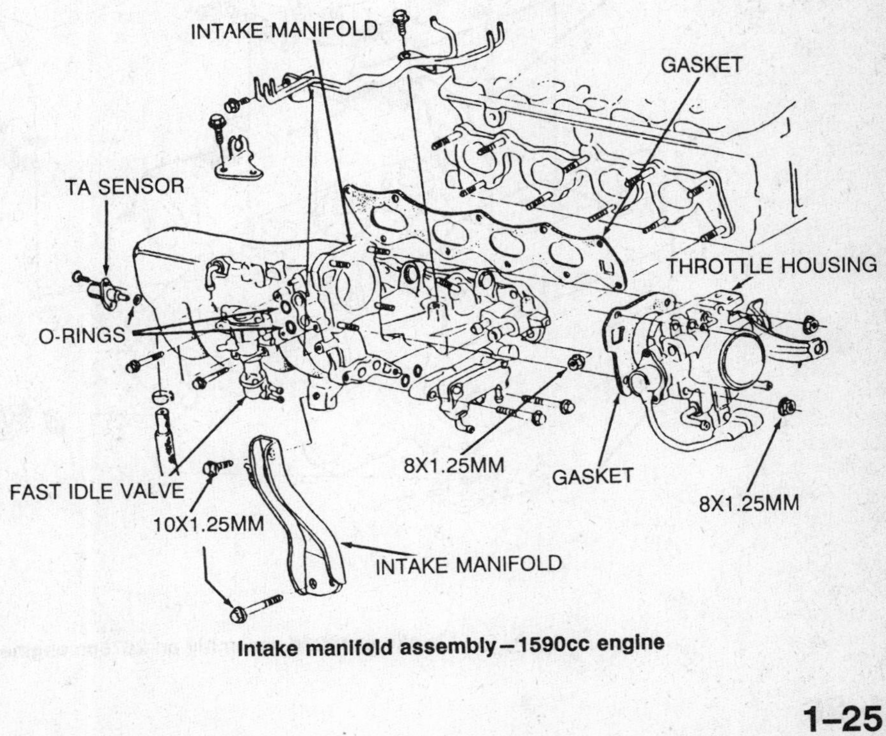

Intake manifold assembly – 1590cc engine

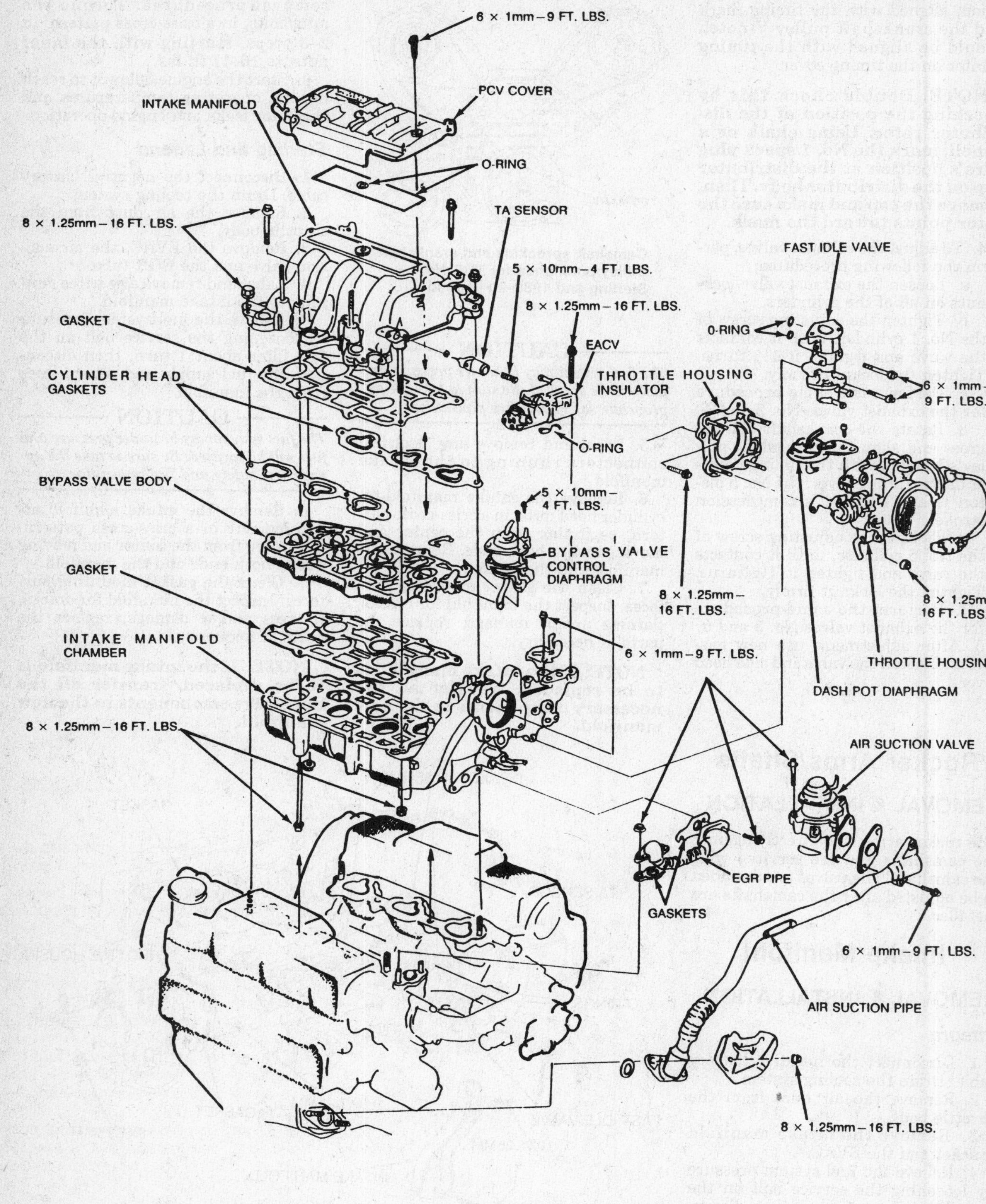

Intake manifold assembly on 2675cc engine

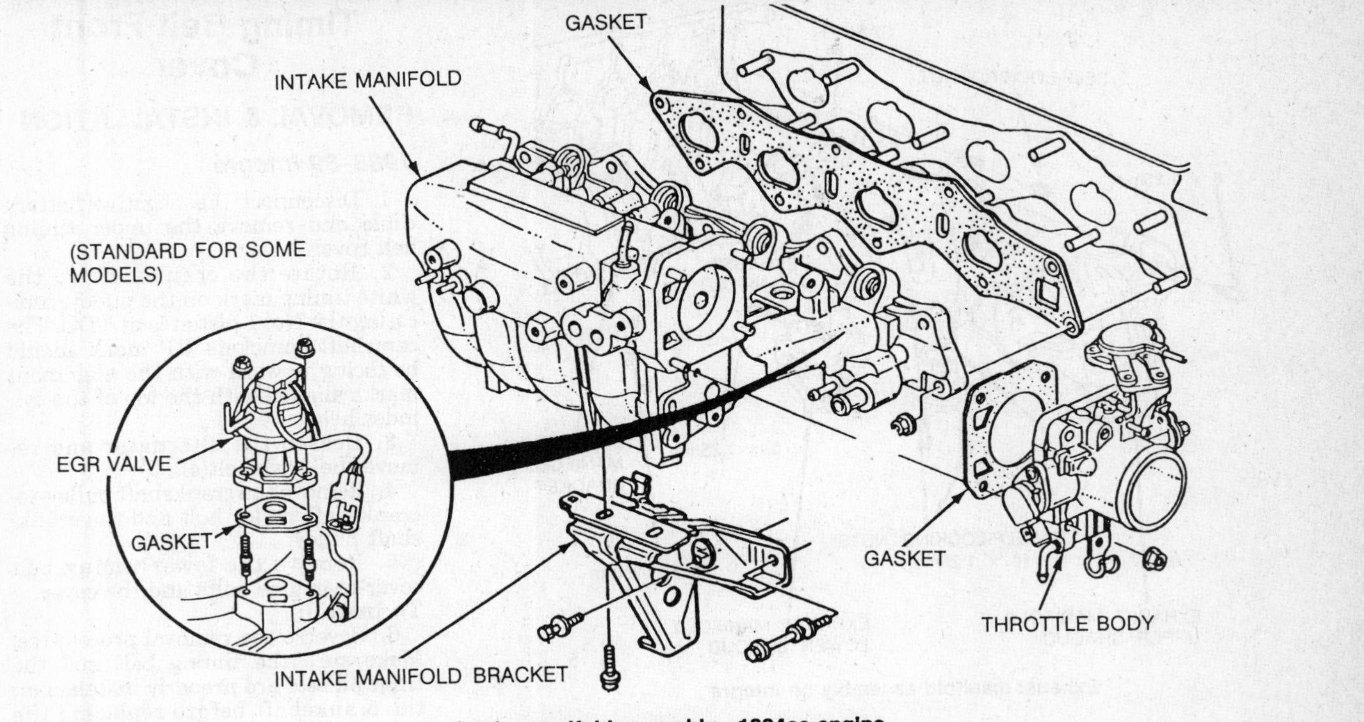

Intake manifold assembly—1834cc engine

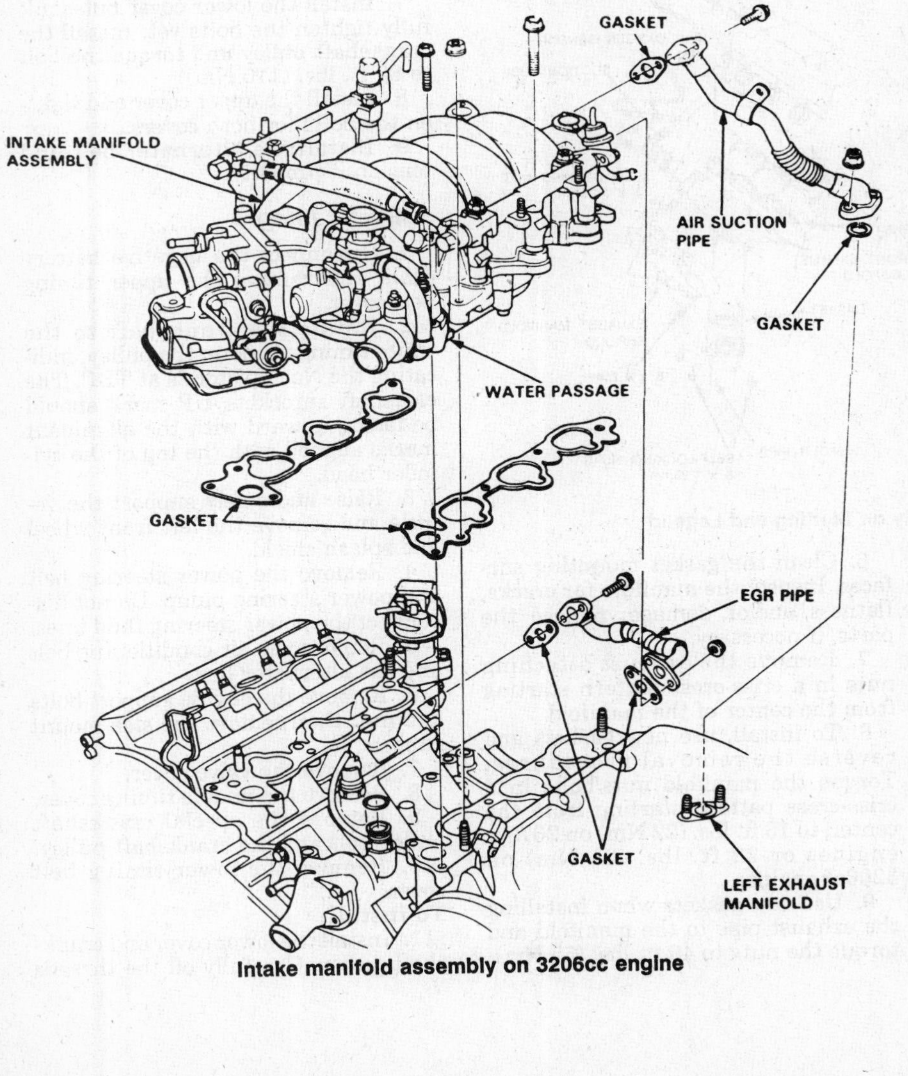

Intake manifold assembly on 3206cc engine

To install:

8. Use new gaskets and reverse the removal procedures. Torque the nuts/bolts, in a criss-cross pattern in 2–3 steps, starting with the inner nuts. Torque the 8mm bolts to 16 ft. lbs. (22 Nm) and the 6mm bolts to 9 ft. lbs. (12 Nm).

9. Start the engine, allow it to reach normal operating temperatures and check for leaks and engine operation.

Exhaust Manifold

REMOVAL & INSTALLATION

Integra

NOTE: Do not perform this operation on a warm or hot engine.

1. Disconnect the negative battery cable. Remove the exhaust manifold shroud.

2. Remove the exhaust pipe-to-exhaust manifold nuts.

3. Remove the oxygen sensor, if equipped.

4. Remove the exhaust manifold bracket bolt.

5. Remove the exhaust manifold-to-cylinder head nuts in a criss-cross pattern starting from the center and the manifold.

6. Clean the gasket mounting surfaces. Inspect the manifold for cracks, flatness and/or damage; replace the parts, if necessary.

To install:

7. Use new gaskets and reverse the removal procedures. Torque the mani-

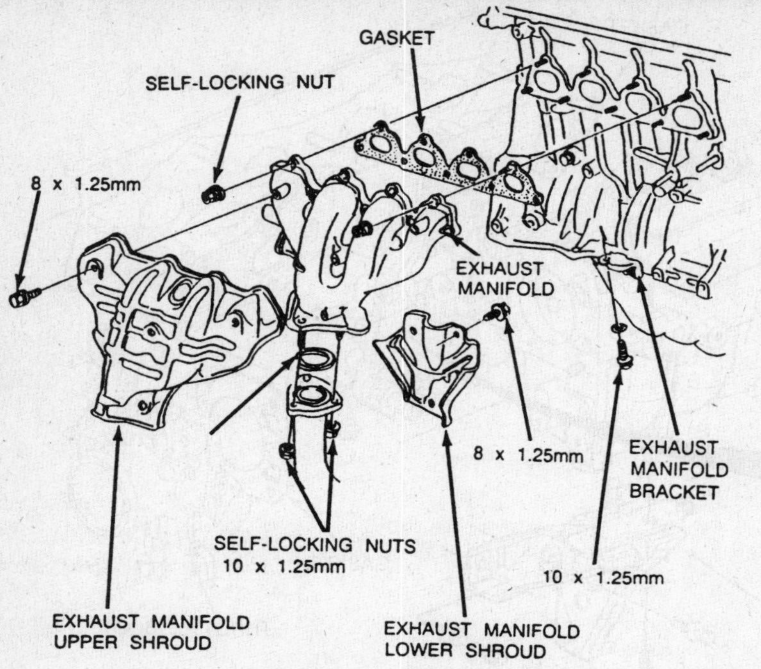

Exhaust manifold assembly on Integra

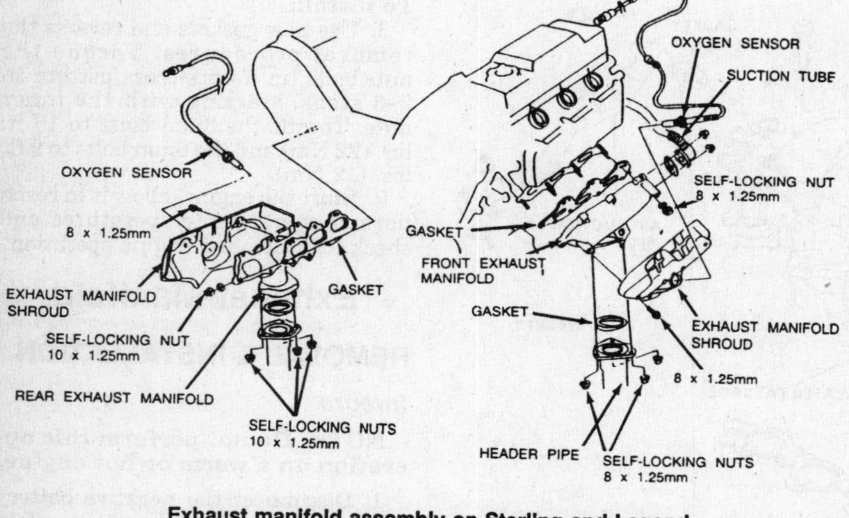

Exhaust manifold assembly on Sterling and Legend

fold nuts in a criss-cross pattern starting from the center, to 23 ft. lbs. (32 Nm) and the exhaust pipe-to-manifold nuts to 40 ft. lbs. (55 Nm).

8. Start the engine and check for leaks.

Sterling, Legend and NSX

1. Disconnect the negative battery cable. Remove the exhaust manifold shrouds.
2. Remove the exhaust pipe-to-exhaust manifold nuts.
3. Remove the oxygen sensors.
4. Remove the air suction tube.
5. Remove the exhaust pipe-to-exhaust manifold nuts.
6. Clean the gasket mounting surfaces. Inspect the manifold for cracks, flatness and/or damage; replace the parts, if necessary.
7. Remove the exhaust attaching nuts in a criss-cross pattern starting from the center of the manifold.
8. To install, use new gaskets and reverse the removal procedures. Torque the manifold nuts/bolts in a criss-cross pattern starting from the center, to 16 ft. lbs. (22 Nm) on 2675cc engines or 22 ft. lbs. (31 Nm) on 3206cc engines.
9. Use new gaskets when installing the exhaust pipe to the manifold and torque the nuts to 40 ft. lbs. (55 Nm).

Timing Belt Front Cover

REMOVAL & INSTALLATION

1988–89 Integra

1. Disconnect the negative battery cable and remove the upper timing belt cover.
2. Rotate the crankshaft to the white timing mark on the pulley, indicating the No. 1 piston is at TDC. The camshaft sprockets UP mark should be facing upward with the alignment marks aligned with the top of the cylinder head.
3. Loosen the alternator and remove the drive belt(s).
4. Remove the crankshaft pulley-to-crankshaft pulley bolt and the crankshaft pulley.
5. Remove the lower timing belt cover-to-engine bolts and the cover.

To install:

6. Reverse the removal procedures. Make sure the timing belt and the front oil seal are properly installed on the crankshaft before replacing the cover.
7. Install the lower cover but don't fully tighten the bolts yet. Install the crankshaft pulley and torque the bolt to 83 ft. lbs. (115 Nm).
8. Install the upper cover and tighten the bolts for both covers.
9. Install the alternator belt and tension it properly.

1990–92 Integra

1. Disconnect the negative battery cable and remove the upper timing belt cover.
2. Rotate the crankshaft to the white timing mark on the pulley, indicating the No. 1 piston is at TDC. The camshaft sprockets UP mark should be facing upward with the alignment marks aligned with the top of the cylinder head.
3. Raise and safely support the vehicle and remove the left front wheel and splash shield.
4. Remove the power steering belt and power steering pump. Do not disconnect the power steering fluid lines.
5. Remove the air conditioning belt and the alternator belt.
6. Remove the engine support bolts and nut, then remove the side mount rubber.
7. Remove the valve cover.
8. Remove the middle timing cover.
9. Remove the special crankshaft pulley bolt and the crankshaft pulley.
10. Remove the lower timing belt cover.

To install:

11. Install the lower cover and crankshaft pulley. Carefully oil the threads

of the pulley bolt without getting oil on the washer.

12. Torque the bolt in 3 steps: to 145 ft. lbs. (200 Nm), then loosen it again, then re-torque to 130 ft. lbs. (180 Nm).

13. The remaining installation is the reverse order of removal.

Sterling and 1988–90 Legend

1. Disconnect the negative battery cable. Rotate the crankshaft to align the crankshaft pulley or flywheel pointer, at Top Dead Center (TDC); the camshaft sprocket notches should align with the marks on the rear timing belt cover.
2. Remove the pulley cover and harness cover from above the timing belt upper cover.
3. Remove the engine sub harness clip.
4. Remove the engine support bolts, loosen the side mount rubber, and raise the side mount bracket. Be sure to use a chain hoist to support the engine.
5. Remove the lower splash guard from under the front of the vehicle.
6. Loosen the air conditioning idler pulley adjusting bolt and remove the air conditioning compressor belt.
7. Remove the alternator adjusting bolt, mounting bolt and remove the alternator with the belt.
8. Remove the power steering pump adjusting bolt, mounting bolt and remove the power steering pump with the belt.
9. Remove the front and rear upper covers.
10. Remove the special crankshaft pulley bolt and remove the crankshaft pulley.
11. Remove the lower timing belt cover.

To install:

12. Installation is the reverse of removal. Make sure the timing belt and oil seal are properly installed before replacing the lower cover.
13. Install the lower belt cover but don't tighten the bolts yet.
14. Install the crankshaft pulley and torque the bolt to 123 ft. lbs. (170 Nm).
15. Install the remaining timing belt covers and tighten all bolts.
16. Install the power steering pump and alternator and properly tension the belts.
17. Install and tension the air conditioner belt.
18. Set the engine in place and install the engine side mount bracket bolts.
19. Secure the wiring in place and install the lower splash shield.

1991–92 Legend

1. Disconnect the negative battery cable.
2. Remove the engine wire harness covers at the front of the engine.
3. Label and disconnect the wiring harness.
4. Remove the breather pipe and vacuum pipe bracket.
5. Remove the alternator, air conditioning and power steering belts.
6. Remove the timing belt upper covers and turn the crankshaft pulley to the TDC No.1 cylinder mark.
7. Remove the crankshaft pulley.
8. Remove the air conditiner belt tensioner pulley.
9. Remove the dip stick tube and the lower timing belt cover.

To install:

10. Make sure the rubber seals and cover are clean and dry and install the lower cover.
11. Carefully oil the threads of the pulley bolt without getting oil on the washer. Install the crankshaft pulley and torque the bolt to 174 ft. lbs. (240 Nm).
12. Use a new O-ring to install the dip stick tube.
13. Install and adjust the drive belts for the air conditioner, power steering and alternator.
14. Install the upper timing belt covers, breather pipe and the wiring harness.

Front Cover Oil Seal

REPLACEMENT

If the proper seal driving tools are available, the front oil seal replacement can be done without removing the oil pump.

1. With camshaft drive belt removed, use a suitable seal removal tool and remove the front oil seal from the oil pump.
2. Apply a light coat of oil to the crankshaft and to the lip of the seal. Make sure the seal contact surface on the oil pump is clean and dry.
3. Using an oil seal driver tool deep enough to fit over the crankshaft, drive in the new seal into the pump until it just bottoms out.

Timing Belt And Tensioner

ADJUSTMENT

Integra

NOTE: Always adjust the timing belt tension with the engine cold. The tensioner is spring-loaded to apply the proper tension to the belt automatically after making the following adjustments.

Timing belt tension adjustment on Integra

1. Turn the crankshaft pulley until the No. 1 piston is at TDC of the compression stroke.
2. Loosen the adjusting bolt on the tensioner pulley.
3. Rotate the crankshaft counterclockwise 3 teeth on the camshaft pulley to create tension on the timing belt.
4. Torque the adjusting bolt on the tensioner pulley to 33 ft. lbs. (45 Nm).
5. If the crankshaft pulley broke loose while turning the crank, torque it as follows:
   a. 1988–83 ft. lbs. (115 Nm)
   b. 1989–119 ft. lbs. (161 Nm)
   c. 1990–92–145–0–130 ft. lbs. (200–0–180 Nm)

NOTE: Place the transaxle in gear and set the parking brake before torquing the crankshaft pulley bolt.

Sterling and Legend

NOTE: Always adjust the timing belt tension with the engine cold. The tensioner is spring-loaded to apply the proper tension to the belt automatically after making the following adjustments.

1. Turn the crankshaft pulley until No. 1 is at TDC of the compression stroke.
2. Rotate the crankshaft clockwise, as viewed from the pulley side of the engine, 9 teeth on the camshaft pulley. The blue mark on the camshaft pulley should match the pointer on the lower cover.
3. Loosen the adjusting bolt about ½ turn.
4. Torque the adjusting bolt to 31 ft. lbs. (43 Nm).

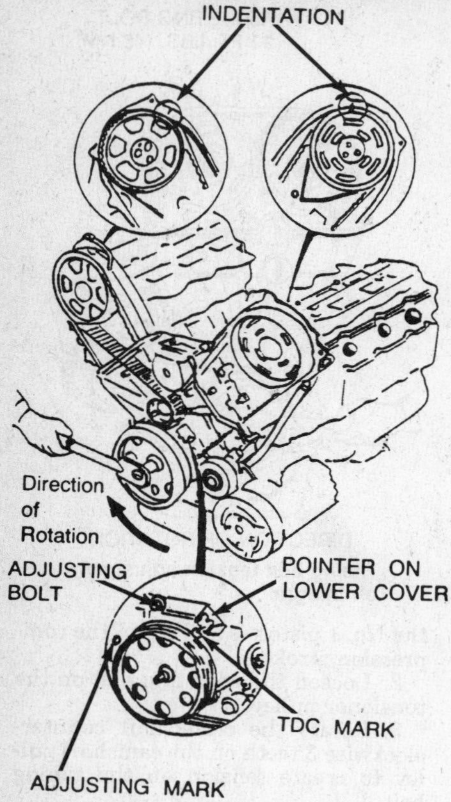

Timing belt tension adjustment on Sterling and Legend

NSX

NOTE: Always adjust the timing belt tension with the engine cold. The tensioner is spring-loaded to apply the proper tension to the belt automatically after making the following adjustments.

1. Turn the crankshaft pulley until No. 1 is at TDC of the compression stroke.
2. Without allowing the crankshaft to turn, turn the exhaust camshaft on the front head counterclockwise just enough to take up the slack in the belt. Repeat this process with the intake camshaft, then the intake and exhaust camshafts on the rear head. All the slack in the belt should end up between the rear exhaust camshaft and the crankshaft.
3. Loosen the adjusting bolt about ½ turn. The slack should be eliminated automatically.
4. Torque the adjusting bolt to 31 ft. lbs. (43 Nm).
5. Rotate the crankshaft clockwise, as viewed from the pulley side of the engine, 9 teeth on the camshaft pulley. The blue mark on the camshaft pulley should match the pointer on the lower cover.
6. Loosen the adjusting bolt about ½ turn. The tension is now correct.
7. Torque the adjusting bolt to 31 ft. lbs. (43 Nm).

REMOVAL & INSTALLATION

1988–89 Integra

1. Rotate the crankshaft pulley until No. 1 is at TDC of the compression stroke.
2. Remove the alternator belt, the power steering belt, and the air conditioning belt, if equipped, crankshaft pulley and camshaft belt cover. Mark the direction of the belt rotation.
3. Loosen the tensioner adjusting bolt ½ turn, push the tensioner pulley to release belt tension, then tighten the bolt again.
4. Slide the timing belt off the camshaft sprockets, crankshaft sprocket and the water pump sprocket; remove it from the engine.
5. Inspect the timing belt; replace it if it is worn. If it is oil soaked, replace it and find the source of the oil leak. Check the condition of the water pump.

To install:

6. Make sure the camshaft and crankshaft are porperly positioned, install the new belt and adjust the tension.
7. Temporarily install the crankshaft pulley and bolt and turn the engine 2 full revolutions. Check that the crankshaft pulley and camshaft sprocket marks are still properly aligned. If not, adjust as necessary.
8. When the timing belt is properly installed and adjusted, remove the crankshaft pulley, instll the timing belt cover and install the pulley. Torque the bolt to 83 ft. lbs (115 Nm) on 1988 models, or to 119 ft. lbs. (161 Nm) on 1989 models.

1990–92 Integra

1. Raise and safely support the vehicle. Remove the left front wheel and remove the wheel well splash guard.
2. Remove the power steering belt and power steering pump. Do not disconnect the power steering fluid lines.
3. Remove the air conditioning belt and the alternator belt.

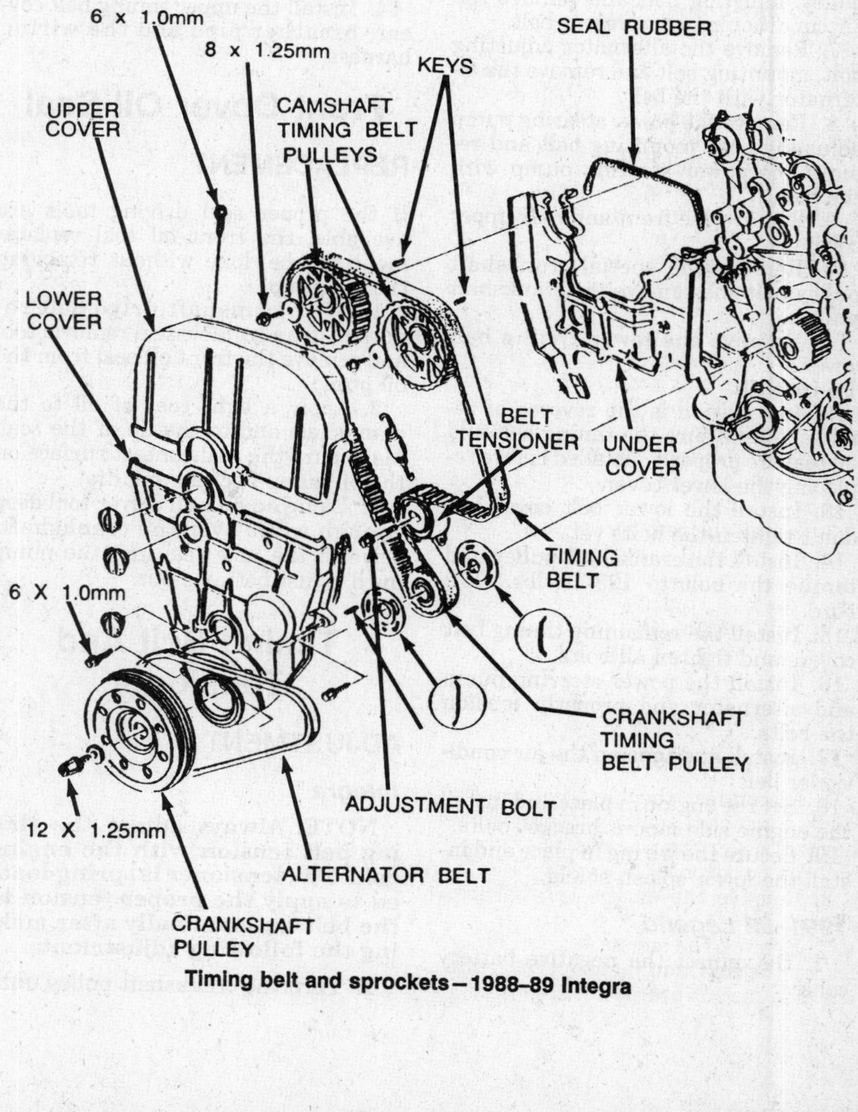

Timing belt and sprockets—1988–89 Integra

4. Remove the left side engine mount.
5. Remove the valve cover.
6. Remove the middle timing cover.
7. Set the crankshaft to TDC of the No. 1 piston: the white crankshaft pulley mark should be aligned with the pointer on the cover and the camshaft sprockets UP mark should be facing upward, sprocket timing marks aligned.
8. Remove the special crankshaft pulley bolt and the crankshaft pulley.
9. Remove the lower timing belt cover.
10. Loosen but do not remove the tensioner adjusting bolt, push the tensioner to slacken the timing belt, then retighten the bolt. If the timing belt is to be reinstalled, mark the direction of rotation. Remove the timing belt.

NOTE: Install the timing belt with the No. 1 piston at TDC on its compression stroke. To set the camshafts to the top dead center position for No. 1 cylinder, align the hole in the camshafts with the holes in the No. 1 camshaft holders and drive a 5.0mm pin punches into the holes.

To install:

11. Make sure the timing belt is properly installed on the crankshaft and that the front oil seal does not leak. Replace the cover and crankshaft pulley and carefully oil the threads of the pulley bolt without getting oil on the washer. Torque the bolt in 3 steps: 145 ft. lbs. (200 Nm), then loosen the bolt completely, then re-torque to 130 ft. lbs. (180 Nm).
12. Make sure the camshaft sprockets and crankshaft pulley are properly aligned with the timing marks. Loosen the tensioner bolt about ½ turn, then torque to 40 ft. lbs. (55 Nm). Tension adjustment is automatically accomplished with the spring on the tensioner. Check timing mark alignment again.
13. Install the remaining parts in reverse order. Torque the engine mount-to-engine bolt and nut to 54 ft. lbs. (75 Nm), torque the mount-to-body bolt to 40 ft. lbs. (55 Nm).

Sterling and 1988–90 Legend

1. Disconnect the negative battery cable. Remove the pulley cover and the harness cover from above the timing belt upper cover.
2. Remove the engine sub-harness clamp.
3. Remove the engine support bolts, loosen the side mount rubber and raise the side mount bracket.

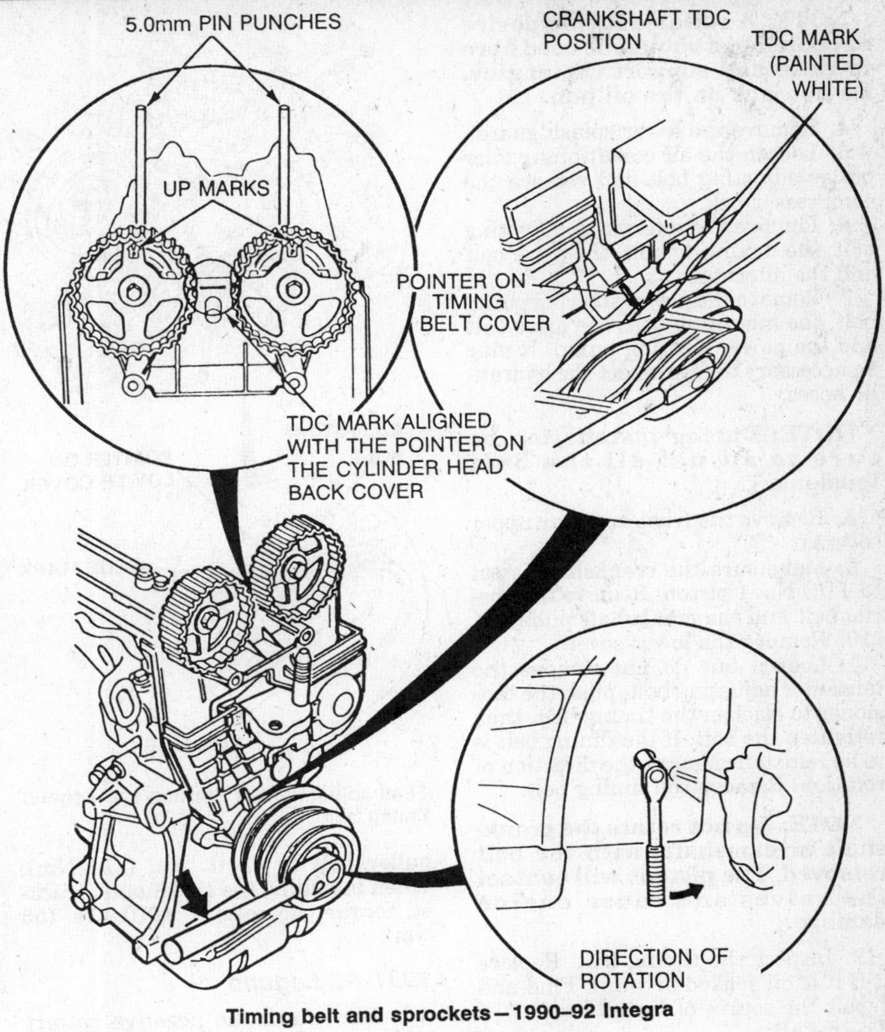

Timing belt and sprockets – 1990–92 Integra

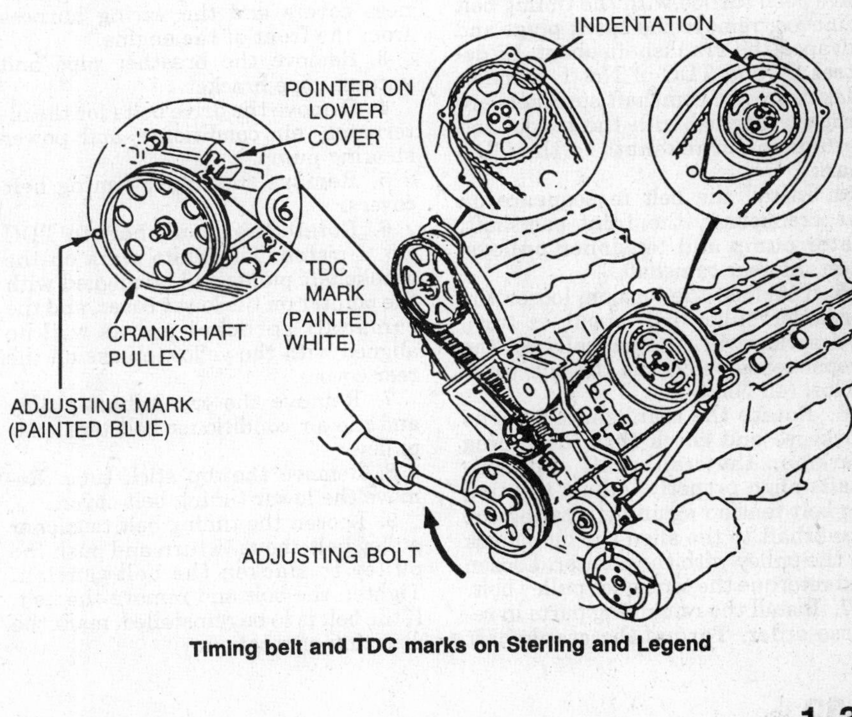

Timing belt and TDC marks on Sterling and Legend

NOTE: A suitable lifting device or chain hoist should be used here to raise and support the engine. Do not jack on the oil pan.

4. Remove the lower splash guard.
5. Loosen the air conditioning idler pulley adjusting bolt and remove the compressor belt.
6. Remove the alternator adjusting bolt, the mounting bolt, the drive belt and the alternator.
7. Remove the power steering pump bolt, the mounting bolt, the drive belt and the power steering pump. It may be necessary to disconnect the hydraulic hoses.

NOTE: During installation be sure to adjust all the belt tensions.

8. Remove the front and rear upper covers.
9. Make sure the crankshaft is set to TDC No. 1 piston. Remove the special bolt and the crankshaft pulley.
10. Remove the lower cover.
11. Loosen but do not remove the tensioner adjusting bolt, push the tensioner to slacken the timing belt, then retighten the bolt. If the timing belt is to be reinstalled, mark the direction of rotation. Remove the timing belt.

NOTE: Do not rotate the crankshaft or camshafts with the belt removed. The pistons will contact the valves and cause engine damage.

12. Inspect the timing belt. Replace it if it is oil soaked or worn. Find and repair the source of the oil leak.

To install:

13. If the crankshaft or camshafts have been turned with the timing belt removed, remove the spark plugs and advance the crankshaft about 15 degrees beyond TDC of No. 1 cylinder. Make sure the camshaft sprockets are properly aligned with the marks and return the crankshaft to the TDC mark.
14. Install the belt in sequence on the crankshaft, the front camshaft, water pump and tensioner pulleys, then the rear camshaft.
15. To adjust the tension, loosen the tensioner pulley bolt about ½ turn. The spring will automatically set the proper tension. Torque the bolt to 31 ft. lbs. (43 Nm).
16. Rotate the crankshaft 6 turns clockwise and check that the timing marks on the crankshaft and camshafts align properly. Adjust the timing belt tension again by rotating the crankshaft to the align the blue mark on the pulley with the pointer. Loosen and retorque the tensioner pulley bolt.
17. Install the remaining parts in reverse order. Torque the crankshaft pulley bolt to 83 ft. lbs. (115 Nm). When installing the side mount bracket, torque the bolts to 40 ft. lbs. (55 Nm).

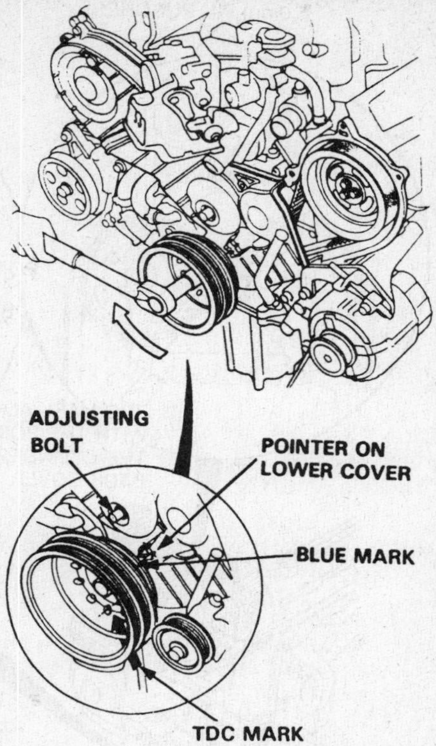

Final adjustment of Sterling and Legend timing belt

1991–92 Legend

1. Disconnect the negative battery cable.
2. Remove the engine wiring harness covers and the wiring harness from the front of the engine.
3. Remove the breather pipe and vacuum pipe bracket.
4. Remove the drive belts for the alternator, air conditioner and power steering pump.
5. Remove the upper timing belt covers.
6. Rotate the crankshaft to TDC No. 1 piston. The white mark on the crankshaft pulley will be aligned with the pointer on the lower cover, and the camshaft sprocket marks will be aligned with the yellow marks on the rear covers.
7. Remove the crankshaft pulley and the air conditioner belt tensioner pulley.
8. Remove the dip stick tube. Remove the lower timing belt cover.
9. Loosen the timing belt tensioner pulley bolt about ½ turn and push the pulley to slacken the belt tension. Tighten the bolt and remove the belt. If the belt is to be reinstalled, mark the direction of rotation.

NOTE: Do not rotate the crankshaft or camshafts with the belt removed. The pistons will contact the valves and cause engine damage.

To install:

10. If the belt is worn or oil soaked, it must be replaced. Find and repair the source of the oil leak before installing a new belt.
11. If the crankshaft or camshafts have been turned with the timing belt removed, remove the spark plugs and advance the crankshaft about 15 degrees beyond TDC of No. 1 cylinder. Make sure the camshaft sprockets are properly aligned with the marks and return the crankshaft to the TDC mark.
12. Install the belt in sequence on the crankshaft, adjuster pulley, the left camshaft, water pump, then the right camshaft.
13. To adjust the tension, loosen the tensioner pulley bolt about ½ turn. The spring will automatically set the proper tension. Torque the bolt to 31 ft. lbs. (43 Nm).
14. Rotate the crankshaft 6 turns clockwise and check that the timing marks on the crankshaft and camshafts align properly. Adjust the timing belt tension again by rotating the crankshaft to the align the blue mark on the pulley with the pointer. Loosen and retorque the tensioner pulley bolt.
15. When installing the crankshaft pulley, oil the threads of the bolt without getting oil on the washer. Torque the bolt to 174 ft. lbs. (240 Nm).
16. Installation of the remaining parts is the reverse of removal.

NSX

1. Disconnect the negative battery cable and remove the timing belt cover. Make sure the crankshaft is at TDC on No. 1 piston. The keyway will be pointed straight up. Also make sure the camshaft sprocket marks are aligned with each other. If the belt is to be reinstalled, mark the direction of rotation.
2. Loosen (do not remove) the belt tensioner pulley bolt about ½ turn, push the pulley to release the tension and tighten the bolt. Remove the belt.

NOTE: Do not rotate the crankshaft or camshafts with the belt removed. The pistons will contact the valves and cause engine damage.

To install:

3. If the belt is worn or oil soaked, it must be replaced. Find and repair the source of the oil leak before installing a new belt.
4. If the crankshaft or camshafts have been turned with the timing belt

removed, remove the spark plugs and advance the crankshaft about 15 degrees beyond TDC of No. 1 cylinder. Make sure the camshaft sprockets are properly aligned with the marks and return the crankshaft to the TDC mark.

5. Install the belt in sequence: on the crankshaft, tension adjuster pulley, front exhaust cam sprocket, front intake cam sprocket, water pump pulley, rear intake cam sprocket, rear exhaust cam sprocket.

6. To adjust the tension, keep the crankshaft from turning and take up the slack in the belt by turning each camshaft in sequence: front exhaust, front intake, rear intake, rear exhaust.

7. Loosen the tensioner pulley bolt to allow the spring to set the tension, then torque the bolt to 31 ft. lbs. (43 Nm). Rotate the crankshaft clockwise about ⅓ turn to align the blue mark on the pulley with the pointer. Loosen and retorque the tensioner pulley bolt again.

8. Rotate the crankshaft 6 full turns clockwise and check for correct crankshaft and camshaft timing mark alignment.

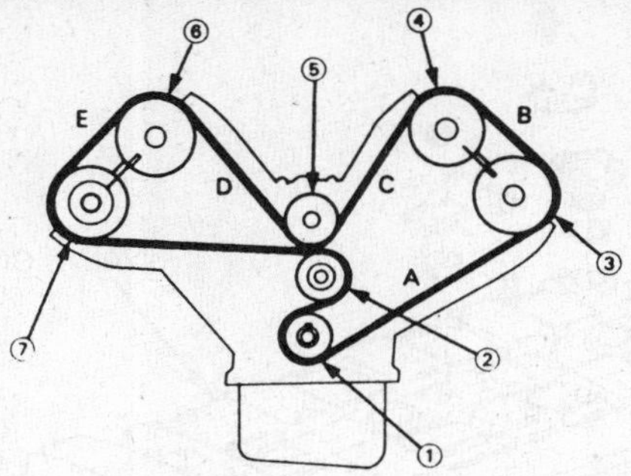

Timing belt installation sequence on NSX

NOTE: Be careful not to loose the Woodruff key.

4. The camshaft oil seal can be replaced without removing the holders. To check camshaft endplay, the rocker arm adjusting screws must be loosened.

5. Installation is the reverse of removal. Torque the camshaft sprocket-to-camshaft bolts to 27 ft. lbs. (38 Nm).

Timing Sprockets

REMOVAL & INSTALLATION

Integra

1. Disconnect the neagtive battery cable. Set the No. 1 cylinder on the TDC of it's compression stroke and remove the timing belt.

2. Align the holes in the No. 1 camshaft bearing holders with the holes in the camshafts. Insert 5.0mm pin punches into the holes to secure the camshafts.

3. Remove the camshaft sprocket-to-camshaft bolts, the washers and the sprockets; remove the sprocket with a pulley remover or a brass hammer.

Sterling, Legend and NSX

1. Disconnect the negative battery cable. Remove the timing belt.

2. Using a camshaft holding tool or equivalent, secure the camshaft sprockets, remove the sprocket-to-camshaft bolts and the sprockets. On the rear camshaft, remove the bolt opposite the locating pin last.

3. On Sterling and 1988–90 Legend, camshaft endplay can be checked without disturbing the valve adjustment. The seal can be replaced without removing the bearing caps.

4. Installation is the reverse of removal. Torque the camshaft sprocket-to-camshaft bolts to 23 ft. lbs. (32 Nm).

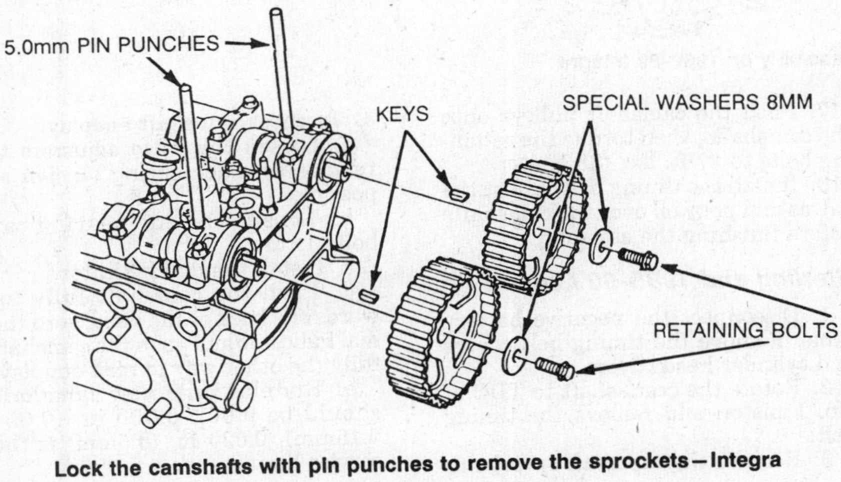

Lock the camshafts with pin punches to remove the sprockets—Integra

Camshaft

REMOVAL & INSTALLATION

Integra

1. Disconnect the negative battery cable. Remove the timing belt cover and cylinder head cover.

2. Rotate the crankshaft to TDC of No. 1 piston and remove the timing belt.

3. Remove the camshaft sprockets.

4. To check camshaft endplay:

a. Loosen the valve adjusters to remove as much spring tension as possible.

b. Loosen the end bearing cap bolts 1 turn.

c. Install the dial indicator.

d. Push the camshaft fully towards the back of the head, zero the dial indicator and push the camshaft fully the other way to read endplay.

e. Endplay on a new camshaft should be 0.002–0.006 in. (0.05–0.15mm), 0.020 in. (0.5mm) is the service limit.

5. To remove the camshaft bearing caps, loosen each bolt 2 turns at a time in a criss-cross pattern to avoid damage to the valves or rockers. Mark the caps so they can be replaced in their original position.

6. Lift the camshafts from the cylinder head, wipe them clean and inspect the lift ramps. Replace the camshafts and rockers if the lobes are pitted, scored or excessively worn.

7. Use Plasti-gauge ® to check bearing clearance: 0.002–0.004 in. (0.050–0.089mm) is standard clearance, 0.006 in. (0.15mm) service limit. If the clearance is too large and a new camshaft does not bring the clearance into specification, the cylinder head must be replaced.

To install:

8. Check the following before installing the camshafts:

a. Be certain the keyways on the camshafts are facing UP (No. 1 cylinder at TDC).

b. The valve locknuts should be loosened and the adjusting screws backed off before installation.

c. Replace the rocker arms in their original positions.

9. Place the rocker arms on the pivot bolts and the valve stems.

10. Install the camshafts with the seals part way on the camshaft, open side (spring) facing in. Lubricate the lip of the seal.

11. Do not apply oil to the holder mating surface of the camshaft seals.

12. Apply liquid gasket to the head mating surfaces of the No. 1 and No. 6 camshaft holders then install them along with the No. 2, 3, 4 and 5.

13. Begin tightening the camshaft

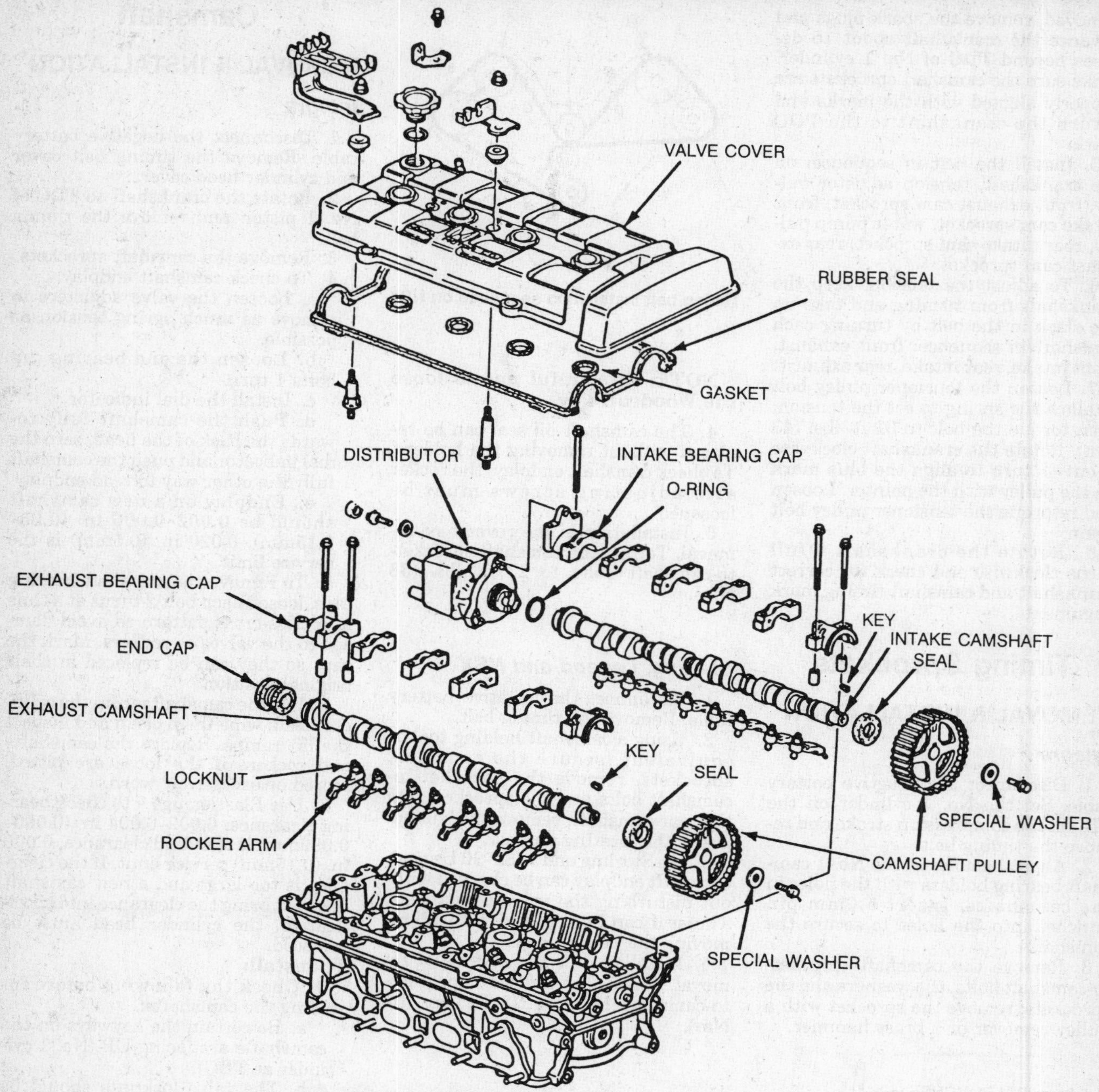

Camshaft assembly on 1988–89 Integra

holders bolts, 2 turns at a time while making sure the rocker arms are positioned on the valve stems. Use a criss-cross pattern when turning the bolts.

14. Using an oil seal driver tool 07947–SB00100 or equivalent, press new oil seals into the No. 1 camshaft holders.

15. Torque the bolts in sequence to 9 ft. lbs. (12 Nm) on 1988–89 models, 7 ft. lbs. (10 Nm) on 1990–92 models. Check that the rockers do not bind on the valves.

16. Install the camshaft pulley keys onto the grooves in the camshafts.

17. Push the camshaft pulleys onto the camshafts, then torque the retaining bolts to 27 ft. lbs. (38 Nm).

18. Install the timing belt, adjust the valves and pour oil over the camshafts before finishing the assembly.

Sterling and 1988–90 Legend

1. Disconnect the negative battery cable. Remove the timing belt covers and cylinder head covers.

2. Rotate the crankshaft to TDC of No. 1 piston and remove the timing belt.

3. Remove the camshaft sprockets.

4. To check camshaft endplay:

a. Loosen the valve adjusters to remove as much spring tension as possible.

b. Loosen the end bearing cap bolts 1 turn.

c. Install the dial indicator.

d. Push the camshaft fully towards the back of the head, zero the dial indicator and push the camshaft fully the other way to read endplay.

e. Endplay on a new camshaft should be 0.002–0.006 in. (0.05–0.15mm), 0.020 in. (0.5mm) is the service limit.

5. To remove the camshaft bearing caps, loosen each bolt 2 turns at a time in a criss-cross pattern to avoid damage to the valves or rockers. Mark the caps so they can be replaced in their original position.

6. Lift the camshafts from the cylinder head, wipe them clean and inspect the lift ramps. Replace the camshafts and rockers if the lobes are pitted, scored or excessively worn.

7. Use Plasti-gauge ® to check bearing clearance: 0.002–0.004 in. (0.050–0.089mm) is standard clearance, 0.006 in. (0.15mm) service limit. If the clearance is too large and a new camshaft does not bring the clearance into specification, the cylinder head must be replaced.

To install:

8. Fill the hydraulic lifter mounting holes with engine oil and install the lifters. Do not rotate the lifters while installing them. Also pour oil into the oil passages and fillers in the cylinder head.

9. With all the adjusting screws loose, install the pushrods and rocker arms in their original positions.

10. Advance the crankshaft 15 degrees, slip the camshaft seals onto the camshafts and lay the camshafts into the heads. Set the front camshaft so both valves on the No. 1 cylinder are be closed. Set the rear camshaft to make sure the No. 4 cylinder exhaust valve is closed.

11. Apply liquid gasket sealer to the camshaft oil seal mounting surface and on the end bearing cap-head contact surface. Install the caps in their original position and tighten each cap bolt 2 turns at a time in sequence to draw the camshaft down evenly against the valve springs.

12. Make sure the oil seal is properly positioned and torque the bolts in 2 steps in sequence to 20 ft. lbs. (28 Nm). Torque the 6mm bolts last to 9 ft. lbs. (12 Nm).

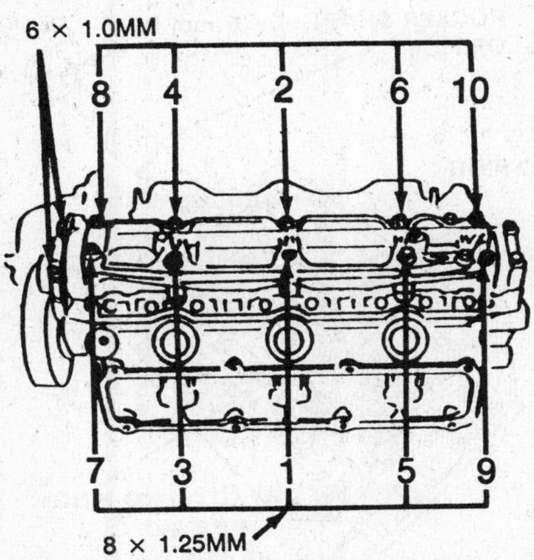

Camshaft tightening sequence—Sterling and 1988–90 Legend

13. Install the upper timing belt cover plate.

14. Install the camshaft pulleys and torque the bolts to 23 ft. lbs. (32 Nm).

15. Install the timing belt, adjust the valves and pour oil over the camshafts before completing the assembly.

1991–92 Legend

1. Disconnect the negative battery cable. Remove the timing belt covers and cylinder head covers.

2. Rotate the crankshaft to TDC of No. 1 piston and remove the timing belt.

3. Remove the camshaft sprockets.

4. When removing the camshaft and rocker arm shaft holder bolts, each bolt must be loosened 2 turns at a time in the proper sequence. This is the only way to avoid damaging the valves or rocker assemblies.

5. After all bolts are loose, the rocker arm shafts can be removed as an assembly if the bolts are not removed from the holders.

6. If the rocker shafts are to be disassembled, note that each rocker arm has a letter **A** or **B** stamped into the side. Before disassembling the rocker arms, make a diagram of the position

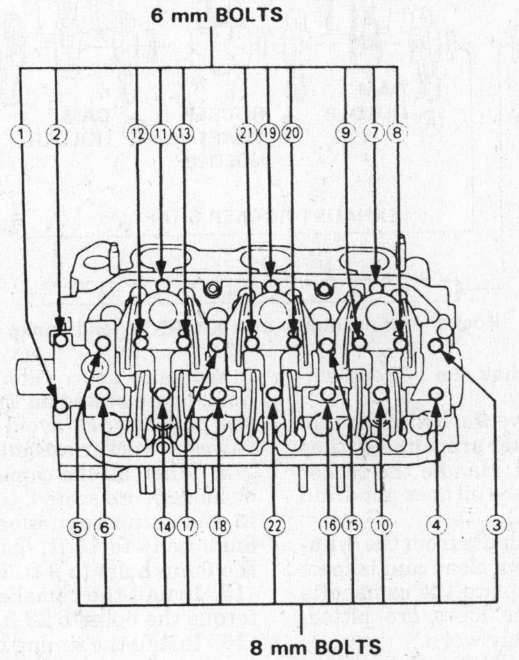

Camshaft holder bolts must be loosened in the proper sequence—1991–92 Legend

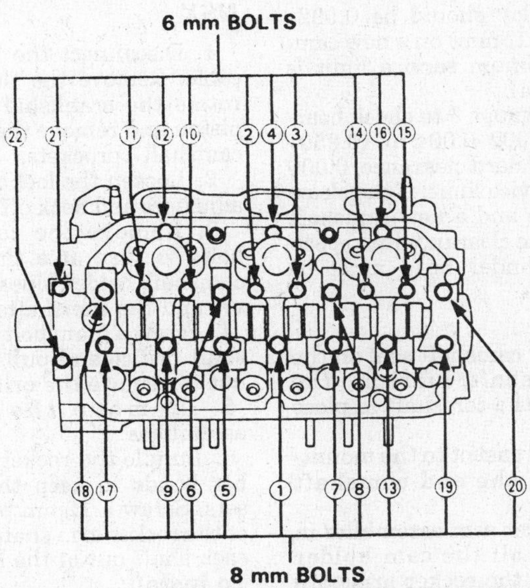

Torque sequence for camshaft and rocker arm holders on 1991–92 Legend

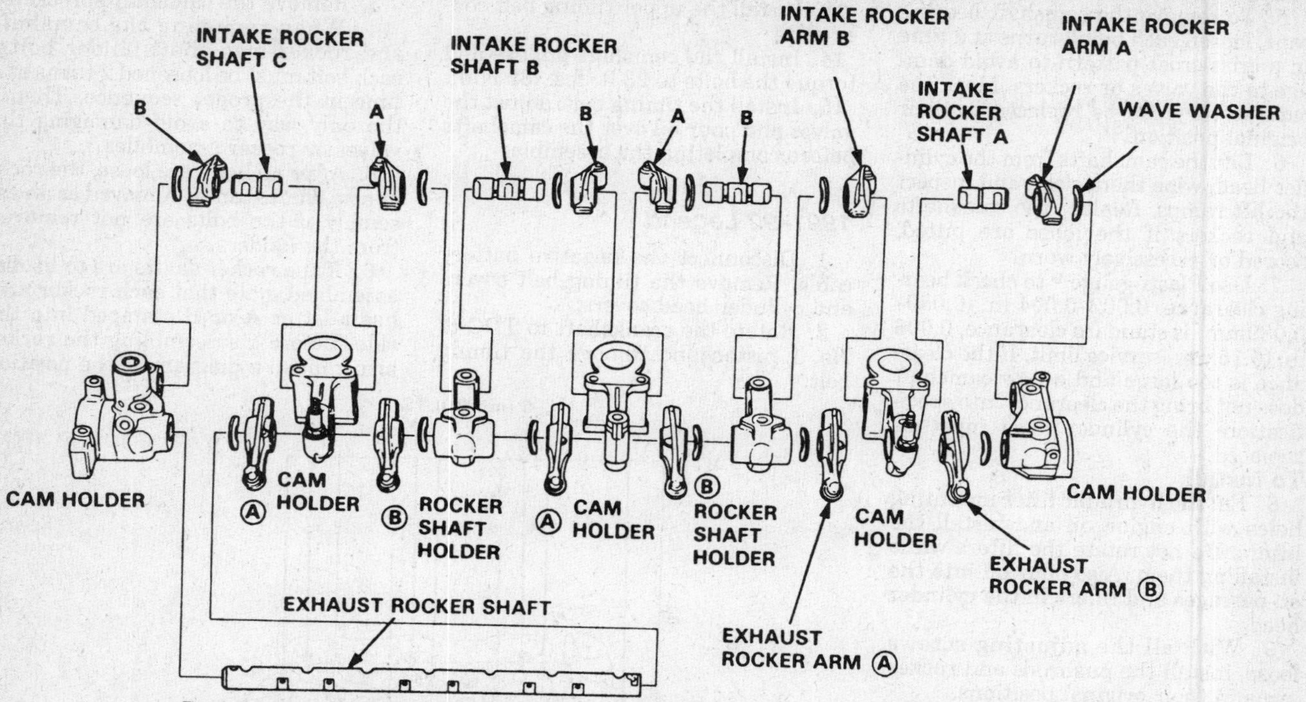

Rocker arm assembly on 1991–92 Legend: make a note of the letters before disassembly

of each letter so they can be reassembled the same way.

7. Do not remove the hydraulic tappets from the rocker arms unless they are to be replaced. Handle the rocker arms carefully so the oil does not drain out of the tappets.

8. Lift the camshafts from the cylinder head, wipe them clean and inspect the lift ramps. Replace the camshafts and rockers if the lobes are pitted, scored or excessively worn.

9. To check camshaft endplay, reinstall the holders without the rocker arm assembly and torque the bolts in sequence. Endplay should be 0.002–0.006 in (0.05–0.15mm) on a new camshaft. The maximum service limit is 0.020 in. (0.5mm).

10. Use Plasti-gauge ® to check bearing clearance: 0.002–0.004 in. (0.050–0.089mm) is standard clearance, 0.006 in. (0.15mm) service limit. If the clearance is too large and a new camshaft does not bring the clearance into specification, the cylinder head must be replaced.

To install:

11. Place a new camshaft seal on the end of the camshaft, lubricate the journals and set the camshaft in place on the head.

12. Apply liquid gasket to the mounting surfaces of the end camshaft holders.

13. Set the rocker arm assemblies in place and start all the cam holder bolts. Make sure the rocker arms are properly positioned and turn each bolt in sequence 2 turns at a time untill the holders are seated on the head. This is the only way to avoid damaging the valves or rocker assemblies.

14. When all the camshaft and rocker holders are seated, torque the bolts in the same sequence. Torque the 8mm bolts to 16 ft. lbs. (22 Nm) and the 6mm bolts to 9 ft. lbs. (12 Nm).

15. Install the camshaft pulleys and torque the bolts to 23 ft. lbs. (32 Nm).

16. Install the timing belt, adjust the valves and pour oil over the camshafts before completing the assembly.

NSX

1. Disconnect the negative battery cable. Remove the timing belt cover, rotate the crankshaft to TDC No. 1 piston and remove the timing belt and camshaft sprockets.

2. Loosen the locknuts on the valve adjusters and back off the adjustment.

3. Remove the camshaft holder bolts ⅓ turn at a time. Remove the camshaft holder pipes, camshaft holders and the camshafts.

4. Screw a 5mm bolt into each rocker shaft orifice and pull the bolt staight out to remove the orifice.

5. Remove the spool valve assemblies.

6. Bundle the rocker arms with rubber bands to keep them together in sets. Screw a 12mm bolt into the end of the rocker arm shaft and slowly pull each shaft out of the head.

To install:

7. Set the O-ring and dowel pin in the oil passage on the front camshaft holder.

8. The camshaft holders are marked for front and rear head. Apply liquid gasket to the sealing surfaces of the end holders and set all the holders onto the head.

9. Place the camshaft holder pipes, oil the bolt threads and start the bolts. Tighten each bolt 2 turns at a time in the sequence shown to draw the camshafts down evenly against the valve springs. Make sure the rockers do not bind on the valves.

10. When all bolts are snug, torque the bolts in sequence. Torque the 8mm bolts to 16 ft. lbs. (22 Nm) and the

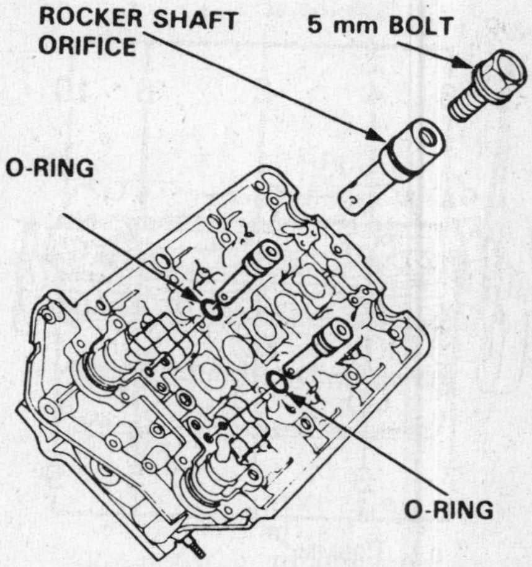

Removing the rocker shaft orifice on NSX

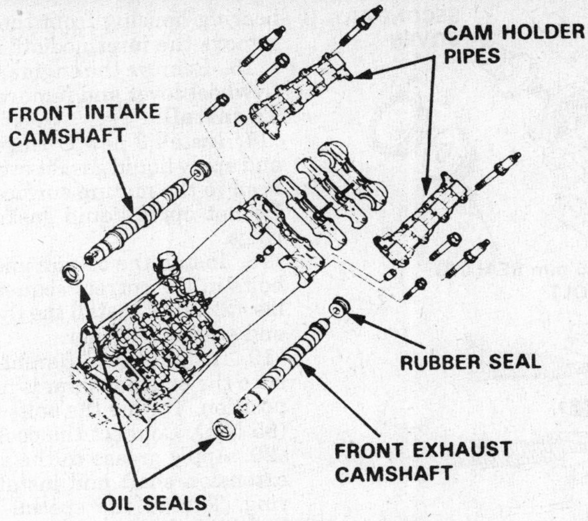

Camshaft removal on NSX

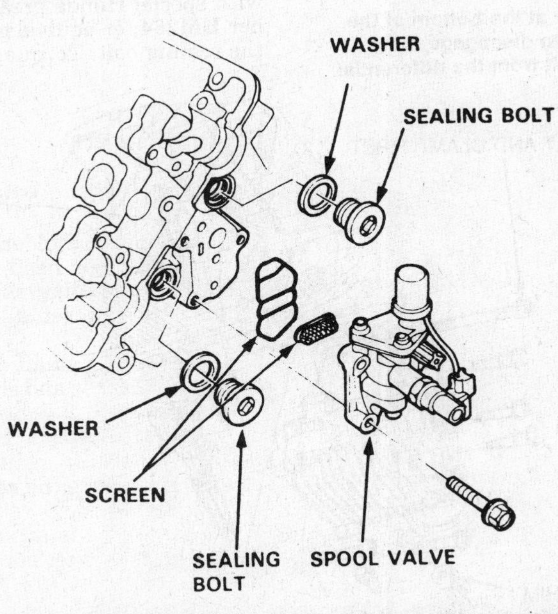

Replace the screen and gaskets when removing the spool valve—NSX

Pistons and Connecting Rods

POSITIONING

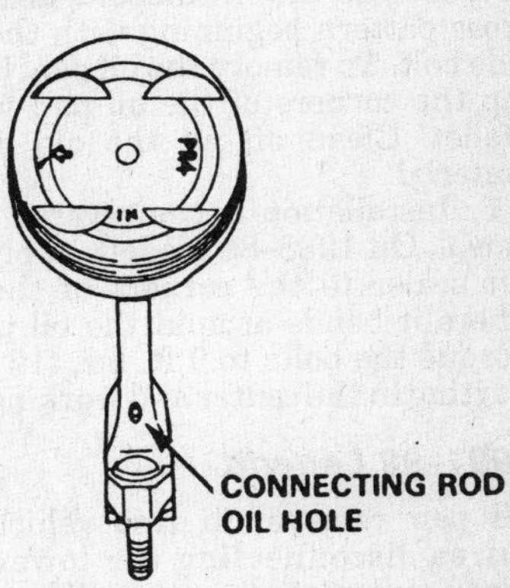

Piston and rod positioning on Integra

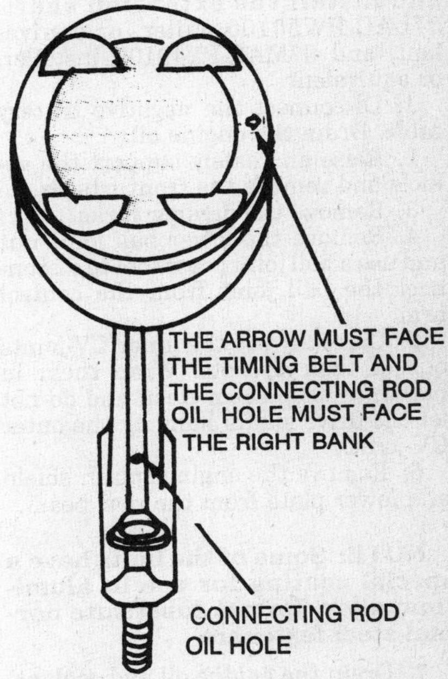

Piston and rod positioning on Sterling, Legend and NSX

6mm bolts to 7 ft. lbs. (10 Nm). Do not over torque the bolts.

11. Install the camshaft sprockets. To hold the camshafts from turning, insert a 5mm punch into the camshaft holder pipe holes and torque the pulley bolts to 51 ft. lbs. (70 Nm).

12. Install the timing belt, set the tension and adjust the valves.

13. Inspect the rocker arms with a mirror. Push the middle arm of each set to make sure it moves freely against the lost motion assembly. If there is any binding or if the arms are locked, remove that arm set and repair or replace as an assembly.

Camshaft Oil Seal

REPLACEMENT

1. Remove the timing sprockets.
2. Using a small prybar, pry the camshaft oil seals from the cylinder heads; be careful not to damage the seal mounting surfaces.
3. Using an oil seal driver tool or equivalent, lubricate the seal lips, apply sealant to the seal housing and drive the new seal into the cylinder head until it seats.
4. To complete the installation, reverse the removal procedures.

ENGINE LUBRICATION

Oil Pan

REMOVAL & INSTALLATION

Integra, Sterling and 1988–90 Legend

1. Disconnect the negative battery cable. Drain the engine oil.

2. Raise and safely support the vehicle. Remove the lower splash pan, if equipped.
3. Remove the exhaust pipe from the manifold and from the catalyst.
4. Loosen the oil pan bolts in a crisscross pattern beginning with the outside bolt. To remove the oil pan, lightly tap the corners of the oil pan with a mallet. Clean off all the old gasket material.
5. Installation is the reverse of removal. On 1988–89 models, apply gasket sealer to the corners of the pan where it bends around the oil pump. Torque the bolts to 9 ft. lbs. (12 Nm), starting in the center and working out.

1991–92 Legend

Oil pan removal in this vehicle requires disconnecting the lower ball joints and removing the differential. Special tools are required to remove and install the extension shaft: 07LAC-PW50100 puller, or equivalent, and 07MAF-PY40100 installer, or equivalent.

1. Disconnect the negative battery cable. Drain the engine oil.
2. Raise and safely support the vehicle and remove the front wheels.
3. Remove the damper forks.
4. Remove the lower ball joint nut and use a ball joint press tool to disconnect the ball joint from the control arm.
5. Carefully pry the inner CV-joints out of their sockets. Wrap them in plastic to keep them clean and do not let the drive shafts hang by the outer CV-joint.
6. Remove the engine splash shield and lower plate from the rear beam.

NOTE: Some of the bolts have a special coating for use in aluminum alloy. Do not substitute normal steel fasteners.

7. Drain the engine oil and coolant.
8. Drain the oil from the differential.
9. Without disconnecting the hoses, remove the power steering speed sensor.
10. Disconnect the differential oil cooler hoses.
11. Shift the transmission into 1st gear (manual) or **P** (automatic) and remove the secondary shaft cover and sealing bolt.
12. Install the puller and disengage the extension shaft from the differential.
13. Remove the mounting bolts and the 26mm shim and remove the differential.
14. Without disconnecting any hoses, remove the air conditioner compressor.
15. Unbolt the intermediate shaft bearing housing from the oil pan and remove the intermediate shaft.
16. Remove the engine stiffener and flywheel cover and remove the oil pan.

To install:

17. Install 3 new O-rings on the pan and apply liquid gasket evenly in a thin bead to the mating surface of the pan. Do not apply liquid gasket to the O-rings.
18. Install the oil pan and torque the bolts in the correct sequence to 16 ft. lbs. (22 Nm). Install the flywheel cover and engine stiffener.
19. Install the differential, making sure the original shim is in the proper position. Torque the bolts to 47 ft. lbs. (65 Nm). Connect the cooling hoses.
20. Apply grease to the spline of the extension shaft and install a new set ring. Thread the special installation tool into the transmission case to install the extension shaft.
21. Pack the extension shaft cavity with special Honda grease, part number UM264, or equivalent, and install the sealing bolt. Torque the bolt to 58

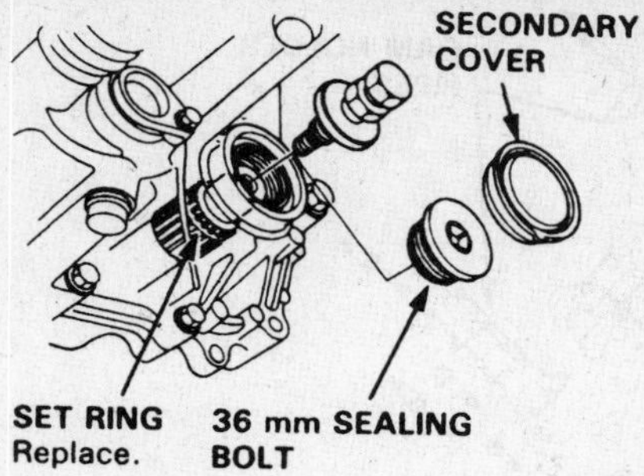

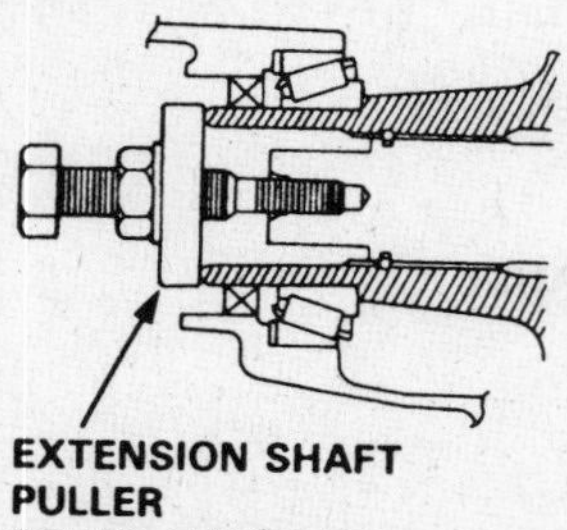

Use the puller at the bottom of the transmission to disengage the extension shaft from the differential

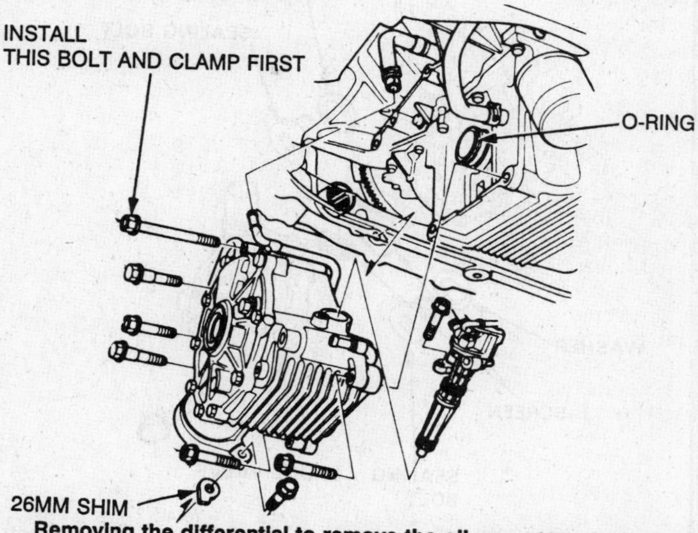

Removing the differential to remove the oil pan—1991–92 Legend

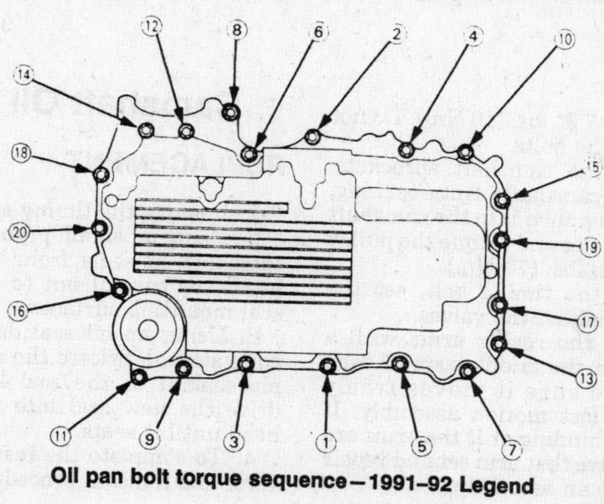

Oil pan bolt torque sequence—1991–92 Legend

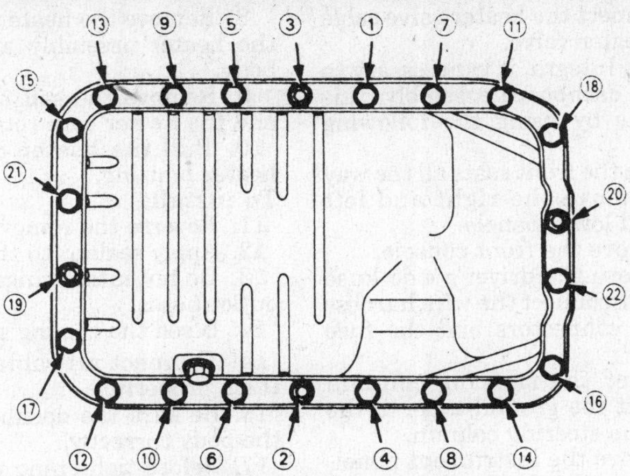

Oil pan bolt torque sequence—NSX

ft. lbs. (80 Nm) and install the secondary cover.

22. Install the air conditioner compressor and adjust the belt tension.
23. Install the intermediate shaft, torque the bolts to 16 ft. lbs. (22 Nm).
24. Install the speed sensor.
25. Install new set rings, then press the CV-joints into their sockets.
26. When installing the lower ball joint nuts, torque them to 51–58 ft. lbs. (70–80 Nm) and install a new cotter pin. Torque the damper fork bolts to 51 ft. lbs. (70 Nm).
27. Install the rear beam and torque the bolts to 44 ft. lbs. (60 Nm). When installing the lower plate, torque the specially coated bolts to 29 ft. lbs. (39 Nm).
28. Install the remaining parts and refill the differential, engine and cooling system. Open the cooling system bleeder at the engine end of the upper radiator hose when filling the system.

NSX

1. Raise and safely support the vehicle and drain the engine oil.
2. Remove the front exhaust pipe and catalytic converter.
3. Remove the oil pan bolts and the oil pan.
4. Installation is the reverse of removal. Use a new gasket and torque the nuts and bolts to 10 ft. lbs. (14 Nm) in the correct sequence.

Oil Pump

REMOVAL & INSTALLATION

1. Disconnect the neagtive battery terminal. Raise and safely support the vehicle, drain the oil and remove the oil pan.
2. Make sure the crankshaft is at TDC on No. 1 cylinder and remove the timing belt cover and the timing belt.
3. Remove the oil pan and the pickup screen. On NSX, remove the baffle plate.
4. Remove the oil pump from the front of the engine. Any time the oil pump is removed, the front oil seal should be replaced.
5. Installation is the reverse of removal. Use new O-rings and apply liquid gasket to the pump mounting face. Torque the 6mm bolts to 9 ft. lbs. (12 Nm) and the 8mm bolts to 17 ft. lbs. (24 Nm).

Front Oil Pump Seal

REPLACEMENT

1. Remove the timing belt.
2. Slide the crankshaft sprocket and belt guides from the crankshaft.
3. Using a small prybar, pry the oil seal from the oil pump housing; be careful not to damage the seal's mounting surface.
4. Using a new oil seal, lubricate the seal lips with engine oil. Using a seal drive tool, drive the new seal into the oil pump housing until it seats.

Rear Main Bearing Oil Seal

REMOVAL & INSTALLATION

1. Remove the oil pan and the transaxle from the vehicle.
2. If equipped with a manual transaxle, perform the following procedures:
   a. Matchmark the pressure plate-to-flywheel.
   b. Insert the clutch alignment tool or equivalent, into the pilot bearing.
   c. Remove the pressure plate-to-flywheel bolts (gradually), the pressure and the clutch plate.
   d. Remove the flywheel-to-crankshaft bolts and the flywheel.
3. If equipped with an automatic transaxle, remove the flexplate-to-flywheel bolts and the flexplate.
4. Remove the rear oil seal housing-to-engine bolts and the gasket.
5. Using a prybar, pry the oil seal from the housing.

To install:

6. Clean the gasket mounting surfaces. Check the flywheel or flexplate for cracks and/or damage; replace it, if necessary.
7. Using an oil seal installation tool or equivalent, drive the new oil seal into the rear oil seal housing until it seats.
8. Apply liquid gasket on the gasket mounting surface. Using oil, lubricate the oil seal lips.
9. Install the oil seal housing and torque the bolts to 9 ft. lbs. (12 Nm); be careful not to damage the oil seal lip.
10. To complete the installation, reverse the removal procedures. Torque the flywheel-to-crankshaft bolts in a cross pattern to 54 ft. lbs. (75 Nm) for automatic transaxles or 76 ft. lbs. (105 Nm) for manual transaxles. Refill the crankcase, start the engine and check for leaks.

ENGINE COOLING

Radiator

REMOVAL & INSTALLATION

Except NSX

1. Disconnect the negative battery cable. Drain the cooling system. On Sterling, it may be necessary to remove the front under protection panel to gain access to the lower radiator and shroud mounting bolts.
2. Disconnect the thermo-switch wire and the fan motor wire. The fans can be removed with the radiator as an assembly.
3. Disconnect the upper and lower hoses at the radiator. If equipped with an automatic transaxle, disconnect and plug the cooling lines at the bottom of the radiator.
4. Remove the hoses to the coolant reservoir.
5. Remove the radiator bolts and the radiator with the fan attached. The fan can be easily unbolted from the back of the radiator.

To install:

6. Install the fan to the radiator and install them as a unit.

7. When filling the system, open the bleeder, which is near where the upper hose connects to the engine.

8. Start the engine and watch the coolant level in the radiator. It will probably require more coolant as the engine warms up.

NSX

1. Turn the ignition switch **ON** and set the heater controls for full temperature. Turn the ignition **OFF** and disconnect the negative battery cable.
2. Remove the expansion tank cap and raise and safely support the vehicle.
3. While the radiator is draining, remove the tunnel cover to gain access to the coolant pipes.
4. Remove the drain bolts to drain the pipes.
5. Install hoses on the drain cocks in the cylinder block (one for each bank) and open them to finish draining the cooling system.
6. When the cooling system is fully drained, disconnect the wiring to the fan, unbolt the radiator brackets and remove the radiator and fan as an assembly.

To install:

7. Use new gaskets when installing the coolant pipe drain bolts.
8. With all the drains closed and the vehicle on the ground, open all 4 bleeders and start filling the system at the expansion tank. Close each bleeder in sequence as coolant flows from it:
   a. on the thermostat housing
   b. at the radiator near the upper hose
   c. in the heater hose pipe near the brake booster
   d. in the pipe at the engine firewall
   e. bleed once more at the thermostat housing
9. The system holds a total of 12.7 US quarts. After all the bleeders have been closed and the system will not take any more coolant, run the engine until the fan runs. Stop the engine and finish filling the system.

Heater Core

REMOVAL & INSTALLATION

Without Air Conditioning

INTEGRA

1. Disconnect the negative battery cable. Drain the cooling system.
2. Disconnect the heater hoses at the firewall.

NOTE: Coolant will run out of the heater hoses when disconnected, place a drain pan under them to catch the coolant.

3. Disconnect the heater valve cable from the heater valve.
4. On the Integra, it is necessary to remove the dashboard assembly; this can be done by using the following procedure:
   a. Slide the front seats all the way back. Remove the right and left dashboard lower panels.
   b. Remove the front console.
   c. Remove the driver's side knee bolster. Disconnect the wire harness from the connectors and the fuse box.
   d. Lower the steering column. Disconnect the ground cable to the right of the steering column.
   e. Remove the instrument panel.
   f. Remove the 4 screws, then pull the gauge assembly out half way and disconnect the speedometer cable and connectors.
   g. Disconnect the antenna cable, wire connector and loosen the 2 screws, then remove the radio assembly.
   h. Disconnect the shift position switch and the shift lock wire connectors from the dashboard wire harness; automatic transaxle vehicles only.
   i. Remove the clock assembly from the top of the dashboard.
   j. Remove the side defroster garnishes from both ends of the dashboard.
   k. Remove the dashboard mounting bolts. Lift up and remove the dashboard assembly.
5. Remove the heater duct. Remove the heater assembly lower mounting nut.
6. Disconnect the air mix cable from the heater.
7. Disconnect the wire harness connector from the function control motor.
8. Remove the heater bolts and pull the heater assembly away from the body.
9. Remove the self tapping screws and the heater core retaining plate.
10. Pull the heater core from the heater housing.

To install:

11. Reverse the removal procedures.
12. Apply sealant to the grommets.
13. Do not interchange the inlet and outlet hoses.
14. Bleed the cooling system.
15. Connect all cables and adjust them properly.
16. Be sure the dashboard fits onto the body correctly.
17. Before tightening the dashboard bolts, make sure the instrument wires are not pinched and that the dashboard is not interfering with the heater control cable. For ease of installation, remove the gauge assembly from the dash.

With Air Conditioning

INTEGRA

It may or may not be necessary to remove the evaporator assembly from the under the instrument panel in order to gain access to the heater core housing. This can be determined by the individual technician performing this operation. The following is a procedure on how to remove the heater core housing if the evaporator housing must be removed.

1. Disconnect the negative battery cable.
2. Properly discharge the refrigerant from the air conditioning system. Disconnect the receiver line and suction hose from the evaporator assembly. Be sure to cap the open fittings to prevent moisture from entering the system.

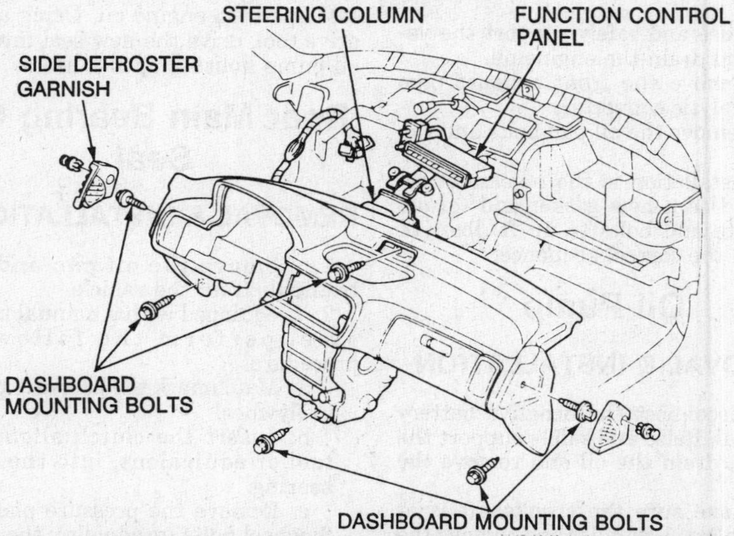

Integra dashboard must be removed to remove the heater core, Sterling and 1988–90 Legend similar

3. Remove the passenger side lower dashboard cover.
4. Remove the glove box assembly.
5. Remove the front console assembly.
6. Remove the passenger side knee bolster panel, located under the glove box frame.
7. Remove the 2 self tapping screws and the air conditioning bands from around the evaporator assembly.
8. Disconnect the wire connector from the thermostat switch and pull off the wire harness from the clamps. Remove the evaporator.
9. Remove the self-tapping screws and remove the heater duct assembly.
10. Remove the heater core housing.

To install:

11. Reverse the removal procedure. Recharge the air conditioning system, paying attention to the following:
   a. When reattaching the actuator, make sure its positioning will not allow the air door to be pulled to far.
   b. Attach the actuator and all linkage, then apply battery voltage and watch the door movement. If necessary, loosen the holding screw and move the actuator up or down.
   c. When adjusting the control rod, connect the recirculation control motor connection to the main wire harness, push **RECIRC** and open the air doors. Then connect the control rod to the arm while holding the air doors open.

STERLING AND 1988–90 LEGEND

CAUTION

This vehicle is equipped with a driver side airbag. To avoid accidental deployment and serious personal injury, the system must be disarmed before beginning this repair procedure. See the disarming procedure in the Chassis Electrical section.

1. Disconnect the negative battery cable. Drain the cooling system.
2. Disconnect the heater hoses at the firewall.

NOTE: Coolant will run out of the heater hoses when disconnected, place a drain pan under them to catch the coolant.

3. Properly discharge the air conditining system. Cap any open fittings to keep moisture out.
4. Remove the dashboard as follows:
   a. Disconnect the negative battery cable. Slide the seat all the way to the rear and remove the lower dashboard panel.
   b. Remove the knee bolster and the left air duct.
   c. Remove the hood opener and the center console.
   d. Disconnect the wire harness from the connector holder.
   e. Remove the dash harness ground bolt from the steering column.
   f. Remove the screws and radio panel assembly, then disconnect the wire connectors, antenna cables and wire tie.
   g. Remove the radio assembly, remove the center dash pocket.
   h. Lower the steering column. Be sure to remove the ignition key from the lock cylinder.
   i. Remove the dashboard mounting bolts.
   j. Pull the dashboard straight back and disconnect the speedometer. Remove the dashboard.
5. Disconnect the wire harness and the vacuum hoses.
6. Remove the heater mounting bolts and pull the heater assembly away from the body.
7. Remove the self-tapping screws and the heater core retaining plate.
8. Pull the heater core from the heater housing.

To install:

19. Apply sealant to the grommets.
10. Do not interchange the inlet and outlet hoses.
11. Bleed the cooling system.
12. Connect all cables and adjust them properly.
13. Be sure the dashboard fits onto the body correctly.
14. Before tightening the dashboard bolts, make sure the instrument wires are not pinched and that the dashboard is not interfering with the heater control cable. For ease of installation, remove the gauge assembly from the dash.
15. Recharge the air conditioning system, paying attention to the following:
   a. When reattaching the actuator, make sure its positioning will not allow the air door to pulled to far.
   b. Attach the actuator and all linkage, then apply battery voltage and watch the door movement. If necessary, loosen the holding screw and move the actuator up or down.
   c. When adjusting the control rod, connect the recirculation control motor connection to the main wire harness, push the **FRESH/RECIRC** switch to **FRESH** and open the air doors. Then connect the control rod to the arm while holding the air doors open.

1991–92 LEGEND

CAUTION

This vehicle is equipped with a driver side airbag and, on LS models, a passenger side airbag. To avoid accidental deployment and serious personal injury, the system must be disarmed before beginning this repair procedure. See the disarming procedure in the Chassis Electrical section.

1. Disconnect the negative battery cable. Drain the cooling system.
2. Properly discharge the air conditioning refrigerant into recovery equipment.
3. Remove the dashboard:
   a. Remove the front seat track end covers and unbolt the seats. Unplug the connectors and remove the front seats.
   b. Unscrew the shift knob and remove the 2 screws directly in front of the shifter boot to remove the console panel.

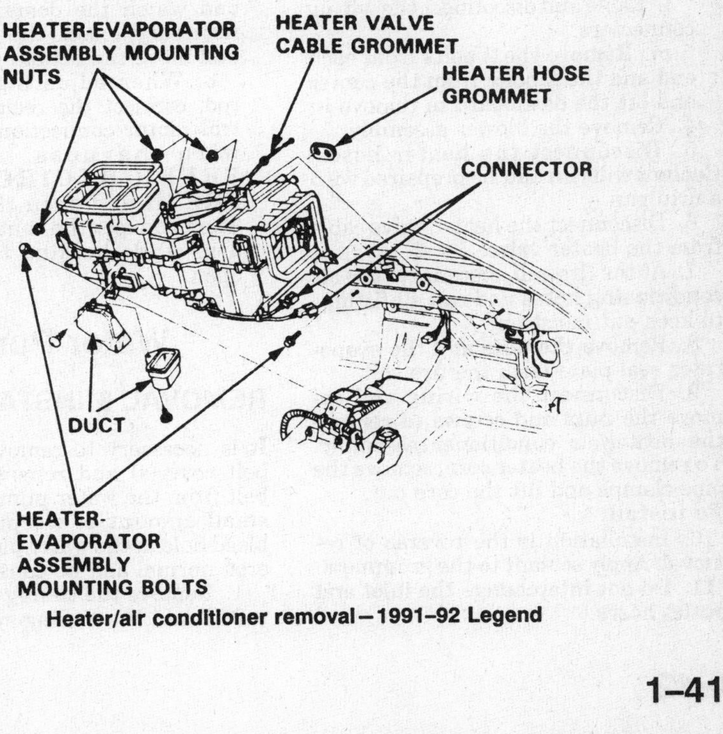

Heater/air conditioner removal—1991-92 Legend

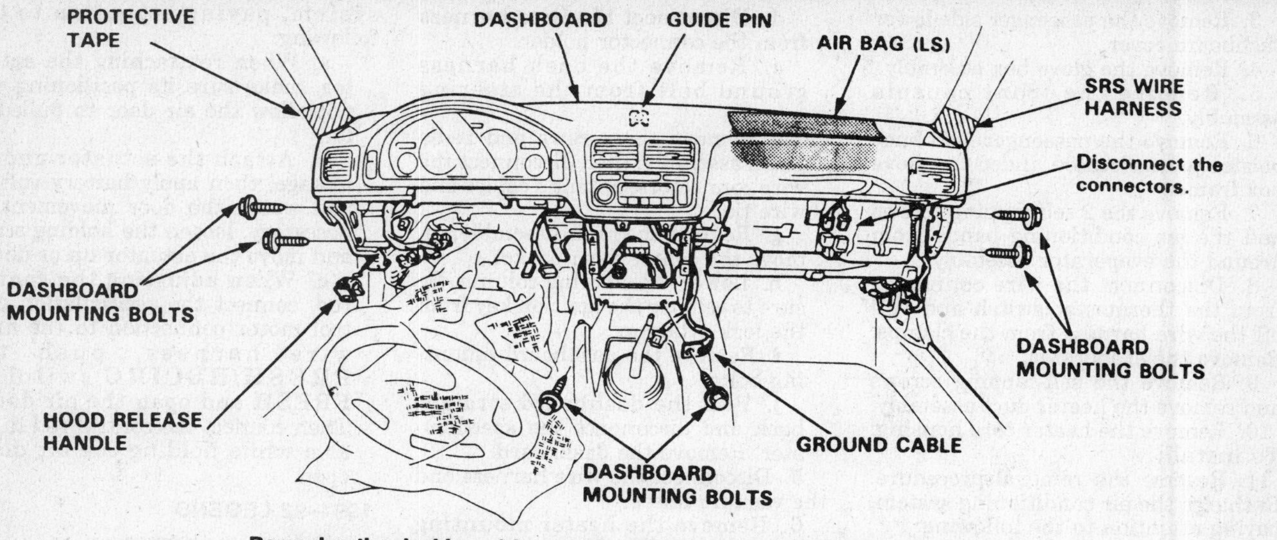

Removing the dashboard from 1991–92 Legend to remove the heater core

c. Remove the center armrest. Be careful not to damage the airbag control unit or wiring below the armrest.
d. Remove the radio.
e. Remove the glove compartment lower panel and the glove compartment.
f. Remove the driver side lower dash panel.
g. Remove the screws securing the center console to the dashboard and remove the center console.
h. Remove the 2 nuts and 2 bolts and lower the steering column.
i. On LS models, remove the passenger side airbag unit.
j. Disconnect the hood release cable from the handle.
k. Remove the side covers from the ends of the dashboard.
l. Lable and disconnect the wiring connectors.
m. Remove the 2 bolts from each end and the 2 bolts from the center and lift the dashboard to remove it.
4. Remove the blower assembly.
5. Disconnect the heater hoses. Coolant will run out, be prepaired with a drip pan.
6. Disconnect the heater valve cable from the heater valve.
7. At the firewall, disconnect the air conditioning hoses and cap all fittings to keep out moisture.
8. Remove the nuts and the evaporator seal plate from the firewall.
9. Disconnect the wiring and remove the nuts and screws to remove the heater/air conditioner assembly. To remove the heater core, remove the pipe clamps and lift the core out.

To install:

10. Installation is the reverse of removal. Apply sealant to the grommets.
11. Do not interchange the inlet and outlet hoses.
12. Bleed the cooling system.
13. Connect all cables and adjust them properly.
14. Be sure the dashboard fits onto the body correctly.
15. Before tightening the dashboard bolts, make sure the instrument wires are not pinched and that the dashboard is not interfering with the control cable. For ease of installation, remove the gauge assembly from the dash.
16. Recharge the air conditioning system, paying attention to the following:
a. When reattaching the actuator, make sure its positioning will not allow the air door to be pulled too far.
b. Attach the actuator and all linkage, then apply battery voltage and watch the door movement. If necessary, loosen the holding screw and move the actuator up or down.
c. When adjusting the control rod, connect the recirculation control motor connection to the main wire harness, push the **FRESH/RECIRC** switch to **FRESH** and open the air doors. Then connect the control rod to the arm while holding the air doors open.

Water Pump

REMOVAL & INSTALLATION

It is necessary to remove the timing belt cover(s) and remove the timing belt from the water pump sprocket. A small amount of weeping from the bleed hole in the water pump is considered normal and no cause for alarm.
1. Remove the timing belt.
2. Drain the cooling system.

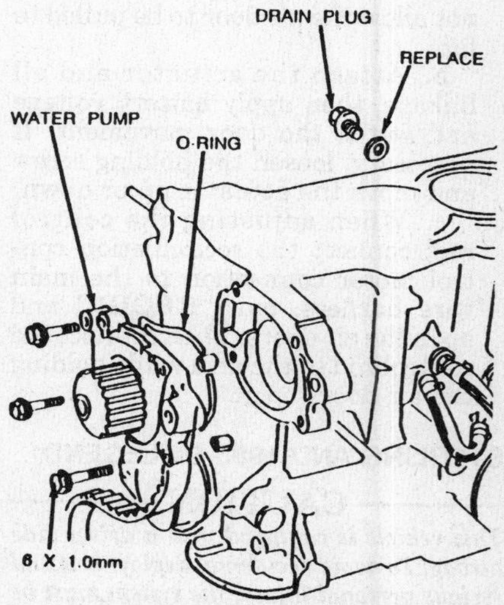

Integra water pump installation

3. Remove the water pump-to-engine bolts and remove together with the drive sprocket.
4. To install, use a new O-ring and reverse the removal procedures. Torque the 6mm bolts to 9 ft. lbs. (12 Nm) and 8mm bolts to 16 ft. lbs. (22 Nm).
5. Refill and bleed the cooling system. Start the engine, allow it to reach normal operating temperatures and check for leaks. Check and/or adjust the engine timing.

Thermostat

REMOVAL & INSTALLATION

1. Drain the cooling system to a level below the thermostat housing. Remove the lower radiator hose from the thermostat housing.
2. Remove the thermostat housing bolts, the housing and the thermostat.

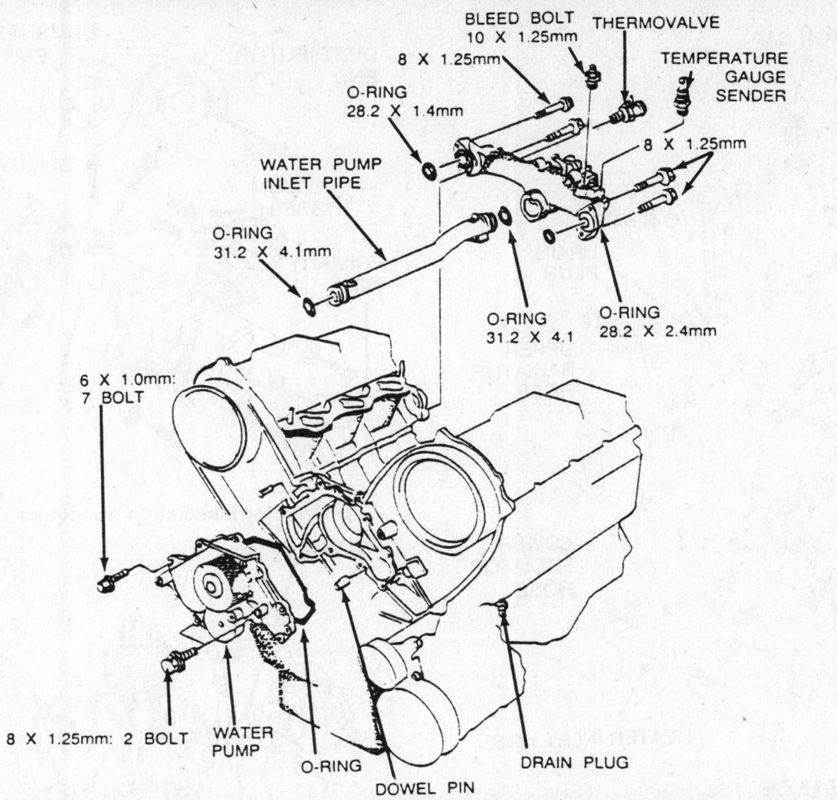

Sterling and 1988–90 Legend water pump installation; later Legend and NSX similar

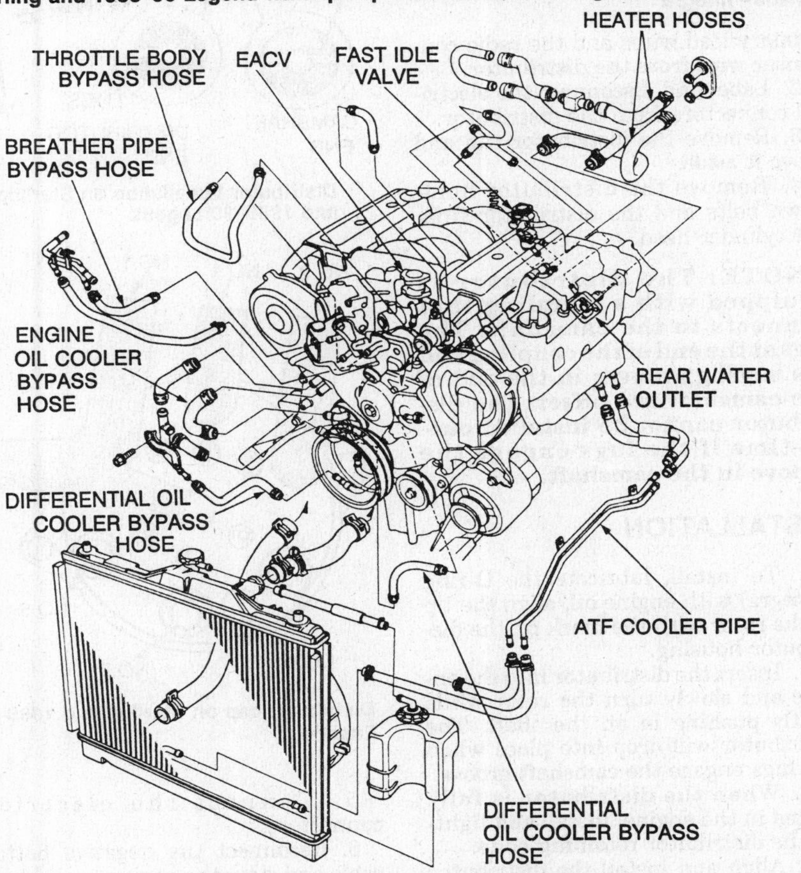

Cooling system schematic—1991–92 Legend

3. Clean the gasket mounting surfaces.

4. To install, use new gaskets and reverse the removal procedures; install the thermostat's spring end toward the engine. Torque the housing bolts to 9 ft. lbs. (12 Nm). Refill and bleed the cooling system.

COOLING SYSTEM BLEEDING

When filling a cooling system, use a 50/50 mixture of antifreeze and clean water. Make sure the antifreeze is suited for use in aluminum cooling system components. Do not mix brands of antifreeze.

Except NSX

1. Set the heater temperature selector for full heat.
2. Fill the coolant reservoir to the **MAX** mark.
3. Loosen the air bleed bolt at the engine end of the upper radiator hose. Fill the radiator to the bottom of the filler neck with the proper coolant mix. Tighten the bleed bolt as soon as the coolant starts to run out in a steady stream without any air bubbles in it.
4. With the radiator cap off, start the engine and allow it to warm up; the cooling fan should go on at least twice. Then, if necessary, add more antifreeze/coolant to bring the level back up to the bottom of the filler neck.
5. Put the radiator cap on, restart the engine and check for any leaks.

NSX

This vehicle has long coolant pipes running under the tunnel between the engine compartment and the radiator. There is a drain in the radiator, in each coolant pipe and in each cylinder bank of the engine. Each must be opened to fully drain the system. When installing the drain plugs in the coolant pipes, be sure to use new sealing washers. There are 4 bleeders which must all be opened when filling the system, then closed in the proper sequence.

1. Make sure the heating system is set for full heat.
2. With the drain plugs all closed, open all 4 bleeders: on the thermostat housing, at the radiator near the upper hose, in the heater hose pipe near the brake booster and in the pipe at the engine firewall.
3. Fill the expansion tank with coolant and close each bleeder when coolant flows from it. Close the bleeders in the following sequence:
   a. on the thermostat housing

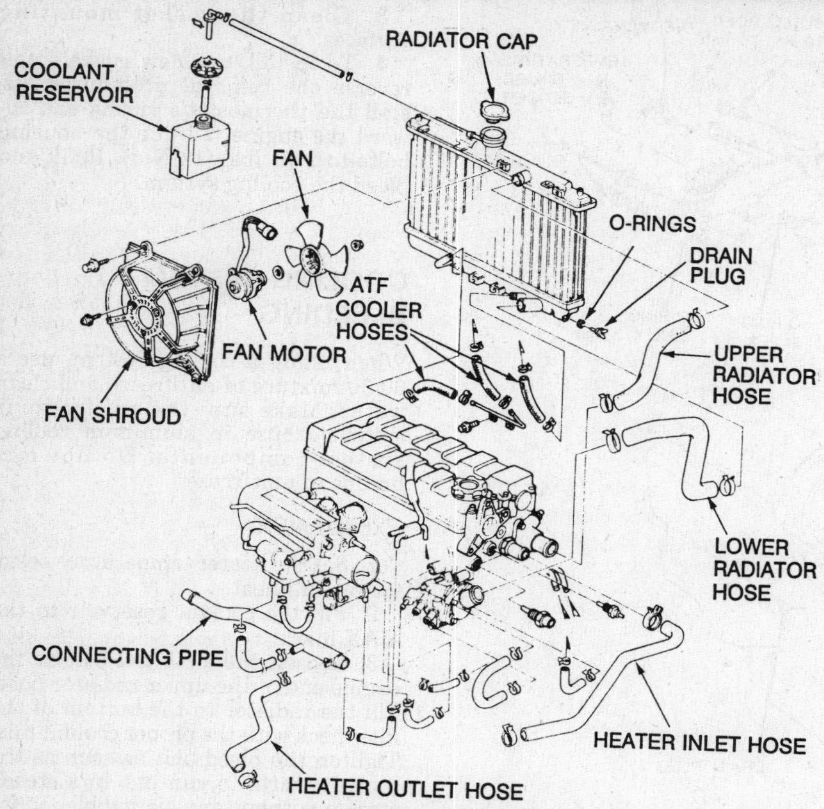

Cooling system schematic—Integra

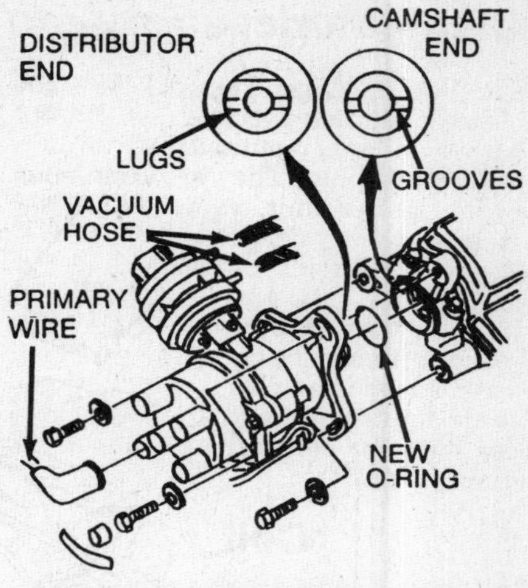

Distributor installation on Integra

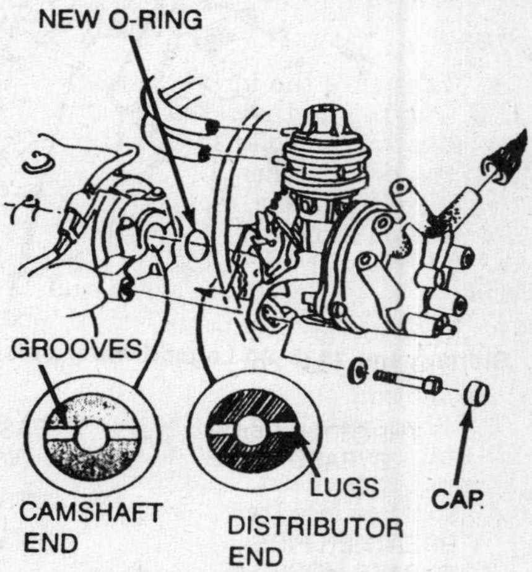

Distributor installation on Sterling and 1988–90 Legend

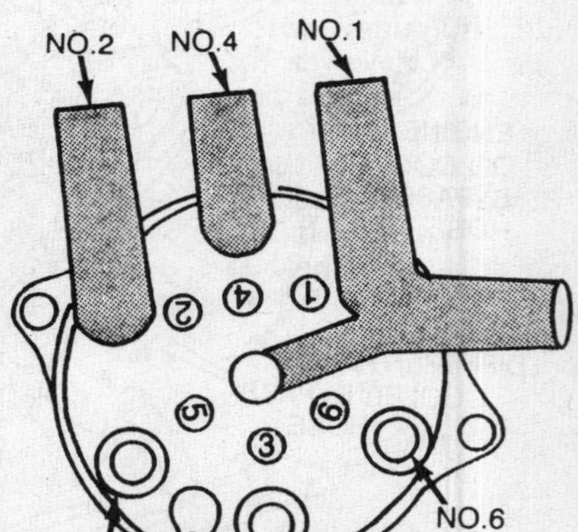

Distributor cap on Sterling and 1988–90 Legend

b. at the radiator near the upper hose

c. in the heater hose pipe near the brake booster

d. in the pipe at the engine firewall

e. bleed once more at the thermostat housing

4. Install the expansion tank cap but do not tighten all the way. Run the engine until the cooling fan runs, stop the engine and check the coolant level.

ENGINE ELECTRICAL

NOTE: Most vehicles are equipped with a theft protected radio. Before disconnecting the battery, obtain the owner's radio activation code. After reconnecting power to the radio, the 5 digit code must be entered to restore operation.

Distributor

REMOVAL

1. Disconnect the negative battery cable. Disconnect the high tension and primary lead wires and the radio condenser wire from the distributor.

2. Label and disconnect the electrical connectors from the distributor.

3. Remove the distributor cap and move it aside.

4. Remove the distributor hold-down bolts and the distributor from the cylinder head.

NOTE: The distributor is equipped with a coupling that connects to the camshaft. The lugs at the end of the coupling and it's mating grooves in the end of the camshaft are offset. The distributor cannot be installed out-of-time if the lugs engage the groove in the camshaft.

INSTALLATION

1. To install, lubricate the O-ring (Integra) with engine oil, align the tip of the rotor with the mark on the distributor housing.

2. Insert the distributor into the engine and slowly turn the rotor while gently pushing in on the shaft. The distributor will drop into place when the lugs engage the camshaft groove.

3. When the distributor is fully seated in the engine, install and tighten the distributor retaining bolts.

4. Align and install the distributor cap, then install the hold-down screws.

5. Connect the electrical connectors.

6. Reconnect the negative battery cable and start the engine to check the ignition timing.

Distributorless Ignition

The 1991–92 Legend and the NSX use a direct ignition system, one coil for each cylinder is mounted directly over each spark plug. They are controlled by an igniter unit, which is a solid state switching device that completes the ground circuit for each coil at the proper time. The igniter unit is controlled by the ECU, which determines ignition timing. To check the timing, a service loop is available that connects to the igniter unit and provides a trigger wire for connecting a standard timing light.

Coil

REMOVAL & INSTALLATION

Legend

1. Make sure the ignition switch is **OFF** and unplug the coil connector.
2. Remove the bolts and lift the coil off of the cylinder head.
3. When installing, torque the bolts to 9 ft. lbs. (12 Nm).

NSX

1. Remove the coil cover.
2. Unplug the connector and remove the coil.
3. When installing, make sure each coil and cover are installed onto the proper cylinder head. They are marked **FF** (front) or **RR** (rear).

Ignition Timing

The timing marks are located on the crankshaft pulley, with a pointer on the timing belt cover. The TDC mark is painted white and the ignition timing mark is painted red.

In all cases, the timing is checked with the engine warmed to operating temperature, allowing the cooling fan to run once, with the engine idling and the transaxle in the **N** position.

ADJUSTMENT

Integra

1. Connect a tachometer according to the manufacturer's instructions.
2. On 1988–89 models, remove the main fuse box lid and connect the BROWN/BLACK and BROWN terminals with a jumper wire.
3. On 1990–92 models, using a suitable jumper wire, connect the GREEN and BROWN terminals of the ignition timing adjusting connector (light gray wire) located under the right side of the dash.
4. Make sure all wires are clear of the cooling fan and hot exhaust manifolds. Start the engine. Point the timing light at the timing mark pointer and the crankshaft pulley.

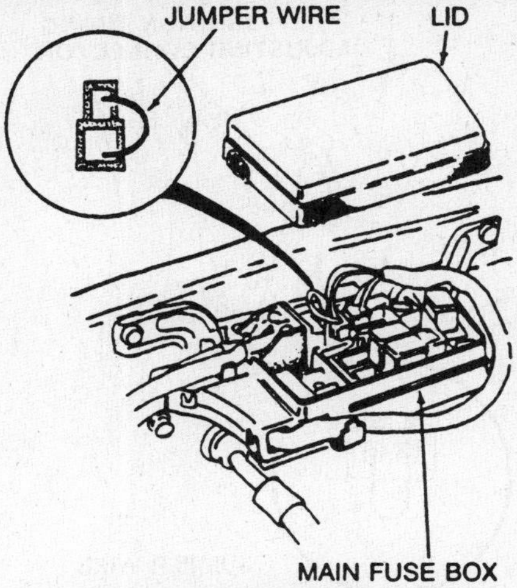

Jumper the brown/black and brown wires in the fuse box to adjust timing on 1988–89 Integra

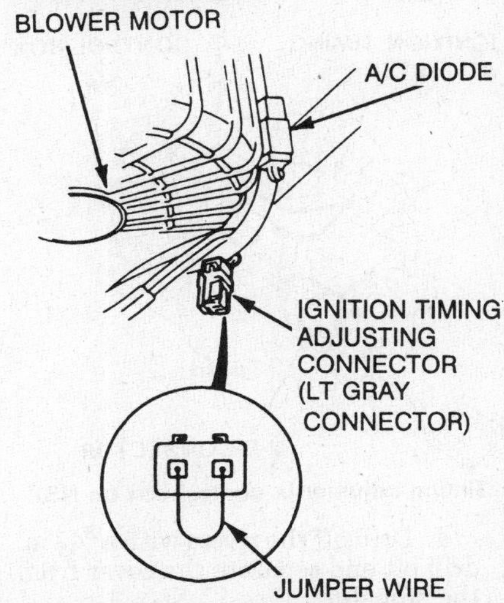

Jumper the timing connector near the blower motor on 1990–92 Integra

5. On 1988–89 models, the pointer should be on the RED mark (12 degrees BTDC) on the crankshaft pulley at 700–800 rpm (manual transaxle) or 650–750 rpm (automatic transaxle).
6. On 1990–92 models, the pointer should be on the RED mark (16 degrees BTDC) on the crankshaft pulley at 700–800 rpm.
7. If necessary, adjust the timing by loosening the distributor adjusting bolts and slowly rotate the distributor in the required direction while observing the timing marks.

CAUTION

Do not grasp the top of the distributor cap while the engine is running as it could give off a shock and cause personal injury. Instead, grab the distributor housing to rotate.

8. After making the necessary adjustment, tighten the hold-down bolts, taking care not to disturb the adjustment.
9. Remove the jumper wire.

Sterling and 1988–90 Legend

Timing is controlled by the ECU according to a very complex program and should only require adjustment due to installation of a major component, such as a new ECU. The adjuster is a potentiometer, in the emission control equipment box mounted on the firewall near the brake booster. It is riveted in place and the rivets must be drilled out to access the adjustment screw.

1. Start the engine and warm to normal operating temperatures; the cooling fan should run once.
2. Stop the engine and connect a timing light and tachometer according to the manufacturer's instructions.
3. Make sure all wires are clear of the cooling fan and hot exhaust manifolds. Start the engine. Point the timing light at the timing mark pointer and the crankshaft pulley.
4. The pointer should be on the RED mark on the crankshaft pulley (15 degrees BTDC) at 680 rpm. If timing is not correct, make sure all other possible faults (vacuum leak, bad con-

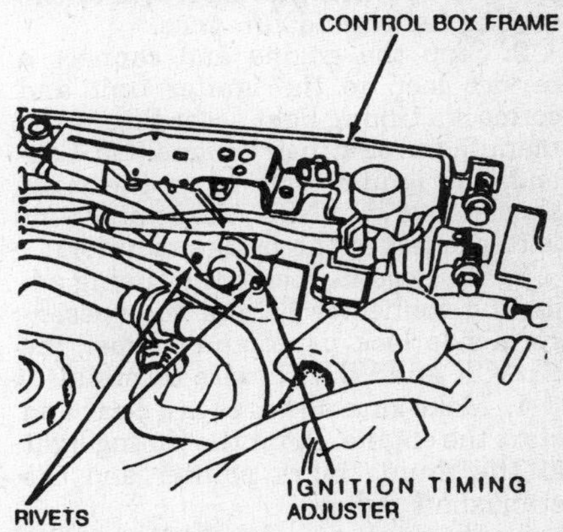

The timing adjuster is riveted in place on Sterling and 1988–90 Legend

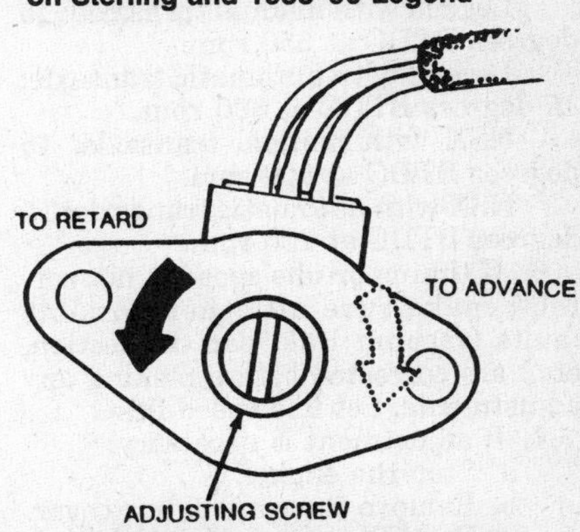

Ignition timing adjuster—Sterling and Legend

nection, etc.) are corrected before adjusting timing.

5. If adjustment is necessary:

a. Stop the engine.

b. Remove the control box upper and lower cover.

c. Drill off the rivets, with a $^3/_{16}$ in. drill bit and separate the stay cover from the adjuster.

d. Start the engine and turn the adjusting screw clockwise to advance or counterclockwise to retard.

e. After adjusting, install the adjuster and stay cover with new rivets.

1991–92 Legend and NSX

These vehicles use a distributorless ignition system. Timing is controlled by the ECU according to a very complex program and should only require adjustment due to installation of a major component, such as a new ECU. The adjuster is a potentiometer, mounted in the emission control equipment box. On Legend, the box is mounted near the left shock tower. On NSX, the box is to the left rear of the engine compartment. The adjuster has a cap that is riveted in place and the rivets must be drilled out to access the adjustment screw.

1. Start the engine and warm to normal operating temperature; the cooling fan should run once.

2. Stop the engine and connect a service loop to the igniter unit and connect a timing light according to the manufacturer's instructions. On Legend, the igniter unit is on the right shock tower. On NSX, the igniter unit is to the left of the intake manifold.

3. On Legend, locate the timing adjusting connector behind the passenger's side kick panel and jumper the BLACK and WHITE wire terminals.

4. Make sure all wires are clear and start the engine. Point the timing light at the timing mark pointer and the crankshaft pulley.

5. The pointer should be on the RED mark on the crankshaft pulley:

Legend with manual transaxle: 15 degrees BTDC at 650 rpm.

Legend with automatic transaxle: 15 degrees BTDC at 600 rpm.

NSX with manual transaxle: 15 degrees BTDC at 800 rpm.

NSX with automatic transaxle: 15 degrees BTDC at 750 rpm.

6. If timing or idle speed is not correct, make sure all other possible faults (vacuum leak, bad connection, etc.) are corrected before making any adjustments. Set idle speed first.

7. If adjustment is necessary:

a. Stop the engine.

b. Remove the control box cover.

c. On NSX, remove the timing adjuster from the control box.

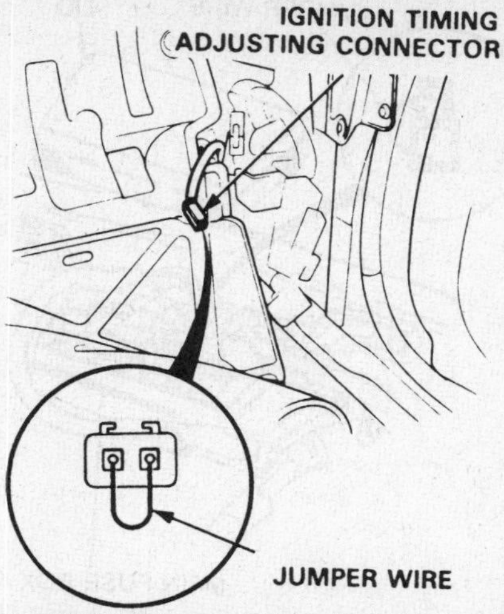

On 1991–92 Legend, the timing adjuster connector is behind the kick panel on the right side

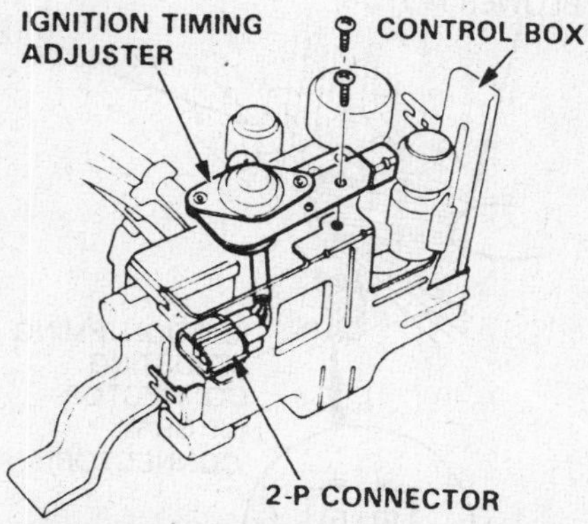

Timing adjuster in control box on NSX

d. Drill off the rivets with a $^3/_{16}$ in. drill bit and separate the cover from the adjuster.

e. On NSX, reconnect the timing adjuster.

f. Start the engine and turn the adjusting screw clockwise to advance or counterclockwise to retard timing.

8. After adjusting, install the cover with new rivets. On NSX, install the adjuster into the control box.

Alternator

PRECAUTIONS

• Observe the proper polarity of the battery connections by making sure the positive (+) and negative (−) terminal connections are not reversed. Mis-connection will allow current to flow in the reverse direction, resulting in damaged diodes and an overheated wire harness.

• Never ground or short out an alternator or regulator terminals.

• Never operate the alternator with it's or the battery's leads disconnected.

• Always remove the battery or disconnect the output lead while charging it.

• Always disconnect the ground cable when replacing any electrical components.

• Never subject the alternator to excessive heat or dampness if the engine is being steam cleaned.

• Never use arc welding equipment with the alternator connected.

BELT TENSION ADJUSTMENT

The initial inspection and adjustment to the alternator drive belt should be performed after the first 3000 miles or if the alternator has been moved for any reason; afterward, inspect the belt tension every 30,000 miles. Before adjusting, inspect the belt for cracks or wear; be sure it's surfaces are free of grease and oil.

1. Push down on the belt halfway between pulleys with a force of about 22 lbs. The belt should deflect:

0.16–0.41 in. (4–11mm)—on Integra

0.43–0.77 in. (11–19.5mm)—on 1988–90 Legend

0.22–0.45 in. (5.5–11.5mm)—on 1991–92 Legend

0.28–0.55 in. (7–14mm)—on NSX

2. To adjust belt tension on Integra, loosen the adjustment lock bolt and move the alternator with a prybar positioned against the front of the alternator housing. Do not apply pressure to any other part of the alternator.

3. To adjust tension on all other models, loosen the adjustment lock bolt and turn the adjusting bolt as required.

4. After obtaining the proper tension, tighten the adjustment lock bolt.

NOTE: Do not over tighten the belt. Damage to the alternator bearings could result.

REMOVAL & INSTALLATION

1988–89 Integra

1. Disconnect the negative battery terminal.

2. Label and disconnect the wires from the plugs on the rear of the alternator.

3. Remove the alternator harness cover, if equipped, the alternator bolts, the V-belt and the alternator.

To install:

4. Position the alternator into the brackets, connect the V-belt and loosely install the bolts.

5. Adjust the alternator belt tension and torque the bolts.

1990–91 Integra

1. Disconnect the negative battery cable.
2. Raise the locking tab on the left front spindle nut and loosen the nut with a $1^{7}/_{16}$ socket.
3. Raise and safely support the vehicle and remove the left front wheel.
4. Remove the damper fork nut and damper pinch bolt. Remove the damper fork.
5. Remove the knuckle-to-lower arm castle nut and separate the lower lower ball joint using a suitable puller with the pawls applied to the lower arm.
6. Pull the knuckle outward and remove the halfshaft outboard CV-joint from the knuckle using a plastic mallet.
7. Carefully pry the inner CV-joint out of the intermediate shaft and remove the halfshaft.

NOTE: Do not pull on the driveshaft, as the CV-joint may come apart. Use care when prying out the assembly and pull it straight to avoid damaging the intermediate shaft seals.

8. Disconnect and tag the alternator wire connection from the alternator. Remove the terminal nut and the white wire from the **B** terminal.
9. Loosen the adjusting nut and remove the alternator nut. Remove the alternator belt from the alternator pulley. Remove the lower through bolt and raise the alternator.
10. Remove the 3 mounting bracket bolts and mounting brackets. Remove the adjusting nut and upper through bolt, pull out the alternator.

To install:

12. Position the alternator and install the through bolt and adjusting nut. Install the mounting brackets and mounting bracket bolts and torque to 33 ft. lbs. (45 Nm).
13. Lower the alternator and install the lower through bolt. Install the alternator belt and adjusting nut. Tension the belt and torque the adjusting nut to 17 ft. lbs. (24 Nm).
14. Reconnect all the wiring.
15. Use a new set ring on the end of the inner CV-joint and slide it into the intermediate shaft. Use the plastic mallet to tap in on the halfshaft and set the ring.
16. Slide the outer CV-joint into place and position the ball joint in the lower arm. Install the nut and torque to 40 ft. lbs. (55 Nm) and tighten as necessary to install a new cotter pin.
17. Install the damper fork, pinch bolt and fork nut. Torque the the pinch bolt to 32 ft. lbs. (44 Nm) and the fork nut to 47 ft. lbs. (65 Nm).
18. Install a new self-locking spindle nut and wheel and place the vehicle on the ground before torquing the spindle nut. Attempting to torque the spindle nut while the vehicle is on jack stands or a lift may cause the vehicle to fall.
19. With the vehicle on the ground, torque the spindle nut to 134 ft. lbs. (185 Nm).
20. Reconnect the negative battery cable.

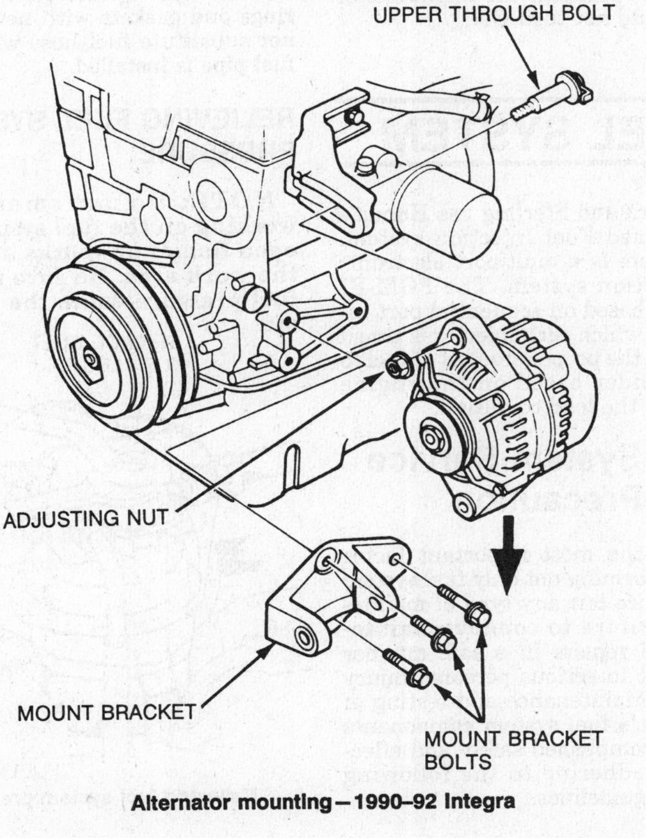

Alternator mounting—1990–92 Integra

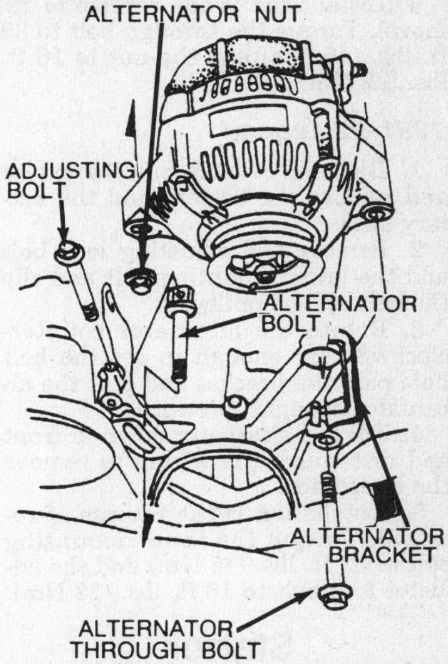

Alternator mounting on Sterling and 1988–90 Legend

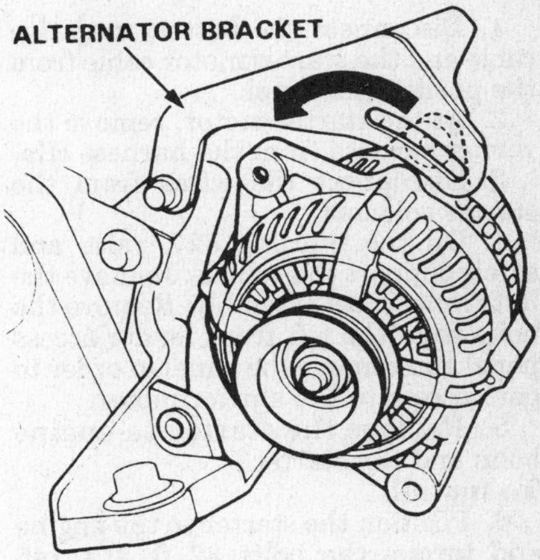

Rotate the alternator to remove it past the bracket

Sterling, 1988–89 Legend and NSX

1. Disconnect the negative battery cable.
2. Remove the alternator belt and wire covers.
3. Disconnect the alternator wiring.
4. Loosen the through bolt and the adjuster bolt and remove the belt.
5. Remove te through bolt and alternator bolt and lift the alternator out.

6. Installation is the reverse of removal. Torque the through bolt to 33 ft. lbs. (45 Nm) and the nut to 16 ft. lbs. (22 Nm).

1991–92 Legend

1. Disconnect both battery cables and remove the battery and the battery base.
2. Remove the adjusting lock bolt and the lower mounting bolt and slip the belt off the pulley.
3. Rotate the alternator counterclockwise far enough to get the bolt hole past the bracket and pull the alternator straight foreward.
4. Tilt the alternator down in front and disconnect the wiring to remove the alternator.
5. Installation is the reverse of removal. Torque the lower mounting bolt to 33 ft. lbs. (45 Nm) and the adjuster lock bolt to 16 ft. lbs. (22 Nm).

Starter

REMOVAL & INSTALLATION

1. Disconnect the battery negative cable and the starter motor cable from the positive terminal.
2. At the starter motor, remove the wiring harness from the harness clip.
3. Disconnect the wires from the starter solenoid.
4. On the Sterling 827, raise and safely support the vehicle. Remove the left front wheel assembly. Remove the bolts from the left front fender access panel and remove the panel in order to gain access to the starter motor.
5. Remove the starter-to-engine bolts and the starter.

To install:

6. Position the starter to the engine and torque the bolts 32 ft. lbs. (45 Nm). Connect the electrical connectors and wiring harness to the starter.
7. Connect the cables to the battery. Check the starter operation.

EMISSION CONTROLS

Please refer to "Emission Controls" in the Unit Repair section for system maintenance procedures. Due to the complex nature of modern electronic engine control systems, comprehensive diagnosis and testing procedures fall outside the confines of this repair manual. For complete information on diagnosis, testing and repair procedures concerning all modern engine and emission control systems, please refer to "Chilton's Guide to Fuel Injection and Electronic Engine Controls".

Maintenance Reminder Lamp

RESETTING

Some 1988–92 Acura vehicles, are equipped with a Scheduled Service Due warning light. This warning light is due to come on at approximately 7500 miles to indicate that an oil and oil filter change is needed. However, if a shorter oil change interval is desired, there are 7 different intervals to choose from all the way down to 1500 miles. To choose a new interval, push the Service Reset button (on the clock control panel) and the Arrow button for approximately 3 seconds, then push the Arrow button until the interval desired appears and then push the Set button. After completing the necessary maintenance service, the warning light must be reset.

In order to reset the maintenance light, open the Service Reset button. With the ignition switch in the **ON** position, hold the reset button in for at least 3 seconds. To verify that the reset is complete, turn the ignition switch to the **OFF** position and back to the **ON** position, the maintenance light should not turn ON.

FUEL SYSTEM

Both Acura and Sterling use Honda's Programmed Fuel Injection system. This system is a multiport electronic fuel injection system. The PGM-FI system is based on sequential port injection by which each injector is timed to provide the proper amount of fuel to each cylinder based on the engine speed and the load condition.

Fuel System Service Precaution

Safety is the most important factor when performing not only fuel system maintenance but any type of maintenance. Failure to conduct maintenance and repairs in a safe manner may result in serious personal injury or death. Maintenance and testing of the vehicle's fuel system components can be accomplished safely and effectively by adhering to the following rules and guidelines.

- To avoid the possibility of fire and personal injury, always disconnect the negative battery cable unless the repair or test procedure requires that battery voltage be applied.
- Always relieve the fuel system pressure prior to disconnecting any fuel system component (injector, fuel rail, pressure regulator, etc.), fitting or fuel line connection. Exercise extreme caution whenever relieving fuel system pressure to avoid exposing skin, face and eyes to fuel spray. Please be advised that fuel under pressure may penetrate the skin or any part of the body that it contacts.
- Always place a shop towel or cloth around the fitting or connection prior to loosening to absorb any excess fuel due to spillage. Ensure that all fuel spillage (should it occur) is quickly removed from engine and paint surfaces. Ensure that all fuel soaked cloths or towels are deposited into a suitable waste container.
- Always keep a dry chemical (Class B) fire extinguisher near the work area.
- Do not allow fuel spray or fuel vapors to come into contact with a spark or open flame.
- Always use a backup wrench when loosening and tightening fuel line connection fittings. This will prevent unnecessary stress and torsion to fuel line piping. Always follow the proper torque specifications.
- Always replace fuel fitting O-rings and gaskets with new ones. Do not substitute fuel hose where metal fuel pipe is installed.

RELIEVING FUEL SYSTEM PRESSURE

NOTE: Do not smoke while working on the fuel system. Keep open flames or sparks away from the work area. Be sure to relieve fuel pressure while the engine is off.

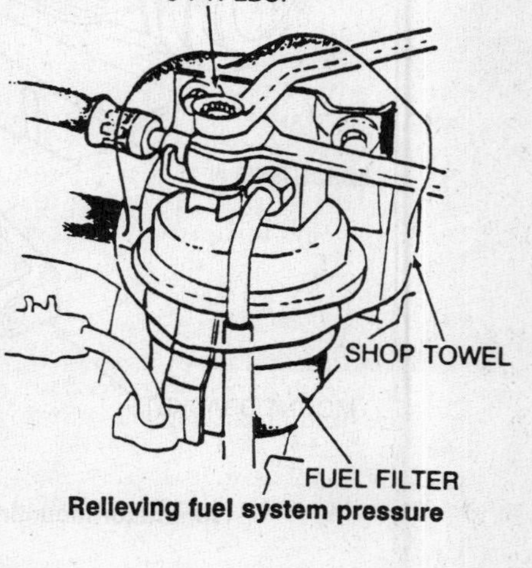

Relieving fuel system pressure

1. Disconnect the negative battery cable and remove the fuel filler cap.
2. Relieve the fuel pressure by slowly loosening the service bolt on the top of the fuel filter about a turn.

NOTE: Always place rag under the filter during this procedure to prevent fuel from spilling onto the engine. Always replace the washer between the service bolt and the banjo bolt, whenever the service bolt has been loosened.

Fuel Tank

REMOVAL & INSTALLATION

1988–89 Integra

1. Raise and safely support the vehicle and remove the rear wheels.
2. Remove the tank drain bolt and drain the fuel into an approved container.
3. Remove the fuel pump cover and disconnect the wiring for the gauge sending unit.
4. When disconnecting the hoses, twist and pull at the same time to work the hose off the tube.
5. Support the tank with a floor jack, remove the straps and lower the tank out of the vehicle. If it sticks on the undercoating, carefully pry it free.
6. Installation is the reverse of removal. Use a new sealing washer on the drain plug and torque to 36 ft. lbs. (50 Nm).

1990–92 Integra

1. Raise and safely support the vehicle and remove the rear wheels.
2. Remove the tank drain bolt and drain the fuel into an approved container.
3. Remove the rear seat and access panel and disconnect the gauge and pump wiring.
4. Remove the fuel hose protector and the 2-way valve. When disconnecting the hoses, twist and pull at the same time to work the hose off the tube.
5. Support the tank with a floor jack, remove the straps and lower the tank out of the vehicle. If it sticks on the undercoating, carefully pry it free.
6. Installation is the reverse of removal. Use a new sealing washer on the drain plug and torque to 36 ft. lbs. (50 Nm).

Legend and Sterling

1. On Sterling and 1988–90 Legend, remove the access panel from the floor of the trunk and disconnect the pump and gauge wiring.
2. On 1991–92 Legend, remove the rear seat and the access panel and disconnect the pump and gauge wiring.
3. Raise and safely support the vehicle and remove the rear wheels.
4. Remove the tank drain bolt and drain the fuel into an approved container.
5. Remove the fuel hose protector on 1991–92 models. When disconnecting the hoses, twist and pull at the same time to work the hose off the tube.
6. Support the tank with a floor jack, remove the straps and lower the tank out of the vehicle. If it sticks on the undercoating, carefully pry it free.
7. Installation is the reverse of removal. Use a new sealing washer on the drain plug and torque to 36 ft. lbs. (50 Nm).

Fuel Filter

REMOVAL & INSTALLATION

The fuel filter is either located on the firewall or inner fender in the engine compartment. The fuel filter should be changed every 4 years or 60,000 miles which ever comes first or whenever the fuel pressure drops below the specified rating (of 36–41 psi with the vacuum pressure hose disconnected) after making sure the fuel pump and pressure regulator are operating properly.

1. Disconnect the negative battery cable and remove the fuel filler cap.
2. Relieve the fuel pressure by slowly loosening the service bolt on the top of the fuel filter about a turn.

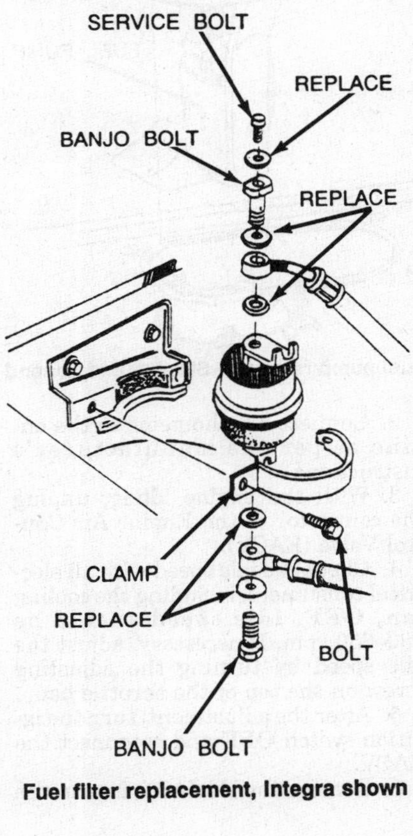

Fuel filter replacement, Integra shown

NOTE: Always place rag under the filter during this procedure to prevent fuel from spilling onto the engine. Always replace the washer between the service bolt and the banjo bolt, whenever the service bolt has been loosened.

3. Remove the fittings from the fuel filter.
4. Remove the fuel filter clamp and the fuel filter.
5. To assemble, use a new filter and washers and reverse the removal procedures.

Electric Fuel Pump

PRESSURE TESTING

On all models, the fuel pump should run for about 2 seconds when the ignition is first turned **ON**. By removing the filler cap and listening at the filler, it should be possible to hear the pump run each time the ignition is switched **ON**. It will then stop again until the starter is activated.

1. Disconnect the negative battery cable and remove the fuel filler cap.
2. Relieve the fuel system pressure.
3. Remove the service bolt and attach a fuel pressure gauge to the top of the fuel filter.
4. Start the engine and measure the fuel pressure with the engine idling and vacuum hose from the pressure regulator disconnected;
   - 1988–89 Integra – 35–41 psi. (240–279 kPa)
   - 1990–92 Integra – 37–44 psi. (255–304 kPa)
   - Sterling and 1988–90 Legend – 36–41 psi. (250–279 kPa)
   - 1991–92 Legend – 38–46 psi. (265–314 kPa)
   - NSX – 46–53 psi. (323–363 kPa)
5. Check to see that the pressure decreases when the vacuum hose is connected.
6. If the pressure is higher than specifications, check for a pinched or clogged fuel return hose or faulty pressure regulator.
7. If the pressure is lower than specifications, check for a clogged filter, defective pressure regulator or leakage in the fuel line.
8. After inspection, remove the pressure gauge.
9. To assemble, use a new filter, washers and reverse the removal procedures.

REMOVAL & INSTALLATION

1988–89 Integra

1. Disconnect the negative battery cable.
2. Relieve the fuel system pressure.
3. Raise and safely support the vehicle and remove the left rear wheel.
4. Remove the fuel pump cover bolts and the cover.
5. Remove the fuel pump mount bolts and the fuel pump with it's mount.
6. Disconnect the fuel hoses and the electrical connectors.
7. Installation is the reverse of removal. Use new sealing washers when connecting the fuel hoses. After the pump is installed, turn the ignition switch **ONN** and **OFF** a number of times to pressurize the system and check for leaks.

1990–92 Integra and NSX

1. On Integra, remove the rear seat and access panel and disconnect the wiring.
2. Remove the fuel tank.
3. Remove the pump flange mounting nuts and lift the pump out.
4. Installation is the reverse of removal. Use new sealing washers when connecting the fuel hoses. After the pump is installed, turn the ignition switch **ON** and **OFF** a number of times to pressurize the system and check for leaks.

Sterling and Legend

1. Disconnect the negative battery cable.
2. Relieve the fuel system pressure.
3. Remove the maintenance access cover in the luggage area.
4. Disconnect the fuel lines and the electrical connectors.
5. Remove the fuel pump from the fuel tank.
6. Installation is the reverse of removal. Use new sealing washers when connecting the fuel hoses. After the pump is installed, turn the ignition switch **ONN** and **OFF** a number of times to pressurize the system and check for leaks.

Fuel Injection

IDLE SPEED ADJUSTMENT

NOTE: The idle mixture is electronically controlled by the fuel injection system and should not be adjusted.

Integra

1. Start the engine and allow it to warm to normal operating temperatures; the cooling fan should turn **ON**.

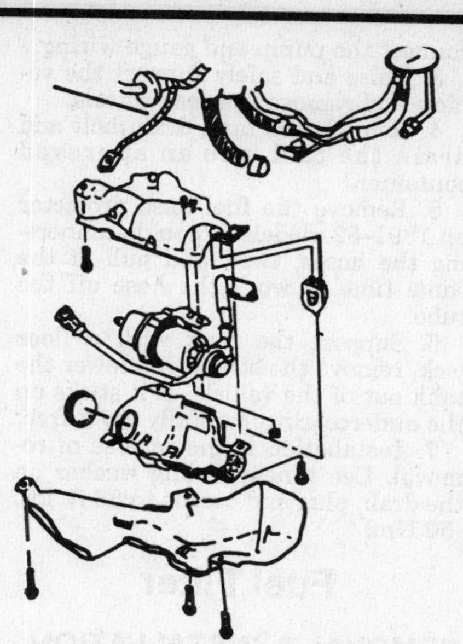

Fuel pump mounting on 1988-89 Integra

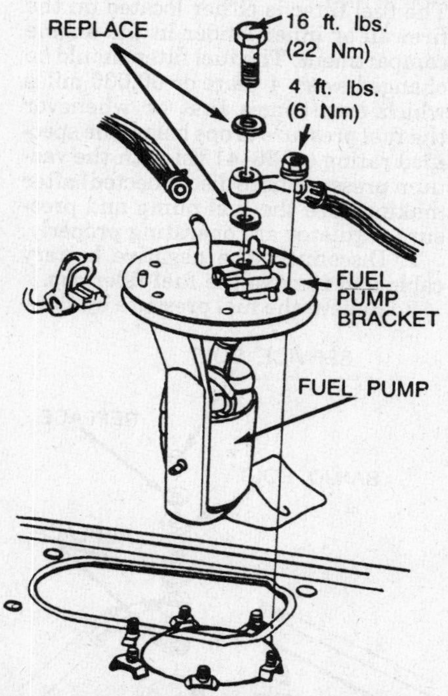

Fuel pump removal—Sterling and Legend

2. Connect a tachometer to the engine as per the manufacturer's instructions.
3. With the engine idling, unplug the connector at the Engine Air Control Valve (EACV).
4. Check the idle speed with all electrical consumers, including the cooling fan, **OFF**. Idle speed should be 500–650 rpm. If necessary, adjust the idle speed by turning the adjusting screw on the top of the throttle body.
5. After the adjustment, turn the ignition switch **OFF** and reconnect the EACV.
6. Remove the HAZARD fuse in the main fuse/relay box at the battery terminal for at least 10 seconds to clear the ECU fault code memory.
7. Start the engine and check idle speed again; it should be 650–750 rpm.
8. Check the idle speed under the following conditions:
   a. With headlights and rear window defogger turned **ON**.
   b. With the air conditioner compressor turned **ON**.
   c. If equipped with an automatic transaxle, shift the transaxle into gear.
9. Under increased electrical load or with the automatic transaxle in gear, idle speed should not decrease more than 50 rpm. With the air conditioner compressor **ON**, idle speed should increase no more than about 50 rpm.

Sterling and 1988–90 Legend

1. Start the engine and allow it to warm to normal operating temperatures; the cooling fan should turn **ON**.
2. Connect a tachometer to the engine, as per the manufacturer's instructions.
3. Make sure the front wheels are pointed straight ahead and check idle speed with all electrical consumers

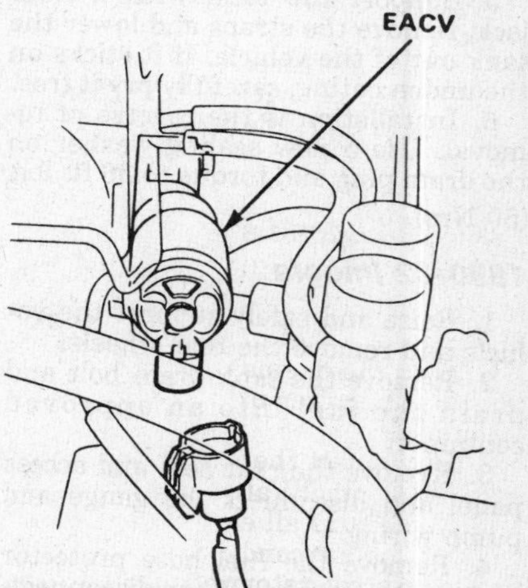

On Integra, disconnect the EACV to check basic idle speed

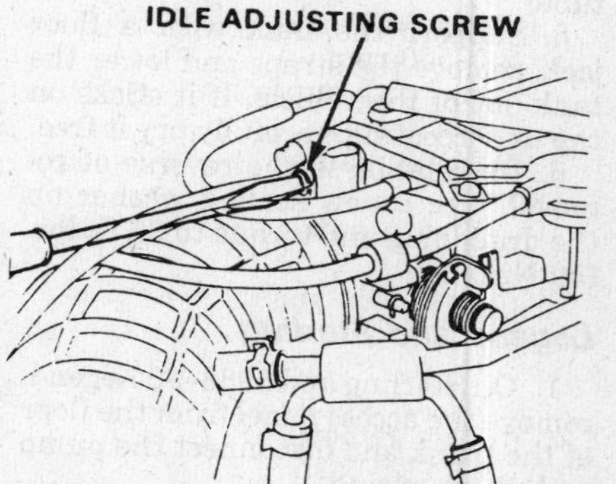

Intergra idle speed adjusting screw on top of the throttle body

OFF, including the cooling fan. Idle speed should be 630–730 rpm.

4. Check the YELLOW LED display at the ECU under the passenger's seat:

a. If the LED is **OFF**, do not adjust the idle speed.

b. If the LED is BLINKING, turn the idle adjusting screw ¼ turn clockwise.

c. If the LED is **ON**, turn the idle adjusting screw ¼ turn counterclockwise.

d. If the vehicle or the ECU has less than about 300 miles (500 km), do not adjust idle speed.

5. Check that the YELLOW LED turns **OFF** after approximately 30 seconds. If it does not turn **OFF**, rotate the idle adjusting screw on the throttle body ¼ turn in the same direction and repeat this operation until the YELLOW LED turns **OFF**.

6. Check the idle speed under the following conditions:

a. With headlights and rear window defogger turned **ON**.

b. Turning the steering wheel.

c. With the air conditioner compressor turned **ON**.

d. If equipped with an automatic transaxle, shift the transaxle into gear.

7. Under increased electrical or hydraulic load or with the automatic transaxle in gear, idle speed should not change.

1991–92 Legend and NSX

1. Start the engine and allow it to warm to normal operating temperatures; the cooling fan should turn **ON**.

2. Stop the engine, install a service loop to the igniter unit and connect a tachometer.

3. Disconnect the wiring to the Engine Air Control Valve (EACV).

4. Make sure all electrical consumers and the air conditioner are **OFF**. Press the accelerator pedal to start the engine, stabilize at about 1000 rpm, then slowly release the pedal.

5. Idle speed should be:

Legend with manual transmission: 400–500 rpm.

Legend with automatic transmission: 430–530 rpm.

NSX with manual transaxle: 600–700 rpm.

NSX with automatic transaxle: 550–650 rpm.

6. Adjust idle speed as required with the idle speed adjusting screw on top of the throttle body.

7. Stop the engine, reconnect the EACV, then remove the No. 15 fuse from the under-dash fuse box for more than 10 seconds to clear the ECU fault code memory.

8. Start the engine. After about 1 minute at no load, idle speed should stabilize at:

Legend with manual transmission: 600–700 rpm.

Legend with automatic transmission: 650–750 rpm in **P** or **N**.

NSX with manual transaxle: 750–850 rpm.

NSX with automatic transaxle: 700–800 rpm in gear.

Fuel Injector

REMOVAL & INSTALLATION

1. Disconnect the negative battery cable.

2. Relieve the fuel pressure from the fuel system.

3. On V6 engines, remove the wiring harness covers and intake manifold covers as required.

4. Disconnect the electrical connectors from the fuel injectors.

5. Disconnect the vacuum hose and the fuel return hose from the fuel pressure regulator. Be sure to place a shop rag or towel over the hose and tube before disconnecting them.

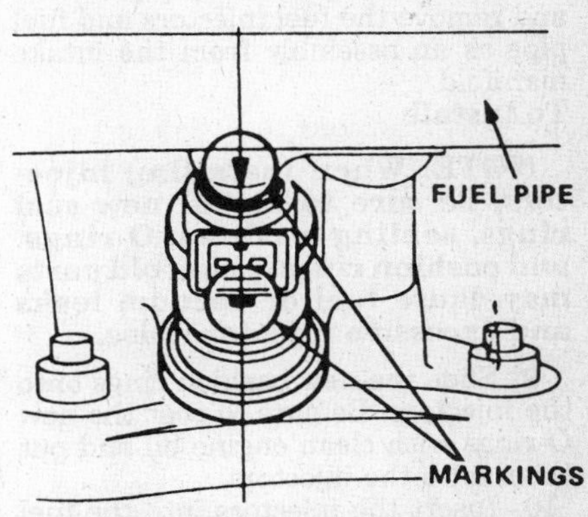

Align the marks on the injector and the fuel pipe

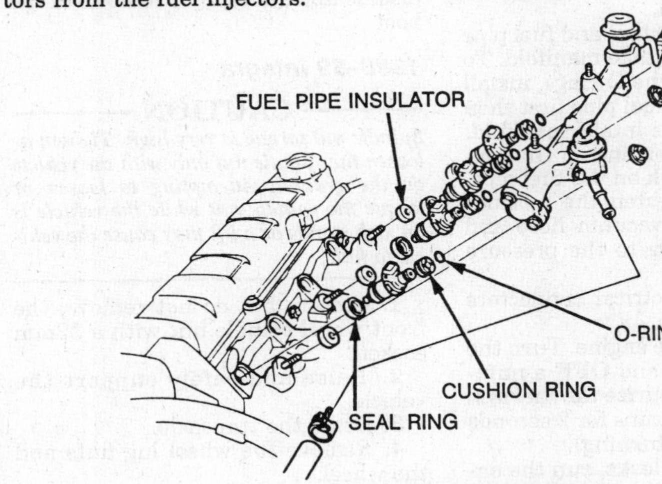

Injector assembly on Integra

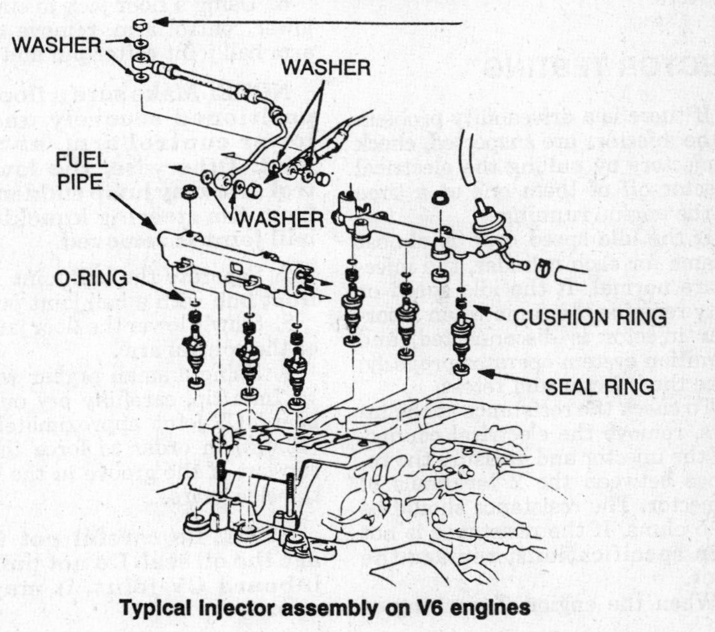

Typical injector assembly on V6 engines

6. Remove the fuel line and the pulsation damper, if equipped.
7. Remove the connector holder and loosen the retainer nuts of the fuel pipe assembly.
8. Disconnect the fuel pipe hoses and remove the fuel injectors and fuel pipe as an assembly from the intake manifold.

To install:

NOTE: When installing injectors, be sure to install new seal rings, sealing washers, O-rings, and cushion rings. Using old parts may cause fuel or vacuum leaks and excessive injector noise.

9. Slide the new cushion rings onto the injectors. Be sure to coat the new O-rings with clean engine oil and put them onto the injectors.
10. Insert the injectors into the fuel pipe first. Coat the new seal rings with clean engine oil and press them into the intake manifold.
11. Install the injectors and fuel pipe assembly into the intake manifold. To prevent damage to the O-rings, install the injectors in the fuel pipe first then install them into the intake manifold.
12. Align the center line on the connector with the mark on the fuel pipe.
13. Install and tighten the retainer nuts. Connect the vacuum hose and the fuel return hose to the pressure regulator.
14. Install the electrical connectors on the fuel injectors.
15. Do not start the engine. Turn the ignition switch **ON** and **OFF** a number of times to pressurize the fuel system (the pump only runs for 2 seconds without the engine turning).
16. If there are no leaks, run the engine to test the injectors, then install the covers.

INJECTOR TESTING

1. If there is a driveability problem and the injectors are suspected, check the injectors by pulling the electrical connector off of them one at a time with the engine running.
2. If the idle speed drop is almost the same for each cylinder, the injectors are normal. If the idle speed or quality remains the same when a particular injector is disconnected, and the ignition system operates properly, replace the injector and retest.
3. To check the resistance at the injectors, remove the electrical connector of the injector and measure the resistance between the 2 terminals of the injector. The resistance should be 1.5–2.5 ohms. If the resistance is not within specifications, replace the injector.
4. When the engine is running at idle, the injector should make a clicking sound as a sign that the injector is working. This can be checked by using a stethoscope to listen to each injector as the engine is idling.

DRIVE AXLE

Halfshaft

REMOVAL & INSTALLATION

The front halfshaft assembly consists of a sub-axle shaft and a halfshaft with 2 constant velocity (CV) joints. The CV-joints are factory packed with special grease and enclosed in sealed rubber boots. The outer joint cannot be disassembled except for removal of the boot.

1988–89 Integra

CAUTION

Spindle nut torque is very high. Tighten or loosen the spindle nut only with the vehicle on the ground. Attempting to loosen or torque the spindle nut while the vehicle is on jack stands or a lift may cause the vehicle to fall.

1. Loosen, but do not remove, the front wheel spindle nut with a 32mm socket.
2. Raise and safely support the vehicle.
3. Drain the transaxle.
4. Remove the wheel lug nuts and the wheel.
5. Remove the spindle nut.
6. Using a floor jack to support the lower control arm, remove the lower arm ball joint cotter pin and nut.

NOTE: Make sure a floor jack is positioned securely under the lower control arm, at the ball joint. Otherwise, the lower control arm may jump suddenly away from the steering knuckle as the ball joint is removed.

7. Separate the ball joint from the front hub with a ball joint puller.
8. Slowly lower the floor jack to lower the control arm.
9. Using a small prybar with a 3.5 × 7mm tip, carefully pry out the inboard CV-joint approximately ½ in. (13mm) in order to force the spring clip out of the groove in the differential side gears.

NOTE: Be careful not to damage the oil seal. Do not pull on the inboard CV-joint, it may come apart.

10. Pull the halfshaft out of the differential or the intermediate shaft.

To install:

11. If either the inboard or outboard joint boot bands have been removed for inspection or disassembly of the joint (only the inboard joint can be disassembled), be sure to repack the joint with a sufficient amount of CV-joint grease.
12. Make sure the CV-joint sub-axle bottoms during installation so the spring clip may hold the halfshaft securely in the differential/intermediate shaft groove. Always replace the spring clip with a new one.
13. Torque the ball joint nut to 29 ft. lbs. (40 Nm), then tighten as required to insert a new cotter pin.
14. With the vehicle on the ground, torque the spindle nut to 134 ft. lbs. (185 Nm) and stake the nut.

Sterling, Legend and 1990–92 Integra

CAUTION

Spindle nut torque is very high. Tighten or loosen the spindle nut only with the vehicle on the ground. Attempting to loosen or torque the spindle nut while the vehicle is on jack stands or a lift may cause the vehicle to fall.

1. With the vehicle on the ground,

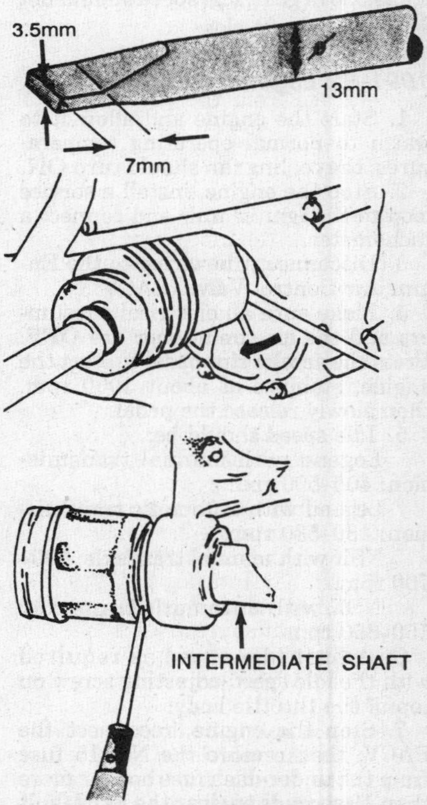

Disengaging the inner CV-joint with a small pry bar

raise the locking tab on the spindle nut and loosen it with a suitable socket.

2. Disconnect the negative battery cable.

3. Raise and safely support the vehicle and remove the spindle nut and front wheels.

4. If removing the right side halfshaft on Legend or Sterling, drain the transaxle.

5. Remove the damper fork nut and damper pinch bolt. Remove the damper fork.

6. Remove the lower ball joint nut and separate the lower ball joint using a suitable puller with the pawls applied to the lower arm.

7. Pull the knuckle outward and remove the halfshaft outboard CV-joint from the knuckle using a suitable plastic mallet.

8. Using a small prybar with a 3.5 × 7mm tip, carefully pry out the inboard CV-joint approximately ½ in. (13mm) in order to force the spring clip out of the groove in the differential side gears.

NOTE: Be careful not to damage the oil seal. Do not pull on the inboard CV-joint, it may come apart.

9. Pull the halfshaft out of the differential or the intermediate shaft.

10. To remove the intermediate shaft, remove the 3 bolts. Lower the bearing support close to the steering gear box and remove the intermediate shaft from the differential. To prevent any damage from the differential oil seal, hold the intermediate shaft horizontal until it is clear of the differential.

To install:

11. If either the inboard or outboard joint boot bands have been removed for inspection or disassembly of the joint (only the inboard joint can be disassembled), be sure to repack the joint with a sufficient amount of the correct CV-joint grease.

12. Always use a new set ring whenever the driveshaft is being installed. Be sure the driveshaft locks in the differential side gear groove and that the CV-joint subaxle bottoms in the differential or the intermediate shaft.

13. Torque the ball joint nut to 40 ft. lbs. (55 Nm) on Integra, 54 ft. lbs. (75 Nm) on Legend and Sterling. Then tighten the nut as required to install a new cotter pin.

14. Torque the lower damper nut and bolt to 47 ft. lbs. (65 Nm) and the upper pinch bolt to 32 ft. lbs. (44 Nm).

15. With the vehicle on the ground, torque the spindle nut to the proper specification, then stake the nut.

Sterling and 1988–90 Legend—180 ft. lbs. (250 Nm)

1991–92 Legend—242 ft. lbs. (335 Nm)

1990–92 Integra— 134 ft. lbs. (185 Nm)

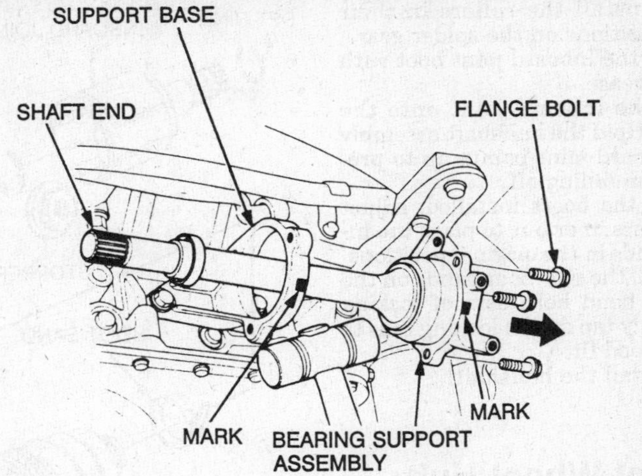

Tap the flange to remove the intermediate shaft from the NSX

Intermediate Shaft

REMOVAL & INSTALLATION

Except NSX

1. Raise and safely support the vehicle and drain the oil from the transaxle.

2. Remove the halfshaft and the bearing heat shield, if equipped.

3. Remove the 3 bearing support bolts.

4. On 1991–92 Legend, slide the shaft out of the oil pan. On all other models, lower the bearing support close to the steering gearbox and remove the intermediate shaft from the differential.

NOTE: To avoid damage to the differential oil seal, hold the intermediate shaft horizontal until it clears the differential.

5. To install, reverse the removal procedure. Torque the bearing support bolts to 16 ft. lbs. (22 Nm) for 8mm bolts, or 29 ft. lbs. (40 Nm) for 10mm bolts.

NSX

1. Raise and safely support the vehicle and drain the oil from the transaxle.

2. Remove the right halfshaft and heat shield.

3. Remove the bearing support flange bolts and tap the flange with a plastic hammer to remove it from the support base.

4. Installation is the reverse of removal. Torque the bolts to 16 ft. lbs. (22 Nm).

CV-Joint Boot

NOTE: The following procedures are for removing the CV-joint boot from the halfshaft once the halfshaft has been removed from the vehicle. If a quick seal boot is to be used, it is recommended that the manufacturers instructions be followed.

REMOVAL & INSTALLATION

NOTE: Be sure to mark the roller grooves during disassembly to ensure proper positioning during reassembly. Before disassembly, mark the spider gear and the driveshaft so they can be installed in their original positions. The inboard joint must be removed to replace the boots.

1. Remove the halfshaft that requires the boot change.

2. Remove the front and rear boot retaining bands and slide the boots off the halfshaft.

To install:

3. Wrap the spline with vinyl tape to prevent damage to the boots. Install the outboard and inboard boots onto the halfshaft, then remove the vinyl tape.

4. Install the stopper ring onto the driveshaft groove. Also install the dynamic damper at this time, if equipped.

5. Install the spider gear onto the halfshaft by aligning the marks and install it in its original position.

6. Fit the snapring into the halfshaft groove.

7. Pack the outboard joint boot with CV-joint grease only. Do not use a substitute or mix types of grease.

8. Fit the rollers to the spider gear with their high shoulders facing out-

ward. Reinstall the rollers in their original positions on the spider gear.

9. Pack the inboard joint boot with CV-joint grease.

10. Fit the inboard joint onto the halfshaft. Hold the halfshaft assembly so the inboard joint points up to prevent it from falling off.

11. With the boots installed, adjust the CV-joints in or out to place the inner boot ends in the original positions.

12. Install the new boot bands on the boots and bend both sets of locking tabs. Lightly tap on the locking tabs to ensure a good fit.

13. Reinstall the halfshaft.

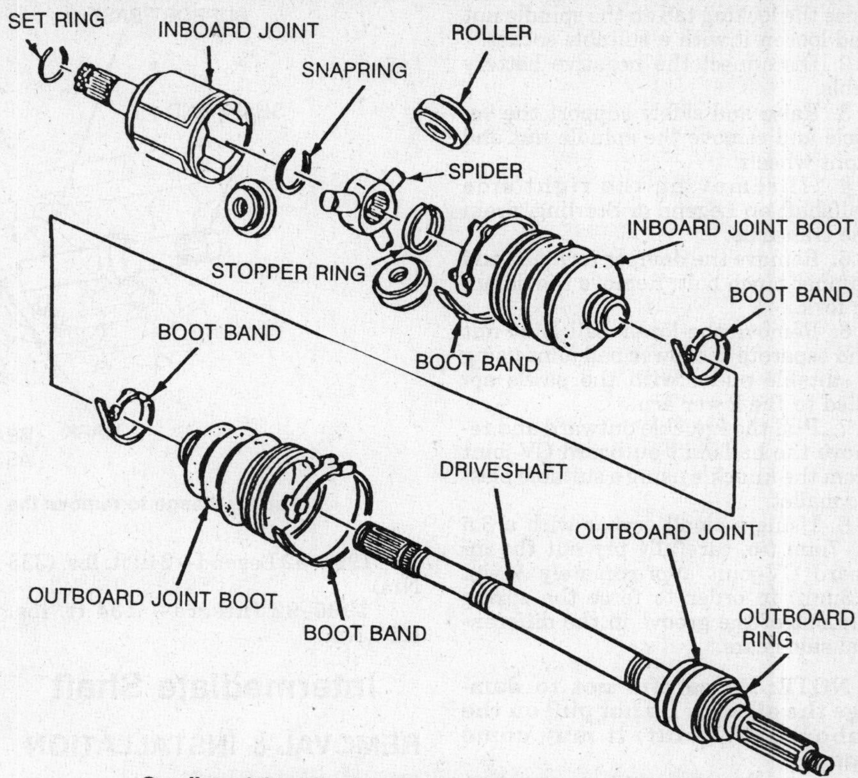

On all models, only the inner CV-joint can be disassembled

Front Wheel Hub, Knuckle and Bearings

REMOVAL & INSTALLATION

NOTE: The following procedures for hub and wheel bearing removal and installation necessitate the use of many special tools and a hydraulic press. Do not attempt this procedure without these special tools.

1988–89 Integra

1. Pry the lock tab away from the spindle and loosen the nut. Slightly loosen the lug nuts.

2. Raise and safely support the vehicle. Remove the front wheel and spindle nut.

3. Remove the brake caliper bolts and the caliper from the knuckle. Do not allow the caliper to hang by the brake hose, support it with a length of wire.

4. Remove the disc brake rotor retaining screws, if equipped. Screw two 8 × 1.25 12mm bolts into the disc brake removal holes and turn the bolts to press the rotor from the hub.

NOTE: Only turn each bolt 2 turns at a time to prevent cocking the disc excessively.

5. Remove the tie rod from the knuckle using a tie rod end removal tool. Use care not to damage the ball joint seals or threads.

6. Use a floor jack to support the lower control arm, then remove the cotter pin and the castle nut from the lower arm ball joint.

NOTE: Be sure to place the jack securely beneath the lower control arm at the ball joint. Otherwise, the tension from the torsion bar may cause the arm to suddenly jump away from the steering knuckle as the ball joint is removed.

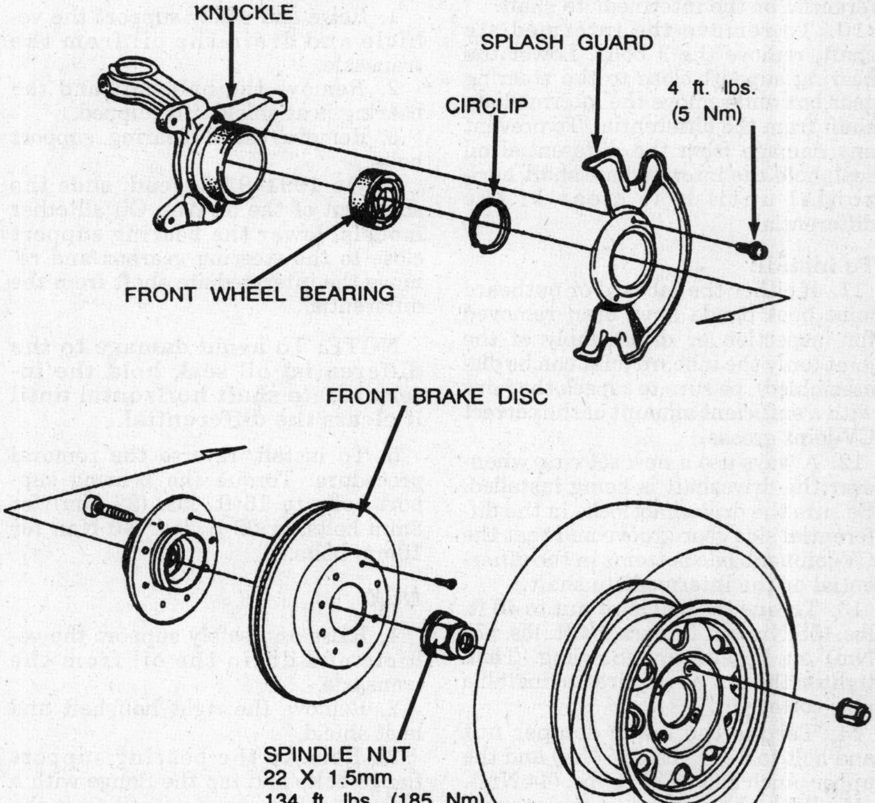

Front knuckle and wheel bearing assembly—1988–89 Integra

7. Remove the lower arm from the knuckle using the ball joint remover.
8. Loosen the pinch bolt which retains the shock in the knuckle. Tap the top of the knuckle with a hammer and slide it off the shock.
9. Remove the knuckle and hub, if still attached, by sliding the assembly off of the halfshaft.
10. Remove the hub from the knuckle a hydraulic press and the proper rams.
11. Remove the splash guard and the snapring.
12. Press the bearing outer race out of the knuckle using a hydraulic press and the proper rams.
13. Remove the outboard bearing inner race from the hub using a bearing puller.

NOTE: Whenever the wheel bearings are removed, always replace them with a new set of bearings and an outer dust seal.

14. Clean all old grease from the halfshaft and spindles on the vehicle.
15. Remove the old grease from the hub and knuckle and thoroughly dry and wipe clean all components.

To install:

16. When installing a new bearing into the knuckle, be sure to press only on the outer race or the bearing will be destroyed. Properly support the knuckle.
17. Install the snapring and the splash guard.
18. Support the hub properly and press on the inner race of the bearing to install the knuckle onto the hub. Do not press on the outer race or on the knuckle or the bearing will be destroyed.
19. The remaining installation is the reverse of the removal procedure. Torque the ball joint and tie rod end nuts to 29 ft. lbs. (40 Nm) and install new cotter pins. Torque the pinch bolt to 47 ft. lbs. (65 Nm).
20. With the vehicle on its wheels, torque the spindle nut to 134 ft. lbs. (185 Nm). Use a new spindle nut and stake it after torquing.

CAUTION

Spindle nut torque is very high. Tighten or loosen the spindle nut only with the vehicle on the ground. Attempting to loosen or torque the spindle nut while the vehicle is on jack stands or a lift may cause the vehicle to fall.

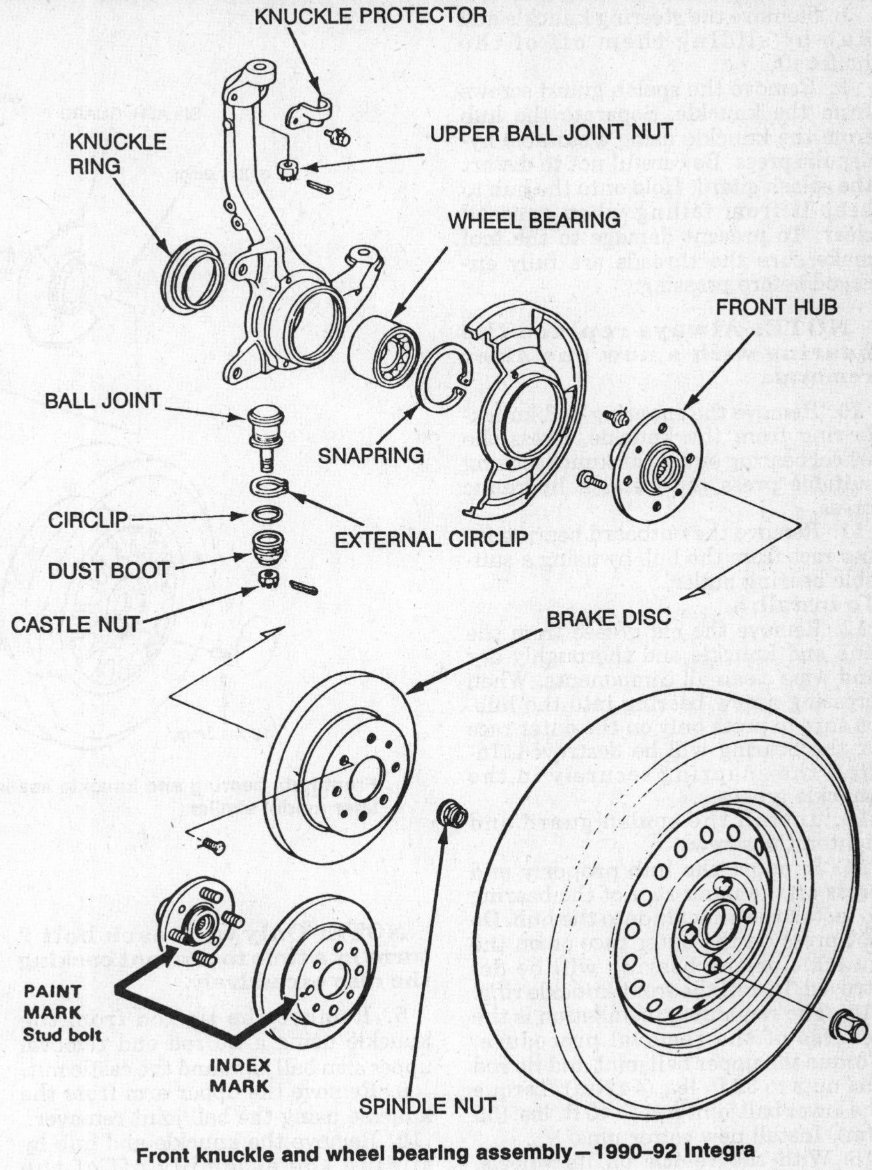

Front knuckle and wheel bearing assembly—1990–92 Integra

1990–91 Integra

1. Pry the lock tab away and loosen the spindle nut. Slightly loosen the lug nuts.
2. Raise and safely support the vehicle. Remove the front wheel and spindle nut.
3. Remove the brake caliper bolts and the caliper from the knuckle. Do not allow the caliper to hang by the brake hose, support it with a length of wire.
4. Remove the disc brake rotor retaining screws, if equipped. Screw two 8 × 1.0mm bolts into the disc brake removal holes and turn the bolts to press the rotor from the hub.

NOTE: Only turn each bolt 2 turns at a time to prevent cocking the disc excessively.

5. Remove the cotter pin from the tool. Use care not to damage the ball joint seals.
6. Remove the cotter pin from the lower arm ball joint and the castle nut.
7. Remove the lower control arm from the knuckle using the ball joint remover.
8. Remove the cotter pin from the tie rod end and remove the castle nut. Break loose the tie rod ball joint using a suitable ball joint removal tool and lift the tie rod out the steering knuckle.
6. Remove the cotter pin and loosen the lower arm ball joint nut half the length of the joint threads. Separate the ball joint and lower arm using a puller with the pawls applied to the lower arm. Avoid damaging the ball joint thread. If necessary, apply a suitable penetrating type lubricant to loosen the ball joint.
7. Remove the knuckle protector. Remove the cotter pin and remove the upper ball pin nut. Separate the upper ball joint and the knuckle using a suitable ball joint removal tool.

8. Remove the steering knuckle and hub by sliding them off of the halfshaft.

9. Remove the spalsh guard screws from the knuckle. Separate the hub from the knuckle using a suitable hydraulic press. Be careful not to distort the splash guard. Hold onto the hub to keep it from falling when pressed clear. To prevent damage to the tool make sure the threads are fully engaged before pressing.

NOTE: Always replace the bearing with a new one after removal.

10. Remove the snapring and knuckle ring from the knuckle. Press the wheel bearing out of the knuckle using suitable press tools and a hydraulic press.

11. Remove the outboard bearing inner race from the hub by using a suitable bearing puller.

To install:

12. Remove the old grease from the hub and knuckle and thoroughly dry and wipe clean all components. When pressing a new bearing into the hub, be sure to press only on the outer race or the bearing will be destroyed. Install the snapring securely in the knuckle groove.

13. Install the spalsh guard and tighten the screws.

14. Support the hub properly and press on the inner race of the bearing to install the knuckle onto the hub. Do not press on the outer race or on the knuckle or the bearing will be destroyed. Install the front knuckle ring.

15. The remaining installation is the reverse of the removal procedure. Torque the upper ball joint and tie rod end nuts to 32 ft. lbs. (44 Nm). Torque the lower ball joint nut to 40 ft. lbs. (55 Nm). Install new cotter pins.

16. With the vehicle on its wheels, torque the spindle nut to 134 ft. lbs. (185 Nm). Use a new spindle nut and stake it after torquing.

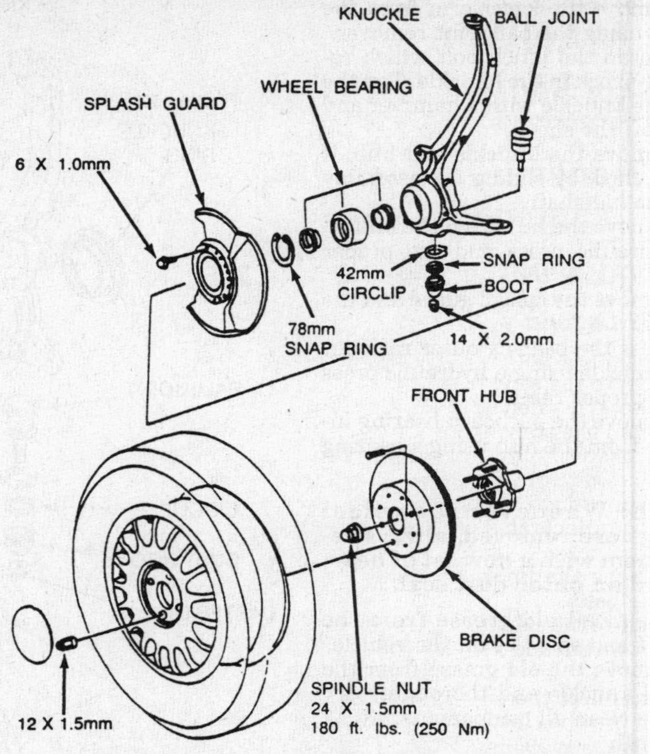

Front hub, bearing and knuckle assembly on Sterling and 1988–90 Legend, later model similar

Sterling and Legend

1. Pry the lock tab away from the spindle and loosen the 36mm nut. Slightly loosen the lug nuts.

2. Raise and safely support the vehicle. Remove the front wheel and spindle nut.

3. Remove the bolts retaining the brake caliper and the caliper from the knuckle. Do not allow the caliper to hang by the brake hose, support it with a length of wire.

4. Remove the disc brake rotor retaining screws if equipped. Screw both 8 × 1.25 × 12mm bolts into the disc brake removal holes and turn the bolts to press the rotor from the hub.

NOTE: Only turn each bolt 2 turns at a time to prevent cocking the disc excessively.

5. Remove the tie rod from the knuckle using a tie rod end removal upper arm ball joint and the castle nut.

9. Remove the upper arm from the knuckle using the ball joint remover.

10. Remove the knuckle and hub by sliding the assembly off of the halfshaft.

NOTE: Any time the hub is removed, the wheel bearing must be replaced with a new one.

11. On Sterling and 1988–90 Legend, properly support the knuckle and press the hub out of the bearing.

12. On 1991–92 Legend, the hub can be removed with a slide hammer. Clamp the knuckle in a vise and secure the slide hammer to the wheel studs.

13. Remove the splash guard and snaprings.

14. Support the knuckle and press the bearing out towards the wheel side.

15. If the inner bearing race stayed on the hub, use a puller to remove it.

To install:

16. Clean all parts and examine for wear. A worn or damged hub will cause premature bearing failure and should be replaced.

17. When pressing in a new bearing, install the inner snapring first and press the bearing in from the wheel side. Be sure to press only on the outer race or the bearing will be damaged.

18. Install the outer snapring and the splash guard.

19. Properly support the knuckle and press the hub into the bearing. Do not press on the wheel studs or they will press out of the hub. Be sure to support the knuckle by the inner race or the bearing will be damaged.

20. Install the knuckle in the reverse order of removal. Torque the lower ball joint nut to 54 ft. lbs. (75 Nm) and tighten as required to install a new cotter pin.

21. Torque the upper ball joint nut to 32 ft. lbs. (44 Nm) and tighten as required to install a new cotter pin. Torque the tie rod end to 36 ft. lbs. (50 Nm) and tighten as required to install a new cotter pin.

22. With the wheel installed and all 4 wheels on the ground, torque the spindle nut and stake it in place.

Sterling and 1988–90 Legend – 180 ft. lbs. (250 Nm)

1991–92 Legend – 242 ft. lbs. (335 Nm)

MANUAL TRANSMISSION

Transmission Assembly

REMOVAL & INSTALLATION

1991–92 Legend

1. Disconnect both battery cables.
2. Remove the strut bar.
3. Raise and safely support the vehicle. Drain the transmission.
4. Remove the emission control equipment box from the firewall without disconnecting the vacuum hoses.
5. Disconnect the transmission wiring.
6. Remove the upper transmission mounting bolts.
7. Remove the exhaust pipe and catalytic converter and remove the heat shield.
8. Disconnect the oil cooler hoses.
9. Remove the clutch slave cylinder without disconnecting the hydraulic hose. Shift into low gear and unbolt the shift rod and torque rod.
10. Remove the lower plate and reinstall the steering gear box mounting bolts.
11. Remove the exhaust pipe bracket and the mount bracket as required.
12. The differential stays on the vehicle. Remove the 36mm sealing bolt and secondary cover and install the extension shaft removal tool. Disconnect the differential extension shaft from the transmission.
13. Remove the flywheel cover.
14. Place a transmission jack securely under the transmission and take the weight off the mounts.
15. Remove the mounts and brackets as required to slide the transmission back away from the engine. Be careful not to lose the shim between the transmission and differential.

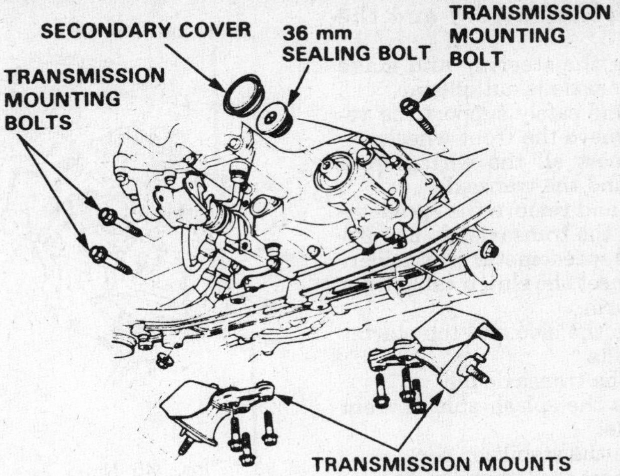

Transmission removal—1991–92 Legend

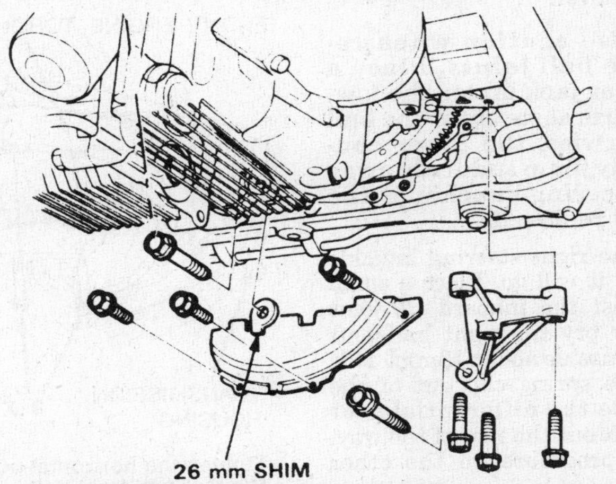

Install the 26mm shim when installing the transmission

To install:

16. Clean and lightly lubricate the release fork contact points with molybdenum grease and install the fork.
17. Install the dowel pins and set the extension shaft in place. Use a new set ring on the shaft and lightly lubricate the splines with molybdenum grease. On automatic transmission, make sure the secondary spring in in the differential side of the extension shaft.
18. Install the transmission and start all of the bolts. Don't forget the transmission-to-differential shim. Torque the 12mm bolts to 54 ft. lbs. (75 Nm).
19. Install all of the engine stiffener 8mm bolts, then torque to 16 ft. lbs. (22 Nm).
20. Install the mounts and brackets. Torque the 10mm bolts to 29 ft. lbs. (39 Nm), 10mm nuts to 36 ft. lbs. (49 Nm), and any remaining 12mm bolts to 54 ft. lbs. (75 Nm).
21. With the transmission in gear, install the extension shaft using the special tool. Make sure the shaft snaps into place on the set ring.
22. Pack the shaft area with molybdenum grease, but keep the thread area clean. Apply liquid gasket to the sealing bolt threads and install the bolt and cover.
23. Install the lower plate and torque the bolts to 28 ft. lbs. (39 Nm). These bolts thread into aluminum and must have the special Dacro® coating to avoid corrosion.
24. Install the slave cylinder and connect the shift linkage and torque rod.
25. Connect the oil cooler hoses, if equipped.
26. Install the heat shield and the exhaust pipe and catalytic converter. Use new locking nuts and gaskets. Torque the exhaust flange nuts to 40 ft. lbs. (55 Nm) and the catalyst flange nuts to 26 ft. lbs. (34 Nm).
27. Install the emission control equipment box and the strut bar.
28. Connect all the wiring and the battery cables and refill the transmission oil.

MANUAL TRANSAXLE

For further information on transmissions/transaxles, please refer to "Chilton's Guide to Transmission Repair".

Transaxle Assembly

REMOVAL & INSTALLATION

1988–89 Integra

1. Disconnect the negative battery

cable from the battery and the transaxle.

2. Unlock the steering and make sure the transaxle is out of gear.

3. Raise and safely support the vehicle and remove the front wheels.

4. Disconnect all the wiring from the starter and the transaxle.

5. Unclip and remove the speedometer cable at the transaxle; do not disassemble the speedometer gear holder.

6. Disconnect the clutch cable from the release arm.

7. Remove the side and top starter mounting bolts.

8. Drain the transaxle oil.

9. Remove the splash shields from the underside.

10. Remove the stabilizer bar.

11. Disconnect the left and right lower ball joints and tie rods ends, using a ball joint remover.

NOTE: Use caution when removing the ball joints. Place a suitable floor jack under the lower control arm securely at the ball joint. Otherwise, the lower control arm may jump suddenly away from the steering knuckle as the ball joint is removed.

12. Turn the right steering knuckle out as far as it will go. Place a small prybar against the inboard CV-joint and carefully pry the right halfshaft out of the transaxle about 13mm. This will force the spring clip out of the groove inside the differential gear splines. Pull it out the rest of the way. Repeat this procedure on the other side.

13. Attach a suitable chain hoist to the engine and lift the engine slightly to take the weight off the mounts.

14. Disconnect the header pipe at the exhaust manifold.

15. Disconnect the shift lever torque rod. To disconnect the shift linkage foreward of the universal joint, slide the pin retainer out of the way and drive out the roll pin.

16. Raise a transaxle jack securely against the transaxle to take up the weight.

17. Remove the bolts from the front transaxle mount at the front engine stiffener.

18. Remove the intake manifold bracket and the rear engine mount bracket.

19. Remove the transaxle housing bolts from the engine torque bracket.

20. Remove the remaining starter mounting bolts and take out the starter.

21. Remove the remaining transaxle mounting bolts.

22. Pull the transaxle away from the engine until it clears the 14mm dowel pins, then lower on the transaxle jack.

23. Separate the mainshaft from the clutch pressure plate and remove the transaxle by lowering the jack.

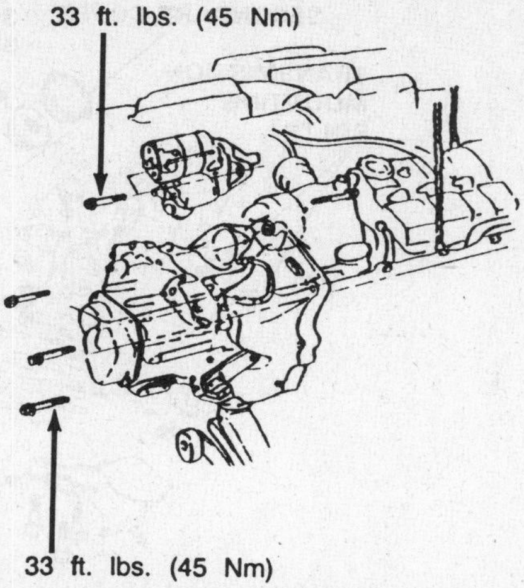

Transaxle removal—1988–89 Integra

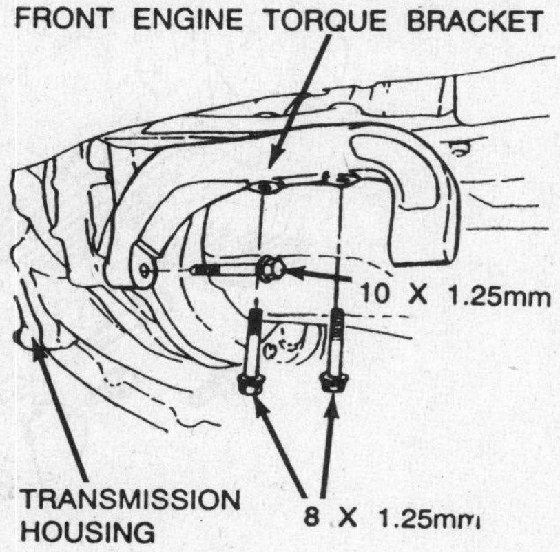

Tighten the horizontal bolt first, then the rear and front bolts

To install:

24. Install the transaxle on a transaxle jack. Clean and lubricate the clutch release bearing surfaces.

25. Make sure both 14mm dowel pins are installed in the clutch housing.

26. Raise the transaxle high enough to align the dowel pins with the matching holes in the block.

27. Roll the transaxle toward the engine and fit the mainshaft into the clutch disc splines. If the driver's side suspension was left in place, install new spring clips on both axles and carefully insert the left axle into the differential when installing the transaxle.

NOTE: Install new 26mm spring clips on both axles. Make sure the axles fully bottom out. Slide the axle in until the spring clip engages the differential.

28. Push and wiggle the transaxle until it fits flush with the flange.

29. Bolt the transaxle to the engine with the mounting bolts from the engine side. Torque the bolts to 50 ft. lbs. (68 Nm).

30. Install the rear mount bracket on the transaxle housing. Torque the mounting bolts to 47 ft. lbs. (65 Nm).

31. Install the engine torque bracket on the transaxle housing. Torque the mounting bolts to 33 ft. lbs. (45 Nm).

32. Loosely install the bolts for the front of the transaxle mount, then torque them to specifications.

33. Install the starter mounting bolts and torque them to 33 ft. lbs. (45 Nm).

34. Install the intermediate shaft, the right and left halfshaft.

35. Turn the right steering knuckle/axle assembly outward far enough to insert the free end of the axle into the transaxle. Repeat this procedure on the other side.

NOTE: Make sure the axles fully bottom out. Slide the axle in until the spring clip engages the differential.

36. Reconnect the shift rod and the shift lever torque rod.

37. Reconnect the lower arm to the ball joints and torque them to 33 ft. lbs. (45 Nm).

38. Reconnect the tie rod end ball joints and torque them to 33 ft. lbs. (45 Nm).

39. Install the engine and wheelwell splash shields.

40. Reconnect the exhaust header pipe.

41. Install the front wheels, lower the vehicle to the ground and tighten the lug nuts to 80 ft. lbs. (110 Nm).

42. Remove the chain hoist from the engine.

43. Install the speedometer cable.

44. Install the transaxle housing bolts and torque them to 33 ft. lbs. (45 Nm).

45. Connect the clutch cable to the release arm, then attach the cable housing end to the transaxle bracket.

46. Connect the wiring.

47. With the ignition key turned **OFF**, connect the ground cable to the battery and the transaxle.

48. Refill the transaxle with the correct oil and adjust the clutch free-play.

49. Check the transaxle for smooth operation.

1990–92 Integra

1. Disconnect the negative first and then the positive battery cables.

2. Remove the 4 battery mounting bolts and remove the battery.

3. Remove the air cleaner case complete with air intake tube. Disconnect the transaxle ground cable.

4. Loosen the clutch cable adjusting nut and disconnect the clutch cable at

the release arm, then disconnect from the clutch cable bracket.

5. Disconnect the electrical connectors for the back-up light switch, oxygen sensor and the starter motor cables and wire harness clamp from the starter.

6. Remove the power steering speed sensor without disconnecting the sensor hose.

7. Disconnect the distributor connectors and remove the distributor mounting bolts. Before removing the distributor, be sure to make some alignment marks on the distributor housing and engine to aid in the installation procedure.

8. Raise and safely support the vehicle. Remove the starter.

9. Drain the transaxle oil into a suitable container.

10. Remove the right front splash shield and splash guard. Remove the center beam bolts and remove the center beam.

11. Remove the cotter pin from the lower right ball joint castle nut, remove the nut and using a ball joint separator, remove the ball joint from the lower arm.

12. Remove the right damper fork. Remove the right radius rod locknut, then the bolts and remove the right radius arm.

13. Remove the right halfshaft assembly.

14. Remove the cotter pin from the lower left ball joint castle nut, remove the nut and using a ball joint separator, remove the ball joint fron the lower arm.

15. Remove the left halfshaft from the intermediate shaft. Remove the intermediate shaft bolts and remove the intermediate shaft.

16. Remove the shift rod and shift lever torque rod.

17. Remove the front engine stiffener and the rear engine stiffener. Remove the 4 bolts from the clutch housing cover and remove the cover.

18. Remove the 2 transaxle mount bolts from the engine side.

19. Remove the 2 transaxle mount bolts from the rear engine mount bracket.

20. Remove the side transaxle mount bolt from the underside. Remove the front transaxle mount bolts and mount.

21. Install the bolts into the cylinder head and attach a suitable lifting device or chain hoist to the bolts. Lift the engine slightly to take the load off of the engine mounts.

22. Place a suitable transaxle jack under the transaxle and raise it enough to take the weight off of the transaxle mounts. Remove the bolts and nuts that attach the brackets to the side transaxle mounts.

23. Remove the 3 transaxle mount bolts from the transaxle side.

24. Pull the transaxle away from the clutch pressure plate until it clears the mainshaft, then remove the transaxle by lowering the jack.

To install:

25. Install the transaxle on a transaxle jack. Clean and lubricate the clutch release bearing surfaces.

26. Make sure both 14mm dowel pins are installed in the clutch housing.

27. Loosely install the transaxle mount bolts, then torque them to 49 ft. lbs. (68 Nm).

28. Secure the transaxle to the engine with the engine side mounting bolt and torque it to 50 ft. lbs. (68 Nm).

29. Install the transaxle to side transaxle mount. Install the transaxle to the front transaxle mount.

30. Install the transaxle to the rear engine mount bracket.

31. Loosely install the bolt in the front stiffener and then torque then to 17 ft. lbs. (24 Nm).

32. Loosely install the bolt in the rear stiffener and then torque then to 17 ft. lbs. (24 Nm).

33. Remove the transaxle jack. Remove the lifting device by removing the hoist bolts from the cylinder head.

34. Reconnect the shift linkage and torque rod.

35. Install the intermediate shaft.

36. Install the left halfshaft, then the left ball joint and lower arm.

37. Install the right halfshaft assembly. Be sure to turn the right steering knuckle fully outward and slide the axle into the differential until the spring clip engages the side gear.

38. Install the right radius arm, damper fork bolt and the right ball joint to the lower arm.

39. Install the center beam. Install the right front splash guard and splash shield.

40. Install the starter motor. Install the distributor, be sure to use the alignment marks made earlier in the removal procedure.

41. Connect the starter motor cables and wire harness clamp. Install the power steering speed sensor.

42. Connect the oxygen sensor connector, back-up light connector and connect the clutch cable to the clutch cable bracket, then connect to the release arm.

43. Connect the transaxle ground. Install the air cleaner case complete with the air intake tube.

44. Install the battery base. Refill the transaxle with the recommended oil.

45. Install the battery and connect the battery cables.

46. Adjust the clutch free-play. Check the ignition timing and road test the vehicle to be sure the transaxle is operating properly.

Sterling and 1988–90 Legend

1. Disconnect the both battery cables from the battery.

2. Disconnect the starter and ground cables.

3. Disconnect the back-up light wires from the engine harness.

4. Loosen the 6mm bolt attaching the harness holder at the side of the transaxle hanger and the release harness from the transaxle.

5. Loosen the 6mm bolts at the side of the battery base and the intake hose band.

6. Remove the air cleaner case assembly along with the intake hose.

7. Remove the 8mm bolts and the clutch slave cylinder with the clutch hose and the pushrod.

NOTE: Do not operate the clutch pedal once the slave cylinder has been removed.

8. Remove the 8mm bolts and clutch damper assembly from the transaxle hanger bracket.

9. Remove the power steering speed sensor without disconnecting the hose.

10. Drain the oil from the transaxle.

11. Remove the halfshafts from the vehicle.

12. Remove the bolts securing the intermediate shaft and remove the shaft.

13. Remove the shift rod and the shift extension.

14. Remove the bolts attaching the torque rod bracket to the clutch case.

NOTE: Replace the torque rod bolts whenever loosened or removed.

15. Place a transaxle jack securely beneath the transaxle.

16. Remove the sub frame center beam.

17. Attach a support chain with two 10mm bolts to the engine block, 1 on each bank. Lift the engine slightly to take the weight off the mounts.

18. Remove the center stop bracket from the transaxle.

19. Remove the flywheel cover. On vehicles with automatic transaxle, unbolt the torque converter from the flywheel.

20. Remove both rear engine mounting bolts from the transaxle.

21. Remove both front engine mounting bolts from the transaxle housing.

22. Remove the starter mounting bolts and the starter assembly.

23. Remove the remaining transaxle mounting bolts.

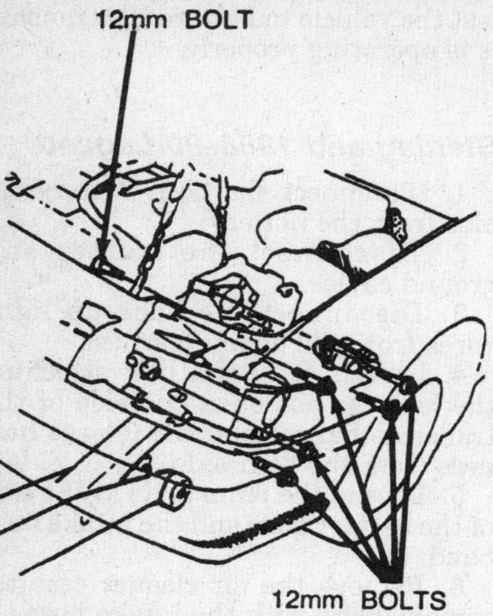

Transaxle removal—Sterling and 1988–90 Legend

24. Pull the transaxle away from the engine until it clears the 14mm dowel pins and lower on the transaxle jack.

To install:

25. Install the transaxle on a transaxle jack; clean and lubricate the clutch release bearing surfaces.
26. Make sure both 14mm dowel pins are installed in the clutch housing.
27. Raise the transaxle high enough to align the dowel pins with the matching holes in the block.
28. Roll the transaxle toward the engine and fit the mainshaft into the clutch disc splines.
29. Install the transaxle mounting bolts and torque to 55 ft. lbs. (75 Nm).
30. Install the starter and torque the mounting bolts.
31. Install the front engine mounting bolts and torque to 29 ft. lbs. (40 Nm).
32. Install the rear engine mounting bolts and torque to 29 ft. lbs. (40 Nm).
33. Install the center stopper bracket bolts and torque to 29 ft. lbs. (40 Nm).
34. Install the flywheel cover.
35. Install the center beam and remove the transaxle jack.
36. Install and torque the new torque rod bracket bolts to 29 ft. lbs. (40 Nm).

NOTE: Replace the torque rod bolts whenever loosened or removed.

37. Remove the engine support chain.
38. Connect the shift linkage or cable.
39. Install the intermediate shaft with the 8mm bolts. Torque the bolts to 29 ft. lbs. (40 Nm).
40. Install the right and left halfshaft.
41. Install the speed sensor.
42. Install the clutch slave cylinder with the 8mm bolts complete with the hose and pushrod. Torque the bolts to 16 ft. lbs. (22 Nm) and adjust.
43. Install the clutch damper assembly and the 8mm bolts to the transaxle hanger bracket. Torque the bolts to 16 ft. lbs. (22 Nm).
44. Install the air cleaner assembly and the air intake hose.
45. Install and torque both 6mm bolts at the side of the battery case and tighten the intake hose band.
46. Tighten the 6mm harness holder bolt at the side of the transaxle hanger.
47. Connect the back-up light switch wire to the engine harness.
48. Connect the starter and ground cables.
49. Connect the both battery cables.
50. Refill the transaxle with the proper fluid.
51. Check the transaxle for smooth operation.

NSX

1. Check and record the rear wheel camber.
2. Disconnect both battery cables.
3. Raise and safely support the vehicle and drain the transaxle oil.
4. Remove the strut bar and the air cleaner assembly.
5. Disconnect the wiring and remove the emission control equipment box without disconnecting the vacuum hoses. Lay the box on the engine..
6. Disconnect all wiring from the transaxle and starter and remove the starter. Remove the back-up light and the neutral switches.
7. Remove the parking brake cable holders from the rear beam rod and remove the rod.
8. Remove the front exhaust pipe and catalytic converter.
9. Remove the parking brake cable holder and wire clamp.
10. Remove the nuts and use the proper joint press to separate the toe control arms.
11. Remove the sway bar links from the bar and from the shock absorber lower mounts.
12. On the right side, remove the lower ball joint nut and use a proper tool to separate the lower ball joint from the control arm.
13. Carefully pry the right inner CV-joint out of the intermediate shaft. Support the halfshaft so it dosen't hang on the outer joint and wrap the inner joint in a plastic bag to keep it clean.
14. Remove the heat shield and pry the intermediate shaft out of the differential.
15. On the left side, make a mark to indicate the position of the lower control arm excentric bolts and remove the bolts. Also remove the rear mounting bolt from the upper control arm front mount.
16. Carefully pry the left halfshaft out of the differential. Support the halfshaft so it dosen't hang on the outer joint and wrap the inner joint in a plastic bag to keep it clean.
17. Remove the shift cable cover and disconnect the shift linkage.
18. Remove the clutch slave cylinder without disconnecting the hydraulic hose. Remove the release fork and hang it on the housing. Remove the flywheel cover.
19. Attach a chain hoist to the transaxle hangers.
20. Place a jack under the transaxle and raise it just enough to take the weight off the mounts.
21. Loosen the front engine mount–to–engine bolt, remove the mount–to–transaxle bolts, then tighten the engine mount bolt again.
22. Remove the transaxle mount bolts and unbolt the transaxle from the engine. Carefully separate the transaxle from the engine and lower it out of the vehicle.

To install:

23. Lightly lubricate the release fork mounting points with molybdenum grease. Set the fork and bearing in place and lightly grease the inside of the release bearing.
24. Make sure the dowel pins are in place and carefully mate the transaxle to the engine. Install 2 of the transaxle–to–engine bolts and torque them to 47 ft. lbs. (65 Nm).
25. Loosen the front engine mount bolt, install the transaxle mount bolts and torque all 3 bolts to 43 ft. lbs. (60 Nm).
26. Install the remaining mounting bolts. Torque the engine–to–transaxle bolts to 47 ft. lbs. (65 Nm) and the mount bolts to 43 ft. lbs. (60 Nm).
27. Install the flywheel cover, connect the shift linkage and install the covers.
28. Install the upper control arm mounting bolt.
29. Install new set rings onto the ends of the halfshafts and the intermediate shaft.
30. Install the left halfshaft into the differential, making sure the set ring clicks into place.
31. Install the lower control arm. Make sure to align the marks on the excentric bolts. Torque the 12mm nut to 69 ft. lbs. (95 Nm) and the 14mm nut to 90 ft. lbs. (125 Nm).
32. Install the intermediate shaft, torque the mounting bolts to 16 ft. lbs.(22 Nm).
33. Install the right side halfshaft and the heat shield.

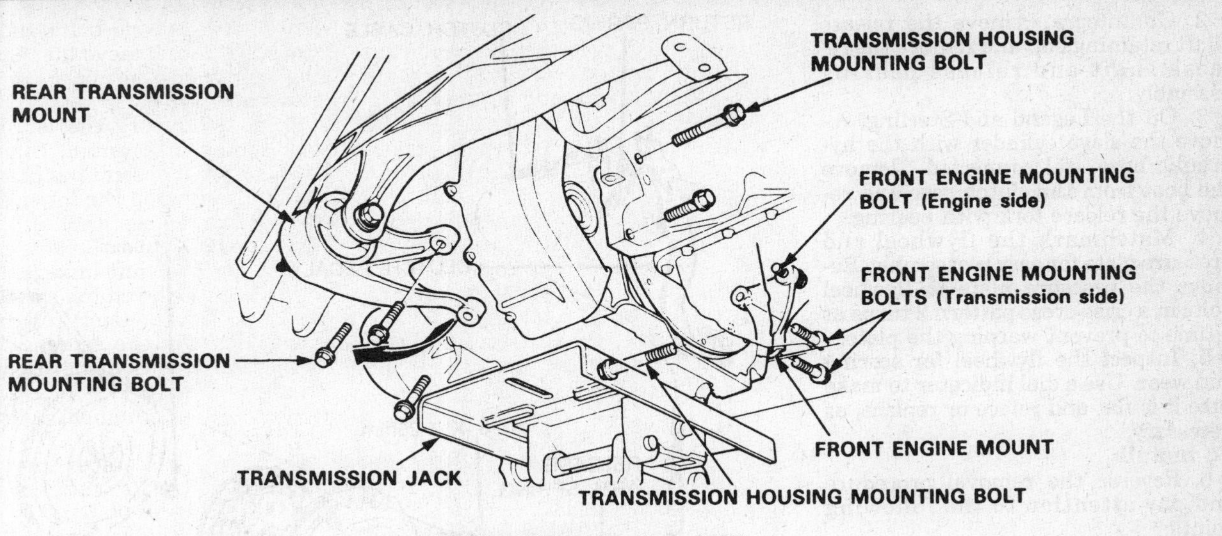

Removing the transaxle on NSX

34. Complete assembly of the rear suspension, brake cables and wiring. Torque the control arm ball joint nut to 43 ft. lbs. (60 Nm) and tighten as required to install a new cotter pin.
35. Replace the toe control link nuts and torque to 32 ft. lbs. (44 Nm) and tighten as required to install a new cotter pin.
36. Use new self locking nuts on the sway bar links and torque to 69 ft. lbs. (95 Nm).
37. Install the exhaust pipe and catalyst. Use new self locking nuts and torque the manifold nuts to 40 ft. lbs. (55 Nm). Torque the catalyst flange nuts to 25 ft. lbs. (34 Nm).
38. Install the rear beam rod. Torque the nuts and bolts to 69 ft. lbs. (95 Nm).
39. Install the starter and connect the wiring.
40. Install the emission control equipment box and connect the wiring.
41. Install the air cleaner assembly and the strut bar. Torque the bolts to 28 ft. lbs. (39 Nm).
42. Refill the fluids and connect the battery.
43. Rear suspension camber alignment must be checked and set.

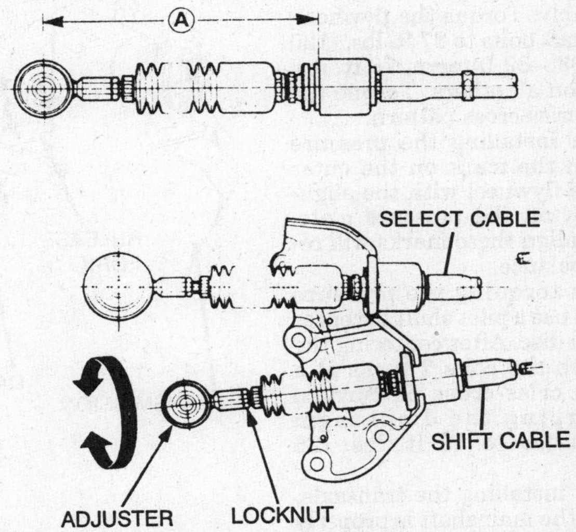

Manual transaxle shift cable adjustment on NSX: measure distance A and adjust at the cable end

LINKAGE ADJUSTMENT

On all models except the NSX, manual shift linkage is not adjustable. If linkage problems are suspected, check the bushings and fastener torque on the shift rod and torque rod.

NSX

There are 2 cables; a shift cable that actually shifts the gears and a selector cable that selects the 1/2, 3/4 or 5/reverse gate. The cables run side by side down the center of the vehicle, with the shift cable on the left, the select cable on the right. Cable adjustment is done by simply measuring cable length and adjusting to specification.

1. Measure and adjust the shift cable first, then the select cable.
2. Measure the cable length from the where the cable housing contacts the bracket to the center of the cable end joint.
3. If adjustment is required, remove the cotter pin and the cable end joint, loosen the locknut and turn the joint on the cable end as required.
4. The shift cable length is 5.53–5.57 in. (140.5–141.5mm). The select cable length is 6.20–6.24 in. (157.5–158.5mm).

CLUTCH

All vehicles except NSX use a single dry disc with a diaphragm spring pressure plate. NSX uses a double disc clutch with a mid-plate. On Integra the clutch is cable operated. All other models use a hydraulic clutch release system with a master and slave cylinder.

Clutch Assembly

REMOVAL & INSTALLATION

Except NSX

1. Disconnect the negative battery cable. Remove the transaxle.

2. On Integra, remove the release shaft retaining bolt and remove the release shaft and release bearing assembly.

3. On the Legend and Sterling, remove the slave cylinder with the hydraulic hose still connected. Remove the boot from the clutch case and remove the release fork with bearing.

4. Matchmark the flywheel and pressure plate for easy reassembly. Remove the pressure plate-to-flywheel bolts in a criss-cross pattern 2 turns at a time to prevent warping the plate.

5. Inspect the flywheel for scoring and wear. Use a dial indicator to make sure it is flat and reface or replace, as necessary.

To install:

6. Reverse the removal procedure and pay attention to the following points:

a. Make sure the flywheel and the end of the crankshaft are clean before assembly. Torque the flywheel-to-crankshaft bolts to 87 ft. lbs. (120 Nm) on 1988–89 Integra, 76 ft. lbs. (105 Nm) on all others. Torque the bolts in a criss-cross pattern.

b. When installing the pressure plate, align the mark on the outer edge of the flywheel with the alignment mark on the pressure plate. Failure to align these marks will result in imbalance.

c. When torquing the pressure plate bolts, use a pilot shaft to center the friction disc. After centering the disc, tighten the bolts 2 turns at a time, in a criss-cross pattern to avoid warping the diaphragm springs; torque to 19 ft. lbs. (26 Nm).

d. When installing the transaxle, make sure the mainshaft is properly aligned with the disc spline and the aligning pins are in place, before torquing the case bolts.

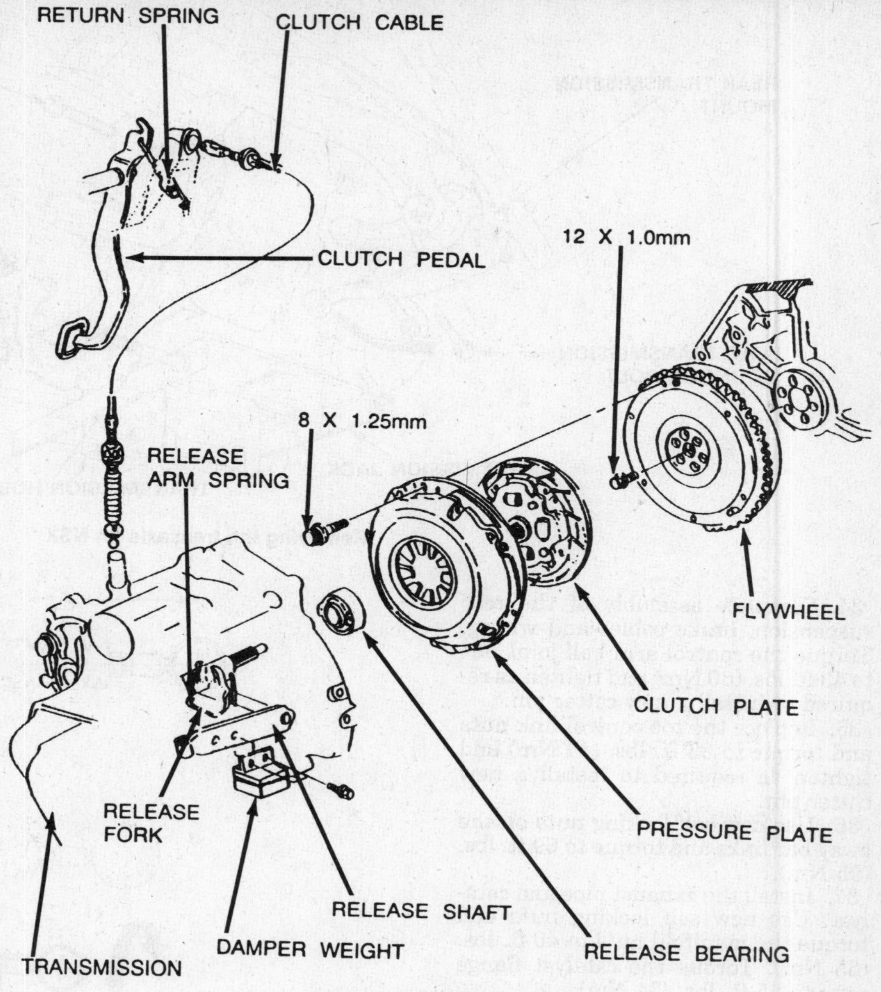

Integra clutch assembly

NSX

1. Remove the transaxle. Install a ring gear holder to prevent the engine from turning.

2. Remove the pressure plate bolts 2 turns at a time in a criss-cross pattern to prevent warping the plate.

3. Remove the release bearing from the pressure plate.

4. Lift off the mid plate and the second disc. Be careful not to lose the spring collar and damper washer that are in the outer edge of the flywheel.

5. Inspect all parts for ware or signs of burning, oil contamination or warping. If any parts are to be replaced, replace the entire assembly as a set, including the flywheel.

To install:

6. If the flywheel has been removed, align the hole in the flywheel with the

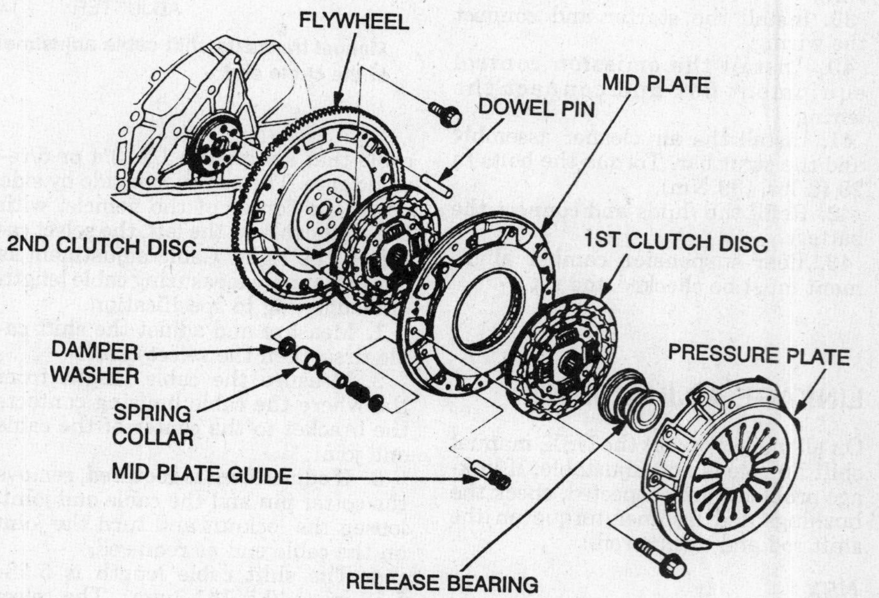

Clutch assembly on NSX

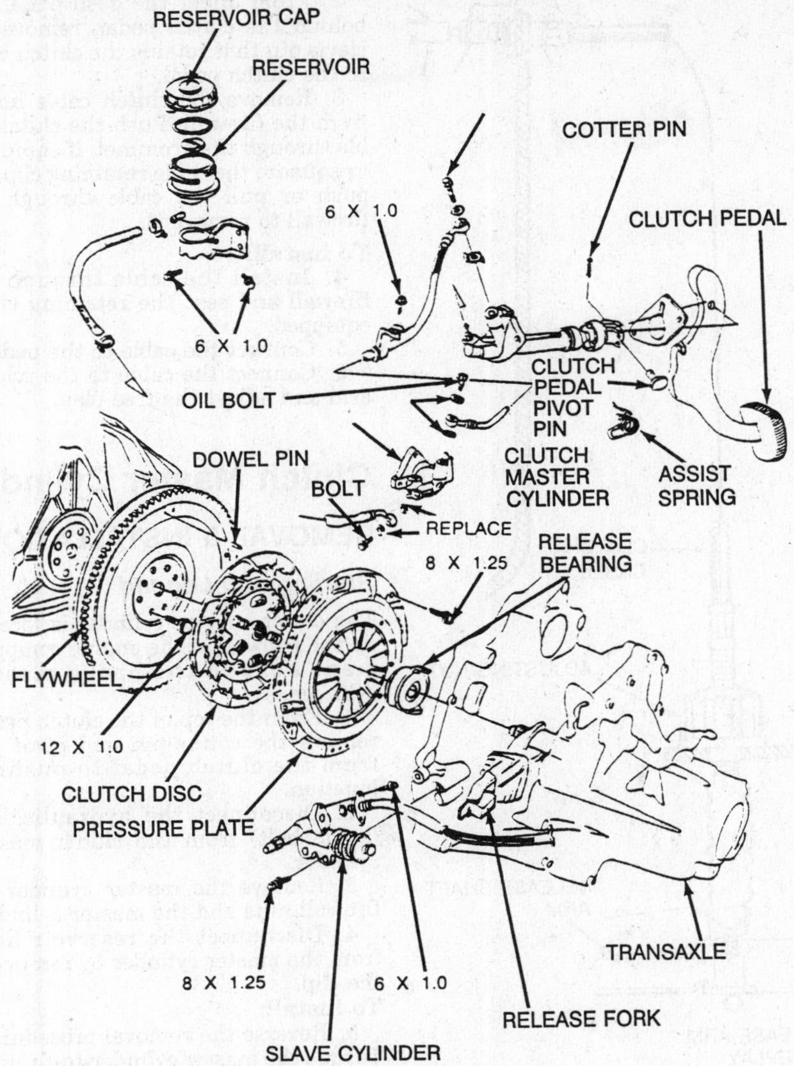

Legend and Sterling clutch assembly

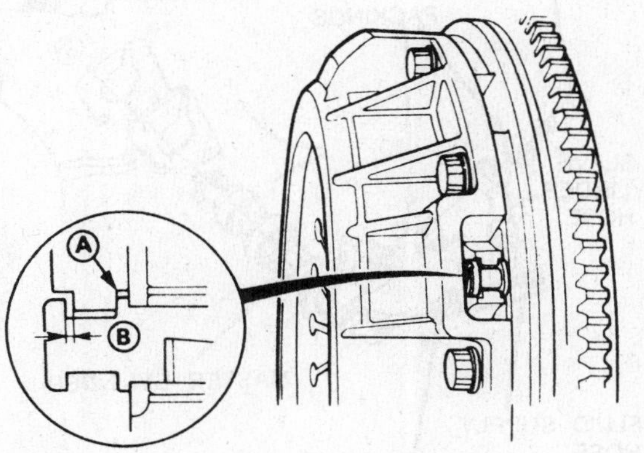

On NSX, screw a 5mm bolt into the rear of the flywheel to set the mid plate guide position

dowel pin on the crankshaft and install the bolts. Torque the bolts in 2 steps in a criss-cross pattern to 76 ft. lbs. (105 Nm).

7. Install the damper washer and spring collar with the concave side of the washer facing in.

8. Be carefull not to mix up the clutch discs. The 2nd disc has a flat plate between the 2 friction surfaces, the first disc has a spring plate. Install the 2nd disc with the index mark away from the flywheel.

9. Install the mid plate with the plate and flywheel marks aligned.

10. Install the 1st clutch disc with the index mark facing out and 180 degrees apart from the mark on the 2nd disc.

11. Install the release bearing onto the pressure plate.

12. Using a pilot tool, install the pressure plate making sure the alignment mark lines up with the mark on the mid plate.

13. Tighten the bolts 2 turns at a time in a criss-cross pattern. Torque to 16 ft. lbs. (22 Nm).

14. After the clutch is assembled, the mid plate guides must be initialized. On the engine side of the flywheel is a small threaded hole. Screw in a 5mm bolt until it just touches the mid plate guide, then turn about another ½ turn.

15. Make sure the mid plate guide touches the back of the mid plate. Use a feeler gauge to measure the gap between the front of the guide and the front of the mid plate. The gap **B** should be 0.016–0.022 in. (0.40–0.55mm).

16. Repeat the process on the other guide.

PEDAL HEIGHT/FREE-PLAY ADJUSTMENT

Integra

1. Adjust the clutch free-play at the release lever by turning the adjusting nut (at the transaxle). The clutch pedal height should be:
   1988–89 – 5.87 (149mm)
   1990–92 – 6.97 (177mm)

2. Make sure there is $^{5}/_{32}$–$^{13}/_{64}$ in. (4.0–5.0mm) of free-play at the tip of the release arm after the adjustment.

3. If equipped with cruise control, turn the adjuster (above the clutch pedal) until the clutch pedal stroke is:
   1988–89 – 5.51–5.71 in. (140–145mm)
   1990–92 – 5.59–5.79 in. (142–147mm)

4. Tighten the locknut securely.

Legend and Sterling

1. Loosen the locknut on the clutch

pedal switch and back off the switch until it does not touch the pedal.

2. Loosen the locknut on the clutch master cylinder pushrod. Turn the pushrod in or out to obtain the correct stroke and height at the clutch pedal.

Stroke at pedal—5.7–5.8 in. (145–148mm)

Clutch pedal free-play—0.040–0.280 in. (1.0–7.0mm)

3. Tighten the locknut on the clutch master cylinder pushrod.

4. Screw the clutch pedal switch until it contacts the pedal.

5. Turn the switch another ¼–½ turn. Tighten the locknut.

NSX

1. Loosen the locknut on the clutch pedal switch and back off the switch until it does not touch the pedal.

2. Loosen the locknut on the clutch master cylinder pushrod. Turn the pushrod in or out to obtain the correct stroke and height at the clutch pedal.

Stroke at pedal—5.12–5.31 in. (130–135mm)

Clutch pedal free-play—0.040–0.280 in. (1.0–7.0mm)

3. Tighten the locknut on the clutch master cylinder pushrod.

4. Screw the clutch pedal switch until it contacts the pedal.

5. Turn the switch another ¼–½ turn. Tighten the locknut.

6. Adjust the safety start switch so the starter will be activated when the pedal is released 0.60–0.78 in. (15–20mm) from the fully depressed position.

Clutch Cable

ADJUSTMENT

Integra

1. Measure the clutch pedal disengagement height.

2. Measure the clutch pedal free-play.

3. Adjust the clutch pedal free-play by turning the adjusting nut. Be sure there is 0.16–0.20 in. (4.0–5.0mm) free-play at the tip of the release arm after the adjustment.

4. Turn the adjusting nut right or left to bring the clutch pedal stroke to the proper specification and then tighten the locknut to 16 ft. lbs. (22 Nm).

REMOVAL & INSTALLATION

1. Release the clutch cable from the release arm by loosening the adjusting nut to allow enough slack to enable the cable to be removed from the elongated hole in the release arm.

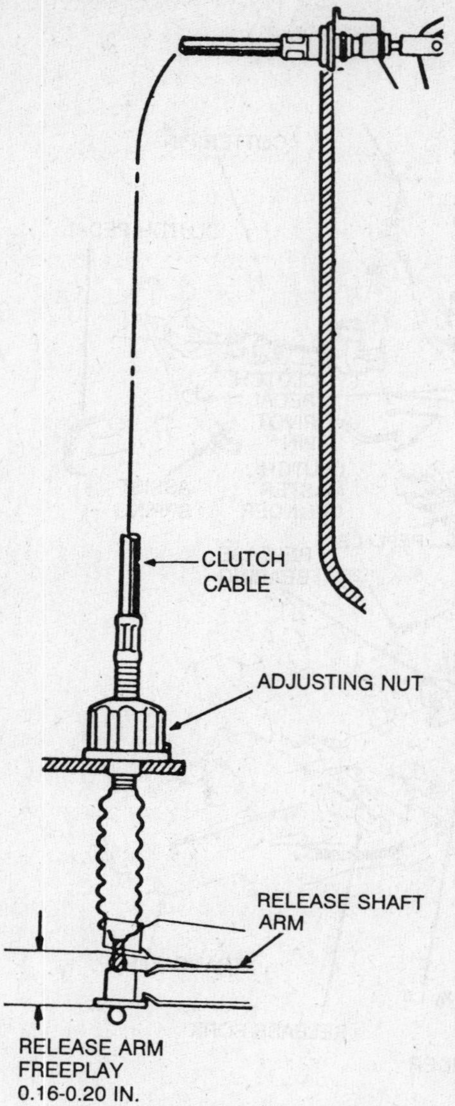

Clutch cable adjustment on Integra

2. From under the dash panel and behind the clutch pedal, remove the clevis pin that retains the clutch cable to the clutch pedal.

3. Remove the clutch cable holder from the firewall. Push the clutch cable through the grommet, if equipped, or squeeze the cable retaining clip and push or pull the cable through the firewall to remove it.

To install:

4. Install the cable through the firewall and seat the retaining clip,if equipped.

5. Connect the cable to the pedal.

6. Connect the cable to the release arm and adjust the free-play.

Clutch Master Cylinder

REMOVAL & INSTALLATION

Sterling and Legend

The clutch master cylinder is located on the firewall in the engine compartment next to the brake master cylinder.

1. From the top of the clutch pedal, remove the cotter pin and pivot pin from the clutch pedal-to-pushrod junction.

2. Disconnect the hydraulic line (banjo bolt) from the clutch master cylinder.

3. Remove the master cylinder-to-firewall nuts and the master cylinder.

4. Disconnect the reservoir hose from the master cylinder by removing the clip.

To install:

5. Reverse the removal procedures. Torque the master cylinder-to-firewall nuts to 16 ft. lbs. (22 Nm). Refill the clutch master cylinder reservoir and bleed the system.

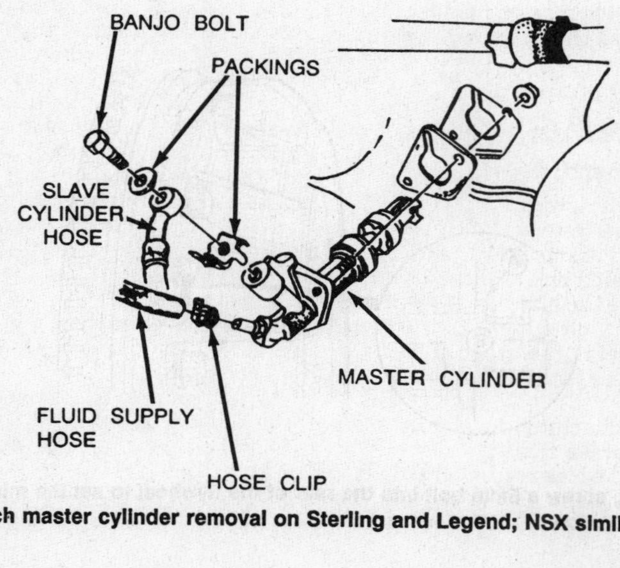

Clutch master cylinder removal on Sterling and Legend; NSX similar

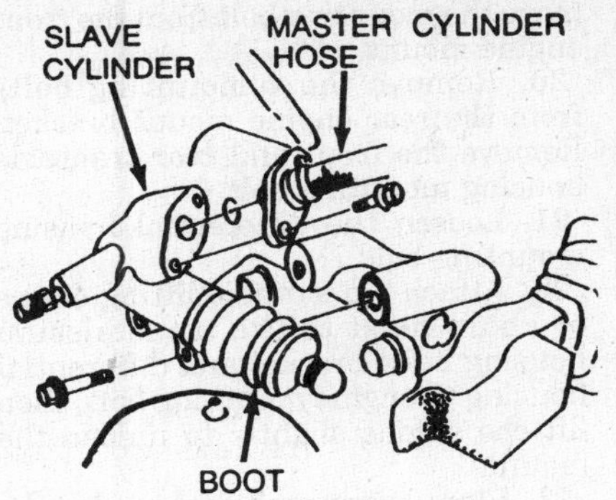

The hydraulic hose is sealed to the slave cylinder with an O-ring

NSX

1. Remove the cotter pin and pedal pin to disconnect the pedal from the master cylinder.
2. Under the front hood, disconnect the pipe from the master cylinder. Be prepared to catch the hydraulic fluid that will leak out.
3. Remove the nuts and unbolt the reservoir bracket to remove the master cylinder with the reservoir attached.
4. Installation is the reverse of removal. Refill the reservoir and bleed the system.

Clutch Slave Cylinder

REMOVAL & INSTALLATION

Sterling and Legend

1. Disconnect and plug the hydraulic line at the slave cylinder.
2. Remove the slave cylinder-to-clutch housing bolts and the slave cylinder.

To install:

3. Reverse the removal procedures. Torque the slave cylinder-to-clutch housing bolts to 16 ft. lbs. (22 Nm). Refill the clutch master cylinder reservoir and bleed the hydraulic system.

Hydraulic Clutch System Bleeding

Sterling, Legend, and NSX

The hydraulic system must be bled whenever the system has been leaking or has been dismantled. The bleed screw is located on the slave cylinder.

1. Remove the bleed screw dust cap.
2. Attach a clear hose to the bleed screw. Immerse the other end of the hose in a clear jar ½ filled with brake fluid.
3. Refill the clutch master cylinder with fresh brake fluid.
4. Open the bleed screw slightly and have an assistant slowly depress the clutch pedal. Close the bleed screw when the pedal reaches the end of it's travel. Allow the clutch pedal to return slowly.
5. Repeat Steps 3–4 until all air bubbles are expelled from the system.
6. Discard the brake fluid in the jar. Replace the dust cap. Refill the master cylinder.

AUTOMATIC TRANSAXLE

For further information on transmissions/transaxles, please refer to "Chilton's Guide to Transmission Repair".

Transaxle Assembly

REMOVAL & INSTALLATION

1988–89 Integra

1. Disconnect the negative battery cable and the ground cable from the transaxle. Raise and safely support the vehicle.
2. Unlock the steering and place the transaxle in **N**.
3. Disconnect the following wires in the engine compartment:
   a. Battery positive cable from the starter.
   b. Black/white wire from the solenoid.
4. Drain the transaxle.
5. Disconnect the speedometer cable.
6. Disconnect and plug the transaxle cooler hoses; wire them up next to the radiator so the automatic transaxle fluid won't drain out.
7. Remove the center console and disconnect the shift cable by removing the adjusting pin.
8. Unscrew the cable guide bolt and pull out the throttle cable.
9. Remove the right and left halfshaft and intermediate shaft.
10. Screw a 10mm bolt at the cylinder head and attach a suitable lifting device or chain hoist and chain to the bolt; attach the other end of the chain to the engine hanger plate and lift the engine slightly to take the weight off the mounts.
11. Remove transaxle stop bracket, if equipped, the undercover and the engine splash shields.
12. Disconnect the header pipe from the exhaust manifold.
13. Using a transaxle jack, place it under the transaxle and raise it enough to take the weight off the mounts.
14. At the front of the engine bracket, remove the bolts from the front transaxle mount.
15. Remove the rear transaxle mount bracket bolts and the mount.
16. From the front transaxle mount bracket, remove the transaxle housing bolts.
17. Remove the torque converter cover. Matchmark the torque converter-to-driveplate location.
18. Rotate the crankshaft pulley and remove the torque converter-to-driveplate bolts.

NOTE: The crankshaft pulley bolt is a right hand thread and may be loosened when the pulley is turned counterclockwise. After removing the driveplate, check that the pulley bolt is torquėd properly.

19. Remove the starter bolts and the starter.
20. Remove the remaining transaxle mounting bolts.
21. Pull the transaxle away from the engine until it clears the 14mm dowel pins and lower it on the transaxle jack.

To install:

22. Install the transaxle on a transaxle jack.
23. Make sure both 14mm dowel pins are installed in the torque converter housing.
24. Raise the transaxle high enough to align the dowel pins with the matching holes in the block. Align the torque converter matchmark and the bolt heads with holes in the driveplate.
25. If the driver's side suspension was left in place, install new spring clips on both axles, then carefully insert the left axle into the differential as the transaxle is raised to the engine.

NOTE: Install new 26mm spring clips on both axles. Make sure the axles fully bottom out. Slide the axle in until the spring clip engages the differential.

26. Push and wiggle the transaxle until it fits flush with the flange.
27. Torque the transaxle-to-engine bolts (from the engine side) to 42 ft. lbs. (58 Nm).
28. Attach the torque converter to the driveplate and torque the bolts in 2 steps in a criss-cross pattern to 9 ft. lbs. (12 Nm). Check for free rotation after torquing the last bolt.
29. Install the shift cable.
30. Remove the transaxle jack.
31. Install the torque converter cover plate.
32. Install the rear mount bracket on the transaxle housing. Torque the bolts to 48 ft. lbs. (65 Nm).

33. Install the front transaxle mount bracket. Torque the bolts to 33 ft. lbs. (45 Nm).
34. Loosely install the bolts for the front of the transaxle mount, then torque them to 18 ft. lbs. (24 Nm).
35. Install the transaxle stop bracket, if equipped, and torque the bolts to 33 ft. lbs. (45 Nm). Install the starter mounting bolts and torque them to 33 ft. lbs. (45 Nm).
36. Install the intermediate shaft, the right and left halfshaft.
37. Turn the right steering knuckle/axle assembly outward far enough to insert the free end of the halfshaft into the transaxle. Repeat this procedure on the other side.

NOTE: Make sure the halfshafts fully bottom out. Slide the halfshaft in until the spring clip engages the differential.

38. Reconnect the lower arm to the ball joints and torque them to 33 ft. lbs. (45 Nm).
39. Reconnect the tie rod end ball joints and torque them to 33 ft. lbs. (45 Nm).
40. Install the engine splash shields.
41. Reconnect the exhaust header pipe.
42. Install the front wheels, lower the vehicle to the ground and torque the lug nuts 80 ft. lbs. (110 Nm).
43. Remove the lifting device from the 10mm bolt on the cylinder head and the engine hanger plate.
44. Install the speedometer cable.
45. Install the 3 top transaxle mounting bolts and torque them to 48 ft. lbs. (65 Nm).
46. Connect the cooler hoses and torque the banjo bolts to 21 ft. lbs. (29 Nm).
47. Attach the shift control cable to the shaft lever with the pin and clip, if removed. Check the adjustment.
48. Reinstall the center console.
49. Connect the engine compartment wiring:
   a. Battery positive cable to the starter.
   b. Black/white wire from the solenoid.
   c. Transaxle ground cable.
50. With the ignition key turned **OFF**, connect the ground cable to the battery and the transaxle.
51. Unscrew the dipstick from the top of the transaxle housing and add 2.5 qts. of Dexron® ATF through the hole. Reinstall the dipstick.

NOTE: If the torque converter was replaced, the transaxle fill quantity is 5.7 qts.

52. Start the engine, set the parking brake and shift the transaxle through all gears 3 times. Check for proper control cable adjustment.
53. Allow the engine to reach operating temperature with the transaxle in **N** or **P**, then turn it off and check the fluid level.
54. Install and adjust the throttle control cable. Road test.

1990–92 Integra

1. Disconnect the negative first and then the positive battery cables.
2. Remove the 4 battery mounting bolts and remove the battery.
3. Remove the air cleaner case complete with air intake tube. Disconnect the transaxle ground cable.
4. Remove the speed sensor, but leave its hoses connected. Be careful not to bend the speedometer cable.
5. Disconnect the speed pulser connector. Disconnect the lock-up control solenoid valve wire connectors.
6. Disconnect the vacuum hose from the vacuum modulator valve. Drain the transaxle fluid into a suitable drain pan.
7. Disconnect the transaxle cooler lines at the joint pipes. Be sure to turn the ends up so as to prevent fluid from flowing out. Be sure to check for leakage at the hose joints at this time.
8. Remove the center beam. Remove the header pipe.
9. Remove the cotter pins from the lower ball joint castle nuts and remove the lower ball joints nuts. Separate the ball joints from the lower arms with a suitable ball joint separator tool.
10. Remove the damper fork.
11. Pry the right and left halfshafts out of the differential and the intermediate shaft. Pull on the inboard CV-joint and remove the right and left halfshafts.
12. Remove the 3 mounting bolts and lower the bearing support. Remove the intermediate shaft from the differential.
13. Remove the engine splash shield and the right wheelwell splash shield.
14. Remove the right damper pinch bolt, then separate the damper fork and damper.
15. Remove the bolts and nut from the right radius arm and remove the radius arm.
16. Remove the front and rear engine stiffners. Remove the torque converter cover and cable holder.
17. Remove the shift control cable by removing the cotter pin, control pin and control lever roller from the control lever.
18. Remove the shift control cable guide, take care not to bend the control cable.
19. Remove the plug, then remove the driveplate bolts 1 at a time while rotating the crankshaft pulley. Remove the mounting bolt from the front engine mount.
20. Remove the 2 mounting bolts from the rear engine mount bracket. Remove the front and rear transaxle housing mounting bolt.
21. Loosen the differential housing mounting bolt.
22. Attach a a suitable lifting device or chain hoist to the transmission housing hoist bracket and differential housing to engine mounting bolt, then lift the engine slightly to unload the mounts.
23. Place a transaxle jack under the transaxle and raise the transaxle enough to take the weight of the mounts.
24. Remove the front engine mount. Remove the 4 transaxle housing mounting bolts and 2 mount bracket bolts.
25. Pull the transaxle away from the engine until it clears the 14mm dowel pins, then lower down the transaxle jack.

To install:

26. Install the transaxle on a transaxle jack.
27. Make sure both 14mm dowel pins are installed in the torque converter housing.
28. Raise the transaxle high enough to align the dowel pins with the matching holes in the block. Align the torque converter matchmark and the bolt heads with holes in the driveplate.
29. Install the 4 transaxle housing mounting bolts, then install the transaxle to the engine block.
30. Install the front engine mount to the front beam. Install the transaxle to the front engine mount.
31. Install the transaxle to the transaxle mount bracket. Remove the transaxle jack.
32. Install the 2 transaxle housing mounting bolts engine side and rear engine mount bracket bolts.
33. Attach the torque converter to driveplate with 8 (6 × 1 × 12mm) bolts and torque to 9 ft. lbs. (12 Nm). Rotate the crankshaft, as necessary, to tighten the bolts half torque, then the final torque in a criss-cross pattern. Check for free rotation after tightening the last bolt.
34. Install the shift control cable and cable guide. Take care not to bend the control cable. Install the torque converter cover and engine stiffners. Loosely install the engine stiffener mounting bolts, then torque them to specifications.
35. Remove the lifting device by removing the hanger plates.
36. Install the radius arm. Be sure to check for deterioration of the radius rod rubber bushings.
37. Install the intermediate shaft.

Install a new set ring on the end of each halfshaft.

38. Install the right and left halfshafts. Be sure to turn the right and left steering knuckle fully outward and slide the axle into the differential until the spring clip engages in the side gear.

39. Install the damper fork. Install the splash shield.

40. Install the damper fork bolts and the ball joint nuts to the lower arms.

41. Install the header pipe and center beam.

42. Install the speed sensor and connect the speed pulser connector.

43. Connect the lock-up control solenoid valve wire connectors. Connect the transaxle oil cooler lines to the joint pipes.

44. Connect the vacuum hose to the modulator. Install the transaxle ground cable.

45. Refill the transaxle with the proper transaxle fluid.

46. Connect the battery cables. Install the air intake hose.

47. Start the engine, set the parking brake and shift the transaxle through all gears 3 times. Check for proper control cable adjustment. Check the ignition timing.

48. Allow the engine to reach operating temperature with the transaxle in **N** or **P**, then turn it off and check the fluid level.

49. Road test the vehicle and make sure the transaxle is operating properly.

Sterling and 1988–90 Legend

1. Disconnect the negative and positive battery cables from the battery.
2. Disconnect the starter motor and ground cables.
3. Drain the transaxle fluid from the transaxle.
4. Remove both 6mm bolts located at the side of the battery base and the intake hose band at the throttle body.
5. Remove the air cleaner assembly along with the intake hose.
6. Remove the speedometer gearbox complete with the power steering speed sensor hose.
7. Disconnect the throttle control cable from the transaxle housing.
8. Disconnect and plug the transaxle cooler hoses at the joint pipes; turn the ends up to prevent the transaxle fluid from flowing out.
9. Near the oil cooler pipe bracket, disconnect the lockup control solenoid valve wire connector and the automatic speed pulser wire connector for 1988–90.
10. Remove the center console, pry off the adjuster pin and disconnect the control cable.
11. Remove the control cable guide

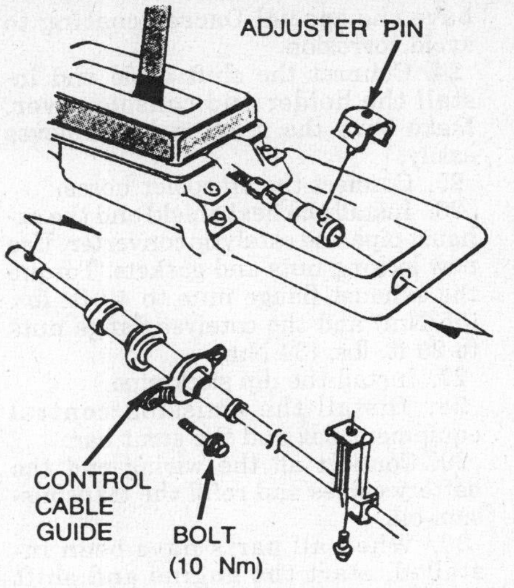

Shift control cable removal on Sterling and 1988–90 Legend

bolts and pull out the cable assembly; be careful not to bend the cable when removing it.

12. Remove the right/left halfshafts and the intermediate shaft.
13. Remove the torque converter case mounting bolts from the torque rod bracket.
14. Attach a chain hoist to the engine at 2 points with bolts and raise the engine slightly to unload the mounts.
15. Remove the front engine mount bolts from the transaxle housing.
16. While holding the locknut, turn off the radius rod.
17. Remove the center beam.
18. Remove the center stopper bracket from the transaxle.
19. Remove the torque converter cover.
20. Place a transaxle jack under the transaxle and raise the transaxle just enough to take the weight off the mounts.
21. Remove both rear engine mount bolts from the transaxle.
22. Matchmark the torque converter-to-driveplate. Remove the plug and the driveplate bolts one at a time while rotating the crankshaft pulley.
23. Remove the starter-to-engine bolts and the starter.
24. Remove the remaining transaxle housing-to-engine bolts.
25. Pull the transaxle away from the engine and lower it from the vehicle.

To install:

26. Install the transaxle on a transaxle jack and raise to engine level.
27. Secure the transaxle to the engine with the mounting bolts.
28. Install the starter motor.
29. Align the matchmarks and install the torque converter-to-driveplate bolts; torque the bolts in 2 steps in a criss-cross pattern to 9 ft. lbs. (12 Nm). Check for free rotation after torquing the last bolt.
30. Install the transaxle to the front engine mount bracket bolts and torque to 29 ft. lbs. (39 Nm).
31. Install the torque converter cover.
32. Install the center stopper bracket to the transaxle.
33. Install the center beam.
34. Connect the radius rod.
35. Remove the chain hoist from the engine.
36. Install the torque rod bracket and torque the bolts to 29 ft. lbs. (40 Nm).

NOTE: Always replace the torque rod bolts with new ones whenever they have been loosened or removed.

37. Remove the transaxle jack.
38. Connect the intermediate shaft, then install the right and left halfshafts.
39. Route the control cables to the center console through the cable guide and secure with the bolt; be careful not to bend the cables.
40. Connect the control cable with the lock pin and reinstall the center console.
41. Connect the lockup control solenoid valve wire connectors.
42. Connect the cooler hoses to the joint pipes.
43. Connect the control cable on the throttle body side.
44. Install the speedometer gearbox.
45. Install the air cleaner assembly and the air intake hose.
46. Install the battery base bolts and tighten the intake hose band on the throttle body.
47. Refill the transaxle with ATF.
48. Connect the starter and ground cables.
49. Connect the battery cables.
50. Start the engine, set the parking brake, then shift the transaxle through all gears 3 times. Check for proper control cable adjustment.
51. Allow the engine reach operating temperature with the transaxle in **N** or **P**, then turn it **OFF** and check the fluid level. Road test the vehicle.

1991–92 Legend

1. Disconnect both battery cables.
2. Remove the strut bar.
3. Drain the transmission.
4. Remove the emission control equipment box from the firewall without disconnecting the vacuum hoses.
5. Disconnect the transmission wiring and the dip stick pipe.
6. Remove the upper transmission mounting bolts.

7. Remove the exhaust pipe and catalytic converter and remove the heat shield.
8. Disconnect the oil cooler hoses.
9. Shift the transmission into **P**. Remove the center console and remove the lock pin to separate the shift cable from the selector lever. The lock pin is to the rear of the selector lever.
10. Remove the lower plate and reinstall the steering gear box mounting bolts.
11. Remove the exhaust pipe bracket and the mount bracket as required.
12. The differential stays on the vehicle. Remove the secondary cover and the 36mm sealing bolt and install the extension shaft removal tool. Disconnect the differential extension shaft from the transmission.
13. Remove the torque converter cover and unbolt the torque converter from the fly wheel. Turn the engine as required.
14. Place a transmission jack securely under the transmission and take the weight off the mounts.
15. Remove the mounts and brackets as required to slide the transmission back away from the engine. Be careful not to lose the shim between the transmission and differential.

To install:

16. Install the dowel pins and set the extension shaft in place. Use a new set ring on the shaft and lightly lubricate the splines with molybdenum grease. Make sure the secondary spring is in the differential side of the extension shaft.
17. Install the transmission and start all of the bolts. Don't forget the transmission-to-differential shim. Torque the 12mm bolts to 54 ft. lbs. (75 Nm).
18. Install all of the engine stiffener 8mm bolts, then torque to 16 ft. lbs. (22 Nm).
19. Install the mounts and brackets. Torque the 10mm bolts to 29 ft. lbs. (39 Nm), 10mm nuts to 36 ft. lbs. (49 Nm), and any remaining 12mm bolts to 54 ft. lbs. (75 Nm).
20. With the transmission in **P**, install the extension shaft using the special tool. Make sure the shaft snaps into place on the set ring.
21. Pack the shaft area with molybdenum grease, but keep the thread area clean. Apply liquid gasket to the sealing bolt threads and install the bolt and cover.
22. Install the torque converter-to-flywheel bolts and torque them in 2 steps in a criss-cross pattern to 9 ft. lbs. (12 Nm). Install the torque converter cover.
23. Install the lower plate and torque the bolts to 28 ft. lbs. (39 Nm). These bolts thread into aluminum and must have the special Dacro® coating to avoid corrosion.
24. Connect the shift cable and install the holder and console cover. Make sure the selector lever moves easily.
25. Connect the oil cooler hoses.
26. Install the heat shield and the exhaust pipe and catalytic converter. Use new locking nuts and gaskets. Torque the exhaust flange nuts to 40 ft. lbs. (55 Nm) and the catalyst flange nuts to 26 ft. lbs. (34 Nm).
27. Install the dip stick pipe.
28. Install the emission control equipment box and the strut bar.
29. Connect all the wiring and the battery cables and refill the transmission oil.
30. When all parts have been installed, start the engine and shift through all the gears 3 times to fill all the passages with fluid and check the shift cable adjustment. When the engine is fully warmed up, stop the engine and check the fluid level.
31. Check the ignition timing.

NSX

1. Check and record the rear wheel camber.
2. Disconnect both battery cables.
3. Drain the transaxle fluid.
4. Remove the strut bar and the air cleaner assembly.
5. Disconnect the wiring and remove the emission control equipment box without disconnecting the vacuum hoses. Lay the box on the engine.
6. Disconnect all wiring from the transaxle and starter and remove the starter. Remove the transaxle fluid cooler.
7. Remove the parking brake cable holders from the rear beam rod and remove the rod.
8. Remove the front exhaust pipe and catalytic converter.
9. Remove the parking brake cable holder and wire clamp.
10. Remove the nuts and use the proper joint press to separate the toe control arms.
11. Remove the sway bar links from the bar and from the shock absorber lower mounts.
12. On the right side, remove the lower ball joint nut and use a proper tool to separate the lower ball joint from the control arm.
13. Carefully pry the right inner CV-joint out of the intermediate shaft. Support the halfshaft so it dosen't hang on the outer joint and wrap the inner joint in a plastic bag to keep it clean.
14. Remove the heat shield and pry the intermediate shaft out of the differential.
15. On the left side, make a mark to indicate the position of the lower control arm eccentric bolts and remove the bolts. Also remove the rear mounting bolt from the upper control arm front mount.
16. Carefully pry the left halfshaft out of the differential. Support the halfshaft so it dosen't hang on the outer joint and wrap the inner joint in a plastic bag to keep it clean.
17. Remove the shift cable cover and disconnect the shift linkage.
18. Remove the flywheel cover and unbolt the torque converter from the flywheel.
19. Attach a chain hoist to the transaxle hangers.
20. Place a jack under the transaxle and raise it just enough to take the weight off the mounts.
21. Loosen the front engine mount-to-engine bolt, remove the mount-to-transaxle bolts, then tighten the engine mount bolt again.
22. Remove the transaxle mount bolts and unbolt the transaxle from the engine. Carefully separate the transaxle from the engine and lower it out of the vehicle.

To install:

23. Make sure the dowel pins are in place and carefully mate the transaxle to the engine. Install 2 of the transaxle-to-engine bolts and torque them to 47 ft. lbs. (65 Nm).
24. Loosen the front engine mount bolt, install the transaxle mount bolts and torque all 3 bolts to 43 ft. lbs. (60 Nm).
25. Install the remaining mounting bolts. Torque the engine-to-transaxle bolts to 47 ft. lbs. (65 Nm) and the mount bolts to 43 ft. lbs. (60 Nm).
26. Install the torque converter-to-flywheel bolts and torque in 2 steps in a criss-cross pattern to 9 ft. lbs. (12 Nm). Do not over torque.
27. Install the flywheel cover, connect the shift linkage and install the covers.
28. Install the upper control arm mounting bolt.
29. Install new set rings onto the ends of the halfshafts and the intermediate shaft.
30. Install the left halfshaft into the differential, making sure the set ring clicks into place.
31. Install the lower control arm. Make sure to align the marks on the eccentric bolts. Torque the 12mm nut to 69 ft. lbs. (95 Nm) and the 14mm nut to 90 ft. lbs. (125 Nm).
32. Install the intermediate shaft, torque the mounting bolts to 16 ft. lbs.(22 Nm).
33. Install the right side halfshaft and the heat shield.
34. Complete assembly of the rear suspension, brake cables and wiring.

Torque the control arm ball joint nut to 43 ft. lbs. (60 Nm) and tighten as required to install a new cotter pin.

35. Replace the toe control link nuts and torque to 32 ft. lbs. (44 Nm) and tighten as required to install a new cotter pin.

36. Use new self locking nuts on the sway bar links and torque to 69 ft. lbs. (95 Nm).

37. Install the exhaust pipe and catalyst. Use new self locking nuts and torque the manifold nuts to 40 ft. lbs. (55 Nm). Torque the catalyst flange nuts to 25 ft. lbs. (34 Nm).

38. Install the rear beam rod. Torque the nuts and bolts to 69 ft. lbs. (95 Nm).

39. Install the starter and connect the wiring. Install the fluid cooler.

40. Install the emission control equipment box and connect the wiring.

41. Install the air cleaner assembly and the strut bar. Torque the bolts to 28 ft. lbs. (39 Nm).

42. Refill the fluids and connect the battery.

43. When assembly is complete, start the engine and shift through all the gears without moving the vehicle, to fill the fluid passages. When normal operating temperature is reached, stop the engine and check the transaxle fluid level.

44. Check the ignition timing.

45. Rear suspension camber alignment must be checked and set.

SHIFT LINKAGE ADJUSTMENT

Except NSX

1. Start the engine and shift into **R**. If the transaxle goes into gear, no adjustment is required.

2. If adjustment is required, stop the engine and remove the center console as required to access the cable lock pin. On 1991–92 Legend, it is to the rear of the shift lever. On all other models it is in front of the lever.

3. Shift the transaxle into:

1988–89 Integra—**D**
1990–92 Integra—**N** or **R**
1988–90 Legend and all Sterling—**R**
1991–92 Legend—**N** or **R**

4. Remove the locking pin from the cable and check the alignment of the hole in the adjuster with the hole in the cable end.

5. Loosen the locknut and turn the adjuster as required to align the holes prefectly. Install the lockpin and test again.

NSX

1. The cable is adjusted at the transaxle end. Start the engine and shift into **R**. If it does not go into gear, stop the engine and put the selector lever into position **1**.

2. Raise and safely support the vehicle and remove the shift cable cover.

3. Note the position of the lockpin so it can be installed the same way. Remove the lockpin and check the alignment of the lockpin hole in the control lever with the hole in the cable end. They must align perfectly.

4. If adjustment is required, loosen the locknut and turn the adjuster. The adjuster has left-hand threads.

5. If any gear does not work properly when the cable is properly adjusted, the problem is inside the transaxle.

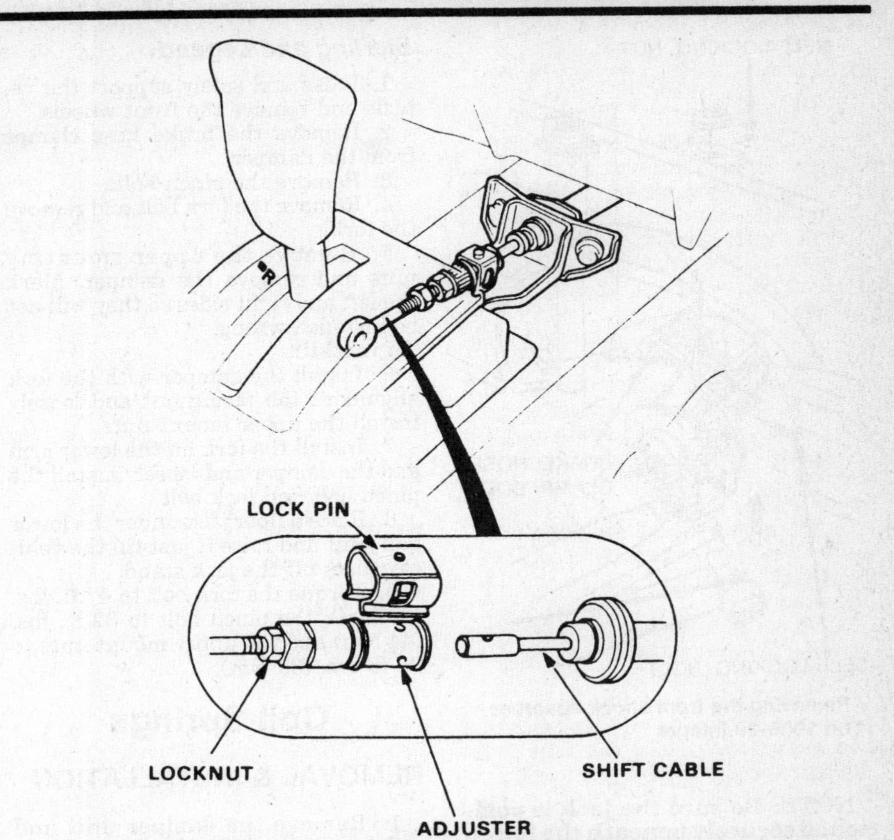

Automatic transaxle shift cable adjustment: align the lockpin holes

THROTTLE CABLE ADJUSTMENT

1. Perform the following checks:

a. Make sure the throttle cable free-play is correct; it should be 0.39–0.47 in. (10–12mm).

b. The engine is operating at normal operating temperatures; the cooling fan turns **ON**.

c. The idle speed is correct.

2. While working the throttle cable by hand, remove the cable free-play.

3. Apply light thumb pressure to the throttle control lever and work the accelerator or throttle linkage; the lever should move as the engine speed increases above idle, if not, adjust the cable.

4. Loosen the control cable nuts at the transaxle, synchronize the control lever to the throttle and tighten the locknuts.

NOTE: To tailor the shift/lock-up characteristics to the driving expectations, adjust the control cable up to 3mm shorter than the synchronized point.

FRONT SUSPENSION

Shock Absorbers

REMOVAL & INSTALLATION

1988–89 Integra

1. Raise and safely support the vehicle. Remove the front wheels.

2. Remove the brake hose clamp bolt.

3. Place a floor jack beneath the lower control arm to support it.

4. Remove the lower shock-to-steering knuckle bolt and slowly lower the jack.

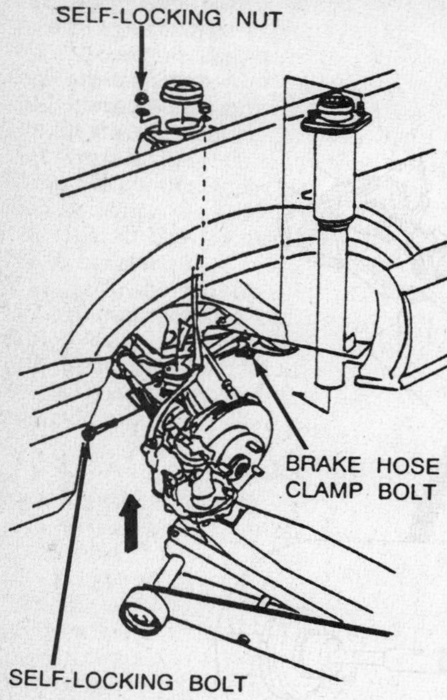

Removing the front shock absorber on 1988–89 Integra

NOTE: Be sure the jack is positioned securely beneath the lower control arm at the ball joint. Otherwise, the tension from the torsion bar may cause the lower control arm to suddenly jump away from the shock absorber as the pinch bolt is removed.

5. Compress the shock absorber by hand, remove the upper locknuts and the shock from the vehicle.

To install:

6. Reverse the removal procedures, taking note of the following:
   a. Use new self locking nuts on the top of the shock assembly and torque to 33 ft. lbs. (45 Nm).
   b. Torque the lower pinch bolt to 47 ft. lbs. (65 Nm).
   c. Install and torque the brake hose clamp to 16 ft. lbs. (22 Nm).

1990–92 Integra

1. Raise and safely support the vehicle and remove the front wheels.
2. Remove the damper pinch bolt.
3. Remove the fork bolt and disengage the fork.
4. Remove the upper mounting nuts and remove the damper unit. Mark the dampers left and right so they will not be installed wrong.
5. Installation is the reverse of removal. Use a new self locking nut on the fork bolt and torque to 47 ft. lbs. (65 Nm). Torque the upper mounting nuts to 29 ft. lbs. (40 Nm) and the pinch bolt to 32 ft. lbs. 44 Nm).

Sterling and Legend

1. Raise and safely support the vehicle and remove the front wheels.
2. Remove the brake hose clamps from the damper.
3. Remove the pinch bolt.
4. Remove the fork bolt and remove the fork.
5. Remove the upper mounting nuts and remove the damper. Mark the left and right sides so they will not be installed wrong.

To install:

6. Install the damper with the fork alignment tab facing out and loosely install the upper mount nuts.
7. Install the fork on the lower arm and the damper and loosely install the pinch bolt and fork bolt.
8. Place a floor jack under the lower ball joint and raise it just till the vehicle raises off the jack stand.
9. Torque the fork bolt to 47 ft. lbs. (65 Nm), the pinch bolt to 32 ft. lbs. (44 Nm) and the upper mount nuts to 28 ft. lbs. (39 Nm).

Coil Springs

REMOVAL & INSTALLATION

1. Remove the damper unit and make a note of the spring seat and bracket positions for reassembly.
2. Install the damper into a spring compressor and tighten the compressor according to manufacturer's instructions.
3. Remove the locking nut from the top of the shock absorber and disassemble the damper and spring as required.

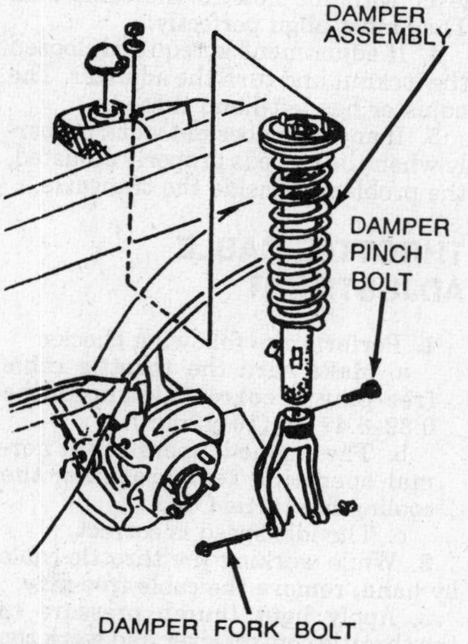

Damper unit removal on Sterling and Legend

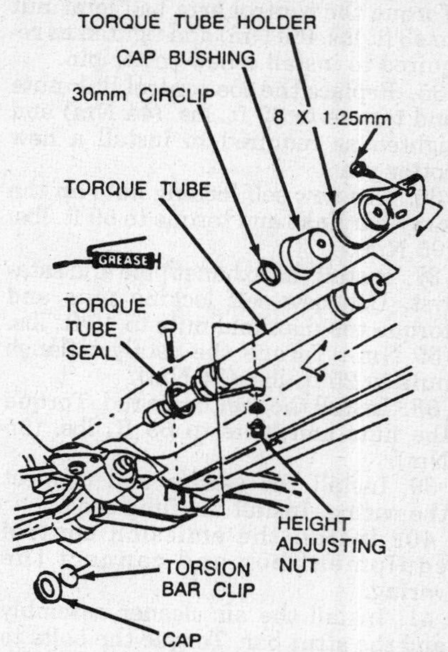

Torsion bar assembly on 1988–89 Integra

4. Installation is the reverse of removal. Be sure to properly position the spring seat and brackets.

Torsion Bar

REMOVAL & INSTALLATION

1988–89 Integra

1. Raise and safely support the vehicle.
2. Remove the height adjusting nut and the torque tube holder.
3. Remove the 30mm circlip.
4. Remove the torsion bar cap and the torsion bar clip by tapping the bar out of the torque tube.

NOTE: The torsion bar will slide easier by moving the lower arm up and down.

5. Tap the torsion bar backward, from the torque tube and remove the torque tube.

To install:

6. Inspect the torsion bar for cracks and/or damage; replace it, if necessary.
7. Install a new seal onto the torque tube. Coat the torque tube seal and tube with grease, install them on the rear beam.
8. Grease the ends of the torsion bar and insert into the torque tube from the rear.
9. Align the projection on the torque tube splines with the cutout in the torsion bar splines and insert the torsion bar approximately 0.394 in. (10mm).

NOTE: The torsion bar will slide easier if the lower arm is moved up and down.

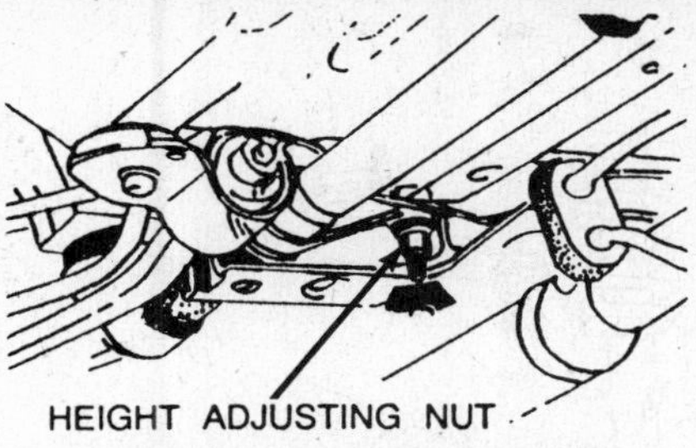

Torsion bar adjustment on 1988–89 Integra

10. Install the torsion bar clip, cap, the 30mm circlip and the torque tube cap.

NOTE: Push the torsion bar forward so there is no clearance between the torque tube and the 30mm circlip.

11. Coat the cap bushing with grease and install it on the torque tube. Install the torque tube holder.
12. Temporarily tighten the height adjusting nut.
13. Lower the vehicle to the ground. Adjust the torsion bar spring height.

ADJUSTMENT

1. Measure the torsion bar spring height between the ground and the highest point of the wheel arch. The measurement should be 25.7 in. (653mm).
2. If the spring height does not meet the specification above, make the following adjustment.
   a. Raise and support the vehicle with the front wheels hanging.
   b. Adjust the spring height by turning the height adjusting nut. Tightening the nut raises the height and loosening the nut lowers the height.

NOTE: The height varies 0.20 in. (5mm)/revolution of the adjusting nut.

3. Lower the front wheels to the ground, then roll and bounce the vehicle up and down several times and recheck the spring height to make sure it is within specifications.

Ball Joints

INSPECTION

Check ball joint play as follows:
1. Raise and safely support the vehicle.
2. Clamp a dial indicator onto the lower control arm and place the indicator tip on the knuckle, near the ball joint.
3. Place a prybar between the lower control arm and the knuckle. Replace the lower control arm if the play exceeds 0.5mm.

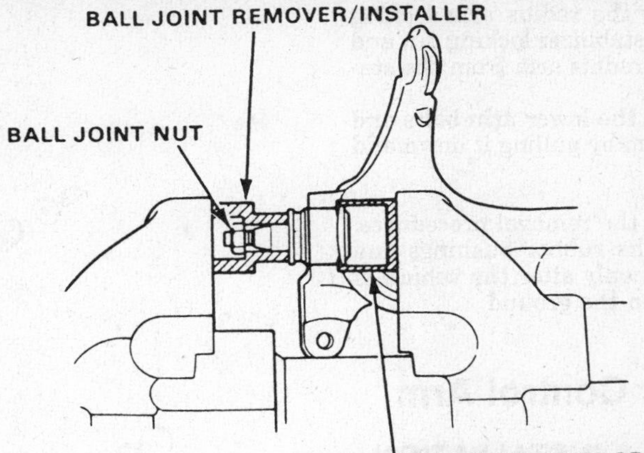

Ball joint removal and installation – 1990–92 Integra

REMOVAL & INSTALLATION

1988–89 Integra

The Integra is equipped with only a lower ball joint. If the lower ball joint play exceeds 0.05mm, replace the lower ball joint and radius arm as an assembly.

1990–92 Integra

1. Raise and support the vehicle safely. Remove the front wheel assemblies. Remove the steering knuckle.
2. Remove the boot by prying off the snap ring. Remove the 40 mm clip.
3. Install the special ball joint removal/installation tool 07965-SB00100 or equivalent, on the ball joint and tighten the ball joint nut.
4. Position the ball joint in this special tool and set this assembly in a vise. Press the ball joint out of the steering knuckle.

To install:

5. Place the ball joint in position by hand. Install the ball joint into the special tool and press in the new ball joint in the vise.
6. Install the 40mm circlip. Adjust the special tool with he adjusting bolt until the end of the tool aligns with the groove on the boot. Slide the clip over the tool and into position.

Sterling and Legend

NOTE: This procedure is performed after the removal of the steering knuckle and requires the use of the following special tools or their equivalent: tool number 07GAF–SD40330 ball joint removal base, 07GAF–SD40320 ball joint installation base and 07GAG–SD40700 clip guide tool.

1. Raise and support the vehicle safely. Remove the front wheel assemblies. Remove the steering knuckle from the vehicle.
2. Position the ball joint removal tool base or equivalent, on the ball joint, position the assembly in a shop press and press the ball joint from the steering knuckle.

To install:

3. Position the new ball joint into the hole of the steering knuckle.
4. Install the ball joint installer tool or equivalent, with the small end facing outward.
5. Position the ball joint installation base tool or equivalent, on the ball joint, position the assembly in a shop press and press the ball joint into the steering knuckle.
6. Seat the snapring in the groove of the ball joint.
7. Install the boot and snapring using the clip guide tool.

Radius Arm

REMOVAL & INSTALLATION

1988–89 Integra

1. Raise and safely support the vehicle. Remove the front wheels.
2. Place a floor jack beneath the lower control arm and remove the ball joint cotter pin/nut.

NOTE: Be sure to place the jack securely beneath the lower control arm at the ball joint. Otherwise, the tension from the torsion bar may cause the arm to suddenly jump away from the steering knuckle as the ball joint is removed.

3. Using a ball joint remover tool or equivalent, remove the ball joint from the steering knuckle.

4. Remove the radius arm locking nuts and the stabilizer locking nut and separate the radius arm from the stabilizer bar.
5. Remove the lower arm bolts and the radius arm by pulling it down and forward.

To install:

6. Reverse the removal procedures. Tighten all the rubber bushings and damper parts only after the vehicle is placed back on the ground.

Upper Control Arm

REMOVAL & INSTALLATION

Sterling and Legend

1. Raise and safely support the vehicle. Remove the front wheel.
2. Remove the cotter pin and the upper control arm-to-steering knuckle nut.
3. Using a ball joint removal tool or equivalent, separate the upper control arm from the steering knuckle.
4. Remove the upper control arm-to-chassis nuts, washers and the upper control arm from the vehicle.

To install:

5. Reverse the removal procedures. Torque the upper control arm-to-chassis nuts to 47 ft. lbs. (65 Nm) and the upper control arm ball joint-to-steering knuckle nut to 32 ft. lbs. (42 Nm).

Lower Control Arm

REMOVAL & INSTALLATION

Sterling and Legend

1. Raise and safely support the vehicle. Remove the front wheels.
2. Disconnect the lower arm ball joint; be careful not to damage the seal.
3. Remove the stabilizer bar brackets, starting with the center brackets.
4. Remove the lower arm ball joint-to-steering knuckle nut. Using a ball joint removal tool, separate the ball joint from the steering knuckle.
5. Remove the radius arm-to-chassis bolt and the arm.

To install:

6. Reverse the removal procedure. Torque the radius arm-to-chassis bolt to 39 ft. lbs. (55 Nm).
7. On Sterling and 1988–90 Legend, torque ball joint nut to 72 ft. lbs. (100 Nm) and tighten as required to insert a new cotter pin.
8. On 1991–92 Legend, torque the ball joint nut to 54 ft. lbs. (75 Nm) and tighten as required to insert a new cotter pin.

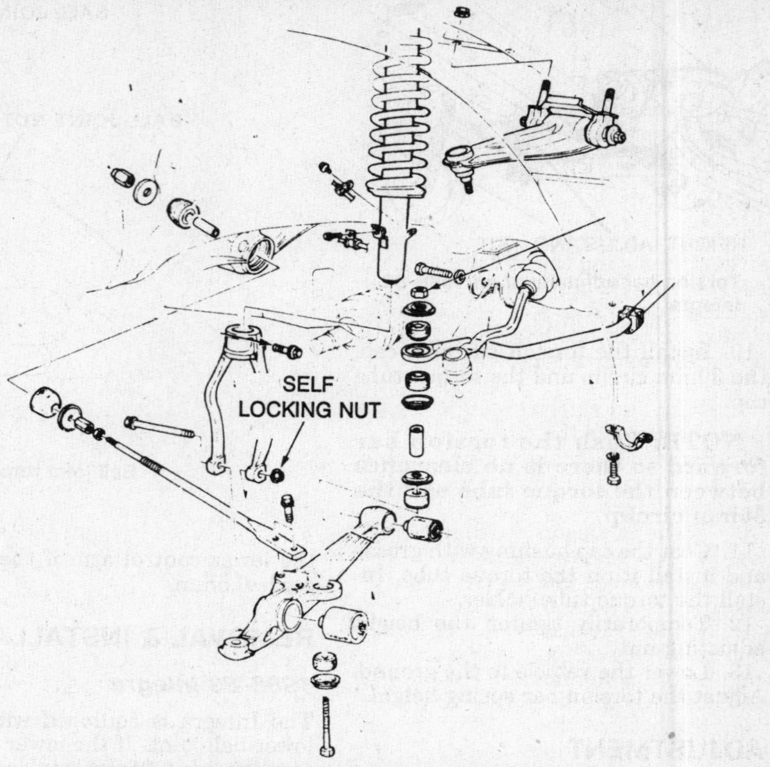

Front suspension on Sterling and 1988–90 Legend

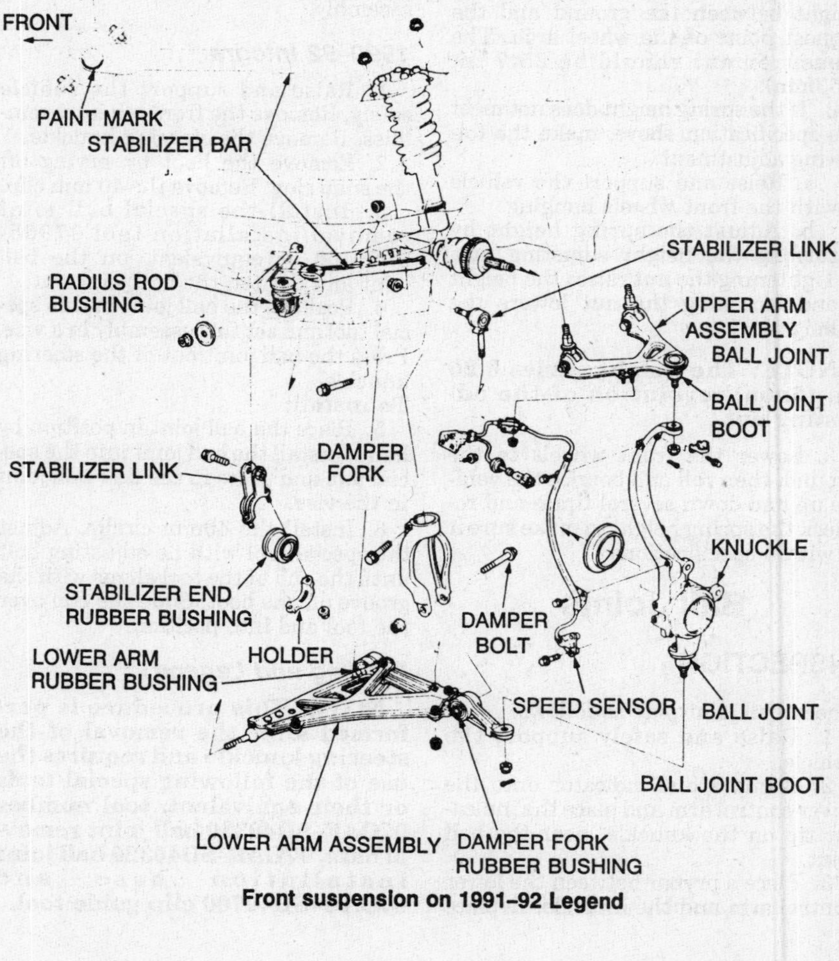

Front suspension on 1991–92 Legend

Front Wheel Bearings

Except NSX

For information on the front wheel bearings, please refer to "Drive Axle" in this section.

NSX

The front bearing hub can be removed without removing the steering knuckle. The spindle nut requires a very high torque and a special fixture is needed to hold the wheel studs in a vice while removing or installing the nut.

1. Raise and safely support the vehicle and remove the front wheel.
2. Remove the brake hose mounting bracket bolts and the speed sensor from the caliper.
3. Without disconnecting the hydraulic hose, remove the brake caliper from the knuckle. Support the caliper so it does not hang by the hose.
4. Remove the disc retaining screws and use a pair of 8mm bolts to press the disc off of the hub.
5. Remove the 4 self locking nuts from the back of the knuckle and remove the hub assembly. The nuts must be replaced.
6. Remove the spindle nut from the back side of the hub assembly and carefully remove the speed sensor pulse wheel with a bearing puller.
7. Press the hub out of the assembly. The bearing race may stay with the hub, but can be removed with a puller.

To install:

8. Press the old bearing out of the carrier. When pressing in a new bearing, be sure to press only on the outer race or the bearing will be destroyed.
9. Press the carrier onto the hub. Be sure to press only on the inner race or the bearing will be destroyed.
10. Install the speed sensor pulser and the spindle nut. Torque the nut to 242 ft. lbs. (335 Nm) and stake the nut.
11. Install the hub assembly and torque the new self locking nuts to 47 ft. lbs. (65 Nm).
12. Install the disc and caliper. Torque the caliper bolts to 80 ft. lbs. (110 Nm).

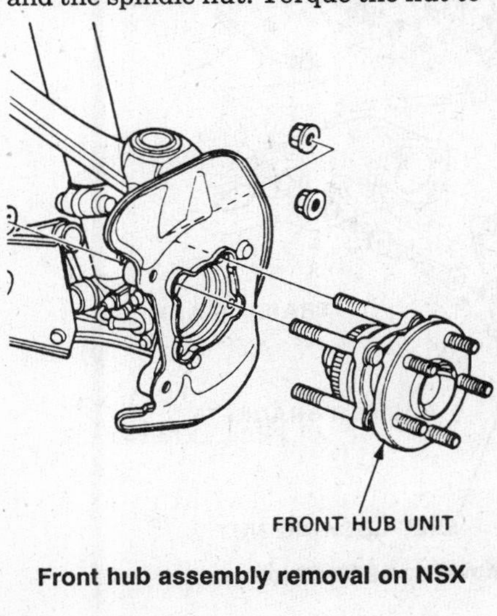

Front hub assembly removal on NSX

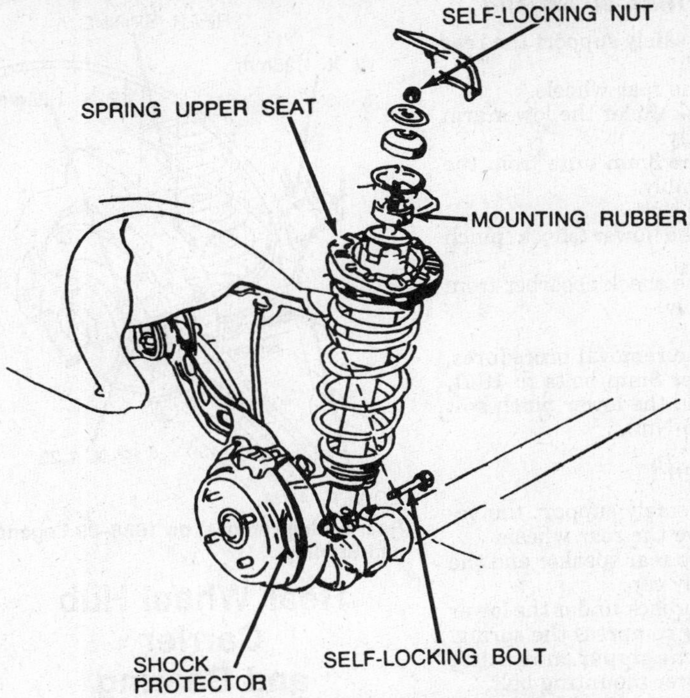

Rear strut assembly—1988–89 Integra

REAR SUSPENSION

Shock Absorber

REMOVAL & INSTALLATION

1988–89 Integra

1. Raise and safely support the rear of the vehicle.
2. Remove the rear wheels.
3. Place a jack under the rear axle beam.
4. Remove the strut maintenance lid and the self locking nut.
5. Lower the jack gradually. Remove the self-locking bolt, rear spring and spring seat.

To install:

6. Fit the upper spring seat into the frame.
7. Install the strut protector on the strut assembly, the dust cover, strut mounting collar and rear spring; temporarily tighten the strut to the axle beam.
8. Fit the inner strut mount rubber into the frame.
9. Raise the axle beam so the damper shaft fits into the frame hole.
10. Install the outer strut mount rubber and washer; torque the self-locking nut to 16 ft. lbs. (22 Nm).
11. Install the strut maintenance lid.
12. Torque the shock on the rear axle beam with the weight of the vehicle placed on the ground. Torque the strut-to-rear axle beam bolt to 40 ft. lbs. (55 Nm).

1990–92 Integra

1. Raise and safely support the vehicle and remove the rear wheels.
2. Remove the upper shock mount cover from the rear panel, just below the speaker.
3. Remove the upper mount nuts.
4. Remove the lower mount bolt and remove the damper assembly (shock and spring) as a unit.
5. Use a spring compressor to remove the spring from the shock absorber.

To install:

6. Installation is the reverse of removal. Torque the upper mount nuts to 29 ft. lbs. (40 Nm). Torque the lower mounting bolt with the weight of the vehicle on the wheels to 40 ft. lbs. (55 Nm).

Sterling and 1988–90 Legend

1. Raise and safely support the rear of the vehicle.
2. Remove the rear wheels.
3. Place a jack under the lower arm and raise slightly.
4. Remove the 8mm nuts from the top of the assembly.
5. Lower the jack.
6. Remove the lower shock pinch bolt.
7. Remove the shock absorber from the hub assembly.

To install:

8. Reverse the removal procedures, torque the upper 8mm bolts to 16 ft. lbs. (22 Nm) and the lower pinch bolt to 47 ft. lbs. (65 Nm).

1990–92 Legend

1. Raise and safely support the vehicle and remove the rear wheels.
2. Remove the rear speaker and the damper assembly cap.
3. Place a floor jack under the lower arm and slightly compress the spring.
4. Remove the upper mounting nuts and the lower mounting bolt.
5. Lower the jack to remove the damper unit.
6. Use a spring compressor to remove the spring from the shock absorber.

To install:

7. Installation is the reverse of removal. Loosley install the mounting nuts and bolt and lower the vehicle onto the wheels to torque them. Torque the upper mounting nuts to 28 ft. lbs. (39 Nm) and the lower mounting bolt to 76 ft. lbs. (105 Nm).

Coil Springs

REMOVAL & INSTALLATION

Sterling and 1988–90 Legend

1. Raise and safely support the rear of the vehicle.
2. Place a floor jack under the lower arm.
3. Pull out the hub carrier lower bolt.
4. Loosen the lower arm outside bolt.
5. Pull out the lower arm inside bolt.
6. Lower the jack gradually and remove the rear spring.

To install:

7. Reverse the removal procedures. Install the rear spring with the lower end of the spring outside.
8. Torque the lower arm-to-hub carrier nut/bolt to 54 ft. lbs. (75 Nm), with the weight of the vehicle on the ground.

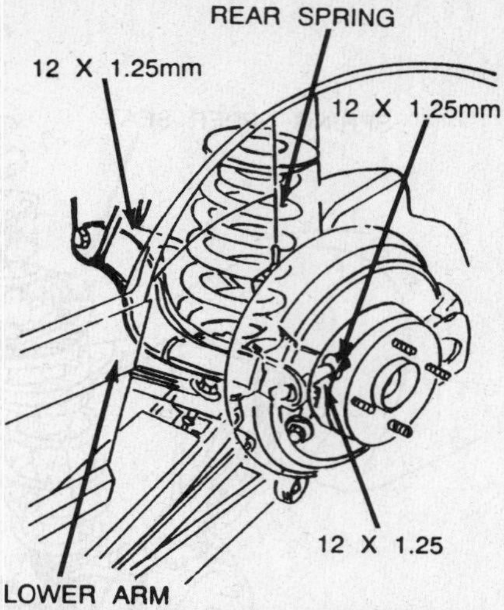

Rear spring removal on 1988–89 Legend and Sterling

Rear Wheel Hub Carrier and Bearing

REMOVAL & INSTALLATION

1990–92 Integra

1. Raise and safely support the vehicle and remove the rear wheels.
2. Remove the brake caliper without disconnecting the hydraulic hose. Support the caliper so it does not hang by the hose.
3. Remove the brake disc.
4. Remove the hub cap and the nut and washer. The torque on the nut is very high, make sure the vehicle is firmly supported and will not fall.
5. Remove the hub/bearing unit from the spindle. The bearing is a sealed unit pressed into the hub.

To install:

6. Installation is the reverse of removal. Torque the hub nut to 134 ft. lbs. (185 Nm) and the caliper mounting bolts to 17 ft. lbs. (23 Nm).

1988–89 Integra

The rear suspension on this vehicle uses a panhard rod and a torsion beam stabilizer bar to control wheel motion. Pressed onto the spindle is a swing bearing unit that allows proper movement of the trailing arm. This procedure is for removing the swing bearing housing to replace the bearing.

1. Raise and safely support the vehicle and remove the rear wheels.
2. Remove the brake caliper without disconnecting the hydraulic hose. Support the caliper so it does not hang by the hose.
3. Remove the brake disc.
4. Remove the hub cap and the nut and washer. The torque on the nut is very high, make sure the vehicle is firmly supported and will not fall.
5. Remove the hub/bearing unit from the spindle. The bearing is a sealed unit pressed into the hub.
6. Remove the hub unit and splash guard.

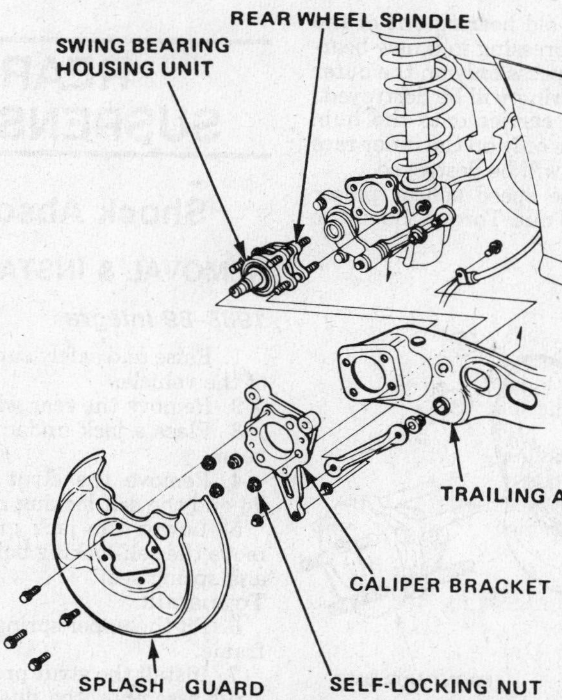

Swing bearing housing removal on 1988–89 Integra

7. Remove the stabilizer control arm and remove the caliper bracket or brake plate.
8. Unbolt the swing bearing housing and remove it.
9. Press the spindle out of the bearing housing, being carefull to not damage the threads.
10. The bearing can be pressed out of the housing but the inner race may stay with the spindle. It can be removed with a bearing puller.

To install:

11. When installing the new bearing, be sure to press only on the outer race or the bearing will be destroyed.
12. When pressing the spindle into the bearing, be sure to support the inner race or the bearing will be destroyed.
13. When installing the swing bearing housing, torque the self-locking nuts to 33 ft. lbs. (45 Nm). Torque the stabilizer control arm nuts to 29 ft. lbs. (40 Nm) with the vehicle is on its wheels.

Sterling and 1988–90 Legend

1. Raise and safely support the vehicle and remove the rear wheels.
2. Remove the caliper without disconnecting the hydraulic hose. Support the caliper so it does not hang by the hose.
3. Remove the disc by pressing it off with a pair of 8mm bolts threaded into the holes between the studs. Turn each bolt 2 turns at a time.
4. Place a floor jack under the lower arm and compress the spring slightly.
5. Remove the hub carrier lower bolt and separate the carrier from the lower arm.
6. Remove the damper assembly pinch bolt and slowly lower the floor jack to remove the hub carrier.
7. To remove the hub and bearing, remove the hub cap from the rear of the carrier and remove the nut. The torque on the nut is very high. Properly secure the hub in a holding fixture and be careful to not damage the hub studs when removing the nut.
8. The hub must be pressed out of the carrier. Be carefull not to damage the spindle threads.
9. Remove the splash guard and the 68mm circlip.
10. Press the bearing out towards the outside of the carrier.
11. The inner race may stay with the hub. It can be removed with a bearing puller.

To install:

12. Press a new bearing into the carrier. Make sure to press only on the outer race or the bearing will be destroyed.
13. Install the circlip and the splash shield.
14. Press the hub into the carrier. Make sure to support the inner race or the bearing will be destroyed.
15. Properly secure the hub in a holding fixture and install the spindle washer and nut. Torque the nut to 180 ft. lbs. (250 Nm). Install the O-ring and cap.
16. Install the carrier in the reverse order of removal. Torque the caliper bolts to 28 ft. lbs. (39 Nm). Torque the lower bolt to 40 ft. lbs. (55 Nm) and the damper pinch bolt to 47 ft. lbs. (65 Nm).

1991–92 Legend

CAUTION

Spindle nut torque is very high. Tighten or loosen the spindle nut only with the vehicle on the ground. Attempting to loosen or torque the spindle nut while the vehicle is on jack stands or a lift may cause the vehicle to fall.

1. With the vehicle on the ground, remove the hub cap and pry the spindle nut lock tab away from the spindle. Loosen the nut.
2. Raise and safely support the vehicle and remove the rear wheels.
3. Remove the caliper without disconnecting the hydraulic hose. Support the caliper so it does not hang by the hose.
4. Remove the disc by pressing it off with a pair of 8mm bolts threaded into the holes between the studs. Turn each bolt 2 turns at a time.
5. Remove the spindle nut and remove the hub/bearing unit from the knuckle. The bearing is a sealed unit pressed into the hub.

To install:

6. Installation is the reverse of removal. Torque the caliper bolts to 28 ft. lbs. (39 Nm). With the vehicle on the ground, torque the spindle nut to 206 ft. lbs. (285 Nm) and stake the nut to the spindle.

STEERING

CAUTION

Some vehicles are equipped with a driver side air bag. To avoid accidental deployment and serious personal injury, the system must be disarmed before beginning any repair procedure. Read the following safety precautions.

- Do not disassemble or tamper with the air bag assembly.
- Always keep the short connector on the air bag connector when the harness is disconnected.
- Be sure to store a removed air bag assembly with the pad surface up. If the air bag is improperly stored face down, accidental deployment could propel the unit with enough force to cause serious injury.
- When rearming the system, connect the battery last with no one in the vehicle.
- Do not install used air bag parts from another vehicle, use only new replacement parts.

DISARMING THE AIR BAG

All Supplemental Restraint System (SRS) wiring is covered with a yellow outer insulation. This wiring harness cannot be repaired and if cut or damaged, the whole harness must be replaced. To disable the air bag:

1. Disconnect the negative battery cable.
2. Remove the access panel from the bottom of the steering wheel. Stored on the panel is a short connector.
3. Unplug the air bag connector and install the short connector on the air bag connector.
4. The air bag can now be safely removed and/or the battery can be reconnected for testing or operating other vehicle systems.
5. To reconnect the air bag, disconnect the battery.
6. Connect the air bag wiring harness and make sure no one is in the vehicle before reconnecting the battery.

Steering Wheel

REMOVAL & INSTALLATION

Without Air Bag

1. Place the steering wheel in the straight ahead position.
2. Disconnect the negative battery cable. Lift off the steering wheel pad.
3. Remove the steering wheel retaining nut and the horn pad.
4. If equipped with cruise control, remove the cruise control set/resume switch.
5. Gently rock the steering wheel from side to side and gently hit the backside of each of the steering wheel spokes with equal force from the palms of your hands. Then pull the steering wheel off of the shaft.

NOTE: Avoid hitting the wheel or the shaft with excessive force. Damage to the shaft could result.

To install:

6. Installation is the reverse of removal. Torque the steering wheel nut to 36 ft. lbs. (50 Nm).

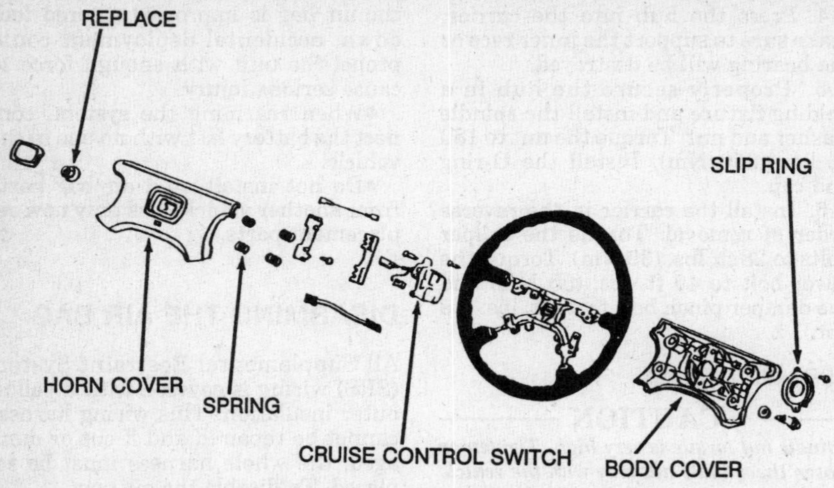

Steering wheel assembly with cruise control

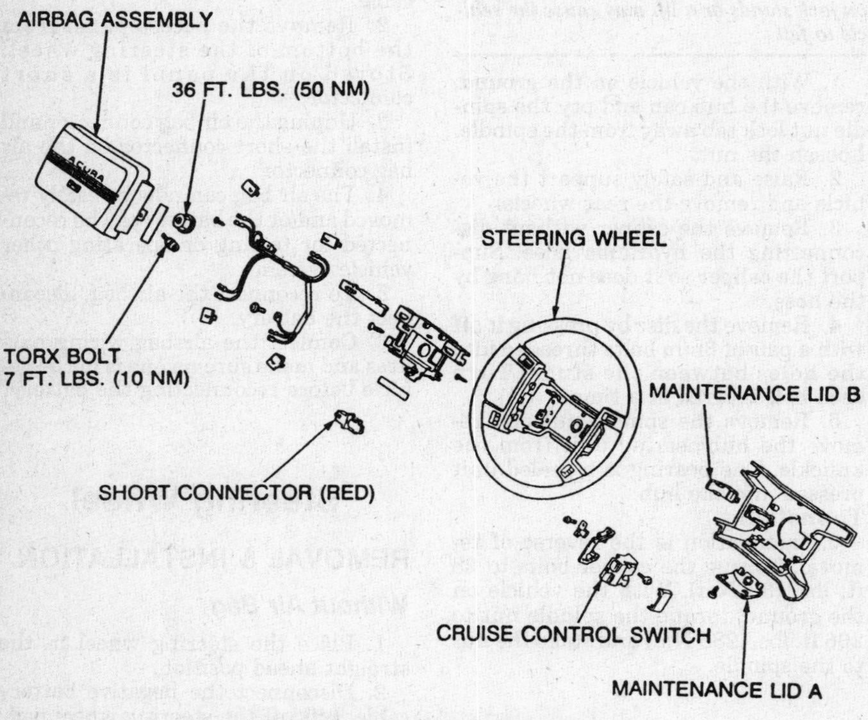

Steering wheel assembly with air bag

With Air Bag

CAUTION

On vehicles equipped with an air bag, the negative battery cable must be disconnected before beginning work. Failure to do so may result in deployment of the air bag and possible injury.

1. Disconnect both the negative and positive battery cable from the battery.
2. Remove the lower maintenance lid below the air bag and then remove the short connector.
3. Disconnect the connector between the air bag and the cable reel.
4. Connect the short connector to the air bag side of the connector.
5. Remove the left side maintenance lid and the cruise control/set resume switch cover.
6. Insert a T30 Torx® bit and remove the Torx® bolts. Remove the air bag assembly.

NOTE: Be sure to store the air bag in a safe place with the pad side facing upwards.

7. Remove the steering wheel retaining nut. Gently hit the backside of each of the steering wheel spokes with equal force with the palms of the hands.

NOTE: Avoid hitting the wheel or the shaft with excessive force. Damage to the shaft could result.

To install:

8. When installing, reverse the order of the removal procedure.
   a. Before installing the steering wheel, the front wheels should be aligned straight forward. Torque the steering wheel nut to 36 ft. lbs. (50 Nm).
   b. Be sure to install the harness wires so they are not pinched or interfering with other parts.
   c. Center the cable reel, by rotating the cable reel clockwise until it stops. Then rotate it counterclockwise (approximately 2 turns) until the yellow gear tooth lines up with the mark on the cover. The arrow on the cable reel lable points straight up.
   d. After reassembly confirm that the wheels are still straight ahead and that the steering wheel spoke angle is correct. If minor spoke angle adjustment is necessary, do so only by adjustment of the tie rods, not by removing and repositioning the steering wheel.
9. Using new Torx® screws, torque the air bag to 7 ft. lbs. (10 Nm). This torque is critical to proper operation of the air bag. Connect the airbag wiring harness and then the battery.
10. After installation, turn the ignition switch **ON**; the instrument panel SRS light should turn **ON** for about 8 seconds and turn **OFF**.
11. Check operation of the horn and the cruise control switches.

Steering Column

REMOVAL & INSTALLATION

Integra

1. Remove the steering wheel center pad. Remove the steering wheel shaft nut.
2. Remove the steering wheel by rocking it slightly from side to side and pull steadily with both hands.
3. Remove the right and left lower instrument panel covers. Remove the front console.
4. Remove the driver's side knee bolster from the steering hanger. Remove the steering joint cover.
5. Remove the steering joint bolts and move the joint toward the steering column. Remove the upper and lower steering column covers.
6. Disconnect each wire coupler from the combination switch. Remove the turn signal canceling sleeve and combination switch assembly.
7. Disconnect each wire coupler

from the fuse box under the left side of the dash.

8. Remove the steering column holder. Remove the attaching nuts and bolts, then remove the steering column assembly.

To install:

9. Install the column in the vehicle. Torque the steering column support bracket bolts to 9 ft. lbs. (13 Nm) and the nuts to 16 ft. lbs. (22 Nm). Torque the lower bolt on the steering joint to 16 ft. lbs. (22 Nm). If equipped with a tilt steering column, use the following procedures.

10. Install the 10mm washer on the tilt lever assembly.

11. Install the bending plate base on the steering column. Insert the tilt lever assembly shaft into the hole in the bending plate base.

12. Install the spin stopper on the shaft. Be sure to apply grease to each sliding surface.

13. Install the stopper collar over the spin stopper. Install the column hanger spring.

14. Torque the tilt locknut to 5 ft. lbs. (7 Nm) and slide the stopper collar to the tilt locknut side.

NOTE: The tilt lock has left hand threads. If the stopper collar cannot be moved, turn the tilt locknut counterclockwise.

15. Pull the tilt lever knob upward and measure the lever preload at 1.38 in. (35.0mm) from the tip of the knob. The lever preload should be 20 lbs.

16. If the preload is out of specifications, adjust it by sliding the stopper collar to the spin stopper side and turning the tilt locknut one turn right or left.

17. Slide the stopper collar to the tilt locknut side.

18. Install the stopper clip between the spin stopper collar and tighten with a 3mm screw.

19. Install the upper column holder and bending plate on the steering column with the rubber bands. Install the bending plate with the arrow toward the gearbox.

20. Install the bending plate guide on the steering column. Loosely install the steering joint on the steering shaft.

21. To finish the installation process, reverse the order of the removal procedure. Be sure to make the adjustment that must be performed on the tilt steering column after installation. The adjustment is as follows:

NOTE: A special tilt steering column adjustment guide tool 07973–6920001 or equivalent, will be needed to perform the following adjustment.

a. Install the adjustment guide tool on the top end of the steering shaft and turn it as far as it will go.

b. Loosely install the upper column holder and bending plate guide attaching bolts and pull the column down to be sure the bending plate is seated snugly against the hook.

c. Loosely install the lower bracket and pull the column down so there is no clearance between the bending plate and hook.

d. Tighten the upper column holder nuts to 9 ft. lbs. (13 Nm). Tighten the lower bracket bolts 16 ft. lbs. (22 Nm).

e. Tighten the bending plate guide bolts to 16 ft. lbs. (22 Nm). Connect the steering joint to the pinion and install and hand tighten the steering joint bolt.

f. Put the steering joint down and tighten the steering joint bolts to 22 ft. lbs. (30 Nm). Make sure the end adjustment guide bottoms against the turn signal switch as shown.

g. Install the column cover, column boot and dashboard lower panel.

Sterling and Legend

1. Disconnect both the negative and positive battery cables from the battery.

2. Remove the lower maintenance lid below the air bag and then remove the short connector.

3. Disconnect the connector between the air bag and the cable reel.

4. Connect the short connector to the air bag side of the connector.

5. Remove the left side maintenance lid and the cruise control/set resume switch cover.

6. Insert a T30 Torx® bit and remove the Torx® bolts. Remove the air bag assembly.

NOTE: Be sure to store the air bag in a safe place with the pad side facing upwards.

7. Remove the steering wheel retaining nut. Gently hit the backside of each of the steering wheel spokes with equal force with the palms of the hands.

NOTE: Avoid hitting the wheel or the shaft with excessive force. Damage to the shaft could result.

8. Disconnect the connectors for the horn, cruise control and release the air bag connector clips. To disconnect the cruise control connector be sure to release both lock tabs on the connector.

9. Remove the yoke joint cover. Remove the instrument panel lower cover then disconnect the wiper control unit connector.

10. Remove the left knee bolster. Remove the column cover's maintenance lid. Disconnect the cable reel harness at the connection located on the underside of the steering column.

11. Remove the rear steering column holder. Remove the steering column mounting nuts.

12. Remove the steering yoke joint bolts and disconnect the yoke joint. Disconnect the harness connectors attached to the column assembly, then remove the column assembly.

To install:

13. Install the column and connect the yoke joint. Make sure the wiring is not pinched and that the yoke joint moves smoothly.

14. Install the cable reel and connect all the wiring.

15. Center the cable reel, by rotating the cable reel clockwise until it stops. Then rotate it counterclockwise (approximately 2 turns) until the yellow gear tooth lines up with the mark on the cover. The arrow on the cable reel lable points straight up.

16. Before installing the steering wheel, the front wheels should be aligned straight forward.

17. After reassembly confirm that the wheels are still straight ahead and that the steering wheel spoke angle is correct. If minor spoke angle adjustment is necessary, do so only by adjustment of the tie rods, not by removing and repositioning the steering wheel.

18. After installation, connect the battery and make sure everything works properly before installing the air bag assembly. Disconnect the battery to install the air bag, then connect the battery with on one in the vehicle.

19. Turn the ignition switch **ON**; the instrument panel SRS light should turn **ON** for about 8 seconds and turn **OFF**.

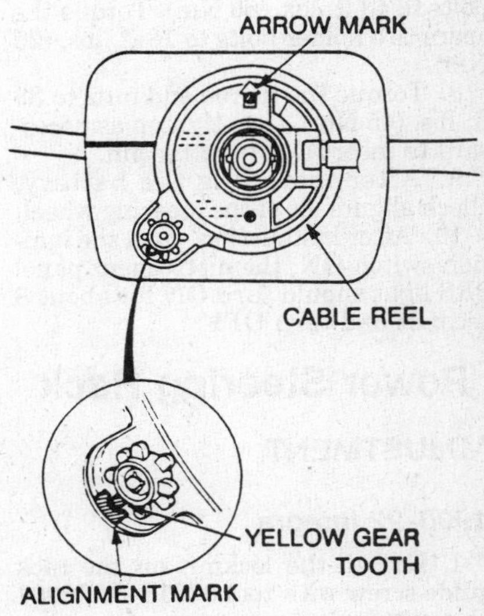

Cable reel alignment on vehicles with air bag

Manual Rack and Pinion

REMOVAL & INSTALLATION

NSX

NOTE: This is an aluminum vehicle. Using normal steel bolts on aluminum will cause an electrolytic reaction: the aluminum around the fastener will corrode and the bolt will loosen. When replacing fasteners, be sure to use bolts that have a Dacro® coating specifically designed for such applications. Dacro® bolts can be identified by a dull grey finish, sometimes with a dull green finish on the threads for more accurate torque wrench readings. These parts should be available at the dealer.

1. Place the front wheels straight ahead. Remove the battery and the spare tire.
2. At the bottom of the steering column, remove the steering joint cover adn remove both bolts from the steering joint.
3. Raise and safely support the vehicle at the proper jacking points. Remove the front wheels.
4. Remove the cotter pins and nuts from the tie rod ends and use a ball joint press to separate the tie rod joints from the steering knuckle. Be carefull to not damage the threads.
5. Remove the spare tire holder.
6. Remove the long bolts from the cross beam and steering rack clamps and lower the rack out of the vehicle.

To install:

7. Installation is the reverse of removal. Torque the long steering rack bolts to 43 ft. lbs. (60 Nm). Torque the spare tire holder bolts to 18 ft. lbs. (25 Nm).
8. Torque the tie rod end nuts to 33 ft. lbs. (45 Nm), then tighten as necessary to insert a new cotter pin.
9. After installing the battery, check alignment of the steering wheel.
10. After installation, turn the ignition switch **ON**; the instrument panel SRS light should turn **ON** for about 8 seconds and turn **OFF**.

Power Steering Rack

ADJUSTMENT

1990–92 Integra

1. Loosen the locknut on the rack guide screw with tool 07916–SA50001 or equivalent.
2. Tighten the guide screw until it compresses the spring against the guide; loosen it, torque it to about 3 ft. lbs. (4 Nm) and back it off about 25 degrees.
3. Torque the locknut to about 18 ft. lbs. (25 Nm) while preventing the guide screw from moving.

Sterling and Legend

1. Loosen the locknut on the rack guide screw with tool 07916–SA50001 or equivalent.
2. Tighten the guide screw until it compresses the spring against the guide; loosen it, torque it to about 2 ft. lbs. (3 Nm) and back it off about 20 degrees.
3. Torque the locknut to about 18 ft. lbs. (25 Nm) while preventing the guide screw from moving.

REMOVAL & INSTALLATION

Integra

1. Remove the steering joint cover, the steering shaft connector bolts and pull the connector up off the pinion shaft. Drain the power steering fluid and remove the gearbox shield.
2. Raise and safely support the vehicle.
3. Remove the front wheels.
4. Remove the cotter pins and unscrew the tie rod end ball joint nuts halfway.
5. Break the tie rod ball joints loose using a tie rod end removal tool or equivalent.
6. Remove the nuts and lift the tie rod ends out of the steering knuckles.
7. If equipped with a manual transaxle, perform the following procedures:
   a. Remove the shift extension from the transaxle case. Slide the boot at the connecting position of the gear shift rod.
   b. Slide the pin retainer out of the way, drive out the spring pin with a punch and disconnect the shift control rod. Note that on reassembly, install the pin retainer back into place after driving the spring pin in.
8. If equipped with an automatic transaxle, remove the shift cable guide from the floor and pull the shift cable down by hand. Remove the shift cable holder and cable from the transaxle case by removing the clamp.
9. On 1988–89 models, remove the self locking nuts connecting the exhaust header pipe to the exhaust pipe. Separate the exhaust pipe from the header pipe.
10. On 1990–92 models, remove the exhaust header pipe from the catalyst.
11. Clean the fluid line connectors of all dirt and oils. Disconnect the fluid lines from the valve body.
12. On 1990–92 modles, remove the center beam.
13. Remove the gearbox mounting bolts. Turn the pinion shaft so the tie rods are all the way to the right side.
14. Drop the gearbox far enough so the end of the pinion shaft comes out of it's hole in the frame channel and rotate it forward until the shaft is pointing to the rear. Slide the gearbox to the right until the tie rod clears the rear beam, lower it from the vehicle to the left.

To install:

15. Install the power steering gear to thechassis and torque the bolts to 32 ft. lbs. (44 Nm). Torque the power steering gear clamp-to-chassis bolts to 29 ft. lbs. (40 Nm).
16. Connect the tie rod ends and toruqe the knuckle nuts to 40 ft. lbs. (55 Nm), then tighten the nuts just enough to install new cotter pins.
17. Connect the shift extension and torque thebolt to 7 ft. lbs. (10 Nm).
18. When everything is assembled, refill the reservoir with new power steering fluid. Start the engine and allow it run at fast idle, turn the steering wheel from lock-to-lock several times to bleed the air out.
19. Check the fluid again and add, if necessary. Check the system for leaks.

Sterling and 1988–90 Legend

1. Remove the steering joint cover and disconnect the steering shaft from the gearbox.
2. Drain the power steering fluid.
3. Remove the gearbox shield.
4. Using cleaning solvent and a brush, clean the control unit, it's lines and the end of the gearbox. Blow dry with compressed air, if possible.
5. Raise and safely support the vehicle.
6. Remove the front wheels.
7. Remove the cotter pins and unscrew the tie rod end ball joint nuts halfway.
8. Break the tie rod ball joints loose, using a tie rod end removal tool or equivalent.
9. Remove the nuts and lift the tie rod ends from the steering knuckles.
10. If equipped with a manual transaxle, perform the following procedures:
    a. Remove the shift extension from the transaxle case.
    b. Disconnect the gearshift rod from the transaxle case by removing the 8mm spring pin.
11. If equipped with an automatic transaxle, remove the shift control cable from the clamp.
12. Remove the center beam bolts and the center beam.

NOTE: Replace the self-locking nuts retaining the center beam, if worn.

13. Disconnect the exhaust header pipe from the manifold. Replace the exhaust gasket and the self-locking nuts when reinstalling the pipe.
14. Remove the header pipe joint nuts and the header pipe.
15. Disconnect the 4 lines from the control unit.
16. Slide the tie rod all the way to the right side.
17. Slide the gear box right so the left tie rod clears the bottom of the rear beam and remove the gearbox.

To install:

18. Reverse the removal procedures. Torque the power steering gear-to-chassis bolts to 28 ft. lbs. (39 Nm), the exhaust pipe-to-exhaust manifold nuts to 40 ft. lbs. (55 Nm), the exhaust pipe-to-muffler nuts to 25 ft. lbs., the center beam-to-chassis bolts to 37 ft. lbs. (51 Nm), the shift extension-to-transaxle bolt to 7 ft. lbs. (10 Nm) and the tie rod end-to-steering knuckle nut to 32 ft. lbs. (44 Nm).
19. Refill the reservoir with new power steering fluid. Start the engine and allow it to run at fast idle, turn the steering wheel from lock-to-lock several times to bled the air out.
20. Check the fluid again and add, if necessary. Check the system for leaks.
21. After installation, turn the ignition switch **ON**; the instrument panel SRS light should turn **ON** for about 8 seconds and turn **OFF**.

1991–92 Legend

1. Disconnect the fluid return hose from the rack and put the end in a container. Start the engine and turn the steering wheel lock–to–lock several times. When fluid stops comming out, stop the engine.
2. Raise and safely support the vehicle and remove the front wheels.
3. Remove the cotter pins and disconnect the tie rod ball joints using a suitable press tool. Be carefull to not damage the threads on the joints.
4. Loosen the steering joint bolt but do not remove it yet.
5. Remove the splash guard. The 2 long bolts also hold the rack in place, and the rack will now be partially hanging on the steering joint.
6. Carefully clean all the hydraulic fitting connections with solvent and a brush and blow them dry.
7. Disconnect the hydraulic fittings and hoses.
8. Remove the hydraulic line mounting clamps from the rack.
9. Place a jack stand under the rack and remove the steering joint bolt. Remove the rack assembly.

To install:

NOTE: Several bolts thread into aluminum. Using normal steel bolts on aluminum will cause an electrolytic reaction: the aluminum around the fastener will corrode and the bolt will loosen. When replacing fasteners, be sure to use bolts that have a Dacro® coating specifically designed for such applications. Dacro® bolts can be identified by a dull grey finish, sometimes with a dull green finish on the threads for more accurate torque wrench readings. These parts should be available at the dealer.

10. Fit the pinion into the steering joint and install the right side mounting rubber and bracket. Do not tighten the bolts yet.
11. Loosely connect the hydraulic lines. Install the hydraulic line cushions and clamps, then tighten the line connections.
12. Install the steering joint bolts, make sure the joint does not bind when turned, then torque the bolts to 16 ft. lbs. (22 Nm).
13. Torque the right side mount bolts to 28 ft. lbs. (39 Nm).
14. Install the splash guard and torque the short bolts to 28 ft. lbs. (39 Nm), the long bolts to 43 ft. lbs. (60 Nm).
15. Connect the tie rod ends and torque the nuts to 40 ft. lbs. (54 Nm), then tighten as necessary to install a new cotter pin.
16. When installation is complete, refill the hydraulic reservoir with new steering fluid, start the engine and turn the steering wheel lock–to–lock several times to bleed the system. Check fluid level again.
17. After installation, turn the ignition switch **ON**; the instrument panel SRS light should turn **ON** for about 8 seconds and turn **OFF**.

Power Steering Pump

REMOVAL & INSTALLATION

1988–89 Integra

1. Disconnect and plug the hoses from the reservoir.
2. Remove the 10mm flange bolts, the belt from the pulley and the pump assembly.

To install:

3. Reverse the removal procedures. Torque the power steering pump-to-engine bolts to 29 ft. lbs. (40 Nm). Be sure to observe the following:
   a. Connect the hoses tightly.
   b. Adjust the belt tension.
   c. Check the fluid level and add, if necessary.
   d. Bleed the air from the system.

1990–92 Integra

1. Drain the power steering fluid. Disconnect the inlet and outlet hoses from the power steering pump and plug them.
2. Remove the belt by loosening the adjusting bolts on the pump bracket.
3. Remove the power steering mounting bolts and remove the power steering pump.

To install:

4. Reverse the removal procedures. Torque the power steering pump-to-engine bolts to 17 ft. lbs. (24 Nm). Be sure to observe the following:
   a. Connect the hoses tightly.
   b. Adjust the belt tension.
   c. Check the fluid level and add, if necessary.
   d. Bleed the air from the system.

NOTE: When installing a new or rebuilt pump, check the power steering pump preload in a vise before installing it on the vehicle. Check the pump preload with a torque wrench: it should take about 3 ft. lbs. (4 Nm) to turn the pump.

Sterling and 1988–90 Legend

1. Remove the belt cover.
2. Drain the fluid from the system.
3. Disconnect and plug the inlet/outlet hoses from the pump.
4. Remove the belt by loosening the pump pivot bolt and adjusting nut.
5. Remove the pump assembly nut/bolt and the assembly.

To install:

6. Reverse the removal procedures. Torque the power steering pump-to-bracket bolt to 28 ft. lbs. (39 Nm) and the power steering pump-to-bracket nut to 16 ft. lbs. (22 Nm). Be sure to observe the following:
   a. Refill the reservoir with new fluid to the Upper Level on the reservoir.
   b. Connect the hoses tightly.
   c. Adjust the belt tension.
   d. Bleed the air from the system.
   e. Check the fluid level and add, if necessary.

1990–92 Legend

1. Disconnect the fluid return hose from the rack and put the end in a container. Start the engine and turn the steering wheel lock–to–lock several times. When fluid stops comming out, stop the engine.
2. Remove the air cleaner cover and duct.
3. Disconnect the hydraulic lines from the pump.
4. Loosen the adjustment and remove the belt.

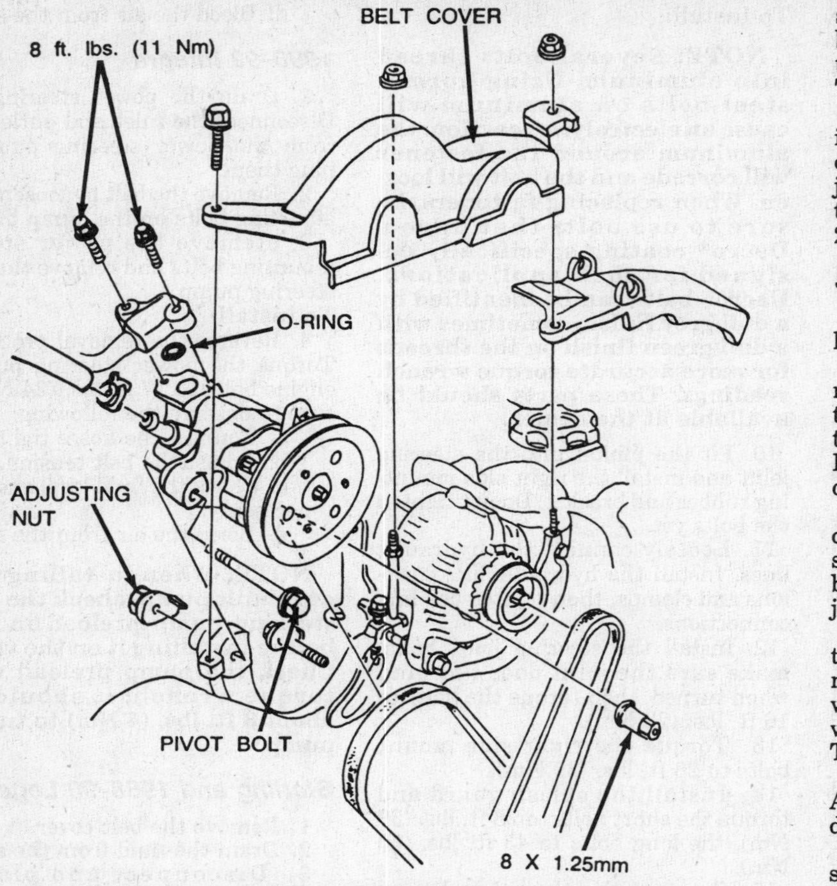

Power steering pump removal—Sterling and 1988–90 Legend

5. Remove the special bolt and nut and remove the pump.

To install:

6. Installation is the reverse of removal. Torque the special bolt to 33 ft. lbs. (45 Nm) and the nut to 16 ft. lbs. (22 Nm).

7. Connect the hydraulic lines, install and adjust the belt and refill the reservoir. To bleed the system, run the engine and turn the steering wheel lock–to–lock several times. Check the fluid level.

BELT ADJUSTMENT

1. Loosen the adjuster arm bolt.

2. Move the pump toward or away from the engine, until the belt can be depressed approximately 19–24mm at the midpoint between both pulleys under moderate thumb pressure. If the tension adjustment is being made on a new belt, the deflection should only be about 11mm, to allow for the initial stretching of the belt.

3. Torque the bolt to 29 ft. lbs. (40 Nm) for Integra or 33 ft. lbs. (45 Nm) for Sterling and Legend. Run the engine, then recheck the adjustment.

SYSTEM BLEEDING

1. Raise and safely support the vehicle.

2. Refill the power steering pump reservoir to the full level.

3. Start the engine and turn the steering wheel from lock-to-lock (several times).

4. After the air bubbles have been eliminated from the system, refill the reservoir and lower the vehicle.

Tie Rod Ends

REMOVAL & INSTALLATION

Integra

1. Raise and safely support the vehicle. Remove the front wheel.

2. Loosen the tie rod end-to-power steering gear jam nut.

3. Remove the tie rod end-to-steering knuckle cotter pin and nut.

4. Using a press type ball joint removal tool, separate the tie rod end from the steering knuckle. Take care to not damage the threads on the joint.

5. While supporting the power steering rod, remove the tie rod end, be sure to count revolutions required to remove the tie rod end.

To install:

6. Install the new tie rod end, turn it the same amount of revolutions necessary to remove it and tighten the jam nut to 42 ft. lbs. (58 Nm) and the tie rod end-to-steering knuckle nut to 29 ft. lbs. (40 Nm).

Sterling and Legend

1. Raise and safely support the vehicle. Remove the front wheels.

2. Remove the cotter pin and the nut from the tie rod end. Use a press type ball joint remover tool, separate the tie rod from the steering knuckle. Be carefull to not damage the threads on the joint.

3. Disconnect the air tube at the dust seal joint. Remove the tie rod dust seal bellows clamps and move the rubber bellows back on the tie rod rack joints.

4. Straighten the tie rod lockwasher tabs at the tie rod-to-rack joint and remove the tie rod by turning it with a wrench. On some models, the lock washer is staked.

To install:

5. Reverse the removal procedure. Always use a new tie rod lockwasher during reassembly.

6. Torque the tie rod end-to-power steering gear to 40 ft. lbs. (55 Nm) and the tie rod end-to-steering knuckle nut to 32 ft. lbs. (44 Nm).

7. Fit the locating lugs into the slots on the rack and bend the outer edge of the washer over the flat part of the rod, after the tie rod nut has been properly tightened.

BRAKES

For all brake system repair and service procedures not detailed below, please refer to "Brakes" in the Unit Repair section.

Master Cylinder

REMOVAL & INSTALLATION

NOTE: Before removing the master cylinder, cover the body surfaces with fender covers and rags to prevent damage to painted surfaces by brake fluid.

1. Disconnect and plug the brake lines at the master cylinder.

2. Remove the master cylinder-to-power booster bolts and the master cylinder from the vehicle.

3. To install, reverse the removal procedure. Torque the master cylinder-to-power booster bolts to 11 ft. lbs. (15 Nm). Bleed the brake system.

Proportioning Valve

REMOVAL & INSTALLATION

1. Disconnect and plug the hydraulic lines from the dual proportioning valve.
2. Remove the proportioning valve-to-bracket bolts and the valve from the vehicle.
3. To install, reverse the removal procedures. Bleed the brake system.

Power Brake Booster

INSPECTION

A preliminary check of the vacuum booster can be made as follows:
1. Depress the brake pedal several times using normal pressure; make sure the pedal height does not vary.
2. Hold the pedal in the depressed position and start the engine. The pedal should drop slightly.
3. Hold the pedal in the above position and stop the engine. The pedal should stay in the depressed position for approximately 30 seconds.
4. If the pedal does not drop when the engine is started or rises after the engine is stopped, the booster is not functioning properly.

REMOVAL & INSTALLATION

1. Disconnect the vacuum hose from the booster.
2. Disconnect and plug the brake lines at the master cylinder.
3. Remove the brake pedal-to-booster link pin and the booster nuts; the pushrod and nuts are located inside the vehicle under the instrument panel.
4. Remove the booster with the master cylinder attached.

To install:

5. Reverse the removal procedure. Torque the power brake booster-to-firewall nuts to 9 ft. lbs. (13 Nm) and the master cylinder-to-power brake booster nuts to 11 ft. lbs. (15 Nm).
6. Check the vacuum booster pushrod-to-master cylinder piston clearance as outlined in the master cylinder removal procedure.
7. Bleed the brake system before operating the vehicle.

Brake Caliper

REMOVAL & INSTALLATION

Front

1. Raise and safely support the vehicle.
2. Remove the front wheel assembly.
3. Remove the banjo bolt and disconnect the brake hose from the caliper.
4. Remove the caliper slide mounting bolts and remove the caliper. Remove the pad spring from the caliper body.
5. To install, reverse the order of the removal procedure. Be sure to properly bleed the brake system after installation.
6. Torque the caliper slide mounting bolts:
   1988–89 Integra—33 ft. lbs. (45 Nm).
   1990–92 Integra—to 24 ft. lbs. (33 Nm).
   Sterling and 1988–90 Legend—24 ft. lbs. (33 Nm).
   1991–92 Legend—36 ft. lbs. (50 Nm).
   NSX—36 ft. lbs. (50 Nm).

Rear

1. Raise and safely support the vehicle.
2. Remove the rear wheel assembly.
3. Remove the caliper shield.
4. Disconnect the parking brake cable from the lever on the caliper by removing the lock pin.
5. Remove the banjo bolt and disconnect the brake hose from the caliper.
6. Remove the 2 caliper slide mounting bolts and remove the caliper from the bracket.

To install:

7. To install, reverse the order of the removal procedure. Be sure to properly bleed the brake system after installation.
8. Torque the caliper slide mounting bolts:
   Integra—28 ft. lbs. (39 Nm).
   1990–92 Integra—16 ft. lbs. (23 Nm).
   Sterling and 1988–90 Legend—20 ft. lbs. (27 Nm).
   1991–92 Legend—17 ft. lbs. (23 Nm).
   NSX—36 ft. lbs. (50 Nm).

Disc Brake Pads

REMOVAL & INSTALLATION

Front

1. Raise and safely support the vehicle. Remove the front wheels.
2. Using a prybar, between the brake pad and the caliper, pry the brake caliper away from the vehicle until the piston is fully seated in the caliper.
3. Remove the lower caliper slide mounting bolt and swing the caliper upward and away from the disc. If necessary, remove both caliper slide mounting bolts without disconnecting the hydraulic line and support the caliper so it does not hang on the line.
4. Remove the brake pad shim, the brake pad retainers and the pad.
5. Install new brake pads, coated with Molykote® M77, between the pads and shims, the shims and retainers.
6. Lower the calipers of the brake pad assemblies and install the caliper slide mounting bolt.

Rear

1. Raise and safely support the rear of the vehicle. Remove the rear wheels.
2. Using a prybar, between the brake pad and the caliper, pry the brake caliper away from the vehicle until the piston is fully seated in the caliper.
3. Remove both caliper slide mounting bolts without disconnecting the hydraulic line and support the caliper so it does not hang on the line.
4. Remove the brake pads and shims.
5. To install, use new brake pads and reverse the removal procedures. Install the caliper slide mounting bolts.

Brake Rotor

REMOVAL & INSTALLATION

1. Raise and safely support the vehicle. Remove the wheel.
2. Remove the brake caliper bolts and the caliper from the knuckle. Do not allow the caliper to hang by the brake hose, support it with a length of wire.
3. Remove the disc brake rotor retaining screws, if equipped. Screw both 8mm into the disc brake removal holes and turn the bolts to press the rotor from the hub.
4. Installation is the reverse order of the removal procedure.

NOTE: Only turn each bolt 2 turns at a time to prevent cocking the disc excessively. Also on the rear brake rotor removal, the rotor should slide off after the rotor retaining bolt is removed.

Parking Brake Cable

ADJUSTMENT

Inspect the following items:
a. Check the ratchet for wear.
b. Check the cables for wear or damage and the cable guide and equalizer for looseness.
c. Check the equalizer cable where it contacts the equalizer and apply grease, if necessary.
d. Check the rear brake adjustment.

1. Block the front wheels. Raise and safely support the rear of the vehicle.
2. Loosen the adjusting nut, located in the console. Make sure the caliper lever is in contact with the pin at both the right and left rear calipers.
3. Pull the parking brake lever up a notch.
4. Tighten the adjusting nut until the rear brakes drag slightly.
5. Release the brake lever and make sure the rear brakes do not drag.
6. The rear brakes should be locked when the hand brake lever is pulled; 4–8 notches for 1988–89 Integra, 6–10 notches for 1990–92 Integra, 7–11 notches for Legend and Sterling, 10–14 notches for NSX.

REMOVAL & INSTALLATION

1. Remove the adjusting nut from the equalizer mounted on the console and separate the cable from the equalizer.
2. Set the parking brake lever to a fully released position. Raise and support the vehicle safely. Remove the cotter pin from the rear caliper assembly.
3. After removing the cotter pin, pull out the pin which connects the cable and the lever.
4. Detach the cable from the guides at the calipers and remove the cable.
5. To install, reverse the removal procedure, making sure grease is applied to the cable and the guides.

Brake System Bleeding

Standard Brakes

NOTE: The master cylinder must be full at the start of the bleeding procedure and checked after bleeding each caliper. Add fluid as required. Use only DOT 3 or 4 brake fluid. If a pressure bleeder is not available it will be necessary to have the aid of an assistant to perform this brake bleeding operation.

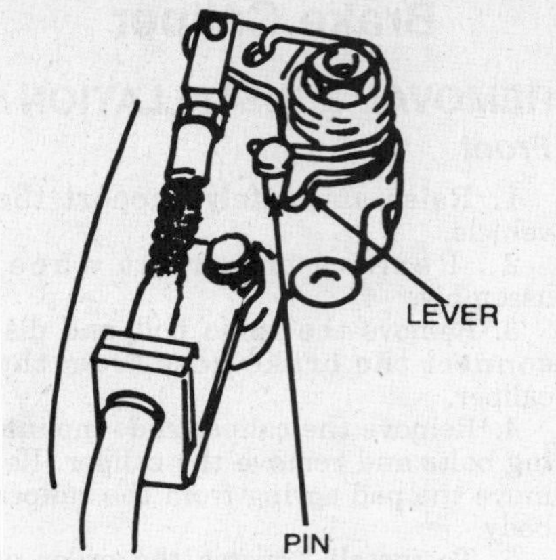

Parking brake cable lever on rear disc brakes

1. Have an assistant slowly pump the brake pedal several times and then apply a steady pressure to the brake pedal.
2. Attach a bleed hose to the bleed screw and place it into a clear container. Loosen the brake bleed screw at the brake caliper furthest away from the master cylinder (passenger's rear) to allow the air to escape from the system.
3. Repeat this procedure for each brake caliper, until no air bubbles appear in the brake fluid. Use the following brake caliper sequence in order to bleed the brake system properly:
a. Right rear, passenger's side brake caliper.
b. Left front, driver's side brake caliper.
c. Left rear, driver's side brake caliper.
d. Right front, passenger's side brake caliper.
4. Check the fluid level in the master cylinder and add, if necessary. Road test the vehicle and check the brake performance.

Anti-Lock Brake System Service

PRECAUTIONS

- The anti-lock brake system accumulator contains a high pressure nitrogen gas, do not puncture, expose to flame or attempt to disassemble the accumulator or it may explode and severe personal injury may result.
- The anti-lock brake system contains brake fluid under extremely high pressure within the power unit pump, accumulator and modulator assembly. Do not disconnect or loosen any lines, hoses, fittings or components without

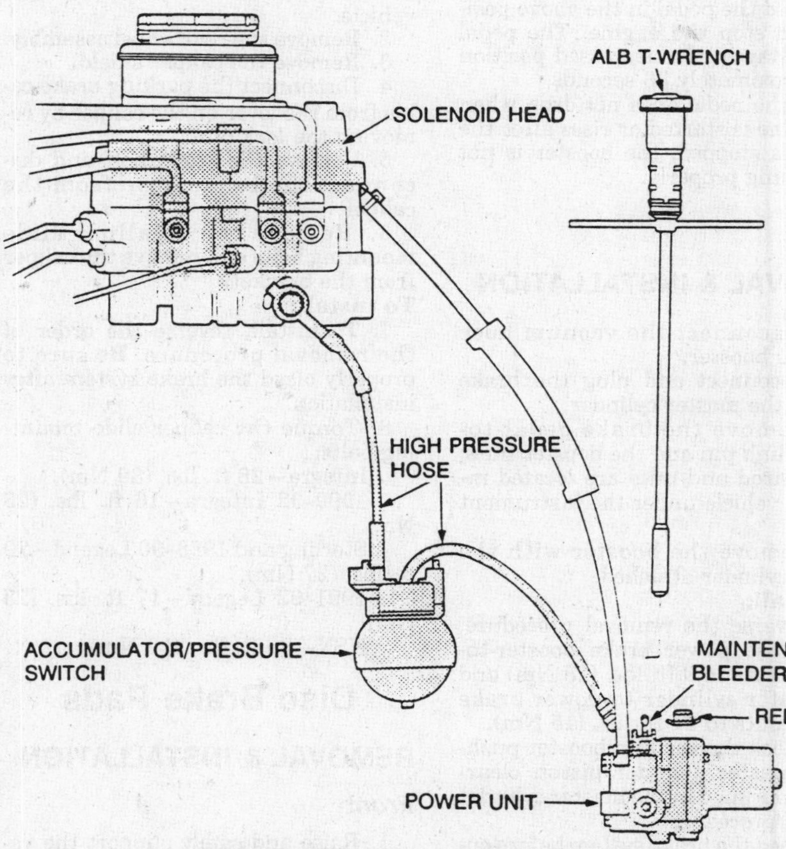

Bleeding the anti-lock brake high pressure system

properly relieving the system pressure. Improper procedures or failure to discharge the system pressure may result in severe or fatal personal injury and/or property damage.

- Use only tool 07HAASG00100 or equivalent to relieve pressure.

RELIEVING ANTI-LOCK BRAKE SYSTEM PRESSURE

1. Insure the ignition switch is **OFF**.
2. Using a syringe or similar device, remove all the fluid from the master cylinder and modulator reservoirs.
3. Remove the red cover from the bleeder port on top of the power unit pump.
4. Install the bleeding tool onto the bleeder. Make certain the reservoir cap on the tool is secured.
5. Using the tool, turn the bleeder about 90 degrees to admit high pressure fluid into the reservoir. As the pressure drops, turn the bleeder open about 1 full turn to completely relieve the system.
6. Retighten the bleeder and remove the tool. Discard the captured brake fluid; do not reuse it. Install the red cap on the bleeder port.

Anti-Lock Brake Modulator

REMOVAL & INSTALLATION

1. Relieve the system pressure.
2. Disconnect and plug the lines from the hydraulic modulator.
3. Remove the mounting bolts and the modulator from the vehicle.
4. To install, reverse the removal procedures. Bleed the brake system.

Accumulator Pressure Switch

REMOVAL & INSTALLATION

NOTE: The anti-lock brake system accumulator contains a high pressure nitrogen gas, do not puncture, expose to flame or attempt to disassemble the accumulator or it may explode and severe personal injury may result.

1. Relieve the accumulator line pressure.
2. Remove the 3 flange bolts, then remove the accumulator from the accumulator bracket.
3. Secure the accumulator in a suitable vise so the relief plug points straight up.
4. Slowly turn the plug 3½ turns and then wait 3 minutes for all pressure to escape.
5. Remove the plug completely and dispose of the accumulator unit.
6. To install, reverse the order of the removal procedure.

Pulsers/Sensors

REMOVAL & INSTALLATION

1. To remove the wheel sensors, it is just a matter of removing the wheel that the sensor must be removed from and remove the the sensor bolts, then remove the sensor.
2. The important part of the wheel sensor removal and installation procedure is the air gap that is necessary when the sensor is installed. Use the following procedure to set the air gap.
   a. Check the pulser/sensor for chipped or damaged teeth.
   b. Measure the air gap between the sensor and pulser all the way around while rotating the wheel by hand. The air gap on the both front and rear sensors should be 0.016–0.039 in. (0.4–1.0mm).
   c. If the gap exceeds 0.039 in. (1.0mm), the probability is a distorted knuckle which should be replaced.
3. Be careful when installing the sensors to avoid twisting the wires. Use a white line on the wires as a guide. After sensor replacement comfirm the proper operation of the ABS system.

CHASSIS ELECTRICAL

Air Bag System

DISARMING

An air bag is a supplimental restraint meant for use with a seat belt. All Supplemental Restraint System (SRS) wiring is covered with a yellow outer insulation. This wiring harness cannot be repaired and if cut or damaged, the whole harness must be replaced. To disable the Air bag:

1. Disconnect the negative battery cable.
2. Remove the access panel from the bottom or side of the steering wheel. Stored on the panel is a red short connector.
3. Unplug the air bag connector and install the red short connector on the air bag connector. This will short the air bag unit electrical terminals and safeguard against static electricity.
4. The 1991–92 Legend LS is also equipped with a passenger side air bag. Remove the glove box and unplug the yellow air bag connector to the right. Plug in the red short connector.
5. The air bag can now be safely removed and/or the battery can be reconnected for testing or operating other vehicle systems.
6. To reconnect the air bag, disconnect the battery.
7. Connect the air bag wiring harness and make sure no one is in the vehicle before reconnecting the battery.

Heater Blower Motor

REMOVAL & INSTALLATION

Without Air Conditioning

1988–89 Integra

1. Disconnect the negative battery cable. Remove the glove box.
2. Remove the frame to the glove box.
3. Remove the blower duct.
4. Disconnect the wire connections from the blower.
5. Remove the screws and remove the blower motor assembly.
6. To install, reverse the removal procedures. Check that there are no air leaks in the blower case.

1990–92 Integra

1. Disconnect the negative battery cable. Remove the passenger side lower dashboard cover.
2. Remove the glove box assembly.
3. Remove the front console assembly.
4. Remove the passenger side knee bolster panel, located under the glove box frame.
5. Remove the self tapping screws and remove the heater duct assembly.
6. Remove the heater blower motor mounting bolts.
7. Disconnect the electrical connectors from the blower motor, resistor and recirculation control motor. Remove the blower motor from the blower motor housing.

To install:

8. Connect the wiring and install the blower motor.
9. When reattaching the actuator, make sure its positioning will not allow the air door to be pulled to far.
10. Attach the actuator and all linkage, connect the battery and operate the system to watch the door movement. If necessary, loosen the holding screw and move the actuator up or down.
11. When adjusting the control rod, connect the recirculation control motor connection to the main wire har-

ness, push **RECIRC** and open the air doors. Then connect the control rod to the arm while holding the air doors open.

12. When all adjustments are completed, disconnect the battery again and complete the dashboard assembly.

With Air Conditioning

1988–89 Integra

1. Disconnect the negative battery cable. Remove the glove box.
2. Remove the frame to the glove box and the side frame.
3. Remove the bolts and the retractor control unit with the bracket.
4. Remove the center console box and the bracket.
5. Unbolt the dashboard lower center bracket and insert a small prybar to pry a 12–15mm clearance, to ease in the removal of the evaporator.
5. Loosen the sealing band toward the right side.
6. Disconnect the wire connections from the blower.
7. Remove the blower bolts and the blower assembly.

To install:

8. Install the blower assembly and connect the wiring. Make sure all wiring is secure and temporarily connect the battery to test the blower. Make sure there are no air leaks in the housing.
9. Disconnect the battery again to complete assembly of the dashboard.

1990–92 Integra

1. Disconnect the negative battery cable.
2. Properly discharge the refrigerant from the air conditioning system into recovery equipment. Disconnect the receiver line and suction hose from the evaporator assembly. Be sure to cap the open fitting to prevent moisture from entering the system.
3. Remove the passenger side lower dashboard cover.
4. Remove the glove box assembly.
5. Remove the front console assembly.
6. Remove the passenger side knee bolster panel, located under the glove box frame.
7. Remove the 2 self tapping screws and the air conditioning bands from around the evaporator assembly.
8. Disconnect the wire connector from the thermostat switch and pull off the wire harness from the clamps. Remove the evaporator.
9. Remove the self tapping screws and remove the heater duct assembly.
10. Remove the heater blower motor mounting bolts.
11. Disconnect the electrical connectors from the blower motor, resistor and recirculation control motor. Remove the blower motor from the blower motor housing.

To install:

12. Reverse the removal procedure.
13. When reattaching the actuator, make sure its positioning will not allow the air door to pulled to far.
14. Attach the actuator and all linkage, then apply battery voltage and watch the door movement. If necessary, loosen the holding screw and move the actuator up or down.
15. When adjusting the control rod, connect the recirculation control motor connection to the main wire harness, push **RECIRC** and open the air doors. Then connect the control rod to the arm while holding the air doors open.

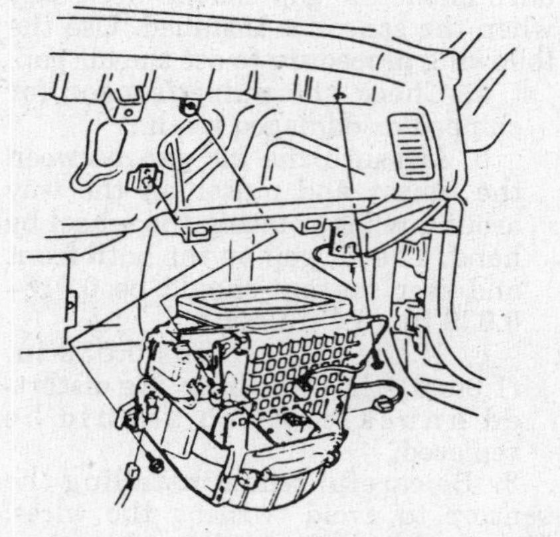

Blower motor removal—1990–92 Integra

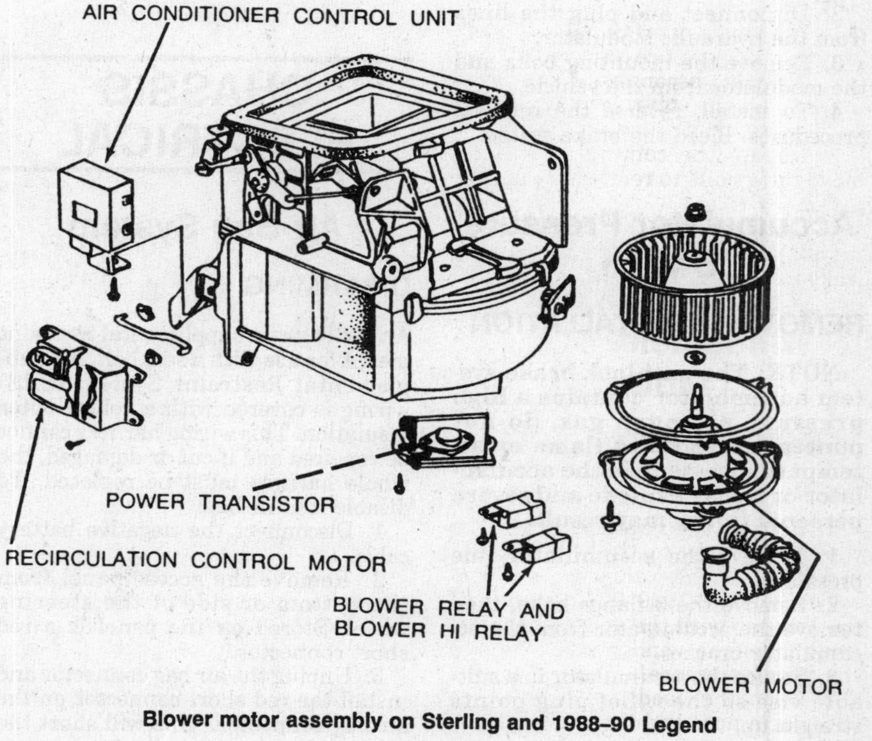

Blower motor assembly on Sterling and 1988–90 Legend

Sterling and 1988–90 Legend

1. Disconnect the negative battery cable.
2. Remove the glove box lower cover screws and the cover.
3. Remove the glove box screws and the glove box.
4. Remove the glove box frame screws, the glove box frame, the clips and the heater duct.
5. Properly discharge the refrigerant from the air conditioning system into recovery equipment.
6. Remove the evaporator as follows:
   a. Disconnect the receiver line and suction hose from the evaporator assembly.
   b. Be sure to cap the open fitting to prevent moisture from entering the system.
   c. Remove the self-tapping screws and the air conditioning bands from around the evaporator assembly.
   d. Disconnect the wire connector from the thermostat switch and pull off the wire harness from the clamps.
   e. Remove the evaporator.
7. Disconnect the wire connectors from the blower.
8. Remove the blower assembly bolts and the assembly.

To install:

9. Install the blower into the case and connect the wiring. Make sure all wiring is secure and temporarily connect the battery to test the blower and

adjust the linkage. Check that there are no air leaks in the blower case.

10. When reattaching the actuator, make sure its positioning will not allow the air door to be pulled too far.

11. Attach the actuator and all linkage, then apply battery voltage and watch the door movement. If necessary, loosen the holding screw and move the actuator up or down.

12. When adjusting the control rod, connect the recirculation control motor connection to the main wire harness, push the **FRESH/RECIRC** switch to **FRESH** and open the air doors. Then connect the control rod to the arm while holding the air doors open.

13. After properly adjusting the linkage, disconnect the battery to complete the dashboard assembly.

1991–92 Legend

1. Disconnect the negative battery cable.
2. Remove the right side lower dashboard panel and unplug the connector.
3. Disconnect the glove box light and remove the glove box.
4. Remove both dashboard end caps.
5. Remove the glove box frame.
6. Unplug the connectors, remove the screws and remove the blower assembly.
7. Installation is the reverse of removal. Make sure there are no air leads in the system.

NSX

1. Working under the front hood, remove the spare tire and the battery.
2. Without disconnecting the wiring, remove the sub-relay box and the water drain duct.
3. Unplug the connectors and remove the 4 bolts to remove the blower motor.
4. Installation is the reverse of removal. Make sure there are no air leaks in the system.

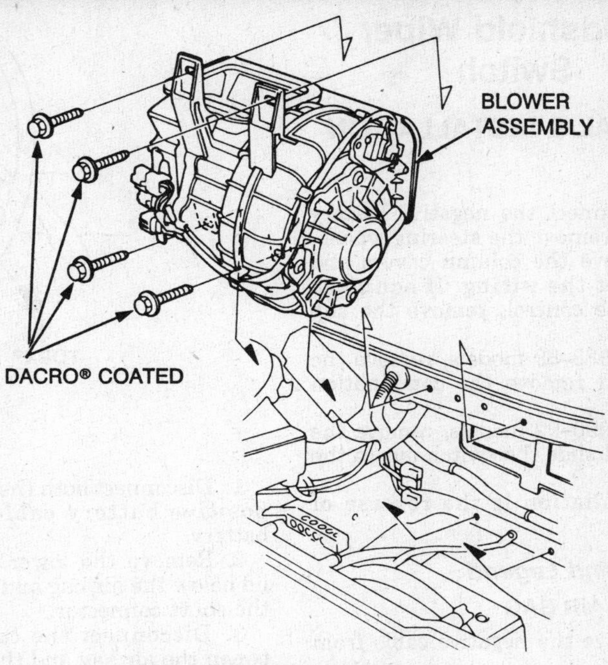

Blower motor removal—NSX

Windshield Wiper Motor

REMOVAL & INSTALLATION

Front

1. Remove the negative cable from the battery. Position the wiper arms in a positioned where they are not concealed. On the 1990–92 Integra, pull the lock tab with the wiper arm lifted away from the windshield to release the spring pressure.
2. Open the hood and remove the wiper arm nuts and the wiper arms.
3. Remove the front air scoop, windshield lower moulding and hood seal, located over the wiper linkage by lightly prying them off the trim clips and removing the retaining screws at the bottom of the windshield.
4. Remove the wiper maintenance grommet. Disconnect the linkage from the wiper motor.
5. Remove the wiper motor water seal cover clamp and the cover, if equipped.
6. Disconnect the wiper motor electrical connector, remove the motor mounting bolts and remove the motor.

To install:

7. Install the motor and connect the wiring. Connect the battery and run the motor in all speeds with the wiper switch to check operation. Turn the switch **OFF** to set the motor in its park position.
8. Lightly c oat the linkage joints with grease and make sure the linkage moves smoothly, then connect the linkage.
9. When installing the wiper arms, be sure to position them on the bottom line of the stopper and then mount the cap nuts.

Rear

INTEGRA AND STERLING

1. Disconnect the negative battery cable. Remove the hatch trim panel.
2. Remove the nut cover, wiper arm nut, wiper arm, cap, special nut, special washer and the cushion rubber.
3. Disconnect the wiper motor electrical connector.

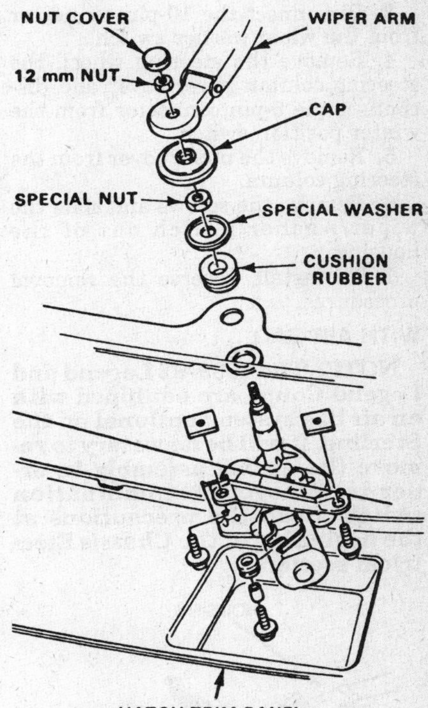

Rear wiper motor assembly

4. Remove the wiper motor mounting nuts with spacers and remove the wiper motor.
5. Installation is the reverse order of the removal procedure.

Windshield Wiper Switch

REMOVAL & INSTALLATION

Integra

1. Disconnect the negative battery cable and remove the steering wheel.
2. Remove the column covers and disconnect the wiring. If equipped with cruise control, remove the slip ring.
3. On 1988–89 models, remove the screws and remove the combination switch.
4. On 1990–92 models, remove the screws and slide the switch out to the side.
5. Installation is the reverse of removal.

Sterling and Legend

WITHOUT AIR BAG

1. Remove the negative cable from the battery.
2. Remove the dashboard lower panel and disconnect the 6-pin and 8-pin connectors from the wiper control unit on the lower panel.
3. Disconnect the 10-pin connector from the wiper/washer switch.
4. Remove the steering wheel, the steering column lower cover and disconnect the 6-pin connector from the winter position switch.
5. Remove the upper cover from the steering column.
6. Remove the screws and slide the wiper/washer switch out of the housing.
7. To install, reverse the removal procedures.

WITH AIR BAG

NOTE: The 1989–92 Legend and Legend Coupe are equipped with an air bag system; optional on the Sterling. It will be necessary to remove the airbag assembly in order to remove the combination switch. Read the precautions at the beginning of the Chassis Electrical section.

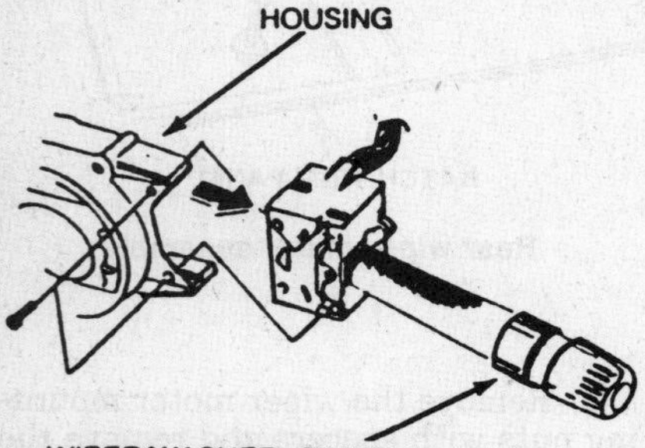

Wiper switch removal—Sterling and Legend

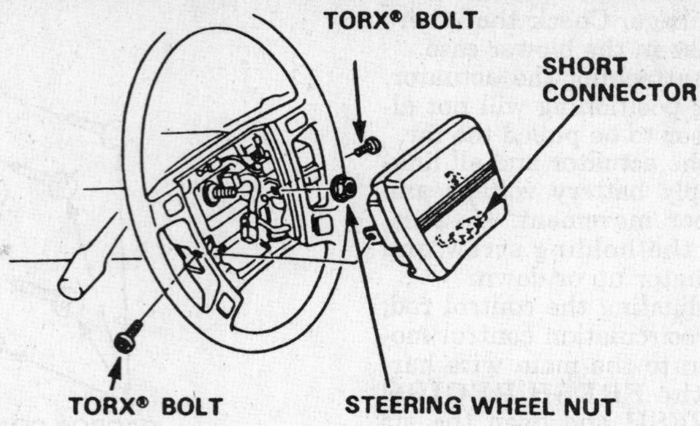

Removing airbag assembly

1. Disconnect both the negative and positive battery cables from the battery.
2. Remove the lower maintenance lid below the air bag and then remove the short connector.
3. Disconnect the connector between the air bag and the cable reel.
4. Connect the short connector to the air bag side of the connector.
5. Remove the left side maintenance lid and the cruise control/set resume switch cover.
6. Insert a T30 Torx® bit and remove the Torx® bolts. Remove the air bag assembly and place it in an out of the way area, such as the back seat, pad side up.
7. Remove the dashboard lower panel and disconnect the 6-pin connector from the wiper position switch and the 6-pin and 8-pin connectors from the wiper control unit on the lower panel.
8. Remove the left knee bolster. Disconnect the combination switch connectors.
9. Remove the upper and lower steering column covers.
10. Remove the lighting and wiper switch mounting screws and remove the switches.

To install:

11. Reverse the removal procedures. Be sure to pay attention to the following:
    a. Be sure to install the airbag wiring so it is not pinched or interfering with other parts. Be sure the battery cables are disconnected.
    b. After installing the airbag assembly, connect the battery and turn the ignition switch to the **ON** position. The instrument panel airbag light should go on for approximately 8 seconds and then go off.
    c. Confirm operations of horn buttons and cruise control set/resume switch.

NSX

NOTE: This vehicle is equipped with an air bag system. It is not necessary to remove the airbag assembly in order to remove the combination switch, but the system should be disarmed before starting work. Read the precautions at the beginning of the Chassis Electrical section.

1. Disconnect both battery cables.
2. Remove the dashboard lower panel and unplug the connectors from the floor wiring harness.
3. Unplug the 18 pin connector from the floor wiring harness.
4. Remove the tilt cover and the upper and lower column covers.
5. Remove the 4 screws and slide the hazzard switch and wiper/washer switch assembly out.
6. Installation is the reverse of removal.

Instrument Cluster

REMOVAL & INSTALLATION

1988–89 Integra

1. Disconnect the negative battery cable. Remove the screws and instrument panel face plate from the dashboard.
2. Remove the right and left switches from the instrument panel then disconnect the wire connecters from the switches.
3. Remove the upper instrument panel caps to access the screws, then remove the panel.
4. Remove the screws under the holes made by removing the switches.
5. Remove the rubber seal, loosen the column cover screws and the instrument panel.
6. Remove the gauge assembly screws and lift out the gauge assembly to disconnect the wire connectors.
7. Disconnect the speedometer cable and remove the gauge assembly.

To install:

8. Install the gauge assembly and connect the speedometer cable and wires.
9. Carefully fit the assembly into place and install the screws.
10. Install the remaining parts and connect the wiring for the switches. Connect the battery and test the switches.

1990–92 Integra

1. Remove the screws and instrument panel from the dsahboard and disconnect the switches.
2. Remove the 4 screws to remove the gauge assembly enough to disconnect the wiring.
3. Installation is the reverse of removal.

Sterling and 1988–89 Legend

NOTE: This vehicle may be equipped with an air bag system. It is not necessary to remove the air bag assembly in order to remove the instrument panel, but the system should be disarmed before starting work.

1. Disconnect the negative battery cable. Remove the 2 screws under the switch pods and 2 screws in front of the gauge panel.
2. Pull the instrument panel straight out and disconnect the wire harness.
3. Remove the gauge assembly screws and lift out the gauge assembly to disconnect the wire connectors. Disconnect the speedometer cable and remove the gauge assembly.
4. On the 827 Sterling use the following procedure:
   a. Remove the instrument cluster cover strip screw.
   b. Remove the cluster screws, 3 off the center of the cowl above the instrument cluster, 2 below the drivers vent, 2 below the air conditioning controls and 2 off the rear access of the cowl.
   c. Remove the dashboard retaining clips and disconnect the switch connectors. Remove the air conditioning mode unit, cowl assembly and drivers vent.
   d. Remove the screws from the control switches, disconnect the electrical connectors and remove the instrument cluster.
5. Once the instrument cluster or panel is removed the speedometer may be removed from the back of the instrument cluster by removing the retaining screws and then removing the speedometer assembly. On some vehicles there may be an electric motor attached to the rear of the speedometer. This can be removed separately or as a complete unit.

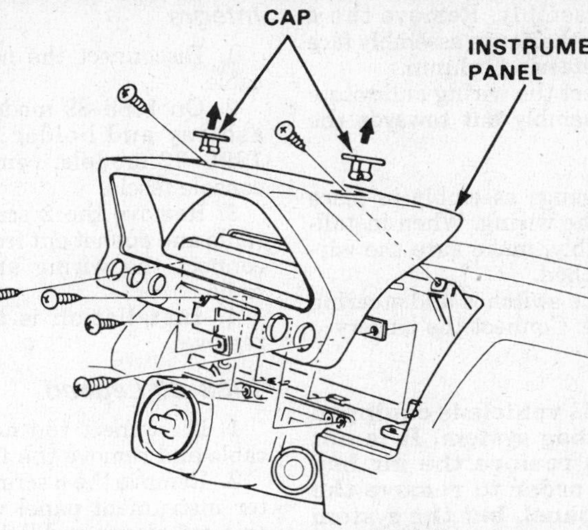

Instrument panel removal—1988–89 Integra

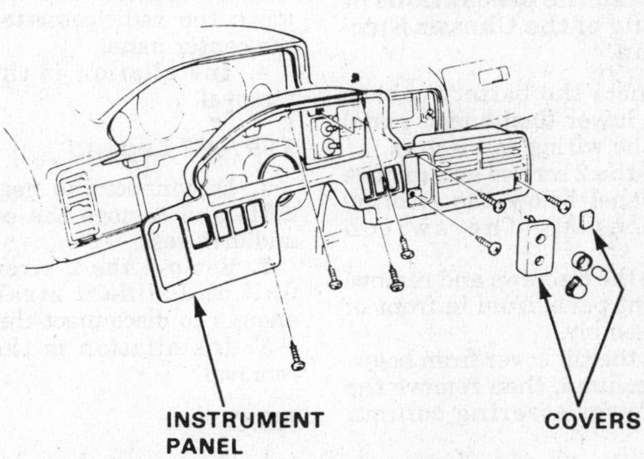

Instrument panel removal—1990–92 Integra

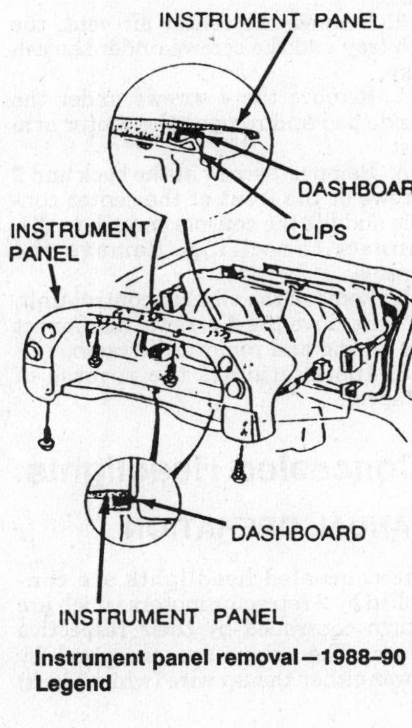

Instrument panel removal—1988–90 Legend

To install:

6. When installing the cluster, be carefull to properly connect and route the wiring and speedometer cable.
7. Set the cluster in place and install the screws.

1991–92 Legend

NOTE: This vehicle is equipped with an air bag system. It is not necessary to remove the air bag assembly in order to remove the instrument panel, but the system should be disarmed before starting work. Read the precautions at the beginning of the Chassis Electrical section.

1. Disconnect the battery cables and remove the lower dashboard panel.
2. Remove the upper and lower steering column covers.
3. Remove the 2 screws, unplug the switch connectors and remove the instrument panel.
4. Place a clean rag over the combination switch to prevent scratching

the gauge assembly. Remove the 4 screws and lay the gauge assembly face down on the steering column.

5. Disconnect the wiring and rotate the gauge assembly out towards the right.

To install:

6. Fit the gauge assembly in place and connect the wiring. When installing the assembly, make sure the wiring is not pinched.

7. Install the switches and steering column covers. Connect the battery.

NSX

NOTE: This vehicle is equipped with an air bag system. It is not necessary to remove the air bag assembly in order to remove the instrument panel, but the system should be disarmed before starting work. Read the precautions at the beginning of the Chassis Electrical section.

1. Disconnect the battery cables. Remove the lower dashboard panel and unplug the wiring connectors.
2. Remove the 2 screws and remove the center panel below the steering column. Unplug the switch connectors.
3. Remove the 6 screws and remove the instrument panel from in front of the gauge assembly.
4. Remove the tilt cover from below the steering column, then remove the upper and lower steering column covers.
5. Unplug the 30 pin connectors from each side of the gauge assembly.
6. Place a clean rag over the combination switch to prevent scratching the gauge assembly. Remove the 4 screws and lay the gauge assembly face down on the steering column.
7. Rotate the gauge assembly out and up towards the right.

To install:

8. Fit the gauge assembly in place and connect the wiring. When installing the assembly, make sure the wiring is not pinched.
9. Install the switches and steering column covers. Connect the battery.

Radio

REMOVAL & INSTALLATION

NOTE: On vehicles equipped with theft protected radios, the word "CODE" will appear in the display when the unit is installed and power turned on. At this time enter the owner's 5 digit code using the numbered preselect buttons.

Integra

1. Disconnect the negative battery cable.
2. On 1988–89 models, remove the ashtray and holder assembly. On 1990–92 models, remove the front console fascia.
3. Remove the 2 screws under the radio and push it out from behind. Disconnect the wiring and remove the radio.
4. Installation is the reverse of removal.

1988–90 Legend

1. Disconnect the negative battery cable and remove the front console.
2. Remove the 6 screws and the center instrument panel with the radio/cassette player and lighter assembly.
3. Disconnect the wiring and remove the radio/cassette player from the center panel.
4. Installation is the reverse or removal.

1991–92 Legend

1. Disconnect the negative battery cable and remove the center console and arm rest.
2. Remove the 2 screws below the unit and slide it straight out far enough to disconnect the wiring.
3. Installation is the reverse of removal.

NSX

1. Disconnect the negative battery cable.
2. Use a clean rag to protect the dashboard and carefully pry out the clock.
3. Remove the center air vent, the ash tray and the screws under the ash tray.
4. Remove the 4 screws under the inside pad and remove the center arm rest.
5. Remove 1 screw at the back and 2 screws at the front of the center console and lift the console enough to disconnect the wiring. Remove the console.
6. Remove the climate control unit.
7. Remove the 4 screws, disconnect the wiring and remove the radio.
8. Installation is the reverse of removal.

Concealed Headlights

MANUAL OPERATION

The concealed headlights are controlled by 2 retractor motors which are inturn controlled by their respective relays. The relays are energized by power either the up wire (white/black) of the down wire (white/yellow), through the slip ring on the retractor motors. The up wire can be powered either by the headlight switch/control unit or by the retractor switch directly. The down wire can be powered by the control unit by either the headlight switch or the retractor switch. The control unit also senses any abnormality in the way the retractor motors operate and warns the driver by illuminating the warning light in the dash assembly.

Each retractor motor has a knob on the motor housing. Turn the motor by hand to manually raise or lower the headlight.

Combination Switch

The headlight switch, dimmer switch, wiper/washer switch and the turn signal switch are all incorporated into the same assembly. On some vehicles, the individual switches can be removed from the combination switch. Otherwise the combination switch must be replaced as a complete unit.

REMOVAL & INSTALLATION

Integra

1. Disconnect the negative battery cable and remove the steering wheel.
2. Remove the column covers and disconnect the wiring. If equipped with cruise control, remove the slip ring.
3. On 1988–89 models, remove the screws and remove the combination switch.
4. On 1990–92 models, remove the screws and slide the switch out to the side.
5. Installation is the reverse of removal.

Sterling and Legend

WITHOUT AIR BAG

1. Remove the negative cable from the battery.
2. Remove the dashboard lower panel and disconnect the 6-pin and 8-pin connectors from the wiper control unit on the lower panel.
3. Disconnect the 10-pin connector from the wiper/washer switch.
4. Remove the steering wheel, the steering column lower cover and disconnect the 6-pin connector from the winter position switch.
5. Remove the upper cover from the steering column.
6. Remove the screws and slide the wiper/washer switch out of the housing.
7. To install, reverse the removal procedures.

WITH AIR BAG

NOTE: The 1989–92 Legend and Legend Coupe are equipped with an air bag system; optional on the Sterling. It will be necessary to remove the air bag assembly in order to remove the combination switch. Read the precautions at the beginning of the Chassis Electrical section.

1. Disconnect both the negative and positive battery cable from the battery.
2. Remove the lower maintenance lid below the air bag and then remove the short connector.
3. Disconnect the connector between the air bag and the cable reel.
4. Connect the short connector to the air bag side of the connector.
5. Remove the left side maintenance lid and the cruise control/set resume switch cover.
6. Insert a T30 Torx® bit and remove the Torx® bolts. Remove the air bag assembly and place it in an out of the way area, such as the back seat, pad side up.
7. Remove the dashboard lower panel and disconnect the 6-pin connector from the wiper position switch and the 6-pin and 8-pin connectors from the wiper control unit on the lower panel.
8. Remove the left knee bolster. Disconnect the combination switch connectors.
9. Remove the upper and lower steering column covers.
10. Remove the lighting and wiper switch mounting screws and remove the switches.

To install:

11. Install the switches and connect the wiring.
12. Install the steering column covers and temporarily connect the battery to test the switches.
13. Disconnect the battery again, assemble the remaining dashboard parts and install the air bag assembly. Be sure to install the air bag wiring so it is not pinched or interfering with other parts. Be sure the battery cables are disconnected.
14. After installing the air bag assembly, connect the battery and turn the ignition switch to the **ON** position. The instrument panel air bag light should go on for approximately 8 seconds and then go off.
15. Confirm operations of horn buttons and cruise control set/resume switch.

NSX

NOTE: This vehicle is equipped with an air bag system. It is not necessary to remove the air bag assembly in order to remove the combination switch, but the system should be disarmed before starting work. Read the precautions at the beginning of the Chassis Electrical section.

1. Disconnect both battery cables.
2. Remove the dashboard lower panel and unplug the connectors from the floor wiring harness.
3. Unplug the 18 pin connector from the floor wiring harness.
4. Remove the tilt cover and the upper and lower column covers.
5. Remove the 4 screws and slide the hazzard switch and wiper/washer switch assembly out.
6. Installation is the reverse of removal.

Ignition Lock/Switch

REMOVAL & INSTALLATION

Integra

1. Disconnect the negative battery cable.
2. Remove the dashboard lower panel, left knee bolster and left kick panel.
3. Remove the steering wheel. Remove the steering column covers.
4. Center punch each of the shear bolts and drill the heads off with a $^{3}/_{16}$ in. drill bit. Be sure to be careful not to damage the switch body when removing the shear head.
5. Remove the shear bolts from the switch body.

To install:

6. Install the new ignition switch without the key inserted. Loosely tighten the new shear bolts. Make sure the projection of the ignition switch is aligned with the hole in the steering column.
7. Insert the ignition key and check for proper operation of the steering wheel lock and that the ignition key turns freely.
8. Tighten the shear bolts until the heads twist off.

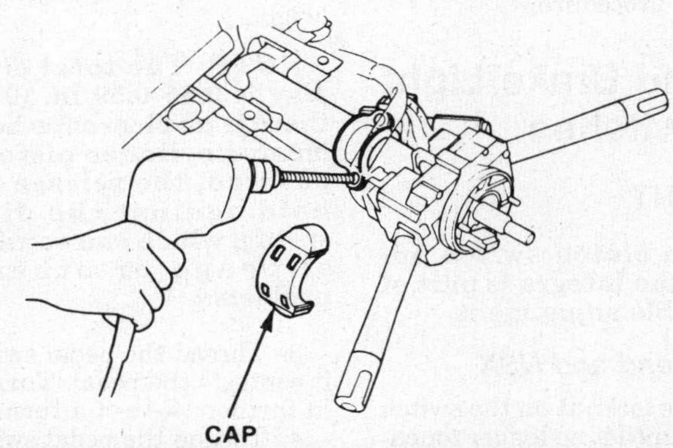

Drill out the shear bolts to replace the ignition lock; Integra shown

Sterling and 1988–90 Legend

1. Disconnect the negative battery cable.
2. Remove the steering column lower cover. Disconnect the ignition switch wire connector from the dash fuse box.
3. Insert the key and place on the **O** position.
4. Remove the 2 screws and replace the base of the switch.

NOTE: The air bag system wire harness is routed near the steering lock assembly. All air bag system wire harness and connectors are colored yellow. Do not use electrical test equipment on these circuits. Be careful not to damage the air bag system wire harness when servicing the steering lock.

5. To remove the lock, remove the steering wheel and the steering column covers.
6. Center punch each of the shear bolts and drill their heads off with a $^{3}/_{16}$ in. drill bit. Be careful not to damage the switch body when removing the shear head.
7. Remove the shear bolts from the switch body.

To install:

8. Install the new ignition switch without the key inserted. Loosely tighten the new shear bolts. Make sure the projection of the ignition switch is aligned with the hole in the steering column.
9. Insert the ignition key and check for proper operation of the steering wheel lock and that the ignition key turns freely.
10. Tighten the shear bolts until the heads twist off.

1991–92 Legend

1. Disconnect the negative battery cable.
2. On Legend, remove the switches from the lower dashboard panel and remove the panel.

3. On NSX, remove the dashboard lower and center panels and the steering column covers. Remove the knee bolster and the lower dash frame.
4. Remove the steering column mounts and lower the column.
5. Disconnect the switch wiring and remove the 2 screws to remove the switch.
6. To remove the lock, grind a slot into the shear bolt head and use a chisel to unscrew and remove the bolt.
7. Insert the key and turn to the first position. Push in the lock pin in the service hole and pull the lock assembly out of the column.

To install:

8. Install the new lock and loosely install the shear bolt. Make sure the switch operates freely before twisting off the head.
9. Install the switch and connect the wiring.
10. Secure the column in place, install the dashboard panels and connect the battery.

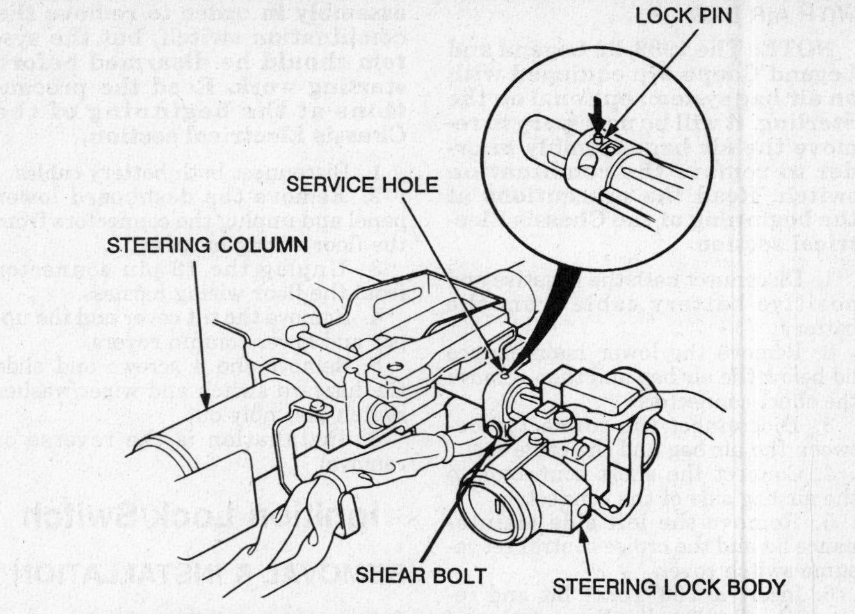

Remove the shear bolt and press the lock pin to remove the assembly—1991–92 Legend and NSX

Stoplight Switch

ADJUSTMENT

1. Loosen the stoplight switch locknut and back off the stoplight switch until it does not touch the brake pedal.
2. If required, adjust the pedal height.
3. Screw in the stoplight switch until the plunger is fully depressed; threaded end touching the pad on the pedal arm.
4. Back off the switch half a turn and tighten the locknut.

REMOVAL & INSTALLATION

1. Disconnect the negative battery cable. Disconnect the stoplight switch electrical connectors.
2. Loosen the stoplight switch locknut and back off the stoplight switch until it is removed from the brake pedal.
3. Installation is the reverse order of the removal procedure.

Clutch and Brake Light Switches

ADJUSTMENT

NOTE: The clutch switch adjustment on the Integra is part of the clutch cable adjustment.

Sterling, Legend and NSX

1. Loosen the locknut on the switch and back it off until it no longer touches the pedal.

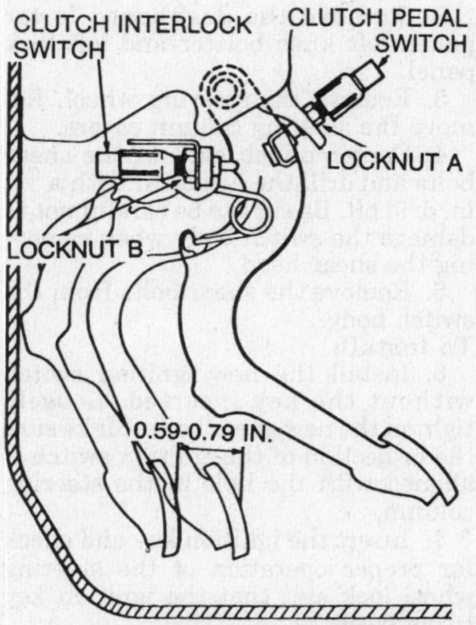

Clutch pedal switch adjustments

2. If required, adjust the pedal height.

NOTE: The total clutch free-play is 0.35–0.59 in. (9–15mm). If there is no clearance between the master cylinder piston and the pushrod, the release bearing is held against the diaphragm spring, which can result in clutch slippage or other clutch problems.

3. Thread the pedal switch in until it contacts the pedal. Turn the switch in further ¼–½ of a turn.
4. Torque the pedal switch locknut to 8 ft. lbs. (10 Nm).
5. To adjust the clutch interlock (starter) switch, loosen the clutch interlock switch locknut. Measure the clearance between the floor board and the clutch pedal with the clutch pedal fully depressed.
6. Release the clutch pedal 0.59–0.79 in. (15–20mm) from the fully depressed position and hold it there. Adjust the position of the clutch interlock switch so the engine will start with the clutch in this position.
7. Thread the clutch interlock switch in further ¼–½ of a turn. Torque the clutch interlock locknut to 8 ft. lbs. (10 Nm).

REMOVAL & INSTALLATION

1. Disconnect the negative battery cable.
2. Remove the instrument panel lower cover and knee bolster, as required.
3. Disconnect the electrical connectors from the switch.
4. Loosen the switch locknut and unscrew the switch from the mounting.
5. To install, reverse the removal procedure and torque the switch locknut to 8 ft. lbs. (10 Nm).

Neutral Safety Switch

ADJUSTMENT

1. The switch is mounted at the base of the automatic transmission shift lever. Remove the console as required to access the switch.

2. Disable the ignition system so the engine will not start.

3. The switch is secured in place by 2 bolts in slotted holes. Place the lever in **N**, loosen the bolts and move the switch as required to allow starter operation in Park and Neutral only.

Fuses Circuit Breakers and Relays

LOCATION

Integra

The Integra fuse/relay box is located in the interior on the drivers side below the dashboard usually next to or behind the left side kick panel. There is also a main fuse box located in the right side of the engine compartment containing two 45 amp and a 65 amp fuse. In addition to these, there is a 10 amp hazard fuse located at the positive battery terminal. There are many relays used on the Integra and some of them are located in the engine fuse box and the interior fuse box. Listed below are some of the other possible relay

Air Conditioning Compressor Relay—located on the left front side of the radiator support.

Air Conditioning Condenser Fan Relay—located on the left front side of the radiator support.

Air Conditioning Radiator Fan Relay—located on the left front side of the radiator support.

Anti-Lock Brake Front Safe Relay—located on the right inner fender panel.

Anti-Lock Brake Motor Relay—located on the right inner fender panel.

Anti-Lock Brake Rear Safe Relay—located on the right inner fender panel.

Day Time Running Light Relay (Canada only)—located behind the right side kick panel.

Fog Light Relay—located on the top of the interior fuse box.

PGM-FI Main Relay—located behind the left side of the instrument panel.

Power Window Relay—located on the top of the interior fuse box.

Rear Defogger Relay—located on the top of the interior fuse box.

Retractable Headlight Relays—located on a relay bracket in the lower right side of the engine compartment.

Starter Relay (Manual Transaxle only)—located behind the radio in the center of the instrument panel.

Sun Roof Relay—located on the top of the interior fuse box.

Legend and Legend Coupe

The fuse box is located in the interior on the drivers side behind the kick panel. There is also a relay box located in the left side of the engine compartment. There are many relays being used on these vehicles and most of them are located in the engine fuse/relay box and the interior fuse/relay box. Listed below are some of the other possible relay locations:

Air Conditioning Compressor Relay—located on the left front side of the inner fender panel.

Air Conditioning Condenser Fan Relay—located on the right front side of the inner fender panel, near the fuse/relay box.

Air Conditioning Condenser Fan Timer Relay—located on the left front side of the inner fender panel.

Air Conditioning Radiator Fan Relay—located on the right front side of the inner fender panel, near the fuse/relay box.

Air Conditioning Radiator Fan Timer Relay—located on the right front side of the inner fender panel, near the fuse/relay box.

Anti-Lock Brake Front Safe Relay—located on the right inner fender panel, near the battery.

Anti-Lock Brake Rear Safe Relay—located on the right inner fender panel, near the battery.

Blower High Relay—located on the blower motor housing.

Blower Motor Relay—located on the blower motor housing.

Day Time Running Light Relay (Canada only)—located under the drivers side seat.

Power Seat Belt Relays—there is one for both front seats, located under the seats.

Sterling

The fuse box is located in the interior on the drivers side below the dashboard usually located next to or behind the left side kick panel. There is also a relay box located in the left side of the engine compartment. There are many relays being used on these vehicles and most of them are located in the engine fuse/relay box and the interior fuse/relay box. Listed below are some of the other possible relay locations:

Air Conditioning Clutch Relay—located in the front corner of the right front inner fender.

Air Conditioning Fan Changeover Relay—located in the engine compartment fuse/relay block.

Anti-Lock Brake System Return Pump Relay—located on the hydraulic modulator.

Anti-Lock Brake System Solenoid Valve Relay—located on the hydraulic modulator.

Blower Motor Relays—located on the blower motor housing.

Cooling Fan Relays—located in the engine compartment fuse/relay block.

Main Relay—located on the left side of the steering wheel, behind instrument panel.

Over-voltage Protection Relay—located in the trunk, next to the ABS control unit.

Power Windows Relay—located in the fuse/relay block located in the passenger compartment.

Wiper Motor Relay—located in the fuse/relay block located in the passenger compartment.

NSX

The main relay box is under the front hood on the right side. Two sub station relay boxes are in the same compartment; one is attached to the rear wall, the other is near the right head light. The main fuse box is in the engine compartment on the left side. A second fuse box in behind the driver side kick panel. Relays located in the rear wall sub station relay box are:

- Air conditioner clutch
- Horns
- Radiator fan low speed
- Radiator fan high speed
- Condenser fans
- Left and right head light retractors
- Blower fan low speed
- Blower fan high speed

Relays located in the other sub station relay box include:

- Windshield wiper low speed
- Intermittent wiper relay
- Windshield wiper high speed
- Washer relay
- ABS rear fail-safe
- ABS front fail-safe
- Power amplifier

The main front relay/fuse box contains relays for:

- Headlight dimmer
- Power windows
- Main lighting relay
- Tail lights
- ABS motor relay

The rear window defogger relay is on the engine compartment fuse panel.

Two head light retractor relays are mounted near the right head light.

The fuel pump relay is behind the driver's seat back.

Computers

LOCATION

Integra

Anti-Lock Brake Control Unit (Hatchback)—located in the top center area of the luggage compartment behind the interior panel.

Anti-Lock Brake Control Unit (Sedan)—located in the left side panel area of the luggage compartment behind the interior panel.

Automatic Shoulder Seat Belt Control Unit—located behind the right side kick panel.

Automatic Transaxle Control Unit—is located behind the left side of the instrument panel.

Cooling Fan Timer Unit—located behind the radio in the center of the instrument panel area.

Cruise Control Unit—located behind the left side of the instrument panel.

Integrated Control Unit—located behind the left side kick panel.

PGM-FI Electronic Control Unit—located on under a protective cover on the drivers side front floorboard area; sometimes under the seat.

Power Door Lock Control Unit—located behind the radio in the center of the instrument panel area.

Retractable Headlight Control Unit—located on the lower left inner fender panel in the engine compartment.

Legend and Legend Coupe

Anti-Lock Brake Control Unit—located behind the drivers side rear seat or the panel next to the rear seat.

Automatic Transaxle Control Unit—located under the right side floor panel.

Cooling Fan Control Unit—located under the instrument panel next to the steering column support.

Cooling Fan Timer Unit—located under the right side floor panel.

Dashlight Brightness Control Unit—located on the left side of the center front console.

Information Control Unit—located behind the upper portion of the glove box assembly.

Integrated Control Unit—located on the interior fuse/relay box located under the left side of the instrument panel.

Interlock Control Unit (Automatic Transaxle)—located on the interior fuse/relay box located under the left side of the instrument panel.

PGM-FI Electronic Control Unit—located under a protective cover located on the right front floor board.

Power Window Control Unit—located inside the driver's inner door panel.

Security Control Unit—located on the right rear shock tower.

Supplemental Restraint System Control Unit—located under the left side of the instrument panel.

Sterling

Anti-Lock Brake Control Unit—located in the luggage compartment, behind the left side panel.

Anti-Theft Control Unit—located in the center console, above the ashtray assembly.

Central Locking Control Unit—located above the steering column.

Cruise Control Unit—located on the left kick panel, behind the fuse/relay box.

Electronic Control Unit—located under the passenger's seat.

Fuel Injection Control Unit—located under the driver's seat.

Ignition Module—located on the side of the distributor assembly.

Power Door Lock Relay Module—there is one located in each of the rear doors.

Remote Door Lock Reciever Control Unit—located in the headliner, next to the rear view mirror.

Vehicle Condition Monitor—located under the passenger's seat.

NSX

Main PGM-FI Computer—behind the passenger side seat back. The main relay for the PGM-FI is next to it.

Cooling Fan Control Unit—behind the passenger side seat back.

Head Light Retractor Control Unit— between the seat backs on the rear wall.

Interlock Control Unit— behind the driver's seat back.

Automatic Transmission Control Unit— behind the driver's seat back. The **Traction Control Unit** is attached to it.

Dash Light Brightness Control Unit—behind the driver's seat back.

Flashers

LOCATION

Integra

The Hazard/Turn Signal Relay, is located either on the interior fuse/relay box or located up behind the left side of the instrument panel.

Sterling and Legend

The Hazard/Turn Signal Relay, is located on the interior fuse/relay box, located under the left side of the instrument panel.

Cruise Control

ADJUSTMENT

Integra

1. Check that the actuator cable operates smoothly with no binding or sticking.
2. Start the engine and let it warm up to reach normal operating temperature.
3. Measure the amount of movement of the output linkage until the engine speed starts to increase.
4. At first, the output linkage should be located at the full closed position. The free-play should be 43 ± 0.06 inches (11 ± 1.5mm).
5. If the free-play is not within specifications, loosen the locknut and turn the adjusting nut as required.
6. Retighten the locknut and recheck the free-play.

Sterling, Legend and NSX

1. Check that the actuator cable operates smoothly with no binding or sticking.
2. Start the engine and let it warm up to reach normal operating temperature.
3. Measure the amount of movement of the actuator rod until the cable pulls on the accelerator lever; engine speed starts to increase.
4. The free-play should be 43 ± 0.06 inches (11 ± 1.5mm).
5. If the free-play is not within specifications, loosen the locknut and turn the adjusting nut, as required.

NOTE: If necessary, check the throttle cable free-play and recheck the actuator free-play.

6. Retighten the locknut and recheck the free-play.

Audi 2

5000, 80, 90, 100, 200, V8, Coupe — All Models

SERIAL NUMBER IDENTIFICATION

Vehicle Identification Plate

The Vehicle Identification Number (VIN) is located on a plate on top of the instrument panel. The VIN number is visible from outside through the left side of the windshield. The VIN number is also stamped into the upper right corner of the firewall. The vehicle identification plate is mounted on the right front wheel housing.

Engine Number

4 and 5 Cylinder Engine

The engine serial number is stamped into the left rear side of the engine block. In addition to the serial number, an engine code number is stamped into the starter end of the engine block, below the cylinder head mounting surface. This number indicates the original cylinder bore size of the engine.

8 Cylinder Engine

The engine serial number is stamped into the left side of the engine block, just above the power steering hydraulic pump.

Transmission/ Transaxle Number

Each transmission or transaxle has 2 series of numbers stamped into the case. The long series includes a 3 digit letter code that also appears on the vehicle model data sticker. The numbers in the long series indicate the date of manufacture. The short series is a 3 digit number code indicating the actual transmission type. That is the number used when buying a replacement unit.

Model Data Sticker

On the inside of the luggage compartment lid is a sticker indicating the model type, chassis number, body type, paint number, engine and transmission codes, and the options package.

Body Panel Identification

The major body panels of all 1991 and later vehicles are marked with the complete VIN and the Audi logo. This is done in accordence with the Motor Vehicle Theft Law Encforcement Act to discourage theft and resale of the vehicle for parts. All authorized replacement parts will also have the Audi logo and a label with the "R DOT" designation in place of the VIN for that vehicle.

SPECIFICATIONS

ENGINE IDENTIFICATION

Year	Model	Engine Displacement cu. in. (cc/liter)	Engine Series Identification	No. of Cylinders	Engine Type
1988	80	121 (1983/2.0)	3A	4	OHC
	80 Quattro	141 (2309/2.3)	NG	5	OHC
	90 ①	121 (1983/2.0)	3A	4	OHC
	90	141 (2309/2.3)	NG	5	OHC
	90 Quattro	141 (2309/2.3)	NG	5	OHC
	5000 CS Turbo	136 (2226/2.2)	MC	5	OHC
	5000 CS Quattro Turbo	136 (2226/2.2)	MC	5	OHC
	5000 CS Quattro Wagon	136 (2226/2.2)	MC	5	OHC
	5000 S	141 (2309/2.3)	NF	5	OHC
	5000 S Wagon	141 (2309/2.3)	NF	5	OHC
	5000 S Quattro	141 (2309/2.3)	NF	5	OHC
1989	80	121 (1983/2.0)	3A	4	OHC
	80 Quattro	141 (2309/2.3)	NG	5	OHC
	90	121 (1983/2.0)	3A	4	OHC
	90	141 (2309/2.3)	NG	5	OHC
	90 Quattro	141 (2309/2.3)	NG	5	OHC
	100	141 (2309/2.3)	NF	5	OHC
	100 Quattro	141 (2309/2.3)	NF	5	OHC
	200	136 (2226/2.2)	MC	5	OHC
	200 Quattro	136 (2226/2.2)	MC	5	OHC

ENGINE IDENTIFICATION

Year	Model	Engine Displacement cu. in. (cc/liter)	Engine Series Identification	No. of Cylinders	Engine Type
1990	80	121 (1983/2.0)	3A	4	OHC
	80 Quattro	141 (2309/2.3)	NG	5	OHC
	90	141 (2309/2.3)	NG	5	OHC
	90 Quattro	141 (2309/2.3)	NG	5	OHC
	100	141 (2309/2.3)	NF	5	OHC
	100 Quattro	141 (2309/2.3)	NF	5	OHC
	200	136 (2226/2.2)	MC	5	OHC
	200 Quattro	136 (2226/2.2)	MC	5	OHC
	V8 Quattro	220 (3562/3.6)	PT	8	OHC
1991-92	80	141 (2309/2.3)	NG	5	OHC
	80 Quattro	141 (2309/2.3)	NG	5	OHC
	90	141 (2309/2.3)	NG	5	OHC
	90 Quattro	141 (2309/2.3)	7A	5	OHC 20V
	Coupe Quattro	141 (2309/2.3)	7A	5	OHC 20V
	100	141 (2309/2.3)	NF	5	OHC
	100 Quattro	141 (2309/2.3)	NF	5	OHC
	200	136 (2226/2.2)	MC	5	Turbo
	200 Quattro	136 (2226/2.2)	3B	5	Turbo 20V
	200 Quattro Wagon	136 (2226/2.2)	3B	5	Turbo 20V
	V8 Quattro	220 (3562/3.6)	PT	8	OHC

OHC—Overhead Camshaft
20V—Twenty Valve

① With automatic transmission

GENERAL ENGINE SPECIFICATIONS

Year	Model	Engine Displacement cu. in. (cc)	Fuel System Type	Net Horsepower @ rpm	Net Torque @ rpm (ft. lbs.)	Bore × Stroke (in.)	Com-pression Ratio	Oil Pressure @ rpm
1988	80	121 (1983)	CIS-M	108 @ 5300	121 @ 3250	3.65 × 3.40	10.4:1	29 @ 2000
	80 Quattro	141 (2309)	CIS-E	130 @ 5700	140 @ 4500	3.25 × 3.40	10.0:1	29 @ 2000
	90 ①	121 (1983)	CIS-M	108 @ 5300	121 @ 3250	3.65 × 3.40	10.4:1	29 @ 2000
	90	141 (2309)	CIS-E	130 @ 5700	140 @ 4500	3.25 × 3.40	10.0:1	29 @ 2000
	90 Quattro	141 (2309)	CIS-E	130 @ 5700	140 @ 4500	3.25 × 3.40	10.0:1	29 @ 2000
	5000 CS Turbo	136 (2226)	CIS-E	158 @ 5500	166 @ 3000	3.19 × 3.40	7.8:1	29 @ 2000
	5000 CS Quattro Turbo	136 (2226)	CIS-E	158 @ 5500	166 @ 3000	3.19 × 3.40	7.8:1	29 @ 2000
	5000 CS Quattro Wagon	136 (2226)	CIS-E	158 @ 5500	166 @ 3000	3.19 × 3.40	7.8:1	29 @ 2000
	5000 S	141 (2309)	KE-III	130 @ 5600	140 @ 4000	3.25 × 3.40	10.0:1	29 @ 2000
	5000 S Wagon	141 (2309)	KE-III	130 @ 5600	140 @ 4000	3.25 × 3.40	10.0:1	29 @ 2000
	5000 S Quattro	141 (2309)	KE-III	130 @ 5600	140 @ 4000	3.25 × 3.40	10.0:1	29 @ 2000
1989	80	121 (1983)	CIS	108 @ 5300	121 @ 3200	3.25 × 3.65	10.5:1	29 @ 2000
	80 Quattro	141 (2309)	CIS-EIII	130 @ 5700	140 @ 4500	3.25 × 3.40	10.0:1	29 @ 2000
	90 ①	121 (1983)	CIS	108 @ 5300	121 @ 3200	3.25 × 3.65	10.5:1	29 @ 2000
	90	141 (2309)	CIS-EIII	130 @ 5700	140 @ 4500	3.25 × 3.40	10.0:1	29 @ 2000
	90 Quattro	141 (2309)	CIS-EIII	130 @ 5700	140 @ 4500	3.25 × 3.40	10.0:1	29 @ 2000
	100	141 (2309)	CIS-EIII	130 @ 5700	140 @ 4500	3.25 × 3.40	10.0:1	29 @ 2000
	100 Quattro	141 (2309)	CIS-EIII	130 @ 5700	140 @ 4500	3.25 × 3.40	10.0:1	29 @ 2000

GENERAL ENGINE SPECIFICATIONS

Year	Model	Engine Displacement cu. in. (cc)	Fuel System Type	Net Horsepower @ rpm	Net Torque @ rpm (ft. lbs.)	Bore × Stroke (in.)	Com-pression Ratio	Oil Pressure @ rpm
1989	200	136 (2226)	CIS	162 @ 5500	177 @ 3000	3.19 × 3.40	7.8:1	29 @ 2000
	200 Quattro	136 (2226)	CIS	162 @ 5500	177 @ 3000	3.19 × 3.40	7.8:1	29 @ 2000
1990	80	121 (1983)	CIS	108 @ 5300	121 @ 3200	3.25 × 3.65	10.5:1	29 @ 2000
	80 Quattro	141 (2309)	CIS-EIII	130 @ 5700	140 @ 4500	3.25 × 3.40	10.0:1	29 @ 2000
	90 ①	121 (1983)	CIS	108 @ 5300	121 @ 3200	3.25 × 3.65	10.5:1	29 @ 2000
	90	141 (2309)	CIS-EIII	130 @ 5700	140 @ 4500	3.25 × 3.40	10.0:1	29 @ 2000
	90 Quattro	141 (2309)	CIS-EIII	130 @ 5700	140 @ 4500	3.25 × 3.40	10.0:1	29 @ 2000
	100	141 (2309)	CIS-EIII	130 @ 5700	140 @ 4500	3.25 × 3.40	10.0:1	29 @ 2000
	100 Quattro	141 (2309)	CIS-EIII	130 @ 5700	140 @ 4500	3.25 × 3.40	10.0:1	29 @ 2000
	200	136 (2226)	CIS	162 @ 5500	177 @ 3000	3.19 × 3.40	7.8:1	29 @ 2000
	200 Quattro	136 (2226)	CIS	162 @ 5500	177 @ 3000	3.19 × 3.40	7.8:1	29 @ 2000
	V8 Quattro	220 (3562)	②	240 @ 5800	245 @ 4000	3.19 × 3.40	10.6:1	NA
1991–92	80	141 (2309)	CIS-EIII	130 @ 5700	140 @ 4500	3.25 × 3.40	10.0:1	29 @ 2000
	80 Quattro	141 (2309)	CIS-EIII	130 @ 5700	140 @ 4500	3.25 × 3.40	10.0:1	29 @ 2000
	90	141 (2309)	CIS-EIII	130 @ 5700	140 @ 4500	3.25 × 3.40	10.0:1	29 @ 2000
	90 Quattro	141 (2309)	MPI	162 @ 6000	162 @ 4500	3.25 × 3.40	10.0:1	29 @ 2000
	Coupe Quattro	141 (2309)	MPI	162 @ 6000	162 @ 4500	3.25 × 3.40	10.0:1	29 @ 2000
	100	141 (2309)	CIS-EIII	130 @ 5600	140 @ 4000	3.25 × 3.40	10.0:1	29 @ 2000
	100 Quattro	141 (2309)	CIS-EIII	130 @ 5600	140 @ 4000	3.25 × 3.40	10.0:1	29 @ 2000
	200	136 (2226)	CIS Turbo	157 @ 5500	177 @ 3000	3.19 × 3.40	8.4:1	29 @ 2000
	200 Quattro	136 (2226)	Motronic	217 @ 5700	228 @ 1950	3.19 × 3.40	9.3:1	29 @ 2000
	200 Quattro Wagon	136 (2226)	Motronic	217 @ 5700	228 @ 1950	3.19 × 3.40	9.3:1	29 @ 2000
	V8 Quattro	220 (3562)	②	240 @ 5800	245 @ 4000	3.19 × 3.40	10.6:1	NA

NA—Not Available
CIS—Continuous Injection System
MPI—Multi Point Injection
① With automatic transmission
KE-III—KE-III Jetronic injection
② Bosch Motronic w/2 Knock Sensors

ENGINE TUNE-UP SPECIFICATIONS

Year	Model	Engine Displacement cu. in. (cc)	Spark Plugs Type	Spark Plugs Gap (in.)	Ignition Timing (deg.) MT	Ignition Timing (deg.) AT	Com-pression Pressure (psi)	Fuel Pump (psi)	Idle Speed (rpm) MT	Idle Speed (rpm) AT	Valve Clearance In.	Valve Clearance Ex.
1988	80	121 (1983)	NA	0.031	6B	6B	NA	88–94	780–900	780–900	Hyd.	Hyd.
	80 Quattro	141 (2309)	N9BYC	0.031	15B	15B	NA	88–94	720–860	720–860	Hyd.	Hyd.
	90 ①	121 (1983)	NA	0.031	6B	6B	NA	88–94	780–900	780–900	Hyd.	Hyd.
	90	141 (2309)	N9BYC	0.031	15B	15B	NA	88–94	720–860	720–860	Hyd.	Hyd.
	90 Quattro	141 (2309)	N9BYC	0.031	15B	15B	NA	88–94	720–860	720–860	Hyd.	Hyd.

ENGINE TUNE-UP SPECIFICATIONS

Year	Model	Engine Displacement cu. in. (cc)	Spark Plugs Type	Spark Plugs Gap (in.)	Ignition Timing (deg.) MT	Ignition Timing (deg.) AT	Compression Pressure (psi)	Fuel Pump (psi)	Idle Speed (rpm) MT	Idle Speed (rpm) AT	Valve Clearance In.	Valve Clearance Ex.
1988	80	121 (1983)	NA	0.031	6B	6B	NA	88–94	780–900	780–900	Hyd.	Hyd.
	80 Quattro	141 (2309)	N9BYC	0.031	15B	15B	NA	88–94	720–860	720–860	Hyd.	Hyd.
	90 ①	121 (1983)	NA	0.031	6B	6B	NA	88–94	780–900	780–900	Hyd.	Hyd.
	90	141 (2309)	N9BYC	0.031	15B	15B	NA	88–94	720–860	720–860	Hyd.	Hyd.
	90 Quattro	141 (2309)	N9BYC	0.031	15B	15B	NA	88–94	720–860	720–860	Hyd.	Hyd.
	100	141 (2309)	N8GY	0.031	15B	15B	NA	88–94	670–770	670–770	Hyd.	Hyd.
	100 Quattro	141 (2309)	N8GY	0.031	15B	15B	NA	88–94	670–770	670–770	Hyd.	Hyd.
	200	136 (2226)	N8GY	0.028	0	0	NA	84–95	750–850	750–850	Hyd.	Hyd.
	200 Quattro	136 (2226)	N8GY	0.028	0	0	NA	84–95	750–850	750–850	Hyd.	Hyd.
	5000 CS Turbo	136 (2226)	N8GY	0.028	0	0	NA	84–95	750–850	750–850	Hyd.	Hyd.
	5000 CS Quattro Turbo	136 (2226)	N8GY	0.028	0	0	NA	84–95	750–850	750–850	Hyd.	Hyd.
	5000 CS Quattro Wagon	136 (2226)	N8GY	0.028	0	0	NA	84–95	750–850	750–850	Hyd.	Hyd.
	5000 S	141 (2309)	N8GY	0.031	15B	15B	NA	88–94	670–770	670–770	Hyd.	Hyd.
	5000 S Wagon	141 (2309)	N8GY	0.031	15B	15B	NA	88–94	670–770	670–770	Hyd.	Hyd.
	5000 S Quattro	141 (2309)	N8GY	0.031	15B	15B	NA	88–94	670–770	670–770	Hyd.	Hyd.
1989	80	121 (1983)	NA	0.031	6B	6B	NA	88–94	780–900	780–900	Hyd.	Hyd.
	80 Quattro	141 (2309)	N9BYC	0.031	15B	15B	NA	88–94	720–860	720–860	Hyd.	Hyd.
	90 ①	121 (1983)	NA	0.031	6B	6B	NA	88–94	780–900	780–900	Hyd.	Hyd.
	90	141 (2309)	N9BYC	0.031	15B	15B	NA	88–94	720–860	720–860	Hyd.	Hyd.
	90 Quattro	141 (2309)	N9BYC	0.031	15B	15B	NA	88–94	720–860	720–860	Hyd.	Hyd.
	100	141 (2309)	N8GY	0.031	15B	15B	160–172	88–94	670–770	670–770	Hyd.	Hyd.
	100 Quattro	141 (2309)	N8GY	0.031	15B	15B	160–172	88–94	670–770	670–770	Hyd.	Hyd.
	200	136 (2226)	N8GY	0.028	①	①	123–144	84–95	750–850	670–770	Hyd.	Hyd.
	200 Quattro	136 (2226)	N8GY	0.028	①	①	123–144	84–95	750–850	670–770	Hyd.	Hyd.

ENGINE TUNE-UP SPECIFICATIONS

Year	Model	Engine Displacement cu. in. (cc)	Spark Plugs Type	Spark Plugs Gap (in.)	Ignition Timing (deg.) MT	Ignition Timing (deg.) AT	Compression Pressure (psi)	Fuel Pump (psi)	Idle Speed (rpm) MT	Idle Speed (rpm) AT	Valve Clearance In.	Valve Clearance Ex.
1990	80	121 (1983)	NA	0.031	6B	6B	NA	88–94	780–900	780–900	Hyd.	Hyd.
	80 Quattro	141 (2309)	N9BYC	0.031	15B	15B	NA	88–94	720–860	720–860	Hyd.	Hyd.
	90 ①	121 (1983)	NA	0.031	6B	6B	NA	88–94	780–900	780–900	Hyd.	Hyd.
	90	141 (2309)	N9BYC	0.031	15B	15B	NA	88–94	720–860	720–860	Hyd.	Hyd.
	90 Quattro	141 (2309)	N9BYC	0.031	15B	15B	NA	88–94	720–860	720–860	Hyd.	Hyd.
	100	141 (2309)	N8GY	0.031	15B	15B	160–172	88–94	670–770	670–770	Hyd.	Hyd.
	100 Quattro	141 (2309)	N8GY	0.031	15B	15B	160–172	88–94	670–770	670–770	Hyd.	Hyd.
	200	136 (2226)	N8GY	0.028	2	2	123–144	84–95	750–850	670–770	Hyd.	Hyd.
	200 Quattro	136 (2226)	N8GY	0.028	2	2	123–144	84–95	750–850	670–770	Hyd.	Hyd.
	V8 Quattro	220 (3562)	②	0.032	—	③	145–189	55–65	—	660–720	Hyd.	Hyd.
1991	80	141 (2309)	W7DCT ③	0.031	15B	15B	145–203	88–94	720–860	720–860	Hyd.	Hyd.
	80 Quattro	141 (2309)	W7DCT ③	0.031	15B	15B	145–203	88–94	720–860	720–860	Hyd.	Hyd.
	90	141 (2309)	W7DCT ③	0.031	15B	15B	145–203	88–94	720–860	720–860	Hyd.	Hyd.
	90 Quattro	141 (2309)	F6DCT ③	0.031	①	①	145–203	55–61	720–860	720–860	Hyd.	Hyd.
	Coupe Quattro	141 (2309)	F6DCT ③	0.031	①	①	145–203	55–61	720–860	720–860	Hyd.	Hyd.
	100	141 (2309)	W7DCT ③	0.031	15B	15B	160–172	88–94	680–770	680–770	Hyd.	Hyd.
	100 Quattro	141 (2309)	W7DCT ③	0.031	15B	15B	160–172	88–94	680–770	680–770	Hyd.	Hyd.
	200	136 (2226)	WR7DS ③	0.028	0①	0①	123–144	84–96 ④	700–740	700–740	Hyd.	Hyd.
	200 Quattro	136 (2226)	F5DPOR ③	0.024	②	②	131–189	43–46	770–830	770–830	Hyd.	Hyd.
	200 Quattro Wagon	136 (2226)	F5DPOR ③	0.024	②	②	131–189	43–46	770–830	770–830	Hyd.	Hyd.
	V8 Quattro	220 (3562)	H6DCO ③	0.032	②	②	145–189	49–54 ④	660–720	660–720	Hyd.	Hyd.
1992	SEE UNDERHOOD SPECIFICATIONS STICKER											

B—Before Top Dead Center
Hyd.—Hydraulic lifters, not adjustable
NA—Not Available
① Top Dead Center, not adjustable
② Controlled by the ECU, not adjustable
③ Bosch number
④ Engine warm and idling, vacuum line connected to regulator on V8.

FIRING ORDERS

NOTE: To avoid confusion, always replace spark plug wires one at a time.

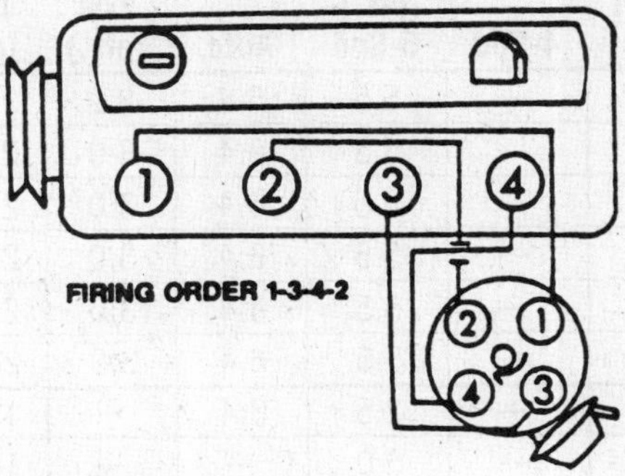

2.0L Engine
Engine Firing Order: 1–3–4–2
Distributor Rotation: Clockwise

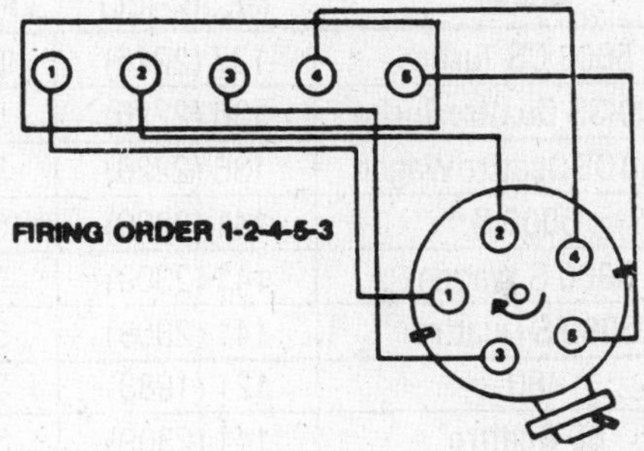

2.2L and 2.3L Engine
Engine Firing Order: 1–2–4–5–3
Distributor Rotation: Clockwise

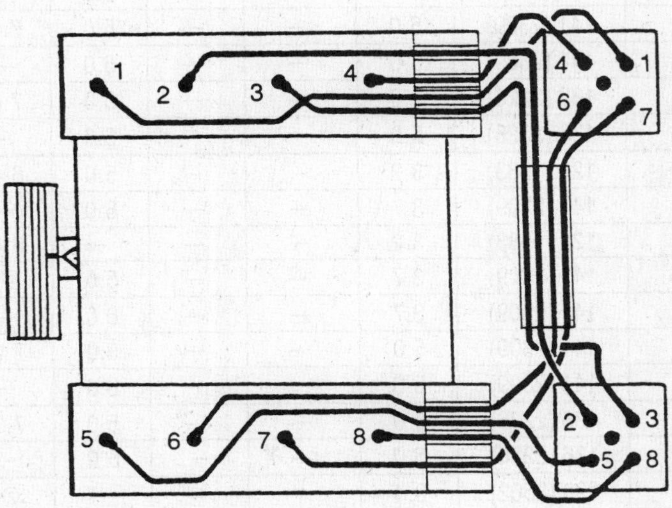

3.6L Engine
Engine Firing Order: 1–5–4–8–6–3–7–2
Distributor Rotation: Clockwise

CAPACITIES

Year	Model	Engine Displacement cu. in. (cc)	Engine Crankcase (qts.) with Filter	Engine Crankcase (qts.) without Filter	Transmission (pts.) 4-Spd	Transmission (pts.) 5-Spd	Transmission (pts.) Auto.	Drive Axle (pts.)	Fuel Tank (gal.)	Cooling System (qts.)
1988	80	121 (1983)	3.2	3.0	—	5.5	6.4	2.2	18.0	7.4
	80 Quattro	141 (2309)	3.7	3.5	—	5.5	6.4	2.2	18.5	8.5
	90 ①	121 (1983)	3.2	3.0	—	—	6.4	2.2	18.0	7.4
	90	141 (2309)	3.7	3.5	—	5.5	6.4	2.2	18.0	8.5
	90 Quattro	141 (2309)	3.7	3.5	—	5.5	6.4	2.2	18.5	7.4
	100	141 (2309)	5.0	4.5	—	5.5	6.4	3.0	21.1	8.5
	100 Quattro	141 (2309)	5.0	4.5	—	5.5	6.4	3.0	20.6	8.5
	200	136 (2226)	5.0	4.5	—	5.5	6.4	3.0	20.6	8.5
	200 Quattro	136 (2226)	5.0	4.5	—	5.5	6.4	3.0	20.6	8.5

CAPACITIES

Year	Model	Engine Displacement cu. in. (cc)	Engine Crankcase (qts.) with Filter	Engine Crankcase (qts.) without Filter	Transmission (pts.) 4-Spd	Transmission (pts.) 5-Spd	Transmission (pts.) Auto.	Drive Axle (pts.)	Fuel Tank (gal.)	Cooling System (qts.)
1988	5000 CS Turbo	136 (2226)	5.3	4.5	—	5.5	6.4	3.0	21.1	8.5
	5000 CS Quattro Turbo	136 (2226)	5.3	4.5	—	5.5	6.4	3.0	21.1	8.5
	5000 CS Quattro Wagon	136 (2226)	5.3	4.5	—	5.5	6.4	3.0	21.1	8.5
	5000 S	141 (2309)	5.3	4.5	—	5.5	6.4	3.0	21.1	8.5
	5000 S Wagon	141 (2309)	5.3	4.5	—	5.5	6.4	3.0	21.1	8.5
	5000 S Quattro	141 (2309)	5.3	4.5	—	5.5	6.4	3.0	21.1	8.5
1989	80	121 (1983)	3.2	—	—	2.5	6.4	—	18	7.4
	80 Quattro	141 (2309)	3.7	—	—	6.0	—	—	18.5	8.5
	90 ①	121 (1983)	3.2	—	—	—	6.4	—	18	7.4
	90	141 (2309)	3.7	—	—	5.0	—	—	18	8.5
	90 Quattro	141 (2309)	3.7	—	—	6.0	—	—	18.5	8.5
	100	141 (2309)	5.0	—	—	5.0	7.0	—	21.1	8.5
	100 Quattro	141 (2309)	5.0	—	—	6.0	—	—	21.1	8.5
	200	136 (2226)	5.0	—	—	5.0	7.0	—	20.6	8.5
	200 Quattro	136 (2226)	5.0	—	—	6.0	—	—	21.1	8.5
1990	80	121 (1983)	3.2	—	—	5.0	6.4	—	18	7.4
	80 Quattro	141 (2309)	3.7	—	—	6.0	—	—	18.5	8.5
	90 ①	121 (1983)	3.2	—	—	—	6.4	—	18	7.4
	90	141 (2309)	3.7	—	—	5.0	—	—	18	8.5
	90 Quattro	141 (2309)	3.7	—	—	6.0	—	—	18.5	8.5
	100	141 (2309)	5.0	—	—	5.0	7.0	—	21.1	8.5
	100 Quattro	141 (2309)	5.0	—	—	6.0	—	—	21.1	8.5
	200	136 (2226)	5.0	—	—	5.0	7.0	—	20.6	8.5
	200 Quattro	136 (2226)	5.0	—	—	6.0	—	—	21.1	8.5
	V8 Quattro	220 (3562)	8.4	—	—	—	②	③	21.0	11.0
1991–92	80	141 (2309)	3.7	—	—	5.0	④	—	15.9	8.6
	80 Quattro	141 (2309)	3.7	—	—	6.0	④	1.6	18.5	8.6
	90	141 (2309)	3.7	—	—	5.0	④	—	15.9	8.6
	90 Quattro	141 (2309)	3.7	—	—	6.0	④	1.6	18.5	8.6
	Coupe Quattro	141 (2309)	3.7	—	—	6.0	④	1.6	18.5	8.6
	100	141 (2309)	5.0	—	—	6.0	⑤	—	21.1	8.5
	100 Quattro	141 (2309)	5.0	—	—	5.5	⑤	1.6	20.6	8.5
	200	136 (2226)	5.0	—	—	5.4	⑤	—	21.1	8.5
	200 Quattro	136 (2226)	5.0	—	—	5.5	⑤	1.6	21.1	8.5
	200 Quattro Wagon	136 (2226)	5.0	—	—	5.5	⑤	1.6	21.1	8.5
	V8 Quattro	220 (3562)	8.5	—	—	6.4	②	③	21.1	11.1

① Automatic transmission
② Initial fill—20.4 pts.
Oil change—8.0 pts.
③ Front final drive—.75 pts. GL5 90 wt.
Rear final drive—3.60 pts. GL5 90 wt.
Both filled for service life. No oil change.
④ Initial fill—12.7 pts.
Oil change—6.3 pts.
⑤ Initial fill—11.4 pts.
Oil change—6.3 pts.

CRANKSHAFT AND CONNECTING ROD SPECIFICATIONS

All measurements are given in inches.

Year	Engine Displacement cu. in. (cc)	Crankshaft Main Brg. Journal Dia.	Crankshaft Main Brg. Oil Clearance	Crankshaft Shaft End-play	Crankshaft Thrust on No.	Connecting Rod Journal Diameter	Connecting Rod Oil Clearance	Connecting Rod Side Clearance
1988	121 (1983)	2.1268–2.1276	NA	NA	3	1.8827–1.8835	NA	NA
	136 (2226)	2.2818–2.2825	0.0006–0.0030	0.003–0.009	4	1.8487–1.8495	0.0004–0.0020	0.016
	141 (2309)	2.2818–2.2825	0.0010–0.0020	0.003–0.009	4	1.8802–1.8810	0.0004–0.0020	0.016
1989	121 (1983)	2.1268–2.1276	NA	NA	3	1.8827–1.8835	NA	NA
	136 (2226)	2.2818–2.2825	0.0006–0.0030	0.003–0.009	4	1.8487–1.8495	0.0004–0.0020	0.016
	141 (2309)	2.2818–2.2825	0.0010–0.0020	0.003–0.009	4	1.8802–1.8810	0.0004–0.0020	0.016
1990	121 (1983)	2.1268–2.1276	NA	NA	3	1.8827–1.8835	NA	NA
	136 (2226)	2.2818–2.2825	0.0006–0.0030	0.003–0.009	4	1.8487–1.8495	0.0004–0.0020	0.016
	141 (2309)	2.2818–2.2825	0.0010–0.0020	0.003–0.009	4	1.8802–1.8810	0.0004–0.0020	0.016
	220 (3562)	NA	NA	NA	NA	NA	NA	NA
1991–92	136 (2226)	2.2818–2.2825	0.0006–0.0030	0.003–0.009	4	1.8487–1.8495	0.0004–0.0020	0.016
	141 (2309)	2.2818–2.2825	0.0010–0.0020	0.003–0.009	4	1.8802–1.8810	0.0004–0.0020	0.016
	220 (3562)	NA	NA	NA	NA	NA	NA	NA

NA—Not available

VALVE SPECIFICATIONS

Year	Engine Displacement cu. in. (cc)	Seat Angle (deg.)	Face Angle (deg.)	Spring Test Pressure (lbs.)	Spring Installed Height (in.)	Stem-to-Guide Clearance (in.) Intake	Stem-to-Guide Clearance (in.) Exhaust	Stem Diameter (in.) Intake	Stem Diameter (in.) Exhaust
1988	121 (1983)	45	45	—	—	0.039	0.051	0.3140	0.3130
	136 (2226)	45	45	—	—	0.039	0.051	0.3140	0.3130
	141 (2309)	45	45	—	—	0.039	0.051	0.3140	0.3130
1989	121 (1983)	45	45	—	—	0.039	0.051	0.3140	0.3130
	136 (2226)	45	45	—	—	0.039	0.051	0.3140	0.3130
	141 (2309)	45	45	—	—	0.039	0.051	0.3140	0.3130
1990	121 (1983)	45	45	—	—	0.039	0.051	0.3140	0.3130
	136 (2226)	45	45	—	—	0.039	0.051	0.3140	0.3130
	141 (2309)	45	45	—	—	0.039	0.051	0.3140	0.3130
	220 (3562)	45	45	—	—	0.039	0.051	0.3140	0.3130
1991–92	136 (2226)	45	45	—	—	0.039	0.051	0.3140	0.3130
	141 (2309)	45	45	—	—	0.039	0.051	0.3140	0.3130
	220 (3562)	45	45	—	—	0.039	0.051	0.3140	0.3130

NOTE: To measure Stem-to-Guide Clearance, insert new valve into guide until end of valve is flush with end of guide. Use dial indicator to measure valve head movement.

PISTON AND RING SPECIFICATIONS

All measurements are given in inches.

Year	Engine Displacement cu. in. (cc)	Piston Clearance	Ring Gap Top Compression	Ring Gap Bottom Compression	Ring Gap Oil Control	Ring Side Clearance Top Compression	Ring Side Clearance Bottom Compression	Ring Side Clearance Oil Control
1988	121 (1983)	0.0011	0.012–0.018	0.012–0.018	0.010–0.018	0.0010–0.0020	0.0010–0.0020	0.0010–0.0020
	136 (2226)	0.0011	0.008–0.020	0.008–0.020	0.010–0.020	0.0010–0.0030	0.0010–0.0030	0.0010–0.0020
	141 (2309)	0.0011	0.008–0.016	0.008–0.016	0.010–0.020	0.0010–0.0030	0.0010–0.0030	0.0010–0.0020
1989	121 (1983)	0.0011	0.012–0.018	0.012–0.018	0.010–0.018	0.0010–0.0020	0.0010–0.0020	0.0010–0.0020
	136 (2226)	0.0011	0.008–0.020	0.008–0.020	0.010–0.020	0.0010–0.0030	0.0010–0.0030	0.0010–0.0020
	141 (2309)	0.0011	0.008–0.016	0.008–0.016	0.010–0.020	0.0010–0.0030	0.0010–0.0030	0.0010–0.0020
1990	121 (1983)	0.0011	0.012–0.018	0.012–0.018	0.010–0.018	0.0010–0.0020	0.0010–0.0020	0.0010–0.0020
	136 (2226)	0.0011	0.008–0.020	0.008–0.020	0.010–0.020	0.0010–0.0030	0.0010–0.0030	0.0010–0.0020
	141 (2309)	0.0011	0.008–0.016	0.008–0.016	0.010–0.020	0.0010–0.0030	0.0010–0.0030	0.0010–0.0020
	220 (3562)	NA	NA	NA	NA	NA	NA	NA
1991–92	136 (2226)	0.0011	0.008–0.020	0.008–0.020	0.010–0.020	0.0010–0.0030	0.0010–0.0030	0.0010–0.0020
	141 (2309)	0.0011	0.008–0.016	0.008–0.016	0.010–0.020	0.0010–0.0030	0.0010–0.0030	0.0010–0.0020
	220 (3562)	NA	NA	NA	NA	NA	NA	NA

NA—Not available

TORQUE SPECIFICATIONS

All readings in ft. lbs.

Year	Engine Displacement cu. in. (cc)	Cylinder Head Bolts	Main Bearing Bolts	Rod Bearing Bolts	Crankshaft Pulley Bolts	Flywheel Bolts	Manifold Intake	Manifold Exhaust	Spark Plugs
1988	121 (1983)	①	48	22③	④	74	15	18	14
	136 (2226)	①	48	22③	258	74	22	26	14
	141 (2309)	①	48	22③	258	74	22	26	14
1989	121 (1983)	①	48	22③	④	74	15	18	14
	136 (2226)	①	48	22③	258	74	22	26	14
	141 (2309)	①	48	22③	258	74	22	26	14
1990	121 (1983)	①	48	22③	④	74	15	18	14
	136 (2226)	①	48	22③	258	74	22	26	14
	141 (2309)	①	48	22③	258	74	22	26	14
	220 (3562)	⑤	NA	NA	258	74	⑥	18	22

TORQUE SPECIFICATIONS

All readings in ft. lbs.

Year	Engine Displacement cu. in. (cc)	Cylinder Head Bolts	Main Bearing Bolts	Rod Bearing Bolts	Crankshaft Pulley Bolts	Flywheel Bolts	Manifold Intake	Manifold Exhaust	Spark Plugs
1991-92	136 (2226)	①	48	22③	258	74	22	26	14
	141 (2309)	①	48	22③	258	74	22	26	14
	220 (3562)	⑤	NA	NA	258	74	②	18	22

NOTE: Always use new rod bearing bolts.
NA—Not available
① In sequence 29 ft. lbs., 43 ft. lbs. and then tighten it a half turn more (180 degrees).
② Lower manifold to block—7 ft. lbs. Lower manifold to upper manifold—11 ft. lbs.
③ Plus a quarter turn (90 degrees).
④ In sequence 66 ft. lbs., then a half turn more (180 degrees).
⑤ In sequence 30 ft. lbs., 44 ft. lbs., and then tighten it a half turn more (180 degrees). It is not necessary to retighten head bolts during maintenance service or after repairs.

BRAKE SPECIFICATIONS

All measurements in inches unless noted

Year	Model	Lug Nut Torque (ft. lbs.)	Master Cylinder Bore	Brake Disc Minimum Thickness	Brake Disc Maximum Runout	Standard Brake Drum Diameter	Minimum Lining Thickness Front	Minimum Lining Thickness Rear
1988	80	81	0.874	0.787④	0.002	—	0.078	0.078⑤
	80 Quattro	81	0.874	0.787④	0.002	—	0.078	0.078⑤
	90	81	0.874	0.787④	0.002	—	0.078	0.078⑤
	90 Quattro	81	0.874	0.787④	0.002	—	0.078	0.078⑤
	5000 CS Turbo	81	0.875	0.787④	0.002	—	③	0.472②
	5000 CS Quattro Turbo	81	0.875	0.787④	0.002	—	③	0.472②
	5000 CS Quattro Wagon	81	0.875	0.787④	0.002	—	③	0.472②
	5000 S	81	0.810	0.787④	0.002	—	③	0.098
	5000 S Wagon	81	0.810	0.787④	0.002	—	③	0.098
	5000 S Quattro	81	0.810	0.787④	0.002	—	③	0.098
1989	80	81	0.874	0.787④	0.002	—	0.078	0.078
	80 Quattro	81	0.874	0.787④	0.002	—	0.078	0.078
	90 ①	81	0.874	0.787④	0.002	—	0.078	0.078
	90	81	0.874	0.787④	0.002	—	0.078	0.078
	90 Quattro	81	0.874	0.787④	0.002	—	0.078	0.078
	100	81	0.874	0.787④	0.002	—	③	0.472②
	100 Quattro	81	0.874	0.787④	0.002	—	③	0.472②
	200	81	0.874	0.905④	0.002	—	③	0.472②
	200 Quattro	81	0.874	0.905④	0.002	—	③	0.472②
1990	80	81	0.874	0.787④	0.002	—	0.078	0.078
	80 Quattro	81	0.874	0.787④	0.002	—	0.078	0.078
	90 ①	81	0.874	0.787④	0.002	—	0.078	0.078
	90	81	0.874	0.787④	0.002	—	0.078	0.078
	90 Quattro	81	0.874	0.787④	0.002	—	0.078	0.078
	100	81	0.874	0.787④	0.002	—	③	0.472②
	100 Quattro	81	0.874	0.787④	0.002	—	③	0.472②
	200	81	0.874	0.905④	0.002	—	③	0.472②
	200 Quattro	81	0.874	0.905④	0.002	—	③	0.472②
	V8 Quattro	81	1.000	⑥	0.002	—	③	0.280③

BRAKE SPECIFICATIONS

All measurements in inches unless noted

Year	Model	Lug Nut Torque (ft. lbs.)	Master Cylinder Bore	Brake Disc Minimum Thickness	Brake Disc Maximum Runout	Standard Brake Drum Diameter	Minimum Lining Thickness Front	Minimum Lining Thickness Rear
1991-92	80	81	0.874	0.787 ④	0.002	—	0.078	0.078
	80 Quattro	81	0.874	0.787 ④	0.002	—	0.078	0.078
	90	81	0.874	0.787 ④	0.002	—	0.078	0.078
	90 Quattro	81	0.874	0.787 ④	0.002	—	0.078	0.078
	100	81	0.874	0.787 ④	0.002	—	③	0.472 ②
	100 Quattro	81	0.874	0.787 ④	0.002	—	③	0.472 ②
	200	81	0.874	0.905 ④	0.002	—	③	0.472 ②
	200 Quattro	81	0.874	0.905 ④	0.002	—	③	0.472 ②
	V8 Quattro	81	1.000	⑥	0.002	—	③	0.280 ③

NOTE: Minimum lining thickness is as recommended by the manufacturer. Due to variations in state inspection regulations, the minimum allowable thickness may be different than recommended by the manufacturer.

① With ventilated discs—0.768 after refinishing—0.728 discard thickness
② Included backing plate
③ Replace the pads when the indicator on the dash turns on
④ All models with rear disc brakes minimum thickness 0.315
⑤ 0.275 in. including backing plate
⑥ Front 0.984 Limit 0.906
Rear 0.787 Limit 0.709

WHEEL ALIGNMENT

Year	Model	Caster Range (deg.)	Caster Preferred Setting (deg.)	Camber Range (deg.)	Camber Preferred Setting (deg.)	Toe-in (in.)	Steering Axis Inclination (deg.)
1988	80	3/4P–13/4P	11/4P	11/4N–1/4N	3/4N	5/64	—
	80 Quattro	3/4P–13/4P	11/4P	113/32N–13/32N	27/32N	5/64	—
	90	3/4P–13/4P	11/4P	11/4N–1/4N	3/4N	5/64	—
	90 Quattro	3/4P–13/4P	11/4P	113/32N–13/32N	27/32N	5/64	—
	5000 CS Turbo	5/32N–11/2P	27/32P	1N–0	1/2N	0	—
	5000 CS Quattro Turbo	11/32P–121/32P	1P	1N–0	1/2N	1/64	—
	5000 CS Quattro Wagon	11/32P–121/32P	1P	1N–0	1/2N	1/64	—
	5000 S	5/32N–11/2P	27/32P	1N–0	1/2N	0	—
	5000 S Wagon	5/32N–11/2P	27/32P	1N–0	1/2N	0	—
	5000 S Quattro	11/32P–121/32P	1P	1N–0	1/2N	1/64	—
1989	80	3/4P–13/4P	11/4P	11/4N–1/4N	3/4N	5/64	—
	80 Quattro	3/4P–13/4P	11/4P	113/32N–13/32N	27/32N	5/64	—
	90	3/4P–13/4P	11/4P	11/4N–1/4N	3/4N	5/64	—
	90 Quattro	3/4P–13/4P	11/4P	113/32N–13/32N	27/32N	5/64	—
	100	5/32N–11/2P	27/32P	1N–0	1/2N	1/16N	—
	100 Quattro	5/32N–11/2P	27/32P	1N–0	1/2N	1/16N	—
	200	5/32N–11/2P	27/32P	1N–0	1/2N	1/16N	—
	200 Quattro	5/32N–11/2P	27/32P	1N–0	1/2N	1/16N	—

WHEEL ALIGNMENT

Year	Model	Caster Range (deg.)	Caster Preferred Setting (deg.)	Camber Range (deg.)	Camber Preferred Setting (deg.)	Toe-in (in.)	Steering Axis Inclination (deg.)
1990	80	3/4P–1 3/4P	1 1/4P	1 1/4N–1/4N	3/4N	5/64	—
	80 Quattro	3/4P–1 3/4P	1 1/4P	1 13/32N–13/32N	27/32N	5/64	—
	90	3/4P–1 3/4P	1 1/4P	1 1/4N–1/4N	3/4N	5/64	—
	90 Quattro	3/4P–1 3/4P	1 1/4P	1 13/32N–13/32N	27/32N	5/64	—
	100	5/32N–1 1/2P	27/32P	1N–0	1/2N	1/16N	—
	100 Quattro	5/32N–1 1/2P	27/32P	1N–0	1/2N	1/16N	—
	200	5/32N–1 1/2P	27/32P	1N–0	1/2N	1/16N	—
	200 Quattro	5/32N–1 1/2P	27/32P	1N–0	1/2N	1/16N	—
	V8 Quattro	5/8N–1 7/8P	1 1/4P	1N–0	1/2N	3/16N	—
1991–92	80	3/4P–1 3/4P	1 1/4P	1 1/4N–1/4N	3/4N	5/64	—
	80 Quattro	3/4P–1 3/4P	1 1/4P	1 13/32N–13/32N	27/32N	5/64	—
	90	3/4P–1 3/4P	1 1/4P	1 1/4N–1/4N	3/4N	5/64	—
	90 Quattro	3/4P–1 3/4P	1 1/4P	1 13/32N–13/32N	27/32N	5/64	—
	100	5/32N–1 1/2P	27/32P	1N–0	1/2N	1/16N	—
	100 Quattro	5/32N–1 1/2P	27/32P	1N–0	1/2N	1/16N	—
	200	5/32N–1 1/2P	27/32P	1N–0	1/2N	1/16N	—
	200 Quattro	5/32N–1 1/2P	27/32P	1N–0	1/2N	1/16N	—
	V8 Quattro	5/8N–1 7/8P	1 1/4P	1N–0	1/2N	3/16N	—

N Negative
P Positive

ENGINE MECHANICAL

NOTE: Most vehicles are equipped with theft protected radios, which cannot be operated if power to the radio is interrupted. Before disconnecting the battery cables, obtain the security code.

Engine Assembly

REMOVAL & INSTALLATION

4 Cylinder Engine

1. Matchmark hood and hinges and remove hood. Disconnect the negative battery cable.
2. Remove the 2 grille retaining clips on the top of the grille. Remove the screw on the bottom and remove the grille.
3. Loosen the right and left sides of the air conditioning condenser. Tie the condenser away from the radiator.
4. Remove the rubber air duct from the throttle valve housing.
5. Remove the hose from the air duct to the auxiliary air regulator.
6. Disconnect the fuel lines from the cold start valve and fuel injectors. Cap the end of the fuel lines. Remove the injectors from the cylinder head.
7. Remove the fuel distributor, air flow sensor, fuel injectors and the air cleaner from the vehicle, as an assembly.
8. Remove the front engine mount-to-chassis bolts and remove the mount.
9. Loosen the nuts on the outer half of the crankshaft pulley and remove the V-belt.

CAUTION

Compressed refrigerant used in the air conditioning system expands and evaporates into the atmosphere at a temperature of −21.7°F or less. This will freeze any surface it comes in contact with, including eyes. In addition, the refrigerant decomposes into a poisonous gas in the presence of flame.

10. Properly discharge the air conditioning system.
11. Remove all air conditioning lines from the compressor and plug the open connections.
12. Remove the crankcase ventilation hose from the valve cover.
13. Support the air conditioning hoses away from the engine.
14. Remove the air conditioning compressor mounting bolts and remove the compressor.
15. Open the heater control valve all the way.
16. Remove the cap on the expansion tank and drain the cooling system.
17. Remove the upper and lower radiator hoses from the radiator.
18. Disconnect and tag the radiator fan wiring and thermo-switch at the radiator. Remove the radiator with the fan and shroud as an assembly.
19. Remove the power steering pump, if equipped, with hoses attached, move aside and secure to body.
20. If equipped with a manual transaxle, disconnect the clutch cable at the release lever.
21. Disconnect and tag the engine wiring.
22. Remove the fuel control pressure regulator (above the oil filter) from the engine, leaving all the fuel lines connected. Support it aside.
23. Remove the air hose from the back of the alternator, if equipped.
24. Disconnect the blue wire from the alternator at the plug located between the battery and the rear of the engine, if equipped.

25. Remove the charcoal filter hose at the intake air duct.
26. Remove the heater hoses from the engine.
27. Remove the throttle cable from the engine.
28. Disconnect and tag all vacuum hoses at the engine.
29. Remove the hose from the auxiliary regulator to the air inlet duct.
30. Remove the 3 upper engine to transaxle mounting bolts.
31. Remove the right and left engine mount nuts.
32. Raise and support the vehicle safely. Disconnect the exhaust pipe from the exhaust manifold.
33. Remove the flywheel cover plate. If equipped with automatic transaxle, remove the torque converter-to-flywheel mounting bolts.

NOTE: Matchmark the converter to flywheel for installation.

34. Remove the front engine mounting bolts and remove the mount.
35. Disconnect and tag the starter wiring and remove the starter.
36. Remove the 2 lower engine to transaxle mounting bolts.
37. Loosen the right and left engine mount nuts on the sub-frame.
38. Remove the bolt from the front exhaust pipe support.
39. Support the transaxle.
40. Lift the engine until the weight is taken off of the engine mounts and carefully separate the engine and transaxle.
41. Remove the engine from the vehicle.

To install:

42. Guide the engine assembly into place and secure the engine mounts.
43. Connect the flexplate to the torque converter, then install the starter.
44. Connect the exhaust, lower the vehicle and reconnect all electrical connections and vacuum lines.
45. Connect fresh air ducts, emission hoses, throttle cable and clutch cable, if equipped.
46. Install the power steering pump, coolant hoses and air conditioning lines.
47. Install the accessory drive belts and connect all fuel lines.
48. Install air conditioning condenser and the grille. Install and align hood.
49. Tighten the engine-to-transaxle bolts to 40 ft. lbs. (54 Nm), starter bolts to 14 ft. lbs. (19 Nm). Tighten the cold start valve, pressure control regulator and radiator mounting bolts to 7 ft. lbs. (10 Nm). Use a new gasket on the cold start valve when installing.

NOTE: To minimize vibration, loosen all the engine and subframe mounting bolts, then tighten while the engine is running at idle. Tighten the front engine mount bolts to 18 ft. lbs. (25 Nm) and the right and left engine mount bolts to 25 ft. lbs. (34 Nm).

5 Cylinder Engine

WITH TURBOCHARGER

NOTE: Tag all hoses and wiring during removal, to use as reference during reassembly.

1. Disconnect the negative battery cable.
2. Open the heater control valve all the way and drain the cooling system.
3. Remove the fuel injector cooling fan blower motor and intake hose from the engine.
4. On 80/90, remove the upper radiator cover, grille, bumper strip. Disconnect the wiring harness in bumper for turn signals and headlights. Remove the bumper.
5. Disconnect the electrical connector from the coolant fan. Remove the upper radiator hose from the engine. Remove the radiator-to-expansion tank hose from the tank and the bleeder hose from the auxiliary radiator.
6. Disconnect the wire from the thermo-switch. Remove the radiator mounting bolts, right-side radiator cover and bottom radiator cover.
7. Remove the windshield washer reservoir from the mount and support it aside.

CAUTION

The compressed refrigerant used in the air conditioning system expands and evaporates into the atmosphere at a temperature of −21.7°F or less. This will freeze any surface it comes in contact with, including eyes. In addition, the refrigerant decomposes into a poisonous gas in the presence of flame.

8. Properly discharge the air conditioning system and disconnect the refrigerant hoses from the air conditioning condenser.
9. Remove the radiator and air conditioning condenser together. Remove the air conditioning compressor and mounting bracket from the engine.
10. Remove the power steering pump drive belt from the pump. Remove the pump from the mounts. Leaving the hoses attached, support it aside.
11. Disconnect the coolant hose from the thermostat housing, wires from the oil pressure switch and temperature sender. Disconnect the wire plugs from the control pressure regulator.
12. Remove the control pressure regulator from the engine leave the fuel lines connected. Support it aside.
13. Remove the throttle rod clips and remove the rod from the engine. Remove the injector line holder and remove the fuel injectors from the cylinder head.
14. Disconnect the electrical connector from the cold start valve and remove the valve from the intake manifold. Leave the fuel line connected.
15. At the throttle body, disconnect the electrical connectors from the throttle valve switches and intake air temperature switch.
16. Disconnect the air intake hose. Disconnect the wire from the auxiliary air regulator, pull off the vacuum hoses and disconnect the breaker hose from the engine.
17. At the 2-way valve, remove and tag the vacuum hoses. Remove the thermo-pneumatic valve. Leave the vacuum lines connected and the rpm sensor.
18. Disconnect the speedometer cable from the transaxle.
19. Remove the distributor from the engine.
20. Disconnect and tag the thermo-time switch and overheating warning lamp connectors. Disconnect the heater hoses from the engine.
21. At the left engine mount, disconnect the brake booster from the firewall, with the reservoir and leave the lines connected. On Quattro vehicles, disconnect the differential lock control lights connector. Disconnect the backup light switch wires.
22. Disconnect the tie rods from the steering rack. Disconnect the steering linkage.
23. If equipped with a manual transaxle, remove the clutch slave cylinder from the bell housing. Leave the line attached. Remove the bracket and pin under the transaxle bracket.
24. Disconnect the left engine mount ground strap. Disconnect the vacuum hose from the auxiliary air valve.
25. Remove the air duct from the intercooler and remove the intercooler.
26. Disconnect and tag the electrical connectors from the alternator. Remove the oil cooler. Leave the lines attached. Disconnect and tag the starter wiring.
27. Disconnect the exhaust pipe at the turbocharger. Remove the transaxle cover plates and the right side transaxle mount. Disconnect the halfshafts from the transaxle. On Quattro vehicles, disconnect the driveshaft from the rear of the transaxle.
28. On Quattro vehicles at the transaxle, disconnect the differential lock, remove the front and rear circlips and push back the boot. Disconnect the cable.
29. Remove the left-side transaxle mounting bolt and mounts from both sides.
30. At both front wheels, remove the

ball joint pinch bolts. At the subframe, remove the mounting bolts and subframe. Separate the ball joints from the steering knuckle.

31. Install an engine lifting device on the engine. Raise the engine slightly and remove the left and right engine mounts. Lower the engine/transaxle assembly from the vehicle.

32. Raise the front of the vehicle and slide the engine/transaxle assembly from under the vehicle.

33. Separate the engine from the transaxle.

To install:

34. Install the engine assembly and temporarily secure the engine mounts.

35. Install the steering joints to the steering knuckles and torque to 22 ft. lbs. (30 Nm). Install the ball joints with the pinch bolts and torque to 44 ft. lbs. (65 Nm).

36. Install the exhaust and 4WD differential lock clips, if equipped.

37. Lower the vehicle and install the electrical connections, oil cooler, air ducts and clutch slave cylinder, if equipped.

38. Install the distributor, speedometer cable and all vacuum lines. Connect the throttle body connectors, throttle linkage and fuel injection pressure regulator.

39. Install the radiator and air conditioning condenser. Connect the hoses.

40. Refill the cooling system. Connect the battery negative cable.

41. To minimize vibration, loosen all the engine and subframe mounting bolts, then torque to 25 ft. lbs. (34 Nm) while the engine is running at idle.

WITHOUT TURBOCHARGER

NOTE: Tag all hoses and wiring during removal to use as reference during reassembly.

1. Disconnect the negative battery cable.
2. Drain the cooling system.
3. Disconnect the radiator and heater hoses from the engine.
4. Remove the control pressure regulator from the engine, without disconnecting the fuel lines.
5. Remove the cold start valve from the intake manifold, without disconnecting the fuel lines.
6. Pull out the fuel injectors from the cylinder head and support the injectors and fuel lines aside.

NOTE: Protect the fuel injectors and the cold start valve with caps.

7. Loosen the air duct and vacuum hoses from the throttle valve assembly.
8. Remove the air box cover and filter.
9. At the top of the grille, pull the hood latch cable guide off of its bracket.
10. If equipped with air conditioning, perform the following procedures:
    a. Remove the 2 clips from the top of the grille and the screw from the bottom. Remove the grille.
    b. Remove the condenser mounting bolts.
    c. Remove the air duct to auxiliary air regulator hose and remove the air duct from the throttle valve housing.
    d. Remove the fuel distributor, air flow sensor, fuel injectors and air box, as a unit.

NOTE: When removing the fuel injectors, leave all of the lines connected and cover the fuel injectors with caps.

    e. Remove the accessory drive belts.

CAUTION

The compressed refrigerant used in the air conditioning system, expands and evaporates into the atmosphere at a temperature of −21.7°F or less. This will freeze any surface it comes in contact with, including eyes. In addition, the refrigerant decomposes into a poisonous gas in the presence of flame.

    f. Properly discharge the refrigerant from the air conditioning system. Remove and plug the air conditioning hoses, move them away from the engine.
    g. Remove the upper/lower compressor mounting bolts and remove the compressor from the engine.

11. Remove the power steering pump from the engine, leaving the hoses connected.
12. Remove the vacuum amplifier.
13. Remove the EGR control valve.
14. Remove the windshield washer reservoir from its holder.
15. Remove the distributor cap and ignition wires. Remove the distributor vacuum hose(s).

NOTE: Tape the distributor dust cap on to prevent it from falling off.

16. Disconnect the throttle linkage from the engine.
17. If equipped with an automatic transaxle, remove the throttle pushrod.
18. Disconnect the oil pressure and water temperature sensor wiring.
19. Remove the exhaust pipe to manifold nuts.
20. Remove the exhaust pipe support bracket from the transaxle.
21. Remove the front engine mount bolts and remove the mount. Disconnect the ground strap on left engine mount, if equipped.
22. Tag and disconnect all wires from the starter and remove the starter.
23. Tag and disconnect all wires leading from the alternator and remove the alternator.
24. If equipped with an automatic transaxle, work through the starter mounting hole to remove the torque converter mounting bolts.
25. Remove the lower engine to transaxle mounting bolts.
26. Support the transaxle and lower the vehicle.
27. Remove the upper engine to transaxle mounting bolts.
28. Remove the left engine support bracket.
29. Loosen the right engine bracket from the right engine mount.
30. Lift the engine until the crankshaft V-belt pulley is behind the grille opening.
31. Carefully detach the engine from the transaxle.
32. Remove the engine assembly by turning it to the right while lifting it out.

To install:

33. Install the engine assembly and secure the mounts.
34. On vehicles with automatic transmission, install the torque converter bolts. On 087 and 089 units, torque the bolts to 22 ft. lbs. (30 Nm). On 097 units, torque the bolts to 44 ft. lbs. (60 Nm).
35. Install the starter and the exhaust system.
36. Lower the vehicle and connect engine compartment wiring including the oil pressure and water temperature sensor wiring.
37. Connect the throttle linkage. Install the distributor cap and wire, the washer reservoir, power steering pump and EGR valve.
38. Install the air conditioning condenser.
39. Install the injectors, control pressure regulator and cold start valve.
40. Install the radiator, refill the cooling system. Connect the battery negative cable.
41. Torque the engine-to-transaxle mounting bolts to 43 ft. lbs. (58 Nm), the starter bolts to 14 ft. lbs. (19 Nm), the air conditioner mounting bolts to 29 ft. lbs. (39 Nm) and the power steering pump and the control pressure regulator mounting bolts to 14 ft. lbs. (19 Nm).

NOTE: To minimize vibration, loosen all the engine and subframe mounting bolts, then tighten while the engine is running at idle. Tighten the engine and subframe mounting bolts to 32 ft. lbs. (43 Nm).

8 Cylinder Engine

The engine is taken out towards the front without the transaxle. All cable ties which have to be released or cut open when removing the engine must be replaced in the same position when the engine is installed.

1. Disconnect the battery negative cable. The battery is under the rear seat.
2. Under the dashboard, remove the retainers and remove left dashboard end cap and the knee protector.
3. Remove the heater duct for the driver-side area, under the dash panel, by removing the right-hand screw and loosening the left-hand clip.
4. Unclip the floor lamp and push it through the opening.
5. Remove the control units with brackets.
6. Open the locking mechanisms on the control unit connectors and disconnect the harness connectors. Lock the connectors so they don't become tangled when removing the cable harness. Pull off connectors 11, 13 and 15. The numbers should be marked on the wiring harness.
7. Remove the connector panel by removing the lower screw, loosening the upper screw and pulling the panel downwards. Press the latch on the butterfly connector lock and slide it sideways to release the connector from the panel.
8. Under the hood, remove the plenum chamber cover and lift up on the rubber grommet on the middle wiring harness. Cut the cable tie and pull the wiring harness carefully out of the passenger compartment and plenum chamber. Open the expansion tank cap for coolant.
9. Raise and safely support vehicle. Remove the sound insulation or under pan.
10. Remove the bolts and nuts holding bumper and bracket, remove the electrical connectors and pull off the bumper towards the front.
11. Drain the coolant from the radiator. Also open the block drains on both sides.
12. Remove the oil cooler hoses from the oil filter housing and the hose from the bottom of the transaxle oil cooler. Remove the bracket for the line on the air conditioner and disconnect the engine ground cable at the engine.
13. Separate the harness for the headlight washing system and air conditioner harness. Disconnect the coolant temperature sensor harness, the cool air duct from the alternator, the coolant hose and hose for headlight washer system.
14. Disconnect the outer half of the air intake elbow for the alternator. Disconnect the wiring from the alternator and starter motor. Unscrew oil filter.
15. Remove the upper bolt on the starter by guiding a 10mm Allen® socket, with extension and flex fitting, through the opening on the transaxle housing over the final drive. Remove starter.
16. Locate and remove the bolts from the front of the subframe on left and right sides. Remove the bolts securing the exhaust system on the left and right sides.
17. Remove the 4 bolts connecting the bottom of the engine and transaxle.
18. Remove the bolts above and below the long member on the left and right sides. Remove the bolts for the cadmium strip and remove the bolts on fender.
19. Remove the harness connector from the temperature sensor at the front of the air conditioning condenser and disconnect the hood release cable.
20. Open the fuse box on the left side behind the hydraulic reservoir. Disconnect the wire to the fan and the ground wire at the top of the suspension strut. Expose the wiring.
21. Remove the air conditioning dryer with the bracket.
22. Remove the intake manifold bracket on the left and right. Pull out the exchanger. Remove the bar-shaped reinforcement strut.
23. Remove the air conditioning condenser bolts and swivel downward. Wiring should remain connected.
24. Remove the wiring bracket from the top of the transaxle cooler. Remove the water hose from both sides of the engine. Remove the bleeder hose from the expansion tank on the radiator. Remove the front apron.
25. Remove the screws securing the upper part of the air cleaner housing. There are 4 screws on the housing and 3 screws at the rear of the housing.
26. Remove the screws securing the lower part of the air cleaner housing. Press toward the rear and lift out. Remove the right-hand stud from the lower air cleaner housing.
27. Disconnect the carbon canister hose from the front of the engine. Cut both cable ties on the fuel injector. Remove the fuel supply and return lines at the right-hand side and release at left. Turn the fuel lines to the left and lay to one side.
28. Remove the coolant hose to expansion tank. Disconnect the high tension wire and connector on the ignition coil at the left and right. Remove the bolts holding the left coil, cable housing and retaining clip. Watch for the spacer sleeve.
29. Remove the left and right heat shields. Remove the housing for the ignition wire retainer. Disconnect the vacuum line.
30. Remove the screws securing the left and right distributor caps. Wiring remains connected. Using a cable tie, secure the 2 distributor caps aside against the PCV hose.
31. Disconnect the throttle cable by unclipping both retaining clips. Remove the screws securing the throttle cable support bracket. Unscrew both coolant hoses. Disconnect the supporting clamp.
32. Separate both harness connectors from the oxygen sensor. Disconnect the vacuum line to the cruise control system.
33. Remove the transaxle oil fill tube bolt from the engine. Pull the wiring harness through the contact plate and place on the engine.
34. Remove the 6 torque converter bolts through the starter opening.
35. Release the tension on the ribbed drive belt by pulling firmly by hand in the middle at the bottom and inserting a pin through the holes provides in the tensioner bracket. Remove the belt.
36. Remove the bolts holding the air conditioning compressor. Lift the compressor over the strut. The lines remain connected.
37. Remove the bolts for the air conditioning compressor bracket and the hydraulic pump. Watch the guide sleeves. Place the bracket with the hydraulic pump on the long member. The lines remain connected. Note that at installation, fit the lower bolts with the guide sleeves in position and tighten lightly. Then the upper bolts can be installed.
38. Remove the nuts from the left and right engine mountings. Remove the bolts securing the engine support at the front. Take note of the shims. The same thickness shims must be used at installation.
39. Disconnect the engine and transaxle at the top by removing 3 of the 4 bolts. Loosen the fourth bolt but do not remove.
40. Install a suitable lifting sling to the front left-hand side and rear right-hand side.
41. Move in hoist, being careful of the air conditioning compressor. Lift engine carefully. Remove last bolt from the top of the engine and transaxle. Unscrew the engine mounting from the left and right sides and pull the engine out from the front. Lift carefully to prevent damage to the transaxle main shaft, clutch and body.
42. Use care when selecting a suitable engine repair stand. Attaching to some types of stands could cause the engine block to distort and cause any cylinder bore measurements to be inaccurate.

To install:

43. Installation is the reverse of the removal procedure. Note that there are guide sleeves in the engine block for centering the engine and transaxle. Make sure they are installed.
44. Install the engine assembly and secure to the transaxle and engine mounts.
45. Install the air conditioning compressor and power steering pump. Fit the lower bolts with the guide sleeves first. Tighten lightly and then install the top bolts.
46. Install all electrical connectors, hoses and sheet metal heat shields.
47. Install the air conditioning condenser and the small brackets that were removed.
48. Install the oil cooler, radiator and other small connections.
49. Refill the cooling system.
50. Install the interior parts removed.
51. Connect battery negative cable.
52. When installing the wiring harness make sure it is first pushed through the contact plate.
53. Always replace all self-locking nuts. Make sure the exhaust system is installed free of strain. Check all fluid levels before starting the engine. Check the fluid level in the automatic transaxle.

Cylinder Head

NOTE: Before removing or installing the cylinder head, align the engine timing marks at TDC. Rotate the crankshaft mark away about ¼ turn (BTDC). This will prevent the valves from hitting the piston heads. Be sure to turn the crankshaft to the proper position after cylinder head installation.

REMOVAL & INSTALLATION

NOTE: Cylinder head removal should not be attempted unless the engine is cold.

Except 8 Cylinder Engine

1. Disconnect the negative battery cable.
2. Drain the cooling system.
3. Disconnect the air duct from the throttle valve assembly on all vehicles except the Turbo and Quattro. On the Turbo and Quattro, remove the hose which runs between the air duct and the turbocharger.
4. Disconnect the throttle cable from the throttle valve assembly.
5. Remove the air duct for the injector cooling fan on the Turbo and Quattro.
6. Clean and remove the fuel injectors and all other fuel lines.

NOTE: Protect the fuel injectors and the cold start valve with caps.

7. Tag and disconnect all vacuum and PCV lines.
8. Remove the hose which runs from the intake manifold to the turbocharger on the Turbo and Quattro.
9. Tag and disconnect all electrical lines leading to the cylinder head.
10. Remove the intake manifold.
11. Disconnect all radiator and heater hoses where they are attached to the cylinder head. Position them aside.
12. Tag and remove all spark plug wires.
13. Remove the distributor. To aid installation, scribe a mark on the body of the distributor and the cylinder head.
14. Separate the exhaust manifold from the exhaust pipe.

NOTE: Exhaust pipe detachment differs slightly on the Turbo and Quattro. First the exhaust pipe must be unbolted from the turbocharger. Second, it must be unbolted from the wastegate at the rear of the engine.

15. Disconnect the EGR valve and oxygen sensor from the exhaust manifold.
16. Remove the heat deflector shield.
17. Remove the oil lines (2) from the turbocharger.
18. Remove the exhaust manifold.

NOTE: When removing the exhaust manifold on the Turbo and Quattro, the manifold, turbocharger and wastegate should all be removed as a unit.

19. Remove the air hose cover from the back of the alternator.
20. Tag and disconnect all wires coming from the back of the alternator and remove the alternator from the engine.
21. Disconnect and plug the hoses coming from the power steering pump.
22. Remove the power steering pump and the V-belt.
23. Remove the timing belt cover and belt.
24. Remove the valve cover.
25. Loosen the cylinder head bolts in the reverse order of the tightening sequence.
26. Remove the bolts and lift the cylinder head off of the engine.

To install:

27. Clean the cylinder head and engine block mating surfaces thoroughly and install the new gasket without any sealing compound. Make sure the words TOP or OBEN are facing up, when the gasket is installed.

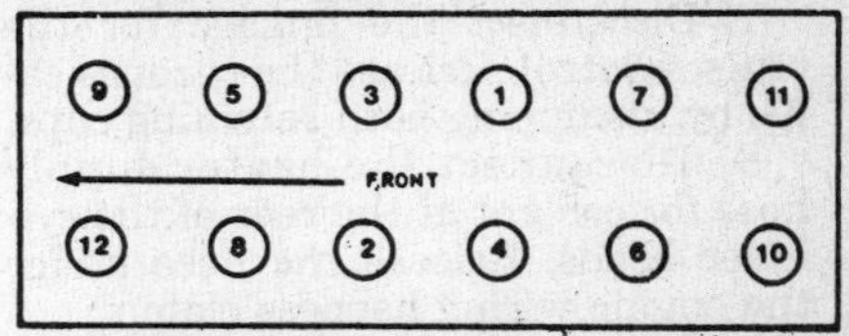

Cylinder head torque sequence for all 5 cylinder engines

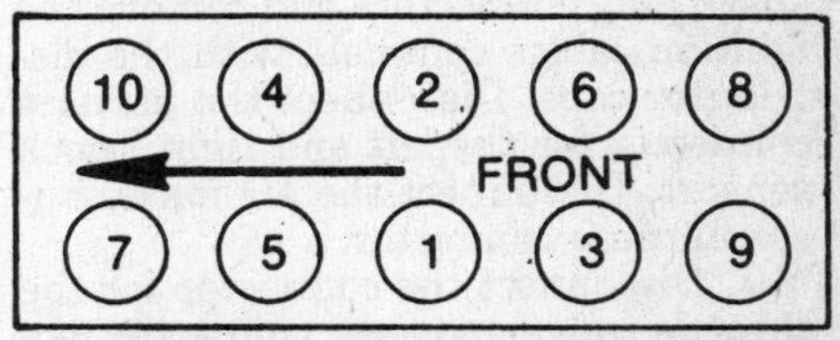

Cylinder head torque sequence for all 4 cylinder engines

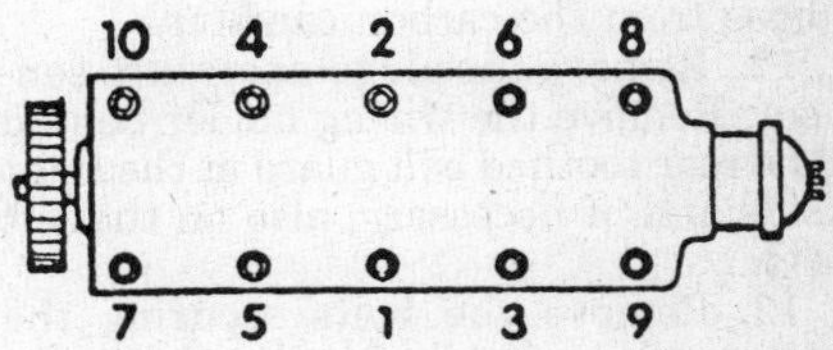

Cylinder head torque sequence for V8 engine

28. Place the cylinder head on the engine block and install bolts No. 8 and 10 first. These holes are smaller and will properly locate the gasket and the head on the engine block.
29. Install the remaining bolts. Torque them in sequence in 3 stages as follows: Step 1 – 29 ft. lbs. (39 Nm), Step 2 – 43 ft. lbs. (58 Nm) and Step 3 – Tighten ½ turn more (180 degrees).
30. Installation of all other components is in the reverse order of removal.

8 Cylinder Engine

1. Disconnect the negative battery cable. The battery is under the rear seat.
2. Remove the toothed cam drive belt. It should not be necessary to loosen or remove the vibration damper.
3. Open the left and right block drains.
4. Remove the bolts holding the exhaust pipe on both sides.
5. Remove the supply line at the fuel manifold. Remove the fuel return line. Remove the bolt holding the pressure regulator to the fuel manifold. At assembly, press the regulator directly into the seat of the O-ring seal. Do not pull in with bolts.
6. Disconnect the breather hose at the rear of the intake manifold, the vacuum hose and the bolt holding the upper part of the supporting clamp for the engine wiring harness. Remove the mounting bracket for ignition wire holder.

7. Disconnect the linkage for the cruise control. Release the throttle cable by unclipping both retaining clips.
8. Disconnect the heater supply hose for coolant at the rear of the cylinder heads. Release the hose under the engine wiring harness clamp.
9. Remove the screws securing the 2 spark plug covers, disconnect the spark plug connectors and remove the ignition cables complete with the distributor caps. Disconnect the harness connector for the left and right knock sensors. Disconnect the air mass sensor harness connector.
10. Disconnect the connector for the throttle valve harness connector potentiometer. Disconnect the idle stabilizer valve harness connector. Remove the idle stabilizer valve. Remove the hose from the carbon canister.
11. Remove the air temperature sensor. Remove the wiring holder behind the rear toothed belt guard at the right side and, if necessary, also on the left side.
12. Remove the bolts securing the coolant hose holder on the right rear cylinder head. Disconnect the heater supply hose from the cylinder heads by pulling towards the rear.
13. Disconnect the harness connector for the Hall sender on the right cylinder head. If the right cylinder head needs to be removed, disconnect the temperature sensor harness connector on the right cylinder head.
14. Remove the breather hose at the top. Turn the hose with pliers until the retaining lug unlatches. Remove the breather hose on the cylinder head, on the left side at the rear.
15. Remove the bolts securing the fuel rails and lift out complete with injectors and place on plenum chamber. At installation, make sure the O-rings are not damaged. Moisten slightly with fuel.
16. Remove the bolts securing the intake manifold and lift out. Watch the front breather hose under the intake manifold.
17. If it is the left hand cylinder head being removed, unscrew the dipstick guide and pull out. Remove the transaxle oil filler tube bolt.
18. Remove the cylinder head cover. Note that the 2 center mounting bolts are different. The bolt with the longer hexagon fitting must be installed at the rear.
19. If the left cylinder head is being removed, remove the bolts securing the cruise control vacuum unit and bracket. Do not forget the deflector plate during installation.
20. Remove the cylinder head bolts in reverse sequence of tightening. Remove the cylinder head.

To install:

21. Installation is the reverse of the removal process, observing the following. Check that the gasket surface is not distorted. Maximum permissible distortion is 0.004 in. (0.1mm).
22. Make sure the gasket surface is clean. Look for the word OBEN on the gasket. This is the top, facing the cylinder head. When installing the cylinder head, watch the centering dowels in the block.
23. Insert the cylinder head bolts hand-tight. Torque in sequence in 3 steps. First to 30 ft. lbs. (41 Nm), then to 44 ft. lbs. (60 Nm). The 3rd step is to give the bolts and additional ½ turn. Use a regular ratchet handle or breaker bar, and turn in one smooth motion without stopping. It should not be necessary to retighten the cylinder head bolts during maintenance or after repairs.
24. When installing the intake manifold, first attach the front breather hose under the intake manifold to the engine.

Valve Lifters

REMOVAL & INSTALLATION

1. Disconnect the negative battery cable. Remove the toothed camshaft drive belt.
2. Remove the cylinder head cover.
3. Remove the camshaft sprocket.
4. On the 8 cylinder engine, on the exhaust side, remove the intermediate flange and bearing cap for the distributor. Loosen the bearing cap in front of the chain, plus caps 2 and 3. Loosen bearing caps 1 and 4 alternately and in a diagonal sequence.
5. On the 8 cylinder engine, on the intake side, remove bearing caps 6 and 7. Loosen bearing caps 5 and 8 alternately and in a diagonal sequence.
6. Remove the cam and lift out the valve lifter. If it is to be reused, it must go in the bore from which it was removed.

To install:

7. Installation is the reverse of removal. Before assembly coat the moving parts with clean oil.
8. On the 8 cylinder engine, install the camshafts with the chain so the markings on the chain sprockets are in alignment. Install the bearing caps so the stamped-on numbers can be read from the intake side.
9. Make sure the bearing caps are installed properly. They will cause cam shaft failure if installed backwards. Tighten bearing caps alternately, in a diagonal sequence to 11 ft. lbs. (15 Nm).

Valve Lash

ADJUSTMENT

All engines are equipped with hydraulic valve lash adjusters that eliminate the need for routine valve lash adjustments. Intermittent valve noise is normal when the engine is cold. If valve noise persists, check the camshaft lobes and/or camshaft followers for wear. Replace if necessary. Do not attempt valve lifter repair. If worn or damaged, replace the complete assembly.

Use care when handling valve lifters. Always place a removed valve lifter on a clean surface with the contact surface or camshaft side facing downward.

After working on the valve train, carefully turn the engine by hand at least 2 turns to make sure the valves do not strike the pistons when the engine is started. Do not start the engine for 30 minutes after installing new valve lifters or the valves may strike the pistons. The lifter must be allowed to bleed down to proper adjustment.

To check a suspect lifter, use the following procedure:

1. Warm the engine to operating temperature until the radiator fan comes on at least once.
2. Bring the engine to approximately 2500 rpm for 2 minutes. If a lifter is still noisy, shut off engine and remove the cylinder head cover.
3. Turn the crankshaft pulley bolt clockwise until the cam lobes of the cylinder to be checked point upward.
4. Push down against the valve lifter with light pressure using a suitable tool. If the valve lifter can be pushed down more than 0.004 in. (0.1mm), replace the lifter.
5. Do not start the engine for 30 minutes after installing new valve lifters or the valves may strike the pistons. The lifter must be allowed to bleed down to proper adjustment.

Intake Manifold

REMOVAL & INSTALLATION

4 Cylinder Engine

1. Disconnect the negative battery cable. Relieve the fuel pressure in the system.
2. Disconnect the throttle cable at the throttle valve housing.
3. Disconnect the wiring for the cold start valve and thermo time switch.
4. Disconnect the ground wire from the intake manifold.
5. Remove the fuel line from the cold start valve. Cap the valve and line.

6. Remove the air boot from the throttle valve housing and sensor plate housing.
7. Disconnect and tag the vacuum hoses at the manifold.
8. Remove the fuel injectors from the cylinder head, without disconnecting the fuel lines from the injectors.
9. Remove the control pressure regulator line and move the regulator aside.
10. Remove the 2 straps that connect the intake and exhaust manifolds.
11. Disconnect the CO percentage check tube from the intake manifold.
12. Remove the intake manifold mounting nuts and remove the manifold from the engine.

NOTE: Before loosening the intake manifold mounting nuts, soak the nuts and studs with lubricant. Studs are very difficult to replace with the cylinder head installed on the engine.

To install:
13. Clean the gasket mating surfaces of the engine and intake manifold.
14. Before installation, hold the intake manifold gasket up to the engine and check for proper fit. Trim, if necessary.
15. Install the intake manifold on the cylinder head. Tighten the 6mm nuts to 7 ft. lbs. (10 Nm) and the 8mm nuts to 18 ft. lbs. (24 Nm).
16. The remainder of the installation is the reverse of the removal procedure.
17. Lubricate the fuel injector O-rings with a drop of engine oil, before installation in the cylinder head.
18. When finished, run the engine and check for leaks.

5 Cylinder Engine

1. Disconnect the negative battery cable.
2. Relieve the fuel system pressure.
3. On non-turbocharged engines, disconnect the air duct from the throttle valve assembly. On turbocharged engines, remove the hose between the air duct and turbocharger.
4. Disconnect the throttle cable and rod from the throttle valve assembly.
5. On turbocharged engines, remove the air duct for the injector cooling fan.
6. Remove the fuel injectors from the cylinder head, with the fuel lines attached.
7. Disconnect the cold start valve wiring and remove the fuel line from the valve.

NOTE: Protect the fuel injectors and cold start valve with caps.

8. Tag and disconnect all vacuum and PCV lines.
9. Tag and disconnect all electrical lines leading to the cylinder head.
10. On turbocharged engines, remove the hose which runs from the intake manifold to the turbocharger; intercooler on the Quattro.
11. Remove the auxiliary air regulator. Remove the air box cover and filter element.
12. Remove the intake manifold mounting nuts and remove the manifold from the engine.

To install:
13. Clean the gasket mating surfaces on the manifold and engine.
14. Using a new gasket, install the manifold on the cylinder head and tighten the nuts to 15 ft. lbs. (20 Nm).
15. The remainder of the installation is the reverse of the removal procedure.
16. Always use new gaskets and O-rings where necessary.
17. Check the engine oil level and correct, if necessary.
18. When finished, run the engine and check for leaks.

8 Cylinder Engine

1. Disconnect the negative battery cable.
2. Relieve the fuel system pressure.
3. Remove the 7 bolts retaining the upper part of the air cleaner housing and the 2 bolts securing the lower part. Press the housing towards the rear and lift out.
4. Remove the fuel supply and return lines. Remove the fuel pressure regulator from the fuel rail.
5. Remove the breather hose at the rear of the intake manifold, remove the vacuum hose and the bolt securing the upper part of the engine wiring harness clamp.
6. Disconnect the harness connectors for the left and right knock sensor, the air mass sensor, the potentiometer, the thermo-switch and idle stabilizer valve. Remove the idle stabilizer valve and disconnect the hose to the carbon canister.
7. Remove the air temperature sensor bolts. Remove the breather hose at the top of the cam cover by using pliers to turn the hose until the retaining lug unlatches. Remove the breather on the cylinder head at the left rear.
8. Remove the bolts securing the fuel rails and lift out the rails with the injectors. Place on the plenum chamber.
9. Remove the intake manifold bolts and lift out the manifold. Watch the front breather hose under the intake manifold.

NOTE: There are 2 oil retention valves that must be replaced if they sound noisy on short drives, although the noise may go away on longer drives. To replace these valves, after the intake manifold has been removed, remove the right hand knock sensor and the bolts securing the engine breather cover and lift out the cover along with the bulkhead panel. It may be easier to lift the right side first. Locate the oil retention valves and remove the circlips. Screw a M6 × 50mm bolt with a large washer into the oil retention valve. Remove the valve by prying evenly under the washer.

To install:
10. Installation is the reverse of the the removal procedure. Use care when installing the intake manifold. First install the front breather hose to the engine under the intake manifold.
11. When reinstalling the fuel system parts, use care not to damage any O-rings.
12. When installing the fuel pressure regulator, first press it directly into the seat of the O-ring seal. Do not pull it in with bolts.

Exhaust Manifold

REMOVAL & INSTALLATION

Non-Turbocharged Engines

EXCEPT 8 CYLINDER ENGINE

NOTE: Although not necessary, more working clearance will be found by removing the intake manifold before removing the exhaust manifold. Before beginning, soak the manifold studs with lubricant to aid in the removal.

1. Disconnect the negative battery cable. Raise and support the vehicle safely. Disconnect the exhaust pipe from the exhaust manifold.
2. Disconnect the EGR valve and oxygen sensor, from the manifold.
3. Remove the heat deflector shield on 4 cylinder engines.
4. Disconnect the CO probe receptacle tube.
5. Remove the exhaust manifold mounting nuts and remove the manifold from the engine.

To install:
6. Clean the gasket mating surfaces of the manifold and engine.
7. Using a new gasket, install the manifold on the engine and tighten the nuts to 22 ft. lbs. (30 Nm).

NOTE: Always replace the old mounting nuts with new brass nuts. Check the condition of the studs before installation. The oxygen sensor, EGR tube, bolts and

nuts exposed to high temperatures should receive a light coating of anti-seize compound on the threads before assembly.

8. Connect the exhaust pipe to the manifold, using a new gasket and tighten nuts to 26 ft. lbs. (35 Nm).
9. Install and tighten the CO measuring tube to 22 ft. lbs. (30 Nm).
10. The remainder of the installation is the reverse of the removal procedure.
11. When finished, run the engine and check for leaks.

8 CYLINDER ENGINE

1. Disconnect negative battery cable.
2. Disconnect the front exhaust pipe at the manifolds.
3. For the left side manifold, unscrew the guide for the dipstick tube and remove.
4. For both the left and right manifolds, remove the bolts securing the exhaust manifolds to the cylinder heads.
5. Remove the manifolds one at a time. Lift up the engine at the edge of the oil pan on the side of the manifold to be removed.

To install:

6. Installation is the reverse of the removal process. Make sure the sealing rings between the exhaust manifold and the cylinder head are installed in such a way that the beading makes contact with the manifold.
7. When installing the dipstick guide, replace the sealing ring.

Turbocharged Engine

1. Disconnect the negative battery cable. Remove the hose which runs between the air duct and the turbocharger.
2. If the intake manifold has not been removed, disconnect the hose which runs from the intake manifold to the turbocharger or intercooler.
3. Disconnect the exhaust pipe from the turbocharger.
4. Disconnect the exhaust pipe from the wastegate on the rear of the manifold.
5. Disconnect the EGR valve and the oxygen sensor, if necessary, from the manifold.
6. Remove the oil lines from the turbocharger.
7. Remove the line from the bottom of the turbocharger to the intercooler, if equipped.

NOTE: The manifold, turbocharger and wastegate are removed as a unit.

8. Remove the manifold assembly.

To install:

9. Installation is the reverse of the removal procedure. Always use new gaskets and O-rings where necessary.

NOTE: The oxygen sensor, EGR tube, bolts and nuts exposed to high temperatures should receive a light coating of anti-seize compound on the threads before assembly.

10. Install and tighten the wastegate nuts to 22 ft. lbs. (30 Nm).
11. Tighten the exhaust manifold mounting nuts to 26 ft. lbs. (35 Nm).
12. Tighten the exhaust pipe nuts to 26 ft. lbs. (35 Nm).
13. Start engine allow to reach normal operating temperature.
14. Check for leaks, and proper engine operation.

Turbocharger

REMOVAL & INSTALLATION

1. Disconnect the negative battery cable.
2. Spray all mounting bolts with a lubricant.
3. Relieve the fuel system pressure.
4. Remove the vacuum tube between the intake air boot and turbocharger.
5. Remove the intake boot and crankcase ventilation hose. Remove the hose assembly between the intake manifold and throttle housing.
6. Remove the air box cover and remove the filter element.
7. Remove the right side engine mount heat shield.
8. Remove the oil supply pipe from the turbocharger. Remove the exhaust pipe from the corrugated pipe. Loosen the exhaust pipe at the transaxle mount and catalytic converter.
9. Remove the retaining clamp from the starter housing and sensor air hose.
10. Remove the exhaust pipe from the turbocharger.
11. Remove the alternator support bolt and position the alternator to the side.
12. Remove the oil return pipe from the turbocharger. Remove mounting bolts and turbocharger.

To install:

13. Install in the reverse order using new gaskets and O-rings, where necessary.

NOTE: Bolts and nuts exposed to high temperatures should receive a light coating of anti-seize compound on the threads before assembly. After servicing the turbocharger, always replace the engine oil along with the turbocharger filter and engine oil filter.

14. Install the turbocharger and torque the mounting nuts to 43 ft. lbs. (60 Nm).
15. Connect the oil supply and return lines using new gaskets.
16. Connect the exhaust pipe and torque the nuts to 25 ft. lbs. (34 Nm).
17. Connect the outlet air hose and the remaining breather hoses.
18. Install the heat shields and turn the ignition switch **ON** to pressurize the fuel system and check for leaks before starting the engine.

Turbocharger Wastegate

REMOVAL & INSTALLATION

NOTE: Although not necessary, more working clearance will be found by removing the intake manifold before removing the wastegate. Before starting, soak the studs with lubricant to aid in the removal.

1. Disconnect the negative battery cable. Remove the wastegate to exhaust pipe connecting tube. There are 3 bolts on the top and on the bottom.
2. Remove the mounting bolt for the tube leading from the wastegate to the exhaust manifold.
3. Remove the vacuum line from the end of the wastegate.
4. Remove the 4 mounting bolts and remove the wastegate from the exhaust manifold.
5. Installation is in the reverse order of removal. Torque the mounting nuts to 18 ft. lbs. (25 Nm).

NOTE: Bolts and nuts exposed to high temperatures should receive a light coating of anti-seize compound on the threads before assembly.

Timing Belt Front Cover

REMOVAL & INSTALLATION

4 Cylinder Engine

UPPER COVER

1. Disconnect the negative battery cable. Loosen the alternator adjusting bolts, pivot the alternator over and slip the drive belt off.
2. Loosen the air conditioning compressor mounting bolts and remove the drive belt.
3. Remove the valve cover nuts and remove the valve cover and retaining straps.
4. Remove the upper timing belt cover nuts. Note the position of the

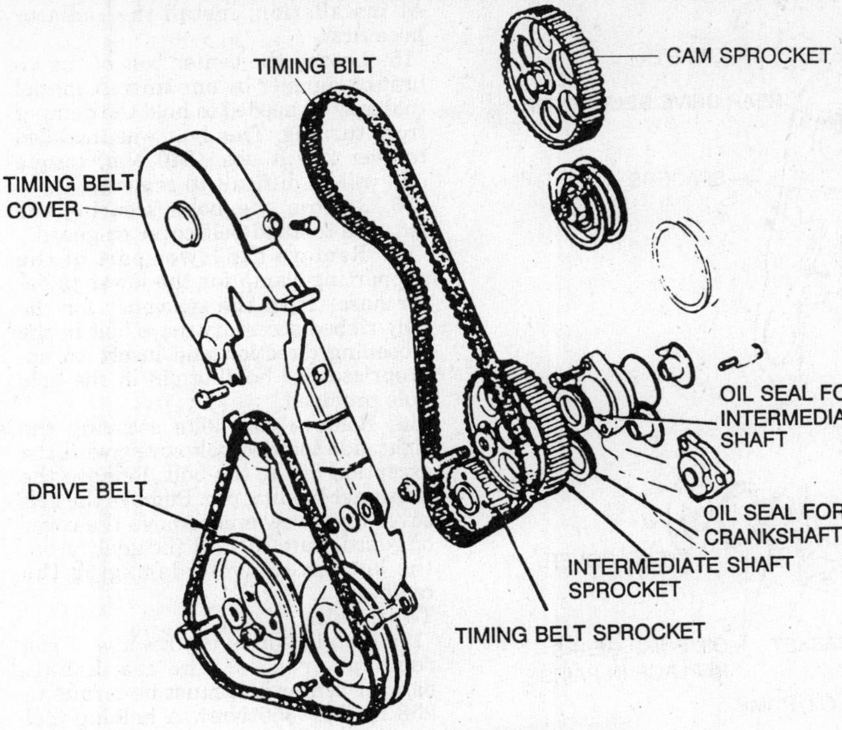

Timing belt on 4 cylinder engine

washers and spacers while removing the cover.

5. Installation is the reverse of the removal procedure.

6. Adjust the drive belt tension when finished.

LOWER COVER

1. Disconnect the negative battery cable. Remove the upper timing belt cover.

2. Using the large bolt on the crankshaft sprocket, rotate the engine until the No. 1 cylinder is at TDC of the compression stroke. At this point, both valves for No. 1 cylinder will be closed and the **0** mark on the flywheel will be aligned with the pointer on the bell housing.

3. Remove the crankshaft pulley retaining bolts and loosen the crankshaft sprocket bolt, if the sprocket or rear cover is to be serviced.

NOTE: To remove the crankshaft sprocket bolt, on manual transaxle vehicles, place the vehicle in 5th gear and have an assistant apply the brake. The will stop the engine from rotating while loosening the bolt. On automatic transaxle vehicles, remove the starter and hold the flywheel from turning, using a flywheel holding tool VW 10–201 or equivalent.

4. Remove the crankshaft pulley.

5. Remove the water pump pulley retaining bolts and remove the pulley.

6. Remove the lower cover retaining nuts and remove the cover. Take care not to lose any of the washers or spacers.

7. Installation is the reverse of the removal procedure.

8. Tighten the crankshaft sprocket bolt to 66 ft. lbs. (89 Nm) plus ½ additional turn.

5 Cylinder Engine

UPPER COVER

1. Disconnect the negative battery cable. Loosen the alternator adjusting bolts and remove the drive belt.

2. Loosen the power steering pump adjusting bolts and remove the drive belt.

3. Remove the retaining nuts and remove the timing belt cover. Take care not to lose any of the washers or spacers.

4. Installation is the reverse of the removal procedure.

5. Adjust the drive belt tension when finished.

LOWER COVER

1. Disconnect the negative battery cable. Remove the upper timing belt cover.

2. Loosen the air conditioning compressor mounting bolts and remove the drive belt.

3. Remove the crankshaft balancer center bolt.

NOTE: To remove the crankshaft balancer bolt, on manual transaxle vehicles, place the vehicle in 5th gear and have an assistant apply the brake. The will stop the engine from rotating while loosening the bolt. On automatic transaxle vehicles, remove the starter and hold the flywheel from turning, using a flywheel holding tool VW 10–201 or equivalent. This bolt is extremely tight.

4. Remove the lower timing belt cover bolts and remove the cover.

To install:

5. Installation is the reverse of the removal procedure.

6. Use the same procedure to install the crankshaft center bolt as when removing the bolt. Apply a locking compound on the bolt threads and tighten the bolt to 258 ft. lbs. (350 Nm) in several steps.

7. Adjust the drive belt tension when finished.

8 Cylinder Engine

1. Disconnect the negative battery cable. The battery is under the rear seat.

2. The toothed cam drive belt must first be removed. Remove the coolant expansion tank cap. Raise and safely support vehicle. Remove the sound insulator or lower pan. Drain coolant from radiator.

3. Remove the alternator cooling air duct. Loosen the clip and disconnect the coolant temperature sensor harness connector. Remove the coolant hose.

4. Disconnect the outer half of the air duct at the alternator. Remove the screws securing the wiring at the alternator. Note that the alternator wiring must be brought out at the side, not downward. Otherwise the alternator air duct cannot be mounted.

5. Release tension on the polyribbed drive belt and remove the belt in a downward direction by placing a 13mm box wrench on the hexagon guide of the tensioner and pressing the wrench slowly upwards.

6. Remove the alternator mounting bolts and remove the alternator.

7. Working from below, remove the 3 bolts for the toothed belt guard.

8. Remove the bolts securing the bracket for the air intake ducts on the left and right and remove the ducts. Remove the bar-shaped strut brace.

9. Remove the 7 screws securing the upper part of the air cleaner housing, then remove the bolts holding the lower part of the housing. Press towards the rear and lift out.

10. Remove the upper radiator hose and clips.

11. Disconnect the electric fan bolts,

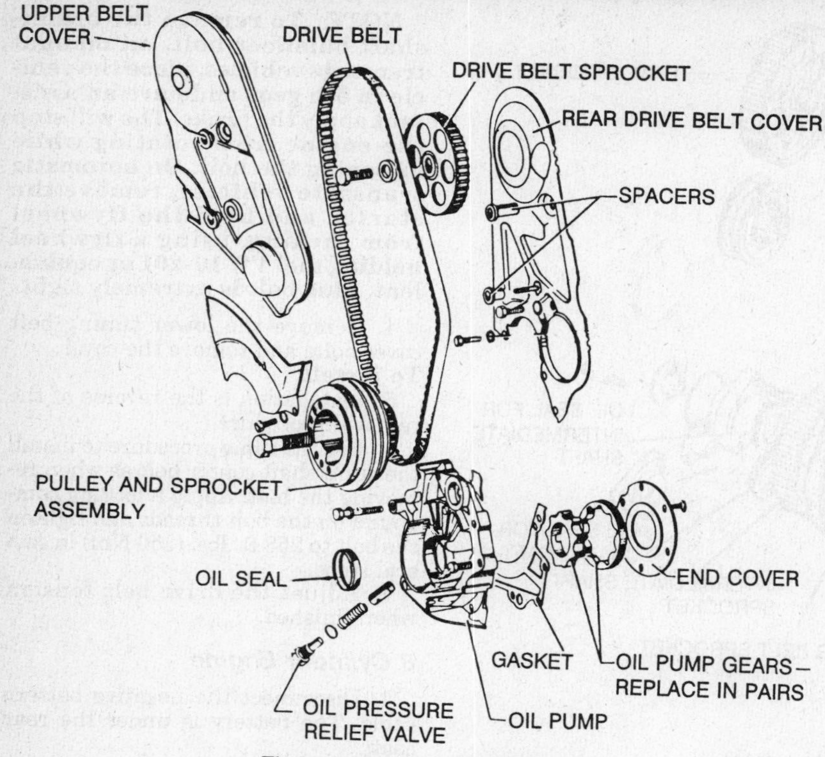

Timing belt on 5 cylinder engine

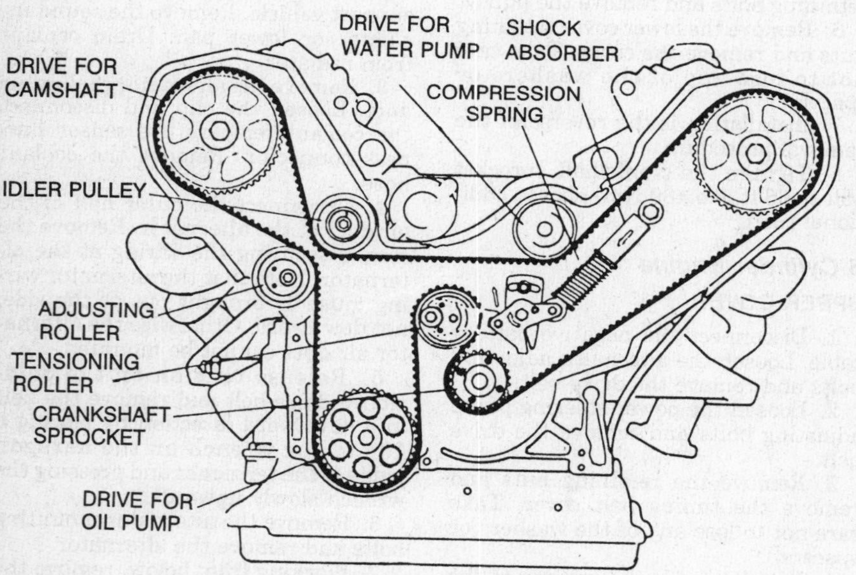

Timing belt layout on V8 engine

lift the fan assembly out and lay to one side, wiring still connected.

12. Remove the bolts securing the supporting clamp for the lower radiator hose and take off the upper part. Disconnect the radiator hose at the thermostat housing. Swing the radiator hose to right at rear.

13. Remove the fan shroud by unscrewing the bolts for the viscous fan at the top. Remove the bolts securing the viscous fan. A 2 pin spanner may be needed to hold the fan hub. Note that this is a left hand thread. Turn to the right to loosen. Lift out the fan and shroud together.

14. Disconnect the engine support at the front. Note any shims. The same thickness shims must be used at installation. Remove the radiator hose. At installation, install the radiator hose first.

15. Loosen the center bolt of the vibration damper by one turn. A special tool may be needed to hold the damper from turning. This bolt was installed to over 250 ft. lbs. (340 Nm) torque and will be difficult to remove.

16. Remove the bolts securing the left side toothed belt cover or guard.

17. Remove the lower part of the supporting clamp for the lower radiator hose. Turn the tensioner for the poly-ribbed accessory drive belt in the loosening direction and insert an appropriate size holding pin in the hole hole provided.

18. Remove the bolts securing the right side toothed belt cover, with the exception of the top bolt. Remove the tensioner holding pin. Remove the belt cover top screw and remove the cover or guard. Carefully lift the guard from the bottom to avoid damaging the radiator.

To install:

19. Installation is the reverse of the removal process. The crankshaft damper center bolt must be torque to 258 ft. lbs. (350 Nm). A holding tool may be needed to keep the crankshaft from turning.

20. The viscous fan hub uses a left hand thread. Turn right to loosen, left to tighten.

OIL SEAL REPLACEMENT

Camshaft Seal

EXCEPT 8 CYLINDER ENGINE

1. Disconnect the negative battery cable. Set engine to TDC. Make sure crankshaft and cam timing marks are aligned.
2. Remove timing belt.
3. Hold camshaft from turning, and remove cam drive sprocket by tapping from behind. Take care not to lose the cam drive key.
4. Pry out the old oil seal. In some cases it may be easier to remove the front camshaft bearing cap.

To install:

5. Installation is the reverse of the removal procedure. Lubricate the seal lip with clean engine oil. Reinstall the front bearing cap if removed. Torque hold-down nuts to 15 ft. lbs. (20 Nm).
6. Use a suitable socket as a driver and tap the new seal into position.
7. Install the cam drive sprocket with the drive key. Check the timing marks. Torque center bolt to 60 ft. lbs. (81 Nm).
8. Install cam belt and cover.

8 CYLINDER ENGINE

1. Disconnect the negative battery cable. Remove the toothed timing belt.

2. Remove the camshaft sprockets.
3. Remove the left and right belt guards.

NOTE: The toothed belt guard on the left-hand side contains 2 sealing rings and 1 shaft seal. The toothed belt guard on the right side contains 1 sealing ring and 1 shaft seal. If one shaft seal is leaking, the seals on both sides must be replaced.

4. Drive out the shaft seal.

To install:

5. When driving in the new seals, make sure the shaft seal is flush with the front edge of the belt guard. Replace the sealing rings.
6. Guide the camshaft sprocket into place.
7. Install the cam belt. Check timing marks on crankshaft and cam sprockets.
8. Clean the sealing surface at the rear of the guard and on the cylinder head. After installing the new sealing rings, apply a thin coat of sealant to both surfaces. Install the rear toothed belt cover and torque the bolts to 7 ft. lbs. (10 Nm).

Crankshaft Seal

4 CYLINDER ENGINE

1. Disconnect the negative battery cable. Remove timing belt.
2. Remove the crankshaft timing belt sprocket assembly.
3. A special oil seal extractor is recommended to pull the seal from the front cover flange.
4. Install the new seal after lubricating with engine oil. Press the seal in to the stop.
5. Reinstall the crankshaft sprocket and timing belt.

5 CYLINDER ENGINE

The oil seal is a part of the oil pump.

1. Disconnect the negative battery cable. Remove the timing belt.
2. Using a small prybar, pry the oil seal from the oil pump.
3. Clean out the seal seat. Using a new seal, lubricate the lip with engine oil. Using a suitable socket, drive the seal in to the seal seat.

NOTE: When installing a new seal, be careful not to damage the lip of the seal.

4. Reinstall the timing belt. Torque the crankshaft pulley bolt to 253 ft. lbs. (343 Nm).

8 CYLINDER ENGINE

1. Disconnect the negative battery cable. Remove the toothed cam drive belt.
2. Remove the vibration damper.
3. A special oil seal extractor is recommended to pull the seal from the front cover flange.
4. Install the new seal after lubricating with engine oil. Press the seal in only until flush. Note that if the crankshaft shows signs of scoring, press the sealing ring in completely.
5. Reinstall the crankshaft sprocket and timing belt.

Timing Belt and Tensioner

ADJUSTMENT

4 Cylinder Engine

1. Holding the large bolt on the tensioner pulley, loosen the small nut and turn the tensioner clockwise to tighten and counterclockwise to loosen.
2. The belt is correctly tensioned when it can be twisted 90 degrees midway between the camshaft and the intermediate shaft drive sprockets.

5 Cylinder Engine

1. Loosen the water pump adjusting bolts and rotate the pump clockwise to tighten and counterclockwise to loosen.
2. The belt is correctly tensioned when it can be twisted 90 degrees midway between the camshaft drive sprocket and the water pump.

8 Cylinder Engine

1. Remove the timing belt covers.
2. Perform the basic setting of the toothed belt tensioner by turning the idler pulley eccentric clockwise. A special turning tool may be needed. Turn the eccentric and measure the shock-absorber shaped damper length. Measure the overall length, not counting the mounting eyes. Measure the length of the barrel. Turn the eccentric until the damper length is 5.11–5.23 in. (130–133mm). Tighten the idler pulley eccentric to 18 ft. lbs. (24 Nm).
3. Remove any camshaft and crankshaft locking tool previously installed. Turn engine at least 2 turns. Tighten the vibration damper center bolt to 258 ft. lbs. (350 Nm). Check the damper length and if necessary, readjust the idler pulley.
4. Reinstall the cam belt cover.

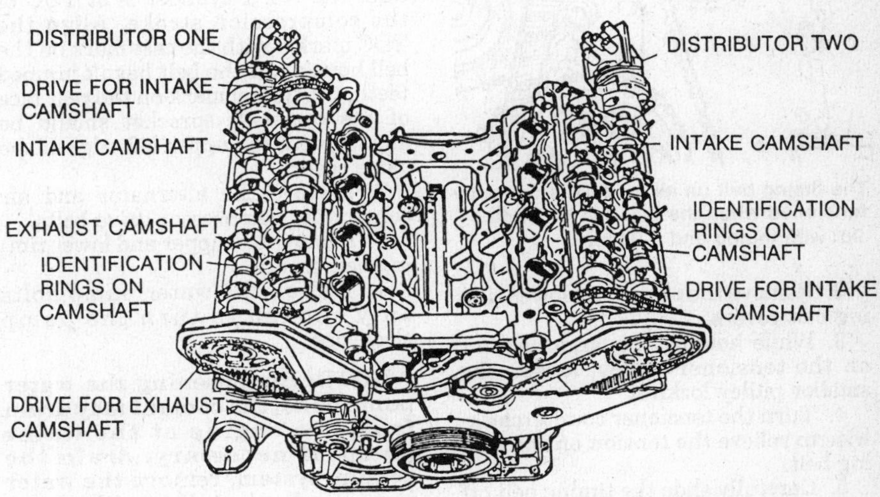

Camshaft layout train on V8 engine

REMOVAL & INSTALLATION

4 Cylinder Engine

1. Disconnect the negative battery cable. Using the large bolt on the crankshaft sprocket, rotate the engine until the No. 1 cylinder is at TDC of the compression stroke. At this point, both valves will be closed and the **0** mark on the flywheel will be aligned with the pointer on the bell housing. If the belt hasn't jumped teeth the timing mark on the rear face of the camshaft sprocket should be aligned with the upper left edge of the valve cover.

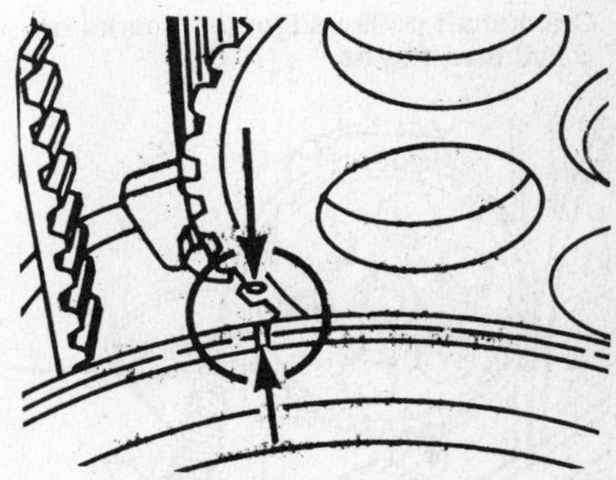

Crankshaft pulley and intermediate sprocket alignment on 4 cylinder engine

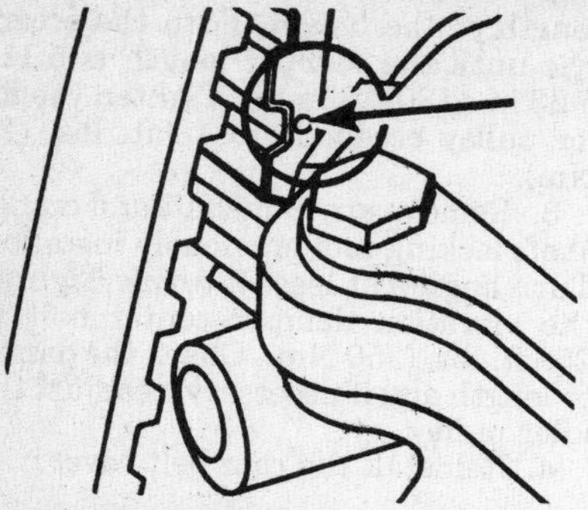

Camshaft sprocket alignment with cylinder head on 4 and 5 cylinder engines

Crankshaft pulley alignment marks on 5 cylinder engine

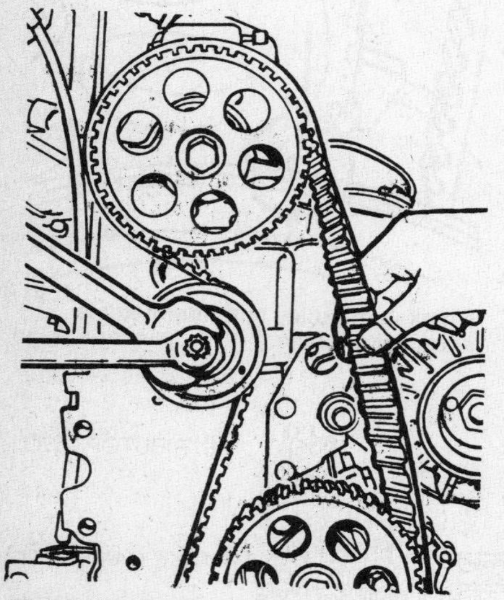

The timing belt on all models is correctly tensioned when the belt can be twisted 90° with thumb and 1 finger

2. Remove the upper and lower timing belt covers.
3. While holding the large hex nut on the tensioner pulley, loosen the smaller pulley locknut.
4. Turn the tensioner counterclockwise to relieve the tension on the timing belt.
5. Carefully slide the timing belt off of the sprockets and remove the belt.

To install:

6. If the engine had moved or jumped timing, use the large bolt on the crankshaft sprocket, rotate the engine until the No. 1 cylinder is at TDC of the compression stroke. At this point, both valves will be closed and the **0** mark on the flywheel will be aligned with the pointer on the bell housing. Rotate the camshaft until the timing mark on the rear face of the camshaft sprocket is aligned with the upper left edge of the valve cover.
7. Install the crankshaft pulley and check that the notch on the pulley is aligned with the mark on the intermediate shaft sprocket. If not, rotate the intermediate shaft until they align.

NOTE: If the timing marks are not correctly aligned with the No. 1 piston at TDC of the compression stroke and the belt is installed, valve timing will be incorrect. Poor performance and possible engine damage can result from the improper valve timing.

8. Remove the crankshaft pulley. Note the pulley location on the crankshaft sprocket so it can be replaced in the same position. Hold the large nut on the tensioner pulley and loosen the smaller locknut. Turn the tensioner counterclockwise to loosen and install the timing belt.
9. Slide the timing belt onto the sprockets and adjust the belt tension. The timing belt tension is correct when the belt can be twisted 90 degrees.
10. The remainder of the installation is the reverse of the removal procedure.

5 Cylinder Engine

1. Disconnect the negative battery cable. Using the large bolt on the crankshaft sprocket, rotate the engine until the No. 1 cylinder is at TDC of the compression stroke. Align the TDC mark **0** with the cast mark on the bell housing. If the belt hasn't jumped teeth, the timing mark on the rear face of the camshaft sprocket should be aligned with the upper left edge of the valve cover.
2. Remove the alternator and air conditioner compressor drive belts.
3. Remove the upper and lower timing belt covers.
4. Loosen the water pump bolts only enough to turn the pump clockwise.

NOTE: By loosening the water pump bolts, the coolant may drain from the engine at the water pump. If necessary, drain the cooling system, remove the water pump and reinstall it with a new O-ring.

5. Slide the timing belt off the sprockets.

To install:

6. If necessary, turn the camshaft until the notch on the back of the sprocket is in line with the left side edge of the cylinder head gasket surface.
7. If necessary, align the TDC **0** mark with the with the lug cast on the bell housing.
8. Install the timing belt and turn the water pump counterclockwise to tighten the belt. Tighten the water pump bolts to 15 ft. lbs. (20 Nm).

NOTE: The timing belt is correctly tensioned when it can be twisted 90 degrees along with the straight run between the camshaft sprocket and water pump. Belt must not be jammed between the oil pump and sprocket when installing the vibration damper.

9. Install the timing belt covers and tighten the bolts to 7 ft. lbs. (10 Nm).
10. Install the alternator and air conditioning compressor belts. These belts are correctly tensioned when they can be depressed ⅜ in. along their longest straight run.

8 Cylinder Engine

1. Disconnect the negative battery cable. The battery is under the rear seat.
2. The toothed cam drive belt must first be removed. Remove the coolant expansion tank cap. Raise and safely support vehicle. Remove the sound insulator or lower pan. Drain coolant from radiator.
3. Remove the alternator cooling air duct. Loosen the clip and disconnect the coolant temperature sensor harness connector. Remove the coolant hose.
4. Disconnect the outer half of the air duct at the alternator. Remove the screws securing the wiring at the alternator.
5. Release tension on the poly-ribbed drive belt and remove the belt in a downward direction, by placing a 13mm box wrench on the hexagon guide of the tensioner and pressing the wrench slowly upwards.
6. Remove the alternator mounting bolts and remove the alternator.
7. Working from below, remove the 3 bolts for the toothed belt guard.
8. Remove the bolts securing the bracket for the air intake ducts on the left and right and remove the ducts. Remove the bar-shaped strut brace.
9. Remove the 7 screws securing the upper part of the air cleaner housing, then remove the bolts holding the lower part of the housing. Press towards the rear and lift out.

10. Remove the upper radiator hose and clips.
11. Disconnect the electric fan bolts, lift the fan assembly out and lay to one side, wiring still connected.
12. Remove the bolts securing the supporting clamp for the lower radiator hose and take off the upper part. Disconnect the radiator hose at the thermostat housing. Swing the radiator hose to the right.
13. Remove the fan shroud by unscrewing the bolts for the viscous fan at the top. Remove the bolts securing the viscous fan. A 2 pin spanner may be needed to hold the fan hub. Note that this is a left hand thread. Turn to the right to loosen. Lift out the fan and shroud together.
14. Disconnect the engine support at the front. Note any shims. The same thickness shims must be used at installation. Remove the radiator hose. At installation, install the radiator hose first.
15. Loosen the center bolt of the vibration damper by one turn. A special tool may be needed to hold the damper from turning. This bolt was installed to over 250 ft. lbs. (340 Nm) torque and will be difficult to remove.
16. Remove the bolts securing the left side toothed belt cover or guard.
17. Remove the lower part of the supporting clamp for the lower radiator hose. Turn the tensioner for the poly-ribbed accessory drive belt in the loosening direction and insert an appropriate size holding pin in the hole provided.
18. Remove the bolts securing the right side toothed belt cover, with the exception of the top bolt. Remove the tensioner holding pin. Remove the belt cover top screw and remove the cover or guard. Carefully lift the guard from the bottom to avoid damaging the radiator.
19. Disconnect the ignition cables at the coils. Remove both distributor caps. Turn the engine to TDC. It may be necessary to temporarily install the damper to align the timing marks. Check that the distributor rotor is pointing to the mark on the housing. If not, turn the crankshaft 1 additional turn. Remove both distributors.
20. Remove the stop plate at the toothed belt tensioner. Disconnect the shock-absorber shaped damper at the top bolt.
21. Remove the belt from the tensioning idler pulley (with eccentric). Pulley is on right side of engine. Take the belt off both camshaft sprockets.
22. A special holding tool is available that is installed on the back of the camshafts. It fits the locating pin on the distributor flanges. If necessary, use a special hook wrench tool to turn the camshaft until the pins latch into the special holding tool. Secure the special tool with the distributor mounting bolts.
23. At the camshaft sprocket end, loosen the mounting bolts 2 turns. Using a plastic hammer, tap the edge of the camshaft sprockets loose.

NOTE: After the sprockets are removed note the grooves machined in the camshaft ends. Woodruff keys must not be installed in the camshaft sprocket/camshaft connection. Unlike the other engines, this engine does not use cam keys.

24. Remove the vibration damper which was previously temporarily reinstalled to line up TDC timing marks. A puller can be used. Unscrew 2 opposing bolts of the 4 bolts connecting the vibration damper and toothed belt sprocket. Use a puller in these holes.
25. Remove the toothed belt.

To install:

26. Installation is the reverse of the removal procedure. Before installing a new belt, the rollers and tensioners must be checked for dirt, rough running and ease of rotation. Clean or replace rollers and tensioners, as necessary.
27. Fit the toothed belt at the crankshaft and install the vibration damper with the belt on the crankshaft. Apply thread locking compound to the center bolt and tighten using hand wrench. A holding tool may be required to keep the crankshaft from turning. It is a must to ensure that the TDC mark on the vibration damper is aligned with the TDC pointer before and after tightening the camshaft sprockets.
28. Guide the toothed belt into position and install the idler pulley with eccentric. Snug the nut enough to hold the eccentric in place but do not fully tighten it yet.
29. Tighten the damper top bolt to 18 ft. lbs. (24 Nm). Engage the damper to the tensioner lever by pressing the lever downward.
30. Perform the basic setting of the toothed belt tensioner by turning the idler pulley eccentric clockwise. A special turning tool may be needed. Turn the eccentric and measure the damper length. Measure the overall length of the barrel not counting the mounting eyes. Turn the eccentric until the damper barrel length is 5.11–5.23 in. (130–133mm). Tighten the idler pulley eccentric to 18 ft. lbs. (24 Nm). Tighten the camshaft sprockets to 33 ft. lbs. (45 Nm).
31. Remove any camshaft and crankshaft locking tool previously installed. Turn engine at least 2 turns. Tighten the vibration damper center bolt to 258 ft. lbs. (350 Nm). Check the damper length and if necessary, readjust the idler pulley.
32. The remaining installation for the toothed cam drive belt is performed in reverse order of removal.

Timing Sprockets

REMOVAL & INSTALLATION

Except 8 Cylinder Engine

All timing belt sprockets except 8 cylinder engines, are located by keys on their respective shafts. Each sprocket is retained by a bolt. To remove any or all of the sprockets, first remove the timing belt cover(s) and timing belt.

1. Disconnect the negative battery cable. Remove the center retaining bolt for the sprocket.
2. Pull the sprocket off of the shaft.
3. If the sprocket is sticking on the shaft, use a gear puller or tap lightly with a plastic hammer. Do not hammer on the sprocket or damage may occur.
4. Remove the sprocket, being careful not to lose the key.
5. Installation is in the reverse order of removal.

NOTE: Always check valve timing after removing the drive sprockets.

8 Cylinder Engine

1. Disconnect the negative battery cable. Remove timing cover, toothed timing belt and both distributors. Note that the timing belt drives the exhaust camshaft. A chain from the exhaust cam drives the intake cam.
2. A special holding tool is available that is installed on the back of the camshafts. It fits the locating pin on the distributor flanges. If necessary, use a special hook wrench tool to turn the camshaft until the pins latch into the special holding tool. Secure the special tool with the distributor mounting bolts.
3. At the camshaft sprocket end, loosen the mounting bolts 2 turns. Using a plastic hammer, tap the edge of the camshaft sprockets slightly loose.

NOTE: After the sprockets are removed, note the grooves machined in the camshaft ends. Woodruff® keys must not be installed in the camshaft sprocket/camshaft connection.

4. When assembling the sprockets to the camshafts, make sure the timing marks found on the backside of the chain sprockets are still aligned. They should align next to each other at the 3 o'clock and 9 o'clock positions.

Camshaft

REMOVAL & INSTALLATION

4 Cylinder Engine

80 AND 90

1. Disconnect the negative battery cable. Remove the upper drive belt, vacuum lines to valve cover and valve cover.
2. Using the large bolt on the crankshaft sprocket, rotate the engine until the No. 1 cylinder is at TDC of the compression stroke. At this point, both valves will be closed and the **0** mark on the flywheel will be aligned with the pointer on the bell housing. If the belt hasn't jumped teeth the timing mark on the rear face of the camshaft sprocket should be aligned with the upper left edge of the valve cover.
3. Remove the timing belt from the camshaft sprocket.
4. Remove the camshaft timing belt sprocket, take care not to loose Woodruff key.
5. First remove bearing caps No. 1 and 3. Bearing cap No. 1 is located at the sprocket. Next remove bearing caps No. 2 and 5, alternately and diagonally. Bearing cap No. 5 is on the opposite end from the cam sprocket. There is no bearing cap No. 4.
6. Remove the camshaft from the cylinder head.

To install:

7. Lubricate the camshaft journals, lobes and contact faces of the caps with assembly lube before reinstallation.
8. Replace the camshaft oil seal in the cylinder head.

NOTE: The bearing caps are offset. Before installing the camshaft, set the bearing caps into position. Check that they are facing in the correct direction. The numbers on the bearing caps are not always on the same side.

9. Install the bearing caps in the proper order, observing the off-center position.
10. Install bearing caps No. 2 and 5 and tighten to 15 ft. lbs. (20 Nm).
11. Install bearing caps No. 1 and 3 and tighten to 15 ft. lbs. (20 Nm).
12. Mount camshaft sprocket and tighten to 59 ft. lbs. (80 Nm).
13. Installation of the remaining components is in the reverse order of removal. When installing the crankshaft timing belt sprocket, be sure the lug on the sprocket is properly installed into the slot.
14. Coat new camshaft seal with oil and press into cylinder head until flush.

NOTE: Always recheck the valve timing and valve clearance after the camshaft has been removed.

5 Cylinder Engine

80 AND 90

1. Disconnect the negative battery cable. Remove the upper drive belt cover, valve cover and upper part of intake manifold, if necessary.
2. Using the large bolt on the crankshaft sprocket, rotate the engine until the No. 1 cylinder is at TDC of the compression stroke. Align the TDC mark **0** with the cast mark on the bell housing. If the belt hasn't jumped teeth the timing mark on the rear face of the camshaft sprocket should be aligned with the upper left edge of the valve cover.
3. Remove the timing belt from the camshaft sprocket. Remove the camshaft sprocket.
4. Remove bearing caps No. 1 and 3.
5. Diagonally loosen bearing caps No. 2 and 4 and remove the bearing caps.
6. Lift the camshaft out of the cylinder head.

To install:

7. When installing, lightly oil the camshaft and bearing journals, with clean engine oil.
8. Position the caps on the same journals from which they were removed.
9. Install bearing caps No. 2 and 4. Tighten alternately and diagonally to 15 ft. lbs. (20 Nm).
10. Install bearing caps No. 1 and 3 and tighten to 15 ft. lbs. (20 Nm).
11. Install the camshaft sprocket and timing belt. Install the valve cover. The camshaft sprocket bolt is tightened to 59 ft. lbs. (80 Nm).

100 AND 200

1. Disconnect the negative battery cable. Remove the upper drive belt cover, valve cover and upper part of intake manifold, if necessary.
2. Using the large bolt on the crankshaft sprocket, rotate the engine until the No. 1 cylinder is at TDC of the compression stroke. Align the TDC mark **0** with the cast mark on the bell housing. If the belt hasn't jumped teeth, the timing mark on the rear face of the camshaft sprocket should be aligned with the upper left edge of the valve cover.
3. Remove the timing belt from the camshaft sprocket. Remove the camshaft sprocket.
4. Diagonally loosen bearing caps No. 2 and 4 and remove the bearing caps.
5. Diagonally loosen bearing caps No. 1 and 3 and remove the bearing caps.

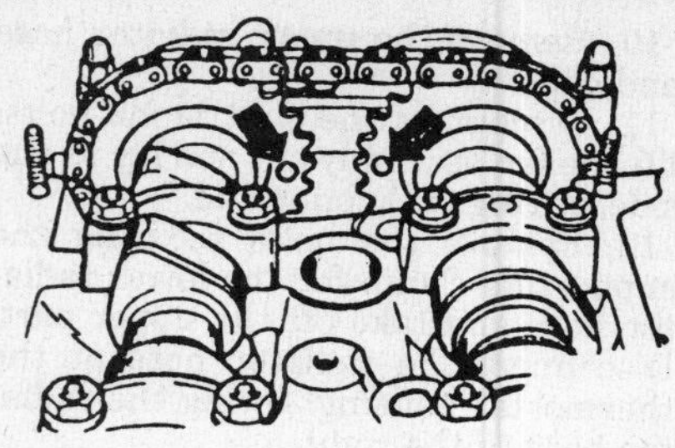

Cam gear alignment marks on V8 engine

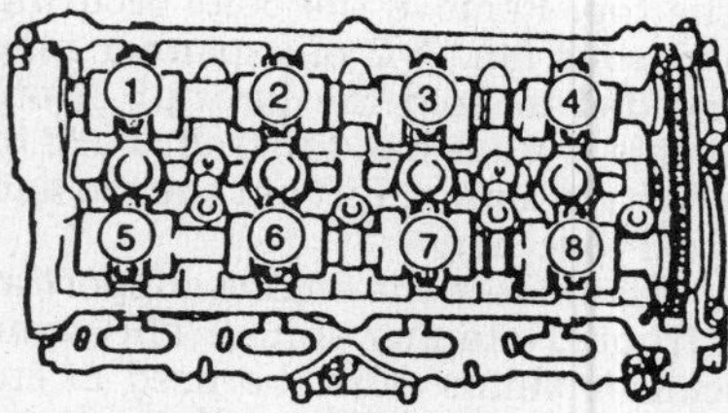

Cam bearing caps are numbered on V8 engine

6. Lift the camshaft out of the cylinder head.

To install:

7. When installing, lightly oil the camshaft and bearing journals, with clean engine oil.
8. Position the caps on the same journals from which they were removed.
9. Tighten the nuts of caps 2 and 4 until snug.
10. Tighten all nuts to 15 ft. lbs. (20 Nm).
11. Install the camshaft sprocket and timing belt. Install the valve cover. The camshaft sprocket bolt is tightened to 58 ft. lbs. (79 Nm).

8 Cylinder Engine

1. Disconnect the negative battery cable. Remove the toothed camshaft drive belt.
2. Remove the cylinder head cover.
3. Remove the camshaft sprocket.
4. On the exhaust side, remove the intermediate flange and bearing cap for the distributor. All bearing caps that are removed should be marked or identified so they can be installed in the same position. Do not mix bearing caps or install then backwards. Loosen the bearing cap in front of the chain, plus caps No. 2 and 3. Loosen bearing caps No. 1 and 4 alternately and in a diagonal sequence.
5. On the intake side, remove bearing caps No. 6 and 7. Loosen bearing caps No. 5 and 8 alternately and in a diagonal sequence.
6. Remove the cam. If a lifter is removed and is to be reused, it must go in the bore from which it was removed.

To install:

7. Installation is the reverse of re-

moval. Before assembly coat the moving parts with clean oil.

8. Install the camshafts with the chain so the markings on the chain sprockets are in alignment. The marks are on the back side of the chain sprocket. They should face each other at the 3 o'clock and 9 o'clock positions. Install the bearing caps so the stamped-on numbers can be read from the intake side.

9. Make sure the bearing caps are installed properly. They will cause parts failure if installed backwards. Torque the bearing caps to 11 ft. lbs. (15 Nm) in the following sequence:

a. caps 5 and 8 in a diagonal sequence.

b. caps 1 and 4 in a diagonal sequence.

c. caps 2 and 3 in a diagonal sequence.

d. caps 6 and 7 in a diagonal sequence.

Piston and Connecting Rod

POSITIONING

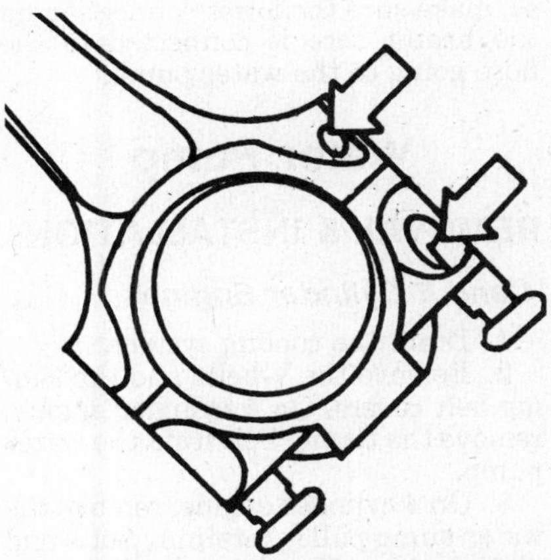

Align the forge marks when assembling the connecting rod caps

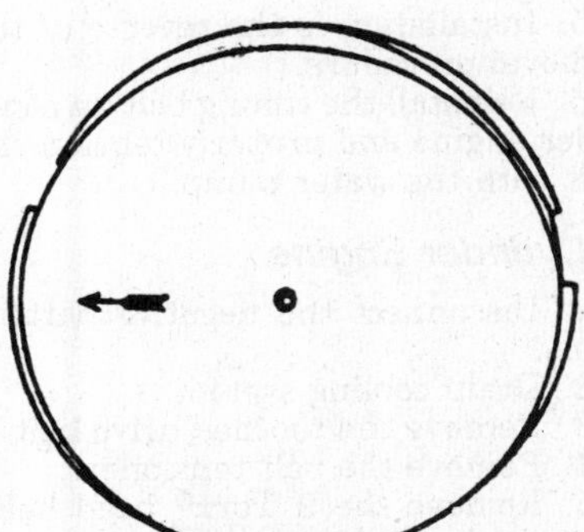

Arrow on piston faces front of vehicle

ENGINE LUBRICATION

Oil Pan

REMOVAL & INSTALLATION

1. Disconnect the negative battery cable. Raise and support the vehicle safely.

2. Drain the oil from the crankcase. Remove the cover plate from under the engine, if equipped.

3. If necessary, remove the 4 bolts from the subframe and lower the subframe. Remove the oil pan bolts while supporting the pan.

4. Lower the pan from the engine. Discard the gasket. Note that the 8 cylinder engine uses a 2 piece oil pan as well as a honeycomb baffle insert. Both an upper and lower pan gasket will be required.

To install:

5. Coat both sides of a new gasket with sealer and install the gasket and oil pan.

6. Tighten the pan bolts to 7 ft. lbs. (10 Nm) on 4 cylinder engines and 15 ft. lbs. (20 Nm) on 5 cylinder engines. On 8 cylinder engines, tighten the lower pan-to-upper pan bolts to 14 ft. lbs. (19 Nm), and the upper pan-to-block bolts to 18 ft. lbs. (24 Nm).

Oil Pump

REMOVAL & INSTALLATION

4 Cylinder Engine

1. Disconnect the negative battery cable. Raise and safely support vehicle. Drain the oil and remove the oil pan.

2. Remove the oil pump mounting bolts and pull the pump down and out of the engine.

3. Unscrew the 2 bolts and separate the pump halves.

4. Clean the lower half in solvent.

5. To remove the oil strainer for cleaning, bend out the metal rim of the oil strainer cover plate and remove it.

6. Examine the gears and the driveshaft for any wear or damage. Replace them, if necessary.

7. Reassemble the pump halves.

8. Prime the pump with oil and install in the reverse order of removal.

5 Cylinder Engine

1. Disconnect the negative battery cable. Loosen and remove the crankshaft pulley bolt.

2. Remove the timing belt covers.

3. Loosen the water pump bolts and turn the pump body clockwise.

4. Remove the timing belt and V-belt pulley with the timing belt sprocket.

5. Remove the dipstick. Raise and safely support vehicle. Drain the engine oil.

6. Remove the front bolts on the subframe and remove the oil pan.

7. Remove the oil suction pipe from the base of the oil pump and bracket to the engine block.

8. Remove the oil pump bolts and remove the oil pump from the front of the engine.

9. Installation is the reverse of the removal procedure.

8 Cylinder Engine

1. Disconnect the negative battery cable. Pull out dipstick. Raise and safely support vehicle.

2. Drain engine oil

3. Remove oil pan bolts and pull off pan assembly.

4. Remove bolts securing oil pump and pull out, disengaging from drive.

5. Installation is the reverse of the removal procedure.

Rear Main Bearing Oil Seal

REMOVAL & INSTALLATION

The rear main oil seal is located at the rear of the engine block. It can be found in a housing or flange behind the flywheel. To replace the seal, remove the transaxle or pull the engine.

1. Disconnect the negative battery cable. Remove the transaxle.

2. Remove the flywheel.

3. Using a suitable tool, pry the old seal out of its housing.

4. To install, lightly oil the replacement seal and press it into place.

NOTE: Be careful not to damage the seal or score the crankshaft.

5. Install the flywheel and the transaxle.

ENGINE COOLING

Radiator

REMOVAL & INSTALLATION

4 Cylinder Engine

NOTE: The 80 and 90 series ve-

hicles use a dual fan electric/belt driven assembly. Replacement is similar to the other 4 cylinder engines.

1. Drain the cooling system.
2. If equipped with air conditioning, remove the grille and detach the condenser from the radiator. Leave refrigerant hoses attached, if possible. If not, properly discharge the air conditioning system.
3. Remove the upper and lower radiator hoses, the expansion tank supply hose and the expansion tank vent hose. Being careful not to crimp them, tie all hoses back aside.
4. Disconnect the wiring at the temperature switch, 2 switches if air conditioned, and the rear of the fan motor.
5. Unscrew the fan shroud retaining bolts and remove the fan, motor and shroud as an assembly.
6. Unscrew the radiator retaining bolts and remove the radiator.
7. Installation is in the reverse order of removal.

Except 4 Cylinder Engine

1. Drain the cooling system.
2. Remove the 3 pieces of the radiator cowl and the fan motor assembly. Take care in removing the fan motor connectors to avoid bending them.
3. Remove the upper and lower radiator hoses and the coolant tank supply hose.
4. Disconnect the coolant temperature switch located on the lower right side of the radiator.
5. Remove the radiator mounting bolts and lift out the radiator.
6. Installation is the reverse of removal. Torque radiator mounting bolts to 14 ft. lbs. and cowl bolts to 7 ft. lbs. (10 Nm).

Electric Cooling Fan

TESTING

Factory procedures call for using the factory tester to check the entire system at the instrument panel multi-pin connectors. It is possible to do a basic check of the sensors with an ohmmeter.

1. Check the radiator coolant temperature sensor. It should have a resistance of approximately 360 ohms when cold and 70 ohms when hot.
2. Check the cylinder head coolant temperature sensor. It should also have a resistance of approximately 360 ohms when cold and 70 ohms when hot.
3. If the sensors check out and the electric cooling fan still does not run, check the wiring and connections. Jumper wires with battery voltage can also be connected to the fan motors to test operation.

REMOVAL & INSTALLATION

1. Disconnect the battery negative cable.
2. Remove the 3 pieces of the radiator cowl.
3. Remove the fan motor assembly. Take care in removing the fan motor connectors to avoid bending them.
4. Installation is the reverse of the removal process.

Heater Core

REMOVAL & INSTALLATION

Except V8 Quattro

1. Disconnect the negative battery cable.
2. At the radiator, pull off the bottom hose and drain the coolant into a container for reuse.
3. Remove the heater hoses from the heat exchanger.
4. At the heater assembly control valve, disconnect the control wire.
5. Remove the console. Remove the left and the right heater covers from below the dashboard. On 80 and 90 model vehicles, remove the instrument panel.
6. Properly discharge the air conditioning system and remove the refrigerant lines from the evaporator.
7. At the heater control unit, pull off the control knobs.
8. Remove the trim plate from the heater control unit.
9. At the heater control unit, remove the retaining screws and the center cover.
10. Remove the heater air ducts and the heater assembly retaining springs.
11. Remove the air plenum from the cowl and the heater assembly from the vehicle.
12. Separate the heater unit and remove the heater core.

To install:

13. Install the heater core and reassemble the heater unit. Make sure the seals are in good condition.
14. Install the assembly into the vehicle and connect the ducts.
15. Connect the hoses and install the controls and instrument panel.
16. Refill the cooling system and evacuate and recharge the air conditioning system, if equipped.

V8 Quattro

1. Disconnect the negative battery cable.
2. Matchmark hood hinges and remove hood.
3. Remove the windshield wiper assembly.
4. Remove the cap from the engine coolant overflow bottle.
5. Remove the heater retaining band.
6. Remove the vacuum hoses from the vacuum servo motors.
7. Clamp the heater hoses and disconnect them from the heater core.
8. Remove the retainers between the body and heater and remove the heater box.
9. Remove the silicone rubber sealant from the heater core inlet/outlet area and separate the housing halves. Remove the heater core from the housing.

To install:

10. Installation is the reverse of the removal procedure. Before installing the fresh air blower guides, lubricate with petroleum jelly.
11. Apply gasket compound around the heater box before installing the heater core to seal box.
12. After installing the heater core, fill the opening between the heater core and the housing with silicone sealant.
13. When connecting the water hoses, make sure the lower connection on the heater core is connected to the hose going to the water pump.

Water Pump

REMOVAL & INSTALLATION

4 and 5 Cylinder Engines

1. Drain the cooling system.
2. Remove the V-belts and the timing belt covers. On 5 cylinder engine, remove the timing belt from the water pump.
3. On 4 cylinder engine, remove the water pump pulley retaining bolts and remove the pulley. Remove the pump retaining bolts. Take note of various lengths and locations. Turn the pump slightly and lift from engine block.
4. Always replace the old gasket or O-ring.
5. Installation is the reverse of the removal procedure.
6. Reinstall the timing belt on 5 cylinder engine and properly tension the belt with the water pump.

8 Cylinder Engine

1. Disconnect the negative battery cable.
2. Drain cooling system.
3. Remove the toothed drive belt.
4. Remove the belt tensioner.
5. Remove the 9 Torx® head bolts and remove water pump.

To install:

6. Installation is the reverse of the

removal process. Always use a new gasket. Torque the water pump bolts to 7 ft. lbs. (10 Nm).

7. When refilling the cooling system, fill the expansion tank with new coolant to the maximum mark. Close the expansion tank and warm the engine until the radiator cooling fan cycles.

8. Check the coolant level and if necessary, top off. When the engine is at normal temperature, the level should be slightly over the maximum mark. When the engine is cold, between the maximum and minimum marks.

Thermostat

REMOVAL & INSTALLATION

4 Cylinder Engine

The thermostat is located in the lower radiator hose neck on the bottom of the water pump housing. The cooling system is drained by removing the thermostat housing.

1. Loosen but do not remove the 2 thermostat housing bolts from the lower water pump neck. Have a large catch pan ready.
2. When the system is drained, remove the bolts.
3. Move the neck, with the hoses attached, aside.
4. Carefully, pry the thermostat out of the water pump housing.

To install:

5. Install the new O-ring and thermostat with the spring towards the engine. Do not use any gasket sealer on rubber gaskets or O-rings.
6. Install the housing, being careful to properly seat the thermostat and O-ring. Torque the bolts to 7 ft. lbs. (10 Nm).
7. Refill the cooling system.

5 Cylinder Engine

The thermostat is located in the lower radiator hose neck, on the left side of the engine block, behind the water pump housing.

1. Drain the cooling system by removing the lower radiator hose.
2. Remove the 2 retaining bolts and remove the thermostat housing. Have a catch pan ready to catch the coolant that is still in the head.
3. Carefully pry the thermostat out of the head.

To install:

4. Install a new O-ring and thermostat with the spring towards the engine. Do not use any gasket sealer on rubber gaskets or O-rings.
5. When installing the housing, be careful to properly seat the O-ring. Torque the bolts to 7 ft. lbs. (10 Nm).
6. Reconnect the hose and refill the system.

8 Cylinder Engine

The thermostat is located at the front right-hand side of the engine below the intake manifold.

1. Drain the cooling system.
2. Remove the 2 retaining bolts from the thermostat housing.

NOTE: It is not necessary to disconnect the lower radiator hose. But removing the hose from the thermostat yoke may ease installation.

3. Remove the thermostat housing.
4. Carefully, pry the thermostat from the engine.

To install:

5. Install the new thermostat with the breather valve at the top or 12 o'clock position. Use a new gasket or O-ring.
6. Install the radiator hose, if removed. Refill with coolant.

Cooling System Bleeding

After working on the cooling system, even to replace the thermostat, the system should be bled. Air trapped in the system will prevent proper filling and leave the radiator coolant level low, causing a risk of overheating.

1. To bleed the system, start with the system cool, the radiator cap off and the radiator filled to about an 1 in. below the filler neck.
2. Start the engine and run it at slightly above normal idle speed. This will insure adequate circulation. If air bubbles appear and the coolant level drops, fill the system with an anti-freeze/water mixture to bring the level back to the proper level.
3. Run the engine this way until the thermostat opens. When this happens, coolant will move abruptly across the top of the radiator and the temperature of the radiator will suddenly rise.
4. At this point, air is often expelled and the level may drop quite a bit. Keep refilling the system until the level is near the top of the radiator and remains constant.
5. If equipped with an overflow tank, fill the radiator right up to the filler neck. Replace the radiator filler cap.

ENGINE ELECTRICAL

NOTE: Most vehicles are equipped with theft protected radios, which cannot be operated if power to the radio is interrupted. Before disconnecting the battery cables, obtain the security code. Never disconnect any electrical connector with the ignition key ON unless specified in the repair procedure, or damage to electronic components may result.

Distributor

REMOVAL

1. Disconnect the wiring harness connector from the distributor cap.

NOTE: V8 Quattro vehicles have 2 distributors, driven off the exhaust camshafts.

2. Unclip and remove the distributor cap and static shield with the spark plug wires still attached.
3. Disconnect and tag the vacuum lines at the distributor, if equipped.
4. Note the position of the rotor in relation to the distributor housing. Scribe a mark on the distributor and engine block for installation. Matchmark the tip of the rotor to the engine. Note the approximate position of the vacuum advance unit in relation to the engine.
5. Remove the distributor hold-down bolt and clamp.
6. Lift the distributor assembly from the engine.

INSTALLATION

Timing Not Disturbed

1. With the rotor pointing in the same direction as when removed, insert the distributor into the engine.
2. Once the distributor is seated into the engine, line up the marks on the distributor and engine with the metal tip of the rotor.
3. Make sure the vacuum advance unit, if equipped, is pointed in the same direction as it was pointed originally. If the marks on the distributor and the engine are lined up properly, this will be done automatically.

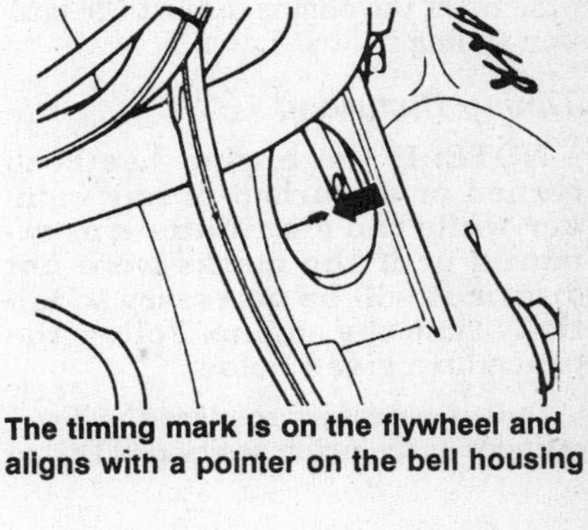

The timing mark is on the flywheel and aligns with a pointer on the bell housing

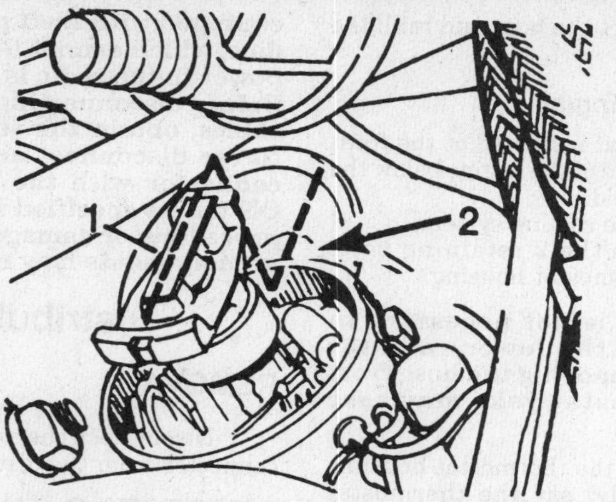

With the cap removed, turn the engine to align the rotor with the mark on the distributor body

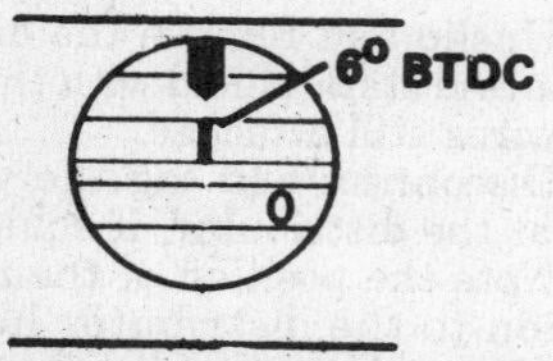

Timing mark alignment on 4 cylinder engine

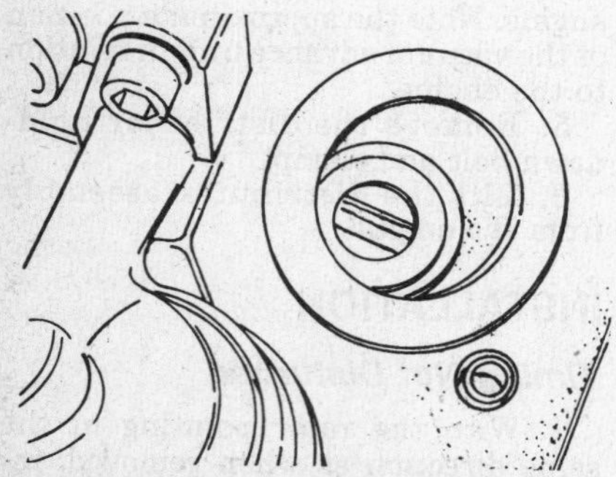

Oil pump dirve shaft must be parallel to the crankshaft on 4 cylinder engine

4. Install the distributor hold-down clamp and bolt.
5. Install the distributor cap and static shield.
6. Install the vacuum lines, if equipped.
7. Install the distributor wiring harness connector.
8. Start the engine. Adjust the ignition timing.

Timing Disturbed

NOTE: If the engine has been turned or disturbed in any manner while the distributor was removed or if the marks were not drawn, it will be necessary to initially time the engine. Follow the procedure given below.

1. It is necessary to place the No. 1 cylinder in the firing position (TDC) to correctly install the distributor. To locate this position, the ignition timing marks on the flywheel and the clutch housing are used.
2. Remove the spark plug from the No. 1 cylinder. Turn the crankshaft until the piston in the No. 1 cylinder is moving up on the compression stroke. This can be determined by placing a finger over the spark plug hole and feeling the air being forced out of the cylinder. Stop turning the engine when the timing mark on the flywheel is aligned with the lug on the flywheel housing.
3. Remove the timing belt cover.
4. Align the mark on the camshaft sprocket with the upper edge of the drive belt cover or with the upper edge of the valve cover gasket mounting surfaces.
5. On 4 cylinder engines, align the oil pump drive pinion lug so it aligns with the threaded hole.
6. Oil the distributor housing lightly where it bears on the cylinder block.
7. Install the distributor so the rotor tip, points to the mark on the distributor housing for the No. 1 cylinder.
8. On 4 cylinder engines, when the distributor shaft has reached bottom, move the rotor back and forth slightly until the drive lug on the oil pump shaft enters the slots cut into the end of the distributor shaft and the distributor assembly slides down into place.
9. Clean the distributor cap and check for signs of cracking or carbon tracks. Install the cap and continue the installation procedure.

Ignition Timing

ADJUSTMENT

Some tachometers, dwell-meters and oscilloscopes will not work with these ignition systems. Some test equipment may be damaged. Consult the manufacturer of the test equipment if there is any doubt.

NOTE: 1988–89 vehicles have an impedance transformer installed on top of the ignition control unit in place of an idle stabilizer as used on previous vehicles. Do not disconnect the transformer when checking the ignition timing.

All engines are timed by aligning the distributor housing with reference marks, no dynamic adjustment is possible. The electronic control unit will retard or advance the timing for each cylinder as required. With suitable equipment, the timing can be checked.

NOTE: V8 Quattro vehicles have 2 ignition coils with power stages and 2 distributors. They are both controlled by the Motronic ECU. Each coil and distributor is responsible for providing spark to 4 cylinders. One distributor is mounted on the back of each cylinder head. Both distributors are driven by lugs on the exhaust camshafts. A Hall sender is installed in the distributor mounted on the right cylinder head. The signal from this Hall unit identifies cylinder No. 1 for the start of the sequential fuel injection and cylinder selective knock regulation. Ignition timing is determined by the electronic control unit and cannot be adjusted.

Alternator

PRECAUTIONS

Several precautions must be observed with alternator equipped vehicles to avoid damage to the unit.

- If the battery is removed for any reason, make sure it is reconnected with the correct polarity. Reversing the battery connections may result in damage to the one-way rectifiers.
- When utilizing a booster battery as a starting aid, always connect the positive to positive terminals and the negative terminal from the booster battery to a good engine ground on the vehicle being started.
- Never use a fast charger as a booster to start vehicles.
- Disconnect the battery cables when charging the battery with a fast charger.
- Never attempt to polarize the alternator.
- Do not use test lamps of more

than 12 volts when checking diode continuity.

- Do not short across or ground any of the alternator terminals.
- The polarity of the battery, alternator and regulator must be matched and considered before making any electrical connections within the system.
- Never separate the alternator on an open circuit. Make sure all connections within the circuit are clean and tight.
- Disconnect the battery ground terminal when performing any service on electrical components.
- Disconnect the battery if arc welding is to be done on the vehicle.

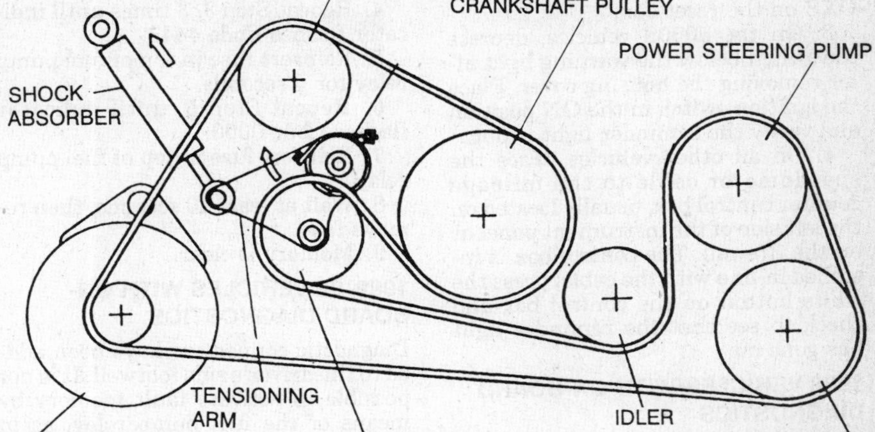

A single belt is used on V8 Quatro

BELT TENSION ADJUSTMENT

The drive belts are correctly tensioned when the longest span of belt between pulleys can be depressed 1/8–1/2 in. (3–13mm) using moderate thumb pressure. To adjust, loosen the slotted adjusting bracket bolt on the alternator. If the alternator hinge bolts are very tight, it may be necessary to loosen them slightly to move the alternator. Move the alternator in or out to obtain the correct tension. Tighten the adjusting bolt when finished.

V-belts under 39 inches in length should deflect about 1/8 inch (3mm). Belts over 40 inches long should deflect about 1/2 inch (13mm).

Poly-ribbed belt alignment is especially important on 8 cylinder engines. Check that the poly-ribbed belt between the air conditioner compressor and the hydraulic pump is in alignment front-to-back to prevent damage to the belt. If the 2 toothed belt pulleys are not in alignment, remove the bolts securing the ribbed belt pulley for the hydraulic pump. Using shims, available in sizes 0.020, 0.040 and 0.060 in. (0.5, 1.0 and 1.5mm) adjust the pulleys until they are in alignment. Note that the poly-ribbed belt used on 8 cylinder engines is designed to last the life of the engine.

REMOVAL & INSTALLATION

NOTE: If equipped with a 4 cylinder engine, the procedure can be done from the top of the vehicle. If equipped with a 5 cylinder engine, remove and install the alternator from below the vehicle.

1. Disconnect the negative battery cable.
2. Disconnect and tag the alternator wiring. On turbocharged engines, the cold air housing must be removed from the back of the alternator.
3. Remove the pivot bolt from the adjusting bracket.
4. Remove the drive belt.
5. Unbolt and remove the alternator.

NOTE: On some 4 cylinder engines, the top alternator mount has a bushing on the engine side of the mount. Check the condition of the bushing and replace if necessary, before installing the alternator.

To install:

6. Hold the alternator in position and install the pivot bolts.
7. Install the drive belt and adjusting bolt.
8. Adjust the belt tension.
9. Connect the electrical connections, making sure they are installed in their original locations.
10. Connect the negative battery cable.

Starter

REMOVAL & INSTALLATION

1. Disconnect the negative battery cable.
2. Raise and safely support the vehicle.
3. Disconnect and tag the starter wiring.
4. On 4 cylinder engine, remove the starter support bracket bolts. Remove the starter mounting bolts from the rear of the starter.
5. On 5 cylinder engine, 1 bolt goes through the transaxle with a nut on the end of the bolt.
6. Remove the starter from the engine.
7. Installation is the reverse of the removal procedure.

EMISSION CONTROLS

Please refer to "Emission Controls" in the Unit Repair section for system maintenance procedures. Due to the complex nature of modern electronic engine control systems, comprehensive diagnosis and testing procedures fall outside the confines of this repair manual. For complete information on diagnosis, testing and repair procedures concerning all modern engine and emission control systems, please refer to "Chilton's Guide to Fuel Injection and Electronic Engine Controls".

Emission Warning Lamps

Resetting

Oxygen Sensor Reminder

VEHICLES WITHOUT ON-BOARD DIAGNOSTICS

Every 30,000 miles a maintenance reminder light in the dashboard will come on. This is an indication that the emission system should be checked and that the oxygen sensor should be replaced.

1. To reset the non-turbocharged vehicles, remove the instrument panel cluster. Remove the switch cover near the **OXS** button. Push the switch to reset the light.

2. On turbocharged vehicles, lift the rear seat and push the button marked **OXS** on the reset box.

3. On the 5000S vehicles, depress the switch below the warning light after removing the housing cover. Place the ignition switch in the **ON** position and verify the reminder light is out.

4. On all other vehicles, trace the speedometer cable to the mileage counter control box, usually located on the left side of the instrument panel or on the firewall. The control box is installed in-line with the cable, press the white button on the control box and check to see that the reminder light has gone out.

1988 VEHICLES WITH ON-BOARD DIAGNOSTICS

The indicator light comes on whenever a fault develops which could cause the vehicle to fail an exhaust emission test. The light will remain on while driving as long as the fault exists. The light will go out after the fault has been repaired or no longer exists. Once the fault has been corrected, the permanent memory can be cleared the with the following procedures:

1. With the ignition **OFF**, insert a fuse in the top of the fuel pump relay.
2. Turn ignition **ON**.

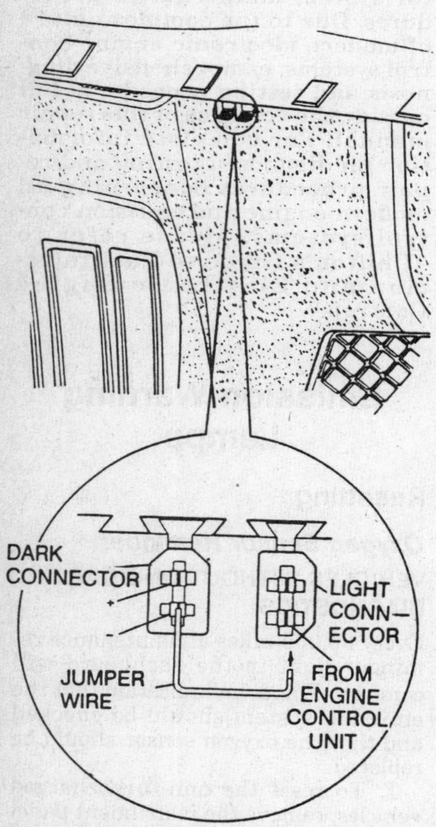

Diagnostic test connector above the pedals

3. Wait at least 4 seconds, then remove the fuse.
4. Repeat Step 3, 3 times until indicator flashes Code 4443.
5. Reinsert fuse in top of fuel pump relay for 4 seconds.
6. Repeat Step 5, until indicator flashes Code 0000.
7. Reinsert fuse in top of fuel pump relay.
8. Wait at least 10 seconds, then remove fuse.
9. Memory is clear.

1989–92 VEHICLES WITH ON-BOARD DIAGNOSTICS

Diagnostic connectors have been added to the driver's side footwell. It is not possible to activate fault memory by means of the fuel pump relay, as in older vehicles. On-board fault memory can only be accessed or cleared with the special tester VAG1551, or equivalent. If the indicator light comes on when a fault develops, the light will remain on while driving as long as the fault exists. The light will go out after the fault has been repaired or no longer exists.

FUEL SYSTEM

Fuel System Service Precaution

Safety is the most important factor when performing not only fuel system maintenance but any type of maintenance. Failure to conduct maintenance and repairs in a safe manner may result in serious personal injury or death. Maintenance and testing of the vehicle's fuel system components can be accomplished safely and effectively by adhering to the following rules and guidelines.

- To avoid the possibility of fire and personal injury, always disconnect the negative battery cable unless the repair or test procedure requires that battery voltage be applied.
- Always relieve the fuel system pressure prior to disconnecting any fuel system component (injector, fuel rail, pressure regulator, etc.), fitting or fuel line connection. Exercise extreme caution whenever relieving fuel system pressure to avoid exposing skin, face and eyes to fuel spray. Please be advised that fuel under pressure may penetrate the skin or any part of the body that it contacts.
- Always place a shop towel or cloth around the fitting or connection prior to loosening to absorb any excess fuel due to spillage. Ensure that all fuel spillage (should it occur) is quickly removed from engine surfaces. Ensure that all fuel soaked cloths or towels are deposited into a suitable waste container.
- Always keep a dry chemical (Class B) fire extinguisher near the work area.
- Do not allow fuel spray or fuel vapors to come into contact with a spark or open flame.
- Always use a backup wrench when loosening and tightening fuel line connection fittings. This will prevent unnecessary stress and torsion to fuel line piping. Always follow the proper torque specifications.
- Always replace worn fuel fitting O-rings with new. Do not substitute fuel hose or equivalent where fuel pipe is installed.

RELIEVING FUEL SYSTEM PRESSURE

Modern fuel injection systems operate under high pressure. This makes it necessary to first relieve the system of pressure before servicing. The pressurized fuel when released may ignite or cause personal injury.

1. Disconnect the power to the fuel pump by removing the relay or the fuel pump fuse, No. 13. (check the list on the fuse box lid to make sure). The fuse can be removed to stop the fuel pump from running. With the engine operating at idle, wait until the engine stalls from fuel starvation.
2. Remove the negative battery cable.
3. Carefully loosen the fuel line on the control pressure regulator or component to be serviced.
4. Wrap a clean rag around the connection, while loosening, to catch any fuel.
5. After service is complete, discard the fuel soaked rag in the proper manner and reconnect negative battery cable, relay or fuses.

Fuel Filter

REMOVAL & INSTALLATION

Most vehicles use a fuel filter mounted under the vehicle, below the fuel tank. An arrow should be on the filter indicating fuel flow direction. Install with arrow pointing to engine. Use care not to mix up fuel supply or return lines. Fuel pressure applied to the return side of the system will cause damage.

In addition, some vehicles use a filter in the engine compartment near the fuel distributor. If equipped, use the following procedure:

1. Make certain to follow precautions and relieve fuel pressure.

2. Disconnect the fuel lines leading into and out of the fuel distributor.
3. Unscrew the filter retaining bracket and remove the filter.
4. Install a new filter in the bracket and reattach bracket to vehicle. Make sure the arrows are pointing in the direction of the fuel flow to the distributor.
5. Reconnect the fuel lines, start the engine and check for leaks.

Electric Fuel Pump

PRESSURE TESTING

NOTE: The fuel tank is pressurized. Open the filler cap carefully. Fuel system pressure is not adjustable.

CIS Systems

1. Using tool VW 1318 or an equivalent 0–100 psi (7 BAR) pressure gauge, connect the gauge in the line to the cold start valve. If using the special tool, position the lever so the valve is closed.
2. Remove the fuel pump relay (No. 10) from the main relay panel and plug a long jumper wire into terminal No. 52. If special tool US 4480/3 or equivalent is available, connect it in place of the fuel pump relay with the switch OFF.
3. Remove the electrical connector from the differential pressure regulator, if equipped.
4. Apply 12 volts to the jumper wire or turn the special tool switch ON to run the fuel pump. The pressure should be 84–96 psi (5.8–6.6 BAR). If the pressure is low, check the fuel pump delivery quantity.
5. If the pressure is higher than specifications, disconnect the fuel tank return line from the diaphragm pressure regulator and repeat the test.
6. If the pressure is within specifications, check for a plugged fuel return line. If the pressure is not within specifications, replace the diaphragm pressure regulator.

Motronic Systems

1. The Motronic system uses electric injectors. Using tool VW 1318 or an equivalent 0–100 psi (7 BAR) pressure gauge, connect the gauge in the system at a convenient place to read the pressure in the supply rail. This can be either at the inlet line or before the pressure regulator. If using the special tool, position the lever so the valve is open.
2. Disconnect the vacuum line to the pressure regulator.
3. Remove the fuel pump relay (No. 10) from the main relay panel and plug a lng jumper wire into terminal No. 52. If special tool US 4480/3 or equivalent is available, connect it in place of the fuel pump relay with the switch OFF.
4. Apply 12 volts to the jumper wire or turn the special tool switch ON to run the fuel pump. The pressure should be 55–61 psi (3.8–4.2 BAR). If the pressure is low, check the fuel pump delivery quantity.
5. If fuel comes out of the regulator at the vacuum hose connection, replace the regulator.
6. Install the fuel pump relay and start the engine. At idle with the engine warm, connect the vacuum line to the pressure regulator. The pressure should decrease by about 8 psi (0.6 BAR). If not, check for blockage in the return lines, no vacuum to the pressure regulator or a bad regulator.

DELIVERY TESTING

1. Remove the fuel pump relay (No. 10) from the main relay panel and plug a long jumper wire into terminal No. 52. If special tool US 4480/3 or equivalent is available, connect it in place of the fuel pump relay with the switch OFF.
2. Remove the fuel pump cover from the floor of the trunk. On 80 and 90 vehicles, raise and safely support vehicle to access the pump.
3. At the fuel pump connector, pull back the rubber cover to expose the terminals, leaveing the plug connected to the pump.
4. Using a voltmeter, connect the probes across the terminals and apply 12 volts to the jumper wire or turn the special tool switch ON to run the pump.
5. Check and note the voltage of the running pump, it should be at least 9.0V. Turn the pump OFF.
6. In the engine compartment, disconnect the fuel line return connection and place it into a graduated container.
7. On 4 cylinder and 10 valve 5 cylinder engines, turn the fuel pump ON for 30 seconds and measure the quantity of fuel collected. Depending on the pump voltage, the delevered quantity should be approximately;
   9 volts – 11 oz. (335cc)
   10 volts – 15 oz. (450cc)
   11 volts – 20 oz. (600cc)
   12 volts – 26 oz. (760cc)
8. On turbo-charged, V8 and 5 cylinder 20 valve engines, run the pump for 15 seconds. Depending on the pump voltage, the delevered quantity should be approximately;
   8 volts – 10 oz. (295cc)
   9 volts – 12 oz. (355cc)
   10 volts – 16 oz. (480cc)
   11 volts – 19 oz. (560cc)
   12 volts – 22 oz. (660cc)
   13 volts – 25 oz. (750cc)
   14 volts – 28 oz. (835cc)

REMOVAL & INSTALLATION

All Except 80/90

1. The fuel pump is located in the fuel tank. Remove the floor cover from the luggage compartment.
2. Disconnect the negative battery cable and the electrical connector from the fuel gauge sender.
3. Mark and remove the hoses from the fuel gauge sender.
4. A special wrench is available to loosen the fuel gauge sender-to-fuel tank retaining ring. Pull out the fuel gauge/fuel pump assembly.
5. From inside the assembly housing, pull off the fuel hoses, detach the electrical connections and remove the gravity vent valve.
6. To install, reverse the removal procedures. Start the engine and check for leaks.

1988–92 80 and 90

The fuel pump is located under the vehicle on a bracket in front of the fuel tank. The fuel pump assembly is located on the right side of front wheel drive vehicles and on the left side on Quattro vehicles. The 80 and 90 vehicles do not use a separate fuel pump filter. The filter is expected to be a lifetime unit, unless the fuel was contaminated.

1. Make certain to follow precautions and relieve fuel pressure.
2. Disconnect the negative battery cable.
3. Raise and safely support vehicle.
4. Carefully loosen fuel line at fuel pump. Catch excess fuel in a container.
5. Remove fuel pump electrical connectors and remove the fuel pump.

To install:

6. Install fuel pump. Connect the fuel lines.
7. Connect the fuel pump electrical connectors.
8. Lower vehicle and connect the negative battery cable.
9. Replace any relays or fuses, that had been removed. Start engine and inspect for fuel leakage.

Fuel Injection

Idle Speed Adjustment

80 and 90

The CIS Motronic system used on 80 and 90 vehicles uses a single Electronic Control Unit (ECU), located behind the air conditioner evaporator assembly. The ECU controls the fuel delivery, ignition system and operation of the emission control components. The CIS Motronic system also incorporates self-diagnostic capabilities.

The idle speed is maintained between 780–900 rpm. No idle speed adjustments are necessary or possible. The idle stabilizer valve is located between the intake air boot and intake manifold.

100, 200 and 5000

The CIS-E system incorporates 2 control units. An Ignition Control Unit (ICU) or Knock Sensor Control Unit (KSCU), on 5000S only, and a Fuel Injection Control Unit (FICU).

The CIS-E system also has self-diagnosis and troubleshooting capabilities. Input and output signals from various sensors, switches and signaling devices are constantly monitored for faults. These faults are stored in the control unit memory. Faults can be displayed by a flashing 4 digit code sequence from an LED light located on the instrument panel.

Idle speed should be 650–790 rpm. The idle speed is not adjustable. If idle speed is not within specification, check for an engine problem such as a vacuum leak, bad connection, etc. Also, check for a defective differential pressure regulator, which is mounted on the fuel distributor.

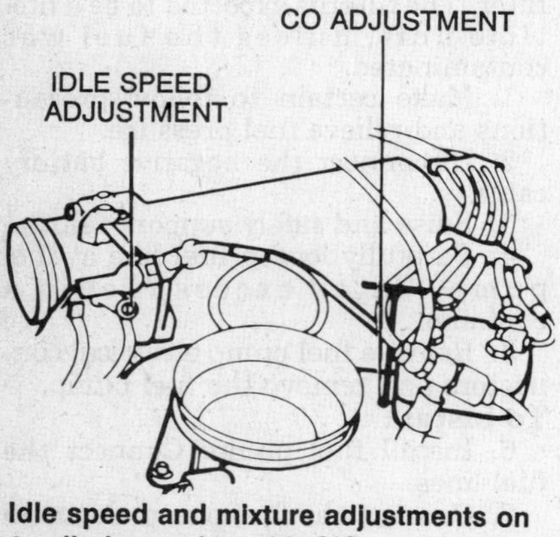

Idle speed and mixture adjustments on 4 cylinder engine with CIS

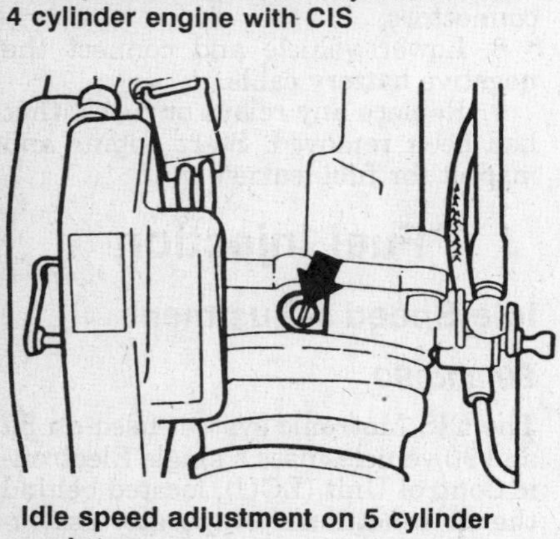

Idle speed adjustment on 5 cylinder engine

V8 Quattro

The Motronic engine management system uses a single ECU to control both ignition and fuel injection. The program for the system has a self-learning adaptive capabilty that continuously learns using a sophisticated feedback system, and constantly readjusts various control settings. These new values are then stored in the ECU memory. The adaptive program allows the system to compensate for changes in the engine's operating conditions, such as intake leaks, altitude changes

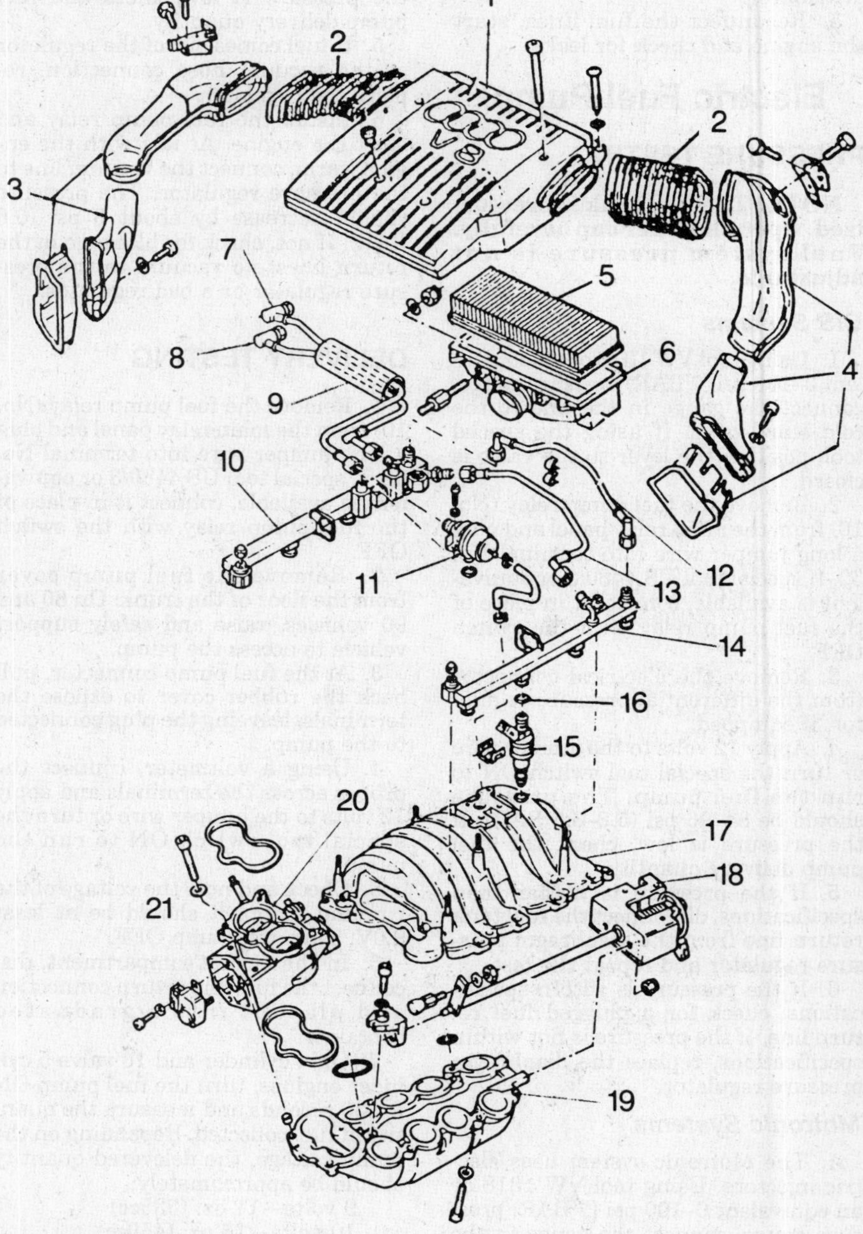

1. Upper air cleaner housing
2. Flexhose
3. Right air intake
4. Left air intake
5. Air cleaner filter element
6. Lower air cleaner housing
7. Return fuel line
8. Supply fuel line
9. Insulator
10. Right fuel manifold
11. Fuel pressure regulator
12. Supply line crossover
13. Return line crossover
14. Left fuel manifold
15. Fuel injector
16. O-ring
17. Upper intake manifold
18. Throttle shaft housing
19. Lower intake manifold
20. Throttle housing
21. Idle and full throttle switch

Injectors and intake system on V8 engine

or any other system malfunction. If the ECU power is interrupted, the vehicle must be driven for a few minutes to re-learn the program.

A Hall effect signal from the right side distributor helps the ECU establish a reference point to start the fuel injection process. After the engine is running, the reference sender and speed sensor provide the necessary information to the ECU for ignition and fuel injection.

Both the idle speed and the ignition timing are controlled by the ECU and cannot be adjusted.

IDLE SPEED PRE-CHECK

The following procedure should be applicable to all vehicles. Before checking idle speed, the following conditions must be met:

A. Start and allow engine to reach normal operating temperature.

B. Check that the throttle valve is in the idle position.

C. All accessories are **OFF**.

D. Fuel pressure test gauge not connected.

1. Connect tester with negative lead to ground, positive lead to the remote positive cable connection and the tachometer lead to the No. 1 terminal on the right side of the ignition coil.
2. Start engine and allow to idle. Check idle speed against specification.
3. If specifications are not as indicated, diagnostic work will need to be performed using the vehicle's fault memory.

Fuel Injector

REMOVAL & INSTALLATION

1. Disconnect the battery negative cable.
2. Remove the left and right support braces, then the intake air duct.
3. Remove the engine compartment support brace.
4. Remove the upper air cleaner attaching bolts, then the upper air cleaner.
5. Remove the lower air cleaner housing attaching bolts, then push the housing back and lift outward.
6. Remove the upper ventilation hose, then the right fuel rail cable tie.

NOTE: When assembling, install a new cable tie in the same position and location as the original.

7. Remove the engine wiring harness support clip attaching bolts, and position the wiring harness aside.
8. Disconnect the throttle valve potentiometer, thermo-switch and idle stabilizer valve electrical connectors.
9. Remove the idle stabilizer valve, then disconnect the vacuum hoses to the carbon canister.
10. Remove the bolts attaching the intake air temperature sensor, then position aside.
11. Disconnect the air mass sensor electrical connectors, then the 2 knock sensor plug connections.
12. Disconnect the fuel return and supply lines, then the pressure regulator vacuum hose.
13. Remove the fuel rail attaching bolts, then the fuel rail and injectors as an assembly.

To install:

14. Use new O-rings and make sure the injectors fit properly onto the fuel rail. Install them as an assembly and torque the bolts to 7 ft. lbs. (10 Nm).
15. Use new gaskets and connect the fuel supply and return lines. Torque the fittings to 18 ft. lbs. (25 Nm).
16. Reconnect all wiring and vacuum hoses.
17. Install the air cleaner housing and ducting and properly secure the wires.
18. Install the support brace.

DRIVE AXLE

Front Halfshaft

REMOVAL & INSTALLATION

Coupe, 80 and 90

NOTE: When loosening or tightening axle nut or bolt, make sure the vehicle is on the ground. Axle nut torque is high enough that attempting to loosen it may cause the vehicle to fall off the support.

1. Loosen the axle nut or bolt and raise and safely support the vehicle.
2. Unbolt and remove the halfshaft-to-transaxle drive flange bolts.
3. Mark the position of the ball joint on the control arm, remove the 2 retaining nuts and disconnect the ball joint.
4. Remove the ball joint-to-steering knuckle bolt and separate the knuckle from the ball joint. Remove the mounting bolts for the control arm/stablizer and push control arm downward, if necessary.
5. Pivot the strut outward and remove the halfshaft.

To install:

6. Installation is the reverse of the removal procedure.
7. When installing the right halfshaft, take care not to damage the boot on the cover plate.
8. Tighten the ball joint-to-control arm/knuckle nuts/bolt to 47 ft. lbs. (64 Nm). Tighten the halfshaft flange bolts to 33 ft. lbs. (45 Nm).
9. Install the wheel and snug the axle nut or bolt. Place the vehicle on the groung and torque the axle nut or bolt to 200 ft. lbs. (270 Nm).
10. Check for and adjust wheel alignment when finished.

5000 Quattro Turbo

1. With the vehicle on the ground, loosen the halfshaft end nut.
2. Raise and support the vehicle safely.
3. Remove the end nut and the wheel assembly. Remove the right backing plate.
4. Remove the halfshaft-to-transaxle flange bolts. Support the halfshaft with wire.
5. Using a halfshaft press tool, attach it to the wheel hub and press the halfshaft from the hub.

To install:

6. Using the locking compound D-6, apply a ¼ in. bead around the front edge of the spline section of the halfshaft.

NOTE: After applying the locking compound, allow it to dry for at least an hour.

7. To install, reverse the removal procedure. Be sure to replace the inner CV-joint gasket. Tighten the

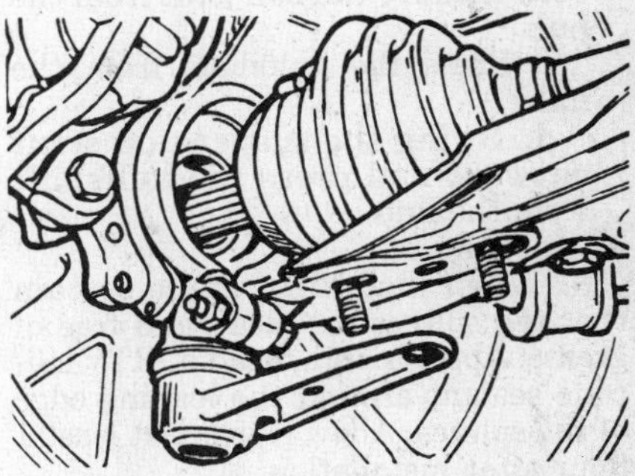

Remove the front halfshaft by pivoting the strut out from the bottom

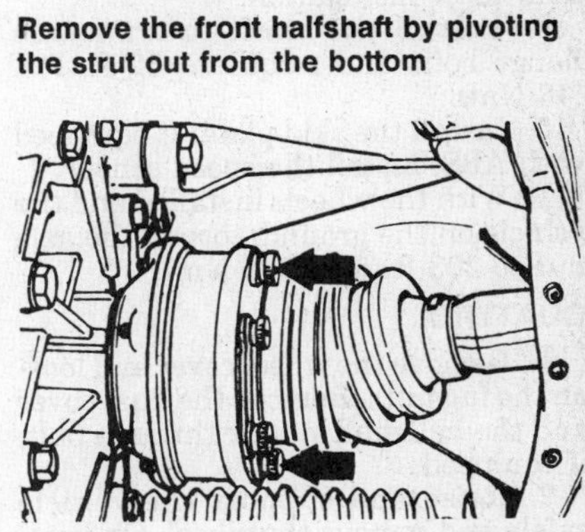

Halfshaft bolts at the transaxle drive flange

halfshaft-to-transaxle flange bolts to 58 ft. lbs. (78 Nm).

8. Snug the axle nut, install the wheel and place the vehicle on the ground to torque the nut to 203 ft. lbs. (275 Nm).

100 and 200

EXCEPT QUATTRO

NOTE: When loosening or tightening axle nuts, make sure the vehicle is on the ground. Axle nut torque is high enouth that attempting to loosen it may cause the vehicle to fall off the support. A puller is required for this procedure.

1. Remove the halfshaft end nut.
2. Raise and support the vehicle safely and remove the wheels. On vehicles equipped with ABS, slide the speed sensor partly out of its mount.
3. On the right side, remove the halfshaft skid plate.
4. Disconnect the halfshaft from the transaxle. Using wire, support the halfshaft.
5. Using a 4-armed puller mounted on the wheel hub, press the halfshaft out of the hub.
6. Guide the inside end of the shaft up over the transaxle and out of the hub.
7. If equipped with an automatic transaxle, perform the following:
   a. Remove the stabilizer bar clamps.
   b. Remove the ball joint-to-hub bolt. Remove the ball joint from the hub.
   c. Press the halfshaft from the hub.
   d. Swing the suspension strut outward and press the halfshaft from the hub.

To install:

8. When installing, make certain that the splines are clean and free of grease. Apply a ¼ in. bead of RTV silicone sealant around the leading edge of the splines. Allow it to set at least 1 hour after installation.
9. Install the shaft-to-transaxle flange bolts and torque to 32 ft. lbs. (43 Nm).
10. Install the skid plate. If equipped with ABS, install the speed sensor.
11. With the wheels installed and the vehicle on the ground, torque the axle nut to 203 ft. lbs. (275 Nm).

QUATTRO

1. Remove the wheel cover and loosen the lug nuts. Remove the dust cover and the halfshaft nut or through bolt, if equipped.
2. Raise and support the vehicle safely and remove the wheel. On vehicles equipped with ABS, slide the speed sensor partly out of its mount. Remove the right backing plate.
3. Disconnect the halfshaft at the transaxle flange and position it aside.
4. Using a suitable puller, press out the stub axle from the hub. Use only a mechanical or hydraulic puller to remove the stub axle. Never use hot air blower or a flame to heat the stub axle.

To install:

5. Replace the gasket on the inner CV-joint.
6. Make sure the splines on the stub axle and the wheel hub are free of oil, grease and old locking compound. Apply a bead of suitable locking compound approximately 13/64 in. wide around the splines and install the stub halfshaft. Allow at least 1 hour for the locking compound to harden after installtion.
7. Install and torque the halfshaft to transaxle bolts to 58 ft. lbs. (79 Nm) and install the wheel.
8. With the vehicle on the ground, torque the halfshaft end nut to 207 ft. lbs. (280 Nm).

V8 Quattro

When loosening or tightening axle nuts, make sure the vehicle is on the ground. Axle nut torque is high enouth that attempting to loosen it may cause the vehicle to fall off the support.

1. Raise and safely support the vehicle. Remove the wheel cover and remove the hexagon through bolt and washer from the halfshaft end.
2. Slightly pull back the wheel speed sensor. Remove both stabilizer bar brackets.
3. Remove the clamp bolt at the steering knuckle and press out the ball joint pivot pin for the lower control arm. Be careful not to damage the joint boot or seal. Never widen the slit in the steering knuckle housing. Use penetrating oil as required.
4. Remove the halfshaft from the transaxle drive flange.
5. Push the strut outward and remove the halfshaft.

To install:

6. Installation is the reverse of the removal procedure. Always use a new gasket at the transaxle drive flange. Pull off the protective film from the replacement gasket and stick to CV-joint.
7. When replacing the ball joint, install the lock bolt with the head facing to the rear of the vehicle. When installing the halfshaft through bolt, always use a new bolt and tighten when the vehicle's weight is on the floor. Torque to 148 ft. lbs. (200 Nm) plus an additional ¼ turn.
8. After installing the halfshaft, insert the wheel speed sensor into the housing as far as possible.

Rear Halfshaft

REMOVAL & INSTALLATION

80 and 90 Quattro

1. With the vehicle resting on the ground, loosen the halfshaft nut.

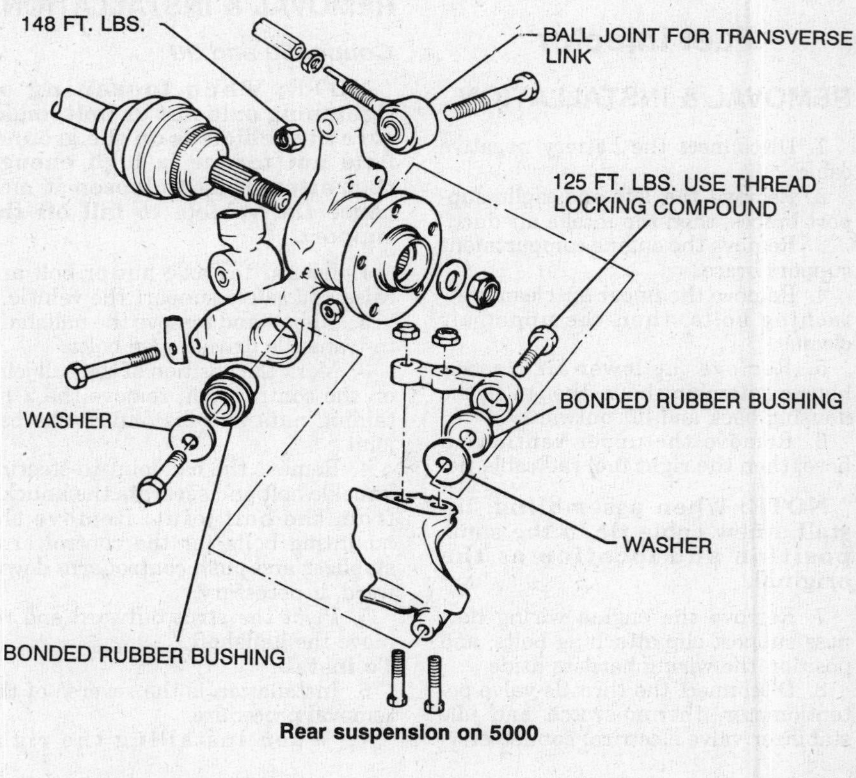

Rear suspension on 5000

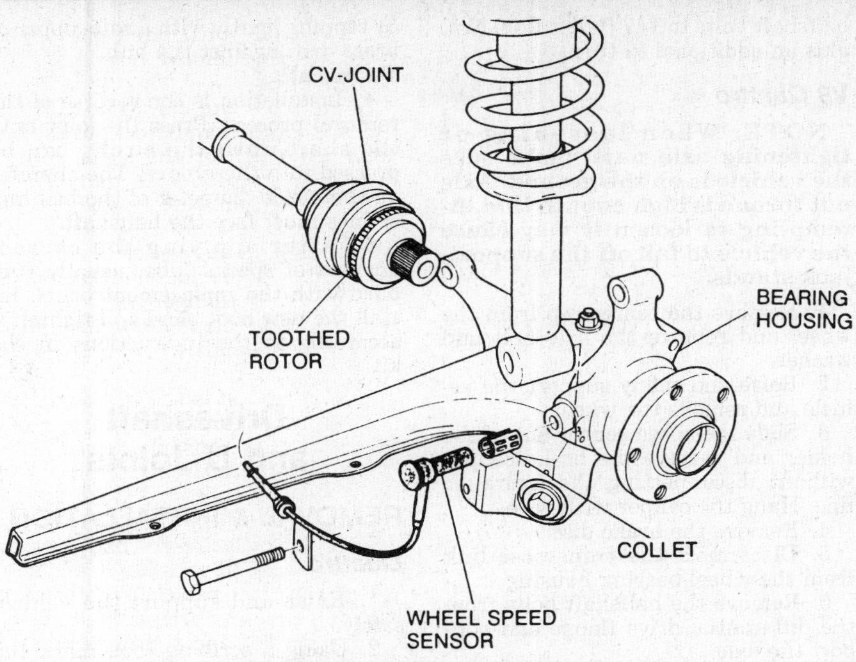

Rear axle speed sensor for ABS 200 shown

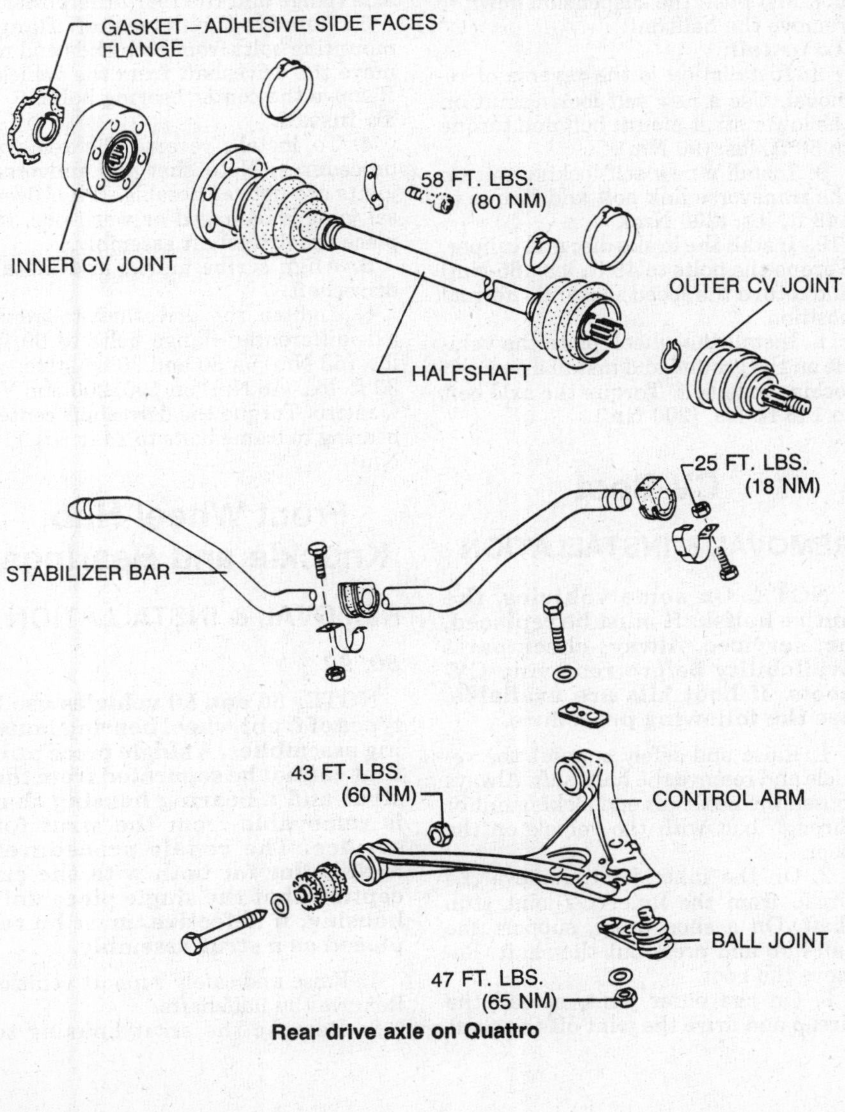

Rear drive axle on Quattro

2. Raise and support the vehicle safely.
3. Remove the halfshaft nut, wheel bolts and wheel assembly.
4. Remove the ball joint nut. Using a ball joint removal tool, separate the ball joint from the strut.
5. Using a suitable tool, pry downward on the lower control arm to remove the ball joint from the control arm. If necessary, loosen lower control arm mounting bolts.
6. Pull the brake hose and parking brake cable, with grommets, from the holding fixture.
7. Remove the inner halfshaft flange bolts. Separate the shaft from the flange and support it.
8. Using an halfshaft pulling tool, attach it to the wheel hub and press the halfshaft out of the hub.

To install:

9. Clean the halfshaft splines of any grease, dirt or locking compound. Using the locking compound D-6, or equivalent, apply a ¼ in. bead around the outer edge of the splines. Allow the locking compound to dry for an hour after installation.
10. When installing, use a new inner flange gasket and reverse the removal procedures. Torque the ball joint nut to 47 ft. lbs. (64 Nm).
11. Tighten the inner halfshaft flange bolts to 58 ft. lbs. (79 Nm) and install the wheel.
12. With the vehicle on the ground, torque the halfshaft to hub nut to 238 ft. lbs. (322 Nm).

5000 Quattro

1. With the vehicle weight on the ground, loosen the halfshaft end nut.
2. Raise and support the vehicle safely.
3. Remove the halfshaft nut, wheel bolts and wheel assembly.
4. Remove the brake caliper-to-strut retaining bolts and remove the caliper, without disconnecting the hydraulic line. Using wire, support the caliper.
5. Remove the brake rotor. Remove the inner halfshaft flange bolts and support the halfshaft.

NOTE: When removing the right side halfshaft, remove the fuel tank cover plate first.

6. Remove the transverse link to wheel bearing housing nut and remove the link.
7. Remove the trapezoidal arm to crossmember nut and bolt. Pry the arm downward.
8. Using a halfshaft pulling tool, attach it to the wheel hub and remove the halfshaft from the hub.

To install:

9. Clean the halfshaft splines of any

grease, dirt or locking compound. Using the locking compound D-6, or equivalent, apply a 1/4 in. bead around the outer edge of the splines. Allow the locking compound to dry for an hour after installation.

10. To install, use a new inner flange gasket and reverse the removal procedures.
11. Tighten the halfshaft flange bolts to 58 ft. lbs. (79 Nm) and snug the axle nut. Install the caliper and torque the bolts to 48 ft. lbs. (65 Nm).
12. Torque the trapezoidal arm-to-crossmember nut to 63 ft. lbs. (85 Nm) and the transverse link-to-wheel bearing housing nut to 148 ft. lbs. (200 Nm).
13. With the vehicle on the ground, torque the halfshaft to hub nut to 266 ft. lbs. (360 Nm).

100 and 200 Quattro

1. With the vehicle weight on the ground, loosen the halfshaft end nut.
2. Raise and support the vehicle safely.
3. Remove the halfshaft nut, wheel bolts and wheel assembly.
4. Remove the brake caliper to strut retaining bolts and remove the caliper, without disconnecting the hydraulic line. Using wire, support the caliper.
5. Remove the brake rotor. Remove the inner halfshaft flange bolts and support the halfshaft.
6. Remove the fuel tank cover plate, if necessary.
7. Remove the transverse link-to-wheel bearing housing nut and remove the link.
8. Remove the trapezoidal arm-to-crossmember nut and bolt. Pry the arm downward.
9. Remove the mounting bolt for suspension strut.
10. Before removing halfshaft, pull speed sensor out of the housing slightly.
11. Press down on wheel bearing housing and remove the halfshaft.
12. Clean the halfshaft splines of any grease, dirt or locking compound.

To install:

13. Use a new inner flange gasket and reverse the removal procedures.
14. Tighten the halfshaft flange bolts to 59 ft. lbs. (80 Nm).
15. Install caliper and torque the bolts to 48 ft. lbs. (65 Nm). Adjustment of parking brake may be necessary.
16. Install the halfshaft bolt and washer assembly, tighten until just snug.
17. Make certain speed sensor sleeve is in place and install speed sensor , by hand, until seated. Install wheels.
18. Lower vehicle and torque halfshaft bolts to 147 ft. lbs. (200 Nm) plus an additional 1/4 turn.

V8 Quattro

NOTE: When loosening or tightening axle nuts, make sure the vehicle is on the ground. Axle nut torque is high enouth that attempting to loosen it may cause the vehicle to fall off the support. jack stands.

1. Remove the center cap from the wheel and remove the axle bolt and washer.
2. Raise and safely support the vehicle and remove the wheel.
3. Slide the speed sensor out of the holder and remove the brake caliper without disconnecting the hydraulic line. Hang the caliper with wire.
4. Remove the brake disc.
5. Disconnect the transverse link from the wheel bearing housing.
6. Remove the halfshaft bolts from the differential drive flange and support the axle.
7. Remove the lower strut mount bolt and push the suspension down to remove the halfshaft.

To install:

8. Installation is the reverse of removal. Use a new self locking nut on the lower strut mount bolt and torque to 66 ft. lbs. (90 Nm).
9. Install a new self locking nut on the transverse link bolt and torque to 148 ft. lbs. (200 Nm).
10. Install the brake disc and caliper. Torque the bolts to 48 ft. lbs. (65 Nm) and return the speed sensor its normal position.
11. Install the wheel, place the vehicle on the ground and install a new self locking axle bolt. Torque the axle bolt to 148 ft. lbs. (200 Nm).

CV-Boot

REMOVAL & INSTALLATION

NOTE: On some vehicles, the entire halfshaft must be replaced, not serviced. Always check parts availability before removing CV-boots. If boot kits are available, use the following procedure.

1. Raise and safely support the vehicle and remove the halfshaft. Always loosen the halfshaft end locking nut or through bolt with the vehicle on the floor.
2. On the inner joint, remove the circlip from the inner CV-joint stub shaft. On a shop press, support the ball hub and press out the shaft. Remove the boot.
3. On the outer joint, spread the circlip and drive the joint off the shaft by tapping lightly with a soft copper or brass drift against the hub.

To install:

4. Installation is the reverse of the removal process. Press the joint onto the shaft until the circlip can be pressed into the groove. The chamfer on the inside diameter of the ball hub splines must face the halfshaft.
5. After applying the correct amount of special lube, usually supplied with the replacement boots, install the new boot band and tighten it according to the instructions in the kit.

Driveshaft and U-Joints

REMOVAL & INSTALLATION

Quattro

1. Raise and support the vehicle safely.
2. Using a scribing tool, mark the position of the driveshaft to the transaxle flange and the rear differential.
3. Remove the driveshaft flange mounting bolts from both ends and remove the driveshaft from the vehicle. Remove the center bearing bolts.

To install:

4. To install, reverse the removal procedures. Note that the universal joints are not replaceable. If a universal joint is damaged or worn out, replace the driveshaft assembly.
5. Align scribe marks and install driveshaft.
6. Tighten the driveshaft-to-transaxle/differential flange bolts to 39 ft. lbs. (53 Nm) on 80 and 90 Quattro, or 33 ft. lbs. (45 Nm) on 100, 200 and V8 Quattro. Torque the driveshaft center bearing to frame bolts to 14 ft. lbs. (19 Nm).

Front Wheel Hub, Knuckle and Bearings

REMOVAL & INSTALLATION

80, 90

NOTE: 80 and 90 vehicles use 2 types of front wheel bearing housing assemblies. A single piece unit that cannot be separated from the strut and a bearing housing that is removable from the strut for service. The repair procedures are similar for both with the exception that the single piece unit housing, if defective, must be replaced as a strut assembly.

1. Raise and safely support vehicle. Remove the halfshafts.
2. Remove the strut housing to

body nuts and remove the strut/hub assembly from the vehicle.

3. Remove the brake disc and splash shield.

4. Using an arbor press with suitable drivers, press the wheel hub from the strut housing.

5. Using snapring pliers, remove the snaprings from both sides of the wheel bearing.

6. Using an arbor press with suitable drivers, press the wheel bearing from the strut housing.

7. Using a wheel puller, pull the wheel bearing race from the wheel hub.

To install:

8. Lightly grease inside the strut housing before installing the new bearing.

9. Be sure to press only on the outer race when pressing the new bearing into the hub. When installing the snaprings, make sure they are properly seated.

10. Install the brake plate onto the bearing housing. Be sure to press only on the inner bearing race when pressing the bearing over the hub.

11. To install the strut and wheel hub assembly, reverse the removal procedures. On 80 and 90 models, torque the upper strut to body nut to 44 ft. lbs. (60 Nm). On all other models, torque the 3 upper strut nuts to 22 ft. lbs. (30 Nm). Check front wheel alignment.

100, 200 and V8 Quattro

1. Raise and safely support vehicle. Remove the halfshafts.

2. Remove the strut to vehicle nuts and remove the strut from the vehicle.

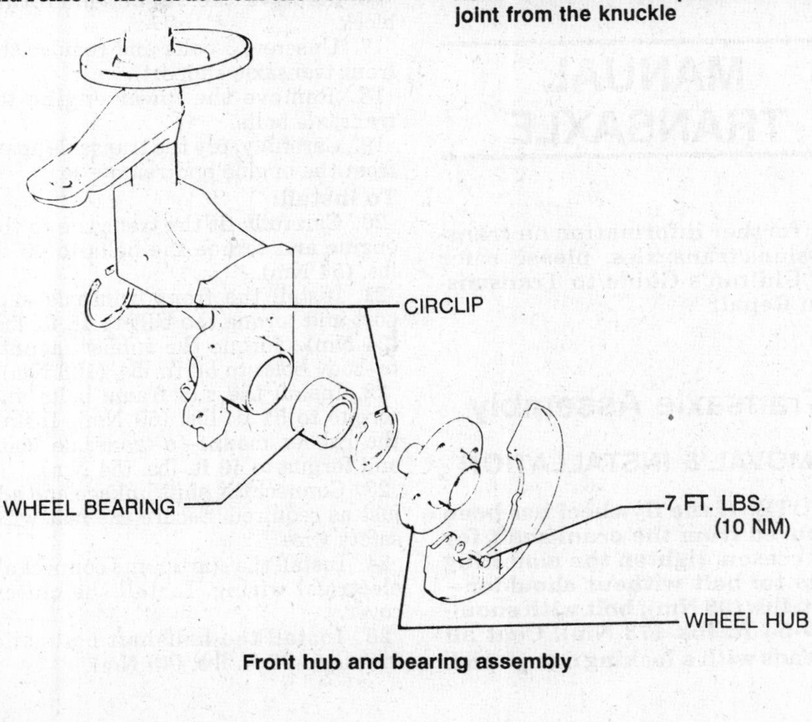

Front hub and bearing assembly

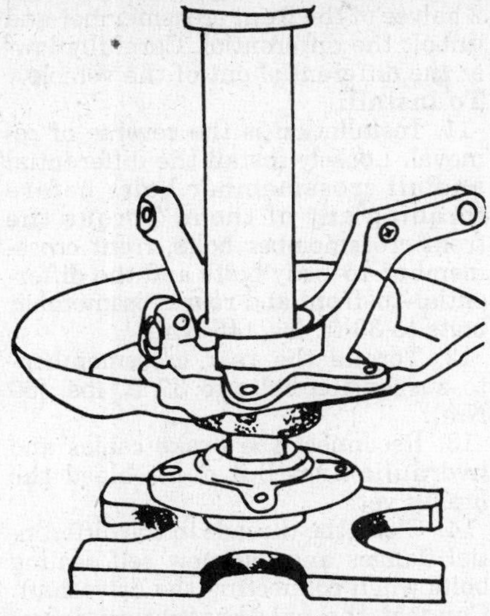

Be sure to press only on the inner race when pressing the bearing and housing onto the hub

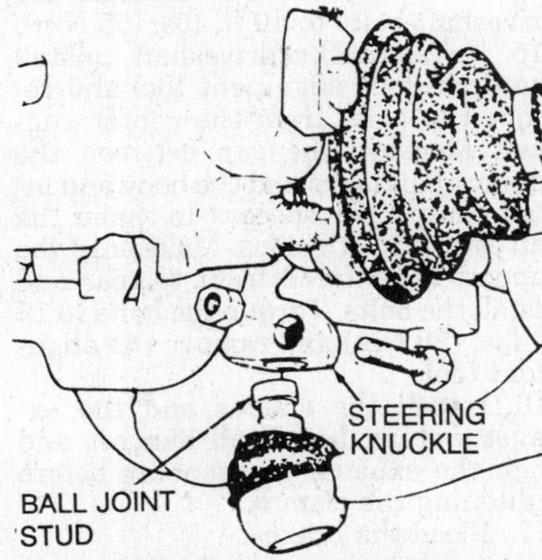

Remove the bolt to separate the ball joint from the knuckle

3. Remove the disc brake rotor and splash shield.

4. Using an arbor press with suitable drivers, press the wheel hub from the strut housing.

5. Using snapring pliers, remove the snaprings from both sides of the wheel bearing.

6. Using an arbor press with suitable driver, press the wheel bearing from strut housing.

7. Using a suitable puller, remove the bearing race from the wheel hub.

To install:

8. Install the outer snapring into the strut housing.

9. Be sure to press only on the outer race when pressing the new bearing into the hub. When installing the snapring, make sure it is properly seated.

10. Install the brake plate onto the bearing housing. Be sure to press only on the inner bearing race when pressing the bearing over the hub.

11. To install the strut and wheel hub assembly, reverse the removal procedures. Tighten the strut to body nuts to 44 ft. lbs. (60 Nm). Check front wheel alignment.

Differential Carrier

REMOVAL & INSTALLATION

80 and 90

1. Raise and safely support the vehicle. Matchmark and disconnect the axle shafts from the flanges and disconnect the drive shaft. Hang them from the body with wire.

2. Remove the differential lock servo and bracket. Disconnect the wiring.

3. Remove the nut from the rear mount bushing.

4. Place a transmission jack under the differential and take the weight of the unit off the mounts. Remove the right side and rear mounts.

5. Pull the differential slightly forward and lower it from the vehicle.

To install:

6. Installation is the reverse of removal. Torque the mount-to-sub frame nuts or bolts to 18 ft. lbs. (25 Nm). Torque the differential-to-mount nuts or bolts to 33 ft. lbs. (45 Nm).

7. Torque the axle flange bolts to 33 ft. lbs. (45 Nm) and the drive shaft bolts to 40 ft. lbs. (55 Nm).

5000, 100 and 200

1. Raise and safely support the vehicle and remove the cover plate.

2. Engage the differential lock and block a wheel or set the parking brake.

3. Matchmark and disconnect the axle shafts and drive shaft and hang them from the body with wire.

4. Release the parking brake and disconnect the cables at the caliper and at the front.
5. Unbolt the left parking brake retainer and remove the heat shield.
6. Support the differential with a transmission jack and take the weight off the mounts.
7. Remove the crossmember and the exhaust pipe.
8. Disconnect the mounts and lower the unit slightly.
9. Note the color coding of the vacuum hoses so they can be properly connected during installation. Disconnect the wiring and vacuum lines and carefully lower the unit out of the vehicle.

To install:

10. Installation is the reverse of removal. Make sure to properly connect the vacuum hoses.
11. Torque the crossmember–to–body bolts and the suspension–to–crossmember bolts to 33 ft. lbs. (45 Nm).
12. Torque the axle shaft 8mm bolts to 33 ft. lbs. (45 Nm) and the 10mm bolts to 60 ft. lbs. (80 Nm).
13. Torque the drive shaft bolts to 40 ft. lbs. (55 Nm).

V8 Quattro

NOTE: Any time the differential is removed from this vehicle, the driveshaft must be properly aligned. This requires the special alignment tool 3139 or equivalent.

1. Raise and safely support the vehicle and remove the rear wheels.
2. Remove the rear muffler system and remove the heat shields, fuel tank shield and the axle shaft joint shield.
3. Disconnect the parking brake cables and the exhaust system support from the crossmember.
4. Matchmark the axle shafts and driveshaft and disconnect them from the differential. Hang them from the body with wire.
5. Detach the parking brake cables from the differential and disconnect them from the calipers.
6. Disconnect the brake hydraulic hoses from the distributor valve on the differential.
7. Use a 15mm socket to remove the mounting nut at the hole in the rear crossmember.
8. Support the rear crossmember and remove the crossmember mounting nuts.
9. At the front crossmember, loosen the left mounting bolt and remove the right mounting bolt to lower the crossmember slightly on the right side. Reposition the parking brake cables to the front.
10. Carefully support the differential with a transmission jack. Separate the 2 halves of the front crossmember and unbolt the differential. Carefully lower the differential out of the vehicle.

To install:

11. Installation is the reverse of removal. Loosely install the differential and all crossmember bolts before torquing any of them. Torque the front crossmember bolts, front crossmember–to–body bolts and the differential–to–front and rear crossmember bolts to 33 ft. lbs. (45 Nm).
12. Torque the rear crossmember–to–suspension bolts to 37 ft. lbs. (50 Nm).
13. Reconnect the brake cables and hydraulic lines but don't bleed the brakes yet.
14. Clean the threads in the differential flanges and use new self sealing bolts when connecting the driveshaft. Connect the axle shafts and driveshaft, making sure to align the matchmarks. Torque the axle shaft bolts to 60 ft. lbs. (80 Nm) and the driveshaft bolts to 40 ft. lbs. (55 Nm).
15. To align the driveshaft, install the driveshaft alignment tool and remove the bolts from the center support. Measure the gap between the support bolt holes and the body and install the proper spacers to make the gap even on both sides. Make sure the support is centered front to rear and install the bolts. Torque the bolts to 15 ft. lbs. (20 Nm) and remove the alignment tool.
16. Install the shields and the exhaust system. Install all hangers and align the exhaust components before tightening the clamps.
17. Bleed the brakes.

MANUAL TRANSAXLE

For further information on transmissions/transaxles, please refer to "Chilton's Guide to Transmission Repair".

Transaxle Assembly

REMOVAL & INSTALLATION

NOTE: If the flywheel has been removed from the crankshaft for any reason, tighten the mounting bolts to: bolt without shoulder–72 ft. lbs. (98 Nm); bolt with shoulder–54 ft. lbs. (73 Nm). Coat all threads with a locking compound.

80 and 90 Except Quattro

1. Disconnect the negative battery cable.
2. Unplug the 2 electrical connectors for the back-up lights. They can be found between the ignition coil and the fuel distributor filter.
3. Remove the upper engine-to-transaxle bolts.
4. Detach the speedometer cable from the transaxle.
5. Detach the clutch cable from the clutch lever.
6. Unbolt the exhaust pipe from the exhaust manifold.
7. Unscrew the 3 mounting bolts and remove the center engine mount.
8. Unbolt the front exhaust pipe from the support bracket and unbolt it from the catalytic converter.
9. Unscrew the 6 screws and remove the left halfshaft from the transaxle. Wire the halfshaft up and aside. Repeat the procedure for the right halfshaft.
10. Remove the clutch cover plate.
11. Tag and disconnect all wires leading to the starter and remove the starter.
12. Remove the bolt from the shift rod coupling.
13. Pry off the linkage coupling with a suitable small prybar.
14. Pull the shift rod coupling off of the shift rod. Place a transaxle jack under the transaxle, support it by lifting up slightly.
15. Loosen the left chassis bolt on the rear transaxle support. Remove the bolts from the right or transaxle side of the support and pivot the support aside.
16. Remove the rubber mounting block.
17. Unscrew 3 bolts and remove the front transaxle support.
18. Remove the lower engine-to-transaxle bolts.
19. Carefully, pry the transaxle apart from the engine and remove it.

To install:

20. Carefully fit the transaxle to the engine and toruqe the bolts to 40 ft. lbs. (54 Nm).
21. Install the front transaxle support and torque the bolt to 18 ft. lbs. (24 Nm). Torque the rubber mount–to–body bolts to 80 ft. lbs. (108 Nm).
22. Install the sub frame bolts and torque to 51 ft. lbs. (69 Nm). Install the rubber mount–to–transaxle bolts and torque to 40 ft. lbs. (54 Nm).
23. Connect the shift linkage and adjust as required. Secure the bolt with safety wire.
24. Install the starter and connect all electrical wiring. Install the clutch cover.
25. Install the halfshaft bolts and torque to 33 ft. lbs. (45 Nm).

26. Reassembly the exhaust system using new self locking nuts. Torque to 25 ft. lbs. (34 Nm).
27. Connect the speedometer and clutch cables and adjust as required.

80 and 90 Quattro

1. Disconnect the negative battery cable.
2. Remove the 3 upper engine-to-transaxle bolts.

NOTE: Tag all bolts during removal, so all bolts can be replaced in their correct locations, as they are not all the same size.

3. Disconnect the ground strap from the transaxle.
4. Remove the wiring connectors from the speedometer sender and the multi-function switch.
5. Disconnect the wiring for the oxygen sensor and oxygen sensor heating element.
6. Remove the engine protection plate.
7. Disconnect the exhaust pipe from the manifold.
8. Separate the exhaust pipe behind the catalyst and remove the pipe and catalyst.
9. Matchmark and remove the driveshaft.
10. Remove the rear crossmember.
11. Remove the shift rod securing bolt at the transaxle and let the shift rod hang.
12. Remove the transaxle cover plate.
13. Remove the right halfshaft shield.
14. Disconnect the left and right halfshafts, turn the steering to the right lock and tie both shafts up.
15. Remove the clutch slave cylinder.
16. Remove the tie rod coupling from the steering rack and turn wheel to the left.
17. Support the engine.
18. Support the transaxle.
19. Remove the transaxle strut at the left rear and front engine mount.
20. Remove the heat shield from the bonded rubber bushing.
21. Remove the bonded rubber bushing support bracket from the transaxle.
22. Remove the bonded rubber bushing.
23. Remove the bolt from the seatbelt tension in cable guide at the left rear of the transaxle. Position the cables and guide aside.
24. Lower the right rear subframe by loosening the mounting bolts.
25. Remove the remaining transaxle to engine bolts.
26. Remove the transaxle.

To install:

27. Installation is the reverse of removal. When installing the transaxle, make certain alignment bushings are in the cylinder block before reassembly.
28. Press clutch master cylinder in with a lever, until retaining bolt can be installed.

NOTE: A replacement bolt for mounting the clutch slave cylinder is available from Audi, with a pointed tip for easier installation.

29. Tighten subframe mounting bolts to 25 ft. lbs. (34 Nm), plus an additional 90 degree turn.
30. Tighten transaxle retaining bolts as follows: Torque the 8mm bolts to 18 ft. lbs. (24 Nm), the 10mm bolts to 33 ft. lbs. (45 Nm) and the 12mm bolts to 48 ft. lbs. (65 Nm).
31. Torque the driveshaft-to-flange bolts to 33 ft. lbs. (45 Nm) and the driveshaft to transaxle and final drive bolts to 40 ft. lbs. (54 Nm).
32. Torque the tie rod coupling-to-steering rack to 33 ft. lbs. (45 Nm).

1988 5000

1. The manual transaxle may be removed with the engine in place. Disconnect the negative battery cable.
2. Remove the air filter, if necessary.
3. Remove the windshield washer bottle.
4. Remove the upper engine-to-transaxle bolts.
5. Raise and safely support the vehicle.
6. Disconnect the speedometer cable from the transaxle.
7. Disconnect all wires and hoses connected to the transaxle.
8. Drive out the clutch slave cylinder lockpin and remove the slave cylinder. Leave the hydraulic line connected.
9. Support the engine, either from above with a hoist or from below with a jack.
10. Remove the heat shield.
11. Remove the lower engine and transaxle splash shield, if equipped.
12. Disconnect the exhaust pipe from the manifold.
13. Remove the right side guard plate.
14. Disconnect the halfshafts from the flanges and support them aside with wires. Disconnect the front-to-rear driveshaft at the rear output shaft on the transaxle and wire it aside.
15. Disconnect the back-up light switch.
16. Pry off the shift and adjusting rods.
17. Remove the lower engine-to-transaxle bolts.
18. Remove the starter.
19. Remove the subframe skid plate.
20. Install a jack under the transaxle and lift it slightly.
21. Remove both transaxle-to-subframe bolts.
22. Remove the right side transaxle bracket.
23. Slide the transaxle back off the locating dowels and remove it from the vehicle.

To install:

24. Carefully fit the transaxle assembly and install the transaxle-to-engine bolts. When installing, place the halfshafts on top of the subframe; tighten the lower bolts first.
25. Torque the transaxle-to-engine bolts to 40 ft. lbs. (54 Nm).
26. Torque the subframe support bolts and the transaxle support bolts to 29 ft. lbs. (39 Nm).
27. Torque the subframe-to-body bolts to 80 ft. lbs. (108 Nm).
28. Install the halfshafts and torque the bolts to 32 ft. lbs. (47 Nm).
29. Install the clutch cylinder and shift rod. Adjust the shift linkage as required.
30. When installing the exhaust pipe, use new self locking nuts, if equipped and torque to 25 ft. lbs. (34 Nm).
31. Connect the wiring and hoses and install the lower body panel.

1988 5000 Quattro

1. Disconnect the negative battery cable and disconnect the rpm sensor.
2. Remove the upper engine-to-transaxle attaching bolts. Disconnect the speedometer.
3. Raise and safely support the vehicle. Disconnect the tie rod coupling from the steering rack, first removing the self locking nuts below the tie rod coupling and second, the mounting bolts.
4. Drive out the clutch slave cylinder lock pin. Remove the clutch slave cylinder leaving the hydraulic lines attached. Disconnect the back-up light switch and shift linkage.
5. Support the engine with a suitable engine hoist. Remove the deflector for the halfshaft and the right transaxle mount. Remove the right transaxle mount.
6. Disconnect the exhaust pipe at the flange and the right halfshaft at the transaxle. Remove the left transaxle mount and disconnect the left halfshaft at the transaxle.
7. On Quattro vehicles, disconnect the driveshaft at the transaxle. Disconnect the differential lock cable and remove the transaxle cover plate.
8. Raise the engine slightly and place a suitable transaxle jack under the transaxle. Remove the lower engine-to-transaxle bolts.
9. Remove the transaxle from un-

der the vehicle. On the Quattro Turbo, remove the transaxle towards the rear of the vehicle, making sure the transaxle clears the halfshafts, tie rods and shift linkage.

To install:

10. Carefully fit the transaxle in place and torque the engine-to-transaxle mounting bolts to 43 ft. lbs. (58 Nm). Torque the transaxle mount-to-subframe attaching bolts to 32 ft. lbs. (43 Nm). Torque the transaxle mount–to–transaxle bolts to 29 ft. lbs. (39 Nm).

11. Torque the halfshaft-to-transaxle mounting bolts to 58 ft. lbs. (79 Nm). Torque the halfshaft attaching bolts to 32 ft. lbs. (43 Nm).

100

1. Disconnect the negative battery cable.
2. Remove the upper engine-to-transaxle bolts.

NOTE: Tag all bolts during removal, so all bolts can be replaced in their correct locations, as they are not all the same size.

3. Disconnect the ground strap from the transaxle, if equipped.
4. Remove the wiring connectors from the speedometer sender and the multi-function switch.
5. Support the engine.
6. Disconnect the wiring for the oxygen sensor and oxygen sensor heating element.
7. Raise and safely support vehicle. Remove the splash shield, if equipped.
8. Disconnect the exhaust pipe from the manifold.
9. Separate the exhaust pipe behind the catalyst and remove the pipe and catalyst.
10. Remove the bolt for the shift rod at the transaxle and separate.
11. Remove the heat shield from the right inner CV-joint.
12. Remove the halfshafts from the flanges and tie aside.
13. Remove the heat shield for the bonded rubber bushing on the right side.
14. Support the transaxle.
15. Remove the strut at the rear of the transaxle.
16. Remove the clutch slave cylinder. Do not remove the hydraulic line from the slave cylinder.
17. Remove the lower transaxle-to-engine bolts.
18. Pry transaxle back and lower assembly.
19. Remove the transaxle.

To install:

20. Installation is the reverse of the removal procedure. When installing the transaxle, make certain alignment **bushings are in the cylinder block before reassembly.**
21. Press clutch master cylinder in with a lever, until retaining bolt can be installed.

NOTE: A replacement bolt for mounting the clutch slave cylinder is available from Audi, with a pointed tip for easier installation.

22. Tighten subframe mounting bolts to 25 ft. lbs. (34 Nm), plus an additional 90 degree turn.
23. Tighten transaxle retaining bolts as follows: Torque the 8mm bolts to 18 ft. lbs. (24), the 10mm bolts to 33 ft. lbs. (45 Nm) and the 12mm bolts to 48 ft. lbs. (65 Nm).
24. Torque the halfshaft-to-flange bolts to 33 ft. lbs. (45 Nm) and the halfshaft-to-transaxle and final drive bolts to 40 ft. lbs. (54 Nm).
25. Torque the tie rod coupling to steering rack to 33 ft. lbs. (45 Nm).

200

1. Disconnect the negative battery cable.
2. Remove the upper engine to transaxle bolts.
3. Remove the connector for the speedometer sender by pressing in the clips. If equipped with a turbocharger, unscrew the cover plate although it cannot be removed yet.
4. Remove the clip from the clutch slave cylinder and drive out spring pin, if equipped. Remove the bolt securing the clutch slave cylinder to the transaxle and remove the cylinder. Leave the hydraulic line connected.
5. Support the engine. Tie up coolant hoses and cables, as needed.
6. Remove the right side guard plate.
7. Disconnect the halfshafts from the flanges and rest both halfshafts on top of the subframe.
8. Tag and disconnect the wire from the back-up light switch. On Quattro vehicles tag and disconnect vacuum hoses at the servo.
9. Pry off the shift and adjusting rods.
10. Remove the lower engine-to-transaxle bolts.
11. Remove the starter.
12. Remove the guard plate from the subframe.
13. With suitable jack, lift transaxle slightly.
14. Remove both rear subframe mounting bolts.
15. Remove both transaxle support bolts from the subframe.
16. Remove the bracket from the transaxle, push tension system cable and bracket off the retainer on transaxle. The retainer can only be removed with the transaxle out of the vehicle.
17. Remove the right side transaxle bracket.
18. Pull transaxle off dowel sleeves.
19. Lower transaxle and take out from below.

To install:

20. Installation is the reverse of the removal procedure. Before installing the transaxle, rest both halfshafts on top of the subframe.
21. Lubricate mainshaft splines.
22. Install transaxle onto dowels and install the lower bolts.
23. Install the tensioning system bracket and cable to the transaxle.
24. Tighten the transaxle bracket and subframe upper bolts to 29 ft. lbs. (39 Nm).
25. Check alignment of transaxle and torque transaxle-to-engine bolts to 40 ft. lbs. (54 Nm).
26. Torque subframe-to-body bolts to 80 ft. lbs. (108 Nm).
27. Torque halfshaft-to-drive flange bolts to 58 ft. lbs. (79 Nm).

V8 Quattro

NOTE: Any time the transaxle is removed from this vehicle, the drive shaft must be properly aligned. This requires the special alignment tool 3139 or equivalent.

1. Disconnect the negative battery cable, located under the rear seat.
2. Remove the strut between the front shock towers.
3. Remove the air cleaner assembly and the ducts.
4. Remove the bracket for the ingition wires. This is held with self locking bolts, which should be replaced.
5. Remove the wiring harness bracket and the wiring retainers from the transaxle. Remove the 2 upper transaxle–to–engine bolts.
6. Remove the bolts for the right side engine mount.
7. Raise and safely support the vehicle and remove the front wheels.
8. Remove the splash shield under the engine and the body crossmember.
9. Remove the front exhaust pipe with the catalytic converter and the transaxle heat sheild.
10. Install the drive shaft alignment tool. Matchmark the drive shaft and remove the drive shaft and heat shield.
11. Remove the transaxle mount–to–frame bolts.
12. Lower the vehicle and install an engine support bridge VAG 10-222A or equivalent across the inner fenders as shown. Connect the bridge to the left engine mount and raise the engine just enough to take the weight off the mounts.
13. Remove the halfshaft shields and disconnect the halfshafts from the

Install the engine support bridge to take the weight off the mounts

flanges. Support the shafts so they do not hang by the outer CV-joints.

14. Disconnect the hydraulic line from the clutch shave cylinder and plug the fittings to keep them clean.

15. Remove the clamp and lock sleeve to disconnect the shift linkage rods.

16. Disconnect the wiring for the back-up lights and speedometer sensor and remove the wiring from the brackets.

17. Remove the cable guide for the seat belt tensioning system and position the cable/guide up out of the way.

18. At the left transaxle mount, pry the wire bracket open to release the oxygen sensor wire.

19. Disconnect the oil lines from the transaxle and plug the fittings. Detatch the lines at the air conditioning compressor.

20. Support the transaxle and subframe with a suitable transmission jack and remove the sub-frame mounting bolts. Lower the sub-frame to give clearance but make sure the radiator fan still turns freely; loosen the engine mounts if needed.

21. Tie the frame in place so the jack can be removed. Secure the halfshafts to the frame.

22. Disconnect the transaxle mounts and remove the engine–to–transaxle bolts. Move the unit back and carefully lower it out of the vehicle.

To install:

23. Installation is the reverse of removal. Check the condition of the dowel sleeves in the engine block and replace if necessary. Observe the following engine–to–transaxle bolt torques:

a. 12mm – 48 ft. lbs. (65 Nm)
b. 10mm – 32 ft. lbs. (45 Nm)
c. 8mm – 18 ft. lbs. (20 Nm)

24. Torque the sub-frame mounting bolts to 48 ft. lbs. (65 Nm) plus an additional ¼ turn.

25. Install all the engine and transaxle mounting bolts loosely before torquing any of them. Torque the engine and transaxle mount bolts and the seat belt tensioning system bolts to 30 ft. lbs. (40 Nm).

26. Connect and adjust the shift linkage before installing the exhaust system. When installing the exhaust, tignten the clamps after making sure everything is properly positioned to minimumize vibration.

27. Install the drive shaft with the matchmarks aligned and torque the flange bolts to 41 ft. lbs. (55 Nm). The shaft must be aligned.

28. To align the drive shaft, install the drive shaft with the alignment tool attached but do not install remove the bolts for the center support. Measure the gap between the support bolt holes and the body and install the proper spacers to make the gap even on both sides. Make sure the support is centered front to rear and install the bolts. Torque the bolts to 15 ft. lbs. (20 Nm) and remove the alignment tool.

29. Torque the halfshaft flange bolts to 58 ft. lbs. (80 Nm). After connecting the clutch slave cylinder line, bleed the system.

LINKAGE ADJUSTMENTS

80, 90 and Coupe, Except Quattro

1. Place the shift lever into the neutral position.

2. Raise and safely support the vehicle and loosen the clamp nut on the shift rod. Be certain that the shift finger slides freely on the shift rod.

3. Working inside the vehicle, remove the gear shift lever knob and the boot.

4. Loosen the shifter base plate bolts slightly. Align the holes in the plate with the holes in the bearing housing and tighten the bolts.

5. Using the alignment tool 3057, or equivalent,slip it over the gearshift lever and make sure the locating pin is in the front centering hole.

6. Position the shift lever to the right detent cut out for 5th and reverse and tighten the lower knurled nut of the tool.

7. At the top of the tool, move the slide with the gear shift lever to the right stop. Tighten the upper knurled nut of the tool.

8. Position the gear shift lever into the left cut-out of the slide. Adjust the shift rod and the shift finger with the transaxle in neutral and tighten the clamp nut.

9. Remove the tool and check the shifting of the gears for smoothness.

5000

1. Remove the gear shift boot.

2. Position the shift lever in neutral.

3. The seam on the plastic stop bracket should line up with the center hole in the curved stop plate. If not, proceed below:

NOTE: On the Quattro, adjust the adjusting rod (center-to-center) to 5.275 in. (134mm) and install the rod.

4. Loosen the 4 bolts at the base of the shifter.

5. Align the holes in the shifter base with the holes in the bearing plate directly below it.

6. Tighten the bolts.

7. Loosen the clamp between the front and rear shift rods; the rear shift rod must move freely.

8. Make certain the front shift rod is in the neutral position.

9. Using the shifter locating tool 3048, or equivalent, place it on the stop plate with the shift lever resting

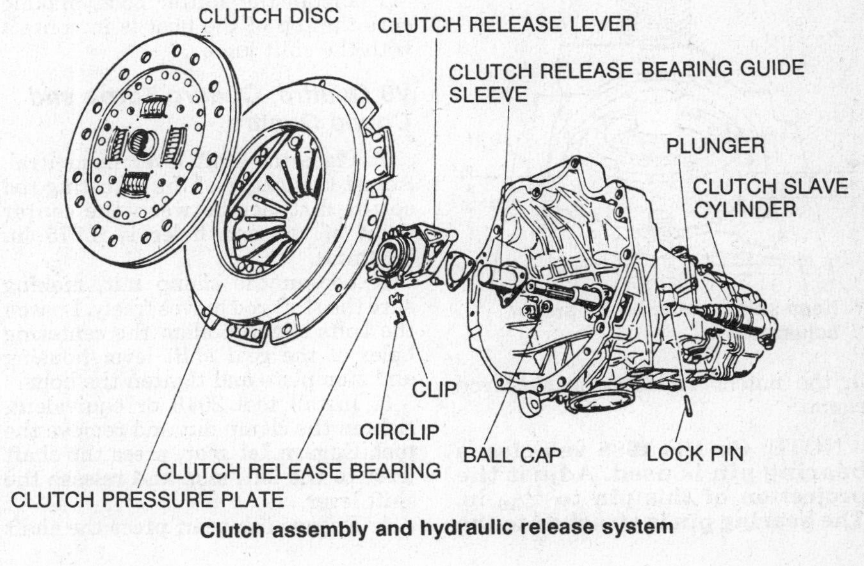

Clutch assembly and hydraulic release system

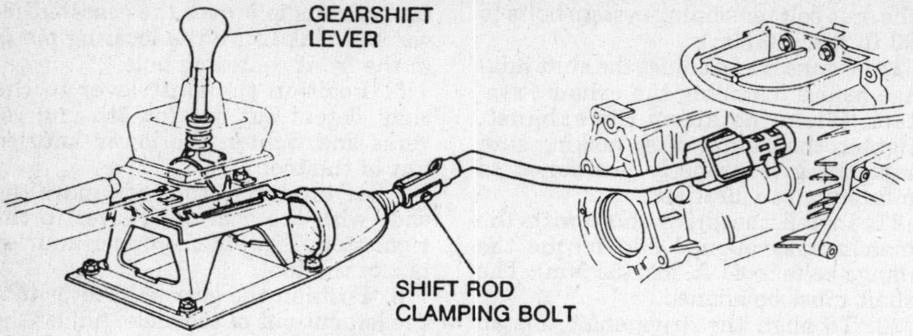

Manual shift linkage—80/90 and 100

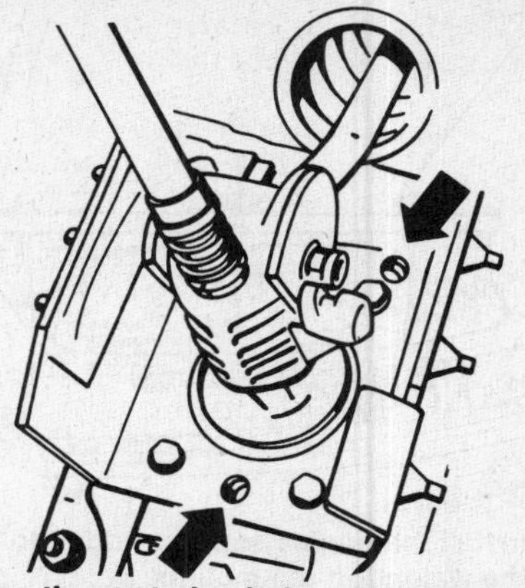

Align the centering holes of the stop lever bearing, then tighten the bolts on 5000

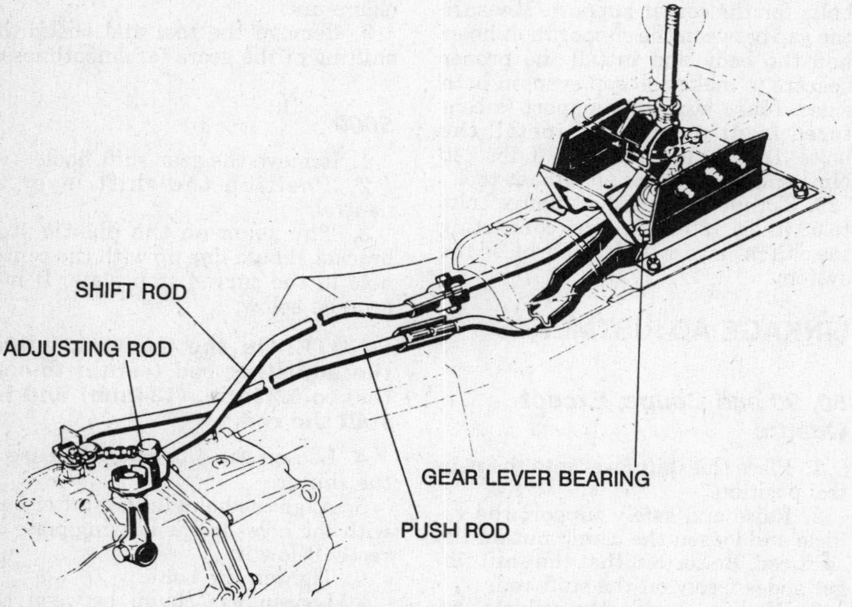

Manual shift linkage—5000, 200 and V8 Quattro

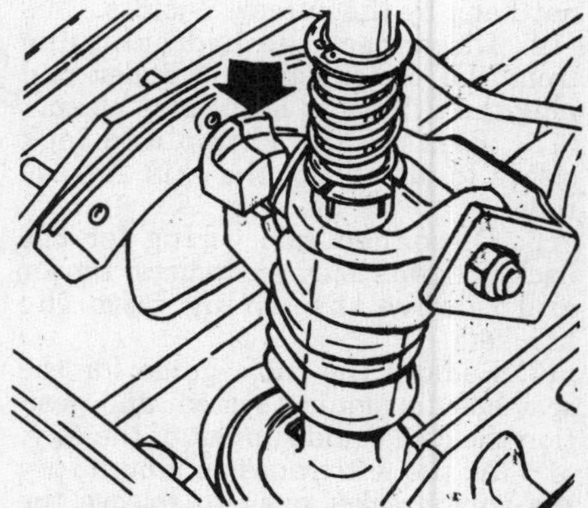

The plastic stop bracket should align with the curved stop plate—5000

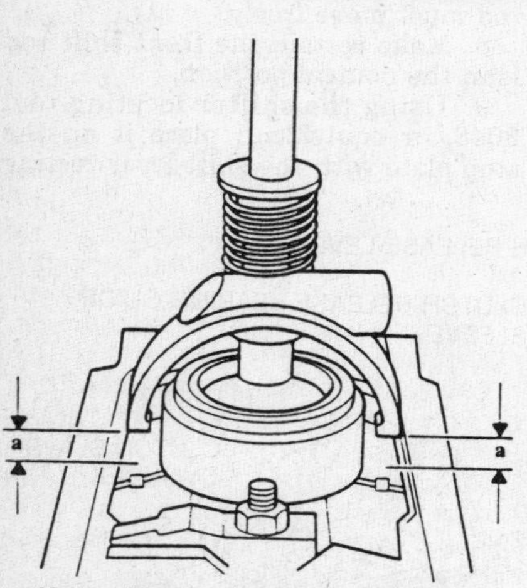

Keep shifter centered for proper adjustment

in the notch. Tighten the shift rod clamp.

NOTE: On the 1988 vehicles, a bearing pin is used. Adjust the projection of this pin to $^{11}/_{16}$ in. The bearing pin is attached to the shift lever lower bearing and faces the rear of the vehicle.

10. Release the shifter and check its operation in all gears.
11. Install the shifter boot, making sure the top of the boot is in contact with the shift knob.

V8 Quattro, Quattro Turbo and Coupe Quattro

1. Place the shift lever in neutral. Adjust the length of the adjusting rod so the distance between the center point of the end holes is 5.275 in. (134mm).
2. Loosen the clamp nut, making sure the shift rod moves freely. Loosen the bolts slightly, align the centering holes of the gear shift lever housing and stop plate and tighten the bolts.
3. Install tool 3048 or equivalent, tighten the clamp nut and remove the tool. Engage 1st gear, press the shaft lever to the left, stop and release the shift lever.
4. Engage 5th gear, press the shaft lever to the right, stop and release the shift lever.
5. If the lever does not spring back approximately the same distance as in Steps 3 and 4, move the gear shift lever housing slightly in the slots sideward.
6. Make sure all gears engage easily without jamming.

100 and 200

1. Loosen the shift rod clamping bolt.
2. Place the gear shift in a vertical position so the dimensions are equal on both sides and retighten shift rod clamp bolt.

CLUTCH

Clutch Assembly

REMOVAL & INSTALLATION

1. Disconnect the negative battery cable. Raise and safely support the vehicle and remove the transaxle.

2. Mark the relationship of the pressure plate to the flywheel, only if it is to be reused.

3. Using a suitable tool, lock the flywheel. Unbolt the pressure plate from the flywheel, loosening the bolts alternately, a little at a time, to prevent warpage.

To install:

4. Installation is the reverse of the removal procedure. Install the clutch with the driven plate on the pressure plate so the spring cage is facing the pressure plate.

5. Hold the clutch assembly against the flywheel, aligning the marks made in Step 2 and the dowel pins on the flywheel with the pressure plate. Insert an alignment shaft tool through the pressure plate and the driven plate into the crankshaft pilot bearing.

6. Install the pressure plate bolts finger tight. Tighten the bolts evenly, in rotation, to avoid distortion. Torque the bolts to 18 ft. lbs. (24 Nm). Remove the alignment shaft.

7. The clutch release bearing in the front of the transaxle should be checked before reassembly. It is retained by 2 springs.

8. Replace the transaxle. Torque the engine-to-transaxle bolts to 40 ft. lbs. (54 Nm) and the halfshaft to 58 ft. lbs. (79 Nm).

FREE-PLAY ADJUSTMENT

All vehicles use a hydraulic clutch release mechanisim. No free play adjustment is required or possible. If the clutch does not release or engage properly and pedal height is correct, try bleeding the system before moving on to more extensive repairs.

PEDAL HEIGHT ADJUSTMENT

The clutch pedal should be about 3/8 in. (10mm) above the brake pedal. To adjust the pedal height, remove the cotter pin holding the clutch master cylinder clevis to the pedal, loosen the locknut on the clevis shaft and turn the shaft to give the required pedal height. Tighten the locknut and install the clevis on the pedal.

Clutch Master Cylinder

REMOVAL & INSTALLATION

NOTE: The use of a pressure bleeder is necessary for this procedure. Before beginning, remove and plug the fluid line from the reservoir to the master cylinder. Empty the fluid in the line into a suitable container.

1. Disconnect the negative battery cable. Locate the master cylinder under the instrument panel, behind the clutch pedal.

2. Remove and plug the line leading to the slave cylinder from the end of the master cylinder.

3. Remove the circlip and the pin which attaches the clevis to the clutch pedal.

4. Remove the 2 master cylinder mounting bolts from the pedal mounting.

5. Remove and plug the reservoir line. Remove the master cylinder.

6. Installation is in the reverse order of removal. Tighten the master cylinder mounting bolts to 15 ft. lbs. (20 Nm).

7. Bleed the clutch system when finished.

Clutch Slave Cylinder

REMOVAL & INSTALLATION

1. Disconnect the negative battery cable. Locate the slave cylinder on top of the transaxle housing.

2. Remove the retaining clip from the pin.

3. Drive out the slave cylinder lock pin, using a small punch.

4. Remove and plug the fluid line at the slave cylinder. This step is necessary only if the cylinder is being replaced.

5. Lightly grease the machined surfaces of the transaxle housing and the slave cylinder.

6. Install the fluid line on the slave cylinder. Install the slave cylinder in the transaxle. Install the retaining pin.

7. The remainder of the installation is the reverse of the removal procedure.

8. If the fluid line was removed, bleed the system.

Hydraulic Clutch System Bleeding

The clutch system should be bled using a pressure bleeder. Follow the instructions that come with the bleeder tank, for the proper bleeding procedure. The maximum line pressure must not exceed 36 psi. (248 KPa).

AUTOMATIC TRANSAXLE

For further information on transmissions/transaxles, please refer to "Chilton's Guide to Transmission Repair".

Transaxle Assembly

REMOVAL & INSTALLATION

089 Series Transaxle

1. Disconnect the negative battery cable.

2. Remove the upper engine-to-transaxle bolts. Raise and support the vehicle safely.

3. Using a suitable engine support tool, secure it to the engine and the vehicle.

4. At the front of the engine, remove both top bolts. Remove the starter.

5. Through the starter opening, remove the torque converter to drive plate bolts and remove torque converter cover plate.

6. Clamp off the coolant hoses at the ATF cooler and remove the hoses from the cooler.

7. Remove the speedometer cable from the transaxle.

8. Remove the inner halfshaft-to-transaxle bolts. Using a wire, tie up the halfshafts.

9. At the left control arm, mark the position of the ball joint and remove the ball joint and the support.

10. Place an oil catch pan under the transaxle, remove the oil filler tube from the oil pan and drain the fluid.

11. Remove the exhaust pipe-to-transaxle bracket.

12. Remove the selector cable bracket from the transaxle. At the transaxle shift lever, remove the selector cable circlip and the cable.

13. At the transaxle, remove the accelerator cable bracket and the cable from the operating lever.

14. From the transaxle mount, remove the center bolt. Using the engine support tool, lift the engine slightly.

15. Remove the throttle cable bracket bolts and the bracket.

16. Support the transaxle and lift it slightly. Remove the lower transaxle-to-engine bolts.

17. Separate the engine from the transaxle and lower it from the vehicle. Be sure to secure the torque converter.

To install:

18. When installing the transaxle, should the torque converter slip off the one-way clutch support, the oil pump shaft could be pulled from the oil pump. This may cause severe damage when bolting the transaxle to the engine. Make sure the torque converter is properly positioned before installing the bolts.

19. Torque the engine-to-transaxle bolts to 41 ft. lbs. (56 Nm), the subframe bolts to 52 ft. lbs. (71 Nm), and the transaxle mount center bolt to 30 ft. lbs. (40 Nm).
20. Install the torque converter-to-driveplate bolts and torque to 22 ft. lbs. (30 Nm). Install the cover plate.
21. Install the halfshaft-to-transaxle bolts and torque to 33 ft. lbs. (45 Nm).
22. Connect the ball joint to the control arm and torque the bolts to 48 ft. lbs. (65 Nm).
23. Connect the hoses to the oil cooler. Install the oil filler tube and refill the transaxle.
24. Connect and adjust the selector cable as required.
25. Connect and adjust the accelerator linkage and align the engine-to-transaxle mounts, if necessary.

ATU 018 Series Transaxle

V8 QUATTRO ALL WHEEL DRIVE

1. Disconnect the negative battery cable. The battery is under the rear bench seat.
2. Raise and safely support vehicle. Remove the front wheels.
3. Remove the cross struts for the spring strut domes.
4. Remove the air cleaner housing and air cleaner assembly.
5. Remove the supports for the ignition cables and the left and right distributor caps. Remove the throttle cable and support.
6. Disconnect the oxygen sensor and probe heater. Remove the cable clamp for the electrical cables running alongside.
7. Disconnect the 2-pin plug connection at the transaxle bell housing. Disconnect the cable clamp for the wire harness next to the 2-pin plug.
8. Disconnect the retaining strap for the ventilation hose at the firewall.
9. Loosen the transaxle filler pipe on the cylinder head
10. Remove the radiator fan. The left fan can be set aside while still connected.
11. Remove the right engine mounting bolts on the body.
12. Raise and safely support vehicle. Remove the lower engine cover, the crossmember, the front exhaust pipe with catalytic converter and the heat deflector for the driveshaft.
13. Remove the driveshaft by loosening the driveshaft mounting bolts on the transaxle, rear final drive and body.

NOTE: A special driveshaft assembly device or aligning jig is required, tool 3213, or equivalent. This tool keeps the multi-piece driveshaft straight and in proper alignment. Attach this jig to the driveshaft and tighten the nuts. Remove the bolts on the transaxle and rear final drive. Support the driveshaft and alignment jig, remove the mounting bolts from the body and carefully lower the the driveshaft from the vehicle. Take note of any shims. Always move and store the driveshaft flat.

14. Detatch the selector lever cable and support bracket at the transaxle, and remove the oil filter. Remove the upper bolt on the starter by guiding a 10mm Allen socket with extension and flex fitting through the opening, on the transaxle housing, over the final drive. Remove the bolt and take out the starter.
15. Working through the starter opening, remove the 6 torque converter bolts.
16. Remove the mounting bolts on both sides of the transaxle mounting.
17. Attach a lifting hoist to the engine mounting on the left side and lift slighty.
18. Under the vehicle, support the subframe using a suitable transaxle jack. Remove the mounting bolts on the subframe and lower the subframe carefully until it hangs freely. Disconnect the halfshafts from the transaxle and tie up.
19. Drain the transaxle and remove the filler pipe. Unscrew the cooling lines, the retaining clip that holds both lines together and tie to the subframe, aside.
20. Remove the left and right transaxle mounts.
21. Disconnect the eight-pin electrical connector on the transaxle by turning counterclockwise.
22. Disconnect the plug on the speed sensor. Push the locking lever down and remove the plug from the multi-function transaxle switch. Remove the retainers for the electrical wires from the transaxle and unclamp the ventilation hose.
23. Remove the tabs on the seat belt tensioning cables and disconnect the cables.
24. Remove the speed and TDC sensor with heat deflector plate on the engine. Remove the 2 upper engine-to-transaxle bolts.
25. Support the transaxle with a suitable transaxle jack. Remove the remaining engine-to-transaxle connecting bolts and remove the transaxle from the engine. Lower the transaxle carefully and secure the torque converter to keep it from falling out. Use caution not to damage the halfshafts, bolt-on parts or the multi-function switch.

To install:

26. Installation is the reverse of the removal process. If a new replacement transaxle is being installed, the following must be carried out. Transfer the catalytic converter mountings on the left and right side of the housing. Transfer the pipes for the transaxle cooler and the support for the seat belt tensioning cables. Check that both guide sleeves are installed in the engine block at the 2 o'clock and 8 o'clock positions, viewed from the flywheel.
27. Make sure the torque converter is secured and install the transaxle in reverse order. Note that if only the torque converter is replaced, it should be carefully positioned on the free-wheel support and should not be tilted. To engage the splines of the pump shaft, rotate the torque converter forward and backward slightly. The torque converter must be inserted onto the free-wheel support up to the stop.
28. Install the transaxle and make sure the halfshafts, any bolt-on parts and the multi-function switch all work freely.
29. Install and adjust the driveshaft. Align with care. Use the following procedure:
   a. With the alignment jig holding the driveshaft straight, set the driveshaft in place.
   b. Carefully, measure the distance between the body pan bolt holes and the mounting ears of the center support bracket. This distance should be the same left to right.
   c. Five different shims are available: 2, 4, 6, 8 and 10mm. Determine the thickness of the required shims.
   d. Push the driveshaft all the way to the rear up to the stop. Mark the position of the center mounting bracket on the floor pan.
   e. Push the driveshaft forward all the way and park the position of the center mounting bracket on the floor pan. The center mounting must be midway between these 2 marks.
   f. Insert the bolts and shims as determined beforehand and tighten to 15 ft. lbs. (20 Nm). Remove the alignment jig.
30. Fill the unit with automatic transaxle fluid.
31. Adjust the selector cable. Check the throttle cable linkage and, if necessary, adjust. Install and tighten all bolts for the lower engine cover. Check engine oil level. Align front end.
32. Tightening torques are:
   a. Cross brace to suspension strut domes-to-17 ft. lbs. (23 Nm).
   b. Engine mounting-to-body to 30 ft. lbs. (45 Nm).
   c. Torque converter-to-driveplate to 26 ft. lbs. (35 Nm).
   d. Transaxle mounting-to-subframe to 30 ft. lbs. (45 Nm).

e. Transaxle support for mounting-to-transaxle to 30 ft. lbs. (45 Nm).

f. Subframe-to-body to 48 ft. lbs. (65 Nm) plus ¼ turn.

g. Halfshaft-to-transaxle to 59 ft. lbs. (80 Nm).

h. Driveshaft-to-transaxle to 41 ft. lbs. (55 Nm).

i. Transaxle-to-engine to 44 ft. lbs. (60 Nm).

j. Support for seat belt cables to 30 ft. lbs. (45 Nm).

k. Side threaded bushings to 111 ft. lbs. (150 Nm).

Automatic Shift Linkage

All vehicles with automatic transaxle covered in this section are equipped with the Audi Shiftlock II transmission control system. The system must be properly adjusted to make sure the transaxle is fully engaged in each shift position. If this is not done, the transaxle may be only partially engaged in a certain range position, causing severe damage due to slippage. Improper adjustment may also make it impossible to shift into or out of Park.

The Shiftlock II system is designed to make it impossible to shift out of Park or Neutral unless certain parameters are met. The system depends on proper operation or condition of fuse S12, the brake light switch circuit, the interior light relay control unit, the driver's door switch circuit and vehicle speed signal.

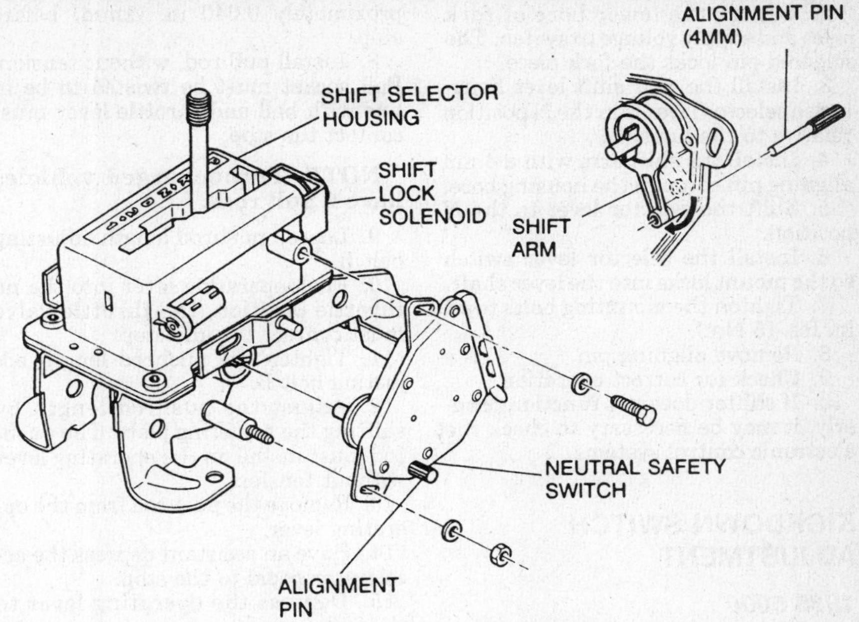

Adjusting neutral safety switch

System Function

With ignition switch **ON**, the selector cannot be shifted out of **P** or **N** unless the brake pedal is depressed. At speeds under 3.7 mph, when shifted into **N** the shifter should lock in **N** after 1 second, unless the brake pedal is depressed. At speeds above 3.7 mph the shifter should not lock.

SYSTEM ADJUSTMENT

1. Adjust the solenoid switch, using a 1mm gauge between the selector lever and solenoid switch. With lever in **R**, push the solenoid against the gauge and tighten to 7 ft. lbs. (9 Nm).

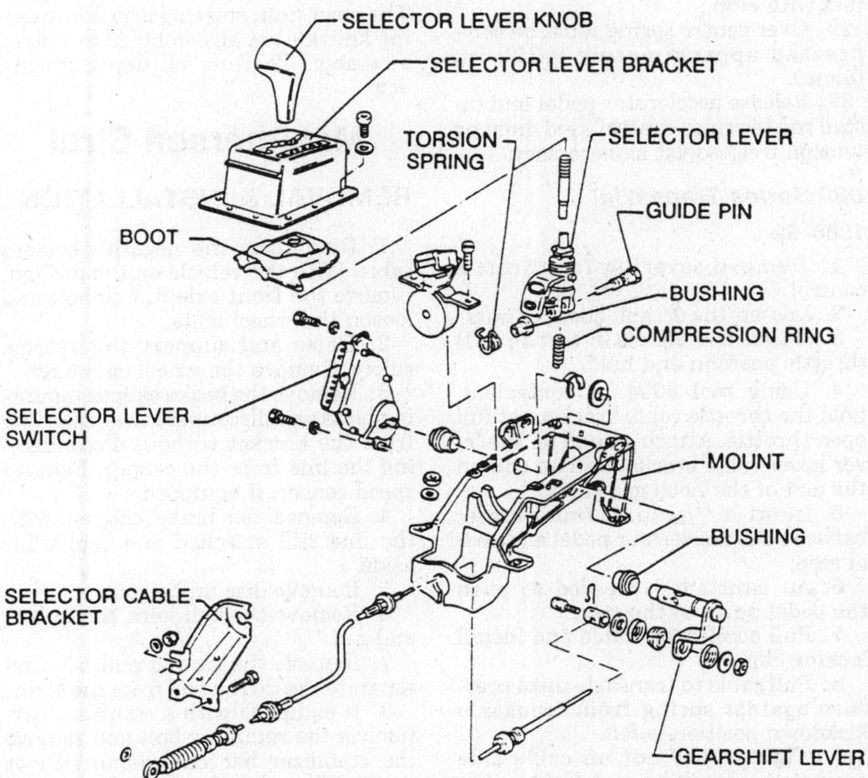

Shiftlock II automatic transaxle shifter assembly

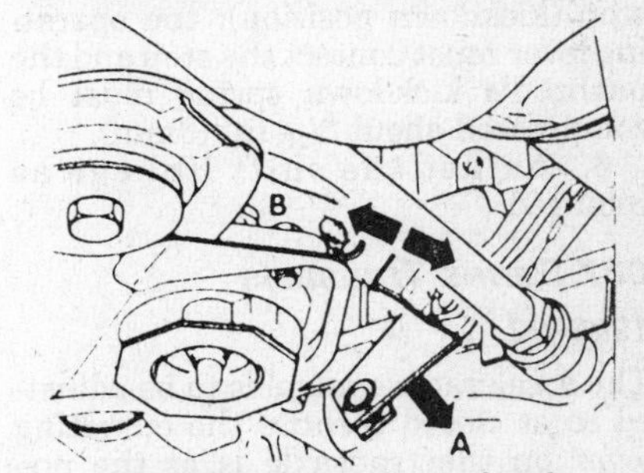

Adjusting transaxle pushrod

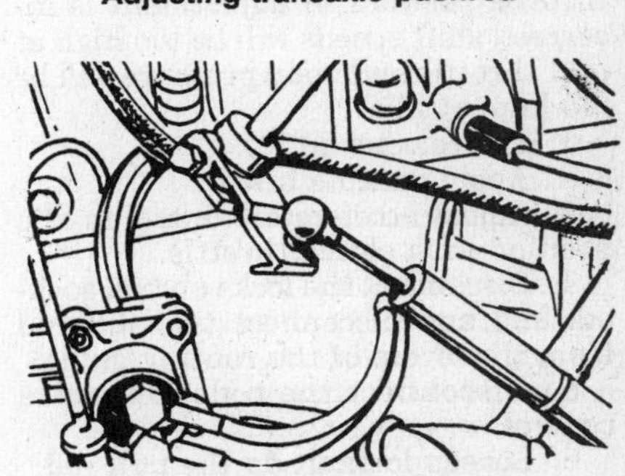
Kickdown detent linkage

2. Center the lower bore of fork piece and supply voltage to switch. The solenoid pin locks the fork piece.
3. Install the gear shift lever housing so selector lever is in the **N** position relative to the housing.
4. Install the shift arm with a 4mm aligning pin through the housing bore.
5. Shift the selector lever to the **N** position.
6. Install the selector lever switch so the mount locks into the lever shaft.
7. Tighten the mounting bolts to 44 in. lbs. (5 Nm).
8. Remove aligning pin.
9. Check for correct operation.
10. If shifter does not function properly, it may be necessary to check the electronic control systems.

KICKDOWN SWITCH ADJUSTMENT

1988 5000

1. Make sure the engine is warmed up. Position the accelerator pedal in the fully released position and make sure the throttle is fully closed.
2. Check the distance between the pedal lower edge and the pedal stop. Clearance should be 3.0 in. (76mm).
3. If not, loosen the lock bolt which holds the cable at the pedal and place the pedal to give the proper clearance. Tighten the lock bolt.
4. Press the pedal to the full throttle position, but not into the kickdown detent. The kickdown take-up spring should not be compressed and the throttle valve should be wide open.
5. Press the accelerator lever to the stop (kickdown position); the operating lever must contact the stop and the pushrod's kickdown spring must be compressed about $^5/_{16}$ in. (8mm).
6. Adjust the shift linkage as required.

087 Series Transaxle

1989–92

The accelerator control is to be adjusted so at closed throttle the operating lever on the transaxle is at the no-throttle position. If adjustment is incorrect, shift speeds will be too high at part throttle and main pressure will be too high at idle.

1. Put selector in **P**.
2. Apply parking brake.
3. Adjust accelerator control in idle position with closed throttle.
4. Disconnect the locks on ball sockets and and disconnect the pull rod from the levers of the routing guide.
5. Disconnect the rods for cruise control.
6. Loosen locknut on the pull rod.
7. Position lever for pull rod approximately 0.040 in. (1mm) before stop.
8. Install pull rod, without tension. Ball socket must be twisted to be in line with ball and throttle lever must contact the stop.

NOTE: Turbocharged vehicles have 2 pull rods.

9. Loosen pushrod length adjusting bolt B.
10. Push operating lever into the no throttle position, the throttle valve must contact the idle stop.
11. Tighten the pushrod length adjusting bolt B.
12. Adjust the pushrod length by shifting the adjusting plate. The pushrod must install on the operating lever without tension.
13. Remove the pushrod from the operating lever.
14. Have an assistant depress the accelerator pedal to the stop.
15. Depress the operating lever to the kickdown stop.
16. Using pliers, pull the accelerator pedal cable back and fasten.
17. Check that the operating lever is in contact with the kickdown stop, readjust the pedal cable, if necessary.
18. Install the pushrod onto operating lever and secure.
19. Check throttle lever operation.
20. Push accelerator cable through full throttle position to kickdown stop.
21. Transaxle lever must be in contact with stop.
22. Over center spring must be compressed approximately 0.320 in. (8mm).
23. Release accelerator pedal and install rod to cruise control. Rod must be tension free, adjust as necessary.

089 Series Transaxle

1988–92

1. Remove covering for throttle control.
2. Loosen the 2 cable locking nuts.
3. Turn the throttle to the full throttle position and hold.
4. Using tool 3004 or equivalent, hold the throttle cable brackets at full open throttle. Attach an end on the lever lower cable bracket and an end on the end of the hood gas strut.
5. Insert a $^{11}/_{16}$ in. (17mm) spacer between the accelerator pedal and pedal stop.
6. An assistant is needed to push the pedal down to the stop.
7. Pull accelerator cable and install locking clip.
8. Pull cable to transaxle until pressure against spring from transaxle kickdown position is felt.
9. Tighten the nut on cable side against the bracket, next tighten the nut on pivot side against bracket.
10. Remove tool.
11. Throttle lever must rest against the idle stop when the accelerator is released.
12. Press accelerator to full throttle position, not kickdown.
13. The pressure point for full throttle position of the accelerator pedal must be approximately ¾ in. (19mm) away from the pedal stop.
14. Press the accelerator to the pedal (kickdown) stop.
15. Transaxle operating lever must contact the kickdown stop.
16. The spring between the cable brackets must be stressed.
17. For vehicles with cruise control, adjust the coupling rod by moving the ball end 0.039–0.059 in. (1–1.5mm).
18. For vehicles with cruise control the air conditioning switch must only switch **ON** at kickdown and not at wide-open throttle, approximately $^3/_{16}$ in. (2mm) between the switch and cable bracket at wide-open throttle (not kickdown).

FRONT SUSPENSION

All vehicles use MacPherson struts. The strut unit, steering arm and steering knuckle are all combined in 1 strut assembly. There is no upper control arm.

MacPherson Strut

REMOVAL & INSTALLATION

1. Disconnect the negative battery cable. With the vehicle on the ground, remove the front axle nut or bolt and loosen the wheel bolts.
2. Raise and support the vehicle safely. Remove the wheel assembly.
3. Remove the brake caliper mounting bolts and disconnect the brake line from the bracket without disconnecting the line from the caliper. Remove speed sensor, if equipped.
4. Remove the brake caliper, with the line still attached and support it aside.
5. Remove disc brake rotor.
6. Remove the ball joint clamp bolt and nut.
7. Remove the tie rod end nut and separate the tie rod end from the strut.
8. If equipped with a stabilizer bar, remove the retaining bolt and remove the stabilizer bar end clamps. Pivot the stabilizer bar downward.
9. Remove the 2 center stabilizer

bar clamps and unbolt it from the lower control arm.

10. Remove the pinch bolt from the steering knuckle and separate the lower ball joint from the knuckle by pushing down on the control arm.

11. Using a suitable hub puller, press the halfshaft out of the hub.

12. On 80 and 90, support the strut assembly, hold the shock absorber piston rod with an internal socket wrench and remove the retaining nut. Remove the strut assembly from the vehicle.

13. On all other models, remove the upper strut cover, support the strut assembly and remove the 3 strut retaining nuts. Remove the strut assembly.

To install:

14. Install the strut in the reverse order. On 80 and 90 models, torque the upper strut retaining nut to 44 ft. lbs. (60 Nm). On other models, torque the 3 upper strut retaining nuts to 22 ft. lbs. (30 Nm).

15. When installing the stabilizer bar, the position is correct if the clamps are difficult to install in the rubber bushings. Attach the clamps loosely, take a short test drive to bring the bushings into the correct position and tighten to 18 ft. lbs. (24 Nm).

16. Tighten the ball joint bolt to 36 ft. lbs. (49 Nm) on 80 and 90 or 47 ft. lbs. on 5000 vehicles or 48 ft. lbs., plus an additional ¼ turn (90 degrees) on 100, 200 and V8 Quattro. Install and seat the speed sensor.

17. When installing the axle shaft, apply a bead of thread locking compound to the splines. When the vehicle is on the ground, torque the center nut or bolt to 207 ft. lbs. (280 Nm) on 5000, 195 ft. lbs. (265 Nm) on 80 and 90, or 148 ft. lbs. (200 Nm) plus ¼ turn on all other models.

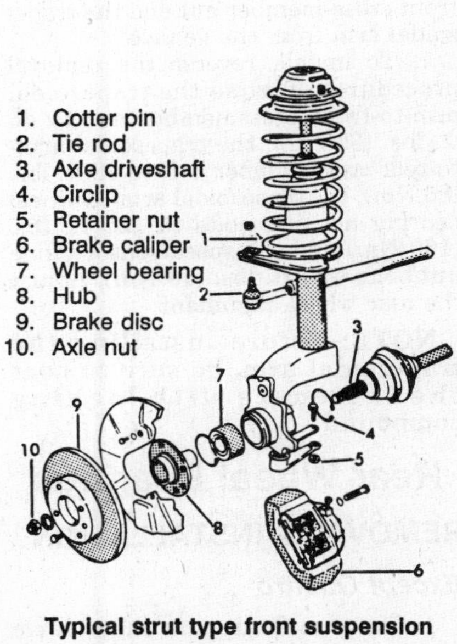

Typical strut type front suspension

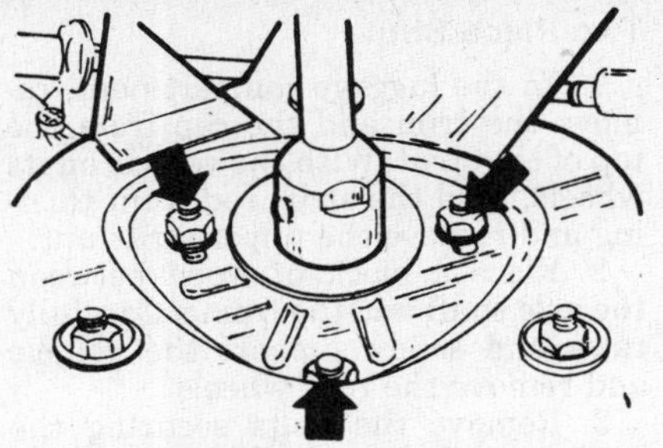

Camber adjustment is at the top of the strut on all except 80/90

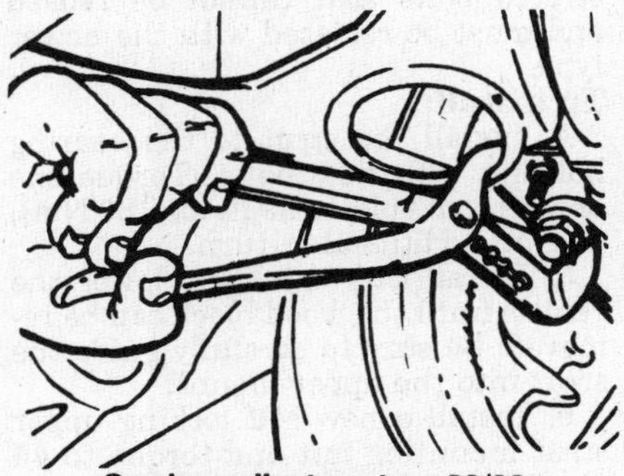

Camber adjustment on 80/90

Lower Ball Joint

REMOVAL & INSTALLATION

Coupe, 80 and 90

1. Raise and safely support the vehicle. Remove the wheel assembly.

2. Remove the lower ball joint clamp nut and bolt and pry the control arm down to disengage the ball joint from the steering knuckle.

3. Matchmark the ball joint to the lower control arm. Remove the ball joint to control arm mounting bolts and remove the ball joint.

To install:

5. Install the new ball joint on the control arm at the matchmarks and tighten the mounting bolts to 46 ft. lbs. (62 Nm).

6. Slowly allow the lower control arm and ball joint to fit into the strut assembly. Install the bolt and tighten to 48 ft. lbs. (65 Nm).

7. Install the wheel assembly and lower the vehicle.

8. Reset the front end alignment when finished.

All Other Models

The ball joint is permenantly assembled to the lower control arm and cannot be replaced separately.

Lower Control Arms

REMOVAL & INSTALLATION

1. Raise and support the vehicle safely. Remove the wheel assembly.

2. Remove the ball joint to strut bolt and nut. Pry and hold the control arm down. Remove the ball joint to control arm mounting bolts and nuts.

3. If equipped with a stabilizer bar, disconnect the end of the stabilizer bar and pull it down.

4. Remove the 2 control arm to subframe bolts and remove the control arm.

NOTE: The ball joint and control arm must be replaced as an assembly.

To install:

5. Installation is in the reverse order of removal. Check control arm bushings for cracking or undue wear. Tighten the control arm-to-subframe bolts to 43 ft. lbs. (58 Nm); the ball joint-to-strut bolt to 46 ft. lbs. (62 Nm) and the stabilizer bar bolts to 18 ft. lbs. (24 Nm).

Sway Bar

REMOVAL & INSTALLATION

1. The sway bar or stabilizer bar is removed and installed with the vehicle standing on its wheels.

2. Remove both sway bar rubber bushing brackets.

3. Disconnect the ends of the sway bar where it goes through the lower control arm and remove.

4. Installation is the reverse of the removal procedure. Use new self-locking nuts and tighten to 80 ft. lbs. (108 Nm).

REAR SUSPENSION

MacPherson Struts

REMOVAL & INSTALLATION

80 and 90

EXCEPT QUATTRO

NOTE: Always remove and install the suspension struts 1 at a time. Do not allow the rear axle to hang in place as this may cause damage to the brake lines.

1. With the vehicle at ground level, open the trunk and remove the trim from around the shock tower.

2. Remove the rubber cap.

3. Hold the strut rod and remove the strut mounting nut.

4. Raise and support the vehicle safely.

5. Remove the lower strut mounting bolt from the axle beam and remove the strut.
6. Installation is the reverse of removal. Torque the upper strut mounting bolt to 14 ft. lbs. (19 Nm) and the lower strut mounting bolt to 43 ft. lbs. (58 Nm).

QUATTRO

Single Piece Strut

1. With the vehicle on the ground, remove the axle nut.
2. Raise and safely support the vehicle and remove the rear wheels.
3. Unbolt the halfshaft from the differential flange.
4. Remove the self locking lower ball joint nut and press the ball joint out of the wheel bearing housing. Use the nut to protect the threads.
5. Loosen the control arm mounting bolts and allow the arm to pivot down out of the way.
6. Remove the tie rod nut and press the tie rod joint out of the bearing housing arm. Use the nut to protect the threads.
7. Pull the parking brake cable out of the strut bracket. Remove the brake caliper with its bracket without disconnecting the hydraulic line. Hang the caliper from the body with wire.
8. Remove the brake disc and slide the wheel speed sensor out of its mounting, if equipped. Press the halfshaft out of the hub.
9. Support the strut from below. In the luggage compartment, remove the trim and the rubber cap at the top of the strut.
10. Hold the top of the strut with an Allen wrench and remove the self locking nut. Lower the strut from the vehicle.

To install:

11. Replace all self locking nuts. Install the strut into the upper mount and torque the new upper nut to 48 ft. lbs. (60 Nm).
12. Make sure the axle shaft splines are clean and apply a bead of threat locking compound to the outer end of the splines. Install the axle shaft but do not torque the center nut until the vehicle is on its wheels. Torque the axle flange bolts to 35 ft. lbs. (45 Nm).
13. Install the brake disc and caliper. Torque the caliper bracket mounting bolts to 92 ft. lbs. (125 Nm).
14. Attach the ball joint to the control arm and the tie rod to the bearing housing and install new self locking nuts. Torque the tie rod nut to 29 ft. lbs. (40 Nm) and the ball joint nut to 54 ft. lbs. (75 Nm).
15. When installation is complete and the vehicle is on the ground, torque the center axle nut to 236 ft. lbs. (320 Nm).

Two Piece Strut

1. In the luggage compartment, remove the trim and the cap from the top of the strut. With the vehicle on its wheels, hold the strut rod from turning and remove the upper strut nut.
2. Place a block of wood between the axle shaft and the frame. Carefully raise and safely support the vehicle and remove the rear wheels.
3. Remove the bolts securing the lower strut to the wheel bearing housing and remove the strut. These are stretch bolts that cannot be reused and must be replaced with the newer type.

To install:

4. Install the strut to the bearing housing with new bolts. Torque the new stretch bolts to 59 ft. lbs. (80 Nm), plus an additional ½ turn.
5. Install the wheel and lower the vehicle until the wood block can be removed. Be sure to carefully guide the strut into the upper mount.
6. Install a new self locking upper strut mounting nut and torque to 44 ft. lbs. (60 Nm). Check the wheel alignment.

5000, 100 and 200

EXCEPT QUATTRO AND V8 QUATTRO

NOTE: The struts must be removed with the weight of the vehicle on the rear wheels. If not, a spring compressor must be used on the rear springs.

1. If the vehicle is not on its wheels, install the spring compressor and compress the spring. Do not attempt to remove the shock with the rear wheels raised without a compressor.
2. Remove the upper strut mounting nut.
3. Remove the lower strut mounting nut.
4. Remove the shock absorber.
5. Installation is the reverse of removal. Torque the lower mounts to 66 ft. lbs. (89 Nm) and the upper to 14 ft. lbs. (19 Nm).

QUATTRO AND V8 QUATTRO

1. Raise and support the vehicle safely. Remove the wheel assembly.
2. Open the trunk and remove the shock absorber covers, the remove the shock absorber-to-body nuts/bolts.
3. Remove the shock absorber-to-rear wheel knuckle assembly. Remove the shock absorber from the vehicle.
4. To install, reverse the removal procedures. Torque the shock absorber-to-body nuts/bolts to 15 ft. lbs. (20 Nm) and the shock absorber-to-rear wheel knuckle assembly bolt to 66 ft. lbs. (89 Nm).

Rear Control Arms

REMOVAL & INSTALLATION

80 and 90 Quattro

1. Raise and support the vehicle safely, under the frame and differential.
2. Using a scribing tool, mark the position of the ball joint carrier with the control arm.
3. Remove the ball joint carrier-to-control arm nuts and the lock plate. Separate the ball joint carrier from the control arm.
4. Remove the control arm-to-subframe bolts and the control arm from the vehicle.
5. To install, reverse the removal procedures. Always using new replacement nuts. Torque the control arm-to-subframe bolts to 43 ft. lbs. (58 Nm) and the ball joint nut to 54 ft. lbs. (75 Nm). Check the rear wheel alignment.

100 and 200 Quattro

On these vehicles, the control arm is called the trapezoidal arm. It is connected to the wheel bearing housing and to 2 separate cross-members. It is recommended to use new self-locking nuts on all applications.

1. Raise and support the vehicle safely, under the frame and differential.
2. Remove the wheel. Along the trapezoidal arm, remove the speed sensor wiring bracket nuts/bolts and the guide.
3. Remove the wheel bearing housing-to-trapezoidal arm front and rear bolts.
4. Remove the trapezoidal arm-to-rear cross-member bolt.
5. At the brake pressure regulator, disconnect the spring.
6. Remove the trapezoidal arm-to-front cross-member nut and the trapezoidal arm from the vehicle.
7. To install, reverse the removal procedures. Torque the trapezoidal arm-to-front crossmember nut to 44 ft. lbs. (60 Nm), the trapezoidal arm-to-rear crossmember bolt to 63 ft. lbs. (85 Nm), the trapezoidal arm-to-wheel bearing housing bolts to 125 ft. lbs. (169 Nm) and the speed sensor guide nut/bolts to 7 ft. lbs. (10 Nm). Adjust the rear wheel alignment.

NOTE: Before installing the trapezoidal arm, be sure to coat the fasteners with locking compound.

Rear Wheel Bearings

REMOVAL & INSTALLATION

Except Quattro

1. Raise and support the vehicle

safely. Remove the wheel assembly.

2. Without disconnecting the hydraulic line, remove caliper assembly from rotor. Suspend caliper from the body with wire, do not let it hang by brake hose.

3. Pry off the grease cap and remove the cotter pin, nut and washer.

4. Remove the outer bearing.

5. Remove the brake rotor.

6. Remove the bearing inner bearing and seal from the rotor hub, using a soft drift or press.

7. Remove the bearing inner and outer race(s) from the rotor, using a soft drift or press.

To install:

8. Clean and inspect mating surfaces for bearing races.

9. Install new races, using soft drift or press.

10. Pack the new bearing with grease and set it into the inner race.

11. Install seal, making sure it is square in the rotor hub.

12. Install rotor, outer bearing, washer, nut and adjust bearing play.

13. Install cotter pin and dust cap.

14. If equipped with disc brakes, install caliper assembly.

15. If hydraulic lines had be removed, install and bleed brakes.

16. If parking brake cable had been remove, install and adjust as necessary.

17. Install the wheel assembly.

18. Lower vehicle and check brakes for proper operation.

ADJUSTMENT

Except Quattro

1. Raise and support the vehicle safely.

2. Remove the grease cap.

3. Remove the cotter pin and the locking nut.

4. While turning the wheel, so the wheel bearing does not jam, tighten the adjusting nut firmly.

5. Back the nut off slightly. The nut is properly adjusted when it is possible to pry the thrust washer side to side with some drag but using light pressure on the tool.

6. Install the locking nut and a new cotter pin. When installing the cap, make sure it is securely in place.

Rear Axle Assembly

REMOVAL & INSTALLATION

Front Wheel Drive Vehicles

1. Raise and support the vehicle safely. Remove the wheel assembly. Detach the muffler hanger bands. Lower and support the muffler and tail pipe.

2. Remove the parking brake cable to equalizer nut. Pry the cable sleeve from the bracket. Remove both parking brake cables at the brackets and disconnect the brake hoses at the brake line brackets. Cap all hoses and lines. Vehicles equipped with anti-lock brakes, disconnect the speed sensor.

3. Remove the nuts from the bolts attaching the trailing arms to the body. Do not remove the bolts at this time. On the right side, disconnect the spring from the brake pressure regulator.

4. Remove the bolts attaching the diagonal arms to the axle and remove the bolts attaching the strut to the axle. Slide out the trailing arm to the body attaching bolts and carefully remove the axle from the vehicle.

NOTE: All bolts through rubber bushings should be tightened with the weight of the vehicle on its wheels. This is done to preset the bushings in a level non-stressed position, to avoid poor handling or tire wear.

To install:

5. After positioning the axle in the vehicle, install both trailing arm bolts finger tight. Attach the lower strut mounts and install the wheel and tire assemblies. Lower the vehicle to the ground.

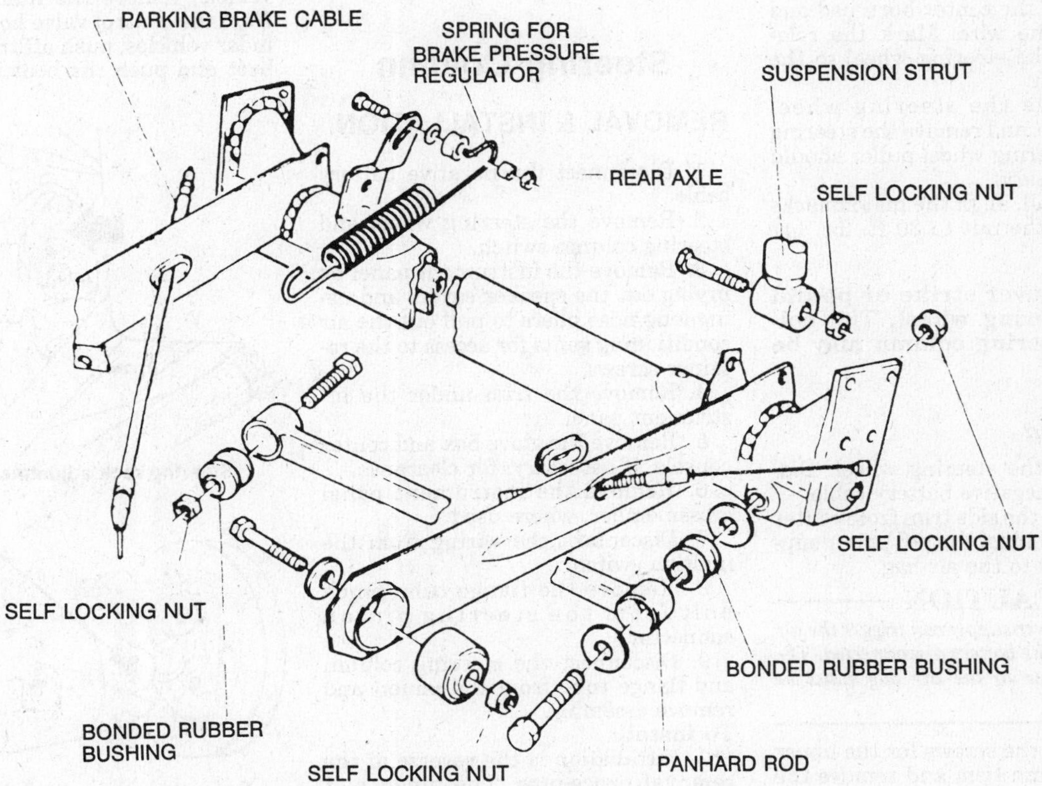

Rear axle attachments on front wheel drive models

6. Torque the trailing arm attaching bolts to 72 ft. lbs. (98 Nm) and the strut bolts to 66 ft. lbs. (90 Nm).
7. Install the diagonal arm and torque the bolt at the axle end to 70 ft. lbs. (95 Nm). Torque the bolt at the body end to 66 ft. lbs. (90 Nm). Raise the vehicle again to install the remaining components.
8. After the installation procedure has been completed, install the speed sensor, if equipped, bleed the brake system and adjust the parking brake, as necessary.

STEERING

Steering Wheel

CAUTION

On vehicles equipped with an air bag, the negative battery cable and reserve power supply must both be disconnected before working on the vehicle. Failure to do so may result in deployment of the air bag and personal injury.

REMOVAL & INSTALLATION

Without Air Bag

1. Center the steering wheel. Disconnect the negative battery cable.
2. Pull off the center horn pad and disconnect the wire. Mark the relationship of the steering wheel to the steering shaft.
3. Remove the steering wheel mounting nut and remove the steering wheel. A steering wheel puller should not be necessary.
4. To install, align the matchmarks and tighten the nut to 30 ft. lbs. (41 Nm).

NOTE: Never strike or pound on the steering wheel. The collapsible steering column may be damaged.

With Air Bag

1. Center the steering wheel. Disconnect the negative battery cable.
2. Remove the side trim from center console, disconnect the red power supply connector to the air bag.

CAUTION

The reserve power supply can trigger the air bag even with the battery disconnected. The power connector to the air bag must be disconnected.

3. Remove the screws for the upper steering column trim and remove the upper trim.
4. Separate the connector for the air bag spiral spring.
5. Remove the air bag Torx® head retaining bolts.
6. Unhook the air bag unit, lift up safety clamp and remove the air bag wiring at the terminal. Place the removed air bag unit face up in a safe place where it will not be disturbed.
7. Remove the steering wheel mounting nut and remove the steering wheel. A steering wheel puller may be necessary.
8. If removing sprial spring, wheel must be in the straight ahead position. Do not twist spring after removing it.

To install:

9. To install, align the matchmarks and tighten the nut to 30 ft. lbs. (41 Nm).

NOTE: Never strike or pound on the steering wheel. The collapsible steering column may be damaged.

10. Reinstall air bag unit, air bag connector and Torx® head screws. Torque the Torx® head screws to 53 in. lbs. (6 Nm).
11. Install steering column upper trim.
12. Connect air bag system power connector.
13. Install the side trim on center console.
14. Connect the negative battery cable.

Steering Column

REMOVAL & INSTALLATION

1. Disconnect the negative battery cable.
2. Remove the steering wheel and steering column switch.
3. Remove the instrument panel by prying out the speaker covers and using long nose pliers to pull out the air conditioning vents for access to the retainer screws.
4. Remove the trim under the instrument panel.
5. Remove the glove box and center console, if necessary, for clearance.
6. Remove the instrument panel crossmember, where used.
7. Disconnect the wiring from the ignition switch.
8. Remove the flange tube pinch bolt from the steering pinion connector.
9. Disconnect the steering column and flange tube from the pinion and remove assembly.

To install:

10. Installation is the reverse of the removal procedure. The pinch bolt should be replaced with a new bolt. Any self-locking nuts should also be replaced with new parts.

Power Steering Rack

ADJUSTMENT

1. Position the wheels in the straight-ahead position.
2. On top of the steering gear, loosen the locknut. Turn the adjusting nut until it just bottoms against the thrust piece. While holding the adjusting screw, tighten the locknut.
3. If the steering rattles, is too tight or does not center, readjust the adjusting screw.

REMOVAL & INSTALLATION

80 and 90

1. Raise and safely support the vehicle.
2. Remove the lower left instrument panel cover, the steering column-to-steering rack clamp bolt and the steering column-to-dash bolts. Remove the steering column from the vehicle.
3. Using a pair of locking pliers, clamp off the fluid return line to the reservoir. Disconnect the fluid pressure line from the steering gear.
4. At the steering column boot, press in on the clips and remove the boot from the panel. From inside the vehicle, remove the fluid return line from the control valve body. On 5 cylinder vehicles, push off the dash panel boot and push the boot into the pas-

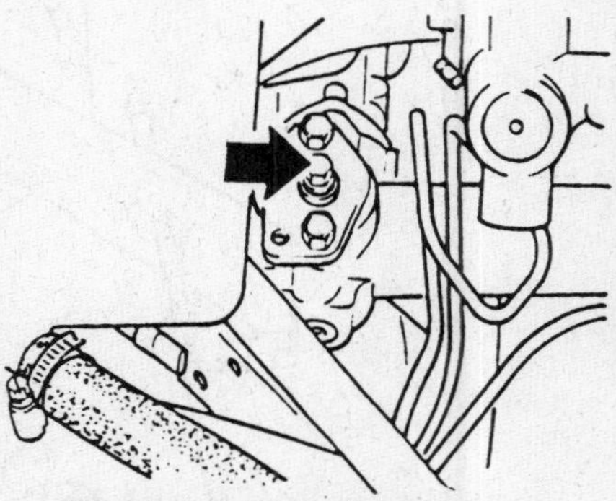

Steering rack adjustment bolt

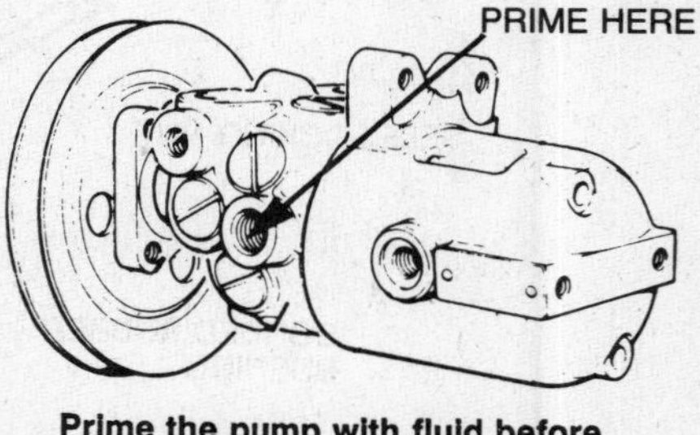

Prime the pump with fluid before installation

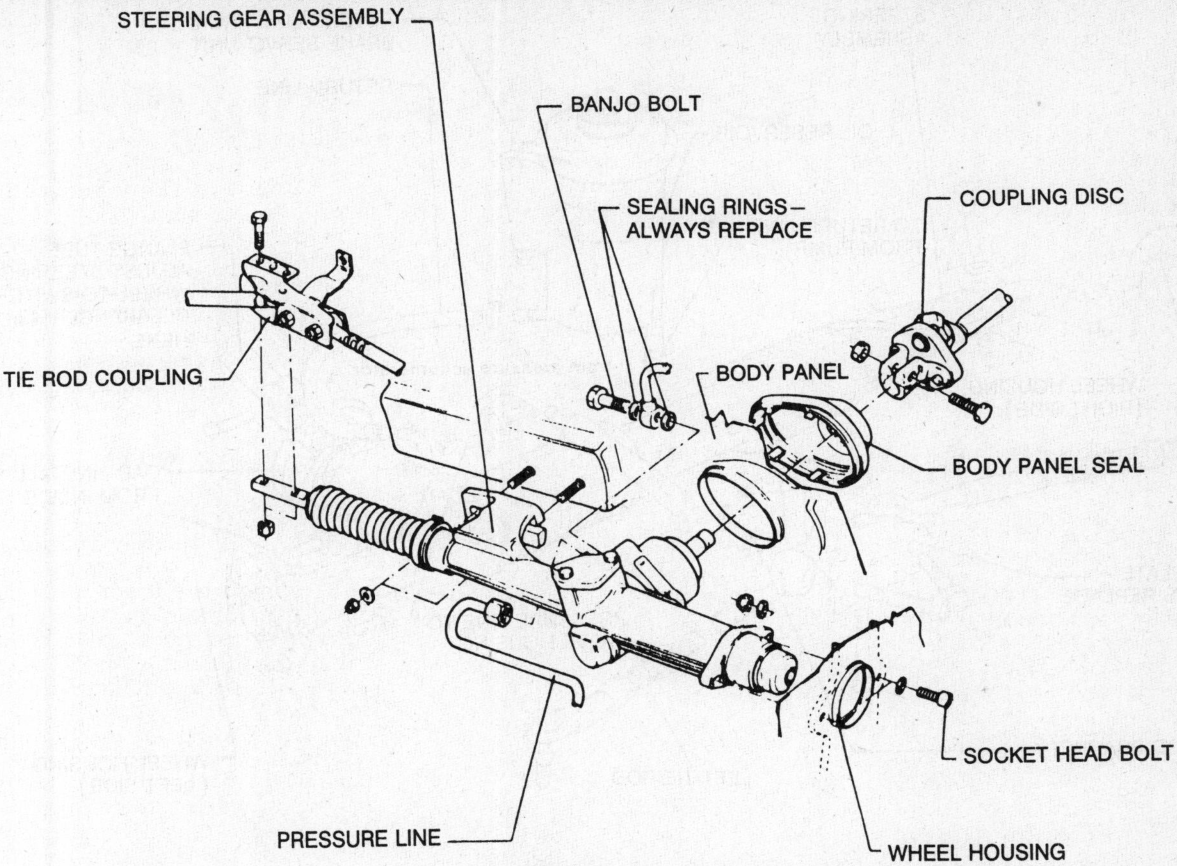

Steering rack assembly removal on 80 and 90

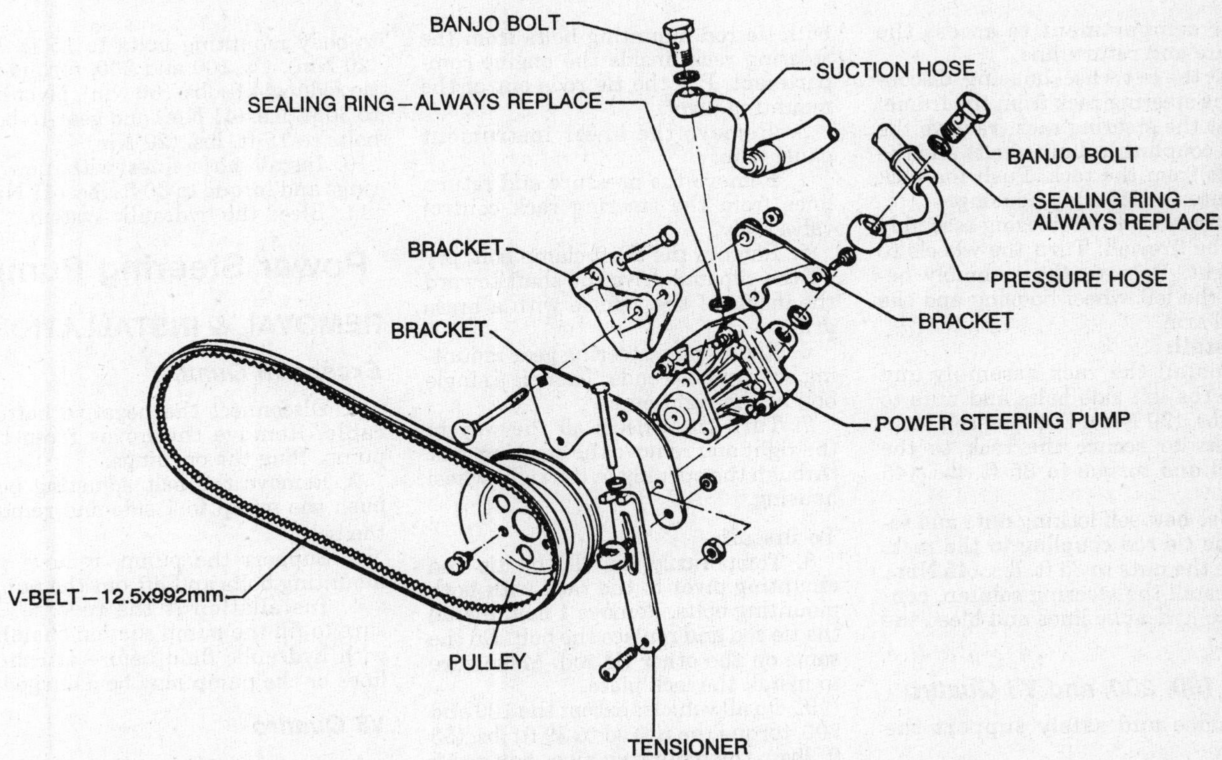

Steering pump assembly on 80 and 90

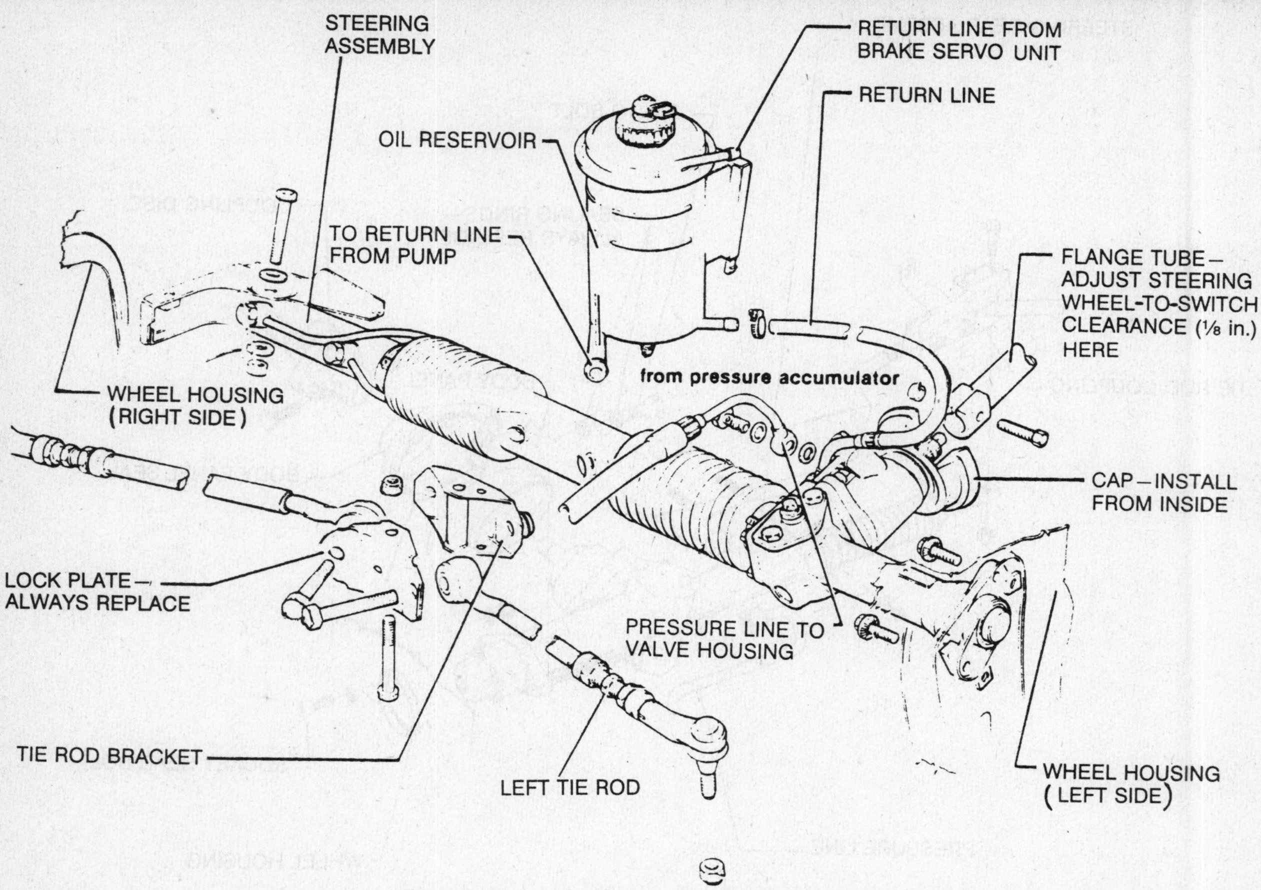

Steering rack assembly on 5000, 100, 200 and V8 Quattro

senger compartment to access the pressure and return line.

5. At the left wheel housing, disconnect the steering rack from the frame.

6. At the steering rack, remove the tie rod coupling locknuts/bolts and the tie rods from the rack. Push the rack back into the steering housing.

7. Disconnect the steering assembly from the firewall. Turn the wheels to the right. Remove the assembly between the left wheel housing and the control arm.

To install:

8. Install the rack assembly and torque the left side bolts and nuts to 14 ft. lbs. (20 Nm). Use new self locking nuts to secure the rack to the firewall and torque to 35 ft. lbs. (45 Nm).

9. Use new self locking nuts and secure the tie rod coupling to the rack. Torque the nuts to 35 ft. lbs. (45 Nm).

10. Install the steering column, connect the hydeaulic lines and bleed the system.

5000, 100, 200, and V8 Quattro

1. Raise and safely support the vehicle.

2. Pry off the lock plate and remove both tie rod mounting bolts from the steering rack, inside the engine compartment. Pry the tie rods out of the mounting pivot.

3. Remove the lower instrument panel trim.

4. Remove the pressure and return lines from the steering rack control valve body.

5. Remove the shaft clamp bolt, pry off the clip and drive the shaft toward the inside of the vehicle with a brass drift.

6. Remove the steering gear mounting bolts at both ends. There is a single bolt at the right end.

7. Turn the wheels all the way to the right and remove the steering gear through the opening in the right wheel housing.

To install:

8. Temporarily install the tie rod mounting pivot to the rack with both mounting bolts. Remove 1 bolt, install the tie rod and replace the bolt. Do the same on the other tie rod. Make sure to install the lock plate.

9. On all vehicles except the 100 and 200, torque the tie rod to 39 ft. lbs. (53 ft. lbs.), the mounting pivot bolt to 15 ft. lbs. (20 Nm) and the steering gear-to-body mounting bolts to 15 ft. lbs. (20 Nm). On 100 and 200, torque the tie rod to 44 ft. lbs. (60 Nm), pivot bolt to 30 ft. lbs. (41 Nm) and gear-to-body bolts to 15 ft. lbs. (20 Nm).

10. Install hose lines with new O-rings and torque to 30 ft. lbs. (41 Nm).

11. Bleed the hydraulic system.

Power Steering Pump

REMOVAL & INSTALLATION

Except V8 Quattro

1. Disconnect the negative battery cable. Remove the hoses from the pump. Plug the openings.

2. Remove the belt adjusting bolt, push the pump to 1 side and remove the belt.

3. Support the pump, remove the mounting bolts and lift out the pump.

4. Installation is the reverse. Be sure to fill the pump suction chamber with hydraulic fluid before attaching lines or the pump may be damaged.

V8 Quattro

1. Disconnect the battery negative cable.

2. Remove the air duct tube bolt near the engine oil dipstick.
3. Remove the retaining screws and remove the air duct elbow.
4. Remove the retaining bolts and move the radiator fan and motor assembly to one side.
5. Loosen the pump pulley mounting bolts.
6. Place a 13mm wrench on the tensioner bolt and move the wrench upward slowly and firmly until tension on the ribbed belt is released. Remove the drive belt from the idler pulley and take the pulley off the pump. Save any adjustment shims.
7. Clamp off the fluid intake and return lines, remove the lines from the pump and seal the openings. Before removing the banjo bolt take note of the proper installation of the pipe, over the corner of the pump support.
8. Remove the mounting bolts and the front support bracket, then remove the rear pump bolts. Push the pump forward from the support and twist the pump for clearance as it is removed.

To install:

9. Installation is the reverse of the removal process. When installing the pump bolts, install all bolts only finger tight until everything is aligned.
10. Before installing the banjo bolt for the hydraulic line, make sure the pipe is properly aligned with the pump support. Use new seals.
11. Check the belt and pulley alignment. Shim the pulley, if necessary.

BELT ADJUSTMENT

1. Loosen the pump mounting bolts.
2. Turn the pump adjusting bolt until the center of the belt can be depressed ⅜ in. (10mm).
3. After adjustment, tighten the mounting bolts.

SYSTEM BLEEDING

1. Fill the reservoir to the **FULL** mark.
2. Raise and safely support the vehicle.
3. Turn the steering wheel with the engine not running from lock to lock several times to remove the air from the system.
4. Add fluid to the reservoir until the level is maintained at $1^{3}/_{16}$ in. (30mm) below the **FULL** mark.
5. Start the engine. As the fluid in the reservoir continues to drop, add fluid to maintain the $1^{3}/_{16}$ in. (30mm) level.
6. Keep bleeding the system until no more air bubbles appear in the reservoir.
7. Turn off the engine and pump the brake pedal at least 20 times.
8. Replenish the fluid to the proper level.

NOTE: When turning the steering wheel, do not use more force than necessary to turn it.

Tie Rod Ends

REMOVAL & INSTALLATION

NOTE: A puller or press is required for this procedure.

1. Raise and support the vehicle safely. Remove the front wheels.
2. Disconnect the outer end of the steering tie rod from the steering knuckle by removing the cotter pin and nut and pressing out the tie rod end. A small puller or press is required to free the tie rod end.
3. Under the hood, pry off the lock plate and remove the mounting bolts from both tie rod inner ends. Pry the tie rod out of the mounting pivot.
4. Install the mounting pivot to the rack with both mounting bolts.
5. Remove 1 bolt, install the tie rod and replace the bolt. Do the same on the other tie rod.
6. **Make sure to install the lock plate. The inner tie rod end bolts should be torqued to 32 ft. lbs. (43 Nm) on 5000, or 44 ft. lbs. (60 Nm) on all other models.**
7. If replacing the adjustable left tie rod, adjust it to the same length as the old one. Check the toe-in.
8. Use new cotter pins or replace self-locking nuts when installing the outer tie rod end.

BRAKES

For all brake system repair and service procedures not detailed below, please refer to "Brakes" in the Unit Repair Section.

Master Cylinder

REMOVAL & INSTALLATION

1. Disconnect the negative battery cable. Have an assistant hold the brake pedal down about 1½ in. Disconnect the brake lines nearest the firewall.
2. Hold a container under the fitting disconnected in Step 1 and have the assistant release the pedal. The contents of the reservoir will drain into the container. Discard the used fluid.
3. Disconnect the other brake line.
4. Disconnect the stoplight switch and any warning switches from the master cylinder.
5. Remove the master cylinder from the power brake unit. Be careful not to loose the sealing ring between the 2 units.

To install:

6. Install the new master cylinder and and torque the mounting bolts to

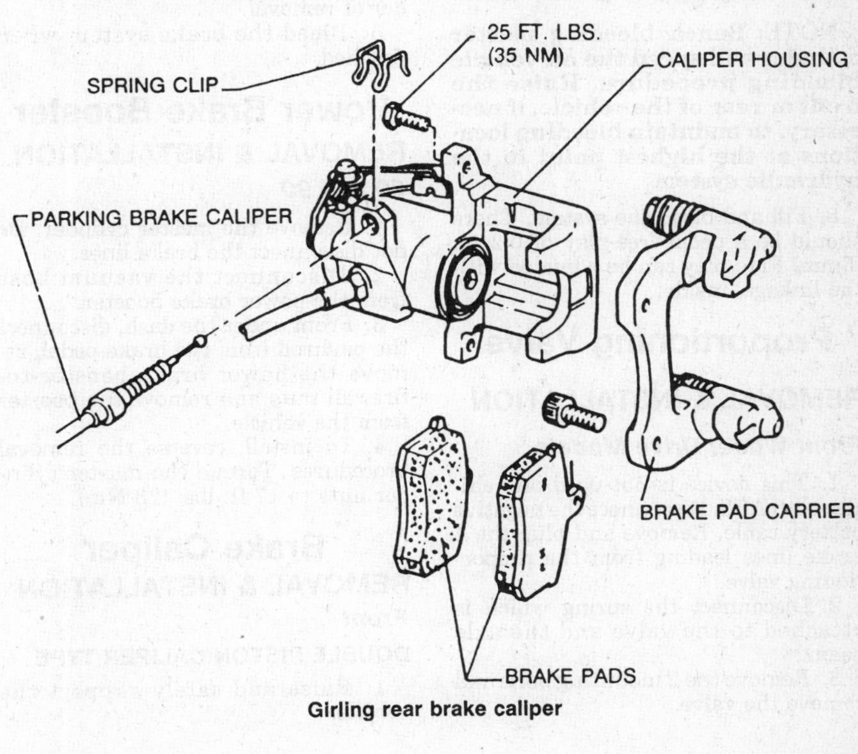

Girling rear brake caliper

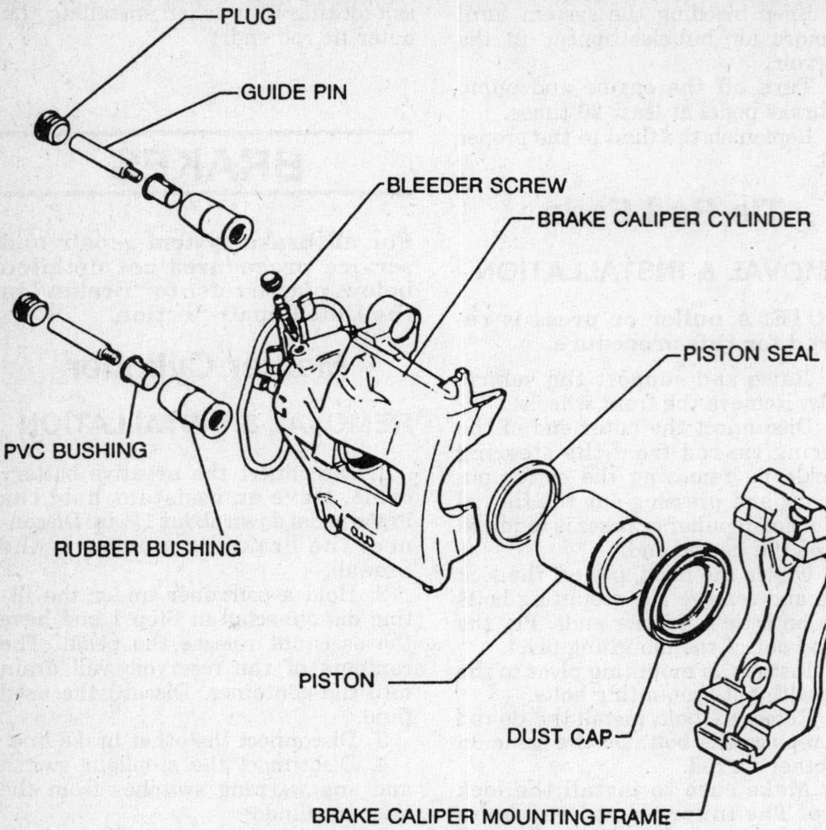

Teves front brake caliper

17 ft. lbs. (23 Nm). Install the switches.

7. Transfer the reservoir from the old master cylinder to the new unit.

NOTE: Bench bleeding master cylinder will speed the on vehicle bleeding procedure. Raise the front or rear of the vehicle, if necessary, to maintain bleeding locations at the highest point in the hydraulic system.

8. Fill and bleed the system. There should be a pedal free-play of 0.2 in. (5mm) Free-play can be adjusted with the linkage, inside.

Proportioning Valve

REMOVAL & INSTALLATION

Front Wheel Drive Models

1. This device is not used on vehicles with ABS. Disconnect the negative battery cable. Remove and plug the 4 brake lines leading from the proportioning valve.
2. Disconnect the spring which is attached to the valve and the axle beam.
3. Remove the 2 mounting bolts and remove the valve.

NOTE: Do not disassemble the valve.

4. Installation is in the reverse order of removal.
5. Bleed the brake system when finished.

Power Brake Booster

REMOVAL & INSTALLATION

80 and 90

1. Remove the master cylinder. Do not disconnect the brake lines.
2. Disconnect the vacuum hose from the power brake booster.
3. From under the dash, disconnect the pushrod from the brake pedal, remove the power brake booster-to-firewall nuts and remove the booster from the vehicle.
4. To install, reverse the removal procedures. Torque the master cylinder nuts to 17 ft. lbs. (23 Nm).

Brake Caliper

REMOVAL & INSTALLATION

Front

DOUBLE PISTON CALIPER TYPE

1. Raise and safely support the vehicle.

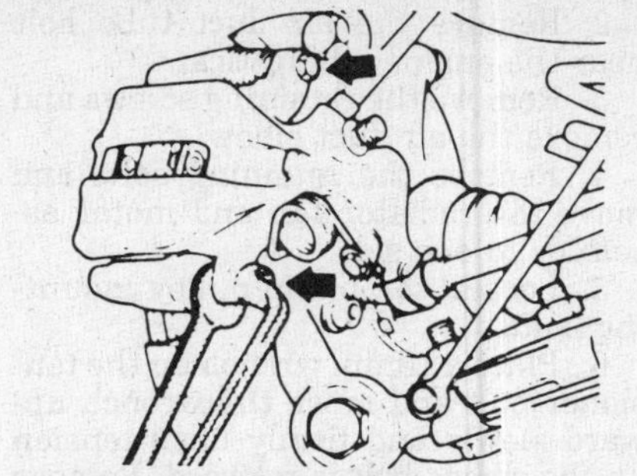
Hold the guide pin with an open end wrench

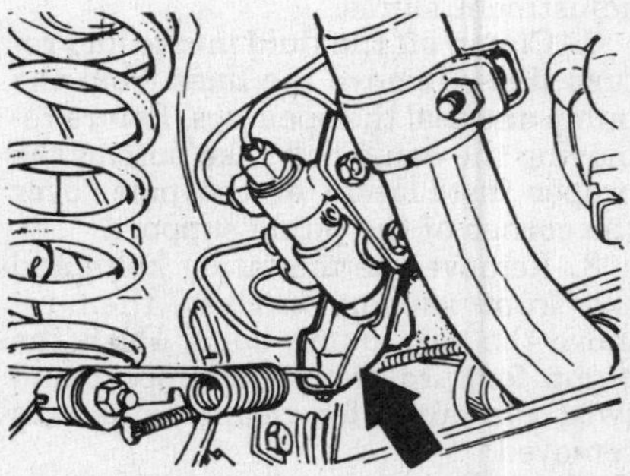
On front wheel drive models without ABS, the proportioning valve on the rear axle should move when the axle moves up and down

2. Remove wheels.
3. Remove the lower caliper bolt, hold guide pin with open end wrench, while loosening. Disconnect the wear indicator, if equipped.
4. Swing brake caliper up and remove brake pads, taking note of spacer shims and heat-shield locations, if equipped.
5. Disconnect brake line and remove caliper.

To install:

6. Installation is the reverse of the removal procedure. Push pistons back into caliper. Place old disc pad on piston side of caliper, using a C-clamp centered on old pad across both piston, push pistons back into bore. Make certain to center C-clamp on pad and caliper to avoid cracking or jamming the pistons in their bores.
7. Install brake pads, shims and heat shield, if equipped. Install brake line.
8. Slide caliper over rotor and align pins. Make sure caliper moves freely on the pins.
9. Install guide pins and torque to 26 ft. lbs. (35 Nm).
10. Connect the wear indicator and install wheels.
11. Lower vehicle and fill master cylinder. Bleed brake system.

GIRLING TYPE

1. Raise and safely support the vehicle.

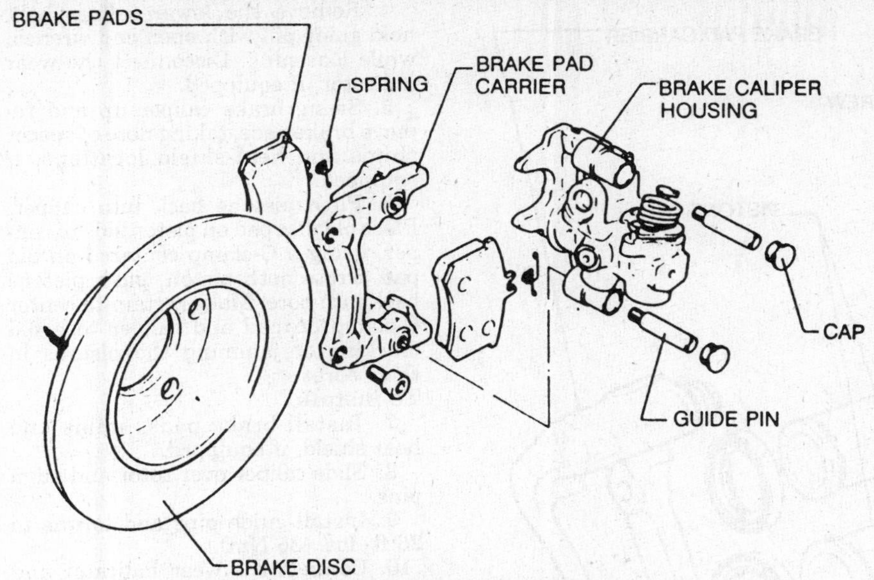

Teves rear caliper assembly

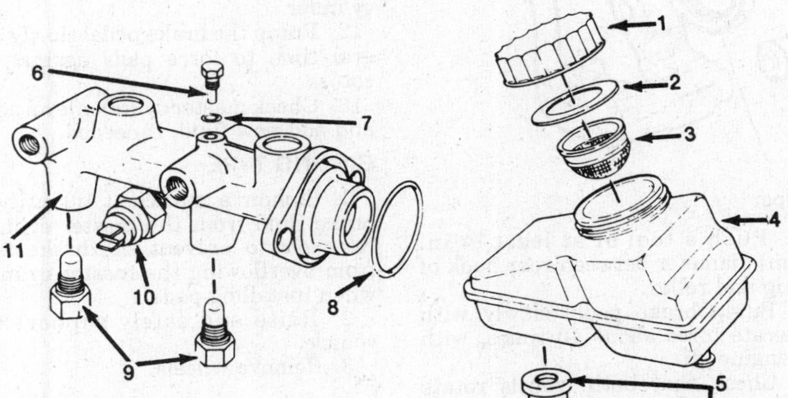

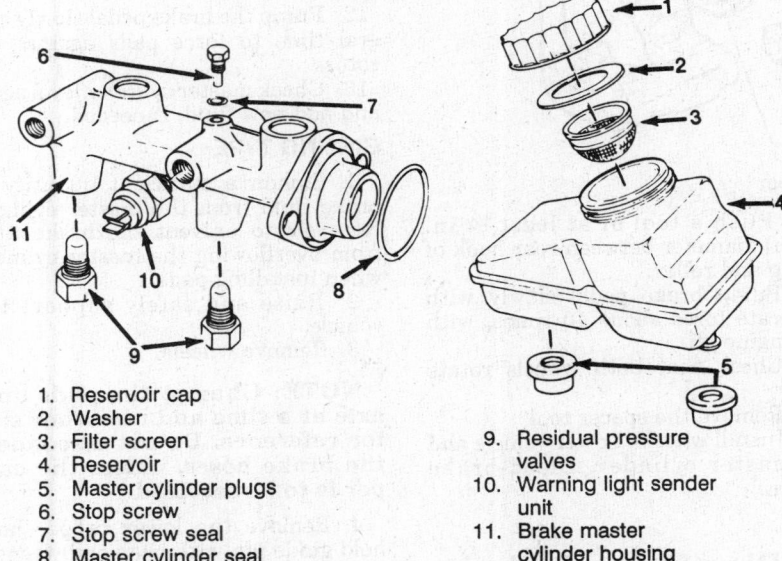

1. Reservoir cap
2. Washer
3. Filter screen
4. Reservoir
5. Master cylinder plugs
6. Stop screw
7. Stop screw seal
8. Master cylinder seal
9. Residual pressure valves
10. Warning light sender unit
11. Brake master cylinder housing

Typical master cylinder assembly

2. Remove wheels.
3. Remove the lower caliper bolt, hold guide pin with open end wrench, while loosening. Disconnect the wear indicator, if equipped.
4. Swing brake caliper up and remove brake pads, taking note of spacer shims and heat-shield locations, if equipped. Remove brake line.

To install:

5. Push piston back into caliper, using a C-clamp in bore of piston. Make certain to center C-clamp on piston and caliper to avoid cracking or jamming the piston in the bore.
6. Install brake pads, shims and heat shield, if equipped. Reconnect brake line.
7. Slide caliper over rotor and align pins.
8. Install guide pins and torque to 25 ft. lbs. (35 Nm).
9. Fill master cylinder and bleed brake system.

TEVES TYPE

1. Raise and safely support the vehicle.
2. Remove wheels.
3. Remove guide pin caps and guide pins.
4. Remove brake hose retaining clip or bracket. Disconnect brake line.
5. Swing caliper up and remove. Remove brake pads, taking note of spacer shims and heat-shield locations. Disconnect the wear indicator, if equipped.

To install:

6. Installation is the reverse of the removal process. Push piston back into caliper, using a C-clamp in bore of piston. Make certain to center C-clamp on piston and caliper to avoid cracking or jamming the piston in the bore.
7. Install brake pads, shims and heat shield, if equipped.
8. Slide caliper over rotor and align pins. Install guide pins and torque to 18 ft. lbs. (25 Nm).
9. Install brake hose and clip. Fill the master cylinder and bleed the brake system.

Rear

GIRLING TYPE

1. Raise and safely support the vehicle.
2. Remove wheels.
3. Remove rear brake caliper housing, hold guide pin with open end wrench while loosening bolts. Disconnect the wear indicator, if equipped.
4. Remove brake pads, taking note of spacer shims and heat-shield locations. Disconnect brake line.

To install:

5. Installation is the reverse of the removal procedure. Screw piston into housing by turning is clockwise with a socket head wrench while pushing in firmly.
6. Install brake pads, shims and heat shield, if equipped.
7. Install caliper on housing. Install new bolts and torque to 25 ft. lbs. (35 Nm).

NOTE: The bolts are self-locking type; it is recommended to always use new bolts.

8. Make certain parking brake is free of tension.
9. Use a prybar to push caliper lever against stop on both sides of vehicle.
10. Parking brake is too tight if the lever of the opposite side caliper is pulled away from the stop.
11. Loosen parking brake adjusting nut and position, as necessary.
12. Push a tool of at least ¼ in. (6mm) diameter between rear hook of spring and roller.
13. Pump brake pedal slowly with moderate force about 40 times, with the engine off.
14. Check that both wheels rotate freely.
15. Remove spacer tool.
16. Install wheels. Lower vehicle and fill master cylinder. Bleed brake system.

TEVES TYPE

1. Raise and safely support the vehicle.

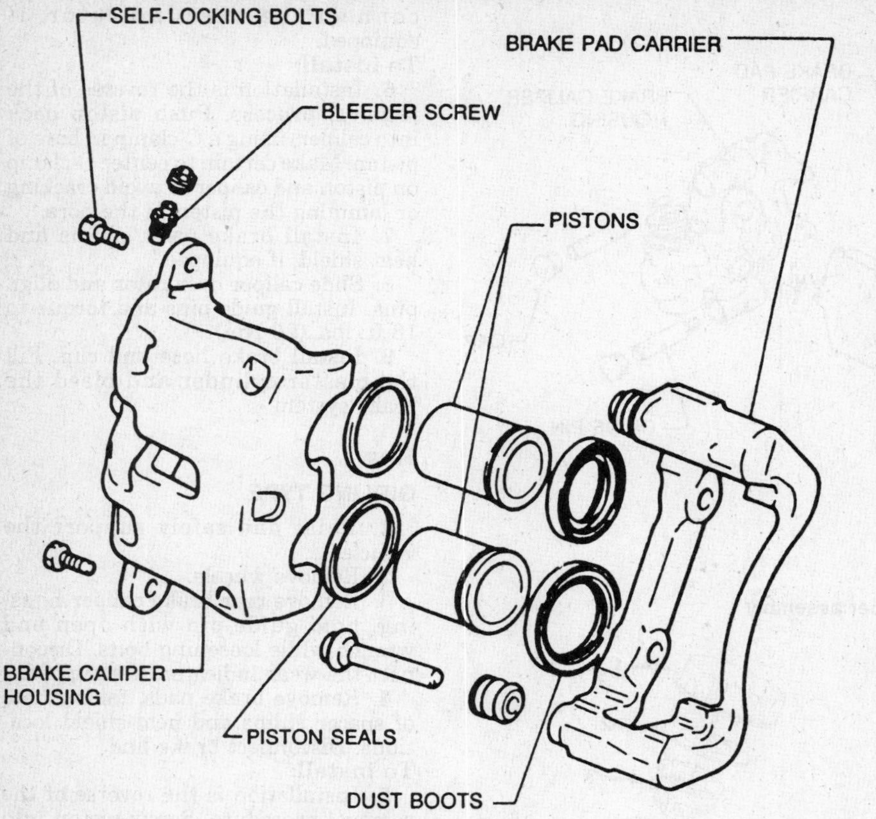

Twin piston caliper

2. Remove rear wheels.
3. Remove both protective caps. Loosen both guide pins but do not pull out of the rubber boots.
4. Pull caliper housing toward the outside of vehicle by hand. Disconnect brake line.
5. Swing the caliper housing to the rear and remove. Disconnect the wear indicator, if equipped.
6. Remove brake pads, taking note of spacer shims and heat-shield locations.

To install:

7. Installation is the reverse of the removal procedure. Push piston back into caliper, using a C-clamp in bore of piston. Make certain to center C-clamp on piston and caliper to avoid cracking or jamming the piston in the bore.
8. Install brake pads, shims and heat shield, if equipped.
9. Slide caliper over rotor and align pins. Install guide pins and torque to 18 ft. lbs. (25 Nm). Install the protective caps. Reconnect brake line.
10. Make certain parking brake is free of tension.
11. Use a prybar to push caliper lever against stop on both sides of vehicle.
12. Parking brake is too tight, if lever of opposite side caliper is pulled away from the stop.
13. Loosen parking brake adjusting nut and position, as necessary.
14. Push a tool of at least ¼ in. (6mm) diameter between rear hook of spring and roller.
15. Pump brake pedal slowly with moderate force about 40 times, with the engine off.
16. Check that both wheels rotate freely.
17. Remove the spacer tool.
18. Install wheels. Lower vehicle and fill master cylinder. Bleed brake system.

Disc Brake Pads

REMOVAL & INSTALLATION

Front

DOUBLE PISTON CALIPER TYPE

1. Siphon a sufficient quantity of brake fluid from the master cylinder reservoir to prevent the brake fluid from overflowing the master cylinder when installing pads.
2. Raise and safely support the vehicle.
3. Remove wheels.

NOTE: Change the pads on 1 axle at a time and use other side for reference. Do not disconnect the brake hoses, unless the caliper is to be serviced.

4. Remove the lower caliper bolt, hold guide pin with open end wrench, while loosening. Disconnect the wear indicator, if equipped.
5. Swing brake caliper up and remove brake pads, taking note of spacer shims and heat-shield locations, if equipped.
6. Push pistons back into caliper. Place old disc pad on piston side of caliper, using a C-clamp centered on old pad across both piston, push pistons back into bore. Make certain to center C-clamp on pad and caliper to avoid cracking or jamming the pistons in their bores.

To install:

7. Install brake pads, shims and heat shield, if equipped.
8. Slide caliper over rotor and align pins.
9. Install guide pins and torque to 26 ft. lbs. (35 Nm).
10. Connect the wear indicator and install wheels.
11. Lower the vehicle and fill master cylinder.
12. Pump the brake pedal slowly several time to force pads against the rotors.
13. Check master cylinder level again and add new fluid, if needed.

GIRLING TYPE

1. Siphon a sufficient quantity of brake fluid from the master cylinder reservoir to prevent the brake fluid from overflowing the master cylinder when installing pads.
2. Raise and safely support the vehicle.
3. Remove wheels.

NOTE: Change the pads on 1 axle at a time and use other side for reference. Do not disconnect the brake hoses, unless the caliper is to be serviced.

4. Remove the lower caliper bolt, hold guide pin with open end wrench, while loosening. Disconnect the wear indicator, if equipped.
5. Swing brake caliper up and remove brake pads, taking note of spacer shims and heat-shield locations, if equipped.
6. Push piston back into caliper, using a C-clamp in bore of piston. Make certain to center C-clamp on piston and caliper to avoid cracking or jamming the piston in the bore.

To install:

7. Install brake pads, shims and heat shield, if equipped.
8. Slide caliper over rotor and align pins.
9. Install guide pins and torque to 25 ft. lbs. (35 Nm).
10. Install wheels.
11. Lower vehicle and fill master cylinder.

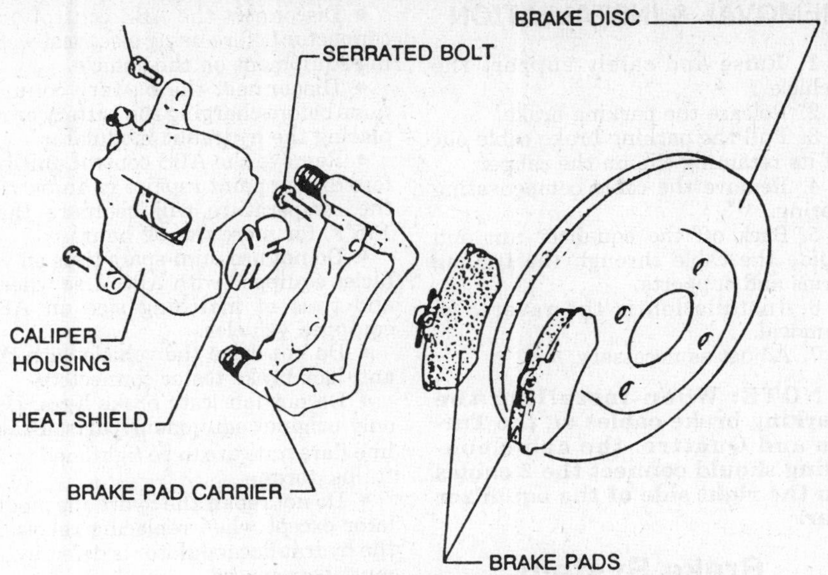

Girling front brake caliper

12. Pump the brake pedal slowly several time to force pads against the rotors.
13. Check master cylinder level again and add new fluid, if needed.

TEVES TYPE

1. Siphon a sufficient quantity of brake fluid from the master cylinder reservoir to prevent the brake fluid from overflowing the master cylinder when installing pads.
2. Raise and safely support the vehicle.
3. Remove wheels.

NOTE: Change the pads on 1 axle at a time and use other side for reference. Do not disconnect the brake hoses, unless the caliper is to be serviced.

4. Remove guide pin caps.
5. Remove guide pins.
6. Remove brake hose retaining clip or bracket.
7. Swing caliper up and secure in position, using a wire. Do not allow caliper to hang from brake hose.
8. Remove brake pads, taking note of spacer shims and heat-shield locations. Disconnect the wear indicator, if equipped.
9. Push piston back into caliper, using a C-clamp in bore of piston. Make certain to center C-clamp on piston and caliper to avoid cracking or jamming the piston in the bore.

To install:

10. Install brake pads, shims and heat shield, if equipped.
11. Slide caliper over rotor and align pins.
12. Install guide pins and torque to 18 ft. lbs. (25 Nm).
13. Install brake hose clip.
14. Install wheels.
15. Lower vehicle and fill master cylinder.
16. Pump the brake pedal slowly several time to force pads against the rotors.
17. Check master cylinder level again and add new fluid if needed.

Rear

GIRLING TYPE

1. Siphon a sufficient quantity of brake fluid from the master cylinder reservoir to prevent the brake fluid from overflowing the master cylinder when installing pads.
2. Raise and safely support the vehicle.
3. Remove wheels.

NOTE: Change the pads on 1 axle at a time and use other side for reference. Do not disconnect the brake hoses, unless the caliper is to be serviced.

4. Remove brake caliper housing, hold guide pin with open end wrench while loosening bolts. Disconnect the wear indicator, if equipped.
5. Remove brake pads, taking note of spacer shims and heat-shield locations.
6. Screw piston into housing by turning is clockwise with a socket head wrench while pushing in firmly.

To install:

7. Install brake pads, shims and heat shield, if equipped.
8. Install caliper on housing.

NOTE: The bolts are self-locking type; it is recommended to always use new bolts.

9. Install new bolts and torque to 25 ft. lbs. (35 Nm).
10. Make certain parking brake is free of tension.
11. Use a prybar to push caliper lever against stop on both sides of vehicle.
12. Parking brake is too tight, if lever of opposite side caliper is pulled away from the stop.
13. Loosen parking brake adjusting nut and position, as necessary.
14. Push a tool of at least ¼ in. (6mm) diameter between rear hook of spring and roller.
15. Pump brake pedal slowly with moderate force about 40 times, with the engine off.
16. Check that both wheels rotate freely.
17. Remove the spacer tool.
18. Install wheels.
19. Lower vehicle and fill master cylinder.

TEVES TYPE

1. Siphon a sufficient quantity of brake fluid from the master cylinder reservoir to prevent the brake fluid from overflowing the master cylinder when installing pads.
2. Raise and safely support the vehicle.
3. Remove wheels.

NOTE: Change the pads on 1 axle at a time and use other side for reference. Do not disconnect the brake hoses, unless the caliper is to be serviced.

4. Remove both protective caps.
5. Loosen both guide pins but do not pull out of the rubber boots.
6. Pull caliper housing toward the outside of vehicle by hand.
7. Swing the caliper housing to the rear and remove. Do not allow caliper to hang from brake hose. Disconnect the wear indicator, if equipped.
8. Remove brake pads, taking note of spacer shims and heat-shield locations.
9. Push piston back into caliper, using a C-clamp in bore of piston. Make certain to center C-clamp on piston and caliper to avoid cracking or jamming the piston in the bore.

To install:

10. Install brake pads, shims and heat shield, if equipped.
11. Slide caliper over rotor and align pins.
12. Install guide pins and torque to 18 ft. lbs. (25 Nm).
13. Install the protective caps.
14. Make certain parking brake is free of tension.

15. Use a prybar to push caliper lever against stop on both sides of vehicle.
16. Parking brake is too tight, if lever of opposite side caliper is pulled away from the stop.
17. Loosen parking brake adjusting nut and position, as necessary.
18. Push a tool of at least ¼ in. (6mm) diameter between rear hook of spring and roller.
19. Pump brake pedal slowly with moderate force about 40 times, with the engine off.
20. Check that both wheels rotate freely.
21. Remove the spacer tool.
22. Install wheels. Refill master cylinder.

Brake Rotor

REMOVAL & INSTALLATION

1. Raise and safely support vehicle.
2. Remove wheels.
3. Remove brake caliper.
4. Remove disc rotor.
5. When replacing rotors, always replace in pairs. If machining, watch wear limit. Always machine both sides, never one side only.

Parking Brake Cable

ADJUSTMENT

NOTE: Because of the self-adjusting rear brakes, adjustment is only necessary after replacement of any of the brake components.

1. Raise and support the vehicle safely.
2. Release the parking brake lever.
3. Check that the parking brake levers at the rear calipers stay on the stop. If not, loosen the adjusting nut.
4. Depress the brake pedal approximately 40 times and pull the parking brake lever to the 3rd tooth.

NOTE: Always make sure basic adjustment on rear brakes is correct first.

5. Tighten the adjusting nut on the parking brake equalizer bar until the rear wheels can just be turned by hand.
6. Release the brake and check that both wheels rotate freely and that the parking brake levers at the rear calipers stay on the stops.
7. Turn the ignition switch **ON** and check that the brake warning light comes **ON** when the parking brake is pulled to the first tooth and goes **OFF** when it is released.

REMOVAL & INSTALLATION

1. Raise and safely support the vehicle.
2. Release the parking brake.
3. Pull the parking brake cable out of its retaining clip on the caliper.
4. Remove the cable compensating spring.
5. Back off the equalizer nut and guide the cable through the trailing arms and supports.
6. Installation is the reverse of removal.
7. Adjust as necessary.

NOTE: When installing the parking brake cables on the Turbo and Quattro, the cable coupling should connect the 2 cables on the right side of the equalizer bar.

Brake System Bleeding

To bleed system, use bottle with transparent hose attached, so brake fluid can be checked for air bubbles. Do not re-use fluid removed from reservoir.

1. Raise and safely support vehicle. Start the engine.
2. Connect bleeder hose and bleed calipers. Have an assistant press the pedal. When in the down position, open the bleeder screw. Close the bleeder screw and press the pedal again. Open the bleeder screw and again allow the air to come out. Repeat at each wheel until the fluid in the container shows no sign of air bubbles.
3. Bleed in the following sequence:
   a. Right rear caliper
   b. Left rear caliper
   c. Right front caliper
   d. Left front caliper
4. Bleed right rear caliper. After bleeding, refill brake fluid reservoir.

Anti-Lock Brake System Service

PRECAUTIONS

The following precautions should be observed when working on the Anti-lock Brake System (ABS).

- Electrical testing should be done using the factory LED tester. This tester must be used to check out the hydraulic modulator, ABS control unit, wheel speed sensors and ABS wiring harness. It is also necessary to perform the test procedure if the brake lines or brake pressure regulators are replaced because of accident damage.
- Switch the ignition OFF before connecting or disconnecting the ABS control unit connector.
- Disconnect the ABS control unit connector before using electrical welding equipment on the vehicle.
- Disconnect the battery connections before charging the battery or replacing the hydraulic modulator.
- Remove the ABS control unit before drying paint repairs in an oven if the temperature will be more than 185°F. for more than 2 hours.
- Do not use mini-spare tires on vehicles equipped with ABS. Use wheels and tires of matching size on ABS equipped vehicles.
- Do not drive the vehicle with the anti-lock brake tester connected.
- Do not fabricate brake lines. Use only original equipment parts. Brake line flare nuts are to be tightened to 11 ft. lbs. torque.
- Do not repair the hydraulic modulator except when replacing relays. If the hydraulic modulator is defective, it must be replaced.

RELIEVING ANTI-LOCK BRAKE SYSTEM PRESSURE

A special factory tool is recommended for ABS system bleed-down. It is a set of high and low pressure gauges that is also used for system pressure testing. It is used as follows:

1. Raise and safely support vehicle. Remove one front wheel.
2. Make sure the ignition switch is **OFF**. Remove the bleeder screw from the caliper.
3. Connect the special tool and install a brake pedal depressor between the brake pedal and the driver's seat.
4. Press the brake pedal until the pressure gauge drops down indicating system pressure has been relieved.

NOTE: If the pressure drops more than 58 psi (4 BAR) in 45 seconds, replace the hydraulic modulator. Press the brake pedal until pressure reads 87 psi (6 BAR). If pressure drops more than 14.5 psi (1 BAR) in 3 minutes, replace the hydraulic modulator.

SYSTEM TEST

The operational ABS system test must be performed after every repair to the service brake components. These repairs include replacement of linings and/or discs, hoses, booster or master cylinder, cables and parking brake components.

This short test insures that nothing within the ABS components was disturbed during the repair. To perform the test:

1. Turn the ignition switch **ON**; the ABS warning lamp should light.

2. Start the engine; the warning lamp should turn off.

3. While the engine is running, switch the ABS **OFF**; the warning lamp(s) should come on.

4. Turn the ignition switch **OFF** and restart it; the ABS warning lamp should go out, showing that the manual switch position has been overridden and the system reset.

5. Drive the vehicle over 20 mph (30 kph); the ABS warning lamp should not come on. The differential locks on Quattro vehicles must not be engaged during this test.

6. In a safe location and under controlled conditions, perform at least one test stop from about 20 mph which engages the anti-lock system. Check system function and vehicle stability.

Hydraulic Modulator

REMOVAL & INSTALLATION

1. Disconnect negative battery cable.
2. Bleed down system pressure.
3. Remove the modulator cover.
4. Remove the harness retainer.
5. Disconnect the hydraulic modulator from the mounting.
6. Unclip the hydraulic fluid reservoir from the mounting.
7. Unscrew the brake lines. Mark the lines so they will be reinstalled in their proper locations. Seal any openings immediately with plugs.

To install:

8. Connect the hydraulic lines to the modulator but do not tighten them yet.
9. Install the modulator, then tighten the lines. Install the reservoir and connect the wiring.
10. Refill with DOT 4 brake fluid only and bleed and test the system.

Wheel Speed Sensors

REMOVAL & INSTALLATION

Front

The front wheel speed sensor is hand pressed into the wheel bearing housing, also called a steering knuckle. A sleeve is inserted into a hole in the housing which retains the speed sensor.

1. Raise and safely support vehicle.
2. Remove wheel and tire assembly.
3. Locate the sensor in the side of the housing and pull out. Left and right are identical.
4. To disconnect the wiring, remove the engine lower cover.
5. Unclip the connector from its mount and disconnect sensor.

To install:

6. Installation is the reverse of the removal procedure. Before inserting the retainer sleeve, the opening in the wheel bearing housing should be greased with brake cylinder paste.
7. Press the sleeve in as far as possible.
8. Push the sensor in to stop by hand.

Rear

The rear wheel speed sensor is hand pressed into the rear wheel bearing housing. A sleeve is inserted into a hole in the front side of the housing which retains the speed sensor.

1. Raise and safely support vehicle.
2. Remove wheel and tire assembly.
3. Locate the sensor in the front side of the rear wheel bearing housing and pull out. Left and right are identical. Unclip the connector and remove sensor.

To install:

4. Installation is the reverse of the removal process. Before inserting the retainer sleeve, the opening in the rear wheel bearing housing should be greased with brake cylinder paste.
5. Press the sleeve in as far as possible.
6. Push the sensor in to stop by hand.

Acceleration Switch

REMOVAL & INSTALLATION

The ABS acceleration switch is located under the left rear seat bench. It is a mercury switch activated during deceleration. When braking with ABS the switch helps the braking system provide additional stabilization during braking.

1. Switch the ignition **OFF**.
2. Remove the mounting screws.
3. Disconnect the wiring.
4. At installation, note that the arrow on the cover must point forward, in the direction of driving.

Electronic Control Unit

REMOVAL & INSTALLATION

The ABS electronic control unit is located under the left side of the rear passenger seat.

1. Switch ignition **OFF**.
2. Raise rear seat and locate ABS control unit. Remove hold down nuts.
3. The ABS control unit plug has locks to secure it. Disconnect by pressing the spring on the narrow end of the plug. To connect, insert the lug of the connector into the adapter and push the connector against the spring. The connector should snap into the lock with a click.

CHASSIS ELECTRICAL

CAUTION

On vehicles equipped with an air bag, the negative battery cable and reserve power supply must both be disconnected before working on the system. Failure to do so may result in deployment of the air bag and possible personal injury. Never disconnect any electrical connector with the ignition switch turned on or damage to electronic controlling devices could occur.

NOTE: All vehicles are equipped with theft protected radios, which cannot be operated if power to the radio is interrupted. Before disconnecting the battery cables, obtain the security code.

Air Bag

DISARMING

1. Disconnect the battery and the red air bag power supply connector. This connector is labled "Air bag". On 80 and 90, the connector is behind a panel under the driver side dashboard. On all other models, the connector is at the front of the center console.

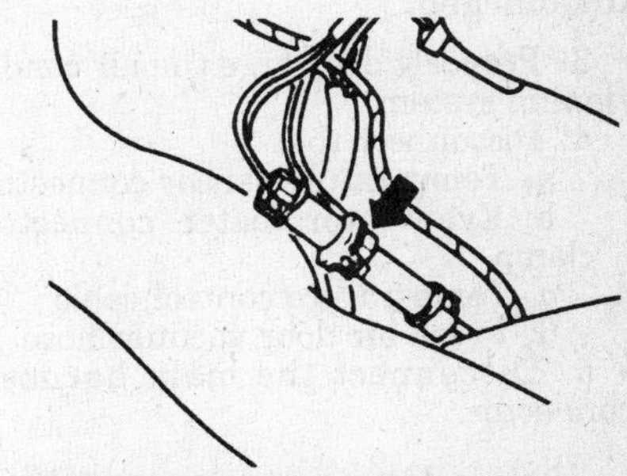

Power supply connector must be disconnected first

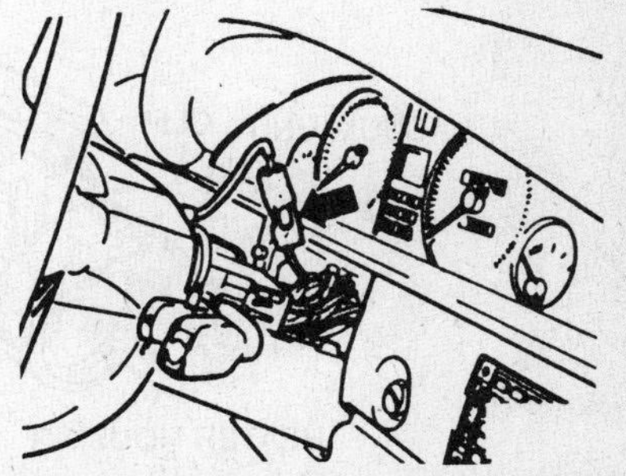

The air bag can be disconnected at the steering column

2. Remove the screws for the upper steering column trim and remove the upper trim.

3. Separate the connector for the air bag spiral spring.

4. If required, a memory saver can now be safely used to preserve the radio code while the battery is disconnected.

Heater Blower Motor

REMOVAL & INSTALLATION

Coupe, 80 and 90

1. The blower is removed from under the hood. Disconnect the negative battery cable.
2. Remove the air plenum from the cowl.
3. Remove the ballast resistor.
4. At the blower motor, disconnect the electrical connector.
5. Remove the blower mounting bolts and the blower from the heater assembly.
6. To install, reverse the removal procedures.

5000, 100 and 200

NOTE: Blower or core removal requires removal and disassembly of the entire unit.

1. Disconnect the negative battery cable.
2. Drain the cooling system.

NOTE: If equipped with air conditioning, the system must be discharged.

3. Properly discharge the air conditioning system.
4. Disconnect the:
   a. Temperature sensor connector
   b. Evaporator/heater connector clamp
   c. Temperature control cable
   d. Fresh air door vacuum hose
5. Disconnect the main harness connector.
6. Loosen the case retaining strap.
7. Remove the coolant hoses at the heater core tubes.
8. Remove the yellow, green and red vacuum hoses from the heater case.
9. Remove the air duct hoses.
10. Remove the heater case mounting screws, 2 in the passenger compartment, 1 in the engine compartment. Remove the 4 evaporator housing mounting screws in the passenger compartment.
11. Support the heater/evaporator unit and pull it away from the firewall.
12. Remove the control cable grommet to facilitate case removal.
13. The case halves may be separated by removing the clips at the top and bottom with a small prybar.
14. Remove the blower motor and the heater core from the unit.

To install:

15. Install the blower motor and heater core into the case. Make sure the case gasket is in good condition and assemble the case. Install the clips.
16. Position the case at the firewall, install the control cable grommet and secure the case in place with the screws.
17. Attach the ducts, wires, control cable, vacuum lines and coolant hoses.
18. Refill the cooling system and leak test and recharge the air conditioner.

CAUTION

Freon will freeze any surface it contacts, including skin and eyes. It also turns into a poisonous gas in the presence of an open flame. Wear eye protection and suitable gloves when working on or around the air conditioning system.

V8 Quattro

1. Disconnect the negative battery cable.
2. Matchmark hood hinges and remove hood.
3. Remove the windshield wiper assembly.
4. Remove the cap from the engine coolant overflow bottle.
5. Remove the heater retaining band.
6. Remove the vacuum hoses from the vacuum servo motors.
7. Clamp off the heater hoses to the heater core and remove the hoses.
8. Remove the retainers between the body and heater. Remove the heater box.
9. After removing the heater box, remove the blower cooling hose.
10. Remove the lock ring, stop washer and grommet. Remove the blower from the housing.

To install:

11. Before installing the fresh air blower guides, lubricate with petroleum jelly. Install the blower motor using the black electrical connection area only.
12. Install the heater unit into the vehicle and connect the vacuum and coolant hoses. When connecting the water hoses, the lower connection on the heater core is connected to the hose going to the water pump.
13. Run the engine and check for leaks before completing the assembly.

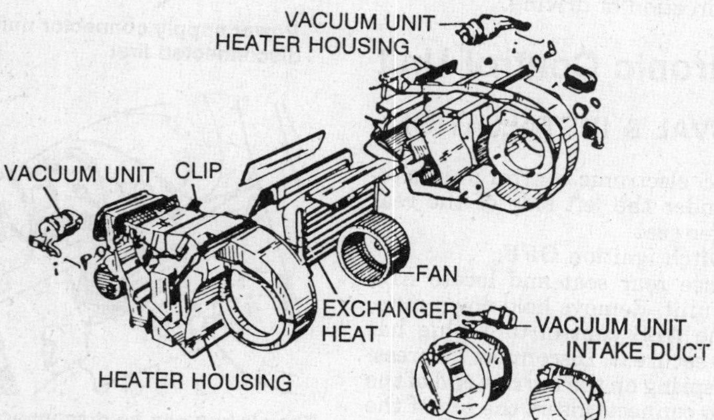

Heater assembly with blower motor

Windshield Wiper Motor

REMOVAL & INSTALLATION

1. Disconnect the negative battery cable and disconnect the wiring harness connector at the motor. Pry off the wiper arms and remove the nuts from the studs in the cowl.
2. Remove the brace-to-body screws. While holding the crank, remove the nut securing the crank to the wiper motor and remove the crank.
3. Remove the bolts securing the wiper motor to the support and remove the motor.
4. Connect the new motor to the wiring harness, run the motor 2 revolutions and turn the wiper switch to the OFF position. The wiper motor should stop in the park position. Remove the linkage followed by the motor.
5. To install, reverse the removal procedures. Make sure the crank is installed in the proper position.

Instrument Cluster

REMOVAL & INSTALLATION

1. Disconnect the negative battery cable. Remove the steering wheel. On V8 Quattro vehicles, the instrument cluster is removed after first removing the instrument panel cover. The screws are under the speaker covers

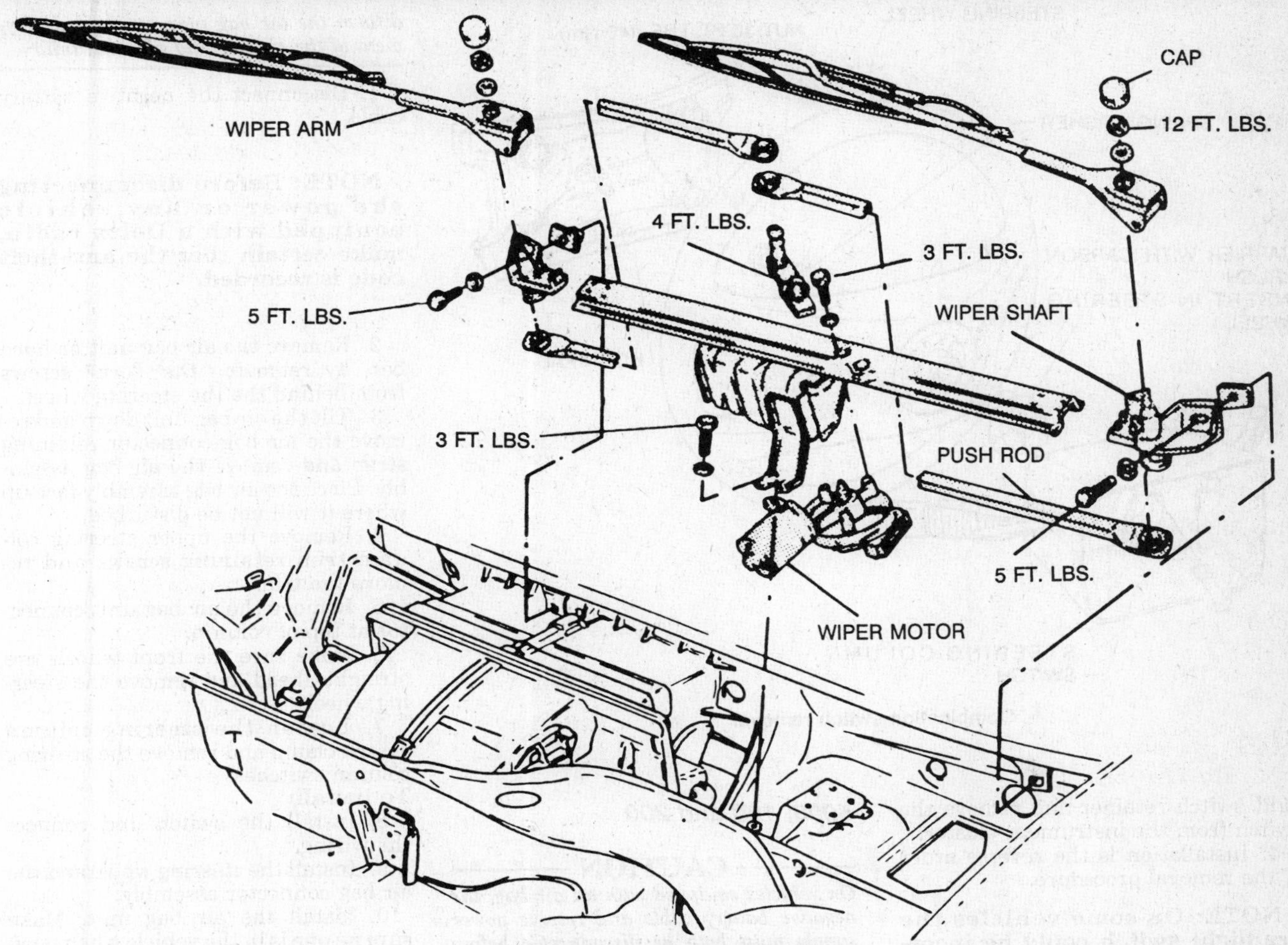

Wiper system assembly 5000 shown

and behind the air conditioning outlets, which are removed with long nose pliers.

2. Loosen the clamp on the steering column switch.
3. Pull forward and remove electrical connector.
4. Remove steering column switches.
5. Tilt instrument cluster back and remove the connector retainers.
6. Remove the electrical connectors.
7. Remove the retaining screws for the instrument cluster and remove the instrument cluster.
8. Installation is in the reverse order of removal.

Headlight Switch

REMOVAL & INSTALLATION

Instrument Panel Mounted

1. Remove the instrument cluster cover.
2. Disconnect the wiring harness connector from the headlight switch.
3. Depress the clips on the head-

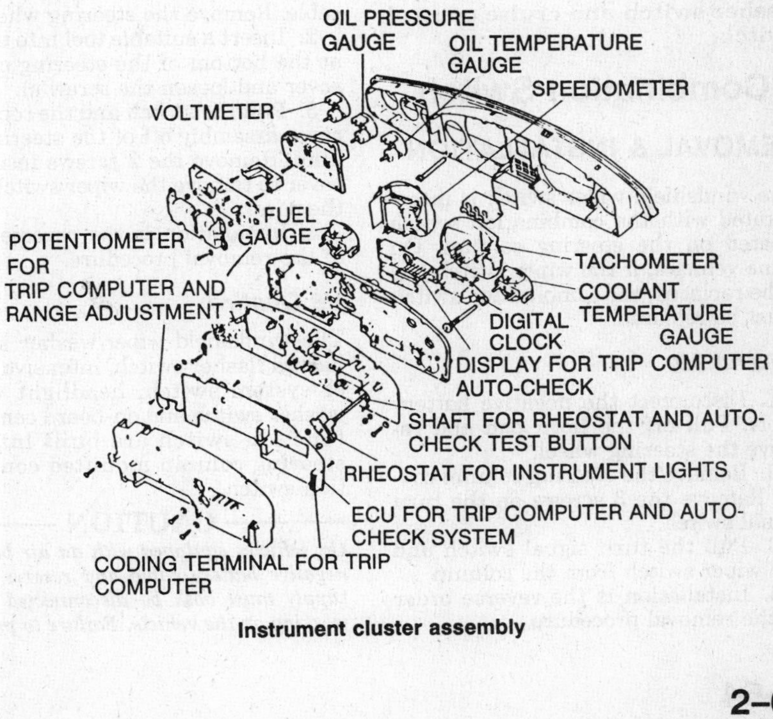

Instrument cluster assembly

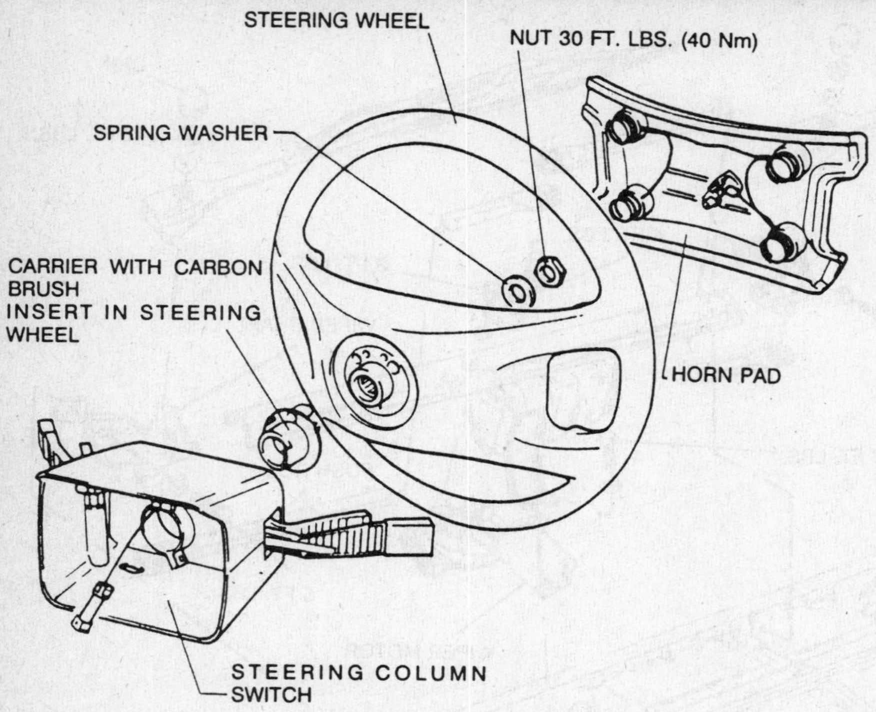

Combination switch removal

light switch retainer and remove the switch from the instrument cluster.

4. Installation is the reverse order of the removal procedure.

NOTE: On some vehicles the headlight switch could be incorporated with the combination switch. This switch can contain the headlight switch, parking light switch, turn signal switch, low/high beam switch, headlight flasher switch and cruise control switch.

Combination Switch

REMOVAL & INSTALLATION

The windshield wiper switch is incorporated with the combination switch located on the steering column. On some vehicles, if the wiper switch has to be replaced, the combination switch must be replaced.

80 and 90

1. Disconnect the negative battery cable. Pull off the horn pad and remove the steering wheel.
2. Remove the steering column cover. Remove the 3 screws on the turn signal switch.
3. Pull the turn signal switch and the wiper switch from the column.
4. Installation is the reverse order of the removal procedure.

5000, 100 and 200

CAUTION

On vehicles equipped with an air bag, the negative battery cable and reserve power supply must both be disconnected before working on the vehicle. Failure to properly disarm the air bag may result in deployment of the air bag and personal injury.

1. Disconnect the negative battery cable. Remove the steering wheel.
2. Insert a suitable tool into the slot at the bottom of the steering column cover and loosen the screw(s).
3. Pull the switch and the top of the cover assembly off of the steering column. Remove the 2 screws inside the cover to remove the wiper switch from the cover.
4. Installation is the reverse order of the removal procedure.

V8 Quattro

The windshield wiper/washer switch, hazard flasher switch, intensive washer system switch, headlight wiper/washer switch and on-board computer function switch are built into one steering column mounted combination switch.

CAUTION

On vehicles equipped with an air bag, the negative battery cable and reserve power supply must both be disconnected before working on the vehicle. Failure to properly disarm the air bag may result in deployment of the air bag and personal injury.

1. Disconnect the negative battery cable.

NOTE: Before disconnecting the power on any vehicle equipped with a Delta radio, make certain that the anti-theft code is recorded.

2. Remove the air bag unit or horn bar, by removing the Torx® screws from behind the the steering wheel.
3. Tilt the air bag unit down and remove the air bag connector retaining strap and remove the air bag assembly. Place the air bag assembly face up where it will not be disturbed.
4. Remove the upper steering column trim retaining screws and remove the trim.
5. Remove the air bag unit connector at top of column.
6. Make sure the front wheels are straight ahead and remove the steering wheel.
7. Loosen the steering column switch clamp and remove the steering column switches.

To install:

8. Install the switch and connect the wiring.
9. Install the steering wheel and the air bag connector assembly.
10. Install the air bag unit. Make sure no one is in the vehicle when power is connected to the unit.

Ignition Lock/Switch

REMOVAL & INSTALLATION

80 and 90

1. Disconnect the negative battery cable.
2. Remove the steering wheel, the steering column covers and the steering column combination switches.
3. Remove instrument cluster retaining screws.
4. Tilt the instrument cluster backwards slightly and remove the electrical connectors.
5. Remove the instrument cluster.
6. Remove the locking compound from the ignition switch screws.
7. Remove the electrical connector from switch.
8. Remove screws and remove the switch.
9. Installation is in the reverse order of removal. Install the ignition switch in the **OFF** position.

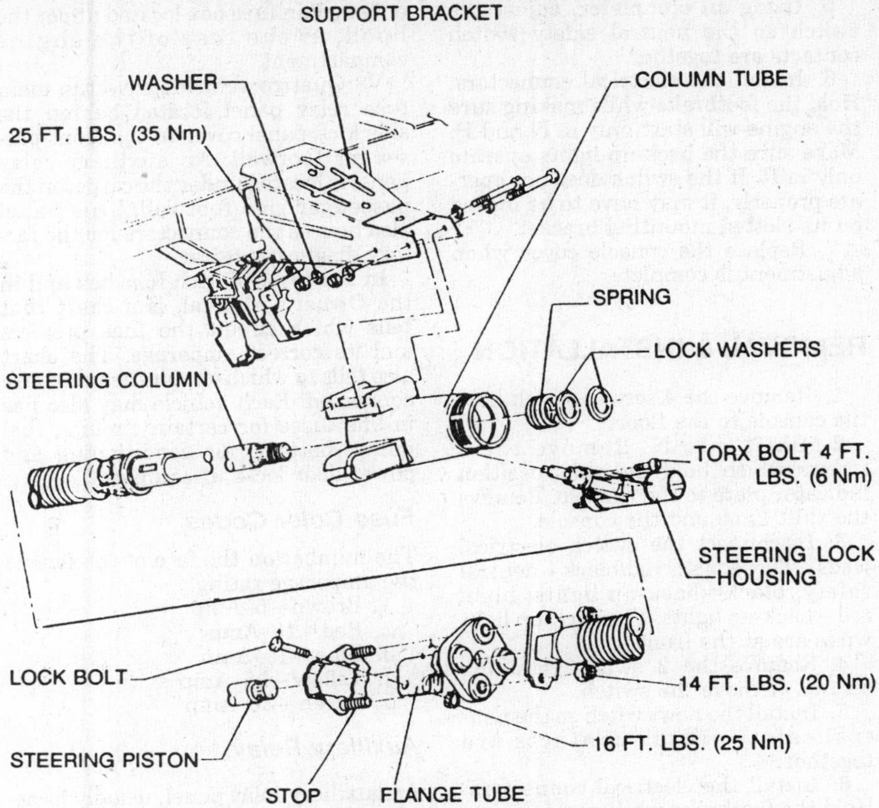

Steering column and lock assembly—80 and 90 shown

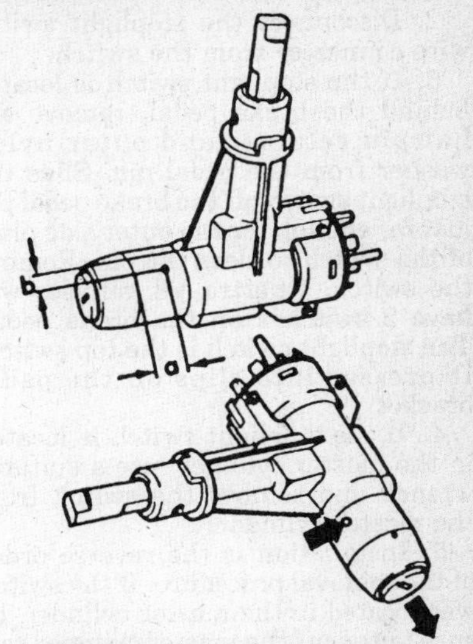

Drill lock housing to remove key cylinder—a = .0.5 in., b = 0.125 in.

5000, 100 and 200

CAUTION

On vehicles equipped with an air bag, the negative battery cable and reserve power supply must both be disconnected before working on the vehicle. Failure to properly disarm the air bag may result in deployment of the air bag and personal injury.

1. Disconnect the negative battery cable. Remove the steering wheel.
2. Remove the instrument cluster attaching bolts. Disconnect the speedometer cable at the transaxle.
3. Pull the instrument cluster forward and detach the speedometer cable at the speedometer. Disconnect the electrical connectors at the instrument cluster and remove the instrument cluster.
4. Remove the locking compound around the ignition switch and remove the switch. To remove the ignition lock cylinder use the following procedure.
5. Support the steering column and drill out the 2 shear bolts using a ⅛ in. drill bit. Loosen the steering column bolts. Remove the left lower dash panel and the left air deflector.
6. Slide the steering column tube with the steering column downward and remove the steering lock from the steering column clamp. Drill a hole in the center of the locking housing using a ⅛ in. drill bit.
7. Push the retaining spring in with a suitable punch and remove the lock cylinder.
8. Installation is the reverse order of the removal procedure.

V8 Quattro

1. Disconnect the negative battery cable.
2. Remove the left and right air bag Torx® bolts from behind the steering wheel.
3. Tilt the air bag unit backward and lift up the safety clamp. Disconnect the wiring from the air bag unit.
4. Remove the steering column trim top section.
5. Separate the air bag connector.
6. Remove the steering wheel.
7. Loosen the steering column switch clamp and remove the column combination switches.
8. Remove the instrument cluster and trim under the left side of the instrument panel.
9. Remove the electrical connector from the ignition switch.
10. Remove the locking compound from the switch mounting screws, loosen the screws slightly and pull the ignition switch assembly from the housing. To remove the ignition lock cylinder:
    a. Remove the glove box.
    b. Remove instrument panel crossmember.
    c. Remove lock housing Torx® screw and remove steering lock housing.
    d. Drill one hole, approx. ⅛ in. (3mm) diameter in steering lock housing only 0.080 in. (2mm) deep, 0.50 in. (12.5mm) from the key end. Do not drill too deeply or the lock cylinder will be damaged.
    e. Remove the lock cylinder by pushing in on the retaining spring.

To install:

11. Installation is the reverse of the removal procedure. When installing the lock cylinder, push into the steering lock housing until the retaining spring engages.
12. After installing the ignition switch, apply locking compound to switch mounting screws.
13. Install the glove box and instrument panel crossmember.
14. Connect the ignition switch wiring and install the instrument cluster.
15. Install the steering column switches and connect all wiring.
16. Install the steering wheel and carefully install the air bag module. Make sure no one is in the vehicle when connecting the battery.

Stoplight Switch

REMOVAL & INSTALLATION

1. Disconnect the negative battery cable.

2. Disconnect the stoplight switch wire connector from the switch.
3. If the stoplight switch is located behind the brake pedal, remove the hairpin retainer and outer nylon washer from the pedal pin. Slide the stoplight switch off the brake pedal pin just far enough for the outer side plate of the switch to clear the pin. Remove the switch. Quattro V8 vehicles will have 2 switches on the brake pedal. The stoplight switch is the top switch. It presses into clips on the pedal bracket.
4. If the stoplight switch is located in the master cylinder, use a suitable wrench and remove the switch from the master cylinder.
5. Installation is the reverse order of the removal procedure. If the switch was located in the master cylinder, be sure to top off the master cylinder reservoir and bleed the brake circuit.

Neutral Safety Switch

The neutral safety switch prevents the engine from being started with the transaxle in any position other than **P** or **N**. It also activates the back-up lights. The switch is at the base of the shift lever, inside the floorshift console except on the V8 Quattro.

On the V8 Quattro, a multi-function switch on the transaxle tells the transaxle ECU which selector lever position has been selected by the driver. It also energizes the back-up light relay and sends a signal to the Automatic Shift Lock (ASL) control unit. This prevents the selector lever from being moved from the **P** or **N** positions unless the driver first steps on the foot brake.

ADJUSTMENT

1. Remove the 4 screws which hold the console to the floor.
2. Shift into **N**. Remove the 2 screws which hold the shift position indicator plate to the console. Remove the shift knob and the console.
3. Disconnect the red/black electrical leads from the switch.
4. Loosen both switch retaining screws.
5. Using an ohmmeter, adjust the switch so the neutral safety switch contacts are together.
6. Install the electrical connectors. Hold the footbrake while making sure the engine will start only in **N** and **P**. Make sure the back-up lights operate only in **R**. If the switch does not operate properly, it may have to be moved on its slotted mounting bracket.
7. Replace the console cover when adjustment is complete.

REMOVAL & INSTALLATION

1. Remove the 4 screws which hold the console to the floor.
2. Shift into **N**. Remove the 2 screws which hold the shift position indicator plate to the console. Remove the shift knob and the console.
3. Disconnect the switch electrical leads. These are: red/black – neutral safety; black – back-up lights; blue/red – back-up lights. The back-up light wires are at the front.
4. Remove the 2 switch retaining screws. Remove the switch.
5. Install the new switch so the neutral safety switch contacts are together.
6. Install the electrical connectors. Hold the footbrake while making sure the engine will start only in **N** and **P**. Make sure the back-up lights operate only in **R**. If the switch does not operate properly, it may have to be moved on its slotted mounting bracket.
7. Replace the console cover when adjustment is complete.

Fuses, Circuit Breakers and Relays

NOTE: Before disconnecting the power on any vehicle equipped with a Delta radio, make certain that the anti-theft code has been recorded.

LOCATION

Several relays are plugged into the fuse box. Most vehicles still have a fuse box mounted under the left side of the dadhboard. Some vehicles may have only a main fuse box located under the hood, at the rear of the engine compartment.

V8 Quattro vehicles have the main fuse relay panel located behind the side kick panel cover in the front passenger footwell. An auxiliary relay panel is located under the carpet in the passenger side footwell. This panel also houses the connectors for the factory diagnostic tester.

In the cover of each fuse box and in the Owner's Manual, is a chart that tells which circuit the fuse protects and its correct amperage. The chart also tells to which circuit the relays are connected. Each vehicle may also use in-line fuses for certain circuits; fuel pump, battery, air conditioning and power door locks, if equipped.

Fuse Color Codes

The number on the face of the fuse is the amperage rating.

1. Brown – 5–Amp
2. Red – 10–Amp
3. Blue – 15–Amp
4. Yellow – 25–Amp
5. Green – 30–Amp

Auxiliary Relay

An auxiliary relay panel, usually located under the left side of the instrument panel may contain additional relays.

Computers

LOCATION

Computers are used in a variety of applications, primarily for engine fuel management. The main computer is located under the instrument panel, generally on the driver's side near the console. A new 4-speed electronically controlled automatic transaxle is now used in the V8 Quattro and is also tied into the computer.

Flashers

LOCATION

Turn signal and hazard flashers are located in the fuse and relay panels.

BMW 3

3-Series, 5-Series, 6-Series, 7-Series—All Models

SERIAL NUMBER IDENTIFICATION

Vehicle Identification Plate

The manufacturer's plate is located in the engine compartment, on the right side inner fender panel or support.

Engine Number

The engine number is located on the left rear side of the engine, above the starter motor.

Vehicle Identification Number

The vehicle identification number is located on a plate, on the drivers side of the instrument panel.

Chassis Number

The chassis number can be found in the engine compartment on the right front inner fender support or facing forward on the right side of the bulkhead. A label is also attached to the upper steering column cover, inside the vehicle.

SPECIFICATIONS

ENGINE IDENTIFICATION

Year	Model	Engine Displacement cu. in. (cc/liter)	Engine Series Identification	No. of Cylinders	Engine Type
1988	325	165 (2693/2.7)	M20B27	6	OHC
	528e	165 (2693/2.7)	M20B27	6	OHC
	325i	152 (2494/2.5)	M20B25	6	OHC
	325iS	152 (2494/2.5)	M20B25	6	OHC
	325iX	152 (2494/2.5)	M20B25	6	OHC
	535i	209 (3428/3.4)	M30B34	6	OHC
	635CSi	209 (3428/3.4)	M30B35MZ	6	OHC
	L6	209 (3428/3.4)	M30B35MZ	6	OHC
	735i	209 (3428/3.4)	M30B35MZ	6	OHC
	M3	140.4 (2302/2.3)	S14	4	DOHC
	M5	210.6 (3453/3.5)	S38Z	6	DOHC
	M6	210.6 (3453/3.5)	S38Z	6	DOHC
	750iL	304.4 (4988/5.0)	M70	12	OHC
1989	325	165 (2693/2.7)	M20B27	6	OHC
	325i	152 (2494/2.5)	M20B25	6	OHC
	325iS	152 (2494/2.5)	M20B25	6	OHC
	325iX	152 (2494/2.5)	M20B25	6	OHC
	525	152 (2494/2.5)	M20B25	6	OHC
	535i	209 (3428/3.4)	M30B34	6	OHC
	635CSi	209 (3428/3.4)	M30B35MZ	6	OHC
	L6	209 (3428/3.4)	M30B35MZ	6	OHC
	735i	209 (3428/3.4)	M30B35MZ	6	OHC
	735iL	209 (3428/3.4)	M30B35MZ	6	OHC
	M3	140.4 (2302/2.3)	S14	4	DOHC
	M5	210.6 (3453/3.5)	S38Z	6	DOHC
	M6	210.6 (3453/3.5)	S38Z	6	DOHC
	750iL	304.4 (4988/5.0)	M70	12	OHC

ENGINE IDENTIFICATION

Year	Model	Engine Displacement cu. in. (cc/liter)	Engine Series Identification	No. of Cylinders	Engine Type
1990	318iS	109.6 (1796/1.8)	M42	4	DOHC
	325i	152 (2494/2.5)	M20B25	6	OHC
	325iS	152 (2494/2.5)	M20B25	6	OHC
	325iX	152 (2494/2.5)	M20B25	6	OHC
	525i	152 (2494/2.5)	M30B25L	6	OHC
	535i	209 (3430/3.5)	M30B34M	6	OHC
	M3	140 (2302/2.3)	S14B23	4	DOHC
	735i	209 (3430/3.5)	M30B35M	6	OHC
	735iL	209 (3430/3.5)	M30B35M	6	OHC
	750iL	304 (4988/5.0)	M70	12	OHC
1991–92	318iS	109.6 (1796/1.8)	M42	4	DOHC
	325i	152 (2494/2.5)	M20B25	6	OHC
	325iX	152 (2494/2.5)	M20B25	6	OHC
	525i	152 (2494/2.5)	M50B25	6	DOHC
	535i	209 (3430/3.5)	M30B35M	6	OHC
	735i	209 (3430/3.5)	M30B35MZ	6	OHC
	735iL	209 (3430/3.5)	M30B35MZ	6	OHC
	750iL	304 (4988/5.0)	M70	12	OHC
	850i	304 (4988/5.0)	M70	12	OHC
	M3	140 (2302/2.3)	S14B23	4	DOHC
	M5	216 (3535/3.6)	S38B36	6	DOHC

OHC—Overhead Camshaft
DOHC—Double Overhead Camshaft

GENERAL ENGINE SPECIFICATIONS

Year	Model	Engine Displacement cu. in. (cc)	Fuel System Type	Net Horsepower @ rpm	Net Torque @ rpm (ft. lbs.)	Bore × Stroke (in.)	Com-pression Ratio	Oil Pressure @ rpm
1988	325	165 (2693)	EFI	121 @ 4250	170 @ 3250	3.307 × 3.189	9.0:1	71 @ 5000
	528e	165 (2693)	EFI	121 @ 4250	170 @ 3250	3.307 × 3.189	9.0:1	71 @ 5000
	325i	152 (2494)	EFI	167 @ 5800	164 @ 4300	3.307 × 2.953	8.8:1	71 @ 6000
	325iS	152 (2494)	EFI	167 @ 5800	164 @ 4300	3.307 × 2.953	8.8:1	71 @ 6000
	325iX	152 (2494)	EFI	167 @ 5800	164 @ 4300	3.307 × 2.953	8.8:1	71 @ 6000
	535i	209 (3428)	EFI	182 @ 5400	213 @ 4000	3.62 × 3.38	8.0:1	71 @ 6100
	635CSi	209 (3428)	EFI	208 @ 5700	225 @ 4000	3.62 × 3.38	9.0:1	64 @ 6200
	L6	209 (3428)	EFI	208 @ 5700	225 @ 4000	3.62 × 3.38	9.0:1	64 @ 6200
	735i	209 (3428)	EFI	208 @ 5700	225 @ 4000	3.62 × 3.38	9.0:1	64 @ 6200
	M3	104.4 (2302)	EFI	194 @ 6750	166 @ 4750	3.67 × 3.30	10.5:1	71 @ 7250
	M5	210.6 (3453)	EFI	256 @ 6500	239 @ 4500	3.67 × 3.30	9.8:1	71 @ 6800
	M6	210.6 (3453)	EFI	256 @ 6500	239 @ 4500	3.67 × 3.30	9.8:1	71 @ 6800

GENERAL ENGINE SPECIFICATIONS

Year	Model	Engine Displacement cu. in. (cc)	Fuel System Type	Net Horsepower @ rpm	Net Torque @ rpm (ft. lbs.)	Bore × Stroke (in.)	Com-pression Ratio	Oil Pressure @ rpm
1989	325	165 (2693)	EFI	121 @ 4250	170 @ 3250	3.307 × 3.189	9.0:1	71 @ 5000
	325i	152 (2494)	EFI	167 @ 5800	164 @ 4300	3.307 × 2.953	8.8:1	71 @ 6000
	325iS	152 (2494)	EFI	167 @ 5800	164 @ 4300	3.307 × 2.953	8.8:1	71 @ 6000
	325iX	152 (2494)	EFI	167 @ 5800	164 @ 4300	3.307 × 2.953	8.8:1	71 @ 6000
	525i	152 (2494)	EFI	167 @ 5800	164 @ 4300	3.31 × 3.295	8.8:1	71 @ 6000
	535i	209 (3428)	EFI	182 @ 5400	213 @ 4000	3.62 × 3.38	8.0:1	71 @ 6100
	635CSi	209 (3428)	EFI	208 @ 5700	225 @ 4000	3.62 × 3.38	9.0:1	64 @ 6200
	L6	209 (3428)	EFI	208 @ 5700	225 @ 4000	3.62 × 3.38	9.0:1	64 @ 6200
	735i	209 (3428)	EFI	208 @ 5700	225 @ 4000	3.62 × 3.38	9.0:1	64 @ 6200
	735iL	209 (3428)	EFI	208 @ 5700	225 @ 4000	3.62 × 3.38	9.0:1	64 @ 6200
	M3	104.4 (2302)	EFI	194 @ 6750	166 @ 4750	3.67 × 3.30	10.5:1	71 @ 7250
	M5	210.6 (3453)	EFI	256 @ 6500	239 @ 4500	3.67 × 3.30	9.8:1	71 @ 6800
	M6	210.6 (3453)	EFI	256 @ 6500	239 @ 4500	3.67 × 3.30	9.8:1	71 @ 6800
1990	318iS	109.6 (1796)	EFI	136 @ 5800	127 @ 4600	3.31 × 3.19	10.0:2	57 @ 6500
	325i	152 (2494)	EFI	168 @ 5800	164 @ 4300	3.31 × 2.95	8.8:1	71 @ 6000
	325iS	152 (2494)	EFI	168 @ 5800	164 @ 4300	3.31 × 2.95	8.8:1	71 @ 6000
	325iX	152 (2494)	EFI	168 @ 5800	164 @ 4300	3.31 × 2.95	8.8:1	71 @ 6000
	525i	152 (2494)	EFI	168 @ 5800	164 @ 4300	3.31 × 2.95	8.8:1	71 @ 6000
	535i	209 (3430)	EFI	208 @ 5700	225 @ 4000	3.62 × 3.39	9.0:1	71 @ 6000
	735i	209 (3430)	EFI	208 @ 5700	225 @ 4000	3.62 × 3.39	9.0:1	64 @ 6200
	735iL	209 (3430)	EFI	208 @ 5700	225 @ 4000	3.62 × 3.39	9.0:1	64 @ 6200
	750iL	304 (4988)	EFI	296 @ 5200	332 @ 4100	3.31 × 2.95	8.8:1	57 @ 6000
	M3	140 (2302)	EFI	192 @ 6750	170 @ 4750	3.68 × 3.31	10.5:1	71 @ 7250
1991-92	318iS	109.6 (1796)	EFI	134 @ 6000	127 @ 4600	3.31 × 3.19	10.0:2	57 @ 6500
	325i	152 (2494)	EFI	168 @ 5800	164 @ 4300	3.31 × 2.95	8.8:1	71 @ 6000
	325iX	152 (2494)	EFI	168 @ 5800	164 @ 4300	3.31 × 2.95	8.8:1	71 @ 6000
	525i	152 (2494)	EFI	189 @ 5900	181 @ 4700	3.31 × 2.95	10.0:1	57 @ 6500
	535i	209 (3430)	EFI	208 @ 5700	225 @ 4000	3.62 × 3.39	9.0:1	71 @ 6000
	735i	209 (3430)	EFI	208 @ 5700	225 @ 4000	3.62 × 3.39	9.0:1	71 @ 6000
	735iL	209 (3430)	EFI	208 @ 5700	225 @ 4000	3.62 × 3.39	9.0:1	71 @ 6000
	750iL	304 (4988)	EFI	296 @ 5200	332 @ 4100	3.31 × 2.95	8.8:1	57 @ 6000
	850i	304 (4988)	EFI	296 @ 5200	332 @ 4100	3.31 × 2.95	8.8:1	57 @ 6000
	M3	140 (2302)	EFI	192 @ 6790	170 @ 4750	3.68 × 3.31	10.5:1	71 @ 7250
	M5	216 (3535)	EFI	310 @ 6400	266 @ 4750	3.68 × 3.39	10.0:1	57 @ 7200

EFI–Electronic Fuel Injection
NA–Not Available

GASOLINE ENGINE TUNE-UP SPECIFICATIONS

Year	Model	Engine Displacement cu. in. (cc)	Spark Plugs Type	Spark Plugs Gap (in.)	Ignition Timing (deg.) MT	Ignition Timing (deg.) AT	Compression Pressure (psi)	Fuel Pump (psi)	Idle Speed (rpm) MT	Idle Speed (rpm) AT	Valve Clearance In.	Valve Clearance Ex.
1988	325	165 (2693)	WR9LS	0.027	①	①	149	36	720	720	0.010	0.010
	528e	165 (2693)	WR9LS	0.027	①	①	149	36	720	720	0.010	0.010
	325i	152 (2494)	WR9LS	0.027	①	①	149	43	760	760	0.010	0.010
	325iS	152 (2494)	WR9LS	0.027	①	①	149	43	760	760	0.010	0.010
	325iX	152 (2494)	WR9LS	0.027	①	①	149	43	760	760	0.010	0.010
	535i	209 (3428)	WR9LS	0.027	①	①	149	43	800	800	0.012	0.012
	635CSi	209 (3428)	WR9LS	0.027	①	①	149	43	800	800	0.012	0.012
	L6	209 (3428)	WR9LS	0.027	①	①	149	43	800	800	0.012	0.012
	M3	140.4 (2302)	WR9LS	0.027	①	①	149	43	—	—	0.012	0.012
	M5	210.6 (3453)	WR9LS	0.027	①	①	149	43	850	—	0.013	0.013
	M6	210.6 (3453)	WR9LS	0.027	①	①	149	43	850	—	0.013	0.013
	735i	209 (3428)	W8LCR	0.027	①	①	149	43	①	①	0.012	0.012
	750i	304 (4988)	F8LCR	0.027	①	①	176	43	700	700	Hyd.	Hyd.
1989	325	165 (2693)	WR9LS	0.027	①	①	149	36	720	720	0.010	0.010
	325i	152 (2494)	WR9LS	0.027	①	①	149	43	760	760	0.010	0.010
	325iS	152 (2494)	WR9LS	0.027	①	①	149	43	760	760	0.010	0.010
	325iX	152 (2494)	WR9LS	0.027	①	①	149	43	760	760	0.010	0.010
	525i	152 (2494)	WR9LS	0.027	①	①	149	43	760	760	0.010	0.010
	535i	209 (3428)	WR9LS	0.027	①	①	149	43	800	800	0.012	0.012
	635CSi	209 (3428)	WR9LS	0.027	①	①	149	43	800	800	0.012	0.012
	L6	209 (3428)	WR9LS	0.027	①	①	149	43	800	800	0.012	0.012
	M3	140.4 (2302)	WR9LS	0.027	①	①	149	43	—	—	0.012	0.012
	M5	210.6 (3453)	WR9LS	0.027	①	①	149	43	850	—	0.013	0.013
	M6	210.6 (3453)	WR9LS	0.027	①	①	149	43	850	—	0.013	0.013
	735i	209 (3428)	W8LCR	0.027	①	①	149	43	①	①	0.012	0.012
	735iL	209 (3428)	W8LCR	0.027	①	①	149	43	①	①	0.012	0.012
	750iL	304 (4988)	F8LCR	0.027	①	①	176	43	700	700	Hyd.	Hyd.
1990	318iS	109.6 (1796)	F03DAR	0.032	①	①	142–156	43	850	—	Hyd.	Hyd.
	325i	152 (2494)	W8LCR	0.028	①	①	142–156	43	760	760	0.010	0.010
	325iS	152 (2494)	W8LCR	0.028	①	①	142–156	43	760	760	0.010	0.010
	325iX	152 (2494)	W8LCR	0.028	①	①	142–156	43	760	760	0.010	0.010
	525i	152 (2494)	W8LCR	0.028	①	①	143–156	43	800	800	③	③
	535i	209 (3430)	W8LCR	0.028	①	①	143–156	43	800	800	③	③
	725i	209 (3430)	W8LCR	0.028	①	①	142–156	43	800	800	③	③
	735iL	209 (3430)	W8LCR	0.028	①	①	142–156	43	800	800	③	③
	750iL	304 (4988)	F8LCR	0.028	①	①	142–170	43	800	800	Hyd.	Hyd.
	M3	140 (2302)	X50L	0.028	①	①	143–156	43	880	880	0.0010–0.0026	0.0010–0.0026

GASOLINE ENGINE TUNE-UP SPECIFICATIONS

Year	Model	Engine Displacement cu. in. (cc)	Spark Plugs Type	Spark Plugs Gap (in.)	Ignition Timing (deg.) MT	Ignition Timing (deg.) AT	Com-pression Pressure (psi)	Fuel Pump (psi)	Idle Speed (rpm) MT	Idle Speed (rpm) AT	Valve Clearance In.	Valve Clearance Ex.
1991	318iS	109.6 (1796)	F03DAR	0.032	①	①	142–156	43	850	—	Hyd.	Hyd.
	325i	152 (2494)	W8LCR	0.028	①	①	142–156	43	760	760	0.010	0.010
	325iS	152 (2494)	W8LCR	0.028	①	①	142–156	43	760	760	0.010	0.010
	325iX	152 (2494)	W8LCR	0.028	①	①	142–156	43	760	760	0.010	0.010
	525i	152 (2494)	W8LCR	0.028	①	①	143–156	43	700	700	Hyd.	Hyd.
	535i	209 (3430)	W8LCR	0.028	①	①	143–156	43	800	800	③	③
	735i	209 (3430)	W8LCR	0.028	①	①	142–156	43	800	800	③	③
	735iL	209 (3430)	W8LCR	0.028	①	①	142–156	43	800	800	③	③
	750iL	304 (4988)	F8LCR	0.028	①	①	142–170	43	800	800	Hyd.	Hyd.
	850i	304 (4988)	F8LCR	0.028	①	①	142–170	43	800	800	Hyd.	Hyd.
	M3	140 (2302)	X5DC	0.024	①	①	143–156	43	880	—	0.0010–0.0026	0.0010–0.0026
	M5	216 (3535)	Y6DC	0.024	①	①	143–156	43	880	—	0.0010–0.0026	0.0010–0.0026
1992	SEE UNDERHOOD SPECIFICATIONS STICKER											

NOTE: The underhood specifications sticker often reflects tune-up changes made in production. Sticker figures must be used if they disagree with those in this chart.

Hyd.—Hydraulic
NA Not available
B Before Top Dead Center

① Motronic injection system—controlled by computer, please refer to the underhood sticker for specifications

② 0.010–0.014 COLD
0.012–0.016 HOT

③ 0.012—COLD
0.014—HOT

FIRING ORDERS

NOTE: To avoid confusion, always replace spark plug wires one at a time.

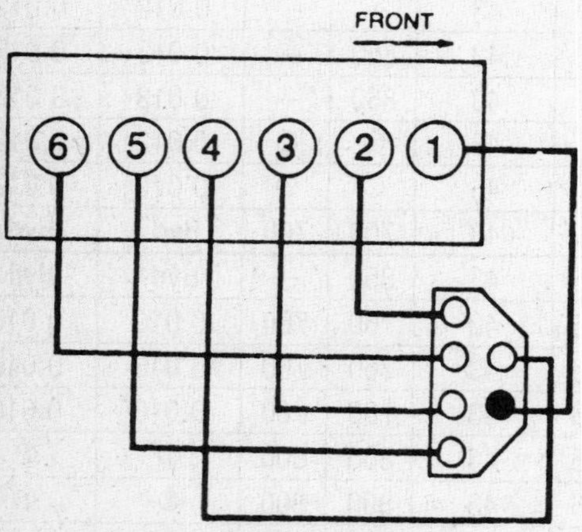

2.5L, 2.7L, 3.4L and 5.0L Engines
Engine Firing Order: 1–5–3–6–2–4
Distributor Rotation: Clockwise

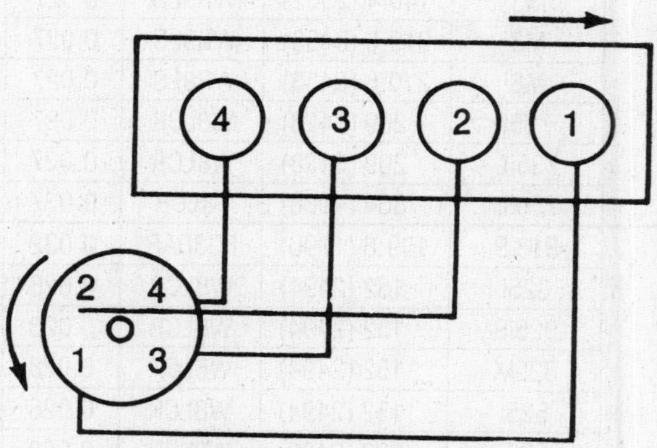

2.3L and 2.7L Engines
Engine Firing Order: 1–3–4–2
Distributor Rotation: Counterclockwise

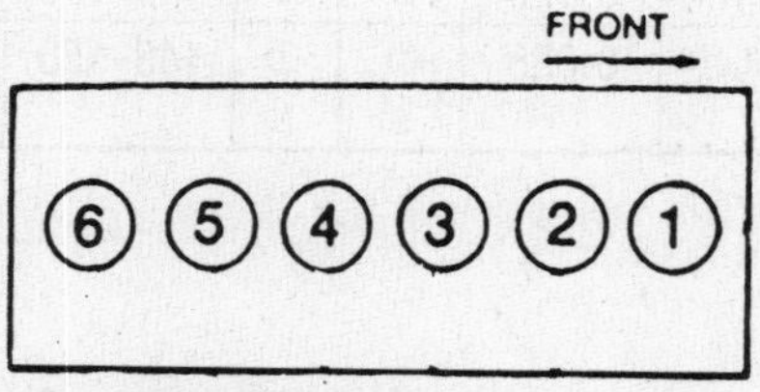

2.5L 24-Valve Engine
Engine Firing Order: 1–5–3–6–2–4
Distributorless Ignition

CAPACITIES

Year	Model	Engine Displacement cu. in. (cc)	Engine Crankcase (qts.) with Filter	Engine Crankcase (qts.) without Filter	Transmission (pts.) 4-Spd	Transmission (pts.) 5-Spd	Transmission (pts.) Auto.	Drive Axle (pts.)	Fuel Tank (gal.)	Cooling System (qts.)
1988	325	165 (2693)	4.5	4.2	—	2.6	6.4	3.6	16.4	11.6
	325i	152 (2494)	5.0	4.75	—	2.6	6.4	3.6	16.4	11.0
	325iS	152 (2494)	5.0	4.75	—	2.6	6.4	3.6	16.4	11.0
	325iX	152 (2494)	5.0	4.75	—	2.6	6.4 ①	3.6 ②	16.4	11.0
	528e	165 (2693)	4.5	4.2	—	3.4	6.4	3.8	16.6	11.6
	535i	209 (3428)	6.1	5.3	—	2.6	6.4	4.0	16.6	11.6
	535iS	209 (3428)	6.1	5.3	—	2.6	6.4	4.0	16.6	12.7
	635CSi	209 (3428)	6.1	5.3	—	3.4	6.4	4.0	16.6	12.7
	L6	209 (3428)	6.1	5.3	—	3.4	6.4	4.0	16.6	12.7
	735i	209 (3428)	6.1	5.3	—	2.6	6.4	4.0	21.4	12.7
	M3	104.4 (2302)	5.0	4.75	—	2.6	6.4	3.6	16.4	12.7
	M5	210.6 (3453)	6.1	5.3	—	2.6	6.4	4.0	14.5	12.7
	M6	210.6 (3453)	6.1	5.3	—	2.6	6.4	4.0	16.6	12.7
	750iL	304 (4988)	7.9	6.8	—	2.6	6.4	4.0	21.4	12.7
1989	325	165 (2693)	4.5	4.2	—	2.6	6.4	3.6	16.4	11.6
	325i	152 (2494)	5.0	4.75	—	2.6	6.4	3.6	16.4	11.0
	325iS	152 (2494)	5.0	4.75	—	2.6	6.4	3.6	16.4	11.0
	325iX	152 (2494)	5.0	4.75	—	2.6	6.4 ①	3.6 ②	16.4	11.0
	525i	152 (2494)	5.0	4.75	—	2.6	6.4	3.6	16.4	11.0
	535i	209 (3428)	6.1	5.3	—	2.6	6.4	4.0	16.6	11.6
	535iS	209 (3428)	6.1	5.3	—	2.6	6.4	4.0	16.6	12.7
	635CSi	209 (3428)	6.1	5.3	—	3.4	6.4	4.0	16.6	12.7
	L6	209 (3428)	6.1	5.3	—	3.4	6.4	4.0	16.6	12.7
	735i	209 (3428)	6.1	5.3	—	2.6	6.4	4.0	21.4	12.7
	735iL	209 (3428)	6.1	5.3	—	2.6	6.4	4.0	21.4	12.7
	M3	104.4 (2302)	5.0	4.75	—	2.6	—	3.6	14.5	12.7
	M5	210.6 (3453)	6.1	5.3	—	2.6	6.4	4.0	16.6	12.7
	M6	210.6 (3453)	6.1	5.3	—	2.6	—	4.0	16.6	12.7
	750iL	304 (4988)	7.9	6.8	—	2.6	6.4	4.0	21.4	12.7
1990	318iS	109.6 (1796)	5.26	5.0	—	2.4	3.2	0.95	14.5	6.9
	325i	152 (2494)	5.0	4.75	—	2.6	6.4	3.6	16.4	11.0
	325iS	152 (2494)	5.0	4.75	—	2.6	6.4	3.6	16.4	11.0
	325iX	152 (2494)	5.0	4.75	—	2.6	6.4 ①	3.6 ②	16.4	11.0
	525i	152 (2494)	5.0	4.75	—	2.6	6.4	3.6	21.0	11.0
	535i	209 (3430)	6.10	5.30	—	2.6	6.4	4.0	21.0	12.7
	725i	209 (3430)	6.10	5.30	—	2.6	6.4	4.0	21.4	12.7
	735iL	209 (3430)	6.10	5.30	—	2.6	6.4	4.0	21.4	12.7
	750iL	304 (4988)	7.90	6.80	—	2.6	7.4	4.0	24	14.8
	M3	140 (2303)	4.5	4.2	—	2.6	—	3.6	14.5	9.5

CAPACITIES

Year	Model	Engine Displacement cu. in. (cc)	Engine Crankcase (qts.) with Filter	without Filter	Transmission (pts.) 4-Spd	5-Spd	Auto.	Drive Axle (pts.)	Fuel Tank (gal.)	Cooling System (qts.)
1991-92	318iS	109.6 (1796)	5.26	5.0	—	2.4	3.2	0.95	14.5	6.9
	325i	152 (2494)	4.46	4.2	—	2.6	6.4	3.6	16.4	11.0
	325iX	152 (2494)	4.46	4.2	—	2.6	6.4 ①	3.6 ②	16.4	11.0
	525i	152 (2494)	6.10	5.30	—	2.2	6.4	3.6	21.4	11.0
	535i	209 (3430)	6.10	5.30	—	2.6	6.4	4.0	21.4	12.7
	735i	209 (3430)	6.10	5.30	—	2.6	6.4	4.0	21.4	12.7
	735iL	209 (3430)	6.10	5.30	—	2.6	6.4	4.0	24	12.7
	750iL	304 (4988)	7.90	6.80	—	2.6	7.4	4.0	24	15.8
	850i	304 (4988)	7.90	6.80	—	4.8	7.4	4.0	24	13.7
	M3	140 (2302)	4.6	4.3	—	2.64	—	3.6	14.5	10.0
	M5	216 (3535)	6.10	5.30	—	2.6	—	4.0	21.4	14

NA Not Available
① 325iX Transfer case—1.1
② 325iX Front drive axle—1.5

CRANKSHAFT AND CONNECTING ROD SPECIFICATIONS

All measurements are given in inches.

Year	Engine Displacement cu. in. (cc)	Crankshaft Main Brg. Journal Dia.	Main Brg. Oil Clearance	Shaft End-play	Thrust on No.	Connecting Rod Journal Diameter	Oil Clearance	Side Clearance
1988	140.4 (2302)	2.1653	0.0012–0.0028	0.0033–0.0068	3	1.88877–1.88940	0.0012–0.0028	0.0016
	152 (2494)	2.3622	0.0012–0.0027	0.0033–0.0068	4	1.7717	0.0012–0.0028	0.0016
	165 (2693)	2.3622	0.0012–0.0027	0.0033–0.0068	4	1.7717	0.0012–0.0028	0.0016
	209 (3428)	2.3622	0.0012–0.0027	0.0033–0.0068	4	1.8898	0.0012–0.0028	0.0016
	210.6 (3453)	2.3622	0.0012–0.0027	0.0033–0.0068	4	1.88877–1.88940	0.0012–0.0028	0.0016
	304 (4988)	2.9521–2.9523	0.0010–0.0030	0.0033–0.0068	—	1.7707–1.7713	0.0006–0.0023	0.0016
1989	140.4 (2302)	2.1653	0.0012–0.0028	0.0033–0.0068	3	1.88877–1.88940	0.0012–0.0028	0.0016
	152 (2494)	2.3622	0.0012–0.0027	0.0033–0.0068	4	1.7717	0.0012–0.0028	0.0016
	165 (2693)	2.3622	0.0012–0.0027	0.0033–0.0068	4	1.7717	0.0012–0.0028	0.0016
	209 (3428)	2.3622	0.0012–0.0027	0.0033–0.0068	4	1.8898	0.0012–0.0028	0.0016
	210.6 (3453)	2.3622	0.0012–0.0027	0.0033–0.0068	4	1.88877–1.88940	0.0012–0.0028	0.0016
	304 (4988)	2.9521–2.9523	0.0010–0.0030	0.0033–0.0068	—	1.7707–1.7713	0.0006–0.0023	0.0016

CRANKSHAFT AND CONNECTING ROD SPECIFICATIONS

All measurements are given in inches.

Year	Engine Displacement cu. in. (cc)	Crankshaft				Connecting Rod		
		Main Brg. Journal Dia.	Main Brg. Oil Clearance	Shaft End-play	Thrust on No.	Journal Diameter	Oil Clearance	Side Clearance
1990	109.6 (1796)	①	NA	0.008–0.016	—	①	0.0008–0.0022	0.0008–0.0020
	140.4 (2302)	2.1653	0.0012–0.0028	0.0033–0.0068	3	1.88877–1.88940	0.0012–0.0028	0.0016
	152 (2494)	2.3622	0.0012–0.0027	0.0033–0.0068	4	1.7717	0.0012–0.0028	0.0016
	209 (3430)	2.3622	0.0012–0.0027	0.0033–0.0068	4	1.88877 1.88940	0.0012–0.0028	0.0016
	304 (4988)	2.9521–2.9523	0.0010–0.0030	0.0033–0.0068	—	1.7707–1.7713	0.0006–0.0023	0.0016
1991–92	109.6 (1796)	2.3615	0.0008–0.0023	0.0031–0.0064	NA	1.7717	0.0008–0.0022	NA
	140.4 (2302)	2.1653	0.0012–0.0028	0.0033–0.0068	3	1.88877–1.88940	0.0012–0.0028	0.0016
	152 (2494)	2.3622	0.0012–0.0027	0.0033–0.0068	4	1.7717	0.0012–0.0028	0.0016
	152 (2494) ②	2.3615	0.0008–0.0023	0.0031–0.0064	NA	1.7717	0.0008–0.0022	NA
	216 (3535)	2.3622	0.0012–0.0027	0.0033–0.0068	4	1.88877 1.88940	0.0012–0.0028	0.0016
	304 (4988)	2.9521–2.9523	0.0010–0.0030	0.0033–0.0068	—	1.7707–1.7713	0.0006–0.0023	0.0016

NA Not Available
① One paint stripe—0.010″
Two paint stripes—0.020″
② 24 Valve Engine

VALVE SPECIFICATIONS

Year	Engine Displacement cu. in. (cc)	Seat Angle (deg.)	Face Angle (deg.)	Spring Test Pressure (lbs.)	Spring Installed Height (in.)	Stem-to-Guide Clearance (in.)		Stem Diameter (in.)	
						Intake	Exhaust	Intake	Exhaust
1988	140.4 (2302)	45	NA	NA	NA	0.025 ②	0.031 ②	0.276	0.276
	152 (2494)	45	NA	NA	NA	0.031 ②	0.031 ②	0.275	0.275
	165 (2693)	45	NA	NA	NA	0.031 ②	0.031 ②	0.275	0.275
	209 (3428)	45	NA	NA	NA	0.031 ②	0.031 ②	0.315	0.315
	210.6 (3453)	45	NA	NA	NA	0.025 ②	0.031 ②	0.276	0.276
	304 (4988)	45	NA	NA	NA	0.020 ②	0.020 ②	0.275	0.275
1989	140.4 (2302)	45	NA	NA	NA	0.025 ②	0.031 ②	0.276	0.276
	152 (2494)	45	NA	NA	NA	0.031 ②	0.031 ②	0.275	0.275
	165 (2693)	45	NA	NA	NA	0.031 ②	0.031 ②	0.275	0.275
	209 (3428)	45	NA	NA	NA	0.031 ②	0.031 ②	0.315	0.315
	210.6 (3453)	45	NA	NA	NA	0.025 ②	0.031 ②	0.276	0.276
	304 (4988)	45	NA	NA	NA	0.020 ②	0.020 ②	0.275	0.275

VALVE SPECIFICATIONS

Year	Engine Displacement cu. in. (cc)	Seat Angle (deg.)	Face Angle (deg.)	Spring Test Pressure (lbs.)	Spring Installed Height (in.)	Stem-to-Guide Clearance (in.)		Stem Diameter (in.)	
						Intake	Exhaust	Intake	Exhaust
1990	109.6 (1796)	NA	NA	NA	NA	NA	NA	NA	NA
	140 (2302)	45	NA	NA	NA	0.025 ②	0.031 ②	0.276	0.276
	152 (2494)	45	NA	NA	NA	0.031 ②	0.031 ②	0.275	0.275
	209 (3430)	45	NA	NA	NA	0.031 ②	0.031 ②	0.275	0.275
	304 (4988)	45	NA	NA	NA	0.031 ②	0.031 ②	0.275	0.275
1991-92	109.6 (1796)	NA	NA	NA	NA	0.020 ②	0.020 ②	0.275	0.275
	140 (2302)	45	NA	NA	NA	0.025 ②	0.031 ②	0.276	0.276
	152 (2494)	45	NA	NA	NA	0.031 ②	0.031 ②	0.275	0.275
	152 (2494) ③	45	NA	NA	NA	0.020 ②	0.020 ②	0.275	0.275
	209 (3430)	45	NA	NA	NA	0.031 ②	0.031 ②	0.275	0.275
	216 (3535)	45	NA	NA	NA	0.025 ②	0.031 ②	0.275	0.275
	304 (4988)	45	NA	NA	NA	0.031 ②	0.031 ②	0.275	0.275

NA Not Available
① A dimension of 1.8110 applies to some springs, depending upon manufacturer. Figure given is free height
② Tilt clearance
③ 24 Valve Engine

PISTON AND RING SPECIFICATIONS

All measurements are given in inches

Year	Engine Displacement cu. in. (cc)	Piston Clearance	Ring Gap			Ring Side Clearance		
			Top Compression	Bottom Compression	Oil Control	Top Compression	Bottom Compression	Oil Control
1988	140 (2302)	0.0012–0.0024	0.0120–0.0220	0.0120–0.0220	0.0100–0.0200	0.0024–0.0035	0.0024–0.0035	0.0008–0.0020
	152 (2494)	0.0004–0.0016	0.0120–0.0200	0.0120–0.0200	0.0100–0.0200	0.0016–0.0028	0.0012–0.0024	0.0008–0.0017
	165 (2693)	0.0004–0.0016	0.0120–0.0200	0.0120–0.0200	0.0100–0.0200	0.0016–0.0028	0.0012–0.0024	0.0008–0.0017
	209 (3428) ①	0.0008–0.0020	0.0120–0.0200	0.0080–0.0160	0.0100–0.0200	0.0200–0.0320	0.0016–0.0028	0.0008–0.0020
	209 (3428) ②	0.0008–0.0020	0.0080–0.0180	0.0160–0.0260	0.0160–0.0240	0.0016–0.0028	0.0012–0.0024	0.0008–0.0022
	210.6 (3453)	0.0012–0.0024	0.0120–0.0220	0.0120–0.0220	0.0100–0.0200	0.0024–0.0035	0.0024–0.0035	0.0008–0.0020
	304 (4988)	0.0004–0.0013	0.0080–0.0160	0.0080–0.0160	0.0100–0.0200	0.0016–0.0025	0.0012–0.0028	0.0008–0.0022
1989	140 (2302)	0.0012–0.0024	0.0120–0.0220	0.0120–0.0220	0.0100–0.0200	0.0024–0.0035	0.0024–0.0035	0.0008–0.0020
	152 (2494)	0.0004–0.0016	0.0120–0.0200	0.0120–0.0200	0.0100–0.0200	0.0016–0.0028	0.0012–0.0024	0.0008–0.0017
	165 (2693)	0.0004–0.0016	0.0120–0.0200	0.0120–0.0200	0.0100–0.0200	0.0016–0.0028	0.0012–0.0024	0.0008–0.0017
	209 (3428) ①	0.0008–0.0020	0.0120–0.0200	0.0080–0.0160	0.0100–0.0200	0.0200–0.0320	0.0016–0.0028	0.0008–0.0020
	209 (3428) ②	0.0008–0.0020	0.0080–0.0180	0.0160–0.0260	0.0160–0.0240	0.0016–0.0028	0.0012–0.0024	0.0008–0.0022

PISTON AND RING SPECIFICATIONS

All measurements are given in inches

Year	Engine Displacement cu. in. (cc)	Piston Clearance	Ring Gap Top Compression	Ring Gap Bottom Compression	Ring Gap Oil Control	Ring Side Clearance Top Compression	Ring Side Clearance Bottom Compression	Ring Side Clearance Oil Control
1989	210.6 (3453)	0.0012–0.0024	0.0120–0.0220	0.0120–0.0220	0.0100–0.0200	0.0024–0.0035	0.0024–0.0035	0.0008–0.0020
	304 (4988)	0.0004–0.0013	0.0080–0.0160	0.0080–0.0160	0.0100–0.0200	0.0016–0.0025	0.0012–0.0028	0.0008–0.0022
1990	109.6 (1796)	0.0004–0.0016	0.0080–0.0160	0.0080–0.0160	0.008–0.018	0.0008–0.0020	0.0008–0.0020	0.0008–0.0022
	140 (2302)	0.0012–0.0024	0.0120–0.0220	0.0120–0.0220	0.0100–0.0200	0.0024–0.0035	0.0024–0.0035	0.0008–0.0020
	152 (2494)	0.0004–0.0016	0.0120–0.0200	0.0120–0.0200	0.0100–0.0200	0.0016–0.0028	0.0012–0.0024	0.0008–0.0017
	209 (3430)	0.0008–0.0020	0.0080–0.0180	0.0160–0.0260	0.0160–0.0240	0.0016–0.0028	0.0012–0.0024	0.0008–0.0020
	304 (4988)	0.0004–0.0013	0.0080–0.0160	0.0080–0.0160	0.0100–0.0200	0.0016–0.0025	0.0012–0.0028	0.0008–0.0022
1991-92	109.6 (1796)	0.0004–0.0016	0.0080–0.0160	0.0080–0.0160	0.008–0.018	0.0008–0.0020	0.0008–0.0020	0.0008–0.0022
	140 (2302)	0.0012–0.0024	0.0120–0.0220	0.0120–0.0220	0.0100–0.0200	0.0024–0.0035	0.0024–0.0035	0.0008–0.0020
	152 (2494)	0.0004–0.0016	0.0120–0.0200	0.0120–0.0200	0.0100–0.0200	0.0016–0.0028	0.0012–0.0024	0.0008–0.0017
	152 (2494) ③	0.0004–0.0016	0.0080–0.0160	0.0080–0.0160	0.008–0.018	0.0008–0.0020	0.0008–0.0020	0.0008–0.0022
	209 (3430)	0.0008–0.0020	0.0080–0.0180	0.0160–0.0260	0.0160–0.0240	0.0016–0.0028	0.0012–0.0024	0.0008–0.0020
	216 (3535)	0.0012–0.0024	0.0120–0.0220	0.0120–0.0220	0.0100–0.0200	0.0024–0.0035	0.0024–0.0035	0.0008–0.0020
	304 (4988)	0.0004–0.0013	0.0080–0.0160	0.0080–0.0160	0.0100–0.0200	0.0016–0.0025	0.0012–0.0028	0.0008–0.0022

NA Not Available
① B34 used in 535i
② B35 used in 6 and 7 series cars
③ 24 Valve Engine

TORQUE SPECIFICATIONS

All readings in ft. lbs.

Year	Engine Displacement cu. in. (cc)	Cylinder Head Bolts	Main Bearing Bolts	Rod Bearing Bolts	Crankshaft Pulley Bolts	Flywheel Bolts	Manifold Intake	Manifold Exhaust	Spark Plugs
1988	140.4 (2302)	⑥	14.5–17.5 ⑤	⑦	311–325	75.5–76.5	6.5–7.0	6.5–7.0	15–21
	152 (2494)	①	42–45	②	283–311	75.5–76.5	22–24	16–18 ⑪	15–21
	165 (2693)	①	42–45	②	283–311	75.5–76.5	22–24	16–18 ⑪	15–21
	209 (3428)	④	42–45	38–41	311–325	75.5–76.5	22–24	16–18 ⑪	15–21
	210.6 (3453)	⑥	14.5–17.5 ⑤	⑦	311–325	75.5–76.5	14–17 ⑩	6.5–7	15–21
	304 (4988)	⑧	⑯	②	311–325	74	16–18	16–18	15–21

TORQUE SPECIFICATIONS

All readings in ft. lbs.

Year	Engine Displacement cu. in. (cc)	Cylinder Head Bolts	Main Bearing Bolts	Rod Bearing Bolts	Crankshaft Pulley Bolts	Flywheel Bolts	Manifold Intake	Manifold Exhaust	Spark Plugs
1989	140.4 (2302)	⑥	14.5–17.5 ⑤	⑦	311–325	75.5–76.5	6.5–7.0	6.5–7.0	15–21
	152 (2494)	①	42–45	②	283–311	75.5–76.5	22–24	16–18 ⑪	15–21
	165 (2693)	①	42–45	②	283–311	75.5–76.5	22–24	16–18 ⑪	15–21
	209 (3428)	④	42–45	38–41	311–325	75.5–76.5	22–24	16–18 ⑪	15–21
	210.6 (3453)	⑥	14.5–17.5 ⑥	⑦	311–325	75.5–76.5	14–17 ⑩	6.5–7	15–21
	304 (4988)	⑮	⑯	②	311–325	74	16–18	16–18	15–21
1990	109.6 (1796)	⑫	③	②	224	87	NA	NA	17
	140.4 (2302)	⑥	14.5–17.5 ⑤	⑦	311–325	71–81	6.5–7.0	6.5–7.0	14–22
	152 (2494)	42–45 ①	②	②	283–311	71–81	22–24	16–18	14–22
	209 (3430)	④	42–45	38–41	311–325	71–81	22–24	16–18	14–22
	304 (4988)	⑧	⑨	②	—	74	14–17	16–18	14–22
1991–92	109.6 (1796)	⑫	③	⑨	217–231	82–94	11	17	14–22
	140.4 (2302)	⑥	③	⑦	82–94	311–325	7.5	7.5	14–22
	152 (2494)	Hex- ① Torx- ⑭	42–46	⑨	82–94	281–309	17	17	14–22
	152 (2494)	⑫	③	⑨	82–94	281–309	11	14	14–22
	209 (3428)	④	42–46	38–41	82–94	311–325	17	17	14–22
	216 (3535)	⑥	③	⑦	82–94	16	⑬	7.5	14–22
	304 (4988)	⑧	⑮	⑨	72	311–325	⑬	17	14–22

NA Not Available

① Step 1—29–33 ft. lbs.
Wait 20 minutes
Step 2—43–47 ft. lbs.
Warm engine fully
Step 3—25°–30° angle torque

② Torque to 14.5 ft. lbs.
Turn 70° angle torque

③ Step 1—17 ft. lbs.
Step 2—50° angle torque

④ Step 1—42–44 ft. lbs.
Wait 20 minutes
Step 2—57–59 ft. lbs.
Run engine warm for 25 minutes
Step 3—30°–40° angle torque

⑤ Step 1—Torque to figure shown
Step 2—47°–53° angle torque

⑥ Step 1—35–37 ft. lbs.
Step 2—57–59 ft. lbs.
Wait 15 minutes
Step 3—71–73 ft. lbs.

⑦ Step 1—7 ft. lbs.
Step 2—21.5 ft. lbs.
Step 3—60°–62° angle torque

⑧ Step 1—22 ft. lbs.
Wait 15 minutes
Step 2—Turn 120° angle torque

⑨ Step 1—17 ft. lbs.
Step 2—Turn 70° angle torque

⑩ Applies to (larger) M8 bolts. Torque M6 (smaller) bolts to 6.5–7.0 ft. lbs.

⑪ Coat the threads of the upper row of bolts with a locking type sealer

⑫ Step 1—24 ft. lbs.
Step 2—90°–95° angle torque
Step 3—90°–95° angle torque

⑬ M6—6.5–8.0 ft. lbs.
M8—14–17 ft. lbs.

⑭ Step 1—22 ft. lbs.
Step 2—90°
Step 3—90°

⑮ Step 1—14.5 ft. lbs.
Step 2—50°–53° angle torque

⑯ Step 1—43 ft. lbs.
Step 2—60° angle torque
Step 3—60° angle torque
Step 4—30° angle torque

BRAKE SPECIFICATIONS

All measurements in inches unless noted

Year	Model	Lug Nut Torque (ft. lbs.)	Master Cylinder Bore	Brake Disc Minimum Thickness	Brake Disc Maximum Runout	Maximum Brake Drum Diameter	Minimum Lining Thickness Front	Minimum Lining Thickness Rear
1988	325 (All)	65–79	—	0.787F/0.315R	0.008	—	0.079	0.079
	528e	65–79	—	0.787F/0.315R	0.008	—	0.079	0.079
	535i	65–79	—	0.787F/0.315R	0.008	—	0.079	0.079
	635CSi	65–79	—	0.906F/0.315R	0.008	—	0.079	0.079

BRAKE SPECIFICATIONS

All measurements in inches unless noted

Year	Model	Lug Nut Torque (ft. lbs.)	Master Cylinder Bore	Brake Disc Minimum Thickness	Brake Disc Maximum Runout	Maximum Brake Drum Diameter	Minimum Lining Thickness Front	Minimum Lining Thickness Rear
1988	M3	65–79	—	0.905F/0.394R	0.008	—	0.079	0.079
	M5	65–79	—	1.102F/0.315R	0.008	—	0.079	0.079
	M6	65–79	—	1.024F/0.315R	0.008	—	0.079	0.079
	735i	65–79	—	0.906F/0.315R	0.008	—	0.079	0.079
	750iL	65–79	—	NA F/0.709R	0.008	—	0.079	0.079
1989	325 (All)	65–79	—	0.787F/0.315R	0.008	—	0.079	0.079
	525i	65–79	—	0.787F/0.315R	0.008	—	0.079	0.079
	535i	65–79	—	0.787F/0.315R	0.008	—	0.079	0.079
	635CSi	65–79	—	0.906F/0.315R	0.008	—	0.079	0.079
	M3	65–79	—	0.905F/0.394R	0.008	—	0.079	0.079
	M5	65–79	—	1.102F/0.315R	0.008	—	0.079	0.079
	M6	65–79	—	1.024F/0.315R	0.008	—	0.079	0.079
	735i	65–79	—	0.906F/0.315R	0.008	—	0.079	0.079
	735iL	65–79	—	0.906F/0.315R	0.008	—	0.079	0.079
	750iL	65–79	—	NA F/0.709R	0.008	—	0.079	0.079
1990	318iS	65–79	—	0.787F/0.315R	0.008	—	0.079	0.079
	325 (All)	65–79	—	0.787F/0.315R	0.008	—	0.079	0.079
	525i	65–79	—	0.787F/0.315R	0.008	—	0.079	0.079
	535i	65–79	—	0.787F/0.315R	0.008	—	0.079	0.079
	M3	65–79	—	0.905F/0.394R	0.008	—	0.079	0.079
	735i	65–79	—	1.024F/0.315R	0.008	—	0.079	0.079
	735iL	65–79	—	1.024F/0.315R	0.008	—	0.079	0.079
	750iL	65–79	—	1.024F/0.709R	0.008	—	0.079	0.079
1991-92	318iS	65–79	—	0.787F/0.315R	0.008	—	0.079	0.079
	325 (All)	65–79	—	0.787F/0.315R	0.008	—	0.079	0.079
	525i	65–79	—	0.787F/0.315R	0.008	—	0.079	0.079
	535i	65–79	—	0.787F/0.315R	0.008	—	0.079	0.079
	735i	65–79	—	1.024F/0.315R	0.008	—	0.079	0.079
	735iL	65–79	—	1.024F/0.315R	0.008	—	0.079	0.079
	750iL	65–79	—	1.024F/0.709R	0.008	—	0.079	0.079
	850i	65–79	—	1.102F/0.709R	0.008	—	0.079	0.079
	M3	65–79	—	0.905F/0.315R	0.008	—	0.079	0.079
	M5	65–79	—	1.039F/0.709R	0.008	—	0.079	0.079

F—Front
R—Rear

WHEEL ALIGNMENT

Year	Model		Caster Range (deg.)	Caster Preferred Setting (deg.)	Camber Range (deg.)	Camber Preferred Setting (deg.)	Toe-in (in.)	Steering Axis Inclination (deg.)
1988	325e	Front	8P–9P	8 1/2P	1 1/10N–1 1/10N	2/3N	0.079	13 5/16P
		Rear	—	—	1 1/3N–2 1/3N	1 5/16N	0.079	—
	325	Front	8P–9P	8 1/2P	1 1/10N–1 1/6N	2/3N	0.079	13 5/16P
		Rear	—	—	1 1/3N–2 1/3N	1 5/16N	0.079	—
	325 ②	Front	8 1/4P–8 3/4P	8 3/4P	1 2/3N–2/3N	1 1/6N	0.079	14 1/3P
		Rear	—	—	1 1/3N–2 1/3P	1 5/16N	0.079	—
	325iX	Front	1P–1 2/3P	1 1/3P	1 1/2N–1/2N	1N	0.024	12 2/3P
		Rear	—	—	1 3/4N–2 1/3P	1 5/16N	0.079	—
	325iX ②	Front	1P–1 2/3P	1 1/3P	1 15/16N–5/16N	1 1/3N	0.024	12 2/3P
		Rear	—	—	1 3/4N–2 1/3P	1 5/16N	0.079	—
	325 Convertible	Front	8P–9P	8 1/2P	1 1/10N–1 1/6N	2/3N	0.079	13 5/16P
		Rear	—	—	1 1/3N–2 1/3N	1 5/6N	0.079	—
	528e	Front	7 3/4P–8 3/4P	8 1/4P	1/2N–1/10P	1/3N	0.079 ①	12 13/16P
		Rear	—	—	—	2 1/3N	0.079 ①	—
	535i	Front	7 3/4P–8 3/4P	8 1/4P	1/2N–1/10P	1/3N	0.079 ①	12 13/16P
		Rear	—	—	—	2 1/3N	0.079 ①	—
	635CSi	Front	7 1/2P–8 1/2P	8P	5/6N–1/6P	1/3N	0.079	12P
		Rear	—	—	2 5/6N–1 5/6N	2 1/3N	0.079 ③	—
	M6	Front	7 3/4P–8 3/4P	8P	5/6N–1/6P	1/3N	0.079 ③	—
		Rear	—	—	2 5/6N–1 5/6N	2 1/3N	0.079 ③	—
	528e	Front	7 1/2P–8 1/2P	8P	5/6N–1/6P	1/3N	0.079 ③	12P
		Rear	—	—	1 5/6N–2 5/6	2 1/3N	0.079 ③	—
	535i	Front	7 1/2P–8 1/2P	8P	5/6N–1/6P	1/3N	0.079 ③	12P
		Rear	—	—	1 5/6N–2 5/6	2 1/3N	0.079 ③	—
	M5	Front	7 1/2P–8 1/2P	8P	5/6N–1/6P	1/3N	0.098 ①	12P
		Rear	—	—	1 5/6N–2 5/6	2 1/3N	0.098 ①	—
	735i	Front	7 1/2P–8 1/2P	8P	3/4N–1/4P	1/4N	0.087 ④	12P
		Rear	—	—	2 5/6N–1 5/6N	2 1/3N	0.087 ①	—
	750iL	Front	7 1/2P–8 1/2P	8P	3/4N–1/4P	1/4N	0.087 ④	12P
		Rear	—	—	2 5/6N–1 5/6N	2 1/3N	0.087 ①	—
1989	325e	Front	8P–9P	8 1/2P	1 1/10N–1 1/10N	2/3N	0.079	13 5/16P
		Rear	—	—	1 1/3N–2 1/3N	1 5/16N	0.079	—
	325	Front	8P–9P	8 1/2P	1 1/10N–1 1/6N	2/3N	0.079	13 5/16P
		Rear	—	—	1 1/3N–2 1/3N	1 5/16N	0.079	—
	325 ②	Front	8 1/4P–8 3/4P	8 3/4P	1 2/3N–2/3N	1 1/6N	0.079	14 1/3P
		Rear	—	—	1 1/3N–2 1/3P	1 5/16N	0.079	—
	325iX	Front	1P–1 2/3P	1 1/3P	1 1/2N–1/2N	1N	0.024	12 2/3P
		Rear	—	—	1 3/4N–2 1/3P	1 5/16N	0.079	—
	325iX ②	Front	1P–1 2/3P	1 1/3P	1 15/16N–5/16N	1 1/3N	0.024	12 2/3P
		Rear	—	—	1 3/4N–2 1/3P	1 5/16N	0.079	—
	325 Convertible	Front	8P–9P	8 1/2P	1 1/10N–1 1/6N	2/3N	0.079	13 5/16P
		Rear	—	—	1 1/3N–2 1/3N	1 5/6N	0.079	—

WHEEL ALIGNMENT

Year	Model		Caster Range (deg.)	Caster Preferred Setting (deg.)	Camber Range (deg.)	Camber Preferred Setting (deg.)	Toe-in (in.)	Steering Axis Inclination (deg.)
1989	525i	Front	7 1/2P–8 1/2P	8P	3/4N–1/4P	1/4N	0.087 ⑥	12P
		Rear	—	—	—	2 1/3N	0.079 ③	—
	535i	Front	7 3/4P–8 3/4P	8 1/4P	1/2N–1/10P	1/3N	0.079 ①	12 13/16P
		Rear	—	—	—	2 1/3N	0.079 ①	—
	635CSi	Front	7 1/2P–8 1/2P	8P	5/6N–1/6P	1/3N	0.079 ⑤	12P
		Rear	—	—	2 5/6N–1 5/6N	2 1/3N	0.079 ⑤	—
	M6	Front	7 3/4P–8 3/4P	8P	5/6N–1/6P	1/3N	0.079 ⑤	12P
		Rear	—	—	2 5/6N–1 5/6N	2 1/3N	0.079 ⑤	—
	528e	Front	7 1/2P–8 1/2P	8P	5/6N–1/6P	1/3N	0.079 ⑤	12P
		Rear	—	—	1 5/6N–2 5/6	2 1/3N	0.079 ⑤	—
	535i	Front	7 1/2P–8 1/2P	8P	5/6N–1/6P	1/3N	0.079 ⑤	12P
		Rear	—	—	1 5/6N–2 5/6	2 1/3N	0.079 ⑤	—
	M5	Front	7 1/2P–8 1/2P	8P	5/6N–1/6P	1/3N	0.098	12P
		Rear	—	—	1 5/6N–2 5/6	2 1/3N	0.098	—
	735iL	Front	7 1/2P–8 1/2P	8P	3/4N–1/4P	1/4N	0.087 ⑥	12P
		Rear	—	—	2 5/6N–1 5/6N	2 1/3N	0.087	—
	750iL	Front	7 1/2P–8 1/2P	8P	3/4N–1/4P	1/4N	0.087 ⑥	12P
		Rear	—	—	2 5/6N–1 5/6N	2 1/3N	0.087	—
1990	318iS	Front	8P–9P	8 1/2P	1 3/16N–3/16N	2/3N	5/32	13 7/8P
		Rear	—	—	2 1/2N–1 1/2N	2N	3/16	—
	325i	Front	8P–9P	8 1/2P	1 3/16N–3/16N	1 11/16N	5/64	13 7/8P
		Rear	—	—	2 13/16N–1 13/16N	2 5/16N	3/32	—
	325i ②	Front	8 1/4–9 1/4P	8 3/4P	1 21/32N–2 1/32N	1 5/32N	—	14 3/8P
		Rear	—	—	3N–2N	2 1/2N	1/8	—
	325iS	Front	8P–9P	8 3/4P	1 3/16N–3/16N	11/16N	5/64	13 7/8P
		Rear	—	—	2 13/16N–1 13/16N	2 5/16N	3/32	—
	325iS ②	Front	8 1/4–9 1/4P	8 3/4P	1 21/32N–2 1/32N	1 5/32N	—	14 3/8P
		Rear	—	—	3N–2N	2 1/2N	1/8	—
	325iX	Front	1 1/16P–2 1/16P	1 9/16P	1 1/2N–1/2N ⑤	1N ⑥	0	12 11/16P
		Rear	—	—	2 1/2N–1 1/2P ②	2N ⑧	3/32	—
	325 Convertible	Front	8P–9P	8 1/2P	1 3/16N–1 3/16N	11/16N	5/64	13 7/8P
		Rear	—	—	2 9/16N–1 1/16N	1 13/16N	5/64	—
	525i	Front	7 3/8P–8 3/8P	7 7/8P	23/32N–9/32P	7/32N	3/32	12 1/16P
		Rear	—	—	2 13/16N–1 13/16N	2 5/16N	3/32	—
	535i	Front	7 3/8P–8 3/8P	7 7/8P	23/32N–9/32P	7/32N	3/32	12 1/16P
		Rear	—	—	2 13/16N–1 13/16	2 5/16N	3/32	—
	M3	Front	9 5/16P–10 5/16P	9 13/16P	1 3/16N–3/16N	11/16N	5/64	14 3/16P
		Rear	—	—	2 13/16N–1 13/16N	2 5/16N	1/8	—
	735i	Front	7 1/2P–8 1/2P	8P	23/32N–9/32N	7/32N	3/32	12 1/16P
		Rear	—	—	2 13/16N–1 13/16N	2 5/16N	3/32	—
	735iL	Front	7 1/2P–8 1/2P	8P	23/32N–9/32N	7/32N	3/32	12 1/16P
		Rear	—	—	2 13/16N–1 13/16N	2 5/16N	3/32	—

WHEEL ALIGNMENT

Year	Model		Caster Range (deg.)	Caster Preferred Setting (deg.)	Camber Range (deg.)	Camber Preferred Setting (deg.)	Toe-in (in.)	Steering Axis Inclination (deg.)
1990	750iL	Front	7 1/2P–8 1/2P	8P	23/32N–9/32N	7/32N	3/32	12 1/16P
		Rear	—	—	2 13/16N–1 13/16N	2 5/16N	3/32	—
1991–92	318iS	Front	8P–9P	8 1/2P	1 3/16N–3/16N	2/3N	5/32	13 7/8P
		Rear	—	—	2 1/2N–1 1/2N	2N	3/16	—
	325i	Front	8P–9P	8 1/2P	1 3/16N–3/16N	1 11/16N	5/64	13 7/8P
		Rear	—	—	2 13/16N–1 13/16N	2 5/16N	3/32	—
	325i ②	Front	8 1/4–9 1/4P	8 3/4P	1 21/32N–21/32N	1 5/32N	—	14 3/8P
		Rear	—	—	3N–2N	2 1/2N	1/8	—
	325iX	Front	1 1/16P–2 1/16P	1 9/16P	1 1/2N–1/2N ⑤	1N ⑥	0	12 11/16P
		Rear	—	—	2 1/2N–1 1/2P ②	2N	3/32	—
	325 Convertible	Front	8P–9P	8 1/2P	1 3/16N–1 3/16N	11/16N	5/64	13 7/8P
		Rear	—	—	2 9/16N–1 1/16N	1 13/16N	5/64	—
	525i	Front	7 3/8P–8 3/8P	7 7/8P	23/32N–9/32P	7/32N	3/32	12 1/16P
		Rear	—	—	2 13/16N–1 13/16N	2 5/16N	3/32	—
	535i	Front	7 3/8P–8 3/8P	7 7/8P	23/32N–9/32P	7/32N	3/32	12 1/16P
		Rear	—	—	2 13/16N–1 13/16	2 5/16N	3/32	—
	735i	Front	7 1/2P–8 1/2P	8P	23/32N–9/32N	7/32N	3/32	12 1/16P
		Rear	—	—	2 13/16N–1 13/16N	2 5/16N	3/32	—
	735iL	Front	7 1/2P–8 1/2P	8P	23/32N–9/32N	7/32N	3/32	12 1/16P
		Rear	—	—	2 13/16N–1 13/16N	2 5/16N	3/32	—
	750iL	Front	7 1/2P–8 1/2P	8P	23/32N–9/32N	7/32N	3/32	12 1/16P
		Rear	—	—	2 13/16N–1 13/16N	2 5/16N	3/32	—
	850i	Front	8 1/2P–7 1/2P	8P	2/3N–1/3P	5/32N	5/32	12P
		Rear	—	—	1 1/2N–1N	1 1/4N	5/32	—
	M3	Front	9 5/16P–10 5/16P	9 13/16P	1 3/16N–3/16N	11/16N	5/64	14 3/16P
		Rear	—	—	2 13/16N–1 13/16N	2 5/16N	1/8	—
	M5	Front	8 2/3P–7 2/3P	8 1/6P	1N–0	1/2N	0	12 2/3P
		Rear	—	—	2 1/2N–1 1/2N	2N	1/8	—

All 300, 500 and 700 series models aligned with 150 lbs. in each front seat, 150 lbs. in rear seat and 46 lbs. in trunk. All 600 series models aligned with 150 lbs. in each front seat and 30 lbs. in trunk on left side.

F Front
R Rear
N Negative
P Positive
① 0.083 with TRX tires
② With "M" suspension
③ 0.083 with TRX 390 rims
④ 0.094 with TRX 415 rims
⑤ MSUSP—1 3/16N–13/16N
⑥ MSUSP—1 5/16N
⑦ MSUSP—3N–2N
⑧ MSUSP—2 1/2N
⑨ MSUSP—1/8 in.

ENGINE MECHANICAL

NOTE: Disconnecting the negative battery cable on some vehicles may interfere with the functions of the on board computer systems and may require the computer to undergo a relearning process, once the negative battery cable is reconnected.

Engine Assembly

REMOVAL & INSTALLATION

325, 325i and 325iS

1. Disconnect the negative battery cable. Remove the transmission.
2. Without disconnecting hoses, loosen and remove the power steering pump bolts and remove the pump and belts and support the pump out of the way.
3. Remove the drain plug and remove the coolant from the radiator. Then, remove the radiator. Unbolt and remove the fan from the engine. Store it in an upright position.
4. Without disconnecting hoses, remove the mounting bolts that run through the compressor body and remove the air conditioner compressor and drive belt and support the compressor out of the way.
5. Remove the through-bolts to disconnect the engine hood supports and

then open the hood and support it securely.

NOTE: The hood must be propped in a secure manner. If it falls during work serious injury could result.

6. Disconnect the accelerator cable. If the vehicle has cruise control, disconnect the cruise control cable. If the vehicle has an automatic transmission, disconnect the throttle cable leading to the transmission.

7. Pull the large, multi-prong plug off the airflow sensor (an integral part of the air cleaner). Loosen the clamp and disconnect the air intake hose at the airflow sensor. Remove the mounting nuts and remove the air cleaner/airflow sensor unit.

8. Disconnect the coolant expansion tank hose. Disconnect the large, multi-prong connector near the air intake hose.

9. The diagnosis plug is a large, screw-on connector located near the thermostat and associated hoses. Unscrew and disconnect this connector.

10. Disconnect the large coolant hoses connecting to the thermostat.

11. Make sure the engine is cold. Place a metal container under the connection to collect fuel; then, disconnect the fuel line at the connection right near the thermostat housing by unscrewing it. Unfasten the fuel line clip about a foot away from this connection.

12. Disconnect the electrical plugs near the diagnosis plug connector. Disconnect the bracket for the dipstick guide tube.

13. Remove the bolts which attach the water pipes going to the engine to mounting brackets.

14. Disconnect the heater hoses at the heater core (near the firewall). Remove the coolant hose running to the top of the block.

15. Place a metal container under the connection to collect fuel; then, disconnect the remaining fuel hose supplying the engine injectors. Disconnect the electrical connectors.

16. Remove the bolt from the mounting brace connecting with the cylinder head.

17. Mark and then disconnect the electrical leads from the starter. Unbolt the starter and lift it out from above.

18. Place a metal container under the connection to collect fuel; then, disconnect the fuel pipe that runs right near the starter.

19. Label electrical connectors on the alternator. Then, pull off the rubber caps for the connectors which are attached with nuts and remove the nuts and any washers. Disconnect the plug-on connector.

20. Disconnect the electrical leads for the coil. Loosen the clips attaching the leads under the distributor and pull the harness away to the left. Disconnect the oil pressure sending unit.

21. Place a drain pan underneath the connections and then disconnect the oil cooler pipes at the crankcase by unscrewing the flare nut fittings.

22. Take the cover off the relay box. Then, lift out the relays and their mounting sockets. Place the relays and associated wiring on top of the engine so they will come out with it.

23. Loosen the mounting clamp and then remove the carbon canister. There is a plate to which a number of electrical leads are connected. Remove the mounting screws and move the plate aside so it will clear the dipstick guide tube when the engine is removed.

24. Remove the 2 bolts that fasten the wiring harness to the firewall. Then, disconnect the engine ground strap.

25. Remove both engine mount through-bolts. Lift out the engine with a suitable hoist, using hooks at front and rear.

To install:

26. Reverse the procedures used for removal and lower the engine into the engine compartment. When the engine is positioned, the guide pin must fit in the bore of the axle carrier. Torque the mounting bolts on the front axle carrier (small bolt) to 18–20 ft. lbs.; the larger bolt to 31–35 ft. lbs. The mount-to-bracket bolts are torqued to 31–35 ft. lbs. Engine-to-bracket mounts are torqued to (small bolt) 16–17 ft. lbs., (large bolt) 31–35 ft. lbs.

27. Connect the fuel lines, use new hose clamps to connect the fuel lines to the fuel filter. Connect all of the multi-prong plugs and all vacuum hoses.

28. Connect the accelerator cable and cruise control cable to the throttle body and adjust the accelerator cable and cruise control cable.

29. Install the coolant recovery tank, use a new hose clamp on the coolant expansion tank.

30. Install the air cleaner and reconnect all electrical plugs. Connect and install the relays in the relay box.

31. Reconnect the wiring to the main control unit and install the idle control unit.

32. Install the air conditioning compressor and power steering pump, properly route the accessory drive belt. Adjust the belt tension.

33. Install the radiator and connect the hoses.

34. Install the transmission.

35. Install the hood support.

36. Make sure all fluid levels are correct before starting the engine. Bleed air from the cooling system.

318iS

1. Disconnect the battery ground cable. Remove the transmission and remove the engine splash guard. Disconnect the gas spring and prop rod and support hood safely in the fully open position.

2. Remove the fan cowl by turning the expansion rivets on the left and right sides. Lift the cowl up and out of the engine compartment.

3. Hold the fan pulley while unscrewing the fan nut from the shaft. The shaft uses lefthand threads; turn the nut counterclockwise to unscrew.

4. Drain the coolant from the engine block. Disconnect the bottom hose from the radiator expansion tank, the engine coolant hoses and the heater hoses from the splash wall. Drain all coolant into clean containers for reuse or proper disposal.

5. Disconnect the air flow meter electrical plug and loosen the hose clamp and mounting screws. Lift the air sensor with the air cleaner up and out of the engine compartment.

6. Unclip the throttle cable and pull the cable out with the rubber holder.

7. Disconnect the fuel lines taking note of their positions. Pull off the vent hose to the filter for tank venting.

8. Disconnect the vacuum fitting at the brake booster.

9. Remove the ignition leads from the coil. Unscrew the connections at the alternator and starter. Disconnect the 2 plugs from the electrical duct. Remove the plug from the throttle valve potentiometer located at the throttle neck. Pull off the tank venting valve plug located next to the air cleaner. Disconnect the fuel injector plug located at the end of the electrical duct near to the fuel pipes. Pull off the idle speed control connector at the rear of the intake manifold. Disconnect the oil pressure switch electrical connection.

10. Unscrew the front and rear intake manifold supports.

11. Remove the electrical duct from the engine. Disconnect the coolant temperature senders for the gauge and the DME.

12. Disconnect the electrical duct and wiring harness on the engine and lay it off to the side of the engine.

13. Use a suitable lifting yoke to attach to the engine lifting eyes. Unscrew the motor mounts and the engine ground strap. Lift out the engine.

To Install:

14. Lower engine into engine compartment. Fasten the motor mounts and the ground strap.

15. Attach the engine wiring harness

and electrical duct. Make sure that the rubber grommets on the duct are clipped in correctly. Connect the the leads to the 2 coolant sensors and the oil pressure switch.

16. Fasten the front and rear intake manifold supports.

17. Connect the idle speed control plug, the fuel injector plug, the tank venting valve plug, the throttle valve potentiometer plug and the electrical lead duct plugs.

18. Reconnect the starter and the alternator. Attach the ignition leads to the coil in the proper order.

19. Refit the vacuum connection to the brake booster. Reconnect the tank vent hose and the fuel hoses. The upper fuel hose is the return line and the bottom is the feed line.

20. Attach the throttle cable and its holder. Replace the air cleaner and air flow meter assembly. Attach the electrical connector to the air flow meter.

21. Connect the heater hoses, engine coolant hoses and the radiator expansion tank hose.

22. Install the fan using tool 11 5 040 or equivalent. Torque the nut to 29 ft. lbs.(40 Nm). If using the tool set the torque wrench to 22 ft. lbs.(30 Nm); the additional length of the tool multiplies the torque to achieve 29 ft. lbs.(40 Nm) at the nut.

23. Replace the fan cowl taking care to engage the tabs at the right and left.

24. Replace the splash guard and the transmission. Reconnect the hood prop rod and gas spring.

25. Add the proper coolant mixture and bleed the cooling system.

26. Connect the battery leads and check all fluid levels before starting the engine.

M3

1. Disconnect the negative battery cable. Remove the transmission.

2. Remove the splash guard from underneath the engine. Put a drain pan underneath and then drain coolant from both the radiator and block.

3. Loosen the hose clamps at either end of the air intake hose leading to the air intake sensor. Pull off the hose. Then, pull both electrical connectors off the air cleaner/airflow sensor unit. Remove both mounting nuts and remove the unit.

4. Disconnect the accelerator and cruise control cables. Unscrew the nuts mounting the cable housing mounting bracket and set the housings and bracket aside.

5. Loosen the clamp and disconnect the brake booster vacuum hose.

6. Loosen the clamp and disconnect the other end of the booster vacuum hose at the manifold. Remove the nut from the intake manifold brace.

7. Loosen the hose clamp and disconnect the air intake hose at the manifold. Then, remove all the nuts attaching the manifold assembly to the outer ends of the intake throttle necks and remove the assembly.

8. Put a drain pan underneath and then loosen the hose clamps and disconnect the coolant expansion tank hoses. Disconnect the engine ground strap.

9. Disconnect the ignition coil high tension lead. Then, label and disconnect the plugs on the front of the block. Remove the nut fastening another lead farther forward of the plugs and move the lead aside so it will not interfere with engine removal.

10. Label and disconnect the plugs from the rear of the alternator. Label the additional leads and then remove the nuts and disconnect those leads. It's best to reinstall nuts once the leads are removed to keep them from being mixed up.

11. Remove the cover for the electrical connectors from the starter. Label the leads and then remove the attaching nuts and disconnect them. Reinstall the nuts.

12. There is a wire running to a connector on the oil pan to warn of low oil level. Pull off the connector, unscrew the carrier for the lead, and then pull the lead out from above. Pull off the connectors near where the lead for the low oil warning system ran and unclip the wires from the carrier.

13. Find the vacuum hose leading to the fuel pressure regulator. Pull it off. Label and then disconnect the plugs. Unscrew the mounting screw for the electrical lead connecting with the top of the block and remove the lead and its carrier.

14. There is a vacuum hose connecting with one of the throttle necks. Disconnect it and pull it out of the intake manifold bracket. Pull off the electrical connector. Pull out the rubber retainer, and then pull the idle speed control out and put it aside. The engine wiring harness is located nearby. Take it out of its carriers.

15. All the fuel injectors are plugged into a common plate. Carefully and evenly pull the plate off the injectors, pull it out past the pressure regulator, and lay it aside.

16. Loosen the clamp and then disconnect the PCV hose. Label and then disconnect the fuel lines connecting the injector circuit. Put a drain pan underneath and then disconnect the heater hose from the cylinder head.

17. Loosen the clamp near the throttle necks and then pull the engine wiring harness out and put it aside. Put a drain pan underneath and then disconnect the heater hose that connects to the block.

18. Loosen the mounting clamp for the carbon canister, slide it out of the clamp, and place it aside with the hoses still connected.

19. Note the routing of the oil cooler lines where they connect at the base of the oil filter. Label them if necessary. Put a drain pan underneath and then unscrew the flared connectors for the lines.

20. Unbolt and remove the fan. Store it in an upright position. Remove the radiator.

21. Support the power steering pump. Remove the adjusting bolt and disconnect and remove the belt. Then, remove the nuts and bolts on which the unit hinges. Pull the unit aside and hang it so there will not be strain on the hoses.

22. Remove the adjusting bolt for the air conditioning compressor and disconnect and remove the belt. Then, remove the nut at one end of the hinge bolt and pull the bolt out, suspending the compressor.

23. Remove the through-bolts to disconnect the engine hood supports and then open the hood and support it securely.

24. Suspend the engine with a suitable lifting device. Then, remove the nuts for the engine mounting bolts. The mounts are on the axle carrier and the nut is at the top on the left and on the bottom on the right. Then, carefully lift the engine out of the compartment, avoiding contact between it and the components remaining in the vehicle.

To install:

25. Keep these points in mind during installation:

a. Torque the engine mounting bolts to 32.5 ft. lbs.

b. Adjust the belt tension for the air conditioning compressor and power steering pump drive belts to give ½–¾ in. deflection.

c. Torque the oil cooler line flare nuts to 25 ft. lbs.

d. When reconnecting the intake manifold to the throttle necks, inspect and, if necessary, replace the O-rings. Torque the mounting nuts to 6.5 ft. lbs.

26. Reverse the procedures used for removal and lower the engine into the engine compartment. When the engine is positioned, the guide pin must fit in the bore of the axle carrier. Torque the mounting bolts on the front axle carrier (small bolt) to 18–20 ft. lbs.; the larger bolt to 31–35 ft. lbs. The mount-to-bracket bolts are torqued to 31–35 ft. lbs.

27. Install the intake manifold assembly and connect the fuel lines, use new hose clamps to connect the fuel lines to the fuel filter. Connect all of

the multi-prong plugs and all vacuum hoses.

28. Connect the accelerator cable and cruise control cable to the throttle body and adjust the accelerator cable and cruise control cable.

29. Install the coolant recovery tank, use a new hose clamp on the coolant expansion tank.

30. Install the air cleaner and reconnect all electrical plugs. Connect and install the relays in the relay box.

31. Reconnect the wiring to the main control unit and install the idle control unit.

32. Install the air conditioning compressor and power steering pump, properly route the accessory drive belt. Adjust the belt tension.

33. Install the radiator and connect the hoses.

34. Install the transmission.

35. Install the hood support and lower the hood.

36. Make sure all fluid levels are correct before starting the engine. Bleed air from the cooling system.

528e

1. Disconnect the negative battery cable. Remove the transmission. Disconnect the exhaust pipe from the exhaust manifold.

2. Remove the splash guard.

3. With the hoses still attached, remove the power steering pump and position it out of the way.

4. Unscrew the drain plug on the engine block, remove the upper and lower radiator hoses and drain the cooling system. After draining, remove the radiator.

5. With the refrigerant hoses still connected, remove the air conditioning compressor and position it out of the way.

6. Disconnect the gas pressure springs, scribe around the hinges and then remove the hood.

7. Disconnect the battery cables, negative cable first and remove the battery.

8. Disconnect the accelerator and cruise control cables. Disconnect and tag all hoses from the throttle housing. Disconnect the air duct.

9. Remove the air filter housing along with the air flow sensor.

10. Tag and disconnect all remaining lines, hoses and wires which may interfere with engine removal.

11. Tag and disconnect all plugs and wires attached to the control unit in the glove box. Unscrew the straps on the firewall and pull the wire harness through to the engine compartment.

12. Disconnect the engine ground strap and then loosen both engine mounts.

13. Attach an engine lifting hoist to the front and rear of the engine, remove the engine mount bolts and then lift out the engine.

To install:

14. Reverse the procedures used for removal and lower the engine into the engine compartment. When the engine is positioned, the guide pin must fit in the bore of the axle carrier. Torque the mounting bolts on the front axle carrier (small bolt) to 18–20 ft. lbs.; the larger bolt to 31–35 ft. lbs. The mount-to-bracket bolts are torqued to 31–35 ft. lbs.

15. Connect all of the electrical wiring and all vacuum hoses.

16. Connect the accelerator cable and cruise control cable to the throttle body and adjust the accelerator cable and cruise control cable.

17. Install the air cleaner and reconnect all electrical plugs.

18. Reconnect the wiring to the main control unit.

19. Install the air conditioning compressor and power steering pump, properly route the accessory drive belt. Adjust the belt tension.

20. Install the radiator and connect the hoses.

21. Install the transmission.

22. Install the hood support.

23. Make sure all fluid levels are correct before starting the engine. Bleed the air from the cooling system.

535i, 635CSi and 1990 525i

1. Disconnect both battery cables, negative first. There is a lead coming from the engine to the positive battery terminal. Disconnect it at the battery. On the 6 Series vehicles, disconnect the ground strap.

2. Unscrew the ground strap for the hood. Support the hood securely and then disconnect the gas props. Then, raise the hood until it is vertical and securely fasten it in place.

3. Remove the transmission. With the engine cool, place a clean container under the coolant drain plug in the side of the block. Remove the plug and drain all coolant from the block. Remove the fan and radiator.

4. Support the power steering pump. Remove the mounting bolts and then hang the pump out of the way in a position that will not put stress on the hoses.

5. Support the air conditioning compressor. Remove the mounting bolts and then hang the compressor out of the way in a position that will not put stress on the hoses.

6. Pull the wire leading to the oxygen sensor out of the clips under the floor. Disconnect the sensor at the exhaust pipe.

7. Pull off the plug at the airflow sensor and remove associated wiring. Remove the hoses and pipes connected to the air cleaner and airflow sensor. Remove the nuts and remove the airflow sensor and air cleaner as an assembly.

8. Pull the large, multi-prong plug off the DME box in the glove compartment. Disconnect the smaller plug that's connected to the same harness and plugged in. Then, run the entire harness back into the engine compartment.

9. Disconnect the engine ground wire located at the rear of the block. Unclip the harness for the DME from the firewall.

10. Disconnect both the low tension and the high tension wire from the coil. Disconnect the wires from the solenoid. Pull the wiring harness out of the holders.

11. Pull off the fuse box cover and the cap. Remove the relays (they have metal covers) on one side of the fuse box. Then, disconnect the wiring harness that leads into the fuse box. On the 635CSi, unclamp the harness where it is clamped to the fender well and remove the diagnosis socket, located right near the fusebox.

12. Disconnect the accelerator and cruise control cables.

13. Unclamp and remove the coolant hose that leads to the expansion tank. Disconnect the fuel return line, collecting any fuel in a metal container for safe disposal. Unclip the wiring harness clips on the wires that run through this area of the engine compartment.

14. Disconnect the fuel supply line, collecting any fuel in a metal container for safe disposal. Disconnect the heater hoses at connections.

15. Pull the main vacuum supply hose off at the intake manifold.

16. Disconnect the remaining main coolant hose and plug it.

17. Install a lifting sling to the hooks on top of the engine. Unbolt the left side engine mount. Remove the main engine ground strap. Unbolt the right side engine mount. Carefully pull the engine out of the compartment.

To install:

18. Reverse the procedures used for removal and lower the engine into the engine compartment. When the engine is positioned, the guide pin must fit in the bore of the axle carrier. Torque the mounting bolts on the front axle carrier (small bolt) to 18–20 ft. lbs.; the larger bolt to 31–35 ft. lbs. The mount-to-bracket bolts are torqued to 31–35 ft. lbs.

19. Install the intake manifold assembly and connect the fuel lines, use new hose clamps to connect the fuel lines to the fuel filter. Connect all of the multi-prong plugs and all vacuum hoses.

20. Connect the accelerator cable and cruise control cable to the throttle body and adjust the accelerator cable and cruise control cable.

21. Install the coolant recovery tank, use a new hose clamp on the coolant expansion tank.

22. Install the air cleaner and reconnect all electrical plugs. Connect and install the relays in the relay box.

23. Reconnect the wiring to the main control unit and install the idle control unit.

24. Install the air conditioning compressor and power steering pump, properly route the accessory drive belt. Adjust the belt tension.

25. Install the radiator and connect the hoses.

26. Install the transmission.

27. Install the hood support and lower the hood.

28. Make sure all fluid levels are correct before starting the engine. Bleed air from the cooling system.

1991–92 525i

1. Disconnect the battery terminals, negative side first, and remove the battery. Unscrew and remove the battery tray. Remove the transmission.

2. Loosen the clamp on the cooling duct to the alternator and remove the duct.

3. Disconnect the plug to the air flow meter and loosen the clamps to the air cleaner duct. Unscrew the mounting bolts and remove the air cleaner assembly.

4. Pull out the expansion rivets that hold the fan cowl. Remove the cowl by pulling up out of the engine compartment.

5. Hold the fan pulley while unscrewing the fan nut from the shaft. The shaft uses lefthand threads; turn the nut counterclockwise to unscrew.

6. Drain the coolant from the block. The drain plug is located between the exhaust manifolds. Disconnect the coolant hoses from the radiator and remove the coolant level switch plug. On automatic transmission equipped vehicles remove the oil lines to the radiator and plug.

7. Disconnect the bottom radiator hose and remove the trim panel from the right side of the engine compartment to expose the side of the radiator and the air conditioner condenser.

8. Pull the plug off of the air conditioner temperature switch.

9. Remove the radiator supporting clips by inserting a small prybar down from above into the slot and pulling back. Pull the radiator free from the clip. Remove the radiator from the vehicle.

10. Disconnect the heater hoses from the heater valve and the heater.

11. Unscrew the fastener from the throttle cable cover and pull the cover forward and off. Unclip the cable and pull the cable out with the rubber holder.

12. Pull the vacuum fitting from the brake booster and plug the openings.

13. Remove the engine and intake manifold covers. Unscrew the bolt holding the ground strap on the front lifting eye. Replace the bolt before lifting the engine.

14. Unscrew the 2 bolts holding the plug plate and pull off the plug plate. Be careful not to damage the rubber seals. Take off the ignition coil electrical plugs. Remove the plug plate complete with the electrical leads.

15. Remove the cylinder head vent hose and pull off the air temperature sensor plug. Remove the tank venting hose and the throttle heating hoses from the throttle body. Remove the throttle valve switch plug. Unclip the idle speed control valve mounted on the manifold. Disconnect the fuel hoses from the pipes.

16. Unscrew the hardware holding the intake manifold to the cylinder head. Remove the intake manifold taking care not to drop anything into the exposed ports.

17. Disconnect the plugs from the temperature sensor, temperature gauge, the oil pressure switch and the idle speed control valve. Disconnect the cylinder identifying sender plug (black) and the pulse sender plug (grey) for the DME. Unscrew the oxygen sensor plug in the holder.

18. Remove the electric leads from the alternator and the starter. Unscrew the electrical lead tray and place the engine wiring harness to the side.

19. Loosen the drive belt for the power steering pump and the air conditioner compressor by turning their respective tensioners clockwise. This will release the tension on the belt and allow the belt to be removed.

20. Unbolt the power steering pump and place to the side without disconnecting the hoses.Unbolt the air conditioner compressor and place to the side without disconnecting the lines.

21. Attach a lifting fixture to the engine lifting hooks. Unscrew the engine mounts and ground strap. Lift the engine out of the vehicle being careful of the front radiator mount.

To Install:

22. Lower the engine into the vehicle and attach the motor mounts and ground strap.

23. Install the power steering pump and the air conditioner compressor. Install the drive belts.

24. Replace the wiring harness and electrical lead tray on the engine. Connect the leads to the starter and alternator. Screw in the plug for the oxygen sensor holder. Connect the leads for the cylinder identifying sender, the DME pulse sender, the temperature sensor, the temperature gauge sender, the oil pressure switch and the idle speed control valve.

25. Install the intake manifold making sure that the intake seals are intact. Replace the intake seals if any signs of deterioration are evident.

26. Attach the fuel lines. The upper line is the return and the lower is the feed.

27. Attach the idle speed control valve hose located on the manifold.

28. Connect the throttle valve switch plug, the throttle valve heating lines and the tank vent line.

29. Connect the air temperature sensor plug and attach the cylinder head venting hoses.

30. Reconnect the plugs for the ignition coils and mount the plug plate. Attach the ground strap to the front lifting eye.

31. Replace the engine and manifold covers. Connect the line to the brake booster.

32. Reconnect the throttle cable and cover.

33. Connect the heater hoses to the valve and inlet. Remount the radiator by pressing down on the mounting clips to fasten. Check that the lower mounts are in place. Connect the temperature switch plug for the air conditioner and replace the trim panel. Connect the cooling system hoses and the automatic transmission lines.

34. Install the fan using tool 11 5 040 or equivalent. Torque the nut to 29 ft. lbs.(40 Nm). If using the tool set the torque wrench to 22 ft. lbs.(30 Nm); the additional length of the tool multiplies the torque to achieve 29 ft. lbs.(40 Nm) at the nut. Replace the radiator cowling.

35. Replace the air cleaner assembly and connect the electrical plug. Install the transmission and fill and bleed the cooling system. Install the battery tray and battery. Check all fluids before starting engine.

M5 and M6

1. Disconnect the battery negative cable. Then, disconnect the positive cable. Scribe matchmarks and then remove the hood.

2. Remove the fan. Remove the drain plugs in the block and radiator. Disconnect the hoses and remove the radiator.

3. Support the power steering pump. Remove the mounting bolts and then hang the pump out of the way in

a position that will not put stress on the hoses.

4. Support the air conditioning compressor. Remove the mounting bolts and then hang the compressor out of the way in a position that will not put stress on the hoses.

5. Remove the transmission.

6. Remove the attaching bolt and, with an appropriate puller, remove the vibration damper from the front of the engine.

7. Remove the bolts at either end and remove the cross brace that runs under the engine. Remove the heat shield.

8. Disconnect the electrical connector going to the airflow sensor. Pull the electrical leads out of the wiring holders. Loosen the hose clamp for the air intake hose. Remove the mounting nut for the air cleaner. Then, remove the air cleaner and airflow sensor as an assembly.

9. Disconnect the large vacuum hose at the bottom of the intake manifold.

10. Disconnect the PCV hoses where they connect to the top of the manifold. Disconnect the throttle cable that runs across the top of the manifold, and the hose running near the front. Remove the bolts fastening the manifold to the outer ends of the intake tubes and remove it.

11. Working inside the glove compartment, disconnect the plug that connects to the DME control. Then, guide the leads through and into the engine compartment. Disconnect the high tension lead and the low tension leads at the coil. Then, unfasten the wiring harness holders for the harness running to the coil where the harness runs along the fender well.

12. Disconnect the fuel hose connection at the rear of the fuel manifold on top of the engine. Disconnect the vacuum hose that runs along the firewall.

13. Disconnect the plugs for the reference mark and speed sensors. Disconnect the hoses on the coolant expansion tank.

14. Working on the fuse box, pull off the large electrical connector. Pull off the diagnosis socket. Disconnect the remaining leads.

15. Disconnect the heater hoses near the firewall. Using a backup wrench, disconnect the lines at the oil cooler. Disconnect the low pressure fuel line at the pressure regulator.

16. Disconnect the starter leads. Cut the straps and remove the solenoid heat shield.

17. Attach a lifting sling to the engine and support the assembly. Disconnect the ground lead. Then, disconnect the left side engine mount, removing the nut from underneath and then unscrewing the bolt out the top. Do the same for the right mount. Carefully lift the engine out of the compartment, tilting the front of the engine upward for clearance.

To install:

18. Keep these points in mind during installation:

a. Torque the engine mounting bolts to 32.5 ft. lbs.

b. Adjust the belt tension for the air conditioning compressor and power steering pump drive belts to give ½–¾ in. deflection.

c. Torque the oil cooler line flare nuts to 25 ft. lbs.

d. When reconnecting the intake manifold to the throttle necks, inspect and, if necessary, replace the O-rings. Torque the mounting nuts to 6.5 ft. lbs.

19. Reverse the procedures used for removal and lower the engine into the engine compartment. When the engine is positioned, the guide pin must fit in the bore of the axle carrier. Torque the mounting bolts on the front axle carrier (small bolt) to 18–20 ft. lbs.; the larger bolt to 31–35 ft. lbs. The mount-to-bracket bolts are torqued to 31–35 ft. lbs.

20. Install the intake manifold assembly and connect the fuel lines, use new hose clamps to connect the fuel lines to the fuel filter. Connect all of the multi-prong plugs and all vacuum hoses.

21. Connect the accelerator cable and cruise control cable to the throttle body and adjust the accelerator cable and cruise control cable.

22. Install the coolant recovery tank, use a new hose clamp on the coolant expansion tank.

23. Install the air cleaner and reconnect all electrical plugs. Connect and install the relays in the relay box.

24. Reconnect the wiring to the main control unit and install the idle control unit.

25. Install the air conditioning compressor and power steering pump, properly route the accessory drive belt. Adjust the belt tension.

26. Install the radiator and connect the hoses.

27. Install the transmission.

28. Install the hood support.

29. Make sure all fluid levels are correct before starting the engine. Bleed air from the cooling system.

735i, 735iL and 750iL

1. Disconnect the negative battery cable and then the positive. Remove the transmission. Scribe hinge locations and remove the hood, or remove support struts and prop it securely all the way up.

2. Remove the splash guard from underneath the engine. Then, with the engine cool, remove the drain plugs in the radiator and block and drain the engine coolant.

3. Loosen the power steering pump bolts from underneath. Turn the adjusting pinion to loosen the belt and remove the belt. Then, remove the mounting bolts and remove the power steering pump without disconnecting the hoses. Support the pump out of the way so as to avoid stressing the hoses.

4. Do the same with the air conditioner compressor, this unit does not have the belt adjusting pinion—it is necessary only to loosen all the bolts and push the compressor toward the engine to remove the belt.

5. Loosen the air intake hose clamp and disconnect the hose. Remove the mounting nut and then remove the air cleaner(s).

6. Unscrew the oil filter cover bolt and disconnect the oil cooler lines and the plug from the oil pressure switch on 750iL.

7. The unit on the opposite side of the intake hose from the air cleaner contains the idle speed control valve, which must be removed next. Loosen the hose clamps and pull off the hoses. Disconnect the electrical connector. Remove the mounting nut and then pull the idle speed control out of the air intake hose.

8. Pull off the retainers for the airflow sensor, and then pull the unit off its mountings, disconnecting the vacuum hose from the PCV system at the same time.

9. Working on the coolant expansion tank, disconnect the electrical connector. Remove the nuts on both sides. Loosen their clamps and then disconnect the hoses and remove the tank.

10. Disconnect the heater hoses at both the control valve and at the heater core.

11. Disconnect the throttle and cruise control cables at the throttle lever. Unbolt the cable housing retainer and remove the housing and cables.

12. Pull off the low amperage starter connectors and disconnect the high amperage connector coming from the battery.

13. Loosen its clamp and then disconnect the coolant hose that runs to the alternator.

14. Disconnect the connecting plug for the oxygen sensor, as well as the other plugs.

15. Loosen the clamps and then disconnect the fuel supply and return pipes.

16. Disconnect the fuel pipe at the injector supply manifold. Disconnect the plug. Disconnect the electrical connector at the throttle body. Lift off the

protective caps and then remove the attaching nuts for the protective cover for the wiring harness for the injectors and remove it.
17. Disconnect the ground strap at the block. Remove the engine mount nut from the top on both sides.
18. Attach a lifting sling to the engine and support the assembly. Disconnect the ground lead. Carefully lift the engine out of the compartment, tilting the front of the engine upward for clearance.

To install:

19. Keep these points in mind during installation:
 a. Torque the engine mounting bolts to 32.5 ft. lbs.
 b. Adjust the belt tension for the air conditioning compressor and power steering pump drive belts to give ½–¾ in. deflection.
 c. Torque the oil cooler line flare nuts to 25 ft. lbs.
 d. When reconnecting the intake manifold to the throttle necks, inspect and, if necessary, replace the O-rings. Torque the mounting nuts to 6.5 ft. lbs.
20. Reverse the procedures used for removal and lower the engine into the engine compartment. When the engine is positioned, the guide pin must fit in the bore of the axle carrier. Torque the mounting bolts on the front axle carrier (small bolt) to 18–20 ft. lbs.; the larger bolt to 31–35 ft. lbs. The mount-to-bracket bolts are torqued to 31–35 ft. lbs.
21. Connect the fuel lines, use new hose clamps to connect the fuel lines to the fuel filter. Connect all of the multi-prong plugs and all vacuum hoses.
22. Connect the accelerator cable and cruise control cable to the throttle body and adjust the accelerator cable and cruise control cable.
23. Install the coolant recovery tank, use a new hose clamp on the coolant expansion tank.
24. Install the air cleaner(s) and reconnect all electrical plugs. Connect and install the relays in the relay box.
25. Install the oil filter cover bolt and connect the oil cooler lines and the plug to the oil pressure switch on the 750iL.
26. Reconnect the wiring to the main control unit and install the idle control unit.
27. Install the air conditioning compressor and power steering pump, properly route the accessory drive belt. Adjust the belt tension.
28. Install the radiator and connect the hoses.
29. Install the transmission.
30. Install the hood support and lower the hood.
31. Make sure all fluid levels are correct before starting the engine. Bleed air from the cooling system.

Engine Mounts

REMOVAL & INSTALLATION

1. Raise and safely support the vehicle.
2. Support the engine using a suitable jacking device. Disconnect the mounting bolts.
3. Remove the ground strap, if equipped. Remove the engine mounts.
4. Install the mount onto the mounting bracket and replace the bolts.
5. Replace the ground strap, if equipped. Remove the jacking device.
6. Lower the vehicle.

Cylinder Head

REMOVAL & INSTALLATION

325, 325i, 325iS, 325iX and 528e

1. Disconnect the negative battery cable. Make sure the engine is cool. Disconnect the exhaust pipes at the manifold and at the transmission clamp. Remove the drain plug at the bottom of the radiator and drain the coolant. Drain the engine oil.
2. Disconnect the accelerator and cruise control cables. If the vehicle has automatic transmission, disconnect the throttle cable that goes to the transmission.
3. Working at the front of the block, disconnect the upper radiator hose, the bypass water hose, and several smaller water hoses. Remove the diagnosis plug located at the front corner of the manifold. Remove the bracket located just underneath. Disconnect the fuel line and drain the contents into a metal container for safe disposal.
4. On the 325, 325i, 325iS and 325iX:
 a. Working on the air cleaner/airflow sensor, disconnect the vacuum hoses, labeling them if necessary. Disconnect all electrical connectors and unclip and remove the wiring harness. There is a relay located in an L-shaped box near the strut tower. Disconnect and remove it. Unclamp and remove the air hose. Remove the mounting nuts and remove the assembly.
 b. Disconnect the hose at the coolant overflow tank. Disconnect the idle speed positioner vacuum hose and then remove the positioner from the manifold.
 c. If equipped with 4 wheel drive, disconnect the vacuum hose from the servo mounted on the manifold.
 d. Place a drain pan underneath and then disconnect the water connections at the front of the intake manifold. Disconnect the electrical connector.
5. On the 528e, perform the following procedures:
 a. Working near the air cleaner/air flow sensor unit, disconnect the vacuum hoses at the intake manifold and at the air intake hose.
 b. Disconnect the electrical connectors and then remove the wiring harness.
 c. Pull off the large hose leading into the unit and loosen the clamp where the air intake hose connects at the intake manifold.
 d. Remove the mounting nuts and remove the air cleaner/airflow sensor.
 e. Disconnect the water hoses at the throttle body.
 f. Disconnect the electrical connector under the throttle body.
 g. Disconnect the bracket under the intake manifold tube.
6. Disconnect the heater water hos-

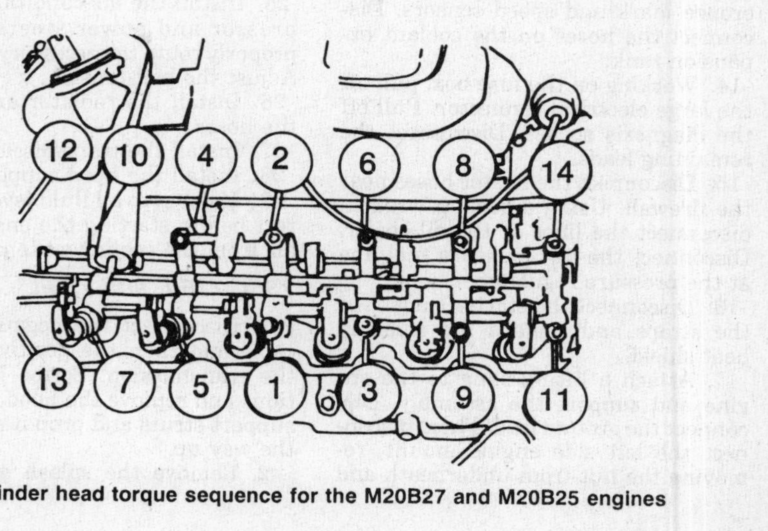

Cylinder head torque sequence for the M20B27 and M20B25 engines es. Press down, in the arrowed direction, on the vent tube collar and install the special tool or a similar device to retain the collar in the unlocked position. Disconnect the vent tube and inspect its O-ring seal, replacing it, if necessary.

7. Unbolt the dipstick tube at the manifold. Remove the fuel hose bracket at the cylinder head. Make sure the engine is cold. Then, place a metal container under the connection and disconnect the fuel hose at the connection.

8. Disconnect the high tension lead from the coil. Disconnect and remove the coolant expansion tank.

9. On the 325, 325i, 325iS and 325iX:

a. If equipped with 4 wheel drive, disconnect the intake manifold vacuum hose leading the the servo that engages 4 wheel drive.

b. Disconnect the fuel injector connectors at all 6 injectors, as well as the 2 additional electrical connectors to sensors on the head. Disconnect the oil pressure sending unit connector. Then, unfasten the carriers and remove this wiring harness toward the left side of the vehicle.

10. On the 528e, perform the following procedures:

a. There is a bracket with various vacuum and electrical fittings that runs from the camshaft cover over toward the intake manifold. Disconnect the electrical connector connected on this bracket and the plug to its left. Remove the nuts fastening the bracket to the camshaft cover and the gasketed flange on the opposite end and remove the bracket. Inspect the gasket and supply a new one for use in installation, if necessary. Unplug the fuel injectors.

b. Disconnect the DME plugs. Disconnect the plugs located near the front 3 fuel injectors and any remaining injector connectors. Unplug the oil pressure sending unit connector. Then, unfasten the mounting clips and pull the wiring harness out toward the left.

11. Disconnect the coil high tension wire and disconnect the high tension wires at the plugs. Then, disconnect the tube in which the wires run at the camshaft cover. Disconnect the PCV hose. Then, remove the retaining nuts and remove the camshaft cover.

12. Turn the crankshaft so the TDC line is aligned with the indicator and the valves of No. 6 cylinder are in overlapping, slightly open position.

13. Remove the distributor cap. Then, unscrew and remove the rotor. Unscrew and remove the adapter just underneath the rotor. Remove the cover underneath the adapter. Check its O-ring and replace it, if necessary.

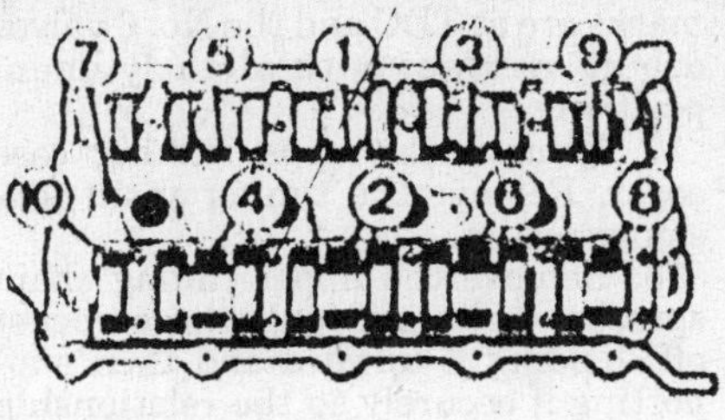

Cylinder head tightening sequence – 318iS

14. Remove the distributor mounting bolts and the protective cover.

15. These engines are equipped with a rubber drive and timing belt. Remove the belt covers. To loosen belt tension, loosen the tension roller bracket pivot bolt and adjusting slot bolt. Push the roller and bracket away from the belt to release the tension, hold the bracket in this position, and retighten the adjusting slot bolt to retain the bracket it this position.

16. Remove the timing belt.

NOTE: Make sure to avoid rotating both the engine and camshaft from this point onward.

17. Remove the cylinder head mounting bolts in exact reverse order of the proper tightening sequence. Then, remove the cylinder head.

To install:

18. Install the head with a new gasket. Check that all passages line up with the gasket holes. Clean the threads on the head bolts and coat with a light coating of oil. Keep oil out of the bolt cavities in the head or the head could be cracked or proper torquing affected.

19. Install the bolts and torque in stages, to the correct specifications. Then, adjust the valves.

20. Clean both cylinder head and block sealing surfaces thoroughly with a hardwood scraper. Inspect the surfaces for flatness. Note that the M20B25 engine gasket is coded 2.5 and the M20B27 engine gasket is coded 2.7.

21. Complete the installation by reversing all removal procedures. Make sure to refill the engine oil pan and cooling system with proper fluids and to bleed the cooling system.

22. Replace the gaskets for the exhaust system connections, if necessary. Coat the studs with the proper sealant. Note that the plugs for the DME reference mark and speed signals should be connected so the gray plug goes to the socket with a ring underneath.

NOTE: Align the timing marks when installing the timing belt. The crankshaft sprocket mark must point at the notch in the flange of the front engine cover. The camshaft sprocket arrow must point at the alignment mark on the cylinder head. Also, the No. 1 piston must be at TDC of the compression stroke. BMW recommends that the timing belt be replaced every time the cylinder head is removed and the belt is disturbed as a consequence. Tension the belt.

23. Start the engine and run it until it is hot. Stop the engine and again remove the camshaft cover. Using an angle gauge, tighten the head bolts 25 degrees farther in numbered order. Reinstall the camshaft cover.

318iS

1. Disconnect the negative battery cable.

2. Remove the ignition coil cover and pull off the spark plug connectors.

3. Remove the complete ignition tackle. Remove the cylinder head cover.

4. Disconnect the coolant hoses and unscrew the temperature sensor.

5. Remove the thermostat housing and thermostat. Unscrew the upper timing case cover.

6. Rotate the engine in the direction of the rotation until the camshaft peaks of the intake and exhaust camshafts for cylinder no.1 face each other. The arrows on the sprocket face up.

7. Remove the chain tensioner. Remove the upper chain guide, chain guide bolt on the right side and the sprockets.

8. Remove the cylinder head bolts from the outside to the inside in several steps using the proper tool.

9. Remove the cylinder head. Clean the sealing surfaces on the cylinder head and the crankcase.

10. The installation is the reverse of the removal procedure

525i(M20B25 Engine); 535i, 635CSi, 735i, 735iL and 750iL

1. Unbolt the exhaust pipe connections at the manifold and at the transmission pipe clamp. Disconnect the negative battery cable.

2. Remove the splash shield from under the engine. With the engine cool, remove the drain plugs from the bottom of the radiator and block. Drain the engine oil.

3. Remove the fan. Lift out the expansion rivets on either side and remove the fan shroud.

4. Loosen the hose clamp and disconnect the air inlet hose. Remove the mounting nut and remove the air cleaner.

5. The unit on the opposite side of the intake hose from the air cleaner

contains the idle speed control valve, which must be removed next. Loosen the hose clamps and pull off the hoses. Disconnect the electrical connector. Remove the mounting nut and then pull the idle speed control out of the air intake hose.

6. Pull off the retainers for the airflow sensor, and then pull the unit off its mountings, disconnecting the vacuum hose from the PCV system at the same time.

7. Working on the coolant expansion tank, disconnect the electrical connector. Remove the nuts on both sides. Loosen their clamps and then disconnect all hoses and remove the tank.

8. Disconnect the heater hoses at both the control valve and at the heater core. Remove the valve, if needed.

9. Disconnect the throttle and cruise control cables at the throttle lever. Unbolt the cable housing retainer and remove the housing and cables.

10. Disconnect the plugs near the thermostat housing. Loosen the hose clamps and pull off the coolant hoses.

11. Disconnect the plug in the line leading to the oxygen sensor. Disconnect the other plugs.

12. Disconnect the fuel supply and return lines, collecting fuel in a metal container for safe disposal.

13. Disconnect the fuel pipe running along the cylinder head, near the manifold. Pull off the electrical connector at the throttle body. Remove the caps, then remove the attaching bolts and remove the wiring harness carrier and harness for the fuel injectors.

14. Disconnect the coil high tension lead. Disconnect the high tension wires at the plugs. Then, remove the mounting nuts and remove the carrier for the high tension wires from the head.

15. Remove the attaching nuts for the camshaft cover and remove it.

16. Turn the engine until the timing

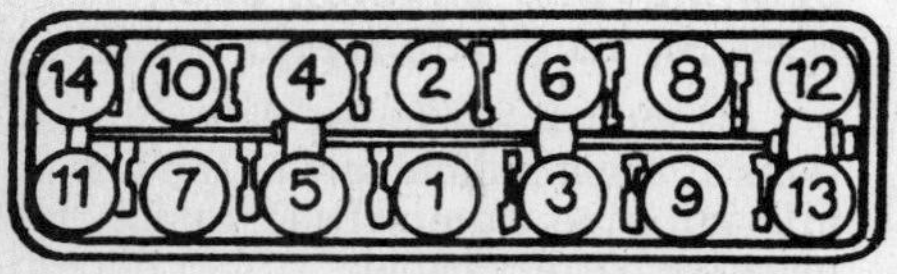

Cylinder head torque sequence for the S38Z engine used in the M5 and M6

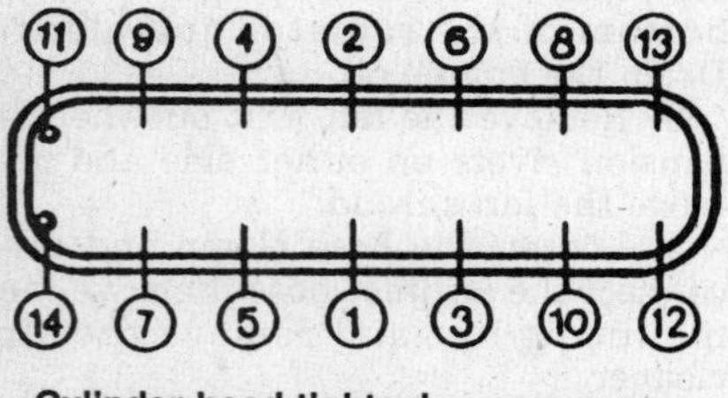

Cylinder head tightening sequence – M30B35 engine

marks are at TDC and the No. 6 valves are at overlap, both slightly open, position.

17. Remove the upper timing case cover. Remove the timing chain tensioner piston.

18. Remove the upper timing chain sprocket bolts and pull the sprocket off, holding it upward and then supporting it securely so the relationship between the chain and sprockets top and bottom will not be lost.

19. Disconnect the upper radiator hose at the thermostat housing. Remove the bolts and remove the support for the intake manifold.

20. Remove the cylinder head bolts in the opposite of numbered order. Then, install 4 special pins part No. 11 1 063 or equivalent. This is necessary to keep the rocker arm shafts from moving. Then, lift off the head.

21. Make checks of the lower cylinder head and block deck surface to make sure they are true. Install a new head gasket, making sure all bolt, oil, and coolant holes line up. Use a gasket marked M30 B35. Use a 0.3mm thicker gasket, if the head has been machined.

To install:

22. Apply a very light coating of oil to the head bolts. Don't let oil get into the bolt holes or apply excessive amounts of oil, or torque could be incorrect and the block could crack. Use the type of bolt without a collar. Install the bolts, finger tight.

23. Torque bolts 1–6 in the correct order to 42–44 ft. lbs. Remove the pins holding the rocker shafts in place. Now, complete the first stage of torquing by torquing bolts 7–14 in the correct order, to the same specification. Adjust the valves after a 15 minute wait. Tighten the bolts, in the correct order, with a torque angle gauge 30–36 degrees, using special tool 11 2 110 or equivalent. Then, reassemble the engine as described below and run it until hot (25 minutes). Then, again remove the valve cover and turn the head bolts, in the correct order, 30–40 degrees.

24. Reinstall the timing sprocket to the camshaft. Make sure the camshaft is in proper time, that new lock plates are used, and that nuts are properly torqued.

25. When reinstalling the timing cover, make sure to apply a liquid sealer to the joints between upper and lower timing covers. The remainder of installation is the reverse of removal. Note these points:

a. Adjust throttle, speed control, and accelerator cables. Inspect and if necessary replace the exhaust manifold gasket.

b. When reinstalling the cylinder block coolant plug, coat it with sealer. Make sure to refill the cooling system and bleed it. Make sure to refill the oil pan with the correct amount of oil.

c. Install the timing chain so the down pin on the camshaft sprocket is at the 8 o'clock when its tapped bores are at right angles to the engine. Torque the sprocket bolts to 6.5–7.5 ft. lbs.

d. Check the camshaft cover gasket, replacing, as necessary. Retighten camshaft cover bolts in the order shown. Torque the bolts to 6.5–7.5 ft. lbs.

e. When reinstalling the fan shroud, make sure all guides are located properly.

f. Coat the tapered portion of the exhaust pipe connection flange with the proper sealant. Torque the attaching nuts to 4.5 ft. lbs. and loosen 1½ turns.

1991-92 525i

1. If engine is not already removed from the vehicle disconnect the negative battery cable and drain the engine coolant. Remove the intake manifold and throttle valve. Disconnect the exhaust pipes and the oxygen sensor wire. Remove the exhaust manifolds. Remove the thermostat housing and engine lifting eye.

2. Pull off the connectors for the ignition coils and remove the coils. Unscrew the cylinder heads cover and remove. Remove the sender from the head and the electrical lead duct.

3. Remove the upper timing case cover and the camshaft cover. Crank the engine in the direction of rotation so that the intake and exhaust camshaft peaks for cylinder No.1 face each other. Hold the camshafts in place with tool 11 3 240 or equivalent. With the camshafts in this alignment the arrows on the sprockets will be facing up. Remove the valve cover mounting studs. Lock the flywheel in place to prevent movement of the crankshaft.

4. Unscrew the chain tensioner and carefully remove. There is a spring contained within the tensioner and may eject out if care is not taken.

5. Press down on the upper chain tensioner and lock it into place using tool 11 3 290 or equivalent. Unscrew the transfer timing chain sprockets and pull the 2 off together with the chain. Remove the upper chain tensioner and the lower chain guide. Pull off the main timing chain sprocket along with the chain. Use a bent piece of wire to hold the chain from falling down into the engine. Do not rotate the engine after this point or the valve timing will be disturbed when the engine is reassembled.

6. Unscrew the bolts on the head at

the ends of the cams. Using a proper sized Torx® bit or tool 11 2 250 loosen the cylinder head bolts in several steps. Use an outside to inside pattern to prevent warpage. On production heads the bolt washers are locked into place while on replacement heads the washers are loose. Keep track of the bolt washers.

To Install:

7. If the camshafts have been removed and reinstalled a waiting period dependent on the ambient temperature is necessary before mounting the cylinder head on the engine. At room temperature wait 4 minutes to allow the lifters to compress fully. At temperatures down to 50° F(10° C) wait 11 minutes. At temperatures lower than 50° F(10° C) wait 30 minutes. This is to prevent contact between the valves and the piston tops. The engine may not be cranked under the same condition for a period of 10 minutes at room temperature; 30 minutes for temperatures down to 50° F(10° C); 75 minutes for temperatures below 50° F(10° C).

8. Clean all mounting surfaces and check the head for warpage. Take care not to drop any pieces of gasket or dirt into the oil or coolant passages. Check the condition of the head locating dowel sleeves.

9. Place a new head gasket on the engine block over the locating dowels and gently place the head on the engine. Align the head with the dowel sleeves and check that the head sits flat on the engine.

10. Cylinder head bolts may only be used once. Lightly oil the threads of the new cylinder head bolts. Check that the head bolt washers are in place and install the bolts. Torque the head bolts in 3 steps; Step 1: 24 ft. lbs.(33 Nm), Step 2 and 3: 93 degree torque angle. Torque the center bolts first and go out in a diagonal pattern.

11. Align main timing chain and sprocket on the can so that the arrow faces up. The bolt holes in the camshaft should be on the left sides of the sprocket slots. This will allow the tensioner to take up the slack in the chain and rotate the gear to the counterclockwise position.

12. Install the upper chain tensioner and the lower chain guide. Install the transfer timing gears and chain on the camshafts with the arrows facing up. Make sure that the pulse sender is installed on the intake cam. Do not tighten the sprocket bolts.

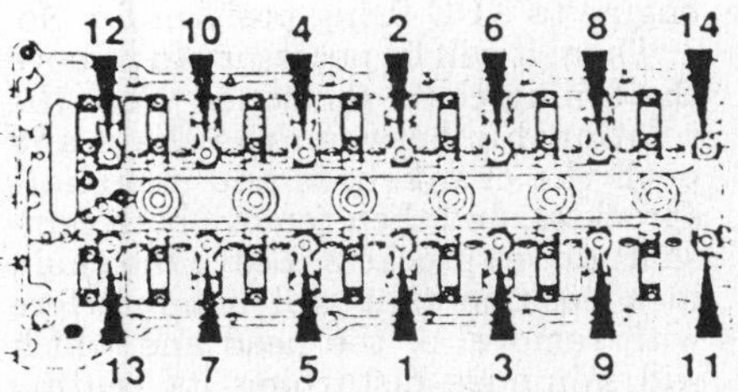

Cylinder head torque sequence—1991–92 525i

13. Install the lower timing chain tensioner with the groove in a vertical position. Use a new sealing ring and torque to 29 ft. lbs.(40 Nm).

14. Release the upper timing chain tensioner and torque the sprocket bolts to 16 ft. lbs.(22 Nm).

15. Install the valve cover mounting studs and remove the flywheel lock. Install the camshaft cover and the timing case covers using new gaskets. Install the lifting eye and the electrical lead duct. Install the camshaft sensor with a new seal if needed.

16. Install the valve cover using new gaskets if necessary. Check that the gasket seats correctly all the way around the seating area.

17. Install the ignition coils and connect the electrical leads to each coil.

18. Install the thermostat with the arrow or vent facing up with a new O-ring. Install the thermostat housing with a new gasket. Install the intake manifold and throttle valve. Install the exhaust manifolds.

M3

1. Disconnect the negative battery cable. Remove the splash guard from under the engine. Put drain pans underneath and remove the drain plugs from both the radiator and block to drain all coolant.

2. Loosen the hose clamps for the air intake hose located next to the radiator and then remove the hose. Disconnect the electrical connectors for the airflow sensor. Then, remove the attaching nuts and remove the air cleaner/ airflow sensor unit.

3. Disconnect the accelerator and cruise control cables. Unbolt the cable mounting bracket and move the cables and bracket aside.

4. Remove the attaching nut, pull off the clamp, and then detach the vacuum hose from the brake booster.

5. Loosen the hose clamp and remove the air intake hose from the intake manifold. Remove the nut from the manifold brace.

6. Loosen the clamp and disconnect the other end of the booster vacuum hose at the manifold. Remove the nut from the intake manifold brace.

7. Loosen the hose clamp and disconnect the air intake hose at the manifold. Then, remove the nuts attaching the manifold assembly to the outer ends of the intake throttle necks and remove the assembly.

8. Put a drain pan underneath and loosen the hose clamps and disconnect the coolant expansion tank hoses. Disconnect the engine ground strap.

9. Disconnect the ignition coil high tension lead. Label and then disconnect the plugs on the front of the block. Remove the nut fastening the lead farther forward of the plugs and move the lead aside so it will not interfere with cylinder head removal.

10. Find the vacuum hose leading to the fuel pressure regulator. Pull it off. Label and then disconnect the 2 plugs. Unscrew the mounting screw for the electrical lead connecting with the top of the block and remove the lead and its carrier.

11. There is a vacuum hose connecting with one of the throttle necks. Disconnect it and pull it out of the intake manifold bracket. Pull off the electrical connector. Pull out the rubber retainer, and then pull the idle speed control out and put it aside. The engine wiring harness is located nearby. Take it out of its carriers.

12. All the fuel injectors are plugged into a common plate. Carefully and evenly pull the plate off the injectors, pull it out past the pressure regulator, and lay it aside.

13. Loosen the clamp and then disconnect the PCV hose. Label and then disconnect the fuel lines connecting with the injector circuit. Put a drain pan underneath and then disconnect the heater hose from the cylinder head.

14. Loosen the clamp near the throttle necks and then pull the engine wiring harness out and put it aside. Put a drain pan underneath and then disconnect the heater hose that connects to the block.

15. Remove the bolts from the flanges connecting the exhaust pipes to the exhaust manifold. Provide new gaskets and self-locking nuts. Disconnect the oxygen sensor plug.

16. Put a drain pan underneath and then disconnect the radiator hoses from the pipe at the front of the block.

17. It is not necessary to remove the timing chain completely, but it is necessary to remove the camshaft cover, front covers for the camshaft drive sprockets, the upper guide rail for the timing chain and then turn the engine to TDC firing position for No. 1. Remove the timing chain tensioner. Note the relationship between the chain and both the crankshaft and camshaft sprockets, and then remove both camshaft drive sprockets. Leave the chain in a position that will not interfere with removal of the head and which will minimize disturbing its routing through the areas on the front of the block.

18. Remove the camshafts.

19. Remove the camshaft followers one at a time, keeping them in exact order for installation in the same positions.

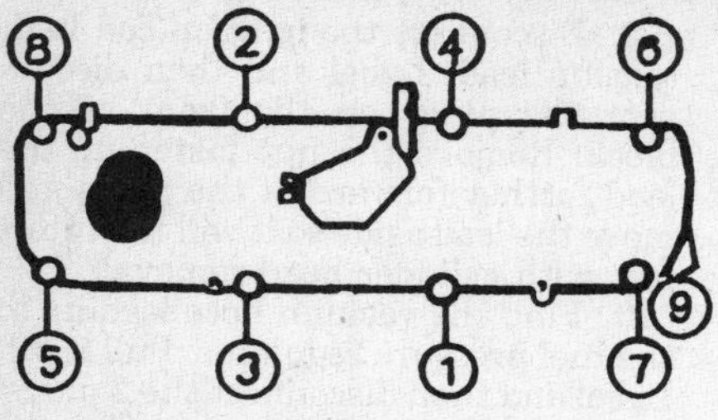

Torque the camshaft cover bolts on the M30, B34 and B35 engines in the order shown

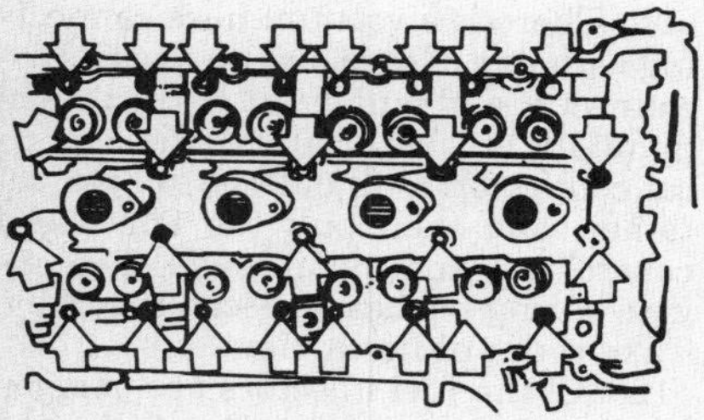

Removing the timing case for the M3. Remove all arrowed bolts

20. Pull off the spark plug connectors. Remove the nuts from the camshaft cover, located just to one side of the row of spark plugs. Remove the ignition lead tube. Remove the remaining nuts and remove the camshaft cover. Provide new gaskets.

21. Remove the bolts, some are accessible from below, that retain the timing case to the head at the front, the timing case houses the lifters and the camshaft lower bearing saddles. Note that one bolt, on the right side of the vehicle, is longer and retains the shaft for the upper timing chain tensioning rail.

22. Remove the coolant pipe that runs along the left/rear of the block. Remove one bolt at the left/front of the block that is located outside the camshaft cover. Then, go along in the area under the camshaft cover and remove all the remaining bolts for the timing case. Remove the timing case.

23. Remove the hex bolts fastening the head to the block at the front. These are located outside the camshaft cover and just behind the water pump drive belt. Then, remove the head bolts located under the camshaft cover in reverse order of the cylinder head torque sequence.

To install:

24. Make checks of the lower cylinder head and block deck surface to make sure they are true. Clean both cylinder head and block sealing surfaces thoroughly. Lubricate the head bolts with a light coating of engine oil. Make sure there is no oil or dirt in the bolt holes in the block. Install a new head gasket, making sure all bolt, oil, and coolant holes line up. Install the bolts as follows:

a. Torque them in the correct order to 35–37 ft. lbs.

b. Then torque them in order to 57–59 ft. lbs.

c. Wait 15 minutes.

d. Torque them, in order, to 71–73 ft. lbs.

e. Remember to reinstall the bolts that go outside the cylinder head cover and fasten the front of the head to the block at front and rear.

25. BMW recommends checking the fit of each tappet in the timing case, by performing the following procedure:

a. Measure a tappet's outside diameter with a micrometer. Then, zero an inside micrometer at this exact dimension.

b. Then, use the inside micrometer to measure the tappet bore that corresponds to this particular tappet. If the resulting measurement is 0.0001–0.0026 in. the tappet may be reused. If it is worn past this dimension, replace it with a new one. If the tappet is being replaced, repeat steps a and b to make sure it will now meet specifications. If the bore were to be worn so much that even a new tappet would not restore clearance to specification, it would be necessary to replace the timing case.

c. Repeat for all the remaining tappets. Make sure to measure each tappet and its corresponding bore only.

26. The remaining steps of installation are the reverse of the removal procedure. Note the following:

a. Before remounting the timing case, replace the O-ring in the oil passage located at the left/front of the block. Also, check the O-rings in the tops of the spark plug bores and replace these as necessary.

b. Install the timing case and torque the bolts in several stages. The smaller (M7) bolts are torqued to 10–12 ft. lbs.; the larger (M8) bolts are torqued to 14.5–15.5 ft. lbs. Install each tappet back into the same bore.

c. When bolting the exhaust pipes to the flange at the manifold, use new gaskets and self-locking nuts and torque the nuts to 36 ft. lbs.

d. When reinstalling the intake manifold, check and, if necessary, replace the O-rings where the manifold tubes connect to the throttle necks. Torque the nuts to 6.5 ft. lbs.

d. Make sure to refill the radiator and bleed the cooling system.

M5 and M6

1. Disconnect the negative battery cable. Scribe matchmarks where the hood hinges attach to the hood. Then, disconnect the support struts, unbolt the hood at the hinges and remove it.

2. Disconnect the electrical connector at the airflow sensor. Loosen the hose clamp at the air intake hose going to the air cleaner, remove the air cleaner attaching nut and remove the air cleaner and airflow sensor.

3. Disconnect the large vacuum hose that connects to the bottom of the intake manifold. Disconnect the PCV hoses where they connect to the top of the manifold. Disconnect the throttle cable that runs across the top of the manifold, and the hose running near the front. Remove the bolts fastening the manifold to the outer ends of the intake tubes and remove it.

4. With the engine cool, drain the coolant from the block. Disconnect the exhaust pipe at the manifold.

5. Working underneath, remove the heat shields. Remove the cross brace and stabilizer bar where they connect to the engine carrier. Remove the exhaust manifold.

6. Disconnect the upper radiator hose. Pull the plugs off the water manifold that connects with the upper radiator hose. Pull off the plug coming from the same harness and connecting to the top of the engine. Then, unclip this harness and pull it out of the way.

7. Loosen the retaining straps and disconnect the electrical connector that runs directly across the front of the block. Disconnect the fuel pipe on the driver's side of the block, collecting fuel in a metal container for safe disposal.

8. Pull the electrical connector off the throttle bypass valve. Disconnect the water hose and remove the bypass valve. Disconnect the large hose just to the right of the throttle bypass valve. Remove the wiring harness clips just to the right.

9. At the rear of the engine, disconnect the fuel return line and collect fuel in a metal container for safe disposal. Disconnect both heater hoses. Remove the conduit for the injector wiring harness from the head. Remove the bolts in the front of the head which run down into the timing cover.

10. It is not necessary to remove the timing chain completely, but it will be necessary to remove the camshaft cover, front covers for the camshaft drive sprockets and the upper guide rail for the timing chain, and then turn the engine to TDC firing position for No. 1. Then, it will be necessary to remove the timing chain tensioner. Note the relationship between the chain and both the crankshaft and camshaft sprockets, and then remove both camshaft drive sprockets. Leave the chain in a position that will not interfere with removal of the head and which will minimize disturbing its routing through the areas on the front of the block.

11. Remove the camshafts as described below.
12. Remove the camshaft followers one at a time, keeping them in exact order for installation in the same positions.
13. Remove the coolant pipe that runs across the front of the block. Remove the bolts (some are accessible from below) that retain the timing case to the head at the front, the timing case houses the lifters and the camshaft lower bearing saddles. Then, go along in the area under the camshaft cover and remove all the remaining bolts for the timing case. Remove the timing case.
14. Loosen the head bolts in reverse of the tightening sequence. Remove the cylinder head.

To install:

15. Make checks of the lower cylinder head and block deck surface to make sure they are true. Lubricate the head bolts with a light coating of engine oil. Make sure there is no oil or dirt in the bolt holes in the block. Install a new head gasket, making sure all bolt, oil, and coolant holes line up. Use a gasket type M6 marked 3.5M 88.3.
16. Replace the O-ring in the head at the right/rear where the coolant pipe comes up from the block. Coat the pipe with a suitable sealer.
17. Install the head onto the block. Install the head bolts and tighten in the correct sequence.
18. When installing the timing case, replace the O-rings in the small oil passages in the ends of the head. Inspect the O-rings in the center of the block and replace them if necessary. Coat all sealing surfaces with a sealer. Tighten the bolts evenly, torquing the smaller (M7) bolts to 10–12 ft. lbs. and the larger (M8) bolts to 14.5–15.5 ft. lbs. Install all lifters back into the same bores.
19. Install the camshafts.
20. Reroute the timing chain, as necessary, and remount the drive sprockets for the camshaft. Install the tensioning rail that goes at the top of the timing chain.
21. Install the front cover.
22. Continue to reverse the removal procedure. Note these points:
    a. When reinstalling the intake manifold, inspect the O-rings and replace, as necessary.
    b. Refill the cooling system with an appropriate anti-freeze/water mix and bleed the cooling system.

Valve Lash

ADJUSTMENT

Except M3, M5 and M6

All engines except the M-Series, dual overhead camshaft designs, are equipped with a single overhead camshaft operating the intake and exhaust valves through rocker arm linkage.

NOTE: The valves must be adjusted cold.

1. Disconnect the negative battery cable. Remove the rocker cover.
2. Rotate the engine until the No. 1 cylinder is at TDC on the compression stroke.

NOTE: Locate No. 1 cylinder firing position by the distributor rotor-to-cap position, or by observing the valve action in the opposite cylinder.

3. Measure the valve clearance between the valve stem end and the rocker arm on the No. 1 cylinder.
4. Adjust the clearance by loosening the locknut on the rocker arm and turning the eccentric with a bent rod inserted through a hole provided on the surface of the eccentric.
5. When the proper clearance is obtained, tighten the locknut and recheck the valve clearance. Complete the adjustment on both valves.
6. Rotate the engine crankshaft to the next cylinder in the firing order, adjust the valves and repeat the procedures until all the valves are adjusted.
7. Replace the rocker cover, using a new gasket.

M3, M5 and M6

NOTE: To perform this procedure, a special tool is needed to depress the valves against spring pressure to gain access to the valve adjusting discs. Use tool 11 3 170 or equivalent. Also needed are: compressed air to lift valve adjusting discs that must be replaced out of the valve tappet; an assortment of adjusting discs of various thicknesses and a precise outside micrometer.

1. Make sure the engine is overnight cold. Disconnect the negative battery cable. Remove the rocker cover.
2. Turn the engine until the No. 1 cylinder intake valve cams, the intake camshaft is labeled **A** on the head, are both straight up.
3. Slide a flat feeler gauge in between each of the camshafts and the adjacent valve tappet. Check to see if the clearance is within the specified range. If it is, proceed with checking the remaining clearances as described starting in Step 8. If not, switch gauges and measure the actual clearance. When actual clearance is achieved, proceed with Steps 4–7.
4. Turn the tappets so the grooves machined into their edges are aligned as shown. Looking at the valves from the center of the engine, the right hand tappet's groove should be at about 5 o'clock and the left hand tappet's groove should be at about 7 o'clock. Use the end of the special tool required for the camshaft involved—in this case the **A** or intake camshaft, the exhaust camshaft end is labeled **E** on the engine and tool. Slide the proper end of the tool, going from the center of the engine outward, under the cam, with the heel of the tool pivoting on the inner side of the camshaft valley. Force the handle downward until the handle rests on the protrusion on the center of the cylinder head.

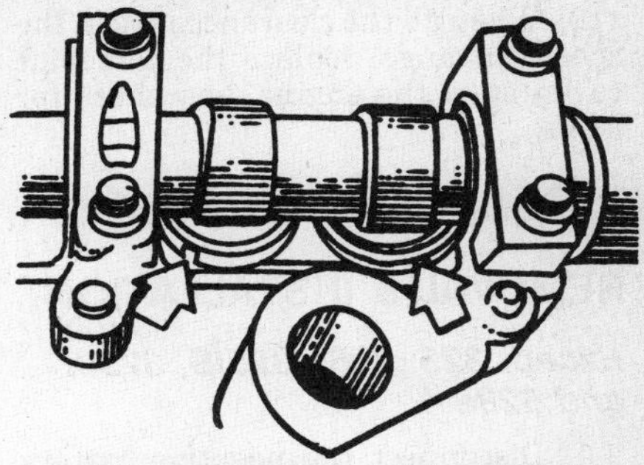

On M5 and M6 DOHC engines, rotate the valve tappets so the grooves machined in the tops are facing as shown before

5. Use compressed air to pop the disc out of the tappet. Read the thickness dimension on the disc.
6. Determine the thickness required as follows:
    a. If the valve clearance is too tight, try the next thinner disc.
    b. If the valve clearance is too loose, try the next thicker disc.
7. Slip the thinner or thicker disc into the tappet with the letter facing downward. Rock the valve spring depressing tool out and remove it. Then, recheck the clearance. Change the disc again, if necessary, until the clearance falls within the specified range.
8. Turn the engine in firing order sequence, turning the crankshaft forward ⅓ of a turn each time to get the intake camshafts to the upward position for each cylinder. Measure the clearance as in Step 3 and, if it is outside the specified range, follow Steps 4–7 to adjust either or both valves. Repeat this for all the intakes, and then turn the engine until No. 1 cylinder exhaust valves are upward.
9. Follow the same sequence for all the exhaust valves, going through the firing order, checking clearance as described in Step 3 and adjusting the valves as in Steps 4–7. Note that it is necessary, however, to use the opposite end of the special tool, the end marked **E** to depress the exhaust valves.

10. When all the clearances are in the specified range, replace the camshaft cover, start the engine, and check for leaks.

Rocker Arms/Shafts

REMOVAL & INSTALLATION

Except 325, 325i, 325iS, 325iX and 528e

1. Disconnect the negative battery cable. Remove the cylinder head.
2. Remove the camshaft.
3. On 6 cylinder engines, remove the retaining bolts and remove the end cover from the rear of the cylinder head. Slide the thrust rings and rocker arms rearward and remove the snaprings from the rocker arm shafts.
4. On 4 cylinder engines:
   a. Remove the distributor flange from the rear of the cylinder head.
   b. Using a long punch, drive the rocker arm shaft from the rear to the front of the cylinder head.

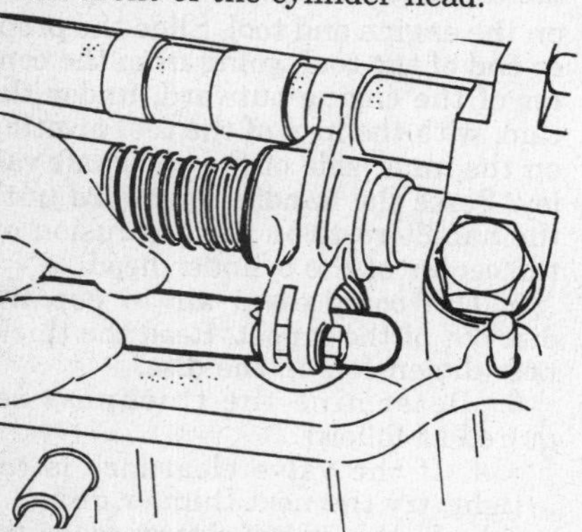

Checking valve clearance with a flat feeler gauge

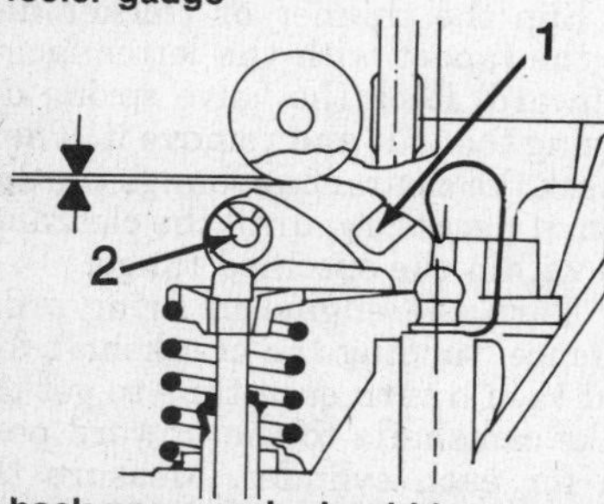

A back-up wrench should be used at (1) when adjusting valves. The locknut that holds the adjusting ecentric is at (2). "V" attempting to measure valve clearance

NOTE: Be sure all snaprings are off the shaft before attempting to drive the shaft from the cylinder head.

   c. The intake rocker shaft is not plugged at the rear, while the exhaust rocker shaft must be plugged. Replace the plug if necessary, during the installation.
5. On 6 cylinder engines:
   a. Install dowel pins part No. 11 1 063 or equivalent to keep the rocker shafts from turning. Then, remove the rocker shaft retaining plugs from the front of the cylinder head. These require a hex head wrench. Then, push back the rocker arms against spring pressure and remove the snaprings retaining the shafts. Remove the dowel pins. If the rocker shafts have welded plugs, the shafts will have to be pressed out of the head with a tool such as 11 3 050 or equivalent.

NOTE: There is considerable force on the springs positioning the rockers. They may pop out. Be cautious and wear safety glasses.

   b. Install a threaded slide hammer into the ends of the rear rocker shafts and remove.

To install:

6. The rocker arms, springs, washers, thrust rings and shafts should be examined and worn parts replaced. Special attention should be given to the rocker arm camshaft followers. If these are loose, replace the arm assembly. The valves can be removed, repaired or replaced, as necessary, while the shafts and rocker arms are out of the cylinder head.
7. Install the rocker arms in position, noting the following procedures:
   a. Design changes of the rocker arms and shafts have occurred with the installation of a bushing in the rocker arm and the use of 2 horizontal oil flow holes drilled into the rocker shaft for improved oil supply. Do not mix the previously designed parts with the later design.
   b. When installing the rocker arms and components to the rocker shafts, install locating pins in the cylinder head bolt bores to properly align the rocker arm shafts. Note that on 6 cylinder engines, the longer rocker shafts go on the chain end of the engine; the openings face the bores for the cylinder head bolts; and the plug threads face outward. The order of installation is: spring, washer, rocker arm, thrust washer, snapring. Note also that newer, short springs may be used with the older design.
   c. Install sealer on the rocker arm shaft retaining plugs and rear cover.
   d. On the 4 cylinder engines, position the rocker shafts so the camshaft retaining plate ends can be engaged in the slots of shafts during camshaft installation.
   e. Adjust the valve clearance.

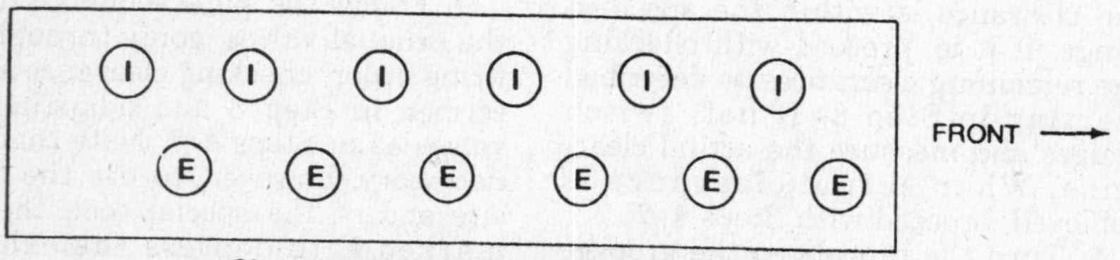

Six cylinder valve location–SOHC engines

325, 325i, 325iS, 325iX and 528e

The cylinder head must be removed before the rocker arm shafts can be removed.

1. Disconnect the negative battery cable. Remove the cylinder head.
2. Mount the head on a suitable holding fixture.
3. Remove the camshaft sprocket bolt and remove the camshaft distributor adapter and sprocket. Reinstall the adapter on the camshaft.
4. Adjust the valve clearance to the maximum allowable on all valves.
5. Remove the front and rear rocker shaft plugs and lift out the thrust plate.
6. Remove the spring-clips from the rocker arms by lifting them off.
7. Remove the exhaust side rocker arm shaft:
   a. Set the No. 6 cylinder rocker arms at the valve overlap position (rocker arms parallel), by rotating the camshaft through the firing order.
   b. Push in on the front cylinder rocker arm and then turn the camshaft in the direction of the intake rocker shaft, using a ½ in. drive breaker bar and a deep well socket to fit over the camshaft adapter. Slide each rocker arm to one side as it develops sufficient clearance away from its actuating camshaft and the valve it actuates. Rotate the camshaft until all of the rocker arms are relaxed.
   c. Remove the rocker arm shaft.
8. Remove the intake side rocker arm shaft:
   a. Turn the camshaft in the direction of the exhaust rocker arm.
   b. Use a deep well socket and ½ in. drive breaker bar on the camshaft adapter to turn the camshaft. Slide each rocker arm to one side as it develops sufficient clearance away from its actuating camshaft and the valve it actuates. Rotate the camshaft until all of the rocker arms are relaxed.
   c. Remove the rocker arm shaft.

9. Install the rocker arm shafts by reversing the removal procedure. Keep the following points in mind:

a. The large oil bores in the rocker shafts must be installed downward, toward the valve guides and the small oil bores and grooves for the guide plate face inward toward the center of the head.

b. The straight sections of the spring clamps must fit into the grooves in the rocker arm shafts.

c. The guide plate must fit into the grooves in the rocker arm shafts.

d. Adjust the valve clearance.

Intake Manifold

REMOVAL & INSTALLATION

M3

NOTE: A Torx® nut driver is needed to perform this operation.

1. Disconnect the negative battery cable. Remove the cap nuts at the outer ends of the 4 throttle necks. Then remove the mounting nuts underneath.
2. Make sure the engine has cooled off. Loosen the hose clamps for the air intake lines and for the fuel lines where they connect with the injection pipe. Collect fuel in a metal container.
3. Disconnect the throttle cable.
4. Pull off the intake manifold. Cut off the crankcase ventilation hose running to it from the crankcase. Then, remove the manifold and place it aside. Supply a new crankcase ventilation hose.
5. Pull off the throttle valve switch plug. Carefully pull the injector plug plate evenly off all 4 injectors.
6. Pull the fuel pressure regulator vacuum hose off the pressure regulator.
7. Remove the 2 mounting bolts for the injector pipe. Then, carefully lift off the pipe and injectors.
8. Unscrew the nut attaching the ball joint at the end of the throttle actuating rod to the throttle linkage. Supply a new self-locking nut.
9. Remove the nuts attaching the throttle necks to the cylinder head. Then, remove the 4 throttle necks as an assembly.
10. Separate the throttle neck assemblies by pulling them apart at the connecting pipe.
11. Inspect the O-rings in the connecting pipe and at the outer ends of the throttle necks. Replace as necessary.
12. Reverse the removal procedure to install. Use the new throttle linkage self-locking nut and the new crankcase ventilation hose.
13. Torque the nuts attaching the throttle necks to the head and the intake manifold to the throttle necks to 6.5–7.0 ft. lbs. Adjust the throttle cable.

6 Cylinder Vehicles Except 1991–92 525i, M5 and M6

1. Disconnect the negative battery cable and drain the cooling system.
2. Disconnect the wire harness at the air flow sensor. Remove the air cleaner and sensor as an assembly. Disconnect the air intake hose running from the air cleaner to the manifold.
3. Remove and tag the vacuum hoses and electrical plugs. Disconnect the accelerator linkage and cruise control linkage, if equipped from the throttle housing.
4. Disconnect the coolant hoses from the throttle housing.
5. Working from the rear of the collector housing, disconnect the vacuum lines, and starting valve connector, fuel line and air line. Tag the hoses and lines for ease of assembly.
6. Remove the EGR valve and line.
7. Remove all intake pipes.
8. Remove the air collector housing from the engine. On vehicles with a single intake manifold casting, remove the nuts and remove the throttle valve body.
9. Disconnect the plugs at the injector valves and remove the valves.
10. Disconnect the wire plugs at the coolant temperature sensor, the temperature time switch and the temperature switch.
11. Pull the wire loom upward through the opening in the intake manifold neck.
12. Remove the coolant hoses from the intake neck.

NOTE: Mark the heater hoses for proper reinstallation.

13. Remove the retaining bolts or nuts and remove either front, rear or both intake manifold necks. On some vehicles, remove the entire assembly.

To install:

14. To install the manifold, use new gaskets and install the manifold to the engine.
15. Install the air intake tubes and the injector valves. Install the collector and bracket.
16. Connect the vacuum line and electrical connections to the timing valve. Install the cold start valve.
17. Connect the line at the EGR valve and the electrical connections at the temperature timing switch.
18. Connect all vacuum, cooling and fuel lines at the throttle housing. Install the accelerator cable and vacuum hoses to the air collector.
19. Install the air cleaner and fill the cooling system. Check all hose connections and fluid levels before operating the engine.

1991–92 525i

1. Disconnect the negative battery cable and drain the coolant to a level below that of the throttle housing. Unscrew the fastener from the throttle cable cover and pull the cover foward and off. Unclip the cable and pull the cable out with the rubber holder.
2. Pull the vacuum fitting from the brake booster and plug the openings.
3. Remove the engine and intake manifold covers. Unscrew the bolt holding the ground strap on the front lifting eye. Replace the bolt before lifting the engine.
4. Unscrew the 2 bolts holding the plug plate and pull off the plug plate. Be careful not to damage the rubber seals. Take off the ignition coil electrical plugs. Remove the plug plate complete with the electrical leads.
5. Remove the cylinder head vent hose and pull off the air temperature sensor plug. Remove the tank venting hose and the throttle heating hoses from the throttle body. Remove the throttle valve switch plug. Unclip the idle speed control valve mounted on the manifold. Disconnect the fuel hoses from the pipes.
6. Unscrew the hardware holding the intake manifold to the cylinder head. Remove the intake manifold taking care not to drop anything into the exposed ports.
7. Installation is reverse of removal. Inspect the sealing rings at the intake port and replace as necessary.

M5 and M6

The M5 and M6 employ a manifold chamber in combination with 6 throttle necks (one for each cylinder), each of which contains its own throttle. The throttle necks are divided into 3 assemblies each containing the necks for 2 adjacent cylinders.

1. Disconnect the negative battery cable. Remove the nuts at the outer ends of the throttle necks, these attach the manifold to the outer ends of the necks. Loosen the hose clamps for the crankcase ventilation hoses and for the air intake hose. Disconnect the accelerator cable.
2. Pull the intake manifold off the throttle necks. Check O-rings and replace any that are hard or cracked.
3. Disconnect the electrical connectors to the cold start valve, throttle bypass valve, and throttle valve switch. Disconnect all the electrical connections going to the fuel injectors and remove the conduit for the injector wires from the throttle necks.
4. Disconnect the vacuum hoses for

the fuel pressure regulator and the heater temperature sensor. Disconnect the fuel return pipe, and collect the fuel in a metal container for safe disposal.

5. Remove the attaching nuts and bolts, and remove the injection pipe and injectors.

NOTE: Clean the throttle shaft thoroughly and be sure not to use pliers on the shaft surface. Otherwise, needle bearings on which the shaft rides may be damaged.

6. Using a center punch, drive out the 4 pins locking the throttle shaft in place. Slide the shaft out of the bearings.
7. Unscrew its mounting nuts and remove the throttle bypass valve. Disconnect the air hoses from this valve.
8. Remove the nuts attaching the throttle valve necks to the head and remove them.
9. Remove the connecting pipes that run between the valve neck units. Replace O-rings, if necessary. Replace all gaskets and make sure gasket surfaces on the head and inner ends of valve necks are clean.
10. Install in reverse order, providing new pins for the throttle shaft and coating its bearing surfaces with a proper lubricant before assembly.
11. Replace the sleeves in the intake manifold, if necessary. Replace the crankcase ventilation hose connecting the intake manifold and crankcase.

318iS

1. Disconnect the negative battery cable. Unscrew the upper manifold section.
2. Disconnect the rear mounting bracket and remove the coolant hose.
3. Loosen the front mounting bracket and disconnect the holder for the preheater.
4. Remove the mounting bolts and lift off the upper manifold section. Pull the hose off the fuel pressure regulator at the same time.
5. Pull the plug plate off the fuel injectors and remove the wire holding clamp.
6. Remove the injection pipe with the fuel injectors attached and remove the lower manifold section.
7. The installation is the reverse of the removal procedure.

750iL

1. Disconnect the negative battery cable. Loosen the clamps for the fuel lines.
2. Pull off the vacuum hoses for the pressure regulators. Lift out the injection pipes with the injectors attached.
3. Remove the distributor caps and the throttle valve necks on the manifolds.
4. Disconnect the spark plug wires and remove the ignition lead pipes.
5. Disconnect the crankcase breather hose and loosen the manifold support nuts.
6. Disconnect the nose guard and remove the intake manifold, using the proper tool.
7. The installation is the reverse of the removal procedure. Tighten the intake manifold mounting bolts to 14–17 ft. lbs.

Exhaust Manifold

REMOVAL & INSTALLATION

Except M3, M5 and M6

The 4 cylinder manifold is a one piece, one outlet unit, while the 6 cylinder manifold assembly consists of a 2 piece, double outlet to the exhaust pipe. One piece can be replaced independently of the other.

1. Disconnect the negative battery cable. Remove the air volume control and if necessary, air cleaner.
2. Disconnect the exhaust pipe at the reactor outlet(s).
3. Remove the guard plate from the reactor(s).
4. Disconnect the air injection pipe fitting, the EGR counter-pressure line, EGR pressure line and any supports.

NOTE: An exhaust filter is used between the reactor and the EGR valve and must be disconnected. Replace the filter if found to be defective.

5. Remove the retaining bolts or nuts at the reactor and remove it from the cylinder head.
6. Install the manifold, using new gaskets. Install the air injection fittings.
7. Connect the exhaust pipe at the reactor. Install the air cleaner.

M3

1. Disconnect the negative battery cable. With the engine cool, remove the drain plug from the block. Remove the 3 electrical connectors from the front of the coolant manifold that runs along the left side of the engine. Disconnect the radiator hose from the front of this pipe. Then, remove all the mounting bolts for this pipe and remove it. Inspect the O-rings and replace any that are worn or damaged.
2. Disconnect the exhaust pipe at the manifold flange. Remove the heat shields from under the engine.
3. Remove the mounting nuts at the cylinder head and remove the manifold.
4. Clean all gasket material from the surfaces of the manifold and head and replace the gaskets.
5. To install, position the manifold on the head, torquing the manifold bolts to 6.5–7.0 ft. lbs. and the coolant pipe mounting bolts to 7.5–8.5 ft. lbs. Torque the bolts at the flange attaching manifold and exhaust pipe first to 22–25 ft. lbs. and then to 36–40 ft. lbs. Make sure to refill the cooling system with fresh anti-freeze/water mix and bleed the cooling system.

M5 and M6

1. Disconnect the negative battery cable. With the engine cool, remove the drain plug from the block. Remove the 3 electrical connectors from the front of the coolant manifold that runs along the left side of the engine. Disconnect the radiator hose from the front of this pipe. Then, remove all the mounting bolts for this pipe and remove it. Inspect the O-rings (one for each cylinder, located in the block), and replace any that are worn or damaged.
2. Disconnect the exhaust pipe at the manifold. Remove the heat shields from under the engine.
3. Remove the crossbrace that runs under the engine by removing the 2 bolts from either end and then removing it.
4. Disconnect the stabilizer bar near both ends where it is bushed to the engine carrier.
5. Attach a lifting sling to the engine. Remove the nut from the right side engine mount and lift the engine slightly for clearance.
6. Remove the mounting bolts and remove the manifold.
7. Clean all gasket material from the surfaces of the manifold and head and replace the gaskets.
8. To install, position the manifold, torquing the manifold bolts to 36–40 ft. lbs. and the coolant pipe mounting bolts to 7.5–8.5 ft. lbs. Make sure to refill the cooling system with fresh anti-freeze/water mix and bleed the cooling system.

Timing Chain Front Cover

REMOVAL & INSTALLATION

318iS and 1991–92 525i

1. Disconnect the negative battery cable. Drain the cooling system and remove the radiator and fan assembly.
2. Remove the drive belts and any accessories that block access to the timing cover. Remove the engine splash shield, if necessary.
3. Remove the vibration damper us-

ing the proper tool. Unscrew the central bolt and remove the vibration damper hub.

4. Remove the timing case cover bolts and remove the timing cover.

NOTE: The timing case cover can be removed without removing the water pump.

5. Reverse the removal procedure to install. Tighten the central hub bolt to 224 ft. lbs. (295 ft. lbs. on the 525i) and the vibration damper bolts to 17 ft. lbs.

M3

1. Disconnect the negative battery cable. Drain the cooling system through the bottom of the radiator. Remove the radiator and fan.
2. Disconnect all electrical plugs, remove the attaching nuts, and remove the air cleaner and airflow sensor.
3. Note and, if necessary, mark the wiring connections. Then, disconnect all alternator wiring. Unbolt the alternator and remove it and the drive belt.
4. Unbolt the power steering pump. Remove the belt and then move the pump aside, supporting it out of the way but in a position where the hoses will not be stressed.
5. Remove the 3 bolts from the bottom of the bell housing and the 2 bolts below it which fasten the reinforcement plate in place.
6. Remove the drain plug and drain the oil from the lower oil pan. Then, remove the lower oil pan bolts and remove the lower pan.
7. Remove the 3 bolts fastening the bottom of the front cover to the front of the oil pan. Loosen all the remaining oil pan bolts so the pan may be shifted downward just slightly to separate the gasket surfaces.
8. Remove the water pump. Remove the center bolt and use a puller to remove the crankshaft pulley.
9. Remove the piston for the timing chain tensioner.
10. Remove the bolts attaching the top of the front cover to the cylinder head. Then, remove all the bolts fastening the cover to the block.
11. Run a sharp bladed tool carefully between the upper surface of the oil pan gasket and the lower surface of the front cover to separate them without tearing the gasket. If the gasket is damaged, remove the oil pan and replace it.

To install:

12. Before reinstalling the cover, use a file to break or file off flashing at the top/rear of the casting on either side so the corner is smooth. Replace all gaskets, coating them with silicone sealer. Where gasket ends extend too far, trim them off. Apply sealer to the area where the oil pan gasket passes the front of the block.
13. Slide the cover straight on to avoid damaging the seal. Install all bolts in their proper positions. Coat the 3 bolts fastening the front cover to the upper oil pan with the proper sealant.
14. Tighten the bolts at the top, fastening the lower cover to the upper cover first. Then, tighten the remaining front cover bolts and, finally, the oil pan bolts to 7 ft. lbs. Inspect the sealing O-rings and replace, as necessary. If it uses the DME distributor with the screw-off type rotor, make sure the bolt at the center of the rotor has its seal in place and that it is installed with a sealer designed to prevent the bolt from backing out.
15. Reverse the remaining portions of the removal procedures, making sure to fill and bleed the cooling system and to refill the oil pan with the correct oil.
16. Torque the oil drain plug to 24 ft. lbs. and both upper and lower oil pan bolts to 7 ft. lbs.

1990 525i, 535i, 635CSi and 735 Series

NOTE: On 535 Series, and 735 Series engines, this procedure requires the use of a special gauge.

1. Disconnect the negative battery cable. Remove the cylinder head cover. Remove the distributor.
2. Drain the coolant to below the level of the thermostat and remove the thermostat housing cover.
3. Remove the mounting bolts and remove the upper timing case cover.
4. Remove the piston which tensions the timing chain, working carefully because of very high spring pressure.
5. Remove the cooling fan and all drive belts. On 6 Series vehicles, the alternator must be swung aside by loosing the front bolt and removing the 2 side bolts. On all vehicles with the M30B35MZ engine, remove its attaching bolts and then remove the drive pulley from the water pump. The power steering pump must be removed, leaving the pump hoses connected and supporting the pump out of the way but so the hoses are not stressed.
6. Remove the flywheel housing cover and lock the flywheel in position with an appropriate tool.
7. Unscrew the nut from the center of the pulley and pull the pulley/vibration damper off the crankshaft.
8. Detach the TDC position transmitter on 635CSi and 7 Series.
9. Loosen all the oil pan bolts, and then unscrew all the bolts from the lower timing case cover, noting their lengths for reinstallation in the same positions. Carefully, use a sharp bladed tool to separate the gasket at the base of the lower timing cover. Then, remove the cover.

To install:

10. To install the lower cover, first coat the surfaces of the oil pan and block with sealer. Put it into position on the block, making sure the tensioning piston holding web (cast into the block) is in the oil pocket. Install all bolts; then tighten the lower front cover bolts evenly; finally, tighten the oil pan bolts evenly.
11. Inspect the hub of the vibration damper. If the hub is scored, install the radial seal so the sealing lip is in front of or to the rear of the scored area. Pack the seal with grease and install it with a sealer installer.
12. Install the pulley/damper and torque the bolt to specifications. When installing, make sure the key and keyway are properly aligned.
13. Remove the flywheel locking tool and reinstall the cover. Reinstall and tension all belts.
14. Before installing the upper cover, use sealer to seal the joint between the back of the lower timing cover and block at the top. On some vehicles, there are sealer wells which are to be filled with sealer. If these are present, fill them carefully. Check the cork seal at the distributor drive coupling and replace it, if necessary.
15. On all but M30B35MZ engines, tighten bolts 1 and 2 on 4 cylinder engines, the lower bolts, slightly. Then, tighten bolts 3–8. Finally, fully tighten the lower bolts.
16. On M30B35MZ engines, note that the top bolt on the driver's side and the bottom bolt on the passenger's side are longer. Tighten the 2 bolts that run down into the lower timing cover first; then tighten the remaining 6 bolts.
17. On the M30B35MZ engine, install the TDC transmitter and its mounting bracket. On remaining engines, install the TDC position transmitter loosely, if so equipped. With the engine at exactly 0 degrees Top Center, as shown by the marker on the front cover, adjust the position of the transmitter, it must fit the curve on the outside of the balancer, and incorporate a notch (for the pin on the balancer) and a ridge against which the transmitter must rest. The straight line distance between the center of the notch and bottom of the ridge must be exactly 37.5mm. Then, tighten the transmitter mounting screw.
18. Just before installing the upper timing case cover, check the condition of that area of the head gasket. It will usually be in good condition. If it should show damage, it must be replaced.
19. Inspect the sealing O-rings and

replace, as necessary. Make sure the bolt at the center of the rotor has its seal in place and that it is installed with a sealer designed to prevent the bolt from backing out.

20. Complete the installation, making sure to bleed the cooling system.

M5 and M6

1. Disconnect the negative battery cable. Pull out the plug and remove the wiring leading to the airflow sensor. Loosen the hose clamp and disconnect the air intake hose. Remove the mounting nut and remove the air cleaner and airflow sensor as an assembly.
2. Remove the radiator and fan. Remove the flywheel housing cover and install a lock, to lock the position of the flywheel. Remove the mounting nut for the vibration damper with a deepwell socket. Pull the damper off with a puller.
3. Remove the pipe that runs across in front of the front cover. Remove the mounting bolts and remove the water pump pulley.
4. Loosen the top/front mounting bolt for the alternator. Remove the lower/front bolt. Loosen the 2 side bolts. Swing the alternator aside.
5. Remove the power steering pump mounting bolts. Make sure to retain the spacer that goes between the pump and oil pan. Swing the pump aside and support it so the hoses will not be under stress.
6. Remove the flywheel housing cover and lock the flywheel in position with an appropriate tool.
7. Unscrew the nut from the center of the pulley and pull the pulley/vibration damper off the crankshaft.
8. Remove the bolts at the top, fastening the lower front cover to the upper front cover. Remove the bolts at the bottom, fastening the lower cover to the oil pan. Loosen the remaining oil pan mounting bolts.
9. Run a sharp bladed tool carefully between the upper surface of the oil pan gasket and the lower surface of the front cover to separate them without tearing the gasket.
10. Loosen and remove the remaining front cover mounting bolts, noting the locations of the TDC sending unit on the upper/right side of the engine and the suspension position sending unit on the upper left. Also, keep track of the bolts that mount these accessories, as their lengths are slightly different. Remove the timing cover, pulling it off squarely.

To install:

11. Before reinstalling the cover, use a file to break or file off flashing at the top/rear of the casting on either side so the corner is smooth. Replace all gaskets, coating them with silicone sealer. Where gasket ends extend too far, trim them off. Apply sealer to the area where the oil pan gasket passes the front of the block.
12. Slide the cover straight on to avoid damaging the seal. Install all bolts in their proper positions. Tighten the bolts at the top, fastening the lower cover to the upper cover first. Then, tighten the remaining front cover bolts and, finally, the oil pan bolts. Inspect the sealing O-rings and replace as necessary. If it uses the DME distributor with the screw-off type rotor, make sure the bolt at the center of the rotor has its seal in place and that it is installed with a sealer designed to prevent the bolt from backing out.
13. Complete the installation procedure, making sure to refill and bleed the cooling system.

750iL

UPPER

1. Disconnect the negative battery cable. Drain the cooling system and remove the fan assembly.
2. Remove both intake manifolds and distributor housings.
3. Disconnect the round rubber mounts, bolts and nuts. Remove both cylinder head covers.
4. Remove the mounting bolts and lift out the timing cover.
5. To install, reverse the removal procedure.

LOWER

1. Disconnect the negative battery cable. Drain the cooling system and remove the fan assembly.
2. Remove the drive belts and the engine splash shield. Remove the tensioning bolt.
3. Unscrew the bolts but do not remove the vibration damper. Remove the central hub bolt with the proper tool.
4. Remove the vibration damper using the proper tool to pull the vibration damper hub from the crankshaft.
5. Drain the engine oil and remove the lower section of the oil pan. Remove the bottom mounting screws from the timing case cover and loosen the adjacent oil pan bolts on both sides.
6. Remove the timing belt tensioner and reference mark sender.
7. Remove the mounting screws and take off the timing case cover.
8. To install, reverse the removal procedure. Tighten the central hub bolt to 318 ft. lbs. and the vibration damper mounting bolts to 17 ft. lbs.

Front Cover Oil Seal

REPLACEMENT

1990 525i, 535i, 635CSi and 735 Series

1. Disconnect the negative battery cable. Position the No. 1 piston at TDC on the beginning of its compression stroke.
2. Remove the flywheel guard and lock the flywheel with a locking tool.
3. Remove the drive belts and the fan.
4. Remove the retaining nut and remove the vibration damper from the crankshaft.

NOTE: The Woodruff® key should be at the 12 o'clock position on the crankshaft.

5. Remove the seal from the timing housing cover with a small pry bar.
6. Using a special seal installer or equivalent, lubricate and install the seal in the cover. This tool is used to press the seal into the bore with even pressure around the entire perimeter.

NOTE: If the balancer hub has serious scoring on the sealing surface, position the seal in the cover so the sealing lip is in front of or behind the scored groove.

7. Lubricate the balancer hub and install it on the crankshaft, being careful not to damage the seal.
8. Complete the assembly, using the reverse of the removal procedure. Be sure to remove the flywheel locking tool before attempting to start the engine.

318iS, 1991–92 525i and 750iL

1. Disconnect the negative battery cable.
2. Remove the vibration damper and hub assembly.
3. Press out the radial oil seal, using the proper tool.
4. Install the new seal flush in conjunction with the central bolt and washer, using the proper seal installer.
5. The remainder of the installation is the reverse of the removal procedure.

Timing Chain and Sprockets

REMOVAL & INSTALLATION

318iS

1. Disconnect the negative battery cable. Remove the vibration damper and hub assembly.
2. Remove the lower timing case

cover. Remove the timing chain tensioner.

3. Unscrew the upper chain guide and top bolt on the right chain guide.

4. Remove the timing chain sprockets and the lift out the chain. Remove the timing chain guide.

5. Remove the tensioning rail, if necessary. Remove the crankshaft sprocket with the proper tool and lift out the Woodruff® key.

6. Remove the reversing roller, if needed.

NOTE: The reversing roller can only be replaced complete with bearings.

7. The installation is the reverse of the removal procedure.

M3

1. Disconnect the negative battery cable. Remove the timing case cover.

2. Remove the camshafts and sprockets. It is not necessary to remove the cover from the rear of the head.

3. Make sure to catch the washer and lock washers which will be released at the front. Remove the 2 mounting bolts for the guide rail, which is located on the left (driver's) side of the engine. These are accessible from the rear.

4. Pull the guide rail forward and then turn it clockwise on its axis, looking at it from above, to free it from the chain.

5. Note the relationships between timing chain and sprocket marks. Remove the chain by separating it from the sprockets at top and bottom.

To install:

6. Engage the timing chain with the crankshaft sprocket so marks line up. Route the chain up through where the guide rail will go. Install the guide rail in reverse of the removal procedure.

7. Then engage the chain with the driver's side (E) sprocket with the marks aligned. Bolt this sprocket and the lock plate onto the front end of the intake camshaft. Use the adapter to keep the sprocket from turning, and torque the bolts to 6–7 ft. lbs. Turn this camshaft in the direction opposite to normal rotation to tension the timing chain on that side.

8. Engage the timing marks with the mark on the passenger's side (A) sprocket and install the sprocket and lock plate onto the front end of the exhaust camshaft. Again, use the adapter to keep the sprocket from turning, and torque the bolts to 6–7 ft. lbs. Make sure the timing chain has stayed in time.

9. Slide the chain tensioner piston into its cylinder. Install a new seal. Now install the spring with the conical end out. Install the cap which retains the spring and torque it to 29 ft. lbs.

10. Turn the engine 1 revolution in the normal direction of rotation. Recheck the timing. With the crankshaft at TDC, one groove on each camshaft faces inward and another on each faces the cast boss on the bearing cap.

11. Install the camshaft and timing chain guides.

12. Install the timing case cover.

M5 and M6

1. Disconnect the negative battery cable. Remove the fan shroud and the fan. Remove the cylinder head cover. Remove the timing cover.

2. Remove the camshaft. Remove the cover from the rear of the head.

3. Remove the water pump.

4. Remove the 2 mounting bolts for the guide rail, which is located on the left (driver's) side of the engine. These are accessible from the rear. Turn the guide rail counterclockwise on it's axis looking at it from above to clear the chain and block and remove it. Be careful to retain all washers.

5. Note the relationships between timing chain and sprocket marks. Remove the timing chain.

To install:

6. Install the timing chain with the marks on all 3 sprockets aligned with marked links on the chain. Make sure the chain runs on the inside of the guide sprocket on the left side of the engine and along the groove in the lower tensioning rail. Install the chain onto the camshaft drive sprockets and then install the sprockets onto the camshafts (note that the exhaust side sprocket is marked **A** and the intake sprocket is marked **E**. Then, install the guide rail with all washers and lock washers by rotating it into position in reverse of the removal procedure.

7. Tighten the camshaft drive sprockets, install the chain tensioner, and install the upper guide rail as described in the camshaft removal and installation procedure. Reverse the remaining removal steps to complete the procedure. Make sure to refill the cooling system with an appropriate antifreeze/water mix and to bleed the cooling system.

535i, 635CSi and 735 Series

1. Disconnect the negative battery cable. Rotate the crankshaft to set the No. 1 piston at TDC, at the beginning of its compression stroke.

2. Remove the distributor.

3. Remove the cylinder head cover, air injection pipe and guard plate.

4. Drain the cooling system and remove the thermostat housing.

5. Remove the upper timing housing cover.

6. Remove the timing chain tensioner piston by unscrewing the cap.

NOTE: The piston is under heavy spring tension.

7. Remove the drive belts and fan.

8. Remove the flywheel guard and lock the flywheel with a locking tool.

9. Remove the vibration damper assembly.

NOTE: The crankshaft Woodruff® key should be in the 12 o'clock position.

10. Remove upper and lower timing covers.

11. Turn the crankshaft so the No. 1 cylinder is at firing position. Open the camshaft lock plates, if equipped, remove the bolts and remove the camshaft sprocket.

12. Remove the chain from the lower sprocket, swing the chain to the right front and out of the guide rail and remove the chain from the engine.

To install:

13. Install the chain in position.

14. Be sure No. 1 piston remains at the top of its firing stroke and the key on the crankshaft is in the 12 o'clock position.

15. Position the camshaft flange so the dowel pin bore is at the 7–8 o'clock position and the upper flange bolt hole is aligned with the cast tab on the cylinder head.

16. Position the chain on the guide rail and swing the chain inward and to the left.

17. Engage the chain on the crankshaft gear and install the camshaft sprocket into the chain.

18. Align the gear dowel pin to the camshaft flange and bolt and sprocket into place. Torque the sprocket bolts to 5 ft. lbs. (6.5–7.5 on the M30B35 engine).

19. Install the chain tensioner piston, spring and cap plug, but do not tighten.

20. To bleed the chain tensioner, fill the oil pocket, located on the upper timing housing cover, with engine oil and move the tensioner back and forth with a screwdriver until oil is expelled at the cap plug. Tighten the cap plug securely.

21. Complete the assembly in the reverse order of removal. Check the ignition timing and the idle speed. Be sure the flywheel holder is removed before any attempt is made to start the engine.

1991–92 525i

1. Disconnect the negative battery cable. Remove the vibration damper and hub assembly.

2. Remove the upper and lower timing case covers. Compress and lock the

upper timing chain tensioner. Unbolt the upper timing chain sprockets and remove from the camshafts along with the chain.

3. Unscrew the upper chain guide and top bolt on the right chain guide.

4. Remove the timing chain sprockets and the lift out the chain. Remove the timing chain guide.

5. Remove the tensioning rail, if necessary. Remove the crankshaft sprocket with the proper tool and lift out the Woodruff® key.

6. Installation is the reverse of removal. Torque the camshaft sprocket bolts to 16 ft. lbs.(22 Nm), the lower timing cover bolts to 6.5–8.0 ft. lbs.(9–11 Nm) for M6 bolts and 16 ft. lbs.(22 Nm) for M8 bolts, the vibration damper hub bolt to 295 ft. lbs.(410 Nm) and the vibration damper bolts to 17 ft. lbs.(23 Nm). Use sealer at the intersections of the timing cover and the pan.

Timing Belt Front Cover

REMOVAL & INSTALLATION

325, 325i, 325iS, 325iX 525i and 528e

The 325 and 528e are equipped with a rubber drive and timing belt and the distributor guard plate is actually the upper timing belt cover.

1. Disconnect the negative battery cable. Remove the distributor cap and rotor. Remove the inner distributor cover and seal.

2. Remove the 2 distributor guard plate attaching bolts and one nut. Remove the rubber guard and take out the guard plate (upper timing belt cover).

3. Rotate the crankshaft to set No. 1 piston at TDC of its compression stroke.

NOTE: At TDC of No. 1 piston compression stroke, the camshaft sprocket arrow should align directly with the mark on the cylinder head.

4. Remove the radiator.

5. Remove the lower splash guard and take off the alternator, power steering and air conditioning belts.

6. Remove the crankshaft pulley and vibration damper.

7. Hold the crankshaft hub from rotating with the proper tool. Remove the crankshaft hub bolt.

8. Install the hub bolt into the crankshaft about 3 turns and use the proper gear puller, to remove the crankshaft hub.

9. Remove the bolt from the engine end of the alternator bracket. Loosen the alternator adjusting bolt and swing the bracket out of the way.

10. Lift out the TDC transmitter and set it aside.

11. Remove the remaining bolt and lift off the lower timing belt cover.

12. The installation is the reverse of the removal procedure.

OIL SEAL REPLACEMENT

1. Disconnect the negative battery cable. Remove the front engine cover.

2. Press the 2 radial oil seals out of the front engine cover.

3. Install the oil seals flush with the front engine cover using the proper tools.

4. Install the front engine cover. Connect the negative battery cable.

Timing Belt and Tensioner

ADJUSTMENT

1. Disconnect the negative battery cable.

2. Remove the front engine cover.

3. Loosen the tensioner bolt. The spring force should be capable of moving the tensioner roller.

4. Rotate the crankshaft in the direction of the rotation to TDC mark. The mark on the camshaft sprocket with the mark on the cylinder head with the crankshaft at TDC mark.

5. Tighten the tensioner bolt and install the front engine cover.

6. Connect the negative battery cable.

REMOVAL & INSTALLATION

1. Disconnect the negative battery cable. Remove the distributor cap and rotor. Remove the inner distributor cover and seal.

2. Remove the 2 distributor guard plate attaching bolts and 1 nut. Remove the rubber guard and take out the guard plate, upper timing belt cover.

3. Rotate the crankshaft to set No. 1 piston at TDC of its compression stroke.

NOTE: At TDC of No. 1 piston compression stroke, the camshaft sprocket arrow should align directly with the mark on the cylinder head.

4. Remove the radiator.

5. Remove the lower splash guard and take off the alternator, power steering and air conditioning belts.

6. Remove the crankshaft pulley and vibration damper.

7. Hold the crankshaft hub from rotating with tool 11 2 150 or equivalent. Remove the crankshaft hub bolt.

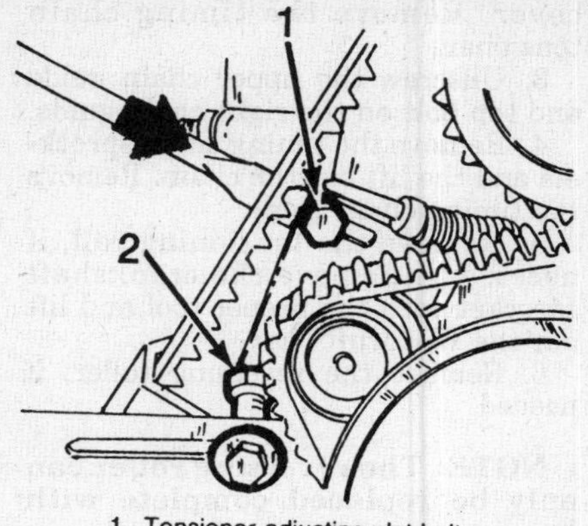

1. Tensioner adjusting slot bolt
2. Tensioner bracket pivot bolt

Releasing the tension on timing belt–528e

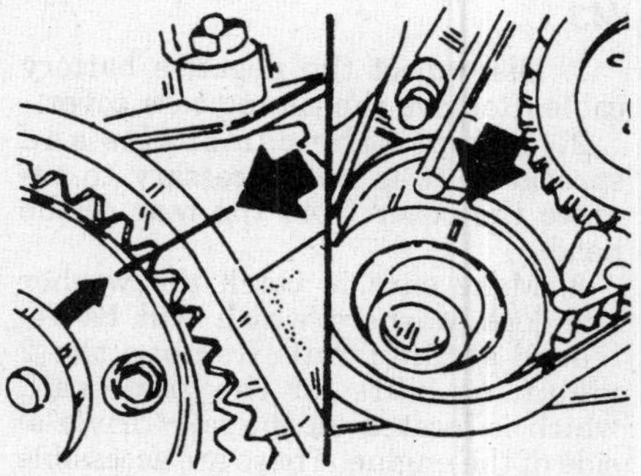

Crankshaft sprocket timing marks aligned for installation of the timing belt–528e

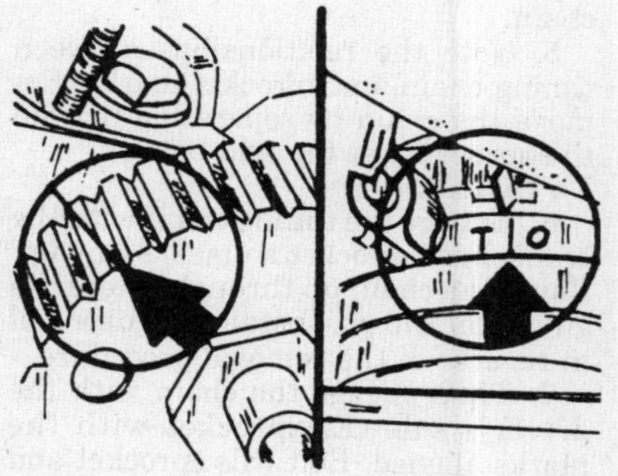

Aligning the marks for timing belt installation–528e

8. Install the hub bolt into the crankshaft about 3 turns and use the proper gear puller, to remove the crankshaft hub.

9. Remove the bolt from the engine end of the alternator bracket. Loosen the alternator adjusting bolt and swing the bracket out of the way.

10. Lift out the TDC transmitter and set it out of the way.

11. Remove the remaining bolt and lift off the lower timing belt cover.

12. Loosen the 2 tensioner pulley bolts and release the tension on the

belt by pushing on the tensioner pulley bracket.

13. Mark the running direction of the timing belt and remove the belt.

14. Remove the 3 bolts across the front of the oil pan and loosen the remaining oil pan bolts. Try not to damage the oil pan gasket. Remove the 6 front engine cover bolts and remove the front engine cover.

To install:

15. Install the cover, noting the following:

a. To tighten the timing belt, turn the engine in the direction of normal engine operation, with a ½ in. drive ratchet wrench on the crankshaft bolt. When the timing belt is tight, torque the 2 tensioner bolts.

b. Align the hub centering pin through the hole in the vibration damper for proper installation.

c. Align the timing marks when installing the timing belt. The crankshaft sprocket mark must point at the notch in the flange of the front engine cover. The camshaft sprocket arrow must point at the alignment mark on the cylinder head. Also, the No. 1 piston must be at TDC of the compression stroke.

d. If the oil pan gasket is damaged, it must be replaced.

e. Check and replace front cover oil seals, if needed.

f. Use tools 11 2 211 (crankshaft seal aligner) and 11 2 212 (intermediate shaft seal aligner) or equivalent, to install the front engine cover without damaging the oil seals.

16. Check the engine oil level.

17. Install engine coolant and bleed the cooling system. Bring the engine up to operating temperature and loosen the bleed screw on top of the thermostat housing. Continue to bleed until escaping coolant is free of bubbles. Add coolant to the expansion tank if needed.

Timing Sprockets

REMOVAL & INSTALLATION

1. Disconnect the negative battery cable. Remove the timing belt.

2. Hold the sprocket with the proper tool. Remove the bolt and take off the collar.

3. Remove the sprocket from the crankshaft with the proper tool.

4. Remove the sprocket from the intermediate shaft in the same way.

5. Reverse the removal procedure to install.

Camshaft

REMOVAL & INSTALLATION

535i, 635CSi and 735 SERIES

1. Disconnect the negative battery cable. Remove the oil line from the top of the cylinder head.

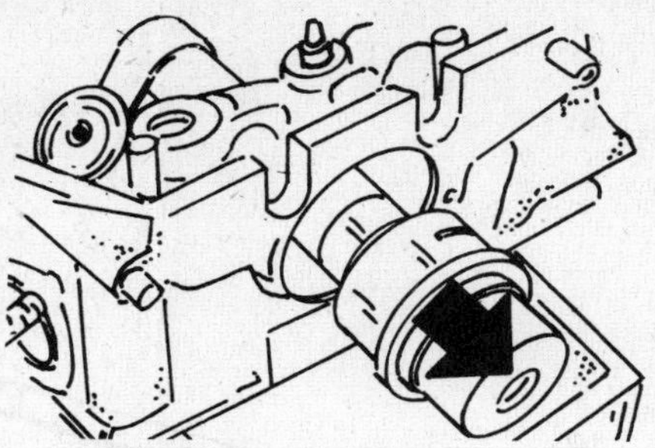

Pulling out the camshaft—528e

NOTE: Observe the location of the seals when removing the hollow oil line studs. Install new seals in the same position.

2. Remove the cylinder head. Support the head in such a way that the valves can be opened during camshaft removal.

3. Adjust the valve clearance to the maximum clearance on all rocker arms.

4. Remove the fuel pump and pushrod on carbureted engines.

5. A tool set 11 1 060 and 00 1 490 or its equivalent, is used to hold the rocker arms away from the camshaft lobes. When installing the tool, move the intake rocker arms of No. 2 and 4 cylinders forward approximately ¼ in. and tighten the intake side nuts to avoid contact between the valve heads. Turn the camshaft 15 degrees clockwise to install the tool. On these engines, to avoid contact between the valve heads, first tighten the tool mounting nuts on the exhaust side to the stop and then tighten the intake side nuts slightly. Reverse this exactly during removal.

6. Remove the camshaft by rotating the camshaft so the 2 cutout areas of the camshaft flange are horizontal and remove the retaining plate bolts.

7. Carefully, remove the camshaft from the cylinder head.

8. The flange and guide plate can be removed from the camshaft by removing the lock plate and nut from the camshaft end.

To install:

9. Install the camshaft and associated components in the reverse order of removal.

10. After installing the camshaft guide plate, the camshaft should turn easily. Measure and correct the camshaft end play.

11. The camshaft flange must be properly aligned with the cylinder head before the sprocket is installed.

12. Install the oil tube hollow stud washer seals properly, 1 above and 1 below the oil pipe. The arrow on the oil line must face forward.

13. Install the cylinder head. Adjust the valves.

325, 325i, 325iS, 325iX, 1990 525i and 528e

The cylinder head and the rocker arm shafts must be removed before the camshaft can be removed.

1. Disconnect the negative battery cable. Remove the cylinder head.

2. Mount the head on a stand. Secure the head to the stand with 1 head bolt.

3. Remove the camshaft sprocket bolt and remove the camshaft distributor adapter and sprocket. Reinstall the distributor adapter on the camshaft.

4. Adjust the valve clearance to the maximum allowable on all valves.

5. Remove the front and rear rocker shaft plugs and lift out the thrust plate.

6. Remove the clips from the rocker arms by lifting them off.

7. Remove the exhaust side rocker arm shaft:

a. Set the No. 6 cylinder rocker arm to the valve overlap position, both rocker arms parallel.

b. Push in on the rocker arm on the front cylinder and turn the camshaft in the direction of the intake rocker shaft, using a ½ in. breaker bar and a deep well socket to fit over the camshaft adapter. Rotate the camshaft until all of the rocker arms are relaxed.

c. Remove the rocker arm shaft.

8. Remove the intake side rocker arm shaft:

a. Turn the camshaft in the direction of the exhaust valves.

b. Use a deep well socket and ½ in. drive breaker bar on the camshaft adapter to turn the camshaft until all of the rocker arms are relaxed.

c. Pull out the rocker arm shaft.

9. Remove the camshaft thrust bearing cover. Check the radial oil seal and round cord seal and replace them if needed.

10. Pull out the camshaft.

To install:

11. Install the camshaft, noting the following:

a. Place the proper tool over the end of the camshaft during installation of the thrust bearing cover; this will protect the oil seals and guide the cover on.

b. The rocker arm thrust plate must be fit into the grooves in the rocker shafts.

c. The straight side of the spring clip must be installed in the groove of the rocker arm shafts.

d. The large oil bores in the rocker shafts must be installed down to the valve guides and the small oil

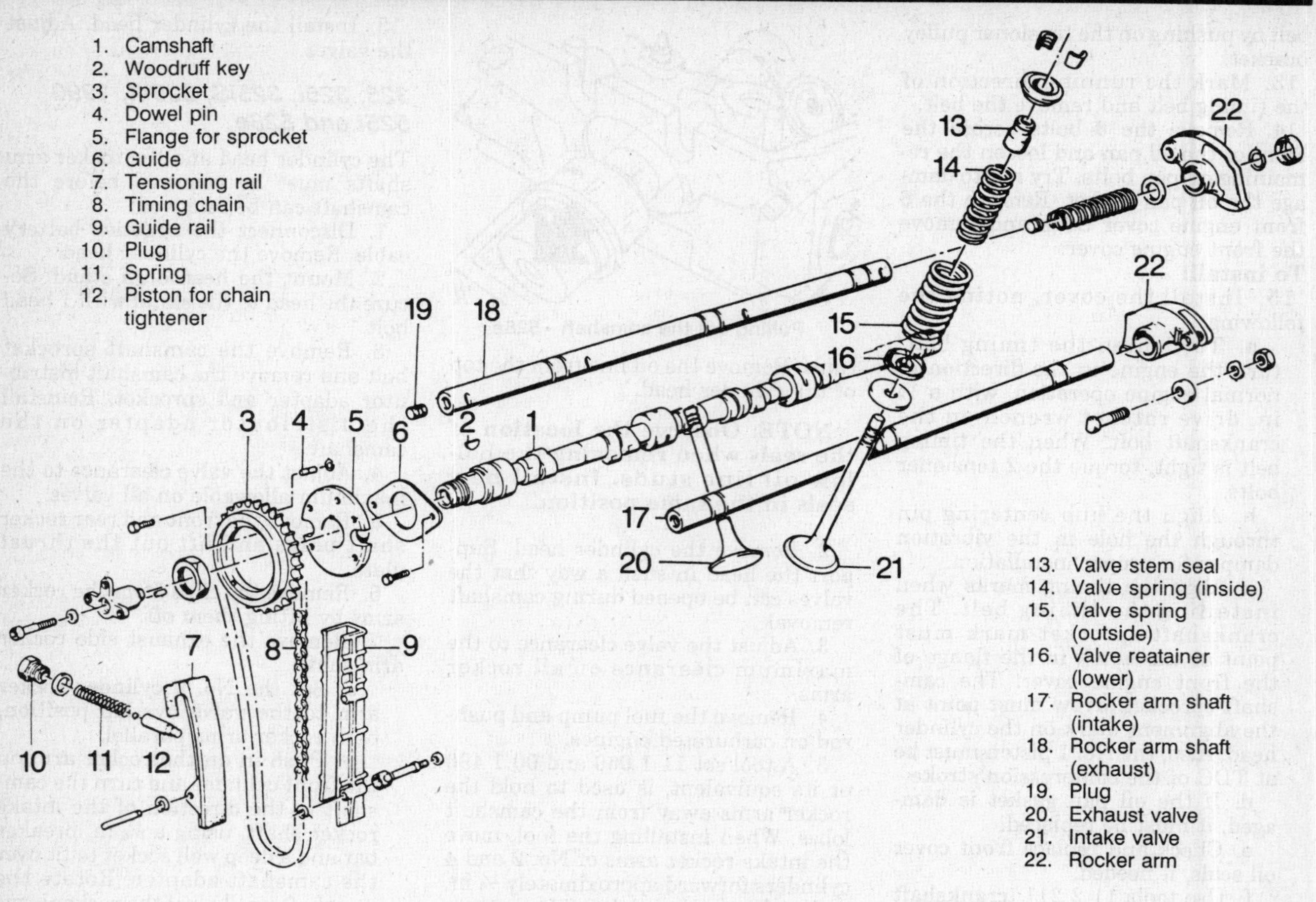

Exploded view of the camshaft assembly—535i

bores must face inward toward the center of the head.

e. Adjust the valve clearance.

12. To complete the installation, reverse the removal procedures.

M3, M5 and M6

NOTE: To perform this operation it is necessary to have a special tool 11 3 010 or equivalent. This is necessary to permit safe removal of the camshaft bearing caps and then safe release of the tension the valve springs put on the camshafts. The job also requires an adapter to keep the camshaft sprockets from turning while loosening and tightening their mounting bolts.

1. Disconnect the negative battery cable. Remove the cylinder head cover. Remove the fan cowl and the fan.

2. Remove the mounting bolts and remove the distributor cap. Remove the mounting screws and remove the rotor. Unscrew the distributor adapter and the protective cover underneath. Inspect the O-ring that runs around the protective cover and replace it, if necessary.

3. Remove the 2 bolts and remove the protective cover from in front of the right side (intake) camshaft. Remove the bolts and remove the distributor housing from in front of the left (exhaust) side cam. Inspect the O-rings, and replace them, if necessary.

4. Remove the 6 mounting bolts from the cover at the rear end of the cylinder head and remove it. Replace the gasket. Note that on the M3, 2 of these bolts are longer. These fit into the 2 holes that are sleeved.

5. Remove the 2 nuts, located at the front of the head, which mount the upper timing chain guide rail. Then, remove the upper guide rail.

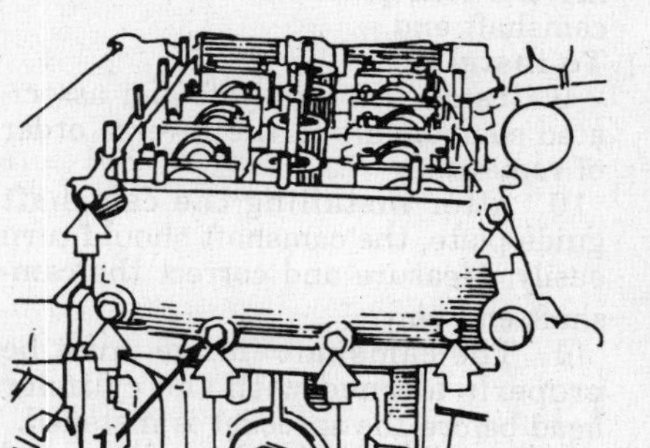

On the rear cover of the M3, install the longer bolts into the 2 holes marked "1"

6. Turn the crankshaft to set the engine at No. 1 cylinder TDC. On the 6 cylinder engine, valves for No. 6 will be at overlap position—both valves just slightly open with timing marks at TDC.

7. Remove the cap for the timing chain tensioner, located on the right side of the front timing cover. Then, slide off the damper housing. Remove the seal, discard it, and supply a new one for re-assembly.

NOTE: The next item to be removed is a plug which keeps the tensioner piston inside its hydraulic cylinder against considerable spring pressure. Use a socket wrench and keep pressure against the outer end of the plug, pushing inward, so spring pressure can be released very gradually once the plug's threads are free of the block.

8. Remove the plug from the tensioner piston, and then release spring tension. Remove the spring and then the piston. Check the length of the

spring. It must be 6.240–6.280 in. in length; otherwise, replace it to maintain stable timing chain tension.

NOTE: The timing chain should remain engaged with the crankshaft sprocket while removing the camshafts. Otherwise, it will be necessary to do additional work to restore proper timing. Keep the timing chain under slight tension by supporting it at the top while removing the camshaft sprockets.

9. Pry open the lock plates for the camshaft sprocket mounting bolts. Install an adapter to hold the sprockets still and remove the mounting bolts.
10. Using an adapter to keep the sprockets from turning and putting tension on the timing chain, loosen and remove the sprocket mounting bolts, keeping the chain supported.
11. Mount the special tool on the timing case, which mounts to the top of the head. Then, tighten the tool's shaft to the stop. This will hold both camshafts down against their lower bearings. Also, mark the camshafts as to which end faces forward.
12. Remove the mounting bolts and remove the camshaft bearing caps. It is possible to save time by keeping the caps in order, although they are marked for installation in the same positions.
13. Once all bearing caps are removed, slowly crank backwards on the tool's shaft to gradually release the tension on the camshafts. Once all tension is released, remove the camshafts.
14. Carefully, remove the camshafts in such a way as to avoid nicking any bearing surfaces or cams.

To install:

15. Oil all bearing and camshaft surfaces with clean engine oil. Carefully, install the camshafts, marked E for intake and A for exhaust, to avoid nicking any wear surfaces. The camshafts should be turned so the groove between the front camshaft and sprocket mounting flange faces straight up. Install the special tool and tighten down on the shaft to seat the camshafts.
16. Install all bearing caps in order or as marked. Torque the attaching bolts to 15–17 ft. lbs. Then, release the tension provided by the tool by turning the bolt and remove the tool.
17. Install the intake sprocket (marked E), install the lock plate, and install the mounting bolts. Use the adapter to keep the sprocket from turning, and torque the bolts to 6–7 ft. lbs. Do the same for the exhaust side sprocket. Make sure the timing chain stays in time.
18. Slide the timing chain tensioner piston into the opening in the cylinder in the block. Install the spring with the conically wound end facing the plug. Install the plug into the end of the sprocket and then install it over the spring. Use the socket wrench to depress the spring until the plug's threads engage with those in the block. Start the threads in carefully and then torque the plug to 27–31 ft. lbs. Install a new seal, connector, damper housing, and the outside cap with a new cap seal. Torque the outside cap to 16–20 ft. lbs. on the engine used in the M5 and M6 and 29 ft. lbs. on the M3 engine.
19. Crank the engine forward just 1 turn in normal direction of rotation. Now, 1 camshaft groove on each side should face toward the center of the head and 1 on each side should face the case boss on the front bearing cap. Lock the sprocket mounting bolts with the tabs on the lockplates.
20. Reverse the remaining removal procedures to complete the installation. Before final tightening of the mounting nuts for the guide rail for the top of the timing chain, go back and forth, measuring the clearance between the sprockets and the center of the guide rail to center it. Then, tighten the mounting nuts.

318iS and 1991–92 525i

1. Disconnect the negative battery cable and strip the head of its covers and related items to expose the camshafts. Remove the timing chains and sprockets.

NOTE: Special tools are required to perform this operation. BMW tools 11 3 260, 11 3 270 and 11 3 250 or their equivalents are required for proper removal and installation of the camshafts and for retention of the valve lash compensators. Without these tools the camshafts will be damaged during removal or installation.

2. Remove the spark plugs and attach the 11 3 260 (plus addition 11 3 270) camshaft removal fixture. Torque the hold down bolts in the spark plug bores to 17 ft. lbs.(23 Nm).
3. Apply load to the bearing caps by rotating the eccentric shaft. This relieves the tension on the bearing cap bolts. Loosen and remove the bearing cap bolts.
4. Remove the camshaft removal fixture after releasing the tension from the eccentric shaft.
5. Remove the camshafts and the bearing caps. Note that the intake camshaft is marked "E" and the exhaust camshaft is marked "A". The camshaft bearing are consecutively numbered with "A" or "E" to designate intake or exhaust side.
6. Hold the valve lash compensators in place using tool 11 3 250 and remove the bearing plate along with the valve plungers.

To Install:

7. Installation is reverse of removal. Inspect the camshafts and valve lash compensators for damage and wear and replace as necessary.
8. When the camshafts have been removed and reinstalled a waiting period dependent on the ambient temperature is necessary before mounting the cylinder head on the engine. At room temperature wait 4 minutes to allow the lifters to compress fully. At temperatures down to 50° F(10° C) wait 11 minutes. At temperatures lower than 50° F(10° C) wait 30 minutes. This is to prevent contact between the valves and the piston tops.
9. The engine may not be cranked under the same conditions as above for a period of 10 minutes at room temperature; 30 minutes for temperatures down to 50° F(10° C); 75 minutes for temperatures below 50° F(10° C).

Auxiliary/Silent Shaft

REMOVAL & INSTALLATION

325, 325i, 325iS, 325iX and 528e

1. Disconnect the negative battery cable. Remove the front cover.
2. Remove the intermediate shaft sprocket.
3. Loosen and remove the 2 retaining screws and then remove the intermediate shaft guide plate.
4. Carefully, slide the intermediate shaft out of the block. Turn the crankshaft, if necessary, to remove it. Inspect the gear on the intermediate shaft, replacing it, if necessary.
5. Install the intermediate shaft to the block. Install the guide plate.
6. Install the front cover.

Piston and Connecting Rod

Positioning

Location of piston in the cylinder bore with ring gaps located 180° apart

ENGINE LUBRICATION

Oil Pan

REMOVAL & INSTALLATION

318iS

1. Disconnect the negative battery cable. Raise and safely support the vehicle.
2. Drain the engine oil.
3. Disconnect the exhaust pipe, if necessary.
4. Remove the lower oil pan mounting bolts and take off the lower oil pan. Remove the upper section oil pan bolts and remove the upper oil pan.
5. Clean the mounting surfaces and install new gaskets.
6. The installation is the reverse of the removal procedure. Tighten the mounting bolts to 6.5 ft. lbs.

325, 325i, 325iS, 325iX and 528e

1. Disconnect the negative battery cable. Raise the vehicle and support it. Drain the engine oil.
2. Remove the front lower splash guard, if necessary.
3. Disconnect the electrical terminal from the oil sending unit.
4. Remove the power steering gear from the front axle carrier, if necessary.
5. Remove the flywheel cover.

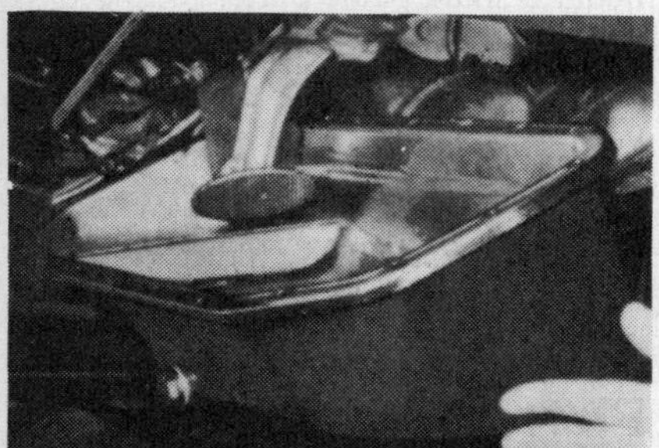

Removal of engine oil pan—typical

6. Remove the oil pan bolts and lower the oil pan. Remove the oil pump bolts and take out the oil pump and oil pan.
7. Install the oil pan, paying attention to the following points:
   a. Clean the gasket surfaces and use a new gasket on the oil pan.
   b. Coat the joints on the ends of the front engine cover with a universal sealing compound.
   c. Install the sending unit wire and the engine oil. If the power steering gear was removed, make sure to refill and bleed this system.

M3

1. Remove the dipstick. Remove the splash guard from underneath the engine. Raise and safely support the vehicle.
2. Remove the drain plug and drain the oil. Unscrew all the bolts for the lower oil pan and remove it.
3. Remove the oil pump.
4. Remove the lower flywheel housing cover by removing the 3 bolts at the bottom of the flywheel housing and the 2 bolts in the cover just ahead of the flywheel housing.
5. Disconnect the oil pressure sending unit plug. Unbolt the oil pan bracket. Disconnect the ground lead. Loosen its clamp and disconnect the crankcase ventilation hose.
6. Remove the oil pan bolts and remove the upper oil pan. Clean all sealing surfaces. Supply a new gasket and the coat the joints where the timing case cover and block meet with a brush-on sealant. Install the pan and torque the bolts evenly to 7 ft. lbs.
7. Reverse the remaining removal procedures to install, cleaning all sealing surfaces and using a new gasket on the lower pan, also. Torque the lower pan bolts, also, to 7 ft. lbs.
8. Install the oil pan drain plug, torquing to 24 ft. lbs. Refill the oil pan with the required amount of approved oil. Start the engine and check for leaks.

Rear main bearing oil seal and end cover housing showing special sealing locations

1990 525i, 535i, 635CSi, M5 and M6

1. Disconnect the negative battery cable. Disconnect the electrical connector and separate the leads from the air cleaner/air flow sensor. Loosen the hose clamp and disconnect the air intake hose. Remove the mounting nut and remove the air cleaner and the airflow sensor as a unit. Remove the fan shroud.
2. Raise and safely support the vehicle. Drain the engine oil. Lower the vehicle.
3. Loosen the belt tension and remove the alternator drive belt. Loosen the upper/front mounting bolt for the alternator and the 2 bolts on the side of the block that mount it at the rear. Remove the lower/front mounting bolt. Then, swing the alternator to the side.
4. Loosen the power steering pump mounts and remove the drive belt. Then, remove the mounting bolts and remove the pump and pump mounting bracket. Make sure to retain spacers. Remove the nuts and bolts that fasten the compressor to the hinge type mounting bracket. Make sure the compressor is suspended so there is no tension on the hoses. Unbolt the hinge type mounting bracket and remove it.
5. Remove the brace plate located under the oil pan. Remove those oil pan bolts that can be reached.
6. Remove the engine ground strap. Remove the engine mount through bolts. Attach a lifting sling to the hooks on top of the engine. Lift the engine slightly for clearance.
7. Shift the power steering pump out of the way and support it so no tension will be placed on the hoses.
8. Remove the remaining oil pan mounting bolts. Turn the crankshaft so the rods for cylinders 5 and 6 are as high as possible. Then, remove the pan.

To install:

9. Clean all sealing surfaces and supply a new gasket. Apply a liquid sealer to the joints between the block and the timing cover on the front and the rear main seal cover at the rear.
10. Install the oil pan in reverse order. Torque the pan bolts to 6.5–7.5 ft. lbs. Make sure to refill the pan with the required amount of the correct oil. Mount all accessories securely and adjust the drive belts.

1991–92 525i

1. Disconnect the negative battery terminal and raise and safely support the vehicle. Drain the engine oil.
2. Loosen the holding bolt for the oil dipstick guide pipe and remove the clamp. Pull the guide tube free of the pan.

3. Remove all the oil pan bolts and remove the pan. Raise the engine slightly if needed for clearance.

To Install:

4. Apply sealer to the joint between the pan, front cover and block.
5. Install new gaskets and install the pan. Torque mounting bolts to 6.5–8.0 ft. lbs.(9–11 Nm).
6. Install the dipstick guide tube using a new base seal and tighten the holding bolt.

735i and 735iL

1. Disconnect the negative battery cable. Loosen the hose clamp for the air intake hose. Remove the mounting nut for the air cleaner, and remove the air cleaner. Remove the fan and shroud.
2. Disconnect the electrical plug and overflow hose from the coolant expansion tank. Be careful not to kink the hose. Remove the mounting nuts and remove the tank.
3. Remove the splash guard for the power steering pump. Loosen the locknut for the pump adjustment and remove the through bolt that mounts the pump lower bracket (which contains the adjustment mechanism) to the block. Swing the bracket aside. Unscrew the bolt attaching the power steering pump lines to the block and shift them aside too.
4. Disconnect the electrical plug for the suspension leveling switch on the left side engine mounting bracket. Raise and safely support the vehicle. Remove the oil pan drain plug and drain the oil.
5. Remove the bracket for the exhaust pipes located near the oil pan.
6. Disconnect the ground strap from the engine. Remove the nuts and washers attaching the engine to the mounts on both sides.
7. Attach an engine lifting sling to the hooks at either end of the cylinder head. Lift the engine as necessary for clearance.
8. Remove all oil pan mounting bolts and remove the pan.

To Install:

9. Clean both sealing surfaces and supply a new gasket. Coat the 4 joints, between the block and timing case cover at the front and the block and rear main seal housing cover at the rear, with a proper sealer. Install the oil pan bolts and torque them to 6.5–7.5 ft. lbs.
10. Reverse the remaining procedures to install the oil pan. Torque the engine mount nuts to 31–34 ft. lbs. Refill the oil pan with the required amount and type of oil.

750iL

1. Disconnect the negative battery cable. Raise and safely support the vehicle.
2. Remove the transmission and the oil pump assembly. Lower the vehicle.
3. Disconnect and remove the windshield washer tank and the coolant expansion tank.
4. Remove the guide tube for the oil dipstick. Disconnect the oil pipe on the tandem pump. Remove the mounting bracket.
5. Unscrew the belt tensioner and remove the oil drain hose.
6. Crank the engine to TDC and unscrew the flywheel using the proper tool.
7. Disconnect the left and right engine mounts at the bottom. Pull off the pipe adapter for oil extraction.

NOTE: Heat the oil pan with a hot air blower to make removing and installing easier.

8. Remove the oil pump consoles. Unscrew the oil pan bolts and remove the oil pan.

To install

9. Clean the mounting surfaces and install a new gasket.
10. Install the oil pan and tighten the mounting bolts to 7 ft. lbs.
11. Connect the left and right engine mounts at the bottom and tighten to 32.5 ft. lbs.
12. Replace the oil consoles and tighten to 25 ft. lbs.
13. Install the flywheel and tighten the bolts to 72 ft. lbs.
14. The remainder of the installation is the reverse of the removal procedure.

Oil Pump

REMOVAL & INSTALLATION

1988-90 525i, 535i, 635CSi, 735i and 750iL

1. Disconnect the negative battery cable and remove the oil pan.
2. Remove the bolts retaining the sprocket to the oil pump shaft and remove the sprocket.
3. Remove the oil pump retaining bolts and lower the oil pump from the engine block. On 6 cyl. engines other than the M30B35, there are 3 bolts at the front and 2 bolts attaching the rear of the oil pickup to the lower end of a support bracket. It is necessary to remove all 5 bolts. On the M30B35, there are only 3 bolts.
4. Do not loosen the chain adjusting shims from the 2 mounting locations.
5. Add or subtract shims between the oil pump body and the engine block to obtain a slight movement of the chain under light thumb pressure.
6. Install the oil pump in position.

NOTE: When used, the 2 shim thicknesses must be the same. Tighten the pump holder at the pick-up end after shimming is completed to avoid stress on the pump.

7. On 6 cylinder engines, other than the M30B35, after the main pump mounting bolts are torqued, loosen the bolts at the bracket on the rear of the pick-up, allowing the pick-up to assume its most natural position. This will relieve tension on the bracket. Tighten the bolts. On the M30B35, torque the oil pump mounting bolts to 16 ft. lbs. and the sprocket bolts to 19 ft. lbs.

318iS

1. Disconnect the negative battery cable.
2. Raise and safely support the vehicle. Drain the engine oil.
3. Remove the timing case cover.
4. Disconnect the oil pump cover mounting bolts and remove the oil pump assembly.
5. Reverse the removal procedure for installation.

325, 325i, 325iS, 325iX, M3 and 528e

1. Raise an safely support the vehicle. Drain the engine oil.
2. Remove the front lower splash guard.
3. Disconnect the electrical terminal from the oil sending unit.
4. Remove the flywheel cover.
5. Remove the oil pan bolts and lower the oil pan. Remove the oil pump bolts and take out the oil pump and oil pan.

To Install:

6. Installation is the reverse of removal. Note the following:
   a. Clean the gasket surfaces and use a new gasket on the oil pan.
   b. Positioning the pump for installation of its mounting bolts, guide the pump driveshaft into the hole in the center of the drive gear.
   c. Coat the joints on the ends of the front engine cover with a universal sealing compound.
   d. Install the sending unit wire and the engine oil.

1991–92 525i

1. Raise and safely support vehicle. Disconnect the negative battery cable. Drain the oil from the engine. Remove the oil pan to access the oil pump drive sprocket.
2. Remove the oil pump drive sprocket nut. Note that it is a left hand thread. Remove the oil pump drive sprocket from the oil pump shaft. Check the shaft splines.
3. Unbolt the oil pump body from the block and remove. Check the condition of the dowel sleeves.

4. Installation is the reverse of removal. Torque the pump mounting bolts to 16 ft. lbs.(22 Nm) and the sprocket nut to 18 ft. lbs.(25 Nm).

CHECKING

318iS

1. Disconnect the negative battery cable.
2. Remove the timing case cover.
3. Measure the play between the pump body and the outer rotor; outer rotor and inner rotor. The distance should be 0.12– 0.20 mm (0.005–0.008 in.).
4. Replace the timing cover assembly and connect the negative battery cable.

Rear Main Bearing Oil Seal

REMOVAL & INSTALLATION

The rear main bearing oil seal can be replaced after the transmission, and clutch/flywheel or the converter/flywheel has been removed from the engine.

Removal and installation, after the seal is exposed, is as follows.

1. Raise and safely support the vehicle. Drain the engine oil and loosen the oil pan bolts. Carefully use a sharp bladed tool to separate the oil pan gasket from the lower surface of the end cover housing.
2. Remove the 2 rear oil pan bolts.
3. Remove the bolts around the outside of the cover housing and remove the end cover housing from the engine block. Remove the gasket from the block surface.
4. Remove the seal from the housing. Coat the sealing lips of the new seal with oil. Install a new seal into the end cover housing with a special seal installer tool. On the 6 cylinder engines, press the seal in until it is about 0.039–0.079 in. deeper than the standard seal, which was installed flush.
5. While the cover is off, check the plug in the rear end of the main oil gallery. If the plug shows signs of leakage, replace it with another, coating it with the proper sealant to keep it in place.

NOTE: Fill the cavity between the sealing lips of the seal with grease before installing.

6. Coat the mating surface between the oil pan and end cover with sealer. Using a new gasket, install the end cover on the engine block and bolt it into place.
7. Complete the installation. If the oil pan gasket has been damaged, replace it. Install the transmission.

ENGINE COOLING

Radiator

REMOVAL & INSTALLATION

1. Disconnect the negative battery cable. Drain the cooling system. On some engines, this requires removing the plug from the bottom radiator tank.
2. If equipped with a coolant expansion tank, remove the cap, disconnect the hose at the radiator and drain the coolant into a clean container. If equipped with a splash guard, remove it.
3. Remove the coolant hoses and disconnect the automatic transmission oil cooler lines and plug their openings as well as the openings in the cooler.
4. Disconnect any of the temperature switch wire connectors.
5. Remove the shroud from the radiator. On some vehicles, this is done by simply pressing plugs toward the rear of the vehicle. On others, there are metal slips that must be pulled upward and off to free the shroud from the radiator. The shroud will remain in the vehicle, resting on the fan on most vehicles. On the 735i and 735iL, remove the fan and shroud together. Make sure to store the fan in a vertical position. The fan must be held stationary with some sort of flat blade cut to fit over the hub and drilled to fit over 2 of the studs on the front of the pulley. Then, unscrew the retaining nut at the center of the fluid drive hub turning it clockwise to remove it because it has left hand threads.
6. If equipped with the M30 B35 engine, remove the fan and shroud; then, spread the retaining clip and pull the oil cooler out to the right. Remove the radiator retaining bolt(s) and lift the radiator from the vehicle.
7. The radiator is installed in the reverse order of removal. Fill and bleed the cooling system.

NOTE: On the M3, there are rubber washers that go on either side of the mounting brackets at the top and that the bottom of the unit is suspended by rubber bushings into which prongs located on the bottom tank will fit. Make sure all parts fit right when the unit is installed.

8. Check that rubber mounts are located so as to effectively isolate the radiator from the chassis, as this will help ensure reliable radiator performance. Note that if the vehicle uses plastic upper and lower radiator tanks and has a radiator drain plug, be careful not to over torque the plugs.
9. Torque engine oil cooler pipes to 18–21 ft. lbs. and transmission cooler pipes to 13–15 ft. lbs.
10. Torque the thermostatic fan hub on the 735i and 735iL to 29–36 ft. lbs.

Heater Core

REMOVAL & INSTALLATION

325, 325i, 325iS, 325iX and M3

1. Disconnect the negative battery cable. Remove the package tray. Remove bolts and remove the left/lower dish trim panel.
2. Drain the coolant, loosen the bolt and remove the clamp bracing the 2 lines going to the heater core.
3. Remove the left side duct carrying air from the heater to the rear seat duct.
4. Unscrew the bolts and remove the lower heater discharge duct.
5. Unscrew the bolts fastening the water lines from the engine compartment to the lines coming down from the heater core. Remove and discard the O-ring seals.
6. Unscrew the bolts, separate the halves of the core housing, and pull the core out of the housing.
7. Installation is the reverse of removal. Replace the O-ring seals for the water lines.

528e, 635CSi, M5 and M6

1. Disconnect the negative battery cable. Remove the instrument panel trim at bottom left. Remove the package tray.
2. Carefully, discharge the air conditioning system through the Schrader® valve, and then cap the valve off.
3. Remove the bolts and remove the trim panel underneath the evaporator unit.
4. Remove the tape type insulation. Get caps for the refrigerant lines. Using a backup wrench, disconnect the low and high pressure lines and cap them.
5. Disconnect the electrical connector for the evaporator. Disconnect the temperature sensor plug, accessible from the outside of the evaporator housing.
6. Remove the bolts and then remove the bracket that braces the housing at the firewall. Remove the mounting bolt from either side of the housing.
7. Unclip both fasteners and remove the housing.
8. Move into the engine compart-

ment and remove the rubber insulator from the cowl.

9. Remove the mounting bolts for the cover which is located under the windshield.

10. Remove the mounting nuts for the heater housing located on either side of the blower.

11. Drain the cooling system and disconnect the hoses at the core.

12. Working inside the vehicle, remove the 3 electrical connectors for the heater housing. Pull off the air ducts.

13. Remove the mounting nuts and remove the heater unit.

14. Remove the air duct connections from the housing. Push the retaining bar back and then split and remove the blower shells.

15. Remove the retaining clips from the housing halves and split the housing. Then, remove the core.

To install:

16. To install, reverse the removal procedure, noting the following points:

a. Cement a new rubber seal on the core.

b. Make sure when reassembling the halves of the housing, all the distributor door flap shafts pass through the holes in the housing.

c. Before reconnecting the refrigerant lines, coat the threads with clean refrigerant oil.

d. Refill the cooling system with clean coolant and bleed it.

e. Properly, evacuate and recharge the air conditioning system.

525i, 535i, 735i, 735iL and 750iL

1. Disconnect the negative battery cable. Drain coolant from the cooling system. Remove the center console.

2. Remove the mounting bolts and remove the right core mounting bracket. Lift out the front blower motor on the 7 series.

3. Remove the core cover screws. Loosen the wire straps and clips and remove the cover.

4. Disconnect the temperature sensor(s).

5. Unscrew the mounting bolts and lift out the heater pipes. Replace the O-rings. Then, lift out the core from the right side.

6. Install in reverse order. Refill and bleed the cooling system.

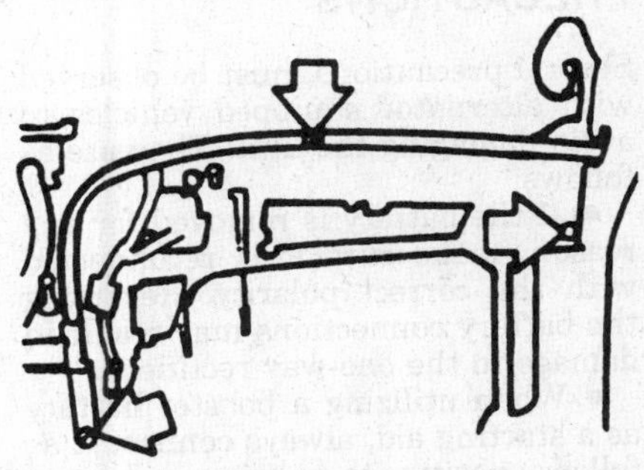

Remove the 2 arrowed screws to remove the heater core—1987-88 735i

Water Pump

REMOVAL & INSTALLATION

525i, 535i, 635CSi, 735i and 750iL

1. Disconnect the negative battery cable. Drain the cooling system.

2. Remove the fan cowl and fan, if necessary.

3. Remove the drive belt and the pulley. Disconnect the bracket, if necessary.

4. Remove the air cleaner with the air flow sensor, if needed.

5. Disconnect the cooling hoses and remove the water pump.

6. The installation is the reverse of the removal procedure. Torque the M8 bolts to 16 ft. lbs. (22 Nm) and the M6 bolts to 6.5 ft. lbs. (9 Nm).

325, 325i, 325iS, 325iX, M3 and 528e

1. Disconnect the negative battery cable. Drain the cooling system.

2. Remove the distributor cap and rotor. Remove the inner distributor cap and rubber sealing ring.

3. The fan must be held stationary with some sort of flat blade cut to fit over the hub and drilled to fit over 2 of the studs on the front of the pulley or use the proper tool. Remove the fan coupling nut; left hand thread—turn clockwise to remove.

4. Remove the belt and pulley.

5. Remove the rubber guard and distributor and or upper timing belt cover.

6. Compress the timing tensioner spring and clamp pin with the proper tool.

NOTE: Observe the installed position of the tensioner spring pin on the water pump housing for reinstallation purposes.

7. Remove the water hoses, remove the 3 water pump bolts and remove the pump.

8. Clean the gasket surfaces and use a new gasket.

9. Install the water pump in position. Note the position of the tensioner spring pin. Torque the M8 bolts to 16 ft. lbs. (22 Nm) and the M6 bolts to 6.5 ft. lbs. (9 Nm).

10. Add coolant and bleed the cooling system.

318iS

1. Disconnect the negative battery cable. Drain the cooling system.

2. Remove the drive belt and the water pump pulley.

3. Remove the pump mounting bolts.

4. Screw 2 bolts into the tapped bores and press the water pump out of the cover uniformly.

5. Lubricate and install a new O-ring.

6. Install the water pump and tighten the bolts to 6.5 ft. lbs.

7. The remainder of the installation is the reverse of the removal procedure.

Thermostat

REMOVAL & INSTALLATION

The thermostat is located near the water pump, either on the cylinder head or intake manifold on some vehicles and is located between 2 coolant hose sections on some vehicles.

Always drain some coolant out and save it in a clean container before removing the thermostat. On the M5, M6 engine, the forward (removable) portion of the housing has a hose connected to it. The hose need not be disconnected to remove the housing. On the engine used in the M5, M6, there is a large O-ring seal for the main portion of the housing and a small, O-ring located above it in a small passage. The M20B25 and MB20B27 engines also use a large O-ring which must be replaced with the thermostat. Replace both these seals on all vehicles.

Note that thermostats for M30B35 engines carry an "A" designation. The thermostat for M20B27 and M20B25 engines is smaller in diameter. On all vehicles, except M3, the thermostat is installed with the thermostatic sensing unit facing inward and the crossband facing outward. Refill and bleed the cooling system.

On the M3, the thermostat is installed in a coolant lines with a 3rd connection that goes to the block. To replace it, first drain coolant and then note the routing of hoses. Loosen all 3 hose clamps and then replace the unit. Refill and bleed the system.

Cooling System Bleeding

With Bleeder Screw On Thermostat Housing

Set the heat valve in the **WARM** position, start the engine and bring it to normal operating temperature. Run the engine at fast idle and open the venting screw on the thermostat housing until the coolant comes out free of air bubbles. Close the bleeder screw and refill the cooling system.

Without Bleeder Screw

Fill the cooling system, place the heater valve in the **WARM** position, close the pressure cap to the second (fully closed) position. Start the engine and bring to normal operating temperature. Carefully release the pressure cap to the first position and squeeze the upper and lower radiator hoses in a pumping action to allow trapped air to escape through the radiator. Recheck the coolant level and close the pressure cap to its second position.

ENGINE ELECTRICAL

NOTE: Disconnecting the negative battery cable on some vehicles may interfere with the functions of the on board computer systems and may require the computer to undergo a relearning process, once the negative battery cable is reconnected.

Distributor

REMOVAL

4 Cylinder Engines Except M3 and 318iS

1. Disconnect the negative battery cable. On all engines so equipped, remove the weather-proof rubber cap protecting the distributor cap and wires from moisture. Prior to removal, using paint, chalk or a sharp instrument, scribe alignment marks showing the relative position of the distributor body to its mount on the rear of the cylinder head.
2. Mark each spark plug wire with a dab of paint or chalk noting its respective cylinder. It will be easier and faster to install the distributor and get the firing order right if the plug wires are left in the cap.
3. Pull up and disconnect the secondary wire (high tension cable leading from the coil to the center of the distributor cap), and remove the spark plug loom retaining nut(s) from the cylinder head cover. Disconnect the vacuum line(s) from the vacuum advance unit.
4. Disconnect the primary wire (low tension wire running from one of the coil terminals to the side of the distributor) at the distributor. On electronic ignition distributors, disconnect the plug.
5. Disconnect the distributor retaining clasps and lift off the cap and wire assembly. On all engines equipped with a dust cap under the rotor, remove the rotor, remove the dust cap and reinstall the rotor.

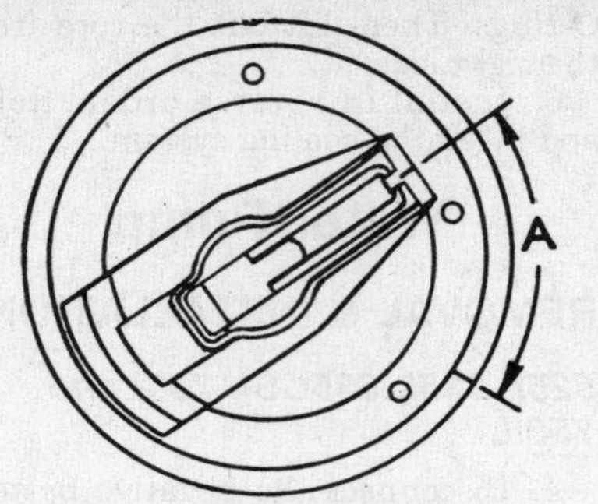

Distance (A) rotor from the housing mark during the removal of the electronic distributor

6. Now, with the aid of a remote starter switch, "bump" the starter a few times until the No. 1 piston is at TDC of its compression stroke. At this time, the notch scribed on the metal tip of the distributor rotor must be aligned with a corresponding notch scribed on the distributor case. Before removing the distributor, make sure these 2 marks coincide.
7. Loosen the clamp bolt at the base of the distributor (where it slides into its mount) and lift the distributor up and out. Notice that the rotor turns clockwise as the distributor is removed. This is because the distributor is gear driven and must be compensated for during installation.

INSTALLATION

Timing Not Disturbed

EXCEPT 6 CYLINDER ENGINES, M3 AND 318iS

1. To install the distributor, position it in the block. Remember to rotate the rotor approximately 1.4 in. counterclockwise from the notch scribed in the distributor body. This will ensure that when the distributor is fully seated in its mount, the marks will coincide. Adjust the ignition timing. Tighten the clamp bolt to 8.0 ft. lbs.
2. Reinstall the cap and wires.
3. Connect the negative battery cable.

EXCEPT 6 CYLINDER ENGINES, M3 AND 318iS

Timing Disturbed

Sometimes, the engine is accidentally turned over while the distributor is removed; in this case, it will be necessary to find TDC position for No. 1 cylinder before installing the distributor. Check the exact position of the crankshaft via the timing marks on the flywheel or front pulley, and obtain exact alignment as indicated by them. Then, proceed to install the distributor.

6 CYLINDER ENGINES, M3 AND 318iS VEHICLES

NOTE: Most 6 cylinders and the 4 cylinder engine used on the M3 and the 318iS use a distributor which is contained within the engine. Other than distributor cap and rotor removal and installation, no service is possible.

Ignition Timing

ADJUSTMENT

1. If equipped with the Motronic control unit, the timing can be checked; however, timing cannot be adjusted. The only cure for improper timing is to replace the control unit. Also, timing must be within a specified range, as the computer changes the timing slightly to allow for various changes in operating condition. In other words, the timing does not have to be right on, but anywhere within the specified range.
2. The engine should be at normal operating temperature and the operation should be performed at normal room temperatures. The engine RPM should be within the specified range under the control of the computer.
3. Look up the control unit number on the unit itself. On 3, 5, and 6 Series vehicles, the unit is in the glove box; on the 7 Series, it is in the right side speaker cutout.
4. Connect a tachometer and a timing light to the engine (the latter to the No. 1 cylinder). Start the engine and check the rpm. If it is not correct, check the idle speed and reset it as necessary. Then, operate the timing light to see if timing is within range. If it is significantly outside the range, the Motronic control unit must be replaced.

Alternator

PRECAUTIONS

Several precautions must be observed with alternator equipped vehicles to avoid damaging the unit. They are as follows:

- If the battery is removed for any reason, make sure it is reconnected with the correct polarity. Reversing the battery connections may result in damage to the one-way rectifiers.
- When utilizing a booster battery as a starting aid, always connect it as follows: positive to positive, and negative (booster battery) to a good ground on the engine.

• Never use a fast charger as a booster to start vehicles with alternating-current (AC) circuits.

• When servicing the battery with a fast charger, always disconnect the battery cables.

• Never attempt to polarize an alternator.

• Avoid long soldering times when replacing diodes or transistors. Prolonged heat is damaging to alternators.

• Do not use test lamps of more than 12 volts for checking diode continuity.

• Do not short across or ground any of the terminals on the alternator.

• The polarity of the battery, alternator, and regulator must be matched and considered before making any electrical connections within the system.

• Never operate the alternator on an open circuit. Make sure all connections within the circuit are clean and tight.

• Turn off the ignition switch and then disconnect the battery terminals when performing any service on the electrical system or charging the battery.

• Disconnect the battery ground cable if arc welding is to be done on any part of the vehicle.

BELT TENSION ADJUSTMENT

The fan belt tension is adjusted by moving the alternator on the slack adjuster bracket. The belt tension is adjusted to a deflection of approximately ½ in. under moderate thumb pressure in the middle of its longest span. On many engines, the position of the top of the alternator is adjusted via a bolt that is geared to the bracket. This bolt is turned to position the alternator and determine tension, and then is locked in position with a lock bolt.

REMOVAL &INSTALLATION

1. Disconnect the negative battery cable.
2. Disconnect the wires from the rear of the alternator, marking them for later installation. Note that there is a ground wire on some vehicles. On the 735i and 735iL, remove the cap and then disconnect the positive terminal at the junction box on the fender well. On the 325, M3, 635CSi, 735i and 735iL, it may be easier to remove the alternator mounting bolts, turn it, and then remove the wires.

On the M5 and M6:

a. Unscrew the nut and loosen the hose clamp. Pull of the plug. Then, lift out the air cleaner and airflow sensor.

b. Make sure the engine is cool. Place a pan under the radiator and disconnect the lower radiator hose.

c. On the 325, 525, M3 and 528e, remove the airflow sensor and air cleaner, if necessary.

d. On the 735i and 735iL, make sure the engine is cool. Place a pan under the radiator and disconnect the lower radiator hose.

3. Loosen the adjusting and pivot bolts, and remove the belt on those vehicles with a standard mounting system. If the alternator has the tensioning bolt described in Step 4, loosen the lock bolt, turn the tensioning bolt so as to eliminate belt tension and then remove the belt. Remove the bolts and remove the alternator. On the 635CSi, 735i and 735iL, it may be necessary to loosen the fan cowl to get at the mounting bolts. On the 535i and 525i, it may be necessary to disconnect a power steering line that runs near the alternator.

To install:

4. Install the alternator in position and install the retaining bolts.
5. Adjust the belt tension to approximately ⅜ in., measured between the balancer and the alternator pulley.
6. The tensioning bolt on the front of the alternator must be turned so as to tension the belt, using a torque wrench, until the torque is approximately 5 ft. lbs. Then, hold the adjustment with one wrench while tightening the locknut at the rear of the unit. Make sure, if the unit has a ground wire on the alternator, it has been reconnected. On the M5 and M6, 735i and 735iL, make sure to reconnect the radiator hose, refill and bleed the cooling system. On the M5, M6 and 528e securely reinstall the air cleaner and airflow sensor. On the 528e, if the power steering line had to be disconnected, reconnect it securely, refill, and bleed the system.

Starter

REMOVAL & INSTALLATION

1. Disconnect the negative battery cable.
2. On 6 cylinder engines with 6 identical intake tubes, it may be necessary to remove No. 6 intake tube for clearance. On 4 cylinder vehicles, remove the intake cowl from the mixture control unit.

On the 325 and the M3:

a. Remove the air cleaner and airflow sensor. Then, remove the mounting bolts for the bracket for the air collector and remove it.

On the 528e:

a. Remove the air cleaner and airflow sensor.

b. Disconnect the electrical leads. Remove the 3 bolts and remove the mounting bracket.

On the 535i and 525i:

a. Make sure the engine is cool. Drain some coolant from the cooling system and then remove the expansion tank.

On the 635CSi:

a. Make sure the engine is cool and drain some coolant out. Disconnect the heater hose that is near the starter.

b. Operate the brake pedal hard 20 times. Disconnect the power steering line that would otherwise prevent access to the starter.

c. Cut off the straps and remove the solenoid switch insulating cover, located right near the solenoid.

On the M5 and M6:

a. Remove the exhaust manifold.

b. Cut off the straps and remove the solenoid switch insulating cover, located right near the solenoid.

3. Remove the starter solenoid wire leads, marking them for later installation, unless they have already been removed. On 4 cylinder vehicles, disconnect the mounting bracket at the block. On the 325 and the M3, drain coolant out of the engine and then disconnect the heater hose located near the starter; also unscrew and remove the coolant pipe if necessary for clearance.

NOTE: Remove the accelerator cable holder on automatic transmission equipped vehicles.

4. Unbolt and remove the starter.

NOTE: On the 525i, 535i, 735i, 735iL, M5 and M6, it may be necessary to use a box wrench with an angled handle to unscrew the main starter mounting bolts. On the 528e and 635CSi, the final mounting bolt must be removed from underneath. On the 325 and M3, the starter must be pulled out from above. On the M3 remove the intake manifold, if necessary.

To install:

5. Install the starter and install the retaining bolts. Install all removed components on all vehicles.
6. Make sure to reconnect all hoses and refill and bleed the cooling system or power steering system.
7. Where the solenoid switch cover has been unstrapped, reinstall it with new straps to locate it properly for electrical safety.

EMISSION CONTROLS

Please refer to "Emission Controls" in the Unit Repair section for system maintenance procedures. Due to the complex nature of modern electronic engine control systems, comprehensive diagnosis and testing procedures fall outside the confines of this repair manual. For complete information on diagnosis, testing and repair procedures concerning all modern engine and emission control systems, please refer to "Chilton's Guide to Fuel Injection and Electronic Engine Controls".

Emission Warning Lamps

RESETTING

Service Interval Reminder Lights

The on board computer is used to evaluate mileage, average engine speed and engine and coolant temperatures as well as other computer input factors that determine maintenance intervals. There are 5 green, a yellow and a red **LED** used to remind the driver of oil changes and other maintenance services.

The green LEDS will be illuminated when the ignition is in the **ON** position and the engine OFF. There will not be as many green LEDS illuminated when as the maintenance time get closer. A yellow LED that is illuminated when the engine is running, will indicate maintenance is now due. The red LED will be illuminated when the service interval has been exceeded by approximately 1000 miles. This is the computers way of saying this is your last warning.

There is a service interval reset tool manufactured by the Assenmacher Tool Company tool 62-1-100. This tool is used to reset BMW 6 cylinder vehicles and the 4 cylinder vehicles. With the aid of an additional adapter tool can also be used on 1988–91 vehicles.

1. Locate the diagnostic connector near the thermostat housing.
2. Plug the special reset tool into the diagnostic connector and place the ignition switch in the **ON** position.
3. Depress the reset button on the tool until all 5 green LEDS are illuminated, showing that the reset has completed.

FUEL SYSTEM

Fuel System Service Precautions

Safety is the most important factor when performing not only fuel system maintenance but any type of maintenance. Failure to conduct maintenance and repairs in a safe manner may result in serious personal injury or death. Maintenance and testing of the vehicle's fuel system components can be accomplished safely and effectively by adhering to the following rules and guidelines.

- To avoid the possibility of fire and personal injury, always disconnect the negative battery cable unless the repair or test procedure requires that battery voltage be applied.
- Always relieve the fuel system pressure prior to disconnecting any fuel system component (injector, fuel rail, pressure regulator, etc.), fitting or fuel line connection. Exercise extreme caution whenever relieving fuel system pressure to avoid exposing skin, face and eyes to fuel spray. Fuel under pressure may penetrate the skin or any part of the body that it contacts.
- Always place a shop towel or cloth around the fitting or connection prior to loosening to absorb any excess fuel due to spillage. Ensure that all fuel spillage (should it occur) is quickly removed from engine surfaces. Ensure that all fuel soaked cloths or towels are deposited into a suitable waste container.
- Always keep a dry chemical (Class B) fire extinguisher near the work area.
- Do not allow fuel spray or fuel vapors to come into contact with a spark or open flame.
- Always use a backup wrench when loosening and tightening fuel line connection fittings. This will prevent unnecessary stress and torsion to fuel line piping. Always follow the proper torque specifications.
- Always replace worn fuel fitting O-rings with new. Do not substitute fuel hose or equivalent where fuel pipe is installed.

RELIEVING FUEL SYSTEM PRESSURE

To relieve the pressure in the system, first find the fuel pump relay plug, located on the cowl. Unplug the relay, leaving it in a safe position where the connections cannot ground. If necessary, tape the plug in place or tape over the connector prongs with electrical tape. Then, start the engine and operate it until it stalls. Crank the engine for 10 seconds after it stalls to remove any residual pressure.

Fuel Filter

REMOVAL & INSTALLATION

Except M5 and M6

On filters that are located near the fuel tank, it is necessary to clamp the fuel lines closed before disconnecting them, or fuel will run out continuously.

1. Disconnect the negative battery cable. Relieve fuel system pressure. Clamp the lines closed if the filter is mounted low, near the fuel tank. Then, loosen the clamps and disconnect the inlet and outlet hoses. Remove the hose clamps or slide them back, well off the connections to make it easier to pull off the hoses, if necessary.
2. The filters will usually be attached to a frame, floor pan or wheel well by a bracket. Loosen the bracket and remove the filter. On the 735 with the M30 B35 engine, remove the Phillips head screw clamping the filter inside the mounting band. Note the direction of flow and then remove the filter.
3. Observe the instructions on the inlet and outlet during installation.

M5 and M6

1. Relieve fuel system pressure. Disconnect the battery cables. Working under the fuel tank, pull back the protective caps and then unscrew the attaching nuts and pull off the electrical connections for the fuel pump.
2. Pinch off the inlet line to the fuel pump and the outlet from the filter. Then, loosen the clamps and disconnect these 2 hoses.
3. Remove the nut that clamps the fuel line near the pump. Then, remove the bolts which mount the pump and filter to the bottom of the body and remove the assembly.
4. Remove the bolt fastening the halves of the bracket together and remove the filter from the bracket. Loosen the clamp on the inlet side of the filter and disconnect the inlet line, noting the direction of flow (arrow). Remove the rubber bushing in which the filter is mounted, and mount it on the new filter.
5. Install the filter in exact reverse order, making sure all clamps are securely tightened. Operate the engine and check for leaks.

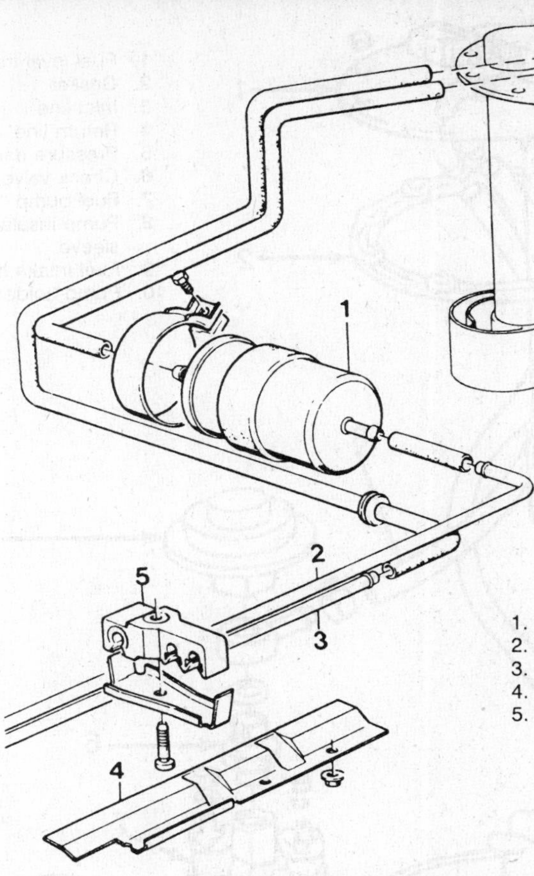

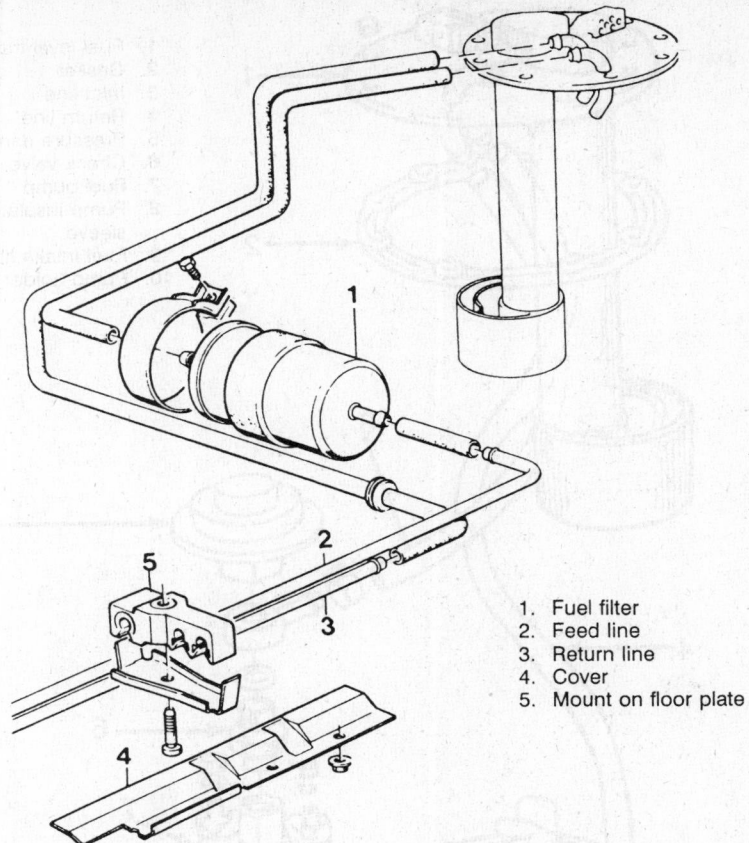

1. Fuel filter
2. Feed line
3. Return line
4. Cover
5. Mount on floor plate

Fuel filter location – 525i and 535i

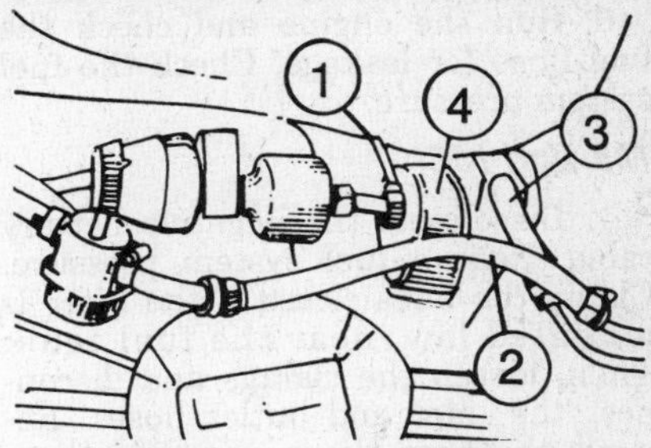

1. Suction line
2. Pressure line
3. Return line
4. Damper

Note the arrangement of the fuel pump, lines and damper – 1988–89 325 series

Electric Fuel Pump

PRESSURE TESTING

318iX, 325, 325i, 325iS and 325iX

1. Relieve fuel system pressure. Tee a pressure gauge into the fuel feed line in front of the pressure regulator.
2. Disconnect the fuel pump relay.
3. Connect a remote starter switch between terminals KL30 and KL87 of the relay. Close the switch and check the pressure. It should be 43 psi. If not, the filter is severely clogged or the fuel pump is defective.

M3

1. Relieve fuel system pressure. Tee a pressure gauge into the fuel return line at the pressure regulator. Then, clamp off the return line so pressure builds up to the maximum level the pump can produce.
2. Remove the trim from the cowl on the right (passenger's side). Then, unplug the fuel pump relay. Connect a remote starter switch between terminals 30 and 87 (left side and top holding the male side of the connector). Energize the switch and check the pressure. It must be 43 psi. Check the filter for excessive clogging. If it is okay, the pump is defective.

525, 525i, 535i, 635CSi, 735i, 735iL, M5 and M6

1. Relieve fuel system pressure. Tee a pressure gauge into the fuel feed line in front of the pressure regulator on M5, M6, tee in between the cold start valve and the fuel rail. Plug the fuel return hose.
2. Pull off the pump relay. Jumper terminals 87 and 30. Measure the delivery pressure. It should be 43 psi on vehicles except the 735i and 735iL. On these vehicles, it should be 48 psi.

REMOVAL & INSTALLATION

325, 325i, 325iS, 325iX and M3

1. Relieve fuel system pressure. Disconnect the negative battery cable. Going to the pump, which is under the vehicle and near the fuel tank, push back any protective caps, note the routing and disconnect the electrical connector(s).
2. Securely clamp the suction hose (coming from the tank) and plug the discharge hose so no fuel can escape.
3. Open the hose clamp connecting the suction hose to the pump and disconnect it.
4. Remove the attaching nuts which mount the pump and bracket to the floor pan and remove both as an assembly.
5. Remove the bolt passing through the 2 parts of the bracket and also mounting the hose attaching strap to the bracket. Then, pull the pump out of the bracket.
6. Loosen the hose clamp for the discharge hose and disconnect it at the pump. Pull the rubber ring off the pump.
7. Note the code number on the pump and make sure to replace it with one of the same number. Inspect all the rubber mounts on the pump mounting bracket and replace any that are cracked or crushed.
8. Install the pump in reverse order. Make sure to unclamp the hoses and then run the engine and check for leaks. Check the fuel system pressure.

All 5 Series Except M5

The fuel pump is an electrical unit, delivering fuel through a pressure regulator, to a fuel distributor or a ringline for the injection valves. The fuel pump is mounted under the vehicle, in the fuel tank, or in the engine compartment.

1. Relieve fuel system pressure. Disconnect the negative battery connector. Push back any protective caps and disconnect the electrical connector(s).
2. If the fuel lines are flexible, pinch them closed with an appropriate tool. Disconnect the fuel lines and plug the ends.
3. Remove the retaining bolts and remove the pump and expansion tank as an assembly.
4. The pump can be separated from the expansion tank after removal.
5. Install the pump in the correct position, be sure to use similar types of hose clamps, if any need replacing. The wrong type clamp can damage the pressure lines.

6. Run the engine and check the fuel lines for leakage. Check the fuel system pressure.

M5 and M6

1. Disconnect the negative battery cable. Relieve fuel system pressure. Clamp the lines closed if the filter is mounted low, near the fuel tank. Then, loosen the clamps and disconnect the inlet and outlet hoses. Remove the hose clamps or slide them back, well off the connections to make it easier to pull off the hoses, if necessary.
2. The filters will usually be attached to a frame, floor pan or wheel well by a bracket. Loosen the bracket and remove the filter. Note the direction of flow and then remove the filter.
3. Observe the instructions on the inlet and outlet during installation.
4. Remove the bolt fastening the halves of the bracket together and remove the filter from the bracket. Loosen the clamp on the outlet side of the fuel pump and disconnect the line. Then, slide off the rubber bushing in which the pump is mounted.
5. Check the code number on the side of the pump and make sure the replacement unit carries the same code.
6. Install the pump in exact reverse order, making sure all clamps are securely tightened. Operate the engine and check for leaks.

635CSi, 735i, 735iL and 750iL

The pump on this vehicle is mounted in the top of the tank along with the fuel level sending unit.

1. Disconnect the negative battery cable. Drain the fuel tank, enough to prevent spillage when removing the pump.
2. Relieve fuel system pressure. Remove the trim panels from the trunk. Then, remove the screws from the cover for the pump/sending unit assembly.
3. Label the fuel hoses connecting at the top of the pump/sending unit assembly. Unclamp and disconnect the fuel hoses and then plug them.
4. Slide the collar for the electrical connector to one side and then unplug the connector.
5. Remove the mounting screws and remove the pump/sending unit assembly. Replace the gasket.
6. Press the retaining locks for the pump unit inward and slide the pump out of the pump/sending unit assembly.
7. Note the routing of the fuel and electrical lines to the pump from the top of the pump/sending unit assembly. Loosen the hose clamp screws and the screws attaching the electrical connectors to the pump. Disconnect the hose and connector.

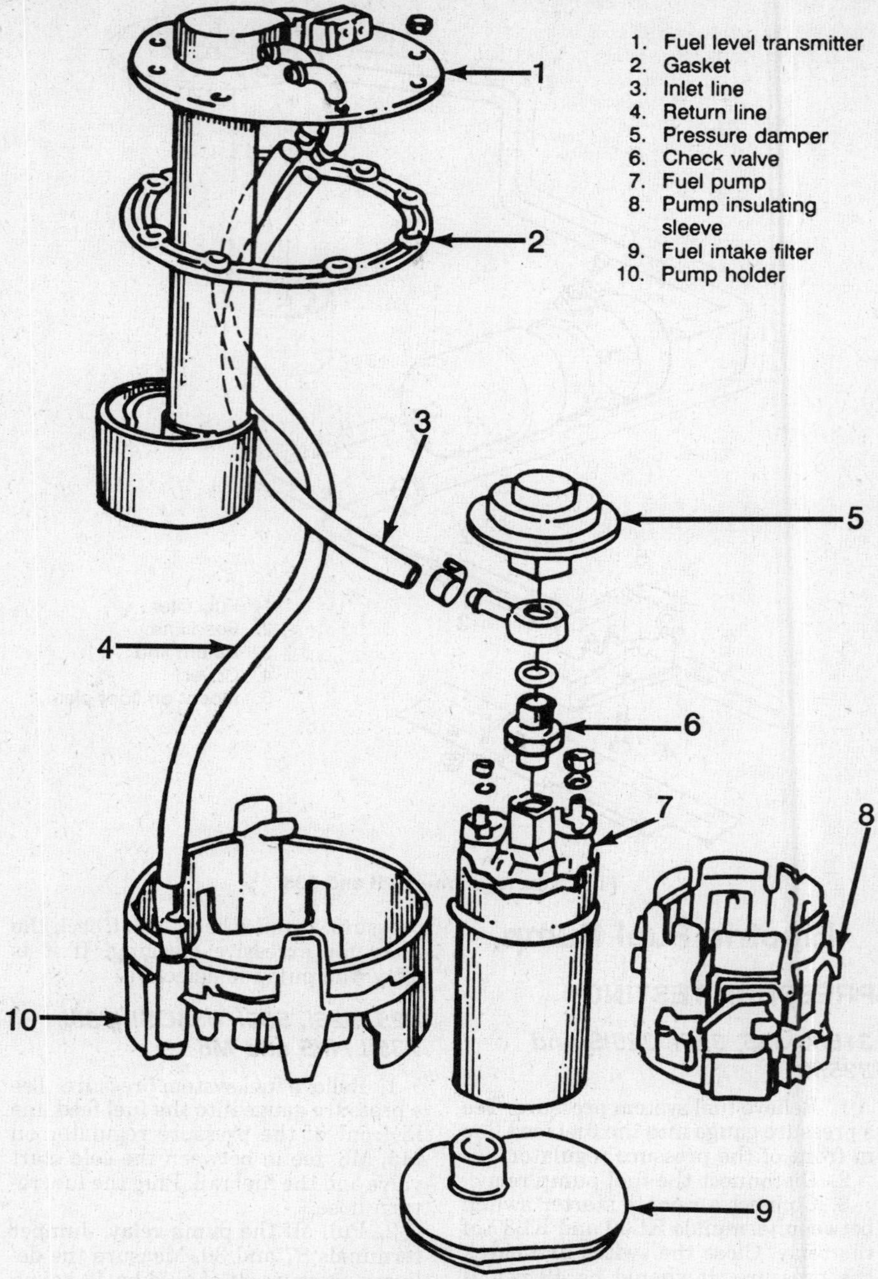

View of the in-tank fuel pump—735i and 735iL

8. Unscrew the pressure regulator from the top of the check valve. Then, unscrew the check valve from the top of the pump.
9. Pull the insulating sleeve off the pump. Then, loosen the retaining screw and slide the filter off the pump.
10. Install the pump in reverse order. Be careful to ensure that the 2 retaining locks fasten the pump in place in a secure manner. Operate the engine and check for leaks.

Fuel Injection

IDLE SPEED AND MIXTURE ADJUSTMENT

NOTE: The idle speed and mixture can be adjusted ONLY with the aid of a CO meter. If this tool is not available, do not attempt any of the following procedures. The idle mixture can be adjusted

ONLY with the aid of a CO meter on most vehicles. Idle speed is not adjustable on any vehicle with the Motronic control unit except the M5 and M6. If idle speed is incorrect, either the idle valve or the idle control unit must be replaced. See the fuel injection unit repair section.

325, 325i and 325iS

1. Disconnect the hose, leading from the throttle housing, that goes to the carbon canister. Do not plug the openings. Remove the bolts on either side of the exhaust manifold plug.
2. Remove the plug in the exhaust manifold, install the test nipple part No. 13 0 100 or equivalent and connect the CO tester 13 0 070 or equivalent, into the open nipple.
3. With the engine valve clearances correctly adjusted, ignition timing correct and the engine at operating temperature, measure the CO percentage at idle speed. The CO nominal value is 0.2–1.2 percent.
4. If the CO level is within the specified range, disconnect the test unit, replace the plug in the exhaust manifold, and conclude the test. If not, adjust the CO as described below.
5. Turn off the engine and then unplug the oxygen sensor plug. Drill a hole in the anti-tamper plug in the throttle body with special tool 13 1092 or equivalent. Then screw the extractor tool 13 1 094 or equivalent into the hole drilled into the plug and draw the plug out. Finally, use an adjustment tool 13 1 060 or 13 1 100 or equivalent to turn the adjustment, with the engine running, until the CO meets nominal values.
6. When the adjustment is complete, install a new anti-tamper plug, and reconnect the oxygen sensor plug and the carbon canister hose. Also, remove the nipple in the exhaust manifold and replace the plug. Reinstall the exhaust manifold bolts.

528e

1. Pull the canister purge hose off the solenoid, leave it unplugged.
2. Connect the CO meter 13 0 070 or equivalent, to the manifold via the nipple 13 0 100.
3. With the engine valve clearances correctly adjusted, ignition timing correct and the engine at operating temperature, measure the CO percentage at idle speed. CO nominal value is 0.2–1.2 percent.
4. If the CO level is within the specified range, disconnect the test unit, replace the plug in the exhaust manifold, and conclude the test. If not, adjust the CO as described below.
5. Turn off the engine and then unplug the oxygen sensor plug. Drill a hole in the anti-tamper plug in the throttle body with special tool No. 13 1 092 or equivalent. Then screw the extractor tool 13 1 094 or equivalent into the hole drilled into the plug and draw the plug out with the impact mass. Finally, use an adjustment tool 13 1 060 or 13 1 100 or equivalent to turn the adjustment, with the engine running, until the CO meets nominal values.
6. When the adjustment is complete, install a new anti-tamper plug, and reconnect the oxygen sensor plug and the carbon canister hose. Also, remove the nipple in the exhaust manifold and replace the plug. Reconnect the hose to the solenoid.

525i, 535i, 635CSi, 735i, 735iL and 750iL

1. Make sure the idle speed is correct. The engine must be hot. Disconnect the evaporative emissions canister purge hose at the bottom of the solenoid mounted on the firewall. Leave the openings unplugged.
2. Unscrew the bolts on the exhaust manifold and install a nipple, part No. 13 0 100 or equivalent and connect the CO test unit 13 0 070 or equivalent. CO should be 0.2–1.2 percent. If CO is not within limits, adjust it as described below.
3. Turn off the engine and unplug the oxygen sensor plug. Then, remove the air flow sensor by removing the air cleaner and removing the mounting bolts to separate the airflow sensor from it.
4. Use special tool 13 1 092 or equivalent to remove the anti-tamper plug. Use this tool to drill a hole in the plug and then use 13 1 094 or equivalent to pull it. The second tool should be screwed into the hole already drilled; use the slide hammer to pull the plug out.
5. Once the plug is removed, install the air flow sensor back onto the air cleaner and reinstall the air cleaner. With the engine idling hot and the oxygen sensor plug still disconnected, measure the CO and adjust it with tool 13 1 060 or 13 1 100 or equivalent. The CO level must meet the nominal value of 0.2–1.2 percent.
6. Once the level is adjusted, stop the engine and reconnect the oxygen sensor plug. Then, remove the air flow sensor, put it on a bench, and install a new anti-tamper plug. Reinstall the airflow sensor and air cleaner. Reconnect the canister purge hose to the solenoid.

M3

NOTE: This test must be performed at essentially sea level altitude. In an area well above sea level, it will be necessary to use a BMW Service tester or equivalent device from another source and run the test with the system's altitude correction box connected.

1. Make sure the engine is at operating temperature and that the air cleaner is in reasonably clean condition. All basic engine tuning factors (spark plug condition and gap, valve adjustment, ignition timing, etc.) must be correct. Turn off all accessories.
2. A special electrical fitting special tool 13 4 010 or equivalent is required to disable the Motronic control system's throttle valve switch. Pull off the electrical connector leading to the throttle valve switch. Then, plug the special tool into the open end of the connector.
3. Adjust the idle speed to specification by turning the screw located just above the "M" on the valve cover Make sure to restore the throttle valve switch when the idle speed is correct.
4. Make sure the engine is in good basic tune, including proper spark plug gap and valve clearances. The engine must be at operating temperature.

On vehicles with no test openings in the exhaust manifold:

a. With a CO meter in the tail pipe, disconnect the oxygen sensor so it no longer influences the mixture produced by the injection system. CO level should be zero. If there is a CO reading, it is necessary to get a special adjusting screw cap remover tool 13 1 011 or equivalent and an adjusting tool 13 1 100 or equivalent.

b. The adjusting screw cap is located on the top surface of the airflow sensor; the adjusting screw is accessible in the aperture underneath. Remove the cap with the cap remover and then turn the adjusting screw slowly to correct the CO value to 0. Reinstall the cap and reconnect the oxygen sensor.

On vehicles with a test opening in the exhaust manifold:

a. Remove the plug in the exhaust manifold and put the probe of the CO meter into the opening. Disconnect the oxygen sensor so it no longer influences the mixture produced by the injection system. CO level should be 0.8–1.2%. If the CO reading is outside the nominal value range, it is necessary to get a special adjusting screw cap remover tool 13 1 011 or equivalent and an adjusting tool 13 1 100 or equivalent.

b. The adjusting screw cap is located on the top surface of the airflow sensor; the adjusting screw is accessible in the aperture underneath. Remove the cap with the cap remover and then turn the adjusting screw slowly to correct the CO value to .8–1.2%. Reinstall the cap and reconnect the oxygen sensor.

M5 and M6

1. Make sure the engine is at operating temperature, and that the air cleaner is in reasonably clean condition. All basic engine tuning factors (spark plug condition and gap, valve adjustment, ignition timing, etc.) must be correct.
2. Adjust the idle speed by turning the screw shown in the illustration.
3. To adjust CO, first remove the cap located at the center of the top surface of the airflow sensor. Use tool 13 1 100 to turn the airflow control screw in the airflow sensor, accessible after the anti-tamper cap is removed. CO must be 0.4–1.2 percent.
4. Install a new cap when CO meets specification.

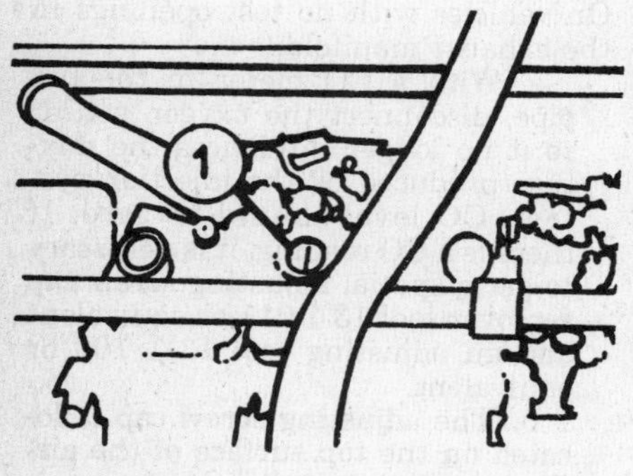

On the M5 and M6 engines, adjust the idle speed by turning this screw (1)

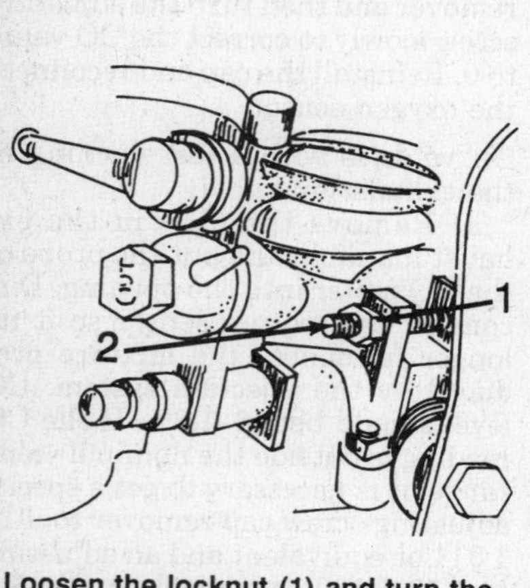

Loosen the locknut (1) and turn the adjusting screw (2) to adjust the idle on the diesel engine

Fuel Injector

REMOVAL & INSTALLATION

Except 318iS and 1991–992 525i

1. Disconnect the negative battery cable. Remove the hose and plastic caps, if necessary.
2. Remove the plugs, plate and screws from the injector pipe.
3. Push up on the injector pipe until the injectors have cleared the guides on the intake manifold or the throttle valve housing.
4. Pull of the plugs and lift up the retainer. Remove the injectors.
5. Check the O-ring and replace, if necessary. Install in the reverse order of the removal. Start the engine and check for leaks.

318iS

1. Disconnect the negative battery cable. Remove the upper section of the collector.
2. Disconnect the rear brace and remove the injector hose.
3. Remove the holder for the preheater.
4. Remove the screws and lift off the upper section of the intake manifold. Disconnect the hose from the fuel pressure regulator at the same time.
5. Remove the plug plate from the fuel injectors and unscrew the clamp.
6. Remove the injection pipe with the injectors attached. Lift off the retainer and remove the fuel injector.
7. Check the O-ring and replace, if necessary. Install in the reverse order of the removal. Start the engine and check for leaks.

1991–92 525i

1. Relieve the fuel pressure and disconnect the negative battery cable.
2. Remove the injector valley cover and disconnect and remove the plug plate.
3. Pull off the vacuum hose from the pressure regulator. Disconnect the fuel feed and reurn hoses from the fuel rail.
4. Unscrew the fasteners holding the fuel rail and remove the fuel rail along with the injectors.
5. Pull off the retainers and remove the injectors. If replacing an injector use the correct style and code number for the application.
6. Inspect the O-rings and replace as necessary. Replace the injectors on the fuel rail and fit the retainers in place.
7. Lubricate the O-rings with vaseline or gear oil prior to installation and replace the fuel rail. The remainder of the installation is reverse of removal.

DRIVE AXLE

Halfshaft

REMOVAL & INSTALLATION

Front

325IX

1. Raise and safely support the vehicle. Remove the front wheels. Remove the drain plug and drain the lube oil from the front axle.
2. Lift out the lock plate in the center of the brake disc with a small prybar. Then, unscrew the collar nut.
3. Remove the attaching nut from each tie rod and then press the rod off the steering knuckle with the proper tool.
4. Remove the retaining nut and then press the control arm off the steering knuckle on either side.
5. Mount the proper tool to the brake disc with 2 wheel bolts. Press the output shaft out of the center of the steering knuckle on that side. Repeat on the other side.
6. To remove the drive axle from the differential on the left side: Install special tool by bolting it together around the axle so the ring on its inner diameter fits into the groove on the shaft. Install the tool onto the shaft so it will rest against the housing and the bolt heads of the tool will rest against it. Screw the 2 bolts in alternately in small increments to get even pressure on the shaft, pulling it out of the differential.
7. To remove the drive axle on the right side: Install the proper tool on the diameter of the shaft directly against the housing. Install the tool by bolting it together around the axle so the ring on its inner diameter fits into the groove on the shaft. Screw the 2 bolts in alternately in small increments to get even pressure on the shaft, pulling it out of the differential.

To install:

8. Install the halfshafts, bearing the following points in mind:
   a. Install the shafts into the housing until the circlip inside engages in the groove of the shaft. It may be necessary to install the removal tool and tap against it with a plastic-headed hammer to drive the shaft far enough into the housing.
   b. Before installing the shafts into the steering knuckle, coat the spline with light oil.
   c. When installing the control arms onto the steering knuckle, torque the nut to 61.5 ft. lbs. and use a new cotter pin. When installing the tie rod onto the steering

knuckle, torque to 61.5 ft. lbs. and use a new self-locking nut.

d. Drive a new lock plate into the brake disc with the proper tool. Torque the nut to 181 ft. lbs.

e. Replace the drain plug and refill the final drive unit with the required lubricant.

Rear

318IS, 325, 325I, 325IS, 325IX AND M3

1. Raise and safely support the vehicle. Remove the rear tire and wheel assembly.
2. Lift out the lock plate and remove the retaining nut from the output flange. Remove the flange.
3. Disconnect the output shaft from the final drive by pressing out with the proper tool and suspend it.
4. Pull out the output shaft with a special tool.
5. Drive out the rear axle shaft with the proper tool.
6. Lift out the snapring. Then, pull out the wheel bearings, using the proper tool.
7. Pull out the seal with the proper tool.
8. If the inner bearing shell is damaged, pull it off with a puller and thrust pad.

To install:

9. To install, pull in the wheel bearing assembly, pull in the seal, insert the circlip and then pull in the rear axle shaft, all in reverse of the removal procedure. Install the axle shaft seal.
10. To install the output shaft, screw the threaded spindle into the shaft all the way, and then use the nut and washer against the outside of the bridge. Reconnect the output shaft to the final drive.
11. Lubricate the bearing surface of the outer nut with oil. Then install and torque the nut.
12. Using the proper installers or equivalent, knock in the lock plate. Use the following torque figures:

Output shaft to drive flange, 42–46 ft. lbs.

Drive flange hub to output shaft,140–152 ft. lbs.

525I, 525IS, 535I, M5, 635CSI, M6, 735I, 735IL AND 750IL

1. Raise and safely support the vehicle.
2. Unscrew the output shaft on the rear drive assembly and the axle shaft end.
3. Remove the output shaft.
4. Install new washers on the rear drive assembly and the axle shaft end.
5. Install the output shaft and tighten to 42–46 ft. lbs.

CV-Boot

REMOVAL & INSTALLATION

Front

325IX

1. Raise and safely support the vehicle. Remove the front wheel assembly.
2. Drain the gear lube and unscrew the collar nut.
3. Disconnect the tie rod ends and the control arm with the proper tool.
4. Remove the output shaft using the proper tool.
5. Loosen both clamps and pull the dust boot off the joint. Clean and remove grease from the joint. Add the required amount of the proper grease and install the boot and tighten the clamps.
6. The remainder of the installation is the reverse of the removal procedure.

Rear

1. Raise and safely support the vehicle.
2. Remove the output shaft and the snapring. Press off the cap and the dust cover.
3. Press the output shaft out of the CV-joint. Clean and remove the grease from the splines of the joint.
4. Install the dust cover with the inside cover on the output shaft.
5. Press on the joint with the cap and install the snapring. Pack the joint and dust cover with the proper grease.
6. Mount he dust cover and install new clamps.
7. Install the output shaft and lower the vehicle.

Driveshaft and U-Joints

REMOVAL & INSTALLATION

318iS, 325, 325i, 325iS and M3

1. Raise and safely support the vehicle. Remove the mufflers. Unscrew and remove the exhaust system heat shield near the fuel tank.
2. Unbolt and remove the cross brace that runs under the driveshaft.
3. Support the transmission. The automatic transmission must be supported by the case and not the pan. Loosen all transmission support bolts and remove. Remove the transmission rear support crossmember.
4. Lower the manual transmission for clearance. Remove the driveshaft bolts from the front coupling.

NOTE: Make sure the drive axle does not rest on the fuel line that runs across under it.

5. Unscrew and remove bolts at the coupling near the final drive.
6. Loosen the threaded sleeve on the driveshaft with a tool such as tool 26 1 040 or equivalent. Unbolt and remove the center mount.
7. Bend the driveshaft downward and remove it, being careful not to allow it to rest on the connecting line on the fuel tank.

To install:

8. Upon installation:

a. Mount the holder for the oxygen sensor plug.

b. Make sure the heat shield clears the fuel tank.

c. Wherever self-locking nuts are used, replace them. On the transmission-end flange, tighten the nuts/bolts only on the flange side, holding the other end stationary.

d. Preload the center mount to 0.157–0.236 in. before tightening the bolts. Torque the mounting bolts to 16 ft. lbs.

e. Lubricate the center bearing with the proper lubricant.

f. Make sure to reinstall the bracket for the oxygen sensor plug.

g. Make sure there is sufficient clearance between the rear heat shield and fuel tank.

325iX

NOTE: Never drive the vehicle with either driveshaft disconnected. This could damage the lockup mechanism in the transfer case.

1. Raise and safely support the vehicle. Remove the exhaust system. Remove both exhaust system heat shields.
2. Loosen the threaded sleeve near the front of the driveshaft with the proper tool. Turn the sleeve several turns outward, but do not disconnect it entirely.
3. Disconnect the driveshaft at the output flange of the transfer case by removing the nuts and the through bolts.
4. Disconnect the driveshaft at the final drive by removing the nuts and the through bolts.

NOTE: Make sure the drive axle does not rest on the fuel line that runs across under it.

5. Slide the sections of the driveshaft together and slide it out of the centering pin on the output flange of the transfer case. Remove it from the vehicle.

To install:

6. Install the driveshaft in reverse order, keeping these points in mind:

a. Whenever self-locking nuts are used, replace them.

b. Hold the nut or bolt in place where it runs through a U-joint and tighten at the opposite end, where the driveshaft flange is located.

c. Check the center bearing for lubrication and if it's dry, lubricate with the proper lubricant.

528e, 525i and 535i

1. Raise and safely support the vehicle.
2. Support the transmission from underneath with the special tools and a floorjack. Remove the nuts and washers from the transmission mounts on top of the rear transmission mounting crossmember. Loosen but do not remove the nuts located underneath which fasten the crossmember to the body. Then, slide this crossmember as far to the rear as it will go.
3. Unscrew the fastening nuts on the forward end of the CV-joint and then discard them.
4. Using a prybar to keep the driveshaft from turning, remove the self locking nuts and bolts fastening the rear of the driveshaft to the final drive.
5. Remove the bolts fastening the center mount to the body. Bend the propshaft down and pull the CV-joint off the transmission flange. Cover the joint to keep it clean.

To install:

6. Replace the gasket that fits between the joint bolts. Install in reverse order, keeping these points in mind:
   a. Replace the self-locking nuts used at either end of the shaft.
   b. Preload the center mount forward by forcing the bracket 0.157–0.197 in. forward from the neutral position.

635CSi

1. Raise and safely support the vehicle. Remove the exhaust system.
3. Remove the heat shield near the fuel tank, if so equipped.
4. Disconnect the driveshaft at the transmission by removing the nuts and bolts from the flexible coupling. If the vehicle has a vibration damper where the shaft connects to the transmission, turn the damper 60 degrees counterclockwise and remove it with the rubber coupling.
5. Loosen the center bearing bolts and remove them.
6. Disconnect the driveshaft at the final drive.
7. Bend the driveshaft down and pull out.

To install:

8. Installation is the reverse of removal.
9. The driveshaft is balanced as an assembly and must only be renewed as a complete assembly.
10. Preload the center bearing in the forward direction to 0.157–0.236 in.
11. Wherever self-locking nuts are used, replace them. Hold the nut or bolt in place where it runs through a U-joint, and torque at the opposite end – where the driveshaft flange is located. Check the center bearing for lubrication and if it's dry, lubricate with the proper lubricant.

735i, 735iL and 750iL

NOTE: If the vehicle has a front universal joint, use tools 24 0 120 and 00 2 020 or equivalent to support the transmission during this operation.

1. Raise and safely support the vehicle. Remove the exhaust system. Remove the heat shield from the floorpan. Remove the nuts and bolts fastening the driveshaft to the transmission at the flexible coupling. Replace the self-locking nuts.
2. If equipped with a front U-joint, support the transmission from underneath with the proper tools. When the transmission is securely supported, remove the 6 bolts and remove the rear transmission mounting crossmember.
3. Remove the self-locking nuts and then the bolts fastening the driveshaft to the final drive. Replace the self-locking nuts. Remove the driveshaft, taking care to keep it protected from dirt.
4. Remove the bolts from the crossbrace and remove the center driveshaft mount. Then, bend the shaft at the middle and remove it from the vehicle by pulling it off the centering pin on the forward end.

To install:

5. Install in reverse order, keeping the following points in mind:
   a. Repack the CV-joint with approved grease and replace the gasket, if necessary.
   b. Check the center bearing for lubrication and if it's dry, lubricate with the proper lubricant.
   c. If the vibration damper at the forward end of the driveshaft must be replaced, turn it 60 degrees to remove it.
   d. When remounting the center mount, preload it forward from its most natural position 0.157–0.236 in.
   e. Torque U-joint bolts to 52 ft. lbs. and CV-joint bolts to 51 ft. lbs.

Front Wheel Hub, Knuckle and Bearings

REMOVAL & INSTALLATION

Except 735i, 735iL and 750iL

1. Raise and safely support the vehicle. Remove the front tire and wheel assembly.

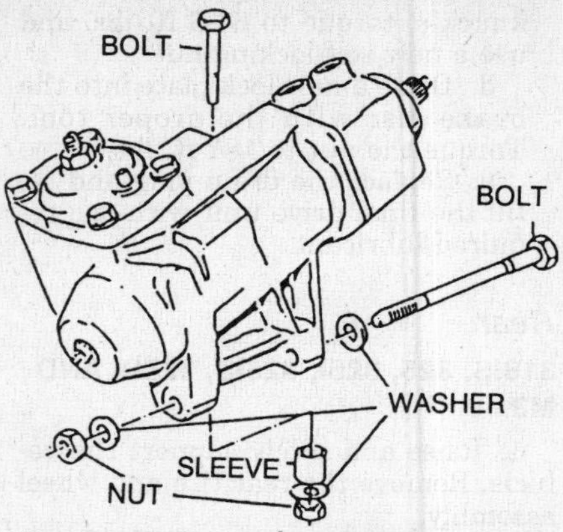

Power steering gear assembly – 5 and 7 Series vehicles

2. Disconnect and suspend the brake caliper from the body without disconnecting the brake line.
2. Remove the setscrew with an Allen® wrench. Pull off the brake disc and pry off the dust cover with a small prybar.
3. Using a chisel, knock the tab on the collar nut away from the shaft. Unscrew and discard the nut.
4. Pull off the bearing with the proper puller set and discard it. On the M3, use a puller set such as 31 2 102/105/106 or equivalent. On the M3, install the main bracket of the puller with wheel bolts.
5. If the inside bearing inner race remains on the stub axle, unbolt and remove the dust guard. Bend back the inner dust guard and pull the inner race off with a special tool capable of getting under the race. Reinstall the dust guard.

NOTE: Do not reuse the bearing unit if removed.

6. If the dust guard has been removed, install a new one. Install a special tool on M3, over the stub axle and screw it in for the entire length of the guide sleeve's threads. Press the bearing on.
7. Reverse the remaining removal procedures to install the disc and caliper. Torque the wheel hub collar nut to 188 ft. lbs. Lock the collar nut by bending over the tab.

735i, 735iL and 750iL

1. Raise and safely support the vehicle. Remove the front tire and wheel assemblies. Remove the attaching bolts and remove and suspend the brake caliper, hanging it from the body so as to avoid putting stress on the brake line.
2. Remove the setscrew with an Allen® wrench. Pull off the brake disc and pry off the dust cover with a small prybar.

3. Using a chisel, knock the tab on the collar nut away from the shaft. Unscrew and discard the nut.

4. Install a puller collar such as 31 2 105 or equivalent to the bearing housing with 3 bolts. Install a puller such as 31 2 102 and 312 2 106 or equivalent and pull off the bearing and discard it.

5. If the inside bearing inner race remains on the stub axle, unscrew and remove the dust guard, using a socket extension. Bend back the inner dust guard and pull the inner race off with a special tool capable of getting under the race. Reinstall the dust guard and install a new dust cover.

6. Then install a special tool over the stub axle and screw it in for the entire length of the guide sleeve's threads. Slide the bearing on and follow it with 31 2 100 or equivalent, and use this tool to press the bearing on.

7. Reverse the remaining removal procedures to install the disc and caliper. Torque the wheel hub collar nut to 210 ft. lbs. Lock the collar nut by bending over the tab.

8. Install a new grease cap coated with the proper lubricant.

Rear Axle Shaft, Bearing and Seal

REMOVAL & INSTALLATION

318iS, 325, 325i, 325iS, 325iX and M3

1. Raise and safely support the vehicle. Remove the rear tire and wheel assembly.

2. Lift out the lock plate and remove the retaining nut from the output flange. Remove the flange.

3. Disconnect the output shaft from the final drive by pressing out with the proper tool and suspend it.

4. Pull out the output shaft with a special tool.

5. Drive out the rear axle shaft with the proper tool.

6. Lift out the snapring. Then, pull out the wheel bearings, using the proper tool.

7. Pull out the seal with a tool such as 33 4 045 or equivalent.

8. If the inner bearing shell is damaged, pull it off with a puller and thrust pad.

To install:

9. Using an appropriate bearing installer, pull in the wheel bearing assembly, pull in the seal, insert the snapring and then pull in the rear axle shaft, all in reverse of the removal procedure. Install the axle shaft seal.

10. To install the output shaft, screw the threaded spindle into the shaft all the way, and then use the nut and washer against the outside of the bridge.

10. Reconnect the output shaft to the final drive.

11. Lubricate the bearing surface of the outer nut with oil. Then install and torque the nut.

12. Using the proper installers or equivalent, knock in the lock plate. Use the following torque figures:

Output shaft to drive flange,42–46 ft. lbs.

Drive flange hub to output shaft,140–152 ft. lbs.

318iS, 325, 325i, 325iS, 325iX and M3

1. Raise and safely support the vehicle. Remove the rear tire and wheel assembly.

2. Disconnect the output shaft at the outer flange and suspend it with wire.

3. Unbolt the caliper and suspend it with the brake line connected. Unbolt and remove the rear disc.

4. Remove the large nut and remove the lock plate. If equipped with ABS, disconnect and then remove the ABS speed sensor by unscrewing it.

5. Unscrew the collar nut. Then, pull off the drive flange with the proper tool(s).

6. Screw on the collar nut until it is just flush with the end of the shaft and use a suitable hammer to knock out the shaft.

7. Remove the snapring. Pull out the wheel bearings, using the proper tool.

8. Pull the inner bearing race off the axle shaft with tool 00 7 500 or equivalent.

To install:

9. Install the new bearing assembly using the proper tools. Then, reinstall the snapring.

10. Install the rear axle shaft with special tools 23 1 300, 33 4 080 and 33 4 020 or equivalent.

11. Install the collar nut and drive in the lock plate with the proper tool(s).

12. Reconnect the output shaft. Remount the brake disc and caliper.

13. Lower the vehicle.

Axle Housing

REMOVAL & INSTALLATION

1. Disconnect the negative battery cable.

2. Raise and safely support the vehicle. Disconnect and remove the rear exhaust.

3. Remove the heat shield, if equipped. Remove the driveshaft and disconnect the center support mount, if necessary.

4. Disconnect the parking brake lever cables. Disconnect and plug the rear brake lines.

5. Support the rear axle housing with a suitable jack device. Disconnect the thrust struts on both sides, if necessary.

6. Disconnect and plug the pulse senders and remove the rear seat cushion, if needed.

7. Remove the rear side trim panel on 3 series convertible.

8. Disconnect the wires at the speedometer pulse sender. Unscrew the rear rubber mounting bolt.

9. Unplug the rear brake pad indicator and disconnect the wire connection.

10. Disconnect the switching valve for the ride control height and the plug on the camber warning sender on the 750iL.

11. Disconnect both rear shock absorbers at the trailing arms and move the brake cables out of the way.

12. Lower the rear axle assembly, using a suitable lifting device.

To install:

13. Install the rear axle assembly, lifting into position with a suitable lifting device.

14. Connect the shock absorbers to the trailing arms. Tighten the bolts when the vehicle is in the normal position to 52–63 ft. lbs. on 3 Series or 90–144 ft. lbs. except 3 Series.

15. Connect the switching valve for the ride control height and the plug on the camber warning sender on the 750iL.

16. Connect the rear brake pad indicator plug and the wire connection.

17. Connect the wires at the speedometer pulse sender. Replace the rear rubber mounting bolt and tighten to 31–35 ft. lbs. except on the 7 series and 36–38 ft. lbs. on the 7 series.

18. Replace the rear side trim panel on 3 series convertible.

19. Connect and plug the pulse senders and replace the rear seat cushion, if needed.

20. Connect the thrust struts on both sides, if needed. Remove the jacking device from under the rear axle assembly.

21. Connect the parking brake cables and the brake lines.

22. Replace the heat shield, if equipped. Replace the driveshaft and connect the center support mount, if necessary.

23. Replace the rear exhaust. Lower the vehicle.

24. Connect the negative battery cable.

MANUAL TRANSMISSION

For further information on transmissions/transaxles, please refer to "Chilton's Guide to Transmission Repair".

Transmission Assembly

REMOVAL & INSTALLATION

318iS, 325, 325i, 325iS, 325iX and M3

1. Disconnect the negative battery cable. Raise and safely support the vehicle. Remove the exhaust system. Remove the cross brace and heat shield. On the 325iX, remove the transfer case.
2. Hold the nuts on the front with one wrench, and remove bolts from the rear with another to disconnect the flexible coupling at the front of the driveshaft. Some vehicles have a vibration damper at this point in the drivetrain. This damper is mounted on the transmission output flange with bolts that are pressed into the damper. On these vehicles, unscrew and remove the nuts located behind the damper.
3. Loosen the threaded sleeve on the driveshaft. Get a special tool to hold the splined portion of the shaft while turning the sleeve.
4. Remove its mounting bolts and remove the center driveshaft mount. Then, bend the driveshaft down at the center and pull it off the transmission output flange. Keep the sections of the driveshaft from pulling apart and suspend it from the vehicle with wire.
5. Remove the retainer and washer, and pull out the shift selector rod.
6. Use a hex-head wrench to remove the self-locking bolts that retain the shift rod bracket at the rear of the transmission and then remove the bracket. If the vehicle has a shift arm, use a screwdriver to pry the spring clip up off the boss on the transmission case and swing it upward. Then, pull out the shift shaft pin.
7. Unscrew and remove the clutch slave cylinder and support it so the hydraulic line can remain connected.
8. The transmission incorporates sending units for flywheel rotating speed and position. Remove the heat shield that protects these from exhaust heat and then remove the retaining bolt for each sending unit. Note that the speed sending unit, which has no identifying ring goes in the bore on the right, and that the reference mark sending unit, which has a marking ring, goes in the bore on the left. If the sending units are installed in reverse positions, the engine will not run at all. Pull these units out of the flywheel housing.
9. Disconnect the wiring connector going to the backup light switch and pull the wires out of the harness.
10. Support the transmission from underneath in a secure manner. Remove mounting bolts and remove the crossmember holding the rear of the transmission to the body. Then, lower the transmission onto the front axle carrier.
11. Using the proper tool, remove the bolts holding the transmission flywheel housing to the engine at the front. Make sure to retain the washers with the bolts. Pull the transmission rearward to slide the input shaft out of the clutch disc and then lower the transmission and remove from the vehicle.

To install:

12. Install the transmission in position under the vehicle. Align the input shaft and install the transmission, note the following points:
   a. Coat the input shaft splines and flywheel housing guide pins with a light coating of suitable grease.
   b. Make sure the front mounting bolts are installed with their washers. Torque them to 46–58 ft. lbs.
   c. Before reinstalling the sending units for flywheel position and speed, make sure their faces are free of either grease or dirt and then coat them with a light coating of a suitable lubricant. Inspect the O-rings and replace them if they are cut, cracked, crushed, or stretched.
   d. When installing the shift rod bracket at the rear of the transmission, use new self-locking bolts and make sure the bracket is level before tightening them. Torque the shift rod bracket bolts to 16.5 ft. lbs. except on the M3, which uses an aluminum bracket. On the M3, torque these bolts to 8 ft. lbs.
   e. Install the clutch slave cylinder with the bleed screw downward.
   f. When installing the driveshaft center bearing, preload it forward 0.157–0.236 in. Check the driveshaft alignment with an appropriate tool such. Replace the nuts and then torque the center mount bolts to 16–17 ft. lbs.
   g. Torque the flexible coupling bolts to 83–94 ft. lbs.

525i, 535i, 635CSi, M5 and M6

1. Disconnect the negative battery cable. Raise and safely support the vehicle. Disconnect and lower the exhaust system to provide clearance for transmission removal. Remove the heat shield brace and transmission heat shield.
2. Support the driveshaft and then unscrew the driveshaft coupling at the rear of the transmission. Use a wrench on both the nut and the bolt.
3. Working at the front of the driveshaft center bearing, unscrew the screw-on type ring type connector which attaches the driveshaft to the center bearing. Then, unbolt the center bearing mount. Bend the driveshaft down and pull it off the centering pin. If the vehicle has a vibration damper, turn it and pull it back over the output flange before pulling the driveshaft off the guide pin. Suspend it from the vehicle.
4. Pull off the wires for the backup light switch. Unscrew the passenger compartment console to disconnect it from the top of the transmission by removing the self-locking bolts. Discard and replace.
5. Pull out the locking clip, and disconnect the shift rod at the rear of the transmission. Take care to keep all the washers.
6. If the transmission is linked to the shift lever with an arm, use a small prybar to lift the spring out of the holder on the bracket and then raise the arm. Pull out the shift shaft bolt.
7. If the vehicle has a flywheel housing cover (semi-circular in shape), remove the mounting bolts and remove the cover.
8. The speed sensor and reference mark sensor on the flywheel housing must be disconnected. Note their locations. The speed sensor goes in the upper bore, marked D. The reference mark sensor, which has a ring, goes in the lower bore, marked B. Check the O-rings for the sensors and install new ones if they are damaged.
9. Support the transmission securely. Then, unbolt and remove the rear transmission crossmember.
10. Remove the upper and lower attaching nuts and remove the clutch slave cylinder, supporting it so the hydraulic line need not be disconnected. Disconnect the reverse gear backup light switch and pull the wires out of the holders.
11. Unscrew the bolts fastening the transmission to the bell housing, using the proper tool. On some vehicles there are Torx® bolts used ; use a Torx® wrench for these. Pull the transmission rearward until the input shaft has disengaged from the clutch disc and then lower and remove it.

To install:

12. Place the transmission in gear. Insert the guide sleeve of the input shaft into the clutch pilot bearing carefully. Turn the output shaft to rotate

the front of the input shaft until the splines line up and it engages the clutch disc.

13. Perform the remaining portions of the procedure in reverse of removal, observing the following points:

a. Make sure the arrows on the rear crossmember point forward.

b. Preload the center bearing mount forward of its most natural position 0.079–0.157 in. On 7 Series vehicles with the M30 B35 engine only, and 6 Series vehicles with the 265/6 transmission (no integral clutch housing), preload the bearing 0.157–0.236 in.

c. In tightening the driveshaft screw on ring, use tool 26 1 040 or equivalent.

d. When reconnecting the nuts and bolt at the transmission coupling, replace the nuts with new ones and turn only the nut, holding the bolts stationary.

e. Make sure DME sensor faces are clean. Coat the sensor outside diameters with the proper lubricant.

f. If equipped with a shift arm, lubricate the bolt with a light layer of a suitable lubricant.

g. Observe these torque figures:

Transmission-to-bell housing–52–58 ft. lbs.

Rear/top transmission Torx® bolts–46–58 ft. lbs.

Center mount-to-body–16–17 ft. lbs.

Front joint-to-transmission–83–94 ft. lbs.

735i and 735iL

1. Disconnect the negative battery cable. Raise and safely support the vehicle. Remove the exhaust system. Remove the attaching bolts and remove the heat shield mounted just to the rear of the transmission on the floorpan.

2. Support the transmission securely from underneath. Then, remove the crossmember that supports it at the rear from the body by removing the mounting bolts on both sides.

3. Using wrenches on both the bolt heads and on the nuts, remove the bolts passing through the vibration damper and front universal joint at the front of the driveshaft.

4. Remove its mounting bolts and remove the center driveshaft mount. Then, bend the driveshaft down at the center and pull it off the transmission output flange. Keep the sections of the driveshaft from pulling apart and suspend it from the vehicle with wire.

5. Pull out the circlip, slide off the washer, and then pull the shift selector rod off the transmission shift shaft. Disconnect the backup light switch.

6. Lower the transmission slightly for access. Then, use a small prybar to lift the spring out of the holder on the bracket and then raise the arm. Pull out the shift shaft bolt.

7. Remove the upper and lower attaching nuts and remove the clutch slave cylinder, supporting it so the hydraulic line need not be disconnected.

8. Unscrew the bolts fastening the transmission to the bell housing. Use a Torx® wrench to remove the bolts. Make sure to retain the washer with each bolt to ensure that they can be readily removed later, if necessary. Pull the transmission rearward until the input shaft has disengaged from the clutch disc and then lower and remove the transmission.

To install:

9. Install the transmission in position under the vehicle. Align the input shaft and install the transmission. Follow these procedures:

a. Preload the center bearing mount forward of its most natural position 0.157–0.236 in.

b. When reconnecting the nuts and bolt at the transmission coupling, replace the nuts with new ones and turn only the nut, holding the bolts stationary.

c. When reconnecting the shift arm, if equipped, lubricate the bolt with a light layer of a suitable lubricant and check the O-ring for crushing, cracks or cuts, replacing it, if damaged.

d. When installing the clutch slave cylinder, make sure the bleeder screw faces downward.

e. Observe these torque figures:

Center mount to body–16–17 ft. lbs.

Front joint-to-transmission–58.5 ft. lbs.

CLUTCH

Clutch Assembly

REMOVAL & INSTALLATION

1. Disconnect the negative battery cable. Raise and safely support the vehicle. Remove the heat shield and then the mounting bolts. Disconnect the speed and reference mark sensors at the flywheel housing. Mark the plugs for reinstallation.

2. Remove the transmission and clutch housing.

3. On vehicles with 6 cylinder engines, a Torx® socket is required. If equipped with a 265/6 transmission (without an integral clutch housing), remove the clutch housing.

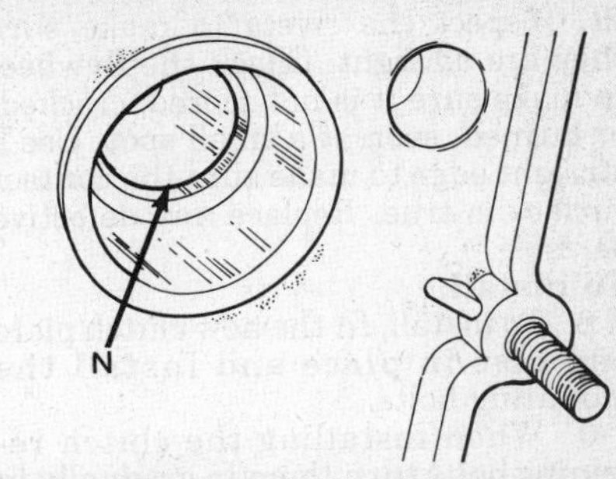

Pack the groove ("N") with the proper grease before installing the transmission on the M6, 635CSi and the 735i models

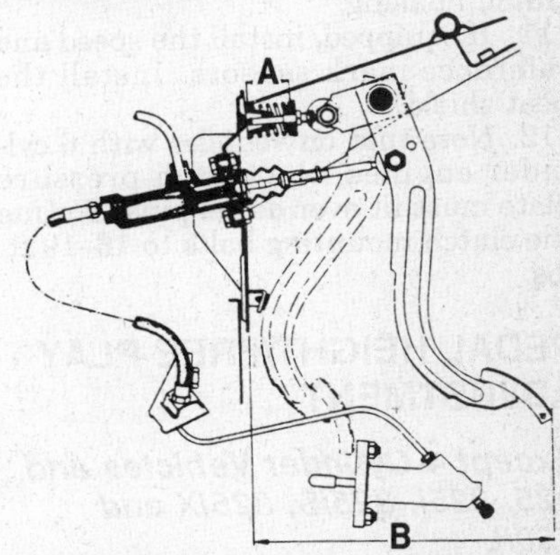

Adjusting the pedal height and overcenter spring on 6 cylinder engine

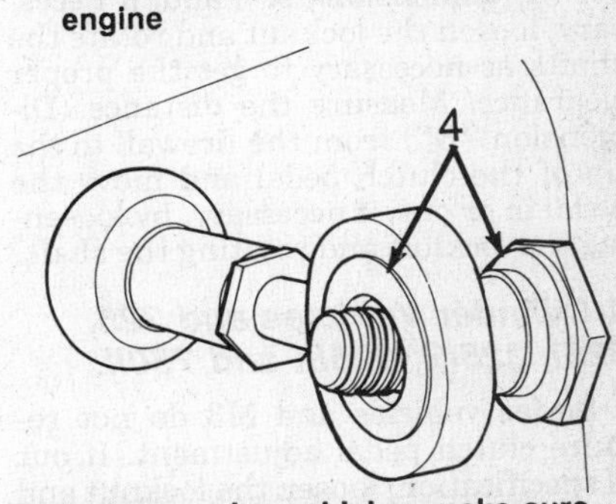

On the 7 series vehicles, make sure the bushing (4) on the clutch master cylinder are still in position

4. Prevent the flywheel from turning, using a locking tool.

5. Loosen the mounting bolts one after another gradually, 1–1½ turns at a time, to relieve tension from the clutch.

6. Remove the mounting bolts, clutch, and drive plate. Coat the splines of the transmission input shaft with Molykote® Longterm 2, Microlube® GL 2611 or equivalent. Make sure the clutch pilot bearing, located in the center of the crankshaft, turns easily.

7. Check the clutch driven disc for excess wear or cracks. Check the integral torsional damping springs, used with lighter flywheels only, for tight

fit. Inspect the rivets to make sure they are all tight. Check the flywheel to make sure it is not scored, cracked, or burned, even at a small spot. Use a straight edge to make sure the contact surface is true. Replace any defective parts.

To install:

8. To install, fit the new clutch plate and disc in place and install the mounting bolts.
9. When installing the clutch retaining bolts turn them in gradually to evenly tighten the clutch disc and to prevent warpage.
10. Install the transmission and the clutch housing.
11. If equipped, install the speed and reference mark sensors. Install the heat shield.
12. Note that on vehicles with 6 cylinder engines, the clutch pressure plate must fit over dowel pins. Torque the clutch mounting bolts to 16–19 ft. lbs.

PEDAL HEIGHT/FREE-PLAY ADJUSTMENT

Except 4 Cylinder Vehicles and 325, 325i, 325iS, 325iX and 750iL

Measure the length of the over-center spring (Dimension "A") and, if necessary, loosen the locknut and rotate the shafts as necessary to get the proper clearance. Measure the distance (Dimension "B") from the firewall to the tip of the clutch pedal and move the pedal in or out, if necessary, by loosening the locknut and rotating the shaft.

4 Cylinder Vehicles and 325, 325i, 325iS, 325iX and 750iL

3 Series vehicles and M3 do not require clutch pedal adjustment. If out of specification, loosen the locknut and rotate the piston rod to correct it.

Clutch Master Cylinder

REMOVAL & INSTALLATION

1. Remove the necessary trim panel or carpet.
2. Disconnect the pushrod at the clutch pedal.
3. Remove the cap on the reservoir tank. On some vehicles, there is a clutch master cylinder reservoir, while on others there is a common reservoir shared with the brake master cylinder. Remove the float container, if equipped. Remove the screen and remove enough brake fluid from the tank until the level drops below the refill line or the connection for the filler pipe, if there is one.
4. Disconnect the coolant expansion tank without removing the hoses on the 735i and 735iL (vehicles with the M30 B35 engine do not require this).
5. Remove the lower/left instrument panel trim. Then, remove the retaining nut from the end of the master cylinder actuating rod where the bolt passes through the pedal mechanism.
6. Disconnect the line to the slave cylinder and the fluid fill line going to the top of the master cylinder. Remove the retaining bolts and remove the master cylinder from the firewall.
7. Install the clutch master cylinder in position. The piston rod bolt should be coated with the proper lubricant. Make sure all bushings remain in position. Bleed the system and adjust the pedal travel with the pushrod to 6 in.

Clutch Slave Cylinder

REMOVAL & INSTALLATION

1. Remove enough brake fluid from the reservoir until the level drops below the refill line connection.
2. Remove the snapring or retaining bolts and pull the unit down.
3. Disconnect the line and remove the slave cylinder.
4. Install the slave cylinder on the transmission. On the 3 Series and M3, if the engine uses the 2-section flywheel, make sure a larger cylinder with a diameter of 0.874 in. is used instead of the usual cylinder (diameter 0.809 in.). Make sure to install the cylinder with the bleed screw facing upward. When installing the front pushrod, coat it with the proper anti-seize compound. Bleed the system.

Hydraulic Clutch System Bleeding

1. Fill the reservoir.
2. Connect a bleeder hose from the bleeder screw to a container filled with brake fluid so air cannot be drawn in during bleeding procedures.
3. Pump the clutch pedal about 10 times and then hold it down.
4. Open the bleeder screw and watch the stream of escaping fluid. When no more bubbles escape, close the bleeder screw and tighten it.
5. Release the clutch pedal and repeat the above procedure until no more bubbles can be seen when the screw is opened.
6. If this procedure fails to produce a bubble-free stream:
   a. Pull the slave cylinder off the transmission without disconnecting the fluid line.

   NOTE: Do not depress the clutch pedal while the slave cylinder is dismounted.

   b. Depress the pushrod in the cylinder until it hits the internal stop. Then, reinstall the cylinder.

AUTOMATIC TRANSMISSION

For further information on transmissions/transaxles, please refer to "Chilton's Guide to Transmission Repair".

Transmission Assembly

REMOVAL & INSTALLATION

Except 325i and 325iX

NOTE: To perform this operation, a support for the transmission, tool 24 0 120 and 00 2 020 or equivalent and a tool for tightening the driveshaft locking ring, tool 26 1 040 or equivalent are required. If the vehicle has the M30 B35 type engine, a special socket, tool 24 1 110 or equivalent will be needed.

1. Disconnect the battery ground cable. Loosen the throttle cable adjusting nuts, release the cable tension, and disconnect the cable at the throttle lever. Then, remove the nuts, and pull the cable housing out of the bracket.
2. Disconnect the exhaust system at the manifold and hangers and lower it out of the way. Remove the hanger that runs across under the driveshaft. Remove the exhaust heat shield from under the center of the vehicle.
3. Support the transmission a a suitable jacking device. Remove the crossmember that supports the transmission at the rear.
4. Remove the driveshaft coupling through bolts and nuts or the CV-joint through bolts and nuts. Either type is located right at the rear of the transmission. Discard used self-locking coupling nuts. Keep the CV-joint clean, and replace its gasket.
5. Unscrew the transmission locking ring at the center mount, if equipped. Then, remove the bolts and remove the center mount. Bend the propshaft downward and pull it off the centering pin. Suspend it with wire from the underside of the vehicle.
6. Drain the transmission oil and discard it. Remove the oil filler neck.

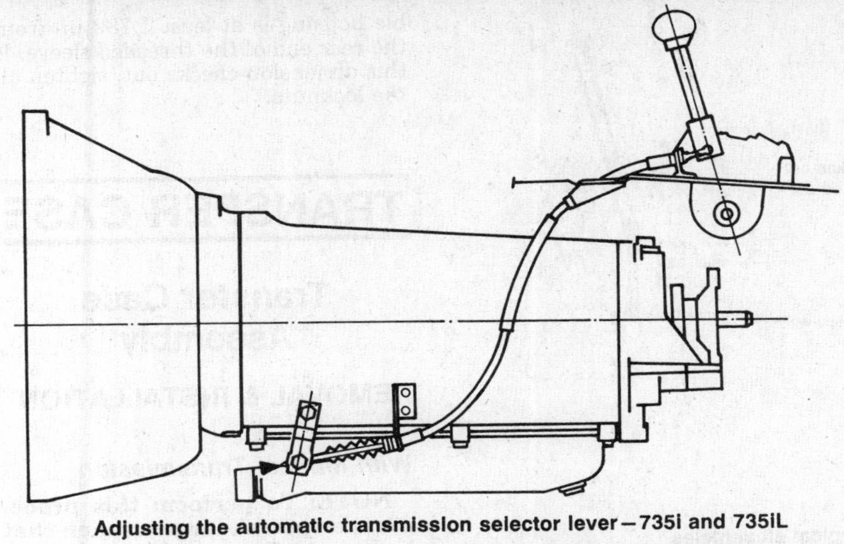

Adjusting the automatic transmission selector lever—735i and 735iL

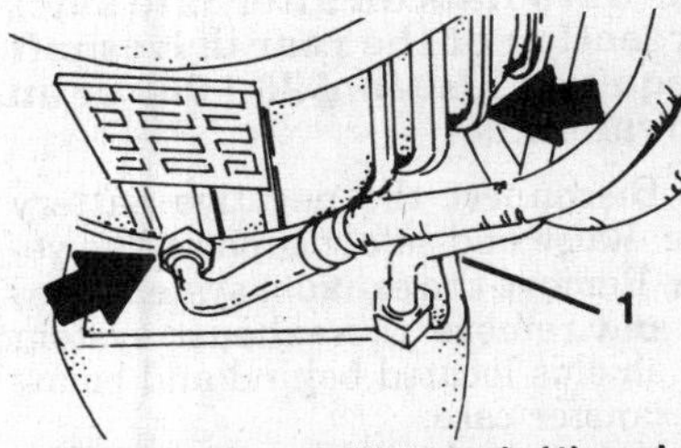

Remove the oil filler neck (1) and disconnect the arrowed hoses to drain fluid from the 4HP-22 transmission

Disconnect the oil cooler lines at the transmission by unscrewing the flare nuts and plug the open connections.

7. If equipped, remove the converter cover by removing the Torx® bolts from behind and the regular bolts from underneath. If equipped with a M30 B35 type engine pull the cover out of the bottom of the transmission housing, just behind the oil pan.

8. Remove the bolts fastening the torque converter to the drive plate, turning the flywheel as necessary to gain access from below. Use a proper socket on vehicles with the M30 B35 type engine.

9. If equipped, remove the guard for the speed and reference mark sensors. Remove the attaching bolt for each and remove each sensor. Keep the sensors clean.

10. Disconnect the shift cable by loosening the locknut fastening it to the shift lever and disconnecting the cable at the cable housing bracket.

11. If the transmission has an electrical connection, turn the bayonet fastener to the left to release the connection, disconnect it, and pull the wire out of the ties.

12. Lower the transmission as far as possible. Then, remove all the Torx® or standard type bolts attaching the transmission to the engine.

13. Remove the small grill from the bottom of the transmission. Then press the converter off with a large prybar through this opening while sliding the transmission out.

To install:

14. Install the transmission under the vehicle and raise it into position. Observe the following points:

a. Make sure the converter is fully installed onto the transmission—so the ring on the front is inside the edge of the case.

b. When reinstalling the driveshaft, tighten the lockring with a special tool.

c. If the driveshaft has a simple coupling, rather than a CV-joint, make sure to replace the self-locking nuts and to hold the bolts still while tightening the nuts to keep from distorting the coupling.

d. When installing the center mount, preload it forward from its most natural position 0.157–0.236 in.

e. Adjust the throttle cables.

325i and 325iX

NOTE: To perform this operation, a support for the transmission, BMW tool 24 0 120 and 00 2 020 or equivalent and a tool for tightening the driveshaft locking ring, BMW tool 26 1 040 or equivalent, are required. If the vehicle the M30 B35 type engine, a special socket, that retains bolts, will also be needed.

1. Disconnect the battery ground cable. Loosen the throttle cable adjusting nuts, release the cable tension, and disconnect the cable at the throttle lever. Then, remove (and retain) the nuts, and pull the cable housing out of the bracket.

2. Disconnect the exhaust system at the manifold and hangers and lower it aside. Remove the hanger that runs across under the driveshaft. Remove the exhaust heat shield from under the center of the vehicle.

3. On the 325iX, remove the transfer case from the rear of the transmission.

4. Drain the transmission oil and discard it. Remove the oil filler neck. Disconnect the oil cooler lines at the transmission by unscrewing the flare nuts and plug the open connections.

5. Support the transmission with the proper tools. Separate the torque converter housing from the transmission by removing the Torx® bolts with the proper tool from behind and the regular bolts from underneath. Retain the washers used with the Torx® bolts.

6. On the 325iX, disconnect the front driveshaft.

7. Remove bolts attaching the torque converter housing to the engine, making sure to retain the spacer used behind one of the bolts. Then, loosen the mounting bolts for the oil level switch just enough so the plate can be removed while pushing the switch mounting bracket to one side.

8. Remove the bolts attaching the torque converter to the drive plate. Turn the flywheel as necessary to gain access to each of the bolts, which are spaced at equal intervals around it. Make sure to re-use the same bolts and retain the washers.

9. To remove the speed and reference mark sensors, remove the attaching bolt for each and remove each sensor. Keep the sensors clean.

10. Turn the bayonet type electrical connector counterclockwise and then pull the plug out of the socket. Then, lift the wiring harness out of the harness bails.

11. Support the transmission using the proper jack. Then, remove the crossmember that supports the transmission at the rear.

12. Disconnect the transmission shift rod. Then, remove the nuts and then the through bolts from the damper-type U-joint at the front of the transmission.

13. Unscrew the transmission locking ring at the center mount, if equipped, using the special tool designed for this purpose. Then, remove the bolts and remove the center mount. Bend the driveshaft downward and pull it off the centering pin. Suspend it with wire from the underside of the vehicle.

14. Lower the transmission as far as possible. Then, remove all the Torx® or standard type bolts attaching the transmission to the engine.

15. Remove the small grill from the bottom of the transmission. Then press the converter off with a large

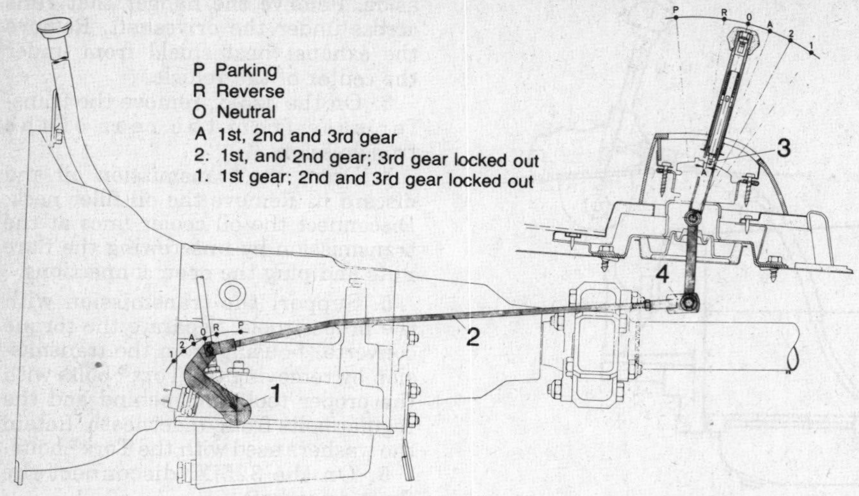

Selector lever adjustment – typical all vehicles

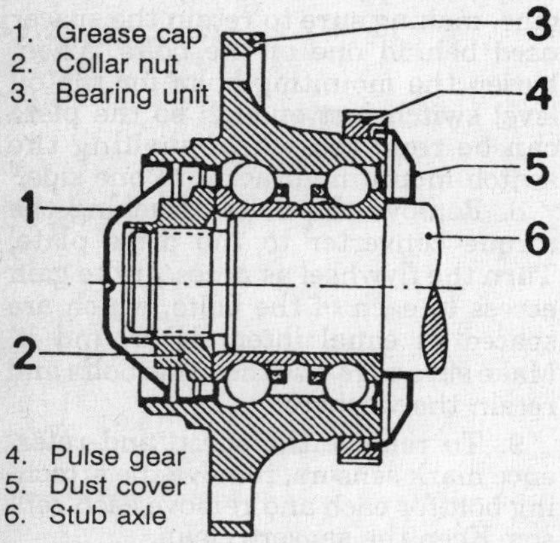

Front wheel bearing assembly – 5 and 7 Series vehicles

prybar passing through this opening while sliding the transmission out.

To install:

16. Install the transmission in position under the vehicle and raise it into position. Observe the following points:

a. Make sure the converter is fully installed onto the transmission – so the ring on the front is inside the edge of the case.

b. When reinstalling the driveshaft, tighten the lockring with the proper tool.

c. Make sure to replace the self-locking nuts on the driveshaft flexible joint and to hold the bolts still while tightening the nuts to keep from distorting it.

d. When installing the center mount, preload it forward from its most natural position 0.157–0.236 in.

e. When reconnecting the bayonet type electrical connector, make sure the alignment marks are aligned after the plug it twisted into its final position.

f. When reinstalling the speed and reference mark sensors, inspect the O-rings used on the sensors and install new ones, if necessary. Make sure to install the speed sensor into the bore marked D and the reference mark sensor, which is marked with a ring, into the bore marked B.

g. Torque the crossmember mounting bolts to 16–17 ft. lbs.

h. If O-rings are used with the transmission oil cooler connections, replace them.

i. Adjust the throttle cables.

SHIFT LINKAGE ADJUSTMENT

1. Move the selector lever to **P** position. Loosen the nut until the cable is free.
2. Push the transmission lever to the **D** or **P** position. Then push the cable rod in the opposite direction.
3. Clamp down the cable rod without tension.
4. Tighten the nut to 7.0–8.5 ft. lbs.

NOTE: Do not bend the cable.

THROTTLE LINKAGE ADJUSTMENT

1. On the injection system throttle body, loosen the 2 locknuts at the end of the throttle cable and adjust the cable until there is a play of 0.010–0.030 in.
2. Loosen the locknut and lower the kickdown stop under the accelerator pedal. Have someone depress the accelerator pedal until the transmission detent can be felt. Then, back the kickdown stop back out until it just touches the pedal.
3. Check that the distance from the seal at the throttle body end of the cable housing is at least 1.732 in. from the rear end of the threaded sleeve. If this dimension checks out, tighten all the locknuts.

TRANSFER CASE

Transfer Case Assembly

REMOVAL & INSTALLATION

With Manual Transmission

NOTE: To perform this procedure, a special, large wrench that locks onto flats on alternate sides of a section of the rear driveshaft is required. Use tool 26 1 060 or an equivalent.

1. Disconnect the negative battery cable. Raise and safely support the vehicle. Remove the exhaust system. Unbolt and remove the exhaust system heat shields located behind and below the transfer case.
2. Unscrew the rear section of the driveshaft at the sliding joint located behind the output flange of the transfer case.
3. Hold the through-bolts stationary and remove the self-locking nuts from in front of the flexible coupling at the transfer case output flange. Discard all the self-locking nuts and replace them.

NOTE: During the next step, be careful not to let the driveshaft rest on the metal fuel line that crosses under it or the line could be damaged.

4. Slide the secti)ns of the driveshaft together at the sliding joint and then pull the front of the driveshaft off the centering pin at the transmission output shaft.

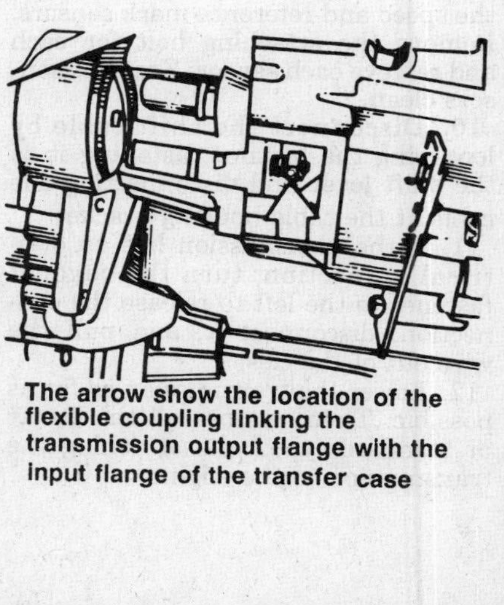

The arrow show the location of the flexible coupling linking the transmission output flange with the input flange of the transfer case

5. Remove the nuts and through bolts from the flexible coupling linking the transmission output flange with the short driveshaft linking the transmission and the transfer case.

6. Support the transmission from underneath in a secure manner. Then, mark each of the 4 bolts fastening the crossmember that supports the transmission at the rear to the body, bolts are of different lengths. Remove the crossmember.

7. Lower the transmission/transfer case unit just enough to gain access to the bolts linking the 2 boxes together. Remove the 2 lower and 2 upper bolts. It is possible to gain access to the upper bolts using a socket wrench with a U-joint and extension.

8. There is a protective cap on the forward driveshaft where it links up with the transfer case. The cap is made of a brittle material, so it must be handled carefully. Gently slide the cap forward until is free of the transfer case.

9. Slide the transfer case to the rear so it can be separated from both the transmission and the forward driveshaft. When it is free, remove it.

To install:

10. Install the transfer case under the vehicle and raise it into position, bearing the following points in mind:

a. Inspect the dowel holes locating the transfer case with the transmission and the guide hole for the output shaft where it slides into the transfer case to make sure these parts will be properly located. Lubricate the guide pin and the splines of the front driveshaft section with grease.

b. When fitting the transfer case onto the transmission, check to make sure the output flange of the transmission is properly aligned with the flexible coupling. Put the through-bolts through the flexible coupling and then install and torque the nuts to 65 ft. lbs. while holding the bolts stationary, rather than turning them.

c. When reconnecting the transfer case to the gearbox, torque the bolts to 30 ft. lbs.

d. Before fitting the driveshaft back onto the rear of the transmission, retain the seal in the protective cap by applying grease to it.

e. Torque the transmission crossmember bolts to 17 ft. lbs.

f. Check the fluid level and fill with the recommended lubricant.

With Automatic Transmission

NOTE: To perform this procedure, a special, large wrench that locks onto flats on alternate sides of a section of the rear driveshaft is required. Use tool 26 1 060 or an equivalent.

1. Disconnect the negative battery cable. Raise and safely support the vehicle. Remove the exhaust system. Unbolt and remove the exhaust system heat shields located behind and below the transfer case.

2. Unscrew the rear section of the driveshaft at the sliding joint located behind the output flange of the transfer case.

3. Hold the through-bolts stationary and remove the self-locking nuts from in front of the flexible coupling at the transfer case output flange. Discard all the self-locking nuts and replace them.

NOTE: During the next step, be careful not to let the driveshaft rest on the metal fuel line that crosses under it or the line could be damaged.

4. Slide the sections of the driveshaft together at the sliding joint and then pull the front of the driveshaft off the centering pin at the transmission output shaft.

5. Remove the nuts and through bolts from the flexible coupling linking the transmission output flange with the short driveshaft linking the transmission and the transfer case.

6. Note the locations of all the washers and then loosen the retaining nut and disconnect the range selector lever cable at the transmission by pulling out the pin. Be careful not to bend the cable in doing this. Then, loosen the nuts that position the cable housing onto the transmission and slide the cable housing backward so it can be separated from the bracket on the transmission housing.

7. There is a protective cap on the forward driveshaft where it links up with the transfer case. The cap is made of a brittle material, so it must be handled carefully. Gently slide the cap forward until is free of the transfer case.

8. Remove the drain plug in the bottom of the pan and drain the transmission fluid.

9. Support the transmission from underneath in a secure manner. Then, mark each of the 4 bolts fastening the crossmember that supports the transmission at the rear to the body, bolts are of different lengths. Remove the crossmember.

10. Remove the 9 nuts fastening the transfer case to the transmission housing. Note the location of the wiring holder so it will be possible to reinstall it on the same bolt.

11. Slide the transfer case to the rear and off the transmission.

To install:

12. Install the transfer case under the vehicle and raise it into position, bearing the following points in mind:

a. Inspect the sealing surfaces as well as the dowel holes in the transfer case to make sure they will seal and locate properly. Clean the sealing surfaces and replace the gasket.

b. When sliding the transfer case back onto the transmission, turn the front driveshaft section slightly to help make the splines mesh.

c. When reconnecting the shift cable, inspect the rubber mounts and replace any that are cut, crushed, or cracked. Adjust the shift cable.

d. Before fitting the driveshaft back onto the rear of the transmission, retain the seal in the protective cap by applying grease to it.

e. When fitting the transfer case onto the transmission, check to make sure the output flange of the transmission is properly aligned with the flexible coupling. Put the through-bolts through the flexible coupling and then install and torque the nuts to 65 ft. lbs. while holding the bolts stationary, rather than turning them.

f. Torque the bolts holding the transfer case to the transmission to 65 ft. lbs.

g. Torque the transmission crossmember bolts to 17 ft. lbs.

h. Check the fluid level and fill with the recommended lubricant.

FRONT SUSPENSION

MacPherson Strut

REMOVAL & INSTALLATION

318iS, 325, 325i, 325iS and M3

1. Disconnect the negative battery cable.

2. Raise and safely support the vehicle. Remove the tire and wheel assembly.

3. Disconnect the brake pad wear indicator plug and ground wire. Pull the wires out of the holder on the strut. Remove the ABS pulse sender, if equipped.

4. Unbolt the caliper and pull it away from the strut, suspending it with a piece of wire from the body. Do not disconnect the brake line.

5. Remove the attaching nut and then detach the push rod on the stabilizer bar at the strut.

6. Unscrew the attaching nut and press off the guide joint with the proper tool.

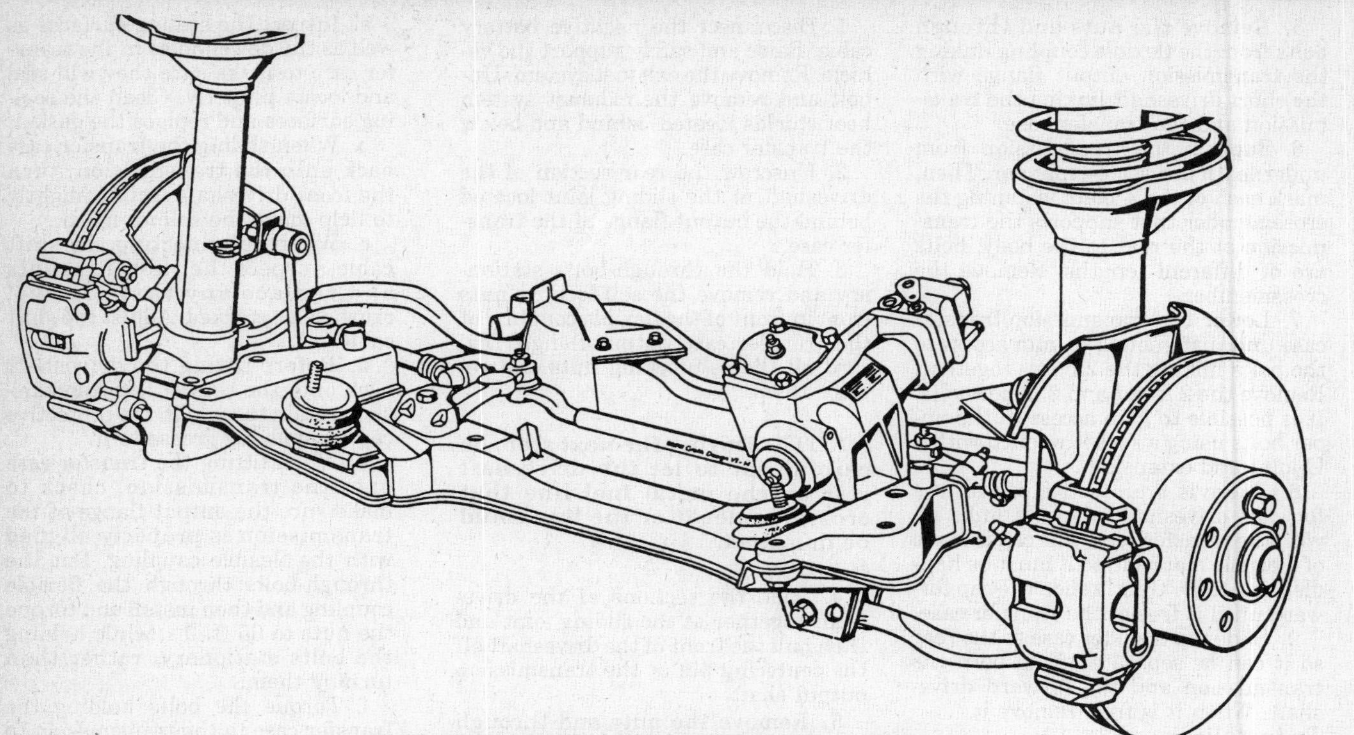

View of the front suspension—5 and 7 Series vehicles

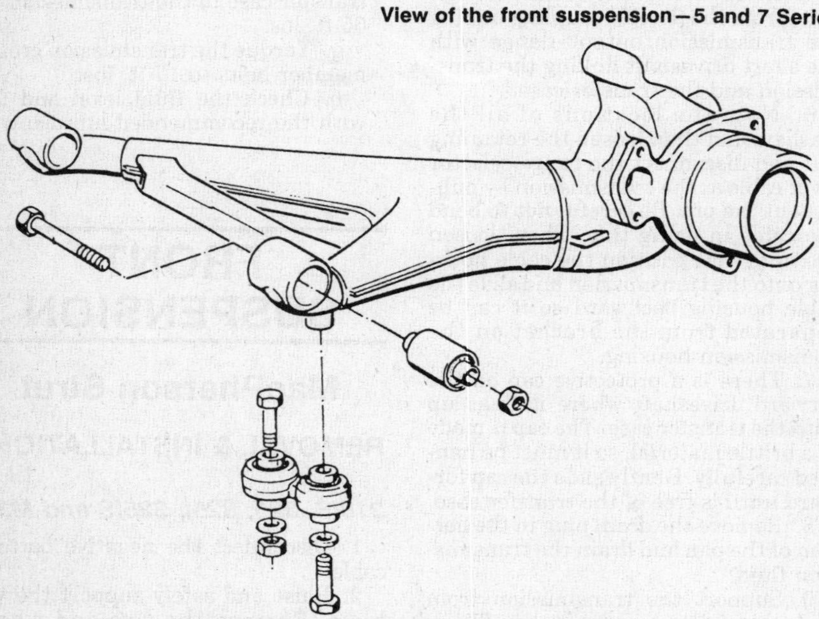

View of the trailing arm assembly—5 and 7 Series Vehicles

7. Unscrew the nut and press off the tie rod joint.
8. Press the bottom of the strut outward and push it over the guide joint pin, using the proper tool. Support the bottom of the strut.
9. Unscrew the nuts at the top of the strut, from inside the engine compartment, then remove the strut.

To install:

10. Install in reverse order, observing the following points:
    a. Replace the self-locking nuts that fasten the top of the strut.
    b. Tie rod and guide joints must have both pins and both bores clean for reassembly. Replace both self-locking nuts.
    c. Torque the control arm to spring strut attaching nut to 43–51 ft. lbs. Torque the spring strut to wheel well nuts to 16–17 ft. lbs.

325iX

1. Disconnect the negative battery cable.
2. Raise and safely support the vehicle. Remove the front tire and wheel assembly. Unplug the ABS pulse transmitter.
3. Lift out the lock plate at the center of the brake disc with a small prybar. Unscrew the collar nut.
4. Disconnect the brake pad wear indicator plug and the ground wire. Pull the wires and brake hose out of the clip on the spring strut. Then, disconnect the small rod at the strut.
5. Remove the brake caliper mounting bolts and support the assembly with a piece of wire, keeping stress off the brake hose.
6. Remove the attaching nut from the tie rod end. Then press the stud off the knuckle with the proper tool.
7. Remove the attaching nut for the control arm and then press the stud off the knuckle with an appropriate tool.
8. Mount the proper tool to the brake disc with 2 of the wheel bolts. Then, press the output shaft out of the center of the knuckle.
9. Support the spring strut from underneath. Remove the cap from the center of the wheel house. Remove the 3 bolts from the upper mount near the wheel housing. Remove the strut.

To install:

10. Install the strut, observing these points:
    a. Torque the nuts attaching the strut to the wheel house to 16 ft. lbs.
    b. Lubricate the splines of the output shaft with oil before pressing it back into the center of the knuckle with the proper tool.

1. Rpring retainer (upper)
2. Ring for hollow piston rod
3. Coil spring
4. Spring retainer (lower)
5. Nut
6. Cap
7. Stop washer
8. Mount
9. Support
10. Rubber mount
11. Protective pipe
12. Shock absorber
13. Screw-on ring
14. Spring strut

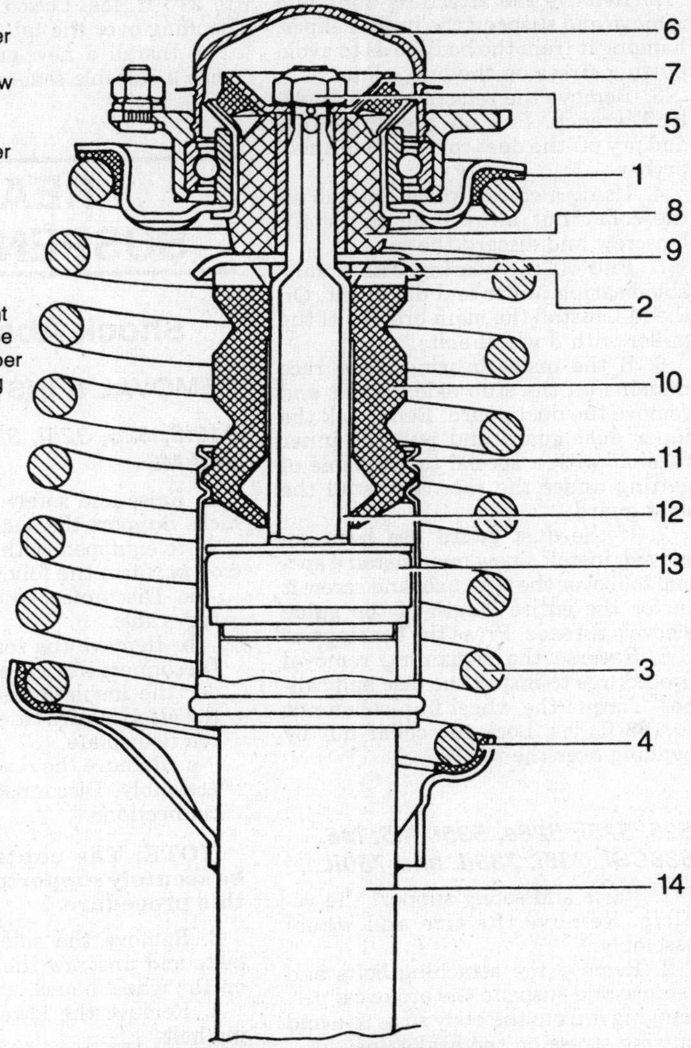

Exploded view of the spring strut assembly

c. Keep grease off the studs for the control arm and tie rod end. Replace the cotter pin on the control arm and the self-locking nut on the tie rod end. Torque the control arm stud nut to 61.5 ft. lbs. Torque the tie rod nut to 61.5 ft. lbs. and then tighten it further to install the cotter pin, if necessary.

d. Replace the lock plate in the center of the disc with the proper tool.

e. Torque the bolts attaching the caliper to the steering knuckle to 63–79 ft. lbs. Lower the vehicle.

f. Connect the negative battery cable.

525, 525i, 528e, 535i, M5, M6, 635CSi, 735i, 735iL and 750iL

1. Disconnect the negative battery cable.

2. Raise and safely support the vehicle. Remove the tire and wheel assembly.

3. Disconnect the brake pad wear indicator plug and ground wire. Pull the wires out of the holder on the strut. Remove the ABS pulse sender, if equipped.

4. Disconnect the stabilizer pushrod with the proper tool.

5. Disconnect the lower strut bolts at the control arm.

6. Support the bottom of the strut and unscrew the nuts at the top of the strut, from inside the engine compartment. Remove the strut.

7. The installation is the reverse of the removal procedure.

Lower Control Arms

REMOVAL & INSTALLATION

318iS, 325, 325i 325iS and M3

1. Raise and safely support the vehicle. Remove the front tire and wheel assembly.

2. Disconnect the rear control arm bracket where it connects to the body by removing the bolts.

3. Remove the nut and disconnect the thrust rod on the front stabilizer bar where it connects to the center of the control arm.

4. Unscrew the nut which attaches the front of the stabilizer bar to the crossmember and remove the nut from above the crossmember. Then, use a plastic hammer to knock this support pin out of the crossmember.

5. Unscrew the nut and press off the guide joint where the control arm attaches to the lower end of the strut, using the proper tool.

To install:

6. Reverse the procedure to install. Keep these points in mind:

a. Replace the self-locking nut that fastens the guide joint to the control arm.

b. Make sure the support pin and the bore in the crossmember are clean before inserting the pin through the crossmember. Replace the original nut with a replacement nut and washer.

c. Torque the control arm-to-spring strut nut to 43–51 ft. lbs. Torque the control arm support to crossmember nut to 29–34 ft. lbs. Torque the push rod on the stabilizer bar to 29–34 ft. lbs.

325iX

1. Raise and safely support the vehicle. Remove the tire and wheel assembly.

2. Disconnect the rear control arm bracket where it connects to the body by removing the bolts.

2. Remove the nut from the top of the stud that attaches to one corner of the control arm and runs through the crossmember.

3. Remove the cotter pin and then remove the nut from the ball joint stud where it passes through the steering knuckle. Then, press the ball joint stud out of the knuckle with the proper tool. Make sure to keep the stud and bore free of grease.

4. Reverse the procedure to install. Keep these points in mind:

a. Make sure the support pin and the bore in the crossmember are clean before inserting the pin through the crossmember. Replace the original nut with a replacement nut and washer equivalent to those shown in the illustration.

b. Torque the control arm-to-spring strut nut to 61.5 ft. lbs. Turn the nut farther, as necessary to align the castellations with the cotter pin hole and install a new cotter pin. Torque the control arm support to crossmember nut to 30 ft. lbs.

525, 525i, 528e, 535i, M5, M6, 635CSi, 735i, 735iL and 750iL

1. Raise and safely support the vehicle. Remove the tire and wheel assembly.
2. Remove the mounting bolts that fasten the bottom of the strut to the steering knuckle.
3. Remove the cotter pin and castellated nut. Use a suitable ball joint remover to press the ball joint end of the control arm off the steering knuckle.
4. Remove the self-locking nut. Then, remove the through bolt and the washers, slide the inner end of the strut and bushing out of the front suspension crossmember.

To install:

5. Install in reverse order, noting the following points:
   a. Make sure both washers are replaced to cushion the bushing where it contacts the suspension crossmember.
   b. Replace the bushing if it is worn or cracked.
   c. Use a new self-locking nut on the bolt fastening the inner end of the strut.
   d. Align the bottom of the strut with the steering knuckle so the tab on the arm fits into the notch on the bottom of the strut. Install the bolts with a locking type sealer.
   e. When installing the arm ball joint onto the steering knuckle, tighten the nut until a castellation lines up with the cotter pin hole and then use a new cotter pin in the nut.
   f. Final tighten the through bolt for the inner end of the arm after the vehicle is on the ground at normal ride height.

Sway Bar

REMOVAL & INSTALLATION

1. Raise and safely support the vehicle.
2. Disconnect the push/thrust rod on both sides.
3. Disconnect the left side control arm bracket on the 3 Series.
4. Disconnect the left and right stabilizer mounts. Remove the stabilizer bar.
5. The installation is the reverse of the removal procedure. Tighten the stabilizer mount bolts to 16 ft. lbs.

Front Wheel Bearings

REMOVAL & INSTALLATION

318iS, 325, 325i, 325iS and M3

1. Raise and safely support the vehicle. Remove the tire and wheel assembly.
2. Remove the attaching bolts and remove and suspend the brake caliper, hanging it from the body so as to avoid putting stress on the brake line.
3. Remove the setscrew with an Allen® wrench. Pull off the brake disc and pry off the dust cover with a small prybar.
4. Using a chisel, knock the tab on the collar nut away from the shaft. Unscrew and discard the nut.
5. Pull off the bearing with a suitable bearing puller and discard it. On the M3, install the main bracket of the puller with 3 wheel bolts.
6. If the inside bearing inner race remains on the stub axle, unbolt and remove the dust guard. Bend back the inner dust guard and pull the inner race off with a special tool capable of getting under the race. Reinstall the dust guard.
7. If the dust guard has been removed, install a new one. Install a special tool over the stub axle and screw it in for the entire length of the guide sleeve's threads. Press the bearing on.
8. Reverse the remaining removal procedures to install the disc and caliper. Torque the wheel hub collar nut to 188 ft. lbs. Lock the collar nut by bending over the tab.

525, 525i, 528e, 535i, M5, M6, 635CSi, 735i, 735iL and 750iL

1. Raise and safely support the vehicle. Remove the tire and wheel assembly.
2. Remove the attaching bolts and remove and suspend the brake caliper, hanging it from the body so as to avoid putting stress on the brake line.
3. Remove the setscrew with an Allen® wrench. Pull off the brake disc and pry off the dust cover with a small prybar.
4. Using a chisel, knock the tab on the collar nut away from the shaft. Unscrew and discard the nut.
5. Using the proper tool, pull off the bearing and discard it.

To Install:

6. If the inside bearing inner race remains on the stub axle, unscrew and remove the dust guard, using a socket extension. Bend back the inner dust guard and pull the inner race off with a special tool capable of getting under the race. Reinstall the dust guard and install a new dust cover.
7. Then install a special tool over the stub axle and screw it in for the entire length of the guide sleeve's threads. Slide the bearing on and follow it with the proper tool and use this tool to press the bearing on.
8. Reverse the remaining removal procedures to install the disc and caliper. Torque the wheel hub collar nut to 210 ft. lbs. Lock the collar nut by bending over the tab.
9. Install a new grease cap coated with a suitable sealer.

REAR SUSPENSION

Shock Absorbers

REMOVAL & INSTALLATION

318iS, 325, 325i, 325iS, 325iX and M3

1. Raise and safely support the vehicle. Support the control arms.
2. If equipped with ride level control, perform the following:
   a. Disconnect the negative battery cable.
   b. Remove the rear seat cushion. Disconnect the left or right plug under the insulation sheet and guide the electric wire through the hole in the floor plate.
   c. Remove the rear tire and wheel assembly. Disconnect the electrical connection.

NOTE: The control arm must be securely supported throughout this procedure.

3. Remove the side backrest, seat belts and unscrew the centering shell on the wheel house.
4. Remove the lower shock retaining bolt.
5. Remove trim inside the trunk, if necessary, and disconnect the upper strut retaining nuts at the wheel arch and remove the assembly.
6. Remove the shock absorber.

To install:

7. Install in reverse order, using new gaskets between the shock and the wheel arch, and new self-locking nuts on top of the strut.
8. Torque the shock-to-body nuts to 16–17 ft. lbs.; spring retainer-to-wheel house nuts – 6 cyl. to 16–17 ft. lbs.; lower bolt to 52–63 ft. lbs. on 4 cylinder engines or 90–103 ft. lbs. on 6 cylinder engines.
9. Final torquing of the lower strut bolt should be done with the vehicle in the normal riding position.

MacPherson Strut

REMOVAL & INSTALLATION

525, 525i, 528e, 535i, M5, M6, 635CSi, 735i, 735iL and 750iL

1. Disconnect the negative battery

cable. Remove the rear seat and back rest.

2. Raise and safely support the vehicle. Support the control arms.

NOTE: The coil spring, shock absorber assembly acts as a strap so the control arm should always be supported.

3. If equipped with automatic ride control perform the following:

a. Diasconnect the low pressure switch electrical connection and turn on the ignition.

b. Disconnect the control rod nut, holding the collar with an 8mm wrench against torque. Don't disconnect the rod at the ball joint.

c. Operate the lever on the control switch in the "discharge" direction for about 20 seconds to discharge fluid from the lines.

d. Disconnect the hydraulic line on the shock absorber.

4. Remove the lower shock retaining bolt.

5. Remove trim if necessary and disconnect the upper strut retaining nuts at the wheel arch and remove the assembly.

To install:

6. Install in reverse order, using new gaskets between the strut and the wheel arch, and new self-locking nuts on top of the strut.

7. Torque the shock-to-body nuts to 16–17 ft. lbs.; spring retainer-to-wheel house nut to 16–17 ft. lbs. Tighten the lower bolt to 52–63 ft. lbs. on the M3 and 90–103 ft. lbs. except on the M3.

8. Replace the gasket that goes between the top of the strut and the lower surface of the wheel well, if necessary

9. Final torquing of the lower strut bolt should be done with the vehicle in the normal riding position.

Coil Springs

REMOVAL & INSTALLATION

318iS, 325, 325i, 325iS, 325iX and M3

1. Raise and safely support the vehicle.

2. Disconnect the rear portion of the exhaust system and hang it from the body.

3. Disconnect the final drive rubber mount, push it down, and hold it down with a wedge.

4. Remove the bolt that connects the rear stabilizer bar to the strut on the side being worked on. Be careful not to damage the brake line.

NOTE: Support the lower control arm securely with a jack or other device that will permit it to be lowered gradually, while maintaining secure support.

5. Then, to prevent damage to the output shaft joints, lower the control arm only enough to slip the coil spring off the retainer.

6. Make sure, in replacing the spring, that the same part number, color code, and proper rubber ring are used.

To Install:

7. Reverse all removal procedures to install, making sure the spring is in proper position, keeping the control arm securely supported until the shock bolt is replaced, and tightening stabilizer bar and lower shock mount bolts with the control arm in the normal ride position.

8. Torque the stabilizer bolt to 22–24 ft. lbs. and the shock bolt to 52–63 ft. lbs.

Trailing Arms

REMOVAL & INSTALLATION

318iS, 325, 325i, 325iS, 325iX and M3

1. Raise and safely support the vehicle. Remove the tire and wheel assembly.

2. Apply the parking brake and disconnect the output shaft at the rear axle shaft, if necessary. Remove the parking brake lever.

3. Remove the brake fluid from the master cylinder reservoir. To do this, it will be necessary to remove the strainer at the top of the reservoir. Disconnect the brake line connection on the rear control arm. Plug the openings.

4. Support the trailing arm securely. Disconnect the shock absorber at the control arm.

5. Remove the nuts and then slide the bolts out of the mounts where the trailing arm is mounted to the axle carrier.

To install:

6. Install in reverse order. Install the bolt that goes into the inner bracket first.

7. Torque the bolts holding the trailing arm to the axle carrier to 48–54 ft. lbs.

8. Make sure the spring is positioned properly top and bottom. Torque the strut bolt to 52–63 ft. lbs.

9. Reinstall the handbrake or reconnect the cable and adjust. Then apply the brake and reconnect the output shaft.

10. Reconnect the brake line, replenish with the proper brake fluid, and bleed the system.

525, 525i, 528e, 535i, M5, M6, 635CSi, 735i, 735iL and 750iL

1. Raise and safely support the vehicle. Remove the tire and wheel assembly.

2. Apply the parking brake to hold the driveshaft stationary. Disconnect the output shaft at the drive flange. Hang the shaft from the body by a piece of wire.

3. Disconnect the parking brake cable at the lever.

4. Remove the float housing from the brake fluid reservoir and then remove as much fluid as possible from the reservoir.

5. Pull the brake cable housing out of the mounting bracket near the control arm. Disconnect the brake line.

6. Pull down and disconnect the plug for the pulse sender, do not damage the rubber grommet. On the 750iL, remove the rear seat to gain access to the pulse sender.

7. Remove the rear pushrod and disconnect the camber warning sender on the rear axle carrier on the 750iL.

8. Support the trailing arm from underneath in a secure manner.

9. Remove the nuts and then remove the bolts to disconnect the control arm from the rear axle carrier.

10. If the vehicle has a stabilizer bar, remove the bolts and remove the attaching bracket for the stabilizer bar.

11. Disconnect the shock absorber and remove the control arm.

To Install:

12. Installation is the reverse of removal. When reattaching the control arm, insert the bolt on the inner bracket first.

13. Attach the shock absorber and the stabilizer bar bracket. On the 750iL attach the camber warning sensor and the pushrod.

14. Connect the pulse sender, the brake line and the brake cable. Connect the drive axle.

15. Tighten all mounting bolts with the vehicle resting on its wheels. Torque the bolts attaching the control arm to the axle carrier to 49–54 ft. lbs. Refill and bleed the brake system.

STEERING

Steering Wheel

CAUTION

On vehicles equipped with an air bag, the negative battery cable must be disconnected, before working on the system. Failure to do so may result in deployment of the air bag and possible personal injury.

REMOVAL & INSTALLATION

1. Disconnect the negative battery cable. Remove steering wheel pad or BMW emblem. Mark the relationship between the steering wheel and shaft for installation in the same position. Unlock the steering wheel lock with the key. Otherwise, the wheel cannot be removed.
2. Unscrew retaining nut and remove the wheel.

NOTE: Be careful not to damage the direction signal cancelling cam, which is right under the steering wheel, in performing this operation. On vehicles equipped with an air bag, it is important to avoid banging on the wheel in any way.

3. Installation is the reverse of removal. Lubricate the direction signal cancelling cam. Replace the self-locking nut on all models, and torque it to 58 ft. lbs.

Steering Column

CAUTION

On vehicles equipped with an air bag, the negative battery cable must be disconnected, before working on the system. Failure to do so may result in deployment of the air bag and possible personal injury.

REMOVAL & INSTALLATION

1. Disconnect the negative battery cable. Remove the steering wheel.
2. Remove the lower instrument trim panel. Disconnect the steering column casing.
3. Disconnect the electrical connections. Press down on the steering spindle, remove the bolt and the spindle.
4. Remove the shear-off bolts with a proper chisel. Remove the mounting nuts/bolts.
5. Press down and remove the steering column.
6. The installation is the reverse of the removal procedure. Install new bolts.

Power Steering Rack

REMOVAL & INSTALLATION

318iS, 325, 325i, 325iS and M3

1. Raise and safely support the vehicle and remove front wheels. Remove the pinch bolt and loosen bolt. Press the spindle off the steering gear.
2. Use a syringe to empty the power steering fluid reservoir. Loosen the clamp and pull off the hydraulic fluid return line from the power steering unit. Discard drained fluid.
3. Disconnect and plug the pressure line.
4. Unscrew left and right side nuts, and press off the tie rods where they connect to the spring struts.
5. Remove the bolts attaching the steering unit to the front axle carrier and remove it.

To install:

6. Install in reverse order, keeping the following points in mind:
   a. The steering unit bolts to the rear holes of the axle carrier. Use new self-locking nuts and torque them to 29–34 ft. lbs.
   b. When reconnecting tie rods to the spring struts, make sure tie rod pins and strut bores are clean. Replace self-locking nut and torque to 40–48 ft. lbs.
   c. Replace the seals on the power steering pump connection, and torque the bolt to 29–32 ft. lbs.
7. Refill the fluid reservoir with specified fluid. Idle the engine and turn the steering wheel back and forth until it has reached right and left lock 2 times each. Then, turn off the engine and refill the reservoir.

325iX

NOTE: To remove the steering gear on 4WD vehicles, use a special tool to support the engine via the body. It is also advisable to use a special tool to support the front axle carrier without damaging it. It is necessary to remove the entire front axle carrier to gain access to the mounting bolts for the steering gear on this vehicle.

1. Raise the vehicle and support it securely. Remove the splash guard. Remove the front wheels.
2. Remove the air cleaner. Use a clean syringe to remove the power steering fluid from the pump reservoir.
3. Attach the support tool and connect it to the engine hooks to be sure the engine is securely supported.
4. Remove the through bolts from the right and left engine mounts.
5. Disconnect both the hydraulic lines running from the power steering pump to the steering gear, and then plug the openings.
6. Loosen both the retaining bolts and then disconnect the steering column spindle off the steering gear.
7. Remove the retaining nuts on both sides and then press the tie rod ends off the steering knuckles with the proper tools. Be careful to keep grease out of the bores and off the tie rod ball studs.
8. Remove the cotter pins, remove the retaining nuts on both sides and then use the proper tool to press the control arm ball joint studs out of the steering knuckles. Be careful to keep grease out of the bores and off the control arm ball studs.
9. Remove the bolts on either side attaching the control arm brackets to the body.
10. Remove the bolts and remove the stabilizer bar mounting brackets from the front axle carrier on both sides.
11. Support the front axle carrier with a suitable jacking device. Then, remove the mounting bolts on either side and remove the axle carrier. Remove the mounting bolts and remove the steering gear from the axle carrier.

To install:

12. Install in reverse order, noting these points:
    a. Clean the bores into which the axle carrier bolts are mounted. Use some sort of locking sealer and torque the bolts to 30 ft. lbs.
    b. Torque the mounting bolts holding the steering gear to front axle carrier to 30 ft. lbs.
    c. Install new cotter pins on the retaining nuts for the control arm ball studs. Torque to 61.5 ft. lbs.
    d. Replace the self-locking nuts on the tie rod end ball studs and connecting the steering column spindle to the steering box. Torque tie rod ball stud nuts to 24–29 ft. lbs.
    e. Replace the gaskets on power steering hydraulic lines.
13. Refill the fluid reservoir with specified fluid. Idle the engine and turn the steering wheel back and forth until it has reached right and left lock 2 times each. Then, turn off the engine and refill the reservoir.

Power Steering Gear

ADJUSTMENT

525i, 535iS, 528e, 535i, M5, M6, 635CSi, 735i, 735iL and 750iL

1. Remove the steering wheel center.
2. With the front wheels in the straight ahead position, remove the cotter pin and loosen the castle nut.
3. Press the center tie rod off the steering drop arm.
4. Turn the steering wheel to the left about 1 turn. Install a friction gauge and turn the wheel to the right, past the point of pressure and the gauge should read 0.72–0.87 ft. lbs.
5. To adjust, turn the steering wheel about 1 turn to the left. Loosen the counter nut and turn the adjusting screw until the specified friction is reached when passing over the point of pressure.

REMOVAL & INSTALLATION

525i, 535iS, 528e, 535i, M5, M6, 635CSi, 735i, 735iL and 750iL

1. Disconnnect the negative battery cable.
2. Remove the steering wheel, if equipped with an air bag (SRS).
3. Discharge the pressure reservoir by pushing in on the brake pedal about 10 times. Draw off hydraulic fluid in the supply tank.
4. Unscrew the bolt and press the tie rod off the steering drop arm with the proper tool.
5. Remove the heat shield on the steering gear and disconnect the ride level height control pipes on the 750iL.
6. Remove the bolt and push the U-joint from the steering gear. Disconnect and plug the hydraulic lines.
7. Unscrew the steering gear mounting bolts and remove the steering gear.

NOTE: If necessary, move the steering drop arm by turning the steering stub to enable the removal of the gear assembly.

To install:

8. Install the steering gear and tighten the mounting bolts.
9. Connect the hydraulic lines, using new seals.
10. Turn the steering wheel counterclockwise or clockwise against the stop and then it back about 1.7 turns until the marks are aligned.
11. Connect the U-joint to the steering gear making sure the bolt is in the locking groove of the steering stub.
12. Install the tie rod to the steering drop arm and replace the self locking nut.
13. Replace the heat shield on the steering gear and connect the ride level height control pipes on the 750iL.
14. Refill the hydraulic fluid and replace the steering wheel, if equipped with an air bag (SRS).
15. Connect the negative battery cable.

Power Steering Pump

REMOVAL & INSTALLATION

1. Disconnect the negative battery cable. Release the pressure from the reservoir.
2. Draw the hydraulic fluid from the pump reservoir. Disconnect and plug the hydraulic lines.
3. Disconnect and plug the hydraulic lines. Remove the bolts and loosen the nuts to turn the adjusting pinion.
4. Remove the drive belt.
5. Remove the bolts from the brackets holding the pump and remove the pump assembly.
6. Reverse the removal procedure for installation. Tighten the adjusting pinion to 6 ft. lbs.
7. If equipped with a tandem pump the removal and installation procedure is the same.

BELT ADJUSTMENT

Tighten the drive belt so when pressure is applied to the belt, the distance between both belt pulleys is 5–10mm of deflection.

1. Disconnect the negative battery cable. Loosen the nuts on the adjusting pinion.
2. Tighten the belt to the recommended specification and tighten the adjusting pinion nuts.
3. Connect the negative battery cable.

SYSTEM BLEEDING

1. Fill the reservoir to the **MAX** mark on the oil stick.
2. Rotate the steering in both directions fully, to each stop, until all the air is removed from the fluid.
3. Check the oil level and fill to the specified mark, if necessary.

Tie Rod Ends

REMOVAL & INSTALLATION

1. Raise and safely support the vehicle. Loosen the clamping bolt that retains the toe-in adjustment by keeping the tie rod end from turning in relation to the tie rod.
2. Remove the cotter pin and castellated or self-locking nut from the bottom of the tie rod end. Then, press the tie rod end out of the steering knuckle with the proper tool. Then, unscrew the tie rod end from the tie rod and remove it, counting the number of turns required.

To Install:

3. Install in reverse order, using a new cotter pin or castellated nut. Recheck the front alignment and reset the toe-in if necessary.
4. Torque the castellated or self-locking nut to 26.5 ft. lbs. and the clamping screw to 10 ft. lbs. Final torque the clamping bolt with the vehicle resting on its wheels.

BRAKES

For all brake system repair and service procedures not detailed below, please refer to "Brakes" in the Unit Repair section.

Master Cylinder

REMOVAL & INSTALLATION

1. Disconnect the negative battery cable. Draw off the brake fluid from the reservoir.
2. Remove the plug and disconnect the hydraulic lines. Remove the hose for the hydraulic clutch, if needed.
3. Remove the reservoir. Remove the mounting bolts and lift out the master cylinder.
4. Install the master cylinder, making sure the rubber ring is making a good seal.
5. The remainder of the installation is the reverse of the removal procedure. Bleed the brake system and connect the negative battery cable.

Proportioning Valve

REMOVAL & INSTALLATION

318iS, 325, 325i, 325iS, 325iX and M3

1. Disconnect the negative battery cable. Draw off hydraulic fluid from the master cylinder with a syringe or hose used only with clean brake fluid.
2. Disconnect the brake lines at the top and bottom of the proportioning valve.
3. Remove the clamp from the valve and disconnect the pressure connection at the union.
4. Check day/year codes, reduction factor, and switch-over pressure to make sure the new valve is identical.
5. Install in reverse order. Bleed the system.

Power Brake Booster

REMOVAL & INSTALLATION

1. Disconnect the negative battery cable. Draw off brake fluid in the reservoir and discard.
2. Remove the reservoir and disconnect the clutch hydraulic hose.
3. Disconnect all brake lines from the master cylinder.
4. Remove the instrument panel trim from the bottom/left inside the passenger compartment.
5. Remove the return spring from the brake pedal. Press off the clip and remove the pin which connects the booster rod to the brake pedal.
6. Remove the 4 nuts and pull the booster and master cylinder off in the engine compartment.
7. If the filter in the brake booster is clogged, it will have to be cleaned. To do this, remove the dust boot, retainer, damper, and filter, and clean the damper and filter. Make sure when re-

installing that the slots in the damper and filter are offset 180 degrees.

8. Install in reverse order. Adjust the stoplight switch for a clearance of 0.197–0.236 in.

9. Inspect the rubber seal between the master cylinder and booster and replace it, if necessary.

Brake Caliper

REMOVAL & INSTALLATION

Front

1. Disconnect the negative battery cable. Draw off brake fluid with a suitable syringe.
2. Disconnect the hydraulic brake lines.
3. Raise and safely support the vehicle. Remove the front tire and wheel assembly.
4. Remove the caliper mounting bolts and disconnect the brake pad wear indicator plug.
5. Remove the caliper assembly.
6. The installation is the reverse of the removal procedure. Bleed the brake system. Tighten the caliper mounting bolts to 63–79 ft. lbs. on the 3 Series and 80–89 ft. lbs. except on the 3 Series. Tighten the guide bolts to 22–25 ft. lbs.

NOTE: Make sure the brake wear indicator wire is held in the correct position by the tab of the dust cap.

Rear

1. Disconnect the negative battery cable. Draw off brake fluid with a suitable syringe.
2. Disconnect the hydraulic brake lines.
3. Raise and safely support the vehicle. Remove the rear tire and wheel assembly.
4. Remove the caliper mounting bolts and disconnect the brake pad wear indicator plug.
5. Remove the caliper assembly by pulling to the rear.
6. The installation is the reverse of the removal procedure. Bleed the brake system. Tighten the mounting bolts to 42–48 ft. lbs. and the guide bolts to 22–25 ft. lbs.

Disc Brake Pads

REMOVAL & INSTALLATION

Front

1. Raise and safely support the vehicle.
2. Remove the tire and wheel assembly.
3. Disconnect the plug for the brake pad wear indicator
4. Remove the caliper guide bolts and the spring clamp.
5. Turn up the caliper and remove the brake pads. The inner pad is located with a spring in the piston.

NOTE: The brake pads on both calipers on 1 axle should be replaced at the same time.

6. Lubricate the mounting pads with a suitable grease.
7. The installation is the reverse of the removal procedure. Bleed the brake system.

Rear

1. Raise and safely support the vehicle.
2. Remove the tire and wheel assembly.
3. Disconnect the plug for the brake pad wear indicator
4. Remove the caliper guide bolts and the spring clamp.
5. Turn up the caliper and remove the brake pads. The inner pad is located with a spring in the piston.

NOTE: The brake pads on both calipers on 1 axle should be replaced at the same time.

6. Lubricate the mounting pads with a suitable grease.
7. The installation is the reverse of the removal procedure. Bleed the brake system.

Brake Rotor

REMOVAL & INSTALLATION

Front

1. Raise and safely support the vehicle. Remove the front tire and wheel assembly.
2. Disconnect the rubber grommet from the bracket, if equipped.
3. Disconnect the plug for the brake pad indicator, if necessary.
4. Disconnect and support the caliper, using a piece of wire.
5. Remove the mounting bolts and remove the brake rotor with the proper tool.

NOTE: The inboard vented discs are balanced. Never remove or reposition the balance clamps.

6. The installation is the reverse of the removal procedure.

Rear

1. Raise and safely support the vehicle. Remove the tire and wheel assembly.
2. Disconnect and support the caliper, using a piece of wire.
3. Remove the mounting bolts and remove the brake rotor with the proper tool.
4. Always replace both discs of 1 axle.

To Install:

5. The installation is the reverse of the removal procedure. Adjust the parking brake.
6. The parking brake must be broken in after replacing the rear brake discs. This is done in the following steps:
   a. Step 1–5 complete stops from 30 mph
   b. Step 2–Allow the brakes to cool off
   c. Step 3–5 complete stops from 30 mph

Parking Brake Cable

ADJUSTMENT

1. The parking brake should be adjusted when the lever can be pulled up more than 8 notches. First, remove the cover pate on the console, the handbrake lever protrudes through this plate. Then, loosen the locknuts and loosen the adjusting nuts for the cables until they are nearly at the ends of the threads.
2. Support the vehicle securely off the rear wheels. Remove 1 wheel bolt from each rear wheel. Then, rotate 1 wheel until the hole left by removing the bolt is about 45 degrees counterclockwise from the 6 o'clock position. This will line the hole in the wheel with the star wheel which adjusts the rear brake shoes which is used for parking only. If it is difficult to turn the star wheel it will help to remove the rear wheel and, if necessary, the brake disc.
3. Turn the adjusting star wheel with a screwdriver until the rear wheel or brake disc can no longer be turned. Then, back it off 4–6 threads.
4. In the passenger compartment, pull the handbrake up 4 notches. Then, turn the cable adjusting nuts in the tightening direction until the rear wheels can just barely be turned. Make sure the adjustment is uniform. Lock the adjusting nuts in position.
5. Release the handbrake and make sure the wheels can now be turned easily, repeating Step 4 to correct a failure to release if necessary.

REMOVAL & INSTALLATION

1. Raise and safely support the vehicle. Remove the rear tire and wheel assembly.
2. Remove the rear brake discs.
2. Using brake spring pliers, disconnect the upper return spring for the

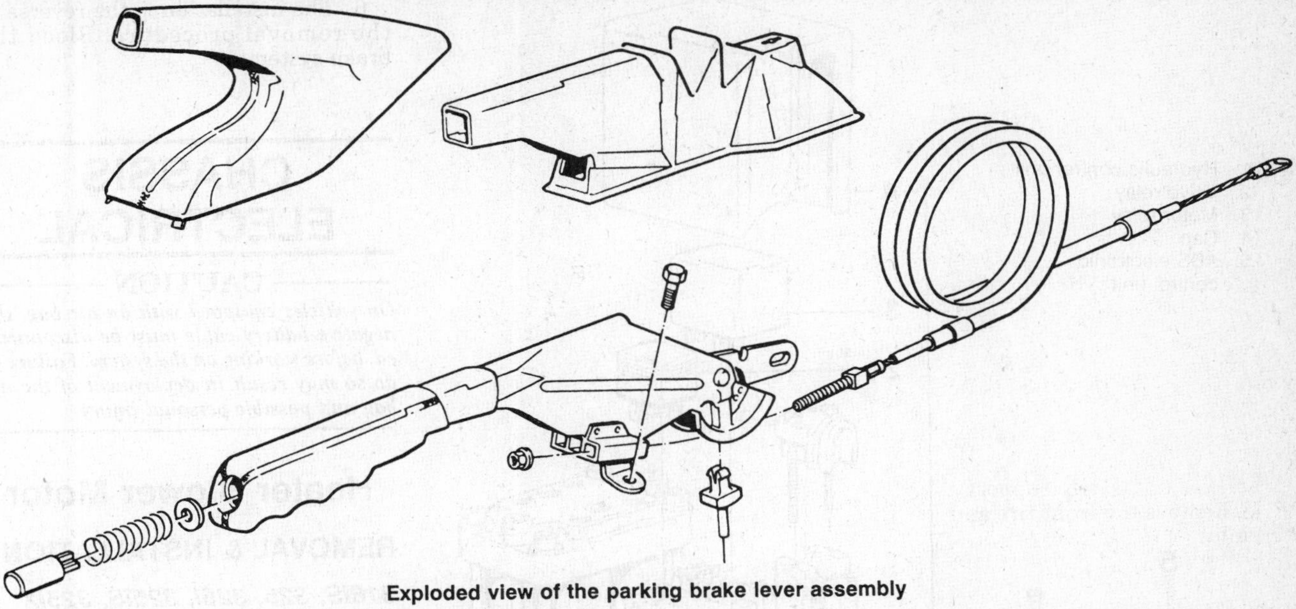

Exploded view of the parking brake lever assembly

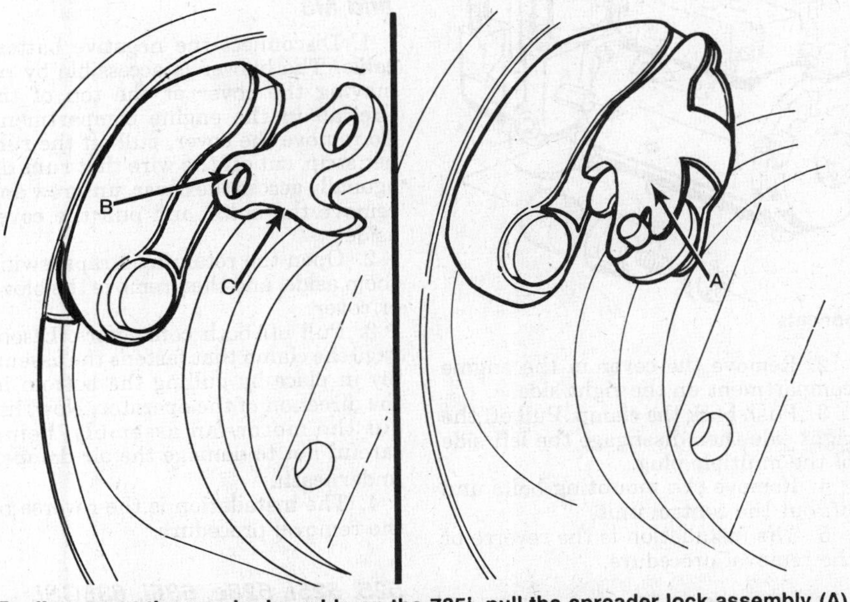

To disconnect the rear brake cable on the 735i, pull the spreader lock assembly (A) out ot the housing and remove the pin at the lower end. Disconnect the cable at (B); pull the inner portion of the spreader lock (C) out of the housing

parking brake shoes. Then, using the proper tool, turn the retaining springs for the parking brake shoes 90 degrees to unlock them and then disconnect them.

3. Separate the parking brake shoes at the top and then remove them from below.

4. Disconnect the spreader locks from the backing plates: first, rock the lower end of the spreader lock outward, and then pull out the pin. Press the cable connection out of the spreader lock. Pull the spreader lock out of the housing. Pull the cable through the backing plate.

5. Disconnect the parking brake cable at the trailing arm.

6. Working inside the vehicle, remove the console cover as follows:

a. Lift out the air grille and remove the nuts underneath.

b. Remove the cap and unscrew the mounting bolt that's located at the forward end of the console on the right side. Lift out the cover that the bolt retains.

c. Remove the bolt on the left side of the forward end of the console. If the vehicle has power windows, disconnect the plugs. Then, lift the console and remove air ducts.

d. Turn the retainer 90 degrees and peel the rubber cover downward. Now, unscrew the adjusting nuts on the parking brake cables and pull them out.

7. Install the cable in position. Adjust the parking brake.

Brake System Bleeding

Note: This procedure is valid for both ABS and non-ABS braking systems. Always use clean, factory approved brake fluid.

1. Fill the master cylinder to the maximum level with the proper brake fluid.

2. Raise and safely support the vehicle. Remove the protective caps from the bleeder screws.

3. The proper bleeding sequence always start with the brake unit farthest from the master cylinder. The proper bleeding sequence is: right rear, left rear, right front and left front.

4. Insert a tight fitting plastic tube over the bleeder screw on the caliper and the other end of the tube in a transparent container partially filled with clean brake fluid.

5. Depress the brake pedal and loosen the bleeder screw to release the brake fluid. Pump the brake pedal to the stop 12 times. Tighten the bleeder screw when the escaping brake fluid is free of air bubbles.

6. Repeat this step on all 4 wheels. Lower the vehicle.

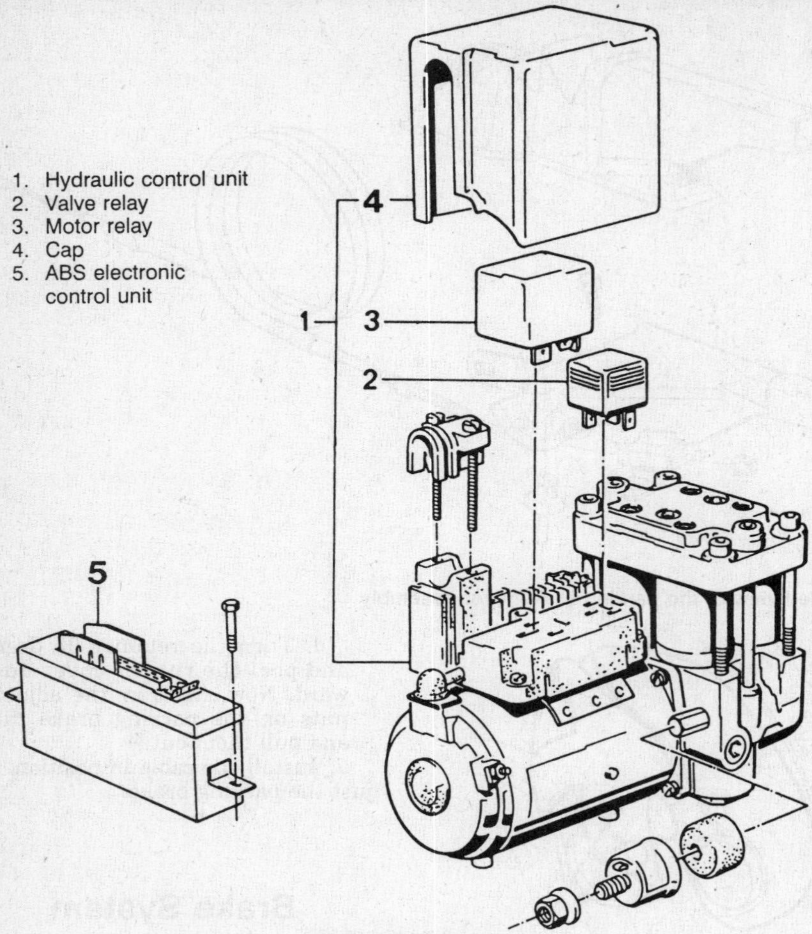

ABS components

Anti-Lock Brake System Service

PRECAUTIONS

- Remove the plugs from the electronic control unit and turn off the ignition when using an electric welder.
- If the battery has been removed, the battery terminals must be tightened on the end poles perfectly after the reinstallation of the battery.
- After the replacement of the hydraulic unit, the control unit, the speed sensors or the wire harness, the entire ABS system has to be checked with the proper tester.
- The brake system must be bled after each job on the brake system.

ABS Electronic Control Unit

REMOVAL & INSTALLATION

1. Disconnect the negative battery cable.
2. Remove the cover in the engine compartment on the right side.
3. Push back the clamp. Pull off the right side then disengage the left side of the multiple plug.
4. Remove the mounting bolts and lift out the control unit.
5. The installation is the reverse of the removal procedure.

Hydraulic Unit

REMOVAL & INSTALLATION

1. Disconnect the negative battery cable.
2. Disconnect and plug the hydraulic brake lines at the unit. Do not mix up the brake lines. The lines are marked as follows:
   a. VL—LF Brake Caliper
   b. VR—RF Brake Caliper
   c. HL—LR Brake Caliper
   d. HR—RR Brake Caliper
3. Remove the cover mounting bolts and lift off the cover.
4. Disconnect the electrical connections and plugs.
5. Loosen the mounting nuts, pull up and remove the hydraulic control unit.
6. The installation is the reverse of the removal procedure. Bleed the brake system.

CHASSIS ELECTRICAL

CAUTION

On vehicles equipped with an air bag, the negative battery cable must be disconnected, before working on the system. Failure to do so may result in deployment of the air bag and possible personal injury.

Heater Blower Motor

REMOVAL & INSTALLATION

318iS, 325, 325i, 325iS, 325iX and M3

1. Disconnect the negative battery cable. The blower is accessible by removing the cover at the top of the firewall in the engine compartment. To remove the cover, pull off the rubber strip, cut off the wire that runs diagonally across the cover, unscrew and remove the bolts, and pull the cover aside.
2. Open the retaining straps, swing them aside, and then remove the blower cover.
3. Pull off both connectors. Disengage the clamp that fastens the assembly in place by pulling the bottom in the direction of the operator. Now, lift out the motor/fan assembly, being careful not to damage the air damper underneath.
4. The installation is the reverse of the removal procedure.

525, 525i, 528e, 535i, 635CSi, M5 and M6

1. Disconnect the negative battery cable.
2. Remove the rubber insulator from the cowl. Remove the mounting bolts for the cover which is located under the windshield.
3. Push back the 3 retaining tabs and remove the 2 shells that cover the blower wheels.
4. Disconnect the electrical connector for the motor. Unclip the retaining strap for the motor and remove the motor and blower wheels.
5. Replace the motor and blower wheels as an assembly (prebalanced). The motor will fit into the housing only one way. Reverse all procedures to install, making sure the flat surface on the inlet cowls face the body.

735i, 735iL and 750iL

1. Disconnect the negative battery cable. Pull the rubber cover off the overflow tank for the cooling system. Disconnect the electrical connector and overflow hose from the overflow tank. Then, remove the mounting nuts and put the tank aside without damaging the hose leading to the radiator.
2. Cut the straps for the wiring harness running across the cowl.
3. Remove the mounting screws and remove the blower cover from the cowl.
4. Disconnect the cable and unclip it where it is clipped to the blower cover. Then, open the plastic retainer and take off the cover.
5. Disconnect the electrical connector. Lift off the metal retainer for the blower motor and remove the blower motor.
6. Install a new, prebalanced motor and blower assembly. Install in reverse order.

Windshield Wiper Motor

The electric wiper motor assembly is located under the engine hood, at the top of the cowl panel. A few vehicles have covers over the wiper motor assembly, while others have the motors exposed. Link rods operate the left and right wiper pivot assemblies from a drive crank bolted to the wiper motor output shaft.

REMOVAL & INSTALLATION

318iS, 325, 325i 325iS, 325iX and M3

1. Disconnect the negative battery cable. Remove the heater motor, as described above. Remove the bracket bracing the windshield wiper motor, which is now visible.
2. Disconnect the electrical connector for the motor.
3. Lift out the grill located at the top of the cowl and disconnect the linkages to both wiper arms at the left side shaft mounts.
4. Disconnect both wiper arms from their shafts by lifting the cover, unscrewing the nut, and pulling the arm off. Then, remove the cover, nut and washer surrounding the shafts and holding the console in place. Now remove the entire console.
5. With the motor still mounted, remove the nut retaining the linkage to the motor shaft. Then, unbolt and remove the motor from the console.
6. Installation is the reverse of the removal procedure.

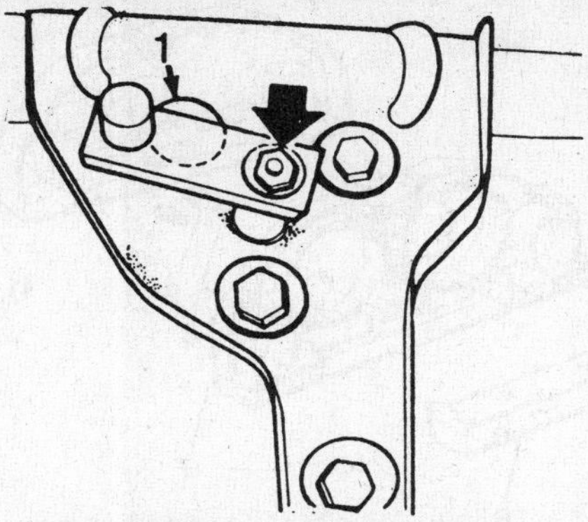

On the 735i and 733i, install the motor crank in the position shown, with the bolt (1) half hidden by the upper edge of the crank

528e, 525i, 535i, 635CSi, M5 and M6

1. Disconnect the negative battery cable. Remove the cowl cover to expose the wiper motor, if equipped.
2. Disconnect the wiper motor crank arm from the motor output shaft by removing the nut and pulling off the crank arm.
3. Remove the motor retaining screws and disconnect the electrical connector.
4. Remove the wiper motor from the vehicle.
5. Reverse the procedure to install the motor.

735i, 735iL and 750iL

1. Disconnect the negative battery cable. Make sure the wipers are in the parked position. Remove the heater blower. Take off the cover near the blower.
2. Disconnect the heater cable and lift out the linkage. Disconnect the temperature sensor.
3. Disconnect the clips, lift the cowl cover slightly and then remove the fresh air inlet cowls on either side. Then, remove the cover.
4. Unscrew bolts and remove the mounting bracket for the wiper housing. Remove the left wiper arm by pulling up the cover, loosening the pinch bolt, and removing it. Remove the right wiper by pulling up the cover, removing the through bolt and then pulling it off.
5. Lift out the clips and remove the cover for the linkage.
6. Unscrew and remove the nuts fastening the linkage to the cowl. Pull the linkage arms downward and out of the cowl.
7. Mark the relationship between the linkage lever and the motor. Remove the nut and disconnect the linkage at the motor shaft. Disconnect the electrical connector and remove the linkage.
8. Remove the mounting bolts and remove the wiper motor. If installing a new motor, connect the motor and operate it until it reaches parked position; then install the linkage so the shaft lever and linkage link are in a straight line.

To install:

9. Perform the remaining portions of the installation in reverse order, noting these points:
   a. Make sure the wiper arms are pressed all the way onto the linkage shafts so the contact pressure control will work.
   b. Make sure the inlet cowling is installed in proper relation to the blower housing and fresh air flap.

Windshield Wiper Switch

REMOVAL & INSTALLATION

1. Disconnect the negative battery cable. Remove the steering wheel. Remove the lower/left instrument panel trim.
2. Remove the screws and remove the lower steering column cover.
3. Push the locking hook for the flasher back and remove the relay, socket facing downward.
4. Take off the upper steering column cover. If the vehicle is equipped with airbags, drive out the pins and lift out the expansion rivet first.
5. Press the retaining hooks inward on both sides, pull the switch out, and then disconnect the electrical connector.
6. Installation is the reverse of removal.

Instrument Cluster

REMOVAL & INSTALLATION

1. Disconnect the negative battery cable. Remove the attaching screws and remove the lower instrument panel trim from under the steering column.
2. Remove the mounting nuts for the trim just under the instrument carrier, and remove it.
3. Unscrew the 4 screws underneath and 2 above the instrument carrier, and remove trim that surrounds the instrument carrier.
4. Remove the 2 screws at the top of the carrier, lift it out of the instrument panel, and then disconnect the plugs. To disconnect the combination plug, first pull the sliding clamp off the center.
5. To replace the speedometer, pull the speedometer from the instrument carrier.

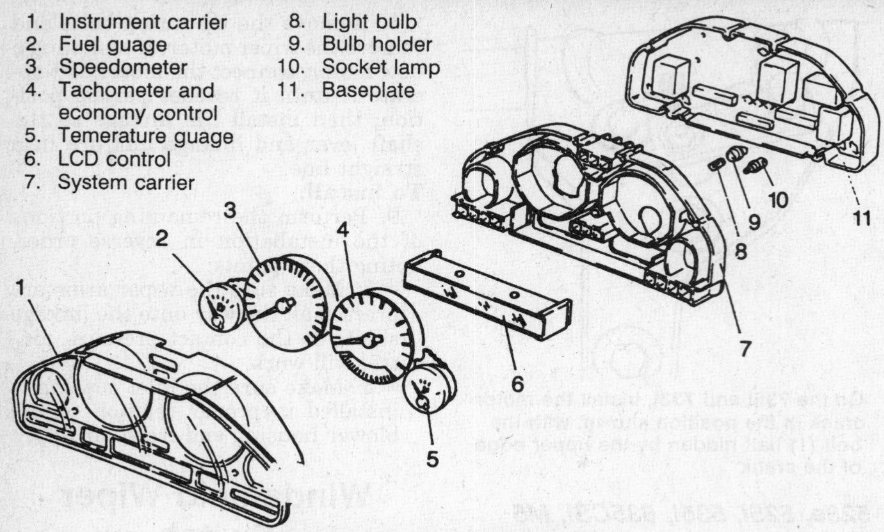

Instrument cluster—525I and 535I

6. Installation is the reverse of removal.

Headlight Switch

REMOVAL & INSTALLATION

1. Disconnect the negative battery cable. Remove the lower/left trim panel screws and remove the panel.
2. Unscrew the knob from the switch.
3. Pull off the connector plug from behind the dash panel. Pull out the switch from behind and remove it.
4. Install in reverse order.

Combination Switch

REMOVAL & INSTALLATION

CAUTION

On vehicles equipped with an air bag, the negative battery cable must be disconnected, before working on the system. Failure to do so may result in deployment of the air bag and possible personal injury.

1. Disconnect the negative battery cable. Remove the steering wheel.
2. Remove the lower instrument panel on the left side.
3. Remove the steering column casing lower section.
4. If equipped with an air bag, drive out the pins and lift out the expansion rivet. Remove the upper section of the steering column casing.
5. Remove the plug and disconnect the electrical connectors.
6. Push in the retaining hooks on both sides and pull out the switch.
7. The installation is the reverse of the removal procedure.

Ignition Switch

REMOVAL & INSTALLATION

1. Disconnect the negative battery cable. Remove the steering wheel.
2. Remove the trim panel on the lower left side.
3. Remove the steering column casing lower section.
4. Press in the retaining hooks on both sides and remove the switch.
5. The installation is the reverse of the removal procedure. Check the position of the ignition switch to the steering wheel lock.

Stoplight Switch

ADJUSTMENT

1. Disconnect the negative battery cable. Disconnect the electrical connector to the switch.
2. Loosen the locknut and then turn the switch outward to remove it.
3. Screw the new switch in. Adjust the gap between the pedal and actuator on the switch to 0.197–0.236 in. on the 3 series and 0.236–0.020 in. except 3 series.
4. Tighten the locknut.

REMOVAL & INSTALLATION

1. Disconnect the negative battery cable.
2. Remove the lower left side trim panel and disconnect the plug.
3. Press down the brake pedal and pull the plunger and sleeve down completely.
4. Press in the retainers and pull back on the switch.
5. The installation is the reverse of the removal procedure

Neutral Safety Switch

ADJUSTMENT

1. Disconnect the leads at the switch terminal.
2. Ground the negative terminal and connect a proper test light to the positive terminal.
3. The test light should light when the gear selector is placed on **P** or **N**.
4. If the switch needs adjusting, unscrew the switch and place thicker shims behind the switch and the transmission housing.

REMOVAL & INSTALLATION

1. Disconnect the negative battery cable. Raise and safely support the vehicle.
2. Disconnect the electrical connection at the switch.
3. Remove the switch using the proper tool.
4. The installation is the reverse of the removal procedure.

Fuses, Circuit Breakers and Relays

LOCATION

The fuse box is located under the engine hood on the left side, near the upper strut housing or near the battery.
Various relays are also mounted on the fuse box for easy accessibility.

Computer

LOCATION

The Digital Motor Electronics (DME) unit is located in the engine compartment on the passenger side by the firewall.

Flashers

LOCATION

The flashers are located behind the bottom center of the lower instrument trim panel.

Cruise Control

ADJUSTMENT

1. Check the distance between the nipple and the nipple holder with the throttle valve closed and the vehicle in the **N** position.
2. Adjust with a knurled nut, if necessary. The distance should be 0.039–0.059 in. exc. 7 series and 0.039–0.079 in. on the 7 series.

Chrysler Corp. Imports 4

Colt, Colt Vista, Conquest—All Models

SERIAL NUMBER IDENTIFICATION

Vehicle Identification Number

The vehicle identification plate is mounted on the left side instrument panel, adjacent to the lower corner of the windshield on the driver's side and is visible through the windshield. A standardized 17 digit Vehicle Identification Number (VIN) is used. The 8th digit designates the engine code and 10th designates model year (J=1988, K=1989, L=1990, M=1991, N=1992).

Vehicle Information Code Plate

The Vehicle Information Code Plate is attached to the bulkhead on the firewall, in the engine compartment. The plate shows model code, engine model, transaxle model and body color code.

Chassis Number

The chassis number plate is stamped on the top center of the firewall, located in the engine compartment.

Engine Number

The engine number is stamped at the right front side, on the top edge of the cylinder block and contains the engine model number and engine serial number.

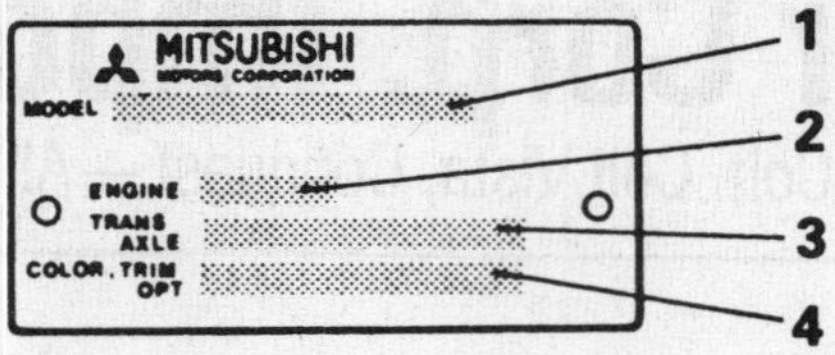

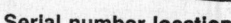

1. Model
2. Engine
3. Transaxle
4. Color, trim option

Vehicle information code plate on engine firewall

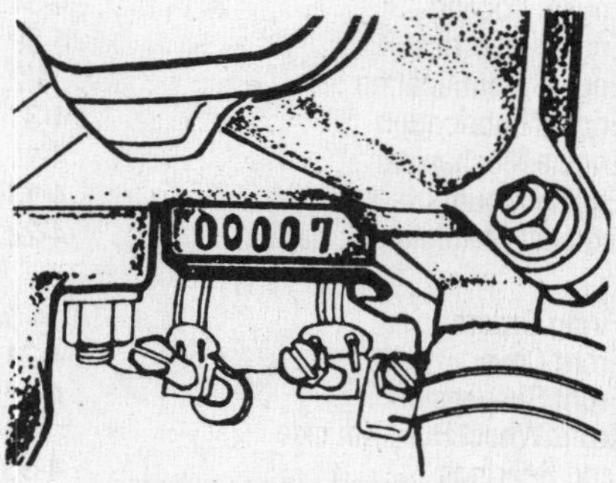

Engine number location

Vehicle Safety Certification Label

The label is located at the lower corner of the driver's door. The label indicates the month and year of manufacture, gross vehicle weight and VIN number.

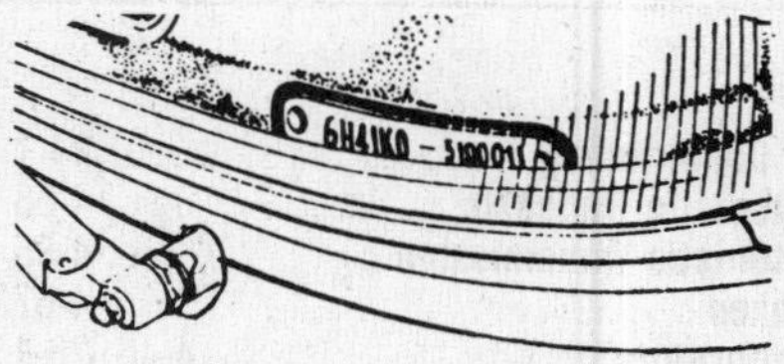

Serial number location

Engine model number

SPECIFICATIONS

ENGINE IDENTIFICATION

Year	Model	Engine Displacement cu. in. (cc/liter)	Engine Series Identification	No. of Cylinders	Engine Type
1988	Colt	89.6 (1468/1.5)	G15B	4	SOHC
	Colt	97.5 (1597/1.6)	G32B	4	SOHC
	Colt Vista	121.9 (1997/2.0)	G63B	4	SOHC
	Conquest	155.9 (2555/2.6)	G54B	4	SOHC
1989	Colt	89.6 (1468/1.5)	4G15	4	SOHC
	Colt	97.3 (1595/1.6)	4G61	4	DOHC
	Colt Wagon	107.1 (1755/1.8)	G37	4	SOHC
	Colt Vista	121.9 (1997/2.0)	G63B	4	SOHC
	Conquest	155.9 (2555/2.6)	G54B	4	SOHC

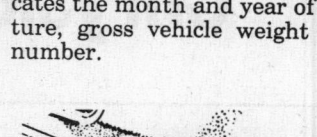

ENGINE IDENTIFICATION

Year	Model	Engine Displacement cu. in. (cc/liter)	Engine Series Identification	No. of Cylinders	Engine Type
1990	Colt	89.6 (1468/1.5)	4G15	4	SOHC
	Colt	97.3 (1595/1.6)	4G61	4	DOHC
	Colt Wagon	107.1 (1755/1.8)	G37	4	SOHC
	Colt Vista	121.9 (1997/2.0)	G63B	4	SOHC
1991-92	Colt	92 (1507/1.5)	4G15	4	SOHC
	Colt Vista	121.9 (1997/2.0)	G63B	4	SOHC

SOHC—Single Overhead Camshaft
DOHC—Double Overhead Camshaft

GENERAL ENGINE SPECIFICATIONS

Year	Model	Engine Displacement cu. in. (cc)	Fuel System Type	Net Horsepower @ rpm	Net Torque @ rpm (ft. lbs.)	Bore × Stroke (in.)	Com-pression Ratio	Oil Pressure @ rpm
1988	Colt	89.6 (1468)	Carb.	68 @ 5000	84 @ 3000	2.99 × 3.23	9.4:1	50–64 @ 750
	Colt	97.5 (1597)	ECI	102 @ 5500	122 @ 3000	3.03 × 3.39	7.6:1	57–71 @ 700
	Colt Vista	121.9 (1997)	MPI	120 @ 5000	108 @ 3500	3.34 × 3.46	8.5:1	57–71 @ 700
	Conquest	155.9 (2555)	EFI/Turbo	170 @ 5000	220 @ 5000	3.59 × 3.86	7.0:1	50–54 @ 850
1989	Colt	89.6 (1468)	MPI	68 @ 5500	82 @ 3500	3.00 × 3.23	9.4:1	50–64 @ 750
	Colt	97.3 (1595)	MPI	102 @ 5500	122 @ 3000	3.24 × 2.95	8.0:1	57–71 @ 700
	Colt Wagon	107.1 (1755)	MPI	87 @ 5000	102 @ 5000	3.17 × 3.39	9.0:1	63 @ 2000
	Colt Vista	121.9 (1997)	MPI	120 @ 5000	116 @ 4500	3.35 × 3.46	8.5:1	42.7 @ 2000
	Conquest	155.9 (2555)	EFI/Turbo	140 @ 5000	317 @ 2500	3.59 × 3.86	7.0:1	56.5 @ 2000
1990	Colt	89.6 (1468)	MPI	68 @ 5500	126 @ 3000	2.99 × 3.23	9.4:1	42 @ 2000
	Colt	97.3 (1595)	MPI	92 @ 6500	134 @ 5000	3.24 × 2.95	8.0:1	42 @ 2000
	Colt Wagon	107.1 (1755)	MPI	65 @ 5000	138 @ 3000	3.17 × 3.39	9.0:1	42 @ 2000
	Colt Vista	121.9 (1997)	MPI	72 @ 5000	153 @ 3500	3.35 × 3.46	8.5:1	42.7 @ 2000
1991-92	Colt	92.0 (1507)	MPI	92 @ 6000	93 @ 3000	2.97 × 3.23	9.2:1	11.5 @ 750
	Colt Vista	121.9 (1997)	MPI	96 @ 5000	113 @ 3500	3.35 × 3.46	8.5:1	11.5 @ @ 700

ECI—Electronic Controlled Injection
EFI—Electronic Fuel Injection
MPI—Multi-Point Injection

GASOLINE ENGINE TUNE-UP SPECIFICATIONS

Year	Model	Engine Displacement cu. in. (cc)	Spark Plugs Type	Spark Plugs Gap (in.)	Ignition Timing (deg.) MT	Ignition Timing (deg.) AT	Compression Pressure (psi)	Fuel Pump (psi)	Idle Speed (rpm) MT	Idle Speed (rpm) AT	Valve Clearance In.	Valve Clearance Ex.
1988	Colt	89.6 (1468)	BPR6ES-11	0.039–0.040	5B	5B	NA	2.7–3.7	700	750	0.006	0.010 ①
	Colt	97.5 (1597)	BPR6ES-11	0.039–0.040	10B	10B	NA	33.6 ②	700	700	0.006	0.010
	Colt Vista	121.9 (1997)	BPR6ES-11	0.039–0.040	5B	5B	NA	36.3	700	700	Hyd. ①	Hyd. ①
	Conquest	155.9 (2555)	BPR6ES-11	0.039–0.040	10B	10B	NA	33.6	850	850	0.006	Hyd. ①
1989	Colt	86.9 (1468)	BPR6ES-11	0.039–0.043	5B	5B	NA	47.6	700	750	0.006	0.010
	Colt	97.3 (1595)	BPR6ES-11	0.028–0.031	5B	5B	NA	47.6 ③	700	750	0.006	0.010
	Colt Wagon	107.1 (1755)	BPR6ES-11	0.039–0.043	5B	5B	NA	47.6	700	700	0.006	0.010
	Colt Vista	121.9 (1997)	WZ0EPR-11	0.039–0.043	5B	5B	NA	47.6	700	700	Hyd.	Hyd.
	Conquest	155.9 (2555)	BPR7EA-11	0.039–0.043	10B	10B	NA	33.6	850	850	Hyd. ①	Hyd. ①
1990	Colt	86.9 (1468)	BPR6ES-11	0.039–0.043	5B	5B	NA	47.6	700	750	0.006	0.010
	Colt	97.3 (1595)	BPR6ES-11	0.039–0.043	5B	5B	NA	47.6	700	750	Hyd.	Hyd.
	Colt Wagon	107.1 (1755)	BPR6ES-11	0.039–0.043	5B	5B	NA	47.6	700	700	0.006	0.010
	Colt Vista	121.9 (1997)	W20EPR-11	0.039–0.043	5B	5B	NA	47.6	700	700	Hyd.	Hyd.
1991	Colt	92.0 (1507)	BPR6ES-11	0.039–0.043	5B	5B	137	47.6	750	750	0.006	0.010
	Colt Vista	121.9 (1997)	BPR6ES-11	0.039–0.043	5B	5B	119	47.6	700	700	Hyd.	Hyd.
1992	SEE UNDERHOOD SPECIFICATIONS STICKER											

NOTE: The Underhood Specifications sticker often reflects tune-up specification changes in production. Sticker figures must be used if they disagree with those in this chart.

NA—Not Available
Hyd.—Hydraulic
① Jet valve clearance—0.010 in.
② MPI 47.6
③ Engines equipped with Turbocharger—36.3 PSI

FIRING ORDERS

NOTE: To avoid confusion, always replace spark plug wires one at a time.

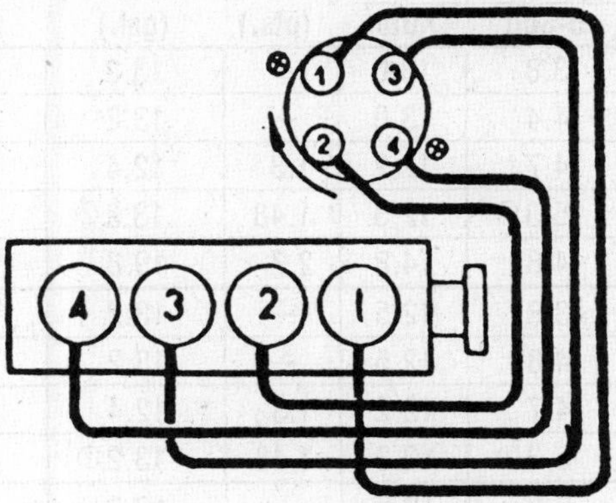

1468cc, 1597cc and 1755cc Engines
Engine Firing Order: 1-3-4-2
Distributor Rotation: Clockwise

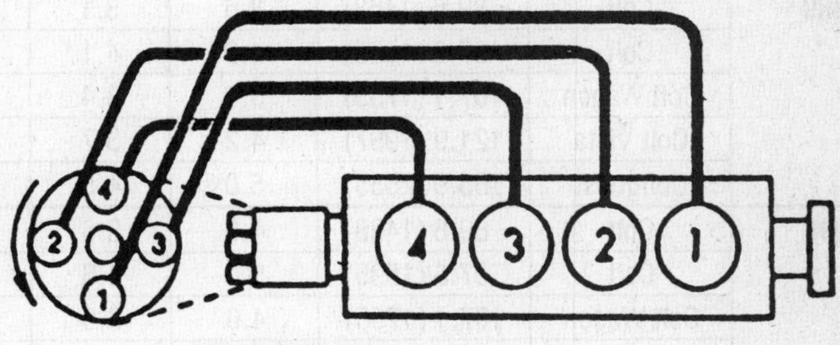

1507cc Engine
Engine Firing Order: 1-3-4-2
Distributor Rotation: Counterclockwise

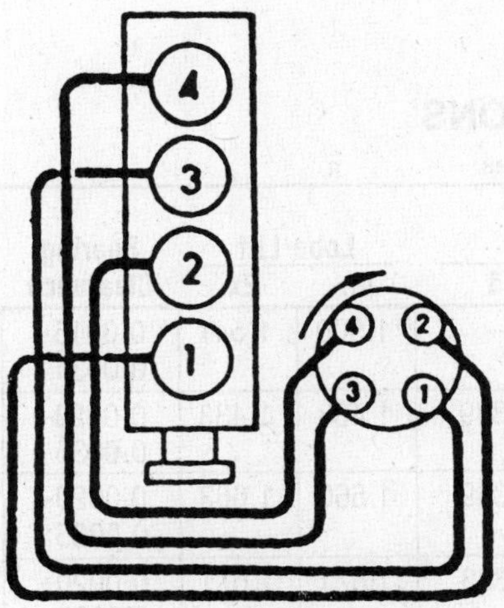

1997cc and 2555cc Engines
Engine Firing Order: 1-3-4-2
Distributor Rotation: Clockwise

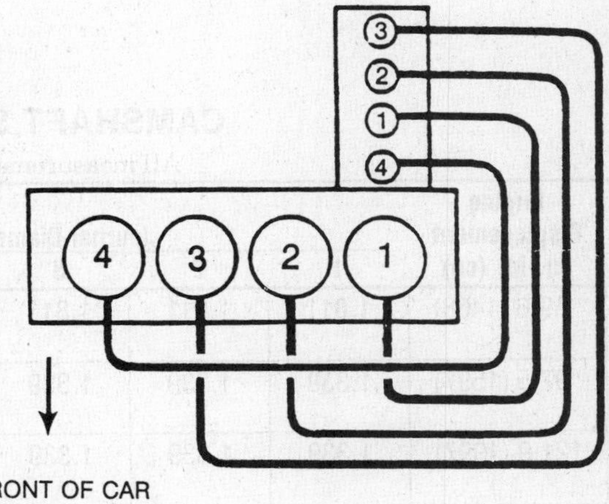

1595cc DOHC Engine
Engine Firing Order: 1-3-4-2
Distributorless Ignition System

CAPACITIES

Year	Model	Engine Displacement cu. in. (cc)	Engine Crankcase (qts.) with Filter	Engine Crankcase (qts.) without Filter	Transmission (pts.) 4-Spd	Transmission (pts.) 5-Spd	Transmission (pts.) Auto.	Drive Axle (pts.)	Fuel Tank (gal.)	Cooling System (qts.)
1988	Colt	89.6 (1468)	3.7	3.2	3.6	3.8	13.0	—	13.2	5.3
	Colt	97.5 (1597)	4.5	3.7	—	4.4	12.3	—	11.9	5.3
	Colt Vista	121.9 (1997)	4.2	3.7	—	5.3③	12.3	1.48	13.2②	7.4
	Conquest	155.9 (2555)	5.0①	4.5	—	2.4	7.4	2.7	19.8	9.2

CAPACITIES

Year	Model	Engine Displacement cu. in. (cc)	Engine Crankcase (qts.) with Filter	Engine Crankcase (qts.) without Filter	Transmission (pts.) 4-Spd	Transmission (pts.) 5-Spd	Transmission (pts.) Auto.	Drive Axle (pts.)	Fuel Tank (gal.)	Cooling System (qts.)
1989	Colt	89.6 (1468)	3.6	3.1	3.6	3.8	13.0	—	13.2	5.3
	Colt	97.3 (1595)	4.6	4.1	—	4.4	13.0	—	13.2	5.3
	Colt Wagon	107.1 (1755)	3.9	3.4	—	4.7	12.2	1.3	12.4	5.3
	Colt Vista	121.9 (1997)	4.2	3.7	—	5.3 ③	12.3	1.48	13.2 ②	7.4
	Conquest	155.9 (2555)	5.0 ①	4.5	—	4.8	14.8	2.7	19.8	9.5
1990	Colt	89.6 (1468)	4.0	3.6	3.6	3.8	12.5	—	13.2	5.3
	Colt	97.3 (1595)	5.0	4.6	—	4.8	12.5	—	13.2	5.3
	Colt Wagon	107.1 (1755)	4.0	3.5	—	4.7	12.2	1.3	12.4	5.3
	Colt Vista	121.9 (1997)	4.2	3.7	—	5.3 ③	12.3	1.48	13.2 ②	7.4
1991-92	Colt	92.0 (1507)	4.0	3.6	3.6	3.8	12.9	—	13.2	5.3
	Colt Vista	121.9 (1997)	4.2	3.7	—	5.2 ③	12.2	1.48	13.2 ②	7.4

① Including oil filter and oil cooler
② 4WD—14.5 gals.
③ 4WD—4.4 pts.

CAMSHAFT SPECIFICATIONS

All measurements given in inches.

Year	Engine Displacement cu. in. (cc)	Journal Diameter 1	Journal Diameter 2	Journal Diameter 3	Journal Diameter 4	Journal Diameter 5	Lobe Lift In.	Lobe Lift Ex.	Bearing Clearance	Camshaft End Play
1988	89.6 (1468)	1.811	1.811	1.811	—	—	1.500	1.541	0.0015–0.0031	0.002–0.008
	97.5 (1597)	1.339	1.339	1.339	1.339	1.339	1.433	1.433	0.0020–0.0035	0.002–0.006
	121.9 (1997)	1.339	1.339	1.339	1.339	1.339	1.660	1.663	0.0020–0.0035	0.004–0.008
	155.9 (2555)	1.339	1.339	1.339	1.339	1.339	1.671	1.671	0.0020–0.0035	0.004–0.008
1989	89.6 (1468)	1.811	1.811	1.811	—	—	1.532	1.534	0.0015–0.0031	0.002–0.008
	97.3 (1595)	1.020	1.020	1.020	1.020	1.020	1.386	1.375	0.0020–0.0035	0.004–0.008
	107.1 (1755)	1.337	1.337	1.337	1.337	1.337	1.414	1.414	0.0020–0.0035	0.002–0.006
	121.9 (1997)	1.339	1.339	1.339	1.339	1.339	1.660	1.663	0.0020–0.0045	0.004–0.008
	155.9 (2555)	1.339	1.339	1.339	1.339	1.339	1.673	1.673	0.0020–0.0045	0.004–0.008
1990	89.6 (1468)	1.811	1.811	1.811	—	—	1.500	1.532	0.0015–0.0032	0.002–0.008
	97.3 (1595)	1.020	1.020	1.020	1.020	1.020	1.386	1.374	0.0020–0.0035	0.004–0.008
	107.1 (1755)	1.337	1.337	1.337	1.337	1.337	1.414	1.414	0.0020–0.0035	0.002–0.008
	121.9 (1997)	1.339	1.339	1.339	1.339	1.339	1.660	1.663	0.0020–0.0035	0.004–0.008

CAMSHAFT SPECIFICATIONS

All measurements given in inches.

Year	Engine Displacement cu. in. (cc)	Journal Diameter					Lobe Lift		Bearing Clearance	Camshaft End Play
		1	2	3	4	5	In.	Ex.		
1991-92	92.0 (1507)	1.811	1.811	1.811	1.811	1.811	1.526	1.539	0.0024–0.0039	0.004–0.008
	121.9 (1997)	1.340	1.340	1.340	1.340	1.340	1.657	1.657	0.0020–0.0035	0.004–0.008

CRANKSHAFT AND CONNECTING ROD SPECIFICATIONS

All measurements are given in inches.

Year	Engine Displacement cu. in. (cc)	Crankshaft				Connecting Rod		
		Main Brg. Journal Dia.	Main Brg. Oil Clearance	Shaft End-play	Thrust on No.	Journal Diameter	Oil Clearance	Side Clearance
1988	89.6 (1468)	1.8898	0.0008–0.0028	0.002–0.007	3	1.6535	0.0004–0.0024	0.004–0.010
	97.5 (1597)	2.2441	0.0008–0.0028	0.002–0.007	3	1.7717	0.0004–0.0028	0.004–0.010
	121.9 (1997)	2.2441	0.0008–0.0028	0.002–0.007	3	1.7717	0.0008–0.0020	0.004–0.010
	155.9 (2555)	2.5984	0.0008–0.0028	0.002–0.007	3	2.0866	0.0008–0.0028	0.004–0.010
	155.9 (2555) Turbo	2.3622	0.0008–0.0020	0.002–0.007	3	2.0866	0.0008–0.0024	0.004–0.010
1989	89.6 (1468)	1.8898	0.0008–0.0028	0.002–0.007	3	1.6535	0.0004–0.0024	0.004–0.010
	97.3 (1595)	2.2441	0.0008–0.0028	0.002–0.007	3	1.7717	0.0004–0.0028	0.004–0.010
	107.1 (1755)	2.2441	0.0008–0.0020	0.002–0.007	3	1.7717	0.0008–0.0020	0.004–0.010
	121.9 (1997)	2.2441	0.0008–0.0028	0.002–0.007	3	1.7717	0.0008–0.0020	0.004–0.010
	155.9 (2555)	2.5984	0.0008–0.0028	0.002–0.007	3	2.0866	0.0008–0.0028	0.004–0.010
	155.9 (2555) Turbo	2.3622	0.0008–0.0020	0.002–0.007	3	2.0866	0.0008–0.0024	0.004–0.010
1990	89.6 (1468)	1.8898	0.0008–0.0018	0.002–0.007	3	1.6535	0.0006–0.0017	0.004–0.010
	97.3 (1595)	2.2441	0.0008–0.0020	0.002–0.007	3	1.7717	0.0008–0.0020	0.004–0.010
	107.1 (1755)	2.2441	0.0008–0.0020	0.002–0.007	3	1.7717	0.0008–0.0020	0.004–0.010
	121.9 (1997)	2.2441	0.0008–0.0020	0.002–0.007	3	1.7717	0.0006–0.0020	0.004–0.010
1991-92	92.0 (1507)	1.8900	0.0008–0.0028	0.002–0.007	3	1.6500	0.0008–0.0024	0.004–0.010
	121.9 (1997)	2.2400	0.0008–0.0020	0.002–0.007	3	1.7700	0.0006–0.0020	0.004–0.010

VALVE SPECIFICATIONS

Year	Engine Displacement cu. in. (cc)	Seat Angle (deg.)	Face Angle (deg.)	Spring Test Pressure (lbs.)	Spring Installed Height (in.)	Stem-to-Guide Clearance (in.)		Stem Diameter (in.)	
						Intake	Exhaust	Intake	Exhaust
1988	89.6 (1468)	45	45	69 @ 1.417	1.417	0.0012–0.0024	0.0020–0.0035	0.3147–0.3153	0.3147–0.3153
	97.5 (1597)	45	45	61 @ 1.470	1.470	0.0012–0.0024	0.0020–0.0035	0.3147–0.3153	0.3147–0.3153
	121.9 (1997)	45	45	40 @ 1.591	1.591	0.0012–0.0024	0.0020–0.0035	0.3147–0.3153	0.3147–0.3153
	155.9 (2555)	45	45	61 @ 1.590	1.590	0.0012–0.0024	0.0020–0.0035	0.3147–0.3153	0.3147–0.3153
1989	89.6 (1468)	44	45	53	—	0.0008–0.0020	0.0020–0.0035	0.2600	0.2600
	97.3 (1595)	44	45	53	—	0.0008–0.0020	0.0020–0.0033	0.2585–0.2586	0.2571–0.2580
	107.1 (1755)	45	45	62	1.470	0.0012–0.0024	0.0020–0.0035	0.3200	0.3200
	121.9 (1997)	45	45	73 @ 1.591	1.591	0.0012–0.0024	0.0020–0.0035	0.3147–0.3153	0.3147–0.3153
	155.9 (2555)	45	45	73 @ 1.591	1.591	0.0012–0.0024	0.0020–0.0035	0.3147–0.3153	0.3147–0.3153
1990	89.6 (1468)	44	45	69 @ 1.417	1.417	0.0008–0.0020	0.0020–0.0035	0.2600	0.2600
	97.3 (1595)	44	45	53	—	0.0008–0.0020	0.0020–0.0033	0.2585–0.2586	0.2571–0.2580
	107.1 (1755)	45	45	62	1.470	0.0012–0.0024	0.0020–0.0035	0.3200	0.3200
	121.9 (1997)	45	45	73 @ 1.591	1.591	0.0012–0.0024	0.0020–0.0035	0.3147–0.3153	0.3147–0.3153
1991–92	92.0 (1507)	44–44.5	45–45.5	①	②	0.0008–0.0020	0.0020–0.0035	0.2585–0.2591	0.2571–0.2579
	121.9 (1997)	45	45	73 @ 1.591	1.591	0.0012–0.0024	0.0020–0.0035	0.3100	0.3100

① Intake—51 @ 1.815
Exhaust—64 @ 1.843
② Intake—1.815
Exhaust—1.843

PISTON AND RING SPECIFICATIONS

All measurements are given in inches.

Year	Engine Displacement cu. in. (cc)	Piston Clearance	Ring Gap			Ring Side Clearance		
			Top Compression	Bottom Compression	Oil Control	Top Compression	Bottom Compression	Oil Control
1988	89.6 (1468)	0.0008–0.0016	0.0080–0.0160	0.0080–0.0160	0.0080–0.0200	0.0012–0.0028	0.0008–0.0024	0.0010–0.0030
	97.5 (1597)	0.0008–0.0016	0.0100–0.0160	0.0080–0.0140	0.0080–0.0280	0.0012–0.0028	0.0008–0.0024	Snug
	121.9 (1997)	0.0008–0.0016	0.0100–0.0180	0.0080–0.0160	0.0080–0.0028	0.0020–0.0040	0.0010–0.0020	Snug
	155.9 (2555)	0.0008–0.0016	0.0098–0.0177	0.0098–0.0177	0.0078–0.0354	0.0024–0.0039	0.0008–0.0024	0.0008–0.0024
	155.9 (2555) Turbo	0.0008–0.0016	0.0120–0.0200	0.0100–0.0160	0.0120–0.0310	0.0020–0.0040	0.0010–0.0020	Snug

PISTON AND RING SPECIFICATIONS

All measurements are given in inches.

Year	Engine Displacement cu. in. (cc)	Piston Clearance	Ring Gap Top Compression	Ring Gap Bottom Compression	Ring Gap Oil Control	Ring Side Clearance Top Compression	Ring Side Clearance Bottom Compression	Ring Side Clearance Oil Control
1989	89.6 (1468)	0.0008–0.0016	0.0080–0.0138	0.0080–0.0138	0.0080–0.0280	0.0012–0.0028	0.0012–0.0028	Snug
	97.3 (1595)	0.0008–0.0016 ①	0.0100–0.0160	0.0140–0.0197	0.0080–0.0280	0.0012–0.0028	0.0008–0.0024	Snug
	107.1 (1755)	0.0008–0.0016	0.0118–0.0140	0.0080–0.0140	0.0080–0.0280	0.0018–0.0033	0.0008–0.0024	Snug
	121.9 (1997)	0.0004–0.0012	0.0100–0.0160	0.0080–0.0140	0.0080–0.0028	0.0012–0.0028	0.0008–0.0024	Snug
	155.9 (2555) Turbo	0.0012–0.0020	0.0120–0.0177	0.0098–0.0177	0.0118–0.0315	0.0020–0.0035	0.0008–0.0024	Snug
1990	89.6 (1468)	0.0008–0.0016	0.0080–0.0138	0.0080–0.0138	0.0080–0.0280	0.0012–0.0028	0.0008–0.0024	Snug
	97.3 (1595)	0.0008–0.0016	0.0100–0.0160	0.0140–0.0197	0.0080–0.0280	0.0012–0.0028	0.0012–0.0028	Snug
	107.1 (1755)	0.0008–0.0016	0.0118–0.0177	0.0080–0.0140	0.0080–0.0280	0.0018–0.0033	0.0008–0.0024	Snug
	121.9 (1997)	0.0004–0.0012	0.0100–0.0160	0.0080–0.0140	0.0080–0.0028	0.0012–0.0028	0.0008–0.0024	Snug
1991-92	92.0 (1507)	0.0008–0.0016	0.0079–0.0157	0.0079–0.0138	0.0079–0.0276	0.0012–0.0028	0.0008–0.0024	Snug
	121.9 (1997)	0.0004–0.0012	0.0098–0.0157	0.0079–0.0138	0.0079–0.0276	0.0012–0.0028	0.0008–0.0024	Snug

① With Turbocharger 0.0012–0.0020

TORQUE SPECIFICATIONS

All readings in ft. lbs.

Year	Engine Displacement cu. in. (cc)	Cylinder Head Bolts	Main Bearing Bolts	Rod Bearing Bolts	Crankshaft Pulley Bolts	Flywheel Bolts	Manifold Intake	Manifold Exhaust	Spark Plugs
1988	89.6 (1468)	50–54	37–39	23–25	57–72	94–101	11–14	11–14	15–21
	97.5 (1597)	50–54	36–40	23–25	80–93	94–101 ①	11–14	11–14	15–21
	121.9 (1997)	65–72	37–39	37–38	80–94	94–101 ①	11–14	11–14	15–21
	155.9 (2555)	65–72	55–61	33–34	80–94	94–101	11–14	11–14	15–21
1989	89.6 (1468)	50–54	37–40	23–25	57–72	94–101	11–14	11–14	15–21
	97.3 (1595)	65–72	47–51	36–38	80–94	94–101 ①	11–14	11–14	15–21
	107.1 (1755)	51–54	37–39	24–25	11–13	94–101	11–14	11–14	15–21
	121.9 (1997)	65–72	37–39	33–35	15–21	94–101	11–14	11–14	15–21
	155.9 (2555)	65–72	55–61	33–34	80–94	94–101	11–14	11–14	15–21
1990	89.6 (1468)	50–54	36–40	14 ②	51–72	94–101	11–14	11–14	15–21
	97.3 (1595)	50–54	47–51	36–38	80–94	94–101 ①	11–14	11–14	15–21
	107.1 (1755)	51–54	37–39	24–25	11–13	94–101	11–14	11–14	15–21
	121.9 (1997)	65–72	37–39	37–38	15–21	94–101 ①	11–14	11–14	15–21

TORQUE SPECIFICATIONS

All readings in ft. lbs.

Year	Engine Displacement cu. in. (cc)	Cylinder Head Bolts	Main Bearing Bolts	Rod Bearing Bolts	Crankshaft Pulley Bolts	Flywheel Bolts	Manifold Intake	Manifold Exhaust	Spark Plugs
1991-92	92.0 (1507)	51-54	47-51	14②	51-72	94-101	11-14	11-14	15-21
	121.9 (1997)	65-72	37-39	37-38	15-21	94-101	11-14	11-14	15-21

① With automatic transmission—84-90 ft. lbs.
② Torque nuts to 14 ft. lbs. and back off. Torque nuts again to 14 ft. lbs., then tighten an additional 1/4 turn.

BRAKE SPECIFICATIONS

All measurements in inches unless noted.

Year	Model	Lug Nut Torque (ft. lbs.)	Master Cylinder Bore	Brake Disc Minimum Thickness	Brake Disc Maximum Runout	Standard Brake Drum Diameter	Minimum Lining Thickness Front	Minimum Lining Thickness Rear
1988	Colt	51-58	0.8125①	0.450④	0.006	7.100	0.080	0.040
	Colt Vista	51-58	0.8750	0.650③	0.006	8.000⑥	0.080	0.040
	Conquest	51-58	0.9400	0.880②	0.006	—	0.080	0.040
1989	Colt	51-58	0.8125①	0.450④	0.006	7.100	0.080	0.040
	Colt Wagon	51-58	0.8700	⑤	0.006	8.000	0.080	0.040
	Colt Vista	51-58	0.8750	0.650③	0.006	8.000⑥	0.080	0.040
	Conquest	51-58	0.9400	0.880②	0.006	—	0.080	0.040
1990	Colt	51-58	0.8125①	0.450	0.006	7.100	0.080	0.040
	Colt Wagon	51-58	0.8700	⑤	0.006	8.000	0.080	0.040
	Colt Vista	51-58	0.8750	0.650③	0.006	8.000	0.080	0.040
1991-92	Colt	51-58	0.8125	0.450	0.006	7.100	0.080	0.040
	Colt Vista	51-58	0.8750	0.650③	0.006	8.000	0.080	0.040

① Colt Turbo—0.8750
② Rear—0.650 in.
③ 4WD—0.880 in.
④ Turbo—0.645 in.
⑤ AD54 type—0.650 in.
PF515 type—0.450 in.
⑥ 4WD—9.000 in.

WHEEL ALIGNMENT

Year	Model	Caster Range (deg.)	Caster Preferred Setting (deg.)	Camber Range (deg.)	Camber Preferred Setting (deg.)	Toe-in (in.)	Steering Axis Inclination (deg.)
1988	Colt	3/16-13/16	11/16	1/2N-1/2P	0	0	NA
	Colt Vista	5/16-15/16	13/16①	1/16N-15/16P	7/16②	0	NA
	Conquest	—	5 13/16	—	0③	0	NA

WHEEL ALIGNMENT

Year	Model	Caster Range (deg.)	Caster Preferred Setting (deg.)	Camber Range (deg.)	Camber Preferred Setting (deg.)	Toe-in (in.)	Steering Axis Inclination (deg.)
1989	Colt	3/16–13/16	11/16	1/2N–1/2P	0	0	NA
	Colt Wagon	3/16–13/16 ①	13/16	1/2N–1/2P ①	0 ②	0	NA
	Colt Vista	5/16–15/16	13/16 ①	1/16N–15/16P	7/16 ②	0	NA
	Conquest		55/16	—	0 ③	0	NA
1990	Colt	3/16–13/16	11/16	1/2N–1/2P	0	0	NA
	Colt Wagon	3/16–13/16 ①	13/16	1/2N–1/2P ①	0 ②	0	NA
	Colt Vista	5/16–15/16	13/16 ①	1/16N–15/16P	7/16 ②	0	NA
1991-92	Colt	—	25/16	—	0 ⑥	0	NA
	Colt Vista	5/16–15/16	13/16	1/16 ④ N–15/16P	7/16 ⑤	0	NA

NA—Not Available

① 4WD Front
13/16 degree caster
13/16 degree camber
4WD Rear
0 degrees caster
0 degrees toe-in

② Rear
5/8 degree camber
0 degrees toe-in

③ Rear
0 degree camber
0 in. toe-in
2WD without Power Steering—3/16–13/16
Preferred setting 11/16

④ 4WD Front
5/16P–15/16P camber
4WD Rear
1/2N–1/2P camber

⑤ 4WD Front
13/16 camber
4WD Rear
0 camber

⑥ Rear
11/16 degrees camber
1/16 degrees toe-in

ENGINE MECHANICAL

NOTE: Disconnecting the negative battery cable on some vehicles may interfere with the functions of the on board computer systems and may require the computer to undergo a relearning process, once the negative battery cable is reconnected.

Engine Assembly

REMOVAL & INSTALLATION

1468cc and 1507cc Engines

The factory recommends that the engine and transaxle be removed as a unit.

1. Disconnect the battery cables, negative cable first. Remove the battery and the tray.
2. Remove the air cleaner assembly. Disconnect the purge control valve. Remove the purge control valve mounting bracket. Remove the windshield washer reservoir, radiator tank and carbon canister.
3. Drain the coolant from the radiator. Remove the radiator assembly with the electric cooling fan attached. Be sure to disconnect the fan wiring harness and the transaxle cooler lines, if equipped.
4. Disconnect the following cables, hoses and wires from the engine and transaxle: clutch, accelerator, speedometer, heater hose, fuel lines, PCV vacuum line, high altitude compensator vacuum hose (California vehicles), bowl vent valve purge hose (U.S.A. vehicles), inhibitor switch (automatic transaxle), control cable (automatic transaxle), starter, engine ground cable, alternator, water temperature gauge, ignition coil, temperature sensor, back-up light (manual transaxle), oil pressure wires and the ISC cable on fuel injected vehicles.

NOTE: On fuel injected vehicles, release fuel system pressure before disconnecting any fuel lines.

5. Remove the ignition coil. Be sure all wires and hoses are disconnected.
6. Raise the vehicle and support safely. Remove the splash shield, if equipped.
7. Drain the lubricant out of the transaxle.
8. Remove the right and left halfshafts from the transaxle and support them with wire. Plug the transaxle case holes so dirt cannot enter.

NOTE: The halfshaft retainer ring should be replaced whenever the shaft is removed.

9. Disconnect the assist rod and the control rod from the transaxle. If the vehicle is equipped with a range selector, disconnect the selector cable.
10. Remove the mounting bolts from the front and rear roll control rods.
11. Disconnect the exhaust pipe from the engine and secure it with wire.
12. Loosen the engine and transaxle mounting bracket nuts. On turbocharged engines, disconnect the oil cooler tube.
13. Lower the vehicle.
14. Attach a lifting device to the engine. Apply slight lifting pressure to the engine. Remove the engine and transaxle mounting nuts and bolts.
15. Make sure the rear roll control rod is disconnected. Lift the engine and transaxle from the vehicle.

NOTE: Make sure the transaxle does not hit the battery bracket when the engine and transaxle are lifted.

To install:

16. Lower the engine and transaxle carefully into position and loosely install the mounting bolts.
17. Temporarily tighten the front and rear roll control rod mounting bolts.
18. Lower the full weight of the engine and transaxle onto the mounts, torque the nuts and bolts to 25 ft. lbs. (34 Nm).
19. Loosen and retighten the roll control rods.
20. Complete the rest of the installation in the reverse order of the removal procedures.
21. Make sure all cables, hoses and wires are connected.
22. Fill the radiator with coolant, the transaxle with lubricant.
23. Adjust the clutch cable and accelerator cable. Adjust the transaxle control rod.
24. Connect the negative battery cable. Start the engine and check for leaks.

1595cc DOHC Engine

1. Disconnect the battery cables, negative cable first. Remove the battery.
2. Raise the vehicle and support it safely. Relieve the fuel system pressure.
3. Drain the cooling system and the engine oil.
4. Disconnect the exhaust pipe from the turbocharger after removing the heat shields.
5. Remove the radiator. Remove the transaxle.
6. Remove the air cleaner. Disconnect the accelerator cable, the vacuum hose from the brake booster and all vacuum hoses. Label the hoses for correct installation.
7. Disconnect the high pressure fuel line and the fuel return hose. Remove their respective mounting O-rings.
8. Disconnect the heater hoses, the oxygen sensor, the coolant temperature sensor and the connection for the engine coolant temperature gauge unit.
9. Disconnect the engine coolant switch for the air conditioner. Disconnect the fuel injector wiring connection, the ignition coil, the power transistor, vacuum lines and the ISC motor.
10. On California vehicles, disconnect the EGR temperature sensor.
11. Disconnect the detonation sensor, the throttle position sensor, the crankshaft angle sensor and the control wiring harness connectors.
12. Disconnect the oil pressure switch for the power steering. Disconnect the alternator wiring. Remove the wiring harness mounting clamps. Disconnect the engine oil pressure switch.
13. Remove the air conditioning compressor, with lines attached and safely wire the assembly out of the way.
14. Remove the power steering pump, with hoses attached and safely wire it out of the way.
15. Connect a chain hoist to the engine with a suitable lifting bracket. Take up slack on the engine. Check to be sure all cables, hoses, harness connectors and vacuum hoses have been disconnected.
16. Remove the engine mounting bracket. Disconnect the front engine roll stopper and the rear roll stopper. Carefully raise and remove the engine assembly.

To install:

17. Installation is the reverse order of the removal procedure.
18. Make sure all cables, vacuum lines, hose and wire connectors are installed or attached.
19. Fill the radiator with the proper coolant mix. Fill the engine with the proper oil.
20. Adjust the clutch and accelerator cables.
21. Connect the negative battery cable. Start the engine and check for leaks.
22. Torque the lower engine mounts to 22–29 ft. lbs. (30–40 Nm), transaxle mount to 29–36 ft. lbs. (40–50 Nm), cylinder head mount to 43–58 ft. lbs. (60–80 Nm) and the exhaust pipe nuts to 14–22 ft. lbs. (20–30 Nm).

1795cc and 1997cc Engines

1. Disconnect the battery cables, negative battery cable first. Remove the battery, battery tray and bracket.
2. Disconnect the engine oil pressure switch and power steering pump connectors.
3. Disconnect the alternator harness.
4. Remove the air cleaner.
5. Remove the high tension cable from the distributor.
6. Disconnect the engine ground wire at the firewall.
7. Remove the windshield washer bottle.
8. Disconnect the brake booster vacuum hose.
9. Disconnect and tag all other vacuum lines connected to the engine.
10. Drain the coolant.
11. Remove the coolant reservoir tank.
12. Remove the radiator.
13. Disconnect the heater hoses at the engine.
14. Disconnect the accelerator cable from the carburetor.
15. Disconnect the speed control cable at the carburetor.
16. Disconnect the speedometer cable at the transaxle.
17. If equipped with air conditioning, the system must be evacuated.
18. Disconnect the hose at the air conditioning compressor and cap all openings immediately.
19. Disconnect the hoses at the power steering pump.
20. Disconnect the fuel return hose, then the fuel inlet hose, at the carburetor.
21. Disconnect the shift control cables at the transaxle.
22. Raise the vehicle and support it safely.
23. Remove the lower cover and skid plate.
24. Drain the transaxle and transfer case.
25. Disconnect the exhaust pipe from the manifold.
26. Remove the clutch slave cylinder.
27. Disconnect the halfshafts at the transaxle.
28. Remove the transfer case extension stopper bracket.

NOTE: The 2 top stopper bracket bolts are easier to get at from the engine compartment, using a T-type box wrench.

29. Lower the vehicle to the ground.
30. Remove the nuts only, from the engine mount-to-body bracket.
31. Remove the range select control valves from the transaxle insulator bracket.
32. Remove the nut only, from the transaxle mounting insulator.
33. Remove the front roll insulator nut.
34. Remove the rear insulator-to-engine nut.

35. Remove the grille and valance panel.
36. Remove the air conditioning condenser.
37. Take up the weight of the engine with a lifting device attached to the lifting eyes.
38. Remove all the mounting bolts.
39. Double check that all wiring, hoses and cables are disconnected from the engine, transaxle and transfer case have been disconnected. Move the assembly forward slightly, to a point at which it will clear the floor pan and lift the whole assembly clear of the vehicle.

To install:

40. Secure the engine to a suitable lifting device.
41. Carefully lower the engine and transaxle into position and loosely install the mounting bolts.
42. Temporarily tighten the front and rear roll control rod mounting bolts.
43. Lower the full weight of the engine and transaxle onto the mounts, tighten the nuts and bolts.
44. Loosen and retighten the roll control rods.
45. Complete the rest of the installation in the reverse order of the removal procedures.
46. Make sure all cables, hoses and wires are connected.
47. Fill the radiator with coolant, the transaxle with lubricant.
48. Adjust the clutch cable and accelerator cable. Adjust the transaxle control rod.
49. Observe the following torques:

Transaxle stopper – 58 ft. lbs. (79 Nm).

Engine-to-body bracket bolts – 47 ft. lbs. (64 Nm).

Rear insulator – 29 ft. lbs. (39 Nm).

Transaxle mount nuts – 58 ft. lbs. (79 Nm).

Heat shield – 7 ft. lbs. (9.5 Nm).

Front roll bracket nuts – 36 ft. lbs. (49 Nm).

50. Connect the negative battery cable. Start the engine and check for leaks.

2555cc Engine

The factory recommends removing the engine and transmission as a unit.

1. Disconnect and remove the battery and tray.
2. Drain the cooling system.
3. Disconnect the coil, throttle positioner solenoid, fuel cut-off solenoid, alternator, starter, transmission switch, backup light switch and temperature and oil pressure gauge sending units.
4. If equipped with air conditioning, the refrigerant must be released from the system. After the system has been drained, disconnect and cap lines at the compressor and condenser.
5. Remove all air cleaners hoses. Remove the wing nut, snap clips and the air cleaner to cover.
6. Remove the 2 retaining nuts and bracket and remove the air cleaner housing.
7. Disconnect the accelerator cable.
8. Remove and plug the radiator hose.
9. Remove the exhaust manifold nuts and drop the pipe down and out of the way.
10. Disconnect and cap the fuel lines at the pump. If equipped with fuel injection, bleed the pressure from the system before disconnecting the lines.
11. Disconnect the vacuum hose from the canister purge valve located on the passenger side firewall. Remove the purge hose which runs from the valve to the intake manifold.
12. Scribe a line around the hood hinges and then remove the hood. Place it away from the work area to avoid it being scratched or dented.
13. Remove the grill, radiator cross panel and the radiator. Disconnect and plug the oil cooler lines, if equipped with automatic transmission. Disconnect and plug the engine oil cooler lines at the engine on turbocharged vehicles. Remove and secure the power steering pump, with hoses connected and place it out of the way.
14. Raise the vehicle and support it safely. Remove the splash shield.
15. Drain the engine oil and the transmission fluid. Remove the driveshaft.
16. Disconnect the speedometer cable and back-up light switch wire. Remove the neutral switch on automatic equipped vehicles.
17. Disconnect the clutch cable from the clutch lever.
18. Remove the control rod and the cross shaft that are located under the transmission.
19. Untie and open the leather shift boot. Pull the rug back. Remove the 4 retaining bolts and remove the shift lever.
20. If equipped with an automatic transmission, disconnect the transmission control rod from the shift linkage.
21. Attach the lifting device to the 2 engine brackets provided by the factory, one near the water neck at the front and the other on the passenger's side at the rear.
22. Raise the engine a slight amount and remove the retaining nuts on the side mounts and the rear crossmember mount.
23. Lift the engine out of the engine compartment.
24. Check the condition of the engine mounts. There are 3: left front, right front and rear. Replace, if required.

To install:

25. Installation is reverse order of the removal procedure.
26. Secure the engine to a suitable lifting device.
27. Carefully lower the engine and transmission into position and loosely install the mounting bolts.
28. Temporarily tighten the front and rear roll control rod mounting bolts.
29. Lower the full weight of the engine and transmission onto the mounts, tighten the nuts and bolts.
30. Loosen and retighten the roll control rods.
31. Make sure all cables, hoses and wires are connected.
32. Fill the radiator with coolant, the transmission with lubricant.
33. Adjust the cruise control and accelerator cable. Adjust the transmission control rod.
34. Observe the following torques:

Front mount-to-crossmember – 15–17 ft. lbs. (20–23 Nm)

Front mount-to-engine bracket nut – 15–17 ft. lbs. (20–23 Nm)

Front engine block-to-bracket bolt – 29–36 ft. lbs. (40–48 Nm)

Rear mount-to-support bracket – 7–8.5 ft. lbs. (10–12 Nm)

Rear mount-to-frame bolt – 15–17 ft. lbs. (20–23 Nm)

Crossmember-to-body bolt – 9–11 ft. lbs. for manual transmission or 7–8.5 ft. lbs. (20–23 Nm) for automatic transmission

35. Connect the negative battery cable. Start the engine and check for leaks.

Engine Mounts

REMOVAL & INSTALLATION

Colt and Colt Vista

1. Disconnect the negative battery cable.
2. Using an engine support fixture tool, center it on the cowl and attach it to the engine. Raise the engine slightly to take the weight off of the engine mounts.
3. From the front of the engine, remove the engine mount bolts and the mount.
4. Inspect the engine mount for deterioration and replace it, if necessary.

To install:

5. To install, support the engine using a engine support fixture tool.
6. Install the engine mounts and the retaining bolts to the engine.
7. Torque the engine mount-to-bracket bolts to 36–47 ft. lbs. (50–65 Nm) and the mount through bolt to 65–80 ft. lbs. (90–110 Nm).

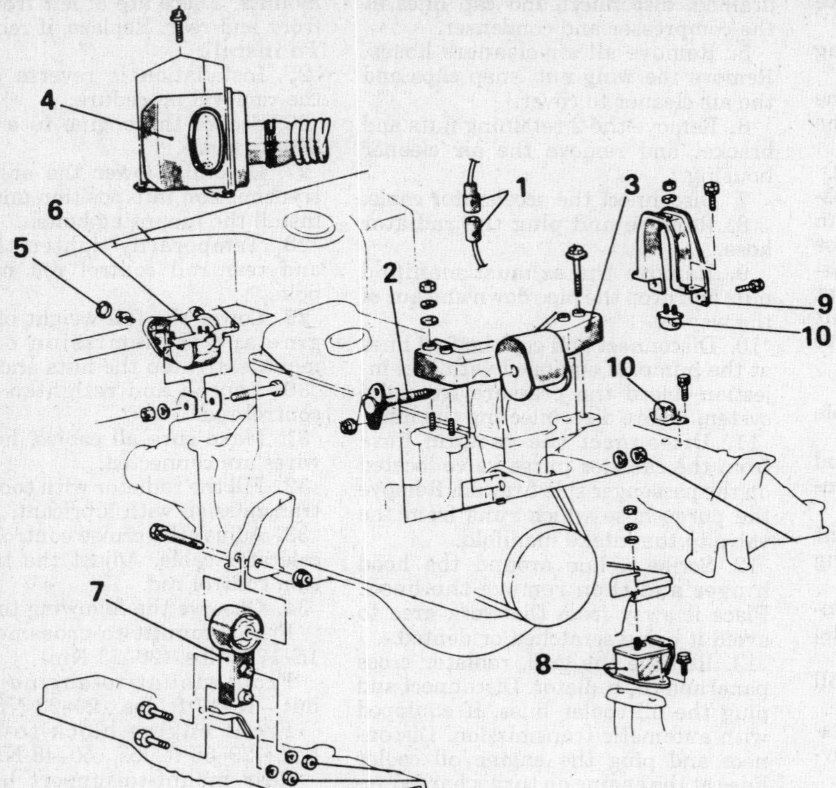

1. Oxygen sensor connector
2. Bolt assembly
3. Engine mount bracket
4. Air filter case
5. Cap
6. Transaxle mount bracket
7. Front roll stopper bracket
8. Rear insulator
9. Transfer extension stopper bracket
10. Transfer extension stopper

Exploded view of the engine mounts

8. Connect the negative battery cable.

Conquest

1. Disconnect the negative battery cable.
2. Raise the vehicle and support it safely.
3. Support the engine and transmission assembly using a suitable jack.
4. Remove the engine mount-to-bracket retaining nuts. Remove the mount-to-frame retaining bolts. Raise the engine slightly, just enough to allow the mount to clear the frame.
5. Remove the mount.

To install:

6. Position the mount between the engine and frame and install the mount-to-bracket retaining nuts.
7. Lower the the engine weight on the engine mounts, but not completely, it may be necessary to move the engine back and forward, order to install the mount-to-frame retaining bolts.
8. Install the mount-to-frame retaining bolts and lower the weight of the engine completely.
9. Torque the mount-to-bracket retaining nuts to 8–12 ft. lbs. (6–9 Nm) and the mount-to-frame retaining bolts to 22–29 ft. lbs. (30–40 Nm).

1. Heat protector
2. Front insulator stopper
3. Engine mounting front insulator

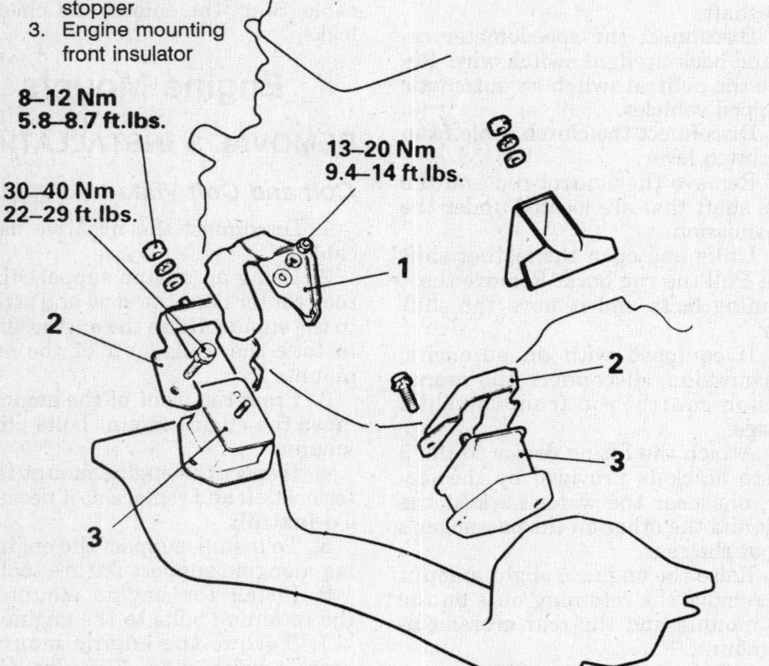

Engine mounts and bolt torques – Conquest

10. Lower the vehicle. Connect the negative battery cable.

Cylinder Head

REMOVAL & INSTALLATION

NOTE: Never remove the cylinder head unless the engine is absolutely cold; the cylinder head could warp.

1988-90 1468cc and 1507cc Engines

1. Disconnect the battery ground cable, remove the air cleaner assembly and the attached hoses.
2. Drain the coolant, remove the upper radiator hose and the heater hoses.

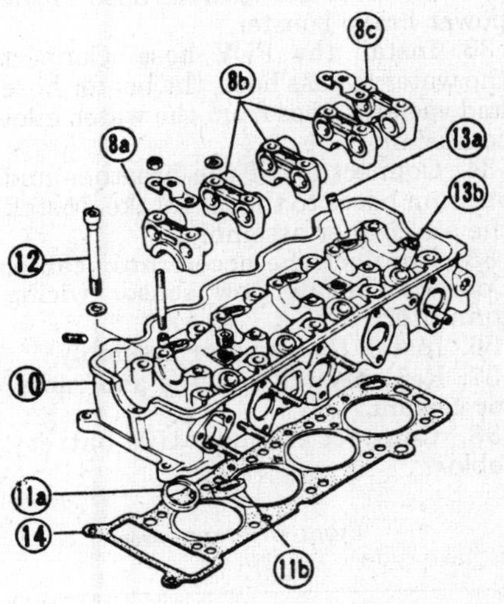

8a. Camshaft bearing cap
8b. No. 2, 3 and 4 caps
8c. Camshaft bearing cap (rear)
10. Cylinder head
11a. Intake valve seat ring
12. Cylinder head bolt
13a. Exhaust valve guide
13b. Intake valve guide
14. Cylinder head gasket

Exploded view of the cylinder head

3. On fuel injected vehicles, release the fuel system pressure. Remove the fuel line, disconnect the accelerator linkage, distributor vacuum lines, purge valve and water temperature gauge wire.
4. Remove the spark plug wires and the fuel pump. Remove the distributor, where necessary.
5. Disconnect the exhaust pipe from the exhaust manifold flange.
6. Remove the exhaust manifold assembly.
7. Remove the intake manifold and carburetor as a unit.
8. Turn the crankshaft to No. 1 piston at TDC on the compression stroke.

NOTE: During the following procedure, do not turn the crankshaft after locating TDC.

9. Remove the timing belt cover. Be sure the knockout pin is at 12 o'clock and the cam sprocket mark and cylinder head pointer are aligned at 3 o'clock. Loosen the timing belt tensioner mounting. Move the tensioner toward the water pump and secure it in that position. Remove the rocker arm cover.

NOTE: The cam pulley need not be removed.

10. Loosen and remove the cylinder head bolts in sequence, 2–3 stages to avoid cylinder head warpage.
11. Remove the cylinder head from the engine block.

To install:

12. Clean the cylinder head and block mating surfaces and install a new cylinder head gasket.
13. Position the cylinder head on the engine block, engage the dowel pins front and rear and install the cylinder head bolts.
14. Tighten the head bolts in, 3 stages, to 50–54 ft. lbs. (68–73 Nm).
15. Locate the camshaft in original position. Pull the camshaft sprocket and belt or chain upward and install on the camshaft.

NOTE: If the dowel pin and the dowel pin hole does not line up between the sprocket and the spacer or camshaft, move the camshaft by bumping either of the 2 projections provided at the rear of the No. 2 cylinder exhaust cam of the camshaft, with a light hammer or other tool, until the hole and pin align. Be certain the crankshaft does not turn.

16. Install the camshaft sprocket bolt and the distributor gear and tighten.
17. Install the timing belt upper front cover and spark plug cable support.
18. Apply sealant to the intake manifold gasket on both sides. Position the gasket and install the intake manifold.

NOTE: Be sure no sealant enters the jet air passages when equipped.

19. Install the exhaust manifold gaskets and the manifold assembly.
20. Connect the exhaust pipe to the exhaust manifold and install the fuel pump. Install the purge valve.
21. Install the water temperature gauge wire, heater hoses and the upper radiator hose.
22. Connect the fuel lines, accelerator linkage, vacuum hoses and the spark plug wires.
23. Fill the cooling system and connect the battery ground cable. Install the distributor.
24. Temporarily adjust the valve clearance to the cold engine specifications.
25. Install the gasket on the rocker arm cover and temporarily install the cover on the engine.
26. Connect the negative battery cable.
27. Start the engine and bring it to normal operating temperature. Stop the engine and remove the rocker arm cover.
28. Adjust the valves to hot engine specifications.
29. Install the rocker arm cover and tighten securely.
30. Install the air cleaner, hoses, purge valve hose and any other removed unit.

COLD ENGINE SPECIFICATIONS

	Inch	mm
Jet valve, if equipped	0.003	0.07
Intake valve	0.003	0.07
Exhaust valve	0.007	0.17

Cold engine specifications

HOT ENGINE SPECIFICATIONS

	Inch	mm
Jet valve, if equipped	0.006	0.15
Intake valve	0.006	0.15
Exhaust valve	0.010	0.25

Hot engine specifications

Crankshaft pulley side

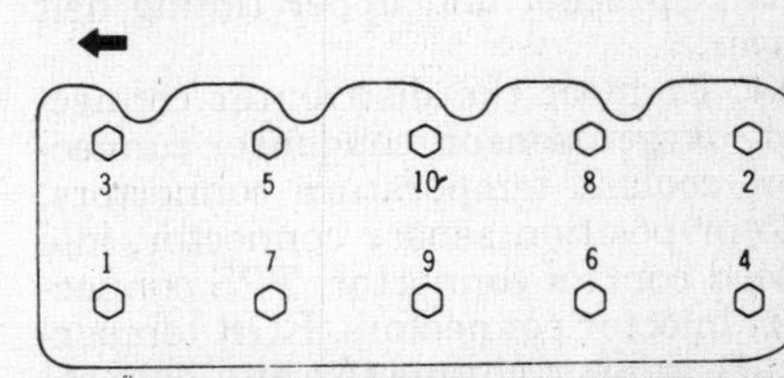

1468cc, 1507cc, 1597cc and 1997cc cylinder head bolt loosening sequence

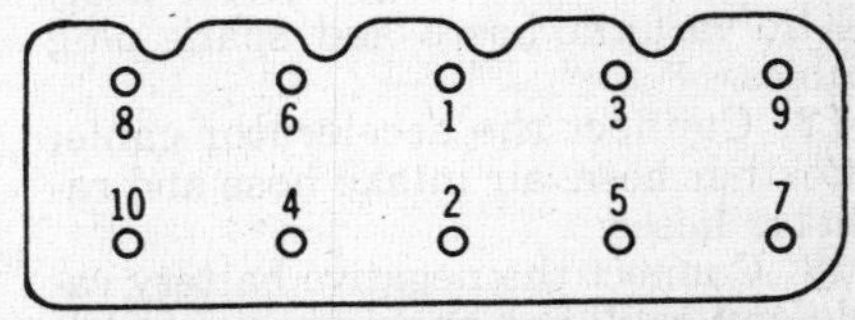

1468cc, 1507cc, 1597cc and 1997cc cylinder head bolt torque sequence

1991-92 1507cc 12 Valve Engine

1. Disconnect the negative battery cable and drain the engine coolant.
2. Disconnect the accelerator cable, breather hose, air intake hose and radiator hoses.
3. Label and disconnect the brake booster hose, EGR vacuum hoses and spark plug cables.
4. Relieve the fuel pressure. Disconnect the fuel hoses.
5. Disconnect the distributor connector, oxygen sensor, condenser connector, coolant temperature connectors, motor position sensor connector, idle speed control connector, TPS connector, injector connectors, EGR temperature sensor connector and control harness.
6. Remove the upper timing belt cover. Rotate the crankshaft clockwise and align the timing marks. Remove the camshaft sprocket and hold the sprocket and timing belt with a piece of wire. Remove the rocker cover and gasket.
7. Remove the exhaust manifold-to-pipe nuts.
8. Remove the cylinder head bolts using a special socket TW-10B or equivalent. Remove the cylinder head and gasket.

To install:

9. Clean the gasket mating surfaces and check for warpage.
10. Match the old gasket with the new and install the gasket with the identification mark facing upward and to the front of the engine.
11. Install the cylinder head and bolts. Using the special head bolt tool TW-10B and torque the bolts in 3 steps and in sequence. 1st step to 20 ft. lbs. (27 Nm), 2nd step to 40 ft. lbs. (54 Nm) and the 3rd step to 51–54 ft. lbs. (70–75 Nm) [Cold].
12. Install the exhaust manifold-to-pipe nuts and torque to 34 ft. lbs. (46 Nm).
13. Install the rocker cover, camshaft sprocket and upper timing belt cover.
14. Connect the distributor connector, oxygen sensor, condenser connector, coolant temperature connectors, motor position sensor connector, idle speed control connector, TPS connector, injector connectors, EGR temperature sensor connector and control harness.
15. Connect the fuel hoses.
16. Connect the brake booster hose, EGR vacuum hoses and spark plug cables.
17. Connect the accelerator cable, breather hose, air intake hose and radiator hoses.
18. Connect the negative battery cable and refill the engine coolant. Adjust the timing and idle if needed.

1595cc DOHC Engine

1. Disconnect the negative battery cable.
2. Drain the engine and radiator coolant.
3. Remove the radiator assembly.
4. Disconnect the accelerator cable. Disconnect the air flow sensor wiring connector.
5. Disconnect all of the breather and vacuum hoses to the air intake. Remove the air cleaner assembly.
6. Remove the PCV hose. Disconnect the water bypass hose, the heater hose and vacuum lines from the water inlet connector.
7. Disconnect the vacuum hose to the power brake booster.
8. Release the fuel system pressure and disconnect the high pressure and fuel return lines. Remove the mounting O-rings.
9. Disconnect the oxygen sensor, engine coolant sensor, temperature gauge connection and the air conditioner coolant temperature switch.
10. Disconnect the fuel injector wiring harness, the ignition coil, power transistor, ISC motor and the EGR sensor connector.
11. Disconnect the detonation sensor, throttle position sensor and crankshaft angle sensor wiring connectors.
12. Remove the center cover and the spark plug wires. Disconnect the control wire harness connector.
13. Remove the timing belt.
14. Remove the rocker cover and rear half moon seal.
15. Remove the heat shield, turbocharger water and oil lines.
16. Disconnect the exhaust pipe from the turbocharger.
17. Remove the turbocharger, exhaust manifold and intake manifold assemblies.
18. Remove the head mounting bolts. Start at the outer ends of the head and loosen, in a criss-cross manner, toward the center of the head. Make 2–3 passes to loosen the bolts, a little at a time, in sequence.
19. Remove the cylinder head. Clean all gasket mounting surfaces. Make sure no gasket material gets into the cylinders, coolant passages or oil passages.

To install:

20. Position the cylinder head, with a new gasket, on the engine. Install and tighten the head mounting bolts.
21. Tighten the head bolts from the center outwards. Tighten in, 3 steps, to 65–72 ft. lbs. (88–98 Nm).
22. Install the turbocharger, exhaust manifold and intake manifold assemblies.
23. Connect the exhaust pipe to the turbocharger.
24. Install the heat shield, turbocharger water and oil lines.
25. Install the rocker cover and rear half moon seal.
26. Install the timing belt.
27. Install the center cover and the spark plug wires. Connect the control wire harness connector.
28. Connect the detonation sensor, throttle position sensor and crankshaft angle sensor wiring connectors.
29. Connect the fuel injector wiring harness, the ignition coil, power transistor, ISC motor and the EGR sensor connector.
30. Connect the oxygen sensor, engine coolant sensor, temperature gauge connection and the air conditioner coolant temperature switch.
31. Connect the high pressure and fuel return lines. Install the new O-rings.
32. Connect the vacuum hose to the power brake booster.
33. Install the PCV hose. Connect the water bypass hose, the heater hose and vacuum lines from the water inlet connector.
34. Connect all of the breather and vacuum hoses to the air intake. Install the air cleaner assembly.
35. Connect the accelerator cable. Connect the air flow sensor wiring connector.
36. Install the radiator assembly.
37. Replenish the engine and radiator coolant.
38. Connect the negative battery cable.

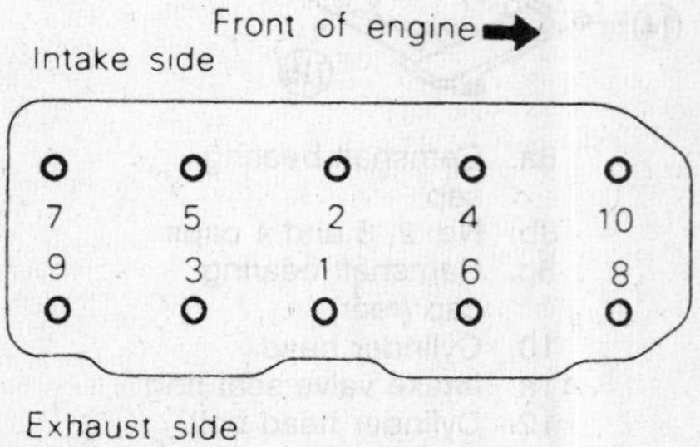

1595cc DOHC cylinder head bolt torque sequence

1597cc, 1755cc and 1997cc Engines

1. Disconnect the negative battery cable. Remove the air cleaner assembly and the attached hoses.
2. Drain the coolant, remove the upper radiator hose and the heater hoses.
3. On fuel injected vehicles, release the fuel system pressure. Remove the fuel line, disconnect the accelerator linkage, distributor vacuum lines, purge valve and water temperature gauge wire.
4. Remove the spark plug wires and

the fuel pump. Remove the distributor, where necessary.

5. Disconnect the exhaust pipe from the exhaust manifold flange.

6. Remove the exhaust manifold assembly.

7. Remove the intake manifold and carburetor as a unit.

8. Turn the crankshaft to No. 1 piston at TDC on the compression stroke.

NOTE: During the following procedure, do not turn the crankshaft after locating TDC.

9. Align the timing mark on the upper under cover of the timing belt with that of the camshaft sprocket. Matchmark the timing belt and the timing mark on the camshaft sprocket with a felt tip pen. Remove the sprocket and insert a 2 in. piece of timing belt or other material between the bottom of the camshaft sprocket and the sprocket holder, on the timing belt lower front cover, to hold the sprocket and belt so the valve timing will not be changed. Remove the timing belt upper under cover and rocker arm cover.

10. Loosen and remove the cylinder head bolts in 2–3 stages to avoid cylinder head warpage.

11. Remove the cylinder head from the engine block.

To install:

12. Clean the cylinder head and block mating surfaces and install a new cylinder head gasket.

13. Position the cylinder head on the engine block, engage the dowel pins front and rear and install the cylinder head bolts.

14. Tighten the head bolts in 3 stages and then torque to:

1597cc engine – 50–54 ft. lbs. (68–73 Nm).

1755cc engine – 50–54 ft. lbs. (68–73 Nm).

1997cc engine – 65–72 ft. lbs. (88–97 Nm).

15. Install the timing belt upper under cover.

16. Locate the camshaft in original position. Pull the camshaft sprocket and belt or chain upward and install on the camshaft.

NOTE: If the dowel pin and the dowel pin hole does not line up between the sprocket and the spacer or camshaft, move the camshaft by bumping either of the 2 projections provided at the rear of the No. 2 cylinder exhaust cam of the camshaft, with a light hammer or other tool, until the hole and pin align. Be certain the crankshaft does not turn.

17. Install the camshaft sprocket bolt and the distributor gear and tighten.

18. Install the timing belt upper front cover and spark plug cable support.

19. Apply sealant to the intake manifold gasket on both sides. Position the gasket and install the intake manifold. Torque the nuts to 11–14 ft. lbs. (15–20 Nm).

NOTE: Be sure no sealant enters the jet air passages when equipped.

20. Install the exhaust manifold gaskets and the manifold assembly. Tighten the nuts to 11–14 ft. lbs. (15–20 Nm).

21. Connect the exhaust pipe to the exhaust manifold and install the fuel pump. Install the purge valve.

22. Install the water temperature gauge wire, heater hoses and the upper radiator hose.

23. Connect the fuel lines, accelerator linkage, vacuum hoses and the spark plug wires.

24. Fill the cooling system and connect the battery ground cable. Install the distributor.

25. Temporarily adjust the valve clearance to the cold engine specifications.

26. Install the gasket on the rocker arm cover and temporarily install the cover on the engine.

27. Start the engine and bring it to normal operating temperature. Stop the engine and remove the rocker arm cover.

28. Adjust the valves to hot engine specifications.

29. Install the rocker arm cover and tighten securely.

30. Install the air cleaner, hoses, purge valve hose and any other removed unit.

2555cc Engine

1. Disconnect the negative battery cable. Remove the air cleaner assembly and the attached hoses.

2. Drain the coolant, remove the upper radiator hose and the heater hoses.

3. Release the fuel system pressure. Remove the fuel line, disconnect the accelerator linkage, distributor vacuum lines, purge valve and water temperature gauge wire.

4. Remove the spark plug wires and the fuel pump. Remove the distributor, where necessary.

5. Disconnect the exhaust pipe from the exhaust manifold flange.

6. Remove the exhaust manifold assembly.

7. Remove the intake manifold.

8. Turn the crankshaft to No. 1 piston at TDC on the compression stroke.

NOTE: During the following procedure, do not turn the crankshaft after locating TDC.

9. Remove the rocker arm cover. Position the camshaft sprocket dowel pin at the 12 o'clock position with the crankshaft pulley notch aligned with the timing mark **T** at the front of the timing chain case. Match the timing chain with the timing mark on the camshaft sprocket. Remove the camshaft sprocket bolt, distributor, gear and the sprocket from the camshaft.

10. Loosen and remove the cylinder head bolts in 2–3 stages to avoid cylinder head warpage.

11. Remove the cylinder head from the engine block.

To install:

12. Clean the cylinder head and block mating surfaces and install a new cylinder head gasket.

13. Position the cylinder head on the engine block, engage the dowel pins front and rear and install the cylinder head bolts.

14. Tighten the head bolts in, 3 stages, to 65–72 ft. lbs. (88–98 Nm).

15. Locate the camshaft in original position. Pull the camshaft sprocket and belt or chain upward and install on the camshaft.

NOTE: If the dowel pin and the dowel pin hole does not line up between the sprocket and the spacer or camshaft, move the camshaft by bumping either of the 2 projections provided at the rear of the No. 2 cylinder exhaust cam of the camshaft, with a light hammer or other tool, until the hole and pin align. Be certain the crankshaft does not turn.

16. Install the camshaft sprocket bolt and the distributor gear and tighten.

17. Install the timing belt upper front cover and spark plug cable support.

18. Apply sealant to the intake manifold gasket on both sides. Position the gasket and install the intake manifold. Tighten nuts to specifications.

NOTE: Be sure no sealant enters the jet air passages when equipped.

19. Install the exhaust manifold gaskets and the manifold assembly.

20. Connect the exhaust pipe to the exhaust manifold and install the fuel pump. Install the purge valve.

21. Install the water temperature gauge wire, heater hoses and the upper radiator hose.

22. Connect the fuel lines, accelerator linkage, vacuum hoses and the spark plug wires.

23. Fill the cooling system and connect the battery ground cable. Install the distributor.

24. Temporarily adjust the valve

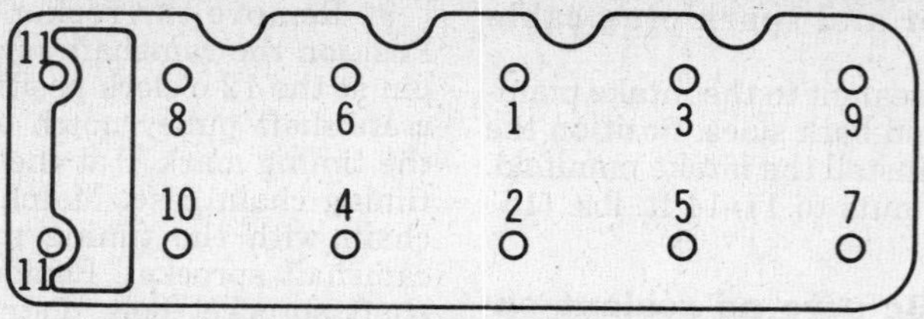

2555cc cylinder head bolt torque sequence

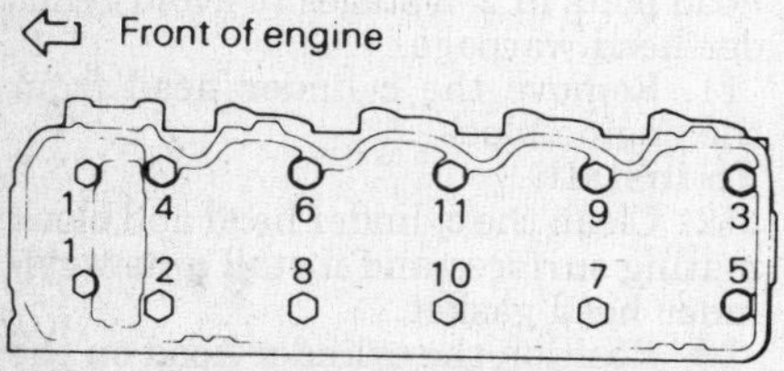

2555cc cylinder head bolt loosening sequence

clearance to the cold engine specifications.

25. Install the gasket on the rocker arm cover and temporarily install the cover on the engine.

26. Connect the negative battery cable.

27. Start the engine and bring it to normal operating temperature. Stop the engine and remove the rocker arm cover.

28. Adjust the valves to hot engine specifications.

29. Install the rocker arm cover and tighten securely.

30. Install the air cleaner, hoses, purge valve hose and any other removed unit.

Automatic Valve Lash Adjusters

REMOVAL & INSTALLATION

1997cc and 2555cc Engines

1. Disconnect the negative battery cable.

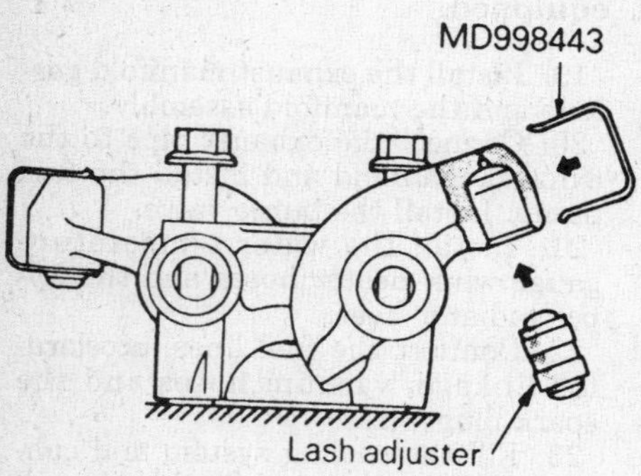

Automatic valve lash adjuster—1997cc and 2555cc engines

2. Disconnect the cables and hoses that route over the rocker cover.

3. Remove the upper timing belt cover, rocker cover and gasket.

4. Remove the camshaft sprocket and timing belt. Secure with a piece of wire.

5. Before removing the rocker arm shaft assembly, use the special lash adjuster retaining tool MD998443 or equivalent, so the adjuster does not fall out. Remove the rocker arm shafts by loosening the bolts gradually from the center and moving outward.

6. Remove the lash adjuster. Mark each adjuster to ensure proper installation.

To install:

7. Insert the adjuster and retaining tool MD998443 or equivalent.

8. Install the rocker arm shaft and torque the large bolts to 14–15 ft. lbs. (19–21 Nm) and the small bolts to 15–19 ft. lbs. (20–27 Nm).

9. Install the camshaft sprocket and timing belt.

10. Install the upper timing belt cover, rocker cover and gasket.

11. Connect the cables and hoses that route over the rocker cover.

12. Connect the negative battery cable.

Valve Lash

ADJUSTMENT

Valve lash must be adjusted on all engines not equipped with automatic lash adjusters. Some engines have a third valve of very small size called a jet valve. The jet valve must be adjusted, whether the engine uses automatic lash adjusters for the normal intake and exhaust valves or not. Thus, on some engines, there are 3 valves per cylinder that must be adjusted.

1. Run the engine until operating temperature is reached.

2. Turn off the engine and block the wheels.

3. Remove all necessary components in order to gain access to the rocker cover.

4. Remove the spark plugs from the cylinder head for easy operations.

a. On some engines it may be necessary to remove the air intake pipe.

b. On some engines it may be necessary to disconnect the oxygen sensor connecting joint. Remove the engine bracket mounting, be sure to place a block of wood on the oil pan and jack it up into to place for the duration of the operation. Remove the upper front timing belt cover, remove the air cleaner assembly on the 1997cc engine and the air intake pipe on the 1755cc engine and remove the rocker cover.

c. On all other vehicles, remove the air cleaner or air intake pipe assembly and remove the rocker cover.

5. Turn each cylinder head bolt, in sequence, back just until it is loose. Torque the cylinder head bolts in the proper sequence to specification.

6. Position the engine crankshaft at TDC with No. 1 cylinder at the firing position. Turn the engine by using a wrench on the bolt in the front of the crankshaft until the 0 degree timing mark on the timing cover lines up with the notch in the front pulley. On some engines, it may be necessary to turn the crankshaft clockwise until the notch on the pulley is lined up with the T mark on the timing belt lower cover.

7. Observe the valve rockers for No. 1 cylinder. If both are in identical positions with the valves up, the engine is in the right position. If not, rotate the engine exactly 360 degrees until the 0 degree timing mark is again aligned. Each jet valve is associated with an intake valve that is on the same rocker lever. In this position you'll be able to adjust all the valves marked A, including associated jet valves which are located on the rockers, on the intake side only.

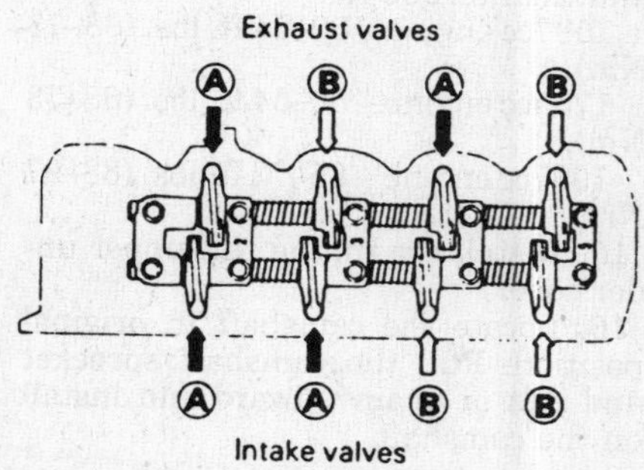

"A" and "B" valve adjusting sequence

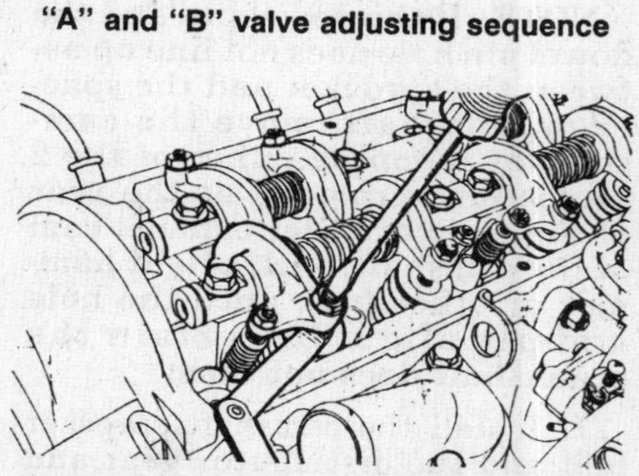

Valve clearance clearance adjustment—all engines

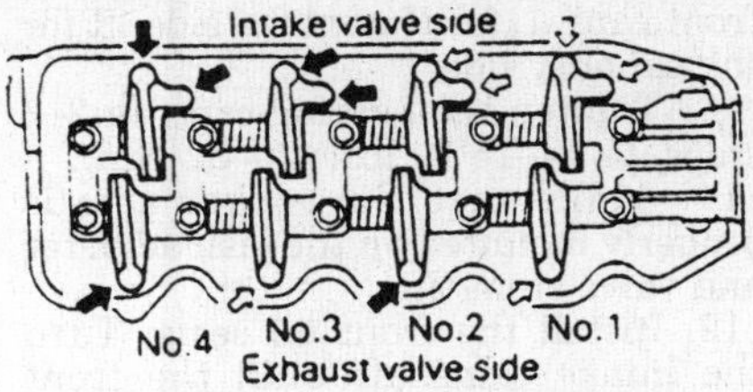

Jet valve adjusting

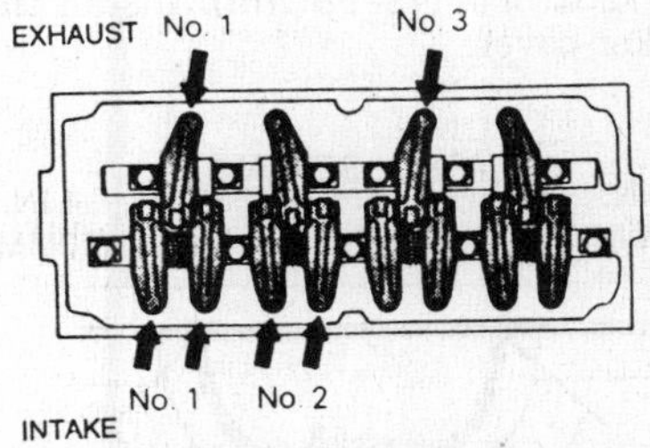

Valve lash adjustment—1507cc 12 valve engine

8. To adjust the appropriate jet valves, first loosen the regular (larger) intake valve adjusting stud by loosening the locknut and backing the stud off 2 turns. Note that this particular step is not required on engines that have automatic lash adjusters.

9. Loosen the jet valve (smaller) adjusting stud locknut, back the stud out slightly and insert the feeler gauge between the jet valve and stud. Make sure the gauge lies flat on the top of the jet valve. Be careful not to twist the gauge or otherwise depress the jet valve spring, rotate the jet valve adjusting stud back in until it just touches the gauge. Tighten the locknut. Make sure the gauge still slides very easily between the stud and jet valve

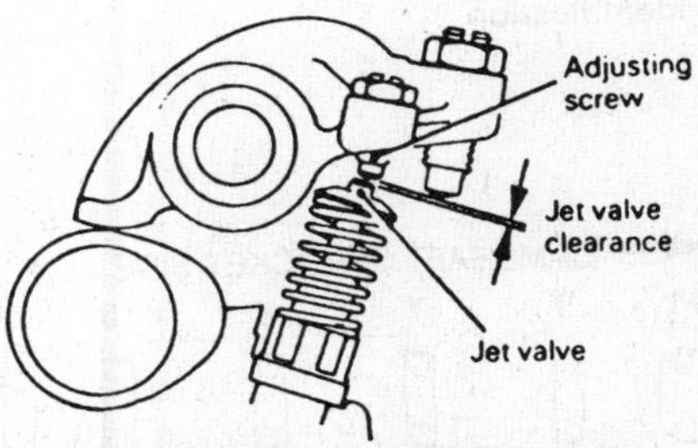

Valve adjustment sequence—typical

and that they both are still just touching the gauge.

NOTE: The clearances must not be too tight.

10. Repeat the entire procedure for the other jet valves associated with rockers labeled A (Dark Arrow).

11. On engines without automatic lash adjusters, repeat the procedure for the intake valves labeled A (Dark Arrow).

12. Repeat the basic adjustment procedure for exhaust valves labeled A on engines without automatic lash adjusters.

13. Turn the engine exactly 360 degrees, until the timing marks are again aligned at 0 degrees BTDC.

14. On engines with automatic lash adjusters, after the jet valves and rockers on the intake side and labeled B (Light Arrow) are adjusted, the valve adjustment procedure is completed. On engines without automatic lash adjusters, adjust the regular intake and exhaust valves labeled B (Light Arrow).

15. Reinstall the cam cover. Run the engine to check for oil leaks.

JET VALVE ADJUSTMENT

NOTE: An incorrect jet valve clearance would affect the emission levels and could also cause engine troubles, so the jet valve clearance must be correctly adjusted. Adjust the jet valve clearance before adjusting the intake valve clearance. The cylinder head bolts should be retorqued before making this adjustment. The jet valve clearance should be adjusted with the adjusting screw on the intake valve side fully loosened.

1. Start the engine and let it run at idle until it reaches normal operating temperature.

2. Remove all spark plugs from the cylinder head for easy operation.

3. On some vehicles it may be necessary to remove the air intake pipe and remove the rocker cover.

4. It may be necessary to disconnect the oxygen sensor. Remove the engine bracket mounting, be sure to place a block of wood on the oil pan and jack it up into to place for the duration of the operation. Remove the upper front timing belt cover, remove the air cleaner assembly on the 1755cc and 1997cc engine.

5. Remove air intake pipe assembly and remove the rocker cover.

6. Set the engine at TDC with No. 1 cylinder at the firing position. Turn the engine by using a wrench on the bolt in the front of the crankshaft until the 0 degree timing mark on the timing cover lines up with the notch in the front pulley. On some engines, it may be necessary to turn the crankshaft clockwise until the notch on the pulley is lined up with the T mark on the timing belt lower cover. This will bring both No. 1 and No. 4 cylinder pistons up to TDC.

NOTE: Never turn the crankshaft counterclockwise.

7. Move the rocker arms on the No. 1 and No. 4 cylinders up and down by hand to determine if the piston in that cylinder is at TDC center on the compression stroke. If the intake and exhaust rocker arms do not move, the piston in that cylinder is not at TDC on the compression stroke.

8. Measure the jet valve clearance at point A.

NOTE: Measure the valve clearance when the No. 1 cylinder or the No. 4 cylinder pistons are at TDC on the compression stroke. Then give the crankshaft 1 clockwise turn to bring the other cylinder piston to TDC on compression stroke.

9. If the jet valve clearance is not as specified (0.010 in. hot and 0.007 in. cold), loosen the rocker arm locknut of the intake valve and loosen the adjusting screw at least 2 turns or more.

10. Loosen the jet valve locknut and adjust the clearance using a feeler gauge while turning the adjusting screw.

NOTE: The jet valve spring has a small tension and the adjustment is somewhat delicate. Be careful not to push in the jet valve by turning the adjusting screw in too much.

11. Tighten the adjusting screw until it touches the feeler gauge. Turn the locknut to secure it, while holding the rocker arm adjusting screw with a suitable tool to keep it from turning.

12. Check the intake and exhaust valve clearance, if it is not within specifications, adjust the valves.

13. Turn the engine by using a wrench on the bolt, in the front of the crankshaft, 360 degrees until the 0 degree timing mark on the timing cover lines up with the notch in the front pulley. On some vehicles turn the crankshaft clockwise until the notch on the pulley is aligned with the T mark on the timing belt lower cover.

14. Repeat Steps 9 through 13 on the other valves marked B for clearance adjustment.

Rocker Arms/Shafts

REMOVAL & INSTALLATION

1468cc, 1507cc, 1597cc and 1755cc Engines

1. Disconnect the negative battery cable. Remove the rocker cover.

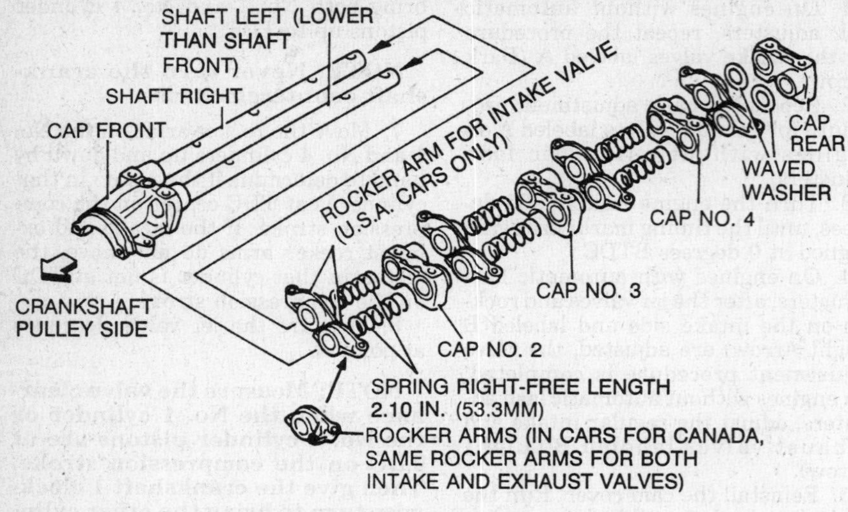

Rocker arm assembly – 1597cc engine

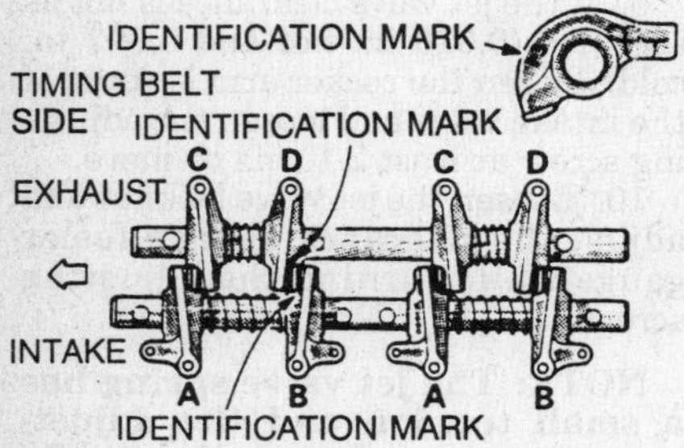

Rocker arm shaft assembly – 1468cc and 1597cc engines

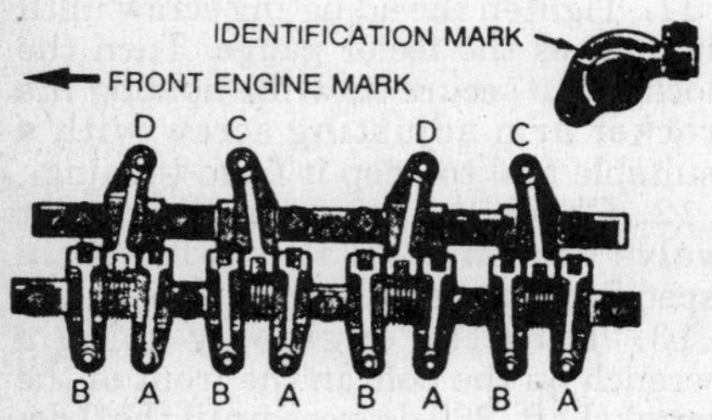

Rocker arm assembly – 1507cc 12 valve engine

Matchmark the camshaft/rocker arm bearing caps to their cylinder head location, except 1468cc and 1507cc engines.

2. Loosen the bearing cap bolts or the rocker shaft bolts on 1468cc and 1507cc engine, from the cylinder head but do not remove them from the caps or shafts. Lift the rocker assembly from the cylinder head as a unit.

3. The rocker arm assembly can be disassembled by the removal of the mounting bolts and dowel pins on some vehicles, from the bearing caps and/or shafts.

NOTE: Keep the rocker arms and springs in the same order as disassembly. The left and right springs have different tension ratings and free length. Observe the location of the rocker arms as they are removed. Exhaust and intake, right and left, are different.

1595cc DOHC Engine

1. Remove the rocker cover, timing cover and cylinder head from the vehicle. Remove the crank angle sensor.
2. Remove both camshaft drive sprockets.
3. Remove both rear (opposite end of the drive sprockets) camshaft bearing caps.
4. Remove both front bearing caps and front oil seals.
5. Remove the remaining camshaft bearing caps alternating from the rear of the head to the front.
6. Remove the camshafts.
7. Remove the rocker arms and the lash adjusters. Remove the valve body assembly from the rear of the cylinder head.
8. Clean and inspect all parts. Check the rollers on the end of the rocker arms. If the rollers are worn or do not rotate smoothly, replace, as necessary.

To install:

9. Install the lash adjusters and rocker arms. Lubricate them prior to installation. Install the valve body. Lubricate the camshafts. Place the camshafts in position. The intake side camshaft has a slit in the rear to drive the crank angle sensor. The bearing caps No. 2–5 are the same shape. When installing them, check the top markings to identify the intake or exhaust side. L or R is marked on the front caps, L for the intake side; R for the exhaust side.
10. Tighten the bearing caps, in 2–3 steps, to 14–15 ft. lbs. (20–22 Nm).
11. Make sure the rocker arm is properly mounted on the lash adjuster and valve stem tip.
12. Install the front oil seals. Turn the intake camshaft until the front dowel pin is facing straight up at the 12 o'clock position. Install the crank angle sensor with the punch mark on the sensor housing aligned with the notch in the plate. Install the drive sprocket and tighten the bolts to 58–72 ft. lbs. (79–98 Nm). Install the rocker cover.

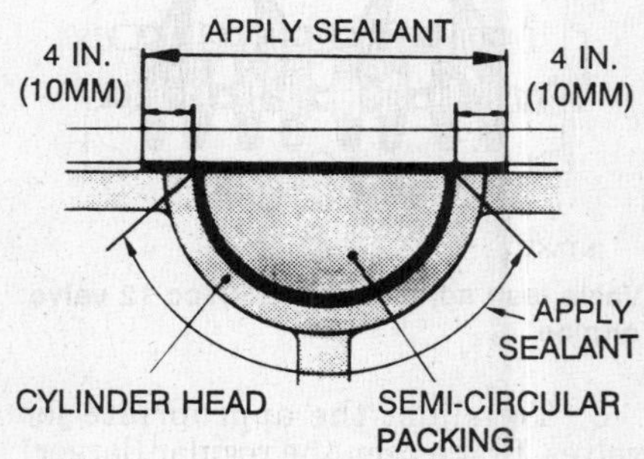

Half moon seal installation

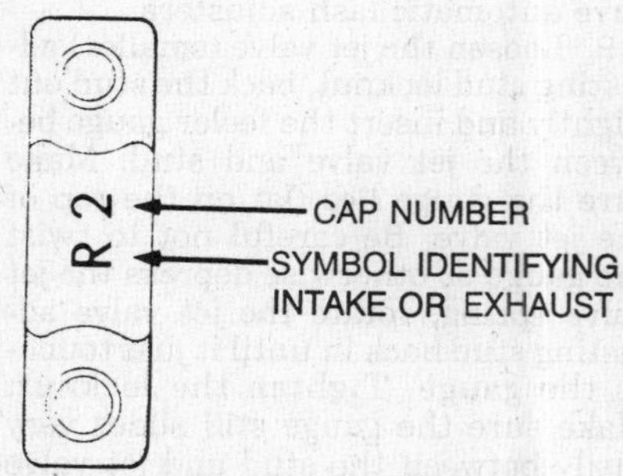

1595cc DOHC camshaft bearing cap identification

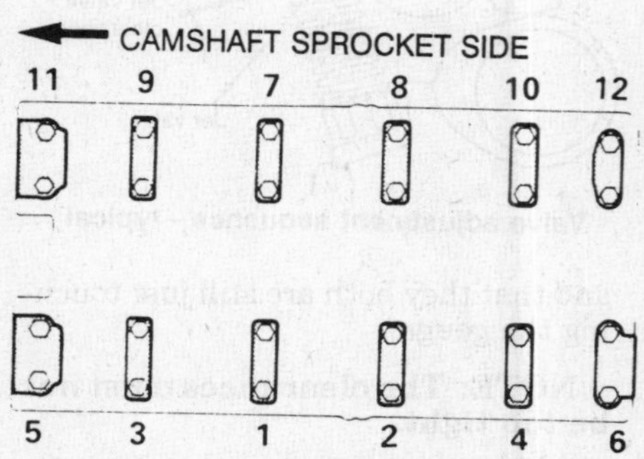

1595cc DOHC camshaft bearing cap installation tightening sequence

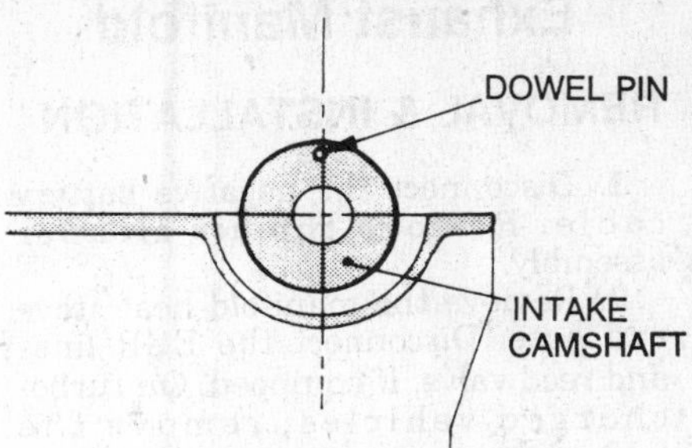

1595cc DOHC intake camshaft dowel pin position

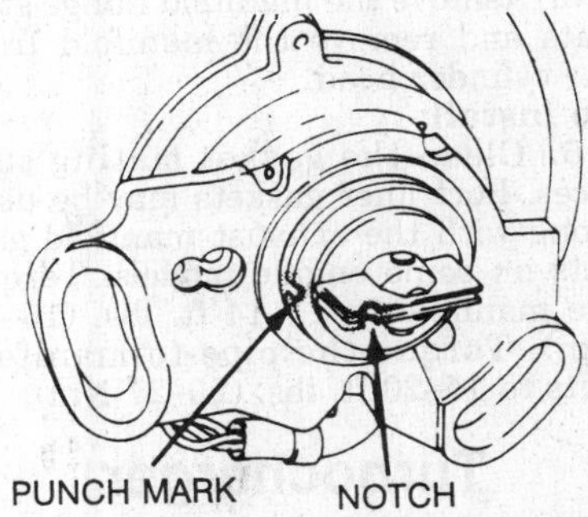

1595cc DOHC crank angle sensor installation alignment

1997cc and 2555cc Engines With Automatic Lash Adjusters

NOTE: A special tool, MD998443 or equivalent, is required for this procedure.

1. Disconnect the negative battery cable. Remove the rocker cover and gasket and the timing belt cover.
2. Turn the crankshaft so the No. 1 piston is a TDC compression. At this point, the timing mark on the camshaft sprocket and the timing mark on the head to the left of the sprocket will be aligned.
3. Remove the camshaft bearing cap bolts.
4. Install the automatic lash adjuster retainer tool MD998443 or equivalent, to keep the adjuster from falling out of the rocker arms.
5. Lift off the bearing caps and rocker arm assemblies.
6. The rocker arms may now be removed from the shaft.

NOTE: Keep all parts in the order in which they were removed. None of the parts are interchangeable. The lash adjusters are filled with diesel fuel, which will spill out if they are inverted. If any diesel fuel is spilled, the adjusters must be bled.

7. Check all parts for wear or damage. Replace any damaged or excessively worn parts.

To install:

8. Assemble all parts in reverse order of the removal procedures. Note the following:
9. The rocker shafts are installed with the notches in the ends facing up.
10. The left rocker shaft is longer than the right.
11. The wave washers are installed on the left shaft.
12. Coat all parts with clean engine oil prior to assembly.
13. Insert the lash adjuster from under the rocker arm and install the special holding tool. If any of the diesel fuel is spilled, the adjuster must be bled.
14. Tighten the bearing cap bolts, working from the center towards the ends to 15 ft. lbs. (20 Nm).
15. Check the operation of each lash adjuster by positioning the camshaft so the rocker arm bears on the low or round portion of the cam pointed part of the can faces straight down. Insert a thin steel wire, or tool MD998442 or equivalent, in the hole in the top of the rocker arm, over the lash adjuster and depress the check ball at the top of the adjuster. While holding the check ball depressed, move the arm up and down. Looseness should be felt. Full plunger stroke should be 0.0866 in. (2.2mm). If not, remove, clean and bleed the lash adjuster.

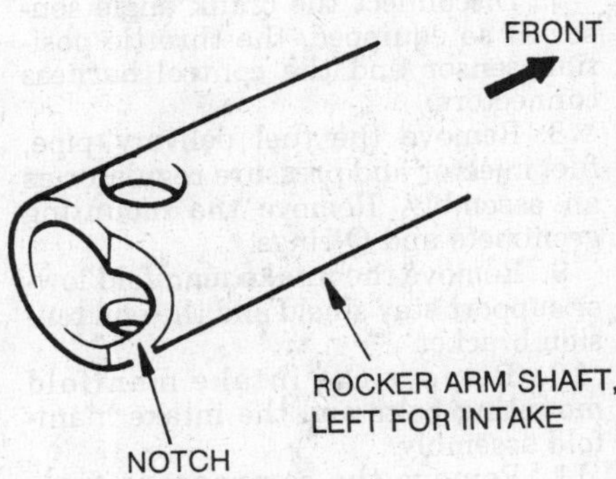

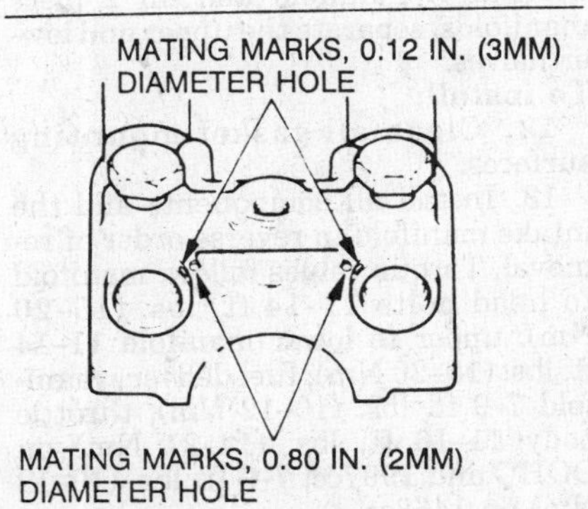

Rocker arm shaft positioning—2555cc engine

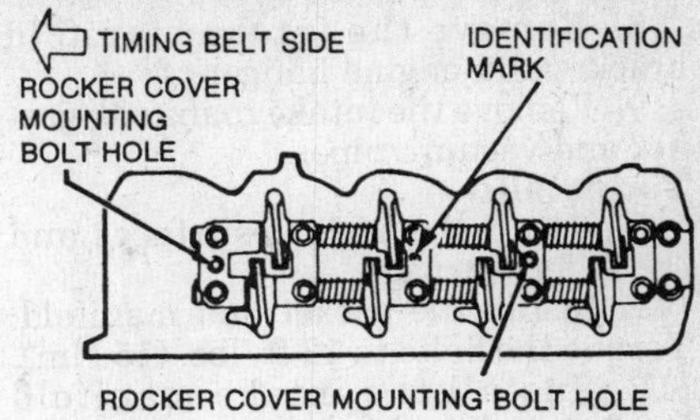

Camshaft bearing cap positioning—2555cc engine

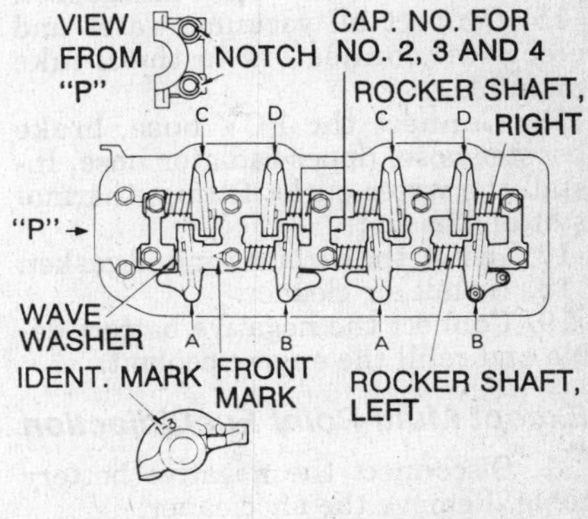

1997cc engine rocker arm shaft assembly

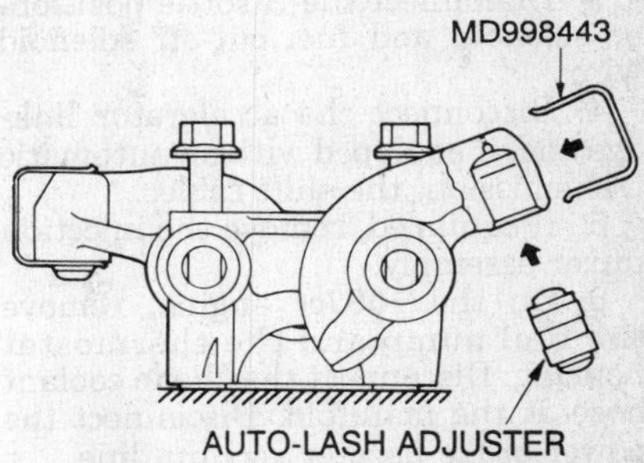

Automatic lash adjuster installation

Intake Manifold

REMOVAL & INSTALLATION

Carbureted Engines

1. Disconnect the negative battery cable and drain the engine coolant.
2. Remove air cleaner.
3. Disconnect, label and remove the carburetor and gasket.
4. Disconnect the PCV hose, brake booster hose, upper radiator hose and remove the water outlet fitting, thermostat and gasket.
5. Disconnect all vacuum, water and electrical connectors from the intake manifold. Label each connection before removal.
6. Remove the EGR valve and gasket.
7. Mark and remove the distributor from the cylinder head.

8. Remove the intake manifold bracket and engine hanger.
9. Remove the intake manifold, gasket and vacuum pipe.

To install:

10. Clean the gasket surfaces and check for warpage.
11. Install the gasket and manifold. Torque the bolts to 12 ft. lbs. (15 Nm).
12. Install the intake manifold bracket and engine hanger.
13. Install the distributor to the cylinder head.
14. Install the EGR valve and gasket.
15. Connect all vacuum, water and electrical connectors to the intake manifold.
16. Connect the PCV hose, brake booster hose, upper radiator hose. Install the water outlet fitting, thermostat and gasket.
17. Install the carburetor and gasket.
18. Install air cleaner.
19. Connect the negative battery cable and refill the engine coolant.

Except Multi-Point Fuel Injection

1. Disconnect the negative battery cable. Remove the air cleaner.
2. Disconnect the fuel line and EGR lines, if equipped.
3. Disconnect the throttle positioner solenoid and fuel cut-off solenoid wires.
4. Disconnect the accelerator linkage and, if equipped with an automatic transmission, the shift cables.
5. If equipped, remove the injection mixer assembly.
6. On the 1597cc engine, remove the fuel pump and the thermostat housing. Disconnect the choke coolant hose at the manifold. Disconnect the power brake booster vacuum line.
7. Drain the coolant.
8. Remove the water hose from throttle body and cylinder head.
9. Remove the heater and water outlet hoses.
10. Disconnect the water temperature sending unit.
11. Remove the manifold.

To install:

12. Clean all mounting surfaces. Before reinstalling the manifold, coat both side with gasket sealer.

NOTE: If the engine is equipped with the jet air system, take care not to get any sealer into the jet air intake passage.

13. Install the manifold and torque the bolts to 11–14 ft. lbs. (15–20 Nm).
14. Install the water temperature sending unit.
15. Install the heater and water outlet hoses.
16. Install the water hose to throttle body and cylinder head.
17. Refill the coolant.
18. On the 1597cc engine, install the fuel pump and the thermostat housing. Connect the choke coolant hose at the manifold. Connect the power brake booster vacuum line.
19. If equipped, install the injection mixer assembly.
20. Connect the accelerator linkage and, if equipped with an automatic transmission, the shift cables.
21. Connect the throttle positioner solenoid and fuel cut-off solenoid wires.
22. Connect the fuel line and EGR lines, if equipped.
23. Connect the negative battery cable and install the air cleaner.

Multi-Point Fuel Injection

1. Disconnect the negative battery cable. Drain the cooling system. Disconnect the air intake hose, accelerator cable and the throttle body stay.
2. Disconnect the water bypass hose and the vacuum hose to the power brake booster.
3. Relieve the fuel system pressure.
4. Disconnect the fuel high pressure and return lines and their mounting O-rings.
5. Disconnect interfering vacuum hose, plug wires and wiring harness connections.
6. Disconnect the oxygen sensor, idle speed control, injector connector, ignition coil and power transistor connectors.
7. Disconnect the crank angle sensor, if so equipped, the throttle position sensor and the control harness connectors.
8. Remove the fuel delivery pipe, fuel injector and pressure regulator as an assembly. Remove the mounting grommets and O-rings.
9. Remove the intake manifold lower support stay shield and the end tension bracket.
10. Remove the intake manifold mounting bolts and the intake manifold assembly.
11. Remove the components from the intake manifold and on 2 piece manifolds, separate the upper and lower halves.

To install:

12. Clean all gasket mounting surfaces.
13. Install all components and the intake manifold in reverse order of removal. Torque values follow: manifold to head bolts 11–14 ft. lbs. (15–20 Nm); upper to lower manifold 11–14 ft. lbs. (15–20 Nm); fuel delivery manifold 7–9 ft. lbs. (10–12 Nm); throttle body 11–16 ft. lbs. (14–21 Nm) on DOHC and 1997cc; 7–9 ft. lbs. (10–12 Nm) on 1468cc.

Exhaust Manifold

REMOVAL & INSTALLATION

1. Disconnect the negative battery cable. Remove the air cleaner assembly.
2. Remove the manifold heat stove and hose. Disconnect the EGR lines and reed valve, if equipped. On turbocharged vehicles, remove the turbocharger.
3. Disconnect the exhaust pipe bracket from the engine block.
4. Remove the exhaust pipe flange bolts, 1 bolt and nut may have to be removed from under the vehicle.
5. Remove the manifold flange stud nuts and remove the manifold from the cylinder head.

To install:

6. Clean the gasket mating surfaces. Port liner gaskets may be used along with the exhaust manifold gaskets on some engine models. Torque the manifold to 11–14 ft. lbs. (15–20 Nm). Torque the pipe-to-manifold nuts to 15–20 ft. lbs. (20–27 Nm).

Turbocharger

REMOVAL & INSTALLATION

NOTE: Make sure the engine and turbocharger are cold, before removing the unit. If replacing the turbocharger, change the engine oil and filter.

1. Drain the cooling system and disconnect the negative battery cable. For clearance on some models, it will be necessary to remove the radiator. Remove the heat shield.
2. Disconnect the air intake hose and vacuum lines. Remove the oxygen sensor from the catalytic converter.
3. Remove the converter-to-turbocharger nuts.
4. Disconnect the hose from the oil return pipe and time chain case.
5. Remove the oil pipe from the turbocharger and oil filter housing.
6. Remove the air intake pipe connecting bolt.
7. Remove the turbocharger mounting nuts and lift the unit off the engine.

To install:

8. Before the oil flare nut is installed at the top of the unit, pour clean engine oil into the turbocharger. Always use new gaskets. Torque the pipe-to-turbocharger nuts to 22–29 ft. lbs. (30–40 Nm), oil flare nuts to 10–13 ft. lbs. (14–19 Nm) and retaining bracket to 9–10 ft. lbs. (12–15 Nm).

Timing Chain and Sprocket

REMOVAL & INSTALLATION

NOTE: The timing chain case is cast aluminum, exercise caution when handling.

2555cc Engines with Silent Shafts

1. Drain the coolant and remove the radiator. Disconnect the battery ground cable.
2. Remove the alternator and accessory belts.
3. Rotate the crankshaft to bring No. 1 piston to TDC, on the compression stroke.
4. Mark and remove the distributor.
5. Remove the crankshaft pulley.
6. Remove the water pump assembly.
8. Raise the vehicle and support it safely.
9. Drain the engine oil and remove the oil pan and screen.
10. Remove the timing case cover.
11. Remove the chain guides. Side (A), top (B), bottom (C), from the outer top (B) chain.
12. Remove the locking bolts from the "B" chain sprockets.
13. Remove the crankshaft sprocket, counter-balance shaft sprocket and the outer chain.
14. Remove the crankshaft and camshaft sprockets and the inner chain.
15. Remove the camshaft sprocket holder and the chain guides, both left and right. Remove the tensioner spring and sleeve from the oil pump.
16. Remove the oil pump by first removing the bolt locking the oil pump driven gear and the right counter-balance shaft and then remove the oil pump mounting bolts. Remove the counter-balance shaft from the engine block.

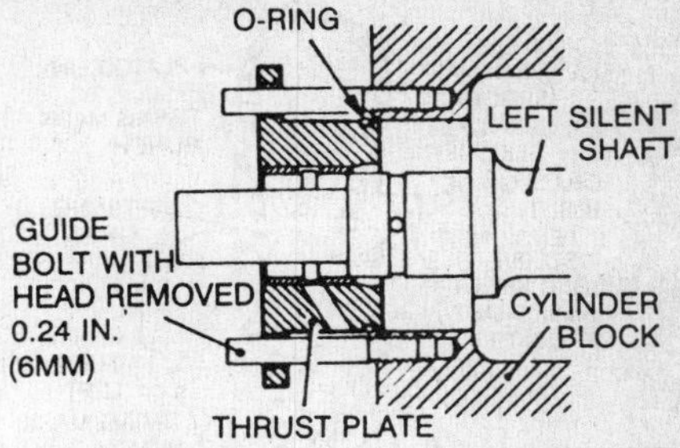

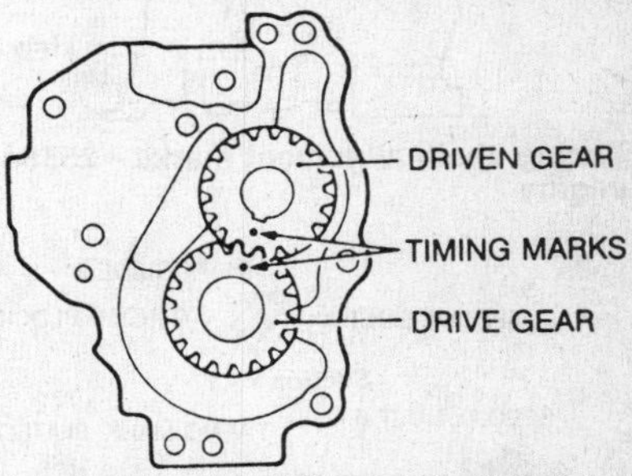

Front timing cover alignment—2555cc engine

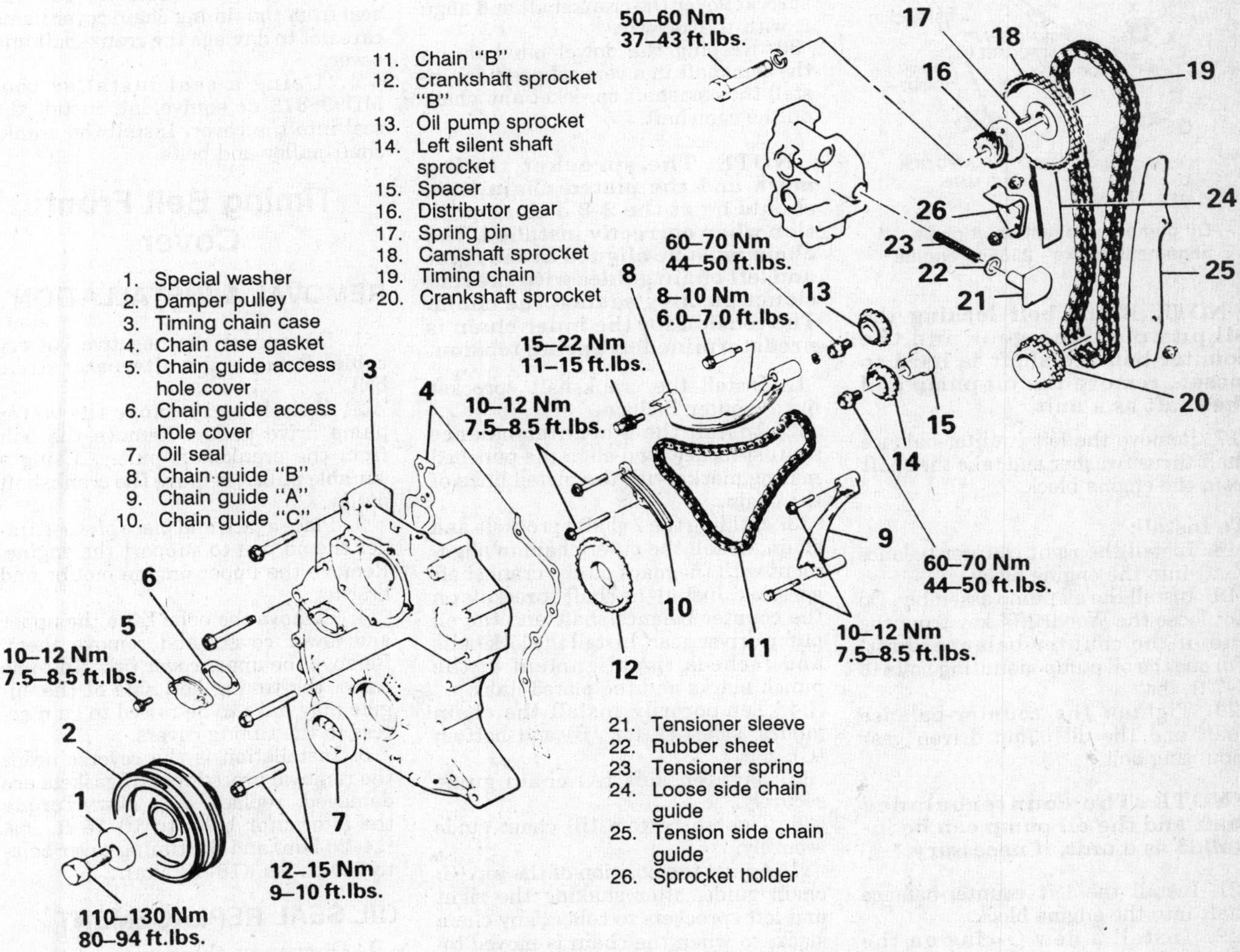

Timing chain and front cover—2555cc engine

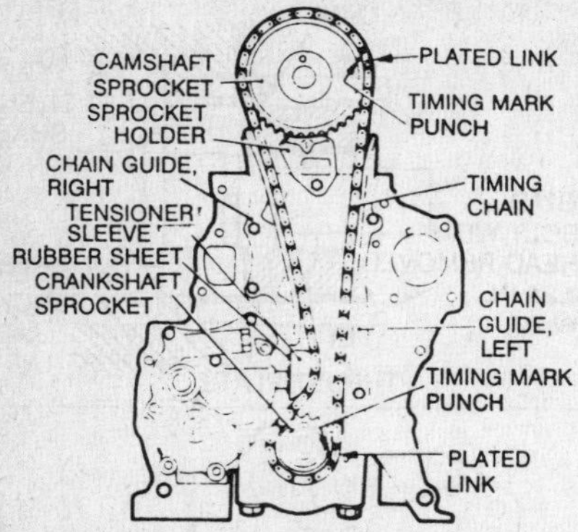

Timing chain alignment marks—2555cc engine

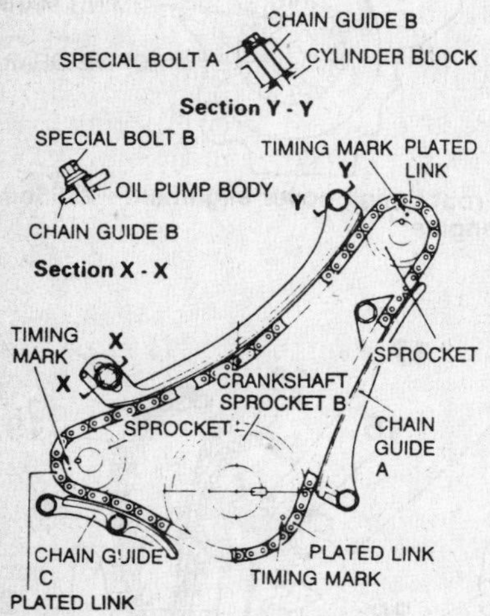

Oil pump and silent shaft chain alignment marks—2555cc engine

NOTE: If the bolt locking the oil pump driven gear and the counter-balance shaft is hard to loosen, remove the oil pump and the shaft as a unit.

17. Remove the left counter-balance shaft thrust washer and take the shaft from the engine block.

To install:

18. Install the right counter-balance shaft into the engine block.
19. Install the oil pump assembly. Do not loose the Woodruff® key from the end of the counter-balance shaft. Torque the oil pump mounting bolts to 6–7 ft. lbs.
20. Tighten the counter-balance shaft and the oil pump driven gear mounting bolt.

NOTE: The counterbalance shaft and the oil pump can be installed as a unit, if necessary.

21. Install the left counter-balance shaft into the engine block.
22. Install a new O-ring on the thrust plate and install the unit into the engine block, using a pair of bolts without heads, as alignment guides.

NOTE: If the thrust plate is turned to align the bolt holes, the O-ring may be damaged.

23. Remove the guide bolts and install the regular bolts into the thrust plate and tighten securely.
24. Rotate the crankshaft to bring No. 1 piston to TDC.
25. Install the cylinder head, if removed.
26. Install the sprocket holder and the right and left chain guides.
27. Install the tensioner spring and sleeve on the oil pump body.
28. Install the camshaft and crankshaft sprockets on the timing chain, aligning the sprocket punch marks to the plated chain links. Both the crankshaft and camshaft sprockets have a small dot on their faces. Assemble the chain and sprockets so that each dot aligns with the plated links on the chain.
29. While holding the sprocket and chain as a unit, install the crankshaft sprocket over the crankshaft and align it with the keyway.
30. Keeping the dowel pin hole on the camshaft in a vertical position, install the camshaft sprocket and chain on the camshaft.

NOTE: The sprocket timing mark and the plated chain link, should be at the 2–3 o'clock position when correctly installed. The chain must be aligned in the right and left chain guides with the tensioner pushing against the chain. The tension for the inner chain is predetermined by spring tension.

31. Install the crankshaft sprocket for the outer B chain.
32. Install the 2 counter-balance shaft sprockets and align the punched mating marks with the plated links of the chain.
33. Holding the 2 shaft sprockets and chain, install the outer chain in alignment with the mark on the crankshaft sprocket. Install the shaft sprockets on the counter balance shaft and the oil pump driver gear. Install the lock bolts and recheck the alignment of the punch marks and the plated links.
34. Temporarily install the chain guides, side (A), top (B) and bottom (C).
35. Tighten side (A) chain guide securely.
36. Tighten bottom (B) chain guide securely.
37. Adjust the position of the top (B) chain guide, after shaking the right and left sprockets to collect any chain slack, so when the chain is moved toward the center, the clearance between the chain guide and the chain links will be approximately $^{9}/_{64}$ in. Tighten the Top (B) chain guide bolts.
38. Install the timing chain cover using a new gasket, being careful not to damage the front seal.
39. Install the oil screen and the oil pan, using a new gasket. Torque the bolts to 4.5–5.5 ft. lbs.
40. Install the crankshaft pulley, alternator and accessory belts and the distributor.
41. Install the oil pressure switch, if removed and install the negative battery cable.
42. Install the fan blades, radiator, fill the system with coolant and start the engine.

Front Cover Oil Seal

REMOVAL & INSTALLATION

1. Disconnect the negative battery cable.
2. Remove the drive belts and crankshaft pulley.
3. Using a seal remover, pull the seal from the timing chain cover using care not to damage the crankshaft and cover.
4. Using a seal installer tool MD998376 or equivalent, install the seal into the cover. Install the crankshaft pulley and belts.

Timing Belt Front Cover

REMOVAL & INSTALLATION

1. Disconnect the negative battery cable. Remove the alternator drive belt.
2. Unbolt and remove the water pump drive pulley. Remove the bolt from the crankshaft pulley. Using a suitable puller, remove the crankshaft pulley.
3. Place a jack and piece of wood under the oil pan to support the engine. Remove the upper engine mount and bracket.
4. Remove the bolts from the upper and lower covers and remove them. Remove the upper cover first. In some cases, the timing belt side of the engine may have to be raised to gain access to the timing covers.
5. Installation is the reverse order the removal procedures. If gaskets are damaged, replace with new. Torque the alternator bolts to 10–15 ft. lbs. (14–20 Nm) and the timing cover bolts to 7–9 ft. lbs. (10–12 Nm).

OIL SEAL REPLACEMENT

1. Disconnect the negative battery cable.

2. Remove the air pump and alternator drive belts. Remove the air pump mounting bracket.
3. Raise and safely support the vehicle. Remove the right inner splash shield.
4. Remove the crankshaft pulley bolt and washer and remove the pulley.
5. Install a seal remover tool over crankshaft nose and turn it tightly into the seal.
6. Tighten the thrust screw to remove the seal.

NOTE: If the front cover is removed from the engine, tap the side of the thrust screw to remove the seal.

To install:
7. Using a oil seal installation tool, drive the new seal into the front cover.
8. Install the crankshaft pulley, washer and retaining bolt.
9. Install the right inner splash shield and lower the vehicle.
10. Install the air pump mounting bracket, air pump and alternator drive belts. Torque the crankshaft pulley bolt to 51–72 ft. lbs. (70–100 Nm).
11. Connect the negative battery cable.

Timing Belt and Tensioner

ADJUSTMENT

1468cc and 1507cc Engines

1. Bring the engine to No. 1 piston at TDC timing marks aligned. Disconnect the negative battery cable.
2. Remove the drive belts, water pump pulley, spacer and timing belt cover.
3. Loosen the tensioner from it's temporary position so the spring pressure will allow it to contact the timing belt.
4. Rotate the crankshaft 2 complete turns in the normal rotation direction to remove any belt slack. Turn the crankshaft until the timing marks are lined up. If the timing has slipped, remove the belt and repeat the procedure.
5. Tighten the tensioner mounting bolts, slotted side (right) first, then the spring side.
6. Once again rotate the engine 2 complete revolutions until the timing marks align. Recheck the belt tension.

NOTE: When the tension side of the timing belt and the tensioner are pushed in horizontally with a moderate force, about 11 lbs. and the cogged side of the belt covers about ¼ in. of the tensioner right side mounting bolt head, the across flats, the tension is correct.

7. Reinstall the timing belt cover, the water pump pulley, spacer, fan blades and drive belt.
8. Connect the negative battery cable.

1595cc DOHC Engine

1. Bring the engine to No. 1 piston at TDC timing marks aligned. Disconnect the negative battery cable.
2. Raise the vehicle and support it safely. Remove the under engine splash shield.
3. Place a piece of wood on a suitable floor jack and support the engine. Remove the engine mount bracket.
4. Remove the alternator and power steering drive belts. Remove the air conditioner drive belt and tensioner assembly.
5. Remove the water pump pulley and the crankshaft pulley.
6. Remove the upper and lower timing belt covers.
7. Lift up the tensioner pulley against the belt and tighten the center bolt to hold it in position.
8. Make sure the timing marks are aligned. Remove the binder clips. Rotate the crankshaft a ¼ turn counterclockwise. Then turn the crankshaft clockwise until the timing marks are aligned.
9. Place special tool MD998752 or equivalent, on a torque wrench. Insert the tool into the place provided on the tension pulley. Loosen the center pulley bolt and apply 2.2 ft. lbs. of pressure against the timing belt with the tension pulley. While holding the required torque, tighten the center bolt. Screw in special tool MD998738 or equivalent, through the left engine support bracket until it contacts the tensioner arm bracket. Turn the tool a little more to secure the tensioner and remove the locking wire placed into the automatic adjuster when it was reset.
10. Remove the special tool. Rotate the crankshaft 2 complete turns clockwise and allow it to set, for about 15 minutes. Then measure the protrusion of the automatic adjuster. It should be 0.015–0.018 in. If the proper amount of protrusion is not present, repeat the tensioning process.
11. Install the upper and lower timing belt covers.
12. Install the crankshaft pulley and water pump pulley.
13. Install the alternator and power steering drive belts. Install the air conditioner drive belt and tensioner assembly.
14. Install the engine mount bracket and lower the engine.
15. Install the under engine splash shield.
16. Connect the negative battery cable.

1597cc and 1755cc Engines

1. Bring the engine to No. 1 piston at TDC, aligned. Disconnect the negative battery cable.
2. Remove the drive belts, water pump pulley, spacer and timing belt cover.
3. Ensure that the sprocket timing marks are aligned, before making the adjustment.
4. Loosen the tensioner mounting bolt and nut and allow the spring tension to move the tensioner against the belt.

NOTE: Make sure the belt comes in complete mesh with the sprocket by lightly pushing the tensioner up by hand toward the mounting nut.

5. Tighten the tensioner mounting nut and bolt.

NOTE: Be sure to tighten the nut before tightening the bolt. Too much tension could result from tightening the bolt first.

6. Recheck all sprocket alignments.
7. Turn the crankshaft through a complete rotation in the normal direction. Do not turn in a reverse direction or shake or push the belt.
8. Loosen the tensioner bolt and nut. Retighten the nut and then the bolt.
9. Reinstall the timing belt covers, the water pump pulley, spacer and drive belts. Connect the negative battery cable.

1997cc Engine

1. Disconnect the negative battery cable. Remove the water pump drive belt and pulley.
2. Remove the crank adapter and crankshaft pulley.
3. Remove the upper and lower timing belt covers.
4. Check the tensioners for a smooth rate of movement.
5. Replace any tensioner that shows grease leakage through the seal.
6. Install the silent shaft belt and adjust the tension, by moving the tensioner into contact with the belt, tighten enough to remove all slack. Tighten the tensioner bolt to 21 ft. lbs.
7. Tighten the silent shaft sprocket bolt to 28 ft. lbs.
8. Install the upper and lower timing belt covers.
9. Install the crank adapter and crankshaft pulley.
10. Install the water pump drive belt

and pulley. Connect the negative battery cable.

REMOVAL & INSTALLATION

NOTE: The timing chain case is cast aluminum, so exercise caution when handling this part.

1468cc and 1507cc Engines

1. Turn the engine until the No. 1 piston is on TDC with the timing marks aligned.
2. Disconnect the negative battery cable.
3. Remove the fan drive belt, the fan blades, spacer and water pump pulley.
4. Remove the timing belt cover.
5. Loosen the timing belt tensioner mounting bolt and move the tensioner toward the water pump. Temporarily secure the tensioner.
6. Remove the crankshaft pulley and slide the belt off of the camshaft and crankshaft drive sprockets.
7. Inspect the drive sprockets for abnormal wear, cracks or damage and replace, if necessary. Remove and inspect the tensioner. Check for smooth pulley rotation, excessive play or noise. Replace tensioner, if necessary.

To install:

8. Reinstall the tensioner, if removed and temporarily secure it close to the water pump.
9. Make sure the timing mark on the camshaft sprocket is aligned with the pointer on the cylinder head and that the crankshaft sprocket mark is aligned with the mark on the engine case.
10. Install the timing belt on the crankshaft sprocket.
11. Install the belt counterclockwise over the camshaft sprocket making sure there is no play on the tension side of the belt. Adjust the belt fore and aft so it is centered on the sprockets.
12. Loosen the tensioner from it's temporary position so the spring pressure will allow it to contact the timing belt.
13. Rotate the crankshaft 2 complete turns in the normal rotation direction to remove any belt slack. Turn the crankshaft until the timing marks are lined up. If the timing has slipped, remove the belt and repeat the procedure.
14. Tighten the tensioner mounting bolts, slotted side (right) first, then the spring side.
15. Once again rotate the engine 2 complete revolutions until the timing marks line up. Recheck the belt tension.

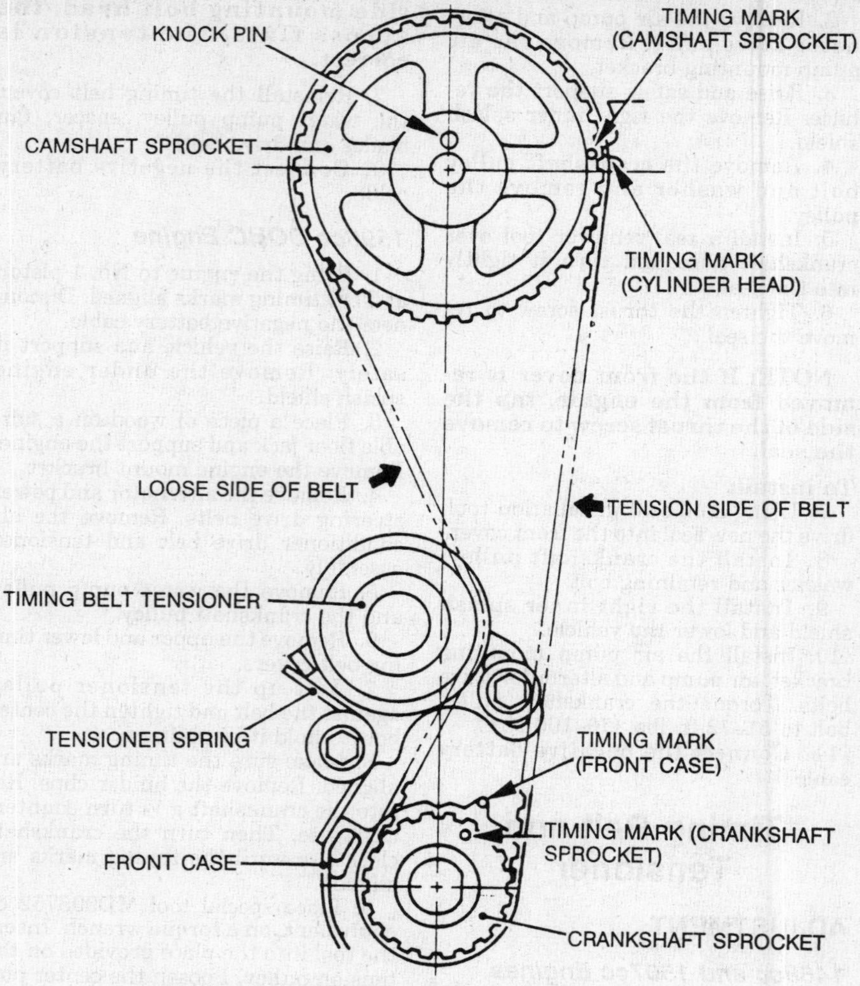

Timing belt installation—1468cc and 1507cc engines

NOTE: When the tension side of the timing belt and the tensioner are pushed in horizontally with a moderate force, about 11 lbs. and the cogged side of the belt covers about ¼ in. of the tensioner right side mounting bolt head the across flats, the tension is correct.

16. Reinstall the timing belt cover, the water pump pulley, spacer, fan blades and drive belt.
17. Connect the negative battery cable.

1595cc DOHC Engine

NOTE: Special tools MD998752 tension pulley torque adapter and MD998738 tension pulley locker or equivalents, are required.

1. Bring the engine to No. 1 piston at TDC (top dead center) timing marks aligned. Disconnect the negative battery cable.
2. Raise the vehicle and support it safely. Remove the under engine splash shield.
3. Place a piece of wood on a suitable floor jack and support the engine. Remove the engine mount bracket.
4. Remove the alternator and power steering drive belts. Remove the air conditioner drive belt and tensioner assembly.
5. Remove the water pump pulley and the crankshaft pulley.
6. Remove the upper and lower timing belt covers.
7. Remove the engine center cover. Remove the breather hose from the rear of the rocker cover. Remove the PCV hose. Disconnect the spark plug cables from the plugs.
8. Remove the rocker cover and rear half-moon seal.
9. Confirm the engine is still at No. 1 TDC. The timing marks on the camshaft sprocket and the upper surface of the cylinder head should coincide. The dowel pin on the front of the camshafts should be in the 12 o'clock position. Remove the automatic belt tensioner. Loosen the tensioner pulley center bolt.

10. If the timing belt is to be reused, mark an arrow, on the belt, in the direction of rotation, for installation reference. Remove the timing belt.

To install:

11. Install the automatic tensioner, after reset.

NOTE: To reset the tensioner: Keep the adjuster level and clamp it in a soft jawed vise. Clamp with the extended adjuster on one side and the end mounting a plug on the other side. If the plug extends out of the adjuster body, place a suitable hole sized washer over the plug so the vise jaw pushes on the washer, not the plug. Close the vise slowly, forcing the adjuster back into the body. When the hole in the adjuster boss aligns with the adjuster rod, insert a snug fitting pin or wire into the holes to keep the rod in the compressed position. With the locking pin or wire in place, install the tensioner.

12. Align the timing marks on the camshaft sprockets. Align the crankshaft timing marks. Align the oil pump timing marks. Place the timing belt around the intake camshaft and secure it to the sprocket with a stationary binder spring clip. Install the timing belt around the exhaust camshaft sprocket, check sprocket marks for alignment and secure the belt with a second binder clip on the exhaust sprocket.

13. Install the timing belt around the idler pulley, oil pump sprocket, crankshaft sprocket and the tensioner pulley.

14. Lift up the tensioner pulley against the belt and tighten the center bolt to hold it in position.

15. Check to see that all of the timing marks are aligned. Remove the binder clips. Rotate the crankshaft a quarter turn counter clockwise. Then turn the crankshaft clockwise until the timing marks are aligned.

16. Place special tool MD998752 or equivalent, on a torque wrench. Insert the tool into the place provided on the tension pulley. Loosen the center pulley bolt and apply 2.2 ft. lbs. of pressure against the timing belt with the tension pulley. While holding the required torque, tighten the center bolt. Screw in special tool MD998738 or equivalent, through the left engine support bracket until it contacts the tensioner arm bracket. Turn the tool a little more to secure the tensioner and remove the locking wire place into the automatic adjuster when it was reset.

17. Remove the special tool. Rotate the crankshaft 2 complete turns clockwise and allow it to sit for about 15 minutes. Then measure the protrusion of the automatic adjuster. It should be 0.015–0.018 in. If the proper amount of protrusion is not present, repeat the tensioning process.

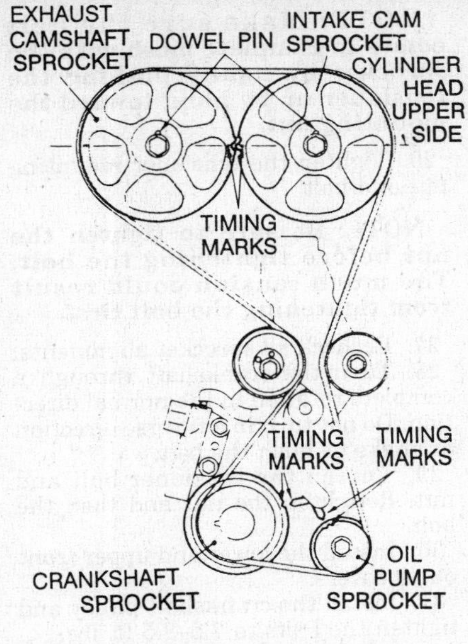

1595cc DOHC timing mark alignment for timing belt installation

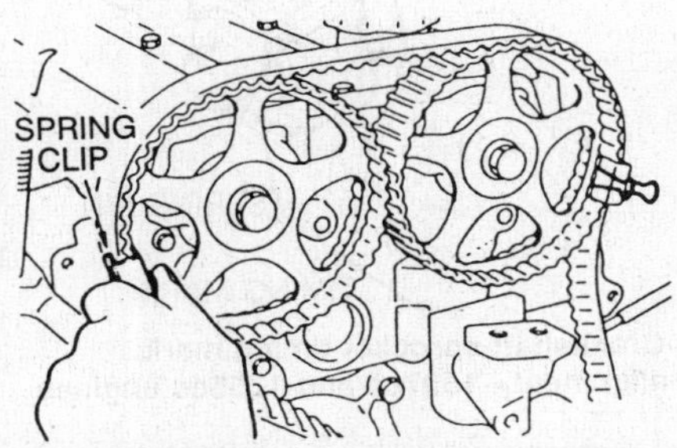

Using binder clips to secure the timing belt

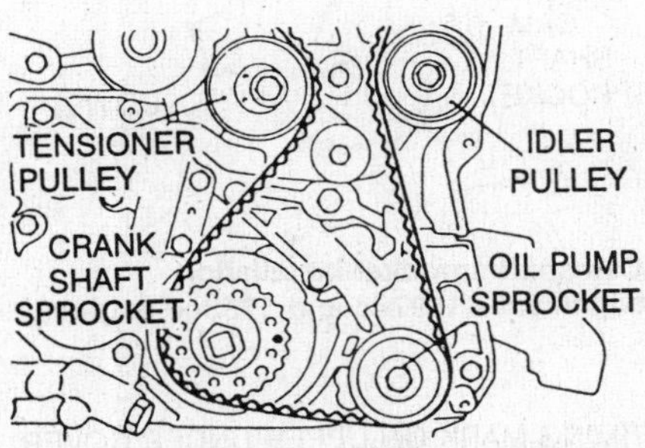

Installing the timing belt around the idler pulley, oil pump sprocket, crankshaft sprocket and tensioner

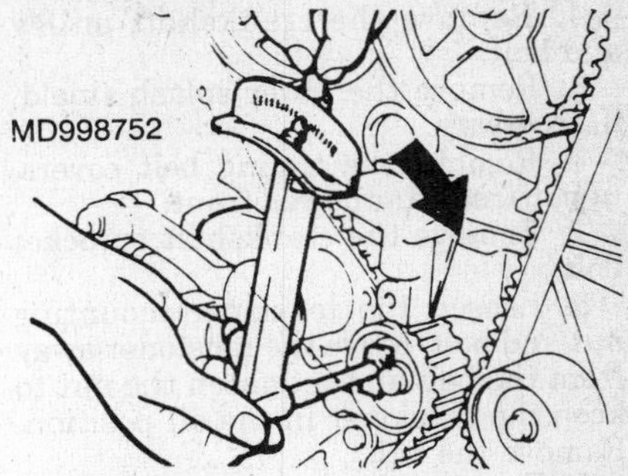

Using special tool MD998752 – 1595cc DOHC

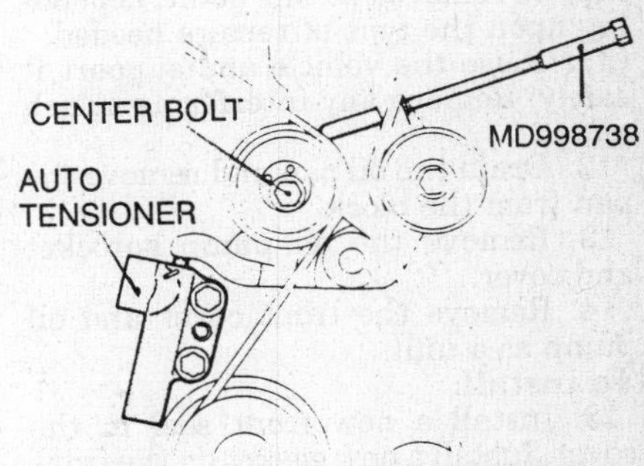

Using special tool MD998752 or equivalent – 1595cc DOHC engine

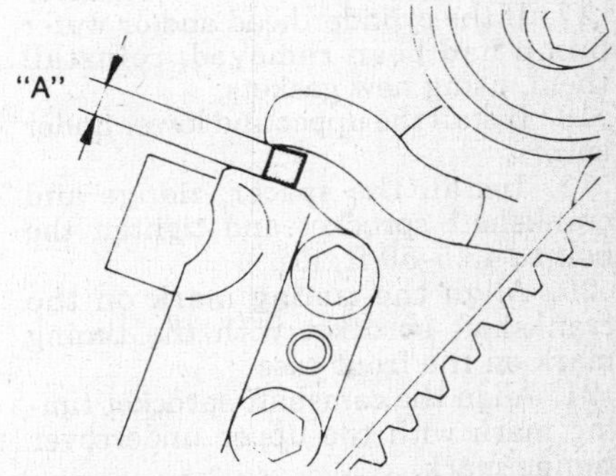

Automatic tensioner extension measurement – 1595cc DOHC engine

1597cc and 1755cc Engines

1. Drain the coolant and remove the radiator on rear wheel drive vehicles only. Disconnect the negative battery cable.

2. Remove the alternator and accessory belts. Remove the belt cover.

3. Rotate the crankshaft to bring No. 1 piston to TDC on the compression stroke. Align the notch on the crankshaft pulley with the T mark on the timing indicator scale and the timing mark on the upper under cover of the timing belt with the mark on the camshaft sprocket. Mark and remove the distributor.

4. Remove the crankshaft pulley and bolt.
5. Remove the lower splash shield, if necessary.
6. Remove the timing belt covers, upper front and lower front.
7. Remove the crankshaft sprocket bolt.
8. Loosen the tensioner mounting nut and bolt. Move the tensioner away from the belt and retighten the nut to keep the tensioner in the off position. Remove the belt.
9. Remove the camshaft sprocket, crankshaft sprocket, flange and tensioner.
10. The water pump or cylinder head may be removed at this point, depending upon the type of repairs needed.
11. Raise the vehicle and support it safely. Remove any interfering splash pans.
12. Drain the oil pan and remove the pan from the block.
13. Remove the oil pump sprocket and cover.
14. Remove the front cover and oil pump as a unit.

To install:

15. Install a new front seal in the cover. Install a new gasket on the front of the cylinder block and install the front cover.
16. Tighten the front cover bolts to 11–13 ft. lbs. Install the oil screen and oil pan. Tighten the bolts to 5 ft. lbs.
17. If the cylinder head and/or water pump had been removed, reinstall them, using new gaskets.
18. Install the upper and lower under covers.
19. Install the spacer, flange and crankshaft sprocket and tighten the bolt to 43.5–50 ft. lbs.
20. Align the timing mark on the crankshaft sprocket with the timing mark on the front case.
21. Align the camshaft sprocket timing mark with the upper undercover timing mark.
22. Install the tensioner spring and tensioner. Temporarily tighten the nut. Install the front end of the tensioner spring (bent at right angles) on the projection of the tensioner and the other end (straight) on the water pump body.
23. Loosen the nut and move the tensioner in the direction of the water pump. Lock it by tightening the nut.
24. Ensure that the sprocket timing marks are aligned and install the timing belt. The belt should be installed on the crankshaft sprocket, the oil pump sprocket and then the camshaft sprocket, in that order, while keeping the belt tight.
25. Loosen the tensioner mounting bolt and nut and allow the spring tension to move the tensioner against the belt.

NOTE: Make sure the belt comes in complete mesh with the sprocket by lightly pushing the tensioner up by hand toward the mounting nut.

26. Tighten the tensioner mounting nut and bolt.

NOTE: Be sure to tighten the nut before tightening the bolt. Too much tension could result from tightening the bolt first.

27. Recheck all sprocket alignments.
28. Turn the crankshaft through a complete rotation in the normal direction. Do not turn in a reverse direction or shake or push the belt.
29. Loosen the tensioner bolt and nut. Retighten the nut and then the bolt.
30. Install the lower and upper front outer covers.
31. Install the crankshaft pulley and tighten the bolts to 7.5–8.5 ft. lbs.

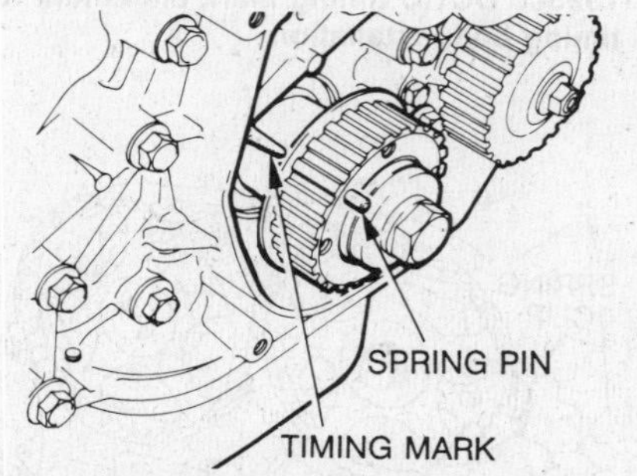

Crankshaft sprocket timing mark alignment—1597cc and 1755cc engines

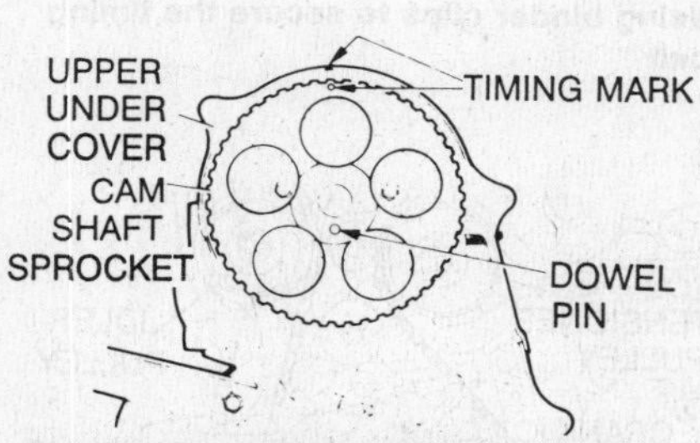

Camshaft sprocket installation alignment—1597cc and 1755cc engines

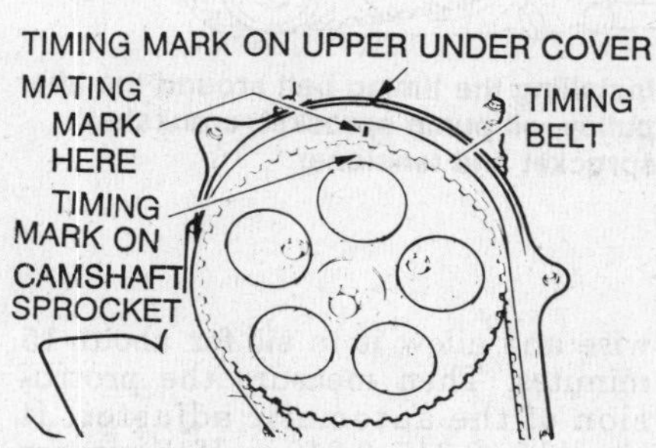

Canshaft timing mark alignment—1597cc and 1755cc engines

32. Install the alternator and belt and adjust. Install the distributor.
33. Install the radiator, fill the cooling system and inspect for leaks.

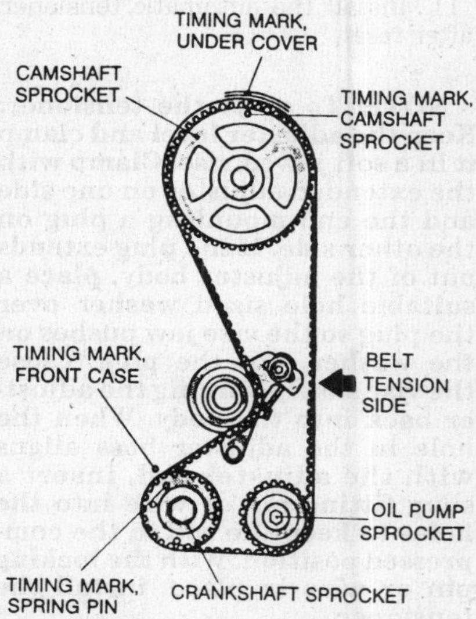

Timing belt installation—1597cc and 1755cc engines

1997cc Engine

NOTE: An 8mm diameter metal bar is needed for this procedure.

1. Disconnect the negative battery cable. Remove the water pump drive belt and pulley.
2. Remove the crank adapter and crankshaft pulley.
3. Remove the upper and lower timing belt covers.
4. Move the tensioner fully in the direction of the water pump and temporarily secure it there.
5. If the timing belt is to be reused, make a paint mark on the belt to indicate the direction of rotation. Slip the belt from the sprockets.
6. Remove the camshaft sprocket bolt and pull the sprocket from the camshaft.
7. Remove the crankshaft sprocket bolt and pull the crankshaft sprocket and flange from the crankshaft.
8. Remove the plug on the left side of the block and insert an 8mm diameter metal bar in the opening to keep the silent shaft in position.
9. Remove the oil pump sprocket retaining nut and remove the oil pump sprocket.
10. Loosen the right silent shaft sprocket mounting bolt until it can be turned by hand.
11. Remove the belt tensioner and remove the timing belt.

NOTE: Do not attempt to turn the silent shaft sprocket or loosen its bolt while the belt is off.

12. Remove the silent shaft belt sprocket from the crankshaft.
13. Check the belt for wear, damage or glossing. Replace it if any cracks, damage, brittleness or excessive wear are found.
14. Check the tensioners for a smooth rate of movement.
15. Replace any tensioner that shows grease leakage through the seal.

To install:

16. Install the silent shaft belt sprocket on the crankshaft, with the flat face toward the engine.
17. Apply light engine oil on the outer face of the spacer and install the spacer on the right silent shaft. The side with the rounded shoulder faces the engine.
18. Install the sprocket on the right silent shaft and install the bolt but do not tighten completely at this time.

NOTE: Align the silent shaft and oil pump sprockets using the timing marks. If the 8mm metal bar can not be inserted into the hole 2.36 in. (60mm), the oil pump sprocket will have to be turned 1 full rotation until the bar can be inserted to the full length. If this procedure is not followed, the engine will run but vibrate at high engine rpm.

19. Install the silent shaft belt and adjust the tension, by moving the tensioner into contact with the belt, tight enough to remove all slack. Tighten the tensioner bolt to 21 ft. lbs.
20. Tighten the silent shaft sprocket bolt to 28 ft. lbs.
21. Install the flange and crankshaft sprocket on the crankshaft. The flange conforms to the front of the silent shaft sprocket and the timing belt sprocket is installed with the flat face toward the engine.

NOTE: The flange must be installed correctly or a broken belt will result.

22. Install the washer and bolt in the crankshaft and torque it to 94 ft. lbs.
23. Install the camshaft sprocket and bolt and torque the bolt to 72 ft. lbs.
24. Install the timing belt tensioner, spacer and spring.
25. Align the timing mark on each sprocket with the corresponding mark on the front case.
26. Install the timing belt on the sprockets and move the tensioner against the belt with sufficient force to allow a deflection of 5–7mm along its longest straight run.
27. Tighten the tensioner bolt to 21 ft. lbs.
28. Install the upper and lower covers, the crankshaft pulley and the crank adapter. Tighten the bolts to 21 ft. lbs.

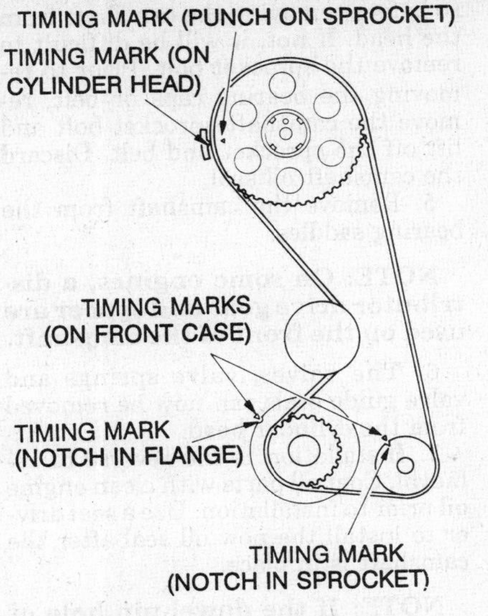

Timing mark alignment—1997cc engine

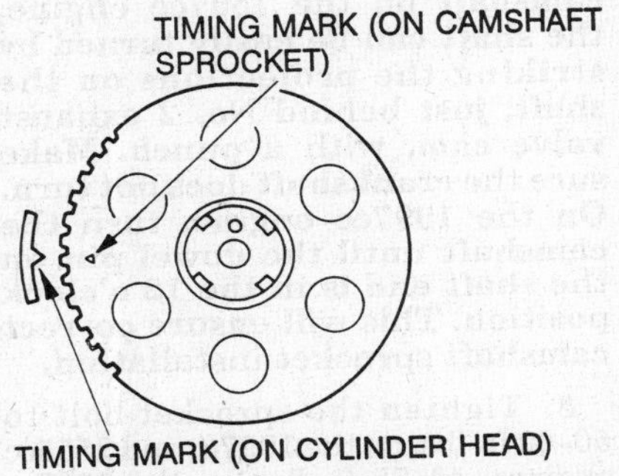

Camshaft timing mark alignment for No. 1 TDC—1997cc engine

29. Remove the 8mm bar and install the plug. Connect the negative battery cable.

SILENT SHAFT BELT REPLACEMENT

NOTE: When replacing the timing belt, the manufacturer recommends to replace the silent shaft belt.

1. After removing the timing belt, mark arrows on the belt to indicate direction of rotation (clockwise).

NOTE: If the silent shaft belt is not installed properly, the engine will run but with a vibration. Align the silent shaft and oil pump sprockets using the timing marks. If the 8mm metal bar can not be inserted into the hole 2.36 inch (60mm), the oil pump sprocket will have to turned 1 full rotation until the bar can be inserted to the full length.

2. Align the timing marks before removal. Loosen the belt tensioner and remove the belt.

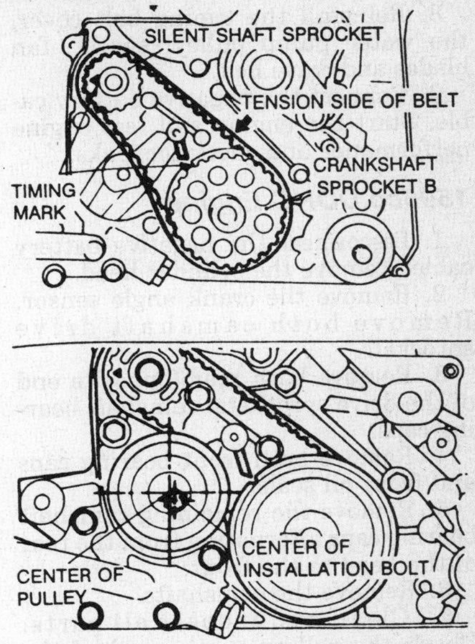

Silent shaft and belt—1997cc engine

To install:

3. Make sure the crankshaft and silent sprocket timing marks are aligned.
4. Fit the belt over the sprockets. Make sure there is no slack in the belt.
5. Temporarily fix the timing belt tensioner so that the center of the tensioner pulley is to the left and above the center of the installation bolt. Temporarily attach the tensioner pulley so that the flange is toward the front of the engine.
6. Adjust the belt so that the slack between the 2 sprockets is within 0.20–0.28 inch (5–7mm).

Camshaft

REMOVAL & INSTALLATION

1468cc and 1507cc Engines

1. Disconnect the negative battery cable.
2. Remove the cylinder head. Remove the cylinder head rear cover.
3. Remove the camshaft thrust case tightening bolt, located on top of the rear mounting boss.
4. Carefully slide the camshaft and thrust case, attached to the rear of the cam, out the rear of the cylinder head.

To install:

5. Carefully slide the camshaft and thrust case into the cylinder head from the front.
6. Install a new cylinder head gasket and install the cylinder head and bolts. Torque the cylinder head bolts to 50–54 ft. lbs. (68–73 Nm).
7. Install the cylinder head rear cover.

8. Reinstall the timing belt cover, the water pump pulley, spacer, fan blades and drive belt.
9. Connect the negative battery cable. Start the engine and test engine performance and check for leaks.

1595cc DOHC Engine

1. Disconnect the negative battery cable. Remove the cylinder head.
2. Remove the crank angle sensor. Remove both camshaft drive sprockets.
3. Remove both rear (opposite end of the drive sprockets) camshaft bearing caps.
4. Remove both front bearing caps and front oil seals.
5. Remove the remaining camshaft bearing caps alternating from the rear of the head to the front.
6. Remove the camshafts.
7. Clean and inspect all parts. Check the rollers on the end of the rocker arms. If the rollers are warn on do not rotate smoothly, replace as necessary.

To install:

8. Lubricate the camshafts. Place the camshafts in position. The intake side camshaft has a slit in the rear to drive the crank angle sensor. The bearing caps No. 2–5 are the same shape. When installing them, check the top markings to identify the intake or exhaust side. Left or Right is marked on the front caps, L for the intake side; R for the exhaust side.
9. Tighten the bearing caps, in 2 or 3 steps, to 14–15 ft. lbs. (20–22 Nm).
10. Make sure the rocker arm is properly mounted on the lash adjuster and valve stem tip.
11. Install the front oil seals. Turn the intake camshaft until the front dowel pin is facing straight up at the 12 o'clock position. Install the crank angle sensor with the punch mark on the sensor housing aligned with the notch in the plate. Install the drive sprocket and tighten the bolts to 58–72 ft. lbs. (79–98 Nm).

1597cc, 1755cc, 1997cc and 2555cc Engines

1. Disconnect the negative battery cable.
2. Remove the rocker cover. Matchmark the rocker arm bearing caps to the cylinder head.
3. Remove the bearing cap bolts from the cylinder head, but do not remove them from the bearing caps and shafts. Lift the rocker arm assembly from the cylinder head.
4. Make sure the timing marks on the camshaft sprocket and head are properly aligned, so No. 1 piston is at TDC of the compression stroke. If the camshaft sprocket is to be removed, do so before removing the camshaft from the head. If not, it will be difficult to remove the sprocket bolt. Prior to removing the bearing caps or belt, remove the camshaft sprocket bolt and lift off the sprocket and belt. Discard the camshaft oil seal.
5. Remove the camshaft from the bearing saddles.

NOTE: On some engines, a distributor drive gear and spacer are used on the front of the camshaft.

6. The valves, valve springs and valve guide seals can now be removed from the cylinder head.
7. Installation is the reverse of removal. Coat all parts with clean engine oil prior to installation. Use a seal driver to install the new oil seal after the camshaft is in place.

NOTE: If the dowel pin hole of the camshaft sprocket will not align with the dowel pin on the camshaft on the 1597cc engine, the shaft can be easily turned by striking the projections on the shaft, just behind No. 2 exhaust valve cam, with a punch. Make sure the crankshaft does not turn. On the 1997cc engine, turn the camshaft until the dowel pin on the shaft end is in the 12 o'clock position. This will ensure correct camshaft sprocket installation.

8. Tighten the sprocket bolt to 50–60 ft. lbs. on the 1997cc and 2555cc engines; 44–55 ft. lbs. on the 1597cc and 1755cc engines. Tighten the rocker cover bolts to 5 ft. lbs.

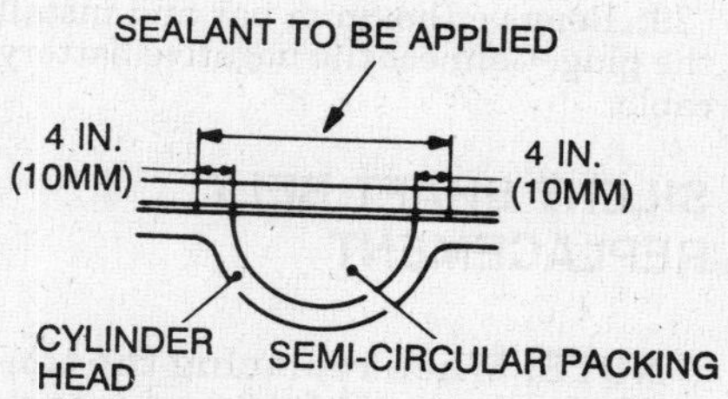

Sealant application on the rocker cover rear seal projections used for turning the shaft in hard-to-turn installations

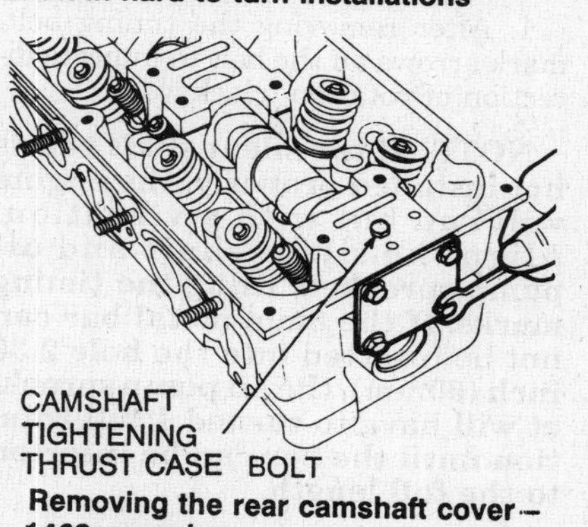

Removing the rear camshaft cover—1468cc engine

Silent Shafts

REMOVAL & INSTALLATION

1997cc Engine

1. Disconnect the negative battery cable. Remove the engine from the vehicle.

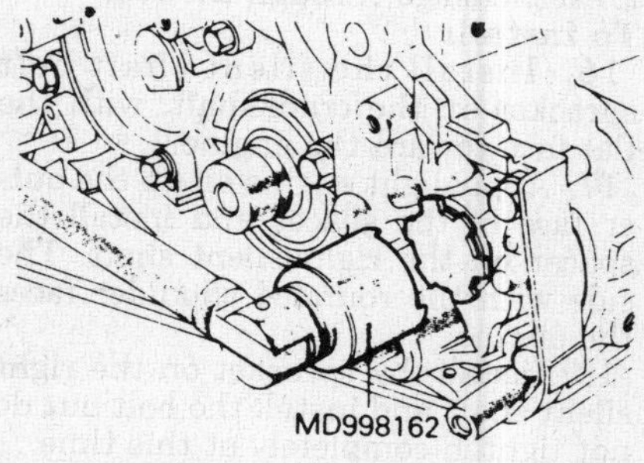

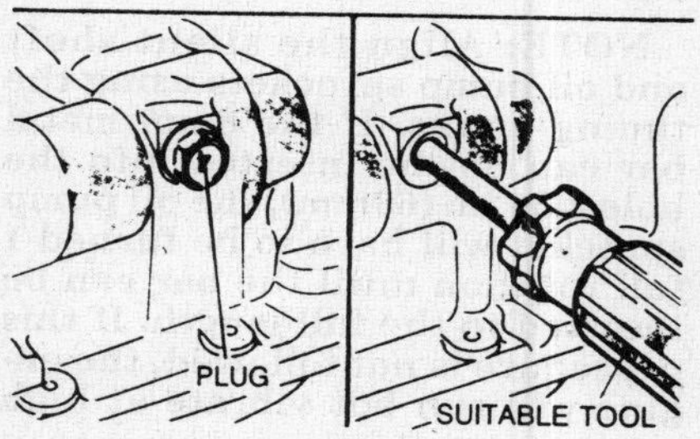

Silent shaft removal—1997cc engine

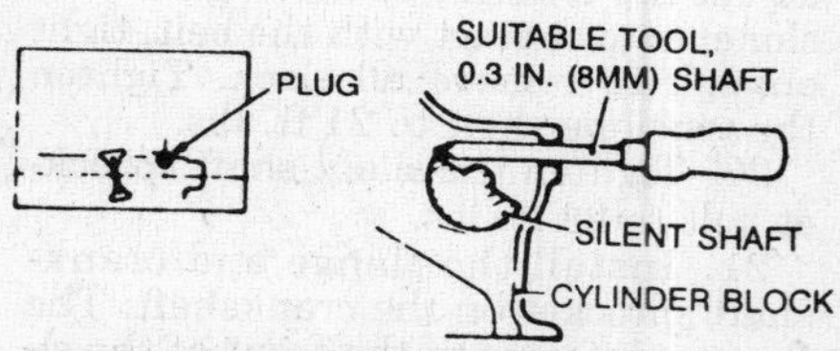

Keeping the left silent shaft in position. The tool must go into the hole at least 2.36 in. (60mm). If not, the silent shaft will be out of time and the engine will vibrate. Rotate the oil pump sprocket 1 full revolution and insert tool—1997cc engine

2. Remove the drive belts, accessories, crankshaft pulley and timing belts.
3. Drain the engine oil and remove the filter. Remove the oil pump sprocket, right silent shaft sprocket and spacer.
4. Remove the oil pan, screen and filter bracket.
5. Using a special tool MD998162 or equivalent, remove the plug cap from the right silent.
6. Remove the plug from the cylinder block left side and insert a suitable tool into the plug hole.
7. Remove the front case, oil pump cover, drive and driven gears.
8. Remove the left and right silent

shafts, being careful not to damage the bearings.

To install:

9. Install the left and right silent shafts, being careful not to damage the bearings.
10. Install the front case, oil pump cover, drive and driven gears.
11. Install the plug to the cylinder block left side silent shaft.
12. Using a special tool MD998162 or equivalent, install the plug cap to the right silent.
13. Install the screen, filter bracket and oil pan.
14. Refill the engine oil and install the filter. Install the oil pump sprocket, right silent shaft sprocket and spacer.
15. Install the drive belts, accessories, crankshaft pulley and timing belts.
16. Install the engine into the vehicle. Connect the negative battery cable.

2555cc Engine

1. Disconnect the negative battery cable. Drain the engine coolant and engine oil.
2. Remove the radiator and air conditioning evaporator, if so equipped. Remove the fan, drive belts and accessories.
3. Remove the oil pump and left silent shaft sprockets.
4. Remove the spacer, oil filter, knock bushing, oil filter bracket, oil pan and oil screen.
5. Remove the oil pump body and gears.
6. Remove the thrust plate by installing 8mm bolts into the holes in the flange and turning the bolts to remove the plate.
7. Remove the right and left silent shafts.

To install:

8. Install the right and left silent shafts.
9. Install the thrust plate using guide pins. Torque the bolts to 8 ft. lbs. (11 Nm).
10. Install the oil pump body and gears. Align the gear timing marks so they facing each other before installation. Pack the gear cavities with grease to hold the gears in place. Torque the bolts to 8 ft. lbs. (11 Nm).
11. Install the spacer, oil filter, knock bushing, oil filter bracket, oil pan and oil screen. Torque the pan bolts to 5 ft. lbs. (7 Nm), filter bracket bolt to 35 ft. lbs. (47 Nm) and oil screen bolts to 15 ft. lbs. (20 Nm).
12. Install the oil pump and left silent shaft sprockets. Torque the bolts to 45 ft. lbs. (61 Nm).
13. Install the radiator and air conditioning evaporator, if so equipped. Remove the fan, drive belts and accessories. Evacuate, recharge and leak test the air conditioning system using environmentally safe equipment.
14. Refill the engine coolant and oil. Connect the negative battery cable and check for leaks.

Piston and Connecting Rod

POSITIONING

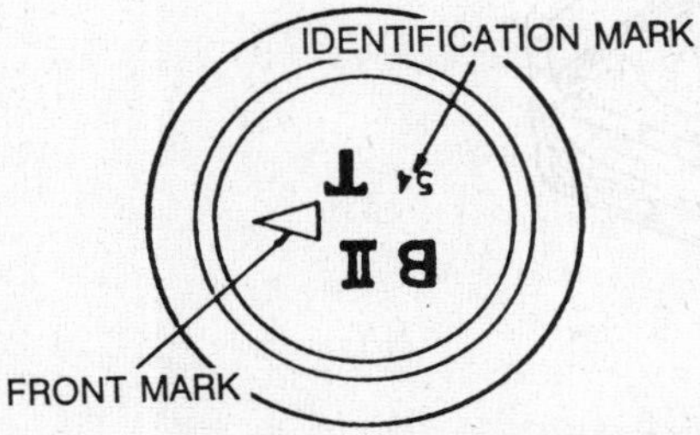

Typical piston identification and direction indicator

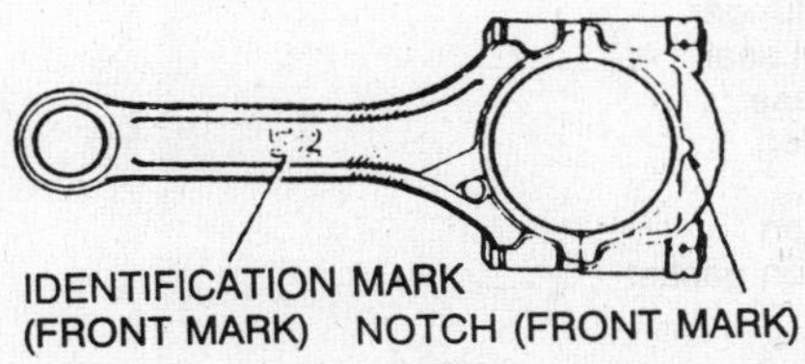

Typical connecting rod identification and front indicator

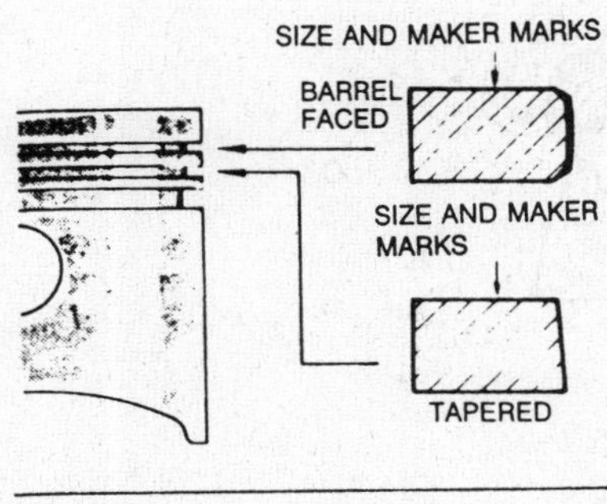

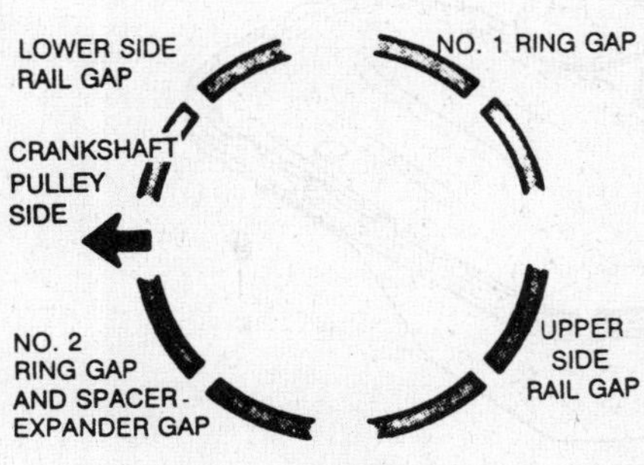

Piston ring positioning

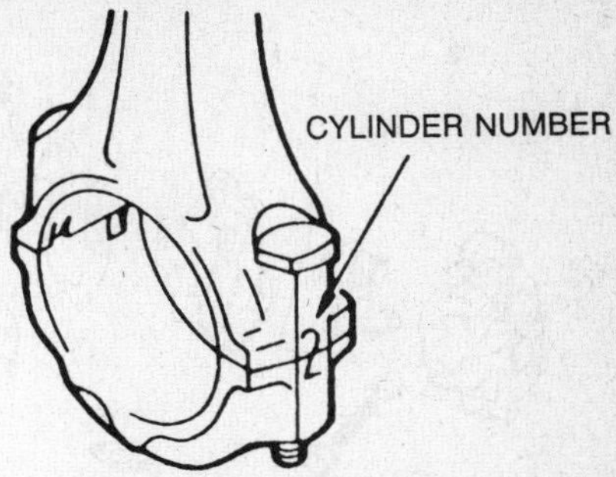

Location of cylinder number on connecting rod

ENGINE LUBRICATION

Oil Pan

REMOVAL & INSTALLATION

The engine may have to be raised off its mount for the pan to clear the suspension crossmember. However, on most front wheel drive vehicles, there is usually enough clearance without raising the engine.

1. Raise the vehicle and support it safely. Remove the underbody splash shield.
2. Unbolt the left and right engine mounts, except on front wheel drive vehicles.
3. On rear wheel drive vehicles, place a jack under the bell housing and raise the engine.
4. Remove the oil pan retaining bolts and remove the oil pan.

To install:

5. Installation is the reverse order of removal procedure.
6. Apply sealant to the front and rear main seal areas. Do not use sealant on the gasket itself. Use a new pan gasket. Torque the oil pan retaining bolts to 4–6 ft. lbs. (6–8 Nm).

Oil Pump

REMOVAL & INSTALLATION

1468cc and 1507cc Engines

1. Disconnect the negative battery cable and drain the oil. Remove the timing belt.
2. Remove the oil pan.
3. Remove the oil screen.
4. Unbolt and remove the front case assembly.
5. Remove the oil pump cover.
6. Remove the inner and outer gears from the front case.

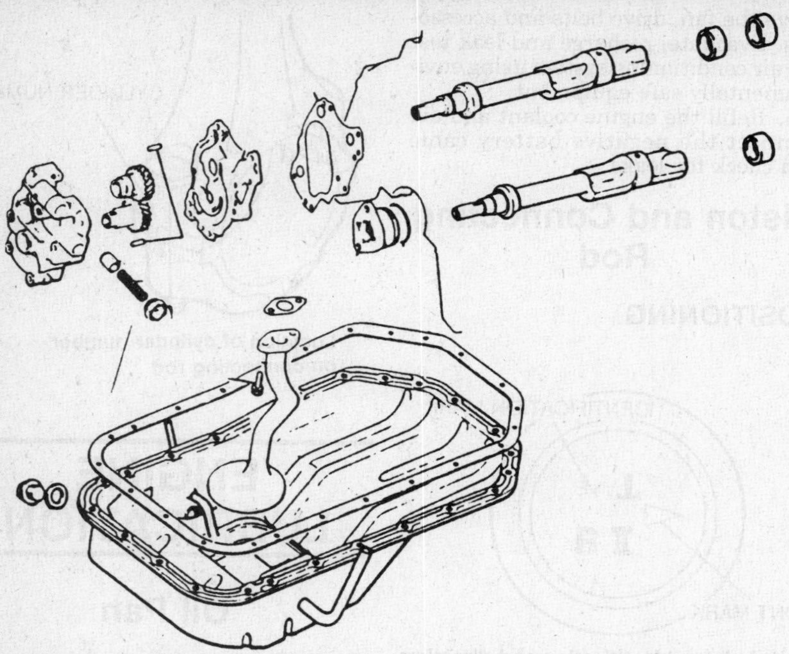

Exploded view of oil pump, oil pan and silent shafts—2555cc engine

1. Front case gasket
2. Oil pump cover
3. Oil pump outer gear
4. Oil pump inner gear
5. Plug
6. Gasket
7. Relief spring
8. Relief plunger
9. Front oil seal
10. Front case
11. Drain plug
12. Oil pan
13. Oil screen
14. Oil screen gasket

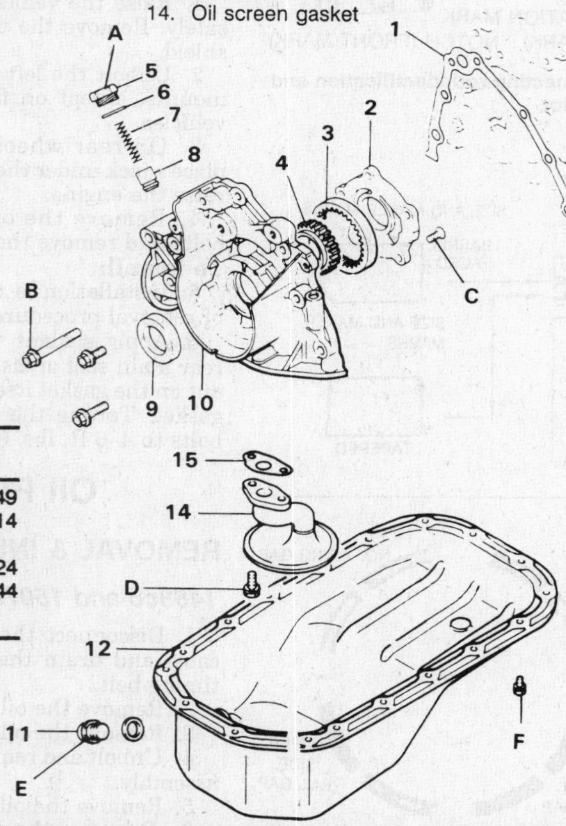

	Nm
A	40–49
B	12–14
C	8–9
D	18–24
E	35–44
F	6–7

Exploded view of oil pump, front cover and oil pan—1468cc engine

NOTE: The outer gear has no identifying marks to indicate direction of rotation. Clean the gear and mark it with an indelible marker.

7. Remove the plug, relief valve spring and relief valve from the case.
8. Check the front case for damage or cracks. Replace the front seal. Replace the oil screen O-ring. Clean all parts thoroughly with a safe solvent.
9. Check the pump gears for wear or damage. Clean the gears thoroughly and place them in position in the case to check the clearances. There is a crescent shaped piece between the 2 gears. This piece is the reference point for 2 measurements. Use the following clearances for determining gear wear:

Outer gear face-to-case—0.0039–0.0079 in.

Outer gear teeth-to-crescent—0.0087–0.0134 in.

Outer gear endplay—0.0016–0.0039 in.

Inner gear teeth-to-crescent—0.0083–0.0126 in.

Inner gear endplay—0.0016–0.0039 in.

10. Check that the relief valve can slide freely in the case.
11. Check the relief valve spring for damage. The relief valve free length should be 1.850 in. load length should be 9.5 lbs. at 1.575 in.
12. Thoroughly coat both oil pump gears with clean engine oil and install them in the correct direction of rotation.
13. Install the pump cover and torque the bolts to 7 ft. lbs.
14. Coat the relief valve and spring with clean engine oil, install them and tighten the plug to 30–36 ft. lbs.
15. Position a new front case gasket, coated with sealer, on the engine and install the front case. Torque the bolts to 10 ft. lbs. Note that the bolts have different shank lengths.
16. Coat the lips of a new seal with clean engine oil and slide it along the crankshaft until it touches the front case. Drive it into place with a seal driver.
17. Install the sprocket, timing belt and pulley.
18. Install the oil screen.
19. Thoroughly clean both the oil pan and engine mating surfaces. Apply a 4mm wide bead of RTV sealer in the groove of the oil pan mating surface.

NOTE: The sealer will set in approximately 15 minutes.

20. Tighten the oil pan bolts to 5–6 ft. lbs. (7–8 Nm). Connect the battery cable and fill the crankcase with oil.

1595cc DOHC, 1597cc and 1755cc Engines

1. Remove the timing belt.
2. Drain the oil.
3. Remove the oil filter, on 1595cc DOHC engines. Remove the oil pan and screen.
4. Remove the oil filter bracket, on 1595cc DOHC engines. Unbolt and remove the front case assembly.

NOTE: On 1597cc and 1755cc engines, if the front case assembly is difficult to remove from the block, there is a groove around the case into which a prybar may be inserted, to aid in removal. Pry slowly and evenly. Don't hammer.

5. On 1597cc and 1755cc engines, remove the oil pressure relief plug, spring and plunger.
6. On 1597cc, remove the nut and pull off the oil pump sprocket.
7. On 1597cc and 1755cc engines, remove the oil pump cover.
8. On 1597cc and 1755cc engines, remove the pump rotor.

To install:

9. Check the case for cracks and damage.
10. Check the oil screen for damage.
11. Replace the oil screen O-ring.
12. Thoroughly clean all parts in a safe solvent.
13. On 1597cc engines, place the rotor back in the case to check clearances.

 Side clearance—0.0024–0.0047 in.

 Tip clearance—0.0016–0.0047 in.

 Body clearance—0.0039–0.0063 in.

 Shaft-to-cover clearance—0.0008–0.0020 in.

14. On 1597cc and 1755cc engines, check that the relief valve plunger slides smoothly in its bore.
15. On 1597cc and 1755cc engines: Check the relief valve spring. The free length should be 1.850 in.; the load length should be 9.5 lbs. at 1.575 in.
16. On 1597cc and 1755cc engines, install a new oil seal, coated with clean engine oil, into the oil pump cover. Drive it into place using a hammer and flat block.
17. On 1597cc and 1755cc, install a new cover gasket in the groove in the case.
18. On 1597cc and 1755cc, coat the rotor with clean engine oil and install it in the cover.
19. On 1597cc and 1755cc, install the cover and tighten the bolts.
20. On 1597cc and 1755cc, install the sprocket and tighten the nut to 28 ft. lbs.
21. On 1597cc and 1755cc, coat the oil relief valve plunger with clean engine oil and install it, along with the spring and plug.
22. Install a new case gasket, coated with sealer, on the block and install the case. Torque the case bolts, on 1597cc engines, 13 ft. lbs. on 1595cc DOHC engines, 20–25 ft. lbs.

NOTE: There are 2 lengths of case bolts.

23. Install the screen. Tighten the bolts, on 1597cc and 1755cc engines, 18 ft. lbs. On 1595cc DOHC engines, 11–16 ft. lbs.
24. Install the oil pan.

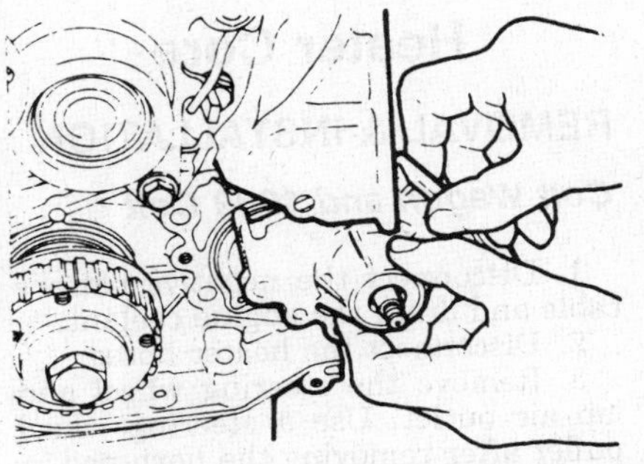

Removing the oil pump cover—1597cc engine

1997cc and 2555cc Engines

1. Remove the timing chain.
2. Remove the oil pump cover and gears.
3. Remove the relief valve plug, spring and plunger.
4. Thoroughly clean all parts in a safe solvent and check for wear and damage.
5. Clean all orifices and passages.

To install:

6. Place the gear back in the pump body and check clearances:

 Gear teeth-to-body—0.0041–0.0059 in.

 Driven gear endplay—0.0024–0.0047 in.

 Drive gear-to-bearing (front end)—0.0008–0.0018 in.

 Drive gear-to-bearing (rear end)—0.0017–0.0026 in.

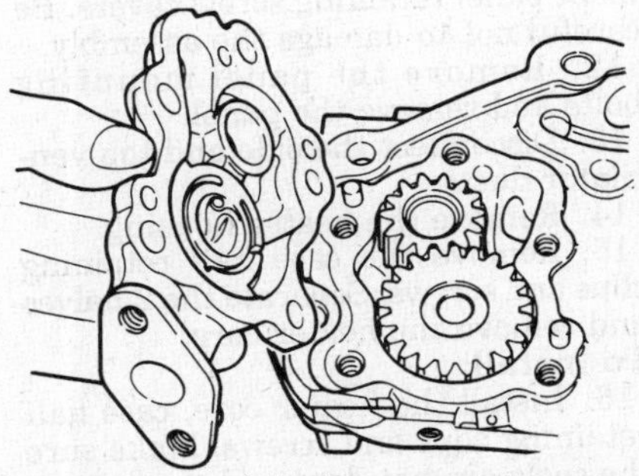

Oil pump cover removal from the 1997cc engine

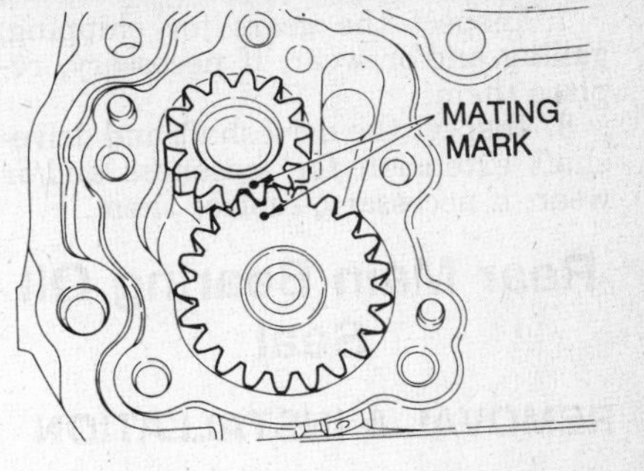

Oil pump gear mating marks on the 1997cc engine

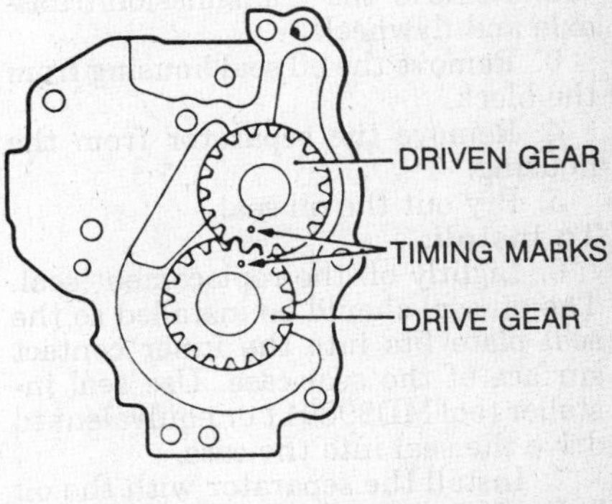

Installation of oil pump drive and driven gears and matching of timing marks, 2555cc engine

NOTE: If gear replacement is necessary, the entire pump body must be replaced.

7. Check the relief valve spring for wear or damage. Free length should be 1.850 in.; load length should be 9.5 lbs. at 1.575 in.
8. Assembly is the reverse of disassembly. Make sure the gears are installed with the mating marks aligned.

CHECKING

1. If foreign matter is present, determine it's source.
2. Check the pump cover and housing for cracks, scoring and/or damage; if necessary, replace the housings.
3. Inspect the idler gear shaft for looseness in the housing; if necessary, replace the pump or timing chain, depending on the model.
4. Inspect the pressure regulator valve for scoring or sticking; if burrs are present, remove them with an oil stone.
5. Inspect the pressure regulator valve spring for loss of tension or distortion, if necessary, replace it.
6. Inspect the suction pipe for looseness, if pressed into the housing and the screen for broken wire mesh; if necessary, replace them.

7. Inspect the gears for chipping, galling and/or wear, if necessary, replace them.
8. Inspect the driveshaft and driveshaft extension for looseness and/or wear; if necessary, replace them.

Rear Main Bearing Oil Seal

REMOVAL & INSTALLATION

The rear main oil seal is located in a housing on the rear of the block.
1. Raise the vehicle and support it safely.
2. Remove the transmission/transaxle and flywheel.
3. Remove the oil seal housing from the block.
4. Remove the separator from the housing.
5. Pry out the oil seal.

To install:

6. Lightly oil the replacement seal. The oil seal should be installed so the seal plate fits into the inner contact surface of the seal case. Use seal installer tool MD998011 or equivalent to drive the seal into the case.
7. Install the separator with the oil holes facing down and install the oil housing.
8. Install the transmission/transaxle. Torque the flywheel bolts to 94–101 ft. lbs. (130–140 Nm) and the transaxle bolts to 20–25 ft. lbs. (27–34 Nm).
9. Lower the vehicle. Refill the the engine with oil.
10. Connect the negative battery cable. Start the engine and check for leaks.

ENGINE COOLING

Radiator

REMOVAL & INSTALLATION

1. Remove the splash panel from the bottom of the vehicle. Drain the radiator by opening the petcock. Remove the shroud, if equipped. On the Conquest, remove the battery.
2. Disconnect the radiator hoses at the engine. On automatic transmission equipped vehicles, disconnect and plug the transmission lines to the bottom of the radiator.
3. Remove the 2 retaining bolts from either side of the radiator. Lift out the radiator. On front wheel drive vehicles, disconnect the electric fan wiring harness. Do not remove the fan motor, blades or bracket—remove as a unit with the radiator.

To install:

4. Installation is the reverse order of the removal procedure.
5. Install the radiator and retaining bolts. Tighten the retaining bolts gradually in a criss-cross pattern.

NOTE: Work around the electric cooling fan when the engine is cold or disconnect the negative battery cable. On some vehicles, the fan will run to cool the engine even when the ignition is off.

Heater Core

REMOVAL & INSTALLATION

Colt Wagon and 1988 Colt

1. Disconnect the negative battery cable and drain the engine coolant.
2. Disconnect the heater hoses.
3. Remove the steering wheel and lap air outlet. Use a steering wheel puller after removing the horn pad.
4. Remove the steering column upper and lower cover, shift knob and floor console assembly.
5. Remove the glove box by opening and quickly pulling up on the right corner of the box and remove the right hinge pin from the half hinge.
6. Remove the defroster duct and side joint from the heater unit.
7. Remove the instrument cluster hood attaching screws and clips. The clips are on the top.
8. Remove the instrument cluster screw and pull the cluster forwards. Disconnect the speedometer cable and electrical connections.
9. Remove the steering column-to-instrument panel bracket bolts.
10. Remove the heater control panel by pushing on the right side from behind the panel cover. Remove the heater control assembly-to-instrument panel retaining screws. Do not remove the control assembly.
11. Using a trim removal tool or equivalent, remove the upper instrument panel retaining screw covers. Be careful not to damage the assembly.
12. Remove the panel mounting bolts and remove the panel.
13. Disconnect the side and top ventilator ducts.
14. Remove the heater housing.
15. Remove the case half retaining clips and screws. Separate the 2 halves and remove the heater core.

To install:

16. Install the heater core, case half retaining clips and screws. Make sure the seals are not damaged.
17. Install the heater housing.
18. Connect the side and top ventilator ducts.
19. Install the instrument panel and mounting bolts.
20. Install the upper instrument panel retaining bolts and covers. Be careful not to damage the assembly.
21. Install the heater control panel.
22. Install the steering column-to-instrument panel bracket bolts.
23. Install the instrument cluster and hood.
24. Install the defroster duct and side joint to the heater unit.
25. Install the glove box.
26. Install the steering column upper and lower cover, shift knob and floor console assembly.
27. Install the steering wheel and lap air outlet.
28. Connect the heater hoses.
29. Connect the negative battery cable and refill the engine coolant.

1989-92 Colt

1. Disconnect the negative battery cable.
2. Drain the cooling system and disconnect the heater hoses.
3. Remove the front seats by removing the covers over the anchor bolts, the underseat tray, the seat belt guide ring, the seat mounting nuts and bolts and disconnect the seat belt switch wiring harness from under the seat. Then lift out the seats.
4. Remove the floor console by first taking out the coin holder and the console box tray. Remove the remote control mirror switch or cover. All of these items require only a plastic trim tool to carefully pry them out.
5. Remove the rear half of the console.
6. Remove the shift lever knob on manual transmission vehicles.
7. Remove the front console box assembly.
8. A number of the instrument panel pieces may be retained by pin type fasteners. They may be removed using the following procedure:
   a. This type of clip is removed by pressing down on the center pin with a suitable blunt pointed tool. Press down a little more than $^1/_{16}$ in. (2mm); this releases the clip. Pull the clip outward to remove it.
   b. Do not push the pin inward more than necessary because it may damage the grommet or the pin may fall in if pushed in too far. Once the clips are removed, use a plastic trim stick to pry the piece loose.
9. Remove both lower cowl trim panels (kick panels).
10. Remove the ashtray.
11. Remove the center panel around the radio.
12. Remove the sunglass pocket at

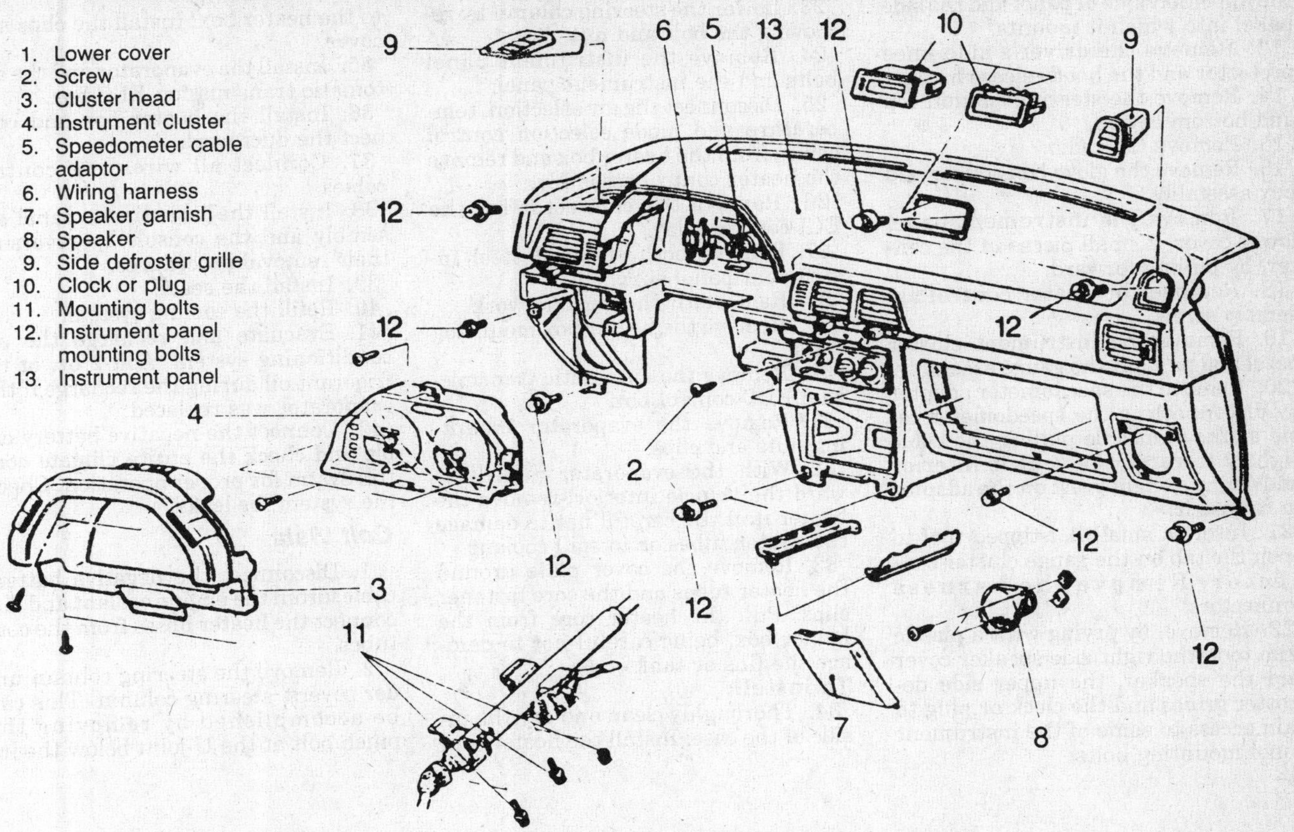

Instrument panel and components—1989-92 Colt

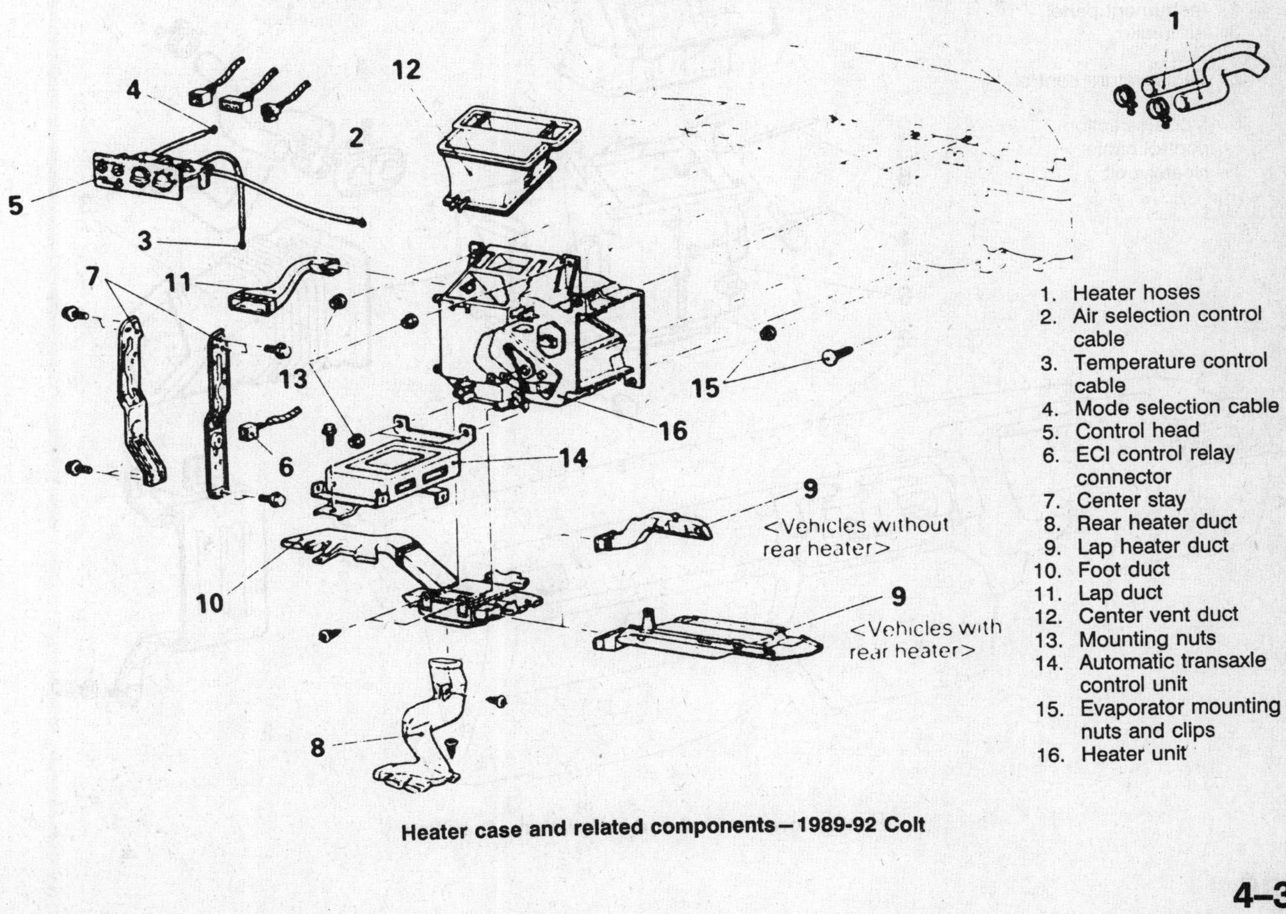

Heater case and related components—1989-92 Colt

the upper left side of panel and the side panel into which it mounts.
13. Remove the driver's side knee protector and the hood release handle.
14. Remove the steering column top and bottom covers.
15. Remove the radio.
16. Remove the glove box striker and box assembly.
17. Remove the instrument panel lower cover, 2 small pieces in the center, by pulling forward.
18. Remove the heater control assembly screw.
19. Remove the instrument cluster bezel and pull out the gauge assembly.
20. Remove the speedometer adapter by disconnecting the speedometer cable at the transaxle pulling the cable sightly towards the vehicle interior and giving a slight twist on the adapter to release it.
21. Insert a small flat-tipped tool to open the tab on the gauge cluster connector. Remove the harness connectors.
22. Remove, by prying with a plastic trim tool, the right side speaker cover and the speaker, the upper side defroster grilles and the clock or plug to gain access to some of the instrument panel mounting bolts.
23. Lower the steering column by removing the bolt and nut.
24. Remove the instrument panel bolts and the instrument panel.
25. Disconnect the air selection, temperature and mode selection control cables from the heater box and remove the heater control assembly.
26. Remove the connector for the ECI control relay.
27. Remove both stamped steel instrument panel supports.
28. Remove the heater ductwork.
29. Remove the heater box mounting nuts.
30. Remove the automatic transmission ELC control box.
31. Remove the evaporator mounting nuts and clips.
32. With the evaporator pulled toward the vehicle interior, remove the heater unit. Be careful not to damage the heater tubes or to spill coolant.
33. Remove the cover plate around the heater tubes and the core fastener clips. Pull the heater core from the heater box, being careful not to damage the fins or tank ends.

To install:

34. Thoroughly clean and dry the inside of the case. Install the heater core to the heater box. Install the clips and cover.
35. Install the evaporator and the automatic transmission ELC box.
36. Install the heater box and connect the duct work.
37. Connect all wires and control cables.
38. Install the instrument panel assembly and the console by reversing their removal procedures.
39. Install the seats.
40. Refill the cooling system.
41. Evacuate and recharge the air conditioning system. Add 2 oz. of refrigerant oil during the recharge if the evaporator was replaced.
42. Connect the negative battery cable and check the entire climate control systm for proper operation. Check the system for leaks.

Colt Vista

1. Disconnect the negative battery cable, drain the engine coolant and disconnect the heater hoses from the core tubes.
2. Remove the steering column under covers, steering column. This can be accomplished by removing the pinch bolt at the U-joint below the in-

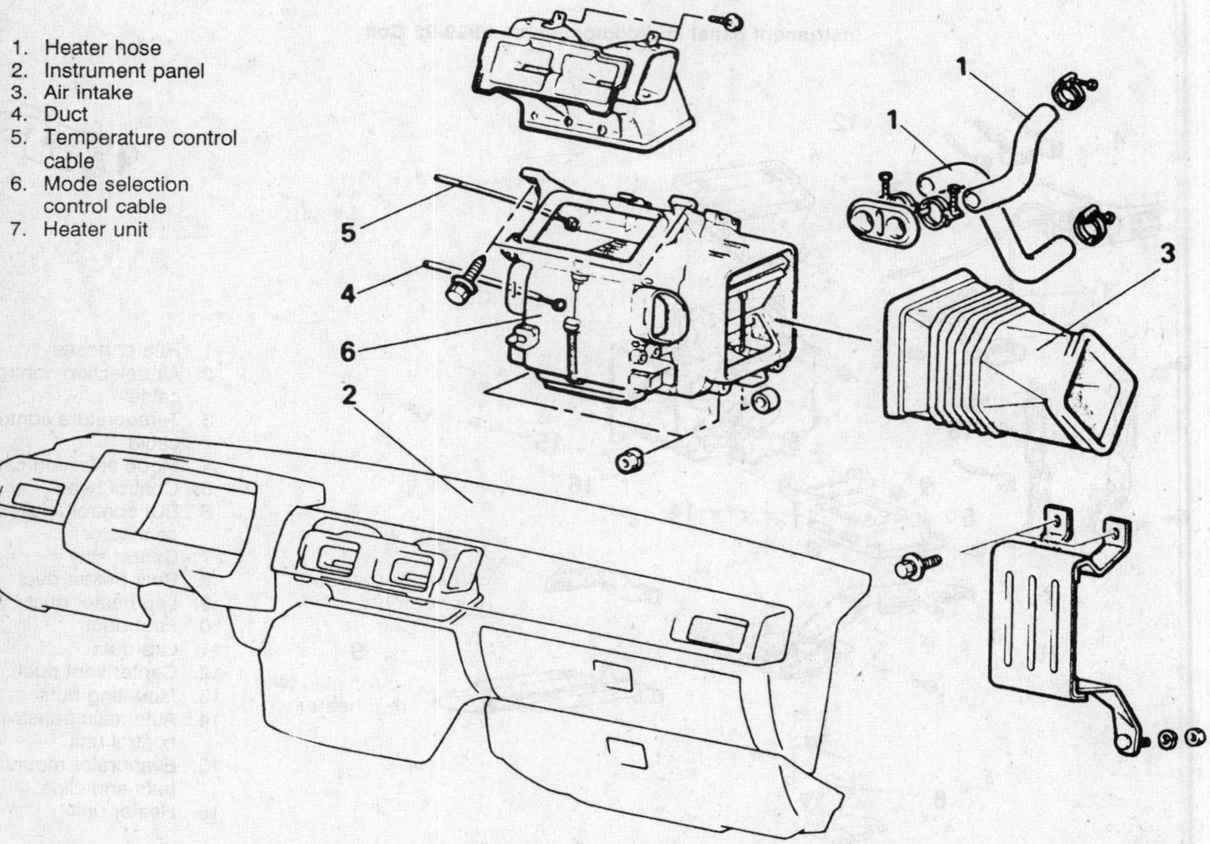

Heater core removal and installation—Colt Vista

strument panel, disconnecting the connectors and pulling the column from the U-joint yoke.

3. Remove the glove box assemblies, lap heater duct, ashtray and hood lock release cable from the instrument panel.

4. Remove the instrument cluster hood covers and hood. Pull the cluster out and disconnect the speedometer cable and electrical connectors.

5. Disconnect the control cables from the heater unit, remove the upper air ducts and disconnect the blower motor harness.

6. Remove the trim panels along the top of the instrument panel, disconnect the antenna feeder wire, remove the instrument panel retaining hardware and remove the assembly.

7. Remove the instrument panel absorber, duct from the right of the heater unit and unit mounting nuts. Remove the unit from the vehicle.

8. Disassemble the heater unit by removing the case clips and screws. Separate the 2 halves and remove the heater core.

To install:

9. Thoroughly clean and dry the inside of the case. Assemble the 2 halves being careful not to damage the seals.

10. Install the heater unit and mounting nuts.

11. Install the right side duct, instrument panel absorber bracket, and instrument panel.

12. Install the hood release cable, ashtray, lap heater duct and glove boxes.

13. Install the steering column and under covers.

14. Connect the heater hoses and refill the cooling system.

15. Connect the negative battery cable and check the system for leaks and proper operation.

Conquest

1. With the engine cold, set the temperature control lever to the extreme right. If equipped with an automatic climate control system, start the engine and use the temperature change switch to select the hottest temperature, then turn the engine off.

2. Disconnect the negative battery cable. Drain the engine coolant.

3. Disconnect the coolant hoses running to the heater pipes at the firewall.

4. Remove the floor console.

5. Remove the steering wheel.

6. Remove the screws holding the hood release handle to the instrument panel.

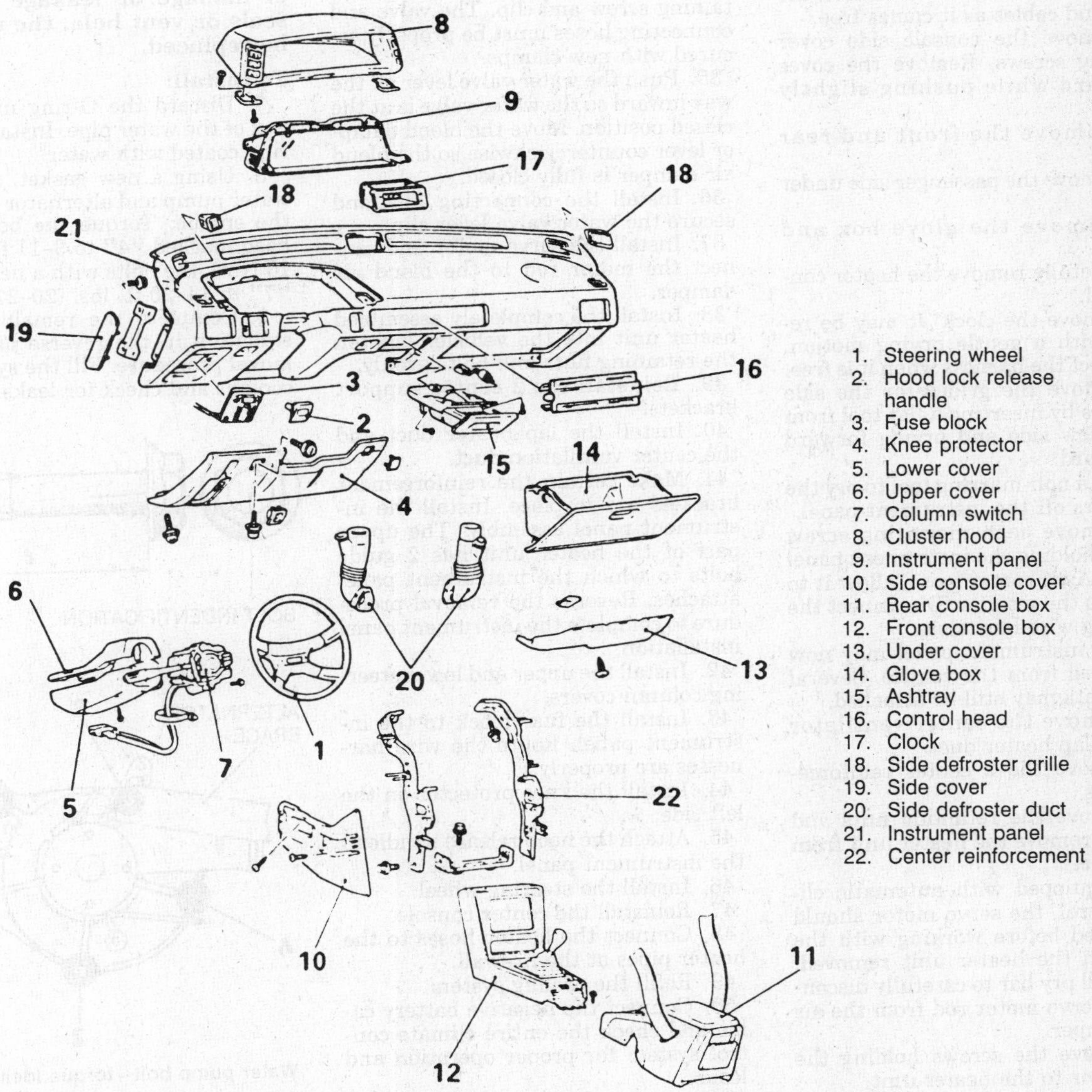

Instrument panel and related components—Conquest

7. Unbolt the fuse box from the instrument panel.
8. Remove the knee protector on the left side. Some of the bolts are hidden behind covers.
9. Remove the screws from the bottom of the steering column cover. Remove both halves of the cover.
10. Remove the attaching screws for the combination switch on the steering column.
11. Disconnect the wiring harnesses and remove the switch.
12. Remove the instrument hood screws. Pull both edges of the bottom of the hood forward; hold it in this position and lift up and out.
13. Disconnect the harness connectors on both sides of the instrument hood.
14. Remove the screws on the bottom and the nuts on the top of the instrument cluster. Pull the bottom edge up and forward to remove. Disconnect the wiring and cables as it comes free.
15. Remove the console side cover mounting screws. Remove the cover downward while pushing slightly forward.
16. Remove the front and rear consoles.
17. Remove the passenger side under cover.
18. Remove the glove box and ashtray.
19. Carefully remove the heater control bezel.
20. Remove the clock. It may be removed with a gentle prying motion. Disconnect the harness when it is free.
21. Remove the grilles for the side defrosters by inserting a flat tool from the window side and prying forward and upward.
22. Use a non-marring tool to pry the side covers off the instrument panel.
23. Remove each mounting screw and bolt holding the instrument panel in place. As it comes loose, allow it to move into the interior. Disconnect the remaining wire harnesses.
24. The instrument panel may now be removed from the vehicle. Several components may still be attached.
25. Remove the center ventilator duct and lap heater duct.
26. Remove the 2 center reinforcement bars.
27. Remove the retaining nuts and bolts and remove the heater unit from the vehicle.
28. If equipped with automatic climate control, the servo motor should be removed before working with the case. With the heater unit removed, use a small pry bar to carefully disconnect the servo motor rod from the air blend damper.
29. Remove the screws holding the servo motor to the heater unit.
30. Carefully unlock the water valve lever clip and disconnect the link between the blend air damper and the water valve lever.
31. Remove the outer clamp from the 2 water tubes.
32. Loosen the clamps on the short joint hoses and disconnect the hoses. Remove the retaining screws holding the water valve in place and remove the water valve.
33. The heater core is held in place by a clip and retaining screw. Once removed, the core should come free of the housing. If the core is blocked by the blend air damper lever, remove the lever. Do not attempt to force the core past the lever.

To install:

34. Thoroughly clean and dry the inside of the case. Reassemble the heater unit, and secure the core with the retaining screw and clip. The valve and connecting hoses must be properly secured with new clamps.
35. Push the water valve lever all the way inward so the water valve is at the closed position. Move the blend damper lever counterclockwise so the blend air damper is fully closed.
36. Install the connecting link and secure the water valve lever clip.
37. Install the servo motor and connect the motor rod to the blend air damper.
38. Install the completely assembled heater unit into the vehicle. Tighten the retaining nuts and bolts evenly.
39. Reinstall the 2 center support brackets.
40. Install the lap heater duct and the center ventilation duct.
41. Make certain the reinforcement brackets are in place. Install the instrument panel assembly. The upper part of the heater unit has 2 guide bolts to which the instrument panel attaches. Reverse the removal procedure to complete the instrument panel installation.
42. Install the upper and lower steering column covers.
43. Install the fuse block to the instrument panel. Route the wire harnesses are properly.
44. Install the knee protector on the left side.
45. Attach the hood release handle to the instrument panel.
46. Install the steering wheel.
47. Reinstall the center console.
48. Connect the heater hoses to the heater pipes at the firewall.
49. Refill the cooling system.
50. Connect the negative battery cable and check the entire climate control system for proper operation and leaks.

Water Pump

REMOVAL & INSTALLATION

Colt and Colt Vista

1. Drain the cooling system.
2. Remove the drive belt and water pump pulley.
3. Remove the timing belt covers and timing belt tensioner.
4. Remove the water pump bolts and alternator bracket.
5. Remove the water pump retaining bolts, it is important to observe the location of each bolt, they are different lengths.
6. Remove the water pump.

NOTE: The pump is not rebuildable. Check for driveshaft side to side play, if excessive replace the pump. If there are signs of damage or leakage from the seals or vent hole, the unit must be replaced.

To install:

7. Discard the O-ring in the front end of the water pipe. Install a new O-ring coated with water.
8. Using a new gasket, mount the water pump and alternator bracket on the engine. Torque the bolts with a head marked "4" to 9–11 ft. lbs. (12–15 Nm); the bolts with a head marked "7" to 14–20 ft. lbs. (20–27 Nm).
9. Complete the remainder of installation in the reverse order of removal procedure. Fill the system with coolant and check for leaks.

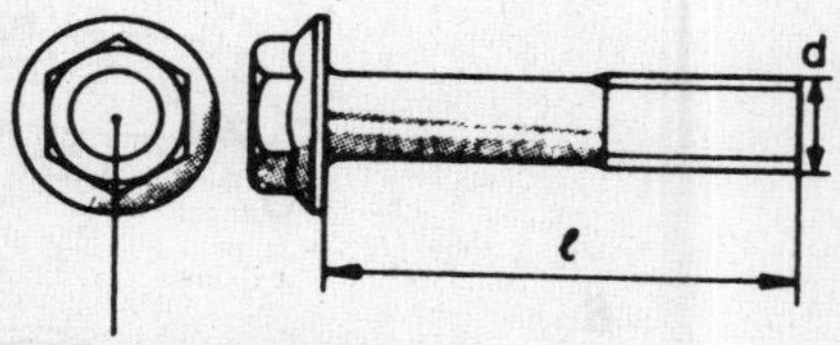

BOLT INDENTIFICATION

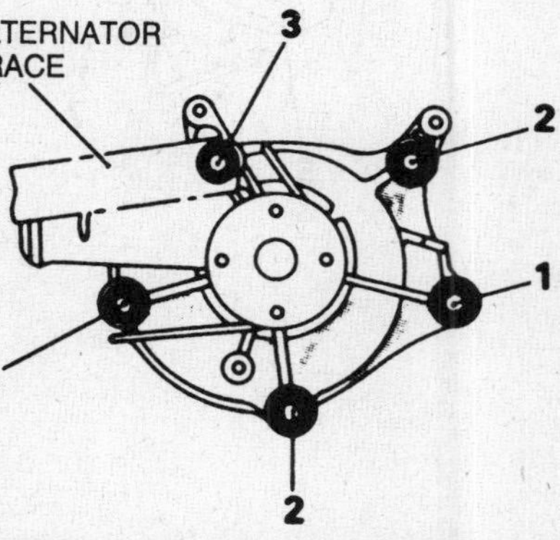

Water pump bolt—torque identification

Conquest

1. Disconnect the negative battery cable.
2. Drain the cooling system.
3. Remove the fan shroud and radiator, if necessary for working room.
4. Remove the alternator belt and accessory belts.
5. Remove the fan blades and/or automatic hub, if equipped.
6. Remove the water pump assembly from the timing chain case or the cylinder block.

To install:

7. Installation is the reverse order of the removal procedures.
8. Fill the radiator with coolant and test for leaks.

Thermostat

REMOVAL & INSTALLATION

Colt and Colt Vista

1. Disconnect the negative battery cable.
2. Drain the cooling system to a point below the thermostat level.
3. Remove the air cleaner.
4. Disconnect the hose at the thermostat water pipe.
5. Remove the water pipe support bracket nut.

NOTE: This nut is also an intake manifold nut. It is very difficult to get to. A deep offset 12mm box wrench is used to remove or replace it.

6. Unbolt and remove the thermostat housing and pipe.
7. Lift out the thermostat. Discard the gasket.

To install:

8. Clean the mating surfaces of the housing and manifold thoroughly.
9. Install the thermostat with the spring facing downward and position a new gasket. The jiggle valve in the thermostat should be on the manifold side.
10. Install the housing and pipe assembly. Torque the housing bolts to 11 ft. lbs. (14 Nm); the intake manifold nut to 14 ft. lbs. (19 Nm).
11. Refill the system with coolant. Connect the negative battery cable.
12. Start the engine and check for leaks.

Conquest

1. Disconnect the negative battery cable.
2. Drain the coolant below the level of the thermostat.
3. Remove the retaining bolts and lift the thermostat housing off the intake manifold with the hose still in position.
4. Raise the small cone shaped cover on the throttle cable to expose the nipple.

NOTE: It is not necessary to remove the upper radiator hose.

5. Lift the thermostat out of the manifold.

To install:

6. Installation is the reverse order of the removal procedures. Use a new gasket and coat the mating surfaces with sealer.
7. Torque the housing retaining bolts to 13–14 ft. lbs. (17–20 Nm).

Cooling System Bleeding

After working on the cooling system, even to replace the thermostat, the system must be bled. Air trapped in the system will prevent proper filling and leave the radiator coolant level low, causing a risk of overheating.

1. To bleed the system, start the system cool, the radiator cap off and the radiator filled to about an inch below the filler neck.
2. Start the engine and run it at slightly above normal idle speed. This will insure adequate circulation. If air bubbles appear and the coolant level drops, fill the system with a mixture of anti-freeze and water to bring the level back to the proper level.
3. Run the engine this way until the thermostat opens. When this happens, the coolant will move abruptly across the top of the radiator and the temperature of the radiator will suddenly rise.
4. At this point, air is often expelled and the level may drop quite a bit. Keep refilling the system until the level is near the top of the radiator and remains constant.
5. If the vehicle has an overflow tank, fill the radiator up to the top of the filler neck.

ENGINE ELECTRICAL

NOTE: Disconnecting the negative battery cable on some vehicles may interfere with the functions of the on board computer systems and may require the computer to undergo a relearning process, once the negative battery cable is reconnected.

Distributor

REMOVAL

Before removing the distributor, position No. 1 cylinder at TDC on the compression stroke and align the timing marks.

1. Disconnect the negative battery cable.
2. Disconnect the spark plug wires from the distributor cap.
3. Disconnect the ignition coil high tension wire from the distributor cap.
4. Remove the vacuum hose from the advance unit.
5. Remove the cap from the distributor.
6. Verify the rotor points to the No. 1 cylinder position and the timing marks on the crankshaft pulley and the timing tab are aligned at TDC.
7. Mark the distributor body to the exact place the rotor points. Matchmark both the distributor mounting flange and the cylinder head.
8. Loosen and remove the retaining nut from the mounting stud. Lift the distributor from the cylinder head. The rotor may turn slightly from the mark on the distributor body. Make note of how far. When the distributor is reinstalled, this is the point to position the rotor.

INSTALLATION

Timing Not Disturbed

1. Position the distributor into the engine while aligning the matchmarks made during removal.
2. Verify the rotor points to the No. 1 cylinder position and the timing marks on the crankshaft pulley and the timing tab are aligned at TDC.
3. Install the distributor retaining nut on the mounting stud and tighten.
4. Install the cap on the distributor.
5. Connect the spark plug wires to the distributor cap.
6. Connect the ignition coil high tension wire to the distributor cap.
7. Connect the negative battery cable to the battery.
8. Start the engine and check the ignition timing whenever the distributor has been removed.

Timing Disturbed

1. With the distributor removed from the engine, turn the crankshaft so the No. 1 piston is on the compression stroke and the timing marks are aligned.
2. Turn the distributor shaft so the rotor points approximately 15 degrees before the rotor position that was marked on the distributor.

3. Insert the distributor, if resistance is met, slight wiggling of the rotor shaft will help seat the distributor.
4. When the distributor seats against the head, align the matchmarks and install the retaining nut. Do not tighten the retaining nut all the way, as the timing must be checked.
5. Reinstall the rotor, cap, plug wires, coil lead, primary lead or harness and connect the vacuum hoses.
6. Connect the negative battery cable. Start the engine, allow it to reach operating temperature and check the ignition timing.

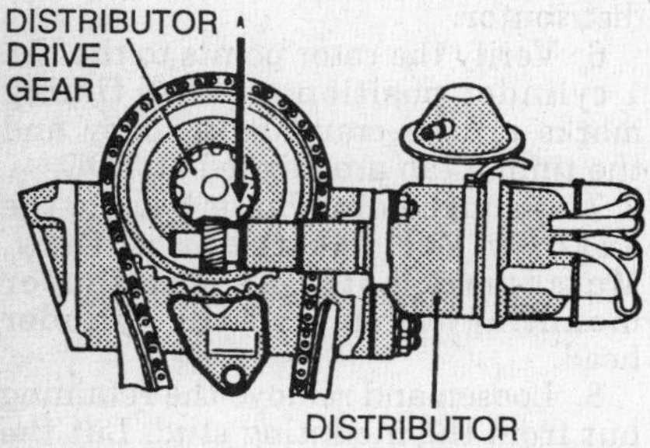

Distributor installation—cylinder head mounted distributors

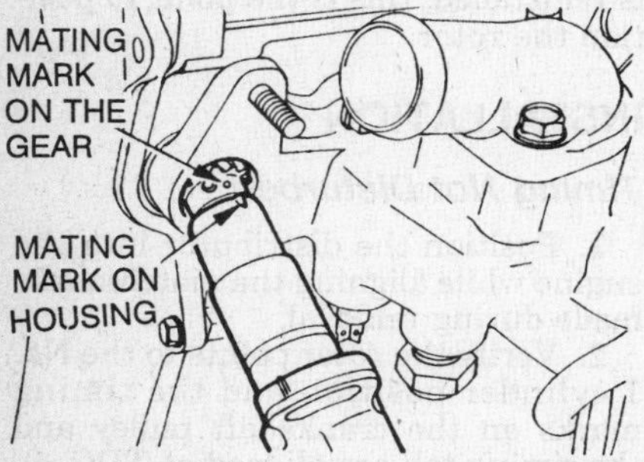

Aligning mating marks for installation of cylinder head mounted distributors

Ignition Timing

ADJUSTMENT

Except 1595cc DOHC Engine

1. Attach the timing light according to the manufacturer's instructions.
2. Locate the timing tab line on the front of the engine and the notch on the crankshaft pulley. Mark them so they are easily recognizable with the timing light. Connect a tachometer by inserting a paper clip into the connector at the distributor and connect the tachometer lead to the paper clip.
3. Start the engine and allow it to reach operating temperature.
4. Point the timing light at the crankshaft pulley marks. The marked line should align with the pulley notch.
5. If the marks do not align, loosen the distributor mounting nut and rotate the distributor slowly, in either direction, to align the timing marks.
6. Tighten the mounting nut when the ignition timing is correct. Stop the engine and remove the timing light.
7. Adjust engine idle if needed.

1595cc DOHC Engine

1. Run the engine until the normal operating temperature is reached. Shut off the engine. Make sure all lights and electrical accessories are off. Make sure the electric cooling fan is not operating when timing. Disconnect the fan harness, if necessary, but take care not to allow the engine to overheat.
2. Connect a timing light, following the light manufacturer's instructions.
3. Insert a paper clip along the terminal surface, of the terminal parallel to the fastener side of the ignition connecter harness, in the engine compartment.
4. Connect a tachometer to the paper clip. Start the engine and check the curb idle speed. The idle speed should be 650–850 rpm.
5. Shut off the engine, connect a lead wire with alligator clips to the terminal for ignition timing adjustment and ground it to a good chassis grounding point.
6. Start the engine and point the timing light at the pulley and timing cover marks. The base timing is 5 degrees before TDC.
7. If timing adjustment is necessary, loosen the crank angle sensor pivot bolt and turn the sensor. Turning the sensor to the right advances the timing, to the left retards it. Tighten the sensor pivot bolt when correct timing is reached. Do not allow the engine to overheat.
8. Stop the engine and disconnect the ground wire. Start the engine and check the curb idle speed. Check the ignition timing, it should now be about 8 degrees before TDC.
9. Timing may vary depending upon the engine control module. If the timing is not about 8 degrees, check the base timing again. If the base timing is still 5 degrees, the ignition timing is functioning normally.

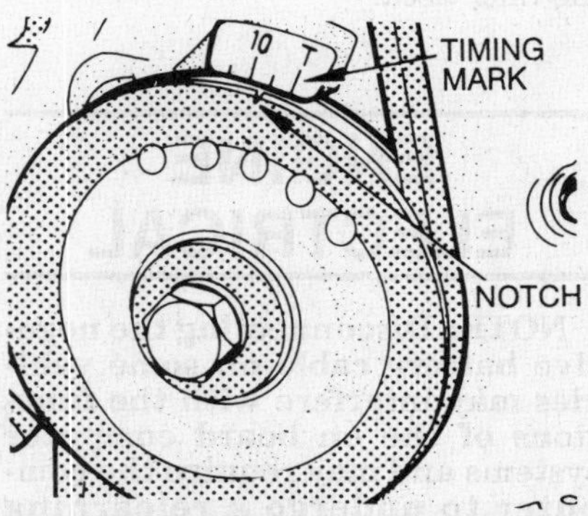

Timing mark positions—1468cc, 1595cc, 1597cc engines

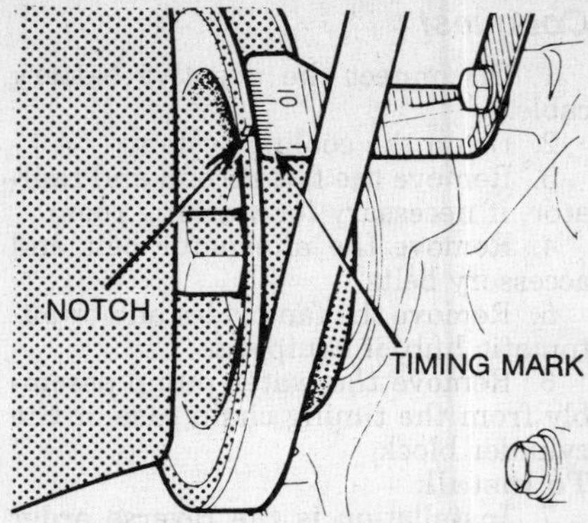

Timing marks—2555cc engine

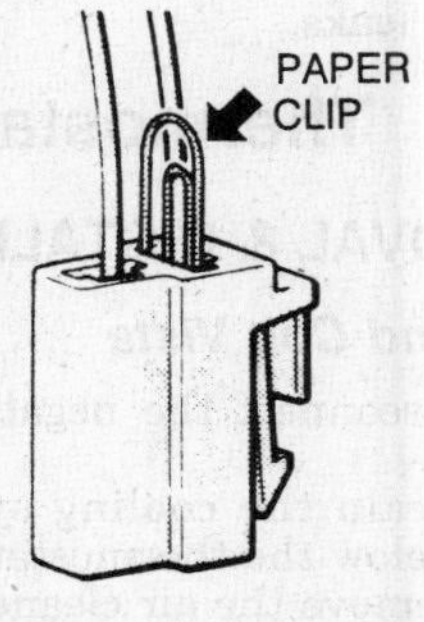

Paper clip installation

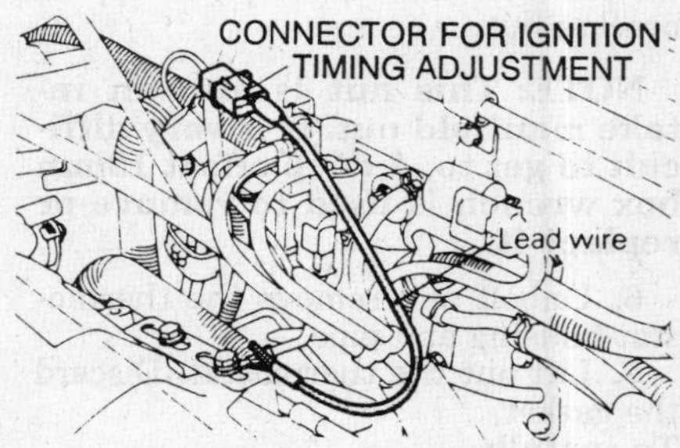

Lead wire connection

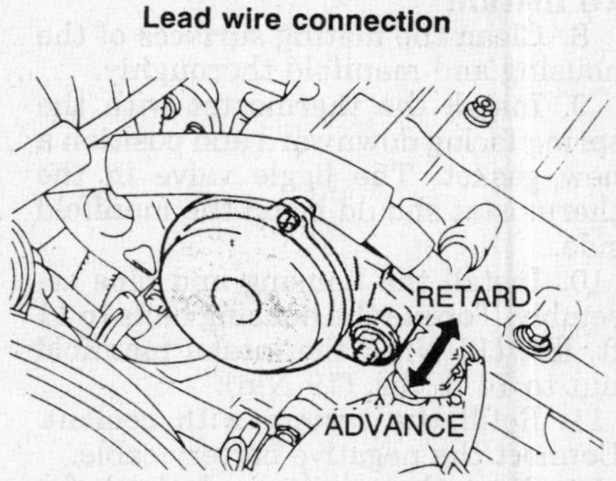

Timing adjustment with crank angle sensor—1595cc DOHC engine

Alternator

PRECAUTIONS

Several precautions must be observed with alternator equipped vehicles to avoid damage to the unit.

- If the battery is removed for any reason, make sure it is reconnected with the correct polarity. Reversing the battery connections may result in damage to the one-way rectifiers.
- When utilizing a booster battery as a starting aid, always connect the positive to positive terminals and the negative terminal from the booster battery to a good engine ground on the vehicle being started.
- Never use a fast charger as a booster to start vehicles.
- Disconnect the battery cables when charging the battery with a fast charger.
- Never attempt to polarize the alternator.
- Do not use test lamps of more than 12 volts when checking diode continuity.
- Do not short across or ground any of the alternator terminals.
- The polarity of the battery, alternator and regulator must be matched and considered before making any electrical connections within the system.
- Never separate the alternator on an open circuit. Make sure all connections within the circuit are clean and tight.
- Disconnect the battery ground terminal when performing any service on electrical components.
- Disconnect the battery if arc welding is to be done on the vehicle.

BELT TENSION ADJUSTMENT

1. Check the drive belts for cracking, fraying and any other deterioration. Replace if necessary.
2. To replace the belt, loosen the stationary mounting bolt and and pivot bolt. If equipped with an adjustment bolt, loosen it to provide the necessary slack for belt removal. Pivot the driven component in its bracket. Remove the old belt and slip the replacement belt over the pulleys.
3. Move the driven component or tighten the adjustment bolt, until the belt can be deflected ¼–⅜ in. at its midpoint.
4. Tighten the mounting and pivot bolts.

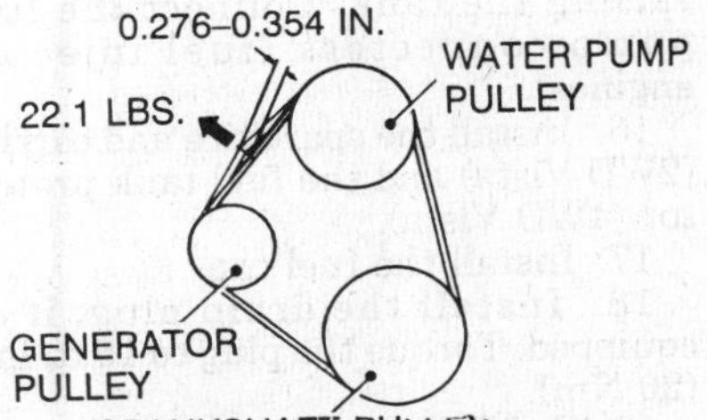

Belt tension adjustment

REMOVAL & INSTALLATION

Colt, Colt Wagon and Colt Vista

1. Disconnect the negative battery cable.
2. Remove the condenser fan motor.
3. Remove the power steering pump from the bracket and support it on the oil reservoir using wire.
4. Remove the power steering pump bracket.
5. Disconnect the wiring connectors from the alternator.
6. Remove the lock bolt and the support bolt.
7. Remove the alternator and the adjusting bolt.

To install:

8. Installation is the reverse order of the removal procedure. Torque alternator brace bolts to to 9–11 ft. lbs. (12–15 Nm) and the pivot bolt to 15–18 ft. lbs. (20–25 Nm).
9. Adjust the belt to proper tension. Connect the negative battery cable.

Conquest

1. Disconnect the negative battery cable.
2. Disconnect the wire connectors from the alternator.
3. If equipped with air conditioning, discharge the air conditioning system and remove the discharge and suction hose connections from the compressor.
4. Remove the compressor mounting bolts and remove the compressor from the engine.
5. Remove alternator mounting bolts and remove the alternator from the engine.

To install:

6. Installation is the reverse order of the removal procedure. Torque the alternator mounting bolts to 14–18 ft. lbs. (20–25 Nm).
7. Torque the compressor mounting bolts to 9–10 ft. lbs. (12–15 Nm) and the compressor lines to 22–26 ft. lbs. (30–34 Nm).
8. Adjust the belt to proper tension. Connect the negative battery cable and recharge the air conditioning system.

Starter

REMOVAL & INSTALLATION

1. Disconnect the battery negative

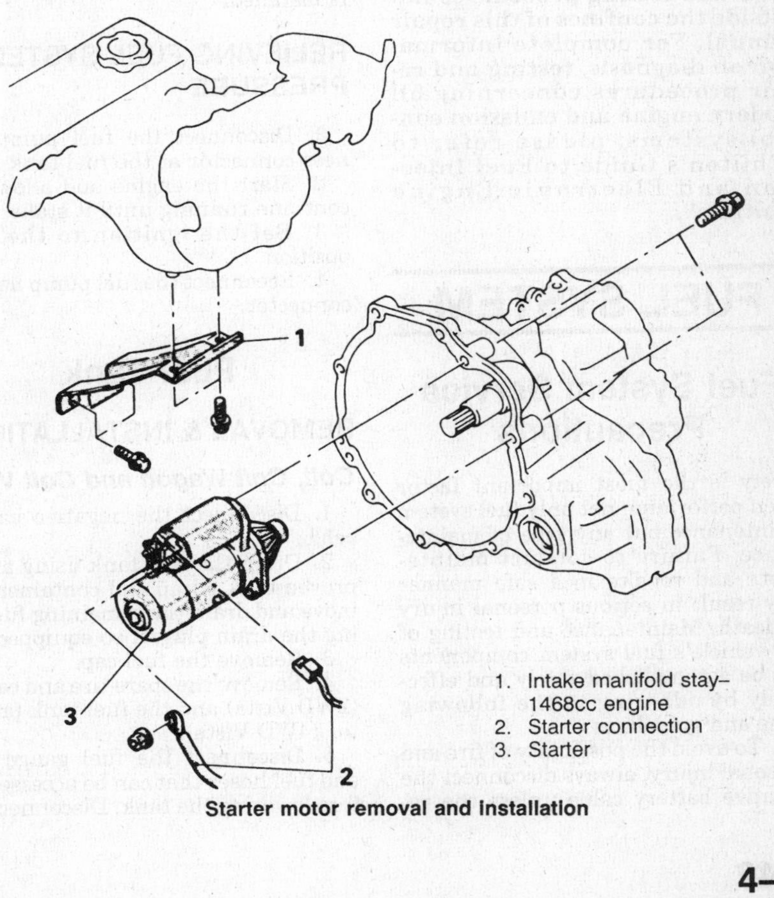

1. Intake manifold stay–1468cc engine
2. Starter connection
3. Starter

Starter motor removal and installation

battery cable and the starter motor wiring.

2. Raise and support the vehicle safely.

3. Remove the intake manifold-to-engine support bracket (Colt and Colt Wagon).

4. Remove the 2 starter attaching bolts and remove the starter motor.

To install:

4. Clean both surfaces of the starter motor flange and the rear plate.

5. Position the starter in the housing opening.

6. Install the manifold-to-engine support bracket (Colt and Colt Wagon).

7. Install the attaching bolts. Tighten evenly to avoid binding.

8. Install the starter wiring and lower the vehicle.

9. Connect the negative battery cable.

EMISSION CONTROLS

Please refer to "Emission Controls" in the Unit Repair section for system maintenance procedures. Due to the complex nature of modern electronic engine control systems, comprehensive diagnosis and testing procedures fall outside the confines of this repair manual. For complete information on diagnosis, testing and repair procedures concerning all modern engine and emission control systems, please refer to "Chilton's Guide to Fuel Injection and Electronic Engine Controls".

FUEL SYSTEM

Fuel System Service Precautions

Safety is the most important factor when performing not only fuel system maintenance but any type of maintenance. Failure to conduct maintenance and repairs in a safe manner may result in serious personal injury or death. Maintenance and testing of the vehicle's fuel system components can be accomplished safely and effectively by adhering to the following rules and guidelines.

- To avoid the possibility of fire and personal injury, always disconnect the negative battery cable unless the repair or test procedure requires that battery voltage be applied.
- Always relieve the fuel system pressure prior to disconnecting any fuel system component (injector, fuel rail, pressure regulator, etc.), fitting or fuel line connection. Exercise extreme caution whenever relieving fuel system pressure to avoid exposing skin, face and eyes to fuel spray. Please be advised that fuel under pressure may penetrate the skin or any part of the body that it contacts.
- Always place a shop towel or cloth around the fitting or connection prior to loosening to absorb any excess fuel due to spillage. Ensure that all fuel spillage (should it occur) is quickly removed from engine surfaces. Ensure that all fuel soaked cloths or towels are deposited into a suitable waste container.
- Always keep a dry chemical (Class B) fire extinguisher near the work area.
- Do not allow fuel spray or fuel vapors to come into contact with a spark or open flame.
- Always use a backup wrench when loosening and tightening fuel line connection fittings. This will prevent unnecessary stress and torsion to fuel line piping. Always follow the proper torque specifications.
- Always replace worn fuel fitting O-rings with new. Do not substitute fuel hose or equivalent where fuel pipe is installed.

RELIEVING FUEL SYSTEM PRESSURE

1. Disconnect the fuel pump harness connector at the fuel tank side.

2. Start the engine and allow it to continue running until it stalls.

3. Set the ignition to the **OFF** position.

4. Reconnect the fuel pump harness connector.

Fuel Tank

REMOVAL & INSTALLATION

Colt, Colt Wagon and Colt Vista

1. Disconnect the negative battery cable.

2. Drain the fuel tank using an approved tank pump and container. Remove and drain the remaining fuel using the drain plug, if so equipped.

3. Remove the fuel cap.

4. Remove the spare tire and carrier (2WD Vista) and the fuel tank protector (4WD Vista).

5. Disconnect the fuel gauge unit and fuel hoses that can be accessed before lowering the tank. Disconnect the fuel pump connectors (fuel injected engines).

6. Disconnect the filler hose.

7. Place a floor jack and a piece of wood under the tank before removing the straps or retaining bolts.

8. Remove the retaining straps (1988 Colt) and the tank retaining bolts.

9. Slowly lower the tank and discon-

Conquest

1. Disconnect the negative battery cable.

2. Drain the fuel tank using an approved tank pump and container. Remove and drain the remaining fuel using the drain plug, if so equipped.

3. Remove the fuel cap and relieve the fuel pressure.

4. Remove the high floor side panel and fuel pipe cover.

5. Place a drain pan under the tank and remove the tank drain plug.

6. Disconnect the fuel and electrical lines that can be accessed before removing the tank.

7. Place a floor jack and a piece of wood under the tank before removing the retaining nuts.

8. Remove the tank retaining nuts.

9. Slowly lower the tank and disconnect any electrical or fuel connections. Be care not to damage the fuel fittings.

nect any electrical or fuel connections. Be care not to damage the fuel fittings.

10. Remove all hardware from the old tank if installing a new tank. Remove the fuel pump nuts and remove the pump.

To install:

11. Install all hardware to the tank, if installing a new tank. Always use new gaskets around the fuel pump and gauge unit. Install the pump and gauge unit. Torque the nuts or bolts to 1.4–2.2 ft. lbs. (2–3 Nm).

12. Slowly raise the tank and connect any electrical or fuel connections. Be care not to damage the fuel fittings.

13. Install the retaining straps (1988 Colt) and the tank retaining bolts. Torque the bolts to 20 ft. lbs. (27 Nm).

14. Connect the filler hose.

15. Connect the fuel gauge unit and fuel hoses that can be accessed after raising the tank. Connect the fuel pump connectors (fuel injected engines).

16. Install the spare tire and carrier (2WD Vista) and the fuel tank protector (4WD Vista).

17. Install the fuel cap.

18. Install the drain plug, if so equipped. Torque the plug to 15 ft. lbs. (20 Nm).

19. Refill the tank and check for leaks. Connect the negative battery cable.

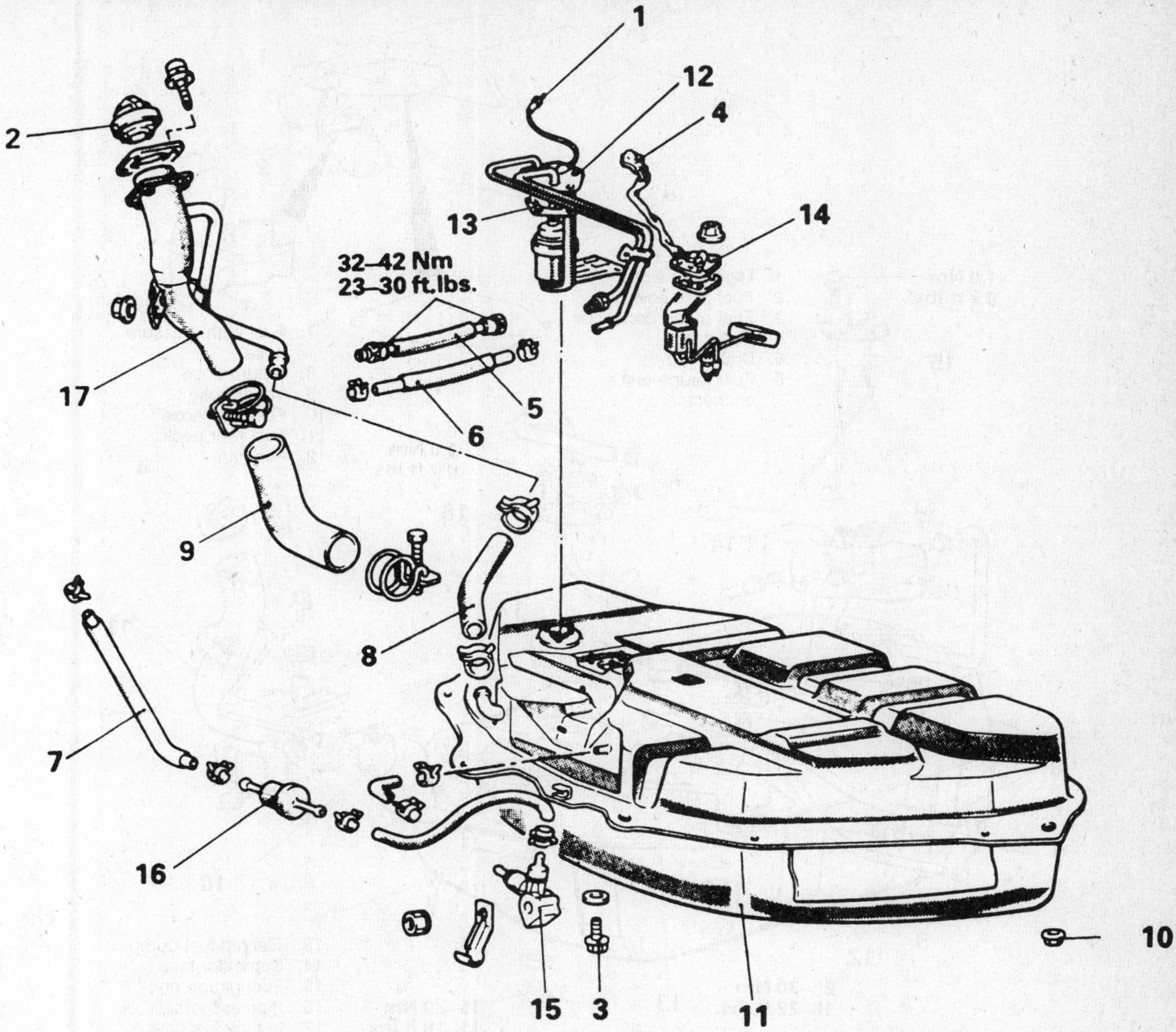

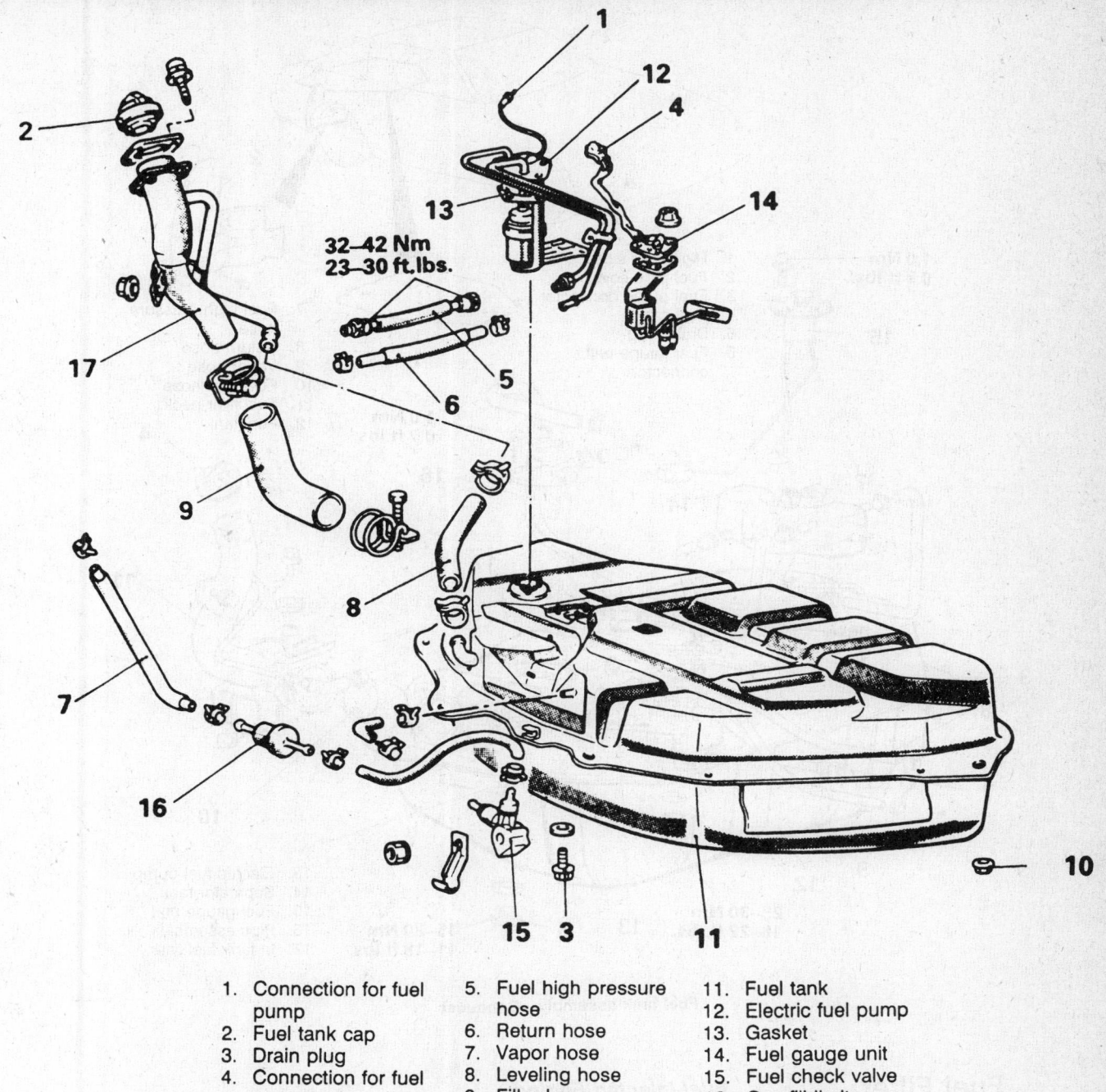

1. Connection for fuel pump
2. Fuel tank cap
3. Drain plug
4. Connection for fuel gauge unit
5. Fuel high pressure hose
6. Return hose
7. Vapor hose
8. Leveling hose
9. Filler hose
10. Self-locking nut
11. Fuel tank
12. Electric fuel pump
13. Gasket
14. Fuel gauge unit
15. Fuel check valve
16. Overfill limiter
17. Filler neck

Fuel tank removal—fuel injected engine shown, others similar

10. Remove all hardware from the old tank if installing a new tank. Remove the fuel pump nuts and remove the pump.

To install:

11. Install all hardware to the new tank, if installing a new tank. Always use new gaskets around the fuel pump and gauge unit. Install the pump and gauge unit. Torque the nuts or bolts to 1.4–2.2 ft. lbs. (2–3 Nm).
12. Slowly raise the tank and connect any electrical or fuel connections. Be care not to damage the fuel fittings.
13. Install the retaining nuts. Torque the bolts to 20 ft. lbs. (27 Nm).
14. Connect the filler hose.
15. Connect the fuel gauge unit and fuel hoses that can be accessed after raising the tank. Connect the fuel pump connectors.
17. Install the fuel cap.
18. Install the drain plug, if so equipped. Torque the plug to 15 ft. lbs. (20 Nm).
19. Install the high floor side panel and fuel pipe cover.
20. Refill the tank and check for leaks. Connect the negative battery cable.

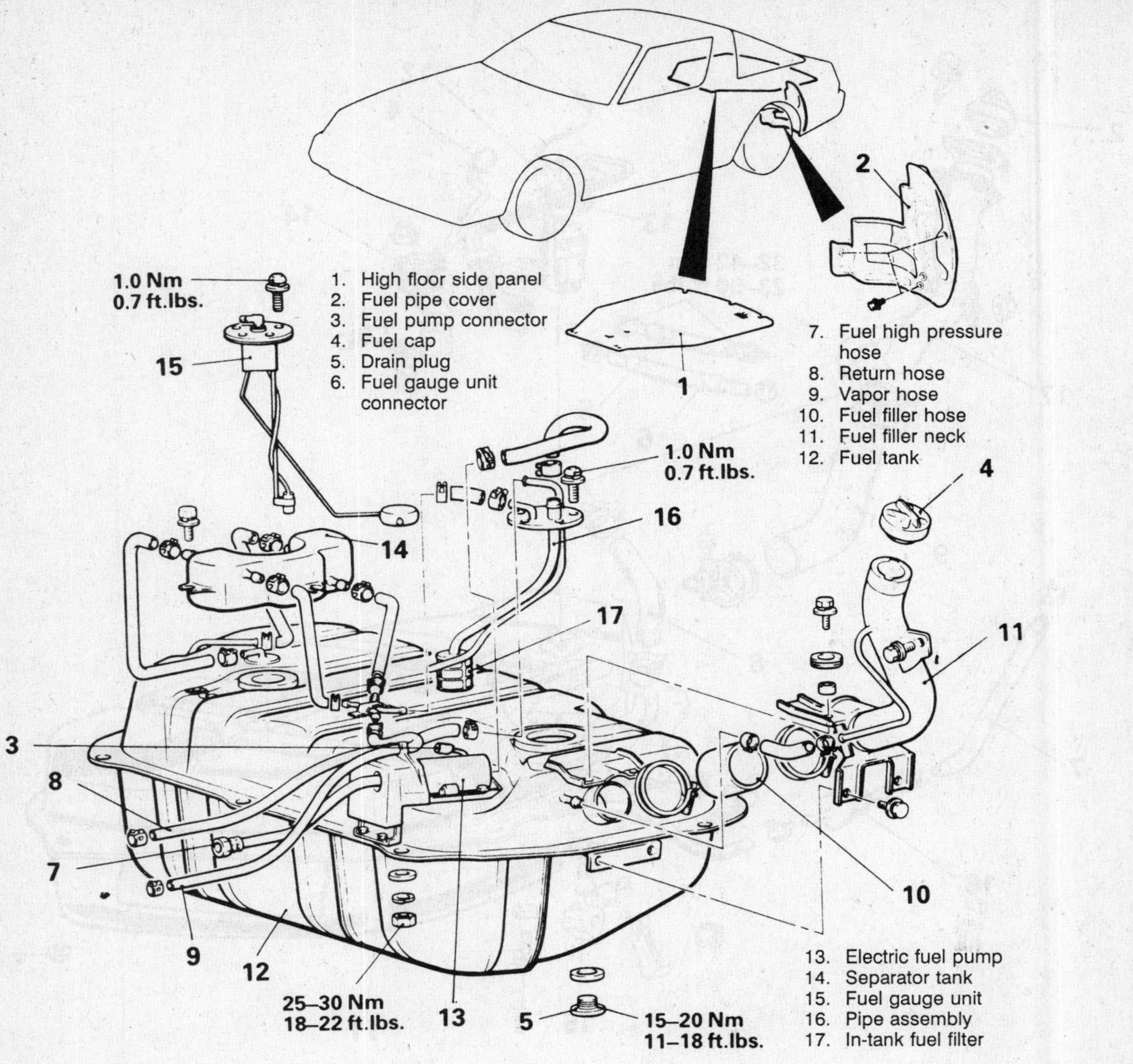

Fuel tank assembly—Conquest

Fuel Filter

REMOVAL & INSTALLATION

Carbureted Engine

1. Disconnect the negative battery cable.
2. Pull the filter from its bracket and discard it.

To install:

3. Snap the replacement filter into the bracket.
4. Install the lines on the filter and tighten the hose clamps.
5. Start the engine and check for leaks.

Fuel Injected Engine

1. Relieve the fuel system pressure.
2. Disconnect the negative battery cable. Remove the air cleaner.
3. Using a backup wrench, remove the fuel line fittings from the fuel filter.

NOTE: Some pressure may still remain in the system, cover the filter connections with a rag to prevent splashing.

4. Remove the fuel filter mounting bolts and the filter from the vehicle.

To install:

5. Installation is the reverse order of the removal procedures. Use a new fuel filter and O-rings. Torque the fuel line-to-filter connectors to 25 ft. lbs. (35 Nm).

Mechanical Fuel Pump

A mechanical fuel pump is used on the 1988 Colt with 1468cc engine.

PRESSURE TESTING

Disconnect the fuel line from the carburetor and attach a pressure tester to the end of the line. Crank the engine. The pressure should be 2.7–3.7 psi.

REMOVAL & INSTALLATION

The pump is mounted on the front side of the engine and is driven by an eccentric on the camshaft.

1. Relieve the fuel system pressure. Disconnect the negative battery cable.
2. Disconnect the fuel lines at the fuel pump.
3. Unbolt the pump mounting bolts, remove the pump, insulator and gasket.
4. Coat both sides of a new insulator and gasket with sealer and install the pump in the reverse order of removal. Torque the mounting bolts and fuel lines to 11–14 ft. lbs. (15–20 Nm).

Electric Fuel Pump

PRESSURE TESTING

Colt and Conquest with Throttle Body Injection

1. Install a suitable fuel pressure gauge to the fuel delivery pipe, be sure to tighten the bolt to 18–25 ft. lbs. (25–34 Nm).
2. Apply voltage to the terminal for the fuel pump drive and activate the fuel pump; then, with fuel pressure applied, check that there is no fuel leakage from the pressure gauge or the special tool connection pipe.
3. Measure the fuel pressure during idling. The standard value is 35–38 psi (240–260 kPa).
4. If the fuel pressure readings are not within specifications, determine the probable cause and make the necessary repairs.
5. Remove all test equipment, use a new gasket and tighten the bolt on the delivery pipe to 18–25 ft. lbs. Start the engine and check for fuel leaks.

Colt and Colt Vista with Multi-Point Injection

1. Install a suitable fuel pressure gauge to the fuel delivery pipe, be sure to tighten the bolt at 18–25 ft. lbs.
2. Apply voltage to the terminal for the fuel pump drive and activate the fuel pump; then, with fuel pressure applied, check that there is no fuel

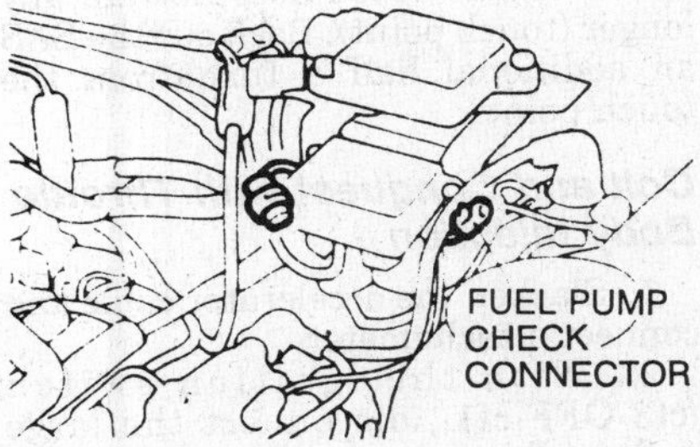

Fuel pump test connector—Colt Turbo

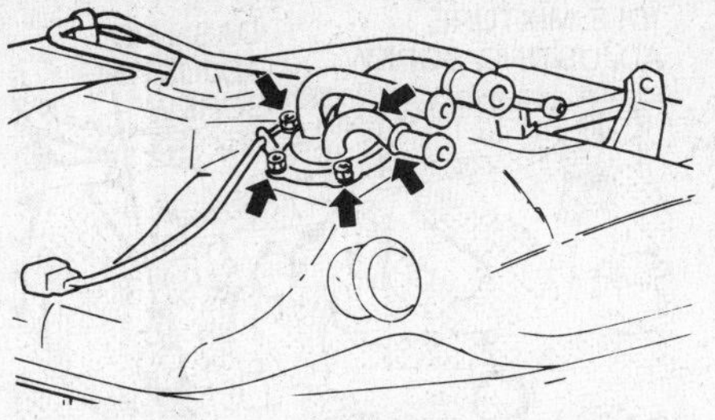

Colt Turbo electric fuel pump location. The arrows indicate the mounting bolts

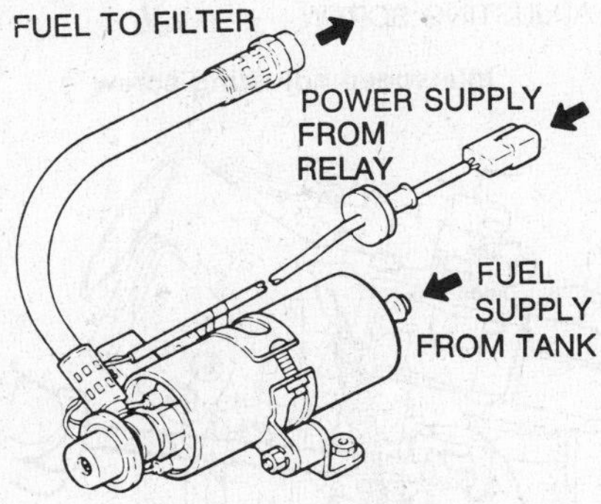

Conquest electric fuel pump

leakage from the pressure gauge or the special tool connection pipe.
3. Disconnect and plug the vacuum hose at the pressure regulator. Measure the fuel pressure during idling. The standard value is 47–50 psi (330–350 kPa).
4. Measure the fuel pressure when the vacuum hose is connected to the pressure regulator. The standard value is 38 psi (270 kPa).
5. If the fuel pressure readings are not within specifications, determine the probable cause and make the necessary repairs.
6. Remove all test equipment, use a new gasket and tighten the bolt on the delivery pipe to 18–25 ft. lbs. Start the engine and check for fuel leaks.

REMOVAL & INSTALLATION

The fuel pump is mounted inside of the fuel tank on Colt and Colt Vista and is mounted externally on top of the fuel tank in the Conquest.

1. Relieve the pressure from the fuel system. Disconnect the negative battery cable.
2. If equipped with a drain plug remove it and drain the fuel into a suitable container. Raise and support the vehicle safely. Remove the left rear wheel.
3. Support the fuel tank with a suitable floor jack. Loosen the fuel tank band mounting nuts and lower the tank for access to the pump support. Then, remove the nut and bolt attaching the pump clamp to the support.
4. Disconnect the fuel lines, noting their locations and remove the pump. If the pump is being replaced, switch the mounting clamp to the new pump and install it at the same angle.

To install:

5. Installation is the reverse order of the removal procedure. Make sure fuel line connections are tight and secure. Operate the pump and check for leaks.

Carburetor

REMOVAL & INSTALLATION

1. Disconnect the negative battery cable.
2. Remove the solenoid valve wiring.
3. Disconnect the air cleaner breather hose, air duct and vacuum tube.
4. Remove the air cleaner.
5. Remove the air cleaner case.
6. If equipped with automatic transaxle, disconnect the accelerator and shift cables, at the carburetor.
7. Disconnect the purge valve hose. Remove the vacuum compensator and fuel lines.
8. Drain the coolant.
9. Remove the water hose between the carburetor and the cylinder head.
10. Remove the carburetor.

To install:

11. Installation is the reverse order of removal procedures.
12. Clean the manifold and carburetor mounting surfaces.
13. Install the carburetor and mounting bolts. Torque the mounting bolts to 11–14 (15–20 Nm).

IDLE SPEED AND MIXTURE ADJUSTMENT

NOTE: The throttle valve adjusting screw should not be tampered with unless the carburetor has been rebuilt. This screw is preset and determines the relationship between the throttle valve and the free lever and has been accurately set at the factory. If this setting is disturbed, the throttle opener adjustment and or dashpot adjustment cannot be done accurately. Also the improper setting (throttle valve opening) will increase the exhaust gas temperature and deceleration, which in turn will reduce the life of the catalyst greatly and deteriorate the exhaust gas cleaning performance. It will also effect the fuel consumption and the engine braking.

1. Disconnect the negative battery cable.
2. Remove the carburetor from the engine.
3. Place the carburetor in a suitable holding fixture with the idle mixture adjusting screw facing up.
4. Drill a 5/64 in. (2mm) hole in the casting surrounding the idle mixture adjusting screw, then redrill the hole to 1/8 in. (3mm).
5. Insert a punch in the hole and drive out plug.
6. Reinstall the carburetor to the engine and connect the negative battery cable.
7. Start the engine and run at fast idle, allow it to reach operating temperature.
8. All lights and accessories must be OFF.
9. Disconnect the electrical connector from the the coolant fan during idle and mixture adjustments.
10. Place the transaxle in **P** or **N**, if equipped with automatic transaxle and neutral, if equipped with manual transaxle.
11. Connect a timing light or tachometer, check and adjust the basic timing, as required.
12. Disconnect the negative battery cable for 3 seconds, then reconnect it.
13. Disconnect the electrical connector from the oxygen sensor.
14. Run the engine at 2000–3000 rpm for 5–10 seconds and allow it to idle for 2 minutes.
15. Adjust the idle CO and engine speed as follows:

Adjust the idle Speed Adjustment Screw No. 1 (SAS–1) to 800 rpm.

Adjust idle Mixture Adjustment Screw (MAS) to 0.1–0.3 percent CO.

NOTE: The idle Speed Adjustment Screw No. 2 (SAS–2) determines the relationship between the throttle valve and free lever. This adjustment is factory preset and therefore should not be disturbed. If the CO adjustment fails, it is likely there is a loss of vacuum at the secondary air hose. Plug the air hose and try the CO adjustment again.

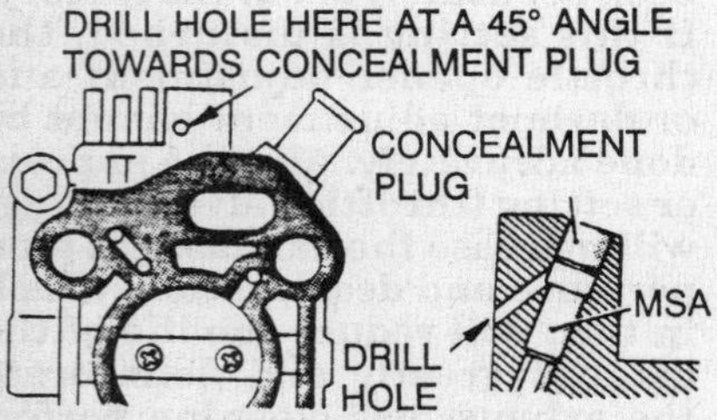

Drilling of carburetor base to remove concelament plug, exposing the mixture adjustment screw

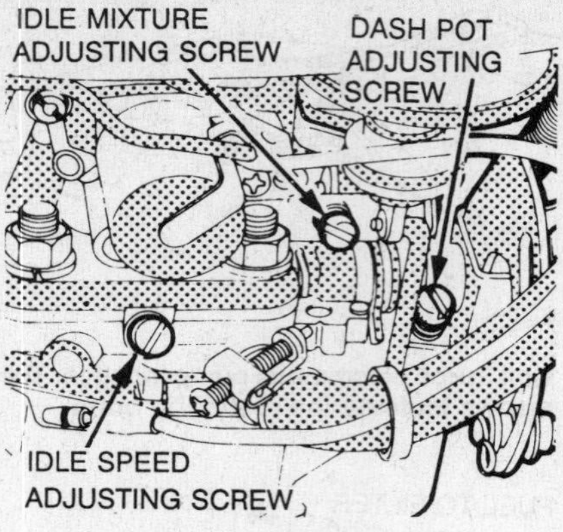

Idle speed adjusting screw

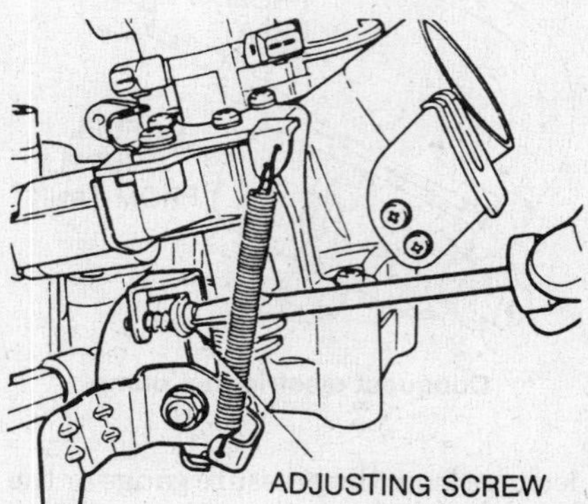

Fuel injection unit idle speed adjusting screw

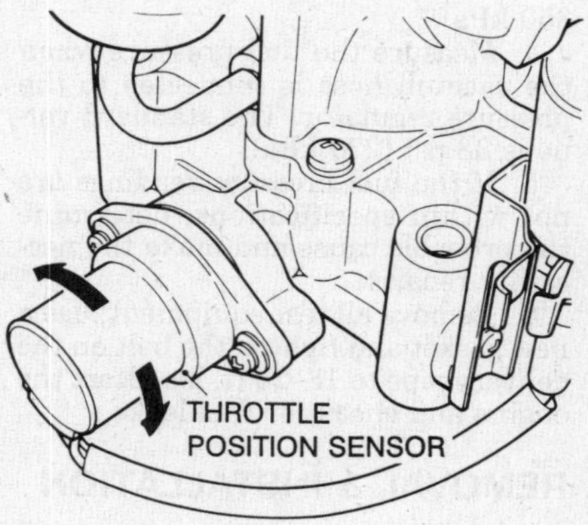

Throttle position sensor adjustment

16. Turn the ignition switch **OFF** and connect the oxygen connector.
17. Install the concealed plug into the hole to seal the idle mixture adjusting screw.
18. Connect the electrical connector to the coolant fan.

Fuel Injection

Idle speed and idle mixture are controlled by the Electronic Control Unit (ECU). Adjustments are therefore not needed.

IDLE SPEED ADJUSTMENT

Colt and Colt Vista

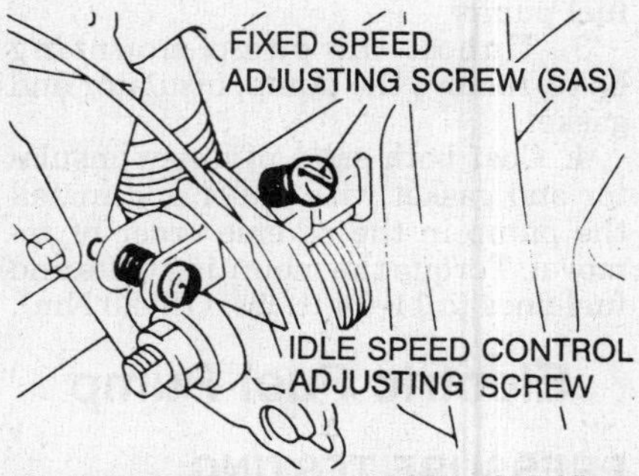

Speed adjusting screw (SAS) – Colt Vista

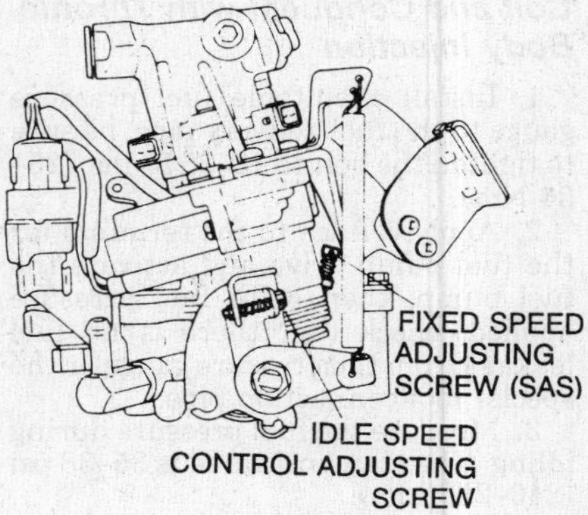

Idle speed adjusting screws

1. Slacken the accelerator cable, connect a tachometer, set the ignition switch to the **ON** position and turn the ignition **OFF**.
2. Disconnect the connectors from the idle speed control servo and lock the idle speed control plunger at the initial position.
3. Back out the speed adjusting screw SAS enough.
4. Start the engine and let idle. Check to see if the idle speed is 750 ± 50 for the Colt and 700 ± 100 for the Colt Vista.
5. If not within specifications, adjust the fixed SAS until the engine speed rises. Then back out the SAS until the engine speed does not fall any longer (touch point). Back out the SAS an additional half a turn from the touch point.

Colt and Conquest with Throttle Body Injection

1. Slacken the accelerator cable and connect a tachometer.
2. With the ignition switch ‹cf3›OFF‹cf1›, disconnect the large and small harness connector from the ECU.

3. Connect a ECI (Electronically Controlled Injection) checker to the ECU harness connector and ECI checker.
4. Connect a digital voltmeter to the extension terminals of the ECI checker.
5. Set the select switch to **A** and check switch to **7**.
6. To prevent throttle valve binding, open the throttle valve by hand 2 or 3 times and then release it to allow it to snap back. Loosen the fixed speed adjusting screw (SAS).
7. Start the engine and allow it to idle.
8. Check that the engine speed and motor position sensor output voltage is within 700 rpm and 0.48–0.52 volts for the Colt and 850 rpm and 0.9 volts for the Conquest.
9. If not within specification, adjust the idle speed control (ISC) adjusting screw.
10. Turn the ignition switch **OFF**.
11. Disconnect the ISC motor connector and fix the ISC motor. Start the engine and idle.
12. Tighten the fixed SAS until the engine speed starts to increase. Then, loosen it until the engine speed ceases to drop and then loosen one turn from the touch point.
13. Stop the engine reconnect any disconnected terminals. Disconnect the negative battery terminal for 6 seconds and reconnect. This erases diagnosis memory during the adjustment.

Fuel Injector

REMOVAL & INSTALLATION

1. Relieve the fuel system pressure.
2. Disconnect the negative battery cable.
3. Disconnect the fuel rail assembly, so the fuel injectors are easily accessible.
4. Remove the injector clip from the fuel rail and injector. Pull the injector straight out of the fuel rail receiver cup.
5. Check the injector O-ring for damage. If the O-ring is damaged, replace it. If the injector is being reused, install a protective cap on the injector tip to prevent damage.

To install:

6. Installation is the reverse order of the removal procedures. Before installing the injector, the rubber O-ring must be lubricated with a drop of clean engine oil to aid in installation.
7. Make sure all fuel connections are tight. Connect the negative battery cable.
8. Start the engine and check for leaks.

DRIVE AXLE

Halfshaft

REMOVAL & INSTALLATION

Except 4WD Colt Vista

1. Remove the hub center cap and loosen the halfshaft axle nut. Loosen the wheel lug nuts.
2. Raise the vehicle and support it safely. Remove the front wheels. Remove the engine splash shield.
3. Remove the lower ball joint and strut bar from the lower control arm.
4. Drain the transaxle fluid.

NOTE: If equipped with a turbocharger, remove the snapring which secures the center bearing.

5. Insert a prybar between the transaxle case, on the raised rib and the halfshaft double offset joint case; do not insert the prybar too deeply or the oil seal will be damaged. Move the bar to the right to withdraw the left driveshaft; to the left to remove the right halfshaft.

NOTE: In the case of the tripod joint driveshaft, be sure to hold the tripod joint case and pull out the shaft straight. Simply pulling the shaft out of position could cause damage to the tripod joint boot or the spider assembly slipping from the case.

6. Plug the transaxle case with a clean rag to prevent dirt from entering the case.
7. Use a special puller driver tool mounted on the wheel studs to push the halfshaft from the front hub. Take care to prevent the spacer from falling out of place.

NOTE: If equipped with a center bearing, after forcing out the halfshaft, remove it by lightly tapping the double offset joint outer race, with a plastic hammer.

To install:

8. Assembly is the reverse of removal. Insert the halfshaft into the hub first, then install the transaxle end. Torque the shaft nut to 180 ft. lbs.

NOTE: Always use a new retaining ring every time the halfshaft is removed.

9. When installing the kit, use the grease supplied with the kit and apply an amount to the inner race and cage.
10. Install the inner race and cock slightly.
11. Apply grease to the balls and install them in the cage.
12. Place the inner race on the halfshaft and install the snapring.
13. Apply grease to the outer race and install.
14. Install the boots and bands.
15. Install the halfshaft using a new retainer ring.
16. Lower the vehicle.

4WD Colt Vista

1. Remove the hub cap and halfshaft nut.
2. Raise and support the vehicle safely.
3. Remove the front wheels.
4. Drain the transaxle fluid.
5. Disconnect the lower ball joint from the knuckle.
6. Remove the strut and stabilizer bar from the lower arm.
7. Remove the center bearing snapring from the bearing bracket.
8. Lightly tap the double offset joint outer race with a wood mallet and disconnect the halfshaft from the Cardan joint.
9. Disconnect the halfshaft from the bearing bracket.

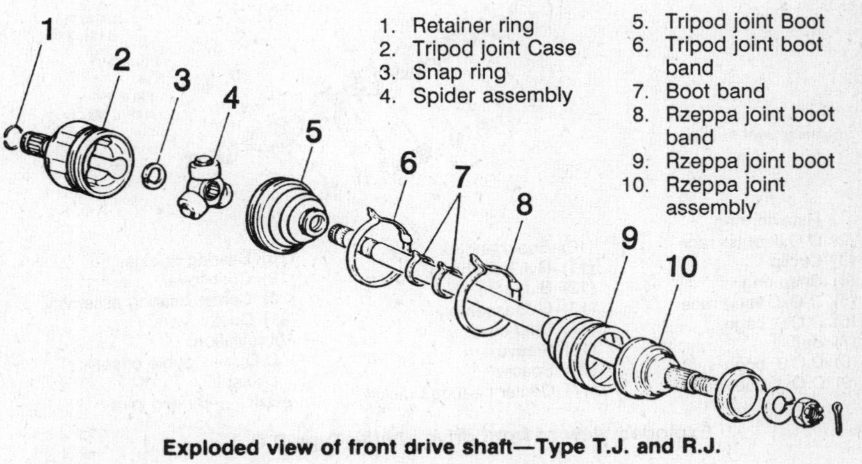

Exploded view of front drive shaft—Type T.J. and R.J.

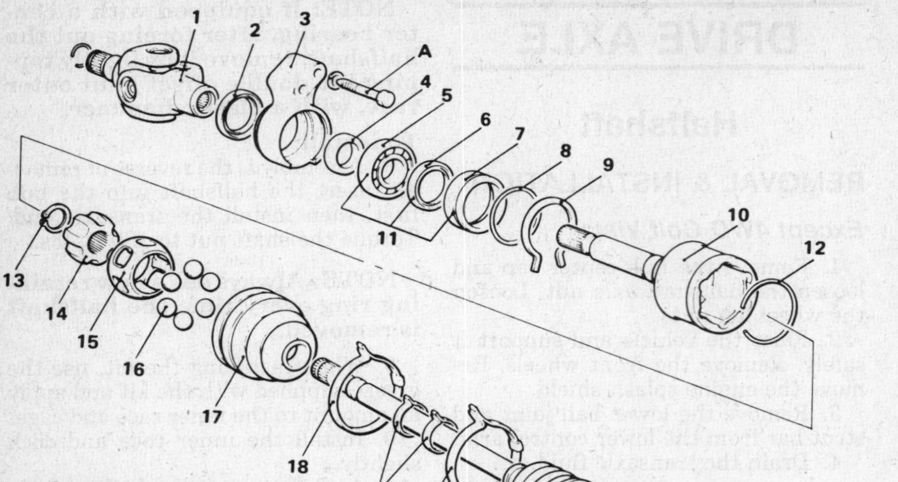

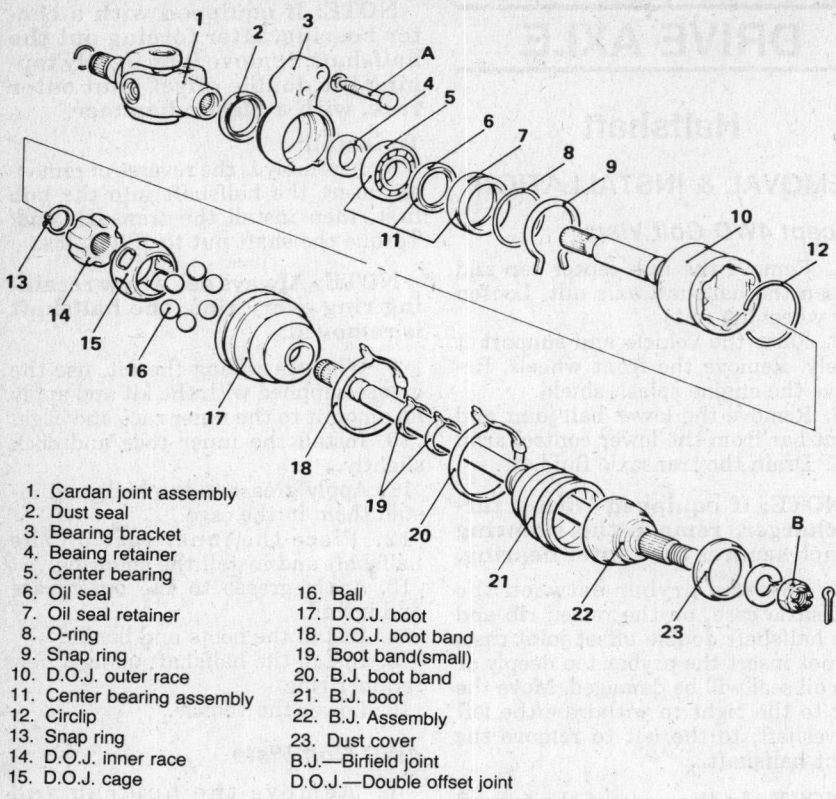

1. Cardan joint assembly
2. Dust seal
3. Bearing bracket
4. Beaing retainer
5. Center bearing
6. Oil seal
7. Oil seal retainer
8. O-ring
9. Snap ring
10. D.O.J. outer race
11. Center bearing assembly
12. Circlip
13. Snap ring
14. D.O.J. inner race
15. D.O.J. cage
16. Ball
17. D.O.J. boot
18. D.O.J. boot band
19. Boot band(small)
20. B.J. boot band
21. B.J. boot
22. B.J. Assembly
23. Dust cover

B.J.—Birfield joint
D.O.J.—Double offset joint

4-wd Vista halfshaft

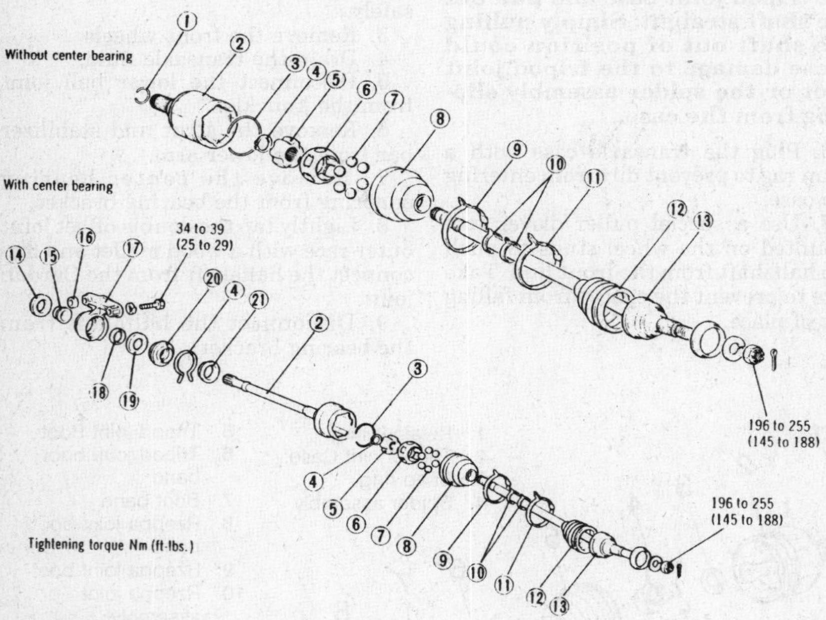

(1) Retainer ring
(2) D.O.J. outer race
(3) Circlip
(4) Snap ring
(5) D.O.J. inner race
(6) D.O.J. cage
(7) Ball
(8) D.O.J. boot
(9) D.O.J. boot band
(10) Boot band
(11) B.J. boot band
(12) B.J. boot
(13) B.J. assembly
(14) Dust cover
(15) Sleeve
(16) Spacer
(17) Center bearing bracket
(18) Bearing retainer
(19) Dust cover
(20) Center bearing assembly
(21) Dust cover

Abbreviation:
D.O.J.—Double offset joint
B.J. —Birfield joint

Exploded view of front drive shaft—Type D.O.J. and B.J.

10. Using a 2-jawed puller secured to the hub lugs, press the halfshaft from the hub.
11. Unbolt and remove the bearing bracket.
12. Using a wood mallet, lightly tap the Cardan joint yoke and remove it from the transaxle. Never pry the Cardan joint from the transaxle. Prying will damage the Cardan joint dust cover.

To install:

13. Install the Cardan joint.
14. Apply a coating of chassis lube on the center bearing.
15. Attach a new O-ring to the oil seal retainer.
16. Install the bearing bracket. Torque the bolts to 40 ft. lbs. (54 Nm).
17. Insert the center bearing in the bearing bracket, making sure it is fully seated, then secure it with the snapring.
18. Coat the halfshaft splines with chassis lube and slide it into the Cardan joint.
19. Slide the halfshaft into the hub and install the nut. Torque the nut to 188 ft. lbs.

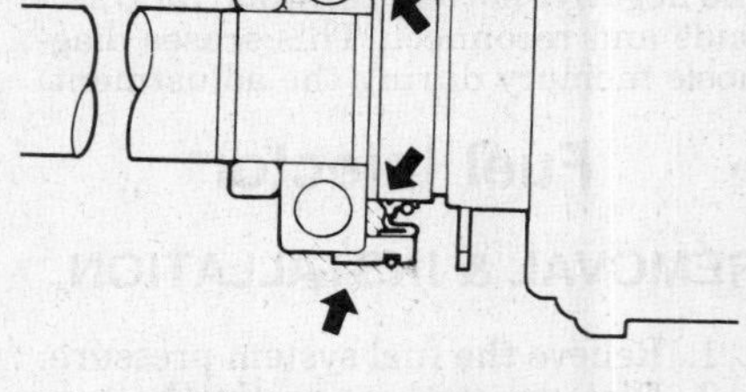

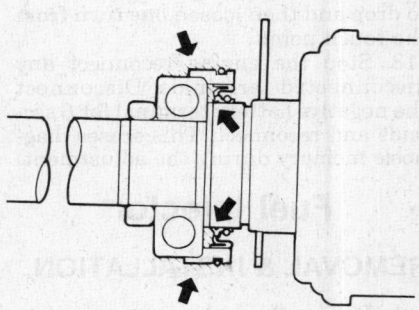

Lubricant application points

CV-Boot

REMOVAL & INSTALLATION

Inner Boot

1. Raise the vehicle and support it safely. Remove the wheel assembly.
2. Remove the halfshaft assembly from the vehicle.
3. Place the assembly in vise. Care must be taken not to crush the tubular shafts.
4. If inner joint needs replacement, cut the small rubber clamp, large metal clamp and remove the rubber boot. These items must be discarded.
5. Inspect for internal wear and/or damage.
6. Clean the grease by hand from inside the joint housing and around

the ball trunnion assembly to inspect it.

7. Mark the tripod and housing for proper reassembly, if it is to be reinstalled.

8. To replace the boot, CV-joint or both, remove the snapring from the groove and tap the trunnion lightly with a brass drift pin. Leave the tripod bearings on the trunnion. Care must be taken to support the bearings as they may fall off.

To install:

9. Installation is the reverse order of the removal procedures.

10. When installing the tripod on the shaft place the chamfer face towards the retainer groove.

NOTE: The grease provided with the repair kit must be used. It cannot be substituted with any other type grease.

Outer Boot

1. Raise the vehicle and support it safely. Remove the wheel assembly.

2. Remove the halfshaft assembly from the vehicle.

3. Place the assembly in vise. Care must be taken not to crush the tubular shafts.

4. At the inner joint, cut the small rubber clamp, large metal clamp and remove the rubber boot. These items must be discarded.

5. Inspect for internal wear and/or damage.

6. Clean the grease by hand from inside the joint housing and around the ball trunnion assembly to inspect it.

7. Mark the tripod and housing for proper reassembly, if it is to be reinstalled.

8. To replace the boot, CV-joint or both, remove the snapring from the groove and tap the trunnion lightly with a brass drift pin. Leave the tripod bearings on the trunnion. Care must be taken to support the bearings as they may fall off.

9. If equipped with a dynamic damper, remove the band and slide the damper from the shaft.

10. Cut both bands from the outer boot and slide the boot from the shaft.

To install:

11. Installation is the reverse order of the removal procedures.

12. When installing the tripod on the shaft place the chamfer face towards the retainer groove.

NOTE: The grease provided with the repair kit must be used. It cannot be substituted with any other type grease.

Driveshaft and U-Joints

REMOVAL & INSTALLATION

Colt Wagon and 4WD Colt Vista

1. Raise and support the vehicle safely.

2. Drain the transfer case.

3. Matchmark the differential companion flange and the driveshaft flange yoke.

4. Unbolt the driveshaft from the differential flange.

5. Remove the 2 center bearing attaching nuts.

NOTE: Make sure the flat washer and the adjusting spacer are not interchanged. Keep them separate for assembly.

6. Pull the driveshaft from the transfer case. Be careful to avoid damaging the transfer case oil seal.

To install:

7. Installation is the reverse order of removal procedure. Torque the center bearing nuts to 25–30 ft. lbs.; the driveshaft-to-differential flange nuts to 20–25 ft. lbs.

Conquest

1. Raise the vehicle and support it safely.

2. Matchmark the rear flange yoke and the differential pinion flange.

3. Remove the driveshaft-to-pinion shaft flange bolts. Remove the driveshaft by pulling it from the rear of the transmission extension housing.

NOTE: Place a container under the transmission extension housing to collect any oil leakage when the driveshaft is removed.

4. To install the shaft, align the front sleeve yoke with the splines of the transmission output shaft and push the driveshaft into the extension housing.

NOTE: Be careful not to damage the rear transmission seal lip upon installation.

5. Align the matchmarks on the rear yokes, install the bolts and tighten securely.

6. Inspect the oil level of the transmission. Lower the vehicle.

Rear Axle Shaft, Bearing and Seal

REMOVAL & INSTALLATION

Conquest

1. Raise and support the vehicle safely.

2. Disconnect the parking brake cable from the rear calipers.

3. Remove the caliper, caliper support and rotor. Do not disconnect the brake line from the caliper, suspend it out of the way.

4. Remove the intermediate shaft and companion flange as described below.

5. Remove the halfshaft housing from the lower control arm.

6. Remove the strut assembly from the halfshaft housing.

7. Loosen the companion flange mounting nut and tap the halfshaft out of the housing with a plastic mallet. Be careful to avoid scratching the oil seal.

8. Remove the spacer and dust covers from inside the housing.

NOTE: Don't remove the bearings unless they are to be replaced, since they will be damaged during removal.

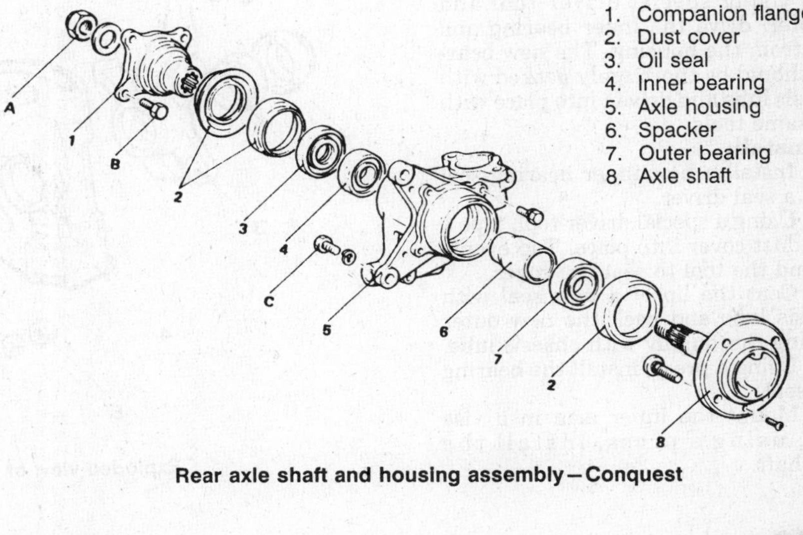

1. Companion flange
2. Dust cover
3. Oil seal
4. Inner bearing
5. Axle housing
6. Spacker
7. Outer bearing
8. Axle shaft

Rear axle shaft and housing assembly—Conquest

9. Remove the outer bearing with a puller.
10. Using a brass drift, drive the inner bearing and seal from the housing.
To install:
11. Press the new outer bearing onto the shaft with the seal side facing the flange side of the shaft.
12. Pack the housing with lithium based wheel bearing grease.
13. Press the inner bearing onto the shaft with the seal side facing the companion flange side of the shaft.
14. Grease the seal bore in the housing and drive the new seal into position.
15. Install the dust covers.
16. Insert the halfshaft and spacer into the housing and attach the companion flange.
17. Place the housing in a vise and install and tighten the companion flange nut to 200–220 ft. lbs.
18. Install all other parts in reverse order of removal. Check halfshaft endplay with a dial indicator. Endplay should be 0.031 in. If endplay exceeds the limit, either the bearing needs replacing or the shaft bearings are not assembled properly.

4WD Colt Vista

1. Raise and support the vehicle safely.
2. Remove the rear wheels.
3. Remove the brake drums.
4. Remove the 3 bolts securing the halfshaft flange to the intermediate shaft flange.
5. Using a special tool, remove the halfshaft flange nut.
6. Using a slide hammer connected to a 2-jawed adapter secured under 2 lug nuts, pull the halfshaft from the housing.
7. Remove the lower arm.
8. Using a special tool, remove the dust cover and the outer wheel bearing and seal from the halfshaft at the same time. Discard the seal.
9. Using special driver tool and adapter, drive the inner bearing and seal from the housing. The new bearing should be thoroughly packed with chassis lube and driven into place with the same tools.
To install:
10. Install a new inner bearing seal with a seal driver.
11. Using a special driver tool, tape a new dust cover into place. Tap evenly around the tool to seat the cover.
12. Coat the lip of a new seal with chassis lube and pack the new outer bearing thoroughly with chassis lube.
13. Using a press, install the bearing and seal.
14. Mount the inner arm in a vise and, using a press, install the halfshaft.
15. Install the halfshaft and inner arm assembly.
16. Torque the halfshaft nut to 160 ft. lbs.
17. Connect the halfshaft and intermediate shaft flanges and torque the bolts to 43 ft. lbs.

Front Wheel Hub, Knuckle and Bearings

REMOVAL & INSTALLATION

Colt and 2WD Colt Vista

NOTE: The following procedure requires the use of several special tools.

1. Remove the halfshaft nut.
2. Raise and support the the vehicle safely. Allow the front suspension to hang freely.
3. Remove the wheels.
4. Remove the caliper and suspend it out of the way, without disconnecting the brake hose.
5. Disconnect the lower ball joint from the knuckle.
6. Disconnect the tie rod end from the knuckle.
7. Using a 2-jawed puller, press the halfshaft from the hub.
8. Unbolt the strut from the knuckle. Remove the hub and knuckle assembly from the vehicle.
9. Install first the arm, then the body of special tool MB991056 (Colt) or MB991001 (Colt Vista) or their equivalent on the knuckle and tighten the nut.
10. Using special tool MB990998 or MB990781 or their equivalent, separate the hub from the knuckle.

NOTE: Prying or hammering will damage the bearing. Use these special tools or their equivalents, to separate the hub and knuckle.

11. Place the knuckle in a vise and separate the rotor from the hub.
12. Using special tools C-293-PA, SP-3183 and MB990781 or their equivalent, remove the outer bearing inner race.
13. Drive the oil seal and inner bearing inner race from the knuckle with a brass drift.
14. Drive out both outer races in a similar fashion.

NOTE: Always replace bearings and races as a set. Never replace just an inner or outer bearing. If either is in need of replacement, both sets must be replaced.

15. Thoroughly clean and inspect all parts. Any suspect part should be replaced.
To install:
16. Pack the wheel bearings with lithium based wheel bearing grease. Coat the inside of the knuckle with similar grease and pack the cavities in the knuckle. Apply a thin coating of grease to the outer surface of the races before installation.
17. Using special tools C-3893 and MB990776 or equivalent, install the outer races.
18. Install the rotor on the hub and torque the bolts to 36–43 ft. lbs. (48–58 Nm).
19. Drive the outer bearing inner race into position.
20. Coat the outer rim and lip of the oil seal and drive the hub side oil seal into place, using a seal driver.

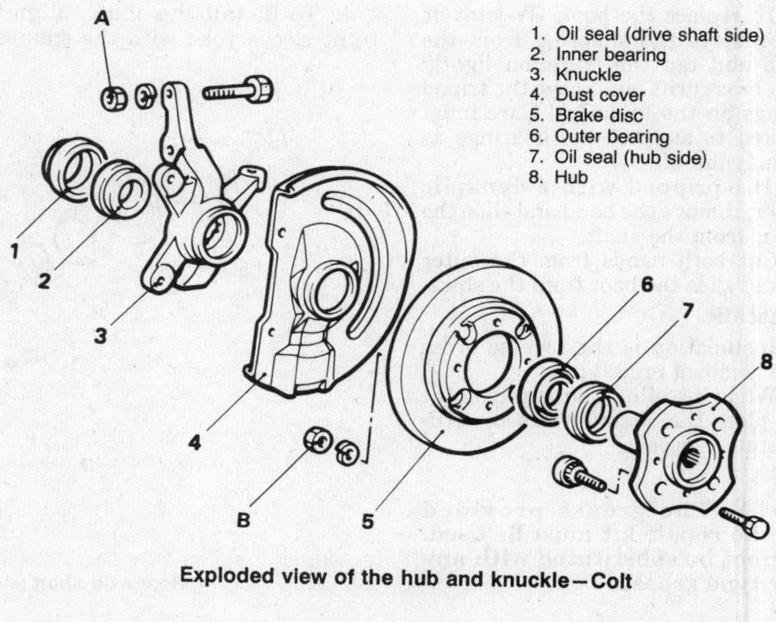

Exploded view of the hub and knuckle – Colt

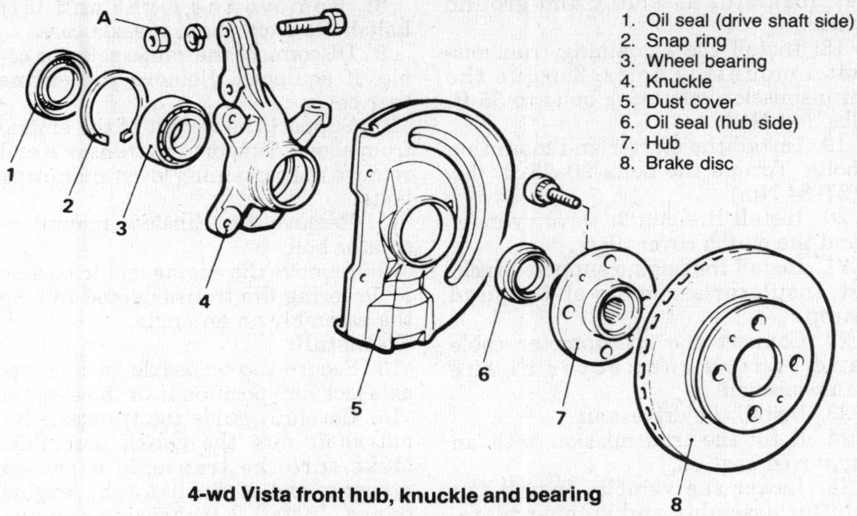

4-wd Vista front hub, knuckle and bearing

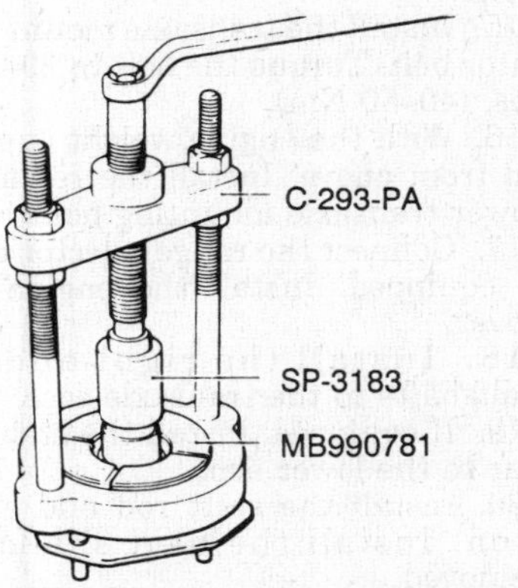

Removing the outer bearing inner race from the hub, using the special tools described

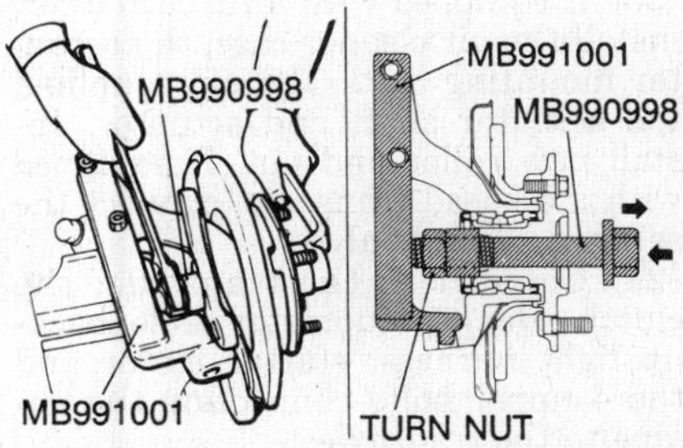

Using special tools to remove the hub from the knuckle

21. Place the inner bearing in the knuckle.
22. Mount the knuckle in a vise. Position the hub and knuckle together. Install tool MB99098 or equivalent and tighten the tool to 147–192 ft. lbs. (190–265 Nm). Rotate the hub to seat the bearing.
23. With the knuckle still in the vise measure the hub starting torque with an inch lb. torque wrench and tool MB990998 or equivalent. Starting torque should be 11.5 inch lbs. or less. If the starting torque is 0, measure the hub bearing axial play with a dial indicator. If axial play exceeds 0.0078 in., while the nut is tightened to 145–192 ft. lbs. (190–265 Nm), the assembly has not been done correctly. Disassemble the knuckle and hub, and start again.
24. Remove the special tool.
25. Place the outer bearing in the hub and drive the seal into place.
26. The remainder of installation is the reverse of removal.

4WD Colt Vista

NOTE: The following procedure requires the use of several special tools.

1. Remove the hub cap and halfshaft nut.
2. Raise the vehicle and support it safely.
3. Remove the front wheels.
4. Drain the transaxle fluid.
5. Disconnect the lower ball joint from the knuckle.
6. Remove the strut and stabilizer bar from the lower arm.
7. Remove the center bearing snapring from the bearing bracket.
8. Lightly tap the double off-set joint outer race with a wood mallet and disconnect the halfshaft from the Cardan joint.
9. Disconnect the halfshaft from the bearing jacket.
10. Using a 2–jawed puller secured to the hub lugs, press the halfshaft from the hub.
11. Unbolt the strut from the hub. Remove the hub and knuckle assembly from the vehicle.
12. Install first the arm, then the body of special tool MB991001 or equivalent on the knuckle and tighten the nut.
13. Using special tool MB9900998 or equivalent, separate the hub from the knuckle and tighten the nut.

NOTE: Prying or hammering will damage the bearing. Use these special tools or equivalent, to separate the hub and knuckle.

14. Matchmark the hub and rotor. The rotor should slide from the hub. If not, insert M8 × 1.25 bolts in the holes between the lugs and tighten them alternately to press the hub from the rotor. Never hammer the rotor to remove it.
15. Using a 2–jawed puller, remove the outer bearing inner race.
16. Remove and discard the outer oil seal.
17. Remove and discard the inner oil seal.
18. Remove the bearing snapring from the knuckle.
19. Using special tools C–4628 and MB991056 or MB991001 or their equivalent, remove the bearing from the knuckle. Using a driver, drive the bearing from the knuckle.

NOTE: Always replace bearings and races as a set. Never replace just an inner or outer bearing. If either is in need of replacement, both sets must be replaced.

20. Thoroughly clean and inspect all parts. Any suspect part should be replaced.
21. Pack the wheel bearings with lithium based wheel bearing grease. Coat the inside of the knuckle with similar grease and pack the cavities in the knuckle. Apply a thin coating of grease to the outer surface of the races before installation.

To install:

22. Using special tools C–4171 and MB990985 or their equivalent, press the bearing into place in the knuckle. Install the snapring.
23. Coat the lips of a new hub side seal with lithium grease. Using a seal driver, install the seal. Make sure it is flush.
24. Install the rotor on the hub.
25. Using special tool MB990998 or equivalent, join the hub and knuckle. Torque the special tool nut to 188 ft. lbs. (260 Nm).
26. Rotate the hub several times to seat the bearing.
27. Mount the knuckle in a vise. Using MB990998 or equivalent and an inch lbs. torque wrench, measure the turning torque. Turning torque should be 15.6 inch lbs. (2.5 Nm) or less. Next, measure the axial play using a dial indicator. Axial play should be 0.008 in. If either the axial play or the turning torque are not within the specified values, the hub and knuckle

have not been properly assembled. Repeat the procedure. If everything checks out okay, go on to the next step.

28. Remove all the special tools.
29. Using a seal driver, drive a new seal coated with lithium grease, into place on the halfshaft side, until it contacts the snapring.
30. The remainder of installation is the reverse order of removal procedures.

MANUAL TRANSMISSION

For further information on transmissions/transaxles, please refer to "Chilton's Guide to Transmission Repair".

Transmission Assembly

REMOVAL & INSTALLATION

1. Raise and safely support the vehicle.
2. Remove the driveshaft.
3. Drain the transmission.
4. Disconnect the speedometer cable and switch connector at the transmission.
5. Remove the clutch slave cylinder.
6. Remove the bell housing cover.
7. Remove the starter.
8. Remove the 2 upper transmission mounting bolts.
9. Support the transmission with a suitable floor jack.
10. Remove the remaining transmission mounting bolts.
11. Remove the engine support bracket, insulator assembly and ground strap.
12. Place the shift lever in the neutral position. Remove the trim plate and unbolt the shifter assembly, removing it and the stopper plate underneath it.
13. Cover the rear of the cylinder head with a heavy cloth to prevent damage from contact with the firewall.
14. Slowly lower the jack, pull it rearward to disengage the transmission from the clutch.

To install:

15. Secure the the transmission on a transmission jack and raise it into position to the clutch assembly.
16. Make sure the transmission is seated properly and is flush to the engine flange. Install 2 mounting bolts on each side of the bell housing.
17. Install the engine support bracket, insulator assembly and ground strap.
18. Install the remaining transmission mounting bolts. Torque the transmission mounting bolts to 35 ft. lbs. (47 Nm).
19. Install the starter and mounting bolts. Torque the bolts 20–25 ft. lbs. (27–34 Nm).
20. Install the clutch slave cylinder and the clutch cover plate.
21. Install the engine support bracket, insulator assembly and ground strap.
22. Connect the speedometer cable and switch connector at the transmission.
23. Install the driveshaft.
24. Refill the transmission with an approved gear oil.
25. Lower the vehicle. Install the shifter assembly and stopper plate. Connect the negative battery cable.

MANUAL TRANSAXLE

For further information on transmissions/transaxles, please refer to "Chilton's Guide to Transmission Repair".

Transaxle Assembly

REMOVAL & INSTALLATION

Colt

1. Disconnect the negative battery cable.
2. Disconnect from the transaxle: the clutch cable, speedometer cable, back-up light harness, starter motor and the 4 upper bolts connecting the engine to the transaxle.
3. If with a turbocharger, remove the air cleaner case, the actuator mounting bolts, the pin coupling, the actuator and shaft and remove the actuator. Discard the collar used with the pin and replace it with a new collar. If equipped with a 5 speed transaxle, disconnect the selector control valve.
4. Raise and support the vehicle safely.
5. Remove the front wheels. Remove the splash shield. Drain the transaxle fluid.
6. Remove the shift rod and extension. It may be necessary to remove any heat shields that interfere.
7. If equipped, remove the stabilizer bar from the lower arm and disconnect the lower arm from the body side.
8. Remove the right and left halfshafts from the transaxle case.
9. Disconnect the range selector cable, if equipped. Remove the engine rear cover.
10. Support the weight of the engine from above. Support the transaxle and remove the remaining lower mounting bolts.
11. Remove the transaxle mount insulator bolt.
12. Remove the engine and transaxle by lowering the transaxle and raising the assembly on an angle.

To install:

13. Secure the transaxle on a transaxle jack and position it to the engine.
14. Carefully guide the transaxle input shaft into the clutch assembly. Make sure the transaxle is seated properly and is flush to the engine flange. Install 2 transaxle-to-engine bolts.
15. Install the transaxle mount insulator bolt. Torque the bolt to 29–36 ft. lbs. (40–50 Nm).
16. With the engine weight supported from above. Install the remaining lower transaxle mounting bolts.
17. Connect the range selector cable, if equipped. Install the engine rear cover.
18. Install the right and left halfshafts to the transaxle case.
19. If equipped, install the stabilizer bar to the lower arm.
20. Install the shift rod and extension. Install the heat shields, if removed.
21. Install the splash shield. Install the front wheels. Refill the transaxle fluid.
22. Lower the vehicle.
23. If equipped with a turbocharger, install the air cleaner case, the actuator mounting bolts, the pin coupling, the actuator shaft and actuator. Install new collar and pin. If equipped with a 5 speed transaxle, connect the selector control valve.
24. Connect to the transaxle: the clutch cable, speedometer cable, back-up light harness, starter motor and the 4 upper bolts connecting the engine to the transaxle.
25. Connect the negative battery cable.

2WD Colt Vista

1. Disconnect the battery cables, negative cable first. Remove the battery and tray.
2. Remove the coolant reservoir.
3. Remove the air cleaner.
4. Disconnect the clutch cable, speedometer cable and backup light wiring from the transaxle.
5. Remove the upper engine-to-transaxle bolts.
6. Disconnect the select control lever and switch harness.

7. Remove the starter.
8. Disconnect and tag all wiring from the transaxle.
9. Raise and support the vehicle safely.
10. Remove the front wheels.
11. Drain the transaxle fluid.
12. Remove the extension and shift rod from the engine compartment.
13. Remove the stabilizer and strut bar from the lower control arm.
14. Remove the left and right halfshafts.
15. Support the transaxle with a suitable floor jack, taking care to avoid damaging the pan.
16. Remove the bell housing cover.
17. Remove the remaining transaxle-to-engine bolts.
18. Remove the transaxle mounting bolt.
19. Lower the jack and slide the transaxle from under the car.

To install:

20. Secure the transaxle on a transaxle jack and position it to the engine.
21. Carefully guide the transaxle input shaft into the clutch assembly. Make sure the transaxle is seated properly and is flush to the engine flange. Install 2 transaxle-to-engine bolts.
22. Install the transaxle mounting bolt. Torque the bolt to 29–36 ft. lbs. (40–50 Nm).
23. Install the lower transaxle-to-engine bolts. Torque the bolts to 32–39 ft. lbs. (43–53 Nm).
24. Remove the floor jack from the transaxle.
25. Install the left and right halfshafts.
26. Install the stabilizer and strut bar to the lower control arm.
27. Install the extension and shift rod.
28. Refill the transaxle with an approved gear oil.
29. Install the front wheels and lower the vehicle.
30. Connect all wiring to the transaxle.
31. Install the starter and mounting bolts. Torque the mounting bolts to 20–24 ft. lbs. (27–34 Nm).
32. Connect the select control lever and switch harness.
33. Install the upper engine-to-transaxle bolts. Torque the bolts to 32–39 ft. lbs. (43–53 Nm).
34. Connect the clutch cable, speedometer cable and backup light wiring to the transaxle.
35. Install the air cleaner.
36. Install the coolant reservoir and refill with coolant.
37. Install the battery tray and battery. Connect the battery cables, positive cable first.

4WD Colt Wagon

1. Disconnect the negative battery cable, drain the transaxle fluid and remove the under cover.
2. Remove the air cleaner, battery and battery tray.
3. Remove the split pin and disconnect the shift and select cables.
4. Disconnect the clutch cylinder line, backup light switch and speedometer cable.
5. Disconnect the stabilizer bar, tie rod ends, lower ball joint and driveshaft from the control arm and strut.
6. Remove the starter motor.
7. Place a jack under the transaxle and support the engine from above.
8. Remove the center crossmember with roll stopper bracket.
9. Remove the rear driveshaft and transfer case.
10. Remove the transaxle mount bracket and bell housing bolts.
11. Using the jack, lower the transaxle and check for interference.

To install:

12. Using the jack, raise the transaxle and check for interference.
13. Install the transaxle mount bracket and bell housing bolts. Torque the bolts to 35 ft. lbs. (47 Nm).
14. Install the rear driveshaft and transfer case. Torque the transfer bolts to 30 ft. lbs. (41 Nm).
15. Install the center crossmember with roll stopper bracket.
16. Remove the jack and engine support.
17. Install the starter motor and torque the bolts to 35 ft. lbs. (41 Nm).
18. Connect the stabilizer bar, tie rod ends, lower ball joint and driveshaft to the control arm and strut.
19. Connect the clutch cylinder line, backup light switch and speedometer cable.
20. Install the split pin and connect the shift and select cables.
21. Install the air cleaner, battery and battery tray.
22. Connect the negative battery cable, refill the transaxle fluid and install the under cover.

4WD Colt Vista

1. Disconnect the battery cables, negative cable first. Remove the battery.
2. Remove the coolant reserve tank.
3. Disconnect the speedometer cable, shift control cable and back-up light harness at the transaxle.
4. Remove the range select control valves and connectors.
5. Tag and disconnect all other wiring attached to the transaxle.
6. Remove the clutch slave cylinder.
7. Remove the vacuum reservoir tank.
8. Disconnect the starter wiring.
9. Remove the upper engine-to-transaxle bolts.
10. Raise and support the vehicle safely.
11. Remove the front wheels, lower engine cover and skid plate.
12. Drain the transaxle and transfer case.
13. Remove the driveshaft.
14. Remove the transfer case extension housing.
15. Remove the left and right halfshafts.
16. Disconnect the right strut from the lower arm.
17. Remove the right fender liner.
18. Take up the weight of the transaxle with a suitable floor jack.
19. Remove the bell housing cover bolts and remove the cover.
20. Remove the remaining engine-to-transaxle bolts.
21. Remove the transaxle mount insulator bolt.
22. Remove the transaxle mounting bracket attaching bolts.
23. Move the transaxle/transfer case assembly to the right. Tilt the right side of the transaxle down, until the transfer case is about level with the upper part of the steering rack tube, then turn it to the left and lower the assembly.

To install:

24. Secure the transaxle/transfer case assembly to a transaxle jack.
25. Raise the assembly in position to the engine. It may be necessary to tilt or angle the assembly in and around the steering rack tube.
26. Once the transaxle/transfer assembly is positioned to the engine, carefully guide the input shaft into the clutch assembly.
27. Install the transaxle mounting bracket attaching bolts. Torque the bolts to 40–43 ft. lbs. (55–60 Nm).
28. Install the transaxle mount insulator bolt. Torque to 40–43 ft. lbs. (55–60 Nm).
29. Install the lower engine-to-transaxle bolts. Torque to 31–40 ft. lbs. (43–55 Nm).
30. Install the bell housing cover bolts and install the cover.
31. Remove the the floor jack from the transaxle.
32. Install the right fender liner.
33. Connect the right strut to the lower arm.
34. Install the left and right halfshafts.
35. Install the transfer case extension housing and the driveshaft.
36. Refill the transaxle and transfer case with an approved gear oil.
37. Install the front wheels, lower engine cover and skid plate.
38. Lower the vehicle.

39. Install the upper engine-to-transaxle bolts. Torque the bolts to 40–43 ft. lbs. (55–60 Nm).
40. Install the starter motor and mounting bolts. Torque the bolts to 22–25 ft. lbs. (30–35 Nm). Connect the starter wiring.
41. Install the vacuum reservoir tank.
42. Install the clutch slave cylinder.
43. Connect all other wiring to the transaxle, as tagged.
44. Install the range select control valves and connectors.
45. Connect the speedometer cable, shift control cable and back-up light harness at the transaxle.
46. Remove the coolant reserve tank.
47. Install the battery. Connect the battery cables, positive cable first.

LINKAGE ADJUSTMENT

1. Disconnect the negative battery cable.
2. Remove the console assembly from the vehicle.
3. At the shift selector assembly, remove the shift cable-to-shift selector cotter pins and disconnect the shift cables from the shift selector assembly.
4. Move the transaxle shift selector and shift selector into the **N** positions.
5. If necessary, turn the shift cable adjuster to adjust the cables length to align with shift lever in the **N** position. Connect the shift cable, flange side of the resin bushing should face cotter pin side of the shift lever, with lever Band install a new cotter pin.
6. At the shift selector, make sure dimensions A and B are equal; if they are not, turn the cable adjuster to make the necessary adjustment.
7. Adjust the other cable in the same fashion.
8. Move the shift selector into each position to make sure it is shifting smoothly.

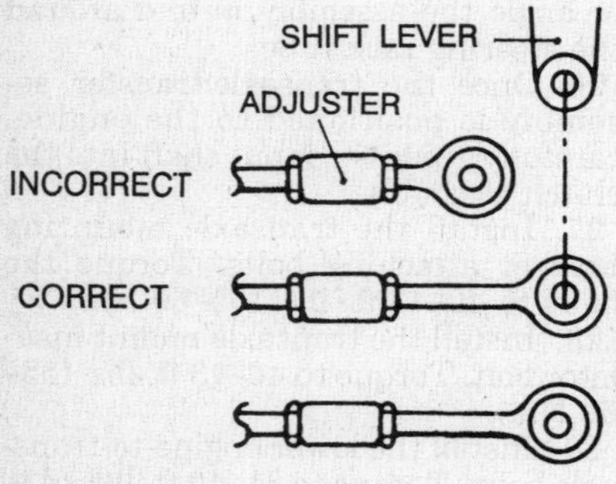

Shift cable position in neutral

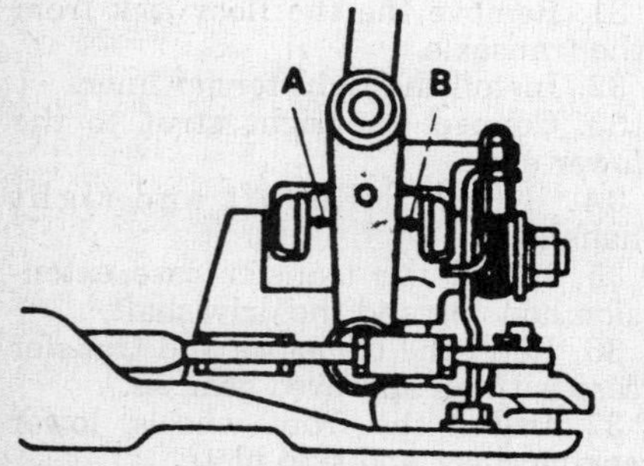

Adjusting the shift cable to make the shift selector's dimension A and B equal

CLUTCH

Clutch Assembly

REMOVAL & INSTALLATION

1. Disconnect the negative battery cable.
2. Raise and support the vehicle safely. On the 4WD Colt Vista, remove the slave cylinder.
3. Remove the transmission or transaxle from the vehicle.
4. Insert a pilot shaft or an old input shaft into the center of the clutch disc, pressure plate and the pilot bearing in the crankshaft.
5. With the pilot tool supporting the clutch disc, loosen the pressure plate bolts gradually and in a criss-cross pattern.
6. Remove the pressure plate and clutch disc.
7. Clean the transmission or transaxle and clutch housing. Clean the flywheel surface with a non-oil based solvent.

NOTE: Before assembly, slide the clutch disc up and down on the input shaft to check for any binding. Remove any rough spots with crocus cloth and then lightly coat the shaft with Lubriplate® or equivalent.

8. To remove the throwout bearing assembly
9. Remove the return clip and take out the throwout bearing carrier and the bearing.
10. To replace the throwout arm use a $^{3}/_{16}$ in. punch, knock out the throwout shaft spring pin and remove the shaft, springs and the center lever.
11. Do not immerse the throwout bearing in solvent; it is permanently lubricated. Blow and wipe it clean. Check the bearing for wear, deterioration, or burning. Replace the bearing if there is any question about its condition.
12. Check the shafts, lever and springs for wear and defects. Replace them if necessary.
13. Examine the clutch disc for the following before reusing it: loose rivets, burned facing, oil or grease on the facing, less than 0.012 in. left between the rivet head and the top of the facing.
14. Check the pressure plate and replace it if any of the following conditions exist: scored or excessive wear, bent or distorted diaphragm, loose rivets.

To install:

15. Insert the control lever into the clutch housing. Install the 2 return springs and the throwout shaft.
16. Lock the shift lever to the shaft with the spring pin.
17. Fill the shaft oil seal with multipurpose grease.
18. Install the throwout bearing carrier and the bearing. Install the return clip.
19. Grease the carrier groove and inner surface.
20. Lightly grease the clutch disc splines.

NOTE: The clutch is installed with the larger boss facing the transmission or transaxle.

21. Support the clutch disc and pressure plate with the pilot tool.
22. Turn the pressure plate so its balance mark aligns with the notch in the flywheel.
23. Install the pressure plate-to-flywheel bolts hand tight. Using a torque wrench and, working in a criss-cross pattern, tighten the bolts to 11–15 ft. lbs. (15–20 Nm).
24. Install the transmission or transaxle.
25. Adjust the clutch free-play.
26. Lower the vehicle and connect the negative battery cable.

PEDAL HEIGHT/FREE-PLAY ADJUSTMENT

1. Measure the distance between the floor and the top of the clutch pedal.
2. The measurement should be as follows:
Colt, Colt Wagon and 2WD Colt Vista – 6.20–6.40 in.
4WD Colt Vista – 7.10–1.30 in.
Conquest – 7.4–7.6 in.
3. If the measurement is not correct, loosen the clutch switch locknut and move the switch in or out as necessary to obtain proper height.
4. Check to see if the clutch pedal free-play is within specifications. Measure the distance from the top of the pedal to the distance the pedal moves before resistance is felt. The distance should be as follows:

1988 Colt with cable type – 0.80–1.20 in.

1988 Colt with hydraulic type – 0.40–1.20 in.

1989–92 Colt, Colt Wagon and Colt Vista – 0.24–0.51 in.

Conquest – 0.20–0.50 in.

5. Adjust the Colt with cable type by turning the outer cable adjusting nut at the bulkhead in the engine compartment. The hydraulic systems are self-adjusting. If the clutch pedal free height is not within specifications, there is probably air in the hydraulic system or a malfunction in the clutch itself. Bleed the air out of the system or diassemble the clutch assembly.

Clutch Cable

ADJUSTMENT

1. Slightly pull the cable out from the firewall.
2. Turn the adjusting wheel on the cable until the play between the wheel and the cable is within the correct dimension.
3. Check the clutch free-play.
   a. Raise and and support the vehicle safely.
   b. Remove the rubber cover from the clutch housing.
   c. Using a 0.030 in. feeler gauge, check the clearance between the pressure plate diaphragm spring and the throwout bearing.
4. If the free travel is not correct, make further adjustments at the cable adjusting wheel.

NOTE: Each turn of the adjusting wheel equals 0.060 in. of adjustment to the wheel and retainer clearance.

5. Lower the vehicle and check the clutch operation.

REMOVAL & INSTALLATION

1. Loosen the cable adjusting wheel inside the engine compartment.
2. Loosen the clutch pedal adjusting bolt locknut and loosen the adjusting bolt.
3. Remove the cable end from the clutch throwout lever.
4. Remove the cable end from the clutch pedal.
5. Installation is the reverse order of removal procedures.

NOTE: Lubricate the cable with engine oil and after installation, install pads isolating the cable from the intake manifold and from the rear side of the engine mount insulator.

Clutch Master Cylinder

REMOVAL & INSTALLATION

1. Loosen the bleeder screw on the slave cylinder and drain the system.
2. Disconnect the pushrod from the clutch pedal.
3. Disconnect the clutch pedal from the pedal bracket.
4. Disconnect the fluid line from the master cylinder.
5. Unbolt and remove the master cylinder and remove.
6. Installation is the reverse order of removal procedures.
7. Install the master cylinder and bleed the system.

NOTE: On the 4WD Colt Vista, the lower master cylinder mounting nut is accessed from inside the vehicle.

Clutch Slave Cylinder

REMOVAL & INSTALLATION

1. Raise and support the vehicle safely.
2. Disconnect the clutch hose from the slave cylinder.
3. Unbolt and remove the cylinder from the clutch housing.
4. Installation is the reverse order of removal procedures. Bleed the system.

Hydraulic Clutch System Bleeding

NOTE: An assistant is needed for the bleeding operation.

1. Raise and support the vehicle safely.
2. Loosen the bleeder screw at the slave cylinder.
3. Make sure the master cylinder is full.
4. Attach a length of rubber hose to the bleeder screw nipple and place the other end in a glass jar half full of clean brake fluid.
5. Have the assistant push the clutch pedal down slowly to the floor. If air is in the system, bubbles will appear in the jar as the pedal is being depressed.
6. When the pedal is at the floor, tighten the bleeder screw.
7. Repeat Steps 5 and 6 until no bubbles are found. Check the master cylinder level frequently to make sure of fluid level.

AUTOMATIC TRANSMISSION

For further information on transmissions/transaxles, please refer to "Chilton's Guide to Transmission Repair".

Transmission Assembly

REMOVAL & INSTALLATION

Conquest

1. Disconnect the negative battery cable.
2. Raise the vehicle and support it safely.
3. Drain the fluid.
4. Remove the dipstick and unbolt the filler tube.
5. Disconnect the control harness, oil feed, oil return tubes, shift control rod and ground cable.
6. Remove the bell housing cover and torque converter bolts.
7. Mark and remove the driveshaft.
8. Place a transmission jack under the transmission oil pan.
9. Remove the rear engine support crossmember and bell housing bolts.
10. Remove the transmission slowly checking for interference during removal.

To install:

11. Check the distance between the bell housing and the torque converter mounting area with a straight edge. The distance should be more than 1.03 in. (26mm). If not, the torque converter is not engaged into the fluid pump.
12. Install the transmission slowly checking for interference during installation. Torque the bellhousing bolts to 25 ft. lbs. (34 Nm).
13. Install the rear engine support crossmember. Lower the transmission onto the rear mount and install bolts.
14. Remove the jack.
15. Install the driveshaft and torque the bolts to 40 ft. lbs. (54 Nm).
16. Install the torque converter bolts and bell housing cover. Use Locktite® or equivalent on the bolts before torquing to 25 ft. lbs. (34 Nm).
17. Connect the control harness, oil feed, oil return tubes, shift control rod and ground cable.
18. Install the dipstick and filler tube.
19. Refill with new transmission fluid (Dexron II).
20. Lower the vehicle.

21. Connect the negative battery cable and check for leaks.

THROTTLE LINKAGE ADJUSTMENT

1. Apply chassis lube to all sliding parts.
2. Place the selector in the **N** position.
3. Turn the adjusting cam until the distance between the adjusting cam and the selector lever end is 15–16mm.

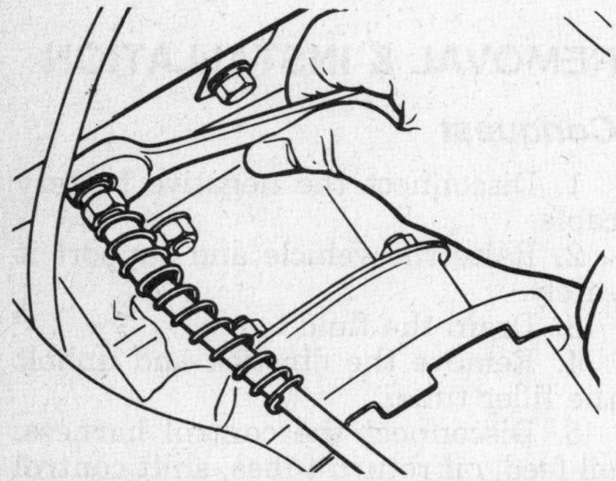

Throttle linkage adjustment (typical)

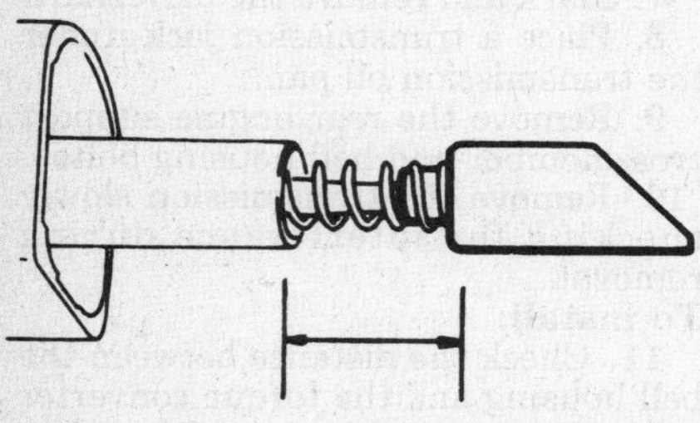

JM600 throttle rod adjustment point

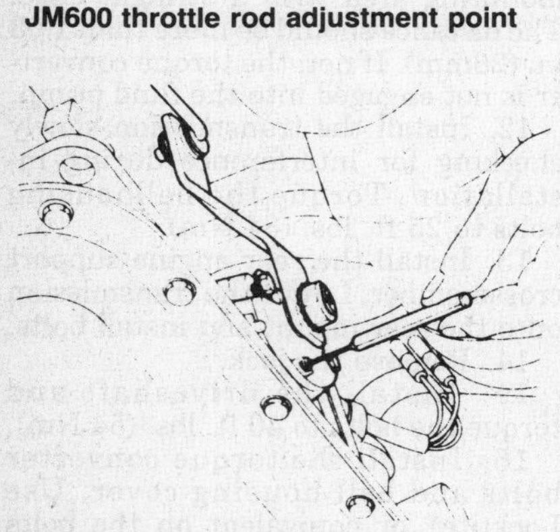

Inhibitor switch adjustment—Conquest

INHIBITOR SWITCH ADJUSTMENT

1. Place the manual valve in the neutral (vertical) position.
2. Remove the lower retaining screw and loosen the retaining bolts.
3. Using an aligning pin 0.079 inch (2.0mm) in diameter, insert the pin into the hole until the pin falls to the bottom of the hole.
4. Torque the retaining bolts to 4–5 ft. lbs. (5–7 Nm).

Measured depth "L" mm (in.)	Rod length mm (in.)	Part number
Under 25.55 (1.0059)	29.0 (1.142)	MD610614
25.65 – 26.05 (1.0098 – 1.0256)	29.5 (1.161)	MD610615
26.15 – 26.55 (1.0295 – 1.0453)	30.0 (1.181)	MD610616
26.65 – 27.05 (1.0492 – 1.0650)	30.5 (1.201)	MD610617
Over 27.15 (1.0689)	31.0 (1.220)	MD610618

Vacuum diaphragm rod selection chart

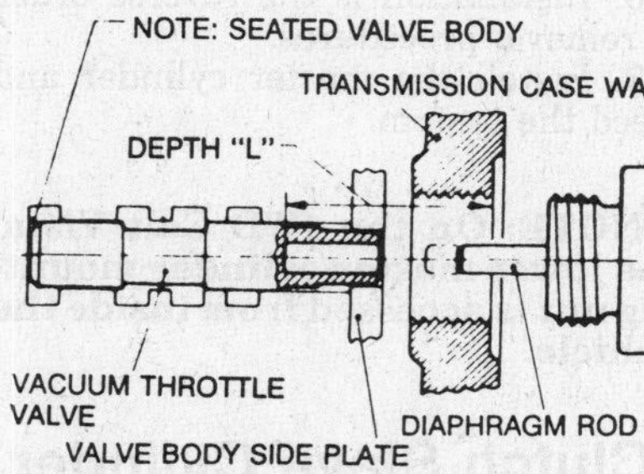

Vacuum diaphragm rod adjustment

KICKDOWN SWITCH ADJUSTMENT

The switch is located at the upper post of the accelerator pedal, inside the vehicle. When the pedal is fully depressed, a click can be heard just before the pedal bottoms out. If the click is not heard, loosen the lock nut and extend the switch until the pedal liver makes contact with the switch and the switch clicks. The switch should not contact too soon.

VACUUM DIAPHRAGM ROD ADJUSTMENT

The diaphragm and length of the rod help determine the shift patterns of the transmission. It is essential that the correct length rod be installed.

1. Disconnect the vacuum hose from the diaphragm and remove the diaphragm from the transmission case.
2. Using a depth gauge, measure the depth of the rod (L measurement). Make sure the throttle valve is pushed into the valve body as far as possible.
3. Check the chart for the proper length rod.
4. Install the rod and diaphragm.

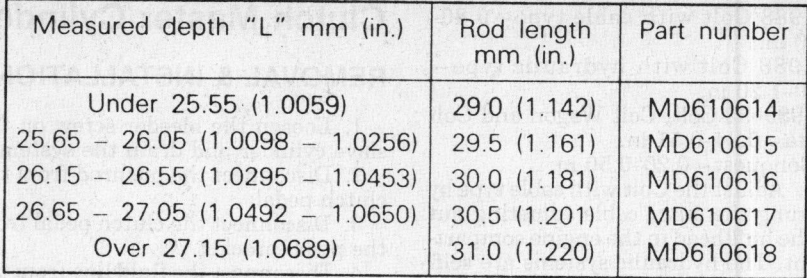

AUTOMATIC TRANSAXLE

For further information on transmissions/transaxles, please refer to "Chilton's Guide to Transmission Repair".

Transaxle Assembly

REMOVAL & INSTALLATION

NOTE: The transaxle and converter must be removed and installed as an assembly.

1. Disconnect the battery cables, negative first. Remove the battery and tray. If equipped with a turbocharger, remove the air cleaner case.
2. Disconnect the throttle control cable at the carburetor. If equipped with fuel injection, disconnect the cable at the throttle body. Disconnect the manual control cable at the transaxle.
3. Disconnect from the transaxle: the inhibitor switch (neutral safety) connecter, fluid cooler hoses and the 4 upper bolts connecting the engine to the transaxle.

NOTE: Cap oil cooler hoses to prevent fluid loss.

4. Raise and support the vehicle safely.
5. Remove the front wheels. Remove the engine splash shield.
6. Drain the transaxle fluid.
7. Disconnect the stabilizer bar at the lower arms and disconnect the control arms from the body. Remove the right and left halfshafts from the transaxle case.
8. Disconnect the speedometer cable. Remove the starter motor.
9. Remove the lower cover from the converter housing. Remove the 3 bolts connecting the converter to the engine driveplate.

NOTE: Never support the full weight of the transaxle on the engine driveplate.

10. Turn and force the converter back and away from the engine driveplate.
11. Support the weight of the engine from above. Support the transaxle and remove the remaining mounting bolts.
12. Remove the transaxle mount insulator bolt.
13. Remove and the transaxle and converter as an assembly.

To install:

14. Secure the transaxle to a transmission jack.
15. Install the torque converter onto the transaxle input shaft. Make sure the converter is fully seated in to the front pump before bolting the transaxle to the engine.
16. Raise the transaxle and position it to the engine. Install the lower transmission-to-engine bolts. Torque the bolts to 31–39 ft. lbs. (43–54 Nm).
17. Install the transaxle mount insulator bolt. Torque the bolts to 31–40 ft. lbs. (43–54 Nm).
18. Install the remaining transaxle-to-engine bolts. Torque the bolts to 31–39 ft. lbs. (43–54 Nm).
19. From the converter housing, install the 3 bolts connecting the converter to the engine driveplate. Torque the bolts to 34–38 ft. lbs. (46–54 Nm).
20. Connect the speedometer cable.
21. Install the starter motor and connect the wiring to it.
22. Connect the stabilizer bar to the lower arms and connect the control arms to the body. Install the right and left halfshafts to the transaxle case.
23. Refill the transaxle fluid.
24. Install the engine splash shield. Install the front wheels.
25. Lower the vehicle.
26. Connect to the transaxle: the inhibitor switch (neutral safety) connecter and the fluid cooler hoses.
27. Connect the throttle control cable to the transaxle.
28. Install the battery and tray. Connect the battery cables, positive first. If equipped with a turbocharger, install the air cleaner case.

SHIFT LINKAGE ADJUSTMENT

NOTE: When it is necessary to disconnect the linkage cable from the lever, which uses plastic grommets as retainers, the grommets should be replaced.

1. Set the parking brake.
2. Move the shift lever into **P**.
3. Loosen the clamp bolt on the gear shift cable bracket.
4. Make sure the preload adjustment spring engages the fork on the transaxle bracket.
5. Pull the shift lever all the way to the front detent position **P** and torque the lock screw to 100 inch lbs. (11 Nm).
6. Check the following conditions:
   a. The detent positions for **N** and **D** should be within limits of hand lever gate stops.
   b. Key start must occur only when the shift lever is in **P** or **N** positions.

THROTTLE LINKAGE ADJUSTMENT

1. Run the engine to normal operating temperature. Shut it off and make sure the throttle plate is closed (curb idle position).
2. Raise the small cone shaped cover on the throttle cable to expose the nipple.
3. Loosen the lower cable bracket bolt.
4. Move the lower cable bracket until the distance between the nipple and the lower cover directly under it is 0.02–0.06 in.
5. Tighten the bracket bolt to 9–11 ft. lbs.

INHIBITOR SWITCH ADJUSTMENT

1. Place the manual valve in the neutral position.
2. Loosen the retaining bolts.
3. Turn the inhibitor switch body until the 0.472 inch (12.0mm) wide end of the manual control lever overlaps the switch body flange.
4. Torque the retaining bolts to 4–5 ft. lbs. (5–7 Nm).

TRANSFER CASE

Transfer Case Assembly

REMOVAL & INSTALLATION

4WD Colt Vista

1. Raise the vehicle and support it safely. Remove the transaxle.
2. Unbolt the transfer case from the transaxle and using a small prybar, separate the two.

To install:

3. Installation is the reverse order of the removal procedure. Torque the attaching bolts to 40–43 ft. lbs. (54–58 Nm).

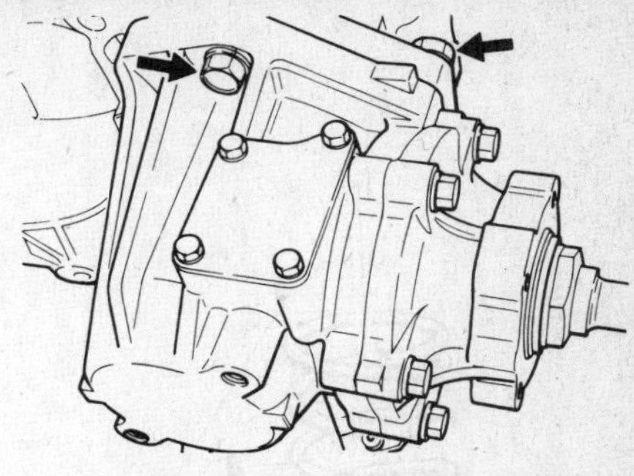

Transfer case attaching bolts

FRONT SUSPENSION

MacPherson Strut

REMOVAL & INSTALLATION

Colt and Colt Vista

1. Raise the vehicle and support it safely, with the wheels hanging.
2. Remove the front wheels.
3. Detach the brake hose from the mounting clip on the strut.
4. Remove the nuts securing the strut to the fender housing.
5. Unbolt the strut from the knuckle.

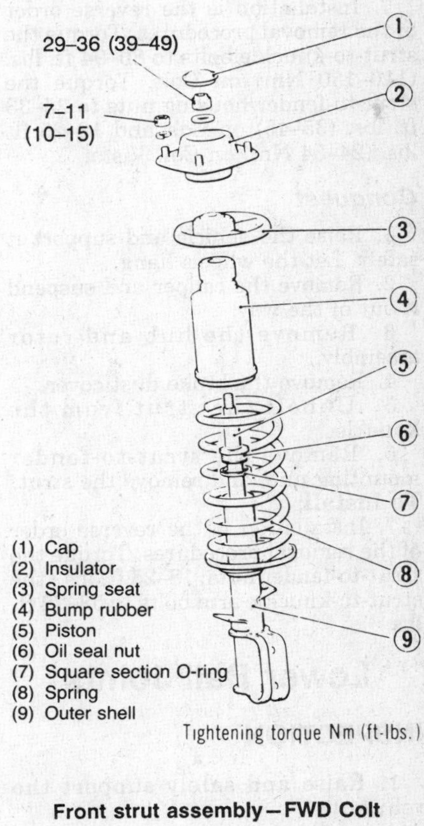

Front strut assembly–FWD Colt

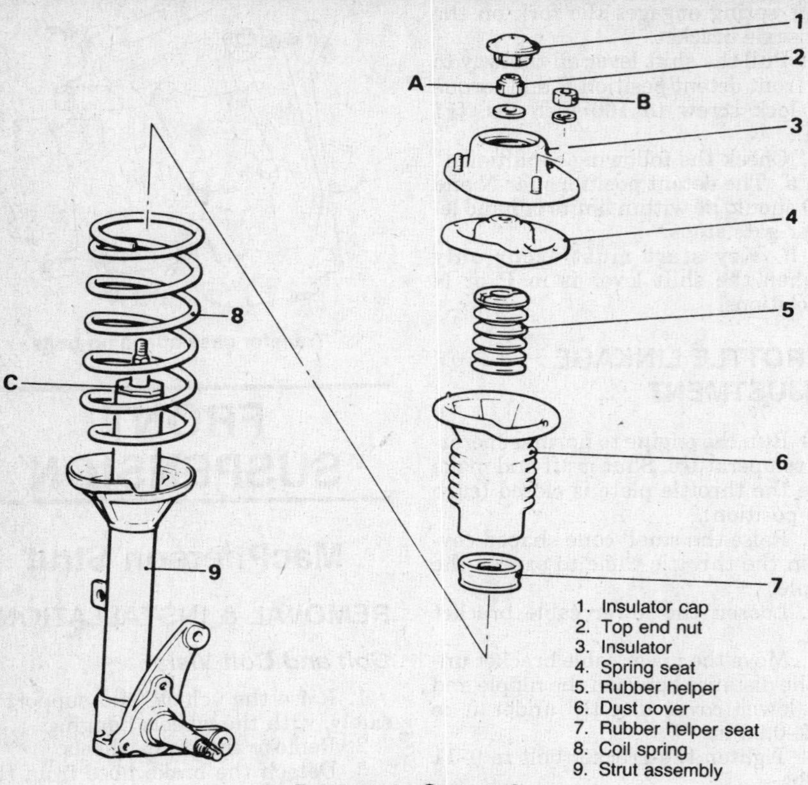

Front strut—Conquest

6. Remove the strut from the vehicle.

To install:

7. Installation is the reverse order of the removal procedures. Torque the strut-to-knuckle bolts to 80–94 ft. lbs. (110–130 Nm) on Colt. Torque the strut-to-fender housing nuts to 25–33 ft. lbs. (35–45) on Colt and 18–25 ft. lbs. (24–34 Nm) on Colt Vista.

Conquest

1. Raise the vehicle and support it safely. Let the wheels hang.
2. Remove the caliper and suspend it out of the way.
3. Remove the hub and rotor assembly.
4. Remove the brake dust cover.
5. Unbolt the strut from the knuckle.
6. Remove the strut-to-fender mounting nuts and remove the strut.

To install:

7. Installation is the reverse order of the removal procedures. Torque the strut-to-fender nuts; 18–25 ft. lbs.; the strut-to-knuckle arm bolts to 58–72 ft. lbs.

Lower Ball Joints

INSPECTION

1. Raise and safely support the vehicle.
2. With the ball joint installed to the steering knuckle, Grasp the top and bottom of the wheel, then, move the wheel using an in and out shaking motion.
3. Observe any movement between the steering knuckle and the control arm. If movement exists, replace the ball joint.

REMOVAL & INSTALLATION

Colt

The ball joint is not replaceable. The ball joint and lower control arm must be replaced as an assembly.

Colt Vista

1. Raise the vehicle and support it safely.
2. Remove the ball joint retaining bolt. Using a special tool, separate the ball joint from the steering knuckle. Remove the control arm from the vehicle.
3. Remove the ball joint dust cover.
4. Using snapring pliers, remove the snapring from the ball joint.
5. Using an adapter plate and driver, press the ball joint from the arm.

To install:

6. Installation is the reverse order of removal procedures. Invert the tool in the press for installation. Coat the lip and interior of the dust cover with lithium based chassis lube. Torque the ball joint nut to 43–52 ft. lbs. (60–72 Nm) and install a new cotter pin, if so equipped.

Conquest

1. Raise the vehicle and support it safely. Remove the tire and wheel assembly.
2. Remove the brake caliper and support from the mounting adapter.
3. Anchor assembly out of the way with wire to the strut spring.
4. Remove the tie rod end nut and separate the tie rod end from the steering knuckle using a removing tool.
5. Remove the bolts securing the strut assembly to the steering knuckle.
6. Tap the connection with a plastic hammer to separate.
7. Remove the ball joint to control arm mounting bolts and remove the ball joint with the knuckle arm attached.
8. Remove the ball joint stud nut and separate the ball joint and knuckle arm.

To install:

9. Installation is the reverse order of the removal procedures.

Torque specifications are:

Ball joint stud nut—43–52 ft. lbs.

Strut-to-knuckle bolts—58–78 ft. lbs.

Ball joint to control arm bolts—43–51 ft. lbs.

NOTE: When self locking nuts are removed, always replace with new self locking nuts.

Lower Control Arms

REMOVAL & INSTALLATION

Colt

1. Raise the vehicle and support it safely. Remove the wheels. Remove the splash shield. If equipped, remove the center crossmember.
2. Disconnect the stabilizer bar from the lower arm.
3. Using a ball joint separator, disconnect the ball joint from the knuckle.
4. Unbolt the lower arm from the body and remove it from the vehicle.

NOTE: The ball joint cannot be separated from the control arm, but must be replaced as an assembly.

5. If the stabilizer bar is to be removed, disconnect the tie rod from the knuckle and unbolt and remove the stabilizer.
6. Check all parts for wear and damage and replace any suspect part.
7. Using an inch lb. torque wrench, check the ball joint starting torque.

Nominal starting effort should be 22–87 inch lbs. Replace if otherwise.

To install:

8. Installation is the reverse order of the removal procedures. Use a new dust cover, the lip and inside of which is coated with lithium based chassis lube. The dust cover should be hammered into place with a driver tool.

9. Install the stabilizer bar so the serrations on the horizontal part protrude 6mm to the inside of the clamp and 23mm of threaded stud appear below the nut at the control arm.

10. The washer on the lower arm shaft should be installed. The left side lower arm shaft has left handed threads. The lower arm shaft nut must be torqued with the wheels hanging freely. Observe the following torques:

Knuckle-to-strut – 54–65 ft. lbs. 1989–90; 80–94 ft. lbs.

Lower arm shaft-to-body – 69–87 ft. lbs.

Stabilizer bar-to-body – 12–20 ft. lbs.

Ball joint-to-knuckle – 44–53 ft. lbs.

Lower arm-to-shaft – 70–88 ft. lbs.

Rear shaft bushing bracket-to-body – 43–58 ft. lbs.

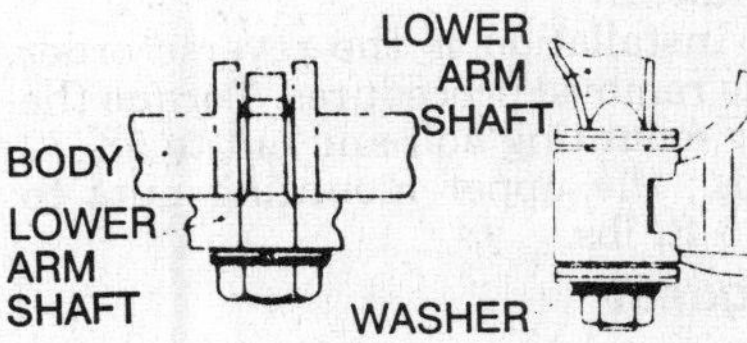

Installation of washer on lower shaft

Colt Vista

1. Raise the vehicle and support it safely.
2. Remove the wheels.
3. Disconnect the stabilizer bar and strut bar from the lower arm.
4. Remove the nut and disconnect the ball joint from the knuckle with a separator.
5. Unbolt the lower arm from the crossmember.
6. Check all parts for wear or damage and replace any suspect part.
7. Using an inch lb. torque wrench, check the ball joint starting torque. Starting torque should be 20–86 inch lbs. If it is not within that range, replace the ball joint.

To install:

8. Installation is the reverse order of the removal procedures. Tighten all fasteners with the wheels hanging freely. Observe the following torques:

Ball joint-to-knuckle – 44–53 ft. lbs.

Arm-to-crossmember-2WD to 90–111 ft. lbs.; 4WD to 58–68 ft. lbs.

Stabilizer bar hanger brackets – 7–9 ft. lbs.

9. When installing the stabilizer bar, the nut on the bar-to-crossmember bolts and the bar-to-lower arm bolts, are not torqued, but turned on until a certain length of thread is exposed above the nut:

2WD stabilizer bar-to-crossmember – 0.31–0.39 in.

2WD stabilizer bar-to-lower control arm – 0.31–0.39 in.

4WD stabilizer bar-to-crossmember – 0.31–0.39 in.

4WD stabilizer bar-to-lower arm – 0.51–0.59 in..

Conquest

1. Raise the vehicle and support it safely.
2. Remove the front wheels.
3. Remove the caliper and suspend it out of the way.
4. Remove the hub and rotor assembly.
5. Disconnect the stabilizer bar and strut bar from the lower arm.
6. Using a separator, remove the attaching nut and disconnect the tie rod from the knuckle arm.
7. Unbolt the strut from the knuckle arm.
8. Unbolt the lower control arm and knuckle arm assembly from the crossmember.
9. Using a special tool, separate the knuckle arm from the lower control arm.

To install:

10. Installation is the reverse order of the removal procedures. Apply sealant to the flange of the knuckle arm where it mates with the strut. Torque the lower control arm shaft bolt to 60–70 ft. lbs.; the ball joint-to-knuckle arm nut to 45–55 ft. lbs.

Sway Bar

REMOVAL & INSTALLATION

1. Raise and safely support the vehicle; allow the suspension to hang free. Remove the left front wheel assembly.
2. Disconnect the stabilizer link bolts and nuts from the control arms. Disconnect the stabilizer shaft from the support assemblies.
3. Loosen the front bolts and remove the bolts from the rear and center of the support assemblies, allowing the supports to be lowered enough to remove the stabilizer bar assembly. Remove the assembly from the vehicle.

To install:

4. Installation is the reverse order of the removal procedures. Loosely assemble all components while insuring that the stabilizer bar is centered, side-to-side. Torque the stabilizer bar support assemblies to 9 ft. lbs. (13 Nm).
5. Lower the vehicle.

Front Wheel Bearings

NOTE: Please refer to the "Drive Axle" Section for FWD and 4WD vehicles.

ADJUSTMENT

Conquest

1. Raise the vehicle and support it safely. Remove the wheel and dust cover. Remove the cotter pin and lock cap from the nut.
2. Torque the wheel bearing nut to 14.5 ft. lbs. (19.6 Nm) and then loosen the nut. Retorque the nut to 3.6 ft. lbs. (4.9 Nm) and install the lock cap and cotter pin.
3. Install the dust cover and the wheel.

REMOVAL & INSTALLATION

Conquest

1. Remove the caliper (pin type) or the caliper and support (sliding type).

NOTE: On sliding type calipers, remove the caliper and support as a unit by unfastening the bolts holding it to the adapter. Support the caliper with wire, do not allow the weight to be supported by the brake hose.

2. Pry off the dust cap. Tap out and discard the cotter pin. Remove the locknut.
3. Being careful not to drop the outer bearing, pull off the brake disc and wheel hub.
4. Remove the grease inside the wheel hub.
5. Using a brass drift, carefully drive the outer bearing race out of the hub.
6. Remove the inner bearing seal and bearing.
7. Check the bearings for wear or damage and replace them, if necessary.

To install:

8. Coat the inner surface of the hub with grease.
9. Grease the outer surface of the bearing race and drift it into place in the hub.
10. Pack the inner and outer wheel bearings with grease.

NOTE: If the brake disc has been removed and/or replaced;

tighten the retaining bolts to 25–29 ft. lbs.

11. Install the inner bearing in the hub. Being careful not to distort it, install the oil seal with its lip facing the bearing. Drive the seal on until its outer edge is even with the edge of the hub.
12. Install the hub/disc assembly on the spindle, being careful not to damage the oil seal.
13. Install the outer bearing, washer and spindle nut. Adjust the wheel bearing.

REAR SUSPENSION

Shock Absorbers

REMOVAL & INSTALLATION

Except 4WD Colt Vista

1. Raise the vehicle and support it safely.
2. Remove the wheel. Remove the upper mounting bolt and nut.
3. While holding the bottom stud mount nut with one wrench, remove the locknut with another wrench.
4. Remove the shock absorber.
5. Check the shock for:
   a. Excessive oil leakage; some minor weeping is permissible.
   b. Bent center rod, damaged outer case, or other defects.
   c. Pump the shock absorber several times, if it offers even resistance on full strokes it may be considered serviceable.

To install:

6. Install the upper shock mounting nut and bolt. Hand tighten the nut.
7. Install the bottom eye of the shock over the spring stud. Tighten the lower nut to 12–15 ft. lbs. on rear wheel drive vehicles; 47–58 ft. lbs. on front wheel drive vehicles.
8. Finally, tighten the upper nut to 47–58 ft. lbs. on all models except station wagons, which are tightened to 12–15 ft. lbs.

4WD Colt Vista

1. Raise the vehicle and support it safely.

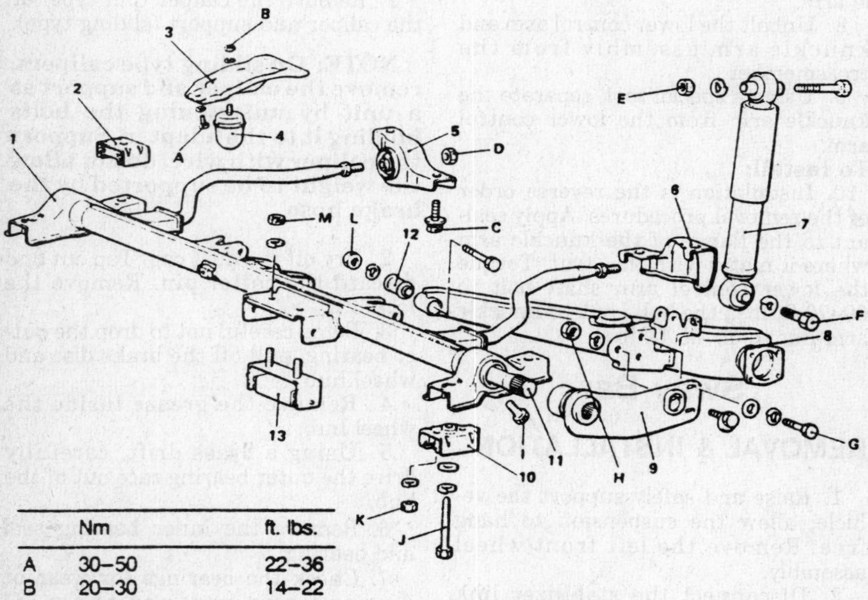

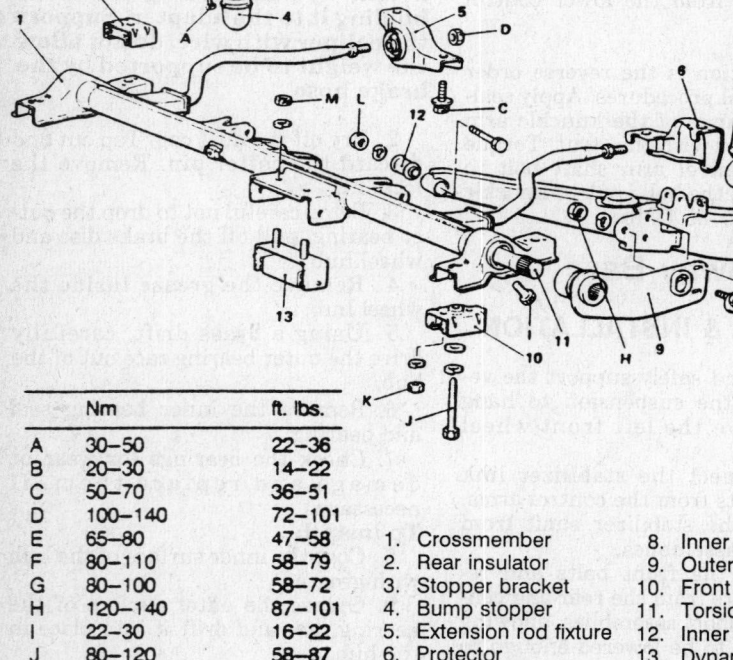

	Nm	ft. lbs.
A	30–50	22–36
B	20–30	14–22
C	50–70	36–51
D	100–140	72–101
E	65–80	47–58
F	80–110	58–79
G	80–100	58–72
H	120–140	87–101
I	22–30	16–22
J	80–120	58–87
K	10–15	7–10
L	70–90	51–65
M	19–28	14–20

1. Crossmember
2. Rear insulator
3. Stopper bracket
4. Bump stopper
5. Extension rod fixture
6. Protector
7. Shock absorber
8. Inner arm
9. Outer arm
10. Front insulator
11. Torsion bar
12. Inner arm bushing
13. Dynamic damper

4wd Vista rear suspension

2. Remove the rear wheels.
3. Using a suitable floor jack, raise the inner control arm slightly.
4. Unbolt the top, then the bottom of the shock absorber. Remove it from the vehicle.
5. Installation is the reverse order of the removal procedures. Torque the top nut to 55–58 ft. lbs.; the bottom bolt to 75–80 ft. lbs.

MacPherson Strut

REMOVAL & INSTALLATION

Colt

1. Raise the vehicle and support it safely. Allow the lower arms and suspension to hang. Remove the wheels.
2. Raise the axle slightly to relax the strut and to support the axle when the strut is removed. Position an additional support under the axle.
3. Take care in jacking that no contact is made on the lateral rod.
4. On hatchback, remove the trunk side trim.
5. Remove the upper dust cover cap.
6. Remove the upper mounting nuts. Remove the lower mounting bolt and nut.
7. Remove the strut.

To install:

8. Installation is the reverse order of the removal procedures. Torque the lower mounting bolt and nut to 58–72 ft. lbs.; the upper mounting nuts to 18–25 ft. lbs.

Conquest

1. Raise the vehicle and support it safely.
2. Remove the rear wheels.
3. Unclip the brake hose at the strut.
4. Unbolt the intermediate shaft from the companion flange.
5. Unbolt the strut assembly from the halfshaft housing. Remove the housing coupling bolts. Separate the strut from the housing by pushing the housing downward while prying open the coupling on the housing.
6. Remove the strut upper end attaching nuts, found under the side trim in the cargo area.
7. Lift out the strut.

To install:

8. Installation is the reverse order of the removal procedures. Torque the upper end nuts to 20–25 ft. lbs.; the strut-to-housing bolts to 50 ft. lbs.; the coupling bolt to 50 ft. lbs.

Coil Springs

REMOVAL & INSTALLATION

Except 4WD Colt Vista

1. Raise the vehicle and support it

safely. Allow the rear wheels to hang.

2. Place a suitable jack under the rear axle and remove the bottom bolts or nuts of the shock absorbers.

3. Lower the rear axle and remove the left and right coil springs.

4. Installation is the reverse order of the removal procedure.

NOTE: When installing the spring, pay attention to the difference in shape between the upper and lower spring seats.

Torsion Bar and Control Arms

Instead of springs, the 4WD Colt Vista uses transversely mounted torsion bars housed inside the rear crossmember, attached to which are inner and outer control arms. The conventional style shock absorbers are mounted on the inner arms.

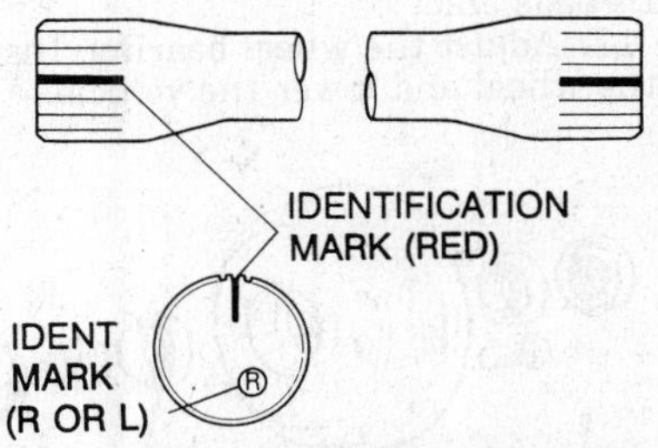

Torsion bar suspension identifying marks

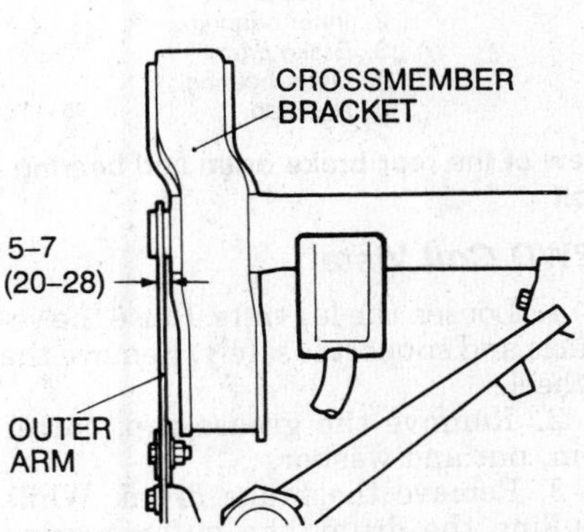

Outer arm-to-crossmember spacing

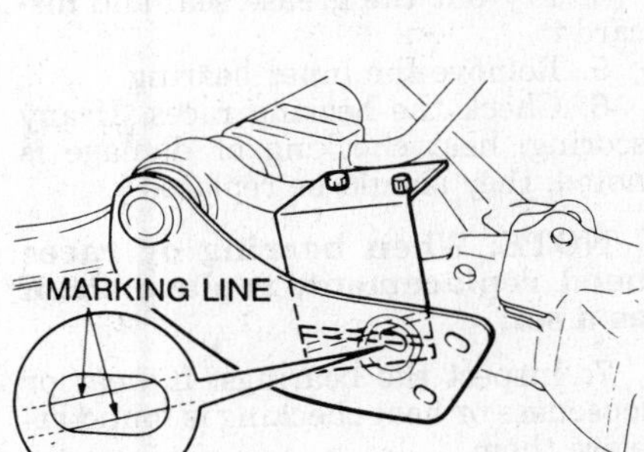

Final alignment of the outer control arm

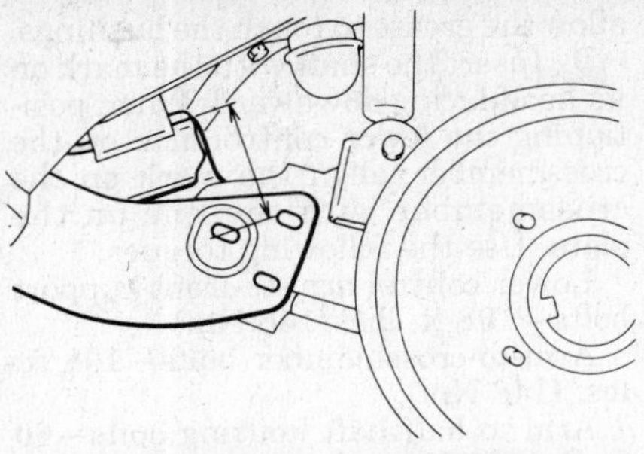

Ride height adjustment point—4WD Colt Vista

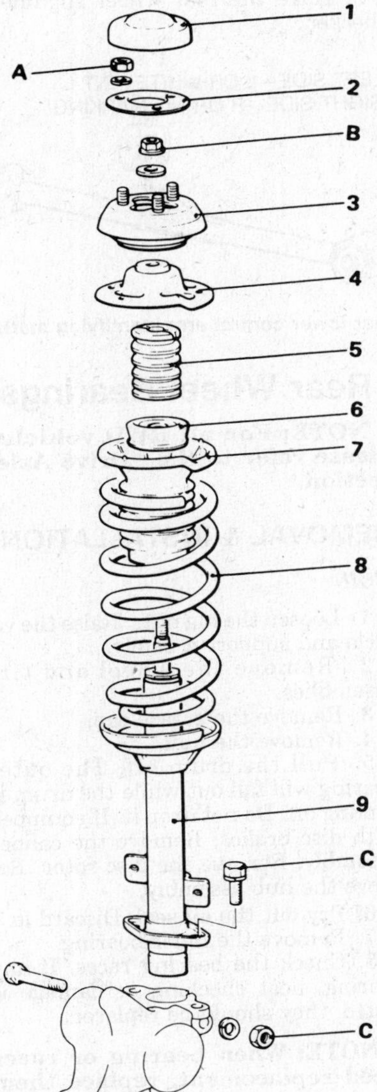

1. Strut house cap
2. Gasket
3. Strut insulator
4. Spring seat
5. Rubber helper
6. Rubber helper seat
7. Dust cover
8. Coil spring
9. Strut

Rear strut—Conquest

4WD Colt Vista

1. Raise the vehicle and support it safely.

2. Remove the differential.

3. Remove the intermediate shafts and halfshafts.

4. Remove the rear brake assemblies.

5. Disconnect the brake lines and parking brake cables from the inner arms.

6. Remove the main muffler.

7. Raise the inner arms slightly with a suitable floor jack and disconnect the shock absorbers.

8. Matchmark, precisely, the upper ends of the outer arms, the torsion bar ends and the top of the crossmember bracket and remove the inner and outer arm attaching bolts.

9. Remove the extension rods fixture attaching bolts.

10. Remove the crossmember attaching bolts and remove the rear suspension assembly from the vehicle.

11. Unbolt and remove the shock from the crossmember.

12. Remove the front and rear insulators from both ends of the crossmember.

13. Loosen but do not remove, the lock bolts securing the outer arm bushings at both ends of the crossmember.

14. Pull the outer arm from the crossmember. The torsion bar will slide out of the crossmember with the outer arm.

15. Remove the torsion bar from either the crossmember or outer arm.

16. Inspect all parts for wear or damage. Inspect the crossmember for bending or deformation.

17. Inner arm bushings may be replaced at this time using a press. The thicker end of the bushing goes on the inner side.

To install:

18. Prior to installation note that the torsion bars are marked with an L or R on the outer end and are not interchangeable.

19. If the original torsion bars are being installed, align the identification marks on the torsion bar end, crossmember and outer arm, install the torsion bar and arm and tighten the lock bolts. Skip Step 20. If new torsion bars are being installed, proceed to Step 20.

20. A special alignment jig must be fabricated. The jig is bolted to the rear insulator hole on the crossmember bracket. Insert the torsion bar into the outer arm, aligning the red identification mark on the torsion bar end with the matchmark made on the outer arm top side. Install the torsion bar and arm so the center of the flanged bolt hole on the arm is 32mm below the lower marking line on the jig. Then,

pull the outer arm off of the torsion bar, leaving the bar undisturbed in the crossmember. Reposition the arm on the torsion bar, 1 serration counterclockwise from its former position. This will make the previously measured dimension, 33mm above the lower line. When the outer arm and torsion bar are properly positioned, the marking lines on the jig will run diagonally across the center of the toe-in adjustment hole. When the adjustment is complete, tighten the lock bolts. The clearance between the outer arm and the crossmember bracket, at the torsion bar, should be 5.0–7.0mm.

21. The remainder of installation is the reverse order of the removal procedure. Observe the following torques:

Extension rod fixture bolts—45–50 ft. lbs.

Extension rod-to-fixture nut—95–100 ft. lbs.

Shock absorber lower bolt—75–80 ft. lbs.

Outer arm attaching bolts—65–70 ft. lbs.

Toe-in bolt—95–100 ft. lbs.

Lock bolts—20–22 ft. lbs.

Crossmember attaching bolts—80–85 ft. lbs.

Front insulator nuts—7–10 ft. lbs.

Inner arm-to-crossmember bolts—60–65 ft. lbs.

Damper-to-crossmember nuts—15–20 ft. lbs.

22. Lower the vehicle to the ground and check the ride height. The ride height is checked on both sides and is determined by measuring the distance between the center line of the toe-in bolt hole on the outer arm and the lower edge of the rebound bumper. The distance on each side should be 4.00–4.11 in. If not or if there is a significant difference between sides, the torsion bars positioning is wrong.

Rear Control Arms

REMOVAL & INSTALLATION

Conquest

1. Raise the vehicle and support it safely. Allow the wheels to hang freely.
2. Remove the rear wheels.
3. Disconnect the parking brake cable from the lower arm.
4. Disconnect the stabilizer bar.
5. Unbolt the lower control arm from the halfshaft housing.
6. Unbolt the lower control arm from the front support.
7. Unbolt the lower control arm from the crossmember and remove it.

To install:

8. Apply a thin coating of chassis lube to the cutout portion of the lower arm-to-halfshaft housing shaft. Do not allow the grease to touch the bushings.
9. Insert the shaft with the mark on its head facing downward. When positioning the lower control arm on the crossmember, align the mark on the crossmember with the line on the plate. Use the following torques:

Lower control arm-to-front support bolts—108 ft. lbs. (148 Nm)

Arm-to-crossmember bolts—108 ft. lbs. (148 Nm)

Arm to halfshaft housing bolts—60 ft. lbs. (81 Nm)

Arm locking pin—15 ft. lbs. (20 Nm)

9. Have the rear wheel alignment checked.

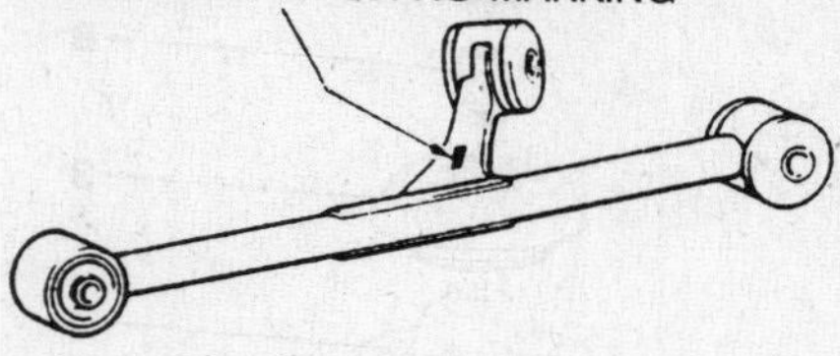

Rear lower control arm identifying marks

Rear Wheel Bearings

NOTE: For all RWD vehicles, please refer to the "Drive Axle" section.

REMOVAL & INSTALLATION

Colt

1. Loosen the lug nuts. Raise the vehicle and support it safely.
2. Remove the wheel and tire assemblies.
3. Remove the grease cap.
4. Remove the nut.
5. Pull the drum off. The outer bearing will fall out while the drum is coming off. Do not drop it. If equipped with disc brakes; Remove the caliper assembly. Remove the disc rotor. Remove the hub assembly.
6. Pry out the oil seal. Discard it.
7. Remove the inner bearing.
8. Check the bearing races. If any scoring, heat checking or damage is noted, they should be replaced.

NOTE: When bearing or races need replacement, replace them as a set.

9. Inspect the bearings. If wear or looseness or heat checking is found, replace them.
10. If the bearings and races are to be replaced, drive out the race with a brass drift.

To install:

11. Before installing new races, coat them with lithium based wheel bearing grease. Drive into place with a brass drift. Make sure they are fully seated.
12. Thoroughly pack the bearings with lithium based wheel bearing grease. Pack the hub with grease.
13. Install the inner bearing and coat the lip and rim of the grease seal with grease. Drive the seal into place with a seal driver.
14. If equipped with drum brakes, place the drum on the hub and install the outer bearing. Do not install the nut at this time.
15. If equipped with drum brakes, use a pull scale attached to one of the lugs to measure the starting force necessary to get the drum to turn. Starting force should be 5 lbs. If the starting torque is greater than specific, replace the bearings.
16. If equipped with drum brakes, install the nut on the halfshaft. Thread the nut on, to a point at which the back face of the nut is 2–3mm from the shoulder of the shaft, where the threads end.
17. Adjust the wheel bearing. Install the wheel and lower the vehicle.

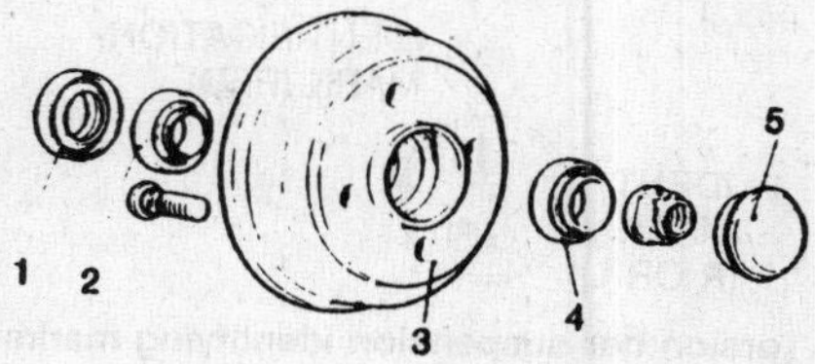

1. Oil seal
2. Inner bearing
3. Brake drum
4. Outer bearing
5. Hub cap

View of the rear brake drum and bearing—Colt

2WD Colt Vista

1. Loosen the lug nuts. Raise the vehicle and support it safely. Remove the wheel.
2. Remove the grease cap, cotter pin, nut and washer.
3. Remove the brake drum. While pulling the drum, the outer bearing will fall out. Do not drop it.
4. Pry out the grease seal and discard it.
5. Remove the inner bearing.
6. Check the bearing races. If any scoring, heat checking or damage is noted, they should be replaced.

NOTE: When bearing or races need replacement, replace them as a set.

7. Inspect the bearings. If wear or looseness or heat checking is found replace them.
8. If the bearings and races are to be replaced, drive out the races with a brass drift.

To install:

9. Before installing new races, coat them with lithium based wheel bearing grease. Drive into place with a brass drift. Make sure they are fully seated.
10. Thoroughly pack the bearings with lithium based wheel bearing grease. Pack the hub with grease.
11. Install the inner bearing and coat the lip and rim of the grease seal with grease. Drive the seal into place with a seal driver.
12. Mount the drum onto the hub, slide the outer bearing into place, install the washer and thread the nut into place.
13. Adjust the wheel bearing, install the wheel and lower the vehicle.

ADJUSTMENT

Colt

DRUM BRAKES

1. Raise the vehicle and support it safely. Remove the rear wheel and wheel bearing cap.
2. Using an inch lb. torque wrench, turn the nut counterclockwise 2–3 turns, noting the average force needed during the turning procedure. Turning torque for the nut should be about 48 inch lbs. If turning torque is not within 5 inch lbs., either way, replace the nut.
3. Tighten the nut to 72–108 ft. lbs. (100–150 Nm) on 1987–88 vehicles and 108–145 ft. lbs. (150–200 Nm) on 1989–91 vehicles.
4. Using a stand mounted gauge, check the axial play of the wheel bearings. Play should be less than 0.0079 in. If play cannot be brought within that figure, the unit is assembled incorrectly.
5. Pack the grease cap with wheel bearing grease and install it.

DISC BRAKES

Install the hub and bearing assembly. Tighten the nut to 108–145 ft. lbs. Install the rotor and caliper assembly.

2WD Colt Vista

1. Raise the vehicle and support it safely. Remove the rear wheel, wheel bearing cap and cotter pin.
2. Install a torque wrench on the nut. While turning the drum by hand, tighten the nut to 15 ft. lbs. Back off the nut until it is loose, then tighten it to 7 ft. lbs.
3. Install the lock cap and insert a new cotter pin. If the lock cap and hole don't align and repositioning the cap can't accomplish alignment, back off the nut no more than 15 degrees. If that won't align the holes either, try the adjustment procedure over again.

Rear Axle Assembly

NOTE: Please refer to the "Drive Axle" Section for RWD vehicles.

REMOVAL & INSTALLATION

Colt

1. Raise and safely support the vehicle. Remove the wheel and tire assemblies.
2. Remove brake fittings and retaining clips holding flexible brake line.
3. Remove parking brake cable adjusting connection nut.
4. Release both parking brake cables from brackets by slipping ball end of cables through brake connectors. Pull parking brake cable through bracket.
5. Pry off grease cap.
6. Remove cotter pin and castle lock.
7. Remove adjusting nut and brake drum.
8. Remove brake assembly and spindle bolts.
9. Set spindle aside and using a piece of wire, hang brake assembly aside.
10. Place supports under rear crossmember to support the rear suspension.
11. Remove shock absorber brackets.
12. Remove trailing arm-to-hanger bracket bolts.
13. Lower jack and remove axle assembly.

To Install:

14. Using a suitable jack, position the rear axle assembly under vehicle.
15. Install trailing arm-to-hanger mounting bracket, finger tighten bolts only.
16. Install shock absorber bolts loosely.
17. Place spindle and brake assembly in position; install bolts, do not tighten at this time.
18. Torque the bolts to 45 ft. lbs. (60 Nm).
19. Install brake drum. Install washer and nut. Adjust wheel bearing. Install the dust cap.
20. Put parking brake cable through the bracket.
21. Slip ball end of parking brake cables through brake connectors on parking brake bracket.
22. Install both retaining clips.
23. Install parking brake cable adjusting connection nut. Tighten until all slack is removed from cables.
24. Install retaining clips and brake tube fittings. Torque fitting to 9 ft. lbs. (12 Nm).
25. Bleed rear brake system and readjust brakes.
26. Install wheel and tire assembly.
27. With vehicle on ground, torque trailing arm-to-hanger bracket mounting bolts to 58–72 ft. lbs. (80–100 Nm).
28. Torque shock absorber mounting bolts to 58–72 ft. lbs. (80–100 Nm).

Colt Vista

1. Raise and safely support the vehicle. Remove the wheel assemblies.
2. Separate the parking brake cable at the connector and cable housing at the floor pan bracket.
3. Separate the brake line at backing plate.
4. Remove the muffler-to-middle exhaust pipe retaining bolts, remove the O-ring hangers and remove the exhaust system.
5. Remove the lower shock absorber through bolts and disconnect the shock at the axle end.
6. Lower the axle until the spring and isolator assemblies can be removed.

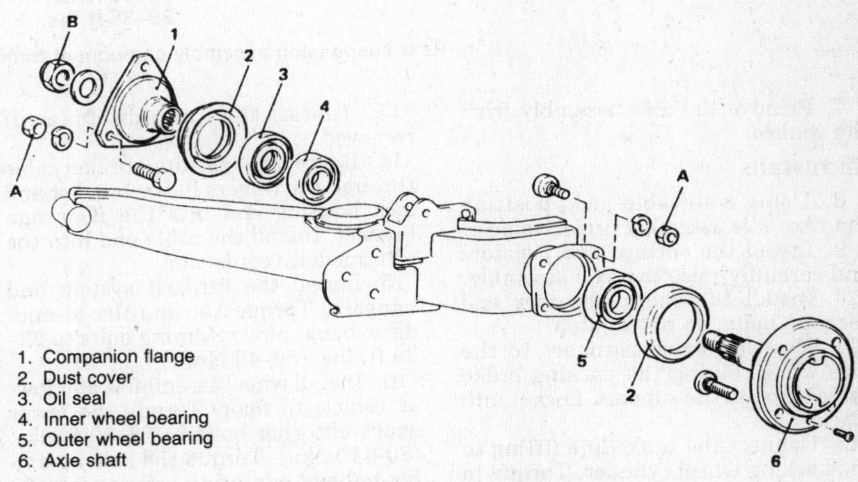

1. Companion flange
2. Dust cover
3. Oil seal
4. Inner wheel bearing
5. Outer wheel bearing
6. Axle shaft

Rear axle shaft—4WD Colt Vista

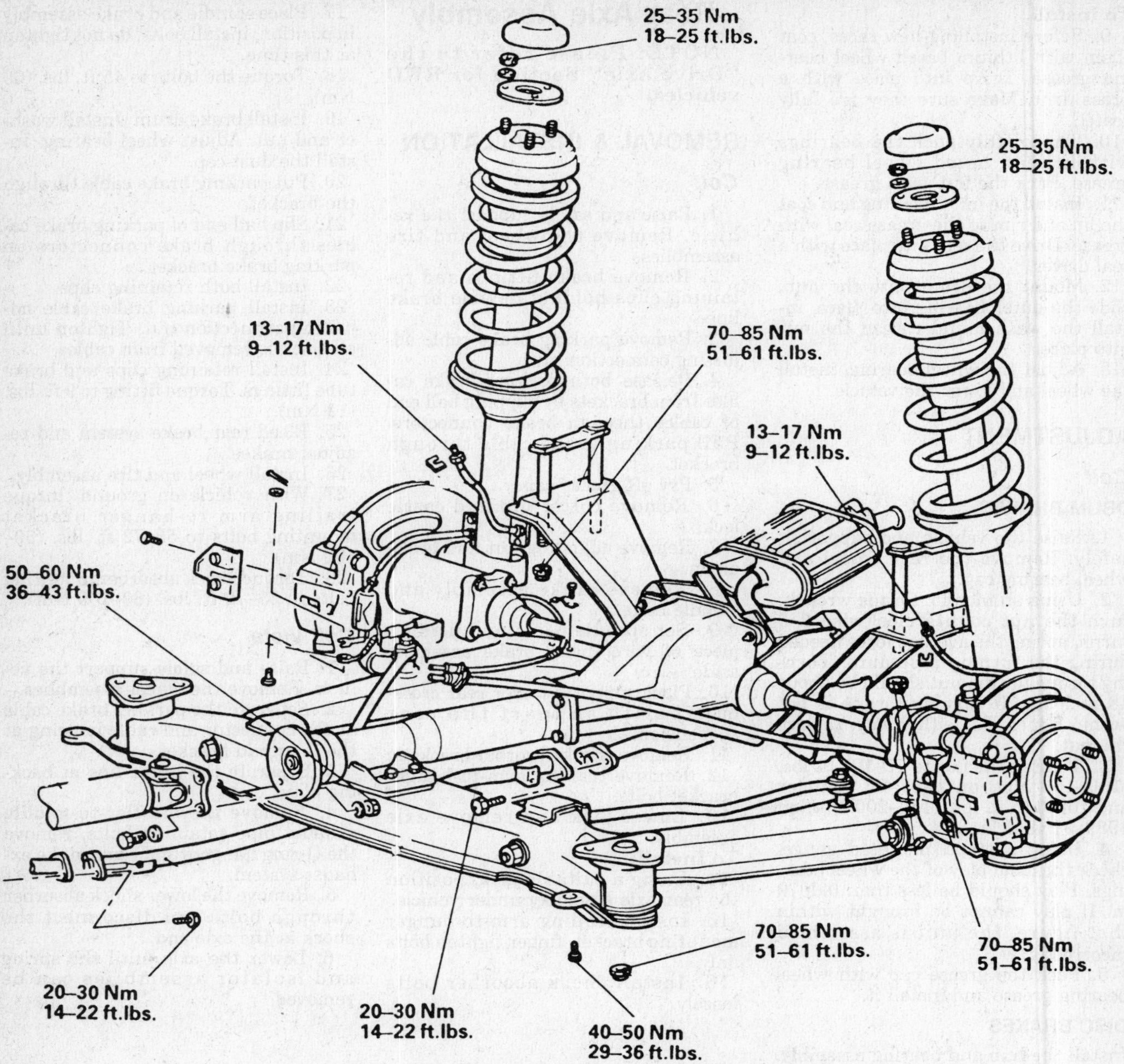

Rear suspension assembly component torques

7. Remove the axle assembly from the vehicle.

To Install:

8. Using a suitable jack, position the rear axle assembly under vehicle.
9. Install the springs and isolators and carefully raise the axle assembly.
10. Install the shock absorber and through bolts; do not tighten.
11. Position brake support to the axle while routing the parking brake cable through the support. Lock it into place.
12. Connect the brake line fitting to the backing wheel cylinder. Torque to 9–12 ft. lbs. (13–17 Nm).
13. Install the hub and drum, if removed.
14. Route the parking brake cable through the fingers in the bracket and lock housing end into the floor pan bracket. Install the cable end into the intermediate connector.
15. Install the exhaust system and hangers. Torque the muffler-to-middle exhaust pipe retaining bolts to 22–29 ft. lbs. (30–40 Nm).
16. Install wheel assemblies and lower vehicle to floor. Torque the lower shock absorber bolts to 58–80 ft. lbs. (80–110 Nm). Torque the axle assembly-to-body mounting bolts to 87–108 ft. lbs. (120–150 Nm).

STEERING

Steering Wheel

REMOVAL & INSTALLATION

1. Pry off the steering wheel center foam pad.
2. Remove the steering wheel retaining nut.
3. Mark the wheel and shaft for proper installation. Using a steering wheel puller, remove the wheel.
4. Be sure the front wheels are in a

straight ahead position. Each spline on the shaft equals about 10 degrees.

5. Installation is the reverse order of removal procedure. Tighten the nut to 30 ft. lbs. (41 Nm).

Steering Column

REMOVAL & INSTALLATION

1. Disconnect the negative battery cable. Remove the steering wheel.
2. Remove the lower column trim cover and the upper column trim cover.
3. Remove the light switch, wiper and washer washer switch. Disconnect the harness connectors.
4. Remove shaft-to-gear retaining bolts and disconnect the shaft at the steering gear.
5. Remove both upper and lower steering column mounting bolts and remove the steering column from the vehicle.

To install:

6. Lower the steering column through the firewall and connect the steering column to the steering gear. Install the retaining bolt. Do not tighten at this time.
7. Install both upper and lower steering column mounting bolts. Torque the mounting bolts to 7–10 ft. lbs. (9–14 Nm).
8. Install the light switch and wiper and washer washer switch. Connect the harness connectors.
9. Install the upper column trim cover and the lower column trim cover.
10. Install the steering wheel. Connect the negative battery cable.

Manual Rack and Pinion

REMOVAL & INSTALLATION

Colt

1. Loosen the lug nuts.
2. Raise and support the vehicle safely.
3. Remove the wheels.
4. Remove the steering shaft-to-pinion coupling bolt.
5. Disconnect the tie rod ends with a separator.
6. Remove the clamps or clamp and bolts securing the rack to the crossmember and remove the unit from the vehicle.

To install:

7. Install the rubber mount for the gear box with the slit on the downside.
8. The remainder of installation is the reverse order of the removal procedures. Torque as follows:

Rack-to-crossmember – 45–60 ft. lbs.
Coupling bolt – 22–25 ft. lbs.
Tie-rod nuts – 11–25 ft. lbs.

9. Road test the vehicle.

2WD Colt Vista

1. Loosen the lug nuts.
2. Raise and support the vehicle safely.
3. Remove the wheels.
4. Remove the steering shaft-to-pinion coupling bolt.
5. Disconnect the tie rod ends with a separator.
6. Remove the crossmember support bracket from the crossmember on the right side of the vehicle.
7. Remove the rear roll stopper-to-center member bolt and move the rear roll stopper forward.
8. Unbolt the rack from the crossmember.

NOTE: The rack is most easily removed using a ratchet and long extension, working from the engine compartment side.

9. Pull the rack out the right side of the vehicle. Pull it slowly to avoid damage.

To install:

10. Installation is the reverse order of the removal procedures. Torque the rack clamp bolts to 43–58 ft. lbs., the tie rod nuts to 17–25 ft. lbs. and the coupling bolt to 22–25 ft. lbs. Fill the system and road test the vehicle.

4WD Colt Vista

1. Remove the steering column.
2. Raise and support the vehicle safely.
3. Remove the front wheels.
4. Using a separator, disconnect the tie rod from the knuckle.
5. Disconnect the steering shaft joint at the rack.
6. Remove the air cleaner.
7. Remove the rack attaching bolts from the rear of the No. 2 crossmember. The bolts are most easily accessed using a long extension and working from the top of the engine compartment.
8. Remove the rear roll stopper-to-center member bolt and move the rear roll stopper forward.
9. From under the vehicle, remove the gear box mounting bolts from the front of the No. 2 crossmember and pull out and to the left on the rack.
10. Lower the rack until the left edge of the left feed tube contacts the lower part of the left fender shield. At this point, remove the left and right feed tubes.
11. Remove the rack from the vehicle.

To install:

12. Position the rack to the No. 2 crossmember and connect the left and right feed tubes.
13. Install the rack, mounting brackets, bushing and retaining bolts. Torque the bolts to 43–58 ft. lbs. (60–80 Nm).
14. Connect the tie rods to the steering knuckle and install the retaining nuts. Torque the tie rod end retaining nuts to 17–25 ft. lbs. (24–34 Nm).
15. Install the wheel assemblies.
16. Install the steering column and connect the steering shaft joint to the rack shaft. Torque the steering shaft-to-rack bolts to 22–25 ft. lbs. (30–35 Nm).
17. Lower the vehicle. Install the air cleaner.
18. Replenish the powering fluid. Start the engine and check for leaks.

Power Steering Gear

ADJUSTMENT

Conquest

NOTE: The steering gear must be disconnected from the steering shaft.

1. Measure the mainshaft preload with an inch lbs. torque wrench. The preload should be 3.5–6.9 inch lbs., with the cross-shaft adjusting bolt backed off.
2. Adjust the valve housing top cover to obtain the proper preload. When correct, lock the top cover with the locking nut.
3. Tighten the cross-shaft adjusting bolt until zero lash is present. Check the total starting torque to rotate the main shaft. The torque should be 5.2–8.7 inch lbs.
4. Adjust the cross-shaft until the required starting torque is obtained and lock the adjusting bolt nut securely.

REMOVAL & INSTALLATION

1. Matchmark and disconnect the steering shaft from the gearbox main shaft.
2. Disconnect the tie rod end and pitman arm from the relay rod.
3. Remove the air cleaner and disconnect the pressure and return lines from the steering gear assembly.
4. Raise the vehicle and support it safely. Remove any interfering splash pans from under the vehicle.
5. If necessary, remove the kickdown linkage splash pan shield and bolts. Move the fuel line aside to avoid damage during removal.
6. Remove the frame bolts from the

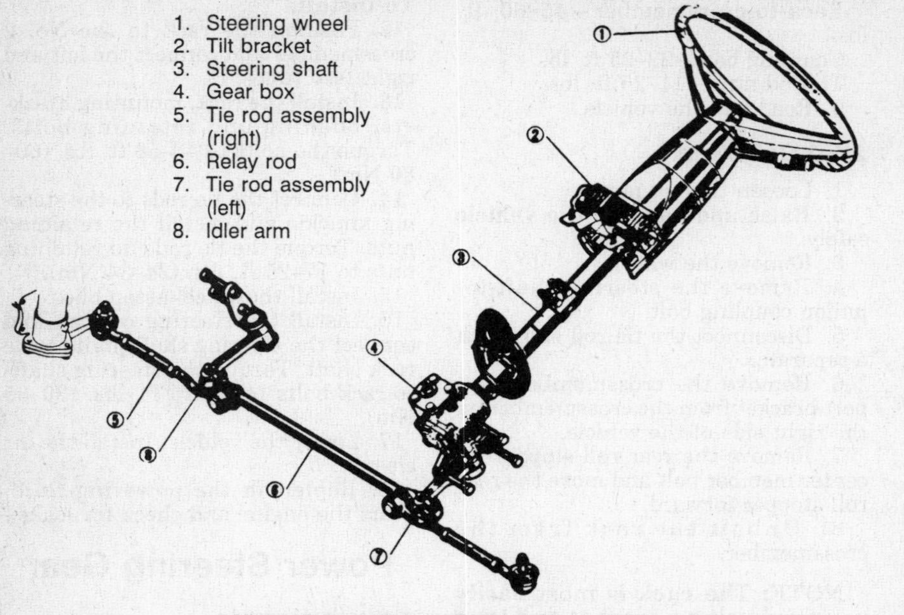

View of the steering system—Conquest

gearbox and lower the unit from the vehicle.

To install:

7. Make sure all matchmarks align. After tightening the pitman arm nut make sure the distance between the centerline of the lowest steering gear mounting bolt and the top of the pitman arm is 19.5mm.

8. Observe the following torques:

Pitman arm nut—94–109 ft. lbs.

Steering gear mounting bolts—40–47 ft. lbs.

Tie rod socket and relay rod—25–33 ft. lbs.

High pressure hose—22–29 ft. lbs.

Return hose—29–36 ft. lbs.

Power Rack and Pinion

REMOVAL & INSTALLATION

Colt

1. Loosen the lug nuts.
2. Raise the vehicle and support it safely.
3. Remove the wheels.
4. Remove the steering shaft-to-pinion coupling bolt.
5. Disconnect the tie rod ends with a separator.
6. Drain the fluid.
7. Disconnect the hoses from the rack.
8. Remove the band from the steering joint cover.
9. Unbolt and remove the stabilizer bar.
10. Remove the rack unit mounting clamp bolts and take the unit out the left side of the vehicle.

To install:

11. Make sure the rubber isolators have their nubs aligned with the holes in the clamps.
12. Apply rubber cement to the slits in the gear mounting grommet.
13. Torque the clamp bolt to 43–58 ft. lbs., the tie rod nuts to 11–25 ft. lbs. and the coupling bolt to 22–25 ft. lbs.
14. Fill the system and road test the car.

2WD Colt Vista

1. Loosen the lug nuts.
2. Raise and support the vehicle safely.
3. Remove the wheels.
4. Remove the steering shaft-to-pinion coupling bolt.
5. Disconnect the tie rod ends with a separator.
6. Disconnect the hoses at the rack.
7. Remove the crossmember support bracket from the crossmember on the right side of the vehicle.
8. Unbolt the rack from the crossmember.

NOTE: The rack is most easily removed using a ratchet and long extension, working from the engine compartment side.

9. Pull the rack out the right side of the vehicle. Pull it slowly to avoid damage.

To install:

10. Installation is the reverse order of the removal procedures. Torque the rack clamp bolts to 43–58 ft. lbs., the tie rod nuts to 17–25 ft. lbs. and the coupling bolt to 22–25 ft. lbs. Fill the system and road test the vehicle.

4WD Colt Vista

1. Remove the steering column.
2. Raise and support the vehicle safely.
3. Remove the front wheels.
4. Using a separator, disconnect the tie rod from the knuckle.
5. Disconnect the steering shaft joint at the rack.
6. Disconnect the fluid lines at the gear box.
7. Remove the air cleaner.
8. Remove the rack attaching bolts from the rear of the No. 2 crossmember. The bolts are most easily accessed using a long extension and working

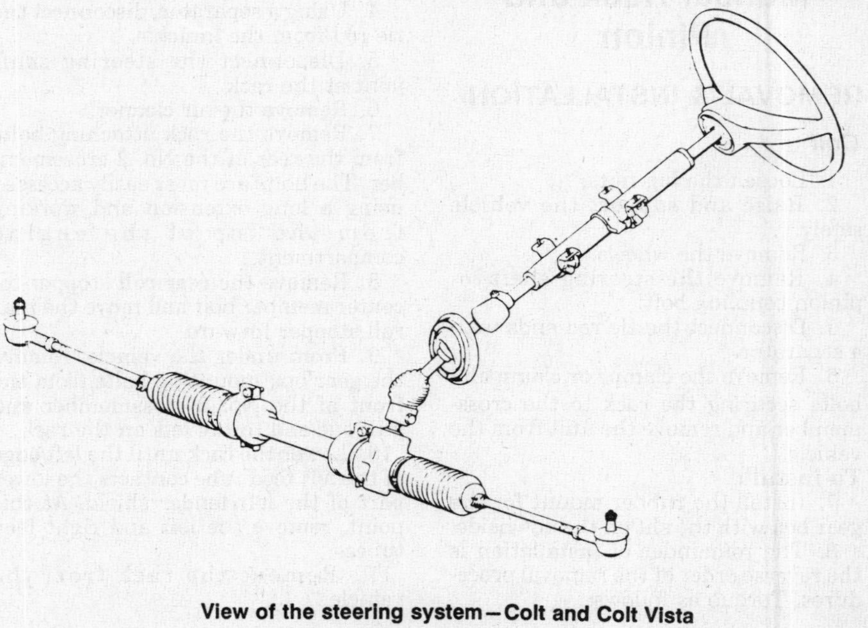

View of the steering system—Colt and Colt Vista

from the top of the engine compartment.

9. From under the vehicle, remove the rack mounting bolts from the front of the No. 2 crossmember and pull out and to the left on the rack.

10. Lower the gear box until the left edge of the left feed tube contacts the lower part of the left fender shield. At this point, remove the left and right feed tubes.

11. Remove the rack from the vehicle.

To install:

12. Installation is the reverse order of the removal procedures. When installing the clamps, make sure the rubber projections are aligned with the holes in the clamps. Install the tie rods so 191–193mm shows between the tie rod end locknut and the beginning of the boot. Torque the gear box mounting bolts to 55–60 ft. lbs.; the tie rod-to-knuckle nut to 20–25 ft. lbs.

Power Steering Pump

REMOVAL & INSTALLATION

1. Remove the drive belt. If the pulley is to be removed, do so now.
2. Disconnect the pressure and return lines. Catch any leaking fluid.
3. Remove the pump attaching bolts and lift the pump from the brackets.

To install:

4. Make sure the bracket bolts are tight and install the pump to the brackets.
5. If the pulley has been removed, install it and tighten the nut securely. Bend the lock tab over the nut.
6. Install the drive belt and adjust to a tension of 22 lbs. at a deflection of 0.28–0.39 in. at the top center of the belt. Tighten the pump bolts securely to hold the tension.
7. Connect the pressure and return lines and fill the reservoir with Dexron® II fluid.
8. Bleed the system.

BELT ADJUSTMENT

1. Press the V-belt by applying pressure of 22 lbs. (100 N) at the center of the belt.
2. Measure the deflection to confirm that it is within the standard range.
   a. Colt, standard value—0.2–0.4 in. (6–9mm).
   b. Colt Vista, standard value—0.3–0.4 in. (7–10mm).
   c. Conquest, standard value—0.35–0.47 in. (9–12mm).
3. To adjust the tension of the belt, loosen the power steering pump mounting bolts, move the power steering pump and then retighten the bolts.

SYSTEM BLEEDING

1. The reservoir should be full of Dexron® II fluid.
2. Raise the vehicle and support it safely.
3. Turn the steering wheel fully to the right and left until no air bubbles appear in the fluid. Maintain the reservoir level.
4. Lower the vehicle and with the engine idling, turn the wheels fully to the right and left. Stop the engine.
5. Install a tube from the bleeder screw on the steering gear box or rack, to the reservoir.
6. Start the engine, turn the steering wheel fully to the left and loosen the bleeder screw.
7. Repeat the procedure until no air bubbles pass through the tube.
8. Tighten the bleeder screw and remove the tube. Refill the reservoir as needed and check that no further bubbles are present in the fluid. An abrupt rise in the fluid level after stopping the engine is a sign of incomplete bleeding. This will cause noise from the pump or control valve.

Tie Rod Ends

REMOVAL & INSTALLATION

NOTE: The following applies to rear wheel drive vehicles. On front wheel drive vehicles, the tie rods are serviced with the steering rack.

1. Raise the vehicle and support it safely.
2. Using a puller, disconnect the tie rod ends from the steering knuckle.
3. Loosen the jam nut and remove the tie rod ends from the tie rod. The outer end is left hand threaded and the inner is right hand threaded.
4. Grease the tie rod threads and install the ends. Turn each end in an equal amount.
5. Install the tie rod assembly on the steering knuckle and relay rod. Tighten the castellated nuts to 29–36 ft. lbs. Use new cotter pins.
6. Adjust the toe-in.

Relay Rod

REMOVAL & INSTALLATION

1. Raise the vehicle and support it safely.
2. Disconnect the tie rod ends from the steering knuckles with a puller.
3. Again using the puller, disconnect the relay rod from the idler arm and the pitman arm.
4. Remove the relay arm.
5. Install the rod in the reverse order of removal. Tighten the tie rod end nuts to 29–36 ft. lbs. Tighten the relay rod-to-pitman arm nut and relay rod-to-idler arm nut to 29–43 ft. lbs. Always use new cotter pins.

Idler Arm

REMOVAL & INSTALLATION

1. Raise the vehicle and support it safely.
2. Disconnect the idler arm from the relay rod using a puller.
3. Remove the retaining bolts and remove the idler arm.
4. Mount the idler arm on the frame and tighten the bolts to 25–29 ft. lbs.
5. Attach the relay rod to the idler arm and tighten the stud nut to 29–43 ft. lbs. Use a new cotter pin.

BRAKES

For all brake system repair and service procedures not detailed below, please refer to "Brakes" in the Unit Repair section.

Master Cylinder

REMOVAL & INSTALLATION

Rear Wheel Drive

NOTE: Be careful not to spill brake fluid on the painted surfaces of the vehicle. The brake fluid will cause damage to the paint.

1. Disconnect all hydraulic lines from the master cylinder. If equipped with a remote reservoir, remove and plug the hoses from the master cylinder caps. If the master cylinder has a fluid level warning device, disconnect the wiring harness.
2. Remove the master cylinder mounting nuts from the power brake booster. Remove the master cylinder.

NOTE: Before installing the master cylinder, make sure there is clearance between the pushrod and master cylinder piston. It should be 0.028–0.043 in.

3. Mount the master cylinder to the power brake booster.
4. Connect all brake lines and wiring harnesses and fill the master cylinder reservoirs with clean fluid.
5. Bleed the brake system.

Front Wheel Drive

1. Disconnect the fluid level sensor.

2. Disconnect the brake tubes from the master cylinder and cap them immediately.
3. If equipped with a turbocharger, remove the reservoir from the reservoir holder.
4. Unbolt and remove the master cylinder from the booster.
5. Installation is the reverse order of the removal procedures. Measure the master cylinder pushrod clearance; it should be 0.016–0.31 in. Torque the mounting bolts to 6–9 ft. lbs. (72–108 inch lbs.).

Proportioning Valve

REMOVAL & INSTALLATION

1. Disconnect the brake lines at the valve.

NOTE: Use a flare wrench, if possible, to avoid damage to the flare nuts and brake lines.

2. Remove the mounting bolts and the valve.
3. Install in the reverse order of the removal procedures. Refill the master cylinder and bleed the brake system.

Power Brake Booster

REMOVAL & INSTALLATION

1. Remove the master cylinder.
2. Disconnect the vacuum line from the booster.
3. Remove the pin connecting the power brake operating rod and the brake lever.
4. Unbolt and remove the booster.
5. Replace the packing on both sides of the booster-to-firewall spacer with new packing.
6. If the check valve was removed, make sure the direction of installation marking on the valve is followed.

To install:

7. Installation is the reverse order of the removal procedures. Torque the booster-to-firewall nuts to 6–9 ft. lbs. (54–108 inch lbs.). Torque the master cylinder-to-booster nuts to 6–9 ft. lbs.
8. Adjust the brake pedal and master cylinder pushrod as explained earlier.

Brake Caliper

REMOVAL & INSTALLATION

Colt and Colt Vista

1. Raise and support the vehicle safely and remove the wheel assembly.

NOTE: On late model vehicles equipped with the PFS15 type front disc brakes, the caliper and pads are retained to the adapter by 2 sleeve pin bolts. Remove the pin bolts and service the assembly as required. Sleeve bolt torque is 16–23 ft. lbs.

2. Disconnect the brake hose from the caliper and plug.
3. Remove the upper and lower pin bolts and remove the caliper and brake pads as an assembly.
4. Remove the inner shim, the anti-squeal shim and the pads from the caliper support assembly.

To install:

5. Installation is the reverse order of the removal procedures. When in-

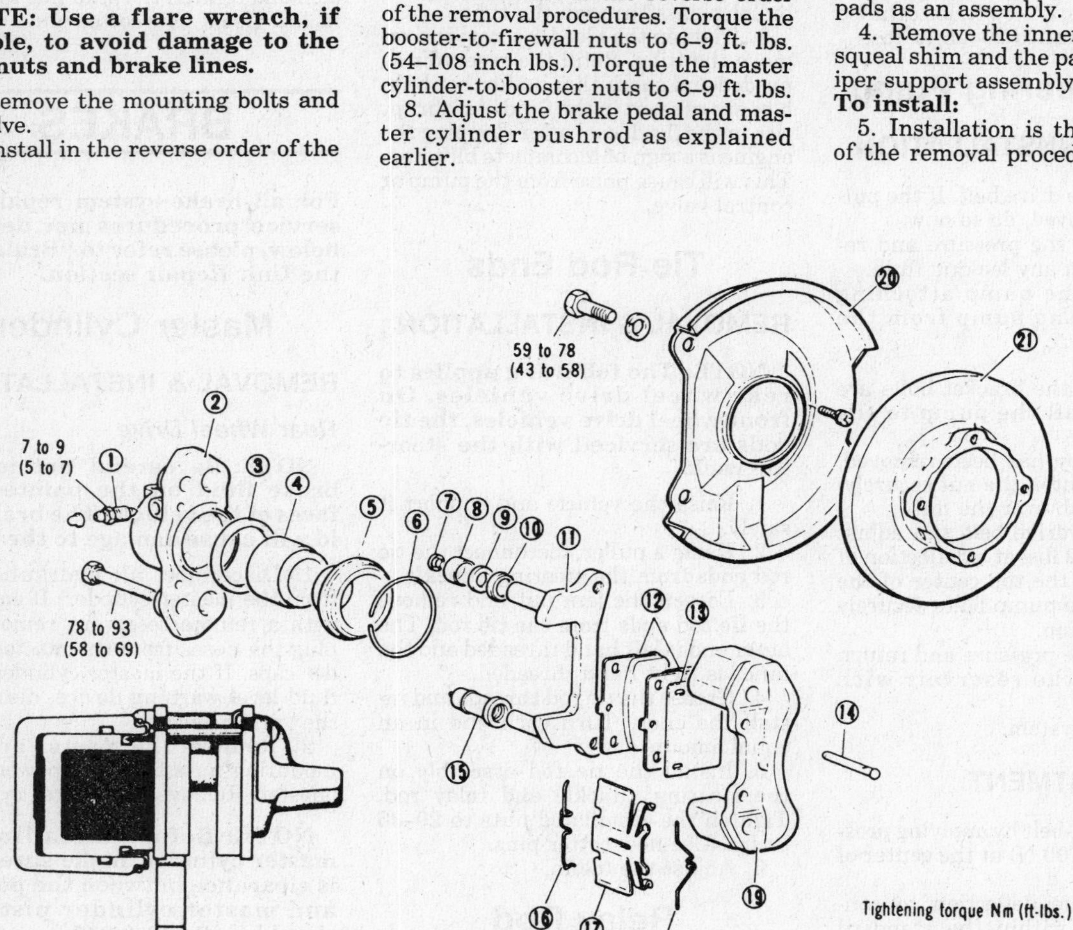

1. Bleeder screw
2. Caliper, inner
3. Piston seal
4. Piston
5. Dust seal
6. Retaining ring
7. Cap plug
8. Torque plate pin cap
9. Oil seal retainer
10. Wiper seal
11. Torque plate
12. Pad assembly
13. Anti-squeak shim
14. Pad retaining pin
15. Torque plate pin bushing
16. K-spring
17. Pad protector
18. M-clip
19. Caliper, outer
20. Dust cover
21. Brake disc

Typical pin type caliper

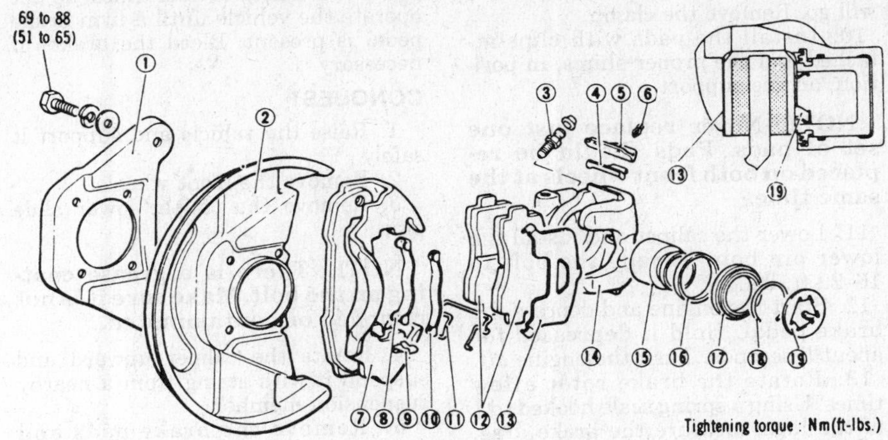

1. Disc brake adapter
2. Dust cover
3. Bleeder screw
4. Pad support plate
5. Stopper plug
6. Spigot pin
7. Caliper support
8. Pad clip (inner)
9. Pad clip B
10. Pad clip (outer)
11. Anti-rattle spring
12. Brake pad
13. Anti-squeak shim (outer)
14. Caliper body
15. Piston
16. Piston seal
17. Dust boot
18. Boot ring
19. Anti-squeak shim (inner)

Exploded view of a typical sliding type caliper

stalling the spacers, apply a coating of an approved grease on the spacers.

6. Torque the caliper retaining bolts to 16–23 ft. lbs. (22–32 Nm). Torque the brake hose fitting to 9–12 ft. lbs. (13–17 Nm).

Disc Brake Pads

REMOVAL & INSTALLATION

Front

COLT

1. Raise the vehicle and support it safely. Remove the wheels.

NOTE: On late vehicles equipped with the PFS15 type front disc brakes; the caliper and pads are retained to the adapter by 2 sleeve pin bolts. Remove the pin bolts and service the assembly as required. Sleeve pin torque is 16–23 ft. lbs.

2. Remove the lower sleeve bolt from the caliper and rotate the caliper upward.

NOTE: There is a grease coating on the bolt. Make sure it is not removed or contaminated.

3. Support the caliper by suspending it with wire or string from a nearby suspension member.
4. Remove the inner, then outer shims from the caliper.
5. Lift out the brake pads.
6. Remove the pad liners.

To install:

7. Clean all parts in solvent made for brake parts.
8. Inspect the dust boot on the caliper piston. If it is torn or brittle, replace it. and consider rebuilding the caliper.
9. Inspect the shims and liners and replace them if damaged.
10. Remove the cap from the master cylinder reservoir and siphon off about ¼ in. of fluid.
11. Using a C-clamp, force the piston back into the caliper as far as it will go. Remove the clamp.
12. Install the liners, pads and inner, then outer shims.

NOTE: Never replace just one set of pads, pads should be replaced on both front wheels at the same time.

13. Lower the caliper and install the lower sleeve bolt. Torque the bolt to 16–23 ft. lbs.
14. Start the engine and depress the brake pedal several times. Hold it depressed for about 5 seconds. Turn the engine off.
15. Rotate the brake rotor a few times. Using a spring scale hooked to 1 of the lugs, measure the brake drag. Remove the pads and perform the spring scale test again. The difference between the drag test with and without the pads should not exceed 15 lbs. If the difference does exceed 15 lbs., the caliper will have to be rebuilt or replaced. When servicing is complete, pump the brakes several times. Do not move the vehicle until a firm brake pedal is present.

COLT AND COLT VISTA

1. Raise the vehicle and support it safely. Remove the front wheels.

NOTE: On late vehicles equipped with the PFS15 type front disc brakes, the caliper and

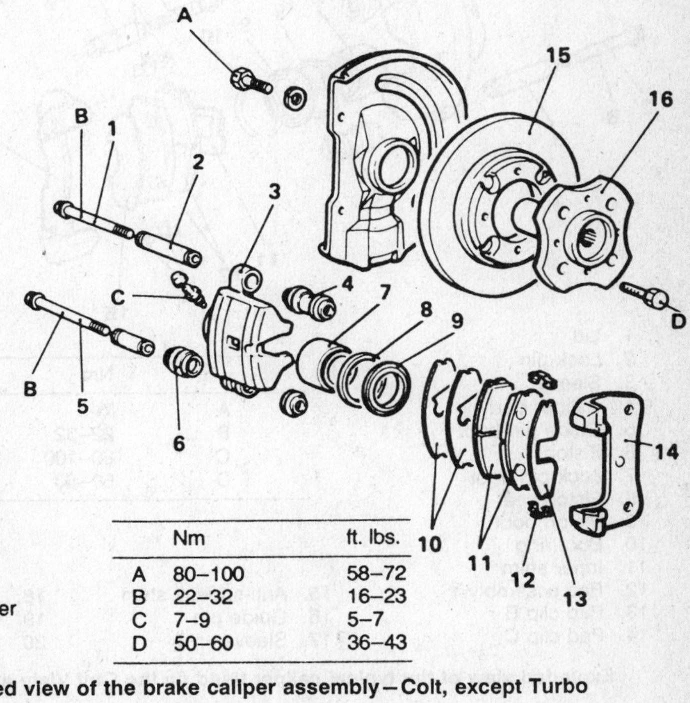

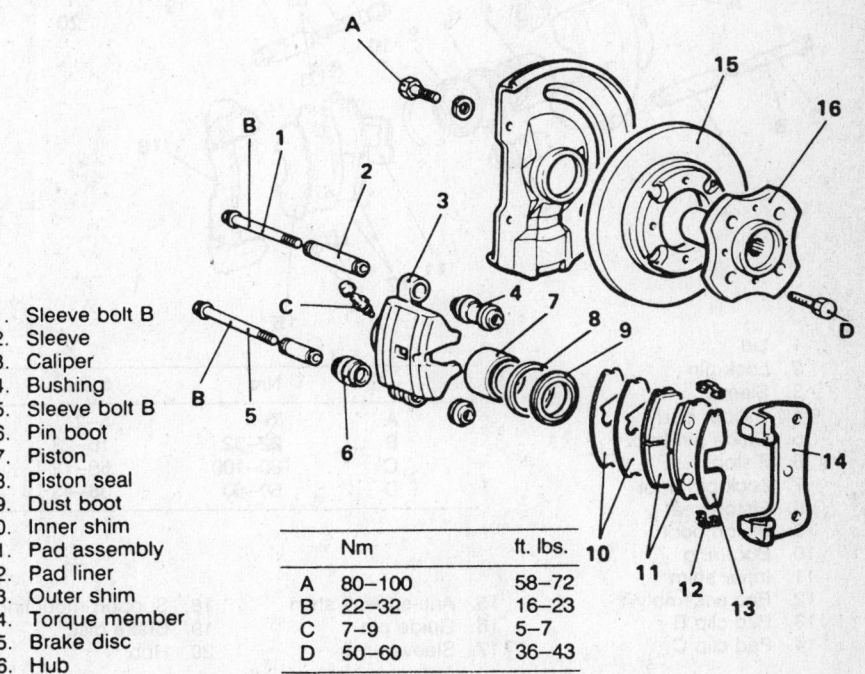

1. Sleeve bolt B
2. Sleeve
3. Caliper
4. Bushing
5. Sleeve bolt B
6. Pin boot
7. Piston
8. Piston seal
9. Dust boot
10. Inner shim
11. Pad assembly
12. Pad liner
13. Outer shim
14. Torque member
15. Brake disc
16. Hub

	Nm	ft. lbs.
A	80–100	58–72
B	22–32	16–23
C	7–9	5–7
D	50–60	36–43

Exploded view of the brake calliper assembly—Colt, except Turbo

pads are retained to the adapter by 2 sleeve pin bolts. Remove the pin bolts and service the assembly as required. Sleeve bolt torque is 16–23 ft. lbs.

2. Remove the lower pin bolt and rotate the caliper upwards. Support the caliper with wire or string from a nearby suspension member.
3. Remove the inner shim, the anti-squeal shim and the pads from the caliper support assembly.
4. Remove the clips from the pads.

To install:

5. Clean all parts in solvent made for brake parts.
6. Inspect the dust boot on the caliper piston. If it is torn or brittle, replace it and consider rebuilding the caliper.
7. Inspect the shims and liners and replace them if damaged.
8. Remove the cap from the master cylinder reservoir and siphon off about ¼ in. of fluid.
9. Using a special tool, force the piston back into the caliper as far as it will go. Remove the clamp.
10. Install the pads with clips attached and the proper shims, in position, on the support.

NOTE: Never replace just one set of pads. Pads should be replaced on both front wheels at the same time.

11. Lower the caliper and install the lower pin bolt. Torque the bolt to 16–23 ft. lbs.
12. Start the engine and depress the brake pedal. Hold it depressed for about 5 seconds. Turn the engine off.
13. Rotate the brake rotor a few times. Using a spring scale hooked to 1 of the lugs, measure the brake drag. Remove the pads and perform the spring scale test again. The difference between the drag test with and without the pads should not exceed 15 lbs. If the difference does exceed 15 lbs., the caliper will have to be rebuilt or replaced.
14. When servicing is complete, pump the brakes several times, do not operate the vehicle until a firm brake pedal is present. Bleed the brakes if necessary.

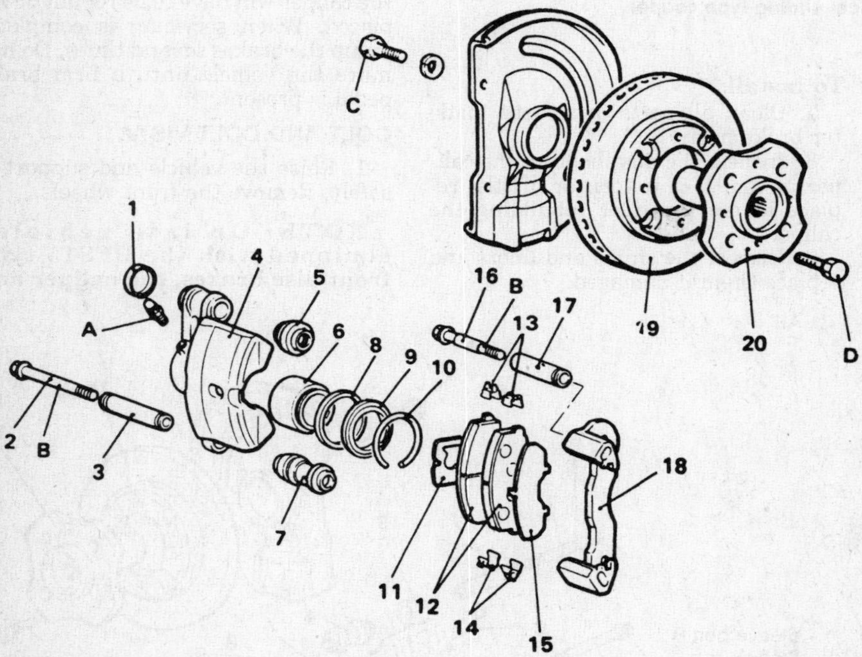

1. Lid
2. Lock pin
3. Sleeve
4. Caliper body
5. Guide pin boot
6. Piston
7. Lock pin boot
8. Piston seal
9. Piston boot
10. Boot ring
11. Inner shim
12. Pad assembly
13. Pad clip B
14. Pad clip C
15. Anti-squeak shim
16. Guide pin
17. Sleeve
18. Support mounting
19. Brake disc
20. Hub

	Nm	ft. lbs.
A	7–9	5–7
B	22–32	16–23
C	80–100	58–72
D	50–60	36–43

Exploded view of the typical caliper used on the Colt Vista and Colt Turbo

CONQUEST

1. Raise the vehicle and support it safely.
2. Remove the front wheels.
3. Remove the caliper lower slide pin.

NOTE: There is a grease coating on the bolt. Make sure it is not removed or contaminated.

4. Rotate the caliper upward and suspend it with string from a nearby suspension member.
5. Remove the brake pads and shims.
6. Remove the clips from the pads.

To install:

7. Clean all parts in solvent made for brake parts.
8. Inspect the dust boot on the caliper piston. If it is torn or brittle, replace it and consider rebuilding the caliper.
9. Inspect the shims and liners and replace them if damaged.
10. Remove the cap from the master cylinder reservoir and siphon off about ¼ in. of fluid.
11. Using a special tool, force the piston back into the caliper as far as it will go. Remove the clamp.
12. Install the pads with clips attached and the proper shims, in position, on the support.

NOTE: Never replace just one set of pads, pads should be replaced on both front wheels at the same time.

13. Rotate the caliper back into position and install the lower slider pin. Torque the pin to 70 ft. lbs.
14. Start the engine and depress the brake pedal. Hold it depressed for about 5 seconds. Turn the engine off.
15. Rotate the brake rotor a few times. Using a spring scale hooked to 1 of the lugs, measure the brake drag. Remove the pads and perform the spring scale test again. The difference between the drag test with and without the pads should not exceed 15 lbs. If the difference does exceed 15 lbs., the caliper will have to be rebuilt or replaced.
16. When servicing is complete, pump the brakes several times, do not operate the vehicle until a firm brake pedal is present. Bleed the brakes if necessary.

Rear

EXCEPT CONQUEST

1. Raise the vehicle and support it

safely. Remove the rear wheel and the caliper dust cover.
2. Disconnect the parking brake cable
3. Remove the spring pin and stopper plug.
4. Move the caliper back and forth to loosen, then remove the caliper from the support.

NOTE: The brake hose need not be disconnected; however, do not suspend the weight of the caliper from the hose.

5. Take time to examine the location of the various clips and springs. Remove the pads from the support. Do not mix up the inner and outer clips, they must be installed in the same location.
6. Seat the caliper piston by pushing in while turning clockwise. When fully seated, 1 of the grooves on the piston must be located vertically at 12 o'clock to accommodate a projection of the brake pad. Install new pads into the support and install the caliper.

CONQUEST

1. Raise the vehicle and support it safely.
2. Remove the wheels.
3. Disconnect the parking brake cable.
4. Remove the lower caliper lockpin and the upper guide pin.

NOTE: There is a grease coating on the bolt. Make sure it is not removed or contaminated.

5. Rotate the caliper upward, if upper guide pin removal is not required. Suspend the caliper with a string from a nearby suspension member.
6. Remove the brake pads and shims.
7. Remove the clips from the pads.
8. Clean all parts in solvent made for brake parts.
9. Inspect the dust boot on the caliper piston. If it is torn or brittle, replace.
10. Inspect the shims and liners and replace them if damaged.
11. Remove the cap from the master cylinder reservoir and siphon off about ¼ in. of fluid.
12. Align the grooves in the caliper and piston and force the piston back into the caliper as far as it will go. Remove the clamp.

To install:

13. Install the pads and shims after attaching the clips in reverse order of removal.

NOTE: Never replace just one set of pads, pads should be replaced on both front wheels at the same time.

14. Rotate the caliper back into position and install the lower slider pin. Torque the pin to 45 ft. lbs.
15. Start the engine and depress the brake pedal. Hold it depressed for about 5 seconds. Turn the engine off.
16. Rotate the brake rotor a few times. Using a spring scale hooked to 1 of the lugs, measure the brake drag. Remove the pads and perform the spring scale test again. The difference between the drag test with and without the pads should not exceed 15 lbs., If the difference does exceed 15 lbs., the caliper will have to be rebuilt or replaced.

Brake Rotor

REMOVAL & INSTALLATION

Front

EXCEPT CONQUEST

1. Raise the vehicle and support it safely.
2. Remove the wheel assembly. Remove the cotter pin and axle nut.
3. Support the caliper by suspending it with wire from a nearby suspension member.
4. Using halfshaft separator tool, separate the halfshaft from the hub assembly.
5. Matchmark the steering knuckle-to-strut bolts and remove the retaining bolts.
6. Remove the lower ball joint retaining nut. Using a ball joint separator tool, separate the ball joint from the knuckle assembly.
7. Remove the knuckle assembly from the vehicle.
8. Secure the knuckle and rotor assembly in a vise. Remove the rotor-to-hub retaining bolts and separate the rotor from the hub.

To install:

9. Installation is the reverse order of the removal procedures. Torque the rotor-hub-bolts to 56–72 ft. lbs. (80–100 Nm). Torque the steering knuckle-to-strut bolts to 54–65 ft. lbs. (75–90 Nm). Torque lower ball joint retaining nut to 43–52 ft. lbs. (60–72 Nm).

CONQUEST

1. Raise the vehicle and support it safely.
2. Remove the wheel assembly and remove the caliper assembly.
3. Support the caliper by suspending it with wire from a nearby suspension member.
4. Remove the axle dust cap, cotter pin and outer wheel bearing.
5. Slide the rotor from the axle.

To install:

6. Installation is the reverse of the removal procedures.
7. Clean all parts and inspect, replace the bearings if showing signs of pitting or overheating. Always install new seals.
8. Adjust the bearings.

Rear

ALL VEHICLES

1. Raise the vehicle and support it safely. Remove the wheel assembly.
2. Remove the caliper. Support the caliper by suspending it with wire from a nearby suspension member.
3. Remove the caliper support retaining bolts and remove the support.
4. Remove the rotor-to-hub retaining screws and slide the rotor from the hub. If equipped with retaining rings, remove the rings from the wheel lugs.

To install:

5. Slide the rotor on the hub and install the rotor-to-hub retaining screws. If equipped with retaining rings, install new rings on the wheel lugs.
6. Install the caliper support and retaining bolts. Torque the bolts to 29–36 ft. lbs. (40–50 Nm).
7. Install the caliper, brake pads and caliper retaining bolts. Torque the bolts to 16–23 ft. lbs. (22–32 Nm).
8. Install the wheel assembly and lower the vehicle.
9. When servicing is complete, pump the brakes several times, do not operate the vehicle until a firm brake pedal is present. Bleed the brakes if necessary.

Brake Shoes

REMOVAL & INSTALLATION

Colt

1. Raise the vehicle and support it safely. Remove rear wheel and brake drum.
2. Remove the lower pressed metal spring clip, the shoe return spring (the large one piece spring between the 2 shoes) and the 2 shoe hold-down springs.
3. Remove the shoes and adjuster as an assembly. Disconnect the parking brake cable from the lever, remove the spring between the shoes and the lever from the rear (trailing) shoe. Disconnect the adjuster retaining spring and remove the adjuster, turn the star wheel in to the adjuster body after cleaning and lubricating the threads.
4. The wheel cylinder may be removed for service or replacement, if necessary.

To install:

5. Clean the backing plate. Install the wheel cylinder if it was removed. Lubricate all contact points on the backing plate, anchor plate, wheel cylinder to shoe contact and parking brake strut joints and contacts. Install

the brake shoes after attaching the parking brake, lever and adjuster assemblies. Install the hold-down and return springs.

6. Pre-adjustment of the brake shoe can be made by turning the adjuster star wheel out until the drum will just slide on over the brake shoes. Before installing the drum make sure the parking brake is not adjusted too tightly, if it is, loosen or the adjustment of the rear brakes will not be correct.

7. If the wheel cylinders were serviced, bleed the brake system. The brake shoes are then adjusted by pumping the brake pedal and applying and releasing the parking brake, Adjust the parking brake stroke. Road test the vehicle.

Colt Vista

1. Raise the vehicle and support it safely.
2. Remove the wheels.
3. Remove the brake drums.
4. Remove the shoe-to-strut spring.
5. Remove the shoe-to-shoe spring.
6. Remove the shoe hold-down spring.
7. Remove the shoe retainer clip.
8. Remove the leading shoe.
9. Remove the brake cable from the lever.
10. Remove the trailing shoe.
11. Remove the brake cable snapring and remove the cable.
12. Inspect all parts for wear or damage.

NOTE: Never replace shoes on one side only. Replace both sets of shoes at the same time.

To install:

13. Assemble the parking brake and adjuster assemblies on the brake shoes.
14. Install the brake shoes and hold-downs. Connect the return springs.
15. Apply a small amount of lithium based grease to the contact pads of the backing plate before installing the shoes.
16. When installing the shoe-to-shoe spring and shoe-to-strut spring, set the adjuster lever all the way back against the shoe.
17. When everything is assembled, pump the pedal several times and adjust the brakes.

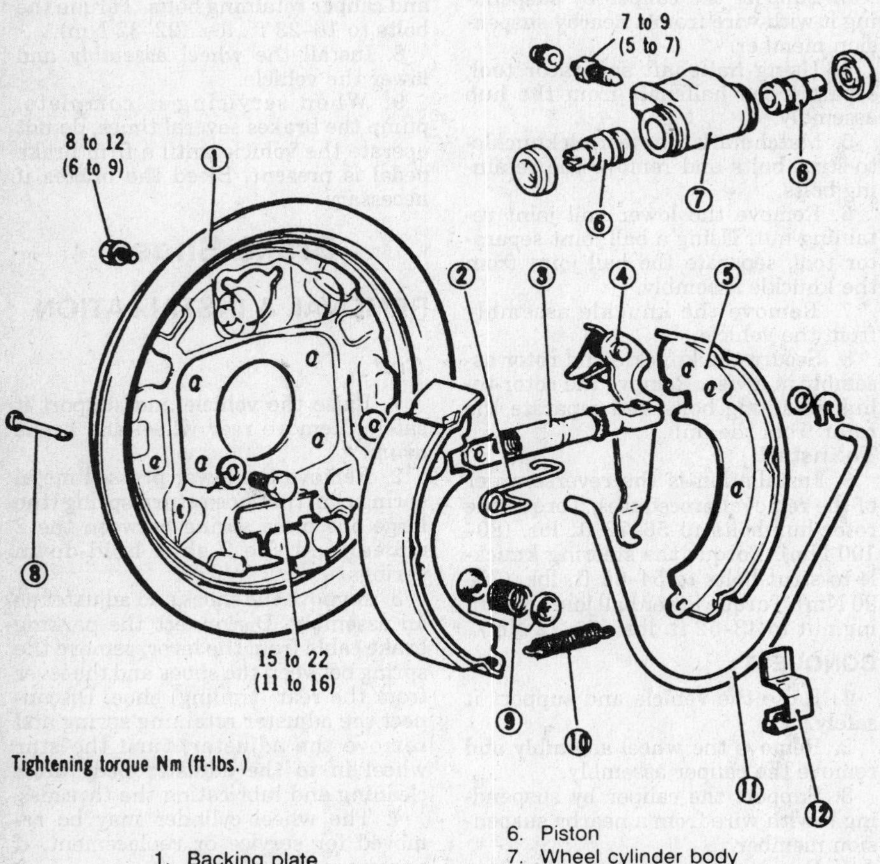

1. Backing plate
2. Spring
3. Adjuster
4. Parking lever
5. Shoe and lining assembly
6. Piston
7. Wheel cylinder body
8. Shoe hold spring pin
9. Shoe hold down spring
10. Shoe to shoe spring
11. Shoe return spring
12. Clip spring

Typical rear drum brake system used on front wheel drive except Vista

Conquest

1. Raise the vehicle and support it safely.
2. Remove the wheel. Make sure the front wheels are blocked securely, release the parking brake and remove the brake drum.

NOTE: The brake drum is retained by 2 small bolts.

3. Disconnect the shoe-to-shoe spring and the strut-to-shoe spring. Disconnect the shoe return spring and remove the brake hold-down assemblies.
4. Disconnect the parking brake cable from the parking brake lever and remove the rear brake shoe.
5. Remove the front brake shoe. Transfer levers and adjusters to the new brake shoes using new U-shaped locks.

To install:

6. Prior to assembly, apply No. 2 brake grease to the contact area of the strut and parking brake lever and strut and adjusting lever. After cleaning the backing plate apply grease to the brake shoe contact points.
7. Connect the parking brake and install the brake shoes with the adjusters and hold-down assemblies. Install the return springs. The lining to drum clearance is automatically adjusted by applying the brakes several times after the drums have been installed. If the wheel cylinders have been rebuilt the brake system must be bled before correct adjustment is possible.

Wheel Cylinder

REMOVAL & INSTALLATION

1. Remove the brake shoes.
2. Place a bucket or some old newspapers under the brake backing plate to catch the brake fluid that will run out of the wheel cylinder.
3. Disconnect the brake line and remove the cylinder mounting bolts.
4. Remove the cylinder from the backing plate.
5. Install the cylinder in the reverse order. Bleed the brake system.

Parking Brake Cable

ADJUSTMENT

Colt and Colt Vista

1. Pull the parking brake lever up

with a force of about 45 lbs. The total number of clicks heard should be 5–7.

2. If the number of clicks was not within that range, release the lever and back off the cable adjuster locknut at the base of the lever and tighten the adjusting nut until there is no more slack in the cable.

3. Operate the lever and brake pedal several times, until no more clicks are heard from the automatic adjuster.

4. Turn the adjusting nut to give the proper number of clicks when the lever is raised full travel.

5. Raise the vehicle and support it safely.

6. Release the brake lever and make sure the rear wheels turn freely. If not, back off on the adjusting nut until they do.

Conquest

1. Pull up on the lever, counting the number of clicks. Total travel should yield 4–5 clicks.

2. If not, remove the center console and turn the adjusting nut on the lever rod to obtain the required travel.

3. Raise the vehicle and support it safely.

4. With the parking brake released, make sure the rear wheels turn freely.

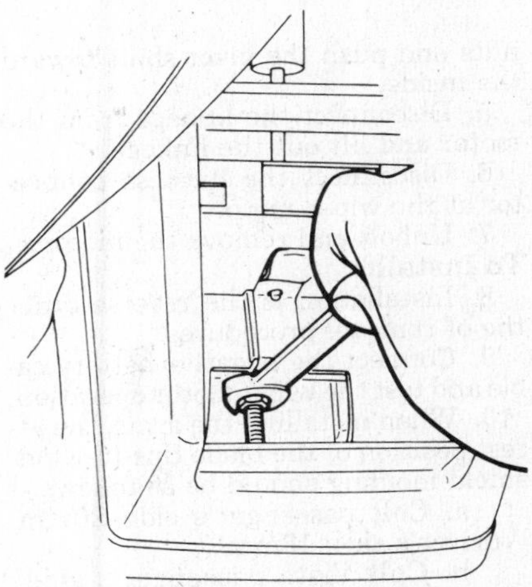

Parking brake cable adjustment—Colt

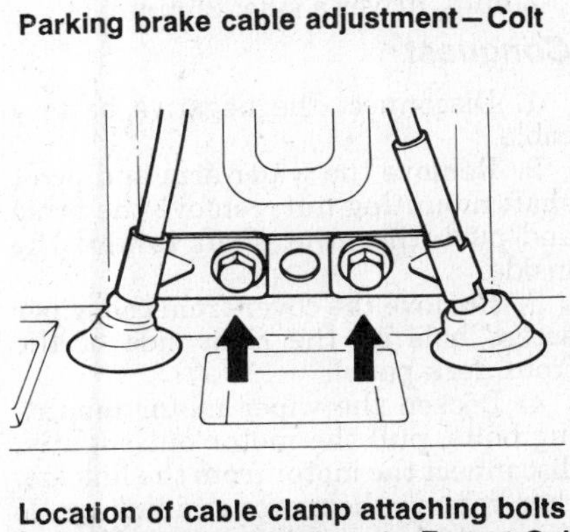

Location of cable clamp attaching bolts in passenger compartment—Except Colt Vista

REMOVAL & INSTALLATION

Colt and Colt Vista

1. Raise the vehicle and support it safely.

2. Disconnect the brake cable at the parking brake lever. Remove the cable clamps inside the driver's compartment. Disconnect the clamps on the rear suspension arm.

3. Remove the rear brake drums and the brake shoes assemblies. Disconnect the parking brake cable from the lever on the rear brake shoe. Remove the brake cables.

4. Install the cable and adjust.

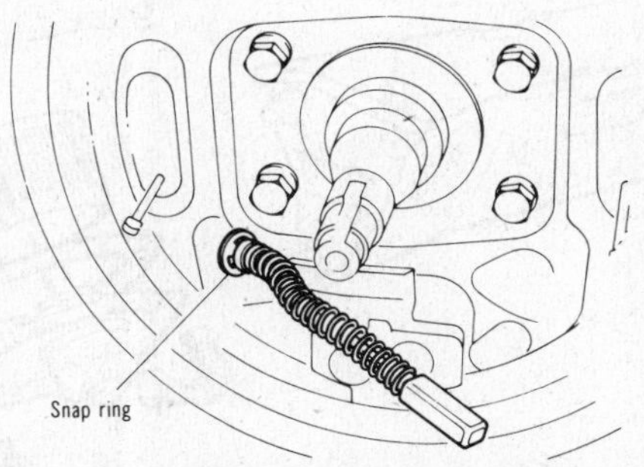

Cable snap ring on backing plate

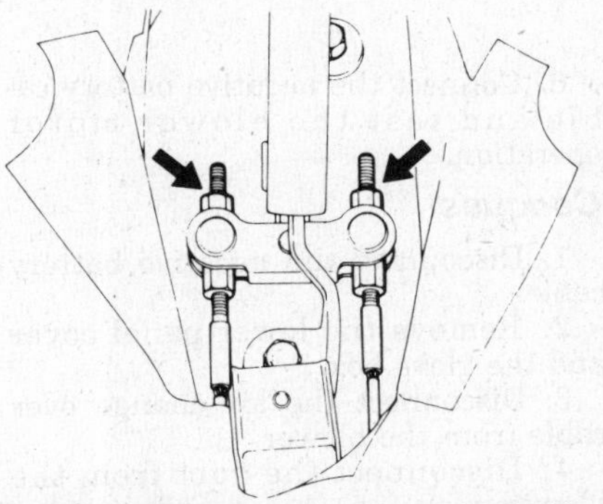

Cable adjusting nuts—front wheel drive models except Vista

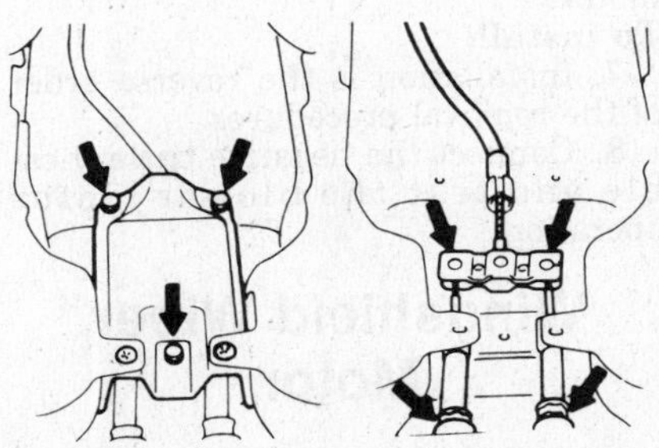

Parking brake equalizer cover bolts and cable coupler—Colt Vista

Conquest

1. Raise the vehicle and support it safely.

2. Release the parking brake. Pull off the clevis pins from both sides of the rear brake. Disconnect the cable from the extension lever.

3. Remove the front cable after disconnecting the parking brake lever. On rear disc brake equipped vehicles, remove the rubber hanger from the center of the axle housing. Remove the parking brake lever and clevis pin linking the lever and cable. Remove the clips under the floor and remove the cable.

4. Install the cable. When installing, make sure the cable clips do not interfere with a rotating part. Adjust the extension lever to stop first. Then adjust the left cable and then the right.

Brake System Bleeding

1. Carefully clean all dirt from around the master cylinder filler cap.

2. If a bleeder tank is used, follow the manufacturer's instructions.

3. Remove the filler cap and refill the master cylinder to the lower edge of the filler neck.

4. Clean off the bleeder connections at all of the wheel cylinders or disc brake calipers. Attach the bleeder hose and fixture to the right rear wheel cylinder bleeder screw and place the end of the tube in a glass jar, submerged in brake fluid.

5. Open the bleeder valve ½–¾ turn. Have an assistant depress the brake pedal and allow it to return slowly. Continue this pumping action to force any air out of the system.

6. When bubbles cease to appear at the end of the bleeder hose, close the bleeder valve and remove the hose. Check the level of the brake fluid in the master cylinder and add fluid, if necessary.

7. After the bleeding operation at each caliper or wheel cylinder has been completed, refill the master cylinder reservoir and replace the filler plug.

NOTE: Never reuse brake fluid which has been removed from the lines through the bleeding process because it contains air bubbles and dirt.

CHASSIS ELECTRICAL

Heater Blower Motor

REMOVAL & INSTALLATION

Colt

1. Disconnect the negative battery cable.

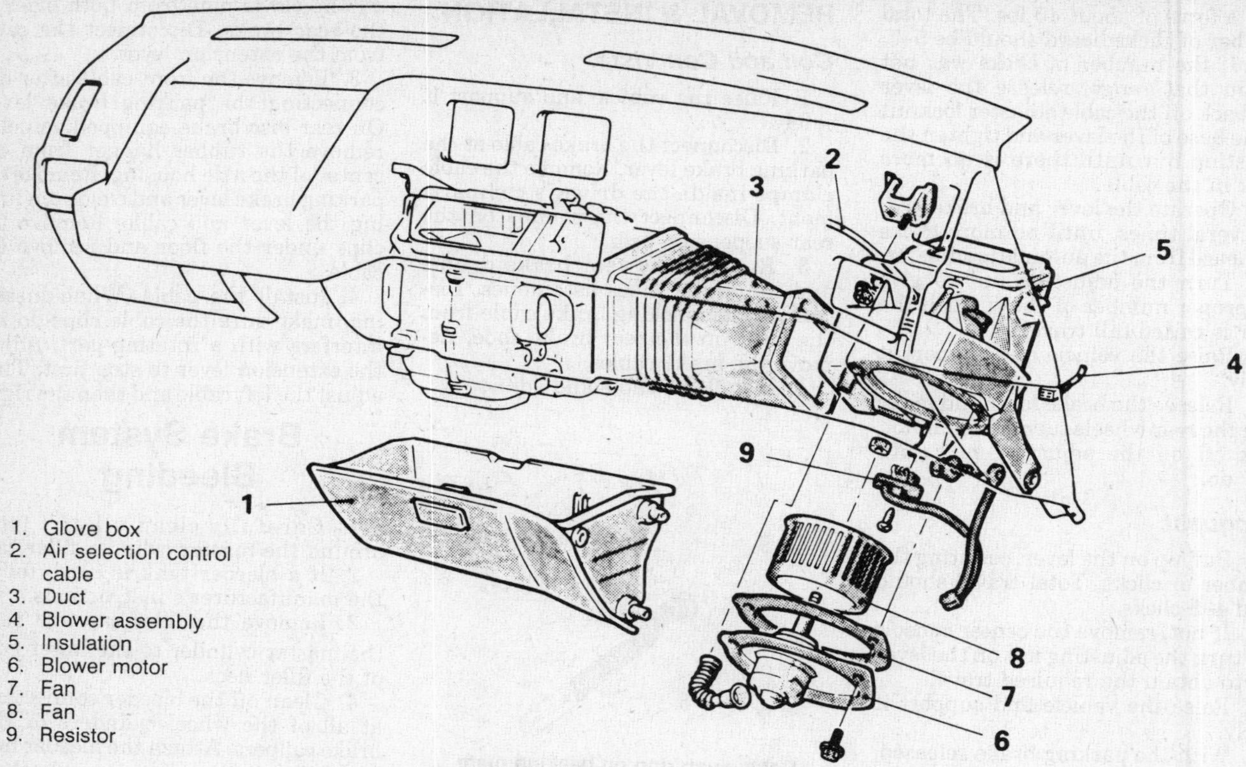

1. Glove box
2. Air selection control cable
3. Duct
4. Blower assembly
5. Insulation
6. Blower motor
7. Fan
8. Fan
9. Resistor

Blower motor—removal and installation

2. Remove the glove box and parcel tray.
3. Disconnect the change over control wire and duct.
4. Disconnect the harness connector at the ECU unit and remove the ECU unit.
5. Remove the blower case assembly from the dash.
6. Unbolt and remove the blower motor from the case.
7. The fan is removable from the motor shaft.

To install:

8. Installation is the reverse of removal.
9. Connect the negative battery cable and test the blower motor operation.

Colt Vista

1. Disconnect the negative battery cable.
2. Remove the upper and lower glove boxes.
3. Disconnect the wiring from the blower assembly.
4. Remove the blower motor mounting bolts and lift out the motor. If the entire blower case is to be removed, the instrument panel will have to be removed first.

To install:

5. Installation is the reverse order of the removal procedures.
6. Connect the negative battery cable and test the blower motor operation.

Conquest

1. Disconnect the negative battery cable.
2. Remove the lower panel cover and the glove box.
3. Disconnect the air change over cable from the blower.
4. Disconnect the duct from the blower.
5. Disconnect the blower wiring.
6. Unbolt and remove the blower motor.

To install:

7. Installation is the reverse order of the removal procedures.
8. Connect the negative battery cable and test the blower motor operation.

Windshield Wiper Motor

REMOVAL & INSTALLATION

Colt and Colt Vista

1. Disconnect the negative battery cable.
2. Remove the wiper arms.
3. Remove the front cowl trim plate.
4. Remove the pivot shaft mounting nuts and push the pivot shaft toward the inside.
5. Disconnect the linkage from the motor and lift out the linkage.
6. Disconnect the harness connector at the wiper motor.
7. Unbolt and remove the motor.

To install:

8. Installation is the reverse order the of removal procedures.
9. Connect the negative battery cable and test the wiper motor operation.
10. When installing the arms, the at-rest position of the blade tips-to-windshield molding should be as follows:
    a. Colt passenger's side–20mm, driver's side: 15mm.
    b. Colt Vista passenger's side–30mm, driver's side: 25mm.

Conquest

1. Disconnect the negative battery cable.
2. Remove the wiper arm and pivot shaft mounting nut, remove the arms and push the pivot shaft toward the inside.
3. Remove the cover from the wiper access hole on the right side of the front deck panel.
4. Loosen the wiper motor mounting bolts, pull the motor out slightly, disconnect the motor from the linkage, then remove the motor and linkage. If the motor's crank arm is to be removed, mark its position first.

To install:

5. Installation is the reverse of order of the removal procedures. Install the wiper arms so the blade tip-to-windshield molding distance, at rest, is ½ in.
6. Connect the negative battery cable and test the wiper motor operation.

Windshield Wiper Switch

REMOVAL & INSTALLATION

NOTE: On Colt Vista and Conquest, the wiper switch is integral with the turn signal switch.

1. Remove the steering wheel.
2. Remove the steering column cover.
3. Pull out and remove the switch knob.
4. Remove the 2 mounting screws and pull the switch out.
5. Installation is the reverse of removal.

Instrument Cluster

REMOVAL & INSTALLATION

1988 Colt

1. Disconnect the negative battery cable.
2. Remove the steering wheel.
3. Remove the glove box.
4. Remove the instrument panel heater duct.
5. Remove the parcel tray.
6. Remove the steering column lower cover.
7. Disconnect the light switch and wiper switch connectors.
8. Remove the steering column upper cover.
9. Remove the instrument cluster trim panel screws and lift off the trim panel.
10. Remove the cluster mounting screws and pull the cluster slightly forward. Disconnect the speedometer cable and electrical connectors. Lift out the cluster.

To install:

11. Installation is the reverse order of the removal procedures.
12. Connect the light switch and wiper switch connectors.
13. Connect the speedometer cable to the speedometer. Connect the negative battery cable.

1989-92 Colt

1. Disconnect the negative battery cable.
2. Remove the heater control cover.
3. Remove the knee protector or lower panel assembly.
4. Remove the cluster bezel and disconnect the electrical connectors.
5. Disconnect the speedometer cable. Remove the cluster retaining screws and pull the cluster out enough to disconnect the speedometer cable and electrical connectors. Release the cable by turning the adapter to the left or right and then remove.

To install:

6. Install the cluster and connect the speedometer and electrical cables.
7. Install the cluster bezel.
8. Install the knee protector or lower panel assembly.
9. Install the heater control cover.
10. Connect the negative battery cable.

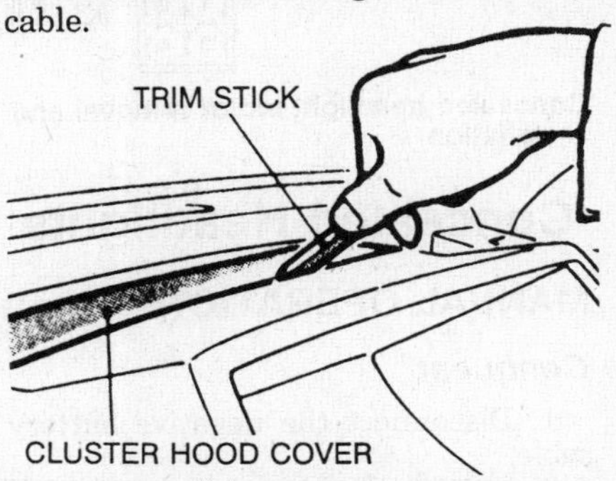

Cluster screw trim panel removal—Colt Vista

Colt Vista

1. Disconnect the negative battery cable.
2. Remove the steering wheel.
3. Remove the ashtray. Remove the cluster screw cover using a wood or plastic trim stick.
4. Remove the cluster trim panel retaining screws.
5. Pull the trim panel slightly toward the front and release the connectors. Lift the trim panel off.
6. Remove the 4 cluster mounting screws, pull the cluster slightly toward the front and disconnect the speedometer cable and electrical connectors.
7. Lift out the cluster.

To install:

8. Installation is the reverse order of the removal procedures.
9. Connect all electrical connectors. Connect the speedometer cable to the speedometer.
10. Connect the negative battery cable.

Conquest

NOTE: The following procedure applies to both the conventional needle type gauge cluster and to the liquid crystal display type. Because the LCD gauges are composed of very delicate components, they must not be subjected to severe shocks. Furthermore, the LCD gauges must not be disassembled.

1. Disconnect the negative battery cable.
2. Remove the cluster hood attaching screws.
3. Pull outward on both bottom side edges of the hood and while holding it in that position, pull it upward and off.
4. Disconnect the wiring from the panel switches.
5. Remove the cluster case attaching screws.
6. Pull both sides of the lower part of the cluster case up and toward the rear of the vehicle.
7. Disconnect the speedometer cable from the back of the case.
8. Disconnect all wiring at the back of the case and lift the case out.

To install:

9. Installation is the reverse order of removal procedure.
10. Connect the electrical wiring at the rear of the instrument cluster.
11. Connect the speedometer cable to the speedometer. Connect the negative battery cable.

Speedometer

REMOVAL & INSTALLATION

1. Disconnect the negative battery cable.

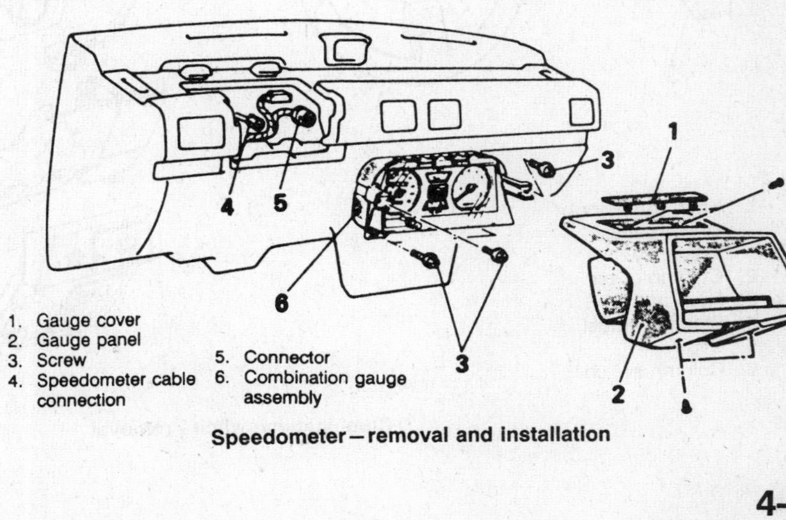

Speedometer—removal and installation

2. Remove the instrument cluster assembly from the dash.
3. Remove the cluster gauge panel retaining screws and remove the gauge panel and lense.
4. Remove the speedometer retaining screws and remove the speedometer from the cluster.

To install:

5. Installation is the reverse order of the removal procedures.
6. Install the instrument cluster assembly to the dash. Connect the negative battery cable.

Radio

REMOVAL & INSTALLATION

1. Disconnect the negative battery cable.
2. Remove the radio/heater control cover (Colt and Vista) or the front console box (Conquest).
3. Remove the radio retaining screws and radio. Pull the assembly out far enough to disconnect the electrical connectors.
4. Remove the radio bracket if installing a new unit.

To install:

5. Install the radio bracket.
6. Install the radio and connect the electrical connectors.
7. Install the radio/heater control cover (Colt and Vista) or the front console box (Conquest).
8. Connect the negative battery cable.

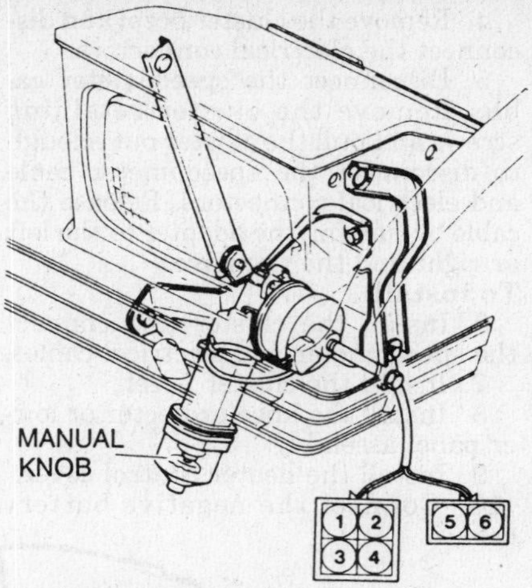

Concealed headlight motor removal and installation

Concealed Headlights

MANUAL OPERATION

Conquest

1. Disconnect the negative battery cable.
2. Lift the hood and look through the access hole in the shield, located behind the bumper.

NOTE: The vehicle is equipped with 2 headlight motors; 1 on each side of the vehicle.

3. Remove the hand wheel cover boot.
4. Turn the manual override hand wheel to raise or lower the headlights.
5. When desired headlight position is achieved, connect the negative battery cable.

Headlight Switch

The headlight switch is an integral part of the combination switch on the Colt Sedan, Colt Vista and Conquest. However, it is separate on the Colt Wagon and is located on left side of the steering column cover.

REMOVAL & INSTALLATION

Colt Wagon

1. Disconnect the negative battery cable.
2. Remove the lower column cover and harness band.
3. Disconnect the electrical connections from the switch.
4. Remove the switch knob retainer screw and knob.
5. Pull the switch out, from behind the upper column cover panel.

To install:

6. Install the switch under and through the upper column cover opening, on the left side of column.
7. Install the knob and retaining screw, securing the switch in place.
8. Connect the electrical connections to the switch. Secure the harness in place with the band.
9. Install the lower column cover and retaining screws. Connect the

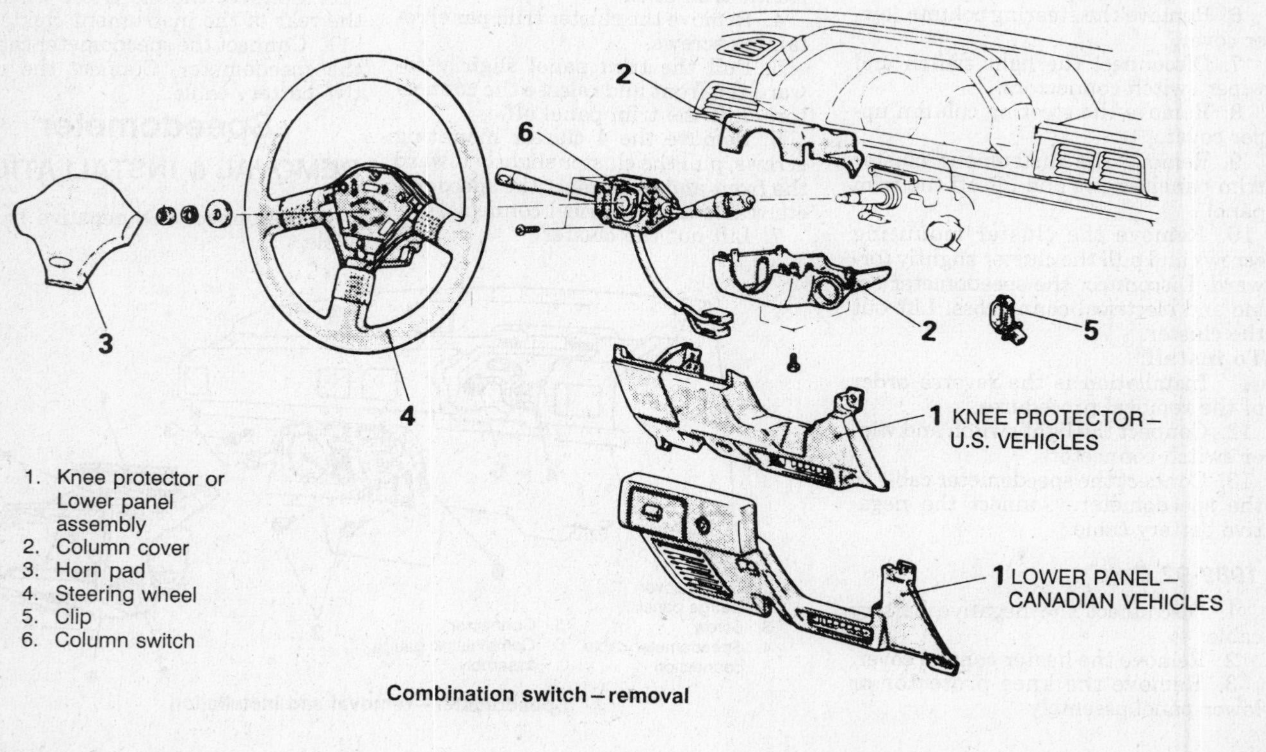

Combination switch – removal

negative battery cable and test the switch operation.

Combination Switch

REMOVAL & INSTALLATION

1. Disconnect the negative battery cable.
2. Remove the steering wheel.
3. Remove heater duct and the lower dash panel assembly.
4. Remove the column covers.
5. Remove the switch retaining screws, disconnect the wiring and remove the switch.

To install:

6. Position the switch to the steering column and secure it in place with the retaining screws. Connect the wiring to the harness connections.
7. Install the heater duct, lower dash panel assembly and the column covers.
8. Install the steering wheel. Connect the negative battery cable.

Ignition Lock/Switch

REMOVAL & INSTALLATION

NOTE: When replacing the ignition switch or key reminder switch only, remove the column cover, remove the screw holding the switch and pull out the switch.

1. Disconnect the negative battery cable.
2. Remove the turn signal switch.
3. Cut a notch in the lock bracket bolt head with a hacksaw.
4. Remove the bolt and lock.
5. Remove the column cover and unbolt and remove the ignition switch.

To install:

6. Install both lock and switch in reverse of removal.

NOTE: When installing the lock, use special break off retaining bolts. The special bolt should be tightened until the head of bolt breaks off. When installing the switch, install the switch bolt loosely and insert and work the key a few times to make sure everything checks out before tightening the bolt.

7. Connect the negative battery cable.

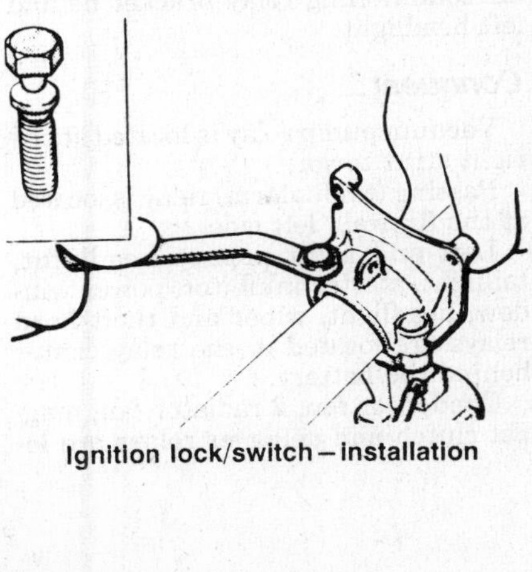

Ignition lock/switch – installation

Stoplight Switch

ADJUSTMENT

1. Disconnect the negative battery cable.
2. Remove the lower dash trim panel.
3. Turn switch adjustment nut until the switch plunger can be pushed in 0.16 in. (4mm), when the pedal is depressed.
4. Tighten the adjustment at that point.
5. Connect the negative battery cable and test the switch operation.

REMOVAL & INSTALLATION

1. Disconnect the negative battery cable.
2. Remove the lower dash trim panel.
3. Disconnect the electrical connections from the switch assembly.
4. Remove the adjustment nut from the switch and pull the switch from the support.

To install:

5. Installation is the reverse order of the removal procedure. Adjust the switch.
6. Connect the negative battery cable and test switch operation.

Clutch Switch

ADJUSTMENT

1. Disconnect the negative battery cable.
2. Remove the lower dash trim panel.
3. Turn switch adjustment nut until the switch plunger can be pushed in 0.16 in. (4mm), when the pedal is depressed.
4. Tighten the adjustment at that point.
5. Connect the negative battery cable.
6. Check the switch for continuity between both terminals when the clutch pedal is depressed.

REMOVAL & INSTALLATION

1. Disconnect the negative battery cable.
2. Remove the lower dash trim panel.
3. Disconnect the electrical connections from the switch assembly.
4. Remove the adjustment nut from the switch and pull the switch from the support.

To install:

5. Installation is the reverse order of the removal procedure. Adjust the switch.
6. Connect the negative battery cable and test switch operation.

Neutral Safety Switch

ADJUSTMENT

Colt and Colt Vista

1. Disconnect the negative battery cable.
2. Place the selector lever in the **N** position.
3. Working under the hood. Place the manual valve in the neutral position.
4. Loosen the switch mounting bolts.
5. Turn the switch until the small end of the the manual lever aligns with the alignment flange on the switch.
6. Tighten the mounting bolts, being careful not to allow the switch to move out of position. Torque the mounting bolts to 7–8 ft. lbs. (10–11 Nm).
7. Connect the negative battery cable.
8. Place the selector lever in **P** and apply the foot brake. Check the starter operation in **P** and **N**.

NOTE: The engine should not start in any other gear range. If it does, repeat adjustment procedure or replace the switch as required.

9. With the ignition switch turned to the **ON** position engine off, place the selector lever in **R** and observe the back-up light operation. The back-up lights should be illuminated.

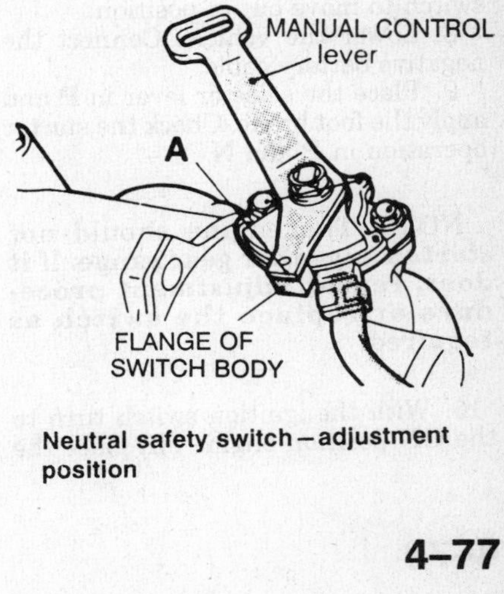

Neutral safety switch – adjustment position

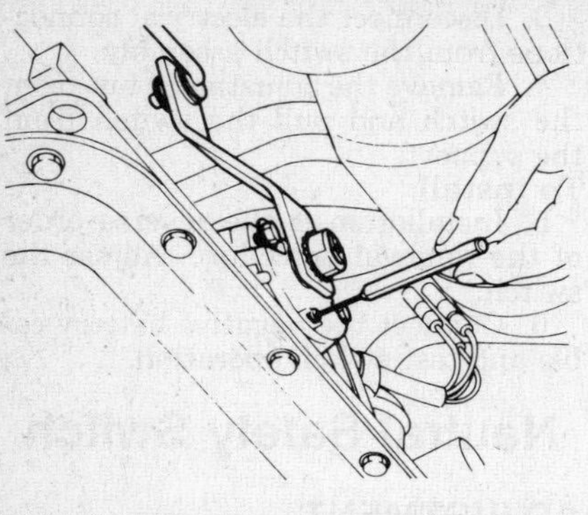

JM600 neutral start switch adjustment

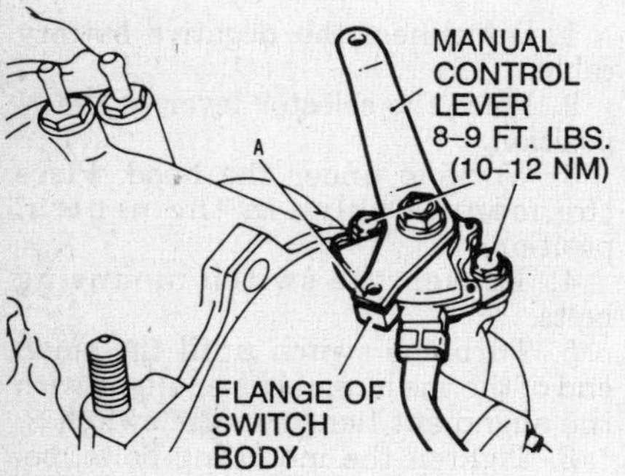

KM series neutral start switch adjustment

Conquest

1. Disconnect the negative battery cable.
2. Place the selector lever in the **N** position.
3. Raise the vehicle and support it safely.
4. Place the manual valve in the neutral position.
5. Remove the small centering screw from the switch and loosen the switch mounting bolts.
6. Using a 0.080 in. (2.0mm) drill or aligning pin, place it in the switch alignment hole.
7. Turn the switch until the pin falls into the hole. Tighten the mounting bolts, being careful not to allow the switch to move out of position.
8. Lower the vehicle. Connect the negative battery cable.
9. Place the selector lever in **P** and apply the foot brake. Check the starter operation in **P** and **N**.

NOTE: The engine should not start in any other gear range. If it does, repeat adjustment procedure or replace the switch as required.

10. With the ignition switch turn to the **ON** position engine off, place the selector lever in **R** and observe the back-up light operation. The back-up lights should be illuminated.

REMOVAL & INSTALLATION

Colt and Colt Vista

1. Disconnect the negative battery cable.
2. Place the selector lever in the **N** position.
3. Working under the hood. Place the manual valve in the neutral position.
4. Remove the manual lever retaining nut and remove the manual lever.
5. Disconnect the electrical connector from the neutral safety switch.
6. Remove the switch mounting bolts and remove the switch from the transaxle.

To install:

7. Place the neutral safety switch on the transaxle manual shaft and install the mounting bolts. Do not tighten at this time.
8. Install the manual lever and retaining nut. Torque the nut to 9–10 ft. lbs. (12–14 Nm).
9. Adjust the switch as required and tighten the mounting bolts to 4–5 ft. lbs. (5–7 Nm).
10. Connect the electrical connector to the switch.
11. Connect the negative battery cable and test the switch operation.

Conquest

1. Disconnect the negative battery cable.
2. Place the selector lever in the **N** position.
3. Raise the vehicle and support it safely.
4. Place the manual valve in the neutral position.
5. Remove the manual lever retaining nut and remove the manual lever.
6. Disconnect the electrical connector from the neutral safety switch.
7. Remove the switch mounting bolts and remove the switch from the transaxle.

To install:

8. Place the neutral safety switch on the transaxle manual shaft and install the mounting bolts. Do not tighten at this time.
9. Install the manual lever and retaining nut. Torque the nut to 9–10 ft. lbs. (12–14 Nm).
10. Adjust the switch as required and tighten the mounting bolts to 4–5 ft. lbs. (5–7 Nm).
11. Connect the electrical connector to the switch.
12. Lower the vehicle and connect the negative battery cable.
13. Test the switch operation.

Fuses

LOCATION

Fuse Block

The main fuse block is located up under the instrument panel on the driver's side of the steering column. The 1989-92 Colt has dedicated fuse blocks behind the right and left headlights. The Colt and Colt Wagon with air conditioning, have an auxiliary fuse located next the receiver/drier. The Colt Vista air conditioning fuse is located behind the left headlight.

Relays

LOCATION

1988 Colt and Colt Wagon

Cold mixture and electric choke relay is located left side firewall next to the ignition coil.

Intermittent wiper relay is located next to the wiper motor.

A/C clutch and condenser fan relay is located in box at the left strut tower.

Radiator fan relay is located in the fuse block.

Transistor relay is located in the harness near the fuse block.

Blower motor and thermo relays are located at the heater unit.

Fuel injection relay is located at the left kick panel.

Seat belt timer and defogger relays are in the fuse block.

Rear intermittent wiper motor is located at the right rear quarter panel (Colt Wagon).

1989-92 Colt

Alternator, radiator fan, power window and headlight relays (in that order) are located in the relay box in the engine compartment behind the battery.

Heater and defogger relays are located in the junction block, left kick panel.

Air Conditioning compressor and condenser fan relays are located in the air conditioning relay bracket behind left headlight.

Conquest

Vacuum pump relay is located at the right strut tower.

Passing (theft alarm) relay is located at the firewall, left side.

Left and right pop-up headlight, taillight, starter inhibitor, power window, headlight, wiper and theft horn relays are located in the relay center behind the battery.

Condenser fan, 2 radiator fan, magnet clutch and defogger relays are lo-

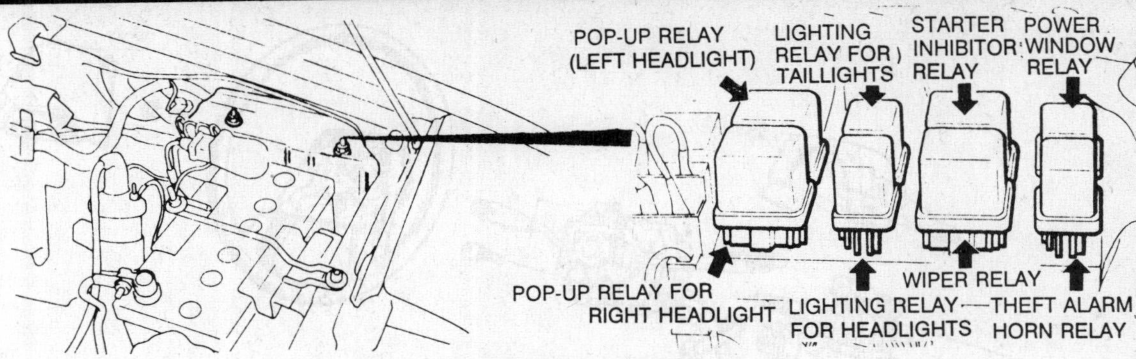

Relay locations—Conquest

cated in the relay center to the side of the battery.

Two Radiator fan motor and fog light relays are located behind the right headlight.

Door lock, overdrive, door unlock and rear brake lockup control relays are located at the left kick panel.

Blower relay is located in the harness at the heater unit.

Blower motor starter cutout and high speed blower relays are located at the heater unit.

Passing control relay is located to the right of the radio.

Power antenna relay is located at the rear compartment panel towards the left side.

Rear intermittent wiper relay is located at the rear compartment panel towards the right side.

Colt Vista

Headlight, intermittent wiper, radiator fan motor, power window and taillight relays are located in the relay block right inner fender.

Alternator and 2 daytime running light relays are located at the left corner of the firewall.

Defogger relay is located at the firewall next to the wiper motor.

Magnet Air Conditioning clutch, condenser fan motor and air conditioning relays are located in front of the windshield washer bottle.

Heater relay is located at the heater control assembly.

Door lock power and door lock control relays are located behind the center console.

Rear intermittent wiper relay is located behind the right rear quarter trim panel.

Automatic seat belt motor relays are located under the seat side panel.

Computers

LOCATION

The Electronic Control Unit (ECU) is located on the passenger's lower kick panel for the 1989-92 Colt and Conquest. The unit is located under the driver's seat or in front of the center console for the Colt Vista. The ECU is located in front of the center console for the 1988 Colt and Colt Wagon.

Circuit Breakers

An automatic seat belt CB for the Conquest is located above the relay block at the left kick panel.

Flashers

LOCATION

The turn signal/hazard flasher is located in the fuse block, located under the instrument panel on the driver's side of the steering column.

Cruise Control

ADJUSTMENT

Colt and Colt Vista

ACCELERATOR CABLE

1. Start the engine and allow it to reach operating temperature and is stable at idle. Shut the engine off.
2. Disconnect the negative battery cable. Remove the air cleaner.
3. Adjust the cable at the throttle side.
4. Loosen the adjustment bolts at the air intake plenum side, freeing the inner cable, use the adjustment bolts to secure the plate so the free-play of the inner cable is 0.04–0.08 in. (1–2mm).

NOTE: If the free-play adjustment is incorrect, either an increase of idle speed or lack of speed control in the high speed range will result.

5. Check and make sure the throttle lever touches the idle position switch.
6. Adjust the accelerator cable at the pedal side.
7. Loosen the adjusting bolt. While keeping the intermediate link of the actuator in close contact with the stopper. Adjust the inner cable free-play to the following:

Manual transalxe—0–0.04 in. (0–1mm)

Automatic transaxle—0.08–0.12 in. (2–3mm).

8. Tighten the adjusting nut.
9. Check the throttle lever at the engine side, it must move 0.04–0.08 in. (1–2mm) when the actuator link is turned.
10. Confirm that the throttle valve fully opens and closes by operating the pedal. Install the air cleaner. Connect the negative battery cable.

Conquest

ACCELERATOR CABLE

1. Start the engine and allow it to reach operating temperature and is stable at idle. Shut the engine off.
2. Disconnect the negative battery cable.
3. Turn the engine on for 15 seconds with the engine not running.
4. Loosen the adjustment nut so the throttle lever is free. Rotate the accelerator nut until the throttle lever just starts to move, then back off ½ turn. Tighten the locknut.
5. Check and make sure the idle position switch touches the stopper after the idle speed control adjustment.

CONTROL CABLE

1. Adjust the accelerator cable.
2. Pull the auto-cruise control cable back until the accelerator pedal just begins to move. Secure the control cable by inserting a clip.
3. Check to ensure that control cable free-play is 0–0.01 in. (0–3mm).

NOTE: If the free-play adjustment is incorrect, either an increase of idle speed or lack of speed control in the high speed range will result.

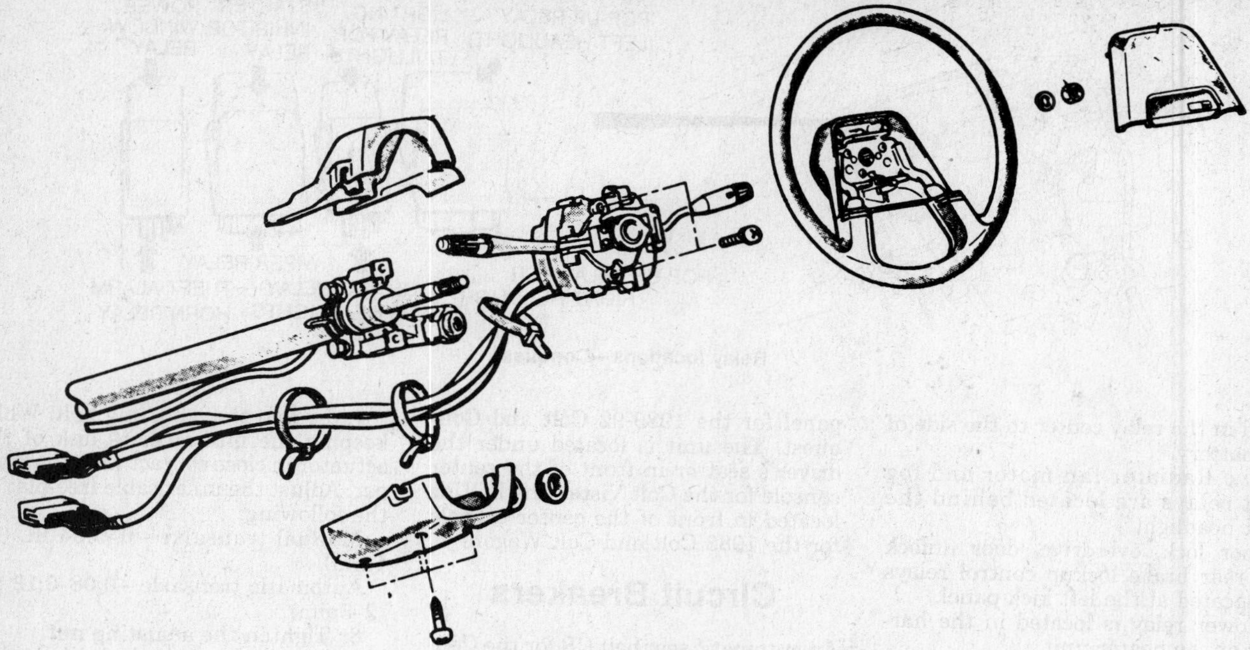

Exploded view of the combination switch

Daihatsu

5

Charade — G100LS, G102LS

SERIAL NUMBER IDENTIFICATION

Vehicle Identification Number

The Vehicle Identification Number (VIN) can be found on the 2 vehicle identification plates. The plates are located at the top left of the instrument panel and at the firewall in the engine compartment.

The VIN is a 17 digit code which identifies the carline, model, year, body type and engine among other items. Character number 8, counting from the left, identifies the type of engine used in the vehicle. A "0" indicates the vehicle is equipped with the the CB EFI, 1.0L 3 cylinder engine. A "2" indicates the vehicle is equipped with the HC EFI, 1.3L 4 cylinder engine.

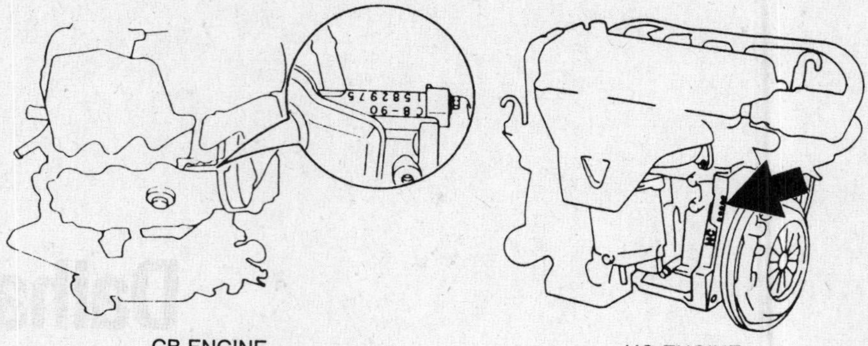

CB ENGINE

HC ENGINE

Engine serial number location

Chassis Number

The chassis number can be found on the manufacturer's plate which is located at the firewall inside the engine compartment.

Engine Number

CB Engine

The engine serial number is stamped on the front side of the cylinder head.

HC Engine

The engine serial number is stamped on the side wall of the cylinder block at the transaxle side.

SPECIFICATIONS

ENGINE IDENTIFICATION

Year	Model	Engine Displacement cu. in. (cc/liter)	Engine Series Identification	No. of Cylinders	Engine Type
1988	G100LS	60.6 (993/1.0)	CB-90	3	OHC
1989	G100LS	60.6 (993/1.0)	CB-90	3	OHC
1990	G100LS	60.6 (993/1.0)	CB-90	3	OHC
	G102LS	79 (1295/1.3)	HC-E	4	OHC 16V
1991-92	G100LS	60.6 (993/1.0)	CB-90	3	OHC
	G102LS	79 (1295/1.3)	HC-E	4	OHC 16V

OHC—Overhead Camshaft

GENERAL ENGINE SPECIFICATIONS

Year	Model	Engine Displacement cu. in. (cc)	Fuel System Type	Net Horsepower @ rpm	Net Torque @ rpm (ft. lbs.)	Bore × Stroke (in.)	Compression Ratio	Oil Pressure @ 3,000 rpm
1988	G100LS	60.6 (993)	MPFI	53 @ 5200	58 @ 3600	2.99 × 2.87	9.5:1	36–71
1989	G100LS	60.6 (993)	MPFI	53 @ 5200	58 @ 3600	2.99 × 2.87	9.5:1	36–71
1990	G100LS	60.6 (993)	MPFI	53 @ 5200	58 @ 3600	2.99 × 2.87	9.5:1	36–71
	G102LS	79 (1295)	MPFI	80 @ 6000	75 @ 4400	2.99 × 2.81	9.5:1	36–71
1991-92	G100LS	60.6 (993)	MPFI	53 @ 5200	58 @ 3600	2.99 × 2.87	9.5:1	36–71
	G102LS	79 (1295)	MPFI	80 @ 6000	75 @ 4400	2.99 × 2.81	9.5:1	36–71

MPFI—Multi point fuel injection

ENGINE TUNE-UP SPECIFICATIONS

Year	Model	Engine Displacement cu. in. (cc)	Spark Plugs Type	Spark Plugs Gap (in.)	Ignition Timing (deg.) MT	Ignition Timing (deg.) AT	Compression Pressure (psi)	Fuel Pump (psi)	Idle Speed (rpm) MT	Idle Speed (rpm) AT	Valve Clearance In.	Valve Clearance Ex.
1988	G100LS	60.6 (993)	①	0.039–0.043	5③	—	178	37–46	800	—	0.006–0.010⑤	0.006–0.010⑤
1989	G100LS	60.6 (993)	①	0.039–0.043	5③	—	178	37–46	800	—	0.006–0.010⑤	0.006–0.010⑤
1990	G100LS	60.6 (993)	①	0.039–0.043	5③	—	178	37–46	800	—	0.006–0.010⑤	0.006–0.010⑤
	G102LS	79 (1295)	②	0.039–0.043	0④	0④	199	37–46	800	850	0.007–0.013⑤	0.010–0.016⑤
1991	G100LS	60.6 (993)	①	0.039–0.043	5③	—	178	37–46	800	—	0.006–0.010⑤	0.006–0.010⑤
	G102LS	79 (1295)	②	0.039–0.043	0④	0④	199	37–46	800	850	0.007–0.013⑤	0.010–0.016⑤
1992	SEE UNDERHOOD STICKER FOR SPECIFICATIONS											

① Champion—RN11YC4
Nippondenso—W16EXR-U11
NGK—BPR5EY-11

② Champion—RC9YC4
Nippondenso—K20PR-U11
NGK—BKR6E-11

③ BTDC with T-E, shorted
④ BTDC with sub advance disconnected
⑤ Hot

FIRING ORDERS

NOTE: To avoid confusion, always replace spark plug wires one at a time.

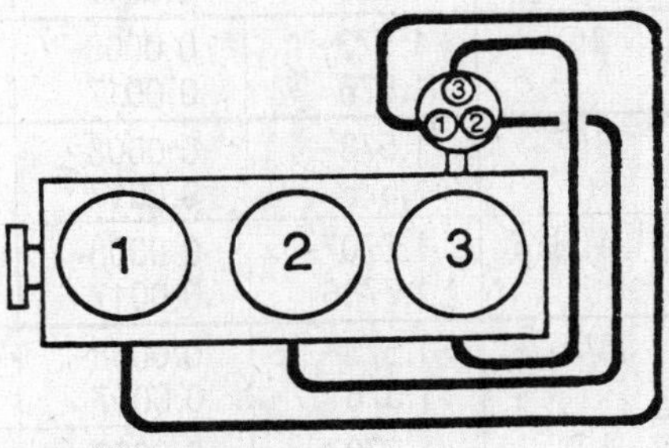

993cc Engine
Engine Firing Order: 1-2-3
Distributor Rotation: Counterclockwise

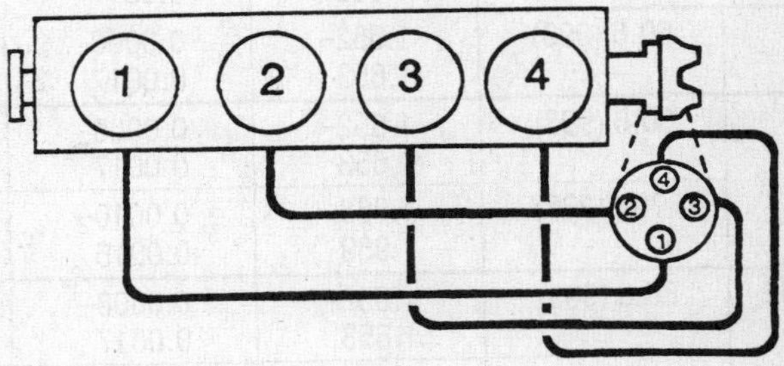

1295cc Engine
Engine Firing Order: 1-3-4-2
Distributor Rotation: Counterclockwise

CAPACITIES

Year	Model	Engine Displacement cu. in. (cc)	Engine Crankcase (qts.) with Filter	Engine Crankcase (qts.) without Filter	Transmission (pts.) 4-Spd	Transmission (pts.) 5-Spd	Transmission (pts.) Auto.	Drive Axle (pts.)	Fuel Tank (gal.)	Cooling System (qts.)
1988	G100LS	60.6 (993)	3.2	2.8	—	2.9	—	—	10.6	4.1①
1989	G100LS	60.6 (993)	3.2	2.8	—	2.9	—	—	10.6	4.1①
1990	G100LS	60.6 (993)	3.2	2.8	—	2.9	—	—	10.6	4.1①
	G102LS	79 (1295)	3.7	3.5	—	2.9	5.3	—	10.6	5.8①
1991-92	G100LS	60.6 (993)	3.2	2.8	—	2.9	—	—	10.6	4.1①
	G102LS	79 (1295)	3.7	3.5	—	2.9	5.3	—	10.6	5.8①

① Including 0.6 for reserve tank

CAMSHAFT SPECIFICATIONS

Year	Engine Displacement cu. in. (cc)	Journal Diameter 1	2	3	4	5	Lobe Lift In.	Ex.	Bearing Clearance	Camshaft End Play
1988	60.6 (993)	1.258–1.259	1.866–1.867	1.906–1.907	—	—	1.574–1.582	1.574–1.582	①	NA
1989	60.6 (993)	1.258–1.259	1.866–1.867	1.906–1.907	—	—	1.574–1.582	1.574–1.582	①	NA
1990	60.6 (993)	1.258–1.259	1.866–1.867	1.906–1.907	—	—	1.574–1.582	1.574–1.582	①	NA
	79 (1295)	NA	NA	NA	—	—	1.314–1.322	1.290–1.298	0.0014–0.0029	0.0040–0.0098
1991–92	60.6 (993)	1.258–1.259	1.866–1.867	1.906–1.907	—	—	1.574–1.582	1.574–1.582	①	NA
	79 (1295)	NA	NA	NA	—	—	1.314–1.322	1.290–1.298	0.0014–0.0029	0.0040–0.0098

NA—Not available
① Front: 0.0016–0.0055
Center: 0.0035–0.0075
Rear: 0.0024–0.0063

CRANKSHAFT AND CONNECTING ROD SPECIFICATIONS

All measurements are given in inches.

Year	Engine Displacement cu. in. (cc)	Crankshaft Main Brg. Journal Dia.	Main Brg. Oil Clearance	Shaft End-play	Thrust on No.	Connecting Rod Journal Diameter	Oil Clearance	Side Clearance
1988	60.6 (993)	1.652–1.653	0.0008–0.0017	0.0008–0.0087	3 ①	1.573–1.576	0.0008–0.0017	0.0059–0.0098
1989	60.6 (993)	1.652–1.653	0.0008–0.0017	0.0008–0.0087	3 ①	1.573–1.576	0.0008–0.0017	0.0059–0.0098
1990	60.6 (993)	1.652–1.653	0.0008–0.0017	0.0008–0.0087	3 ①	1.573–1.576	0.0008–0.0017	0.0059–0.0098
	79 (1295)	1.968–1.969	0.0010–0.0016	0.0008–0.0086	4 ①	1.7707–1.7716	0.0008–0.0017	0.0060–0.0150
1991–92	60.6 (993)	1.652–1.653	0.0008–0.0017	0.0008–0.0087	3 ①	1.573–1.576	0.0008–0.0017	0.0059–0.0098
	79 (1295)	1.968–1.969	0.0010–0.0016	0.0008–0.0086	4 ①	1.7707–1.7716	0.0008–0.0017	0.0060–0.0150

① Thrust washers fitted in cylinder block

VALVE SPECIFICATIONS

Year	Engine Displacement cu. in. (cc)	Seat Angle (deg.)	Face Angle (deg.)	Spring Test Pressure (lbs.)	Spring Installed Height (in.) ⑤	Stem-to-Guide Clearance (in.) Intake	Exhaust	Stem Diameter (in.) Intake	Exhaust
1988	60.6 (993)	45 ①	45.5	56.7 ③	1.705	0.0016–0.0028	0.0018–0.0030	NA	NA
1989	60.6 (993)	45 ①	45.5	56.7 ③	1.705	0.0016–0.0028	0.0018–0.0030	NA	NA
1990	60.6 (993)	45 ①	45.5	56.7 ③	1.705	0.0016–0.0028	0.0018–0.0030	NA	NA
	79 (1295)	45 ②	45.5	58.2 ④	1.78–1.81	0.0008–0.0023	0.0010–0.0025	NA	NA

VALVE SPECIFICATIONS

Year	Engine Displacement cu. in. (cc)	Seat Angle (deg.)	Face Angle (deg.)	Spring Test Pressure (lbs.)	Spring Installed Height (in.) ⑤	Stem-to-Guide Clearance (in.)		Stem Diameter (in.)	
						Intake	Exhaust	Intake	Exhaust
1991-92	60.6 (993)	45 ①	45.5	56.7 ③	1.705	0.0016–0.0028	0.0018–0.0030	NA	NA
	79 (1295)	45 ②	45.5	58.2 ④	1.78–1.81	0.0008–0.0023	0.0010–0.0025	NA	NA

NA—Not available
① Valve contacting angle—45°
Refacing angle—intake: 20° 45° 60°
exhaust: 30° 45° 70°
② Valve contacting angle—45°
Refacing angle—intake: 30° 45° 70°
exhaust: 20° 45° 70°
③ Installed tension at 1.374 in.
④ Installed tension at 1.50 in.
⑤ Spring free length

PISTON AND RING SPECIFICATIONS

All measurements are given in inches.

Year	Engine Displacement cu. in. (cc)	Piston Clearance	Ring Gap			Ring Side Clearance		
			Top Compression	Bottom Compression	Oil Control	Top Compression	Bottom Compression	Oil Control
1988	60.6 (993)	0.0014–0.0022	0.0079–0.0157	0.0079–0.0138	0.0079–0.0315	0.0012–0.0028	0.0008–0.0024	0.0004–0.0012
1989	60.6 (993)	0.0014–0.0022	0.0079–0.0157	0.0079–0.0138	0.0079–0.0315	0.0012–0.0028	0.0008–0.0024	0.0004–0.0012
1990	60.6 (993)	0.0014–0.0022	0.0079–0.0157	0.0079–0.0138	0.0079–0.0315	0.0012–0.0028	0.0008–0.0024	0.0004–0.0012
	79 (1295)	0.0018–0.0026	0.0110–0.0160	0.0140–0.0200	0.0080–0.0270	0.0012–0.0027	0.0008–0.0023	NA
1991-92	60.6 (993)	0.0014–0.0022	0.0079–0.0157	0.0079–0.0138	0.0079–0.0315	0.0012–0.0028	0.0008–0.0024	0.0004–0.0012
	79 (1295)	0.0018–0.0026	0.0110–0.0160	0.0140–0.0200	0.0080–0.0270	0.0012–0.0027	0.0008–0.0023	NA

NA—Not available

TORQUE SPECIFICATIONS

All readings in ft. lbs.

Year	Engine Displacement cu. in. (cc)	Cylinder Head Bolts ②	Main Bearing Bolts	Rod Bearing Bolts	Crankshaft Pulley Bolts ①	Flywheel Bolts	Manifold		Spark Plugs
							Intake	Exhaust	
1988	60.6 (993)	39.8–47.0	39.1–47.7	17.4–22.4	65.1–72.0	28.9–36.2	10.8–15.9	21.7–32.5	10.8–15.9
1989	60.6 (993)	39.8–47.0	39.1–47.7	17.4–22.4	65.1–72.0	28.9–36.2	10.8–15.9	21.7–32.5	10.8–15.9
1990	60.6 (993)	39.8–47.0	39.1–47.7	17.4–22.4	65.1–72.0	28.9–36.2	10.8–15.9	21.7–32.5	10.8–15.9
	79 (1295)	43.4–49.2	32.5–39.8	25.3–32.5	65.1–72.0	57.9–72.0	10.8–15.9	21.7–32.5	10.8–15.9

TORQUE SPECIFICATIONS

All readings in ft. lbs.

Year	Engine Displacement cu. in. (cc)	Cylinder Head Bolts ②	Main Bearing Bolts	Rod Bearing Bolts	Crankshaft Pulley Bolts ①	Flywheel Bolts	Manifold Intake	Manifold Exhaust	Spark Plugs
1991-92	60.6 (993)	39.8–47.0	39.1–47.7	17.4–22.4	65.1–72.0	28.9–36.2	10.8–15.9	21.7–32.5	10.8–15.9
	79 (1295)	43.4–49.2	32.5–39.8	25.3–32.5	65.1–72.0	57.9–72.0	10.8–15.9	21.7–32.5	10.8–15.9

① Crankshaft timing belt attaching bolt
② Tighten cylinder head bolts in 2–3 steps.

BRAKE SPECIFICATIONS

All measurements in inches unless noted.

Year	Model	Lug Nut Torque (ft. lbs.)	Master Cylinder Bore	Brake Disc Minimum Thickness	Brake Disc Maximum Runout	Standard Brake Drum Diameter	Minimum Lining Thickness Front	Minimum Lining Thickness Rear
1988	G100LS	NA	0.81	0.39	0.04	7.126	0.12	0.04
1989	G100LS	NA	0.81	0.39	0.04	7.126	0.12	0.04
1990	G100LS	NA	0.81	0.39	0.04	7.126	0.12	0.04
	G102LS	NA	0.81	0.67	0.04	7.913	0.12	0.04
1991-92	G100LS	NA	0.81	0.39	0.04	7.126	0.12	0.04
	G102LS	NA	0.81	0.67	0.04	7.913	0.12	0.04

NA—Not available

WHEEL ALIGNMENT

Year	Model		Caster Range (deg.)	Caster Preferred Setting (deg.)	Camber Range (deg.)	Camber Preferred Setting (deg.)	Toe-in (in.)	Steering Axis Inclination (deg.)
1988	G100LS	Front	2⁷⁄₁₆P–3⁷⁄₁₆P	2¹⁵⁄₁₆P	0P–1P	¹¹⁄₃₂P	0	12
		Rear	—	—	1⁹⁄₃₂N–³⁄₃₂N	¹¹⁄₁₆N	1/4	—
1989	G100LS	Front	2⁷⁄₁₆P–3⁷⁄₁₆P	2¹⁵⁄₁₆P	0P–1P	¹¹⁄₃₂P	0	12
		Rear	—	—	1⁹⁄₃₂N–³⁄₃₂N	¹¹⁄₁₆N	1/4	—
1990	G100LS	Front	2⁷⁄₁₆P–3⁷⁄₁₆P	2¹⁵⁄₁₆P	0P–1P	¹¹⁄₃₂P	0	12
		Rear	—	—	1⁹⁄₃₂N–³⁄₃₂N	¹¹⁄₁₆N	1/4	—
	G102LS	Front	2⁷⁄₁₆P–3⁷⁄₁₆P	2¹⁵⁄₁₆P	0P–1P	¹¹⁄₃₂P	0	12
		Rear	—	—	1⁹⁄₃₂N–³⁄₃₂N	¹¹⁄₁₆N	1/4	—
1991-92	G100LS	Front	2⁷⁄₁₆P–3⁷⁄₁₆P	2¹⁵⁄₁₆P	0P–1P	¹¹⁄₃₂P	0	12
		Rear	—	—	1⁹⁄₃₂N–³⁄₃₂N	¹¹⁄₁₆N	1/4	—
	G102LS	Front	2⁷⁄₁₆P–3⁷⁄₁₆P	2¹⁵⁄₁₆P	0P–1P	¹¹⁄₃₂P	0	12
		Rear	—	—	1⁹⁄₃₂N–³⁄₃₂N	¹¹⁄₁₆N	1/4	—

N—Negative
P—Positive

ENGINE MECHANICAL

NOTE: Disconnecting the negative battery cable on some vehicles may interfere with the functions of the on board computer systems and may require the computer to undergo a relearning process, once the negative battery cable is reconnected.

Engine Assembly

REMOVAL & INSTALLATION

1.0L Engine

1. Disconnect the negative and positive battery cables and remove the battery.
2. Disconnect the windshield washer hose, matchmark the hood with the hinges and remove the hood.
3. Drain the engine coolant by removing the plug from the bottom of the radiator.
4. Drain the engine oil.
5. Drain the transaxle oil.
6. Remove the 2 wiring clamps from the battery tray.
7. Remove the 4 bolts retaining the battery tray and remove the tray.
8. Disconnect the accelerator cable from the throttle body. Remove the accelerator cable from the clamp. Position the cable aside.
9. Label and disconnect the hoses attached to the air cleaner assembly.
10. Disconnect the air cleaner hose at the air cleaner side.
11. Remove the air cleaner by removing the 4 bolts.
12. Remove the PCV hose.
13. Remove the air cleaner hose from the throttle body.
14. Label and disconnect the brake booster, air conditioning fast idle VSV (air conditioning equipped vehicles only) and barometric pressure VSV vacuum hoses.
15. Disconnect the spark plug wires.
16. Disconnect the engine-to-fuse block wiring harness connector.
17. Disconnect the transaxle ground cable at the transaxle side.
18. Disconnect the engine wire clamp at the lower side of the ignitor.
19. Disconnect the ignitor connector.
20. Disconnect the engine wire and cowl wire.
21. Remove the clutch cable weight. Disconnect the clutch cable.
22. Disconnect the speedometer cable from the transaxle.
23. Disconnect the ground cable between the engine and the radiator.
24. Disconnect the ground cable to the alternator.
25. If equipped with air conditioning, disconnect the compressor clutch wire.
26. Relieve the fuel system pressure. Disconnect the fuel hose at the upper side of the fuel filter.
27. Remove the engine harness clamp bolt at the lower side of the pressure sensor. Disconnect the pressure sensor connector.
28. Remove the upper and lower radiator hoses.
29. Remove the heater inlet pipe hose.
30. Remove the heater hose from the air plenum.
31. Remove the vacuum hose between the BVSV and the charcoal canister.
32. Disconnect the radiator fan motor connector.
33. Remove the radiator.
34. Remove the compressor drive belt. Remove the 4 compressor retaining bolts from the bracket. Remove the compressor from the bracket and set aside supporting, compressor with a length of mechanics wire.
35. Remove the glove box door subassembly.
36. Disconnect the ECU connector from the engine wiring harness.
37. Remove the 2 engine harness clamps. Draw the engine harness toward the engine compartment.
38. Raise and safely support the vehicle.
39. Remove the right and left engine under covers.
40. Disconnect the exhaust pipe from the exhaust manifold by removing the 3 nuts.
41. Remove the exhaust pipe support No. 1.
42. Remove the stabilizer bar. Remove the lower arm from the lower arm bracket.
43. Disconnect the control rod.
44. Remove the halfshaft.
45. Disconnect the engine mounting front stopper and engine mounting rear bracket from the engine lower member.
46. Remove the damper weight.
47. Install a suitable engine lifting device and support the engine.
48. Remove the engine mounting upper right insulator and front mounting bracket attaching nuts.
49. Disconnect the engine mounting lower left insulator and rear mounting attaching bolts.
50. Remove the engine front mounting bracket from the cylinder head.
51. Remove the engine from the vehicle.

To install:

52. Lower engine into vehicle and install the engine front mounting bracket to the cylinder head.
53. Tighten the engine mounting lower left bracket to transaxle. Tighten to 22–33 ft. lbs. (30–44 Nm).
54. Fit the front mounting bracket to the engine mounting upper right insulator. Secure the front mounting bracket. Tighten to 29–40 ft. lbs. (39–54 Nm).
55. Remove the lifting device.
56. Install the engine mounting front stopper and engine mounting rear bracket to the engine lower mounting member. Tighten the front stopper to 54–76 ft. lbs. (74–103 Nm). Tighten the rear bracket to 54–76 ft. lbs. (74–103 Nm).

NOTE: The front stopper should be mounted at the center of the lower engine mounting stopper.

57. Install the damper weight to the engine mounting rear bracket.
58. Install the halfshaft using a new snapring at the end of the halfshaft.
59. Install the transaxle control rod. Tighten to 12–22 ft. lbs. (17–30 Nm).
60. Install the lower arm to the lower arm bracket. Tighten to 58–76 ft. lbs. (79–103 Nm).
61. Install the stabilizer bar. Tighten stabilizer bracket to 29–43 ft. lbs. (39–58 Nm). Using a new nut, tighten stabilizer nut to 54–80 ft. lbs. (74–108 Nm).
62. Install the exhaust pipe using a new gasket. Tighten 3 nuts to 36–55 ft. lbs. (49–75 Nm).
63. Install the exhaust pipe support. Tighten to 25–36 ft. lbs. (34–49 Nm).
64. Install the right and left engine undercovers.
65. Lower the vehicle.
66. Connect the ECU wiring harness connector. Fit the rubber grommet into the dash panel.
67. Install the 2 clamps retaining the engine wiring harness.
68. Connect the ECU connector to the engine harness.
69. Install the glove box door subassembly.
70. Install the air conditioning compressor. Tighten the 4 bolts to 18 ft. lbs. (25 Nm).
71. Install the compressor drive belt and adjust tension.
72. Connect the alternator ground cable.
73. Connect the compressor clutch wiring connector.
74. Install the radiator. Ensure that the ground cable is connected to the right side bolt.
75. Connect the radiator cooling fan motor connector.
76. Connect the heater hose to the air plenum.
77. Install the heater hose to the inlet pipe.

78. Install the upper and lower radiator hoses.
79. Connect the pressure sensor connector and install the wiring harness clamp and retaining bolt.
80. Install the fuel line hose to the fuel filter, using the union bolt and a new gasket. Tighten to 25–33 ft. lbs. (34–44 Nm).
81. Connect the ground cable connecting the exhaust manifold with the radiator attaching bolt.
82. Connect the speedometer cable.
83. Connect the clutch cable. Install the clutch cable weight.
84. Connect the engine-to-cowl wiring harness connector.
85. Connect the ignitor connector.
86. Install the engine wire clamp under the ignitor.
87. Connect the transaxle ground cable.
88. Connect the engine-to-fuse block connector.
89. Connect the spark plug wires.
90. Connect the vacuum hoses for the brake booster and air conditioning VSV, if equipped with air conditioning. Connect the barometric pressure VSV.
91. Install the air cleaner hose to the throttle body.
92. Install the PCV hose to the cylinder head cover.
93. Install the air cleaner assembly.
94. Connect the air cleaner hose. Connect the air cleaner rubber hoses.
95. Connect the vacuum hose between the charcoal canister and the BVSV.
96. Connect the accelerator cable to the throttle body.

NOTE: Adjust play in accelerator cable to 0.118–0.315 in. (3.0–8.0mm).

97. Install battery tray with the 4 mounting bolts. Install wire harness clamps.
98. Fill the engine and transaxle with oil.
99. Fill cooling system.
100. Install the battery.
101. Connect the alternator wire and terminal of the relay block assembly to the positive battery terminal.
102. Connect negative battery cable.
103. Install the hood.
104. Start the engine and check for coolant, fuel and oil leaks. Stop engine and check oil level. Add, as necessary.

1.3L Engine

1. Disconnect the positive and negative battery cables and remove the battery.
2. Disconnect the windshield washer hose, matchmark the hood with the hinges and remove the hood.
3. Drain the engine coolant by removing the plug from the bottom of the radiator.
4. Drain the engine oil. Remove the oil filter.
5. Drain the transaxle oil.
6. Remove the wiring clamp and washer hose clamp from the battery tray.
7. Remove the 4 bolts retaining the battery tray and remove the tray.
8. Disconnect the accelerator cable from the throttle body. Remove the accelerator cable from the clamp. Position the cable aside.
9. Remove the power steering pump drive belt. Remove the power steering pump and set aside, supporting with a length of mechanics wire.
10. Remove the air conditioning belt. Remove the belt tensioner assembly.
11. If equipped with air conditioning, disconnect the compressor clutch wire.
12. Remove the compressor drive belt. Remove the 4 compressor retaining bolts from the bracket. Remove the compressor from the bracket and set aside, supporting compressor with a length of mechanics wire.
13. Remove the glove box door subassembly.
14. Disconnect the ECU connector from the engine wiring harness.
15. Remove the air cleaner by removing the 4 bolts.
16. Remove the PCV hose.
17. Remove the air cleaner hose from the throttle body.
18. Label and disconnect the brake booster and air conditioning fast idle VSV, if equipped with air conditioning. Disconnect the barometric pressure VSV vacuum hoses.
19. Disconnect the spark plug wires.
20. Disconnect the engine-to-fuse block wiring harness connector.
21. Disconnect the transaxle ground cable at the transaxle side.
22. Disconnect the engine wire clamp at the lower side of the ignitor.
23. Disconnect the engine-to-cowl wiring harness connector.
24. Remove the clutch cable weight. Disconnect the clutch cable.
25. Disconnect the speedometer cable from the transaxle.
26. Disconnect the ground cable between the engine and the radiator.
27. Disconnect the ground cable to the alternator.
28. Relieve the fuel system pressure. Disconnect the fuel hose at the upper side of the fuel filter.
29. Remove the engine harness clamp bolt at the lower side of the pressure sensor. Disconnect the barometric pressure sensor connector.
30. Remove the upper and lower radiator hoses.
31. Remove the heater hoses.
32. Remove the vacuum hose between the BVSV and the charcoal canister.
33. Disconnect the radiator fan motor connector.
34. Remove the radiator.
35. Remove the 2 engine harness clamps. Draw the engine harness toward the engine compartment.
36. Raise and safely support the vehicle.
37. Remove the engine under cover.
38. Disconnect the exhaust pipe from the exhaust manifold by removing the 3 nuts.
39. Remove the exhaust pipe support No. 1.
40. Remove the stabilizer bar. Remove the lower arm from the lower arm bracket.
41. Disconnect the control rod.
42. Remove the halfshafts at the transaxle end.
43. Remove the attaching bolt at the exhaust manifold side of the exhaust pipe.
44. Remove the attaching bolt of the exhaust pipe support bracket clamp to the exhaust front pipe.
45. Install a suitable engine lifting device and support the engine.
46. Remove the bolts connecting the engine mounting rear brackets.
47. Remove the engine mounting left support attaching bolts from the engine mounting lower left insulator and the engine mounting front left bracket. Remove support.
48. Remove the bolt and nut connecting the engine mounting front left bracket to the insulator.
49. Lower the engine slightly. Remove the engine mounting bracket from the vehicle.
50. Remove the engine from the vehicle.

To install:

51. Lower engine into vehicle and install the engine front mounting bracket to the cylinder head.
52. Tighten the engine mounting front insulator attaching bolts. Tighten bolt to 29–40 ft. lbs. (39–54 Nm). Tighten nut to 11–17 ft. lbs. (15–23 Nm).
53. Connect the left front mounting bracket to the body. Tighten to 29–40 ft. lbs. (39–54 Nm).
54. Install engine mounting lower left insulator to the engine mounting lower left bracket. Tighten to 29–40 ft. lbs. (39–54 Nm).
55. Install the engine mounting left support. Tighten to 29–40 ft. lbs. (39–54 Nm).
56. Adjust engine height so the engine rear bracket is aligned with the engine mounting bracket.
57. Connect the engine mounting rear bracket to the engine mounting bracket. Tighten to 40–51 ft. lbs. (54–69 Nn).

58. Remove the lifting device.
59. Install the transaxle control rod. Tighten to 12–22 ft. lbs. (17–30 Nm).
60. Install the stabilizer bar. Tighten stabilizer bracket to 29–43 ft. lbs. (39–58 Nm). Tighten stabilizer nut, using a new nut, to 54–80 ft. lbs. (74–108 Nm).
61. Install the engine undercover.
62. Install the exhaust pipe using a new gasket. Tighten 3 nuts to 36–55 ft. lbs. (49–75 Nm).
63. Install the exhaust pipe support. Tighten to 25–36 ft. lbs. (34–49 Nm).
64. Lower the vehicle.
65. Connect the ECU wiring harness connector. Fit the rubber grommet into the dash panel.
66. Install the 2 clamps retaining the engine wiring harness.
67. Connect the ECU connector to the engine harness.
68. Install the glove box door subassembly.
69. Install the air conditioning compressor. Tighten the 4 bolts to 18 ft. lbs. (25 Nm).
70. Install the compressor drive belt and adjust tension.
71. Connect the compressor clutch wiring connector.
72. Install the vane pump. Install power steering pump belt and adjust.
73. Connect the alternator ground cable.
74. Install the radiator. Ensure that the radiator is properly seated in the mounting grommets.
75. Connect the radiator cooling fan motor connector.
76. Connect the heater hoses.
77. Install the upper and lower radiator hoses.
78. Connect the barometric pressure sensor connector and install the wiring harness clamp and retaining bolt.
79. Install the fuel line hose to the fuel filter, using the union bolt and a new gasket. Tighten to 25–33 ft. lbs. (34–44 Nm).
80. Connect the ground cable connecting the exhaust manifold with the radiator attaching bolt.
81. Connect the speedometer cable.
82. Connect the clutch cable. Install the clutch cable weight.
83. Connect the engine-to-cowl wiring harness connector.
84. Connect the transaxle ground cable.
85. Connect the engine-to-fuse block connector.
86. Connect the spark plug wires.
87. Connect the vacuum hoses for the brake booster, air conditioning VSV, if equipped with air conditioning, and the barometric pressure VSV.
88. Install the air cleaner hose to the throttle body.
89. Install the PCV hose to the cylinder head cover.
90. install the air cleaner assembly.
91. Connect the air cleaner hose. Connect the air cleaner rubber hoses.
92. Connect the vacuum hose between the charcoal canister and the BVSV.
93. Connect the accelerator cable to the throttle body.

NOTE: Adjust play in accelerator cable to 0.118–0.315 in. (3.0–8.0mm).

94. Install battery tray with the 4 mounting bolts. Install wire harness clamps.
95. Fill the engine and transaxle with oil.
96. Fill cooling system.
97. Install the battery.
98. Connect the alternator wire and terminal of the relay block assembly to the positive battery terminal.
99. Connect negative battery cable.
100. Install the hood.
101. Start the engine and check for coolant, fuel and oil leaks. Stop engine and check oil level. Add, as necessary.

Engine Mounts

REMOVAL & INSTALLATION

1. Support the engine.
2. Remove the bolts attaching the mount bracket to the engine and body.
3. Remove insulator attaching bolts.
4. Remove insulator.

To install:

5. Install insulator attaching bolts.
6. Install mount bracket and attach bracket-to-body bolts.
7. Remove engine support.

Cylinder Head

REMOVAL & INSTALLATION

1.0L Engine

1. Disconnect the negative battery cable.
2. Drain the coolant.
3. Drain the engine oil.
4. Remove the timing belt.
5. Label and disconnect all vacuum hoses necessary to remove cylinder head.
6. Relieve the system fuel pressure, disconnect the fuel hose at the upper side of the fuel filter and plug the hose to prevent the entry of dirt.
7. Disconnect the radiator lower hose from the inlet pipe.
8. Remove the heater hose from the inlet pipe.
9. Disconnect the radiator upper hose at the cylinder head.
10. Disconnect the heater hose from the intake manifold.
11. Remove the EGR pipe.
12. Label and disconnect all electrical connectors necessary to remove cylinder head.
13. Disconnect the fuel hose between the cold start injector and the fuel delivery pipe.
14. Remove the air plenum support. Disconnect the fuel hose clamp at the back of the air plenum. Remove the air plenum retaining bolts and nuts and remove the air plenum.
15. Remove the fuel return hose from the pressure regulator. Remove the fuel hose from the delivery pipe. Remove the fuel delivery pipe from the intake manifold.
16. Remove the 3 fuel delivery pipe insulators from the intake manifold. Remove the fuel injectors.

NOTE: Label the fuel injectors so they may be reinstalled in their original positions.

17. Disconnect the radiator fan motor connector.
18. Remove the engine wire harness clamp under the cold start injector time switch.
19. Remove the engine wire harness from the water outlet pipe bracket clamp.
20. Disconnect the oil cooler hose from the intake manifold.
21. Remove the intake manifold retaining bolts and nuts and remove the intake manifold. Remove the intake manifold gasket and discard.
22. Label and remove the spark plug wires. Remove the spark plugs.
23. Remove the exhaust pipe-to-exhaust manifold retaining nuts. Remove the exhaust pipe clamp. Disconnect the exhaust pipe from the exhaust manifold. Remove the exhaust manifold cover. Remove the exhaust manifold support attaching bolts. Remove the oxygen sensor. Remove the exhaust manifold attaching bolts and nuts and remove the exhaust manifold.
24. Remove distributor attaching bolt and remove distributor.
25. Remove the alternator bracket attaching bolt from the head.
26. Remove the cylinder head cover.
27. Loosen the cylinder head bolts over 2–3 stages, in sequence. Remove the cylinder head.

To install:

28. Install a new cylinder head gasket on the engine block.

NOTE: Before installing the cylinder head, ensure that the cylinder head bolt holes are free of oil and coolant.

29. Rotate the camshaft until the key groove is at the top position. If this position is not maintained, the valves

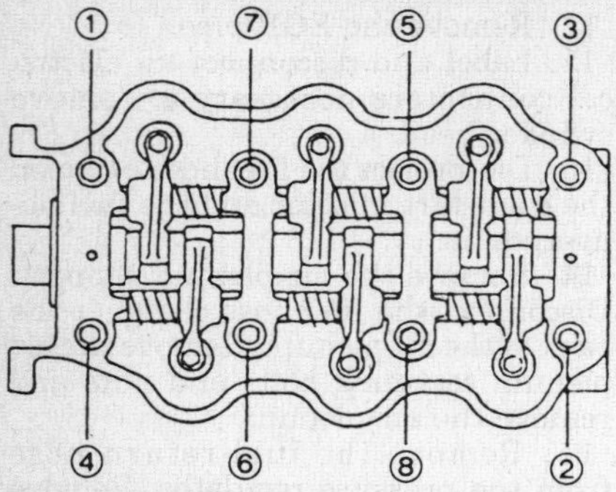

Cylinder head bolt removal sequence – 1.0L engine

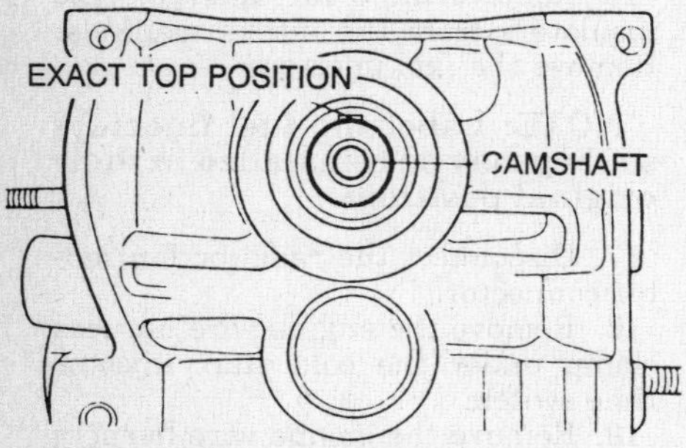

Proper camshaft position prior to cylinder head installation – 1.0L engine

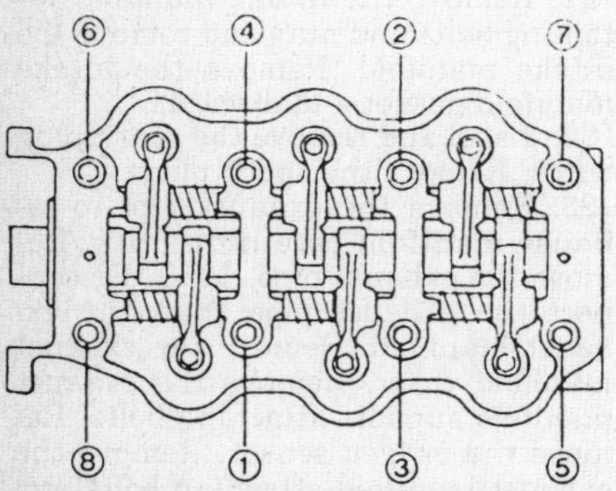

Cylinder head bolt installation sequence 1.0L engine

will interfere with the piston causing damage.

30. Place the cylinder head on the engine block.

31. Ensure that the threads of the cylinder head bolts are dry and install them into the engine block.

32. Tighten the cylinder head bolts, in sequence, over 2–3 stages. Tighten to 40–47 ft. lbs. (54–64 Nm).

33. Fill the camshaft chamber of each cylinder with approximately 1.84 cu. in. (30cc) of oil.

34. Fill the distributor drive gear chamber with approximately 1.84 cu. in. (30cc) of oil.

35. Replace the O-ring on the distributor body. Remove the distributor cap. Set the distributor shaft so drilled mark at the end of the distributor drive gear is aligned with the drilled mark on the distributor body. Insert the distributor assembly into the distributor housing so the split line of the distributor body is aligned with the embossed line of the distributor housing.

36. Tighten the distributor attaching bolts. Install the distributor cap on the distributor.

37. Install a new exhaust manifold gasket and install the exhaust manifold. Tighten bolts to 11–16 ft. lbs. (15–22 Nm) starting in the center and working toward the outside.

38. Install the exhaust manifold support. Tighten to 22–33 ft. lbs. (29–44 Nm).

39. Install the oxygen sensor using a new gasket. Tighten to 22–29 ft. lbs. (30–39 Nm).

40. Install the exhaust manifold cover. Tighten to 4–7 ft. lbs. (6–9 Nm).

41. Install the exhaust pipe using an new gasket. Tighten to 36–55 ft. lbs. (49–75 Nm). Install the exhaust pipe clamp. Tighten to 25–36 ft. lbs. (34–49 Nm).

42. Install the spark plugs. Connect the spark plug wires.

43. Install the intake manifold using a new gasket. Connect the engine wiring harness to the 2 stud bolts on the cylinder head. Tighten the bolts to 11–16 ft. lbs. (15–22 Nm) starting in the center and working toward the outside.

44. Connect the oil cooler hose.

45. Connect the radiator fan motor connector.

46. Install the 3 injector insulators to the intake manifold.

47. Install a new O-ring on the injectors.

48. Install the injectors into the insulators.

49. Install the fuel delivery pipe heat insulator to the intake manifold. Apply a small amount of silicone or gasoline to the O-ring of the injector. Install the fuel delivery pipe to the intake manifold. Secure the fuel delivery pipe using the retaining nuts. Tighten to 7–12 ft. lbs. (10–16 Nm).

50. Connect the fuel return hose to the pressure regulator.

51. Connect the fuel hose union bolt using a new gasket. Tighten to 25–33 ft. lbs. (34–44 Nm).

52. Connect the injector connector.

53. Install the air plenum using a new gasket. Tighten to 11–16 ft. lbs. (15–22 Nm).

54. Install the fuel hose clamp to the back of the air plenum.

55. Connect the fuel hose to the cold start injector and fuel delivery pipe using a new gasket. Tighten the union bolt to 8–13 ft. lbs. (12–18 Nm).

56. Install the engine wiring harness to the bolt at the back of the air plenum.

57. Install the EGR pipe.

58. Connect all wiring connectors that were disconnected for cylinder head removal.

59. Connect all vacuum lines that were disconnected for cylinder head removal.

60. Install the heater hose to the intake manifold.

61. Install the radiator upper hose to the cylinder head.

62. Connect the heater hose to the water inlet pipe.

63. Install the radiator lower hose to the inlet pipe.

64. Connect the fuel hose to the fuel filter using a new gasket. Tighten to 25–33 ft. lbs. (34–44 Nm).

65. Connect the pressure sensor connector.

66. Install the alternator bracket to the cylinder head.

67. Install the timing belt.

68. Adjust the valve clearance.

69. Install the cylinder head cover. Tighten to 6–9 ft. lbs. (8–12 Nm).

70. Connect the accelerator cable to the throttle body. Adjust the accelerator cable freeplay to 3.0–8.0mm. Secure it to the throttle body with the locknut. Tighten to 7–12 ft. lbs. (10-16 Nm). Secure the accelerator cable wire to the clamp.

71. Install the air plenum support.

72. Fill the engine with oil.

73. Connect negative battery cable.

74. Fill the cooling system and check for leaks. Start engine and allow engine to come to normal operating temperature. Recheck for coolant leaks. Allow engine to warm up sufficiently to confirm operation of cooling fan.

75. Check engine timing.

1.3L Engine

1. Disconnect the negative battery cable.

2. Drain the coolant into a clean container for reuse.

3. Relieve fuel system pressure.

4. Remove the air cleaner. Disconnect the PCV.

5. Remove the accelerator cable.

6. If equipped with automatic transaxle, remove the throttle valve (TV) cable.

7. Label and remove the rubber hoses from the air plenum.

8. Remove the modulator.

9. Remove the EGR valve assembly.

10. Label and disconnect electrical connectors required for removal of the head.

11. Label and disconnect injector connectors.

12. Remove the engine wire clamp and clamp bolt.

13. Disconnect the throttle body coolant circulating hose.
14. Remove the throttle body attaching nuts and bolts and the air plenum support. Remove the throttle body.
15. Remove the vacuum pipe assembly.
16. Remove the air plenum supports.
17. Remove the air plenum attaching nuts and bolts, loosening evenly, in 2–3 steps, in sequence. Remove the air plenum.
18. Disconnect the fuel supply and return hoses from the fuel rail.
19. Remove the fuel rail.
20. Label and remove the fuel injectors.
21. Remove the injector and fuel rail insulators.
22. Disconnect the radiator thermo control switch from the water inlet and remove the water inlet from the cylinder block.
23. Remove the thermostat and water inlet hose from the water inlet.
24. Disconnect the water by-pass inlet hose. Disconnect the heater inlet hose.
25. Remove the timing belt.

NOTE: Do not rotate the camshaft independently of the crankshaft with the timing belt removed as damage to the valves may result.

26. Remove the intake manifold by loosening the bolts, in sequence, over 2–3 stages.
27. Remove the radiator and cooling fan assembly.
28. Disconnect the oxygen sensor. Remove the wire from the clamp in the cylinder head cover. Remove the oxygen sensor.
29. Remove the dipstick.
30. Remove the engine ground cable from the cylinder head cover.
31. Remove the exhaust manifold top and side covers.
32. If equipped with air conditioning, remove the exhaust manifold lower cover.
33. Remove the dipstick tube clamp attaching screw and remove the tube.
34. Remove the exhaust pipe clamp bolt and remove the clamp.
35. Remove the nuts attaching the exhaust pipe to the exhaust manifold.
36. Remove the exhaust manifold support.
37. Remove the exhaust manifold bolts and nuts, in sequence, by loosening them evenly, over 2–3 stages. Remove the exhaust manifold gasket and discard.
38. Label and remove the spark plug wires. Remove the spark plugs.
39. Remove the distributor vacuum advance hose. Remove the distributor.
40. Disconnect the appropriate radiator and coolant hoses for removal of the head.

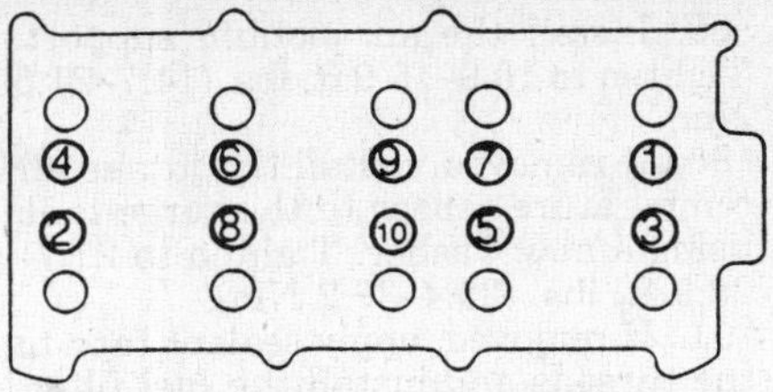

Cylinder head bolt removal sequence—1.3L engine

41. Remove the cylinder head cover.
42. Remove the camshaft pulley attaching bolt from the camshaft and remove the pulley.

NOTE: Do not allow the camshaft to turn during removal.

43. Loosen the 10 bolts retaining the rocker arm shaft, in sequence, over 2–3 stages. Remove the valve rocker shaft together with the rocker arms from the cylinder head.

NOTE: Remove the rocker arm spacers and wave washers from the shaft. Arrange the spacers and washers in order so they may be reinstalled in their original positions.

44. Remove the 2 attaching bolts from camshaft cap No. 5. Remove the remaining camshaft bearing cap bolts and caps.
45. Remove the camshaft from the cylinder head.
46. Loosen the cylinder head bolts, in sequence, over 2–3 stages.

NOTE: Head bolts 1 and 3 are shorter than the other bolts. Bolts 1 and 3 are 4.41 in. (112mm) and the other bolts are 6.10 in. (155mm).

47. Remove the cylinder head.

To install:

48. Clean the cylinder bolt holes.
49. Clean the cylinder block upper gasket surface. Install the cylinder head gasket using the aligning pins for reference.
50. Turn the crankshaft so the crankshaft key groove is at the top position.

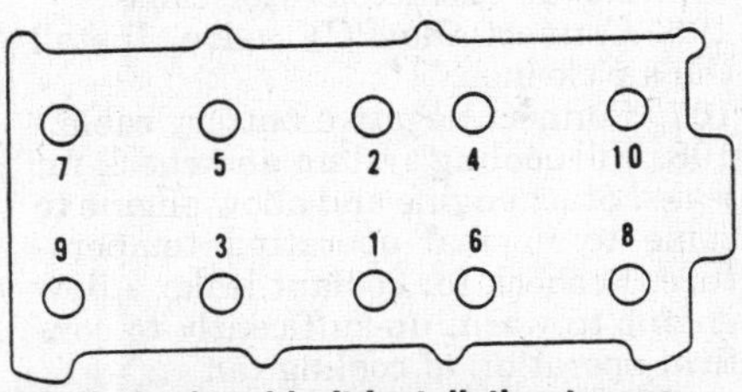

Cylinder head bolt installation torque sequence—1.3L engine

51. Install the cylinder head on the engine block.
52. Coat each cylinder head bolt with a thin film of engine oil and install in the engine block. Tighten the cylinder head bolts, in sequence, over 2–3 stages. Tighten to 43.4–49.2 ft. lbs. (58.8–66.7 Nm).
53. Clean camshaft and rocker arm bolts holes.
54. Apply engine oil to the camshaft journals and thrust bearing sections.
55. Install the camshaft in the cylinder head so the locating pin for the camshaft timing belt pulley is at the top position.
56. Apply Three Bond 1104 sealant or equivalent, to No. 1 camshaft bearing cap attaching section. Wipe off any excess bond that oozes from No. 1 cap. Install the camshaft bearing caps in sequence embossed on the caps.
57. Assemble the rocker arms and wave washers onto the valve rocker shafts in the same order as when disassembled.
58. Install the rocker shaft on the camshaft caps.
59. Ensure that the rocker shaft retaining bolts are clean and install in the cylinder head. Tighten evenly, over 2–3 stages, to 26.7 ft. lbs. (36.2 Nm) for M10 bolts and 12.29 ft. lbs. (16.6 Nm) for M8 bolts.
60. Install the spacers on the rocker shaft between the intake valve rocker arms.
61. Apply sealant tape to the threads and install the coolant temperature sensor. Tighten to 18.1–25.3 ft. lbs. (24.5–34.3 Nm).
62. Connect the bypass hoses and bypass pipe to the water outlet. Install the water outlet using a new gasket.
63. Apply sealant tape to the threads and install the Bi-metal Vacuum Switching Valve (BVSV). Tighten to 18.1–25.3 ft. lbs. (24.5–34.3 Nm).
64. Apply sealant tape to the threads and install the air conditioning water temperature switch. Tighten to 18.1–25.3 ft. lbs. (24.5–34.3 Nm).
65. Apply engine oil to the bore of the T-type camshaft seal and install using a suitable seal installer.
66. Install the camshaft timing belt pulley aligning the pulley with the locating pin on the camshaft and the **F** mark faces out. Install the retaining bolts and tighten to 10.9–15.9 ft. lbs. (14.7–21.5 Nm).

NOTE: Prevent the camshaft from turning while torquing the camshaft timing belt pulley.

67. Replace the O-ring on the distributor body. Align the cut-out section of the distributor body with the cut-out groove of the cup ring. Apply a few drops of engine oil to the O-ring and install the distributor in the cylinder

head with the cut-out sections facing the top of the engine.
68. Center the distributor mounting bolt holes, in the elongated holes, on the distributor mounting flange and install the mounting bolts. Tighten to 10.9–15.9 ft. lbs. (14.7–21.5 Nm).
69. Connect the distributor connector, the water temperature sender gauge, the water temperature sensor and the air conditioning water temperature switch connectors.
70. Fill the oil well of each cylinder with 30cc of engine oil.
71. Install the exhaust manifold gasket on the cylinder head. Ensure that the side of the gasket where the grommet bulges out faces the exhaust manifold.
72. Connect exhaust manifold intermediate pipe to exhaust manifold using a new gasket. Tighten to 14.5–21.6 ft. lbs. (19.6–29.4 Nm).
73. Install exhaust manifold to cylinder head. Tighten, in sequence, over 2–3 stages, to 21.7–32.5 ft. lbs. (29.4–44.1 Nm).
74. Install exhaust manifold support to the cylinder block. Tighten 21.7–32.5 ft. lbs. (29.4–44.1 Nm).
75. Connect the exhaust pipe to the exhaust manifold intermediate pipe using a new gasket. Tighten to 36.2–55.0 ft. lbs. (49.0–74.5 Nm).
76. Install the exhaust pipe clamp. Tighten to 25.3–36.2 ft. lbs. (34.3–49.0 Nm).
77. Replace the O-ring on the dipstick tube and install dipstick tube in cylinder block. Install the dipstick.
78. Install the exhaust manifold covers.
79. Connect the water outlet hose to the water outlet.
80. Install the intake manifold using a new gasket. Install nuts hand tight.
81. Connect the heater inlet hose to the cylinder head.
82. Install the water bypass pipe to the adjacent intake manifold stud bolt.
83. Tighten intake manifold bolts in sequence to 10.9–15.9 ft. lbs. (14.7–21.5 Nm).
84. Connect the water bypass inlet hose.
85. Install the thermostat so the jiggle pin faces up. Install the water inlet. Tighten to 4.4–6.5 ft. lbs. (5.9–8.8 Nm).
86. Connect the thermo control switch connector.
87. Install the air plenum. Tighten, in sequence, over 2–3 stages, to 21.7–32.5 ft. lbs. (29.4–44.1 Nm).
88. Install the engine hanger and air plenum support to the stud bolt on the cylinder head. Tighten the nut and bolt to 13.8–22.4 ft. lbs. (18.6–30.3 Nm).
89. Install the air plenum support. Tighten to 10.9–15.9 ft. lbs. (14.7–21.5 Nm).
90. If removed, install the intake air temperature sensor to the surge tank using a new washer. Tighten to 21.7–28.9 ft. lbs. (29.4–39.2 Nm).
91. If removed, apply sealant tape to the threads and install the fuel filter. Tighten to 8.7–14.4 ft. lbs. (11.8–19.6 Nm).
92. Install the injector vibration insulator to the intake manifold.
93. Inspect injector grommets and replace, as necessary. Install new O-rings. Insert the injector into the vibration insulator hole of the intake manifold.
94. Install the delivery pipe insulator to the stud bolt of the intake manifold.
95. Apply a light film of clean engine oil to the O-ring of the injector and install the delivery pipe. Install the delivery pipe attaching nuts and tighten to 10.8–15.9 ft. lbs. (14.7–21.6 Nm).

NOTE: After torquing the delivery pipe attaching nuts, ensure that the injectors can be turned by hand. If not, check for damaged O-rings.

96. Install the fuel hose to the delivery pipe using a new gasket. Tighten to 25.4–32.5 ft. lbs. (23.2–44.1 Nm). Connect the fuel return hose to the pressure regulator.
97. Apply a suitable thread sealer to each throttle body attaching bolt. Install the throttle body to the air plenum using a new gasket. Install the air plenum support. Tighten to 10.9–15.9 ft. lbs. (14.7–21.5 Nm).
98. Install the vacuum pipe assembly to the air plenum.
99. Connect the coolant hoses to the throttle body.
100. Install and connect wiring harness connectors.
101. Install the EGR valve to the intake manifold using a new gasket.
102. Install the modulator.
103. Connect vacuum hoses removed for disassembly.
104. If equipped with automatic transaxle, install the TV cable.
105. Install the accelerator cable.
106. Connect the PCV valve. Install the air cleaner.
107. Connect negative battery cable.
108. Fill cooling system and check for leaks. Start engine and allow engine to come to normal operating temperature. Recheck for coolant leaks. Allow engine to warm up sufficently to confirm operation of cooling fan.

Valve Lash

ADJUSTMENT

1.0L Engine

NOTE: The valve clearance adjustment is performed with the engine in the hot condition: coolant temperature above 176°F (80°C) and the oil temperature above 140°F (60°C).

1. Disconnect the negative battery cable.
2. Disconnect the spark plug wires and accelerator cable from the retaining clamps.
3. Remove the air plenum.
4. Remove the oil filler cap. Disconnect the crankcase ventilation hose. Remove the cylinder head cover.
5. Turn the crankshaft until the timing mark on the flywheel is aligned with the timing mark on the bell housing. Ensure that the rocker arms for No. 1 cylinder are up, indicating that both valves are closed.
6. With No. 1 cylinder at TDC of its compression stoke, check clearance of: No. 1 cylinder intake and exhaust valves; No. 2 cylinder exhaust valve and No. 3 cylinder intake valve.

Valve clearances (hot): Intake and Exhaust – 0.006–0.010 in. (0.15–0.25mm).

Reference clearances (cold): Intake and Exhaust – 0.004–0.008 in. (0.10–0.020mm).

7. Rotate crankshaft 360 degrees.
8. With No. 1 cylinder at TDC of its exhaust stroke (valves in overlap), check clearance of: No. 2 cylinder intake valve and No. 3 cylinder exhaust valve.
9. Adjust valves by loosening locknut on the rocker arm and turning adjusting screw.

NOTE: A stubby prybar of no more than 2 in. in length is required to adjust the intake valves.

10. Install the cylinder head cover using a new gasket. Tighten bolts to 6–9 ft. lbs. (7.8–11.8 Nm).
11. Install the air plenum.
12. Attach the spark plug wires and accelerator cable to clamps.
13. Connect negative battery cable.

1.3L Engine

NOTE: The valve clearance adjustment is performed with the engine in the hot condition: coolant temperature at 167–185°F (75–85°C) and the oil temperature above 149°F (65°C).

1. Disconnect the negative battery cable.

2. Disconnect the spark plug wires from the retaining clamps and the accelerator cable from the throttle body.

3. Disconnect the crankcase ventilation hose. Detach the oxygen sensor harness from the clamp. Detach the engine ground wire from the cylinder head cover. Detach the accelerator cable from the clamps.

4. Remove the cylinder head cover.

5. Turn the crankshaft until the recessed mark on the crankshaft pulley is aligned with the timing mark timing belt cover. Ensure that the rocker arms of No. 1 cylinder are up, indicating that both valves are closed.

6. With No. 1 cylinder at TDC of its compression stoke, check clearance of: No. 1 cylinder intake and exhaust valves; No. 2 cylinder intake valve and No. 3 cylinder exhaust valve.

Valve clearances (hot): Intake–0.007–0.011 in. (0.20–0.30mm), Exhaust–0.011–0.014 (0.28–0.38mm).

Reference clearances (cold): Intake–0.007 in. (0.18mm), Exhaust–0.10 in. (0.25mm).

7. Rotate crankshaft 360 degrees.

8. With No. 4 cylinder at TDC of its compression stroke, check clearance of: No. 2 cylinder exhaust valve; No. 3 cylinder intake valve and No. 4 intake and exhaust valves.

9. Adjust valves by loosening locknut on the rocker arm and turning adjusting screw.

10. Install the cylinder head cover using a new gasket. Inspect spark plug rubber grommets for wear. Replace, as required. Ensure that the grommets are seated properly.

11. Tighten the cylinder head cover bolts, over 2–3 stages, in sequence, to 2–4 ft. lbs. (3–5 Nm).

12. Attach the accelerator cable to clamp. Install the oxygen sensor harness. Install the engine ground wire to the cylinder head. Connect the PCV hoses to the cylinder head cover. Install the spark plug wires and position wires in retainers.

13. Connect negative battery cable.

Rocker Arms/Shafts

REMOVAL & INSTALLATION

1.0L Engine

1. Disconnect the negative battery cable.

2. Drain the cooling system.

3. Remove the cylinder head and place in a suitable holding fixture.

4. Loosen the locknut and loosen the adjusting screw of each valve rocker arm.

5. Use a suitable puller to remove the rocker arm shafts, making sure to hold the rocker arm compression springs as the shaft is removed.

NOTE: Be sure to keep all parts in order so they may be reassembled in the same order.

To install:

6. Apply engine oil to the valve rocker shaft, valve rocker arm, compression spring and valve rocker shaft hole of the cylinder block.

7. Fit the rocker arms and compression springs onto the shaft while inserting into position.

NOTE: The valve rocker shafts differ in length between the intake and exhaust sides. The intake rocker shaft is 11.004 in. (279.5mm) in length. The exhaust rocker shaft is 11.201 in. (284.5mm) in length.

8. Install the cylinder head on the engine block.

9. Adjust the valve clearances to cold specifications.

10. Connect negative battery cable.

11. Fill cooling system and check for leaks. Start engine and allow engine to come to normal operating temperature. Recheck for coolant leaks. Allow engine to warm up sufficiently to confirm operation of cooling fan.

12. Readjust valve clearances to hot specifications.

1.3L Engine

1. Disconnect the negative battery cable.

2. Drain the cooling system.

3. Remove the cylinder head cover and timing belt.

4. Loosen the 10 rocker arm retaining bolts evenly, over 2–3 stages, and remove.

5. Remove the rocker arm shaft together with the rocker arms from the cylinder head.

NOTE: When disassembling the rocker arms, spacers and wave washers from the rocker arm shaft, keep all disassembled parts in order so they can be reassembled in the same position.

To install:

6. Assemble the valve rocker arms and wave washers onto the valve rocker shaft. The intake rocker shaft is identified by recessed sections along the shaft.

7. Install the valve rocker shaft on the camshaft caps. For ease of installation, insert the rocker arm onto the camshaft side first.

8. Install the 10 rocker arm retaining bolts and tighten, evenly, over 2–3 stages. Tighten the M10 bolts to 27 ft. lbs. (36 Nm). Tighten the M8 bolts to 13 ft. lbs. (17 Nm).

9. Install the spacers between the rocker arms on the intake valve rocker arm shaft.

10. Install the timing belt.

11. Adjust the valve clearances to the cold specification.

12. Connect negative battery cable.

13. Fill cooling system and check for leaks. Start engine and allow engine to come to normal operating temperature. Recheck for coolant leaks. Allow engine to warm up sufficiently to confirm operation of cooling fan.

14. Readjust the valves to the hot specifications.

Intake Manifold

REMOVAL & INSTALLATION

1.0L Engine

1. Disconnect the negative battery cable.

2. Drain the cooling system to a point where the heater hose to the intake manifold can be removed.

3. Remove the heater hose from the intake manifold.

4. Remove the EGR pipe.

5. Disconnect the idle-up Vacuum Switching Valve (VSV) connector.

6. Disconnect the EGR VSV connector.

7. Disconnect the intake air temperature sensor connector.

8. Disconnect the wire clamp attached to the back of the air plenum.

9. Disconnect the fuel line to the cold start injector.

10. Disconnect the throttle body and cold start injector connectors.

11. Remove the air plenum support.

12. Disconnect the vacuum hose connecting the air plenum and oil filler cap.

13. Disconnect the fuel hose clamp at the back of the air plenum.

14. Remove the bolts and nuts attaching the air plenum and remove the air plenum and gasket.

15. Disconnect the injector connectors.

16. Remove the fuel return hose from the pressure regulator.

17. Remove the fuel hose from the delivery pipe.

18. Remove the fuel delivery pipe from the intake manifold.

19. Remove the 3 fuel delivery pipe insulators from the intake manifold. Remove the fuel injectors.

20. Disconnect the engine ground terminal at the intake manifold.

21. Disconnect the water temperature switch, water temperature sender, cold start injector time switch and radiator fan motor switch connectors, as required.

22. Remove the engine wire clamp under the cold start injector time switch.
23. Remove the clamp from the water outlet pipe bracket.
24. Disconnect the oil cooler hose from the intake manifold.
25. Remove the bolts and nuts attaching the intake manifold.
26. Remove the vacuum hose between the Bimetal Vacuum Switching Valve (BVSV) and the charcoal cannister.
27. Remove the intake manifold and gasket.

To install:

28. Ensure that the gasket mounting surfaces are free of gasket material and install a new intake manifold gasket.
29. Position the intake manifold in place and loosely install the attaching bolts and nuts.
30. Position the engine wiring harness clamps onto the intake manifold studs.
31. Tighten the nuts to 11–16 ft. lbs. (15–22 Nm).
32. Connect the vacuum hose between the BVSV and the charcoal cannister.
33. Connect the oil cooler hose to the intake manifold.
34. Install the wire clamp for the intake manifold water temperature sensor.
35. Connect the engine ground terminal to the intake manifold.
36. Connect the water temperature switch, water temperature sender, cold start injector time switch and radiator fan motor switch connectors, if removed.
37. Install the clamp to the water outlet pipe bracket.
38. Install the engine wire clamp to the cold start injector time switch.
39. Install the 3 fuel delivery pipe insulators to the intake manifold. Install the injectors.
40. Connect the fuel delivery pipe to the intake manifold.
41. Connect the fuel hose to the fuel delivery pipe.
42. Connect the fuel return hose to the pressure regulator.
43. Connect the fuel injector connectors.
44. Install a new air plenum gasket and position the air plenum on the intake manifold and install the attaching nuts and bolts.
45. Connect the fuel hose clamp at the back of the air plenum.
46. Connect the air plenum support.
47. Connect the throttle body and cold start injector connectors.
48. Connect the fuel line to the cold start injector.
49. Connect the wire clamp to the back of the air plenum.
50. Connect the intake air temperature sensor connector.
51. Connect the EGR VSV connector.
52. Connect the idle-up VSV connector.
53. Connect the heater hose to the intake manifold.
54. Connect negative battery cable.
55. Fill cooling system and check for leaks. Start engine and allow engine to come to normal operating temperature. Recheck for coolant leaks. Check for vacuum leaks. Allow engine to warm up sufficiently to confirm operation of cooling fan.

1.3L Engine

1. Disconnect the negative battery cable.
2. Remove the air cleaner.
3. Disconnect the coolant circulating hose from the throttle body. Disconnect the accelerator cable from the throttle body.
4. Remove the nuts and bolts retaining the air plenum supports and remove the supports.
5. Remove the air plenum attaching nuts and bolts, evenly, over 2–3 stages, and remove the air plenum.
6. Remove the intake manifold attaching nuts and bolts, evenly, over 2–3 stages, and remove the intake manifold.

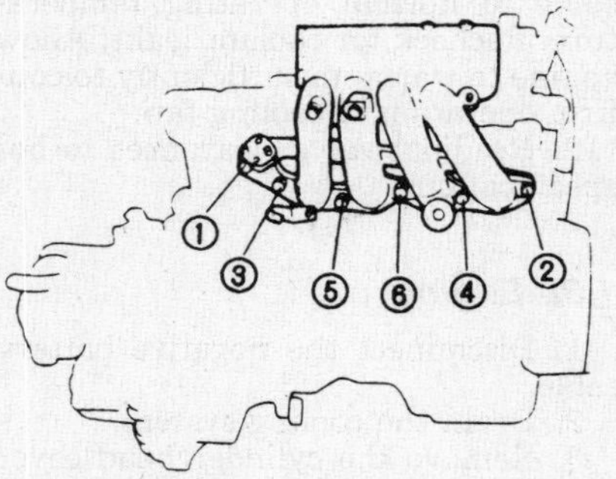

Air plenum removal sequence – 1.3L engine

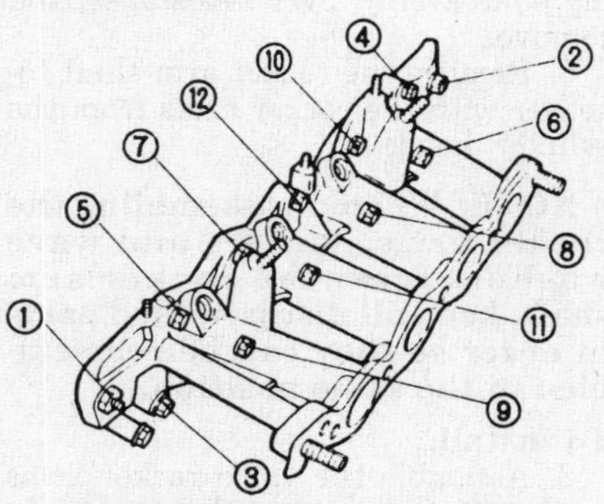

Intake manifold removal sequence – 1.3L engine

To install:

7. Position the intake manifold in place using a new gasket and install the attaching nuts and bolts, in sequence, evenly over 2–3 stages. Tighten to 11–16 ft. lbs. (15–22 Nm).
8. Position the air plenum in place and install the attaching nuts and bolts. Tighten evenly, over 2–3 stages, in sequence, to 22–33 ft. lbs. (29–44 Nm).
9. Install the air plenum supports.
10. Connect the coolant circulating hose and the accelerator cable to the throttle body.
11. Install the air cleaner.
12. Connect negative battery cable.

Exhaust Manifold

REMOVAL & INSTALLATION

1.0L Engine

1. Disconnect the negative battery cable.
2. Disconnect the oxygen sensor connector.
3. Disconnect the distributor connector.
4. Raise and safely support the vehicle.
5. Remove the nuts attaching the exhaust pipe to the exhaust manifold.
6. Remove the exhaust pipe clamp.
7. Disconnect the exhaust pipe from the exhaust manifold.
8. Lower vehicle.
9. Remove the exhaust manifold cover.

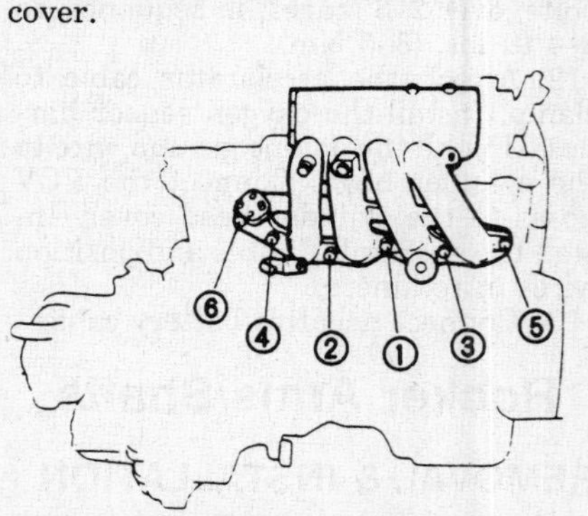

Air plenum bolt installation torque sequence – 1.3L engine

Intake manifold bolt installation torque sequence – 1.3L engine

10. Remove the exhaust manifold support attaching bolts.
11. Remove the exhaust manifold.

To install:

12. Install a new exhaust manifold gasket and position the exhaust manifold in place. Install the attaching bolts and tighten to 11-16 ft. lbs. (15-22 Nm).
13. Install the exhaust manifold support and tighten to 22-33 ft. lbs. (30-44 Nm).
14. Install the oxygen sensor, if removed, using a new gasket. Tighten to 22-29 ft. lbs. (29-39 Nm).
15. Install the exhaust manifold cover. Be sure to connect the engine ground cable with the proper attaching bolt. Tighten to 4-7 ft. lbs. (6-9 Nm).
16. Raise and safely support the vehicle.
17. Connect the exhaust pipe to the exhaust manifold using a new gasket. Tighten bolts to 36-55 ft. lbs. (49-75 Nm).
18. Install the exhaust pipe clamp. Tighten to 25-36 ft. lbs. (34-49 Nm).
19. Lower vehicle.
20. Install the distributor connector.
21. Install the oxygen sensor connector.
22. Connect negative battery cable.

1.3L Engine

1. Disconnect the negative battery cable.
2. Drain the cooling system.
3. Remove the radiator and cooling fan assembly.
4. Disconnect the oxygen sensor wire. Remove the wire from the clamp on the cylinder head cover. Remove the oxygen sensor.
5. Remove the oil dipstick.
6. If equipped with radio, disconnect the engine ground cable from the cylinder head.
7. Remove the exhaust manifold top and side covers. If equipped with air conditioning, remove the exhaust manifold lower covers.
8. Remove the dipstick tube.
9. Raise and safely support the vehicle.
10. Remove the exhaust pipe clamp.
11. Disconnect the exhaust pipe from the exhaust manifold.
12. Remove the exhaust manifold support.
13. Lower vehicle.
14. Remove the exhaust manifold attaching bolts, evenly, over 2-3 stages, and remove the exhaust manifold and gasket.

To install:

15. If separated, connect the 2 sections of the exhaust manifold using a new gasket. Tighten bolts to 15-22 ft. lbs. (20-29 Nm).
16. Install a new exhaust manifold gasket, making sure gasket mating surfaces are free of gasket material.
17. Position the exhaust manifold and install the attaching bolts. Tighten, in sequence, evenly, over 2-3 stages, to 22-33 ft. lbs. (29-44 Nm).
18. Raise and safely support the vehicle.
19. Install the exhaust manifold support and tighten to 22-33 ft. lbs. (29-44 Nm).
20. Connect the exhaust pipe to the exhaust manifold using a new gasket. Tighten to 36-55 ft. lbs. (49-75 Nm).
21. Install the exhaust pipe clamp. Tighten to 25-36 ft. lbs. (34-49 Nm).
22. Lower vehicle.
23. Insert the dipstick tube using a new O-ring. Install the dipstick tube attaching bolt and tighten a few turns. Do not tighten completely at this time.
24. Install the exhaust manifold top cover.
25. Tighten the dipstick tube. Install the dipstick.
26. Install the exhaust manifold side cover and, if equipped with air conditioning, install the exhaust manifold lower cover.
27. Install the oxygen sensor using a new gasket. Connect the oxygen sensor wire connector. Install the wire from the clamp to the cylinder head cover.
28. Install the radiator and cooling fan assembly.
29. Connect negative battery cable.
30. Fill cooling system and check for leaks. Start engine and allow engine to come to normal operating temperature. Recheck for coolant leaks. Allow engine to warm up sufficiently to confirm operation of cooling fan.

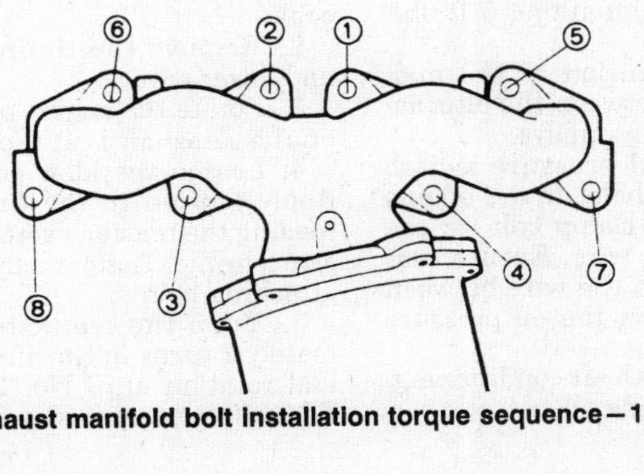

Exhaust manifold bolt installation torque sequence – 1.3L engine

Timing Belt Front Cover

REMOVAL & INSTALLATION

1.0L Engine

1. Disconnect the negative battery cable.
2. Loosen the accelerator cable locknut from the throttle body support. Remove the accelerator cable from the throttle lever. Remove the accelerator cable from the clamp.
3. Remove the PCV hose.
4. Label and remove the air cleaner hoses. Remove the air cleaner assembly.
5. Disconnect the alternator ground cable.
6. If equipped with air conditioning, perform the following:
   a. Loosen the idler pulley mount nut.
   b. Loosen the adjusting bolt.
   c. Remove the drive belt.
7. Loosen the alternator lock bolts, relieve the tension and remove the alternator drive belt.
8. Remove the right front engine under cover.
9. Using a supporting pad under the oil pan, slightly raise the engine to release the tension on the engine mount.
10. Remove the nuts attaching the front engine mount bracket.
11. Remove the 3 bolts attaching the front engine mount to the cylinder head.
12. Lower the lifting device slightly and remove the front engine mount bracket.
13. Remove the water pump pulley.
14. Set the engine to TDC of No. 1 cylinder on the compression stroke.
15. Prevent the crankshaft from turning by inserting a suitable tool into the ring gear.
16. Remove the crankshaft pulley.
17. Remove the timing belt upper and lower covers.

To install:

18. Install the timing belt upper and lower covers. Tighten to 1.4-2.9 ft. lbs. (2.0-3.9 Nm).
19. Remove the crankshaft pulley bolt. Insert the crankshaft pulley with the key groove on the crankshaft pulley aligned with the crankshaft key. Install the crankshaft pulley bolt and tighten to 65-72 ft. lbs. (88-98 Nm).
20. Install the water pump pulley. Tighten to 4-7 ft. lbs. (6-9 Nm).
21. Install the front engine mount bracket to the cylinder head. Tighten

M12 bolt to 36–51 ft. lbs. (49–67 Nm). Tighten M10 bolt to 22–33 ft. lbs. (29–44 Nm).

22. Tighten the front engine mount bracket to 29–40 ft. lbs. (39–54 Nm).
23. Install the front right engine under cover.
24. Remove the lifting device.
25. Install the alternator drive belt. Adjust the alternator drive belt tension to 22 lbs. (10 kg) with 0.197–0.276 in. (5–7mm) deflection applied at the midpoint between the alternator and water pump pulleys.
26. If equipped with air conditioning, install the air conditioning drive belt. Adjust the air conditioning compressor drive belt tension to 22 lbs. (10 kg) with 0.16–0.24 in. (4.0–6.0mm) deflection at the midpoint between the crankshaft and compressor pulleys.
27. Tighten the idler pulley mounting nut. Tighten to 23–35 ft. lbs. (31–47 Nm).
28. Connect the alternator ground cable.
29. Install the air cleaner assembly. Tighten bolts to 7.2–11.6 ft. lbs. (9.8–15.7 Nm). Install the rubber hoses to the air cleaner.
30. Install the PCV hose.
31. Install the accelerator cable to the throttle lever.
32. Tighten the locknut to the throttle body support. Tighten to 7.2–11.6 ft. lbs. (9.8–15.7 Nm).
33. Connect negative battery cable.

1.3L Engine

1. Disconnect the negative battery cable.
2. If equipped with air conditioning, perform the following:
   a. Drain the cooling system into a clean container for reuse.
   b. Remove the radiator and cooling fan assembly..
   c. Remove the air conditioning belt by loosening the locknut and tensioner bolt on the air conditioning idler pulley.
   d. Remove the idler pulley assembly.
   e. Remove the air conditioning compressor and set aside. Support the compressor using a length of mechanics wire.
3. If equipped with power steering, perform the following:
   a. Loosen the power steering belt tensioner locknut and tensioner bolt. Remove the power steering belt.
   b. Remove the power steering pump and set aside. Support the power steering pump using a length of mechanics wire.
4. Remove the hoses from the air cleaner case. Remove the attaching bolts and hose bands and remove the air cleaner case.
5. Disconnect the oil pressure switch connector.
6. Remove the bolts attaching the oil pressure switch wire. Pull wire through the hole in the engine mount and set wire aside.
7. Using the alternator belt tension, loosen the bolts on the water pump pulley.
8. Loosen the adjusting bolt and 2 alternator mounting bolts to slacken tension on the alternator belt and remove the water pump pulley and alternator belt.
9. Using a supporting pad under the oil pan, slightly raise the engine to release the tension on the engine mount.
10. Remove the 3 attaching bolts and 1 nut for the right engine mount.
11. Remove the 3 attaching bolts for the engine mount front insulator.
12. Remove the right front engine mount bracket.
13. Remove the service hole cover in the fender well, just forward of the right front wheel.
14. Remove the crankshaft pulley, accessing the 4 bolts through service hole in the fender well.
15. Remove the timing belt upper and lower covers.

To install:

16. Install the timing belt upper and lower covers. Tighten to 1.4–3.9 ft. lbs. (2–4 Nm).
17. Install the crankshaft timing belt pulley. Tighten bolts to 15–22 ft. lbs. (20–29 Nm). Install the service hole cover.
18. Install the right front engine mount bracket. Tighten bolt to 29–40 ft. lbs. (39–54 Nm).
19. Install the right side engine mount bracket and front engine mount insulator. Tighten bolt to 29–40 ft. lbs. (39–54 Nm). Tighten nut to 11–17 ft. lbs. (15–23 Nm).
20. Remove lifting device from under oil pan.
21. Temporarily attach the water pump pulley.
22. Install the alternator belt and adjust tension using the alternator adjuster. Tighten alternator locknut .
23. Tighten the water pump pulley attaching bolts, using belt tension to hold the pulley. Tighten to 4–7 ft. lbs. (4–9 Nm).
24. Check belt tension at the midpoint of the belt between the alternator and water pump pulleys.
25. Route the oil pressure switch wire through the hole in the engine mount. Install the clamp bolt for the oil pressure switch wire. Ensure that there is no slack in the wire between the clamps. Connect the oil pressure switch connector.
26. If equipped with air conditioning, perform the following:
   a. Install the air conditioning compressor to the engine block.
   b. Install the air conditioning belt tensioner pulley assembly to the cylinder block. Install the air conditioner belt and adjust tension.
   c. Install the radiator and cooling fan assembly.
27. If equipped with power steering, perform the following:
   a. Install the power steering pump to the cylinder head.
   b. Install the power steering belt and adjust tension.
28. Connect negative battery cable.
29. Fill cooling system and check for leaks. Start engine and allow engine to come to normal operating temperature. Recheck for coolant leaks. Allow engine to warm up sufficiently to confirm operation of cooling fan.

OIL SEAL REPLACEMENT

1. Disconnect the negative battery cable.
2. Remove the timing belt.
3. Remove the idler pulley and return spring.
4. Remove the crankshaft timing belt sprocket.
5. Remove the crankshaft timing belt sprocket flange.
6. Remove the front oil seal using an appropriate pry tool.

To install:

7. Install a new front oil seal using an appropriate seal driver.
8. Install the crankshaft timing belt sprocket flange.
9. Install the crankshaft timing belt sprocket.
10. Install the idler pulley and return spring.
11. Install the timing belt.
12. Connect negative battery cable.

Timing Belt and Tensioner

ADJUSTMENT

1.0L Engine

1. Disconnect the negative battery cable.
2. Remove the timing belt upper and lower covers.
3. Rotate the engine so the **F** mark on the camshaft is at the top position.
4. Loosen the idler pulley locknut. Apply tension to the timing belt exceeding the tension exerted by the tension spring. Temporarily tighten the attaching bolt.
5. Turn the crankshaft approximately 2 turns in the direction of normal rotation until No. 1 piston is at TDC on its compression stroke. The **F**

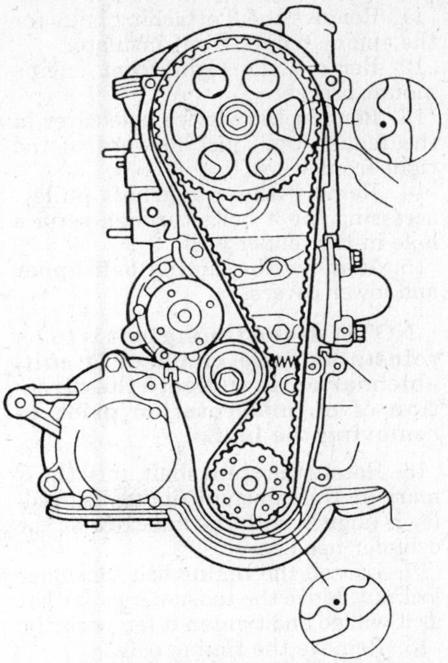

Timing mark locations—1.0L engine

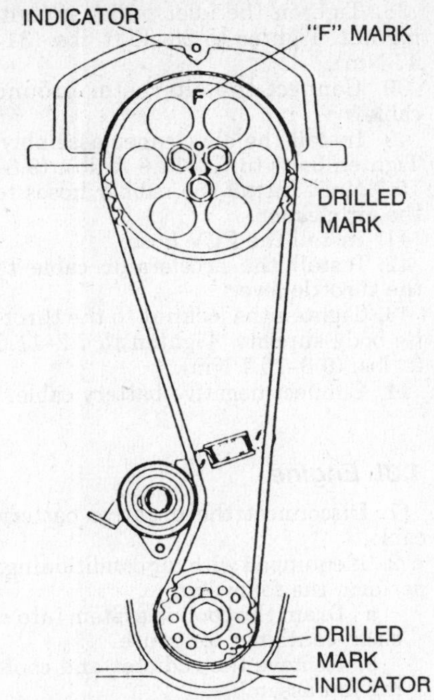

Timing mark locations—1.3L engine

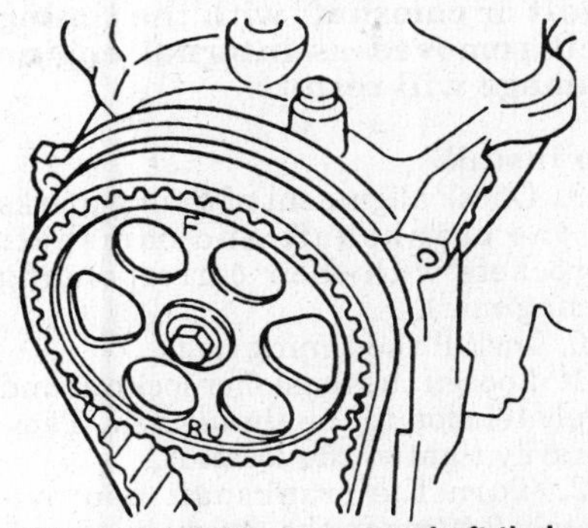

"F" mark position on camshaft timing belt sprocket—1.0L engine

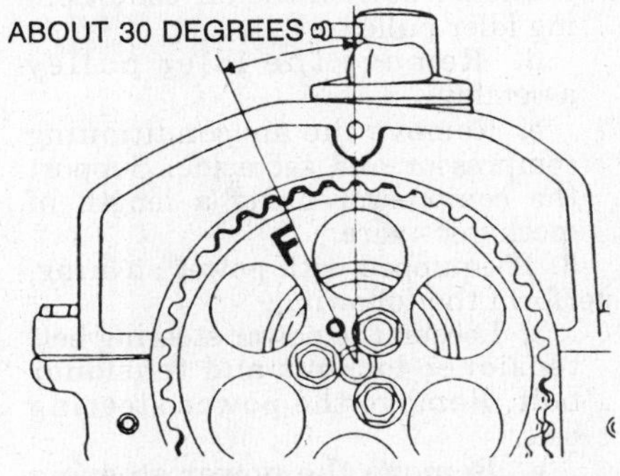

"F" mark position on camshaft timing belt sprocket for adjusting timing belt tension—1.3L engine

mark on the camshaft should be at the top position.

6. Loosen the idler pulley locknut.
7. Tighten the timing belt tensioner locknut to 25–33 ft. lbs. (33–44 Nm).
8. Ensure that the timing marks on the camshaft and crankshaft timing belt sprockets are aligned with the corresponding marks on the timing belt.
9. Install the timing belt upper and lower covers.
10. Connect negative battery cable.

1.3L Engine

1. Disconnect the negative battery cable.
2. Remove the timing belt upper and lower covers.
3. Rotate the engine so the **F** mark on the camshaft is at the top position. Ensure that the drilled timing mark on the crankshaft timing belt sprocket is aligned with the mark on the engine block.
4. Loosen the idler pulley locknut. Apply tension to the timing belt exceeding the tension exerted by the tension spring. Temporarily tighten the attaching bolt.
5. Turn the crankshaft approximately 2 turns in the direction of normal rotation until No. 1 piston is approaching TDC on its compression stroke. The **F** mark on the camshaft should be at the 11 o'clock position or 3 camshaft sprocket teeth before the reference mark on the valve cover.
6. Loosen the idler pulley locknut.
7. Turn the crankshaft in the direction of rotation until the **F** mark is at the 12 o'clock position and aligned with the reference mark on the valve cover.
8. Tighten the timing belt tensioner locknut to 22–33 ft. lbs. (29–44 Nm).
9. Ensure that the timing marks on the camshaft and crankshaft timing belt sprockets are aligned with the corresponding marks on the timing belt.
10. Install the timing belt upper and lower covers.
11. Connect negative battery cable.

REMOVAL & INSTALLATION

1.0L Engine

1. Disconnect the negative battery cable.
2. Loosen the accelerator cable locknut from the throttle body support. Remove the accelerator cable from the throttle lever. Remove the accelerator cable from the clamp.
3. Remove the PCV hose.
4. Label and remove the air cleaner hoses. Remove the air cleaner assembly.
5. Disconnect the alternator ground cable.
6. If equipped with air conditioning, perform the following:
   a. Loosen the idler pulley mount nut.
   b. Loosen the adjusting bolt.
   c. Remove the drive belt.
7. Loosen the alternator lock bolts, relieve the tension and remove the alternator drive belt.
8. Remove the right front engine under cover.
9. Using a supporting pad under the oil pan, slightly raise the engine to release the tension on the engine mount.
10. Remove the nuts attaching the front engine mount bracket.
11. Remove the 3 bolts attaching the front engine mount to the cylinder head.
12. Lower the lifting device slightly and remove the front engine mount bracket.
13. Remove the water pump pulley.
14. Set the engine to TDC of No. 1 cylinder on the compression stroke.
15. Prevent the crankshaft from turning by inserting a suitable tool into the ring gear.
16. Remove the crankshaft pulley.
17. Remove the timing belt upper and lower covers.
18. Remove the crankshaft timing belt pulley flange.

NOTE: If the timing belt is to be reinstalled, use chalk or a suitable marker to indicate the direction of normal rotation prior to removing the belt.

19. Loosen the tensioner pulley bolt. Push the bolt to the left as far as it will go and tighten it temporarily.

NOTE: Do not rotate the crankshaft or camshaft with the timing belt removed as internal engine damage will result.

20. Remove the timing belt.

To install:

21. Ensure that the timing marks on the camshaft and crankshaft are aligned with corresponding marks on the engine.
22. Install the timing belt on the camshaft and crankshaft timing sprockets.

NOTE: When installing a new belt, align the mark on the back side of the belt with the corresponding mark on each pulley.

23. Loosen the tensioner pulley locknut. Apply tension to the timing belt exceeding the tension exerted by the tension spring. Temporarily tighten the attaching bolt.
24. Turn the crankshaft approximately 2 turns in the direction of normal rotation until No. 1 piston is at TDC on its compression stroke. The **F** mark on the camshaft should be at the top position.
25. Loosen the tensioner pulley locknut.
26. Tighten the timing belt tensioner locknut to 25–33 ft. lbs. (33–44 Nm).
27. Ensure that the timing marks on the camshaft and crankshaft timing belt sprockets are aligned with the corresponding marks on the timing belt.
28. Install the crankshaft timing belt pulley flange. Ensure that the crankshaft timing belt pulley flange with the protruding side faces toward the crankshaft timing belt sprocket.
29. Install the timing belt upper and lower covers. Tighten to 1.4–2.9 ft. lbs. (2.0–3.9 Nm).
30. Remove the crankshaft pulley bolt. Insert the crankshaft pulley with the key groove on the crankshaft pulley aligned with the crankshaft key. Install the crankshaft pulley bolt and tighten to 65–72 ft. lbs. (88–98 Nm).
31. Install the water pump pulley. Tighten to 4–7 ft. lbs. (6–9 Nm).
32. Install the front engine mount bracket to the cylinder head. Tighten M12 bolt to 36–51 ft. lbs. (49–67 Nm). Tighten M10 bolt to 22–33 ft. lbs. (29–44 Nm).
33. Tighten the front engine mount bracket to 29–40 ft. lbs. (39–54 Nm).
34. Install the front right engine under cover.
35. Remove the lifting device.
36. Install the alternator drive belt. Adjust the alternator drive belt tension to 22 lbs. (10 kg) with 0.197–0.276 in. (5–7mm) deflection applied at the midpoint between the alternator and water pump pulleys.
37. If equipped with air conditioning, install the air conditioning drive belt. Adjust the air conditioning compressor drive belt tension to 22 lbs. (10 kg) with 0.16–0.24 in. (4.0–6.0mm) deflection at the midpoint between the crankshaft and compressor pulleys.
38. Tighten the idler pulley mounting nut. Tighten to 23–35 ft. lbs. (31–47 Nm).
39. Connect the alternator ground cable.
40. Install the air cleaner assembly. Tighten bolts to 7.2–11.6 ft. lbs. (9.8–15.7 Nm). Install the rubber hoses to the air cleaner.
41. Install the PCV hose.
42. Install the accelerator cable to the throttle lever.
43. Tighten the locknut to the throttle body support. Tighten to 7.2–11.6 ft. lbs. (9.8–15.7 Nm).
44. Connect negative battery cable.

1.3L Engine

1. Disconnect the negative battery cable.
2. If equipped with air conditioning, perform the following:
   a. Drain the cooling system into a clean container for reuse.
   b. Remove the radiator and cooling fan assembly.
   c. Remove the air conditioning belt by loosening the locknut and tensioner bolt on the air conditioning idler pulley.
   d. Remove the idler pulley assembly.
   e. Remove the air conditioning compressor and set aside. Support the compressor using a length of mechanics wire.
3. If equipped with power steering, perform the following:
   a. Loosen the power steering belt tensioner locknut and tensioner bolt. Remove the power steering belt.
   b. Remove the power steering pump and set aside. Support the power steering pump using a length of mechanics wire.
4. Remove the hoses from the air cleaner case. Remove the attaching bolts and hose bands and remove the air cleaner case.
5. Disconnect the oil pressure switch connector.
6. Remove the bolts attaching the oil pressure switch wire. Pull wire through the hole in the engine mount and set wire aside.
7. Using the alternator belt tension, loosen the bolts on the water pump pulley.
8. Loosen the adjusting bolt and 2 alternator mounting bolts to slacken tension on the alternator belt and remove the water pump pulley and alternator belt.
9. Using a supporting pad under the oil pan, slightly raise the engine to release the tension on the engine mount.
10. Remove the 3 attaching bolts and 1 nut for the right engine mount.
11. Remove the 3 attaching bolts for the engine mount front insulator.
12. Remove the right front engine mount bracket.
13. Remove the service hole cover in the fender well, just forward of the right front wheel.
14. Remove the crankshaft pulley, accessing the 4 bolts through service hole in the fender well.
15. Remove the timing belt upper and lower covers.

NOTE: If the timing belt is to be reinstalled, use chalk or a suitable marker to indicate the direction of normal rotation prior to removing the belt.

16. Rotate the crankshaft until the **F** mark on the camshaft timing belt pulley is aligned with the indicator or the cylinder head cover.
17. Loosen the timing belt tensioner locknut. Move the tensioner as far left as it will go and tighten it temporarily.
18. Remove the timing belt.

NOTE: Do not rotate the crankshaft or camshaft with the timing belt removed as internal engine damage will result.

To install:

19. Check alignment of timing marks of the crankshaft and camshaft sprockets with their corresponding timing marks.
20. Install the timing belt.
21. Loosen the tensioner locknut and apply tension to the timing belt. Temporarily tighten the locknut.
22. Turn the crankshaft approximately 2 turns in the direction of normal rotation until No. 1 piston is approaching TDC on its compression stroke. The **F** mark on the camshaft should be at the 11 o'clock position or 3 camshaft sprocket teeth before the reference mark on the valve cover.
23. Loosen the idler pulley locknut.
24. Turn the crankshaft in the direction of rotation until the **F** mark is at the 12 o'clock position and aligned with the reference mark on the valve cover.
25. Tighten the timing belt tensioner locknut to 22–33 ft. lbs. (29–44 Nm).
26. Ensure that the timing marks on the camshaft and crankshaft timing belt sprockets are aligned with the corresponding marks on the timing belt.
27. Install the timing belt upper and lower covers. Tighten to 1.4–3.9 ft. lbs. (2–4 Nm).
28. Install the crankshaft timing belt pulley. Tighten bolts to 15–22 ft. lbs. (20–29 Nm). Install the service hole cover.
29. Install the right front engine mount bracket. Tighten bolt to 29–40 ft. lbs. (39–54 Nm).

30. Install the right side engine mount bracket and front engine mount insulator. Tighten bolt to 29–40 ft. lbs. (39–54 Nm). Tighten nut to 11–17 ft. lbs. (15–23 Nm).
31. Remove lifting device from under oil pan.
32. Temporarily attach the water pump pulley.
33. Install the alternator belt and adjust tension using the alternator adjuster. Tighten alternator locknut .
34. Tighten the water pump pulley attaching belts, using the belt tension. Tighten to 4–7 ft. lbs. (4–9 Nm).
35. Check belt tension at the mid-point of the belt between the alternator and water pump pulleys.
36. Route the oil pressure switch wire through the hole in the engine mount. Install the clamp bolt for the oil pressure switch wire. Ensure that there is no slack in the wire between the clamps. Connect the oil pressure switch connector.
37. If equipped with air conditioning, perform the following:
   a. Install the air conditioning compressor to the engine block.
   b. Install the air conditioning belt tensioner pulley assembly to the cylinder block. Install the air conditioner belt and adjust tension.
   c. Install the radiator and cooling fan assembly.
38. If equipped with power steering, perform the following:
   a. Install the power steering pump to the cylinder head.
   b. Install the power steering belt and adjust tension.
39. Connect negative battery cable.
40. Fill cooling system and check for leaks. Start engine and allow engine to come to normal operating temperature. Recheck for coolant leaks. Allow engine to warm up sufficiently to confirm operation of cooling fan.

Timing Sprockets

REMOVAL & INSTALLATION

1.0L Engine

1. Disconnect the negative battery cable.
2. Remove the timing belt.
3. Remove the timing belt tensioner pulley and return spring.
4. Remove the crankshaft timing belt sprocket.
5. If required, remove the crankshaft timing belt sprocket flange.
6. Prevent the camshaft from turning by inserting a rod through a hole in the sprocket into the rib in the cylinder head and remove the set bolt.
7. Remove the camshaft timing belt sprocket using a suitable puller.

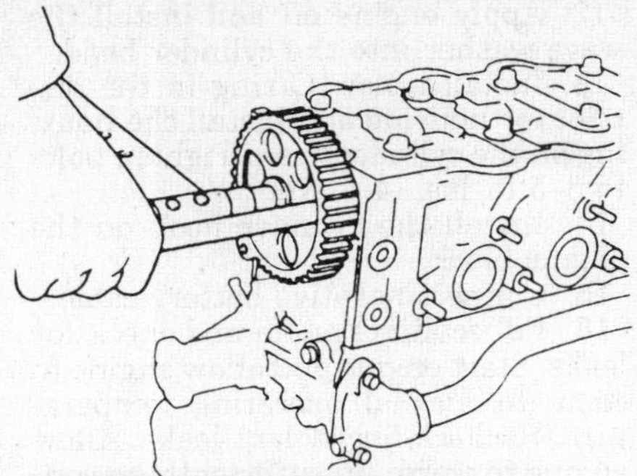

Removal of camshaft timing belt sprocket – 1.0L engine

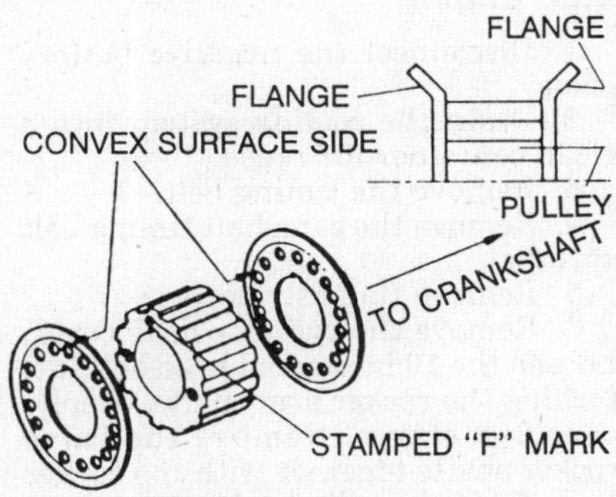

Crankshaft pulley and flange position – 1.0L engine

To install:

8. Align the groove in the camshaft timing belt sprocket with the key in the camshaft and push the sprocket onto the camshaft.
9. Prevent the camshaft from turning and tighten the set bolt to 22–33 ft. lbs. (29–44 Nm).
10. If removed, install the crankshaft timing belt sprocket flange with its recessed side facing the sprocket.
11. Align the groove in the crankshaft timing belt sprocket with the key in the crankshaft and push the sprocket onto the crankshaft.
12. Install the timing belt tensioner pulley and spring. Push the pulley assembly to the left and temporarily tighten the locknut.
13. Install the timing belt.
14. Connect negative battery cable.

1.3L Engine

1. Disconnect the negative battery cable.
2. If equipped with air conditioning, remove the radiator and cooling fan assembly. Properly discharge the A/C system and remove the air conditioning compressor.
3. If equipped with power steering, remove the power steering pump.

NOTE: Do not rotate the crankshaft or camshaft with the timing belt removed as internal engine damage will result.

4. Remove the timing belt.
5. Remove the oxygen sensor wire from its clamp.
6. Label and disconnect the spark plug wires from the spark plugs.
7. Disconnect the PCV hoses.
8. Disconnect the accelerator cable from its clamp.
9. If equipped with a radio, remove the engine ground wire from the cylinder head cover.
10. Remove the cylinder head cover loosening the 8 attaching bolts, evenly, in 2–3 stages.
11. Prevent the camshaft timing belt sprocket from turning by inserting a rod through one of the holes in the sprocket.

NOTE: Be careful not to allow the rod to rest on the gasket surface while holding the camshaft timing sprocket.

12. Remove the bolts attaching the camshaft timing sprocket and remove the sprocket.
13. Prevent the crankshaft from turning.
14. Remove the crankshaft timing belt sprocket set bolt.
15. Remove the crankshaft timing belt sprocket. If any difficulty is encountered in removing the sprocket reinstall the set bolt a few turns and use a suitable gear puller to remove the sprocket.

To install:

16. Install the crankshaft timing belt sprocket flange with its recessed side facing the oil pump.
17. Install the crankshaft timing belt sprocket, aligning it with the key groove.
18. Install the crankshaft timing belt sprocket set bolt. Prevent the crankshaft from turning and tighten the set bolt to 65–72 ft. lbs. (88–98 Nm).
19. Install the camshaft timing belt sprocket on the camshaft so the **F** mark faces away from the cylinder head and the locating pin hole is aligned with the pin.
20. Install the camshaft timing belt sprocket attaching bolts. Prevent the

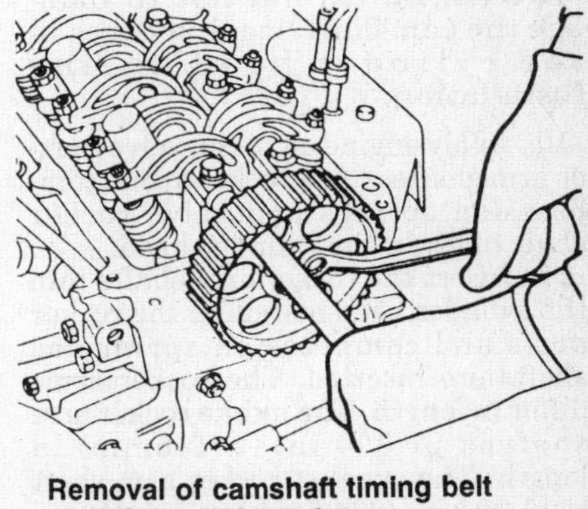

Removal of camshaft timing belt sprocket – 1.3L engine

camshaft from turning and tighten the bolts to 11–16 ft. lbs. (15–22 Nm).

21. Install the cylinder head cover and tighten the attaching bolts, evenly, over 2–3 stages, to 2.2–3.6 ft. lbs. (3.0–4.9 Nm).
22. If equipped with a radio, install the engine ground wire to the cylinder head cover.
23. Connect the accelerator cable to its clamp.
24. Connect the PCV hoses.
25. Connect the spark plug wires.
26. Connect the oxygen sensor wire to its clamp.
27. Install the timing belt.
28. If equipped with power steering, install the power steering pump and drive belt.
29. If equipped with air conditioning, install the air conditioning compressor and drive belt. Properly charge the A/C system.
30. Connect negative battery cable.

Camshaft

REMOVAL & INSTALLATION

1.0L Engine

1. Disconnect the negative battery cable.
2. Drain the cooling system into a clean container for reuse.
3. Remove the cylinder head and place in a suitable fixture.
4. Remove the distributor housing.
5. Remove the wave washer.
6. Loosen each valve rocker arm adjusting screw.
7. Remove the valve rocker arm shafts using a suitable puller.
8. Remove the camshaft, pulling it toward the rear of the cylinder head.

NOTE: Be careful not to damage the camshaft bearing bores in the cylinder head during removal.

To install:

9. Apply engine oil to the camshaft bearing bores in the cylinder head and the camshaft bearing journals.

NOTE: Be careful not to damage the camshaft bearing bores in the cylinder head during installation.

10. Apply engine oil to the valve rocker arm shafts, valve rocker arms, compression springs and valve rocker shaft holes in the cylinder head.
11. Insert the rocker arm shafts into the cylinder head installing the rocker arms and compression springs as shafts are inserted. The rocker arms differ in length. The intake rocker arm shaft is 11.004 in. (279.5mm) in length. The exhaust rocker arm shaft is 11.201 in. (284.5mm) in length.
12. Apply engine oil and install the wave washer into the cylinder head.
13. Install a new O-ring in the distributor housing and install the housing on the cylinder head. Tighten bolts to 3–5 ft. lbs. (4–7 Nm).
14. Install the cylinder head on the engine block.
15. Connect negative battery cable.
16. Fill cooling system and check for leaks. Start engine and allow engine to come to normal operating temperature. Recheck for coolant leaks. Allow engine to warm up sufficiently to confirm operation of cooling fan.

1.3L Engine

1. Disconnect the negative battery cable.
2. Drain the cooling system into a clean container for reuse.
3. Remove the timing belt.
4. Remove the camshaft timing belt sprocket.
5. Remove the distributor.
6. Remove the cylinder head cover. Loosen the 10 hexagonal head bolts attaching the rocker arm shafts, evenly, over 2–3 stages. Remove the valve rocker shafts together with the rocker arms from the cylinder head.
7. Remove the 2 bolts attaching camshaft bearing cap No. 5.
8. Remove the remaining camshaft bearing caps.
9. Remove the camshaft from the cylinder head.

To install:

10. Apply engine oil to the camshaft bearing saddles in the cylinder head and the thrust surfaces. Apply engine oil to the bearing journals and lobes on the camshaft. Set the camshaft on the cylinder head so the locating pin for the camshaft timing belt pulley is at the top position.
11. Install the camshaft bearing caps in the sequence embossed on the bearing caps.
12. Assemble the valve rocker arms and wave washers onto the valve rocker arm shafts. Apply engine oil while assembling the rocker arms and wave washers.
13. Install the valve rocker arm shafts on the camshaft bearing caps.
14. Ensure that the attaching bolts and the mounting holes in the cylinder head are clean and free of oil and dirt.
15. Tighten the bolts, evenly, in 2–3 stages. Tighten the M10 bolts to 27 ft. lbs. (36 Nm). Tighten the M8 Bolts to 12 ft. lbs. (17 Nm).
16. Install the spacers between the intake valve rocker arms on the rocker shaft.

NOTE: Do not allow the camshaft to turn independently of the crankshaft with the timing belt removed as internal engine damage will result.

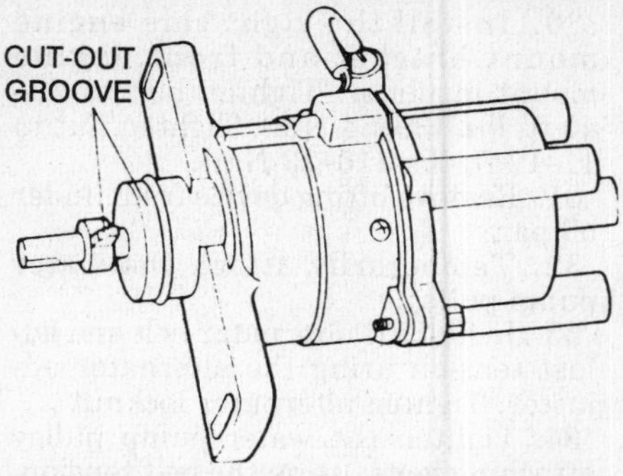

Distributor shaft alignment – 1.3L engine

17. Install the camshaft timing belt sprocket to the end of the camshaft. Prevent the camshaft from turning and tighten the bolts to 11–16 ft. lbs. (15–21 Nm).
18. Replace the O-ring at the base of the distributor. Align the cut-out section of the distributor properly with the cut-out groove of the cup ring. Install the distributor in the cylinder head, aligning the protrusion of the distributor with the camshaft groove. The aligned cut-out sections must be at the top position on the engine.
19. Center the threaded holes in the cylinder in the elongated holes on the distributor flange and install the attaching bolts. Tighten to 11–16 ft. lbs. (15–22 Nm).
20. Install the cylinder head cover.
21. Install the timing belt.
22. Connect negative battery cable.
23. Fill cooling system and check for leaks. Start engine and allow engine to come to normal operating temperature. Recheck for coolant leaks. Allow engine to warm up sufficiently to confirm operation of cooling fan.
24. Check ignition timing.

Balance Shaft

REMOVAL & INSTALLATION

1.0L Engine

1. Disconnect the negative battery cable.
2. Drain the cooling system into a clean container for reuse.
3. Drain the engine oil.
4. Remove the engine assembly from the vehicle and place in a suitable holding fixture.

NOTE: Do not allow the camshaft to turn independently of the crankshaft with the timing belt removed as internal engine damage will result.

5. Remove the timing belt.

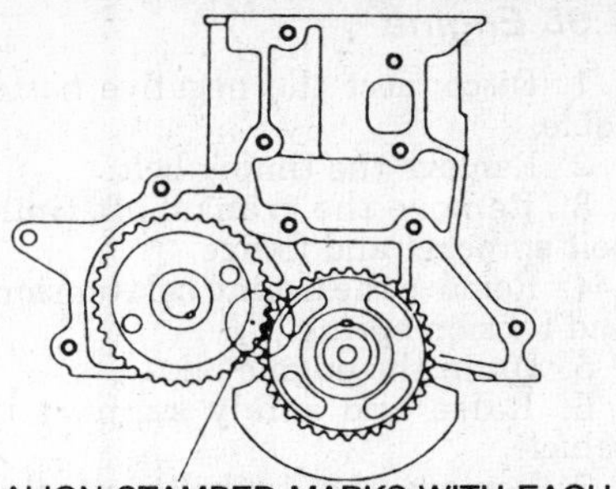

Balance shaft gear alignment marks—1.0L engine

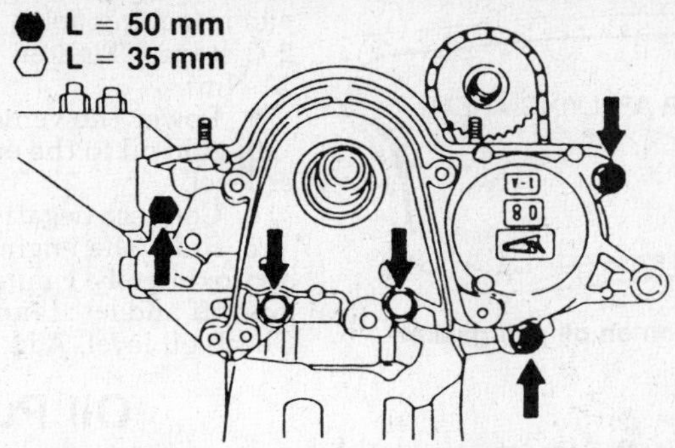

Balance shaft gear cover bolt pattern—1.0L engine

6. Remove the crankshaft timing belt sprocket.
7. Remove the oil pan and gasket.
8. Remove the balance shaft gear cover and gasket.
9. Remove the oil pump driven sprocket and drive chain.
10. Remove the oil pump and oil pump outlet pipe.
11. Remove the balance weight.
12. Remove the oil pump sprocket and drive chain.
13. Align the stamped mark on the crankshaft gear with the stamped mark on the balance shaft gear.
14. Remove the hexagonal socket head cap bolt, using a hexagonal wrench key (5mm).
15. Pull the balance shaft toward the front of the cylinder block.

To install:

16. Apply engine oil to the bearing bores in the engine block and the bearing journals on the balance shaft.
17. Slide the balance shaft into its bore making sure the alignment marks on the balance shaft and crankshaft gears line up.
18. Secure the thrust plate of the balance shaft by tightening the hexagonal bolt to 7–11 ft. lbs. (10–15 Nm).
19. Install the oil pump drive sprocket to the balance shaft.
20. Install the balance weight to the balance shaft with the key groove aligned. Insert the washer. Tighten the bolt to 22–33 ft. lbs. (29–44 Nm).
21. Replace the O-ring on the oil pump outlet. Connect the outlet pipe and tighten the bolts temporarily.
22. Install the oil pump and outlet pipe to the cylinder block.
23. Install the oil pump drive chain to the oil pump sprocket on the balance shaft.
24. Install the oil pump drive sprocket to the oil pump with the drive chain installed. Ensure that the stamped marks "CB OUTSIDE" face the front of the engine. Tighten the attaching bolt.
25. Apply engine oil to the balance shaft gear cover oil seal.
26. Install the balance shaft gear cover using a new gasket. Tighten to 7–12 ft. lbs. (10–16 Nm).
27. Install the oil pan using a new gasket.
28. Install the crankshaft timing belt sprocket.
29. Install the engine assembly in the vehicle.
30. Install the timing belt.
31. Add oil to the crankcase.
32. Connect negative battery cable.
33. Fill cooling system and check for leaks. Start engine and allow engine to come to normal operating temperature. Recheck for coolant leaks. Allow engine to warm up sufficiently to confirm operation of cooling fan.

Piston and Connecting Rod

POSITIONING

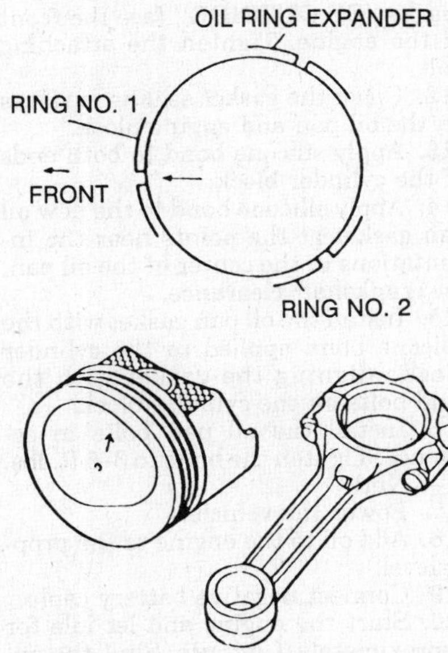

Piston and connecting rod for 1.0L and 1.3L engines

ENGINE LUBRICATION

Oil Pan

REMOVAL & INSTALLATION

1.0L Engine

1. Disconnect the negative battery cable.
2. Drain the engine oil.
3. Raise and safely support the vehicle.
4. Remove the 26 bolts attaching the oil pan to the engine block. Remove the oil pan and gasket.

To install:

5. Clean the gasket sealing surfaces on the oil pan and engine block.
6. Apply silicone bond to both ends of the cylinder block.
7. Apply silicone bond to the new oil pan gasket at the points near the indentations in the center of the oil pan for crankshaft clearance.
8. Install the oil pan gasket with the silicone bond applied to the cylinder block, aligning the gasket with the stud bolts on the cylinder block.
9. Install the oil pan bolts in sequence. Tighten the bolts to 3–5 ft. lbs. (4–7 Nm).
10. Lower the vehicle.
11. Add oil to the engine to the proper level.
12. Connect negative battery cable.
13. Start the engine and let idle for approximately 1 minute. Shut the engine off and let stand for 3–5 minutes. Check oil level. Add, as necessary.

1.3L Engine

1. Disconnect the negative battery cable.
2. Drain the engine oil.
3. Raise and safely support the vehicle.
4. Remove the 14 bolts attaching

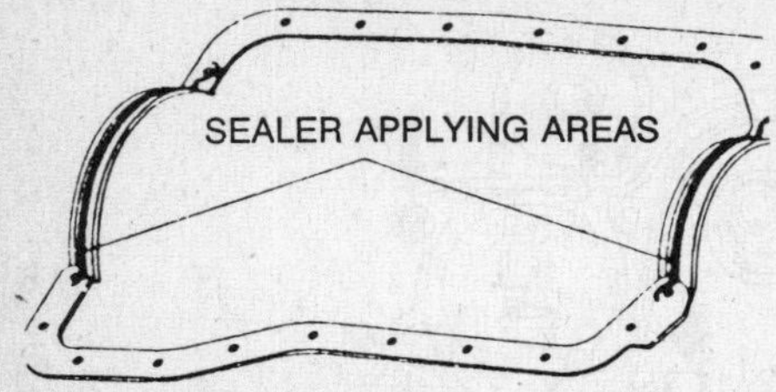

Sealer application on oil pan gasket—1.0L engine

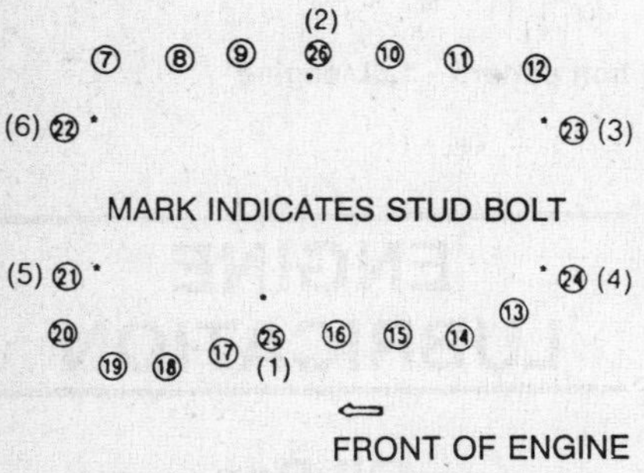

Oil pan bolt tightening pattern—1.0L engine. The numerals in parentheses denote the sequence for tightening the stud bolt nuts.

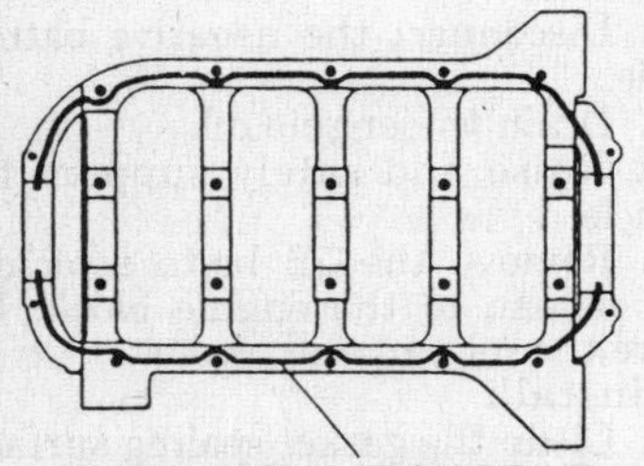

Oil pan sealer application—1.3L engine

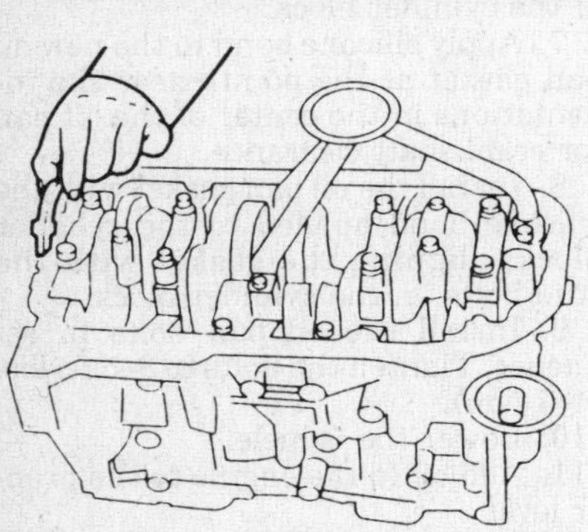

Oil pan gasket application—1.3L engine

the oil pan to the engine block. Remove the oil pan and gasket.

To install:

5. Clean the gasket sealing surfaces on the oil pan and engine block.
6. Apply silicone bond to the gasket sealing surface on the cylinder block.
7. Install a new semi-circular gasket at the front of the engine.
8. Install the oil pan. Tighten the oil pan attaching bolts, in sequence, over 2–3 stages. Tighten to 5–9 ft. lbs. (7–11 Nm).
9. Lower the vehicle.
10. Add oil to the engine to the proper level.
11. Connect negative battery cable.
12. Start the engine and let idle for approximately 1 minute. Shut the engine off and let stand for 3–5 minutes. Check oil level. Add, as necessary.

Oil Pump

REMOVAL & INSTALLATION

1.0L Engine

1. Disconnect the negative battery cable.
2. Drain the engine oil.
3. Raise and safely support the vehicle.
4. Remove the 26 bolts attaching the oil pan to the engine block. Remove the oil pan and gasket.
5. Remove the oil pump driven sprocket and drive chain.
6. Remove the oil pump and oil pump outlet pipe.

To install:

7. Replace the oil pump outlet pipe O-ring. Apply engine oil to the new O-ring.
8. Connect the outlet pipe to the oil pump tightening the bolts temporarily.
9. Install the oil pump with the outlet pipe to the cylinder block Tighten the attaching bolts.
10. Install the oil pump drive chain to the balance shaft oil pump sprocket.
11. Install the oil pump drive sprocket to the oil pump with the drive chain installed. Ensure that the stamped marks "CB OUTSIDE" face the front of the engine. Tighten the attaching bolt.
12. Clean the gasket sealing surfaces on the oil pan and engine block.
13. Apply silicone bond to both ends of the cylinder block.
14. Apply silicone bond to the new oil pan gasket at the points near the indentations in the center of the oil pan, for crankshaft clearance.
15. Install the oil pan gasket with the silicone bond applied to the cylinder block, aligning the gasket with the stud bolts on the cylinder block.
16. Install the oil pan bolts in sequence. Tighten the bolts to 3–5 ft. lbs. (4–7 Nm).
17. Lower the vehicle.
18. Add oil to the engine to the proper level.
19. Connect negative battery cable.
20. Start the engine and let idle for approximately 1 minute. Shut the engine off and let stand for 3–5 minutes. Check oil level. Add, as necessary.

1.3L Engine

1. Disconnect the negative battery cable.
2. Remove the timing belt.
3. Remove the crankshaft timing belt sprocket and flange.
4. Remove the timing belt tensioner and tension spring.
5. Drain the engine oil.
6. Raise and safely support the vehicle.
7. Remove the 14 bolts attaching the oil pan to the engine block. Remove the oil pan and gasket.
8. Remove the oil pump strainer.
9. Lower the vehicle.
10. Remove the oil pump.

To install:

11. Apply a suitable sealer to the oil pump installation surface on the cylinder block.
12. Replace the oil pump O-ring.
13. Apply engine oil to the inner surface of the oil seal. Install the oil pump to the cylinder block. Tighten the oil pump attaching bolts to 4–7 ft. lbs. (6–9 Nm).
14. Raise and safely support the vehicle.
15. Install the oil strainer using a new gasket.
16. Clean the gasket sealing surfaces on the oil pan and engine block.
17. Apply silicone bond to the gasket sealing surface on the cylinder block.
18. Install a new semi-circular gasket at the front of the engine.
19. Install the oil pan. Tighten the oil pan attaching bolts, in sequence, over 2–3 stages. Tighten to 5–9 ft. lbs. (7–11 Nm).
20. Lower the vehicle.
21. Add oil to the engine to the proper level.
22. Connect negative battery cable.
23. Start the engine and let idle for approximately 1 minute. Shut the engine off and let stand for 3–5 minutes. Check oil level. Add, as necessary.

Rear Main Bearing Oil Seal

REMOVAL & INSTALLATION

1.0L Engine

1. Disconnect the negative battery cable.
2. Remove the transaxle assembly from the vehicle.
3. If equipped with manual transaxle, remove the clutch assembly from the flywheel.
4. Prevent the crankshaft from turning. Remove the flywheel or drive plate.
5. Remove the rear main oil seal retainer.

6. Remove the oil seal from the retainer.

To install:

7. Install a new rear main oil seal using a suitable installer.
8. Apply engine oil to the oil seal. Install the retainer assembly on the engine block.
9. Install the flywheel with locating pins aligned or the drive plate.
10. Prevent the crankshaft from turning. Tighten the bolts evenly, over 2–3 stages, in a star pattern. On the 1.0L engine, tighten the bolts to 29–36 ft. lbs. (39–49 Nm). On the 1.3L engine, tighten the bolts to 33–47 ft. lbs. (44–64 Nm).
11. If equipped with manual transaxle, install the clutch assembly.
12. Install the transaxle assembly.
13. Connect the negative battery cable.

ENGINE COOLING

Radiator

REMOVAL & INSTALLATION

1.0L Engine

1. Disconnect the negative battery cable.
2. Drain the cooling system into a clean container for reuse.
3. Disconnect the coolant reservoir hose.
4. Disconnect the upper and lower radiator hoses.
5. Disconnect the fan motor connector.
6. Remove the 2 radiator retaining bolts. Note position of ground cable on right side bolt and set aside.
7. Remove the radiator with fan shroud and fan motor assembly attached from the vehicle.

To install:

8. Install fan shroud and fan motor assembly to the radiator.
9. Install radiator in vehicle and install 2 retaining bolts.

NOTE: Ensure that ground cable is reconnected with right side retaining bolt.

10. Connect fan motor connector.
11. Connect upper and lower radiator hoses.
12. Connect coolant reservoir hose.
13. Fill cooling system.
14. Connect negative battery cable.
15. Start engine and check for leaks. Allow engine to come to normal operating temperature and recheck for leaks. Allow engine to warm up sufficiently to confirm operation of cooling fan.

1.3L Engine

1. Disconnect the negative battery cable.
2. Remove the engine under cover.
3. Drain the cooling into a clean container for reuse.
4. If equipped with automatic transaxle, disconnect the 2 oil cooler hoses for the automatic transaxle.

NOTE: Plug oil lines to prevent excessive loss of transaxle fluid.

5. Disconnect the coolant reservoir hose and radiator upper and lower hoses.
6. If equipped with manual transaxle, disconnect the oil cooler hose from the radiator lower tank.
7. Disconnect the fan motor connector.
8. Disconnect the upper radiator hose from the clamp.
9. Remove the radiator upper bracket.
10. Remove the radiator together with the fan shroud from the vehicle.
11. Remove the fan shroud from the radiator.

To install:

12. If equipped with automatic transaxle, replace the automatic transaxle oil hoses and hose bands with new ones.
13. Install the fan shroud together with the fan motor on the radiator.
14. Inspect the rubber mounting grommet and replace if cracked or damaged.
15. Ensure that the radiator is positioned on the grommet when installing.
16. Install the radiator bracket.

NOTE: If equipped with radio, ensure that the ground cable is installed together with bracket on the air cleaner side.

17. Install the radiator hose to the fan shroud clamp.
18. Connect the fan motor connector.
19. Connect the oil cooler hose to radiator. Install new hose bands.
20. Connect the radiator upper and lower hoses to the radiator. Install new hose bands.
21. Connect the coolant reservoir hose. Install new hose bands.
22. If equipped with automatic transaxle, connect the oil cooler hoses.

NOTE: Never reuse oil cooler hoses and hose bands.

23. Install the engine under cover.
24. Connect negative battery cable.
25. If equipped with automatic transaxle, add automatic transaxle fluid.
26. Fill cooling system.
27. Start engine and check for leaks. Allow engine to come to normal operating temperature and recheck for leaks. Allow engine to warm up sufficiently to confirm operation of the cooling fan.

Heater Core

REMOVAL & INSTALLATION

1. Disconnect the negative battery cable.
2. Drain the cooling system into a clean container for reuse.

NOTE: The instrument panel must be removed to gain access to heater assembly containing the heater core.

3. Remove the 4 screws retaining the instrument cluster finish panel subassembly.
4. Pull subassembly out slightly to allow room to disconnect the rear window defogger switch and rear wiper switch couplers.
5. Remove instrument cluster finish panel subassembly.
6. Remove 4 instrument cluster assembly retaining screws.
7. Pull instrument cluster assembly out slightly to allow room to disconnect the speedometer cable and the wiring harness connector.

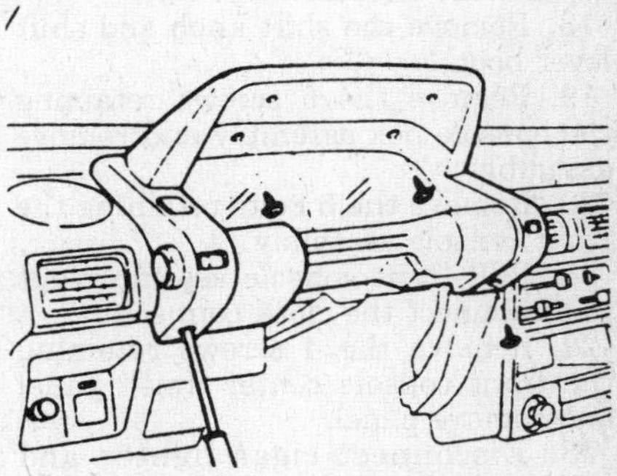

Removal of instrument cluster finish panel

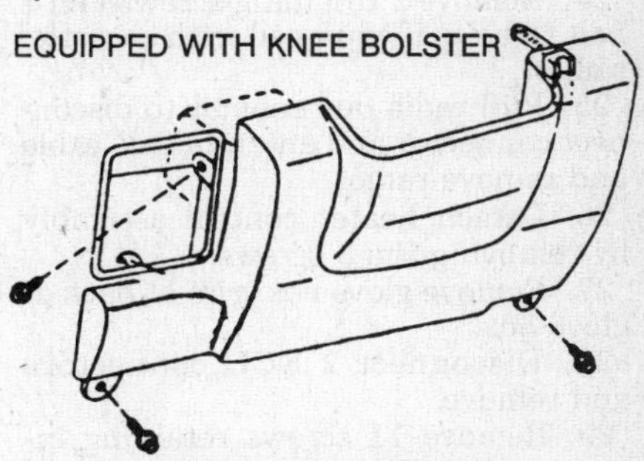

Removal of knee bolster

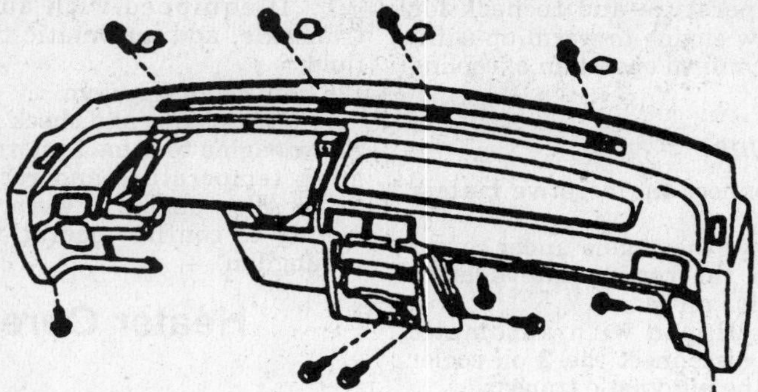

Removal of instrument panel

8. Remove the instrument cluster assembly.
9. Remove the switch base plate from the driver's side knee bolster by pushing it out from behind.
10. Remove the 3 bolts retaining the knee bolster.
11. Detach the clip located at the right side of the knee bolster by reaching through opening in instrument panel for instrument cluster.
12. Remove the knee bolster.
13. Remove the 4 instrument panel lower reinforcement retaining bolts.
14. Disconnect the steering column wiring harness connector.
15. Remove the steering shaft universal joint connector bolts.
16. Remove the 2 steering column attaching bolts and 3 nuts and remove the steering column.
17. Remove the 2 screws retaining the glove box door subassembly and remove the subassembly.
18. Remove the shift knob and shift lever boot.
19. Remove the 6 screws retaining the console box assembly and remove assembly.
20. Remove the 6 bolts retaining the front console assembly.
21. Pull front console assembly out to disconnect the clock connector.
22. Remove the 4 screws retaining the front console center finish panel and remove panel.
23. Disconnect cigar lighter and ashtry illumination lamp connectors and remove ashtray.
24. Remove 2 retaining screws and 1 bolt (2 bolts if equipped with cassette/radio).
25. Pull radio out enough to disconnect connector and antenna feed cable and remove radio.
26. Detach heater control assembly by removing the 3 screws.
27. Remove glove box trim at back of glove box.
28. Disconnect 2 ECU connectors and remove.
29. Remove 11 screws retaining instrument panel.
30. Pull out instrument panel enough to disconnect front speakers at the right and left sides, rheostat and door control switch connectors.
31. Remove the instrument panel assembly.
32. Disconnect the inside/outside air switching cable from the blower assembly.
33. Disconnect the 2 water hoses from the heater assembly.
34. If not equipped with air conditioning, remove the air duct.
35. If equipped with air conditioning, remove duct band connecting the heater assembly to the air conditioning cooling unit.
36. Remove the heater assembly by removing the 2 nuts and 2 bolts.

To install:

37. Install the heater assembly and attach with 2 nuts and 2 bolts.
38. If not equipped with air conditioning, install the air duct.
39. If equipped with air conditioning, install the duct band connecting the heater assembly to the cooling unit.
40. Connect the 2 water hoses to the heater assembly.
41. Install the inside/outside air switching cable to the blower assembly and adjust as follows:
    a. Set the inside/outside air switching lever of the heater control to the **RECIRC** position.
    b. Insert and clamp the cable securely.
42. Place the instrument panel in vehicle and connect front left and right speakers, rheostat and door control switch connectors.
43. Place the instrument panel in position and install the 11 retaining screws.
44. Connect 2 ECU connectors and install ECU in position.
45. Install glove box trim.
46. Install heater control assembly with 3 retaining screws.
47. Connect radio connector and antenna feed cable to back of radio.
48. Install radio with 2 retaining screws and 1 bolt (2 bolts if equipped with cassette/radio).
49. Install front console center finish panel with 4 retaining screws.
50. Connect clock connector to front console assembly.
51. Install front console assembly with 6 bolts.
52. Install console box assembly with 6 retaining screws.
53. Install the shift lever boot and shift knob.
54. Install glove box subassembly with 2 screw.
55. Install the steering shaft universal joint connecting bolts.
56. Install steering column with 2 attaching bolts and 3 nuts.
57. Install the 4 instrument panel lower reinforcement retaining bolts.
58. Install the knee bolster with 3 bolts, making sure right side clip is in place.
59. Install the switch base plate in knee bolster.
60. Install instrument cluster enough to connect wiring harness connector and speedometer cable and attach with 4 screws.
61. Install instrument cluster finish panel enough to connect rear window defogger switch and rear wiper switch connectors.
62. Install instrument cluster finish panel with 4 screws.
63. Connect negative battery cable.
64. Fill cooling system and check for leaks. Start engine and allow engine to come to normal operating temperature. Recheck for coolant leaks. Allow engine to warm up sufficently to confirm operation of cooling fan.
65. Operate heating system to insure proper installation and function of system.

Water Pump

REMOVAL & INSTALLATION

1.0L Engine

1. Disconnect the negative battery cable.
2. Drain coolant into a clean container for reuse.
3. Remove the timing belt.
4. Remove the water inlet.
5. Remove the throttle body hose from water pump.
6. Remove the water pump by removing the 3 bolts.

To install:

7. Replace the O-ring for the water inlet pipe.
8. Install the water pump to the cylinder block with new gasket. Tighten water pump to 11–16 ft. lbs. (15–22 Nm).
9. Install the water inlet to the cylinder block with an new gasket. Insert

the water inlet pipe into the water inlet at the same time.

10. Connect the throttle body hose to the water pump.
11. Install the timing belt.
12. Install the timing belt cover.
13. Install the water pump pulley.
14. Connect negative battery cable.
15. Fill cooling system and check for leaks. Start engine and allow engine to come to normal operating temperature. Recheck for coolant leaks. Allow engine to warm up sufficently to confirm operation of cooling fan.

1.3L Engine

1. Disconnect the negative battery cable.
2. Drain coolant into a clean container for reuse.
3. Remove the timing belt.
4. Remove the water pump by removing the attaching bolts and nuts of the water pump.

To install:

5. Remove the gasket material from the water pump installing surface of the cylinder block.
6. Install the water pump to the cylinder block with new gasket. Tighten water pump to 11–16 ft. lbs. (15–22 Nm).
7. Install the timing belt.
8. Install the timing belt cover.
9. Install the water pump pulley.
10. Connect negative battery cable.
11. Fill cooling system and check for leaks. Start engine and allow engine to come to normal operating temperature. Recheck for coolant leaks. Allow engine to warm up sufficently to confirm operation of cooling fan.

Thermostat

REMOVAL & INSTALLATION

1.0L Engine

1. Disconnect the negative battery cable.
2. Drain the coolant into a clean container for reuse.
3. Remove the water outlet hose.
4. Remove the water outlet.
5. Remove the thermostat.

To install:

6. Install the thermostat to the intake manifold.
7. Install the water inlet with the 2 nuts, using a new gasket.
8. Connect the water outlet pipe.
9. Connect negative battery cable.
10. Fill cooling system and check for leaks. Start engine and allow engine to come to normal operating temperature. Recheck for coolant leaks. Allow engine to warm up sufficently to confirm operation of cooling fan.

1.3L Engine

1. Disconnect the negative battery cable.
2. Drain the coolant into a clean container for reuse.
3. Disconnect the radiator thermo control switch connector.
4. Remove the water inlet and thermostat.

To install:

5. Install the thermostat so the jiggle pin faces up.
6. Install the water inlet. Tighten to 4–7 ft. lbs. (6–9 Nm).
7. Connect the radiator thermo control switch connector.
8. Connect negative battery cable.
9. Fill cooling system and check for leaks. Start engine and allow engine to come to normal operating temperature. Recheck for coolant leaks. Allow engine to warm up sufficently to confirm operation of cooling fan.

Cooling System Bleeding

The cooling system is self-bleeding due to a jiggle pin bleeder valve designed into the thermostat. The jiggle pin allows air bubbles to pass through the thermostat and out of the system. It is essential that the thermostat be installed with the jiggle pin facing up in order for the self-bleeding capability to function properly. Failure to due so could lead to air being trapped in the cooling system resulting in a dangerous overheating condition.

ENGINE ELECTRICAL

NOTE: Disconnecting the negative battery cable on some vehicles may interfere with the functions of the on board computer systems and may require the computer to undergo a relearning process, once the negative battery cable is reconnected.

Distributor

REMOVAL

1.0L Engine

1. Disconnect the negative battery cable.
2. Disconnect the spark plug and coil wires from the distributor cap.
3. Disconnect the distributor connector.
4. Remove the distributor set bolt.
5. Pull out the distributor from the distributor housing.
6. Remove the O-ring from the distributor housing and discard.

1.3L Engine

1. Disconnect the negative battery cable.
2. Disconnect the spark plug wire from the distributor cap.
3. Disconnect the distributor connector.
4. Disconnect the vacuum advance hoses.

NOTE: Tag the hoses prior to disconnecting so they can be reinstalled in their original positions.

5. Remove the distributor set bolt.
6. Pull out the distributor from the cylinder head.

NOTE: Insert a suitable cloth under the distributor opening to catch the oil.

7. Remove O-ring from the distributor housing and discard.

INSTALLATION

Timing Not Disturbed

1.0L ENGINE

1. Install a new O-ring to the distributor.
2. Remove the distributor cap.
3. Position the distributor shaft so the drilled mark at the forward end of the distributor drive gear is aligned with the drilled mark on the distributor body.
4. Insert the distributor assembly into the distributor housing so the split line of the distributor body is aligned with the embossed line of the distributor housing.
5. Temporarily tighten the distributor attaching bolt.
6. Install the distributor cap.
7. Connect distributor connector.
8. Connect spark plug wires.
9. Connect negative battery cable.
10. Start engine and allow to come to normal operating temperature. Check timing.
11. Check idle speed. Tighten distributor attaching bolt.

1.3L ENGINE

1. Align the cut-out section of the distributor housing with the cut-out section of the coupling.
2. Install the distributor. Ensure that the distributor attaching bolt holes in the cylinder head are in the center of the elongated holes in the base of the distributor housing.
3. Tighten the distributor attaching bolts temporarily.

4. Connect the vacuum advance hoses.
5. Connect the distributor connector and install it to the clamp.
6. Connect the spark plug wires.
7. Install the clamp to spark plug wire.
8. Connect negative battery cable.
9. Start engine and allow to come to normal operating temperature. Check timing.

Timing Disturbed

1.0L ENGINE

1. Remove the oil filler cap.
2. Observe the cam lobes of No. 1 cylinder through the oil fill hole while rotating the engine in the direction of normal rotation until the lobes are on their base circles (valves closed) indicating that cylinder No. 1 is on its compression stroke. Confirm that No. 1 cylinder is on its compression stroke by feeling for play in rocker arms with fingers.
3. Continue rotating engine in direction of normal rotation until timing marks are aligned.
4. Install a new O-ring to the distributor.
5. Remove the distributor cap.
6. Position the distributor shaft so the drilled mark at the forward end of the distributor drive gear is aligned with the drilled mark on the distributor body.
7. Insert the distributor assembly into the distributor housing so the split line of the distributor body is aligned with the embossed line of the distributor housing.
8. Temporarily tighten the distributor attaching bolt.
9. Install the distributor cap.
10. Connect distributor connector.
11. Connect spark plug wires.
12. Connect negative battery cable.
13. Start engine and allow to come to normal operating temperature. Check timing.
14. Check idle speed. Tighten distributor attaching bolt.

1.3L ENGINE

1. Remove the oil filler cap.
2. Observe the cam lobes of No. 1 cylinder through the oil fill hole, while rotating the engine in the direction of normal rotation until the lobes are on their base circles (valves closed), indicating that cylinder No. 1 is on its compression stroke. Confirm that No. 1 cylinder is on its compression stroke by feeling for play in rocker arms with fingers. Reinstall oil filler cap.
3. Continue rotating engine in direction of normal rotation until timing marks are aligned.
4. Align the cut-out section of the distributor housing with the cut-out section of the coupling.
5. Install the distributor. Ensure that the distributor attaching bolt holes in the cylinder head are centered in the elongated holes in the base of the distributor housing.
6. Tighten the distributor attaching bolts temporarily.
7. Connect the vacuum advance hoses.
8. Connect the distributor connector and install it to the clamp.
9. Connect the spark plug wires.
10. Install the clamp to spark plug wire.
11. Connect negative battery cable.
12. Start engine and allow to come to normal operating temperature. Check timing.

Ignition Timing

ADJUSTMENT

1.0L Engine

1. Connect positive lead of timing light to positive battery terminal and negative lead of timing light to a solid engine ground. Clamp timing light inductive pick-up to No. 1 spark plug wire (at timing belt end of engine).
2. Connect tachometer positive lead to negative terminal on ignition coil and tachometer negative lead to a solid engine ground.

NOTE: If the tachometer does not have a 3 cylinder setting, set the tachometer to the 6 cylinder range and multiply the reading by 2.

3. Start engine and let idle until engine reaches normal operating temperature and cooling fan cycles on and off.
4. Connect a jumper wire between terminals 11 (brown) and ground terminal (black) of the check connector.
5. Remove rubber cap from the throttle body. Adjust idle speed by turning the idle screw.

NOTE: Check idle speed with all accessories OFF and cooling fan not running.

6. Check and adjust ignition timing.
7. Tighten distributor attaching bolt if timing was adjusted.
8. Recheck timing after tightening bolt.
9. Remove jumper wire from the check connector.
10. Install cap to the check connector.
11. Check and readjust engine idle speed.
12. Install the rubber cap to the throttle body.
13. Disconnect timing light and tachometer.

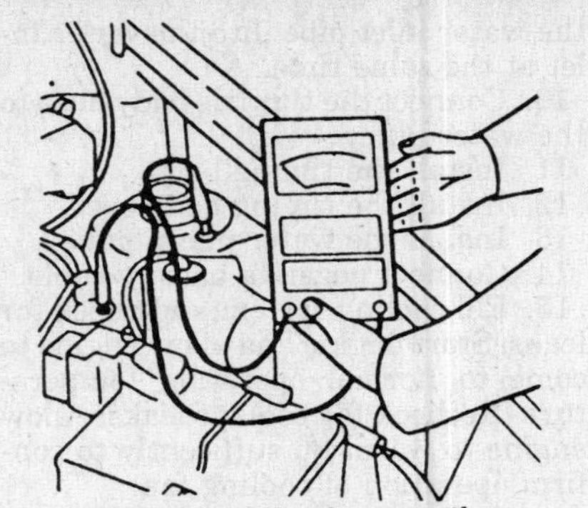

Tachometer lead connection—1.0L engine

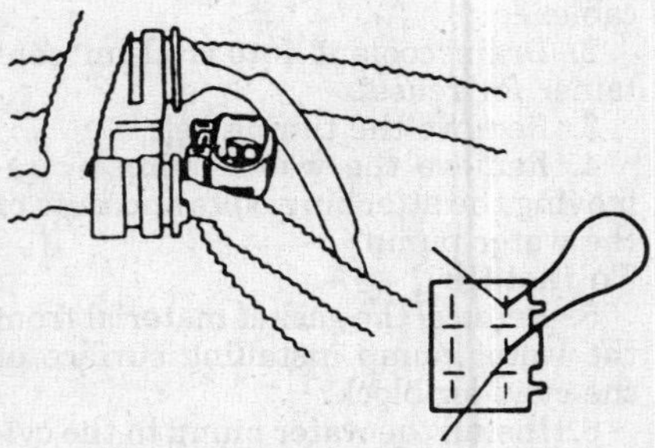

Connect jumper lead to check connector-1.0L engine

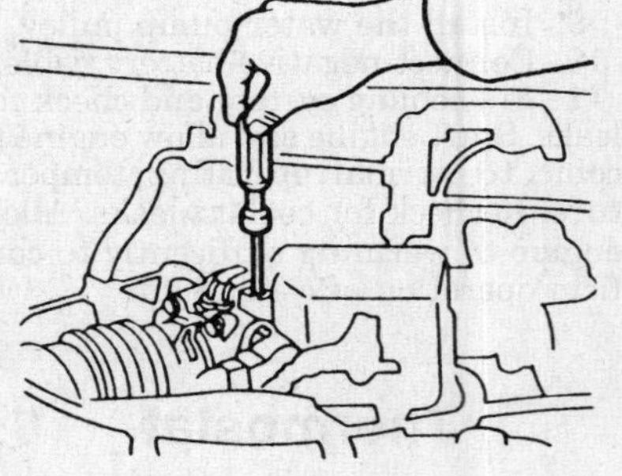

Idle speed setting—1.0L engine

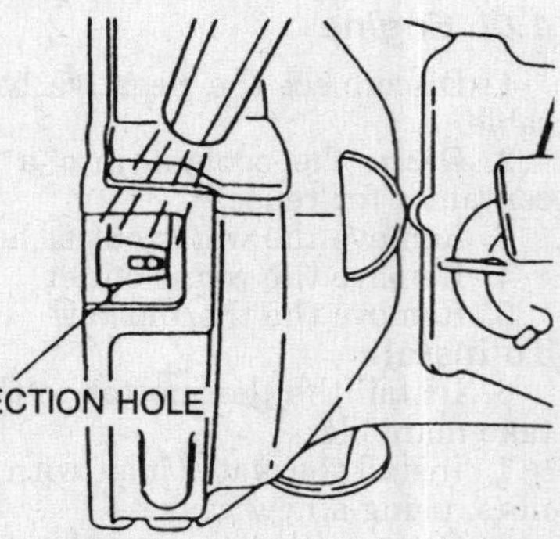

Timing mark location—1.0L engine

1.3L Engine

1. Connect positive lead of timing light to positive battery terminal and negative lead of timing light to a solid engine ground. Clamp timing light in-

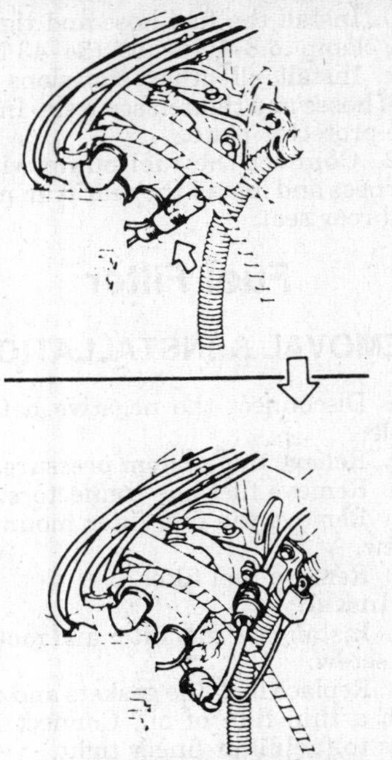

Distributor connector adapter installation – 1.3L engine

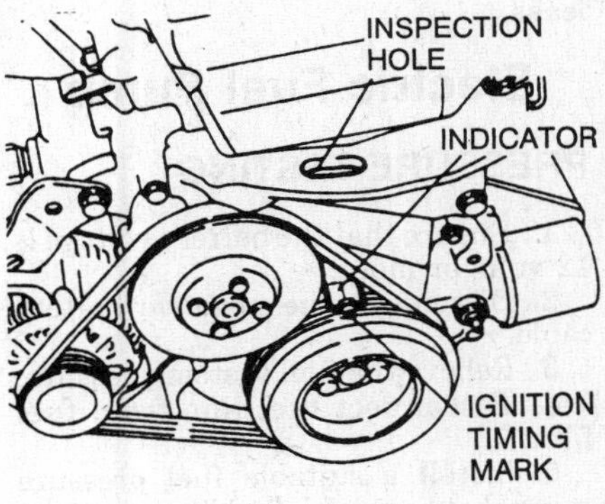

Timing mark location – 1.3L engine

ductive pick-up to No. 1 spark plug wire (at timing belt end of engine).

2. Connect adapter wire to distributor connector. Connect tachometer positive lead to distributor connector adapter and tachometer negative lead to a solid engine ground.
3. Disconnect and plug the vacuum advance hose.
4. Ensure engine idle speed is under 1000 rpm.
5. Check and adjust ignition timing.
6. Reconnect the vacuum hose to the distributor.
7. Check and readjust engine idle speed.
8. Disconnect timing light and tachometer. Remove distributor connector adapter and connect distributor connector.

Alternator

PRECAUTIONS

Several precautions must be observed with alternator equipped vehicles to avoid damage to the unit.

- If the battery is removed for any reason, make sure it is reconnected with the correct polarity. Reversing the battery connections may result in damage to the one-way rectifiers.
- When utilizing a booster battery as a starting aid, always connect the positive to positive terminals and the negative terminal from the booster battery to a good engine ground on the vehicle being started.
- Never use a fast charger as a booster to start vehicles.
- Disconnect the battery cables when charging the battery with a fast charger.
- Never attempt to polarize the alternator.
- Do not use test lamps of more than 12 volts when checking diode continuity.
- Do not short across or ground any of the alternator terminals.
- The polarity of the battery, alternator and regulator must be matched and considered before making any electrical connections within the system.
- Never separate the alternator on an open circuit. Make sure all connections within the circuit are clean and tight.
- Disconnect the battery ground terminal when performing any service on electrical components.
- Disconnect the battery if arc welding is to be done on the vehicle.

BELT TENSION ADJUSTMENT

1. Disconnect the negative battery cable.
2. Check tension of the alternator belt by applying 22 lbs. (10 kg) of pressure to the belt at the midpoint between the alternator and water pump pulleys. Deflection should not exceed 0.163–0.204 in. (4.0–5.0mm) for a new belt and 0.204–0.245 in. (5.0–6.0mm) for a used belt.
3. To adjust, loosen alternator mounting bolts and adjuster locknut.
4. Adjust belt tension, as necessary.
5. Tighten adjuster locknut and mounting bolts.
6. Connect negative battery cable.

REMOVAL & INSTALLATION

1.0L Engine

1. Disconnect the negative battery cable.
2. Remove alternator connector cover plate and disconnect the connector.
3. Remove the nut and wires from the alternator.
4. Remove alternator drive belt.
5. Remove right rear engine undercover.
6. Remove alternator mounting and lock bolts and remove alternator from the underside of the vehicle.
7. Complete the installation of the alternator by reversing the removal procedure.

1.3L Engine

1. Disconnect the negative battery cable.
2. Disconnect the alternator wire from terminal B. Disconnect the alternator connector.
3. Remove the accessory drive belts.
4. Remove the alternator locking plate. Loosen the attaching and adjusting bolts. Remove the alternator.
5. Remove the engine under rear cover. Remove the alternator attaching bolts. Remove the alternator from the engine compartment.
6. Complete the installation of the alternator by reversing the removal procedure.

Starter

REMOVAL & INSTALLATION

1.0L Engine

1. Disconnect the negative battery cable.
2. Disconnect the wires from the starter.
3. Remove the 2 starter retaining bolts and remove the starter.
4. Complete the installation of the starter by reversing the removal procedure.

1.3L Engine

1. Disconnect the negative battery cable.
2. Raise and safely support the vehicle.
3. Remove the crossmember.
4. Disconnect the wires from the starter.
5. Remove the starter motor attaching bolts.
6. Remove the starter from the vehicle.
7. Complete the installation of the starter by reversing the removal procedure.

EMISSION CONTROLS

Please refer to "Emission Controls" in the Unit Repair section for system maintenance procedures. Due to the complex nature of modern electronic engine control systems, comprehensive diagnosis and testing procedures fall outside the confines of this repair manual. For complete information on diagnosis, testing and repair procedures concerning all modern engine and emission control systems, please refer to "Chilton's Guide to Fuel Injection and Electronic Engine Controls".

FUEL SYSTEM

Fuel System Service Precautions

Safety is the most important factor when performing not only fuel system maintenance but any type of maintenance. Failure to conduct maintenance and repairs in a safe manner may result in serious personal injury or death. Maintenance and testing of the vehicle's fuel system components can be accomplished safely and effectively by adhering to the following rules and guidelines.

- To avoid the possibility of fire and personal injury, always disconnect the negative battery cable unless the repair or test procedure requires that battery voltage be applied.
- Always relieve the fuel system pressure prior to disconnecting any fuel system component (injector, fuel rail, pressure regulator, etc.), fitting or fuel line connection. Exercise extreme caution whenever relieving fuel system pressure to avoid exposing skin, face and eyes to fuel spray. Please be advised that fuel under pressure may penetrate the skin or any part of the body that it contacts.
- Always place a shop towel or cloth around the fitting or connection prior to loosening to absorb any excess fuel due to spillage. Ensure that all fuel spillage (should it occur) is quickly removed from engine surfaces. Ensure that all fuel soaked cloths or towels are deposited into a suitable waste container.
- Always keep a dry chemical (Class B) fire extinguisher near the work area.
- Do not allow fuel spray or fuel vapors to come into contact with a spark or open flame.
- Always use a backup wrench when loosening and tightening fuel line connection fittings. This will prevent unnecessary stress and torsion to fuel line piping. Always follow the proper torque specifications.
- Always replace worn fuel fitting O-rings with new. Do not substitute fuel hose or equivalent where fuel pipe is installed.

RELIEVING FUEL SYSTEM PRESSURE

1. Disconnect the negative battery cable.
2. Place a suitable container near the fuel filter fitting.
3. Put a wrench on the fitting and a backup wrench on the filter fitting boss and wrap a shop towel around the connection.
4. Slowly, loosen the connection allowing the shop towel to absorb as much of the fuel as possible.
5. After repairs have been completed, tighten the fitting using the backup wrench.
6. Connect negative battery cable.
7. Start the engine and check for fuel leaks.

Fuel Tank

REMOVAL & INSTALLATION

1. Raise and support the vehicle safely.
2. Drain the fuel from the tank by removing the drain plug located at the left side of the tank. After all fuel has been drained, replace the plug using a new gasket. Tighten plug to 8–9 ft. lbs. (34–39 Nm).
3. Remove the rear seat back and cushion.
4. Remove the right rear quarter trim panel and taping to access the connector.
5. Remove the fuel sender and wiring as a unit.
6. Remove the protector plate. Label and disconnect all fuel and emission hoses.
7. Support the fuel tank with a floorjack and remove the 4 fuel tank attaching bolts.
8. Lower the fuel tank from the vehicle.

To install:

9. Raise the tank into position and tighten the 4 attaching bolts. Make sure the fuel gage harness is routed through to the inside of the vehicle.
10. Install the fuel hose and tighten the clamp to 8–10 ft. lbs. (34–43 Nm).
11. Install all other emissions and fuel hoses and tighten securely. Install the protector plate.
12. Connect the fuel pump wiring harness and install the rear trim panel and rear seat.

Fuel Filter

REMOVAL & INSTALLATION

1. Disconnect the negative battery cable.
2. Relieve fuel system pressure.
3. Remove fuel line connectors.
4. Remove the fuel filter mounting screw.
5. Remove fuel filter.

To install:

6. Install new fuel filter and mounting screw.
7. Replace fuel line gaskets and coat with a thin film of oil. Connect fuel lines to fuel filter finger tight.
8. Using a backup wrench to hold fuel filter, tighten fuel line nut to 26–32 ft. lbs. (34–43 Nm).
9. Connect negative battery cable.
10. Start engine and check for fuel leaks.

Electric Fuel Pump

PRESSURE TESTING

1. Ensure that the battery voltage is 12 volts or more.
2. Disconnect the negative battery cable.
3. Relieve the fuel system pressure.
4. Disconnect fuel line from fuel filter.
5. Install a suitable fuel pressure gauge between the fuel line and the fuel filter using a new gasket.
6. Connect negative battery cable.
7. Remove the cap on the check terminal.
8. Connect a jumper wire between fuel pump terminal F (white/black) and the ground terminal (black).
9. Turn the ignition switch to the **ON** position.
10. Check fuel pressure. Specified fuel pressure: 37–46 psi. (2.6–3.2 kg/cm^2).
11. Turn ignition switch to the **OFF** position. Let stand for 3 minutes.
12. Check residual pressure. Specified fuel pressure: 35.5 psi. (2.5 kg/cm^2) or more.
13. Disconnect the negative battery cable.
14. Relieve fuel system pressure.
15. Remove fuel pressure gauge from the system and reconnect fuel line to fuel filter replacing old gaskets. Coat new gaskets with a thin film of oil and

install finger tight. Using a backup wrench to hold fuel filter, tighten fuel line nut to 26–32 ft. lbs. (34–43 Nm).
16. Remove jumper wire and install cap on check terminal.
17. Connect negative battery cable.
18. Start engine and check for fuel leaks.

REMOVAL & INSTALLATION

1.0L Engine

1. Disconnect the negative battery cable.
2. Drain fuel from the fuel tank.
3. Remove the tank from the vehicle.
4. Remove the fuel main hose and fuel return pipe from the fuel pump bracket.
5. Disconnect the wire from the fuel pump bracket.
6. Remove the fuel pump bracket attaching bolts.
7. Remove the fuel pump bracket from the fuel tank.
8. Remove the nut retaining the fuel pump to the fuel pump bracket.
9. Disconnect the 2 lines from the fuel pump bracket.
10. Detach the fuel hose clip from the fuel hose.
11. Remove the ground cable attaching screw. Remove the ground wire.
12. Remove the fuel pump from the fuel pump bracket.
13. Remove the 2 nuts and disconnect the wires.
14. Remove the fuel hose from the fuel pump.

To install:

15. Connect the fuel hose to the fuel pump. Attach the hose clip.
16. Connect the 2 wires to the fuel pump. Ensure that the blue cable is on the positive terminal.
17. Install the rubber cushion on the fuel pump and install fuel pump to the fuel pump bracket.
18. Install the rubber cushion to the fuel pump bracket. Connect the ground cable. Secure the fuel pump by tightening the attaching bolts.
19. Attach the fuel hose clip to the fuel hose.
20. Install the positive wire of the fuel pump to the fuel pump bracket.
21. Install the fuel pump bracket with a new gasket to the fuel tank. Tighten the bolts in stages to 2–4 ft. lbs. (1.5–4.9 Nm).
22. Connect the wire to the fuel pump bracket.
23. Connect the hose and pipe to the fuel pump bracket. Coat a new gasket with a thin film of oil and install flare nut finger tight. Tighten to 25–32 ft. lbs. (34.3–43.1 Nm).
24. Install the fuel tank.
25. Fill fuel tank with fuel.
26. Connect negative battery cable.
27. Start engine and check for leaks.

Fuel Injection

IDLE SPEED ADJUSTMENT

1. Ensure that engine is at normal operating temperature.
2. Connect a timing light to verify proper ignition timing. If timing is not within specification, loosen distributor holddown and rotate distributor to correct timing. Tighten distributor holddown and recheck timing.
3. Connect a tachometer and adjust engine idle speed screw until proper idle speed is obtained.
4. Disconnect tachometer and timing light.
5. Test drive vehicle.

IDLE MIXTURE ADJUSTMENT

The idle mixture is controlled automatically by the Electronic Control Unit (ECU).

Fuel Injector

REMOVAL & INSTALLATION

1. Disconnect the negative battery cable.
2. Relieve fuel system pressure.
3. Remove the pressure regulator.
4. Remove the injector.

To install:

5. Check the insulator and grommet of each injector for damage.
6. Install the insulator on the manifold section.
7. Install the grommet on the injector.
8. Replace the injector O-ring.
9. Insert the injector into the insulator.
10. Install the delivery pipe.
11. Connect negative battery cable.
12. Start the engine and check for fuel leaks.

DRIVE AXLE

Halfshaft

REMOVAL & INSTALLATION

1. Disconnect the negative battery cable.
2. Raise and safely support the vehicle.
3. Remove the front wheels.
4. Drain the transaxle oil.
5. Remove the cotter pin and front wheel adjusting lock cap.
6. Remove the stub shaft nut.
7. Disconnect the outer tie rod from the steering knuckle.
8. Remove the engine undercovers.
9. Remove the stabilizer bar.
10. Remove the lower control arm from the bracket.
11. Separate the extension rod subassembly and shift and selector subassembly from the transaxle.
12. Remove the outer stub shaft from the steering knuckle.
13. Remove the inner stub shaft from the differential fitting.

NOTE: Be careful not to damage oil seals during removal.

To install:

14. Inspect the differential oil seal prior to installation of the halfshaft. Replace as needed.
15. Apply grease to the serrated section of the outboard joint stub shaft. Insert the outboard joint section into the knuckle. Temporarily, install the plate washer and nut.
16. Apply grease to the lip of the oil seal. Insert the inboard joint stub shaft into the differential.
17. Install the shift and selector shaft and extension rod. Tighten to 12–22 ft. lbs. (9–16 Nm).
18. Install the lower control arm to the bracket. Tighten to 54–76 ft. lbs. (40–56 Nm).
19. Install the stabilizer bar to the lower control arm assembly. Tighten to 54–76 ft. lbs. (40–56 Nm).
20. Install the stabilizer lower bracket to the body. Tighten to 43–61 ft. lbs. (32–45 Nm).
21. Install the engine undercovers.
22. Attach the tie rod end to the steering knuckle and tighten the castellated nut to 22–33 ft. lbs. (16–24 Nm). Install a new cotter pin.
23. Install the stub shaft nut and tighten to 130–166 ft. lbs. (96–123 Nm). Install the front wheel locknut and insert a new cotter pin.
24. Install the front wheels.
25. Connect negative battery cable.

CV-Boot

REMOVAL & INSTALLATION

1. Disconnect the negative battery cable.
2. Raise and safely support the vehicle.
3. Remove the halfshaft assembly from the vehicle and place in a vise.
4. Pry up the boot band clip and detach the boot.
5. Paint a match mark on the inboard joint and shaft prior to disas-

sembly to ensure reassembly in the same position. Do not use a punch to make match marks on the joint or shaft. Remove the front axle inboard joint subassembly.

6. Remove the snapring retaining the inboard tripod joint.
7. Punch a match mark on the tip of the tripod joint and shaft.
8. Use a brass drift to remove the tripod joint from the shaft.

NOTE: Be sure to place the end of the brass drift against the boss section of the joint, not the roller section.

9. Pry up the inner boot band. Remove the front halfshaft joint inboard boot.
10. Pry up the outboard boot band.
11. Remove the front halfshaft joint outboard boot.

To install:

12. Wind vinyl tape or equivalent around the splined tip of the shaft so boot will not be damaged during installation.
13. Fit the boot and a new boot band (smaller diameter) on the outboard joint.
14. Pack the outboard joint with grease.
15. Fit a new band (large diameter) in place.
16. Temporarily install the outboard joint boot onto the halfshaft.
17. Remove the vinyl tape that was wound around the splined portion.
18. Face the non-splined side of the inboard tripod joint toward the outboard joint.
19. Align the match marks.
20. Drive the tripod assembly onto the shaft lightly using the brass drift.
21. Attach the snapring onto the shaft.
22. Pack the inboard joint with grease.
23. Install the inboard join, aligning the match marks.
24. Fit new boot bands in place.
25. Install the halfshaft assembly in the vehicle.
26. Lower the vehicle.
27. Connect the negative battery cable.

Front Wheel Hub, Spindle and Bearings

REMOVAL & INSTALLATION

1. Disconnect the negative battery cable.
2. Raise and safely support the vehicle.
3. Remove the front wheel.
4. Disconnect the disc brake caliper

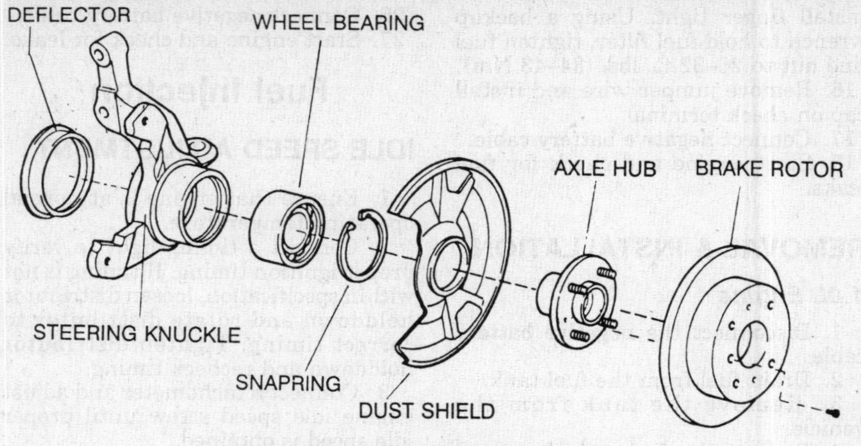

Steering knuckle and hub assembly

from the steering knuckle and support using a length of mechanics wire.

5. Remove the front disc attaching screws and remove the disc from the front axle hub.
6. Remove the cotter pin and front wheel adjusting lock cap. Remove the nut, using a 30mm socket and a suitable tool to prevent the axle from turning.
7. Use a slide hammer or equivalent to remove the axle hub.
8. Separate the outer tie rod from the steering knuckle.
9. Remove the lower ball joint attaching bolt and nut.
10. Remove the steering knuckle attaching nuts. Leave the bolts in place at this time.
11. Support the steering knuckle and draw out the attaching bolts for the strut lower bracket.

NOTE: Protect the outer CV-joint boot during removal. Be careful not to distort the disc brake outer cover during hub removal.

12. Place the hub in a vise and remove the dust seal from the axle hub.
13. Remove the bearing inner race from the axle hub.
14. Remove the snapring.
15. Press the bearing out of the hub.

To install:

16. Press the new bearing into the hub.
17. Install a new snapring.
18. Press the front axle hub into position.

NOTE: Be sure to press only on the inner race.

19. Insert the steering knuckle into the halfshaft.
20. Connect the steering knuckle to the ball joint. Connect the steering knuckle to the strut lower bracket. Tighten to 65–94 ft. lbs. (48–69 Nm).
22. Install the lower ball joint. Tighten to 58–76 ft. lbs. (43–56 Nm).
23. Install the plate washer in the proper direction. Install the nut temporarily.
24. Install the front disc.
25. Attach the tie rod end to the steering knuckle and tighten the castellated nut to 22–33 ft. lbs. (16–24 Nm). Install a new cotter pin.
26. Install the disc pad guide plates to the steering knuckle. Tighten the caliper attaching bolts to 23–30 ft. lbs. (17–22 Nm).
27. Prevent the halfshaft from turning and tighten the axle nut to 130–166 ft. lbs. (96–121 Nm).
28. Install the front wheel adjusting lock cap to the nut. Install a new cotter pin.
30. Install the front wheel.
31. Connect negative battery cable.
32. Check front wheel alignment.

MANUAL TRANSAXLE

For further information on transmissions/transaxles, please refer to "Chilton's Guide to Transmission Repair".

Transaxle Assembly

REMOVAL & INSTALLATION

1.0 Engine

1. Disconnect the negative and positive battery cables.
2. Remove the battery hold-down clamp.

3. Disconnect the harness clamp and washer hose.
4. Remove the battery tray.
5. Remove the clutch cable weight. Loosen the adjusting nut. Disconnect the clutch cable from the clutch release lever. Remove the clutch cable through the bracket hole.
6. Disconnect the starter harness.
7. Remove the 2 starter attaching bolts.
8. Disconnect the speedometer cable.
9. Disconnect the check engine terminal, cowl harness clamp, backup lamp connector and transaxle ground cable.
10. Detach the 3 transaxle wiring harness clamps.
11. Remove the transaxle attaching bolts except for 1 bolt in the front center.
12. Raise and safely support the vehicle.
13. Remove the engine undercovers.
14. Disconnect the front exhaust pipe from the manifold and the front bracket.
15. Remove the stiffener attaching bolt and exhaust pipe bracket support attaching bolts.
16. Disconnect the shift and select shaft and extension rod connectors.
17. Remove the stabilizer bar.
18. Remove the lower control arm installing nut from the bracket and remove the lower arm.
19. Remove the tie rod cotter pin and castellated nut.
21 Separate the tie rod end from the steering knuckle. Remove the halfshafts.
22. Remove the 2 attaching bolts and detach the left hand steering rack cover.
23. Support the lower part of the transaxle. Remove the lower left engine mount bracket attaching bolts.
24. Remove the 2 remaining transaxle attaching bolt at the front center of the housing. Slowly lower the transaxle and remove from under the vehicle.

To install:

25. Ensure that the clutch disc is centered.
26. Raise the transaxle and position it to the engine, making sure the locating pins are properly aligned.
27. Install and temporarily tighten the attaching bolts, making sure not to cause the clutch disc to slip.
28. Attach the clutch housing undercover. Tighten the undercover together with the exhaust pipe bracket support to 11–16 ft. lbs. (15–22 Nm).
29. Tighten the transaxle assembly attaching bolts to 36–51 ft. lbs. (49–67 Nm).
30. Tighten the lower left engine mount bolts to 22–33 ft. lbs. (29–44 Nm).
31. Install the halfshafts.
32. Attach the lower control arm bracket and temporarily tighten the installing nut.
33. Temporarily tighten the stabilizer bar end nut. Tighten the cushion and stabilizer bar bracket. Tighten to 29–43 ft. lbs. (39–59 Nm).
34. Attach the tie rod end to the steering knuckle and tighten the castellated nut to 22–32 ft. lbs. (29–44 Nm). Install a new cotter pin.
35. Connect the shift and select shaft connectors. Tighten to 12–22 ft. lbs. (17–30 Nm).
36. Connect the extension rod connector. Tighten to 12–22 ft. lbs. (17–30 Nm).
37. Install the left hand steering rack cover.
38. Install the front exhaust pipe to the exhaust manifold and tighten to 36–51 ft. lbs. (49–69 Nm). Install the front support bracket and tighten to 25–36 ft. lbs. (34–49 Nm).
39. Install the engine under covers.
40. Lower the vehicle.
41. Rock the vehicle to settle the suspension.
42. Tighten the stabilizer bar installing nut to 54–80 ft. lbs. (74–108 Nm).
43. Tighten the lower control arm bolt at the bracket to 51–72 ft. lbs. (69–98 Nm).
44. Fill the transaxle with 2.4 qts. (2.24L) of transaxle oil. Install the filler plug.
45. Attach the 3 wiring harness clamps to the transaxle.
46. Connect the speedometer transaxle ground cables.
47. Connect the cowl harness clamp and check engine terminal to the bracket.
48. Install the starter and tighten the 2 bolts to 29–40 ft. lbs. (39–54 Nm). Install the starter harness. Connect the backup lamp connector.
49. Connect the clutch cable to the clutch release lever through the bracket hoe. Adjust clutch cable free-play. Attach the clutch cable weight.
50. Install the battery tray and battery. Secure the hold-down clamp. Connect the positive and negative battery cables.

1.3L Engine

1. Disconnect the negative and positive battery cables. Remove the battery and battery tray. Disconnect the wire harness clamp with the EFI check terminal from the upper part of the transaxle. Disconnect the backup lamp switch.
2. Drain the cooling system into a clean container for reuse.
3. Drain the transaxle oil if the transaxle is to be disassembled.
4. Raise and safely support the vehicle.
5. Remove the engine undercover.
6. Remove the lower suspension brace from the lower control arm brackets.
7. Remove the extension rod and shift and select shaft from the transaxle case.
8. Remove the exhaust pipe.
9. Remove the front stabilizer bar.
10. Disconnect the lower control arms from the bracket ends.
11. Remove the exhaust support bracket from the engine stiffener.
12. Remove the halfshafts on the right and left sides.
13. Remove the speedometer cable and transaxle ground cable.
14. Disconnect the clutch cable from

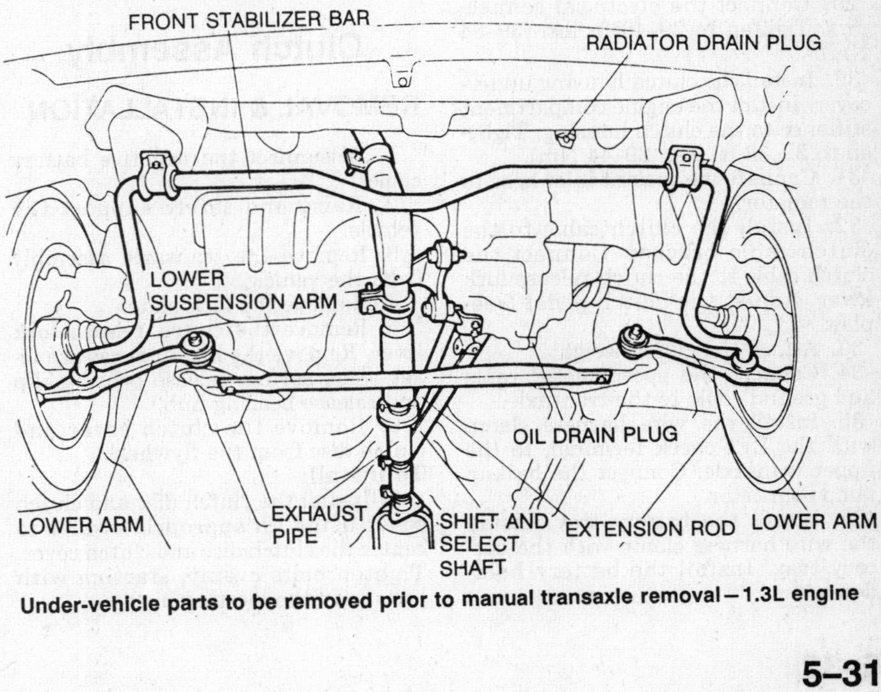

Under-vehicle parts to be removed prior to manual transaxle removal—1.3L engine

the clutch release fork lever and clutch cable bracket.

15. Disconnect the coolant inlet hose from the radiator.

16. Remove the engine compartment stiffener from the engine and clutch housing.

17. Remove the starter.

18. Support the engine and transalxe.

19. Remove 3 of the 4 bolts connecting the transaxle to the engine. Leave 1 in the front loose.

20. Remove the engine mount bolts. from the rear engine mount bracket and the lower left transaxle mount. Remove the support bolt from the upper side of the transaxle case.

21. Remove the remaining transaxle-to-engine bolt.

22. Slowly lower the transaxle and remove from under the vehicle.

To install:

23. Ensure that the clutch disc is centered.

24. Attach the transaxle to the engine block. Tighten 2 bolts at the front , 1 bolt at the rear and 2 upper bolts to 36–51 ft. lbs. (49–69 Nm).

NOTE: Installing the transaxle may require lowering the engine slightly.

25. Install the lower left engine mount-to-transaxle case bracket bolts. Install the upper transaxle case support bolt. Tighten to 36–51 ft. lbs. (49–69 Nm).

26. Install the rear engine mount insulator bolts to the transaxle case bracket. Tighten to 29–40 ft. lbs. (39–54 Nm).

27. Remove the engine and transmission lifting devices.

28. Install the starter motor assembly to the engine.

29. Connect the electrical connectors. Tighten to 29–40 ft. lbs. (39–54 Nm).

30. Install the clutch housing undercover. Install the engine compartment stiffener to the clutch housing. Tighten to 22–33 ft. lbs. (29–44 Nm).

31. Connect the coolant inlet hose to the radiator.

32. Install the clutch cable to the clutch cable bracket. Connect the clutch cable to the clutch release fork lever. Adjust the clutch pedal freeplay.

33. Attach the clutch weight.

34. Connect the speedometer cable and ground cable to the transaxle.

35. Install the wire harness clamp with the EFI check terminal, to the upper transaxle. Connect the backup lamp connector.

36. Install the battery tray. Install the wire harness clamp with the battery tray. Install the battery hold-down.

37. Connect the fusible link to the battery.

38. Connect the halfshafts.

39. Install the exhaust support bracket to the engine stiffener. Tighten to 22–33 ft. lbs. (29–44 Nm).

40. Install the lower control arms to the brackets. Tighten to 51–72 ft. lbs. (69–98 Nm).

41. Install the front stabilizer bar to the lower control arm and stabilizer bracket. Tighten lower control arm to 54–80 ft. lbs. (84–108 Nm). Tighten stabilizer bracket to 29–43 ft. lbs. (39–59 Nm).

42. Connect the exhaust pipe.

43. Install the extension rod and shift and select shaft to the transaxle case, using new attaching bolts. Tighten extension rod to 12–22 ft. lbs. (17–30 Nm). Tighten shift and select shaft to 12-22 ft. lbs. (17–30 Nm).

44. Install the lower suspension brace to the lower arm brackets. Tighten to 29–40 ft. lbs. (39–54 Nm).

45. Lower the vehicle.

46. If drained, fill the transaxle with 2.4 qts. (2.25L of oil).

47. Fill the radiator with coolant and check for leaks.

48. Install the engine undercover.

49. Connect the positive and negative battery cables.

50. Start engine and allow engine to come to normal operating temperature. Recheck for coolant leaks. Allow engine to warm up sufficiently to confirm operation of cooling fan.

CLUTCH

Clutch Assembly

REMOVAL & INSTALLATION

1. Disconnect the negative battery cable.

2. Raise and safely support the vehicle.

3. Remove the transaxle assembly from the vehicle.

4. Remove the lock plate.

5. Remove the clutch release fork lever. Remove the bushing, release lever yoke, spring, release bearing clip and release bearing hub.

6. Remove the clutch cover and clutch disc from the flywheel.

To install:

7. Install the clutch disc and clutch cover using an appropriate pilot to center the clutch disc and clutch cover. Tighten bolts evenly, starting with bolts near the locating pin.

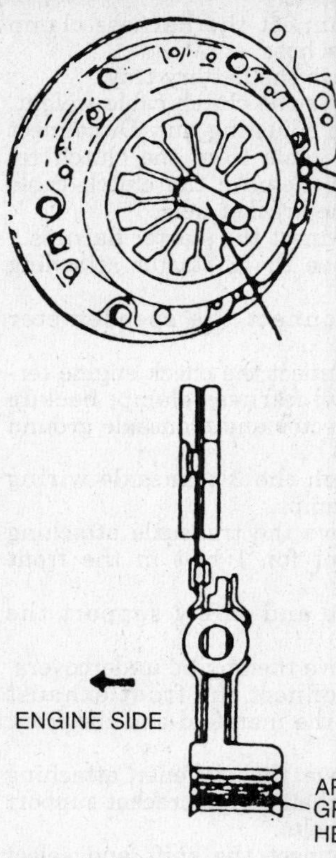

Proper clutch disc installation—1.0L and 1.3L engines

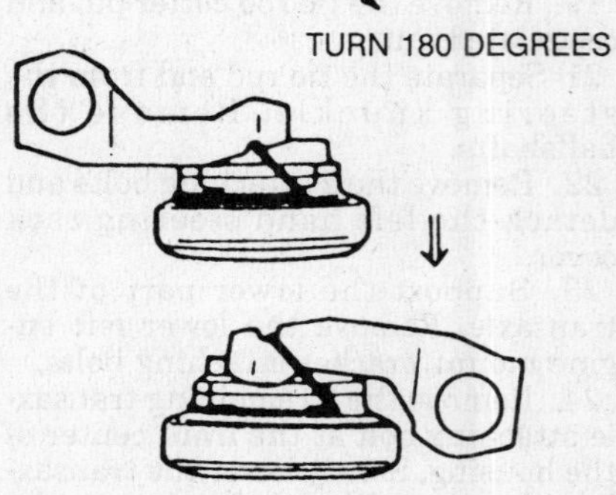

Attaching release lever yoke to clip—1.0L and 1.3L engines

NOTE: Ensure that clutch disc is installed in the proper direction. Apply a long-life grease to the clutch disc splines.

8. Assemble the release bearing and clip to the clutch release lever yoke by placing clip on the release bearing. Place the cut-out section of the release lever yoke face down on the clip. Pivot the release lever yoke 180 degrees over the clip to connect the yoke to the clip.

9. Assemble the bushing, dust seal, torsion spring and clutch release lever in position.

1. Bolt
2. Lock plate
3. Clutch release fork lever
4. Bushing
5. Clutch release lever yoke
6. Torsion spring
7. Bushing
8. Dust seal
9. Release bearing clip
10. Clutch release bearing hub
11. Clutch cover
12. Clutch disc

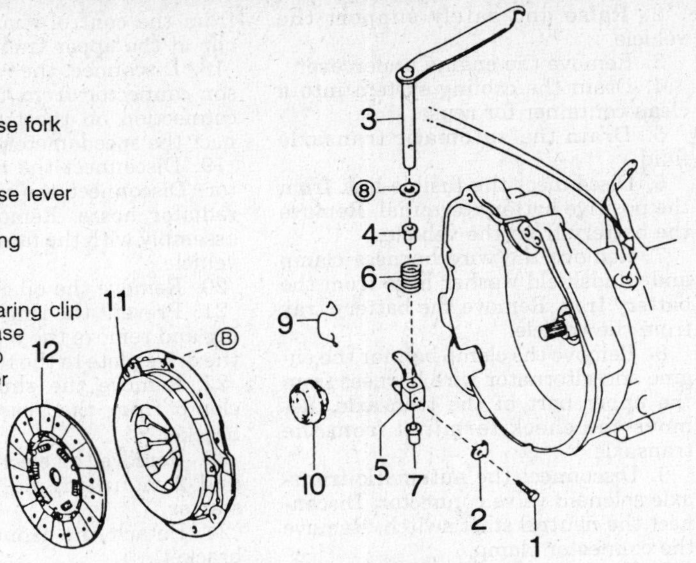

Clutch assembly—1.0L and 1.3L engines

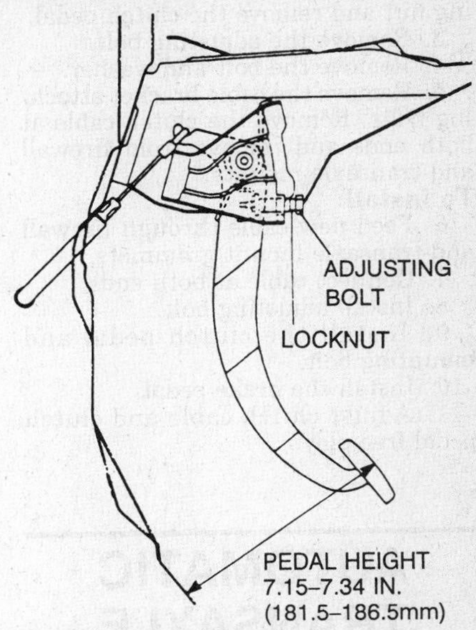

Clutch pedal height adjustment

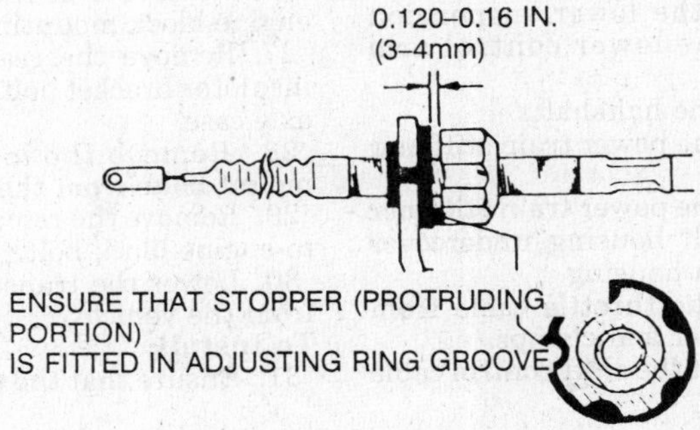

Clutch cable adjustment

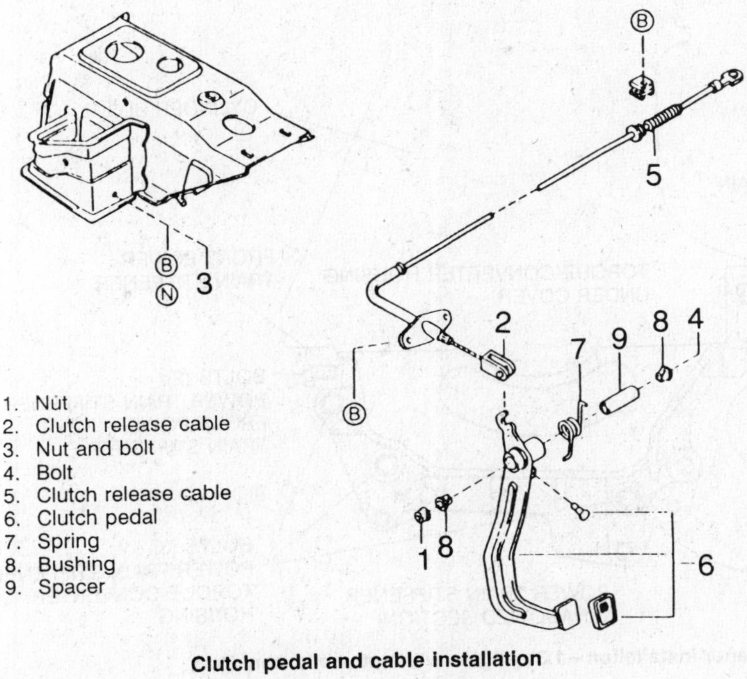

1. Nut
2. Clutch release cable
3. Nut and bolt
4. Bolt
5. Clutch release cable
6. Clutch pedal
7. Spring
8. Bushing
9. Spacer

Clutch pedal and cable installation

10. Assemble the lock plate bolt and lever.
11. Check the operation of the release hub and yoke.
12. Install the transaxle assembly to the vehicle.
13. Lower the vehicle.
14. Connect negative battery cable.

PEDAL HEIGHT/FREE-PLAY ADJUSTMENT

1. Slacken the clutch cable locknut. Turn the stopper bolt until the installation height conforms to specification. Clutch pedal free travel should be: 0.59–1.18 in. (15–30mm).
2. Tighten the locknut.

Clutch Cable

ADJUSTMENT

1. Pull the outer cable lightly until resistance from the clutch is felt.
2. Turn the adjusting ring until it lightly touches the protruding portion of the rubber grommet. Clutch cable endplay should be: 0.120–0.160 in. (3–4mm).

NOTE: Ensure that the stopper (protruding portion) is fitted securely in the adjusting groove. If the stopper is not aligned in the adjusting groove, turn the adjusting ring counterclockwise to fit the stopper.

REMOVAL & INSTALLATION

1. Remove the brake pedal.
2. Remove the clutch pedal mount-

ing nut and remove the clutch pedal.

3. Remove the adjusting bolt.
4. Remove the bolt and washer.
5. Remove the cable bracket attaching bolts. Remove the clutch cable at both ends and remove from firewall and transaxle mounts.

To install:

6. Feed new cable through firewall and transaxle mount grommets.
7. Connect cable at both ends.
8. Install adjusting bolt.
9. Install the clutch pedal and mounting bolt.
10. Install the brake pedal.
11. Adjust clutch cable and clutch pedal free-play.

AUTOMATIC TRANSAXLE

For further information on transmissions/transaxles, please refer to "Chilton's Guide to Transmission Repair".

Transaxle Assembly

REMOVAL & INSTALLATION

1. Disconnect the negative battery cable.
2. Raise and safely support the vehicle.
3. Remove the engine undercover.
4. Drain the cooling system into a clean container for reuse.
5. Drain the automatic transaxle fluid.
6. Disconnect the fusible link from the positive battery terminal. Remove the battery from the vehicle.
7. Remove the wire harness clamp and windshield washer hose from the battery tray. Remove the battery tray from the vehicle.
8. Remove the clamp bolt for the engine and alternator wire harness from the upper part of the transaxle. Remove the check terminal from the transaxle.
9. Disconnect the automatic transaxle solenoid valve connector. Disconnect the neutral start switch. Remove the connector clamp.
10. Disconnect the transaxle ground cable.
11. Remove the front stabilizer bar.
12. Remove the lower suspension brace from the lower control arm brackets.
13. Remove the halfshafts.
14. Remove the power train stiffener attaching bolts.
15. Remove the power train stiffener with the clutch housing undercover from the clutch housing.
16. Remove the throttle cable from the throttle lever and clamps.
17. Disconnect the shift control cable from the control shaft lever and the clip at the upper transaxle.
18. Disconnect the vehicle speed sensor connector from the speedometer connection on the transaxle. Disconnect the speedometer cable.
19. Disconnect the radiator fan motor. Disconnect the upper and lower radiator hoses. Remove the radiator assembly with the fan shroud from the vehicle.
20. Remove the oil cooler hoses.
21. Prevent the ring gear from turning and remove the 6 bolts connecting the drive plate to the torque converter.
22. Remove the shift control cable clamp from the rear engine mount bracket.
23. Remove the electrical connectors and 2 mounting bolts and remove starter.
24. Detach the front left stabilizer bracket.
25. Support the engine and transaxle separately.
26. Remove 3 of the 4 transaxle-to-engine block mounting bolts.
27. Remove the rear engine mount insulator bracket bolts from the transaxle case.
28. Remove the lower left engine mount bolts from the transaxle case.
29. Remove the remaining transaxle-to-engine block bolt.
30. Lower the transaxle and remove from the vehicle.

To install:

31. Ensure that the torque converter

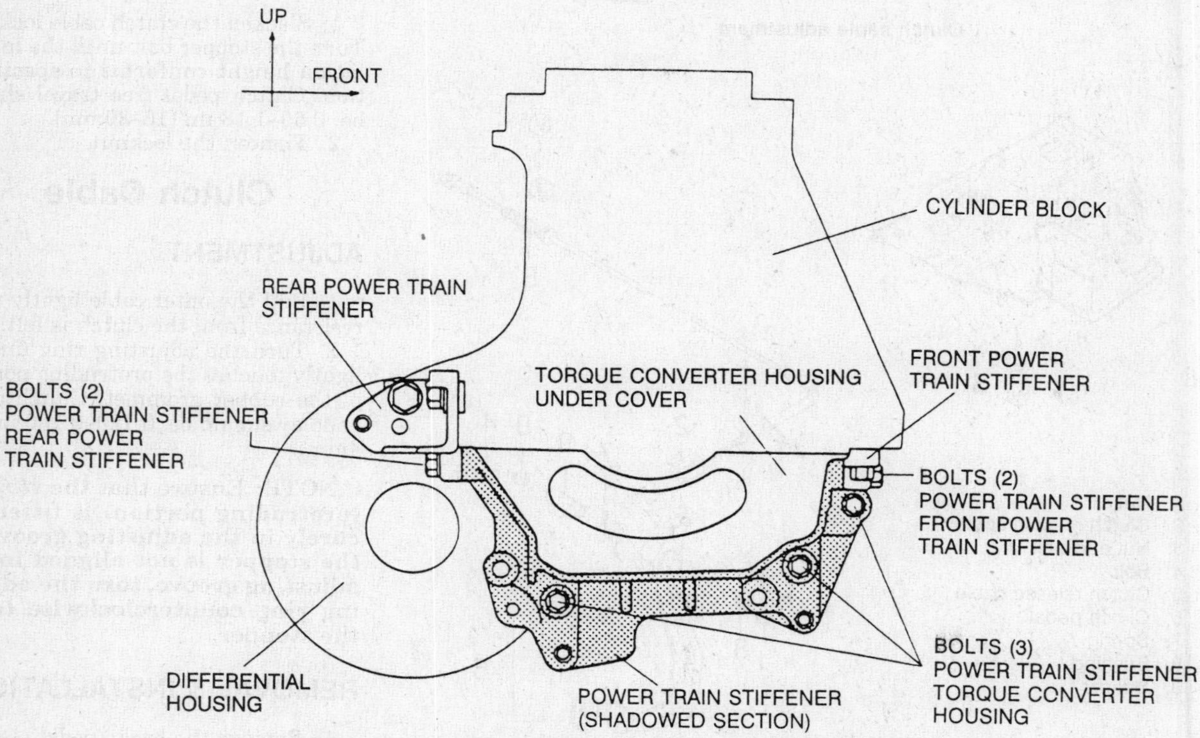

Power train stiffener installation—1.3L engine with automatic transaxle

is fitted positively with the automatic transaxle. Apply grease to the crankshaft fitting on the torque converter.

32. Install the automatic transaxle in the vehicle, making sure not to pry on the torque converter or drive plate during installation. Install the 4 attaching bolts. Tighten to 36–51 ft. lbs. (49–68 Nm).

33. Install the lower left engine mount bolts to the transaxle bracket. Tighten to 21–32 ft. lbs. (29–44 Nm).

34. Install the rear engine mount insulator bracket bolts to the transaxle. Tighten to 40–51 (54–69 Nm).

35. Remove the engine and transaxle support devices.

36. Install the front stabilizer bracket. Tighten to 29–44 ft. lbs. (39–59 Nm).

37. Install the starter motor and connectors. Tighten bolts to 36–51 ft. lbs. (49–69 Nm).

38. Install the shift control cable clamp to the rear engine mount bracket. Tighten to 10–18 ft. lbs. (13–24 Nm).

39. Prevent the ring gear from turning and install the 6 drive plate-to-converter bolts. Tighten to 17–23 ft. lbs. (23–32 Nm).

40. Install the oil cooler lines using new line.

41. Install the radiator fan motor and shroud to the radiator. Connect the upper and lower radiator hoses. Connect the radiator fan motor.

42. Connect the speedometer cable. Connect the vehicle speed sensor.

43. Attach the clip for the shift control cable to the upper part of the transaxle.

44. Position the shift control shaft to the **N** position. Tighten the adjusting bolt while the shift lever is being pulled slightly from the **N** position to the **R** position with the shift control cable at the transaxle.

45. Ensure that the shift lever can be moved to each of the drive ranges with the proper amount of detent resistance. Tighten the adjusting bolt to 12–18 ft. lbs. (16–24 Nm).

46. Connect the throttle cable from the transaxle to the throttle lever. Secure the cable clamps.

47. Install the power train stiffener to the torque converter housing undercover temporarily.

48. Install the bolts connecting the power train stiffener to the front and rear stiffeners, tightening the power train stiffener-to-torque converter housing bolts first. Tighten bolts to 21–32 ft. lbs. (29–44 Nm).

49. Install the halfshafts.

50. Install the lower control arms to the brackets. Tighten to 55–76 ft. lbs. (74–103 Nm).

51. Install the front stabilizer bar. Tighten to 29–44 ft. lbs. (39–59 Nm).

52. Connect the exhaust pipe.

53. Install the lower suspension brace to the lower control arm brackets. Tighten bolts to 29–44 ft. lbs. (39–54 Nm).

54. Install the transaxle ground cable. Install the neutral start switch connector. Connect the solenoid valve connector.

55. Install the engine and alternator wire harness clamp bolt to the upper side of the transaxle.

56. Install the battery tray. Attach the windshield washer hose and wire harness clamp to the battery tray. Install the battery and the battery holddown clamp. Connect the fusible link and positive battery cable.

57. Fill the automatic transaxle with fluid.

58. Fill the radiator with coolant.

59. Install the engine undercover.

60. Lower the vehicle. Tighten lower control arm-to-bracket bolts to 51–72 ft. lbs. (69–98 Nm). Tighten stabilizer bar-to-lower control arm bolts to 55–80 ft. lbs. (74–108 Nm).

61. Connect negative battery cable.

62. Start engine and allow engine to come to normal operating temperature. Recheck for coolant leaks. Allow engine to warm up sufficiently to confirm operation of cooling fan.

63. Check automatic transaxle fluid level.

SHIFT LINKAGE ADJUSTMENT

1. Disconnect the control cable from the transaxle control shaft lever.
2. Ensure that the shift lever can be moved smoothly while operating the shift lever button. If shift lever action is difficult, replace the shift lever assembly.
3. Align the control shaft lever with the **N** range. With the shift lever in **N** and pulled lightly toward **R** at the transaxle, install the control cable on the control shaft lever. Tighten the adjusting bolt.
4. Ensure that the shift lever can be moved smoothly.

THROTTLE LINKAGE ADJUSTMENT

1. Ensure that the throttle valve cable can be moved smoothly when operated independently.
2. Connect the cable to the throttle lever. Turn the adjusting nut so the gap between the cable and the adjusting nut (dimension A) is 0.0–0.020 in. (0.0–0.5mm).

FRONT SUSPENSION

MacPherson Strut

REMOVAL & INSTALLATION

1. Disconnect the negative battery cable.
2. Raise and safely support the vehicle.
3. Remove the front wheel.
4. Remove the clip retaining the flexible brake hose to the strut housing. Disconnect the hose from the strut housing.
5. Remove the nuts attaching the strut to the steering knuckle.

NOTE: Before removing the left strut, remove the disc brake caliper attaching bolt from the upper side.

6. Working in the engine compartment, remove the 2 nuts attaching the suspension support.
7. Remove the bolts attaching the strut to the steering knuckle.
8. Remove the strut from the vehicle.

To install:

9. Working in the engine compartment, install the suspension support on the fender apron using new nuts. Tighten to 15–22 ft. lbs. (20–30 Nm).
10. Mount the lower strut bracket on

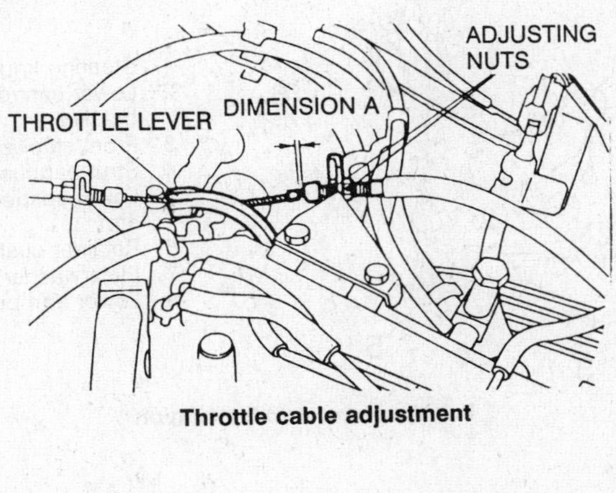

Throttle cable adjustment

the steering knuckle using new nuts and bolts. Tighten to 80–116 ft. lbs. (108–158 Nm).

NOTE: If the left strut was removed, install the disc brake caliper upper attaching bolt to the knuckle and tighten to 23–30 ft. lbs. (31–41 Nm).

11. Install the flexible hose to the strut bracket.
12. Install the flexible hose clip.
13. Install the wheels.
14. Lower the vehicle.
15. Check the front end alignment.

Lower Ball Joints

INSPECTION

With the lower ball joint separated from the steering knuckle, check for looseness and excessive play in the ball joint.

REMOVAL & INSTALLATION

The lower ball joint is integral to the lower control arm. If the lower ball joint is determined to be defective, replace the lower control arm.

Lower Control Arms

REMOVAL & INSTALLATION

1. Disconnect the negative battery cable.
2. Raise and safely support the vehicle.
3. Remove the front wheel.
4. Remove the end nut from the stabilizer bar.
5. Remove the ball joint nut and bolt.
6. Remove the nut attaching the lower control arm to the body.
7. On G102 vehicles, remove the lower suspension brace attaching bolts.
8. Remove the engine undercovers.
9. Remove the lower control arm bracket.
10. Remove the lower control arm pivot bolt and then remove the lower control arm.

To install:

11. Install and temporarily tighten the lower control arm ball joint and stabilizer bar end nut.
12. Install the ball joint nut and bolt. Tighten to 58–76 ft. lbs. (79–103 Nm).
13. Install the lower control arm bracket. Tighten to 54–76 ft. lbs. (73–103 Nm).
14. Tighten the lower control arm attaching nut temporarily.
15. Install the engine undercovers.
16. On G102 vehicles, install the lower suspension brace. Tighten to 29–40 ft. lbs. (39–54 Nm).
17. Install the front wheel.
18. Lower the vehicle.
19. Rock the front suspension several times to settle the suspension. Tighten the stabilizer bar and lower control arm bolts and nuts to 54–76 ft. lbs. (73–103 Nm).
20. Check front end alignment.

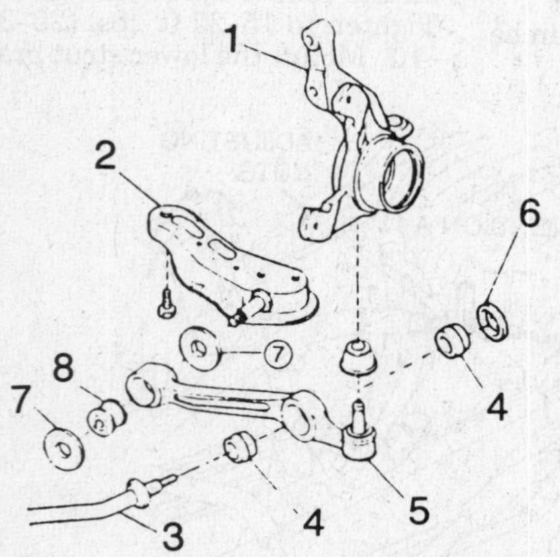

1. Steering knuckle
2. Lower control arm bracket
3. Front stabilizer bar
4. Strut bar cushion
5. Lower suspension arm
6. Retainer cushion
7. Plate washer
8. Lower arm bushing

Lower control arm installation

Stabilizer Bar

REMOVAL & INSTALLATION

1. Disconnect the negative battery cable.
2. Raise and safely support the vehicle.
3. Remove the engine undercovers.
4. Remove the bolts attaching the stabilizer bar bracket to the body.
5. Remove the nuts attaching the stabilizer bar to the lower control arm.
6. Remove the stabilizer bar from the vehicle.

To install:

7. Install the stabilizer bar in the lower control arm. Tighten the bolts to 54–76 ft. lbs. (40–56 Nm).
8. Install the stabilizer bar brackets to the body. Tighten to 43–61 ft. lbs. (32–45 Nm).
9. Install the engine undercovers.
10. Lower the vehicle.
11. Connect negative battery cable.

REAR SUSPENSION

MacPherson Strut

REMOVAL & INSTALLATION

1. Disconnect the negative battery cable.
2. Raise and safely support the vehicle.
3. Remove the rear wheel.
4. Remove the brake tube from the bracket.
5. Detach the clip and disconnect the flexible hose from the strut.
6. Remove the clamp retaining the brake tube located at the back of the strut housing.
7. Remove the nuts attaching the strut to the axle carrier. Do not remove the bolts at this time.
8. On G100 and G102 Hatchback, working inside the vehicle, perform the following:
   a. Remove the package tray.
   b. Tilt the seat back.
   c. Remove the package tray side trim.
   d. Disconnect the speaker.
   e. Loosen, but do not remove, the nut attaching the strut to the suspension support.
   f. Remove the nuts attaching the suspension support to the body.
9. On G102 Sedan, working inside the vehicle, perform the following:
   a. Remove the rear seat back hinge bolts.
   b. Remove the screw attaching the rear wheel quarter trim.
   c. Remove the rear quarter trim.
10. Remove the bolts attaching the axle carrier to the strut. Remove the strut from the body.

To install:

11. Install the suspension support to the body. Tighten to 7–12 ft. lbs. (9–16 Nm).
12. Mount the strut to the axle carrier and install new nuts and bolts. Push the axle carrier to the lower (positive) side and tighten to 80–116 ft. lbs. (108–158 Nm).
13. Install the flexible hose to the strut housing. Secure the clip.
14. Install the brake tube to the fitting.
15. Install the brake tube retaining

clamp located at the back of the strut housing.

16. On G100 and G102 Hatchback, working inside the vehicle, perform the following:

a. Tighten the strut-to-suspension support attaching nut to 25–40 ft. lbs. (34–54 Nm).

b. Connect the speaker.

c. Install the package tray side trim.

d. Raise the seatback.

e. Install the package tray.

17. On G102 Sedan, working inside the vehicle, perform the following:

a. Install the rear wheel quarter trim.

b. Install the rear seat back hinge.

18. Bleed the rear brakes.
19. Install the rear wheel.
20. Lower the vehicle.
21. Connect negative battery cable.

Rear Control Arms

REMOVAL & INSTALLATION

1. Disconnect the negative battery cable.
2. Raise and safely support the vehicle.
3. Remove the wheel.
4. Remove the bolt and nut attaching the stabilizer link to suspension arm No. 1.
5. Remove the bolt and nut attaching the suspension arm to the axle carrier.
6. Put a match mark on the body bracket and toe adjusting cam as a guide during reinstallation. Remove the bolt attaching the suspension arm to the adjusting cam on the body. Remove suspension arm No. 1.
7. Remove the rear stabilizer bar bracket from the suspension arm.
8. Remove the bolt and nut attaching suspension arm No. 2 to the axle carrier.
9. Remove the bolt and nut attaching suspension arm No. 2 to the body. Remove the suspension arm.

To install:

10. Install suspension arm No. 1 to the body temporarily.
11. Install suspension arm No. 1 to the axle carrier temporarily.
12. Install the rear stabilizer bar bracket to the suspension arm. Tighten to 7–12 ft. lbs. (9–16 Nm).
13. Mount the suspension arm to the body and tighten temporarily.
14. Align the match marks on the toe adjuster and body.
15. Tighten the suspension arm to the axle carrier temporarily.
16. Tighten the stabilizer link attaching bolt to 14–22 ft. lbs. (19–30 Nm).
17. Install the rear wheel.
18. Lower the vehicle. Connect negative battery cable.
19. Rock the vehicle up and down several times to settle the suspension.
20. With the vehicle weight applied to the suspension, tighten the bolts for suspension arm No. 1 to 80–101 ft. lbs. (108–137 Nm).
21. Tighten the bolts for suspension arm No. 2 to 80–101 ft. lbs. (108–137 Nm) for the bolts to the axle carrier and 51–64 ft. lbs. (69–87 Nm) for the bolts to the body.

Rear Wheel Bearings

REMOVAL & INSTALLATION

G100 and G102 Hatchback

1. Disconnect the negative battery cable.
2. Raise and safely support the vehicle.
3. Remove the rear wheel.
4. Remove the grease cap. Remove the cotter pin and castellated nut and plate washer.
5. Remove the brake drum using a suitable puller.
6. Place the brake drum on the work bench. Using a brass drift, drive out the inner bearing. With the drum in this position, drive out the outer bearing retainer.
7. Invert the brake drum and drive out the outer bearing.

To install:

8. Pack the hub and bearing with grease.
9. Install the outer bearing, outer retainer and inner bearing using a suitable seal/bearing installer.
10. Install the brake drum.
11. Install the rear wheel.
12. Lower the vehicle.
13. Connect the negative battery cable.

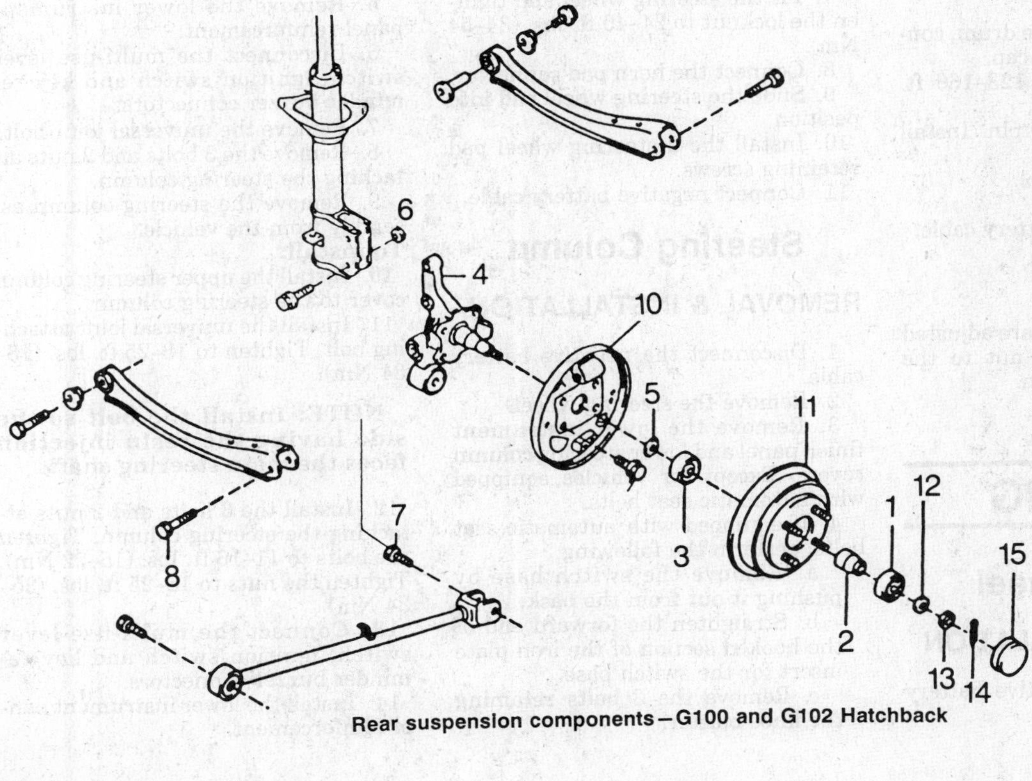

1. Radial ball bearing
2. Rear axle bearing outer retainer
3. Radial ball bearing
4. Rear axle carrier
5. Rear axle bearing inner retainer
6. Nut
7. Bolt
8. Bolt
9. Bolt
10. Rear brake assemlby
11. Rear brake drum
12. Plate washer
13. Castellated nut
14. Cotter pin
15. Grease retainer cap

Rear suspension components—G100 and G102 Hatchback

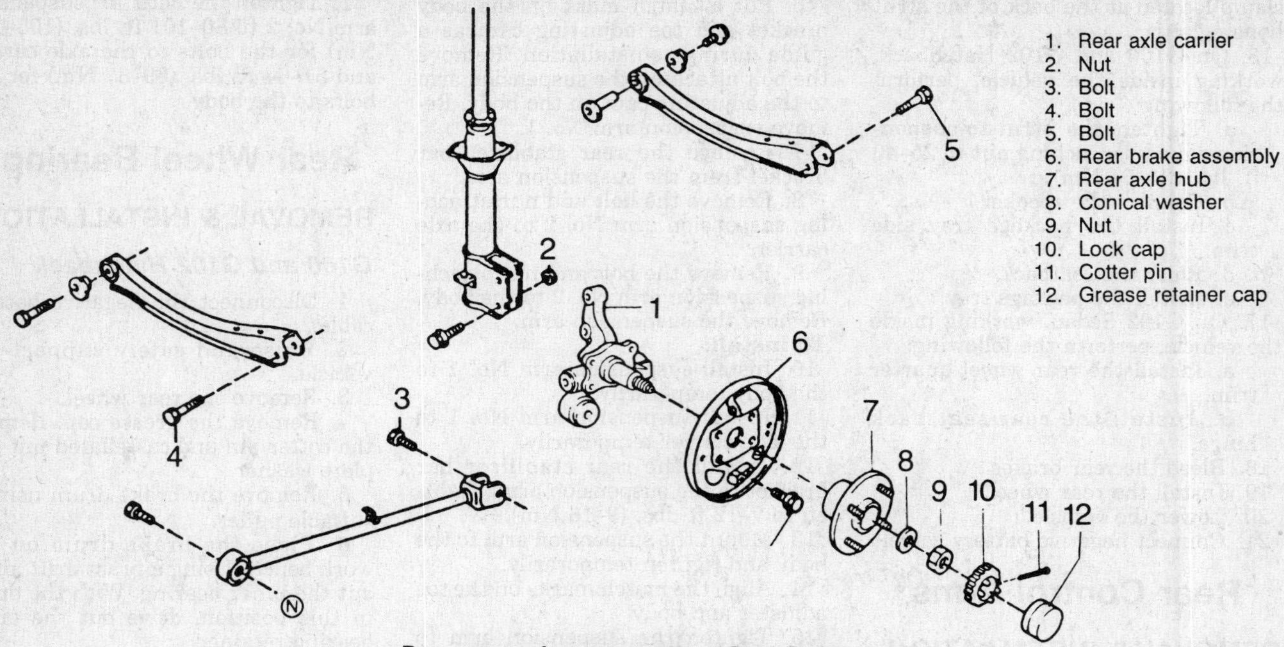

Rear suspension components – G102 Sedan

G102 Sedan

1. Disconnect the negative battery cable.
2. Raise and safely support the vehicle.
3. Remove the rear wheel.
4. Remove the grease retainer cap, cotter pin, lock cap, nut, conical washer, brake drum and hub.
5. The unit wheel bearing is integral to the hub.

To install:

6. Install the hub, brake drum, conical washer, nut and lock cap.
7. Tighten the nut to 123–166 ft. lbs. (167–225 Nm).
8. Install a new cotter pin. Install the grease retainer cap.
9. Install the rear wheel.
10. Lower the vehicle.
11. Connect negative battery cable.

ADJUSTMENT

The rear wheel bearing(s) are adjusted by tightening the wheel nut to the proper torque specification.

STEERING

Steering Wheel

REMOVAL & INSTALLATION

1. Disconnect the negative battery cable.
2. Remove the 2 screws retaining the steering wheel pad.
3. Remove the steering wheel pad by pushing it upward.
4. Disconnect the horn pad connector.
5. Remove the steering wheel locknut.
6. Remove the steering wheel using a suitable puller.

To install:

7. Fit the steering wheel and tighten the locknut to 24–40 ft. lbs. (34–54 Nm).
8. Connect the horn pad connector.
9. Slide the steering wheel pad into position.
10. Install the 2 steering wheel pad retaining screws.
11. Connect negative battery cable.

Steering Column

REMOVAL & INSTALLATION

1. Disconnect the negative battery cable.
2. Remove the steering wheel.
3. Remove the lower instrument finish panel and lower steering column covers. Except for vehicles equipped with automatic seat belts.
4. If equipped with automatic seat belts, perform the following:
   a. Remove the switch base by pushing it out from the back.
   b. Straighten the forward end of the hooked section of the iron plate insert for the switch base.
   c. Remove the 3 bolts retaining the knee bolster.
   d. Remove the 4 screws retaining the instrument cluster.
   e. Remove the 4 screws retaining the instrument cluster and remove the cluster.
   f. Detach the clip on the right side of the knee bolster by inserting finger through the aperture left by the removed cluster.
   g. Remove the lower steering column cover.
5. Remove the lower instrument panel reinforcement.
6. Disconnect the multi-use lever switch, ignition switch and key reminder buzzer connectors.
7. Remove the universal joint bolt.
8. Remove the 3 bolts and 2 nuts attaching the steering column.
9. Remove the steering column assembly from the vehicle.

To install:

10. Install the upper steering column cover to the steering column.
11. Install the universal joint attaching bolt. Tighten to 18–25 ft. lbs. (25–34 Nm).

NOTE: Install the bolt so the side having the resin injection faces the main steering shaft.

12. Install the 3 bolts and 2 nuts attaching the steering column. Tighten the bolts to 11–16 ft. lbs. (15–22 Nm). Tighten the nuts to 18–25 ft. lbs. (25–34 Nm).
13. Connect the multi-use lever switch, ignition switch and key reminder buzzer connectors.
14. Install the lower instrument panel reinforcement.

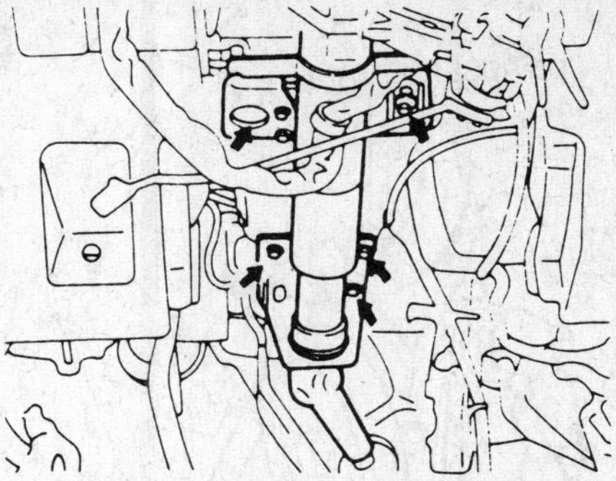

Steering column mounting bolts

15. Install the lower instrument finish panel and lower steering column cover. Except for vehicles equipped with automatic seat belts.
16. If equipped with automatic seat belts, perform the following:
   a. Install the lower steering column cover screws.
   b. Insert the clip for the knee bolster to the instrument panel.
   c. Bend the forward end of the hook section of the iron plate.
   d. Tighten the 3 bolts attaching the knee bolster.
   e. Install the switch base by pushing it in from the front.
17. Install the steering wheel.
18. Connect negative battery cable.

Manual Steering Rack

ADJUSTMENT

With the steering rack removed from the vehicle and installed in a vise, measure the rack preload and the steering pinion preload.

NOTE: Steps 1 and 2 below should be carried out with the steering rack in the neutral position (equivalent to the wheel straight ahead).

1. Tighten the rack guide spring cap. Tighten to 5 ft. lbs. (7 Nm).
2. Move the steering rack back and for the about 15 times so as to settle the rack. Tighten the rack guide spring cap again to 9 ft. lbs. (12 Nm).
3. Back off the rack guide spring cap 35–55 degrees.
4. Using an inch pound torque wrench, measure the steering pinion preload. Specified value (starting torque) should be: 4.8–5.9 inch lbs. (0.6–1.1 Nm).
5. Measure the starting force of the steering rack, using a spring scale with a rope attached to the end of the steering rack. Specified value: not to exceed 14 kg.
6. If the preload does not fall within specification, repeat Steps 1 through 5.

REMOVAL & INSTALLATION

1. Disconnect the negative battery cable.
2. Raise and safely support the vehicle.
3. Remove the front wheels.
4. Remove the rear engine undercovers.
5. Remove the bolt from the universal joint at the manual rack input shaft. Disconnect the universal joint.
6. Remove the cotter pins and castellated nuts from the outer tie rods. Disconnect the tie rod ends from the steering knuckles.
7. Remove the 2 bolts attaching the damper weights.
8. Remove the 4 steering rack housing bracket set bolts.
9. Remove the steering gear assembly from the vehicle.

To install:

10. Install the grommets to the steering rack assembly. Insert the rack unit into the vehicle.
11. Install the steering rack universal joint to the pinion. Install the bolt and tighten temporarily.
12. Connect the steering rack to the body. Tighten the bolts to 29–40 ft. lbs. (39–54 Nm.).

NOTE: Install the steering rack housing brackets so the end with the elongated hole faces up.

13. Tighten the universal joint attaching bolt to 18–25 ft. lbs. (25–34 Nm).

NOTE: When tightening the bolt, be sure to limit the length of exposed splines on the pinion shaft to 0.200 in. (5mm) or less.

14. Install the damper weight.
15. Install the rear undercover.
16. Connect the tie rod end to the steering knuckle. Tighten the castellated nut to 21–32 ft. lbs. (29–44 Nm). Install a new cotter pin.
17. Install the wheels.
18. Lower the vehicle.
19. Connect negative battery cable.
20. Check alignment.

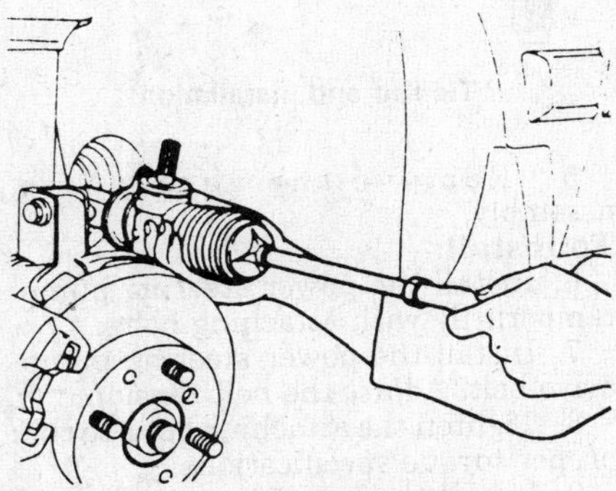

Installation of steering gear

Power Steering Rack

ADJUSTMENT

The procedure for checking the rack preload and steering pinion preload is the same as manual steering gear.

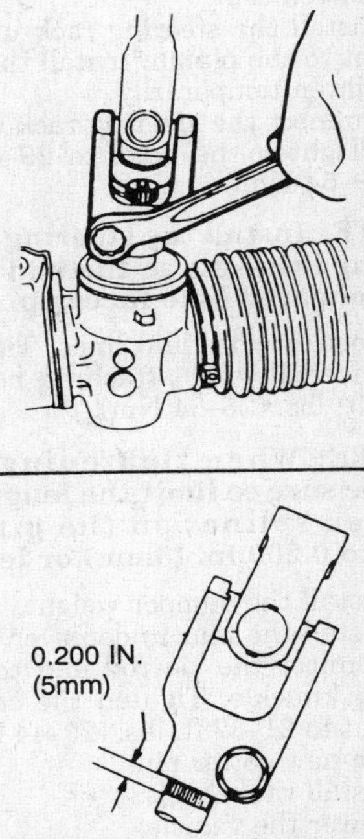

Proper depth of pinion shaft into universal joint

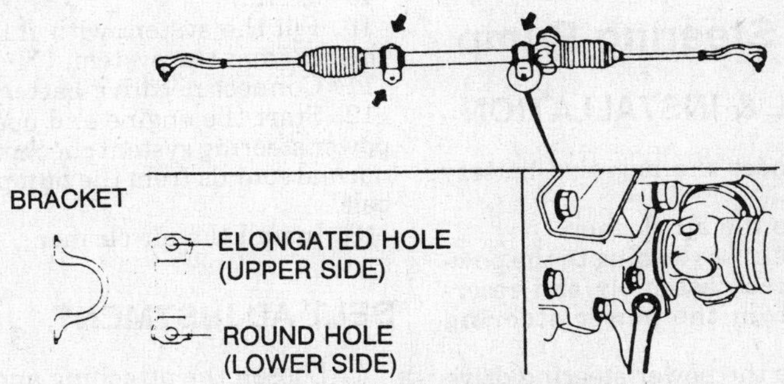

Positioning of steering gear brackets

REMOVAL & INSTALLATION

1. Disconnect the negative battery cable.
2. Raise and safely support the vehicle.
3. Remove the front wheels.
4. Remove the rear engine undercovers. Disconnect the fluid lines from the rack and drain the fluid.
5. Remove the bolt from the universal joint at the rack input shaft. Disconnect the universal joint.
6. Remove the cotter pins and castellated nuts from the outer tie rods. Disconnect the tie rod ends from the steering knuckles.
7. Remove the 2 bolts attaching the damper weights.
8. Remove the 4 steering rack housing bracket set bolts.
9. Remove the steering gear assembly from the vehicle.

To install:

10. Install the grommets to the steering rack assembly. Insert the rack unit into the vehicle.
11. Install the steering rack universal joint to the pinion. Install the bolt and tighten temporarily.
12. Connect the steering rack to the body. Tighten the bolts to 29–40 ft. lbs. (39–54 Nm.).

NOTE: Install the steering rack housing brackets so the end with the elongated hole faces up.

13. Connect the fluid lines. Tighten the universal joint attaching bolt to 18–25 ft. lbs. (25–34 Nm).

NOTE: When tightening the bolt, be sure to limit the length of exposed splines on the pinion shaft to 0.200 in. (5mm) or less.

14. Install the damper weight.
15. Install the rear undercover.
16. Connect the tie rod end to the steering knuckle. Tighten the castellated nut to 21–32 ft. lbs. (29–44 Nm). Install a new cotter pin.
17. Install the wheels.
18. Lower the vehicle.
19. Connect negative battery cable.
20. Fill the steering system with fluid and bleed. Check alignment.

Power Steering Pump

REMOVAL & INSTALLATION

1. Disconnect the negative battery cable.
2. Remove the air cleaner.
3. Remove hoses leading to the power steering rack assembly and reservoir tank from the power steering pump.
4. Remove the power steering drive belt.

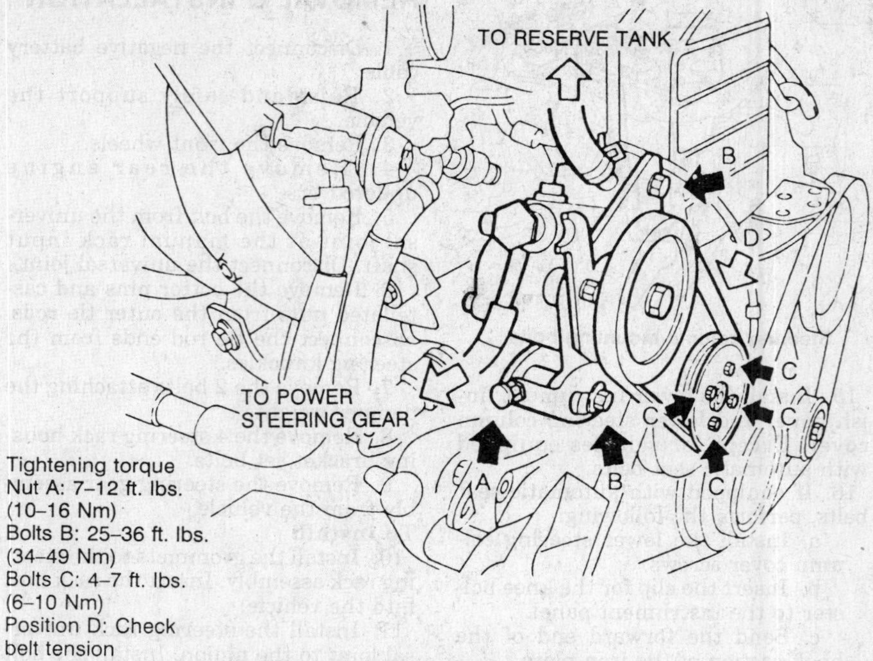

Power steering pump installation

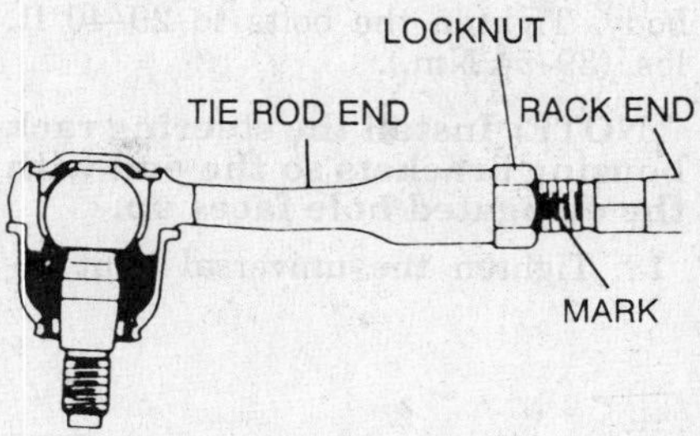

Tie rod end installation

5. Remove the vane pump assembly.

To install:

6. Install the power steering pump temporarily with attaching bolts.
7. Install the power steering pump drive belt. Adjust the belt tension.
8. Tighten the attaching bolts to the proper torque specifications.
9. Install the hoses from the power steering gear and the reservoir tank. Tighten the union bolt to 36–44 ft. lbs. (49–59 Nm).
10. Fill the system with fluid. Bleed the air from the system.
11. Connect negative battery cable.
12. Start the engine and operate the power steering system checking for abnormal sounds from the pump or drive belt.
13. Install the air cleaner.

BELT ADJUSTMENT

1. Loosen the attaching and adjusting bolts.
2. Check the belt tension at midpoint between the pulleys. For a new belt: tension should be 55–88 lbs. (25–40 kg) at 0.315–0.394 in. (8–10mm) deflection. For a used belt: tension should be 33–55 lbs. (15–25 kg) at 0.394–0.551 in. (10–14mm).
3. Tighten the attaching and adjusting bolts.

SYSTEM BLEEDING

1. Check the fluid level in the reservoir tank, add as necessary.
2. With the engine running at idle speed, turn the steering wheel to either lock position. Hold there for 2–3 seconds. Turn the steering wheel to the opposite lock position and hold for 2–3 seconds. Repeat sequence 2–3 times or until all air is bled from the system.
3. Check the fluid level.

Tie Rod Ends

REMOVAL & INSTALLATION

1. Disconnect the negative battery cable.
2. Raise and safely support the vehicle.
3. Remove the front wheel.
4. Mark the position of the tie rod end on the rack end.
5. Loosen the locknut.
6. Remove the cotter pin and castellated nut attaching tie rod to the steering knuckle.
7. Remove the tie rod end.

To install:

8. Install the new tie rod end on the rack end. Thread the tie rod end up to the mark made on rack end before removal of the old tie rod end.
9. Connect new tie rod end to the steering knuckle.
10. Install castellated nut and tighten to 22–33 ft. lbs. (29–44 Nm). Install a new cotter pin.
11. Install the front wheel.
12. Lower the vehicle.
13. Connect negative battery cable.
14. Check alignment.

BRAKES

For all brake system repair and service procedures not detailed below, please refer to "Brakes" in the Unit Repair section.

Master Cylinder

REMOVAL & INSTALLATION

1. Disconnect the negative battery cable.
2. Disconnect the level switch connector.
3. Disconnect the brake lines from the master cylinder.

NOTE: Do not allow brake fluid to touch painted surfaces. If a spill occurs, wipe up immediately.

4. Remove the 2 attaching nuts and remove the master cylinder and gasket from the brake booster.
5. Remove the level switch connector from the bracket.

To install:

6. Install the master cylinder using a new gasket.
7. Connect the brake tubes to the master cylinder.
8. Connect the level switch connector and install it to the bracket.
9. Fill the master cylinder with brake fluid.
10. Bleed the system.
11. Check the brake system for leaks.
12. Connect negative battery cable.
13. Test drive the vehicle to ensure proper brake operation.

Proportioning Valve

REMOVAL & INSTALLATION

1. Remove the brake lines connected to the proportioning valve.
2. Remove the proportioning valve.
3. Install the new proportioning valve.
4. Bleed the system.

Power Brake Booster

REMOVAL & INSTALLATION

1. Disconnect the negative battery cable.
2. Remove the master cylinder.
3. Disconnect the vacuum hose.
4. Remove the ignition coil wire and the clutch cable.
5. Working from under the dash panel, remove the clip and pin. Separate the master cylinder push rod clevis from the brake pedal.
6. Remove the brake booster assmbly and gasket from the vehicle.

To install:

7. Install the brake booster assembly using a new gasket.
8. Attach the master cylinder push rod clevis to the brake pedal using the pin and clip.
9. Attach the vacuum hose.
10. Install the clutch cable.
11. Install the master cylinder.
12. Install the ignition coil wire.
13. Connect negative battery cable.

Brake Caliper

REMOVAL & INSTALLATION

1. Disconnect the negative battery cable.
2. Raise and safely support the vehicle.
3. Remove the front wheel.
4. Separate the flexible hose from the brake tube.
5. Detach the clip from the strut.
6. Disconnect the flexible hose from the strut bracket.
7. Disconnect the flexible hose from the disc brake caliper.
8. Remove the attaching bolts and remove the caliper from the vehicle.
9. Remove the disc brake pad with the anti-squeal shims from the caliper.

To install:

10. Install the disc brake pad guide plate on the knuckle.
11. Install the brake pad in the caliper, with the anti-squeal shims.
12. Install the disc brake front caliper assembly on the steering knuckle. Tighten to 23–33 ft. lbs. (31–41 Nm).
13. Attach the flexible hose to the front disc brake caliper.
14. Attach the flexible hose to the bracket on the strut using the clip.
15. Temporarily install the flexible hose and brake tube by hand.
16. Tighten the flexible hose and brake tube.
17. Attach the clip at the bracket on the body.

NOTE: After installation of the flexible brake hose, turn the wheel from lock to lock to ensure free and unobstructed movement of the brake hose.

18. Bleed the air from the system.
19. Install the front wheel.
20. Lower the vehicle.
21. Test drive the vehicle to ensure proper brake operation.

Disc Brake Pads

REMOVAL & INSTALLATION

1. Disconnect the negative battery cable.
2. Raise and safely support the vehicle.
3. Remove the front wheel.
4. Remove the caliper attaching bolts and remove the caliper from the brake disc.
5. Remove the disc brake pad with anti-squeal shims.
6. Support the caliper using a length of mechanics wire. Do not allow the caliper to hang by the flexible brake hose unsupported.
7. Detach the disc brake pad guide plate.

To install:

8. Drain a small amount of brake fluid from the master cylinder. Use a clamp to press the caliper piston into the bore.

NOTE: Remove only 1 caliper at a time. If both calipers are removed at the same time, the hydraulic pressure created by pressing one piston into its bore may force the piston of the other caliper out of its bore.

9. Install a new disc brake pad guide plate on the knuckle.
10. Install the new brake pad in the caliper, with anti-sqeal shims.
11. Install the front disc brake caliper assembly on the knuckle. Tighten to 23–30 ft. lbs. (31–41 Nm).
12. Install the front wheel.
13. Lower the vehicle.
14. Connect negative battery cable.
15. Check the brake fluid level in the master cylinder reservoir and add, as necessary.
16. Before starting the engine, slowly press the brake pedal until the pedal rises to the normal position and a steady pedal is felt.
17. Recheck the brake fluid level in the master cylinder reservoir.
18. Start the engine and test drive to ensure proper brake operation.

Brake Rotor

REMOVAL & INSTALLATION

1. Disconnect the negative battery cable.

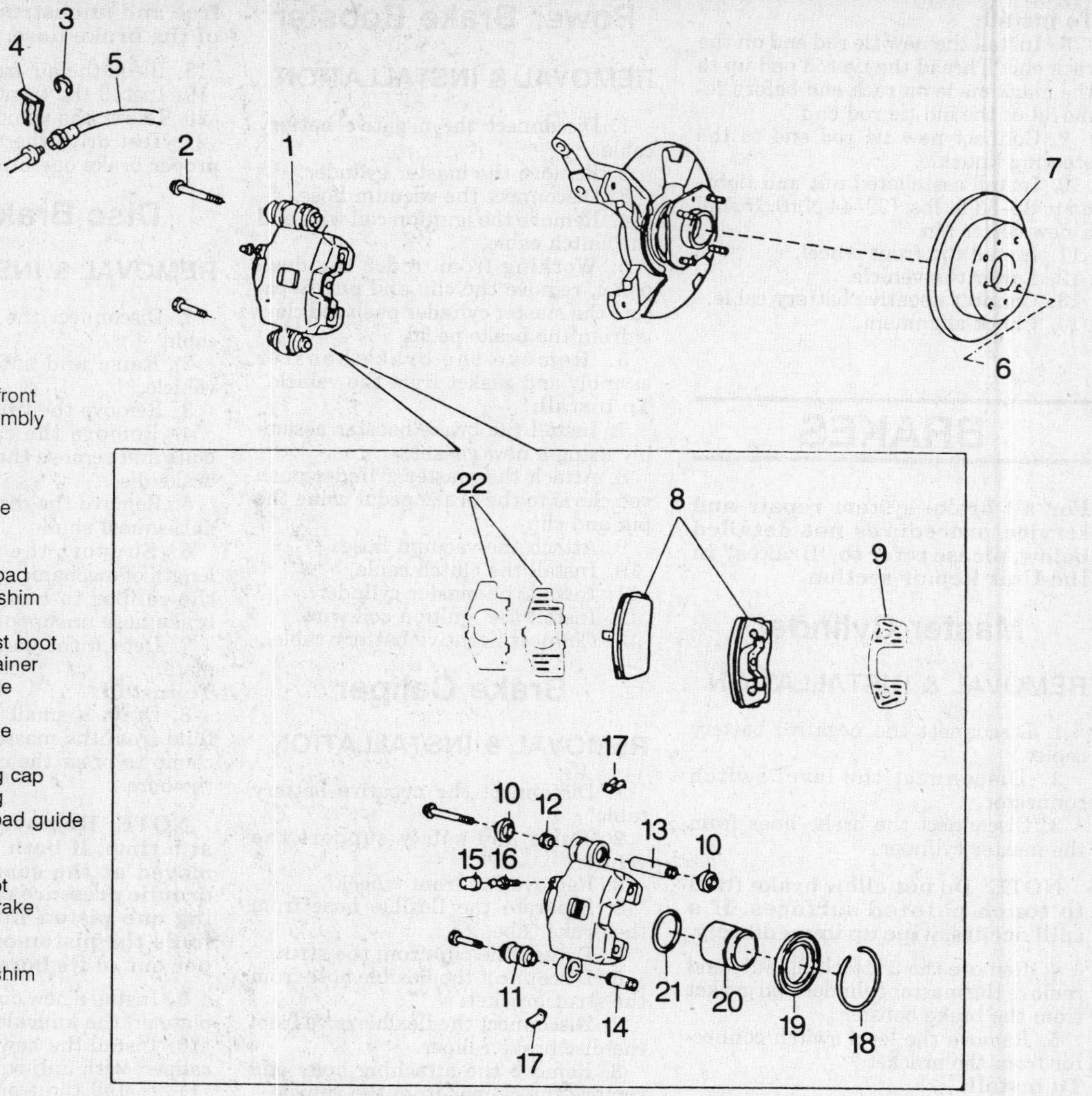

1. Disc brake front caliper assembly
2. Bolt
3. "E" ring
4. Tube clamp
5. Flexible hose
6. Screw
7. Front disc
8. Disc brake pad
9. Anti-squeal shim
10. Pin boot
11. Bushing dust boot
12. Bushing retainer
13. Cylinder slide bushing
14. Cylinder slide bushing
15. Bleeder plug cap
16. Bleeder plug
17. Disc brake pad guide plate
18. Set ring
19. Cylinder boot
20. Front disc brake piston
21. Piston seal
22. Anti-squeal shim

Front disc brake installation

2. Raise and safely support the vehicle.
3. Remove the front wheel.
4. Remove the bolts attaching the brake caliper and remove the caliper from the brake rotor and support the caliper using a length of mechanics wire. Do not allow the caliper to hang by the flexible brake hose unsupported.
5. Remove the screw attaching the brake rotor to the wheel hub. Remove the rotor.

To install:

6. Install the rotor on the wheel hub and attach with screw.
7. Install the brake caliper on the steering knuckle and tighten to 23–30 ft. lbs. (31–41 Nm).

NOTE: If installing a new rotor, it may be necessary to press the caliper piston into its bore a small amount to allow for the additional thickness of the new rotor.

8. Install the front wheel.
9. Lower the vehicle.
10. Connect negative battery cable.
11. Before starting the engine, slowly press the brake pedal until the pedal rises to the normal position and a steady pedal is felt.
12. Test drive the vehicle to ensure proper brake operation.

Brake Drums

REMOVAL & INSTALLATION

1. Disconnect the negative battery cable.
2. Raise and safely support the vehicle.

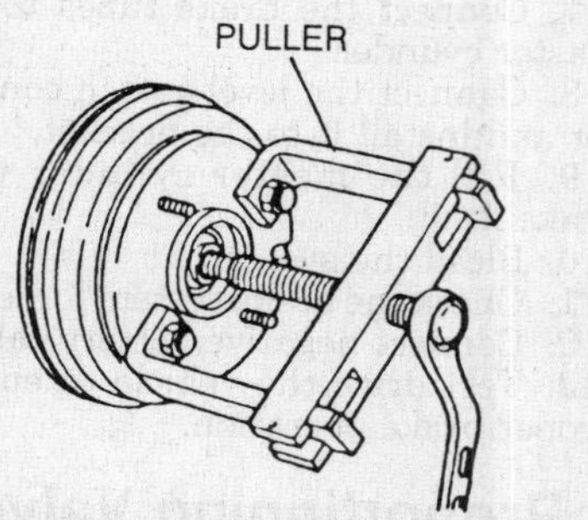

Removal of brake drum using puller

3. Remove wheel covers and rear wheels.
4. Remove the grease cap, cotter pin, castellated nut and plate washer.
5. Remove the brake drum using an appropriate puller.

To install:

6. Install the brake drum and install the plate washer and castellated nut. Tighten the nut to 43–72 ft. lbs. (49–98 Nm).
7. Install a new cotter pin. Install the grease cap.
8. Install the rear wheels and wheel covers.
9. Lower the vehicle.
10. Before starting the engine, slowly press the brake pedal until the pedal rises to the normal position and a steady pedal is felt.
11. Connect negative battery cable.
12. Test drive the vehicle to ensure proper brake operation.

Brake Shoes

REMOVAL & INSTALLATION

1. Disconnect the negative battery cable.
2. Raise and safely support the vehicle.
3. Remove wheel covers and rear wheels.
4. Remove the grease cap, cotter pin, castellated nut and plate washer.
5. Remove the brake drum using an appropriate puller.
6. Remove the tension spring.
7. Remove the leading shoe hold-down spring and pin.
8. Remove the brake shoe, parking brake shoe strut and tension spring from the leading shoe.
9. Remove the trailing shoe hold-down spring and pin.
10. Remove the parking brake cable from the parking brake shoe lever.
11. Detach the "C" ring from the parking brake lever pin on the trailing shoe and automatic adjusting lever parts.

To install:

12. Install the parking brake shoe lever and automatic adjusting lever parts to the trailing brake shoe.
13. Apply grease to the brake shoe contact points on the backing plate.
14. Connect the parking brake cable to the parking brake shoe lever.
15. Connect the brake shoe to the backing plate. Install the shoe hold-down spring and pin.
16. Connect the leading brake shoe to the rear brake backing plate. Install the shoe hold-down spring and pin.
17. Install the tension spring.
18. Install the brake drum.
19. Adjust the brake shoes.
20. Install the rear wheel and wheel cover.
21. Bleed the brake system.
22. Connect negative battery cable.
23. Test drive the vehicle to ensure proper brake operation.

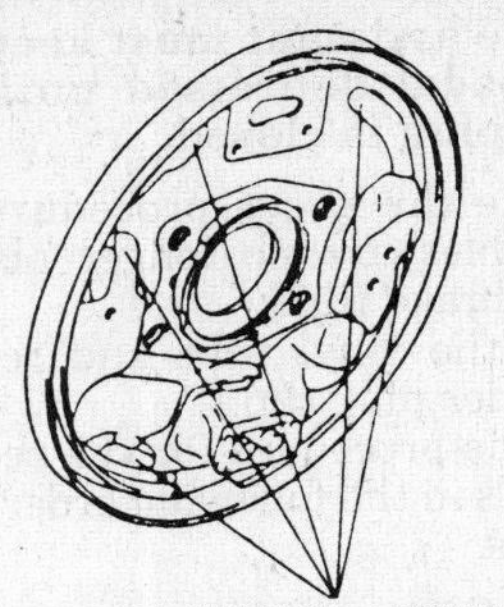

Grease points on backing plate

Wheel Cylinder

REMOVAL & INSTALLATION

1. Disconnect the negative battery cable.
2. Raise and safely support the vehicle.
3. Remove wheel covers and rear wheels.
4. Remove the grease cap, cotter pin, castellated nut and plate washer.
5. Remove the brake drum using an appropriate puller.
6. Remove the tension spring.
7. Remove the leading shoe hold-down spring and pin.
8. Remove the brake shoe, parking brake shoe strut and tension spring from the leading shoe.
9. Remove the trailing shoe hold-down spring and pin.
10. Remove the parking brake cable from the parking brake shoe lever.
11. Detach the "C" ring from the parking brake lever pin on the trailing shoe and automatic adjusting lever parts.
12. Disconnect the brake tube from the wheel cylinder.
13. Remove the wheel cylinder attaching bolts from behind the brake backing plate and remove the wheel cylinder.

To install:

14. Install the wheel cylinder on the backing plate. On the G102 vehicles, tighten the 2 bolts to 7–9 ft. lbs. (10–13 Nm). On the G100 vehicles, tighten the 2 bolts to 6–9 ft. lbs. (8–12 Nm).
15. Connect the brake tube to the wheel cylinder. Tighten nut to 9–13 ft. lbs. (13–18 Nm).
16. Install the parking brake shoe lever and automatic adjusting lever parts to the trailing brake shoe.
17. Apply grease to the brake shoe contact points of the backing plate.
18. Connect the parking brake cable to the parking brake shoe lever.
19. Connect the brake shoe to the rear brake backing plate. Install the shoe hold-down spring and pin.
20. Connect the leading brake shoe to the rear brake backing plate. Install the shoe hold-down spring and pin.
21. Install the tension spring.
22. Install the brake drum.
23. Adjust the brake shoes.
24. Install the rear wheel and wheel cover.
25. Bleed the brake system.
26. Connect negative battery cable.
27. Test drive the vehicle to ensure proper brake operation.

Parking Brake Cable

ADJUSTMENT

1. Ensure that the rear drum service brakes are in proper adjustment.
2. If equipped with automatic seat belts, remove the front seats.
3. Remove the coin box and console box from the vehicle.
4. Loosen parking brake cable adjusting nut.
5. Turn adjusting nut until the handle moves 4–7 notches when pulled by a force of 44 lbs. (20 kg).
6. Confirm operation of the parking brake indicator lamp.
7. Install the console box and coin box.
8. If equipped with automatic seat belts, install the front seats.

REMOVAL & INSTALLATION

1. If equipped with automatic seat belts, remove the front seats.
2. Remove the coin box and console box from the vehicle.
3. Remove the parking brake cable adjusting nut and connector.
4. Remove the parking brake handle.
5. Remove the parking brake tube protector.
6. Remove the parking brake cable from the parking brake pull rod.
7. Raise and safely support the vehicle.
8. Remove the exhaust pipe.
9. Remove the brake shoe.
10. Remove the parking brake cable from the rear brake backing plate.

To install:

11. Install the parking brake cable to the rear brake backing plate.
12. Install the brake shoe-related parts.
13. Install the under-body parking brake cable clamps.
14. Install the exhaust pipe.
15. Lower the vehicle.
16. Attach the parking brake cable to the parking brake pull rod.
17. Temporarily install the adjusting nut.
18. Install the parking brake tube protector.
19. Depress the brake pedal 4–5

times to adjust the rear brake clearance.

20. Turn adjusting nut until the handle moves 4–7 notches when pulled by a force of 44 lbs. (20 kg).

21. Confirm operation of the parking brake indicator lamp.

22. Install the console box and coin box.

23. If equipped with automatic seat belts, install the front seats.

Brake System Bleeding

1. Fill the brake master cylinder reservoir with brake fluid.

2. Raise and safely support the vehicle.

3. Connect a length of vinyl hose to the bleeder plug of the right rear wheel cylinder.

4. Submerge one end of the vinyl hose in a container filled with brake fluid. Connect the other end of the vinyl hose to the wheel cylinder bleeder plug.

5. Have an assistant slowly depress the brake pedal and hold it.

6. Open the bleeder plug of the right rear wheel cylinder ⅓–½ turn until the bubbles stop coming out of the tube. Close the bleeder plug.

7. Have the assistant release the brake pedal.

NOTE: The assistant must keep the brake pedal depressed until the bleeder plug is closed.

8. Continue the above procedure until air bubbles are no longer observed in the brake fluid.

9. Remove the vinyl tube and replace the bleeder plug cap.

10. Repeat the procedure for the remaining wheels in the following order:
    Left front
    Left rear
    Right front

11. Check the brake fluid level in the master cylinder reservoir frequently during the bleeding operation.

CHASSIS ELECTRICAL

Heater Blower Motor

REMOVAL & INSTALLATION

1. Disconnect the negative battery cable.

2. Remove the glove box door assembly.

3. Remove glove box/instrument panel reinforcement.

4. Remove glove box trim piece.

5. Remove blower motor connector and 3 blower motor retaining screws.

6. Remove blower motor.

7. Complete the installation of the blower motor by reversing the removal procedure.

Windshield Wiper Motor

REMOVAL & INSTALLATION

Front Motor

1. Disconnect the negative battery cable.

2. Remove the front wiper arm covers and wiper arm retaining nuts. Remove wiper arm and blade assemblies.

3. Remove the cowl top ventilator louver.

4. Remove the hood-to-cowl top seal.

5. Disconnect the wiper motor connector.

6. Remove the set bolt.

7. Disconnect the motor from the link. Remove the motor.

8. Complete the installation of the front wiper motor by reversing the removal procedure.

Rear Motor

1. Disconnect the negative battery cable.

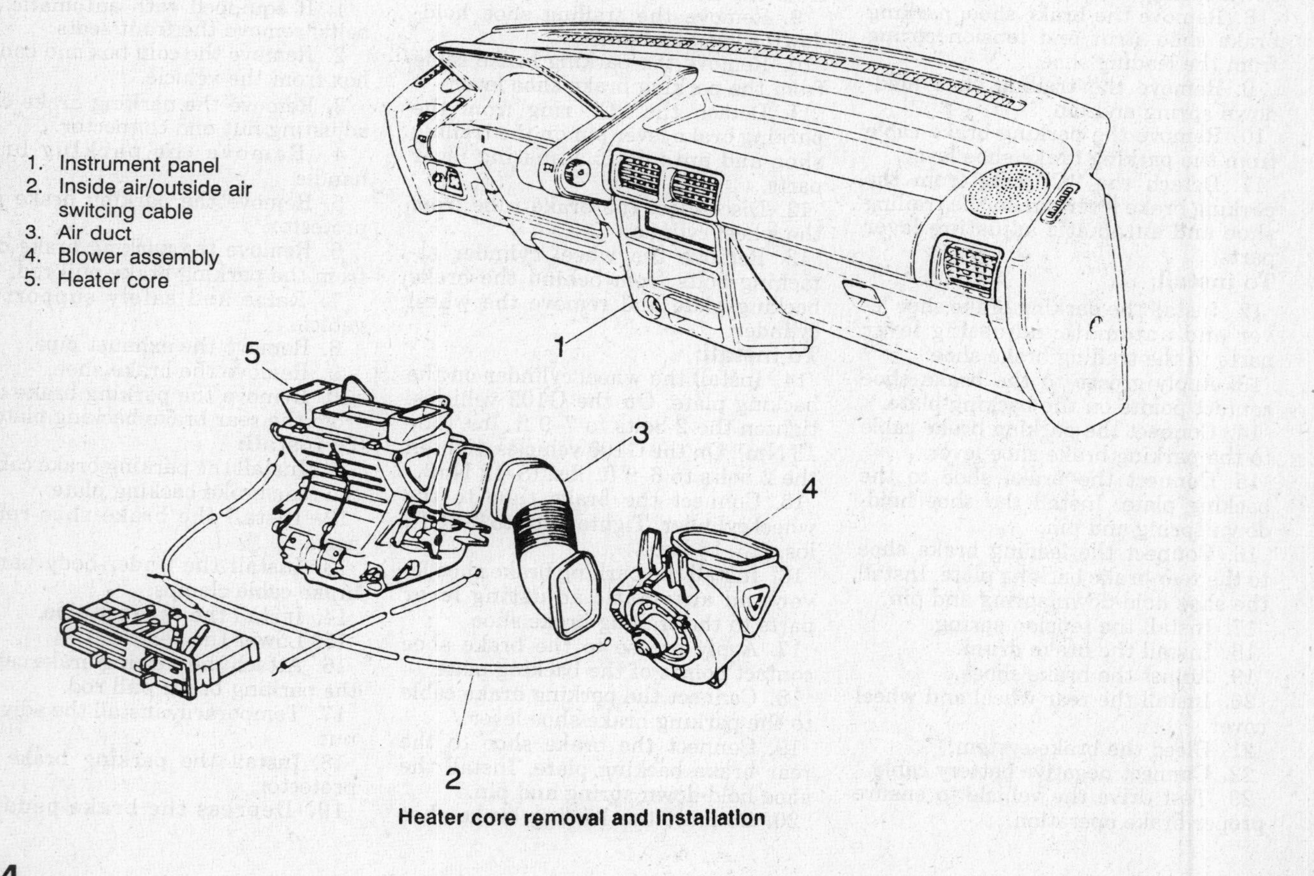

Heater core removal and installation

2. Remove the wiper arm cover.
3. Remove the wiper arm and blade by removing the nut.
4. Remove the wiper link cap.
5. Remove the washer and wiper link packing by removing the hexagonal nut.
6. Open the hatch. Remove back door trim board by pushing the center section of the clip and detaching.
7. Disconnect the connector. Remove the rear wiper motor assembly.
8. Complete the installation of the rear wiper motor by reversing the removal procedure.

Instrument Cluster

REMOVAL & INSTALLATION

1. Disconnect the negative battery cable.
2. Remove the finish panel sub-assembly from the instrument panel.
3. Disconnect the speedometer cable.
4. Disconnect electrical connectors from the instrument panel.
5. Remove the combination meter assembly from the instrument panel.
6. Remove speedometer assembly from instrument panel.
7. Complete the installation of the speedometer and instrument panel by reversing the removal procedure.

Radio

REMOVAL & INSTALLATION

1. Remove the negative battery cable.
2. Remove the console.
3. Remove the instrument cluster finish center panel.
4. Remove the radio assembly.
5. Installation is the reverse of removal.

Combination Switch

REMOVAL & INSTALLATION

1. Disconnect the negative battery cable.
2. Remove 2 screws retaining the steering wheel pad.
3. Remove the steering wheel pad by pushing it upward.
4. Disconnect the horn switch connector.
5. Remove the steering wheel locknut.
6. Install a suitable steering wheel puller and remove steering wheel.
7. Remove the steering column lower cover. Disconnect the combination switch connector.
8. Remove the combination switch.
9. Complete the installation of the combination switch by reversing the removal procedure.

Ignition Lock Cylinder

REMOVAL & INSTALLATION

1. Disconnect the negative battery cable.
2. Remove the steering column lower cover.
3. Fabricate a rod of approximately 0.078 in. (2mm) in diameter and 4 in. (100mm) in length with a 90 degree turn of 0.591 in. (15mm).
4. Set the ignition key to the **ACC** position.
5. Push down on the stop pin with the rod and draw out the cylinder.
6. Complete the installation of the ignition key cylinder by reversing the removal procedure.

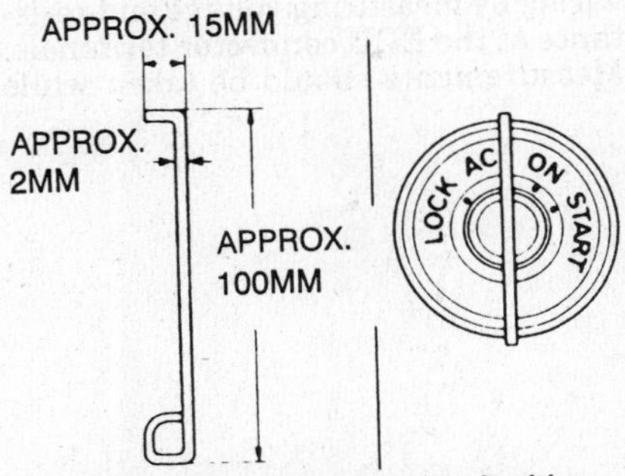

Special tool for removing ignition key cylinder

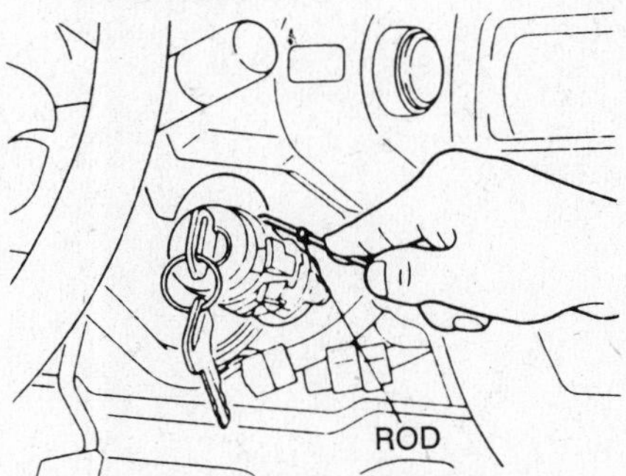

Removal of ignition key cylinder

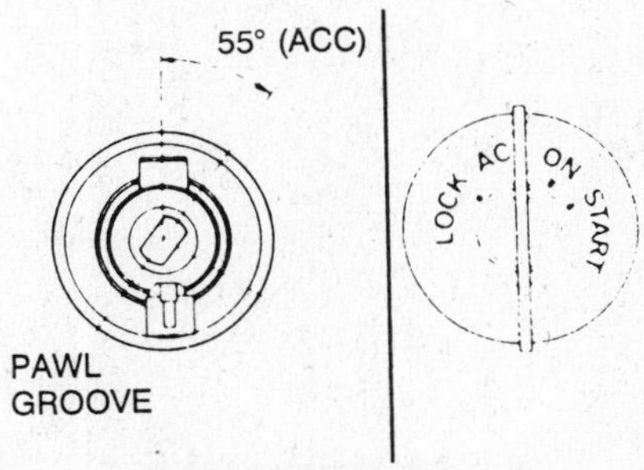

Key position to remove ignition key cylinder

Stoplight Switch

ADJUSTMENT

Loosen the locknut and turn the switch until the pedal cushion comes in contact with the edge of the threaded portion of the stoplight switch. Tighten the locknut.

REMOVAL & INSTALLATION

1. Disconnect the negative battery cable.
2. Disconnect the stoplight switch electrical connector.
3. Loosen the locknut and remove the switch.

To install:

4. Install new switch and turn until the pedal cushion comes in contact with the edge of the threaded portion of the stoplight switch. Tighten the locknut.
5. Connect the electrical connector.
6. Connect negative battery cable.

Neutral Safety Switch

ADJUSTMENT

1. Disconnect neutral safety switch connector.
2. Set the control shaft lever to the **N** position. Check continuity between terminals N (neutral) and E (ground) to determine proper installation angle of neutral safety switch.
3. Secure the neutral safety switch at the center position of the range where continuity exists between terminals N and E.
4. Connect neutral safety switch connector.

REMOVAL & INSTALLATION

1. Disconnect the negative battery cable.
2. Disconnect neutral safety switch connector.
3. Remove the shift shaft lever.
4. Remove the neutral safety switch retaining bolt and the neutral safety switch connector clamp.
5. Remove the neutral safety switch.

To install:

6. Install neutral safety switch and loosely install retaining bolt and connector clamp.
7. Install the shift shaft lever.
8. Set the control shaft lever to the **N** position. Check continuity between terminals N (neutral) and E (ground) to determine proper installation angle of neutral safety switch.
9. Secure the neutral start switch at the center position of the range where

continuity exists between terminals N and E.

10. Tighten the neutral safety switch retaining bolt to 31 ft. lbs. (22.6 Nm).

11. Connect neutral safety switch connector.

12. Connect negative battery cable.

Fuses, Circuit Breakers and Relays

LOCATION

Fuses

The fuse panels are located under the instrument panel near the steering column and in the engine compartment, next to the battery. When replacing a blown fuse, ensure that the fuse is of the proper value for that circuit. Never substitute a fuse of higher value. A fuse that blows repeatedly, is evidence of a grounded or shorted circuit.

Circuit Breakers

The circuit breakers are located on the sub-fuse block near the main fuse block. When testing a circuit breaker, ensure that there is continuity between the terminals.

Relays

The relays are found in various locations near the circuits they control and in a relay box in the engine compartment, next to the battery. Relays are remote switches that control high amperage circuits. Problems in a relay-controlled circuit can result from the control circuit, the load circuit, the electrical consumer or the relay itself.

Computers

LOCATION

Electronic Fuel Injection (EFI) Electronic Control Unit (ECU)

The ECU is located behind the glove box. The ECU monitors and controls engine functions. Check EFI circuit wiring by measuring voltage and resistance at the ECU connector terminals. Measurements should be taken while all connectors are connected. Ensure that battery voltage is 11 volts or more.

Automatic Transaxle ECU

The automatic transaxle ECU is located behind and to the left of the instrument cluster.

Cruise Control ECU

The cruise control ECU is located to the driver's side of the center console, adjacent to the radio.

Automatic Seat Belt Control ECU

The belt control ECU is located behind the left front kick panel, below the instrument panel.

Flashers

LOCATION

The turn signal/hazard flasher is located on the main fuse block.

Chevrolet Imports/Geo

6

Metro, Nova, Prizm, Spectrum, Sprint/Firefly, Storm — All Models

SERIAL NUMBER IDENTIFICATION

Vehicle Identification Plate

The code plate is attached to the bulkhead on the firewall, in the engine compartment. The plate shows model code, engine family, transaxle code and body color.

Vehicle Identification Number

The VIN plate is attached to the instrument panel on the drivers side, near the windshield. The VIN number is a 17 digit number, the 8th digit denotes engine family and the 10th digit denotes vehicle model year.

SPECIFICATIONS

ENGINE IDENTIFICATION

Year	Model	Engine Displacement cu. in. (cc/liter)	Engine Series Identification	No. of Cylinders	Engine Type
1988	Sprint/Firefly	61 (1000/1.0)	5	3	SOHC
	Sprint/Firefly	61 (1000/1.0)	2	3	SOHC
	Spectrum	90 (1471/1.5)	7	4	SOHC
	Spectrum	90 (1471/1.5)	9	4	SOHC
	Nova	97 (1600/1.6)	4	4	SOHC
	Nova	97 (1600/1.6)	5	4	DOHC
1989	Metro	61 (1000/1.0)	6	3	SOHC
	Spectrum	90 (1471/1.5)	7	4	SOHC
	Prizm	97 (1600/1.6)	6	4	SOHC
1990	Metro	61 (1000/1.0)	6	3	SOHC
	Prizm	97 (1600/1.6)	5	4	DOHC
	Storm	97 (1600/1.6)	6	4	SOHC
	Storm	97 (1600/1.6)	5	4	DOHC
1991-92	Metro	61 (1000/1.0)	6	3	SOHC
	Prizm	97 (1600/1.6)	6	4	DOHC
	Prizm	97 (1600/1.6)	5	4	DOHC
	Storm	97 (1600/1.6)	6	4	SOHC
	Storm	97 (1600/1.6)	5	4	DOHC

SOHC—Single overhead cam engine
DOHC—Dual overhead cam engine

GENERAL ENGINE SPECIFICATIONS

Year	Model	Engine Displacement cu. in. (cc.)	Fuel System Type	Net Horsepower @ rpm	Net Torque @ rpm (ft. lbs.)	Bore × Stroke (in.)	Compression Ratio	Oil Pressure @ 3000 rpm
1988	Sprint	61 (1000)	2 bbl	48 @ 5100	77 @ 3200	2.91 × 3.03	9.5:1	48
	Sprint	61 (1000)	2 bbl	46 @ 4700	78 @ 3200	2.91 × 3.03	9.8:1	48
	Sprint	61 (1000)	EFI	70 @ 5500	107 @ 3500	2.91 × 3.03	8.3:1	48
	Spectrum	90 (1471)	2 bbl	70 @ 5400	87 @ 3400	3.03 × 3.11	9.6:1	49 @ 5200
	Spectrum	90 (1471)	Turbo	110 @ 5400	120 @ 3400	3.03 × 3.11	8.0:1	49 @ 5200
	Nova	97 (1600)	2 bbl	74 @ 5200	85 @ 2800	3.19 × 3.03	9.0:1	34 @ 2000
	Nova	97 (1600)	EFI	110 @ 6600	98 @ 4800	3.19 × 3.03	9.4:1	56 @ 3000

GENERAL ENGINE SPECIFICATIONS

Year	VIN	No. Cylinder Displacement cu. in. (liter)	Fuel System Type	Net Horsepower @ rpm	Net Torque @ rpm (ft. lbs.)	Bore × Stroke (in.)	Compression Ratio	Oil Pressure @ rpm
1989	Metro ②	61 (1000)	EFI	55 @ 5700	58 @ 3300	2.91 × 3.03	9.5:1	39 @ 4000
	Metro	61 (1000)	EFI	49 @ 4700	58 @ 3300	2.91 × 3.03	9.5:1	39 @ 4000
	Spectrum	90 (1471)	2 bbl	70 @ 5400	87 @ 3400	3.03 × 3.11	9.6:1	49 @ 5200
	Prizm	97 (1600)	MFI	102 @ 5800	101 @ 4800	3.19 × 3.03	9.5:1	56 @ 3000
1990	Metro	61 (1000)	EFI	55 @ 5700	58 @ 3300	2.91 × 3.03	9.5:1	39 @ 4000
	Prizm	97 (1600)	MFI	102 @ 5800	101 @ 4800	3.15 × 3.11	9.1:1	56 @ 3000
	Prizm ③	97 (1600)	MFI	130 @ 7000	102 @ 5800	3.15 × 3.11	9.8:1	56 @ 3000
	Storm ③	97 (1600)	MFI	102 @ 5800	101 @ 4800	3.20 × 3.00	9.5:1	—
	Storm ③	97 (1600)	MFI	130 @ 6800	105 @ 6000	3.20 × 3.00	10.3:1	—
1991–92	Metro	61 (1000)	EFI	55 @ 5700	58 @ 3300	2.91 × 3.03	9.5:1	39 @ 4000
	Prizm	97 (1600)	MFI	102 @ 5800	101 @ 4800	3.15 × 3.11	9.1:1	56 @ 3000
	Prizm ③	97 (1600)	MFI	130 @ 7000	102 @ 5800	3.15 × 3.11	9.8:1	56 @ 3000
	Storm ③	97 (1600)	MFI	102 @ 5800	101 @ 4800	3.20 × 3.00	9.5:1	—
	Storm ③	97 (1600)	MFI	130 @ 6800	105 @ 6000	3.20 × 3.00	10.3:1	—

EFI—Electronic Fuel Injection
MFI—Multi-Point Fuel Injection
① ER model
② LSI model
③ DOHC

ENGINE TUNE-UP SPECIFICATIONS

Year	Model	Engine Displacement cu. in. (cc)	Spark Plugs		Ignition Timing (deg.)		Compression Pressure (psi)	Fuel Pump (psi)	Idle Speed (rpm)		Valve Clearance	
			Type	Gap (in.)	MT	AT			MT	AT	In.	Ex.
1988	Sprint	61 (1000)	R43CXLS	0.039–0.043	10	6	199	4.0	750 ⑪	850	0.006	0.008
	Sprint	61 (1000) ⑨	R43CXLS	0.039–0.043	12	—	199	2.5–3.3	750	—	0.006	0.008
	Spectrum	90 (1471)	BPR6ES-11	0.040	15 ①	10 ③	128–179	3.8–4.7	700	950	0.006	0.010
	Spectrum	90 (1471) ⑨	BPR6ES-11	0.040	15 ②	NA	128–179	28.4 ⑤	950	NA	0.006	0.010
	Nova	97 (1600)	⑥	0.043	0	0	128–178	3.5	650	750	0.008	0.012
	Nova	97 (1600) ⑩	BCPR5EP11	0.043		10B ④	142–179	NA	800	800	⑦	⑧
1989	Metro	61 (1000)	R43CXLS	0.039–0.043	⑬	⑬	199	26	750	850	Hyd.	Hyd.
	Spectrum	90 (1471)	R42XLS	0.040	15 ①	10 ③	128–179	3.8–4.7	750	1000	0.006	0.010
	Prizm	97 (1600)	BCPRSEY	0.031	10B	10B	142–191	⑫	700	700	0.006–0.010	0.008–0.012
1990	Metro	61 (1000)	R43CXLS	0.039–0.043	⑬	⑬	199	26	750	850	Hyd.	Hyd.
	Prizm	97 (1600)	BCPRSEY	0.031	10B	10B	142–191	⑫	700	700	0.006–0.010	0.008–0.012
	Storm	97 (1600)	R42XLS	0.040	10B ⑬	10B ⑬	—	NA	700	700	0.006–0.010	0.008–0.0012
	Storm ⑩	97 (1600)	BKR6E-11	0.041	10B ⑬	10B ⑬	—	NA	700	700	0.004–0.008	0.008–0.0012

GASOLINE ENGINE TUNE-UP SPECIFICATIONS

Year	Model	Engine Displacement cu. in. (cc)	Spark Plugs Type	Spark Plugs Gap (in.)	Ignition Timing (deg.) MT	Ignition Timing (deg.) AT	Compression Pressure (psi)	Fuel Pump (psi)	Idle Speed (rpm) MT	Idle Speed (rpm) AT	Valve Clearance In.	Valve Clearance Ex.
1991	Metro	61 (1000)	R43CXLS	0.039–0.043	⑬	⑬	199	26	750	850	Hyd.	Hyd.
	Prizm	97 (1600)	BCPRSEY	0.031	10B	10B	142–191	⑫	700	700	0.006–0.010	0.008–0.012
	Storm	97 (1600)	R42XLS	0.040	10B ⑬	10B ⑬	—	NA	700	700	0.006–0.010	0.008–0.012
	Storm ⑩	97 (1600)	BKR6E-11	0.041	10B ⑬	10B ⑬	—	NA	700	700	0.004–0.008	0.008–0.012
1992	SEE UNDERHOOD SPECIFICATIONS STICKER											

NOTE: The underhood specifications sticker often reflects tune-up specification changes made in production. Sticker figures must be used if they disagree with those in this chart.

NA—Not available
Hyd.—Hydraulic
① @ 750 rpm
② @ 950 rpm
③ @ 1000 rpm
④ Use jumper wire to short circuit both terminals of the check engine connector located near the wiper motor. When the jumper wire is removed and the transaxle is in Neutral, the ignition timing should be more than 16 degrees BTDC (manual) or 12 degrees BTDC (automatic)
⑤ @ 900 rpm with pressure regulator connected
⑥ USA—BPR5EY11
Calif—BPR4EY11
⑦ Cold—0.006–0.010 in.
Hot—0.008–0.012 in.
⑧ Cold—0.008–0.012 in.
Hot—0.010–0.014 in.
⑨ Turbo engine
⑩ Twin Cam engine
⑪ ER Model—700 rpm
⑫ See "Pressure Testing" in text
⑬ See Underhood Sticker

FIRING ORDERS

NOTE: To avoid confusion, always replace spark plug wires one at a time.

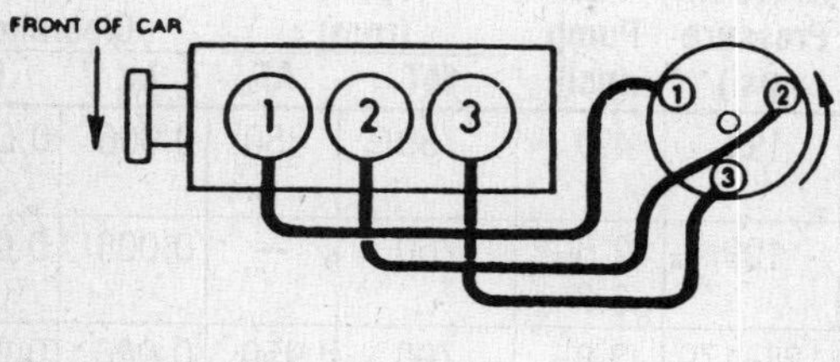

1.0L Engine
Engine Firing Order: 1-3-2
Distributor Rotation: Counterclockwise

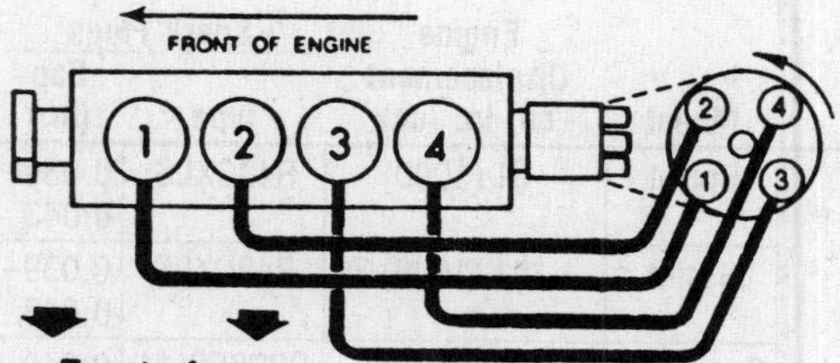

1.5L Engine
Engine Firing Order: 1-3-4-2
Distributor Rotation: Counterclockwise

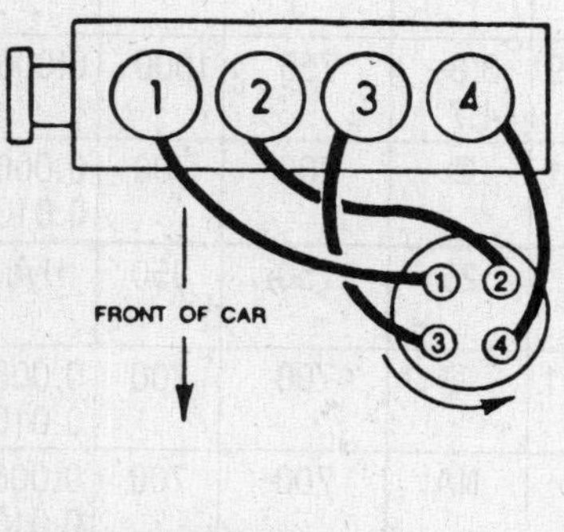

1.6L SOHC Engine
Engine Firing Order:1-3-4-2
Distributor Rotation: Counterclockwise

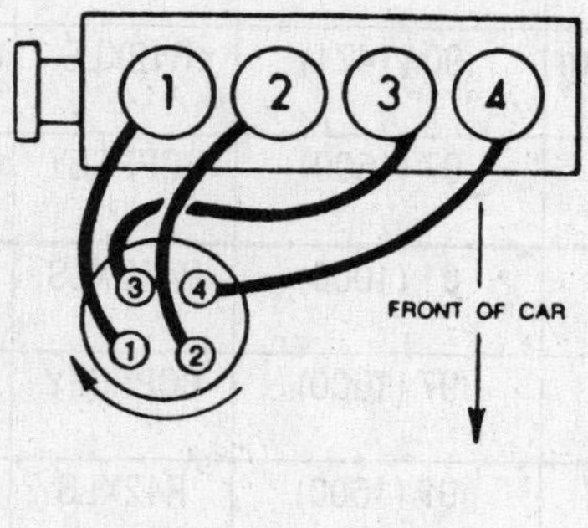

1.6L DOHC Engine
Engine Firing Order: 1-3-4-2
Distributor Rotation: Clockwise

CAPACITIES

Year	Model	Engine Displacement cu. in. (cc.)	Engine Crankcase (qts.) with Filter	Engine Crankcase (qts.) without Filter	Transmission (pts.) 4-Spd	Transmission (pts.) 5-Spd	Transmission (pts.) Auto.	Drive Axle (pts.)	Fuel Tank (gals.)	Cooling System (qts.)
1988	Sprint	61 (1000/1.0)	3.5	3.5	—	4.8	9.5	—	8.3	4.5
	Sprint	61 (1000/1.0)	3.5	3.5	—	4.8	9.5	—	8.3	4.5
	Spectrum	90 (1471/1.5)	3.4	3.0	—	5.8	12.2	—	11.0	7.5
	Spectrum	90 (1471/1.5)	3.4	3.0	—	5.8	12.2	—	11.0	7.5
	Nova	97 (1600/1.6)	3.5	3.2	—	5.4	11.6	—	13.2	6.4
	Nova	97 (1600/1.6)	3.9	3.6	—	5.4	16.6	—	13.2	6.3
1989	Metro	61 (1000/1.0)	3.7	3.7	—	4.8	9.6	—	8.7	4.5
	Spectrum	90 (1471/1.5)	3.4	3.0	—	4.0	13.8	—	11.0	6.8
	Prizm	97 (1600/1.6)	3.4	3.2	—	5.8	11.6	—	13.0	6.3
1990	Metro	61 (1000/1.0)	3.7	3.7	—	5.0	10.4	—	10.0	①
	Prizm	97 (1600/1.6)	3.9	3.6	—	12.0	②	3.0⑦	13.2	③
	Storm	97 (1600/1.6)	3.5	3.1	—	4.0	13.6	3.0	12.4	④
	Storm	97 (1600/1.6)	4.2	3.8	—	5.4	⑥	3.0	12.4	⑤
1991-92	Metro	61 (1000/1.0)	3.7	3.7	—	5.0	10.4	—	10.0	①
	Prizm	97 (1600/1.6)	3.4	3.2	—	12.0	②	3.0⑦	13.2	③
	Prizm	97 (1600/1.6)	3.9	3.6	—	12.0	②	3.0⑦	13.2	③
	Storm	97 (1600/1.6)	3.5	3.1	—	4.0	13.6	3.0	12.4	④
	Storm	97 (1600/1.6)	4.2	3.8	—	5.4	⑥	3.0	12.4	⑤

① 4.1 qts Manual Transaxle
4.2 qts Automatic Transaxle
② 11.6 pts 3 speed Automatic
12.2 pts 4 speed Automatic
③ 6.0 qts Manual Transaxle
5.8 qts Automatic Transaxle
④ 7.1 qts Manual Transaxle
7.6 qts Automatic Transaxle
⑤ 7.3 qts Manual Transaxle
7.8 qts Automatic Transaxle
⑥ 11.6 pts 3 speed Automatic
15.2 pts 4 speed Automatic
⑦ 3 Speed Automatic Transaxle only

CAMSHAFT SPECIFICATIONS

All measurements given in inches.

Year	Engine Displacement cu. in. (cc)	Journal Diameter 1	Journal Diameter 2	Journal Diameter 3	Journal Diameter 4	Journal Diameter 5	Lobe Lift In.	Lobe Lift Ex.	Bearing Clearance	Camshaft End Play
1988	1.0 (1000)	1.7372–1.7381	1.7451–1.7460	1.7530–1.7539	1.7609–1.7618	—	1.512	1.512	0.0029	—
	1.5 (1471)	1.0210–1.0220	1.0210–1.0220	1.0210–1.0220	1.0210–1.0220	1.021–1.022	1.426	1.426	0.0024–0.0044	0.0039–0.0071
	1.6 (1600)	1.1015–0.1022	1.1015–0.1022	1.1015–0.1022	1.1015–0.1022	—	1.541 ①	1.541 ①	0.0015–0.0029	0.0031–0.0071
	1.6 (1600)	1.0610–0.0616	1.0610–0.0616	1.0610–0.0616	1.0610–0.0616	1.061–0.062	1.399–1.400	1.399–1.400	0.0014–0.0028	0.0031–0.0075
1989	1.0 (1000)	1.0220–1.0228	1.1795–1.1803	1.1795–1.1803	—	—	1.560 2.566	1.560 2.566	0.0008–0.0024	—
	1.5 (1471)	1.0210–1.0220	1.0210–1.0220	1.0210–1.0220	1.0210–1.0220	1.021–1.022	1.426	1.426	0.0024–0.0044	0.0039–0.0071
	1.6 (1600)	0.9822	0.9035	0.9035	0.9035	—	1.370	1.359	0.0014–0.0028	0.0043
1990	1.0 (1000)	1.0220–1.0228	1.1795–1.1803	1.1795–1.1803	—	—	1.560 2.566	1.560 2.566	0.0008–0.0024	—
	1.6 (1600) ②	0.9822	0.9035	0.9035	0.9035	—	1.370	1.359	0.0014–0.0028	0.0043

CAMSHAFT SPECIFICATIONS

All measurements given in inches.

Year	Engine Displacement cu. in. (cc)	Journal Diameter 1	2	3	4	5	Lobe Lift In.	Ex.	Bearing Clearance	Camshaft End Play
1990	1.6 (1600) ③	1.0610–1.0616	1.0610–1.0616	1.0610–1.0616	1.0610–1.0616	—	1.394–1.398	1.394–1.398	0.0014–0.0028	0.0031–0.0075
	1.6 (1600) ③	1.0157	1.0157	1.0157	1.0157	—	1.426	1.426	0.0059	0.0080
1991-92	1.0 (1000)	1.0220–1.0228	1.1795–1.1803	1.1795–1.1803	—	—	1.560 2.566	1.560 2.566	0.0008–0.0024	—
	1.6 (1600) ②	0.9822	0.9035	0.9035	0.9035	—	1.370	1.359	0.0014–0.0028	0.0043
	1.6 (1600) ③	1.0610–1.0616	1.0610–1.0616	1.0610–1.0616	1.0610–1.0616	—	1.394–1.398	1.394–1.398	0.0014–0.0028	0.0031–0.0075
	1.6 (1600) ③	1.0157	1.0157	1.0157	1.0157	—	1.426	1.426	0.0059	0.0080

① Minimum lobe height
② Prizm
③ Storm

CRANKSHAFT AND CONNECTING ROD SPECIFICATIONS

All measurements are given in inches.

Year	Engine Displacement cu. in. (cc.)	Crankshaft Main Brg. Journal Dia.	Main Brg. Oil Clearance	Shaft End-play	Thrust on No.	Connecting Rod Journal Diameter	Oil Clearance	Side Clearance
1988	1.0 (1000)	①	0.0012	0.0044–0.0122	3	1.6532	0.0012–0.0019	NA
	1.0 (1000) ④	①	0.0012	0.0044–0.0122	3	1.6532	0.0012–0.0019	0.0039–0.780
	1.5 (1471)	1.8865–1.8873	0.0008–0.0020	0.0024–0.0095	2	1.5720–1.5726	0.0010–0.0023	0.0079–0.0138
	1.6 (1600) ⑤	1.8865–1.8873	0.0008–0.0020	0.0024–0.0095	2	1.5720–1.5726	0.0010–0.0023	0.0079–0.0138
1989	1.0 (1000)	①	0.0012	0.0044–0.0122	3	1.6529–1.6535	0.0012–0.0019	NA
	1.5 (1471)	1.8865–1.8873	0.0008–0.0020	0.0024–0.0095	2	1.5526	0.0009–0.0023	0.0079–0.0138
	1.6 (1600)	1.8865–1.8873	0.0006–0.0013	0.0008–0.0073	3	1.5420–1.5748	0.0008–0.0020	0.0059–0.0098
1990	1.0 (1000)	①	0.0012	0.0044–0.0122	3	1.6529–1.6535	0.0012–0.0019	NA
	1.6 (1600) ③	1.8891–1.8898	0.0006–0.0013	0.0008–0.0073	3	1.5742–1.5748	0.0006–0.0013	0.0059–0.0098
	1.6 (1600) ②	2.0440–2.0448	0.0008–0.0020	0.0024–0.0095	2	1.5724–1.5728	0.0007–0.0018	0.0079–0.0138
1991-92	1.0 (1000)	①	0.0012	0.0044–0.0122	3	1.6529–1.6535	0.0012–0.0019	NA
	1.6 (1600) ③	1.8891–1.8898	0.0006–0.0013	0.0008–0.0073	3	1.6529–1.6535	0.0006–0.0013	0.0059–0.0098
	1.6 (1600) ②	2.0440–2.0448	0.0008–0.0020	0.0024–0.0095	2	1.5722–1.5728	0.0010–0.0023	0.0059–0.0078

NA Not available
① Bearing cap stamped
No. 1—1.7710–1.7712
No. 3—1.7712–1.7714
No. 2—1.7714–1.7716
No. 4—1.7710–1.7712
② Storm
③ Prizm
④ Fuel injected
⑤ Single overhead camshaft

VALVE SPECIFICATIONS

Year	Engine Displacement cu. in. (cc.)	Seat Angle (deg.)	Face Angle (deg.)	Spring Test Pressure (lbs.)	Spring Installed Height (in.)	Stem-to-Guide Clearance (in.)		Stem Diameter (in.)	
						Intake	Exhaust	Intake	Exhaust
1988	61 (1000)	45	45	60	1.63	0.0014	0.0020	0.2745	0.2740
	61 (1000)	45	45	60	1.63	0.0014	0.0020	0.2745	0.2740
	90 (1471)	45	45	47 @ 1.57	1.57	0.0009–0.0022	0.0012–0.0025	0.2745–0.2750	0.2740–0.2744
	90 (1471)	45	45	47 @ 1.57	1.57	0.0009–0.0022	0.0012–0.0025	0.2745–0.2750	0.2740–0.2744
	97 (1600)	45	44.5	32.2	1.366	0.0010–0.0024	0.0012–0.0026	0.2350–0.2356	0.2348–0.2354
	97 (1600) ②	45	44.5	46.3	1.52	0.0010–0.0024	0.0012–0.0026	0.2744–0.2750	0.2742–0.2748
1989	61 (1000)	45	45	44 @ 1.28	—	0.0008–0.0021	0.0014–0.0024	0.2148–0.2151	0.2146 0.2151
	90 (1471)	45	45	47 @ 1.57	1.57	0.0009–0.0022	0.0118–0.0025	0.2740–0.2750	0.2740–0.2744
	97 (1600)	45	45.5	32.2	1.36	0.0031	0.0039	0.2350–0.2356	0.2348–0.2354
1990	61 (1000)	45	45	44 @ 1.28	—	0.0008–0.0021	0.0014–0.0024	0.2148–0.2151	0.2146–0.2151
	97 (1600) ①	45	45.5	32.2	1.36	0.0031	0.0039	0.2350–0.2356	0.2348–0.2354
	97 (1600) ①	45	45.5	—	—	0.0010–0.0024	0.0012–0.0026	0.2350–0.2356	0.2348–0.2354
	97 (1600) ②	45	45.5	—	—	0.0009	0.0018	0.2335	0.2335
	97 (1600) ②	45	45	—	—	0.0009	0.0018	0.2335	0.2335
1991–92	61 (1000)	45	45	44 @ 1.28	—	0.0008–0.0021	0.0014–0.0024	0.2148–0.2151	0.2146–0.2151
	97 (1600) ①	45	45.5	32.2	1.36	0.0031	0.0039	0.2350–0.2356	0.2348–0.2354
	97 (1600) ①	45	45.5	—	—	0.0010–0.0024	0.0012–0.0026	0.2350–0.2356	0.2348–0.2354
	97 (1600) ②	45	45.5	—	—	0.0009	0.0018	0.2335	0.2335
	97 (1600) ②	45	45	—	—	0.0009	0.0018	0.2335	0.2335

① Prizm ② Storm ③ Single overhead camshaft

PISTON AND RING SPECIFICATIONS

All measurements are given in inches.

Year	Engine Displacement cu. in. (cc.)	Piston Clearance	Ring Gap			Ring Side Clearance		
			Top Compression	Bottom Compression	Oil Control	Top Compression	Bottom Compression	Oil Control
1988	61 (1000)	0.0008–0.0015	0.0079–0.0129	0.0079–0.0137	0.0079–0.0275	0.0012–0.0027	0.0008–0.0023	—
	61 (1000)	0.0008–0.0015	0.0079–0.0119	0.0079–0.0119	0.0079–0.0237	0.0012–0.0030	0.0008–0.0023	—
	61 (1000) ①	0.0008–0.0015	0.0079–0.0157	②	0.0079–0.0275	0.0012–0.0027	—	—
	90 (1471)	0.0011–0.0019	0.0098–0.0138	②	0.0039–0.0236	0.0010–0.0026	—	—

PISTON AND RING SPECIFICATIONS

All measurements are given in inches.

Year	Engine Displacement cu. in. (cc.)	Piston Clearance	Ring Gap Top Compression	Ring Gap Bottom Compression	Ring Gap Oil Control	Ring Side Clearance Top Compression	Ring Side Clearance Bottom Compression	Ring Side Clearance Oil Control
1988	90 (1471)	0.0011–0.0019	0.0106–0.0153	0.0098–0.0145	0.0039–0.0236	0.0010–0.0026	0.0008–0.0024	—
	97 (1600) ⑤	0.0035–0.0043	0.0098–0.0185	0.0059–0.0165	0.0118–0.0402	0.0160–0.0031	0.0012–0.0028	Snug
	97 (1600)	0.0039–0.0047	0.0098–0.0138	0.0078–0.0118	0.0059–0.0031	0.0016–0.0031	0.0012–0.0028	Snug
1989	61 (1000)	0.0008–0.0015	0.0079–0.0129	0.0079–0.0137	0.0079–0.0275	0.0012–0.0027	0.0008–0.0023	—
	90 (1471)	0.0011–0.3500	0.0025–0.0035	—	—	0.0009–0.0026	—	—
	97 (1600)	0.0024–0.0031	0.0098–0.0138	0.0059–0.0118	0.0039–0.0236	0.0016–0.0031	0.0012–0.0028	—
1990	61 (1000)	0.0008–0.0015	0.0079–0.0129	0.0079–0.0137	0.0079–0.0275	0.0012–0.0027	0.0008–0.0023	—
	97 (1600) ④	0.0011–0.0019	0.0110–0.0157	0.0177–0.0236	0.0039–0.0236	0.00117–0.00315	0.00078–0.00236	—
	97 (1600) ④	0.0024–0.0031	0.0110–0.0157	0.0177–0.0236	0.0039–0.0236	0.0012–0.0032	0.0008–0.0024	—
	97 (1600) ③	0.0039–0.0047	0.0098–0.0185	0.0079–0.0165	0.0059–0.0205	0.0012–0.0028	0.0012–0.0028	0.0039–0.0236
	97 (1600) ③	0.0024–0.0031	0.0098–0.0138	0.0059–0.0118	0.0039–0.0236	0.0016–0.0031	0.0012–0.0028	—
1991-92	61 (1000)	0.0008–0.0015	0.0079–0.0129	0.0079–0.0137	0.0079–0.0275	0.0012–0.0027	0.0008–0.0023	—
	97 (1600) ④	0.0011–0.0019	0.0110–0.0157	0.0177–0.0236	0.0039–0.0236	0.00117–0.00315	0.00078–0.00236	—
	97 (1600) ④	0.0024–0.0031	0.0110–0.0157	0.0177–0.0236	0.0039–0.0236	0.0012–0.0032	0.0008–0.0024	—
	97 (1600) ③	0.0039–0.0047	0.0098–0.0185	0.0079–0.0165	0.0059–0.0205	0.0012–0.0028	0.0012–0.0028	0.0039–0.0236
	97 (1600) ③	0.0024–	0.0098–0.0031	0.0059–0.0138	0.0039–0.0118	0.0016–0.0236	0.0012–0.0031	— 0.0028

① ER model
② ER model has only one compression ring
③ Prizm
④ Storm
⑤ Double overhead camshaft

TORQUE SPECIFICATIONS

All readings in ft. lbs.

Year	Engine Displacement cu. in. (cc.)	Cylinder Head Bolts	Main Bearing Bolts	Rod Bearing Bolts	Crankshaft Pulley Bolts	Flywheel Bolts	Manifold Intake	Manifold Exhaust	Spark Plugs
1988	61 (1000)	48	38	25	50	44	17	17	20
	90 (1471)	②	65	25	108	22 ①	17	17	18
	97 (1600)	43	43	29	87	58	20	18	13
	97 (1600) ⑧	③	44	36	101	58	20	18	13
1989	61 (1000)	54	40	26	8	45	17	17	18
	90 (1471)	②	65	25	108	22 ①	17	17	18
	97 (1600)	44	44	36	87	58 ④	14	18	20

TORQUE SPECIFICATIONS

All readings in ft. lbs.

Year	VIN	No. Cylinder Displacement cu. in. (liter)	Cylinder Head Bolts	Main Bearing Bolts	Rod Bearing Bolts	Crankshaft Pulley Bolts	Flywheel Bolts	Manifold Intake	Manifold Exhaust	Spark Plugs
1990		61 (1000)	54	40	26	8	45	17	17	18
		97 (1600) ⑤	44	44	36	87	58 ④	14	18	20
		97 (1600) ⑥	58	44	11 ⑦	87	22 ⑦	17	30	21
1991-92		61 (1000)	54	40	26	8	45	17	17	18
		97 (1600) ⑤	44	44	36	87	58 ④	14	18	20
		97 (1600) ⑥	58	44	11 ⑦	87	22 ⑦	17	30	21

① Tighten an additional 45 degrees after torquing
② 1st step—29 ft. lbs.; 2nd step—58 ft. lbs.
③ 1st—Torque in sequence to 22 ft. lbs.
2nd—Torque in sequence another 1/4 turn
3rd—Torque in sequence another 1/4 turn
④ With automatic transaxle, 47 ft. lbs.
⑤ Prizm
⑥ Storm
⑦ Plus 45–60 Degree turn
⑧ Double overhead camshaft

BRAKE SPECIFICATIONS

All measurements in inches unless noted.

Year	Model	Lug Nut Torque (ft. lbs.)	Master Cylinder Bore	Brake Disc Minimum Thickness	Brake Disc Maximum Runout	Standard Brake Drum Diameter	Minimum Lining Thickness Front	Minimum Lining Thickness Rear
1988	Sprint	29–50	0.825	0.315	0.0028	7.09	0.315 ①	0.110 ①
	Spectrum	65	0.810	0.378	0.0059	7.09	0.039	0.039
	Nova	76	NA	0.492	0.0059	7.87	0.039	0.039
1989	Metro	41	0.825	0.315	0.0040	7.09	0.315	0.110 ①
	Spectrum	65	0.810	0.378	0.0059	7.09	0.039	0.039
	Prizm	76	NA	0.669	0.0035	7.87	0.030	0.039
1990	Metro	41	0.825	0.315	0.0040	7.09	0.315	0.110 ①
	Prizm	76	NA	0.669	0.0035	7.87	0.030	0.039
	Storm	87	NA	0.669	0.0035	7.87	0.030	0.039
1991-92	Metro	41	0.825	0.315	0.0040	7.09	0.315	0.110 ①
	Prizm	76	NA	0.669	0.0035	7.87	0.030	0.039
	Storm	87	NA	0.669	0.0035	7.87	0.030	0.039

NA Not available
① Lining plus shoe rim

WHEEL ALIGNMENT

Year	Model	Caster Range (deg.)	Caster Preferred Setting (deg.)	Camber Range (deg.)	Camber Preferred Setting (deg.)	Toe-in (in.)	Steering Axis Inclination (deg.)
1988	Sprint	—	3 3/16	—	1/4	0	12 3/16
	Spectrum	1 3/4P–2 3/4P	2 1/4P	7/16N–1 1/16P	11/32P	0 + 1/16	① ②
	Nova	1/8–1 2/3P	9/10P	3/4N–1/4P	1/4N	0–0.078	—
	Nova Twincam	1N–1 1/2P	1/4P	3/4N–1/4P	1/4N	0–0.078	—

WHEEL ALIGNMENT

Year	Model	Caster Range (deg.)	Caster Preferred Setting (deg.)	Camber Range (deg.)	Camber Preferred Setting (deg.)	Toe-in (in.)	Steering Axis Inclination (deg.)
1989	Metro	1P–5P	3P	1N–1P	0	0	25 11/16
	Spectrum	1 3/4P–2 3/4P	2 1/4P	11/16N–1 5/16P	5/16	0	16
	Prizm	11/16P–2 3/16P ③	1 7/16P	9/6N–1 5/16P	3/16	3/64	NA
	Prizm	9/16P–2 1/8P ④	1 5/16P	1/2N–1P	1/4	3/64	NA
1990	Metro	1P–5P	3P	1N–1P	0	0	25 11/16
	Prizm	11/16P–2 3/16P ③	1 7/16P	9/6N–1 5/16P	3/16	3/64	NA
	Prizm	9/16P–2 1/8P ④	1 5/16	1/2N–1P	1/4	3/64	NA
	Storm	2P–4P	3P	1N–1/4P	3/8N	0	10 3/16
1991–92	Metro	1P–5P	3P	1N–1P	0	0	25 11/16
	Prizm	11/16P–2 3/16P ③	1 7/16P	9/6N–1 5/16P	3/16	3/64	NA
	Prizm	9/16P–2 1/8P ④	1 5/16	1/2N–1P	1/4	3/64	NA
	Storm	2P–4P	3P	1N–1/4P	3/8N	0	10 3/16

① Inside—37°40′ full lock
② Outside—32°30′ full lock
③ Manual transaxle
④ Automatic transaxle

ENGINE MECHANICAL

NOTE: Disconnecting the negative battery cable on some vehicles may interfere with the functions of the on board computer systems and may require the computer to undergo a relearning process, once the negative battery cable is reconnected.

Engine Assembly

REMOVAL & INSTALLATION

Metro

1. Relieve the fuel pressure from the fuel system.
2. Using a scratch awl, scribe the hood hinge-to-hood outline, then, using an assistant remove the hood.
3. Disconnect the negative battery cable. Drain the cooling system.
4. Remove the air cleaner assembly. Remove the radiator assembly along with the cooling fan.
5. Disconnect and tag all the necessary electrical connections.
6. Disconnect and tag all the necessary vacuum lines.
7. Disconnect the fuel return hose and fuel feed hose from the throttle body.
8. Disconnect the heater inlet and outlet hoses.
9. Disconnect the following cables:
   a. The accelerator cable from the throttle body.
   b. The clutch cable from the transaxle (for M/T models).
   c. The gear select cable and the oil pressure control cable from the transaxle (for A/T models).
   d. The speedometer cable from the transaxle.
10. Raise and safely support the vehicle.
11. Disconnect the exhaust pipe from the exhaust manifold.
12. Disconnect the gear shift control shaft and the extension to the transaxle for (M/T model).
13. Drain the engine oil and transaxle oil.
14. Remove the drive axles from the differential side gears of the transaxle. For the engine/transaxle removal, it is not necessary to remove the drive axle from the steering knuckle.
15. Remove the engine rear torque rod bracket from the transaxle (for A/T models).
16. Lower the vehicle.
17. Install a suitable chain hoist to the lifting device on the engine.
18. Remove the right side engine mounting from its bracket.
19. On vehicles equipped with a automatic transaxle, remove the transaxle rear mounting nut.
20. On vehicles equipped with a manual transaxle, remove the transaxle rear mounting from the body.
21. Remove the transaxle left side mounting bracket.
22. Lift the engine and transaxle assembly out from the vehicle. Separate the transaxle from the engine.

To install:

31. Install the transaxle to the engine, then a suitable hoist onto the engine lifting brackets.
32. Install engine and transaxle into vehicle and leave the hoist connected to the lifting device.
33. On vehicles equipped with a automatic transaxle, install the transaxle rear mounting nut.
34. On vehicles equipped with a manual transaxle, install the transaxle rear mounting from the body.
35. Install the transaxle left side mounting bracket.
36. Install the transaxle right side engine mounting to its bracket.
37. Tighten all the bolts and nuts.
38. Remove the lifting device.
39. To complete the installation procedure, reverse the removal procedure.
40. Adjust the clutch pedal free-play.
41. Adjust the gear select cable, and oil pressure control cable.
42. Adjust the accelerator cable play.
43. Refill the transaxle with the recommended fluid. Do the same for the engine oil and engine coolant.
44. Reconnect the negative battery cable. Start the engine and check for

leakage of any kind. Make all necessary repairs and adjustments.

45. Torque the transaxle-to-engine bolts and nuts to 37 ft. lbs. (50 Nm).

46. Torque the engine mounting nuts to 37 ft. lbs. (50 Nm).

47. Torque the engine mounting left hand bracket bolts to 37 ft. lbs. (50 Nm).

48. Torque the exhaust pipe to manifold bolts to 37 ft. lbs. (50 Nm).

49. Torque the flywheel retaining bolts to 47 ft. lbs. (64 Nm).

Nova

EXCEPT TWINCAM ENGINE

1. Disconnect the negative terminal from the battery. Drain the cooling system. Drain the engine oil.

2. Properly discharge the air conditioning system.

3. Using a scratch awl, scribe the hood hinge-to-hood outline, then, using an assistant remove the hood.

4. Remove the air cleaner assembly and associated ducts.

5. From the radiator, remove the upper coolant hose and the overflow hose. Disconnect the coolant hose from the coolant pipe at the rear of the cylinder head and the coolant hose from the thermostat housing.

6. Disconnect the fuel hoses from the fuel pump.

7. Remove the drive belt from the alternator, the power steering pump, if equipped, and the air conditioning compressor, if equipped. If equipped with power steering, remove the power steering pump-to-engine bolts and move the pump aside, do not disconnect the pressure hoses. If equipped with air conditioning, remove the compressor-to-engine bolts and move it aside; do not disconnect the pressure hoses.

8. Label and disconnect the electrical connectors that will interfere with the engine removal.

9. Label and disconnect the vacuum hoses running between the engine and firewall or fender well mounted accessories.

10. Label and disconnect the electrical connectors from the transaxle.

11. Disconnect the speedometer cable from the transaxle.

12. Raise and support the vehicle safely. Drain the engine oil and the transaxle fluid.

13. Disconnect the exhaust pipe-to-exhaust manifold bolts and separate the exhaust pipe from the manifold.

14. Disconnect the air hose from the catalytic converter, if equipped.

15. If equipped with an automatic transaxle, disconnect and plug the oil cooler tubes from the radiator.

16. Remove the undercovers from both sides of the vehicle.

17. Disconnect the cable and the bracket from the transaxle.

18. Disconnect the steering knuckles from the lower control arms.

19. Disconnect the halfshafts from the transaxle.

20. Remove the flywheel cover. If equipped with an automatic transaxle, mark the torque converter-to-flexplate, then, remove the torque converter-to-flexplate bolts and move the torque converter back into the transaxle.

21. Disconnect the front and rear engine mounts from the center member.

22. Lower the vehicle.

23. Remove the radiator-to-chassis bolts and the radiator (with the fans) from the vehicle.

24. Using a overhead lift, attach it to and support the engine.

25. Remove the through bolt from right-side engine mount, then, the left-side transaxle mount bolt and the mount.

26. Remove the engine/transaxle assembly from the vehicle. Remove the transaxle-to-engine bolts and separate the transaxle from the engine, then, secure the engine to a workstand.

To install:

27. Attach a suitable chain hoist to the lifting device on the engine. Slowly lift the engine/transaxle assembly into the vehicle.

28. Tilt the transaxle downward to clear the left mount.

29. Install the transaxle mounts. Remove the engine lifting device.

30. Install the radiator with fan assemblies.

31. Raise and safely support the vehicle.

33. Install the center member and connect the speedometer cable.

34. Connect the front and rear mounting at the center member.

35. Install the flywheel to torque converter. Torque the bolts in 3 passes to 58 ft. lbs. (78 Nm).

36. Install the flywheel cover.

37. Connect the halfshaft at the transaxle.

38. Connect the steering knuckles at the lower control arm.

39. Connect the cable and bracket at the transaxle.

40. Place the air conditioning compressor in position (if so equipped), and tighten the bolts.

41. Place the power steering pump in position (if so equipped), and tighten the bolts.

42. Install the left and right hand under covers.

43. Connect the transaxle cooler lines at the radiator.

44. Connect the air hose at the converter.

45. Connect the exhaust pipe to the manifold.

46. lower the vehicle.

47. Connect the speedometer cable at the transaxle.

48. Connect the wires at the transaxle.

49. Connect all disconnected vacuum lines and electrical connectors.

50. Install and adjust the alternator, power steering and the air conditioning compressor belts.

51. Connect the fuel lines at the fuel pump.

52. Install the coolant hoses to their respective placed on the engine and the radiator.

53. Install the air cleaner assembly.

54. With the aid of an assistant, install the hood assembly.

55. Refill the transaxle with the recommended transaxle fluid and the same for the engine oil and the cooling system.

56. Reconnect the negative battery cable. Start the engine and check for leakage of any kind. Make all necessary repairs and adjustments.

57. Torque the engine-to-crossmember mount bolts to 29 ft. lbs. (39 Nm).

58. Torque the exhaust pipe-to-exhaust manifold to 46 ft. lbs. (62 Nm).

59. Torque the power steering pump-to-bracket bolt to 29 ft. lbs. (39 Nm).

60. Torque the power steering adjusting bolt to 32 ft. lbs. (43 Nm).

TWINCAM ENGINE

1. Disconnect the negative terminal from the battery.

2. Properly relieve the fuel system pressure.

3. Drain the cooling system.

4. Using a scratch awl, scribe the hood hinge-to-hood outline, then, using an assistant remove the hood.

5. Remove the air cleaner assembly, the coolant tank reservoir and the PVC hose.

6. Disconnect the heater hoses from the water inlet housing and the fuel hose from the fuel filter.

7. If equipped with a manual transaxle, remove the clutch slave cylinder-to-transaxle bolts and the slave cylinder, then, move the cylinder aside.

8. Disconnect the vacuum hose from the charcoal canister.

9. Disconnect the speedometer cable from the transaxle and the accelerator cable from the throttle body.

10. If equipped with cruise control, perform the following procedures:

a. Remove the cables from the throttle body.

b. Disconnect the vacuum hose from the actuator.

c. Remove the actuator cover bolts and the cover.

d. Disconnect the actuator connector, then, remove the actuator.

11. Remove the ignition coil.

12. To remove the main wiring har-

ness, perform the following procedures:

a. Remove the right side of the cowl panel and disconnect the No. 4 junction block connectors.

b. Remove the ECM cover and disconnect the ECM connectors, then, pull the main wiring harness into the engine compartment.

13. Disconnect the No. 2 junction block connectors and the ground strap terminals.

14. Disconnect the windshield washer change valve connector, the battery cable from the starter, the cruise control vacuum pump and switch connectors.

15. Disconnect the vacuum hose from the power brake booster.

16. If equipped with air conditioning, perform the following procedures:

a. Remove the vane pump pulley nut.

b. Loosen the idler pulley adjusting and pulley nuts.

c. Remove the compressor-to-bracket bolts, then, move the compressor aside and secure it.

d. Disconnect the oil pressure connector.

e. Remove the compressor bracket bolts, the vane pump bolts, then, move the vane pump and bracket aside and suspend it.

17. Raise and support the vehicle safely. Drain the engine crankcase and transaxle fluid.

18. Remove the splash shields.

19. If equipped with an automatic transaxle, disconnect and plug the oil cooler from the radiator.

20. Remove the exhaust pipe-to-exhaust manifold bolts and separate the pipe from the manifold. Disconnect the oxygen sensor connector.

21. Remove the flywheel housing cover.

22. Remove the front and rear engine mounts from the center member, then, the center member.

23. Disconnect the right-side control arm from the steering knuckle and halfshafts from the transaxle.

24. Lower the vehicle.

25. Using a vertical hoist, secure the engine to it and support the engine; secure the engine wiring and hoses to the lift chain.

26. Remove the right-side engine mount, then, the left-side engine mount from the transaxle bracket.

NOTE: When lifting the engine be careful not to damage the throttle position sensor or the power steering gear housing.

27. Lift the engine/transaxle assembly from the vehicle.

28. To separate the transaxle from the engine, perform the following procedures:

a. Remove the radiator fan temperature switch connector and the start injector time switch connector.

b. Disconnect the vacuum hoses from the BVSV's.

c. Remove the No. 1 and 2 hoses from the water bypass pipes.

d. Disconnect the electrical connector from the back-up switch, the water temperature sensor and the water temperature switch.

e. If equipped with an automatic transaxle, disconnect the neutral start switch connector and the transaxle solenoid connector, then, remove the torque converter-to-flexplate bolts; be sure to push the torque converter back into the transaxle.

f. Remove the starter, the transaxle-to-engine bolts and the transaxle.

To install:

29. Reassemble the transaxle to the engine as follows:

a. Install the transaxle to the engine. Install the transaxle-to-engine bolts and torque the bolts to 47 ft. lbs. (64 Nm).

b. Install the starter assembly and torque the retaining bolts to 29 ft. lbs. (39 Nm).

c. If equipped with an automatic transaxle, reconnect the neutral start switch connector and the transaxle solenoid connector. Install the torque converter and the converter-to-flexplate bolts (be sure to push the torque converter back into the transaxle) torque the bolts to 27 ft. lbs. (20 Nm).

d. Reconnect the electrical connector to the back-up switch, the water temperature sensor and the water temperature switch.

e. Install the No. 1 and 2 hoses to the water bypass pipes.

f. Reconnect the vacuum hoses to the BVSV's.

g. Install the radiator fan temperature switch connector and the start injector time switch connector.

30. Attach a suitable chain hoist to the lifting device on the engine. Slowly lift the engine/transaxle assembly into the vehicle.

31. Tilt the transaxle downward to clear the mounts. When installing the engine, be careful not to hit tje steering gear housing and the throttle position sensor.

32. With the engine assembly sitting in the proper position, install the left and right hand motor mount and through bolts.

33. Remove the chain hoist from the vehicle. Raise and support the vehicle safely.

34. Install the halfshafts to the transaxle, torque the nuts to 27 ft. lbs. (36 Nm).

35. Install the right hand control arm to the steering knuckle and torque the nuts to 64 ft. lbs. (47 Nm).

36. Install the flywheel housing under cover. Install the engine mounting center member. Torque the center member bolt to 29 ft. lbs. (39 Nm).

37. Install the front and rear mount onto the center member and install the 2 hole covers. Torque the mount bolts to 35 ft. lbs. (48 Nm). Torque the front and rear mount through bolts to 58 ft. lbs. (78 Nm).

38. Install the exhaust pipe with a new gasket onto the exhaust manifold. Torque the exhaust pipe nuts to 46 ft. lbs. (62 Nm).

39. Reconnect the oxygen sensor. Install the oil cooler lines and lower the vehicle.

40. Install the air conditioning bracket, compressor and drive belt. Torque the vane pump pulley to 28 ft. lbs. (38 Nm). Install and adjust the drive belts.

41. Install the brake booster vacuum hose.

42. Install the No. 2 junction block connectors, the starter cable to the battery positive terminal and the ground strap terminals.

43. Connect the water change valve connector, the cruise control vacuum pump connector and the vacuum switch connector.

44. Connect the main engine wire harness as follows:

a. Push the ECM and No. 4 junction connectors wiring harness from the engine compartment into the interior.

b. Connect the ECM connectors and install the cover.

c. Connect the No. 4 junction block connectors and install the right cowl (kick) panel.

45. Install the ignition coil.

46. Install the cruise control actuator, if so equipped. Install the accelerator cables and cruise control cables.

47. Install the speedometer cable and the shift control cable. Install the charcoal canister vacuum hose.

48. Install the clutch release cylinder.

49. Install the fuel return hose to the pressure regulator. Install the heater and air hoses to the air valve.

50. Install the fuel hose to the fuel filter. Install the heater hoses to the water inlet housing. Install the PCV hose.

51. Install the coolant recovery tank. Install the air cleaner assembly.

52. Refill the transaxle with the recommended transaxle fluid and the same for the engine oil and the cooling system.

53. Reconnect the negative battery cable. Start the engine and check for leakage of any kind. Make all neces-

sary repairs and adjustments.

54. Raise and support the vehicle safely. Install the splash shields and lower the vehicle.

55. With the aid of an assistant, install the hood assembly.

1989-90 Prizm

1. Remove the hood, with the aid of an helper.2. Disconnect the negative battery cable, then the positive battery cable and remove the battery.

3. Raise the vehicle and safely support it.

4. Remove the left and right splash shields.

5. Drain the engine oil and the transmission oil.

6. Drain the engine coolant and save it in closed containers for reuse.

7. Remove the air cleaner hose and the air cleaner assembly.

8. Remove the coolant reservoir. Remove the radiator and fan assembly.

9. Disconnect the accelerator cable and if equipped with automatic transaxle, the throttle cable.

10. Disconnect and remove the cruise control actuator.

11. Label and disconnect the main engine wiring harness from its related sensors and switches.

12. Remove the ground strap connector and its bolt. Disconnect the wiring to the vacuum sensor, the oxygen sensor and the air conditioning compressor.

13. Label and disconnect the brake booster vacuum hose, the power steering vacuum hose, the charcoal canister vacuum hose and the vacuum switch vacuum hose.

14. Carefully disconnect the fuel inlet and return lines.

CAUTION

The fuel system is under pressure. Release pressure slowly and contain spillage. Observe no smoking/no open flame precautions. Have a Class B-C (dry powder) fire extinguisher within arm's reach at all times.

15. Disconnect the heater hoses.

16. Loosen the power steering pump mounting bolt and through bolt. Remove the drive belt.

17. Remove the 4 bolts holding the air conditioner compressor and remove the compressor. Do not loosen or remove any lines or hoses. Move the compressor out of the way and hang it from a piece of stiff wire.

18. Disconnect the speedometer cable from the transaxle.

19. On vehicles with manual transaxles, unbolt the clutch slave cylinder from the bell housing and move the cylinder out of the way. Don't disconnect any lines or hoses. Disconnect the shift control cables by removing the 2 clips, the washers and retainers.

20. On vehicles with automatic transaxles, remove the clip and retainer and separate the control cable from the shift lever.

21. Elevate the vehicle and support it safely.

22. Remove the 2 bolts from the exhaust pipe flange and separate the pipe from the exhaust manifold.

23. Remove the nuts and bolts and separate the halfshafts from the transaxle.

24. Remove the through bolt from the rear transaxle mount.

25. Remove the nuts from the center transaxle mount and the rear mount.

26. Lower the vehicle to the ground and attach the lifting equipment to the brackets on the engine. Take tension on the hoist line or chain just enough to support the motor but no more. Hang the engine wires and hoses on the chain or cable.

27. Remove the 3 exhaust hanger bracket nuts and the hanger. Remove the 2 center crossmember-to-main crossmember bolts. Remove the 3 center crossmember-to-radiator support bolts.

CAUTION

Support the crossmembers with a jack when loosening the bolts. The pieces are heavy and could fall.

28. Remove the 8 crossmember-to-body bolts, then remove the 2 bolts holding the control arm brackets to the underbody. Remove the 2 center mount-to-transaxle bolts and remove the mount. Carefully lower the center mount and crossmember and remove from under the vehicle.

29. At the left engine mount, remove the 3 bolts and the bracket, then remove the bolt, 2 nuts, through bolt and mounting. Remove the 3 bolts and the air cleaner bracket.

30. Loosen and remove the 5 bolts and disconnect the mounting bracket from the transaxle bracket. Remove the through bolt and mounting.

31. Carefully and slowly raise the engine and transaxle assembly out of the vehicle. Tilt the transaxle down to clear the right engine mount. Be careful not to hit the steering gear housing. Make sure the engine is clear of all wiring, lines and hoses.

32. Support the engine assembly on a suitable stand; do not allow it to remain on the hoist for any length of time.

33. With the engine properly supported, disconnect the reverse light switch and the neutral safety switch (automatic transaxle).

34. Remove the rear end cover plate.

35. For automatic transaxles, remove the 6 torque converter mounting bolts.

36. Remove the starter.

37. Support the transaxle, remove the retaining bolts in the case and remove the transaxle from the engine. Pull the unit straight off the engine; do not allow it to hang partially removed on the shaft. Keep the automatic transaxle level; if it tilts forward the converter may fall off.

To install:

38. Before reinstalling the engine in the car, several components must be reattached or connected. Install the transaxle to the engine; tighten the 12mm bolts to 47 ft. lbs. (64 Nm) and the 10mm bolts to 34 ft. lbs. (46 Nm).

39. Install the starter, its cable and connector. Tighten the mounting bolts to 29 ft. lbs. (39 Nm).

40. Install the 6 torque converter-to-flexplate bolts on automatic transaxles. Tighten the bolts to 20 ft. lbs. (27 Nm).

41. Install the rear cover plate and connect the wiring to the reverse light switch and the neutral safety switch (automatic transaxle).

42. Attach the chain hoist or lift apparatus to the engine and lower it into the engine compartment. Before it is completely in position, attach the power steering pump and its through bolt to the motor.

NOTE: Tilt the transaxle downward and lower the engine to clear the left motor mount. As before, be careful not to hit the power steering housing (rack) or the throttle position sensor. Be sure that the engine is clear of all wiring, hoses and cables.

43. Level the engine and align each mount with its bracket.

44. Install the right mounting insulator (bushing) to the engine bracket with the 2 nuts and bolt. Tighten the bolt temporarily.

45. Align the right insulator with the body bracket and install the through bolt and nut. Temporarily tighten the nut and bolt.

46. Align the left mounting insulator with the transaxle case bracket. Temporarily install the 3 bracket bolts.

47. With the engine held in place by these mounts, repeat steps 44, 45 and 46, tightening the bolts to the following tightness. Step 44: 38 ft. lbs. (52 Nm); step 45: 64 ft. lbs. (87 Nm); step 46: 35 ft. lbs. (47 Nm).

48. Install the left side mounting support with its 2 bolts; tighten them to 15 ft. lbs. (20 Nm).

49. With the engine securely

mounted in the vehicle, the lifting equipment may be removed. Elevate the vehicle and support it safely.

50. Install the center mount to the transaxle with its 2 bolts and tighten them to 45 ft. lbs. (61 Nm).

51. Position the center mount over the front and rear studs and start 2 nuts on the center mount only. Loosely install the 3 center support-to-radiator support bolts.

52. Loosely install the 2 front mount bolts. Raise the main crossmember into place over the rear studs and align all the underbody bolts.

53. Install the 2 rear mount nuts; leave them loose.

54. Loosely install the 8 underbody bolts, the lower control arm bracket bolts, the 2 center support-to-crossmember bolts and the exhaust hanger bracket and nuts.

55. With everything loose, but in place, make a second pass over all the nuts and bolts tightening them to the following specifications:

Crossmember-to-underbody bolts: 152 ft. lbs. (206 Nm).

Lower control arm bracket-to-underbody bolts: 94 ft. lbs. (127 Nm).

Center support-to-radiator support: 45 ft. lbs. (61 Nm).

Center support-to-crossmember: 45 ft. lbs. (61 Nm).

Front, center and rear mount bolts: 45 ft. lbs. (61 Nm).

Exhaust hanger bracket nuts: 9 ft. lbs. (12 Nm).

56. Install the rear transaxle mount and tighten its bolt to 64 ft. lbs. (87 Nm).

57. Install the nuts on the center transaxle mount and tighten them to 45 ft. lbs. (61 Nm).

58. Reconnect the halfshafts to the transaxle.

59. Using a new gasket, connect the exhaust pipe to the manifold and install the exhaust pipe bolts, tightening them to 18 ft. lbs. (24 Nm).

60. Lower the vehicle to the ground.

61. Either connect the control cables to the shift outer lever and selector lever and attach the control cables to manual transaxles or reconnect the control cable to the shift lever and install the clip and retainer on automatic transaxles. If equipped with manual transaxle, reattach the clutch slave cylinder to its mount.

62. Attach the speedometer cable to the transaxle.

63. Install the air conditioning compressor and drive belt if so equipped.

64. Install the power steering pump, pivot bolt and drive belt if so equipped. Adjust the belts to the correct tension.

65. Install the fuel inlet and outlet lines.

66. Connect the heater hoses. Make sure they are in the correct positions and that the clamps are in sound condition.

67. Connect the vacuum hoses to the vacuum switch, the charcoal canister, the vacuum sensor, the power steering and the brake booster.

68. Connect the wiring to the air conditioning, the oxygen sensor and the vacuum sensor.

69. Observing the labels made at the time of disassembly, reconnect the main engine harness to its sensors and switches. Work carefully and make sure each connector is properly matched and firmly seated.

70. Install the ground strap connector and its bolt; connect the wiring at the No. 2 junction block in the engine compartment.

71. Install the cruise control actuator if so equipped.

72. Connect the accelerator cable and throttle cable (automatic) to their brackets.

73. Install the radiator and cooling fan assembly. Install the overflow reservoir.

74. Install the air cleaner assembly and the air intake hose.

75. Install the battery. Connect the positive cable to the starter terminal, then to the battery. Do not connect the negative battery cable at this time.

76. Fill the tansaxle with the correct amount of fresh fluid.

77. Refill the engine coolant.

78. Fill the engine with the correct amount of fresh oil.

79. Double check all installation items, paying particular attention to loose hoses or hanging wires, untightened nuts, poor routing of hoses and wires (too tight or rubbing) and tools left in the engine area.

80. Connect the negative battery cable. Start the engine and allow it to idle. As the engine warms up, shift the automatic tansaxle into each gear range allowing it to engage momentarily. After each gear has been selected, put the shifter in **P** and check the tansaxle fluid level.

81. Shut the engine off and check the engine area carefully for leaks, particularly around any line or hose which was disconnected during removal.

82. Elevate and support the vehicle. Replace the left and right splash shields and lower the vehicle.

83. With the help of an assistant, reinstall the hood. Adjust the hood for proper fit and latching.

1991 Prizm

1. Disconnect the negative terminal from the battery.

2. Using a scratch awl, scribe the hood hinge-to-hood outline. Disconnect the windshield washer fluid lines from the hood. Remove the 4 hood hinge bolts (2 on each side) disengaging the hood from the hinge. Using an assistant remove the hood.

3. Properly relieve the fuel system pressure.

4. Raise and safely support the vehicle.

5. Remove the left and right stone shields.

6. Drain the engine oil and the cooling system.

7. Drain the transaxle fluid. Lower the vehicle.

8. Remove the air cleaner hose and air cleaner assembly. Remove the radiator, cooling fan and cooling overflow reservoir.

9. Disconnect the accelerator cable and throttle cable from the bracket.

10. Remove the cruise control actuator.

11. Disconnect and tag all electrical component connectors.

12. Disconnect and tag all the necessary vacuum lines.

13. Remove the fuel inlet and outlet lines.

14. Remove the heater hoses from the engine.

15. Remove the power steering pump and bracket from the engine and set it off to the side. Leave the power steering hydraulic lines hook up to the pump.

16. Remove the air conditioning compressor and bracket bolts from the engine and set it off to the side. Leave the freon lines hooked up to the compressor.

17. Remove the speedometer cable from the transaxle.

18. Disconnect the clutch release cylinder without disconnecting the line and hose (manual transaxles only).

19. On the manual transaxle models, disconnect the transaxle control cables by removing the 2 clips, washers and retainers. Disconnect the control cables from the shift outer lever and select outer lever.

20. Disconnect the control cable from the shift lever (automatic transaxle only).

21. Raise and safely support the vehicle.

22. Disconnect the front exhaust pipe from the exhaust manifold.

23. Remove the transaxle mounts. Lower the vehicle.

24. Remove the engine mounts.

25. Attach a suitable chain hoist to the lifting device on the engine. Slowly lift the engine/transaxle assembly out of the vehicle and place it in a suitable engine stand.

To install:

26. Attach a suitable chain hoist to the lifting device on the engine. Slowly lift the engine/transaxle assembly into the vehicle.

26. Tilt the transaxle downward to clear the left mount.
27. Raise and safely support the vehicle. Install the engine and transaxle mounts.
28. Install the drive axles.
29. Install the exhaust pipe to the exhaust manifold.
30. Install the right and left and stone shields. Lower the vehicle.
31. Install the control cable to the shift lever (automatic transaxle).
32. On vehicles equipped with manual transaxles, install the control cables to the shift outer lever and select outer lever. Install the clutch release cylinder.
33. Install the air conditioner compressor bracket, compressor and bolts.
34. Install the power steering pump and bracket to the engine.
35. Install the heater hoses to the engine.
36. Install the fuel inlet and outlet lines.
37. Reconnect the disconnected vacuum lines.
38. Reconnect the disconnected electrical connectors.
39. Install the cruise control actuator.
40. Install the accelerator and throttle cable to the bracket.
41. Install the radiator, cooling fan and overflow reservoir. Refill the cooling system.
42. Install the air cleaner hose and air cleaner assembly.
43. Refill the transaxle.
44. Refill the engine with the recommended oil.
45. Reconnect the battery.
46. Install the hood assembly.
47. Start the engine and check for leakage of any kind. Make all necessary repairs and adjustments.
48. Torque the center transaxle mount-to-center crossmember nuts 45 ft. lbs. (61 Nm)
49. Torque the rear transaxle mount-to-main crossmember nuts 45 ft. lbs. (61 Nm).
50. Torque the right engine mount through bolt 64 ft. lbs. (87 Nm).
51. Torque the right engine mount support bolts 35 ft. lbs. (47 Nm).
52. Torque the right engine mount support bolts 35 ft. lbs. (47 Nm).
53. Torque the engine mount-to-engine bracket nuts 45 ft. lbs. (61 Nm).
54. Torque the engine mount-to-bracket bolt 45 ft. lbs. (61 Nm).
55. On vehicles equipped with an automatic transaxle, torque the flywheel retaining bolts to 47 ft. lbs. (64 Nm).
56. On vehicles equipped with a manual transaxle, torque the flywheel retaining bolts to 58 ft. lbs. (78 Nm).

Storm

1. Disconnect the battery cables, then remove battery and battery tray from vehicle.
2. Scribe match marks on the hood hinge-to-hood, then remove the hood.
3. Discharge air conditioning system.
4. Drain cooling system, then disconnect the accelerator cable from the throttle valve.
5. Disconnect the breather hose from the intake air duct, then remove the intake air duct from the throttle valve.
6. Remove the air cleaner cover, filter and the body from the vehicle.
7. Remove the MAP sensor hose from the MAP sensor, then the brake booster vacuum hose.
8. Disconnect the 2 canister hoses from the pipes on the intake manifold (common chamber). If equipped with twincam engine, remove the canister pipe support bracket.
9. Disconnect the 2 cable harness connectors, located near the left shock tower.
10. Disconnect the ignition coil ground cable from the terminal on the thermostat housing flange.
11. Disconnect the ignition coil-to-distributor wire, the 2 primary wires from the ignition coil, then remove the ignition coil and bracket from the vehicle.
12. Disconnect the engine harness ground cable from the left inner fender.
13. Disconnect the harness terminal from the relay and fuse box.
14. Disconnect the cooling fan electrical connector and the 2 battery cable connectors.
15. Disconnect the oxygen sensor electrical connector, then the ground cable terminals from the left front side of the common chamber and the rear of the cylinder head cover.
16. If equipped with automatic transaxle, disconnect the electrical connectors from transaxle.
17. If equipped with manual transaxle, loosen the tow adjusting nuts and disconnect the clutch cable. Disconnect the 2 transaxle shaft cables by removing the cotter pin and clip from the shaft cable bracket.
18. If equipped with automatic transaxle, disconnect the shift cable by removing the cotter pin from the shaft cable lever.
19. On all vehicles, remove the 2 heater hoses from the engine.
20. Disconnect the speedometer from the transaxle, then the upper radiator hose from the radiator.
21. Disconnect the fuel feed line and fuel return hose, near the filter, then the coolant recovery tank with bracket.
22. Remove the power steering belt, power steering pump and bracket from the vehicle.
23. Remove the cooling fan and shroud, then raise and support the vehicle safely.
24. Remove the right and left under covers and the lower radiator hose from the engine.
25. If equipped with automatic transaxle, disconnect the oil cooler, lines from the transaxle.
26. Remove the air conditioning compressor bracket bolts and position compressor aside.
27. Remove the front tire and wheel assemblies, then the halfshafts.
28. Remove the front exhaust pipe from the exhaust manifold.
29. Lower the vehicle and install a suitable engine hoist onto the lifting brackets on the engine.
30. Remove the engine mounts and transaxle mounts, then lift the engine and transaxle assembly out from the vehicle. Separate the transaxle from the engine.

To install:

31. Install the transaxle to the engine, then a suitable hoist onto the engine lifting brackets.
32. Install engine and transaxle into vehicle, then the transaxle mounts and engine mounts.
33. Remove the engine hoist from the vehicle, then raise and support vehicle safely.
34. Install the front exhaust pipe to the exhaust manifold, then the halfshafts.
35. Install the tire and wheel assemblies, then the 2 air conditioning bracket mounting bolts.
36. If equipped with automatic transaxle, connect the 2 cooler lines to the transaxle.
37. Install the lower radiator hose, then the right and left undercovers. Lower the vehicle.
38. Install the cooling fan and shroud, then the power steering pump and belt onto the engine.
39. Install the coolant recovery tank with bracket, then connect the fuel feed line and fuel return hose.
40. Install the upper radiator hose, then connect the speedometer cable to the transaxle.
41. Connect the 2 heater hoses to the engine, then the transaxle shift cables, if equipped.
42. If equipped with manual transaxle, connect the clutch cable and adjust.
43. If equipped with automatic transaxle, connect the electrical connectors to the transaxle.
44. Connect the ground cable terminal to the rear of the cylinder head cover.
45. Connect the 2 ground terminals to the right side of the intake manifold (common chamber). If equipped with

twincam engine, install the canister pipe support bracket.

46. Connect the oxygen sensor electrical connector and the 2 battery cable harness connectors.

47. Connect the cooling fan harness connector, then the chassis harness terminal connector to the relay and fuse box.

48. Connect the engine ground cable to the left inner fender, then install the ignition coil and bracket.

49. Connect the 2 primary wires to the ignition coil and the ignition coil-to-distributor wire.

50. Connect the ignition coil ground wire to the terminal at the thermostat housing.

51. Connect the 2 cable harness connectors, located near the left shock tower.

52. Connect the 2 canister hoses to the intake manifold (common chamber).

53. Connect the brake booster vacuum hose, then the MAP sensor hose to the MAP sensor.

54. Install the air cleaner body, air cleaner filter and the cover.

55. Connect the air intake duct to the throttle body and the breather hose to the air intake duct.

56. Connect the accelerator cable to the throttle valve.

57. Fill the cooling system, engine oil and transaxle with suitable fluids. Install the battery tray and battery.

58. Align match marks on the hood hinge during removal and install the hood. Connect the battery negative cable. Start the engine and check for leaks.

Spectrum

1. Remove the hood, relieve the fuel system pressure and disconnect the negative battery cable.
2. Drain the cooling system.
3. If equipped with carburetor, remove the air cleaner and the throttle cable at the carburetor.
4. Disconnect the heater hoses at the intake manifold, the coolant hose at the thermostat housing and the thermostat housing at the cylinder head.
5. On turbocharged models, remove the throttle cable, fuel lines and connectors, and also remove turbocharger vacuum, oil and water lines.
6. Remove the distributor from the cylinder head.
7. Disconnect the oxygen sensor electrical connector.
8. Support the engine using a vertical lift and remove the right motor mount.
9. Disconnect the necessary electrical connectors and vacuum hoses.
10. Disconnect the flex hose at the exhaust manifold and the lower radiator hose at the block.
11. Remove the upper air conditioning compressor bolt and remove the belt.
12. Disconnect the power steering bracket at the block and remove the belt.
13. Disconnect the fuel lines from the fuel pump and the electrical connectors from under the carburetor, if equipped.
14. Remove the upper starter bolt. Raise and support the vehicle safely.
15. Drain the oil from the crankcase and remove the oil filter.
16. Disconnect the oil temperature switch connector.
17. Disconnect the exhaust pipe bracket at the block and the exhaust pipe at the manifold.
18. Remove the air conditioning compressor and move to one side. Do not disconnect the air conditioning refrigerant lines. Remove the alternator wires.
19. Remove the flywheel cover and the converter bolts, then install a flywheel holding tool J-35271 or equivalent.
20. Disconnect the starter wires and remove the starter.
21. Remove the front right wheel and inner splash shield.
22. Lower the engine by lowering the crossmember enough to gain access to the crankshaft pulley bolts, then remove the pulley.
23. Raise the engine and crossmember. Remove the engine support.
24. Lower the vehicle and support the transaxle.
25. Remove the transaxle to engine bolts. Remove the engine.

To install:

26. Reverse the procedure to install.
27. Adjust the drive belts and refill the fluids.
28. Connect all electrical connectors and vacuum lines.
29. Connect the battery negative cable. Start engine and check for leaks.

Sprint

1. Properly relieve the fuel system pressure. Remove the battery cables.
2. Remove the hood, battery, battery tray, air cleaner and the outside air duct.
3. Drain the cooling system.
4. Disconnect and tag the radiator, heater and vacuum hoses from the engine.
5. Disconnect the cooling fan wiring.
6. Remove the cooling fan, shroud and radiator as an assembly.
7. Remove the fuel hoses from the fuel pump.
8. Remove the brake booster hose from the intake manifold, accelerator cable from the carburetor and speed control cable from the transaxle.
9. Remove the clutch cable and bracket from the transaxle.
10. Disconnect and tag the necessary wiring from the engine and transaxle.
11. Remove the air conditioning compressor adjusting bolt and drive belt splash shield.
12. Raise and support the vehicle safely. Drain the engine oil and the transaxle fluid.
13. Disconnect the exhaust pipe from the exhaust manifold.
14. Remove the air conditioning pivot bolt, the drive belt and the mounting bracket.
15. Disconnect the gear shift control shaft and extension rod at the transaxle.
16. Disconnect the ball joints.
17. Remove the halfshafts from the transaxle.
18. Remove the engine torque rods and the transaxle mount nut.
19. Lower the vehicle.
20. Remove the engine side mount and the mount nuts.
21. Connect a vertical hoist to the engine, then lift the engine and transaxle assembly from the vehicle.

To install:

22. Lower the engine and transmission assembly into the vehicle.
23. Install the rear mount.
24. Raise and safely support the vehicle.
25. Raise the engine. Remove the front mount and bracket at the frame.
26. Align the stud of the motor mount with the hole in the engine.
27. Lower the engina and install the bolt into the frame bracket.
28. To complete the installation procedure, reverse the removal procedure.
29. Connect all electrical connectors and vacuum hoses.
30. Refill the engine, the transaxle and the cooling system.
31. Connect the battery negative cable. Start the engine and check for leaks.

Engine Mounts

REMOVAL & INSTALLATION

Nova

1. Disconnect the negative terminal from the battery.
2. Raise and support the vehicle safely.
3. Using an engine support fixture tool No. J-28467 or equivalent, attach it to the engine and support it.
4. Remove the center member hole covers, loosen the engine center mount and lower the vehicle.

5. Remove the 2 bolts/nuts from the front, rear and right-side engine mounts.
6. Lower the vehicle.
7. From the left-side engine mount, remove the 2 bolts on manual transaxle or 3 bolts on automatic transaxle.
8. Remove the through bolts from the engine mounts. Raise the engine in order to relieve engine weight from the mount. Remove the mounts from the vehicle.
9. To install, reverse the removal procedures. Torque the engine-to-cross member bolts to 35 ft. lbs. (47 Nm) and the engine mount through bolts to 58 ft. lbs. (79 Nm).

Prizm

1. Disconnect the negative terminal from the battery.
2. Raise and support the vehicle safely.
3. Using an engine support fixture tool No. J-28467 or equivalent, attach it to the engine and support it.
4. Turn the front wheels all the way to the right and remove the 6 bolts from the lower right stone shield.
5. Remove the 2 lower engine mount-to-engine bracket nuts and lower the vehicle.
6. Remove the 3 bolts and the right engine mount support.
7. Remove the 1 upper bolt from the engine mount-to-engine bracket.
8. Remove the windshield washer reservoir.
9. Remove the right engine mount through bolt and engine mount.

To install:

10. Install the right engine mount and through bolt. Torque the through bolt to 64 ft. lbs. (87 Nm).
11. Install the windshield wiper reservoir.
12. Install the 1 upper bolt to engine mount-to-engine bracket. Torque the bolt to 45 ft. lbs. (61 Nm).
13. Raise and safely support the vehicle.
14. Torque the engine mount-to-engine bracket nuts to 45 ft. lbs. (61 Nm).
15. Install the right side stone shield.
16. Install the 3 bolts and the right engine mount support. Torque the support bolts to 35 ft. lbs. (47 Nm).
17. Reconnect the negative battery cable.

NOTE: On vehicles equipped with manual transaxles, make sure the insulator space is equal all the way around the insulator. On vehicles equipped with automatic transaxles check that the space above and below the insulator is equal.

Storm

RIGHT

1. Disconnect the battery negative cable.
2. Support the engine using support support tool J–28467-Aor equivalent.
3. Raise and support vehicle safely, then remove the nut and bolt from the mounting bracket.
4. Remove the through bolt from the engine mount, then the engine mount from the vehicle.
5. Reverse procedure to install. Torque bridge bracket bolt to 29 ft. lbs. (40 Nm) and the right mount through bolt to 50 ft. lbs. (68 Nm).

LEFT

1. Disconnect the battery negative and positive cables, then remove the battery.
2. Remove the cover plate.
3. Remove the left engine mount bolts from the transaxle case.
4. Remove the left engine mount center bolt, then the left engine mount from the vehicle.
5. Reverse procedure to install. Torque left engine mount center bolt to 50 ft. lbs (68 Nm) and the engine mount-to-transaxle case bolts to 35 ft. lbs. (48 Nm).

REAR

1. Disconnect the battery negative cable.
2. Support the engine using engine support tool J–28467–A or equivalent.
3. Raise and support vehicle safely.
4. Remove the dampener weight from the engine mounting, then the bolt from the transaxle case.
5. Remove the center bolts from the center beam and the rear engine mount from the vehicle.
6. Reverse procedure to install. Torque rear mount center bolts to 76 ft. lbs. (102 Nm) and the rear mount transaxle case bolt to 37 ft. lbs. (50 Nm).

TORQUE ROD

1. Disconnect the battery negative cable.
2. Support the engine using engine support tool J–28467–A or equivalent.
3. Raise and support vehicle safely.
4. Remove the nut and 2 bolts from the torque rod bracket at the cylinder block.
5. Remove the center bolt and nut from the center beam.
6. Remove the torque rod with the bracket.
7. Installation is the reverse order of the removal procedure. Torque the torque rod center bolt to 50 ft. lbs. (68 Nm). Torque the rod bracket bolts and nuts to 28 ft. lbs. (38 Nm).

Spectrum

RIGHT HAND ENGINE MOUNT

1. Disconnect the negative battery terminal from the battery.
2. Raise and support the vehicle safely.
3. Support the engine.
4. Remove the 2 bolts and the plate on the engine side.
5. Remove the bolt and the mounting rubber.
6. Installation is the reverse order of the removal procedure. Be sure to install the right-hand engine mounting rubber and temporarily tighten the new bolts. Torque the 2 engine side bolts first to a torque of 45 ft. lbs. (61 Nm). Torque the 1 body side bolts first to a torque of 37 ft. lbs. (50 Nm).

TORQUE ROD

1. Disconnect the negative battery terminal from the battery.
2. Remove the engine side bolt and th body side nut and bolt.
3. Remove the torque rod.
4. Installation is the reverse order of the removal procedure. Torque the engine side bolt first to a torque of 57 ft. lbs. (76 Nm). Torque the body side nut and bolt to 42 ft. lbs. (57 Nm).

FRONT AND REAR ENGINE MOUNTS

1. Disconnect the negative battery terminal from the battery.
2. Raise and support the vehicle safely.
3. Support the engine.
4. Remove the through nuts and bolts from the front and rear engine mounts.
5. Remove the 4 bolts attaching the beam and remove beam.
6. Remove the front and or rear engine mounting rubbers.

NOTE: Install new bolts for mounts and torque engine mounting nut and bolts to 51 ft. lbs. (69 Nm).

7. Installation is the reverse order of the removal procedure.

Sprint and Metro

FRONT

1. Disconnect the negative battery cable.
2. Remove the motor mount nut.
3. Raise the and support the vehicle safely. Support the engine.
4. Remove the mount and frame bracket. Separate and remove the mount from the bracket.
5. Installation is the reverse of the removal procedure. Torque all mounting bolts to 41 ft. lbs. (55 Nm).

REAR

1. Disconnect the negative battery cable.
2. Raise and support the vehicle safely.
3. Remove the motor mount to body bracket nut.
4. Support the engine.
5. Remove the mount and frame bracket. Separate and remove the mount from the bracket.
6. Installation is the reverse of the removal procedure. Torque all mounting bolts to 41 ft. lbs. (55 Nm).

Cylinder Head

REMOVAL & INSTALLATION

Nova

EXCEPT TWINCAM ENGINE

1. Disconnect the negative terminal from the battery.
2. Drain the engine coolant into a clean container, opening both the radiator and cylinder block drain cocks.
3. Remove the air cleaner. Label and disconnect all vacuum hoses.
4. Raise and support the vehicle safely. Drain the engine oil. Remove the exhaust pipe-to-exhaust manifold nuts and separate the exhaust pipe from the manifold. Remove the exhaust pipe bracket from the engine. Remove the hose from the catalytic converter pipe.
5. If equipped with power steering, loosen the power steering pump pivot bolt. Lower the vehicle.
6. Disconnect the accelerator and throttle cables from the carburetor and cable bracket.
7. Disconnect electrical harness from the cowl, the oxygen sensor and the distributor.
8. Disconnect the fuel hoses from the fuel pump.
9. Disconnect the upper radiator hose from the water outlet, then, remove the water outlet from the cylinder head. Remove the heater hose.
10. If equipped with power steering, remove the adjusting bracket from the engine.
11. Remove the PCV valve and the wiring harness that passes over the valve cover.
12. Label and disconnect the spark plug wires, the electrical connector and the vacuum hoses from the distributor.
13. Remove the upper timing belt cover-to-cylinder head bolts and the cover.
14. Remove the cylinder head cover-to-cylinder head bolts, the cover and the gasket.
15. Remove the alternator drive belt. Remove the water pump pulley-to-water pump bolts and the pulley.
16. Using a socket wrench on the crankshaft pulley bolt, rotate the crankshaft to position the No. 1 cylinder on the TDC of its compression stroke; the crankshaft pulley notch is aligned with the **0** degrees mark on the timing plate and the No. 1 cylinder rocker arms are loose.
17. Remove the distributor-to-cylinder head hold-down bolts and the distributor.
18. Matchmark the timing belt and sprocket for reassembly in the same position; mark an arrow on the timing belt for rotation direction.
19. Loosen the idler pulley bolt. Move it so as to release the timing belt tension and snug the idler pulley bolt. Remove the timing belt; avoid twisting or bending it.
20. Loosen the head bolts, in sequence, in 3 stages, then, remove them. Lift the head directly off the block. If it is necessary to pry the head off the block, use a bar between the head and the projection provided on top of the block.

NOTE: Do not pry except at the projection provided. Be careful not to damage the block or cylinder head sealing surface.

To install:

21. Using the proper tool, clean the gasket mounting surfaces. Using wire brush, clean the cylinder head chambers.
22. Use new gaskets, sealant, if necessary, and reverse the removal procedures.

NOTE: When installing the cylinder head gasket, position the side with the sealer facing upwards.

23. Torque the cylinder head-to-engine bolts, in sequence, using 3 passes, to 43 ft. lbs. (58 Nm), the camshaft sprocket-to-camshaft bolt to 34 ft. lbs. (46 Nm) and the timing belt idler bolt to 27 ft. lbs. (37 Nm).
24. Rotate the crankshaft through 2 complete revolutions and check the timing belt tension; the tension should be 0.024–0.28 in.
25. Adjust the valves with the engine cold. Operate the engine until normal operating temperatures are reached and check for leaks. Readjust the valves with the engine hot. Set the ignition timing.

1988 TWINCAM ENGINE

1. Disconnect the negative terminal from the battery. Relieve the fuel system pressure.
2. Drain the engine coolant. Remove the air cleaner assembly.

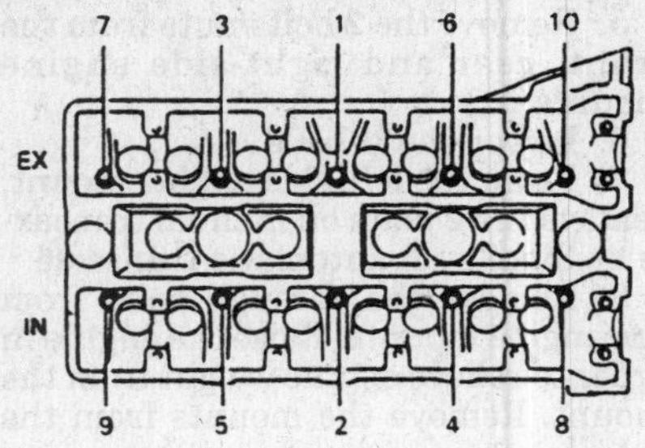

Cylinder head bolt torqueing sequence, twincam engine – Nova

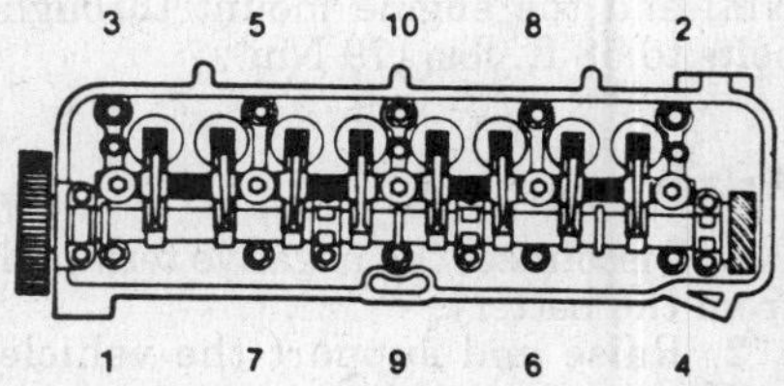

Cylinder head bolt loosening sequence except twincam engine – Nova

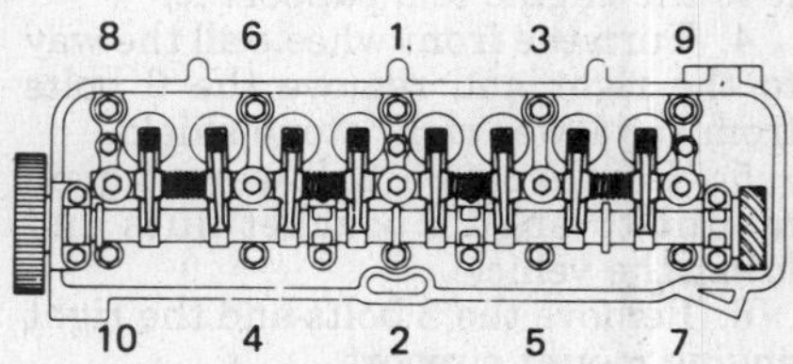

Cylinder head bolt torqueing sequence except twincam engine – Nova

3. Disconnect the throttle cable and the cruise control cable from the throttle linkage. Remove the ignition coil.
4. From the rear of the cylinder head, remove the heater hose. Remove the vacuum hoses from the throttle body. Remove the water outlet hose from cylinder head and the radiator.
5. If equipped with cruise control, remove the actuator and the bracket assembly.
6. Remove the hoses from the PCV valve and the power brake booster.
7. Remove the pressure regulator, the EGR valve (with lines) and the cold start injector hose.
8. Disconnect and remove the No. 1 fuel line.
9. From the auxiliary air valve, remove the No. 1 and No. 2 water bypass hoses.
10. Remove the vacuum pipe and the cylinder head rear cover.
11. Label and disconnect the electrical harness connectors. Remove the distributor-to-cylinder hold-down bolt and the distributor.
12. Remove the exhaust manifold-to-cylinder head bolts and separate the exhaust manifold from the cylinder head.

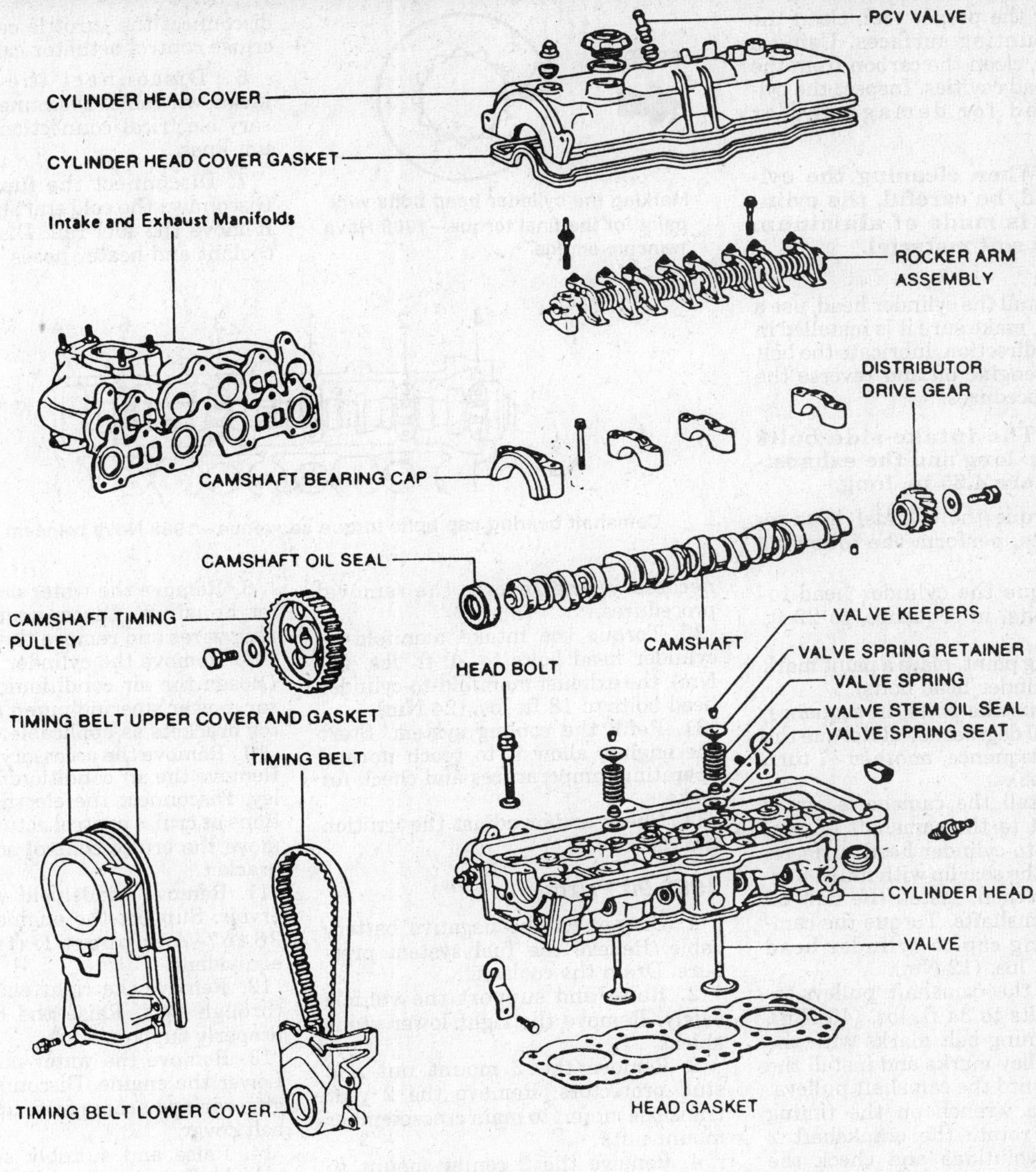

Exploded view of the cylinder head assembly – 1988 Nova overhead cam engine

13. Remove the fuel delivery pipe-to-engine bolts and the delivery pipe with the injectors; do not drop the fuel injectors.
14. Remove the intake manifold bracket, the intake manifold-to-cylinder head bolts (in sequence), the intake manifold and the intake air control valve.
15. If equipped with power steering, remove the drive belt. Remove the alternator drive belt and the cylinder head covers.
16. Remove the water outlet with the No. 1 bypass pipe and drive belt adjusting bar assembly.
17. To position the No. 1 cylinder on the TDC of its compression stroke, perform the following procedures:

a. Remove the spark plugs.

b. Using a socket wrench on the crankshaft pulley, rotate the crankshaft to align the notch in the crankshaft pulley with the idler pulley bolt.

c. The valve lifters of the No. 1 cylinder should be loose; if not, rotate the crankshaft 1 complete revolution.

18. Remove the right side engine mount, the right-side engine mount bracket, then, the upper and middle timing belt covers.
19. Using chalk or paint, place matchmarks on the timing belt and the timing belt pulleys, then, remove the timing belt from the timing belt pulleys.

NOTE: When removing the timing belt, be sure to support it so the meshing with the timing belt pulleys does not change. Do not allow it to come in contact with oil or water.

20. While securing the camshafts, remove each camshaft pulley-to-camshaft bolt, washer and pulley. Remove the inner timing belt cover.
21. Remove the camshaft bearing cap-to-cylinder head bolts, the caps (keep them in order) and the camshafts (keep them in order).
22. Using the cylinder head bolt removal sequence, remove the cylinder head bolts and lift the cylinder head from the engine.

23. Using the proper tool, clean the gasket mounting surfaces. Using a wire brush, clean the carbon from the cylinder head cavities. Inspect the cylinder head for damage and/or warpage.

NOTE: When cleaning the cylinder head, be careful, the cylinder head is made of aluminum which is a soft material.

To install:

24. To install the cylinder head, use a new gasket, make sure it is installed in the correct direction, lubricate the bolt threads in engine oil and reverse the removal procedures.

NOTE: The intake-side bolts are 3.45 in. long and the exhaust-side bolts are 4.25 in. long.

25. To torque the cylinder head-to-engine bolts, perform the following procedures:

a. Torque the cylinder head-to-engine bolts, in sequence, to 22 ft. lbs.

b. Using paint, place a paint mark on the cylinder head bolts.

c. Torque the bolts, in sequence, ¼ turn (90 degrees).d. Retorque the bolts, in sequence, another ¼ turn (90 degrees).

26. To install the camshafts, apply RTV sealant to the camshaft oil seal bearing cap-to-cylinder head surfaces, lightly coat the seal lip with multi-purpose grease, then, install the new oil seals and camshafts. Torque the camshaft bearing cap-to-cylinder head bolts to 9 ft. lbs. (12 Nm).

27. Install the camshaft pulleys-to-camshaft bolts to 34 ft. lbs. (46 Nm). Align the timing belt marks with the camshaft pulley marks and install the timing belt onto the camshaft pulleys.

28. Using a wrench on the timing belt pulley, rotate the crankshaft 2 complete revolutions and check the timing belt alignment points.

29. To complete the installation, use new O-rings, new gaskets, sealant, if necessary, and reverse the removal procedures.

30. Torque the intake manifold-to-cylinder head bolts to 20 ft. lbs. (27 Nm), the exhaust manifold-to-cylinder head bolts to 18 ft. lbs. (24 Nm).

31. Refill the cooling system. Start the engine, allow it to reach normal operating temperatures and check for leaks.

32. Check and/or adjust the ignition timing.

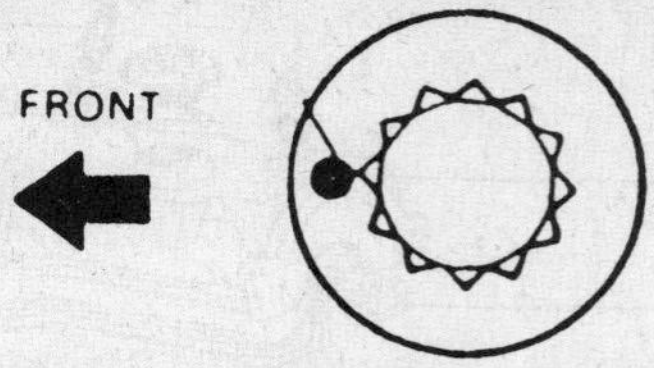

Marking the cylinder head bolts with paint for the final torque – 1988 Nova twincam engine

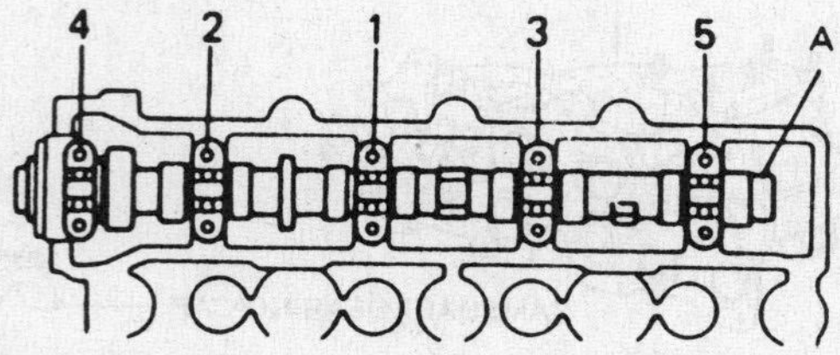

Camshaft bearing cap bolts torque sequence – 1988 Nova twincam engine

1989-90 Prizm

1. Disconnect the negative battery cable. Relieve the fuel system pressure. Drain the coolant.

2. Raise and support the vehicle safely. Remove the right lower stone shield.

3. Remove the 2 mount nut and stud protectors. Remove the 2 rear transaxle mount to main crossmember mount nuts.

4. Remove the 2 center mount to center crossmember nuts. Lower the vehicle.

5. Remove the air cleaner assembly, disconnect the throttle cable and the cruise control actuator cable.

6. Disconnect the transaxle kickdown cable. Disconnect all necessary electrical connections and vacuum lines.

7. Disconnect the fuel inlet line. Disconnect the cold start injector pipe. Remove the fuel rail. Disconnect the coolant and heater hoses.

8. Remove the water outlet and inlet housings. Disconnect the spark plug wires and remove the PCV valve.

9. Remove the cylinder head cover. Loosen the air conditioning compressor, power steering pump and generator brackets as applicable.

10. Remove the accessory drive belts. Remove the air conditioning idler pulley. Disconnect the electrical connections at cruise control actuator and remove the cruise control actuator and bracket.

11. Remove windshield washer reservoir. Support the engine with a J-28467-A support fixture or its equivalent.

12. Remove the right engine mount through bolt. Raise the engine and properly support it.

13. Remove the water pump pulley. Lower the engine. Disconnect the engine wiring harness from upper timing belt cover.

14. Raise and suitably support the vehicle. Remove the cylinder head-to-cylinder block bracket.

15. Remove the exhaust manifold support bracket. Disconnect the exhaust pipe from exhaust manifold.

16. Remove the upper and center timing belt cover. Remove the right engine mount bracket. Remove the distributor.

17. Set the No. 1 cylinder at TDC on its compression stroke. Turn the crankshaft pulley and align its groove with the **0** mark of the timing belt cover.

18. Check that the camshaft gear hole is aligned with the exhaust camshaft cap mark. Remove the plug from the lower timing belt cover.

19. Place alignment marks on the camshaft timing gear and belt.

20. Loosen the idler pulley mount bolt and push the idler pulley toward

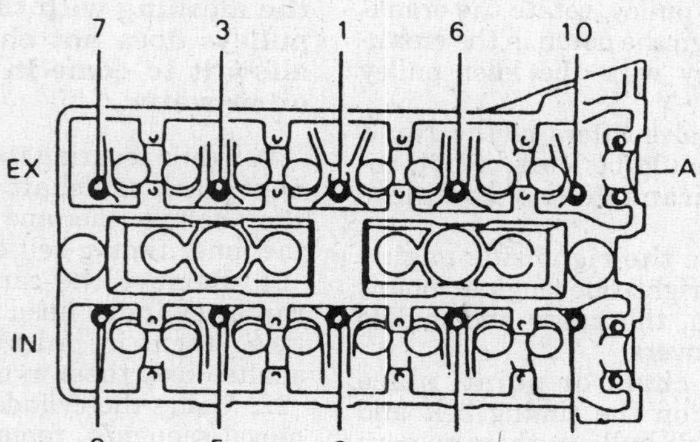

Cylinder head bolt torque sequence – 1988 Nova twincam engine

the left as far as it will go, then tighten temporarily.

21. Remove the timing belt from the camshaft timing gear after marking its position in relation to the camshaft timing gear.

22. Hold the timing belt with a cloth.

NOTE: Support the belt so the meshing of the crankshaft timing gear and timing belt does not shift. Be careful not to drop anything inside the timing belt cover. Do not allow the belt to come in contact with oil, water or dust.

23. Remove the cylinder head bolts, in sequence, using a 10mm, 12 point deep well socket.

NOTE: Head warping or cracking could result from incorrect removal.

24. Remove the cylinder head with intake and exhaust manifolds. If the head is difficult to lift off, carefully pry with a bar between the cylinder head and a cylinder block projection.

NOTE: Be careful not to damage the cylinder head and block mating surface. Lift the cylinder head from the dowels on the cylinder block and place it on wooden blocks on a bench.

25. Remove the intake and exhaust manifolds.

26. Remove as necessary the camshafts, the valve lifters and shims, the spark plug tubes, the valves using a J-8062 spring compressor and a J-37979–A adapter, the valve stem oil seals and the half circle plug.

To install:

27. Install the half circle plug to the cylinder head. Apply GM No. 1052751 or equivalent, sealant to the plug.

28. Install the valves. Install the new oil seals on the valves using a J-38232 seal installer.

NOTE: The intake valve oil seal is brown and the exhaust valve oil seal is black.

29. Install the spring seat, spring, and spring retainer on the cylinder head. Using a J-8062 spring compressor and a J-37979–A adapter, compress the valve springs and place the 2 keepers around valve stem. Remove the J 8062 spring compressor and the J-37979–A adapter.

30. Apply GM No. 1052751 sealant to the spark plug tube hole of the cylinder head and using a press, install a new spark plug tube to a protrusion height of 1.835 – 1.866 in. (46.6 – 47.4mm).

31. Install the engine hangers to the cylinder head. Install the valve lifters and shims. Install the camshafts. Install the intake manifold.

32. Carefully install the cylinder head in position on the cylinder head gasket.

NOTE: Apply a light coating of engine oil on the bolt threads and under the bolt head before installation.

33. Install the 10 cylinder head bolts, in several passes and in sequence. Tighten the cylinder head bolts to 44 ft. lbs. (60 Nm).

34. Install the timing belt. Install the distributor. Install the engine mount bracket. Install the air conditioning idler pulley.

35. Install the center and upper timing belt covers. Raise and suitably support the vehicle. Install the exhaust manifold to exhaust pipe with a new gasket.

36. Install the 2 new exhaust pipe bolts.Tighten the exhaust pipe bolts to 18 ft. lbs. (25 Nm). Install the exhaust manifold support bracket.

37. Install the cylinder head-to-cylinder block bracket. Lower the vehicle.

38. Connect the engine harness to upper timing belt cover. Raise and properly support the engine.

39. Install the water pump pulley. Lower engine. Install the right engine mount through bolt. Tighten to 64 ft. lbs. (87 Nm).

40. Remove the J–28467-A support fixture. Install the windshield washer reservoir.

41. Install the cruise control actuator and bracket and connect the electrical connector.

42. Install the accessory drive belts and adjust to the proper tensions, as applicable.

43. Install the cylinder head cover. Install the PCV valve and spark plug wires. Install the water inlet and outlet housings.

44. Install the heater hoses and all coolant hoses. Install the fuel rail. Install the cold-start injector pipe.

45. Install the fuel inlet line. Connect all necessary electrical connections and vacuum lines.

46. Install the transaxle kickdown cable, connect the cruise control actuator cable and the throttle cable. Install the air cleaner assembly.

47. Raise and safely support the vehicle. Install the 2 rear mount-to-main crossmember nuts. Install the 2 center transaxle mount-to-center crossmember nuts.

48. Tighten the rear transaxle mount-to-main crossmember nuts to 45 ft. lbs. (61 Nm). Tighten the center transaxle mount-to-center crossmember nuts to 45 ft. lbs. (61 Nm).

49. Install both mount nut and stud protectors. Install the right lower stone shield.

50. Lower the vehicle. Refill coolant and install the battery negative cable.

1991-92 Prizm

1. Disconnect the negative battery cable

2. Open the draincocks on the engine and radiator. Collect the coolant in clean containers.

3. Remove the air cleaner assembly.

4. Disconnect the cruise control cable if so equipped.

5. Disconnect the throttle cable from the throttle linkage.

6. Remove the heater hose from the cylinder head rear cover.

7. Label and remove the vacuum hoses from the throttle body.

8. If equipped with cruise control, remove the actuator and bracket assembly.

9. Remove the ignition coil.

10. Remove the upper radiator hose from the cylinder head and the radiator.

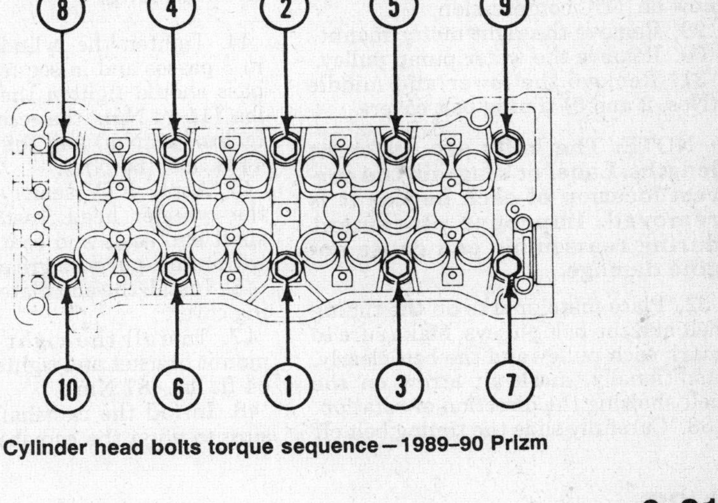

Cylinder head bolts torque sequence—1989–90 Prizm

11. Remove the brake booster vacuum hose.
12. Remove the PCV hose.
13. Unbolt and remove as a unit the fuel pressure regulator.
14. Unbolt and remove the EGR valve with the lines attached.
15. Remove the cold start injector hose.

CAUTION

The fuel system is under pressure. Release pressure slowly and contain spillage. Observe no smoking/no open flame precautions. Have a Class B-C (dry powder) fire extinguisher within arm's reach at all times.

16. Remove the No. 1 fuel line.
17. Remove the first and second water bypass hoses from the auxiliary air valve.
18. Remove the vacuum pipe and the cylinder head rear cover.
19. Remove, disconnect or reposition the wiring harness(es) around the head as necessary.
20. Remove the distributor.
21. Remove the exhaust manifold and its gaskets.
22. Remove the fuel delivery pipe and the injectors. Do not drop the injectors.
23. Remove the intake manifold support bracket; remove the intake manifold and the intake air control valve.
24. Remove the power steering drive belt.
25. Remove the upper timing belt cover and the valve covers.
26. Remove the water outlet fitting with the bypass pipe and the belt adjusting bar.
27. Remove the spark plugs.
28. Turn the crankshaft clockwise, stopping so that the groove in the crank pulley aligns with the idler pulley bolt. Additionally, check that the valve lifters on No. 1 cylinder are loose (the cam lobes are not depressing the lifters). If the valves are under tension, rotate the crank one full revolution and check again. The engine is now on TDC/compression.
29. Remove the right motor mount.
30. Remove the water pump pulley.
31. Remove the lower and middle (Nos. 2 and 3) timing belt covers.

NOTE: The bolts are different lengths. Label or diagram the correct location of each bolt as it is removed. Improper placement during reassembly can cause engine damage.

32. Place matchmarks on the timing belt and the belt pulleys. Make sure to mark each pulley and the belt clearly. Additionally, mark an arrow on the belt showing the direction of rotation.
33. Carefully slide the timing belt off the camshaft pulleys. Do not pry on the belt with tools. Keep the belt under light upward tension so that the bottom (crankshaft) end doesn't shift position on its pulley.
34. Remove the camshaft pulleys. Use an adjustable wrench to counterhold the cams during removal. Look for the flats on the cam and fit the wrench to them.
35. With the pulleys removed, the end plate (otherwise called No. 4 timing cover) may be removed.
36. Loosen the head bolts in the proper sequence. Make 3 complete passes, loosening them slowly, evenly and in order.
37. Remove the cylinder head. If it is difficult to remove, it may be pried up gently with a suitable tool. Be very careful not to scratch or gouge the mating surfaces when prying the head up.
38. Keeping the head upright, place it on wooden blocks on the workbench. If the head is to receive further work, the various components will need to be removed. If the head is not to be worked on, the mating surface must be cleaned of all gasket and sealant material before reinstallation.
39. Clean the engine block mating surface of all gasket and sealant material. Use plastic or wooden scrapers so as not gouge the metal. Remove all traces of liquids from the surface and clean out the bolt holes.
40. Install the new head gasket on the block. Make sure it is properly placed and that all the holes and passages in the block line up with the holes in the gasket.
41. Place the head in position and make sure it is properly seated and aligned.
42. Apply a light coat of oil to the threads of the cylinder head bolts.
43. Install the 10 cylinder head bolts.

NOTE: The bolts for the exhaust side are 108mm long; the bolts for the intake side are 87.5mm long.

44. Tighten the cylinder head bolts in 3 passes and in sequence. The first pass should tighten them to 8-10 ft. lbs. (11-14 Nm), the second pass to 16 ft. lbs. (22 Nm) and the third pass to 44 ft. lbs. (60 Nm).
45. Apply RTV sealer or similar to the cylinder head. Install new camshaft end seals and coat them lightly with multi-purpose grease.
46. Install the end plate or No. 4 timing cover.
47. Install the right side engine mount bracket and tighten its bolts to 64 ft. lbs. (87 Nm).
48. Install the camshaft pulleys. Be sure to align the camshaft knock pin and the camshaft pulley. Tighten the pulley bolts to 34 ft. lbs. (47 Nm).
49. Install the lower and middle (No. 2 and 3) timing belt covers. Remember that the bolts are different lengths; make certain the correct bolt is in the correct location.
50. Install the water pump pulley.
51. Install the right engine mount. Tighten the through bolt to 64 ft. lbs. (87 Nm).
52. Install the spark plugs. Correct tightness is 13 ft. lbs. (18 Nm).
53. Reinstall the water outlet with the bypass pipe and the belt adjusting bar.
54. Install the valve covers.
55. Install the alternator and power steering drive belts. Adjust the belts to the correct tension.
56. Install the intake manifold and intake air control valve. Tighten the bolts to 20 ft. lbs. (27 Nm).
57. Install the bracket and support for the intake manifold.
59. Install the fuel delivery pipe and the injectors. Make sure the insulators and spacers have been placed properly. Make sure the injectors rotate smoothly in their seats.
62. Tighten the delivery pipe retaining bolts to 13 ft. lbs. (18 Nm).
63. Install the exhaust manifold tighten its nuts and bolts to 18 ft. lbs. (25 Nm).
64. Reinstall the distributor.
65. Attach, reposition or connect the wiring harness(es) around the head. Make sure that all retaining clips are used and are secure. Double check the wiring to eliminate any contact with moving parts.
66. Install the vacuum pipe and the cylinder head rear cover with a new gasket.
67. Connect the first and second bypass hoses to the auxiliary air valve.
68. Connect the No. 1 fuel line.
69. Use new gaskets and connect the cold start injector line. Tighten the bolts to 13 ft. lbs. (18 Nm).
70. Replace the EGR valve and use a new gasket.
71. Use a new o-ring and attach the fuel pressure regulator. Tighten the regulator bolts to 6.8 ft. lbs (82 in. lbs.)
72. Install the PCV hose and the brake vacuum hose.
73. Install the radiator hose at the radiator and the cylinder head.
74. Install the ignition coil.
75. Install the cruise control actuator and bracket assembly if so equipped.
76. Observing the labels made earlier, connect the vacuum hose to the throttle body.
77. Attach the heater hose to the cylinder head rear cover.
78. Connect the throttle valve and accelerator cables.

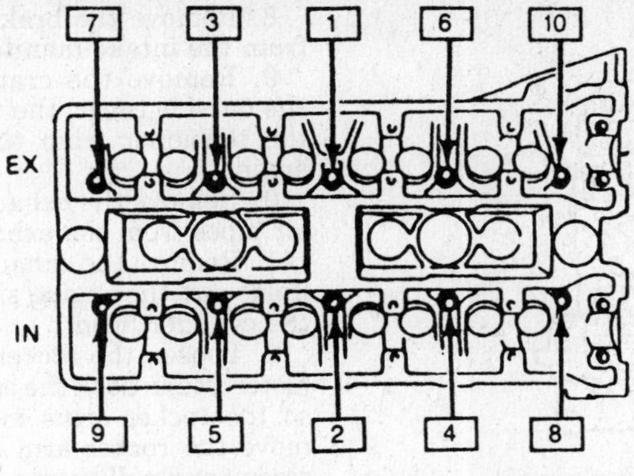

Cylinder head bolts torque sequence – 1991–92 Prizm

79. Connect the cruise control cable, if so equipped.
80. Install the air cleaner assembly.
81. Confirm that the drain cocks on the radiator and engine block are closed. Fill the cooling system with coolant.
82. Double check all installation items, paying particular attention to loose hoses or hanging wires, untightened nuts, poor routing of hoses and wires (too tight or rubbing) and tools left in the engine area.
83. Connect the negative battery cable.
84. Start the engine. During the warm up period, check for any sign of leakage or overheating. Check engine timing and adjust the idle speed if necessary.
85. After the engine is shut off, check the drive belts and adjust the tension if necessary.

Storm

1. Disconnect the battery negative cable.
2. Drain cooling system.
3. Disconnect the accelerator cable from the throttle valve, then the breather hose from the intake air duct.
4. Disconnect the intake air duct from the throttle valve, then the MAP sensor hose from the MAP sensor.
5. Disconnect the brake booster hose, then the 2 canister hoses from the pipes on the intake manifold (common chamber).
6. Disconnect the EGR vacuum hoses, then the oxygen sensor harness electrical connector.
7. Disconnect the ignition coil ground cable from the thermostat housing flange.
8. Disconnect the coolant temperature sensor and thermo unit harness connector from the thermostat housing.
9. Remove the cable harness clip from the coolant outlet pipe bracket.
10. Disconnect the 2 cable harness electrical connectors, located near the left strut tower.
11. Disconnect the engine heater hoses, then the upper radiator hose from the radiator.
12. Disconnect the fuel feed and return hoses, then raise and support vehicle safely.
13. Remove the right undercover, then the front exhaust pipe from the exhaust manifold. Lower the vehicle.
14. Remove the right engine mount, then the alternator drive belt.
15. Remove the power steering belt.
16. Remove the engine mounting bracket from the timing case cover, then remove the timing belt.
17. Remove the cylinder head center cover, cylinder head bolts, then the cylinder head.

To install:

18. Clean the cylinder head gasket mounting surfaces, then install the cylinder head with a new gasket.
19. First torque cylinder head bolts in sequence to 29 ft. lbs. (40 Nm), then a final torque in sequence of 58 ft. lbs. (79 Nm).
20. Install the timing belt and the cylinder head cover, then the engine mounting bracket onto the timing case cover.
21. Raise and safely support the vehicle, then install the right engine mount.
22. Install the front exhaust pipe to the exhaust manifold, then right side under cover. Lower the vehicle.
23. Connect the fuel feed line and fuel return hose.
24. Connect the coolant bypass pipe bracket to the cylinder head.
25. Install the upper radiator hose, then the 2 heater hoses onto the engine.
26. Connect the 2 cable harness connectors, located near the left strut tower.
27. Connect the 2 ground cable terminals, located to the right side of the intake manifold (common chamber).
28. Connect the coolant temperature sensor and thermo unit electrical connectors on the thermostat housing.
29. Connect the ignition coil ground cable to the terminal on the thermostat housing flange.
30. Connect the oxygen sensor electrical connector.
31. Connect the canister hoses to the canister pipes on the intake manifold (common chamber).
32. Connect the brake booster vacuum hose and the MAP sensor vacuum hose.
33. Install the intake air duct to the throttle valve, then the PCV hose to the intake air duct.
34. Connect the accelerator cable to the throttle valve.
35. Connect the battery negative cable and fill cooling system. Start engine and check for leaks.

Spectrum

1. Relieve the fuel system pressure. Disconnect the negative battery terminal from the battery.
2. Drain the cooling system.
3. Remove the air cleaner. Remove the suction pipe and clips for the air induction system.
4. Disconnect the flex hose and oxygen sensor at the exhaust manifold.
5. Disconnect the exhaust pipe bracket at the block and the exhaust pipe at the manifold. On turbocharged

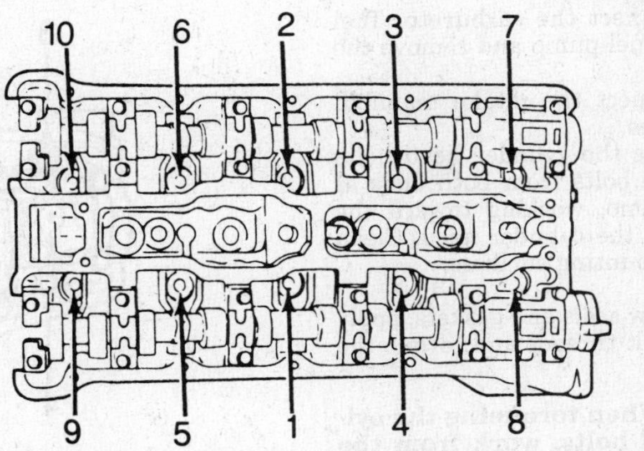

Cylinder head bolt torqueing sequence – Storm with twincam engine

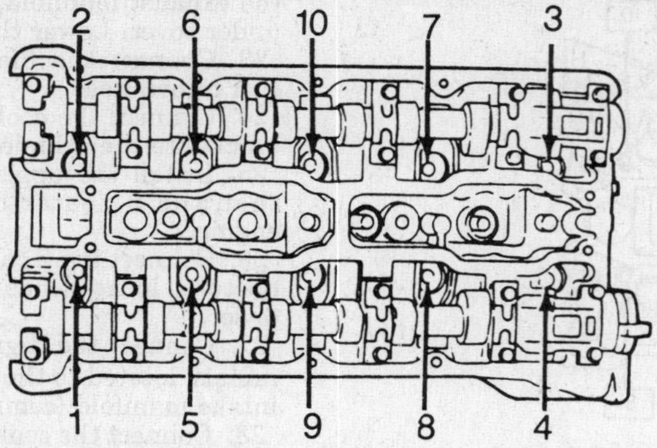

Cylinder head bolt removing sequence – Storm with twincam engine

engine disconnect exhaust pipe at wastegate manifold and disconnect vacuum line for turbocharger control.

6. Disconnect the spark plug wires.
7. Remove the thermostat housing, the distributor, the vacuum advance hoses and the ground cable at the cylinder head.
8. Disconnect the fuel hoses at the fuel pump on non-turbocharged engine.
9. From the carburetor, if equipped, remove the necessary hoses and the throttle cable.
10. Remove engine harness assembly from fuel injectors and fuel line from fuel injector pipe on turbocharged model.
11. Disconnect the vacuum switching valve electrical connector and the heater hoses.
12. Remove the alternator, power steering and air conditioning adjusting bolts, brackets and drive belts.
13. Support the engine using a vertical hoist. Remove the right hand motor mount and the bracket at the front cover.
14. Rotate the engine to align the timing marks, then remove the timing gear cover.
15. Loosen the tension pulley and remove the timing belt from the camshaft timing pulley.
16. Disconnect the carburetor fuel line at the fuel pump and remove the fuel pump.
17. Disconnect the intake manifold coolant hoses.
18. Remove the cylinder head bolts (remove the bolts from both ends at the same time, working toward the middle) and the cylinder head. Clean all of the mounting surfaces.

To install:

20. Use new seals and gaskets, apply oil to the bolt threads and torque the head bolts.

NOTE: When torqueing the cylinder head bolts, work from the middle toward both ends, alternating from one side to the other. First, torque the bolts to 29 ft. lbs. (40 Nm) and then final torque them to 58 ft. lbs. (79 Nm).

21. After torqueing, adjust the valve clearance and complete the installation procedures, by reversing the removal procedures.

Sprint

1. Disconnect the negative battery cable and relieve the fuel system pressure.
2. Drain the cooling system.
3. Remove the air cleaner and the cylinder head cover.
4. Remove the distributor cap, then mark the position of the rotor and the distributor housing with the cylinder head. Remove the distributor and the case from the cylinder head.
5. Remove the accelerator cable from the carburetor. Remove the emission control and the coolant hoses from the carburetor/intake manifold.
6. Remove the electrical lead connectors from the carburetor/intake manifold and the lead wire from the oxygen sensor.
7. Remove the fuel hoses from the fuel pump and the pump from the cylinder head.
8. Remove the brake vacuum hose from the intake manifold.
9. Remove the crankshaft pulley, the outside cover, the timing belt and the tensioner from the front of the engine.
10. Remove the exhaust and the 2nd air pipes from the exhaust manifold.
11. Remove the exhaust/intake manifolds and the engine side mount from the cylinder head.
12. Loosen the rocker arm valve adjusters, turn back the adjusting screws so the rocker arms move freely. Remove the rocker arm shaft retaining screws and pull out the shafts. Remove the rocker arms and springs from the cylinder head.

NOTE: Make a note of the differences between the rocker arm shafts. The intake shaft's stepped end is 0.55 in., which faces the camshaft pulley; the exhaust shaft's stepped end is 0.59 in., which faces the distributor.

13. Remove the mounting bolts and the cylinder head from the engine.

To install:

14. Use new gaskets and reverse the removal procedures. Torque the cylinder head bolts to 46–51 ft. lbs. (63-70 Nm) and the rocker arm shaft screws to 7–9 ft. lbs. (9-12 Nm). Adjust the valve clearances. Refill the cooling system. Check and/or adjust the ignition timing.

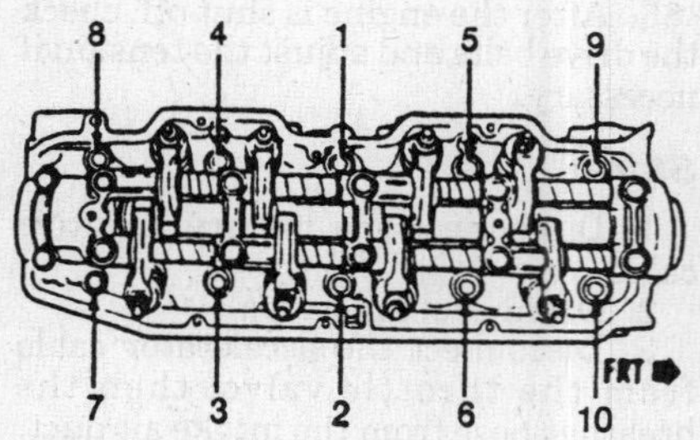

Cylinder head bolt torque sequence – Spectrum

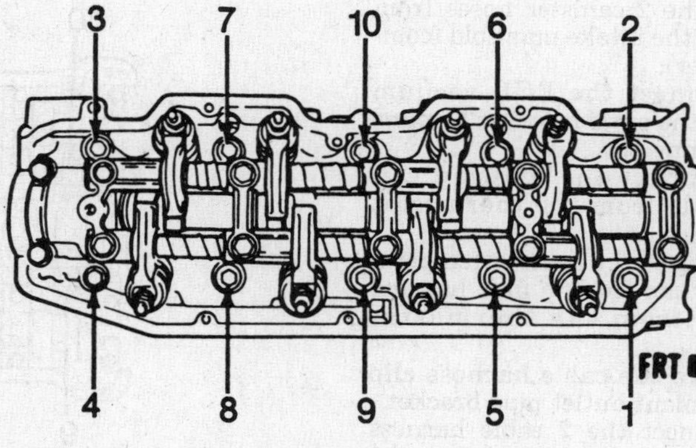

Cylinder head bolt removal sequence – Spectrum

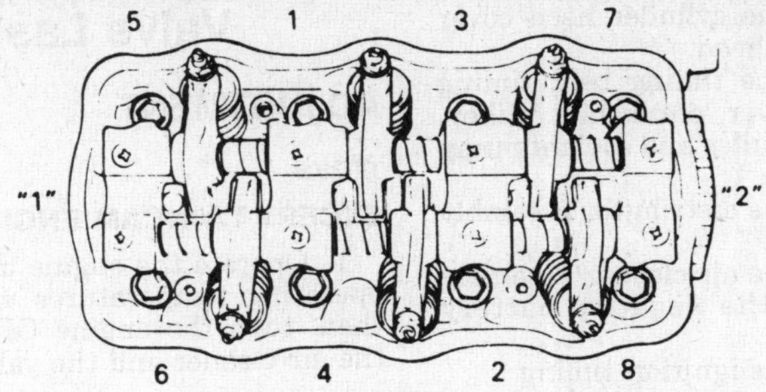

Cylinder head bolts torque sequence – Sprint

Metro

1. Disconnect the negative battery cable and relieve the fuel system pressure.
2. Drain the cooling system.
3. Remove the air cleaner and the cylinder head cover.
4. Remove the distributor cap, then mark the position of the rotor and the distributor housing with the cylinder head. Remove the distributor and the case from the cylinder head.
5. Disconnect and tag all the necessary electrical connectors. Disconnect and tag all necessary vacuum lines.
6. Disconnect the throttle switch or throttle position solenoid.
7. Remove the fuel injectors.
8. Remove the vacuum switching valve for the EGR valve.
9. Disconnect the idle speed control solenoid valve.
10. Remove the ground wires from the intake manifold.
11. Disconnect the oxygen sensor and release the wire harness above from the clamps.
12. Disconnect the heater and coolant hoses.
13. Disconnect the fuel return hose and fuel feed hose from the throttle body.
14. Disconnect the accelerator cable from the throttle body.
15. Raise and safely support the vehicle.
16. Disconnect the No. 1 exhaust pipe from the exhaust manifold.
17. Remove the cylinder head cover.
18. Remove the cylinder head bolts. The cylinder head with throttle body, intake manifold and exhaust manifold. With the camshaft still mounted to the cylinder, 1 of the 6 valves is out toward combustion chamber (it is open). Be sure not to place the cylinder head on any flat surface with its mating surface with the cylinder block facing down. It will cause damage to such an open valve.

To install:

19. Install the new cylinder head gasket with the top mark provided on the gasket on the top side. (toward the cylinder head) and on the crankshaft pulley side.
20. Install the cylinder head. Apply clean engine oil to the threads of the cylinder head bolts. Install the bolts and gradually tighten them with a torque wrench to 54 ft. lbs. (73 Nm).
21. Install the rubber seal between the water pump and the cylinder head.
22. Install the timing belt.
23. Install the distributor assembly.
24. To complete the installation procedure, reverse the removal procedure.
25. Adjust the water pump belt.
26. Adjust the accelerator cable play.
27. Refill the cooling system. Reconnect the negative battery cable.

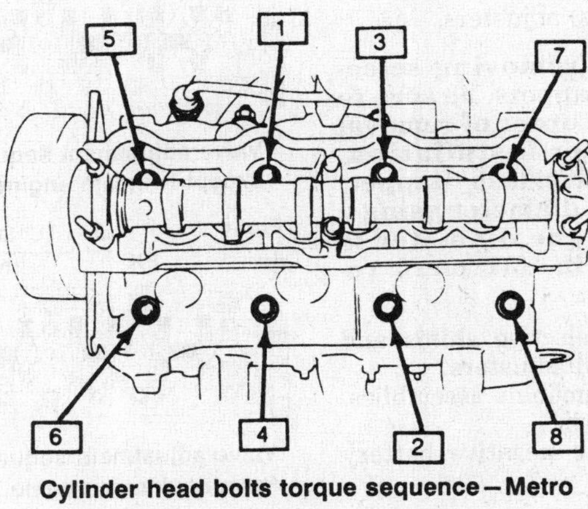

Cylinder head bolts torque sequence – Metro

28. Start the vehicle check for leaks of any kind. Make all repairs and adjustments as necessary.

Valve Lifters

REMOVAL & INSTALLATION

NOTE:The Metro 1.0L engine and the Storm 1.6L twin cam are the only engines using hydraulic valve adjusters. All the other engines that are being used in these vehicles are of the single or dual overhead cam design without valve adjusters.

Valve Adjuster

REMOVAL & INSTALLATION

Metro

1. Disconnect the negative battery cable.
2. Remove the air cleaner assembly.
3. Remove the cylinder head cover assembly.
4. Set the engine up on top dead center of the compression stroke on the No. 1 cylinder. Make an alignment mark on the distributor cap and engine block and remove the distributor assembly.
5. Remove the crankshaft pulley, timing belt outside cover and the timing belt.

NOTE: After removing the timing belt, set the key on the crankshaft in position by turning the crankshaft. This is to prevent interference between the valves and the piston when reinstalling the camshaft.

6. Remove the camshaft timing belt gear. Lock the camshaft with a proper size rod inserted into the hole 0.39 in. (10mm) in it. Loosen the camshaft timing belt gear bolt.

NOTE: The mating surface of the cylinder head and cover must not be damaged in this work. So, put a clean shop cloth between the rod and mating surfaces and use care not to bump the rod against the mating surfaces hard when loosening.

7. Remove the camshaft housing from the cylinder head.
8. Remove the camshaft from the cylinder head.
9. Remove the valve lash adjuster from the cylinder head.

NOTE: Never disassemble the hydraulic valve lash adjuster. Do not apply force to the body of the adjuster, or oil in high pressure

chamber. Immerse the removed adjuster in clean engine oil until it is reinstalled. If the adjuster is left in the air, place it with its valve lash adjuster body facing down. Do not place it on its side or with the valve lash adjuster body facing up.

10. Check the adjuster for pitting, scratches or damage. If any of these conditions are found.

To install:

11. Before installing the valve lash adjuster to the cylinder head, fill the oil passage of the cylinder head with the engine oil according to the following procedure.
   a. Pour the engine oil through the camshaft journal oil holes and check that the oil comes out from the oil holes in the sliding part of the valve lash adjuster.
12. Apply the engine oil around the valve lash adjuster and then install it to the cylinder head.
13. Install the camshaft to cylinder head. After applying engine oil to the camshaft journal and all around the cam, then position the camshaft into the cylinder head so that the camshaft timing belt gear pin hole in camshaft is at the lower position.
14. Install the camshaft housing to the camshaft and the cylinder head.
15. Apply the engine oil to the sliding surface of each housing against the camshaft journal.
16. Apply the sealant to the mating surface of the No. 1 and No. 3 housing which will mate with the cylinder head.
17. There are marks provided on each camshaft housing indicating position and direction for installation. Install the housing as indicated by these marks.
18. As the camshaft housing No. 1 retains the camshaft in the proper position as to the thrust direction, make sure to first fit the No. 1 housing to the No. 1 journal of the camshaft securely.
19. After applying the engine oil to the housing bolts, tighten them temporarily. Then tighten in the proper sequence. Tighten the bolts a little at a time and evenly among bolts, repeat the tightening sequence 3 to 4 times before they are tightened to the proper torque of 8 ft. lbs. (11 Nm).
20. Install the camshaft oil seal. After applying engine oil to the oil seal lip, press-fit the camshaft oil seal until the oil seal surface becomes flush with the housing surfaces.
21. Install the camshaft timing belt gear to the camshaft after installing the dwell pin to the camshaft. While locking the camshaft, install the camshaft pulley and retaining bolt and torque the bolt to 44 ft. lbs. (60 Nm).
22. Install the cylinder head cover to the cylinder head.
23. Install the timing belt, timing belt outside cover, crankshaft pulley, coolant pump pulley and coolant pump belt.
24. Install the distributor assembly into the engine.
25. Install the air cleaner assembly and reinstall the negative battery cable.
26. Adjust the ignition timing.

NOTE: Do not turn the camshaft or start the engine (valves should be operated) for about a half an hour after reinstalling the hydraulic valve lash adjusters and camshaft. As it takes time for valves to settle in place, operating engine within a half an hour after their installation may cause interference to occur between the valves and piston.

27. If air is trapped in the valve lash adjuster, the valve may make tapping sound when engine is operated after valve lash adjuster is installed. In such a case, run the engine for about an half hour at approximately 2000 rpm, and then the air will be purged and the tapping sound cease. Should the tapping sound not cease, it is possible that the valve lash adjuster is defective. among the 6 of them, if a defective adjuster can not be located, check as follows:
   a. Stop the engine and remove the cylinder head cover.
   b. Push the adjuster downward by hand (with less than 33 lbs. of force) when the cam lobe is not on the adjuster to be checked and check if the clearance exists between the cam and the adjuster. If it does, the adjuster is defective and needs to be replaced.

Storm Twin Cam Engine

1. Disconnect the negative battery cable.
2. Remove the camshaft assemblies.
3. Remove the selective shims and valve lash (tappets) adjusters.

NOTE: When removing selective shims and tappets, be sure to arrange them in order of removal to ensure proper installation. Measure the valve lash (tappet) adjuster outside diameter using a micrometer. If the diameter is less than 1.2l8 in. (31 mm), replace the tappet.

4. Install the selective shims and valve lash (tappets) adjusters.
5. Install the camshaft assemblies. Adjust the valve lash.
6. Reconnect the negative battery cables.

Valve Lash

ADJUSTMENT

Nova

EXCEPT TWINCAM ENGINE

1. Operate the engine until normal operating temperatures are reached, then, turn the engine OFF. Remove the air cleaner and the valve cover.

NOTE: If clearances are being set because parts have been disassembled, adjust the valves COLD, then, reset them with the engine HOT.

2. Using a socket wrench on the crankshaft pulley bolt, turn the crankshaft until the No. 1 cylinder is positioned to the TDC of its compression stroke; the rocker arms of the No. 1 cylinder should be loose.

NOTE: The notch on the crankshaft pulley should align with the 0 degrees mark on the timing plate. Make sure the rocker arms on the No. 1 cylinder are loose and the rockers on the No. 4 cylinder are tight. If not, turn the crankshaft one complete revolution and align the marks again.

3. Using a 0.008 in. feeler gauge, adjust the intake valve clearance of cylinder No. 1 and 2. Using a 0.012 in. feeler gauge, adjust the exhaust valve clearance of cylinders No. 1 and 3.
4. To adjust each valve, perform the following procedures:
   a. Loosen the rocker arm adjusting nut; it may be necessary to back-off the adjusting screw.
   b. Slide the feeler gauge between the rocker arm and valve tip. The surfaces will just touch, giving a very slight pull on the gauge.
   c. Using a adjusting tool and a wrench, to hold the rocker arm lock-

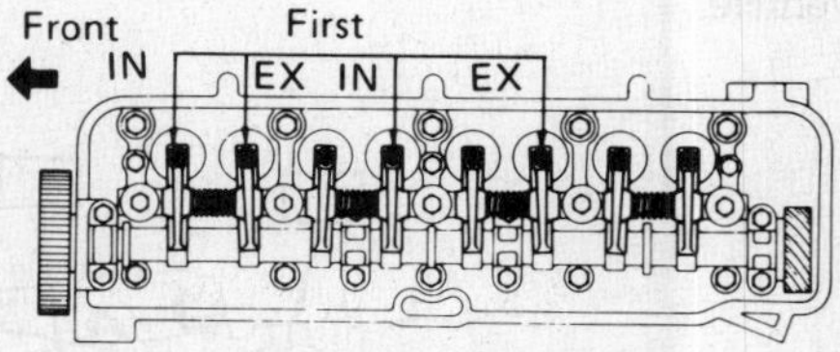

Valve adjustment sequence—step 1 except twincam engine in Nova

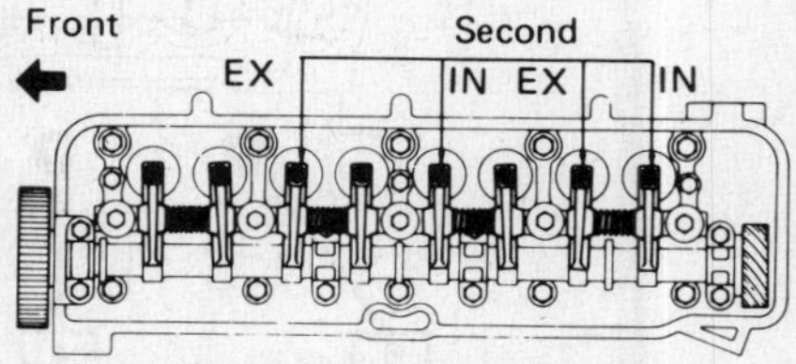

Valve adjustment sequence—step 2 except twincam engine inNova

Measured Clearance (mm)	Installed Shim Thickness (mm)																																																							
	2 500	2 525	2 550	2 575	2 600	2 620	2 625	2 640	2 650	2 660	2 675	2 680	2 700	2 720	2 725	2 740	2 750	2 760	2 775	2 780	2 800	2 820	2 825	2 840	2 850	2 860	2 875	2 880	2 900	2 920	2 925	2 940	2 950	2 960	2 975	2 980	3 000	3 020	3 025	3 040	3 050	3 060	3 075	3 080	3 100	3 120	3 125	3 140	3 150	3 160	3 175	3 180	3 200	3 225	3 250	3 275	3 300
0 000 - 0 009											02	02	02	04	04	04	04	04	06	06	06	08	08	08	08	08	10	10	10	12	12	12	12	12	14	14	14	16	16	16	16	16	18	18	18	20	20	20	20	20	22	22	22	24	24	26	28
0 010 - 0 025										02	02	02	02	04	04	04	04	06	06	06	06	08	08	08	08	10	10	10	10	12	12	12	12	14	14	14	14	16	16	16	16	18	18	18	18	20	20	20	20	22	22	22	22	24	24	26	26
0 026 - 0 029									02	02	02	02	04	04	04	04	06	06	06	06	08	08	08	08	10	10	10	10	12	12	12	12	14	14	14	14	16	16	16	16	18	18	18	18	20	20	20	20	22	22	22	22	24	24	26	26	28
0 030 - 0 040								02	02	02	02	02	04	04	04	04	06	06	06	06	08	08	08	08	10	10	10	10	12	12	12	12	14	14	14	14	16	16	16	16	18	18	18	18	20	20	20	20	22	22	22	22	24	24	26	26	28
0 041 - 0 050								02	02	02	02	02	04	04	04	06	06	06	06	06	08	08	08	10	10	10	10	10	12	12	12	14	14	14	14	14	16	16	16	18	18	18	18	18	20	20	20	22	22	22	22	22	24	24	26	26	28
0 051 - 0 070						02	02	02	02	02	04	04	04	06	06	06	06	06	08	08	08	10	10	10	10	10	12	12	12	14	14	14	14	14	16	16	16	18	18	18	18	18	20	20	20	22	22	22	22	22	24	24	24	26	26	28	28
0 071 - 0 075					02	02	02	02	02	04	04	04	04	06	06	06	06	08	08	08	08	10	10	10	10	12	12	12	12	14	14	14	14	16	16	16	16	18	18	18	18	20	20	20	20	22	22	22	22	24	24	24	24	26	26	28	28
0 076 - 0 090					02	02	02	02	04	04	04	04	06	06	06	06	08	08	08	08	10	10	10	10	12	12	12	12	14	14	14	14	16	16	16	16	18	18	18	18	20	20	20	20	22	22	22	22	24	24	24	24	26	26	28	28	30
0 091 - 0 100					02	02	02	04	04	04	04	04	06	06	06	08	08	08	08	08	10	10	10	12	12	12	12	12	14	14	14	16	16	16	16	16	18	18	18	20	20	20	20	20	22	22	22	24	24	24	24	24	26	26	28	28	30
0 101 - 0 120				02	02	04	04	04	04	04	06	06	06	08	08	08	08	08	10	10	10	12	12	12	12	12	14	14	14	16	16	16	16	16	18	18	18	20	20	20	20	20	22	22	22	24	24	24	24	24	26	26	26	28	28	30	30
0 121 - 0 125			02	02	02	04	04	04	04	06	06	06	06	08	08	08	08	10	10	10	10	12	12	12	12	14	14	14	14	16	16	16	16	18	18	18	18	20	20	20	20	22	22	22	22	24	24	24	24	26	26	26	26	28	28	30	30
0 126 - 0 140			02	02	04	04	04	04	06	06	06	06	08	08	08	08	10	10	10	10	12	12	12	12	14	14	14	14	16	16	16	16	18	18	18	18	20	20	20	20	22	22	22	22	24	24	24	24	26	26	26	26	28	28	30	30	32
0 141 - 0 149			02	02	04	04	04	06	06	06	06	06	08	08	08	10	10	10	10	10	12	12	12	14	14	14	14	14	16	16	16	18	18	18	18	18	20	20	20	22	22	22	22	22	24	24	24	26	26	26	26	26	28	28	30	30	32
0 150 - 0 250																																																									
0 251 - 0 270	04	06	06	08	08	10	10	10	10	10	12	12	12	14	14	14	14	14	16	16	16	18	18	18	18	18	20	20	20	22	22	22	22	22	24	24	24	26	26	26	26	26	28	28	28	30	30	30	30	30	32	32	32	34	34		
0 271 - 0 275	04	06	06	08	08	10	10	10	10	12	12	12	12	14	14	14	14	16	16	16	16	18	18	18	18	20	20	20	20	22	22	22	22	24	24	24	24	26	26	26	26	28	28	28	28	30	30	30	30	32	32	32	32	34	34		
0 276 - 0 290	06	06	08	08	10	10	10	10	12	12	12	12	14	14	14	14	16	16	16	16	18	18	18	18	20	20	20	20	22	22	22	22	24	24	24	24	26	26	26	26	28	28	28	28	30	30	30	30	32	32	32	32	34	34			
0 291 - 0 300	06	06	08	08	10	10	10	12	12	12	12	12	14	14	14	16	16	16	16	16	18	18	18	20	20	20	20	20	22	22	22	24	24	24	24	24	26	26	26	28	28	28	28	28	30	30	30	32	32	32	32	32	34	34			
0 301 - 0 320	06	08	08	10	10	12	12	12	12	12	14	14	14	16	16	16	16	16	18	18	18	20	20	20	20	20	22	22	22	24	24	24	24	24	26	26	26	28	28	28	28	28	30	30	30	32	32	32	32	32	34	34	34				
0 321 - 0 325	06	08	08	10	10	12	12	12	12	14	14	14	14	16	16	16	16	18	18	18	18	20	20	20	20	22	22	22	22	24	24	24	24	26	26	26	26	28	28	28	28	30	30	30	30	32	32	32	32	34	34	34	34				
0 326 - 0 340	08	08	10	10	12	12	12	12	14	14	14	14	16	16	16	16	18	18	18	18	20	20	20	20	22	22	22	22	24	24	24	24	26	26	26	26	28	28	28	28	30	30	30	30	32	32	32	32	34	34	34	34					
0 341 - 0 350	08	08	10	10	12	12	12	14	14	14	14	14	16	16	16	18	18	18	18	18	20	20	20	22	22	22	22	22	24	24	24	26	26	26	26	26	28	28	28	30	30	30	30	30	32	32	32	34	34	34	34						
0 351 - 0 370	08	10	10	12	12	14	14	14	14	14	16	16	16	18	18	18	18	18	20	20	20	22	22	22	22	22	24	24	24	26	26	26	26	26	28	28	28	30	30	30	30	30	32	32	32	34	34	34	34	34							
0 371 - 0 375	08	10	10	12	12	14	14	14	14	16	16	16	16	18	18	18	18	20	20	20	20	22	22	22	22	24	24	24	24	26	26	26	26	28	28	28	28	30	30	30	30	32	32	32	32	34	34	34	34								
0 376 - 0 390	10	10	12	12	14	14	14	14	16	16	16	16	18	18	18	18	20	20	20	20	22	22	22	22	24	24	24	24	26	26	26	26	28	28	28	28	30	30	30	30	32	32	32	32	34	34	34	34									
0 391 - 0 400	10	10	12	12	14	14	14	16	16	16	16	16	18	18	18	20	20	20	20	20	22	22	22	24	24	24	24	24	26	26	26	28	28	28	28	28	30	30	30	32	32	32	32	32	34	34	34										
0 401 - 0 420	10	12	12	14	14	16	16	16	16	16	18	18	18	20	20	20	20	20	22	22	22	24	24	24	24	24	26	26	26	28	28	28	28	28	30	30	30	32	32	32	32	32	34	34	34												
0 421 - 0 425	10	12	12	14	14	16	16	16	16	18	18	18	18	20	20	20	20	22	22	22	22	24	24	24	24	26	26	26	26	28	28	28	28	30	30	30	30	32	32	32	32	34	34	34	34												
0 426 - 0 440	12	12	14	14	16	16	16	16	18	18	18	18	20	20	20	20	22	22	22	22	24	24	24	24	26	26	26	26	28	28	28	28	30	30	30	30	32	32	32	32	34	34	34	34													
0 441 - 0 450	12	12	14	14	16	16	16	18	18	18	18	18	20	20	20	22	22	22	22	22	24	24	24	26	26	26	26	26	28	28	28	30	30	30	30	30	32	32	32	34	34	34	34	34													
0 451 - 0 470	12	14	14	16	16	18	18	18	18	18	20	20	20	22	22	22	22	22	24	24	24	26	26	26	26	26	28	28	28	30	30	30	30	30	32	32	32	34	34	34	34	34															
0 471 - 0 475	12	14	14	16	16	18	18	18	18	20	20	20	20	22	22	22	22	24	24	24	24	26	26	26	26	28	28	28	28	30	30	30	30	32	32	32	32	34	34	34	34																
0 476 - 0 490	14	14	16	16	18	18	18	18	20	20	20	20	22	22	22	22	24	24	24	24	26	26	26	26	28	28	28	28	30	30	30	30	32	32	32	32	34	34	34	34																	
0 491 - 0 500	14	14	16	16	18	18	18	20	20	20	20	20	22	22	22	24	24	24	24	24	26	26	26	28	28	28	28	28	30	30	30	32	32	32	32	32	34	34	34																		
0 501 - 0 520	14	16	16	18	18	20	20	20	20	20	22	22	22	24	24	24	24	24	26	26	26	28	28	28	28	28	30	30	30	32	32	32	32	32	34	34	34																				
0 521 - 0 525	14	16	16	18	18	20	20	20	20	22	22	22	22	24	24	24	24	26	26	26	26	28	28	28	28	30	30	30	30	32	32	32	32	34	34	34	34																				
0 526 - 0 540	16	16	18	18	20	20	20	20	22	22	22	22	24	24	24	24	26	26	26	26	28	28	28	28	30	30	30	30	32	32	32	32	34	34	34	34																					
0 541 - 0 550	16	16	18	18	20	20	20	22	22	22	22	22	24	24	24	26	26	26	26	26	28	28	28	30	30	30	30	30	32	32	32	34	34	34	34	34																					
0 551 - 0 570	16	18	18	20	20	22	22	22	22	22	24	24	24	26	26	26	26	26	28	28	28	30	30	30	30	30	32	32	32	34	34	34	34	34																							
0 571 - 0 575	16	18	18	20	20	22	22	22	22	24	24	24	24	26	26	26	26	28	28	28	28	30	30	30	30	32	32	32	32	34	34	34	34																								
0 576 - 0 530	18	18	20	20	22	22	22	22	24	24	24	24	26	26	26	26	28	28	28	28	30	30	30	30	32	32	32	32	34	34	34	34																									
0 591 - 0 600	18	18	20	20	22	22	22	24	24	24	24	24	26	26	26	28	28	28	28	28	30	30	30	32	32	32	32	32	34	34	34																										
0 601 - 0 620	18	20	20	22	22	24	24	24	24	24	26	26	26	28	28	28	28	28	30	30	30	32	32	32	32	32	34	34	34																												
0 621 - 0 625	18	20	20	22	22	24	24	24	24	26	26	26	26	28	28	28	28	30	30	30	30	32	32	32	32	34	34	34	34																												
0 626 - 0 640	20	20	22	22	24	24	24	24	26	26	26	26	28	28	28	28	30	30	30	30	32	32	32	32	34	34	34	34																													
0 641 - 0 650	20	20	22	22	24	24	24	26	26	26	26	26	28	28	28	30	30	30	30	30	32	32	32	34	34	34	34	34																													
0 651 - 0 670	20	22	22	24	24	26	26	26	26	26	28	28	28	30	30	30	30	30	32	32	32	34	34	34	34	34																															
0 671 - 0 675	20	22	22	24	24	26	26	26	26	28	28	28	28	30	30	30	30	32	32	32	32	34	34	34	34																																
0 676 - 0 690	22	22	24	24	26	26	26	26	28	28	28	28	30	30	30	30	32	32	32	32	34	34	34	34																																	
0 691 - 0 700	22	22	24	24	26	26	26	28	28	28	28	28	30	30	30	32	32	32	32	32	34	34	34																																		
0 701 - 0 720	22	24	24	26	26	28	28	28	28	28	30	30	30	32	32	32	32	32	34	34	34																																				
0 721 - 0 725	22	24	24	26	26	28	28	28	28	30	30	30	30	32	32	32	32	34	34	34	34																																				
0 726 - 0 740	24	24	26	26	28	28	28	28	30	30	30	30	32	32	32	32	34	34	34	34																																					
0 741 - 0 750	24	24	26	26	28	28	28	30	30	30	30	30	32	32	32	34	34	34	34																																						
0 751 - 0 770	24	26	26	28	28	30	30	30	30	30	32	32	32	34	34	34	34	34																																							
0 771 - 0 775	24	26	26	28	28	30	30	30	30	32	32	32	32	34	34	34	34																																								
0 776 - 0 790	26	26	28	28	30	30	30	30	32	32	32	32	34	34	34	34																																									
0 791 - 0 800	26	26	28	28	30	30	30	32	32	32	32	32	34	34	34																																										
0 801 - 0 820	26	28	28	30	30	32	32	32	32	32	34	34	34																																												
0 821 - 0 825	26	28	28	30	30	32	32	32	32	34	34	34	34																																												
0 826 - 0 840	28	28	30	30	32	32	32	32	34	34	34	34																																													
0 841 - 0 850	28	28	30	30	32	32	32	34	34	34	34	34																																													
0 851 - 0 870	28	30	30	32	32	34	34	34	34	34																																															
0 871 - 0 875	28	30	30	32	32	34	34	34	34																																																
0 876 - 0 890	30	30	32	32	34	34	34	34																																																	
0 891 - 0 900	30	30	32	32	34	34	34																																																		
0 901 - 0 925	30	32	32	34	34																																																				
0 926 - 0 950	32	32	34	34																																																					
0 951 - 0 975	32	34	34																																																						
0 976 - 1 000	34	34																																																							
1 001 - 1 025	34																																																								

AVAILABLE SHIMS

Shim No.	Thickness	Shim No.	Thickness
02	2.500 (0.0984)	20	2.950 (0.1161)
04	2.550 (0.1004)	22	3.000 (0.1181)
06	2.600 (0.1024)	24	3.050 (0.1201)
08	2.650 (0.1043)	26	3.100 (0.1220)
10	2.700 (0.1063)	28	3.150 (0.1240)
12	2.750 (0.1083)	30	3.200 (0.1260)
14	2.800 (0.1102)	32	3.250 (0.1280)
16	2.850 (0.1122)	34	3.300 (0.1299)
18	2.900 (0.1142)		

Intake valve clearance (cold):
0.15 – 0.25 mm (0.006 – 0.010 in.)

Example: **A 2.800 mm shim is installed and the measured clearance is 0.450 mm. Replace the 2.800 mm shim with shim No. 24 (3.050 mm).**

Intake valve shim size chart—Nova twincam engine and Prizm

Measured Clearance (mm)	Installed Shim Thickness (mm)																											
	2.500	2.525	2.550	2.575	2.600	2.620	2.625	2.640	2.650	2.660	2.675	2.680	2.700	2.720	2.725	2.740	2.750	2.760	2.775	2.780	2.800	2.820	2.825	2.840	2.850	2.860	2.875	2.880
0.000 - 0.009														02	02	02	02	02	04	04	04	06	06	06	06	06	08	08
0.010 - 0.025														02	02	02	02	04	04	04	06	06	06	06	06	08	08	08
0.026 - 0.040													02	02	02	02	04	04	04	04	06	06	06	06	08	08	08	08
0.041 - 0.050												02	02	02	02	04	04	04	04	06	06	06	06	08	08	08	08	10
0.051 - 0.070											02	02	02	04	04	04	04	04	06	06	06	08	08	08	08	08	10	10
0.071 - 0.090									02	02	02	02	04	04	04	04	06	06	06	06	08	08	08	08	10	10	10	10
0.091 - 0.100								02	02	02	02	04	04	04	04	06	06	06	06	08	08	08	08	10	10	10	10	12
0.101 - 0.120						02	02	02	02	02	04	04	04	06	06	06	06	06	08	08	08	10	10	10	10	10	12	12
0.121 - 0.140					02	02	02	02	04	04	04	04	06	06	06	06	08	08	08	08	10	10	10	10	12	12	12	12
0.141 - 0.150					02	02	02	04	04	04	04	06	06	06	06	08	08	08	08	10	10	10	10	12	12	12	12	14
0.151 - 0.170				02	02	04	04	04	04	04	06	06	06	08	08	08	08	08	10	10	10	12	12	12	12	12	14	14
0.171 - 0.190			02	02	04	04	04	04	06	06	06	06	08	08	08	08	10	10	10	10	12	12	12	12	14	14	14	14
0.191 - 0.199			02	02	04	04	04	06	06	06	06	08	08	08	08	10	10	10	10	12	12	12	12	14	14	14	14	16
0.200 - 0.300																												
0.301 - 0.320	04	06	06	08	08	10	10	10	10	10	12	12	12	14	14	14	14	14	16	16	16	18	18	18	18	18	20	20
0.321 - 0.325	04	06	06	08	08	10	10	10	10	12	12	12	12	14	14	14	14	16	16	16	16	18	18	18	18	20	20	20
0.326 - 0.340	06	06	08	08	10	10	10	10	12	12	12	12	14	14	14	14	16	16	16	16	18	18	18	18	20	20	20	20
0.341 - 0.350	06	06	08	08	10	10	10	12	12	12	12	12	14	14	14	16	16	16	16	16	18	18	18	20	20	20	20	20
0.351 - 0.370	06	08	08	10	10	12	12	12	12	12	14	14	14	16	16	16	16	16	18	18	18	20	20	20	20	20	22	22
0.371 - 0.375	06	08	08	10	10	12	12	12	12	14	14	14	14	16	16	16	16	18	18	18	18	20	20	20	20	22	22	22
0.376 - 0.390	08	08	10	10	12	12	12	12	14	14	14	14	16	16	16	16	18	18	18	18	20	20	20	20	22	22	22	22
0.391 - 0.400	08	08	10	10	12	12	12	14	14	14	14	14	16	16	16	18	18	18	18	18	20	20	20	22	22	22	22	22
0.401 - 0.420	08	10	10	12	12	14	14	14	14	14	16	16	16	18	18	18	18	18	20	20	20	22	22	22	22	22	24	24
0.421 - 0.425	08	10	10	12	12	14	14	14	14	16	16	16	16	18	18	18	18	20	20	20	20	22	22	22	22	24	24	24
0.426 - 0.440	10	10	12	12	14	14	14	14	16	16	16	16	18	18	18	18	20	20	20	20	22	22	22	22	24	24	24	24
0.441 - 0.450	10	10	12	12	14	14	14	16	16	16	16	16	18	18	18	20	20	20	20	20	22	22	22	24	24	24	24	24
0.451 - 0.470	10	12	12	14	14	16	16	16	16	16	18	18	18	20	20	20	20	20	22	22	22	24	24	24	24	24	26	26
0.471 - 0.475	10	12	12	14	14	16	16	16	16	18	18	18	18	20	20	20	20	22	22	22	22	24	24	24	24	26	26	26
0.476 - 0.490	12	12	14	14	16	16	16	16	18	18	18	18	20	20	20	20	22	22	22	22	24	24	24	24	26	26	26	26
0.491 - 0.500	12	12	14	14	16	16	16	18	18	18	18	18	20	20	20	22	22	22	22	22	24	24	24	26	26	26	25	26
0.501 - 0.520	12	14	14	16	16	18	18	18	18	18	20	20	20	22	22	22	22	22	24	24	24	26	26	26	26	26	28	28
0.521 - 0.525	12	14	14	16	16	18	18	18	18	20	20	20	20	22	22	22	22	24	24	24	24	26	26	26	26	28	28	28
0.526 - 0.540	14	14	16	16	18	18	18	18	20	20	20	20	22	22	22	22	24	24	24	24	26	26	26	26	28	28	28	28
0.541 - 0.550	14	14	16	16	18	18	18	20	20	20	20	20	22	22	22	24	24	24	24	24	26	26	26	28	28	28	28	28
0.551 - 0.570	14	16	16	18	18	20	20	20	20	20	22	22	22	24	24	24	24	24	26	26	26	28	28	28	28	28	30	30
0.571 - 0.575	14	16	16	18	18	20	20	20	20	22	22	22	22	24	24	24	24	26	26	26	26	28	28	28	28	30	30	30
0.576 - 0.590	16	16	18	18	20	20	20	20	22	22	22	22	24	24	24	24	26	26	26	26	28	28	28	28	30	30	30	30
0.591 - 0.600	16	16	18	18	20	20	20	22	22	22	22	22	24	24	24	26	26	26	26	26	28	28	28	30	30	30	30	30
0.601 - 0.620	16	18	18	20	20	22	22	22	22	22	24	24	24	26	26	26	26	26	28	28	28	30	30	30	30	30	32	32
0.621 - 0.625	16	18	18	20	20	22	22	22	22	24	24	24	24	26	26	26	26	28	28	28	28	30	30	30	30	32	32	32
0.626 - 0.640	18	18	20	20	22	22	22	22	24	24	24	24	26	26	26	26	28	28	28	28	30	30	30	30	32	32	32	32
0.641 - 0.650	18	18	20	20	22	22	22	24	24	24	24	24	26	26	26	28	28	28	28	28	30	30	30	32	32	32	32	32
0.651 - 0.670	18	20	20	22	22	24	24	24	24	24	26	26	26	28	28	28	28	28	30	30	30	32	32	32	32	32	34	34
0.671 - 0.675	18	20	20	22	22	24	24	24	24	26	26	26	26	28	28	28	28	30	30	30	30	32	32	32	32	34	34	34
0.676 - 0.690	20	20	22	22	24	24	24	24	26	26	26	26	28	28	28	28	30	30	30	30	32	32	32	32	34	34	34	34
0.691 - 0.700	20	20	22	22	24	24	24	26	26	26	26	26	28	28	28	30	30	30	30	30	32	32	32	34	34	34	34	34
0.701 - 0.720	20	22	22	24	24	26	26	26	26	26	28	28	28	30	30	30	30	30	32	32	32	34	34	34	34	34		
0.721 - 0.725	20	22	22	24	24	26	26	26	26	28	28	28	28	30	30	30	30	32	32	32	32	34	34	34	34			
0.726 - 0.740	22	22	24	24	26	26	26	26	28	28	28	28	30	30	30	30	32	32	32	32	34	34	34	34				
0.741 - 0.750	22	22	24	24	26	26	26	28	28	28	28	28	30	30	30	32	32	32	32	32	34	34	34					
0.751 - 0.770	22	24	24	26	26	28	28	28	28	28	30	30	30	32	32	32	32	32	34	34	34							
0.771 - 0.775	22	24	24	26	26	28	28	28	28	30	30	30	30	32	32	32	32	34	34	34	34							
0.776 - 0.790	24	24	26	26	28	28	28	28	30	30	30	30	32	32	32	32	34	34	34	34								
0.791 - 0.800	24	24	26	26	28	28	28	30	30	30	30	30	32	32	32	34	34	34	34	34								
0.801 - 0.820	24	26	26	28	28	30	30	30	30	30	32	32	32	34	34	34	34	34										
0.821 - 0.825	24	26	26	28	28	30	30	30	30	32	32	32	32	34	34	34	34											
0.826 - 0.840	26	26	28	28	30	30	30	30	32	32	32	32	34	34	34	34												
0.841 - 0.850	26	26	28	28	30	30	30	32	32	32	32	32	34	34	34													
0.851 - 0.870	26	28	28	30	30	32	32	32	32	32	34	34	34															
0.871 - 0.875	26	28	28	30	30	32	32	32	32	34	34	34	34															
0.876 - 0.890	28	28	30	30	32	32	32	32	34	34	34	34																
0.891 - 0.900	28	28	30	30	32	32	32	34	34	34	34	34																
0.901 - 0.925	28	30	30	32	32	34	34	34	34																			
0.926 - 0.950	30	30	32	32	34	34	34																					
0.951 - 0.975	30	32	32	34	34																							
0.976 - 1.000	32	32	34	34																								
1.001 - 1.025	32	34	34																									
1.026 - 1.050	34	34																										
1.051 - 1.075	34																											

Measured Clearance (mm)	Installed Shim Thickness (mm)																												
	2.900	2.920	2.925	2.940	2.950	2.960	2.975	2.980	3.000	3.020	3.025	3.040	3.050	3.060	3.075	3.080	3.100	3.120	3.125	3.140	3.150	3.160	3.175	3.180	3.200	3.225	3.250	3.275	3.300
0.000 - 0.009	08	10	10	10	10	10	12	12	12	14	14	14	14	14	16	16	16	18	18	18	18	18	20	20	20	22	22	24	24
0.010 - 0.025	10	10	10	10	10	12	12	12	14	14	14	14	14	16	16	16	18	18	18	18	18	20	20	20	22	22	22	24	26
0.026 - 0.040	10	10	10	10	12	12	12	12	14	14	14	14	16	16	16	16	18	18	18	18	20	20	20	20	22	22	24	24	26
0.041 - 0.050	10	10	10	12	12	12	12	14	14	14	14	16	16	16	16	18	18	18	18	20	20	20	20	22	22	22	24	24	26
0.051 - 0.070	10	12	12	12	12	12	14	14	14	16	16	16	16	16	18	18	18	20	20	20	20	20	22	22	22	24	24	26	26
0.071 - 0.090	12	12	12	12	14	14	14	14	16	16	16	16	18	18	18	18	20	20	20	20	22	22	22	22	24	24	26	26	28
0.091 - 0.100	12	12	12	14	14	14	14	16	16	16	16	18	18	18	18	20	20	20	20	22	22	22	22	24	24	24	26	26	28
0.101 - 0.120	12	14	14	14	14	14	16	16	16	18	18	18	18	18	20	20	20	22	22	22	22	22	24	24	24	26	26	28	28
0.121 - 0.140	14	14	14	14	16	16	16	16	18	18	18	18	20	20	20	20	22	22	22	22	24	24	24	24	26	26	28	28	30
0.141 - 0.150	14	14	14	16	16	16	16	18	18	18	18	20	20	20	20	22	22	22	22	24	24	24	24	26	26	26	28	28	30
0.151 - 0.170	14	16	16	16	16	16	18	18	18	20	20	20	20	20	22	22	22	24	24	24	24	24	26	26	26	28	28	30	30
0.171 - 0.190	16	16	16	16	18	18	18	18	20	20	20	20	22	22	22	22	24	24	24	24	26	26	26	26	28	28	30	30	32
0.191 - 0.199	16	16	16	18	18	18	18	20	20	20	20	22	22	22	22	24	24	24	24	26	26	26	26	28	28	28	30	30	32
0.200 - 0.300																													
0.301 - 0.320	20	22	22	22	22	22	24	24	24	26	26	26	26	26	28	28	28	30	30	30	30	30	32	32	32	34	34		
0.321 - 0.325	20	22	22	22	22	24	24	24	24	26	26	26	26	28	28	28	28	30	30	30	30	32	32	32	32	34	34		
0.326 - 0.340	22	22	22	22	24	24	24	24	26	26	26	26	28	28	28	28	30	30	30	30	32	32	32	32	34	34			
0.341 - 0.350	22	22	22	24	24	24	24	24	26	26	26	28	28	28	28	28	30	30	30	32	32	32	32	32	34	34			
0.351 - 0.370	22	24	24	24	24	24	26	26	26	28	28	28	28	28	30	30	30	32	32	32	32	32	34	34	34				
0.371 - 0.375	22	24	24	24	24	26	26	26	26	28	28	28	28	30	30	30	30	32	32	32	32	34	34	34	34				
0.376 - 0.390	24	24	24	24	26	26	26	26	28	28	28	28	30	30	30	30	32	32	32	32	34	34	34	34					
0.391 - 0.400	24	24	24	26	26	26	26	26	28	28	28	30	30	30	30	30	32	32	32	34	34	34	34	34					
0.401 - 0.420	24	26	26	26	26	26	28	28	28	30	30	30	30	30	32	32	32	34	34	34	34	34							
0.421 - 0.425	24	26	26	26	26	28	28	28	28	30	30	30	30	32	32	32	32	34	34	34	34								
0.426 - 0.440	26	26	26	26	28	28	28	28	30	30	30	30	32	32	32	32	34	34	34	34									
0.441 - 0.450	26	26	26	28	28	28	28	28	30	30	30	32	32	32	32	32	34	34	34										
0.451 - 0.470	26	28	28	28	28	28	30	30	30	32	32	32	32	32	34	34	34												
0.471 - 0.475	26	28	28	28	28	30	30	30	30	32	32	32	32	34	34	34	34												
0.476 - 0.490	28	28	28	28	30	30	30	30	32	32	32	32	34	34	34	34													
0.491 - 0.500	28	28	28	30	30	30	30	30	32	32	32	34	34	34	34	34													
0.501 - 0.520	28	30	30	30	30	30	32	32	32	34	34	34	34	34															
0.521 - 0.525	28	30	30	30	30	32	32	32	32	34	34	34	34																
0.526 - 0.540	30	30	30	30	32	32	32	32	34	34	34	34																	
0.541 - 0.550	30	30	30	32	32	32	32	32	34	34	34																		
0.551 - 0.570	30	32	32	32	32	32	34	34	34																				
0.571 - 0.575	30	32	32	32	32	34	34	34	34																				
0.576 - 0.590	32	32	32	32	34	34	34	34																					
0.591 - 0.600	32	32	32	34	34	34	34	34																					
0.601 - 0.620	32	34	34	34	34	34																							
0.621 - 0.625	32	34	34	34	34																								
0.626 - 0.640	34	34	34	34																									
0.641 - 0.650	34	34	34																										
0.651 - 0.670	34																												
0.671 - 0.675	34																												

AVAILABLE SHIMS mm (in.)

Shim No.	Thickness	Shim No.	Thickness
02	2.500 (0.0984)	20	2.950 (0.1161)
04	2.550 (0.1004)	22	3.000 (0.1181)
06	2.600 (0.1024)	24	3.050 (0.1201)
08	2.650 (0.1043)	26	3.100 (0.1220)
10	2.700 (0.1063)	28	3.150 (0.1240)
12	2.750 (0.1083)	30	3.200 (0.1260)
14	2.800 (0.1102)	32	3.250 (0.1280)
16	2.850 (0.1122)	34	3.300 (0.1299)
18	2.900 (0.1142)		

Exhaust valve clearance (cold):
0.20 – 0.30 mm (0.008 – 0.012 in.)

Example: **A 2.800 mm shim is installed and the measured clearance is 0.450 mm. Replace the 2.800 mm shim with shim No. 22 (3.000 mm).**

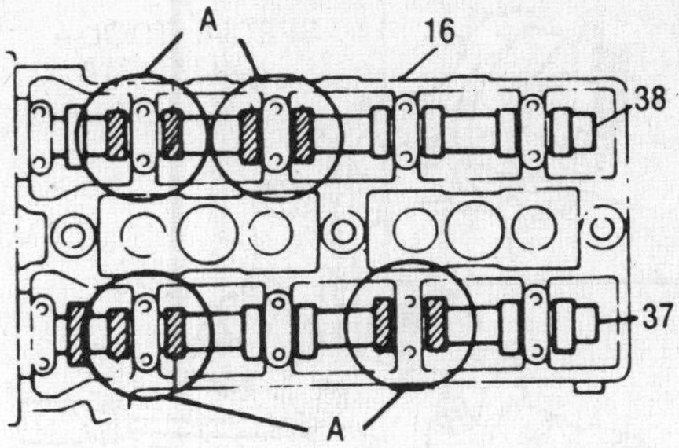

A. Adjust valves (1 & 2 intake) (1 & 3 exhaust) 16. Cylinder head 37. Exhaust valve camshaft 38. Intake valve camshaft

Measuring the valve clearance step 1—twincam engine in Nova

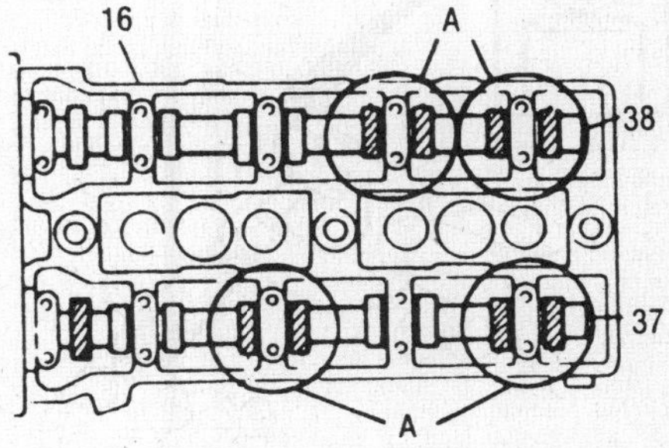

A. Adjust valves (3 & 4 intake) (2 & 4 exhaust) 16. Cylinder head 37. Exhaust valve camshaft 38. Intake valve camshaft

Measuring the valve clearance step 2—twincam engine in Nova

nut, adjust the valve clearance, then, tighten the rocker arm locknut.

d. Recheck the clearance and readjust, if necessary.

5. Rotate the crankshaft 1 complete revolution (360 degrees), then, realign the crankshaft pulley notch with the **0** degrees mark on the timing plate.

6. Using a 0.008 in. feeler gauge, adjust the intake valve clearance of cylinder No. 3 and 4. Using a 0.012 in. feeler gauge, adjust the exhaust valve clearance of cylinders No. 2 and 4.

7. To install, use a new gasket, sealant, if necessary, and reverse the removal procedures. Install the air cleaner. Adjust the engine timing and idle speed.

TWINCAM ENGINE

1. With the engine COLD, remove the valve covers.

2. To inspect the valve clearances, perform the following procedures:

a. Using a socket wrench on the crankshaft pulley, rotate the crankshaft until the No. 1 cylinder is positioned to the TDC of its compression stroke; the valve lifters of the No. 1 cylinder should be loose.

NOTE: The crankshaft pulley notch will align with the 0 degrees mark on the timing belt cover.

b. Using a feeler gauge, measure and record (valves not within specifications) the intake valve-to-lifter clearances of cylinders No. 1 and 2; the exhaust valve-to-lifter clearances of cylinders No. 1 and 3.

c. Rotate the crankshaft 1 complete revolution (360 degrees) and realign the crankshaft pulley notch with the **0** degrees mark on the timing plate; the valve lifters of the No. 4 cylinder should be loose.

d. Using a feeler gauge, measure and record (valves not within specifications) the intake valve-to-lifter clearances of cylinders No. 3 and 4; the exhaust valve-to-lifter clearances of cylinders No. 2 and 4.

3. Rotate the crankshaft pulley until the cam lobe, valve being worked on, is positioned in the upward direction.

4. Using the valve clearance adjustment tool set No. J-37141 or equivalent, press the valve lifter downward, then, secure it in downward position (using another tool) and remove the first tool.

5. Using a prybar or a magnetic finger, remove the adjusting shim.

6. To select the correct valve shim(s), perform the following procedures:

a. Using a micrometer, measure the thickness of the old shim.

b. Using the valve clearance measurement (already acquired), subtract 0.008 in. (intake valve) or 0.010 in. (exhaust valve) from it; the new calculation is the difference between the old shim and the new shim.

c. Using the difference (just calculated), add it to the old shim thickness, then, select (from the chart) a new shim with the thickness closest to the new calculation.

7. Install the new shim and remove the hold-down tool.

8. After all valves have met specifications, use new gaskets, sealant (if necessary) and reverse the removal procedures. Adjust the engine timing and idle speed.

Prizm

NOTE: The use of the correct special tools or their equivalent is required for this procedure. The valve adjustment requires removal of the adjusting shims (Tool kit J-37141 or equivalent) and accurate measurement of the shims with a micrometer. A selection of replacement shims is also required.

1. Remove the valve cover.

2. Turn the crankshaft to align the groove in the crankshaft pulley with the **0** mark on the timing belt cover. Removing the spark plugs makes this easier, but is not required.

3. Check that the lifters on No. 1 cylinder are loose and those on No. 4 are tight. If not, turn the crankshaft pulley one full revolution (360°).

4. Using the feeler gauge, measure the clearance on the 4 valves. Make a written record of any measurements which are not within specification.

5. Rotate the crankshaft pulley one full turn (360°) and check the clearance on the other 4 valves. Any measurements not within specification should be recorded.

6. For any given valve needing adjustment:

a. Turn the crankshaft pulley un-

1. Valves—No. 1 cylinder
2. Valves—No. 2 cylinder
3. Intake camshaft
4. Exhaust camshaft
5. Valves—No. 3 cylinder

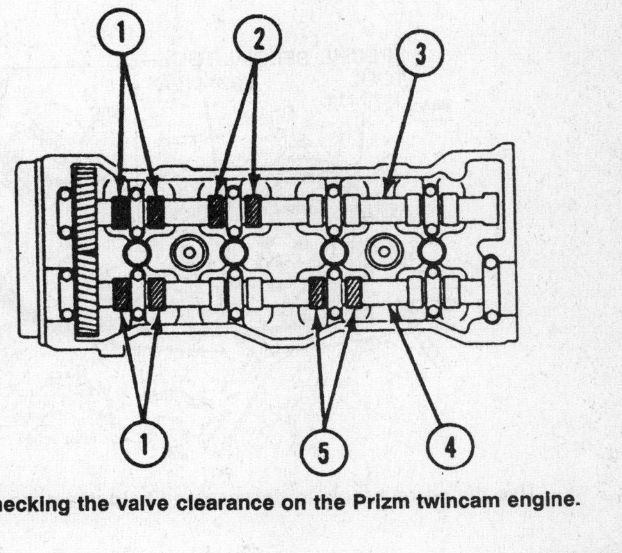

First pass when checking the valve clearance on the Prizm twincam engine.

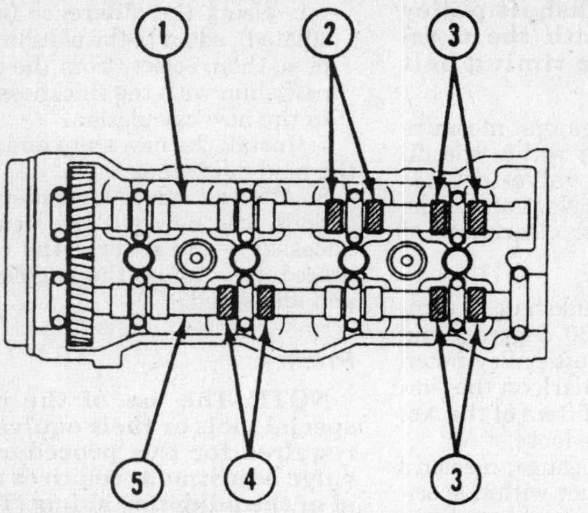

1. Intake camshaft
2. Valves—No. 3 cylinder
3. Valves—No. 4 cylinder
4. Valves—No. 2 cylinder
5. Exhaust camshaft

Second pass when checking the valve clearance on the Prizm twincam engines requires the crankshaft to be rotated 360 degrees

til the camshaft lobe points upward over the valve. This takes the tension off the valve and spring.

b. Using the forked tool, press the valve lifter downward and hold it there.

Some tool kits require a second tool for holding the lifter in place, allowing the first to be removed

c. Using small magnetic tools, remove the adjusting shim from the top of the lifter.

d. Use the micrometer and measure the thickness of the shim removed. Determine the thickness of the new shim using the formula below. For the purposes of the following formula, T = Thickness of the old shim; A = Valve clearance measured; N = Thickness of the new shim.

For the intake side (camshaft nearest to the intake manifold):
N = T + (A ‹minus› 0.008 in. or 0.20mm)

For the exhaust side (camshaft nearest to the exhaust manifold):
N = T + (A ‹minus› 0.10 in. or 0.25mm)

e. Select a shim closest to the calculated thickness. Use the lifter depressor tool to press down the lifter and install the shim. Shims are available in 17 sizes from 0.0984 in. (2.50mm) to 0.1299 (3.30mm). The standard increment is 0.002 in. (0.05mm)

f. Repeat steps a through e for each valve needing adjustment.

7. Reinstall the valve cover.

8. Check and adjust the timing and idle speed.

Spectrum and Storm

EXCEPT TWINCAM ENGINE

1. Remove the cylinder head cover.
2. Rotate the engine until the notched line on the crankshaft pulley aligns with the **0** degree mark on the timing gear case. The position of the No. 1 piston should be at TDC of the compression stroke.

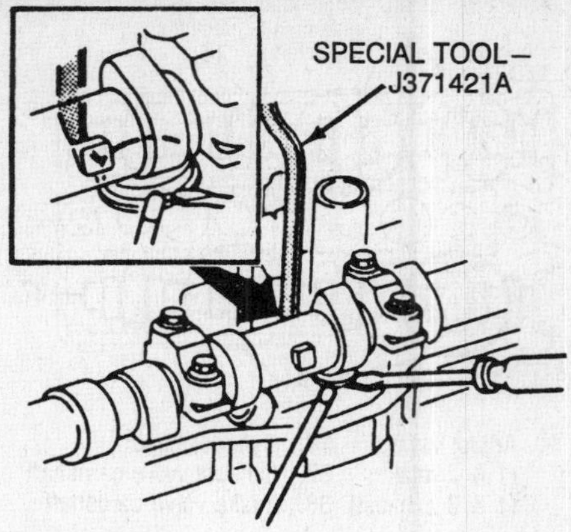

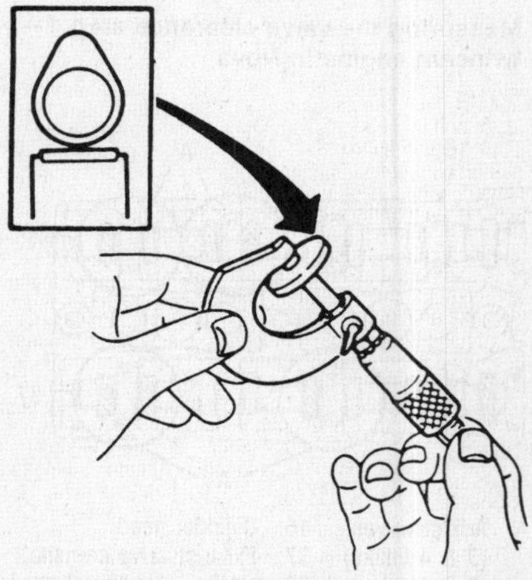

Carefully remove the shim to be replaced (above) with a small suitable tool and magnetic tools. Clean and dry the shim before measuring with the micrometer

NOTE: The notch on the crankshaft pulley should align with the 0 degrees mark on the timing gear case. Make sure the rocker arms on the No. 1 cylinder are loose and the rockers on the No. 4 cylinder are tight. If not, turn the crankshaft one complete revolution and align the marks again.

3. The valve lash between the rocker arm the valve stem on the intake and exhaust valves on the number 1 cylinder, the intake valves on the number 2 cylinder and the exhaust valves on the number 3 cylinder. If the valve lash on the intake valves is not 0.006 in. (0.15mm), adjust the valve lash. If the valve lash on the exhaust valve is not 0.010 in. (0.25mm), adjust the valve lash.
4. Adjust the valve lash by loosening the adjusting screw locknut and turning the adjusting screw until the proper specification is obtained. Tighten the adjusting screw locknut.
5. Set the intake valve to 0.006 in.

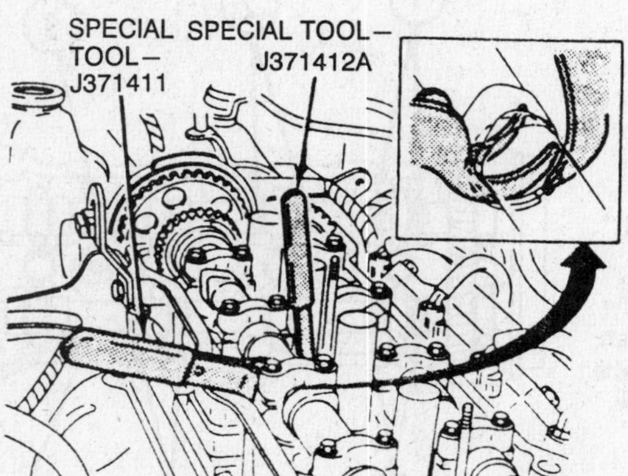

Use the correct tools to depress and hold down the valve lifter

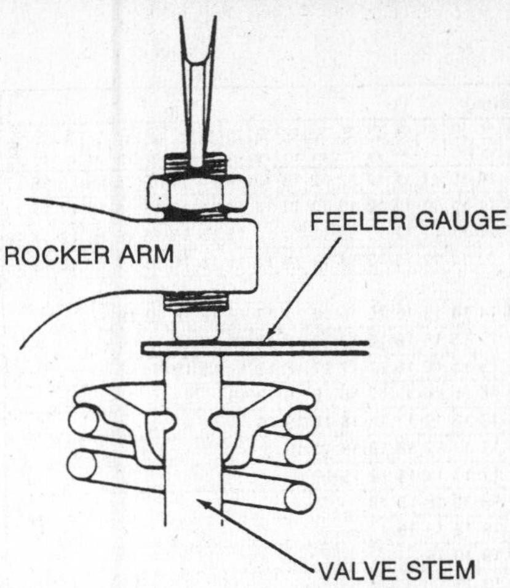

Valve lash adjustment—Spectrum

(cold) for No. 1 and 2 cylinders; exhaust valves to 0.010 in. (cold) for No. 1 and 3 cylinders.

6. When piston in No. 4 cylinder is at TDC on compression stroke set the intake valves to 0.006 in. (cold) for No. 3 and 4 cylinders; exhaust valves to 0.010 in. (cold) for No. 2 and 4 cylinders.

7. After the adjustment has been completed, replace the head cover.

Storm

TWINCAM ENGINE

1. Disconnect the negative battery cable.
2. Remove the cylinder head cover.
3. Position the No. 1 cylinder at TDC on its compression stroke.

NOTE: The notch on the crankshaft pulley should align with the 0 degrees mark on the timing gear case. Make sure the rocker arms on the No. 1 cylinder are loose and the rockers on the No. 4 cylinder are tight. If not, turn the crankshaft one complete revolution and align the marks again.

5. Using a feeler gauge, measure the clearance between the cam lobe and the selective shim on the intake and exhaust valves on the No. 1 cylinder, then the intake valves on the No. 2 cylinder and the exhaust valves on the No. 3 cylinder. Note readings.
6. Rotate the crankshaft 360 degrees. Using a feeler gauge, measure the clearance between the cam lobe and the selective shim on the intake and exhaust valves on the No. 4 cylinder, then the intake valves on the No. 3 cylinder and the exhaust valves on the No. 2 cylinder. Note readings.
7. The valve clearance obtained on the exhaust valves should be between 0.008–0.012 in. (0.20‹n-dash›0.30mm). If not, replace the selective shim by turning the camshaft lobe downward and installing tool J-38413-2 or J-38413-3 between the camshaft journal and the cam lobe next to the selective shim. Turn cam lobe upward and remove the selective shim. Install new shim using the selective shim chart.
8. The valve clearance obtained on the intake valves should be between 0.004–0.008 in. (0.10‹n-dash›0.20mm). If not, replace the selective shim by turning the camshaft lobe downward and installing tool J-38413-2 or J-38413-3 between the camshaft journal and the cam lobe next to the selective shim. Turn cam lobe upward and remove the selective shim. Install new shim using the selective shim chart.
9. Install the cylinder head covers and connect the battery negative cable. Start the engine and check for leaks.

Sprint

1. Remove the air cleaner on carbureted engine. Remove the rocker cover.
2. Rotate the crankshaft clockwise and align the **V** mark on the crankshaft pulley with the **0** mark on the timing tab.
3. Remove the distributor cap and make sure the rotor is facing the fuel pump. If not, rotate the crankshaft 360 degrees.

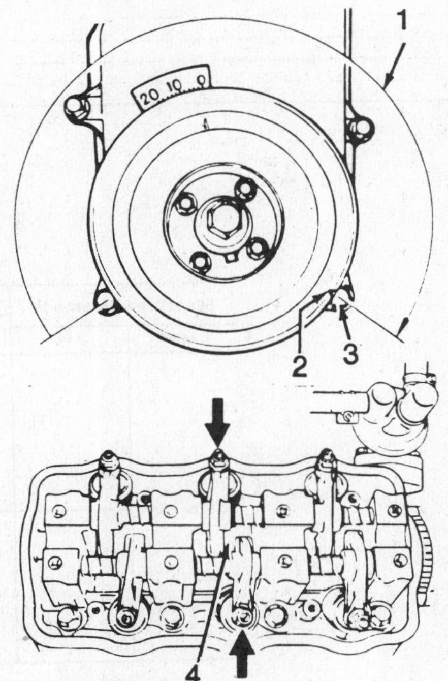

1. 240 degrees
2. Timing notch
3. Right mounting bolt
4. No. 2 cylinder

Valve identification in cylinder head—Sprint

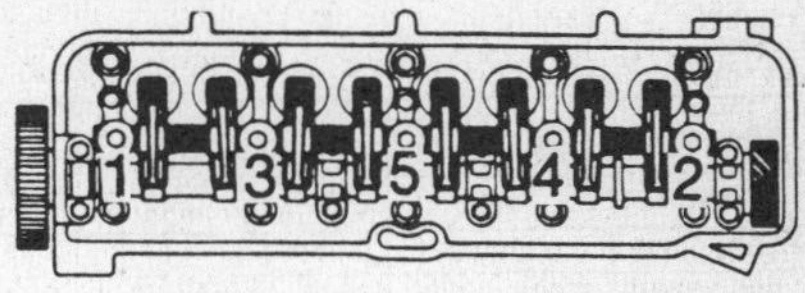

Valve lash adjusting screw—Sprint

4. Check and/or adjust the valves of the No. 1 cylinder.

NOTE: On a COLD engine, adjust the valves to 0.006 in. (intake) and 0.008 in. (exhaust). On a HOT engine, adjust the valves to 0.010 in. (intake) and 0.012 in. (exhaust).

5. After adjusting the valves of the No. 1 cylinder, rotate the crankshaft pulley 240 degrees, the **V** mark should align with the lower left oil pump mounting bolt, when facing the crankshaft pulley, then adjust the valves of the No. 3 cylinder.
6. After adjusting the valves of the No. 3 cylinder, rotate the crankshaft pulley 240 degrees, the **V** mark should align with the lower right oil pump mounting bolt, when facing the crankshaft pulley, then adjust the valves of the No. 2 cylinder.
7. After the valves have been adjusted, install the removed items by reversing the removal procedures. Torque the valve adjustment locknuts to 11–13 ft. lbs. (15-18 Nm).

Rocker Arms/Shafts

REMOVAL & INSTALLATION

Nova

EXCEPT TWINCAM ENGINE

1. Disconnect the negative battery cable. Remove the air cleaner and valve cover.
2. Remove the 5 rocker shaft assembly retaining bolts in several stages—note that they must be loosened in the correct sequence: Front bolt 1st, rear bolt 2nd, forward—center bolt 3rd, rearward-center bolt 4th and the center bolt 5th.
3. Remove the rocker arm/shaft assembly.
4. Inspect for wear by attempting to rock the rocker levers on the shaft. If negligible motion is felt, wear is acceptable. If there is noticeable wear, note the order of assembly and the fact that there are 2 types of rockers. Remove the bolts and slide the rockers, springs and pedestals from the shaft.
5. Using an internal dial indicator, measure the inside diameter of each rocker lever; using a micrometer, measure the shaft diameter at the rocker

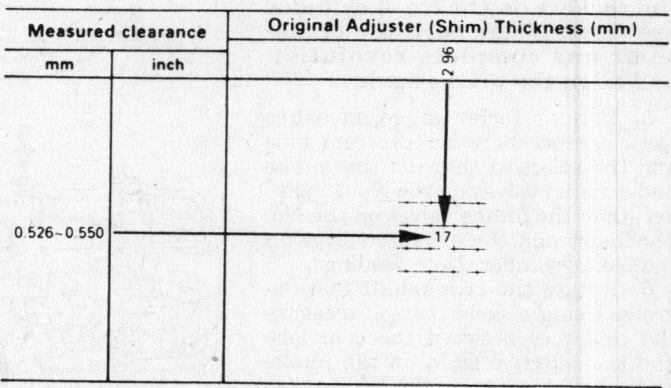

Measured clearance		Original Adjuster (Shim) Thickness (mm)																																														
mm	inch	2.52	2.54	2.56	2.58	2.60	2.62	2.64	2.66	2.68	2.70	2.72	2.74	2.76	2.78	2.80	2.82	2.84	2.86	2.88	2.90	2.92	2.94	2.96	2.98	3.00	3.02	3.04	3.06	3.08	3.10	3.12	3.14	3.16	3.18	3.20	3.22	3.24	3.26	3.28	3.30	3.32	3.34	3.36	3.38	3.40	3.42	3.44	3.46	3.48
0.000~0.025	0.000~0.001									1	1	2	2	2	3	3	4	4	4	5	5	6	6	6	7	7	8	8	8	9	9	10	10	10	11	11	12	12	12	13	13	14	14	14	15	15	16	16	16	17
0.026~0.050	0.001~0.002							1	1	1	2	2	3	3	3	4	4	5	5	5	6	6	7	7	7	8	8	9	9	9	10	10	11	11	11	12	12	13	13	13	14	14	15	15	15	16	16	17	17	17
0.051~0.075	0.002~0.003						1	1	1	2	2	3	3	3	4	4	5	5	5	6	6	7	7	7	8	8	9	9	9	10	10	11	11	11	12	12	13	13	13	14	14	15	15	15	16	16	17	17	17	18
0.076~0.100	0.003~0.004					1	1	2	2	2	3	3	4	4	4	5	5	6	6	6	7	7	8	8	8	9	9	10	10	10	11	11	12	12	12	13	13	14	14	14	15	15	16	16	16	17	17	18	18	18
0.101~0.200	0.004~0.008	Replacement not to be required																																																
0.201~0.225	0.008~0.009	2	2	2	3	3	4	4	4	5	5	6	6	6	7	7	8	8	8	9	9	10	10	10	11	11	12	12	12	13	13	14	14	14	15	15	16	16	16	17	17	18	18	18	19	19				
0.226~0.250	0.009~0.010	2	3	3	3	4	4	5	5	5	6	6	7	7	7	8	8	9	9	9	10	10	11	11	11	12	12	13	13	13	14	14	15	15	15	16	16	17	17	17	18	18	19	19	19					
0.251~0.275	0.010~0.011	3	3	3	4	4	5	5	5	6	6	7	7	7	8	8	9	9	9	10	10	11	11	11	12	12	13	13	13	14	14	15	15	15	16	16	17	17	17	18	18	19	19	19						
0.276~0.300	0.011~0.012	3	4	4	4	5	5	6	6	6	7	7	8	8	8	9	9	10	10	10	11	11	12	12	12	13	13	14	14	14	15	15	16	16	16	17	17	18	18	18	19	19								
0.301~0.325	0.012~0.013	4	4	4	5	5	6	6	6	7	7	8	8	8	9	9	10	10	10	11	11	12	12	12	13	13	14	14	14	15	15	16	16	16	17	17	18	18	18	19	19									
0.326~0.350	0.013~0.014	4	5	5	5	6	6	7	7	7	8	8	9	9	10	10	11	11	11	11	12	12	13	13	14	14	14	15	15	15	16	16	17	17	17	18	18	19	19	19										
0.351~0.375	0.014~0.015	5	5	5	6	6	7	7	7	8	8	9	9	9	10	10	11	11	11	12	12	13	13	13	14	14	15	15	15	16	16	17	17	17	18	18	19	19	19											
0.376~0.400	0.015~0.016	5	6	6	6	7	7	8	8	8	9	9	10	10	10	11	11	12	12	12	13	13	14	14	14	15	15	16	16	16	17	17	18	18	18	19	19													
0.401~0.425	0.016~0.017	6	6	6	7	7	8	8	8	9	9	10	10	10	11	11	12	12	12	13	13	14	14	14	15	15	16	16	16	17	17	18	18	18	19	19														
0.426~0.450	0.017~0.018	6	7	7	7	8	8	9	9	9	10	10	11	11	11	12	12	13	13	13	14	14	14	15	15	16	16	17	17	17	18	18	19	19	19															
0.451~0.475	0.018~0.019	7	7	7	8	8	9	9	9	10	10	11	11	12	12	12	13	13	13	14	14	15	15	15	16	16	17	17	17	18	18	19	19	19																
0.476~0.500	0.019~0.020	7	8	8	8	9	9	10	10	10	11	11	12	12	12	13	13	14	14	14	15	15	16	16	16	17	17	18	18	18	19	19																		
0.501~0.525	0.020~0.021	8	8	8	9	9	10	10	10	11	11	12	12	12	13	13	14	14	14	15	15	16	16	16	17	17	18	18	18	19	19																			
0.526~0.550	0.021~0.022	8	9	9	9	10	10	11	11	11	12	12	13	13	13	14	14	15	15	15	16	16	17	17	17	18	18	19	19	19																				
0.551~0.575	0.022~0.023	9	9	9	10	10	11	11	11	12	12	13	13	13	14	14	15	15	15	16	16	17	17	17	18	18	19	19	19																					
0.576~0.600	0.023~0.024	9	10	10	10	11	11	12	12	12	13	13	14	14	14	15	15	16	16	16	17	17	18	18	18	19	19																							
0.601~0.625	0.024~0.025	10	10	10	11	11	12	12	12	13	13	14	14	14	15	15	16	16	16	17	17	18	18	18	19	19																								
0.626~0.650	0.025~0.026	10	11	11	11	12	12	13	13	13	14	14	15	15	15	16	16	17	17	17	18	18	19	19	19																									
0.651~0.675	0.026~0.027	11	11	11	12	12	13	13	13	14	14	15	15	15	16	16	17	17	17	18	18	19	19	19																										
0.676~0.700	0.027~0.028	11	12	12	12	13	13	14	14	14	15	15	16	16	16	17	17	18	18	18	19	19																												
0.701~0.725	0.028~0.029	12	12	12	13	13	14	14	14	15	15	16	16	16	17	17	18	18	18	19	19																													
0.726~0.750	0.029~0.030	12	13	13	13	14	14	15	15	15	16	16	17	17	17	18	18	19	19	19																														
0.751~0.775	0.030~0.031	13	13	13	14	14	15	15	15	16	16	17	17	17	18	18	19	19	19																															
0.776~0.800	0.031~0.032	13	14	14	14	15	15	16	16	16	17	17	18	18	18	19	19																																	
0.801~0.825	0.032~0.033	14	14	14	15	15	16	16	16	17	17	18	18	18	19	19																																		
0.826~0.850	0.033~0.034	14	15	15	15	16	16	17	17	17	18	18	19	19	19																																			
0.851~0.875	0.034~0.035	15	15	15	16	16	17	17	17	18	18	19	19	19																																				
0.876~0.900	0.035~0.036	15	16	16	16	17	17	18	18	18	19	19																																						
0.901~0.925	0.036~0.037	16	16	16	17	17	18	18	18	19	19																																							
0.926~0.950	0.0365~0.0374	16	17	17	17	18	18	19	19	19																																								
0.951~0.975	0.037~0.038	17	17	17	18	18	19	19	19																																									
0.976~1.000	0.038~0.039	17	18	18	18	19	19																																											
1.001~1.025	0.039~0.040	18	18	18	19	19																																												
1.026~1.050	0.040~0.041	18	19	19	19																																													
1.051~1.075	0.041~0.042	19	19	19																																														
1.076~1.100	0.042~0.043	19																																																

Thickness of available adjuster (Shim)

NO in Chart	Thickness (mm)	NO in Chart	Thickness (mm)
1	2.55	11	3.05
2	2.60	12	3.10
3	2.65	13	3.15
4	2.70	14	3.20
5	2.75	15	3.25
6	2.80	16	3.30
7	2.85	17	3.35
8	2.90	18	3.40
9	2.95	19	3.45
10	3.00		

How to use the chart

[Example]
Measured clearance; C.550mm
Original adjuster thickness; 2.96mm
(Thickness mark (2.96) is printed on the adjuster surface)

1. Draw straight lines as shown in the chart.
2. Select No.17 available adjuster to be replaced by finding cross point of straight lines.
3. Replace the 2.96mm adjuster with No.17 (3.35mm) adjuster.

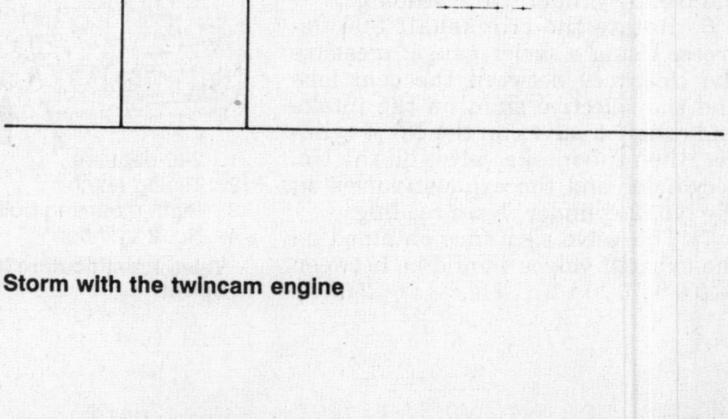

Intake valve shim size chart—Storm with the twincam engine

Measured clearance		Original Adjuster (Shim) Thickness (mm)																																																
mm	inch	2.52	2.54	2.56	2.58	2.60	2.62	2.64	2.66	2.68	2.70	2.72	2.74	2.76	2.78	2.80	2.82	2.84	2.86	2.88	2.90	2.92	2.94	2.96	2.98	3.00	3.02	3.04	3.06	3.08	3.10	3.12	3.14	3.16	3.18	3.20	3.22	3.24	3.26	3.28	3.30	3.32	3.34	3.36	3.38	3.40	3.42	3.44	3.46	3.48
0.000~0.025	0.000~0.001														1	1	2	2	2	3	3	4	4	4	5	5	6	6	6	7	7	8	8	8	9	9	10	10	10	11	11	12	12	12	13	13	14	14	14	15
0.026~0.050	0.001~0.002												1	1	1	2	2	3	3	3	4	4	5	5	5	6	6	7	7	7	8	8	9	9	9	10	10	11	11	11	12	12	13	13	13	14	14	14	15	15
0.051~0.075	0.002~0.003											1	1	1	2	2	3	3	3	4	4	5	5	5	6	6	7	7	7	8	8	9	9	9	10	10	11	11	11	12	12	13	13	13	14	14	15	15	15	16
0.076~0.100	0.003~0.004										1	1	2	2	2	3	3	4	4	4	5	5	6	6	6	7	7	8	8	8	9	9	10	10	10	11	11	12	12	12	13	13	14	14	14	15	15	16	16	16
0.101~0.125	0.004~0.005									1	1	2	2	2	3	3	4	4	4	5	5	6	6	6	7	7	8	8	8	9	9	10	10	10	11	11	12	12	12	13	13	14	14	14	15	15	16	16	16	17
0.126~0.150	0.005~0.006							1	1	1	2	2	3	3	3	4	4	5	5	5	6	6	7	7	7	8	8	9	9	9	10	10	11	11	11	12	12	13	13	13	14	14	15	15	15	16	16	17	17	17
0.151~0.175	0.006~0.007						1	1	1	2	2	3	3	3	4	4	5	5	5	6	6	7	7	7	8	8	9	9	9	10	10	11	11	11	12	12	13	13	13	14	14	15	15	15	16	16	17	17	17	18
0.176~0.200	0.007~0.008					1	1	2	2	2	3	3	4	4	4	5	5	6	6	6	7	7	8	8	8	9	9	10	10	10	11	11	12	12	12	13	13	14	14	14	15	15	16	16	16	17	17	18	18	18
0.201~0.300	0.008~0.012	Replacement not to be required
0.301~0.325	0.012~0.013	2	2	2	3	3	4	4	4	5	5	6	6	6	7	7	8	8	8	9	9	10	10	10	11	11	12	12	12	13	13	14	14	14	15	15	16	16	16	17	17	18	18	18	19	19
0.326~0.350	0.013~0.014	2	3	3	3	4	4	5	5	5	6	6	7	7	7	8	8	9	9	9	10	10	11	11	11	12	12	13	13	13	14	14	15	15	15	16	16	17	17	17	18	18	19	19	19
0.351~0.375	0.014~0.015	3	3	3	4	4	5	5	5	6	6	7	7	7	8	8	9	9	9	10	10	11	11	11	12	12	13	13	13	14	14	15	15	15	16	16	17	17	17	18	18	19	19	19
0.376~0.400	0.015~0.016	3	4	4	4	5	5	6	6	6	7	7	8	8	8	9	9	10	10	10	11	11	12	12	12	13	13	14	14	14	15	15	16	16	16	17	17	18	18	18	19	19
0.401~0.425	0.016~0.017	4	4	4	5	5	6	6	6	7	7	8	8	8	9	9	10	10	10	11	11	12	12	12	13	13	14	14	14	15	15	16	16	16	17	17	18	18	18	19	19
0.426~0.450	0.017~0.018	4	5	5	5	6	6	7	7	7	8	8	9	9	9	10	10	11	11	11	12	12	13	13	13	14	14	15	15	15	16	16	17	17	17	18	18	19	19	19
0.451~0.475	0.018~0.019	5	5	5	6	6	7	7	7	8	8	9	9	9	10	10	11	11	11	12	12	13	13	13	14	14	15	15	15	16	16	17	17	17	18	18	19	19	19
0.476~0.500	0.019~0.020	5	6	6	6	7	7	8	8	8	9	9	10	10	10	11	11	12	12	12	13	13	14	14	14	15	15	16	16	16	17	17	18	18	18	19	19
0.501~0.525	0.020~0.021	6	6	6	7	7	8	8	8	9	9	10	10	10	11	11	12	12	12	13	13	14	14	14	15	15	16	16	16	17	17	18	18	18	19	19
0.526~0.550	0.021~0.022	6	7	7	7	8	8	9	9	9	10	10	11	11	11	12	12	13	13	13	14	14	14	15	15	16	16	17	17	17	18	18	19	19	19
0.551~0.575	0.022~0.023	7	7	7	8	8	9	9	9	10	10	11	11	12	12	12	13	13	13	14	14	15	15	15	16	16	17	17	17	18	18	19	19	19
0.576~0.600	0.023~0.024	7	8	8	8	9	9	10	10	10	11	11	12	12	12	13	13	14	14	14	15	15	16	16	16	17	17	18	18	18	19	19
0.601~0.625	0.024~0.025	8	8	8	9	9	10	10	10	11	11	12	12	12	13	13	14	14	14	15	15	16	16	16	17	17	18	18	18	19	19
0.626~0.650	0.025~0.026	8	9	9	9	10	10	11	11	11	12	12	13	13	13	14	14	15	15	15	16	16	17	17	17	18	18	19	19	19
0.651~0.675	0.026~0.027	9	9	9	10	10	11	11	11	12	12	13	13	13	14	14	15	15	15	16	16	17	17	17	18	18	19	19	19
0.676~0.700	0.027~0.028	9	10	10	10	11	11	12	12	12	13	13	14	14	14	15	15	16	16	16	17	17	18	18	18	19	19
0.701~0.725	0.028~0.029	10	10	10	11	11	12	12	12	13	13	14	14	14	15	15	16	16	16	17	17	18	18	18	19	19
0.726~0.750	0.029~0.030	10	11	11	11	12	12	13	13	13	14	14	15	15	15	16	16	17	17	17	18	18	19	19	19
0.751~0.775	0.030~0.031	11	11	11	12	12	13	13	13	14	14	15	15	15	16	16	17	17	17	18	18	19	19	19
0.776~0.800	0.031~0.032	11	12	12	12	13	13	14	14	14	15	15	16	16	16	17	17	18	18	18	19	19
0.801~0.825	0.032~0.033	12	12	12	13	13	14	14	14	15	15	16	16	16	17	17	18	18	18	19	19
0.826~0.850	0.033~0.034	12	13	13	13	14	14	15	15	15	16	16	17	17	17	18	18	19	19	19
0.851~0.875	0.034~0.035	13	13	13	14	14	15	15	15	16	16	17	17	17	18	18	19	19	19
0.876~0.900	0.035~0.036	13	14	14	14	15	15	16	16	16	17	17	18	18	18	19	19
0.901~0.925	0.036~0.037	14	14	14	15	15	16	16	16	17	17	18	18	18	19	19
0.926~0.950	0.0365~0.0374	14	15	15	15	16	16	17	17	17	18	18	19	19	19
0.951~0.975	0.037~0.038	15	15	15	16	16	17	17	17	18	18	19	19	19
0.976~1.000	0.038~0.039	15	16	16	16	17	17	18	18	18	19	19
1.001~1.025	0.039~0.040	16	16	16	17	17	18	18	18	19	19
1.026~1.050	0.040~0.041	16	17	17	17	18	18	19	19	19
1.051~1.075	0.041~0.042	17	17	17	18	18	19	19	19
1.076~1.100	0.042~0.043	17	18	18	18	19	19
1.101~1.125	0.043~0.044	18	18	18	19	19
1.126~1.150	0.044~0.045	18	19	19	19
1.151~1.175	0.045~0.046	19	19	19
1.176~1.200	0.046~0.047	19

Thickness of available adjuster (Shim)

NO in Chart	Thickness (mm)	NO in Chart	Thickness (mm)
1	2.55	11	3.05
2	2.60	12	3.10
3	2.65	13	3.15
4	2.70	14	3.20
5	2.75	15	3.25
6	2.80	16	3.30
7	2.85	17	3.35
8	2.90	18	3.40
9	2.95	19	3.45
10	3.00		

Note; Thickness mark is printed on the surface to be contacted with tappet.

How to use the chart

[Example]
Measured clearance; 0.550mm
Original adjuster thickness; 2.96mm
(Thickness mark (2.96) is printed on the adjuster surface)

1. Draw straight lines as shown in the chart.
2. Select No.15 available adjuster to be replaced by finding cross point of straight lines.
3. Replace the 2.96mm adjuster with No.15 (3.25mm) adjuster.

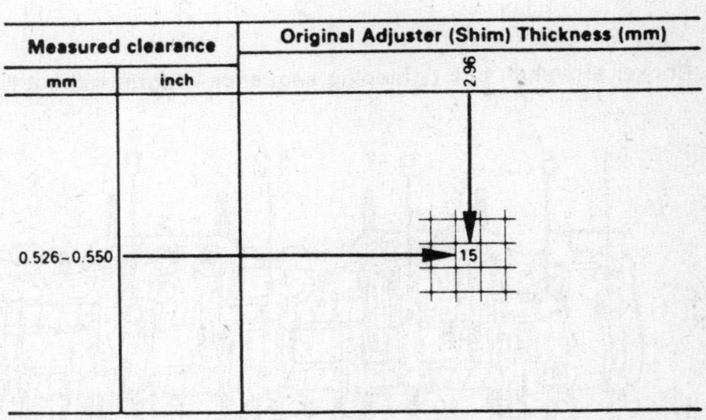

Exhaust valve shim size chart—Storm with the twincam engine

wear areas. Subtract the shaft shaft diameter from the rocker arm inside diameter; the difference must not exceed 0.0024 in. If necessary, replace the rockers and/or the shaft to correct the clearance problems.

To install:

6. Assemble the pedestals, rockers, springs and bolts in reverse order of disassembly. Using clean engine oil, lubricate wear surfaces thoroughly. Install the rocker arm shaft with the oil holes facing downward.

7. Loosen the valve adjusting screw locknuts. Install the rocker arm assembly onto the cylinder head and start the bolts, tightening them finger tight. Torque the rocker arm assembly-to-cylinder head bolts (in sequence) using 3 passes to 18 ft. lbs. (24 Nm): center bolt – first, center/rearward bolt – second, center/forward bolt – third, rear bolt – fourth and front bolt – last. Perform the valve adjustment.

8. To complete the installation, use new gasket(s), sealant, if necessary, and reverse the removal procedures.

Nova and Prizm Twincam Engines

The twin camshaft motors (one cam for the intake valves and one for the exhaust valves) use direct-acting cams; that is, the lobes of the camshaft act directly on the valve mechanism. These engines do not have rocker arms.

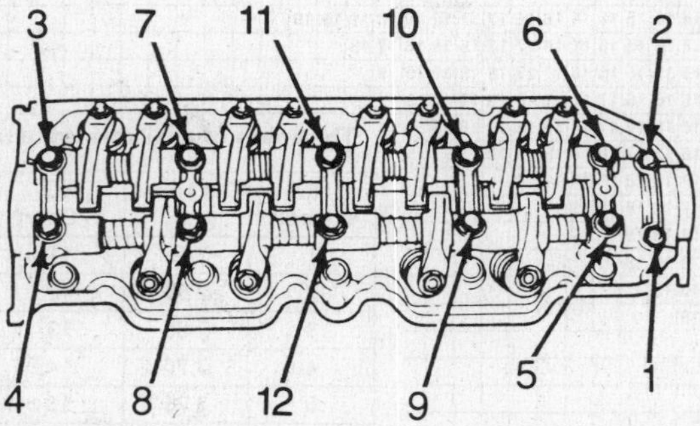

Rocker arm/shaft bolt removal sequence – Storm without twincam engine

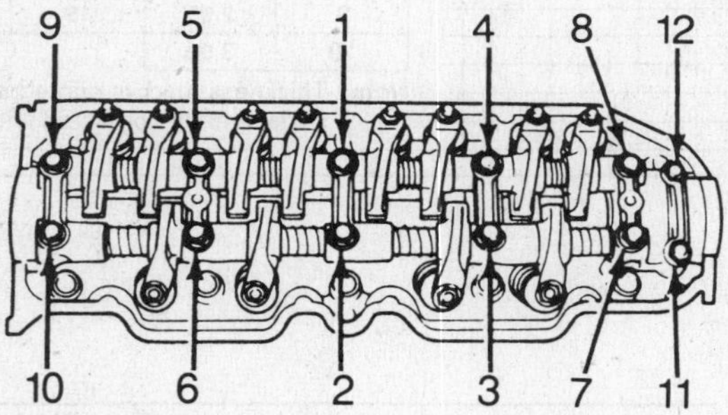

Rocker arm/shaft bolt tightening sequence – Storm without twincam engine

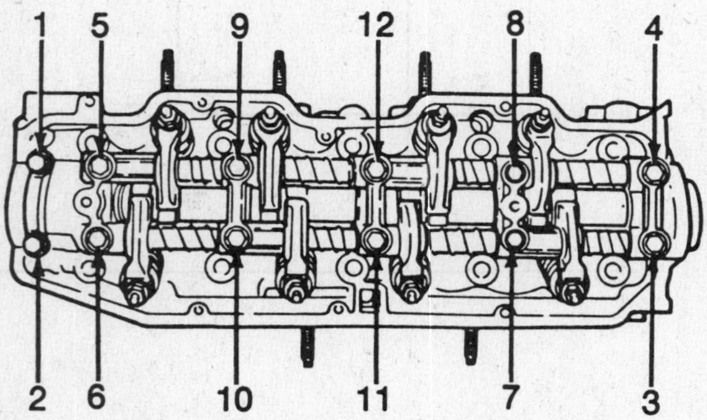

Rocker arm/shaft bolt removal sequence – Spectrum

Storm

1. Disconnect the battery negative cable.
2. Remove the cylinder head valve cover.
3. Remove the rocker arm bracket bolts in the proper order.
4. Remove the rocker shaft/arm assembly from the vehicle.
5. Remove the rocker arms from the rocker shaft.

To install:

6. Apply a light coat of engine oil to the rocker arm brackets.
7. Install the rocker arms on the rocker arm shafts.
8. Install the rocker arm shafts on the engine with the ID marks toward the front of the engine.
9. Apply a suitable silicone sealant to the number 1 and number 5 rocker brackets.
10. Install the rocker arm brackets. Torque the rocker arm bracket bolts in sequence to 16 ft. lbs. (22 Nm).
11. Adjust the valves and install the cylinder head cover. Reconnect the battery negative cable.

Spectrum

1. Disconnect the negative battery terminal from the battery. Remove the PCV hoses.
2. Remove the spark plug wires from the mounting clip.
3. Remove the ground wire from the right rear side of the head cover.
4. Support the engine and remove the right side engine mounting rubber, bolts and plate.
5. Remove the mounting bracket on the timing cover.
6. Remove the 4 bolts holding the timing cover and the 2 bolts holding the cylinder head cover.
7. Loosen the timing cover and remove the cylinder head cover.

NOTE: If the cylinder head cover sticks, strike the end of the cover with a rubber mallet.

8. Remove the rocker arm bracket bolts in sequence, work from both ends equally, toward the middle.
9. Remove the rocker arm shafts and then the rocker arms from the shafts.
10. Using the proper tool, clean the sealing surfaces of the cover and the cylinder head.

To install:

11. To install, apply sealer to the sealing surfaces and reverse the removal procedures.

NOTE: The rocker arm shafts are different from each other, make sure they are installed in the same position that they were removed. Install the rocker arms

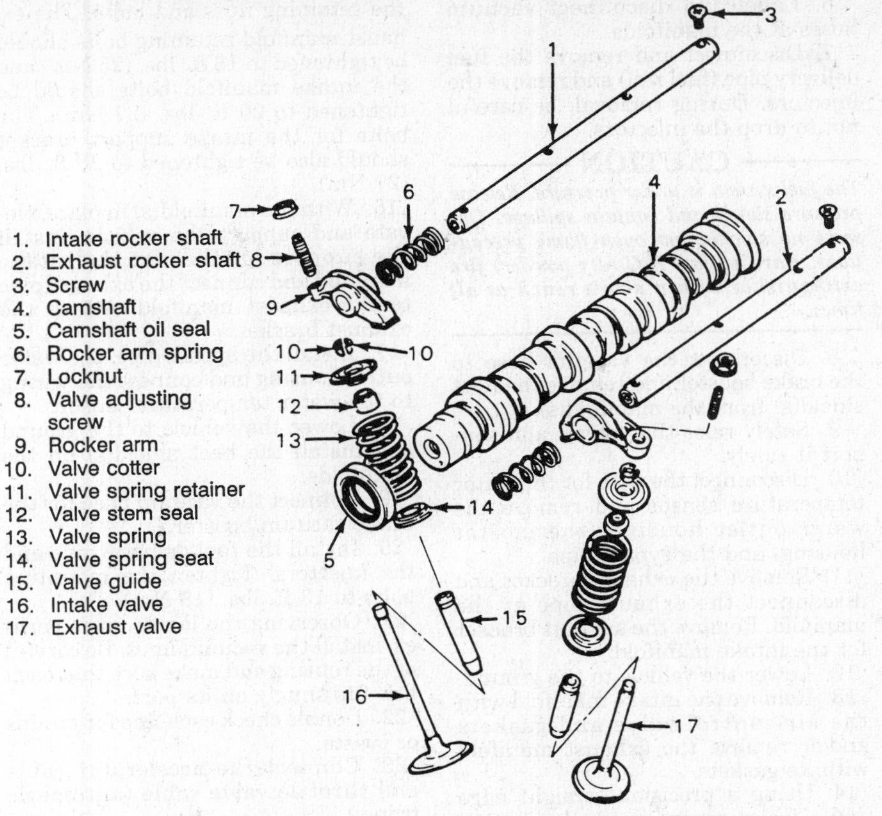

Exploded view of the rocker arm assembly and camshaft—Sprint

with the identification marks toward the front of the engine. Apply sealant to the bracket and cylinder head mating surfaces of the front and rear rocker brackets.

12. To complete the installation, mount the rocker assemblies securely to the dowel pins on the cylinder head. Torque the rocker arm bolts to 16 ft. lbs. (22 Nm). Start the engine and check for leaks.

Sprint

1. Disconnect the negative battery cable.
2. Remove the air cleaner and the cylinder head cover.
3. Remove the distributor cap, then mark the position of the rotor and the distributor housing with the cylinder head. Remove the distributor and the case from the cylinder head.
4. Loosen the rocker arm valve adjusters, turn back the adjusting screws so the rocker arms move freely.
5. Remove the rocker arm shaft retaining screws and pull out the shafts. Remove the rocker arms and springs from the cylinder head.

NOTE: Make a note of the differences between the rocker arm shafts. The intake shaft's stepped end is 0.55 in., which faces the camshaft pulley; the exhaust shaft's stepped end is 0.59 in., which faces the distributor.

6. To install, use new gaskets and reverse the removal procedures. Torque the rocker arm shaft screws to 7–9 ft. lbs. (9-12 Nm). Adjust the valve clearances. Check and/or adjust the ignition timing.

Intake Manifold

REMOVAL & INSTALLATION

Nova

TWINCAM ENGINE

1. Disconnect the negative terminal from the battery. Remove the air cleaner assembly.
2. Drain the cooling system. Remove the upper radiator hose.
3. Disconnect the accelerator and throttle valve cable from the throttle body.
4. Label and disconnect the necessary vacuum hoses. Disconnect the brake vacuum hose from the intake manifold.
5. Relieve the fuel pressure, then, disconnect and remove the fuel delivery pipe with the fuel injectors.
6. Raise and support the vehicle safely.
7. Disconnect the temperature sensor connector from the water outlet housing. Remove the water outlet housing-to-engine bolts with the No. 1 bypass pipe.
8. Remove the intake manifold bracket, the intake manifold-to-engine bolts, the intake manifold, with the air control valve, and gaskets from the cylinder head.

To install:

9. Using the proper tool, clean the gasket mounting surfaces. Inspect the intake manifold and air control valve for damage and/or warpage; maximum warpage for both is 0.002 in., if the warpage is greater, replace the intake manifold or air control valve.
10. To install, use new gaskets and reverse the removal procedures. Torque the intake manifold-to-cylinder head bolts to 20 ft. lbs. (27 Nm), the intake manifold bracket-to-engine bolts to 20 ft. lbs. (27 Nm) and the fuel delivery pipe-to-engine bolts to 13 ft. lbs. (18 Nm). Start the engine and check for leaks.

1989-90 Prizm

1. Disconnect the negative battery cable.
2. Drain the cooling system.
3. Remove the air cleaner assembly.
4. Label and disconnect the vacuum hoses at the manifold.
5. Label and disconnect the wiring to the throttle position sensor, the cold start injector, the injector connectors, the air control valve and the vacuum sensor.

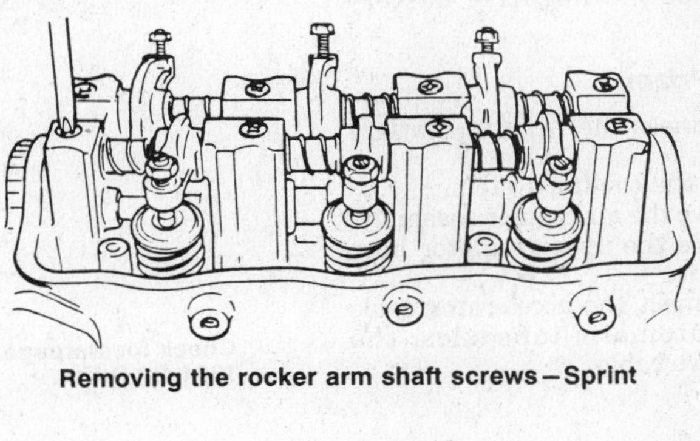

Removing the rocker arm shaft screws—Sprint

6. Disconnect the cold start injector pipe.

CAUTION

The fuel system is under pressure. Release pressure slowly and contain spillage. Observe no smoking/no open flame precautions. Have a Class B-C (dry powder) fire extinguisher within arm's reach at all times.

7. Disconnect the water hose from the air valve.
8. Raise and safely support the vehicle.
9. Remove the intake manifold support bracket. Lower the vehicle to the ground.
10. Remove the 7 bolts, 2 nuts and the ground cable. Remove the intake manifold and its gaskets.

To install:

11. Measure the intake manifold mating surface with a precision straight edge and a feeler gauge. If the warpage exceeds 0.2mm, the manifold must be replaced.
12. To reassemble, install the manifold with new gaskets in position. Attach the 7 bolts, 2 nuts and the ground cable connector.
13. Raise and safely support the car; install the manifold support bracket and its bolts. Lower the vehicle to the ground.
14. Tighten the manifold mounting nuts and bolts evenly to 14 ft. lbs. (19 Nm).
15. Connect the water hose to the air valve.
16. Connect the fuel line to the cold-start injector.
17. Connect the wiring to the throttle position sensor, the cold start injector, the injector connectors, the air control valve and the vacuum sensor.
18. Observing the labels made earlier, install the vacuum lines. Be careful of the routing and make sure that each line fits snugly on its port. Double check each line for crimps or twists.
19. Connect the accelerator and throttle valve (automatic transaxle) cables to their brackets.
20. Refill the coolant.
21. Install the air cleaner assembly and connect the negative battery cable.

1991-92 Prizm

1. Disconnect the negative battery cable.
2. Drain the cooling system.
3. Remove the air cleaner assembly.
4. Remove the upper radiator hose at the engine.
5. Disconnect the accelerator cable and, on automatic tansaxles, the throttle valve cable.
6. Label and disconnect vacuum hoses at the manifolds.
7. Disconnect and remove the fuel delivery pipe (fuel rail) and remove the injectors. During removal, be careful not to drop the injectors.

CAUTION

The fuel system is under pressure. Release pressure slowly and contain spillage. Observe no smoking/no open flame precautions. Have a Class B-C (dry powder) fire extinguisher within arm's reach at all times.

8. Disconnect the vacuum hose to the brake booster and remove the heat shield(s) from the manifold(s).
9. Safely raise the vehicle and support it safely.
10. Disconnect the wire for the water temperature sensor and remove the water outlet housing (thermostat housing) and the bypass pipe.
11. Remove the exhaust bracket and disconnect the exhaust pipe at the manifold. Remove the support bracket for the intake manifold.
12. Lower the vehicle to the ground.
13. Remove the intake manifold with the air control valve and gaskets and/or remove the exhaust manifold with its gaskets.
14. Using a precision straight edge and a feeler gauge, check the mating surfaces of the manifolds for warpage. If the warpage is greater than the maximum allowable specification, replace the manifold.

Maximum allowable warpage:
Intake Manifold: 0.05mm
Exhaust Manifold: 0.3mm
Air Control Valve: 0.05mm.

To install:

15. When reinstalling, always use new gaskets and make sure they are properly positioned. Place the manifold(s) in position and loosely install the nuts and bolts until all are just snug. Double check the placement of the manifold(s) and in 2 passes tighten the retaining nuts and bolts. The exhaust manifold retaining bolts should be tightened to 18 ft. lbs. (25 Nm) and the intake manifold bolts should be tightened to 20 ft. lbs. (27 Nm). The bolts for the intake support bracket should also be tightened to 20 ft. lbs. (27 Nm).
16. With the manifold(s) in place, elevate and support the vehicle; install the support bracket for the intake manifold and connect the exhaust pipe to the exhaust manifold. Attach the exhaust bracket.
17. Install the bypass pipe, the water outlet housing and connect the wiring to the water temperature sensor.
18. Lower the vehicle to the ground and install the heat shield(s) on the manifolds.
19. Connect the vacuum hose for the brake vacuum booster.
20. Install the fuel delivery pipe and the injectors. Tighten the mounting bolts to 13 ft. lbs. (18 Nm).
21. Observing the labels made earlier, install the vacuum lines. Be careful of the routing and make sure that each line fits snugly on its port.
22. Double check each line for crimps or twists.
23. Connect the accelerator cable and throttle valve cable (automatic trans.)
24. Reconnect the upper radiator hose.
25. Install the air cleaner assembly.
26. Refill the coolant.
27. Connect the negative battery cable.

Storm

EXCEPT TWINCAM ENGINE

1. Disconnect the battery negative cable and the ignition coil wire.
2. Disconnect the accelerator cable from the throttle valve and the intake manifold (common chamber).

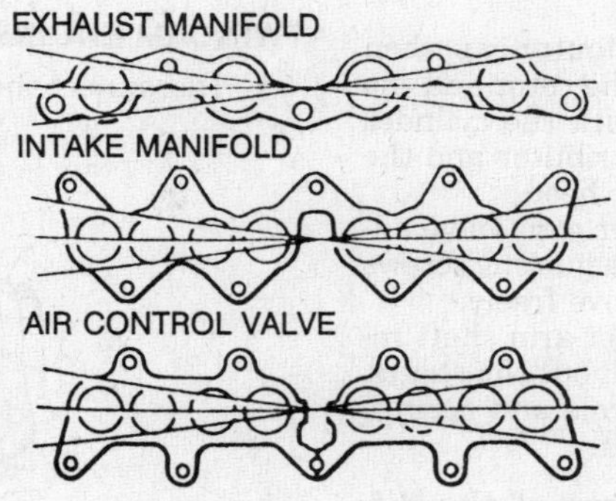

Check for warpage in 3 positions on the manifolds and air control valve—1991–92 Prizm.

3. Disconnect the 2 cable harness connectors, located near the left shock tower.

4. Disconnect the cable harness from from the MAT sensor, then the TPS sensor and the intake air control valve.

5. Remove the intake air duct from the throttle valve, then the PCV hose from the cylinder head cover.

6. Disconnect the EGR valve and canister vacuum hoses from the throttle valve.

7. Disconnect the EGR pipe from the EGR valve and the exhaust manifold.

8. Remove the intake manifold (common chamber) bracket bolt, then the throttle vale assembly bolts and the throttle valve.

9. Remove the coolant bypass pipe clip bolt, then disconnect the MAP sensor from the intake manifold (common chamber).

10. Disconnect the brake booster vacuum hose, then the canister vacuum hose from the intake manifold (common chamber) and the throttle valve.

11. Disconnect the pressure regulator vacuum hose from the common chamber, then the EGR vacuum hose.

12. Remove the engine hanger bolt, then the intake manifold (common chamber) attaching nuts and bolts. Remove the intake manifold (common chamber) from the vehicle.

To install:

13. Install the intake manifold (common chamber) and a new gasket onto the induction port. Torque nuts and bolts to 17 ft. lbs. (23 Nm).

14. Install the coolant bypass pipe clip bolts.

15. Install the throttle valve with a new gasket. Torque throttle valve bolts to 17 ft. lbs. (23 Nm).

16. Connect the EGR pipe flange to the exhaust manifold. Torque flange bolts to 17 ft. lbs. (23 Nm).

17. Connect the EGR vacuum hose, then the pressure regulator vacuum hose to the intake manifold (common chamber).

18. Connect the canister vacuum hose to the common chamber and the throttle valve, then the breather hose to the intake air duct.

19. Connect the cable harness to the MAT sensor, TPS and the intake air control valve.

20. Connect the 2 cable harness connectors, located near the left shock tower.

21. Connect the accelerator cable to the throttle valve and to the intake manifold (common chamber).

22. Connect the ignition coil wire. Connect the battery negative cable and fill cooling system. Start engine and check for leaks.

TWINCAM ENGINE

1. Disconnect the battery negative cable.

2. Disconnect the accelerator cable clip and the PCV hose from the intake air duct.

3. Disconnect the accelerator cable from the throttle valve.

4. Disconnect the MAP sensor hose from the MAP sensor, then the vacuum hose form the brake vacuum booster.

5. Disconnect the 2 canister hoses from intake manifold (common chamber).

6. Remove the canister pipe bracket from the intake manifold (common chamber).

7. Disconnect the vacuum hose from the fuel pressure regulator, then the vacuum hose from the induction port.

8. Remove the throttle valve from the intake manifold (common chamber).

9. Remove the alternator harness clip, then the 3 fuel injector harness cable clips.

10. Loosen the EGR pipe bracket on the exhaust manifold, then loosen the from the thermostat housing.

11. Remove the 2 intake manifold (common chamber) bracket attaching bolts, located on the left side of the engine and the engine hanger bolt, located on the right side of the intake manifold (common chamber).

12. Remove the intake manifold (common chamber) attaching nuts and bolts, then remove the intake manifold (common chamber) from the induction port assembly.

To install:

13. Install the intake manifold (common chamber) with a new gasket onto the induction port assembly. Torque nuts and bolts to 17 ft. lbs. (23 Nm).

14. Install the intake manifold (common chamber) bracket bolts, located on the left side of the engine and the hanger bolt, located on the right side of the engine.

15. Install the EGR pipe bracket bolts on the exhaust manifold and the EGR clip on the thermostat housing. Torque EGR bracket bolts to 32 ft. lbs. (44 Nm).

16. Install the alternator harness cable clip, then the 3 fuel injector harness cable clips.

17. Install the throttle valve onto the intake manifold (common chamber), then connect the induction control valve vacuum hose.

18. Connect the fuel pressure regulator vacuum hose to the intake manifold (common chamber).

19. Connect the canister pipe bracket onto the intake manifold (common chamber).

20. Connect the brake booster vacuum hose, then the MAP sensor hose to the MAP sensor.

21. Connect the accelerator cable clip to the intake air duct.

22. Connect the battery negative cable and fill cooling system. Start engine and check for leaks.

Spectrum

NON-TURBOCHARGED ENGINE

1. Disconnect the negative battery terminal from the battery. Drain the cooling system.

2. Remove the bolt securing the alternator adjusting plate to the engine.

3. Disconnect and label all of the hoses attached to the air cleaner and remove the air cleaner.

4. Disconnect the air inlet temperature switch wiring connector.

5. Disconnect and label the hoses, electrical connectors and control cable attached to the carburetor.

6. If equipped with air conditioning, disconnect the FIDC vacuum hose, the pressure tank control valve hose, the distributor 3-way connector hose and the VSV wiring connector.

7. Remove the carburetor attaching bolts, located beneath the intake manifold, then remove the carburetor and the EFE heater.

8. At the intake manifold, remove the PCV hose, the water bypass hose, the heater hoses, the EGR valve/canister hose, the distributor vacuum advance hose and the ground wires.

9. Disconnect the thermometer unit switch wiring connector.

10. Remove the intake manifold attaching nuts/bolts and the intake manifold.

11. Clean the sealing surfaces of the intake manifold and cylinder head.

To install:

12. Use new gaskets and reverse the removal procedures.

13. Torque the intake manifold to 17 ft. lbs. (23 Nm); then adjust the engine control cable and the alternator belt tension. Refill the engine with coolant and check for leaks.

TURBOCHARGED ENGINE

1. Relieve the fuel system pressure. Disconnect the negative battery terminal from the battery.

2. Remove pressure regulator and oil separator.

3. Disconnect vacuum line from VSV and remove vacuum switching valve from bracket.

4. Remove oil separator/VSV bracket and hanger as an assembly.

5. Remove throttle valve assembly and engine harness assembly (mark or tag harness connections if necessary).

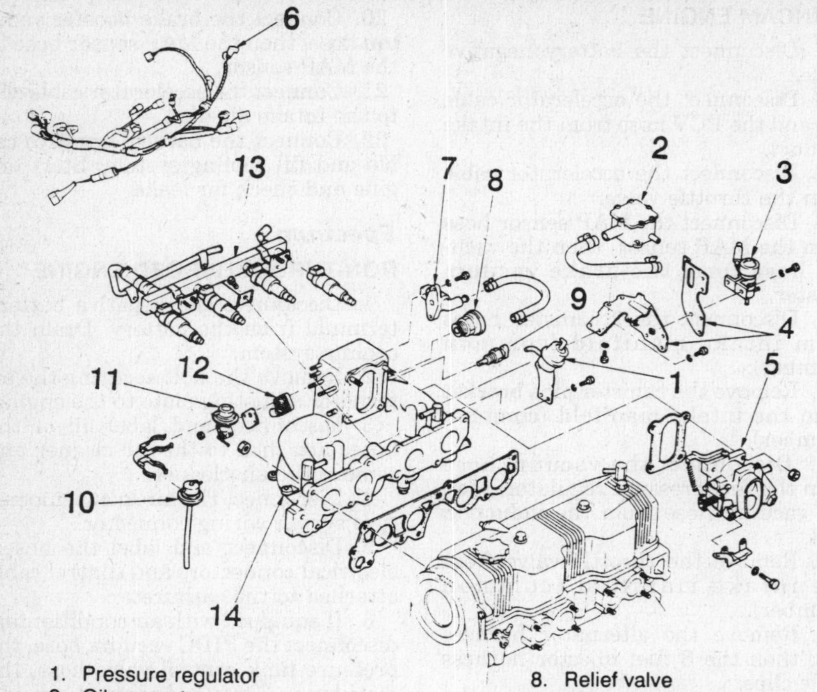

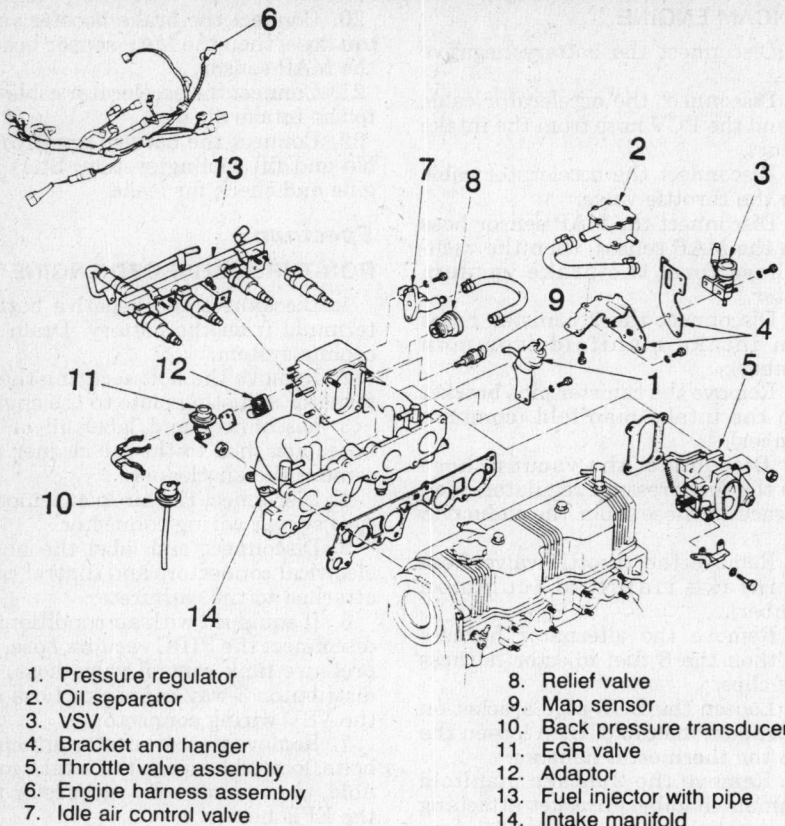

1. Pressure regulator
2. Oil separator
3. VSV
4. Bracket and hanger
5. Throttle valve assembly
6. Engine harness assembly
7. Idle air control valve
8. Relief valve
9. Map sensor
10. Back pressure transducer
11. EGR valve
12. Adaptor
13. Fuel injector with pipe
14. Intake manifold

Intake manifold turbocharged model – Spectrum

6. Remove idle air control valve, relief valve and MAP sensor.
7. Disconnect vacuum line from back pressure transducer and unclip transducer from hold-down bracket.
8. Remove EGR valve and adaptor plate.
9. Remove fuel injectors and fuel pipe connected to rail as one unit and position out of way then remove intake manifold with common chamber
10. To install, use new gaskets and reverse the removal procedures. Torque the intake manifold to 17 ft. lbs. (23 Nm). Start and run engine check for leaks.

Sprint

NON-TURBOCHARGED ENGINE

1. Relieve the fuel system pressure. Disconnect the negative battery cable.
2. Drain the cooling system.
3. Disconnect the air cleaner element, the EGR modulator, the warm air, the cool air, the 2nd air and the vacuum hoses from the air cleaner case.
4. Remove the air cleaner case, the electrical lead wires and the accelerator cable from the carburetor.
5. Disconnect the emission control and the fuel hoses from the carburetor.
6. Remove the water hoses from the choke housing.
7. Remove the electrical lead wires, the emission control, the coolant and the brake vacuum hoses from the intake manifold.
8. Remove the intake manifold from the cylinder head.
9. Clean the mating gasket surfaces.
10. To install, use new gaskets and reverse the removal procedures. Torque the intake manifold-to-cylinder head bolts to 14–20 ft. lbs. (19-27 Nm). Refill the cooling system.

TURBOCHARGED ENGINE

1. Disconnect the negative battery cable.
2. Drain the cooling system when the engine is cool.

NOTE: The fuel delivery pipe is under high pressure even after the engine is stopped, direct removal of the fuel line may result in dangerous fuel spray. Make sure to release the fuel pressure according to the procedure outlined under Fuel System in this section.

3. Remove the surge tank together with the throttle body.
4. Disconnect the fuel injector couplers.
5. Disconnect the fuel hoses from the delivery pipe.
6. Remove the delivery pipe together with the injectors.
7. Disconnect the water temperature gauge wire (yellow/white).
8. Disconnect the starter injector time switch coupler (brown).

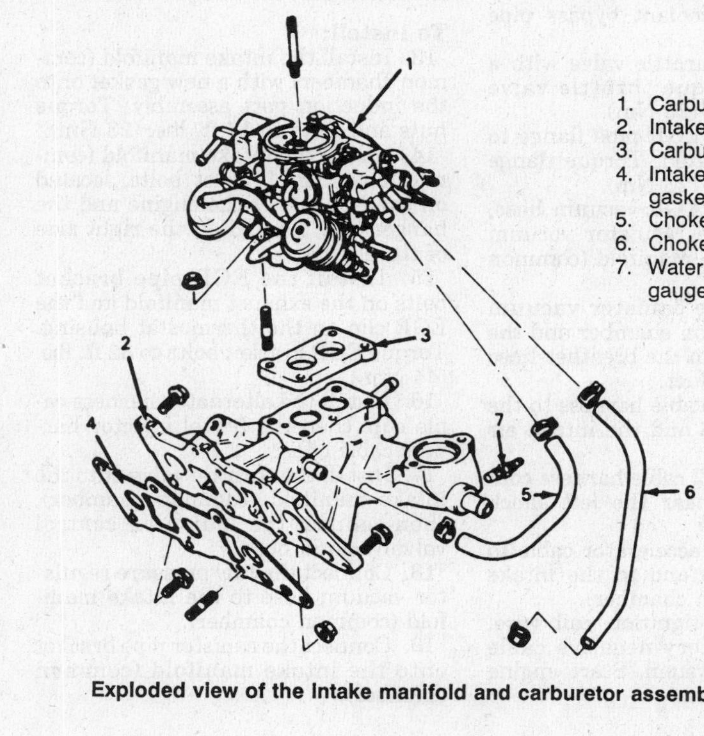

1. Carburetor
2. Intake manifold
3. Carburetor insulator
4. Intake manifold gasket
5. Choke No. 1 hose
6. Choke No. 2 hose
7. Water temperature gauge

Exploded view of the Intake manifold and carburetor assembly – Sprint

9. Disconnect the water temperature sensor coupler (green).
10. Disconnect the radiator fan switch coupler.
11. Disconnect the water hoses and the EGR vacuum hoses.
12. Remove the intake manifold retaining bolts then remove the manifold from the vehicle.**To install:**
13. Installation is the reverse of removal, with the following precautions:
a. Use a new intake manifold gasket.
b. If an injector was removed from the delivery pipe a new O-ring should be used.
c. Torque the intake manifold retaining bolts to 17 ft. lbs. (23 Nm).
d. After the ignition is turned on check for fuel leaks.

Metro

1. Relieve the fuel system pressure.
2. Disconnect the negative battery cable.
3. Drain the cooling system.
4. Remove the air cleaner assembly.
5. Disconnect the following wires:
a. Vacuum switching valve for EGR valve
b. Water temperature sensor
c. Idle speed control solenoid valve
d. Ground wires from the intake manifold
e. Fuel injector
f. Throttle switch or throttle position sensor
g. Water temperature gauge
6. Disconnect the fuel return and feed hoses from the throttle body.
7. Disconnect the water hoses from the throttle body and the intake manifold.
8. Disconnect the following hoses:
a. Canister purge hose from the intake manifold
b. Canister hose from its pipe
c. Pressure sensor hose from the intake manifold
d. Brake booster hose from the intake manifold
9. Disconnect the PCV hose from the cylinder head cover.
10. Disconnect the accelerator cable from the throttle body.
11. Disconnect any other lines and cables, as necessary.
12. Remove the intake manifold with the throttle body from the cylinder head.

To install:

13. Install the intake manifold to the cylinder head, using a new gasket, install the clamps and tighten the intake manifold retaining bolts to 17 ft. lbs. (23 Nm).
14. Reinstall all vacuum and water hoses.
15. Install the fuel feed and return hoses.
16. Install all electrical lead wires.
17. Install the accelerator cable to the throttle body.
18. Install the air cleaner assembly.
19. Fill the cooling system and reconnect the negative battery cable.

Exhaust Manifold

REMOVAL & INSTALLATION

1988 Nova

TWINCAM ENGINE

1. Disconnect the negative terminal from the battery.
2. Remove the exhaust manifold heat shield.
3. Raise and support the vehicle safely.
4. Remove the exhaust pipe-to-exhaust manifold nuts and separate the pipe from the manifold. Remove the exhaust manifold bracket from the exhaust manifold, the exhaust manifold-to-engine bolts, the exhaust manifold and gasket (discard the gasket) from the cylinder head.
5. Using the proper tool, clean the gasket mounting surfaces. Inspect the exhaust manifold for damage and/or warpage; maximum warpage is 0.012 in., if the warpage is greater, replace the exhaust manifold.
6. To install, use a new gasket and reverse the removal procedures. Torque the exhaust manifold-to-cylinder head bolts to 18 ft. lbs. (25 Nm). Start the engine and check for leaks.

1989-90 Prizm

1. Disconnect the negative battery cable.
2. Remove the 5 bolts and remove the upper heat shield (insulator) from the manifold.
3. Raise and safely support the vehicle.
4. Disconnect the exhaust pipe from the exhaust manifold and remove the manifold support and its 2 bolts.
5. Lower the vehicle to the ground. Disconnect the oxygen sensor wire.
6. Remove the 2 nuts and 3 bolts holding the manifold to the engine. Remove the manifold and its gaskets. When the manifold is clear of the car, remove the lower heat shield.
7. Measure the exhaust manifold mating surface with a precision straight edge and a feeler gauge. If the warpage exceeds 0.28mm, the manifold must be replaced.

To install:

8. Before reinstalling, attach the lower heat shield to the manifold with the 3 bolts. Tighten the bolts to 18 ft. lbs. (25 Nm).
9. Install the manifold with new gaskets and tighten its bolts and nuts to 18 ft. lbs. (25 Nm).
10. Raise the vehicle and safely support. Install the manifold support and tighten the bolts to 18 ft. lbs. (25 Nm).
11. Connect the exhaust pipe to the manifold with new gaskets and tighten the bolts to 18 ft. lbs. (25 Nm).
12. Lower the vehicle to the ground. Install the upper heat shield on the manifold and tighten its 5 bolts to 18 ft. lbs. (25 Nm).
13. Connect the wiring to the oxygen sensors.
14. Connect the negative battery cable.

1991-92 Prizm

1. Disconnect the negative battery cable.
2. Drain the cooling system.
3. Remove the air cleaner assembly.

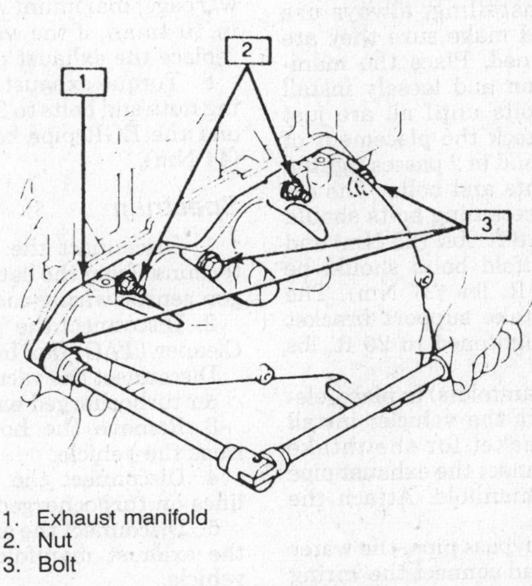

1. Exhaust manifold
2. Nut
3. Bolt

Location of the attaching hardware, on the exhaust manifold – 1989-90 Prizm

4. Remove the upper radiator hose at the engine.
5. Disconnect the accelerator cable and, on automatic tansaxles, the throttle valve cable.
6. Label and disconnect vacuum hoses at the manifolds.
7. Disconnect and remove the fuel delivery pipe (fuel rail) and remove the injectors. During removal, be careful not to drop the injectors.

CAUTION

The fuel system is under pressure. Release pressure slowly and contain spillage. Observe no smoking/no open flame precautions. Have a Class B-C (dry powder) fire extinguisher within arm's reach at all times.

8. Disconnect the vacuum hose to the brake booster and remove the heat shield(s) from the manifold(s).
9. Safely raise the vehicle and support it.
10. Disconnect the wire for the water temperature sensor and remove the water outlet housing (thermostat housing) and the bypass pipe.
11. Remove the exhaust bracket and disconnect the exhaust pipe at the manifold. Remove the support bracket for the intake manifold.
12. Lower the vehicle to the ground.
13. Remove the intake manifold with the air control valve and gaskets and/or remove the exhaust manifold with its gaskets.**To install:**
14. Using a precision straight edge and a feeler gauge, check the mating surfaces of the manifolds for warpage. If the warpage is greater than the maximum allowable specification, replace the manifold.
Maximum allowable warpage: Intake Manifold: 0.05mm; Exhaust Manifold: 0.3mm; Air Control Valve: 0.05mm.
15. When reinstalling, always use new gaskets and make sure they are properly positioned. Place the manifold(s) in position and loosely install the nuts and bolts until all are just snug. Double check the placement of the manifold(s) and in 2 passes tighten the retaining nuts and bolts. The exhaust manifold retaining bolts should be tightened to 18 ft. lbs. (25 Nm) and the intake manifold bolts should be tightened to 20 ft. lbs (27 Nm). The bolts for the intake support bracket should also be tightened to 20 ft. lbs (27 Nm).
16. With the manifold(s) in place, elevate and support the vehicle; install the support bracket for the intake manifold and connect the exhaust pipe to the exhaust manifold. Attach the exhaust bracket.
17. Install the bypass pipe, the water outlet housing and connect the wiring to the water temperature sensor.
18. Lower the vehicle to the ground and install the heat shield(s) on the manifolds.
19. Connect the vacuum hose for the brake vacuum booster.
20. Install the fuel delivery pipe and the injectors. Tighten the mounting bolts to 13 ft. lbs. (18 Nm).
21. Observing the labels made earlier, install the vacuum lines. Be careful of the routing and make sure that each line fits snugly on its port.
22. Double check each line for crimps or twists.
23. Connect the accelerator cable and throttle valve cable (automatic trans.)
24. Reconnect the upper radiator hose.
25. Install the air cleaner assembly.
26. Refill the coolant.
27. Connect the negative battery cable.

Storm

1. Disconnect the battery negative cable.
2. Remove the heat protector, then disconnect the oxygen sensor electrical connector.
3. If equipped with twincam engine, Disconnect the EGR pipe clip from the thermostat housing.
4. Disconnect the EGR pipe from the exhaust manifold and the EGR valve.
5. Remove the front exhaust pipe from the exhaust manifold.
6. Remove the exhaust manifold attaching nuts and bolts, then remove the exhaust manifold from the engine.
To install:
7. Reverse procedure to install. Using the proper tool, clean the gasket mounting surfaces. Inspect the exhaust manifold for damage and/or warpage; maximum warpage is 0.0157 in. (0.4mm), if the warpage is greater, replace the exhaust manifold.
8. Torque exhaust manifold attaching nuts and bolts to 30 ft. lbs. (41 Nm) and the EGR pipe bolts to 32 ft. lbs. (44 Nm).

Spectrum

1. Disconnect the negative battery terminal from the battery and the oxygen sensor wiring connector.
2. Disconnect the Thermostatic Air Cleaner (TAC) flex hose.
Disconnect the vacuum and oil lines on turbocharged engines.
3. Remove the hot air cover and raise the vehicle.
4. Disconnect the vacuum and oil lines on turbocharged engines.
5. Disconnect the exhaust pipe from the exhaust manifold and lower the vehicle.
6. Remove the nuts and bolts securing the exhaust manifold to the cylinder head. Clean the gasket mounting surfaces.
7. To install, use new ga002ts and reverse the removal procedures. Torque the exhaust manifold to 17 ft. lbs. (23 Nm) or 21 ft. lbs. (28 Nm) turbocharged model then start the engine and check for leaks.

Sprint

NON-TURBOCHARGED ENGINE

1. Disconnect the negative battery cable.
2. Raise and support the vehicle safely.
3. Remove the exhaust pipe at the exhaust manifold.
4. Remove the lower heat shield bolt and the 2nd air pipe at the exhaust manifold.
5. If equipped, remove the air conditioning drive belt and the lower adjusting bracket.
6. Lower the vehicle.
7. Remove the spark plug and the oxygen sensor wires.
8. Remove the hot air shroud from the exhaust manifold.
9. Remove the 2nd air valve hoses, the valve and the pipe from the exhaust manifold.
10. Remove the mounting bolts and the exhaust manifold.**To install:**
11. Clean the gasket mating surfaces.
12. To install, use a new gasket and reverse the removal procedures. Torque the exhaust manifold fasteners to 14–20 ft. lbs. (19-27 Nm) and the exhaust pipe to 30–43 ft. lbs. (41-58 Nm).

TURBOCHARGED ENGINE

1. Disconnect the battery negative cable and drain the cooling system.
2. Remove the hood.
3. Remove the front grille attaching screws, then the front grille from the vehicle.
4. Remove the intercooler, then the radiator hoses.
5. Disconnect the radiator fan motor electrical connector, then remove the front upper member.
6. If equipped with air conditioning, discharge the air conditioning system, then remove the air conditioning condensor.
7. Remove the radiator from the vehicle.
8. Disconnect the front bumper from the damper flange. Place a stand under the front bumper to prevent it from dropping, then remove the coupler clamps and bolts and pull the bumper outward.
9. Remove the exhaust pipe bolts.
10. If equipped with air conditioning, remove the air conditioning compressor.

11. Remove the turbocharger cover, then disconnect the oxygen sensor electrical connector.
12. Remove the turbocharger side cover, then the lower exhaust pipe support bracket bolt.
13. Remove the upper exhaust pipe, along the lower exhaust pipe.
14. Disconnect the air outlet pipe, then the air inlet hose clamp bolt on the cylinder head.
15. Remove the air inlet pipe, then disconnect the oil drain pipe.
16. Disconnect the water pipe cylinder head clamp bolt, then the water hoses.
17. Remove the turbocharger attaching bolts, then the turbocharger from the vehicle.
18. Remove the exhaust pipe from the exhaust manifold. Remove the manifold retaining bolts and remove the exhaust manifold and gaskets.**To install:**
20. Reverse procedure to install. Install a new manifold gasket and torque the bolts to 17 ft. lbs. (23 Nm).
21. Connect the battery negative cable and recharge air conditioning system.
22. Check and refill fluids as necessary.
23. Start engine and check for leaks.

Metro

1. Disconnect the negative battery cable.
2. Remove the exhaust pipe from the exhaust manifold.
3. Remove the manifold retaining bolts and remove the exhaust manifold and gaskets.
4. Reverse procedure to install. Install a new manifold gasket and torque the bolts to 17 ft. lbs. (23 Nm). Reconnect the oxygen sensor coupler and the battery negative cable.

Combination Manifold

REMOVAL & INSTALLATION

1988 Nova

EXCEPT TWINCAM ENGINE

The intake and exhaust manifolds on the this engine are a one-piece or combination design. They can not be separated from each other or removed individually.

1. Disconnect the negative battery cable.
2. Remove the air cleaner assembly.
3. Label and disconnect all vacuum hoses at the carburetor.
4. Disconnect the accelerator cable and, for automatic tansaxles, the throttle cable.
5. Label and disconnect the electrical connections at the carburetor.
6. Disconnect the fuel line at the fuel pump.
7. Carefully loosen and remove the carburetor mounting bolts and remove the carburetor.

CAUTION

The carburetor bowls contain gasoline which may spill or leak during removal. Observe no smoking/no open flame precautions. Have a Class B-C (dry powder) fire extinguisher within arm's reach at all times.

NOTE: Keep the carburetor level during removal and handling. As soon as it is off the car, wrap or cover it with a clean towel to keep dirt out.

8. Remove the Early Fuel Evaporation (EFE) gasket.
9. Remove the vacuum line and dashpot bracket.
10. Carefully remove the heat shields on the manifold; don't break the bolts.
11. Safely elevate and support the vehicle.
12. Disconnect the exhaust pipe at the manifold and exhaust bracket at the engine.
13. Remove the hose at the converter pipe.
14. Lower the vehicle to the ground and remove or disconnect the vacuum hose to the brake booster.
15. Remove the bracket for the accelerator and throttle cables.
16. Evenly loosen and then remove the bolts and nuts holding the manifold to the engine. Remove the manifold and its gaskets.
17. Using a precision straight edge and a feeler gauge, check the mating surfaces of the manifold for warpage. If the warpage is greater than the maximum allowable specification, replace the manifold.

Maximum permitted warpage: Intake 0.2mm; Exhaust 0.3mm.

To install:

18. When reinstalling, always use new gaskets and make sure they are properly positioned. Place the manifold in position and loosely install the nuts and bolts until all are just snug. Double check the placement of the manifold and in 2 passes tighten the retaining nuts and bolts to 18 ft .lbs. (25 Nm).
19. Reinstall the bracket for the accelerator and throttle cables and connect the vacuum hose to the brake vacuum booster.
20. Elevate and safely support the vehicle.
21. Reconnect the hose at the converter pipe. Install the exhaust bracket at the engine and, using new gaskets, connect the exhaust pipe to the manifold.
22. Lower the vehicle to the ground. Install the heat shield onto the manifold.
23. Reinstall the vacuum line and the dashpot bracket.
24. Install the EFE gasket.
25. Reinstall the carburetor. Slowly and evenly tighten the mounting nuts and bolts to 8-10 ft. lbs. (11-14 Nm). Connect the electrical connectors to the carburetor.
26. Connect and secure the fuel line to the fuel pump.
27. Attach the accelerator cable and throttle valve cable (automatic trans.).
28. Observing the labels made earlier, install the vacuum lines. Be careful of the routing and make sure that each line fits snugly on its port. double check each line for crimps or twists.
29. Install the air cleaner assembly and connect the negative battery cable.

Turbocharger

REMOVAL & INSTALLATION

Spectrum

1. Disconnect the battery negative cable.
2. Remove the upper and lower heat protector shields from the turbocharger assembly.
3. Remove the manifold heat protector, then disconnect the oxygen sensor electrical connector.
4. Disconnect the vacuum pipe wastegate and position aside.
5. Disconnect the water lines, then disconnect the return and delivery oil lines.
6. Disconnect the exhaust pipe from the wastegate manifold.

NOTE: The exhaust manifold studs should be soaked with CRC or equivalent to prevent studs from breaking before removal.

7. Remove the turbocharger and wastegate from the vehicle as an assembly.
8. To install, use a new gasket on the exhaust manifold-to-turbocharger housing and reverse the removal procedure. Connect the battery negative cable and refill all fluids. Start the engine and check for leaks.

Sprint

1. Disconnect the battery negative cable and drain the cooling system.
2. Remove the hood.
3. Remove the front grille attaching screws, then the front grille from the vehicle.

4. Remove the intercooler, then the radiator hoses.
5. Disconnect the radiator fan motor electrical connector, then remove the front upper member.
6. If equipped with air conditioning, discharge the air conditioning system, then remove the air conditioning condensor.
7. Remove the radiator from the vehicle.
8. Disconnect the front bumper from the damper flange. Place a stand under the front bumper to prevent it from dropping, then remove the coupler clamps and bolts and pull the bumper outward.
9. Remove the exhaust pipe bolts.
10. If equipped with air conditioning, remove the air conditioning compressor.
11. Remove the turbocharger cover, then disconnect the oxygen sensor electrical connector.
12. Remove the turbocharger side cover, then the lower exhaust pipe support bracket bolt.
13. Remove the upper exhaust pipe, along the lower exhaust pipe.
14. Disconnect the air outlet pipe, then the air inlet hose clamp bolt on the cylinder head.
15. Remove the air inlet pipe, then disconnect the oil drain pipe.
16. Disconnect the water pipe cylinder head clamp bolt, then the water hoses.
17. Remove the turbocharger attaching bolts, then the turbocharger from the vehicle.

To install:

18. Reverse procedure to install.
19. Connect the battery negative cable and recharge air conditioning system.
20. Check and refill fluids as necessary. Start engine and check for leaks.

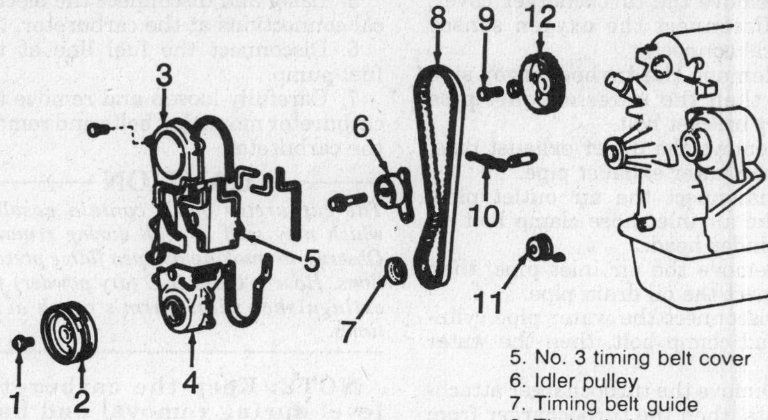

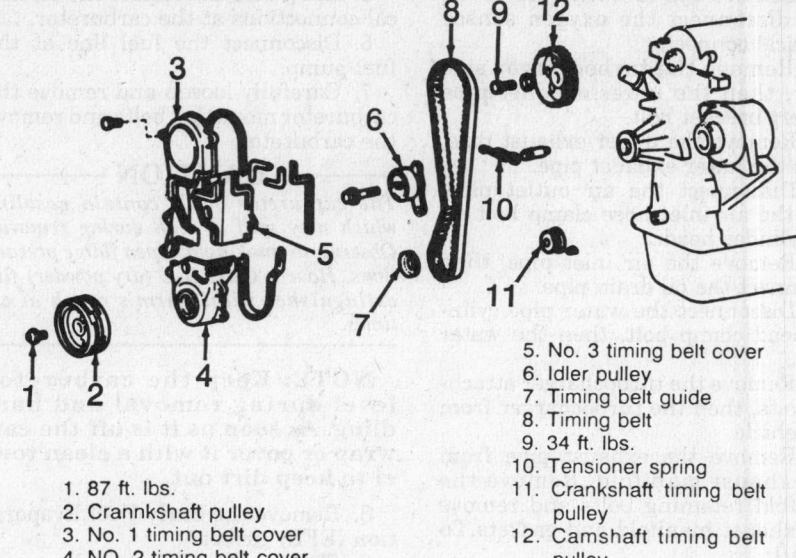

1. 87 ft. lbs.
2. Crankshaft pulley
3. No. 1 timing belt cover
4. NO. 2 timing belt cover
5. No. 3 timing belt cover
6. Idler pulley
7. Timing belt guide
8. Timing belt
9. 34 ft. lbs.
10. Tensioner spring
11. Crankshaft timing belt pulley
12. Camshaft timing belt pulley

Exploded view of the timing belt assembly—except twincam engine—Nova

Timing Belt Front Cover

REMOVAL & INSTALLATION

Nova—Except Twincam Engine

This engine uses a 3 piece timing belt cover assembly; any individual cover can be removed by performing one of the following procedures.

UPPER

1. Disconnect the negative terminal from the battery.
2. Loosen the water pump pulley bolts and remove the alternator/water pump drive belt. If equipped with power steering, remove the power steering pump drive belt.
3. Remove the water pump pulley bolts and pulley. Drain the cooling system.
4. Disconnect the upper radiator hose from the water pump outlet. Label and disconnect all vacuum hoses that may be in the way.
5. Remove the upper timing belt front cover-to-engine bolts.

NOTE: To remove the lower timing belt cover-to-engine bolts, it may be necessary to raise and support the vehicle, then, remove them from underneath.

6. Remove the upper timing belt front cover and gasket.
7. Using the proper tool, clean the gasket mounting surfaces.
8. To install, use a new gasket, sealant, if necessary, and reverse the removal procedures. Adjust the drive belts. Refill the cooling system. Start the engine, allow it to reach normal operating temperatures and check for leaks.

MIDDLE

1. Remove the upper timing belt front cover.
2. If equipped with air conditioning, loosen the idler pulley mounting bolt. Loosen the adjusting nut, then, remove the air conditioning drive belt, the idler pulley, with adjusting bolt.
3. Remove the alternator bolts and move it aside.
4. Remove the middle timing belt front cover-to-engine bolts, the cover and gasket.
5. Using the proper tool, clean the gasket mounting surfaces.
6. To install, use a new gasket, sealant, if necessary, and reverse the removal procedures. Adjust the drive bolts. Refill the cooling system. Start the engine, allow it to reach normal operating temperatures and check for leaks.

LOWER

1. Disconnect the negative terminal from the battery.
2. Loosen the alternator adjusting bolts and remove the drive belt.
3. If equipped with air conditioning, remove the drive belt.
4. Raise and support the vehicle safely.
5. Remove the right-side under cover, the flywheel cover, the crankshaft pulley-to-crankshaft bolt and the crankshaft pulley.
6. Remove the lower timing belt front cover-to-engine bolts, the cover and gasket.
7. Using the proper tool, clean the gasket mounting surfaces.
8. To install, use a new gasket, sealant, if necessary, and reverse the removal procedures. Torque the crankshaft pulley-to-crankshaft bolt to 80–94 ft. lbs. (108-127 Nm). Adjust the drive belt(s).

Nova—With Twincam Engine

This engine uses a 3 piece timing belt front cover assembly of an interlocking design. To removal any portion of the cover, disassembly must start from the top and work to the bottom.

1. Disconnect the negative terminal from the battery.
2. Raise and support the vehicle safely, then, remove the right side wheel assembly.

3. Remove the under carriage splash shield and drain the cooling system.
4. Disconnect the accelerator cable, the cruise control cable, if equipped, the cruise control actuator, if equipped, and the ignition coil.
5. Remove the water outlet housing-to-engine bolts and the housing.
6. Remove the drive belt from the power steering pimp, if equipped, and the alternator. Disconnect the spark plug wires from the spark plugs and the spark plugs from the engine.
7. To position the No. 1 cylinder on the TDC of its compression stroke, perform the following procedures:
   a. Using a socket wrench on the crankshaft pulley bolt, rotate the crankshaft to align the notch on the crankshaft pulley with the idler pulley bolt.
   b. Remove the oil filler cap and look for the hole in the camshaft; if it cannot be seen, rotate the crankshaft 1 complete revolution and check for it again.
8. Remove the right-side engine mount, the water pump pulley bolts and the pulley.
9. To remove the crankshaft pulley, perform the following procedures:
   a. Using a crankshaft pulley holding tool, secure and hold the pulley while removing the crankshaft pulley-to-crankshaft bolt.
   b. Using the crankshaft pulley puller tool, press the crankshaft pulley from the crankshaft.
10. Remove the timing belt front covers-to-engine bolts, the covers and the gaskets.**To install:**
11. Using the proper tool, clean the gasket mounting surfaces.
12. To install, use new gaskets, sealant (if necessary) and reverse the removal procedures.
13. Using the crankshaft pulley holding tool, secure and hold the pulley while installing the crankshaft pulley-to-crankshaft bolt.
14. To complete the installation, reverse the removal procedures. Adjust the drive belts. Refill the cooling system. Start the engine, allow it to reach normal operating temperatures and check for leaks.

1989-90 Prizm

1. Disconnect the negative battery cable.
2. Elevate the vehicle and safely support it.
3. Remove the right splash shield under the vehicle.
4. Lower the vehicle. Remove the wiring harness from the upper timing belt cover.
5. Depending on equipment, loosen the air conditioner compressor, the power steering pump and the alternator on their adjusting bolts. Remove the drive belts.
6. Remove the crankshaft pulley. The use of a counterholding tool such as J 8614-01 or similar is highly recommended.
7. Remove the valve cover.
8. Remove the windshield washer reservoir.
9. Elevate and safely support the vehicle.
10. Support the engine either from above (Tool 28467-A or chain hoist) or below (floor jack and wood block) and remove the through bolt at the right engine mount.
11. Remove the protectors on the mount nuts and studs for the center and rear transaxle mounts.
12. Remove the 2 rear transaxle mount-to-main crossmember nuts. Remove the 2 center transaxle mount-to-center crossmember nuts.
13. Carefully elevate the engine enough to gain access to the water pump pulley.
14. Remove the water pump pulley. Lower the engine to its normal position.
15. Remove the 4 bolts and the lower timing cover. Remove the center timing cover and its bolt, then the upper cover with its 4 bolts.
16. If further work is to be done, the vehicle may be lowered to the ground but the engine must remain supported until the mount(s) are reinstalled.

To install:

17. When reinstalling, make certain that the gaskets and their mating surfaces are clean and free from dirt and oil. The gasket itself must be free of cuts and deformations and must fit securely in the grooves of the covers.
18. Install the covers and the bolts; tighten the bolts to 3.6 ft. lbs. (44 in. lbs.)
19. Elevate the engine and install the water pump pulley.
20. Lower the engine to its normal position. Install the through bolt in the right engine mount and tighten it to 64 ft. lbs. (87 Nm) with the bolt secure, the engine lifting apparatus may be removed.
21. Install the valve cover.
22. Install the crankshaft pulley and tighten its bolt to 87 ft. lbs. (118 Nm).
23. Reinstall the air conditioning compressor, the power steering pump and the alternator. Install their belts and adjust them to the correct tension.
24. Reconnect the wiring harness to the upper timing belt cover.
25. Raise the vehicle and safely support it.
26. Install the 2 nuts on the center transaxle mount and the rear transaxle mount. Tighten all the nuts to 45 ft. lbs. (61 Nm).
27. Install the protectors on the nuts and studs.
28. Install the splash shield under the vehicle.
29. Lower the vehicle to the ground.
30. Install the windshield washer reservoir and connect the negative battery cable.

1991-92 Prizm

1. Disconnect the negative battery cable.

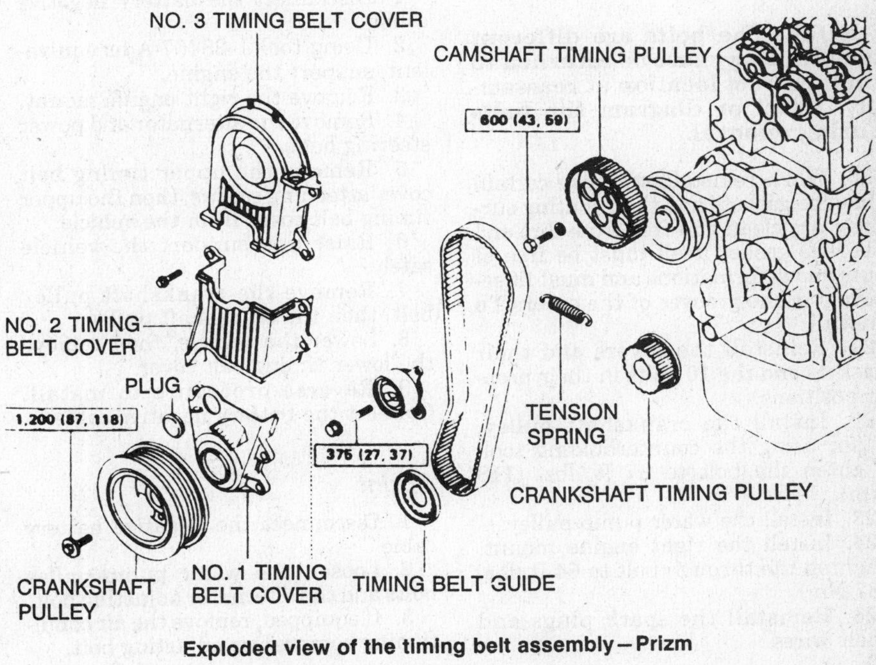

Exploded view of the timing belt assembly—Prizm

2. Elevate the vehicle and safely support it on jackstands.
3. Remove the right front wheel.
4. Remove the splash shield from under the vehicle.
5. Drain the coolant into clean containers. Close the draincocks when the system is empty.
6. Lower the vehicle to the ground. Disconnect the accelerator cable and, if equipped, the cruise control cable.
7. Remove the cruise control actuator if so equipped.
8. Carefully remove the ignition coil.
9. Disconnect the radiator hose at the water outlet.
10. Remove the power steering drive belt and the alternator drive belt.
11. Remove the spark plugs.
12. Rotate the crankshaft clockwise and set the engine to TDC/compression on No. 1 cylinder. Align the crankshaft marks at zero; look through the oil filler hole and make sure the small hole in the end of the camshaft can be seen.
13. Raise and safely support the vehicle. Disconnect the center engine mount.
14. Lower the vehicle to the ground.
15. Support the engine either from above or below. Disconnect the right engine mount from the engine.
16. Raise the engine and remove the mount.
17. Remove the water pump pulley.
18. Remove the crankshaft pulley. The use of a counterholding tool such as J-8614-01 or similar is highly recommended.
19. Remove the 10 bolts and remove the timing belt covers with their gaskets.

NOTE: The bolts are different lengths; they must be returned to their correct location at reassembly. Label or diagram the bolts during removal.

20. When reinstalling, make certain that the gaskets and their mating surfaces are clean and free from dirt and oil. The gasket itself must be free of cuts and deformations and must fit securely in the grooves of the covers.**To install:**
21. Reinstall the covers and their gaskets and the 10 bolts in their proper positions.
22. Install the crankshaft pulley, again using the counterholding tool. Tighten the bolt to 87 ft. lbs. (118 Nm).
23. Install the water pump pulley.
24. Install the right engine mount. Tighten the through bolt to 64 ft. lbs. (87 Nm).
25. Reinstall the spark plugs and their wires.
26. Install the alternator drive belt and the power steering drive belt. Adjust the belts to the correct tension.
27. Connect the radiator hose to the water outlet port.
28. Install the ignition coil.
29. Install the cruise control actuator and the cruise control cable if so equipped.
30. Connect the accelerator cable.
31. Refill the cooling system with the correct amount of anti-freeze and water.
32. Connect the negative battery cable.
33. Start the engine and check for leaks. Allow the engine to warm up and check the work areas carefully for seepage.
34. Install the splash shield under the vehicle.
35. Install the right front wheel.

Storm

EXCEPT TWINCAM ENGINE

1. Disconnect the battery negative cable.
2. Remove the alternator belt, then the power steering belt.
3. Using tool J–28467‹n-dash›A or equivalent, support the engine.
4. Remove the right engine mount.
5. Remove the timing belt cover attaching screws, then the timing belt cover from the vehicle.
6. Reverse procedure to install. Connect the battery negative cable.

TWINCAM ENGINE

1. Disconnect the battery negative cable.
2. Using tool J-28467-A or equivalent, support the engine.
3. Remove the right engine mount.
4. Remove the alternator and power steering belts.
5. Remove the upper timing belt cover attaching screws, then the upper timing belt cover from the vehicle.
6. Raise and support the vehicle safely.
7. Remove the crankshaft pulley bolt, then the crankshaft pulley.
8. Lower the vehicle, then remove the lower timing belt cover.
9. Reverse procedure to install. Connect the battery negative cable.

Sprint

1. Disconnect the negative battery cable.
2. Loosen the water pump pulley bolts and the alternator adjusting bolt.
3. If equipped, remove the air conditioning compressor adjusting bolt.
4. Raise and support the vehicle safely.
5. Remove the drive belt splash shield, the right fender plug and the drive belts.
6. Remove the crankshaft and the water pump pulleys.
7. Remove the bolts from the bottom of the belt cover.
8. Lower the vehicle.
9. Remove the bolts from the top of the belt cover, then the cover from the vehicle.
10. Reverse procedure to install. Connect the battery negative cable.

Spectrum

1. Disconnect negative battery cable.
2. Support the engine.
3. Remove the front mount bracket attached to the front cover.
4. Remove front cover.
5. To install, reverse the removal procedures.

Metro

1. Disconnect the negative battery cable.
2. Raise and support the vehicle safely.
3. Remove the fender apron extension on the right side. Remove the clip after pushing the center pin.
4. Remove the water pump belt and its pulley. Loosen the generator pivot bolts and remove the water bolt.
5. Remove the crankshaft pulley by removing the pulley bolts. It is not necessary to loosen the crankshaft timing belt pulley bo0lt at the center.
6. Remove the timing belt outside cover.
7. To install, reverse the removal procedures.

Front Cover Oil Seal

REMOVAL & INSTALLATION

All Vehicles

The front cover oil seal replacement can and should be done when the front timing belt cover has been removed.

1. With the timing cover containing the oil seal removed, use a suitable seal removal tool and remove the front oil seal from the cover.
2. To install, apply a light coat of oil to the crankshaft and the lip of the new oil seal.
3. Using a suitable seal driver tool, drive in the new oil seal into the cover until the end of the seal sits squarely with the cover.
4. Reinstall the cover on the vehicle.

Timing Belt and Tensioner

ADJUSTMENT

Nova and Prizm

1. Remove the front cover assembly.
2. Using finger pressure on the longest span between pulleys (except twincam) or between the camshaft pulleys (twincam), measure the timing belt deflection; 4.4 lbs. at 0.24‹n-dash›0.28 in. (except twincam) or 0.16 in. (twincam).
3. If adjustment is not correct, loosen the idler pulley bolt and correct the belt tension.
4. To install, the front covers, reverse the removal procedures.

Spectrum and Storm

1. Remove the front cover.
2. Loosen the timing belt tension pulley bolt.

NOTE: If the belt has been removed or replaced with a new one, perform the following procedures to stretch the belt.

3. Using an Allen wrench, insert it into the hexagonal hole of the tension pulley. Hold the pulley stationary and temporarily tighten the tension pulley-to-engine bolt.
4. Rotate the crankshaft 2 complete revolutions and align the crankshaft timing pulley groove with the mark on the oil pump.
5. Loosen the tension pulley-to-engine bolt.
6. Using the Allen wrench and a timing belt tension gauge, apply 38 ft. lbs. (52 Nm) of tension to the timing belt on Spectrum and 31 ft. lbs. (42 Nm) on tension on the timing belt on Storm. Hold the pulley stationary and torque the tension pulley-to-engine bolt to 37 ft. lbs. (50 Nm) on Spectrum and 31 ft. lbs. (42 Nm) on Storm.
7. To complete the adjustment, install the front cover.

Sprint

1. Disconnect the battery negative cable.
2. Remove the front cover.
3. Turn the camshaft pulley clockwise and align the mark on the pulley with the **V** mark on the inside cover.
4. Using a 17mm wrench, turn the crankshaft clockwise and align the punch mark on the crankshaft pulley with the arrow mark on the oil pump.
5. With the timing marks aligned, install the timing belt so there is no belt slack on the right side (facing the engine) of the engine, apply belt tension with the tensioner pulley.
6. Turn the crankshaft 1 rotation clockwise to remove the belt slack. Torque the tensioner stud, first, and then the tensioner bolt to 17‹n-dash›21 ft. lbs. (23-28 Nm).

Metro

1. Disconnect the battery negative cable.
2. Remove the front cover.
3. Take up the slack of the timing belt by turning the crankshaft 2 rotations clockwise after installing it. After making sure that the belt is free from slack, tighten the tensioner stud first to 8 ft. lbs. (11 Nm) and the tensioner bolt to 20 ft. lbs. (27 Nm). Then confirm again that the 2 sets of marks are aligned respectively.
4. Install the timing belt outside cover. Before installing, make sure that the seal is between the water pump and oil pump case. Tighten the timing belt cover bolts to 8 ft. lbs. (11 Nm).

REMOVAL & INSTALLATION

NOTE: Timing belts must always be handled carefully and kept completely free of dirt, grease, fluids and lubricants. This includes any accidental contact from spillage, fingerprints, rags, etc. These same precautions apply to the pulleys and contact surfaces on which the belt rides.

The belt must never be crimped, twisted or bent. Never use tools to pry or wedge the belt into place. Such actions will damage the structure of the belt and possibly cause breakage.

Nova

EXCEPT TWINCAM ENGINE

1. Remove the timing belt covers.
2. If not done as part of the cover removal, rotate the crankshaft clockwise to the TDC/compression position for No. 1 cylinder. Insure that the crankshaft marks align at zero and that the rocker arms on No. 1 cylinder are loose.
3. Loosen the timing belt idler pulley to relieve the tension on the belt.
4. Make matchmarks on the belt and both pulleys showing the exact placement of the belt. Mark an arrow on the belt showing its direction of rotation.
5. Carefully slip the timing belt off the pulleys.

NOTE: Do not disturb the position of the camshaft or the crankshaft during removal.

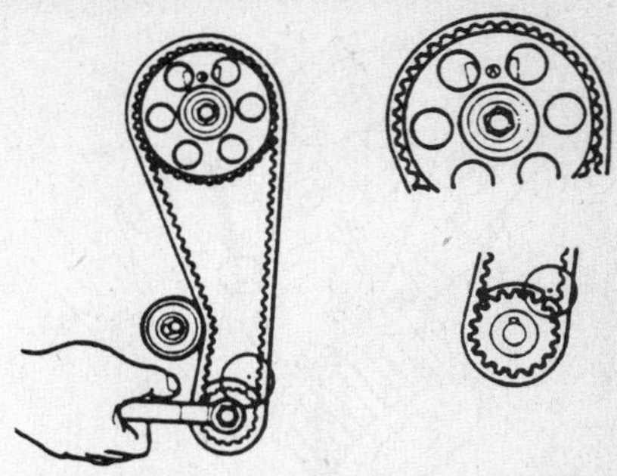

Aligning the valve timing marks—except twincam engine—Nova

6. Remove the idler pulley bolt, pulley and return spring.
7. Use an adjustable wrench mounted on the flats of the camshaft to hold the cam from moving. Loosen the center bolt in the camshaft pulley and remove the pulley.
8. Check the timing belt carefully for any signs of cracking or deterioration. Pay particular attention to the area where each tooth or cog attaches to the backing of the belt. If the belt shows signs of damage, check the contact faces of the pulleys for possible burrs or scratches.
9. Check the idler pulley by holding it and spinning it. It should rotate freely and quietly. Any sign of grinding or abnormal noise indicates replacement of the pulley.
10. Check the free length of the tension spring. Correct length is 38.5mm measured at the inside faces of the hooks. A spring which has stretched during use will not apply the correct tension to the pulley; replace the spring.
11. Test the tension of the spring, look for 8.4 lbs. of tension at 50mm of length. If in doubt, replace the spring.**To install:**
12. Reinstall the camshaft timing belt pulley, making sure the pulley fits properly on the shaft and that the timing marks align correctly. Tighten the center bolt to 34 ft. lbs. (46 Nm).

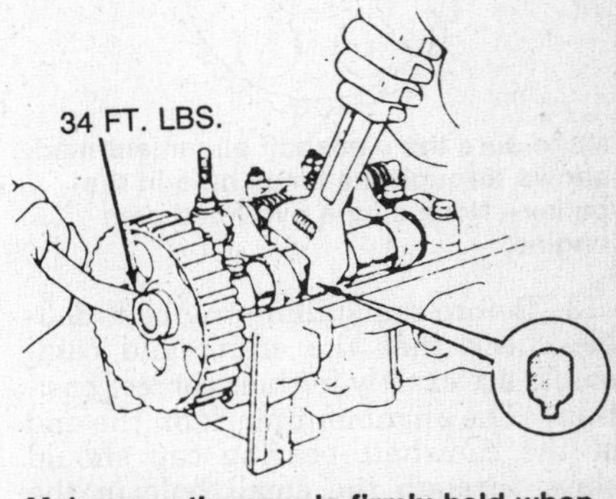

Make sure the cam is firmly held when removing the timing belt sprocket—Nova single overhead cam engine.

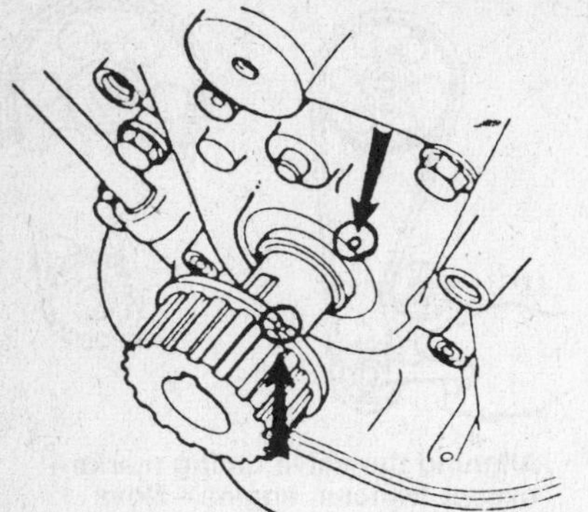

The crankshaft and oil pump marks must be aligned before reinstalling the timing belt—Nova single overhead cam engine.

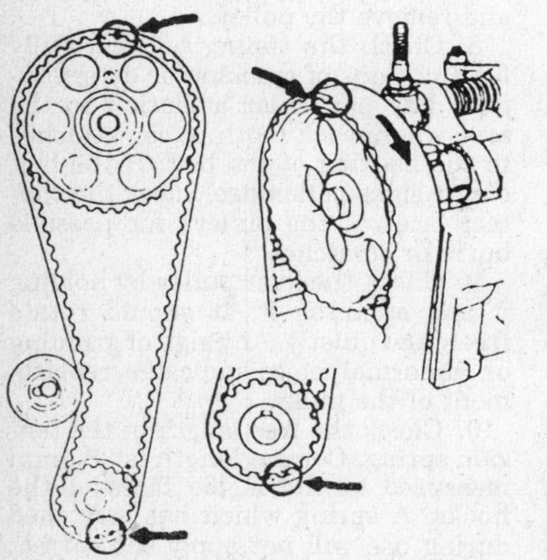

Always matchmark the belt and both pulleys before removing the timing belt. Mark and arrow on the belt showing the direction—Nova single overhead cam engine.

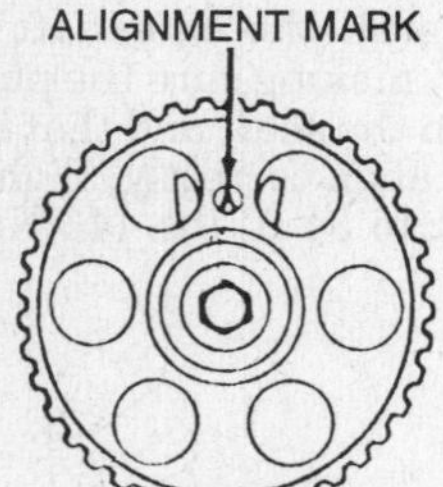

Make sure the camshaft alignment mark shows through the small hole in the pulley—Nova single overhead cam engine.

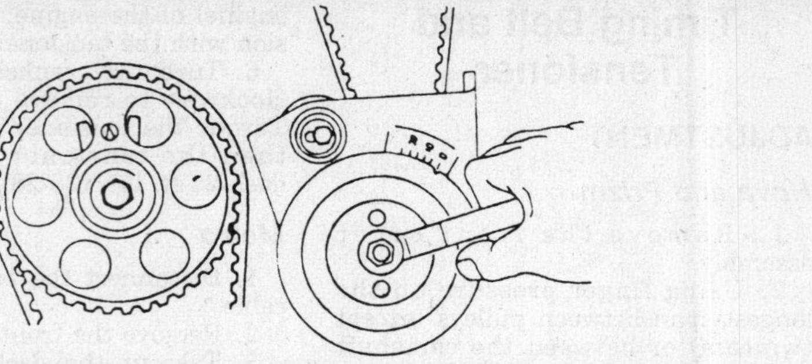

Timing mark alignment setting—Nova single overhead cam engine.

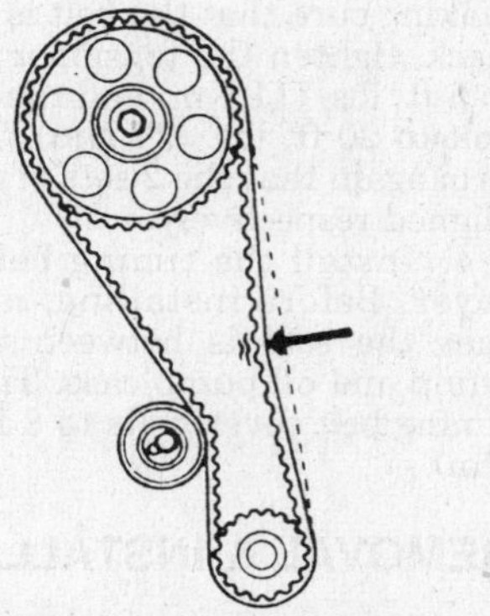

After the timinmg beltis installed, always check for the proper tension—Nova single overhead cam.

13. Before reinstalling the belt, double check that the crank and camshafts are exactly in their correct positions. The alignment mark on the end of the camshaft bearing cap should show through the small hole in the camshaft pulley and the small mark on the crankshaft timing belt pulley should align with the mark on the oil pump.

14. Reinstall the timing belt idler pulley and the tension spring. Pry the pulley to the left as far as it will go and temporarily tighten the retaining bolt. This will hold the pulley in its loosest position.

15. Install the timing belt, observing the matchmarks made earlier. Make sure the belt is fully and squarely seated on the upper and lower pulleys.

16. Using the equipment installed during the removal of the timing covers, elevate the engine enough to gain access to the work area.

17. Loosen the retaining bolt for the timing belt idler pulley and allow it to tension the belt.

18. Temporarily install the crankshaft pulley bolt and turn the crank clockwise 2 full revolutions from TDC to TDC. Insure that each timing mark realigns exactly.

19. Tighten the timing belt idler pulley retaining bolt to 27 ft. lbs. (37 Nm).

20. Measure the timing belt deflection (Tool 23600 B or similar), looking for 6–7mm of deflection at 4.4 pounds of pressure. If the deflection is not correct, readjust the idler pulley by repeating steps 15 through 18.

21. Remove the bolt from the end of the crankshaft.

22. Lower the engine into position and install the right engine mount.

23. Install the timing belt guide onto the crankshaft and install the lower timing belt cover.

24. Continue reassembly of the timing belt covers as outlined previously in this chapter.

1989-90 Prizm

1. Remove the timing belt covers.
2. If not done as part of the cover removal, rotate the crankshaft clockwise to the TDC/compression position for No. 1 cylinder.
3. Loosen the timing belt idler pulley to relieve the tension on the belt, move the pulley away from the belt and temporarily tighten the bolt to hold it in the loose position.
4. Make matchmarks on the belt and both pulleys showing the exact placement of the belt. Mark an arrow on the belt showing its direction of rotation.
5. Carefully slip the timing belt off the pulleys.

NOTE: Do not disturb the position of the camshafts or the crankshaft during removal.

6. Remove the idler pulley bolt, pulley and return spring.
7. Use an adjustable wrench mounted on the flats of the camshaft to hold the cam from moving. Loosen the center bolt in the camshaft timing pulley and remove the pulley.
8. Check the timing belt carefully for any signs of cracking or deterioration. Pay particular attention to the area where each tooth or cog attaches to the backing of the belt. If the belt shows signs of damage, check the contact faces of the pulleys for possible burrs or scratches.
9. Check the idler pulley by holding it and spinning it. It should rotate freely and quietly. Any sign of grinding or abnormal noise indicates replacement of the pulley.
10. Check the free length of the tension spring. Correct length is 38.5mm measured at the inside faces of the hooks. A spring which has stretched

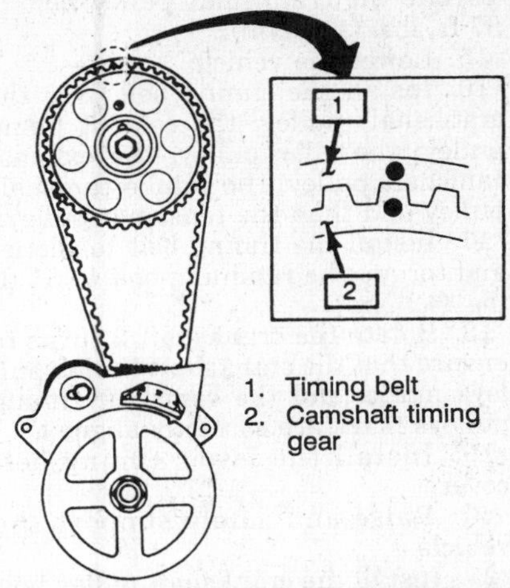

Mark the timing belt and camshaft timing gear – 1989–90 Prizm

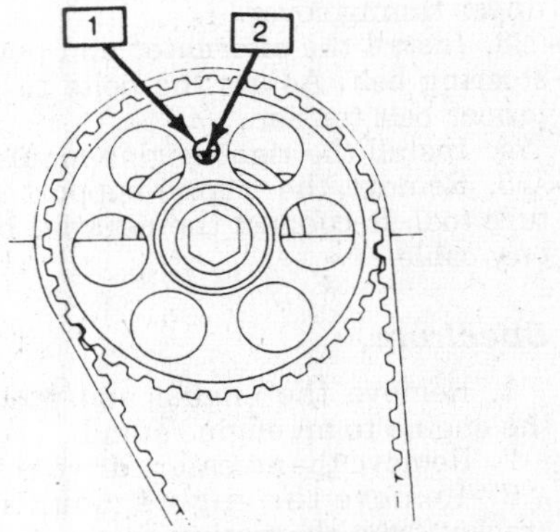

Aligning the camshaft gearhole and exhaust camshaft cap mark – 1989–90 Prizm

during use will not apply the correct tension to the pulley; replace the spring.

11. Test the tension of the spring, look for 8.4 lbs. of tension at 50mm of length. If in doubt, replace the spring.

12. Reinstall the camshaft timing belt pulley, making sure the pulley fits properly on the shaft and that the timing marks align correctly. Tighten the center bolt to 43 ft. lbs. (58 Nm).

13. Before reinstalling the belt, double check that the crank and camshafts are exactly in their correct positions. The alignment mark on the end of the camshaft bearing cap should show through the small hole in the camshaft pulley and the small mark on the crankshaft timing belt pulley should align with the mark on the oil pump.

14. Reinstall the timing belt idler pulley and the tension spring. Pry the pulley to the left as far as it will go and temporarily tighten the retaining bolt. This will hold the pulley in its loosest position.

15. Install the timing belt, observing the matchmarks made earlier. Make sure the belt is fully and squarely seated on the upper and lower pulleys.

16. Loosen the retaining bolt for the timing belt idler pulley and allow it to tension the belt.

17. Temporarily install the crankshaft pulley bolt and turn the crank clockwise 2 full revolutions from TDC to TDC. Insure that each timing mark realigns exactly.

18. Tighten the timing belt idler pulley retaining bolt to 27 ft. lbs. (37 Nm).

19. Measure the timing belt deflection (Tool 23600 B or similar), looking for 5–6mm of deflection at 4.4 pounds of pressure. If the deflection is not correct, readjust the idler pulley by repeating steps 15 through 18.

20. Remove the bolt from the end of the crankshaft.

21. Install the timing belt guide onto the crankshaft and install the lower timing belt cover.

22. Continue reassembly of the timing belt covers.

1988 Nova and 1991–92 Prizm

TWINCAM ENGINE

1. Remove the timing belt covers.
2. Remove the timing belt guide from the crankshaft pulley.
3. Loosen the timing belt idler pulley, move it to the left (to take tension off the belt) and tighten its bolt.
4. Make matchmarks on the belt and all pulleys showing the exact placement of the belt. Mark an arrow on the belt showing its direction of rotation.
5. Carefully slip the timing belt off the pulleys.

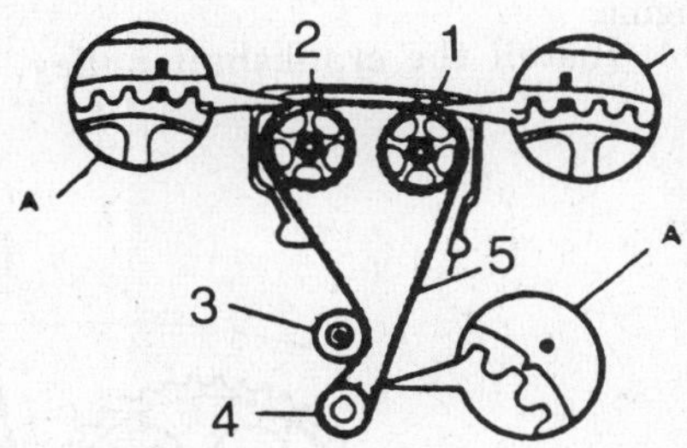

Aligning the valve timing marks – twincam engine Nova and 1991–92 Prizm

NOTE: Do not disturb the position of the camshafts or the crankshaft during removal.

6. Remove the idler pulley bolt, pulley and return spring.
7. Remove the PCV hose and the valve covers.
8. Use an adjustable wrench to counterhold the camshaft. Be careful not to damage the cylinder head. Loosen the center bolt in each camshaft pulley and remove the pulley. Label the pulleys and and keep them clean.
9. Check the timing belt carefully for any signs of cracking or deterioration. Pay particular attention to the area where each tooth or cog attaches to the backing of the belt. If the belt shows signs of damage, check the contact faces of the pulleys for possible burrs or scratches.
10. Check the idler pulley by holding it and spinning it. It should rotate freely and quietly. Any sign of grinding or abnormal noise indicates replacement of the pulley.
11. Check the free length of the tension spring. Correct length is 43.5mm measured at the inside faces of the hooks. A spring which has stretched during use will not apply the correct tension to the pulley; replace the spring.
12. Test the tension of the spring, look for 22 lbs. of tension at 50mm of length. If in doubt, replace the spring.**To install:**
13. Align the camshaft knock pin and the pulley. Reinstall the camshaft timing belt pulleys, making sure the pulley fits properly on the shaft and that the timing marks align correctly. Tighten the center bolt on each pulley to 34 ft. lbs. (46 Nm). Be careful not to damage the cylinder head during installation.
14. Before reinstalling the belt, double check that the crank and camshafts are exactly in their correct positions. The alignment marks on the pulleys should align with the cast marks on the head and oil pump.

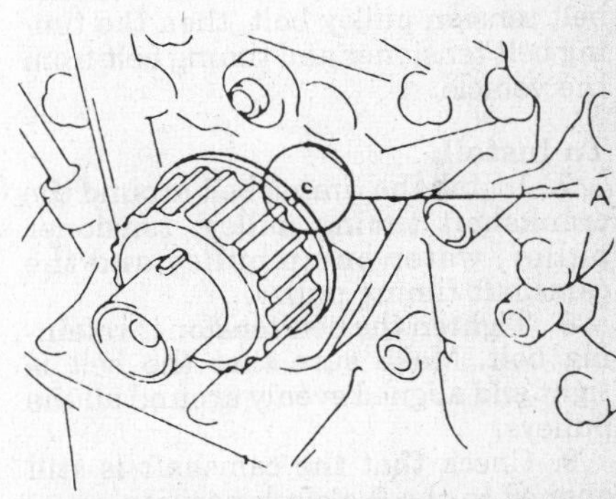

Aligning the oil pump and crankshaft timing marks – twincam engine Nova and 1991–92 Prizm

15. Reinstall the valve covers and the PCV hose.
16. Install the timing belt idler pulley and its tensioning spring. Move the idler to the left and temporarily tighten its bolt.
17. Carefully observing the matchmarks made earlier, install the timing belt onto the pulleys.
18. Slowly release tension on the idler pulley bolt and allow the idler to take up tension on the timing belt. Do not allow the idler to slam into the belt; the belt may become damaged.
19. Temporarily install the crankshaft pulley bolt. Turn the engine clockwise through 2 complete revolutions, stopping at TDC. Check that each pulley aligns with its marks.
20. Using tool J 23600-B or similar, check the tension of the timing belt at a point halfway between the 2 camshaft sprockets. The correct deflection is 4mm at 4.4 lbs. pressure. If the belt tension is incorrect, readjust it by repeating steps 19 and 20. If the tension is correct, tighten the idler pulley bolt to 27 ft. lbs. (37 Nm).
21. Remove the crankshaft pulley bolt.
22. Install the timing belt guide onto the crankshaft timing pulley.
23. Reinstall the timing belt covers, following procedures outlined previously in this chapter.

Storm

SINGLE OVERHEAD CAM ENGINE

1. Disconnect the battery negative cable.
2. Remove the front cover(s).
3. Rotate the crankshaft so that the No. 4 cylinder is at TDC on its compression stroke by aligning the camshaft pulley timing mark to the 9 o'clock position.
4. Raise and support the vehicle safely.
5. Remove the crankshaft pulley bolt, the 4 crankshaft pulley side bolts and the crankshaft pulley from the vehicle.
6. Lower the vehicle. Loosen the belt tension pulley bolt, then the timing belt tensioner and timing belt from the vehicle.

To install:

7. Install the timing belt around the crankshaft timing pulley, tensioner pulley, water pump pulley and the camshaft timing pulley.
8. Tighten the belt tensioner retaining bolt. Make sure that the belt is tight and aligned evenly around all the pulleys.
9. Check that the camshaft is still aligned to the 9 o'clock position.
10. Install the timing belt cover, secure it with the 2 lower mounting bolts and torque the bolts to 89 in. lbs. (10 Nm).
11. Install the crankshaft pulley to the crankshaft dampner, secure the center bolt and the 4 side bolts. Torque the center bolt to 87 ft. lbs. (118 Nm) and the 4 side bolts to 17 ft. lbs. (23 Nm).
12. Install the right side under cover. Lower the vehicle. Install the 4 upper timing belt cover retaining bolts.
13. Install the right engine mount.
14. Install the power steering pump drive belt, alternator drive belt and any other drive belts that have been removed. Adjust them to the proper belt tension.
15. Remove engine support fixture tool J-28467-A or equivalent from the engine and reconnect the negative battery cable.

TWINCAM ENGINE

1. Disconnect the negative battery cable.
2. Install engine support fixture tool J-28467-A or equivalent.
3. Remove the right engine mount.
4. Remove the power steering belt and alternator belt.
5. Remove the upper timing belt cover.
6. Raise and safely support the vehicle.
7. Remove the crankshaft pulley.
8. Lower the vehicle.
9. Remove the lower timing belt cover.
10. Align the crankshaft pulley to the TDC mark.
11. Loosen the tensioner pulley retaining belt ½ turn and remove the timing belt.

To install:

12. Align the camshaft pulleys timing mark.
13. Raise and safely support the engine.
14. Install the crankshaft pulley. Torque the crankshaft pulley bolt to 87 ft. lbs. (118 Nm).
15. Lower the vehicle.
16. Install the timing belt over the crankshaft pulley, the coolant pump pulley, the idler pulley, the exhaust camshaft pulley, the intake camshaft pulley and then the tensioner pulley.
17. Install the timing belt tensioner and torque the retaining bolt to 31 ft. lbs. (42 Nm).
18. Rotate the crankshaft 2 turns to ensure that the crankshaft timing pulleys marks and the camshaft timing pulleys mark are correctly aligned.
19. Install the lower timing belt cover.
20. Raise and safely support the vehicle.
21. Install the crankshaft pulley bolt and torque the crankshaft pulley bolt to 87 ft. lbs. (118 Nm).
22. Lower the vehicle and install the upper timing cover.
23. Install the alternator and power steering belt. Adjust the belts to the proper belt tension.
24. Install the right engine mount.
25. Remove the engine support fixture tool. Reconnect the negative battery cable.

Spectrum

1. Remove the engine and mount the engine to an engine stand.
2. Remove the accessory drive belts.
3. Remove the engine mounting bracket from the timing cover.
4. Rotate the crankshaft until the notch on the crankshaft pulley aligns with the **0** degree mark on the timing cover and the No. 4 cylinder is on TDC of the compression stroke.
5. Remove the starter and install the flywheel holding tool No. J<n-dash>35271 or equivalent.
6. Remove the crankshaft bolt, boss and pulley.

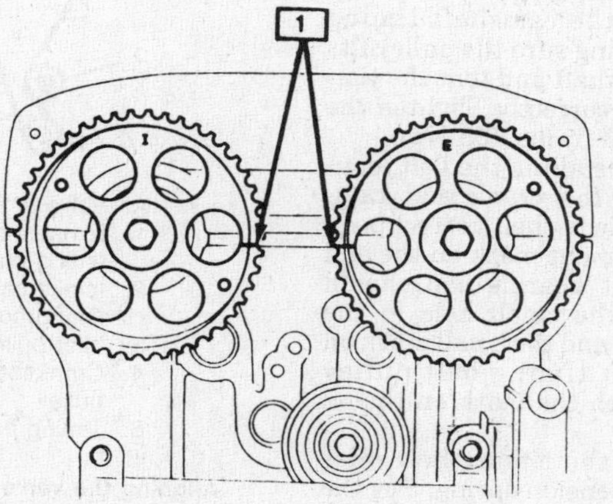

Camshaft pulley timing marks—twincam engine Storm

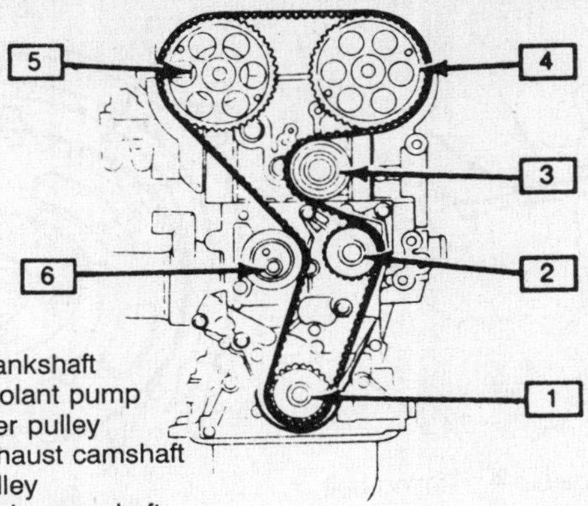

1. Crankshaft
2. Coolant pump
3. Idler pulley
4. Exhaust camshaft pulley
5. Intake camshaft pulley
6. Belt tensioner

Timing belt installation sequence—twincam engine Storm

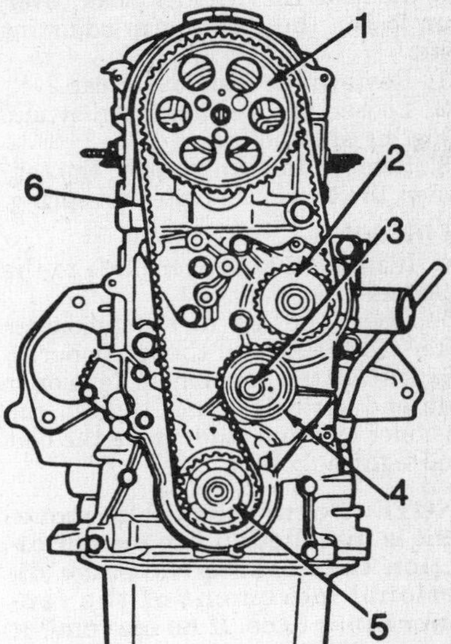

1. Camshaft timing pulley
2. Water pump timing pulley
3. Bolt
4. Tension pulley
5. Crankshaft timing pulley
6. Timing belt

View of the timing belt assembly—Spectrum and Storm

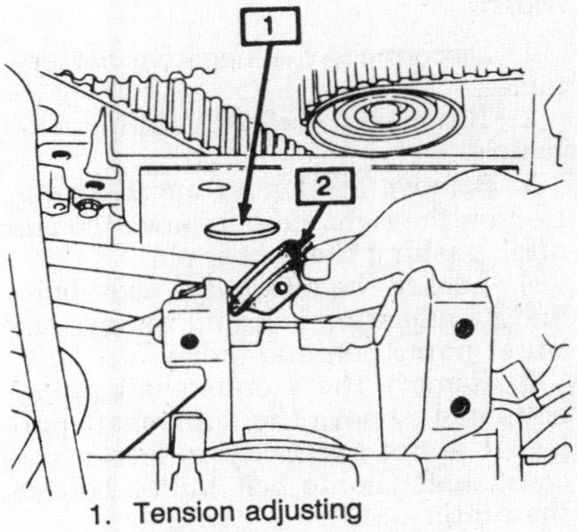

1. Tension adjusting hole
2. Tension adjusting hole cover

Timing belt tension adjusting hole cover—twincam engine Storm

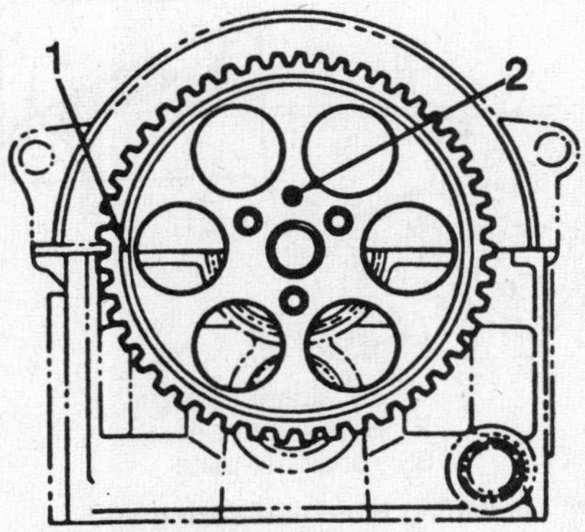

1. Alignment mark
2. Dowel

Aligning the camshaft pulley—Spectrum

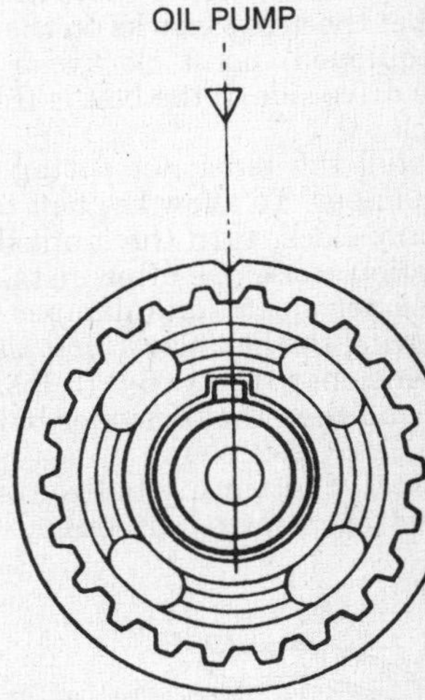

Crankshaft pulley timing mark alignment—Spectrum

7. Remove the timing cover bolts and the timing cover.
8. Loosen the tension pulley bolt.
9. Insert an Allen wrench into the tension pulley hexagonal hole and loosen the timing belt by turning the tension pulley clockwise.
10. Remove the timing belt.
11. Remove the head cover.

NOTE: Inspect the timing belt for signs of cracking, abnormal wear and hardening. Never expose the belt to oil, sunlight or heat. Avoid excessive bending, twisting or stretching.

To install:

12. Position the Woodruff key on the crankshaft followed by the crankshaft timing gear. Align the groove on the timing gear with the mark on the oil pump.
13. Align the camshaft timing gear mark with the upper surface of the cylinder head and the dowel pin in its uppermost position.
14. Place the timing belt arrow in the direction of the engine rotation and install the timing belt. Tighten the tension pulley bolt.
15. Turn the crankshaft 2 complete revolutions and realign the crankshaft timing gear groove with the mark on the oil pump.
16. Loosen the tension pulley bolt and apply tension to the belt with an Allen wrench. Torque the pulley bolt to 37 ft. lbs. (50 Nm) while holding the pulley stationary.
17. Adjust the valve clearances.
18. To complete the installation, reverse the removal procedures. Torque the crankshaft pulley-to-crankshaft bolt to 109 ft. lbs. (148 Nm).

Sprint

1. Disconnect the negative battery cable.
2. Loosen the water pump pulley bolts and the alternator adjusting bolt.
3. If equipped, remove the air conditioning compressor adjusting bolt.
4. Raise and support the vehicle safely.
5. Remove the drive belt splash shield, the right fender plug and the drive belts.
6. Remove the crankshaft and the water pump pulleys.
7. Remove the bolts from the bottom of the belt cover.
8. Lower the vehicle.
9. Remove the bolts from the top of the belt cover, then the cover from the vehicle.

10. Remove the cylinder head cover, then loosen the rocker arm adjusting bolts.
11. Remove the distributor cap.
12. Loosen the tensioner pulley and adjusting stud bolt.
13. Remove the timing belt, the tensioner, the tensioner plate and spring.

To install:

14. Install the tensioner plate to the tensioner.
15. Insert the lug of the tensioner plate into the hole of the tensioner.
16. Install the tensioner, tensioner plate and spring. Do not tighten the tensioner bolt and stud, make the bolt hand tight only.

NOTE: Be sure that plate movement is installed in the proper direction that causes the same directional movement of the tensioner inner race. If no movement between the plate and inner race occurs remove the tensioner and plate again and reinsert the plate lug into the tensioner hole.

17. Turn the camshaft pulley clockwise and align the timing mark on the camshaft pulley with the **V** mark on the belt inside cover.
18. Turn the crankshaft clockwise, using a 17mm wrench. Align the punch mark on the timing belt pulley with the arrow mark on the oil pump.
19. With the 4 marks aligned, install the timing belt on the 2 pulleys in such a way that the arrow marks on the belt (if so equipped) point clockwise and that the drive side of the belt is free of any slack.
20. Install the tensioner spring and spring damper. To allow the belt to be free of any slack, turn the crankshaft one rotation clockwise after installing the tensioner spring and damper. After removing the belt slack, first tighten the tensioner stud to 6-9 ft. lbs. (8-12 Nm) and then the tensioner bolt to 13-20 ft. lbs. (18-28 Nm).
21. Install the timing belt outside cover and the crankshaft pulley. Fit the keyway on the pulley to the key on the crank timing belt pulley and tighten the 4 bolts to 7-9 ft. lbs. (10-13 Nm).
22. Adjust the intake and exhaust valves.
23. Install the water pump pulley and water pump belt. Adjust the belts to the proper belt tension.
24. Install the cylinder head cover and air cleaner.
25. Install the negative battery cable.

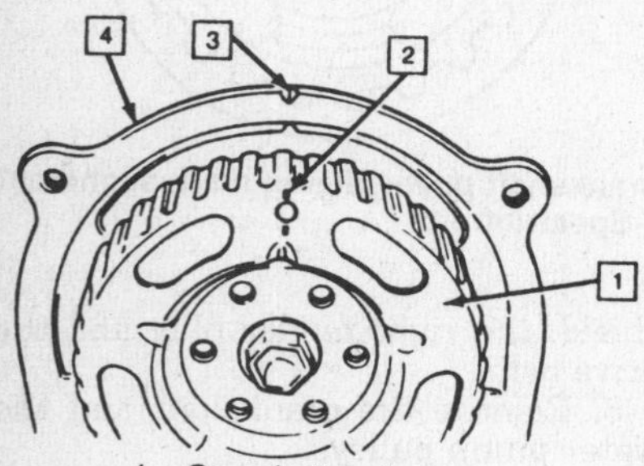

1. Camshaft pulley
2. Timing mark
3. V mark
4. Belt inside cover

Camshaft sprocket and inner belt cover timing mark alignment – Sprint

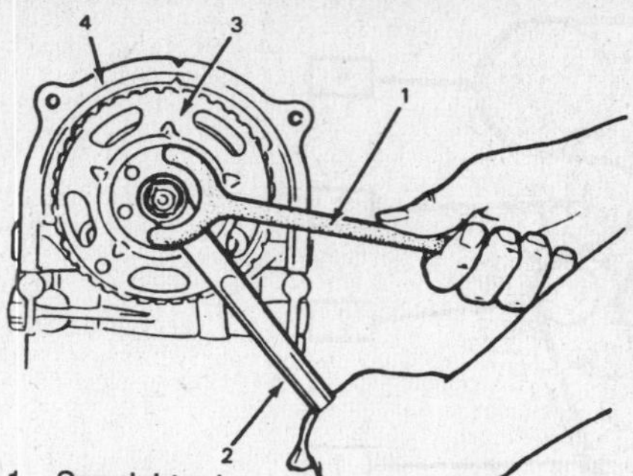

1. Special tool
2. Wrench
3. Camshaft timing belt pulley
4. Timing belt cover

Removing the camshaft pulley bolt using a lock holder tool – Sprint

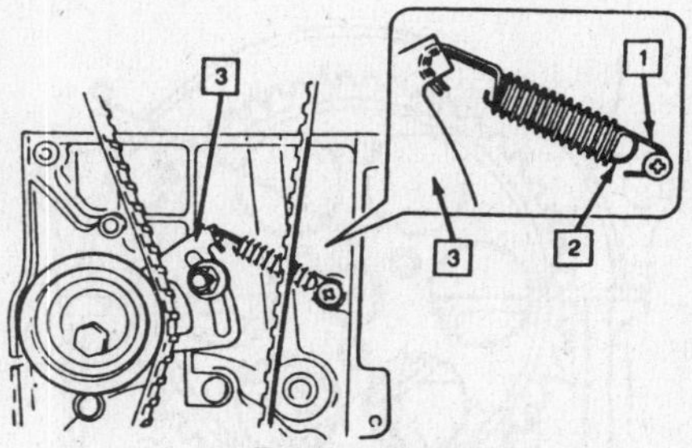

1. Tensioner spring
2. Spring damper
3. Tensioner plate

Installing tensioner spring and dampener – Sprint and Metro

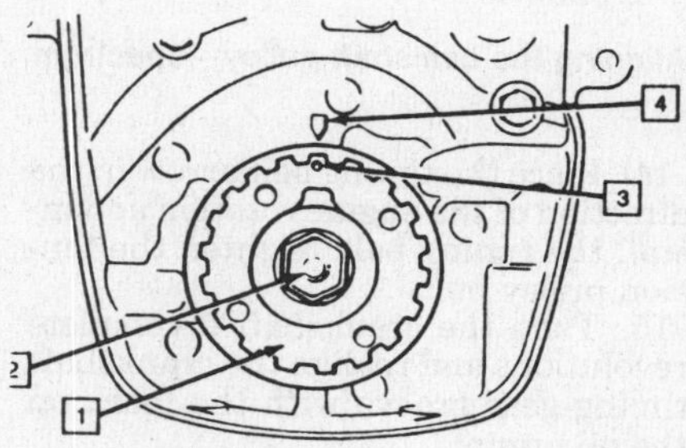

1. Crankshaft timing belt pulley
2. Pulley bolt
3. Punch Mark
4. Arrow Mark

Crankshaft sprocket and oil pump timing mark alignment – Sprint and Metro

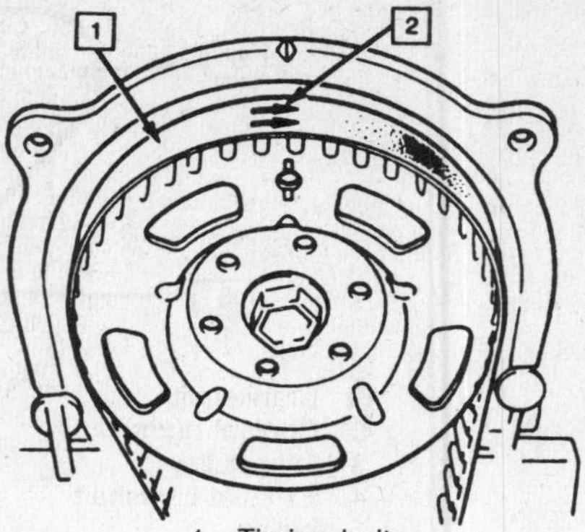

1. Timing belt
2. Arrow mark

Arrow marks on the timing belt show direction of rotation – Sprint and Metro

Metro

1. Disconnect the negative battery cable.
2. Raise and safely support the vehicle.
3. Remove the fender apron extension on the right side. Remove the clip after pushing the center pin.
4. Loosen the alternator pivot bolts and its adjusting bolt and remove the water pump belt and pulley.
5. Remove the 4 crankshaft pulley bolts and remove the crankshaft pulley. It is not necessary to loosen the crankshaft timing belt pulley bolt at the center.
6. Remove the timing belt cover. Before removing the timing belt, align the 4 timing marks by turning the crankshaft.
7. Remove the timing belt tensioner, tensioner plate, tensioner spring, spring damper and timing belt.

NOTE: After the timing belt is removed, never turn th camshaft or crankshaft independently. If turned, interference may occur among the pistons, and valves and parts related to the pistons and valves may be damaged.

8. Inspect the timing belt for wear or cracks and replace as necessary. Check the tensioner for smooth rotation.

To install:

9. Install the tensioner plate to the tensioner.
10. Insert the lug of the tensioner plate into the hole of the tensioner.
11. Install the tensioner, tensioner plate and spring. Do not tighten the tensioner bolt and stud, make the bolt hand tight only.

NOTE: Be sure that plate movement is installed in the proper direction that causes the same directional movement of the ten-

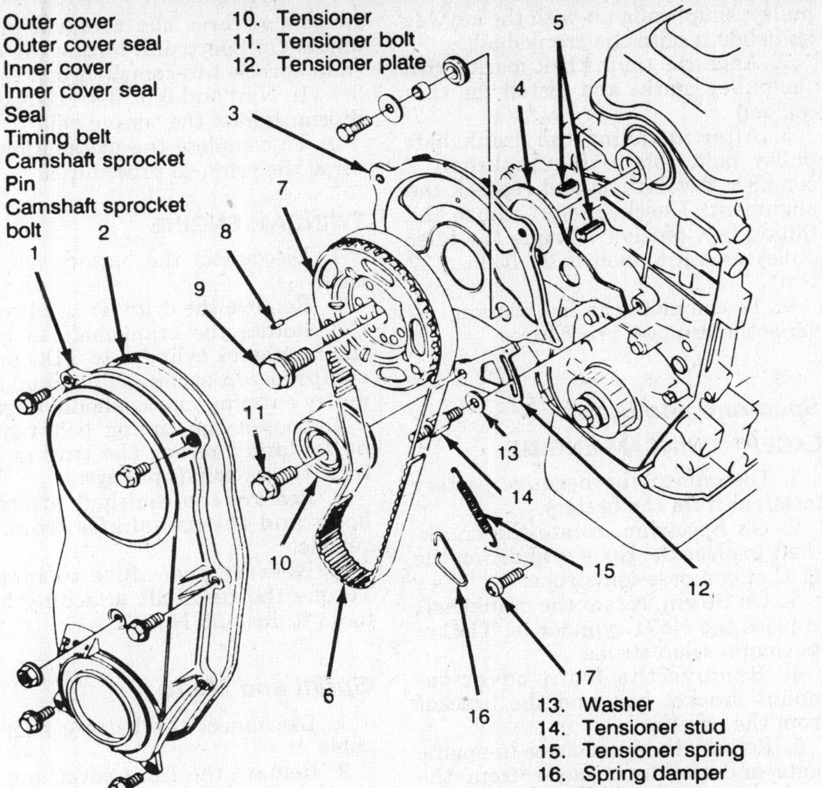

Timing belt, tensioner and sprockets exploded view—Sprint and Metro

sioner inner race. If no movement between the plate and inner race occurs remove the tensioner and plate again and reinsert the plate lug into the tensioner hole.

12. Check that the timing mark on the camshaft timing pulley is aligned with the **V** mark on the cylinder head cover. If not align the 2 marks by turning the camshaft by be careful not to turn it more than the allowable range.

13. Check that the punch mark on the crankshaft timing belt pulley is aligned with the arrow mark on the oil pump case. If not align the 2 marks by turning the camshaft by be careful not to turn it more than the allowable range.

14. With the 4 marks aligned, install the timing belt on the 2 pulleys in such a way that the arrow marks on the belt (if so equipped) coincide with the rotating direction of the crankshaft. In this state the No. 1 piston is at top dead center of its compression stroke.

15. Install the tensioner spring and spring damper. To allow the belt to be free of any slack, turn the crankshaft 2 rotations clockwise after installing the tensioner spring and damper. After removing the belt slack, first tighten the tensioner stud to 8 ft. lbs. (11 Nm) and then the tensioner bolt to 20 ft. lbs. (27 Nm).

21. Install the timing belt outside cover. Before installing the cover, make sure that seal is between the water pump and the oil pump case. Torque the timing belt cover bolts to 8 ft. lbs. (11 Nm).

22. Install the crankshaft pulley. Fit the keyway on the pulley to the key on the crank timing belt pulley and tighten the 4 bolts to 8 ft. lbs. (11 Nm).

24. Install the water pump belt pul-

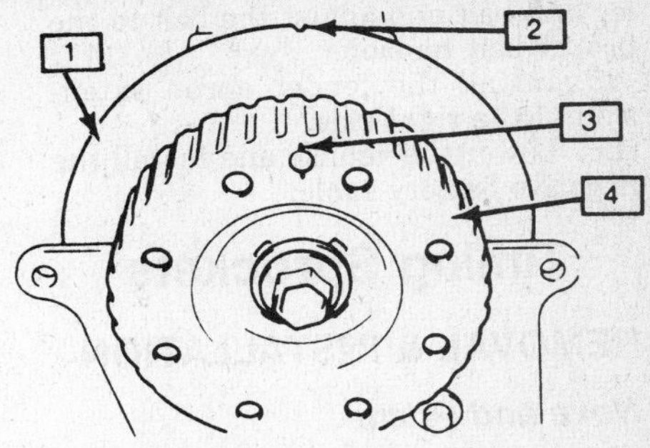

1. Cylinder head cover
2. V mark
3. Camshaft gear timing mark
4. Camshaft timing belt gear

Camshaft timing marks—Metro

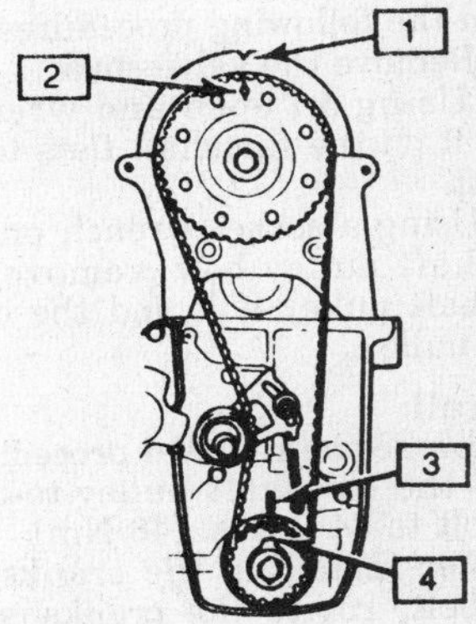

1. V mark on cylinder head cover
2. Timing mark on camshaft timing belt gear
3. Arrow amrk on oil pump case
4. Punch mark on crankshaft timing belt gear

Aligning the timing marks—Metro

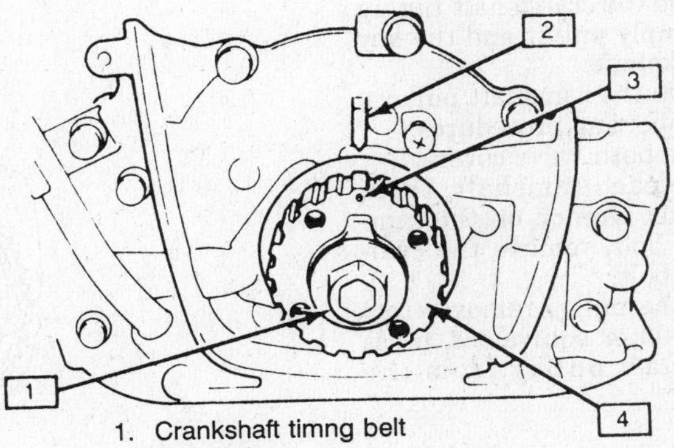

1. Crankshaft timng belt gear bolt
2. Arrow mark
3. Punch mark
4. Crankshaft timing belt gear

Crankshaft timing marks—Metro

ley and belt and adjust the belt to the proper belt tension.

25. Install the fender apron extension of the right side.

26. Lower the vehicle and install the negative battery cable.

Timing Sprockets

REMOVAL & INSTALLATION

Nova and Prizm

EXCEPT TWINCAM ENGINE

1. Disconnect the negative battery cable.
2. Remove the timing belt.
3. To remove the crankshaft timing belt pulley, simply pull it and the key from the crankshaft.
4. To remove the camshaft pulley, perform the following procedures:
   a. Remove the valve cover.
   b. Using an open end wrench, place it on the camshaft flats to secure it.
   c. Using a socket wrench on the camshaft pulley bolt, remove the camshaft pulley bolt and the camshaft pulley.

To install:

5. Reverse the removal procedures. Torque the camshaft pulley-to-camshaft bolt to 34 ft. lbs. (46 Nm).
6. After installing the crankshaft pulley bolt, rotate the crankshaft 2 complete revolutions and recheck the alignment. Check and/or adjust the timing belt tension. Torque the idler pulley mounting bolt to 27 ft. lbs. (37 Nm).
7. To complete the installation, reverse the removal procedures.

TWINCAM ENGINE

1. Disconnect the negative battery cable.
2. Remove the timing belt.
3. To remove the crankshaft timing belt pulley, simply pull it and the key from the crankshaft.
4. To remove the camshaft pulleys, perform the following procedures:
   a. Remove both valve covers.
   b. Secure each camshaft, then, using a socket wrench on the camshaft pulley bolt, remove the camshaft pulley bolt.
   c. Using the pulley remover tool No. J-1859-03 or equivalent, press each camshaft pulley from the camshafts.

To install:

5. To install the camshaft pulleys, align each with the knock pin and reverse the removal procedures. Torque the camshaft pulley-to-camshaft bolt to 34 ft. lbs. (47 Nm).
6. To install the crankshaft timing pulley, simply align it with the keyway and slide it onto the crankshaft.
7. Align the timing belt marks with the pulley marks and install the timing belt.
8. After installing the crankshaft pulley bolt, rotate the crankshaft 2 complete revolutions and recheck the alignment. Check and/or adjust the timing belt tension. Torque the idler pulley mounting bolt to 27 ft. lbs. (37 Nm).
9. To complete the installation, reverse the removal procedures.

Spectrum and Storm

EXCEPT TWINCAM ENGINE

1. Disconnect the negative battery terminal from the battery.
2. On Spectrum, rotate the crankshaft to place the No. 4 cylinder on the TDC of compression stroke.
3. On Storm, rotate the crankshaft to place the No. 1 cylinder on TDC of its compression stroke.
4. Remove the front cover-to-mount bracket bolt and the bracket from the vehicle.
5. Remove the front cover-to-engine bolts and the front cover from the engine.

NOTE: Make sure the camshaft dowel pin is positioned at the top and the mark on the cam sprocket is aligned with the upper cylinder head surface.

6. Loosen the timing belt tension pulley-to-engine bolt, then remove the timing belt.
7. Remove the camshaft sprocket-to-camshaft bolts, the camshaft sprocket; allow the timing belt to hang.
8. If the engine has not been disturbed, reverse the removal procedures. On Spectrum, torque the camshaft sprocket-to-camshaft bolt to 7 ft. lbs. (10 Nm) and 9 ft. lbs. (12 Nm) on Storm. Adjust the timing belt.
9. To complete the installation, reverse the removal procedures.

TWINCAM ENGINE

1. Disconnect the battery negative cable.
2. Remove the cylinder head cover.
3. Rotate the crankshaft to bring the number 1 cylinder to TDC on its compression stroke. The camshaft pulley's timing marks should align.
4. Loosen the timing belt tension pulley and remove the timing belt from the camshaft pulleys.
5. Remove the camshaft attaching bolts and the camshafts from the vehicle.
6. Reverse procedure to install. Torque the camshaft attaching bolts to 43 ft. lbs. (59 Nm).

Sprint and Metro

1. Disconnect the battery negative cable.
2. Remove the front cover and the timing belt.
3. Using tool J-34836 or equivalent, hold the camshaft sprocket and remove the retaining bolt, camshaft sprocket, alignment pin and cover.
4. Using a suitable spanner wrench, remove the crankshaft pulley bolt and the crankshaft pulley.
5. Reverse procedure to install. Align the "V" mark on the timing belt cover with the timing mark on the camshaft sprocket and the arrow mark on the engine block with the punch mark on the crankshaft timing sprocket. Torque the camshaft bolt to 44 ft. lb. (60 Nm) and the crankshaft bolts to 8 ft. lbs. (10 Nm).

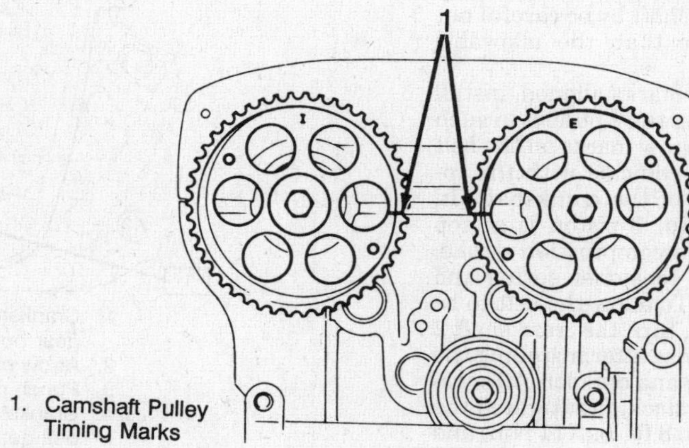

Camshaft pulley timing marks – Storm

Camshaft

REMOVAL & INSTALLATION

Nova

EXCEPT TWINCAM ENGINE

1. Remove the upper timing belt front cover and the valve cover; do not remove the timing belt.
2. Disconnect the negative battery cable and the spark plug wires, then, remove the distributor-to-engine hold-down bolt, the distributor and the distributor gear bolt.
3. Disconnect the hoses from the fuel pump, then, remove the fuel pump.
4. Using a socket wrench on the crankshaft pulley bolt, rotate the crankshaft (clockwise) to position the No. 1 cylinder on the TDC of its compression stroke; the rocker arms of the No. 1 cylinder will be loose, if not, rotate the crankshaft 1 complete revolution.
5. Loosen the rocker arm adjusting nuts and back off the adjusting screw. Remove the rocker shaft-to-cylinder head assembly.
6. Place alignment marks on the timing belt and the timing pulleys; also, mark the direction of timing belt rotation.
7. Loosen the idler pulley bolt and push the pulley as far left as possible, then, retighten the bolt. Remove the timing belt from the camshaft timing pulleys, support it so it will remain in mesh with the crankshaft pulley; be careful not to get oil on the timing belt.
8. Use a large open-end wrench, secure the camshaft (on the flats), then remove the camshaft pulley-to-camshaft bolt; the camshaft flats are located between the first and second cam lobes. Remove the camshaft pulley.
9. Remove the camshaft bearing cap bolts, the caps and the camshaft; keep the caps in order for reinstallation purposes.
10. Remove the distributor drive gear.

To install:

11. Inspect the camshaft for damage and/or wear; if necessary, replace the camshaft.
12. Insert the distributor drive gear, plate washer and bolt.
13. Using clean engine oil, coat all bearing surfaces, then, install the camshaft and No. 2, 3 and 4 bearing caps, in their proper positions and direction.
14. To install a new camshaft oil seal, apply grease the oil seal lips and sealant to the outside edge, then, slip the seal onto the camshaft; make sure it is on straight, as a crooked seal will leak.
15. Using sealant, apply it to the bottom surfaces of the No. 1 bearing cap and install it. Install all bearing cap bolts finger tight.
16. Torque the bearing cap bolts, alternately and evenly, to 8–10 ft. lbs. (11-14 Nm).
17. Using a dial indicator, inspect the camshaft thrust clearance, front-to-rear movement; it should be 0.0031–0.0071 in. with a limit of 0.0098 in. Torque the distributor drive gear bolt to 22 ft. lbs. (29 Nm).
18. To complete the installation, adjust the valves, use new gaskets, sealant, if necessary, and reverse the removal procedures. Start the engine, allow it to reach normal operating temperatures and check for leaks.

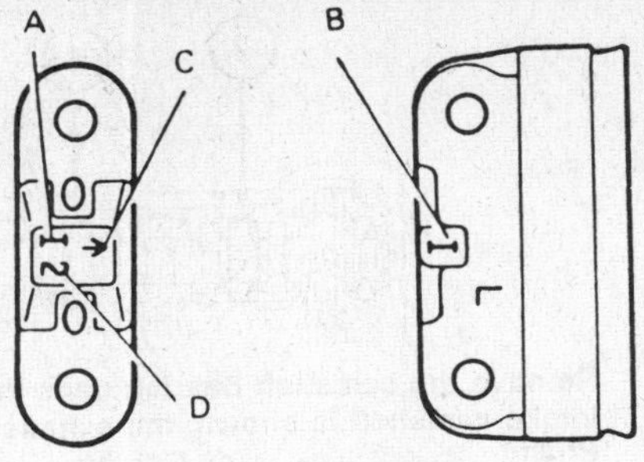

A. I = intake; E = exhaust
B. I = intake; E = exhaust
C. Front Mark
D. I.D. for bearings No. 2–5

View of the camshaft bearing caps twincam engine – Nova

TWINCAM ENGINE

1. Disconnect the negative battery cable.
2. Remove the cylinder head covers and the camshaft pulleys.
3. Loosen and remove the camshaft bearing caps-to-cylinder head bolts in sequence. Remove the camshaft bearing caps and camshafts; be sure to keep the parts in order for reinstallation purposes.
4. Using the proper tool, clean the gasket mounting surfaces. Inspect the camshaft for wear and/or damage, if necessary, replace the camshaft.

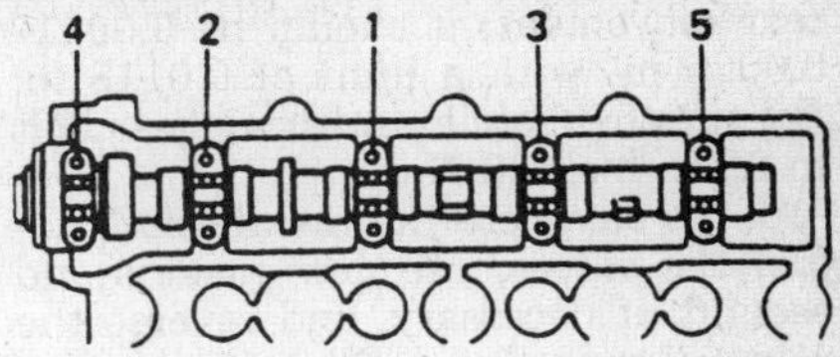

Camshaft bearing cap torque sequence twincam engine – Nova

To install:

5. Using clean engine oil, coat all bearing surfaces, then, install the camshaft and No. 2, 3 and 4 bearing caps, in their proper positions and direction.
6. To install a new camshaft oil seal, apply grease the oil seal lips and sealant to the outside edge, then, slip the seal onto the camshaft; make sure it is on straight, as a crooked seal will leak.
7. Using sealant, apply it to the bottom surfaces of the No. 1 bearing cap and install it. Install all bearing cap bolts finger tight.
8. Torque the bearing cap bolts, alternately and evenly, to 8–10 ft. lbs. (11-14 Nm).
9. Using a dial indicator, inspect the camshaft thrust clearance, front-to-

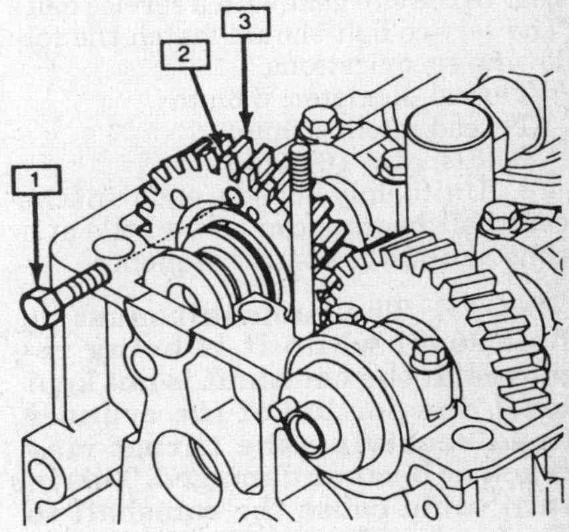

1. Service bolt
2. Sub-gear
3. Main gear

Always install a service bolt to lock the intake cam together – 1989-90 Prizm

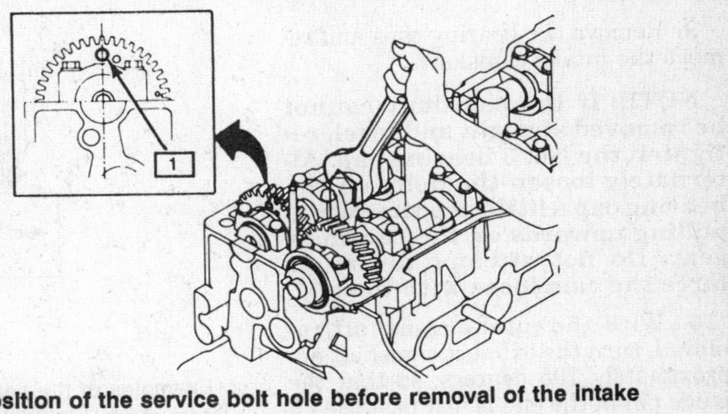

The correct position of the service bolt hole before removal of the intake camshaft – 1989-90 Prizm

rear movement; it should be 0.0031–0.0075 in. with a limit of 0.0118 in. Torque the distributor drive gear bolt to 22 ft. lbs. (30 Nm).

10. To complete the installation, adjust the valves, use new gaskets and sealant, if necessary, and reverse the removal procedures. Start the engine, allow it to reach normal operating temperatures and check for leaks.

1989-90 Prizm

1. Remove the valve cover.
2. Remove the timing belt covers.
3. Remove the timing belt and idler pulley.
4. Hold the exhaust camshaft with an adjustable wrench and remove the camshaft timing belt gear. Be careful not to damage the head or the camshaft during this work.
5. Gently turn the camshafts with an adjustable wrench until the service bolt hole in the intake camshaft end gear is straight up or in the "12 o'clock" position.
6. Alternately loosen the bearing cap bolts in the number one (closest to the pulleys) intake and exhaust bearing caps.
7. Attach the intake camshaft end gear to the sub gear with a service bolt. The service bolt should match the following specifications:
   Thread diameter: 6.0mm
   Thread pitch: 1.0mm
   Bolt length: 16-20mm
8. Uniformly loosen each intake camshaft bearing cap bolt a little at a time and in the correct sequence.

NOTE: The camshaft must be held level while it is being removed. If the camshaft is not kept level, the portion of the cylinder head receiving the thrust may crack or become damaged. This in turn could cause the camshaft to bind or break.

Before removing the intake camshaft, make sure the rotational force has been removed from the sub gear; that is, the gear should be in a neutral or "unloaded" state.

9. Remove the bearing caps and remove the intake camshaft.

NOTE: If the camshaft cannot be removed straight and level, retighten the No. 3 bearing cap. Alternately loosen the bolts on the bearing cap a little at a time while pulling upwards on the camshaft gear. Do not attempt to pry or force the cam loose with tools.

10. With the intake camshaft removed, turn the exhaust camshaft approximately 105 degrees, so that the guide pin in the end is just past the "5 o'clock" position. This puts equal loadings on the camshaft, allowing easier and safer removal.
11. Loosen the exhaust camshaft bearing cap bolts a little at a time and in the correct sequence.
12. Remove the bearing caps and remove the exhaust camshaft.

NOTE: If the camshaft cannot be removed straight and level, retighten the No. 3 bearing cap. Alternately loosen the bolts on the bearing cap a little at a time while pulling upwards on the camshaft gear. Do not attempt to pry or force the cam loose with tools.

13. When reinstalling, remember that the camshafts must be handled carefully and kept straight and level to avoid damage. **To install:**
14. Place the exhaust camshaft on the cylinder head so that the cam lobes press evenly on the lifters for cylinders Nos. 1 and 3. This will put the guide pin in the "just past 5 o'clock" position.
15. Place the bearing caps in position according to the number cast into the cap. The arrow should point towards the pulley end of the motor.
16. Tighten the bearing cap bolts gradually and in the proper sequence to 9.5 ft. lbs. (13 Nm).
17. Apply multi-purpose grease, such as GM 1051344 or similar, to a new exhaust camshaft oil seal.
18. Install the exhaust camshaft oil seal using tool J 35403 or similar. Be very careful not to install the seal on a slant or allow it to tilt during installation.
19. Turn the exhaust cam until the cam lobes of No. 4 cylinder press down

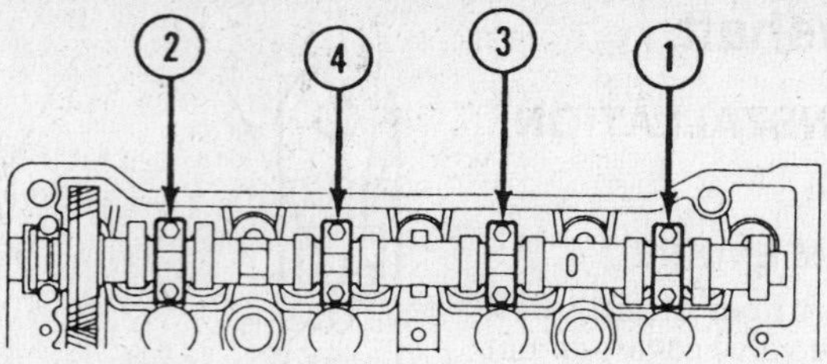

Remove the camshaft bearing caps in the order that is shown in this figure. The intake camshaft is shown, the exhaust cam uses the identical order—1989-90 Prizm

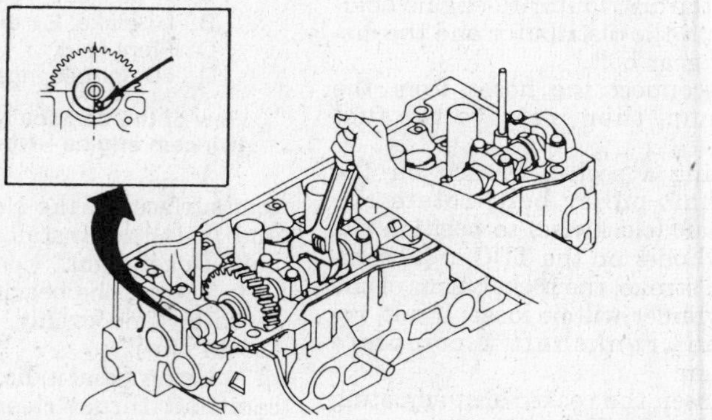
Setting the exhaust cam guide pin to the just past 5 o'clock position

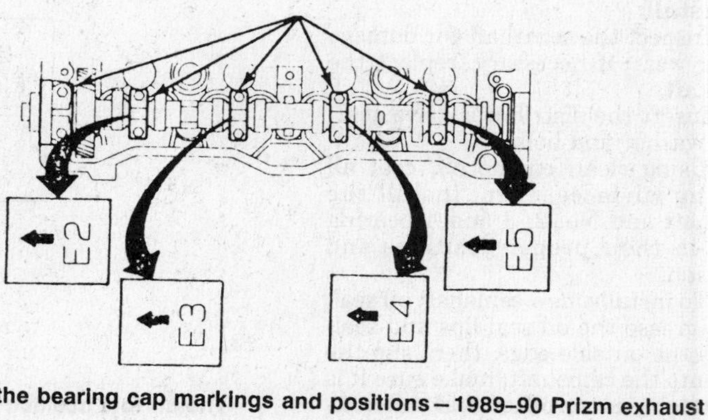

Examples of the bearing cap markings and positions—1989-90 Prizm exhaust cam shown

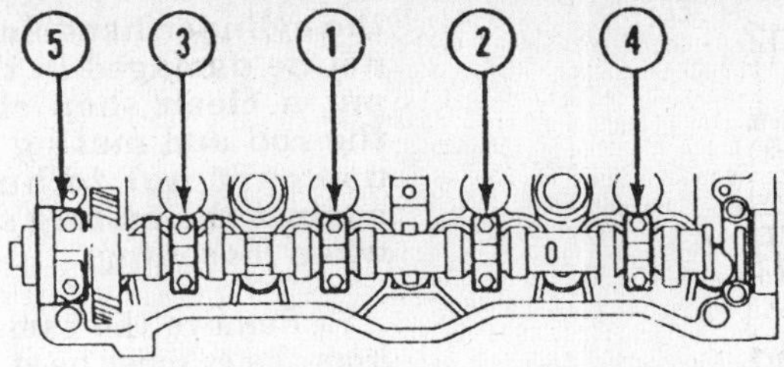

Torque the camshaft bearing caps in the order that is shown in this figure. The exhaust camshaft is shown, the intake cam uses the identical order—1989–90 Prizm

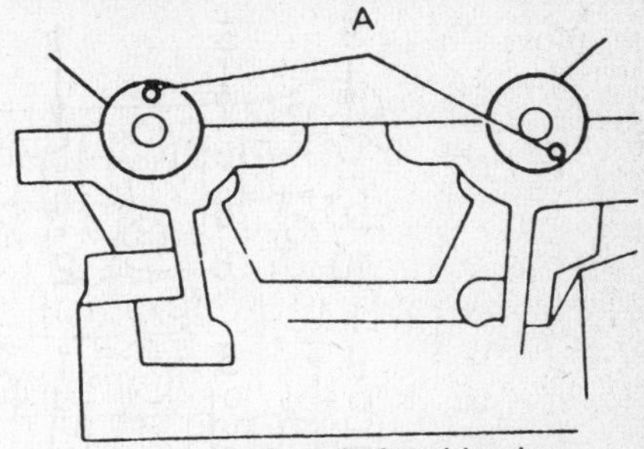

A. Camshaft guide pins
B. Exhaust camshaft
C. Intake camshaft

Correct position of guide pins on the camshaft—Nova and the 1991–92 Prizm twincam engine

on their lifters.

20. Hold the intake camshaft next to the exhaust camshaft and engage the gears by matching the alignment marks on each gear.

21. Keeping the gears engaged, roll the intake camshaft down and into its bearing journals.

22. Place the bearing caps for Nos. 2,3,4 and 5 in position. Observe the numbers on each cap and make certain the arrows point to the pulley end of the motor.

23. Gradually tighten each bearing cap bolt in the same order as the exhaust camshaft bolts. Tighten each bolt to 9.5 ft. lbs. (13 Nm)

24. Remove any retaining pins or bolts in the intake camshaft gears.

25. Install the number one bearing cap for the intake camshaft.

NOTE: If the No. 1 bearing cap does not fit properly, gently push the cam gear towards the rear of the engine by levering between the gear and the head.

26. Turn the exhaust camshaft one full revolution from TDC/compression on No. 1 cylinder to the same position. Check that the mark on the exhaust camshaft gear matches exactly with the mark on the intake camshaft gear.

27. Counterhold the exhaust camshaft and install the timing belt pulley. Tighten the bolt to 43 ft. lbs. (59 Nm).

28. Double check both the crankshaft and camshaft positions, insuring that they are both set to TDC/compression for No. 1 cylinder.

29. Install the timing belt.

30. Install the timing belt covers and the valve cover.

1991–92 Prizm

1. Remove the valve covers and the timing belt cover.

2. Make certain the engine is set to TDC/compression on No. 1 cylinder. Remove the timing belt.

3. Remove the crankshaft pulley if so desired.

4. Remove the camshaft timing belt pulleys.

5. Loosen and remove the camshaft bearing caps in the proper sequence. It it recommended that the bolts be loosened in 2 or 3 passes.

6. With the bearing caps removed, the camshaft(s) may be lifted clear of the head. If both cams are to be removed, label them clearly—they are not interchangable.

To install:

7. When reinstalling, place the camshaft(s) in position on the head. The exhaust cam has the distributor drive gear on it. Observe the markings on the bearing caps and place them according to their numbered positions. The arrow should point to the front of the engine.

8. Tighten the bearing cap bolts in the correct sequence and in 3 passes to a final tightness of 9 ft. lbs. (12 Nm.).

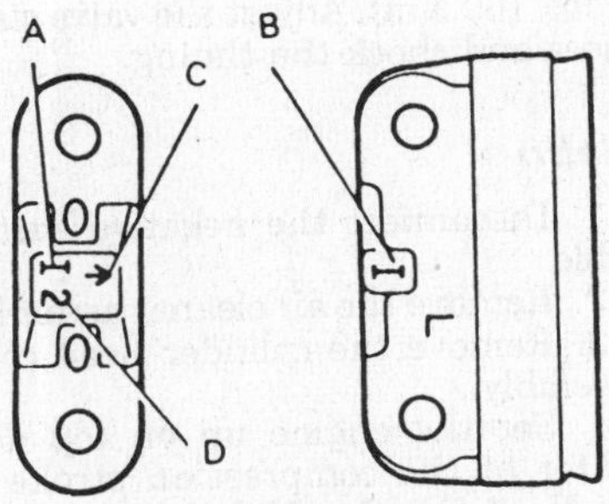

A. I = intake; E = exhaust
B. I = intake; E = exhaust
C. Front mark
D. I.D. for bearing No. 2 thru No. 5

Examples of the bearing cap markings. Note that the bearing on the right is used for position No. 1 only—Nova 92 Prizm twincam engine

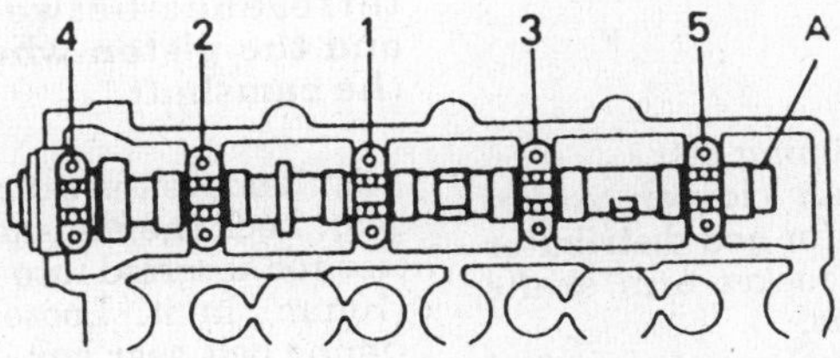

Install the camshaft bearing caps in the order shown—Nova and the 1991–92 Prizm twincam engine

9. Position the camshafts so that the guide pins (knock pins) are in the proper position. This step is critical to the correct valve timing of the engine.

10. Install the camshaft timing pulleys and tighten the bolts to 34 ft. lbs. (47 Nm).

11. Double check the positioning of the camshaft pulleys and the guide pin.

12. Install the crankshaft pulley if it was removed. Tighten its bolt to 87 ft. lbs. (118 Nm) and double check its position to be on TDC.

13. Install the timing belt and tensioner. Adjust the belt according to procedures outlined previously in this chapter.

14. Install the timing belt covers.

15. Install the valve covers.

Storm

1. Disconnect the battery negative cable.

2. Remove the cylinder head cover.

3. Rotate the crankshaft to position No. 1 at TDC on its compression stroke by aligning the camshaft(s) pulley timing mark with the cylinder head cover.

4. Loosen the timing belt tensioner.

5. Remove the timing belt pulley(s) from the camshaft(s).

6. Remove the distributor.

7. Remove the camshaft(s) bearing cap bolts, then remove the camshaft

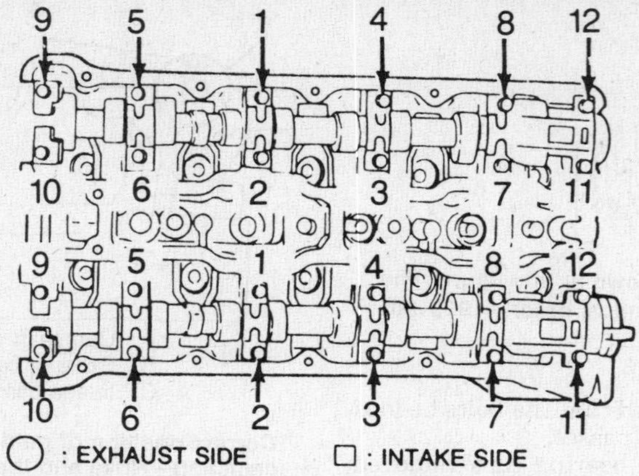

Camshaft bearing cap torque sequence twincam engine – Storm

and seal from the vehicle. Discard seal.

8. Reverse procedure to install. Torque camshaft bearing cap bolts to 8 ft. lbs. (10 Nm). Connect the battery negative cable. Start the engine and check for leaks.

Spectrum

1. Disconnect the negative battery terminal from the battery.
2. Align the crankshaft pulley notch with the **0** degree mark on the timing cover.
3. Remove the cylinder head cover.
4. Remove the timing cover.
5. Loosen the camshaft timing gear bolts; Do not rotate the engine.
6. Loosen the timing belt tensioner and remove the timing belt from the camshaft timing gear.
7. Remove the rocker arm shaft/rocker arm assembly.
8. Remove the distributor bolt and the distributor.
9. Remove the camshaft and the camshaft seal.
10. To install, drive a new camshaft seal on the camshaft using the seal installation tool No. J-35268 or equivalent, reverse the removal procedures, adjust the valves and the timing belt.

Sprint

1. Remove the timing belt.
2. Remove the air cleaner, rocker arm cover, distributor and distributor case. Remove the rocker arm shafts and the rocker arms.
3. Remove the fuel pump and fuel pump pushrod from the cylinder head.
4. Using a spanner wrench tool J-34836 to hold the camshaft pulley, remove the camshaft pulley bolt, the pulley, the alignment pin and the inside cover.
5. Carefully slide the camshaft from the rear of the cylinder head.
6. Clean the gasket mounting surfaces. Check for wear and/or damage, replace the parts as necessary.
7. To install, use new gaskets/seals and reverse the removal procedures. Torque the camshaft pulley bolt to 44 ft. lbs. (60 Nm). Adjust the valve clearances and check the timing.

Metro

1. Disconnect the negative battery cable.
2. Remove the air cleaner assembly.
3. Remove the cylinder head cover assembly.
4. Set the engine up on top dead center of the compression stroke on the No. 1 cylinder. Make an alignment mark on the distributor cap and engine block and remove the distributor assembly.
5. Remove the crankshaft pulley, timing belt outside cover and the timing belt.

NOTE: After removing the timing belt, set the key on the crankshaft in position by turning the crankshaft. This is to prevent interference between the valves and the piston when reinstalling the camshaft.

6. Remove the camshaft timing belt gear. Lock the camshaft with a proper size rod inserted into the hole 0.39 in. (10mm) in it. Loosen the camshaft timing belt gear bolt.

NOTE: The mating surface of the cylinder head and cover must not be damaged in this work. So, put a clean shop cloth between the rod and mating surfaces and use care not to bump the rod against the mating surfaces hard when loosening.

7. Remove the camshaft housing from the cylinder head.
8. Remove the camshaft from the cylinder head.

To install:

9. Install the camshaft to cylinder head. After applying engine oil to the camshaft journal and all around the cam, the position the camshaft into the cylinder head so that the camshaft timing belt gear pin hole in camshaft is at the lower position.
10. Install the camshaft housing to the camshaft and the cylinder head.
11. Apply the engine oil to the sliding surface of each housing against the camshaft journal.
12. Apply the sealant to the mating surface of the No. 1 and No. 3 housing which will mate with the cylinder head.
13. There are marks provided on each camshaft housing indicating position and direction for installation. Install the housing as indicated by these marks.
14. As the camshaft housing No. 1 retains the camshaft in the proper position as to the thrust direction, make sure to first fit the No. 1 housing to the No. 1 journal of the camshaft securely.
15. After applying the engine oil to the housing bolts, tighten them temporarily. Then tighten in the proper sequence. Tighten the bolts a little at a time and evenly among bolts, repeat the tightening sequence 3 to 4 times before they are tighten to the proper torque of 8 ft. lbs. (11 Nm).
16. Install the camshaft oil seal. After applying engine oil to the oil seal lip, press-fit the camshaft oil seal until the oil seal surface becomes flush with the housing surfaces.
17. Install the camshaft timing belt gear to the camshaft after installing the dwell pin to the camshaft. While locking the camshaft, install the camshaft pulley and retaining bolt and torque the bolt to 44 ft. lbs. (60 Nm).
18. Install the cylinder head cover to the cylinder head.
19. Install the timing belt, timing belt outside cover, crankshaft pulley, coolant pump pulley and coolant pump belt.
20. Install the distributor assembly into the engine.
21. Install the air cleaner assembly and reinstall the negative battery cable.
22. Adjust the ignition timing.

Piston and Connecting Rod

POSITIONING

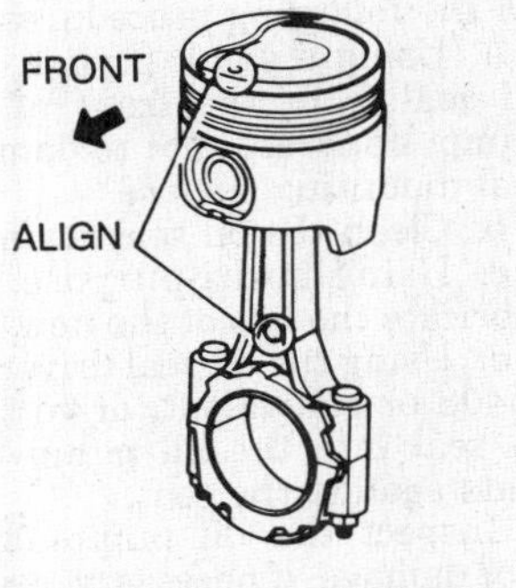

Typical piston alignment marks

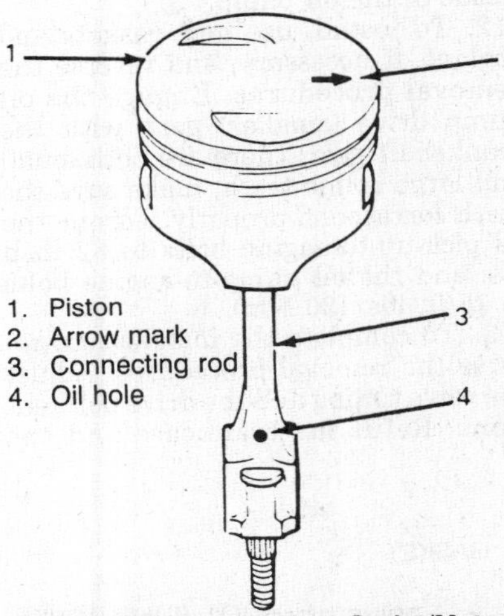

Piston alignment marks—Sprint/Metro

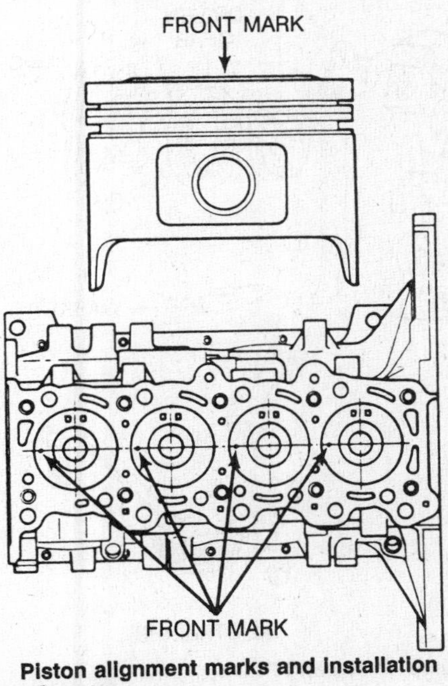

Piston alignment marks and installation—Spectrum

ENGINE LUBRICATION

Oil Pan

REMOVAL & INSTALLATION

Nova

1. Disconnect the negative battery cable.
2. Safely raise and support the vehicle.
3. Drain the engine oil and replace the drain plug when the pan is empty.
4. Remove the right splash shield.
5. On the twincam engines, remove the front exhaust pipe. Disconnect it at the manifold and the catalytic converter. Remove the exhaust bracket.
6. Remove the oil pan bolts.
7. Remove the oil pan. If the pan is difficult to remove, tap it gently with a rubber or plastic mallet. Do not use a pry bar to release it.

To install:

8. Clean the pan thoroughly and remove all sludge and solid matter. Clean the mating surfaces of the pan and the engine, removing all traces of old gasket material and sealer. During the cleaning, remove the drain bolt and clean the threads. Install a new gasket on the bolt and install the bolt in the pan.
9. Apply a new gasket and/or sealant to the pan and install the pan to the engine.
10. Install and tighten the pan bolts to 4 ft. lbs. (48 in. lbs.)
11. Install the right splash shield.
12. Lower the vehicle to the ground.
13. Refill the engine oil and connect the negative battery cable.
14. Start the engine and check for leaks.

1989-90 Prizm

1. Disconnect the negative battery cable.
2. Elevate and safely support the vehicle.
3. Remove the right and left splash shields.
4. Drain the oil.
5. Disconnect the sensor to the oxygen sensor.
6. Remove the front exhaust pipe. Disconnect the pipe at the catalytic converter and the exhaust manifold.
7. Remove the 2 nuts and the 19 bolts from the oil pan. Remove the oil pan from the engine.

NOTE: Use caution when removing the oil pan. The oil pump at the front of the engine may be damaged during removal. Use caution not to damage the edge of the pan.

8. Remove the 2 nuts and 2 bolts holding the oil pick-up and strainer to the engine. Remove the pick-up assembly and its gasket.

To install:

9. Clean the mating surfaces of the pan and cylinder block of all sealant and gasket material. Remove any traces of oil on these surfaces.
10. Thoroughly clean the pick-up and screen assembly and allow it to air dry.
11. Install the pick-up and strainer assembly with a new gasket. Tighten the nuts and bolts to 7.5 ft. lbs. (90 in. lbs.).
12. Apply a continuous bead of GM Sealant 1050026 or similar to both sides of the new oil pan gasket and place the gasket in position on the pan.
13. Install the pan onto the engine with all the nuts and bolts finger tight.
14. Tighten the nuts and bolts to 3.6 ft. lbs. (44 in. lbs.).
15. Connect the exhaust pipe to the catalytic converter and to the exhaust manifold.
16. Connect the wiring to the oxygen sensor.
17. Install the left and right splash shields.
18. Lower the vehicle to the ground.
19. Refill the engine oil and connect the negative battery cable.
20. Start the engine and check for leaks.

1991–92 Prizm

1. Disconnect the negative battery cable.
2. Safely raise and support the vehicle on jackstands.
3. Drain the engine oil and replace the drain plug when the pan is empty.
4. Remove the right splash shield.
5. Remove the front exhaust pipe. Disconnect it at the manifold and the catalytic converter.
6. Remove the exhaust bracket.
7. Remove the oil pan bolts. Remove the oil pan. If the pan is difficult to remove, tap it gently with a rubber or plastic mallet. Do not use a pry bar to release it.

NOTE: Do not bend or deform the edge (flange) of the oil pan during removal.

8. Clean the pan thoroughly and remove all sludge and solid matter. Clean the mating surfaces of the pan and the engine, removing all traces of old gasket material and sealer. During the cleaning, remove the drain bolt and clean the threads.

To install:

9. Install a new gasket on the bolt

and install the bolt in the pan. Apply a new gasket and/or sealant to the pan and install the pan to the engine.

10. Install and tighten the pan bolts to 4 ft. lbs. (48 in. lbs.).

11. Install the front exhaust pipe and its bracket.

12. Install the right splash shield.

13. Lower the vehicle to the ground.

14. Refill the engine oil and connect the negative battery cable.

15. Start the engine and check for leaks.

Storm

1. Disconnect the battery negative cable.
2. Raise and support the vehicle safely, then drain engine oil.
3. Remove the right undercover.
4. Remove the front exhaust pipe from the exhaust manifold, then remove the torque rod.
5. Remove the flywheel dust cover. Remove the stiffener from the cylinder head (if so equipped).
6. Remove the oil pan attaching bolts, then remove the oil pan.
7. Reverse procedure to install. Apply suitable sealant to the oil pan gasket. Torque bolts to 89 in. lbs (10 Nm). Connect battery negative cable. Start engine and check for leaks.

Spectrum

1. Disconnect the negative battery terminal from the battery.
2. Raise and support the vehicle safely, then drain the crankcase.
3. Disconnect the exhaust pipe bracket from the block and the exhaust pipe at the manifold.
4. Disconnect the right hand tension rod located under the front bumper.
5. Remove the oil pan bolts and oil pan, then clean the sealing surfaces.
6. To install, use a new gasket, apply sealant to the oil pump housing and the rear retainer housing, reverse the removal procedures. Torque the oil pan bolts to 7 ft. lbs. (10 Nm). Torque the exhaust pipe to manifold nuts to 42 ft. lbs. (57 Nm). Torque the exhaust pipe to converter bolts to 20 ft. lbs. (28 Nm).

Sprint and Metro

1. Remove the negative battery cable.
2. Raise and support the vehicle safely.
3. Drain the engine oil.
4. Remove the flywheel dust cover.
5. Remove the exhaust pipe at the exhaust manifold.
6. Remove the oil pan bolts, the pan and the oil pump strainer.
7. Clean the gasket mating surfaces.
8. To install, use new gaskets and reverse the removal procedures. Torque the oil pan bolts to 9 ft. lbs. (11 Nm). Refill the engine oil.

Oil Pump

REMOVAL & INSTALLATION

Nova

1. Raise and support the vehicle safely. Drain the engine oil and remove the oil pan. Remove the timing belt cover assembly.
2. Remove the oil pickup-to-engine brace bolts and the oil pickup.
3. Attach a lifting sling to the engine lift points and securely suspend the engine.
4. Mark the timing belt alignment between the camshaft and the crankshaft pulleys; also, mark the timing belt's direction of rotation. Loosen the idler pulley bolt, relieve the timing belt tension and remove the timing belt from the crankshaft sprocket; keep it engaged with the upper pulley.
5. Remove the crankshaft timing belt pulley and the timing belt idler pulley.
6. Remove the dipstick and dipstick tube.
7. Remove the oil pump-to-engine bolts and the oil pump; it may be necessary to tap lightly on the lower rear surface of the oil pump to loosen it.

To install:

8. Using the proper tool, clean the gasket mounting surfaces.
9. To replace the oil pump seal, perform the following procedures:
   a. Using a small prybar, pry the oil seal from the front of the oil pump; be careful not to damage the seal mounting surface.
   b. Clean the oil seal surface.
   c. Using multi-purpose grease, lubricate the lips of the new oil seal.
   d. Using the oil seal driver tool J-35403 or equivalent, drive the new oil seal into the oil pump until it seats against the seat.
10. Inspect the oil pump for wear and/or damage; if necessary, replace or repair the oil pump.
11. Using petroleum jelly, pack the inside of the oil pump.
12. To install, use new gaskets and sealant, if necessary, and reverse the removal procedures. Engage the oil pump drive (smaller) gear with the crankshaft gear; there are both small and large spline teeth, make sure the teeth correspond properly. Torque the oil pick-up-to-engine bolts to 82 inch lbs. and the oil pump-to-engine bolts to 15 ft. lbs. (20 Nm).
13. To complete the installation, reverse the removal procedures. Adjust the valve timing and the drive belt tensions. Refill the crankcase and the

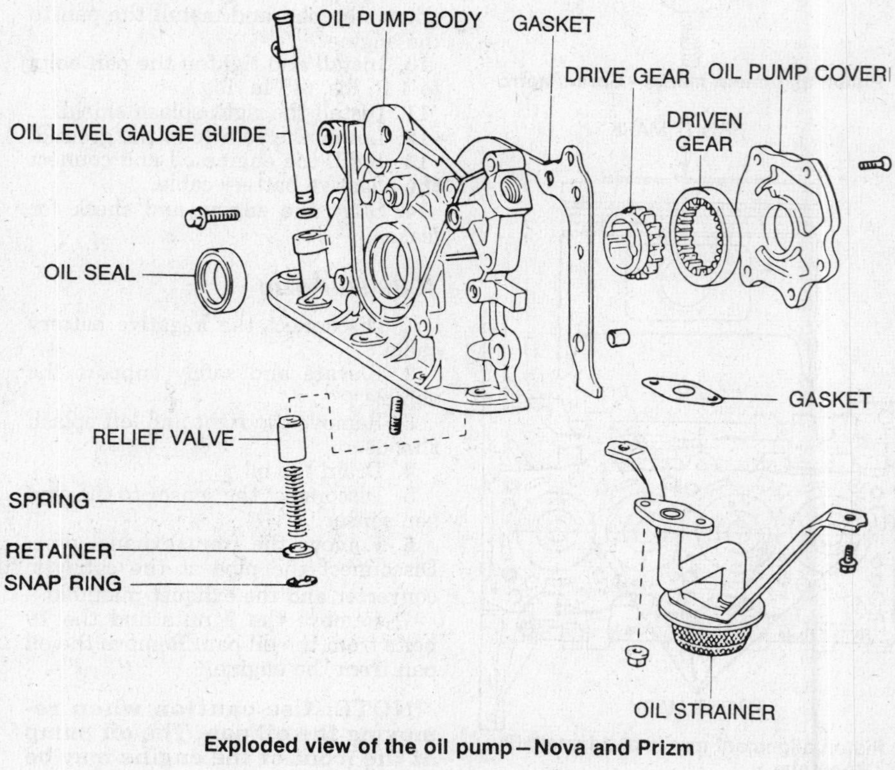

Exploded view of the oil pump—Nova and Prizm

cooling system. Start the engine, allow it to reach normal operating temperatures and check for leaks.

1989–90 Prizm

1. Disconnect the negative battery cable and elevate the vehicle. Safely support it. Remove the splashshield(s).
2. Remove the protectors from the 2 center engine mount nuts and studs.
3. Remove the 2 center transaxle mount-to-center crossmember nuts.
4. Remove the 2 rear transaxle mount-to-main crossmember nuts.
5. Drain the engine oil.
6. Remove the oil pan. Remove the oil pick-up and strainer assembly.
7. Lower the vehicle to the ground.
8. Depending on equipment, loosen the air conditioning compressor bracket, the power steering pump bracket and the alternator bracket as applicable. Remove the drive belts.
9. Remove the alternator from its mounts and place it out of the way. The wiring may be left attached.
10. Lift out the windshield washer fluid reservoir.
11. Support the engine. This may be done from above with tool J 28467-A or a chain hoist or from below with a floor jack. Be very careful of the jack placement (the oil pan is removed); use a piece of wood to distribute the load and protect the engine.
12. Remove the through bolt in the right engine mount.
13. Remove the water pump pulley.
14. Lower the engine to its normal position.
15. Remove the crankshaft pulley.
16. Remove the timing belt covers.
17. Remove the timing belt guide from the crank pulley.
18. Loosen the idler pulley bolt, push it all the way to the left and tighten the bolt. This removes tension from the belt.
19. Matchmark the belt and all the pulleys so that the belt may be reinstalled exactly as it was before. Mark an arrow on the belt showing direction of rotation.
20. Remove the timing belt from the lower pulley (crankshaft timing pulley). If careful, the belt may remain undisturbed on the camshaft pulleys.
21. Remove the idler pulley and spring.
22. Remove the dipstick and dipstick tube.
23. Remove the crankshaft timing pulley.
24. Raise the vehicle and safely support it.
25. Remove the 7 bolts holding the oil pump.
26. Remove the 7 bolts in the oil pump and carefully remove the pump. If it is difficult to remove, tap it lightly with a plastic or rubber mallet. Do not pry it off or strike it with a metal hammer.
27. When reinstalling, place a new gasket on the block. Install the oil pump to the crankshaft with the spline teeth to the drive gear engaged with the large teeth of the crankshaft.
28. Install the 7 retaining bolts and tighten them to 16 ft. lbs. (22 Nm).
29. Lower the vehicle to the ground.
30. Install the timing belt idler pulley.
31. Install the dipstick tube and dipstick.
32. Install the timing belt.
33. Install the timing belt guide. It should install with the cupped side facing outward.
34. Make sure the gaskets are properly seated in the timing belt covers and reinstall the covers. Make sure each bolt is in the correct hole.
35. Install the crankshaft pulley. Tighten the bolt to 87 ft. lbs. (118 Nm).
36. Elevate the motor to gain access to the water pump.
37. Install the water pump pulley.
38. Install the right engine mount through bolt. Tighten the through bolt to 64 ft. lbs. (87 Nm). When the bolt is secure, the engine lifting apparatus may be removed.
39. Position and install the alternator.
40. Reinstall the drive belts for the alternator, power steering and air conditioning as applicable. Adjust the belts to the correct tension.
41. Raise the vehicle and safely support it.
42. Install the oil pick-up and strainer assembly.
43. Apply a continuous bead of sealer (GM 1050026 or similar) to both sides of the new pan gasket.
44. Place the gasket on the pan and install the pan to the block. Tighten the bolts and nuts to 4 ft. lbs. (44 in. lbs.)
45. Install the 2 rear transaxle mount-to-main crossmember nuts and tighten them to 45 ft. lbs. (61 Nm). Install the 2 center transaxle mount-to-center crossmember nuts and tighten them to 45 ft. lbs. (61 Nm).
46. Install the protectors over the nuts and studs for the mounts.
47. Lower the vehicle to the ground.
48. Install the windshield washer fluid reservoir.
49. Refill the engine with the correct amount of fresh oil.
50. Connect the negative battery cable.
51. Start the engine and check for leaks. Allow the engine to warm up to normal operating temperature and check the work area carefully for signs of seepage.
52. With the engine shut off, check the tension of the drive belts and adjust if necessary. Reinstall the splash guard(s) under the vehicle.

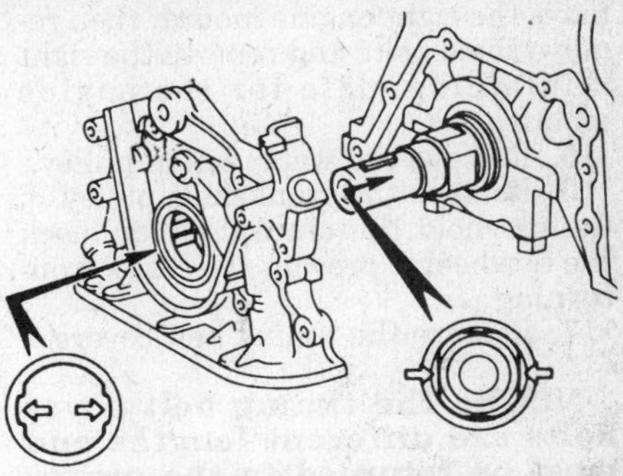

Correct positions for installation of the oil pump on the Prizm

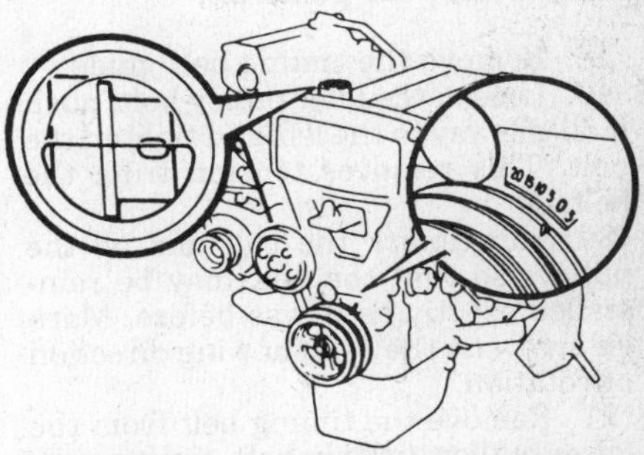

Aligning the motor at TDC/compression for No. 1 cylinder – Nova and the 1991–92 Prizm

1991-92 Prizm

1. Disconnect the negative battery cable.
2. Remove the oil pan.
3. Remove the oil pick-up and strainer.
4. Remove the oil pan baffle plate.
5. Drain the cooling system.
6. Disconnect the accelerator cable or linkage.
7. Remove the cruise control actuator if so equipped.
8. Remove the washer tank.
9. Remove the upper radiator hose at the engine block.
10. Remove the power steering and/or the air conditioning drive belt(s).
11. Loosen the bolts to the water pump pulley and then remove the alternator drive belt.
12. Remove the spark plugs.
13. Rotate the crankshaft and position the engine at TDC/compression. The crankshaft mark aligns at zero and the camshaft, when viewed through the oil filler cap, has a small cavity pointing upward.
14. Use a floor jack and a piece of wood to slightly elevate the engine. Re-

move the right engine mount; then remove the 3 bolts and remove the right reinforcing plate for the engine mount.

15. Remove the water pump pulley.
16. Remove the crankshaft pulley. Counterhold the crankshaft or block the flywheel to prevent the crank from turning.
17. Remove the timing belt covers.

NOTE: The timing belt cover bolts are different lengths and must be returned to the proper hole at reassembly. During removal, diagram or label each bolt and its correct position.

18. Remove the timing belt guide.
19. Loosen the idler pulley bolt, push it all the way to the left and tighten the bolt. This removes tension from the belt.
20. Matchmark the belt and all the pulleys so that the belt may be reinstalled exactly as it was before. Mark an arrow on the belt showing direction of rotation.
21. Remove the timing belt from the lower pulley (crankshaft timing pulley). If careful, the belt may remain undisturbed on the camshaft pulleys.
22. Remove the idler pulley and spring.
23. Remove the crankshaft timing pulley.
24. Remove the PCV hose.
25. Remove the dipstick and tube.
26. Remove the 7 bolts in the oil pump and carefully remove the pump. If it is difficult to remove, tap it lightly with a plastic or rubber mallet. Do not pry it off or strike it with a metal hammer.**To install:**
27. When reinstalling, place a new gasket on the block. Install the oil pump to the crankshaft with the spline teeth to the drive gear engaged with the large teeth of the crankshaft.
28. Install the 7 retaining bolts and tighten them to 7 ft. lbs. (10 Nm).
29. Install the dipstick tube and dipstick.
30. Install the crankshaft timing pulley.
31. Install the timing belt idler pulley.
32. Install the timing belt.
33. Install the timing belt guide. It should install with the cupped side facing outward.
34. Make sure the gaskets are properly seated in the timing belt covers and reinstall the covers. Make sure each bolt is in the correct hole.
35. Install the crankshaft pulley. Again using a counterholding device, tighten the bolt to 87 ft. lbs. (118 Nm).
36. Install the water pump pulley and tighten the bolts finger tight. Install the valve covers.
37. Install the right side engine mount. Tighten the through bolt to 64 ft. lbs. (87 Nm).
38. Install the reinforcement for the right motor mount and tighten the bolts to 31 ft. lbs. (43 Nm).
39. Install the spark plugs.
40. Install the alternator drive belt and tighten the water pump pulley bolts.
41. Install and adjust the power steering and/or the air conditioner drive belts.
42. Connect the upper radiator hose.
43. Install the windshield washer reservoir.
44. If equipped with cruise control, reinstall the cruise control actuator.
45. Connect the accelerator cable or linkage.
46. Install the oil pan baffle plate. Clean the contact surfaces thoroughly, apply a bead of sealer to the baffle plate and press the baffle plate into position. Be very careful not to get any sealant into the oil passages.
47. Install the oil pick-up and strainer assembly. Install the PCV hose.
48. Install the oil pan.
49. Using new gaskets, reconnect the exhaust pipe. Tighten the bolts to the exhaust manifold 46 ft. lbs. (62 Nm).
50. Install the flywheel cover.
51. Install the stiffener plate and the center engine mount.
52. Refill the engine with oil.
53. Refill the coolant system.
54. Start the engine and check for leaks. Allow the engine to warm up to normal operating temperature and check the work area carefully for signs of seepage.
55. With the engine shut off, check the tension of the drive belts and adjust if necessary. Reinstall the splash guards under the vehicle.

Storm

1. Disconnect the battery negative cable.
2. Remove the power steering and alternator belts.
3. Remove the timing belt.
4. Raise and support the vehicle safely.
5. Remove the crankshaft pulley.
6. Remove the oil pump attaching bolts, then the oil pump from the vehicle.
7. Reverse procedure to install. Apply suitable sealant to the oil pump gasket. Install bolts and torque to 89 inch lbs. (10 Nm). Connect the battery negative cable. Start engine and check for leaks.

Spectrum

1. Remove the engine from the vehicle. Drain the engine oil from the crankcase.
2. Remove the alternator belt. Remove the starter assembly.
3. Install the flywheel holding tool J-35271 or equivalent, to secure the flywheel.
4. Remove the crankshaft pulley and boss.
5. Remove the timing cover bolts and the timing cover.
6. Loosen the tension pulley and remove the timing belt.
7. Remove the crankshaft timing gear and the tension pulley.
8. Remove the oil pan bolts, oil pan, oil strainer fixing bolt and the oil strainer assembly.
9. Remove the oil pump bolts and the oil pump assembly.
10. Remove the sealing material from the oil pump and engine block sealing surfaces.
11. To install, lubricate the oil pump, use new gaskets, apply sealant to the sealing surfaces and reverse the removal procedures.

NOTE: Before installing the oil pump it would be a good idea to check the oil pressure relief valve and spring incorporated into the oil pump assembly. Remove the relief valve retaining plug along with the spring. Clean or replace the valve and spring assembly as necessary and reinstall it. Torque the retaining plug to 27 ft. lbs. (37 Nm). Torque the oil pump mounting bolts to 7 ft. lbs. (10 Nm). Be careful no to accidentally force the garter spring out of position during the oil pump assembly.

Sprint

1. Remove the timing belt.
2. Raise and support the vehicle safely. Drain the engine oil and remove the oil pan.
3. Use a suitable tool to hold the crankshaft timing belt pulley, remove the crankshaft bolt and pull the timing pulley from the shaft.
4. Remove the alternator mounting

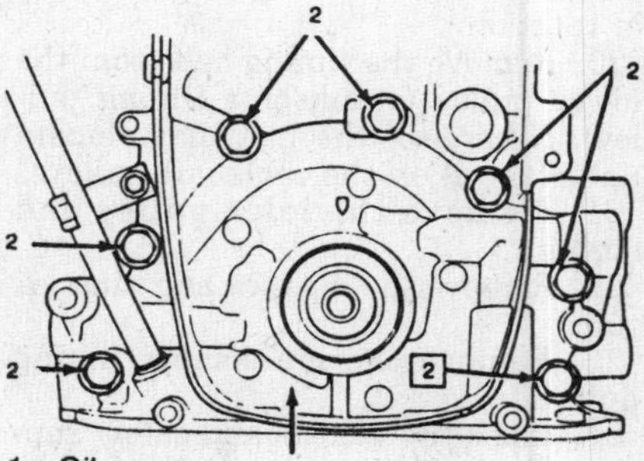

1. Oil pump
2. Oil pump mounting bolts

Oil pump mounting—Sprint and Metro

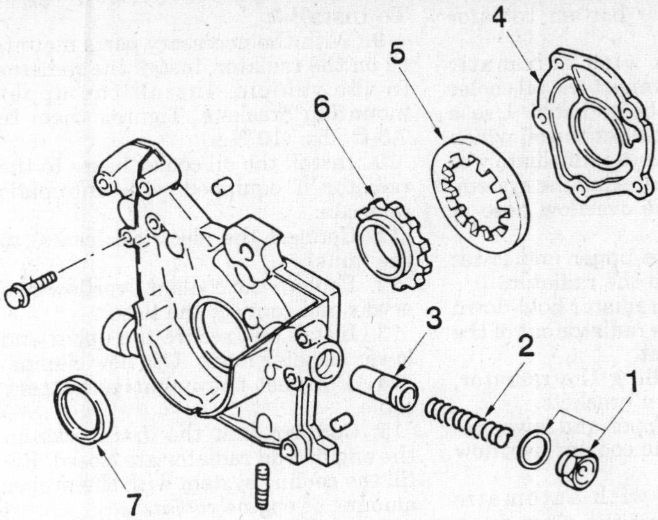

Exploded view of the oil pump—Spectrum

bracket and the air conditioning compressor bracket, if equipped.
5. Remove the alternator adjusting bolt and the upper cover bolt.
6. Remove the oil pump mounting bolts and the oil pump.
7. Pry the crankshaft oil seal from the oil pump.
8. Clean the gasket mounting surfaces. Remove the gear plate from the back of the oil pump and pack the oil pump gears with petroleum jelley.
9. To install, use new gaskets/seals and reverse the removal procedures. Torque the oil pump bolts to 7‹n-dash›9 ft. lbs. (10-12 Nm)and the crankshaft timing pulley bolt to 47–54 ft. lbs. (65-75 Nm). Adjust the valve clearances and check the timing.

NOTE: To install the oil pump to the engine, place the oil seal guide tool J-34853 on the crankshaft and slide the oil pump onto the alignment pins. After installing the oil seal housing, trim the gasket flush with the bottom of the case. Install the 4 shorter bolts into the top of the oil pump and the 3 longer ones into the bottom portion of the oil pump assembly.

Metro

1. Remove the negative battery cable.
2. Raise and support the vehicle safely.
3. Drain the engine oil.
4. Remove the water pump belt, pulley, alternator, alternator bracket and air conditioning mounting bracket, if equipped.
5. Remove the crankshaft pulley, timing belt outside cover, timing belt and tensioner.
6. Disconnect the engine oil level gauge.
7. Remove the crankshaft timing belt gear and timing belt guide. with eh crankshaft locked, remove the crankshaft timing belt pulley bolt.
8. Remove the oil pan bolts, oil pan, oil strainer fixing bolt and the oil strainer assembly.
9. Remove the oil pump bolts and the oil pump assembly.
10. Remove the sealing material from the oil pump and engine block sealing surfaces.
11. To install, lubricate the oil pump, use new gaskets, apply sealant to the sealing surfaces and reverse the removal procedures. Torque the 8 ft. lbs. (11 Nm). Torque the crankshaft timing belt gear bolt to 81 ft. lbs. (110 Nm). Install the rubber seal between the oil pump and the water pump.

NOTE: The edge of the oil pump gasket could bulge out; if it does, cut the bulge off, making the edge smooth and flush with end faces of pump case and cylinder block.

Rear Main Bearing Oil Seal

REMOVAL & INSTALLATION

Nova and Prizm

1. Remove the transaxle from the vehicle.
2. If equipped with a manual transaxle, perform the following procedures:
   a. Matchmark the pressure plate-to-flywheel.
   b. Remove the pressure plate-to-flywheel bolts and the clutch assembly from the vehicle.
   c. Remove the flywheel-to-crankshaft bolts and the flywheel.
3. If equipped with an automatic transaxle, perform the following procedures:
   a. Matchmark the flywheel-to-crankshaft.
   b. Remove the torque converter driveplate-to-crankshaft bolts and the torque converter driveplate.
4. Remove the rear end plate-to-engine bolts and the rear end plate.
5. If removing the rear oil seal retainer, perform the following procedures:
   a. Remove the rear oil seal retainer-to-engine bolts, rear oil seal retainer to oil pan bolts and the rear oil seal retainer.
   b. Using a small prybar, pry the rear oil seal retainer from the mating surfaces.
   c. Using a drive punch, drive the oil seal from the rear bearing retainer.
   d. Using the proper tool, clean the gasket mounting surfaces.

6. To remove the rear oil seal, with the rear oil seal retainer installed, use a small prybar and pry the seal from the rear oil seal retainer.

NOTE: When removing the rear oil seal, be careful not to damage the seal mounting surface.

To install:

7. Clean the oil seal mounting surface.
8. Using multi-purpose grease, lubricate the new seal lips.
9. Using an rear oil seal installation tool J-35388 or equivalent, tap the seal straight into the bore of the retainer.
10. If the rear oil seal retainer was removed from the vehicle, use a new gasket and sealant, if necessary, and reverse the removal procedures; be careful when installing the oil seal over the crankshaft.
11. To complete the installation, reverse the removal procedures. Torque the flywheel-to-crankshaft bolts to:
   a. 58 ft. lbs. (78 Nm) on vehicles equipped with automatic transaxles.
   b. 47 ft. lbs. (64 Nm) on vehicles equipped with manual001856axles.
11. Torque the converter drive plate-to-crankshaft bolts to 61 ft. lbs. (83 Nm).

Spectrum and Storm

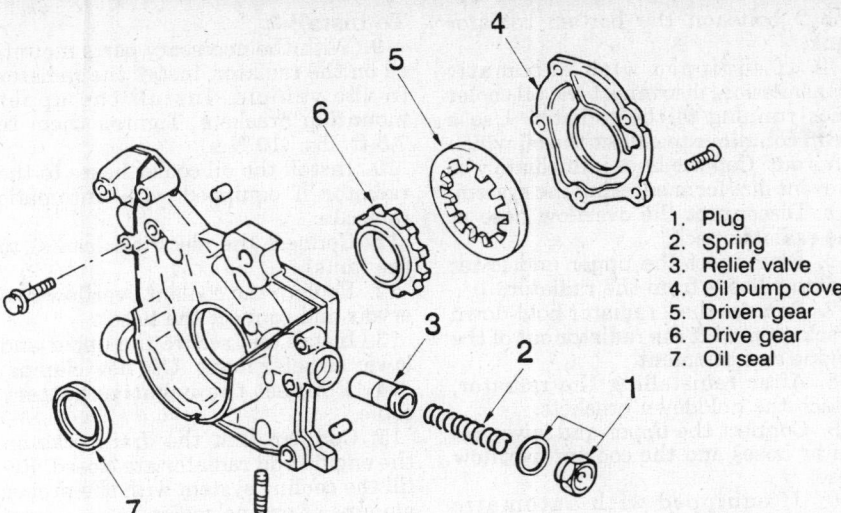

1. Remove the transaxle.
2. Remove the oil pan. On Storm, remove the right and left undercovers.
3. Remove the pressure plate and clutch on manual transaxle or the torque converter on automatic trans-

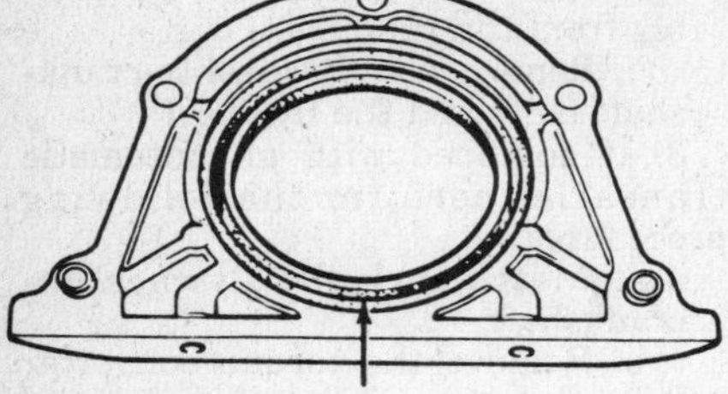

Rear main oil seal installed in seal housing – Sprint and Metro

axle, the flywheel bolts and the flywheel from the crankshaft.

4. Remove the rear oil seal retainer and remove the oil seal from the retainer. Clean the sealing surfaces.
5. Using a new oil seal, install the new seal in the oil seal retainer.
6. To install, use new gaskets, apply sealer to the mounting surfaces, apply oil to the seal lips, align the dowel pins of the retainer with the engine block and reverse the removal procedures.

Sprint and Metro

1. Remove the transaxle.
2. Raise and support the vehicle safely. Drain the oil and remove the oil pan.
3. Remove the pressure plate, the clutch plate and the flywheel.
4. Remove the mounting bolts and the rear seal housing.
5. Pry the oil seal from the oil seal housing.
6. To install, use new gaskets/seals and reverse the removal procedures. Torque the oil seal housing to 7–9 ft. lbs. (10-12 Nm) and the flywheel to the proper specification.

NOTE: After installing the oil seal housing, trim the gasket flush with the bottom of the case.

ENGINE COOLING

Radiator

REMOVAL & INSTALLATION

Nova

1. Drain the coolant by opening the engine block and radiator drain cocks.
2. Unplug the wiring to the cooling fan(s).
3. Remove the fan shroud, the 4 bolts on the top of the radiator tank and, if equipped with air conditioning, the 2 bolts on the bottom radiator tank.
4. If equipped with automatic transmission, disconnect the oil cooler lines running to the radiator. Use a clean container to collect the oil which runs out. Cap the lines immediately to prevent dirt form entering the system.
5. Disconnect the overflow hose at the radiator neck.
6. Disconnect the upper and lower radiator hoses from the radiator.
7. Remove the 2 radiator hold-down brackets and lift the radiator out of the engine compartment.
8. After reinstalling the radiator, attach the holddown brackets.
9. Connect the upper and lower radiator hoses and the coolant overflow hose.
10. If equipped with automatic transmission, remove the plugs from the lines and connect the oil cooler lines to the radiator.

To install:

11. Reinstall the 4 bolts at the top tank and, on automatics, install the 2 bolts at the lower tank.
12. Connect the wiring for the fan(s).
13. Confirm that the draincocks on the engine and radiator are closed. Refill the cooling system with the proper amount of engine coolant.
14. Add automatic tansaxle fluid in an amount equal to that lost from the oil cooler during removal.
15. Start the engine and check hose and line connections for leaks. Allow the engine to warm up to normal operating temperature. Check carefully for leaks under both cold and hot conditions.
16. Check the automatic transmission fluid and add if necessary.

Prizm

1. Disconnect the negative battery cable.
2. Drain the coolant by opening the engine block and radiator drain cocks.
3. Remove the coolant overflow reservoir.
4. Disconnect the upper radiator hose from the radiator and the lower radiator hose from the thermostat housing.
5. If equipped with automatic transmission, disconnect the oil cooler lines running to the radiator.
6. Remove the upper radiator brackets.
7. Disconnect the wiring running to the fan(s).
8. Remove the radiator and fan assembly from the vehicle. If the radiator is to be worked on or replaced, the fan and shroud assembly, the lower hose and the lower rubber mounts must be removed. Don't forget to install them on the new unit before installing it.

To install:

9. With the necessary parts mounted on the radiator, install the radiator in the vehicle. Install the upper mounting brackets. Tighten them to 7.5 ft. lbs. (10 Nm)
10. Install the oil cooler hoses to the radiator if equipped with automatic tansaxle.
11. Connect the electrical lead(s) to the fan(s).
12. Replace the coolant overflow reservoir and connect the hose.
13. Install and secure the upper and lower radiator hoses. Use new clamps.
14. Connect the negative battery cable.
15. Confirm that the draincocks on the engine and radiator are closed. Refill the cooling system with the proper amount of engine coolant.
16. Add automatic tansaxle fluid in an amount equal to that lost from the oil cooler during removal.
17. Start the engine and check hose and line connections for leaks. Allow the engine to warm up to normal operating temperature. Check carefully for leaks under both cold and hot conditions.
18. Check the automatic transmission fluid and add if necessary.

Spectrum

1. Disconnect the battery negative cable.
2. Drain the cooling system.
3. Remove the air intake duct.
4. Remove the fan motor cable from the fan motor socket and disconnect the thermo-switch cable.
5. Remove the fan motor assembly.
6. Remove the radiator hoses at the radiator, the coolant recovery hose at the filler neck and the oil cooler lines, if equipped with automatic transaxle.
7. Remove the radiator attaching bolts and the radiator.
8. Reverse procedure to install. Connect the battery negative cable.

Storm

1. Disconnect the battery negative cable.
2. Drain cooling system.
3. If equipped with automatic transaxle, disconnect the oil cooler lines from the radiator.
4. Remove the upper and lower radiator hoses.
5. Disconnect the fan motor cable from the rear of the fan motor socket.
6. Remove the coolant recovery hose from the radiator filler neck.
7. Remove the radiator attaching bolts and the radiator with fan and motor assembly from the vehicle.
8. Reverse procedure to install. Connect the battery negative cable and refill system. start engine and check for leaks.

Sprint and Metro

1. Disconnect the battery negative cable.
2. Drain the cooling system.
3. Disconnect the cooling fan motor electrical connector and the air inlet hose.
4. Remove the inlet, outlet and the reservoir tank hoses from the radiator.
5. If equipped with automatic transaxle, disconnect the oil cooler lines from the radiator.
6. Remove the mounting bolts, cooling fan motor, shroud and the radiator from the vehicle.
7. Reverse procedure to install. Connect the battery negative cable and refill system.

REMOVAL & INSTALLATION

All Vehicles

1. Disconnect the negative terminal from the battery.
2. Label and disconnect the electrical connector from the cooling fan motor.
3. Remove the fan shroud-to-radiator frame bolts and the fan/shroud assembly from the vehicle.
4. Remove the fan blade-to-motor nut, fan blade and washer.
5. Remove the fan-to-shroud bolts and the fan motor from the shroud.
6. Test the fan motor and replace it, if necessary.
7. To install, reverse the removal procedures and check the fan operation.

Heater Core

REMOVAL & INSTALLATION

Nova

1. Disconnect the negative battery cable. Drain the cooling system.
2. In the engine compartment, disconnect the heater hoses from the heater unit.
3. From inside the vehicle (under the dash), remove the lower heater unit case-to-heater case clips and remove the lower case.
4. Using a medium prybar, separate and remove the lower portion of the case from the heater case.
5. Remove the heater core from the heater case.
6. Inspect the heater hoses for cracking and deterioration, then, the heater core for leakage and corrosion; replace the items, if necessary.
7. To install, reverse the removal procedures. Refill the cooling system. Start the engine, allow it to reach normal operating temperatures and check for leaks. Turn the heater controls to max heat and check the heater operation.

Prizm

The heater case and core are located directly behind the center console. The access the case and core, the entire console must be removed as well as most of the instrument panel assembly.

1. Disconnect the negative battery cable. Remove the steering wheel.
2. Remove the trim bezel from the instrument panel.
3. Remove the cup holder from the console.
4. Remove the radio.
5. Remove the instrument panel assembly, cluster assembly, center console and all console trim, lower dash trim, side window air deflectors, and all instrument panel wiring harnesses.
6. Drain the coolant from the cooling system.
7. Disconnect all cables and ducts from the heater case.
8. Disconnect the blower switch harness and the heater control assembly.
9. Disconnect the 2 center console support braces.
10. Remove all mounting bolts, nuts and clips from the heater and air distribution cases.
11. Remove the heater and air distribution cases.
12. Remove the screws and clips from the case, separate the case halves, and remove the core from the case.
13. Installation is the reverse of removal. Fill the cooling system.

Spectrum

1. Disconnect the negative battery cable. Disconnect the heater hoses in the engine compartment.
2. At the lower part of the heater unit case, remove the 6 retaining clips.
3. Using a small prybar, pry open the lower part of the case and remove it.
4. Remove the core assembly insulator and the core assembly.
5. To install, reverse the removal procedures.

Storm

1. Disconnect the battery negative cable.
2. Disconnect heater hoses from inside the engine compartment.
3. Remove the instrument panel assembly.
4. If equipped with air conditioning, remove the evaporator assembly.
5. Remove the duct between the blower motor and the heater unit.
6. Remove the center ventilation duct.
7. Remove the 4 nuts attaching the heater unit and remove the heater unit from the vehicle.
8. Remove the 5 screws attaching the mode control case to the heater core case, then the mode control case. Do not remove the link assembly as this time.
9. Remove 5 screws to separate the 2 halves of the heater core case and remove the heater core from the case.
10. Reverse procedure to install. Connect the battery negative cable.

Sprint

1. Disconnect the negative battery cable. Drain the cooling system.
2. Disconnect the 2 water hoses from the radiator at the heater unit.
3. Remove the glove box from the upper instrument panel.
4. Remove the defroster hoses from the heater case.
5. Disconnect the electrical connectors from the blower motor and the heater resistor.
6. Disconnect the 3 control cables from the heater case side levers.
7. Pull out the center vent louver.
8. Disconnect both side vent ducts from the center duct vent.
9. Remove the center duct vent and the ashtray's upper plate.
10. Remove the instrument member stay and the heater assembly mounting nuts.
11. Loosen the 3 heater case top mounting bolts through the glove box opening.
12. Raise the dash panel and remove the heater control assembly.
13. Separate the heater case into 2 sections by removing the clips.
14. Pull the heater core from the heater unit.
15. To install, reverse the removal procedures. Refill the cooling system. Start the engine, bring it to normal operating temperature and check for leaks.

Metro

1. Disconnect the negative battery cable. Drain the cooling system.
2. Remove the clamps and heater core hoses from the heater core.
3. Remove the instrument panel carrier assembly.
4. Remove the heater control assembly from the instrument support member.
5. Remove the temperature and mode control cables from the heater case.
6. Remove the 2 mounting bolts and 2 mounting nuts from the heater case. Remove the heater case from the vehicle.
7. Remove all the heater case re-

1. Car heater assembly
2. Blower motor
3. Seal
4. Blower fan
5. Resistor
6. Resistor plate
7. Case clamp
8. Defroster damper
9. Temp damper
10. Vent damper
11. Heater pipe cover
12. Heater core
13. Heater left case
14. Heater right case
15. Duct
16. Vent link plate
17. Temp lever
18. Temp plate
19. Link lever
20. Mode lever
21. Link No. 2 lever
22. Defroster link plate
23. Vent link shaft
24. Defroster link shaft
25. Heater control lever assembly
26. Control lever knob
27. Air control cable
28. Heat control cable
29. Fresh air control cable
30. Heater grommet
31. Defroster link spring
32. defroster link spring washer

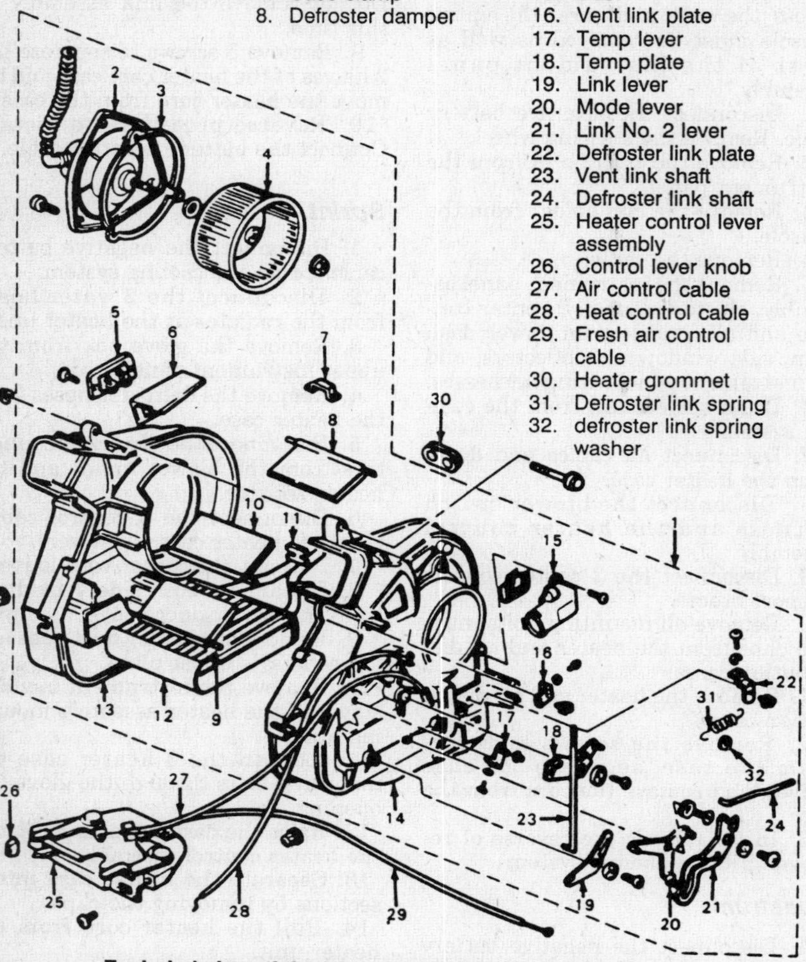

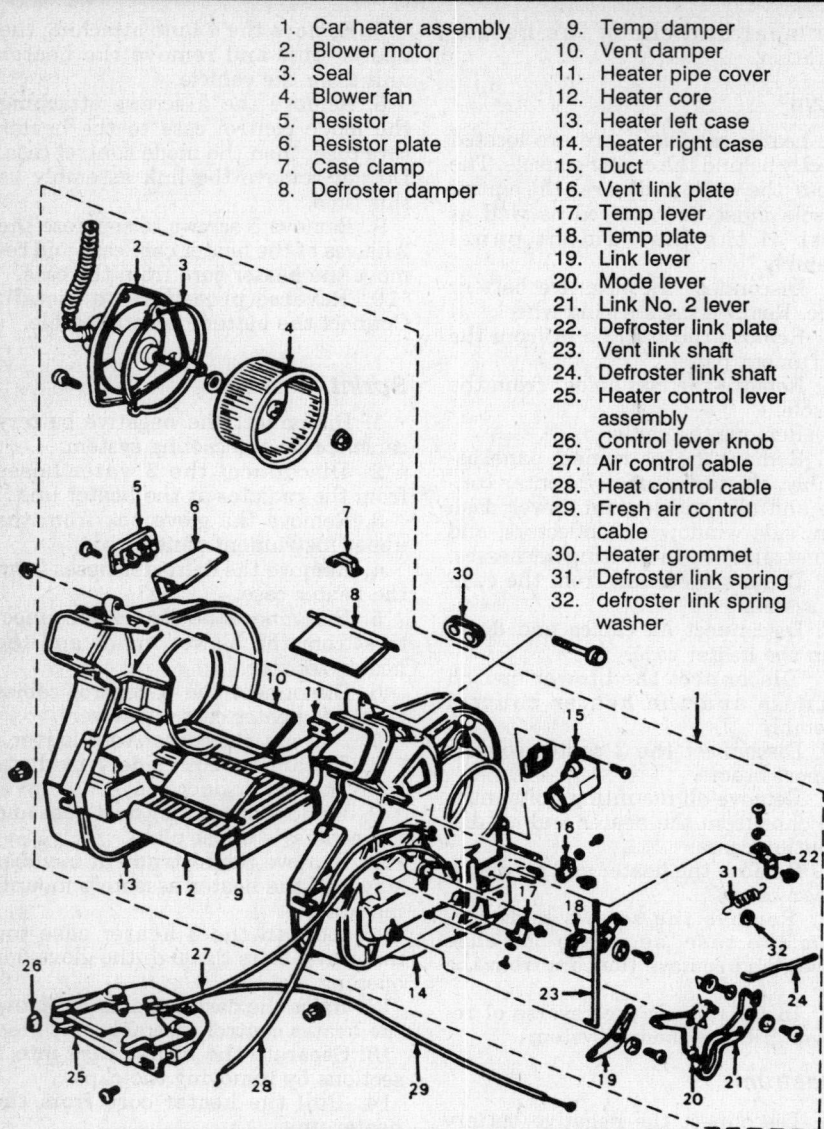

Exploded view of the heater assembly—Sprint and Metro

taining clips and 2 attaching screws from the heater case.

8. Separate the heater case halves.
9. Carefully pull the heater core from the heater case.
10. To install, reverse the removal procedures. Refill the cooling system. Start the engine, bring it to normal operating temperature and check for leaks.

Water Pump

REMOVAL & INSTALLATION

Nova and 1991-92 Prizm

1. Disconnect the negative battery cable.
2. Drain the cooling system.
3. Loosen the water pump pulley-to-water pump bolts. If equipped with power steering, remove the power steering pump drive belt.
4. Loosen the alternator adjusting and mounting bolts, move the alternator to relieve the belt tension, then remove the alternator/water pump drive belt.
5. Remove the water pump pulley-to-water pump bolts and the pulley.
6. Remove the water pump inlet-to-engine bolts (from the side of the block), the inlet pipe-to-water pump nuts and the inlet pipe, discard the O-ring.
7. Remove the dipstick tube bracket bolt and the dipstick tube; be sure to plug the hole in the block with a clean rag.
8. For the non-twincam engine, remove the upper timing belt front cover. For the twincam engine, remove the upper and middle timing belt front covers.
9. Remove the water pump-to-engine bolts and the water pump from the engine; discard the water pump-to-engine O-ring.

To install:

10. Using the proper tool, clean the gasket mounting surfaces.
11. To install the water pump, use a new O-ring and reverse the removal procedures. Torque the water pump-to-engine bolts to 11 ft. lbs. (15 Nm).
12. When installing the oil dipstick tube, use a new O-ring and coat it with engine oil.
13. To complete the installation, use a new O-ring and reverse the removal procedures. Refill the cooling system. Start the engine, allow it to reach normal operating temperatures and check for leaks.

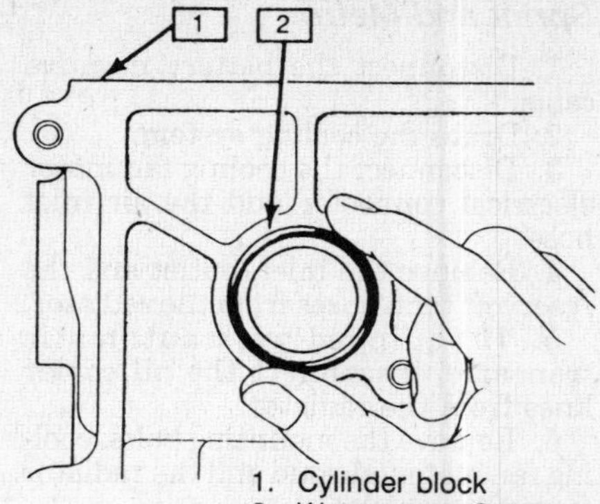

1. Cylinder block
2. Water pump O-ring

Always use a new gasket (O-ring) when installing a new water pump.

1989–90 Prizm

1. Remove the radiator cap. Drain the engine coolant by opening the radiator and engine block draincocks.
2. Raise and safely support the vehicle on jackstands.
3. Remove the 2 nuts for the rear motor mount.
4. Lower the vehicle to the ground.
5. Remove the windshield washer fluid container.
6. If so equipped, remove the cruise control bracket with the control module.
7. Remove the through bolt for the right motor mount.
8. Place a jack under the engine. Use a piece of wood between the engine and the jack.
9. Raise the engine slowly and carefully. Keep a close watch on lines and cables. The engine need only be raised enough to gain access to various nuts and bolts.
10. Loosen the water pump pulley bolts, but leave them in place.
11. Loosen the alternator lock bolt and pivot nut; swing the alternator towards the engine and remove the drive belt.

12. Loosen the pivot bolts and the lock bolt for the power steering pump and move it towards the engine; remove the power steering belt.
13. With the belts removed, the water pump pulley may be removed from the pump.
14. Lower the jack, allowing the engine to return to place.
15. Remove the water inlet and water bypass hoses from the water inlet pipe.
16. Remove the clamp holding the water inlet pipe to the engine. Loosen and remove the 2 nuts holding the inlet pipe to the water pump. Remove the pipe and its o-ring.
17. Remove the mounting bolt for the dipstick tube; remove the tube and dipstick. Immediately plug the hole in block to prevent fluid from polluting the oil.

NOTE: During the following steps, if coolant should get by the plug in the dipstick hole and run into the motor, the engine oil must be changed before starting the motor. Failure to do so can damage the engine bearings.

18. Remove the upper timing belt cover.

NOTE: The timing belt is exposed with the cover(s) removed. Do not allow oil or coolant to contact the belt. This includes any fluid which may be accidentally transferred through rags, fingerprints, etc.

19. Remove the 3 water pump bolts and remove the water pump.
20. When reinstalling, position a new o-ring on the engine and fit the water pump. Tighten the 3 bolts to 11 ft. lbs. (15 Nm).

To install:

21. Install the upper timing belt cover.
22. Install a new o-ring on the dipstick tube.
23. Remove the plug in the engine and install the tube. Tighten the mounting bolt.
24. Using a new o-ring, install the water inlet pipe at the back of the water pump. Tighten the 2 nuts to 14 ft. lbs. (19 Nm).
25. Connect the water inlet and water bypass hoses to the water inlet pipe.
26. Following the same jacking procedure as before, elevate the motor with the floor jack.
27. Install the water pump pulley and tighten the bolts finger tight. It will be easier to do the final tightening when the belts are installed.
28. Install the power steering belt and adjust its tension.
29. Install the alternator drive belt and adjust its tension.
30. Tighten the water pump pulley bolts to 17 ft. lbs. (23 Nm).
31. Lower the jack, allowing the engine to return to its normal place.
32. Install the through bolt for the right motor mount. Tighten it to 64 ft. lbs. (86 Nm).
33. Reinstall the cruise control module and the bracket, if so equipped.
34. Raise and safely support the vehicle.
35. Install the 2 nuts for the rear mount and tighten them to 38 ft. lbs. (52 Nm).
36. Lower the vehicle to the ground.
37. Replace the washer fluid container.
38. Confirm that the draincocks on the radiator and engine block are closed. Refill the cooling system with coolant.
39. Start the engine and check for leaks. Pay particular attention to any hose or fitting which was disassembled during the repair.
40. After the engine is shut off, double check the drive belts for proper tension and adjust as necessary.

Storm

EXCEPT TWINCAM ENGINE

1. Disconnect the battery negative cable and drain cooling system.
2. Remove the power steering pump belt and the timing belt.
3. Remove the belt tension pulley.
4. Remove the water pump-to-block attaching bolts and the water pump from the vehicle.
5. Reverse procedure to install. Clean mating surfaces of all gasket material. Torque water pump bolts to 17.4 ft. lbs. (23.5 Nm). Connect battery negative cable and refill cooling system.

TWINCAM ENGINE

1. Disconnect the battery negative cable and drain cooling system.
2. Support the engine using a suitable floor jack and remove the right front engine mount.
3. Remove the engine mount bridge and the upper timing belt cover.
4. Remove the power steering belt and the lower timing belt cover.
5. Loosen the timing belt tension pulley and remove the timing belt.
6. Remove the power steering pump and bracket.
7. Remove the water pump attaching bolts, then the water pump from the vehicle.
8. Reverse procedure to install. Connect the battery negative cable and refill cooling system. Start engine and check for leaks.

Spectrum

1. Disconnect the negative battery cable and drain the cooling system.
2. Loosen the power steering pump adjustment bolts and remove the belt.
3. Remove the timing belt.
4. Remove the tension pulley and spring.
5. Remove the water pump mounting bolts, the water pump and gasket. Clean the mounting surfaces of all gasket material.
6. To install, reverse the removal procedures. Torque the water pump to 17 ft. lbs. (23 Nm) and the tension pulley to 30 ft. lbs. (41 Nm).

Sprint and Metro

1. Disconnect the negative battery cable.
2. Drain the cooling system.
3. Remove the water pump belt and pulley.
4. Remove the crankshaft pulley, the timing belt outside cover, the timing belt and the tensioner.
5. Remove the mounting bolts and the water pump.
6. Clean the gasket mating surfaces.
7. To install, use a new gasket/sealer and reverse the removal procedures. Torque the water pump bolts to 7.5–9 ft. lbs. (10-12 Nm). Adjust the water pump belt deflection to ¼–⅜ in. between the water pump and the crankshaft pulleys.

Thermostat

REMOVAL & INSTALLATION

Nova and Prizm

NOTE: A thermostat with an internal by-pass should never be removed as a countermeasure to overheating. Removing the thermostat actually makes the problem worse because more coolant bypasses the radiator, thereby reducing cooling even more.

1. Drain the cooling system.
2. Remove the water inlet housing and remove the thermostat. Carefully observe the positioning of the thermostat within the housing.
3. Install the new thermostat in the housing, making sure it is in correctly. It is possible to install it backwards. Additionally, make certain that the air bleed valve aligns with the protrusion on the water inlet housing. Failure to observe this placement can result in poor air bleeding and possible overheating.
4. Install the water inlet housing cover with a new gasket. Install the 2 hold down bolts and tighten them to

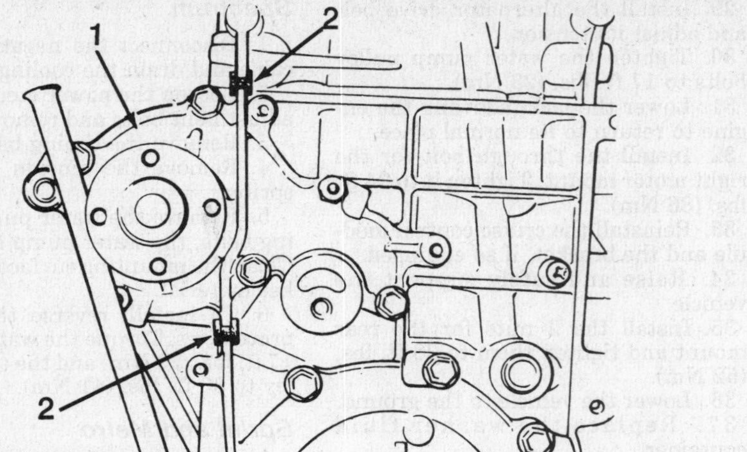

1. Water pump
2. Rubber seals

Water pump mounting – Sprint and Metro

20 ft. lbs. (27 Nm). Do not overtighten these bolts!

5. Refill the cooling system with coolant.
6. Start the engine. During the warm up period, observe the temperature gauge for normal behavior. Also during this period, check the water inlet housing area for any sign of leakage. Remember to check for leaks under both cold and hot conditions.

Spectrum and Storm

1. Disconnect the battery negative cable. Drain cooling system.
2. Remove the top radiator hose from from the outlet pipe.
3. Remove the outlet pipe bolts, outlet pipe, gasket and thermostat from the thermostat housing.
4. Reverse procedure to install. Refill cooling system. Connect battery negative cable. Start engine and check for leaks.

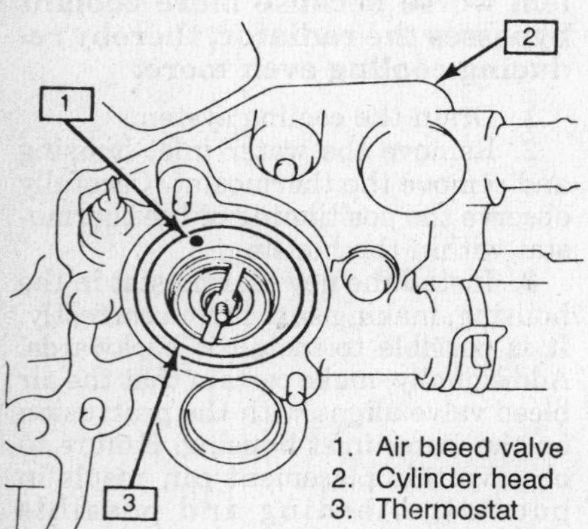

1. Air bleed valve
2. Cylinder head
3. Thermostat

The correct placement of the air bleed valve during the thermostat replacement – Prizm

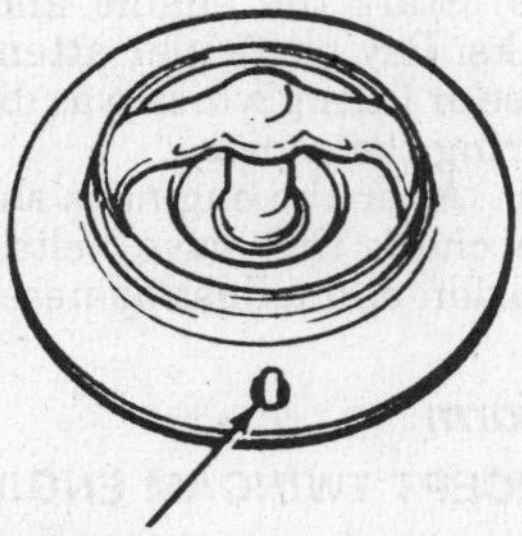

AIR BLEED VALVE

Typical air bleed valve type thermostat

Sprint and Metro

1. Disconnect the battery negative cable.
2. Drain cooling system to a level below the thermostat.
3. Remove the air cleaner.
4. Disconnect the electrical connectors at the thermostat cap.
5. Remove the inlet hose, cap mounting bolts and the thermostat from the themostat housing.
6. Clean the gasket mounting surfaces. Ensure that the thermostat air bleed hose is clear.
7. Reverse procedure to install. Install thermostat into housing with the spring side down. Fill cooling system and connect the battery negative cable. Start engine and check for leaks.

Cooling System Bleeding

After performing any repairs on the cooling system, it must be bled. Air trapped in the system will prevent proper filling and leave the radiator coolant level low, causing a risk of overheating.

1. To bleed the system, start with the system cool, radiator cap off and the radiator filled to about an inch below the filler neck.
2. Start the engine and run it at slightly above normal idle speed. This will ensure adequate circulation. If air bubbles appear and the coolant level drops, fill the cooling system to bring the level back to the proper level.
3. Run the engine until the thermostat opens. When this happens, coolant will will move abruptly across the top of the radiator and the temperature of the radiator will suddenly rise.
4. At this point, air is often expelled and the level may drop quite a bit. Keep refilling the system until the level is near the top of the radiator and remains constant.
5. If the vehicle has an overflow tank, fill the radiator right up to the filler neck. Replace the radiator filler cap.

ENGINE ELECTRICAL

NOTE: Disconnecting the negative battery cable on some vehicles may interfere with the functions of the on board computer systems and may require the computer to undergo a relearning process, once the negative battery cable is reconnected.

Distributor

REMOVAL

Nova and 1991–92 Prizm

EXCEPT TWINCAM ENGINE

1. Disconnect the negative battery cable.
2. Disconnect the distributor wire at its connector.
3. Label and disconnect the vacuum hoses running to the vacuum advance unit on the side of the distributor.
4. Remove the distributor cap (leave the spark plug wires connected) and swing it out of the way.
5. Carefully note the position of the distributor rotor relative to the distributor housing; a mark made on the casing will be helpful during reassembly. Use a marker or tape so the mark doesn't rub off during the handling of the case.
6. Remove the distributor hold-down bolts.
7. Carefully pull the distributor out until it stops turning counterclockwise.
8. If the engine has not been moved out of position, align the rotor with the

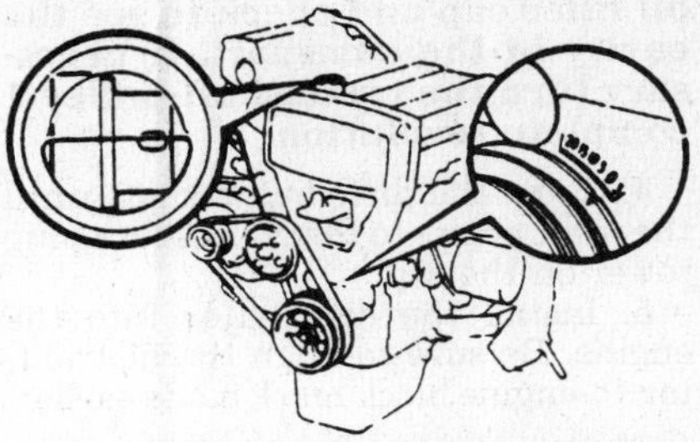

Aligning the crankshaft pulley and the camshaft cavity, twincam engine – Nova and Prizm

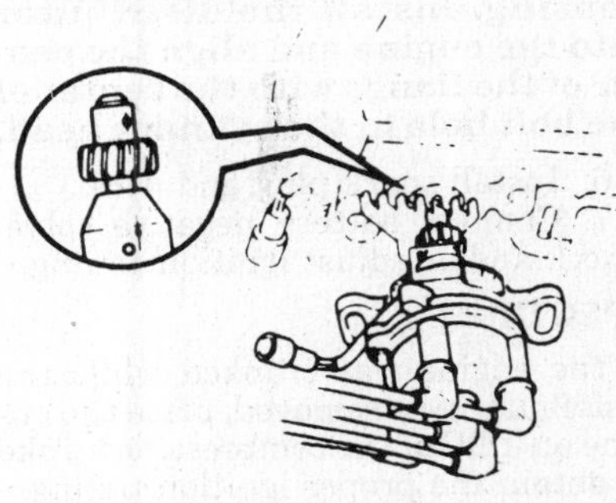

Aligning the distributor driveshaft with the housing twincam engine – Nova and Prizm

mark made earlier and reinstall the distributor. Position it carefully and make sure the drive gear engages properly within the engine. Install the holding bolts.

9. Install the distributor cap and reattach the vacuum lines to their correct ports.

10. Install the wiring to the distributor, and connect the battery cable.

11. Check and adjust the timing as necessary.

TWINCAM ENGINE

1. Disconnect the negative battery cable.

2. Label and disconnect the coil and spark plug wiring at the distributor cap.

3. Disconnect the distributor wire at its connector.

4. Remove the distributor hold-down bolts. Before moving or disturbing the distributor, mark the position of the distributor relative to the engine and the position of the rotor relative to the case. Use a marker or tape so the mark doesn't rub off during the handling of the case.

5. Remove the distributor from the engine.

6. Remove the O-ring from the distributor shaft.

7. If the engine has not been moved out of position, align the rotor with the mark made earlier and reinstall the distributor. Install a new O-ring and lubricate it with clean engine oil. Position it carefully and make sure the drive gear engages properly within the engine. Install the holding bolts.

8. Reconnect the wiring and the spark plug wires. Connect the negative battery cable.

9. Check and adjust the timing as necessary.

Storm and 1989-90 Prizm

1. Disconnect the negative battery cable.

2. Disconnect all electrical connections at the distributor, including the plug wires.

3. Remove the distributor cap.

4. Mark the position of the distributor case relative to the engine. Use a marker or tape so the mark doesn't rub off during the handling of the case. Also mark the position of the distributor rotor relative to the case.

5. Remove the distributor mounting bolts.

6. Remove the distributor from the engine and remove the O-ring from the distributor shaft.

7. If the engine has not been moved out of position, align the rotor with the mark made earlier and reinstall the distributor. Position it carefully and make sure the drive gear engages properly within the engine. Install the holding bolts.

8. Reconnect the wiring and the spark plug wires. Connect the negative battery cable.

9. Check and adjust the timing as necessary.

Spectrum

1. Disconnect the negative battery terminal from the battery.

2. Remove the distributor cap.

3. Mark and remove all electrical leads and vacuum lines connected to the distributor assembly.

4. Mark the relationship of the rotor to the distributor housing and the distributor housing to the engine.

5. Remove the hold-down bolt, clamp and distributor.

Sprint and Metro

1. Disconnect the negative battery cable.

2. Disconnect the wiring harness at the distributor and the vacuum line at the distributor vacuum unit.

3. Remove the distributor cap.

NOTE: Mark the distributor body in reference to where the ro-

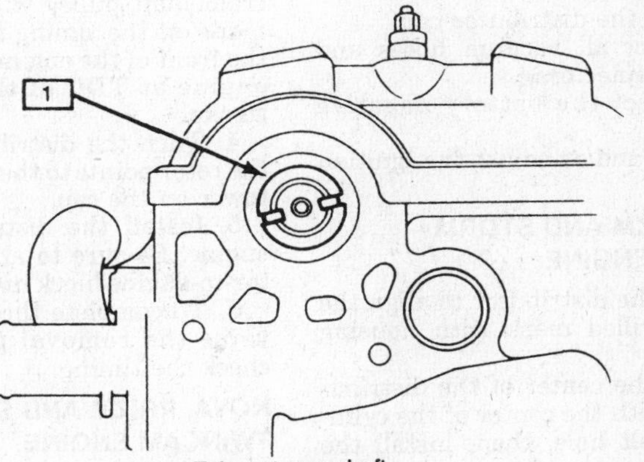

Positioning the camshaft for top dead center – 1989–90 Prizm

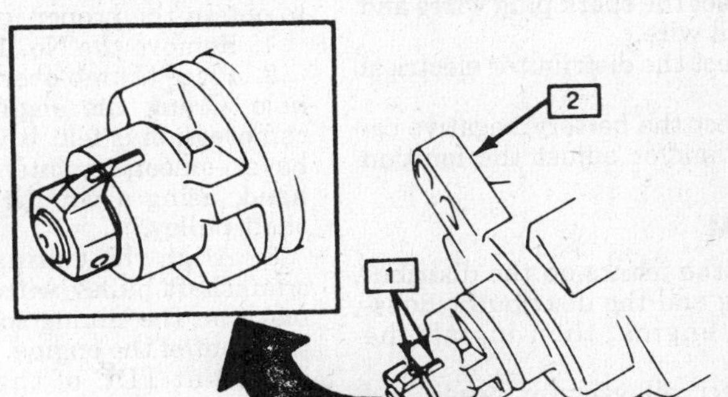

Positioning the distributor alignment marks – 1989–90 Prizm

tor is pointing. Mark the distributor hold-down bracket and cylinder head for a reinstallation location point.

4. Remove the hold-down bolt and the distributor from the cylinder head. Do not rotate the engine after the distributor has been removed. On the Sprint models, when the distributor assembly has been removed, the engine oil may start to flow out, it would be wise to install a drain pan under the distributor before removing it.

INSTALLATION

Timing Not Disturbed

NOVA, PRIZM AND STORM—EXCEPT TWINCAM ENGINE

1. Use a new O-ring on the distributor housing, lubricate the drive gear teeth with engine oil, align the protrusion at the bottom of the distributor housing with the pin on the side of the distributor drive gear, mesh the gears and install the distributor.

NOTE: Ensure that alignment marks on the distributor housing-to-engine and the rotor-to-distributor housing align before installing distributor.

2. Install the distributor cap.
3. Connect all vacuum hoses and electrical connectors.
4. Connect the battery negative cable.
5. Check and/or adjust the ignition timing.

NOVA, PRIZM AND STORM—TWINCAM ENGINE

1. Turn the distributor to align the halfshaft drilled mark with housing cavity.
2. Align the center of the distributor flange with the center of the cylinder head bolt hole, then, install the distributor.
3. Install the hold-down bolt and torque it to 14 ft. lbs. (19 Nm).
4. Connect the spark plug wires and ignition coil wire.
5. Connect the distributor electrical connector.
6. Connect the battery negative cable. Check and/or adjust the ignition timing.

SPECTRUM

1. Align the marks on the distributor housing and the distributor housing-to-the engine, then install the distributor.
2. Connect all vacuum hoses and electrical connectors.
3. Install the distributor cap and spark plug wires.
4. Connect the battery negative cable. Check and/or adjust ignition timing.

SPRINT AND METRO

1. Align the reference marks on the distributor housing to the distributor hold-down bracket.
2. Install the distributor into the offset slot in the camshaft, then the hold-down bolt.
3. Connect vacuum hoses and electrical connectors to the distributor.
4. Install the distributor cap, then connect the battery negative cable. Check and/or adjust the ignition timing.

Timing Disturbed

NOVA, PRIZM AND STORM—EXCEPT TWINCAM ENGINE

If the engine was cranked while the distributor was removed, place the engine on TDC of the compression stroke to obtain the proper ignition timing.

1. Remove the No. 1 spark plug.
2. Place finger over the spark plug hole. Crank the engine slowly until compression is felt. It would be easier to have someone rotate the engine by hand, using a wrench on the crankshaft pulley.
3. Align the timing mark on the crankshaft pulley with the **0** degrees mark on the timing scale attached to the front of the engine. This places the engine at TDC of the compression stroke.
4. Turn the distributor shaft until the rotor points to the No. 1 spark plug tower on the cap.
5. Install the distributor into the engine. Be sure to align the distributor-to-engine block mark made earlier.
6. To complete the installation, reverse the removal procedures and check the timing.

NOVA, PRIZM AND STORM—TWINCAM ENGINE

If the engine was cranked while the distributor was removed, place the engine on TDC of the compression stroke to obtain the proper ignition timing.

1. Remove the No. 1 spark plug.
2. Place thumb over the spark plug hole. Crank the engine slowly until compression is felt. It will be easier to have someone rotate the engine by hand, using a wrench on the crankshaft pulley.
3. Align the timing mark on the crankshaft pulley with the **0** degrees mark on the timing scale attached to the front of the engine. This places the engine at TDC of the compression stroke.

NOTE: On the Prizm models, verify that the timing marks are aligned properly by, removing the oil filler cap and check to see the cavity in the camshaft. If necessary turn the crankshaft pulley 1 complete revolution.

4. Turn the distributor shaft until the rotor points to the No. 1 spark plug tower on the cap.
5. Install the distributor into the engine. Be sure to align the distributor-to-engine block mark made earlier.

NOTE: On the Prizm models, align the drilled marks on the driven gear with the cavity of the housing. Install the distributor into the engine and align the center of the flange with the center of the bolt hole in the cylinder head.

6. Install spark plug and wire.
7. Connect battery negative cable. Check and/or adjust ignition timing.

SPECTRUM

If the engine was cranked while the distributor was removed, place the engine on TDC of the compression stroke to obtain the proper ignition timing.

1. Remove the No. 1 spark plug.
2. Place thumb over the spark plug hole. Crank the engine slowly until compression is felt. It will be easier to have someone rotate the engine by hand, using a wrench on the crankshaft pulley.
3. Align the timing mark on the crankshaft pulley with the **0** degrees mark on the timing scale attached to the front of the engine. This places the engine at TDC of the compression stroke.
4. Turn the distributor shaft until the rotor points to the No. 1 spark plug tower on the cap.
5. Install the distributor into the engine. Be sure to align the distributor-to-engine block mark made earlier.
6. Install the No. 1 spark plug and connect all vacuum hoses and electrical connectors.
7. Install distributor cap, then connect the battery negative cable.
Check and/or adjust the ignition timing.

SPRINT AND METRO

If the engine was cranked while the distributor was removed, place the engine on TDC of the compression stroke to obtain the proper ignition timing.

1. Remove the No. 1 spark plug.
2. Place thumb over the spark plug hole. Crank the engine slowly until compression is felt. It will be easier to have someone rotate the engine by hand, using a wrench on the crankshaft pulley.
3. Align the timing mark on the crankshaft pulley with the **0** degrees mark on the timing scale attached to the front of the engine. This places the

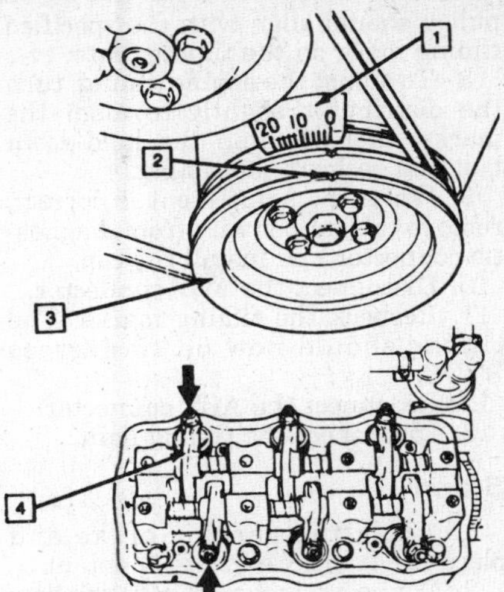

1. Timing tab
2. Timing notch
3. Crankshaft pulley
4. No. 1 cylinder

Timing mark and zero mark—Sprint and Metro

engine at TDC of the compression stroke.

4. Turn the distributor shaft until the rotor points to the No. 1 spark plug tower on the cap.
5. Install the distributor into the engine. Be sure to align the distributor-to-engine block mark made earlier.
6. Install the No. 1 spark plug. Connect all vacuum hoses and electrical connectors to the distributor.
7. Install distributor cap, then connect the battery negative cable.
8. Check and/or adjust ignition timing.

Ignition Timing

ADJUSTMENT

NOTE: When connecting the tachometer in the following procedures, be sure not to ground the tachometer terminal. Grounding the tachometer could result in damage to the igniter or ignition coil.

Be sure that the tachometer being used is compatible with the ignition system before installation. Incompatible equipment installation could cause damage to the ignition system.

Nova

EXCEPT TWINCAM ENGINE

1. Set the parking brake and place the transaxle in **N**. Run the engine until normal operating temperatures are reached, then, turn off the engine.
2. Install a timing light to the No. 1 spark plug wire according to the manufacturer's instructions.

NOTE: For inductive timing lights, the induction clip can simply be installed over the plug wire. For other lights, the pick-up wire must be connected between the spark plug boot and the spark plug. Connect a tachometer according to the manufacturer's instructions.

3. Disconnect and plug the distributor-to-intake manifold vacuum hoses.
4. Loosen the distributor flange hold-down bolt to finger tight.
5. Start the engine, then, check and/or adjust the engine rpm as necessary.
6. Aim the timing light at the scale on the timing cover near the front pulley. If the timing is not correct, turn the distributor slightly to correct it. Once the reading is correct, tighten the hold-down bolt and recheck the timing. Readjust the idle rpm as necessary.
7. Stop the engine, remove the timing light, then, unplug and reconnect the distributor vacuum hoses.

TWINCAM ENGINE

1. Firmly apply the parking brake and place the transaxle in the **N** detent.
2. Run the engine until normal operating temperatures are reached, then, stop the engine.
3. Using a jumper wire, connect it to the check engine connector located near the wiper motor.
4. Using a timing light, connect it to the No. 1 spark plug wire. Loosen the distributor hold-down bolt until it is finger tight.

NOTE: For inductive timing lights, the induction clip can simply be installed over the plug wire. For other lights, the pick-up wire must be connected between the spark plug boot and the spark plug. Connect a tachometer according to the manufacturer's instructions.

5. Start the engine, then, check and/or adjust the idle speed as necessary.
6. Aim the timing light at the timing cover plate near the crankshaft pulley.
7. If the timing is not correct, adjust the engine timing, turn the distributor slightly to align the marks, then, tighten the hold-down bolt and recheck the timing.
8. When the adjustment is correct, remove the jumper wire from the check engine connector and recheck the timing marks. Readjust the idle speed rpm as necessary.

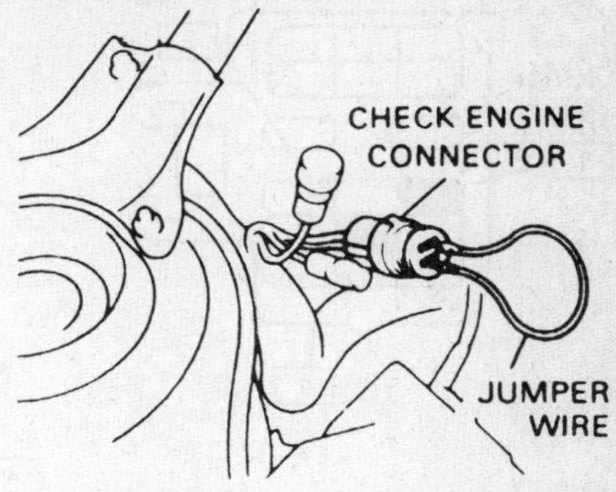

Installing the jumper wire in the check engine connector—Nova twincam engine

Prizm

1. Firmly apply the parking brake and place the transaxle in the **N** detent.
2. Connect a tachometer to the battery and the diagnostic connector. Do not ground the tachometer terminal.
3. Run the engine until normal operating temperatures are reached, then stop the engine.
4. Remove the diagnostic check connector cap (located on the left inner fender) and insert a jumper wire between terminals **E1** and **T**(of the diagnostic check connector. This will disconnect the ECM's control of the ignition timing, leaving the vehicle to operate on base timing.

NOTE: On the 1991-92 Prizm Base and LSi models, insert the jumper wire between terminal E1 and TE1 of the diagnostic connector.

5. Using a timing light, connect it to the No. 1 spark plug wire. Loosen the distributor hold-down bolt until it is finger tight.
6. Start the engine, then, check and/or adjust the idle speed; it should be 700 rpm.
7. Aim the timing light at the timing cover plate near the crankshaft pulley; the notch on the crankshaft

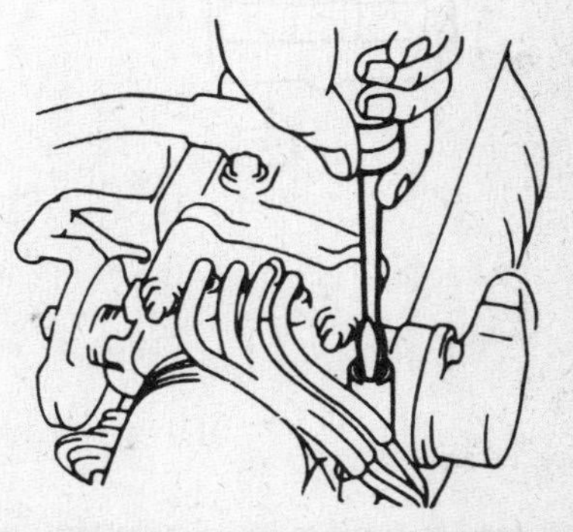

Adjusting the idle speed screw–twincam engine—Nova and Prizm

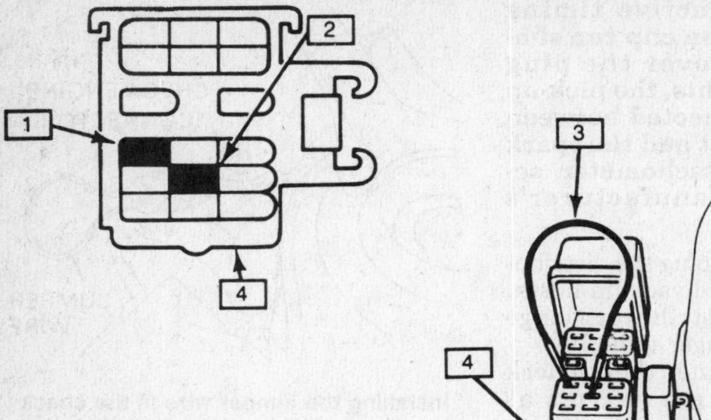

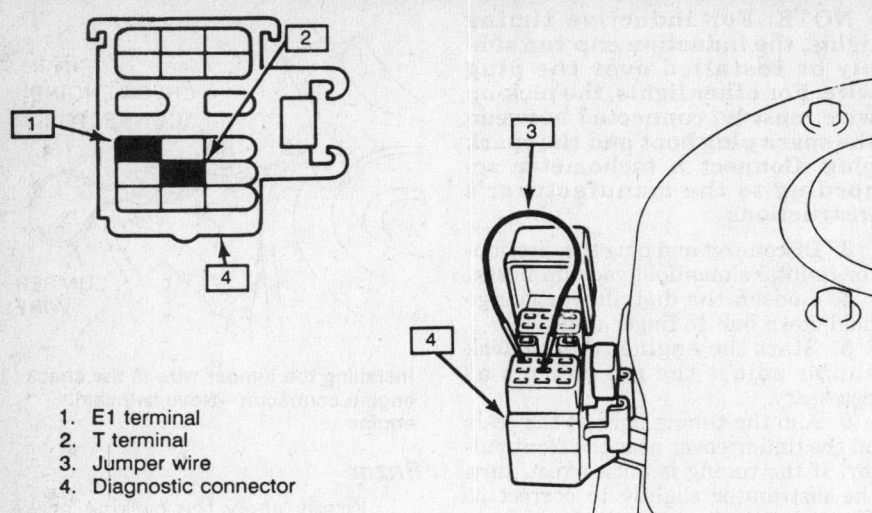

1. E1 terminal
2. T terminal
3. Jumper wire
4. Diagnostic connector

Diagnostic connector with jumper wire – Prizm

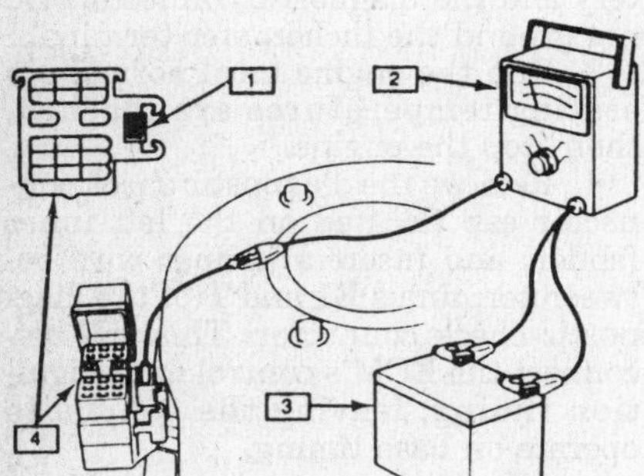

1. IG terminal
2. Tachometer
3. Battery
4. Diagnostic connector

Tachometer connections – Prizm

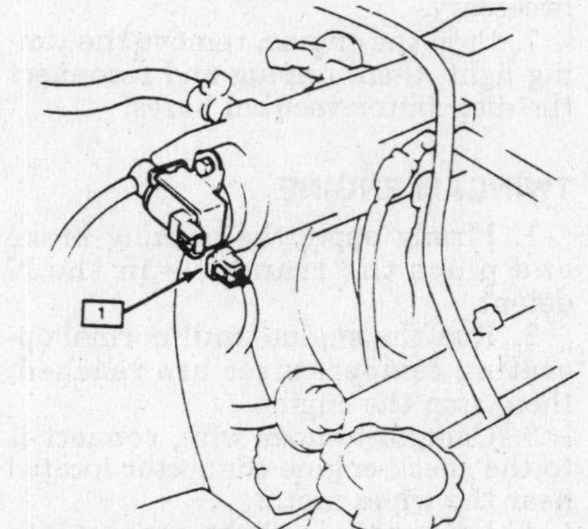

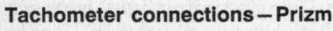

1. AVC connector

Removing the AVC connector – Prizm

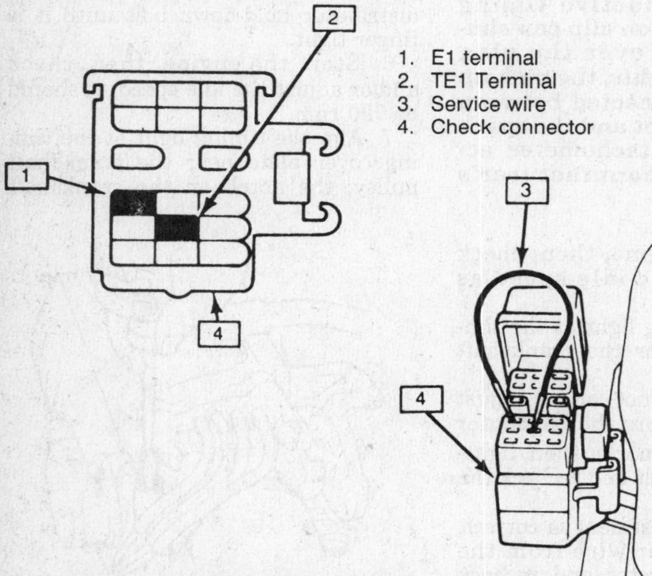

1. E1 terminal
2. TE1 Terminal
3. Service wire
4. Check connector

Installing the jumper wire in the check engine connector – 1991–92 Prizm (Base and LSi models)

pulley should align with the specified timing mark on the timing plate.

8. To adjust the engine timing, turn the distributor slightly to align the marks, then, tighten the hold-down bolt and recheck the timing.

9. When the adjustment is correct, remove the jumper wire from diagnostic connector and install the cap.

10. Disconnect the ACV connector.

11. Recheck the timing marks. The timing should now be 10 degrees BTDC.

12. Reconnect the ACV connector.

13. Disconnect the timing light.

Storm

1. Apply the parking brake and place the transaxle in the **N** detent.

2. Connect a tachometer to the battery and the diagnostic connector. Do not ground the tachometer terminal.

3. Run the engine until normal operating temperatures are reached, then stop the engine.

4. Connect a fused jumper wire between terminals 1 and 3 on the ALDL connector (located under the right hand instrument panel).

5. Using a timing light, connect it to the No. 1 spark plug wire. Loosen the distributor hold-down bolt until it is finger tight.

6. Start the engine, then, check and/or adjust the idle speed; it should be 700 rpm.

7. Aim the timing light at the timing cover plate near the crankshaft pulley; the notch on the crankshaft pulley should align with the specified timing mark on the timing plate.

8. To adjust the engine timing, turn the distributor slightly to align the marks, then, tighten the hold-down bolt and recheck the timing.

9. When the adjustment is correct, remove the jumper wire from the ALDL connector.

10. Disconnect the timing light.

Spectrum

1. Set the parking brake and block the wheels.

2. Place the manual transaxle in **N** or the automatic transaxle in the **P** detent.

3. Allow the engine to reach normal operating temperature. Make sure the choke valve is open. Turn off all of the accessories.

4. If equipped with power steering, place the front wheels in a straight line.

5. Disconnect and plug the distributor vacuum line, the canister purge line, the EGR vacuum line and the ITC valve vacuum line at the intake manifold.

6. Connect a timing light to the No. 1 spark plug wire and a tachometer to

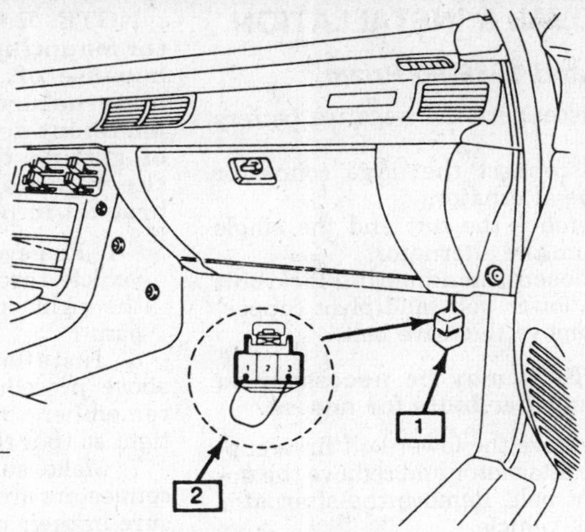

1. ALDL Connector C–52
2. Jumper Wire

ALDL connector location—Storm

the tachometer filter connector on the coil, tachometer filter is mounted near distributor hold-down bolt.

NOTE: Check the idle speed and adjust as needed.

7. Loosen the distributor flange bolt.
8. Using the timing light, align the notch on the crankshaft pulley with the mark on the timing cover by turning the distributor.

NOTE: Adjust the timing to 15 degrees BTDC at 750 rpm for manual transaxle or 10 degrees BTDC at 1000 rpm for automatic transaxle.

9. After the timing marks have been aligned, tighten the distributor flange bolt, then reinstall all vacuum lines.

Sprint and Metro

Before setting timing , make sure the headlights, heater fan, engine cooling fan and any other electrical equipment is turned off. If any current drawing systems are operating, the idle up system will operate and cause the idle speed to be higher than normal.

1. Connect a tachometer to the negative terminal of the ignition coil. Connect a timing light to the No. 1 spark plug wire. Refer to the underhood sticker.

NOTE: On the Metro vehicles equipped with the Electronic Spark Control Distributor (Base and Xfi models), remove the diagnostic check connector cap (this connector is located next to the ignition coil). Insert a fused jumper wire between terminals C and D of the diagnostic connector.

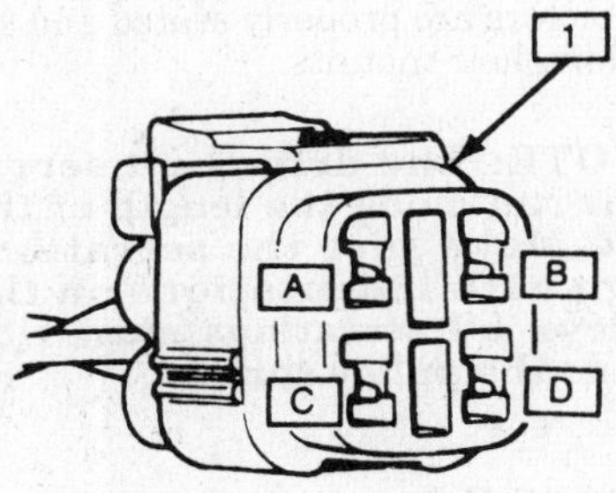

A. A/F duty check terminal
B. Diagnosis switch terminal
C. Ground
D. Test switch terminal
1. Diagnsotic connector

Exploded view of the check engine connector terminal identification—Metro (with Electronic Spark Control Distributor)

2. Start and run the engine until it reaches normal operating temperature.
3. Check and/or adjust the idle speed. Correct speed should be 750 rpm for engines with manual transaxle and 850 rpm on engines with automatic transaxles.

NOTE: To adjust the idle speed, turn the throttle adjustment screw on the carburetor.

4. With the engine at the proper idle speed, aim the timing light at the crankshaft pulley and timing marks.
5. To adjust the ignition timing, loosen the distributor hold-down bolt and rotate the distributor until the correct timing marks are aligned. Tighten the distributor hold-down bolt and recheck the timing.
6. With the timing adjusted, stop the engine and remove the testing equipment.

Alternator

PRECAUTIONS

Several precautions must be observed with alternator equipped vehicles to avoid damage to the unit.

- If the battery is removed for any reason, make sure it is reconnected with the correct polarity. Reversing the battery connections may result in damage to the one-way rectifiers.
- When utilizing a booster battery as a starting aid, always connect the positive to positive terminals and the negative terminal from the booster battery to a good engine ground on the vehicle being started.
- Never use a fast charger as a booster to start vehicles.
- Disconnect the battery cables when charging the battery with a fast charger.
- Never attempt to polarize the alternator.
- Do not use test lamps of more than 12 volts when checking diode continuity.
- Do not short across or ground any of the alternator terminals.
- The polarity of the battery, alternator and regulator must be matched and considered before making any electrical connections within the system.
- Never separate the alternator on an open circuit. Make sure all connections within the circuit are clean and tight.
- Disconnect the battery ground terminal when performing any service on electrical components.
- Disconnect the battery if arc welding is to be done on the vehicle.

BELT TENSION ADJUSTMENT

Nova

The belt tension on most components is adjusted by moving the components, alternator, within the range of the slotted bracket. Check the belt tension every 12 months or 10,000 miles. Push in on the drive belt about midway between the crankshaft pulley and the driven component. If the belt deflects more than $^{9}/_{16}$ in., adjustment is required.

1. Loosen the adjustment nut and bolt in the slotted bracket. Slightly loosen the pivot bolt.
2. Pull the component outward to increase tension. Push inward to re-

duce tension. Tighten the adjusting nut/bolt and the pivot bolt.

3. Recheck the drive belt tension and readjust, if necessary.

Spectrum

NOTE: The following procedure requires the use of GM belt tension gauge tool BT-33-95-ACBN, regular V-belts, or BT-33-97M, poly V-belts.

1. If the belt is cold, operate the engine, at idle speed, for 15 minutes; the belt will seat itself in the pulleys allowing the belt fibers to relax or stretch. If the belt is hot, allow it to cool, until it is warm to touch.

NOTE: A used belt is one that has been rotated at least 1 complete revolution on the pulleys. This begins the belt seating process and it must never be tensioned to the new belt specifications.

2. Loosen the component-to-mounting bracket bolts.
3. Using a belt tension gauge, place the gauge at the center of the belt between the longest span.
4. Applying belt tension pressure on the component, adjust the drive belt tension to the correct specification. Belt tension should deflect about ¼ in. over a 7–10 in. span or ½in. over a 13–16 in. span.5. While holding the correct tension on the component, tighten the component-to-mounting bracket bolt.
6. When the belt tension is correct, 70–110 inch lbs., remove the tension gauge.

Prizm and Storm

1. Disconnect the battery negative cable.
2. Place a suitable belt tension gauge on the belt.
3. Belt tension should be 110–150 inch lbs. on a old belt and 140–180 inch lbs. on a new belt.
4. If specifications are not as indicated, loosen the upper and lower alternator belt and adjust as necessary.
5. Connect the battery negative cable.

Sprint and Metro

1. Disconnect the negative battery cable.
2. Place a suitable belt tension gauge on the belt.
3. Adjust the drive belt to have ¼–⅜in. play on the longest run of the drive belt.

REMOVAL & INSTALLATION

Nova and 1989-90 Prizm

1. Disconnect the negative battery cable.
2. Disconnect the large connector from the alternator.
3. Remove the nut and the single wire from the alternator.
4. Loosen the adjusting lock bolt (Prizm, lower bolt) and pivot (upper) bolt. Remove the drive belt.

NOTE: It may be necessary to remove other belts for access.

5. Remove the lower bolt first, support the alternator and remove the upper pivot bolt. Remove the alternator from the vehicle.
6. Installation is reverse of the above procedure. When reinstalling, remember to leave the bolts finger tight so that the belt may be adjusted.
7. Make sure that the plugs and connectors are properly seated and secure in their mounts.

NOTE: The drive belt serrations run along the length of the belt. Make sure the serrations align with indentations on the pulleys; all serrations must ride inside the pulley surface.

1991-92 Prizm

1. Disconnect the negative battery terminal from the battery.

NOTE: Failure to disconnect the negative cable may result in injury from the positive battery lead at the alternator and may short the alternator and regulator during the removal process.

2. Remove the rubber protector, nut and wire from the battery B terminal on the alternator.
3. Remove the 3-wire alternator connector.
4. On the Prizm Base and LSi models use the following steps;
   a. Remove the upper and lower alternator bolts and nuts.
   b. Remove generator.
5. On the Prizm GSi models use the following steps;
   a. Raise and support the vehicle safely.
   b. Remove the power steering and air conditioning drive belts.
   c. Remove the lower mounting bolt and nut from the alternator bracket.
   d. Remove the alternator drive belt from the pulley.
   e. Remove the lower alternator mounting bracket (1 nut and 2 bolts).

NOTE: Removal of the alternator mounting bracket is necessary because of the extremely proximate nature of the engine compartment design. There is no way of getting the alternator out of the vehicle with the mounting bracket in place.

   f. Remove the alternator from the vehicle through the space between the right splash shield and the oil pan.
6. Installation is reverse of the above procedure. When reinstalling, remember to leave the bolts finger tight so that the belt may be adjusted.
7. Make sure that the plugs and connectors are properly seated and secure in their mounts.

Spectrum

1. Disconnect the negative battery terminal from the battery.

NOTE: Failure to disconnect the negative cable may result in injury from the positive battery lead at the alternator and may short the alternator and regulator during the removal process.

2. Disconnect and label the 2 terminal plug and the battery leads from the rear of the alternator.
3. Loosen the mounting bolts. Push the alternator inwards and slip the drive belt off the pulley.
4. Remove the mounting bolts and remove the alternator.

To install:

5. Place the alternator in its brackets and install the mounting bolts. Do not tighten them yet.
6. Slip the belt back over the pulley. Pull outwards on the unit and adjust the belt tension. Tighten the mounting and adjusting bolts.
7. Install the electrical leads and the negative battery cable.

Storm

1. Disconnect the battery negative cable and remove the belt adjusting bolt from the alternator.
2. Raise and support vehicle and remove the right undercover.
3. Loosen the lower retaining bolt. Remove the alternator belt, then tag and disconnect electrical leads from the alternator.
4. Remove alternator mounting bracket-to-engine block attaching bolts, then remove the alternator and bracket from the vehicle.
5. Installation is reverse of the above procedure. When reinstalling, remember to leave the bolts finger tight so that the belt may be adjusted.
6. Make sure that the plugs and connectors are properly seated and secure in their mounts.

Sprint and Metro

1. Disconnect the negative battery cable.
2. Disconnect the wiring connectors from the back of the alternator.
3. Remove the adjusting arm mounting bolt, the lower pivot bolt and the drive belt.
4. Remove the alternator.
5. To install, reverse the removal procedures. Adjust the drive belt to have ¼–⅜ in. play on the longest run of the drive belt.

Starter

REMOVAL & INSTALLATION

Nova, Storm and 1989-90 Prizm

1. Disconnect the negative terminal from the battery.
2. Disconnect the electrical connectors from the starter terminals.
3. Remove the transaxle cable and bracket from the transaxle.
4. Remove the starter-to-engine bolts and the starter from the vehicle.
5. To install, reverse the removal procedures. Torque the starter-to-engine bolts to 29 ft. lbs. (39 Nm).

1991-92 Prizm

BASE AND LSI MODELS

1. Disconnect the negative battery cable.
2. Disconnect the positive battery cable from the starter solenoid by removing the nut from terminal 30 stud.
3. Disconnect the ignition switch lead from terminal 50.
4. Remove the 2 starter assembly retaining bolts and remove the starter assembly.
5. To install, reverse the removal procedures. Torque the starter-to-engine bolts to 26 ft. lbs. (35 Nm).

GSi MODEL

1. Disconnect the negative battery cable.
2. Remove the rear cooling fan assembly.
3. Raise and safely support the vehicle safely.
4. Remove the right and left splash shields.
5. Disconnect the oxygen sensor connector from the front exhaust pipe.
6. Disconnect the front exhaust pipe.
7. Remove the starter assembly mounting bolts.
8. Disconnect the positive battery cable from the starter solenoid by removing the nut from terminal 30 stud.
9. Disconnect the ignition switch lead from terminal 50. Remove the starter assembly from the vehicle.
10. To install, reverse the removal

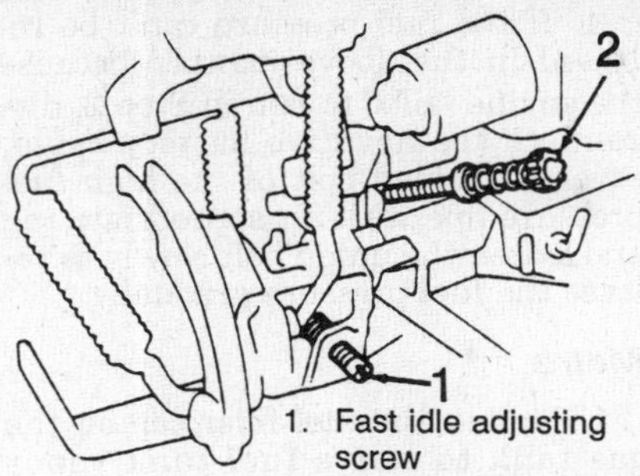

Starter motor mounting—Sprint and Metro

procedures. Torque the starter-to-engine bolts to 29 ft. lbs. (39 Nm).

Spectrum

1. Disconnect the negative battery terminal from the battery.
2. Disconnect the ignition switch lead wire and the battery cable from the starter motor terminal.
3. Remove the 2 mounting bolts from the starter and remove the starter.
4. To install, reverse the removal procedures.

Sprint and Metro

1. Disconnect the negative battery cable.
2. Disconnect the ignition switch wire and the battery cable from the starter.
3. Remove the 2 engine-to-starter mounting bolts and remove the starter.
4. To install, reverse the removal procedures.

EMISSION CONTROLS

Please refer to "Emission Controls" in the Unit Repair section for system maintenance procedures. Due to the complex nature of modern electronic engine control systems, comprehensive diagnosis and testing procedures fall outside the confines of this repair manual. For complete information on diagnosis, testing and repair procedures concerning all modern engine and emission control systems, please refer to "Chilton's Guide to Fuel Injection and Electronic Engine Controls".

FUEL SYSTEM

Fuel System Service Precautions

Safety is the most important factor when performing not only fuel system maintenance but any type of maintenance. Failure to conduct maintenance and repairs in a safe manner may result in serious personal injury or death. Maintenance and testing of the vehicle's fuel system components can be accomplished safely and effectively by adhering to the following rules and guidelines.

- To avoid the possibility of fire and personal injury, always disconnect the negative battery cable unless the repair or test procedure requires that battery voltage be applied.
- Always relieve the fuel system pressure prior to disconnecting any fuel system component (injector, fuel rail, pressure regulator, etc.), fitting or fuel line connection. Exercise extreme caution whenever relieving fuel system pressure to avoid exposing skin, face and eyes to fuel spray. Please be advised that fuel under pressure may penetrate the skin or any part of the body that it contacts.
- Always place a shop towel or cloth around the fitting or connection prior to loosening to absorb any excess fuel due to spillage. Ensure that all fuel spillage (should it occur) is quickly removed from engine surfaces. Ensure that all fuel soaked cloths or towels are deposited into a suitable waste container.
- Always keep a dry chemical (Class B) fire extinguisher near the work area.
- Do not allow fuel spray or fuel vapors to come into contact with a spark or open flame.
- Always use a backup wrench when loosening and tightening fuel line connection fittings. This will prevent unnecessary stress and torsion to fuel line piping. Always follow the proper torque specifications.
- Always replace worn fuel fitting O-rings with new. Do not substitute fuel hose or equivalent where fuel pipe is installed.

RELIEVING FUEL SYSTEM PRESSURE

Nova and Prizm

NOTE: Make sure the engine is cold before disconnecting any portion of the fuel system.

1. Disconnect the negative battery cable.

NOTE: On the Prizm models, disconnect the circuit opening relay located under the dash panel, near the ECM. Crank the engine and allow it to run for until it stalls, crank the engine for an additional 30 seconds. Then disconnect the negative battery.

2. Loosen the fuel filler cap to relieve tank vapor pressure.
3. Using a shop rag, wrap it around the fuel line fitting.
4. Slowly loosen and remove the cold start injector valve fuel line at the fuel rail. Absorb any excess fuel remaining in the line.
5. When replacing the be sure to install a new O-ring.

Spectrum

1. Remove the fuel pump fuse from the fuse block or disconnect the harness connector at the tank.
2. Start the engine. It should run and then stall when the fuel in the lines is exhausted. When the engine stops, crank the starter for about 3 seconds to make sure all pressure in the fuel lines is released.
3. Disconnect the negative battery cable.
4. Install the fuel pump fuse after repair is made and reconnect the battery cable.

Storm

1. Remove the fuel cap from the fuel tank.
2. Remove the fuel pump relay from the underhood relay center.
3. Start the engine and allow it to stall. Crank the engine for an additional 30 seconds.
4. Reconnect the negative battery cable.

Sprint

1. Release the fuel vapor pressure in the fuel tank by removing the fuel tank cap then reinstalling it.
2. With the engine running, remove the connector of the fuel pump relay and wait until the engine stops itself.

NOTE: The main relay and fuel pump relay are identical. Which one to connect to the fuel lead wire is not specified. Identify the fuel pump relay by the color of the lead wire. The fuel pump relay lead wire is pink, pink/white, white/blue, white/blue.

3. Once the engine is stopped, crank it a few times with the starter for about 3 seconds each time, with the relay connector disconnected.
4. If the fuel pressure can't be released in the above manner because the engine failed to run in Step 2, disconnect the negative battery cable, cover the union bolt of the high fuel pressure line with an appropriate rag and loosen the union bolt slowly to release the fuel pressure gradually.

Metro

1. Loosen the fuel filler cap on the gas tank to relive fuel tank vapor pressure.
2. Place the transaxle gear shift lever in the **N** or **P** position. Set the parking brake and block the drive wheels.
3. Remove the main fuse block cover and engine coolant reservoir from its bracket.
4. Remove the screws holding the main fuse block.
5. Remove the coupler holding the fuel pump relay and disconnect it from the fuel pump relay.
6. Start the engine and let it run until it stalls. Crank the engine for 3 seconds to insure there is no fuel left in the lines.
7. After completing the repairs. Reverse Steps 2-4.

Fuel Tank

REMOVAL & INSTALLATION

NOTE: Before removing fuel system parts, clean them with a spray-type engine cleaner. Follow the instructions on the cleaner. Do not soak fuel system parts in liquid cleaning solvent.

CAUTION

The fuel injection system is under pressure. Release pressure slowly and contain spillage. Observe no smoking/no open flame precautions. Have a Class B-C (dry powder) fire extinguisher within arm's reach at all times.

Nova and Prizm

1. Remove the filler cap.
2. Using a siphon or pump, drain the fuel from the tank and store it in a proper metal container with a tight cap.
3. Remove the rear seat cushion to gain access to the electrical wiring.
4. Disconnect the fuel pump and sending unit wiring at the connector.
5. Raise the vehicle and safely support it on jackstands.
6. Loosen the clamp and remove the filler neck and overflow pipe from the tank.
7. Remove the supply hose from the tank. Wrap a rag around the fitting to collect escaping fuel. Disconnect the breather hose from the tank, again using a rag to control spillage.
8. Cover or plug the end of each disconnected line to keep dirt out and fuel in.
9. Support the fuel tank with a floor jack or tansaxle jack. Use a broad piece of wood to distribute the load. Be careful not to deform the bottom of the tank.
10. Remove the fuel tank support strap bolts.
11. Swing the straps away from the tank and lower the jack.
12. Remove the fuel filler pipe extension, the breather pipe assembly and the sending unit assembly. Keep these items in a clean, protected area away from the vehicle.
13. While the tank is out and disassembled, inspect it for any signs of rust, leakage or metal damage. If any problem is found, replace the tank. Clean the inside of the tank with water and a light detergent and rinse the tank thoroughly several times.
14. Inspect all of the lines, hoses and fittings for any sign of corrosion, wear or damage to the surfaces. Check the pump outlet hose and the filter for restrictions.
15. When reassembling, always replace the sealing gaskets with new ones. Also replace any rubber parts showing any sign of deterioration.

To install:

16. Connect the breather pipe assembly and the filler pipe extension.

NOTE: Tighten the breather pipe screw to 17 in. lbs. (23 Nm) and all other attaching screws to 30 in. lbs. (41 Nm).

17. Place the fuel tank on the jack and elevate it into place within the vehicle. Attach the straps and install the strap bolts, tightening them to 29 ft. lbs. (40 Nm).
18. Connect the breather hose to the tank pipe, the return hose to the tank pipe and the supply hose to its tank pipe. tighten the supply hose fitting to 21 ft. lbs. (28 Nm).
19. Connect the filler neck and overflow pipe to the tank. Make sure the clamps are properly seated and secure.
20. Lower the vehicle to the ground.
21. Connect the pump and sending unit electrical connectors to the harness.
22. Install the rear seat cushion.
23. Using a funnel, pour the fuel that was drained from its container into the fuel filler.
24. Install the fuel filler cap.
25. Start the engine and check carefully for any sign of leakage around the tank and lines.

Storm

1. Relieve the fuel system pressure and disconnect the negative battery cable.
2. Remove the fuel filler cap from the fuel tank.
3. Have a Class B fire extinguisher near the work area. Use a hand operated pump device to drain as much fuel through the fuel filler neck as possible.
4. Use a siphon to remove the remainder of the fuel in the tank by connecting the siphon to the fuel pump outlet fitting.

CAUTION

Never drain or store fuel in an open container because there is the possibility of a fire or an explosion.

5. Reinstall the fuel pump outlet hose to its fitting, install the fuel filler cap.
6. Raise and safely support the vehicle.
7. Remove the exhaust pipe from the catalytic converter.
8. Loosen the clamp on the fuel filler neck hose from the fuel tank.
9. Loosen the clamp for the fuel overflow pipe hose from the fuel tank.
10. Disconnect the fuel gauge sending unit and fuel pump electrical connector.
11. Spread the clamp for the fuel vapor hose on the fuel tank.
12. Remove the fuel pump supply hose from the fuel pipe.
13. Remove the fuel pump return hose.
14. Using a suitable tansaxle jack, support the fuel tank.
15. Remove the 6 fuel tank retaining bolts and lower the tank.

To install:

16. Place the fuel tank in its proper space using the tansaxle jack, Install the 6 fuel tank retaining bolts and torque them to 14 ft. lbs. (19 Nm).
17. Remove the tansaxle jack. Install the fuel return hose.
18. Install the fuel pump supply hose, vapor hose and fuel gauge and fuel pump electrical connector. Torque the fuel pump supply hose fitting nut to 25 ft. lbs. (34 Nm).
19. Install the fuel overflow pipe hose.
20. Install the exhaust pipe to the catalytic converter. Install the fuel filler neck hose.
21. Lower the vehicle. Refill the fuel tank and install the fuel filler cap. Install the negative battery cable.

Metro

1. Relieve the fuel system pressure and disconnect the negative battery cable.
2. Remove the fuel filler cap from the fuel tank.
3. Have a Class B fire extinguisher near the work area. Use a hand operated pump device to drain as much fuel through the fuel filler neck as possible.
4. Use a siphon to remove the remainder of the fuel in the tank by connecting the siphon to the fuel pump outlet fitting.

CAUTION

Never drain or store fuel in an open container because there is the possibility of a fire or an explosion.

5. Reinstall the fuel filler cap.
6. Remove the rear seat cushion from vehicle (if so equipped).
7. Disconnect the fuel pump motor connector and sending unit electrical connectors.
8. Disconnect the wire harness grommet and harness through the vehicle floor pan.
9. Raise and safely support the vehicle.
10. Remove the fuel filler hose clamp from the filler neck assembly.
11. Remove the fuel breather hose clamp from the filler neck assembly.
12. Remove the fuel filter inlet hose clamp and hose from the filter.

NOTE: A small amount of fuel may be released after the fuel hose is disconnected. In order to reduce the chance of personal injury, cover the fitting to be disconnected with a shop towel.

13. Remove the fuel vapor clamp and hose, and fuel return hose clamps and hose from the respective fuel lines.
14. Using a suitable tansaxle jack, support the fuel tank.
15. Remove the fuel tank retaining bolts and lower the tank.

To install:

16. Reconnect all fuel hoses, lines and the fuel breather hose to the fuel tank.
17. Place the fuel tank in its proper space using the tansaxle jack, Install the fuel tank retaining bolts.
18. Remove the tansaxle jack.
19. Reconnect all fuel hoses, lines and the fuel breather hose to the fuel tank.
20. Install the fuel breather hose and hose clamp to the filler neck assembly.
21. Install the fuel vapor hose and hose clamps. Install the fuel return hose and hose clamp to the respective fuel lines.
22. Install the fuel filter inlet hose and hose clamp to the filter.
23. Lower the vehicle. Install the wire harness and grommet through the vehicle floor pan.
24. Connect the fuel pump motor and the sending unit connectors.
25. Install the rear seat cushion. Reconnect the negative battery cable.

Sprint

1. Relieve the fuel system pressure and disconnect the negative battery cable.
2. Remove the fuel filler cap from the fuel tank.
3. Have a Class B fire extinguisher near the work area. Use a hand operated pump device to drain as much fuel through the fuel filler neck as possible.
4. Use a siphon to remove the remainder of the fuel in the tank by connecting the siphon to the fuel pump outlet fitting.

CAUTION

Never drain or store fuel in an open container because there is the possibility of a fire or an explosion.

5. Reinstall the fuel filler cap.
6. Remove the rear seat cushion from vehicle (if so equipped).
7. Raise and safely support the vehicle.
8. Remove the 3 fuel lines from the fuel pipes.
11. Remove the fuel filler hose and the breather hose.
12. Using a suitable tansaxle jack, support the fuel tank.
13. Remove the fuel tank retaining bolts and lower the tank.

To install:

14. Place the fuel tank in its proper space using the tansaxle jack, Install the fuel tank retaining bolts.
15. Remove the tansaxle jack.
16. Reconnect all fuel hoses, lines and the fuel breather hose to the fuel tank.
17. Lower the vehicle.
18. Install the rear seat cushion to the vehicle (if so equipped).
19. Reconnect the negative battery cable.

Spectrum

1. Relieve the fuel system pressure and disconnect the negative battery cable.
2. Remove the fuel filler cap from the fuel tank.
3. Have a Class B fire extinguisher near the work area. Use a hand operated pump device to drain as much fuel through the fuel filler neck as possible.
4. Use a siphon to remove the remainder of the fuel in the tank by connecting the siphon to the fuel pump outlet fitting.

CAUTION

Never drain or store fuel in an open container because there is the possibility of a fire or an explosion.

5. Reinstall the fuel filler cap.

6. Disconnect the sending unit wire from the fuel tank terminal.
7. Unfasten the clips and disconnect the 3 rubber hoses on the rear side of the fuel tank.
8. Unfasten the clips and disconnect the fuel delivery and fuel return hose from the side of the fuel tank.
9. Remove the fuel filler pipe bracket bolts.
10. Using a suitable tansaxle jack, support the fuel tank.
13. Remove the fuel tank retaining bolts and lower the tank.

To install:

14. Place the fuel tank in its proper space using the tansaxle jack, Install the fuel tank retaining bolts.
15. Install the fuel filler pipe bracket bolts.
16. Install the clips and reconnect the fuel delivery and fuel return hose to the side of the fuel tank.
17. Install the clips and reconnect the 3 rubber hoses on the rear side of the fuel tank.

Fuel Filter

REMOVAL & INSTALLATION

Nova and Prizm

The fuel filter is located on the firewall. The element inside the filter reduces the flow speed of the fuel causing dirt particles that are heavier than gasoline to settle to the bottom. The lighter particles are filtered by the element.

Carbureted Engine

CAUTION

Gasoline is hazardous and extremely flammable. Remove the hoses slowly and contain spillage. Observe no smoking/no open flame precautions. Have a Class B-C (dry powder) fire extinguisher within arm's reach at all times.

1. Remove the fuel filler cap.
2. Note the routing of inlet and outlet lines and the direction of flow as marked by the arrow on the filter. With a pair of pliers, shift the clips on the inlet and outlet hoses back and well away from the connections on the filter.
3. Disconnect the fuel lines, using a twisting motion to break them loose. Pull the filter out of its retaining clip. Plug the lines immediately to prevent spillage.
4. Install the filter in the holder. When reconnecting the fuel lines, make sure the hoses are connected to the correct ports on the filter.(The arrow on the filter indicates direction of flow and points towards the carburetor).

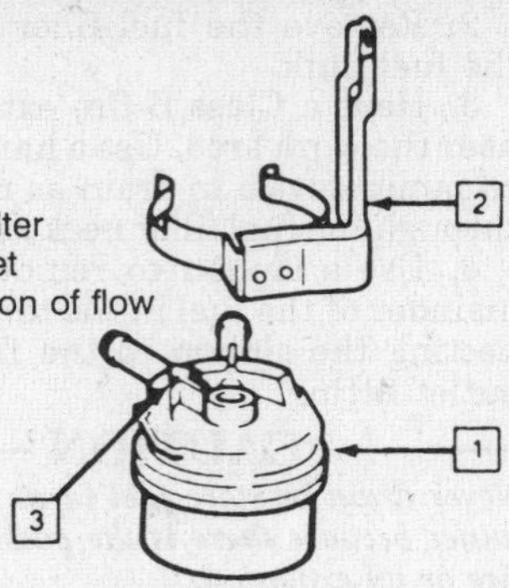

1. Fuel filter
2. Bracket
3. Direction of flow

Fuel filter and mounting bracket—Nova carbureted models

5. Install the hose clips to the inside of the bulged sections of the fuel filter connections, and not right at the ends of the fuel lines. Start the engine and check for leaks.
5. Install the fuel filler cap.

Fuel Injected Engines

CAUTION

The fuel system is under pressure. Release pressure slowly and contain spillage. Observe no smoking/no open flame precautions. Have a Class B-C (dry powder) fire extinguisher within arm's reach at all times. Perform this procedure only on a cold motor.

NOTE: There is also a sock type fuel filter attached to the fuel pump which is located in the fuel tank. This filter does not require routine maintenance, but when ever the fuel pump is replaced it would be wise to change the sock type fuel filter.

1. Disconnect the negative battery cable.
2. Relieve the pressure in the fuel line as follows:
   a. Fit a wrench of the exact size onto the upper line connection and fit a second wrench on the flats of the filter. This counterholding arrangement will prevent the filter and lines from turning during removal.
   b. Wrap the upper filter area in layers of rags to absorb spray and spillage.
   c. Slowly loosen the bolt holding the line to the filter. When the pressure is equalized, remove the bolt and the line. Plug the line immediately.
3. Again using 2 wrenches, disconnect the lower line to the filter. Be prepared to deal with fuel spillage. Be very careful not to crimp or bend the fuel line. Plug the line immediately.
4. Remove the filter from its holder and install a new one.

To install:

5. Apply a thin coat of oil to the threads and connect the bottom fuel line to the filter. Always begin the threads by hand, then tighten to 22 ft. lbs. (30 Nm).
6. Install the upper line with new gaskets. Make sure the line is held within the forked bracket on the filter and begin the bolt by hand to insure proper threading.
7. Tighten the bolt to 22 ft. lbs. (30 Nm). Reconnect the negative battery cable.
8. Start the engine and check for leaks. (The engine may require a longer period of cranking until it starts due to loss of fuel pressure.)

Spectrum

1. Relieve the pressure in the fuel system. Remove the fuel filler cap.
2. Note the routing of inlet and outlet lines and the direction of flow as marked by the arrow on the filter. With a pair of pliers, shift the clips on the inlet and outlet hoses back and well away from the connections on the filter.

NOTE: The fuel filter is located under the power brake booster.

3. Disconnect the fuel lines, using a twisting motion to break them loose. Pull the filter out of its retaining clip. Plug the lines immediately to prevent spillage.

To install:

4. Install the filter in the holder. When reconnecting the fuel lines, make sure the hoses are connected to the correct ports on the filter. The arrow on the filter indicates direction of flow and points towards the carburetor.
6. Install the hose clips to the inside of the bulged sections of the fuel filter connections, and not right at the ends of the fuel lines. Start the engine and check for leaks.
7. Install the fuel filler cap.

Storm

1. Relieve the pressure in the fuel system.
2. Disconnect the negative battery cable.
3. Remove the air intake duct.
4. Disconnect the fuel filter outlet pipe from the fuel filter upper fitting.
5. Raise and support the vehicle safely. Be sure to cap the ends of the fuel pipe fittings to prevent both the entry of dirt and the spillage of fuel.
6. Loosen the fuel filter bracket clamp bolt.
7. Remove the fuel filter from the bracket as follows.
   a. Fit a wrench of the exact size onto the upper line connection and

fit a second wrench on the flats of the filter. This counterholding arrangement will prevent the filter and lines from turning during removal.

b. Wrap the upper filter area in layers of rags to absorb spray and spillage.

c. Slowly loosen the bolt holding the line to the filter. When the pressure is equalized, remove the bolt and the line. Plug the line immediately.

d. Again using 2 wrenches, disconnect the lower line to the filter. Be prepared to deal with fuel spillage. Be very careful not to crimp or bend the fuel line. Plug the line immediately.

8. Remove the filter from its holder and install a new one.

9. Installation is the reverse order of the removal procedure. Torque the fuel fitting nuts to 25 ft. lbs. (34 Nm).

Sprint

CARBURETED ENGINE

1. Remove and replace the fuel tank cap. This releases the fuel pressure within the fuel system.
2. Disconnect the battery negative cable.
3. Remove the clamps from the inlet and outlet hoses and remove the hoses from the fuel filter.
4. Remove the fuel filter from the bracket.
5. Reverse procedure to install. When reconnecting the fuel lines, make sure the hoses are connected to the correct ports on the filter. The arrow on the filter indicates direction of flow and points towards the carburetor.
6. Start engine and check for leaks.

FUEL INJECTED ENGINE

1. Relieve the fuel system pressure.
2. Place an appropriate container under the fuel filter.
3. Use a 17mm and a 19mm wrench to loosen the inlet and outlet pipes, then remove the fuel filter.
4. Reverse procedure to install. Use new gaskets and ensure that end marked out on the filter is placed up. Torque inlet and outlet pipe bolts to 22–28 ft. lbs. (30-34 Nm). Start engine and check for leaks.

Metro

1. Relieve the pressure in the fuel system.
2. Disconnect the negative battery cable.
3. Remove and replace the fuel tank cap. This releases the fuel pressure within the fuel system.
4. Raise and support the vehicle safely.
5. Place a drain pan under the fuel filter. Disconnect the fuel filter inlet hose clamp and the hose from the filter. Disconnect the fuel filter outlet hose clamp and the hose from the fuel feed line.

CAUTION

A small amount of fuel may be released after the fuel hose is disconnected. In order to reduce the chance of personal injury, cover the fittings to be disconnected with a shop towel.

6. Remove the fuel filter, mounting bracket, and outlet hose as an assembly from the vehicle.
7. Disconnect the fuel filter outlet hose clamp and hose from the filter. Remove the mounting bracket from the filter.
8. Reverse procedure to install. Start engine and check for leaks.

Mechanical Fuel Pump

PRESSURE TESTING

Nova

1. Using a fuel pressure gauge, connect it between the fuel pump and the carburetor; using a pair of vise grips or equivalent, squeeze off the return hose.
2. Operate the engine at idle and note the reading on the gauge.
3. Fuel pump pressure should be 7.6 psi. If specification obtained is not as specified, the fuel pump should be replaced.
4. Remove the gauge, reconnect the fuel lines. Start the engine and check for leaks.

Spectrum

1. Disconnect the fuel line from the carburetor. Install a rubber hose about 10 inches long. Attach a low reading pressure gauge.
2. Hold the gauge at least 16 inches above the fuel pump. If equipped, pinch the fuel return line.
3. Start the engine and run it at slow idle, using the fuel that is left in the carburetor.
4. Fuel pump pressure should be 3.8–4.7 psi. If specification obtained is not as indicated, replace the fuel pump.

REMOVAL & INSTALLATION

Nova

1. Disconnect the negative battery cable.
2. Note the routing of the fuel lines and label, if necessary.
3. Remove the fuel line clips, then disconnect the fuel lines.

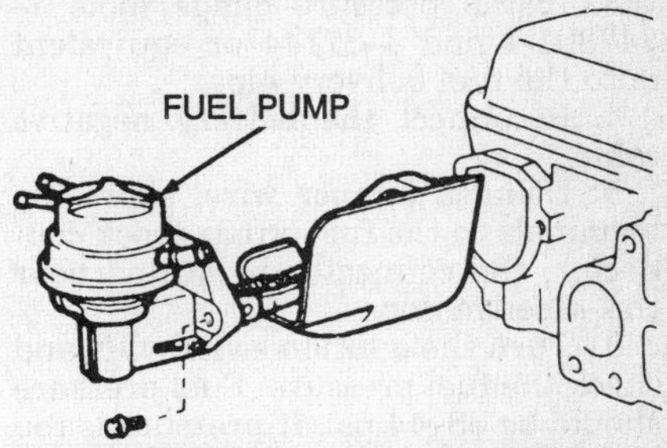

Removing the fuel pump assembly—Nova single overhead cam engine

4. Remove the fuel pump-to-cylinder head bolts, then the pump, gasket and heat shield. Note position of the heat shield.
5. Clean the gasket mounting surface. Ensure not to scratch the aluminum surface of the cylinder head.
6. Reverse procedure to install. Use a new gasket. When reconnecting the fuel lines, ensure that the clips are installed to the inside of the bulged sections of the fuel pump connections. Start the engine and check for leaks.

Spectrum

1. Disconnect the fuel and return hoses from the fuel pump.
2. Remove the bolts, fuel pump and heat insulator assembly.
3. After removing the fuel pump, cover the mounting face of the cylinder head to prevent oil discharge.
4. Reverse procedure to install. Replace the heat insulator assembly.

Sprint

1. Remove and replace the fuel tank cap, this procedure releases the pressure within the fuel system.
2. Disconnect the battery negative cable.
3. Remove the air cleaner from the carburetor.
4. Remove the fuel inlet, outlet and return hoses from the fuel pump.
5. Remove the fuel pump mounting bolts, the pump and the pump rod from the cylinder head.
6. Reverse procedure to install. Use a new gasket and lubricate the pump rod.

Electric Fuel Pump

PRESSURE TESTING

Nova

1. Disconnect the battery negative cable and the electrical connector from the cold start injector.
2. Using a shop cloth, wrap it around the cold start injector pipe, then slowly loosen the union bolt and remove. Discard the gaskets.

3. Install pressure gauge tools J–347301-1 and J-37144 or equivalent onto the fuel delivery pipe.

4. Reconnect the battery negative cable.

5. Using a jumper wire, short the terminals to the fuel pump check connector; the connector is located near the wiper motor.

6. Turn the ignition switch **ON** and check the fuel pressure. Fuel pressure should be 38–44 psi. If pressure is too high, replace the pressure regulator. If the pressure is low, check the fuel pump, fuel filter, pressure regulator, hoses and connections.

7. Remove the service wire from the check connector. Start the engine.

8. Disconnect the vacuum sensing hose from the pressure regulator and pinch it off.

9. The fuel pressure at idle should be 38–44 psi.

10. Reconnect the vacuum sensing hose to the pressure regulator and allow the engine to idle for one and a half minutes. Recheck the fuel pressure. Fuel pressure should be 30–33 psi; if no pressure, check the vacuum sensing hose and the pressure regulator.

11. Stop the engine and inspect the fuel pressure for 5 minutes; it should remain above 21 psi. If the pressure does not remain high, check the fuel pump, pressure regulator and/or the injectors.

12. After checking the fuel pressure, disconnect the battery negative cable and remove the fuel pressure gauge.

13. Using new gaskets, reconnect the cold start injector hose to the delivery pipe and the battery negative cable. Start the engine and check for leaks.

Prizm

1. Disconnect the battery negative cable.

2. Using a shop cloth, wrap it around the cold start injector pipe and slowly loosen the union bolt and remove. Discard the gaskets.

3. Remove the delivery pipe to cold start injector banjo fitting bolt and the 2 gaskets. Discard the gaskets.

4. Install pressure gauge tools J–34370-1 and J-38347 onto the fuel delivery pipe. Close the gauge valve.

5. Reconnect the battery negative cable.

6. Open the gauge valve and turn the ignition switch to the **RUN** position.

7. Install a jumper wire between the **+B** and **FP** terminals of the fuel pump check connector. The fuel pump connector is located near the wiper motor.

8. Open the gauge air bleed valve and purge the air from the gauge. Use a suitable container to catch the fuel from the gauge air bleed valve.

9. The gauge reading should be 38–44 psi. If the pressure is too high, replace the pressure regulator or a restricted fuel return line. If the pressure is low, check the fuel pump, fuel filter, pressure regulator, the hoses and connections.

10. Pinch the fuel pressure regulator return hose. The pressure should be 57 psi.

11. Remove the fuel pump jumper wire and observe the pressure gauge. A very slow drop in pressure is normal.

12. The fuel pressure at idle should be 30–33 psi.

13. Remove the fuel pressure regulator sensing hose and plug the hose end. The fuel pressure gauge should read 38–44 psi.

14. Connect the vacuum sensing hose and quickly open and close the throttle valve for the air induction system. The fuel pressure gauge should show quick increases in pressure as the throttle is snapped open and decreases as it returns to idle.

15. Stop the engine. If improper pressure readings are shown, check the fuel pump, pressure regulator and/or the injectors, restricted fuel lines or filter and for improper injector control, thermal timer/ECM.

16. Connect the battery negative cable.

17. Use the fuel pressure gauge air bleed valve to relieve the system pressure.

18. Remove the fuel pressure gauge and adapter.

19. Using new gaskets, reconnect the cold start injector fitting bolt and torque to 13 ft. lbs. (18 Nm). Check for leaks using the fuel jumper wire.

Storm

1. Relieve fuel system pressure.

2. Disconnect the fuel pressure line from the fuel filter. Wrap a shop towel around the line catch any fuel leakage.

3. Install fuel gauge adapter J‹n-dash›35957-10 between the fuel pressure line and the fuel filter, located on the fuel rail.

4. Install fuel gauge J‹n-dash›34730-1 to the adapter.

5. Disconnect the vacuum hose from the pressure regulator.

6. Turn the ignition switch to the **ON** position. Fuel pressure should be 35–42 psi.

7. If pressure is with specification but not holding, proceed as follows:

a. Turn the ignition switch to the **OFF** position.

b. Using battery voltage, connect a jumper wire between the BLK/RED wire and the RED/WHT wire on the fuel pump relay connector, located in the engine compartment.

c. Pinch the pressure line and remove the jumper wire from the test connector. If pressure held, check for a leaking coupling hose or a faulty in-tank pump. If pressure did not hold, repeat Step 7b pinching the fuel return line.

d. If pressure holds, replace the pressure regulator assembly. If pressure does not hold, check for leaking injector and spark plugs.

8. If pressure obtained is below specification, proceed as follows:

a. Check for a restricted in-line fuel filter; replace, if necessary.

b. Turn the ignition switch to the **OFF** position.

c. Apply battery voltage to the fuel pump relay connector with the fuel pump removed.

d. Pinch the fuel return line and note pressure. Pressure should be above 65 psi.

e. If not, check for a faulty in-tank pump. If pressure is above 65 psi, replace the pressure regulator.

9. If pressure obtained is above specification, proceed as follows:

a. Disconnect the vacuum line to the pressure regulator and the fuel return line hose.

b. Install a $^5/_{16}$ in. hose to the pressure regulator side of the return line and insert the other end into a suitable container.

c. Turn the ignition switch to the **ON** position and check the fuel pressure. Fuel pressure should be 35–42 psi. If not, replace the pressure regulator.

d. If fuel pressure is 35–42 psi, check for a restriction in the fuel return line.

10. If no pressure was obtained on the pressure gauge, proceed as follows:

a. Remove the gas cap and turn the ignition switch to the **ON** position. Fuel pump should operate for approximately 2 seconds.

b. If fuel pump operates, check for faulty wiring and connectors.

c. If fuel pump does not operate, remove the fuel pump relay from the relay center. Using a suitable test light, check RED/WHT wire on the fuel pump relay connector for power.

d. If test light does not light, check for an open circuit in the RED/WHT wire. If test light illuminates, install a jumper wire between the RED/WHT and BLK/RED wire on the fuel pump relay connector. Fuel pump should operate.

e. If fuel pump does not operate, check the fuel pump connector at the fuel tank using the test light. Test light should illuminate. If not, check the BLK/RED wire for an open circuit. If Test light illuminates, check for an open circuit in the fuel ground and/or a faulty fuel pump.

f. If the fuel pump operates, turn the ignition switch **OFF** for approximately 10 seconds. Connect a test light between the BLK wire and the PNK/WHT wire on the fuel pump relay connector. Turn the ignition switch **ON**, test light should light for approximately 2 seconds.

g. If test light illuminates, check for a faults fuel pump relay. If test light does not illuminate, turn the ignition switch **OFF** for approximately 10 seconds. Using the test light check the A1 terminal of the ECM. Turn the ignition switch **ON**. Test light should light for approximately 2 seconds. If not, replace the ECM. If test light illuminates, check for an open circuit in the PNK/WHT wire and/or the BLK wire.

Spectrum

1. Relieve the fuel system pressure.
2. Disconnect the fuel hoses between the pressure regulator and the fuel distributor pipe.
3. Connect fuel pressure gauge J–33945 between the pressure regulator and the fuel distributor pipe.
4. Start the engine and check the fuel pressure under the 2 different conditions:

   a. Vacuum hose of the pressure regulator disconnected, intake manifold side end of hose must be plugged. The pressure should read 35.6 psi.

   b. Vacuum hose of the pressure regulator connected, at engine speed of 900 rpm. The pressure should read 28.4 psi.

5. After on-vehicle inspection is made, remove the fuel pressure gauge, reconnect fuel line and check for leaks.

Sprint and Metro

1. Relieve fuel system pressure.
2. Disconnect the fuel inlet line at the fuel filter and install a suitable pressure gauge on the line.
3. Jump the fuel pump relay, using a suitable jumper wire, and check the system pressure on the gauge.
4. Fuel pressure should be 25‹n-dash›33 lbs.
5. As the pressure reaches 33 lbs., the relief valve in the pump should pulsate the pressure so it is always within specification.
6. If the pressure is not within specification, check for restrictions in the fuel lines or replace the in-tank fuel pump.
7. Before removing the pressure gauge, relieve the fuel pressure again.
8. Reconnect the fuel line. Start the engine and check for leaks.

REMOVAL & INSTALLATION

CAUTION

The fuel injection system is under pressure. Release pressure slowly and contain spillage. Observe no smoking/no open flame precautions. Have a Class B-C (dry powder) fire extinguisher within arm's reach at all times.

Nova and Prizm

1. Remove the filler cap.
2. Using a siphon or pump, drain the fuel from the tank and store it in a proper metal container with a tight cap.
3. Remove the rear seat cushion to gain access to the electrical wiring.
4. Disconnect the fuel pump and sending unit wiring at the connector.
5. Raise the vehicle and safely support it.
6. Loosen the clamp and remove the filler neck and overflow pipe from the tank.
7. Remove the supply hose from the tank. Wrap a rag around the fitting to collect escaping fuel. Disconnect the breather hose from the tank, again using a rag to control spillage.
8. Cover or plug the end of each disconnected line to keep dirt out and fuel in.
9. Support the fuel tank with a floor jack or tansaxle jack. Use a broad piece of wood to distribute the load. Be careful not to deform the bottom of the tank.
10. Remove the fuel tank support strap bolts.
11. Swing the straps away from the tank and lower the jack. Balance the tank with a hand or have a helper assist. The tank is bulky and may have some fuel left in it. If its balance changes suddenly, the tank may fall.
12. Remove the fuel filler pipe extension, the breather pipe assembly and the sending unit assembly. Keep these items in a clean, protected area away from the vehicle.
13. To remove the electric fuel pump:

    a. Disconnect the 2 pump-to-harness wires.

    b. Loosen the pump outlet hose clamp at the bracket pipe.

    c. Remove the pump from the bracket and the outlet hose from the bracket pipe.

    d. Separate the outlet hose and the filter from the pump.

14. While the tank is out and disassembled, inspect it for any signs of rust, leakage or metal damage. If any problem is found, replace the tank. Clean the inside of the tank with water and a light detergent and rinse the tank thoroughly several times.
15. Inspect all of the lines, hoses and fittings for any sign of corrosion, wear or damage to the surfaces. Check the pump outlet hose and the filter for restrictions.**To install:**
16. When reassembling, always replace the sealing gaskets with new ones. Also replace any rubber parts showing any sign of deterioration.
17. Assemble the outlet hose and filter onto the pump; then attach the pump to the bracket.
18. Connect the outlet hose clamp to the bracket pipe and connect the pump wiring to the harness wire.
19. Install the fuel pump and bracket assembly onto the tank.
20. Install the sending unit assembly.
21. Connect the breather pipe assembly and the filler pipe extension.

NOTE: Tighten the breather pipe screw to 17 in. lbs. and all other attaching screws to 30 in. lbs.

22. Place the fuel tank on the jack and elevate it into place within the vehicle. Attach the straps and install the strap bolts, tightening them to 29 ft. lbs. (39 Nm).
23. Connect the breather hose to the tank pipe, the return hose to the tank pipe and the supply hose to its tank pipe. tighten the supply hose fitting to 21 ft. lbs. (28 Nm).
24. Connect the filler neck and overflow pipe to the tank. Make sure the clamps are properly seated and secure.
25. Lower the vehicle to the ground.
26. Connect the pump and sending unit electrical connectors to the harness.
27. Install the rear seat cushion.
28. Using a funnel, pour the fuel that was drained from its container into the fuel filler.
29. Install the fuel filler cap.
30. Start the engine and check carefully for any sign of leakage around the tank and lines.

Storm

1. Relieve the fuel system pressure, then disconnect the battery negative cable.
2. Drain the fuel from the fuel tank, then remove the fuel tank assembly.
3. Remove 8 fuel pump attaching screws, then the fuel pump/bracket from the fuel tank.
4. Remove the fuel pump supply hose extension from the bracket.
5. Remove the gasket from the fuel pump bracket and/or tank.
6. Remove the filter from the pump pickup tube.
7. Disconnect the electrical connectors from the fuel pump, outlet hose from pump and remove the pump from the bracket.

8. Reverse procedure to install. Connect battery negative cable. Start engine and check for leaks.

Spectrum

1. Relieve fuel pressure then disconnect negative battery cable.
2. Raise and support the vehicle safely. Drain fuel tank.
3. Remove all gas line hose connections and fuel pump ground wire.
4. Remove filler neck hose and clamp.
5. Remove breather hose and clamp.
6. Disconnect fuel tank hose to evaporator pipe.
7. Remove fuel tank mounting bolts and lower tank from vehicle. At this point remove hose from pump to fuel filter.
8. Remove fuel pump bracket plate and fuel pump as an assembly.
9. Remove pump bracket, rubber cushion and fuel pump filter.
10. To install, reverse the removal procedures. Be careful, to push the lower side of the fuel pump, together with the rubber cushion, into the fuel pump bracket.

Sprint

NOTE: The fuel tank must be lowered to gain access to the fuel pump which is located in the fuel tank.

1. Relieve the fuel system pressure.
2. Disconnect the negative battery cable.
3. Remove the rear seat cushion and disconnect the fuel gauge and fuel pump lead wires.
4. Raise and support the vehicle safely.
5. Drain the fuel tank by pumping or siphoning the fuel out through the fuel feed line (tank to fuel filter line).
6. Remove the tank from the vehicle.
7. Remove the fuel pump and fuel level gauge bracket from the fuel tank.
8. Remove the fuel pump from the fuel tank.
9. Installation is the reverse of removal. Clean the fuel filter in the tank.

Metro

1. Relieve the fuel system pressure.
2. Disconnect the negative battery cable.
4. Raise and support the vehicle safely.
5. Drain the fuel tank by pumping or siphoning the fuel out through the fuel feed line (tank to fuel filter line).
6. Remove the tank from the vehicle.
7. Fuel feed and return clamps and hoses from the fuel pump assembly.
8. Remove the 12 attaching screws from the fuel pump assembly and remove the assembly with the gasket from the fuel tank.
9. Remove the 1 mounting screw from the fuel pump motor assembly. Remove the 2 fuel pump motor connectors and remove the fuel pump motor from the fuel pump assembly.
10. Installation is the reverse order of the removal procedure.

Carburetor

REMOVAL & INSTALLATION

Nova

1. Disconnect the negative battery cable. To remove the air cleaner, perform the following procedures:
   a. Disconnect the air intake hose.
   b. Label and disconnect the emission control hoses from the air cleaner.
   c. Remove the wing nut, mounting bolts and the air cleaner from the carburetor.
2. Disconnect the accelerator cable from the carburetor. If equipped with an automatic transaxle, disconnect the transaxle throttle linkage from the carburetor.
3. Disconnect the wiring connector form the carburetor solenoid valve(s).
4. Label and disconnect the emission control hoses from the carburetor. Disconnect and drain the fuel inlet hose. Disconnect the evaporative emissions canister hose.
5. Remove the cold mixture heater wire clamp and the EGR vacuum control bracket.
6. Remove the carburetor-to-intake manifold nuts, the carburetor and gasket from the intake manifold. Using a clean shop rag, seal off the intake manifold opening.
7. Using the proper tool, clean the gasket mounting surfaces of the carburetor and manifold.
8. To install, use a new gasket and reverse the removal procedures. Install and adjust the throttle and transaxle linkages to the carburetor. Start the engine and check for fuel leaks.

Spectrum

1. Disconnect the negative battery terminal from the battery.
2. Remove the air cleaner.
3. Disconnect the harness connector and hoses.
4. Remove the accelerator cable from the carburetor.
5. Remove the bolts securing the carburetor to the intake manifold. Remove the carburetor and place a cover over the intake manifold.
6. To install, reverse the removal procedures and torque carburetor fixing bolts to 7.2 ft. lbs. (9 Nm) then start the engine and check for leaks.

Sprint

1. Remove and replace the fuel tank cap; this procedure releases the pressure within the fuel system.
2. Disconnect the negative battery cable.
3. Disconnect the warm air, the cold air, the second air, the vacuum and the EGR modulator hoses from the air cleaner case.
4. Remove the air cleaner case from the carburetor.
5. Disconnect the accelerator cable and the electrical wiring from the carburetor.
6. Remove the emission control and the fuel hoses from the carburetor.
7. Disconnect the No. 1 and No. 2 choke hoses from the carburetor.
8. Remove the mounting bolts and the carburetor from the intake manifold.
9. To install, use new gaskets and reverse the removal procedures. Torque the carburetor mounting bolts to 18 ft. lbs. (25 Nm).

IDLE SPEED ADJUSTMENT

Nova

1. Turn off all of the accessories, firmly set the parking brake and position the transaxle in **N**.
2. Check and/or adjust the ignition timing.
3. Start the engine and allow it to reach normal operating temperatures; make sure the choke is in the wide open position.
4. Inspect the fuel level sight glass on the carburetor to make sure the fuel is at the correct level.
5. At the distributor, locate the service engine connector, remove the rubber cap from it. Using a tachometer, connect the positive terminal to the service connector.

NOTE: When using a tachometer, consult the manufacturer's information to be sure it is compatible with the system.

6. Adjust the idle speed screw to 650 rpm for manual transaxles and 750 rpm for automatic transaxles.

Spectrum

1. Set the parking brake and block the wheels.
2. Place the manual transaxle in **N** or the automatic transaxle in **P**. Check the float level. Establish a normal operating temperature and make sure the choke plate is open.
3. Turn off all of the accessories and

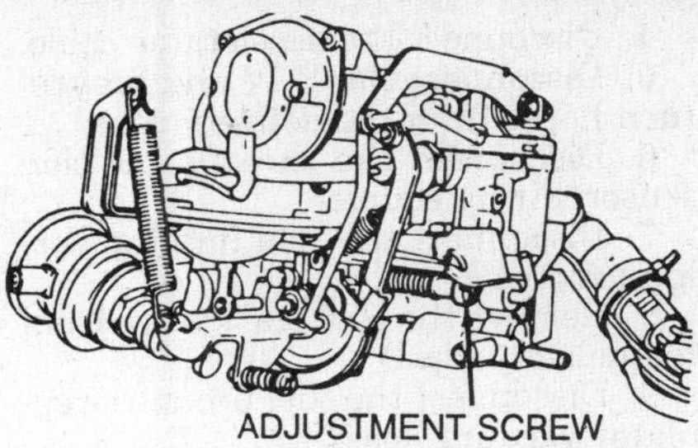

Idle speed adjustment screw location – Spectrum

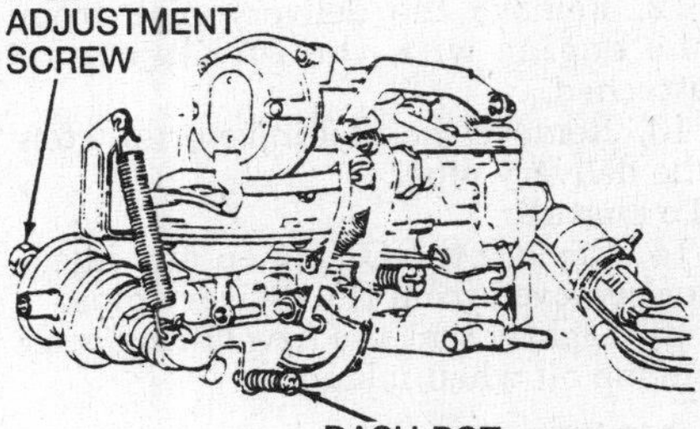

Fast idle control device adjustment screw location – Spectrum

wait until the cooling fan is not operating.

4. If equipped with power steering, place the wheels in the straight forward position. Remove the air filter.
5. Disconnect and plug the distributor vacuum line, canister purge line, EGR vacuum line and ITC valve vacuum line.
6. Connect a tachometer to the coil tachometer connector and a timing light to the No. 1 spark plug wire. Check the timing and idle speed.
7. If the idle speed needs adjusting, turn the idle speed adjusting screw.
8. If equipped with air conditioning, adjust the system to **MAX/COLD** and place the blower on **HIGH** position. Set the fast idle speed by turning the adjust bolt of the fast idle control diaphragm to 850 rpm on automatic transaxle or 980 rpm on manual transaxle.
9. When adjustment is completed, turn the engine off, remove the test equipment, install the air filter and vacuum lines.

Sprint

Check and/or adjust the accelerator cable free-play, ignition timing, valve lash and the emission control wiring and hoses. Make sure the headlights, heater fan, engine cooling fan and any other electrical equipment is turned **OFF**. If any current drawing system is operating, the idle up system will operate and cause the idle speed to be higher than normal.

1. Connect a tachometer to the primary negative terminal of the ignition coil and refer to specifications.
2. Place the transaxle in **N**, set the parking brake and block the wheels.
3. Start and run the engine until it reaches normal operating temperature.
4. Check and/or adjust the idle speed, it should be 700–800 rpm with manual transaxles, 800–900 rpm with automatic transaxles.

NOTE: To adjust the idle speed, turn the throttle adjustment screw on the carburetor.

5. With the engine at the proper idle speed, check and/or adjust the idle up speed.
6. Stop the engine and remove the tachometer.

SERVICE ADJUSTMENTS

For all carburetor service adjustment procedures and specifications, please refer to "Carburetor Service" in the Unit Repair section.

Fuel Injection

IDLE SPEED ADJUSTMENT

Nova, Prizm and Storm

1. Make sure the ignition timing is correct.
2. Using a jumper wire, connect it to the check engine connector located near the wiper motor.
3. Start the engine, then, check and/or adjust the idle speed; it should be 800 rpm on Nova and 700 rpm on Prizm and Storm.
4. When the adjustment is correct, remove the jumper wire from the check engine connector and recheck the timing marks.

Spectrum

1. Set the parking brake.
2. Block the front wheels.

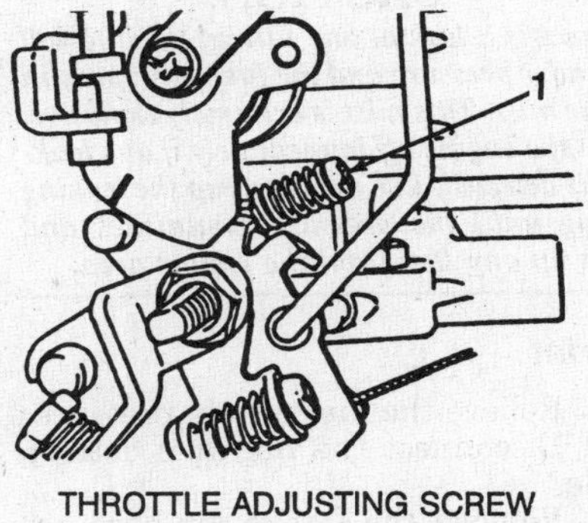

Idle speed adjustment screw – Sprint

3. Place the select lever in **N**.
4. Make the idling speed adjustment with the engine at normal operating temperature, with air conditioning off and front wheels facing straight ahead.

NOTE: All electrical equipment (lights, rear defogger, heater, etc.) should be turned off.

5. Make sure check engine light is not on.
6. Ground test terminal (ALDL connector).
7. Increase engine speed over 2000 rpm to reset the position of idle air control valve.
8. Set idle adjust screw to 950 rpm.
9. Remove test terminal ground and clear ECM trouble code.

Sprint and Metro

1. With a tachometer attached, run the engine until it reaches normal operating temperature.
2. Turn off all electrical loads such as lights, air conditioning etc.
3. Turn the idle speed adjusting screw until the idle speed is at 700–800 rpm on Sprint and Metro equipped with manual transaxle or 800–900 rpm on Metro equipped with automatic transaxle.

Fuel Injector

REMOVAL & INSTALLATION

CAUTION

The fuel system is under pressure. Release pressure slowly and contain spillage. Observe no smoking/no open flame precautions. Have a Class B-C (dry powder) fire extinguisher within arm's reach at all times.

Nova and 1991-92 Prizm

1. Disconnect the negative battery cable.
2. Disconnect the PCV hose from the valve cover.
3. Remove the vacuum sensing hose from the pressure regulator.
4. Disconnect the fuel return hose from the pressure regulator.
5. Place a towel or container under the cold start injector pipe. Loosen the 2 union bolts at the fuel line and remove the pipe with its gaskets.
6. Remove the fuel inlet pipe mounting bolt and disconnect the fuel inlet hose by removing the fuel union bolt, the 2 gaskets and the hose.
7. Disconnect the injector electrical connections.
8. At the fuel delivery pipe (rail), remove the 3 bolts. Lift the delivery pipe and the injectors free of the engine. Don't drop the injectors.

9. Remove the 4 insulators and 3 collars from the cylinder head.
10. Pull the injectors free of the delivery pipe.**To install:**
11. Before installing the injectors back into the fuel rail, install a new O-ring on each injector.
12. Coat each O-ring with a light coat of gasoline (never use oil of any sort)and install the injectors into the delivery pipe. Make certain each injector can be smoothly rotated. If they do not rotate smoothly, the O-ring is not in its correct position.
13. Install the insulators into each injector hole. Place the 3 spacers on the delivery pipe mounting holes in the cylinder head.
14. Place the delivery pipe and injectors on the cylinder head and again check that the injectors rotate smoothly. Install the 3 bolts and tighten them to 13 ft. lbs. (18 Nm).
15. Connect the electrical connectors to each injector.
16. Install 2 new gaskets and attach the inlet pipe and fuel union bolt. Tighten the bolt to 22 ft. lbs. (30 Nm). Install the mounting bolt.
17. Install new gaskets and connect the cold start injector pipe to the delivery pipe and cold start injector. Install the fuel line union bolts and tighten them to 13 ft. lbs. (18 Nm).
18. Connect the fuel return hose and the vacuum sensing hose to the pressure regulator. Attach the PCV hose to the valve cover.
19. Connect the battery cable to the negative battery terminal. Start the engine and check for leaks.

CAUTION

If there is a leak at any fitting, the line will be under pressure and the fuel may spray in a fine mist. This mist is extremely explosive. Shut the engine off immediately if any leakage is detected. Use rags to wrap the leaking fitting until the pressure diminishes and wipe up any fuel from the engine area.

1989-90 Prizm

1. Disconnect the negative battery cable.
2. Disconnect the PCV hoses from the valve cover and the vacuum sensing hose from the fuel pressure regulator.
3. Disconnect the fuel return hose from the fuel pressure regulator.
4. Remove the wiring connectors from the injectors.
5. Remove the pressure regulator by loosening the 2 bolts and pulling the regulator from the delivery pipe.
6. Label and remove the 4 vacuum hoses running to the EGR vacuum modulator. Remove the nut and bracket with the modulator.
7. Disconnect the fuel union bolt at the inlet pipe. Remove the pipe and the 2 gaskets.
8. Remove the 2 bolts holding the delivery pipe and then remove the delivery pipe and the injectors. Don't drop the injectors.
9. Remove the 2 spacers and the 4 insulators from the cylinder head.
10. Pull the injectors free of the delivery pipe.**To install:**
11. Before installing the injectors back into the fuel rail, install a new O-ring on each injector.
12. Coat each O-ring with a light coat of gasoline (never use oil of any sort)and install the injectors into the delivery pipe. Make certain each injector can be smoothly rotated. If they do not rotate smoothly, the O-ring is not in its correct position.
13. Install the 4 insulators and two spacers in place.
14. Place the delivery pipe and injectors on the cylinder head and again check that the injectors rotate smoothly. Install the 2 bolts and tighten them to 11 ft. lbs. (15 Nm).
15. Install 2 new gaskets and attach the inlet pipe and fuel union bolt. Tighten the bolt to 22 ft. lbs. (30 Nm).
16. Install the EGR vacuum modulator with its bracket and nut. Connect the 4 vacuum hoses to their proper ports.
17. Install new gaskets and connect the cold start injector pipe to the delivery pipe and cold start injector. Install the fuel line union bolts and tighten them to 13 ft. lbs. (18 Nm).
18. Install a new O-ring on the pressure regulator. Push the regulator into the delivery pipe and install the 2 bolts. Tighten the bolts to 5.5 ft. lbs. (65 in. lbs.)
19. Connect the injector wiring connectors.
20. Connect the fuel return hose and the vacuum sensing hose to the pressure regulator. Attach the PCV hoses to the valve cover.
21. Connect the battery cable to the negative battery terminal. Start the engine and check for leaks.

CAUTION

If there is a leak at any fitting, the line will be under pressure and the fuel may spray in a fine mist. This mist is extremely explosive. Shut the engine off immediately if any leakage is detected. Use rags to wrap the leaking fitting until the pressure diminishes and wipe up any fuel from the engine area.

Sprint

1. Relieve the fuel system pressure.
2. Disconnect the negative battery cable.
3. Remove the intake air hose between the throttle body and intercooler.
4. Disconnect the accelerator cable.
5. Disconnect the PCV and fuel return hoses from the delivery pipe.
6. Disconnect the throttle position sensor wiring coupler.
7. Disconnect the fuel injector wiring harness.
8. Remove the cold start valve from the delivery pipe.
9. Disconnect the fuel pressure regulator vacuum hose.
10. Disconnect the fuel feed hose from the delivery pipe.
11. Remove the fuel pressure regulator from the delivery pipe.
12. Remove the delivery pipe from the engine with the fuel injectors attached.
13. Remove the fuel injector(s) from the delivery pipe.

To install:

14. Replace the O-ring on any injector removed from the delivery pipe.
15. Lightly coat the new O-ring with engine oil when installing.

NOTE: After the injectors are installed in the fuel delivery pipe, make sure they rotate freely in the pipe before installation.

16. Check the insulator in the engine and replace, if necessary.
17. Install the delivery pipe and injector assembly in the engine.
18. Continue the installation in the reverse of the removal procedure.
19. When finished, run the engine and check for fuel leaks.

Metro

1. Relieve the fuel system pressure.
2. Disconnect the negative battery cable.
3. Remove the air cleaner assembly.
4. Remove the injector cover and upper insulator.
5. Disconnect the electrical connector and remove the injector.

To install:

6. Apply a thin coat of transaxle fluid or gasoline to the new upper and lower O-rings and install them on the injectors.
7. Install a new lower injector insulator into the injector cavity.
8. Push the injector straight into the fuel injection cavity. Do not twist.
9. Install a new upper insulator and the injection cover.
10. Install the injector cover screws to 31 inch lbs.
11. Connect the electrical connector to the injector, facing it lug side upward on clamp subwire securely.
12. Connect the negative battery cable.

Storm

1. Relieve the fuel system pressure.

2. Disconnect the negative battery cable.
3. Remove the air duct assembly.
4. Disconnect the throttle cable at the throttle body, the throttle cable from the bracket and the TPS electrical connector.
5. Disconnect the idle air control valve and the intake air temperature sensor electrical connectors.
6. Disconnect the 2 vacuum hose and the 2 coolant hoses at the throttle body.
7. Remove the 4 retaining bolts at the throttle body assembly.
8. Remove the throttle body.
9. Remove the 4 vacuum hoses at the plenum.
10. Disconnect the PCV valve at the plenum. Disconnect the ECM ground wires.
11. Remove the engine lift hook brackets.
12. Remove the plenum support brackets.
13. Remove the 9 plenum retaining nuts and bolts. Remove the plenum assembly. Remove the 2 fuel rail retaining bolts.
14. Remove the 4 injector electrical connectors.
15. Slide the injector retaining clip along the fuel rail ledge away from the injector to disengage, then spread the clip open while pulling downward. Discard the clip after it has been removed.
16. Remove the injector assembly.
17. Remove the O-ring from the top and bottom of the injectors and discard them.

To install:

NOTE: Different injectors are calibrated for different flow rates. When ordering new fuel injectors, be sure to order the identical part number that is inscribed on the old injector.

18. Install new O-rings to the top and bottom of the injectors and lubricate them with clean engine oil.
19. Install the fuel injectors into the fuel rail. Install the new injector retaining clip. Be sure that the clip is parallel to the injector electrical connector.

NOTE: Be sure to install the injector assembly into the fuel rail injector socket, with the electrical connector and retaining clip facing outward. Line up the injector with the injector socket, push upward slowly to engage the retaining clip on the fuel rail ledge then push the injector all the way in to firmly seat it in th socket.

20. Install the 4 injector electrical connectors.
21. Install the 2 fuel rail retaining bolts and torque them to 28 ft. lbs. (38 Nm).
22. Install the plenum assembly. Install the 9 plenum retaining nuts and bolts.
23. Install the plenum support brackets.
24. Install the engine lift hook brackets.
25. Reconnect the PCV valve at the plenum and the ECM ground wires.
26. Install the 4 vacuum hoses at the plenum.
27. Install the throttle body.
28. Install the 4 retaining bolts at the throttle body assembly. Torque the bolts 16 ft. lbs. (22 Nm).
29. Reconnect the 2 vacuum hose and the 2 coolant hoses at the throttle body.
30. Reconnect the idle air control valve and the intake air temperature sensor electrical connectors.
31. Reconnect the throttle cable at the throttle body, the throttle cable from the bracket and the TPS electrical connector.
32. Install the air duct assembly.
33. Reconnect the negative battery cable.

DRIVE AXLE

Halfshaft

REMOVAL & INSTALLATION

Nova

EXCEPT TWINCAM ENGINE

1. Remove the wheel cover.
2. Remove the cotter pin, hub nut cap, hub nut and washer.
3. Loosen the wheel nuts.
4. Elevate and safely support the vehicle.
5. Remove the front wheel.
6. Remove the lower control arm to ball joint attaching nuts and bolts.
7. Use a ball joint separator such as GM J-24319-01 or equivalent to remove the tie rod ball joint from the knuckle.
8. Remove the bolts holding the brake caliper bracket to the steering knuckle. Use stiff wire to suspend the caliper out of the way; do not let the caliper hang by its hose. Remove the brake disc.
9. Using a puller such as GM J-25287 or equivalent, push the axle from the hub.
10. Use a slide hammer and appropriate end fitting (GM J-2619-01 and J 35762 or equivalents) to pull the halfshaft from the transaxle. Remove the shaft from the vehicle.

To install:

11. When reinstalling, install shaft into transaxle. If necessary, use a long brass drift and a hammer to drive the housing ribs onto the inner joint.
12. Install the shaft into the wheel hub.
13. Install the lower control arm to the lower ball joint. Tighten the nuts and bolts to 59 ft. lbs. (80 Nm).
14. Install the tie rod end to the steering knuckle and tighten the nut to 36 ft. lbs. (49 Nm).
15. Install the brake disc; install the brake caliper and tighten the bolts to 65 ft. lbs. (88 Nm).
16. Install the wheel.
17. Install the hub nut and washer.
18. Lower the vehicle to the ground.
19. Tighten the wheel lugs to 76 ft. lbs. (103 Nm). Tighten the hub nut to 137 ft. lbs. (186 Nm).
20. Install the nut, cap, cotter pin and washer. Install the wheel cover.

TWINCAM ENGINE

1. Raise the vehicle and support it safely. Remove the tires.
2. Remove the cotter pin and locknut cap.
3. Have an assistant step on the brake pedal and at the same time, loosen and remove the bearing locknut.
4. While the assistant is still depressing the brake pedal, loosen and remove the 6 nuts which connect the halfshaft to the differential side gear shaft.
5. Remove the brake caliper and position it out of the way. Remove the brake disc.
6. Remove the 2 retaining nuts and then disconnect the lower arm from the steering knuckle.
7. Use a two-armed puller and remove the axle hub from the outer end of the halfshaft.
8. Remove the halfshaft.

To install:

9. To reinstall, place the outboard side of the shaft into the axle hub and then insert the inner end into the differential. Finger tighten the 6 nuts.

NOTE: Be careful not to damage the boots during installation.

10. Install the lower arm to the steering knuckle and tighten the bolts to 59 ft. lbs. (80 Nm).
11. Install the disc and reinstall the caliper and bracket. tighten the bolts to 65 ft. lbs. (88 Nm).
12. While an assistant depresses the brake pedal, tighten the 6 nuts holding the axle to the side gear shaft to 27 ft. lbs. (36 Nm).
13. Install the hub nut and washer.
14. Install the wheel.
15. Lower the vehicle to the ground.

16. Tighten the wheel lugs to 76 ft. lbs. (103 Nm). Tighten the hub nut to 137 ft. lbs. (186 Nm).
17. Install the nut, cap, cotter pin and washer. Install the wheel cover.

Prizm

1. Remove the wheel cover.
2. Remove the cotter pin, hub nut cap, hub nut and washer.
3. Loosen the wheel nuts.
4. Elevate and safely support the vehicle.
5. Remove the front wheel.
6. Remove the lower control arm to ball joint attaching nuts and bolts.
7. Use a ball joint separator such as GM J-24319-01 or equivalent to remove the tie rod ball joint from the knuckle.
8. Remove the bolts holding the brake caliper bracket to the steering knuckle. Use stiff wire to suspend the caliper out of the way; do not let the caliper hang by its hose. Remove the brake disc.
9. Push the axle from the hub using a brass or plastic hammer.

NOTE: If the axle can not be separated from the hub using a brass or plastic hammer. Use a puller such as GM J-25287 or equivalent, to push the axle from the hub.

10. Use a slide hammer and appropriate end fitting (GM J-2619-01 and J 35762 or equivalents) to pull the halfshaft from the transaxle. Remove the shaft from the vehicle.

To install:

11. When reinstalling, install shaft into transaxle. If necessary, use a long brass drift and a hammer to drive the housing ribs onto the inner joint.
12. Install the shaft into the wheel hub.
13. Install the lower control arm to the lower ball joint. Tighten the nuts and bolts to 105 ft. lbs. (142 Nm).
14. Install the tie rod end to the steering knuckle and tighten the nut to 36 ft. lbs. (49 Nm).
15. Install the brake disc; install the brake caliper and tighten the bolts to 65 ft. lbs. (88 Nm).
16. Install the wheel.
17. Install the hub nut and washer.
18. Lower the vehicle to the ground.
19. Tighten the wheel lugs to 76 ft. lbs. (103 Nm). Tighten the hub nut to 137 ft. lbs. (186 Nm).
20. Install the nut, cap, cotter pin and washer. Install the wheel cover.

Spectrum

1. Raise and support the vehicle safely, allowing the wheels to hang.
2. Remove the front wheel assem-

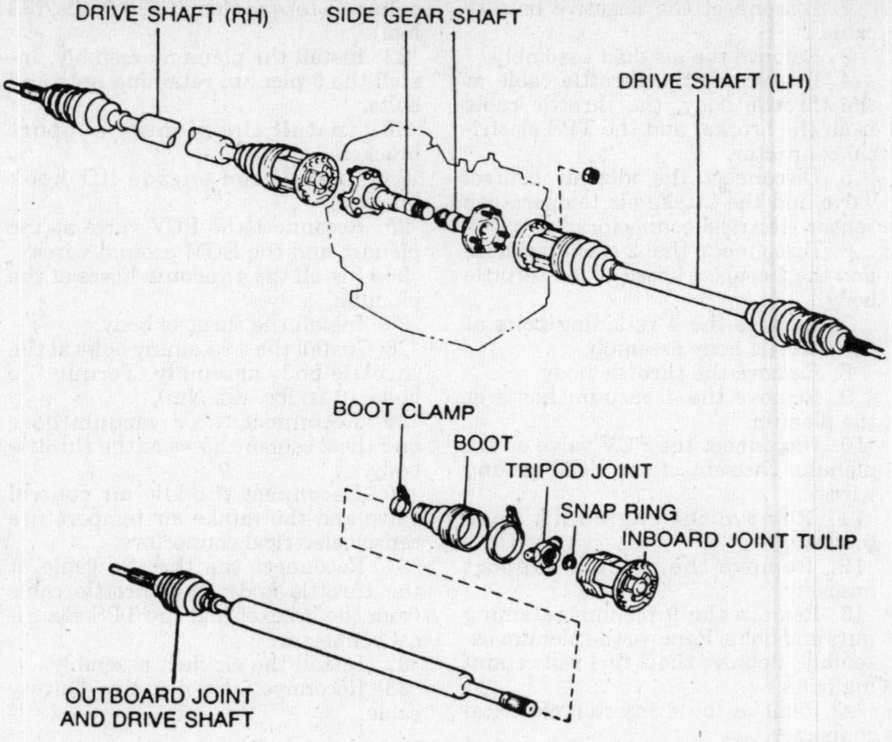

Drive axle assembly—Nova and Prizm

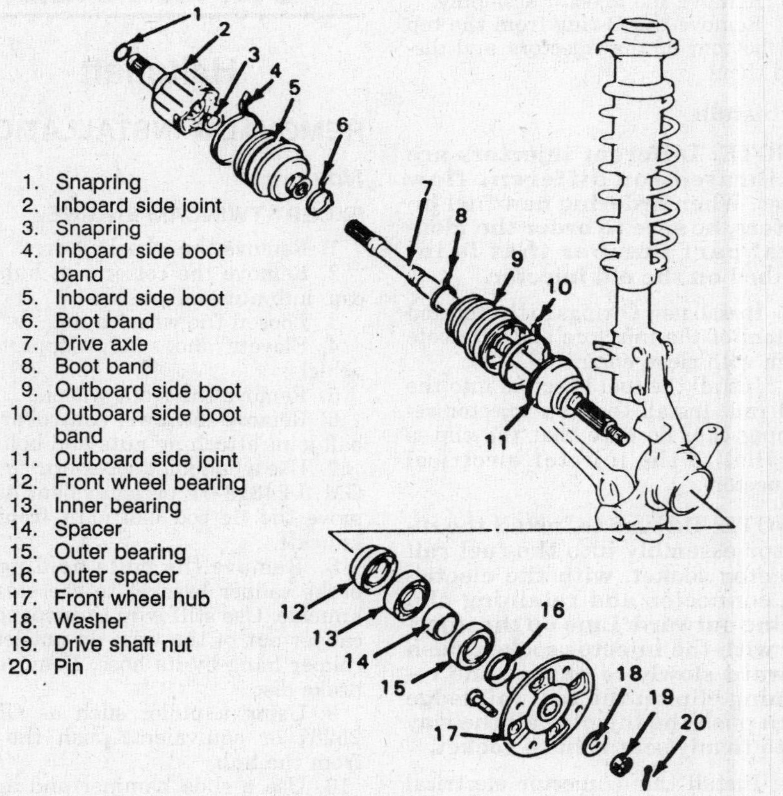

Exploded view of the halfshaft—Sprint and Metro

1. Deflector Ring
2. Constant Velocity Joint Housing
3. Haflshaft Snapring
4. Large Boot Clamp
5. Outer Boot
6. Small Boot Clamp
7. Left Hand Drive Axle Shaft
8. Halfshaft Damper
9. Inner Boot
10. Circular Clip
11. Ball Guide
12. Ball Joint
13. Ball Retainer
14. Double Offset Joint Outer Race
15. Halfshaft Retaining Clip
16. Halfshaft Snapring
17. Tri-Pot Joint Housing
18. Tri-Pot Joint Ball
19. Tri-Pot Joint Ball And Bearing Retainer
20. Tri-Pot Joint Needle Bearings
21. Tri-Pot Joint Spider

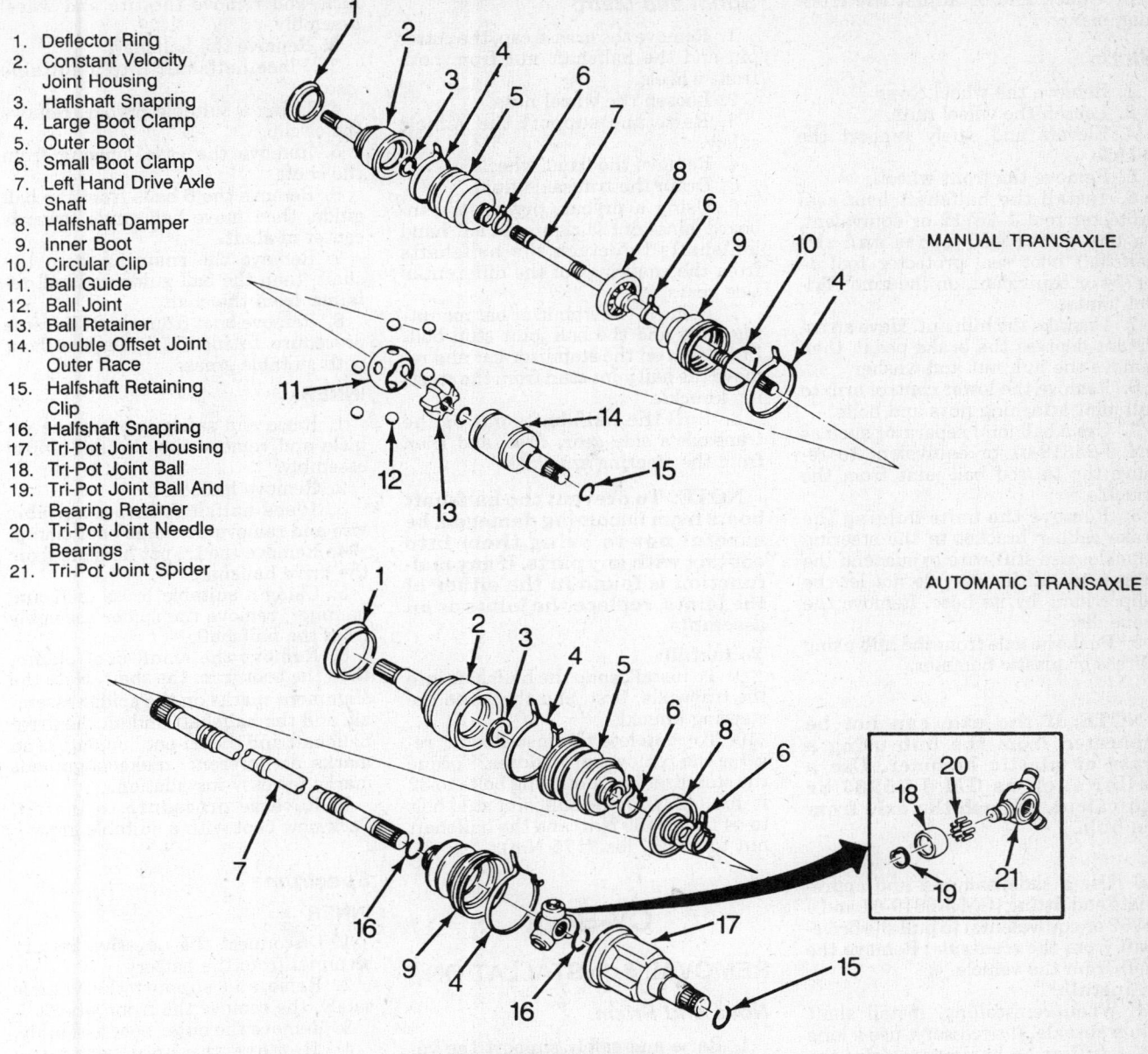

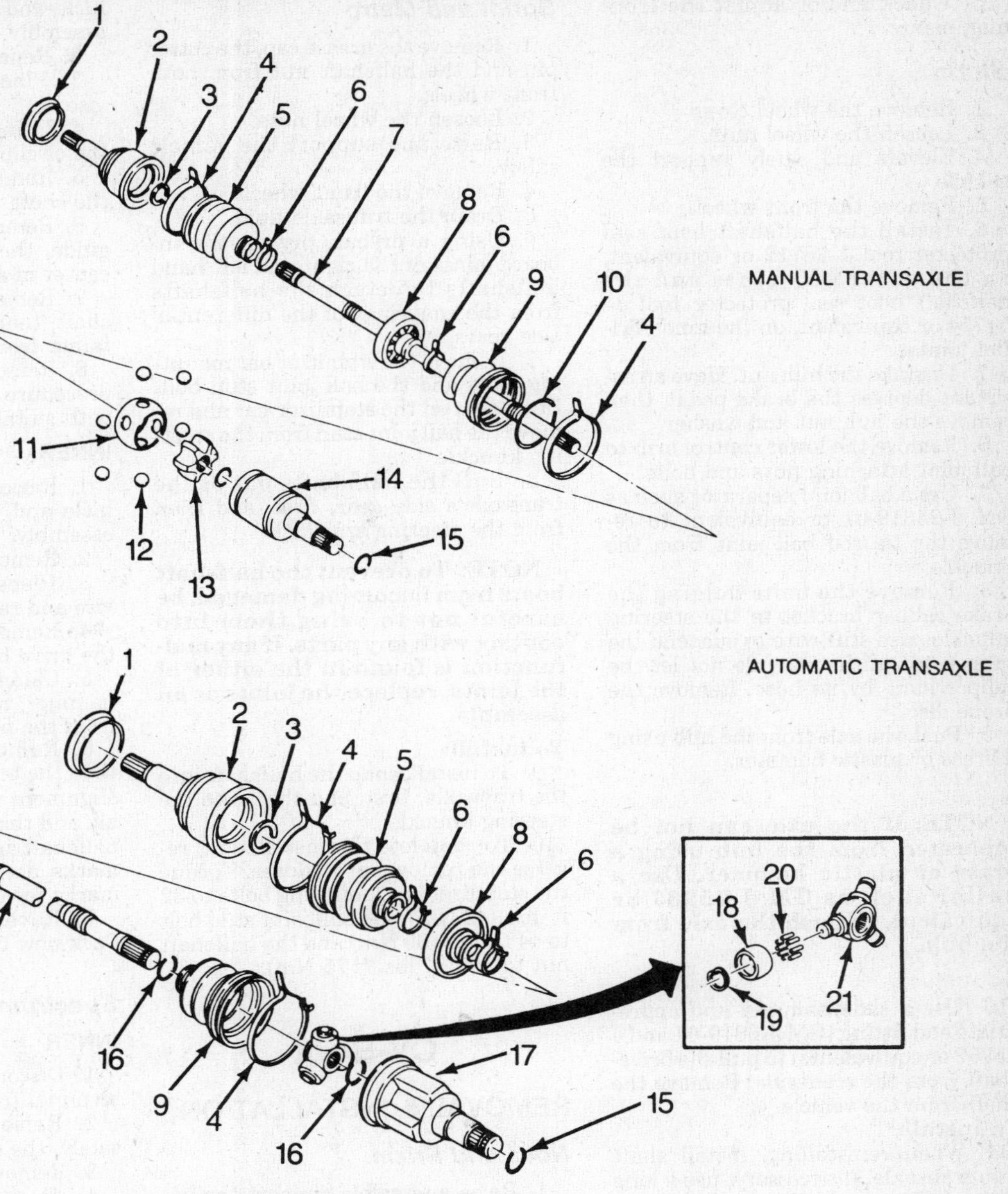

Exploded view of the halfshafts—Storm

blies, the hub grease caps, the hub nuts and the cotter pins.

3. Install the halfshaft boot seal protector tool J-28712 or equivalent, on the outer CV-joints and the halfshaft boot seal protector tool J-34754 or equivalent, on the inner Tri-Pot joints.

NOTE: Clean the halfshaft threads and lubricate them with a thread lubricant.

4. Have an assistant depress the brake pedal, then remove the hub nut and washer.

5. Remove the caliper-to-steering knuckle bolts and support the caliper, on a wire, out of the way.

6. Remove the rotor. Remove the drain plug and drain the oil from the transaxle.

7. Using a slide hammer puller and the puller attachment tool J-34866 or equivalent, pull the hub from the halfshaft.

8. Remove the tie rod-to-steering knuckle cotter pin and the nut. Using the ball joint separator tool J-21687-02 or equivalent, press the tie rod ball joint from the steering knuckle.

9. Remove the lower ball joint-to-control arm nuts/bolts.

10. Swing the steering knuckle assembly outward and slide the halfshaft from the steering knuckle.

11. Place a large prybar between the differential case and the inboard constant velocity joint. Pry the halfshaft from the differential case.

12. Remove the halfshaft assembly.

NOTE: When installing the halfshaft, press it into the differential case until it locks with with snapring.

To install:

13. Use new cotter pins and reverse the removal procedures.

14. Torque the ball joint-to-control arm nuts/bolts to 80 ft. lbs. (108 Nm), the caliper-to-steering knuckle bolts to 41 ft. lbs. (55 Nm) and the halfshaft-to-hub nut to 137 ft. lbs. (186 Nm).

15. Check and/or adjust the front alignment.

Storm

1. Remove the wheel cover.
2. Loosen the wheel nuts.
4. Elevate and safely support the vehicle.
5. Remove the front wheel.
6. Install the halfshaft boot seal protector tool J-28712 or equivalent, on the outer CV-joints and the halfshaft boot seal protector tool J-34754 or equivalent, on the inner Tri-Pot joints.
7. Unstake the hub nut. Have an assistant depress the brake pedal, then remove the hub nut and washer.
6. Remove the lower control arm to ball joint attaching nuts and bolts.
7. Use a ball joint separator such as GM J-24319-01 or equivalent to remove the tie rod ball joint from the knuckle.
8. Remove the bolts holding the brake caliper bracket to the steering knuckle. Use stiff wire to suspend the caliper out of the way; do not let the caliper hang by its hose. Remove the brake disc.
9. Push the axle from the hub using a brass or plastic hammer.

NOTE: If the axle can not be separated from the hub using a brass or plastic hammer. Use a puller such as GM J-25287 or equivalent, to push the axle from the hub.

10. Use a slide hammer and appropriate end fitting (GM J-2619-01 and J 35762 or equivalents) to pull the driveshaft from the transaxle. Remove the shaft from the vehicle.

To install:

11. When reinstalling, install shaft into transaxle. If necessary, use a long brass drift and a hammer to drive the housing ribs onto the inner joint.
12. Install the shaft into the wheel hub.
13. Install the lower control arm to the lower ball joint. Tighten the nuts and bolts to 115 ft. lbs. (156 Nm).
14. Install the tie rod end to the steering knuckle and tighten the nut to 40 ft. lbs. (54 Nm).
15. Install the brake disc; install the brake caliper and tighten the bolts to 65 ft. lbs. (88 Nm).
16. Install the wheel.
17. Install the hub nut and washer.
18. Lower the vehicle to the ground.
19. Tighten the wheel lugs to 76 ft. lbs. (103 Nm). Tighten the hub nut to 137 ft. lbs. (186 Nm).
20. Install the nut, cap, cotter pin and washer. Install the wheel cover.

Sprint and Metro

1. Remove the grease cap, the cotter pin and the halfshaft nut from both front wheels.
2. Loosen the wheel nuts.
3. Raise and support the vehicle safely.
4. Remove the front wheels.
5. Drain the transaxle fluid.
6. Using a prybar, pry on the inboard joints of the right and left hand halfshafts to detach the halfshafts from the snaprings of the differential side gears.
7. Remove the stabilizer bar mounting bolts and the ball joint stud bolt. Pull down on the stabilizer bar and remove the ball joint stud from the steering knuckle.
8. Pull the halfshafts out of the transaxle's side gear, first, and then from the steering knuckles.

NOTE: To prevent the halfshaft boots from becoming damaged, be careful not to bring them into contact with any parts. If any malfunction is found in the either of the joints, replace the joints as an assembly.

To install:

9. To install, snap the halfshaft into the transaxle, first, and then into the steering knuckle.
10. To complete the installation, reverse the removal procedures. Torque the stabilizer bar mounting bolts to 32 ft. lbs. (43 Nm); the ball joint stud bolt to 44 ft. lbs. (60 Nm) and the halfshaft nut to 129 ft. lbs. (175 Nm).

CV-Boot

REMOVAL & INSTALLATION

Nova and Prizm

1. Raise and safely support the vehicle and remove the tire and wheel assembly.
2. Remove the halfshaft assembly.
3. Remove the boot retaining clamps.
4. Remove the inboard joint tulip.
5. Remove the tripod joint snapring and the tripod joint from the halfshaft.
6. Remove the inboard and outboard joint boots.
7. To install, reverse the removal procedures; the inboard joint and clamp is larger than the outboard clamp. Face the beveled side of the tripod axial spline towards the outboard joint.

Storm

OUTER

1. Raise and safely support the vehicle and remove the tire and wheel assembly.
2. Remove the halfshaft.
3. Place halfshaft into a suitable vise.
4. Using a suitable prybar, remove the circlip.
5. Remove the case housing from the shaft.
6. Remove the 6 balls from the ball guide, then move ball guide towards center of shaft.
7. Remove the snapring from the shaft, then the ball guide and ball retainer from the shaft.
8. Remove boot from shaft. Reverse procedure to install. Pack new boot with suitable grease.

INNER

1. Raise and safely support the vehicle and remove the tire and wheel assembly.
2. Remove halfshaft.
3. Place halfshaft into a suitable vise and remove the large boot clamp.
4. Remove the tri-pot housing from the drive halfshaft.
5. Using a suitable brass drift and hammer, remove the spider assembly from the halfshaft.
6. Remove the small boot clamp, then the boot from the shaft. Note the alignment marks on the spider assembly and the halfshaft and on the drive halfshaft and the tri-pot housing. If no marks are present, make alignment marks for easy installation.
7. Reverse procedure to install. Pack new boot with a suitable grease.

Spectrum

INNER

1. Disconnect the negative battery terminal from the battery.
2. Raise and support the vehicle safely, the remove the front wheels.
3. Remove the outer boot assembly.
4. Remove the boot retaining clamps and the spacer ring.
5. Slide the axle and the spider bearing assembly out of the tri-pot housing. Install the spider retainer onto the spider bearing assembly.
6. Remove the spider assembly and the boot from the axle.
7. To install, pack the new boot with grease and reverse the removal procedures.

OUTER

1. Disconnect the negative battery terminal from the battery.
2. Raise and support the vehicle safely, then remove the front wheels.
3. Remove the brake caliper and support on a wire, then remove the rotor.
4. Slide the outer CV-joint assembly off the halfshaft.

5. Remove the bearing retaining ring, the boot retainer, the clamp and the outer boot.
6. To install, pack the new boot with grease and reverse the removal procedures.

Sprint and Metro

NOTE: Do not disassemble the wheel side joint (outboard). Replace if found to be defective. Do not disassemble the spider of the differential side joint. If the spider is found to be defective, replace the differential side joint assembly.

1. With the axleshaft removed from the vehicle, remove the boot band from the differential side joint.
2. Remove the housing from the differential side joint.
3. Remove the snapring and spider from the shaft.
4. Remove the inside and outside boots from the shaft.

To install:

5. Liberally apply the joint grease to the wheel side joint. Use the black joint grease in the tube included in the wheel side boot set or wheel side joint assembly.
6. Fit the wheel side boot on the shaft. Fill the inside of the boot with the joint grease, approximately 80 grams and fix the boot bands.
7. Fit the differential side boot on the shaft.
8. Liberally apply the joint grease to the differential side joint on the shaft. Use the yellow grease in the tube included in the differential side boot set or differential side joint assembly.
9. Install the spider of the differential side joint on the shaft, facing its chamfered side to the wheel side joint.
10. After installing the spider, fit the snapring in the groove on the shaft.
11. Fill the inside of the differential side boot with the joint grease, approximately 130 gram and install the housing. Fix the boot to the housing with a boot band.
12. Correct any distortions or bends in the boots.
13. Install the halfshaft in the vehicle.

Front Wheel Hub, Knuckle and Bearings

REMOVAL & INSTALLATION

Nova

1. Loosen the wheel nuts and the center axle nut.
2. Raise the vehicle and safely support it.
3. Remove the wheel.
4. Remove the brake hose retaining clip at the strut.
5. Disconnect the brake flex hose from the metal brake line. Use a small pan to collect spillage and plug the lines as soon as possible.
6. Remove the bolts holding the brake caliper to the knuckle; support the caliper with a piece of stiff wire out of the way.
7. Remove the brake disc.
8. Remove the drive axle nut. Use GM Tool J-25287 or equivalent to push out the drive axle.
9. Remove the cotter pin and the tie rod (steering rod) nut at the knuckle. Use a tie rod separator (GM Tool J-24329-01 or equivalent) to separate the joint.
10. Remove the nuts and bolt holding the ball joint to the control arm.
11. Matchmark the camber adjusting cam and the strut.
12. Remove the 2 bolts holding the knuckle to the strut. Remove the knuckle. The ball joint may be removed from the knuckle if desired.
13. Mount the knuckle securely in a vise. Use a screwdriver to remove the outer dust cover.
14. Using a slide hammer and puller (GM J-26941 or equivalent), remove the inner grease seal from the knuckle.
15. Remove the inner bearing snapring.
16. Remove the brake splash shield.
17. Remove the hub using an extractor such as GM Tools J-25287 and J-35378 or equivalent.

NOTE: Whenever the hub is removed, the inner and outer grease seals MUST be replaced with new seals. The seals are not reusable.

18. Use the same extractor to remove the outer bearing race from the hub.
19. Remove the outer grease seal with the slide hammer and puller.
20. Using a bearing driver of the correct size (GM Tools J-35399 and J-35379 or equivalent), remove the bearing assembly.
21. Clean and inspect all parts but do not wash or clean the wheel bearing; it cannot be repacked. If the bearing is damaged or noisy, it must be replaced.

To install:

22. Using a bearing driver of the correct size (GM J-8092 and J-35411 or equivalent), install the bearing into the hub.
23. Use a seal driver to install a new outer grease seal. (GM J-35737-1 or equivalent).
24. Apply sealer to the brake splash shield and install it to the knuckle.
25. Apply a layer of multi-purpose grease to the seal lip, seal and bearing. Install the hub with GM Tools J-8092 and J-35399 or equivalent.
26. Install the snapring.
27. Install a new inner grease seal using the correct size driver. (GM J-35737-2 or equivalent).
28. Install the outer dust cover (open end down) with GM Tool J-35379 or equivalent.
29. Install the lower ball joint to the control arm and tighten the nuts and bolt to 59 ft. lbs. (80 Nm).
30. If the ball joint was removed from the knuckle, reinstall it. Install a new nut and temporarily tighten it to 14 ft. lbs. (19 Nm). Back off the nut until clear of the knuckle and then retighten it to 82 ft. lbs. (111 Nm).
31. Install the camber adjusting cam into the knuckle. Connect the knuckle to the strut lower bracket.
32. Insert the bolts from rear to front and align the matchmarks of the camber adjusting cam and the strut. Tighten the nuts to 105 ft. lbs. (142) (single overhead cam engine) or 166 ft. lbs. (225 Nm) twincam engine).
33. Connect the tie rod to the knuckle. Install the nut and tighten to 36 ft. lbs. (49 Nm). Install the cotter pin.
34. Install the halfshaft into the hub.
35. Double check the nuts and bolt holding the ball joint to the lower control arm. Correct torque is 59 ft. lbs. (80 Nm).
36. Install the brake disc.
37. Reinstall the brake caliper. Tighten the bolts to 65 ft. lbs. (88 Nm).
38. Connect the brake flex hose to the metal brake line. Correct tightness is 11 ft. lbs. (15 Nm). Do not overtighten this connection.
39. Install the wheel
40. Lower the vehicle to the ground.
41. Tighten the wheel nuts to 76 ft. lbs. (103 Nm).
42. Install the washer and nut onto the halfshaft end. Tighten the bolt to 137 ft. lbs. (186 Nm) and install the cap and cotter pin.
43. Bleed the brake system.

Prizm

1. Loosen the wheel nuts and the center axle nut.
2. Raise the vehicle and safely support it.
3. Remove the wheel.
4. Remove the center axle nut.
5. Remove the brake caliper and hang it out of the way on a piece of stiff wire. Do not disconnect the brake line; do not allow the caliper to hang by the hose.
6. Remove the brake disc.
7. Remove the cotter pin and nut from the tie rod end.
8. Remove the tie rod end from the knuckle using a joint separator (GM Tool J-6627-A or equivalent).
9. Remove the bolt and 2 nuts holding the bottom of the ball joint to the

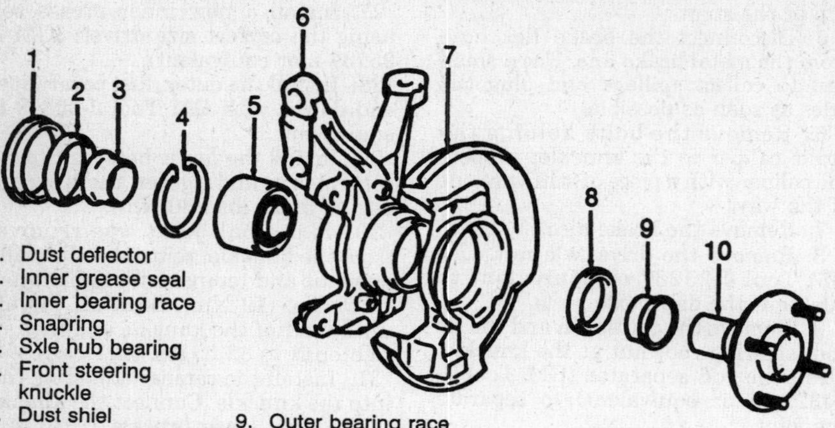

1. Dust deflector
2. Inner grease seal
3. Inner bearing race
4. Snapring
5. Sxle hub bearing
6. Front steering knuckle
7. Dust shiel
8. Outer grease seal
9. Outer bearing race
10. Drive axle hub

Exploded view of the hub/bearing assembly—Nova and Prizm

control arm and remove the arm from the knuckle.

10. Remove the 2 nuts from the steering knuckle. Place a protective cover or shield over the CV boot on the halfshaft.

11. Using a plastic mallet, tap the halfshaft free of the hub assembly.

12. Remove the 2 bolts and remove the axle hub assembly.

13. Clamp the knuckle in a vise with protected jaws.

14. Remove the dust deflector. Loosen the nut holding the ball joint to the knuckle. Use a ball joint separator (GM Tool J-35413 or equivalent) to loosen and remove the joint

15. Use a slide hammer/extractor (GM Tool J-26941 or equivalent) to remove the outer oil seal.

16. Remove the snapring.

17. Using a hub puller and pilot (GM Tools J-25287 and J-35378 or equivalents), pull the axle hub from the knuckle.

18. Remove the brake splash shield (3 bolts).

19. Use a split plate bearing remover, puller pilot and a shop press, remove the inner bearing race from the hub.

20. Remove the inner oil seal with the same tools used to remove the outer seal.

21. Place the inner race in the bearing. Support the knuckle and use an axle hub remover (GM Tool J-35399) with a plastic mallet to drive out the bearing.

22. Clean and inspect all parts but do not wash or clean the wheel bearing; it cannot be repacked. If the bearing is damaged or noisy, it must be replaced.**To install:**

23. Press a new bearing race into the steering knuckle using a bearing driver of the correct size (GM J-8092 and J-37777 or equivalent).

24. Place a new bearing inner race on the hub bearing.

25. Insert the side lip of a new oil seal into the seal installer (GM J-35737-01 or equivalent) and drive the oil seal into the steering knuckle.

26. Apply multi-purpose grease to the oil seal lip.

27. Apply sealer to the brake splash shield and install the shield.

28. Use a hub installer (GM J-35399) to press the hub into the steering knuckle.

29. Install a new snapring into the hub.

30. Using a seal installer of the correct size, install a new outer oil seal into the steering knuckle.

31. Apply multi-purpose grease to the seal surfaces which will contact the halfshaft.

32. Support the knuckle and drive in a new dust deflector.

33. Install the ball joint into the knuckle and tighten the nut to 94 ft. lbs. (127 Nm).

34. Temporarily install the hub assembly to the lower control arm and fit the driveaxle into the hub.

35. Install the knuckle to strut bolts, then install the tie rod end to the knuckle.

36. Tighten the steering knuckle to strut assembly nuts to 194 ft. lbs. (263 Nm) and the strut assembly bracket to body nuts to 29 ft. lbs. (39 Nm) and tighten the tie rod end nut to 36 ft. lbs. (49 Nm). Install the cotter pin.

37. Remove the old nut from the lower ball joint and install a new castle nut. Tighten the nut to 94 ft. lbs. (127 Nm) and install a new cotter pin. (The old nut was used to draw the joint into the knuckle; the new nut assures retention.)

38. Connect the ball joint to the lower control arm and tighten the nuts to 105 ft. lbs. (142 Nm).

39. Install the brake disc.

40. Install the brake caliper and tighten the bolts to 65 ft. lbs. (88 Nm).

41. Install the center nut and washer on the drive axle.

42. Install the wheel

43. Lower the vehicle to the ground.

44. Tighten the wheel nuts to 76 ft. lbs. (103 Nm) and the axle bolt to 137 ft. lbs. (186 Nm). Install the cap and cotter pin.

45. Remove the protective cover from the CV boot.

Spectrum and Storm

1. Raise and support the vehicle safely, allowing the wheels to hang.

2. Remove the front wheel assemblies, the hub grease caps and the cotter pins.

3. Install the halfshaft boot seal protector tool No. J–28712 or equivalent, on the outer CV-joints and the halfshaft boot seal protector tool No. J–34754 or equivalent, on the inner Tri-Pot joints.

NOTE: Clean the halfshaft threads and lubricate them with a thread lubricant.

4. Remove the hub nut and washer.

5. Remove the caliper-to-steering knuckle bolts and support the caliper, on a wire, out of the way.

6. Remove the rotor.

7. Using a slide hammer puller and the puller attachment tool J‹n-dash›34866 or equivalent, pull the hub from the halfshaft.

8. Remove the tie rod-to-steering knuckle cotter pin and the nut. Using the ball joint separator tool J‹n-dash›21687–02 or equivalent, press the tie rod ball joint from the steering knuckle.

9. To remove the steering knuckle from the vehicle, perform the following procedures:

a. Remove the lower ball joint-to-control arm nuts/bolts.

NOTE: Before separating the steering knuckle from the strut, be sure to scribe matchmarks on each component.

b. Remove the steering knuckle-to-strut nuts/bolts and the steering knuckle from the vehicle.

10. Using a medium prybar, pry the grease seals from the steering knuckle. Using a pair of internal snapring pliers, remove the internal snaprings from the steering knuckle.

11. Support the steering knuckle (face down) on an arbor press (on 2 press blocks). Position tool J‹n-dash›35301 or equivalent, on the rearside of the hub bearing, then, press the bearing from the steering knuckle.

12. Using an arbor press, the wheel puller tool J–35893 or equivalent and a piece of bar stock, press the bearing inner race from the wheel hub.
13. Clean the parts in solvent and blow dry with compressed air.

To Install:

14. Using wheel bearing grease, lubricate the inside of the steering knuckle.
15. Using the internal snapring pliers, install the outer snapring into the steering knuckle.
16. Position the steering knuckle on a arbor press (outer face down), a new wheel bearing and the bearing installation tool J–35301 or equivalent, then press the bearing inward until it stops against the snapring.
17. Using the internal snapring pliers, install the inner snapring into the steering knuckle.
18. Using the grease seal installation tool J–35303 or equivalent, drive new grease seals into both ends of the steering knuckle until they seat against the snaprings.
19. To install the hub into the steering knuckle, perform the following procedure:
   a. Position tool J–35302 or equivalent, facing upward on a arbor press.
   b. Position the steering knuckle (facing upward) on the tool J‹n-dash›35302, the wheel hub and a piece of bar stock (on top).
   c. Press the assembly together until the hub bottoms out on the wheel bearing.
20. To install steering knuckle, use new cotter pins and reverse the removal procedures. Lubricate the new bearing seal with wheel bearing grease.
21. On Spectrum, torque the steering knuckle-to-strut bolts to 87 ft. lbs. (118 Nm), the ball joint-to-control arm nuts/bolts to 80 ft. lbs. (108 Nm), caliper-to-steering knuckle bolts to 41 ft. lbs. (55 Nm) and the halfshaft-to-hub nut to 137 ft. lbs. (186 Nm).
22. On Storm, torque steering knuckle nuts and bolts to 115 ft. lbs. (156 Nm), ball joint pinch bolt to 48 ft. lbs. (66 Nm), tie rod-to-steering knuckle nut to 40 ft. lbs. (54 Nm), brake caliper-to-steering knuckle bolts to 72 ft. lbs. (98 Nm) and hub nut to 137 ft. lbs. (186 Nm).
23. Check and/or adjust the front end alignment.

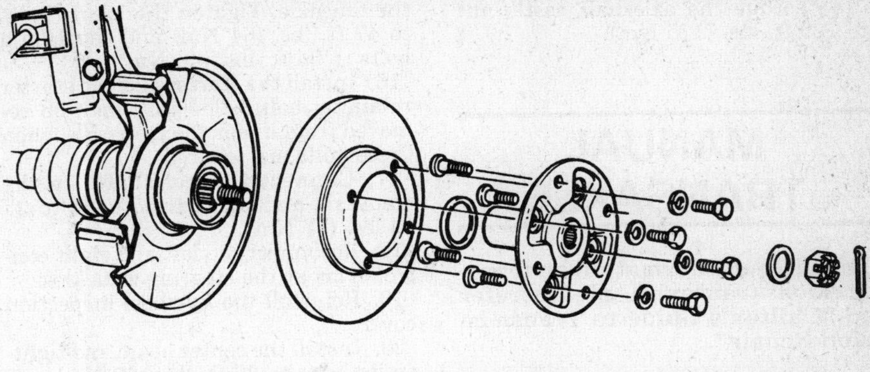

Exploded view of the front wheel hub assembly—Sprint and Metro

Sprint and Metro

1. Raise and support the vehicle safely. Remove the front wheel assembly. Remove the hub from the steering knuckle.
2. Remove the tie rod end cotter pin and nut.
3. Using the ball joint removal tool J–21687–02, remove the ball joint from the steering knuckle.
4. Remove the ball stud bolt from the steering knuckle.
5. Remove the strut-to-steering knuckle bolts.
6. Remove the steering knuckle and support the axleshaft.
7. Using a brass drift, drive the inner and outer wheel bearings from the steering knuckle.
8. Remove the spacer and clean the steering knuckle cavity.

To install:

9. Lubricate the new bearings and the steering knuckle cavity.
10. Using the installation tool J–34856, drive the new bearings, with the internal seals facing outward, into the steering knuckle.
11. Using the seal installation tool J–34881, drive the new seal into the steering knuckle, grease the seal lip.
12. To complete the installation, reverse the removal procedures.
13. Torque the strut-to-steering knuckle bolts to 59 ft. lbs. (80 Nm). Torque the ball joints-to-steering knuckle nuts to 44 ft. lbs. (60 Nm).

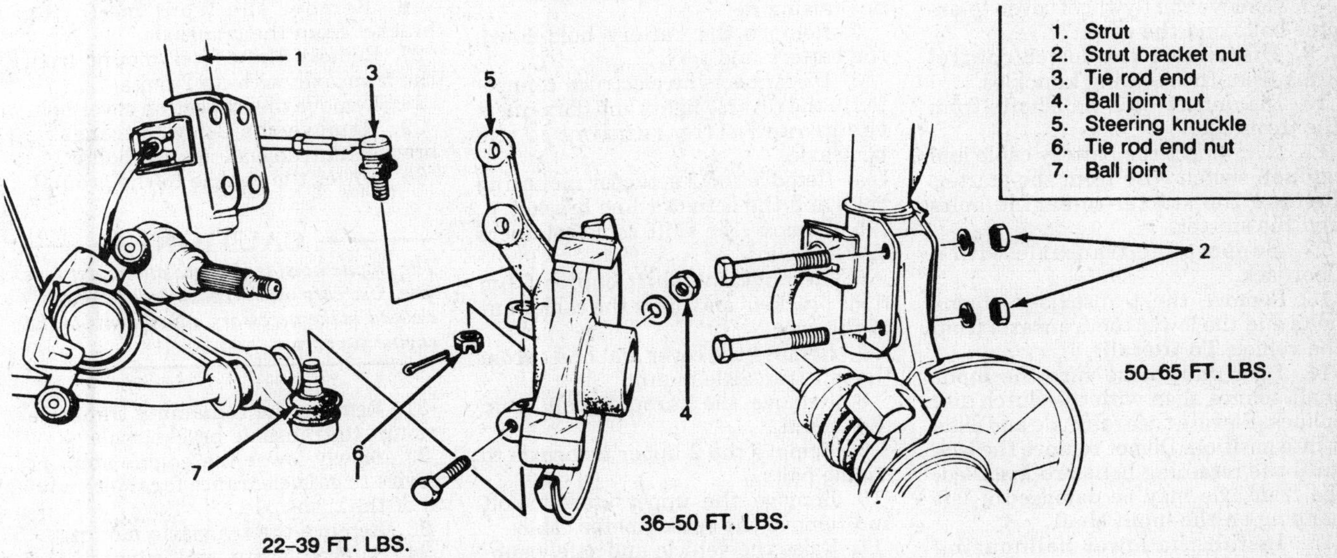

Front steering mounting knuckle—Sprint and Metro

14. Torque the axleshaft castle nut to 129 ft. lbs. (175 Nm).

MANUAL TRANSAXLE

For further information on transmissions/transaxles, please refer to "Chilton's Guide to Transmission Repair".

Transaxle Assembly

REMOVAL & INSTALLATION

Nova

1. Disconnect the negative battery terminal.
2. Remove the air cleaner and inlet duct.
3. From the transaxle, disconnect the back-up light switch connector, the speedometer cable, the thermostat housing and the ground wire.
4. Remove the 4 shift cable-to-transaxle clips, the clutch slave cylinder-to-transaxle bolts and the slave cylinder.
5. Remove the (2) upper transaxle-to-engine bolts and the upper transaxle mount bolt.
6. Attach an engine support tool or equivalent, and support the engine. Raise and support the front of the vehicle safely.
7. Remove the left wheel. From under the vehicle, remove the left, right and center splash shields. Remove the center beam-to-chassis bolts and the center beam.
8. Remove the flywheel cover-to-engine bolts and the cover.
9. Disconnect the lower control arms from the steering knuckles.
10. Disconnect both halfshafts from the transaxle.
11. Disconnect the battery cable and ignition switch wire from the starter. Remove the starter-to-engine bolts and the starter.
12. Support the transaxle with a floor jack.
13. Remove the transaxle-to-engine bolts and the lower the transaxle from the vehicle.**To install:**
14. To install, make sure the input shaft splines align with the clutch disc splines. Elevate the transaxle and slide it into position. Do not remove the jack until the retaining bolts are installed; the transaxle may be damaged if left hanging on the input shaft.
15. Install the lower bellhousing bolts and the bolts at the back side of the tansaxle. Tighten the 12mm bolts to 47 ft. lbs. (64 Nm) and the 10mm bolts to 34 ft. lbs. (46 Nm).
16. Install the starter and the starter retaining bolts. The jack may be removed from under the transaxle when these bolts are secured.
17. Connect the halfshafts to the transaxle and tighten the bolts to 27 ft. lbs. (37 Nm).
18. Reconnect the left and right control arms at the steering knuckles.
19. Reinstall the flywheel inspection cover.
20. Install the center beam and tighten its bolts to 29 ft. lbs. (39 Nm).
21. Install the splash shields and the left front wheel.
22. Lower the vehicle from the stands. Install the 2 upper bellhousing bolts and the transaxle mount upper bolt. The engine support devices may be removed from the engine.
23. Connect the backup light switch.
24. Reinstall the shift cables and their clips. Install the clutch slave cylinder.
25. Connect the ground wire to the tansaxle and attach the wiring to the starter.
26. Connect the thermostat housing and the speedometer cable to the transaxle.
27. Install the air cleaner and intake duct.
28. Connect the negative battery cable.
29. Check the level of the fluid in the transaxle and fill or top off as needed.

Prizm

1. Install an engine support and tension it to support the engine without raising it.
2. Remove the battery hold-down, the battery and tray.
3. Disconnect the electrical connector to the reverse lights and disconnect the ground strap running to the transaxle.
4. Remove the 2 actuator mounting bolts and the actuator line bracket.
5. Remove the shift cable retainers and end clips.
6. Remove the shift cables from their brackets and place the cables out of the way.
7. Remove the cover and brace from the left transaxle mount.
8. Remove the through bolt from the mount.
9. Remove the 2 upper transaxle to engine bolts.
10. Remove the upper starter bolt and remove the speedometer cable.
11. Raise the vehicle and safely support it.

CAUTION

The engine hoist is in place and under tension. Use care when repositioning the vehicle and make necessary adjustments to the engine support.

12. Remove the splash shields. Drain the transaxle oil.
13. Disconnect the electrical connections at the starter.
14. Remove the bottom starter bolt and the starter.
15. Remove the drive axles.
16. Remove the 3 bolts holding the center crossmember to the radiator support.
17. Remove the covers from the front and center mount bolts.
18. Remove the 2 front mount bolts, then the center mount bolts and then the 2 rear mount bolts.
19. Remove the 2 bolts holding the center crossmember to the main crossmember.
20. Remove the 3 exhaust hanger bracket nuts and the exhaust hanger.
21. Use a floor jack and a wide piece of wood to support the main crossmember.
22. Remove the 8 bolts holding the main crossmember to the body.
23. Remove the 2 bolts holding the lower control arm brackets to the body.

CAUTION

The crossmembers are loose and free to fall. Make sure they are properly supported.

24. Slowly lower the main crossmember while holding onto the center crossmember.
25. At the front transaxle mount, remove the through bolt and mount.
26. Remove the front mounting bracket from the transaxle.
27. Remove the center mount from the transaxle with its 2 bolts.
28. Remove the inspection cover bolt.
29. Remove the 2 lower transaxle bracket to transaxle mount bolts.
30. Lower the vehicle to the ground.

CAUTION

The engine hoist is in place and under tension. Use care when repositioning the vehicle and make necessary adjustments to the engine support.

31. Remove the remaining transaxle mount to transaxle bracket bolt.
32. Slowly lower the engine support device to gain clearance for the removal of the transaxle.
33. Remove the transaxle mount.
34. Safely elevate and support the vehicle.

CAUTION

The engine hoist is in place and under tension. Use care when repositioning the vehicle and make necessary adjustments to the engine support.

35. Support the transaxle with a floor jack, making sure it is properly placed and balanced.
36. Remove the lower front and rear bolts holding the transaxle to the engine.
37. Remove the transaxle assembly from the engine and lower it slowly on the floor jack.

To install:

38. Elevate the transaxle into position, making sure the input shaft aligns with the clutch splines.
39. Install the lower front and rear bolts holding the transaxle to the engine.
40. Remove the floor jack from under the transaxle.
41. Attach the electrical wiring to the starter.
42. Install the lower starter bolt snugly.
43. Lower the vehicle to the ground

CAUTION

The engine hoist is in place and under tension. Use care when repositioning the vehicle and make necessary adjustments to the engine support.

44. Install the left transaxle mount upper bolt and 2 lower bolts snugly.
45. Take tension on the engine support device and raise the transaxle. Install the through bolt loosely in the mount.
46. Tighten the through bolt to 69 ft. lbs. (87 Nm). Tighten the upper transaxle mount bolt to 45 ft. lbs. (61 Nm).
47. Install the mount cover and bracket. Tighten the cover bolts to 45ft. lbs. (61 Nm).
48. Install the 2 upper transaxle to engine bolts and tighten them to 34 ft. lbs. (46 Nm).
49. Install the upper starter bolt and tighten it to 29 ft. lbs. (39 Nm).
50. Connect the speedometer cable.
51. Position the shift cables into the brackets and connect the cable retainers and end clips.
52. Install the 2 actuator mounting bolts snugly, then install the actuator line bracket and bolt. Tighten the actuator line and mounting bolts to 15 ft. lbs. (20 Nm).
53. Connect the ground strap and the electrical connector for the reverse lights.
54. Install the air cleaner assembly.
55. Elevate and safely support the vehicle.

CAUTION

The engine hoist is in place and under tension. Use care when repositioning the vehicle and make necessary adjustments to the engine support.

56. Tighten the 2 remaining lower transaxle mount bolts to 45 ft. lbs. (61 Nm). (These are the bolts installed in step 39.).
57. Tighten the lower starter bolt to 29 ft. lbs. (39 Nm).
58. Install the center mount and its 2 bolts and tighten them to 45 ft. lbs. (61 Nm). Install the inspection cover bolt and tighten it to 45 ft. lbs. (61 Nm).
59. Install the front mount bracket on the transaxle.
60. Install the front mount and through bolt loosely.

NOTE: When installing the front mount, the weight on the mount must go toward the transaxle for proper mount alignment.

61. Position the center crossmember over the center and rear transaxle mount studs; start 2 nuts on the center mount.
62. Loosely install the 3 bolts holding the center crossmember to the radiator support.
63. Loosely install the 2 front mount bolts.
64. Raise the main crossmember into position over the rear mount studs and align all underbody bolts. Install the 2 rear mount nuts loosely.
65. Install the 8 main crossmember to underbody bolts loosely.
66. Install the 2 lower control arm bracket bolts loosely.
67. Loosely install the 2 bolts holding the center crossmember to the main crossmember.
68. Install the exhaust hanger bracket and its 3 nuts. The crossmembers, mounts, through bolts and brackets should now all be in place and held loosely by their nuts and bolts. If any repositioning is necessary, do so now.
69. Tighten the components below in the order listed to the correct torque specification:
   a. Main crossmember to underbody bolts: 152 ft. lbs. (206 Nm).
   b. Lower control arm bolts: 94 ft. lbs. (127 Nm).
   c. Center crossmember to radiator support bolts: 45 ft. lbs. (61 Nm).
   d. Front, center and rear mount bolts: 45 ft. lbs. (61 Nm).
   e. Exhaust hanger bracket nuts: 9.5 ft. lbs. (12.5 Nm)
   f. Front mount through bolt: 69 ft. lbs. (87 Nm).
70. Install the covers on the front and center mount bolts.
71. Reinstall the halfshafts.
72. Fill the transaxle with the correct amount of oil.
73. Install the splash shields.
74. Lower the vehicle to the ground. Remove the engine support apparatus.
75. Install the battery tray, battery and hold-down clamp.

Storm

1. Disconnect the battery cables, then remove the battery and tray from the vehicle.
2. Drain the transaxle fluid.
3. Remove the air cleaner assembly.
4. Disconnect the electrical connectors from the transaxle.
5. Disconnect the ground cable and engine wiring harness clamp from the transaxle.
6. Disconnect the ignition coil ground cable from the engine, then the ground cable at the ignition coil.
7. Disconnect the speedometer cable, clutch cable and shifter cables.
8. Support the engine using engine support tool J-28467-A or equivalent.
9. Raise and support the vehicle safely, then remove the tire and wheel assemblies.
10. Remove the front undercovers, then the ball joints from the steering knuckles.
11. Remove the left and right halfshaft, then the front exhaust pipe.
12. Remove the torque rod and bracket, then the left transaxle mount.
13. Remove the front transaxle through bolt, then the center beam with the rear transaxle mount.
14. Remove the engine stiffener attaching bolts, then the engine stiffener from the vehicle.
15. Remove the flywheel dust cover from the clutch housing. Support the transaxle using a suitable jack.
16. Remove the transaxle-to-engine attaching bolts, then the transaxle from the vehicle.

To install:

17. Install the transaxle, then the transaxle-to-engine attaching bolts. Torque bolts to 55 ft. lbs. (75 Nm).
18. Remove the engine support tool J-28467-A or its equivalent..
19. Install the flywheel dust cover, engine stiffener and center beam with the rear transaxle mount.
20. Install the left transaxle mount, then the front transaxle mount through bolt.
21. Torque the center crossmember bolts to 45 ft. lbs. (61 Nm). Torque the left transaxle mount bolts to 29 ft. lbs. (39 Nm) and the front transaxle through bolt to 64 ft. lbs. (87 Nm).
22. Install the torque rod and bracket, then the front exhaust pipe.
23. Install the right and left halfshafts, then the ball joints onto the steering knuckles.
24. Install the front undercovers, then the front tire and wheel assemblies.

25. Lower the vehicle and remove the engine support tool.
26. Connect the shift cable, clutch cable and speedometer cable.
27. Install the battery bracket and the ignition coil assembly.
28. Install the ignition coil ground cable onto the engine.
29. Install the engine harness wiring clamp, then the ground cable.
30. Connect all electrical connectors to the transaxle. Install the battery and battery tray.
31. Connect the battery cables. Adjust the shift cables, if necessary. Start engine and check for leaks.

Spectrum

1. Disconnect the negative battery terminal from the battery and the transaxle.
2. Disconnect the wiring connectors, speedometer cable, clutch cable and shift cables from the transaxle.
3. Remove the air cleaner heat tube.
4. Remove the upper transaxle-to-engine bolts.
6. Raise and support the vehicle safely. Drain the oil from the transaxle.
7. Remove the left front wheel assembly and splash shield.
8. Disconnect the left tie rod at the steering knuckle and the left tension rod.
9. Disconnect the halfshafts and remove the shafts by pulling them straight out from the transaxle, avoid damaging the oil seals.
10. Remove the dust cover at the clutch housing.
11. Using a floor jack, support the transaxle, then remove the transaxle-to-engine retaining bolts.
12. While sliding the transaxle away from the engine, carefully lower the jack, guiding the right halfshaft out of the transaxle.

NOTE: The right halfshaft must be installed into the transaxle when the transaxle is being mated to the engine.

To install:

12. When installing the transaxle, guide the right halfshaft into the shaft bore as the transaxle is being raised.
13. Install the transaxle-to-engine mounting bolts. Torque bolts to 55 ft. lbs. (75 Nm).
14. Install the left halfshaft into its bore on the transaxle.
15. Install the left tension rod and torque bolts to 80 ft. lbs. (108 Nm).
16. Install the tie rod to the steering knuckle.
17. Install the clutch housing dust cover bolts and the splash shield.
18. Install the tire and wheel assembly and lower the vehicle.
19. Install the remaining transaxle-to-engine attaching bolts. Torque bolts to 55 ft. lbs. (75 Nm).
20. Connect the ground cable at transaxle, clutch cable, speedometer cable and the battery negative cable.
21. Start engine and check for leaks.

Sprint

1. Disconnect the negative battery cable and the ground strap at the transaxle.
2. Remove the air cleaner and air pipe.
3. Remove the clutch cable from the clutch release lever.
4. Remove the starter and speedometer cable. Disconnect and tag the electrical wires and wiring harness from the transaxle.
5. Remove the front and rear torque rod bolts at the transaxle.
6. Raise and support the vehicle safely.
7. Drain the transaxle fluid.
8. Remove the exhaust pipe at the exhaust manifold and at the 1st exhaust hanger.
9. Remove the clutch housing lower plate.
10. Disconnect the gear shift control shaft and the extension rod at the transaxle.
11. Remove the left front wheel.
12. Using a prybar, pry on the inboard joints of the right and left hand halfshafts. This will detach the halfshafts from the snaprings of the differential side gears.
13. On the left side, remove the stabilizer bar mounting bolts and ball joint stud bolt. Push down on the stabilizer bar and remove the ball joint stud from the steering knuckle.
14. Pull the left halfshaft out of the transaxle.
15. Remove the front torque rod.
16. Secure and support the transaxle case with a jack.
17. Remove the transaxle-to-body mounting bolts and the mounts.
18. Remove the transaxle-to-engine mounting bolts.
19. Disconnect the transaxle from the engine by sliding it to the left side and lower the jack.

NOTE: When removing the transaxle, support the right halfshaft, so it does not become damaged.

To install:

20. Guide the right halfshaft into the transaxle and reverse the removal procedures. Torque the transaxle-to-engine bolts to 35 ft. lbs. (47 Nm).
21. Torque the transaxle-to-mount bolts to 34 ft. lbs. (46 Nm). Torque the mounting member bolts to 40 ft. lbs. (54 Nm).
22. Torque the stabilizer bar bolts to 30 ft. lbs. (41 Nm). Torque the ball joint stud bolt to 44 ft. lbs. (60 Nm). Adjust the clutch cable and refill the transaxle.
23. Install the clutch housing lower plate.
24. Install the exhaust pipe to the exhaust manifold and the hanger.
25. Connect the gear shift control shaft and the extension rod on the transaxle.
26. Connect the speedometer cable and connect all electrical connectors to the transaxle.
27. Install the air cleaner and air pipe. Connect the battery negative cable.

Metro

1. Disconnect the negative battery cable and the ground strap at the transaxle.
2. Remove the clutch cable adjusting nuts, retaining clip from the cable and cable from the bracket.
3. Disconnect and tag all the wiring harness clamps and connectors involved with the transaxle removal.
4. Remove the speedometer cable boot, speedometer case clip and speedometer cable from the case.
5. Remove the transaxle retaining bolts.
6. Remove the starter assembly and starter motor plate.
7. Remove the vacuum hose from the pressure sensor.
8. Install the engine support to prevent the engine from lowering excessively.
9. Raise and support the vehicle safely. Drain the transaxle oil.
10. Remove the gear shaft control shaft bolt and nut and detach the control shaft from the gear shift shaft.
11. Extension rod nut and remove the rod with washers.
12. Remove the exhaust pipe front and rear flange bolts.
13. Remove the clutch housing lower plate.
14. Remove the left front wheel.
15. Remove the left tie rod end.
16. Remove the left ball joint by removing the joint stud bolt.
17. Remove both halfshafts at the transaxle.
18. Support the transaxle with a suitable jack and remove the transaxle retaining bolts and nuts.
19. Remove the 2 rear engine mounting bolts.
20. Remove the 3 bolts and 2 nuts from the transaxle mounting left hand bracket, remove the left hand bracket.
21. Lower the transaxle with the engine attached in order to detach it from the stud bolt at the engine rear mounting portion. Pull the transaxle

straight out toward the left side to disconnect the input shaft from the clutch cover, lower and remove the transaxle assembly.

To install:

22. While the transaxle is being raised into its correct position, install the right hand halfshaft into the differential.
23. Install the transaxle along with the transaxle to engine nuts and bolts. Install the left hand bracket with its 3 bolts and 2 nuts. Torque them to 37 ft. lbs. (50 Nm).
24. Install the 2 rear engine mounting nuts and torque them to 37 ft. lbs. (50 Nm).
25. Lower the transaxle supporting jack. Torque the transaxle to engine bolt and nut to 37 ft. lbs. (50 Nm).
26. Install the left hand halfshaft to the transaxle. Be sure to push each driveaxle in fully to engage the snaprings with the differential gear.
27. Install the left ball joint and ball joint stud bolt. Torque the ball joint bolt and nut to 44 ft. lbs. (60 Nm).
28. Install the left tie rod end, castle nut and cotter pin. Torque the castle nut to 32 ft. lbs. (43 Nm).
29. Install the left front wheel.
30. Install the clutch housing lower plate.
31. Install the exhaust pipe front and rear flange nuts.
32. Install the extension rod nut and washers. Torque the rod nut to 24 ft. lbs. (33 Nm).
33. Install the control shaft to gear shift and install the gear shift control shaft bolt and nut. Torque the gear shift control shaft bolt and nut to 13 ft. lbs. (18 Nm).
34. Refill the transaxle with the recommended lubricant.
35. Lower the vehicle.
36. Remove the engine support fixture.
37. Install the vacuum hose to the pressure sensor.
38. Install the starter, starter motor plate and 2 bolts.
39. Install the transaxle retaining bolts. Torque the retaining bolts to 37 ft. lbs. (50 Nm).
40. Install the speedometer cable to case, speedometer case clip and speedometer cable boot.
41. Install the clutch cable bracket, retaining clip to cable and clutch cable adjusting nut. Adjust the clutch free-play as necessary.
42. Install the negative battery cable and the ground strap to the transaxle.

LINKAGE ADJUSTMENT

Nova

Adjustment of shift lever free-play is accomplished through the use of a selective shim installed in the bottom of the lower shift lever seat.

Select a shim of a thickness that allows a preload of 0.1–0.2 lbs. at the top of the lever and install it in the shift lever seat.

To install the shim, perform the following procedures:

1. Disconnect the negative terminal from the battery.
2. Remove the console and the shifter boot.
3. Remove the shifter cover, shift support and cap.
4. Remove the shifter spacer, shifter seat and the shim.
5. Install the new shim and reassemble the shifter.
6. Check the shifter free-play, using a pull scale, for the proper preload.
7. Repeat the procedure, if necessary.

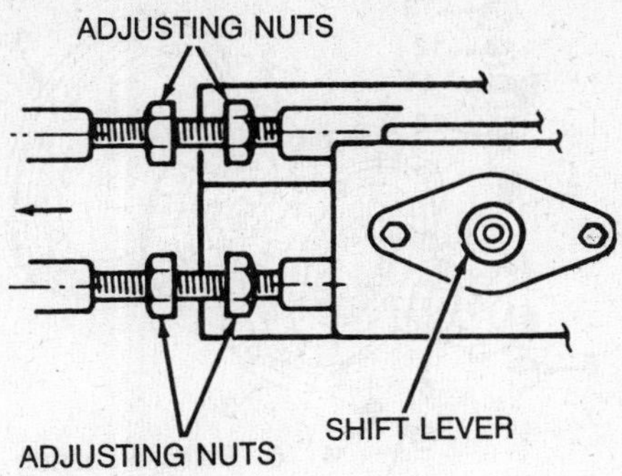

Adjusting the shift linkage of the transaxle – Spectrum

Spectrum and Storm

1. Loosen the adjusting nuts.
2. Place the transaxle and the shift lever in the **N** position.
3. Turn the adjusting nuts until the shift lever is in the vertical position.
4. Tighten the adjusting nuts.

Sprint and Metro

1. At the console, loosen the gear shift control housing nuts and the guide plate bolts.
2. Adjust the guide plate by displacing it toward the front and rear so that the gear shift control lever is brought in the middle of the guide plate at the right angle.
3. Once the guide plate is positioned properly, tighten the guide plate bolts to 7 ft. lbs. (9 Nm) and then the housing nuts to 4 ft. lbs. (5 Nm).

CLUTCH

Clutch Assembly

REMOVAL & INSTALLATION

Nova and Prizm

NOTE: Do not allow grease or oil to contaminate any of the disc, pressure plate or flywheel friction surfaces.

1. Remove the transaxle from the vehicle.
2. Remove the clutch cover and disc from the bellhousing.
3. Unfasten the release fork bearing clips. Withdraw the release bearing hub, complete with the release bearing.
4. Remove the tension spring from the clutch linkage.
5. Remove the release fork and support.
6. Punch matchmarks on the clutch cover (pressure plate) and flywheel so that the pressure plate can be returned to its original position during installation.
7. Slowly unfasten the screws which attach the retracting springs. Loosen each screw one turn at a time until the tension is released.

CAUTION

If the screws are released too quickly, the clutch assembly will fly apart, causing possible injury.

8. Separate the pressure plate from the clutch cover/spring assembly.
9. Inspect the parts for wear or deterioration. It is strongly recommended that all 3 components of the clutch system–disc, pressure plate and bearing–be replaced as a unit if any part is worn.
10. Inspect the flywheel for any signs of cracking, bluing in the steel (a sign of extreme heat) or scoring. Any bluing or cracks which are found require replacement of the flywheel. While the flywheel will not be perfectly smooth, it should be free of all but the slightest ridges and valleys or scores. A scored flywheel will immediately attack a new clutch disc, causing slippage and vibration.

To install:

11. When reassembling, apply a thin coating of multipurpose grease to the release bearing hub and release fork contact points. Also, pack the groove inside the clutch hub with multipurpose grease and lubricate the pivot points of the release fork.
12. Align the matchmarks on the clutch cover and flywheel which were

made during disassembly. Install the clutch and pressure plate assembly and tighten the retaining bolts just finger tight.

13. Center the clutch disc by using a clutch pilot tool or an old input shaft. Insert the pilot into the end of the input shaft front bearing, wiggle it gently to align the clutch disc and pressure plate and tighten the retaining bolts. The bolts should be tightened in 2 or 3 steps, gradually and evenly. Final bolt torque is 14 ft. lbs. (19 Nm).

14. Install the release bearing, fork and boot.

15. Reinstall the transaxle.

Spectrum and Storm

1. Remove the transaxle.
2. Install the pilot shaft tool J-35282 or equivalent, into the pilot bearing to support the clutch assembly during the removal procedures.

NOTE: Observe the alignment marks on the clutch and the clutch cover and pressure plate assembly. If the markings are not present, be sure to add them.

3. Loosen the clutch cover and pressure plate assembly retaining bolts evenly, one at a time, until the spring pressure is released.
4. Remove the clutch cover and pressure plate assembly and clutch plate.

NOTE: Check the clutch disc, flywheel and pressure plate for wear, damage or heat cracks. Replace all damaged parts.

5. Before installation, lightly lubricate the pilot shaft splines, pilot bearing and pilot release bearing surface with grease.
6. To install, reverse the removal procedures. Torque the clutch cover/pressure plate-to-flywheel bolts evenly to 13 ft. lbs. (18 Nm), to avoid distortion.

Sprint and Metro

1. Remove the transaxle.
2. Install tool J-34860 (J-37761 on the Metro) into the pilot bearing to support the clutch assembly.

NOTE: On the Sprint models, look for the X mark or white painted number on the clutch cover and the X mark on the flywheel. If there are no markings, mark the clutch cover and the flywheel for reassembly purposes.

3. Loosen the clutch cover-to-flywheel bolts, one turn at a time (evenly) until the spring pressure is released.
4. Remove the clutch cover and clutch disc.

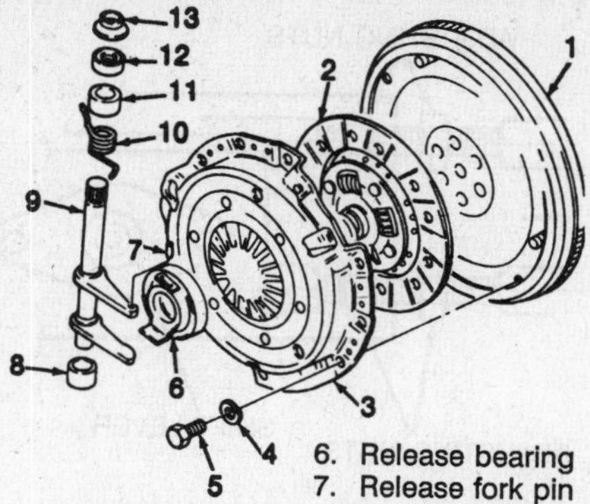

1. Flywheel
2. Disc
3. Clutch cover
4. Lock washer
5. Bolt
6. Release bearing
7. Release fork pin
8. No. 2 bushing
9. Release shaft
10. Return spring
11. No. 1 bushing
12. Shaft seal
13. Shaft cover

Exploded view of the clutch assembly – Sprint and Metro

5. Inspect the parts for wear, if necessary, replace the parts.
6. To install, reverse the removal procedures. Torque the clutch cover bolts to 18 ft. lbs. (23 Nm).

PEDAL HEIGHT/FREE-PLAY ADJUSTMENT

1. Check pedal height as measured from the insulating sheet on the floor to the front-center of the pedal; it should be 5.65–6.043 in. If the height is not correct, perform the following procedures:
   a. Remove the lower instrument finish panel and air duct.
   b. Loosen the locknut on the pedal stopper bolt, located at the top of the pedal.
   c. Turn the stopper bolt inward (to decrease) or outward (to increase) until it is within specifications.
   d. Tighten the locknut, recheck and readjust, if necessary.
2. To check and/or adjust the clutch pedal free-play, perform the following procedures:
   a. Measure the clutch pedal height.
   b. Push the pedal until increased resistance is felt as the clutch pressure plate springs begin to be compressed. Measure the pedal at this point, then, subtract the smaller figure from the larger one; this is the free-play dimension.

NOTE: The free-play dimension should be 0.20–0.59 in.

   c. If necessary to adjust the free-play, loosen the pushrod locknut, located between the pedal and the clutch master cylinder. Turn the pushrod, clockwise to decrease or counter-clockwise to increase, until the dimension is within specifications.
   d. Tighten the locknut, recheck and readjust, if necessary.

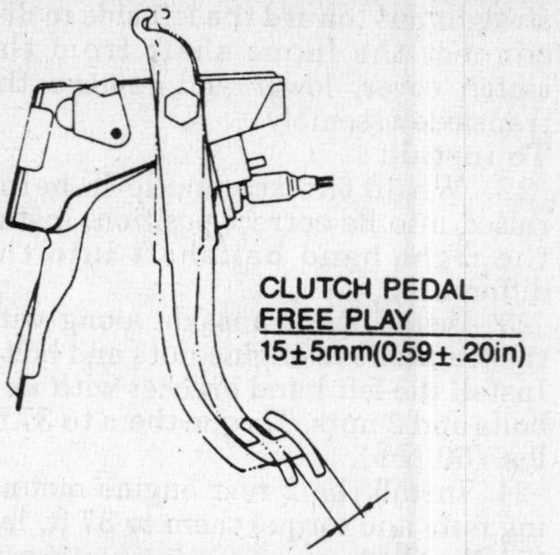

Clutch pedal free-play adjustment – Spectrum and Storm

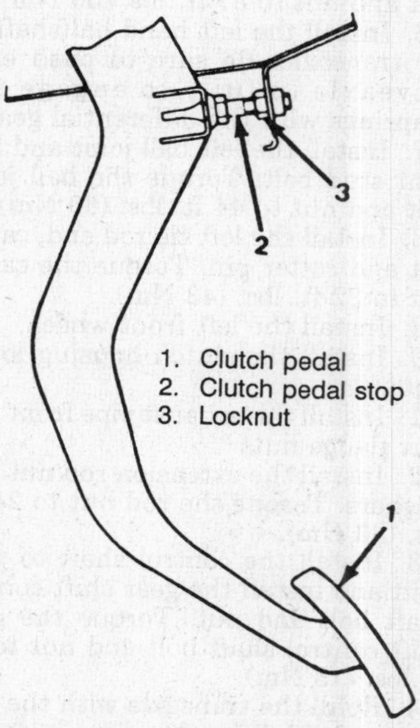

Clutch pedal height adjustment – Sprint and Metro

Spectrum and Storm

1. Disconnect the negative battery terminal from the battery.
2. Loosen the adjusting nut and pull the cable to the rear until it turns freely.
3. Adjust the cable length by turning the adjusting nut.
4. When the clutch pedal free-play travel reaches 0.39–0.79 in. (0.20-0.59 in. on the Storm) release the cable.
5. When the adjustment has been completed, tighten the locknut.

Sprint and Metro

1. At the transaxle, move the clutch release arm to check the free-play, it should be 0.08–0.16 in. (0.06-0.08 in. Metro).
2. If necessary, turn the clutch cable joint nut to adjust the cable length.

NOTE: The clutch pedal height should be adjusted so the clutch pedal is the exact same height as the brake pedal. The pedal is adjusted at the stop bolt on the upper end of the pedal pivot.

Clutch Cable

REMOVAL & INSTALLATION

Spectrum and Storm

1. Disconnect the negative battery terminal from the battery.
2. Loosen the clutch cable adjusting nuts. Disconnect the cable from the release arm and cable bracket.
3. At the clutch pedal, remove the cable retaining bolt.
4. Disconnect the cable from the front of the dash.
5. Remove the clutch cable from the vehicle.
6. To install, grease the clutch cable pin and reverse the removal procedures.
7. Adjust the pedal free-play.

Sprint and Metro

1. Disconnect the negative battery cable.
2. Remove the clutch cable joint nut and disconnect the cable from the release arm.
3. Remove the clutch cable bracket mounting nuts and remove the bracket from the cable.
4. Remove the cable retaining bolts at the clutch pedal.
5. Remove the cable from the vehicle.

To install:

6. Before installation, apply grease to the hook and pin end of the cable.
7. Connect the cable to the clutch pedal and install the retaining bolts.
8. Install the clutch cable bracket on the cable.
9. Position the bracket on the transaxle and install the mounting bolts.
10. Connect the cable to the release lever and install the joint nut on the cable.
11. Adjust the pedal free-play as previously outlined and connect the negative battery cable.

Clutch Master Cylinder

REMOVAL & INSTALLATION

Nova and Prizm

1. Drain or siphon the fluid from the master cylinder.
2. Disconnect the hydraulic line to the clutch from the master cylinder.

NOTE: Do not spill brake fluid on the painted surfaced of the vehicle.

3. Inside the car, remove the underdash panel and the air duct.
4. Remove the pedal return spring.
5. Remove the spring clip and clevis pin.
6. Unfasten the bolts which secure the master cylinder to the firewall. Withdraw the assembly from the firewall side.

To install:

7. Install the master cylinder with its retaining nuts to the firewall.
8. Connect the line from the clutch to the master cylinder.
9. Connect the clevis and install the clevis pin and spring clip.
10. Install the pedal return spring.
11. Fill the reservoir with clean, fresh brake fluid and bleed the system.
12. Check the cylinder and the hose connection for leaks.
13. Adjust the clutch pedal.
14. Reinstall the air duct and underdash cover panel.

Clutch Slave Cylinder

REMOVAL & INSTALLATION

Nova and Prizm

NOTE: Do not spill brake fluid on the painted surface of the vehicle.

1. Raise the front of the vehicle and support it safely.
2. If necessary, remove the rear splash shield to gain access to the release cylinder.
3. Remove the clutch fork return spring.
4. Unfasten the hydraulic line from the release cylinder by removing its retaining nut.
5. Remove the release cylinder retaining nuts and remove the cylinder.

To install:

6. Reinstall the cylinder to the clutch housing and tighten the bolts to 9 ft. lbs. (12 Nm).
7. Connect the hydraulic line and tighten it to 11 ft. lbs. (15 Nm).
8. Install the clutch release spring.
9. Bleed the system and remember to top up the fluid in the master cylinder when finished.
10. Install the splash shield if it was removed.
11. Lower the vehicle to the ground.

Hydraulic Clutch System Bleeding

Nova and Prizm

1. Fill the clutch master cylinder reservoir with brake fluid.

NOTE: Do not spill brake fluid on the painted surface of the vehicle.

2. Fit a vinyl bleeder tube over the bleeder screw at the front of the slave cylinder and place the other end in a clean jar half filled with brake fluid.
3. Have an assistant depress the clutch pedal several times. Loosen the bleeder screw and allow the fluid to flow into the jar.
4. Tighten the screw and have the assistant release the clutch pedal.
5. Repeat bleeding procedure until no air bubbles are present in the fluid.
6. Refill the master cylinder to the specified level, check the system for leaks, then, adjust the clutch pedal height and free-play.

AUTOMATIC TRANSAXLE

For further information on transmissions/transaxles, please refer to "Chilton's Guide to Transmission Repair".

Transaxle Assembly

REMOVAL & INSTALLATION

Nova

WITH SINGLE OVERHEAD CAM ENGINE AND A131L TRANSAXLE

1. Disconnect the negative battery cable
2. Remove the air intake tube.
3. Disconnect the speedometer cable and neutral start wiring connector at the tansaxle. Loosen the lug nuts on the left front wheel.
4. Disconnect the thermostat housing at the tansaxle.
5. Disconnect the ground cable at the tansaxle.
6. Remove one upper mount to bracket bolt.
7. Label and disconnect any interfering wiring in the area of the transmission case.

8. Disconnect the throttle valve cable at the carburetor.
9. Remove 2 upper bellhousing bolts.
10. Support the engine using either GM tool J-28467-A or equivalent.
11. Raise the vehicle and support it safely.
12. Remove the left front wheel.
13. Remove the splash shields under the vehicle.
14. Remove the center beam.
15. Disconnect the shift cable at the tansaxle.
16. Remove the shift cable bracket.
17. Remove the cooler bracket and disconnect the cooler lines at the transaxle. Plug the lines to prevent leakage.
18. Remove the inspection cover on the bell housing and remove the torque converter bolts through the inspection hole.
19. Disconnect the left control arm at the ball joint.
20. Disconnect the right control arm at the ball joint.
21. Remove the right and left halfshafts from the transaxle.
22. Disconnect the starter bolts
23. Remove the 3 rear transaxle bolts.
24. Support the transaxle with a floor jack.
25. Remove the remaining mount bolts and remove the remaining bellhousing bolts.
26. Remove the transaxle from the vehicle. Handle the unit carefully and do not allow it to tip; the torque converter may fall from the unit and become damaged.**To install:**
27. Before reinstallation, install the torque converter in the tansaxle. If the converter has been drained, refill it with 2.3 US qts. of Dexron® II ATF.
28. Check the torque converter installation. Use a straightedge and calipers to check the distance between the face of the converter and the surface of the transaxle housing. Correct free space is more than 20mm.
29. Apply multi-purpose grease to the contact points of the torque converter and the crankshaft.
30. Place the transaxle on the jack and lift into position. Align the guide pin with one of the holes in the driveplate (flywheel) and align the 2 pins on the block with the converter housing holes. Temporarily install one bolt to hold everything together.
31. Install all of the tansaxle housing bolts and tighten them. The larger bolts (12mm) should be tightened to 47 ft. lbs. (64 Nm) and the smaller (10mm) bolts tightened to 34 ft. lbs. (46 Nm).
32. Install the left engine mount and tighten the bolts to 38 ft. lbs. (52 Nm).
33. Install the 6 torque converter bolts. Remove the guide pin and install the white bolt, then install the 5 yellow bolts. Tighten the bolts evenly and to 13 ft. lbs. (18 Nm).
34. Install the inspection plate (engine rear cover plate).
35. Reinstall the starter.
36. Connect the halfshafts to the transaxle.
37. Install the center engine support and tighten the bolts to 29 ft. lbs. (39 Nm).
38. Install the front and rear mounts, tightening the 4 bolts to 29 ft. lbs. (39 Nm).
39. Connect the left and right control arms to the ball joints.
40. Reinstall the thermostat housing.
41. Connect the oil cooler lines to the transaxle.
42. Connect and adjust the shift control cable.
43. Install the speedometer cable.
44. Install and adjust the throttle valve cable.
45. Connect the neutral start switch connector. Connect or reposition any wiring which was moved to gain access during removal.
46. Install the air cleaner.
47. Fill the transaxle with the correct amount of Dexron® II ATF. If the transaxle has been disassembled and drained, install 5.8 US qts. If the transaxle was not disassembled and still contains fluid, add 2.4 US qts.
48. Double check all installation items, paying particular attention to loose hoses or hanging wires, untightened nuts, poor routing of hoses and wires (too tight or rubbing) and tools left in the engine area.
49. Install the splash shields below the engine. Install the left front wheel and lower the vehicle to the ground.
50. Connect the negative battery cable.
51. Start the engine and allow it to idle. With the wheels blocked front and rear and the brakes engaged, engage each gear and check that the transxle engages properly. Check that the vehicle will not roll in **PARK**.
52. Allow the engine to idle with the transaxle in **PARK**; check the fluid level on the dipstick. Check all assembly points for any sign of leakage.

WITH TWINCAM ENGINE AND A240E TRANSAXLE

1. Disconnect the negative battery cable.
2. Remove the air cleaner assembly.
3. Disconnect the neutral start switch.
4. Disconnect the solenoid valve wiring connector.
5. Disconnect the speed sensor.
6. Disconnect the speedometer cable.
7. Remove the water inlet assembly.
8. Remove the throttle cable from the throttle linkage.
9. Remove the transaxle dipstick tube.
10. Disconnect the battery ground cable from the transaxle housing.
11. Loosen, but do not remove the lug nuts on the left front wheel. Raise the vehicle and safely support it.
12. Drain the transaxle oil.
13. Remove the splash shields from under the engine.
14. Remove the shift control cable and brackets.
15. Disconnect the oil cooler lines at the transaxle. Plug the lines to prevent leakage.
16. Remove the front and rear engine mount bolts.
17. Remove the center engine mount member.
18. Carefully disconnect the oxygen sensor connector and remove the front exhaust pipe.
19. Disconnect and remove the left and right halfshafts from the transaxle.
20. Remove the left front wheel.
21. Remove the cotter pin, lock nut cap, lock nut and washer.
22. Remove the brake caliper assembly and support it with stiff wire out of the way. Do not disconnect the hose. Do not allow the caliper to hang by the hose.
23. Remove the brake disc.
24. Remove the steering knuckle from the lower control arm.
25. Remove the left halfshaft.
26. Remove the starter motor and the bracket (stiffener plate).
27. Remove the engine rear cover plate (inspection plate).
28. Remove the torque converter mounting bolts.
29. Lower the vehicle to the ground.
30. Support the engine using GM tool J-28467 or equivalent.
31. Remove the 3 bolts from the rear engine mount.
32. Raise and safely support the vehicle. Remember that the engine support is in place; make necessary adjustments during jacking.
33. Support the transaxle with a jack. Remove the transaxle, keeping it level.
34. Remove the torque converter.

To install:

35. Before reinstallation, install the torque converter in the tansaxle. If the converter has been drained, refill it with 2.3 US qts. of Dexron® II ATF.
36. Check the torque converter installation. Use a straightedge and calipers to check the distance between the face of the converter and the surface of the transaxle housing. Correct free space is more than 20mm.

37. Apply multi-purpose grease to the contact points of the torque converter and the crankshaft.
38. Place the transaxle on the jack and lift into position. Use a guide pin to align the holes in the driveplate (flywheel) and align the 2 pins on the block with the converter housing holes. Temporarily install one bolt to hold everything together.
39. Install all of the tansaxle housing bolts and tighten them. The larger bolts (12mm) should be tightened to 47 ft. lbs. (64 Nm) and the smaller (10mm) bolts tightened to 34 ft. lbs. (46 Nm).
40. Install the 3 engine rear mount bolts and tighten them to 38 ft. lbs. (52 Nm).
41. Install the 6 torque converter bolts. Remove the guide pin and install the white bolt, then install the 5 yellow bolts. Tighten the bolts evenly and to 20 ft. lbs. (27 Nm).
42. Install the inspection plate (engine rear cover plate).
43. Reinstall the stiffener plate and the starter.
44. Install the left halfshaft with its washer, lock nut, lock nut cap and cotter pin.
45. Reconnect the steering knuckle to the lower control arm.
46. Install the brake disc. Install the caliper and tighten the mounting bolts to 18 ft. lbs. (24 Nm).
47. Install the left front wheel; tighten the wheel nuts to 76 ft. lbs. (103 Nm).
48. Connect both halfshafts to the transaxle.
49. Reinstall the front exhaust pipe using new gaskets. Connect the wiring to the oxygen sensor.
50. Install the engine center mounting member. Tighten the bolts to 29 ft. lbs. (39 Nm).
51. Install the bolts for the front and rear engine mounts. Correct torque is 29 ft. lbs. (39 Nm). When the mount bolts are secure, the engine support apparatus may be removed.
52. Connect the oil cooler lines to the transaxle.
53. Install the shift control cable and brackets.
54. Replace the splash shields under the engine.
55. Lower the vehicle to the ground.
56. Connect the ground strap to the transaxle housing.
57. Install the transaxle dipstick tube.
58. Install the water inlet assembly.
59. Attach the throttle cable to the linkage and adjust as necessary.
60. Install the speedometer cable and the speed sensor.
61. Connect the wiring to the solenoid valve.
62. Connect the wiring to the neutral start switch.
63. Install the air cleaner.
64. Fill the transaxle with the correct amount of Dexron® II ATF. If the transaxle has been disassembled and drained, install 7.6 US qts. If the transaxle was not disassembled and still contains fluid, add 3.1 US qts.
65. Double check all installation items, paying particular attention to loose hoses or hanging wires, untightened nuts, poor routing of hoses and wires (too tight or rubbing) and tools left in the engine area.
66. Connect the negative battery cable.
67. Start the engine and allow it to idle. With the wheels blocked front and rear and the brake depresses, engage each gear and check that the transxle engages properly. Check that the vehicle will not roll in **PARK**.
68. Allow the engine to idle with the transaxle in **PARK**; check the fluid level on the dipstick. Check all assembly points for any sign of leakage.

Prizm

1. Support the engine with GM tool J-28467-A or equivalent.
2. Remove the battery hold-down, battery and tray.
3. Remove the air cleaner assembly
4. Disconnect the connector at the neutral start switch.
5. Disconnect the ground strap from the transaxle case.
6. Remove the throttle cable from the throttle body.
7. Disconnect the cooling lines at the transaxle and plug the lines to prevent leakage.
8. Remove the shift cable retainer and end clip.
9. Remove the shift cable from its brackets and place the cable out of the way.
10. Remove the upper shift cable bracket from the transaxle housing.
11. Remove the left transaxle mount brace.
12. Remove the through bolt in the mount.
13. Remove the 2 upper transaxle-to-engine bolts.
14. Remove the upper starter bolts.
15. Disconnect the speedometer cable.
16. Raise the vehicle and safely support it.
17. Remove the splash shields from below the vehicle.
18. Drain the transaxle oil.
19. Disconnect the electrical connections to the starter. Remove the bottom starter bolt and remove the starter.
20. Remove the halfshafts from the transaxle.
21. Remove the 3 bolts holding the center crossmember to the radiator support.
22. Remove the covers from the front and center mount bolts; remove the 2 front mount bolts, then the 2 center mount bolts.
23. Remove the 2 nuts from the rear mount.
24. Remove the 2 bolts holding the center crossmember to the main crossmember.
25. Remove the 3 nuts holding the exhaust hanger bracket and remove the bracket.
26. Use a floor jack to support the main crossmember. Check its placement carefully so that the crossmember will be balanced during removal.
27. Remove the 8 bolts holding the main crossmember to the underbody.
28. On the left and right sides, remove the 2 bolts holding the control arm bracket to the underbody.

CAUTION

Lower the main crossmember slowly while holding onto the center crossmember. The center crossmember is free to fall at this point and could cause serious injury or damage.

29. At the front mount, remove the through bolt and the mount.
30. Remove the front mount bracket from the transaxle.
31. Remove the center mount from the transaxle with its 2 bolts.
32. Remove the torque converter bolt shield and remove the bolts holding the torque converter to the flywheel.
33. Remove the 2 bolts holding the lower transaxle mount to the bracket.
34. Lower the vehicle to the ground.
35. Remove the remaining transaxle mount-to-bracket bolt.
36. Adjust the engine support apparatus to lower the transaxle and gain clearance for removal.
37. Disconnect the transaxle mount.
38. Raise and safely support the vehicle.
39. Use a floor jack to support the transaxle. Make sure it is properly positioned.
40. Remove the front and rear lower bolts holding the transaxle to the engine.
41. Carefully remove the transaxle, keeping it as level as possible. Remove the torque converter.**To install:**
42. Before installing the transaxle, make sure that the torque converter is all the way into the pump and bushing. Use a straightedge across the front of the transaxle case and measure the inset to the converter. Clearance must be at least 20mm.
43. Lightly coat the converter to flywheel contact point with multi-purpose grease.

44. Support the transaxle on the floor jack and move it into place. Install the transaxle and hold it in place with the front and rear lower transaxle-to- engine bolts. Tighten the bolts snugly; they will be final tightened later.
45. Remove the jack from under the transaxle.
46. Install the starter and its lower bolt snugly. Connect the starter wiring.
47. Lower the vehicle to the ground.

NOTE: The engine support apparatus has been repositioned and the crossmembers are removed. Be very careful when moving the vehicle.

48. Install the left transaxle mount.
49. Install the bolt for the upper mount bracket and the 2 bolts for the lower bracket snugly.
50. Adjust the engine support apparatus to raise the transaxle into position.
51. Install the through bolt for the transaxle mount and tighten it to 64 ft. lbs. Tighten the upper transaxle mounting bracket bolt to 45 ft. lbs.
52. Install the brace at the mount and tighten the bolts to 13 ft. lbs.
53. Install the 2 upper transaxle-to-engine bolts snugly.
54. Install the upper starter bolt snugly.
55. Tighten all the transaxle-to-engine bolts to 34 ft. lbs. (46 Nm).
56. Tighten the upper starter bolt to 29 ft. lbs. (40 Nm).
57. Install the speedometer cable.
58. Install the upper shift cable bracket, position the shift cable into the bracket and install the cable retainer and end clips.
59. Install the oil cooling lines.
60. Connect the throttle cable at the throttle body.
61. Install the ground strap to the transaxle case.
62. Install the connector to the neutral start switch.
63. Install the air cleaner assembly.
64. Raise and safely support the vehicle.
65. Tighten the 2 lower transaxle mount bolts to 45 ft. lbs. (61 Nm).
66. Tighten the lower starter bolt to 29 ft. lbs. (40 Nm).
67. Install the torque converter to flywheel bolts and tighten them to 31 ft. lbs. (42 Nm).
68. Install the converter bolt shield.
69. Install the center mount with two bolts to the transaxle; tighten the bolts to 45 ft. lbs. (61 Nm).
70. Install the front mount bracket on the transaxle and tighten the bolts to 13 ft. lbs. (18 Nm).
71. Install the front mount. Loosely install the through bolt.

NOTE: As the front mount is installed, the weight on the mount must go towards the transaxle; this allows the mount to align properly.

72. Position the center crossmember over the center and rear transaxle mount studs. Start 2 nuts on the center mount.
73. Loosely install the 3 bolts which hold the center crossmember to the radiator support.
74. Loosely install the 2 front mount bolts.
75. Raise the main crossmember into position over the rear mounting studs.
76. Align all the under body bolts and bolt holes.
77. Install the 2 rear mount nuts loosely.
78. Loosely install the 8 bolts to hold the main crossmember to the underbody.
79. On each side, loosely install the 2 bolts holding the control arm brackets to the underbody.
80. Loosely install the 2 bolts holding the center crossmember to the main crossmember.
81. Install the exhaust hanger bracket and its 3 nuts. The crossmembers, mounts, through bolts and brackets should now all be in place and held loosely by their nuts and bolts. If any repositioning is necessary, do so now.
82. Tighten the components below in the order listed to the correct torque specification:
 a. Main crossmember to underbody bolts: 152 ft. lbs. (206 Nm).
 b. Lower control arm bolts: 94 ft. lbs. (127 Nm).
 c. Center crossmember to radiator support bolts: 45 ft. lbs. (61 Nm).
 d. Front, center and rear mount bolts: 45 ft. lbs. (61 Nm).
 e. Exhaust hanger bracket nuts: 9.5 ft. lbs. (13 Nm).
 f. Front mount through bolt: 64 ft. lbs. (87 Nm).
83. Install the covers on the front and center mount bolts.
84. Reinstall the drive axles.
85. Fill the transaxle with the correct amount of Dexron® II ATF. If the transaxle has been disassembled and drained, install 5.8 US qts. If the transaxle was not disassembled and still contains fluid, add 2.4 US qts.
86. Double check all installation items, paying particular attention to loose hoses or hanging wires, untightened nuts, poor routing of hoses and wires (too tight or rubbing) and tools left in the engine area.
87. Install the splash shields below the engine. Install the left front wheel and lower the vehicle to the ground.
88. Remove the engine support apparatus.
89. Install the battery tray, battery and hold-down.
90. Start the engine and allow it to idle. With the wheels blocked front and rear and the brake depressed, engage each gear and check that the transaxle engages properly. Check that the vehicle will not roll in **PARK**.
91. Allow the engine to idle with the transaxle in **PARK**; check the fluid level on the dipstick. Check all assembly points for any sign of leakage.

Storm

1. Disconnect the battery cables, then remove the battery and tray from the vehicle.
2. Remove the intake air duct and breather tube from the air cleaner assembly.
3. Disconnect the electrical connectors, shift cable, shift cable bracket, breather hose and speedometer cable from the transaxle.
4. Disconnect the vacuum diaphragm hose from the vacuum diaphragm, if equipped.
5. Install engine support tool J–28467–A or equivalent, then remove the left transaxle through bolt.
6. Remove 4 transaxle-to-engine attaching bolts, then raise and support the vehicle safely.
7. Remove the right and left undercovers, then the front wheel and tire assemblies.
8. Disconnect the left control arm from the steering knuckle, then drain the transaxle fluid.
9. Remove the halfshafts and suspend on a wire.
10. Remove the front transaxle mount through bolt, then the dampener from the rear mount through bolt.
11. Remove the rear transaxle mount through bolt, then the 2 front center crossmember mounting bolts.
12. Remove the front exhaust pipe-to-exhaust manifold attaching nuts, then 2 rear front pipe bolts. Disconnect the front pipe from the exhaust manifold.
13. Remove the 2 rear center crossmember mounting bolts, then the crossmember from the vehicle. Lower the vehicle.
14. Lower the engine using engine support tool J–28467 or equivalent.
15. Raise and safely support the vehicle, then position a suitable jack under the transaxle.
16. Remove the rear mount to transaxle case bolt, the front mount attaching bolt.
17. Remove the front mounting bracket attaching bolts, then the front mounting bracket from the engine.
18. Remove the flywheel cover, then the flywheel-to-torque converter attaching bolts.

19. Disconnect the oil cooler lines from the transaxle, then remove the 2 rear transaxle mount bolts. Lower the transxale from the vehicle.

To install:

20. Raise the transaxle into position, then install the 2 rear transaxle mounting bolts. Remove transaxle jack.
21. Lower the vehicle, then install the 4 transaxle-to-engine mounting bolts.
22. Using engine support tool J–28467–A or equivalent, raise the engine slightly.
23. Install the through bolt on the left transaxle mount. Torque through bolt to 64 ft. lbs. (87 Nm).
24. Raise and safely support the vehicle, then install the flywheel-to-torque converter attaching bolts. Torque converter bolts to 31 ft. lbs. (42 Nm).
25. Install the flywheel cover, then connect the oil cooler lines to the transaxle. Torque oil cooler line trunion bolts tov 10 ft. lbs. (15 Nm).
26. Install the front transaxle mount nut and bolt. Torque bolt to 45 ft. lbs. (61 Nm).
27. Install the rear mount through bolt. Torque rear transaxle mount attaching bolts to 29 ft. lbs. (39 Nm) and the rear mount through bolt to 64 ft. lbs. (87 Nm).
28. Install the center crossmember. Torque crossmember bolts to 45 ft. lbs. (61 Nm).
29. Install the front mount through bolt. Torque bolt to 64 ft. lbs. (87 Nm).
30. Install the dampener on the rear mount through bolt.
31. Install the front pipe to the exhaust manifold, the 2 rear front pipe attaching bolts.
32. Install the halfshafts.
33. Install the left control arm onto the steering knuckle, then the right and left under covers.
34. Install the tire and wheel assemblies, then lower the vehicle. Torque the engine-to-transaxle attaching bolts to 31 ft. lbs. (42 Nm).
35. Remove the engine support tool, then install the speedometer to the transaxle.
36. Install the breather hose, shaft cable bracket and shift cable to the transaxle.
37. Install the vacuum hose to the vacuum diaphragm, if equipped.
38. Connect the electrical connectors to the transaxle.
39. Install the air breather tube and the air intake duct onto the air cleaner assembly.
40. Install the battery and battery tray. Connect the battery cables and check transaxle fluid. Start engine and check for leaks.

Spectrum

1. Disconnect the negative battery terminal from the battery.
2. Remove the air duct tube from the air cleaner.
3. From the transaxle, disconnect the shift cable, speedometer cable, vacuum diaphragm hose, engine wiring harness clamp and the ground cable.
4. At the left-fender, disconnect the inhibitor switch and the kickdown solenoid wiring connectors.
5. Disconnect the oil cooler lines from the transaxle.
6. Remove the 3 upper transaxle-to-engine mounting bolts. Raise and support the vehicle safely.
7. Remove both front-wheels and the left-front fender splash shield.
8. Disconnect both tie rod ends at the steering knuckles.
9. Remove both front tension rod brackets and disconnect the rods from the control arms.
10. Disengage the halfshafts from the transaxle.
11. Remove the flywheel dust cover and the converter-to-flywheel attaching bolts.
12. Remove the transaxle rear mount through bolt.
13. Disconnect the starter wiring and the starter. Support the transaxle.
14. Remove the lower transaxle-to-engine mounting bolts and remove the transaxle.

To install:

15. Install transaxle onto engine and converter to the flywheel. Torque the converter-to-flywheel at 30 ft. lbs. (41 Nm), the transaxle-to-engine at 56 ft. lbs. (76 Nm).
16. Install the flywheel dust cover.
17. Install both halfshafts into the transaxle.
18. Install the tension rod brackets and both tie rod ends to the steering knuckle.
19. Install the splash shield to the left front fender and the tire and wheel assemblies.
20. Connect the transaxle cooler lines.
21. Connect all electrical connectors and hoses that were disconnected during the removal.
22. Connect the speedometer cable and shift cable to the transaxle.
23. Connect the battery negative cable, adjust the shift linkage and fill the transaxle with Dexron® II automatic tansaxle fluid. Start engine and check for leaks.

Sprint and Metro

1. Disconnect the air suction guide from the air cleaner.
2. Disconnect the negative and the positive battery cables.
3. Remove the battery and the battery bracket tray.
4. Remove the negative cable from the transaxle.
5. Disconnect the solenoid wire coupler and the shift lever switch wire couplers.
6. Remove the wiring harness from the transaxle.
7. Remove the speedometer cable from the transaxle.
8. Disconnect the oil pressure control cable from the accelerator cable and the accelerator cable from the transaxle.
9. Remove the select cable from the transaxle.
10. Remove the starter motor.
11. Drain the transaxle fluid.
12. Disconnect the oil outlet and inlet hoses from the oil pipes. After disconnecting, plug the 2 oil hoses to prevent fluid in the hoses and oil cooler from draining.
13. Raise the vehicle and support it safely.
14. Remove the exhaust No. 1 pipe.
15. Remove the clutch housing lower plate.
16. Remove the 6 driveplate bolts. To lock the driveplate, engage a prybar with the driveplate gear through the notch provided at the under side of the transaxle case.
17. Remove the left hand front halfshaft.
18. Detach the inboard joint of the right hand halfshaft from the differential.
19. Disconnect the transaxle mounting member.
20. Securely support the transaxle with a suitable jack for removal.
21. Remove the transaxle left mounting.
22. Remove the bolts fastening the engine and the transaxle.
23. Disconnect the transaxle from the engine by sliding towards the left side and carefully lower the jack.

NOTE: When removing the transaxle assembly from the engine, move it in parallel with the crankshaft and use care so as not to apply excessive force to the driveplate and torque converter. After removing the transaxle assembly, be sure to keep it so the oil pan is at the bottom. If the transaxle is tilted, fluid in it may flow out.

To install:

24. Reverse the removal procedure noting the following important steps.
25. Before installing the transaxle assembly apply grease around the cup at the center of the torque converter. Then measure the distance between the torque converter flange nut and

the transaxle case housing. The distance should be more than 0.85 in. (21.4mm). If the distance is less than 0.85 in. (21.4mm), the torque converter has been installed incorrectly and must be removed and reinstalled correctly.

26. When installing the transaxle, guide the right halfshaft into the differential side gear as the transaxle is being raised.

27. After inserting the inboard joints of the right hand and left hand halfshafts into the differential side gears, push the inboard joints into the side gears until the snaprings on the halfshafts engage the side gears.

28. After connecting the oil pressure control cable to the accelerator cable, check the oil pressure control cable play and adjust if necessary.

29. Install the select cable.

30. Refill the transaxle and check the fluid level.

31. Tighten the following bolts and nuts to specifications.

a. Drive plate bolts—14 ft. lbs. (19 Nm).

b. Mounting member bolts‹m-dash›40 ft. lbs. (55 Nm).

c. Transaxle mounting nuts‹m-dash›33 ft. lbs. (45 Nm).

d. Transaxle mounting bolts (8mm)—40 ft. lbs. (55 Nm).

e. Transaxle mounting nuts (8mm)—40 ft. lbs. (55 Nm).

f. Stabilizer shaft mounting bolts—31 ft. lbs. (42 Nm).

g. Ball stud bolt—44 ft. lbs. (60 Nm).

h. Wheel nuts—40 ft. lbs. (55 Nm).

SHIFT LINKAGE ADJUSTMENT

Nova and Prizm

1. Loosen the swivel nut on the lever.

2. Push the manual lever fully towards the right side of the vehicle.

3. Return the lever 2 notches to the **N** position.

4. Set the shift lever in the **N** position.

5. While holding the lever lightly towards the **R** position, tighten the swivel nut.

6. On the 1991-92 Prizm, us the following procedure;

a. Loosen the cap nut on the manual lever.

b. Position the manual lever in the **NEUTRAL** range.

c. Position the shift selector lever in the **NEUTRAL** range.

d. Tighten the cap nut to 89 in. lbs. (10 Nm).

Spectrum

1. Loosen the 2 adjusting nuts at the control rod link and connect the shift cable to the link on the transaxle.

2. Shift the transaxle into the **N** detent.

3. Place the shifter lever into the **N** position.

4. Rotate the link assembly clockwise to remove slack in the cable.

5. Tighten the rear adjusting nut until it makes contact with the link. Tighten the front adjusting nut until it makes contact with the link and tighten the adjusting nuts.

Storm

1. Place the ignition switch in the **LOCK** position.

2. Set the selector lever in the **PARK** position.

3. Loosen the adjuster nuts at the transaxle.

4. Be sure that the shift lever at the transaxle is in the **PARK** position.

5. Pull the cable forward and tighten the forward adjuster nut until it contacts the shift lever.

6. Tighten the rear nut until it comes in contact with the shift lever.

7. Tighten both adjuster nuts.

8. To adjust the brake drive cable, use the following procedure:

a. Place the shift lever to the **PARK** position.

b. Place the ignition switch to the **LOCK** position.

c. Pull the cable forward at the shift lever bracket and tighten the forward adjuster nut until it makes contact with the bracket.

d. Tighten the rear nut until it makes contact with the shift lever bracket and then tighten both the adjuster nuts.

Sprint and Metro

1. Place the shift lever in the **NEUTRAL** position.

2. Turn the adjusting nut in until it contacts the manual select cable joint.

3. Tighten the locknut.

4. To adjust the interlock cable (back drive cable), use the following procedure.

a. Shift the selector to the **PARK** position.

b. Loosen the adjusting and locknut on the interlock cable.

c. Pull the outer wire (interlock cable) forward so there is no deflection on the inner wire, tighten the adjusting nut hand tight only and then tighten the locknut.

NOTE: After tightening the nuts, make sure that with the shifter lever shifted to the PARK position, the ignition key can be turned from the ACC to the LOCK position and the key can be removed from the ignition switch. With the selector lever shifted to any range other than the PARK position, the ignition key can not be turned from the ACC to LOCK position.

5. On the Metro adjust the shift lock solenoid so that it will operate as follows:

a. When the ignition switch is turned **OFF**, the solenoid is not operated.

b. When the ignition switch is turned **ON**, and the brake pedal is depressed, the solenoid is operated and the lock plate is positioned properly.

c. There is no clearance between the lock plate and the guide plate.

d. If the manual release knob is pulled when the ignition switch is turned **OFF**, the selector lever can be shifted from the **P** range to any other range.

NOTE: After tightening the solenoid retaining nuts, make sure that with the shifter lever shifted to the PARK position, the ignition key can be turned from the ACC to the LOCK position and the key can be removed from the ignition switch.

THROTTLE LINKAGE ADJUSTMENT

Nova and Prizm

1. With the ignition **OFF**, depress the accelerator pedal all the way. On carbureted vehicles, check that the throttle plates are fully open. If they are not, adjust the throttle linkage.

2. Peel the rubber dust boot back from the throttle valve cable.

3. Loosen the adjustment nuts on the throttle cable bracket (rocker cover) just enough to allow cable housing movement.

4. Have an assistant depress the accelerator pedal fully.

5. Adjust the cable housing so that the distance between its end and the cable stop collar is 0-0.4 in. (0‹n-dash›1.0mm.

6. Tighten the adjustment nuts. Make sure that the adjustment hasn't changed. Install the dust boot.

Storm

1. Position the ignition switch/key in the **LOCK** position.

2. Place the selector lever in the **P** position.

3. Loosen the adjuster nuts on the transaxle. Ensure that the shift lever on the transaxle is in the **P** position.

4. Pull the cable forward, then tighten the forward adjuster nut until it contacts the shift lever. Tighten the rear nut until it contacts the shift lever.
5. Tighten both adjuster nuts.

FRONT SUSPENSION

MacPherson Strut

REMOVAL & INSTALLATION

Nova and Prizm

1. Under the hood, remove the 3 or 4 (4A-GE) small nuts holding the top of the strut to the shock tower. Do not loosen the larger center nut.
2. Loosen the wheel lug nuts at the appropriate wheel.
3. Raise the vehicle and safely support it. It need not be any higher than the distance necessary to separate the tire from the ground. Do not place the jackstands under the control arms.
4. Remove the wheel. Install a cover over the halfshaft boot to protect it from fluid and impact damage.
5. On Nova vehicles:
   a. Remove the brake flex hose clip at the strut bracket.
   b. Disconnect the brake flex hose from the brake pipe at the strut. Remove the brake hose clips. Use a small pan to catch any leakage.
   c. Pull the brake hose back through the opening in the strut bracket. Plug the lines to prevent any dirt form entering.
   d. Remove the 2 brake caliper mounting bolts and remove the caliper. Hang it out of the way with a piece of wire. Do not allow it to hang by the flex hose and do not disconnect the hose from the caliper.
   e. Mark the position of the adjusting cam for reassembly.
6. On Prizm vehicles:
   a. Disconnect the brake hose from the brake caliper and drain the fluid into a small pan.
   b. Remove the clip from the brake hose and remove the hose from the bracket.
   c. Use a sharp instrument or scribing tool to make matchmarks in all 3 dimensions on the steering knuckle. The strut must be reinstalled in its exact previous position.
7. Remove the 2 bolts which attach the shock absorber to the steering knuckle. The steering knuckle bolt holes have collars that extend about 5mm. Be careful to clear them when separating the steering knuckle from the strut assembly.

NOTE: Press down on the lower suspension arm in order to remove the strut assembly. This must be done to clear the collars on the steering knuckle bolt holes when removing the strut assembly.

8. Remove the strut assembly. Remember that the spring is still under tension. It will stay in place as long as the top nut on the shock piston shaft is not loosened. Handle the strut carefully and do not allow the coating on the spring to become damaged.

To install:

9. Loosely assemble all components onto the strut assembly. Make sure the mark on the upper spring seat is facing the outside of the vehicle.

NOTE: Never reuse a self-locking nut. Always replace self-locking nuts with new hardware.

10. Compress the spring, carefully aligning the shaft guide rod with the hole in the upper mount. Align the lower spring seat. Do not over-compress the spring; compress it just enough to allow installation of the shaft nut.
11. Install the shaft nut and tighten it until the strut shaft begins to rotate.
12. Double check that the spring is correctly seated in the upper and lower mounts and reposition it as needed. Slowly release the tension on the spring compressor and remove it from the strut assembly.
13. Tighten the shaft nut to 34 ft. lbs. (46 Nm).
14. On Nova vehicles, install the camber adjusting cam into the knuckle, observing the matchmarks made during disassembly.
15. Place the strut assembly in position and install the strut to knuckle attaching bolts. Tighten the bolts to the correct torque:
   Nova with single overhead cam engine: 105 ft. lbs. (142 Nm).
   Nova with twincam engine: 166 ft. lbs. (225 Nm).
   Prism 194 ft. lbs. (263 Nm).
8. Using a floor jack and a piece of wood, gently elevate the control arm to the point that the upper mount can be aligned with the holes in the shock tower. Insert the bolts into the upper holes and install the nuts. On the Nova equipped with the single overhead cam engine, tighten the nuts to 23 ft. lbs. (46 Nm) or 29 ft. lbs. (39 Nm) on the Nova and Prizm with the twincam engine.
9. Pack the shaft nut area with grease and install the dust cover.
10. On Nova vehicles, install the brake caliper and tighten the bolts to 65 ft. lbs. (88 Nm). Pull the brake hose through the strut bracket opening and connect the fitting. Tighten the fitting to 11 ft. lbs. (15 Nm). Install the flex hose clip at the strut bracket.
11. On Prizm vehicles, install the brake hose in the bracket and install the clip. Connect the hose to the caliper and tighten the fitting to 22 ft. lbs. (30 Nm).
12. Install the wheel and install the lug nuts snugly.
13. Lower the vehicle to the ground. Tighten the wheel lug nuts to 76 ft. lbs. (103 Nm).
14. Bleed the brake system and top up the brake fluid level.

Spectrum and Storm

1. From in the engine compartment, remove the nuts attaching the strut to the body.
2. Loosen the wheel nuts. Raise and support the vehicle safely. Lower the vehicle slightly so the weight of the vehicle rests on the support and not the control arms.
3. Remove the wheel and tire assembly.
4. Remove the brake hose clip at the strut bracket.
5. Disconnect the brake hose at the brake caliper. Cap all openings.
6. Pull the brake hose through the opening in the strut bracket.
7. Remove the nuts attaching the strut to the steering knuckle, then the strut assembly from the vehicle.
8. Reverse procedure to install.
9. On the Spectrum, torque the strut-to-body nuts to 80 ft. lbs. (108 Nm) and the upper strut retaining nuts to 41 ft. lbs. (55 Nm).
10. On the Storm, torque the upper strut retaining nuts to 58 ft. lbs. (78 Nm). Torque the steering knuckle to strut assembly nuts to 116 ft. lbs. (157 Nm).

Sprint and Metro

1. Raise and support the vehicle safely.
2. Remove the tire and wheel assembly.
3. Remove the brake hose clip, then the hose from the strut.
4. Remove the upper strut support nuts from the engine compartment.
5. Remove the strut-to-steering knuckle bolts, then the strut assembly from the vehicle.
6. Reverse procedure to install. Torque upper mounting nuts to 20 ft. lbs. (27 Nm) and the strut-to-steering knuckle bolts to 59 ft. lbs. (80 Nm).

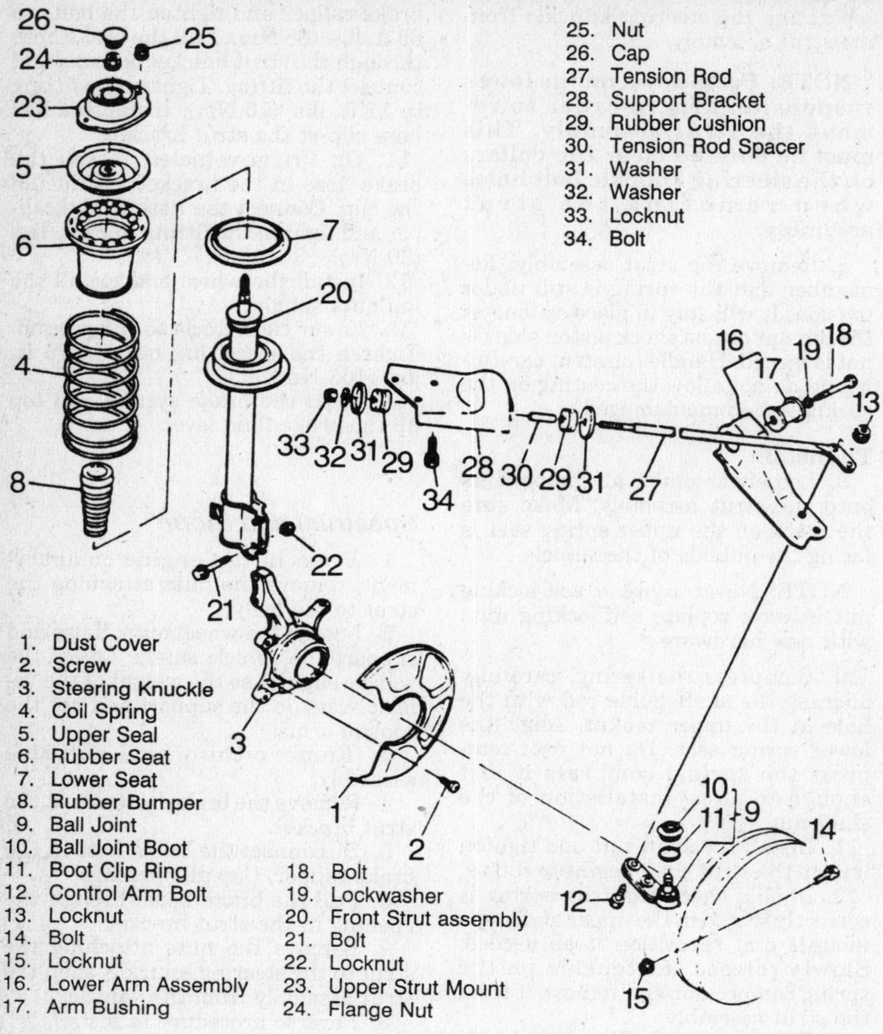

Exploded view of the front suspension assembly—Spectrum

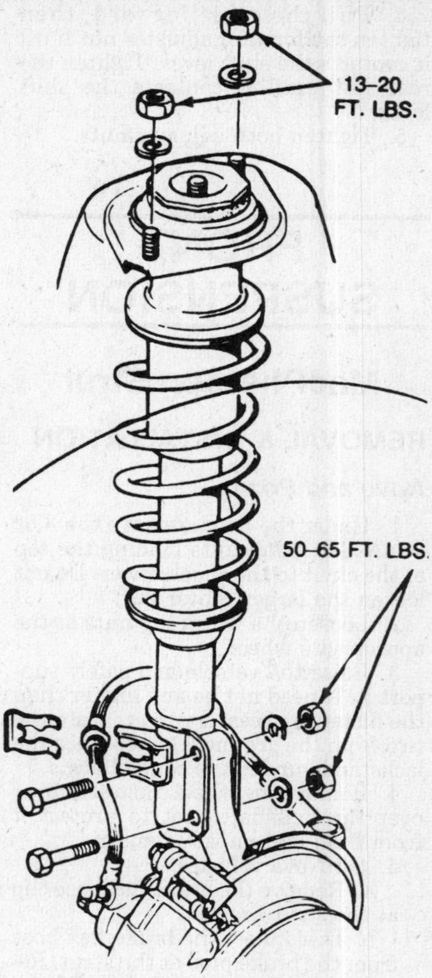

Front suspension strut mounting—Sprint and Metro

Torsion Bars

REMOVAL & INSTALLATION

Spectrum

1. Raise and support the vehicle safely.
2. If equipped with a stabilizer bar, remove the nuts, bolts and insulators attaching it to the tension rod.
3. Remove the nut and washer attaching the tension rod to the body.
4. Remove the nuts and bolts attaching the tension rod to the control arm, then the tension rod from the vehicle.
5. Reverse procedure to install. Torque tension rod-to-body nuts to 72 ft. lbs. (98 Nm) and the tension rod-to-control arm nuts to 80 ft. lbs. (108 Nm).

Lower Ball Joints

INSPECTION

Nova and Prizm

1. Turn the front wheels so they are in a straight ahead position, then chock the rear wheels.
2. Raise and support the front of the vehicle, then place a wooden block approximately 7–8 in. under it. Lower the vehicle onto the block until the spring is compressed to only about half its compression when the vehicle is resting on it.
3. Attempt to move the lower arm up and down. There should be no noticeable play.
4. If the ball joint is removed from the vehicle, check the required rotating torque with an inch lb. torque wrench. Flip the ball joint back and forth several times. Install the nut and turn the stud with a torque wrench at a rate of about 1 turn in 3 seconds. At the fifth turn, measure the required torque; it should be 9–30 inch lbs. If specifications are not as indicated, replace the ball joint.

Spectrum and Storm

Raise the front of the vehicle and safely support it on stands. Do not place stands under the control arms; the arms must hang free. Grasp the tire at the top and bottom and move the top of the tire through an in-and-out motion. Look for any horizontal motion in the steering knuckle relative to the control arm. Such motion is an indication of looseness within the ball joint. If the joint is checked while disconnected from the knuckle, it should have minimal or no free-play and should not twist in its socket under finger pressure. Replace any joint showing looseness or free play.

Sprint and Metro

The lower ball is part of the lower control arm. Therefore, if ball joint replacement is necessary, it must be replace as an assembly.

REMOVAL & INSTALLATION

Nova and Prizm

1. Elevate and safely support the front of the vehicle. Do not place the support under the control arms; they must hang free.

NOTE: Do not allow the driveshaft joints to over-extend. The CV joints can become disconnected under extreme extension.

2. Install a protective cover over the CV boot.
3. Remove the wheel.
4. Remove the cotter pin from the ball joint nut.
5. Loosen the castle nut but do not remove it. Unscrew it just to the top of the threads and install the ball joint separator. (GM Tool 34754 or equivalent) Use the nut to bear on the tool; this protects the threaded shaft from damage during removal.
6. Use the separator to loosen the ball joint from the steering knuckle.
7. Remove the nuts and bolt holding the ball joint to the control arm.
8. Remove the ball joint from the control arm and steering knuckle.

To install:

9. When reinstalling, attach the ball joint to the control arm and tighten the 2 bolts and 1 nut to 59 ft. lbs. (80 Nm) on the Nova and 105 ft. lbs. (142 Nm) on the Prizm.
10. Carefully install the ball joint to the steering knuckle. Use a new castle nut and tighten it to 82 ft. lbs. (111 Nm) on the Nova and 94 ft. lbs. (127 Nm) on the Prizm.
11. Install the cotter pin through the castle nut and stud.
12. Remove the protector from the CV boot.
13. Install the wheel.
14. Lower the vehicle to the ground.

Spectrum and Storm

1. Loosen the wheel nuts.
2. Loosen the wheel nuts. Raise and support the vehicle safely. Lower the vehicle slightly so the weight of the vehicle rests on the supports and not the control arms.
3. Remove the tire and wheel assembly.

NOTE: Care must be exercised to prevent the CV-joints from being overextended. When either end of the halfshaft is disconnected, overextension of the joint will result in separation of the internal components and possible joint failure. Drive axle joint seal protectors should be used any time service is performed on or near the drive axles. Failure to observe this could result in interior joint or seal damage and possible joint failure.

4. Install a inner drive joint seal protector (J-34754 or equivalent) to the drive axle boot.
5. Remove the 2 nuts attaching the ball joint to the tension rod and the control arm assembly.
5. Remove the pinch bolt attaching the ball joint to the steering knuckle, then the ball joint from the vehicle.
6. Installation is the reverse order of the removal procedure.On Spectrum, torque the knuckle-to-ball joint nut to 51 ft. lbs. (69 Nm) and the tension rod-to-control arm nuts to 72 ft. lbs. (98 Nm). On Storm, torque the ball joint-to-control arm bolts to 115 ft. lbs. (156 Nm) and the pinch bolt to 49 ft. lbs. (66 Nm).

Lower Control Arms

REMOVAL & INSTALLATION

Nova

1. Raise and safely support the vehicle. Do not place the supports under the control arms.
2. Remove the nuts and bolts holding the ball joint to the lower control arm.
3. On vehicles with the twincam engine, remove the nut holding the sway bar link to the control arm and disconnect the link and bar from the control arm.
4. Remove the nuts and bolts holding the control arm to the body.
5. Remove the control arm from the vehicle and check it carefully for cracks, bends or crimps in the metal or corrosion damage. Check the rubber bushing and replace it if any sign of damage or deformation is found.
6. Before reinstalling the arm, install or replace the bushing and tighten its retaining nut to 76 ft. lbs. (103 Nm).

NOTE: Never reuse a self-locking nut; always replace the removed nut with a new one.

7. Position the control arm to the body and install the nuts and bolts. The front nut and bolt should be tightened to 105 ft. lbs. (142 Nm) and the rear to 72 ft. lbs. (98 Nm).
8. On vehicles with the twincam engine, reinstall the sway bar and link; tighten the nut to 13 ft. lbs. (18 Nm).
9. Install the ball joint attaching nuts and bolts and tighten them to 59 ft. lbs. (80 Nm).
10. Check the wheel alignment.

Prizm

NOTE: Both lower control arms and the suspension crossmember must be removed as a unit even if only 1 arm is damaged.

1. Raise and safely support the vehicle. Do not place the stands under the control arms or the suspension crossmember.
2. Remove the nuts and bolts holding the ball joints to the lower control arms.
3. Remove the nut and bolt holding the control arm rear brackets to the crossmember.
4. Place a floor jack under the suspension crossmember. Use a broad piece of wood between the jack and crossmember to evenly distribute the loading.
5. Remove the 6 bolts and 2 nuts holding the suspension crossmember. Carefully lower the crossmember (with the control arms attached) and remove from the vehicle.
6. Remove the mounting bolt holding the control arm to the crossmember and remove the arm. Inspect the arm and bushing for damage, deformation or corrosion damage.

To install:

7. Install the control arm(s) to the crossmember and partially tighten the bolts. They should be tight enough to hold firmly, yet still be able to pivot when moderate force is applied.
8. Install the suspension crossmember and control arms to the body of the vehicle and tighten the 6 nuts and 2 nuts.
9. Install the bolts holding the rear control arm brackets and partially tighten them.
10. Connect the ball joints to each arm and tighten the bolts to 105 ft. lbs. (142 Nm).
11. Lower the vehicle to the ground. Bounce the front end up and down several times to stabilize the suspension. The partially tightened joints will flex and seek a "normal" position.
12. With the vehicle on the ground (don't raise it or the suspension position will be lost), tighten the bolt holding the arm to the crossmember to 152 ft. lbs. (206 Nm). Tighten the nut and bolt holding the rear bracket to the crossmember to 14 ft. lbs. (19 Nm). and the rear bracket bolts to 94 ft. lbs. (127 Nm). On the 1991-92 Prizm torque the lower control arm nut to 152 ft. lbs. (206 Nm).
13. Check the alignment.

Storm

1. Raise and support the vehicle safely.
2. Remove the tire and wheel assembly.
3. Remove the stabilizer bar from the control arm.
4. Remove the ball stud from the steering knuckle.
5. Remove the front bushing-to-body attaching bolt, then the rear bushing-to-crossmember attaching bolts.
6. Remove the control arm from the vehicle.
7. Reverse procedure to install. Torque the rear bushing-to-crossmember bolt to 51 ft. lbs. (69 Nm); front bushing-to-body attaching bolt to 95 ft. lbs. (129 Nm) and the ball stud-to-knuckle pinch bolt to 46 ft. lbs.

Spectrum

1. Raise and safely support the vehicle.
2. Remove the tire and wheel assembly.
3. Remove the control arm-to-tension arm attaching nuts and bolts.
4. Remove the nut and bolt attaching the control arm to the body, then the control arm from the vehicle.
5. Reverse procedure to install.

NOTE: Raise the control arm to a distance of 15 inches from the top of the wheel well to the center of the hub. Retorque the control arm-to-body bolts to 41 ft. lbs. (55 Nm) and the control arm-to-tension rod bolts to 80 ft. lbs. (108 Nm). This procedure aligns the bushing arm to the body.

Sprint

1. Raise and support the front of the vehicle safely.
2. Remove the front tire and wheel assembly.
3. Remove the cotter pin, castle nut, washer and the bushing from the stabilizer bar.
4. Remove the stabilizer bar-to-body mounting bracket bolts.
5. Remove the ball studs and the control arm bolts, then the control arm from the vehicle.
6. Reverse procedure to install. Torque the control arm-to-body bolt to 36–50 ft. lbs. (50-70 Nm); ball stud-to-steering knuckle to 36‹n-dash›50 ft. lbs. (50-70 Nm); stabilizer bar-to-control arm nut to 29‹n-dash›65 ft. lbs. (40-90Nm) and the stabilizer bar-to-body nuts to 22‹n-dash›39 ft. lbs. (30-50Nm).

Metro

1. Raise and support the vehicle safely.
2. Remove the wheel assemblies.
3. Remove the ball stud bolts.
4. Remove the suspension arm bracket nut.
5. Remove the suspension arm bracket bolts.
6. Remove the rear bracket and suspension arm.
7. Installation is the reverse order of the removal procedure. Torque the ball stud nut to 44 ft. lbs. (60 Nm). Torque the suspension arm rear bracket bolts to 32 ft. lbs. (43 Nm).

Stabilizer Bar

REMOVAL & INSTALLATION

Nova

1. Raise and safely support the vehicle.
2. Disconnect the stabilizer-to-lower control arm bolts.
3. Remove the stabilizer bar-to-chassis brackets.
4. Disconnect the exhaust pipe from the exhaust manifold.
5. Remove the stabilizer bar from the vehicle.

NOTE: Never reuse a self locking nut, always use a new one.

6. Reverse procedure to install. Torque the exhaust pipe-to-exhaust manifold bolts to 46 ft. lbs. (62 Nm), the stabilizer bar-to-body bolts to 14 ft. lbs. (19 Nm) and the stabilizer bar-to-lower control arm bolts to 13 ft. lbs. (18 Nm).

1991–92 Prizm

GSI MODELS

1. Raise and safely support the vehicle.
2. Remove the left stabilizer link from the stabilizer bar.
3. Remove the right stabilizer link from the stabilizer bar.
4. Remove the catalytic converter from the front exhaust pipe.
5. Remove the right stabilizer bracket from the floor pan.
6. Remove the left stabilizer bracket from the floor pan.
7. Remove the stabilizer bar from the vehicle. Remove the bushing brackets and retainers from the stabilizer bar.

To install:

8. Installation is the reverse of the removal procedure. Use the following torque specifications:
   a. Torque the bushing bracket retaining bolts to 29 ft. lbs. (39 Nm).
   b. Torque the left and right stabilizer link nuts to 26 ft. lbs. (35 Nm).
   c. Torque the left and right stabilizer bracket bolts to 94 ft. lbs. (127 Nm).
   d. Torque the left and right stabilizer bracket nut and bolt to 14 ft. lbs. (19 Nm).
9. To remove the stabilizer link, use the following procedure.
   a. With the vehicle still raise and safely supported, remove the stabilizer link from the stabilizer bar.
   b. Remove the stabilizer link from the control arm.
   c. Remove the stabilizer link from the vehicle.
10. Torque the stabilizer link nuts to 26 ft. lbs. (35 Nm).

Storm

1. Install engine support tool J-28467–A onto the engine.
2. Raise the vehicle, then position suitable support under the suspension supports. Lower the vehicle onto the support, not the control arms.
3. Remove the tire and wheel assemblies.
4. Disconnect the front exhaust pipe from the exhaust manifold.
5. Disconnect the power steering lines from the rack and pinion gear.
6. From inside the vehicle, remove the boot from the steering shaft, then the steering shaft.
7. Remove the ball joints and tie rods from the steering knuckles.
8. Remove the engine torque rod from the center beam.
9. Remove the engine rear mount.
10. Remove the 2 bolts from the center beam and the 4 bolts from the crossmember, then the center beam and crossmember assembly with the rack and pinion gear and stabilizer bar.
11. Remove the rack and pinion assembly from the crossmember, then the stabilizer bar from the crossmember.

To install:

12. Install the stabilizer bar and rack and pinion assembly to the crossmember.
13. Install the center beam and crossmember assembly with the rack and pinion gear and stabilizer bar.
14. Install the 2 bolts at the center beam and the 4 bolts at the crossmember.
15. Install the rear engine mount.
16. Install the engine torque rod at the center beam, then the tie rods to the steering knuckle.
17. From inside the vehicle, install the steering shaft and boot.
18. Install the front exhaust pipe, then the tire and wheels.
19. Torque ball joint bolt to 48 ft. lbs. (66 Nm); tie rod-to-knuckle bolt 40 ft. lbs. (54 Nm); rack and pinion-to-cross-

member nuts 65 ft. lbs.; stabilizer bar-to-control arm bolts 19 ft. lbs. (26 Nm); stabilizer bar-to-crossmember bolts to 12 ft. lbs. (16 Nm) and the crossmember-to-body bolts to 137 ft. lbs. (186 Nm).

20. Lower the vehicle and remove the engine support tool.

Sprint

1. Raise and safely support the vehicle.
2. Remove the tire and wheel assemblies.
3. Remove the stabilizer bar-to-body attaching bolts.
4. Remove the cotter pin, castle nut, washer and bushing, then the stabilizer bar from the lower control arms.
5. Reverse procedure to install. Torque stabilizer bar-to-control arm nuts to 29-60 ft. lbs. (40-90 Nm) and the stabilizer bar-to-body bolts to 22-39 ft. lbs. (30-55 Nm).

Sway Bar

REMOVAL & INSTALLATION

Nova With Twincam Engine

1. Raise and support the vehicle. Disconnect the sway bar links from the lower control arms.
2. Disconnect the sway bar brackets from the body.

NOTE: Check the bushings inside the brackets for wear or deformation. A worn bushing can cause a distinct bang or "thunk" as the bar twists during cornering.

3. Disconnect the exhaust pipe from the exhaust manifold.
4. Remove the sway bar from the vehicle. Examine the insulators (bushings) carefully for any sign of wear and replace them if necessary.

To install:

5. To reinstall, place the bar in position and reconnect the exhaust system using new nuts. Tighten the nuts to 46 ft. lbs. (62 Nm).
6. Install both stabilizer bar brackets and tighten the bolts to 14 ft. lbs. (19 Nm).
7. Connect the sway bar links to the control arms with the bolts, insulators and new nuts. Tighten the nuts to 13 ft. lbs. (18 Nm).

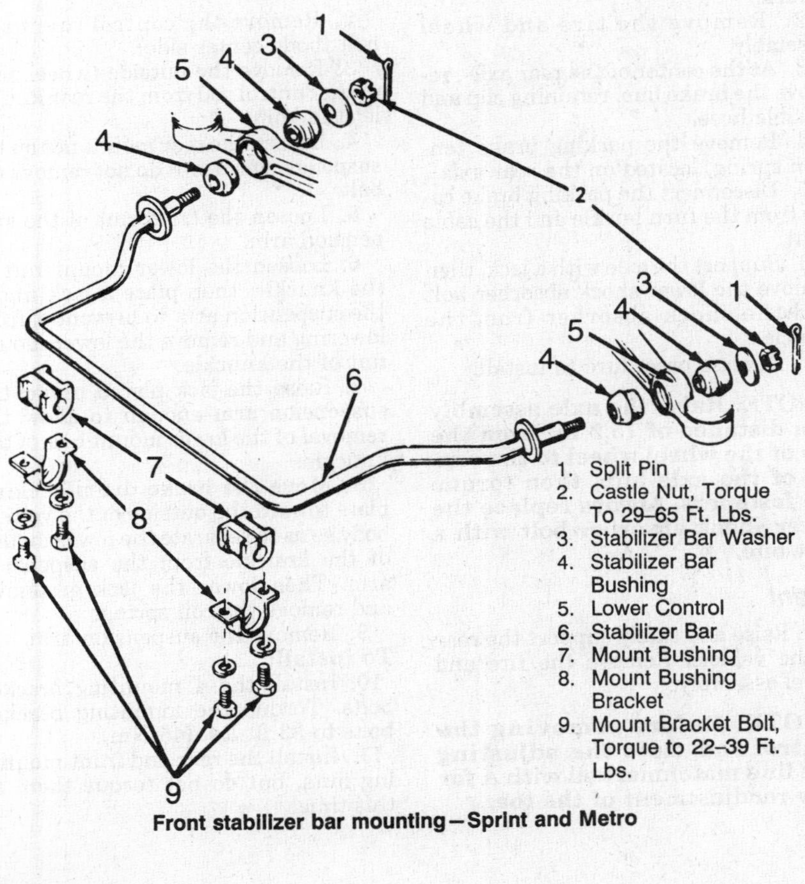

Front stabilizer bar mounting—Sprint and Metro

REAR SUSPENSION

Shock Absorbers

REMOVAL & INSTALLATION

Spectrum

1. Open the trunk and lift off the trim cover, hatch back vehicles only. Remove the upper shock absorber nut and raise and safely support the vehicle.
2. Remove the shock absorber lower attaching bolt.
3. Remove the shock absorber from the vehicle.
4. Reverse procedure to install.

NOTE: When replacing the shock absorber, never reuse the old lower bolt, always use a new one.

Sprint

1. Raise and safely support the vehicle.
2. Remove the lower mounting nut, lock washer and the outer washer.
3. Remove the upper mounting bolt, lock washer and nut.
4. Remove the shock absorber from the vehicle.
5. Reverse procedure to install. Torque upper mounting bolt to 13-20 ft. lbs. (18-28 Nm) and the lower mounting nut to 32-50 ft. lbs. (45-70 Nm).

Metro

1. Raise and safely support the vehicle. Remove the wheels.
2. Place a support under the suspension arm for support when it lowers.
3. Remove the shock (strut) support nuts and push the shock down.
4. Remove the lower shock mount bolt.
5. Remove the shock from the knuckle by pulling the upper side of the knuckle. Compress the shock as short as possible for removal. If the shock is hard to remove, open the slit of the knuckle by inserting a wedge.

NOTE: Do not open the knuckle slit wider than necessary. Do not lower the jack more than necessary during the strut removal to prevent the coil spring from coming off, or a brake flexible hose from stretching.

6. Installation is the reverse order of the removal procedure. Torque the shock lower mounting bolt to 44 ft. lbs.

(60 Nm). Torque the upper shock to vehicle bolts to 24 ft. lbs. (30 Nm).

MacPherson Strut

REMOVAL & INSTALLATION

Nova

1. Working inside the rear of the vehicle, remove the rear quarter window garnish molding, back window panel and the speaker grille, if necessary.
2. Raise and safely support the vehicle by placing support under the rear of the vehicle. Remove the tire and wheel assembly.
3. Disconnect the flexible brake hose from the strut. Remove the flexible hose and clip from the mounting point on the strut, then reconnect the brake line to the flex hose to prevent an excessive amount of brake fluid from draining.

NOTE: Before removing the strut from the hub carrier, be sure to mark the location of the strut to the hub carrier for reinstallation purposes.

4. Remove the strut to hub attaching nuts and bolts, then separate the strut from the hub carrier.
5. Remove the strut-to-chassis nuts, then carefully remove the strut assembly from the vehicle.
6. Reverse procedure to install. Torque the strut-to-chassis nuts to 17 ft. lbs. (23 Nm) and the strut to hub carrier nuts and bolts to 105 ft. lbs. (142 Nm) Bleed the brake system.

Prizm

1. On Sedans, remove the seat back side cushion. On Hatchbacks, remove the rear sill side panel.
2. Raise and safely support the vehicle, then place a suitabl support under the suspension support.
3. Lower the vehicle slightly so the weight rest on the jackstands and not the suspension arms.
4. Remove the tire and wheel assembly.
5. Disconnect the brake line hose and backing plate, then the brake hose rom the brake hose bracket.
6. Disconnect the stabilizer bar link from the strut assembly, then the strut assembly mounting nuts and bolts.
7. Remove the strut assembly mounting nuts holding the top of the strut support, then the strut assembly from the vehicle.
8. Reverse procedure to install. Torque the strut to body nuts to 29 ft. lbs. (39 Nm); strut assembly to knuckle mounting bolts to 105 ft. lbs. (142 Nm) and the stabilizer bar link to the strut assembly to 26 ft. lbs. (35 Nm). Bleed the brake system.

Storm

1. Raise and safely support the rear of the vehicle, then place a supportunder the suspension support. Remove the tire and wheel assembly.
2. Lower the vehicle slightly so the weight rest on the support and not the suspension arms.
3. Remove the flexible bake hose clip at the strut and the brake hose from the brake pipe.
4. Disconnect the stabilizer bar link from the strut assembly, then the strut assembly mounting nuts and bolts.
5. Open the rear hatch, then remove the trim cover panel from the strut tower.
6. Remove the strut assembly mounting nuts holding the top of the strut support, then the strut assembly from the vehicle.
7. Reverse procedure to install. Torque strut tower nuts to 50 ft. lbs. (68 Nm) and the strut to knuckle bolts to 116 ft. lbs. (157 Nm).

Coil Springs

REMOVAL & INSTALLATION

Spectrum

1. Raise and safely support the vehicle.
2. Remove the tire and wheel assembly.
3. At the center of the rear axle , remove the brake line, retaining clip and flexible hose.
4. Remove the parking brake tension spring, located on the rear axle.
5. Disconnect the parking brake cable from the turn buckle and the cable joint.
6. Support the axle with a jack, then remove the lower shock absorber bolt and the shock absorber from the vehicle.
7. Reverse procedure to install.

NOTE: Raise the axle assembly to a distance of 15.2 in. from the top of the wheel wheel to the center of the axle hub, then torque the fasteners. Always replace the lower shock absorber bolt with a new one.

Sprint

1. Raise and safely support the rear of the vehicle. Remove the tire and wheel assembly.

NOTE: Before removing the control rod, note the adjusting bolt line matchmarked with A for easy readjustment of the toe.

2. Remove the control rod-to-knuckle nut/bolt, the control rod-to-chassis nut/bolt, then the control rod from the vehicle.
3. Loosen both suspension arm-to-chassis mount nuts and the suspension arm-to-knuckle nut. Using a floor jack, position it under the suspension arm to prevent it from lowering.
4. Remove the suspension arm-to-knuckle nut and raise the jack slightly to allow removal of the bolt.
5. Pull the brake drum/backing plate assembly outward to disengage knuckle's lower mount from the suspension arm.
6. Lower the jack and remove the coil spring.
7. Reverse procedure to install. Torque the front suspension arm-to-chassis nut to 44 ft. lbs. (60 Nm). and the rear to 37 ft. lbs. (50 Nm).
8. Check and/or adjust the toe.

Metro

1. Raise and support the vehicle safely.

NOTE: To facilitate the toe adjustment after reinstallation, confirm which one of the lines stamped on the washer is in the closest alignment with the stamped line on the control rod. If not markes, add the alignment marks.

2. Remove the control rod inside bolt (body center side).
3. Remove the outside (wheel side) of the control rod from the rear knuckle stud bolt.
4. Loosen the rear mount not on the suspension arm, but do not remove the bolt.
5. Loosen the front nut of the suspension arm.
6. Loosen the lower mount nut of the knuckle, then place a jack under the suspension arm to prevent it from lowering and remove the lower mount nut of the knuckle.
7. Raise the jack placed under the suspension arm enough to allow the removal of the lower mount bolt of the knuckle.
8. Move the brake drum/backing plate toward the outside of the vehicle body so as to separate the lower mount of the knuckle from the suspension arm. Then lower the jack gradually and remove the coil spring.
9. Remove the suspension arm.

To install:

10. Install the 4 mounting bracket bolts. Torque the mounting bracket bolts to 33 ft. lbs. (45 Nm).
11. Install the rear and front mounting nuts, but do not torque them at this time.

NOTE: Make sure that the front mounting washer is installed in the proper direction.

12. Place the jack under the suspension arm.
13. Install the coil spring on the spring seat of the suspension arm then raise the suspension arm. When seating the coil spring, mate the spring end with the stepped part of the suspension arm spring.
14. Install the lower knuckle mount bolt. Torque the bolt to 37 ft. lbs. (50 Nm).
15. Remove the jack from under the suspension arm.
16. Install the inside and outside control rod bolts, but do not tighten them at this time.
17. Install the wheel assemblies and lower the vehicle.
18. Install the control rod inside and outside nut and torque them to 59 ft. lbs. (80 Nm).

NOTE: When tightening the nuts, it is most desirable to have the vehicle off the hoist and in a non-loaded state. Also when tightening the inside nut, align the line stamped on the body with the line on the washer as confirmed before removal or align the matchmarks if marked.

19. Install the suspension arm front and rear nuts. Torque the front nuts to 44 ft. lbs. (60 Nm) and the rear nuts to 37 ft. lbs. (50 Nm). After tightening the suspension arm outer nut, make sure that the washer is not tilted.
20. Check the rear wheel alignment.

Rear Control Arms

REMOVAL & INSTALLATION

Nova

FRONT

1. Raise and support the vehicle by placing supports under the frame, then remove the rear wheel.
2. If equipped with a twincam engine, remove the rear suspension arm-to-stabilizer bar nut, retainer and cushion.
3. Remove the front suspension arm-to-hub carrier nut/bolt.
4. Remove the front suspension arm to chassis nut/bolt, then the suspension arm from the vehicle.
5. Reverse procedure to install. On twincam, torque the stabilizer bar-to-front suspension arm nut/bolt to 11 ft. lbs. (15 Nm).
6. Lower the vehicle, then bounce the vehicle to stabilize the suspension. Torque the front suspension arm bolts to 64 ft. lbs. (87 NM).

REAR

1. Raise and support the vehicle under the frame, then remove the rear tire and wheel assembly.
2. Remove the rear suspension arm-to-hub carrier nut and bolt.

NOTE: Before removing the rear cam bolt, mark the alignment of the cam bolt for reinstallation purposes.

3. Remove the rear suspension arm-to-chassis cam bolt, then the suspension arm.
4. Reverse procedure to install. Torque rear suspension arm-to-chassis cam bolt to 64 ft. lbs. (87 Nm).

Prizm

1. Raise and safely support the vehicle.
2. Remove the rear suspension arm to body attaching nut, then suspension knuckle bolt.
3. Remove the rear suspension arm to rear knuckle mounting bolts.
4. Remove the rear suspension arm from the vehicle.

To install:

5. When installing the rear suspension arm, align the matchmarks on the cam and body. Lower the vehicle and bounce the vehicle up and down to stabilize the suspension. Torque suspension arm mounting bolts to 87 ft. lbs. (118 Nm).
6. Check the rear wheel alignment.
7. To remove the strut rod, use the following procedure.
   a. Raise and safely support the vehicle.
   b. Remove the strut rod to body mounting bolt and nut.
   c. Remove the strut rod to rear suspension knuckle mounting bolt and nut.
   d. Remove the strut rod.
8. Installation is the reverse order of the removal procedure. Torque the strut rod to rear suspension knuckle bolt to 87 ft. lbs. (118 Nm). Torque the strut rod to body bolt and nut to 87 ft. lbs. (118 Nm).

Storm

1. Raise and safely support the vehicle.
2. Remove the tire and wheel assembly.
3. Remove trailing control arm nuts, then the trailing control arm from the vehicle.
4. Remove the right and left lateral control arms as follows:
   a. Remove the right lateral control arm attaching bolts from the rear crossmember and the rear suspension knuckle, then the right lateral control arm from the vehicle.
   b. Remove the left lateral control arms by loosening the 2 rear crossmember-to-body attaching bolts/nuts, then push crossmember down as far as possible and support with a jack. Remove the bolt from the rear crossmember and rear suspension knuckle, then the left lateral control arms from the vehicle.
5. Reverse procedure to install. Torque all nuts and bolts to 94 ft. lbs. (128 Nm).
6. When removing the stabilizer bar use the following procedure:
   a. With the vehicle still in the raised and supported position and the wheel assemblies removed.
   b. Remove the bolts on the shackle at both strut sides.
   c. Remove the mounting bolts and brackets from stabilizer bar on the crossmember.
   d. Remove the stabilizer bar.
7. Torque the link retaining bolts to 19 ft. lbs. (26 Nm) and torque the stabilizer mounting bracket bolts to 71 ft. lbs. (96 Nm).

Sprint

1. Raise and safely support the rear of the vehicle. Remove the tire and wheel assembly.

NOTE: Before removing the control rod, note the adjusting bolt line matchmarked with A for easy readjustment of the toe.

2. Remove the control rod-to-knuckle nut/bolt, the control rod-to-chassis nut/bolt, then the control rod from the vehicle.
3. Loosen both suspension arm-to-chassis mount nuts and the suspension arm-to-knuckle nut. Using a floor jack, position it under the suspension arm to prevent it from lowering.
4. Remove the suspension arm-to-knuckle nut and raise the jack slightly to allow removal of the bolt.
5. Pull the brake drum/backing plate assembly outward to disengage knuckle's lower mount from the suspension arm.
6. Lower the jack and remove the coil spring.
7. Remove the suspension arm-to-chassis nuts and bolts, then the suspension arm.
8. Reverse procedure to install. Torque the suspension arm to knuckle nut/bolt to 37 ft. lbs. (50 Nm); front suspension arm-to-chassis nut to 44 ft. lbs. (60 Nm) and the rear to 37 ft. lbs. (50 Nm).
9.Check and/or adjust the toe.

Metro

1. Raise and support the vehicle safely.

NOTE: To facilitate the toe adjustment after reinstallation, confirm which one of the lines stamped on the washer is in the closest alignment with the stamped line on the control rod. If not markes, add the alignment marks.

2. Remove the brake hose from the control rod buy pulling off the E-ring.
3. Remove the outside (wheel side) of the control rod from the rear knuckle stud bolt.
4. Remove the control rod inside bolt (body center side). Hold the inside bolt with another wrench to prevent it from turning as the nut is turned.
5. Remove the control rod inside bolt and the control rod.
6. Installation is the reverse order of the removal procedure. To determine the installing direction of the control rod, note the angle of the brake flexible hose mounting brackets welded to the control rod. Torque the control rod outside and inside nut to 59 ft. lbs. (80 Nm).

Rear Axleshaft, Bearing and Seal

REMOVAL & INSTALLATION

Nova

1. Raise and safely support the vehicle.
2. Remove the wheel.
3. Remove the brake drum or brake caliper and brake disc. If the caliper is removed, suspend it out of the way with a piece of stiff wire; do not let it hang by the hose.
4. On drum brake models, disconnect the brake line from the wheel cylinder.
5. Remove the 4 bolts holding the axle hub/bearing assembly to the axle carrier and remove the hub and bearing assembly.
6. Remove the splash shield (if so equipped) or the rear brake shoes (drum brake models), then remove the O-ring. It must be replaced at reassembly.
7. Place the hub/bearing assembly in a vise with protected jaws. Do not clamp the hub any tighter than needed to hold it securely.
8. Remove the hub nut.
9. Using an extractor such as GM Tool J-25287 or equivalent, remove the bearing case from the axle hub.
10. Remove the inner race, the inboard bearing and the outboard bearing.
11. Use the extractor again to remove the inner race from the outboard bearing.
12. Remove the seal from the axle hub.
13. To remove the outer bearing race, install the outboard bearing inner race so that it pushes against the outer race. Use Tool J-35440 or similar to press the race free.

NOTE: Whenever the bearing assembly is removed, it must be replaced with a new bearing assembly. Reuse of the bearing after disassembly is not recommended.

To install:

14. To reassemble, apply grease around the bearing outer race.
15. Press the new bearing outer race into the bearing case with Tool J-35400 or equivalent.
16. Install new bearings (inner and outer) and the inner races into the bearing case.
17. Lightly coat a new bearing seal with multi-purpose grease and install with a seal driver such as GM Tool J-35736 or equivalent.
18. Install the bearing case onto the hub with GM Tool J-35440 or equivalent.
19. Tighten the rear nut to 90 ft. lbs. (122 Nm) and stake the nut with a chisel.
20. Install a new O-ring to the axle carrier.
21. Assemble the hub to either the brake backing plate or the splash shield and install. Tighten the bolts to 59 ft. lbs. (80 Nm).
22. On drum brake models, reinstall the rear brake shoes and connect the brake line to the wheel cylinder.
23. Reinstall either the rear brake drum or the brake disc and caliper.
24. Install the wheel and lower the vehicle to the ground.
25. Refill the master cylinder and bleed the brake system.

Prizm

1. Raise and safely support the vehicle.
2. Remove the wheel.
3. Remove the brake drum or brake caliper and brake disc. If the caliper is removed, suspend it out of the way with a piece of stiff wire; do not let it hang by the hose.
4. Remove the 4 bolts holding the axle hub/bearing assembly to the axle carrier and remove the hub and bearing assembly.
5. Remove the O-ring from the backing plate.
6. Mount the hub/bearing assembly in a vise with protected jaws.
7. Use a hammer and chisel to unstake the wheel bearing lock nut.
8. Remove the lock nut.
9. Using a split plate bearing remover, a race puller pilot and an halfshaft puller (GM Tools J-22912, J-38278 and J-8433 respectively), separate the halfshaft from the hub.
10. Using the same tools, remove the inner bearing race.
11. Use a slide hammer and seal puller to remove the oil seal from the halfshaft.
12. Press the inner race off with a bearing race remover such as GM Tool J-35400 or equivalent.
13. Clean all components thoroughly and examine for any signs of cracking, abrasion or corrosion. Whenever the wheel bearings are removed or disassembled, they must be replaced with new bearings. Reuse of the old bearings is not recommended.

To install:

14. To reassemble, apply multi-purpose grease around the outer race of a new bearing and install the bearing into the inner race.
15. Using a seal installer and slide hammer, install a new oil seal onto the halfshaft.
16. Using the proper size installation tool, press the bearing and bearing outer race into the axle hub.
17. Press the bearing inner race into the axle hub.
18. Press the outer race onto the halfshaft.
19. Install a new wheel bearing lock nut and tighten it to 90 ft. lbs. (123 Nm). Use a hammer and chisel the stake the nut in place.
20. Install a new O-ring onto the knuckle.
21. Place the axle hub/bearing assembly in position, install the bolts and tighten them to 59 ft. lbs. (80 Nm).
22. Reinstall the brake drum.
23. Install the wheel and lower the vehicle to the ground.
24. Although this repair should not have affected the rear alignment, it is recommended that the alignment be checked and adjusted if necessary.

Spectrum

1. Raise and safely support the vehicle.
2. Remove the tire and wheel assemblies, then the brake line, retaining clip and flexible hose from the center of the rear axle.
3. Remove the tension spring from the rear axle, disconnect the parking brake cable from the turn buckle and the cable joint.
4. Using a suitable jack, support the lower side of the rear axle and remove the lower shock absorber bolt, then the shock absorber from the vehicle.
5. Carefully lower the jack and remove the coil spring.
6. Remove the bolts attaching the

1. Fuel tank band
2. Strut tower cover
3. Strut rod piston nut
4. Strut assembly
5. Rear toe adjustment bolt
6. No. 2 suspension arm-to-body bolt
7. Stabilizer bar link-tostrut assembly nut
8. No. 2 suspension arm
9. Strut assembly-to-knuckle nut
10. No. 2 suspension arm-to-knuckle bolt
11. Brake line
12. Brake drum
13. Brake assembly
14. Strut rod
15. Strut rod-to-knuckle bolt
16. Strut rod-to-body bolt
17. No. 1 suspension arm-to-knuckle bolt
18. No. 1 suspension arm
19. No. 1 suspension arm-to-body bolt
20. Stabilizer bracket bolt
21. Stabilizer bar bracket
22. Bushing
23. Stabilizer bar link
24. Stabilizer bar link nut
25. Stabilizer bar

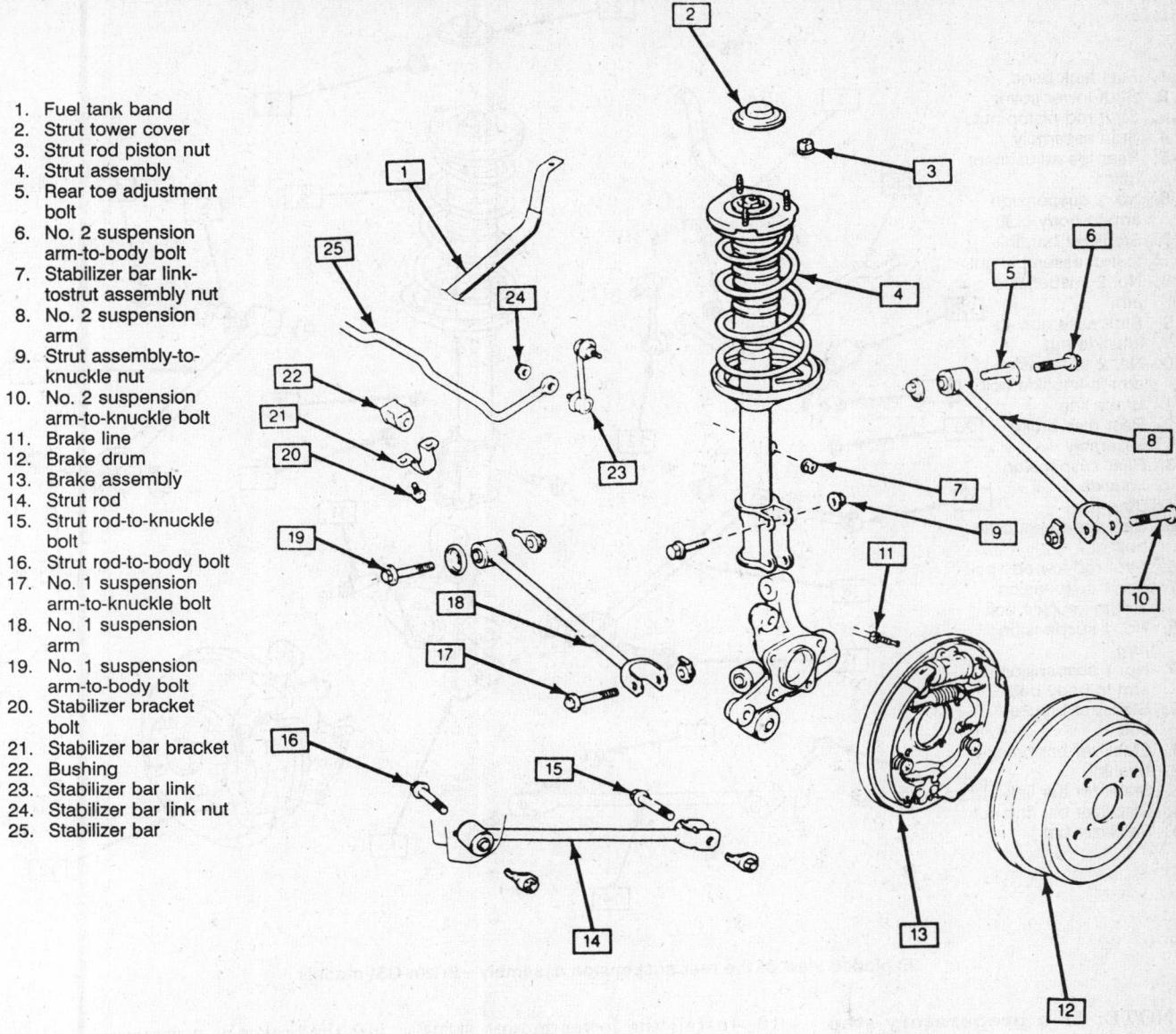

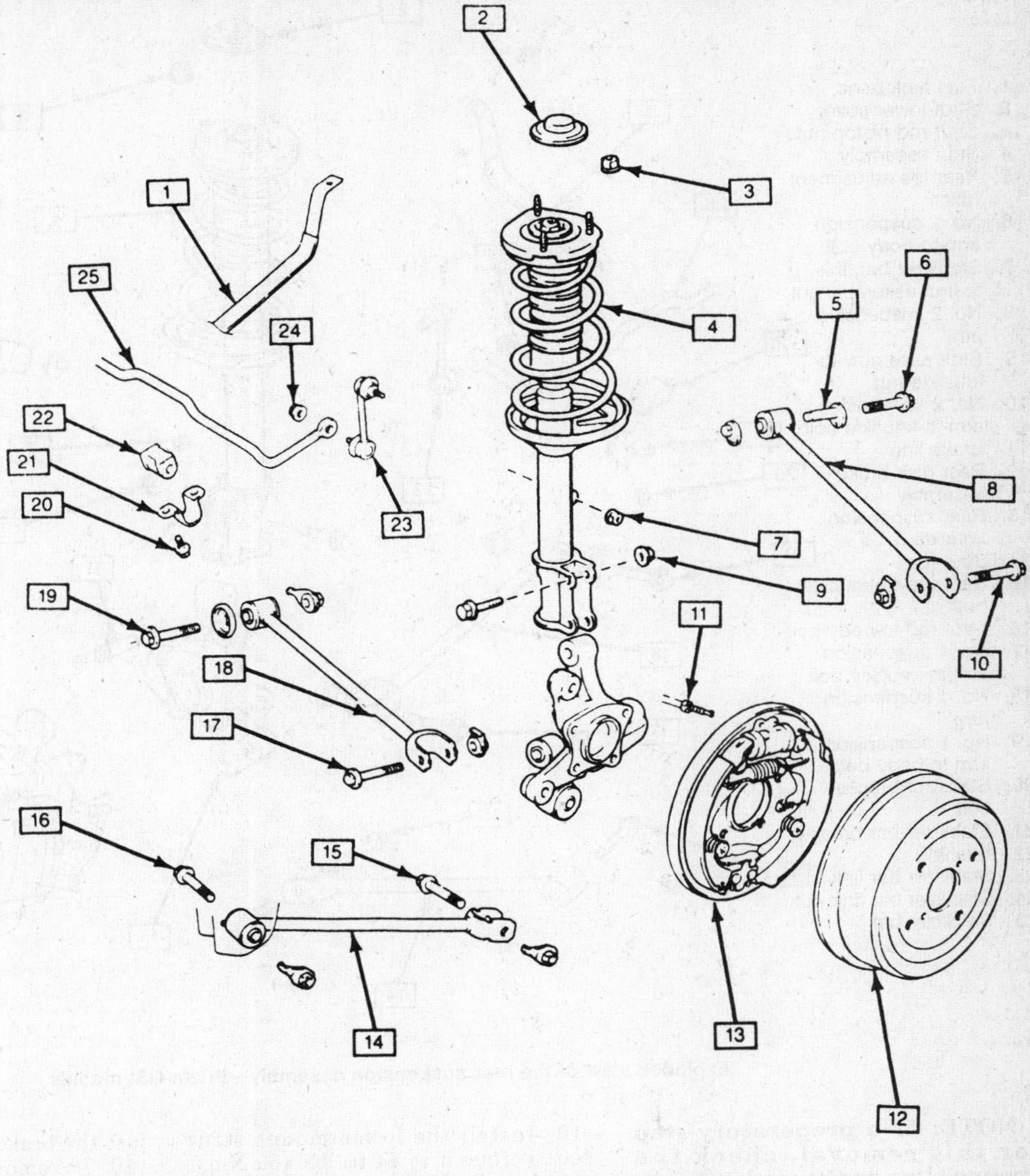

Exploded view of the rear suspension assembly – Prizm Base and LSi models

rear axle to the body, then the rear axle assembly from the vehicle.

7. Reverse procedure to install. Set rear trim height.

8. If removing the wheel bearing use the following procedure:

a. Raise and support the vehicle safely.

b. Remove the rear wheel assemblies.

c. Remove the hub cap, cotter pin, hub nut, washer and outer bearing.

d. Remove the hub.

e. Using a slide hammer puller and attachment, pull the oil seal from the hub. Remove the inner bearing.

f. Using a brass drift and a hammer, drive both bearing races from the hub.

g. Clean, inspect and/or replace all parts.

h. To install, pack the bearings with grease, coat the oil seal lips with grease and reverse the removal procedures. Torque hub nut to 22 ft. lbs. (29 Nm).

NOTE: If the cotter pin holes are out of alignment upon reassembly, use a wrench to tighten the nut until the hole in the shaft and a slot of the nut align.

Metro

1. Raise and safely support the vehicle.

2. Remove the tire and wheel assemblies, then the brake line, retaining clip and flexible hose from the center of the rear axle.

3. Remove the tension spring from the rear axle, disconnect the parking brake cable from the turn buckle and the cable joint.

4. Remove the brake backing plate from the knuckle after removing the 4 plate retaining bolts.

5. Using a suitable jack, support the under suspension are to prevent it from lowering.

1. Fuel tank band
2. Strut tower cover
3. Strut rod piston nut
4. Strut assembly
5. Rear toe adjustment bolt
6. No. 2 suspension arm-to-body bolt
7. Stabilizer bar link-tostrut assembly nut
8. No. 2 suspension arm
9. Strut assembly-to-knuckle nut
10. No. 2 suspension arm-to-knuckle bolt
11. Brake line
12. Rear disc brake assembly
13. Rear suspension knuckle
14. Strut rod
15. Strut rod-to-knuckle bolt
16. Strut rod-to-body bolt
17. No. 1 suspension arm-to-knuckle bolt
18. No. 1 suspension arm
19. No. 1 suspension arm-to-body bolt
20. Stabilizer bracket bolt
21. Stabilizer bar bracket
22. Bushing
23. Stabilizer bar link
24. Stabilizer bar link nut
25. Stabilizer bar

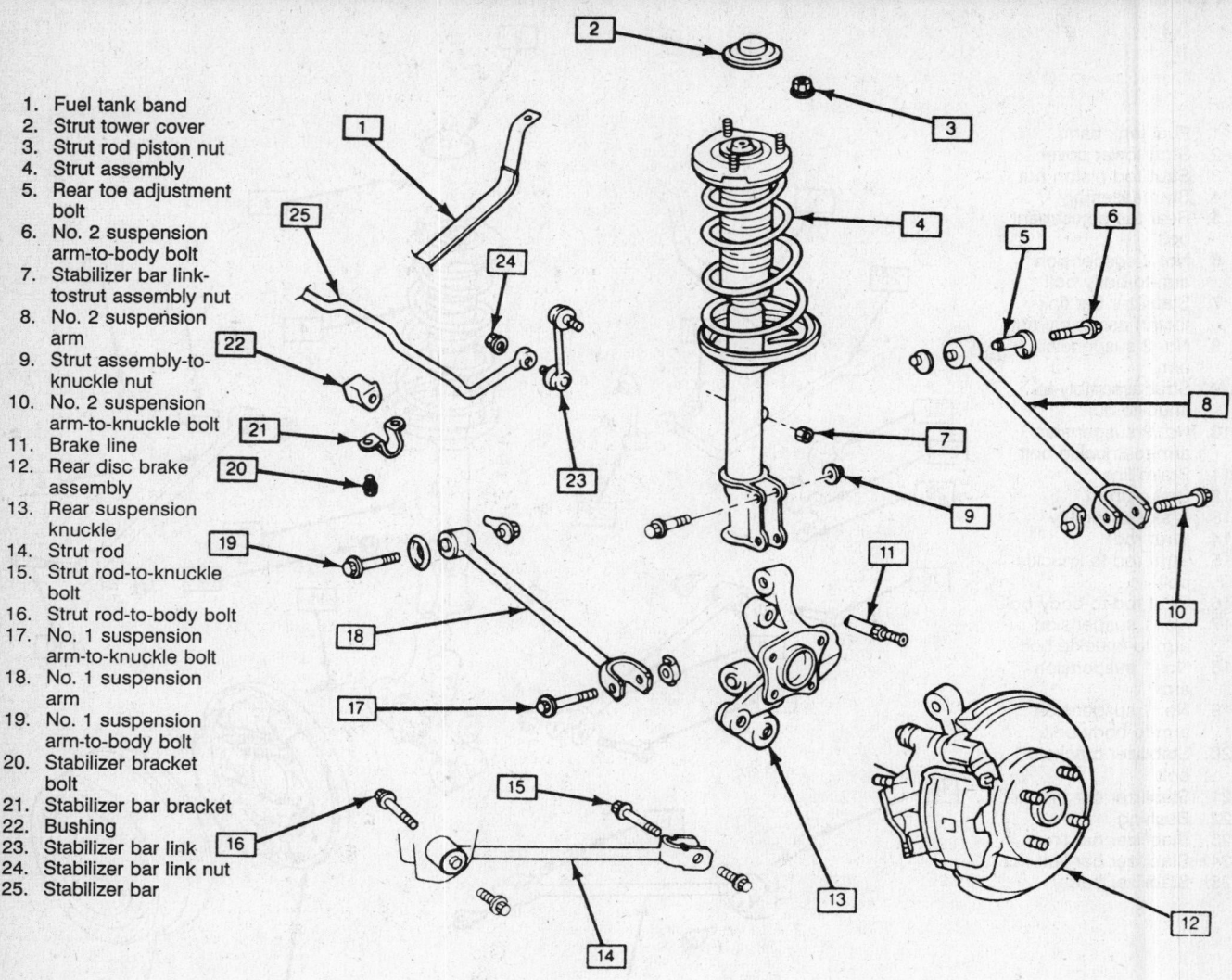

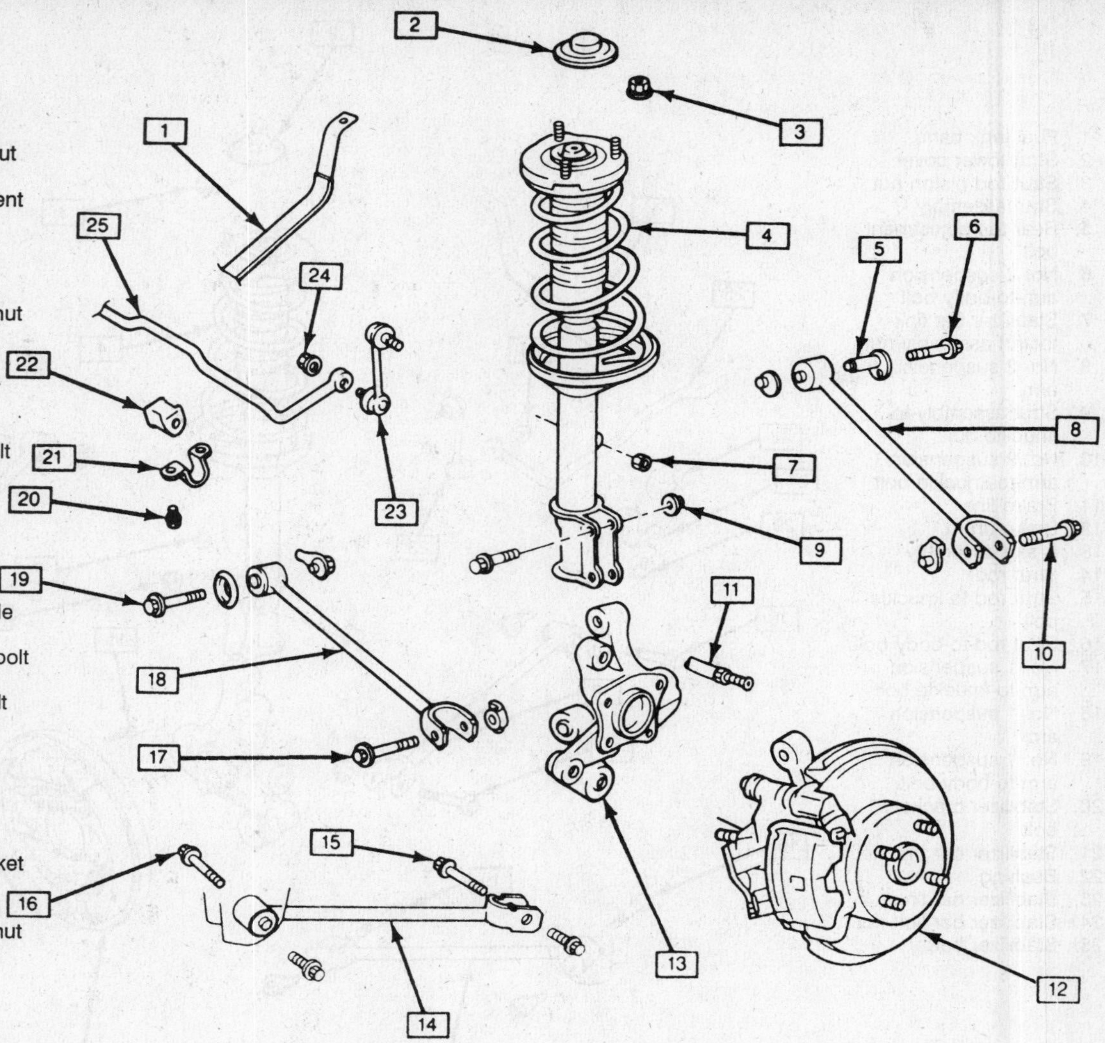

Exploded view of the rear suspension assembly—Prizm GSi models

NOTE: As a preparatory step for this removal, check the stamped line on the washer to use for a guide in the reinstallation.

6. Remove the lower strut mounting bolt.
7. Remove the lower knuckle mounting bolt.
8. Remove the knuckle (spindle) from the suspension arm and knuckle from the strut.

NOTE: If it is hard to remove the knuckle from the strut, open the slit in the knuckle by inserting a wedge. Do not open the slit wider thean necessary.

To install:

9. Install the knuckle to the strut. Align the projection on the strut against the slit of the knuckle and push the strut into the knuckle until it is properly positioned.
10. Install the lower mount strut bolt. Torque it to 44 ft. lbs. (60 Nm).
11. Install the lower mount end of the knuckle to suspension arm. Torque the nut to 44 ft. lbs. (60 Nm).
12. Remove the jack from under the suspension arm.
13. Torque the knuckle lower mount nut to 37 ft. lbs. (50 Nm).
14. Install the backing plate. Torque the retaining bolts to 17 ft. lbs. (23 Nm).
15. Install the brake hose bracket to the knuckle. Install the brake line to the wheel cylinder and tighten the line to 12 ft. lbs. (16 Nm).
16. Install the breather plug cap to the breather plug. Install the brake drum assembly.
17. Install the control rod and control rod nuts. Torque the nuts to 59 ft. lbs. (80 Nm).
18. Bleed the brake system and adjust the brakes as necessary.
19. If removing the wheel bearing use the following procedure:
    a. Raise and support the vehicle safely.
    b. Remove the wheel assembly.
    c. Remove the dust cap, the cotter pin, the castle nut and the washer.
    d. Loosen the adjusting nuts of the parking brake cable.
    e. Remove the plug from the rear of the backing plate. Insert a suitable tool through the hole, making contact with the shoe hold-down spring, then push the spring to release the parking brake shoe lever.
    f. Using a slide hammer tool and a brake drum remover tool, pull the brake drum from the halfshaft.
    g. Using a brass drift and a hammer, drive the rear wheel bearings from the brake drum.

NOTE: When installing the wheel bearings, face the sealed sides (numbered sides) outward. Fill the wheel bearing cavity with bearing grease.

h. Drive the new bearings into the brake drum with the bearing installation tool.

i. To install, use a new seal and reverse the removal procedures. Torque the hub castle nut to 41 ft. lbs. (55 Nm). Bleed the brake system. Operate the brakes 3‹n-dash›5 times to obtain the proper drum-to-shoe clearance. Adjust the parking brake cable.

Storm

1. Raise and safely support the vehicle.
2. Remove the tire and wheel assemblies, then the brake line, retaining clip and flexible hose from the center of the rear axle.
3. Remove the tension spring from the rear axle, disconnect the parking brake cable from the turn buckle and the cable joint.
4. Using a suitable jack, support the under suspension are to prevent it from lowering.
5. Before removing the hub assembly, check the endplay as follows:
   a. Using a suitable dial indicator tool (J-8001 or equivalent) along with the magnetic base (J-26900-1 or equivalent) measure the hub assembly endplay.
   b. If the endplay exceeds 0.020 in. (0.05mm), replace the axle hub bearing.
   c. Press the axle hub bearing out of the hub assembly or if possible use a brass drift and a hammer and drive the rear wheel bearings from the axle hub.
   d. Drive the new bearings into the axle hub assembly with a bearing installation tool.
6. Remove the 4 retaining bolts holding the hub assembly to the knuckle.
7. Remove the hub assembly from the knuckle.
8. Remove the brake backing plate from the knuckle after removing the 4 plate retaining bolts.
9. Remove the through bolt from the trailing link at the axle side.
10. Remove the through bolt from the lateral link at the axle side.
11. Remove the knuckle (spindle) assembly from the strut.

To install:

12. Install the knuckle (spindle) assembly to the strut.
13. Install the through bolt to the lateral link at the axle side. Torque the bolt to 94 ft. lbs. (128 Nm).
14. Install the through bolt from the trailing link at the axle side. Torque the bolt to 94 ft. lbs. (128 Nm).
15. Install the brake backing plate to the knuckle along with the 4 plate retaining bolts. Torque the bolts to 12 ft. lbs. (16 Nm).
16. Install the rear suspension knuckle mounting bolts and nuts to the strut assembly.
17. Install the lateral link mounting bolt to the rear suspension and partially tighten.
18. Install the trailing link mounting bolt to the rear suspension and partially tighten.
19. Install the hub assembly to the knuckle. Install the retaining bolts and torque them to 49 ft. lbs. (66 Nm).
20. Torque the following:
    a. Lateral link bolt to knuckle to 94 ft. lbs. (128 Nm).
    b. Trailing link bolt to knuckle to 94 ft. lbs. (128 Nm).
    c. Strut to knuckle nuts and bolts to 116 ft. lbs. (157 Nm).
21. Check the rear wheel alignment.
22. Install the tire and wheel assemblies.
23. Raise the vehicle and remove the jack. Lower the vehicle and bounce the vehicle up and down to stabilize the suspension.

Sprint

1. Raise and safely support the rear of the vehicle.
2. Remove the brake drum, then the bake cable clip from the trailing arm bracket.
3. Disconnect the brake line from the wheel cylinder, then remove the brake line from the bracket on the trailing arm.
4. Remove the shoe hold-down spring, then disconnect the parking brake cable from the shoe.
5. Remove the brake shoes.
6. Remove the parking brake cable attaching clip, then the parking brake cable from the backing plate and trailing arm hole.
7. Remove brake backing plate attaching bolts, then the brake backing plate.
8. Support the axle using a suitable jack, then remove the lateral arm attaching bolts and the lower shock absorber nut.
9. Lower the rear axle slowly and remove the coil spring.
10. Remove the trailing arm bolts, then remove the rear axle assembly from the vehicle.
11. Reverse procedure to install. Torque backing plate bolts to 13.5–20 ft. lbs. (18–28 Nm), trailing arm bolts to 50.5–65 ft. lbs. (70–90 Nm), lateral rod bolts to 32–50 ft. lbs. (45–70 Nm) and shock absorber bolt to 32–50 ft. lbs. (45-70 Nm).
12. If removing the wheel bearing use the following procedure:
    a. Raise and support the vehicle safely.
    b. Remove the wheel assembly.
    c. Remove the dust cap, the cotter pin, the castle nut and the washer.
    d. Loosen the adjusting nuts of the parking brake cable.
    e. Remove the plug from the rear of the backing plate. Insert a suitable tool through the hole, making contact with the shoe hold-down spring, then push the spring to release the parking brake shoe lever.
    f. Using a slide hammer tool and a brake drum remover tool, pull the brake drum from the halfshaft.
    g. Using a brass drift and a hammer, drive the rear wheel bearings from the brake drum.

NOTE: When installing the wheel bearings, face the sealed sides (numbered sides) outward. Fill the wheel bearing cavity with bearing grease.

h. Drive the new bearings into the brake drum with the bearing installation tool.

i. To install, use a new seal and reverse the removal procedures. Torque the hub castle nut to 58–86 ft. lbs. (80-120 Nm). Bleed the rear brake system. Operate the brakes 3–5 times to obtain the proper drum-to-shoe clearance. Adjust the parking brake cable.

STEERING

Steering Wheel

CAUTION

On vehicles equipped with an air bag, the negative battery cable must be disconnected, before working on the system. Failure to do so may result in deployment of the air bag and possible personal injury.

REMOVAL & INSTALLATION

Nova and Prizm

1. Disconnect the negative terminal from the battery.
2. Remove the screws from the bottom of the steering wheel pad and pull the pad upward and off the steering wheel.
3. Remove the steering wheel-to-steering column nut. Matchmark the steering wheel-to-steering column relationship.
4. Using a steering wheel puller tool, screw the bolts into both sides of

the steering column, turn the puller center bolt to press the steering wheel from the steering shaft.

NOTE: When working on the steering column, be careful not to strike the column in any way, for it is constructed of a collapsible design and will not withstand major shock.

5. To install, align the matchmarks and reverse the removal procedures. Torque the steering wheel-to-steering column nut to 25 ft. lbs. (34 Nm).

Storm

1. If equipped, with an air bag system, disable the air bag system as follows:
   a. Turn the ignition switch to the **OFF** position.
   b. Disconnect the battery negative cable.
   c. Remove fuses C-22 and C-23 from the fuse box.
   d. Disconnect the orange 3-way connector at the base of the steering column.
2. Remove the air bag module attaching screws, then the air bag assembly from the vehicle.

CAUTION

When carrying a live air bag module, make sure that the bag and trim cover are pointed away from you. Never carry the air bag module by the wires or the connector on the underside of the air bag module. In case of an accidental deployment, the bag will then deploy with minimal chance of injury. When placing a live air bag module on a bench or other surface, always face the bag and trim cover in the up position, away from the surface. Never rest a steering column assembly on the steering wheel with the air bag module face down and the column vertical. This is necessary so that a free space is provided to allow the air bag to expand in the unlikely event of accidental deployment. Otherwise, personal injury could result.

3. Remove the steering wheel attaching nut, then using a suitable puller, remove the steering wheel.
4. Disconnect the electrical connectors from the steering wheel, then the rear steering wheel cover.

To install:

5. Reverse procedure to install. Torque steering nut to 25 ft. lbs. (34 Nm) and the air bag module attaching screws to 44 inch lbs. (5 Nm).
6. Reactivate the air bag system as follows:
   a. Turn the ignition switch to the **OFF** position.
   b. Connect the orange 3-way connector at the base of the steering column.
   c. Install fuses C-22 and C-23 to the fuse box.
   d. Connect the battery negative cable.

Spectrum

1. Disconnect the negative battery terminal from the battery.
2. Using a suitable tool, remove the shroud screws from the rear-side of the steering wheel (Type 1) or pry the shroud from the steering wheel (Type 2).
3. Disconnect the horn connector and remove the shroud.
4. Remove the nut/washer retaining the steering wheel to the steering shaft.
5. Using a steering wheel puller, remove the steering wheel.

NOTE: The steering shaft and the steering column are designed to absorb impact from collision. Be careful not to severely jar the steering column and shaft during the removal and installation.

6. To install, reverse the removal procedures.

Sprint

1. Disconnect the negative battery cable.
2. Loosen the pad screws and remove the pad.
3. Remove the steering wheel nut.
4. Scribe a matchmark line on the steering wheel and the shaft.
5. Using a steering wheel puller, pull the steering wheel from the steering shaft.
6. To install, reverse the removal procedures. Torque the steering wheel nut to 19–29 ft. lbs. (26-39 Nm).

Metro

1. If equipped, with an air bag system, disable the air bag system as follows:
   a. Turn the ignition switch to the **OFF** position.
   b. Remove the **SIR IG** fuse in the supplemental inflatable restraint fuse block.
   c. Remove the rear plastic access cover to the air bag module.
   d. Disconnect the yellow 2-way connector and connector position assurance (CPA) inside the inflator module housing.
2. Disconnect the negative battery cable.
3. Remove the air bag module attaching screws, then the air bag assembly from the vehicle.

CAUTION

When carrying a live air bag module, make sure that the bag and trim cover are pointed away from you. Never carry the air bag module by the wires or the connector on the underside of the air bag module. In case of an accidental deployment, the bag will then deploy with minimal chance of injury. When placing a live air bag module on a bench or other surface, always face the bag and trim cover in the up position, away from the surface. Never rest a steering column assembly on the steering wheel with the air bag module face down and the column vertical. This is necessary so that a free space is provided to allow the air bag to expand in the unlikely event of accidental deployment. Otherwise, personal injury could result.

NOTE: The air bag system coil assembly is easily damaged if the correct steering wheel puller tools are not used.

4. Remove the steering wheel attaching nut, then using a suitable puller, remove the steering wheel.
5. Disconnect the electrical connectors from the steering wheel, then the rear steering wheel cover.

To install:

6. Reverse procedure to install. Torque steering nut to 25 ft. lbs. (34 Nm) and the air bag module attaching screws to 44 inch lbs. (5 Nm).
7. Reactivate the air bag system as follows:
   a. Turn the ignition switch to the **OFF** position.
   b. Connect the yellow 2-way connector and connector position assurance (CPA) inside the inflator module housing.
   c. Install the **SIR IG** fuse in the supplemental inflatable restraint fuse block.
   d. Install the rear plastic access cover to the air bag module.
   e. Turn the ignition switch to the **RUN** position. Observe the **INFLATABLE RESTRAINT** indicator lamp.If the lamp does not flash 7 to 9 times and then remain off, there is a problem in the air bag system and further diagnostic testing of the system is needed.

Steering Column

REMOVAL & INSTALLATION

Nova

1. Disconnect the negative battery cable. Remove the steering wheel.
2. Remove the combination switch.
3. Loosen the hole cover clamp screw.
4. Remove the pinch bolt from the yoke.

5. Remove the yoke from the steering gear.
6. Remove the bolts holding the lower column mounting brackets.
7. Remove the bolts holding the upper column to the instrument panel.
8. Remove the column from the vehicle.

To install:

9. When reinstalling, place the column assembly into position and install the upper and lower bracket nuts and bolts finger tight.
10. Position the column assembly so the end of the lower support holes touch the mounting bolts.
11. Tighten the upper and lower support nuts and bolts to 19 ft. lbs. (26 Nm).
12. Install the yoke and tighten the pinch bolt to 26 ft. lbs. (35 Nm).
13. Install the hole cover clamp.
14. Install the combination switch.
15. Install the steering wheel.

Prizm

NOTE: While it is possible to remove the column with the steering wheel attached, it is much easier if the wheel is removed.

1. Disconnect the negative battery cable
2. Remove the left side lower dash trim panel; let it rest on the floor.
3. Remove the upper and lower column covers.
4. Disconnect the wiring harnesses to the combination switch and the ignition switch.
5. Disconnect the park/lock cable from the lock cylinder housing. (automatic tansaxle only)
6. Remove the air filter assembly to gain access to the steering gear.
7. Disconnect the yoke from the steering gear.
8. Remove the lower column mounting bolts.
9. Remove the upper column mounting bolts.
10. Remove the steering column from the vehicle.

To install:

11. When reinstalling, place the column in position and tighten the upper and lower mounting bolts to 19 ft. lbs. (26 Nm).
12. Connect the shaft yoke to the steering gear and tighten the bolt to 26 ft. lbs. (35 Nm).
13. Install the air filter assembly.
14. Connect the wiring harnesses to the combination switch and the ignition switch.
15. Connect the park/lock cable to the lock cylinder housing. Make certain the clip is firmly seated.
16. Install the upper and lower column covers.
17. Reinstall the left side lower dash trim panel.
18. Connect the negative battery cable.

Storm

1. If equipped, with an air bag system, disable the air bag system as follows:
   a. Turn the ignition switch to the **OFF** position.
   b. Disconnect the battery negative cable.
   c. Remove fuses C-22 and C-23 from the fuse box.
   d. Disconnect the orange 3-way connector at the base of the steering column.
2. Remove the air bag module attaching screws, then the air bag assembly from the vehicle.

CAUTION

When carrying a live air bag module, make sure that the bag and trim cover are pointed away from you. Never carry the air bag module by the wires or the connector on the underside of the air bag module. In case of an accidental deployment, the bag will then deploy with minimal chance of injury. When placing a live air bag module on a bench or other surface, always face the bag and trim cover in the up position, away from the surface. Never rest a steering column assembly on the steering wheel with the air bag module face down and the column vertical. This is necessary so that a free space is provided to allow the air bag to expand in the unlikely event of accidental deployment. Otherwise, personal injury could result.

3. Remove the steering wheel attaching nut, then using a suitable puller, remove the steering wheel.
4. Disconnect the electrical connectors from the steering wheel, then the rear steering wheel cover.
5. Remove the lower switch panel and the dash lighter panel.
6. Disconnect the hood release cable, then remove the lap air deflector.
7. Remove the left lower trim panel, then the upper steering column attaching bolts and lower the column.
8. Remove the 2 piece steering column attaching screws, then the 2 piece steering column cover.
9. Disconnect the electrical connectors, then the back drive cable from the ignition switch.
10. Remove the pinch bolt from the steering column knuckle, then the 2 lower column mount attaching nuts. Remove the column from the vehicle.

To install:

11. Install steering column assembly, then the 2 rear column mount nuts. Torque nuts to 18 ft. lbs. (24 Nm).
12. Install intermediate shaft to column, then the pinch bolt. Torque pinch bolt to 30 ft. lbs. (40 Nm).
13. Connect the back drive cable to the ignition switch.
14. Connect all the electrical connectors, then install the 2 piece steering column cover.
15. Install the upper column bolts. Torque bolts to 18 ft. lbs. (24 Nm).
16. Install the lower dash trim panel and the air conditioning lap deflector.
17. Connect the hood release cable, then install the lighter panel and the lower switch panel.
18. Install the steering wheel. Connect the battery negative cable.
19. Reactivate the air bag system as follows:
   a. Turn the ignition switch to the **OFF** position.
   b. Connect the orange 3-way connector at the base of the steering column.
   c. Install fuses C-22 and C-23 to the fuse box.
   d. Connect the battery negative cable.

Spectrum

1. Disconnect the negative battery cable.
2. From under the dash, remove the steering column protector nut, clip and protector.
3. Remove the pinch bolt between the intermediate shaft and the steering shaft.
4. Remove the mounting bracket bolts from the lower column.
5. Remove the steering column-to-instrument panel mounting bolts.
6. Remove the electrical connectors and park lock cable at the ignition switch.

NOTE: If equipped with an automatic transaxle, remove the park lock cable bracket.

7. Remove the steering column assembly.
8. To install, reverse the removal procedures. Torque the steering shaft pinch bolts to 19 ft. lbs. (26 Nm).

Sprint

1. Disconnect the negative battery cable. Remove the steering wheel and combination switch.
2. Remove the lower pinch bolt and separate the lower steering column shaft from the steering column. Disconnect and tag the electrical connectors from the steering column.
3. Remove the upper steering column mounting bolts and the column from the dash.
4. To install, reverse the removal procedure. Torque the lower bracket

bolts to 8–12 ft. lbs. (11-17 Nm), the upper bracket bolts to 10 ft. lbs. (14 Nm) and the steering shaft bolt to 15–22 ft. lbs. (20-30 Nm).

Metro

NOTE: The steering column is very susceptible to damage once it has been removed from the vehicle.

The vehicle's wheels must be in a straight ahead position and the key must be in the LOCK position when removing or installing the steering column. Failure to do so may cause the air bag coil assembly to become uncentered and may result in unneeded air bag system damage. In the event air bay deployment has occurred, inspect the coil assembly for any signs of scorching, melting or other damage due to excessive heat. If the coil has been damaged, replace it. The steering column should never be supported by only the lower support bracket, damage to the column lower bearing adapter could result.

1. If equipped, with an air bag system, disable the air bag system as follows:
   a. Turn the ignition switch to the **OFF** position.
   b. Remove the **SIR IG** fuse in the supplemental inflatable restraint fuse block.
   c. Remove the rear plastic access cover to the air bag module.
   d. Disconnect the yellow 2-way connector and connector position assurance (CPA) inside the inflator module housing.
2. Disconnect the negative battery cable.
3. Remove the air bag module attaching screws, then the air bag assembly from the vehicle.

CAUTION

When carrying a live air bag module, make sure that the bag and trim cover are pointed away from you. Never carry the air bag module by the wires or the connector on the underside of the air bag module. In case of an accidental deployment, the bag will then deploy with minimal chance of injury. When placing a live air bag module on a bench or other surface, always face the bag and trim cover in the up position, away from the surface. Never rest a steering column assembly on the steering wheel with the air bag module face down and the column vertical. This is necessary so that a free space is provided to allow the air bag to expand in the unlikely event of accidental deployment. Otherwise, personal injury could result.

NOTE: The air bag system coil assembly is easily damaged if the correct steering wheel puller tools are not used.

4. Remove the steering wheel attaching nut, then using a suitable puller, remove the steering wheel.
5. Disconnect the electrical connectors from the steering wheel, then the rear steering wheel cover.
6. Remove the steering shaft trim panel. Remove the lower steering column trim panel and the steering column reinforcement plate.
7. Disconnect the steering column electrical connectors. Remove the brake transaxle shift interlock cable from the ignition switch.
8. Remove the steering column to steering shaft joint pinch bolt. Remove the upper and lower steering column mounting nuts and remove the steering column from the vehicle.

To install:

9. Installation is the reverse order of the removal procedure. Torque the steering column upper and lower mounting nuts to 10 ft. lbs. (14 Nm). Torque the steering column to steering shaft joint pinch bolt to 18 ft. lbs. (25 Nm).
10. Reactivate the air bag system as follows:
   a. Turn the ignition switch to the **OFF** position.
   b. Connect the yellow 2-way connector and connector position assurance (CPA) inside the inflator module housing.
   c. Install the **SIR IG** fuse in the supplemental inflatable restraint fuse block.
   d. Install the rear plastic access cover to the air bag module.
   e. Turn the ignition switch to the **RUN** position. Observe the **INFLATABLE RESTRAINT** indicator lamp.If the lamp does not flash 7 to 9 times and then remain off, there is a problem in the air bag system and further diagnostic testing of the system is needed.

Manual Rack and Pinion

ADJUSTMENT

Nova and Prizm

NOTE: To perform the adjustment procedure, the steering rack and pinion assembly should be removed from the vehicle.

1. Remove the cover from the intermediate shaft.
2. Loosen the upper pinch bolt. Remove the lower pinch bolt at the pinion shaft.
3. Loosen the wheel lug nuts.
4. Elevate and safely support the vehicle.
5. Remove both front wheels.
6. Remove the cotter pins from both ball joints and remove the nuts.
7. Using a tie rod separator, remove both tie rod joints from the knuckles.
8. Remove the nuts and bolts attaching the steering rack to the body.
9. Remove the rack through the access hole.

To install:

10. Install the rack through the access hole. Secure it with the retaining bolts and nuts and tighten them to 43 ft. lbs. (58 Nm).
11. Connect the tie rods to each knuckle. Tighten the nuts to 36 ft. lbs. (49 Nm) and install new cotter pins.
12. Install the front wheels.
13. Lower the vehicle to the ground.
14. Install the lower pinch bolt at the pinion shaft. Tighten the upper and lower bolts to 26 ft. lbs. (35 Nm).
15. Install the cover on the intermediate shaft.

Prizm

1. Remove the cover from the intermediate shaft.
2. Loosen the upper pinch bolt. Remove the lower pinch bolt at the pinion shaft.
3. Loosen the wheel lug nuts.
4. Elevate and safely support the vehicle.
5. Remove both front wheels.
6. Install an engine support and tension it to support the engine without raising it.

CAUTION

The engine hoist is in place and under tension. Use care when repositioning the vehicle and make necessary adjustments to the engine support.

7. Remove the 3 bolts holding the center crossmember to the radiator support.
8. Remove the covers from the front and center mount bolts.
9. Remove the 2 front mount bolts, then the center mount bolts.
10. Support the crossmember and remove the 2 rear mount bolts.
11. Remove the 2 bolts holding the center crossmember to the main crossmember.
12. Use a floor jack and a wide piece of wood to support the main crossmember.
13. Remove the 8 bolts holding the main crossmember to the body.
14. Remove the 2 bolts holding the lower control arm brackets to the body.

CAUTION

The crossmembers are loose and free to fall. Make sure they are properly supported.

15. Slowly lower the main crossmember while holding onto the center crossmember.
16. Remove the cotter pins from both ball joints and remove the nuts.
17. Using a tie rod separator, remove both tie rod joints from the knuckles.
18. Remove the nuts and bolts attaching the steering rack to the body.
19. Remove the rack through the right side wheel well.

To install:

20. To reinstall the rack, place it in position through the right wheel well and tighten the bracket bolts to 45 ft. lbs. (59 Nm).
21. Attach the tie rods to the knuckles. Tighten the nuts to 36 ft. lbs. (49 Nm) and install new cotter pins.
22. Position the center crossmember over the center and rear transaxle mount studs; start 2 nuts on the center mount.
23. Loosely install the 3 bolts holding the center crossmember to the radiator support.
24. Loosely install the 2 front mount bolts.
25. Raise the main crossmember into position over the rear mount studs and align all underbody bolts. Install the 2 rear mount nuts loosely.
26. Install the 8 main crossmember to underbody bolts loosely.
27. Install the 2 lower control arm bracket bolts loosely.
28. Loosely install the 2 bolts holding the center crossmember to the main crossmember.
29. The crossmembers, bolts and brackets should now all be in place and held loosely by their nuts and bolts. If any repositioning is necessary, do so now.
30. Tighten the components below in the order listed to the correct torque specification:

Main crossmember to underbody bolts: 152 ft. lbs. (206 Nm).

Lower control arm bolts: 94 ft. lbs. (127 Nm).

Center crossmember to radiator support bolts: 45 ft. lbs. (59 Nm).

Front, center and rear mount bolts: 45 ft. lbs. (59 Nm).

31. Install the covers on the front and center mount bolts.
32. Install the front wheels.
33. Lower the vehicle to the ground.
34. Connect the yoke to the pinion and tighten both the upper and lower bolts to 26 ft. lbs. (35 Nm).
35. Install the yoke cover.

Spectrum

1. Remove the intermediate shaft cover.
2. Loosen the upper pinch bolt. Remove the lower pinch bolt at the pinion shaft.
3. Raise and safely support the vehicle. Remove the wheel assemblies.
4. Remove both tie rod ends from the steering knuckles and the left inner tie rod from the rack.
5. Remove the steering gear-to-body attaching nuts and the rack and pinion assembly from the vehicle.
6. Installation is the reverse order of the removal procedure.

Sprint and Metro

1. Slide the driver's seat back as far as possible.
2. Pull off the front part of the floor mat on the driver's side and remove the steering shaft joint cover.
3. Loosen the steering shaft upper joint bolt, but do not remove.
4. Remove the steering shaft lower joint bolt and disconnect the lower joint from the pinion.
5. Raise and support the vehicle safely.
6. Remove the tie rod ends from the steering knuckles. Mark the left and right tierods accordingly.
7. From under the dash, remove the steering joint cover.
8. Remove the lower steering shaft-to-steering gear clinch bolt and separate the steering shaft from the steering gear.
9. Remove the steering gear mounting bolts, the brackets and the steering gear case from the vehicle.
10. Installation is the reverse order of the removal procedure. On the Sprint, torque the steering gear case bolts to 14–22 ft. lbs. (20-30 Nm); the steering gear-to-steering shaft bolt to 14–22 ft. lbs. (20-30 Nm) and the tie rod end-to-steering knuckle nut to 22–40 ft. lbs. (30-55 Nm).
11. On the Metro, torque the steering gear case bolts to 18 ft. lbs. (25 Nm); the steering gear-to-steering shaft bolt to 18 ft. lbs. (25 Nm) and the tie rod end-to-steering knuckle nut to 32 ft. lbs. (43 Nm).

ADJUSTMENT

Nova

1. Remove the steering rack and pinion and place it in a vise.
2. To adjust the pinion bearing turning torque, perform the following procedures:
   a. Using the pinion bearing locknut wrench tool No. J–35415 or equivalent, loosen the pinion bearing locknut.
   b. Using a torque wrench, pinion spanner wrench tool No. J‹n-dash›35416 or equivalent, and socket tool No. J–35422 or equivalent, adjust the pinion bearing screw torque to 3.2 inch lbs.
   c. Loosen the adjusting screw until the turning torque is 2‹n-dash›2.9 inch lbs.
   d. Using sealant, coat the pinion locknut threads, then, torque the nut to 83 ft. lbs. (113).
   e. Recheck the turning torque, if it is incorrect, repeat this procedure.
3. To adjust the rack guide screw, perform the following procedure:
   a. At the rear of the steering rack, loosen the rack guide spring cap locknut.
   b. Remove the rack guide adjusting plug.
   c. Install the rack guide adjusting plug and count the number of rotations, then, back-off the plug ½the number of turns.d. Using a socket wrench and socket tool J–35423 or equivalent, hold the rack guide adjusting plug. Using a torque wrench with a locknut wrench adapter tool J–35692 or equivalent, torque the rack guide locknut to 18 ft. lbs. (25 Nm), then, back-off the nut 25 degrees (use a compass to measure the position in degrees).
   e. Using a torque wrench and socket tool J–35422 or equivalent, check the pinion shaft preload; it should be 6.9–11.3 inch lbs.

NOTE: If the preload is insufficient, retorque the locknut and back it off 12 degrees, then, recheck the preload.

4. Using sealant, coat the locknut threads and torque the locknut to 51 ft. lbs. (69 Nm).
5. To install the steering gear, reverse the removal procedures.

Power Rack and Pinion

REMOVAL & INSTALLATION

Nova

1. Remove the intermediate shaft cover.
2. Loosen the upper pinch bolt and remove the lower pinch bolt.
3. Place a drain pan below the power steering rack assembly. Clean the area around the line fittings on the rack.
4. Loosen the front wheel lug nuts.
5. Safely elevate and support the vehicle.
6. Remove the front wheels.
7. Remove the cotter pins and nuts from both tie rod joints. Separate the

joints from the knuckle using a tie rod joint separator.

8. Support the transaxle with a jack.

9. Remove the rear bolts holding the engine crossmember to the body.

10. Remove the nut and bolt holding the rear engine mount to the mount bracket.

11. Label and disconnect the fluid pressure and return lines at the rack.

12. Remove the 4 bolts and nuts holding the rack brackets to the body. It will be necessary to slightly raise and lower the rear of the transaxle to gain access to the bolts.

13. Remove the rack through the access hole.

To install:

14. When reinstalling, place the rack in position through the access hole and install the retaining brackets to the body. Tighten the nuts and bolts to 39 ft. lbs. (53 Nm).

15. Connect the fluid lines to the rack.

16. Install the nut and bolt holding the rear engine mount to the mount bracket. Tighten the nut and bolt to 29 ft. lbs. (39 Nm).

17. Reinstall the engine crossmember bolts and tighten them to 29 ft. lbs. (39 Nm).

18. Remove the jack from the transaxle.

19. Connect the tie rod ends to the knuckles. Tighten the nuts to 36 ft. lbs. (49 Nm) and install new cotter pins.

20. Install the wheels and lower the vehicle to the ground.

21. Connect the intermediate shaft to the steering rack. Install the lower bolt; tighten both the upper and lower bolts to 26. ft. lbs. (35 Nm). Install the intermediate shaft cover.

22. Add fluid and bleed the system.

Prizm

1. Place a drain pan under the steering rack.

2. Remove the cover from the intermediate shaft.

3. Loosen the upper pinch bolt. Remove the lower pinch bolt at the pinion shaft.

4. Loosen the wheel lug nuts.

5. Elevate and safely support the vehicle.

6. Remove both front wheels.

7. Install an engine support and tension it to support the engine without raising it.

CAUTION

The engine hoist is now in place and under tension. Use care when repositioning the vehicle and make necessary adjustments to the engine support.

8. Remove the 3 bolts holding the center crossmember to the radiator support.

9. Remove the covers from the front and center mount bolts.

10. Remove the 2 front mount bolts, then the center mount bolts and then the 2 rear mount bolts.

11. Remove the 2 bolts holding the center crossmember to the main crossmember.

12. Use a floor jack and a wide piece of wood to support the main crossmember.

13. Remove the 8 bolts holding the main crossmember to the body.

14. Remove the 2 bolts holding the lower control arm brackets to the body.

CAUTION

The crossmembers are loose and free to fall. Make sure they are properly supported.

15. Slowly lower the main crossmember while holding onto the center crossmember.

16. Remove the cotter pins from both tie rod ball joints and remove the nuts.

17. Using a tie rod separator, remove both tie rod joints from the knuckles.

18. Label and disconnect the fluid pressure and return lines from the rack.

19. Remove the nuts and bolts attaching the steering rack to the body.

20. Remove the rack through the right side wheel well.

To install:

21. To reinstall the rack, place it in position through the right wheel well and tighten the bracket bolts to 43 ft. lbs. (58 Nm).

22. Connect the fluid lines to the rack and tighten the fittings to 33 ft. lbs. (44 Nm). Make certain the fittings are correctly threaded before tightening them.

23. Attach the tie rods to the knuckles. Tighten the nuts to 36 ft. lbs. (49 Nm) and install new cotter pins.

24. Position the center crossmember over the center and rear transaxle mount studs; start 2 nuts on the center mount.

25. Loosely install the 3 bolts holding the center crossmember to the radiator support.

26. Loosely install the 2 front mount bolts.

27. Raise the main crossmember into position over the rear mount studs and align all underbody bolts. Install the 2 rear mount nuts loosely.

28. Install the 8 main crossmember to underbody bolts loosely.

29. Install the 2 lower control arm bracket bolts loosely.

30. Loosely install the 2 bolts holding the center crossmember to the main crossmember.

31. The crossmembers, bolts and brackets should now all be in place and held loosely by their nuts and bolts. If any repositioning is necessary, do so now.

32. Tighten the components below in the order listed to the correct torque specification:

Main crossmember to underbody bolts: 152 ft. lbs. (206 Nm).

Lower control arm bolts: 94 ft. lbs. (127 Nm).

Center crossmember to radiator support bolts: 45 ft. lbs. (61 Nm).

Front, center and rear mount bolts: 45 ft. lbs. (61 Nm).

33. Install the covers on the front and center mount bolts.

34. Install the front wheels.

35. Lower the vehicle to the ground.

36. Connect the yoke to the pinion and tighten both the upper and lower bolts to 26 ft. lbs. (35 Nm).

37. Install the yoke cover.

38. Add power steering fluid to the reservoir and bleed the system.

Spectrum

1. Raise and support the vehicle safely.

2. Remove both tie rod ends from the steering knuckles and the right inner tie rod from the rack.

3. Place a drain pan under the rack assembly and clean around the pressure lines at the rack valve.

4. Cut the plastic retaining straps at the power steering lines and hose.

5. Remove the power steering pump lines, the rack valve and drain the fluid into the pan.

6. Remove the rack and pinion.

7. To install, reverse the removal procedures, add fluid, bleed the system and check the toe-in.

Storm

1. Remove the steering shaft cover and 3 retaining nuts on the dust boot retaining ring.

2. Raise and safely support the vehicle. Remove the tires and wheels.

3. Remove the dust boot from the bulkhead. Place a drain pan beneath the vehicle.

4. Remove the pinch bolt from the intermediate steering shaft. Disconnect the tierod ends from the steering knuckles.

5. Remove the hold down bracket from the steering lines.

6. Remove the high pressure line from the steering rack and pinion. Remove the rack to crossmember bolts.

7. Remove the retaining brackets from the rack mounts. Position the rack away from the mounts.

8. Remove the return line from the rack. Remove the intermediate steering shaft knuckle.

9. Remove the rack and pinion from the right side of the vehicle.
10. Installation is the reverse order of the removal procedure. Torque the rack to crossmember bolts to 41 ft. lbs. (59 Nm). Torque the high pressure and return lines to 20 ft. lbs. (27 Nm). Torque the tierod end nuts 40 ft. lbs. (54 Nm).

ADJUSTMENT

Nova

NOTE: To perform the adjustment procedure, the steering rack and pinion assembly should be removed from the vehicle.

1. Remove the steering rack and place it in a vise.
2. To adjust the pinion bearing turning torque, perform the following procedures:
   a. Using the socket wrench and socket tool No. J–35428 or equivalent, loosen the pinion bearing locknut.
   b. Using a socket wrench and socket tool No. J-35428 or equivalent, hold the pinion from turning. Using a torque wrench and socket, torque the lower pinion locknut to 48 ft. lbs. (65 Nm).
3. To adjust the rack guide cap, perform the following procedure:
   a. At the rear of the steering gear, loosen the rack guide spring cap locknut.
   b. Using a socket wrench and socket tool No. J-35423 or equivalent, torque the rack guide locknut to 18 ft. lbs. (24 Nm), then, back-off the nut 12 degrees, use a compass to measure the position in degrees.
   c. Using a torque wrench and socket tool No. J-35428 or equivalent, check the pinion shaft preload; it should be 7–11 inch lbs.
4. Using sealant, coat the locknut threads and torque the locknut to 33 ft. lbs. (45 Nm).
5. To install the steering rack, reverse the removal procedures. Bleed the power steering system.

Power Steering Pump

REMOVAL & INSTALLATION

Nova

EXCEPT TWINCAM ENGINE

1. Place a drain pan below the pump.
2. Remove the air cleaner assembly
3. Remove the clamp from the fluid return hose. Disconnect the pressure and return hoses at the pump. Plug the hoses and suspend them with the ends upward to prevent leakage.
4. Loosen the pump pulley nut. Push down on the belt to keep the pulley from turning.
5. Remove the adjusting bolt.
6. Remove the pivot bolt and remove the drive belt.
7. Remove the pump assembly.
8. Remove the pump bracket.
9. Remove the pulley. Be careful not to lose the small woodruff key between the pulley and the shaft.

To install:

10. To reinstall, place the pump in position and temporarily install the 2 mounting bolts.
11. Install the pump bracket and tighten the bolts to 29 ft. lbs. (39 Nm).
12. Install the pump pulley and the woodruff key. Tighten the pulley nut to 32 ft. lbs. (43 Nm).
13. Install the drive belt, making certain that all the grooves of the belt are engaged on the pulley. Adjust the belt to the proper tension.
14. Connect the pressure and return lines to the pump. Tighten the fittings to 33 ft. lbs. (44 Nm). Install the clamp on the return hose.
15. Install the air cleaner assembly.
16. Fill the reservoir to the proper level with power steering fluid and bleed the system.
17. After the vehicle has been driven for about an hour, double check the belt adjustment.

TWINCAM ENGINE

1. Place a drain pan below the pump.
2. Remove the air cleaner assembly.
3. Disconnect the return hose from the pump, then disconnect the pressure hose. Plug the lines immediately to prevent fluid loss and contamination.
4. Remove the splash shield under the engine.
5. Remove the pulley nut. Push down on the drive belt to prevent the pulley from turning.
6. Loosen the idler pulley nut and loosen the adjusting bolt.
7. Remove the drive belt.
8. Loosen the pump pulley and woodruff key. Don't lose the woodruff key.
9. Remove the upper mounting bolt.
10. Loosen the lower mounting bolt and pivot the pump downward.
11. Disconnect the oil pressure switch connector.
12. Remove the pump bracket mounting bolts; remove the pump from the engine with the bracket attached.
13. Remove the pulley from the pump and the pump from the bracket.
14. Remove the idler pulley bracket.

To install:

15. When reinstalling, mount the pump on the bracket and loosely install the lower mounting bolt.
16. Temporarily insert the pulley onto the pump shaft without the woodruff key. The pulley cannot be installed after the pump is installed on the engine.
17. Install the pump and bracket onto the engine. Tighten the upper mounting bolts to 29 ft. lbs. (39 Nm).
18. Connect the oil pressure switch connector.
19. Install the idler pulley bracket; tighten the 3 mounting bolts to 29 ft. lbs. (39 Nm).
20. Tighten the lower mounting bolts to 29 ft. lbs. (39 Nm).
21. Install the woodruff key into the pulley and install the drive belt. Make certain the ribs of the belt are properly placed on all the pulleys.
22. Tighten the pulley nuts on the pump and idler to 28 ft. lbs. (38 Nm).
23. Connect the pressure hose and tighten its fitting to 33 ft. lbs. (45 Nm).
24. Connect the return hose.
25. Install the air cleaner and install the lower splash shield.
26. Adjust the belt to the proper tension.
27. Fill the reservoir to the proper level with power steering fluid and bleed the system.
28. After the vehicle has been driven for about an hour, double check the belt adjustment.

1989-90 Prizm

1. Place a drain pan below the pump.
2. Elevate and safely support the vehicle.
3. Remove the right front wheel.
4. Place a floor jack under the engine block and support it. Use a broad piece of wood to spread the load evenly and prevent damage.
5. Remove the bolt from the right side engine mount and lower the engine about 50mm to gain access to the lower power steering pump throughbolt.
6. Working through the right wheel well, remove the lower pump throughbolt.
7. Disconnect the fluid lines from the pump and plug them immediately.
8. Remove the upper mounting bolt from the pump and remove the pump.

To install:

9. When reinstalling, place the pump in position and install the mounting bolts. Tighten them to 29 ft. lbs. (39 Nm).
10. Raise the engine to its normal position and install the engine mount bolt, tightening it to 69 ft. lbs. (93 Nm).
11. Connect the fluid lines to the

pump and tighten the pressure hose fitting to 34 ft. lbs. (46 Nm).

12. Install the belt and adjust it to the proper tension.

13. Install the right front wheel.

14. Remove the jack and drain pan from under the engine.

15. Lower the vehicle to the ground.

16. Fill the reservoir to the proper level with power steering fluid and bleed the system.

17. After the vehicle has been driven for about an hour, double check the belt adjustment.

NOTE: If replacing the pump, switch the pulley and the mounting nut to the new pump.

1991-92 Prizm

BASE AND LSI MODELS

1. Disconnect the negative battery cable. Remove the air cleaner.

2. Remove the power steering pump drive belt.

3. Loosen the power steering hose retaining clip and remove the return hose from the power steering pump.

4. Remove the pressure hose from the pump using a injection line wrench J-29698 or equivalent.

5. Remove the upper mounting bolt from the pump.

6. Remove the vacuum hoses from the power steering switch. Remove the 3 bolts retaining the pump bracket mount and the 2 from the engine block and the one from the engine mount.

7. Remove the lower mounting bolt and the power steering pump from the vehicle.

8. Installation is the reverse order of the removal procedure. Torque the power steering pump mounting brackets to 29 ft. lbs. (39 Nm). Connect the pressure hose to the power steering pump union bolt and torque it to 32 ft. lbs. (44 Nm). Adjust the belt tension and bleed the system.

GSi MODELS

1. Disconnect the negative battery cable. Place a drain pan under the vehicle.

2. Remove the windshield washer reservoir retaining bolt; position the reservoir out of the way.

3. Loosen the drive pulley retaining nut. Use a J-35416 or equivalent adjusting screw wrench to keep the pulley from turning while loosening the retaining nut.

4. Raise and support the vehicle safely. Loosen the jam nut on the idler pulley. Lower the vehicle and remove the power steering pump drive belt.

5. Remove the drive pulley retaining nut from the shaft and remove the pulley from the pump.

6. Remove the power steering pressure lines from the pump.

7. Remove the 2 power steering pump retaining bolts and remove the pump from the bracket.

8. Installation is the reverse order of the removal procedure. Torque the power steering pump mounting brackets to 29 ft. lbs. (39 Nm). Connect the pressure hose to the power steering pump union bolt and torque it to 32 ft. lbs. (44 Nm). Adjust the belt tension and bleed the system.

Spectrum and Storm

1. Disconnect the negative battery cable.

2. Place a drain pan below the pump.

3. Remove the pressure hose clamp, pressure hose and return hose. Drain the fluid from the pump and reservoir.

4. Remove the adjusting bolt, pivot bolt and drive belt.

5. Remove the pump assembly.

6. To install, reverse the removal procedures, tighten the pressure hose to 20 ft. lbs. (27 Nm), and the pivot bolt to 15 ft. lbs. (20 Nm), adjust the drive belt, fill the reservoir and bleed the system.

BELT ADJUSTMENT

Nova

1. Using a belt tension gauge tool BT–33–73F or equivalent, position it on the drive belt, between the longest span of 2 pulleys.

2. Loosen the power steering adjusting and pivot bolts.

3. Move the pump to adjust the drive belt tension.

NOTE: The belt deflection should be 0.31–0.39 in. with moderate thumb pressure (about 20 lbs.) applied in the center of the span.

4. Torque the power steering pump pivot/adjusting bolts to 29 ft. lbs. (40 Nm).

1989-90 Prizm

1. Install a J-23600-B or equivalent belt tension gauge on the pump belt. Loosen the 2 pump attaching bolts and nuts.

2. Using a suitable belt tension adjusting tool (between the pump and the engine block), push the pump away from the engine until the belt tension is set to specifications.

a. If a new belt is being used, tighten the belt tensioner to 100-150 lbs.

b. If a used belt is being installed, tighten the belt tensioner to 60-100 lbs.

3. Torque the upper bolt and then the lower through bolt to 29 ft. lbs. (39 Nm).

1991-92 Prizm

1. Disconnect the negative battery cable. Loosen the upper and lower alternator mounting bolts.

2. Remove the alternator drive belt.

3. Loosen the upper and lower power steering mounting bolts.

4. Make sure that the power steering drive belt is properly aligned in the grooves on the pulleys. Install a J-23600-B or equivalent belt tension gauge on the pump belt.

5. Using a suitable belt tension adjusting tool (between the pump and the engine block), push the pump away from the engine until the belt tension is set to specifications.

a. If a new belt is being used, tighten the belt tensioner to 30-46 lbs.

b. If a used belt is being installed, tighten the belt tensioner to 18-30 lbs.

6. Torque the upper bolt and then the lower through bolt to 29 ft. lbs. (39 Nm).

7. Make sure that the alternator drive belt is properly aligned in the grooves on the pulleys. Install a J-23600-B or equivalent belt tension gauge on the pump belt.

8. Using a suitable belt tension adjusting tool apply tension on the alternator until the belt tension is set to specifications.

a. If a new belt is being used, tighten the belt tensioner to 141-182 lbs.

b. If a used belt is being installed, tighten the belt tensioner to 111-152 lbs.

9. Torque the upper bolt to 17 ft. lbs. (23 Nm) and then the lower through bolt and nut to 26 ft. lbs. (35 Nm).

Storm

1. Install belt tension gauge J–23600–B or equivalent, onto the pump drive belt.

2. Loosen the pump attaching nuts and bolts.

3. Using a ½ in. drive ratchet or breaker bar, move pump until the correct belt tension is obtained.

4. Belt tension should be 90 lbs. If equipped with air conditioning, belt tension should be 145 lbs.

5. Adjust the pivot and adjusting bolts to 15 ft. lbs. (20 Nm).

Spectrum

NOTE: The following procedures require the use of GM belt tension gauge BT–33–95–ACBN (regular V-belts) or BT–33–97M (poly V-belts).

1. If the belt is cold, operate the en-

gine, at idle speed, for 15 minutes; the belt will seat itself in the pulleys allowing the belt fibers to relax or stretch. If the belt is hot, allow it to cool, until it is warm to the touch.

NOTE: A used belt is one that has been rotated at least 1 complete evolution on the pulleys. This begins the belt seating process and it must never be tensioned to the new belt specifications.

2. Loosen the component-to-mounting bracket bolts.
3. Using a GM belt tension gauge No. BT-33-95-ACBN (standard V-belts) or BT-33-97M (poly V-belts), place the tension gauge at the center of the belt between the longest span.
4. Applying belt tension pressure on the component, adjust the drive belt tension to the correct specifications. The belt tension should deflect about ¼ in. over a 7–10 in. span or ½in. over a 13–16 in. span.5. While holding the correct tension on the component, tighten the component-to-mounting bracket bolt.
6. When the belt tension is correct (70–110 inch lbs.), remove the tension gauge.

SYSTEM BLEEDING

Nova and Prizm

1. With the engine running, turn the wheel all the way to the left and shut off the engine.
2. Add power steering fluid to the **COLD** or **MIN** mark on the indicator.
3. Start the engine and run at fast idle for about 15 seconds. Stop the engine and recheck the fluid level. Add to the **COLD** mark as needed.
4. Start the engine and bleed the system by turning the wheels from left to right 3 or 4 times.
5. Stop the engine and check the fluid level and condition. Fluid with air in it is a light tan color. This air must be eliminated from the system before normal operation can be obtained. Repeat Steps 3 and 4 until the correct fluid color and fluid level is obtained.

Spectrum and Storm

1. Turn the wheels to the extreme left.
2. With the engine stopped, add power steering fluid to the min mark on the fluid indicator.
3. Start the engine and run it for 15 seconds at fast idle.
4. Stop the engine, recheck the fluid level and refill to the min mark.
5. Start the engine and turn the wheels from side to side, 3 times.
6. Stop the engine check the fluid level.

NOTE: If air bubbles are still present in the fluid, the procedures must be repeated.

Tie Rod Ends

REMOVAL & INSTALLATION

Nova and Prizm

1. Raise the front of the vehicle and support it safely. Remove the wheel.
2. Remove the cotter pin and nut holding the tie rod to the steering knuckle.
3. Using a tie rod separator, press the tie rod out of the knuckle.

NOTE: Use only the correct tool to separate the tie rod joint. Replace the joint if the rubber boot is cracked or ripped.

4. Matchmark the inner end of the tie rod to the end of the steering rack.
5. Loosen the locknut and remove the tie rod from the steering rack.
6. Install the tie rod ends onto the rack ends and align the matchmarks made earlier.
7. Tighten the locknuts to 35 ft. lbs. (47 Nm).
8. Connect the tie rod joint to the knuckle. Tighten the nut to 36 ft. lbs.(49 Nm) and install a new cotter pin.
9. Install the wheel and lower the vehicle to the ground.

Spectrum and Storm

1. Raise and safely support the vehicle. Remove the tire and wheel assemblies.
2. Remove the castle nut from the ball joint. Using a ball joint removal tool, separate the tie rod from the steering knuckle.
3. Disconnect the retaining wire from the inner boot and pull back the boot.
4. Using a chisel, straighten the staked part of the locking washer between the tie rod and the rack.
5. Remove the tie rod from the rack.
6. Reverse procedure to install.

Sprint and Metro

1. Raise and safely support the front of the vehicle. Remove the front wheel assembly.
2. Remove the cotter pin and the castle nut from the tie rod end.
3. Using the ball joint removal tool J-21687-02, remove the tie rod end ball joint from the steering knuckle.
4. Loosen the locknut on the tie rod end.
5. Unscrew the tie rod end from the tie rod, count the number of revolutions necessary to remove the tie rod end, for installation purposes.
6. At the steering gear, remove the boot clamps and pull the boot back over the tie rod.
7. Using a pair of pliers, bend the lockwasher back from the tie rod joint.
8. Using 2 wrenches, hold the steering rack and unscrew the tie rod end. Remove the tie rod and slide the boot from the tie rod.
9. Installation is the reverse order of the removal procedure, reverse the removal procedure. On the Sprint, torque the tie rod-to-steering rack to 51–72 ft. lbs. (69-98 Nm), the tie rod end locknut to 26–44 ft. lbs. (35-60 Nm) and the tie rod end to steering knuckle to 22–40 ft. lbs. (30-54 Nm).
10. On the Metro, the tie rod end locknut to 33 ft. lbs. (45 Nm) and the tie rod end to steering knuckle nut to 32 ft. lbs. (43 Nm).
11. With the tie rod secured to the steering gear, bend the lockwasher over the flat spot on the tie rod ball end.

BRAKES

For all brake system repair and service procedures not detailed below, please refer to "Brakes" in the Unit Repair section.

Master Cylinder

REMOVAL & INSTALLATION

NOTE: Be careful not to spill brake fluid on the painted surfaces of the vehicle; it will damage the finish.

Nova and Prizm

1. Disconnect the negative battery cable.
2. Clean the area at the reservoir and brake lines to prevent entry of dirt into the system.
3. Disconnect the wiring to the brake fluid level switch. Release the wiring from any clips.
4. On Prizm vehicles, remove the air intake duct.
5. Use a syringe to remove the fluid from the reservoir. Store the fluid in a clean glass jar with a lid.
6. Disconnect the brake lines from the master cylinder. Plug or tape the lines immediately to keep dirt and moisture out of the system.
7. Remove the retaining nuts holding the master cylinder to the brake booster.

8. Remove the 3-way union from the booster stud.
9. Remove the master cylinder from the studs.
10. Remove the seal or gasket from the booster.

To install:

11. When reinstalling, always use a new gasket or seal and install the master cylinder to the booster. On Nova vehicles, confirm that the "UP" mark on the master cylinder boot is in the correct position.
12. Install the 3-way union bracket over the stud and install the retaining nuts finger tight.
13. Connect the brake lines to the master cylinder. Make certain each fitting is correctly threaded and tighten each fitting 1–2 turns. The job is made easier by having a small amount of movement available at the master cylinder mounting studs.
14. Tighten the master cylinder retaining nuts to 9.5 ft. lbs. (13 Nm).
15. Tighten the brake line fittings to 11 ft. lbs. (15 Nm). Do not overtighten these fittings.
16. On Prizm vehicles, install the air intake duct.
17. Connect the wiring to the brake fluid sensing switch and attach any wiring clips.
18. Fill the master cylinder reservoir.
19. Bleed the brake system.
20. Connect the negative battery cable

Spectrum and Storm

1. Disconnect the negative battery cable. Remove some brake fluid from the master cylinder with a syringe.
2. On Storm, remove the top of the air cleaner and the air duct.
3. Disconnect and cap or tape the openings of the brake tube.
4. Disconnect the brake fluid level warning switch connector.
5. Remove the 2 nuts securing the master cylinder to the power brake booster.
6. Remove the master cylinder from the power brake booster.
7. To install, reverse the removal procedures, add fluid to the reservoir and bleed the brake system.

NOTE: It may be necessary to adjust the brake booster push ros length. Adjuster the push rod length until the push rod lightly touches the pin head.

Sprint and Metro

1. Disconnect the negative battery cable. Clean around the reservoir cap and take some of the fluid out with a syringe.
2. Disconnect and plug the brake tubes from the master cylinder.
3. Remove the mounting nuts and washers.
4. Remove the master cylinder.
5. To install, reverse the removal procedures. Torque the mounting bolts to 8–12 ft. lbs. (11-16 Nm). Bleed the brake system.

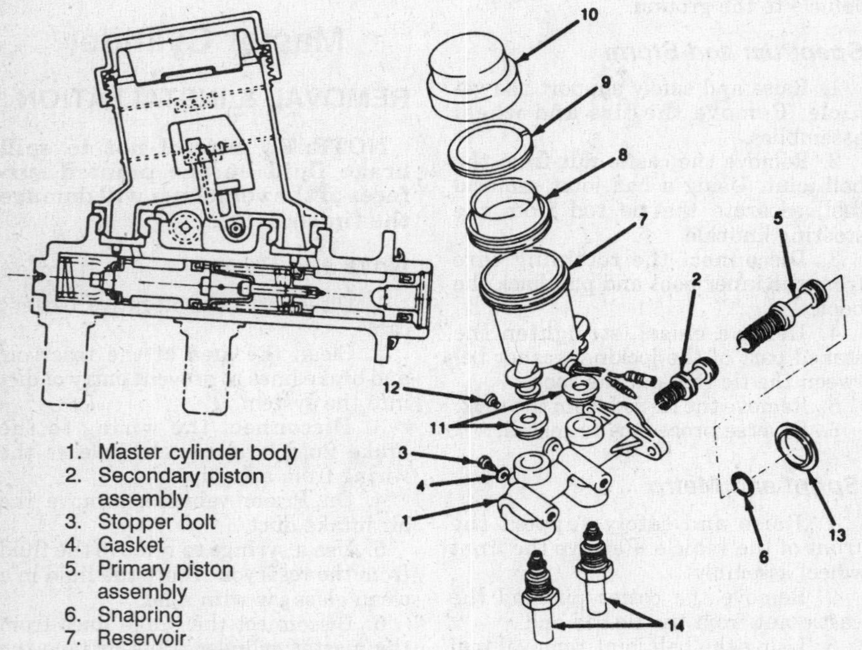

1. Master cylinder body
2. Secondary piston assembly
3. Stopper bolt
4. Gasket
5. Primary piston assembly
6. Snapring
7. Reservoir
8. Reservoir diaphragm
9. Diaphragm retainer
10. Reservoir cap
11. Reservoir grommets
12. Reservoir screw
13. Dust seal
14. Proportioning valve

Exploded view of the master cylinder assembly—Spectrum

Proportioning Valve

REMOVAL & INSTALLATION

Nova and Prizm

The proportioning valve (if so equipped) is located on the center of the firewall under the hood. Except for leakage or impact damage, it rarely needs replacement. If it must be removed, all 5 brake lines must be labeled and removed and the valve removed from its mount. Clean the fittings before removal to prevent dirt from entering the ports.

1. Disconnect and plug the brake lines from the proportioning valve unions.
2. Remove the proportioning valve-to-bulkhead bolts, then the valves.

NOTE: If the proportioning valve is defective, it must be replaced as an assembly; it cannot be rebuilt.

3. To install, reverse the removal procedure. Bleed the brake system and check for leaks.

Spectrum

1. Clean the area around the reservoir and the brake pipe connections.
2. Remove the brake fluid from the master cylinder reservoir using a syringe.
3. Disconnect the brake pipes from the proportioning valve. Cap all openings.
4. While holding the master cylinder, use a box wrench and remove the proportioning valves from the master cylinder.

NOTE: It may be necessary to remove the master cylinder and place it into a suitable vise to sufficiently hold it while removing the proportioning valves.

5. To install, reverse the removal procedure. Fill the reservoir and bleed the system.

Power Brake Booster

REMOVAL & INSTALLATION

Nova and Prizm

NOTE: To perform this procedure, use a booster pushrod gauge GM tool No. J-34873‹n-dash›A or equivalent, to set the booster pushrod length.

1. Disconnect the negative battery cable.

2. On Prizm vehicles, remove the top of the air cleaner and the intake duct. Remove the charcoal canister mounting nuts.
3. Remove the brake master cylinder from the booster.
4. Remove the vacuum hose from the booster.
5. Inside the vehicle, disconnect the pedal return spring. Disconnect the clip and the clevis pin.
6. Remove the brake booster retaining nuts. It will be helpful to have a helper support the booster while the nuts are loosened.
7. Remove the booster from the engine compartment.
8. To adjust the power brake booster pushrod, perform the following procedures:
a. Using the push rod gauge tool J-34873-A or equivalent, set the short-side on the booster.

NOTE: The head of the pin sits near the end of the booster pushrod.

b. Check the gap between the head of the tool's pin and the pushrod; it should be 0. If necessary, adjust the pushrod by turning it until the pushrod just touches the pin.

To install:

8. When reinstalling, have a helper hold the booster in position while installing the retaining nuts. Tighten the nuts to 9.5 ft. lbs. (13 Nm).
9. Install the clevis pin and clip, then install the pedal return spring.
10. Connect the vacuum hose to the booster.
11. Install the master cylinder onto the booster and tighten the nuts to 9.5 ft. lbs. (13 Nm.)
12. On Prizm vehicles, install the charcoal canister mounting bolts and install the air cleaner top and intake duct.
13. Connect the negative battery cable.
14. Bleed the brake system.

Spectrum and Storm

1. Disconnect the negative battery cable.
2. On Storm, remove the top of the air cleaner and the air cleaner duct.
3. Remove the master cylinder.
4. Remove the vacuum hose from the vacuum servo.
5. Remove the clevis pin from the brake pedal.
6. Remove the 4 nuts from the brake assembly under the dash and remove the power booster from the engine compartment.
7. To install, reverse the removal procedures.

Sprint and Metro

1. Disconnect the negative battery cable.
2. Remove the master cylinder.
3. Disconnect the pushrod clevis pin from the brake pedal arm.
4. Disconnect the vacuum hose from the brake booster.
5. Remove the mounting nuts from under the dash and the booster.
6. To install, reverse the removal procedures. Torque the booster-to-cowl nuts to 14–20 ft. lbs. (19-27 Nm). Bleed the brake system, if necessary.

Brake Caliper

REMOVAL & INSTALLATION

1. Raise and safely support the front of the vehicle. Set the parking brake and block the rear wheels.
2. Siphon a sufficient quantity of brake fluid from the master cylinder reservoir to prevent the brake fluid from overflowing the master cylinder when removing or installing the calipers. This is necessary as the piston must be forced into the cylinder bore to provide sufficient clearance to install the caliper.
3. Remove the wheel, then reinstall 2 lug nuts finger tight to hold the disc in place.

NOTE: Disassemble brakes one wheel at a time. This will prevent parts confusion and also prevent the opposite caliper piston from popping out during installation. On Nova and Prizm, mark the relationship between the wheel and the axle hub before removing the tire and wheel assembly.

4. Disconnect the hose union at the caliper. Use a pan to catch any spilled fluid and immediately plug the disconnected hose.
5. Remove the 2 caliper mounting bolts and then remove the caliper from the mounting bracket.

To install:

6. Use a caliper compressor, a C-clamp or large pair of pliers to slowly press the caliper piston back into the caliper.
7. Install the caliper assembly to the mounting plate. Before installing the retaining bolts, apply a thin, even coating of anti-seize compound to the threads and slide surfaces. Don't use grease or spray lubricants; they will not hold up under the extreme temperatures generated by the brakes. Tighten the bolts.
8. Install the brake hose to the caliper. Always use a new gasket and tighten the union.
9. Bleed the brake system.
10. Remove the 2 lugs holding the disc in place and install the wheel.
11. Lower the vehicle to the ground. Check the level of the brake fluid in the master cylinder reservoir; it should be at least to the middle of the reservoir.

Rear Brake Caliper

REMOVAL & INSTALLATION

1991-92 Prizm

1. Raise and safely support the vehicle safely.
2. Remove the rear wheel assemblies.
3. Disconnect the brake line union bolt.
4. Place the brake line a suitable container to catch the brake fluid.
5. Remove the 2 caliper retaining bolts.
6. Remove the brake caliper from the vehicle.
7. Installation is the reverse order of the removal procedure. Install the caliper retaining bolts and torque them to 14 ft. lbs. (20 Nm). Torque the brake line union bolt to 22 ft. lbs. (30 Nm).

Disc Brake Pads

REMOVAL & INSTALLATION

1. Raise and safely support the front of the vehicle. Set the parking brake and block the rear wheels.
2. Siphon a sufficient quantity of brake fluid from the master cylinder reservoir to prevent the brake fluid from overflowing the master cylinder when removing or installing the brake pads. This is necessary as the piston must be forced into the cylinder bore to provide sufficient clearance to install the pads.
3. Remove the wheel, then reinstall 2 lug nuts finger tight to hold the disc in place.

NOTE: Disassemble brakes one wheel at a time. This will prevent parts confusion and also prevent the opposite caliper piston from popping out during pad installation.

4. Remove the 2 caliper mounting bolts and then remove the caliper from the mounting bracket. Position the caliper out of the way and support it with wire so it doesn't hang by the brake line.

NOTE: It may be necessary to rock the caliper back and forth a bit in order to reposition the piston so it will clear the brake pads.

5. Remove the brake pads, the wear indicators, the anti-squeal shims, the support plates and the anti squeal springs (if so equipped). Disassemble slowly and take note of how the parts fit together. This will save much time during reassembly.

6. Inspect the brake disc (both sides) for scoring or gouging. Measure the disc for both thickness and run-out. Complete inspection procedures are given later in this section.

7. Inspect the pads for remaining thickness and condition. Any sign of uneven wear, cracking, heat checking or spotting is cause for replacement. Compare the wear of the inner pad to the outer pad. While they will not wear at exactly the same rate, the remaining thickness should be about the same on both pads. If one is heavily worn and the other is not, suspect either a binding caliper piston or dirty slides in the caliper mount.

8. Examine the 2 caliper retaining bolts and the slide bushings in which they run. Everything should be clean and dry. If cleaning is needed, use spray solvents and a clean cloth. Do not wire brush or sand the bolts–this will cause grooves in the metal which will trap more dirt. Check the condition of the rubber dust boots and replace them if damaged.

To install:

9. Install the pad support plates onto the mounting bracket.

10. Install new pad wear indicators onto each pad, making sure the arrow on the tab points in the direction of disc rotation.

11. Install new anti-squeal pads to the back of the pads.

12. Install the pads into the mounting bracket and install the anti-squeal springs.

13. Use a caliper compressor, or a C-clamp to slowly press the caliper piston back into the caliper. If the piston is frozen, or if the caliper is leaking hydraulic fluid, the caliper must be overhauled or replaced.

14. Install the caliper assembly to the mounting plate. Before installing the retaining bolts, apply a thin, even coating of anti-seize compound to the threads and slide surfaces. Don't use grease or spray lubricants; they will not hold up under the extreme temperatures generated by the brakes. Tighten the bolts to specification.

15. Remove the 2 lugs holding the disc in place and install the wheel.

16. Lower the vehicle to the ground. Check the level of the brake fluid in the master cylinder reservoir; it should be at least to the middle of the reservoir.

17. Depress the brake pedal several times and make sure that the movement feels normal. The first brake pedal application may result in a very "long" pedal due to the pistons being retracted. Always make several brake applications before starting the vehicle. Bleeding is not usually necessary after pad replacement.

18. Recheck the fluid level and add to the "MAX" line if necessary.

NOTE: Braking should be moderate for the first 5 miles or so until the new pads seat correctly. The new pads will bed best if put through several moderate heating and cooling cycles. Avoid hard braking until the brakes have experienced several long, slow stops with time to cool in between. Taking the time to properly bed the brakes will yield quieter operation, more efficient stopping and contribute to extended brake life.

Rear Brake Caliper

REMOVAL & INSTALLATION

1991-92 Prizm

1. Raise and safely support the vehicle safely.

2. Remove the rear wheel assemblies.

3. Remove the 2 caliper retaining bolts.

4. Remove the brake caliper from the vehicle. Position the caliper out of the way and support it with wire so it doesn't hang by the brake line.

5. Remove the 2 brake pads, 2 anti-squeal shims and 4 pad guide plates.

6. Installation is the reverse order of the removal procedure. Torque the caliper retaining bolts to 14 ft. lbs. (20 Nm).

Brake Rotor

REMOVAL & INSTALLATION

Front

EXCEPT 1991–92 PRIZM

1. Elevate and safely support the vehicle. If only the front end is supported, set the parking brake and block the rear wheels.

2. Remove the wheel.

3. Remove the brake caliper from its mount and suspend it out of the way. Don't disconnect the hose and don't let the caliper hang by the hose. Remove the brake pads with all the clips, shims, etc.

4. Install all the lug nuts to hold the rotor in place. If the nuts are open at both ends, it is helpful to install them backwards (tapered end out) to secure the disc. Tighten the nuts a bit tighter than finger tight, but make sure all are at approximately the same tightness.

5. Check the run-out and thickness measurements of the rotor.

6. Remove the 2 bolts holding the caliper mounting bracket to the steering knuckle. These bolts will be tight. Remove the 4 lug nuts holding the rotor.

7. Remove the bracket from the knuckle. Before removing the rotor, make a mark on the rotor indexing one wheel stud to one hole in the rotor. This assures the rotor will be re-installed in its original position, serving to eliminate minor vibrations in the brake system.

To install:

8. When reinstalling, make certain the rotor is clean and free of any particles of rust or metal from resurfacing. Observe the index mark made earlier and fit the rotor over the wheel lugs. Install 2 lug nuts to hold it in place.

9. Install the caliper mounting bracket in position and tighten the bolts.

10. Install the brake pads and the hardware.

11. Install the caliper. Tighten the mounting bolts.

12. Install the wheel and lower the vehicle to the ground.

Rear

1991-92 Prizm

1. Raise and safely support the vehicle safely.

2. Remove the rear wheel assemblies.

3. Remove the 2 caliper retaining bolts.

4. Remove the brake caliper from the vehicle. Position the caliper out of the way and support it with wire so it doesn't hang by the brake line.

5. Remove the 2 brake pads, 2 anti-squeal shims and 4 pad guide plates.

6. Remove the rotor assembly.

7. Installation is the reverse order of the removal procedure. Torque the caliper retaining bolts to 14 ft. lbs. (20 Nm).

Brake Drums

REMOVAL & INSTALLATION

Nova, Prizm and Storm

1. Raise and support the vehicle safely.

2. Remove the tire and wheel assembly.

3. Insert a suitable tool through the hole in the backing plate and hold the automatic adjusting lever away from the adjusting bolt.

4. Using another suitable tool, turn the adjusting bolt to reduce the brake shoe adjustment.

5. Remove the brake drum.

6. Reverse procedure to install. Adjust brakes.

Spectrum

1. Raise and support the vehicle safely.
2. Remove the tire and wheel assembly.
3. Remove the cotter pin, nut and washer, then the hub and drum.
4. Reverse procedure to install.

Sprint and Metro

1. Raise and support the vehicle safely.
2. Remove the tire and wheel assembly.
3. Remove the spindle cap without damaging the sealing portion of the cap.
4. Unfasten the staked portion of the nut using a suitable chisel.
5. Remove the castle nut and washer.
6. Slacken the parking brake cable by loosening its adjusting nuts.
7. Remove the backing plate plug, located on the back side of the backing plate.
8. Insert a suitable tool into the plug until its tip contacts the shoe hold-down and push the spring in the direction of the leading shoe. The allows a greater clearance between the shoes and the drum.
9. Using slide hammer J‹n-dash›2619-01 and drum remover J–34866 or equivalent, remove the drum.
10. Reverse procedure to install.

Brake Shoes

REMOVAL & INSTALLATION

Nova, Prizm and Storm

1. Disconnect the battery negative cable.
2. Raise and support the vehicle safely.
3. Remove the brake drums.
4. Remove the return spring, retainers, hold-down springs and pins.
5. Remove the anchor spring, then disconnect the parking brake cable from the parking brake lever.
6. Remove the adjuster spring, then the shoes and adjuster.
7. Reverse procedure to install. Adjust brakes.

Spectrum

1. Raise and support the vehicle safely.
2. Remove the brake drums.
3. Remove the return spring and the automatic adjuster spring.
4. Remove the leading shoe holding pin, then the shoe and adjuster.
5. Remove the trailing shoe holding pin, then disconnect the parking brake cable from the lever and remove the trailing shoe. Remove the lever from the shoe.
6. Reverse procedure to install. Adjust brakes.

Sprint and Metro

1. Disconnect the battery negative cable.
2. Raise and support the vehicle safely.
3. Remove the brake drum.
4. Remove the brake shoe hold-down spring, then disconnect the parking brake shoe lever and remove the parking brake shoes.
5. Disconnect the bottom return spring, then remove the strut and upper return spring from the shoe.
6. Remove the parking brake shoe lever from the shoe.
7. Reverse procedure to install. Adjust brakes.

Wheel Cylinder

REMOVAL & INSTALLATION

Nova, Prizm and Storm

1. Raise and safely support the rear of the vehicle.
2. Remove the rear tire and wheel assembly.
3. Disconnect and plug the brake line at the wheel cylinder to prevent hydraulic fluid from leaking.
4. Remove the brake drums and shoes.
5. Remove the wheel cylinder-to-backing plate attaching bolts, then the wheel cylinder from the vehicle.
6. Reverse procedure to install. Torque the wheel cylinder-to-backing plate bolts to 7 ft. lbs. (10 Nm). Bleed system and adjust brakes.

Spectrum

1. Disconnect the battery negative cable.
2. Raise and support the vehicle safely.
3. Remove the brake drums and shoes.
4. Clean the area around the brake pipe, then disconnect the brake pipe from the wheel cylinder. Cap all openings.
5. Remove the 2 wheel cylinder attaching bolts, then the wheel cylinder from the vehicle.
6. Reverse procedure to install. Torque the wheel cylinder attaching bolts to 7 ft. lbs. (10 Nm). Bleed system and adjust brakes.

Sprint and Metro

1. Disconnect the battery negative cable.
2. Raise and support the vehicle safely, then remove the tire and wheel assembly.
3. Remove the brake drum and shoes.
4. Remove the bleeder screw from the wheel cylinder.
5. Loosen the brake pipe flare nut, then remove the wheel cylinder attaching bolts. Disconnect the brake pipe from the wheel cylinder. Cap all openings.
6. Reverse procedure to install. Fill the master cylinder with brake fluid and bleed the system.

Parking Brake Cable

ADJUSTMENT

Nova and Prizm

1. Release the parking brake (all the way). Using 44 lbs. of pulling pressure, slowly pull the lever upward and count the number of clicks; 4–7 clicks (except twincam engine) or 5–8 (twincam engine).
2. If the number of clicks is incorrect, adjust the parking brake cable by performing the following procedures:
   a. Remove the console box.
   b. At the rear of the parking brake handle, loosen the cable nut, then, turn the adjusting nut.
3. Secure the adjusting nut position when tighten the locknut. Check the adjustment and repeat Steps 2 and 3, if necessary. Tighten the adjusting nut securely and ensure that the adjustment is correct.

Prizm Rear Wheel Disc Brake

1. Pull up the parking brake lever and count the number of clicks. The brake lever should click 5 to 8 clicks.
2. If less than 5 or more than 8 clicks are heard, the parking brake requires adjustment.
3. Before adjusting the parking brake lever in the console, depress the brake pedal several times and recheck the number of clicks at the parking brake lever.
4. If the parking brake is still less than 5 or more than 8 clicks adjust the parking brake lever as follows:
   a. Remove the center console box.
   b. Loosen the jam nut.
   c. Adjust the parking brake cable.
   d. Tighten the jam nut. Torque the jam nut to 53 in. lbs. (6 Nm).
5. Install the center console box.

Spectrum and Storm

The parking brake adjustment is normal when the lever moves 7–9 notches at 66 lbs. If it is not within limits, adjust the rear brakes. If this adjustment does not affect the specifications, adjust the parking brake turn buckle.

Sprint and Metro

1. Remove both door seal plates and the seat belt buckle bolts at the floor.
2. Disconnect the shoulder harness bolts at the floor and the interior, bottom trim panels.
3. Raise the rear seat cushion.
4. Pull up the carpet to gain access to the parking brake lever.
5. Loosen the parking brake cable adjusting nuts.
6. Adjust the parking brake cables, so they work evenly.
7. Adjust the cable, so when the parking brake handle is pulled, its travel is between 5–8 notches, with 44 lbs. of force.
8. After adjustment, reverse the removal procedures.

REMOVAL & INSTALLATION

Nova and 1989-90 Prizm

1. Elevate and safely support the vehicle. If only the rear wheels are elevated, block the front wheels. Release the parking brake after the vehicle is supported.
2. Remove the rear wheel(s).
3. If equipped with drum brakes, remove the brake drum and remove the brake shoes.
4. If equipped with disc brakes, remove the clip from the parking brake cable and remove the cable from the caliper assembly.
5. If equipped with drum brakes, remove the parking brake retaining bolts at the backing plate.
6. Remove any exhaust heat shields which interfere with the removal of the cable.
7. Remove the 2 cable clamps.
8. Disconnect the cable retainer.
9. Remove the cable from the equalizer (yoke).

To install:

10. When reinstalling, fit the end of the new cable into the equalizer and make certain it is properly seated.
11. Install the cable retainer, and, working along the length of the cable, install the clamps.

NOTE: Make certain the cable is properly routed and does not contain any sharp bends or kinks.

12. Feed the cable through the backing plate and install the retaining bolts.
13. If equipped with disc brakes, connect the cable to the arm and install the clip.
14. If equipped with drum brakes, reinstall the shoes. The cable will be connected to the shoes during the installation process.
15. Reinstall the wheel(s) and lower the vehicle to the ground.

1991-92 Prizm

1. Raise and support the vehicle safely.
2. Remove the tire and wheel assemblies.
3. Remove the brake pad assemblies.
4. Remove the parking brake cable retaining bolts.
5. Remove the exhaust shields as necessary.
6. Remove the 2 parking brake cable clamps.
7. Remove the parking brake cable retainer and remove the cable from the equalizer.
8. Installation is the reverse order of the removal procedure. Be sure to readjust the parking brake cable.

Spectrum and Storm

1. Remove the parking brake lever assembly as follows:
   a. Remove the console box.
   b. Disconnect the parking brake switch electrical connector.
   c. Remove the lever-to-chassis bolts, then the lever.
   d. Remove the parking brake lever and switch from the lever.
2. Raise and safely support the vehicle.
3. Remove the parking brake cable as follows:
   a. Separate the front cable from the rear cable.
   b. Remove the tension spring from the rear axle housing assembly.
   c. Remove the rear wheel assembly, hub and drum.
   d. Disconnect the parking brake cable from the rear brake lever.
   e. Remove the parking brake cable-to-chassis bolt, then the cable(s) from the vehicle.
4. Reverse procedure to install. Lubricate the cable(s) with a suitable grease. Torque the backplate-to-chassis bolt to 30 ft. lbs. (41 Nm) and the lever-to-chassis bolts to 10 ft. lbs. (14 Nm). Adjust the parking brake.

Sprint and Metro

1. Raise and support the vehicle safely, then remove the tire and wheel assembly.
2. Remove the brake drum.
3. Disconnect the parking cable from the brake shoe lever and the backing plate.
4. Remove the cable(s) from the chassis mounts, then the cable from the vehicle.
5. Installation is the reverse of the removal procedure.

Brake System Bleeding

1. Clean the bleeder screw at each wheel.
2. Start with the wheel farthest from the master cylinder (right rear).
3. Attach a rubber hose to the bleeder screw and place the end in a clear container of brake fluid.
4. Fill the master cylinder with brake fluid. Have an assistant slowly pump up the brake pedal and hold the pressure.
5. Open the bleed screw about ¼ turn, press the brake pedal to the floor, close the bleed screw and slowly release the pedal. Continue until no more air bubbles are forced from the cylinder on application of the brake pedal.
6. Repeat procedure on remaining wheel cylinders and calipers, still working from the cylinder/caliper farthest from the master cylinder.

NOTE: Master cylinders equipped with bleed screws may be bled independently. When bleeding the master cylinder, it is necessary to cap 1 reservoir section while bleeding the other to prevent pressure loss through the cap vent hole.

CHASSIS ELECTRICAL

Air Bag

DISARMING

Storm

1. Disable the air bag system as follows:
   a. Turn the ignition switch to the **OFF** position.
   b. Disconnect the battery negative cable.
   c. Remove fuses C-22 and C-23 from the fuse box.
   d. Disconnect the orange 3-way connector at the base of the steering column.
2. Reactivate the air bag system as follows:
   a. Turn the ignition switch to the **OFF** position.
   b. Connect the orange 3-way con-

nector at the base of the steering column.

c. Install fuses C-22 and C-23 to the fuse box.

d. Connect the battery negative cable.

Metro

1. If equipped, with an air bag system, disable the air bag system as follows:

a. Turn the ignition switch to the **OFF** position.

b. Remove the **SIR IG** fuse in the supplemental inflatable restraint fuse block.

c. Remove the rear plastic access cover to the air bag module.

d. Disconnect the yellow 2-way connector and connector position assurance (CPA) inside the inflator module housing.

2. Disconnect the negative battery cable.

3. Reactivate the air bag system as follows:

a. Turn the ignition switch to the **OFF** position.

b. Connect the yellow 2-way connector and connector position assurance (CPA) inside the inflator module housing.

c. Install the **SIR IG** fuse in the supplemental inflatable restraint fuse block.

d. Install the rear plastic access cover to the air bag module.

e. Turn the ignition switch to the **RUN** position. Observe the **INFLATABLE RESTRAINT** indicator lamp.If the lamp does not flash 7 to 9 times and then remain off, there is a problem in the air bag system and further diagnostic testing of the system is needed.

Heater Blower Motor

REMOVAL & INSTALLATION

Nova

The heater blower motor is located inside the vehicle, behind the glove box.

1. Disconnect the negative battery cable. Remove the 3 screws attaching the retainer.
2. Remove the glove box assembly.
3. Remove the air duct between the heater case and blower assembly.
4. Disconnect the blower motor wire connector at the motor case.
5. Disconnect the air source selector control cable at the blower assembly.
6. Loosen the 2 nuts and the bolt attaching the blower assembly, then remove the blower.
7. With the blower removed, check the case for any debris or signs of fan contact. Inspect the fan for wear spots, cracked blades or hub, loose retaining nut or poor alignment.
8. To reinstall, place the blower in position, making sure it is properly aligned within the case. Install the 2 bolts and the nut and tighten them.
9. Connect the selector control cable at the blower assembly.
10. Connect the wire harness to the motor and install the ductwork between the heater case and the blower assembly.
11. Install the glove box assembly and install the retainer with its 3 screws.

Prizm

1. Disconnect the negative battery cable.
2. Remove the rubber duct running between the heater case and the blower.
3. Disconnect the wiring from the motor.
4. Remove the 3 screws holding the motor and remove the blower motor.
5. With the blower removed, check the case for any debris or signs of fan contact. Inspect the fan for wear spots, cracked blades or hub, loose retaining nut or poor alignment.
6. To reinstall, place the blower in position, making sure it is properly aligned within the case. Install the three screws and tighten them.
7. Connect the wiring to the motor.
8. Install the rubber air duct and connect the negative battery cable.

Storm

The blower motor is located under the instrument panel at the far right side of the vehicle. It is accessible from below the instrument panel.

1. Disconnect the negative battery cable.
2. Disconnect the rubber air duct between the motor and the heater assembly.
3. Disconnect the electrical connector from the motor.
4. Remove the 3 screws retaining the motor and remove the motor.
5. Installation is the reverse of removal.

Spectrum

1. Disconnect the negative battery cable. Disconnect the blower motor electrical connector at the motor case.
2. If equipped with air conditioning, remove the rubber hose from the blower case.
3. Rotate the blower motor case counterclockwise and remove the blower motor assembly.
4. To install, reverse the removal procedures.

Sprint

1. Disconnect the negative battery cable.
2. Disconnect the defroster hose on the steering column side.
3. Disconnect the blower motor electrical connector.
4. Remove the 3 mounting screws and the blower motor.
5. To install, reverse the removal procedures.

Metro

1. Disconnect the negative battery cable.
2. Remove the 2 attaching screws from the glove box striker and remove the striker.
3. Remove the 1 attaching screw from the rear of the glove box upper panel and remove the panel.
4. Remove the blower motor and blower resistor electrical connections.
5. Remove the fresh/recirculate control cable from the blower case assembly.
6. Remove the 3 blower case mounting bolts and case from the vehicle.
7. Remove the air hose from the blower case. Remove the 3 blower motor mounting screws and remove the blower motor from the case.
8. Installation is the reverse order of the removal procedure.

Windshield Wiper Motor

REMOVAL & INSTALLATION

Nova and Prizm

FRONT

1. Disconnect the negative battery terminal.
2. Disconnect the electrical connector from the wiper motor.
3. Remove the mounting bolts and remove the motor from the firewall.
4. Remove the wiper linkage from the wiper motor assembly.
5. Installation is the reverse of removal.

REAR

The rear wiper motor is located in the rear hatch.

1. Disconnect the negative terminal from the battery.
2. Remove the rear wiper arm-to-wiper motor nut and wiper arm.
3. From inside the rear hatch, remove the rear wiper cover, then, disconnect the electrical connector from the rear wiper motor.
4. Remove the wiper motor-to-hatch screws and the wiper motor from the hatch.
5. To install, reverse the removal

procedures. Check the operation of the rear wiper motor.

Storm

FRONT

1. Disconnect the negative terminal from the battery.
2. Remove the cowl vent grille. From the engine compartment, disconnect the electrical connector from the windshield wiper motor.
3. Disconnect the wiper motor from the windshield wiper crank arm; be careful not to bend the linkage.
4. Remove the 2 retaining bolts from the charcoal canister mounting bracket, allowing the canister to slip down and provide access to the wiper motor mounting bolts.
5. Remove the wiper motor-to-chassis screws.
6. To install, reverse the removal procedures. Check the operation of the front windshield wiper motor.

REAR

The rear wiper motor is located in the rear hatch.

1. Disconnect the negative terminal from the battery.
2. Remove the rear wiper arm-to-wiper motor nut and wiper arm.
3. From inside the rear hatch, remove the rear wiper cover (hatch trim panel), then, disconnect the electrical connector from the rear wiper motor.
4. Remove the wiper motor-to-hatch screws and the wiper motor from the hatch.
5. To install, reverse the removal procedures. Check the operation of the rear wiper motor.

Spectrum

FRONT

1. Disconnect the negative battery terminal from the battery.
2. Remove the locknuts retaining the wiper arms and the wiper arms.
3. Remove the cowl cover, wiper motor cover and the electrical connector.
4. Disconnect the drive arm from the wiper link.
5. Remove the mounting bolts and the wiper motor.
6. To install, reverse the removal procedures.

REAR

1. Disconnect the negative battery terminal from the battery.
2. Remove the trim pad and the wiper arm assemblies.
3. Remove the mounting bolts and the motor assembly.
4. Disconnect the electrical connector.
5. To install, reverse the removal procedures.

Sprint

FRONT

1. Disconnect the negative battery cable.
2. Remove the wiper motor retaining bolts.
3. Disconnect the crank arm from the wiper motor.
4. Remove the wiper motor from the vehicle.
5. To install, reverse the removal procedures.

REAR

1. Disconnect the negative battery cable.
2. Remove the electrical connector from the rear wiper motor.
3. Remove the rear motor mounting bracket.
4. Disconnect the motor from the wiper linkage.
5. Remove the motor from the vehicle.
6. To install, reverse the removal procedures.

Metro

FRONT

1. Disconnect the negative battery cable.
2. Remove the wiper motor retaining bolts.
3. Disconnect the crank arm from the wiper motor.
4. Remove the wiper motor from the vehicle.
5. To install, reverse the removal procedures.

REAR

1. Disconnect the negative battery cable.
2. Remove the right and left speakers from the hatchback door inner trim panel, if so equipped.
3. Remove the retaining clips and hatchback door inner trim panel from the vehicle.
4. Remove the wiper motor electrical connector and wiper motor ground screw.
5. Remove the wiper cranking arm retaining nut from the wiper motor shaft.
6. Remove the 3 wiper motor mountin screws and the wiper motor assembly from the vehicle.
7. Installation is the reverse order of the removal procedure.

Windshield Wiper Switch

REMOVAL & INSTALLATION

Nova

FRONT

1. Disconnect the negative battery cable. Remove the steering wheel.
2. Remove the lower left dashboard trim panel.
3. Disconnect the air duct from the vent in the lower panel.
4. Remove the upper and lower steering column covers.
5. Disconnect the wiring from the combination switch to the dashboard wiring harness.
6. Remove the mounting bolts and remove the combination switch.

To install:

7. When reinstalling, position the switch carefully onto the column and secure the mounting screws.
8. Connect the wiring harness(es) from the switch to the dashboard harness.
9. Install the upper and lower column covers.
10. Install the air duct to the lower trim panel and install the panel.
11. Install the steering wheel.

REAR

1. Disconnect the negative battery cable.
2. Remove the hood release lever.
3. Remove the 4 screws from the lower left dash trim and pull the trim out.
4. Disconnect the wiring from the radio speaker.
5. If equipped with air conditioning, remove the ductwork from the lower air outlet and remove the trim panel from the vehicle.
6. Remove the steering wheel.
7. Remove the upper and lower steering column covers.
8. Remove the 2 screws from the trim panel (bezel).
9. Pull the panel out, releasing the spring clips behind the dash. With the panel loose, disconnect the wiring to the dash switches.
10. Remove the switch from the panel.

To install:

11. To reinstall, press the switch into place in the lower trim bezel and connect the wiring to the switches.
12. Install the lower trim bezel, making sure all the clips engage.
13. Install the steering column upper and lower covers.
14. Connect the air conditioning ductwork to the lower air outlet if so equipped.

15. Attach the wiring to the radio speaker.
16. Install the lower left dashboard trim panel and its 4 screws.
17. Install the hood release lever.
18. Install the steering wheel.
19. Connect the negative battery cable.

Prizm

FRONT

1. Disconnect the negative battery cable.

NOTE: If either the front wiper or washer switch is malfunctioning, replace the multi-function switch as a unit.

2. Remove the steering wheel assemblies.
3. Remove the upper and lower steering column covers.
4. Disconnect the multi-function switch electrical electrical.
5. Remove the multi-function switch.
6. Installation is the reverse order of the removal procedure.

REAR

1. Disconnect the negative battery cable.
2. Remove the trim bezel.
3. Disconnect the electrical connector.
4. Remove the rear wiper-washer switch.
5. Installation is the reverse order of the removal procedure.

Storm

1. Disconnect the negative battery cable.
2. Remove the cluster switch panel.
3. Remove the 2 screws securing the wiper/washer switch assembly to the cluster switch panel.
4. Remove the wiper/washer switch assembly from the cluster switch panel.
5. Installation is the reverse order of the removal procedure.

Spectrum

FRONT

1. Disconnect the negative battery cable. Remove the instrument cluster bezel.
2. Remove the wiper switch electrical connector, attaching nuts and bracket.
3. Remove the wiper switch.
4. To install, reverse the removal procedures.

REAR

1. Disconnect the negative battery cable. Using a small tool, pry the switch panel from the dash.
2. Pull the switch out and disconnect the electrical connector.
3. To install, reverse the removal procedures.

Sprint

1. Disconnect the negative battery cable.
2. Remove the steering column trim panel.
3. Lower the steering column.
4. Remove the cluster bezel and the bezel.
5. Disconnect the wiper switch connector.
6. Remove the wiper switch.
7. To install, reverse the removal procedures.

Metro

1. Disconnect the negative battery cable.
2. Remove the cluster switch panel.
3. Remove the 2 screws securing the wiper/washer switch assembly to the cluster switch panel.
4. Remove the wiper-washer switch assembly from the cluster switch panel.
5. Remove the 2 screws securing the wiper-washer switch housing to the wiper-washer switch assembly.
6. Remove the wiper-washer switch housing from the wiper-washer switch assembly.
7. Installation is the reverse order of the removal procedure.

Instrument Cluster

REMOVAL & INSTALLATION

Nova

1. Disconnect the negative battery cable.
2. Remove the steering wheel.
3. Remove the left side speaker grille. The grille may be attached with a clip (which must be pulled loose to remove the grille) or with a screw.
4. Remove the lower trim cover from the steering column.
5. Remove the hood release lever.
6. Remove the heater duct assembly.
7. Remove the instrument hood. Remove the air conditioner outlet register, then remove the 4 screws and remove the hood.
8. Remove the 6 screws from the instrument cluster, and disconnect the speedometer cable and the wiring connectors. Remove the meter assembly from the instrument panel.

To install:

9. To reinstall, connect the wiring harnesses and the speedometer cable to the cluster. Place the cluster in the dash and secure the 6 screws. Make certain the wiring is properly placed so as not to become pinched or crushed.
10. Install the air conditioning outlet then install the hood and its screws.
11. Install the heater duct assembly.
12. Install the hood release lever and the lower steering column trim.
13. Reinstall the speaker grille.
14. Install the steering wheel and connect the negative battery cable.

Prizm

NOTE: Removing the steering wheel is not required, but may make the job easier.

1. Disconnect the negative battery cable.
2. Remove the hood release lever.
3. Remove the 4 screws from the lower left dash trim and pull the trim out.
4. Disconnect the wiring from the radio speaker.
5. If equipped with air conditioning, remove the ductwork from the lower air outlet.
6. Remove the trim panel from the vehicle.
7. Remove the upper and lower steering column covers.
8. Remove the 2 screws from the trim panel (bezel).
9. Pull the panel out, releasing the spring clips behind the dash. With the panel loose, disconnect the wiring to the dash switches.
10. Remove the switches from the panel.
11. Remove the 2 electrical connectors and the cigarette lighter from the trim bezel and remove the bezel from the vehicle.
12. Remove the 4 screws holding the instrument cluster trim.
13. Disconnect the wiring from the hazard (4-way) flasher and dimmer switches.
14. Remove the cluster trim panel.
15. Remove the 4 attaching screws holding the cluster, move it away from the dash and disconnect the wiring harnesses and the speedometer cable.
16. Remove the instrument cluster from the vehicle.

To install:

17. When reinstalling, connect the speedometer and electrical cables to the cluster. Install the cluster and the 4 retaining screws.
18. Attach the wiring connectors for the hazard flasher and the dimmer switches.
19. Install the cluster trim bezel.
20. Place the dash switches in place on the lower trim bezel and connect the wiring to the switches.
21. Install the lower trim bezel, making sure all the clips engage.
22. Install the steering column upper and lower covers.

23. Connect the air conditioning ductwork to the lower air outlet if so equipped.
24. Attach the wiring to the radio speaker.
25. Install the lower left dashboard trim panel and its 4 screws.
26. Install the hood release lever.
27. Connect the negative battery cable.

Storm

1. Disconnect battery negative cable.
2. Remove the cluster switch panel.
3. Remove the 4 screws securing the instrument cluster. Disconnect the electrical connectors and speedometer from the back of the instrument cluster.
4. Remove the cluster assembly.
5. Installation is the reverse order of the removal procedure. Connect battery negative cable.

Spectrum

1. Disconnect the negative battery terminal from the battery.
2. Remove the instrument cluster bezel retaining screws and bezel.
3. Disconnect the windshield wiper and lighting switch connectors.
4. Remove the instrument cluster retaining screws and pull out the assembly.
5. Remove the trip reset knob and the assembly glass.
6. Remove the buzzer, sockets and bulbs.
7. Remove the speedometer assembly, fuel and temperature gauge.
8. Remove the tachometer, if equipped.
9. To install, reverse the removal procedures.

Sprint

1. Disconnect the negative battery cable.
2. Remove the steering column trim panel.
3. Lower the steering column.
4. Remove the cluster lens and the cluster mounting screws.
5. Disconnect the speedometer cable at the transaxle and at the instrument cluster.
6. Disconnect and mark the electrical connectors at the instrument cluster.
7. Remove the instrument cluster from the vehicle.
8. To install, reverse the removal procedures.

Metro

1. Disconnect battery negative cable.
2. Remove the cluster switch panel.
3. Remove the 4 screws securing the instrument cluster. Disconnect the electrical connectors and speedometer from the back of the instrument cluster.
4. Remove the cluster assembly.
5. Installation is the reverse order of the removal procedure. Connect battery negative cable.

Speedometer

REMOVAL & INSTALLATION

All Models

1. Disconnect the negative battery cable.
2. Remove the instrument cluster from the instrument panel.
3. Remove the cluster lens and retainer from the cluster.
4. Remove the speedometer retaining screws from the rear of the cluster.
5. Remove the speedometer-odometer from the cluster.
6. Installation is the reverse order of the removal procedure. Connect battery negative cable.

Radio

REMOVAL & INSTALLATION

Nova

1. Disconnect the negative battery cable.
2. Remove the steering wheel and then the upper and lower steering column covers.
3. Use a suitable tool to gently pry the switches from the lower dash trim panel. Disconnect the wiring and remove the switches.
4. Remove the 2 screws under the panel and pull out the center cluster finish panel. It is also secured by spring clips behind the dash—pull straight out away from the dash so as not to break the plastic.
5. Remove the 7 screws holding the upper trim panel.
6. Remove the 8 screws holding the radio and its accessories trim cover.
7. Slide the radio out of the dash. Disconnect the antenna lead and the wiring connector.

To install:

8. When reinstalling, connect the wiring and antenna leads and place the radio in the dash. Make sure the wiring is not crushed or pinched.
9. Install the 8 screws, making sure the radio is straight and in position.
10. Install the upper trim panel.
11. Route the switch cables through the openings in the center trim panel. Install the center cluster finish panel, making sure all the clips engage properly. Install the 2 screws.
12. Connect the switches to the wiring harnesses and push the switches firmly into place in the trim panel.
13. Install the upper and lower steering column covers and reinstall the steering wheel.
14. Connect the negative battery cable.

Prizm

1. Remove the 7 screws from the steering column covers and remove the covers.
2. Remove the 2 attaching screws from the lower part of the trim panel.
3. Remove the trim panel, being careful of the concealed spring clips behind the panel.
4. Disconnect the wiring from the switches mounted in the trim panel.
5. Remove the 4 mounting screws from the radio.
6. Remove the radio from the dash until the wiring connectors are exposed.
7. Disconnect the 2 electrical connectors and the antenna cable from the body of the radio and remove the radio from the vehicle.

To install:

8. When reinstalling, connect all the wiring and antenna cable first, then place the radio in position within the dash.
9. Install the 4 attaching screws.
10. Reconnect the wiring harnesses to the switches in the trim panel and make sure the switches are secure in the panel.
11. Install the trim panel (make sure all the spring clips engage) and install the 2 screws.
12. Install the steering column covers and the 7 screws.
13. Install the steering wheel.

Storm

1. Disconnect battery negative cable.
2. Remove 4 heater control knobs.
3. Remove 4 radio bezel attaching screws and the bezel.
4. Disconnect the illumination light and harness from the radio bezel.
5. Remove the 3 left hand console side panel attaching screws, then the left hand console side panel.
6. Remove the 3 right hand console side panel attaching screws, then the right hand console side panel.
7. Remove the radio attaching screws and the radio from the instrument panel.
8. Disconnect electrical connectors and antenna. Reverse procedure to install and connect the battery negative cable.

Spectrum

1. Disconnect battery negative cable.
2. Remove the radio cover attaching screws and the radio cover.
3. Remove the radio and bracket.
4. Disconnect electrical connectors and antenna from the radio.
5. Reverse procedure to install. Connect battery negative cable.

Sprint

1. Disconnect battery negative cable.
2. Remove the ashtray and radio knobs.
3. Remove the ashtray assembly.
4. Remove the radio mounting nuts.
5. Remove the radio and disconnect the electrical connectors and antenna.
6. Reverse procedure to install. Connect battery negative cable.

Metro

1. Disconnect the negative battery cable.
2. Open the glove box. Remove the 1 screw and the glove box open panel.
3. Remove the 1 screw securing the radio to mounting bracket. Access through the glove box opening.
4. Push the radio out of the instrument panel.
5. Disconnect the electrical connector and antenna lead.
6. Remove the radio from the vehicle.
7. To install, reverse the removal procedures.

Headlight Switch

REMOVAL & INSTALLATION

Storm

The headlight control switch is located at the left hand side of the instrument panel on the meter hood.

1. Disconnect the battery negative cable.
2. Remove the meter hood.
3. Remove the instrument cluster from the meter hood.
4. Remove the 2 clips attaching headlight control harness.
5. Remove the 4 screws attaching the headlight switch to the meter hood.
6. Disconnect electrical connectors from the switch and remove the switch.
7. Reverse procedure to install. Connect battery negative cable.

Spectrum

The headlight control switch is a 3 position, push type switch which is located at the left side of the instrument panel.

1. Disconnect the negative battery cable. Remove the instrument cluster bezel retaining screw and the bezel.
2. Disconnect the electrical connectors.
3. Place the bezel on a bench and remove the 2 nuts securing the headlight control switch.
4. Remove the headlight control switch.
5. To install, reverse the removal procedures.

Sprint

1. Disconnect the negative battery cable.
2. Remove the steering column trim panel.
3. Lower the steering column.
4. Remove the cluster bezel and the bezel.
5. Disconnect the headlight switch connector.
6. Remove the headlight switch.
7. To install, reverse the removal procedures.

Metro

1. Disconnect the negative battery cable.
2. Remove the instrument cluster switch panel.
3. Remove the 2 screws securing the light switch assembly to the cluster switch panel.
4. Remove the light switch assembly from the cluster switch panel.
5. Remove the 2 screws securing the light switch housing to the light switch assembly.
6. Remove the light switch housing from the light switch assembly.
7. To install, reverse the removal procedures.

Combination Switch

REMOVAL & INSTALLATION

Nova

1. Disconnect the negative battery cable. Remove the steering wheel.
2. Remove the lower instrument finish panel, air duct and column lower cover.
3. From the base of the steering column shroud, disconnect the ignition switch and turn signal electrical connector.
4. Remove the combination switch-to-steering column screws and the combination switch with the upper column cover.
5. To install, reverse the removal procedures.

Prizm

1. Disconnect the negative battery cable. Remove the steering wheel.
2. Remove the lower left side instrument finish panel.
3. Remove the upper and lower steering column covers.
4. Disconnect the combination switch connector.
5. Disconnect the 4 bolts holding the combination switch and remove the switch from the vehicle.
6. To install, reverse the removal procedures.

Spectrum

1. Disconnect the negative battery terminal from the battery.
2. Remove the horn shroud, steering wheel nut/washer and steering wheel assembly.
3. Remove the steering cowl attaching screw and steering cowl.
4. Disconnect the

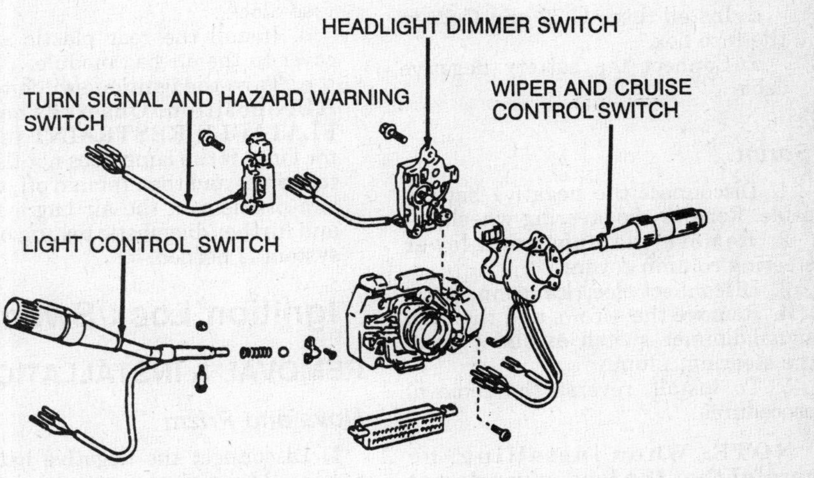

Exploded view of the combination switch—Nova and Prizm

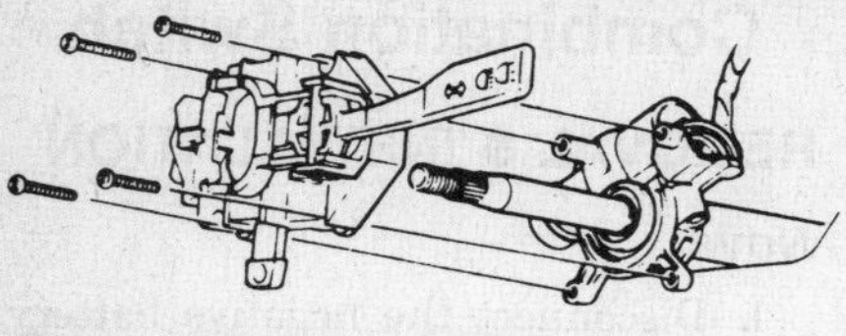

Turn signal/dimmer switch mounting—Sprint and Metro

combination/starter switch connector.
5. Remove the turn signal/dimmer switch attaching screw and switch.
6. To install, reverse the removal procedures.

Storm

1. Disable the air bag system as follows:
   a. Turn the ignition switch to the **OFF** position.
   b. Disconnect the battery negative cable.
   c. Remove fuses C-22 and C-23 from the fuse box.
   d. Disconnect the orange 3-way connector at the base of the steering column.
2. Remove the 4 bolts attaching the inflator module to the steering wheel. Disconnect the inflator module electrical connector.
3. Disconnect the electrical connector from the horn.
4. Remove the steering wheel nut and the steering wheel.
5. Remove the steering wheel cowl assembly and harness.
6. Remove the combination switch attaching screws, disconnect electrical connectors and the combination switch from the vehicle.
7. Reverse procedure to install. Reactivate the air bag system as follows:
   a. Turn the ignition switch to the **OFF** position.
   b. Connect the orange 3-way connector at the base of the steering column.
   c. Install fuses C-22 and C-23 to the fuse box.
   d. Connect the battery negative cable.

Sprint

1. Disconnect the negative battery cable. Remove the steering wheel.
2. Remove the upper and lower steering column covers.
3. Disconnect electrical connectors.
4. Remove the screws and the turn signal/dimmer switch assembly from the steering column.
5. To install, reverse the removal procedures.

NOTE: When installing, be careful that the lead wires do not get caught by the lower cover.

Metro

1. Place the ignition switch in the **LOCK** position. If equipped, with an air bag system, disable the air bag system as follows:
   a. Turn the ignition switch to the **OFF** position.
   b. Remove the **SIR IG** fuse in the supplemental inflatable restraint fuse block.
   c. Remove the rear plastic access cover to the air bag module.
   d. Disconnect the yellow 2-way connector and connector position assurance (CPA) inside the inflator module housing.
2. Disconnect the negative battery cable.
3. Remove the 6 screws retaining upper and lower column covers.
4. Remove the steering column covers from the steering cover.
5. Remove the 4 electrical connectors, 2 from the fuse block, 1 from the air bag wire harness and 1 from the main wire harness.
6. Remove the 4 screws retaining the air bag coil and turn signal/dimmer switch assembly.
7. Remove the switch assembly from the steering column.

NOTE: The coil assembly will become uncentered if the steering column is separated from the steering gear and it is allowed to rotate.

8. Installation is the reverse order of the removal procedure.
9. Reactivate the air bag system as follows:
   a. Turn the ignition switch to the **OFF** position.
   b. Connect the yellow 2-way connector and connector position assurance (CPA) inside the inflator module housing.
   c. Install the **SIR IG** fuse in the supplemental inflatable restraint fuse block.
   d. Install the rear plastic access cover to the air bag module.
   e. Turn the ignition switch to the **RUN** position. Observe the **INFLATABLE RESTRAINT** indicator lamp.If the lamp does not flash 7 to 9 times and then remain off, there is a problem in the air bag system and further diagnostic testing of the system is needed.

Ignition Lock/Switch

REMOVAL & INSTALLATION

Nova and Prizm

1. Disconnect the negative battery cable. Remove the combination switch.
2. If equipped with a tilt steering column, perform the following procedures:
   a. Remove the tension springs and grommets, the tilt lever (the bolt has left-hand threads), the adjusting nut/washer.
   b. Pull out the lock bolt, then, remove the upper and lower column supports.
3. From the lower steering column, disconnect the ignition switch electrical connector.
4. Remove the retainer-to-upper bracket screws and the retainer from the upper bracket.
5. Using snapring pliers, remove the snapring from the upper bracket.
6. Insert the key into the ignition switch and release the steering lock.
7. Using a hammer and a pin punch, drive the tapered bolt from the upper bracket.
8. Remove the upper bracket-to-steering column tube bolts and the upper bracket.

To install:

9. Release the steering lock and install the upper bracket-to steering column bolts, tighten the bolts finger tight. Torque the upper bracket-to-steering column bolts to 14 ft. lbs. (19 Nm).
10. If installing the tilt steering mechanism, perform the following procedures:
   a. Apply grease to the bushings and the O-rings, then, install the lower support-to-tube.
   b. Using multi-purpose grease, apply it to the tilt bracket-to-steering column mating surfaces, then, install the upper support and lock bolt.

NOTE: If there is any play in the adjusting support, snug-up the adjusting nut.

   c. Install the tilt lever. Move the lever to loosen the bracket-to-column bolt, adjust the column height and move the lever to lock the column position; if the lever is out of position, reposition the adjusting nut.
   d. Install the tilt lever retaining screw (left-hand thread) and torque it. Install the tension springs and grommets.
11. To complete the installation, reverse the removal procedures.

Spectrum

1. Disconnect the negative battery cable. Remove the combination switch.
2. Insert the key into the ignition and place the key in the **ON** position, the lock bar must be pulled all the way in.

3. Remove the snapring and rubber cushion from the steering shaft.
4. Disconnect the switch wires at the connectors.
5. Remove the 2 screws retaining the ignition/starter switch and remove the switch.
6. To install, reverse the removal procedures.

Storm

1. Disable the air bag system as follows:
   a. Turn the ignition switch to the **OFF** position.
   b. Disconnect the battery negative cable.
   c. Remove fuses C-22 and C-23 from the fuse box.
   d. Disconnect the orange 3-way connector at the base of the steering column.
2. Remove the 4 bolts attaching the inflator module to the steering wheel. Disconnect the inflator module electrical connector.
3. Disconnect the electrical connector from the horn.
4. Remove the steering wheel nut and the steering wheel.
5. Remove the steering wheel cowl assembly and harness.
6. With the key in the **OFF** position, depress the retaining pin and remove the ignition lock cylinder from the ignition switch. Disconnect the electrical connector from the switch.
7. Remove the ignition switch snapring, rubber seal and spacer collar from the steering shaft.
8. Remove the back drive cable from the ignition switch. Remove the ignition switch retaining screws and remove the switch from the steering column.

To install:

9. Installation is the reverse order of the removal procedure.
10. Reactivate the air bag system as follows:
    a. Turn the ignition switch to the **OFF** position.
    b. Connect the orange 3-way connector at the base of the steering column.
    c. Install fuses C-22 and C-23 to the fuse box.
    d. Connect the battery negative cable.

Sprint

1. Disconnect the negative battery cable. Remove the steering column.
2. Place the column on a bench.
3. Using a sharp point center punch and a hammer, remove the steering lock mounting bolts.
4. Turn the ignition key to **ACC** or **ON** positions and remove the lock assembly from the steering column.
5. To install, reverse the removal procedures. After installing the lock, turn the key to **LOCK** position and pull out the key. Turn the steering shaft to make sure the shaft is locked. Install new mounting bolts to the lock housing, tighten until the bolt heads break off. Torque the lower bracket bolts to 8–12 ft. lbs. (11-16 Nm); the upper bracket bolts to 10 ft. lbs. (13 Nm) and the steering shaft bolt to 15–22 ft. lbs. (20-30 Nm).

Metro

1. Place the ignition switch in the **LOCK** position. If equipped, with an air bag system, disable the air bag system as follows:
   a. Turn the ignition switch to the **OFF** position.
   b. Remove the **SIR IG** fuse in the supplemental inflatable restraint fuse block.
   c. Remove the rear plastic access cover to the air bag module.
   d. Disconnect the yellow 2-way connector and connector position assurance (CPA) inside the inflator module housing.
2. Disconnect the negative battery cable.
3. Remove the 6 screws retaining upper and lower column covers.
4. Remove the steering column covers from the steering cover.
5. Remove the 4 electrical connectors, 2 from the fuse block, 1 from the air bag wire harness and 1 from the main wire harness.
6. Remove the 4 screws retaining the air bag coil and turn signal/dimmer switch assembly.

NOTE: The coil assembly will become uncentered if the steering column is separated from the steering gear and it is allowed to rotate.

7. Remove the 1 screws and ignition switch assembly from the steering column. Remove the 2 screws and ignition key warning switch from the steering column.
8. Pull the ignition switch harness free from the instrument panel and remove it from the vehicle.
9. Installation is the reverse order of the removal procedure.
10. Reactivate the air bag system as follows:
    a. Turn the ignition switch to the **OFF** position.
    b. Connect the yellow 2-way connector and connector position assurance (CPA) inside the inflator module housing.
    c. Install the **SIR IG** fuse in the supplemental inflatable restraint fuse block.
    d. Install the rear plastic access cover to the air bag module.
    e. Turn the ignition switch to the **RUN** position. Observe the **INFLATABLE RESTRAINT** indicator lamp.If the lamp does not flash 7 to 9 times and then remain off, there is a problem in the air bag system and further diagnostic testing of the system is needed.

Stoplight Switch

REMOVAL & INSTALLATION

Nova and Prizm

1. Disconnect the negative battery cable. Remove the lower instrument panel cover.
2. Disconnect the electrical connector from the stoplight switch.
3. Remove the stoplight switch-to-bracket nut and the switch.
4. To install, reverse the removal procedures. Adjust the switch so the stoplights turn with slight movement of the brake pedal.

Spectrum and Storm

1. Disconnect the negative battery cable. Remove stoplight switch locknut.
2. Remove switch by pulling straight out of pedal assembly.
3. To install push switch straight in, push the brake pedal by turning the stoplight switch, so free-play in the brake pedal is eliminated, then tighten the stoplight switch locknut.

Sprint and Metro

1. Disconnect the negative battery cable. Disconnect the stoplight switch wiring at the brake pedal.
2. Remove the switch from the plate and install the new one.
3. Adjust the switch so there is 0.02–0.04 in. clearance between the contact plate and the end of the threads on the switch. Tighten the locknut and check the clearance again.
4. Connect the battery cable and check that the stoplights are not on with the pedal in the resting position.

Clutch Switch

ADJUSTMENT

The clutch start switch clearance is adjusted by loosening the front locknut and depressing the clutch pedal fully. Adjust the clutch start switch clearance to specification. Tighten the locknut.

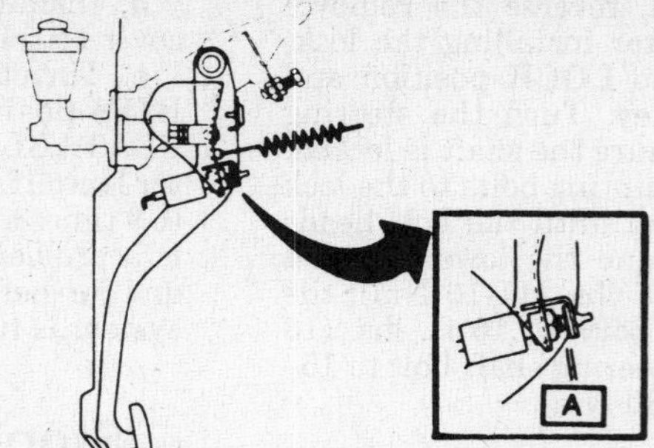

Clutch start switch clearance

REMOVAL & INSTALLATION

Nova and Prizm

1. Disconnect the negative terminal from the battery.
2. Remove the necessary trim panels in order to gain access to the clutch switch retaining screws.
3. Remove the clutch switch retaining screws and front locknut.
4. Pull the switch down from its retainer and disconnect the electrical connections.
5. To install, reverse the removal procedures.

Spectrum

1. Disconnect the negative battery terminal from the battery.
2. Disconnect the electrical connector from the switch.
3. Remove the clutch start switch-to-brake pedal stop bracket screw and the switch from the clutch pedal.
4. To install, reverse the removal procedures

Sprint, Metro and Storm

1. Disconnect the negative battery terminal from the battery.
2. Disconnect the electrical connector from the switch.
3. Loosen the locknut then unscrew the clutch start switch from the clutch pedal.
4. To install, reverse the removal procedures

Neutral Safety Switch

ADJUSTMENT

Nova and Prizm

1. Loosen the neutral start switch bolts and set the shifter lever in the **N** position.
2. Disconnect the neutral start switch electrical connector.
3. Connect an ohmmeter between the terminals on the switch.
4. Adjust the switch to the point where there is continuity between the terminals.
5. Connect the neutral switch electrical connector. Torque switch bolts to 48 inch lbs. (5.4 Nm). Check switch operation.

Storm

1. Remove the intake air duct and air breather tube from the air cleaner assembly.
2. Loosen the neutral safety switch attaching screws.
3. Place selector lever in the **N** position.
4. Install a pin into the adjustment holes in both the neutral safety switch and the switch lever.
5. Torque the retaining screws to 26 inch lbs. (3 Nm).
6. Install the air breather tube and intake air duct onto the air cleaner assembly.

Metro

1. Place selector lever in the **N** position.
2. Using a suitable tool, turn the neutral safety switch joint clockwise or counterclockwise to position the this slot at the 11 o'clock position. Check that a click is heard from the switch in this position.
3. Now turn the switch by hand and stop at the position where a click is heard from the joint.
4. Torque the neutral safety switch retaining bolts to 17 ft. lbs. (23 Nm).

REMOVAL & INSTALLATION

Nova, Prizm and Storm

1. Disconnect the negative battery cable.
2. On Storm equipped with 4-speed transaxle, proceed as follows:
   a. Remove the intake air duct and breather tube from the air cleaner assembly.
   b. Disconnect the shift control cable from the selector cable.
   c. Remove the neutral safety switch attaching screws and disconnect electrical connectors.
   d. Remove the neutral safety switch. Reverse procedure to install. Adjust neutral safety switch, if necessary.
3. Disconnect the electrical connector at the switch.
4. Raise the vehicle and support it safely.
5. Remove the switch retaining bolts and remove the switch.
6. Installation is the reverse of removal. Align the groove and neutral basic line. Hold the switch in position and tighten the bolts to 48 inch lbs.
7. Check and make sure the engine starts in only the **P** and **N** detents.

Spectrum

1. Disconnect the negative battery cable.
2. Disconnect the electrical connector for the switch at the left fender.
3. Raise the vehicle and support it safely.
4. Remove the switch retaining bolts and remove the switch.
5. Installation is the reverse of removal. Add transaxle fluid as necessary.
6. Check and make sure the engine starts in only the **P** and **N** detents.

Sprint

This vehicle uses a back drive system (solenoid) to keep the selector lever always in the **P** detent position when starting the engine with the ignition key.

1. Disconnect the negative battery cable.
2. Remove the solenoid to housing attaching screws.
3. Disconnect the solenoid wire and remove the solenoid.

To install:

4. Shift the selector lever to the **P** range.
5. Apply grease to the upper and lower edges of the lock plate before installing the back drive solenoid.
6. Position the solenoid to the housing and install the retaining screws hand tight.
7. Connect the solenoid wire.
8. Adjust the solenoid position so there is no clearance between the lock plate and the guide plate and tighten the retaining screws.

Metro

1. Disconnect the negative battery cable.

2. Raise and support the vehicle safely.

3. Disconnect the neutral safety switch couplers.

4. Remove the neutral safety switch retaining screws. Remove the switch from the transaxle.

5. Installation is the reverse order of the removal procedure. Adjust the switch as necessary.

Fuses, Circuit Breakers and Relays

LOCATION

Fusible Links

Fusible links are located in the engine harness at the starter solenoid and the left hand front of the dash at the battery junction block. On the Prizm there is a fusible link box located on the positive battery cable.

Circuit Breakers

The circuit breakers can be found incorporated in the switch it represents or they can be found on the fuse and relay boards, mounted to the boards with blades similar to the fuses. Before removing a breaker, always disconnect the negative battery cable to prevent potentially damaging electrical "spikes" within the system. Simply remove the breaker by pulling straight out from the relay board. Do not twist the relay; damage may occur to the connectors inside the housing.

NOTE: Some circuit breakers on the Nova and Prizm do not reset automatically. Once tripped, they must be reset by hand. Use a small screwdriver or similar tool; insert it in the hole in the back of the breaker and push gently. Once the breaker is reset, either check it for continuity with an ohmmeter or reinstall it and check the circuit for function.

Reinstall the circuit breaker by pressing it straight in to its mount. Make certain the blades line up correctly and that the circuit breaker is fully seated. Reconnect the negative battery cable and check the circuit for function.

Fuse Panel

NOVA AND PRIZM

On the Nova the main fuse panel is located behind the left hand kick panel. On the Prizm, the main fuse panel is attached to the number 1 junction block on the lower left hand shroud. There is also a ECM-IG fuse attached to the number 1 junction block.

STORM

The main fuel panel is attached to the junction block behind the left hand kick panel inside the vehicle. There is also a fuse/relay box located on the left hand side of the engine compartment near the battery. There is also a relay box located on the right hand side of the engine compartment near the strut tower.

SPECTRUM

The fuse panel is located at the lower left hand side of the instrument panel, concealed by a cover.

SPRINT

The fuse panel is located on the underside of the steering column.

METRO

The main fuse/relay box is located on the left hand side of the engine compartment near the battery.

Various Relays

NOVA AND PRIZM

Starter Relay—located at the front left side of the engine compartment.

Air Conditioning Relay—located at the front left side of the engine compartment.

Cooling Fan Relay—located at the front left side of the engine compartment.

Main Engine Relays—located at the front left side of the engine compartment.

Defogger Relay—located on the passenger side of the vehicle under the dash assembly.

Heater Relay—located on the passenger side of the vehicle under the dash assembly.

Charge Light Relay—located on the passenger side of the vehicle under the dash assembly.

Seat Belt Relay—located on the passenger side of the vehicle under the dash assembly.

Tailight Relay—located behind the driver's side kick panel.

SPECTRUM AND STORM

Various relays are attached to the brackets under the left-side of the dash. All units are easily replaced with plug-in modules.

SPRINT AND METRO

A/C Condenser Fan Relay—located on the left side of the firewall above the shock tower.

A/C Magnetic Clutch Relay—located on the left side of the firewall above the shock tower.

Computers

LOCATION

Nova and Prizm

The Electronic Control Module (ECM) is located towards the center of the vehicle under the instrument panel. In order to gain access to the electronic control module, it will first be necessary to remove various trim panel assemblies.

Storm

The Electronic Control Module (ECM) is located under the dash on the left hand side of the steering column.

Spectrum

The Electronic Control Module (ECM) controls the operation of the engine and is located under the dash on the right side of the vehicle.

Sprint and Metro

The Electronic Control Module (ECM) is located inside the left hand instrument panel.

Flashers

LOCATION

Nova and Prizm

The turn signal flasher is located on the driver's-side kick panel. In order to gain access to the unit, it will be necessary to first remove certain under dash padding.

Storm

The turn signal and hazard flasher are located above the fuse box, under the left hand dash panel.

Spectrum

The turn signal flasher is located behind the instrument panel, on the left-hand side of the steering column. Replacement is accomplished by unplugging the old flasher and inserting a new one.

The hazard flasher is located behind the instrument panel, on the left-hand side of the steering column. Replacement is accomplished by unplugging the old flasher and inserting a new one.

Sprint and Metro

The flasher is located near the junction/fuse block, under the left hand side of the instrument panel.

Cruise Control

ADJUSTMENT

Nova and Prizm

1. Check that the actuator cables are properly installed.
2. Measure the cable stroke to where the throttle valve begins to open.
3. If the cable stroke is not approximately 0.39 in. (10 mm) with a slight amount of free-play.
4. If the cable stroke is not within specification, adjust the cable to the correct specification.

Spectrum and Sprint

1. Position the accelerator pedal so that the engine runs at a normal idling speed. Hold the accelerator pedal in this position.
2. Lightly secure the cable casing to the bracket using the cable clamp and self locking hex nut.
3. Adjust the cable casing so that there is a light tension on the servo cable. Make sure not to pull the throttle off the idle setting.
4. Tighten the clamp securely.

7

Honda

Accord, Civic, CRX, Prelude—All Models

SERIAL NUMBER IDENTIFICATION

Vehicle Identification Plate

The vehicle identification plate is mounted on the top edge of the instrument panel and is visible from the outside of the vehicle.

Engine Number

The engine number is stamped into the side of the cylinder block, near the transaxle. The first 5 digits indicate engine model identification. The remaining numbers refer to emission equipment and production sequence.

Vehicle Identification Number

The vehicle identification number is located on the vehicle identification plate and on the engine cowl under the hood. The vehicle identification number is also located on the driver's side door jam, near the lock striker plate.

Transaxle Number

The transaxle serial number is stamped on the top of the transaxle/clutch case.

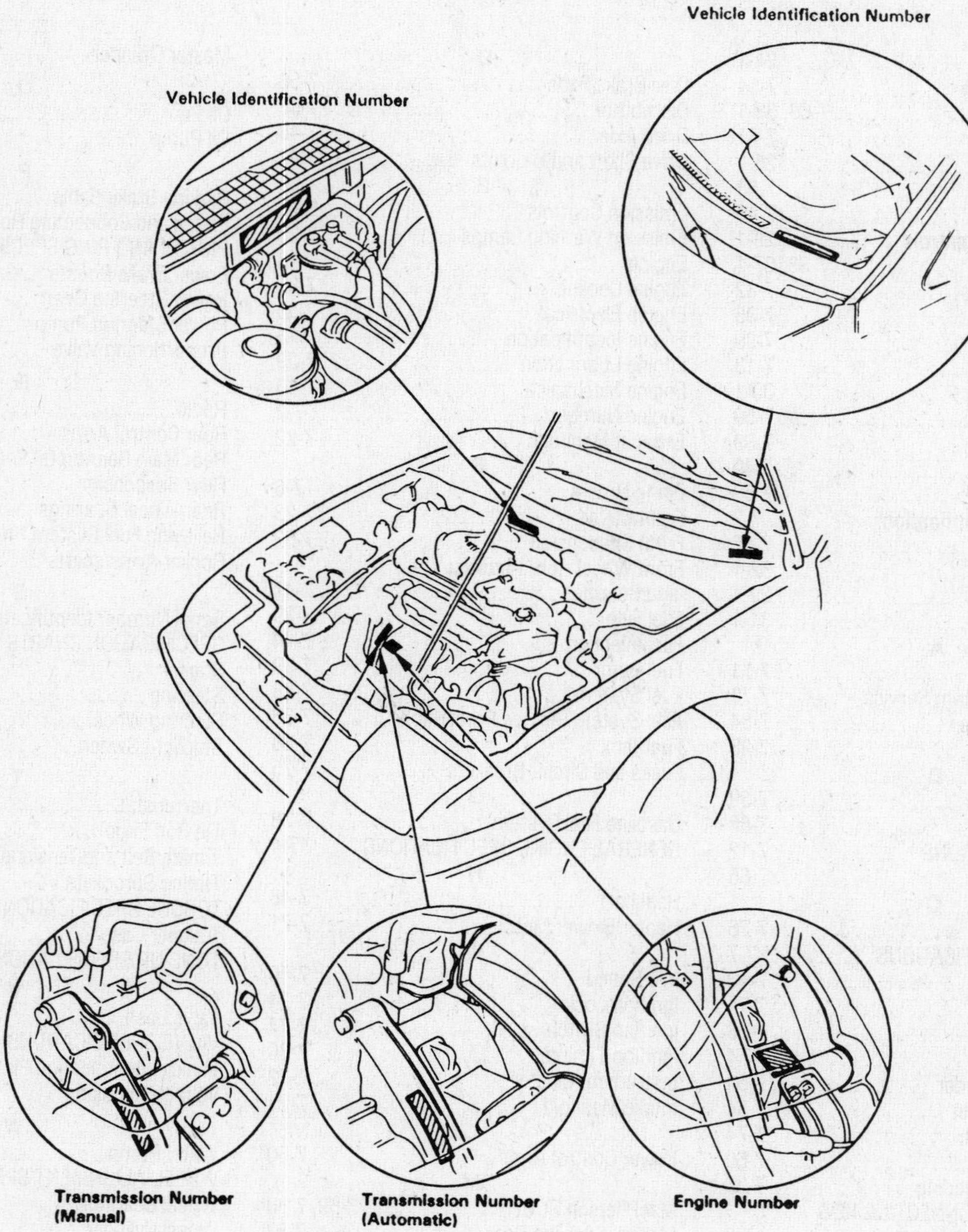

Vehicle, engine and transaxle identification numbers

SPECIFICATIONS

ENGINE IDENTIFICATION

Year	Model	Engine Displacement cu. in. (cc/liter)	Engine Series Identification	No. of Cylinders	Engine Type
1988	Civic	91.0 (1493/1.5)	D15B1	4	SOHC 16-valve
	Civic/CRX	91.0 (1493/1.5)	D15B2	4	SOHC 16-valve
	Civic/CRX, HF	91.0 (1493/1.5)	D15B6	4	SOHC 8-valve
	Civic/CRX, Si	97.0 (1590/1.6)	D16A6	4	SOHC 16-valve
	Accord, DX/LX	119.0 (1955/2.0)	A20A1	4	SOHC 12-valve
	Accord LX-i	119.0 (1955/2.0)	A20A3	4	SOHC 12-valve
	Prelude	119.0 (1955/2.0)	B20A3	4	SOHC 12-valve
	Prelude Si	119.0 (1955/2.0)	B20A5	4	SOHC 12-valve
1989	Civic	91.0 (1493/1.5)	D15B1	4	SOHC 16-valve
	Civic/CRX	91.0 (1493/1.5)	D15B2	4	SOHC 16-valve
	Civic/CRX, HF	91.0 (1493/1.5)	D15B6	4	SOHC 8-valve
	Civic/CRX, Si	97.0 (1590/1.6)	D16A6	4	SOHC 16-valve
	Accord, DX/LX	119.0 (1955/2.0)	A20A1	4	SOHC 12-valve
	Accord LX-i	119.0 (1955/2.0)	A20A3	4	SOHC 12-valve
	Prelude	119.0 (1955/2.0)	B20A3	4	SOHC 12-valve
	Prelude Si	119.0 (1955/2.0)	B20A5	4	DOHC 16-valve
1990	Civic	91.0 (1493/1.5)	D15B1	4	SOHC 16-valve
	Civic/CRX	91.0 (1493/1.5)	D15B2	4	SOHC 16-valve
	Civic/CRX, HF	91.0 (1493/1.5)	D15B6	4	SOHC 8-valve
	Civic/CRX, Si	97.0 (1590/1.6)	D16A6	4	SOHC 16-valve
	Accord, DX/LX	132.0 (2156/2.2)	F22A1	4	SOHC 16-valve
	Accord EX	132.0 (2156/2.2)	F22A4	4	SOHC 16-valve
	Prelude S	119.0 (1955/2.0)	B20A3	4	SOHC 12-valve
	Prelude Si	119.0 (1955/2.0)	B20A5	4	DOHC 16-valve
	Prelude Si	125.0 (2056/2.1)	B21A1	4	DOHC 16-valve
1991-92	Civic	91.0 (1493/1.5)	D15B1	4	DOHC 16-valve
	Civic/CRX	91.0 (1493/1.5)	D15B2	4	DOHC 16-valve
	Civic/CRX, HF	91.0 (1493/1.5)	D15B6	4	SOHC 8-valve
	Civic/CRX, Si	97.0 (1590/1.6)	D16A6	4	SOHC 16-valve
	Accord, DX/LX	132.0 (2156/2.2)	F22A1	4	SOHC 16-valve
	Accord EX	132.0 (2156/2.2)	F22A4	4	SOHC 16-valve
	Accord SE	132.0 (2156/2.2)	F22A6	4	SOHC 16-valve
	Prelude Si	119.0 (1958/2.0)	B20A5	4	DOHC 16-valve
	Prelude Si	125.0 (2056/2.1)	B21A1	4	DOHC 16-valve

SOHC—Single Overhead Cam
DOHC—Double Overhead Cam

GENERAL ENGINE SPECIFICATIONS

Year	Model	Engine Displacement cu. in. (cc)	Fuel System Type	Net Horsepower @ rpm	Net Torque @ rpm (ft. lbs.)	Bore × Stroke (in.)	Com-pression Ratio	Oil Pressure @ rpm
1988	Civic	91.0 (1493)	DP-FI	70 @ 5500	83 @ 3000	2.95 × 3.33	9.6:1	50 @ 2000
	Civic/CRX	91.0 (1493)	DP-FI	92 @ 6000	89 @ 4500	2.95 × 3.33	9.2:1	50 @ 2000
	Civic/CRX, HF	91.0 (1493)	MP-FI	62 @ 4500	90 @ 2000	2.95 × 3.33	9.6:1	50 @ 2000
	Civic/CRX, Si	97.0 (1590)	MP-FI	105 @ 6000	98 @ 5000	2.95 × 3.54	9.1:1	50 @ 2000
	Accord DX/LX	119.0 (1955)	2 bbl	98 @ 5500	109 @ 3500	3.25 × 3.58	9.1:1	55 @ 2000
	Accord LX-i	119.0 (1955)	MP-PFI	110 @ 5500	114 @ 4500	3.25 × 3.58	9.3:1	55 @ 2000
	Prelude	119.0 (1955)	Dual Sidedraft	100 @ 5500	107 @ 4000	3.19 × 3.74	9.1:1	50 @ 2000
	Prelude Si	119.0 (1955)	FI	110 @ 5500	114 @ 4500	3.18 × 3.74	9.0:1	50 @ 2000
1989	Civic	91.0 (1493)	DP-FI	70 @ 5500	83 @ 3000	2.95 × 3.33	9.2:1	74 @ 3000
	Civic/CRX	91.0 (1493)	DP-FI	92 @ 6000	89 @ 4500	2.95 × 3.33	9.2:1	74 @ 3000
	Civic/CRX, HF	91.0 (1493)	MP-FI	62 @ 4500	90 @ 2000	2.95 × 3.33	9.2:1	74 @ 3000
	Civic/CRX, Si	97.0 (1590)	MP-FI	108 @ 6000	100 @ 5000	2.95 × 3.54	9.1:1	74 @ 3000
	Accord DX/LX	119.0 (1955)	2 bbl	98 @ 5500	109 @ 3500	3.26 × 3.58	9.1:1	55–65 @ 3000
	Accord LX-i/SE-i	119.0 (1955)	MP-PFI	120 @ 5800	122 @ 4000	3.26 × 3.58	9.3:1	55–65 @ 3000
	Prelude	119.0 (1955)	Dual Sidedraft	①	111 @ 4800	3.19 × 3.74	9.1:1	75–87 @ 3000
	Prelude Si	110.3 (1955)	MP-PFI	135 @ 6200	127 @ 4500	3.19 × 3.74	9.0:1	75–87 @ 3000
1990	Civic	91.0 (1493)	DP-FI	70 @ 5500	83 @ 3000	2.95 × 3.33	9.2:1	50 @ 3000
	Civic/CRX	91.0 (1493)	DP-FI	92 @ 6000	89 @ 4500	2.95 × 3.33	9.2:1	50 @ 3000
	Civic/CRX, HF	91.0 (1493)	MP-FI	62 @ 4500	90 @ 2000	2.95 × 3.33	9.6:1	50 @ 3000
	Civic/CRX, Si	97.0 (1590)	MP-FI	108 @ 6000	100 @ 5000	2.95 × 3.54	9.1:1	50 @ 3000
	Accord DX/LX	132.0 (2156)	MP-FI	125 @ 5200	137 @ 4000	3.35 × 3.74	8.8:1	50 @ 3000
	Accord EX	132.0 (2156)	MP-FI	130 @ 5200	142 @ 4000	3.35 × 3.74	8.8:1	50 @ 3000
	Prelude S	119.0 (1955)	Dual Sidedraft	①	111 @ 4000	3.19 × 3.74	9.1:1	50 @ 3000
	Prelude Si	119.0 (1955)	MP-FI	135 @ 6200	127 @ 4000	3.19 × 3.74	9.0:1	50 @ 3000
	Prelude Si	125.0 (2056)	MP-FI	140 @ 5800	135 @ 5000	3.27 × 3.74	9.4:1	50 @ 3000
1991–92	Civic	91.0 (1493)	DP-FI	70 @ 5500	83 @ 3000	2.95 × 3.33	9.2:1	50 @ 3000
	Civic/CRX	91.0 (1493)	DP-FI	92 @ 6000	89 @ 4500	2.95 × 3.33	9.2:1	50 @ 3000
	Civic/CRX, HF	91.0 (1493)	MP-FI	62 @ 4500	90 @ 2000	2.95 × 3.33	9.6:1	50 @ 3000
	Civic/CRX, Si	97.0 (1590)	MP-FI	108 @ 6000	100 @ 5000	2.95 × 3.54	9.1:1	50 @ 3000
	Accord DX/LX	132.0 (2156)	MP-FI	125 @ 5200	137 @ 4000	3.35 × 3.74	8.8:1	50 @ 3000
	Accord EX	132.0 (2156)	MP-FI	130 @ 5200	142 @ 4000	3.35 × 3.74	8.8:1	50 @ 3000
	Accord SE	132.0 (2156)	MP-FI	140 @ 5600	142 @ 4500	3.35 × 3.74	8.8:1	50 @ 3000
	Prelude Si	119.0 (1958)	MP-FI	135 @ 6200	127 @ 4000	3.19 × 3.74	9.0:1	50 @ 3000
	Prelude Si	125.0 (2056)	MP-FI	140 @ 5800	135 @ 5000	3.27 × 3.74	9.4:1	50 @ 3000

DP-FI Dual Point Fuel Injected
MP-FI Multipoint Fuel Injected
MP-PFI Multipoint Port Fuel Injected
① Manual transaxle—104 @ 5800
Automatic transaxle—105 @ 5800

ENGINE TUNE-UP SPECIFICATIONS

Year	Model	Engine Displacement cu. in. (cc)	Spark Plugs Type	Spark Plugs Gap (in.)	Ignition Timing (deg.) MT	Ignition Timing (deg.) AT	Compression Pressure (psi)	Fuel Pump (psi)	Idle Speed (rpm) MT	Idle Speed (rpm) AT	Valve Clearance ⑤ In.	Valve Clearance ⑤ Ex.
1988	Civic/CRX HF	91.0 (1493)	BCPR6E-11	0.042	14B	14B	185	35	600–700	700–800	0.005–0.007	0.007–0.009
	Civic/CRX Std.	91.0 (1493)	BCPR6E-11	0.042	18B	18B	185	35	600–700	700–800	0.007–0.009	0.009–0.011
	Civic/CRX Si	97.0 (1590)	BCPR6E-11	0.042	18B	18B	185	35	700–800	700–800	0.007–0.009	0.009–0.011
	Accord DX/LX	119.0 (1955)	BPR5EY-11	0.042	24B ⑥	15B ①	171	3.0	800–850	700–800	0.005–0.007	0.010–0.012
	Accord LX-i	119.0 (1955)	BPR5EY-11	0.042	15B ①	15B ①	178	35	750–800	750–800	0.005–0.007	0.010–0.012
	Prelude	119.0 (1955)	BCPR5E-11	0.042	20B	12B	156	2.5	800–850	750–800	0.005–0.007	0.001–0.001
	Prelude Si	119.0 (1955)	BCPR6E-11	0.042	15B	15B	178	35	750–800	750–800	0.003–0.005	0.006–0.008
1989	Civic/CRX	91.0 (1493)	BCPR6E-11	0.042	18B	18B	185	36	700–800	700–800	0.007–0.009	0.009–0.011
	Civic HF	91.0 (1493)	BCPR6E-11	0.042	18B	18B	185	36	600–650	600–650	0.005–0.007	0.007–0.009
	Civic/CRX Si	97.0 (1590)	BCPR6E-11	0.042	18B	18B	185	36	700–800	700–800	0.007–0.009	0.009–0.011
	Accord DX/LX	119.0 (1955)	BPR5EY-11	0.042	24B ②	15B	171	2.6–3.3	750–850	680–790	0.005–0.007	0.010–0.012
	Accord LX-i	119.0 (1955)	BPR5EY-11	0.042	15B	15B	178	33–39	700–800	700–800	0.005–0.007	0.010–0.012
	Prelude	119.0 (1955)	BCPR5E-11	0.042	20B ③	15B ④	171	1.3–2.1	750–850	700–800	0.005–0.007	0.010–0.012
	Prelude Si	119.0 (1955)	BCPR6E-11	0.042	15B	15B	178	36	700–800	700–800	0.003–0.005	0.006–0.008
1990	Civic	91.0 (1493)	BCPR6E-11	0.042	18B	18B	185	36	700–800	700–800	0.007–0.009	0.009–0.011
	Civic/CRX	91.0 (1493)	BCPR6E-11	0.042	18B	18B	185	36	700–800	700–800	0.007–0.009	0.009–0.011
	Civic/CRX, HF	91.0 (1493)	BCPR5E-11	0.042	14B	14B	185	36	600–700 ⑪	700–800	0.005–0.007	0.007–0.009
	Civic/CRX, Si	97.0 (1590)	BCPR6E-11	0.042	18B	18B	185	36	700–800	700–800	0.007–0.009	0.009–0.011
	Accord DX/LX	132.0 (2156)	ZFR5F-11	0.042	15B	15B	178	36	650–750	650–750	0.0094–0.011	0.011–0.012
	Accord EX	132.0 (2156)	ZFR5F-11	0.042	15B	15B	178	36	650–750	650–750	0.0094–0.011	0.0110–0.0126
	Prelude S	119.0 (1955)	BCPR5E-11	0.042	20B ③	15B ④	178	1.3–2.1	750–850	700–800	0.005–0.007	0.010–0.012
	Prelude Si	119.0 (1955)	ZFR5F-11	0.042	15B	15B	178	36	700–800	700–800	0.003–0.005	0.006–0.008
	Prelude Si	125.0 (2056)	ZFR5F-11	0.042	15B	15B	178	36	700–800	700–800	0.003–0.005	0.006–0.008

ENGINE TUNE-UP SPECIFICATIONS

Year	Model	Engine Displacement cu. in. (cc)	Spark Plugs Type	Spark Plugs Gap (in.)	Ignition Timing (deg.) MT	Ignition Timing (deg.) AT	Com-pression Pressure (psi)	Fuel Pump (psi)	Idle Speed (rpm) MT	Idle Speed (rpm) AT	Valve Clearance ⑤ In.	Valve Clearance ⑤ Ex.
1991	Civic	91.0 (1493)	BCPR6E-11	0.042	18B	18B	185	36	700–800	700–800	0.007–0.009	0.009–0.011
	Civic/CRX	91.0 (1493)	BCPR6E-11	0.042	18B	18B	185	36	700–800	700–800	0.007–0.009	0.009–0.011
	Civic/CRX, HF	91.0 (1493)	BCPR5E-11	0.042	14B	14B	185	36	550–650	550–650	0.005–0.007	0.007–0.009
	Civic/CRX, Si	97.0 (1590)	BCPR6E-11	0.042	18B	18B	185	36	700–800	700–800	0.007–0.009	0.009–0.011
	Accord DX/LX	132.0 (2156)	ZFR5F-11	0.042	15B	15B	178	36	700–800	700–800	0.0094–0.011	0.011–0.012
	Accord EX	132.0 (2156)	ZFR5F-11	0.042	15B	15B	178	36	700–800	700–800	0.0094–0.011	0.011–0.012
	Accord SE	132.0 (2156)	ZFRSF-11	0.042	15B	15B	178	36	700–800	700–800	0.0094–0.011	0.011–0.012
	Prelude Si	119.0 (1958)	ZFR5F-11	0.042	15B	15B	178	36	700–800	700–800	0.003–0.005	0.006–0.008
	Prelude Si	125.0 (2056)	ZFR5F-11	0.042	15B	15B	178	36	700–800	700–800	0.003–0.005	0.006–0.008
1992	SEE UNDERHOOD SPECIFICATIONS STICKER											

NOTE: The underhood specifications sticker often reflects tune-up changes made in production. Sticker figures must be used if they disagree with those in this chart.

B Before top dead center

—Not applicable

① Aim timing light at red mark on flywheel or torque converter drive plate with the distributor vacuum hose connected at the specified idle speed.

② California—20B

③ California—15B

④ Califorina—10B

⑤ Jet valve adjustment
Except 1342cc and 1488cc—0.005- 0.007
1342cc and 1488cc—0.007- 0.009

⑥ California—20B

FIRING ORDERS

NOTE: To avoid confusion, always replace spark plug wires one at a time.

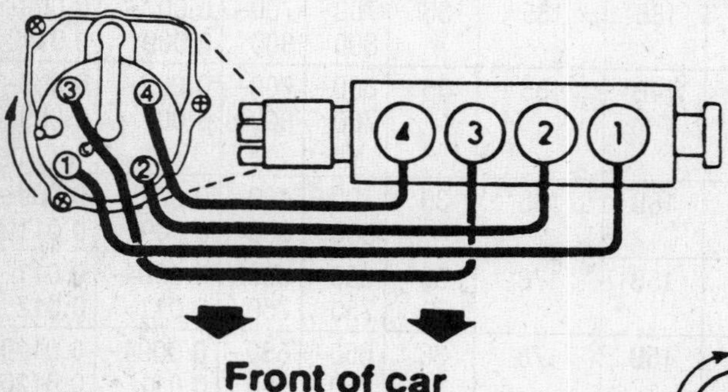

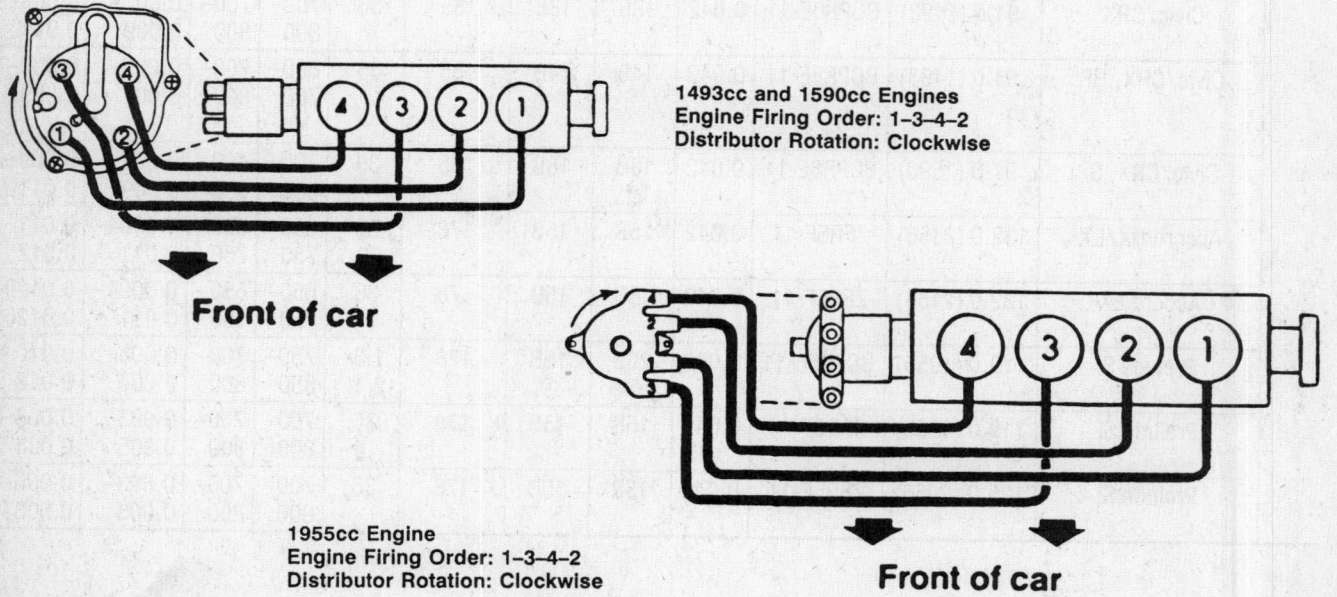

1493cc and 1590cc Engines
Engine Firing Order: 1–3–4–2
Distributor Rotation: Clockwise

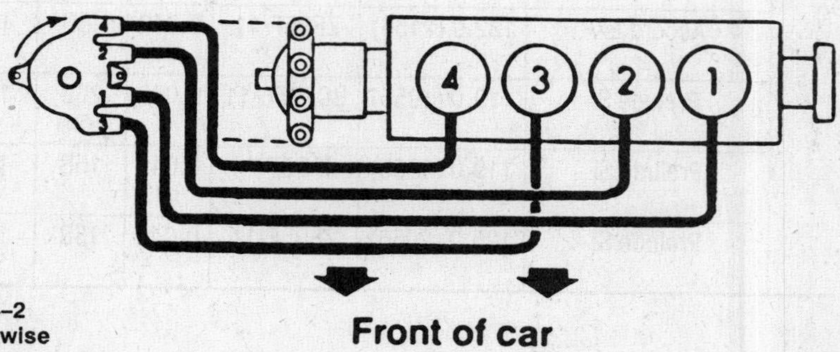

1955cc Engine
Engine Firing Order: 1–3–4–2
Distributor Rotation: Clockwise

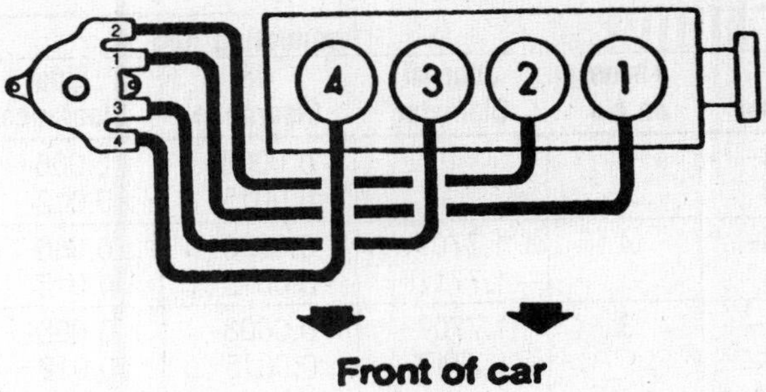

2155cc (1988–90) and 2056cc Engines
Engine Firing Order: 1–3–4–2
Distributor Rotation: Clockwise

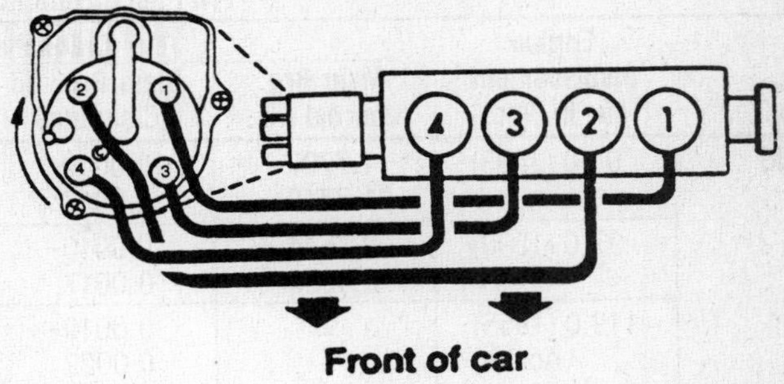

2156cc Engine
Engine Firing Order: 1–3–4–2
Distributor Rotation: Clockwise

CAPACITIES

Year	Model	Engine Displacement cu. in. (cc)	Engine Crankcase (qts.) with Filter	Engine Crankcase (qts.) without Filter	Transmission (pts.) 4-Spd	Transmission (pts.) 5-Spd	Transmission (pts.) Auto.	Drive Axle (pts.)	Fuel Tank (gal.)	Cooling System (qts.)
1988	Civic/CRX	91.0 (1493)	3.7	3.2	—	5.0	5.0	1.4⑨	11.9	5.8
	Civic/CRX Si	97.0 (1590)	3.7	3.2	—	5.0	5.0	1.4⑨	11.9	5.8
	Accord	119.0 (1955)	3.7	3.2	—	5.0	6.0	—	15.9	5.8
	Prelude	119.0 (1955)	4.1	3.6	4.0	4.0	6.0	—	15.9	8.2
1989	Civic/CRX	91.0 (1493)	3.7	3.2	—	4.0①	5.0②	1.4⑨	11.9③	④
	Civic/CRX	97.0 (1590)	3.7	3.2	—	4.0①	5.0②	1.4⑨	11.9③	④
	Accord	119.0 (1955)	3.7	3.2	—	4.8	6.4	—	15.9	⑥
	Prelude	119.0 (1955)	4.1	3.6	4.0	4.0	6.0	—	15.9	⑤
1990	Civic/CRX	91.0 (1493)	3.7	3.2	—	4.0①	5.0②	1.4⑨	11.9③	④
	Civic/CRX	97.0 (1590)	3.7	3.2	—	4.0①	5.0②	1.4⑨	11.9③	④
	Accord	132.0 (2156)	4.0	3.7	—	4.0	5.0	—	17.0	⑦
	Prelude	119.0 (1955)	4.0	3.6	—	4.4	6.0	—	15.9	⑧
	Prelude	125.0 (2056)	4.0	3.6	—	4.4	6.0	—	15.9	8.2
1991-92	Civic/CRX	91.0 (1493)	3.7	3.2	—	4.0⑨	5.0②	1.4⑨	11.9③	④
	Civic/CRX	97.0 (1590)	3.7	3.2	—	4.0⑨	5.0④	1.4⑨	11.9③	④
	Accord	132.0 (2156)	4.0	3.7	—	4.0	5.0	—	17.0	⑦
	Prelude	119.0 (1958)	4.0	3.6	—	4.4	6.0	—	15.9	8.2
	Prelude	125.0 (2056)	4.0	3.6	—	4.4	6.0	—	15.9	8.2

① 4WD—5.0
② 4WD—6.0
③ HF—10.6
④ Automatic transaxle—5.7
Manual transaxle—5.8
⑤ Fuel injection—8.2
Carbureted with manual transaxle—7.2
Carbureted with automatic transaxle—7.9
⑥ Manual transaxle—6.9
Automatic transaxle—7.2
⑦ Manual transaxle—7.0
Automatic Transaxle—7.5
⑧ Prelude 2.0S—Manual transaxle—7.2
Automatic transaxle—7.9
⑨ 4WD Rear differential

CRANKSHAFT AND CONNECTING ROD SPECIFICATIONS

All measurements are given in inches.

Year	Engine Displacement cu. in. (cc)	Crankshaft Main Brg. Journal Dia.	Crankshaft Main Brg. Oil Clearance	Crankshaft Shaft End-play	Crankshaft Thrust on No.	Connecting Rod Journal Diameter	Connecting Rod Oil Clearance	Connecting Rod Side Clearance
1988	91.0 (1493)	1.7707–1.7718	0.0010–0.0017	0.004–0.014	4	③	0.0008–0.0015	0.006–0.012
	97.0 (1590)	2.1644–2.1654	0.0010–0.0017	0.004–0.014	4	1.7707–1.7717	0.0008–0.0015	0.006–0.012
	119.0 (1955) Accord	②	0.0010–0.0022 ①	0.004–0.014	3	1.7707–1.7717	0.0008–0.0015	0.006–0.012
	119.0 (1955) Prelude	2.1644–2.1654	0.0010–0.0017 ④	0.004–0.014	3	1.7707–1.7717 ⑤	0.0010–0.0017	0.006–0.012
1989	91.0 (1493)	1.7707–1.7718	⑥	0.004–0.014	4	1.6526–1.6535	0.0008–0.0015	0.006–0.012
	97.0 (1590)	2.1644–2.1654	⑦	0.004–0.014	4	1.7707–1.7717	0.0008–0.0015	0.006–0.012
	119.0 (1955) Accord	②	0.0010–0.0022 ①	0.004–0.014	3	1.7707–1.7717	0.0008–0.0015	0.006–0.012
	119.0 (1955) Prelude	2.1644–2.1654	0.0010–0.0017 ④	0.004–0.014	3	1.7707–1.7717 ⑤	0.0010–0.0017 ⑧	0.006–0.012
	119.0 (1955) Prelude Si	2.1644–2.1654	0.0010–0.0017 ④	0.004–0.014	3	1.8888–1.8900 ⑤	0.0010–0.0017	0.006–0.012
1990	91.0 (1493)	1.7707–1.7718	⑥	0.004–0.014	4	1.6526–1.6535	0.0008–0.0015	0.006–0.012
	97.0 (1590)	2.1644–2.1654	⑦	0.004–0.014	4	1.7707–1.7717	0.0008–0.0015	0.006–0.012
	132.0 (2156)	⑨	⑩	0.004–0.014	4	1.7710–1.7717	0.0008–0.0017	0.006–0.012
	119.0 (1955)	2.1644–2.1654 ⑫	⑦	0.004–0.014	3	1.7707–1.7717	⑪	0.006–0.012
	125.0 (2056)	2.1644–2.1654 ⑫	⑦	0.004–0.014	3	1.8888–1.8900	0.0010–0.0017	0.006–0.012

CRANKSHAFT AND CONNECTING ROD SPECIFICATIONS

All measurements are given in inches.

Year	Engine Displacement cu. in. (cc)	Crankshaft Main Brg. Journal Dia.	Crankshaft Main Brg. Oil Clearance	Crankshaft Shaft End-play	Crankshaft Thrust on No.	Connecting Rod Journal Diameter	Connecting Rod Oil Clearance	Connecting Rod Side Clearance
1991-92	91.0 (1493)	1.7707–1.7718	⑥	0.004–0.014	4	1.6526–1.6535	0.0008–0.0015	0.006–0.012
	97.0 (1590)	2.1644–2.1654	⑦	0.004–0.014	4	1.7707–1.7717	0.0008–0.0015	0.006–0.012
	132.0 (2156)	⑨	⑩	0.004–0.014	4	1.7710–1.7717	0.0008–0.0017	0.006–0.012
	119.0 (1955)	2.1644–2.1654 ⑫	⑦	0.004–0.014	3	1.7707–1.7717	⑪	0.006–0.012
	125.0 (2056)	2.1644–2.1654 ⑫	⑦	0.004–0.014	3	1.8888–1.8900	0.0010–0.0017	0.006–0.012

① No. 3—0.0013-0.0024
② Accord:
No. 1—1.9676-1.9685
No. 3—1.9671-1.9680
No. 2, 4, 5:1.9673-1.9683
③ HF—1.4951-1.4961
Std.—1.5739-1.5748
④ No. 3—0.0012-0.0019
⑤ Prelude Si—1.8888-1.8900
⑥ No. 2, 3, 4—0.0010-0.0017
No. 1, 5—0.0007-0.0014
⑦ No. 1, 5—0.0007-0.0014
No. 2, 4—0.0010-0.0017
No. 3—0.0012-0.0019
⑧ Prelude with fuel injection—0.0010-0.0017
⑨ No. 1, 2—1.9676-1.9685
No. 3—1.9674-1.9683
No. 4, 5—1.9655-1.9688
⑩ No. 1, 2—0.0009-0.0018
No. 3—0.0014-0.0017
No. 4, 5—0.0005-0.0015
⑪ Prelude 2.0S—0.0008-0.0015
Prelude 2.0Si—0.0010-0.0017
⑫ No. 3—2.1642-2.1651

VALVE SPECIFICATIONS

Year	Engine Displacement cu. in. (cc)	Seat Angle (deg.)	Face Angle (deg.)	Spring Test Pressure (lbs.)	Spring Installed Height (in.)	Stem-to-Guide Clearance (in.) Intake	Stem-to-Guide Clearance (in.) Exhaust	Stem Diameter (in.) Intake	Stem Diameter (in.) Exhaust
1988	91.0 (1493)	45	45	NA	②	0.001–0.002	0.002–0.003	0.2157–0.2161	0.2147–0.2150
	97.0 (1590)	45	45	NA	②	0.001–0.002	0.002–0.003	0.2157–0.2161	0.2147–0.2150
	119.0 (1955) Accord Prelude	45	45	NA	①	0.001–0.002	0.002–0.004	0.2591–0.2594	0.2732–0.2736
	119.0 (1955) Prelude Si	45	45	NA	1.683	0.001–0.002	0.002–0.003	0.2591–0.2594	0.2579–0.2583
1989	91.0 (1493)	45	45	NA	③	0.001–0.002	0.002–0.003	0.2157–0.2161	0.2147–0.2150
	97.0 (1590)	45	45	NA	③	0.001–0.002	0.002–0.003	0.2157–0.2161	0.2147–0.2150
	119.0 (1955) Accord Prelude	45	45	NA	①	0.001–0.002	0.002–0.004	0.2591–0.2594	0.2732–0.2736
	119.0 (1955) Prelude Si	45	45	NA	1.683	0.001–0.002	0.002–0.003	0.2591–0.2594	0.2579–0.2583

VALVE SPECIFICATIONS

Year	Engine Displacement cu. in. (cc)	Seat Angle (deg.)	Face Angle (deg.)	Spring Test Pressure (lbs.)	Spring Installed Height (in.)	Stem-to-Guide Clearance (in.)		Stem Diameter (in.)	
						Intake	Exhaust	Intake	Exhaust
1990	91.0 (1493)	45	45	NA	NA	0.001–0.002	0.002–0.003	0.2157–0.2161	0.2147–0.2150
	97.0 (1590)	45	45	NA	NA	0.001–0.002	0.002–0.003	0.2157–0.2161	0.2147–0.2150
	132.0 (2156)	45	45	NA	NA	0.0009–0.0019	0.002–0.003	0.2157–0.2161	0.2146–0.2150
	119.0 (1955)	45	45	NA	NA	0.001–0.002	④	0.2591–0.2594	⑤
	125.0 (2056)	45	45	NA	NA	0.001–0.002	0.002–0.003	0.2591–0.2594	0.2579–0.2583
1991–92	91.0 (1493)	45	45	NA	NA	0.001–0.002	0.002–0.003	0.2157–0.2161	0.2147–0.2150
	97.0 (1590)	45	45	NA	NA	0.001–0.002	0.002–0.003	0.2157–0.2161	0.2147–0.2150
	132.0 (2156)	45	45	NA	NA	0.0009–0.0019	0.002–0.003	0.2157–0.2161	0.2146–0.2150
	119.0 (1955)	45	45	NA	NA	0.001–0.002	④	0.2591–0.2594	⑤
	125.0 (2056)	45	45	NA	NA	0.001–0.002	0.002–0.003	0.2591–0.2594	0.2579–0.2583

NA Not Available
① Intake—1.913
Exhaust—1.876
② 1493cc:
Intake—1.8498–1.8880
Exhaust—1.9278–1.9463
1590cc:
Intake—1.8498–1.8683
Exhaust—1.9278–1.9263
③ Intake—1.8498–1.8683
Exhaust—1.9278–1.9263
④ Prelude 2.0S—0.002–0.004
Prelude 2.0Si—0.002–0.003
⑤ Prelude 2.0S—0.2732–0.2736
Prelude 2.0Si—0.2579–0.2583

PISTON AND RING SPECIFICATIONS

All measurements are given in inches

Year	Engine Displacement cu. in. (cc)	Piston Clearance	Ring Gap			Ring Side Clearance		
			Top Compression	Bottom Compression	Oil Control	Top Compression	Bottom Compression	Oil Control
1988	91.0 (1493)	0.0004–0.0016	0.006–0.014	0.006–0.014	0.008–0.024	0.0012–0.0024	0.0012–0.0022	Snug
	97.0 (1590)	0.0004–0.0016	0.006–0.014	0.006–0.014	0.008–0.024	0.0012–0.0024	0.0012–0.0022	Snug
	119.0 (1955)	0.0008–0.0016	0.008–0.014	0.016–0.022	0.008–0.028 ①	0.0012–0.0024 ②	0.0012–0.0024 ②	Snug
1989	91.0 (1493)	0.0004–0.0016	0.006–0.012	0.012–0.018	0.008–0.024	0.0012–0.0024	0.0012–0.0022	Snug
	97.0 (1590)	0.0004–0.0016	0.006–0.012	0.012–0.018	0.008–0.024	0.0012–0.0024	0.0012–0.0022	Snug
	119.0 (1955)	0.0008–0.0016	0.008–0.014	0.016–0.022 ③	0.008–0.028 ①	0.0012–0.0024 ②	0.0012–0.0024	Snug

PISTON AND RING SPECIFICATIONS

All measurements are given in inches

Year	Engine Displacement cu. in. (cc)	Piston Clearance	Ring Gap Top Compression	Ring Gap Bottom Compression	Ring Gap Oil Control	Ring Side Clearance Top Compression	Ring Side Clearance Bottom Compression	Ring Side Clearance Oil Control
1990	91.0 (1493)	0.0004–0.0016	0.006–0.012	0.012–0.018	0.008–0.031	0.0012–0.0024	0.0012–0.0022	Snug
	97.0 (1590)	0.0004–0.0016	0.006–0.012	0.012–0.018	0.008–0.031	0.0012–0.0024	0.0012–0.0022	Snug
	132.0 (2156)	0.0008–0.0016	0.008–0.014	0.016–0.022	0.007–0.027	0.0014–0.0024	0.0011–0.0022	Snug
	119.0 (1955)	0.0008–0.0016	0.008–0.014	0.016–0.022	0.008–0.020	0.0012–0.0022	0.0012–0.0022	Snug
	125.0 (2056)	0.0004–0.0013	0.010–0.014	0.018–0.022	0.008–0.020	0.0014–0.0026	0.0012–0.0024	Snug
1991-92	91.0 (1493)	0.0004–0.0016	0.006–0.012	0.012–0.018	0.008–0.031	0.0012–0.0024	0.0012–0.0022	Snug
	97.0 (1590)	0.0004–0.0016	0.006–0.012	0.012–0.018	0.008–0.031	0.0012–0.0024	0.0012–0.0022	Snug
	132.0 (2156)	0.0008–0.0016	0.008–0.014	0.016–0.022	0.007–0.027	0.0014–0.0024	0.0011–0.0022	Snug
	119.0 (1955)	0.0008–0.0016	0.008–0.014	0.016–0.022	0.008–0.020	0.0012–0.0022	0.0012–0.0022	Snug
	125.0 (2056)	0.0004–0.0013	0.010–0.014	0.018–0.022	0.008–0.020	0.0014–0.0026	0.0012–0.0024	Snug

① Prelude equipped with carburetor—0.008–0.020
② Prelude—0.0012-0.0022
③ Prelude—0.016-0.022

TORQUE SPECIFICATIONS

All readings in ft. lbs.

Year	Engine Displacement cu. in. (cc)	Cylinder Head Bolts	Main Bearing Bolts	Rod Bearing Bolts	Crankshaft Pulley Bolts	Flywheel Bolts	Manifold Intake	Manifold Exhaust	Spark Plugs
1988	91.0 (1493)	49	48	23	83	87①	25	25	13
	97.0 (1590)	49	48	23	83	87①	25	25	13
	119.0 (1955)	49	49	23②	83	76①	20③	23④	13
1989	91.0 (1493)	47	⑤ ⑦ ⑧	23	119⑤	87①	16	23	13
	97.0 (1590)	47	⑤ ⑦ ⑧	23	119⑤	87①	16	23	13
	119.0 (1955)	49	49⑤	23	108⑤	76①	16	23⑥	13
1990	91.0 (1493)	47	33⑤	23	119⑤	87①	16	23	13
	97.0 (1590)	47	47⑤	23	119⑤	87①	16	23	13
	132.0 (2156)	78	52⑤	34	159⑤	76①	16	23	13
	119.0 (1955)	49⑤	49⑤	23⑨	108⑤	76①	16	22	13
	125.0 (2056)	49⑤	49⑤	34	108⑤	76①	16	22	13
1991-92	91.0 (1493)	47	33⑤	23	119⑤	87①	16	23	13
	97.0 (1590)	47	47⑤	23	119⑤	87①	16	23	13
	132.0 (2156)	78	52⑤	34	159⑤	76①	16	23	13
	119.0 (1955)	49⑤	49⑤	23⑨	108⑤	76①	16	22	13
	125.0 (2056)	49⑤	49⑤	34	108⑤	76①	16	22	13

① Auto Transaxle—54 ft. lbs.
② Fuel injected engine—33 ft. lbs.
③ Prelude—16 ft. lbs.
④ Prelude—26 ft. lbs.
⑤ Dip bolts in clean engine oil
⑥ Prelude with fuel injection—26 ft. lbs.
⑦ Civic and CRX except Si—33 ft. lbs. Civic and CRX Si—47 ft. lbs.
⑧ Station Wagon 2WD—33 ft. lbs. Station Wagon 4WD—47 ft. lbs.
⑨ Fuel injected engine—34 ft. lbs.

BRAKE SPECIFICATIONS

Year	Model	Lug Nut Torque (ft. lbs.)	Master Cylinder Bore	Brake Disc Minimum Thickness	Brake Disc Maximum Runout	Standard Brake Drum Diameter	Minimum Lining Thickness Front	Minimum Lining Thickness Rear
1988	Civic/CRX	80	NA	③	0.004	7.90 ②	0.120	0.080
	Accord	80	NA	① ④	0.006	7.850	0.120	0.080
	Prelude	80	NA	① ④	0.004	7.850	0.120	0.080
1989	Civic	80	NA	0.67	⑫	7.90	0.120	0.080
	Civic SW	80	NA	0.67	⑫	7.870	⑥	0.080
	CRX	80	NA	⑤	⑫	7.90	0.120	0.080
	Accord	80	NA	⑩	⑫	7.870 ⑨	0.120	0.031
	Prelude	80	NA	⑦	⑧	⑨	0.120	0.080
1990	Civic	80	NA	0.67 ⑪	0.004	7.90	0.120	0.079
	Civic SW	80	NA	0.67 ⑬	0.004	7.90	0.120	0.080
	CRX	80	NA	0.67	0.004	7.90	0.120	0.080
	Accord	80	NA	0.83	0.004	8.660	0.063	0.080
	Prelude	80	NA	⑦	⑧	⑨	0.120	0.080
1991-92	Civic	80	NA	0.67 ⑪	0.004	7.90	0.120	0.079
	Civic Wagon	80	NA	0.75 ⑭	0.004	7.90	0.120	0.080
	CRX	80	NA	0.67 ⑬	0.004	7.90	0.120	0.080
	Accord	80	NA	0.83F 0.32R	0.004	8.660 ⑮	0.120	0.080
	Prelude	80	NA	0.75F 0.31R	0.004	—	0.120	0.080

NA—Not Available
① Rear disc—0.310
② Civic Wagon—7.850
③ Civic/CRX HF—0.590
All others—0.670
④ Accord DX/LX and Prelude—0.670
Accord LX-i and Prelude Si—0.750
⑤ Std and Si—0.670
HF—0.590
⑥ 4WD—0.120
Except 4WD—0.060
⑦ Si—0.750
S—0.670
⑧ Front—0.004
Rear—0.006
⑨ Rear disc rotor—0.390
⑩ DX, LX—0.670
LXi, SEi—0.750
⑪ Civic EX—0.750
⑫ Civic Wagon DX—0.750
⑬ CRX Std—0.750
CRX HF—0.590
⑭ 4WD—0.67
⑮ Parking Brake

WHEEL ALIGNMENT

Year	Model	Caster Range (deg.)	Caster Preferred Setting (deg.)	Camber Range (deg.)	Camber Preferred Setting (deg.)	Toe-in (in.)	Steering Axis Inclination (deg.)
1988	Civic exc. SW	2P–4P	3P	1N–1P	0	0	7 5/16
	Civic SW	1 15/16P–3 15/16P	2 15/16P	11/16N–1 5/16P	5/16P	0	7 1/4
	Civic 4WD	1 15/16P–3 15/16P	2 15/16P	7/16N–1 9/16P	9/16P	0	6 15/16
	Accord	1/2N–1 1/2P	1/2P	1N–1P	0	0	6 13/16
	Prelude	1 3/16P–2 13/16P	2 3/8P	1N–1P	0	0	6 13/16
1989	Civic exc. SW	2P–4P	3P	1N–1P	0	0	—
	Civic SW	1 15/16P–3 15/16P	2 15/16P	11/16N–1 5/16P	5/16P	0	—
	Civic 4WD	1 15/16P–3 15/16P	2 15/16P	7/16N–1 9/16P	9/16P	0	—
	Accord	1/2N–1 1/2P	1/2P	1N–1P	0	0	—
	Prelude	1 13/16P–2 13/16P	2 3/8P	1N–1P	0	0	—

WHEEL ALIGNMENT

Year	Model	Caster Range (deg.)	Caster Preferred Setting (deg.)	Camber Range (deg.)	Camber Preferred Setting (deg.)	Toe-in (in.)	Steering Axis Inclination (deg.)
1990	Civic exc. SW	2P–4P	3P	1N–1P	0	0	—
	Civic SW	1 15/16P–3 15/16P	2 15/16P	11/16N–1 5/16P	5/16P	0	—
	Civic 4WD	1 15/16P–3 15/16P	2 15/16P	7/16N–1 9/16P	9/16P	0	—
	Accord	2P–4P	3P	1N–1P	0	0	—
	Prelude	1 13/16P–2 13/16P	2 3/8P	1N–1P	0	0	—
1991–92	Civic exc. SW	2P–4P	3P	1N–1P	0	0	—
	Civic SW	1 15/16P–3 15/16P	2 15/16P	11/16N–1 5/16P	5/16P	0	—
	Civic 4WD	1 15/16P–3 15/16P	2 15/16P	7/16N–1 9/16P	9/16P	0	—
	Accord	2P–4P	3P	1N–1P	0	0	—
	Prelude	1 13/16P–2 13/16P	2 3/8P	1N–1P	0	0	—

SW—Station Wagon
4WD—Four Wheel Drive
P—Positive
N—Negative

ENGINE MECHANICAL

Engine Assembly

REMOVAL & INSTALLATION

Civic and CRX

1. Disconnect the negative battery cables. Remove the battery and the battery tray.
2. Apply the parking brake and place blocks behind the rear wheels. Raise the vehicle and support it safely.
3. Scribe a line where the hood brackets meet the inside of the hood.
4. Disconnect the windshield washer fluid tubes. Unbolt and remove the hood.
5. Remove the engine and wheelwell splash shields.
6. Drain the oil from the engine, the coolant from the radiator and the transaxle oil from the transaxle.
7. Remove the air intake duct and the front air intake duct.
8. Relieve the fuel pressure from the fuel system, by slowly loosening the banjo bolt on the fuel filler approximately 1 turn.

NOTE: Keep any and all open flames away from the work area. Before disconnecting any fuel lines, the fuel pressure should be relieved. Place a suitable shop towel over the fuel filler to prevent the pressurized fuel from spraying over the engine.

9. Disconnect and tag the engine compartment harness connectors, battery wires and transaxle ground cable.
10. Remove the throttle cable by loosening the lock nut and the throttle cable adjust nut, then slip the throttle cable end out of the throttle bracket and accelerator linkage. Be sure not to bend the cable when removing it. Do not use pliers to remove the cable from the linkage. Always replace a kinked cable with a new one.
11. Disconnect and tag the engine wire connectors and spark plug wires. Bring the engine up to TDC on the No. 1 cylinder. Mark the position of the distributor rotor in relation to the distributor housing and the distributor in relation to the engine block. Remove the distributor assembly from the cylinder head.
12. Disconnect the radiator hoses and heater hoses. Disconnect the transaxle fluid cooler lines. Remove the speedometer cable.

NOTE: Do not remove the speedometer cable holder, because the speedometer gear may fall into the transaxle housing.

13. If equipped with power steering, remove the mounting bolts and power steering belt, then without disconnecting the hoses, pull the pump away from it's mounting bracket and lay aside.
14. Disconnect and tag the alternator wiring, remove the alternator adjusting bolts, mounting bolts and belt. Remove the alternator from the vehicle.
15. Loosen the air conditioning belt adjust bolt and the idler puller nut. Remove the compressor mounting bolts. Disconnect the air conditioning suction and discharge lines, only if it is necessary. Lift the compressor out of the bracket with the air conditioning hoses attached and wire the compressor to the front beam of the vehicle.

NOTE: If it is necessary to remove the air conditioning suction and discharge lines, discharge the refrigerant from the air conditioning system. Be sure to properly discharge the refrigerant into a suitable container and be sure to wear safety goggles and gloves.

16. If equipped with an automatic transaxle, proceed as follows:
    a. Remove the header pipe, header pipe bracket, torque converter cover and shift control cable holder.
    b. Remove the shift control cable by removing the cotter pin, control pin and control lever roller from the control lever.
17. If equipped with manual transaxles, remove the shift lever torque rod, shift rod and clutch cable. On reassembly, slide the retainer back into place after driving in the spring pin.

18. Remove the wheelwell splash shields and engine splash shields. Remove the right and left halfshafts from the transaxle and cover the shafts with a plastic bag so as to prevent the oil from spilling over the work area. Be sure to coat all precision finished surfaces with clean engine oil or grease.
19. On 4WD vehicles equipped with automatic transaxles, remove the cable clip and the control pin. Loosen the shift control cable nut and then remove the control cable.
20. On 4WD vehicles equipped with manual transaxles, remove the cotter pins and the 3 cable bracket mounting bolts. Remove the cable bracket from the rear of the transaxle mount bracket.
21. Attach a suitable chain hoist to the engine block hoist brackets and raise the hoist just enough to remove the slack from the chain. To attach the rear engine chain, remove the plastic radiator hose bracket and hook the chain to the top of the clutch cable bracket.
22. Remove the rear transaxle mount bracket. Remove the bolts from the front transaxle bolt mount. Remove the bolts from the engine side mount. Remove the bolts from the engine side transaxle mounts.
23. Check that the engine/transaxle assembly are completely free of vacuum, fuel, coolant hoses and electrical wires.
24. Slowly raise the engine approximately 6 in. and stop. Check again that the engine/transaxle assembly are completely free of vacuum, fuel, coolant hoses and electrical wires.
25. Raise the engine/transaxle assembly all the way up and out of the vehicle, once it is clear from the vehicle, lower the assembly into a suitable engine stand.

To install:

26. Installation is the reverse order of the removal procedure. Use the following steps to aid in the installation procedure.
27. Torque the engine mount bolts in the following sequence; be sure to replace the rear transaxle bolt and the front transaxle bolt with new bolts:
    a. Side transaxle mount—40 ft. lbs. (54 Nm).
    b. Rear transaxle mount bracket—43 ft. lbs. (58 Nm).
    c. Front transaxle mount—43 ft. lbs. (58 Nm).
    d. Engine side mount—40 ft. lbs. (54 Nm).

NOTE: Failure to tighten the bolts in the proper sequence can cause excessive noise and vibration and reduce bushing life. Be sure to check that the bushings are not twisted or offset.

28. Check that the spring clip on the end of each driveshaft clicks into place. Be sure to use new spring clips on installation.
29. After assembling the fuel line parts, turn the ignition switch (do not operate the starter) to the **ON** position so the fuel pump is operated for approximately 2 seconds so as to pressurize the fuel system. Repeat this procedure 2–3 times and check for a possible fuel leak.
30. Bleed the air from the cooling system at the bleed bolt with the heater valve open.
31. Adjust the throttle cable tension, install the air conditioning compressor and belt and adjust all belt tensions. Adjust the clutch cable free-play and check that the transaxle shifts into gear smoothly.
32. Check the ignition timing.
33. Install the speedometer cable, be sure to align the tab on the cable end with the slot holder. Install the clip so the bent leg is on the groove side. After installing, pull the speedometer cable to make sure it is secure.

Prelude and 1988–89 Accord

1. Raise the vehicle and support it safely.
2. Disconnect both battery cables from the battery. Remove the battery, and then remove the battery tray from the engine compartment.
3. On Prelude, remove the knob caps covering the headlights' manual retracting knobs, then turn the knobs to bring the headlights to the **ON** position.
4. On Prelude, remove the 5 screws retaining the grille and remove the grille.
5. Remove the splash guard from under the engine. Unbolt and remove the hood.
6. Remove the oil filler cap and drain the engine oil.
7. Remove the radiator cap, then open the radiator drain petcock and drain the coolant from the radiator.
8. Remove the transaxle filler plug, then remove the drain plug and drain the transaxle.
9. On carbureted vehicles:
    a. Label and then remove the wires at the coil and the engine secondary ground cable located on the valve cover.
    b. Remove the air cleaner cover and filter.
    c. Remove the air intake ducts. Remove the 2 nuts and 2 bolts from the air cleaner, remove the air control valve. Remove the air cleaner as required.
    d. Loosen the locknut on the throttle cable and loosen the cable adjusting nut, then slip the cable end out of the carburetor linkage.

NOTE: Be careful not to bend or kink the throttle cable. Always replace a damaged cable.

    e. Disconnect the No. 1 control box connector. Remove the control box from its bracket and let it hang next to the engine.
    f. Disconnect the fuel line at the fuel filter and remove the solenoid vacuum hose at the charcoal canister.
    g. On California and high altitude vehicles, remove the air jet controller.
10. On fuel injected vehicles:
    a. Remove the air intake duct. Disconnect the cruise control vacuum tube from the air intake duct and remove the resonator tube.
    b. Remove the secondary ground cable from the top of the engine.
    c. Disconnect the air box connecting tube. Unscrew the tube clamp bolt and disconnect the emission tubes.
    d. Remove the air cleaner case mounting nuts and remove the air cleaner case assembly.
    e. Loosen the locknut on the throttle cable and loosen the cable adjusting nut, then slip the cable end out of the bracket and linkage.

NOTE: Be careful not to bend or kink the throttle cable. Always replace a damaged cable.

    f. Disconnect the following wires, the ground cable at the fuse box. The engine compartment sub-harness connector and clamp. The high tension wire and ignition primary leads at the coil. The radio condenser connector at the coil.
    g. Using the following procedures relieve the fuel system pressure by placing a shop rag over the fuel filter to absorb any gasoline which may be sprayed on the engine while relieving the pressure. Slowly loosen the service bolt approximately 1 full turn. This will relieve any pressure in the system. Using a new sealing washer, retighten the service bolt.
    h. Disconnect the fuel return hose from the pressure regulator. Remove the banjo nut and then remove the fuel hose.
    i. Disconnect the vacuum hose from the brake booster.
11. Disconnect the radiator and heater hoses at the engine. Label the heater hoses so they can be installed correctly.
12. On automatic transaxle equipped vehicles, disconnect the transaxle oil cooler hoses at the transaxle, let the fluid drain from the hoses, then hang the hoses up near the radiator.
13. On manual transaxle equipped vehicles, loosen the clutch cable ad-

justing nut and remove the clutch cable from the release arm.

14. Disconnect the battery cable at the transaxle and the starter cable at the starter motor terminal.

15. Disconnect both engine harness connectors.

16. Remove the speedometer cable clip, then pull the cable out of the holder.

NOTE: Do not remove the holder as the speedometer gear may drop into the transaxle.

17. If equipped with power steering:
   a. Remove the speed sensor complete with hoses.
   b. Remove the adjusting bolt and the drive belt.
   c. Without disconnecting the hoses, pull the pump away from its mounting bracket and position it out of the way.
   d. Remove the power steering hose bracket from the cylinder head.

18. Remove the center beam beneath the engine. On Accord loosen the radius rod nuts to aid in the removal of the driveshafts.

19. If equipped with air conditioning:
   a. Remove the compressor clutch lead wire.
   b. Loosen the belt adjusting bolt.

NOTE: Do not remove the air conditioner hoses. The air conditioner compressor can be moved without discharging the air conditioner system.

   c. Remove the compressor mounting bolts, then lift the compressor out of the bracket with the hoses attached, and hang it on the front bulkhead with a piece of wire.

20. If equipped with manual transaxle, remove the shift rod yoke attaching bolt and disconnect the shift lever torque rod from the clutch housing.

21. If equipped with automatic transaxle:
   a. Remove the center console.
   b. Place the shift lever in reverse, then remove the lock pin from the end of the shift cable.
   c. Unscrew the cable mounting bolts and remove the shift cable holder.
   d. Remove the throttle cable from the throttle lever. Loosen the lower locknut, then remove the cable from the bracket.

NOTE: Do not loosen the upper locknut as it will change the transaxle shift points.

22. Disconnect the right and left lower ball joints and the tie rod ends.

23. Remove the halfshafts as follows:
   a. Lower the vehicle. Loosen the 32mm spindle nuts with a socket. Raise and support the vehicle safely.
   b. Remove the front wheel, and the spindle nut.
   c. Remove the damper fork and the damper pinch bolts.
   d. Remove the ball joint bolt and separate the ball joint from the front hub (Accord) or lower arm control (Prelude).
   e. Disconnect the tie rods from the steering knuckles.
   f. On Accord, remove the sway bar bolts.
   g. Pull the front hub outward and off the halfshafts.
   h. Using a small pry bar, pry out the inboard CV-joint approximately ½ in. in order to release the spring clip from the differential, then pull the halfshaft out of the transaxle case.

NOTE: When installing the halfshaft, insert the shaft until the spring clip clicks into the groove. Always use a new spring clip when installing driveshafts.

24. On fuel injected vehicles, disconnect the sub-engine harness connectors and clamp.

25. Remove the exhaust header pipe.

26. Attach a chain hoist to the engine and raise it just enough to remove the slack.

27. Disconnect the No. 2 control box connector, lift the control box off its bracket, and let it hang next to the engine.

28. If equipped with air conditioning, remove the idle control solenoid valve.

29. Remove the air chamber (if equipped).

30. Remove the 3 engine mount bolts located under the air chamber, then push the engine mount into the engine mount tower.

31. Remove the front engine mount nut, then remove the rear engine mount nut.

32. Loosen and remove the alternator belt. Disconnect the alternator wire harness and remove the alternator.

33. Remove the bolt from the rear torque rod at the engine, then loosen the bolt in the frame mount and swing the rod up and out of the way.

34. Raise the engine carefully from the vehicle, checking that all wires and hoses have been removed from the engine/transaxle. Raise the engine all the way up and remove it from the vehicle.

To install:

35. Installation is the reverse of the removal procedure. Use the following steps to aid in the installation procedure.

36. Tighten the engine mount bolts on 1988–89 Accord in the following sequence:
   a. Tighten the side engine mount through bolt snug only.
   b. Tighten the side engine mount mounting bolts to 40 ft. lbs. (55 Nm).
   c. Tighten the front engine mount nut to 28 ft. lbs. (39 Nm).
   d. Tighten the rear engine mount nut to 28 ft. lbs. (39 Nm).
   e. Tighten the side engine mount through bolt to 28 ft. lbs. (39 Nm).
   f. Tighten the rear torque rod-to-frame mount bolt snug only.
   g. Tighten the rear torque rod-to-engine mount bolt to 54 ft. lbs. (75 Nm).
   h. Tighten the rear torque rod-to-frame mount bolt to 54 ft. lbs. (75 Nm).
   i. Check that the rubber damper on the center beam is centered in it's mount on the transaxle. If not, loosen the bolts for the center beam and insulator and adjust as necessary. Tighten the center beam mounting bolts to 37 ft. lbs. (51 Nm) and the center beam transaxle mount nuts to 40 ft. lbs. (55 Nm).

37. Tighten the engine mount bolts on Prelude in the following sequence:
   a. Replace the 3 rear engine mount bolts with new ones and tighten to 40 ft. lbs. (55 Nm).
   b. Replace the rear engine mount-to-frame bolt with a new one and tighten temporarily.
   c. Replace the front engine mount-to-frame bolt with a new one and tighten temporarily.
   d. Tighten the transaxle mount bolts to 28 ft. lbs. (39 Nm) for the vertical bolt and 40 ft. lbs. (55 Nm) for the horizontal bolt. Tighten the transaxle mount-to-frame bolt to 54 ft. lbs. (75 Nm).
   e. Tighten the side engine mount bolts to 28 ft. lbs. (39 Nm) and the side engine mount through bolt to 54 ft. lbs. (75 Nm).
   f. Tighten the rear engine mount-to-frame bolt to 51 ft. lbs. (70 Nm).
   g. Tighten the front engine mount-to-frame bolt to 51 ft. lbs. (70 Nm).

NOTE: Failure to tighten the bolts in the proper sequence can cause excessive noise and vibration and reduce bushing life. Check that the bushings are not twisted or offset.

38. Check that the spring clip on the end of each halfshaft clicks into the differential. Use new clips on installation.

39. After assembling the fuel line parts, turn the ignition switch (do not operate the starter) to the **ON** position, so the fuel pump is operated for approximately 2 seconds so as to pressurize the fuel system. Repeat this

procedure 2–3 times and check for a possible fuel leak.

40. Bleed the air from the cooling system at the bleed bolt, with the heater valve open.

41. Adjust the throttle cable tension, install the air conditioning compressor and belt and adjust all belt tensions. Adjust the clutch cable free-play and check that the transaxle shifts into gear smoothly.

42. Check the ignition timing.

1990–92 Accord

1. Disconnect the battery cables and remove the battery and battery case.
2. Raise and safely support the vehicle.
3. Place the hood in a vertical position and safely support it in place. Do not remove the hood.
4. Remove the engine splash shield. Drain the engine oil, coolant and transaxle fluid.
5. Remove the air intake duct and the air cleaner case.
6. Relieve the fuel system pressure by slowly loosening the service bolt on the fuel pipe about 1 turn.
7. Remove the fuel feed hose from the fuel pipe and the return hose from the pressure control valve.
8. Disconnect the 2 connectors and remove the control box from the firewall.

NOTE: Do not disconnect the vacuum hoses.

9. Disconnect the vacuum hose from the charcoal canister and the charcoal canister hose from the throttle body.
10. Remove the ground cable from the transaxle.
11. Remove the throttle cable by loosening the locknut, then slip the cable end out of the throttle bracket and accelerator linkage.

NOTE: Be careful not to bend the cable when removing. Do not use pliers to remove the cable from the linkage. Always replace a kinked cable with a new one.

12. Disconnect the connector and the vacuum hose, then remove the cruise control actuator.
13. Remove the brake booster vacuum hose and mount vacuum hose from the intake manifold.
14. Disconnect the 3 engine wire harness connectors from the main wire harness at the right side of the engine compartment and remove the engine wire harness terminal and the starter cable terminal from the underhood relay box and clamps. Then remove the transaxle ground terminal.
15. Disconnect the 2 engine wire harness connectors from the main harness and the resistor at the left side of the engine compartment.
16. Remove the engine ground wire from the cylinder head cover and power steering pump bracket.
17. Remove the mounting bolts and the power steering belt from the power steering pump, then without disconnecting the hoses, pull the pump away from it's mounting bracket. Support the pump out of the way.
18. Remove the mounting bolts and belt from the air conditioning compressor, then without disconnecting the hoses, pull the compressor away from it's mounting bracket. Support the compressor out of the way.
19. Disconnect the heater hoses. Disconnect the radiator hoses, automatic transaxle cooler hoses and the cooling fan motor connectors. Remove the radiator/cooling fan assembly.
20. Remove the speed sensor without disconnecting the hoses or connector.
21. Remove the center beam.
22. Remove the exhaust pipe nuts and bracket mounting bolts.
23. Remove the halfshafts as follows:
    a. Remove the wheel and tire assemblies.
    b. Raise the locking tab on the spindle nut and remove it.
    c. Remove the damper fork nut and damper pinch bolt and remove the damper fork.
    d. Remove the cotter pin and castle nut from the lower ball joint.
    e. Using a suitable puller, separate the lower control arm from the knuckle.
    f. Pull the knuckle outward and remove the halfshaft outboard CV-joint from the knuckle using a suitable plastic hammer.
    g. Using a suitable pry bar, pry the halfshaft out to force the set ring at the end of the halfshaft past the groove.
    h. Pull the inboard CV-joint and remove the halfshaft and CV-joint out of the differential case or intermediate shaft as an assembly.

NOTE: Do not pull on the halfshaft, as the CV-joint may come apart. Tie plastic bags over the halfshaft ends to protect them.

24. On manual transaxle equipped vehicles, remove the clutch release hose from the clutch damper on the transaxle housing. Remove the shift cable and the select cable with the cable bracket from the transaxle.

NOTE: Be careful not to bend the cable when removing. Do not use pliers to remove the cable. Always replace a kinked cable with a new one.

25. On automatic transaxle equipped vehicles, remove the engine stiffener, then remove the torque converter cover. Remove the cable holder, then remove the shift control lever bolt and shift control cable.

NOTE: Be careful not to bend the cable when removing. Do not use pliers to remove the cable. Always replace a kinked cable with a new one.

26. Attach a suitable lifting device to the engine. Raise the engine to unload the engine mounts.
27. Remove the front and rear engine mounting bolts.
28. Remove the engine side mount and mounting bolt and the side transaxle mount and mounting bolt.
29. Make sure the engine/transaxle assembly is completely free of vacuum hoses, fuel and coolant hoses and electrical wires.
30. Slowly raise the engine approximately 6 in. Check again that all hoses and wires have been disconnected from the engine/transaxle assembly.
31. Raise the engine/transaxle assembly all the way and remove it from the vehicle.

To install:

32. Installation is the reverse of the removal procedure. Attention to the following steps will aid installation.
33. Tighten the engine mounting bolts in the following sequence:
    a. Tighten the rear engine mount-to-frame bolts snug only.
    b. Replace the rear engine mount through bolt with a new one and tighten snug only.
    c. Replace the front engine mount through bolt with a new one and tighten snug only.
    d. Tighten the side transaxle mount through bolt snug only.
    e. Tighten the engine side mount through bolt snug only.
    f. Tighten the side transaxle mount-to-block nuts to 28 ft. lbs. (39 Nm).
    g. Tighten the engine side mount-to-block bolt and nut to 40 ft. lbs. (55 Nm).
    h. Tighten the rear engine mount through bolt to 47 ft. lbs. (65 Nm).
    i. Tighten the rear engine mount-to-frame bolts to 40 ft. lbs. (55 Nm).
    j. Tighten the front engine mount through bolt to 47 ft. lbs. (65 Nm).
    k. Tighten the side transaxle mount through bolt to 40 ft. lbs. (55 Nm).
    l. Tighten the engine side mount through bolt to 40 ft. lbs. (55 Nm).

NOTE: Failure to tighten the bolts in the proper sequence can cause excessive noise and vibration and reduce bushing life.

Check that the bushings are not twisted or offset.

34. Make sure the spring clip on the end of each halfshaft clicks into place. Use new clips when installing.
35. Bleed the air from the cooling system at the bleed bolt with the heater valve open.
36. Adjust the throttle cable tension and check the clutch pedal free-play.
37. Check that the transaxle shifts into gear smoothly.
38. Adjust the tension of the accessory drive belts.
39. After assembling the fuel line parts, turn the ignition switch **ON**, but do not operate the starter, so the fuel pump is operated for approximately 2 seconds and the fuel is pressurized. Repeat 2–3 times and check for fuel leakage.
40. Check the ignition timing.

Cylinder Head

REMOVAL & INSTALLATION

Civic and CRX

1. Be sure the engine is cold. Disconnect the negative battery cable. Drain the cooling system.
2. Remove the brake booster vacuum hose from the brake master cylinder power booster. Remove the engine secondary ground cable from the valve cover.
3. Remove the air intake hose and the air chamber. Relieve the fuel pressure. Disconnect the fuel hoses and fuel return hose.
4. Remove the air intake hose and resonator hose. Disconnect the throttle cable at the throttle body. On vehicles equipped with automatic transaxles, disconnect the throttle control cable at the throttle body.
5. Disconnect the charcoal canister hose at the throttle valve.
6. Disconnect the following engine wire connectors from the cylinder head and the intake manifold:
   a. 14 prong connector from the main wiring harness
   b. EACV connector
   c. Intake air temperature sensor connector
   d. Throttle angle sensor connector
   e. Injector connectors
   f. Ignition coil from the distributor
   g. Top dead center/crank sensor connector from the distributor.
   h. Coolant temperature gauge sender connector.
   i. Coolant temperature sensor connector.
   j. Oxygen sensor.
7. Disconnect the vacuum hoses and the water bypass hoses from the intake manifold and throttle body.
8. Remove the upper radiator hose and the heater hoses from the cylinder head.
9. Remove the PCV hose, charcoal canister hose and vacuum hose from the intake manifold, and remove the vacuum hose from the brake master cylinder power booster.
10. Loosen the air conditioning idler pulley and remove the air conditioning belt. Remove the alternator belt. If equipped with power steering, remove the power steering belt and pump bracket.
11. Remove the intake manifold bracket. Remove the exhaust manifold bracket, then remove the header pipe.
12. Remove the exhaust manifold shroud, then remove the exhaust manifold.
13. Mark the position of the distributor in relation to the engine block, remove and tag the spark plug wires and remove the distributor assembly.
14. Remove the valve cover. Remove the timing belt cover.
15. Mark the direction of rotation on the timing belt. Loosen the timing belt adjuster bolt, then remove the timing belt from the camshaft pulley.

NOTE: Do not crimp or bend the timing belt more than 90 degrees or smaller than 1 in. (25mm) in diameter.

16. Remove the cylinder head bolts. Once the bolts are all removed, remove the cylinder head along with the intake manifold from the engine. Remove the intake manifold from the cylinder head.

To install:

17. Install the cylinder head in the reverse order of the removal procedure.
18. Always use a new head gasket.
19. Be sure the cylinder head and the engine block surfaces are clean, level and straight.
20. Be sure the **UP** mark on the timing belt pulley is at the top.
21. Install the intake manifold and tighten the nuts in a criss-cross pattern in 2–3 steps to 17 ft. lbs. (23 Nm) starting with the inner nuts.
22. Be sure the cylinder head dowel pins and control jet are aligned.
23. Install the bolts that secure the intake manifold to its bracket but do not tighten them at this point.
24. Position the cam correctly and install the cylinder head bolts.
25. Tighten the cylinder head bolts in 2 steps. On the first step tighten all the bolts, in sequence, to 22 ft. lbs. (30 Nm). On the final step, using the same sequence, tighten the bolts to 47 ft. lbs. (65 Nm).

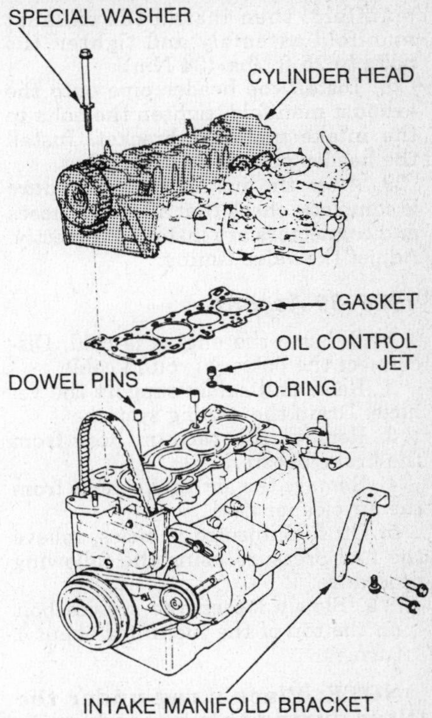

Cylinder head installation on all Civic and CRX

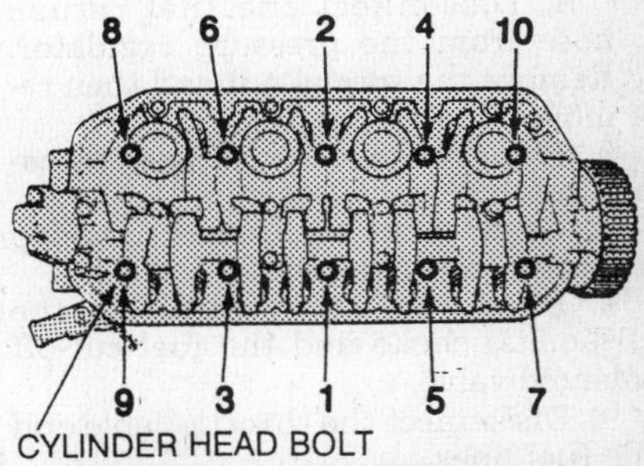

Cylinder head bolt torque sequence on Civic and CRX except CRX HF

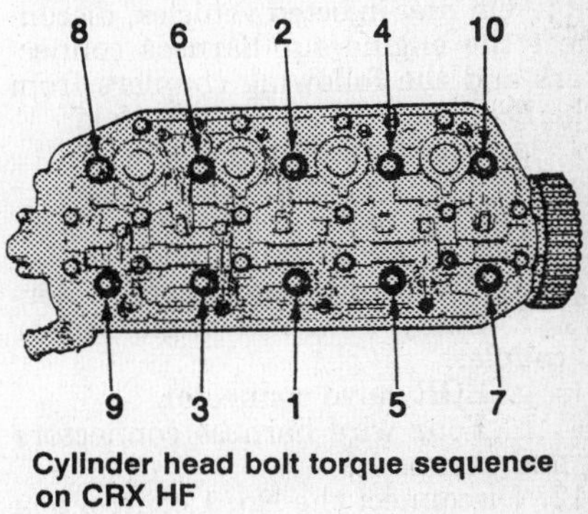

Cylinder head bolt torque sequence on CRX HF

26. On the Standard and Si vehicles, install the exhaust manifold and tighten the nuts in a criss-cross pattern in 2 or 3 steps to 25 ft. lbs. (34 Nm) starting with the inner nuts.
27. On the CRX HF vehicles, install the catalytic converter to the exhaust

manifold, then install the exhaust manifold assembly and tighten the bolts to 25 ft. lbs. (34 Nm).

28. Install the header pipe onto the exhaust manifold. tighten the bolts to the intake manifold bracket. Install the header pipe on to its bracket.

29. After the installation procedure is complete, check that all tubes, hoses and connectors are installed correctly. Adjust the valve timing.

1988–89 Accord

1. Be sure the engine is cold. Disconnect the battery ground cable.
2. Raise and safely support the vehicle. Drain the cooling system.
3. Remove the vacuum hose from the brake booster.
4. Remove the air intake ducts from the air cleaner case.
5. On fuel injected vehicles, relieve the fuel pressure using the following procedure:
   a. Slowly loosen the service bolt on the top of the fuel filter about 1 turn.

NOTE: Place a rag under the filter during this procedure to prevent fuel from spilling onto the engine.

   b. Disconnect the fuel return hose from the pressure regulator. Remove the special nut and then remove the fuel hose.

6. Remove the secondary ground cable from the valve cover.
7. Remove the air cleaner, tagging all hoses for installation.
8. Disconnect the wires from the automatic choke and the fuel cut-off solenoid valve.
9. Disconnect the throttle cable and the fuel lines.
10. Disconnect the connector and hoses from the distributor.
11. On fuel injected vehicles, disconnect the engine sub-harness connectors and the following couplers from the head and the intake manifold:
    a. The 4 injector couplers
    b. The TA sensor connector
    c. The ground connector
    d. The TW sensor connector
    e. The throttle sensor connector
    f. The crankshaft angle sensor coupler
    g. EGR valve connector
    h. Four wire harness connectors and clamps
12. Disconnect the No. 1 control box hoses from the tubing manifold.
13. On California and high altitude vehicles, disconnect the air jet controller hoses.
14. Disconnect the oxygen sensor coupler.
15. Disconnect the cooling system hoses at the cylinder head.
16. Remove the power steering pump, if equipped. Do not disconnect the pump hoses. Also, remove the hose clamp bolt on the cylinder head.
17. Remove the power steering pump bracket.
18. Remove the cruise control actuator, if equipped.
19. If equipped with air conditioning, disconnect the idle boost solenoid hoses.
20. Remove the engine splash guard from under the vehicle, if equipped.
21. Remove the exhaust header pipe and pull it clear of the exhaust manifold.
22. Remove the air cleaner base mount bolts and disconnect the hose from the intake manifold to the breather chamber.
23. Remove the valve cover, upper timing belt cover and then loosen the belt tensioner to remove the belt.
24. Remove the cylinder head bolts and remove the head.

NOTE: Loosen the cylinder head bolts in the reverse order of the torque sequence, ⅓ turn at a time to prevent warpage to the cylinder head.

To install:

25. Installation is the reverse of the removal procedure.
26. Make sure the cylinder head gasket surfaces are clean.
27. Make sure the **UP** mark on the timing belt pulley is at the top.
28. Install the intake and exhaust manifolds and tighten the nuts in a criss-cross pattern in 2–3 steps, beginning with the inner nuts.
29. Make sure the head dowel pins and oil control jet are aligned.
30. Install the bolts that secure the intake manifold to it's bracket, but do not tighten them yet.
31. Position the cam correctly.
32. Tighten the cylinder head bolts in 2 steps. Tighten all bolts in sequence to 22 ft. lbs. (30 Nm) and then to 49 ft. lbs. (68 Nm) in the same sequence.
33. Install the exhaust pipe on the exhaust manifold. Tighten the bolts for the intake manifold bracket.
34. Install the exhaust pipe on it's bracket.

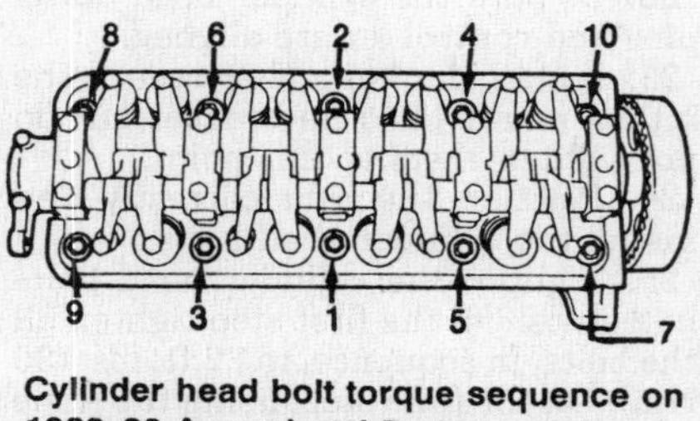

Cylinder head bolt torque sequence on 1988–89 Accord and Prelude with SOHC engine

35. After installation, check that the tubes, hoses and connectors are installed correctly.
36. Adjust the valve timing.

Prelude

SOHC Engine

1. Disconnect the negative battery cable.
2. Bring the No. 1 cylinder to TDC.
3. Drain the cooling system.
4. Remove the brake booster vacuum hose from the tubing manifold.
5. Remove the engine secondary ground cable from the valve cover.
6. Disconnect the radio condenser connector, ignition coil wire and ignition primary connector.
7. Remove the air cleaner cover.
8. Remove the fuel tube from the fuel filter.
9. Disconnect the throttle cable from the carburetor.
10. Disconnect the engine wire harness connectors from the cylinder head.
11. Remove the emission control box and vacuum tank, then disconnect the 2 connectors. Do not remove the emission hoses.
12. Disconnect the charcoal canister vacuum hoses and the upper radiator, heater and bypass hoses.
13. If equipped with cruise control, remove the actuator.
14. Remove the power steering and alternator belts.
15. Disconnect the inlet hose of the power steering pump, then remove the pump from the cylinder head.

NOTE: When the hose is disconnected fluid will run out. Protect the alternator by covering it with a shop towel. Plug the inlet hose.

16. Remove the alternator.
17. Remove the intake manifold bracket.
18. Remove the exhaust manifold bracket and the exhaust pipe.
19. Remove the valve cover and the timing belt upper cover.
20. Remove the crankshaft pulley, then remove the timing belt lower cover.
21. Loosen the timing belt adjust bolt, then remove the timing belt.
22. Remove the cylinder head bolts and remove the cylinder head.
23. Remove the EGR pipe, if equipped, and air suction pipe from the intake and exhaust manifolds.
24. Remove the exhaust manifold shroud, oxygen sensor and exhaust manifold from the cylinder head.
25. Remove the intake manifold from the cylinder head.

To install:

26. Install the cylinder head in the reverse order of removal. Always use a

new head gasket and make sure the cylinder head and block surfaces are clean. Check the cylinder head surface for warpage. If warpage is less than 0.002 in. (0.05mm), resurfacing is not required. The maximum resurface limit is 0.008 in. (0.2mm) based on a cylinder head height of 3.54 in. (90.0mm).

27. The **UP** mark on the timing belt pulley should be at the top.

28. Install the intake and exhaust manifolds and tighten the nuts in a criss-cross pattern in 2–3 steps, beginning with the inner nuts.

29. Make sure the head dowel pins and oil control jet are aligned.

30. Install the bolts that secure the intake manifold to it's bracket, but do not tighten.

31. Position the cam correctly.

32. Tighten the cylinder head bolts in 2 steps. Tighten all bolts in sequence to 22 ft. lbs. (30 Nm) and then to 49 ft. lbs. (68 Nm) in the same sequence.

33. Install the exhaust pipe on the exhaust manifold. Tighten the bolts for the intake manifold bracket.

34. Install the exhaust pipe on it's bracket.

35. After installation, check that the tubes, hoses and connectors are installed correctly.

36. Adjust the valve timing.

DOHC ENGINE

1. Be sure the engine is cold. Disconnect the negative battery cable.

2. Drain the cooling system.

3. Remove the brake booster vacuum hose from the intake manifold.

4. Remove the engine secondary ground cable from the valve cover. Disconnect the radio condenser connector and the ignition coil wire.

5. Remove the air cleaner assembly. Relieve the fuel system pressure.

6. Disconnect the fuel lines. Remove the air intake hose and the resonator hose. Disconnect the throttle cable at the throttle body.

7. Disconnect the throttle control cable at the throttle body, if equipped with automatic transaxle. Disconnect the charcoal canister hose at the throttle valve.

8. Disconnect and tag all the necessary wire harness connectors from the cylinder head. Remove the emission control box and vacuum tank, then disconnect the 2 connectors. Do not remove the emission hoses.

9. Remove the upper radiator hose. Remove the heater hoses from the cylinder head. Remove the water bypass hoses from the water pump inlet pipe.

10. If equipped with cruise control, remove the actuator.

11. Remove the power steering pump belt and the alternator belt. Also remove the air conditioning belt if equipped.

12. Disconnect the inlet hose from the power steering pump and remove the power steering pump from the cylinder head. Remove the alternator assembly as well.

13. Remove the intake manifold bracket. Remove the exhaust manifold bracket and then the header pipe.

14. Mark the position of the distributor rotor in relation to the distributor housing and the distributor housing in relation to the cylinder head. Remove and tag the ignition wires and then remove the distributor assembly.

15. Remove the cylinder sensor. Remove the valve cover. Remove the timing belt middle cover.

16. Remove the crankshaft pulley and then remove the lower timing belt cover. Loosen the timing belt adjusting bolt and then remove the timing belt. Be sure to mark the rotation of the timing belt, if the belt is to be used again.

NOTE: Do not crimp or bend the timing belt more than 90 degrees or smaller than 1 in. (25mm) in diameter.

17. Remove the camshaft holders, camshafts and rocker arms. Remove the cylinder head bolts taking notice of the bolt holes that the 2 longer bolts come out of and remove the cylinder head.

18. Remove the exhaust manifold shroud and EGR pipe, then remove the exhaust manifold from the cylinder head. Remove the intake manifold from the cylinder head.

To install:

19. Installation is the reverse of the removal procedure. Attention to the following steps will aid installation.

20. Thoroughly clean the mating surfaces of the head and block.

21. Always use a new gasket.

22. Make sure the head dowel pins and oil control jet are aligned. Make sure the **UP** marks or cut-out on the timing belt pulleys are at the top. Tighten the cylinder head bolts in 2 equal steps. Apply engine oil to all the cylinder head bolts and the washers. Place the 2 longer bolts in the No. 1 and No. 2 positions in the cylinder head torque sequence. Tighten all bolts to 22 ft. lbs. (30 Nm), in sequence, and then to 49 ft. lbs. (68 Nm) in sequence.

23. Install the intake manifold and tighten the nuts in a criss-cross pattern in 2–3 steps to 16 ft. lbs. (21 Nm).

24. Install the exhaust manifold and bracket and tighten the nuts in a criss-cross pattern in 2–3 steps to 26 ft. lbs. (35 Nm).

25. After the installation procedure is complete, check that all tubes, hoses and connectors are installed correctly. Adjust the valve timing.

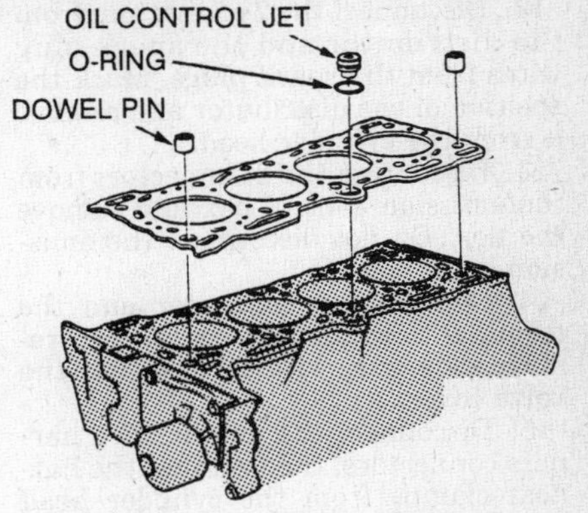

Cylinder head gasket installation on Prelude with DOHC engine

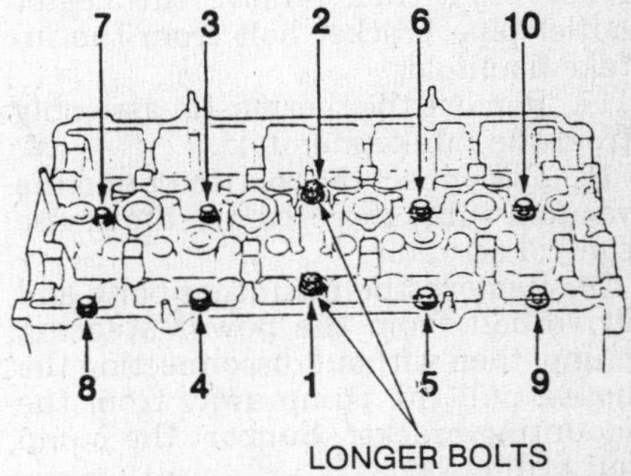

Cylinder head bolt torque sequence on Prelude with DOHC engine

1990–92 Accord

1. Disconnect the negative battery cable.

2. Bring the No. 1 cylinder to TDC.

3. Drain the cooling system.

4. Relieve the fuel system pressure.

5. Remove the fuel feed and return hose.

6. Remove the vacuum hose, breather hose and air intake duct.

7. Remove the water by pass hose from the cylinder head.

8. Remove the charcoal canister hose from the throttle body.

9. Remove the brake booster vacuum hose from the intake manifold. On automatic transaxle equipped vehicles, remove the vacuum hose mount.

10. Remove the cruise control vacuum hose.

11. Remove the throttle cable from the throttle body. On automatic transaxle equipped vehicles, remove the throttle control cable at the throttle body.

NOTE: Be careful not to bend the cable when removing. Do not use pliers to remove the cable from the linkage. Always replace a kinked cable with a new one.

12. Disconnect the 2 connectors from the distributor and the spark plug wires from the spark plugs. Mark the position of the distributor and remove it from the cylinder head.
13. Disconnect the 2 connectors from the emission control box and remove the box. Do not disconnect the emission hoses.
14. Remove the connector and the terminal from the alternator, then remove the engine wire harness from the valve cover.
15. Disconnect the engine wire harness connectors, then remove the harness clamps from the cylinder head and the intake manifold.
16. Remove the upper radiator hose and the heater inlet hose from the cylinder head, then remove the heater outlet pipe bracket bolt from the intake manifold.
17. Remove the thermostat assembly from the intake manifold.
18. Disconnect the connector and the vacuum tube, then remove the cruise control actuator.
19. Remove the mounting bolts and drive belt from the power steering pump, then without disconnecting the hoses, pull the pump away from the mounting bracket. Support the pump out of the way.
20. Raise and safely support the vehicle.
21. Remove the front wheel and tire assemblies.
22. Remove the splash shield.
23. Remove the intake manifold bracket bolts.
24. Remove the exhaust manifold and the exhaust manifold heat insulator.
25. Remove the intake manifold.
26. Remove the valve cover and engine ground wire.
27. Remove the side engine mount bracket stay, then remove the timing belt upper cover.
28. Mark the rotation of the timing belt if it is to be used again. Loosen the timing belt adjusting bolt and then release the timing belt.

NOTE: Push the tensioner to release tension from the belt, then retighten the adjusting bolt.

29. Remove the timing belt from the driven pulley.
30. Remove the cylinder head bolts, then remove the cylinder head.

NOTE: To prevent warpage, unscrew the bolts in sequence 1/3 turn at a time. Repeat the sequence until all bolts are loosened.

To install:

31. Installation is the reverse of the removal procedure. Attention to the following steps will aid installation.

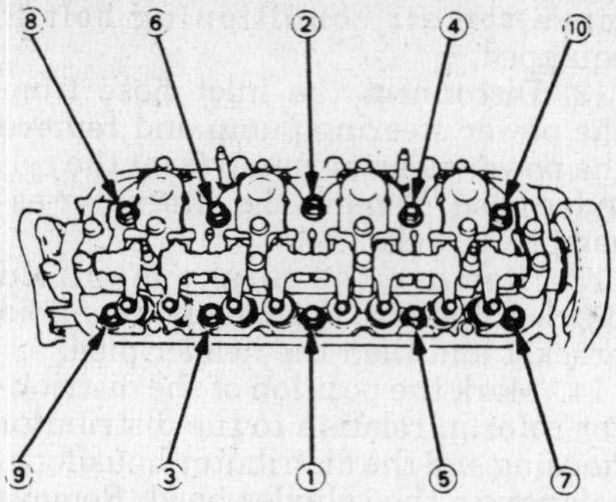

Cylinder head bolt torque sequence on 1990–92 Accord

32. Make sure all cylinder head and block gasket surfaces are clean. Check the cylinder head for warpage. If warpage is less than 0.002 in. (0.05mm), cylinder head resurfacing is not required. Maximum resurface limit is 0.008 in. (0.2mm) based on a cylinder head height of 3.935 in. (99.95mm).
33. Always use a new head gasket.
34. The **UP** mark on the camshaft pulley should be at the top.
35. Make sure the No. 1 cylinder is at TDC.
36. The cylinder head dowel pins and oil control jet must be aligned.
37. Install the bolts that secure the intake manifold to it's bracket but do not tighten them.
38. Position the cam correctly.
39. Tighten the cylinder head bolts sequentially in 3 steps:
 Step 1 – 29 ft. lbs. (40 Nm).
 Step 2 – 51 ft. lbs. (70 Nm).
 Step 3 – 78 ft. lbs. (108 Nm).
40. Install the intake manifold and tighten the nuts in a criss-cross pattern, in 2–3 steps, beginning with the inner nuts. Final torque should be 16 ft. lbs. (22 Nm). Always use a new intake manifold gasket.
41. Install the heat insulator to the cylinder head and the block.
42. Install the exhaust manifold and tighten the nuts in a criss-cross pattern in 2–3 steps, beginning with the inner nut. Final torque should be 23 ft. lbs. (32 Nm). Always use a new exhaust manifold gasket.
43. Install the exhaust manifold bracket, then install the exhaust pipe, the bracket and upper shroud.
44. Check the ignition timing.

Valve Lash

ADJUSTMENT

NOTE: The valves should be adjusted cold. Cylinder head temperature must be less than 100°F (38°C). Adjustment is the same for intake and exhaust valves.

1. Disconnect the negative battery cable and remove the valve cover.
2. Bring the No. 1 piston to TDC. The **UP** mark on the pulley(s) should be at the top and the TDC grooves on the back side of the pulley(s) should align with the cylinder head surface. The distributor rotor must be pointing towards the No. 1 spark plug wire.
3. Adjust the valves on the No. 1 cylinder. Valve clearance is as follows:
 Civic and CRX except CRX HF: Intake, and auxiliary, if equipped – 0.007–0.009 in. (0.17–0.22mm). Exhaust – 0.009–0.011 in. (0.22–0.27mm).
 1988–92 CRX HF: Intake – 0.005–0.007 in. (0.12–0.17mm). Exhaust – 0.007–0.009 in. (0.17–0.22mm).
 1988–89 Accord and Prelude with SOHC engine: Intake – 0.005–0.007 in. (0.12–0.17mm). Exhaust – 0.010–0.012 in. (0.25–0.30mm).
 1990–92 Accord: Intake – 0.009–0.011 in. (0.24–0.28mm). Exhaust – 0.011–0.013 in. (0.28–0.32mm).
 1988–92 Prelude with DOHC engine: Intake – 0.003–0.005 in. (0.08–0.12mm). Exhaust – 0.006–0.008 in. (0.16–0.20mm).
4. Loosen the locknut and turn the adjusting screw until the feeler gauge slides back and forth with a slight amount of drag.
5. Tighten the locknut and check the clearance again. Repeat adjustment if necessary.
6. Rotate the crankshaft 180 degrees counterclockwise. The cam pulley(s) will turn 90 degrees counterclockwise. The **UP** mark should align with the cylinder head surface and the distributor rotor should point to the No. 3 spark plug wire. Adjust the valves on the No. 3 cylinder.
7. Rotate the crankshaft 180 degrees counterclockwise to bring the No. 4 piston to TDC. The **UP** mark should be at the bottom and the TDC

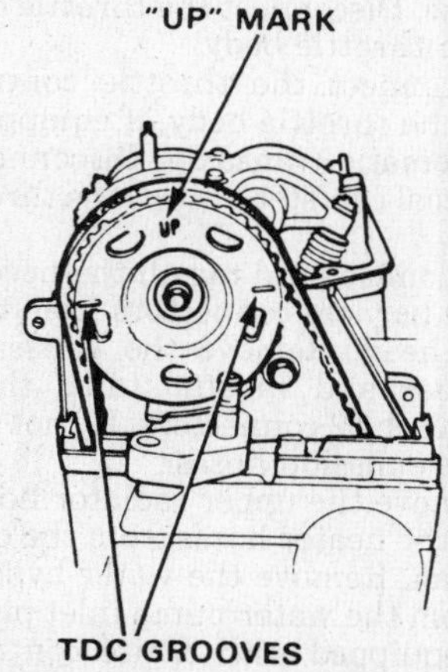

Position of the sprocket marks for No. 1 TDC on Civic and CRX with single camshaft engine

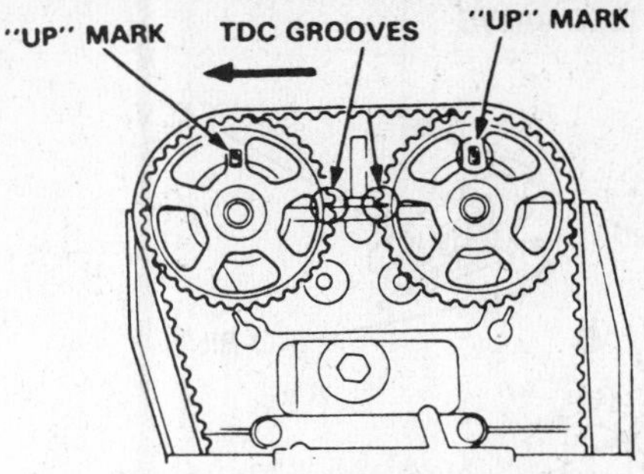

Position of the sprocket marks for No. 1 TDC on dual camshaft engines

grooves should again be aligned with the cylinder head surface. The distributor rotor should point to the No. 4 spark plug wire. Adjust the valves on the No. 4 cylinder.

8. Rotate the crankshaft 180 degrees counterclockwise to bring the No. 2 piston to TDC. The **UP** mark should align with the cylinder head surface and the distributor should point to the No. 2 spark plug wire. Adjust the valves on the No. 2 cylinder.

9. Replace the valve cover.

Rocker Arms/Shafts

REMOVAL & INSTALLATION

Except DOHC Engine

1. Disconnect the negative battery cable.

2. Remove the valve cover and bring the No. 1 cylinder to TDC.

3. Remove the rocker arm bolts. Unscrew the bolts 2 turns at a time, in a criss-cross pattern, to prevent damaging the valves or rocker assembly.

4. Remove the rocker arm/shaft assemblies. Leave the rocker arm bolts in place as the shafts are removed to keep the bearing caps, springs and rocker arms in place on the shafts.

5. If the rocker arms or shafts are to be replaced, identify the parts as they are removed from the shafts to ensure reinstallation in the original location.

To install:

6. Lubricate the camshaft journals and lobes.

7. Set the rocker arm assembly in place and loosely install the bolts. Tighten each bolt 2 turns at a time in the proper sequence to ensure that the rockers do not bind on the valves. Tighten the rocker arm bolts to 16 ft. lbs. (22 Nm) except on 1990–92 Accord. On 1990–92 Accord tighten the 6mm bolts to 9 ft. lbs. (12 Nm) and the 8mm bolts to 16 ft. lbs. (22 Nm).

8. Replace the valve cover and connect the negative battery cable.

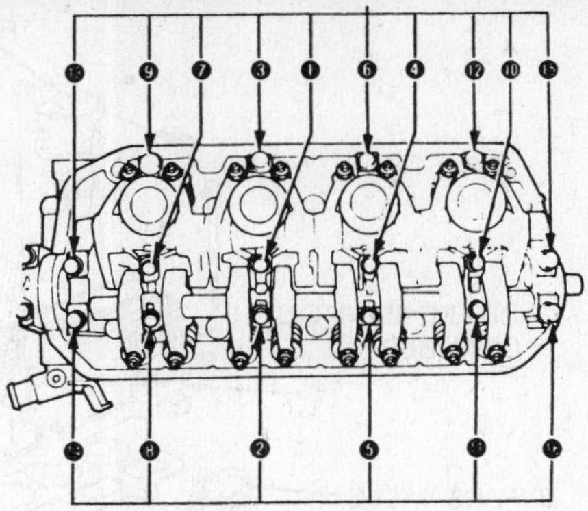

Rocker arm shaft torque sequence—Civic and CRX with single camshaft engine

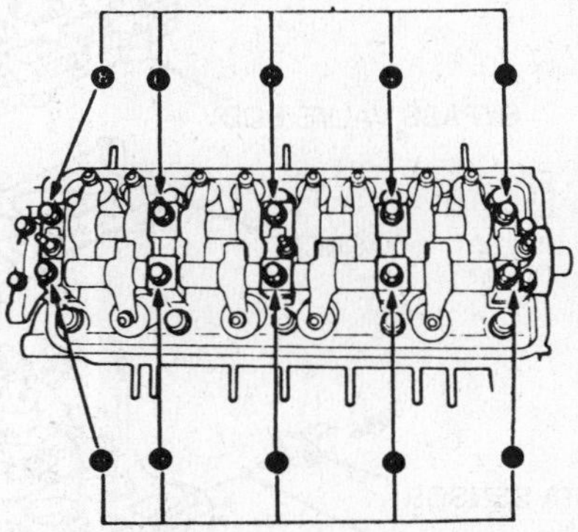

Rocker arm shaft torque sequence—1988–89 Accord and Prelude with single camshaft engine

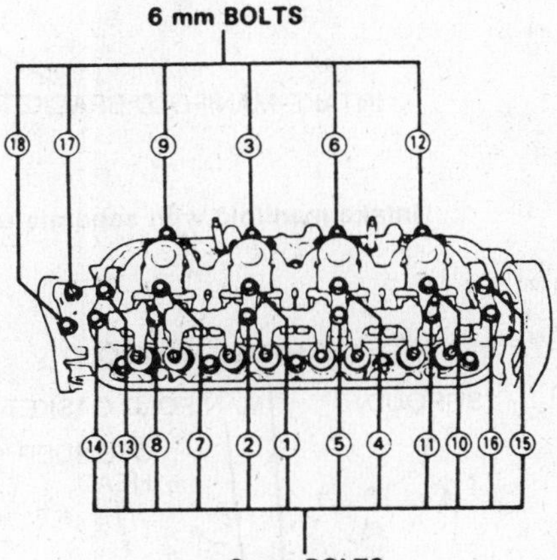

Rocker arm shaft torque sequence—1990–92 Accord and Prelude with dual camshaft engine

DOHC Engine

1. Disconnect the negative battery cable.

2. Remove the valve cover and bring the No. 1 cylinder to TDC.

3. Remove the timing belt cover and the timing belt.

4. Remove the camshaft bearing caps and remove the camshafts.

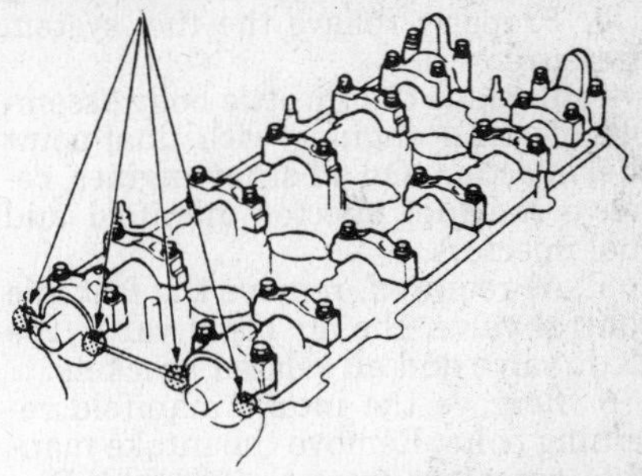

Use non-hardening sealant when installing the valve cover on dual camshaft engines

5. Remove the rocker arms.

6. Installation is the reverse of the removal procedure.

7. Apply liquid gasket to the No. 1 and No. 6 camshaft bearing caps and install them with the rest of the caps. Make sure the caps are installed in their proper positions as indicated by their markings.

8. Tighten each camshaft bearing cap bolt gradually, to prevent binding. Tighten the bolts to 9 ft. lbs. (12 Nm).

Intake Manifold

REMOVAL & INSTALLATION

Carbureted Engine

1. Disconnect the negative battery cable. Drain the coolant from the radiator.

2. Remove the air cleaner and case from the carburetor(s).

3. Remove the air valve, EGR valve, air suction valve and air chamber if equipped.

4. Label and remove all wires and vacuum hoses running to the intake manifold.

5. Remove the intake manifold attaching nut in a criss-cross pattern, beginning from the center. Then remove the manifold.

To install:

6. Installation is the reverse of the removal procedure.

7. Clean all the old gasket material from the manifold and the cylinder head.

8. Always use a new gasket.

9. Tighten the nuts in a criss-cross pattern in 2–3 steps, starting with the inner nuts. Tighten the nuts to 16 ft. lbs. (22 Nm).

10. Be sure all hoses and wires are connected properly.

Fuel Injected Engine

1. Disconnect the negative battery cable. Drain the cooling system.

2. Label and disconnect all required electrical connectors and vacuum lines.

3. Properly relieve the fuel system pressure.
4. Remove the throttle body assembly on 1.5L engines with dual-point fuel injection. On all other engines, remove the fuel injector manifold and fuel injectors.
5. As required, remove the fast idle control valve, the air bleed valve, the EGR valve and all related brackets.
6. Remove the intake manifold retaining bolts. Remove the intake manifold assembly from the vehicle. Discard the gaskets.

NOTE: Some Accord and Prelude engines have an upper and lower manifold chamber. Separate the upper chamber from the lower manifold before removing the assembly from the vehicle.

To install:

7. Installation is the reverse of the removal procedure.
8. Make sure all gasket mating surfaces are clean prior to installation.
9. Always use a new gasket.
10. Tighten the intake manifold nuts in 2–3 steps in a criss-cross pattern starting with the inside nuts. Tighten the nuts to 16 ft. lbs. (22 Nm).
11. Make sure that all hoses and wires are connected properly.

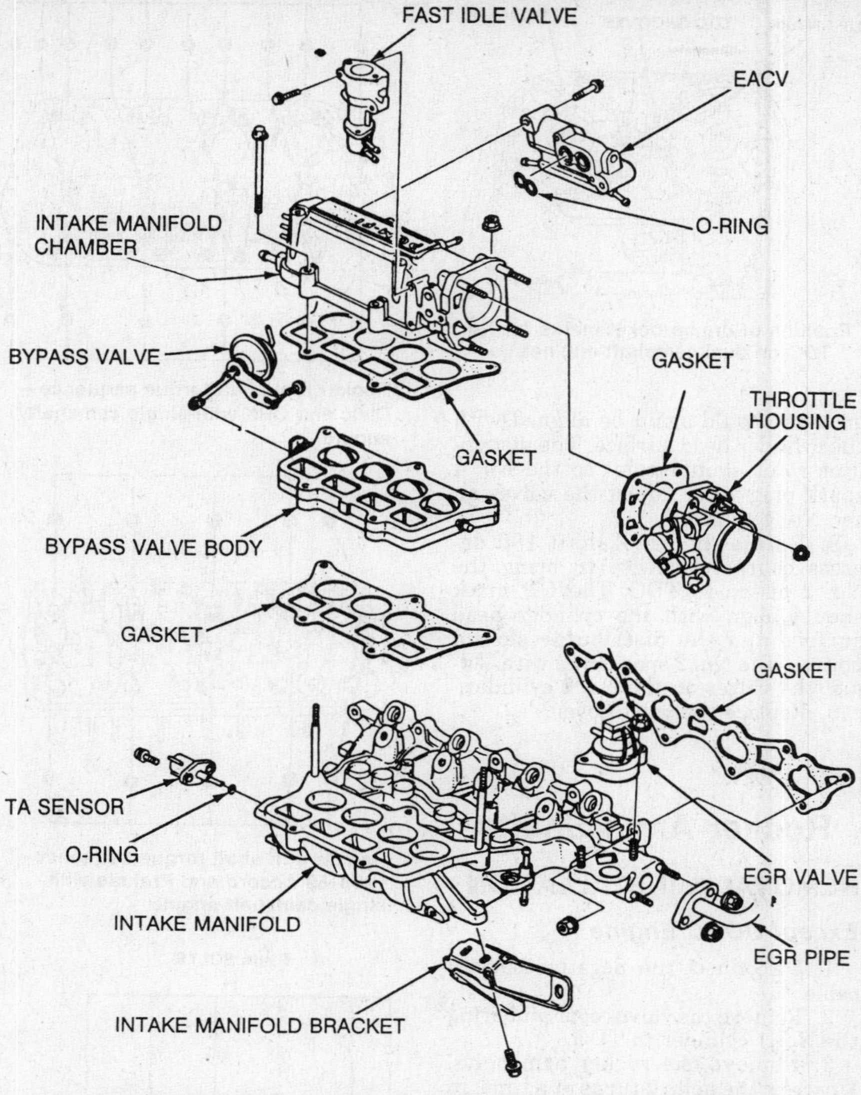

Intake manifold with separate upper chamber—1989 Accord shown

Exhaust Manifold

REMOVAL & INSTALLATION

1. Disconnect the negative battery cable.
2. Raise and safely support the vehicle.
3. Remove the engine splash shield, as necessary.
4. Disconnect the exhaust pipe or catalytic converter, as required.
5. Lower the vehicle.
6. Remove the exhaust manifold heat shield.

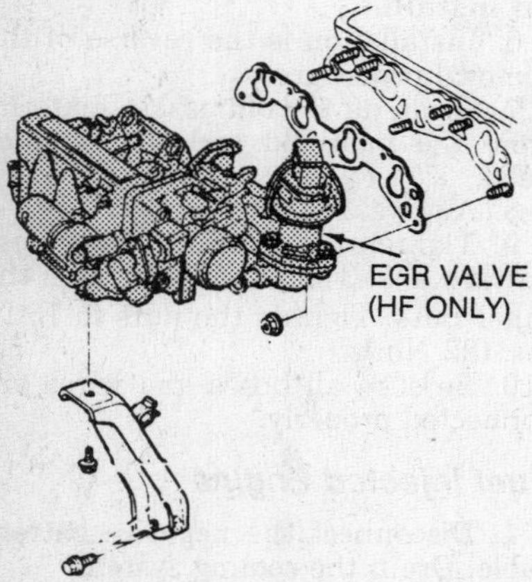

Intake manifold installation on Civic and CRX with multi-point fuel injection

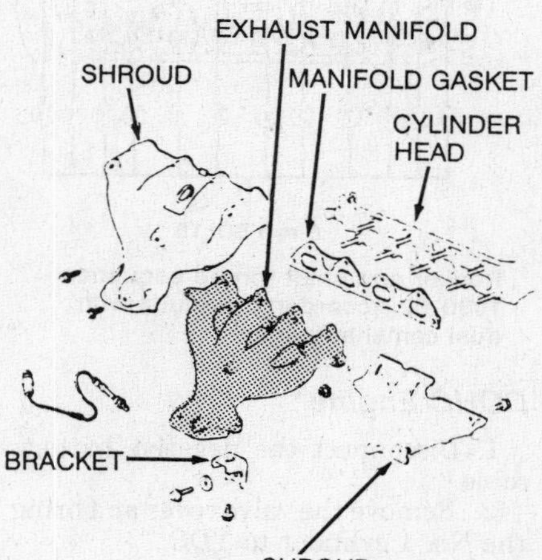

Typical exhaust manifold installation—Civic shown

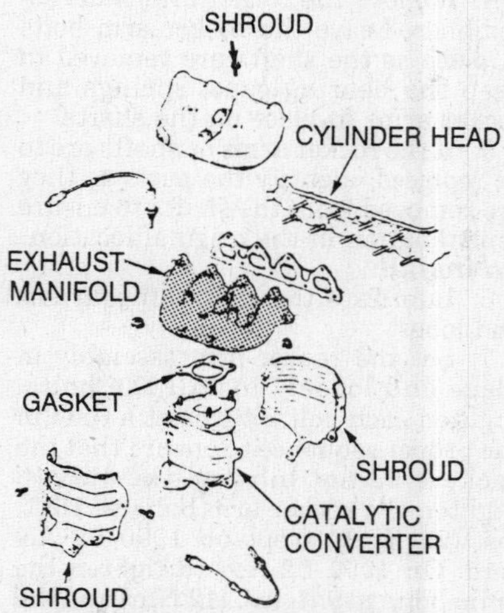

Exhaust system on CRX HF has the catalytic converter close coupled to the manifold

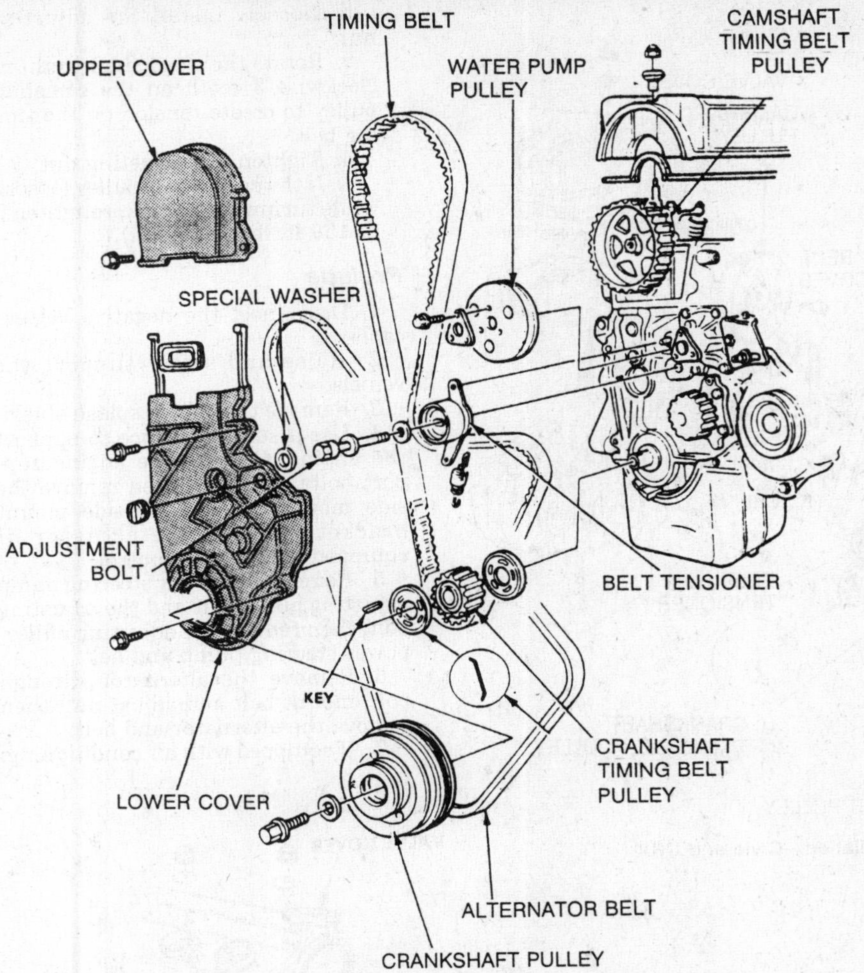

Timing belt and cover installation—1988–89 Accord

7. Disconnect the exhaust manifold brackets, EGR tube and oxygen sensor, as required.
8. Remove the exhaust manifold nuts and the exhaust manifold.

To install:

9. Installation is the reverse of the removal procedure.
10. Make sure all gasket mating surfaces are clean prior to installation. Always use a new gasket.
11. Tighten the exhaust manifold nuts in 2–3 steps in a criss-cross pattern, beginning with the inner nuts. Tighten the nuts to 22 ft. lbs. (32 Nm).

Timing Belt Front Cover

REMOVAL & INSTALLATION

1988–89 Accord

1. Disconnect the negative battery cable.
2. Bring the piston of the No. 1 cylinder to TDC.
3. Remove all accessory drive belts.
4. Remove the water pump pulley.
5. Remove the crankshaft pulley bolt and using a suitable puller, remove the crankshaft pulley.
6. Remove the timing belt cover bolts and the upper and lower covers.
7. Installation is the reverse of the removal procedure. Tighten the crankshaft pulley bolt to 108 ft. lbs. (150 Nm).

1988–92 Civic and CRX

1. Disconnect the negative battery cable.
2. Raise and safely support the vehicle.
3. Remove the left front wheel and tire assembly.
4. Remove the left front wheel well splash shield.
5. If equipped, remove the power steering belt and pump.
6. If equipped with air conditioning, remove the adjust pulley with bracket and the belt.
7. Remove the power steering bracket, loosen the alternator adjust bolt and through bolt and remove the alternator belt.
8. Use a suitable device to support the engine. Remove the engine side support bolts and nut and then remove the side mount rubber.
9. Remove the valve cover, the crankshaft pulley bolt and the crankshaft pulley.
10. Remove the timing belt upper and lower cover.
11. Installation is the reverse of the removal procedure. Apply engine oil to the crankshaft pulley. On 1988 models, torque the pulley bolt to 83 ft. lbs. (115 Nm). On 1989–92 models, torque the pulley bolt to 119 ft. lbs. (165 Nm)

1990–92 Accord

1. Disconnect the negative battery cable.
2. Raise and safely support the vehicle.
3. Remove the engine splash shield.
4. Disconnect the connector, then remove the cruise control actuator. Do not disconnect the control cable.
5. Remove the mounting bolt, nut and drive belt from the power steering pump, then without disconnecting the hoses, pull the pump away from the mounting bracket.
6. Disconnect the alternator terminal and the connector, then remove the engine wire harness from the valve cover.
7. Loosen the alternator mounting bolt, nut and adjusting nut, then remove the alternator belt or if equipped, the air conditioning belt.
8. Remove the valve cover. Remove the side engine mount bracket stay, if equipped.
9. Remove the upper timing belt cover.
10. Use a suitable device to support the engine, then remove the side engine bolt.
11. Remove the dipstick and pipe and remove the timing belt tensioner adjusting nut.
12. Remove the crankshaft pulley bolt and the crankshaft pulley. Remove the 2 rear bolts from the center beam, to allow the engine to drop down and give clearance to remove the lower timing belt cover. Remove the lower cover.

To install:

13. Installation is the reverse of the removal procedure.
14. Apply oil to the threads of the crankshaft pulley bolt and tighten it to 159 ft. lbs. (220 Nm).
15. After installing the lower cover, the timing belt and balancer belt tension must be adjusted as follows:
    a. Make sure the No. 1 cylinder is at TDC.

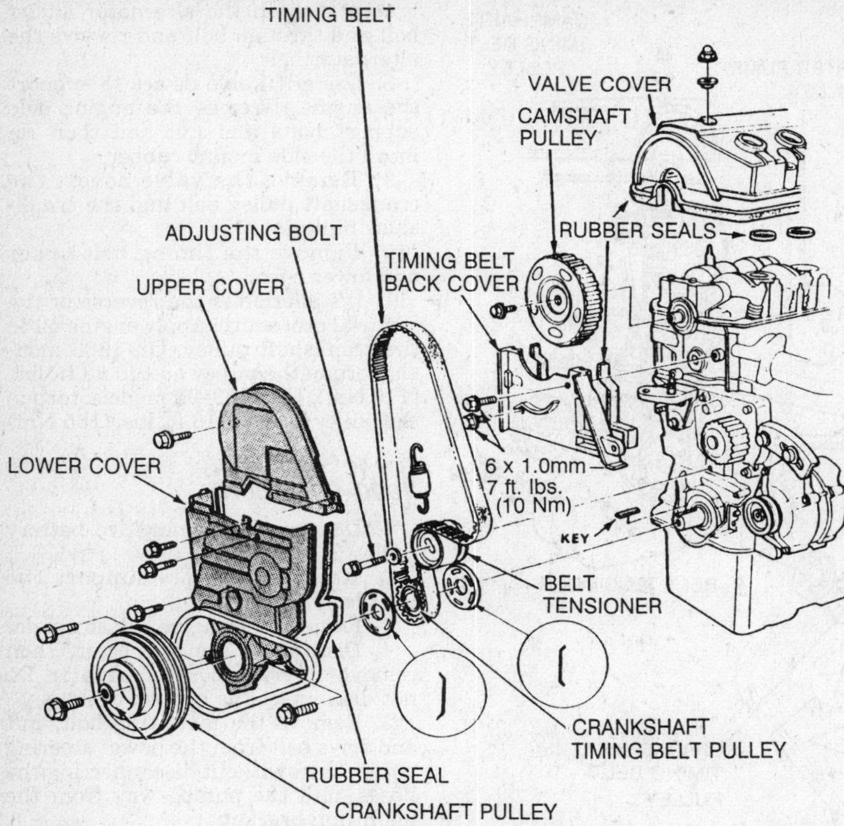

Timing belt and cover installation—Civic and CRX

b. Loosely install the adjusting nut.

c. Rotate the crankshaft counterclockwise 3 teeth on the camshaft pulley to create tension on the timing belt.

d. Tighten the adjusting nut.

e. If the crankshaft pulley loosens while turning the crank, retighten it to 159 ft. lbs. (220 Nm).

Prelude

1. Disconnect the negative battery cable.
2. Raise and safely support the vehicle.
3. Remove the engine splash shield.
4. Use a suitable device to support the engine. Remove the engine support bolts and nuts, then remove the side mount rubber and side mount bracket. Remove the actuator, if equipped with cruise control.
5. Remove the power steering pump adjusting pulley nut and the adjusting bolt, then remove the adjusting pulley, power steering pump and belt.
6. Remove the alternator through bolt, mount bolt and adjust nut, then remove the alternator and belt.
7. If equipped with air conditioning,

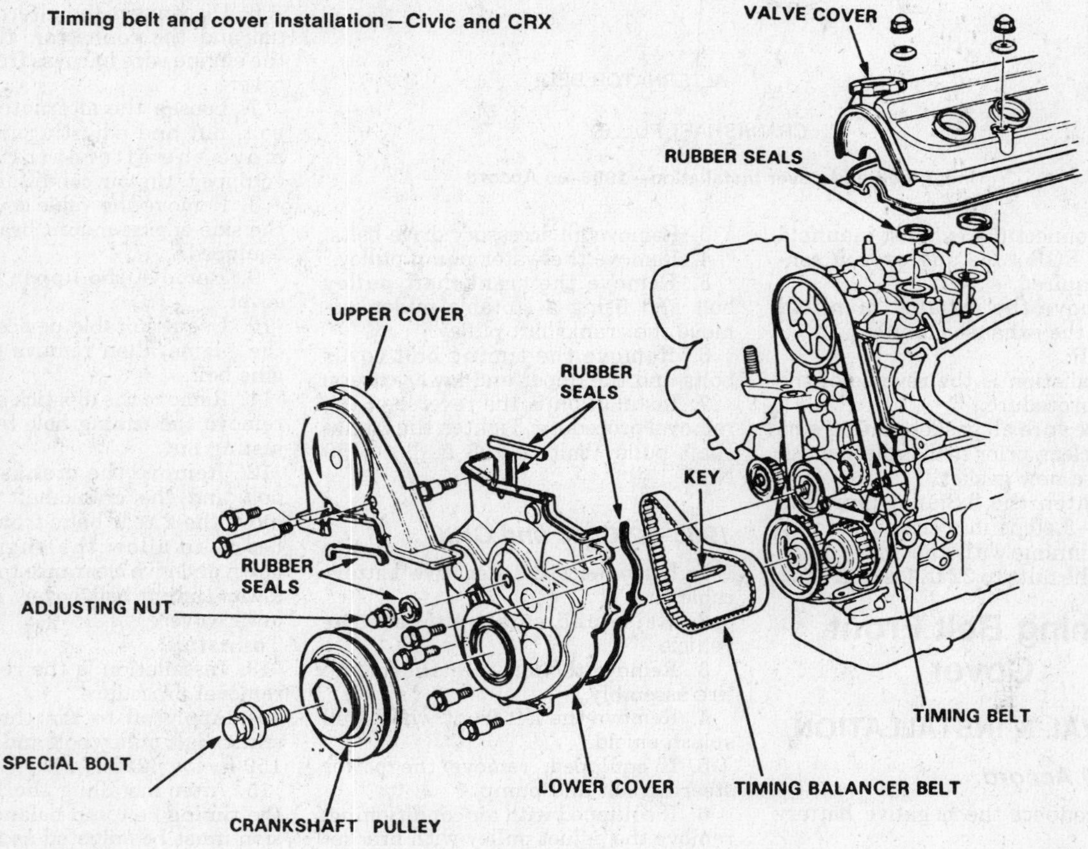

Timing belt and cover installation—1990–92 Accord

remove the air conditioning compressor mount bolts, then remove the air conditioning compressor and the belt.

8. Remove the ignition wire and the engine wire harness protector from the valve cover, if necessary.

9. Remove the valve cover.

10. Remove the crankshaft pulley bolt and the crankshaft pulley.

11. Remove the timing belt covers.

12. Installation is the reverse of the removal procedure. Apply engine oil to the threads of the crankshaft pulley bolt and tighten it to 83 ft. lbs. (115 Nm) on 1988 Prelude and 108 ft. lbs. (150 Nm) on 1989–91 Prelude.

OIL SEAL REPLACEMENT

1. Disconnect the negative battery cable.

2. Remove the timing belt cover and the timing belt.

3. Remove the crankshaft timing sprocket.

4. Using a suitable seal removal tool, remove the seal from the front of the engine.

5. Installation is the reverse of the removal procedure. Place a thin coat of oil on the seal lip prior to installation. Use a suitable seal driver to install the seal. Be sure to install the seal with the open (spring) side facing the inside of the engine.

Timing Belt and Tensioner

ADJUSTMENT

NOTE: The timing belt tensioner is spring loaded, to apply proper tension to the belt automatically, after making the following adjustment.

1. Disconnect the negative battery cable.

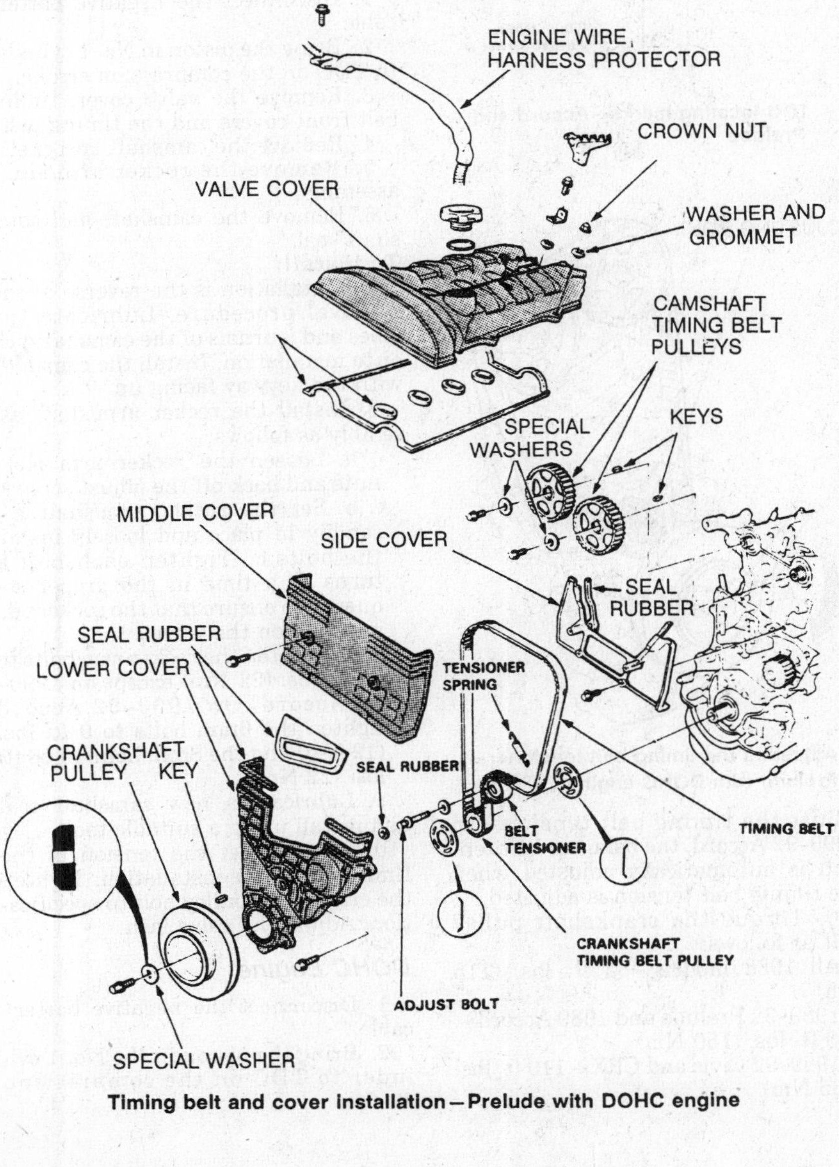

Timing belt and cover installation—Prelude with DOHC engine

2. Remove the valve cover or upper timing belt cover.

3. Set the piston in No. 1 cylinder at TDC.

4. Loosen, but do not remove, the adjusting bolt.

5. Rotate the crankshaft counterclockwise 3 teeth on the camshaft pulley to create tension on the timing belt.

6. Tighten the adjusting bolt.

7. If the crankshaft pulley broke loose while turning the crankshaft, retighten it to specification.

8. Reinstall or connect the remaining components after adjustment is completed.

REMOVAL & INSTALLATION

1. Disconnect the negative battery cable.

2. Bring the piston in No. 1 cylinder to TDC on the compression stroke.

3. Remove the valve cover and timing belt front covers.

4. Mark the direction of timing belt rotation. On 1990–92 Accord, mark the direction of timing balancer belt rotation.

5. On all except 1990–92 Accord, loosen the adjusting bolt and remove the timing belt. On 1990–92 Accord, push the timing balancer belt tensioner and the timing belt tensioner to remove tension on the belts, then reinstall and tighten the adjusting nut. Remove the timing balancer belt and the timing belt.

To install:

6. Align the camshaft sprocket(s) and crankshaft sprocket as follows:

a. Make sure the **UP** mark on the camshaft sprocket(s) is at the top most position. The sprocket timing marks should be aligned with the cylinder head upper surface.

b. On Accord and Prelude, remove the timing inspection hole cover at the rear of the engine block. Make sure the TDC mark on the flywheel, indicated by a white painted mark, is aligned with the pointer in the inspection hole.

c. On Civic and CRX, temporarily reinstall the lower timing belt cover and crankshaft pulley. Make sure the TDC mark on the crankshaft pulley, indicated by a white painted mark, is aligned with the pointer on the timing cover. Remove the lower timing belt cover and crankshaft pulley.

7. Install the timing belt. If the old timing belt is reused, install the belt in the same rotational direction, as indicated by the mark that was made during removal.

8. On 1990–92 Accord, align the timing belt balancer pulleys and install the balancer belt as follows:

a. The timing belt balancer drive pulley should already be at TDC, if the timing belt is installed correctly.

b. Align the groove on the front timing balancer belt driven pulley with the pointer on the oil pump body.

c. Remove the bolt from the maintenance hole on the cylinder block, next to the rear balancer shaft. Align the rear timing balancer belt driven pulley using a 6 × 100mm bolt, or equivalent. Mark a line at the 74mm length of the bolt. Align the pulley by inserting the bolt through the maintenance hole. The bolt must be inserted to a depth where the line is flush with the maintenance hole surface.

d. Install the timing balancer belt. If the old balancer belt is reused, install the belt in the same rotational direction, as indicated by the mark that was made during removal. After the balancer belt is installed, remove the rear balancer belt driven pulley alignment bolt and reinstall the original bolt in the maintenance hole. Tighten the bolt to 22 ft. lbs. (30 Nm).

9. Installation of the remaining components is the reverse of the removal procedure. Be sure to properly adjust the timing belt tension. On 1990–92 Accord, the balancer belt tension is automatically adjusted when the timing belt tension is adjusted.

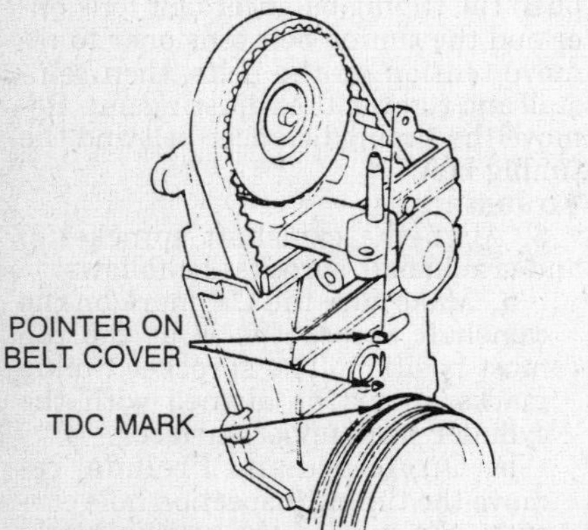

TDC locating marks on single camshaft engines

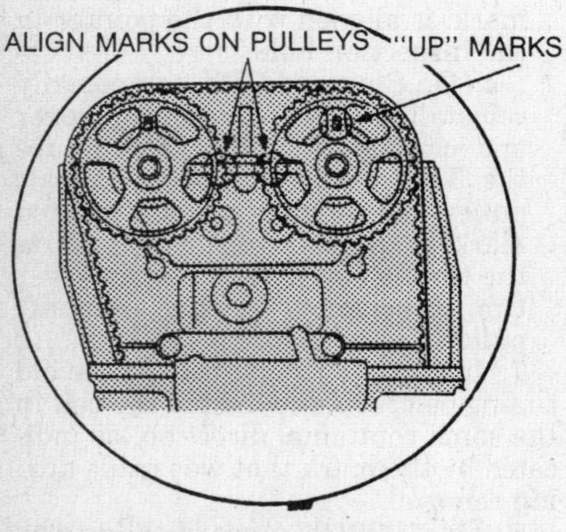

Aligning the timing marks on dual camshaft engines

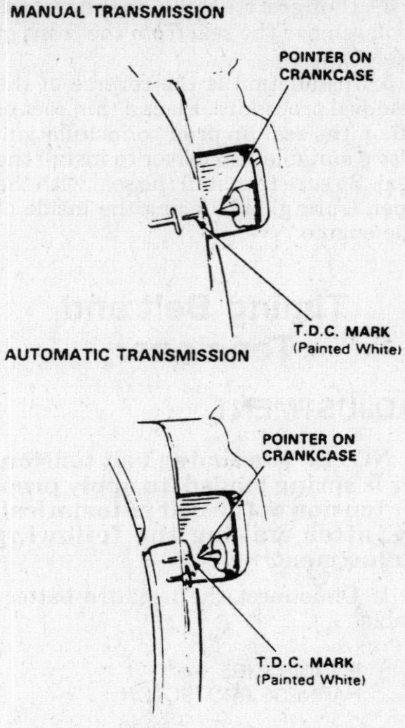

TDC locating marks—Accord and Prelude

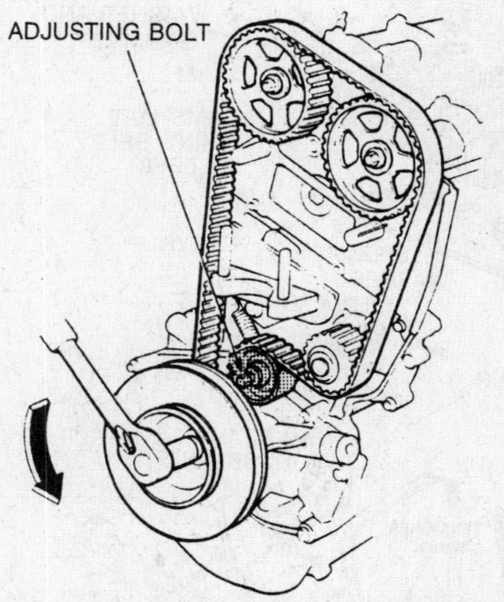

Adjusting the timing belt tensioner—Prelude with DOHC engine

10. Torque the crankshaft pulley bolt as follows:

All 1988 models—83 ft. lbs. (115 Nm)

1989–92 Prelude and 1989 Accord—108 ft. lbs. (150 Nm)

1989–92 Civic and CRX—119 ft. lbs. (165 Nm)

1990–92 Accord—159 ft. lbs. (220 Nm)

Timing Sprockets

REMOVAL & INSTALLATION

1. Disconnect the negative battery cable.
2. Remove the valve cover, timing belt covers and the timing belt.
3. Remove the crankshaft and camshaft timing sprockets.
4. Installation is the reverse of the removal procedure. Tighten the camshaft sprocket retaining bolts to 27 ft. lbs. (38 Nm).

Camshaft

REMOVAL & INSTALLATION

Except DOHC Engine

1. Disconnect the negative battery cable.
2. Bring the piston in No. 1 cylinder to TDC on the compression stroke.
3. Remove the valve cover, timing belt front covers and the timing belt.
4. Remove the camshaft sprocket.
5. Remove the rocker arm/shaft assembly.
6. Remove the camshaft and camshaft seal.

To install:

7. Installation is the reverse of the removal procedure. Lubricate the lobes and journals of the camshaft prior to installation. Install the camshaft with the keyway facing up.
8. Install the rocker arm/shaft assembly as follows:

a. Loosen the rocker arm locknuts and back off the adjust screws.

b. Set the rocker arm/shaft assembly in place and loosely install the bolts.c. Tighten each bolt 2 turns at a time in the proper sequence to ensure that the rockers do not bind on the valves.

d. Tighten the rocker arm bolts to 16 ft. lbs. (22 Nm) except on 1990–92 Accord. On 1990–92 Accord tighten the 6mm bolts to 9 ft. lbs. (12 Nm) and the 8mm bolts to 16 ft. lbs. (22 Nm).

9. Lubricate a new camshaft seal and install using a suitable tool.
10. Properly set the tension of the timing belt after installation. Tighten the crankshaft pulley bolt to specification. Adjust the valve lash.

DOHC Engine

1. Disconnect the negative battery cable.
2. Bring the piston in the No. 1 cylinder to TDC on the compression stroke.

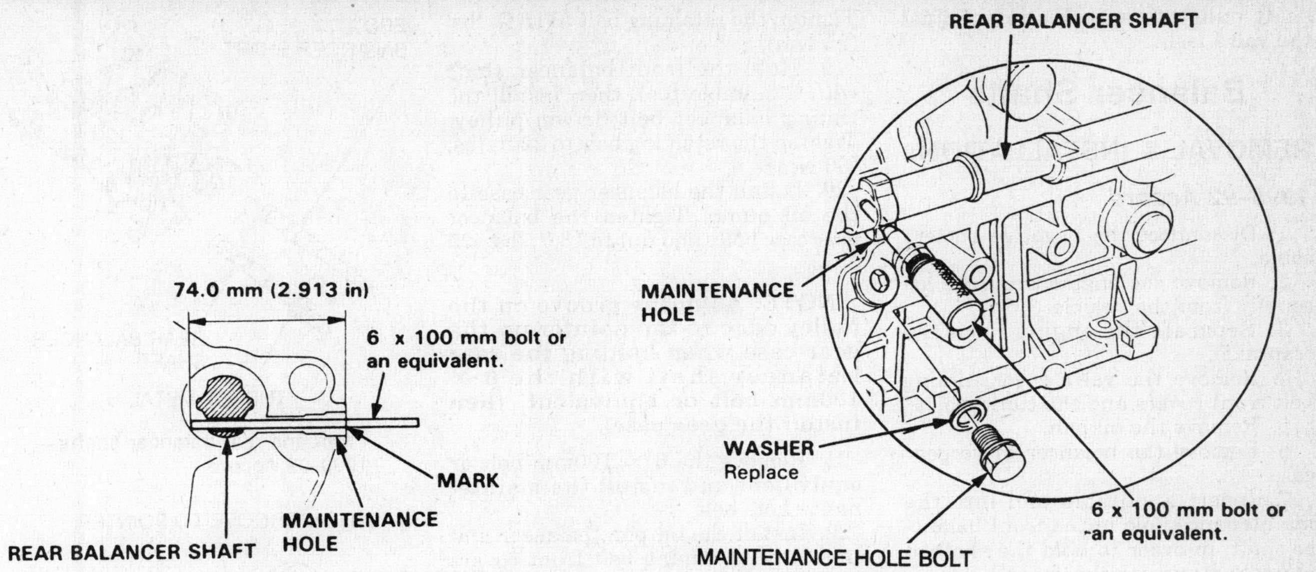

Aligning the belt driven rear timing balancer—1990–92 Accord

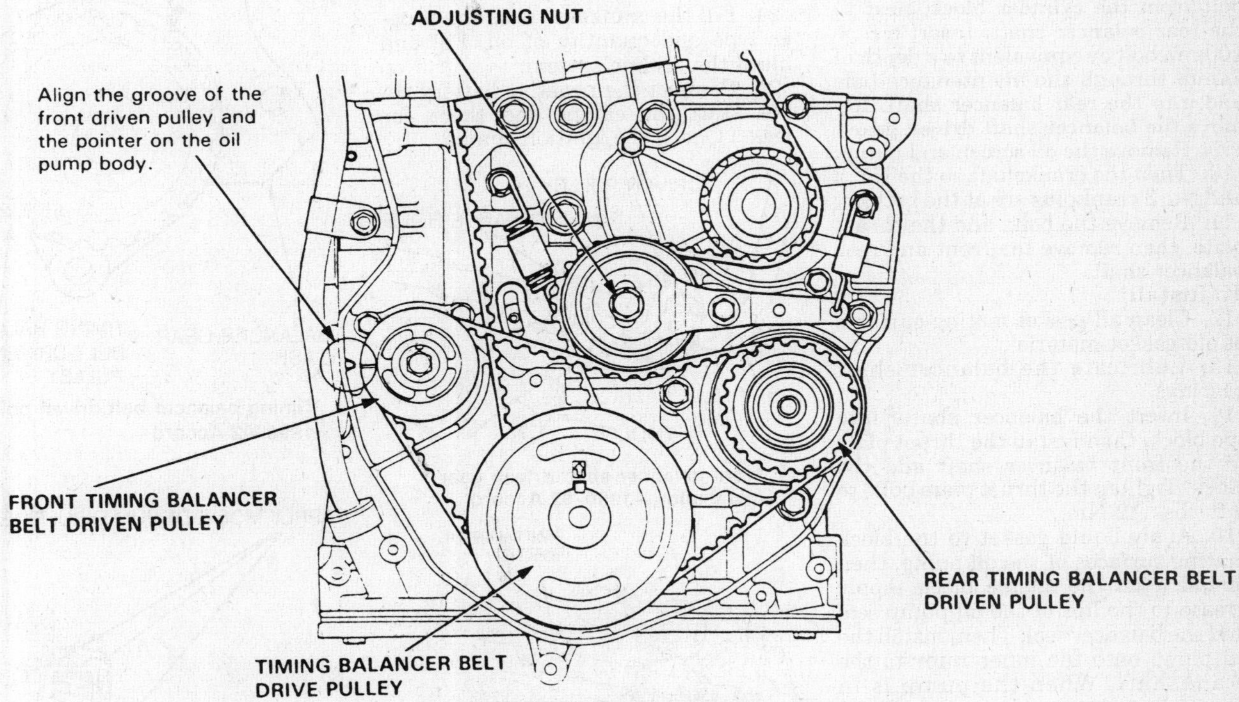

Timing balancer belt installation—1990–92 Accord

3. Remove the valve cover, timing belt front covers and the timing belt.
4. Remove the camshaft sprockets.
5. Remove the camshaft bearing caps and the camshafts and camshaft seals.

To install:

6. Installation is the reverse of the removal procedure. Inspect the rocker arms for wear or damage and replace as necessary prior to installation of the camshafts. The rocker arm locknuts and adjust screws should be backed off before installation of the camshafts. Lubricate the lobes and journals of the camshafts before installing. Install the camshafts with keyways facing UP.
7. Apply liquid gasket to the No. 1 and No. 6 camshaft bearing caps and install them with the rest of the caps. Make sure the caps are installed in their proper positions as indicated by their markings.
8. Tighten each camshaft bearing cap bolt gradually, to prevent binding. Tighten the bolts to 9 ft. lbs. (12 Nm).
9. Lubricate new camshaft seals and install using a suitable tool.
10. Properly tension the timing belt after installation. Tighten the crank-

shaft pulley to specification. Adjust the valve lash.

Balancer Shafts

REMOVAL & INSTALLATION

1990–92 Accord

1. Disconnect the negative battery cable.
2. Remove the engine/transaxle assembly from the vehicle.
3. Separate the engine from the transaxle.
4. Remove the valve cover, timing belt front covers and the timing belt.
5. Remove the oil pan.
6. Remove the balancer drive gear case.
7. Insert a suitable tool into the maintenance hole in the front balancer shaft, in order to hold the shaft in place and remove the front balancer driven pulley.
8. Remove the maintenance hole bolt from the cylinder block, next to the rear balancer shaft. Insert a 6 × 100mm bolt or equivalent to a depth of 74mm through the maintenance hole and into the rear balancer shaft. Remove the balancer shaft driven gear.
9. Remove the oil screen and pump.
10. Turn the crankshaft so the No. 2 and No. 3 crankpins are at the bottom.
11. Remove the bolts and the thrust plate, then remove the front and rear balancer shaft.

To install:

12. Clean all gasket mating surfaces of old gasket material.
13. Lubricate the balancer shaft bearings.
14. Insert the balancer shafts into the block, then install the thrust plate to the front balancer shaft and the block. Tighten the thrust plate bolts to 9 ft. lbs. (12 Nm).
15. Apply liquid gasket to the block mating surfaces of the oil pump, then install it on the engine block. Apply grease to the lips of the oil pump seal and the balancer seal. Then, install the oil pump onto the inner rotor to the crankshaft. When the pump is in place, clean any excess grease off the crankshaft and the balancer shaft, then check that the oil seal lips are not distorted. Apply liquid gasket to the oil pump retaining bolts and tighten the bolts to 9 ft. lbs. (12 Nm).
16. Install the oil screen.
17. Apply molybdenum disulfide to the thrust surfaces of the balancer gears, before installing the balancer driven gear and the balancer drive gear case.
18. Hold the rear balancer shaft with the 6 × 100mm bolt or equivalent, then install the balancer driven gear and the balancer drive belt pulley. Tighten the retaining bolt to 18 ft. lbs. (25 Nm).
19. Hold the front balancer shaft with a suitable tool, then install the timing balancer belt driven pulley. Tighten the retaining bolt to 22 ft. lbs. (30 Nm).
20. Install the balancer gear case to the oil pump. Tighten the balancer gear case bolts and nut to 18 ft. lbs. (25 Nm).

NOTE: Align the groove on the pulley edge to the pointer on the gear case when holding the rear balancer shaft with the 6 × 100mm bolt or equivalent, then install the gear case.

21. Remove the 6 × 100mm bolt or equivalent and install the maintenance hole bolt.
22. Install the oil pan, balancer and timing belts, timing belt front covers and the valve cover.
23. Install the engine assembly in the vehicle.
24. Fill the crankcase with the proper type and quantity of oil. Fill and bleed the cooling system.
25. Connect the negative battery cable, start the engine and check for leaks. Check the ignition timing.

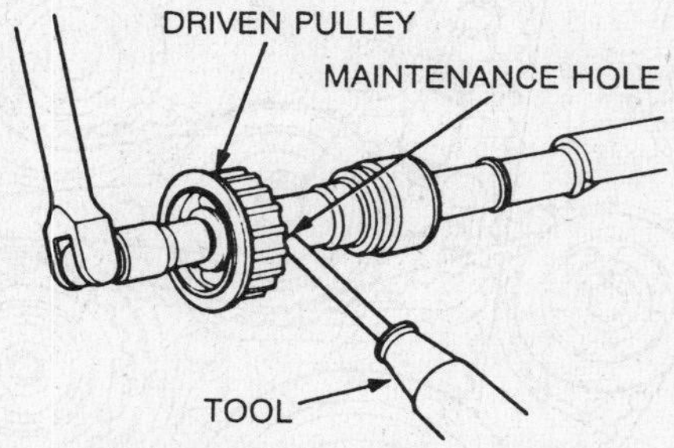

Front balancer shaft driven gear installation – 1990–92 Accord

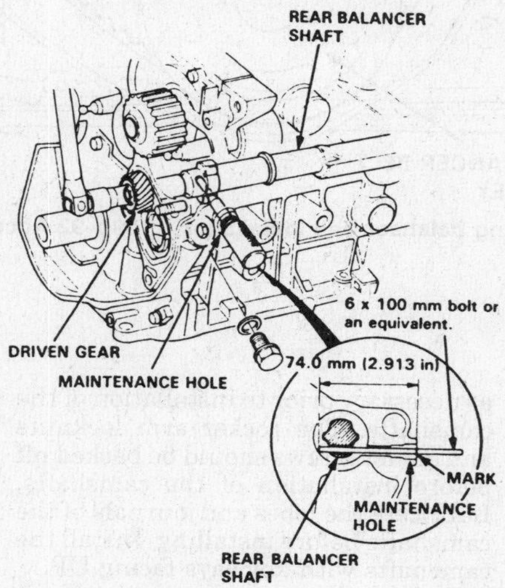

Rear balancer shaft driven gear installation – 1990–92 Accord

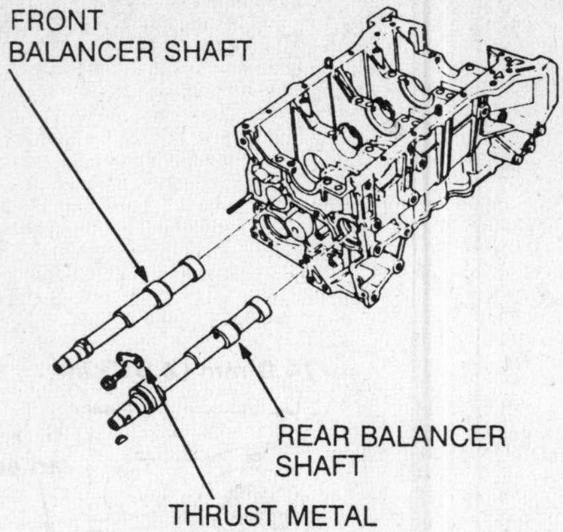

Front and rear balancer shafts – 1990–92 Accord

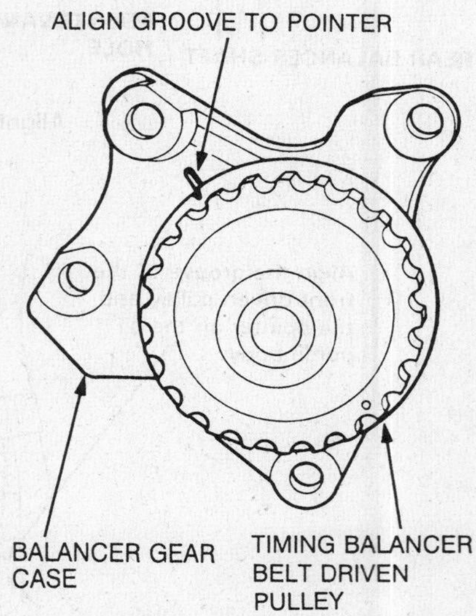

Timing balancer belt driven pulley – 1990–92 Accord

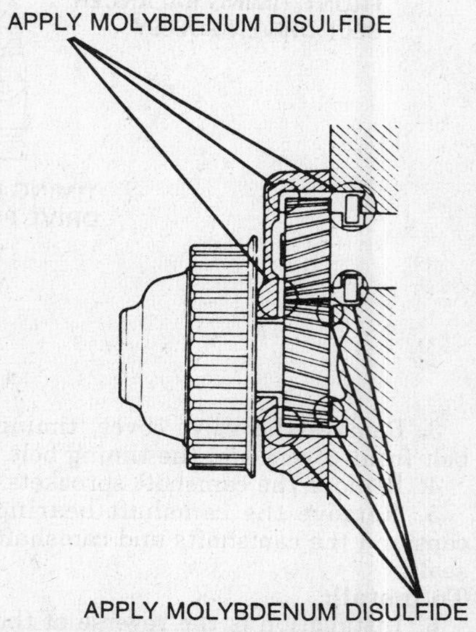

Apply molybdenum disulfide to the thrust surfaces of the balancer gears

Piston and Connecting Rod

POSITIONING

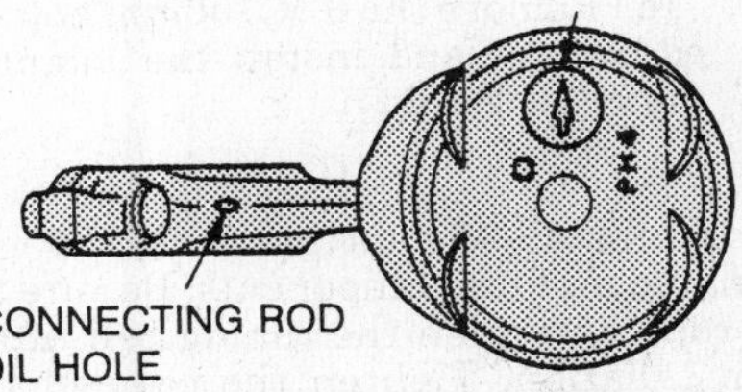

Piston and connecting rod positioning: The arrow must face the timing belt end of the engine and the connecting rod oil hole must face the intake manifold

ENGINE LUBRICATION

Oil Pan

REMOVAL & INSTALLATION

Except 4WD Civic Wagon

1. Disconnect the negative battery cable.
2. Raise and safely support the vehicle.
3. Remove the engine splash shield, if equipped.
4. Drain the engine oil.
5. Remove the exhaust header pipe, if necessary.
6. Remove the oil pan bolts and nuts and the oil pan.
7. Installation is the reverse of the removal procedure. Make sure all gasket mating surfaces are clean prior to installation.
8. Apply a coat of sealant to both sides of the oil pan gasket. Tighten the bolts and nuts in 2 steps in a criss-cross pattern beginning at the center of the pan. The final torque should be 9 ft. lbs. (12 Nm). Tighten the oil pan drain plug to 33 ft. lbs. (45 Nm).

4WD Civic Wagon

MANUAL TRANSAXLE

1. Disconnect the negative battery cable.
2. Raise and safely support the vehicle.
3. Remove the engine and transaxle splash shield.
4. Drain the engine and transaxle oil.
5. Remove the exhaust header pipe.
6. Mark the position of the driveshaft flange in relation to the companion flange and remove the driveshaft.
7. Remove the left side cover from the transfer case.

NOTE: Be careful not to damage the thrust shim and mating surface.

8. Remove the driven gear from the transfer case.

NOTE: Be careful not to damage the thrust shim and mating surface.

9. Remove the transfer case from the clutch housing.
10. Remove the clutch case cover.
11. Remove the oil pan by removing the bolts and nuts.

To install:

12. Clean all gasket mating surfaces.
13. Apply sealant to both sides of a new oil pan gasket and install the gasket and the oil pan. Tighten the bolts and nuts in 2 steps in a criss-cross pattern starting at the center of the pan. The final torque should be 9 ft. lbs. (12 Nm).
14. Install and tighten the oil drain plug to 33 ft. lbs. (45 Nm).
15. Apply liquid gasket to the clutch housing mating surface of the transfer case. Install the transfer case on the clutch housing. Tighten the transfer case bolts to 33 ft. lbs. (45 Nm).
16. Install the drive gear thrust shim on the transfer shaft. Lubricate the drive gear and install it on the transfer shaft. Install the transfer thrust shim and left side cover on the transfer case. Apply liquid gasket to the side cover bolts and tighten them to 33 ft. lbs. (45 Nm).
17. Apply a thin film of sealant at the top and bottom of the transfer case opening and install the driven gear thrust shim and the driven gear. Tighten the mouning bolts to 19 ft. lbs. (26 Nm).
18. Install the driveshaft, aligning the marks that were made during the removal procedure. Tighten the bolts to 24 ft. lbs. (33 Nm).
19. Install the exhaust header pipe and the engine and transaxle splash shields. Install and tighten the transaxle drain plug to 30 ft. lbs. (40 Nm).
20. Fill the transaxle with the proper type of oil, to the required level.
21. Lower the vehicle and fill the crankcase with the proper type of oil, to the required level.
22. Connect the negative battery cable, start the engine and check for leaks.

AUTOMATIC TRANSAXLE

Disconnect the negative battery cable.
2.Raise and safely support the vehicle.
3.Remove the engine and transaxle splash shield.
4.Drain the engine and transaxle oil.
5.Remove the exhaust header pipe.
6.Mark the position of the driveshaft flange in relation to the companion flange and remove the driveshaft.
7.Remove the driven gear assembly from the transfer case.
8.Remove the left side cover and then the drive gear from the transfer case. Rotate the cover using the bolt closest to the front of the vehicle as the axis. This bolt is not removed from the cover.

NOTE: Be careful not to damage the thrust shim and mating surface.

9. Remove the transfer case from the clutch housing.
10. Remove the clutch case cover.
11. Remove the oil pan by removing the bolts and nuts.

To install:

12. Clean all gasket mating surfaces.
13. Apply sealant to both sides of a new oil pan gasket and install the gasket and the oil pan. Tighten the bolts and nuts in 2 steps in a criss-cross pattern starting at the center of the pan. The final torque should be 9 ft. lbs. (12 Nm).
14. Install and tighten the oil drain plug to 33 ft. lbs. (45 Nm).
15. Apply liquid gasket to the clutch housing mating surface of the transfer case. Attach a new O-ring to the groove in the transfer left side cover.
16. Install the transfer case on the clutch housing. Install the bolt that remained in the left side cover, in the transfer case before installing the case on the clutch housing. Tighten the transfer case bolts to 33 ft. lbs. (45 Nm).
17. Install the drive gear thrust shim on the transfer shaft. Lubricate the drive gear and install it on the transfer shaft. Install the transfer thrust shim and left side cover on the transfer case. Apply liquid gasket to the side cover bolts and tighten them to 33 ft. lbs. (45 Nm).
18. Apply a thin film of sealant at the top and bottom of the transfer case opening and install the driven gear thrust shim and the driven gear. Tighten the mounting bolts to 19 ft. lbs. (26 Nm).
19. Install the driveshaft, aligning the marks that were made during the removal procedure. Tighten the bolts to 24 ft. lbs. (33 Nm).
20. Install the exhaust header pipe and the engine and transaxle splash shields. Install and tighten the transaxle drain plug to 29 ft. lbs. (40 Nm).
21. Lower the vehicle and fill the crankcase with the proper type of oil, to the required level. Fill the transaxle

with the proper type and quantity of oil.

22. Connect the negative battery cable, start the engine and check for leaks.

Oil Pump

REMOVAL & INSTALLATION

Except 1988–89 Accord

1. Disconnect the negative battery cable.
2. Raise and safely support the vehicle.
3. Drain the engine oil.
4. Bring the No. 1 cylinder to TDC. On Civic and CRX, the mark on the crankshaft pulley should align with the index mark on the timing cover. On Accord and Prelude, the mark on the flywheel should align with the pointer in the inspection hole.
5. Remove the necessary accessory drive belts and the crankshaft pulley.
6. Remove the valve cover and the timing belt covers.
7. On 1990–92 Accord, remove the following:
   a. Timing balancer belt
   b. Timing belt
   c. Timing belt tensioner
   d. Timing balancer belt tensioner
   e. Timing belt drive pulley
   f. Timing balancer belt driven pulley. Insert a suitable tool into the maintenance hole in the front balancer shaft in order to hold the shaft in place and remove the front balancer driven pulley.
   g. Balancer drive gear case
   h. Balancer driven gear. Remove the maintenance hole bolt from the cylinder block next to the rear balancer shaft. Insert a 6 × 100mm bolt or equivalent, to a depth of 74mm through the maintenance hole and into the rear balancer shaft. Remove the balancer shaft driven gear.
8. On all other vehicles, remove the following:
   a. Timing belt tensioner
   b. Timing belt
   c. Timing belt drive pulley
9. Remove the oil pan and oil screen.
10. Remove the oil pump mount bolts and the oil pump assembly.

To install:

11. Installation is the reverse of the removal procedure. Make sure all gasket mating surfaces are clean prior to installation.
12. Inspect the crankshaft oil seal and replace as necessary prior to installing the oil pump.
13. Apply liquid gasket to the cylinder block mating surface of the block. Apply a light coat of oil to the crankshaft seal lip. Install a new O-ring on the cylinder block and install the oil pump. Apply liquid gasket to the threads of the oil pump mounting bolts and tighten them to 9 ft. lbs. (12 Nm).
14. Install the oil screen.
15. On 1990–92 Accord, perform the following procedure:
   a. Apply molybdenum disulfide to the thrust surfaces of the balancer gears, before installing the balancer driven gear and the balancer drive gear case.
   b. Hold the rear balancer shaft with the 6 × 100mm bolt or equivalent, then install the balancer driven gear and the balancer drive belt pulley. Tighten the retaining bolt to 18 ft. lbs. (25 Nm).
   c. Hold the front balancer shaft with a suitable tool, then install the timing balancer belt driven pulley. Tighten the retaining bolt to 22 ft. lbs. (30 Nm).
   d. Install the balancer gear case to the oil pump. Align the groove on the pulley edge to the pointer on the gear case when holding the rear balancer shaft with the 6 × 100mm bolt or equivalent, then install the gear case. Tighten the balancer gear case bolts to 18 ft. lbs. (25 Nm).
   e. Remove the 6 × 100mm bolt or equivalent and install the maintenance hole bolt.
16. Install the oil pan and the remainder of the components. Be sure to properly tension the timing belt after installation. Tighten the crankshaft pulley bolt to specification.

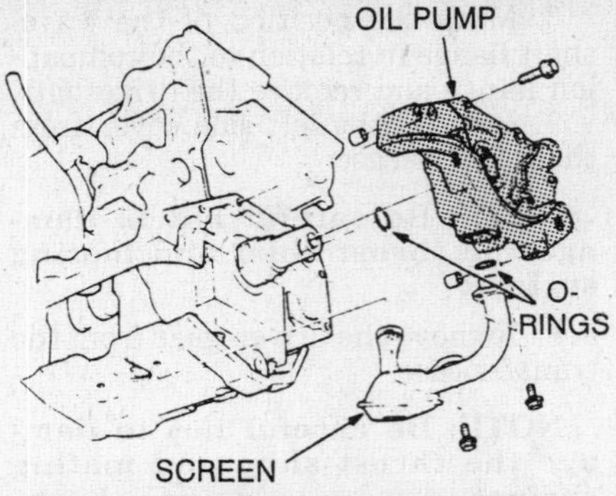

Oil pump installation on Prelude, Civic and CRX similar

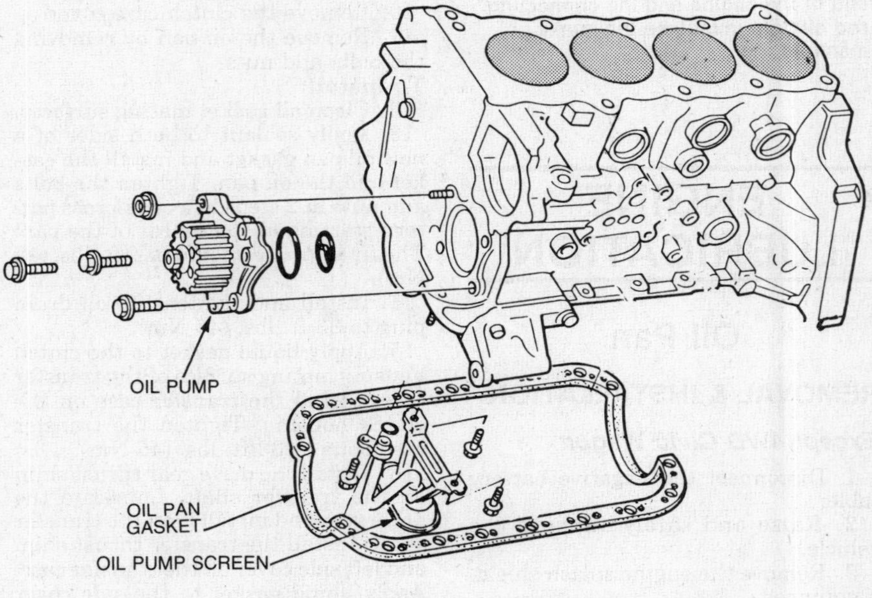

Oil pump installation on 1988–89 Accord

1988–89 Accord

1. Disconnect the negative battery cable.
2. Remove the necessary accessory drive belts.
3. Remove the crankshaft pulley, timing belt covers and timing belt.
4. Remove the 3 bolts and 1 nut retaining the oil pump to the cylinder block and remove the pump.
5. Installation is the reverse of the removal procedure. Make sure all gasket mating surfaces are clean prior to installation.
6. Apply liquid gasket around the O-ring groove, then install the new O-ring. Install a new gasket to the pump housing and install the pump. Tighten the pump bolts and nut to 9 ft. lbs. (12 Nm).
7. Install the remainder of the components. Be sure to properly tension the timing belt after installation. Tighten the crankshaft pulley bolt to specification.

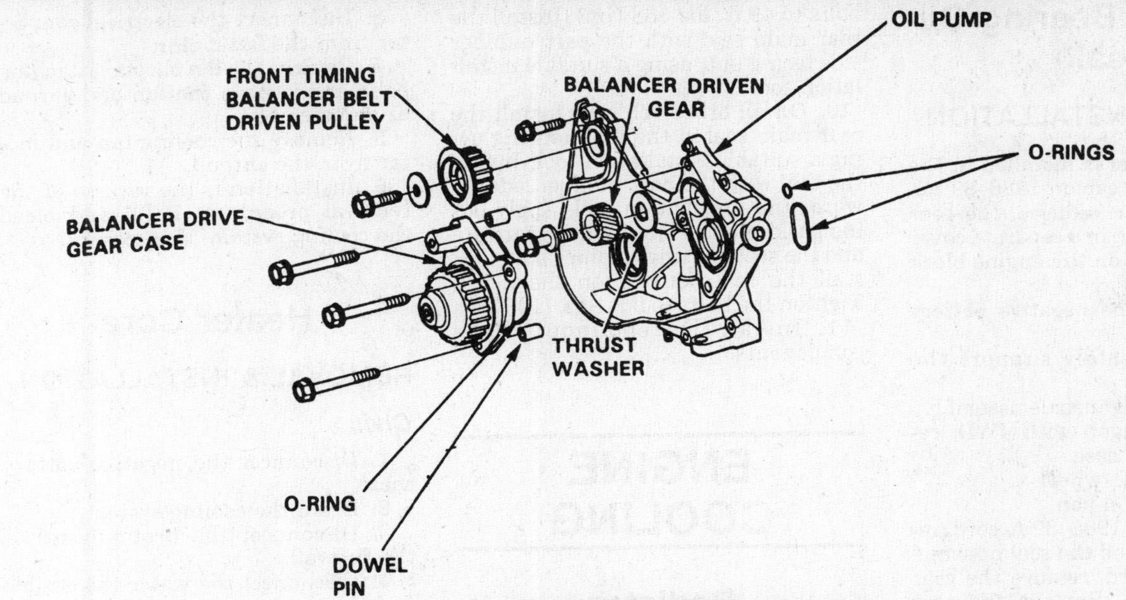

Oil pump installation on 1990–92 Accord

CHECKING

1. Remove the oil pump.
2. Check the rotor radial clearance using a feeler gauge. The clearance must be less than 0.008 in. (0.2mm).
3. Check the housing-to-rotor axial clearance using a feeler gauge on 1988–89 Accord, and a feeler gauge and straight-edge on all others. The clearance must be less than 0.005 in. (0.12mm) on Prelude and 1990–92 Accord, and less than 0.006 in. (0.15mm) on all others.
4. Check the housing-to-rotor radial clearance using a feeler gauge. The clearance must be less than 0.008 in. (0.2mm). It is necessary to remove the 2 screws and disassemble the pump housing on 1988–89 Accord to perform this procedure.
5. Inspect both rotors and pump housing for scoring or other damage.
6. Replace components that are not within specification or appear worn.
7. Reassemble the oil pump on 1988–89 Accord, using thread locking fluid on the 2 screws.
8. Install the oil pump.

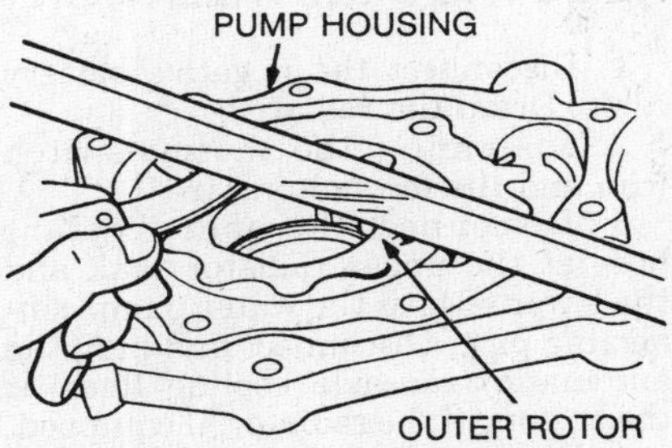

Checking housing-to-rotor axial clearance – except 1988–89 Accord

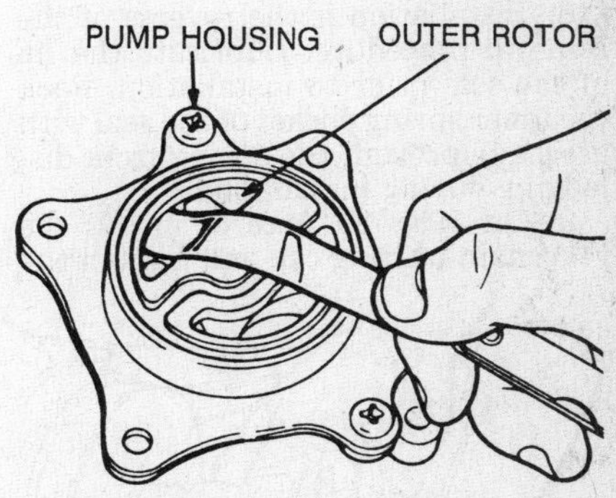

Checking housing-to-rotor axial clearance – 1988–89 Accord

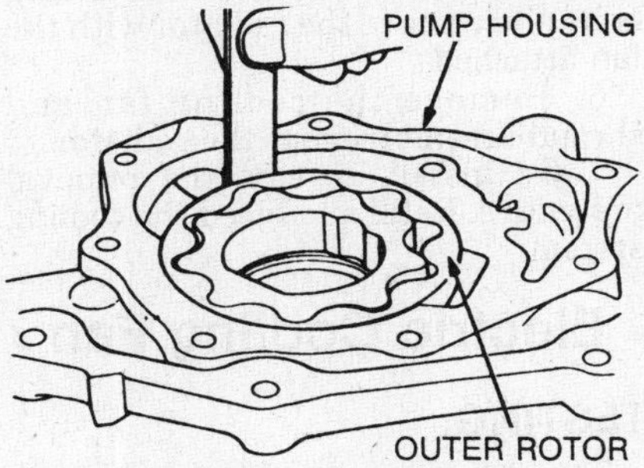

Checking housing-to-rotor radial clearance – except 1988–89 Accord

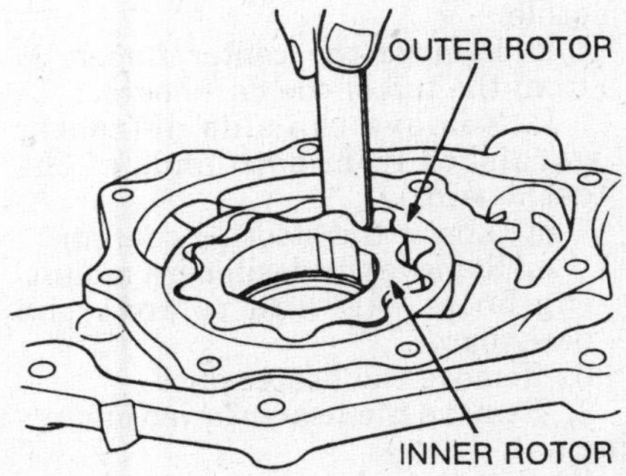

Checking rotor radial clearance – except 1988–89 Accord

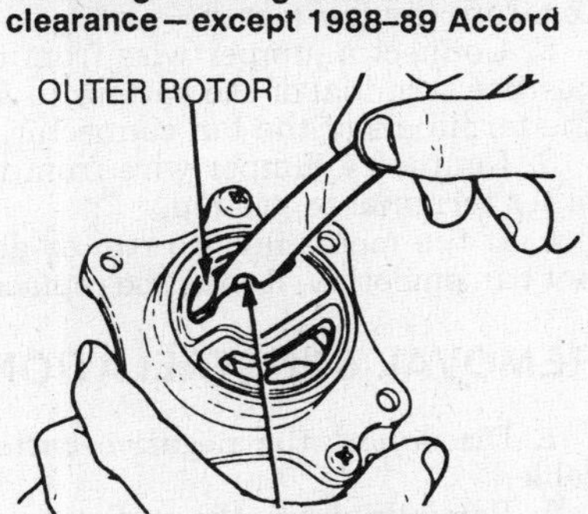

Checking rotor radial clearance – 1988–89 Accord

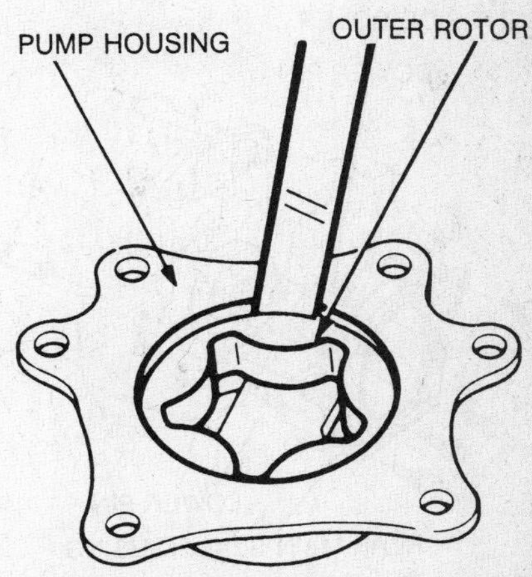

Checking housing-to-rotor radial clearance – 1988–89 Accord

Rear Main Bearing Oil Seal

REMOVAL & INSTALLATION

The rear main seal is installed in the rear main bearing cap on 1988–89 Accord. On all other vehicles, the rear main seal is housed in a separate cover which is mounted on the engine block with 4 bolts.

1. Disconnect the negative battery cable.
2. Raise and safely support the vehicle.
3. Remove the transaxle assembly.
4. On Civic Wagon with 4WD, remove the transfer case.
5. Remove the flywheel.
6. Remove the oil pan.
7. On all except 1988–89 Accord, remove the 4 bolts and the seal housing. On 1988–89 Accord, remove the rear main bearing cap. Remove the rear main seal.

To install:

8. Installation is the reverse of the removal procedure. Lubricate the lip of the seal prior to installation. Pack the inner spring pocket of the seal with grease to prevent the spring from dislodging during installation.
9. On 1988–89 Accord, install the rear main bearing cap and tighten the bolts to 49 ft. lbs. (68 Nm). Install the rear main seal with the part number side facing out, using a suitable installation tool.
10. On all other vehicles, install the rear main seal in the seal housing using a suitable installation tool. Install the seal with the part number side towards the installation tool. Apply liquid gasket to the block mating surface and the seal housing retainer bolts. Install the seal housing on the block. Tighten the bolts to 9 ft. lbs. (12 Nm).
11. Install the remainder of the components.

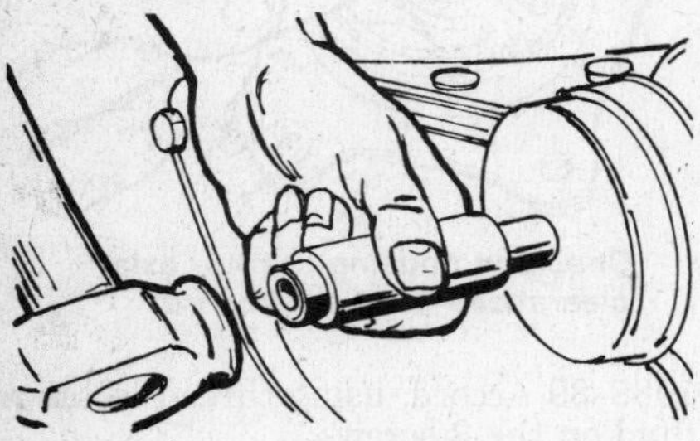

Installing the rear main seal on 1988–89 Accord

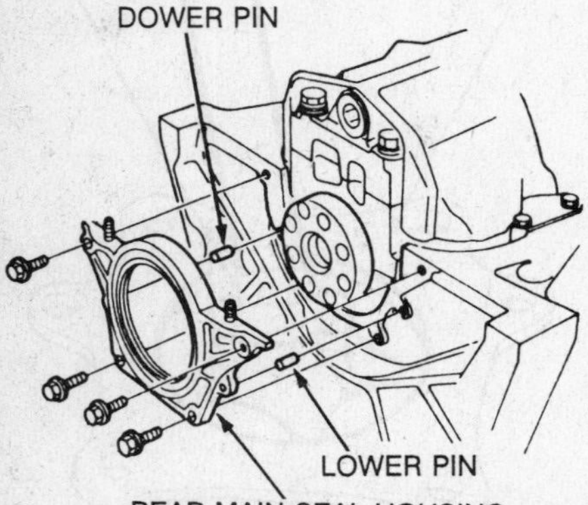

Installing the rear main seal housing—except 1988–89 Accord

ENGINE COOLING

Radiator

REMOVAL & INSTALLATION

1. Disconnect the negative battery cable. Drain the radiator.
2. Disconnect the thermo-switch wire and the fan motor wire.
3. Disconnect the upper coolant hose at the upper radiator tank and the lower hose at the water pump connecting pipe. Disconnect and plug the automatic transaxle cooling lines at the bottom of the radiator, if equipped.
4. Remove the hoses to the coolant reservoir, if equipped.
5. Detach the radiator mounting bolts and remove the radiator with the fan attached.
6. Remove the cooling fan and shroud assembly from the radiator.
7. To install, reverse the removal procedure. Refill and bleed the cooling system.

Electric Cooling Fan

TESTING

1. Disconnect the electrical connector from the fan motor.
2. Connect a jumper wire from the positive terminal of the battery to 1 of the terminals of the fan connector.
3. Connect a jumper wire from the other terminal to ground.
4. If the motor fails to run or does not run smoothly, it must be replaced.

REMOVAL & INSTALLATION

1. Disconnect the negative battery cable.
2. Partially drain the radiator and remove the upper radiator hose, if necessary.
3. Disconnect the electrical connector from the fan motor.
4. Remove the fan shroud attaching bolts and remove the fan and shroud as an assembly.
5. Remove the cooling fan and motor from the shroud.
6. Installation is the reverse of the removal procedure. Refill and bleed the cooling system, if necessary.

Heater Core

REMOVAL & INSTALLATION

Civic

1. Disconnect the negative battery cable.
2. Drain the cooling system.
3. Disconnect the heater hoses at the firewall.
4. Disconnect the water valve cable from the water valve.
5. Remove the dashboard by performing the following procedure:
   a. Slide the seats back fully and remove the dashboard center panel on 1988–89 vehicles or the center console on 1990–91 vehicles.
   b. Remove the fuse lid.
   c. Remove the knee bolster on 1990–91 vehicles.
   d. Disconnect the wire harnesses from the connector holder and disconnect the sunroof switch connector, if equipped. Remove the fuse box mounting nuts and lower the fuse box, if necessary.
   e. Disconnect the ground cable at the right of the steering column and the power door mirror switch connector.
   f. Remove the knob, then remove the side air vent face plate.
   g. Remove the 2 screws attaching the side air vent control lever.
   h. Remove the center panel and radio, then remove the 3 screws attaching the heater control panel to the dashboard.
   i. Remove the gauge upper panel or instrument panel, as necessary.
   j. Disconnect the speedometer cable.
   k. Remove the center upper lid from the top of the dashboard.
   l. Remove the side defroster garnishes from both ends of the dashboard.
   m. Lower the steering column.
   n. Remove the dashboard mounting bolts, lift and remove the dashboard.
6. Remove the heater duct.
7. Remove the heater lower mounting nut.
8. Remove the steering column bracket and duct assembly.
9. Remove the 2 heater mounting

bolts and the clip, then remove the heater assembly.

10. Remove the tapping screws and heater core cover.

11. Remove the tapping screw and clamp and remove the heater core.

To install:

12. Install the heater core into the housing and make sure the cover is properly sealed.

13. Install the heater assembly and connect the ducting.

14. Fit the dashboard into place and secure the steering column.

15. Install the instruments and connect the wiring.

16. Install the center console and connect the wiring and heater controls.

17. Install the fuse box and connect the wiring.

18. Connect the heater hoses and fill and bleed the cooling system. Adjust the controls as required.

CRX

1. Disconnect the negative battery cable.
2. Drain the cooling system.
3. Disconnect the heater hoses at the firewall.
4. Disconnect the water valve cable from the water valve.
5. Remove the dashboard by performing the following procedure:
   a. Slide the seats back fully and remove the front console.
   b. Remove the fuse lid. Disconnect the wire harnesses from the connector holder and disconnect the sunroof switch connector, if equipped. Remove the fuse box mounting nuts and lower the fuse box, if necessary.
   c. Disconnect the ground cable at the right of the steering column.
   d. Remove the coin box.
   e. Remove the knob, then remove the side air vent face plate.
   f. Remove the 2 screws attaching the side air vent control lever.
   g. Remove the 3 screws attaching the heater control panel to the dashboard.
   h. Remove the instrument panel.
   i. Disconnect the speedometer cable.
   j. Remove the center upper lid from the top of the dashboard.
   k. Remove the side defroster garnishes from both ends of the dashboard.
   l. Lower the steering column.
   m. Remove the dashboard mounting bolts, lift and remove the dashboard.
6. Remove the heater duct.
7. Remove the heater lower mounting nut.
8. Remove the steering column bracket and duct assembly.
9. Remove the 2 heater mounting bolts, disconnect the wire harness connector from the function control motor and then remove the heater assembly.
10. Remove the tapping screws and heater core cover.
11. Remove the tapping screw and clamp and remove the heater core.

To install:

12. Install the heater core into the housing and make sure the cover is properly sealed.
13. Install the heater assembly and connect the ducts and wiring.
14. Fit the dashboard into place and secure the steering column.
15. Install the instruments and connect the wiring.
16. Install the center console and connect the wiring and heater controls.
17. Install the fuse box and connect the wiring.
18. Connect the heater hoses and fill and bleed the cooling system. Adjust the controls as required.

Accord

1988–89

1. Disconnect the negative battery cable.
2. Drain the cooling system.
3. Disconnect the heater hoses at the firewall.
4. Disconnect the heater valve cable from the heater valve.
5. Remove the 2 heater lower mounting nuts.
6. Remove the dashboard by performing the following procedure:
   a. Slide the seats back fully and remove the front console.
   b. Remove the dashboard lower panel and lower the steering column.
   c. Remove the hood release handle but do not disconnect the cable.
   d. Disconnect the wire harnesses from the connector holders at the fuse area.
   e. Remove the ashtray, ashtray holder and the radio.
   f. Remove the heater control knobs, the heater control cover, disconnect the heater control connectors and remove the heater control.
   g. Remove the instrument cluster and the clock.
   h. Remove the dashboard mounting bolts. Lift the dashboard so as it is removed so it will slide up and off the guide pin at the middle. Support it from underneath so it will not fall when it comes off the pin.
7. Disconnect the cables from the heater assembly.
8. Remove the heater duct.
9. Remove the 4 heater mounting bolts, then pull the heater away from the body.
10. Remove the tapping screws and retaining plates and the heater core from the heater housing.

To install:

11. Install the heater core into the housing and make sure the cover is properly sealed.
12. Install the heater assembly and connect the ducts and control cables.
13. Carefully fit the dashboard into place and install the bolts.
14. Install the instruments and connect the wiring and speedometer cable.
15. Install the heater controls, ash tray and radio.
16. Connect the wiring at the fuse box.
17. Connect the heater hoses and fill and bleed the cooling system. Adjust the controls as required.

1990–92

1. Disconnect the negative battery cable.
2. Drain the cooling system.
3. Disconnect the heater hoses at the heater.
4. Disconnect the heater valve cable from the heater valve.
5. Remove the dashboard by performing the following procedure:
   a. Slide the seats back fully and remove the console.
   b. Remove the knee bolster, lower panel and steering column.
   c. Disconnect the dashboard wire harness from the connectors and fuse box.
   d. Remove the carpet clips and disconnect the antenna lead.
   e. Disconnect the heater control cable and function control cable.
   f. Remove the caps from both sides of the dash and the clock.
   g. Remove the 7 dashboard mounting bolts, lift and remove the dashboard.
6. Remove the heater duct.
7. Remove the instrument sub-pipe.
8. Remove the 4 heater mounting nuts and the heater assembly.
9. Remove the air mix rod from the clip, the self-tapping screws and heater core cover and the self tapping screw and clamp. Remove the heater core from the heater housing.

To install:

10. Install the heater core into the housing and make sure the cover is properly sealed. Install the air mix door rod.
11. Install the heater assembly and connect the ducts.
12. Install the instrument sub-pipe.
13. Carefully fit the dashboard into place and install the bolts.
14. Connect the heater controls and wiring.

15. Install the console and knee bolster and secure the steering column in place.
16. Connect the heater hoses and fill and bleed the cooling system. Adjust the controls as required.

Prelude

1. Disconnect the negative battery cable.
2. Drain the cooling system.
3. Disconnect the heater hoses at the heater.
4. Disconnect the heater valve cable from the heater.
5. Remove the dashboard by performing the following procedure:
   a. Slide the seats back fully and remove the dashboard lower panel and the front and rear consoles.
   b. Disconnect the wire harnesses from the connector holder and fuse box.
   c. Remove the 6 screws and radio panel, then disconnect the wire connectors and antenna cable.
   d. Remove the radio assembly.
   e. Disconnect the heater control cable and the connector and wire harnesses from the heater control unit.
   f. Remove the clock from the top of the dashboard.
   g. Lower the steering column.
   h. Remove the dashboard mounting bolts. Lift the dashboard as it is removed, so it will slide up and off the guide pin in the middle. Hold the dashboard from underneath so it will not fall when it comes off of the pin.
6. Remove the heater duct.
7. Remove the heater lower mounting nuts.
8. Remove the steering column bracket.
9. Remove the 2 heater mounting bolts, disconnect the wire harness connector from the function control motor and then remove the heater assembly.
10. Remove the integrated control unit and bracket from the heater assembly.
11. Remove the 2 tapping screws, bracket, set plate and heater core cover.
12. Remove the 2 tapping screws, heater core setting plate and clamp.
13. Pull the heater core from the heater housing.

To install:

14. Install the heater core into the housing and make sure the cover is properly sealed. Complete the heater assembly.
15. Install the heater assembly and connect the ducts and wiring.
16. Install the instrument sub-pipe.
17. Carefully fit the dashboard into place and install the bolts.

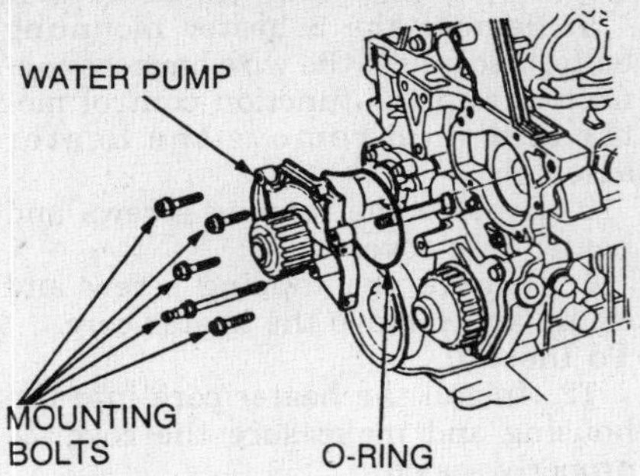

Water pump installation on 1990–92 Accord

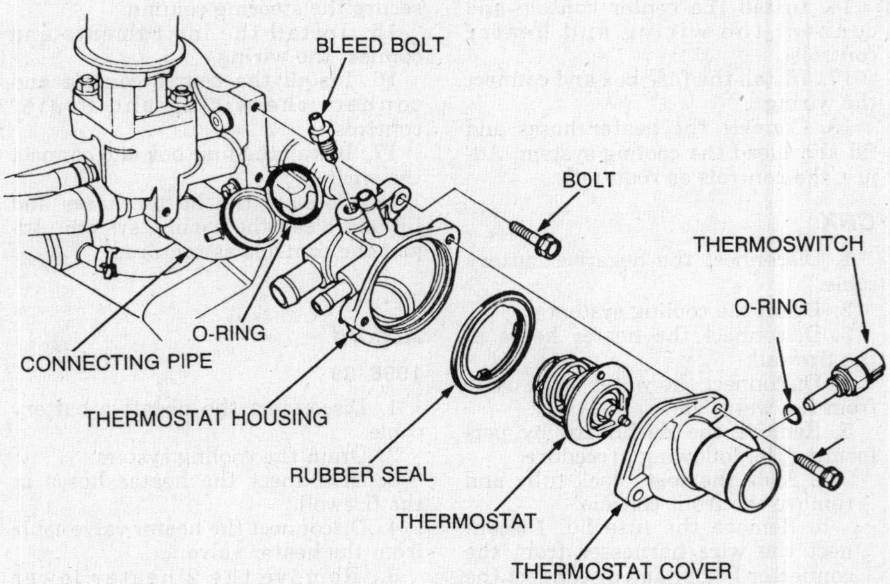

Thermostat housing assembly on 1990–92 Accord

18. Connect the heater controls and wiring.
19. Install the radio and connect the wiring.
20. Connect the wiring to the fuse box.
21. Connect the heater hoses and fill and bleed the cooling system. Adjust the controls as required.

Water Pump

REMOVAL & INSTALLATION

Except 1988–89 Accord

1. Disconnect the negative battery cable.
2. Drain the cooling system.
3. Remove the accessory drive belts.
4. Remove the timing belt cover and the timing belt.
5. Remove the water pump mounting bolts and the water pump.
6. Installation is the reverse of the removal procedure. Use a new O-ring seal when installing the pump. Tighten the water pump mounting bolts to 9 ft. lbs. (12 Nm).
7. Fill and bleed the cooling system.

1988–89 Accord

1. Disconnect the negative battery cable.
2. Drain the cooling system.
3. Remove the accessory drive belts.
4. Remove the timing belt cover.
5. Remove the water pump pulley and the water pump mounting bolts.
6. Remove the water pump and the O-ring seal.
7. Installation is the reverse of the removal procedure. Use a new O-ring seal when installing the pump. Tighten the water pump mounting bolts and the pulley bolts to 9 ft. lbs. (12 Nm).
8. Fill and bleed the cooling system.

Thermostat

The thermostat housing is located on the cylinder head with the exception of Civic and CRX and 1990–91 Accord, where it is located at the end of the water pump inlet tube.

REMOVAL & INSTALLATION

1. Disconnect the negative battery cable.
2. Drain the cooling system.
3. Disconnect the radiator hose from the thermostat housing outlet.
4. Remove the thermostat housing outlet and remove the thermostat.
5. Installation is the reverse of the removal procedure. Use new gaskets and O-rings. Install the thermostat

with the pin towards the thermostat housing outlet. Tighten the thermostat housing outlet bolts to 9 ft. lbs. (12 Nm).

6. Fill and bleed the cooling system.

Cooling System Bleeding

1. Loosen the air bleed bolt in the water outlet and fill the radiator to the bottom of the filler neck with antifreeze/coolant. Tighten the bleed bolt as soon as the coolant starts to run out in a steady stream without any air bubbles in it.
2. With the radiator cap off, start the engine and allow it to warm up (the cooling fan should go on at least twice). Then if necessary add more antifreeze/coolant to bring the level back up to the bottom of the filler neck.
3. Put the radiator cap on, restart the engine and check for any leaks.

ENGINE ELECTRICAL

Distributor

REMOVAL

1. Disconnect the negative battery cable.
2. Disconnect the spark plug wires and tag for reassembly in the same positions.
3. Disconnect the coil wire and primary lead wire, if equipped.
4. Disconnect the vacuum advance hoses, if equipped.
5. Disconnect the necessary electrical connectors.
6. Remove the distributor cap.
7. Using a suitable marking tool, mark the position of the distributor rotor in relation to the distributor housing and mark the position of the distributor housing in relation to the cylinder head.
8. Remove the distributor hold-down bolts and remove the distributor. Remove and discard the distributor O-ring.

INSTALLATION

Timing Not Disturbed

1. Coat a new distributor O-ring with engine oil and install on the distributor. Install the distributor, aligning the distributor housing and distributor rotor with the marks that were made during the removal procedure.

NOTE: The distributor is equipped with locating lugs which mesh with corresponding grooves in the end of the camshaft. The lugs and grooves are both offset to prevent installing the distributor 180 degrees out of time.

2. Install the distributor hold-down bolts and tighten temporarily.
3. Install the distributor cap.
4. Connect the electrical connectors.
5. Connect the vacuum hoses, if equipped.
6. Connect the coil wire and the primary lead wire, if equipped.
7. Connect the spark plug wires in their original positions.
8. Check the ignition timing and tighten the distributor hold-down bolts to 16 ft. lbs. (22 Nm). Recheck the ignition timing.

Timing Disturbed

1. Disconnect the spark plug wire from the No. 1 cylinder spark plug and remove the spark plug.
2. Place a finger over the spark plug hole and turn the engine over slowly, by hand, until compression is felt.
3. On Civic vehicles, align the **RED** timing mark on the crankshaft pulley with the pointer on the timing belt cover. On Accord and Prelude vehicles, remove the rubber cap from the inspection window at the rear of the cylinder block. Align the **RED** timing mark on the driveplate (automatic transaxle) or flywheel (manual transaxle) with the pointer on the cylinder block.
4. Coat a new distributor O-ring with engine oil and install on the distributor. Install the distributor with the distributor rotor pointing to the No. 1 spark plug tower on the distributor cap.

NOTE: The distributor is equipped with locating lugs which mesh with corresponding grooves in the end of the camshaft. The lugs and grooves are both offset to prevent installing the distributor 180 degrees out of time.

5. Install the distributor hold-down bolts and tighten temporarily.
6. Install the distributor cap.
7. Connect the electrical connectors.
8. Connect the vacuum hoses, if equipped.
9. Connect the coil wire and the primary lead wire, if equipped.
10. Install the No. 1 spark plug and tighten to 13 ft. lbs. (18 Nm). Connect the No. 1 spark plug wire to the spark plug.
11. Connect the spark plug wires to the distributor cap in their original positions.
12. Set the ignition timing and tighten the distributor hold-down bolts to 16 ft. lbs. (22 Nm). Recheck the ignition timing.

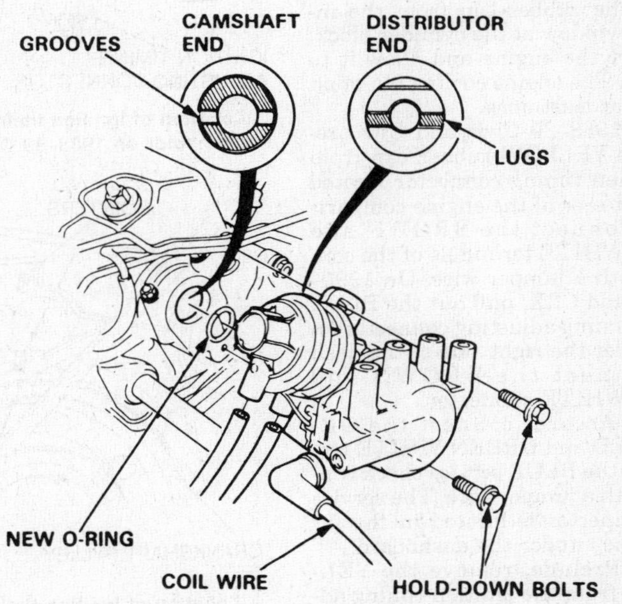

Typical distributor installation

Ignition Timing

ADJUSTMENT

1988–89 Accord and Carburetor Equipped Prelude

1. Remove the rubber cap from the inspection window of the cylinder block.
2. Start the engine and allow it to warm up. The cooling fan must come on at least once.
3. Disconnect the vacuum hoses from the vacuum advance diaphragm and plug them.
4. Connect the pick-up lead from a suitable timing light to the No. 1 spark plug wire. Make the remaining timing light connections according to the manufacturers instructions.
5. With the engine idling, aim the timing light at the pointer on the engine block and the driveplate (automatic transaxle) or flywheel (manual transaxle).
6. Adjust the ignition timing according to the specification listed on the underhood emission label. Adjust as necessary by loosening the distributor adjusting bolts and turning the distributor housing counterclockwise to advance the timing, or clockwise to retard the timing.
7. Tighten the adjusting bolts to 16 ft. lbs. (22 Nm) and recheck the timing.
8. Connect the vacuum advance hoses and replace the rubber cap to the inspection window.

Civic, CRX, 1990–92 Accord and Prelude with Fuel Injection

1. On Accord and Prelude vehicles, remove the rubber cap from the inspection window of the cylinder block.
2. Start the engine and allow it to warm up. The engine cooling fan must come on at least once.
3. On 1988–89 Civic and CRX, remove the YELLOW rubber cap from the ignition timing connector located in the left rear of the engine compartment. Connect the BROWN and GREEN/WHITE terminals of the connector with a jumper wire. On 1990–92 Civic and CRX, pull out the BLUE ignition timing adjusting connector located under the right side of the dash and connect the BROWN and GREEN/WHITE connector.
4. On Accord, connect the ORANGE/RED and GREEN/WHITE terminals of the BLUE service check connector with a jumper wire. The service check connector is located in the far right corner under the dashboard.
5. On Prelude, remove the YELLOW cap from the ignition timing adjusting connector located behind the ignition coil and connect the BROWN and GREEN/WHITE terminals with a jumper wire.
6. Connect the pick-up lead of a suitable timing light to the No. 1 spark plug wire. Make the other timing light connections according to the manufacturers instructions.
7. With the engine idling, aim the timing light at the pointer on the timing belt cover and the crankshaft pulley on Civic and CRX. On Accord and Prelude, aim the timing light at the pointer on the cylinder block and the driveplate (automatic transaxle) or flywheel (manual transaxle).
8. Adjust the ignition timing to the specification listed on the underhood vehicle emission label. Adjust as necessary by loosening the distributor adjusting bolts and turning the distributor housing counterclockwise to advance the timing, or clockwise to retard the timing.
9. Tighten the adjusting bolts to 16 ft. lbs. (22 Nm) and recheck the timing.
10. Remove the jumper wire from the ignition timing adjusting connector on Civic, CRX and Prelude, and install the YELLOW rubber cap, if equipped. On Accord, remove the jumper wire from the BLUE service check connector and install the rubber cap to the inspection window.

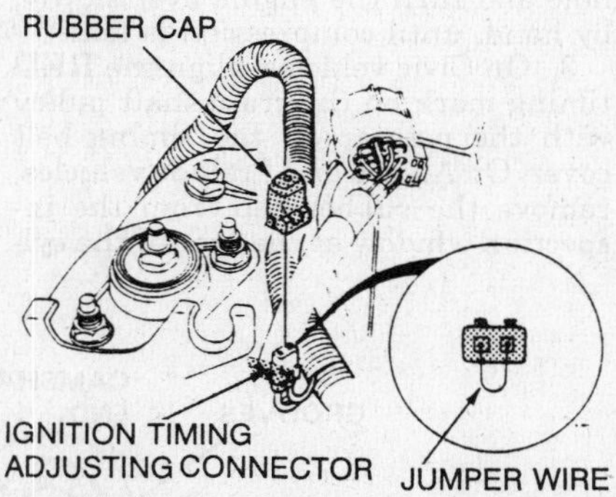

Location of ignition timing adjusting connector on 1988–89 Civic and CRX

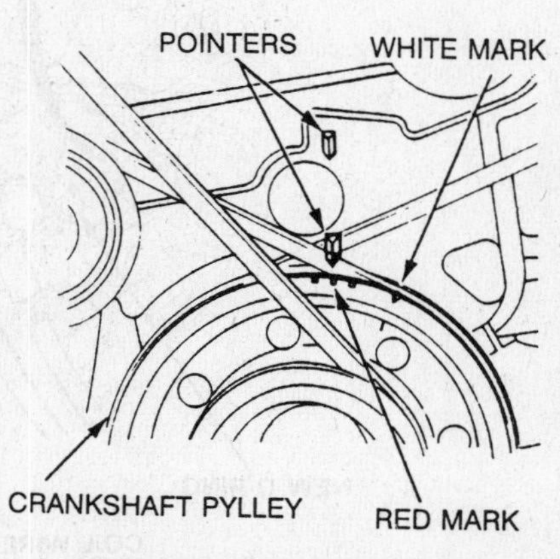

Location of ignition timing marks—Civic and CRX

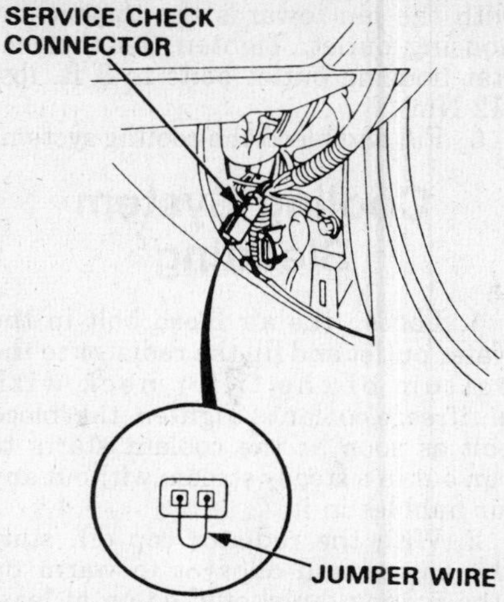

Location of service check connector on 1990–92 Accord

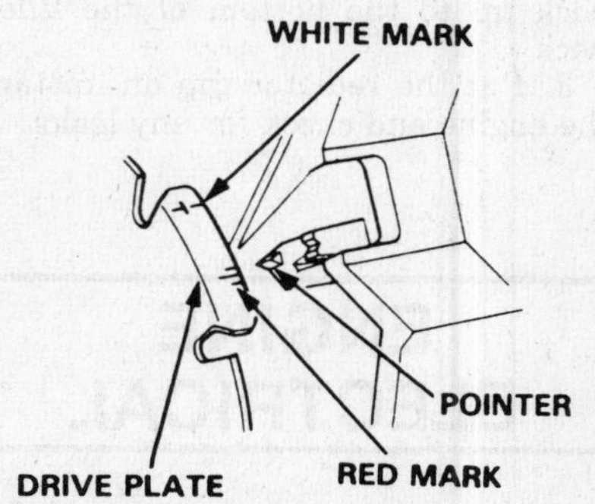

Timing pointer and timing marks on Accord and Prelude with carburetors

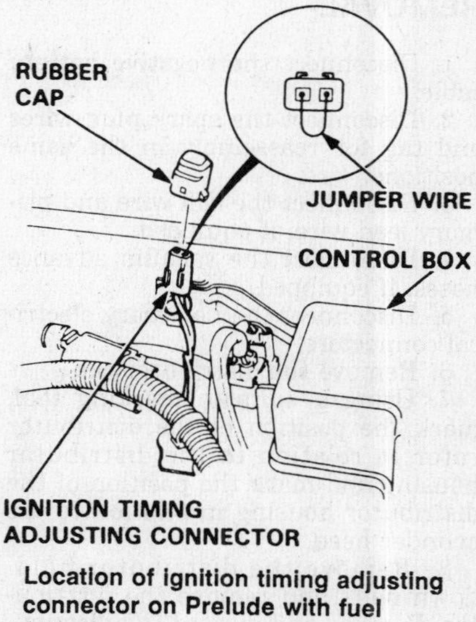

Location of ignition timing adjusting connector on Prelude with fuel injection

Alternator

PRECAUTIONS

Several precautions must be observed

with alternator equipped vehicles to avoid damage to the unit.

- If the battery is removed for any reason, make sure it is reconnected with the correct polarity. Reversing the battery connections may result in damage to the one-way rectifiers.
- When utilizing a booster battery as a starting aid, always connect the positive to positive terminals and the negative terminal from the booster battery to a good engine ground on the vehicle being started.
- Never use a fast charger as a booster to start vehicles.
- Disconnect the battery cables when charging the battery with a fast charger.
- Never attempt to polarize the alternator.
- Do not use test lamps of more than 12 volts when checking diode continuity.
- Do not short across or ground any of the alternator terminals.
- The polarity of the battery, alternator and regulator must be matched and considered before making any electrical connections within the system.
- Never separate the alternator on an open circuit. Make sure all connections within the circuit are clean and tight.
- Disconnect the battery ground terminal when performing any service on electrical components.
- Disconnect the battery if arc welding is to be done on the vehicle.

BELT TENSION ADJUSTMENT

1. On all vehicles except 1988–89 Accord, apply a force of 22 lbs. and measure the deflection of the alternator belt between the alternator and the crankshaft pulley. On 1988–89 Accord, apply a force of 22 lbs. and measure the deflection of the alternator belt between the alternator and the water pump pulley.
2. On a belt in service, the deflection should be as follows:
   Civic and CRX: 0.35–0.43 in. (9–11 mm)
   1988–89 Accord: 0.24–0.35 in. (6–9mm)
   1990–92 Accord: 0.39–0.47 in. (10–12mm)
   Prelude: 0.39–0.47 in. (10–12mm)
3. If the belt deflection is not as specified, loosen the alternator pivot bolt. On Civic and CRX, loosen the alternator adjusting bolt and move the alternator using a suitable prybar positioned against the front of the alternator housing. Tighten the adjusting bolt when the proper tension is obtained. On Accord and Prelude, loosen the alternator nut or bolt and turn the adjusting nut or bolt until the proper tension is obtained. Tighten the alternator nut or bolt and the pivot bolt. Recheck the belt deflection.

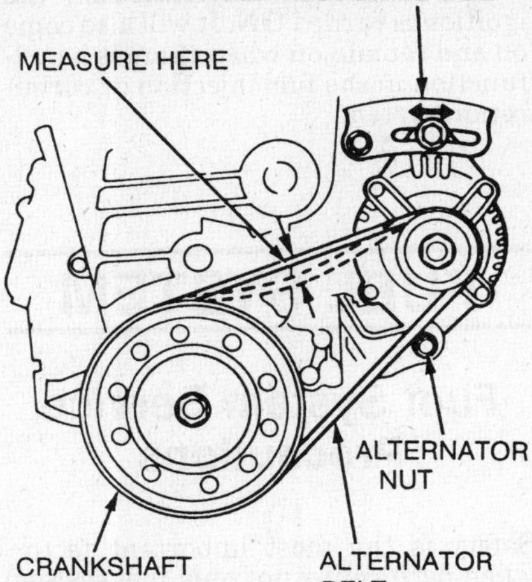

Alternator belt adjustment—Civic and CRX

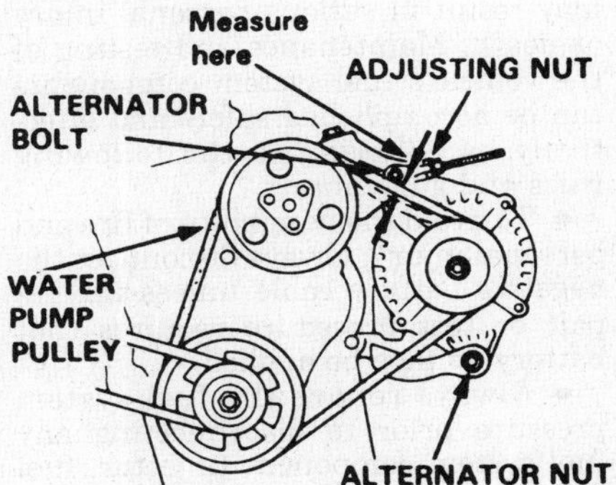

Alternator belt adjustment—1989–89 Accord

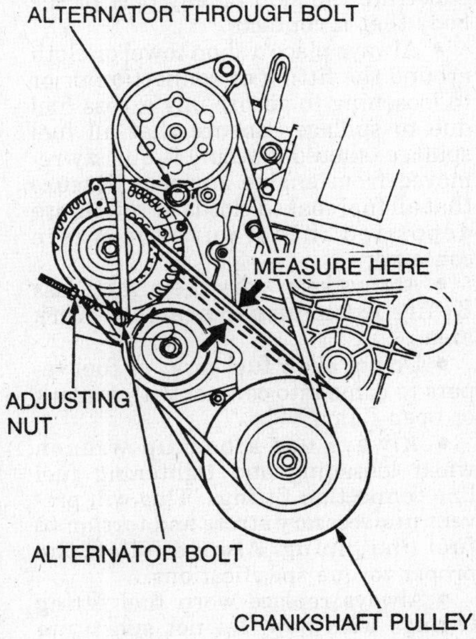

Alternator belt adjustment—Prelude

4. If a new belt is installed, the deflection should be as follows when first measured:
   Civic and CRX: 0.25–0.35 in. (7–9mm)
   1988–89 Accord: 0.16–0.24 in. (4–6mm)
   1990–92 Accord without air conditioning: 0.33–0.43 in. (8.5–11mm)
   1990–92 Accord with air conditioning: 0.18–0.28 in. (4.5–7mm)
   Prelude: 0.31–0.39 in. (8–10mm)

REMOVAL & INSTALLATION

Except 1990–92 Accord

1. Disconnect the negative battery cable.
2. Remove the air cleaner assembly if necessary.
3. on 1988–89 Accord.
4. On 1988–89 Accord, disconnect the left driveshaft from the steering knuckle.
4. Disconnect the electrical connectors from the alternator.
5. Loosen the alternator adjusting bolt or nut and through bolt and remove the alternator belt.
6. Remove the alternator adjusting bolt or nut and through bolt and remove the alternator. If necessary, remove the mount bracket bolts and the upper and lower mount brackets.
7. Installation is the reverse of the removal procedure. Tighten the alternator through bolt or through bolt nut to 33 ft. lbs. (45 Nm). Tighten the alternator adjusting bolt on Civic and CRX and the alternator bolt on Prelude and Accord vehicles to 17 ft. lbs. (24 Nm).

1990–92 Accord

1. Disconnect the negative battery cable.
2. Remove the power steering pump and the cruise control actuator. Do not disconnect the actuator cable.
3. Disconnect the electrical connectors from the alternator.
4. Loosen the adjusting bolt, then remove the alternator nut. Remove the alternator belt from the alternator pulley.
5. Remove the adjusting bolt, the lower through bolt and the stay.
6. Remove the upper through bolt and the alternator. If necessary, remove the 4 mount bracket bolts, the mount bracket and the heat insulator.
7. Installation is the reverse of the removal procedure. Tighten the upper through bolt to 33 ft. lbs. (45 Nm) and the alternator nut to 18 ft. lbs. (26 Nm). If the mount bracket was re-

moved, apply liquid gasket to the mount bracket bolt threads and tighten to 36 ft. lbs. (50 Nm).

Voltage Regulator

The voltage regulator on all vehicles is now an internal part of the alternator. It can still usually be replaced separately but the alternator must be removed and partially disassembled.

Starter

REMOVAL & INSTALLATION

1. Disconnect the negative battery cable.
2. Disconnect the starter cable from the starter motor.
3. Remove the engine compartment sub wire harness from the harness clip on the starter motor, if equipped.
4. Disconnect the wire from the starter solenoid.
5. Remove the 2 bolts retaining the starter motor and remove the starter motor.
6. Installation is the reverse of the removal procedure. Tighten the starter motor retaining bolts to 32 ft. lbs. (45 Nm).

EMISSION CONTROLS

Please refer to "Emission Controls" in the Unit Repair section for system maintenance procedures. Due to the complex nature of modern electronic engine control systems, comprehensive diagnosis and testing procedures fall outside the confines of this repair manual. For complete information on diagnosis, testing and repair procedures concerning all modern engine and emission control systems, please refer to "Chilton's Guide to Fuel Injection and Electronic Engine Controls".

Emission Warning Lamps

There are no dashboard warning lamps indicating periodic maintenance or component replacement is necessary. All fuel injected and some carbureted vehicles are equipped with a CHECK ENGINE light. This light comes on momentarily each time the ignition is turned **ON**. It will also come on and remain on when there is a malfunction in the fuel injection or carburetion system.

FUEL SYSTEM

Fuel System Service Precautions

Safety is the most important factor when performing not only fuel system maintenance but any type of maintenance. Failure to conduct maintenance and repairs in a safe manner may result in serious personal injury or death. Maintenance and testing of the vehicle's fuel system components can be accomplished safely and effectively by adhering to the following rules and guidelines.

- To avoid the possibility of fire and personal injury, always disconnect the negative battery cable unless the repair or test procedure requires that battery voltage be applied.
- Always relieve the fuel system pressure prior to disconnecting any fuel system component (injector, fuel rail, pressure regulator, etc.), fitting or fuel line connection. Exercise extreme caution whenever relieving fuel system pressure to avoid exposing skin, face and eyes to fuel spray. Please be advised that fuel under pressure may penetrate the skin or any part of the body that it contacts.
- Always place a shop towel or cloth around the fitting or connection prior to loosening to absorb any excess fuel due to spillage. Ensure that all fuel spillage (should it occur) is quickly removed from engine surfaces. Ensure that all fuel soaked cloths or towels are deposited into a suitable waste container.
- Always keep a dry chemical (Class B) fire extinguisher near the work area.
- Do not allow fuel spray or fuel vapors to come into contact with a spark or open flame.
- Always use a backup wrench when loosening and tightening fuel line connection fittings. This will prevent unnecessary stress and torsion to fuel line piping. Always follow the proper torque specifications.
- Always replace worn fuel fitting O-rings with new. Do not substitute fuel hose or equivalent where fuel pipe is installed.

RELIEVING FUEL SYSTEM PRESSURE

1. Make sure the engine is cold.
2. Disconnect the negative battery cable.
3. Remove the fuel filler cap.
4. Use a suitable box end wrench on the 6mm service bolt at the top of the fuel filter (fuel pipe on 1990–92 Accord), while holding the special banjo bolt with another wrench.
5. Place a rag or shop towel over the 6mm service bolt.
6. Slowly loosen the 6mm service bolt 1 complete turn.

Fuel Tank

REMOVAL & INSTALLATION

1. Raise and safely support the vehicle and remove the rear wheels.
2. Remove the drain bolt and drain the fuel from the tank. Be sure to take the appropriate fire safety precautions.
3. In the luggage area on 2 wheel drive vehicles, remove the access covers and disconnect the fuel pump and gauge unit wiring. On 4 wheel drive vehicles, the access cover is under the rear seat.
4. To disconnect the hoses, loosen the clamps and slide them back. Carefully twist the hose while pulling it off the fitting to avoid damage to the hose or the flared fittings.
5. On 4 wheel drive vehicles, remove the exhaust pipe and muffler and the rear propeller shaft.
6. Place a jack under the tank and remove the nuts to allow the tank straps to fall free. The tank may stick to the vehicle's undercoating but it can be pried out of the mounts. Be careful not to damage the fittings.

To install:

7. Position the tank under the vehicle and install the straps, washers and nuts onto the hooks.
8. Wiggle the tank while tightening the nuts a few turns at a time to make sure the tank seats properly into the mount. Torque the strap nuts to 16 ft. lbs. (22 Nm).
9. Make sure the drain plug in installed with a new gasket and torque to 36 ft. lbs. (50 Nm).
10. Connect the hoses and wiring.
11. On 4 wheel drive vehicles, install the drive shaft and torque the bolts to 24 ft. lbs. (33 Nm). Use new self locking nuts when attaching the exhaust pipe to the catalytic converter and torque to 25 ft. lbs. (34 Nm).

Fuel Filter

REMOVAL & INSTALLATION

Carbureted Vehicles

FRONT FILTER

1. Disconnect the negative battery cable.
2. Use suitable fuel line clamps to pinch off the fuel lines and prevent fuel from leaking.
3. Slide the fuel line retaining clamps back. Remove the fuel lines from the filter by using a twisting motion as the line is pulled off.
4. Remove the fuel filter.
5. Install the replacement filter, connect the fuel lines and position the fuel line retaining clamps.
6. Remove the fuel line clamps and connect the negative battery cable.
7. Start the engine and check for fuel leaks.

REAR FILTER

1. Disconnect the negative battery cable.
2. Raise and safely support the vehicle.
3. Remove the fuel filter and holder.
4. Push in the tab of the fuel filter to release the holder, then remove the filter from the holder.
5. Attach suitable fuel line clamps to pinch off the fuel lines and prevent fuel from leaking.
6. Slide the fuel line retaining clamps back. Remove the fuel lines from the filter by using a twisting motion as the line is pulled off.
7. Remove the fuel filter.
8. Install the replacement filter, connect the fuel lines and position the fuel line retaining clamps.
9. Remove the fuel line clamps and install the holder on the fuel filter.
10. Install the fuel filter and holder.
11. Lower the vehicle, connect the negative battery cable, start the engine and check for fuel leaks.

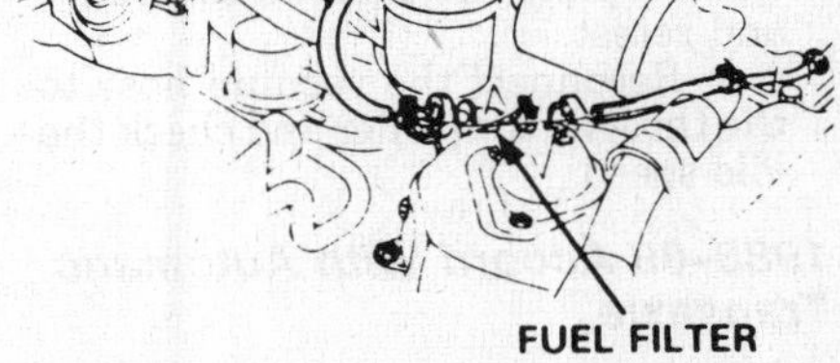

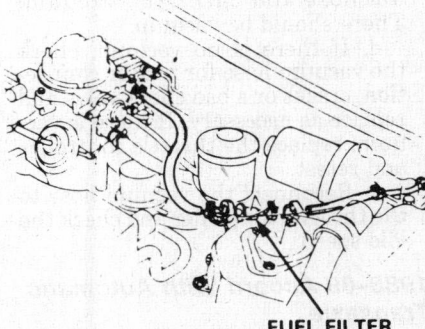

Front fuel filter location on Accord with carburetor

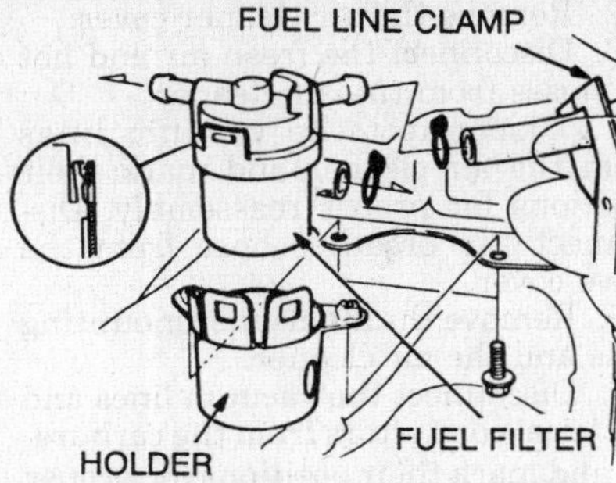

Rear fuel filter location on vehicles with carburetor

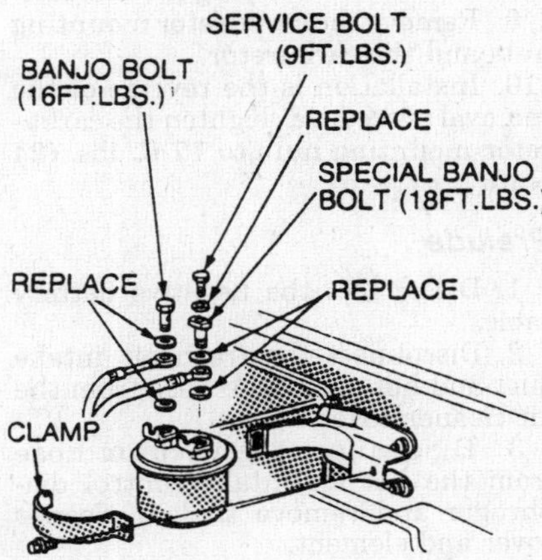

Fuel filter installtion on fuel injected vehicles

Fuel Injected Vehicles

1. Disconnect the negative battery cable.
2. Relieve the fuel system pressure.
3. Remove the banjo bolt(s) and the fuel lines from the fuel filter.
4. Remove the fuel filter clamp and the fuel filter.
5. Installation is the reverse of the removal procedure. Always use new washers during installation. Tighten the banjo bolts to 16 ft. lbs. (22 Nm) and the service bolt to 9 ft. lbs. (12 Nm).

Electric Fuel Pump

PRESSURE TESTING

Carbureted Vehicles

1. Remove the fuel cut-off relay from the underdash fuse panel.
2. Using a jumper wire, connect the No. 1 terminal to the No. 2 terminal located at the fuse box side of the fuel cut-off relay.
3. Disconnect the fuel line at the fuel filter in the engine compartment and connect a suitable fuel pressure gauge.
4. Turn the ignition **ON** until the pressure stabilizes, then turn the key **OFF**. The fuel pressure should be as follows:
   1988 Prelude and 1988–89 Accord–2.6–3.3 psi.
   1989–91 Prelude–1.3–2.1 psi.
5. If the fuel pressure is less than specified, replace the fuel pump and retest.
6. If the fuel pressure is correct, remove the pressure gauge and hold a graduated container under the hose.
7. Turn the ignition **ON** for 1 minute, then turn the ignition **OFF** and measure the amount of fuel flow.
8. The fuel flow should be as follows:
   1988 Prelude and 1988–89 Accord–25.7 oz. (760cc) or more.
   1989–91 Prelude–20 oz. (600cc) or more.
9. If the fuel flow is as specified, reconnect the fuel cut-off relay and the fuel line. If the fuel flow is less than specified, replace the fuel pump and retest.

NOTE: Check for a clogged fuel filter and/or fuel line before replacing the pump.

Fuel Injected Vehicles

1. Disconnect the negative battery cable.
2. Relieve the fuel system pressure.
3. Attach a suitable fuel pressure gauge to service port of the fuel filter (fuel pipe on 1990–92 Accord).
4. Connect the negative battery cable and disconnect the vacuum hose at the pressure regulator.
5. Start the engine and allow it to idle. The fuel pressure should be 36–41 psi.
6. If the fuel pressure is not as specified, check if there is battery voltage available at the fuel pump. If battery voltage is available, replace the fuel pump. If there is no voltage, check the main relay and wire harness.
7. If the fuel pump is OK, check the following:
   a. If the fuel pressure is higher than specified, inspect for a pinched or clogged fuel return hose or pipe or a faulty pressure regulator.
   b. If the fuel pressure is lower than specified, inspect for a clogged fuel filter, pressure regulator failure or leakage in the fuel line.

REMOVAL & INSTALLATION

Prelude and 1988–89 Accord

The fuel pump is located in the fuel tank.

1. Disconnect the negative battery cable.
2. Relieve the fuel system pressure on fuel injected vehicles.
3. Remove the left maintenance access cover in the trunk.

4. Disconnect the fuel lines.
5. Remove the fuel pump mounting bolts and remove the fuel pump from the fuel tank. If the pump is hard to remove, slightly lower the fuel tank by loosening the fuel tank mounting nuts.
6. Installation is the reverse of the removal procedure. Use a new O-ring when installing the pump. Before installing the maintenance access cover, turn the ignition switch **ON** and check for fuel leaks.

NOTE: When installing the maintenance access cover, make sure the seal is attached to the cover.

Civic, CRX and 1990–92 Accord

The fuel pump is located in the fuel tank.

1. Disconnect the negative battery cable.
2. Relieve the fuel system pressure.
3. Raise and safely support the vehicle.
4. Remove the fuel tank drain bolt and drain the fuel into a suitable container.
5. On Civic and Accord, disconnect the fuel pump electrical connector in the trunk. On CRX, remove the storage compartment and disconnect the fuel pump electrical connector. On Civic Wagon, remove the rear seat and disconnect the fuel pump electrical connector.
6. On Civic Wagon with 4WD, remove the driveshaft from the rear differential and the exhaust pipe and muffler.
7. Remove the 2-way valve cover and fuel hose protector.
8. Disconnect the fuel lines.
9. Place a suitable support under the fuel tank.
10. Remove the fuel tank strap nuts and let the straps hang.
11. Remove the fuel tank.
12. Remove the fuel pump mounting nuts and the fuel pump.

To install:

13. Install the pump assembly into the tank and make sure the unit is properly sealed.
14. Install the tank into the vehicle and tighten the strap nuts. Connect the hoses and wiring.
15. On 4WD vehicles, connect the driveshaft and the exhaust system.
16. Put about 3 gallons of fuel into the tank and test the system.

Carburetor

REMOVAL & INSTALLATION

1988–89 Accord

1. Disconnect the negative battery cable.
2. Remove the air cleaner cover.
3. Disconnect the fresh air and hot air hoses from the air cleaner.
4. Disconnect the vacuum lines from the air cleaner and mark their positions for proper reassembly. Disconnect the breather hose from the valve cover.
5. Remove the air cleaner mounting nuts and the air cleaner.
6. Disconnect the vacuum lines and electrical connectors from the carburetor and mark their positions for proper reassembly.
7. Disconnect the throttle cable.
8. Disconnect and plug the fuel line.
9. Remove the carburetor mounting nuts and the carburetor.
10. Installation is the reverse of the removal procedure. Tighten the carburetor mounting nuts to 17 ft. lbs. (24 Nm).

Prelude

1. Disconnect the negative battery cable.
2. Disconnect the fresh air intake duct and hot air intake hose from the air cleaner cover.
3. Disconnect the vacuum hose from the hot air intake control diaphragm and remove the air cleaner cover and element.
4. Disconnect the breather hose from the valve cover. Disconnect the vacuum lines from the air cleaner base and mark their positions for proper reassembly.
5. Disconnect the electrical connectors from the air cleaner base and remove the bolts from the air cleaner base.
6. Remove the retaining nuts, air screens, flanges and the air cleaner base.
7. Disconnect all vacuum hoses and electrical connectors and mark their position for proper reassembly.
8. Disconnect the throttle cable and the fuel line.
9. Loosen the insulator bands and remove the carburetor.
10. Installation is the reverse of the removal procedure.

IDLE SPEED ADJUSTMENT

Except 1988–89 Accord With Automatic Transaxle

NOTE: The carburetors must be properly synchronized on Prelude before making idle speed adjustments.

1. Start the engine and warm it up to normal operating temperature. The cooling fan must come on at least once.
2. Disconnect the vacuum hose from the intake air control diaphragm and clamp the hose end.
3. Connect a tachometer According to the manufacturers instructions.
4. On 1988–91 Prelude, make sure the fast idle lever is not seated against the fast idle cam. If it is, replace the left carburetor.
5. Check the idle speed with the headlights, heater blower, rear window defroster, cooling fan and air conditioner **OFF**. Check the underhood emission label for idle speed specification.
6. Adjust the idle speed, if necessary, by turning the throttle stop screw.
7. If the idle speed is excessively high on Accord with manual transaxle, check the dashpot system as follows:
   a. Disconnect the vacuum hose from the throttle controller and check for vacuum. If there is no vacuum, check the vacuum line for leaks or blockage and replace as necessary. If there is vacuum, replace the throttle controller and retest.
   b. With the engine idling, disconnect the vacuum hose from the throttle controller. The engine speed should rise to 1400–2000 rpm.
   c. If the engine speed is not within 1400–2000 rpm, loosen the locknut and adjust by turning the adjusting nut.
   d. If the engine speed does not change, check the throttle controller linkage for free movement. If there is no problem, replace the throttle controller and retest.
9. If the engine speed is excessively high on 1988–91 Prelude, check the throttle control as follows:
   a. Disconnect the vacuum hose from the throttle controller and check the engine speed. The engine speed should be 1700–2700 rpm on manual transaxle equipped vehicles and 1400 – 2400 rpm on automatic transaxle equipped vehicles.
   b. If the engine speed is excessively high, adjust the engine speed by bending the tab.
   c. If the engine speed does not change, connect a suitable vacuum pump to the throttle control vacuum hose and check the vacuum. There should be vacuum.
   d. If there is no vacuum, check the vacuum hose for proper connection, cracks or a bad check valve and replace as necessary. If there is vacuum, replace the throttle controller and retest.
   e. Reconnect the vacuum hose to the throttle controller and check the idle speed.

1988–89 Accord With Automatic Transaxle

1. Start the engine and warm it up to normal operating temperature. The cooling fan must come on at least once.

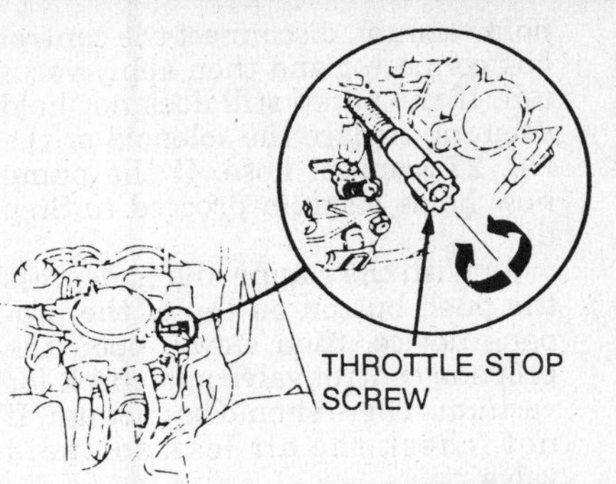

Throttle stop screw location – 1988–89 Accord

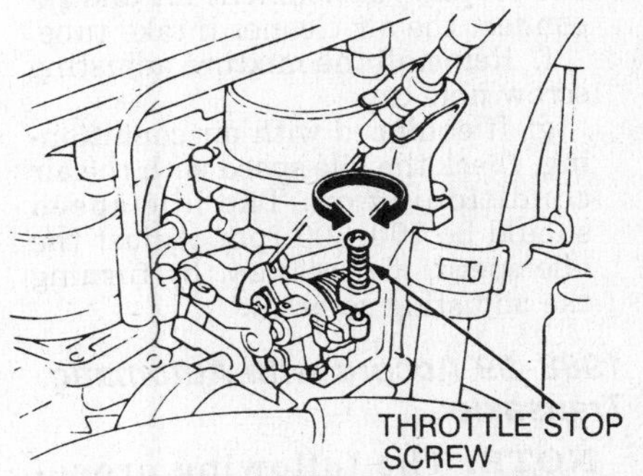

Throttle stop screw location – 1988–91 Prelude

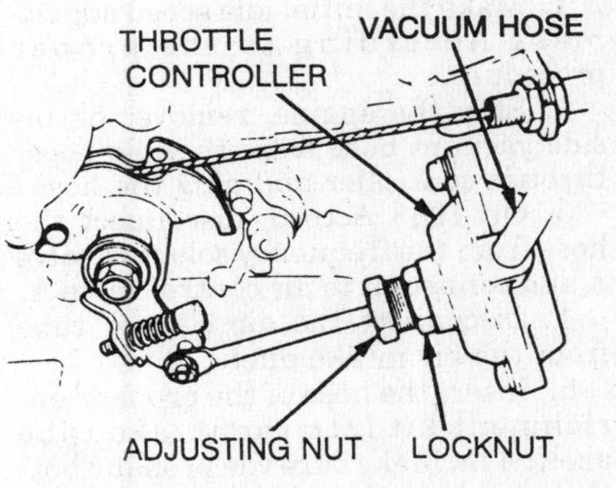

Throttle controller – 1988–89 Accord

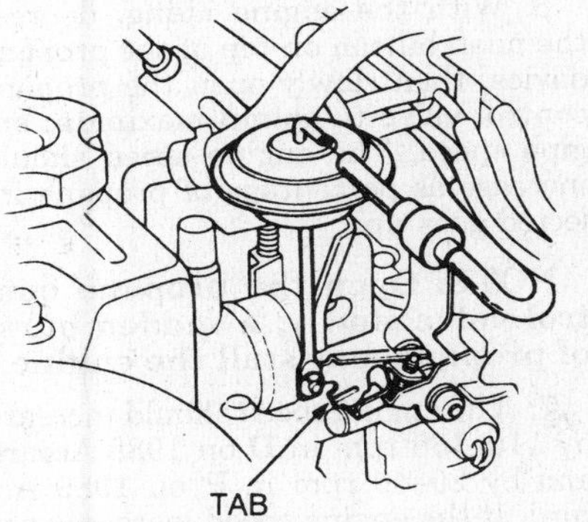

Throttle controller adjusting tab – 1988–91 Prelude

2. Disconnect the vacuum hose from the intake air control diaphragm and clamp the hose end.

3. Connect a tachometer according to the manufacturers instructions.

4. On 1988 Accord, remove the air filter from frequency solenoid valve C and plug the opening in the solenoid valve.

5. On 1989 Accord, disconnect the vacuum hose from the 2-way joint between the frequency solenoid valve C and the vacuum hose manifold and plug the vacuum hose. Disconnect the vacuum hose from the 2-way joint between the frequency solenoid valve C and the throttle controller and plug the hose of the frequency solenoid valve side. Connect a vacuum pump to the hose of the throttle controller and apply 20 in. Hg.

6. Turn back the idle control screw until the end is flush with the bracket. With the headlights, heater blower, rear window defogger, cooling fan and air conditioner off and the transaxle in **N** or **P**, lower the idle speed as much as possible by turning the throttle stop screw.

7. Adjust the idle speed by turning the idle control screw to 580–680 rpm.

8. Adjust the idle speed by turning the throttle stop screw to 650–750 rpm.

9. With the transaxle in any gear except **P** or **N**, adjust the idle speed by turning adjusting screw A. The idle speed should be 625–725 rpm at high altitude and 650–750 rpm at low altitude.

10. Shift the transaxle to **N** or **P**.

11. If equipped with air conditioning, adjust the idle speed by turning adjusting screw B to 650–750 rpm with the air conditioning on.

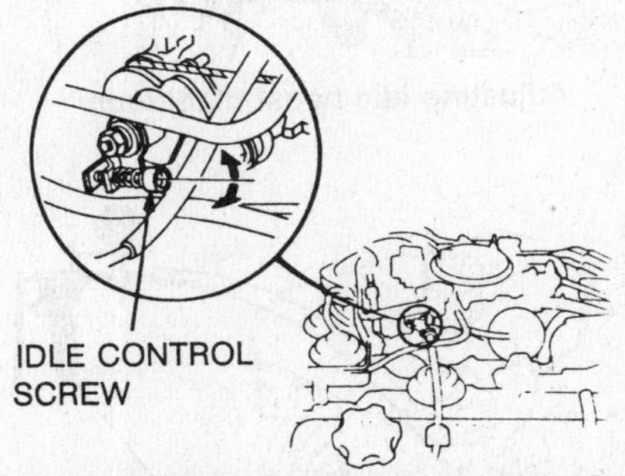

Idle control screw location – 1988–89 Accord with automatic transaxle

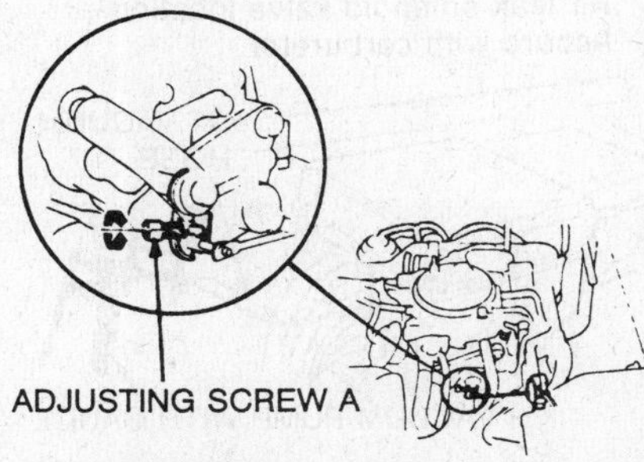

Adjusting screw A location – 1988–89 Accord with automatic transaxle

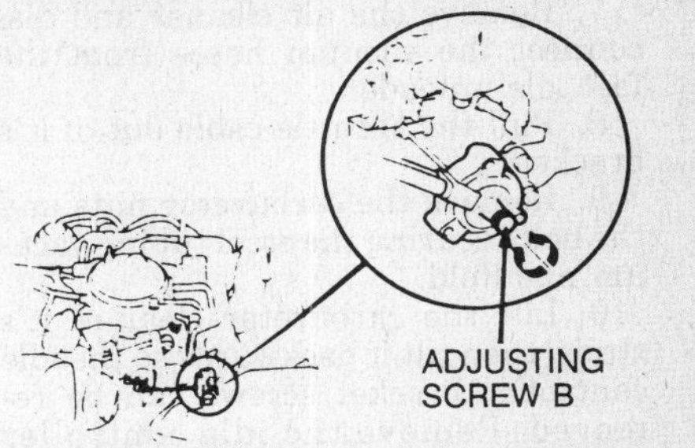

Adjusting screw B location – 1988–89 Accord with automatic transaxle

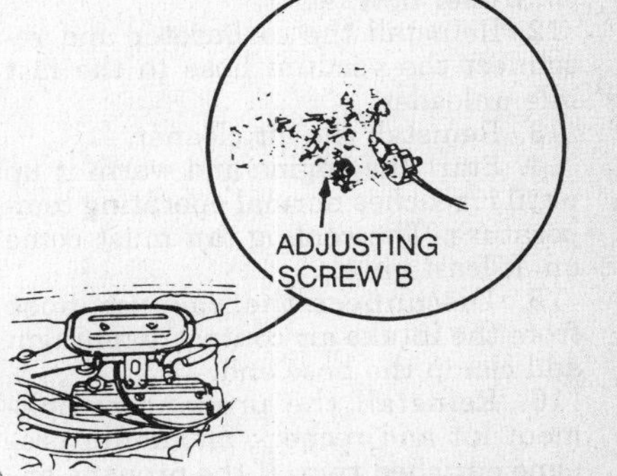

Adjusting screw B location – 1988–89 Accord with manual transaxle

IDLE MIXTURE ADJUSTMENT

1988–89 Accord With Manual Transaxle

NOTE: The following procedure requires a propane enrichment kit.

1. Make the initial idle speed adjustment according to the proper procedure.

2. Disconnect the air cleaner tube from the air intake duct.

3. Insert the hose of the propane enrichment kit into the intake tube about 4 inches. Make sure the propane bottle has adequate gas before beginning the test.

4. With the engine idling, depress the button on top of the propane device, then slowly open the propane control valve to obtain maximum engine speed. Engine speed should increase as percentage of propane injected goes up.

NOTE: Open the propane control valve slowly; a sudden burst of propane may stall the engine.

5. The engine speed should increase by 39–70 rpm. If not, the mixture is improperly adjusted. Proceed to Step 6. If the engine speed increases as specified, proceed to Step 17.

6. Close the propane control valve and shut **OFF** the engine.

7. Remove the air cleaner and disconnect the vacuum hoses from the fast idle unloader.
8. Pull the throttle cable out of it's bracket.
9. Remove the carburetor nuts and the bolt securing the steel tubing vacuum manifold.
10. Lift the carburetor clear of it's studs, then tilt it backwards so the idle controller bracket screws can be removed. Remove the idle controller bracket.
11. Remove the mixture adjusting screw hole cap, then reinstall the idle controller bracket.
12. Reinstall the carburetor and reconnect the vacuum hose to the fast idle unloader.
13. Reinstall the air cleaner.
14. Start the engine and warm it up until it reaches normal operating temperature. The cooling fan must come on at least once.
15. Disconnect the vacuum hose from the intake air control diaphragm and clamp the hose end.
16. Reinstall the propane enrichment kit and recheck maximum propane enriched rpm. If the propane enriched speed is too low, the mixture is too rich: turn the mixture screw ¼ turn clockwise and recheck. If the propane enriched speed is too high, the mixture is too lean: turn the mixture screw ¼ turn counterclockwise and recheck.
17. Close the propane control valve and recheck the idle speed. It should be as specified on the underhood emission label.

NOTE: Run the engine at 2500 rpm for 10 seconds to stabilize condition.

18. If the idle speed is as specified, proceed to Step 20. If the idle speed is not as specified, proceed to Step 19.
19. Recheck the idle speed and if necessary, adjust by turning the throttle stop screw. Repeat Steps 16 and 17.
20. Finish adjustments through the following procedure:
   a. The intake air temperature must be above 149°F (65°C).
   b. Disconnect the vacuum hose from the air suction valve and plug the hose.
   c. On 1988 Accord, open the control box lid. Disconnect the lower vacuum hose of the air leak solenoid valve, located between the solenoid valve and the air filter, from the air filter and connect a vacuum gauge to the hose. On 1989 Accord, disconnect the No. 27 vacuum hose from the pipe and plug the pipe. Attach a vacuum pump/gauge to the hose and apply vacuum. The vacuum pump should hold vacuum. If it does, proceed to Step d. If the pump does not

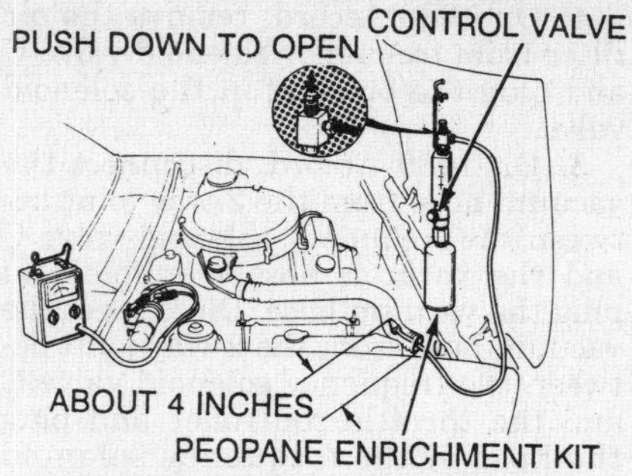

Mixture adjustment using propane enrichment

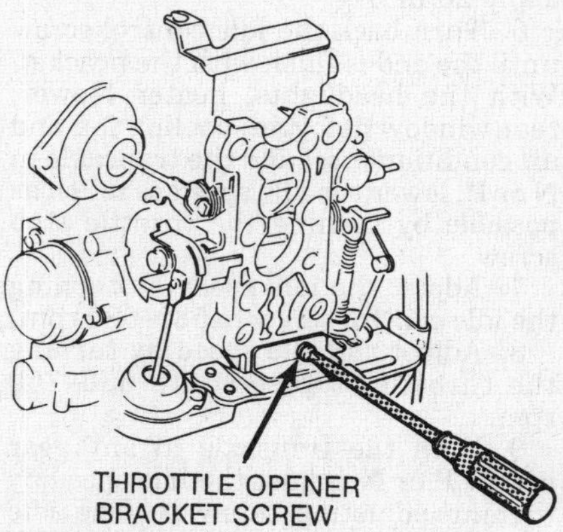

Throttle opener bracket—Accord with carburetor

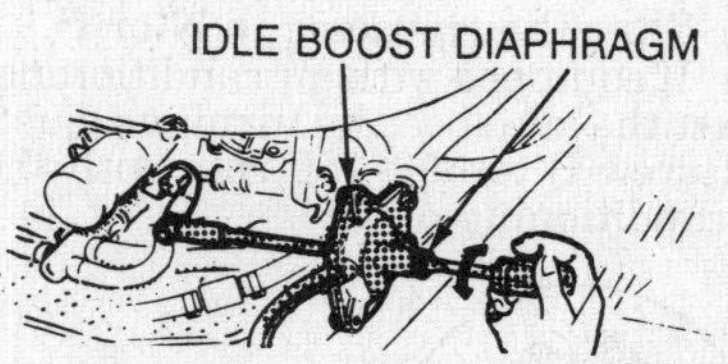

Adjusting idle boost diaphragm

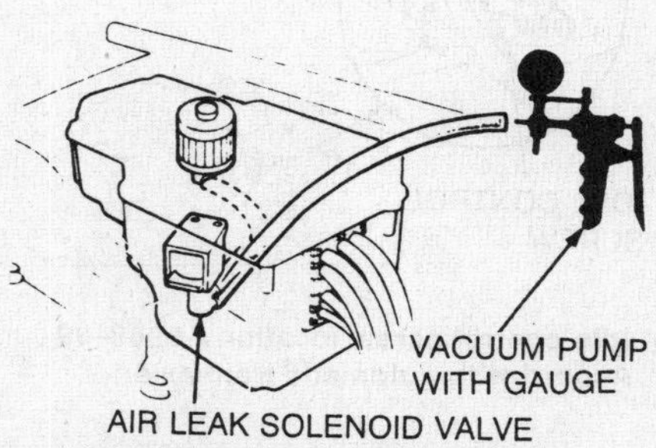

Air leak solenoid valve location—Accord with carburetor

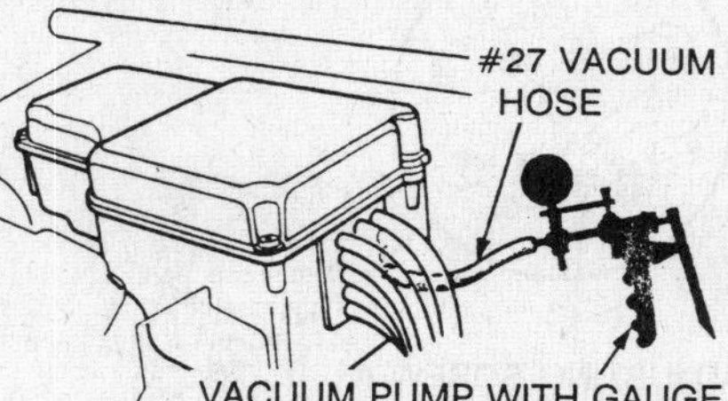

No. 27 vacuum hose location—1989 Accord with carburetor

hold vacuum, disconnect the control box connector and then apply vacuum. If the pump still does not hold vacuum, replace the solenoid or the No. 27 vacuum hose. If the pump now holds vacuum proceed to Step d.
   d. With the engine idling, depress the push button on top of the propane device, then slowly open the propane control valve and check for vacuum. There should be vacuum. If not, check the air leak solenoid valve.
   e. Reconnect all hoses, remove the propane enrichment kit and reconnect the air cleaner intake tube.
   f. Reinstall the mixture adjusting screw hole cap.
   g. If equipped with air conditioning, check the idle speed with the air conditioning on. The idle speed should be 700–800 rpm. Adjust the idle speed, if necessary, by turning the adjusting screw B.

1988–89 Accord with Automatic Transaxle

NOTE: The following procedure requires a propane enrichment kit.

1. Make the initial idle speed adjustment According to the proper procedure.
2. Stop the engine, remove the inside vacuum hose from the idle boost throttle controller and plug the hose.
3. On 1988 Accord, disconnect the hose from the frequency solenoid valve A and connect it to air control valve A.
4. Disconnect the air cleaner tube from the air intake duct.
5. Insert the hose of the propane enrichment kit into the intake tube about 4 in. Make sure the propane bottle has adequate gas before beginning the test.
6. With the engine idling, depress the push button on top of the propane device, then slowly open the propane control valve to obtain maximum engine speed. The engine speed should increase as percentage of propane injected goes up.

NOTE: Open the propane control valve slowly; a sudden burst of propane may stall the engine.

7. The engine speed should increase by 110–160 rpm in **D** on 1988 Accord and by 20–40 rpm in **P** on 1989 Accord. If the engine speed increases per specification, proceed to Step 19. If the engine speed does not increase per specification, proceed to Step 8.
8. Close the propane control valve and shut **OFF** the engine.
9. Remove the air cleaner and disconnect the vacuum hoses from the fast idle unloader.

10. Pull the throttle cable out of it's bracket.
11. Remove the carburetor nuts and the bolt securing the steel tubing vacuum manifold.
12. Lift the carburetor clear of it's studs, then tilt it backwards so the idle controller bracket screws can be removed. Remove the idle controller bracket.
13. Remove the mixture adjusting screw hole cap, then reinstall the idle controller bracket.
14. Reinstall the carburetor and reconnect the vacuum hose to the fast idle unloader.
15. Reinstall the air cleaner.
16. Start the engine and warm it up until it reaches normal operating temper ature. The cooling fan must come on at least once.
17. Disconnect the vacuum hose from the intake air control diaphragm and clamp the hose end.
18. Reinstall the propane enrichment kit and recheck maximum propane enriched rpm. If the propane enriched speed is too low, the mixture is too rich: turn the mixture screw ¼ turn clockwise and recheck. If the propane enriched speed is too high, the mixture is too lean; turn the mixture screw ¼ turn counterclockwise and recheck.
19. Stop the engine, close the propane control valve, remove all plugs and reconnect all hoses.
20. Restart the engine and recheck the idle speed.

NOTE: Raise the engine speed to 2500 rpm 2–3 times in 10 seconds and then check the idle speed.

21. The idle speed should be 680–780 rpm in **N** or **P**. If the idle speed is as specified, proceed to Step 22. If the idle speed is not as specified, repeat the initial idle setting procedure.
22. The intake air temperature must be above 149°F (65°C).
23. Disconnect the vacuum hose from the air suction valve and plug the hose.
24. On 1988 Accord, open the control box lid. Disconnect the lower vacuum hose of the air leak solenoid valve, located between the solenoid valve and the air filter, from the air filter and connect a vacuum gauge to the hose. On 1989 Accord, disconnect the No. 27 vacuum hose from the pipe and plug the pipe. Attach a vacuum pump/gauge to the hose and apply vacuum. The vacuum pump should hold vacuum. If it does, proceed to Step 25. If the pump does not hold vacuum, disconnect the control box connector and then apply vacuum. If the pump still does not hold vacuum, replace the solenoid or the No. 27 vacuum hose. If the pump now holds vacuum proceed to Step 25.
25. With the engine idling, depress the push button on top of the propane device, then slowly open the propane control valve and check for vacuum. There should be vacuum. If not, check the air leak solenoid valve.
26. Reconnect all hoses, remove the propane enrichment kit and reconnect the air cleaner intake tube.
27. Reinstall the mixture adjusting screw hole cap.
28. Recheck the idle speed with the automatic transaxle shift lever in gear. The idle speed should be 680–780 rpm.
29. Recheck the idle speed with the air conditioning on and with the shift lever in **P** or **N** position. The idle speed should be 700–800 rpm.
30. Recheck the idle speed with the air conditioning on and the shift lever in gear. The idle speed should be 700–800 rpm.

1988–91 Prelude with Carburetors

NOTE: The following procedure requires a propane enrichment kit. The carburetor must be properly adjusted before making mixture adjustments.

1. Make the initial idle speed adjustment According to the proper procedure.
2. Disconnect the 2P connector from the EACV and disconnect the hose from the vacuum hose manifold, then cap the hose end.

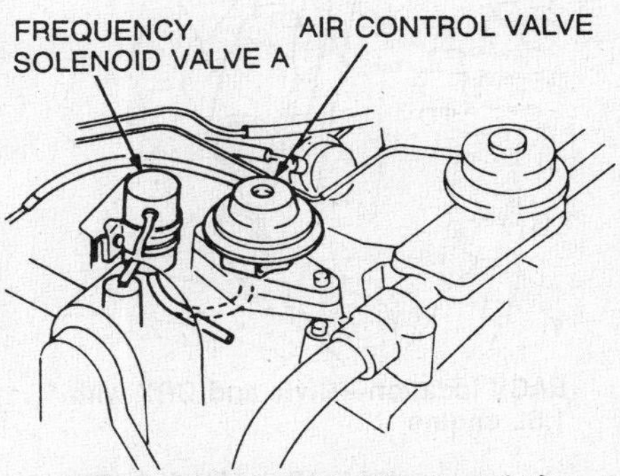

Frequency solenoid valve A and air control valve A location—1988 Accord

3. Disconnect the cap from the vacuum hose manifold. If equipped with air conditioning, disconnect the vacuum hose from the vacuum hose manifold. Disconnect the air cleaner intake tube from the air intake duct. Note the engine speed when starting the engine and that the idle speed is stable.
4. Insert the propane enrichment hose into the opening of the intake tube about 4 in.
5. With the engine idling, depress the push button on top of the propane device, then slowly open the propane control valve to obtain maximum engine speed. The engine speed should increase as the percentage of propane goes up.

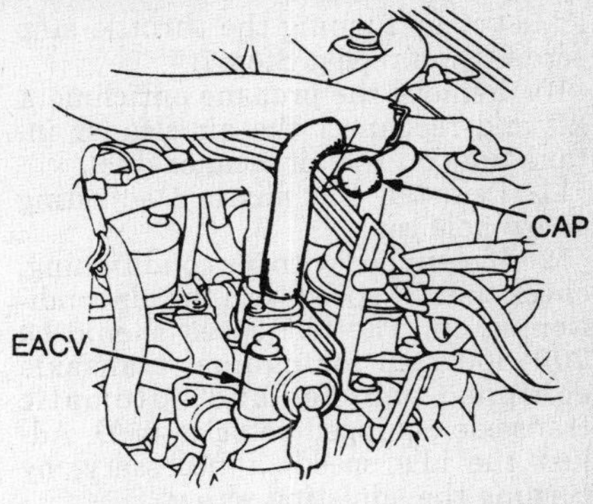

EACV location—1988–91 Prelude

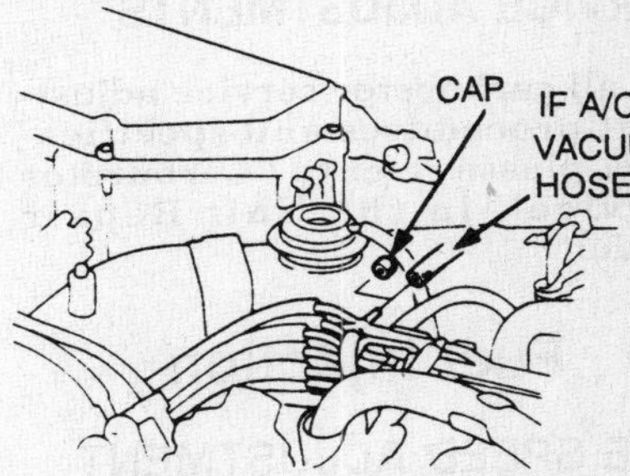

Vacuum hose manifold location—1988–91 Prelude

NOTE: Open the propane control valve slowly; a sudden burst of propane may stall the engine.

6. The idle rpm should increase by 150–190 rpm on vehicles with manual transaxle and 40–60 rpm on automatic transaxle vehicles in **D**. If the engine speed does not increase per specification, the mixture is improperly adjusted. Proceed to Step 7. If the engine speed increases as specified, proceed to Step 8.
7. Remove the mixture adjusting screw hole caps and recheck the maximum propane enriched rpm. If the propane enriched speed is too low, the mixture is too rich; turn both mixture screws ¼ turn clockwise and recheck. If the propane enriched speed is too high, the mixture is too lean; turn both mixture screws ¼ turn counterclockwise and recheck.
8. Reconnect the connector and cap or hose. Close the propane control valve.
9. Remove the EFI-ECU fuse for 10 seconds to reset the control unit and recheck the idle speed. The idle speed should be as specified on the underhood emission label. If the idle speed is as specified, proceed to Step 10. If the idle speed is not as specified,

adjust it by turning the throttle stop screw, then repeat Step 7.

10. Remove the propane enrichment kit and reconnect the air cleaner intake tube on the air cleaner duct.

11. Reinstall the mixture adjusting screw hole cap.

12. If equipped with air conditioning, check the idle speed with the air conditioning on. The idle speed should be 700–800 rpm for manual transaxle equipped vehicles and automatic transaxle equipped vehicles in **D**. Adjust the idle speed, if necessary, by turning the adjusting screw.

SERVICE ADJUSTMENTS

For all carburetor service adjustment procedures and specifications, please refer to "Carburetor Service" in the Unit Repair section.

Fuel Injection

IDLE SPEED ADJUSTMENT

1. Start the engine and warm it up to normal operating temperature. The cooling fan must come on at least once.

2. Connect a tachometer according to the manufacturers instructions.

3. Disconnect the 2P connector from the Electronic Air Control Valve (EACV).

4. Set the front wheels in the straight ahead position and check the idle speed in no-load conditions in which the headlights, blower fan, rear defroster, cooling fan and air conditioner are not operating. Vehicles with automatic transaxle should be checked with the transaxle in **N** or **P**.

5. The idle speed should be as follows:

Civic and CRX with 1.5L engine – 575–675 rpm.

Civic and CRX with 1.6L engine – 500–600 rpm.

CRX HF – 450–550 rpm.

Prelude – 600–700 rpm.

1988–89 Accord – 600–700 rpm.

1990–92 Accord – 550–650 rpm.

6. Adjust the idle speed, if necessary, by turning the adjusting screw. If the idle speed is excessively high on Civic and CRX, check the throttle control system.

7. Turn the ignition switch **OFF**.

8. Reconnect the 2P connector on the EACV.

9. On Civic and CRX, remove the HAZARD fuse in the main fuse box for 10 seconds to reset the ECU. On 1988–89 Accord, remove the No. 11 (10A) fuse in the underhood relay box for 10 seconds to reset the ECU. On 1990–92 Accord, remove the BACK-UP fuse in the underhood relay box for 10 sec-

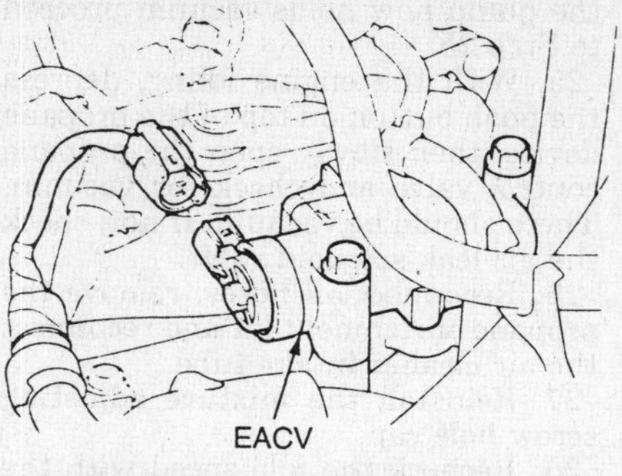

EACV location – Civic and CRX with 1.5L engine

Idle adjusting screw location – Civic and CRX with 1.5L fuel injected engine

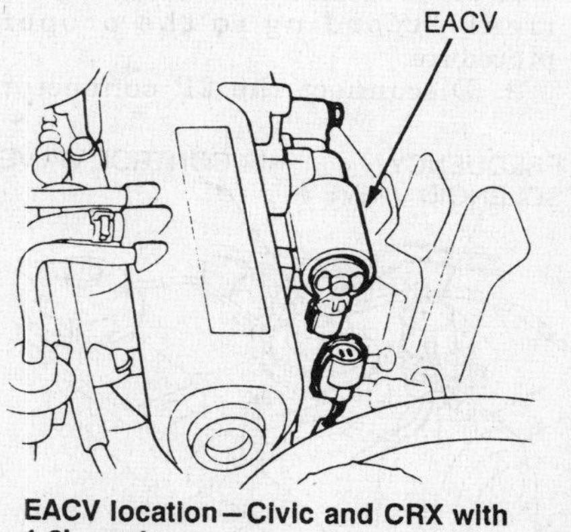

EACV location – Civic and CRX with 1.6L engine

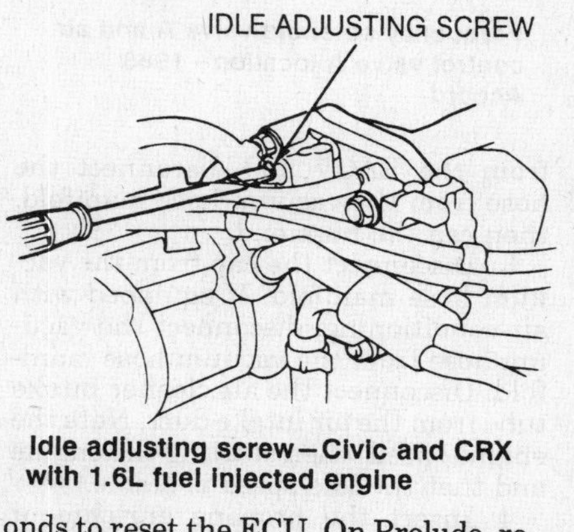

Idle adjusting screw – Civic and CRX with 1.6L fuel injected engine

onds to reset the ECU. On Prelude, remove the CLOCK (10A) fuse from the underhood relay box for 10 seconds to reset the ECU.

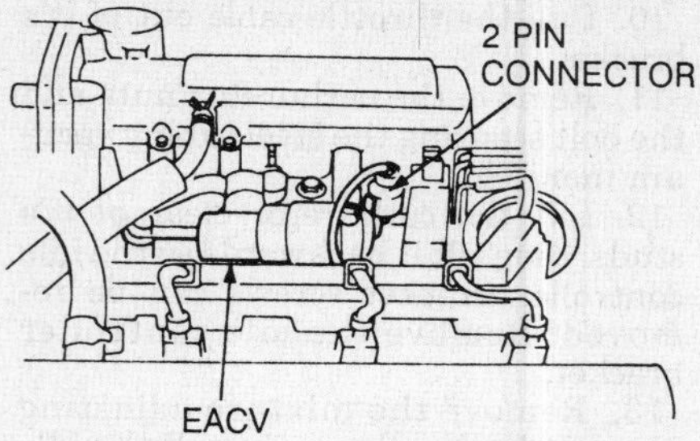

EACV location – 1988–89 Accord

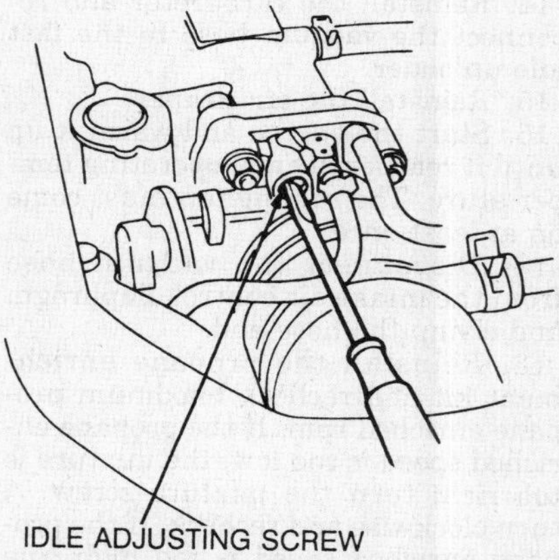

Idle adjusting screw location – 1988–89 Accord with fuel injection

10. Set the front wheels in the straight ahead position and restart the engine. Idle the engine with no-load conditions in which the headlights, blower fan, rear window defogger, cooling fan and air conditioner are not operating for 1 minute. Recheck the idle speed. Vehicles with automatic transaxle should be checked with the transaxle in **N** or **P**.

11. The idle speed should be 700–800 rpm for all except CRX HF and 1990–92 Accord. The idle speed on CRX HF should be 550–650 rpm for 49 state vehicles and 600–700 rpm for California vehicles. The idle speed for 1990–92 Accord should be 650–750 rpm.

12. Idle the engine for 1 minute with the headlights on high and the rear window defogger on. On all vehicles with automatic transaxle except 1990–92 Accord, place the transaxle in gear. The idle speed should be as follows:

1988 Civic and CRX – 730 – 830 rpm

1989–92 Civic and CRX standard – 750–850 rpm.

1989–92 Civic Si and CRX Si – 730–830 rpm.

1989–92 CRX HF 49 state – 600–700 rpm.

1988–92 CRX HF California – 700–800 rpm.

1988–89 Accord – 700–800 rpm

1990–92 Accord – 720–820 rpm, automatic transaxle in **N** or Prelude – 700–800 rpm.
P.

13. Idle the engine for 1 minute with the heater fan switch on high and the air conditioner on, then check the idle speed. Vehicles with automatic transaxle should be checked with the transaxle in **N** or **P**.

14. The idle speed should be as follows:

1988 Civic and CRX except HF–730–830 rpm.

1988–92 CRX HF–700–800 rpm.

1989–92 Civic, CRX except HF, 1.5L Civic Wagon and 1.6L Civic Wagon with manual transaxle–750–850 rpm.

1.6L Civic Wagon with automatic transaxle–770–870 rpm.

1988–89 Accord and Prelude–700–800 rpm.

1990–92 Accord and Prelude–720–820 rpm.

Fuel Injector

REMOVAL & INSTALLATION

Except 1.5L Engine with Dual Point Fuel Injection

1. Disconnect the negative battery cable.
2. Relieve the fuel pressure.
3. Disconnect the electrical connectors from the fuel injectors.
4. Disconnect the vacuum hose and fuel return hose from the fuel pressure regulator.

NOTE: Place a rag or shop towel over the hose and tube before disconnecting them.

5. Disconnect the fuel line from the fuel pipe.
6. On 1988–91 Prelude, disconnect the EACV from the intake manifold.
7. Remove the fuel pipe retainer nuts and the fuel pipe.
8. Remove the injectors from the intake manifold.

To install:

9. Slide new cushion rings onto the injectors.
10. Coat new O-rings with clean engine oil and install them on the injectors.
11. Insert the injectors into the fuel pipe first.
12. Coat new seal rings with clean engine oil and press them into the intake manifold.
13. Install the injectors and fuel pipe assembly in the intake manifold.

NOTE: To prevent damage to the O-rings, install the injectors in the fuel pipe first, then install them in the intake manifold.

14. Align the center line marking on the fuel injector with the mark on the fuel pipe.
15. Install and tighten the fuel pipe retainer nuts.

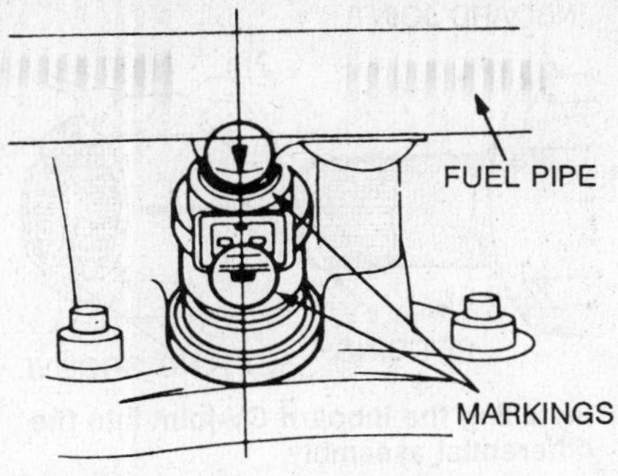

Aligning the fuel injector and fuel pipe marks

16. Connect the fuel line to the fuel pipe and the vacuum hose and fuel return line to the pressure regulator.
17. Connect the electrical connectors to the injectors.
18. Connect the negative battery cable and turn the ignition switch **ON** for 2 seconds, but do not operate the starter. Repeat 2–3 times and check for fuel leaks.

1.5L Engine with Dual Point Fuel Injection

1. Disconnect the negative battery cable.
2. Relieve the fuel pressure.
3. Remove the air intake chamber.
4. Disconnect the electrical connector from the fuel injector.
5. Loosen the screws and remove the injector from the throttle body. Place a rag or shop towel over the throttle body after removal.
6. Installation is the reverse of the removal procedure. Use new O-rings and coat them with engine oil prior to installation. After the injector is inserted, make sure it turns smoothly approximately 30 degrees.
7. Before installing the air intake chamber, connect the negative battery cable and turn the ignition switch **ON** for approximately 2 seconds. Repeat 2–3 times and check for fuel leaks.

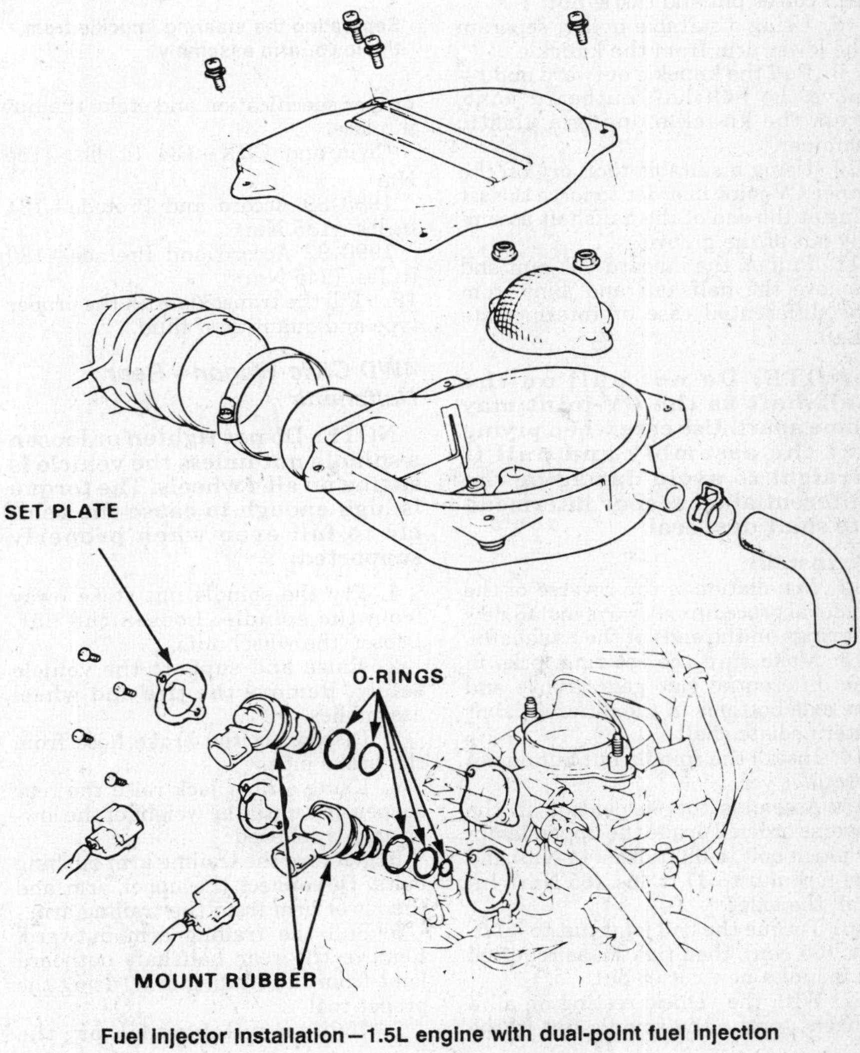

Fuel injector installation–1.5L engine with dual-point fuel injection

DRIVE AXLE

Halfshaft

REMOVAL & INSTALLATION

Front Halfshafts

NOTE: Do not tighten or loosen a spindle nut unless the vehicle is sitting on all 4 wheels. The torque is high enough to cause the vehicle to fall even when properly supported.

1. Loosen the front spindle nut.
2. Raise and safely support the vehicle.
3. Remove the front wheel and tire assemblies and the spindle nut.
4. Drain the transaxle fluid and replace the drain plug.
5. Remove the damper fork nut and damper pinch bolt.
6. Remove the damper fork.
7. Remove the knuckle-to-lower arm cotter pin and castle nut.
8. Using a suitable puller, separate the lower arm from the knuckle.
9. Pull the knuckle outward and remove the halfshaft outboard joint, from the knuckle, using a plastic hammer.
10. Using a suitable tool, pry on the inner CV-joint in order to force the set ring at the end of the halfshaft assembly out of the groove.
11. Pull on the inboard CV-joint and remove the halfshaft and joint from the differential case or intermediate shaft.

NOTE: Do not pull on the halfshaft as the CV-joint may come apart. Use care when prying out the assembly and pull it straight to avoid damaging the differential oil seal or intermediate shaft dust seal.

To install:

12. Installation is the reverse of the removal procedure. Always install new set rings on the ends of the halfshafts.
13. Make sure the set ring locks in the differential side gear groove and the axle bottoms in the differential or intermediate shaft.
14. Install the spindle nut but do not torque it yet.
15. Assemble the suspension in the reverse order. Torque the upper damper pinch bolt to 32 ft. lbs. (44 Nm) and the fork nut to 47 ft. lbs. (65 Nm). Install the axle.
16. Torque the ball joint nut to 40 ft. lbs. (55 Nm), then tighten as required to install a new cotter pin.
17. With the vehicle resting on all 4 wheels, torque the spindle nut to the proper specification and stake the nut in place.

Civic and CRX – 134 ft. lbs. (185 Nm)
1988–89 Accord and Prelude – 134 ft. lbs. (185 Nm)
1990–92 Accord and Prelude – 180 ft. lbs. (245 Nm)

18. Fill the transaxle with the proper type and quantity of fluid.

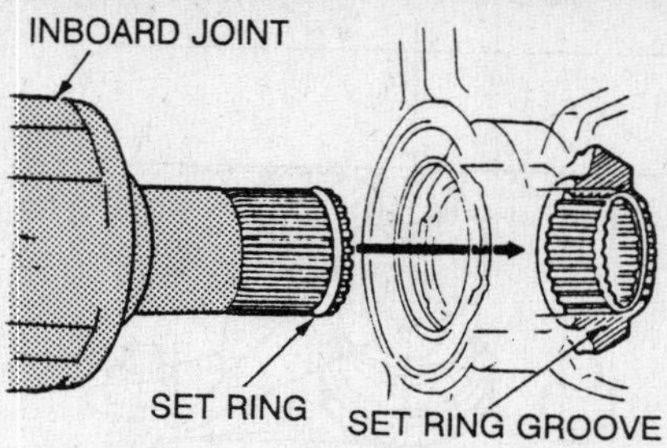

Installing the inboard CV-joint into the differential assembly

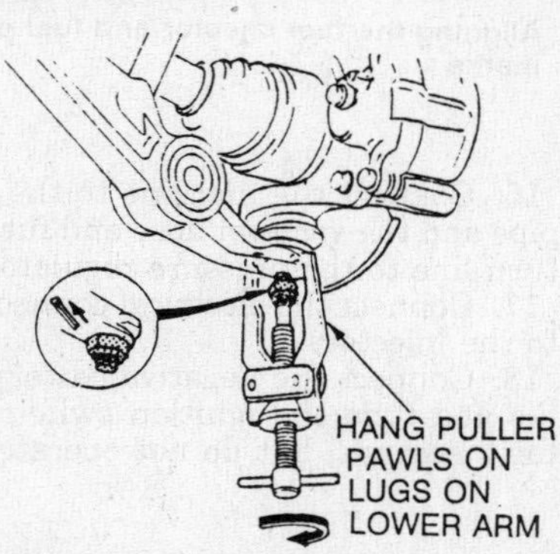

Separating the steering knuckle from the lower arm assembly

4WD Civic Wagon – Rear Halfshafts

NOTE: Do not tighten or loosen a spindle nut unless the vehicle is sitting on all 4 wheels. The torque is high enough to cause the vehicle to fall even when properly supported.

1. Pry the spindle nut stake away from the spindle. Loosen the nut. Loosen the wheel nuts.
2. Raise and support the vehicle safely. Remove the tire and wheel assemblies.
3. Disconnect the brake hose from the brake pipe.
4. Using a floor jack raise the rear suspension until the weight of the lower arm is relieved.
5. Remove the trailing arm bushing bolts. Disconnect the upper arm and the lower arm from the trailing arm.
6. Pull the trailing arm outward. Remove the rear halfshaft outboard joint from the trailing arm, using the proper tool.
7. Using a suitable tool, pry the halfshaft assembly to force the set ring at the halfshaft end past the groove.
8. Pull the inboard joint and remove the halfshaft and the CV-joint from the differential case as an assembly.

To install:

9. Always install a new set ring on the end of the axle. Make sure the set ring locks in the differential side gear groove and the CV-joint subaxle bottoms in the differential.
10. Reassemble the suspension and torque the upper and lower arm bushing bolts to 40 ft. lbs. (55 Nm) and the trailing arm bushing bolts to 47 ft. lbs. (65 Nm).
11. When the vehicle is resting on all 4 wheels, torque the spindle nut to 134 ft. lbs. (185 Nm).

CV-Boot

REMOVAL & INSTALLATION

1. Raise and safely support the vehicle.
2. Remove the halfshaft.
3. If replacing the inboard CV-joint boot, perform the following procedure:
   a. Place the halfshaft in a suitable holding fixture where it will remain in position during disassembly.
   b. Remove the boot bands. If the boot bands are the welded type, they must be cut to be removed. After removing the bands, push the CV-joint boot away from the end of the halfshaft to gain access to the CV-joint.
   c. Remove the inboard CV-joint. Mark the components during disassembly to ensure proper positioning during reassembly.
   d. Remove the CV-joint boot.
   e. Installation is the reverse of the removal procedure. Check the CV-joint components for wear prior to installation and replace as necessary.
   f. Thoroughly pack the inboard CV-joint and boot with molybdenum disulfide grease. Always install new boot bands.
4. If replacing the outboard CV-joint boot, perform the following procedure:
   a. Place the halfshaft in a suitable holding fixture where it will remain in position during disassembly.
   b. Remove the inboard CV-joint and boot. Do not try to remove or disassemble the outboard CV-joint.
   c. Remove the boot bands and the outboard CV-joint boot.
   d. Installation is the reverse of the removal procedure. Thoroughly pack the outboard CV-joint boot with molybdenum disulfide grease. Always install new boot bands.
5. Install the halfshaft in the reverse order of the removal procedure.

Driveshaft and U-Joints

REMOVAL & INSTALLATION

Intermediate Shaft

4WD CIVIC WAGON, PRELUDE AND 1990–92 ACCORD

1. Raise the vehicle and support it safely.
2. Drain the transaxle fluid and replace the drain plug.
3. Remove the driver's side halfshaft.
4. Remove the three 10mm bolts.
5. Lower the bearing support close to the steering gearbox and remove the intermediate shaft from the differential.

NOTE: To avoid damage to the differential oil seal, hold the intermediate shaft in a horizontal position until it is clear of the differential.

6. Installation is the reverse of the removal procedure. Tighten the three 10mm bolts to 29 ft. lbs. (40 Nm). Fill the transaxle with the proper type and quantity of fluid.

Rear Driveshafts

4WD CIVIC WAGON

1. Raise the vehicle and support it safely.
2. Mark the position of the driveshafts in relation to the flanges for reassembly.
3. Remove the No. 1 driveshaft protector.
4. Disconnect the No. 1 driveshaft and viscous coupling.
5. Remove the front center bearing support from the body.
6. Remove the No. 1 driveshaft by disconnecting the U-joint.
7. Remove the No. 3 driveshaft protector.
8. Disconnect the No. 3 driveshaft and rear differential.
9. Remove the rear center bearing support from the body, then remove the viscous coupling and No. 3 driveshaft.

To install:

10. Install the rear center bearing support on the frame and tighten the bolts to 29 ft. lbs. (40 Nm).
11. Temporarily connect the No. 3 driveshaft and rear differential using the 12-point bolts and yoke nuts. Tighten the bolts to 24 ft. lbs. (33 Nm).
12. Install the No. 3 driveshaft protector and tighten the bolts to 29 ft. lbs. (40 Nm).
13. Temporarily connect the No. 1 driveshaft and front differential using the 12-point bolts and yoke nuts. Tighten the bolts to 24 ft. lbs. (33 Nm).
14. Install the front center bearing support on the frame and tighten the bolts to 29 ft. lbs. (40 Nm). Temporarily connect the No. 1 driveshaft and viscous coupling using 12-point bolts. Tighten the bolts to 24 ft. lbs. (33 Nm).
15. Install the No. 1 driveshaft protector and tighten the bolts to 29 ft. lbs. (40 Nm).

Front Wheel Hub, Knuckle and Bearings

REMOVAL & INSTALLATION

Civic and CRX

NOTE: Do not tighten or loosen a spindle nut unless the vehicle is sitting on all 4 wheels. The torque is high enough to cause the vehicle to fall even when properly supported.

1. Pry the spindle nut stake away from the spindle, then loosen the nut.
2. Raise and safely support the vehicle.
3. Remove the wheel and tire assembly and the spindle nut.
4. Remove the caliper mounting bolts and the caliper. Support the caliper out of the way with a length of wire. Do not let the caliper hang from the brake hose.
5. Remove the 6mm brake disc retaining screws. Screw two 8 × 1.25 × 12mm bolts into the disc to push it away from the hub.

NOTE: Turn each bolt 2 turns at a time to prevent cocking the brake disc.

6. Remove the cotter pin from the tie rod castle nut, then remove the nut. Break the tie rod ball joint using a suitable ball joint remover, then lift the tie rod out of the knuckle.
7. Remove the cotter pin and loosen the lower arm ball joint nut half the length of the joint threads.
8. Separate the ball joint and lower arm using a suitable puller with the pawls applied to the lower arm.

NOTE: Avoid damaging the ball joint boot. If necessary, apply penetrating type lubricant to loosen the ball joint.

9. Remove the knuckle protector.
10. Remove the cotter pin and remove the upper ball pin nut.
11. Separate the upper ball joint and knuckle using a suitable tool.
12. Remove the knuckle and hub by sliding them off the halfshaft.
13. Remove the splash guard screws from the knuckle.
14. Position the knuckle/hub assembly on a hydraulic press. Press the hub from the knuckle using a suitable driver while supporting the knuckle with a suitable base.

NOTE: The bearing must be replaced with a new one after removal.

15. Remove the 76mm snapring and knuckle ring from the knuckle.
16. Press the wheel bearing out of the knuckle using a suitable driver while supporting the knuckle with a suitable base.
17. Remove the outboard bearing inner race from the hub using a suitable bearing puller.

To install:

18. Clean the knuckle and hub thoroughly before reassembly.
19. Press a new wheel bearing into the hub using a suitable driver while supporting the knuckle with a suitable base.
20. Install the 76mm snapring securely in the knuckle groove.
21. Install the splash guard and tighten the screws to 7 ft. lbs. (10 Nm).
22. Place the knuckle into position on the hydraulic press and press onto the hub using a suitable driver. The maximum press load should be 2 tons.
23. Install the front knuckle ring on the knuckle.
24. Install the knuckle/hub assembly onto the vehicle in the reverse order of the removal procedure. Tighten the upper ball pin nut and tie rod nut to 32 ft. lbs. (44 Nm) and the lower ball joint castle nut to 40 ft. lbs. (55 Nm).
25. With all 4 wheels resting on the ground, torque the spindle nut to 134 ft. lbs. (185 Nm).

Prelude and 1988–89 Accord

NOTE: Do not tighten or loosen a spindle nut unless the vehicle is sitting on all 4 wheels. The torque is high enough to cause the vehicle to fall even when properly supported.

1. Pry the spindle nut stake away from the spindle and loosen the nut.
2. Raise and safely support the vehicle.
3. Remove the wheel and tire assembly and the spindle nut.
4. Remove the caliper mounting bolts and the caliper. Support the caliper out of the way with a length of wire. Do not let the caliper hang from the brake hose.
5. Remove the 6mm brake disc retaining screws. Screw two 8 × 1.25 × 12mm bolts into the disc to push it away from the hub.

NOTE: Turn each bolt 2 turns at a time to prevent cocking the brake disc.

6. Remove the cotter pin from the tie rod castle nut, then remove the nut. Break the tie rod ball joint using a suitable ball joint remover, then lift the tie rod out of the knuckle.
7. Remove the cotter pin and loosen the lower arm ball joint nut half the length of the joint threads.
8. Separate the ball joint and lower arm using a suitable puller with the pawls applied to the lower arm.

NOTE: Avoid damaging the ball joint boot. If necessary, apply penetrating type lubricant to loosen the ball joint.

9. Remove the upper ball joint shield, if equipped.
10. Pry off the cotter pin and remove the upper ball joint nut.
11. Separate the upper ball joint and knuckle using a suitable tool.
12. Remove the knuckle and hub by sliding them off the halfshaft.
13. Remove the splash guard screws from the knuckle.
14. Position the knuckle/hub assembly in a hydraulic press. Press the hub from the knuckle using a suitable driver while supporting the knuckle.

NOTE: The bearing must be replaced with a new one after removal.

15. Remove the splash guard and snapring from the knuckle.
16. Press the wheel bearing out of the knuckle using a suitable driver while supporting the knuckle.
17. Remove the outboard bearing inner race from the hub using a suitable bearing puller.

To install:

18. Clean the knuckle and hub thoroughly.
19. Press a new wheel bearing into the knuckle using a suitable driver while supporting the knuckle.
20. Install the snapring.
21. Install the splash shield and tighten the screws to 4 ft. lbs. (5 Nm).
22. Press the knuckle onto the hub using a suitable fixture.
23. Install the front knuckle ring on the knuckle.
24. Install the knuckle/hub assembly on the vehicle in the reverse of the removal procedure. Tighten the upper ball joint nut and tie rod end nut to 32 ft. lbs. (44 Nm). Install new cotter pins. Tighten the lower ball joint nut to 40 ft. lbs. (55 Nm) and install a new cotter pin.
25. With all 4 wheels resting on the ground, torque the spindle nut to 134 ft. lbs. (185 Nm) except 1990–92 Prelude, where it is tightened to 180 ft. lbs. (250 Nm).

1990–91 Accord

NOTE: Do not tighten or loosen a spindle nut unless the vehicle is sitting on all 4 wheels. The torque is high enough to cause the vehicle to fall even when properly supported.

1. Pry the spindle nut stake away from the spindle, then loosen the nut.
2. Raise and safely support the vehicle.
3. Remove the wheel and tire assembly and the spindle nut.
4. Remove the caliper mounting bolts and the caliper. Support the caliper out of the way with a length of wire. Do not let the caliper hang from the brake hose.
5. Remove the cotter pin from the tie rod castle nut, then remove the nut. Break loose the tie rod ball joint using a suitable ball joint remover, then lift the tie rod out of the knuckle.
6. Remove the cotter pin and loosen the lower arm ball joint nut half the length of the joint threads.
7. Separate the ball joint and lower arm using a suitable puller with the pawls applied to the lower arm.

NOTE: Avoid damaging the ball joint boot. If necessary, apply penetrating type lubricant to loosen the ball joint.

8. Pull the knuckle outward and remove the halfshaft outboard joint from the knuckle using a suitable tool.
9. Remove the cotter pin and the upper ball joint nut. Break loose the upper ball joint using a suitable tool.

NOTE: Avoid damaging the ball joint boot. If necessary, apply penetrating type lubricant to loosen the ball joint.

10. Remove the 4 bolts and remove the knuckle from the hub unit.
11. Remove the splash guard screws and the splash guard from the knuckle.
12. Remove the 4 bolts, then separate the hub unit from the brake disc.
13. Position the hub in a suitable hydraulic press. Press the wheel bearing from the hub while adequately supporting the hub.
14. Remove the outboard bearing inner race from the hub using a suitable bearing puller.

NOTE: The wheel bearing must be replaced with a new one after removal.

To install:

15. Clean the knuckle and hub thoroughly.
16. Position the hub in a suitable hydraulic press. Press a new wheel bearing into the hub using a suitable driver.
17. Install the hub on the brake disc and tighten the bolts to 40 ft. lbs. (55 Nm).
18. Install the splash guard and tighten the screws to 7 ft. lbs. (10 Nm).
19. Install the knuckle on the hub and tighten the bolts to 33 ft. lbs. (45 Nm).
20. Installation of the knuckle/hub assembly on the vehicle is the reverse of the removal procedure. Tighten the upper ball joint nut and the tie rod nut to 32 ft. lbs. (44 Nm) and install new cotter pins. Tighten the lower ball joint nut to 40 ft. lbs. (55 Nm) and install a new cotter pin.
21. With all 4 wheels resting on the ground, install a new spindle nut and torque to 180 ft. lbs. (245 Nm). After tightening, use a suitable drift to stake the spindle nut shoulder against the spindle.

Pinion Seal

REMOVAL & INSTALLATION

4WD Civic Wagon

1. Raise and safely support the vehicle.
2. Remove the rear wheel and tire assemblies.
3. Keep the rear halfshafts in their normal horizontal position by raising the lower arms to their normal road level position.
4. Mark the position of the driveshaft in relation to the pinion flange, then disconnect the driveshaft from the pinion flange.
5. Using a suitable inch lb. torque wrench, turn the pinion flange by the pinion nut and record the reading. This is the total bearing preload.
6. Hold the pinion flange using a

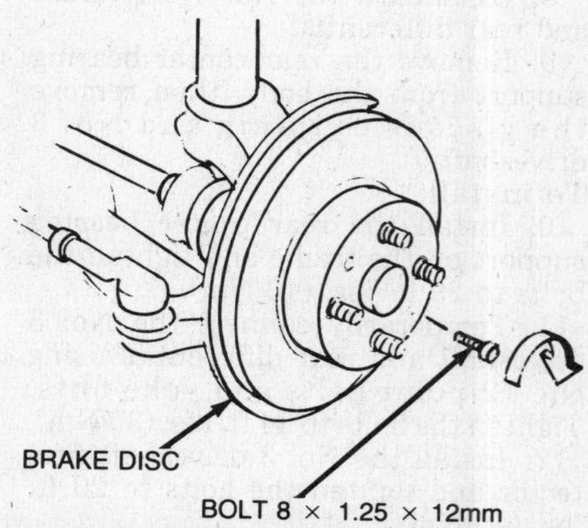

Disc brake rotor removal—except 1990–92 Accord

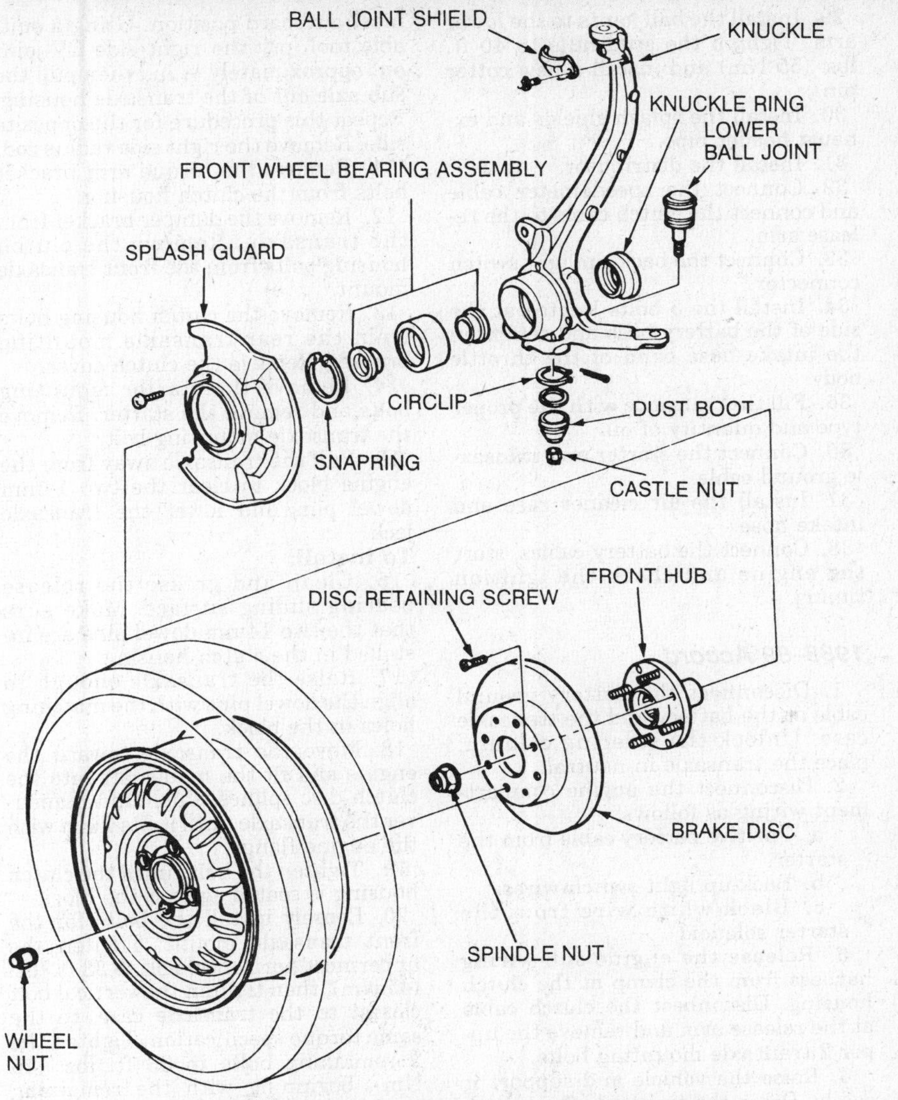

Front steering knuckle, hub and bearing—1988–89 Accord

suitable holding tool and remove the pinion nut.

7. Remove the pinion flange and use a suitable removal tool to remove the pinion seal.

To install:

8. Lubricate the lip of a new pinion seal.
9. Use a suitable installation tool to install the pinion seal.
10. Install the pinion flange and pinion nut. Tighten the pinion nut until the reading on an inch lb. torque wrench is the same as that recorded prior to removal.
11. Connect the driveshaft to the pinion flange, aligning the marks that were made during the removal procedure. Tighten the bolts to 24 ft. lbs. (33 Nm).
12. Install the wheel and tire assemblies.
13. Lower the vehicle.

Differential Carrier

REMOVAL & INSTALLATION

4WD Civic Wagon

1. Raise and safely support the vehicle.
2. Drain the oil from the differential and replace the drain bolt.
3. Remove the driveshaft and the right and left rear halfshafts.
4. Remove the differential mounting bolts from the lower differential mounting bracket.
5. Remove the 2 bolts from the upper differential bracket and remove the differential and bracket as an assembly.
6. Remove the upper differential bracket from the differential and remove the differential carrier from the differential housing.

To install:

7. Clean the mating surfaces of the differential carrier and differential housing.
8. Apply sealant to the mating surfaces and install the carrier to the housing, aligning the dowel pin. Apply sealant to the bolt threads and tighten them to 16 ft. lbs. (22 Nm).
9. Install the differential bracket on the differential and tighten the bolts to 40 ft. lbs. (55 Nm).
10. Install the differential and upper bracket and tighten the upper bracket bolts to 43 ft. lbs. (59 Nm).
11. Install and tighten the lower differential bracket bolts to 40 ft. lbs. (55 Nm).
12. Installation of the remainder of the components is the reverse of the removal procedure.

MANUAL TRANSAXLE

Transaxle Assembly

REMOVAL & INSTALLATION

Civic and CRX

1. Disconnect the battery cables from the battery.
2. Remove the 3 mount bolts and loosen the 1 bolt located at the side of the battery base. Remove the intake hose band at the throttle body.
3. Remove the air cleaner case complete with the intake hose. Disconnect the starter and transaxle ground cables.
4. Disconnect the speedometer, but be sure not to disassemble the speedometer gear holder.
5. Disconnect the back-up light switch connector and the clutch cable release arm.
6. Drain the transaxle fluid into a suitable drain pan. Disconnect the connectors and remove the mount bolts.
7. Remove the distributor assembly, as required.
8. Remove the starter mounting bolts and remove the starter assembly. Remove the engine splash shield and the right wheelwell splash shield.
9. Remove the header pipe. Remove the cotter pin and the lower arm ball joint nut, separate the ball joint and lower arm.
10. Remove the bolts and nut, then remove the right radius rod. Remove the right and left halfshafts. Remove the header pipe bracket. Remove the

shift lever torque rod and shift rod from the clutch housing.

11. On 4WD vehicles, remove the driveshaft and the intermediate shaft. Remove the cable bracket and the side transaxle mount from the transaxle housing and body.

12. Install a bolt at the cylinder head and attach a suitable chain hoist to the bolt and the other end to the engine hanger plate. Lift the engine slightly to unload the mounts.

13. Place a suitable transaxle jack under the transaxle and raise it just enough to take the weights off of the mounts.

14. Remove the front transaxle mounting bolts. Remove the rear transaxle mounting bolts. Remove the side transaxle mount, remove the 5 remaining transaxle mounting bolts and pull the transaxle assembly far enough away from the engine to clear the 14mm dowel pins.

15. Separate the mainshaft from the clutch pressure plate and remove the transaxle by lowering the jack.

To install:

16. Make sure the two 14mm dowel pins are installed in the clutch housing.

17. Raise the transaxle into position with the transaxle jack. Loosely install the transaxle mount bolts and then tighten to 43 ft. lbs. (60 Nm).

18. Install the engine side mount bolt and tighten to 50 ft. lbs. (68 Nm).

19. Install the transaxle-to-rear transaxle mount bracket. Tighten the bolts to 40 ft. lbs. (55 Nm).

20. Install the transaxle-to-front transaxle mount and tighten the bolts to 29 ft. lbs. (40 Nm).

21. Install the transaxle-to-side transaxle mount and tighten the bolts to 43 ft. lbs. (60 Nm).

22. Install the starter and tighten the starter bolts to 33 ft. lbs. (45 Nm).

23. Remove the transaxle jack and the chain hoist and bolts.

24. Install the shift lever torque rod and shift rod. After reassembly, slide the retainer back into place after driving in the spring pin.

25. Install the header pipe bracket and tighten the bolts to 16 ft. lbs. (22 Nm).

26. On 4WD vehicles, install the cable bracket to the rear transaxle mount bracket. Install the select and shift cables. Install the intermediate shaft and the driveshaft.

27. Install a new set ring on the end of each halfshaft. Install the right and left halfshafts. Turn the right steering knuckle fully outward and slide the axle into the differential, until the spring clip is felt to engage the differential side gear.

28. Install the damper fork and radius rod. Tighten the radius rod nut to 39 ft. lbs. (44 Nm).

29. Install the ball joints to the lower arm. Tighten the stud nuts to 40 ft. lbs. (55 Nm) and install a new cotter pin.

30. Install the splash shields and exhaust header pipe.

31. Install the distributor.

32. Connect the speedometer cable and connect the clutch cable to the release arm.

33. Connect the back-up light switch connector.

34. Install the 3 bolts located at the side of the battery base and retighten the intake hose band of the throttle body.

35. Fill the transaxle with the proper type and quantity of oil.

36. Connect the starter and transaxle ground cable.

37. Install the air cleaner case and intake hose.

38. Connect the battery cables, start the engine and check the ignition timing.

1988–89 Accord

1. Disconnect the battery ground cable at the battery and the transaxle case. Unlock the steering column; place the transaxle in neutral.

2. Disconnect the engine compartment wiring as follows:

a. Positive battery cable from the starter.

b. Back-up light switch wires.

c. Black/white wire from the starter solenoid

3. Release the engine sub wiring harness from the clamp at the clutch housing. Disconnect the clutch cable at the release arm and remove the upper 2 transaxle mounting bolts.

4. Raise the vehicle and support it safely. Drain the transaxle fluid.

5. Remove the front wheels. Place a suitable transaxle jack into position under the transaxle.

6. Disconnect the speedometer cable.

NOTE: When removing the speedometer cable from the transaxle, it is not necessary to remove the entire cable holder. Remove the end boot, gear holder seal, the cable retaining clip and then pull the cable out of the holder.

7. Disconnect the shift lever torque rod from the clutch housing. Remove the bolt from the shift rod clevis.

8. Disconnect the tie rod ball joints and remove them using a suitable ball joint remover tool.

9. Remove the lower arm ball joint bolt from the right side lower control arm, then using a puller disconnect the ball joint from the knuckle. Remove the damper fork bolt.

10. Turn each steering knuckle to its most outboard position. Using a suitable tool, pry the right side CV-joint out approximately ½ in., then pull the sub axle out of the transaxle housing. Repeat this procedure for the opposite side. Remove the right side radius rod.

11. Remove the torque arm bracket bolts from the clutch housing.

12. Remove the damper bracket from the transaxle. Remove the clutch housing bolts from the front transaxle mount.

13. Remove the clutch housing bolts from the rear transaxle mounting bracket. Remove the clutch cover.

14. Remove the starter mounting bolts and remove the starter. Remove the transaxle mounting bolt.

15. Pull the transaxle away from the engine block to clear the two 14mm dowel pins and lower the transaxle jack.

To install:

16. Clean and grease the release bearing sliding surface. Make sure that the two 14mm dowel pins are installed in the clutch housing.

17. Raise the transaxle enough to align the dowel pins with the matching holes in the block.

18. Move the transaxle toward the engine and fit the mainshaft into the clutch disc splines. Push and maneuver the transaxle until it fits flush with the engine flange.

19. Tighten the bolts until the clutch housing is seated against the block.

20. Loosely install the bolts for the front transaxle mount. Tighten the uppermost horizontal bolt to 33 ft. lbs. (45 Nm), then tighten the vertical bolt closest to the transaxle case, to the same torque specification. Tighten the 2 remaining bolts to 33 ft. lbs. (45 Nm), beginning with the remaining horizontal bolt.

21. Install the upper torque arm and it's bracket. Tighten the bracket mounting bolts to 47 ft. lbs. (65 Nm) and the torque arm bolts to 54 ft. lbs. (75 Nm).

22. Remove the transaxle jack.

23. Install the starter and tighten the mounting bolts to 33 ft. lbs. (45 Nm).

24. Install new spring clips on the end of each stub axle. Turn the right steering knuckle/axle assembly outward enough to insert the free end of the axle into the transaxle. Repeat on the opposite side. Slide the axle in until the spring clips are felt engaging the differential.

25. Install the lower ball joints and tie rods. Tighten the tie rod nuts to 32 ft. lbs. (44 Nm) and the ball joint nuts to 40 ft. lbs. (55 Nm). Install new cotter pins. Install the damper fork bolt.

26. Connect the shift linkage.

27. Connect the shift lever torque rod to the clutch housing and tighten the bolt to 16 ft. lbs. (22 Nm).

28. Install the front wheel and tire assemblies. Install the transaxle drain plug and tighten to 29 ft. lbs. (40 Nm).
29. Lower the vehicle.
30. Install the clutch cable at the release arm.
31. Coat a new O-ring with oil, put it on the speedometer gear holder, then install the holder in the transaxle housing and secure with the hold-down tab and bolt.
32. Install the engine sub-wire harness in the clamp at the clutch housing.
33. Connect all engine compartment wiring.
34. Fill the transaxle with the proper type and quantity of oil.
35. Connect the negative battery cable.

1990–92 Accord

1. Disconnect the battery cables and remove the battery.
2. Raise and safely support the vehicle.
3. Remove the air intake hose and battery base.
4. Disconnect the starter wires and remove the starter.
5. Disconnect the transaxle ground cable and the back-up light switch wire.
6. Remove the cable stay and then disconnect the cables from the top housing of the transaxle. Remove both cables and the stay together.
7. Disconnect the connector and remove the speed sensor, but leave it's hoses connected.
8. Remove the front wheel and tire assemblies.
9. Remove the engine splash shield and drain the transaxle fluid.
10. Remove the mounting bolts and clutch slave cylinder with the clutch pipe and pushrod.
11. Remove the mounting bolt and clutch hose joint with the clutch pipe and clutch hose.

NOTE: Do not operate the clutch pedal once the slave cylinder has been removed. Be careful not to bend the pipe.

12. Remove the center beam and the header pipe.
13. Remove the cotter pins and lower arm ball joint nuts. Separate the ball joints and lower arms using a suitable tool.
14. Remove the damper fork bolt.
15. Use a suitable tool to pry the right and left halfshafts out of the differential and the intermediate shaft. Pull on the inboard joint and remove the right and left halfshafts.
16. Remove the 3 mounting bolts and lower the bearing support.
17. Remove the intermediate shaft from the differential.
18. Remove the right damper pinch bolt, then separate the damper fork and damper. Remove the bolts and nut, then remove the right radius rod.
19. Remove the engine stiffener and the clutch cover.
20. Remove the intake manifold bracket.
21. Remove the rear engine mount bracket stay and remove the 3 rear engine mount bracket mounting bolts.
22. Remove the transaxle housing mounting bolt on the engine side. Swing the right halfshaft to the inner fender.
23. Place a suitable jack under the transaxle and raise the transaxle just enough to take the weight off the mounts.
24. Remove the transaxle mount bolt and loosen the mount bracket nuts.
25. Remove the 3 transaxle housing mounting bolts.
26. Remove the transaxle from the vehicle.

To install:

27. Make sure the 4 dowel pins are installed.
28. Raise the transaxle into position.
29. Install the 3 transaxle mounting bolts and tighten to 47 ft. lbs. (65 Nm).
30. Install the transaxle mount and mount bracket. Install the through bolt and tighten temporarily. Make sure the engine is level and tighten the 3 mount bracket nuts to 40 ft. lbs. (55 Nm). Tighten the through bolt to 47 ft. lbs. (65 Nm).
31. Install the transaxle housing mounting bolts on the engine side and tighten to 47 ft. lbs. (65 Nm).
32. Install the 3 rear engine bracket mounting bolts and tighten to 40 ft. lbs. (55 Nm).
33. Install the rear engine mount bracket stay. Tighten the mounting bolt to 28 ft. lbs. (39 Nm) and then tighten the mounting nut to 15 ft. lbs. (21 Nm).
34. Install the intake manifold bracket and tighten the bolts to 16 ft. lbs. (22 Nm).
35. Install the clutch cover and tighten the bolts to 9 ft. lbs. (12 Nm).
36. Install the engine stiffener and loosely install the mounting bolts. Tighten the stiffener-to-transaxle case mounting bolt to 28 ft. lbs. (39 Nm), then tighten the 2 stiffener-to-cylinder block mounting bolts to 28 ft. lbs. (39 Nm) beginning with the bolt closest to the transaxle.
37. Install the radius rod. Tighten the radius rod mounting bolts to 76 ft. lbs. (105 Nm) and the radius rod nut to 32 ft. lbs. (44 Nm).
38. Install the damper fork. Tighten the damper pinch bolt to 32 ft. lbs. (44 Nm).
39. Install the intermediate shaft.
40. Install a new set ring on the end of each halfshaft. Install the right and left halfshafts. Turn the right and left steering knuckle fully outward and slide the axle into the differential, until the spring clip is felt engaging the differential side gear.
41. Install the damper fork bolt and ball joint nut to the lower arms. Tighten the nut while holding the damper fork bolt to 40 ft. lbs. (55 Nm). Tighten the ball joint nut to 40 ft. lbs. (55 Nm). Install a new cotter pin.
42. Install the header pipe and center beam. Tighten the center beam bolts to 28 ft. lbs. (39 Nm).
43. Install the clutch hose joint and clutch slave cylinder to the transaxle housing. Tighten the slave cylinder mounting bolts to 16 ft. lbs. (22 Nm).
44. Install the speed sensor. Tighten the mounting bolt to 13 ft. lbs. (18 Nm).
45. Install the shift cable and select cable to the shift arm lever and to the select lever respectively. Tighten the cable bracket mounting bolts to 16 ft. lbs. (22 Nm). Install new cotter pins.
46. Connect the back-up light switch coupler.
47. Install the starter. Tighten the 10 × 1.25mm bolt to 32 ft. lbs. (45 Nm) and the 12 × 1.25mm bolt to 54 ft. lbs. (75 Nm). Connect the starter wires.
48. Install the transaxle ground cable.
49. Install the front wheel and tire assemblies.
50. Fill the transaxle with the proper type and quantity of oil.
51. Lower the vehicle.
52. Install the battery and connect the battery cables.
53. Check the clutch pedal free-play.
54. Start the vehicle and check the transaxle for smooth operation.

Prelude

1. Disconnect the negative battery cable at the battery and the transaxle. Raise and safely support the vehicle.
2. Disconnect the wiring for the starter and the back-up light switch.
3. On fuel injected vehicles, remove the air cleaner case.
4. Remove the power steering speed sensor from the transaxle, without removing the power steering hoses.
5. Remove the shift cable and the select cable from the top cover of the transaxle. Remove the mounting bolt from the cable stay. Remove the cables and the stay together.
6. Remove the upper transaxle mounting bracket.
7. Remove the 4 transaxle-to-block attachment bolts, that must be removed from the engine compartment.
8. Raise and safely support the vehicle.

9. Remove the front wheel and tire assemblies.
10. Remove the engine splash shield.
11. Drain the transaxle oil.
12. Remove the clutch slave cylinder and the center beam.
13. Remove the right radius rod. Remove the right and left halfshafts.
14. Remove the intermediate shaft.
15. Remove the engine stiffener and the clutch cover.
16. Support the transaxle with a suitable jack.
17. Remove the 3 lower bolts from the rear engine mounting bracket. Loosen but do not remove the top bolt. This bolt will support the weight of the engine.
18. Remove the 2 remaining engine-to-transaxle mounting bolts.
19. Pull the transaxle away from the engine and disengage the input shaft from the clutch disc. Lower the transaxle and remove it from the vehicle.

To install:

20. Make sure the 14mm dowel pin is installed in the transaxle.
21. Raise the transaxle into position.
22. Install the engine side transaxle mount bolts and tighten to 47 ft. lbs. (65 Nm). Install the transaxle side transaxle mount bolts and tighten to 47 ft. lbs. (65 Nm).
23. Attach the transaxle mounting bracket. Tighten the transaxle mount bolts to 28 ft. lbs. (39 Nm) and the mount through bolt to 54 ft. lbs. (75 Nm).
24. Install and tighten the rear engine mounting bracket bolts to 54 ft. lbs. (75 Nm).
25. Attach the clutch cover and tighten the bolts to 9 ft. lbs. (12 Nm).
26. Attach the engine stiffener and tighten the mounting bolts to 28 ft. lbs. (39 Nm). The bolts should be tightened in sequence beginning with the top most stiffener-to-transaxle case bolt, followed by the remaining stiffener-to-transaxle bolt, then the stiffener-to-cylinder block bolt closest to the transaxle and finally the remaining stiffener-to-cylinder block bolt.
27. Install the intermediate shaft and the left and right halfshaft.
28. Install the center beam and tighten the bolts to 37 ft. lbs. (51 Nm).
29. Install the clutch slave cylinder with the clutch hose and pushrod. Tighten the slave cylinder bolts to 16 ft. lbs. (22 Nm).
30. Attach the transaxle side shift cable and select cable to the shift arm lever and to the select lever respectively. Tighten the bracket mounting bolts to 16 ft. lbs. (22 Nm).
31. Connect the back-up light switch coupler.
32. Attach the right and left front damper forks.
33. Install the speed sensor assembly and the air cleaner case.
34. Connect the starter cable and connect the ground cable to the transaxle.
35. Install the wheel and tire assemblies.
36. Fill the transaxle with the proper type and quantity of oil.
37. Lower the vehicle and connect the negative battery cable. Check the clutch pedal free-play.

LINKAGE ADJUSTMENT

Prelude, 1990–92 Accord and Civic Wagon with 4WD, feature cable operated gear shift mechanisms that are adjustable. All other vehicles have non-adjustable, rod operated, gear shift linkage.

Prelude and 4WD Civic Wagon

SELECT CABLE

1. Disconnect the negative battery cable.
2. Remove the console.
3. With the transaxle in neutral, check that the groove in the lever bracket is aligned with the index mark on the selector cable.
4. If the index mark is not aligned with the groove in the cable, loosen the locknuts and turn the adjuster as necessary.

NOTE: After adjustment, check the operation of the gear shift lever. Make sure the threads of the cables do not extend out of the cable adjuster by more than 0.4 in. (10mm).

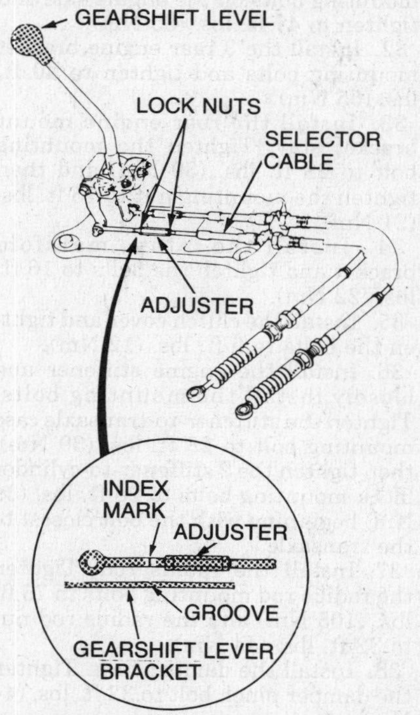

Selector cable adjustment – 4WD Civic Wagon and Prelude

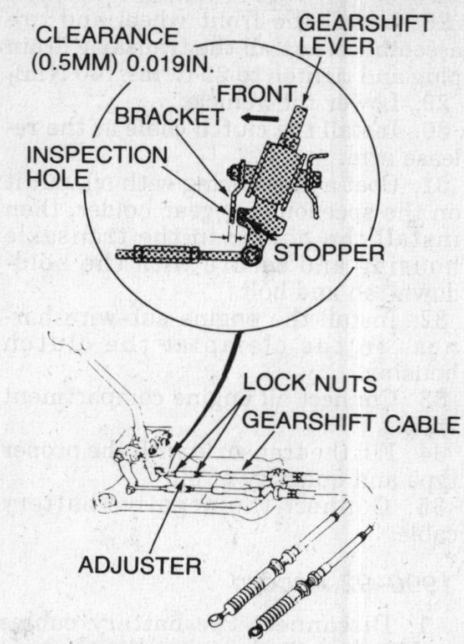

Gear shift cable adjustment – 4WD Civic Wagon and Prelude

5. Replace the console and connect the negative battery cable.

SHIFT CABLE

1. Disconnect the negative battery cable.
2. Remove the console.
3. Place the transaxle in 4th gear.
4. Measure the clearance between the gear shift lever bracket and stopper, while pushing the lever forward. The clearance should be 0.24–28 in. (6.0–7.0mm).
5. If the clearance is outside specification, loosen the locknuts and turn the adjuster in or out until the correct clearance is obtained.

NOTE: After adjustment, check the operation of the gear shift lever. Make sure the threads of the cables do not extend out of the cable adjuster by more than 0.4 in. (10mm).

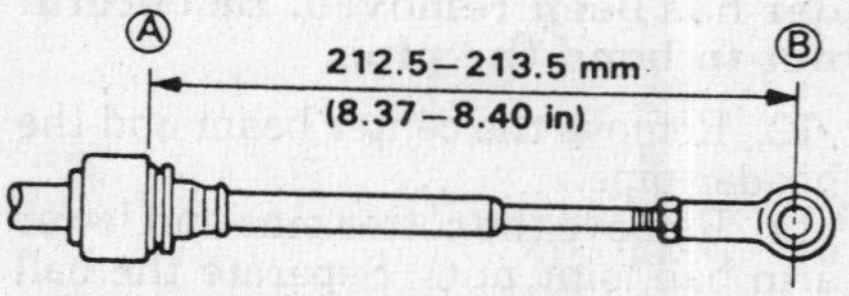

Select cable clearance measurement – 1990–92 Accord

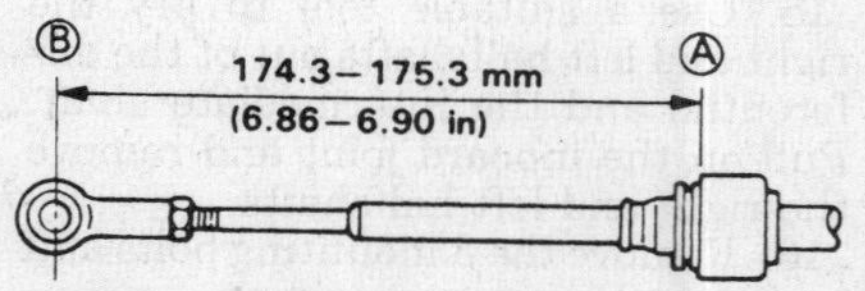

Shift cable clearance measurement – 1990–92 Accord

6. Replace the console and connect the negative battery cable.

1990–92 Accord

SELECT CABLE

1. Disconnect the negative battery cable.
2. Remove the console.
3. Place the shift lever in the neutral position.
4. Measure the clearance between (A) and (B). It should be 8.37–8.40 in. (212.5–213.5mm).
5. If the clearance is incorrect, disconnect the select cable from the linkage, loosen the locknut and turn the adjuster, as necessary.
6. Tighten the locknut and connect the select cable to the linkage. Install a new cotter pin.

NOTE: Make sure the new cotter pin is seated firmly. After adjustment, check the operation of the shift lever.

7. Install the console and connect the negative battery cable.

SHIFT CABLE

1. Disconnect the negative battery cable.
2. Remove the console.
3. Place the shift lever in the neutral position.
4. Measure the clearance between (A) and (B). The clearance should be 6.86–6.90 in. (174.3–175.3mm).
5. If the clearance is incorrect, disconnect the shift cable from the change lever, loosen the locknut and turn the adjuster, as necessary.
6. Tighten the locknut and connect the shift cable to the change lever. Install a new cotter pin.

NOTE: Make sure the new cotter pin is seated firmly. After adjustment, check the operation of the gear shift lever.

7. Install the console and connect the negative battery cable.

CLUTCH

Clutch Assembly

REMOVAL & INSTALLATION

1. Disconnect the negative battery cable. Raise and safely support the vehicle. Remove the transaxle from the vehicle. Matchmark the flywheel and clutch for reassembly.
2. Hold the flywheel ring gear with a tool made for this purpose, remove the retaining bolts and remove the pressure plate and clutch disc. Remove the bolts 2 turns at a time working in a criss-cross pattern, to prevent warping the pressure plate.
3. At this time, inspect the flywheel for wear, cracks or scoring and resurface or replace, as necessary.
4. If the clutch release bearing is to be replaced, perform the following procedure on all except Prelude and 1990–92 Accord:
   a. Remove the 8mm special bolt.
   b. Remove the release shaft and the release bearing assembly.
   c. Separate the release fork from the bearing by removing the release fork spring from the holes in the release bearing.
5. To remove the release bearing on Prelude and 1990–92 Accord, perform the following procedure:
   a. Remove the boot from the clutch housing.
   b. Remove the release fork from the clutch housing by squeezing the release fork set spring with a suitable tool.
   c. Remove the release bearing from the release fork.
6. Check the release bearing for excessive play by spinning it by hand. Replace if there is excessive play.

To install:

7. If the flywheel was removed, make sure the flywheel and crankshaft mating surfaces are clean. Align the hole in the flywheel with the crankshaft dowel pin and install the flywheel bolts finger tight. Install the ring gear holder and tighten the flywheel bolts in a criss-cross pattern. Tighten the flywheel bolts to 76 ft. lbs. (105 Nm) on Prelude and Accord, 87 ft. lbs. (120 Nm) on Civic and CRX.
8. Install the clutch disc and pressure plate by aligning the dowels on the flywheel with the dowel holes in the pressure plate. If the same pressure plate is being installed that was removed, align the marks that were made during the removal procedure. Install the pressure plate bolts finger tight.
9. Insert a suitable clutch disc alignment tool into the splined hole in the clutch disc. Tighten the pressure plate bolts in a criss-cross pattern 2 turns at a time to prevent warping the pressure plate. The final torque should be 19 ft. lbs. (26 Nm).
10. Remove the alignment tool and ring gear holder.
11. If the release bearing was removed, replace it in the reverse order of the removal procedure. Place a light coat of molybdenum disulfide grease on the inside diameter of the bearing prior to installation.
12. Install the transaxle, making sure the mainshaft is properly aligned with the clutch disc splines and the transaxle case is properly aligned with the cylinder block, before tightening the transaxle case bolts.
13. Adjust the clutch pedal free-play and connect the negative battery cable.

PEDAL HEIGHT/FREE-PLAY ADJUSTMENT

Civic, CRX and 1988–89 Accord

1. Measure the clutch pedal disengagement height.
2. Measure the clutch pedal free play.
3. Adjust the clutch pedal free-play by turning the clutch cable adjusting nut, found at the end of the clutch cable housing near the release shaft.
4. Turn the adjusting nut until the clutch pedal free-play is as follows:
   Civic and CRX – 0.6–0.8 in. (15–20mm).
   1988–89 Accord – 0.6–1.0 in. (15–25mm).
5. After adjustment, make sure the free-play at the tip of the release arm is as follows:
   Civic and CRX – 0.12–0.16 in. (3.0–4.0mm).
   1988–89 Accord – 0.20–0.25 in. (5.2–6.4mm).

Prelude and 1990–92 Accord

NOTE: The clutch is self adjusting to compensate for wear. The total clutch pedal free-play is 0.35–0.59 in. (9–15mm). If there is no clearance between the master cylinder piston and pushrod, the release bearing is held against the diaphragm spring of the pressure plate, which can result in clutch slippage or other clutch malfunction.

1. Loosen the locknut on clutch pedal switch A and back off the pedal switch until it no longer touches the clutch pedal. Clutch pedal switch A is the switch that contacts the clutch pedal below the clutch pedal pivot.
2. Loosen the locknut on the clutch master cylinder pushrod and turn the pushrod in or out to get the specified stroke and height at the clutch pedal. The pedal stroke should be 5.3–5.5 in. (135–140mm) on Prelude and 5.6 in. (142mm) on Accord. The pedal height should be 8.1 in. (207mm) on Prelude and 8.27 in. (210mm) on Accord.
3. Tighten the pushrod locknut.
4. Thread in clutch pedal switch A until it contacts the clutch pedal, then turn it in ¼–½ turn further.
5. Tighten the locknut on clutch pedal switch A.

Clutch Cable

REMOVAL & INSTALLATION

1. Disconnect the negative battery cable.
2. Disconnect the cable end from the brake pedal.
3. Remove the adjuster nut assembly from its mounting.
4. Raise and support the vehicle safely.
5. Disconnect the cable end from the release arm. Remove the cable from the vehicle.
6. Installation is the reverse of the removal procedure. Adjust the cable to specification.

Clutch Master Cylinder

REMOVAL & INSTALLATION

1. Disconnect the negative battery cable. Pry out the cotter pin and pull the pedal pin out of the yoke.
2. Remove the nuts and bolts retaining the clutch master cylinder and remove the cylinder from the engine compartment.
3. Disconnect and plug the hydraulic lines from the master cylinder.
4. Installation is the reverse order of the removal procedure. Bleed the clutch hydraulic system.

Clutch Slave Cylinder

REMOVAL & INSTALLATION

1. Disconnect the negative battery cable.
2. Disconnect and plug the clutch hose from the slave cylinder.
3. Remove the 2 retaining bolts and remove the slave cylinder.
4. Installation is the reverse of the removal procedure. Bleed the clutch hydraulic system.

Hydraulic Clutch System Bleeding

The hydraulic system must be bled whenever the system has been leaking or dismantled. The bleed screw is located on the slave cylinder.

1. Remove the bleed screw dust cap.
2. Attach a clear hose to the bleed screw. Immerse the other end of the hose in a clear jar, half filled with brake fluid.
3. Fill the clutch master cylinder with fresh brake fluid.
4. Open the bleed screw slightly and have an assistant slowly depress the clutch pedal. Close the bleed screw when the pedal reaches the end of its travel. Allow the clutch pedal to return slowly.
5. Repeat Steps 3–4 until all air bubbles are expelled from the system.
6. Discard the brake fluid in the jar. Replace the dust cap. Refill the master cylinder.

AUTOMATIC TRANSAXLE

For further information on transmissions/transaxles, please refer to "Chilton's Guide to Transmission Repair".

Transaxle Assembly

REMOVAL & INSTALLATION

Civic and CRX

1. Disconnect the battery cables and remove the battery.
2. Raise and safely support the vehicle.
3. Remove the 3 mount bolts and loosen the 1 bolt located at the side of the battery base and intake hose band of the throttle body.
4. Remove the air cleaner case, complete with the intake hose.
5. Remove the starter and the transaxle ground cable.
6. Disconnect the lock-up control solenoid valve wire connector. On 4WD Civic Wagon, disconnect the lock-up control solenoid valve and shift control solenoid valve wire connectors and disconnect the automatic transaxle speed pulser connector.
7. Disconnect the control cable at the control lever.
8. Drain the transaxle fluid. Remove the filler plug to speed draining. After draining reinstall the drain plug with a new washer.
9. Disconnect and plug the cooler hoses at the joint pipes.
10. Mark the position of the distributor housing in relation to the cylinder head. Disconnect the connectors and remove the mount bolts, then remove the distributor from the cylinder head.
11. Remove the shift cable by removing the cotter pin, control pin, control lever roller and loosening the locknut.
12. Disconnect the speedometer cable. Do not disassemble the speedometer gear holder.
13. On 4WD Civic Wagon, remove the driveshaft.
14. Remove the torque converter cover. Remove the plug, then remove the driveplate bolts 1 at a time while rotating the crankshaft pulley.
15. Remove the engine splash shield and the right wheel well splash shield.
16. Remove the header pipe.
17. Remove the cotter pins and lower arm ball joint nuts and then separate the ball joints using a suitable tool.
18. Remove the bolts and nut, then remove the right radius rod.
19. Remove the right and left halfshafts. On 4WD Civic Wagon, remove the intermediate shaft.
20. On 4WD Civic Wagon, remove the 9 mounting bolts, then remove the transaxle cover.
21. Install bolts in each end of the cylinder head and attach a suitable hoist to the bolts. Lift the engine slightly to unload the mounts.
22. Place a suitable jack under the transaxle and raise the transaxle just enough to take weight off of the mounts.
23. On 4WD Civic Wagon, remove the transaxle mounting bolt on the engine side.
24. Remove the bolts from the front transaxle mount.
25. Remove the rear transaxle mount bracket by removing the 4 mounting bolts.
26. Remove the 4 mounting bolts. Remove the side transaxle mount.
27. Remove the transaxle-to-cylinder block mounting bolts.
28. On 4WD Civic Wagon, loosen the side engine mounting bolt and tilt the engine.
29. Pull the transaxle away from the engine until it clears the 14mm dowel pins, then lower the transaxle jack.

To install:

30. Make sure the two 14mm dowel pins are installed in the torque converter housing.
31. Raise the transaxle into position and loosely install the transaxle-to-cylinder block bolts. Tighten them to 43 ft. lbs. (60 Nm).
32. Install the engine side mounting bolt (12 × 1.25 × 70mm) and tighten to 50 ft. lbs. (68 Nm).
33. Install the transaxle to rear transaxle mount bracket and tighten the bolts to 40 ft. lbs. (55 Nm).
34. Install the transaxle to the front and side transaxle mounts. Tighten the mount retaining bolts to 29 ft. lbs. (40 Nm) and the mount through bolts to 40 ft. lbs. (55 Nm).
35. Remove the transaxle jack. On 4WD Civic Wagon, tighten the side engine mount bolt to 40 ft. lbs. (55 Nm).
36. Remove the hoist chain and bolts from the cylinder head.
37. Attach the torque converter to the driveplate with the eight 12mm bolts. Tighten the bolts in 2 steps, first to 4.5 ft. lbs. (6 Nm) in a criss-cross

pattern and finally to 9 ft. lbs. (12 Nm) in the same pattern.

38. Install the shift cable and cable holder.

39. Install the torque converter cover and header pipe bracket.

40. On 4WD Civic Wagon, install the transaxle cover and the intermediate shaft.

41. Install a new set ring on the end of each halfshaft.

42. Turn the right steering knuckle fully outward and slide the axle into the differential, until the spring clip can be felt engaging the differential side gear. Repeat the procedure on the left side.

43. Install the damper fork bolt and radius rod.

44. Install the ball joints to the lower arms and tighten the nuts to 40 ft. lbs. (55 Nm). Install new cotter pins.

45. On 4WD Civic Wagon, install the driveshaft.

46. Install the splash shields and header pipe.

47. Install the distributor.

48. Connect the lock-up control solenoid valve wire connector. On 4WD Civic Wagon, connect the lock-up control solenoid valve and shift control solenoid valve wire connectors and connect the automatic transaxle speed pulser connector.

49. Connect the transaxle cooler hoses to the joint pipes and the control cable to the control lever.

50. Connect the speedometer cable.

51. Install the starter and tighten the bolts to 33 ft. lbs. (45 Nm).

52. Connect the positive battery cable to the starter and connect the ground cable at the transaxle.

53. Install the air intake case and hose.

54. Install the 3 bolts located at the side of the battery base and retighten the intake hose band of the throttle body.

55. Lower the vehicle, install the battery and connect the battery cables.

56. Fill the transaxle with the proper type and quantity of fluid.

57. Start the engine, set the parking brake and shift through all gears 3 times. Check for proper control cable adjustment. Check the ignition timing.

58. Let the engine reach operating temperature with the transaxle in **N** or **P**, then turn the engine off and check the fluid level.

1988-89 Accord

1. Disconnect the negative battery cable at the battery and the transaxle.

2. Unlock the steering and place the transaxle in **N**.

3. Disconnect the positive battery cable from the starter and the wire from the starter solenoid.

4. Disconnect and plug the transaxle cooler hoses.

5. Remove the starter and the top transaxle mounting bolt.

6. Raise and safely support the vehicle.

7. Remove the front tire and wheel assemblies.

8. Drain the transaxle and reinstall the drain plug with a new washer.

9. Remove the throttle control cable by removing the cable end from the throttle lever, loosening the locknut on the cable end side of the bracket and removing the cable from the bracket.

10. Remove the power steering speed sensor complete with speedometer cable and hoses.

11. Remove 2 upper transaxle mounting bolts.

12. Place a suitable transaxle jack securely beneath the transaxle. Attach a suitable hoist to the engine and raise the engine to just take the weight of the engine off the mounts.

13. Remove the subframe center beam and splash pan.

14. Remove the ball joint pinch bolt from the right side lower control arm, then use a suitable puller to disconnect the ball joint from the knuckle. Remove the damper fork bolt.

15. Turn the right side steering knuckle to it's most outboard position. Using a suitable tool, pry the CV-joint out approximately ½ in., then pull the CV-joint out of the transaxle housing.

NOTE: Do not pull on the halfshaft or knuckle since this may cause the inboard CV-joint to separate. Pull on the inboard CV-joint.

16. Remove the transaxle damper bracket located in front of the torque converter cover plate.

17. Remove the torque converter cover plate.

18. Remove the center console and shift indicator.

19. Remove the lock pin from the adjuster and shift cable. Remove both bolts and pull the shift cable out of the housing.

20. Unbolt the torque converter assembly from the driveplate by removing the 8 bolts.

21. Remove the 3 rear engine mounting bolts from the transaxle housing. Remove the rear engine mount.

22. Remove the front transaxle mount's 2 bolts.

23. Remove the lower transaxle mounting bolt.

24. Pull the transaxle away from the engine to clear the two 14mm dowel pins. Pry the left side CV-joint out approximately ½ in. Pull the transaxle out and lower the transaxle.

To install:

25. Attach the shift cable to the shift arm with the pin, then secure the cable to the edge of the housing with the cable holder and bolt. Tighten the bolt to 9 ft. lbs. (12 Nm).

26. Make sure the two 14mm dowel pins are installed in the transaxle housing.

27. Install new spring clips on the end of each axle.

28. Raise the transaxle into position, aligning the dowel pins with the holes in the block and the torque converter bolt heads with the holes in the driveplate. Fit the left axle into the differential as the transaxle is raised up to the engine.

29. Install the 2 lower transaxle mounting bolts but do not torque at this time.

30. Install the rear engine mounts on the transaxle housing and tighten the bolts to 28 ft. lbs. (39 Nm).

31. Install the front transaxle mount bolts and tighten to 28 ft. lbs. (39 Nm).

32. Attach the torque converter to the driveplate with the eight 12mm bolts. Tighten the bolts in 2 steps, first to 4.5 ft. lbs. (6 Nm) in a criss-cross pattern and finally to 9 ft. lbs. (12 Nm) in the same pattern.

33. Remove the transaxle jack.

34. Install the torque converter cover plate and tighten the bolts to 9 ft. lbs. (12 Nm).

35. Install the wind stop rubber on the center beam and tighten the nuts to 40 ft. lbs. (55 Nm). Install the wind stop bracket on the transaxle housing and tighten the 3 bolts to 22 ft. lbs. (31 Nm).

36. Remove the hoist.

37. Install the starter and tighten the bolts to 33 ft. lbs. (45 Nm).

38. Install the rear torque rod and brackets. Tighten the bracket mounting bolts to 46 ft. lbs. (65 Nm) and the torque rod bolts to 54 ft. lbs. (75 Nm).

39. Turn the right steering knuckle fully outward and slide the axle into the differential until the spring clip is felt engaging the differential side gear.

40. Reconnect the ball joint to the knuckle, then tighten it's bolt to 40 ft. lbs. (55 Nm). Reinstall the damper fork and tighten it's bolt to 32 ft. lbs. (44 Nm).

41. Install the speedometer cable. Align the tab on the cable end with the slot in the holder. Install the clip so the bent leg is on the groove side.

NOTE: After installing, pull the speedometer cable to see if it is secure.

42. Install the wheel and tire assemblies and lower the vehicle.

43. Install the remaining transaxle mounting bolts. Tighten all the trans-

axle mounting bolts to 33 ft. lbs. (45 Nm).

44. Connect the transaxle cooler hoses and tighten the banjo bolts to 21 ft. lbs. (29 Nm).

45. Connect the positive battery cable to the starter, the solenoid wire to the solenoid, the wire to the water temperature sending unit and the wires to the ignition timing thermosensor.

46. Connect the negative battery cable to the transaxle.

47. Unscrew the dipstick from the top of the transaxle end cover and add 3.2 quarts of Dexron® ATF through the hole. Reinstall the dipstick.

NOTE: If the transaxle and torque converter have been disassembled, add a total of 6.3 quarts.

48. Install and reconnect the shift cable. Install the console.

49. Connect the negative battery cable, start the engine, set the parking brake and shift the transaxle through all gears 3 times. Check for proper shift cable adjustment.

50. Let the engine reach operating temperature with the transaxle in **N** or **P**, then turn the engine off and check the fluid level.

51. Install the throttle control cable and adjust.

1990–92 Accord

1. Disconnect the battery cable and remove the battery.
2. Raise and safely support the vehicle.
3. Remove the air intake hose air cleaner case and battery base.
4. Disconnect the throttle cable from the throttle control lever.
5. Disconnect the transaxle ground cable and the speed sensor connectors.
6. Disconnect the starter cables and remove the starter.
7. Remove the rear mount bracket stay nut first. Remove the bolt, then remove the rear mount bracket stay.
8. Remove the speed sensor, but leave it's hoses connected.
9. Disconnect the lock-up control solenoid valve and shift control solenoid valve connectors.
10. Drain the transaxle fluid and reinstall the drain plug with a new washer.
11. Disconnect the transaxle cooler hoses from the joint pipes. Plug the hoses.
12. Remove the center beam.
13. Disconnect the oxygen sensor connector.
14. Remove the exhaust header pipe and the splash shield.
15. Remove the cotter pins and lower arm ball joint nuts, then separate the ball joints from the lower arms using a suitable tool.
16. Using a suitable tool, pry the right and left halfshafts out of the differential. Pull on the inboard CV-joints and remove the right and left halfshafts.
17. Remove the right damper pinch bolt, then separate the damper fork and damper.
18. Remove the bolts and nut, then remove the right radius rod.
19. Tie plastic bags over the halfshaft ends.
20. Remove the torque converter cover and control cable holder.
21. Remove the shift control cable by removing the cotter pin, control pin and control lever roller from the control lever.
22. Remove the plug, then remove the driveplate bolts 1 at a time while rotating the crankshaft pulley.
23. Remove the rear, engine side transaxle housing mounting bolts.
24. Remove the mounting bolts from the rear engine mount bracket.
25. Attach a suitable hoist to the transaxle housing hoisting brackets, then lift the engine slightly.
26. Place a suitable jack under the transaxle and raise the jack just enough to take weight off of the mounts.
27. Remove the 4 transaxle housing mounting bolts and 3 mount bracket nuts.
28. Pull the transaxle away from the engine until it clears the 14mm dowel pins, then lower it on the transaxle jack.

To install:

29. Make sure the two 14mm dowel pins are installed in the torque converter housing.
30. Raise the transaxle into position and install the 4 transaxle housing mounting bolts. Tighten the bolts to 47 ft. lbs. (65 Nm).
31. Install the transaxle to transaxle mount bracket and tighten the nuts to 28 ft. lbs. (39 Nm).
32. Remove the transaxle jack.
33. Install the 2 engine side transaxle housing mounting bolts and tighten to 47 ft. lbs. (65 Nm). Install the rear engine mount bracket bolts and tighten to 40 ft. lbs. (55 Nm).
34. Attach the torque converter to the driveplate with 8 bolts. Tighten the bolts in 2 steps, first to 4.5 ft. lbs. (6 Nm) in a criss-cross pattern and finally to 9 ft. lbs. (12 Nm) in the same pattern. Check for free rotation after tightening the last bolt.
35. Install the shift control cable and control cable holder. Tighten the control cable holder bolts to 13 ft. lbs. (18 Nm).
36. Install the torque converter cover and tighten the bolts to 9 ft. lbs. (12 Nm).
37. Remove the hoist.
38. Install the radius rod. Tighten the mounting bolts to 76 ft. lbs. (105 Nm) and the nut to 32 ft. lbs. (44 Nm).
39. Install the damper fork. Tighten the damper pinch bolt to 32 ft. lbs. (44 Nm).
40. Install a new set ring on the end of each halfshaft.
41. Turn the right steering knuckle fully outward and slide the axle into the differential until the spring clip is felt engaging the differential side gear. Repeat the procedure on the left side.
42. Install the damper fork bolts and ball joint nuts to the lower arms. Tighten the nut to 40 ft. lbs. (55 Nm) while holding the damper fork bolt. Tighten the ball joint nut to 40 ft. lbs. (55 Nm) and install a new cotter pin.
43. Install the splash shield, the center beam and the exhaust header pipe. Tighten the center beam bolts to 28 ft. lbs. (39 Nm).
44. Connect the oxygen sensor connector.
45. Install the speed sensor and tighten the bolt to 13 ft. lbs. (18 Nm).
46. Install the rear mount bracket stay. Tighten the mounting bolt first, to 28 ft. lbs. (39 Nm) and then tighten the nut to 15 ft. lbs. (21 Nm).
47. Install the starter and tighten the bolts to 33 ft. lbs. (45 Nm). Connect the cables to the starter.
48. Connect the lock-up control solenoid valve and shift control solenoid valve connectors.
49. Connect the speed sensor connectors and the transaxle ground cable.
50. Connect the transaxle cooler hoses to the joint pipes.
51. Install the battery base, air cleaner case and air intake hose. Install the battery.
52. Lower the vehicle and connect the battery cables at the battery.
53. Fill the transaxle with the proper type and quantity of fluid.
54. Start the engine, set the parking brake and shift the transaxle through all gears 3 times. Check for proper control cable adjustment.
55. Let the engine reach operating temperature with the transaxle in **N** or **P**, then turn the engine off and check the fluid level.

Prelude

1. Disconnect the negative battery cable at the battery and the transaxle.
2. Drain the transaxle fluid and replace the drain plug.
3. Disconnect the wiring for the starter, lock-up control solenoids, shift control solenoids and speed pulser.
4. On fuel injected vehicles, remove the air inlet hose and the air cleaner case.

5. Remove the speed sensor from the transaxle without removing the hoses.
6. Disconnect the throttle control cable at the transaxle bracket.
7. Disconnect the transaxle cooler hoses at the joint pipes and cap the joint pipes.
8. Remove the upper transaxle mounting bracket.
9. Remove the transaxle-to-cylinder block attachment bolts that must be removed from the engine compartment.
10. Raise and safely support the vehicle.
11. Remove the front wheel and tire assemblies.
12. Remove the splash shield and the center beam.
13. Remove the right radius rod completely.
14. Remove the right and left halfshafts and the intermediate shaft.
15. Remove the engine stiffener and the torque converter cover.
16. Remove the shift cable from the transaxle.
17. Remove the bolts from the driveplate.
18. Support the transaxle with a suitable jack.
19. Remove the lower bolt from the rear engine mounting bracket. Loosen, but do not remove the top bolt. This bolt will support the weight of the engine.
20. Remove the remaining engine-to-transaxle mounting bolts.
21. Separate the transaxle from the engine block. Disengage the two 14mm dowel pins and lower the transaxle.

To install:

22. Raise the transaxle into position and install the mounting bolts. Tighten the bolts to 47 ft. lbs. (65 Nm).
23. Attach the torque converter to the driveplate with the mounting bolts. Tighten the bolts in 2 steps, first to 4.5 ft. lbs. (6 Nm) in a criss-cross pattern and finally to 9 ft. lbs. (12 Nm) in the same pattern. Check for free rotation after tightening the last bolt.
24. Install the transaxle to the rear engine mount bracket with the mounting bolts. Tighten the bolts to 54 ft. lbs. (75 Nm).
25. Install the shift cable with the control pin and a new cotter pin.
26. Install the torque converter cover and the cable holder.
27. Install the engine stiffener. The engine stiffener bolts must be tightened to 28 ft. lbs. (39 Nm) in their proper order. First tighten the uppermost stiffener-to-transaxle housing bolt followed by the remaining stiffener-to-transaxle housing bolt. Next tighten the stiffener-to-cylinder block bolt closest to the transaxle followed by the remaining stiffener-to-cylinder block bolt.
28. Install the intermediate shaft and the halfshafts.
29. Install the center beam and the right and left front damper fork.
30. Install the radius rod on the transaxle side.
31. Install the transaxle mounting bracket and tighten the bolts to 28 ft. lbs. (39 Nm).
32. Connect the lock-up control solenoid valve connector, the shift control solenoid valve coupler and the connector of the speed pulser.
33. Connect the throttle control cable to the throttle control lever.
34. Install the speed sensor assembly and the air cleaner case.
35. Connect the oil cooler hoses and connect the starter and ground cables.
36. Connect the battery cables to the battery.
37. Start the engine, set the parking brake and shift the transaxle through all gears 3 times. Check for proper control cable adjustment.
38. Let the engine reach operating temperature with the transaxle in **N** or **P**, then turn the engine off and check the fluid level.

SHIFT CABLE ADJUSTMENT

1. Start the engine. Shift the transaxle to **R**, to see if the reverse gear engages.
2. Shut the engine off and disconnect the negative battery cable.
3. Remove the console.
4. On 1988 Civic, CRX and Accord, place the selector lever in **D**. On 1988 Prelude, place the selector lever in **R**. On 1989 Accord and Prelude and 1989–92 Civic and CRX, place the selector lever in **N** or **R**. On 1990–92 Accord and Prelude, place the selector lever in **N**. Remove the lock pin from the cable adjuster.
5. Check that the hole in the adjuster is perfectly aligned with the hole in the shift cable.

NOTE: There are 2 holes in the end of the shift cable. They are positioned 90 degrees apart to allow cable adjustments in ¼ turn increments.

6. If not perfectly aligned, loosen the locknut on the shift cable and adjust as required.
7. Tighten the locknut and install the lock pin on the adjuster.

NOTE: If the lock pin feels like it is binding when being installed, the cable is still out of adjustment and must be adjusted again.

8. Connect the negative battery cable, start the engine and check the shift lever in all gears. Install the console.

THROTTLE LINKAGE ADJUSTMENT

Carbureted Engine

THROTTLE CONTROL CABLE BRACKET

1. Disconnect the negative battery cable.
2. Disconnect the throttle control cable from the throttle control lever.
3. Bend down the lock tabs of the lock plate and remove the two 6mm bolts to free the bracket.
4. Loosely install a new lock plate.
5. Adjust the position of the bracket by measuring the distance between the cable housing side of the bracket and the bracket side edge of the throttle control lever. Measure between the same points that the cable would pass through the bracket and lever.
6. Tighten the two 6mm bolts when the measurement is 3.287 in. (83.5mm) 1988–89 Accord. Tighten the two 6mm bolts when the measurement is 6.18 in. (157.0mm) on Prelude. The bolts should be tightened to 9 ft. lbs. (12 Nm).

NOTE: Make sure the control lever does not get pulled toward the bracket side as the bolts are tightened.

7. Bend up the lock plate tabs against the bolt heads, connect the throttle control cable and connect the negative battery cable.

THROTTLE CONTROL CABLE – 1988

1. Start the engine and warm it up to normal operating temperature. The cooling fan must come on at least once.
2. Make sure the throttle cable play, idle speed and automatic choke operation are correct.
3. Check the distance between the throttle control lever and the throttle control bracket and adjust as necessary.
4. Turn the engine off and disconnect the negative battery cable.
5. Disconnect the throttle control cable from the control lever.
6. If the vehicle is equipped with a dash pot, disconnect the vacuum hose from the dash pot, connect a vacuum pump and apply vacuum. This simulates a normal operating amount of pull by the dash pot, as if the engine were running.
7. Attach a weight of about 3 lbs. to the accelerator pedal. Raise the pedal, then release it. This will allow the weight to remove the normal free-play from the throttle cable.

8. Secure the throttle cable with clamps.
9. Lay the end of the throttle cable aside on Civic and CRX and on the shock tower on Accord and Prelude.
10. Adjust the distance between the throttle control cable end and the locknut closest to the cable housing to 3.366 in. (85.5mm) on all except 1988 Prelude. The distance on 1988 Prelude should be 6.22 in. (158.0mm).
11. Insert the end of the throttle control cable in the groove of the throttle control lever. Insert the throttle control cable in the bracket and secure with the other locknut.

NOTE: Make sure the cable is not kinked or twisted.

12. Check that the cable moves freely by depressing the accelerator.
13. Remove the weight on the accelerator pedal and push the pedal to make sure there is at least 0.08 in. (2.0mm) play at the throttle control lever.
14. Connect the negative battery cable and the vacuum hose to the dash pot.
15. Start the engine and check the synchronization between the carburetor and the throttle control cable. The throttle control lever should start to move as the engine speed increases.
16. If the throttle control lever starts to move before the engine speed increases, turn the cable locknut closest to the cable housing counterclockwise and retighten the locknut closest to the cable end.
17. If the throttle control lever moves after the engine speed increases, turn the locknut closest to the cable housing clockwise and retighten the locknut closest to the cable end.

THROTTLE CONTROL CABLE—1989–92

1. Start the engine and bring it to normal operating temperature. The cooling fan must come on at least once.
2. Make sure the throttle cable free-play and idle speed are correct.
3. Check the distance between the throttle control lever and throttle control bracket and adjust, as necessary.
4. On 1990–92 Prelude, disconnect the vacuum hose from the throttle controller and connect a vacuum pump to the controller and apply vacuum.
5. Apply light thumb pressure to the throttle control lever. Have an assistant depress the accelerator. The lever should move just as the engine speed increases above idle. If not, proceed to Step 6.
6. Loosen the nuts on the control cable at the transaxle end and synchronize the control lever to the throttle.

NOTE: The shift/lock-up characteristics can be tailored to the driver's expectations by adjusting the control cable up to 3mm shorter than the synchronized point.

Fuel Injected Engine

THROTTLE CONTROL CABLE—1988

1. Loosen the locknuts on the throttle control cable.
2. Press down on the throttle control lever until it stops.
3. While pressing down on the throttle control lever, pull on the throttle linkage to check the amount of throttle control cable free-play.
4. Remove all throttle control cable free-play by gradually turning the locknut closest to the cable housing. Keep turning the locknut until no movement can be felt in the throttle link, while continuing to press down on the throttle control lever, pull open the throttle link. The control lever should begin to move at precisely the same time as the link.

NOTE: The adjustment of the throttle control cable is critical for proper operation of the transaxle and lock-up torque converter.

5. Have an assistant depress the accelerator to the floor. While depressed, check that there is at least 0.08 in. (2.0mm) play in the throttle control lever. Check that the cable moves freely by depressing the accelerator.

THROTTLE CONTROL CABLE—1989–92

1. Start the engine and bring it up to operating temperature. The cooling fan must come on at least once.
2. Make sure the throttle cable free-play and idle speed are correct.
3. On dash pot equipped vehicles, disconnect the vacuum hose from the dash pot, connect a vacuum pump and apply vacuum. This simulates a normal operating amount of pull by the dash pot as if the engine were running.
4. Remove the throttle cable free-play.
5. Apply light thumb pressure to the throttle control lever, then work the throttle linkage. The lever should move just as the engine speed increases above idle. If not, proceed to Step 6.
6. Loosen the nuts on the control cable at the transaxle end and synchronize the control lever to the throttle.

NOTE: The shift/lock-up characteristics can be tailored to the driver's expectations by adjusting the control cable up to 3mm shorter than the synchronized point.

7. Remove the vacuum pump and connect the vacuum hose to the dash pot.

TRANSFER CASE

Transfer Case Assembly

REMOVAL & INSTALLATION

Manual Transaxle

1. Disconnect the negative battery cable.
2. Raise and safely support the vehicle.
3. Drain the transaxle fluid and replace the drain plug.
4. Remove the exhaust header pipe.
5. Disconnect the driveshaft from the transfer case.
6. Remove the transaxle splash shield.
7. Remove the left side cover from the transfer case.
8. Remove the driven gear from the transfer case.

NOTE: Be careful not to damage the thrust shim and mating surface.

9. Remove the transfer case from the clutch housing.
10. Remove the clutch case cover.

To install:

11. Make sure all mating surfaces are clean prior to installation.
12. Apply liquid gasket to the clutch housing mating surface of the transfer case and install the transfer case on the clutch housing. Tighten the bolts to 33 ft. lbs. (45 Nm).
13. Lubricate the drive gear with oil and install the drive gear thrust shim and the drive gear on the transfer shaft. Install the transfer thrust shim and left side cover on the transfer case. Apply liquid gasket to the threads and tighten the bolts to 33 ft. lbs. (45 Nm).
14. Apply liquid gasket to the mating surface on the top and bottom of the transfer case opening, install a new O-ring and install the driven gear thrust shim and driven gear in the transfer case. Tighten the bolts to 19 ft. lbs. (26 Nm).
15. Installation of the remaining components is the reverse of the removal procedure. Lower the vehicle and fill the transaxle with the proper type and quantity of fluid.
16. Connect the negative battery ca-

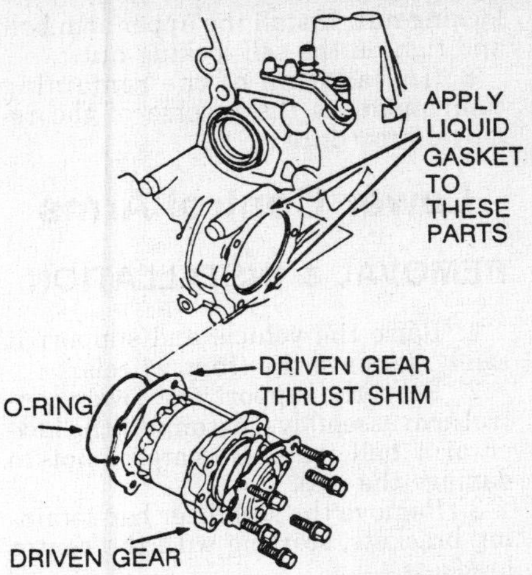

Transfer case driven gear installation—4WD Civic Wagon with

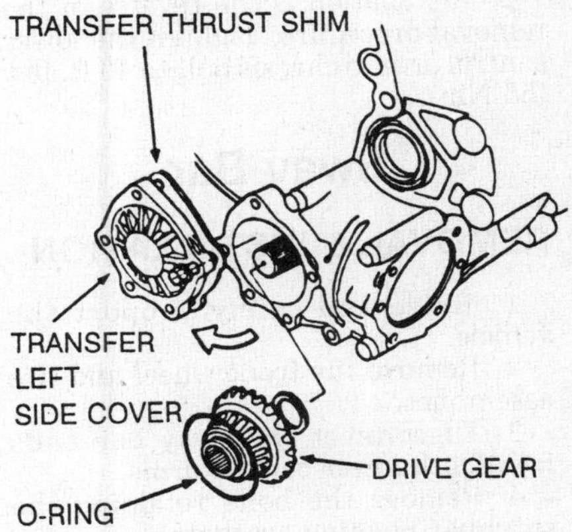

Transfer case drive gear and left side cover installation—4WD Civic Wagon with automatic transaxle

ble, start the vehicle and check for leaks.

Automatic Transaxle

1. Disconnect the negative battery cable.
2. Raise and safely support the vehicle.
3. Drain the transaxle fluid and replace the drain plug.
4. Remove the exhaust header pipe.
5. Disconnect the driveshaft from the transfer case.
6. Remove the transaxle splash shield.
7. Remove the driven gear assembly from the transfer case.
8. Remove the drive gear from the transfer case. In this procedure, remove all of the left side cover bolts except the 1 closest to the front of the vehicle. Loosen this bolt and use it as the axis, about which the cover can be rotated to gain access to the drive gear. Leave this bolt in the cover and transfer case.

NOTE: Be careful not to damage the thrust shim and mating surface.

9. Remove the transfer case from the clutch housing.
10. Remove the clutch case cover.

To install:

11. Make sure all mating surfaces are clean prior to installation.
12. Apply liquid gasket to the clutch housing mating surface of the transfer case. Attach a new O-ring to the groove on the left side cover.
13. Install the transfer case with the 1 bolt and left side cover still attached. Tighten the transfer case bolts to 33 ft. lbs. (45 Nm).
14. Lubricate the drive gear with oil and install the drive gear thrust shim and the drive gear on the transfer shaft. Install the transfer thrust shim and left side cover on the transfer case. Apply liquid gasket to the threads and tighten the bolts to 33 ft. lbs. (45 Nm).
15. Apply liquid gasket to the mating surface on the top and bottom of the transfer case opening, install a new O-ring and install the driven gear thrust shim and driven gear in the transfer case. Tighten the bolts to 19 ft. lbs. (26 Nm).
16. Installation of the remaining components is the reverse of the removal procedure. Lower the vehicle and fill the transaxle with the proper type and quantity of fluid.
17. Connect the negative battery cable, start the vehicle and check for leaks.

FRONT SUSPENSION

MacPherson Strut

REMOVAL & INSTALLATION

1. Raise and safely support the vehicle.
2. Remove the front tire and wheel assembly.
3. Remove the damper fork bolts and remove the damper fork.
4. Remove the flange nuts and remove the strut assembly.
5. Installation is the reverse of the removal procedure. Tighten the damper fork nut to 47 ft. lbs. (65 Nm) while holding the damper fork bolt. Tighten the damper fork pinch bolt to 32 ft. lbs. (44 Nm).
6. The flange nuts should not be tightened until the strut is under vehicle load. Tighten them to 28 ft. lbs. (39 Nm).

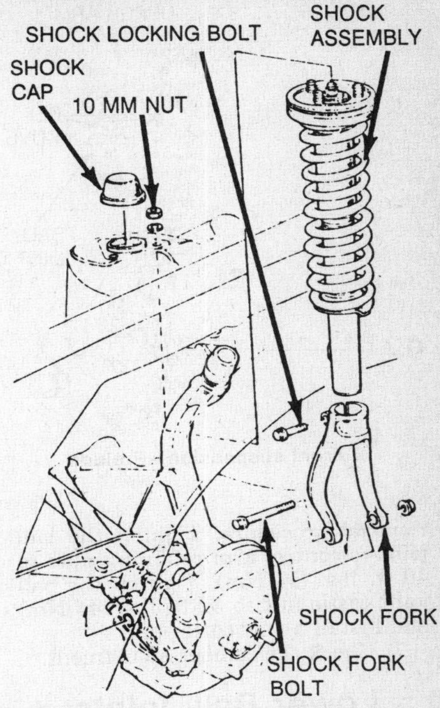

MacPherson strut installation—Accord and Prelude

Upper Ball Joints

INSPECTION

1. Raise and safely support the vehicle.
2. Remove the front wheel and tire assembly.
3. Grasp the steering knuckle and move it back and forth.
4. Replace the upper control arm on all except Prelude if any play is detected. On Prelude, replace the upper ball joint.

REMOVAL & INSTALLATION

Except Prelude

The upper ball joint is an integral component of the upper control arm. If the ball joint is defective the entire upper control arm must be replaced.

Prelude

1. Raise and safely support the vehicle.
2. Remove the front wheel and tire assembly.
3. Remove the cotter pin and castle nut from the upper ball joint.
4. Using a suitable tool, separate the upper ball joint from the steering knuckle.
5. Remove the 2 retaining nuts and the ball joint.
6. Installation is the reverse of the

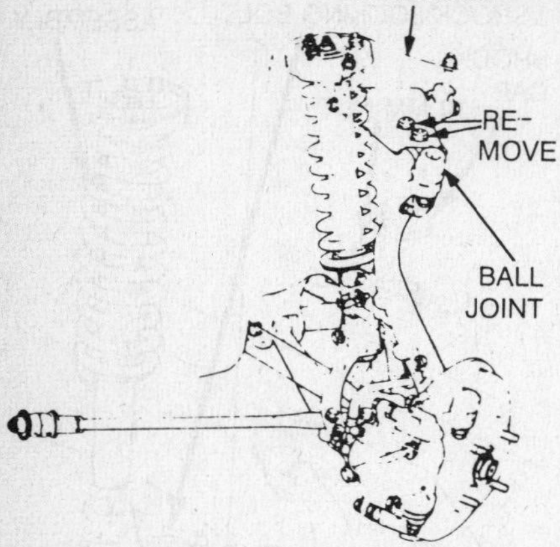

Front suspension—Prelude

removal procedure. Tighten the ball joint-to-control arm retaining nuts to 40 ft. lbs. (55 Nm). Tighten the ball joint castle nut to 32 ft. lbs. (44 Nm) and install a new cotter pin.

7. Check the camber adjustment.

Lower Ball Joints

INSPECTION

1. Raise and safely support the vehicle.
2. Remove the wheel and tire assembly.
3. Grasp the steering knuckle close to the lower ball joint and move it back and forth. Replace the ball joint, if any movement is detected.

REMOVAL & INSTALLATION

1. Raise and safely support the vehicle.
2. Remove the wheel and tire assembly.
3. Remove the steering knuckle.
4. Pry off the snapring and remove the boot.
5. Remove the circlip.
6. Install a suitable ball joint remover/installer 07965–SB00100 or equivalent, on the ball joint and tighten the ball joint nut.
7. Position ball joint remover base 07JAF–SH20200 or equivalent, over the ball joint and then set the assembly in a suitable vise. Press the ball joint out of the knuckle.

To install:

8. Place the ball joint in position by hand.
9. Install a suitable ball joint remover/installer 07965–SB00100 or equivalent, and ball joint installer base 07965–SB00200 or equivalent, over the ball joint and position in a suitable vise. Press the ball joint in.
10. Installation of the remaining components is the reverse of the removal procedure.

Upper Control Arms

REMOVAL & INSTALLATION

Civic, CRX and 1990–92 Accord

1. Raise and support the vehicle safely.
2. Remove the front wheels. Properly support the lower control arm assemblies.
3. Remove the self locking nuts, upper control arm bolts and upper control anchor bolts. Separate the upper ball joint using a suitable ball joint separator tool.
4. Place the upper control arm assembly into a suitable holding fixture and drive out the upper arm bushing.

To install:

5. Drive the new upper arm bushing into the the upper arm anchor bolts. On Civic and CRX, center the bushing so 0.3543 in. (9mm) protrudes from each side of the anchor bolt. On Accord, drive in the bushing so the leading edges are flush with the anchor bolt.
6. Install the upper control arm assembly and install the upper arm bolts, then tighten the self locking nuts. Be sure to align the upper arm anchor bolt with the mark on the upper arm.
7. Installation of the remaining components is the reverse of the removal procedure.

Prelude and 1988–89 Accord

1. Raise and support the vehicle safely.
2. Remove the front wheels. Properly support the lower control arm assemblies.
3. Remove the self locking nuts, upper control arm bolts and upper control anchor bolts. Separate the upper ball joint using a suitable ball joint separator tool.
4. Place the upper control arm assembly into a suitable holding fixture and remove the self locking nut, upper arm bolt, upper arm anchor bolts and housing seals.
5. Remove the upper arm collar. Drive out the upper arm bushing, using a suitable drift.
6. Replace the upper control arm bushings, bushing seals and upper control arm collar with new ones. Be sure to coat the ends and the insides of the upper control arm bushings, and the sealing lips of the upper control arm bushing with grease.
7. After Step 6 is completed, apply sealant to the threads and underside of the upper arm bolt heads and self locking nut. Install the upper arm bolt and tighten the self locking nut.
8. Installation of the remaining components is the reverse of the removal procedure.

Lower Control Arms

REMOVAL & INSTALLATION

1. Raise the vehicle and support it safely. Remove the front wheels.
2. Properly support the lower control arm assembly. Disconnect the lower arm ball joint. Be careful not to damage the seal.
3. Remove the stabilizer bar retaining brackets, starting with the center brackets.
4. Remove the lower arm pivot bolt.
5. Disconnect the radius rod and remove the lower arm.
6. Installation is the reverse of the removal procedure. Tighten the lower control arm to chassis bolt to 40 ft. lbs. (55 Nm).

Sway Bar

REMOVAL & INSTALLATION

1. Raise and safely support the vehicle.
2. Remove the front wheel and tire assemblies.
3. Disconnect the sway bar ends from both lower control arms.
4. Remove the bolts retaining the sway bar bushing brackets.
5. Remove the sway bar.
6. Installation is the reverse of the removal procedure.

REAR SUSPENSION

MacPherson Strut

REMOVAL & INSTALLATION

1. On Civic, CRX and 1988–89 Accord, remove the strut upper cover from inside the vehicle and remove the upper strut retaining nuts.
2. On all other vehicles, remove the strut upper cover from inside the trunk and remove the upper strut retaining nuts.
3. Raise and safely support the vehicle.
4. Remove the rear wheel and tire assembly.
5. Remove the strut mounting bolt,

lower the suspension and remove the strut.

6. Installation is the reverse of the removal procedure. Tighten the strut lower mounting bolt to 40 ft. lbs. (55 Nm) with the strut under vehicle load.

Rear Control Arms

REMOVAL & INSTALLATION

Trailing Arm

EXCEPT ACCORD

1. Raise and safely support the vehicle.
2. Remove the rear wheel and tire assembly.
3. On all except Civic Wagon with 4WD, perform the following procedure:
   a. Remove the brake drum or rotor.
   b. Remove the spindle nut and hub unit.
   c. Disconnect the parking brake cable and the brake line from the wheel cylinder. Plug the line.
   d. Remove the brake backing plate.
4. On 4WD Civic Wagon, remove the brake drum and spindle nut and disconnect the parking brake cable.
5. Disconnect the brake hose from the brake line. Plug the line.
6. Support the lower control arm or beam axle with a suitable jack.
7. Remove the trailing arm bushing mounting bolts.
8. Disconnect the upper arm and compensator arm from the trailing arm, if equipped.
9. On Civic Wagon with 4WD, remove the rear halfshaft outboard CV-joint from the trailing arm using a suitable puller.
10. Remove the trailing arm from the vehicle.
11. Installation is the reverse of the removal procedure. Tighten all bolts and nuts with the vehicle on the ground. Bleed the brake system.

ACCORD

1. Raise and safely support the vehicle.
2. Remove the rear wheel and tire assembly.
3. Support the lower control arm using a suitable jack.
4. Remove the bolt from the trailing arm bushing.
5. Remove the mounting nuts from the knuckle and remove the trailing arm.
6. Installation is the reverse of the removal procedure.

Upper Control Arm

1. Raise and safely support the vehicle.

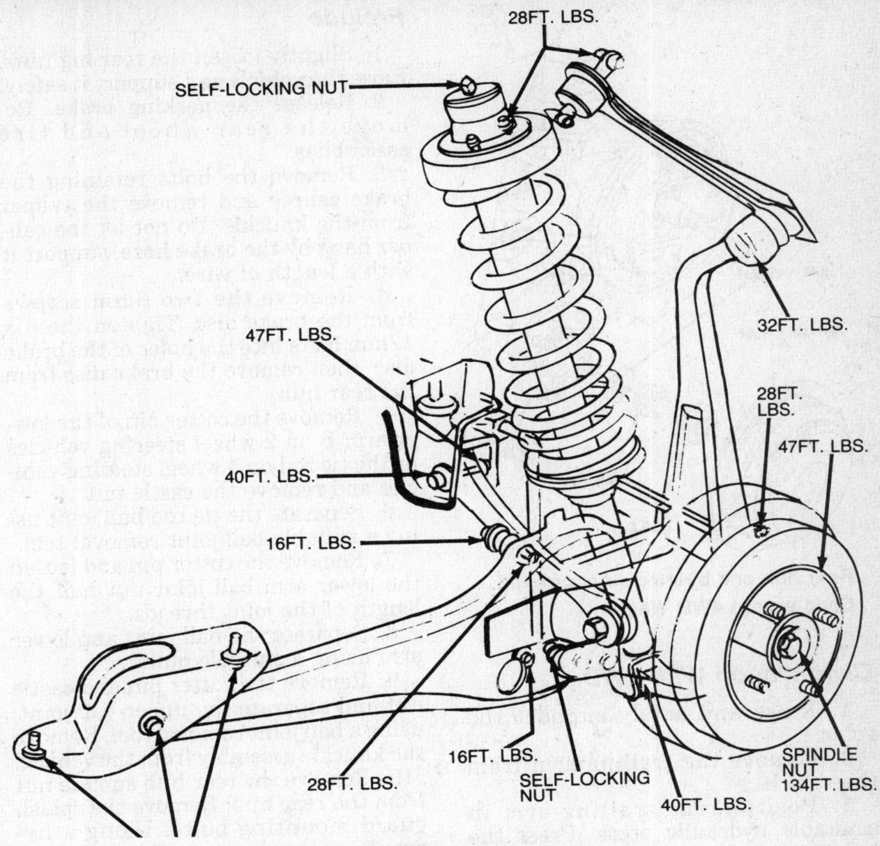

Rear suspension – 1988–89 Accord

2. Remove the rear wheel and tire assembly.
3. Support the lower control arm or rear axle, as necessary.
4. On Accord and Prelude, remove the cotter pin and castle nut from the upper ball joint and use a suitable tool to separate the ball joint from the knuckle.
5. Remove the upper control arm mounting bolts and the upper control arm.
6. Installation is the reverse of the removal procedure.

Lower Control Arm

1. Raise and safely support the vehicle.
2. Remove the rear wheel and tire assembly.
3. Remove the strut and/or radius rod mounting bolts from the lower control arm, if necessary.
4. On Prelude, remove the cotter pin and castle nut from the ball joint and separate the ball joint from the knuckle using a suitable tool.
5. Remove the lower arm mounting bolts and remove the lower control arm.
6. Installation is the reverse of the removal procedure.

Rear Wheel Bearings

REMOVAL & INSTALLATION

Accord and Civic Except Wagon With 4WD

NOTE: Do not tighten or loosen a spindle nut unless the vehicle is sitting on all 4 wheels. The torque is high enough to cause the vehicle to fall even when properly supported.

1. Loosen the rear lug nuts and the spindle nut. Raise the vehicle and support it safely.
2. Release the parking brake. Remove the rear wheel and the brake drum.
3. Remove the rear bearing hub cap and nut.
4. Pull the hub unit off of the spindle.
5. Installation is the reverse order of removal. With the vehicle on the ground, torque the new spindle nut to 134 ft. lbs. (185 Nm), then stake the nut.

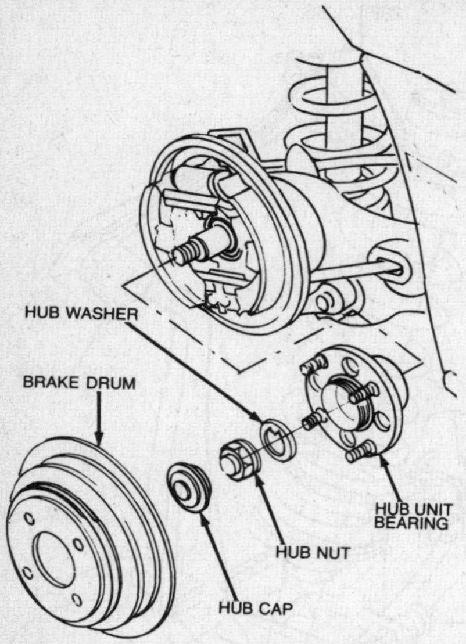

Rear hub and bearing—Accord and Civic except 4WD Wagon

Civic Wagon With 4WD

1. Raise and safely support the vehicle.
2. Remove the trailing arm from the vehicle.
3. Position the trailing arm in asuitable hydraulic press. Press the hub from the trailing arm using a suitable driver while supporting the trailing arm.

NOTE: Be careful not to distort the brake backing plate. Hold onto the rear hub and trailing arm to keep it from falling when pressed clear.

4. Remove the outboard bearing inner race from the hub using a suitable bearing puller.
5. Remove the 64mm snapring.
6. Remove the bolts and the backing plate.
7. Remove the O-ring from the groove of the bearing holder plate.
8. Press the wheel bearing out of the trailing arm, using a suitable driver, while supporting the trailing arm.

To install:

9. Clean the trailing arm and hub thoroughly.
10. Press a new wheel bearing into the trailing arm, using a suitable driver, while supporting the trailing arm.
11. Install the O-ring on the groove of the bearing holder plate.
12. Install the backing plate and the snapring.
13. Press the trailing arm onto the hub, using a suitable guide and driver, while supporting the hub.
14. Installation of the remaining components is the reverse of the removal procedure.

Prelude

1. Slightly loosen the rear lug nuts. Raise the vehicle and support it safely.
2. Release the parking brake. Remove the rear wheel and tire assemblies.
3. Remove the bolts retaining the brake caliper and remove the caliper from the knuckle. Do not let the caliper hang by the brake hose, support it with a length of wire.
4. Remove the two 6mm screws from the brake disc. Tighten the 8 x 12mm bolts into the holes of the brake disc, then remove the brake disc from the rear hub.
5. Remove the cotter pin of the lower arm B on 2 wheel steering vehicles or the tie rod on 4 wheel steering vehicles and remove the castle nut.
6. Separate the tie rod ball joint using a suitable ball joint removal tool.
7. Remove the cotter pin and loosen the lower arm ball joint nut half the length of the joint threads.
8. Separate the ball joint and lower arm using a suitable puller.
9. Remove the cotter pin and castle nut and separate the upper ball joint, using a ball joint removal tool. Remove the knuckle assembly from the vehicle.
10. Remove the rear hub spindle nut from the rear hub. Remove the splash guard mounting bolts. Using a hydraulic press, separate the hub from the knuckle.

NOTE: Set the rear hub at the hub/disc assembly base firmly, so the knuckle will not tilt the assembly in the press. Take care not to distort the splash guard. Hold onto the hub to keep it from falling after it is pressed out.

11. Remove the splash guard and 68mm circlip from the knuckle.
12. Using a hydraulic press and suitable press tools, press the wheel bearing out of the knuckle.
13. Remove the bearing inner race using a suitable bearing remover.

To install:

14. Place the rear wheel bearing in the tool fixture, then set the knuckle into position and apply downward pressure with a hydraulic press. Fit the 68mm circlip into the groove of the knuckle.
15. Install the splash guard. Place the hub in the tool fixture, then set the knuckle into position and apply downward pressure with a hydraulic press. Install the rear hub nut and torque the spindle nut to 180 ft. lbs. (250 Nm).
16. Install the knuckle assembly onto the vehicle and install all nuts and bolts loosely. Torque the lower ball joint nut to 40 ft. lbs. (55 Nm), then tighten as required to install a new cotter pin.
17. On vehicles with 4 wheel steering, torque the tie rod end joint to 32 ft. lbs. (44 Nm), then tighten as required to install a new cotter pin.
18. When everything is assembled, lower the vehicle and tighten the rubber bushing nuts and bolts with the weight of the vehicle on the wheels. Torque the upper bushing nut to 32 ft. lbs. (44 Nm) and the lower bushing bolt to 40 ft. lbs. (55 Nm).

STEERING

CAUTION

The Accord Wagon is equipped with a driver's side air bag. To avoid accidental deployment and serious personal injury, the system must be disarmed before beginning any repair procedure.

Steering Wheel

REMOVAL & INSTALLATION

1. Disconnect the negative battery cable. Disarm the air bag, if equipped, and remove the air bag unit or the steering wheel pad. Disconnect the necessary electrical connections under the steering wheel pad.
2. Remove the steering wheel retaining nut. Remove the steering wheel by rocking it from side to side, as it is pulled up steadily by hand.
3. Installation is the reverse of the removal procedure. Be sure to tighten the steering wheel nut to 36 ft. lbs. (50 Nm).

Steering Column

REMOVAL & INSTALLATION

1. Disconnect the negative battery cable.
2. Remove the steering wheel.
3. Remove the lower cover panel. Remove the driver's knee bolster, if equipped.
4. Remove the upper and lower column covers.
5. Disconnect the wire couplers from the combination switch. Remove the turn signal cancelling sleeve and the combination switch.
6. Remove the steering joint cover and remove the steering joint bolt(s).
7. Disconnect each wire coupler from the fuse box under the left side of the dashboard.
8. Remove the steering column retaining brackets.
9. Remove the nuts attaching the bending plate guide and bending plate and remove the steering column assembly.

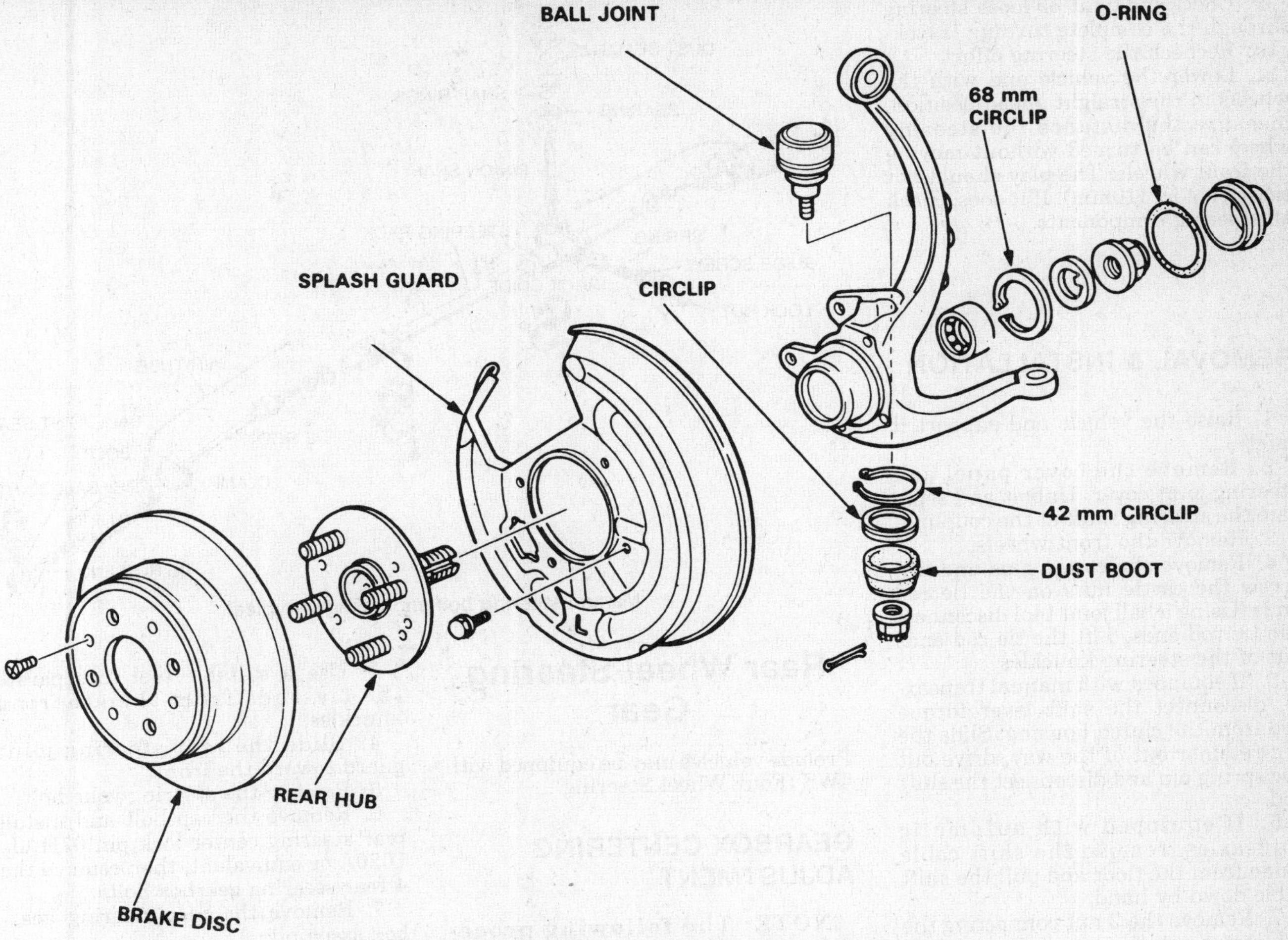

Rear wheel bearing and hub assembly on Prelude

To install:

10. Fit the column into place and secure the bracket, bending plate and guide. Torque the nuts to 16 ft. lbs. (22 Nm).
11. Install the steering joint bolts and torque to 16 ft. lbs. (22 Nm).
12. Connect the wiring at the fuse box.
13. Install the switches and connect the wiring.
14. Install the knee bolster and steering wheel. Torque the wheel nut to 36 ft. lbs. (50 Nm).

Manual Steering Rack and Pinion

ADJUSTMENT

1. Raise and safely support the vehicle.
2. Turn the steering wheel with a suitable spring gauge and check the reading.
3. If the reading exceeds 3.3 lbs., adjust the steering gear as oulined in the following Steps.
4. Place the front wheels in the straight ahead position.
5. Loosen the rack screw locknut.
6. On 1988–92 Civic Si, perform the following procedure:
   a. Tighten, loosen and retighten the rack guide screw 2 times, to 3.6 ft. lbs. (5 Nm), then back it off 15 degrees.
   b. Tighten the locknut on the rack guide screw to 49 ft. lbs. (68 Nm).
7. On 1988–92 CRX Si, perform the following procedure:
   a. Retighten the rack guide screw until it compresses the spring and seats against the rack guide.
   b. Back off the rack guide screw 15 degrees and install the locknut on the rack guide screw. Tighten the locknut to 18 ft. lbs. (25 Nm).
8. On all other vehicles, perform the following procedure:
   a. Tighten the rack screw until it compresses the spring and seats against the rack guide.
   b. Back the rack guide screw off 40–60 degrees and tighten the locknut on the rack guide screw to 18 ft. lbs. (25 Nm).

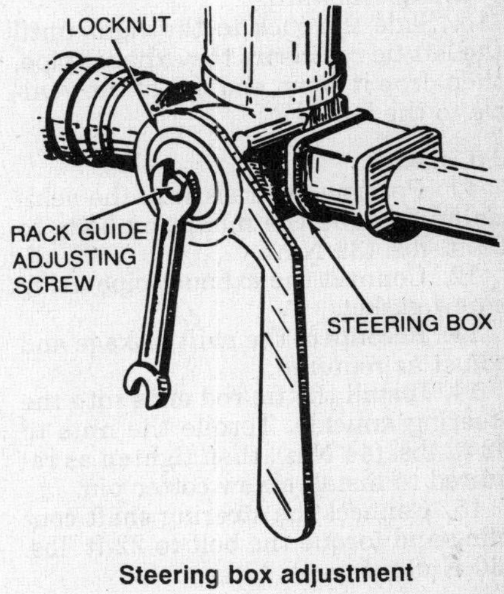

Steering box adjustment

9. Check for tight or loose steering through the complete turning travel.
10. Recheck the steering effort.
11. Lower the vehicle and with the wheels in the straight ahead position, measure the distance the steering wheel can be turned without moving the front wheels. The play should not exceed 0.4 in. (10mm). If it does, check all steering components.

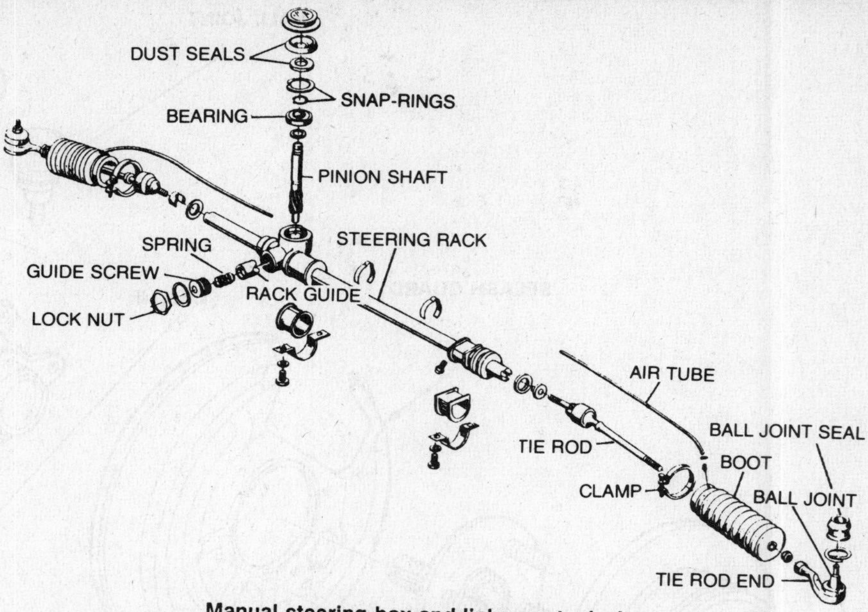

Manual steering box and linkage—typical

REMOVAL & INSTALLATION

1. Raise the vehicle and support it safely.
2. Remove the cover panel and steering joint cover. Unbolt and separate the steering shaft at the coupling.
3. Remove the front wheels.
4. Remove the cotter pins and unscrew the castle nuts on the tie rod ends. Using a ball joint tool disconnect the tie rod ends. Lift the tie rod ends out of the steering knuckles.
5. If equipped with manual transaxle, disconnect the shift lever torque rod from the clutch housing. Slide the pin retainer out of the way, drive out the spring pin and disconnect the shift rod.
6. If equipped with automatic transaxles, remove the shift cable guide from the floor and pull the shift cable down by hand.
7. Remove the 2 nut connecting the exhaust header pipe to the exhaust pipe and move the exhaust pipe out of the way.
8. Push the rack all the way to the right and remove the brackets or mounting bolts. Slide the tie rod ends all the way to the right.
9. Drop the rack far enough to permit the end of the pinion shaft to come out of the hole in the frame channel, then rotate it forward until the shaft is pointing rearward.
10. Slide the rack to the right until the left tie rod clears the exhaust pipe, then drop it down and out of the vehicle to the left.

To install:

11. Position the rack into the vehicle and torque the mounting bolts to 29 ft. lbs. (39 Nm).
12. Connect the exhaust pipe using a new gasket.
13. Reconnect the shift linkage and adjust as required.
14. Install the tie rod ends into the steering knuckle. Torque the nuts to 32 ft. lbs. (44 Nm), then tighten as required to install a new cotter pin.
15. Connect the steering shaft coupling and torque the bolt to 22 ft. lbs. (30 Nm).

Rear Wheel Steering Gear

Prelude vehicles may be equipped with 4WS (Four Wheel Steering).

GEARBOX CENTERING ADJUSTMENT

NOTE: The following procedure must be used after reassembling/replacing the steering gearbox components or in preparing to solve complaints of mis-adjusted steering wheel angle.

1. Center the steering and steering wheel by eye.
2. Raise and safely support the vehicle.
3. Remove the gearbox cap bolt using a TORX® T-40 bit.
4. Insert rear steering center lock pin 07HAJ–SF1020A or equivalent, in the gearbox.
5. Turn the steering wheel right or left slightly until the pin of the tool seats fully. The red mark on the pin should not be visible.

NOTE: Do not turn the steering wheel quickly when the tool is seated and do not force past the locking point after the pin is seated or the pin may be damaged.

REMOVAL & INSTALLATION

1. Raise and safely support the vehicle.
2. Remove the rear wheel and tire assemblies.
3. Use a suitable tool to separate the tie-rods from the steering knuckles.
4. Slide the rear steering joint guard toward the front.
5. Remove the steering yoke bolt.
6. Remove the cap bolt and install rear steering center lock pin 07HAJ–1020A or equivalent, then remove the 4 rear steering gearbox bolts.
7. Remove the rear steering gearbox assembly.
8. Installation is the reverse of the removal procedure. Tighten the 4 gearbox mounting bolts to 29 ft. lbs. (40 Nm). Tighten the steering yoke bolt to 22 ft. lbs. (30 Nm). Tighten the tie-rod nuts to 32 ft. lbs. (44 Nm) and install new cotter pins.

Power Steering Rack and Pinion

ADJUSTMENT

Civic and Accord

1. Remove the steering gear splash shield, if equipped.
2. Loosen the rack guide adjusting locknut.
3. Tighten the adjusting screw until it compresses the spring and seats against the guide, then loosen it. Retorque it to 35 inch lbs., then back it off 20 degrees on 1988–92 Civic, 25 degrees on 1988–89 Accord and 35 degrees on 1990–92 Accord.
4. Hold it in that position while adjusting the locknut to 18 ft. lbs.
5. Recheck the play, and then move the wheels lock-to-lock, to make sure the rack moves freely.

Prelude

1. Make sure the rack is well lubricated.
2. Loosen the rack guide adjusting locknut.
3. Tighten the adjusting screw until it compresses the spring and seats against the guide, then loosen it. Retorque it to 24 inch lbs., then back it off 20–30 degrees (15–25 degrees on 1990–92 vehicles) on 2 wheel steering vehicles. Retorque it to 3 ft. lbs. and 30–40 degrees on 4 wheel steering vehicles).
4. Hold it in that position while adjusting the locknut to 18 ft. lbs.
5. Recheck the play, and then move the wheels lock-to-lock, to make sure the rack moves freely.

REMOVAL & INSTALLATION

Civic

1. Disconnect the negative battery cable. Raise the vehicle and support it safely.
2. Remove the cover panel and steering joint cover. Unbolt and separate the steering shaft at the coupling.
3. Drain the power steering fluid by disconnecting the return hose at the box and running the engine while turning the steering wheel lock to lock until fluid stops draining. Remove the gearbox shield, if equipped. Remove the front wheels.
4. Remove the cotter pins and unscrew the castle nuts on the tie rod ends. Using a ball joint tool, disconnect the tie rod ends. Lift the tie rod ends out of the steering knuckles.
5. If equipped with manual transaxle, disconnect the shift lever torque rod from the clutch housing. Slide the pin retainer out of the way, drive out the spring pin and disconnect the shift rod.
6. If equipped with automatic transaxle, remove the shift cable guide from the floor and pull the shift cable down by hand.
7. Remove the 2 nuts connecting the exhaust header pipe to the exhaust pipe and remove the exhaust header pipe. Disconnect the 3 hydraulic lines from the control unit.
8. Slide the tie rod ends all the way to the right and remove the steering rack mounting bolts.
9. Drop the gearbox far enough to permit the end of the pinion shaft to come out of the hole in the frame channel, then rotate it forward until the shaft is pointing rearward.
10. Slide the gearbox to the right until the left tie rod clears the beam, then drop it down and out of the vehicle to the left.

To install:

11. Installation is the reverse of removal. Torque the mounting bracket bolts to 29 ft. lbs. (40 Nm). If equipped with a manual transaxle, reinstall the pin retainer after driving in the pin and be sure the projection on the pin retainer is in the hole.
12. Fill the system with fluid and bleed the air from the system.

Accord and Prelude

1. Disconnect the negative battery cable. Raise the vehicle and support it safely.
2. Remove the steering shaft joint cover and disconnect the steering shaft at the coupling.
3. Drain the power steering fluid by disconnecting the return hose at the box and running the engine, while turning the steering wheel lock to lock until fluid stops draining.
4. Remove the gearbox shield.
5. Remove the front wheels.
6. Using a ball joint tool, disconnect the tie rods from the knuckles.
7. If equipped with manual transaxle, remove the shift extension from the transaxle case. Disconnect the gear shift rod from the transaxle case by removing the 8mm bolt.
8. If equipped with automatic transaxle, remove the control cable clamp.
9. Remove the center beam.
10. On the 4 wheel steering vehicles, separate the joint guard cap and the joint guard. Remove the joint bolt from the driven pinion side. Remove the joint bolt from the center steering shaft side, then slide the joint back to disconnect it from the driven pinion.
11. Remove the exhaust header pipe.
12. Disconnect the hydraulic lines at the steering control unit. On 4 wheel steering vehicles, disconnect the driven drain hose.
13. On Prelude, remove the mounting bolts and lower the front sway bar.
14. Shift the tie rods all the way right.
15. Remove the gearbox mounting bolts.
16. Slide the gearbox right so the left tie rod clears the bottom of the rear beam. Remove the gearbox.

To install:

17. Position the gear box in the vehicle and torque the clamp bolts to 16 ft. lbs. (22 Nm) on 1988–89 Accord and 29 ft. lbs. (40 Nm) on all others.
18. On Prelude, install the sway bar.
19. Connect the hydraulic lines and the exhaust pipe.
20. Connect the shift linkage.
21. Connect the tie rod ends to the steering knuckles. Torque the nuts to 32 ft. lbs. (44 Nm) and tighten as requied to install a new cotter pin.
22. Connect the steering shaft coupling and torque to 22 ft. lbs. (30 Nm).
23. Fill the system with fluid and bleed the air from the system.

Power Steering Pump

REMOVAL & INSTALLATION

1. Drain the fluid from the system as follows:
   a. Disconnect the cooler return hose from the reservoir and place the end in a large container.
   b. Start the engine and allow it to run at fast idle. Turn the steering wheel from lock to lock several times, until fluid stops running from the hose. Shut off the engine and discard the fluid.
   c. Reattach the hose on all vehicles with a separate reservoir.
2. Disconnect the inlet and outlet hoses at the pump. Remove the drive belt.
3. Remove the bolts and remove the pump.
4. Installation is the reverse of the removal procedure. Adjust the belt tension, fill the reservoir and bleed the air from the system.

BELT ADJUSTMENT

1. Push on the belt mid way between the pulleys with a force of about 22 lbs. The belt deflection should be as follows:

1988–92 Civic – 0.35–0.47 in. (9–12mm).

1988–89 Accord – 0.55–0.67 in. (14–17mm).

1990–92 Accord – 0.50 – 0.62 in. (12.5–16mm).

1988–92 Prelude – 0.43–0.51 in. (11–13mm).

2. If belt deflection is not as specified, adjust as follows:
   a. 1990–92 Accord – Loosen the pivot bolt and mounting nut. Turn the adjusting bolt to get the proper tension. Tighten the pivot bolt to 33 ft. lbs. (45 Nm) and the mounting nut to 16 ft. lbs. (22 Nm).
   b. 1988–92 Prelude – Loosen the adjusting pulley bolt and turn the adjusting bolt to get the proper tension. Tighten the pulley bolt to 35 ft. lbs. (49 Nm).
   c. All others – Loosen the pump pivot bolt and the adjusting nut or bolt. Pry the pump away from the engine to get the proper tension. Tighten the pivot bolt to 28 ft. lbs. (39 Nm). Tighten the adjusting nut or bolt to 28 ft. lbs. (39 Nm) except 1988–89 Accord. Tighten the adjusting nut on Accord to 16 ft. lbs. (22 Nm).

SYSTEM BLEEDING

1. Make sure the reservoir is filled to the full mark.
2. Start the engine and allow it to idle.
3. Turn the steering wheel from side to side several times, lightly contacting the stops.
4. Turn the engine off.
5. Check the fluid level in the reservoir and add if necessary.

Tie Rod Ends

REMOVAL & INSTALLATION

1. Raise the vehicle and support it safely. Remove the front wheels.
2. Remove the cotter pins and castle nuts from the tie rod ends. Use a ball joint remover to remove the tie rod from the knuckle.
3. Disconnect the air tube at the dust seal joint. Remove the tie rod dust seal bellows clamps and move the rubber bellows on the tie rod rack joints.
4. Straighten the tie rod lock washer tabs at the tie rod-to-rack joint and remove the tie rod by turning it with a wrench.
5. To install, reverse the removal procedure. Always use a new tie rod lock washer during reassembly. Install the locating lugs into the slots on the rack and bend the outer edge of the washer over the flat part of the rod, after the tie rod nut has been properly tightened.
6. Check the toe setting of the front end alignment.

BRAKES

For all brake system repair and service procedures not detailed below, please refer to "Brakes" in the Unit Repair section.

Master Cylinder

REMOVAL & INSTALLATION

1. Disconnect the negative battery cable. Disconnect and plug the brake lines at the master cylinder.
2. Remove the master cylinder-to-vacuum booster attaching bolts and remove the master cylinder from the vehicle.
3. To install, reverse the removal procedure. Before operating the vehicle, bleed the brake system.

Proportioning Valve

REMOVAL & INSTALLATION

1. Disconnect the negative battery cable.
2. Disconnect and plug the brake lines at the valve.
3. Remove the valve mounting bolts and the valve.
4. Installation is the reverse of the removal procedure. Bleed the brake system.

Power Brake Booster

REMOVAL & INSTALLATION

1. Disconnect the negative battery cable. Disconnect the vacuum hose at the booster.
2. It may be possible to remove the master cylinder to brake booster retaining nuts and than position the master cylinder assembly to the side on some vehicles. If not, the master cylinder will have to be removed from the vehicle.
3. Remove the brake pedal-to-booster link pin and the 4 nuts retaining the booster. The pushrod and nuts are located inside the vehicle under the instrument panel.
4. Remove the booster assembly from the vehicle.
5. To install, reverse the removal procedure. If the master cylinder was removed, bleed the brake system.

Brake Caliper

REMOVAL & INSTALLATION

Front

1. Raise and safely support the vehicle.
2. Remove the front wheel and tire assembly.
3. Remove the banjo bolt and disconnect the brake hose from the caliper. Plug the hose.
4. Remove the mounting bolt(s) and the caliper.

To install:

5. Installation is the reverse of the removal procedure. Use new gaskets on the banjo bolt and torque to 25 ft. lbs. (35 Nm).
6. On vehicles that have long pins below the threads of the caliper bolt, torque the bolt to 54 ft. lbs. (75 Nm).
7. On vehicles that have short bolts with no pin beyond the threads, torque the bolt to 36 ft. lbs. (50 Nm).
8. Bleed the brakes.

Rear

1. Raise and safely support the vehicle.
2. Remove the rear wheel and tire assembly.
3. Remove the caliper shield.
4. Disconnect the parking brake cable from the lever on the caliper by removing the lock pin.
5. Remove the banjo bolt and disconnect the brake hose from the caliper. Plug the hose.
6. Remove the 2 caliper mounting bolts and the caliper from the bracket.

To install:

7. Installation is the reverse of the removal procedure. Use new gaskets on the banjo bolt and torque to 25 ft. lbs. (35 Nm).
8. Torque the caliper bracket bolts to 28 ft. lbs. (39 Nm).
9. Bleed the brakes.

Disc Brake Pads

REMOVAL & INSTALLATION

Front

1. Remove the master cylinder cover and remove half the quantity of brake fluid in the master cylinder.
2. Raise and support the vehicle safely. Remove the tire and wheel assemblies.
3. As required, separate the brake hose clamp from the knuckle by removing the retaining bolts.
4. Remove the lower caliper retaining bolt and pivot the caliper out of the way.
5. Remove the pad shim and pad retainers. Remove the disc brake pads from the caliper.

To install:

6. Clean the caliper thoroughly; remove any rust. Check the brake rotor for grooves or cracks and machine or replace, as necessary.
7. Install the pad retainers. Apply a suitable disc brake pad lubricant to both surfaces of the shims and the back of the disc brake pads. Do not get any lubricant on the braking surface of the pad.
8. Install the pads and shims.
9. Use a suitable tool to push in the caliper piston so the caliper will fit over the pads.
10. Pivot the caliper down into position and tighten the mounting bolts.
11. Connect the brake hose to the knuckle, if removed. Install the wheel and tire assembly and lower the vehicle.
12. Check the master cylinder and add fluid as required, then replace the master cylinder cover. Depress the brake pedal several times to seat the pads.

Rear

1. Remove the master cylinder cov-

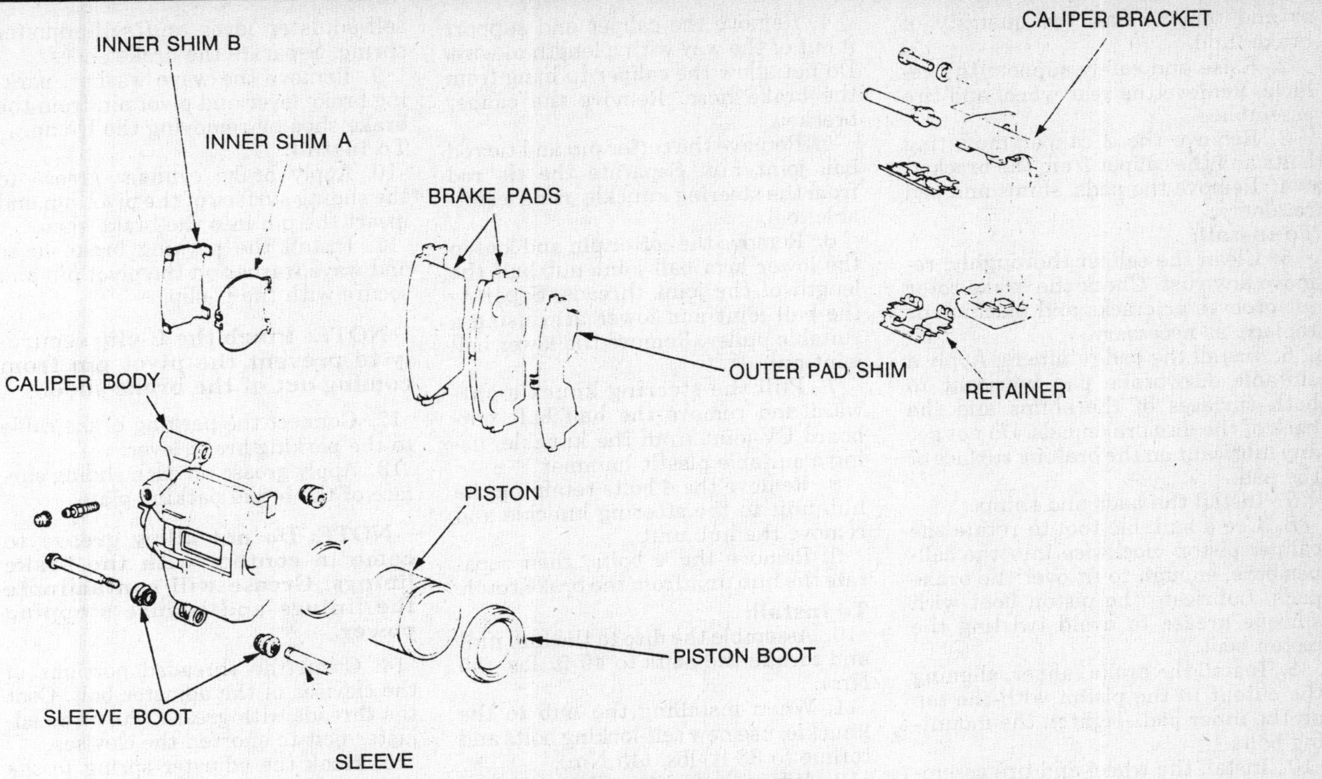

Typical front disc brake assembly

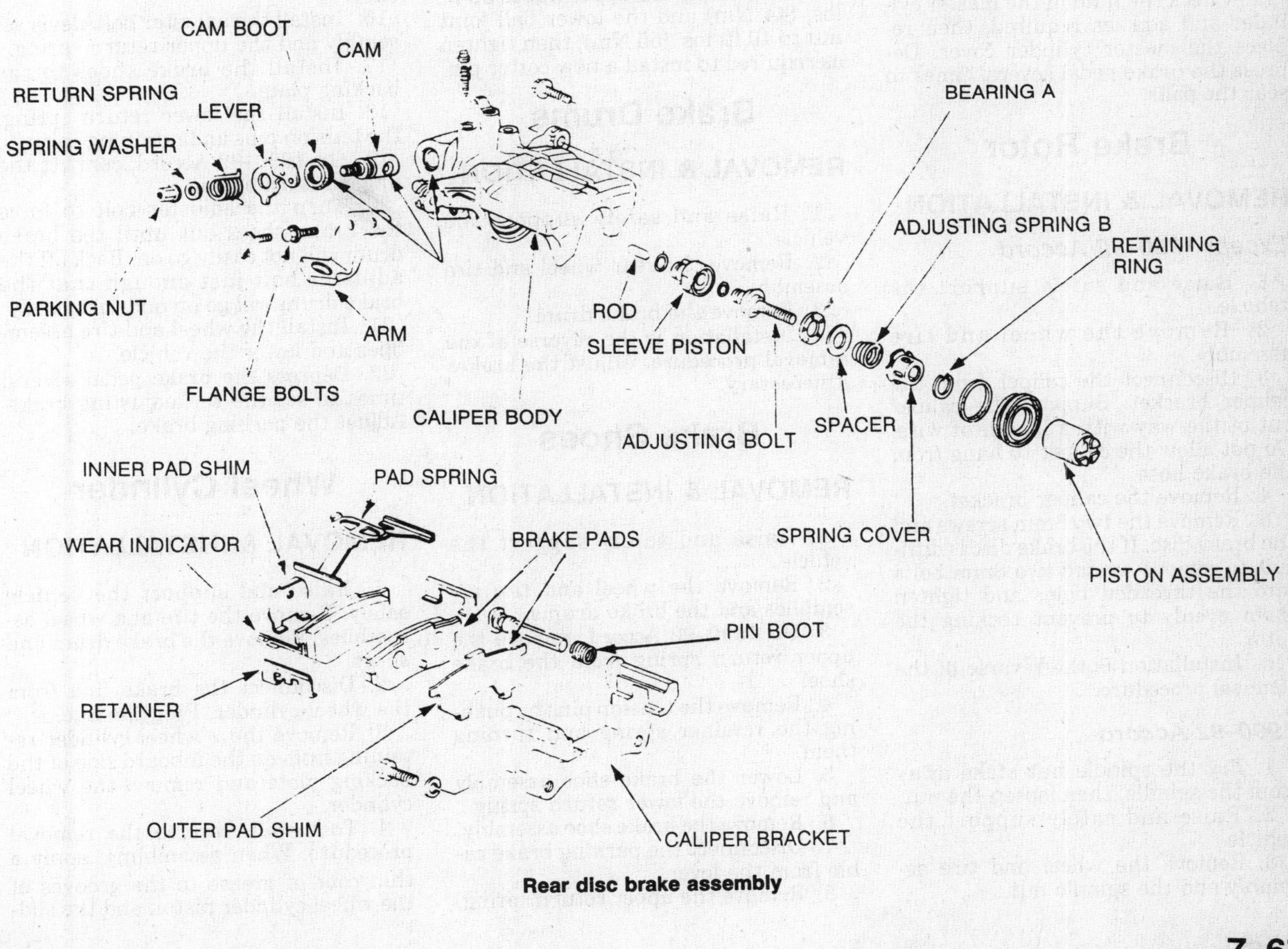

Rear disc brake assembly

er and remove half the quantity of brake fluid.

2. Raise and safely support the vehicle. Remove the rear wheel and tire assemblies.

3. Remove the 2 caliper mounting bolts and the caliper from the bracket.

4. Remove the pads, shims and pad retainers.

To install:

5. Clean the caliper thoroughly; remove any rust. Check the brake rotor for grooves or cracks and machine or replace, as necessary.

6. Install the pad retainers. Apply a suitable disc brake pad lubricant to both surfaces of the shims and the back of the disc brake pads. Do not get any lubricant on the braking surface of the pad.

7. Install the pads and shims.

8. Use a suitable tool to rotate the caliper piston clockwise into the caliper bore, enough to fit over the brake pads. Lubricate the piston boot with silicone grease to avoid twisting the piston boot.

9. Install the brake caliper, aligning the cutout in the piston with the tab on the inner pad. Tighten the mounting bolts.

10. Install the wheel and tire assemblies and lower the vehicle.

11. Check the fluid in the master cylinder and add, as required, then replace the master cylinder cover. Depress the brake pedal several times to seat the pads.

Brake Rotor

REMOVAL & INSTALLATION

Except 1990–92 Accord

1. Raise and safely support the vehicle.

2. Remove the wheel and tire assembly.

3. Disconnect the caliper from the caliper bracket. Support the caliper out of the way with a length of wire. Do not allow the caliper to hang from the brake hose.

4. Remove the caliper bracket.

5. Remove the two 6mm screws and the brake disc. If the brake disc is difficult to remove, install two 8mm bolts into the threaded holes and tighten them evenly to prevent cocking the rotor.

6. Installation is the reverse of the removal procedure.

1990–92 Accord

1. Pry the spindle nut stake away from the spindle, then loosen the nut.

2. Raise and safely support the vehicle.

3. Remove the wheel and tire assembly and the spindle nut.

4. Remove the caliper and support it out of the way with a length of wire. Do not allow the caliper to hang from the brake hose. Remove the caliper bracket.

5. Remove the cotter pin and tie rod ball joint nut. Separate the tie rod from the steering knuckle using a suitable tool.

6. Remove the cotter pin and loosen the lower arm ball joint nut half the length of the joint threads. Separate the ball joint and lower arm using a suitable puller. Remove the lower ball joint nut.

7. Pull the steering knuckle outward and remove the halfshaft outboard CV-joint from the knuckle, using a suitable plastic hammer.

8. Remove the 4 bolts retaining the hub unit to the steering knuckle and remove the hub unit.

9. Remove the 4 bolts, then separate the hub unit from the brake rotor.

To install:

10. Assemble the disc to the hub unit and torque the bolts to 40 ft. lbs. (55 Nm).

11. When installing the hub to the knuckle, use new self-locking bolts and torque to 33 ft. lbs. (45 Nm).

12. When installing the steering knuckle, torque the upper nut to 32 ft. lbs. (44 Nm) and the lower ball joint nut to 40 ft. lbs. (55 Nm), then tighten as required to install a new cotter pin.

Brake Drums

REMOVAL & INSTALLATION

1. Raise and safely support the vehicle.

2. Remove the rear wheel and tire assembly.

3. Remove the brake drum.

4. Installation is the reverse of the removal procedure. Adjust the brakes if necessary.

Brake Shoes

REMOVAL & INSTALLATION

1. Raise and safely support the vehicle.

2. Remove the wheel and tire assemblies and the brake drums.

3. On 1990–92 Accord, remove the upper return spring from the brake shoe.

4. Remove the tension pins by pushing the retainer spring and turning them.

5. Lower the brake shoe assembly and remove the lower return spring.

6. Remove the brake shoe assembly.

7. Disconnect the parking brake cable from the lever.

8. Remove the upper return spring, self-adjuster lever and self-adjuster spring. Separate the brake shoes.

9. Remove the wave washer, parking brake lever and pivot pin from the brake shoe by removing the U-clip.

To install:

10. Apply brake cylinder grease to the sliding surface of the pivot pin and insert the pin into the brake shoe.

11. Install the parking brake lever and wave washer on the pivot pin and secure with the U-clip.

NOTE: Pinch the U-clip securely to prevent the pivot pin from coming out of the brake shoe.

12. Connect the parking brake cable to the parking brake lever.

13. Apply grease on each sliding surface of the brake backing plate.

NOTE: Do not allow grease to come in contact with the brake linings. Grease will contaminate the linings and reduce stopping power.

14. Clean the threaded portions of the clevises of the adjuster bolt. Coat the threads with grease. Turn the adjuster bolt to shorten the clevises.

15. Hook the adjuster spring to the adjuster lever first, then to the brake shoe.

16. Install the adjuster bolt/clevis assembly and the upper return spring.

17. Install the brake shoes to the backing plate.

18. Install the lower return spring, the tension pins and retaining springs.

19. On 1990–92 Accord, connect the upper return spring.

20. Turn the adjuster bolt to force the brake shoes out until the brake drum will not easily go on. Back off the adjuster bolt just enough that the brake drum will go on and turn easily.

21. Install the wheel and tire assemblies and lower the vehicle.

22. Depress the brake pedal several times to set the self adjusting brake. Adjust the parking brake.

Wheel Cylinder

REMOVAL & INSTALLATION

1. Raise and support the vehicle safely. Remove the tire and wheel assemblies. Remove the brake drum and shoes.

2. Disconnect the brake line from the wheel cylinder. Plug the line.

3. Remove the 2 wheel cylinder retaining nuts on the inboard side of the backing plate and remove the wheel cylinder.

4. To install, reverse the removal procedure. When assembling, apply a thin coat of grease to the grooves of the wheel cylinder piston and the slid-

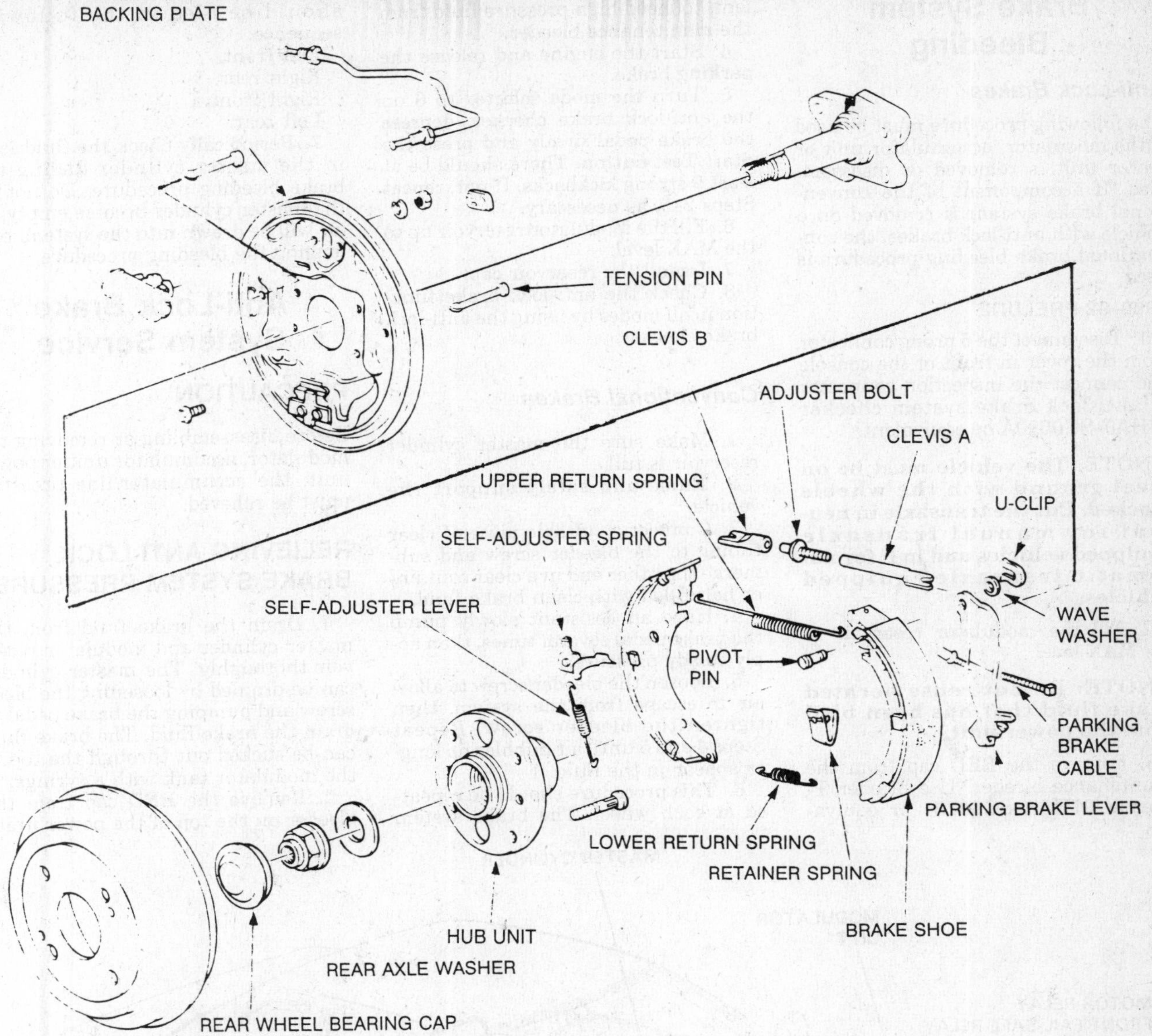

Rear brake assembly

ing surfaces of the backing plate. Bleed the brakes.

Parking Brake Cable

ADJUSTMENT

1. Raise and safely support the vehicle.
2. On rear disc brake equipped vehicles, make sure the lever of the rear brake caliper contacts the brake caliper pin.
3. On drum brake equipped vehicles, make sure the rear brakes are properly adjusted.
4. Pull the parking brake lever up 1 notch.
5. Remove the access cover at the rear of the console and tighten the adjusting nut until the rear wheels drag slightly when turned.
6. Release the parking brake lever and check that the rear wheels do not drag when turned. Readjust if necessary.
7. With the equalizer properly adjusted, the parking brake should be fully applied when the parking brake lever is pulled up 6–10 clicks on Civic and CRX, 7–11 clicks on 1988–89 Accord and 1988–92 Prelude and 4–8 clicks on 1990–92 Accord.

REMOVAL & INSTALLATION

1. Remove the access cover at the rear of the console. Loosen the adjusting nut until the cable ends can be disconnected from the equalizer.
2. Raise and safely support the vehicle.
3. Remove the rear wheel and tire assemblies.
4. On disc brake equipped vehicles, pull out the lock pin, remove the clevis pin and remove the clip.
5. On drum brake equipped vehicles, remove the brake drum and brake shoes. Disconnect the cable from the backing plate.
6. Detach the cables from the cable guides and remove the cables from the vehicle.
7. Installation is the reverse of the removal procedure. Adjust the parking brake.

Brake System Bleeding

Anti-Lock Brakes

The following procedure must be used if the modulator, accumulator unit or power unit is removed or disassembled. If a component of the conventional brake system is removed on a vehicle with anti-lock brakes, the conventional brake bleeding procedure is used.

1990–92 PRELUDE

1. Disconnect the 6 prong connector from the cover in front of the console and connect the inspection connector to anti-lock brake system checker 07HAJ–SG0010A or equivalent.

NOTE: The vehicle must be on level ground with the wheels blocked. Put the transaxle in neutral for manual transaxle equipped vehicles and in P for automatic transaxle equipped vehicles.

2. Fill the modulator reservoir to the MAX level.

NOTE: Do not reuse aerated brake fluid that has been bled from the power unit.

3. Remove the RED cap from the maintenance bleeder. Use bleeder T-wrench 07HAA–SG00101 or equivalent, to bleed high pressure fluid from the maintenance bleeder.
4. Start the engine and release the parking brake.
5. Turn the mode selector to **6** on the anti-lock brake checker, depress the brake pedal firmly and press the Start Test button. There should be at least 2 strong kickbacks. If not, repeat Steps 2–5, as necessary.
6. Fill the modulator reservoir up to the MAX level.
7. Install the reservoir cap.
8. Check the anti-lock brake function in all modes by using the anti-lock brake checker.

Conventional Brakes

1. Make sure the master cylinder reservoir is full.
2. Raise and safely support the vehicle.
3. Connect a suitable piece of clear tubing to the bleeder screw and submerge the other end in a clear container half filled with clean brake fluid.
4. Have an assistant slowly pump the brake pedal several times, then apply steady pressure.
5. Loosen the bleeder screw to allow air to escape from the system, then tighten the bleeder screw. Repeat Steps 4 and 5 until air bubbles no longer appear in the fluid.
6. This procedure should be repeated at each wheel. The brake system should be bled in the following sequence:
   Left front.
   Right rear.
   Right front.
   Left rear.
7. Periodically check the fluid level in the master cylinder during the brake bleeding procedure. Do not let the master cylinder become empty, as air will be drawn into the system, prolonging the bleeding procedure.

Anti-Lock Brake System Service

PRECAUTION

Before disassembling or removing the modulator, accumulator unit or power unit, the accumulator/line pressure must be relieved.

RELIEVING ANTI-LOCK BRAKE SYSTEM PRESSURE

1. Drain the brake fluid from the master cylinder and modulator reservoir thoroughly. The master cylinder can be drained by loosening the bleed screw and pumping the brake pedal to drain the brake fluid. The brake fluid can be sucked out through the top of the modulator tank with a syringe.
2. Remove the RED cap from the bleeder on the top of the power unit.

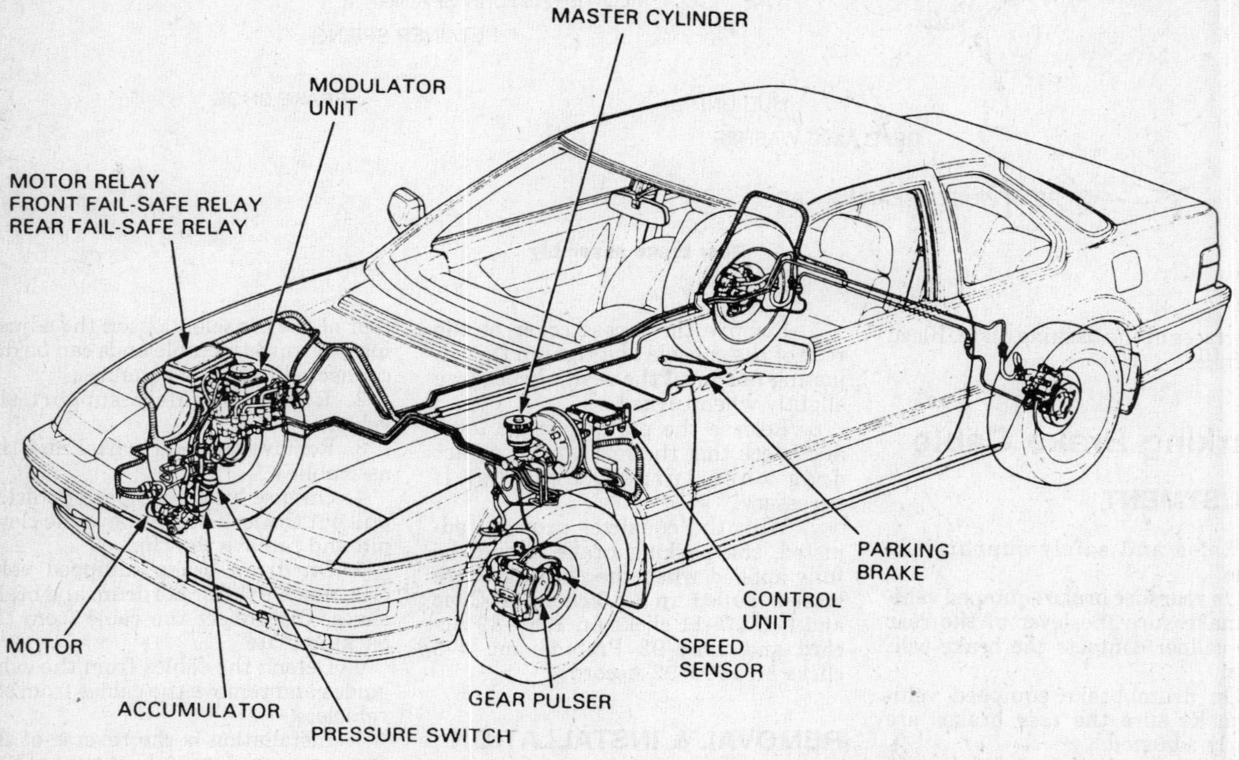

Anti-lock brake system components – 1990–92 Prelude

3. Install bleeder T-wrench 07HAA-SG00101 or equivalent on the bleeder screw and turn it out slowly 90 degrees to collect high pressure fluid into the reservoir. Turn the T-wrench out 1 complete turn to drain the brake fluid thoroughly.
4. Retighten the bleeder screw and discard the fluid. Reinstall the RED cap.

Anti-Lock Brake System Control Unit

REMOVAL & INSTALLATION

1. Disconnect the negative battery cable.
2. Remove the cover that is mounted in front of the center console.
3. Remove the control unit mounting bolts, then remove the control unit.
4. Disconnect the electrical connectors and remove the control unit from the vehicle.

NOTE: When the control unit mounting bolts are removed, the control unit's memory is cleared.

5. Installation is the reverse of the removal procedure.

Modulator

REMOVAL & INSTALLATION

1. Disconnect the negative battery cable.
2. Relieve the anti-lock brake system pressure.
3. Disconnect the hydraulic lines from the modulator and plug the lines.
4. Disconnect the electrical connectors.
5. Remove the modulator mounting bolts and remove the modulator.
6. Installation is the reverse of the removal procedure. Be sure to bleed the anti-lock brake hydraulic system According to the proper procedure.

Accumulator

REMOVAL & INSTALLATION

1. Disconnect the negative battery cable.
2. Relieve the anti-lock brake system pressure.
3. Disconnect the hydraulic lines from the accumulator and plug the lines.
4. Disconnect the electrical connectors.
5. Remove the accumulator mounting bolts and remove the accumulator.
6. Installation is the reverse of the removal procedure. Be sure to bleed the anti-lock brake hydraulic system, According to the proper procedure.

Power Unit

REMOVAL & INSTALLATION

1. Disconnect the negative battery cable.
2. Relieve the anti-lock brake system pressure.
3. Disconnect the hoses from the accumulator and plug them.
4. Disconnect the electrical connectors.
5. Remove the power unit mounting bolts and remove the power unit.
6. Installation is the reverse of the removal procedure. Be sure to bleed the anti-lock brake hydraulic system According to the proper procedure.

Speed Sensor

REMOVAL & INSTALLATION

1. Raise and safely support the vehicle.
2. Disconnect the sensor electrical connectors.
3. Remove the sensor and wire guide mounting bolts and remove the sensor.
4. Installation is the reverse of the removal procedure.

CHASSIS ELECTRICAL

Air Bag

CAUTION

To avoid accidental deployment and serious personal injury, the air bag system must be disarmed before beginning any repair procedure. Read the following safety precautions.

- Do not disassemble or tamper with the air bag assembly.
- Be sure to store a removed air bag assembly with the pad surface up. If the air bag is improperly stored face down, accidental deployment could propel the unit with enough force to cause serious injury.
- When rearming the system, connect the battery last with no one in the vehicle.
- Do not install used air bag parts from another vehicle, use only new replacement parts.

DISARMING THE AIR BAG

All Supplemental Restraint System (SRS) wiring is covered with a yellow outer insulation. This wiring harness cannot be repaired and if cut or damaged, the whole harness must be replaced. To disable the air bag:

1. Disconnect the negative battery cable.
2. Remove the access panel from the bottom of the steering wheel. Installed on the panel is a red short connector.
3. Unplug the air bag connector and install the short connector on the air bag connector.
4. The air bag can now be safely removed and/or the battery can be reconnected for testing or operating other vehicle systems.
5. To reconnect the air bag, disconnect the battery.
6. Connect the air bag wiring harness and make sure no one is in the vehicle before reconnecting the battery.

Heater Blower Motor

REMOVAL & INSTALLATION

Without Air Conditioning

1. Disconnect the negative battery cable.
2. Remove the glove box and glove box frame.
3. Remove the heater duct.
4. Disconnect the electrical connectors from the blower motor.
5. Remove the blower motor mounting bolts and the blower motor.
6. Installation is the reverse of the removal procedure.

With Air Conditioning

EXCEPT 1990–92 ACCORD

1. Disconnect the negative battery cable.
2. Properly discharge the air conditioning system.
3. Disconnect the receiver line and suction hose from the evaporator.

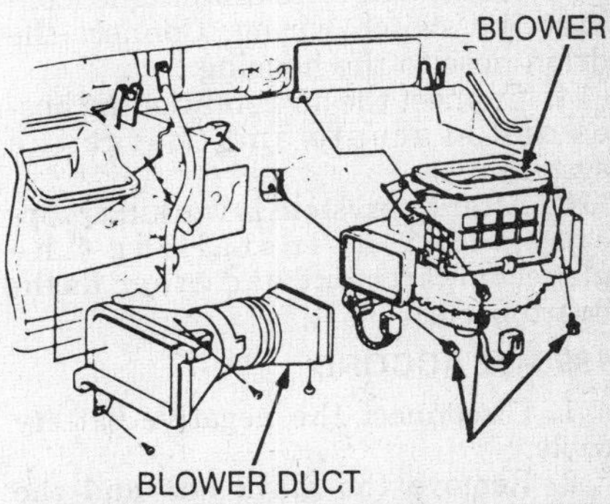

Heater blower motor installation—1988–89 Accord without air conditioning shown

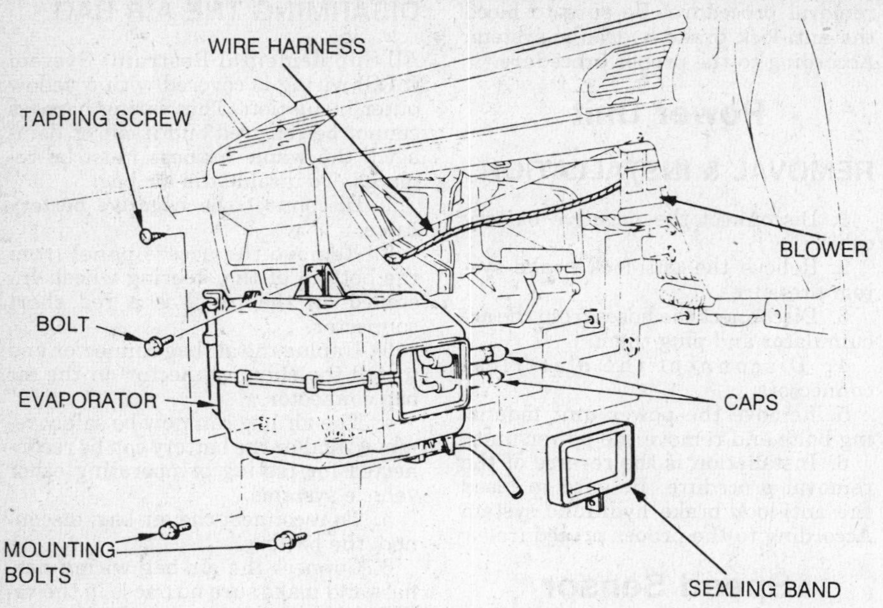

Evaporator installation – 1988–89 Accord shown

NOTE: Cap the open fittings immediately to keep moisture out of the system.

4. Remove the glove box and glove box frame.
5. Disconnect the drain hose from the evaporator lower housing, if equipped.
6. Loosen the sealing band, if equipped, and slide it to the right.
7. Disconnect the thermostat switch wire connector and pull the wire from the clamps.
8. Remove the evaporator and blower retaining bands, as necessary.
9. Remove the evaporator mounting bolts and remove the evaporator.
10. Disconnect the blower motor electrical connectors.
11. Remove the blower motor mounting bolts and remove the blower motor.

To install:

12. Install the blower motor and connect the wiring.
13. Install the evaporator and connect the switch wiring. Connect the drain hose to the housing.
14. Connect the air conditioning hoses and evacuate and charge the system.
15. After the system is working properly, finish installing the glovecompartment and other dashboard parts.

1990–92 ACCORD

1. Disconnect the negative battery cable.
2. Remove the glove box and the glove box frame.
3. Turn over the carpet and remove the side cover. Remove the control unit bracket mounting nuts. Disconnect the connectors and remove the control unit bracket.
4. Remove the retaining band and remove the blower undercover.

NOTE: Be careful not to break the tabs while removing the blower undercover.

5. Remove the blower mounting nuts, disconnect the electrical connectors and remove the blower.
6. Installation is the reverse of the removal procedure. When installing the glove box frame, the face which covers the dashboard is installed with double-sided adhesive tape.

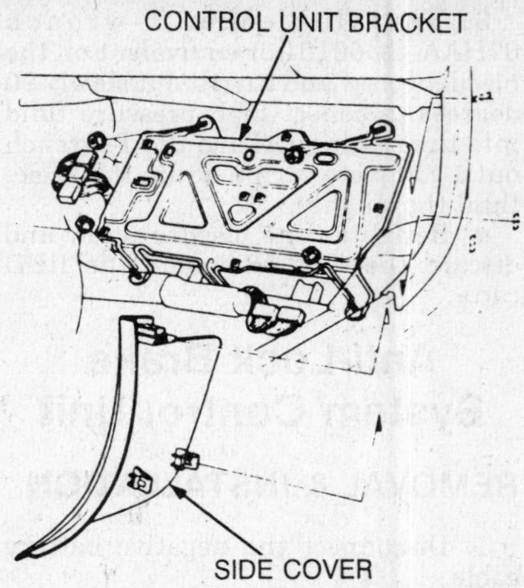

Control unit location – 1990–92 Accord

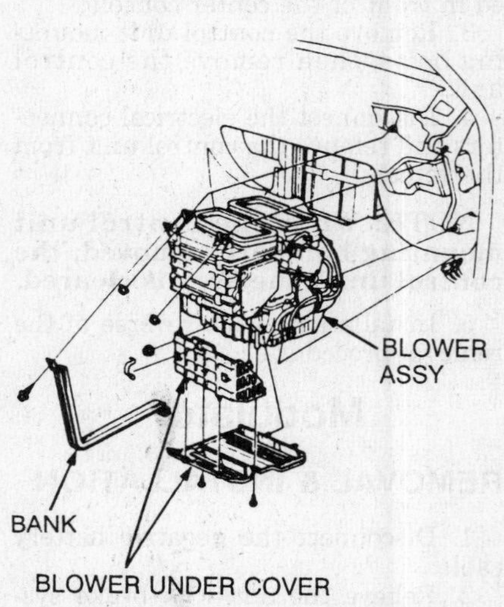

Blower installation – 1990–92 Accord

Windshield Wiper Motor

REMOVAL & INSTALLATION

1. Disconnect the negative battery cable.
2. Remove the wiper arm retaining nuts and remove the wiper arms.
3. Remove the front air scoop, if equipped, and hood seal located over the wiper linkage at the bottom of the windshield.
4. Disconnect the linkage from the wiper motor.
5. Remove the wiper motor water seal cover clamp and remove the cover, if equipped.
6. Disconnect the wiper motor electrical connector, remove the motor mounting bolts and remove the motor.
7. Installation is the reverse of the removal procedure. Coat the linkage joints with grease and make sure the linkage moves smoothly.

Instrument Cluster

REMOVAL & INSTALLATION

Civic

1988–89

1. Disconnect the negative battery cable.
2. Remove the dashboard lower panel.
3. Remove the dashlight brightness controller and rear window defogger switch by pushing them out and then disconnecting the connectors.
4. Remove the 4 instrument panel retaining screws and remove the instrument panel from the dashboard.
5. Remove the caps positioned over the 2 screws retaining the gauge visor.
6. Remove the gauge visor retaining screws and remove the gauge visor.

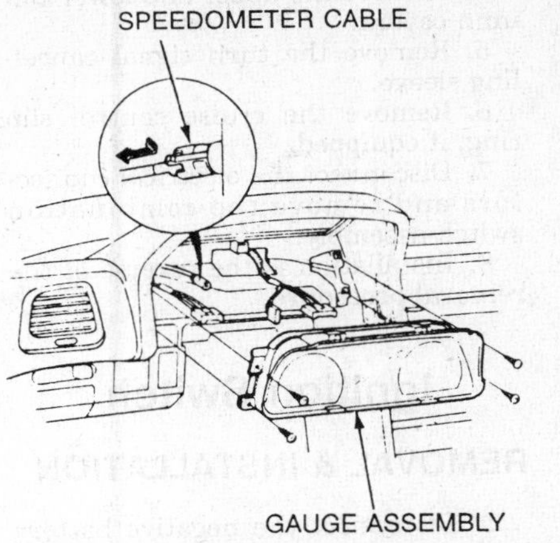

Instrument cluster installation—1990–92 Civic

7. Remove the 4 screws retaining the gauge assembly and pull out the assembly to gain access to the speedometer cable and electrical connectors.
8. Disconnect the electrical connectors and the speedometer cable. Remove the gauge assembly.
9. Installation is the reverse of the removal procedure.

1990–92

1. Disconnect the negative battery cable.
2. Remove the caps positioned over the 2 gauge visor retaining screws.
3. Remove the gauge visor retaining screws and remove the gauge visor.
4. Remove the 4 screws retaining the instrument panel to the dashboard, disconnect the switch connectors and remove the instrument panel.
5. Remove the 4 screws retaining the gauge assembly and pull out the assembly to gain access to the speedometer cable and the electrical connectors.
6. Disconnect the speedometer cable and electrical connectors and remove the gauge assembly.
7. Installation is the reverse of the removal procedure.

Civic Wagon

1988–89

1. Disconnect the negative battery cable.
2. Remove the dashlight brightness controller and rear window defogger switch by pushing them out and disconnecting the connectors.
3. Remove the caps positioned over the upper instrument panel retaining screws.
4. Remove the upper and lower instrument panel retaining screws and remove the instrument panel.
5. Remove the gauge assembly retaining screws and pull the assembly out to gain access to the speedometer cable and electrical connectors.
6. Disconnect the speedometer cable and electrical connectors and remove the gauge assembly.
7. Installation is the reverse of the removal procedure.

1990–92

1. Disconnect the negative battery cable.
2. Remove the caps positioned over the upper instrument panel retaining screws.
3. Remove the upper and lower instrument panel retaining screws.
4. Disconnect the switch connectors and remove the instrument panel.
5. Remove the 4 gauge assembly retaining screws and pull out the assembly to gain access to the speedometer cable and electrical connectors.
6. Disconnect the speedometer cable and electrical connectors and remove the gauge assembly.
7. Installation is the reverse of the removal procedure.

CRX

1988–89

1. Disconnect the negative battery cable.
2. Remove the dashlight brightness controller and rear window defogger switch by pushing them out and disconnecting the connectors.
3. Remove the caps positioned over the upper instrument panel retaining screws.
4. Remove the upper and lower instrument panel retaining screws and remove the instrument panel.
5. Remove the gauge assembly retaining screws and pull the assembly out to gain access to the speedometer cable and electrical connectors.
6. Disconnect the speedometer cable and electrical connectors and remove the gauge assembly.
7. Installation is the reverse of the removal procedure.

1990–92

1. Disconnect the negative battery cable.
2. Remove the caps positioned over the upper instrument panel retaining screws.
3. Remove the upper and lower instrument panel retaining screws.
4. Disconnect the switch connectors and remove the instrument panel.
5. Remove the 4 gauge assembly retaining screws and pull out the assembly to gain access to the speedometer cable and electrical connectors.
6. Disconnect the speedometer cable and electrical connectors and remove the gauge assembly.
7. Installation is the reverse of the removal procedure.

Accord

1988–89

1. Disconnect the negative battery cable.
2. Remove the switches from the instrument panel by inserting a suitable prying tool under the bottom center of the switch and prying it loose. Pull the switch straight back and disconnect the connectors.
3. Use a suitable prying tool to remove the upper lid.
4. Remove the 2 upper and 4 lower instrument panel screws and the instrument panel.
5. Remove the 4 gauge assembly screws and pull the assembly back to gain access to the speedometer cable and electrical connectors.
6. Disconnect the speedometer cable and the electrical connectors and remove the gauge assembly.
7. Installation is the reverse of the removal procedure.

1990–92

1. Disconnect the negative battery cable.
2. Remove the front console mounting screws. On manual transaxle equipped vehicles, remove the shift lever knob.
3. Remove the front console.
4. Remove the ashtray and ashtray holder.
5. Loosen the 2 screws retaining the radio and disconnect the wire harness connector and the antenna lead. Remove the radio.
6. Remove the coin box, cruise control master switch, sunroof switch and panel brightness controller.
7. Remove the side and center air vents.
8. Remove the 12 mounting screws and disconnect the electrical connectors.

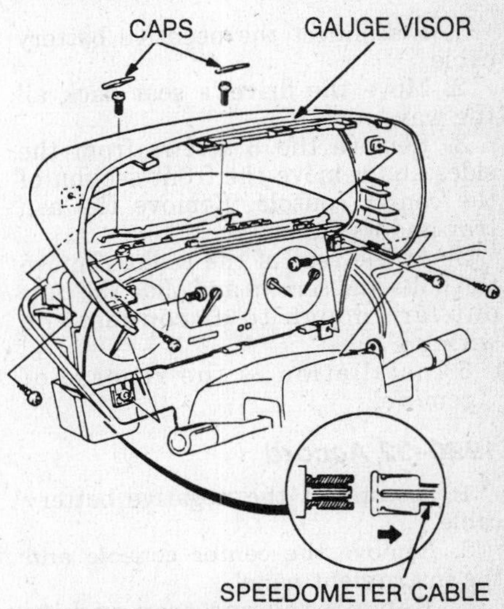

Instrument cluster installation—1988–92 Prelude

9. Remove the instrument panel.
10. Remove the 4 gauge assembly screws and the gauge assembly. Disconnect the electrical connectors from the gauge assembly.
11. Installation is the reverse of the removal procedure.

Prelude

1. Disconnect the negative battery cable.
2. Remove the dashlight brightness controller and retractor/fog light switch from the instrument panel and disconnect the connectors.
3. Remove the 5 retaining screws and the instrument panel from the gauge visor.
4. Remove the 4 screws and pull out the gauge assembly from the dashboard to gain access to the electrical connectors.
5. Disconnect the electrical connectors and remove the gauge assembly.
6. Installation is the reverse of the removal procedure.

Radio

REMOVAL & INSTALLATION

Civic and CRX

1. Disconnect the negative battery cable.
2. Remove the screws from the center instrument panel and lift the panel off. On 1988 models, disconnect the cigarette lighter wires.
3. Remove the 2 screws from the bottom of the rear of the radio and slide it out far enough to disconnect the wiring.
4. Installation is the reverse of removal.

1988–89 Accord

1. Disconnect the negative battery cable.
2. Move the driver's seat back all the way.
3. Remove the 8 screws from the sides and remove the front section of the center console. Remove the ash tray and its frame.
5. At the rear of the radio, remove the bolts and screws and slide the unit out far enough to disconnect the wiring.
6.Installation is the reverse of removal.

1990–92 Accord

1. Disconnect the negative battery cable.
2. Remove the center console and the instrument panel.
3. Remove the ash tray and its frame.
4. Remove the 2 screws from the bottom of the rear of the radio and slide it out far enough to disconnect the wiring.
5. Installation is the reverse of removal.

Prelude

1. Disconnect the negative battery cable.
2. Remove the front console.
3. Remove the 6 screws holding the center of the instrument panel and pull the panel out far enough to disconnect the wiring for the radio and the cigarette lighter.
4. Remove the radio from the panel.
5. Installation is the reverse of removal.

Concealed Headlights

MANUAL OPERATION

1. Remove the cover from the engine compartment fuse box and remove the fuse for the headlight motor that does not work.

NOTE: Always remove the fuse before manually operating a headlight motor, otherwise the motor may suddenly activate.

2. Remove the cap from the top of the headlight motor, then turn the knob in the direction of the arrow (clockwise) until the headlight is as far up or down as it will go.
3. Replace the cap and reinstall the fuse.

Combination Switch

REMOVAL & INSTALLATION

1. Disconnect the negative battery cable.
2. Remove the steering wheel.
3. Remove the lower cover and driver's knee bolster, if equipped.

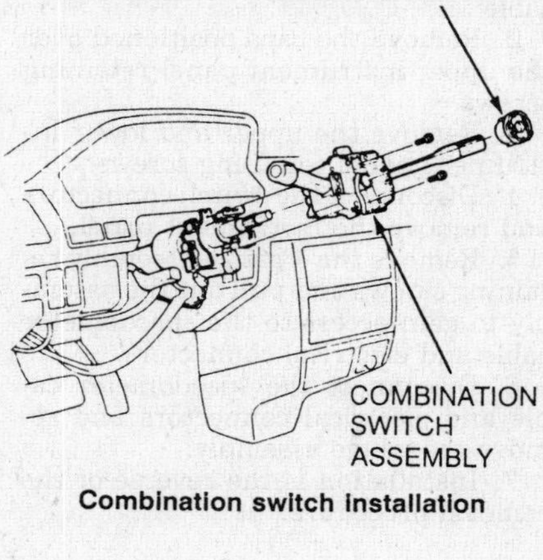

Combination switch installation

4. Remove the upper and lower column covers.
5. Remove the turn signal cancelling sleeve.
6. Remove the cruise control slip ring, if equipped.
7. Disconnect the electrical connectors and remove the combination switch assembly.
8. Installation is the reverse of the removal procedure.

Ignition Switch

REMOVAL & INSTALLATION

1. Disconnect the negative battery cable.
2. Remove the steering wheel and steering column covers as necessary, to gain access to the switch.
3. Disconnect the electrical connector.
4. Insert the ignition key and turn it to the **O** position.
5. Remove the 2 screws and remove the base of the switch.
6. Installation is the reverse of the removal procedure.

Ignition Lock

REMOVAL & INSTALLATION

1. Disconnect the negative battery cable.
2. Remove the steering wheel and the steering column covers.
3. Disconnect the ignition switch connector.
4. Center punch each of the 2 shear bolts and drill their heads off with a suitable drill bit.

NOTE: Do not damage the switch body when removing the shear bolt heads.

5. Remove the shear bolts from the switch body and remove the switch.
To install:
6. Install the new ignition switch without the key inserted.
7. Loosely tighten the new shear bolts.

NOTE: Make sure the projection on the ignition switch is aligned with the hole in the steering column.

8. Insert the ignition key and check for proper operation of the steering wheel lock and that the ignition key turns freely.
9. Tighten the shear bolts until the hex heads twist off.

Stoplight Switch

ADJUSTMENT

1. Loosen the stoplight switch locknut and back off the stoplight switch until it is no longer touching the brake pedal.
2. Loosen the brake pushrod locknut and screw the pushrod in or out until the pedal height from the floor is as follows:
1988–92 Civic and CRX – 6.02 in. (153mm).
1988–89 Accord – 8.07 in. (205mm).
1990–92 Accord with manual transaxle – 7.48 in. (190mm).
1990–92 Accord with automatic transaxle – 7.68 in. (195mm).
1988–92 Prelude with manual transaxle – 7.0 in. (176mm).
1988–92 Prelude with automatic transaxle – 7.2 in. (183mm).
3. After adjustment, tighten the brake pushrod locknut.
4. Screw in the stoplight switch until it's plunger is fully depressed, threaded end touching the pad on the pedal arm, then back off the switch ½ turn and tighten the locknut.

NOTE: Make sure the brake lights go off when the pedal is released.

REMOVAL & INSTALLATION

1. Disconnect the negative battery cable.
2. Disconnect the electrical connector.
3. Loosen the locknut and unscrew the switch from it's mounting.
4. Installation is the reverse of the removal procedure. The switch must be adjusted after installation.

Clutch Switch

ADJUSTMENT

1. Loosen the locknut on the clutch interlock switch.
2. Depress the clutch pedal fully and then release 0.59–0.79 in. (15–20mm) from the fully depressed position and hold. Adjust the position of the clutch switch so the engine will start with the clutch pedal in this position.
3. Thread the clutch switch in further ¼–½ turn and tighten the locknut.

REMOVAL & INSTALLATION

1. Disconnect the negative battery cable.
2. Disconnect the electrical connector from the switch.

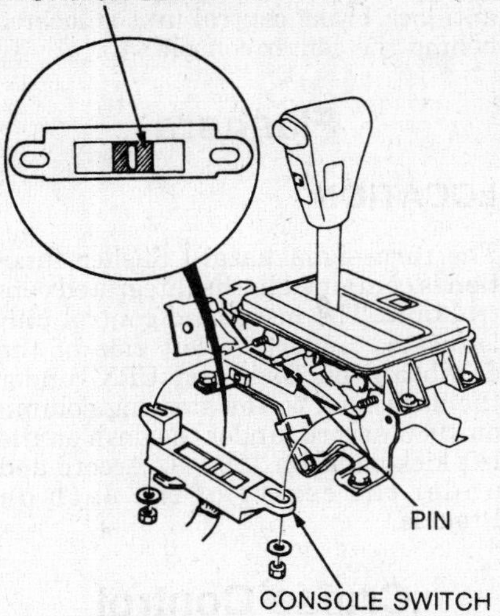

Neutral safety switch installation

3. Loosen the locknut on the clutch switch and unscrew the switch from it's mounting.
4. Installation is the reverse of the removal procedure. Adjust the switch after installation.

Neutral Safety Switch

REMOVAL & INSTALLATION

1. Disconnect the negative battery cable.
2. Remove the console and disconnect the electrical connectors from the neutral safety switch.
3. Remove the 2 neutral safety switch mounting nuts and remove the neutral safety switch.

To install:

4. Position the switch slider to the **N** position.
5. Shift the selector lever to **N**, then slip the neutral safety switch into position.
6. Attach the switch with the 2 nuts.
7. Connect the electrical connectors and the negative battery cable. Make sure the engine starts when the shift lever is in the **N** position in the range of free-play in the switch.
8. Install the console.

Fuses and Relays

LOCATION

Fuses

The fuse panel is located under the left side of the dashboard on all vehicles. All vehicles also have a fuse box in the right side of the engine compartment.

Relays

Air Conditioner Condenser Fan Relay – Civic and CRX – located in right front of engine compartment. Accord – located in left front of engine compartment. Prelude – located in underhood relay box.

Air Conditioner Clutch Relay – Civic, CRX and Prelude – located in right front of engine compartment. Accord – located in left front of engine compartment.

Radiator Cooling Fan Relay – Civic and CRX – located in right front of engine compartment. Accord and Prelude – located in underhood relay box.

Turn Signal and Hazard Relay – Civic and CRX – located on dashboard fuse panel. Accord – located on underdash fuse panel. Prelude – located on underdash relay panel.

Intermittent Windshield Wiper Relay – 1988–89 Accord – located on underdash fuse panel. 1990–92 Accord – located under underhood relay box.

Rear Window Defogger Relay – Civic and CRX – located on dashboard fuse panel. Accord and Prelude – located on underdash fuse panel.

Power Window Relay – Civic – located on dashboard fuse panel. Accord – located in underhood relay box. Prelude – located on underdash relay panel.

ECM Main Relay – All models – under left side of dashboard. **Starter Relay** – Civic and CRX – located under left side of dashboard. Accord – located under left side of dashboard. Prelude – located on underdash relay box.

Cigarette Lighter Relay – 1990–92 Civic and CRX – located under dashboard to the right of steering column. Accord – located under left side of dashboard.

Heater Motor Relay – 1990–92 Civic Wagon – located under dashboard to the right of steering column.

Sunroof Relay – 1988–92 CRX – located behind underdash fuse panel. 1990–92 Accord – located on underdash fuse panel. Prelude – located on underdash relay panel.

Blower Motor Relay – 1990–92 Accord – located on underdash fuse panel.

Headlight Retractor Relays – 1988–89 Accord – located in right front of engine compartment. Prelude – right retractor relay located in right front of engine compartment, left retractor relay located in left front of engine compartment.

Dimmer Relay – Prelude and 1990–92 Accord – located in underhood relay box.

Lighting Relay – Prelude and

1990–92 Accord – located in underhood relay box.

Fuel Cut-Off Relay – 1988–89 Accord with carbureted engine – located on underdash fuse panel.

Computers

LOCATION

Civic and CRX

The main ECM is located under a panel on the right front floor. The automatic transaxle control unit on 1989–92 Civic Wagon is located under the driver's seat.

Accord

The engine Electronic Control Module (ECM) is located under the driver's seat on 1988–89 Accord. The ECM is located under a panel on the right front floor on 1990–92 Accord. The automatic transaxle control unit on 1990–92 Accord is located next to the engine ECM.

Prelude

The ECM is located under a panel on the right front floor. The automatic transaxle control unit is located next to the ECM on 1988–89 Prelude and 1990–92 Prelude equipped with the 2.1L engine. The automatic transaxle control unit is located behind the center console on 1990–92 Prelude equipped with the 2.0L engine. The anti-lock brake control unit is located behind the center console.

Flashers

LOCATION

The turn signal/hazard flasher function is contained in the integrated control unit. The integrated control unit is located under the left side of the dashboard on Civic and CRX, under the dash next to the steering column on 1989 Accord, under the dash on the left kick panel on 1990–92 Accord and under the center of the dash on Prelude.

Cruise Control

ADJUSTMENT

1. Make sure the actuator cable operates smoothly, with no binding or sticking.
2. Start the engine and warm it up to normal operating temperature.
3. Measure the amount of movement of the actuator rod or output linkage on 1990–92 Civic Sedan, until the engine speed starts to increase. At first, the actuator linkage should be located at the full close position. Freeplay should be 0.37–0.49 in. (9.5–12.5mm).

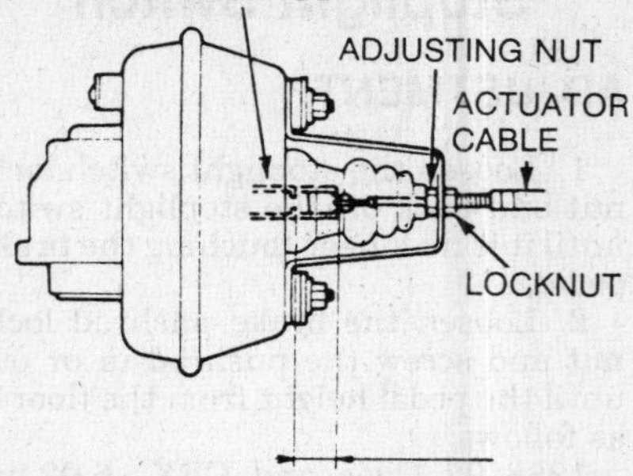

Cruise control actuator cable adjustment – Accord and Prelude

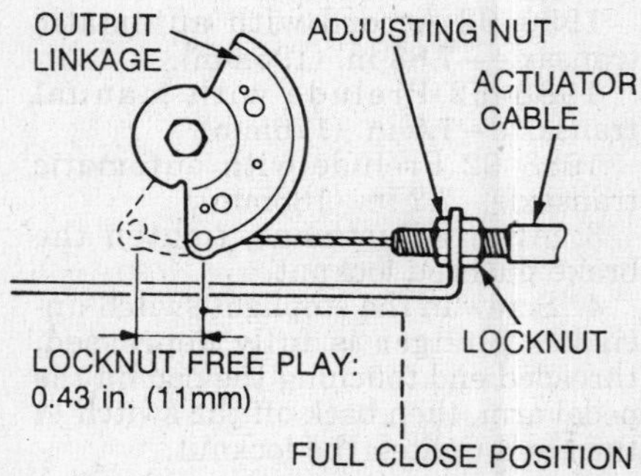

Cruise control actuator cable adjustment – 1990–92 Civic Sedan

4. If free-play is not within specification, loosen the locknut and turn the adjusting nut as required.

8 Hyundai

Excel, Scoupe, Sonata—All Models

SERIAL NUMBER IDENTIFICATION

Vehicle Identification Plate

The Vehicle Identification Number (VIN) is located on a plate attached to the left front of the dash panel, so it can be seen through the windshield when standing beside the vehicle, in front of the driver's door.

The letters and numbers in the VIN digits can be interpreted according to their positions in the sequence as follows:

1. Manufacturing country
2. Make
3. Vehicle type
4. Type of seat belt system
5. Vehicle line
6. Trim code/price class
7. Body type
8. Engine displacement: J=1468cc (L4), S=2351cc (L4) and T=2972cc (V6)
9. Check digit: a special letter or number code is used to verify the serial number. This contains no useful information for the owner.
10. Vehicle year: J=1988, K=1989, L=1990, M=1991 and N=1992
11. Plant where the vehicle was built
12. Transaxle code

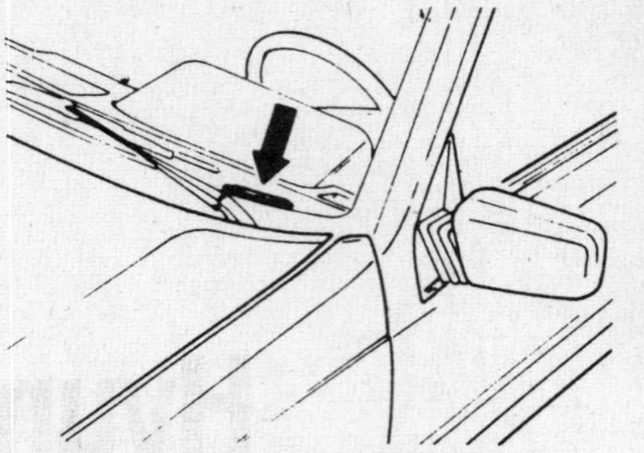

Serial number location

Engine Number

The engine model and serial numbers in all cases are stamped on the top edge of the block near the front of the engine. In most cases, they are located on the right side of the engine.

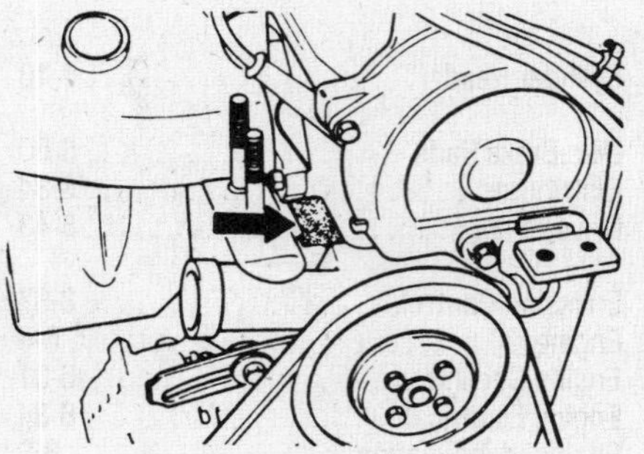

Engine number location

Engine model number

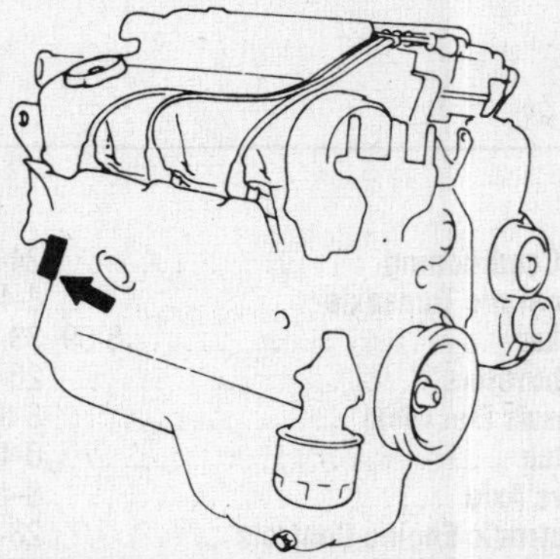

Engine number location—6 cylinder engine

Vehicle Identification Number

The Vehicle Identification Label (VIN) is located on the top center of the firewall in the engine compartment.

SPECIFICATIONS

ENGINE IDENTIFICATION

Year	Model	Engine Displacement cu. in. (cc/liter)	Engine Series Identification	No. of Cylinders	Engine Type
1988	Excel	89.6 (1468/1.5)	—	4	OHC
1989	Excel	89.6 (1468/1.5)	—	4	OHC
	Sonata	143.5 (2351/2.4)	—	4	OHC
1990	Excel	89.6 (1468/1.5)	—	4	OHC
	Sonata	143.5 (2351/2.4)	—	4	OHC
	Sonata	181.4 (2972/3.0)	—	6	OHC
1991-92	Excel	89.6 (1468/1.5)	—	4	OHC
	Sonata	143.5 (2351/2.4)	—	4	OHC
	Sonata	181.4 (2972/3.0)	—	6	OHC
	Scoupe	89.6 (1468/1.5)	—	4	OHC

OHC—Overhead Camshaft

GENERAL ENGINE SPECIFICATIONS

Year	Model	Engine Displacement cu. in. (cc)	Fuel System Type	Net Horsepower @ rpm	Net Torque @ rpm (ft. lbs.)	Bore × Stroke (in.)	Com-pression Ratio	Oil Pressure @ rpm
1988	Excel	89.6 (1468)	2 bbl	77 @ 5300	84 @ 3000	2.97 × 3.23	9.4:1	45 @ 2000
1989	Excel	89.6 (1468)	2 bbl	77 @ 5300	84 @ 3000	2.97 × 3.23	9.4:1	45 @ 2000
	Sonata	143.5 (2351)	MPI	126 @ 5100	180 @ 2600	3.41 × 3.94	8.5:1	45 @ 2000
1990	Excel	89.6 (1468)	2 bbl	77 @ 5300	84 @ 3000	2.97 × 3.23	9.4:1	45 @ 2000
	Excel	89.6 (1468)	MPI	77 @ 5300	84 @ 3000	2.97 × 3.23	9.4:1	45 @ 2000
	Sonata	143.5 (2351)	MPI	126 @ 5100	180 @ 2600	3.41 × 3.94	8.5:1	45 @ 2000
	Sonata	181.4 (2972)	MPI	142 @ 5000	168 @ 2500	3.59 × 2.99	8.9:1	30–80 @ 3000
1991–92	Excel	89.6 (1468)	2 bbl	77 @ 5300	84 @ 3000	2.97 × 3.23	9.4:1	45 @ 2000
	Excel	89.6 (1468)	MPI	81 @ 5500	91 @ 3000	2.97 × 3.23	9.4:1	45 @ 2000
	Sonata	143.5 (2351)	MPI	116 @ 4500	142 @ 3500	3.41 × 3.94	8.5:1	45 @ 2000
	Sonata	181.4 (2972)	MPI	142 @ 5000	168 @ 2500	3.59 × 2.99	8.9:1	30–80 @ 3000
	Scoupe	89.6 (1468)	MPI	81 @ 5500	91 @ 3000	2.97 × 3.23	9.4:1	11 @ 750

MPI Multi-Port Fuel Injection

ENGINE TUNE-UP SPECIFICATIONS

Year	Model	Engine Displacement cu. in. (cc)	Spark Plugs Type	Spark Plugs Gap (in.)	Ignition Timing (deg.) MT	Ignition Timing (deg.) AT	Com-pression Pressure (psi)	Fuel Pump (psi)	Idle Speed (rpm) MT	Idle Speed (rpm) AT	Valve Clearance In.	Valve Clearance Ex.
1988	Excel	89.6 (1468)	RN9YC4	0.039–0.043	5B	5B	164	2.8–3.6	750	750	0.006	0.010
1989	Excel	89.6 (1468)	RN9YC4	0.039–0.043	5B	5B	164	2.8–3.6	750	750	0.006	0.010
	Sonata	143.5 (2351)	RN9YC4	0.039–0.043	5B	5B	160	48.0	750	750	Hyd.	Hyd.
1990	Excel	89.6 (1468)	RN9YC4	0.039–0.043	5B	5B	164	2.8–3.6	700	700	0.006	0.010
	Excel	89.6 (1468)	RN9YC4	0.039–0.043	5B	5B	164	2.8–3.6	700	700	0.006	0.010
	Sonata	143.5 (2351)	RN9YC4	0.039–0.043	5B	5B	160	48.0	750	750	Hyd.	Hyd.
	Sonata	181.4 (2972)	PGR5A11	0.039–0.043	12B	12B	178	48.0	750	750	Hyd.	Hyd.
1991	Excel	89.6 (1468)	RN9YC4	0.039–0.043	5B	5B	164	2.8–3.6	700	700	0.006	0.010
	Excel	89.6 (1468)	RN9YC4	0.039–0.043	5B	5B	164	2.8–3.6	700	700	0.006	0.010
	Sonata	143.5 (2351)	RN9YC4	0.039–0.043	5B	5B	160	48.0	750	750	Hyd.	Hyd.
	Sonata	181.4 (2972)	PGR5A11	0.039–0.043	12B	12B	178	48.0	750	750	Hyd.	Hyd.
	Scoupe	89.6 (1468)	RN9YC4	0.039–0.043	5B	5B	192	②	①	①	0.006	0.010
1992	SEE UNDERHOOD SPECIFICATIONS STICKER											

NOTE: Valve clearance is set with the engine at normal operating temperature. On USA engines, the jet valve must be set before the intake valve. Jet valve clearance is 0.010 in.

① 700 ± 100 rpm

② 46–49 Fuel pressure regulator vacuum hose disconnected
39 Fuel pressure regulator vacuum hose connected

FIRING ORDERS

NOTE: To avoid confusion, always replace spark plug wires one at a time.

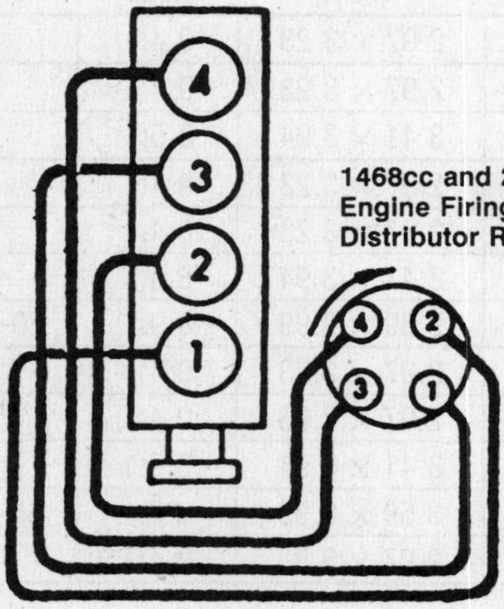

1468cc and 2351cc Engines
Engine Firing Order: 1–3–4–2
Distributor Rotation: Clockwise

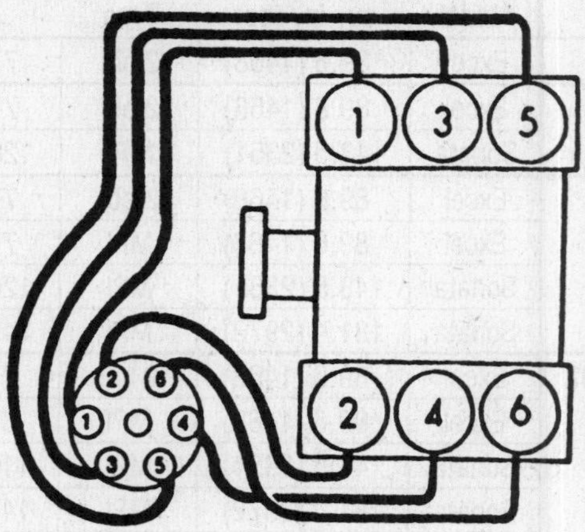

2972cc Engine
Engine Firing Order: 1–2–3–4–5–6
Distributor Rotation: Counterclockwise

CAPACITIES

Year	Model	Engine Displacement cu. in. (cc)	Engine Crankcase (qts.) with Filter	Engine Crankcase (qts.) without Filter	Transmission (pts.) 4-Spd	Transmission (pts.) 5-Spd	Transmission (pts.) Auto.	Drive Axle (pts.)	Fuel Tank (gals.)	Cooling System (qts.)
1988	Excel	89.6 (1468)	3.6	—	4.4	4.4	12.2	—	10.6 ①	5.6
1989	Excel	89.6 (1468)	3.6	—	4.4	4.4	12.2	—	10.6 ①	5.6
	Sonata	143.5 (2351)	4.0	—	—	5.3	12.3	—	16.0	7.4
1990	Excel	89.6 (1468)	3.6	—	4.4	4.4	12.2	—	10.6 ①	5.6
	Sonata	143.5 (2351)	4.0	—	—	5.3	12.3	—	16.0	7.4
	Sonata	181.4 (2972)	4.0	—	—	5.3	12.3	—	17.0	7.4
1991–92	Excel	89.6 (1468)	3.6	—	4.4	4.4	12.2	—	10.6 ①	5.6
	Sonata	143.5 (2351)	4.0	—	—	5.3	12.3	—	16.0	7.4
	Sonata	181.4 (2972)	4.0	—	—	5.3	12.3	—	17.0	7.4
	Scoupe	89.6 (1468)	3.6	—	3.8	3.8	12.8	—	11.9 ①	5.6

① Optional 13.2 gal. tank

CAMSHAFT SPECIFICATIONS

All measurements given in inches.

Year	Engine Displacement cu. in. (cc)	Journal Diameter 1	Journal Diameter 2	Journal Diameter 3	Journal Diameter 4	Journal Diameter 5	Lobe Lift In.	Lobe Lift Ex.	Bearing Clearance	Camshaft End Play
1988	89.6 (1468)	1.338	1.338	1.338	—	—	1.500	1.504	0.00197–0.00354	0.002–0.008
1989	89.6 (1468)	1.338	1.338	1.338	—	—	1.500	1.504	0.00197–0.00354	0.002–0.008
	143.5 (2351)	1.336	1.336	1.336	1.336	1.336	1.753	1.758	0.00200–0.00350	0.004–0.008

CAMSHAFT SPECIFICATIONS

All measurements given in inches.

Year	Engine Displacement cu. in. (cc)	Journal Diameter 1	2	3	4	5	Lobe Lift In.	Ex.	Bearing Clearance	Camshaft End Play
1990	89.6 (1468)	1.338	1.338	1.336	—	—	1.500	1.504	0.00197–0.00354	0.002–0.008
	143.5 (2351)	1.336	1.336	1.336	1.336	1.336	1.753	1.758	0.00200–0.00350	0.004–0.008
	181.4 (2972)	1.336	1.336	1.336	1.336	—	1.628	1.628	0.00200–0.00350	0.004–0.008
1991-92	89.6 (1468)	1.338	1.338	1.336	—	—	1.500	1.504	0.00197–0.00354	0.004–0.008
	143.5 (2351)	1.336	1.336	1.336	1.336	1.336	1.753	1.758	0.00200–0.00350	0.004–0.008
	181.4 (2972)	1.336	1.336	1.336	1.336	—	1.628	1.628	0.00200–0.00350	0.004–0.008

CRANKSHAFT AND CONNECTING ROD SPECIFICATIONS

All measurements are given in inches.

Year	Engine Displacement cu. in. (cc)	Crankshaft Main Brg. Journal Dia.	Main Brg. Oil Clearance	Shaft End-play	Thrust on No.	Connecting Rod Journal Diameter	Oil Clearance	Side Clearance
1988	89.6 (1468)	1.8898	0.0008–0.0028	0.002–0.007	3	1.6535	0.0004–0.0024	0.004–0.010
1989	89.6 (1468)	1.8898	0.0008–0.0028	0.002–0.007	3	1.6535	0.0004–0.0024	0.004–0.010
	143.5 (2351)	2.2436	0.0008–0.0020	0.002–0.007	3	1.7709–1.7715	0.0004–0.0024	0.004–0.010
1990	89.6 (1468)	1.8898	0.0008–0.0028	0.002–0.007	3	1.6535	0.0004–0.0024	0.004–0.010
	143.5 (2351)	2.2436	0.0008–0.0020	0.002–0.007	3	1.7709–1.7715	0.0004–0.0024	0.004–0.010
	181.4 (2972)	2.3622	0.0008–0.0020	0.002–0.009	3	1.9685	0.0006–0.0018	0.004–0.010
1991-92	89.6 (1468)	1.8898	0.0008–0.0028	0.002–0.007	3	1.6535	0.0004–0.0024	0.004–0.010
	143.5 (2351)	2.2436	0.0008–0.0020	0.002–0.007	3	1.7709–1.7715	0.0004–0.0024	0.004–0.010
	181.4 (2972)	2.3622	0.0008–0.0020	0.002–0.009	3	1.9685	0.0006–0.0018	0.004–0.010

VALVE SPECIFICATIONS

Year	Engine Displacement cu. in. (cc)	Seat Angle (deg.)	Face Angle (deg.)	Spring Test Pressure (lbs. @ in.)	Spring Installed Height (in.)	Stem-to-Guide Clearance (in.) Intake	Exhaust	Stem Diameter (in.) Intake	Exhaust
1988	89.6 (1468)	45	45	53 @ 1.07 ①	1.42 ②	0.0012–0.0024	0.0020–0.0035	0.2598	0.2598

VALVE SPECIFICATIONS

Year	Engine Displacement cu. in. (cc)	Seat Angle (deg.)	Face Angle (deg.)	Spring Test Pressure (lbs. @ in.)	Spring Installed Height (in.)	Stem-to-Guide Clearance (in.) Intake	Stem-to-Guide Clearance (in.) Exhaust	Stem Diameter (in.) Intake	Stem Diameter (in.) Exhaust
1989	89.6 (1468)	45	45	53 @ 1.07 ①	1.42 ②	0.0012–0.0024	0.0020–0.0035	0.2598	0.2598
	143.5 (2351)	45	45	73 @ 1.591 ①	1.591 ②	0.0012–0.0024	0.0020–0.0035	0.3150	0.3150
1990	89.6 (1468)	45	45	53 @ 1.07 ①	1.42 ②	0.0012–0.0024	0.0020–0.0035	0.2598	0.2598
	143.5 (2351)	45	45	73 @ 1.591 ①	1.591 ②	0.0012–0.0024	0.0020–0.0035	0.3150	0.3150
	181.4 (2972)	45	45	74 @ 1.591	1.59	0.0012–0.0024	0.0020–0.0035	0.3150	0.3134
1991-92	89.6 (1468)	45	45	53 @ 1.07 ①	1.42 ②	0.0012–0.0024	0.0020–0.0035	0.2598	0.2598
	143.5 (2351)	45	45	73 @ 1.591 ①	1.591 ②	0.0012–0.0024	0.0020–0.0035	0.3150	0.3150
	181.4 (2972)	45	45	74 @ 1.591	1.59	0.0012–0.0024	0.0020–0.0035	0.3150	0.3134

① Jet valve—7.7 @ 0.846
② Jet valve—0.846

PISTON AND RING SPECIFICATIONS

All measurements are given in inches.

Year	Engine Displacement cu. in. (cc)	Piston Clearance	Ring Gap Top Compression	Ring Gap Bottom Compression	Ring Gap Oil Control	Ring Side Clearance Top Compression	Ring Side Clearance Bottom Compression	Ring Side Clearance Oil Control
1988	89.6 (1468)	0.0008–0.0016	0.008–0.014	0.008–0.014	0.008–0.028	0.0012–0.0028	0.0008–0.0024	Snug
1989	89.6 (1468)	0.0008–0.0016	0.008–0.014	0.008–0.014	0.008–0.028	0.0012–0.0028	0.0008–0.0024	Snug
	143.5 (2351)	0.0004–0.0012	0.010–0.016	0.008–0.014	0.008–0.028	0.0012–0.0028	0.0008–0.0024	Snug
1990	89.6 (1468)	0.0008–0.0016	0.008–0.014	0.008–0.014	0.008–0.028	0.0012–0.0028	0.0008–0.0024	Snug
	143.5 (2351)	0.0004–0.0012	0.010–0.016	0.008–0.014	0.008–0.028	0.0012–0.0028	0.0008–0.0024	Snug
	181.4 (2972)	0.0008–0.0016	0.012–0.018	0.010–0.016	0.008–0.028	0.0012–0.0035	0.0008–0.0024	Snug
1991-92	89.6 (1468)	0.0008–0.0016	0.008–0.014	0.008–0.014	0.008–0.028	0.0012–0.0028	0.0008–0.0024	Snug
	143.5 (2351)	0.0004–0.0012	0.010–0.016	0.008–0.014	0.008–0.028	0.0012–0.0028	0.0008–0.0024	Snug
	181.4 (2972)	0.0008–0.0016	0.012–0.018	0.010–0.016	0.008–0.028	0.0012–0.0035	0.0008–0.0024	Snug

TORQUE SPECIFICATIONS

All readings in ft. lbs.

Year	Engine Displacement cu. in. (cc)	Cylinder Head Bolts	Main Bearing Bolts	Rod Bearing Bolts	Crankshaft Pulley Bolts	Flywheel Bolts	Manifold Intake	Manifold Exhaust	Spark Plugs
1988	89.6 (1468)	①	36–39	23–25	9–11④	94–101	12–14	12–14	18
1989	89.6 (1468)	①	36–39	23–25	9–11④	94–101	12–14	12–14	18
	143.5 (2351)	②	36–40	36–38	14–22③	94–100	11–14	11–14	18
1990	89.6 (1468)	①	36–39	23–25	9–11④	94–101	12–14	12–14	18
	143.5 (2351)	②	36–40	36–38	14–22③	94–100	11–14	11–14	18
	181.4 (2972)	②	55–61	36–38	109–115	65–70	11–14	11–16	18
1991-92	89.6 (1468)	①	36–39	23–25	9–11④	94–101	12–14	12–14	18
	143.5 (2351)	②	36–40	36–38	14–22③	94–100	11–14	11–14	18
	181.4 (2972)	②	55–61	36–38	109–115	65–70	11–14	11–16	18

① Cold—51–54 ft. lbs.; warm—58–61 ft. lbs.
② Cold—65–72; warm—72–80
③ Sprocket bolt—94
④ Sprocket—72

BRAKE SPECIFICATIONS

All measurements in inches unless noted.

Year	Model	Lug Nut Torque (ft. lbs.)	Master Cylinder Bore	Brake Disc Minimum Thickness	Brake Disc Maximum Runout	Standard Brake Drum Diameter	Minimum Lining Thickness Front	Minimum Lining Thickness Rear
1988	Excel	50–58①	0.8125	②	0.006	7.086	0.04	0.04
1989	Excel	50–58①	0.8125	②	0.006	7.086	0.04	0.04
	Sonata	50–60	1.0000	0.787	0.006	9.000	0.04	0.04
1990	Excel	50–58①	0.8125	②	0.006	7.086	0.04	0.04
	Sonata	50–60	1.0000	0.866	0.006	9.000	0.04	0.04
1991-92	Excel	50–58①	0.8125	②	0.006	7.086	0.04	0.04
	Sonata	50–60	1.0000	0.866	0.006	9.000	0.04	0.04
	Scoupe	65–80	0.8750	0.670	0.006	7.165	0.04	0.04

① Aluminum wheels—58–72
② Sumitomo—0.449
Tokico—0.675

WHEEL ALIGNMENT

Year	Model	Caster Range (deg.)	Caster Preferred Setting (deg.)	Camber Range (deg.)	Camber Preferred Setting (deg.)	Toe-in (in.)	Steering Axis Inclination (deg.)
1988	Excel	½P–1⅙P	⅚P	0–1P	½P	1/16 in–5/64 out	①
1989	Excel	½P–1⅙P	⅚P	0–1P	½P	1/16 in–5/64 out	①
	Sonata	½P–2½P	2P	0–1P	½P	1/8 in–5/64 out	②
1990	Excel	½P–1⅙P	⅚P	0–1P	½P	1/16 in–5/64 out	①
	Sonata	½P–2½P	2P	0–1P	½P	1/8 in–5/64 out	②
1991-92	Excel	½P–1⅙P	⅚P	0–1P	½P	1/16 in–5/64 out	①
	Sonata	½P–2½P	2P	0–1P	½P	1/8 in–5/64 out	②
	Scoupe	⑥	③	⅔N–⅓P	⅙N	⑤	④

① Inside wheel—35⅔; outside wheel—29 9/32
② King pin angle—13°25′
③ Manual steering 1½P
Power steering 1P
④ King pin angle—13°14′
⑤ 0.157 in–0.079 out
⑥ ½P–1½P

ENGINE MECHANICAL

NOTE: Disconnecting the negative battery cable on some vehicles may interfere with the functions of the on board computer systems and may require the computer to undergo a relearning process, once the negative battery cable is reconnected.

Engine Assembly

REMOVAL & INSTALLATION

4 Cylinder Engine

NOTE: The factory recommends that the engine and transaxle be removed as a unit.

1. Disconnect the negative battery cable. Remove the air cleaner assembly. Disconnect the purge control vacuum hose from the purge valve. Remove the purge control valve mounting bracket. Remove the windshield washer reservoir, radiator tank and carbon canister.
2. Drain the coolant from the radiator. Disconnect the upper and lower radiator hoses and then remove the radiator assembly with the electric cooling fan attached. Be sure to disconnect the fan wiring harness.
3. Disconnect the electrical connectors for the back-up lights and engine harness, located near the battery tray. If equipped with a 5 speed transaxle, disconnect the select control valve connector. Disconnect the alternator harness connectors and the oil pressure sending unit.
4. Label and disconnect the automatic transaxle oil cooler hoses. Avoid spilling oil and cap the openings.
5. Label and disconnect all low tension wires and the one high tension wire going to the coil from the distributor. Disconnect the engine ground.
6. Disconnect the brake booster vacuum hose at the intake manifold.
7. Disconnect the fuel supply, return and vapor hoses at the side of the engine. Avoid spilling fuel.
8. Disconnect the heater hoses from the side of the engine. Disconnect the accelerator cable at the engine side.
9. Remove the clutch control cable for manual transaxle or transaxle shifter control cable for automatic transaxles from the transaxle.
10. Unscrew and disconnect the speedometer cable at the transaxle. Disconnect the air conditioning compressor mounting bracket.
11. Raise and safely support the vehicle. Remove the splash shield. Remove the drain plug and drain the transaxle fluid. Disconnect the exhaust pipe at the manifold. Then, suspend the pipe securely with wire.
12. If equipped with a manual transaxle, remove the shift control rod and extension rod.
13. Disconnect the stabilizer bar at both lower control arms. Remove the bolts that attach the lower control arms to the body on either side. Support the arms from the body.

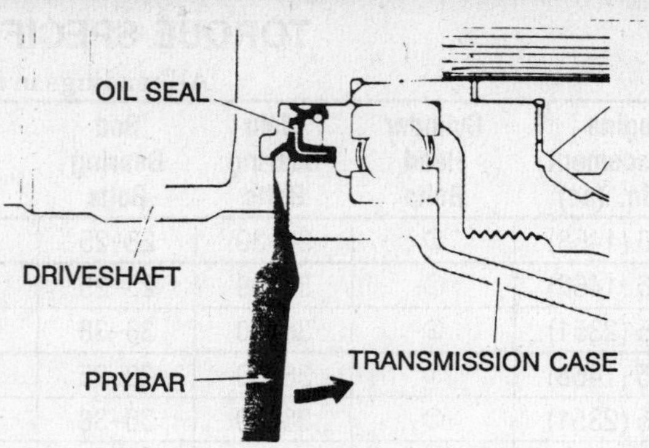

Removing the driveshafts

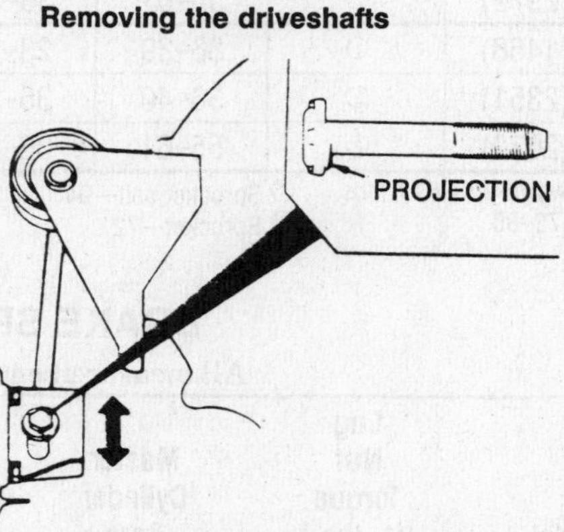

Installing the lower bolt on the roll stopper

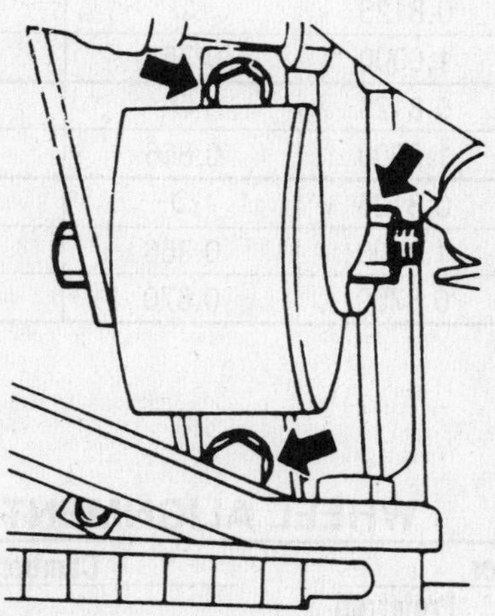
Remove the front roll stop nut and the bolt; or, remove the attaching bolts from the engine damper

14. Disconnect the halfshafts at the transaxle on both sides. Then, seal off the openings in the transaxle. Be sure to replace the circlips holding the halfshafts in the transaxle. Support the halfshafts from the body.
15. Attach a suitable lift, via chains or cables, to both the engine lifting hooks. Put just a little tension on the cables. Then, remove the nut and bolt from the front roll stopper; unbolt the brace from the top of the engine damper.
16. Separate the rear roll stopper from the No. 2 crossmember. Remove the attaching nut from the left mount insulator bolt, but do not remove the bolt.
17. Raise the engine just enough that the crane is supporting its weight. Check that everything is disconnected from the engine.
18. Remove the blind cover from the inside of the right fender inner shield. Remove the transaxle mounting bracket bolts.
19. Remove the left mount insulator bolt. Then, press downward on the transaxle while lifting the engine/transaxle assembly to guide it up and out of the vehicle.

NOTE: Make sure the transaxle does not hit the battery bracket during engine and transaxle removal.

To install:

20. Using a lifting device, lower the engine and transaxle carefully into position and loosely install the mounting bolts. Temporarily tighten the front and rear roll control rods mounting bolts. Lower the full weight of the engine and transaxle onto the mounts and tighten the nuts and bolts. Loosen and retighten the roll control rods.
21. Install the transaxle mounting bracket bolts. Install the blind cover to the inside of the right fender inner shield.
22. Assemble the rear roll stopper to the No. 2 crossmember. Install retaining nut and bolt.
23. Install new circlips on the halfshafts prior to installing. Connect the halfshafts at the transaxle on both sides.
24. Attach the lower control arms to the body on either side. Connect the stabilizer bar to both lower control arms.
25. If equipped with a manual transaxle, install the shift control rod and extension rod.
26. Raise the vehicle and support it safely. Connect the exhaust pipe at the manifold. Install the splash shield.
27. Connect the speedometer cable at the transaxle. Connect the air conditioning compressor mounting bracket.
28. Install the clutch control cable, for manual transaxle, or shifter control cable, for automatic transaxle, to the transaxle.
29. Lower the vehicle. Connect the heater to the engine. Connect the accelerator cable at the engine side.
30. Connect the fuel supply, return and vapor hoses at the side of the engine.
31. Connect the brake booster vacuum hose at the intake manifold.
32. Connect all low tension wires and the high tension wire going to the coil from the distributor. Connect the engine ground.
33. Connect the automatic transaxle oil cooler hoses.
34. Connect the electrical connectors for the back-up lights and engine harness, located near the battery tray. If equipped with 5 speed, connect the select control valve connector. Connect the alternator harness connectors and the oil pressure sending unit.
35. Install the radiator and electric cooling fan assembly. Connect the upper and lower radiator hoses. Observe the following torque values:

Left mount large insulator nut—65–80 ft. lbs. (88–108 Nm)

Left mount small insulator nut—22–29 ft. lbs. (29–40 Nm)

Left mount bracket-to-engine nuts/bolts—36–47 ft. lbs. (48–63 Nm)

Transaxle mount insulator nut—65–80 ft. lbs. (88–108 Nm)

Transaxle insulator bracket-to-side frame bolts—22–29 ft. lbs. (29–40 Nm)

Transaxle bracket assembly-to-automatic transaxle nuts—65–80 ft. lbs. (88–108 Nm)

Transaxle mount bracket to manual transaxle bolts—40–43 ft. lbs. (54–58 Nm)

Rear roll insulator nut—33–43 ft. lbs. (45–58 Nm)

Rear roll stopper bracket-to-crossmember assembly bolts—22–29 ft. lbs. (29–40 Nm)

Front roll insulator nut—33–43 ft. lbs. (45–58 Nm)

Front roll stopper bracket-to-crossmember assembly bolts—33–40 ft. lbs.

Lower roll insulator-to-roll damper bracket bolt—22–29 ft. lbs. (29–40 Nm)

Center crossmember-to-body bolts—43–58 ft. lbs. (58–77 Nm)

36. Replenish all fluids. Adjust the transaxle and accelerator linkages. Start the engine and check for leaks as well as proper gauge operation. Replace the hood and the air conditioning system.

6 Cylinder Engine

NOTE: The factory recommends that the engine and transaxle be removed as a unit.

1. Disconnect the negative battery cable. Remove the air cleaner assembly.
2. Disconnect the backup light, engine, alternator and oil pressure harnesses.
3. Drain the engine coolant into a suitable container.
4. Label and disconnect the transaxle oil cooler lines, radiator hoses and remove the radiator assembly.
5. Disconnect the brake booster, fuel, evaporative canister and heater hoses.
6. Disconnect the accelerator, transaxle, cruise control and speedometer cables.
7. Detach the air conditioning compressor from the mounting bracket and hange out of the way with a piece of wire. Do not disconnect the refrigerant lines.
8. Remove the power steering pump.
9. Raise the vehicle and support safely.
10. Remove the oil pan shield and drain the transaxle.
11. Disconnect the front exhaust pipe.
12. Remove the lower arm ball joint and stabilizer bar at the point where it is mounted to the lower arms.
13. Remove the halfshaft from the housing by prying against the transaxle housing with a suitable prybar.
14. Hang the lower arm and driveshafts from the body with wire.
15. Attach an engine lifting device to the engine and raise the engine enough to take the tension off the engine mounts.
16. Remove the front roll stopper, engine damper and rear roll stopper.
17. Remove the engine mount bolts.
18. Slowly raise the engine and transaxle and hold in the raised position. Disconnect any connected cables or harnesses.
19. Remove the blind plugs from the inside of the right fender shield and remove the transaxle mounting bracket bolts.
20. Remove the left mount insulator bolt.
21. While directing the transaxle side downward, lift the engine and transaxle assembly up and out of the vehicle.

To install:

22. While directing the transaxle side downward, direct the engine and transaxle assembly into the vehicle. Observe the following torque values:

Left mount large insulator nut—65–80 ft. lbs. (88–108 Nm)

Left mount small insulator nut—22–29 ft. lbs. (29–40 Nm)

Left mount bracket-to-engine nuts/bolts—36–47 ft. lbs. (48–63 Nm)

Transaxle mount insulator nut—65–80 ft. lbs. (88–108 Nm)

Transaxle insulator bracket-to-side frame bolts—22–29 ft. lbs. (29–40 Nm)

Transaxle bracket assembly-to-automatic transaxle nuts—65–80 ft. lbs. (88–108 Nm)

Transaxle mount bracket to manual transaxle bolts—40–43 ft. lbs. (54–58 Nm)

Rear roll insulator nut—33–43 ft. lbs. (45–58 Nm)

Rear roll stopper bracket-to-crossmember assembly bolts—22–29 ft. lbs. (29–40 Nm)

Front roll insulator nut—33–43 ft. lbs. (45–58 Nm)

Front roll stopper bracket-to-crossmember assembly bolts—33–40 ft. lbs. (45–58 Nm)

Lower roll insulator-to-roll damper bracket bolt—22–29 ft. lbs. (29–40 Nm)

Center crossmember-to-body bolts—43–58 ft. lbs. (58–77 Nm)

23. Install the left mount insulator bolt.
24. Install the tranaxle mounting bracket bolts and blind plugs the inside of the right fender shield.
25. Slowly lower the engine and transaxle into position.
26. Install the engine mount bolts.
27. Install the front roll stopper, engine damper and rear roll stopper.
28. Remove the engine lifting device.

29. Connect the lower arm and halfshafts.
30. Install the halfshafts to the transaxle housing. Make sure the C-clips are fully engaged into the differential assembly.
31. Install the lower arm ball joint and stabilizer bar at the point where it is mounted to the lower arms.
32. Connect the front exhaust pipe. Refill the transaxle with fluid.
33. Install the oil pan shield.
34. Lower the vehicle.
35. Install the power steering pump.
36. Install the air conditioning compressor to the mounting bracket.
37. Connect the accelerator, transaxle, cruise control and speedometer cables.
38. Connect the brake booster, fuel, evaporative canister and heater hoses.
39. Install the radiator. Connect the transaxle oil cooler lines and radiator hoses.
40. Refill the engine coolant.
41. Connect the backup light, engine, alternator and oil pressure harnesses.
42. Connect the negative battery cable. Install the air cleaner assembly.
43. Start the engine and check for leaks.

Engine Mounts

REMOVAL & INSTALLATION

The engine mounts can be removed and installed by supporting the engine and transaxle assembly from below.

1. Disconnect the negative battery cable.
2. Remove the upper mount-to-body bracket bolts.
3. Using an engine holding fixture tool, support the engine.
4. Raise and safely support the vehicle.
5. Remove the left side inner fender shield.
6. Remove the lower engine mount-to-body bracket bolt.
7. Raise the engine, slightly, and remove the engine mount through bolt.
8. Remove the lower engine mount-to-engine bracket bolt and the mount.

To install:

9. Support the engine from below. To protect the engine, place a board between the support and the engine.
10. Install the lower engine mount and mount-to-engine bracket bolt.
11. Raise or lower the engine slightly, to install engine mount through bolt.
12. Install the left side inner fender shield.
13. Install the upper mount and mount-to-body bracket bolts.
14. Lower the vehicle and connect the negative battery cable.

Cylinder Head

REMOVAL & INSTALLATION

NOTE: Never remove the cylinder head unless the engine is cold, a hot cylinder head will warp.

1468cc Engine

1. Disconnect the negative battery cable.
2. Drain the cooling system and then disconnect the upper radiator hose. Remove the PCV hose that runs between the air cleaner and the rocker cover.
3. Remove the air cleaner. Disconnect the fuel lines. Label and disconnect any vacuum lines running to the cylinder head, manifold or carburetor from other parts of the engine compartment. Disconnect the heater hoses going to the head.
4. Label and disconnect the spark plug wires. Remove the rocker cover. Turn the crankshaft over until the TDC timing marks align and both No. 1 cylinder valves are closed, both rockers are off the cams. Then, remove the distributor.
5. Remove the carburetor, if equipped. Remove the intake manifold. Remove the exhaust manifold.
6. Remove the timing belt cover. Note the location of the camshaft sprocket timing mark. Loosen both timing belt tensioner mounting bolts and then lever it over toward the water pump as far as it will go. Retighten the adjusting bolt to hold the tensioner in this position. Pull the timing belt off the camshaft sprocket but leave it engaged with the other sprockets.
7. Using a hex type wrench, loosen the head bolts in the proper sequence. When all have been loosened, remove them. Then, pull the head off the en-

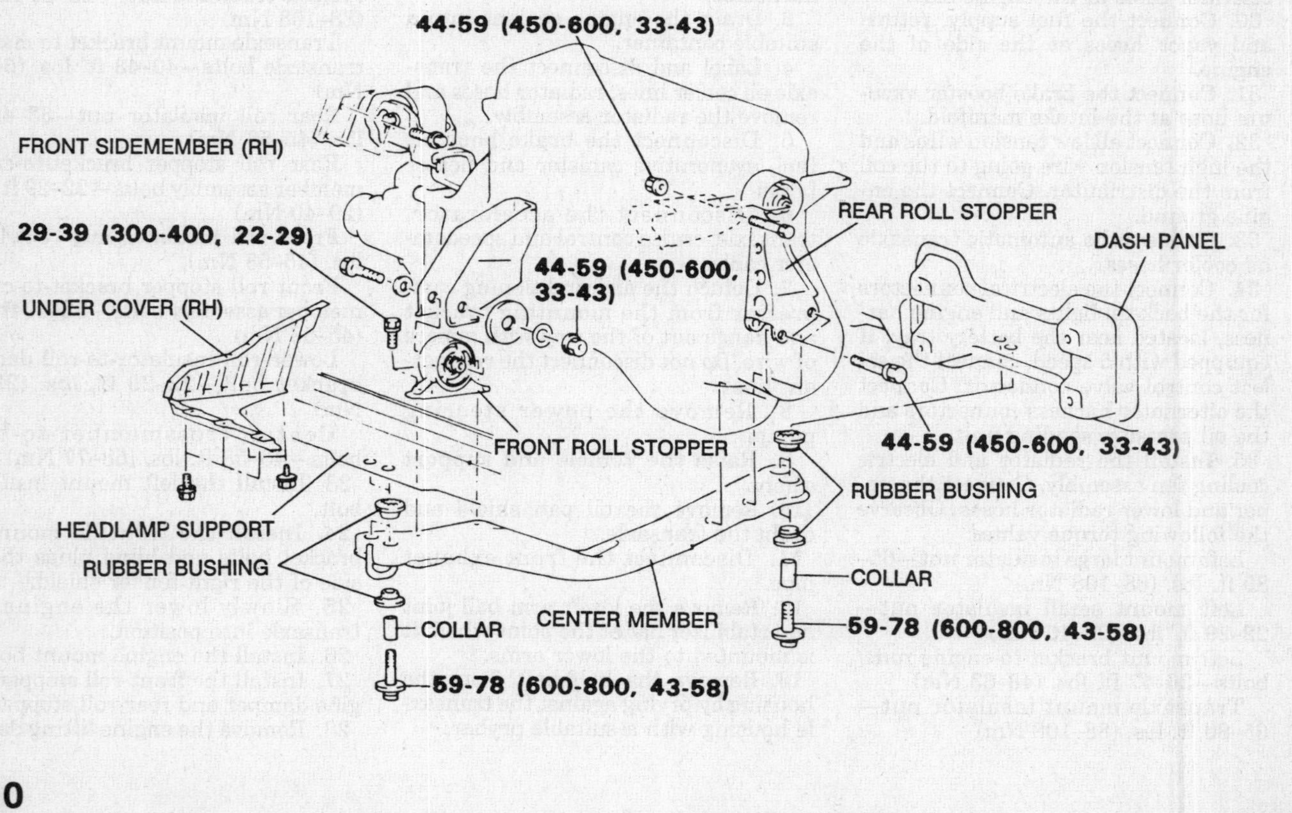

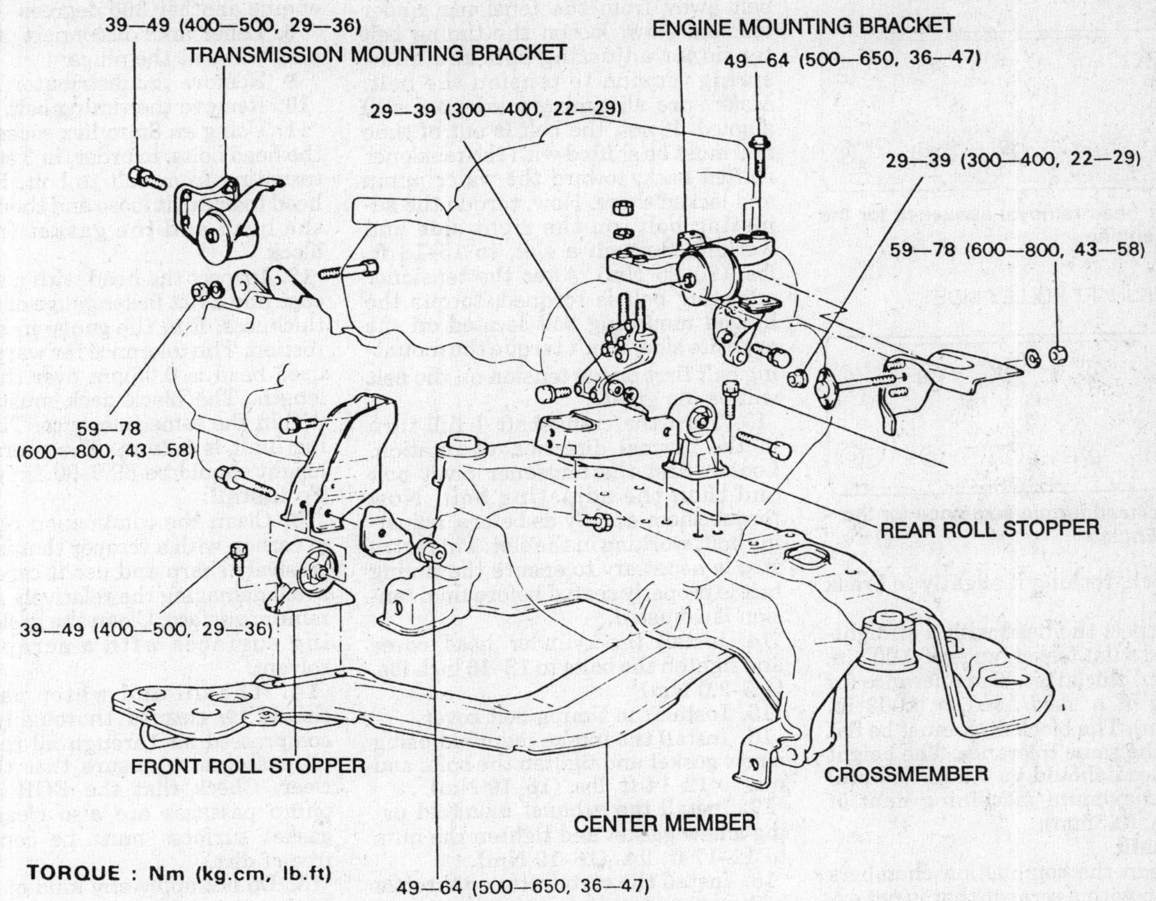
39—49 (400—500, 29—36)
TRANSMISSION MOUNTING BRACKET
29—39 (300—400, 22—29)
ENGINE MOUNTING BRACKET
49—64 (500—650, 36—47)
29—39 (300—400, 22—29)
59—78 (600—800, 43—58)
59—78
(600—800, 43—58)
REAR ROLL STOPPER
39—49 (400—500, 29—36)
FRONT ROLL STOPPER
CROSSMEMBER
CENTER MEMBER
TORQUE : Nm (kg.cm, lb.ft)
49—64 (500—650, 36—47)

Sonata engine mounts

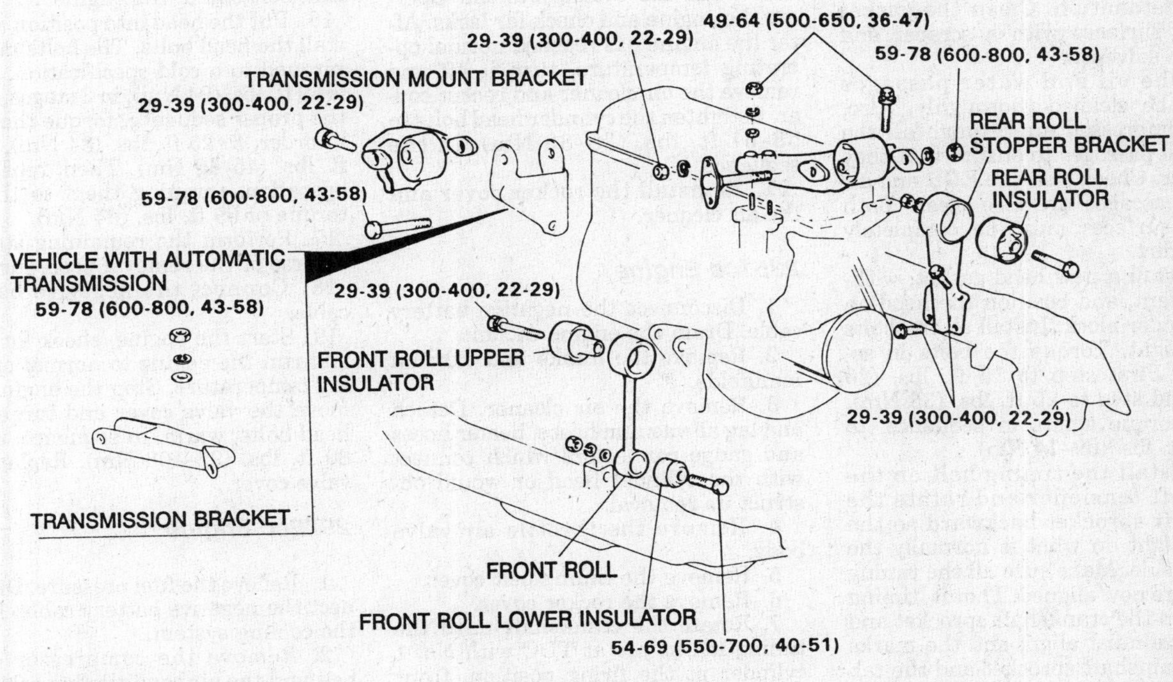
29-39 (300-400, 22-29)
49-64 (500-650, 36-47)
59-78 (600-800, 43-58)
TRANSMISSION MOUNT BRACKET
29-39 (300-400, 22-29)
REAR ROLL
STOPPER BRACKET
REAR ROLL
INSULATOR
59-78 (600-800, 43-58)
VEHICLE WITH AUTOMATIC
TRANSMISSION
59-78 (600-800, 43-58)
29-39 (300-400, 22-29)
FRONT ROLL UPPER
INSULATOR
29-39 (300-400, 22-29)
TRANSMISSION BRACKET
FRONT ROLL
FRONT ROLL LOWER INSULATOR
54-69 (550-700, 40-51)
TORQUE : Nm (kg·cm, lb·ft)

Excel engine mounts

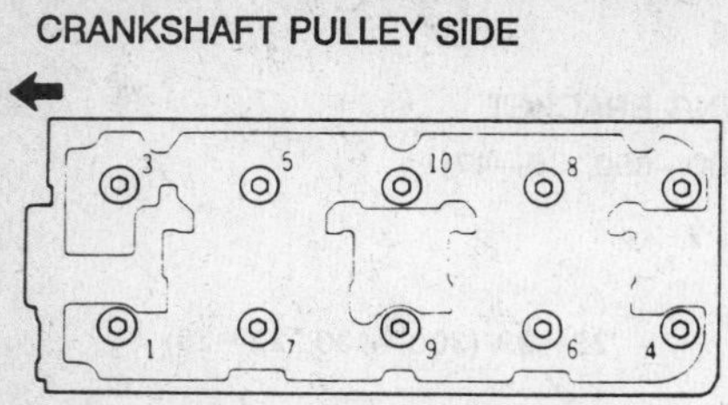

Cylinder head removal sequence for the 1468cc engine

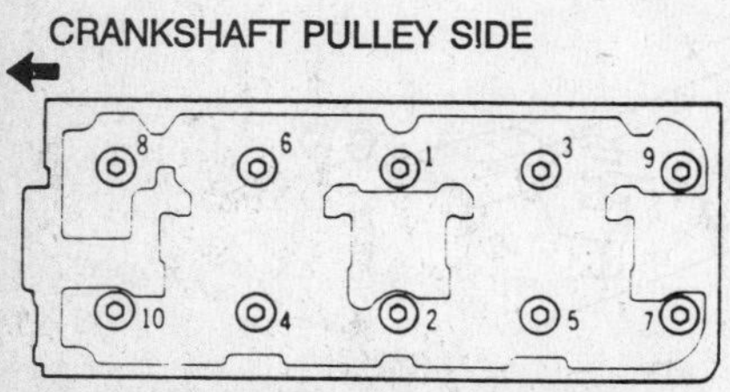

Cylinder head torque sequence for the 1468cc Engine

gine block, rocking it slightly to break it loose.

8. Inspect the head with a straightedge and a flat feeler gauge of 0.002 in. (0.05mm) thickness. The tolerance for warping of a used head is 0.002 in. (0.05mm). The block deck must be flat within the same tolerance. The height of the head should be 3.5 in. (89mm) with a maximum machining limit of 0.012 in. (0.3mm).

To install:

9. Clean the combustion chambers of carbon with a scraper that is not excessively sharp, being careful not to damage the aluminum surface. Sharp edges in the combustion chambers can cause detonation. Clean the gasket mating surfaces with a scraper and suitable solvent.

10. The oil and water passages should be cleaned thoroughly. Also, blow compressed air through all the small oil passages to ensure that they are clear. Check that the EGR and air pump passages are also clear. Both gasket surfaces must be completely free of dirt.

11. Install a new head gasket, without sealant, and position the head on the cylinder block. Install all the bolts finger tight. Torque the bolts in sequence. First step to 15 ft. lbs. (20 Nm), 2nd step to 25 ft. lbs. (35 Nm). Then, torque again, in sequence, to 51–54 ft. lbs. (69–74 Nm).

12. Install the timing belt on the camshaft tensioner and rotate the camshaft sprocket backward so the belt is tight on what is normally the tension side. Make sure all the timing marks are now aligned. That is, timing marks on the crankshaft sprocket and front case must align; and the marks on the camshaft sprocket and the tab on the cylinder head must be simultaneously aligned with the side of the belt away from the tensioner under tension. Now, loosen the timing belt tensioner adjusting bolt and allow spring tension to tension the belt. Make sure all timing marks are still aligned. If not, the belt is out of time and must be shifted with the tensioner shifted back, toward the water pump and locked there. Now, torque the adjusting bolt, on the right side and working through a slot, to 15–18 ft. lbs. (20–25 Nm). After the tensioner adjusting bolt is torqued, torque the hinged mounting bolt located on the opposite side. Don't torque the mounting bolt first or the tension on the belt will be too great.

13. Turn the crankshaft 1 full turn in the normal direction of rotation. Loosen first the tensioner pivot bolt and then the adjusting bolt. Now torque them exactly as before, adjusting bolt, working in the slot. This extra step is necessary to ensure the timing belt is properly seated before final tension is adjusted.

14. Install the cylinder head cover and tighten the bolts to 13–16 inch lbs. (1.5–2.0 Nm).

15. Install the timing belt cover.

16. Install the intake manifold using a new gasket and tighten the bolts and nut to 12–14 ft. lbs. (16–19 Nm).

17. Install the exhaust manifold using a new gasket and tighten the nuts to 12–17 ft. lbs. (16–19 Nm).

18. Install the carburetor or throttle body assembly.

19. Install the distributor. Connect all hoses, lines and the air cleaner.

20. Refill the cooling system. Operate the engine and check for leaks. After the engine has reached normal operating temperature, turn it off and remove the air cleaner and rocker cover. Retighten the cylinder head bolts to 58–61 ft. lbs. (78–83 Nm), in the sequence.

21. Reinstall the rocker cover and the air cleaner.

2351cc Engine

1. Disconnect the negative battery cable. Drain the engine coolant.

2. Remove the intake and exhaust manifolds.

3. Remove the air cleaner. Detach and tag all vacuum hoses, heater hoses and gauge connectors which connect with the cylinder head or would obstruct its removal.

4. Remove the throttle air valve body.

5. Remove the timing belt cover.

6. Remove the rocker cover.

7. Rotate the crankshaft until the timing marks are at TDC with No. 1 cylinder at the firing position, front valves closed fully. If the rockers are not all the way off the cams, turn the engine another 360 degrees.

8. Label and disconnect all spark plug wires at the plugs.

9. Remove the distributor.

10. Remove the timing belt.

11. Using an 8mm hex socket loosen the head bolts, in order, in 3 stages, alternating from bolt to bolt. Rock the head to break it loose and then remove the head and the gasket from the block.

12. Inspect the head with a straightedge and a flat feeler gauge of 0.10mm thickness. Run the gauge in every direction. The tolerance for warping of a used head is 0.05mm over the entire length. The block deck must be flat within the same tolerance. The refacing limit is 0.2mm. The overall head height should be 89.9–90.1mm.

To install:

13. Clean the combustion chambers of carbon with a scraper that is not excessively sharp and use it carefully to avoid damaging the relatively soft aluminum surface. Clean the gasket mating surfaces with a scraper and solvent.

14. The oil and water passages should be cleaned thoroughly, Blow compressed air through all the small oil passages to ensure that they are clear. Check that the EGR and air pump passages are also clear. Both gasket surfaces must be completely free of dirt.

15. Do not apply any kind of sealant to the gasket. Install the head gasket on the block deck, with the identification mark facing upward and on the cam belt end of the engine.

16. Put the head into position and install the head bolts. The bolts must be torqued to a cold specification, which is 69 ft. lbs. (94 Nm), in 3 stages. Using the proper sequence, torque the bolts, in order, to 25 ft. lbs. (34 Nm), 33-36 ft. lbs. (45–48 Nm). Then, repeat the operation, torquing them to the full torque of 69 ft. lbs. (94 Nm).

17. Perform the remaining steps in reverse of the removal procedure.

18. Connect the negative battery cable.

19. Start the engine, check for leaks and run the engine to normal operating temperature. Stop the engine. Remove the valve cover and torque the head bolts, warm, in sequence, to 72–80 ft. lbs. (96–108 Nm). Replace the valve cover.

2972cc Engine

1. Relieve the fuel pressure. Disconnect the negative battery cable. Drain the cooling system.

2. Remove the compressor drive belt and the air conditioning compressor from its mount and support it

8a. Camshaft bearing cap
8b. No. 2, 3 and 4 caps
8c. Camshaft bearing cap (rear)
10. Cylinder head
11a. Intake valve seat ring
11b. Exhaust valve seat ring
12. Cylinder head bolt
13a. Exhaust valve guide
13b. Intake valve guide
14. Cylinder head gasket

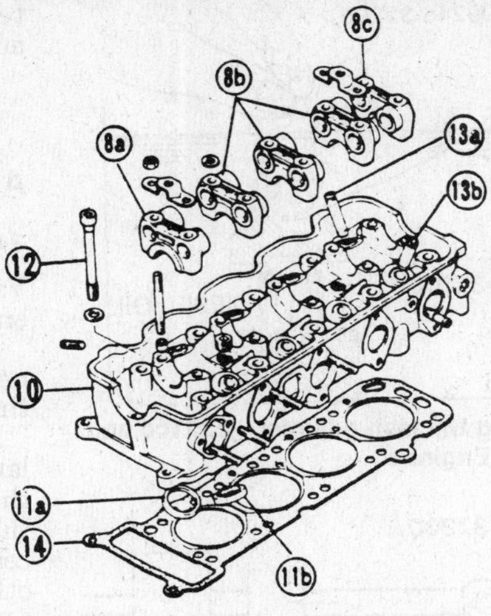

Exploded view of the cylinder head—2351cc Engine

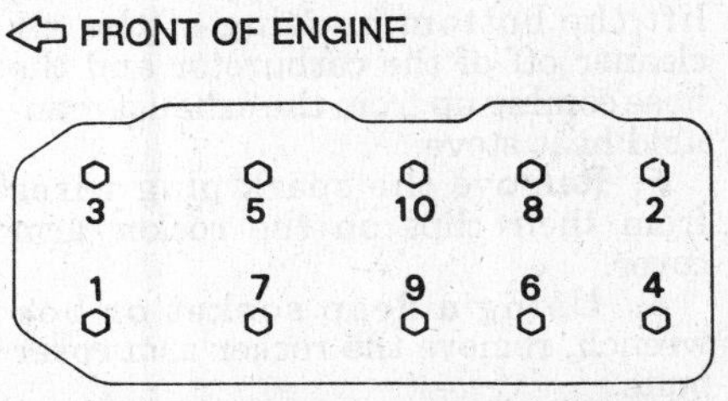

2351cc Engine cylinder head loosening sequence

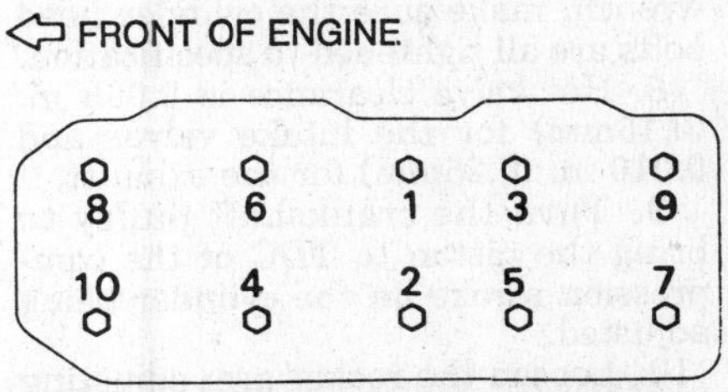

2351cc Engine cylinder head tightening sequence

aside. Using a ½ in. drive breaker bar, insert it into the square hole of the serpentine drive belt tensioner, rotate it counterclockwise to reduce the belt tension and remove the belt. Remove the alternator and power steering pump from the brackets and move them aside.

3. Raise the vehicle and support safely. Remove the right front wheel and the inner splash shield.

4. Remove the crankshaft pulleys and the torsional damper.

5. Lower the vehicle. Using a floor jack and a block of wood positioned under the oil pan, raise the engine slightly. Remove the engine mount bracket from the timing cover end of the engine and the timing belt covers.

6. To remove the timing belt, perform the following procedures:

a. Rotate the crankshaft to position the No. 1 cylinder on the TDC of its compression stroke; the crankshaft sprocket timing mark should align with the oil pan timing indicator and the camshaft sprockets timing marks (triangles) should align with the rear timing belt covers timing marks.

b. Mark the timing belt in the direction of rotation for reinstallation purposes.

c. Loosen the timing belt tensioner and remove the timing belt.

NOTE: When removing the timing belt from the camshaft sprocket, make sure the belt does not slip off of the other camshaft sprocket. Support the belt so it can not slip off of the crankshaft sprocket and opposite side camshaft sprocket.

7. Remove the air cleaner assembly. Label and disconnect the spark plug wires and the vacuum hoses.

8. Remove the valve cover.

9. Install auto lash adjuster retainer tools MD998443 or equivalent, on the rocker arms.

10. If removing the front cylinder head, matchmark the distributor rotor

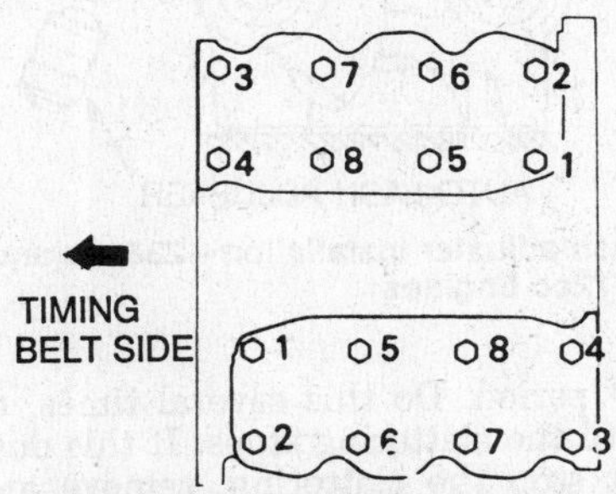

Cylinder head bolt loosening sequence—2972cc Engine

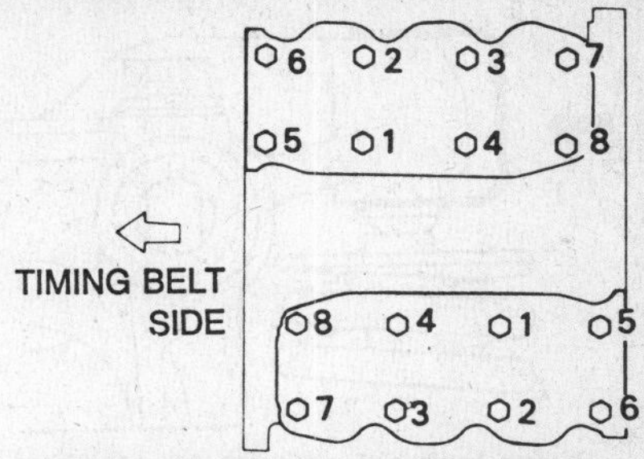

Cylinder head bolt torque sequence—2972cc Engine

to the distributor housing and the housing to distributor extension locations. Remove the distributor and the distributor extension.

11. Remove the camshaft bearing assembly to cylinder head bolts but do not remove the bolts from the assembly. Remove the rocker arms, rocker shafts and bearing caps as an assembly, as required. Remove the camshafts from the cylinder head and inspect them for damage.

12. Remove the intake manifold assembly.

13. Remove the exhaust manifold.

14. Remove the cylinder head bolts, starting from the outside and working inward. Remove the cylinder head from the engine.

15. Clean the gasket mounting surfaces and check the heads for warpage; maximum warpage is 0.008 in. (0.20mm).

To install:

16. Install the new cylinder head gasket over the dowels on the engine block.

17. Install the cylinder head(s) on the engine and torque the cylinder head bolts, in sequence, using 3 even steps, to 70 ft. lbs. (95 Nm).

18. Install or connect all items that were removed or disconnected during the removal procedure.

19. When installing the timing belt over the camshaft sprocket, use care not to allow the belt to slip off the opposite camshaft sprocket.

20. Make sure the timing belt is installed on the camshaft sprocket in the same position as when removed.

21. Refill the cooling system. Connect the negative battery cable. Start the engine and check for leaks using the DRB I or II to activate the fuel pump.

22. Adjust the timing, as required.

Valve Lash Adjuster

The lash adjuster is similar to a hydraulic lifter that maintains a zero lash adjustment. The lash adjuster is incorporated in the rocker arm and is removable. Though the lash adjuster is non-adjustable, if an abnormal tap-

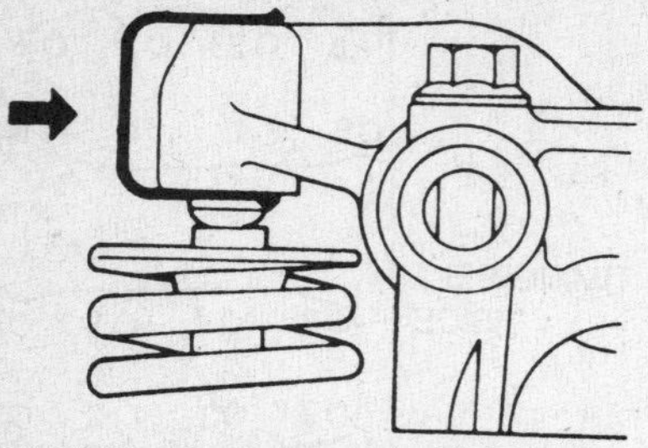

Valve lash adjuster retainer tool MD998443

ping noise is heard during idling, air in the lash adjuster may be the problem. The adjuster can be bled and rechecked for noise.

REMOVAL & INSTALLATION

2351cc and 2972cc Engines

1. Disconnect the negative battery cable. Remove the air cleaner assembly.
2. Remove the valve cover.

NOTE: The lash adjuster is incorporated in the rocker arm on the side of the valve spring. When removing the rocker arm, a special tool must be used the prevent the lash adjuster from falling out.

3. Using the lash adjuster retainer tools MD998443 or equivalent, install them on the rocker arms.
4. On the right side cylinder head, remove the distributor extension.
5. Hold the rear end of the camshaft down. If the rear of the camshaft cannot be held down, the belt will dislodge and the valve timing will be lost. Loosen the camshaft cap bolts but do not remove them from the caps. Remove the caps, arms, shafts and bolts all as an assembly.
6. Remove the lifter(s) from the rocker arm(s).
7. Lubricate the lifter(s) and their bore(s) with clean engine oil.
8. The installation is the reverse of the removal procedure.
9. Connect the negative battery cable and check the lifters for proper operation.

BLEEDING THE LASH ADJUSTERS

If the lash adjuster is removed and the diesel fuel contained inside is spilled, submerge the adjuster in clean diesel fuel and compress it several times to expel the air. If air is still trapped after assembly and installation and a clattering noise is heard when the engine is started, the air can be bled by increasing engine speed from idle to 3000 rpm and back to idle over a minute period. Do this several times, or until the clattering stops. If this does not stop the clattering, remove and submerge the lifter in clean diesel fuel, compressing it several times. If clattering continues, replace the lash adjuster.

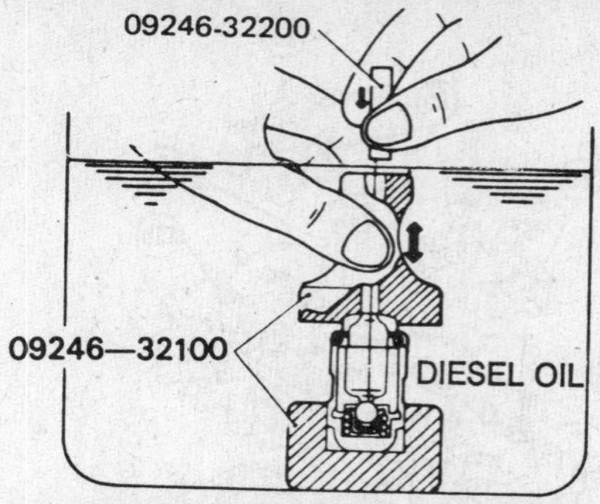

Bleeding the lash adjuster—2351cc and 2972cc Engines

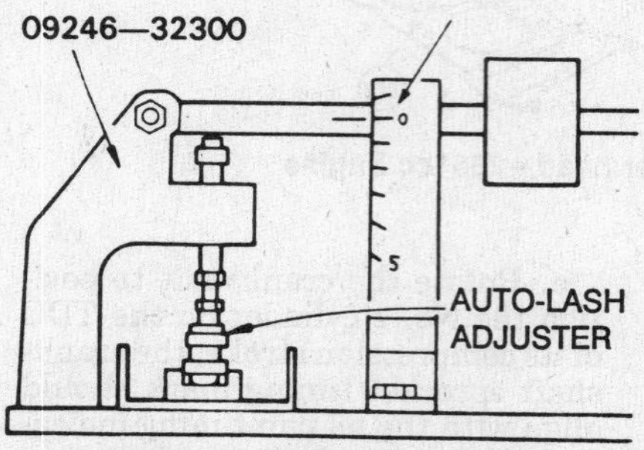

Lash adjuster leak-down—2351cc and 2972cc Engines

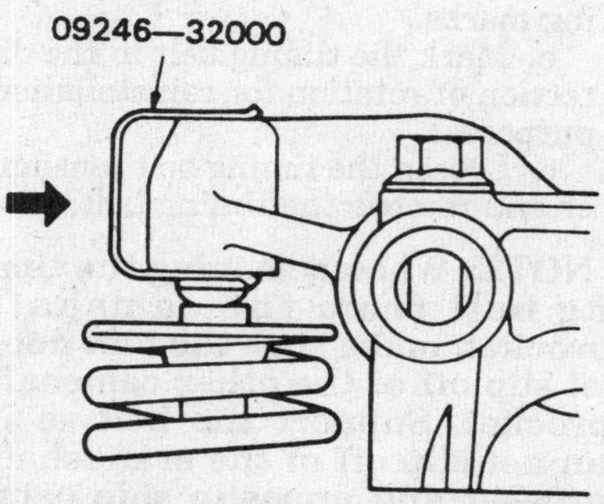

Lash adjuster holding tool installed—2351cc and 2972cc Engines

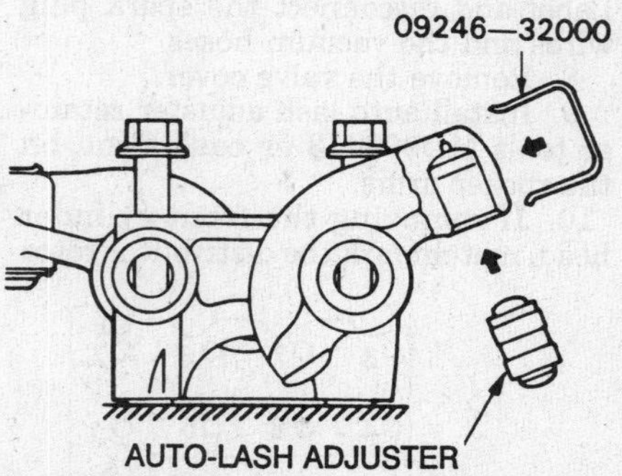

Lash adjuster installation—2351cc and 2972cc Engines

Valve Lash

ADJUSTMENT

1468cc Engine

Valve clearance is adjusted with the engine off.

1. Run the engine until it reaches normal operating temperature and then turn it off.
2. Remove the air cleaner. Pull the large crankcase ventilation hose off the front of the air cleaner. Disconnect the 2 smaller hoses, one goes to the rear of the rocker arm cover and the other to the intake manifold.
3. Loosen and remove the nuts and bracket which attach the air cleaner to the rocker arm cover.
4. On carburetor equipped vehicles, lift the bottom housing of the air cleaner off of the carburetor and the hose coming up from the exhaust manifold heat stove.
5. Remove the spark plug wires from their clips on the rocker arm cover.
6. Using a deep socket or box wrench, remove the rocker arm cover bolts.
7. Carefully lift the rocker arm cover off the cylinder head. Using an $^5/_{16}$ in. (8mm) Allen socket and a torque wrench, make sure the cylinder head bolts are all tightened to specification.
8. Hot valve clearance is 0.006 in. (0.15mm) for the intake valves and 0.010 in. (0.25mm) for the exhaust.
9. Turn the crankshaft pulley to bring the piston to TDC of the compression stroke on the cylinder being adjusted.
10. Loosen the rocker arm adjusting screw locknuts.
11. Using the correct thickness feeler gauge, turn the adjusting screw until the gauge just snaps through the valve stem and the rocker arm.
12. Repeat the procedure to adjust the valves of each cylinder.

NOTE: Loose valve clearances will result in excessive wear and valve train chatter. Tight valve clearance will result in burnt valve seats. Make sure to set the valve clearance to the exact specifications.

13. Apply non-hardening sealer to the rocker arm cover gasket. Always use a new gasket.
14. Install the cover, hoses, spark plug wires and the air cleaner in the reverse order of removal. Tighten the rocker arm cover bolts to 48–60 inch lbs.

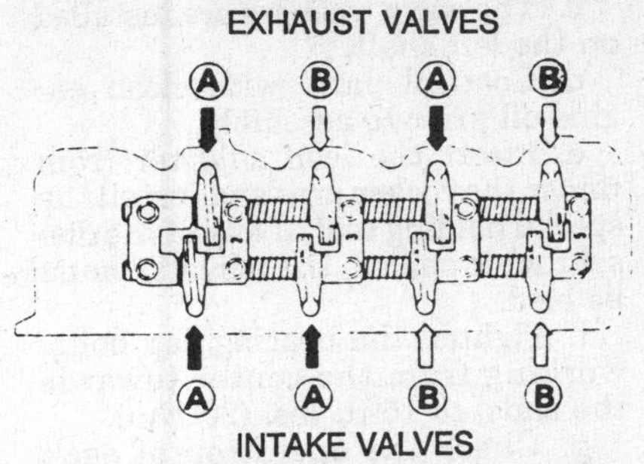

"A" and "B" valve adjusting positions—2351cc Engine

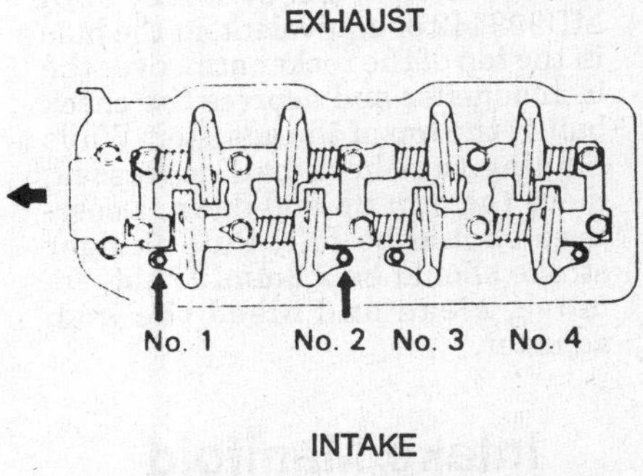

Jet valve adjustment sequence

15. Start the engine and check for leaks.

Jet Valve

ADJUSTMENT

1468cc and 2351cc Engine

The jet valve clearance is normally adjusted with the engine stopped and at normal operating temperature. However, after the engine has been rebuilt or a valve job done, the valves should first be set and adjusted with the engine cold. The basic procedure is the same, hot or cold, the only differences being the engine temperature and the jet valve clearance setting. The cold setting is 0.007 in. (0.17mm). The hot setting is 0.010 in. (0.25mm).

1. Start the engine and allow it to reach normal operating temperature.
2. Turn the engine off and disconnect the negative battery cable. Remove the air cleaner. Disconnect the large crankcase ventilation hose from the front of the air cleaner. Disconnect the smaller hoses from the rear of the rocker arm cover and the intake manifold.
3. Loosen and remove the nuts and bracket which secure the air cleaner to the rocker arm cover.
4. Lift the bottom housing of the air cleaner off of the carburetor, with the hose from the exhaust manifold heat stove attached.
5. Unsnap the spark plug wires from their clips on the rocker arm cover.
6. Remove the rocker arm cover bolts.
7. Carefully lift the rocker arm cover off the cylinder head. Using an $\frac{5}{16}$ in. (8mm) Allen socket and a torque wrench, torque the cylinder head bolts. Turn each bolt, in sequence, to specifications. After the first bolt, in sequence, has been torqued, move on to the second one, repeating the procedure. Continue, in order, until all the bolts have been torqued. Make sure the cylinder head bolts are all tightened, in sequence, to specification.
8. Remove the spark plugs.
9. Remove the distributor cap.

NOTE: A crankshaft pulley access hole is located on the left side frame member. Remove the covering plug and use a ratchet extension to turn the crankshaft when adjusting the valves.

10. Rotate the crankshaft until the No. 1 cylinder is at TDC of the compression stroke. Turn the engine by using a wrench on the bolt in the front of the crankshaft until the **TDC** or **0** timing mark on the timing cover lines up with the notch in the front pulley. Observe the valve rockers for No. 1 cylinder. If both are in identical positions with the valves up, the engine is in the right position. If not, rotate the engine exactly 360 degrees until the **TDC** or **0** degree timing mark is again aligned. Each jet valve is associated with an intake valve that is on the same rocker lever. In this position, adjust No. 1 and No. 2 jet valves, which are located on the rockers on the intake side only.
11. To adjust the appropriate jet valves, first loosen the regular (larger) intake valve adjusting stud by, loosening the locknut and backing the stud off 2 turns (1468cc only). Now, loosen the jet valve (smaller) adjusting stud locknut, back the stud out slightly and insert an 0.010 in. (0.25mm) feeler gauge between the jet valve and stud. Make sure the gauge lies flat on the top of the jet valve. Being careful not to twist the gauge or otherwise depress the jet valve spring, rotate the jet valve adjusting stud back in until it just touches the gauge. Now, tighten the locknut. Make sure the gauge still slides very easily between the stud and jet valve and they both are still just touching the gauge. Readjust, if necessary. Note that, especially with the jet valve, the clearance must not be too tight. Repeat entire the procedure for the other jet valves associated with rockers labeled No. 1 and 2.
12. Turn the engine exactly 360 degrees, until the timing marks are again aligned at **TDC** or **0**. First, perform the adjustment procedure for all the jet valves on rockers labeled No. 3 and 4, intake side only.

NOTE: Loose valve clearances will result in excessive wear and valve train chatter; tight valve clearance will result in burnt valve seats.

13. Apply non-hardening sealer to the rocker arm cover gasket. Always use a new gasket.
14. Install the cover, hoses, spark plug wires and the air cleaner in the reverse order of removal. Tighten the rocker arm cover bolts to 48–60 inch lbs.
15. Start the engine and check for leaks. It's best to install new gaskets and seals wherever they are used and to observe torque specifications for the cam cover bolts.

2972cc Engine

The 2972cc engine uses hydraulic lash adjusters, no valve adjustment is necessary.

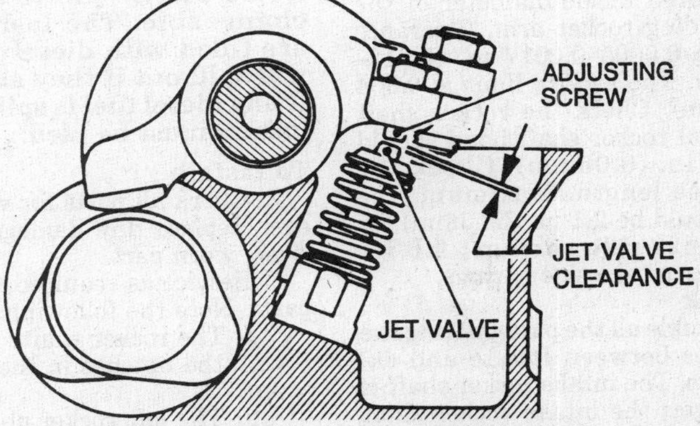

Jet valve clearance adjustment—2351cc Engine

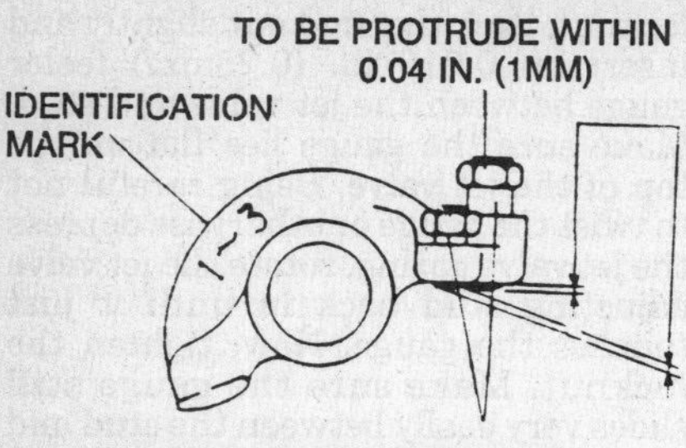

Rocker arm adjusting screw

Rocker Arms/Shafts

REMOVAL & INSTALLATION

1468cc Engine

1. Disconnect the negative battery cable. Remove the PCV hose running from the rocker cover and the air cleaner. Remove the air cleaner.
2. Remove the upper timing belt cover. Remove the rocker cover.
3. Loosen the bearing cap bolts or the rocker shaft mounting bolts but do not remove them and remove each rocker shaft, rocker arms and springs as an assembly. Disassemble the whole assembly by progressively removing each bolt and then the associated springs and rockers, keeping all parts in the exact order of disassembly. The left and right springs have different tension ratings and free length. Observe the location of the rocker arm as they are removed. Exhaust and intake, right and left are different. Do not mix them up.
4. Check the rocker arm face contacting the cam lobe and the adjusting screw that contacts the valve stem for excess wear. Inspect the fit of the rockers on the shaft. Replace adjusting screws, rockers, and/or shafts that show excessive wear. Pay special attention to the contact pad ends of the rocker arms and the ball surface of the adjustings studs. Check the diameter of the shaft at the rocker mounting points and subtract that number from the measured inside diameter of the corresponding rocker arm. Clearance should be 0.0005–0.0017 in. (0.013–0.043mm). The service limit is 0.004 in. (0.1mm). Check the rocker shaft bend. Total rocker shaft bend should be 0.002 in. (0.05mm). Check the spring free length. Maximum free length should be 2.1 in. (53.3mm) for the exhaust side springs; 2.6 in. (66mm) for intake side springs.

To install:

5. Assemble all the parts, noting the differences between intake and exhaust parts. The intake rocker shaft is much longer; the intake rocker shaft springs are over 3 in. long, while those for the exhaust side are less than 2 in. long; intake rockers have the extra adjusting screw for the jet valve; rockers are labeled 1–3 and 2–4 for the cylinder with which they are associated. Torque the rocker shaft mounting bolts to 15–19 ft. lbs. (20–26 Nm).
6. Adjust the valve clearances. This step may be omitted only if all parts are being reused.
7. Install the rocker cover with a new gasket, torquing the bolts to 12–18 inch lbs. (1.5–2.0 Nm).
8. Install the air cleaner and PCV valve. Connect the battery cable.
9. Run the engine at idle speed until it is hot. Then, unless valves did not require adjustment, remove the valve cover again and adjust the valve clearances with the engine hot.
10. Replace the rocker cover and timing belt cover, air cleaner and PCV valve.

2351cc Engine

NOTE: A special tool 09246-32000 (MD998443) or equivalent, is required to retain the automatic lash adjusters in this procedure.

1. Disconnect the negative battery cable. Remove the rocker cover and gasket and the timing belt cover.
2. Turn the crankshaft so the No. 1 piston is at TDC compression. At this point, the timing mark on the camshaft sprocket and the timing mark on the head to the left of the sprocket will be aligned.
3. Remove the camshaft bearing cap bolts.
4. Install the automatic lash adjuster retainer tool 09246-32000 (MD998443) or equivalent, to keep the adjuster from falling out of the rocker arms.
5. Lift off the bearing caps and rocker arm assemblies.
6. The rocker arms may now be removed from the shafts.

NOTE: Keep all parts in the order in which they were removed. None of the parts are interchangeable. The lash adjusters are filled with diesel fuel, which will spill out if they are inverted. If any diesel fuel is spilled, the adjusters must be bled.

To install:

7. Check all parts for wear or damage. Replace any damaged or excessively worn part.
8. Service as required, assemble all parts. Note the following:
   a. The rocker shafts are installed with the notches in the ends facing up.
   b. The left rocker shaft is longer than the right.
   c. The wave washers are installed on the left shaft.
   d. Coat all parts with clean engine oil prior to assembly.
   e. Insert the lash adjuster from under the rocker arm and install the special holding tool. If any of the diesel fuel is spilled, the adjuster must be bled.
   f. Tighten the bearing cap bolts, working from the center towards the ends, to 15 ft. lbs. (20 Nm).
   g. Check the operation of each lash adjuster by positioning the camshaft so the rocker arm bears on the low or round portion of the cam. Insert a thin steel wire, tool MD998442 or equivalent, in the hole in the top of the rocker arm, over the lash adjuster and depress the check ball at the top of the adjuster. While holding the check ball depressed, move the arm up and down. Looseness should be felt. Full plunger stroke should be 2.2mm. If not, remove, clean and bleed the lash adjuster.

Intake Manifold

REMOVAL & INSTALLATION

1468cc Engine

1. Disconnect the negative battery cable. Remove the air cleaner assembly.
2. Disconnect the fuel line and the EGR lines, if equipped with EGR. Tag and disconnect all vacuum hoses.
3. Disconnect the throttle positioner and fuel cut-off solenoid wires.
4. Disconnect the throttle linkage.
5. If equipped with an automatic transaxle, disconnect the shift cable linkage.
6. Disconnect the power brake booster vacuum line.
7. Drain the cooling system.
8. Disconnect the choke water hose at the manifold.
9. Remove the heater and water outlet hoses, disconnect the water temperature sending unit.
10. Remove the mounting nuts that hold the manifold to the cylinder head. Remove the intake manifold.

To install:

11. Clean all mounting surfaces. Before installing the manifold, coat both sides of a new gasket with a gasket sealer.

NOTE: If equipped with jet air system, take care not to get any sealer into the jet air intake passage.

12. Install the intake manifold assembly to the engine block.
13. Reconnect the heater and water

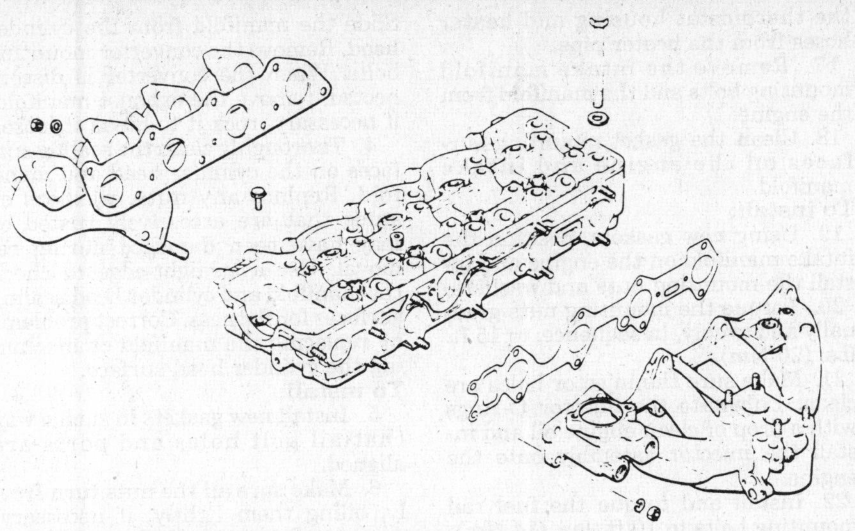

1468cc Engine intake and exhaust manifolds—carburetor shown, fuel injected similar

outlet hoses. Connect the water temperature sending unit.

14. Connect the brake booster vacuum line. Connect the choke hose at the manifold.

15. Connect the throttle and shift cable linkages, the fuel lines and all vacuum hoses. Install the air cleaner.

16. Refill the engine with coolant. Connect the negative battery cable.

2351cc Engine

1. Disconnect the negative battery cable. Remove the air cleaner assembly.
2. Release fuel system pressure. Disconnect the fuel line and the EGR lines. Tag and disconnect all vacuum hoses.
3. Disconnect the throttle positioner and fuel cut-off solenoid wires.
4. Disconnect the throttle linkage. If equipped with automatic transaxle, disconnect the shift cable linkage.
5. Remove the heater and water outlet hoses, disconnect the water temperature sending unit. Disconnect the oxygen sensor connecter, power transistor connecter, ISC connector, ignition coil connector, etc. and the distributor.
7. Remove the mounting nuts that hold the manifold to the cylinder head. Remove the intake manifold lower and upper sections with injector assembly as a unit.
8. Clean all mounting surfaces. Before installing the manifold, coat both sides with a gasket sealer.

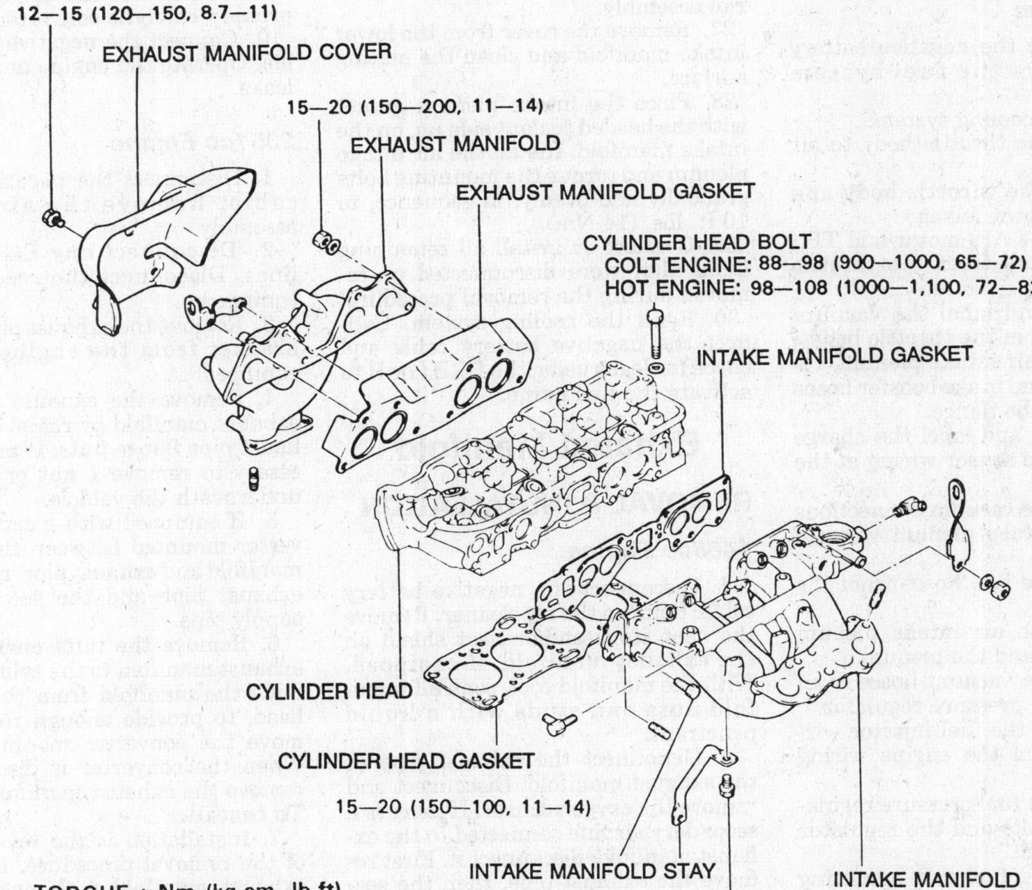

2351cc Engine intake and exhaust manifolds

To install:

9. Clean all mounting surfaces. Before installing the manifold, coat both sides of a new gasket with a gasket sealer.

NOTE: If equipped with jet air system, take care not to get any sealer into the jet air intake passage.

10. Install the intake manifold lower and upper sections with injector assembly as a unit. Install the mounting nuts that hold the manifold to the cylinder head.
11. Install the heater and water outlet hoses, disconnect the water temperature sending unit. Connect the oxygen sensor connecter, power transistor connecter, ISC connector, ignition coil connector, etc. and the distributor.
12. Connect the throttle linkage. If equipped with automatic transaxle, connect the shift cable linkage.
13. Connect the throttle positioner and fuel cut-off solenoid wires.
14. Connect the fuel line the EGR lines. Connect all vacuum hoses.
15. Refill the engine with coolant. Connect the negative battery cable.

2972cc Engine

1. Disconnect the negative battery cable. Relieve the fuel system pressure.
2. Drain the cooling system.
3. Remove the throttle body to air cleaner hose.
4. Remove the throttle body and transaxle kickdown linkage.
5. Remove the AIS motor and TPS wiring connectors from the throttle body.
6. Remove and label the vacuum hose harness from the throttle body.
7. From the air intake plenum, remove the PCV and brake booster hoses and the EGR tube flange.
8. Disconnect and label the charge and temperature sensor wiring at the intake manifold.
9. Remove the vacuum connections from the air intake plenum vacuum connector.
10. Remove the fuel hoses from the fuel rail.
11. Remove the air intake plenum mounting bolts and the plenum.
12. Remove the vacuum hoses from the fuel rail and pressure regulator.
13. Disconnect the fuel injector wiring harness from the engine wiring harness.
14. Remove the fuel pressure regulator mounting bolts and the regulator from the fuel rail.
15. Remove the fuel rail mounting bolts and the fuel rail from the intake manifold.
16. Separate the radiator hose from the thermostat housing and heater hoses from the heater pipe.
17. Remove the intake manifold mounting bolts and the manifold from the engine.
18. Clean the gasket mounting surfaces on the engine and intake manifold.

To install:

19. Using new gaskets, position the intake manifold on the engine and install the mounting nuts and washers.
20. Torque the mounting nuts gradually and evenly, in sequence, to 15 ft. lbs. (20 Nm).
21. Make sure the injector holes are clean. Lubricate the injector O-rings with a drop of clean engine oil and install the injector assembly onto the engine.
22. Install and torque the fuel rail mounting bolts to 10 ft. lbs. (14 Nm).
23. Install the fuel pressure regulator onto the fuel rail.
24. Install the fuel supply and return tube and the vacuum crossover hold-down bolt.
25. Connect the fuel injection wiring harness to the engine wiring harness.
26. Connect the vacuum harness to the fuel pressure regulator and fuel rail assembly.
27. Remove the cover from the lower intake manifold and clean the mating surface.
28. Place the intake plenum gasket with the beaded sealant side up, on the intake manifold. Install the air intake plenum and torque the mounting bolts gradually and evenly, in sequence, to 10 ft. lbs. (14 Nm).
29. Connect or install all remaining items that were disconnected or removed during the removal procedure.
30. Refill the cooling system. Connect the negative battery cable and check for leaks using the DRB I or II to activate the fuel pump.

Exhaust Manifold

REMOVAL & INSTALLATION

1468cc Engine

1. Disconnect the negative battery cable. Remove the air cleaner. Remove the heat stove and/or heat shield on the exhaust manifold, if equipped. With the manifold cool, soak all manifold nuts and studs with a liquid penetrant.
2. Disconnect the exhaust pipe at the exhaust manifold. Disconnect and remove the oxygen sensor. If there is a secondary air line connected to the exhaust manifold, disconnect it. First remove the exhaust pipe, then the secondary air supply pipe.
3. Support the manifold and remove all attaching nuts and washers. Slide the manifold from the cylinder head. Remove the converter mounting bolts. When the converter is disconnected, remove the exhaust manifold, if necessary, rock it to break it loose.
4. Thoroughly clean the sealing surfaces on the cylinder head and manifold. Replace any nuts, washers or studs that are excessively rusted or may have been damaged during removal. Use a straight-edge to check the manifold and cylinder head sealing surfaces for flatness. Correct problems by replacing the manifold or machining the cylinder head surface.

To install:

5. Install new gaskets in such a way that all bolt holes and ports are aligned.
6. Make sure all the nuts turn freely, oiling them lightly, if necessary. Also, make sure all the studs are screwed all the way into the block.
7. Place the manifold in position and support it, install all washers and nuts hand tight.
8. Torque the exhaust manifold-to-cylinder head nuts to 14 ft. lbs. (20 Nm), alternately and in several stages.
9. Install piping, heat stoves and shields. Connect the exhaust pipe or primary catalytic converter.
10. Connect the negative battery cable. Operate the engine and check for leaks.

2351cc Engine

1. Disconnect the negative battery cable. Remove the air cleaner assembly.
2. Disconnect any EGR or heat lines. Disconnect the reed valve, if equipped.
3. Remove the exhaust pipe support bracket from the engine block, if equipped.
4. Remove the exhaust pipe from exhaust manifold by removing the exhaust pipe flange nuts. It may be necessary to remove 1 nut or bolt from underneath the vehicle.
5. If equipped with a catalytic converter mounted between the exhaust manifold and exhaust pipe, remove the exhaust pipe and the secondary air supply pipe.
6. Remove the nuts mounting the exhaust manifold to the cylinder head. Slide the manifold from the cylinder head, to provide enough room to remove the converter mounting bolts. When the converter is disconnected, remove the exhaust manifold.

To install:

7. Installation is the reverse order of the removal procedure. Install the exhaust manifold with new gaskets. New gaskets should be used and on some engines, port liner gaskets are used.

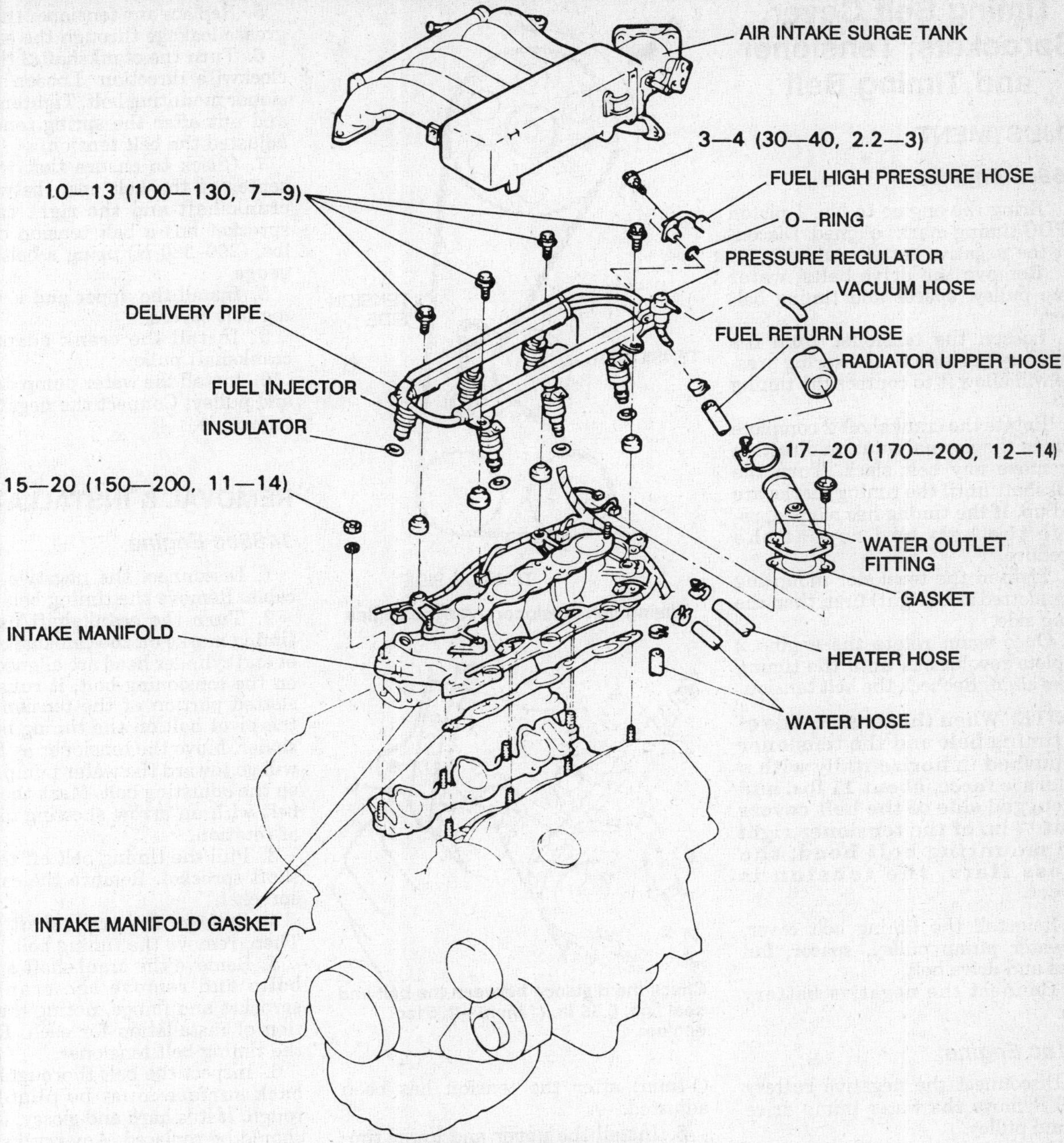

2972cc engine intake manifold

8. Torque the exhaust manifold-to-cylinder head bolts to 14 ft. lbs. (20 Nm).

2972cc Engine

1. Disconnect the negative battery cable. Raise the vehicle and support safely.
2. Disconnect the exhaust pipe from the rear exhaust manifold at the articulated joint.
3. Disconnect the EGR tube from the rear manifold and unplug the oxygen sensor wire.
4. Remove the crossover pipe to manifold bolts.
5. Remove the rear manifold to cylinder head nuts and the manifold.
6. Lower the vehicle and remove the heat shield from the manifold.
7. Remove the front manifold-to-cylinder head nuts and the manifold.
8. Clean the gasket mounting surfaces. Inspect the manifolds for cracks, flatness and/or damage.

To install:

9. When installing, the numbers 1–3–5 on the gaskets are used with the rear cylinders and 2–4–6 are on the gasket for the front cylinders. Torque the manifold-to-cylinder head nuts to 14 ft. lbs. (20 Nm).
10. Install the crossover pipe to the manifold.
11. Connect the EGR tube and oxygen sensor wire.
12. Connect the exhaust pipe to the rear exhaust manifold, at the articulated joint.
13. Connect the negative battery cable and check the manifolds for leaks.

Timing Belt Cover, Sprockets, Tensioner and Timing Belt

ADJUSTMENT

1468cc Engines

1. Bring the engine to No. 1 piston at TDC timing marks aligned. Disconnect the negative battery cable.
2. Remove the drive belts, water pump pulley, spacer and timing belt cover.
3. Loosen the tensioner from it's temporary position so the spring pressure will allow it to contact the timing belt.
4. Rotate the crankshaft 2 complete turns in the normal rotation direction to remove any belt slack. Turn the crankshaft until the timing marks are lined up. If the timing has slipped, remove the belt and repeat the procedure.
5. Tighten the tensioner mounting bolts, slotted side (right) first, then the spring side.
6. Once again rotate the engine 2 complete revolutions until the timing marks align. Recheck the belt tension.

NOTE: When the tension side of the timing belt and the tensioner are pushed in horizontally with a moderate force, about 11 lbs. and the cogged side of the belt covers about 1/4 in. of the tensioner right side mounting bolt head, the across flats, the tension is correct.

7. Reinstall the timing belt cover, the water pump pulley, spacer, fan blades and drive belt.
8. Connect the negative battery cable.

2351cc Engine

1. Disconnect the negative battery cable. Remove the water pump drive belt and pulley.
2. Remove the crank adapter and crankshaft pulley.
3. Remove the upper and lower timing belt covers.
4. Check the tensioners for a smooth rate of movement.
5. Replace any tensioner that shows grease leakage through the seal.
6. Loosen the tensioner mounting nut and bolt. Tighten the bolt and nut after the spring tension has adjusted the belt tension.
7. Check to ensure that when the center of the belt span on the tension side and seal line of the under cover are held between the thumb and forefinger. The clearance "C", between the belt and seal line should be 0.55 in. (14mm) after the tension has been adjusted.
8. Install the upper and lower timing belt covers.
9. Install the crank adapter and crankshaft pulley.
10. Install the water pump drive belt and pulley. Connect the negative battery cable.

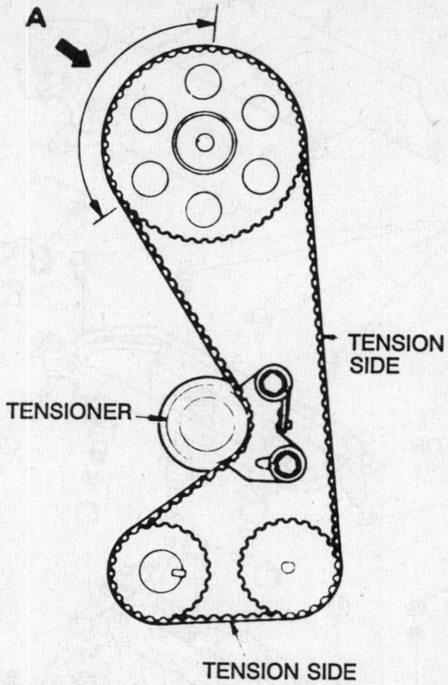

Timing belt tensioner—2351cc engine

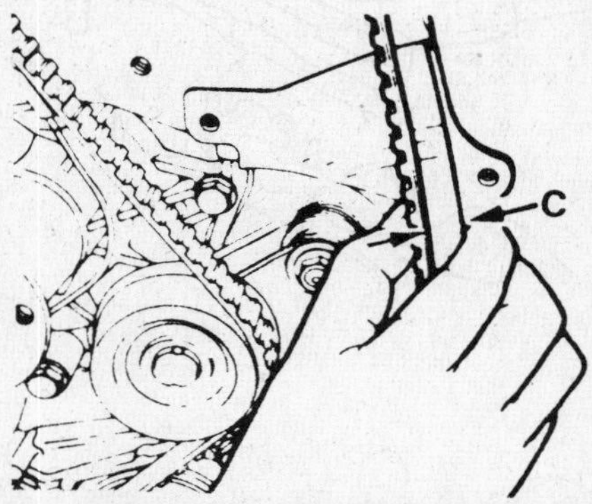

Check the distance between the belt and seal line, 0.55 in. (14mm)—2351cc engine

2972cc Engine

1. Disconnect the negative battery cable. Remove the water pump drive belt and pulley.
2. Remove the crank adapter and crankshaft pulley.
3. Remove the upper and lower timing belt covers.
4. Check the tensioners for a smooth rate of movement.
5. Replace any tensioner that shows grease leakage through the seal.
6. Turn the crankshaft 2 times in a clockwise direction. Loosen the tensioner mounting bolt. Tighten the bolt and nut after the spring tension has adjusted the belt tension.
7. Check to ensure that when the center of the belt span between the crankshaft and the right camshaft sprocket has a belt tension of 57–84 lbs. (260–380 N) using a belt tension gauge.
8. Install the upper and lower timing belt covers.
9. Install the crank adapter and crankshaft pulley.
10. Install the water pump drive belt and pulley. Connect the negative battery cable.

REMOVAL & INSTALLATION

1468cc Engine

1. Disconnect the negative battery cable. Remove the timing belt cover.
2. Turn the crankshaft until the timing marks on the camshaft sprocket and cylinder head are aligned. Loosen the tensioning bolt, it runs in the slotted portion of the tensioner, and the pivot bolt on the timing belt tensioner. Move the tensioner as far as it will go toward the water pump. Tighten the adjusting bolt. Mark the timing belt with an arrow showing direction of rotation.
3. Pull the timing belt off the camshaft sprocket. Remove the camshaft sprocket.
4. Remove the crankshaft pulley. Then, remove the timing belt.
5. Remove the crankshaft sprocket bolts and remove the crankshaft sprocket and flange, noting the direction of installation for each. Remove the timing belt tensioner.
6. Inspect the belt thoroughly. The back surface must be pliable and rough. If it is hard and glossy, the belt should be replaced. Any cracks in the belt backing or teeth or missing teeth mean the belt must be replaced. The canvas cover should be intact on all the teeth. If rubber is exposed anywhere, the belt should be replaced.
7. Inspect the tensioner for grease leaking from the grease seal and any roughness in rotation. Replace a tensioner for either defect.
8. The sprockets should be inspected and replaced, if there is any sign of damaged teeth or cracking anywhere. Do not immerse sprockets in solvent, as solvent that has soaked into the metal may cause deterioration of the timing belt later. Do not clean the tensioner in solvent either, as this may wash the grease out of the bearing.

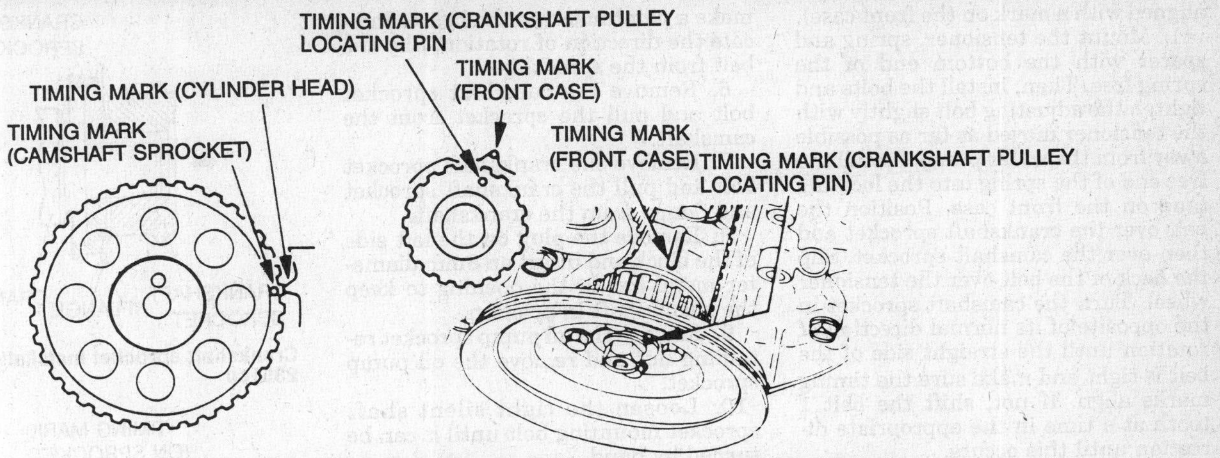

Installing the crankshaft and camshaft sprockets on the 1468cc Engine

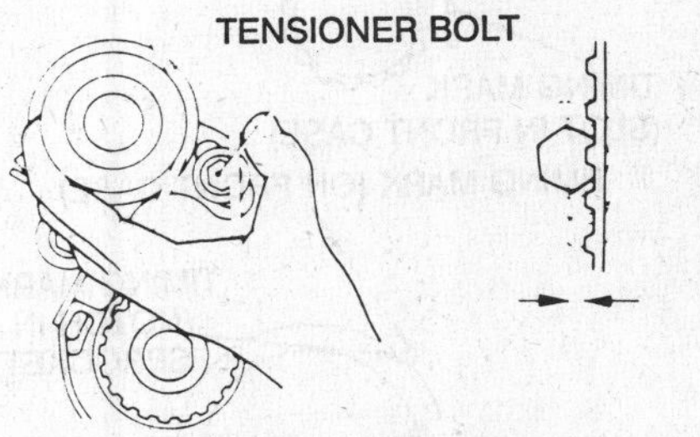

Checking belt tension on the 1468cc Engine

Installing the belt tensioner spring on the 1468cc Engine

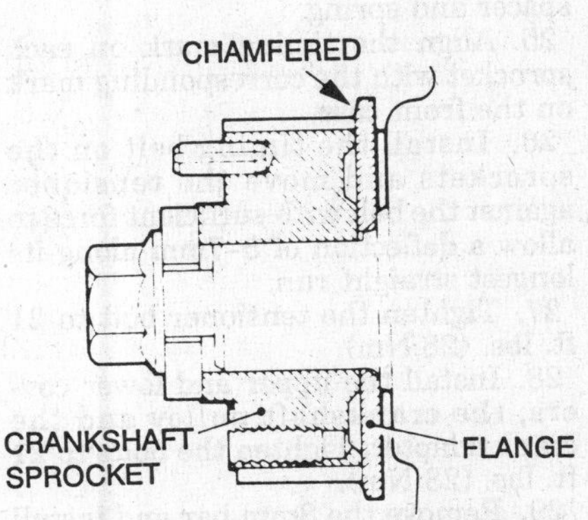

Installing the crankshaft sprocket on the 1468cc Engine

To install:

9. Install the flange and crankshaft sprocket. The flange must go on first with the chamfered area outward. The sprocket is installed with the boss forward and the studs for the fan belt pulley outward. Install and torque the crankshaft sprocket bolt to 51–72 ft. lbs. (69–98 Nm). Install the camshaft sprocket and bolt, torquing it to 47–54 ft. lbs. (64–74 Nm).

10. Align the timing marks of the camshaft sprocket. Check that the crankshaft timing marks are still in alignment (the locating pin on the front of the crankshaft sprocket is

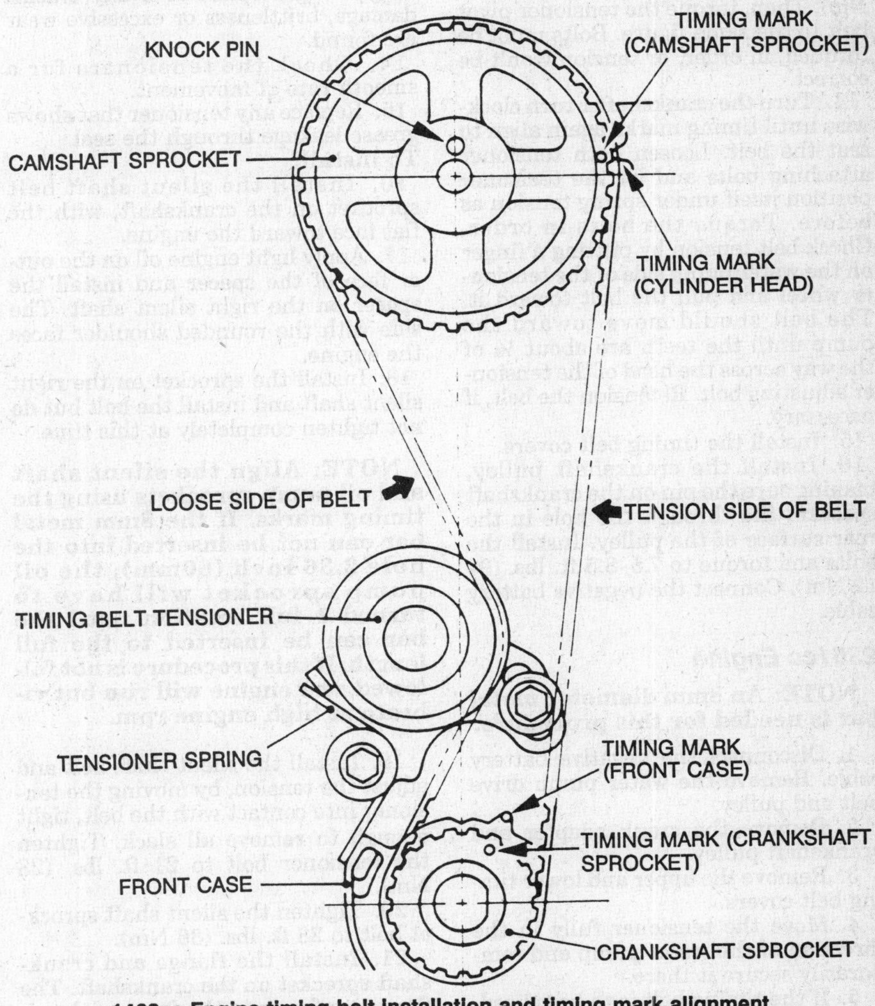

1468cc Engine timing belt installation and timing mark alignment

aligned with a mark on the front case).

11. Mount the tensioner, spring and spacer with the bottom end of the spring free. Then, install the bolts and tighten the adjusting bolt slightly with the tensioner moved as far as possible away from the water pump. Install the free end of the spring into the locating tang on the front case. Position the belt over the crankshaft sprocket and then over the camshaft sprocket. Slip the back of the belt over the tensioner wheel. Turn the camshaft sprocket in the opposite of its normal direction of rotation until the straight side of the belt is tight and make sure the timing marks align. If not, shift the belt 1 tooth at a time in the appropriate direction until this occurs.

12. Loosen the tensioner mounting bolts so the tensioner works, without the interference of any friction, under spring pressure. Make sure the belt follows the curve of the camshaft pulley so the teeth are engaged all the way around.

13. Correct the path of the belt, if necessary. Torque the tensioner adjusting bolt to 15–18 ft. lbs. (20–26 Nm). Then, torque the tensioner pivot bolt to the same figure. Bolts must be torqued, in order, or tension won't be correct.

14. Turn the crankshaft 1 turn clockwise until timing marks again align to seat the belt. Loosen both tensioner attaching bolts and let the tensioner position itself under spring tension as before. Torque the bolts in order. Check belt tension by putting a finger on the water pump side of the tensioner wheel and pull the belt toward it. The belt should move toward the pump until the teeth are about ¼ of the way across the head of the tensioner adjusting bolt. Retension the belt, if necessary.

15. Install the timing belt covers.

16. Install the crankshaft pulley, making sure the pin on the crankshaft sprocket fits through the hole in the rear surface of the pulley. Install the bolts and torque to 7.5–8.5 ft. lbs. (9–12 Nm). Connect the negative battery cable.

2351cc Engine

NOTE: An 8mm diameter metal bar is needed for this procedure.

1. Disconnect the negative battery cable. Remove the water pump drive belt and pulley.
2. Remove the crank adapter and crankshaft pulley.
3. Remove the upper and lower timing belt covers.
4. Move the tensioner fully in the direction of the water pump and temporarily secure it there.
5. If the timing belt is to be reused, make a paint mark on the belt to indicate the direction of rotation. Slip the belt from the sprockets.
6. Remove the camshaft sprocket bolt and pull the sprocket from the camshaft.
7. Remove the crankshaft sprocket bolt and pull the crankshaft sprocket and flange from the crankshaft.
8. Remove the plug on the left side of the block and insert an 8mm diameter metal bar in the opening to keep the silent shaft in position.
9. Remove the oil pump sprocket retaining nut and remove the oil pump sprocket.
10. Loosen the right silent shaft sprocket mounting bolt until it can be turned by hand.
11. Remove the belt tensioner and remove the timing belt.

NOTE: Do not attempt to turn the silent shaft sprocket or loosen its bolt while the belt is off.

12. Remove the silent shaft belt sprocket from the crankshaft.
13. Check the belt for wear, damage or glossing. Replace it if any cracks, damage, brittleness or excessive wear are found.
14. Check the tensioners for a smooth rate of movement.
15. Replace any tensioner that shows grease leakage through the seal.

To install:

16. Install the silent shaft belt sprocket on the crankshaft, with the flat face toward the engine.
17. Apply light engine oil on the outer face of the spacer and install the spacer on the right silent shaft. The side with the rounded shoulder faces the engine.
18. Install the sprocket on the right silent shaft and install the bolt but do not tighten completely at this time.

NOTE: Align the silent shaft and oil pump sprockets using the timing marks. If the 8mm metal bar can not be inserted into the hole 2.36 inch (60mm), the oil pump sprocket will have to turned 1 full rotation until the bar can be inserted to the full length. If this procedure is not followed, the engine will run but vibrate at high engine rpm.

19. Install the silent shaft belt and adjust the tension, by moving the tensioner into contact with the belt, tight enough to remove all slack. Tighten the tensioner bolt to 21 ft. lbs. (28 Nm).
20. Tighten the silent shaft sprocket bolt to 28 ft. lbs. (36 Nm).
21. Install the flange and crankshaft sprocket on the crankshaft. The flange conforms to the front of the silent shaft sprocket and the timing belt sprocket is installed with the flat face toward the engine.

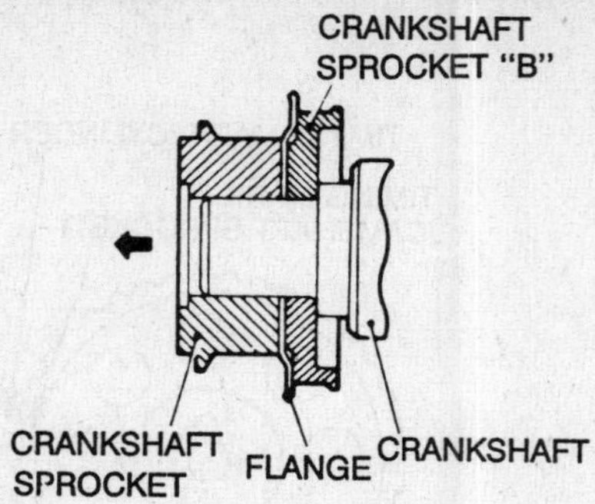

Crankshaft sprocket installation on the 2351cc

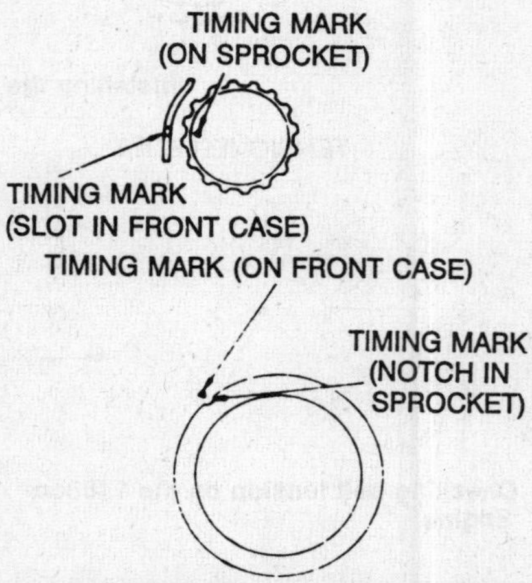

2351cc right counterbalance shaft timing mark alignment

NOTE: The flange must be installed correctly or a broken belt will result.

22. Install the washer and bolt in the crankshaft and torque it to 94 ft. lbs. (130 Nm).
23. Install the camshaft sprocket and bolt and torque the bolt to 72 ft. lbs. (96 Nm).
24. Install the timing belt tensioner, spacer and spring.
25. Align the timing mark on each sprocket with the corresponding mark on the front case.
26. Install the timing belt on the sprockets and move the tensioner against the belt with sufficient force to allow a deflection of 5–7mm along its longest straight run.
27. Tighten the tensioner bolt to 21 ft. lbs. (28 Nm).
28. Install the upper and lower covers, the crankshaft pulley and the crank adapter. Tighten the bolts to 21 ft. lbs. (28 Nm).
29. Remove the 8mm bar and install the plug. Connect the negative battery cable.

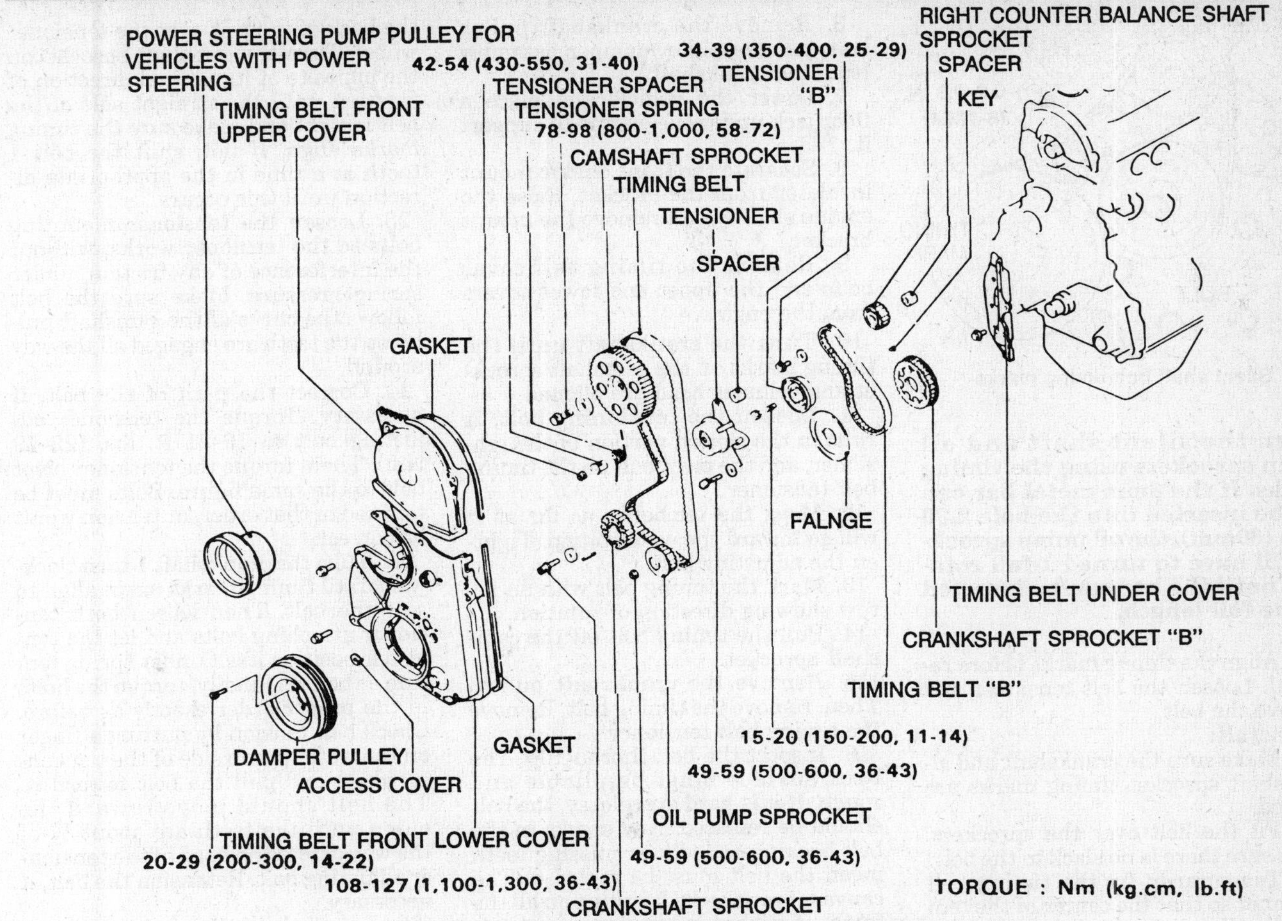

2351cc right counterbalance shaft seal installation

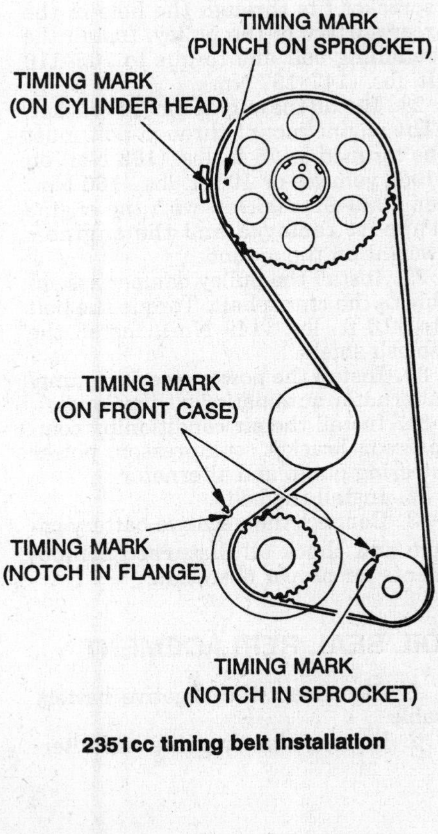

2351cc timing belt installation

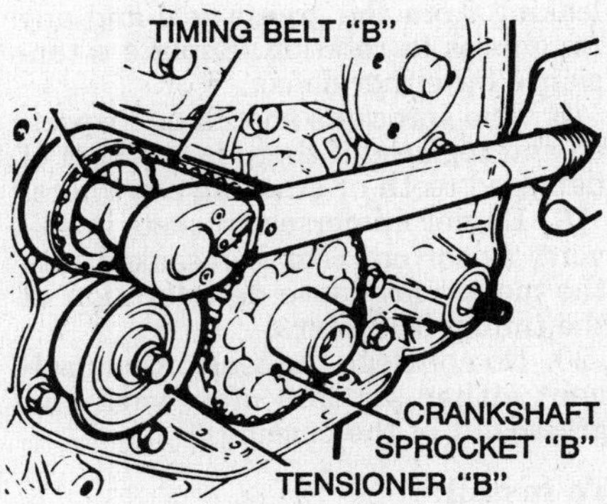

Removing tensioner "B"

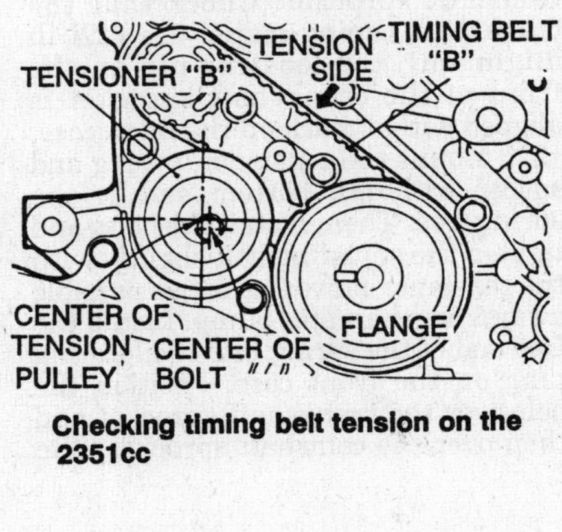

Checking timing belt tension on the 2351cc

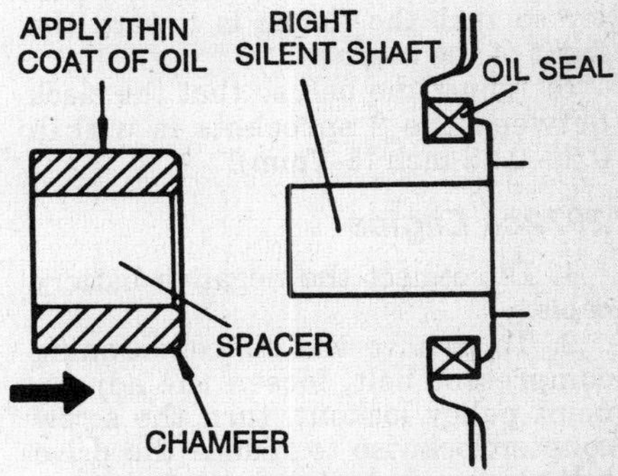

2351cc right counterbalance shaft seal installation

SILENT SHAFT BELT

NOTE: When replacing the timing belt, the manufacturer recommends to replace the silent shaft belt.

1. After removing the timing belt, mark arrows on the belt to indicate direction of rotation (clockwise).

NOTE: If the silent shaft belt is not installed properly, the engine will run but with a vibration.

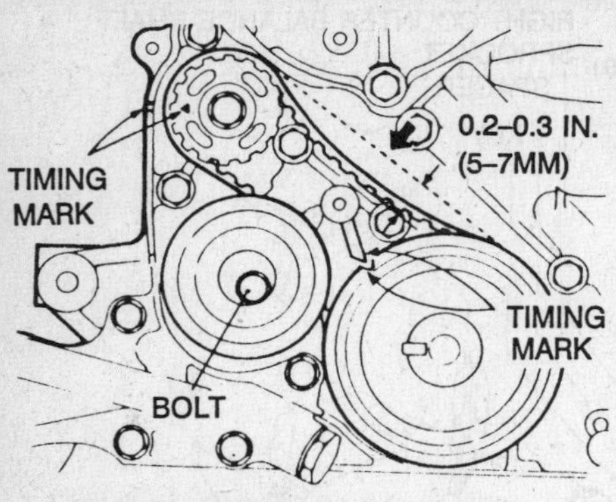

Silent shaft belt timing marks

Align the silent shaft and oil pump sprockets using the timing marks. If the 8mm metal bar can not be inserted into the hole 2.36 inch (60mm), the oil pump sprocket will have to turned 1 full rotation until the bar can be inserted to the full length.

2. Align the timing marks before removal. Loosen the belt tensioner and remove the belt.

To install:

3. Make sure the crankshaft and silent shaft sprocket timing marks are aligned.
4. Fit the belt over the sprockets. Make sure there is no slack in the belt.
5. Temporarily fix the timing belt tensioner so that the center of the tensioner pulley is to the left and above the center of the installation bolt. Temporarily attach the tensioner pulley so that the flange is toward the front of the engine.
6. Adjust the belt so that the slack between the 2 sprockets is within 0.20–0.28 inch (5–7mm).

2972cc Engine

1. Disconnect the negative battery cable.
2. To remove the air conditioning compressor belt, loosen the adjustment pulley locknut, turn the screw counterclockwise to reduce the drive belt tension and remove the belt.
3. To remove the serpentine drive belt, insert a ½ in. breaker bar in to the square hole of the tensioner pulley, rotate it counterclockwise to reduce the drive belt tension and remove the belt.
4. Remove the air conditioning compressor and the air compressor bracket, power steering pump and alternator from the mounts and support them to the side. Remove power steering pump/alternator automatic belt tensioner bolt and the tensioner.
5. Raise the vehicle and support safely. Remove the right inner fender splash shield.
6. Remove the crankshaft pulley bolt and the pulley/damper assembly from the crankshaft.
7. Lower the vehicle and place a floor jack under the engine to support it.
8. Separate the front engine mount insulator from the bracket. Raise the engine slightly and remove the mount bracket.
9. Remove the timing belt cover bolts and the upper and lower covers from the engine.
10. Turn the crankshaft until the timing marks on the camshaft sprocket and cylinder head are aligned.
11. Loosen the tensioning bolt, it runs in the slotted portion of the tensioner, and the pivot bolt on the timing belt tensioner.
12. Move the tensioner as far as it will go toward the water pump. Tighten the adjusting bolt.
13. Mark the timing belt with an arrow showing direction of rotation.
14. Pull the timing belt off the camshaft sprocket.
15. Remove the crankshaft pulley. Then, remove the timing belt. Remove the timing belt tensioner.
16. Inspect the belt thoroughly. The back surface must be pliable and rough. If it is hard and glossy, the belt should be replaced. Any cracks in the belt backing or teeth or missing teeth mean the belt must be replaced. The canvas cover should be intact on all the teeth. If rubber is exposed anywhere, the belt should be replaced.
17. Inspect the tensioner for grease leaking from the grease seal and any roughness in rotation. Replace a tensioner for either defect.
18. The sprockets should be inspected and replaced if there is any sign of damaged teeth or cracking anywhere.
19. Do not immerse sprockets in solvent, as solvent that has soaked into the metal may cause deterioration of the timing belt later.
20. Do not clean the tensioner in solvent either, as this may wash the grease out of the bearing.

To install:

21. Align the timing marks of the camshaft sprocket. Check that the crankshaft timing marks are still in alignment, the locating pin on the front of the crankshaft sprocket is aligned with a mark on the front case.
22. Mount the tensioner, spring and spacer with the bottom end of the spring free. Then, install the bolts and tighten the adjusting bolt slightly with the tensioner moved as far as possible away from the water pump. Install the free end of the spring into the locating tang on the front case. Position the belt over the crankshaft sprocket and then over the camshaft sprocket. Slip the back of the belt over the tensioner wheel. Turn the camshaft sprocket in the opposite of its normal direction of rotation until the straight side of the belt is tight and make sure the timing marks align. If not, shift the belt 1 tooth at a time in the appropriate direction until this occurs.
23. Loosen the tensioner mounting bolts so the tensioner works, without the interference of any friction, under spring pressure. Make sure the belt follows the curve of the camshaft pulley so the teeth are engaged all the way around.
24. Correct the path of the belt, if necessary. Torque the tensioner adjusting bolt to 16–21 ft. lbs. (22–29 Nm). Then, torque the tensioner pivot bolt to the same figure. Bolts must be torqued in that order, or tension won't be correct.
25. Turn the crankshaft 1 turn clockwise until timing marks again align to seat the belt. Then loosen both tensioner attaching bolts and let the tensioner position itself under spring tension as before. Finally, torque the bolts in the proper order exactly as before. Check belt tension by putting a finger on the water pump side of the tensioner wheel and pull the belt toward it. The belt should move toward the pump until the teeth are about ¼ of the way across the head of the tensioner adjusting bolt. Retension the belt, if necessary.
26. Install the timing belt covers.
27. Install the crankshaft pulley, making sure the pin on the crankshaft sprocket fits through the hole in the rear surface of the pulley. Install the retaining bolt and torque to 108–116 ft. lbs. (147–157 Nm).
28. Install the engine mount bracket. The engine mount through bolt must be torqued to 75 ft. lbs. (102 Nm) on 1988 vehicles or 100 ft. lbs. (136 Nm) on 1989–92 vehicles, with the engine support removed and the engine's weight on the mount.
29. Install the pulley damper assembly to the crankshaft. Torque the bolt to 110 ft. lbs. (149 Nm). Install the splash shield.
30. Install the power steering pump/alternator automatic belt tensioner.
31. Install the air conditioning compressor bracket, compressor, power steering pump and alternator.
32. Install the belts.
33. Connect the negative battery cable and check all disturbed components for proper operation.

OIL SEAL REPLACEMENT

1. Disconnect the negative battery cable.
2. Remove the air pump and alter-

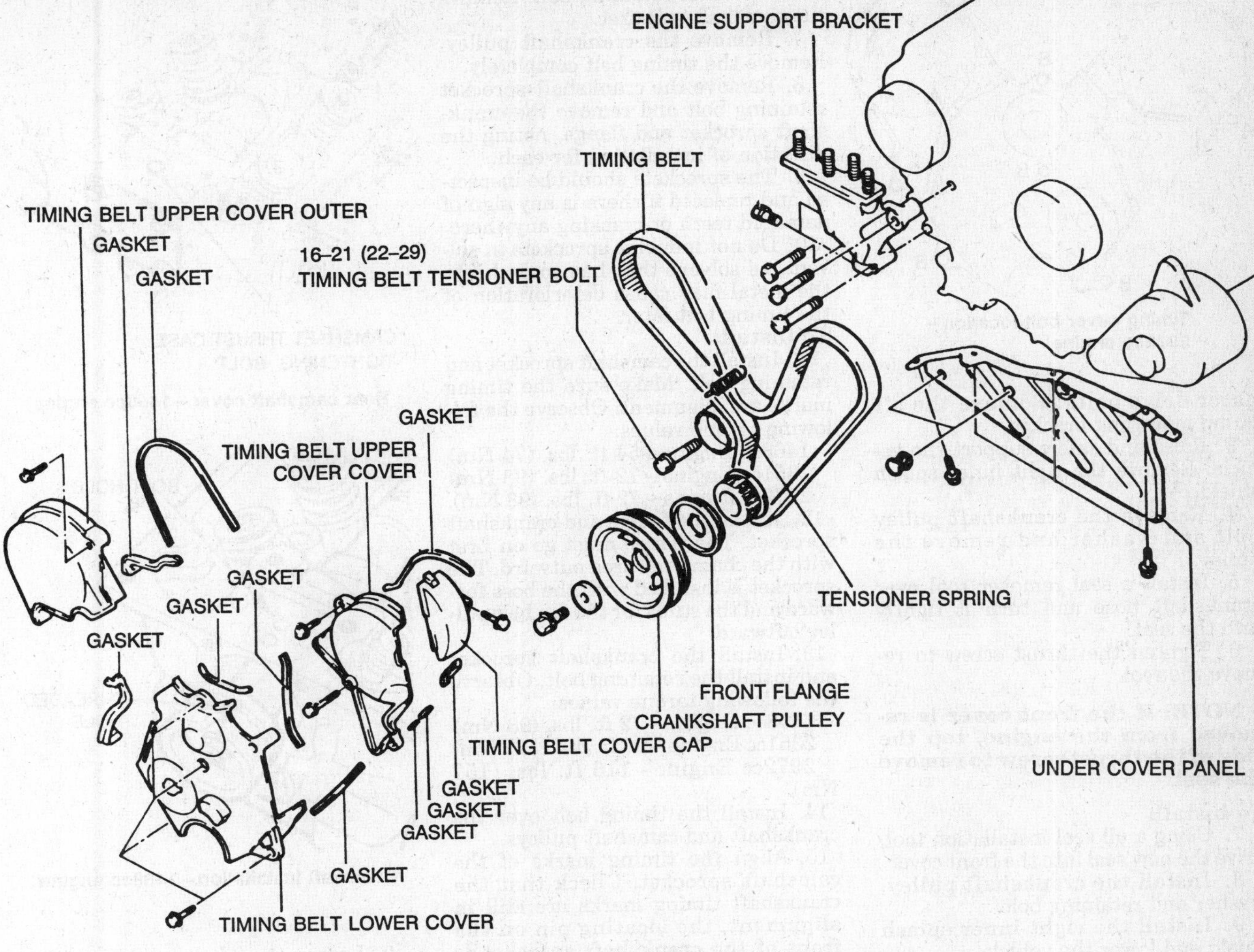

Timing belt cover and related components—2972cc engine

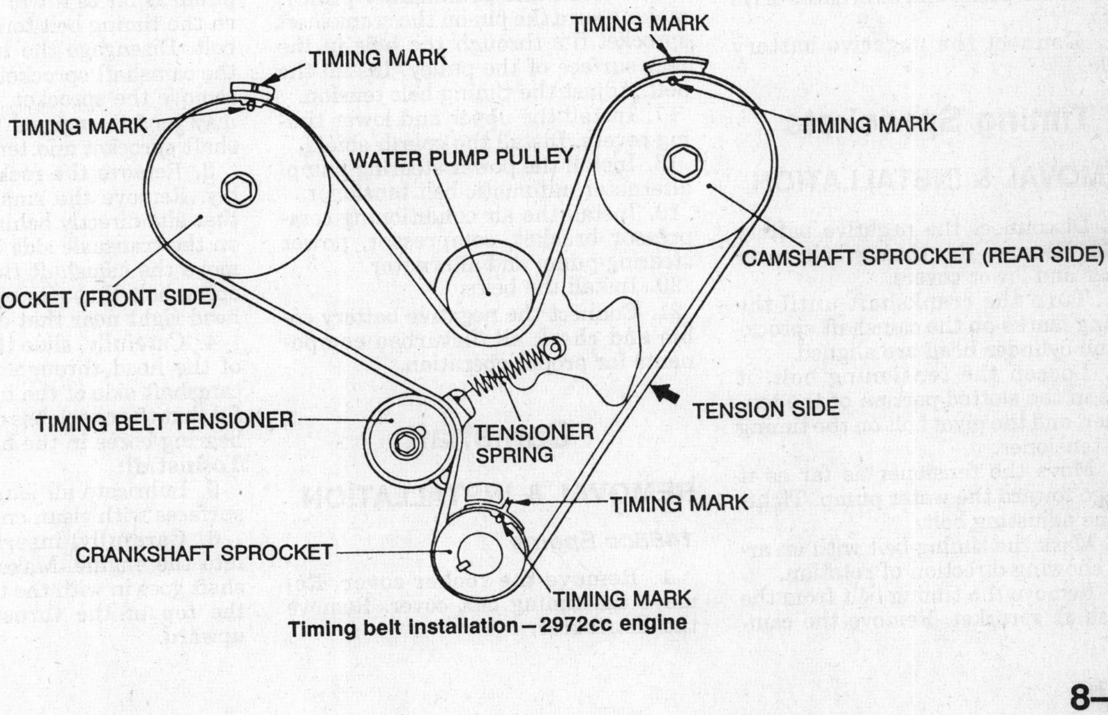

Timing belt installation—2972cc engine

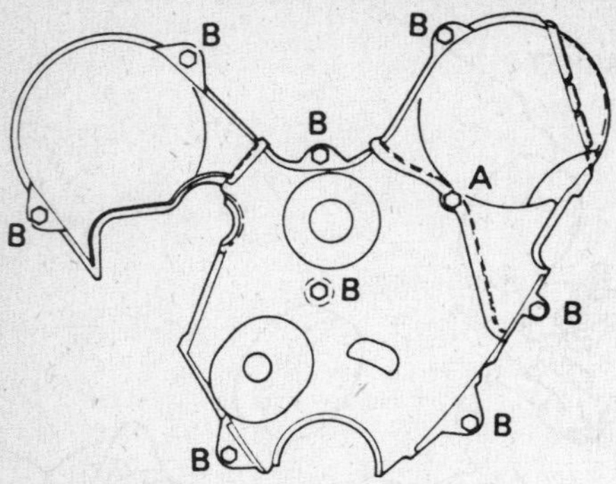

Timing cover bolt location—2972cc engine

nator drive belts. Remove the air pump mounting bracket.

3. Raise and safely support the vehicle. Remove the right inner splash shield.

4. Remove the crankshaft pulley bolt and washer and remove the pulley.

5. Install a seal remover tool over crankshaft nose and turn it tightly into the seal.

6. Tighten the thrust screw to remove the seal.

NOTE: If the front cover is removed from the engine, tap the side of the thrust screw to remove the seal.

To install:

7. Using a oil seal installation tool, drive the new seal into the front cover.

8. Install the crankshaft pulley, washer and retaining bolt.

9. Install the right inner splash shield and lower the vehicle.

10. Install the air pump mounting bracket, air pump and alternator drive belts.

11. Connect the negative battery cable.

Timing Sprockets

REMOVAL & INSTALLATION

1. Disconnect the negative battery cable. Remove the timing belt cover upper and lower covers.

2. Turn the crankshaft until the timing marks on the camshaft sprocket and cylinder head are aligned.

3. Loosen the tensioning bolt, it runs in the slotted portion of the tensioner, and the pivot bolt on the timing belt tensioner.

4. Move the tensioner as far as it will go toward the water pump. Tighten the adjusting bolt.

5. Mark the timing belt with an arrow showing direction of rotation.

6. Remove the timing belt from the camshaft sprocket. Remove the camshaft sprocket retaining bolt. Remove the camshaft sprocket.

7. Remove the crankshaft pulley. Remove the timing belt completely.

8. Remove the crankshaft sprocket retaining bolt and remove the crankshaft sprocket and flange, noting the direction of installation for each.

9. The sprockets should be inspected and replaced if there is any sign of damaged teeth or cracking anywhere.

10. Do not immerse sprockets in solvent, as solvent that has soaked into the metal may cause deterioration of the timing belt later.

To install:

11. Install the camshaft sprocket and retaining bolt. Make sure the timing marks are alignment. Observe the following torque values:

1468cc Engine—54 ft. lbs. (74 Nm)
2351cc Engine—72 ft. lbs. (98 Nm)
2972cc Engine—72 ft. lbs. (98 Nm).

12. Install the flange and crankshaft sprocket. The flange must go on first with the chamfered area outward. The sprocket is installed with the boss forward and the studs for the fan belt pulley outward.

13. Install the crankshaft sprocket and install the retaining bolt. Observe the following torque values:

1468cc Engine—72 ft. lbs. (98 Nm)
2351cc Engine—94 ft. lbs. (127 Nm)
2972cc Engine—116 ft. lbs. (157 Nm).

14. Install the timing belt over the crankshaft and camshaft pulleys.

15. Align the timing marks of the camshaft sprocket. Check that the crankshaft timing marks are still in alignment, the locating pin on the front of the crankshaft sprocket is aligned with a mark on the front case.

16. Install the crankshaft pulley, making sure the pin on the crankshaft sprocket fits through the hole in the rear surface of the pulley. Install the bolt. Adjust the timing belt tension.

17. Install the upper and lower timing covers. Install the splash shield.

18. Install the power steering pump/alternator automatic belt tensioner.

19. Install the air conditioning compressor bracket, compressor, power steering pump and alternator.

20. Install the belts.

21. Connect the negative battery cable and check all disturbed components for proper operation.

Camshaft

REMOVAL & INSTALLATION

1468cc Engine

1. Remove the rocker cover. Remove the timing belt cover. Remove the distributor.

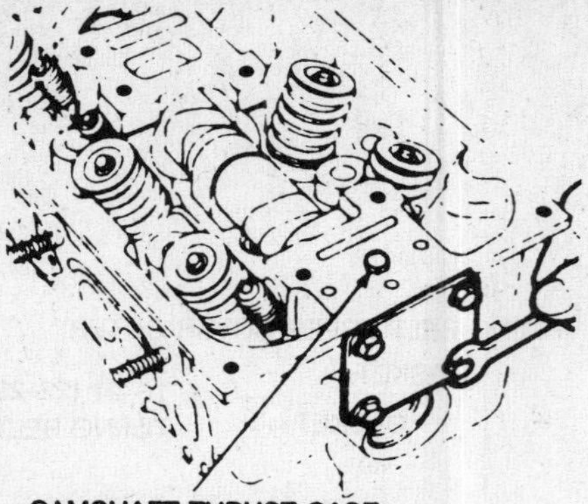

Rear camshaft cover—1468cc engine

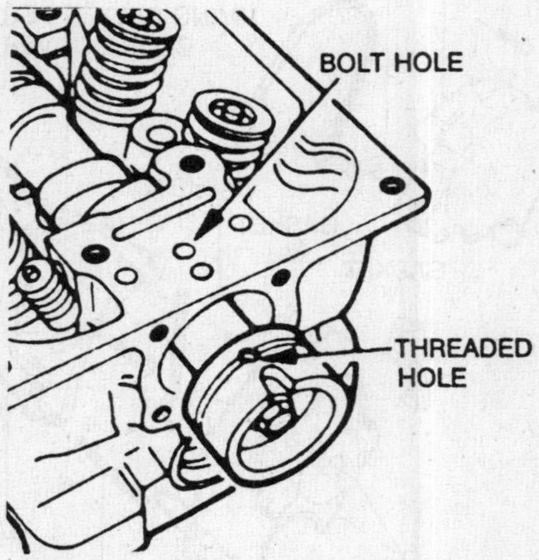

Camshaft installation—1468cc engine

2. Loosen the 2 bolts and move the timing belt tensioner toward the water pump as far as it will go, then retighten the timing belt tensioner adjusting bolt. Disengage the timing belt from the camshaft sprocket and unbolt and remove the sprocket. The timing belt may be left engaged with the crankshaft sprocket and tensioner.

3. Remove the rocker shaft assembly. Remove the small, square cover that sits directly behind the camshaft on the transaxle side of the head. Remove the camshaft thrust case tightening bolt that sits on the top of the head right near that cover.

4. Carefully, slide the camshaft out of the head through the hole in the camshaft side of the head, being careful that the cam lobes do not strike the bearing bores in the head.

To install:

5. Lubricate all journal and thrust surfaces with clean engine oil.

6. Carefully, insert the camshaft into the engine. Make sure the camshaft goes in with the threaded hole in the top of the thrust case straight upward.

7. Align the bolt hole in the trust case and the cylinder head surface.
8. Install the thrust case bolt and tighten firmly.
9. Install the rear cover with a new gasket and install and tighten the bolts.
10. Coat the external surface of the front oil seal with engine oil.
11. Using special installer tool MD 998306-01 or equivalent, drive the a new front camshaft oil seal into the clearance between the cam and head at the forward end, making sure the seal seats fully.
12. Install the camshaft sprocket and torque the bolt to 47–54 ft. lbs. (64–74 Nm).
13. Reconnect the timing belt, check the timing and adjust the belt tension.
14. Reinstall the rocker shaft assembly. Adjust the valves and install the rocker and timing belt covers.

2351cc Engine

NOTE: A special tool 09246-32000 (MD998443) or equivalent, is required to retain the automatic lash adjusters in this procedure.

1. Remove the rocker cover and gasket and the timing belt cover.
2. Remove the timing belt and camshaft sprocket.
3. Turn the crankshaft so the No. 1 piston is at TDC compression. At this point, the timing mark on the camshaft sprocket and the timing mark on the head to the left of the sprocket will be aligned.
4. Remove the camshaft bearing cap bolts.
5. Install the automatic lash adjuster retainer tool 09246-32000 or equivalent, to keep the adjuster from falling out of the rocker arms.
6. Lift off the bearing caps and rocker arm assemblies.
7. Lift out the camshaft.

NOTE: Keep all parts in the order in which they were removed. None of the parts are interchangeable. The lash adjusters are filled with diesel fuel, which will spill out if they are inverted. If any diesel fuel is spilled, the adjusters must be bled.

To install:

8. Check all parts for wear or damage. Replace any damaged or excessively worn part.
9. Coat the camshaft with clean engine oil and place it on the head.
10. Assemble all parts. Note the following:
   a. The rocker shafts are installed with the notches in the ends facing up.
   b. The left rocker shaft is longer than the right.
   c. The wave washers are installed on the left shaft.
   d. Coat all parts with clean engine oil prior to assembly.
   e. Insert the lash adjuster from under the rocker arm and install the special holding tool. If any of the diesel fuel is spilled, the adjuster must be bled.
   f. Tighten the bearing cap bolts, working from the center towards the ends, to 15 ft. lbs. (20 Nm).
   g. Check the operation of each lash adjuster by positioning the camshaft so the rocker arm bears on the low or round portion of the cam, pointed part of the can faces straight down. Insert a thin steel wire, or tool MD998442 or equivalent, in the hole in the top of the rocker arm, over the lash adjuster and depress the check ball at the top of the adjuster. While holding the check ball depressed, move the arm up and down. Looseness should be felt. Full plunger stroke should be 2.2mm. If not, remove, clean and bleed the lash adjusters.

2972cc Engine

1. Disconnect the negative battery cable. Remove the air cleaner assembly and valve covers.
2. Install auto lash adjuster retainer tools MD998443 or equivalent on the rocker arms.
3. If removing the right side (front) camshaft, remove the distributor extension.
4. Remove the camshaft bearing caps but do not remove the bolts from the caps.
5. Remove the rocker arms, rocker shafts and bearing caps, as an assembly.
6. Remove the camshaft from the cylinder head.
7. Inspect the bearing journals on the camshaft, cylinder head and bearing caps.

To install:

8. Lubricate the camshaft journals and camshaft with clean engine oil and install the camshaft in the cylinder head.
9. Align the camshaft bearing caps with the arrow mark depending on cylinder numbers and install in numerical order.
10. Apply sealer at the ends of the bearing caps and install the assembly.
11. Torque the rocker arm and shaft assembly bolts to 15 ft. lbs. (21 Nm).
12. Install the distributor extension, if removed.
13. Install the valve cover and all related parts. Torque the valve cover retaining bolts to 7 ft. lbs. (10 Nm).
14. Connect the negative battery cable and road test the vehicle.

Counterbalance Shafts

REMOVAL & INSTALLATION

2351cc Engine

1. Disconnect the negative battery cable.
2. Remove the alternator and accessory belts. Remove the belt cover.
3. Rotate the crankshaft to bring No. 1 piston to TDC on the compression stroke. Align the notch on the crankshaft pulley with the T mark on the timing indicator scale and the timing mark on the upper under cover of the timing belt with the mark on the camshaft sprocket.
4. Remove the crankshaft pulley and bolt.
5. Remove the timing belt covers, upper front and lower front.
6. Remove the crankshaft sprocket bolt.
7. Loosen the tensioner mounting nut and bolt. Move the tensioner away from the belt and retighten the nut to keep the tensioner in the off position. Remove the tensioner.
8. Remove the camshaft sprocket, crankshaft sprocket, flange and tensioner.
9. Loosen the counterbalance shaft sprocket mounting bolt.
10. Remove the belt tensioner and remove the timing belt.
11. Remove the crankshaft sprocket (inner) and counterbalance shaft sprocket.
12. Remove the upper and lower under timing belt covers.
13. Remove the oil pump sprocket and cover.
14. Remove the plug at the bottom of the left side of the cylinder block and insert an 8mm metal bar to keep the left counter balance shaft in position while removing the sprocket nut.
15. Remove the front cover and oil pump as a unit, with the left counter shaft attached.
16. Remove the oil pump gear and left counterbalance shaft.

NOTE: To aid in removal of the front cover, a driver groove is provided on the cover, above the oil pump housing. Avoid prying on the thinner parts of the housing flange or hammering on it to remove the case.

17. Remove the right counterbalance shaft from the engine block.

To install:

18. Install a new front seal in the cover. Install the oil pump drive and

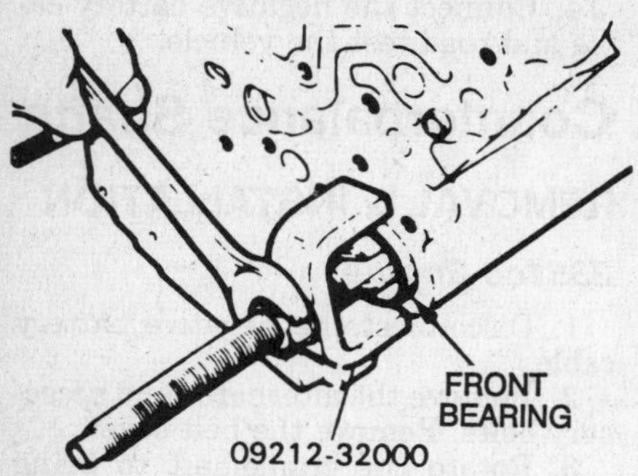

Removing the right counterbalance shaft

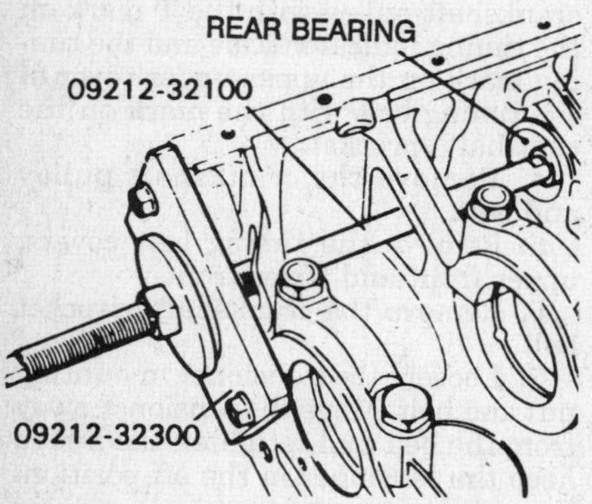

Removing the left counterbalance shaft

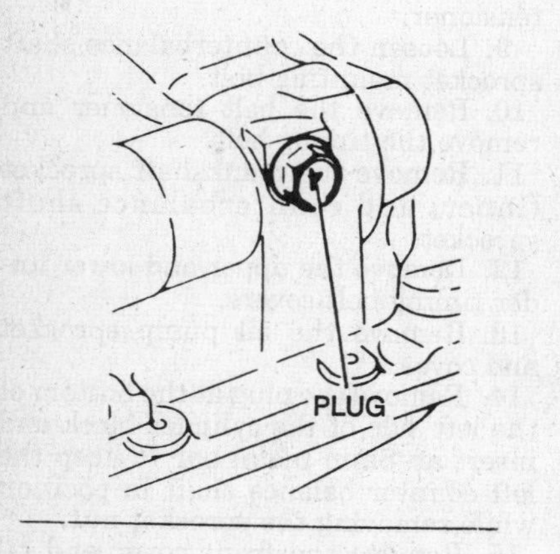

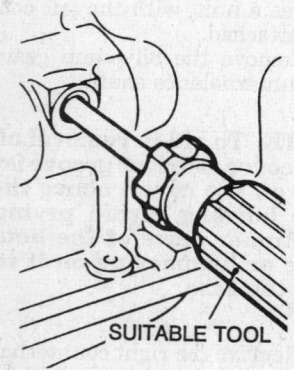

Remove the plug insert a 8mm bar to position the left counterbalance shaft –2351cc engine

driven gears in the front case, aligning the timing marks on the pump gears.

19. Install the left counterbalance shaft in the driven gear and temporarily tighten the bolt.

20. Install the right counterbalance shaft into the cylinder block.

21. Install an oil seal guide on the end of the crankshaft and install a new gasket on the front of the engine block for the front cover.

22. Install a new front case packing.

23. Insert the left counterbalance shaft into the engine block and at the same time, guide the front cover into place on the front of the engine block.

24. Insert an 8mm metal bar at the bottom of the left side of the block and hold the left counterbalance shaft and tighten the bolt. Install the hole plug.

25. Install an O-ring on the oil pump cover and install it on the front cover.

26. Tighten the oil pump cover bolts and the front cover bolts to 11–13 ft. lbs. (15–18 Nm).

27. Install the upper and lower under covers.

28. Install the spacer on the end of the right counterbalance shaft, with the chambered edge toward the rear of the engine.

29. Install the counterbalance shaft sprocket and temporarily tighten the bolt.

30. Install the inner crankshaft sprocket and align the timing marks on the sprockets with those on the front case.

31. Install the inner tensioner (B) with the center of the pulley on the left side of the mounting bolt and with the pulley flange toward the front of the engine.

32. Lift the tensioner by hand, clockwise, to apply tension to the belt. Tighten the bolt to secure the tensioner.

33. Check that all alignment marks are in their proper places and the belt deflection is approximately ¼–½ in. on the tension side.

NOTE: When the tensioner bolt is tightened, make sure the shaft of the tensioner does not turn with the bolt. If the belt is too tight there will be noise and if the belt is too loose, the belt and sprocket may come out of mesh.

34. Tighten the counterbalance shaft sprocket bolt to 22–28.5 ft. lbs. (29–40 Nm).

35. Install the flange and crankshaft sprocket. Tighten the bolt to 43–50 ft. lbs. (58–67 Nm).

36. Install the camshaft spacer and sprocket. Tighten the bolt to 44–57 ft. lbs. (61–75 Nm).

37. Align the camshaft sprocket timing mark with the timing mark on the upper inner cover.

38. Install the oil pump sprocket, tightening the nut to 25–28 ft. lbs. (34–39 Nm). Align the timing mark on the sprocket with the mark on the case.

NOTE: To be assured that the phasing of the oil pump sprocket and the left counterbalance shaft is correct, a metal rod should be inserted in the plugged hole on the left side of the cylinder block. If it can be inserted more than 60mm, the phasing is correct. If the tool can only be inserted approximately 25mm, turn the oil pump sprocket through 1 turn and realign the timing marks. Keep the metal rod inserted until the installation of the timing belt is completed. Remove the tool from the hole and install the plug, before starting the engine.

39. Install the tensioner spring and tensioner. Temporarily tighten the nut. Install the front end of the tensioner spring (bent at right angles) on the projection of the tensioner and the other end (straight) on the water pump body.

40. If the timing belt is correctly tensioned, there should be about 12mm clearance between the outside of the belt and the edge of the belt cover. This is measured about halfway down the side of the belt opposite the tensioner.

41. Complete the assembly by installing the upper and lower front covers.

42. Install the crankshaft pulley, alternator and accessory belts and adjust to specifications.

43. Install the radiator, fill the cooling system and start the engine.

Piston and Connecting Rod

POSITIONING

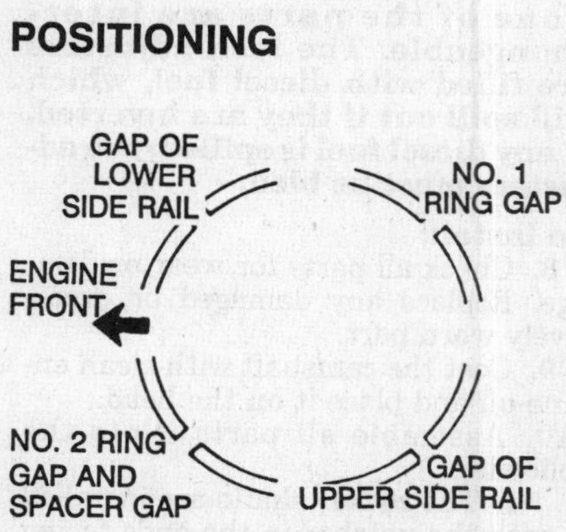

Piston ring positioning

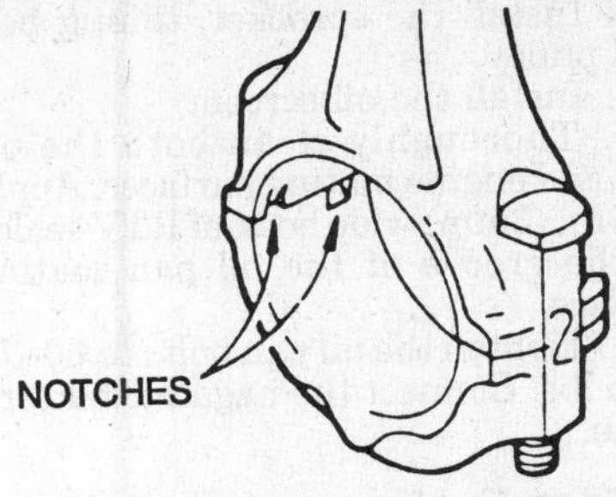

Connecting rod cap installation

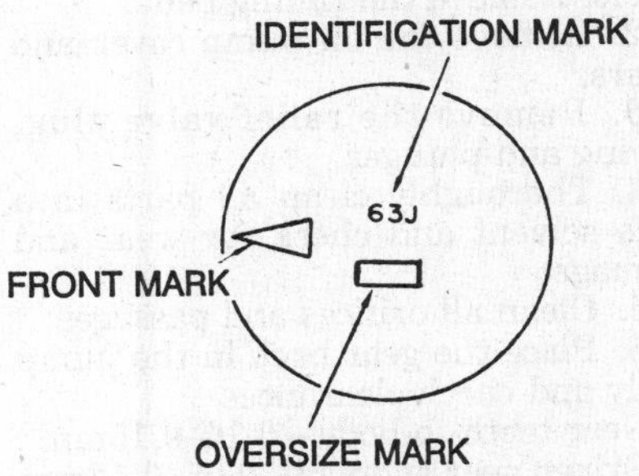

Piston installation

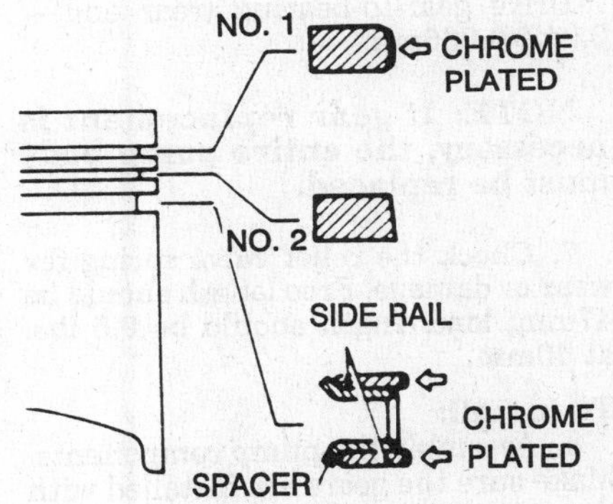

Piston ring installation

1. Front case gasket
2. Oil pump cover
3. Oil pump outer gear
4. Oil pump inner gear
5. Plug
6. Gasket
7. Relief spring
8. Relief plunger
9. Front oil seal
10. Front case
11. Drain plug
12. Oil pan
13. Oil screen
14. Oil screen gasket

	Nm
A	40–49
B	12–14
C	8–9
D	18–24
E	35–44
F	6–7

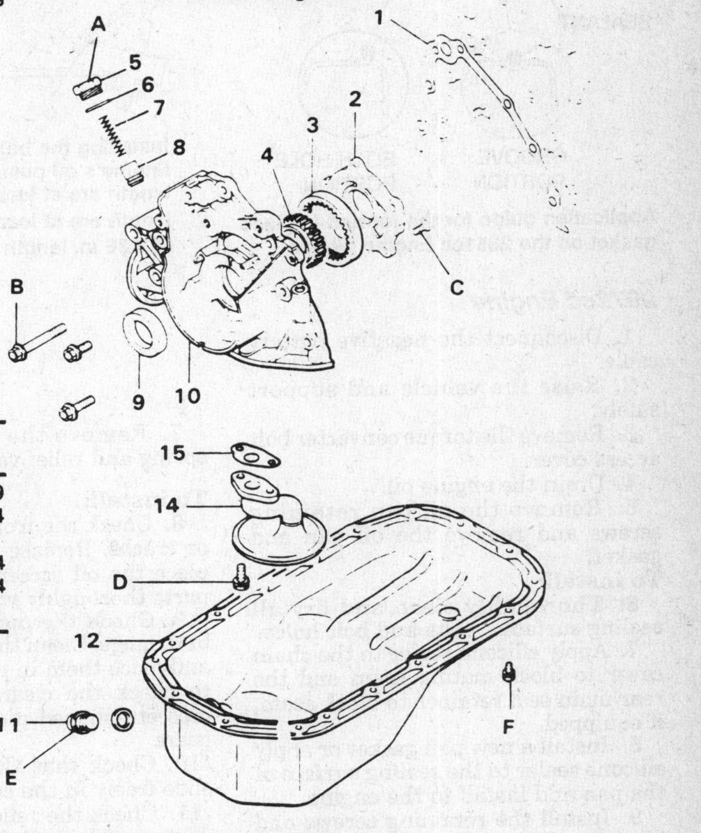

1468cc Engine pan and front case

ENGINE LUBRICATION

Oil Pan

REMOVAL & INSTALLATION

1468cc Engine

1. Disconnect the negative battery cable. Raise the vehicle and support it safely.
2. Drain the oil.
3. Remove the underbody splash shield.
4. Remove the oil pan bolts, drop the pan and slide it out from under the vehicle.
5. Clean the mating surfaces of the oil pan and the engine block.
6. Apply a ⅛ in. (3mm) bead of RTV sealer along the groove in the oil pan.

To install:

7. Using non-hardening sealer, glue a new gasket to the oil pan.
8. Install the oil pan. Hand tighten the retaining bolts.
9. Starting at one end of the pan, gradually tighten the retaining bolts to 48–72 inch lbs. in a criss-cross pattern.
10. Lower the engine and tighten the mount retaining nuts.
11. Install the oil pan drain plug.
12. Install the splash shield and lower the vehicle.
13. Refill the crankcase with oil. Start the engine and check for leaks.

2351cc Engine

1. Disconnect the negative battery cable. Raise the vehicle and support it safely.
2. Drain the oil.
3. Remove the underbody splash shield.
4. Remove the oil pan bolts, drop the pan and slide it out from under the vehicle.
5. Clean the mating surfaces of the oil pan and the engine block.

To install:

6. Apply sealer to the engine block at the block-to-chain case and block-to-rear oil seal case joint faces.
7. Use a non-hardening sealer and secure a new gasket to the oil pan.
8. Install the oil pan. Hand-tighten the retaining bolts.
9. Starting at one end of the pan, tighten the pan bolts to 48–72 inch lbs. in a criss-cross pattern.
10. Install the oil pan drain plug.
11. Install the splash shield and lower the vehicle.
12. Fill the crankcase with the proper amount of oil. Connect the negative battery cable. Start the engine and check for leaks.

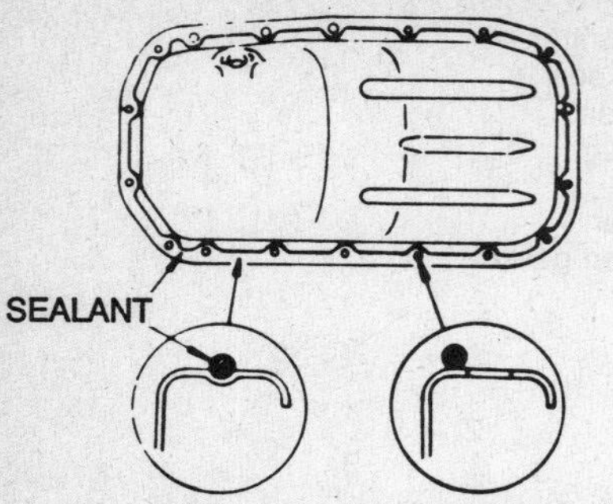

Application guide for the formed-in-place gasket on the 2351cc Engine oil pan

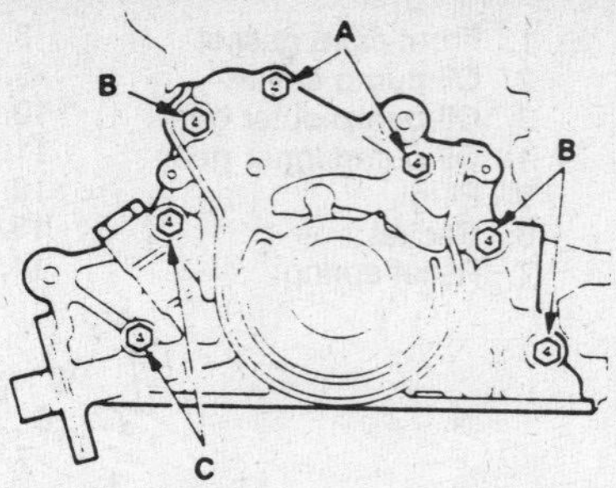

Installing the bolts for the 1468cc Engine's oil pump. Bolts of 0.79 in. length are at location A; of 1.18 in. length are at location B; and those of 2.36 in. length are at C.

2972cc Engine

1. Disconnect the negative battery cable.
2. Raise the vehicle and support safely.
3. Remove the torque converter bolt access cover.
4. Drain the engine oil.
5. Remove the oil pan retaining screws and remove the oil pan and gasket.

To install:

6. Thoroughly clean and dry all sealing surfaces, bolts and bolt holes.
7. Apply silicone sealer to the chain cover to block mating seam and the rear main seal retainer to block seam, if equipped.
8. Install a new pan gasket or apply silicone sealer to the sealing surface of the pan and install to the engine.
9. Install the retaining screws and torque to 50 inch lbs. (6 Nm).
10. Install the torque converter bolt access cover, if equipped. Lower the vehicle.
11. Install the dipstick. Fill the engine with the proper amount of oil.
12. Connect the negative battery cable and check for leaks.

Oil Pump

REMOVAL & INSTALLATION

1468cc Engine

1. Disconnect the negative battery cable. Remove the timing belt.
2. Remove the oil pan.
3. Remove the oil screen.
4. Unbolt and remove the front case assembly.
5. Remove the oil pump cover.
6. Remove the inner and outer gears from the front case.

NOTE: The outer gear has no identifying marks to indicate direction of rotation. Clean the gear and mark it with an indelible marker.

7. Remove the plug, relief valve spring and relief valve from the case.

To install:

8. Check the front case for damage or cracks. Replace the front seal. Replace the oil screen O-ring. Clean all parts thoroughly with a safe solvent.
9. Check the pump gears for wear or damage. Clean the gears thoroughly and place them in position in the case to check the clearances. There is a crescent-shaped piece between the 2 gears.
10. Check that the relief valve can slide freely in the case.
11. Check the relief valve spring for damage. The relief valve free length should be 1.8 in. (47mm). Load length should be 13.4 lbs. at 1.6 in. (40mm).
12. Thoroughly coat both oil pump gears with clean engine oil and install them in the correct direction of rotation.
13. Install the pump cover and torque the bolts to 6–7 ft. lbs. (8–10 Nm).
14. Coat the relief valve and spring with clean engine oil, install them and tighten the plug to 30–36 ft. lbs (39–49 Nm).
15. Position a new front case gasket, coated with sealer, on the engine and install the front case. Torque the bolts to 10 ft. lbs. (14 Nm). Note that the bolts have different shank lengths. Use the following guide to determine which bolts go where.
Bolts marked:
A: 0.08 in. (20mm)
B: 1.2 in. (30mm)
C: 2.4 in. (60mm)
16. Coat the lips of a new seal with clean engine oil and slide it along the crankshaft until it touches the front case. Drive it into place with a seal driver.
17. Install the sprocket, timing belt and pulley.
18. Install the oil screen.
19. Thoroughly clean both the oil pan and engine mating surfaces. Apply a ⅛ in. (3mm) wide bead of RTV sealer in the groove of the oil pan mating surface.
20. Tighten the oil pan bolts to 60–72 inch lbs. Connect the negative battery cable.

2351cc Engine

1. Disconnect the negative battery cable. Remove the timing belt.
2. Remove the oil pump cover and gears.
3. Remove the relief valve plug, spring and plunger.
4. Thoroughly clean all parts in a safe solvent and check for wear and damage.
5. Clean all orifices and passages.
6. Place the gear back in the pump body and check clearances.
Gear teeth-to-body – 0.10–0.15mm
Driven gear endplay – 0.06–0.12mm
Drive gear-to-bearing (front end) – 0.020–0.045mm
Drive gear-to-bearing (rear end) – 0.043–0.066mm.

NOTE: If gear replacement is necessary, the entire pump body must be replaced.

7. Check the relief valve spring for wear or damage. Free length should be 47mm; load/length should be 9.5 lbs. at 40mm.

To install:

8. Assembly the pump components. Make sure the gears are installed with the mating marks aligned.
9. Install the timing belt. Connect the negative battery cable.

2972cc Engine

1. Disconnect the negative battery cable. Remove the dipstick.
2. Raise the vehicle and support safely. Remove the timing belt, drain the engine oil and remove the oil pan from the engine. Remove the oil pickup.
3. Remove the oil pump mounting bolts and remove the pump from the front of the engine. Note the different length bolts and their position in the pump for installation.

To install:

4. Clean the gasket mounting surfaces of the pump and engine block.
5. Prime the pump by pouring fresh oil into the pump and turning the rotors. Using a new gasket, install the oil pump on the engine and torque all bolts to 11 ft. lbs. (15 Nm).

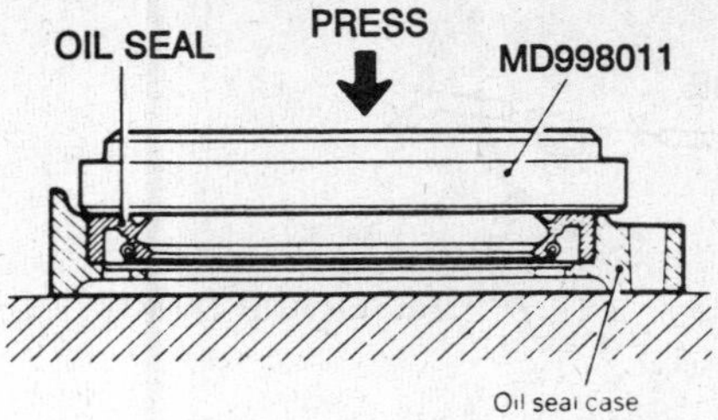

Pressing a new rear main seal into place

6. Install the balancer and crankshaft sprocket to the end of the crankshaft.
7. Clean out the oil pickup or replace, if necessary. Replace the oil pickup gasket ring and install the pickup to the pump.
8. Install the timing belt, oil pan and all related parts.
9. Install the dipstick. Fill the engine with the proper amount of oil.
10. Connect the negative battery cable and check the oil pressure.

Rear Main Bearing Oil Seal

REMOVAL & INSTALLATION

NOTE: The rear main seal is located in a housing on the rear of the block. To replace the seal, it is necessary to remove the transaxle and perform the work from underneath the vehicle or remove the engine and perform the work on an engine stand.

1. Raise the vehicle and support it safely. Remove the transaxle from the vehicle.
2. Unscrew the retaining bolts and remove the housing from the cylinder block. Remove the separator from the housing.
3. Using a small prybar, pry out the old seal.
4. Clean the housing and the separator.
5. Lightly oil the replacement seal. Tap the seal into the housing. The oil seal should be installed so the seal plate fits into the inner contact surface of the seal case.
6. Install the separator into the housing so the oil hole faces down.
7. Oil the lips of the seal and install the housing on the rear of the engine block.
8. Install the transaxle in the vehicle.
9. Lower the vehicle. Start the engine and check for leaks.

ENGINE COOLING

Radiator

REMOVAL & INSTALLATION

1. Disconnect the negative battery cable.
2. On Scoupe, set the warm water flow control knob of the heater control to the hot position.
3. Drain the radiator. Raise the vehicle and support it safely. Remove the splash shield from under the vehicle.
4. Remove the fan shroud and disconnect the fan motor wiring harness.
5. Disconnect the radiator hoses and, if equipped, the automatic transaxle cooler hoses.
6. Disconnect the expansion tank hose.
7. Remove the radiator mounting bolts and lift out the radiator and fan assembly. The fan and motor may be left attached to the radiator and removed with the radiator as one unit.

To install:

8. Install the radiator. Tighten the retaining bolts gradually in a crisscross pattern.
9. Connect the expansion tank hose, the radiator hoses and the automatic transaxle oil cooler lines.
10. Install the fan shroud and connect the fan wiring.
11. Install the splash shield and refill the engine with coolant.
12. Connect negative battery cable.

Heater Core

REMOVAL & INSTALLATION

Excel

1. Disconnect the negative battery cable.
2. Set the heater control to **HOT** and drain the cooling system.
3. Disconnect the coolant hoses at the heater core tubes, in the engine compartment.
4. Remove the lower instrument panel section.
5. Remove the center console and on-board computer.
6. Loosen the heater duct mounting screw.
7. Pushing downward and pull, remove the heater ducts.
8. Disconnect the heater control cable.
9. Disconnect the wiring at the motor.
10. Remove the heater case mounting bolts and remove the heater case from under the dash.
11. Separate the case halves and remove the blower.

To install:

12. Assemble the case halves and install the blower.
13. Install the heater case and mounting bolts.
14. Connect the wiring at the motor.
15. Connect the heater control cable.
16. Install the heater ducts.
17. Tighten the heater duct mounting screw.
18. Install the center console and on-board computer.
19. Install the lower instrument panel section.
20. Connect the coolant hoses at the heater core tubes, in the engine compartment.
21. Set the heater control to **HOT** and refill the cooling system.
22. Connect the negative battery cable.

Sonata

1. Disconnect the negative battery cable.
2. Place the control in the HOT position.
3. Drain the cooling system.
4. Remove the heater hoses from the core tubes.
5. Discharge the air conditioning system.
6. Disconnect the suction and liquid refrigerant lines at the firewall connectors. Always use back-up wrenches. Cap all openings at once.
7. Remove the front and rear center consoles.
8. Remove the heater side covers.
9. Remove the glove box, center crash pad cover, center crash pad and the radio.
10. Remove the lower crash pad.
11. Remove the console mounting bracket and center support.
12. Remove the left and right rear heat duct assemblies and the rear heating joint duct.
13. Remove the control unit.
14. Disconnect the blower speed actuator connector and, on Canadian vehicles, disconnect the blend door actuator connector.
15. Remove the heater/air conditioning unit.
16. Remove the blower motor from the case.
17. Separate the case halves and lift out the core.

To install:

18. Assemble the case halves.
19. Install the blower motor to the case.
20. Install the heater/air conditioning unit.
21. Connect the blower speed actua-

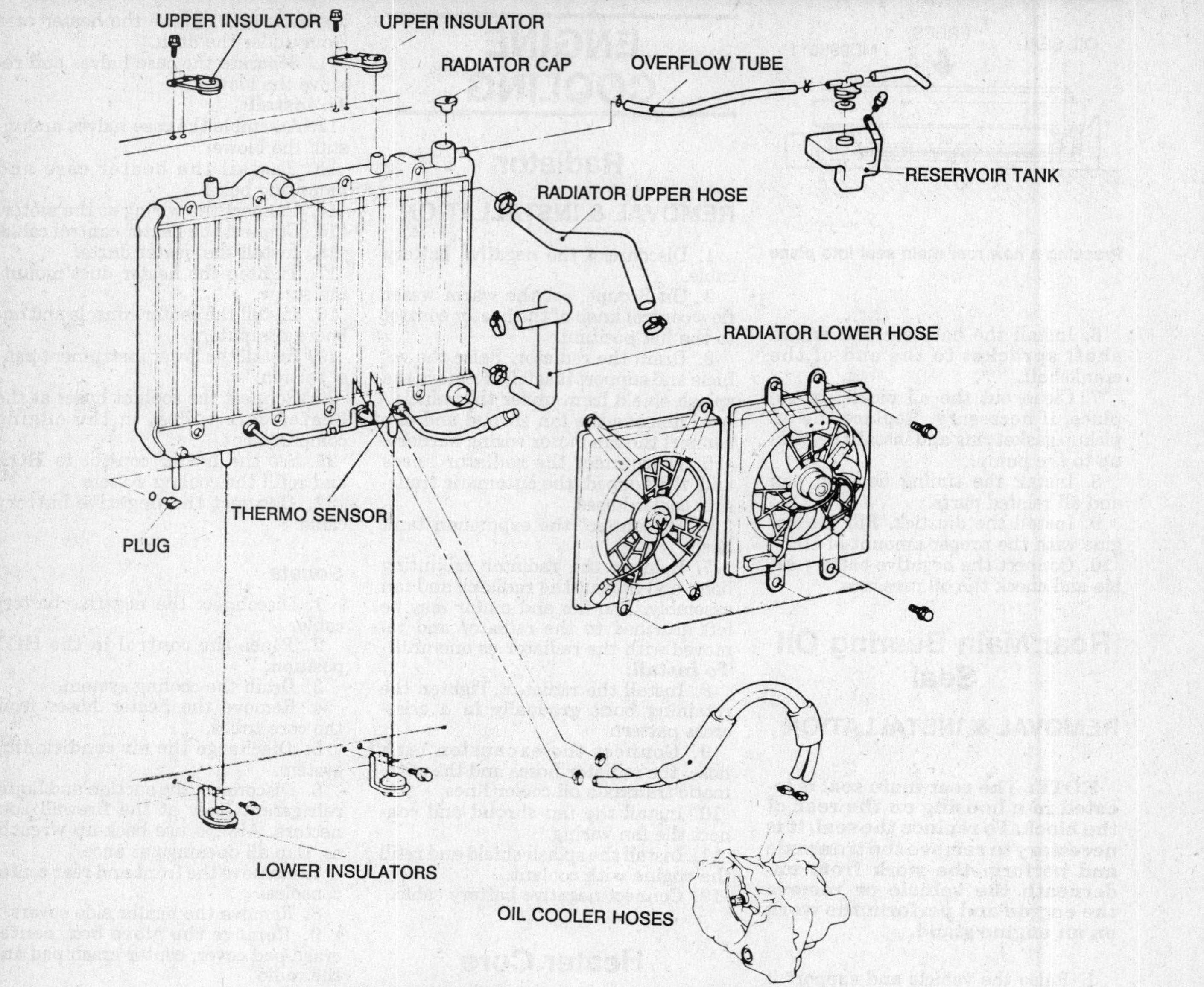

Typical radiator and cooling fan—Scoupe shown, others similar

tor connector and, on Canadian vehicles, connect the blend door actuator connector.

22. Install the control unit.
23. Install the left and right rear heat duct assemblies and the rear heating joint duct.
24. Install the console mounting bracket and center support.
25. Install the lower crash pad.
26. Install the glove box, center crash pad cover, center crash pad and the radio.
27. Install the heater side covers.
28. Install the front and rear center consoles.
29. Connect the suction and liquid refrigerant lines at the firewall connectors. Always use back-up wrenches.
30. Recharge the air conditioning system.
31. Install the heater hoses to the core tubes.
32. Refill the cooling system.
33. Connect the negative battery cable.

Scoupe

1. Disconnect the negative battery cable.
2. Set the heater control to **HOT** and drain the cooling system.
3. Disconnect the coolant hoses at the heater core tubes, in the engine compartment.
4. Remove the console assembly, cluster facia panel and lower crash pad center skin.
5. Loosen the heater duct mounting screw.
6. Remove the heater ducts.
7. Disconnect the heater control cable.
8. Disconnect the wiring at the motor.
9. Remove the heater case mounting bolts and remove the heater case from under the dash.
10. Separate the case halves and remove the blower.

To install:

11. Assemble the case halves.
12. Install the heater case and mounting bolts.
13. Connect the wiring at the motor.
14. Connect the heater control cable.
15. Install the heater ducts.

16. Tighten the heater duct mounting screw.
17. Install the console assembly, cluster facia panel and lower crash pad center skin.
18. Connect the coolant hoses at the heater core tubes, in the engine compartment.
19. Refill the cooling system.
20. Connect the negative battery cable.

Water Pump

REMOVAL & INSTALLATION

1468cc Engine

1. Disconnect the negative battery cable.
2. Loosen the 4 bolts attaching the water pump pulley to the pulley flange. Loosen the alternator mounting bolts, slide the alternator toward the engine and remove the belt. Remove the radiator cap, open the drain cock at the bottom of the radiator and drain the coolant from the radiator into a clean container.
3. Remove the timing belt covers, timing belt and tensioner.
4. Remove the water pump mounting bolts, noting the 3 different lengths and locations. Remove the pump and gasket, disconnecting the outlet at the water pipe (don't lose the O-ring).

To install:

5. Clean gasket surfaces and coat a new gasket with sealer. Then, position the gasket on the front of the block with all bolt holes aligned. Replace the O-ring for the outlet water pipe.
6. Install the pump connecting the outlet water pipe. Install the bolts with the shortest at the bottom; 2 just slightly longer at the 1 and 4 o'clock positions on the right side of the pump; next-to-longest bolt at the 8 o'clock position, just under the outlet; and the longest bolt at the 11 o'clock position and also attaching the alternator brace. Torque the bolts with a head mark, 4, to 9–11 ft. lbs. (12–15 Nm); those with a head mark 7, to 14–20 ft. lbs. (20–26 Nm).
7. Install the remaining parts in reverse order. Final tightening of the water pump pulley bolts is done after the V-belt has been installed and tensioned. Recheck tension after the pulley bolts are tightened. Close the radiator drain and refill the system. Run the engine until the thermostat opens and then add coolant until the level stabilizes before replacing the radiator cap. Check for leaks. Connect the negative battery cable.

2351cc Engine

1. Disconnect the negative battery cable. Drain the cooling system.
2. Remove all drive belts and water pump pulley. It may be necessary to remove an engine mount, the power steering pump and alternator brace to gain necessary clearance, depending the vehicle.
3. Remove the timing belt covers, timing belt tensioner and timing belt.
4. Remove the water pump bolts.
5. Remove the water pump.

NOTE: The pump is not rebuildable. If there are signs of damage or leakage from the seals or vent hole, the unit must be replaced.

6. Discard the O-ring in the front end of the water pipe. Install a new O-ring coated with water.
7. Using a new gasket, mount the water pump. Torque the bolts with a head marked 4, to 10 ft. lbs. (14 Nm) or the bolts with a head marked 7, to 20 ft. lbs. (27 Nm).
8. Assemble the remaining components, install the timing belt. Fill the system with a 50 percent mix of antifreeze.

2972cc Engine

1. Disconnect the negative battery cable.
2. Drain the cooling system.
3. Remove the timing cover. If the same timing belt will be reused, mark the direction of the timing belt's rotation, for installation in the same direction. Make sure the engine is positioned so the No. 1 cylinder is at the TDC of its compression stroke and the sprockets timing marks are aligned with the engine's timing mark indicators.
4. Loosen the timing belt tensioner bolt and remove the belt. Position the tensioner as far away from the center of the engine as possible and tighten

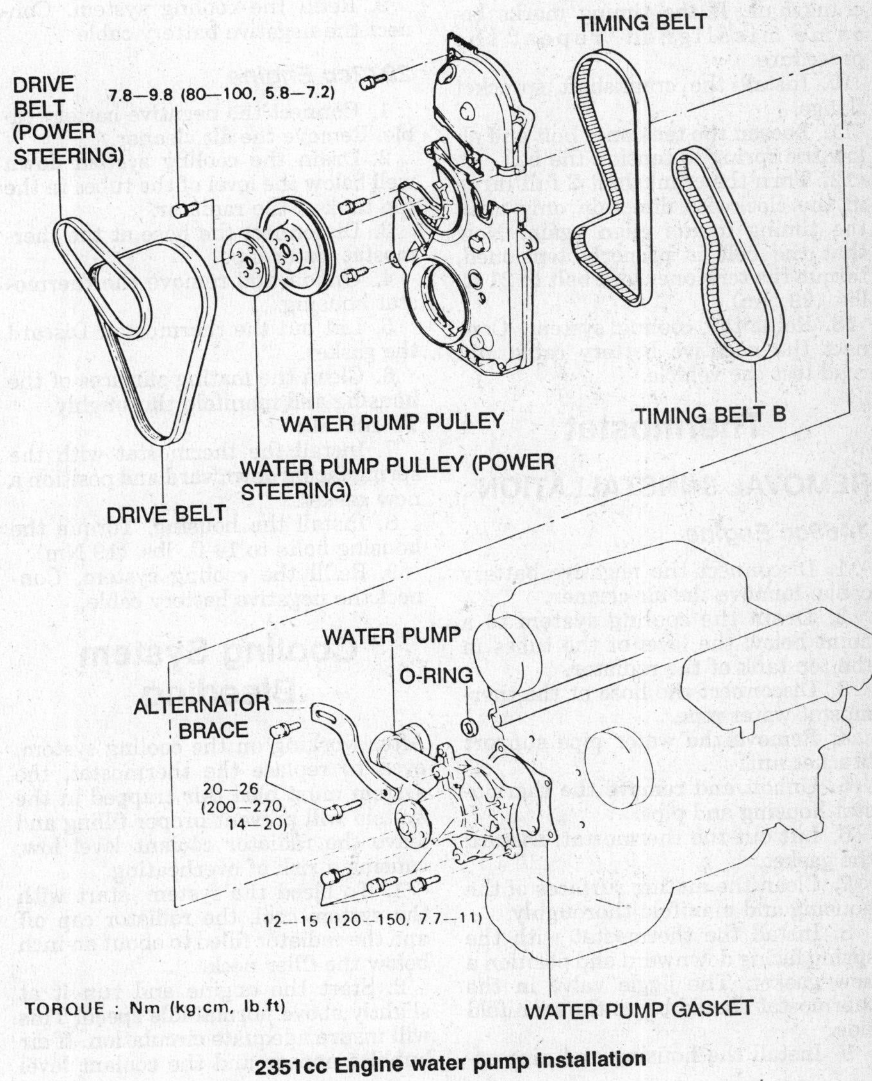

2351cc Engine water pump installation

the bolt. Remove the water pump mounting bolts, separate the pump from the water inlet pipe and remove the pump from the engine.

To install:

5. Install the pump with a new gasket to the engine. Torque the water pump mounting bolts to 20 ft. lbs. (27 Nm).
6. If not already done, position both camshafts so the marks align with those on the alternator bracket (rear bank) and inner timing cover (front bank). Rotate the crankshaft so the timing mark aligns with the mark on the oil pump.
7. Install the timing belt on the crankshaft sprocket and while keeping the belt tight on the tension side (right side), install the belt on the front camshaft sprocket.
8. Install the belt on the water pump pulley, then the rear camshaft sprocket and the tensioner.
9. Rotate the front camshaft counterclockwise to tension the belt between the front camshaft and the crankshaft. If the timing marks became misaligned, repeat the procedure.
10. Install the crankshaft sprocket flange.
11. Loosen the tensioner bolt and allow the spring to tension the belt.
12. Turn the crankshaft 2 full turns in the clockwise direction only until the timing marks align again. Now that the belt is properly tensioned, torque the tensioner lock bolt to 21 ft. lbs. (29 Nm).
13. Refill the cooling system. Connect the negative battery cable and road test the vehicle.

Thermostat

REMOVAL & INSTALLATION

1468cc Engine

1. Disconnect the negative battery cable. Remove the air cleaner.
2. Drain the cooling system to a point below the level of the tubes in the top tank of the radiator.
3. Disconnect the hose at the thermostat water pipe.
4. Remove the water pipe support bracket nut.
5. Unbolt and remove the thermostat housing and pipe.
6. Lift out the thermostat. Discard the gasket.
7. Clean the mating surfaces of the housing and manifold thoroughly.
8. Install the thermostat with the spring facing downward and position a new gasket. The jiggle valve in the thermostat should be on the manifold side.
9. Install the housing and pipe assembly. Torque the housing bolts to 10 ft. lbs. (14 Nm); the intake manifold nut to 14 ft. lbs. (19 Nm).
10. Refill the cooling system. Connect the negative battery cable.

2351cc Engine

1. Connect the negative battery cable. Remove the air cleaner.
2. Drain the cooling system down well below the level of the tubes in the top tank of the radiator.
3. Disconnect the hose at the thermostat water pipe.
4. Unbolt and remove the thermostat housing.
5. Lift out the thermostat. Discard the gasket.
6. Clean the mating surfaces of the housing and manifold thoroughly.

To install:

7. Install the thermostat with the spring facing downward and position a new gasket.
8. Install the housing. Torque the housing bolts to 14 ft. lbs. (19 Nm).
9. Refill the cooling system. Connect the negative battery cable.

2972cc Engine

1. Connect the negative battery cable. Remove the air cleaner.
2. Drain the cooling system down well below the level of the tubes in the top tank of the radiator.
3. Disconnect the hose at the thermostat water pipe.
4. Unbolt and remove the thermostat housing.
5. Lift out the thermostat. Discard the gasket.
6. Clean the mating surfaces of the housing and manifold thoroughly.

To install:

7. Install the thermostat with the spring facing downward and position a new gasket.
8. Install the housing. Torque the housing bolts to 14 ft. lbs. (19 Nm).
9. Refill the cooling system. Connect the negative battery cable.

Cooling System Bleeding

After working on the cooling system, even to replace the thermostat, the system must bled. Air trapped in the system will prevent proper filling and leave the radiator coolant level low, causing a risk of overheating.

1. To bleed the system, start with the system cool, the radiator cap off and the radiator filled to about an inch below the filler neck.
2. Start the engine and run it at slightly above normal idle speed. This will insure adequate circulation. If air bubbles appear and the coolant level drops, fill the system with a mixture of anti-freeze and water to bring the level back to the proper level.
3. Run the engine this way until the thermostat opens. When this happens, the coolant will move abruptly across the top of the radiator and the temperature of the radiator will suddenly rise.
4. At this point, air is often expelled and the level may drop quite a bit. Keep refilling the system until the level is near the top of the radiator and remains constant.
5. If the vehicle has an overflow tank, fill the radiator to the top of the filler neck.

ENGINE ELECTRICAL

NOTE: Disconnecting the negative battery cable on some vehicles may interfere with the functions of the on board computer systems and may require the computer to undergo a relearning process, once the negative battery cable is reconnected.

Distributor

REMOVAL

1. Disconnect the battery ground cable. Remove the spark plug wires from the spark plugs and the coil wire from the coil. Then, disconnect the retaining clips or unfasten the 2 screws that hold on the distributor cap and pull the distributor cap and seal off the distributor. Locate the cap and wires away from the distributor. Disconnect the vacuum advance line. Disconnect the distributor wiring connector.
2. Turn the engine until the rotor points to the No. 1 cylinder position and the timing marks on the crankshaft pulley and the timing tab are aligned at **TDC** or **0**.
3. Mark the distributor body to the exact place the rotor points. Matchmark both the distributor mounting flange and the cylinder head.
4. Carefully pull the distributor out of the engine, noting the direction and degree to which the rotor turns. Mark the location of the rotor after it has turned.

INSTALLATION

Timing Not Disturbed

1. Install the distributor while

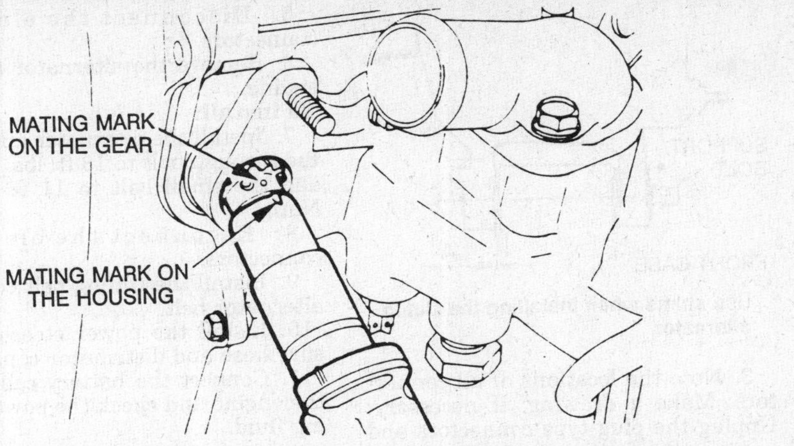

Aligning mating marks for installation of cylinder head mounted distributors

aligning the matchmarks made during removal.

2. Position the rotor so it is aligned with the matchmark on the distributor body after the distributor was pulled part way out.

3. Carefully work the distributor into the engine until the gears at the bottom engage and then begin turning the rotor. If there is resistance, turn the rotor back and forth slightly so the gears mesh. Once the gears engage and inserting the distributor causes the rotor to turn.

4. Push the distributor in until it seats and the rotor is aligned with the first mark made on the body.

5. Install the distributor hold-down bolt and reconnect the electrical connections.

6. Check and/or adjust the ignition timing.

Timing Disturbed

If the engine was cranked with the distributor removed, it will be necessary to position the No. 1 cylinder at TDC on the compression stroke. Follow the procedure listed here. This will enable the proper setting of the ignition timing.

1. Remove the No. 1 spark plug.
2. Place a finger over the spark plug hole. Crank the engine slowly until compression is felt.
3. Align the timing mark on the crankshaft pulley with the **0** degree mark on the timing scale attached to the front of the engine. This places the No. 1 cylinder at the TDC of the compression stroke.
4. Turn the distributor shaft until the rotor points to the No. 1 spark plug tower on the cap.
5. Install the distributor into the engine.
6. Tighten the distributor hold-down bolt and reconnect the electrical connections. Check the timing and adjust, as necessary.

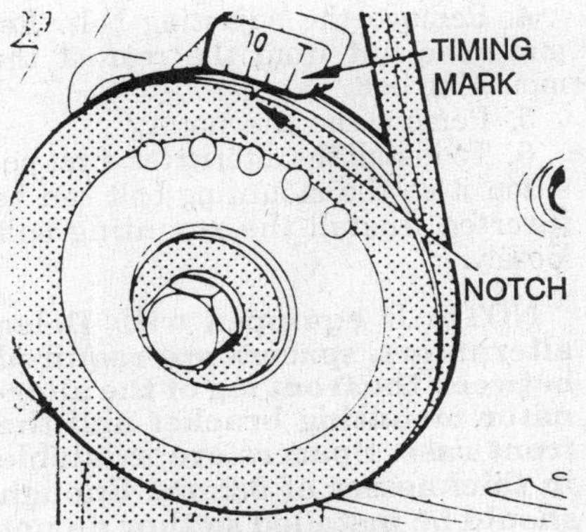

Timing marks

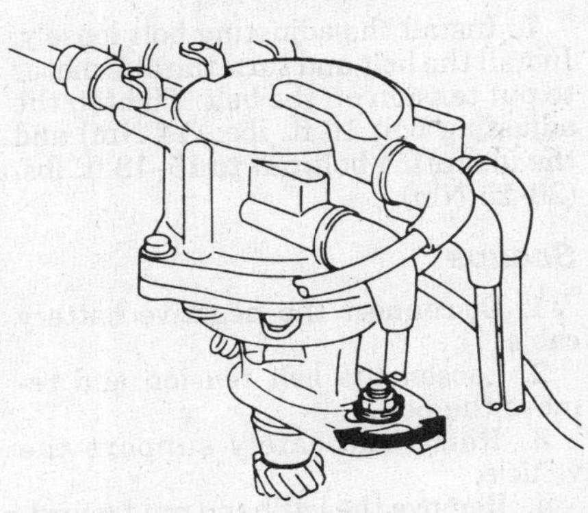

Adjusting ignition timing—2972cc engine

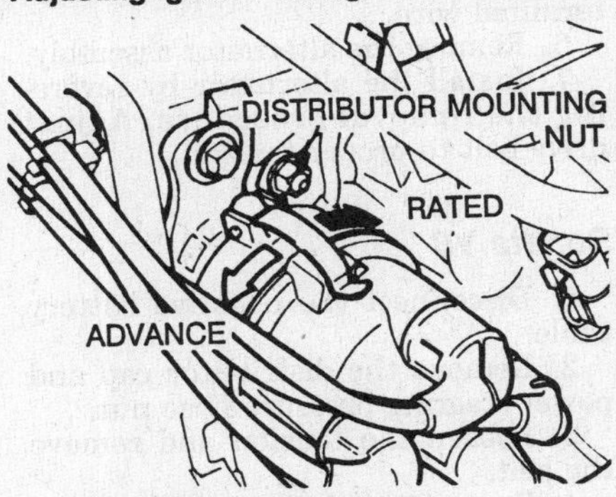

Adjusting ignition timing—4 cylinder engine

Ignition Timing

ADJUSTMENT

1. Locate the timing tab line on the front of the engine and the notch on the crankshaft pulley. Mark them with chalk.
2. Run the engine until it is at normal operating temperature.
3. Leave the engine idling, apply the hand brake and put the transaxle in neutral (manual) or **PARK** (automatic). Turn off all accessories.

NOTE: On high altitude engines equipped with vacuum advance, the distributor vacuum hoses must be disconnected and plugged.

4. Install a tachometer, according to the manufacturer's instructions.
5. Verify that the engine idle speed is correct. If not adjust it, because incorrect idle speed will change the timing.
6. Connect a timing light according to the manufacturer's instructions.
7. Direct the timing light at the crankshaft pulley marks. The marked line should align with the pulley notch.
8. If the marks do not align, loosen the distributor mounting nut and rotate the distributor slowly in either direction to align the timing marks. When the timing is correct, tighten the distributor hold-down bolt. Recheck the timing.
9. Turn the engine off and disconnect the timing light and tachometer.

Alternator

PRECAUTIONS

Several precautions must be observed with alternator equipped vehicles to avoid damage to the unit.

- If the battery is removed for any reason, make sure it is reconnected with the correct polarity. Reversing the battery connections may result in damage to the one-way rectifiers.
- When utilizing a booster battery as a starting aid, always connect the positive to positive terminals and the negative terminal from the booster battery to a good engine ground on the vehicle being started.
- Never use a fast charger as a booster to start vehicles.
- Disconnect the battery cables when charging the battery with a fast charger.
- Never attempt to polarize the alternator.
- Do not use test lamps of more than 12 volts when checking diode continuity.

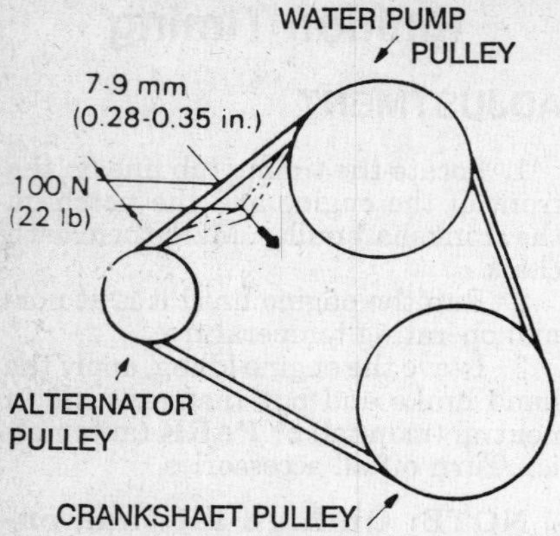

Adjusting the tension on the drive belts

- Do not short across or ground any of the alternator terminals.
- The polarity of the battery, alternator and regulator must be matched and considered before making any electrical connections within the system.
- Never separate the alternator on an open circuit. Make sure all connections within the circuit are clean and tight.
- Disconnect the battery ground terminal when performing any service on electrical components.
- Disconnect the battery if arc welding is to be done on the vehicle.

BELT TENSION ADJUSTMENT

The alternator drive belt is correctly tensioned when the longest span of belt between pulleys can be depressed ⅛–½ in. by moderate thumb pressure. To adjust, loosen the slotted adjusting bracket bolt on the alternator. If the alternator hinge bolts are very tight, it may be necessary to loosen them slightly to move the alternator. Move the alternator in or out by hand to get the correct tension, then tighten the adjusting bolt.

V-belts under 39 in. (100cm) in length should deflect about ⅛ in. (3mm). Belts over 40 in. (101cm) long should deflect about ½ in. (13mm).

NOTE: Be careful not to overtighten the belt, as this may damage the alternator bearings.

REMOVAL & INSTALLATION

Excel and Sonata with 4 Cylinder Engine

1. Turn off the ignition switch and disconnect both battery cables.
2. Loosen the support bolt and adjusting bolt and then shift the alternator toward the engine so belt tension is relieved. Remove the belt.

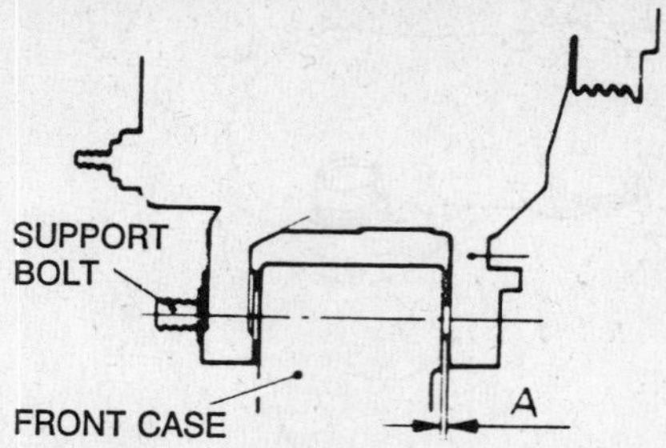

Use shims when installing the Delco alternator

3. Note the locations of all connectors. Make a drawing, if necessary. Unplug the plug type connectors and unscrew the fastening nuts for terminal type connectors. Clean any dirty connections.
4. Remove the adjusting bolt. Remove the nut from the rear of the mounting bolt.
5. Remove the alternator.
6. To install the alternator, first position it so the mounting bolt can be inserted. Install the mounting bolt loosely.

NOTE: If equipped with Delco alternators, spacers are required between the front leg of the alternator mounting bracket and the front case. Spacers are available in thicknesses of 0.2mm. Enough should be installed so they do not fall out when removed.

7. Install the adjusting bolt loosely. Install the belt and turn the alternator to put tension on the belt. Tighten the adjusting bolt 10 ft. lbs. (14 Nm) and the mounting bolt/nut to 15–18 ft. lbs. (20–25 Nm).

Scoupe

1. Disconnect the negative battery cable.
2. Loosen the belt tension and remove the belt.
3. Raise and safely support the vehicle.
4. Remove the left hand mud guard.
5. Disconnect the alternator B+ terminal wire.
6. Remove the alternator assembly.
7. Install the alternator by reversing the removal procedure. Adjust drive belt to proper tension.

Sonata V6

1. Disconnect the negative battery cable.
2. Remove the distributor cap and power steering pressure hose nut.
3. Loosen the tension and remove the belt.
4. Remove the timing belt cover cap and timing belt upper cover.
5. Disconnect the electrical connectors.
6. Remove the alternator from the engine.

To install:

7. Install the alternator and torque the through bolt to 18 ft. lbs. (25 Nm) and the small bolt to 11 ft. lbs. (15 Nm).
8. Reconnect the electrical connectors.
9. Install the timing belt cover and alternator belt.
10. Install the power steering pressure hose and distributor cap.
11. Connect the battery cable, start the engine and check the power steering fluid.

Starter

REMOVAL & INSTALLATION

Except Scoupe

1. Disconnect the negative battery cable. Then, mark and disconnect all wiring connectors at the starter.

NOTE: It will be helpful to remove the battery and battery tray, as well as the engine underside shield.

2. Raise and support the vehicle safely. Then remove the starter mounting bolts and remove the starter.
3. Clean the surfaces of the starter motor flange and the flywheel housing where the starter attaches. Then, install the starter motor. Tighten the bolts to 16–23 ft. lbs. (21–30 Nm) for the 4 cylinder engine and 20–25 ft. lbs. (27–34 Nm) for the V6 engine.
4. Install the battery and tray, if removed.
5. Reconnect the negative battery cable.

Scoupe

1. Disconnect the negative battery cable.
2. Remove the EGR valve assembly.
3. Remove the speedometer cable and the heater valve.
4. Disconnect the starter motor connector and terminal.
5. Remove the starter motor assembly.
6. Install the starter motor by reversing the removal procedure.

EMISSION CONTROLS

Please refer to "Emission Controls" in the Unit Repair section for system maintenance procedures. Due to the complex nature of modern electronic engine control systems, comprehensive diagnosis and testing procedures fall outside the confines of this repair manual. For complete information on diagnosis, testing and repair procedures concerning all modern engine and emission control systems, please refer to "Chilton's Guide to Fuel Injection and Electronic Engine Controls".

FUEL SYSTEM

Fuel System Service Precautions

Safety is the most important factor when performing not only fuel system maintenance but any type of maintenance. Failure to conduct maintenance and repairs in a safe manner may result in serious personal injury or death. Maintenance and testing of the vehicle's fuel system components can be accomplished safely and effectively by adhering to the following rules and guidelines.

- To avoid the possibility of fire and personal injury, always disconnect the negative battery cable unless the repair or test procedure requires that battery voltage be applied.
- Always relieve the fuel system pressure prior to disconnecting any fuel system component (injector, fuel rail, pressure regulator, etc.), fitting or fuel line connection. Exercise extreme caution whenever relieving fuel system pressure to avoid exposing skin, face and eyes to fuel spray. Please be advised that fuel under pressure may penetrate the skin or any part of the body that it contacts.
- Always place a shop towel or cloth around the fitting or connection prior to loosening to absorb any excess fuel due to spillage. Ensure that all fuel spillage (should it occur) is quickly removed from engine surfaces. Ensure that all fuel soaked cloths or towels are deposited into a suitable waste container.
- Always keep a dry chemical (Class B) fire extinguisher near the work area.
- Do not allow fuel spray or fuel vapors to come into contact with a spark or open flame.
- Always use a backup wrench when loosening and tightening fuel line connection fittings. This will prevent unnecessary stress and torsion to fuel line piping. Always follow the proper torque specifications.
- Always replace worn fuel fitting O-rings with new. Do not substitute fuel hose or equivalent where fuel pipe is installed.

RELIEVING FUEL SYSTEM PRESSURE

1. Remove the fuel pump fuse from the fuse block.
2. Start the engine and let it run until the remaining fuel in the lines is consumed.
3. Crank engine again to make sure any fuel in the lines has been removed.
4. With the ignition **OFF** replace the fuel pump fuse.
5. Disconnect the negative battery cable.

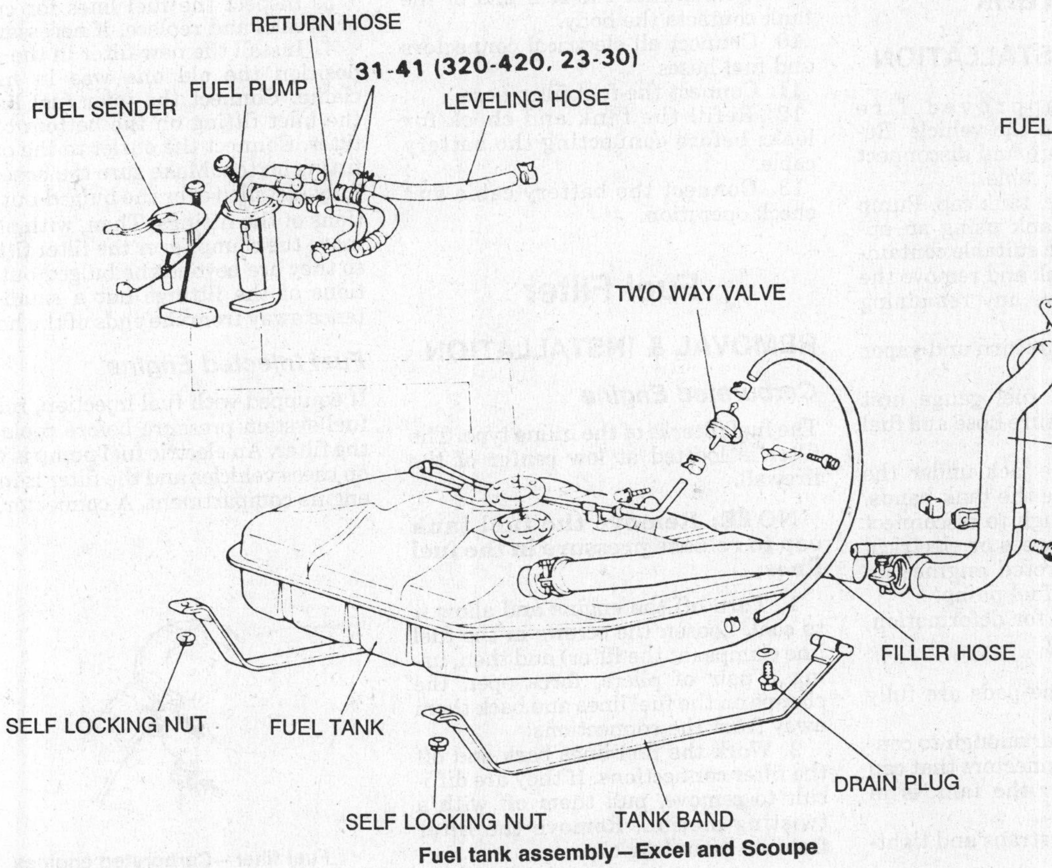

Fuel tank assembly—Excel and Scoupe

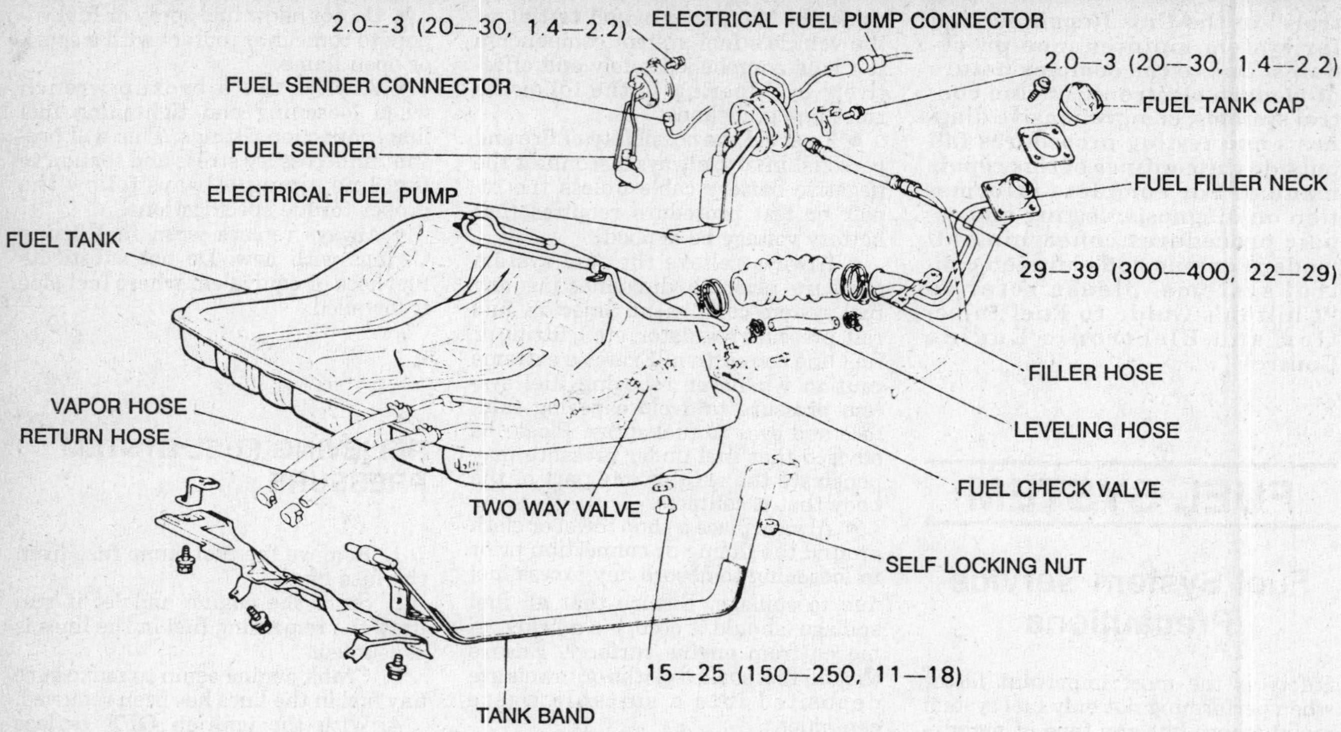

Fuel tank assembly—Sonata

Fuel Tank

REMOVAL & INSTALLATION

1. Have an approved fire extinquisher close to the vehicle. Relieve the fuel pressure and disconnect the negative battery cable.
2. Remove the fuel tank cap. Pump the fuel from the tank using an approved pump. Place a suitable container under the fuel tank and remove the drain plug to remove any remaining fuel.
3. Disconnect the return and vapor hoses.
4. Disconnect the fuel gauge unit connector, high pressure hose and fuel filler hose.
5. Place a suitable jack under the fuel tank and remove the tank bands. Lower the tank enough to disconnect any remaining fuel hoses or electrical connections. Carbureted engines do not have an in-tank fuel pump.
6. Check the tank for deformation, corrosion or cracking.

To install:

7. Confirm that the pads are fully bonded to the tank.
8. Raise the tank far enough to connect the hoses and connectors that can not be accessed after the tank is in place.
9. Install the tank straps and tighten the nuts until the rear end of the tank contacts the body.
10. Connect all electrical connectors and fuel hoses.
11. Connect the fuel filler hose.
12. Refill the tank and check for leaks before connecting the battery cable.
13. Connect the battery cable and check operation.

Fuel Filter

REMOVAL & INSTALLATION

Carbureted Engine

The fuel filter is of the inline type. The filter is located at low center of the firewall.

NOTE: Remove the fuel tank cap to release pressure in the fuel lines.

1. Turn off the engine and allow it to cool. Loosen the screws in the fuel line clamps (at the filter) and then, using a pair of pliers, force open the clamps on the fuel lines and back them away from the connections.
2. Work the fuel lines back and off the filter connections. If they are difficult to remove, pull them off with a twisting motion. Remove the filter from its mounting clip.
3. Inspect the fuel lines for cracks or breaks and replace, if necessary.
4. Install the new filter in the same position the old one was in in the clamp. Connect the inlet fuel line to the inlet fitting on the bottom of the filter. Connect the outlet to the outlet fitting on top. Make sure the hoses are fully installed over the bulged-out portions of the fittings. Then, with pliers, move the clamps over the filter fittings so they are beyond the bulged-out sections of the fittings but a small distance away from the ends of the hoses.

Fuel Injected Engine

If equipped with fuel injection, relieve fuel system pressure before replacing the filter. An electric fuel pump is used on these vehicles and the filter is in the engine compartment. A connector, for

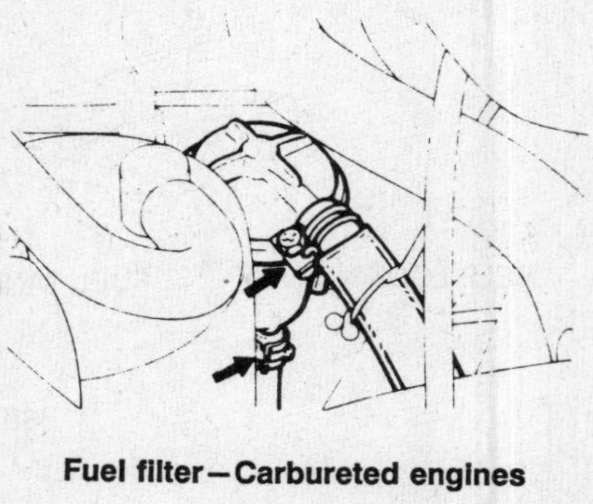

Fuel filter—Carbureted engines

checking the fuel function, is located under the battery tray. With the engine running, disconnect the connector, when the engine stops no pressure will remain in the system.

NOTE: Cover the fuel line fittings with a shop towel before loosening the eye bolts securing the fuel lines to the fuel filter.

1. Hold the fuel filter nuts securely and loosen the eye bolts securing the fuel lines to the fuel filter and remove the fuel lines from the filter ends.
2. Unclip the filter from the mounting bracket.
3. Installation is the reverse order of the remove procedure.
4. Install a new filter. Start the engine and check for leaks.
5. If a clogged filter is suspected, remove the filter and blow compressed air through the inlet and outlet fittings, reinstall the filter. Replace the old filter. Torque the filter fitting bolts to 18–25 ft. lbs. (25–34 Nm).

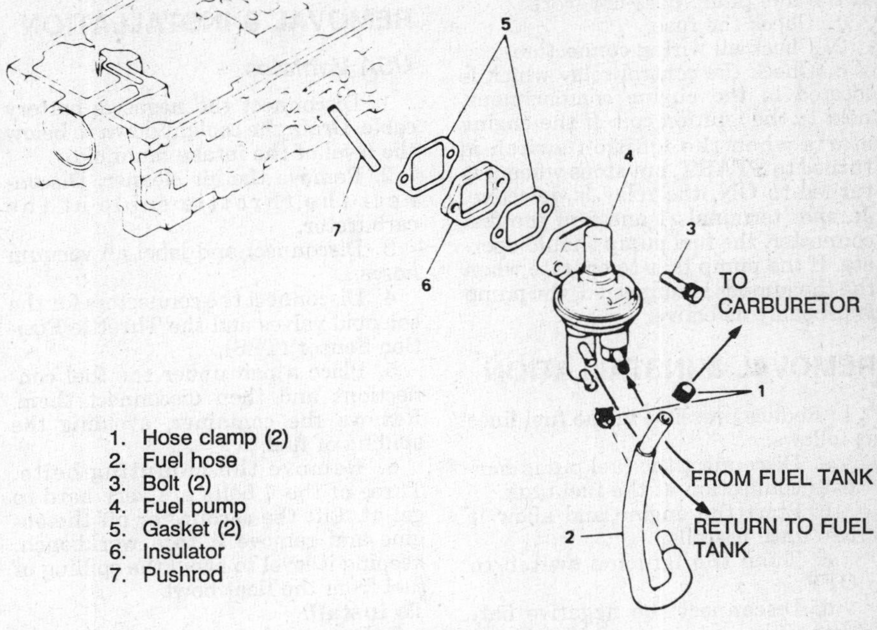

Mechanical fuel pump

Mechanical Fuel Pump

PRESSURE TESTING

1. Disconnect the inlet line, coming from the filter, at the pump.
2. Connect a vacuum gauge to the pump nipple.
3. Remove the high tension cable at the coil. Crank the engine and observe the gauge reading.
4. A vacuum of 2.7–3.7 psi should be produced. If there is a blow-back of pressure, the inlet valve on the pump is leaking and the unit must be replaced.
5. Disconnect the hose at the carburetor and connect a fuel pressure gauge.
6. Disconnect the return hose at the pump and plug the fitting at the pump.
7. Check the pressure while the engine is idling. The pressure should be 2.76–3.63 psi.
8. Check the pump volume by disconnecting the carburetor fuel hose and insert the end into a graduated container. Start the engine and measure the amount of fuel pumped within 1 minute. The volume should be 0.85 pint (0.60 liters).
9. Reconnect all disconnected hoses and check for leaks.

REMOVAL & INSTALLATION

The mechanical fuel pump operates directly off of a camshaft eccentric. A fuel return valve is located in the upper body of the pump. If the fuel temperature rises above 122°F (50°C), the valve opens and routes fuel back to the tank, preventing percolation.

1. Disconnect the negative battery cable. Remove the 2 screws and remove the plastic heat shield.
2. Disconnect the 3 fuel pump lines.
3. Unscrew the retaining nuts and remove the fuel pump and pushrod.
4. Remove the gasket, insulator and gasket.
5. Clean the fuel pump.
6. Apply non-hardening sealer to both sides of the gaskets. Position a gasket, the insulator and the other gasket on the head studs.
7. Set the No. 1 piston at TDC of the compression stroke and insert the pushrod into the head. Install the pump and torque the nuts to 25 ft. lbs. (34 Nm).
8. Connect the negative battery cable.

Electric Fuel Pump

All fuel injected vehicles are equipped with an electric fuel pump. The fuel pump is in the gas tank.

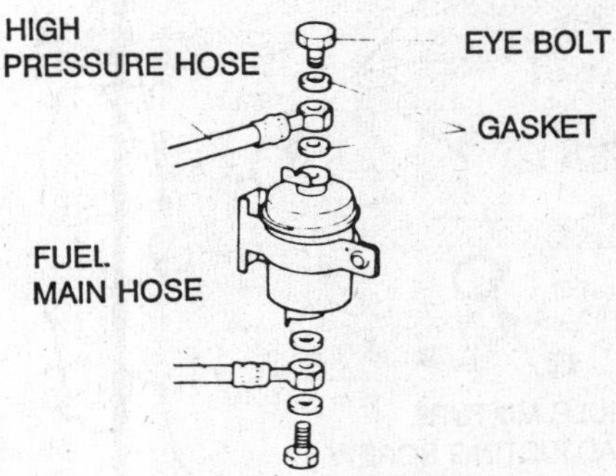

Fuel filter—Injected engine

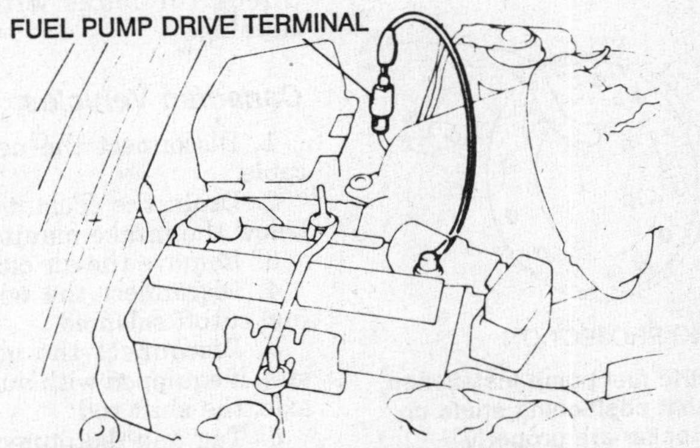

Fuel pump drive terminal—Injected engines

PRESSURE TESTING

If the fuel pump does not work:

1. Check the fuse.
2. Check all wiring connections.
3. Check the control relay which is located in the engine compartment, next to the ignition coil. If the engine starts when the ignition switch is turned to **START**, but stops when it is turned to **ON**, the relay is defective. Jumper terminals 1 and 2 of the test connecter, the fuel pump should operate. If the pump fails to operate when the the jumper is connected, the pump is probably defective.

REMOVAL & INSTALLATION

1. Reduce pressure in the fuel lines as follows:
   a. Disconnect the fuel pump harness connection at the fuel tank.
   b. Start the engine and allow it run until it stalls.
   c. Turn the ignition switch to **OFF**.
   d. Disconnect the negative battery cable.
2. Raise the vehicle and support it safely. Remove the fuel tank.
3. Disconnect the hoses at the pump.
4. Unbolt and remove the pump.
5. Install the new pump. Use a new gasket. Install the fuel tank. Connect the fuel lines and the electrical connectors to the fuel pump. Connect the negative battery cable.

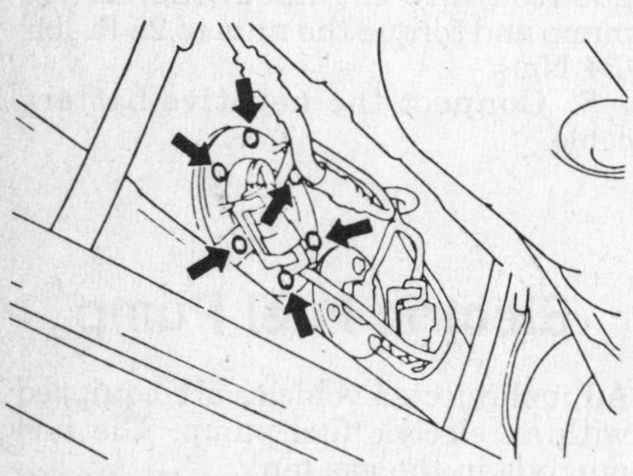

Electric fuel pump attaching screws

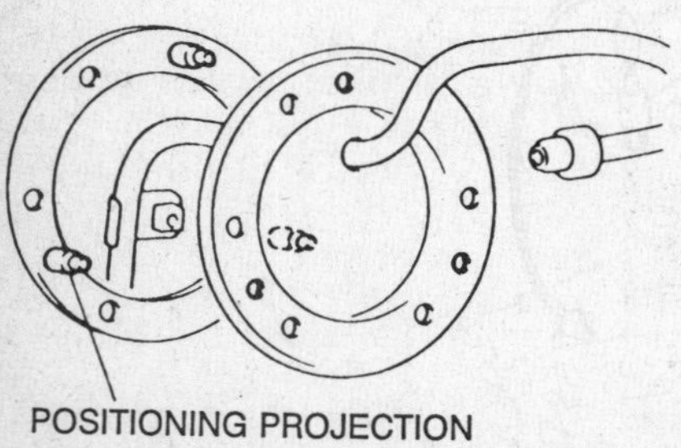

During electric fuel pump installation, make sure that positioning studs on the packing collar are properly positioned

Carburetor

REMOVAL & INSTALLATION

USA Vehicles

1. Disconnect the negative battery cable. Drain the coolant down to below the level of the intake manifold.
2. Remove the air cleaner. Disconnect the throttle cable at the carburetor.
3. Disconnect and label all vacuum hoses.
4. Disconnect the connectors for the solenoid valves and the Throttle Position Sensor (TPS).
5. Place a pan under the fuel connections and then disconnect them. Remove the container, avoiding the spilling of fuel.
6. Remove the mounting bolts. Three of the 4 bolts are very hard to get at. Lift the carburetor off the engine and remove it to a workbench, keeping it level to avoid the spilling of fuel from the float bowl.

To install:

7. Inspect the mating surfaces of the carburetor and manifold. They should be clean and free of nicks or burrs. Clean and, if necessary, remove any slight imperfections with crocus cloth. Put a new carburetor gasket on the surface of the manifold.
8. Position the carburetor on top of the gasket with all holes aligned. Install the carburetor bolts and tighten them alternately and evenly.
9. Connect the throttle linkage. Depress the accelerator pedal and make sure the throttle blade opens all the way. Adjust, if necessary.
10. Connect the vacuum hoses. Make sure all are soft and free of cracks to make a good seal. Replace hoses that are hard and cracked. Reconnect the fuel hoses.
11. Install the remaining parts in reverse order. Connect the negative battery cable. To start the engine, set the choke and operate the starter. Do not attempt to prime the engine by pouring gas into the carburetor inlet. Check for leaks with the engine running.

Canadian Vehicles

1. Disconnect the negative battery cable.
2. Drain the coolant to a level just below the intake manifold.
3. Remove the air cleaner.
4. Disconnect the wiring from the fuel cutoff solenoid.
5. Disconnect the accelerator rod and, if equipped with automatic transaxle, the shift rod.
6. Tag and disconnect the vacuum hoses from the carburetor.
7. Place suitable rags or a container under the fuel inlet and return hoses to catch any leaking fuel and disconnect the hoses.
8. Disconnect the water hose which runs between the carburetor and the cylinder head.
9. Unscrew the 4 retaining nuts and remove the carburetor. Hold the carburetor level to avoid a fuel spill.
10. Mount the carburetor on the intake manifold and attach the choke water hose.
11. Reconnect the fuel lines and vacuum hoses to the carburetor.
12. Connect the accelerator or shift rod.
13. Reconnect the fuel cut-off solenoid and replace the air cleaner.
14. Refill the system with coolant. Connect the negative battery cable.

IDLE SPEED ADJUSTMENT

The idle speed is adjusted periodically to compensate for engine wear or after engine work is performed. Idle mixture adjustments are not required as a matter of routine but only when major carburetor work is required. The emission control system compensates as required to ensure a stable idle mixture. If a rough idle is apparent, check the idle mixture with a CO meter, if one is available.

Also, Hyundais have an idle-up solenoid that operates to prevent stalling under certain conditions. This does not require adjustment as a matter of routine either but may be adjusted, if the system is not functioning properly.

NOTE: The idle mixture adjustment is preset at the factory. The mixture adjusting screw is inaccessible without removing and modifying the carburetor. Since this adjustment is preset, it should not be changed unless major, unscheduled maintenance has been performed on the carburetor. This adjustment can only be made with a CO meter.

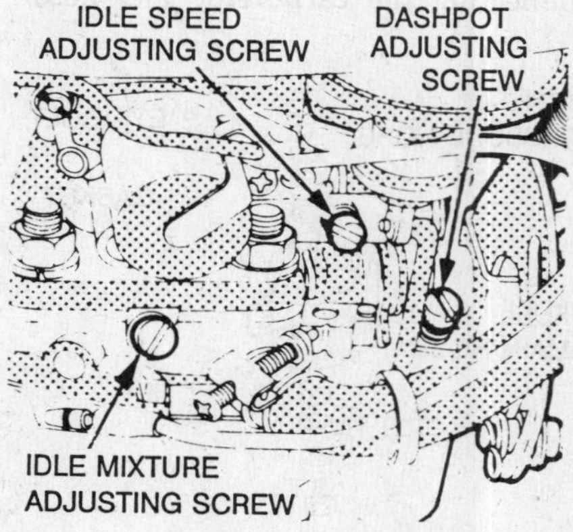

Carburetor adjustment points

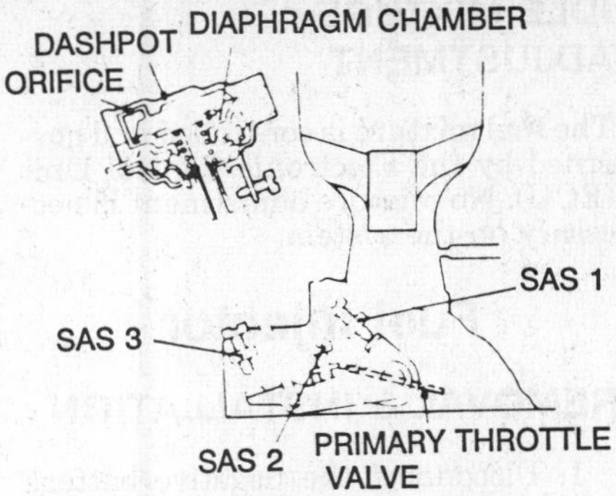

Idle speed adjustment points for carbureted engines

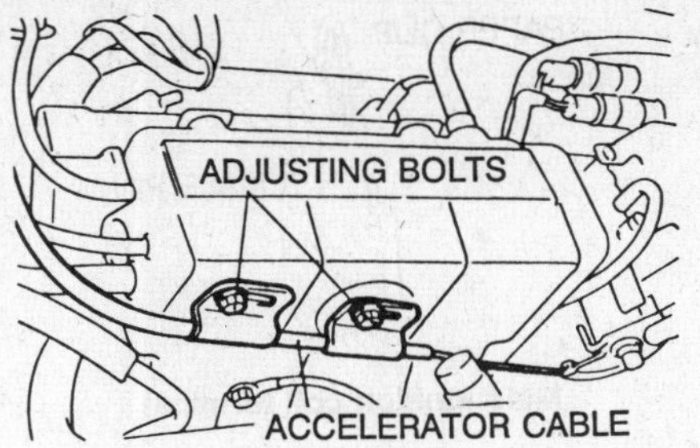

ISC control cable adjusting points

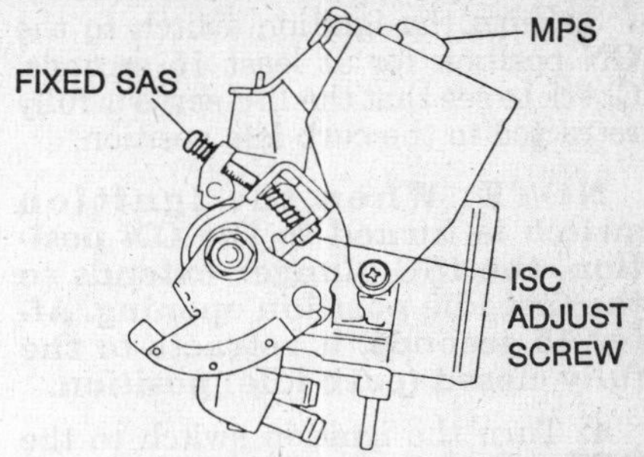

Speed control screws for fuel injected engines

The idle adjustment is made under the following conditions:

Lights Off
All accessories Off
Electric cooling fan Off
Transaxle in neutral and parking brake applied

1. Start and run the engine at idle until normal operating temperature is reached.
2. Check the underhood decal or the tune-up specification charts for the correct curb idle speed.
3. Connect a tachometer, according to the manufacturer's instructions and adjust the idle speed screw until the correct rpm is reached.
4. Run the engine for at least 5 seconds at 2000–3000 rpm. Then, reduce the speed to idle rpm for at least 2 minutes.
5. If the idle speed is not at the specified rpm, turn the idle speed adjusting screw until the proper rpm is reached.
6. Check the underhood specifications sticker for the correct idle speed rpm.

SERVICE ADJUSTMENTS

For all carburetor service adjustment procedures and specifications, please refer to "Carburetor Service" in the Unit Repair section.

Fuel Injection

IDLE SPEED ADJUSTMENT

NOTE: This adjustment must be made any time the Idle Speed Control (ISC) servo, Throttle Position Sensor (TPS), mixing body or throttle body has been removed. A digital voltmeter is essential for this operation.

Except Scoupe and Sonata V6

1. Run the engine to normal operating temperature, then turn it off.

Multi-tester connections for the ISC adjustment

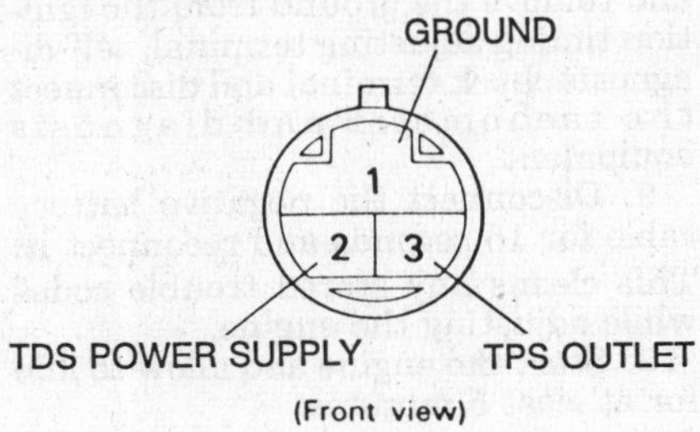

TPS connector

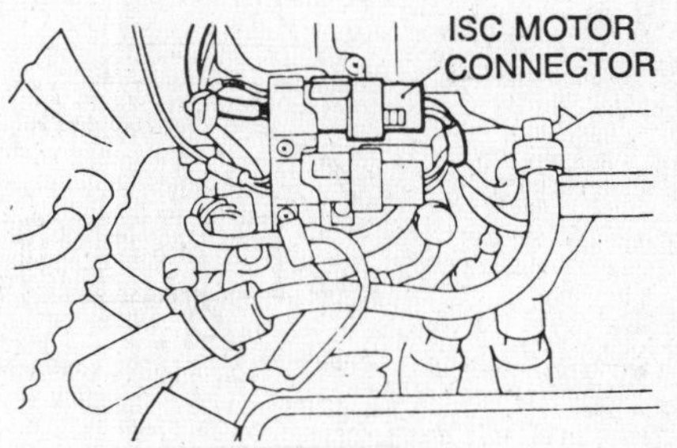

ISC motor connector

2. Disconnect the accelerator cable at the throttle lever on the mixing body.
3. Loosen the screws holding the TPS and turn it fully clockwise. Tighten the screws.
4. Turn the ignition switch to the **ON** position for at least 15 seconds, then turn it **OFF**. This will automatically set the ISC servo to the proper position.
5. Disconnect the ISC servo wiring connector.

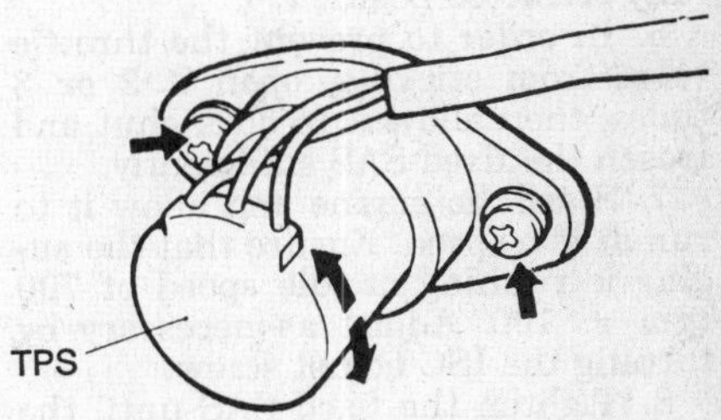

TPS adjustment

6. Start the engine and check the idle speed with a tachometer. Idle speed should be 600 rpm. If not, adjust it with the adjusting screw.
7. Insert the digital voltmeter test probes in the TPS connector GW and B holes.
8. Turn the ignition switch to the **ON** position. Do not start the engine.
9. Read the voltage. If indicated voltage is not 0.45–0.51 volts, loosen the TPS mounting screws and turn the sensor clockwise or counterclockwise until the indicated voltage is 0.48 volts. Tighten the screws and apply a thread-locking sealant.
10. Open the throttle valve fully and allow it close. Recheck the indicated voltage. Adjust, if necessary.
11. Remove the voltmeter and reconnect the wiring.
12. Start the engine. Recheck the idle speed. Adjust, if necessary and stop the engine.
13. Turn the ignition switch to the **ON** position for at least 15 seconds, then turn it **OFF**.
14. Reconnect the cable and adjust, if necessary, to remove any slack, using the adjusting nut at the throttle lever.

Scoupe

1. Loosen the accelerator cable.
2. Connect the multi-use tester to the diagnosis connector in the fuse box. If using a digital voltmeter, connect the voltmeter between terminals 1 and 2 of the throttle position sensor.

NOTE: Do not disconnect the TPS connector from the throttle body.

3. Turn the ignition switch to the **ON** position for at least 15 seconds. Check to see that the ISC servo is fully retracted to the curb idle position.

NOTE: When the ignition switch is turned to the ON position, the ISC plunger extends to the fast idle position opening. After 15 seconds, it retracts to the fully closed (curb idle) position.

4. Turn the ignition switch to the **OFF** position.
5. Disconnect the ISC motor connector and secure the ISC motor at the fully retracted position.
6. In order to prevent the throttle valve from sticking, open it 2 or 3 times, then allow it to click shut and loosen the fixed SAS sufficiently.
7. Start the engine and allow it to run at idle speed. Ensure that the engine is running at idle speed of 700 rpm ± 100. Adjust as necessary by turning the ISC adjust screw.
8. Tighten the fixed SAS until the engine speed starts to increase. Then, loosen the screw until the engine speed ceases to drop (touch point) and loosen an additional ½ turn.
9. Turn the ignition switch to the **OFF** position.
10. Turn the ignition switch to the **ON** position and check TPS output voltage. The output voltage should be 0.48–0.52 volts.
11. If the voltage is out of specification, loosen the TPS mounting screws and adjust by turning the TPS.
12. Turn the ignition switch to the **OFF** position.
13. Adjust the accelerator cable to 0–0.04 in. (0–1mm) for a manual transaxle and 0.08–0.12 in. (2–3mm) for an automatic transaxle.
14. Connect the ISC motor connector.
15. Disconnect the voltmeter and connect the TPS connector.
16. Start the engine and check to be sure that the idle speed is correct.
17. Turn the ignition switch to the **OFF** and disconnect the negative battery cable for 15 seconds and reconnect.

Sonata V6

NOTE: Adjustment of the idle speed is usually unnecessary because the idle speed control servo adjusts the idle speed.

1. Warm up the engine, turn all accessories **OFF**, place the transaxle in **P** and center the steering wheel.
2. Loosen the accelerator cable and connect a tachometer if not using a multi-use tester. Insert a paper clip into **pin 1** of the primary side of the ignition coil terminal and connect the tachometer.

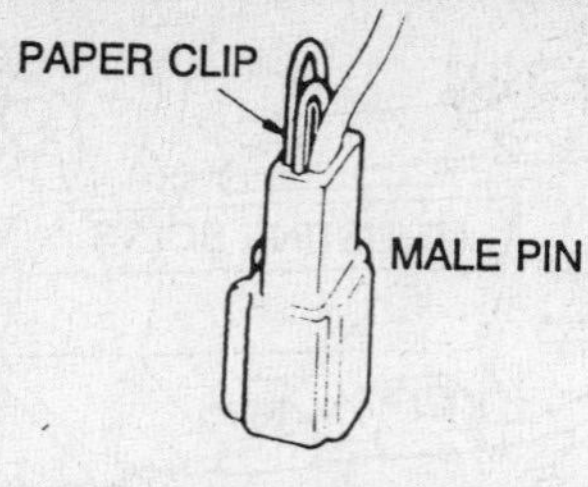

Male ignition coil terminal

3. Ground the terminal of the ignition timing adjusting connector.
4. Start the engine and let idle. Adjust the ignition timing if needed.
5. Turn the ignition switch OFF and ground the self-diagnosis check terminal.
6. Run the engine for more than 5 seconds at 3000 rpm. Then run at idle for 2 minutes.
7. If the idle speed is not 700±100 rpm, turn the speed adjusting screw (SAS) to the specified rpm.
8. Turn the ignition switch OFF and remove the ground from the ignition timing adjusting terminal, self-diagnosis check terminal and disconnect the tachometer and diagnosis equipment.
9. Disconnect the negative battery cable for 15 seconds and reconnect it. This cleans any stored trouble codes while adjusting the engine.
10. Start the engine and allow to idle for at least 5 minutes.

IDLE MIXTURE ADJUSTMENT

The fuel mixture is controlled and governed by the Electronic Control Unit (ECU). No mixture adjustment is necessary on the system.

Fuel Injector

REMOVAL & INSTALLATION

1. Disconnect the negative battery cable.
2. Relieve the fuel pressure.
3. Remove the air cleaner-to-throttle body hose.
4. Disconnect the throttle cable from the throttle body and disconnect the kickdown linkage. Remove the throttle cable bracket attaching bolts.
5. Disconnect the connectors to the throttle body.
6. Matchmark and carefully remove the vacuum hoses from the throttle body.
7. Remove the PCV and brake booster hoses from the air intake plenum.
8. Remove the ignition coil from the intake plenum, if mounted there.
9. Remove the EGR tube flange from the intake plenum, if equipped.
10. Unplug the coolant temperature sensor and charge temperature sensor, if equipped.
11. Remove the vacuum connection from the air intake plenum vacuum connector.

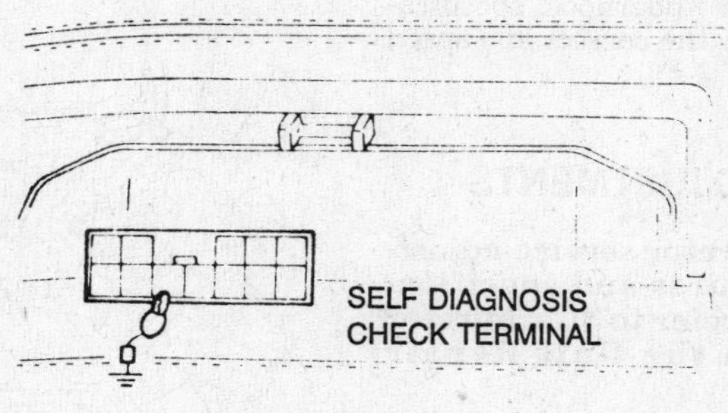

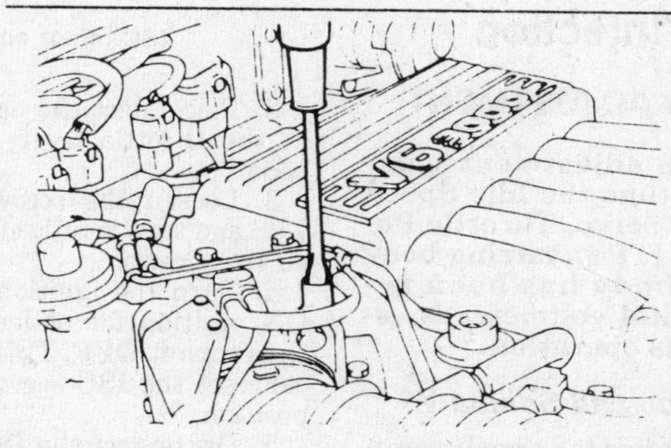

Diagnosis connector and idle speed adjusting screw—Sonata V6

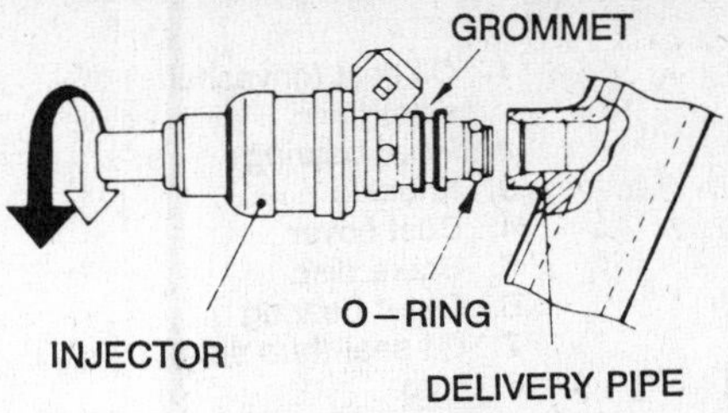

Fuel Injector

12. Remove the fuel hoses from the fuel rail and plug them.
13. Remove the air intake plenum-to-intake manifold bolts, the plenum and gaskets. Cover the intake manifold openings.
14. Remove the vacuum hoses from the fuel rail.
15. Disconnect the fuel injector wiring harness.
16. Remove the fuel rail attaching bolts and the fuel rail with the wiring harness from the vehicle. Position the rail upside down so the injectors are easily accessible.
17. Remove the small connector retainer clip and unplug the injector. Remove the injector clip off the fuel rail and injector. Pull the injector straight out of the rail.

To install:

18. Lubricate the rubber O-ring with clean oil and install to the rail receiver cap. Install the injector clip to the top slot of the injector, plug in the connector and install the connector clip.
19. Install the fuel rail to the vehicle and plug in the injector harness. Connect the vacuum hoses to the fuel rail.
20. Install new intake plenum gaskets with the beaded sealer side up and install the intake plenum. Torque the attaching bolts and nuts to 115 inch lbs. (13 Nm).
21. Install the fuel hoses to the fuel rail.
22. Install or connect all items that were removed or disconnected from the intake plenum and throttle body.
23. Connect the negative battery cable and check for leaks using the DRB I or II to activate the fuel pump.

DRIVE AXLE

Halfshaft

REMOVAL & INSTALLATION

1. Remove the hub center cap and loosen the halfshaft end nut.
2. Loosen the wheel lug nuts.
3. Raise the vehicle and support it safely.

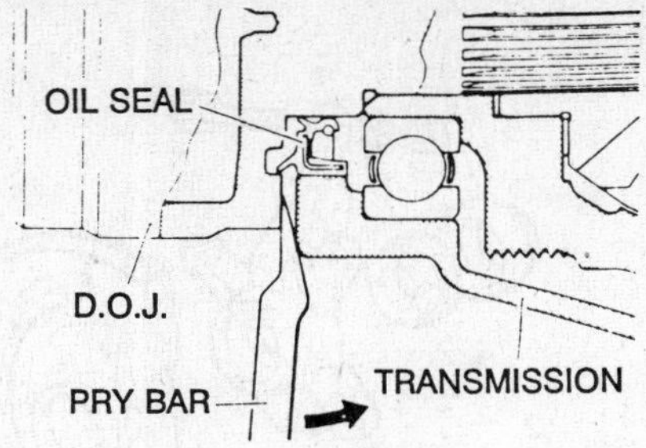

Halfshaft removal

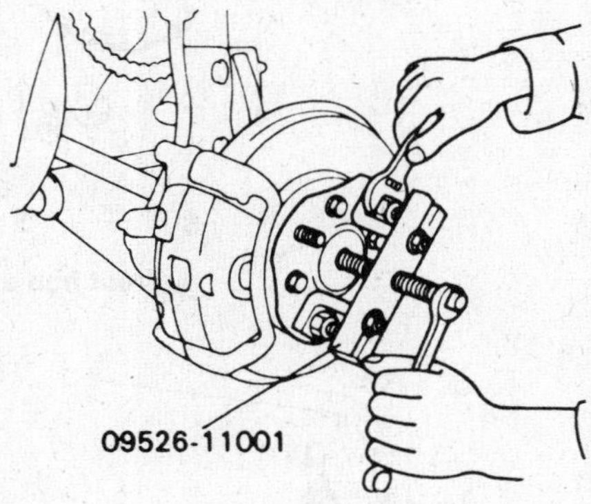

Halfshaft removal

4. Remove the front wheels.
5. Remove the engine splash shield.
6. Remove the lower ball joint and strut bar from the lower control arm.

NOTE: Place the lower arm ball joint on the lower arm to prevent damage to the ball joint dust boot.

7. Drain the transaxle fluid.
8. Insert a prybar between the transaxle case, on the raised rib, and the halfshaft inner joint case. Move the bar to the right to withdraw the left halfshaft; to the left to remove the right halfshaft.

NOTE: Do not insert the prybar too deeply (7mm) or the oil seal will be damaged.

9. Plug the transaxle case with a clean rag to prevent dirt from entering the case.
10. Use a puller/driver mounted on the wheel studs to push the halfshaft from the front hub. Take care to prevent the spacer shims from falling out of place.

To install:

11. Insert the halfshaft into the hub, first, then install the transaxle end.
12. Install the hub nut washer.
13. Torque the halfshaft end nut to 185 ft. lbs. (260 Nm); the lower arm-to-ball joint nuts to 87 ft. lbs. (120 Nm) [42–53 ft. lbs. (56–70 Nm) for Sonata, 1990-92 Excel and Scoupe]; the lower arm-to-strut bar nuts to 87 ft. lbs. (120 Nm).

NOTE: Always use a new inner joint retaining ring.

CV-Boot

REMOVAL & INSTALLATION

Inner Boot

1. Raise the vehicle and support it safely. Remove the wheel assembly.
2. Remove the halfshaft assembly from the vehicle.
3. Place the assembly in vise. Care must be taken not to crush the tubular shafts.
4. If inner joint needs replacement, cut the small clamp, large metal clamp and remove the rubber boot. These items must be discarded.
5. Inspect for internal wear and/or damage.
6. Clean the grease by hand from inside the joint housing and around the ball trunnion assembly to inspect it.
7. Mark the tripod and housing for proper reassembly, if it is to be reinstalled.
8. To replace the boot, CV-joint or both, remove the snapring from the groove and tap the trunnion lightly with a brass drift. Leave the tripod bearings on the trunnion. Care must be taken to support the bearings as they may fall off.

To install:

9. Installation is the reverse order of the removal procedures.
10. When installing the tripod on the shaft place the chamfer face towards the retainer groove.

NOTE: The grease provided with the repair kit must be used. It cannot be substituted with any other type grease.

Outer Boot

1. Raise the vehicle and support it safely. Remove the wheel assembly.
2. Remove the halfshaft assembly from the vehicle.
3. Place the assembly in vise. Care must be taken not to crush the tubular shafts.
4. At the inner joint, cut the small clamp, large metal clamp and remove the rubber boot. These items must be discarded.
5. Inspect for internal wear and/or damage.
6. Clean the grease by hand from inside the joint housing and around the ball trunnion assembly to inspect it.
7. Mark the tripod and housing for proper reassembly, if it is to be reinstalled.
8. To replace the boot, CV-joint or both, remove the snapring from the

groove and tap the trunnion lightly with a brass drift. Leave the tripod bearings on the trunnion. Care must be taken to support the bearings as they may fall off.

9. If equipped with a dynamic damper, remove the band and slide the damper from the shaft.

10. Clean any dirt or corrosion from the shaft and lightly oil the shaft. Cut both bands from the outer boot and slide the boot from the shaft.

To install:

11. Installation is the reverse order of the removal procedures. Keep the shaft oiled to aid in boot installation.

12. When installing the tripod on the shaft place the chamfer face towards the retainer groove.

NOTE: The grease provided with the repair kit must be used. It cannot be substituted with any other type grease.

Front Wheel Hub, Knuckle and Bearings

REMOVAL & INSTALLATION

NOTE: The following procedure requires the use of a number of special tools. Always replace bearings and races as a set. Never replace just an inner or outer bearing. If either is in need of replacement, both sets must be replaced.

1. Remove the center hub cap and halfshaft nut. Raise the vehicle and support it safely, positioned so the wheels hang freely. Remove the front wheel and tire assembly.

2. Remove the brake caliper without disconnecting the hydraulic line and suspend it out of the way with a piece of wire.

3. Disconnect the stabilizer bar and strut bar from the lower control arm. Disconnect the ball joint at the steering knuckle. Disconnect the tie rod end ball joint at the steering knuckle as well.

4. Remove the halfshaft from the transaxle and press the halfshaft out of the hub with a 2 jawed puller.

5. Unbolt and remove the hub and knuckle from the bottom of the strut and remove the hub and knuckle assembly from the vehicle.

6. Several special tools are required to press the hub and disc from the steering knuckle and to remount them. Use 09517–21600 or equivalent. Do not attempt to hammer the parts apart, or damage to the bearing may result. Install the arm of the special tool then the body onto the knuckle

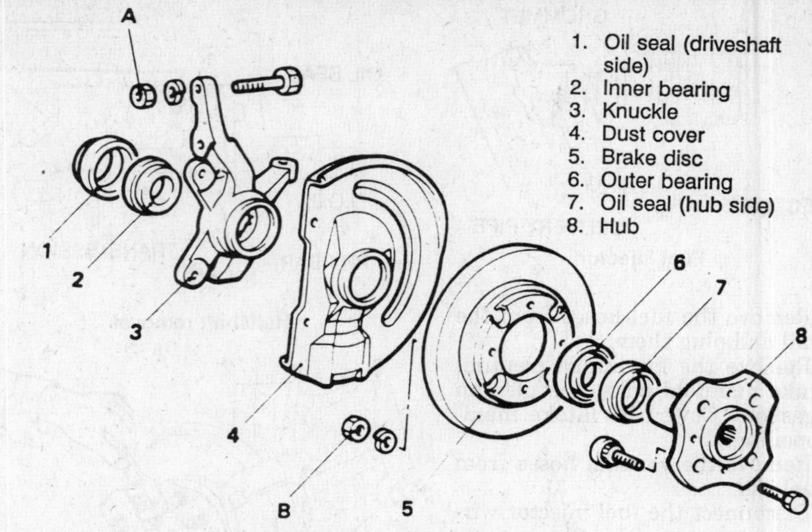

Front hub and knuckle

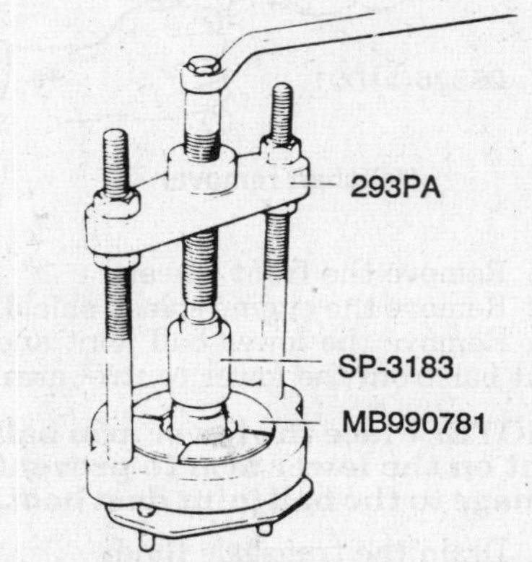

Removing the outer bearing inner race from the hub

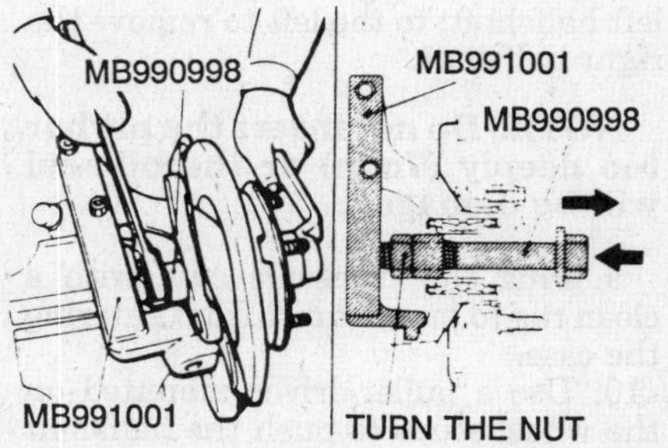

Removing the hub from the knuckle

and tighten the nut manually. Using special tool 09517–21500, separate the hub from the knuckle. Pull the bearings out, noting their positions and direction of installation (smaller diameter inward).

7. Matchmark the relationship between the brake disc and hub. Then place the knuckle in a vise and separate the rotor from the hub by removing the 4 attaching bolts.

8. Special tools 09532–1100, 09532–11301 (pulling ring and pulling collar) and 09517–21100 (stepped plate) or equivalent are needed. Fit the pulling lips of the collets onto the inner race and secure the pulling collar to the collets with the bolts provided. Then, attach the stepped plate to the hub.

9. Attach the pulling ring to the assembly, turning it and moving it up and down so the top of the pulling collar fits into the groove on the ring. Then, use an open-end wrench to keep the special tools from turning, while removing the bolt at the top of the assembly downward with another wrench. This will press the inner race out of the hub. Do this for both inner races.

10. Drive the oil seal and inner bearing inner race from the knuckle with a brass drift.

11. Using a brass drift and a hammer, tap the bearing outer races out of the knuckle.

12. Thoroughly clean and inspect all parts. Any suspect part should be replaced.

To install:

13. Apply multi-purpose grease to the outside surfaces of the bearing outer races to make them easy to press in. Using special tools 09500–21000, 09517–21300 and 09517–21200 or equivalents, install the outer races.

14. Install rotor on the hub and torque the mounting bolts to 36–43 ft. lbs. (49–59 Nm).

15. Drive the outer bearing inner race into position. Coat the outer ring and lip of the oil seal and drive the hub side oil seal into place, using a seal driver.

16. Mount the knuckle in a vise. Place the hub and knuckle together.

Then use 09517–21500 or equivalent, to tighten the hub to the knuckle, torquing to 147–192 ft. lbs. (260 Nm) [167 ft. lbs. (250 Nm) for Scoupe]. Then, rotate the hub to seat the bearing.

17. With the hub still in the vise, check the turning torque of the bearing with an inch lbs. torque wrench and tool 09517-21500. Starting torque should be 11.5 inch lbs. If the starting torque is 0, measure the hub bearing axial play with a dial indicator. If axial play exceeds 0.004 in. (0.1mm), while the nut is tightened, the parts have been assembled incorrectly.

18. Remove the special tool. Place the outer bearing in the hub and drive the seal into place. Lower strut-to-knuckle mounting bolts are torqued to 54–65 ft. lbs. (75–88 Nm). The remaining procedures are the reverse of removal except that the final torquing of the lower arm-to-ball joint connecting bolt should be accomplished after the vehicle is on the ground.

MANUAL TRANSAXLE

For further information on transmissions/transaxles, please refer to "Chilton's Guide to Transmission Repair".

Transaxle Assembly

REMOVAL & INSTALLATION

1. Disconnect the negative battery cable. Remove the battery and battery tray.
2. On 5 speed transaxles, disconnect the electrical connector for the selector control valve.

NOTE: The actuator-to-shaft coupling pin collar is not reusable; replace it.

3. Disconnect and remove the speedometer cable. If equipped with a cable operated clutch, disconnect the clutch cable. If equipped with a hydraulically operated clutch, remove the clevis pin connecting the slave cylinder to the release fork shaft and remove the slave cylinder mounting bolts and the bolts attaching the hydraulic line support bracket to the transaxle. Support the slave cylinder assembly with a length of mechanics wire.
4. Disconnect the back-up lamp electrical connector. Remove the starter motor electrical harness.
5. Remove the 6 transaxle mounting bolts accessible from the top side of the transaxle.
6. Unbolt and remove the starter motor.
7. Raise the vehicle and support it safely. Then, remove the splash shield from under the engine. Drain the transaxle fluid.
8. Disconnect the extension rod and the shift rod at the transaxle end and lower them.
9. Disconnect the stabilizer bar at the lower control arm.
10. Remove the halfshafts.
11. Support the transaxle from below with a floor jack. Make sure the support is widely enough spread that the transaxle pan will not be damaged. Then, remove the attaching bolts and remove the bell housing cover.
12. Remove the lower bolts attaching the transaxle to the engine.
13. Remove the transaxle insulator mount bolt. Remove the cover from inside the right fender shield and remove the transaxle support bracket.
14. Remove the transaxle mount bracket.
15. Pull the assembly away from the engine and lower it from the vehicle.

To install:

16. Installation is the reverse of removal. Observe the following torques:
    - M10–7T engine-to-transaxle bolts—35 ft. lbs. (48 Nm)
    - M8–10T engine-to-transaxle bolts—25 ft. lbs. (34 Nm)
    - M8–20T bell housing cover bolts—15 ft. lbs. (20 Nm)
    - M8–14T bell housing cover bolts—9 ft. lbs. (12 Nm)
    - Transaxle mounting bracket bolt—40 ft. lbs. (54 Nm)
17. Refill the transaxle with the specified fluid to the level of the filler plug.
18. If equipped, adjust the clutch cable. If equipped with a hydraulic clutch, install the slave cylinder mounting bolts and the bolts attaching the hydraulic line support bracket. Install the clevis pin.
19. Make sure the gearshift lever works correctly.

Upper transaxle bolt locations

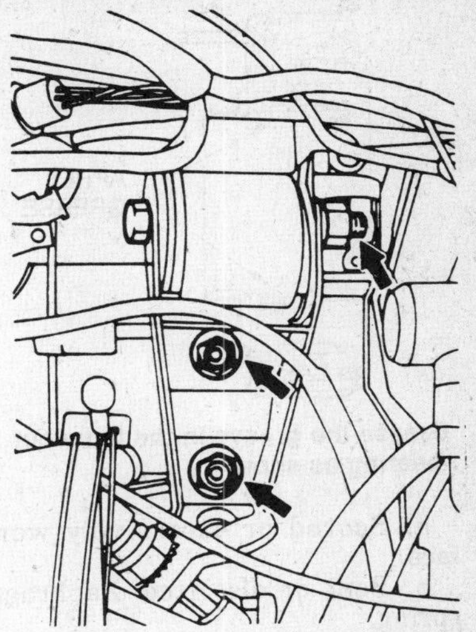

Transaxle mount insulator bolt locations

CLUTCH

Clutch Assembly

REMOVAL & INSTALLATION

1. Remove the transaxle. Insert the forward end of an old transaxle input shaft or a clutch disc guide tool into the splined center of the clutch disc, pressure plate and the pilot bearing in the crankshaft. This will keep the disc from dropping when the pressure plate is removed from the flywheel.
2. Loosen the clutch mounting bolts alternately and diagonally in very small increments, no more than 2 turns at a time, so as to avoid warping the cover flange. When all bolts are free, remove the pressure plate and disc.
3. Remove the snapring. Remove the clevis pin.
4. Remove the return clip and then remove the release bearing.
5. Use a center punch and hammer to remove the spring pins from the clutch release fork and shaft. Discard the spring pins.
6. Remove the release shaft. Remove the release fork, seals and return spring.

If the clutch disc is not being replaced, examine it for the following before re-using it.

a. Loose rivets
b. Burned facing
c. Oil or grease on the facing
d. Less than 0.3mm left between the rivet head and the top of the facing.

Check the pressure plate and replace it, if any of the following conditions exist:

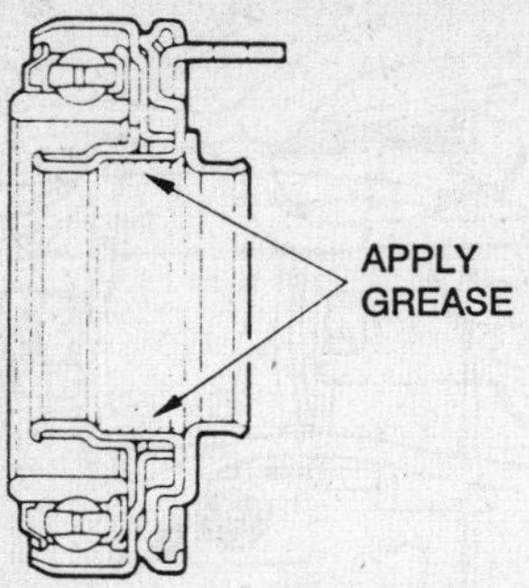

Grease the groove in the throwout bearing as shown

a. Scored or excessively worn face.
b. Bent or distorted diaphragm spring.
c. Loose rivets.

To install:

7. Grease the bearing areas for the release shaft. Install the release shaft, seals, return spring and the release fork. Apply grease to the throw-out bearing contacting surfaces of the release fork.
8. Align the lock pin holes of the release fork and shaft and drive in 2 new spring pins. Make sure the spring pin slot is at right angles to the centerline of the control shaft.
9. Apply grease into the groove in the release bearing and install it into the front bearing retainer in the transaxle. Install the return clip to the release bearing and fork.
10. Make sure the surfaces of the pressure plate and flywheel are wiped clean of grease and lightly sand them with crocus cloth. Lightly grease the clutch disc and transaxle input shaft splines.
11. Locate the clutch disc on the flywheel with the stamped mark facing outward. Use a clutch disc guide or old input shaft to center the disc on the flywheel and then install the pressure plate over it. Install the bolts and tighten them evenly. Tighten them in increments of 2 turns or less to avoid warping the pressure plate. Torque to 11–15 ft. lbs. (15–21 Nm).
12. Remove the clutch disc centering tool. Install the transaxle. Adjust the clutch free-play.

PEDAL HEIGHT ADJUSTMENT

Excel

Measure the pedal height from the top of the pedal pad to the closest point on the floor. The distance should be 7.3–7.6 in. (185–192mm). Loosen the clutch switch locknut and move the pedal stop bolt. Then, tighten the locknut.

Sonata

Measure the pedal height from the face of the pedal to the floorboard. Pedal height should be 6.97–7.17 in. (177–182mm). If not:

1. Disconnect the clutch switch wiring.
2. Loosen the locknut and turn the switch, as required.
3. Tighten the locknut.

Scoupe

Measure the pedal height from the top of the pedal pad to the closest point on the floor. The distance should be 7 in. (178mm). Loosen the clutch switch locknut and move the pedal stop bolt. Then, tighten the locknut.

CLEVIS PIN PLAY ADJUSTMENT

Sonata

Clevis pin play is measure at the pedal while observe the pin. Play should be 0.04–0.12 in. (1–3mm). If not, loosen the locknut and turn the pushrod, as required. Tighten the locknut.

FREE-PLAY ADJUSTMENT

Excel

Slightly pull the cable away from the firewall. Turn the adjusting wheel on the cable until the play between the wheel and the cable retainer is 0.20–0.25 in. (5–6mm). Release the cable and make sure the end of the tension spring engages the adjusting wheel, so the wheel won't turn. Check the clutch pedal free-play. Free-play should be 0.8–1.2 in. (20–30mm). If it is outside specification, adjust it by means of the adjusting wheel on the cable.

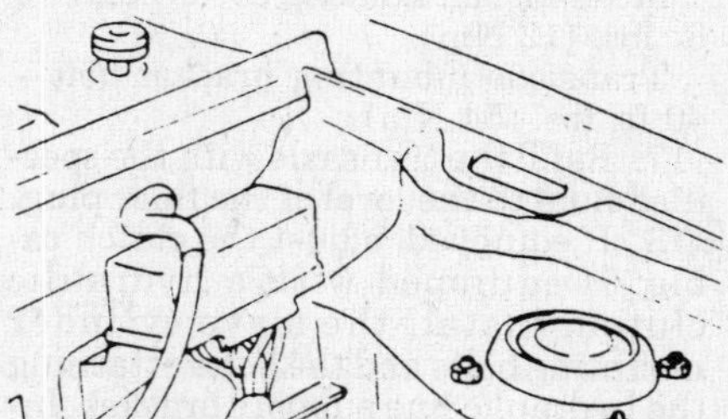

Clutch pedal height adjustment

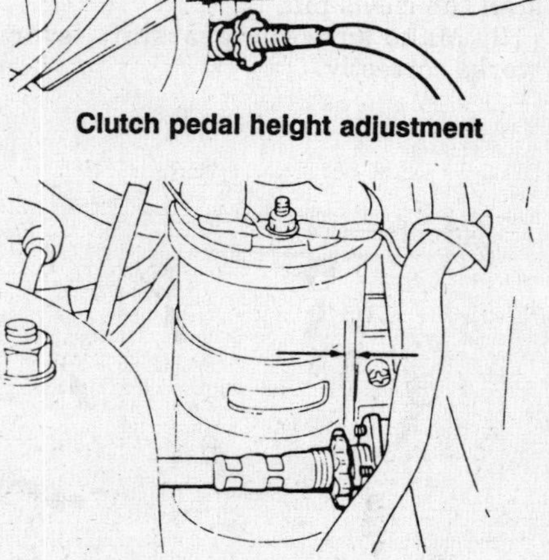

Clutch cable adjustment

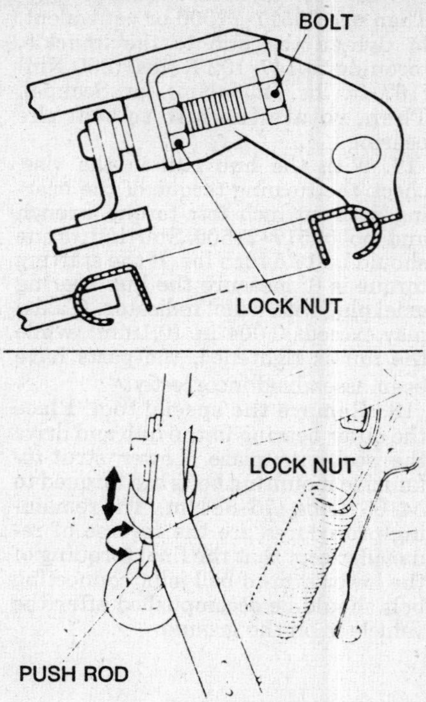

Clutch pedal height and pedal clevis pin adjustment

Sonata and Scoupe

Check this adjustment after checking pedal height and clevis play. Free-play should be 0.24–0.51 in. (6–13mm). If not, there is air in the system which must be bled. If air persists, there is a leak and the system must be repaired.

Clutch Cable

REMOVAL & INSTALLATION

Excel

1. Completely back-off the cable adjusting wheel in the engine compartment.
2. Raise the vehicle and support it safely.
3. Pull out the split pin from the end of the clutch control lever and disconnect the cable from the lever.
4. Disconnect the cable at the clutch pedal.

To install:

5. Installation is the reverse of removal. Make sure the cable does not touch any hot or moving parts. Lubricate the cable with clean engine oil. Adjust the clutch.

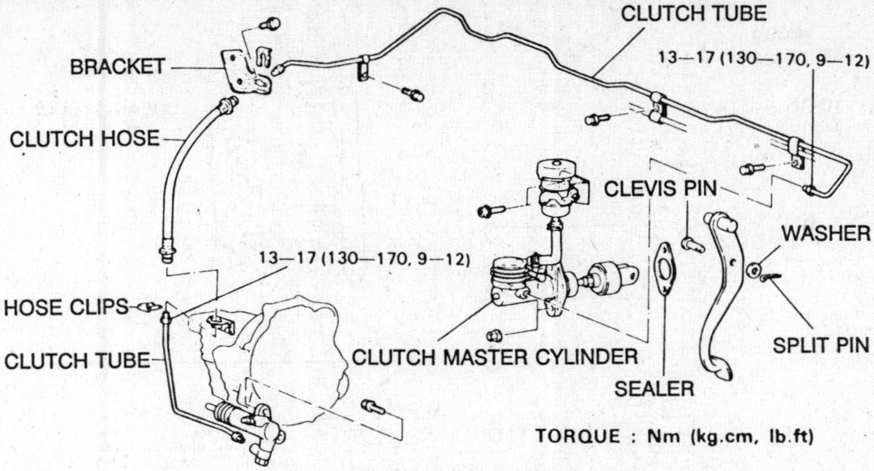

Sonata hydraulic clutch components—Scoupe similar

Clutch Master Cylinder

REMOVAL & INSTALLATION

1. Loosen the bleeder screw on the slave cylinder and drain the system.
2. Disconnect the pushrod from the clutch pedal.
3. Disconnect the clutch pedal from the pedal bracket.
4. Disconnect the fluid line from the master cylinder.
5. Unbolt and remove the master cylinder.
6. Install the master cylinder and bleed the system. Torque the mounting nuts to 7–10 ft. lbs. (9–14 Nm) and the hydraulic fitting to 7–10 ft. lbs. (13–17 Nm).

Clutch Slave Cylinder

REMOVAL & INSTALLATION

1. Disconnect the clutch hose from the slave cylinder.
2. Unbolt and remove the cylinder from the clutch housing.
3. Install the slave cylinder and bleed the system. Torque the bolts to 11–16 ft. lbs. (15–22 Nm) and the hydraulic fitting to 7–10 ft. lbs. (13–17 Nm).

Hydraulic Clutch System Bleeding

1. Raise the vehicle and support it safely.
2. Loosen the bleeder screw at the slave cylinder.
3. Make sure the master cylinder is full.
4. Attach a length of rubber hose to the bleeder screw nipple and place the other end in a glass jar half full of clean brake fluid.
5. Have an assistant push the clutch pedal down slowly to the floor. If air is in the system, bubbles will appear in the jar as the pedal is being depressed.
6. When the pedal is at the floor, have an assistant hold it there. Tighten the bleeder screw.
7. Repeat Steps 5 and 6 until no bubbles are found. Check the master cylinder level frequently to make sure you don't run low on fluid.

AUTOMATIC TRANSAXLE

For further information on transmissions/transaxles, please refer to "Chilton's Guide to Transmission Repair".

Transaxle Assembly

REMOVAL & INSTALLATION

NOTE: The transaxle and torque converter must be removed and installed as an assembly.

1. Disconnect the negative battery cable. Remove the battery and battery tray. Remove the air cleaner and housing.
2. Disconnect the throttle control cable and the manual control cable. Loosen the locknut which uses a star washer and locates the cable housing on the bracket. Also, remove the locknut at the very end of the cable, where it connects with the neutral safety switch.
3. Disconnect the inhibitor switch connector, pulse generator connector, oil cooler hoses, solenoid valve connector and speedometer cable from the transaxle. Immediately install clean caps in the open ends of the hoses. Keep the hoses pointed up so fluid will not escape until the caps are installed.
4. Remove the 5 bolts attaching the converter housing to the engine that are accessible from above.
5. Raise the vehicle and support it safely. Label and then disconnect the starter motor wiring. Then, remove the mounting bolts and the starter.
6. Remove the front wheels. Remove the engine splash shield.
7. Drain the transaxle fluid by removing the drain plug. Remove the transaxle pan bolts and remove the pan and drain it.
8. Remove the undercover. Disconnect the strut bars and stabilizer bar from the lower control arm. Disconnect the lower arm at the crossmember.
9. Disconnect both halfshafts from the transaxle and suspend them in a secure manner.
10. Remove the bell housing cover. Turn the engine for access and remove the bolts connecting the drive plate to the front of the torque converter. Make sure to push the converter as far as it will go toward the transaxle after the bolts have been removed.
11. Support the weight of the engine from above, using a chain hoist. Support the transaxle from underneath with a floor jack in such a way that the support will be spread out and will not dent the transaxle pan. Remove the remaining bolts connecting the transaxle to the engine.

NOTE: Never support the full weight of the transaxle on the engine driveplate.

12. Remove the transaxle mount insulator bolts. Remove the transaxle insulator mount bracket from the transaxle.
13. Slide the transaxle and converter away from the engine, to the right and then lower and remove it as an assembly.

To install:

14. Observe the following torques:

Torque converter-to-driveplate mounting bolts—34–38 ft. lbs. (47–52 Nm) Transaxle assembly mounting bolts—35 ft. lbs. (48 Nm)

Engine-to-transaxle bolts—35 ft. lbs. (48 Nm)

Engine-to-transaxle bolts—25 ft. lbs. (34 Nm)

15. Adjust the throttle control cable and neutral safety switch. Test the neutral safety switch.

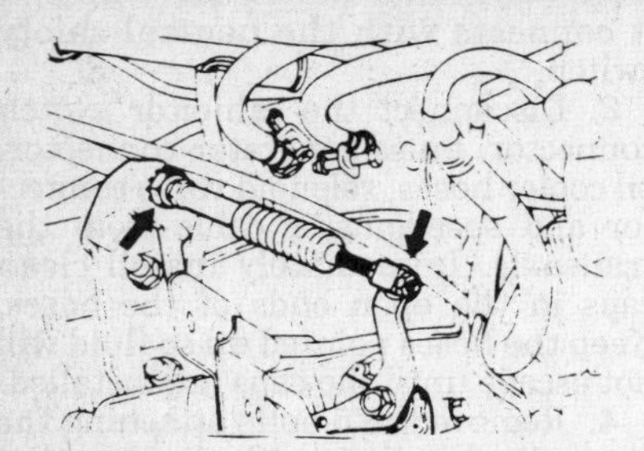

Disconnecting the throttle control cable at the transaxle

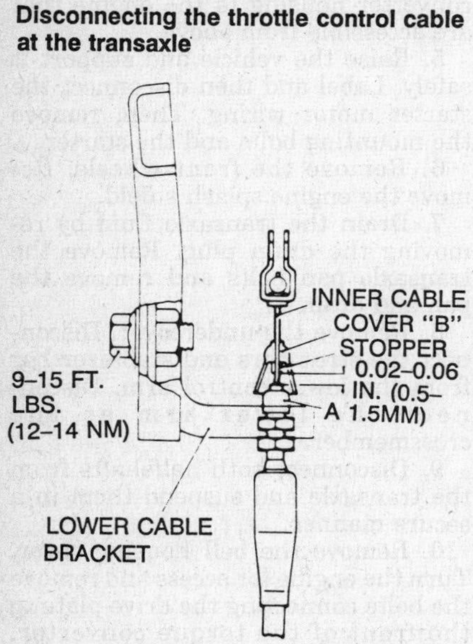

Excel throttle cable adjustment

16. Refill the transaxle to the proper level. Make sure the neutral safety switch wiring does not rub against the insulator mount bracket.

17. Connect the negative battery cable.

18. Start the engine and allow to come to normal operating temperature. Check fluid level in the transaxle.

THROTTLE CONTROL CABLE ADJUSTMENT

Excel

1. Make sure the engine is warm with the throttle in normal idling position.

2. Loosen the lower cable bracket mounting bolt. Pull the small rubber cover located near the transaxle back toward the housing to expose the nipple. Now, move the cable bracket until the distance between the nipple and the outer end of the cover next to the bracket is 0.02–0.06 in. (0.5–1.5mm). Then, torque the bracket mounting bolt to 9–10.5 ft. lbs. (12–14 Nm).

3. With the engine off, open the throttle all the way and hold it there. Then, pull the cable further upward to make sure it still has freedom of movement; that it has not bottomed out. If necessary, repeat the adjustment.

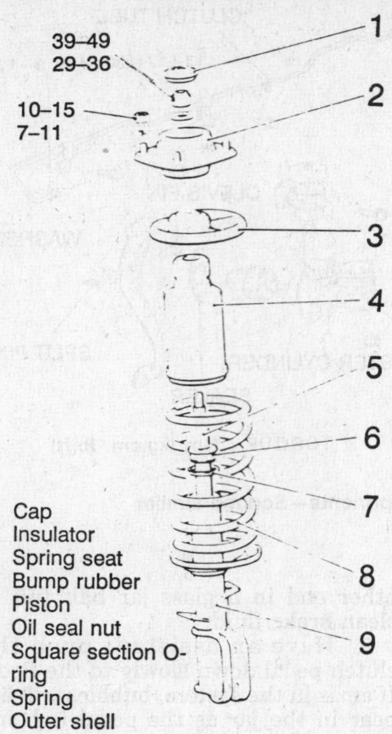

1. Cap
2. Insulator
3. Spring seat
4. Bump rubber
5. Piston
6. Oil seal nut
7. Square section O-ring
8. Spring
9. Outer shell

Excel and Scoupe front strut

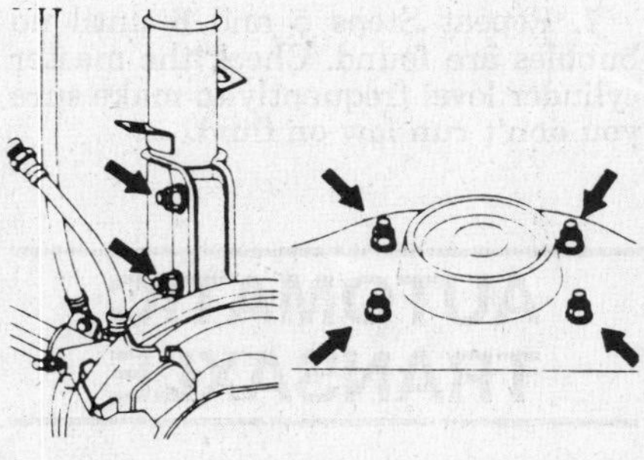

Excel and Scoupe strut attachment points

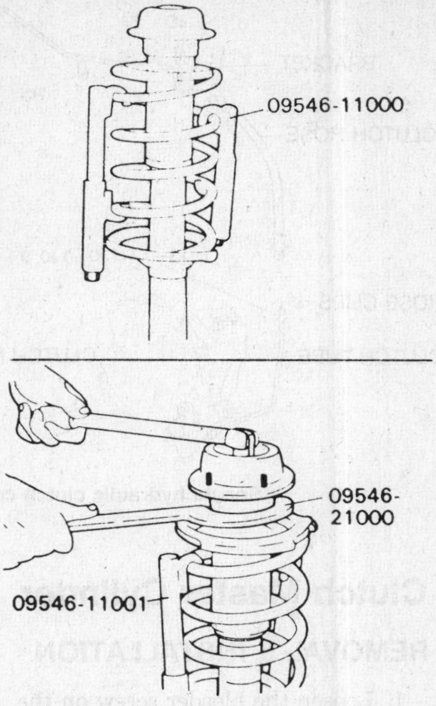

A MacPherson strut spring compressor must be used

FRONT SUSPENSION

MacPherson Strut

REMOVAL & INSTALLATION

1. Raise the vehicle and support it safely. Remove the front wheels. Detach the brake hose bracket at the strut.

2. Remove the nuts securing the strut to the fender well.

3. Unbolt the strut lower end from the knuckle.

4. Remove the strut from the vehicle.

5. To install, reposition the strut and tighten the bolts. Observe the following torques:

Strut-to-knuckle bolts – 55–65 ft. lbs. (75–88 Nm) for Excel.

Strut-to-knuckle bolts – 65–76 ft. lbs. (88–103 Nm) for Sonata.

Strut-to-knuckle bolts – 54–64 ft. lbs. (74–87 Nm) for Scoupe.

Strut-to-fender well nuts – 7–11 ft. lbs. (10–15 Nm) for Excel.

Strut-to-fender well nuts – 18–25 ft. lbs. (25–34 Nm) for Sonata.

Strut-to-fender well nuts – 11–14 ft. lbs. (15–19 Nm) for Scoupe.

Lower Ball Joints

INSPECTION

Raise the vehicle and support it safely. Disconnect the ball joint at the lower end of the strut. Install the nut back onto the ball stud. Then, with an inch lbs., torque wrench, measure the torque required to start the ball joint rotating. The figures are 26–86 inch lbs. If the figures are within specification, the ball joint is satisfactory. If the figure is too high, the joint should be replaced. If the figure is too low. reuse

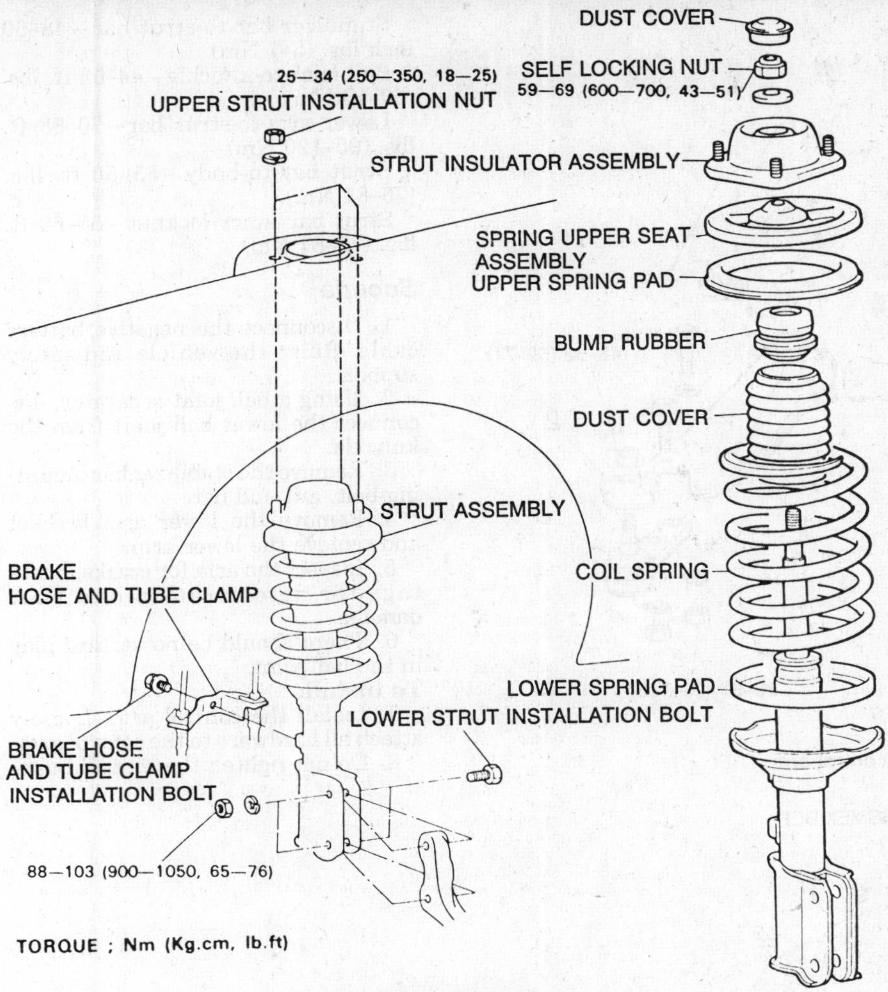

Sonata front strut

the joint, provided its rotation is smooth and even. If there is roughness or play, it must be replaced.

REMOVAL & INSTALLATION

Excel and Scoupe

1. Raise the vehicle and support it safely.
2. Unbolt the ball joint from the control arm.
3. Remove the stud retaining nut.
4. Use a ball joint removing tool and separate the ball joint from the steering knuckle.
5. Replace the ball joint and tighten the ball joint to control arm to 69–87 ft. lbs. (99–118 Nm). Torque the ball joint stud nut 43–52 ft. lbs. (59–71 Nm).

Sonata

1. Raise the vehicle and support it safely. Remove the tire and wheel assemblies.
2. Remove the lower control arm from the vehicle.
3. Remove the ball joint dust cover. Using a special tool, press the ball joint from the control arm.

To install:

4. Apply grease to the lip of the control arm and to the ball joint contact surfaces.
5. Place the ball joint in the control arm. Using a special tool, press the ball joint into the control arm. The ball joint must be pressed evenly into the control arm.
6. Install a new dust cover on the ball joint. Install the control arm assembly into the vehicle. Torque the ball joint-to-steering knuckle retaining nut to 52 ft. lbs. (71 Nm).
7. Install the wheel and tire assembly. Lower the vehicle.

Lower Control Arms

REMOVAL & INSTALLATION

Excel

1. Raise the vehicle and support it safely. Remove the front wheel and tire assembly.
2. Remove the undercover.
3. Disconnect the stabilizer bar from the lower arm. Remove the nut from under the control arm and take off the washer and spacer.
4. Remove the ball joint stud nut and press the tool off with tool MB991113 or equivalent.
5. Remove the bolts which retain the spacer at the rear and the nut and washers on the front of the lower arm shaft (at the front). Slide the arm forward, off the shaft and out of the bushing.
6. Replace the dust cover on the ball joint. The new cover must be greased on the lip and inside with 2 EP multipurpose grease or equivalent, and pressed on, with special tool MB990800 or equivalent, until it is fully seated.

To install:

7. When installing the control arm, the nut on the stabilizer bar bolt must be torqued until the link shows 0.8–0.9 in. (21–23mm) of threads below the bottom of the nut.
8. The washer for the lower arm must be installed as shown. The left side arm shaft has a left hand thread.
9. Torque the all fasteners with the vehicle on the ground. Observe the following torques:

Knuckle-to-strut – 54–65 ft. lbs. (75–88 Nm)

Lower arm shaft-to-body – 69–87 ft. lbs. (94–120 Nm)

Stabilizer bar-to-body – 22–29 ft. lbs. (29–38 Nm)

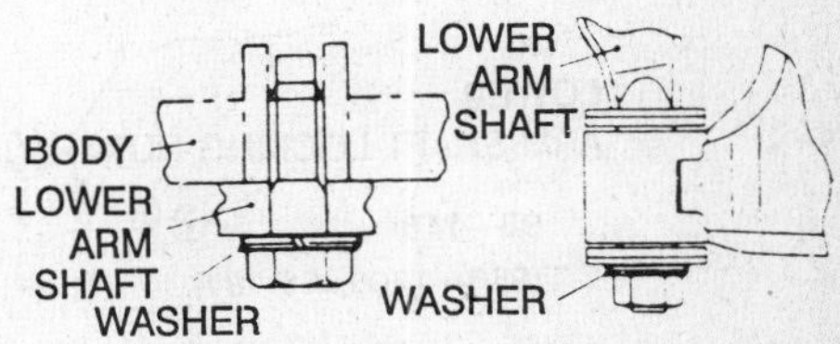

Lower control arm installation on the Excel

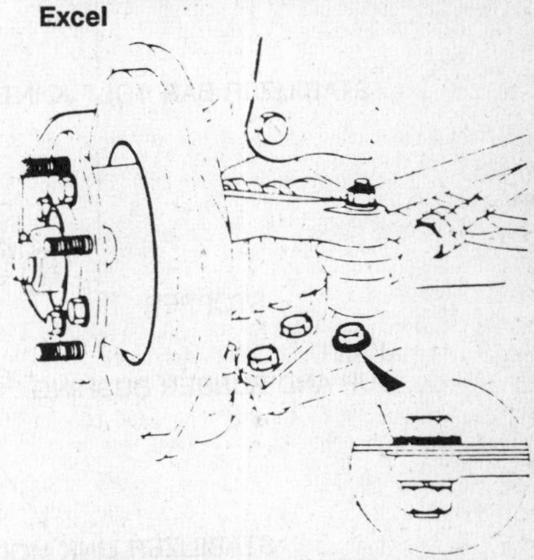

On the Excel, install a flat washer after the ball joint is mounted in the control arm

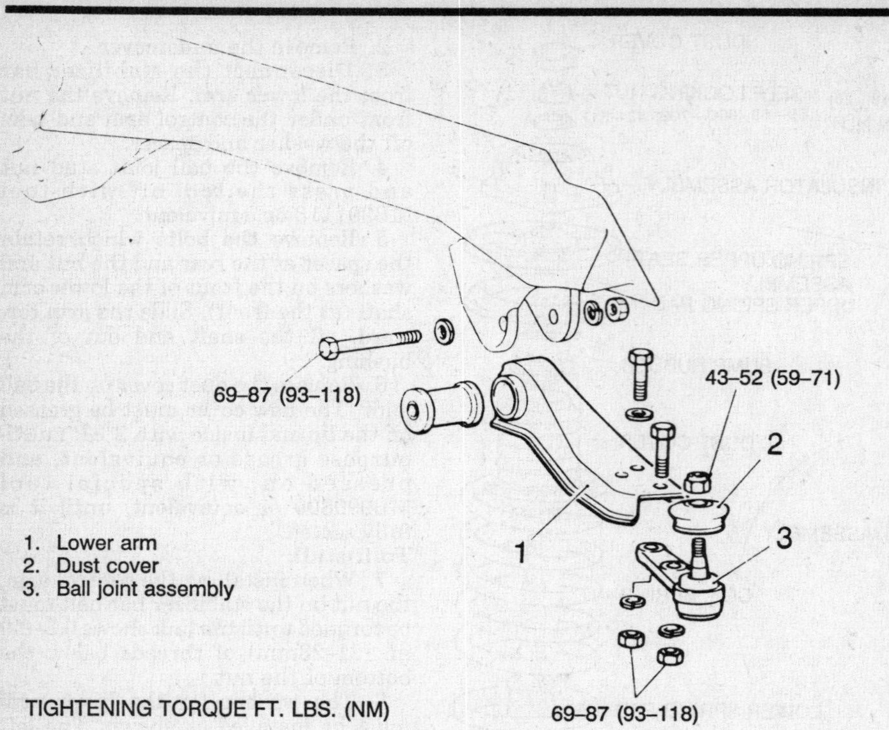

Excel lower control arm

Stabilizer bar-to-strut bar – 48–60 inch lbs. (5–7 Nm)

Ball joint-to-knuckle – 44–53 ft. lbs. (60–73 Nm)

Lower arm-to-strut bar – 70–88 ft. lbs. (95–120 Nm)

Strut bar-to-body – 55–60 ft. lbs. (75–81 Nm)

Strut bar inner locknut – 55–60 ft. lbs. (75–81 Nm)

Scoupe

1. Disconnect the negative battery cable. Raise the vehicle and safely support.
2. Using a ball joint separator, disconnect the lower ball joint from the knuckle.
3. Remove the stabilizer bar mounting bolt, nut and bar.
4. Remove the lower arm bracket and remove the lower arm.
5. Inspect the arm for cracks, bushings for deterioration and boot damage.
6. There should be no vertical play in the ball joint.

To install:

7. Install the control arm. Loosely attach all hardware to the control arm.
8. Do not tighten the control bush-

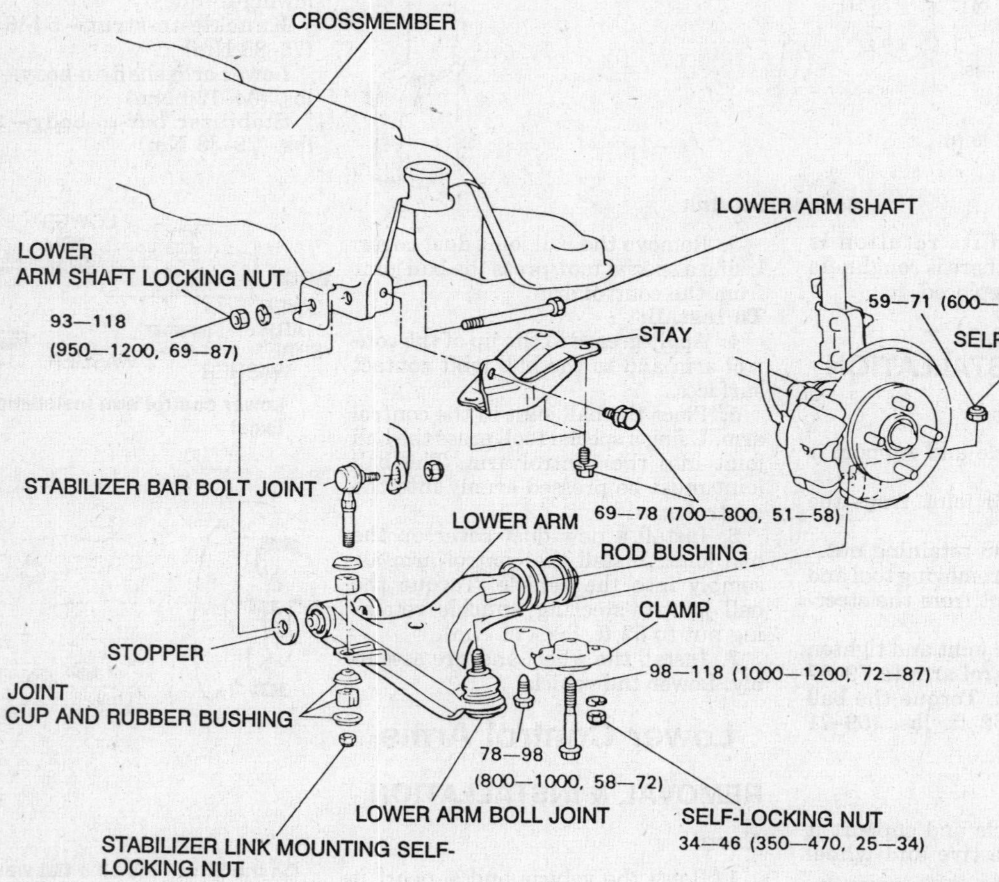

Sonata lower control arm and related components

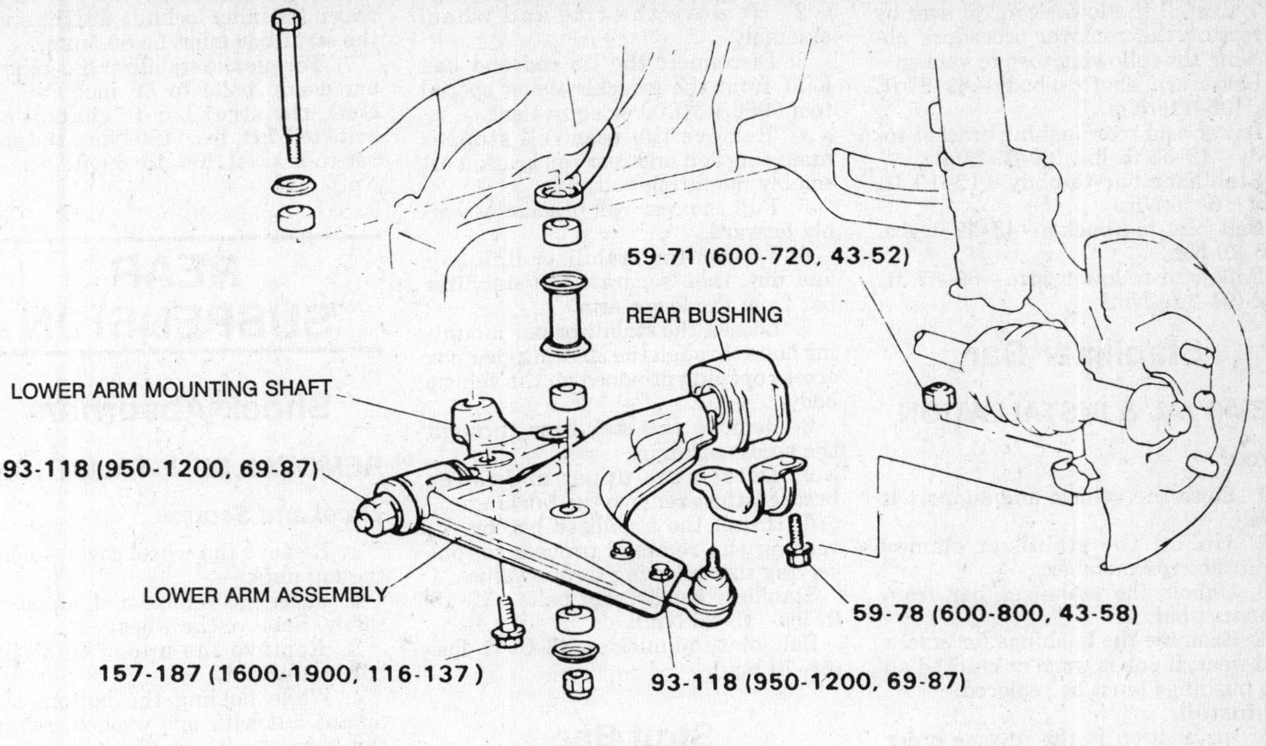

Scoupe lower control unit

ing nuts before the suspension is loaded. Torque the ball joint to 43–52 ft. lbs. (59–71 Nm), lower arm mounting shaft to 116–137 ft. lbs. (157–187 Nm) and the stabilizer shaft nut to 43–58 ft. lbs. (59–78 Nm).

9. Load the suspension and torque the lower bushing nuts to 69–87 ft. lbs. (93–117 Nm).

Sonata

1. Raise the vehicle and support it safely.
2. Loosen the ball joint nut and, using special tool 09568-3100 or equivalent. Disconnect the joint from the knuckle. Be sure to secure the tool's cord to a nearby part.
3. Once the ball joint is free, remove the nut.
4. Remove the stabilizer bar.
5. Remove the control arm-to-crossmber bolt.

To install:

6. Installation is the reverse of removal. Torque the control arm-to-crossmember bolt to 70–87 ft. lbs. (95–110 Nm). Install the stabilizer bar link nut until 5–7mm of threads show beneath the nut. Torque the ball joint stud nut to 50 ft. lbs. (68 Nm).

Scoupe

1. Raise and safely support the vehicle.
2. Remove the tire and wheel assembly.
3. Using special tool 09545–21000 or equivalent, disconnect the lower arm ball joint from the knuckle.
4. Remove the stabilizer bar mounting bolt and nut and detach the stabilizer bar from the lower arm.
5. Detach the lower arm bracket. Remove the lower arm mounting shaft bolts and separate.
6. Remove the lower arm.

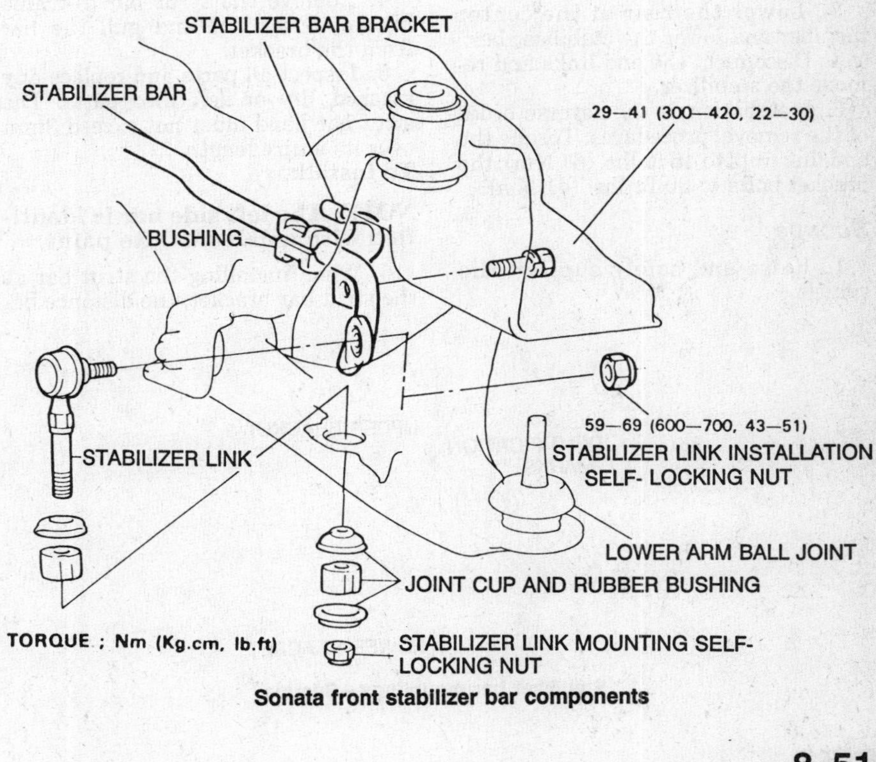

Sonata front stabilizer bar components

7. Install the lower control arm by reversing the removal procedure observing the following torque values:
Lower arm shaft-to-body—69–87 ft. lbs. (95–115 Nm)
Lower arm rear bushing bracket-to-body—43–58 ft. lbs. (60–79 Nm)
Stabilizer bar-to-body—12–19 ft. lbs. (16–26 Nm)
Ball joint-to-knuckle—43–52 ft. lbs. (60–70 Nm)
Ball joint-to-lower arm—69–87 ft. lbs. (94–115 Nm)

Stabilizer Bar

REMOVAL & INSTALLATION

Excel

1. Raise the vehicle and support it safely.
2. Unbolt the stabilizer clamps from the crossmember.
3. Unbolt the stabilizer bar from the strut bar.
4. Examine the bushings for cracks and wear, if one is worn or cracked all the bushings must be replaced.

To install:

5. Installation is the reverse order of the removal procedures. Reposition the stabilizer bar and tighten the chassis clamp bolts to 29 ft. lbs. (39 Nm); the strut bar clamps to 50 inch lbs. (67 Nm).

Sonata

1. Raise the vehicle and support it safely.
2. Remove the stabilizer bar brackets from the crossmember.
3. Lower the rear of the center member and lower the stabilizer bar.
4. Disconnect the end links and remove the stabilizer.
5. Installation is the reverse order of the removal procedures. Torque the end link nuts to 45 ft. lbs. (61 Nm); the bracket bolts to 30 ft. lbs. (41 Nm).

Scoupe

1. Raise and safely support the vehicle.
2. Remove the tire and wheel assembly.
3. Disconnect the tie rod end ball joint from the knuckle using special tool 09568–31000 or equivalent.
4. Remove the rear roll stopper mounting bolt and rear roll bracket assembly mounting bolt.
5. Pull the rear roll bracket assembly forward.
6. Loosen the stabilizer link bolt and nut, then separate the stabilizer bar from the lower arm.
7. Loosen the stabilizer bar mounting bolts through the steering gear box access opening provided on the vehicle body.
8. Remove the stabilizer through the access opening.
9. Detach the upper and lower bracket, then remove the bushing.
10. Install the stabilizer bar by reversing the removal procedure, observing the following torque values:
Stabilizer bar bracket bolts—12–19 ft. lbs. (15–26 Nm).
Ball joint-to-knuckle—43–52 ft. lbs. (58–70 Nm).

Strut Bar

REMOVAL & INSTALLATION

Excel

1. Raise the vehicle and support it safely.
2. Unbolt the stabilizer bar from the strut bar.
3. Remove the bolts securing the strut bar to the control arm.
4. Remove the strut bar-to-frame bracket outer nut and pull the bar from the bracket.
5. Inspect all parts and replace any cracked, dry or deformed parts. The strut bar bend must not exceed 3mm over its entire length.

To install:

NOTE: The left side bar is identified with a dab of white paint.

6. When installing the strut bar at the strut bar bracket, the distance between the inner locknut and the end of the strut bar must be 80.5mm.
7. Torque the stabilizer bar-to-strut bar clamp bolts to 50 inch lbs. (5.5 Nm); the strut bar-to-control arm bolts to 87 ft. lbs. (120 Nm); the strut bar-to-bracket nut to 60 ft. lbs. (81 Nm).

REAR SUSPENSION

Shock Absorbers

REMOVAL & INSTALLATION

Excel and Scoupe

1. Remove the wheel cover. Loosen the lug nuts.
2. Raise the vehicle and support it safely. Remove the wheel.
3. Remove the upper mounting bolt/nut or nut.
4. While holding the bottom stud mount nut with one wrench, remove the locknut with another wrench, or on some vehicles, remove the nut and bolt from the mounting bracket.
5. Remove the shock absorber.
6. Check the shock for:
a. Excessive oil leakage, some minor weeping is permissible;
b. Bent center rod, damaged outer case, or other defects;
c. Pump the shock absorber several times, if it offers even resistance on full strokes it may be considered serviceable.

To install:

7. Install the upper shock mounting nut and bolt. Hand-tighten the nut.
8. Install the bottom eye of the shock over the spring stud or into the mounting bracket and insert the bolt and nut. Tighten the nut to 47–58 ft. lbs. (64–78 Nm).
9. Tighten the upper fasteners to 47–58 ft. lbs. (64–78 Nm).

MacPherson Strut

REMOVAL & INSTALLATION

Sonata

1. Raise the vehicle and support it safely. Allow the lower arms and suspension to hang. Remove the wheels.
2. Place a block of wood on a floor jack and position the jack under the axle beam. Raise the axle slightly to relax the strut and to support the axle when the strut is removed. Position an additional support under the axle.
3. Take care in jacking that no contact is made on the lateral rod.

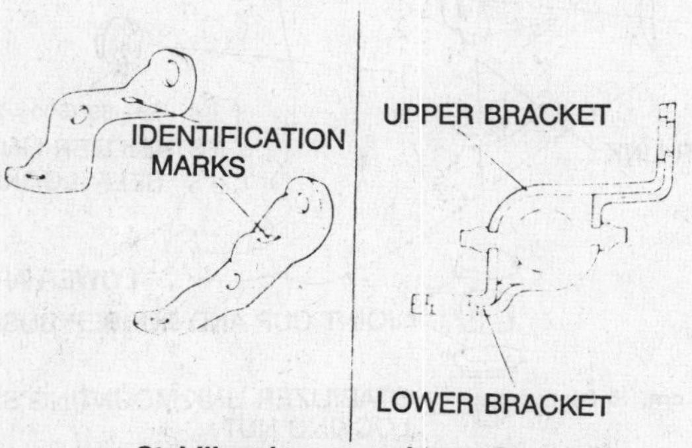

Stabilizer bar mountings—Scoupe

1. Suspension arm (R.H.)
2. Dust cover
3. Clamp
4. Bushing A
5. Bushing B
6. Rubber stopper
7. Suspension arm (L.H.)
8. Rubber bushing (inner)
9. Rubber stopper
10. Fixture
11. Rubber bushing (outer)
12. Washer
13. Stabilizer bar
14. Spring seat
15. Coil spring
16. Shock absorber
17. Bump stopper

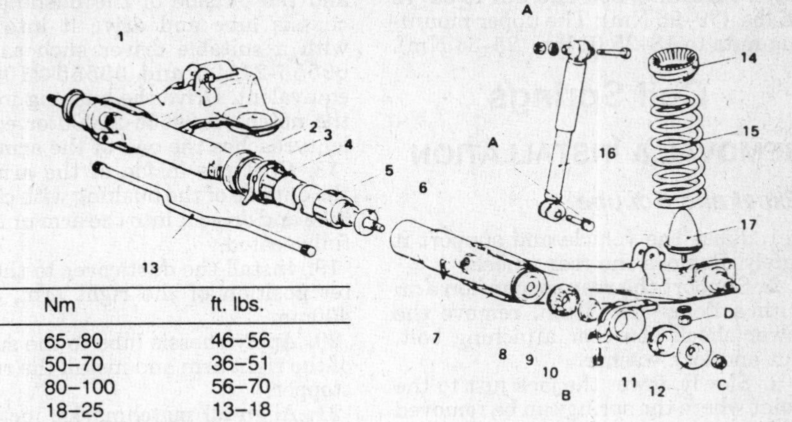

	Nm	ft. lbs.
A	65–80	46–56
B	50–70	36–51
C	80–100	56–70
D	18–25	13–18

Excel and Scoupe rear suspension

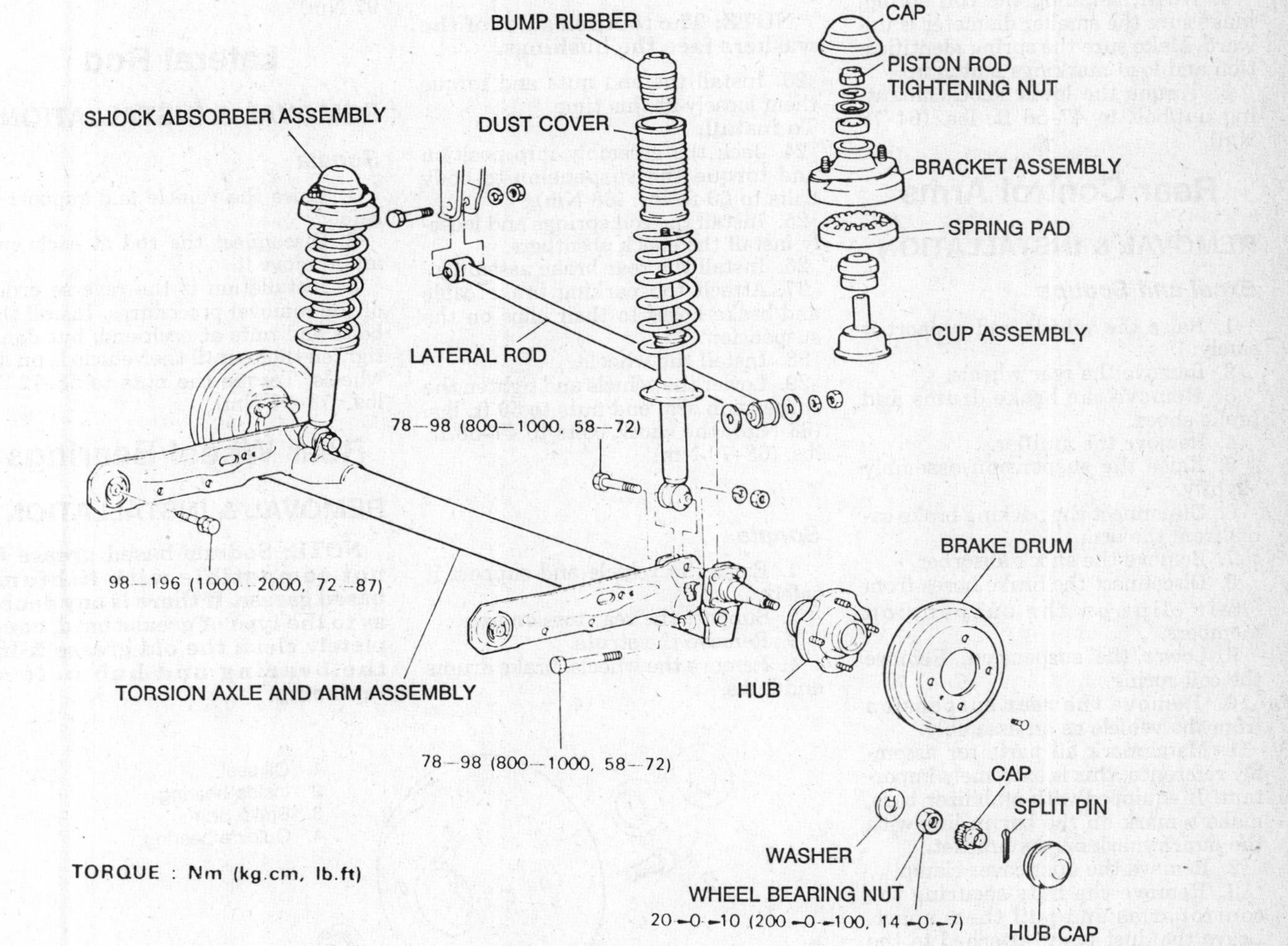

Sonata rear suspension components

4. Remove the upper dust cover cap from the strut assembly.
5. Remove the upper mounting nuts. Remove the lower mounting bolt and nut.
6. Remove the strut.

To install:

7. Installation is the reverse order of the removal procedures. Torque the

lower mounting bolt and nut to 58–72 ft. lbs. (79–96 Nm); The upper mounting nuts to 18–25 ft. lbs. (25–34 Nm).

Coil Springs

REMOVAL & INSTALLATION

Excel and Scoupe

1. Raise the vehicle and support it safely. Remove the rear wheels.
2. Support the rear suspension arm with a floor jack. Then, remove the lower shock absorber attaching bolt, nut and lock washer.
3. Slowly, lower the jack just to the point where the spring can be removed and remove the spring. If the spring is being replaced, transfer the spring seat to the new spring.

To install:

4. When installing the coil spring, make sure the smaller diameter is upward. Make sure the spring identification and load markings match up.
5. Torque the lower shock mounting nut/bolt to 47–58 ft. lbs. (64–78 Nm).

Rear Control Arms

REMOVAL & INSTALLATION

Excel and Scoupe

1. Raise the vehicle and support it safely.
2. Remove the rear wheels.
3. Remove the brake drums and brake shoes.
4. Remove the muffler.
5. Raise the suspension assembly slightly.
6. Disconnect the parking brake cable from the arm.
7. Remove the shock absorber.
8. Disconnect the brake hoses from their clips on the suspension members.
9. Lower the suspension. Remove the coil spring.
10. Remove the rear suspension from the vehicle as an assembly.
11. Matchmark all parts for assembly reference; this is extremely important! If equipped with stabilizer bars, make a mark on the bar in line with the punch mark on the bracket.
12. Remove the dust cover clamp.
13. Remove the nuts securing the control arms and pull them apart. Leave the dust cover attached to the right arm.
14. Remove the rubber stopper from the right arm.
15. Using a flat bladed chisel, drive bushing from right the arm.
16. Using a brass drift, drive bushing out from the left arm.
17. Coat the inside of the left arm and the outside of the bushing with chassis lube and drive it into place with a suitable driver such as tools 09555-21100 and 09555-21000 or equivalent. Drive the bushing in until the notch on 09555-21000 or equivalent, reaches the end of the arm.
18. Coat the inside of the arm and the outside of the bushing with chassis lube and drive it into the arm until it is fully seated.
19. Install the dust cover to the center position of the right arm, about 400mm.
20. Apply chassis lube to the surface of the right arm and install the rubber stopper.
21. Align all matchmarks, including the stabilizer bar and slowly push the suspension halves together.
22. Install all remaining bushing, washers and attaching parts.

NOTE: The toothed sides of the washers face the bushings.

23. Install the end nuts and torque them loosely at this time.

To install:

24. Jack the assembly into position and torque the suspension-to-body bolts to 50 ft. lbs. (68 Nm).
25. Install the coil springs and loosely install the shock absorbers.
26. Install the rear brake assembly.
27. Attach the parking brake cable and brake hoses to their clips on the suspension.
28. Install the wheels.
29. Lower the vehicle and tighten the suspension arm end nuts to 50 ft. lbs. (68 Nm); the shock bolts to 47–58 ft. lbs. (63–79 Nm).

Sonata

1. Raise the vehicle and support it safely.
2. Support the rear torsion axle.
3. Remove the struts.
4. Remove the wheels, brake drums and hubs.
5. Disconnect and cap the brake lines.
6. Disconnect the parking brake cables.
7. Disconnect the lateral rod at the axle.
8. Remove the control arm-to-frame bolts and lower the assembly from the vehicle.

To install:

9. Installation is the reverse order of the removal procedures. Install all fasteners securely but don't tighten them to the final torque until the vehicle is resting on its wheels.
10. Torque the fasteners to the following torque:

Control arm-to-frame bolts – 72–87 ft. lbs. (96–120 Nm)

Lateral rod bolt – 58–72 ft. lbs. (78–97 Nm)

Strut lower bolt – 58–72 ft. lbs. (78–97 Nm)

Lateral Rod

REMOVAL & INSTALLATION

Sonata

1. Raise the vehicle and support it safely.
2. Disconnect the rod at each end and remove it.
3. Installation is the reverse order of the removal procedures. Install the bolts and nuts at each end, but don't tighten them until the vehicle is on its wheels. Torque the nuts to 58–72 ft. lbs. (78–97 Nm).

Rear Wheel Bearings

REMOVAL & INSTALLATION

NOTE: Sodium-based grease is not compatible with lithium-based grease. If there is any doubt as to the type of grease used, completely clean the old grease from the bearing and hub before replacing.

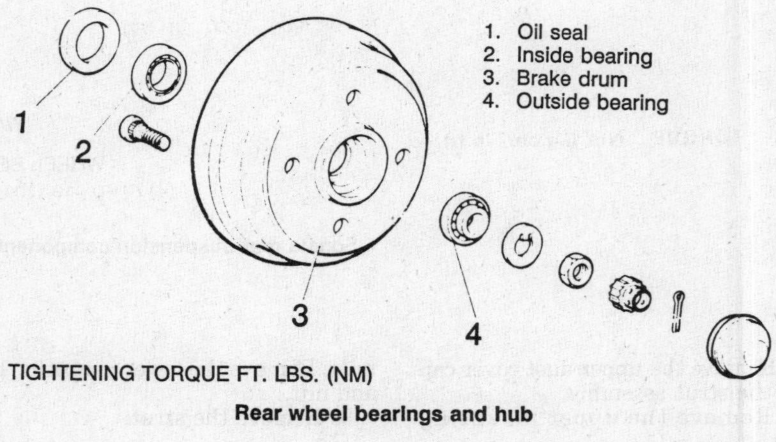

Rear wheel bearings and hub

1. Loosen the lug nuts. Safely raise and support the rear of the vehicle.
2. Remove the grease cap, cotter pin serrated nut cap, axle shaft nut and washer.
3. Pull outward on the brake drum slightly to remove the bearing. Pull the drum off completely.
4. Pry the inner grease seal from the hub and discard it.
5. Remove the inner bearing. If the bearings are being replaced, drive the bearing races from the hub.
6. Clean all old grease from the hub and bearings. If the old bearing are beings reused, clean them in a safe solvent and inspect them thoroughly.
7. If the bearings are being replaced, coat the new races with EP lithium wheel bearing grease and drive them into the hub, making sure they are fully and squarely seated.
8. Pack the hub cavity with new EP lithium wheel bearing grease, until the cavity is full.
9. Pack the bearings completely.
10. Install the inner bearing and drive a new grease seal into place.
11. Install the drum on the spindle and install the outer bearing, washer and shaft nut. Tighten the nut to 15 ft. lbs. (20 Nm) while turning the drum, to seat the bearings. Back off on the nut until it is loose, then torque it to 48 inch lbs.
12. Install the serrated nut cap and a new cotter pin. If the cotter pin holes have to be aligned, back off on the nut no more than 15 degrees; if not, repeat the adjustment procedure.

ADJUSTMENT

Excel and Scoupe

1. Raise the vehicle and support it safely. Remove the rear wheel assembly.
2. Remove dust cover from the hub. Remove excess grease from adjustment nut.
3. Loosen the self locking nut. Torque the nut to 145 ft. lbs. (196 Nm).
4. Fill the dust cap with grease and install.
5. Using a cable and tension gauge tool, check the brake drum turning force. The service limit is 11 inch lbs. before the drum starts to move. If the turning force exceeds the limit, loosen the adjustment nut and retighten it.
6. Reheck the turning force, if it is not within the specifications, replace the wheel bearings.

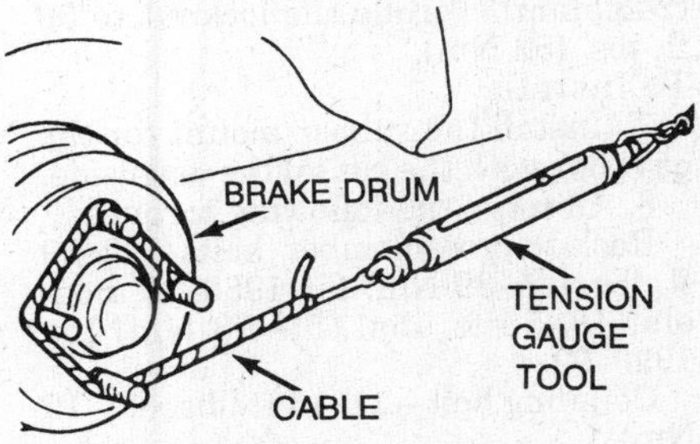

Checking brake drum turning force with a tension gauge tool

Sonata

1. Raise the vehicle and support safely.
2. Remove the rear wheel assembly.
3. Remove the dust cap.
4. Remove the cotter pin, nut lock and nut.
5. Remove the thrust washer and the outer wheel bearing.
6. Remove the drum with the inner wheel bearing and the grease seal.
7. Remove the grease seal and remove the inner bearing.

To install:

8. Lubricate the inner bearing and install to the drum.
9. Install a new grease seal.
10. Install the drum to the vehicle.
11. Lubricate and install the outer wheel bearing to the spindle.
12. Install the thrust washer.
13. Install and tighten the wheel bearing nut to 14 ft. lbs. (20 Nm) while rotating the drum.
14. Back off the adjusting nut ¼ turn and retorque to 7 ft. lbs. (10 Nm).
15. Install the nut lock and a new cotter pin.

Rear Axle Assembly

REMOVAL & INSTALLATION

1. Raise and safely support the vehicle. Remove the wheel assemblies.
2. Separate the parking brake cable at the connector and cable housing at the floor pan bracket.
3. Separate the brake line at backing plate.
4. Remove the muffler-to-middle exhaust pipe retaining bolts, remove the O-ring hangers and remove the exhaust system.
5. Remove the lower shock absorber through bolts and disconnect the shock at the axle end.
6. Lower the axle until the spring and isolator assemblies can be removed.
7. Remove the axle assembly from the vehicle.

To Install:

8. Using floor jacks, position the rear axle assembly under vehicle.
9. Install the springs and isolators and carefully raise the axle assembly.
10. Install the shock absorber and through bolts, do not tighten.
11. Position brake support to the axle while routing the parking brake cable through the support. Lock it into place.
12. Connect the brake line fitting to the backing wheel cylinder. Torque to 9–12 ft. lbs. (13–17 Nm).
13. Install the hub and drum, if removed.
14. Route the parking brake cable through the fingers in the bracket and lock housing end into the floor pan bracket. Install the cable end into the intermediate connector.
15. Install the exhaust system and hangers. Torque the muffler-to-middle exhaust pipe retaining bolts to 22–29 ft. lbs. (30–40 Nm).

STEERING

Steering Wheel

REMOVAL & INSTALLATION

Excel and Scoupe

1. Disconnect the negative battery cable. Pull off the horn cover at the center of the wheel by grasping the upper edge for Excel and prying off at the lower edge for Scoupe. Then, disconnect the horn wire connector.
2. Remove the steering wheel retaining nut. Matchmark the relationship between the wheel and shaft.

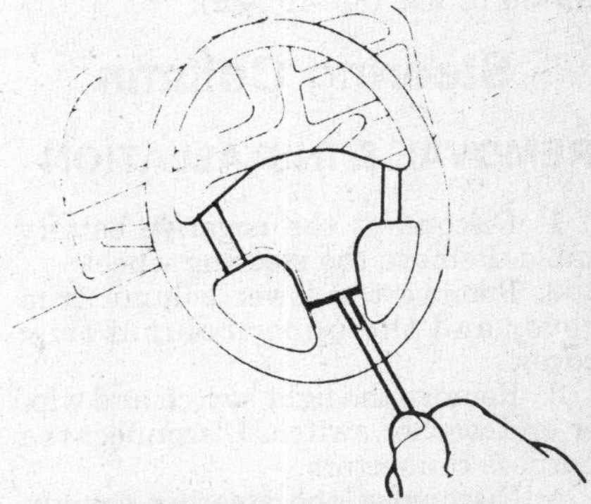
Removing horn pad—Scoupe

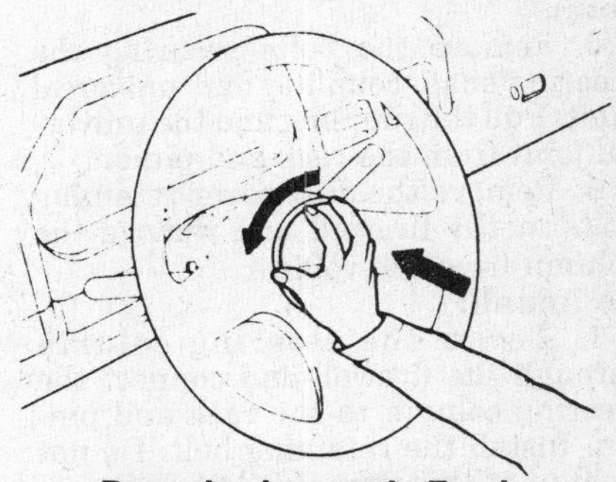
Removing horn pad—Excel

3. Remove the steering wheel dynamic dampener.
4. Screw the 2 bolts of a steering wheel puller into the wheel. Then, turn the bolt at the center of the puller to force the wheel off the steering shaft. Do not pound on the wheel to remove it or the collapsible steering shaft may be damaged.

To install:

5. The steering wheel can be pushed onto the shaft splines by hand far enough to start the retaining nut. Install the retaining nut and torque it to 26–32 ft. lbs. (34–44 Nm).

Sonata

1. Disconnect the negative battery cable.
2. Remove the screws from the back of the horn pad and lift it off.
3. Disconnect the horn wire connector.
4. Pull the dynamic damper forward and off.
5. Remove the steering wheel retaining nut. Matchmark the relationship between the wheel and shaft.
6. Screw the 2 bolts of a steering wheel puller into the wheel. Then, turn the bolt at the center of the puller to force the wheel off the steering shaft. Do not pound on the wheel to remove it or the collapsible steering shaft may be damaged.

To install:

7. The steering wheel can be pushed onto the shaft splines by hand far enough to start the retaining nut. Install the retaining nut and torque it to 29–36 ft. lbs. (39–49 Nm).

Steering Column

REMOVAL & INSTALLATION

1. Disconnect the negative battery cable. Remove the steering wheel.
2. Remove the lower column trim cover and the upper column trim cover.
3. Remove the light switch and wiper and washer switch. Disconnect the harness connectors.
4. Disconnect the steering column mounting bolts and lay the column down.
5. Remove the bolts securing the steering shaft coupling and universal joint. Pull the coupling and the universal joint from the rack and pinion.
6. Remove the dust cover retaining bolts at the firewall and remove the column from the vehicle.

To install:

7. Lower the steering column through the firewall and connect the steering column to the rack and pinion. Install the retaining bolt. Do not tighten at this time.
8. Install both upper and lower steering column mounting bolts. Tighten the steering column mounting bolts as follows:
   a. On Sonata, tighten the upper column to column member mounting bolts to 9–13 ft. lbs. (13–18 Nm) and the lower column to column member mounting bolts to 6–9 ft. lbs. (8–12 Nm).
   b. On 1988–89 Excel, tighten the column and shaft assembly mounting bracket bolts to 7 ft. lbs. (10 Nm) and 1990–92 Excel and Scoupe to 9–13 ft. lbs. (13–18 Nm).
9. Tighten the steering shaft and joint bolt to 11–14 ft. lbs. (15–20 Nm).
10. Install the light switch and wiper and washer washer switch. Connect the harness connectors.
11. Install the upper column trim cover and the lower column trim cover.
12. Install the steering wheel. Connect the negative battery cable.

Manual Rack and Pinion

ADJUSTMENT

1. Mount the rack and pinion assembly in a soft jawed vise, clamping on the rack mounting area, only.
2. Using a spline adapter on an inch-pound torque wrench, turn the pinion shaft at the rate of 1 full turn every 4–6 seconds, turning the steering from lock-to-lock. Measure the total preload lock-to-lock. Preload should be 3.6–9.6 inch lbs.
3. Place a pull scale on each tie rod end, in turn and pull straight away. The rack starting force should be 11–66 lbs.
4. If the specifications in either Steps 2 or 3 are not met, the rubber cushion and yoke spring behind the pinion shaft nut will have to be replaced.

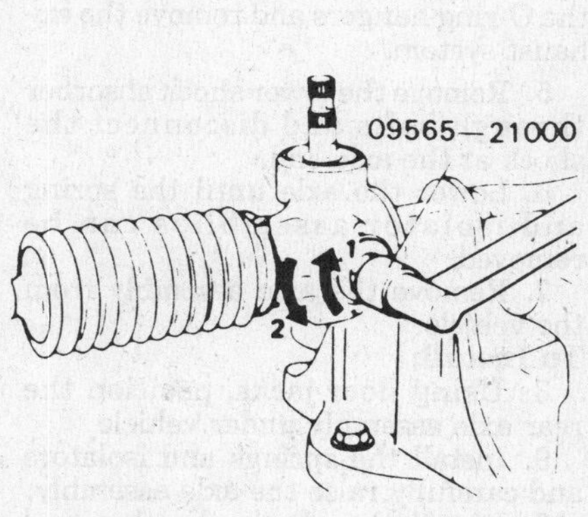

Pinion preload adjustment—Excel and Scoupe with manual steering

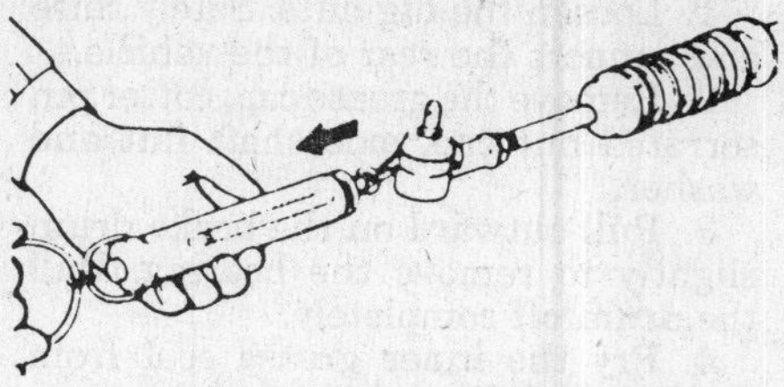
Measuring the pinion preload on the manual steering rack—Excel

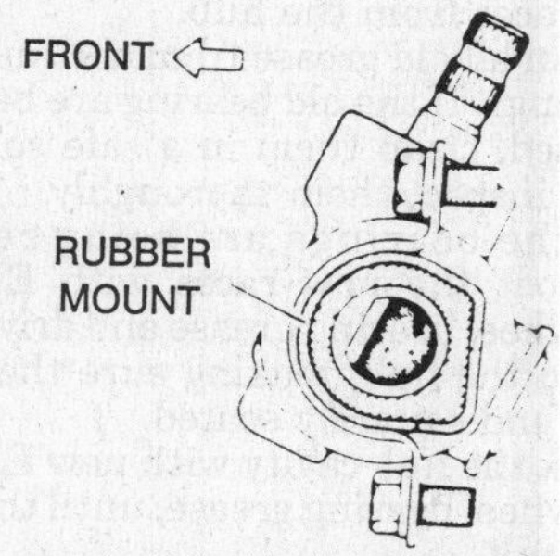

Install the rubber gearbox mount as shown—Excel

REMOVAL & INSTALLATION

Excel and Scoupe

1. Loosen the lug nuts.
2. Raise the vehicle and support it safely.
3. Remove the wheels.
4. Remove the steering shaft-to-pinion coupling bolt.
5. Disconnect the tie rod ends with a separator.
6. Removing the clamps securing the rack to the crossmember and remove the unit from the vehicle. The tie rod ends can now be removed. Prior to removal, count the exact number of exposed threads on the tie rod ends, then loosen the locknut and unscrew the tie rod end. When installing new tie rod ends, oil the threads and screw them into place so the previously noted number of threads is visible with the locknut tight. As a further reference, the distance between the end of the tie rod boot and the centerline of the tie rod ball stud should be 9.6 in. (243.5mm). Torque the locknut to 38 ft. lbs. (52 Nm).

To install:

7. Install the rubber mount for the gear box with the slit on the downside.
8. Observe the following torques:

Rack-to-crossmember bolts—22–29 ft. lbs. (25–39 Nm) for 1988–89 models, 43–58 ft. lbs. (58–78 Nm) for 1990–92

Coupling bolt—11–14 ft. lbs. (15–19 Nm)

Tie rod end slotted nuts—11–25 ft. lbs. (15–34 Nm)

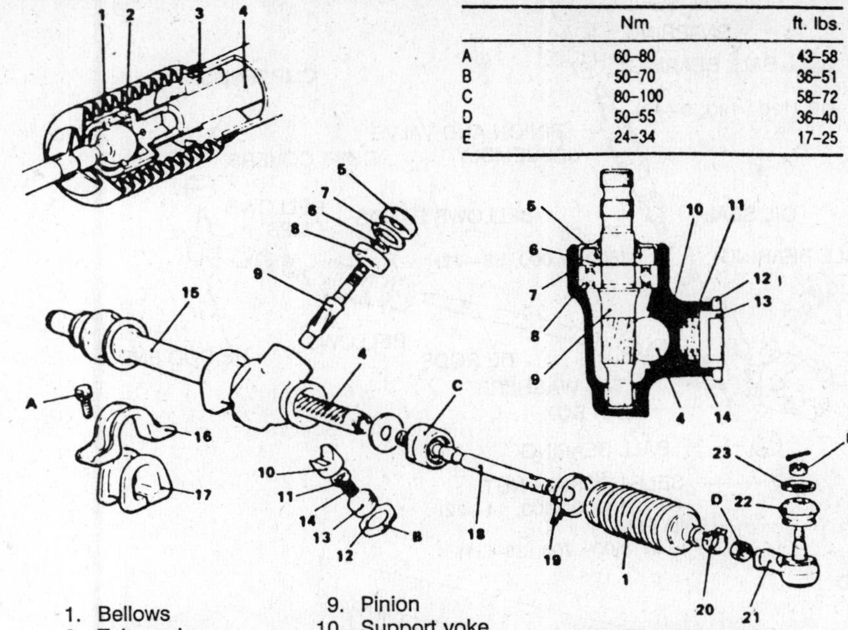

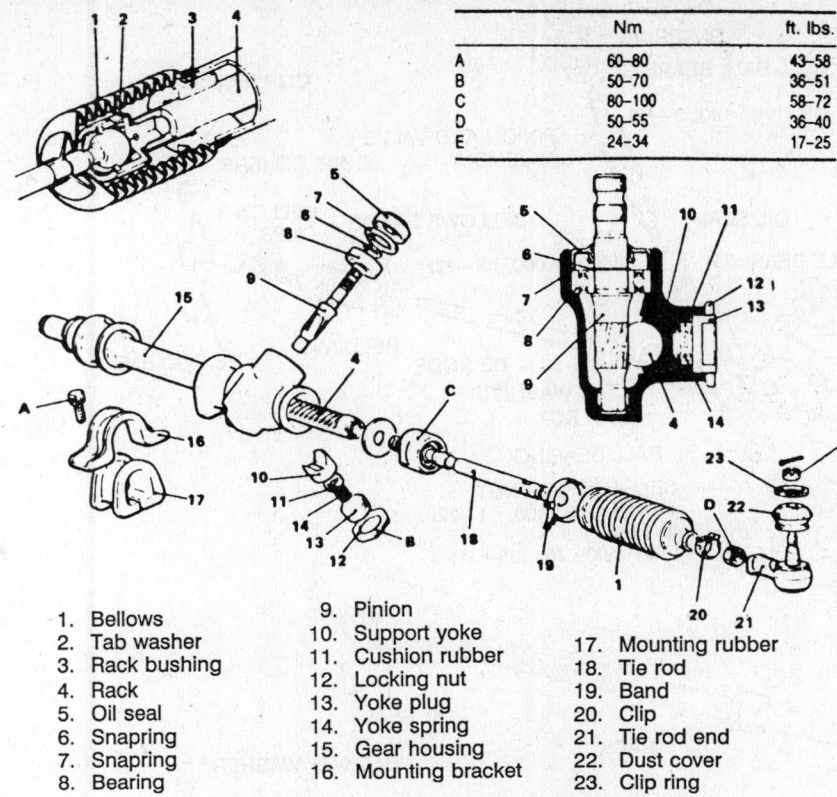

	Nm	ft. lbs.
A	60–80	43–58
B	50–70	36–51
C	80–100	58–72
D	50–55	36–40
E	24–34	17–25

1. Bellows
2. Tab washer
3. Rack bushing
4. Rack
5. Oil seal
6. Snapring
7. Snapring
8. Bearing
9. Pinion
10. Support yoke
11. Cushion rubber
12. Locking nut
13. Yoke plug
14. Yoke spring
15. Gear housing
16. Mounting bracket
17. Mounting rubber
18. Tie rod
19. Band
20. Clip
21. Tie rod end
22. Dust cover
23. Clip ring

Manual rack and pinion steering assembly

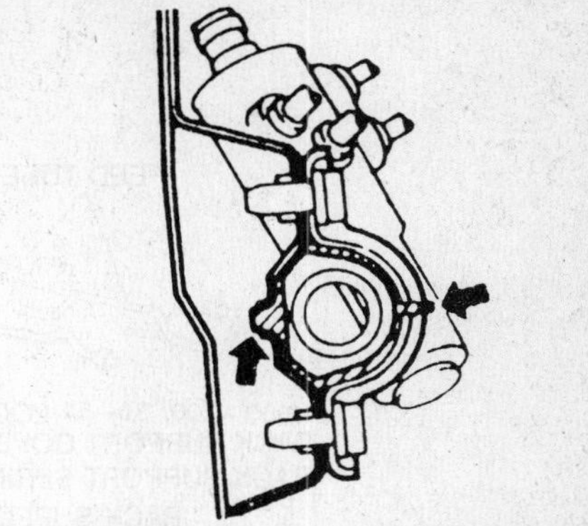

Rubber isolator alignment on the power steering rack—Excel

Power Rack and Pinion

ADJUSTMENT

1. Mount the rack in a soft jawed vise, clamping the vise on the rack mounting areas, only.
2. Using a spline adapter on an inch-pound torque wrench, rotate the pinion shaft several times, lock-to-lock and note the total pinion preload. Preload should be 5–11 inch lbs.
3. If the preload is note within specifications, adjust the position of the rack support cover and recheck the preload. If it does not work, the rack support cover components are defective.

REMOVAL & INSTALLATION

Excel and Scoupe

1. Loosen the lug nuts.
2. Raise the vehicle and support it safely.
3. Remove the wheels.
4. Remove the steering shaft-to-pinion coupling bolt.
5. Disconnect the tie rod ends with a separator.
6. Drain the fluid.
7. Disconnect the hoses from the gear box.
8. Remove the band from the steering joint cover.
9. Unbolt and remove the stabilizer bar.
10. Remove the rear roll stopper-to-center member bolt and move the rear roll stopper forward, as required.
11. Remove the rack unit mounting clamp bolts and take the unit out the left side of the vehicle. The tie rod ends can now be removed. Prior to removal, count the exact number of exposed threads on the tie rod ends, then loosen the locknut and unscrew the tie rod end.

To install:

12. When installing new tie rod ends, oil the threads and screw them into place so the previously noted number of threads is visible with the locknut tight. As a further reference, the distance between the end of the tie rod boot and the point at which the locknut touches the tie rod ball socket body should be 6.1–6.2 in. (155.5–157.5mm); 6.9–7.0 in. (174.3–176.3mm) for Scoupe. Torque the locknut to 38 ft. lbs. (52 Nm).
13. When installing the power steering rack, make sure the rubber isolators have their nubs aligned with the holes in the clamps.
14. Apply rubber cement to the slits in the gear mounting grommet.
15. Tighten the clamp bolt to 43–58 ft. lbs. (58–78 Nm), the tie rod end slotted nuts to 11–25 ft. lbs. (14–34 Nm) and the coupling bolt to 11–14 ft. lbs. (14–19 Nm).
16. Fill the system with Dexron®II ATF.

Sonata

1. Loosen the lug nuts.
2. Raise the vehicle and support it safely.
3. Remove the wheels.
4. Remove the steering shaft-to-pinion coupling bolt.
5. Disconnect the tie rod ends with a separator.
6. Drain the fluid.
7. Disconnect the hoses from the gear box.
8. Remove the center member and temporarily retighten the front muffler.
9. Unbolt and remove the stabilizer bar.
10. Remove the rack unit mounting clamp bolts and take the unit out the right side of the vehicle. The tie rod ends can now be removed. Prior to removal, count the exact number of exposed threads on the tie rod ends, then loosen the locknut and unscrew the tie rod end.

To install:

11. When installing new tie rod ends, oil the threads and screw them into place so the previously noted number of threads is visible with the locknut tight. As a further reference, the distance between the end of the tie rod boot and the point at which the locknut touches the tie rod ball socket body should be 187.4mm. Torque the locknut to 38 ft. lbs. (52 Nm).
12. When installing the power steering rack, make sure the rubber

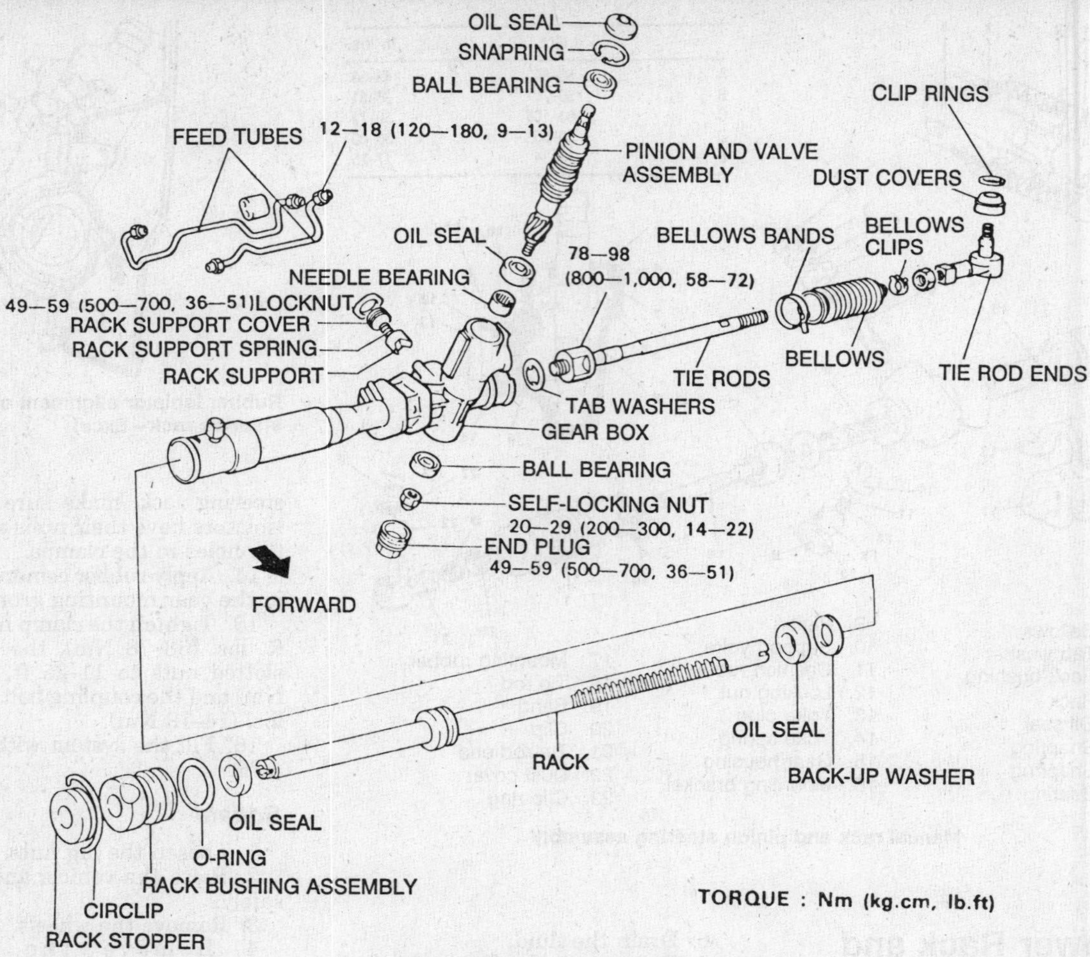

Sonata steering components

isolators have their nubs aligned with the holes in the clamps.

13. Apply rubber cement to the slits in the gear mounting grommet. Tighten the clamp bolt to 43–58 ft. lbs. (58–78 Nm), the tie rod nuts to 11–25 ft. lbs. (14–34 Nm) and the coupling bolt to 22–25 ft. lbs. (29–34 Nm).

14. Fill the system with Dexron®II ATF.

Power Steering Pump

REMOVAL & INSTALLATION

1. Place a drain pan under the pump.
2. Disconnect the pressure hose from the pump.
3. Disconnect the return hose from the pump.
4. Loosen the pump mounting bolts and remove the drive belt.
5. Remove the pump-to-mounting bracket bolts and lift out the pump.

To install:

6. Install the power steering pump. When installing the return line, push it at least 1.2 in. (30mm) onto the return tube. Fill the system with Dexron®II ATF, start the engine and turn the steering lock-to-lock several times to bleed any trapped air.

BELT ADJUSTMENT

1. Press the V-belt by applying pressure of 22 lbs. (98 N) at the center of the belt.
2. Deflection of the belt should be 0.28–0.39 in. (7–10mm).
3. To adjust the tension of the belt, loosen the oil pump mounting bolts, move the oil pump and then retighten the bolts.

SYSTEM BLEEDING

1. Ensure that the reservoir is full of Dexron II® automatic transmission fluid.
2. Raise and safely support the front wheels of the vehicle.
3. Turn the steering wheel from lock to lock 5 or 6 times.
4. Disconnect the coil wire and connect to a solid ground. Operate the starter motor intermittently for 15 to 20 seconds and turn the steering wheel from lock to lock 5 or 6 times.

NOTE: Ensure that the reservoir is full during air bleeding to prevent the fluid level from falling below the lower position of the filter.

5. Connect the coil wire and start the engine.
6. Turn the steering wheel from lock to lock until no more air bubbles are visible in the reservoir.
7. Confirm that the oil is not milky and that the fluid level is correct.
8. Confirm that there is little change in the fluid level when the steering wheel is turned to the left and right.

NOTE: An abrupt rise in the fluid level after stopping the engine is a sign of incomplete bleed-

ing. If this occurs, repeat the bleeding procedure.

Tie Rod Ends

REMOVAL & INSTALLATION

1. Raise the vehicle and support it safely. Remove the front wheels.
2. Remove the cotter pin and then remove the ball stud retaining nut. Use a vise-like tool MB991113 or equivalent, to press the ball stud down and out of the steering knuckle.
3. Using a back-up wrench on the flats at the inner end of the tie rod end, loosen the nut that retains the end to the tie rod coming out of the steering box. Now, unscrew the tie rod end, counting the turns required to remove it.

To install:

4. Install the new tie rod end in reverse order. Torque the castellated nut retaining the ball stud to 11–25 ft. lbs. (15–33 Nm). Then, turn it just far enough to align the castellations with the hole in the stud and install a new cotter pin. Torque the inner nut to 36–40 ft. lbs. (49–54 Nm).

BRAKES

For all brake system repair and service procedures not detailed below, please refer to "Brakes" in the Unit Repair section.

Master Cylinder

REMOVAL & INSTALLATION

1. Disconnect the negative battery cable. Disconnect the fluid level sensor.
2. Disconnect the brake tubes from the master cylinder and cap them immediately.

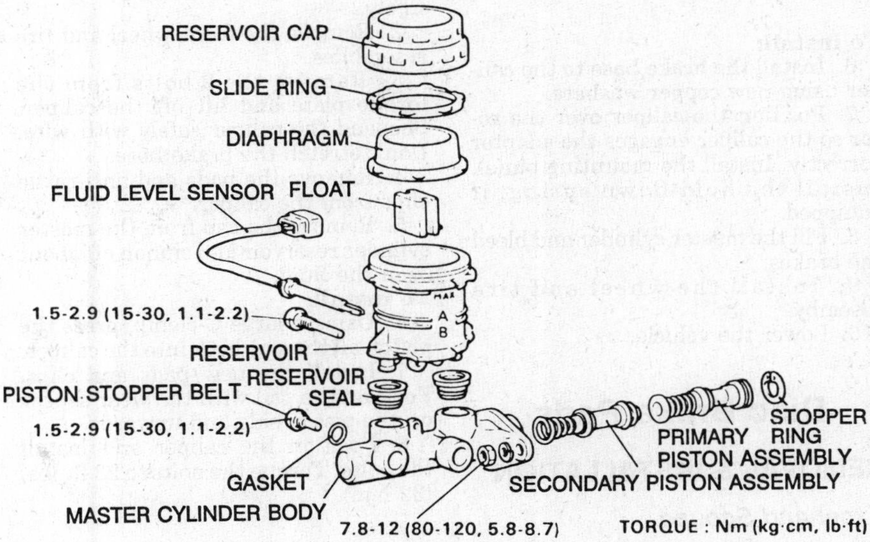

Excel master cylinder

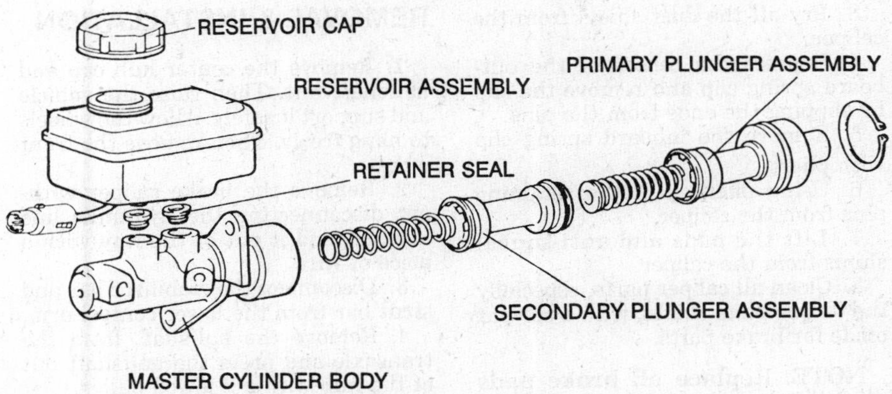

Sonata master cylinder

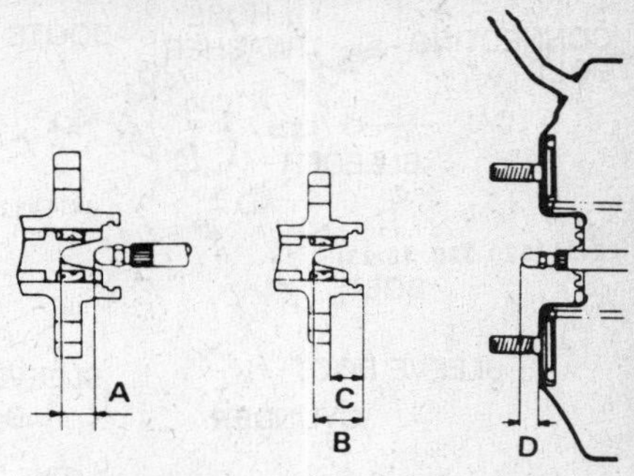

Determining brake booster pushrod clearance

3. Unbolt and remove the master cylinder from the booster.

To install:

4. Position the master cylinder to the booster and install the mounting bolts. Torque the mounting bolts to 6–9 ft. lbs. (8–12 Nm); 10–16 ft. lbs. (14–22 Nm) for Scoupe.
5. Connect the brake tubes to the master cylinder. Torque the tubes to 9–12 ft. lbs. (13–17 Nm).
6. Bleed the brake system. Connect the negative battery cable. Road test the vehicle.

Proportioning Valve

On Excel and Sonata, the proportioning valve is located under the master cylinder and supported by the brake lines to which it is connected and a through bolt. On Scoupe and Sonata V6 the proportioning valves are threaded into the master cylinder where the hydraulic fittings connect. It does not require routine check or adjustment.

REMOVAL & INSTALLATION

1. Disconnect the negative battery cable. Disconnect the brake lines at the valve.

NOTE: Use a flare nut wrench to avoid damage to the lines and fittings.

2. Remove the mounting bolts and remove the valve.

NOTE: If the proportioning valve is found to be defective, it must be replaced.

3. Install the proportioning valve and tighten the mounting bolts to 15 ft. lbs. (20 Nm).
4. Refill the system with fluid and bleed the brakes. Connect the negative battery cable.

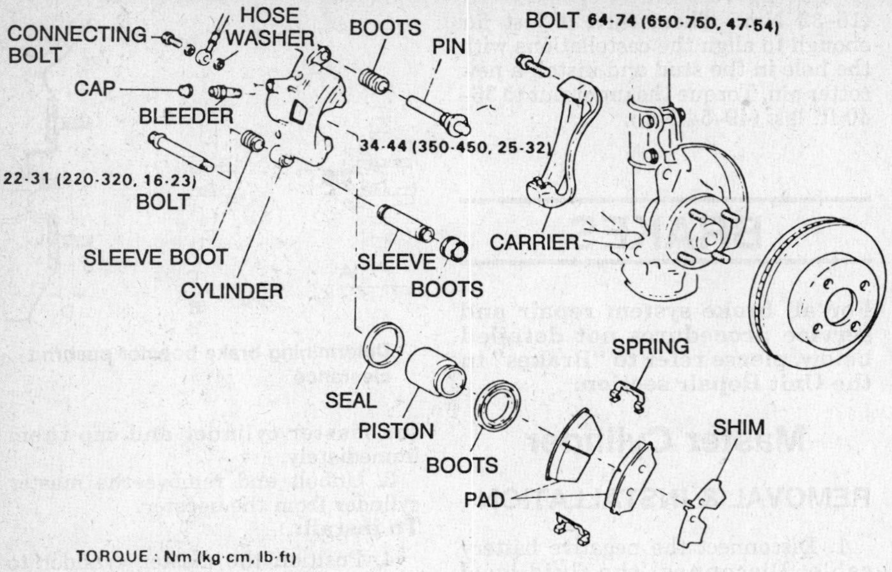

Excel front disc brake components

Power Brake Booster

REMOVAL & INSTALLATION

1. Disconnect the negative battery cable. Slide back the clip and disconnect the vacuum supply line at the brake booster. Pull gently in order to avoid damaging the check valve.
2. Remove the master cylinder.
3. Disconnect the pushrod at the brake pedal. This requires pulling the lockpin out of the pedal clevis pin and then pulling the latter out of the pedal lever and clevis rod.
4. Remove the mounting bolts and nuts from the firewall and remove the booster.

To install:

5. Installation is the reverse order of the removal procedures. Install the brake booster on the firewall and tighten the mounting nuts to 72–108 inch lbs. (8–12 Nm); 10–12 ft. lbs. (13–17 Nm) for Scoupe. Bleed the system.

Brake Caliper

REMOVAL & INSTALLATION

1. Raise the vehicle and support safely.
2. Remove the tire and wheel assembly.
3. Remove the caliper mounting pin(s).
4. Lift the caliper off of the rotor. Remove the outer pad from the caliper.
5. Remove the brake hose retaining bolt from the caliper.

To install:

6. Install the brake hose to the caliper using new copper washers.
7. Position the caliper over the rotor so the caliper engages the adaptor correctly. Install the mounting pin(s). Install the hold-down spring, if equipped.
8. Fill the master cylinder and bleed the brakes.
9. Install the wheel and tire assemby.
10. Lower the vehicle.

Disc Brake Pads

REMOVAL & INSTALLATION

Excel and Scoupe

1. Raise the vehicle and support it safely.
2. Remove the front wheels.
3. Pry off the dust shield from the caliper.
4. Depress the center of the outboard spring clip and remove the clip by slipping the ends from the pins.
5. Remove the inboard spring clip with pliers.
6. Using pliers, pull the retaining pins from the caliper.
7. Lift the pads and anti-squeal shims from the caliper.
8. Clean all caliper parts, especially the torque plate shafts, with a solvent made for brake parts.

NOTE: Replace all brake pads at the same time. Never replace the pads on 1 wheel only.

9. If the dust protector or spring clips are weak, damaged or deformed, replace them.
10. Remove the cap from the master cylinder reservoir and, using a clean suction gun, remove about 6mm of fluid.

To install:

11. Using a C-clamp, force the caliper piston back into the caliper as far as it will go.
12. Install the inboard pad and anti-squeal shim.
13. Install the outboard pad and anti-squeal shim.
14. Install the pins.
15. Install the spring clips.
16. Install the dust shield.
17. Install the wheels and lower the vehicle. Depress the brake pedal a few times. The first couple of strokes on the pedal will feel overly long. However, the pads will set themselves and the stroke will return to normal.

Sonata

1. Raise the vehicle and support it safely.
2. Remove the front wheel and tire assemblies.
3. Remove the 2 bolts from the torque plate and lift off the caliper. Suspend the caliper safely with wire. Don't stretch the brake hose.
4. Remove the pads and anti-rattle clips from the caliper.
5. Remove the cap from the master cylinder reservoir and siphon off about ⅓ of the fluid.

To install:

6. Using a large C-clamp, press the piston all the way back into the caliper.
7. Install the new pads and clips. Position the pad with the wear sensor on the piston side and upwards.
8. Position the caliper and install the bolts. Torque the bolts to 23 ft. lbs. (32 Nm).

Brake Rotor

REMOVAL & INSTALLATION

1. Remove the center hub cap and halfshaft nut. Then raise the vehicle and support it safely. Allow the wheels to hang freely. Then remove the front wheel.
2. Remove the brake caliper without disconnecting the hydraulic line and suspend it out of the way with a piece of wire.
3. Disconnect the stabilizer bar and strut bar from the lower control arm.
4. Remove the halfshaft from the transaxle and press the halfshaft out of the hub with a 2 jawed puller.
5. Unbolt and remove the hub and knuckle from the bottom of the strut

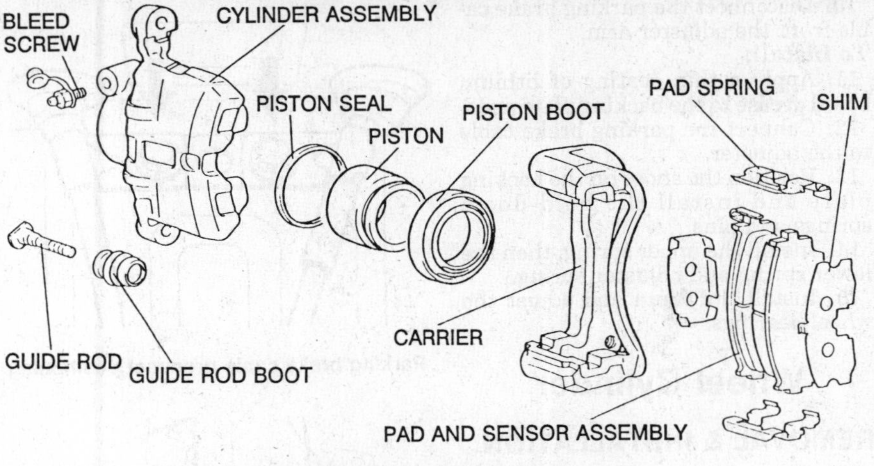

Sonata front disc brake components

and remove the hub and knuckle assembly from the vehicle.

6. Several special tools are required to press the hub and disc from the steering knuckle and to remount them. Use 09517–21600 or equivalent. Do not attempt to hammer the parts apart, or the bearing will be damaged. Install the arm of the special tool then the body onto the knuckle and tighten the nut manually. Using special tool 09517–21500, separate the hub from the knuckle. Pull the bearings out, noting their positions and direction of installation (smaller diameter inward).

7. Matchmark the relationship between the brake disc and hub. Then place the knuckle in a vise and separate the rotor from the hub by removing the attaching bolts.

To install:

8. Install the hub to the rotor and attaching bolts.

9. Install the hub and knuckle to the bottom of the strut.

10. Install the halfshaft to the transaxle.

11. Connect the stabilizer bar and strut bar to the lower control arm.

12. Install the brake caliper.

13. Install the halfshaft nut and center hub cap and halfshaft nut.

14. Install the front wheel and lower the vehicle.

15. Torque the hub nut to 188 ft. lbs. (255 Nm).

Brake Drums

REMOVAL & INSTALLATION

1. Raise the vehicle and support safely.

2. Remove the wheel and tire assembly.

3. Remove the dust cap.

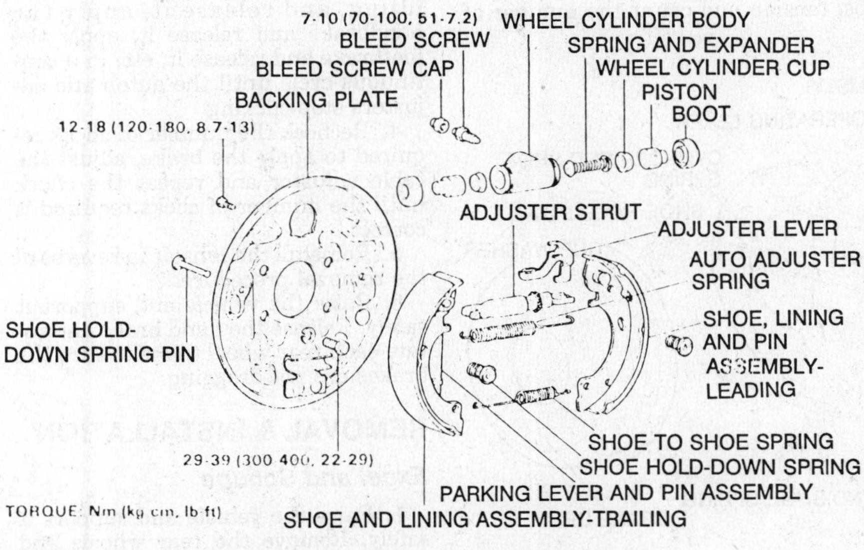

Excel and Scoupe rear brake components

4. Remove the cotter pin and nut lock.

5. Remove the wheel bearing nut and washer from the spindle.

6. Remove the outer wheel bearing.

7. Remove the drum with the inner wheel bearing from the spindle. If the drum is difficult to remove, remove the plug from the rear of the backing plate and push the self adjuster lever away from the star wheel. Rotate the star wheel to retract the shoes. Remove the grease seal.

To install:

8. Lubricate and install the inner wheel bearing. Install a new grease seal.

9. Install the drum to the spindle.

10. Lubricate and install the outer wheel bearing, washer and nut. When the bearing preload is properly set, install the nut lock and a new cotter pin.

11. Install the grease cap.

12. Install the wheel and tire assembly. Adjust the rear brakes as required.

Brake Shoes

REMOVAL & INSTALLATION

Excel and Scoupe

1. Raise the vehicle and support it safely.

2. Remove the rear wheels.

3. Pry off the hub grease cap.

4. Remove the cotter pin, lock cap and bearing adjusting nut.

5. Remove the outer bearing and pull the hub and drum assembly from the spindle.

6. Thoroughly clean the spindle.

7. Remove the lower pressed metal spring clip, the shoe return spring, the large 1 piece spring between the 2 shoes, and the shoe hold-down springs.

8. Remove the shoes and adjuster as an assembly.

9. Disconnect the parking brake cable from the lever.

10. Remove the spring between the shoes and the lever from the rear (trailing) shoe.

11. Disconnect the adjuster retaining spring and remove the adjuster, turn the star wheel in to the adjuster body after cleaning and lubricating the threads.

To install:

12. Clean the backing plate with solvent made for cleaning brakes.

13. Lubricate all contact points on the backing plate, anchor plate, wheel cylinder to shoe contact and parking brake strut joints and contacts with lithium based grease.

14. Installation of the brake shoes, from this point, is the reverse of re-

moval after the lever has been transferred to the new rear (trailing) shoe.

15. Pre-adjustment of the brake shoe can be made by turning the adjuster star wheel out until the drum will just slide on over the brake shoes. Before installing the drum make sure the parking brake is not adjusted too tightly, if it is, loosen it, or the adjustment of the rear brakes will not be correct.

16. Position the hub assembly on the spindle.

17. Insert the packed outer bearing, retaining washer and adjusting nut.

18. While rotating the hub, tighten the adjusting nut to 15 ft. lbs. (20 Nm), then back it off until loose.

19. Torque the nut again, this time to 48 inch lbs.

20. Install the lock cap and turn the nut counterclockwise just enough to align the cap and cotter pin hole. Insert a new cotter pin. Never back the nut off more than 15 degrees to align the cotter pin hole. If more is necessary, repeat the entire adjustment procedure.

21. The brakes shoes are adjusted by pumping the brake pedal and applying and releasing the parking brake. Adjust the parking brake stroke. Road test the vehicle.

Sonata

1. Raise the vehicle and support it safely.
2. Remove the wheels.
3. Remove the hub nut, outer bearing and brake drum.
4. Thoroughly clean the spindle.
5. Clean the brake shoes and backing plate with a commercially available solvent.
6. Remove the lower spring.
7. Remove the upper spring.
8. Remove the hold-down springs.
9. Remove the shoes and adjuster.
10. Disconnect the parking brake cable from the adjuster arm.

To install:

11. Apply a thin coating of lithium based grease to the backing plate pads.
12. Connect the parking brake cable to the adjuster.
13. Position the shoes on the backing plate and install the hold-down springs and pins.
14. Install the upper spring, then the lower spring and adjuster spring.
15. Install the drum and adjust the wheel bearings.

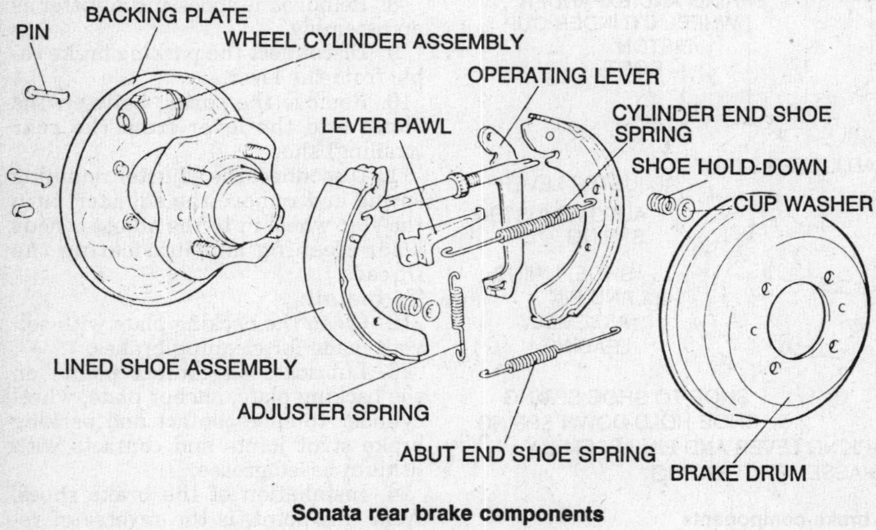

Sonata rear brake components

Wheel Cylinder

REMOVAL & INSTALLATION

1. Raise the vehicle and support it safely. Remove the tire and wheel assembly.
2. Remove the brake drums and shoes.
3. Disconnect the brake line where it connects to the wheel cylinder behind the brake backing plate and plug the open end of the brake line.
4. Remove the bolts that fasten the wheel cylinder to the backing plate from behind it and remove the wheel cylinder.
5. Install the wheel cylinder and tighten the mounting bolts for the wheel cylinder to 72–108 inch lbs. (8–12 Nm); 9–13 ft. lbs. (12–18 Nm) for Scoupe. Bleed the system. Make sure the self-adjusters have taken up play so the brakes actuate normally before operating the vehicle.

Parking Brake Cable

ADJUSTMENT

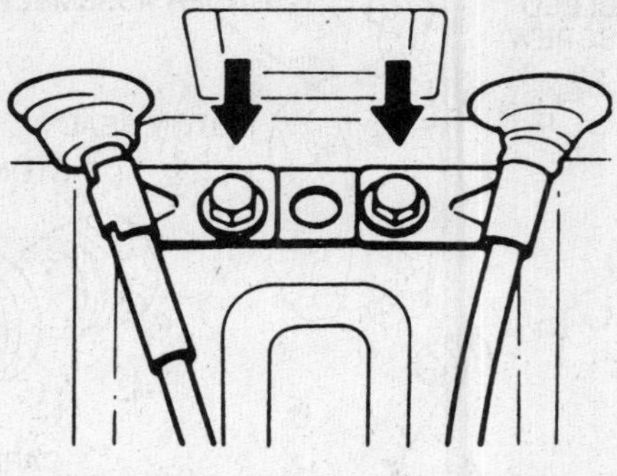

Parking brake cable grommet positioning

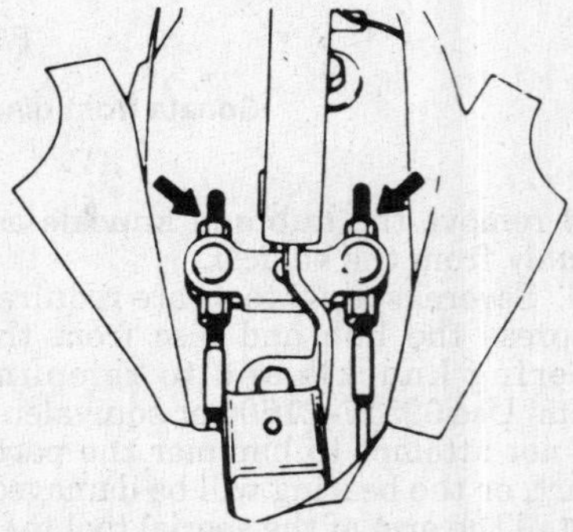

Parking brake cable adjusting nuts

1. Apply the brake with about 45 lbs. tension and count the number of clicks required. 5–7 clicks should be required on the Excel; 8–9 clicks on the Sonata; 6–7 clicks on Scoupe. If the number of clicks is incorrect, proceed with the remaining steps.
2. Remove the rear console box. Remove the parking brake cover and the ashtray. Then, remove the console mounting screws and remove the console.
3. Release the brake and then adjust the cable adjusting nuts until all cable slack is just removed. Then, apply the footbrake, the engine should be idling, and release it, apply the handbrake and release it, apply the footbrake and release it, etc. in a continuous cycle until the automatic adjusters stop clicking.
4. Recheck the number of clicks required to apply the brake, adjust the cable adjuster and repeat the check until the number of clicks required is correct.
5. Reinstall the console in reverse of the removal procedure.
6. Raise the vehicle and support it safely. Release the hand brake and rotate each rear wheel to make sure the brakes are not dragging.

REMOVAL & INSTALLATION

Excel and Scoupe

1. Raise the vehicle and support it safely. Remove the rear wheels and brake drums.

2. Remove the console box and rear seat.

3. Release the hand brake and then disconnect the cable connectors at the equalizer. It may be necessary to loosen the cable adjusting nuts to do this.

4. Disconnect all cable clamps from the body. Remove the mounting bolts for the large mounting clamp located just forward of where the cables pass through the body grommets.

5. Pull the cables and grommets out of the body.

6. Disconnect the cables at the rear brakes.

To install:

7. When installing the cables, make sure the grommets are installed in the body completely and that the concave side faces to the rear. Adjust the hand brake mechanism. Adjust the switch for the indicator light so the light comes on when the lever is pulled 1 notch.

Sonata

1. Raise the vehicle and support it safely.

2. Remove the console.

3. Remove the cable adjuster, pin, equalizer and nut holder.

4. Remove the parking brake switch.

5. Remove the parking brake lever.

6. Remove the rear seat cushion and roll back the carpet.

7. Remove the brake cable clamp and grommet.

8. Remove the brake drums and shoes.

9. Disconnect the brake cables from the adjusting arms.

10. Remove the retaining clips and push the cable out of the backing plates.

To install:

11. Connect the brake cables to the adjusting arms.

12. Install the brake shoes and drums.

13. Install the rear seat cushion and roll back the carpet.

14. Install the parking brake lever.

15. Install the parking brake switch.

16. Install the cable adjuster, pin, equalizer and nut holder.

17. Install the console.

18. Lower the vehicle and test the brakes.

Brake System Bleeding

1. Fill the master cylinder with fresh brake fluid. Check the level often during the procedure.

2. Starting with the right rear wheel, remove the protective cap from the bleeder, if equipped, and place where it will not be lost. Clean the bleed screw.

CAUTION

When bleeding the brakes, keep face away from the brake area. Spewing fluid may cause facial and/or visual injury. Do not allow brake fluid to spill on the vehicle's finish; it will remove the paint.

3. If the system is empty, the most efficient way to get fluid down to the wheel is to loosen the bleeder about ½–¾ turn, place a finger firmly over the bleeder and have a helper pump the brakes slowly until fluid comes out the bleeder. Once fluid is at the bleeder, close it before the pedal is released inside the vehicle.

NOTE: If the pedal is pumped rapidly, the fluid will churn and create small air bubbles, which are almost impossible to remove from the system. These air bubbles will eventually congregate and a spongy pedal will result.

4. Once fluid has been pumped to the caliper or wheel cylinder, open the bleed screw again, have an assistant press the brake pedal to the floor, lock the bleeder and have an assistant slowly release the pedal. Wait 15 seconds and repeat the procedure, including the 15 second wait, until no more air comes out of the bleeder upon application of the brake pedal. Remember to close the bleeder before the pedal is released inside the vehicle each time the bleeder is opened. If not, air will be induced into the system.

5. If a helper is not available, connect a small hose to the bleeder, place the end in a container of brake fluid and proceed to pump the pedal from inside the vehicle until no more air comes out the bleeder. The hose will prevent air from entering the system.

6. Repeat the procedure on remaining wheel cylinders in order:

a. Left rear
b. Right front
c. Left front

7. Hydraulic brake systems must be totally flushed, if the fluid becomes contaminated with water, dirt or other corrosive chemicals. To flush, bleed the entire system until all fluid has been replaced with the correct type of new fluid.

8. Install the bleeder cap(s) on the bleeder to keep dirt out. Always road test the vehicle.

CHASSIS ELECTRICAL

Heater Blower Motor

REMOVAL & INSTALLATION

NOTE: On Excel and Sonata, in order to remove either the blower or core, the heater case must be removed.

1988-89 Excel

1. Disconnect the negative battery cable.

2. Place the control in the **HOT** position.

3. Drain the cooling system.

4. Remove the heater hoses from the core tubes.

5. Remove the lower instrument panel section.

6. Remove the center console and on-board computer.

7. Loosen the heater duct mounting screw. Then, pushing downward and pulling, remove the heater ducts.

8. Disconnect the heater control cable.

9. Disconnect the wiring at the motor.

10. Remove the heater case mounting bolts and remove the heater case from under the dash.

11. Separate the case halves and remove the blower.

12. Installation is the reverse order of the removal procedures. Adjust the control cable and refill the cooling system.

Sonata

CAUTION

When discharging the refrigerant from the system, extreme care should be observed. Always wear protective eye wear, freon coming in contact with the eyes can cause permanent blindness. Freon coming in contact with the skin can cause frost bite. Always cover fittings with a clean towel and open the fitting slowly.

1. Disconnect the negative battery cable.

2. Place the control in the **HOT** position.

3. Drain the cooling system.

4. Remove the heater hoses from the core tubes.

5. Discharge the air conditioning system.

6. Disconnect the suction and liquid refrigerant lines at the firewall connectors. Always use back-up wrenches. Cap all openings at once.

7. Remove the front and rear center consoles.

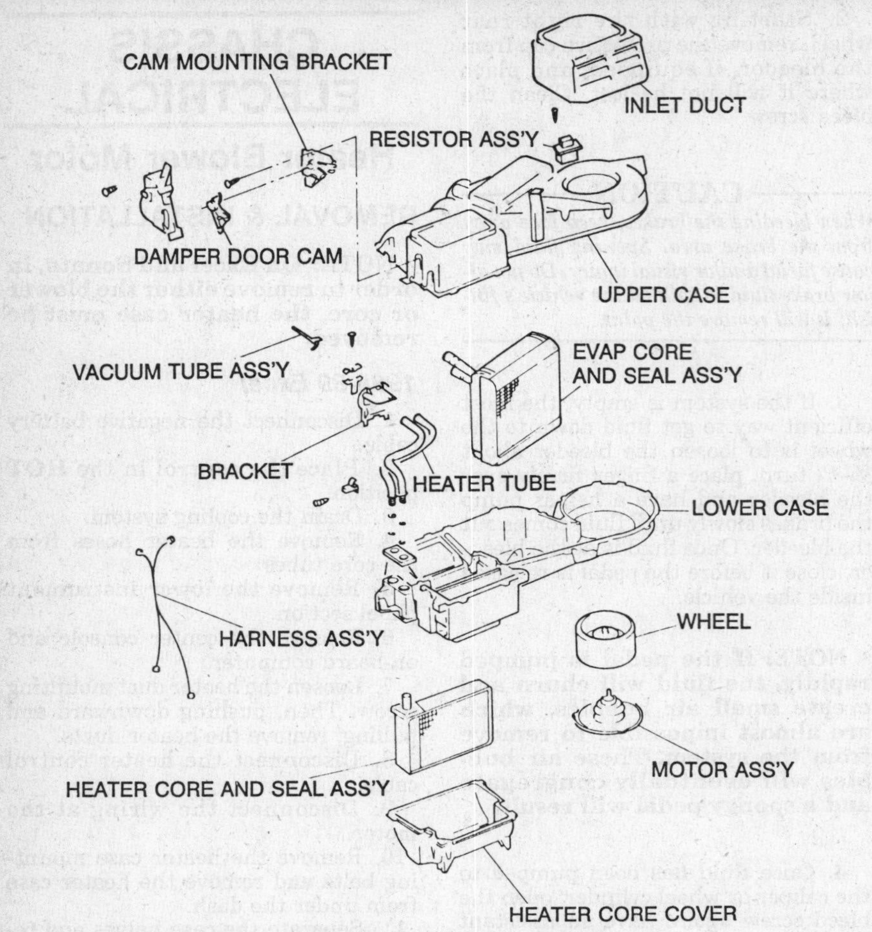

Heater-A/C components for USA Sonatas

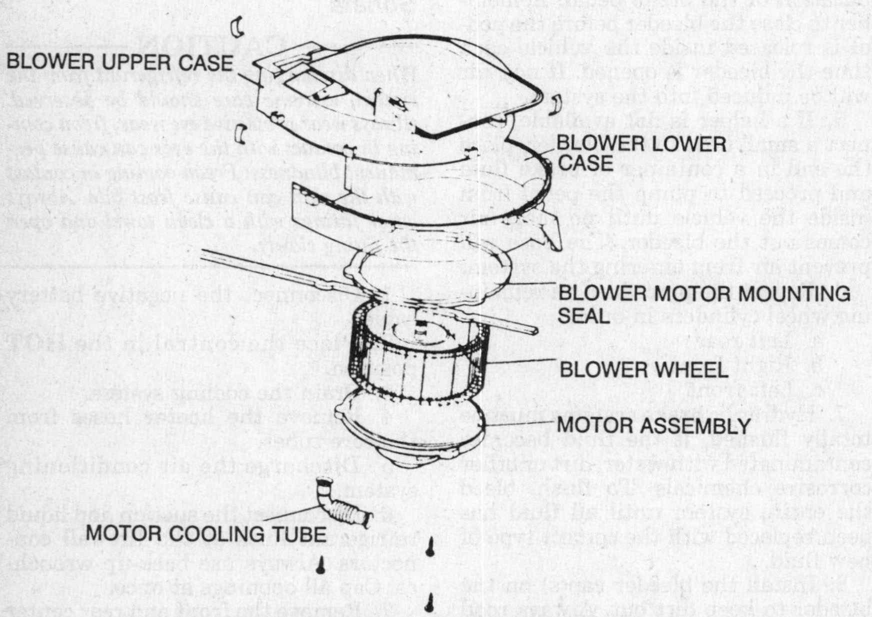

Blower motor assembly – Excel and Scoupe

8. Remove the heater side covers.
9. Remove the glove box, center crash pad cover, center crash pad and the radio.
10. Remove the lower crash pad.
11. Remove the console mounting bracket and center support.
12. Remove the left and right rear heat duct assemblies and the rear heating joint duct.
13. Remove the control unit.
14. Disconnect the blower speed actuator connector and, in Canada, disconnect the blend door actuator connector.
15. Remove the heater/air conditioning unit.
16. Remove the blower motor from the case.
17. Installation is the reverse order of the removal procedures. Adjust the control cable and refill the cooling system.

Scoupe and 1990–92 Excel

1. Disconnect the negative battery cable.
2. Remove the glove box housing cover assembly.
3. Disconnect the resistor and blower motor connector.
4. Pull out the blower unit and disconnect the fresh/recirc vacuum connector.
5. Install the blower motor by reversing the removal procedure.

Windshield Wiper Motor

REMOVAL & INSTALLATION

Front

1. Disconnect the negative battery cable. Remove the air inlet and cowl front center trim panels. Remove the 3 pivot shaft mounting nuts and push the pivot shafts into the area under the cowl.
2. Remove the motor mounting bolts. Pull the motor into the best possible position for access and use prybar, to pry the linkage off the motor crank arm. Remove the motor and then the linkage.
3. If the motor is being replaced, matchmark the position of the crank arm of the motor shaft of the new motor and then remove the nut and crank arm, transferring both to the new motor.

To install:

4. Torque the pivot shaft nuts to 4.3–5.8 ft. lbs. (5–8 Nm).
5. Position the wiper arms so, on the Excel, the blades are about 15mm above the lower windshield molding, on the driver's side and 20mm above it, on the passenger's side.

6. On the Sonata, each blade tip should be about 30mm above the moulding. On Scoupe, the driver's side blade tip should be about 50mm above the lower windshield moulding and the passenger side blade tip should be about 30mm above the lower windshield moulding. Torque the wiper arm mounting nuts to 7–12 ft. lbs. (10–15 Nm).

7. Make sure the wiper motor is securely grounded. Connect the negative battery cable.

Rear

1. Remove the wiper blade and arm by lifting the wiper blade locknut cover and removing the locknut. Then, pull the arm from the shaft.
2. Remove the lift gate trim panel and disconnect the wiring harness connector.
3. Matchmark the relationship of the crank arm to the motor and remove the crank arm.
4. Remove the inside and outside motor mounting nuts and remove the motor.
5. Installation is the reverse of the removal procedures.

NOTE: When installing the wiper arm, the distance between the tip of the blade and the lower window molding should be 40mm.

Windshield Wiper Switch

REMOVAL & INSTALLATION

Rear

1. Pry the switch bezel from the instrument panel.
2. Reach behind the panel and disconnect the wiring from the switch.
3. Depress the 2 retainers and pull the switch from the panel.
4. Installation is the reverse order of the removal procedures.

Instrument Cluster

REMOVAL & INSTALLATION

Excel

1. Disconnect the negative battery cable. Remove the meter hood attaching screws, located at the bottom and tilt the lower meter hood outward. Pull the hood downward to release the locking tangs at the top and remove it.
2. Remove the meter assembly mounting screws and pull the unit outward. Disconnect the speedometer cable and all connectors. Remove the unit.
3. Installation is the reverse order of the removal procedures.

Sonata

1. Disconnect the negative battery cable. Remove the steering column support bolts and carefully lower the column on the front seat.
2. Remove the cluster trim panel.
3. Remove the cluster mounting screws and slowly pull the cluster outward. Disconnect the wires.
4. Installation is the reverse of removal. Torque the steering column bolts to 20 ft. lbs. (27 Nm).

Scoupe

1. Disconnect the negative battery cable.
2. Remove the ashtray.
3. Remove the lower crash pad center facia panel.
4. Remove the digital clock and remote mirror switch.
5. Remove the 7 screws at the cluster facia panel.
6. Remove the 4 screws retaining the cluster and connectors
7. Install the instrument cluster by reversing the removal procedure.

Speedometer

REMOVAL & INSTALLATION

1. Disconnect the negative battery cable.
2. Remove the instrument cluster from the instrument panel.
4. Place the instrument cluster on clean work area.
5. Remove the speedometer lens retaining screws from the side of the cluster and remove the lens.
6. From the rear side of the cluster, remove the speedometer retaining screws and carefully remove the speedometer from the cluster assembly.

To install:

7 Install the speedometer into the cluster and install the retaining screws.
8. Position the cluster assembly to the dash, while inserting the speedometer cable into the speedometer, push in securely.
9. Install the instrument cluster face plate and lens to the panel.
10. Connect the negative battery cable.

Combination Switch

REMOVAL & INSTALLATION

1. Disconnect the negative battery cable. Remove the steering wheel. Remove the steering column cover.

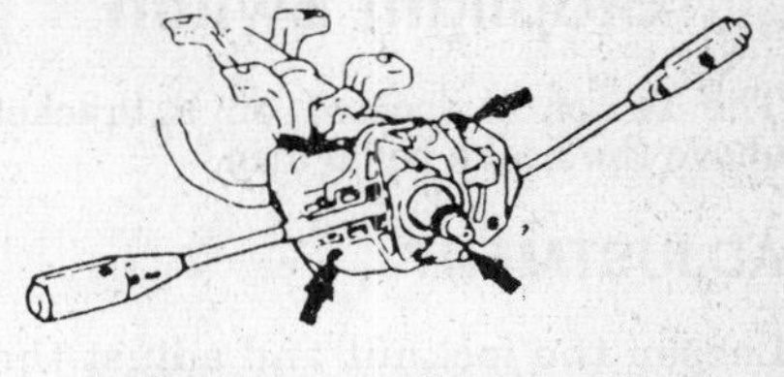

Windshield wiper switch mounting

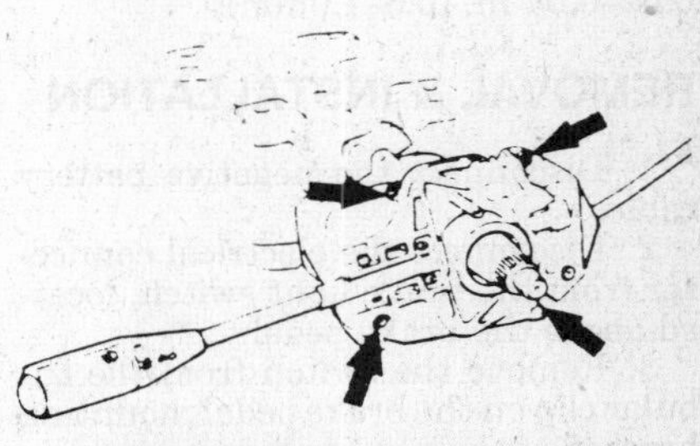

Combination switch removal

2. Unplug the electrical connectors. If necessary, remove the harness retainer.
3. Remove the retaining screws and slide the switch off the steering column.
4. Installation is the reverse order of the removal procedures.

Ignition Lock/Switch

REMOVAL & INSTALLATION

1. Disconnect the negative battery cable. Remove the steering wheel. Remove the steering column cover.
2. Remove the combination switch.
3. Unplug the electrical connector for the ignition lock.
4. Use a hacksaw to cut a slit in the top of each of the fastening bolts. Unscrew the bolts and remove the switch.
5. When installing the new switch, align the halves of the assembly around the steering column, align the assembly with the column boss and then install the special new installation bolts just loosely. Verify that the ignition switch works and then tighten the bolts until their heads break off.

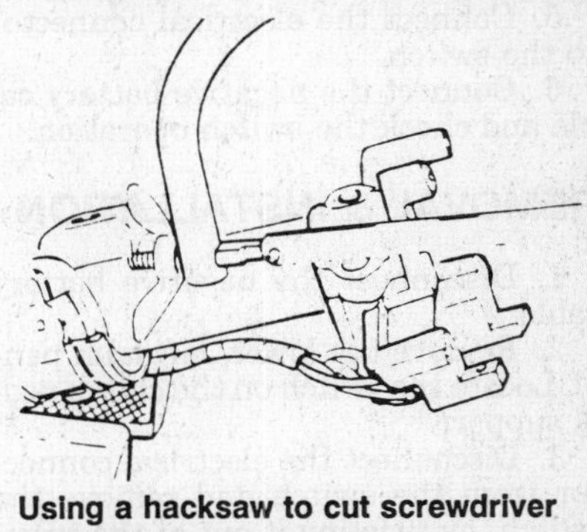

Using a hacksaw to cut screwdriver grooves into the ignition lock mounting bolts

Stoplight Switch

The switch is located on a bracket above the brake pedal arm.

ADJUSTMENT

Loosen the locknut and adjust the switch so the distance between the switch outer case and the pedal arm is 0.02–0.04 in. (0.5–1.0mm).

REMOVAL & INSTALLATION

1. Disconnect the negative battery cable.
2. Disconnect the electrical connector from the brake light switch, located above the brake pedal.
3. Remove the switch from the tubular clip on the brake pedal mounting bracket.

To install:

4. Insert the switch into the clip until the switch body seats on the clip.
5. Pull the brake pedal rearward against the internal pedal stop. The switch will be moved in the tubular clip providing the proper adjustment.
6. Connect the electrical connector to the switch.
7. Connect the negative battery cable and check the switch operation.

Clutch Switch

ADJUSTMENT

1. Disconnect the negative battery cable.
2. Remove the lower, left trim panel and locate the switch on the clutch pedal support.
3. Disconnect the electrical connector from the switch and remove the switch, by twisting it out of the tubular retaining clip.
4. Pull back on the clutch pedal and push the switch through the retaining clip noting the clicks; repeat this procedure until no more clicks can be heard.
5. Connect the electrical connector to the switch.
6. Connect the negative battery cable and check the switch operation.

REMOVAL & INSTALLATION

1. Disconnect the negative battery cable.
2. Remove the lower, left trim panel. Locate the switch on the clutch pedal support.
3. Disconnect the electrical connector from the switch and remove the switch, by twisting it out of the tubular retaining clip.

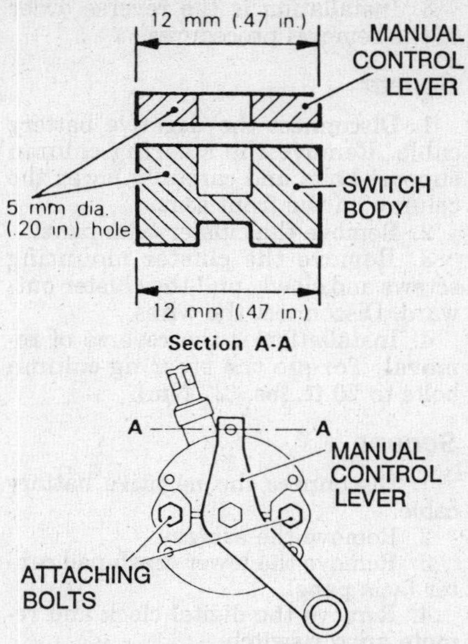

Neutral start switch adjustments on the Sonata

To install:

4. Using a new retaining clip, install the switch and connect the electrical connector.
5. To adjust the switch, pull back on the clutch pedal, push the switch through the retaining clip, noting the clicks; repeat this procedure until no more clicks can be heard.
6. Connect the negative battery cable and check the switch operation.

Neutral Safety Switch

ADJUSTMENT

Excel and Scoupe

1. Apply the parking brake. Place the gearshift lever in **N** position.
2. Loosen the mounting screws of the neutral switch so it can be rotated. Now, rotate it so the end of the operating lever (A) is directly over the flange on the switch body and the holes in that flange and the outer end of the lever are aligned.
3. Hold the switch securely in place while torquing the mounting screws to 90–100 inch lbs.
4. Recheck the function of the switch by attempting to start the engine in all selector positions. It should start only in **P** and **N**.

Sonata

1. Place the shifter in the **N** position.
2. Loosen the control cable coupler and free the cable.
3. Place the control lever in the neutral position.
4. Turn the switch body until the wide (12mm) end of the control lever aligns with the switch body's widest part or turn the switch body until; the 5mm hole in the control lever aligns with the 5mm hole in the switch body. Tighten the nuts.

REMOVAL & INSTALLATION

1. Disconnect the negative battery cable.
2. Raise the vehicle and support it safely. Disconnect the shift linkage from the transaxle.
3. Disconnect the electrical connector from the switch.
4. Remove the switch to transaxle bolts and the switch from the vehicle.
5. To install, position the shifter shaft in the **N** position.
6. Align the shifter shaft flats with the switch and assemble the mounting bolts loosely.
7. Adjust the switch. Connect the negative battery cable and test operation.

Fuses and Relays

LOCATION

Fuse Block

The fuse block is located up under the instrument panel on the driver's side of the steering column.

Fusible Link

The main fusible link is located under the hood, inline of the positive battery cable.

Relays

There are 2 relay boxes located under the hood, on the right front apron. On Scoupe, the relay box is found in the passenger compartment on the firewall below the steering column.

Computers

LOCATION

The Electronic Control Unit (ECU) is located on the passengers lower kick panel. On Scoupe, the ECU is located behind the dash panel on the drivers side, below the air outlet duct.

Flashers

LOCATION

The turn signal and hazard flashers are located in the fuse block, located under the instrument panel on the driver's side of the steering column.

Infiniti

G20, M30, Q45 — All Models

9

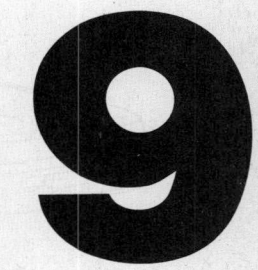

SERIAL NUMBER IDENTIFICATION

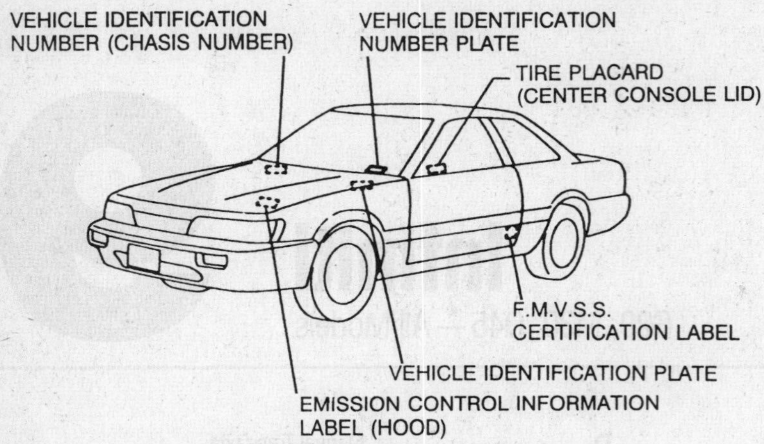

Vehicle identification plate locations – M30

Vehicle Identification Number

The vehicle identification number plate is located at the upper left corner of the dash panel, as viewed through the windshield.

Engine Number

The engine number is located at the rear of the engine block. It is centrally positioned on Q45, slightly right of center on M30 and at the left side on the G20.

Chassis Number

The chassis number plate is located in the engine compartment, at the upper right portion of the firewall.

Transaxle Number

The G20 manual transaxle identification number is located at the top of the case near the bellhousing. The G20 automatic transaxle identification number is located at the top of the case on the govenor cap.

Transmission Number

The automatic transmission identification number for both the M30 and Q45 is located at the right rear of the case on the tailshaft.

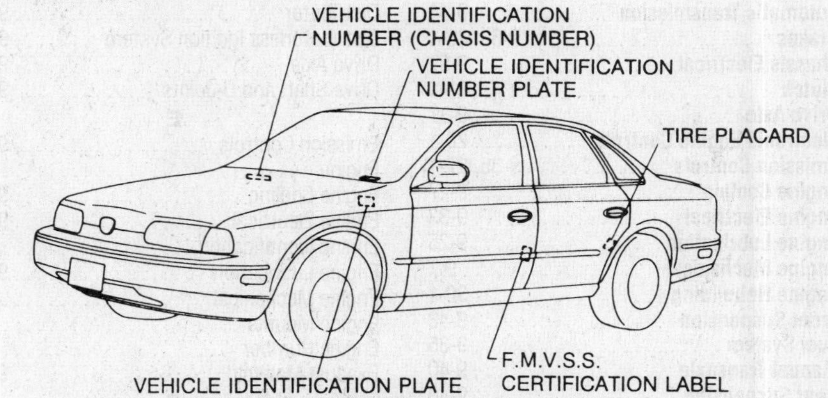

Vehicle identification plate locations – Q45

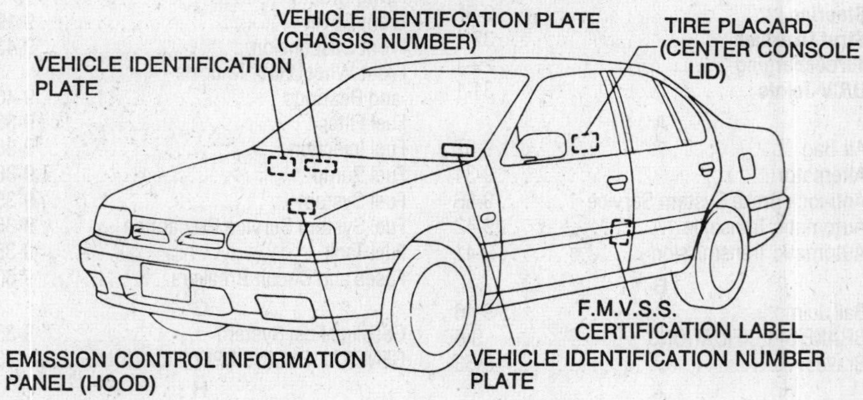

Vehicle identification plate locations – G20

SPECIFICATIONS

ENGINE IDENTIFICATION

Year	Model	Engine Displacement cu. in. (cc/liter)	Engine Series Identification	No. of Cylinders	Engine Type
1990	Q45	274 (4494/4.5)	VH45DE	8	DOHC
	M30	181 (2960/3.0)	VG30	6	SOHC

ENGINE IDENTIFICATION

Year	Model	Engine Displacement cu. in. (cc/liter)	Engine Series Identification	No. of Cylinders	Engine Type
1991-92	Q45	274 (4494/4.5)	VH45DE	8	DOHC
	M30	181 (2960/3.0)	VG30	6	SOHC
	G20	122 (1998/2.0)	SR20DE	4	DOHC

SOHC—Single Overhead Camshaft
DOHC—Double Overhead Camshaft

GENERAL ENGINE SPECIFICATIONS

Year	Model	Engine Displacement cu. in. (cc)	Fuel System Type	Net Horsepower @ rpm	Net Torque @ rpm (ft. lbs.)	Bore × Stroke (in.)	Com-pression Ratio	Oil Pressure @ rpm
1990	Q45	274 (4494)	MPFI	278 @ 6000	280 @ 4000	3.66 × 3.26	10.2:1	67–81 @ 3000
	M30	181 (2960)	MPFI	162 @ 5200	180 @ 3600	3.43 × 3.27	9.0:1	53–67 @ 3200
1991-92	Q45	274 (4494)	MPFI	278 @ 6000	280 @ 4000	3.66 × 3.26	10.2:1	67–81 @ 3000
	M30	181 (2960)	MPFI	162 @ 5200	180 @ 3600	3.43 × 3.27	9.0:1	53–67 @ 3200
	G20	122 (1998)	MPFI	140 @ 6400	132 @ 4800	3.38 × 3.38	10.0:1	46–57 @ 3200

MPFI—Multi-Point Fuel Injection

ENGINE TUNE-UP SPECIFICATIONS

Year	Model	Engine Displacement cu. in. (cc)	Spark Plugs Type	Spark Plugs Gap (in.)	Ignition Timing (deg.) MT	Ignition Timing (deg.) AT	Com-pression Pressure (psi)	Fuel Pump (psi)	Idle Speed (rpm) MT	Idle Speed (rpm) AT	Valve Clearance In.	Valve Clearance Ex.
1990	Q45	274 (4494)	PFR6B-11	0.040	—	15B	185 @ 300 rpm	34	—	750	Hyd.	Hyd.
	M30	181 (2960)	PFR6B-11	0.040	—	15B	173 @ 300 rpm	43	—	800	Hyd.	Hyd.
1991	Q45	274 (4494)	PFR6B-11	0.040	—	15B	185 @ 300 rpm	34	—	750	Hyd.	Hyd.
	M30	181 (2960)	PFR6B-11	0.040	—	15B	173 @ 300 rpm	43	—	800	Hyd.	Hyd.
	G20	122 (1998)	PFR6B-11	0.040	15B	15B	178 @ 300 rpm	43	800	800	Hyd.	Hyd.
1992	SEE UNDERHOOD STICKER											

Hyd.—Hydraulic

FIRING ORDERS

NOTE: To avoid confusion, always replace spark plug wires one at a time.

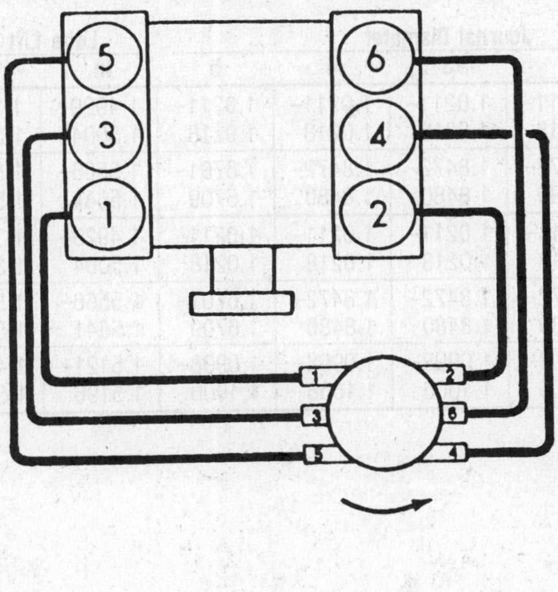

3.0L Engine
Engine Firing Order: 1-2-3-4-5-6
Distributor Rotation: Counterclockwise

FIRING ORDERS

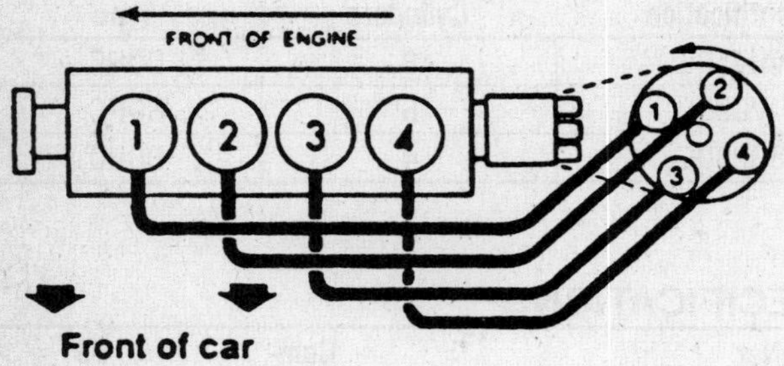

2.0L Engine
Engine Firing Order: 1-3-4-2
Distributor Rotation: Counterclockwise

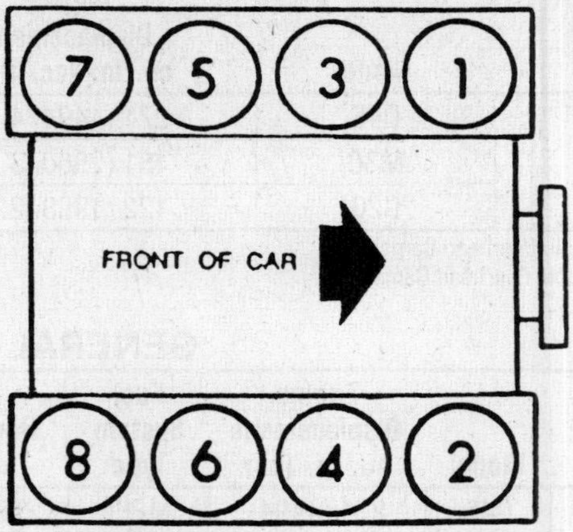

4.5L Engine
Engine Firing Order: 1-8-7-3-6-5-4-2
Distributorless Ignition System

CAPACITIES

Year	Model	Engine Displacement cu. in. (cc)	Engine Crankcase (qts.) with Filter	Engine Crankcase (qts.) without Filter	Transmission (pts.) 4-Spd	Transmission (pts.) 5-Spd	Transmission (pts.) Auto.	Drive Axle (pts.)	Fuel Tank (gal.)	Cooling System (qts.)
1990	Q45	274 (4494)	6½	6	—	—	18¼	3⅛	22½	11
	M30	181 (2960)	4½	4¼	—	—	17½	2¾	16	9½
1991-92	Q45	274 (4494)	6½	6	—	—	18¼	3⅛	22½	11
	M30	181 (2960)	4½	4¼	—	—	17½	2¾	16	9½
	G20	122 (1998)	3⅝	3⅝	—	7½	17¾	—	16	①

① 6½ qts.—Manual transaxle
6⅞ qts.—Automatic transaxle

CAMSHAFT SPECIFICATIONS

Year	Engine Displacement cu. in. (cc)	Journal Diameter 1	Journal Diameter 2	Journal Diameter 3	Journal Diameter 4	Journal Diameter 5	Lobe Lift ① In.	Lobe Lift ① Ex.	Bearing Clearance	Camshaft End Play
1990	274 (4494)	1.0211–1.0218	1.0211–1.0218	1.0211–1.0218	1.0211–1.0218	1.0211–1.0218	1.4929–1.5004	1.3889–1.3964	0.0018–0.0059	0.0028–0.0079
	181 (2960)	1.8866–1.8874	1.8472–1.8480	1.8472–1.8480	1.8472–1.8480	1.6701–1.6709	1.5566–1.5641	1.5566–1.5641	0.0024–0.0041	0.0012–0.0024
1991-92	274 (4494)	1.0211–1.0218	1.0211–1.0218	1.0211–1.0218	1.0211–1.0218	1.0211–1.0218	1.4929–1.5004	1.3889–1.3964	0.0018–0.0059	0.0028–0.0079
	181 (2960)	1.8866–1.8874	1.8472–1.8480	1.8472–1.8480	1.8472–1.8480	1.6701–1.6709	1.5566–1.5641	1.5566–1.5641	0.0024–0.0041	0.0012–0.0024
	122 (1998)	1.0998–1.1006	1.0998–1.1006	1.0998–1.1006	1.0998–1.1006	1.0998–1.1006	1.5121–1.5196	1.4929–1.5004	0.0018–0.0034	0.0022–0.0055

CRANKSHAFT AND CONNECTING ROD SPECIFICATIONS

All measurements are given in inches.

Year	Engine Displacement cu. in. (cc)	Crankshaft Main Brg. Journal Dia.	Crankshaft Main Brg. Oil Clearance	Crankshaft Shaft End-play	Crankshaft Thrust on No.	Connecting Rod Journal Diameter	Connecting Rod Oil Clearance	Connecting Rod Side Clearance
1990	274 (4494)	①	0.0005–0.0020	0.0039–0.0118	3	②	0.0008–0.0026	0.0079–0.0157
	181 (2960)	③	0.0011–0.0035	0.0020–0.0118	4	1.9667–1.9675	0.0006–0.0035	0.0079–0.0157
1991-92	274 (4494)	①	0.0005–0.0020	0.0039–0.0118	3	②	0.0008–0.0026	0.0079–0.0157
	181 (2960)	③	0.0011–0.0035	0.0020–0.0118	4	1.9667–1.9675	0.0006–0.0035	0.0079–0.0157
	122 (1998)	④	0.0002–0.0009	0.0039–0.0102	3	⑤	0.0008–0.0018	0.0079–0.0138

① Grade No. 0—2.5180-2.5183
Grade No. 1—2.5178-2.5180
Grade No. 2—2.5176-2.5178
Grade No. 3—2.5173-2.5176
② Grade No. 0—2.0460-2.0462
Grade No. 1—2.0457-2.0460
Grade No. 2—2.0455-2.0457
③ Grade No. 0—2.4790-2.4793
Grade No. 1—2.4787-2.4790
Grade No. 2—2.4784-2.4787
④ Grade No. 0—2.1643-2.1646
Grade No. 1—2.1641-2.1643
Grade No. 2—2.1639-2.1641
Grade No. 3—2.1636-2.1639
⑤ Grade No. 0—1.8885-1.8887
Grade No. 1—1.8883-1.8885
Grade No. 2—1.8880-1.8883

VALVE SPECIFICATIONS

Year	Engine Displacement cu. in. (cc)	Seat Angle (deg.)	Face Angle (deg.)	Spring Test Pressure (lbs.)	Spring Installed Height (in.)	Stem-to-Guide Clearance (in.) Intake	Stem-to-Guide Clearance (in.) Exhaust	Stem Diameter (in.) Intake	Stem Diameter (in.) Exhaust
1990	274 (4494)	45.25	44.75	120.4	1.86 ①	0.0017 ②	0.0018 ②	0.275	0.314
	181 (2960)	45.25	45	③	④	0.0039	0.0039	0.274	0.314
1991-92	274 (4494)	45.25	44.75	120.4	1.86 ①	0.0017 ②	0.0018 ②	0.275	0.314
	181 (2960)	45.25	45	③	④	0.0039	0.0039	0.274	0.314
	122 (1998)	45.15–45.45	44.53–45.07	128–144	1.18	0.0008–0.0021	0.0016–0.0029	0.235	0.234

① Free height
② Maximum
③ Inner—57.3
Outer—117.7
④ Inner—1.74
Outer—2.02

PISTON AND RING SPECIFICATIONS

All measurements are given in inches.

Year	Engine Displacement cu. in. (cc)	Piston Clearance	Ring Gap Top Compression	Ring Gap Bottom Compression	Ring Gap Oil Control	Ring Side Clearance Top Compression	Ring Side Clearance Bottom Compression	Ring Side Clearance Oil Control
1990	274 (4494)	0.0004–0.0012	0.0106–0.0390	0.0154–0.0390	0.0079–0.0390	0.0016–0.0040	0.0012–0.0040	—
	181 (2960)	0.0010–0.0018	0.0083–0.0390	0.0071–0.0390	0.0079–0.0390	0.0016–0.0040	0.0012–0.0040	—
1991-92	274 (4494)	0.0004–0.0012	0.0106–0.0390	0.0154–0.0390	0.0079–0.0390	0.0016–0.0040	0.0012–0.0040	—
	181 (2960)	0.0010–0.0018	0.0083–0.0390	0.0071–0.0390	0.0079–0.0390	0.0016–0.0040	0.0012–0.0040	—
	122 (1998)	0.0004–0.0012	0.0079–0.0118	0.0138–0.0197	0.0079–0.0236	0.0018–0.0031	0.0012–0.0026	—

TORQUE SPECIFICATIONS

All readings in ft. lbs.

Year	Engine Displacement cu. in. (cc)	Cylinder Head Bolts	Main Bearing Bolts	Rod Bearing Bolts	Crankshaft Pulley Bolts	Flywheel Bolts	Manifold Intake	Manifold Exhaust	Spark Plugs
1990	274 (4494)	①	②	③	260–275	61–69	12–15	20–23	15–20
	181 (2960)	④	67–74	③	90–98	61–69	12–14	13–16	15–20
1991-92	274 (4494)	①	②	③	260–275	61–69	12–15	20–23	15–20
	181 (2960)	④	67–74	③	90–98	61–69	12–14	13–16	15–20
	122 (1998)	⑤	⑥	③	105–112	61–69	13–15	13–15	15–20

NOTE: Always tighten bolts in specified sequence.

① ⓐ Tighten bolts to 22 ft. lbs. (29 Nm)
ⓑ Tighten bolts to 69 ft. lbs. (93 Nm)
ⓒ Loosen bolts completely.
ⓓ Tighten bolts to 18–25 ft. lbs. (25–34 Nm)
ⓔ Turn bolts 90 to 95 degrees clockwise or if angle wrench is not available, tighten bolts to 69–72 ft. lbs. (93–98 Nm)

② See text

③ Step 1: 10–12 ft. lbs.
Step 2: An additional 60–65 degrees or tighten to a final torque of 28–33 ft. lbs.

④ Step 1: 22 ft. lbs.
Step 2: 43 ft. lbs.
Step 3: Loosen all bolts completely
Step 4: 22 ft. lbs.
Step 5: An additional 60–65 degrees or tighten to a final torque of 40–47 ft. lbs.

⑤ ⓐ Tighten all bolts to 29 ft. lbs. (39 Nm)
ⓑ Tighten all bolts to 58 ft. lbs. (78 Nm)
ⓒ Loosen all bolts completely
ⓓ Tighten all bolts to 25–33 ft. lbs. (34–44 Nm)
ⓔ Turn all bolts 90 degrees–100 degrees clockwise
ⓕ Turn all bolts an additional 90 degrees–100 degrees clockwise
• Do not turn any bolt 180 degrees–200 degrees clockwise all at once

⑥ ⓐ Tighten bolts to 24–28 ft. lbs. (32–38 Nm)
ⓑ Turn all bolts 45 degrees–50 degrees clockwise

BRAKE SPECIFICATIONS

All measurements in inches unless noted.

Year	Model	Lug Nut Torque (ft. lbs.)	Master Cylinder Bore	Brake Disc Minimum Thickness	Brake Disc Maximum Runout	Standard Brake Drum Diameter	Minimum Lining Thickness Front	Minimum Lining Thickness Rear
1990	Q45	72–87	1.06	①	0.003	—	0.080	0.080
	M30	76–90	1.00	②	0.003	—	0.080	0.080
1991-92	Q45	72–87	1.06	①	0.003	—	0.080	0.080
	M30	76–90	1.00	②	0.003	—	0.080	0.080
	G20	76–90	0.937	③	0.003	—	0.080	0.080

① Front: 1.024
Rear: 0.315

② Front: 0.787
Rear: 0.354

③ Front: 0.790
Rear: 0.310

WHEEL ALIGNMENT

Year	Model		Caster Range (deg.)	Caster Preferred Setting (deg.)	Camber Range (deg.)	Camber Preferred Setting (deg.)	Toe-in (in.)	Steering Axis Inclination (deg.)
1990	Q45	front	5¾P–7¼P	6½P	1½N-0	¾N	1/32	12–13½
		rear	—	—	1½N–½N	1N	1/8	—
	M30	front	4P–5½P	4¾P	½N–1P	¾P	0	12–13½
		rear	—	—	1N–¼P	⅜N	1/16	—
1991-92	Q45	front	5¾P–7¼P	6½P	1½N-0	¾N	1/32	12–13½
		rear	—	—	1½N–½N	1N	1/8	—
	M30	front	4P–5½P	4¾P	½N–1P	¾P	0	12–13½
		rear	—	—	1N–¼P	⅜N	1/16	—
	G20	front	1–2½P	2P	¾N–¾P	0	0–¼P	13¾–15¼
		rear	—	—	1¾N–¼P	1¼N	¼N–¼P	—

ENGINE MECHANICAL

NOTE: Disconnecting the negative battery cable on some vehicles may interfere with the functions of the on board computer systems and may require the computer to undergo a relearning process, once the negative battery cable is reconnected.

Engine Assembly

REMOVAL & INSTALLATION

G20

1. Disconnect the negative battery cable. Raise and support the vehicle safely. Remove the engine under cover. Matchmark the hood with the hood hinges and remove.
2. Drain the coolant from both the cylinder block and radiator.
3. Drain the engine oil.
4. Release fuel system pressure and remove fuel line.
5. Label and remove all vacuum lines and wiring harness connectors.
6. Remove exhaust tubes, ball joints and drive shafts.
7. Remove the radiator and fans.
8. Remove the drive belts.
9. Remove the alternator, compressor and power steering pump from the engine and lay them aside. do not disconnect the compressor or power steering pump lines.
10. Support the engine with a hoist and the transmission with a suitable jack. Raise the engine and transaxle slightly and remove the center member.
11. Remove the bolts from the rear engine mount and slowly lower the hoist and transaxle jack.
12. Remove the engine and transaxle from beneath the vehicle.

To install:

13. Install the center member on the engine (if removed). Ensure that all insulators are correctly positioned on the brackets. Tighten bolts to 57–72 ft. lbs. (77–98 Nm).
14. If equipped with manual transaxle, ensure that the distance between the center of the insulator through-bolt and the center member is 2.28–2.36 in. (58–60mm). Tighten through-bolt to 46–58 ft. lbs. (62–78 Nm).
15. Carefully install the engine and tighten the center member-to-frame bolts. Install the rear engine mount.
16. Install the alternator, compressor and power steering pump.
17. Connect all vacuum hoses and wiring harness connectors. Connect the fuel line.
18. Install the exhaust tubes, ball joints, drive shafts, the radiator and fans, the drive belts.
19. Fill the coolant system with anti-freeze and the crankcase with oil.
20. Install the engine under cover and hood.

M30

1. Mark the hood hinge relationship and remove the hood and engine undercover.
2. Release the fuel system pressure and disconnect the negative battery cable. Raise and support the vehicle safely.
3. Drain the cooling system and the oil pan.
4. Remove the air cleaner and disconnect the throttle cable.
5. Disconnect or remove the following:
   - Drive belts
   - Ignition wire from the coil to the distributor
   - Ignition coil ground wire and the engine ground cable
   - Block connector from the distributor
   - Fusible links
   - Engine harness connectors
   - Fuel and fuel return hoses
   - Upper and lower radiator hoses
   - Heater inlet and outlet hoses
   - Engine vacuum hoses
   - Carbon canister hoses and the air pump air, cleaner hose
   - Any interfering engine accessory: power steering pump, air conditioning compressor or alternator
6. Remove the air pump air cleaner.
7. Remove the carbon canister.
8. Remove the auxiliary fan, washer tank, grille and radiator (with fan assembly).
9. Disconnect the speedometer cable.
10. Remove the spring pins from the transaxle gear selector rods.
11. Install engine slingers to the block and connect a suitable lifting device to the slingers.
12. Disconnect the exhaust pipe at both the manifold connection and the clamp holding the pipe to the engine.
13. Drain the transaxle gear oil.
14. Lower the shifter and selector rods and remove the bolts from the motor mount brackets. Remove the nuts holding the front and rear motor mounts to the frame.
15. Lift the engine/transaxle assembly up and away from the vehicle.

To install:

16. Lower the engine and transaxle assembly into the vehicle. When lowering the engine onto the frame, make sure to keep it as level as possible.
17. Check the clearance between the frame and transaxle and make sure the engine mount bolts are seated in the groove of the mounting bracket.
18. After installing the motor mounts, adjust and install the buffer rods. The front should be 3.50–3.58 in. (89–91mm), and the rear, 3.90–3.98 in. (99–101mm).
19. Raise the shifter and selector rods to their normal operating positions.
20. Connect the exhaust pipe to the manifold connection and the clamp holding the pipe to the engine.
21. Disconnect the lifting device and remove the engine slingers.
22. Insert the spring pins into the transaxle gear selector rods.
23. Connect the speedometer cable.
24. Install the auxiliary fan, washer tank, grille and radiator (with fan assembly).
25. Install the carbon canister.
26. Install the air pump air cleaner.
27. Install or connect all hoses, belts, harnesses, connectors and components that were necessary to remove the engine.
28. Connect the throttle cable and install the air cleaner.
29. Fill the transaxle and cooling system to the proper levels.
30. Install the hood and connect the negative battery cable.
31. Make all the necessary engine adjustments. Charge the air conditioning system.
32. Tighten the following to the proper torque:
   - Front engine mount bracket-to-engine bolts to 33–43 ft. lbs.
   - Front engine mount-to-bracket bolts to 29–36 ft. lbs.
   - Rear engine mount-to-crossmember bolts to 16–21 ft. lbs.
   - Rear crossmember-to-transaxle bolts to 32–41 ft. lbs.

Q45

1. Disconnect the negative battery cable.
2. Relieve the pressure from the fuel system.
3. Mark the relation of the hood to the hinge brackets and remove the hood.
4. Raise and safely support the vehicle.
5. Remove the engine splash shield.
6. Drain the coolant and the engine oil.
7. Disconnect the transmission cooler lines from the radiator.
8. Remove the radiator hoses and remove the radiator and shroud.
9. Tag and disconnect all vacuum hoses, fuel lines and electrical connectors.

10. Disconnect the exhaust pipes from the exhaust manifolds.
11. Mark the position of the driveshaft on the flanges and remove the driveshaft.
12. Remove the accessory drive belts.
13. Remove the alternator, air conditioning compressor and power steering pump.
14. Remove the lower steering joint.
15. Remove the sway bar, transverse link and tension rod with bracket.
16. Place a suitable jack under the transmission and disconnect the transmission rear mount.
17. Remove the suspension member attaching bolts.
18. Remove the engine mounting bolts.
19. Attach a suitable hoist to the engine. Lower the transmission jack and the hoist and lower the engine and transmission from under the vehicle.

To install:

20. Raise the engine and transmission into position.
21. Install the engine mounting nuts and bolts and tighten to 41 ft. lbs. (55 Nm).
22. Install the suspension member bolts.
23. Install and tighten the rear mount-to-body bolts to 41 ft. lbs. (55 Nm).
24. Remove the hoist and the transmission jack.
25. Install the sway bar, transverse link and the tension rod with bracket.
26. Install the lower steering joint.
27. Install the alternator, air conditioning compressor and power steering pump.
28. Install and adjust the accessory drive belts.
29. Install the radiator and shroud. Install the radiator hoses.
30. Install the driveshaft, aligning the marks that were made during the removal procedure.
31. Connect the exhaust pipes to the exhaust manifolds.
32. Connect all electrical connectors, fuel lines and vacuum hoses.
33. Fill the crankcase with the proper type of engine oil to the required level. Fill the cooling system with the proper type and quantity of coolant.
34. Install the hood, aligning the marks that were made during the removal procedure.
35. Connect the negative battery cable, start the engine and check for leaks.

Engine Mount

REMOVAL & INSTALLATION

G20

1. Disconnect the negative battery cable.
2. Matchmark the engine mount to its frame mounting location.
3. Raise the vehicle and support safely, if necessary. Using the proper equipment, support the weight of the engine.
4. Inspect all mounts to determine which is defective. A defective mount will have the rubber portion of the mount separated from the metal backing or stud.
5. Remove all bolts and nuts that attach the mount to the engine, transaxle or frame and remove the mount assembly from the vehicle.
6. Remove the through bolt and separate the insulator from the bracket, as required.

To install:

7. Installation is the reverse of removal.
8. If equipped with manual transaxle, ensure that the distance between the center of the front mounting bracket through-bolt and the center member is 2.28–2.36 in. (58–60mm). Tighten through-bolt to 46–58 ft. lbs. (62–78 Nm).
9. Tighten center member bolts to 57–72 ft. lbs. (77–98 Nm); rear fluid engine mount to 32–41 ft. lbs. (43–55 Nm); front fluid engine mount to 36–43 ft. lbs. (49–59 Nm).

M30

1. Disconnect the negative battery cable.

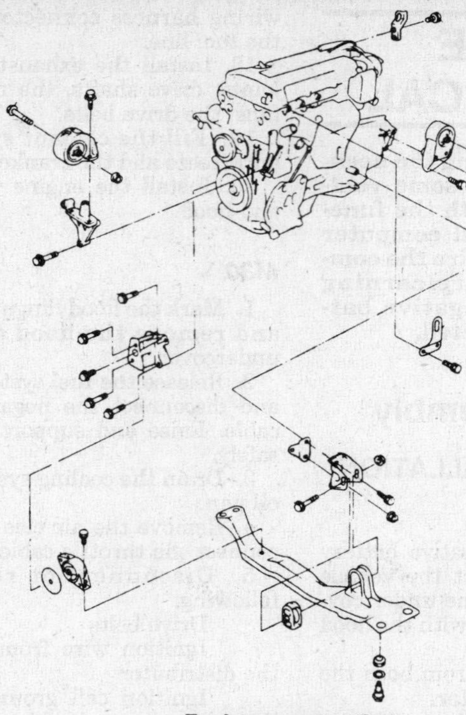

Engine mounts – G20

2. Matchmark the engine mount to its frame mounting location.
3. Raise the vehicle and support safely, if necessary. Using the proper equipment, support the weight of the engine.
4. Inspect all mounts to determine which is defective. A defective mount will have the rubber portion of the mount separated from the metal backing or stud.
5. Remove all bolts and nuts that attach the mount to the engine, transmission or frame and remove the mount assembly from the vehicle.
6. Remove the through bolt and separate the insulator from the bracket, as required.
7. Installation is the reverse of removal.
8. Tighten bolts as follows: engine mount-to-frame 29–36 ft. lbs. (39–49 Nm); engine mount-to-engine 33–43 ft. lbs. (44–59 Nm); transmission mount-to-crossmember 16–21 ft. lbs. (22–28 Nm); transmission mount-to-transmission 21–41 ft. lbs. (43–55 Nm).

Q45

1. Disconnect the negative battery cable.
2. Matchmark the engine mount to its frame mounting location.
3. Raise the vehicle and support safely, if necessary. Using the proper equipment, support the weight of the engine.

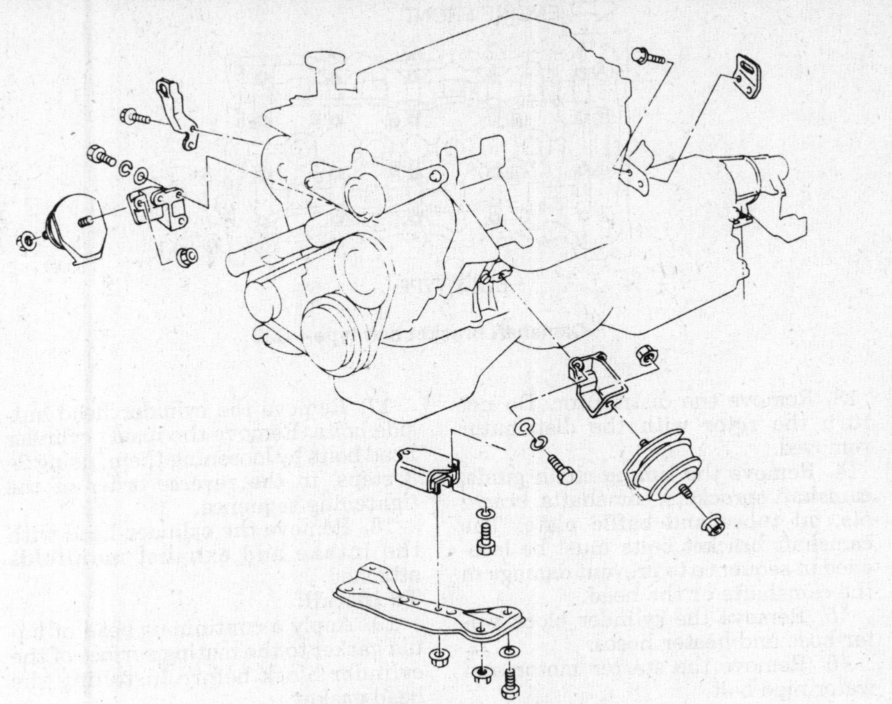

Engine mounts—M30

4. Inspect all mounts to determine which is defective. A defective mount will have the rubber portion of the mount separated from the metal backing or stud.

5. Remove all bolts and nuts that attach the mount to the engine, transmission or frame and remove the mount assembly from the vehicle.

6. Remove the through bolt and separate the insulator from the bracket, as required.

7. Installation is the reverse of removal.

8. Tighten bolts as follows: engine mount-to-frame 41–49 ft. lbs. (55–67 Nm); engine mount-to-engine 32–41 ft. lbs. (43–55 Nm); transmission mount-to-crossmember 16–21 ft. lbs. (22–28 Nm); transmission mount-to-transmission 32–41 ft. lbs. (43–55 Nm).

Cylinder Head

REMOVAL & INSTALLATION

G20

1. Relieve the fuel system pressure

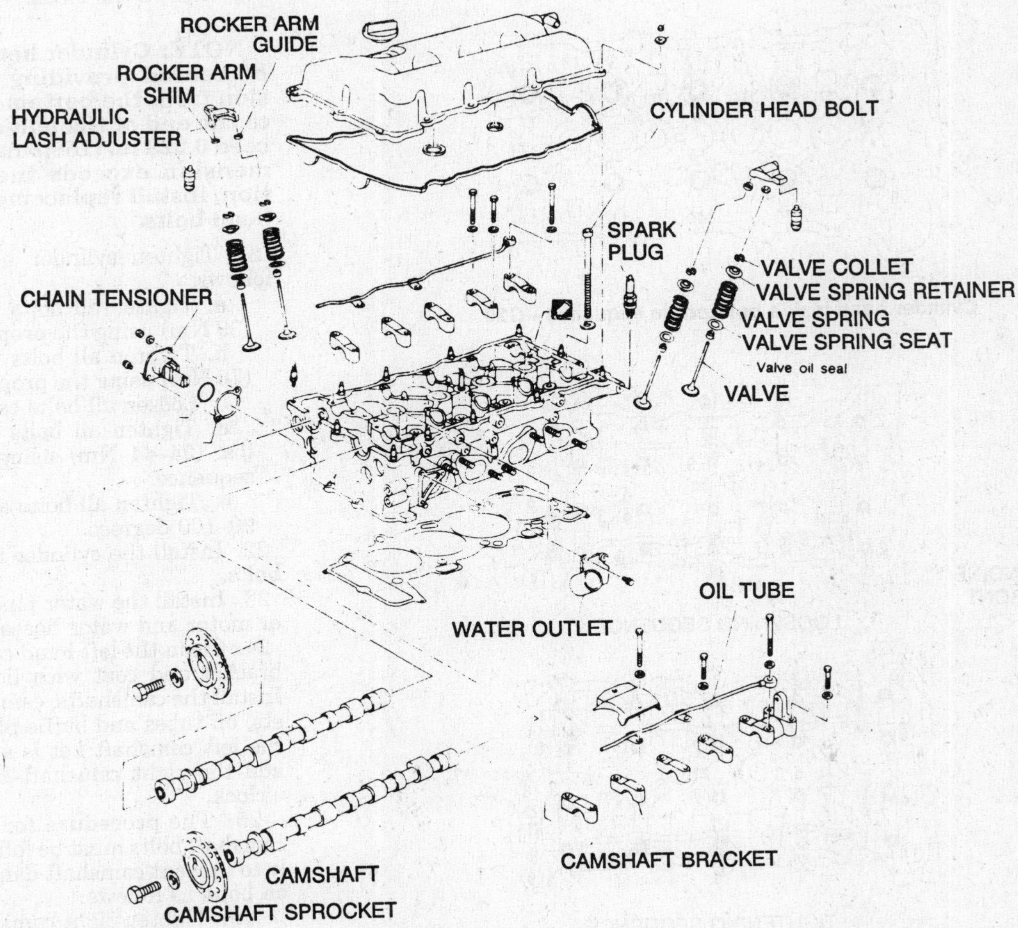

Cylinder head components—G20

and Disconnect the negative battery cable.

2. Drain the coolant from the radiator and engine block. Remove the radiator.

3. Remove the right front wheel and engine side cover.

4. Remove the air duct to the intake manifold.

5. Remove the drive belts, water pump pulley, alternator and power steering pump.

6. Label and remove the vacuum hoses, fuel hoses and wire harness connectors.

7. Remove all the spark plugs, the AIV valve and resonator.

8. Remove the rocker cover and oil separator. Loosen rocker cover bolts, using 2–3 steps, in the opposite sequence of tightening

9. Remove the intake manifold supports.

10. Remove the oil filter bracket and power steering oil pump bracket.

11. Set No. 1 piston at TDC on the compression stroke by rotating the crankshaft.

12. Remove the chain tensioner.

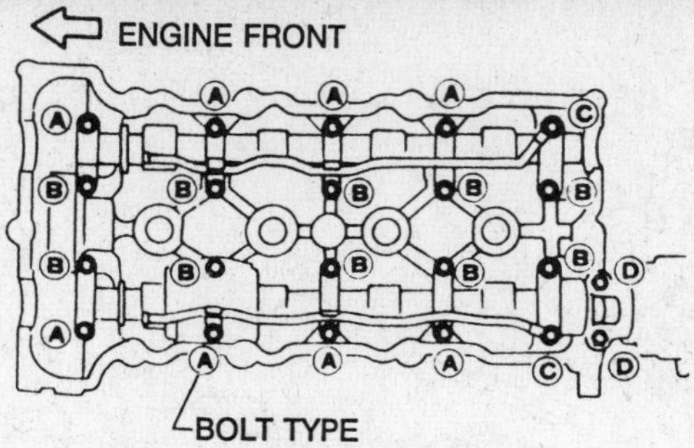

Camshaft bracket bolt type – G20

13. Remove the distributor. Do not turn the rotor with the distributor removed.

14. Remove the timing chain guide, camshaft sprockets, camshafts, brackets, oil tubes and baffle plate. The camshaft bracket bolts must be loosened in sequence to prevent damage to the camshafts or the head.

15. Remove the cylinder block water hose and heater hoses.

16. Remove the starter motor and water pipe bolt.

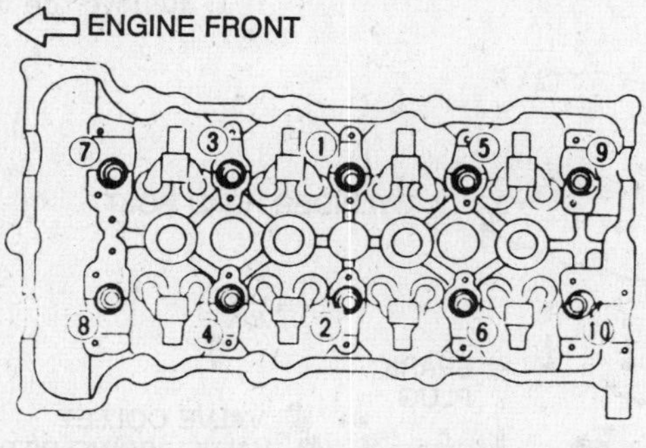

Cylinder head inside bolt torque sequence – G20

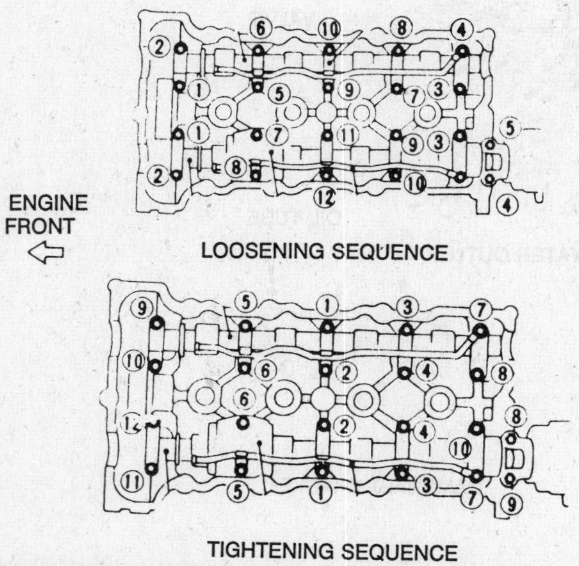

Camshaft bracket loosening and tightening sequence – G20

17. Remove the cylinder head outside bolts. Remove the inside cylinder head bolts by loosening them, using 2–3 steps, in the reverse order of the tightening sequence.

18. Remove the cylinder head with the intake and exhaust manifolds attached.

To install:

19. Apply a continuous bead of liquid gasket to the mating surface of the cylinder block before installing the head gasket.

20. Install the the gasket and cylinder head on the block.

NOTE: Cylinder head bolts may be reused providing the dimension from the bottom of the head to the end of the bolt does not exceed 6.228 in. (158.2mm). If the dimension exceeds the specification, install replacement cylinder head bolts.

21. Tighten cylinder head bolts as follows:

a. Tighten all bolts to 29 ft. lbs. (39 Nm) using the proper sequence.

b. Tighten all bolts to 58 ft. lbs. (78 Nm) using the proper sequence.

c. Loosen all bolts completely.

d. Tighten all bolts to 25–33 ft. lbs. (34–44 Nm) using the proper sequence.

e. Tighten all bolts an additional 90–100 degrees.

22. Install the cylinder head outside bolts.

23. Install the water pipe bolt, starter motor and water hoses.

24. Clean the left hand camshaft end bracket and coat with liquid gasket. Install the camshafts, camshaft brackets, oil tubes and baffle plate. Ensure the left camshaft ket is at 12 o'clock and the right camshaft key is at 10 o'clock.

25. The procedure for tightening camshaft bolts must be followed exactly to prevent camshaft damage. Tighten bolts as follows:

a. Tighten right camshaft bolts 9 and 10 (in that order) to 1.5 ft. lbs. (2 Nm) then tighten bolts 1 through

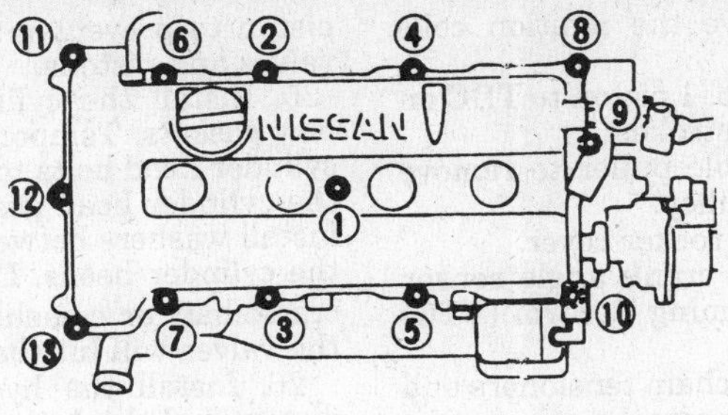

Rocker cover tightening sequence – G20

8 (in that order) to the same specification.

b. Tighten left camshaft bolts 11 and 12 (in that order) to 1.5 ft. lbs. (2 Nm) then tighten bolts 1 through 10 (in that order) to the same specification.

c. Tighten all bolts in sequence to 4.5 ft. lbs. (6 Nm).

d. Tighten all bolts in sequence to 6.5–8.5 ft. lbs. (9–12 Nm) for type A, B and C bolts, and 13–19 ft. lbs. (18–25 Nm) for type D bolts.

26. Line up the mating marks on the timing chain and camshaft sprockets and install the sprockets. Tighten sprocket bolts to 101–116 ft. lbs. (137–157 Nm).

27. Install the timing chain guide, distributor (ensure that rotor head is at 5 o'clock position), chain tensioner, oil filter bracket and power steering oil pump bracket.

28. Install intake manifold supports. Clean the rocker cover and mating surfaces and apply a continious bead of liquid gasket to the mating surface.

29. Install the rocker cover and oil separator. Tighten the rocker cover bolts as follows:

a. Tighten nuts 1, 10, 11, and 8 in that order to 3 ft. lbs. (4 Nm).

b. Tighten nuts 1 through 13 as indicated in the figure to 6–7 ft. lbs. (8–10 Nm).

30. Install the AIV and resonator, spark plugs, power steering pump, alternator water pump pulley and drive belts, air duct to the intake manifold and the radiator.

31. Install all vacuum and fuel hoses, and reconnect all electrical connections.

32. Install the engine side cover, right front wheel and engine under cover.

33. Refill the cooling system.

M30

1. Relieve the fuel system pressure and disconnect the negative battery cable.

2. Drain the cooling system.

3. Remove the timing belt.

NOTE: do not rotate either the crankshaft or camshaft from this point onward or the valves could be bent by hitting the tops of the pistons.

4. Disconnect and tag all vacuum and water hoses connected to the intake collector.

5. Remove the distributor, ignition wires and disconnect the accelerator and cruise control (ASCD) cables from the intake manifold collector.

6. Remove the collector cover and the collector from the intake manifold. Disconnect and tag all harness connectors and vacuum lines to gain access to the cover retaining bolts on these models.

7. Remove the intake manifold and fuel tube assembly. Loosen the intake manifold bolts starting from the front of the engine and proceed in criss-cross pattern towards the center.

8. Remove the exhaust collector bracket.

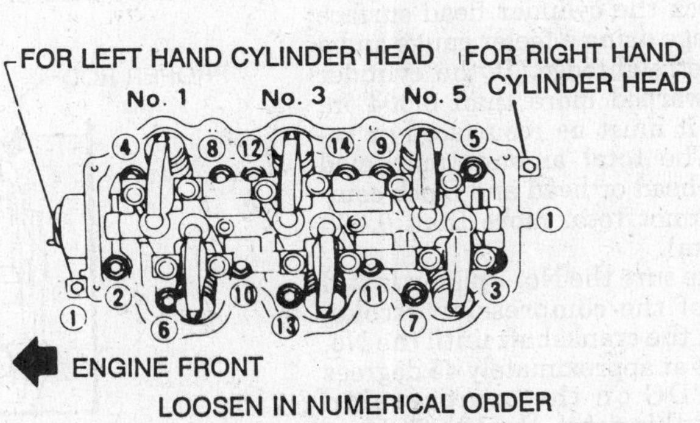

Cylinder head bolt removal sequence – M30

9. Remove the exhaust manifold covers.

10. Disconnect the exhaust manifold from the exhaust pipe.

11. Remove the camshaft pulleys and the rear timing cover securing bolts. Remove the rocker arm covers.

12. Separate the air conditioning compressor and alternator from the their mounting brackets. Remove the mounting brackets. Do not disconnect the refrigerant lines from the compressor or serious injury will result.

13. Remove the cylinder head bolts in the correct sequence. Lift the cylinder head off the engine block with the exhaust manifolds attached. It may be necessary to tap the head lightly with a rubber mallet to loosen it.

To install:

14. Make sure the No. 1 cylinder is set at TDC on its compression stroke as follows:

a. Align the crankshaft timing mark with the mark on the oil pump housing.

b. The knock pin in the front end of the camshaft should be facing upward.

NOTE: do not rotate crankshaft and camshaft separately because valves will hit piston head.

15. Install the cylinder head with a new gasket. Apply clean engine oil to the threads and seats of the bolts and install the bolts with washers in the correct position. Note that bolts 4, 5, 12, and 13 are 4.95 in. (127mm) long. The other bolts are 4.13 in. (106mm) long.

16. Torque the bolts in the proper sequence as follows:

a. Torque all bolts, in sequence, to 22 ft. lbs. (29 Nm).

b. Torque all bolts, in sequence, to 43 ft. lbs. (58 Nm).

c. Loosen all bolts completely.

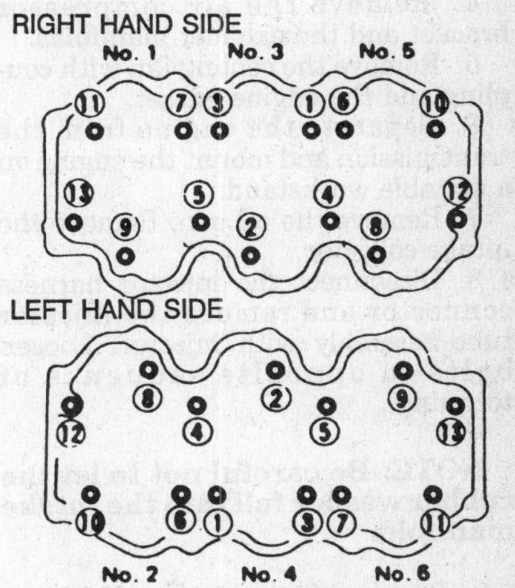

Cylinder head bolt tightening sequence – M30

d. Torque all bolts, in sequecne, to 22 ft. lbs. (29 Nm).

e. Torque all bolts, in sequence, to 40–47 ft. lbs (54–64 Nm). Using an angle torque wrench, torque them 60–65 degrees tighter rather than going to 40–47 ft. lbs. (54–64 Nm)

17. Install the alternator and air conditioner compressor mounting brackets. Mount the compressor and alternator.
18. Install the rear timing cover bolts. Install the camshaft pulleys. Make sure the pulley marked R3 goes on the right and that marked L3 goes on the left. Align the timing marks if necessary and then install the timing belt and adjust the belt tension.
19. Connect the exhaust manifold to the exhaust pipe.
20. Install the exhaust manifold covers.
21. Install the exhaust collector bracket.
22. Install the intake manifold and fuel tube assembly.
23. Install the intake manifold collector cover.
24. Connect the accelerator and cruise control cables to the intake manifold and install the distributor and ignition wires.
25. Connect the vacuum and water hoses to the intake collector.
26. Install and tension the timing belt.
27. Fill the cooling system and connect the negative battery cable.
28. Make all the necessary engine adjustments.

Q45

1. Disconnect the negative battery cable.
2. Remove the engine and transmission assembly from the vehicle.
3. Remove the suspension member and engine mounts from the engine.
4. Remove the air compressor bracket and the exhaust manifolds.
5. Remove the cooling fan with coupling and the engine gusset.
6. Separate the engine from the transmission and mount the engine on a suitable workstand.
7. Remove the oil pan. Remove the intake collector.
8. Disconnect the injector harness connector and remove the injector tube assembly with injector. Loosen bolts in opposite sequence of torquing..

NOTE: Be careful not to let the rubber washer fall into the intake manifold.

9. Remove the intake manifold.
10. Remove the ornamental rocker cover and remove the ignition coils and spark plugs.
11. Bring the No. 1 piston to TDC on the compression stroke.
12. Use a suitable puller to remove the crankshaft pulley.
13. Remove the rocker cover.
14. Remove the crank angle sensor and the Valve Timing Control (VTC) solenoid.
15. Remove the chain tensioners and the upper front covers.
16. Remove the front timing chain cover.

NOTE: The timing chain will not be disengaged or dislocated from the crankshaft sprocket unless the front cover is removed. The cast portion of the front cover is located on the lower side of the crankshaft sprocket so the timing chain is not disengaged from the sprocket.

17. Remove the VTC assembly and the camshaft sprocket.
18. Remove the oil pump chain and the timing chains.

NOTE: do not attempt to disassemble the VTC assembly since they are difficult to reassemble accurately in the field. If it should be disassembled, the VTC assembly must be replaced with a new one.

19. Remove the camshaft brackets in the reverse order of torquing sequence. Use 2–3 steps. Remove the camshafts. Mark the parts so they can be reinstalled in their original positions.
20. Remove the rocker arm and hydraulic lash adjuster. Be sure to identify each adjuster so it can be reinstalled in it's original position.
21. Remove the cylinder head and gasket. Loosen the head bolts in 2–3 steps working from the outside bolts in towards the center bolts.

To install:

22. Make sure all mating surfaces are clean before installation.
23. Check the cylinder head surface for warpage using a feeler gauge and a suitable straightedge. If the cylinder head is warped more than 0.004 in. (0.1mm) it must be resurfaced or replaced. The total amount machined from the head or head and block combined, cannot total more than 0.008 in. (0.2mm).
24. Make sure the No. 1 piston is still at TDC of the compression stroke, then turn the crankshaft until the No. 1 piston is at approximately 45 degrees before TDC on the compression stroke. At this point, the No. 3 piston will be at the same height as the No. 1 piston to prevent interference of the valves and pistons.
25. Install the cylinder heads with new gaskets. Temporarily tighten the cylinder head bolts to avoid damaging the cylinder head gaskets. Be sure to install washers between the bolts and the cylinder heads. Do not rotate the crankshaft or camshaft separately or the valves will hit the pistons.
26. Install the hydraulic lash adjusters and check them as follows:

a. When the rocker arm can be moved at least 0.04 in. (1.0mm) by pushing at the hydraulic lash adjuster location, it indicates that there is air in the high pressure chamber. Noise will be emitted from the hydraulic lash adjuster if the engine is started without bleeding the air.

b. Remove the hydraulic lash adjuster and dip in a container filled with engine oil. While pushing the top of the plunger down, insert a suitable thin rod through the hole in the top of the plunger and lightly push the check ball. Air is completely bled when the plunger no longer moves.

NOTE: Air cannot be bled from the lash adjusters by running the engine.

27. Install the rocker arms, camshafts and camshaft brackets on the right bank and tighten in the proper sequence to 9–10 ft. lbs (12–14 Nm).
28. Install the VTC assembly and the exhaust cam sprocket on the right bank.
29. After making sure the camshafts are still correctly positioned, turn the crankshaft clockwise to bring the No. 1 piston to TDC on the compression stroke.
30. Install the timing chain on the right bank, aligning the mating marks on the chain with those on the crankshaft and camshaft sprockets.
31. Install the chain tensioner on the right bank.
32. Turn the crankshaft approximately 120 degrees clockwise from the point where the No. 1 piston is at TDC

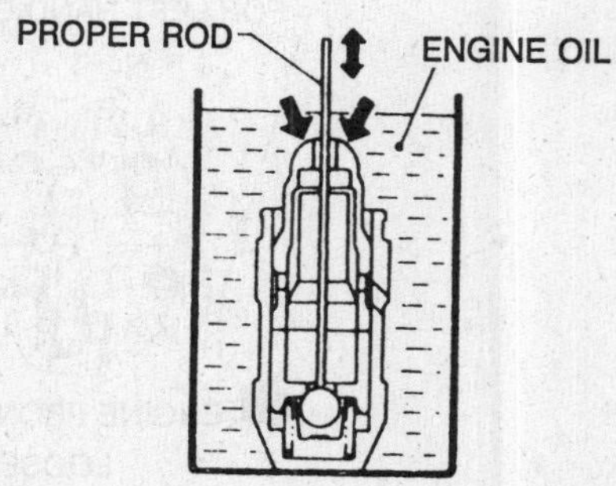

Hydraulic lash adjuster bleeding—Q45

on the compression stroke. At this point, the valves on the left bank still remain unlifted.

33. Correctly position the camshafts for the left cylinder head and tighten brackets in the proper sequence to 9–10 ft. lbs (12–14 Nm).

Install the VTC assembly and the exhaust cam sprocket.

34.
Install the timing chain on the left bank, aligning the mating marks on the chain with those on the crankshaft and camshaft sprockets.

35.
Install the oil pump chain and sprockets.

36.
Install the oil pump chain guides. Place a 0.04 in. (1.0mm) feeler gauge between the upper chain guide and chain before assembling the chain guides. The force applied to the chain is equivalent to the upper chain guide weight.

37.
Apply suitable sealer and install the front covers.

38.
Install the chain tensioner for the left bank.

39.
Apply suitable sealer to the rubber plugs and install them on the cylinder head.

40.
Install the crank angle sensor, VTC solenoid, rocker cover and crank pulley.

41.
Bring the piston in No. 1 cylinder to TDC on the compression stroke.

42. Tighten the cylinder head bolts

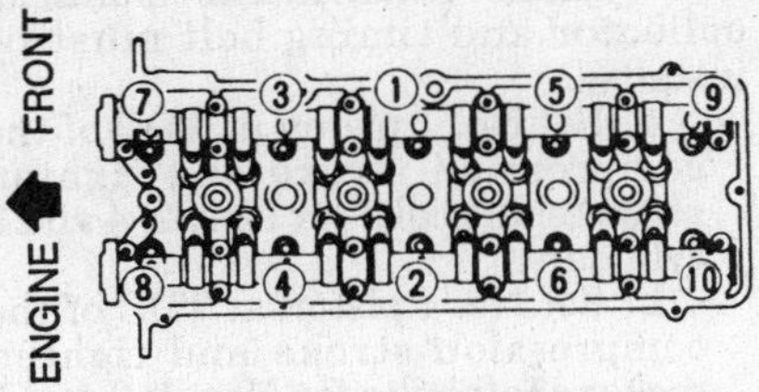

Cylinder head torque sequence—Q45

in the proper torque sequence as follows:

a. Tighten the bolts in sequence to 22 ft. lbs. (29 Nm).

b. Tighten the bolts in sequence to 69 ft. lbs. (93 Nm).

c. Loosen the bolts completely.

d. Tighten the bolts in sequence to 18–25 ft. lbs. (25–34 Nm).

e. Turn the bolts in sequence 90–95 degrees or 69–72 ft. lbs. (93–98 Nm).

43. Install the intake manifold bolts in their proper positions on the cylinder head and lightly tighten the mounting bolts.

44. Connect the injector tube assemblies, including the fuel injectors, to the intake manifolds and lightly tighten the mounting bolts.

NOTE: Be careful not to let the rubber washer fall into the intake manifold.

45. Install the intake collector and lightly tighten the mounting bolts.

46. Tighten the intake manifold mounting bolts at the cylinder head, remove the intake collectors and tighten the intake manifolds to 12–15 ft. lbs. (16–21 Nm).

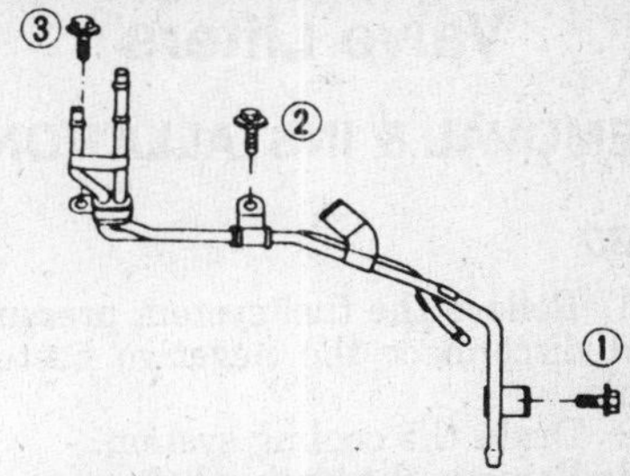

Sub-fuel tubes torque sequence

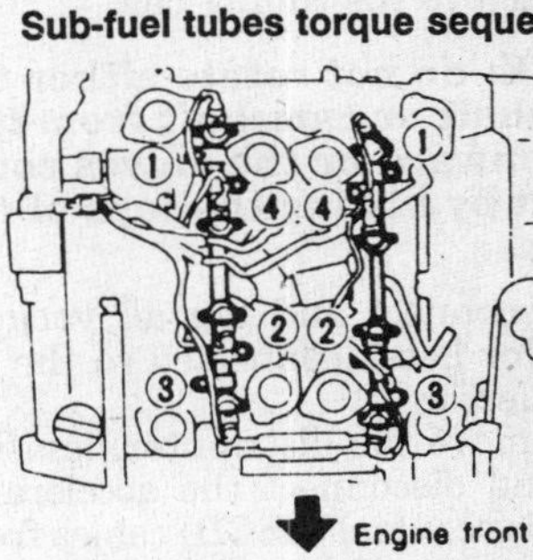

Injector tube torque sequence

47. Tighten the sub-fuel tubes, in sequence, first to 3.1–4.3 ft. lbs. (4.2–5.9 Nm) and then to 6.2–8.0 ft. lbs. (8.4–10.8 Nm).

48. Tighten the injector tube assemblies, in sequence, first to 6.9–8.0 ft. lbs. (9.3–10.8 Nm) and then to 15–20 ft. lbs. (21–26 Nm).

49. Install the intake collectors and tighten to 9–11 ft. lbs. (12–15 Nm).

50. Install the exhaust manifolds.

51. Install the rocker covers and tighten in the proper sequence to 5–7 ft. lbs. (7–10 Nm).

52. Installation of the remaining components is the reverse of the removal procedure.

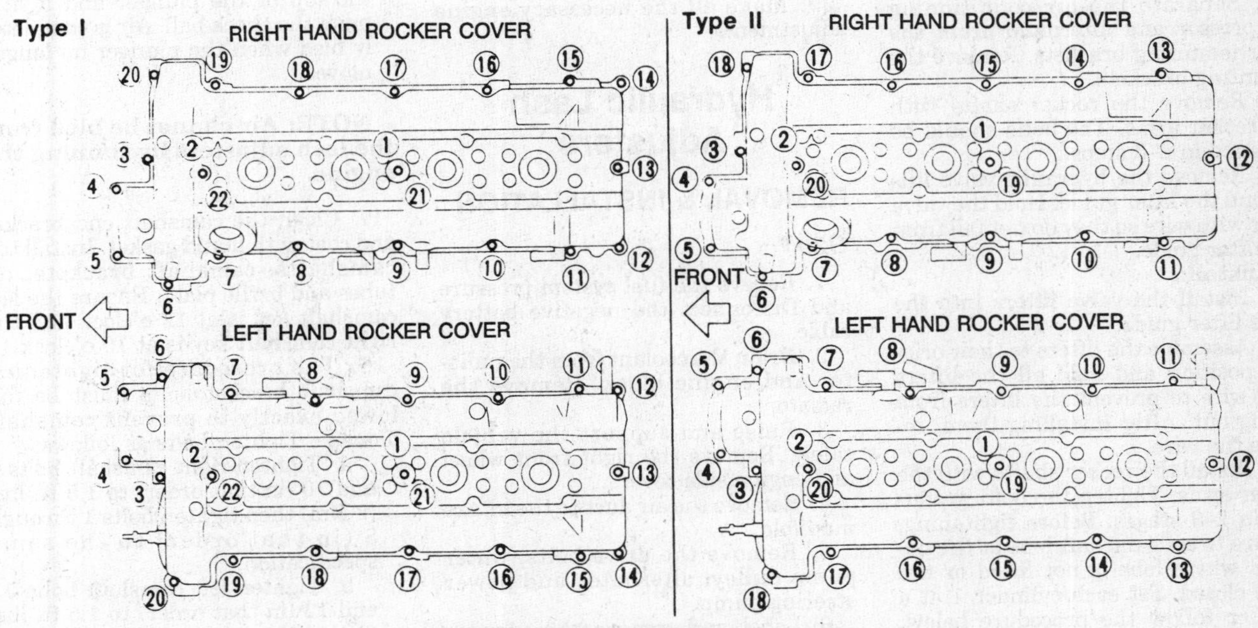

Rocker cover bolt torque sequence—Q45

Valve Lifters

REMOVAL & INSTALLATION

M30

1. Relieve the fuel system pressure and disconnect the negative battery cable.
2. Drain the cooling system.
3. Remove the timing belt.

NOTE: do not rotate either the crankshaft or camshaft from this point onward or the valves could be bent by hitting the tops of the pistons.

4. Disconnect and tag all vacuum and water hoses connected to the intake collector.
5. Remove the distributor, ignition wires and disconnect the accelerator and cruise control (ASCD) cables from the intake manifold collector.
6. Remove the collector cover and the collector from the intake manifold. Disconnect and tag all harness connectors and vacuum lines to gain access to the cover retaining bolts on these models.
7. Remove the intake manifold and fuel tube assembly. Loosen the intake manifold bolts starting from the front of the engine and proceed in criss-cross pattern towards the center.
8. Remove the exhaust collector bracket.
9. Remove the exhaust manifold covers.
10. Disconnect the exhaust manifold from the exhaust pipe.
11. Remove the camshaft pulleys and the rear timing cover securing bolts. Remove the rocker arm covers.
12. Separate the air conditioning compressor and alternator from the their mounting brackets. Remove the mounting brackets.
13. Remove the rocker shafts with the rocker arms. The bolts should be loosened in 2–3 steps.
14. Remove the hydraulic valve lifters and the lifter guide. Hold the valve lifter with wire so they do not fall from the lifter guide.

To install:

15. Install the valve lifters into the valve lifter guide.
16. Assemble the lifters to their original position and hold all the lifters with wire to prevent the lifters from falling out. After installing them, remove the wire.
17. Install the rocker shafts with the rocker arms. Tighten the bolts gradually in 2–3 stages. Before tightening, be sure to set camshaft lobe at the position where lobe is not lifted or the valve closed. Set each cylinder 1 at a time or follow the procedure below. The cylinder head, intake manifold, collector and timing belt must be installed:
    a. Set No. 1 piston at TDC of the compression stroke and tighten rocker shaft bolts for Nos. 2, 4 and 6 cylinders.
    b. Set No. 4 piston at TDC of the compression stroke and tighten rocker shaft bolts for Nos. 1, 3 and 5 cylinders.
    c. Torque specification for the rocker shaft retaining bolts is 13–16 ft. lbs. (18–22 Nm).
18. Install the alternator and air conditioner compressor mounting brackets. Mount the compressor and alternator.
19. Install the rear timing cover bolts. Install the camshaft pulleys. Make sure the pulley marked R3 goes on the right and that marked L3 goes on the left. Align the timing marks if necessary and then install the timing belt and adjust the belt tension.
20. Connect the exhaust manifold to the exhaust pipe.
21. Install the exhaust manifold covers.
22. Install the exhaust collector bracket.
23. Install the intake manifold and fuel tube assembly.
24. Install the intake manifold collector cover.
25. Connect the accelerator and cruise control cables to the intake manifold and install the distributor and ignition wires.
26. Connect the vacuum and water hoses to the intake collector.
27. Install and tension the timing belt.
28. Fill the cooling system and connect the negative battery cable.
29. Make all the necessary engine adjustments.

Hydraulic Lash Adjusters

REMOVAL & INSTALLATION

G20

1. Relieve the fuel system pressure and Disconnect the negative battery cable.
2. Drain the coolant from the radiator and engine block. Remove the radiator.
3. Raise and support the vehicle safely. Remove the right front wheel and engine side cover.
4. Remove the air duct to the intake manifold.
5. Remove the drive belts, water pump pulley, alternator and power steering pump.
6. Label and remove the vacuum hoses, fuel hoses and wire harness connectors.
7. Remove all the spark plugs, the AIV valve and resonator.
8. Remove the rocker cover and oil separator. Loosen rocker cover bolts, using 2–3 steps, in the opposite sequence of tightening
9. Remove the intake manifold supports.
10. Remove the oil filter bracket and power steering oil pump bracket.
11. Set No. 1 piston at TDC on the compression stroke by rotating the crankshaft.
12. Remove the chain tensioner.
13. Remove the distributor. Do not turn the rotor with the distributor removed.
14. Remove the timing chain guide, camshaft sprockets, camshafts, brackets, oil tubes and baffle plate. The camshaft bracket bolts must be loosened in sequence to prevent damage to the camshafts or the head.
15. Remove the hydraulic lash adjuster and rocker arm assembly.

To install:

16. Install the hydraulic lash adjusters and check them as follows:
    a. When the rocker arm can be moved at least 0.04 in. (1.0mm) by pushing at the hydraulic lash adjuster location, it indicates that there is air in the high pressure chamber. Noise will be emitted from the hydraulic lash adjuster if the engine is started without bleeding the air.
    b. Remove the hydraulic lash adjuster and dip in a container filled with engine oil. While pushing the top of the plunger down, insert a suitable thin rod through the hole in the top of the plunger and lightly push the check ball. Air is completely bled when the plunger no longer moves.

NOTE: Air cannot be bled from the lash adjusters by running the engine.

17. Clean the camshaft end bracket and coat with liquid gasket. Install the camshafts, camshaft brackets, oil tubes and baffle plate. Ensure the left camshaft ket is at 12 o'clock and the right camshaft key is at 10 o'clock.
18. The procedure for tightening camshaft bracket bolts must be followed exactly to prevent camshaft damage. Tighten bolts as follows:
    a. Tighten right camshaft bolts 9 and 10 (in that order) to 1.5 ft. lbs. (2 Nm) then tighten bolts 1 through 8 (in that order) to the same specification.
    b. Tighten left camshaft bolts 11 and 12 (in that order) to 1.5 ft. lbs. (2 Nm) then tighten bolts 1 through

10 (in that order) to the same specification.

c. Tighten all bolts in sequence to 4.5 ft. lbs. (6 Nm).

d. Tighten all bolts in sequence to 6.5–8.5 ft. lbs. (9–12 Nm) for type A, B and C bolts, and 13–19 ft. lbs. (18–25 Nm) for type D bolts.

19. Line up the mating marks on the timing chain and camshaft sprockets and install the sprockets. Tighten sprocket bolts to 101–116 ft. lbs. (137–157 Nm).

20. Install the timing chain guide, distributor (ensure that rotor head is at 5 o'clock position) and chain tensioner.

21. Install intake manifold supports. Clean the rocker cover and mating surfaces and apply a continious bead of liquid gasket to the mating surface.

22. Install the rocker cover and oil separator. Tighten the rocker cover bolts as follows:

a. Tighten nuts 1, 10, 11, and 8 in that order to 3 ft. lbs. (4 Nm).

b. Tighten nuts 1 through 13 in the proper sequence to 6–7 ft. lbs. (8–10 Nm).

23. Installation of the remaining components is the reverse of removal procedures.

Q45

1. Disconnect the negative battery cable.

2. Remove the engine and transmission assembly from the vehicle.

3. Remove the suspension member and engine mounts from the engine.

4. Remove the air compressor bracket.

5. Remove the cooling fan with coupling and the engine gusset.

6. Separate the engine from the transmission and mount the engine on a suitable workstand.

7. Remove the oil pan.

8. Remove the ornamental rocker cover and remove the ignition coils and spark plugs.

9. Bring the No. 1 piston to TDC on the compression stroke.

10. Use a suitable puller to remove the crankshaft pulley.

11. Remove the rocker cover.

12. Remove the crank angle sensor and the Valve Timing Control (VTC) solenoid.

13. Remove the chain tensioners and the upper front covers.

14. Remove the front timing chain cover.

NOTE: The timing chain will not be disengaged or dislocated from the crankshaft sprocket unless the front cover is removed. The cast portion of the front cover is located on the lower side of the crankshaft sprocket so the timing chain is not disengaged from the sprocket.

15. Remove the VTC assembly and the camshaft sprocket.

16. Remove the oil pump chain and the timing chains.

NOTE: do not attempt to disassemble the VTC assembly since they are difficult to reassemble accurately in the field. If it should be disassembled, the VTC assembly must be replaced with a new one.

17. Remove the camshaft brackets and the camshafts. Mark the parts so they can be reinstalled in their original positions.

18. Remove the rocker arm and hydraulic lash adjuster. Be sure to identify each adjuster so it can be reinstalled in it's original position.

To install:

19. Make sure all mating surfaces are clean before installation.

20. Install the hydraulic lash adjusters and check them as follows:

a. When the rocker arm can be moved at least 0.04 in. (1.0mm) by pushing at the hydraulic lash adjuster location, it indicates that there is air in the high pressure chamber. Noise will be emitted from the hydraulic lash adjuster if the engine is started without bleeding the air.

b. Remove the hydraulic lash adjuster and dip in a container filled with engine oil. While pushing the top of the plunger down, insert a suitable thin rod through the hole in the top of the plunger and lightly push the check ball. Air is completely bled when the plunger no longer moves.

NOTE: Air cannot be bled from the lash adjusters by running the engine.

21. Install the rocker arms, camshafts and camshaft brackets on the right bank.

22. Install the VTC assembly and the exhaust cam sprocket on the right bank.

23. Make sure the camshafts are still correctly positioned and the piston in the No. 1 cylinder is still at TDC.

24. Install the timing chain on the right bank, aligning the mating marks on the chain with those on the crankshaft and camshaft sprockets.

25. Install the chain tensioner on the right bank.

26. Turn the crankshaft approximately 120 degrees clockwise from the point where the No. 1 piston is at TDC on the compression stroke. At this point, the valves on the left bank still remain unlifted.

27. Correctly position the camshafts for the left cylinder head. Install the VTC assembly and the exhaust cam sprocket.

28. Install the timing chain on the left bank, aligning the mating marks on the chain with those on the crankshaft and camshaft sprockets.

29. Install the oil pump chain and sprockets.

30. Install the oil pump chain guides. Place a 0.04 in. (1.0mm) feeler gauge between the upper chain guide and chain before assembling the chain guides. The force applied to the chain is equivalent to the upper chain guide weight.

31. Apply suitable sealer and install the front covers.

32. Install the chain tensioner for the left bank.

33. Apply suitable sealer to the rubber plugs and install them on the cylinder head.

34. Install the crank angle sensor, VTC solenoid, rocker cover and crank pulley.

35. Installation of the remaining components is the reverse of the removal procedure.

Rocker Arms

REMOVAL & INSTALLATION

G20

1. Relieve the fuel system pressure and Disconnect the negative battery cable.

2. Drain the coolant from the radiator and engine block. Remove the radiator.

3. Raise and support the vehicle safely. Remove the right front wheel and engine side cover.

4. Remove the air duct to the intake manifold.

5. Remove the drive belts, water pump pulley, alternator and power steering pump.

6. Label and remove the vacuum hoses, fuel hoses and wire harness connectors.

7. Remove all the spark plugs, the AIV valve and resonator.

8. Remove the rocker cover and oil separator. Loosen rocker cover bolts, using 2 through 3 steps, in the opposite sequence of tightening

9. Remove the intake manifold supports.

10. Remove the oil filter bracket and power steering oil pump bracket.

11. Set No. 1 piston at TDC on the compression stroke by rotating the crankshaft.

12. Remove the chain tensioner.

13. Remove the distributor. Do not

turn the rotor with the distributor removed.

14. Remove the timing chain guide, camshaft sprockets, camshafts, brackets, oil tubes and baffle plate. The camshaft bracket bolts must be loosened in sequence to prevent damage to the camshafts or the head.

15. Remove rocker arm assembly.

To install:

16. Check the hydraulic lash adjusters to ensure they did not bleed down during disassembly. If bleed down has occured, remove the lash adjuster and reprime.

NOTE: Air cannot be bled from the lash adjusters by running the engine.

17. Clean the camshaft end bracket and coat with liquid gasket. Install the camshafts, camshaft brackets, oil tubes and baffle plate. Ensure the left camshaft ket is at 12 o'clock and the right camshaft key is at 10 o'clock.

18. The procedure for tightening camshaft bracket bolts must be followed exactly to prevent camshaft damage. Tighten bolts as follows:

a. Tighten right camshaft bolts 9 and 10 (in that order) to 1.5 ft. lbs. (2 Nm) then tighten bolts 1 through 8 (in that order) to the same specification.

b. Tighten left camshaft bolts 11 and 12 (in that order) to 1.5 ft. lbs. (2 Nm) then tighten bolts 1 through 10 (in that order) to the same specification.

c. Tighten all bolts in sequence to 4.5 ft. lbs. (6 Nm).

d. Tighten all bolts in sequence to 6.5–8.5 ft. lbs. (9–12 Nm) for type A, B and C bolts, and 13–19 ft. lbs. (18–25 Nm) for type D bolts.

19. Line up the mating marks on the timing chain and camshaft sprockets and install the sprockets. Tighten sprocket bolts to 101–116 ft. lbs. (137–157 Nm).

20. Install the timing chain guide, distributor (ensure that rotor head is at 5 o'clock position) and chain tensioner.

21. Install intake manifold supports. Clean the rocker cover and mating surfaces and apply a continious bead of liquid gasket to the mating surface.

22. Install the rocker cover and oil separator. Tighten the rocker cover bolts as follows:

a. Tighten nuts 1, 10, 11, and 8 in that order to 3 ft. lbs. (4 Nm).

b. Tighten nuts 1 through 13 in the proper sequence to 6–7 ft. lbs. (8–10 Nm).

32. Installation of the remaining components is the reverse of removal procedures.

M30

1. Relieve the fuel system pressure and disconnect the negative battery cable.

2. Drain the cooling system.

3. Remove the timing belt.

NOTE: do not rotate either the crankshaft or camshaft from this point onward or the valves could be bent by hitting the tops of the pistons.

4. Disconnect and tag all vacuum and water hoses connected to the intake collector.

5. Remove the distributor, ignition wires and disconnect the accelerator and cruise control (ASCD) cables from the intake manifold collector.

6. Remove the collector cover and the collector from the intake manifold. Disconnect and tag all harness connectors and vacuum lines to gain access to the cover retaining bolts on these models.

7. Remove the intake manifold and fuel tube assembly. Loosen the intake manifold bolts starting from the front of the engine and proceed in criss-cross pattern towards the center.

8. Remove the exhaust collector bracket.

9. Remove the exhaust manifold covers.

10. Disconnect the exhaust manifold from the exhaust pipe.

11. Remove the camshaft pulleys and the rear timing cover securing bolts. Remove the rocker arm covers.

12. Separate the air conditioning compressor and alternator from the their mounting brackets. Remove the mounting brackets.

13. Remove the rocker shafts with the rocker arms. The bolts should be loosened in 2–3 steps.

To install:

14. Install the rocker shafts with the rocker arms. Tighten the bolts gradually in 2–3 stages. Before tightening, be sure to set camshaft lobe at the position where lobe is not lifted or the valve closed. Set each cylinder 1 at a time or follow the procedure below. The cylinder head, intake manifold, collector and timing belt must be installed:

a. Set No. 1 piston at TDC of the compression stroke and tighten rocker shaft bolts for Nos. 2, 4 and 6 cylinders.

b. Set No. 4 piston at TDC of the compression stroke and tighten rocker shaft bolts for Nos. 1, 3 and 5 cylinders.

c. Torque specification for the rocker shaft retaining bolts is 13–16 ft. lbs. (18–22 Nm).

15. Install the alternator and air conditioner compressor mounting brackets. Mount the compressor and alternator.

16. Install the rear timing cover bolts. Install the camshaft pulleys. Make sure the pulley marked R3 goes on the right and that marked L3 goes on the left. Align the timing marks, if necessary, and then install the timing belt and adjust the belt tension.

17. Connect the exhaust manifold to the exhaust pipe.

18. Install the exhaust manifold covers.

19. Install the exhaust collector bracket.

20. Install the intake manifold and fuel tube assembly.

21. Install the intake manifold collector cover.

22. Connect the accelerator and cruise control cables to the intake manifold and install the distributor and ignition wires.

23. Connect the vacuum and water hoses to the intake collector.

24. Install and tension the timing belt.

25. Fill the cooling system and connect the negative battery cable.

26. Make all the necessary engine adjustments.

Q45

1. Disconnect the negative battery cable.

2. Remove the engine and transmission assembly from the vehicle.

3. Remove the suspension member and engine mounts from the engine.

4. Remove the air compressor bracket.

5. Remove the cooling fan with coupling and the engine gusset.

6. Separate the engine from the transmission and mount the engine on a suitable workstand.

7. Remove the oil pan.

8. Remove the ornamental rocker cover and remove the ignition coils and spark plugs.

9. Bring the No. 1 piston to TDC on the compression stroke.

10. Use a suitable puller to remove the crankshaft pulley.

11. Remove the rocker cover.

12. Remove the crank angle sensor and the Valve Timing Control (VTC) solenoid.

13. Remove the chain tensioners and the upper front covers.

14. Remove the front timing chain cover.

NOTE: The timing chain will not be disengaged or dislocated from the crankshaft sprocket unless the front cover is removed. The cast portion of the front cover is located on the lower side of the crankshaft sprocket so the

timing chain is not disengaged from the sprocket.

15. Remove the VTC assembly and the camshaft sprocket.
16. Remove the oil pump chain and the timing chains.

NOTE: do not attempt to disassemble the VTC assembly since they are difficult to reassemble accurately in the field. If it should be disassembled, the VTC assembly must be replaced with a new one.

17. Remove the camshaft brackets and the camshafts. Mark the parts so they can be reinstalled in their original positions.
18. Remove the rocker arms. Be sure to identify each rocker arm so it can be reinstalled in it's original position.

To install:

19. Make sure all mating surfaces are clean before installation.
20. Install the rocker arms, camshafts and camshaft brackets on the right bank. Properly lubricate the rocker arms and camshafts prior to installation.
21. Install the VTC assembly and the exhaust cam sprocket on the right bank.
22. Make sure the camshafts are still correctly positioned and the piston in the No. 1 cylinder is still at TDC.
23. Install the timing chain on the right bank, aligning the mating marks on the chain with those on the crankshaft and camshaft sprockets.
24. Install the chain tensioner on the right bank.
25. Turn the crankshaft approximately 120 degrees clockwise from the point where the No. 1 piston is at TDC on the compression stroke. At this point, the valves on the left bank still remain closed.
26. Correctly position the camshafts and rocker arms for the left cylinder head. Properly lubricate the rocker arms and camshafts prior to installation. Install the VTC assembly and the exhaust cam sprocket.
27. Install the timing chain on the left bank, aligning the mating marks on the chain with those on the crankshaft and camshaft sprockets.
28. Install the oil pump chain and sprockets.
29. Install the oil pump chain guides. Place a 0.04 in. (1.0mm) feeler gauge between the upper chain guide and chain before assembling the chain guides. The force applied to the chain is equivalent to the upper chain guide weight.
30. Apply suitable sealer and install the front covers.
31. Install the chain tensioner for the left bank.
32. Apply suitable sealer to the rubber plugs and install them on the cylinder head.
33. Install the crank angle sensor, VTC solenoid, rocker cover and crank pulley.
34. Installation of the remaining components is the reverse of the removal procedure.

Intake Manifold

REMOVAL & INSTALLATION

G20

1. Disconnect the negative battery cable.
2. Properly relieve the fuel system pressure.
3. Drain the cooling system.
4. Tag and disconnect the fuel lines, vacuum hoses and electrical connectors. Disconnect the throttle linkage.
5. Remove the intake manifold collector. Loosen bolts in the reverse order of the torquing sequence.
6. Remove the injector tube assembly and remove the intake manifolds. Loosen bolts in the reverse order of the torquing sequence.

To install:

7. Make sure all mating surfaces are clean prior to installation.
8. Install the intake manifold bolts, in their proper positions, on the cylinder head and lightly tighten the mounting bolts.
9. Connect the injector tube assemblies, including the fuel injectors, to the intake manifolds and lightly tighten the mounting bolts.
10. Install the intake collector and lightly tighten the mounting bolts.
11. To tighten the intake manifold mounting bolts at the cylinder head, remove the intake collector bolts and tighten the intake manifolds in sequence to 13–15 ft. lbs. (18–21 Nm).
12. Tighten the injector tube assemblies, in sequence, first to 6.9–8.0 ft. lbs. (9.3–10.8 Nm) and then to 15–20 ft. lbs. (21–26 Nm).
13. Install the intake collector and tighten the bolts in sequence to 12–15 ft. lbs. (16–21 Nm).
14. Reconnect the fuel lines, vacuum hoses and electrical connectors. Disconnect the throttle linkage.
15. Refill the cooling system, connect the negative battery cable and start engine and test for leaks.

M30

1. Relieve the fuel system pressure, disconnect the negative battery cable and drain the cooling system.
2. Remove the distributor and the ignition wires.
3. Disconnect the ASCD and accelerator wires from the intake manifold collector.

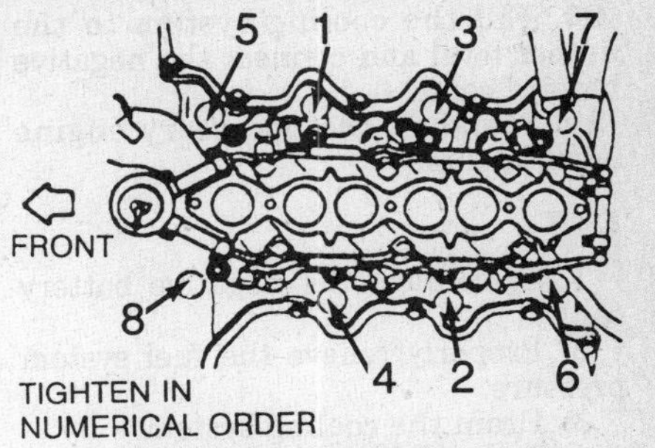

Intake manifold installation torque sequence – M30

4. Disconnect the harness connectors for the AAC valve, throttle sensor and idle switch.
5. Disconnect the air cut out valve water hose.
6. Disconnect the PCV valve hoses.
7. Disconnect the vacuum hoses from the vacuum gallery, swirl control valve, master brake cylinder, EGR control valve and EGR flare tube.
8. Loosen the upper collector cover bolts in proper sequence and remove the upper intake manifold collector from the engine. Remove the collector gasket.
9. Disconnect the engine ground harness.
10. Loosen the lower collector bolts, in sequence, and remove the lower intake manifold collector from the engine.
11. Disconnect the harness connectors for all injectors, engine temperature switch and sensor, power valve control solenoid valve, EGR control solenoid valve, EGR. temperature sensor (California only).
12. Disconnect the vacuum gallery hoses.
13. Disconnect the pressure regulator valve vacuum hose, heater hose, fuel feed and return hose.
14. Remove the intake manifold and fuel tube assembly. Loosen intake manifold bolts in numerical order.

To install:

15. Install the intake manifold and fuel tube assembly with a new gasket. Tighten the manifold bolts and nuts, in 2–3 stages, in sequence.
16. Connect the hoses and electrical wires to the intake manifold and fuel tube.
17. Install the upper and lower collector and collector cover with new gaskets. Tighten collector to intake manifold bolts, in 2–3 stages, by reversing the removal sequence.
18. Connect the vacuum lines, hoses, cables and brackets to the collector cover and collector assembly.
19. Install the distributor and ignition wires.

20. Fill the cooling system to the proper level and connect the negative battery cable.
21. Make all the necessary engine adjustments.

Q45

1. Disconnect the negative battery cable.
2. Properly relieve the fuel system pressure.
3. Drain the cooling system.
4. Tag and disconnect the fuel lines, vacuum hoses and electrical connectors. Disconnect the throttle linkage.
5. Remove the intake manifold collector.
6. Remove the injector tube assembly and remove the intake manifolds.

To install:

7. Make sure all mating surfaces are clean prior to installation.
8. Install the intake manifold bolts, in their proper positions, on the cylinder head and lightly tighten the mounting bolts.
9. Connect the injector tube assemblies, including the fuel injectors, to the intake manifolds and lightly tighten the mounting bolts.

NOTE: Be careful not to let the rubber washer fall into the intake manifold.

10. Install the intake collector and lightly tighten the mounting bolts.
11. Tighten the intake manifold mounting bolts at the cylinder head, remove the intake collectors and tighten the intake manifolds to 12–15 ft. lbs. (16–21 Nm).
12. Tighten the sub-fuel tubes, in sequence, first to 3.1–4.3 ft. lbs. (4.2–5.9 Nm) and then to 6.2–8.0 ft. lbs. (8.4–10.8 Nm).
13. Tighten the injector tube assemblies, in sequence, first to 6.9–8.0 ft. lbs. (9.3–10.8 Nm) and then to 15–20 ft. lbs. (21–26 Nm).
14. Install the intake collector and tighten to 9–11 ft. lbs. (12–15 Nm).
15. Install the remaining components in the reverse order of their removal.

Exhaust Manifold

REMOVAL & INSTALLATION

G20

1. Disconnect the negative battery cable. Raise and support the vehicle safely.
2. Remove the undercover and dust covers, if equipped. Disconnect the exhaust pipe at the manifold flange.
3. Remove the AIV, AIV tube and attaching bracket.
4. Disconnect the exhuast gas sensor electrical connection and remove the sensor.
5. Remove the exhaust manifold cover.
6. Remove the exhaust manifold nuts in reverse order of torquing sequence.
7. Remove the exhaust manifold and gasket.

To install:

8. Clean the gasket mating surface and intall a new exhaust manifold gasket.
9. Install the exhaust manifold and tighten the manifold nuts, in sequence, to 27–35 ft. lbs. (37–48 Nm).
10. Install the exhaust manifold cover and exhaust gas sensor. Reconnect the sensor electrical connection.
11. Install the AIV, AIV tube and attaching bracket.
12. Install the exhaust pipe to the manifold flange and tighten the nuts to 30– 35 ft. lbs. (41–48 Nm).
13. Lower the vehicle, start the engine and check for leaks.

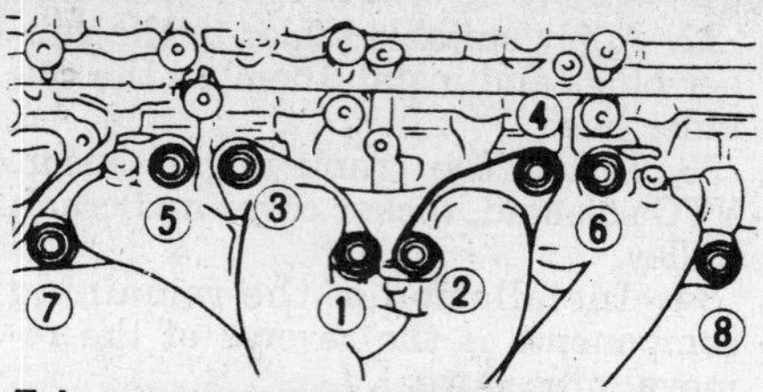

Exhaust manifold tightening sequence – G20

M30

1. Disconnect the negative battery cable. Raise and support the vehicle safely.
3. Remove the air cleaner or collector assembly, if necessary for access.
4. Remove the heat shield(s), if equipped.
5. Disconnect the exhaust pipe from the exhaust manifold.
6. Remove or disconnect the temperature sensors, oxygen sensors, air induction pipes, bracketry and other attachments from the manifold.
7. Loosen and remove the exhaust manifold attaching nuts and remove the manifold(s) from the block. Discard the exhaust manifold gaskets and replace with new.
8. Clean the gasket surfaces and check the manifold for cracks and warpage.

To install:

9. Install the exhaust manifold with a new gasket. Torque the manifold fasteners from the center outward in several stages.
10. Install or connect the temperature sensors, oxygen sensors, air induction pipes, brackets and other attachments to the manifold.

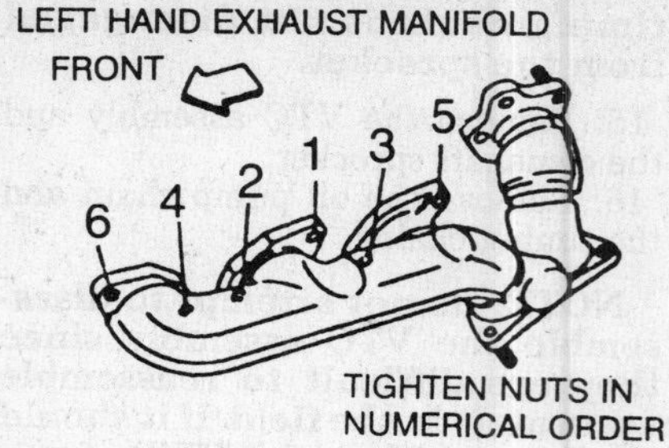

Left hand exhaust manifold installation torque sequence – M30

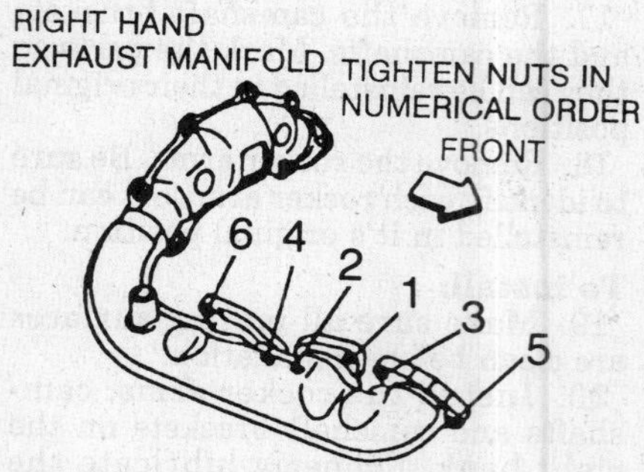

Right hand exhaust manifold installation torque sequence – M30

11. Connect the exhaust pipe to the manifold or turbo outlet using a new gasket.
12. Install the heat shields.
13. Install the air cleaner or collector assembly.
14. Install the under covers and dust covers.
15. Connect the negative battery cable.

Q45

1. Disconnect the negative battery cable. Raise and support the vehicle safely.
2. Remove the undercover and dust covers, if equipped. Disconnect the exhaust pipe at the manifold flange.
3. Disconnect the exhuast gas sensor electrical connection and if necessary, remove the sensor.
4. Remove the exhaust manifold nuts in reverse order of torquing sequence.
5. Remove the exhaust manifold and gasket.

To install:

6. Clean the gasket mating surface and intall a new exhaust manifold gasket.
7. Install the exhaust manifold and tighten the manifold nuts, in sequence, to 20–23 ft. lbs. (27–31 Nm).
8. Install exhaust gas sensor (if removed) and tighten to 30–37 ft. lbs. (40–50 Nm. Reconnect the sensor electrical connection.
9. Install the exhaust pipe to the manifold flange and tighten the nuts

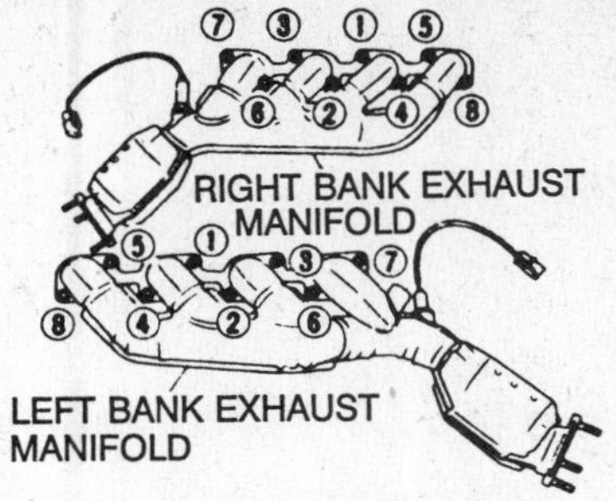

Exhaust manifold torque sequence—Q45

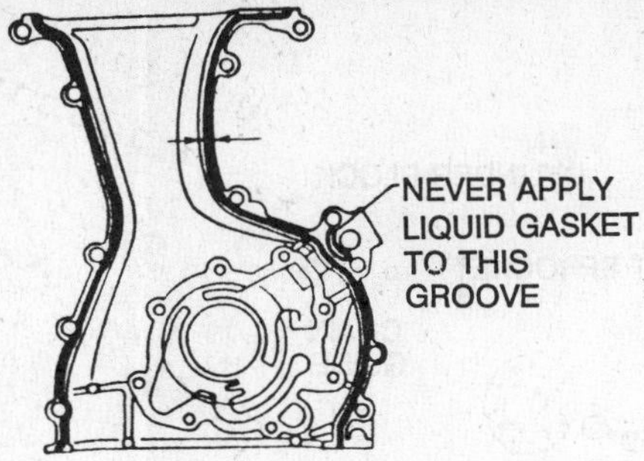

Front timing cover gasket sealing surface—G20

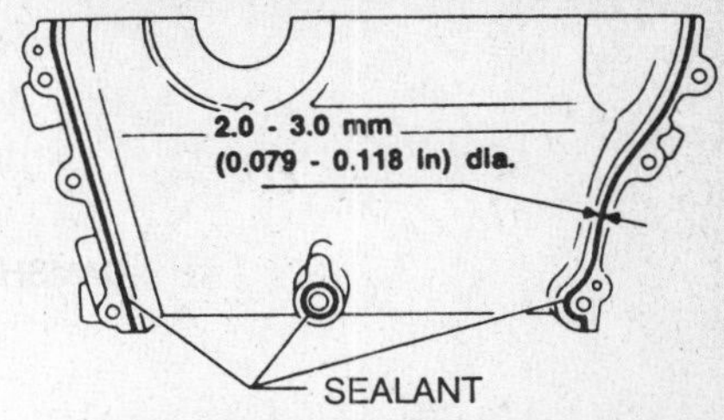

Upper front right cover sealant application areas—Q45

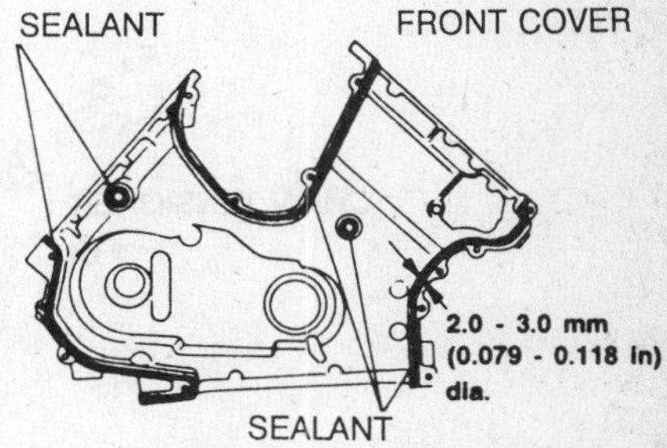

Front cover sealant application areas—Q45

to 33– 44 ft. lbs. (45–60 Nm).

13. Lower the vehicle, start the engine and check for leaks.

Timing Chain Front Cover

REMOVAL & INSTALLATION

G20

1. Disconnect the negative battery cable.
2. Drain the engine oil and coolant.
3. Remove the cylinder head.
4. Raise and support the vehicle safely. Remove the oil pan, oil strainer and baffle plate.
5. Remove the crankshaft pulley using a suitable puller. Removal of the radiator may be necessary to gain clearance.
6. Place a suitable jack under the main bearing beam. Remove the front engine mount.
7. Loosen the front cover bolts in 2–3 steps and remove the front cover.

To install:

8. Clean all mating surfaces of liquid gasket material.
9. Apply a continious bead of liquid gasket to the mating surface of the timing cover. Install the oil pump drive spacer and front cover. Tighten front cover bolts to 5–6 ft. lbs. (6–8 Nm). Wipe excess liquid gasket material.
10. Install front engine mount.
11. Install crankshaft pulley and tighten bolt to 105–112 ft. lbs.(142–152 Nm). Set No. 1 piston at TDC on the compression stroke.
12. Install the oil strainer and baffle. Install the oil pan.
13. Before installing the cylinder head, place a bead of liquid gasket at the parting line between the front cover and the engine block.
14. Install the cylinder head.
15. Lower the vehicle, connect the negative battery cable, start the engine and check for leaks.

Q45

1. Disconnect the negative battery cable.
2. Remove the engine and transmission assembly from the vehicle.
3. Remove the suspension member and engine mounts from the engine.
4. Remove the air compressor bracket.
5. Remove the cooling fan with coupling and the engine gusset.
6. Separate the engine from the transmission and mount the engine on a suitable workstand.
7. Remove the oil pan.
8. Remove the ornamental rocker cover and remove the ignition coils and spark plugs.
9. Bring the No. 1 piston to TDC on the compression stroke.
10. Use a suitable puller to remove the crankshaft pulley.
11. Remove the rocker cover.
12. Remove the crank angle sensor and the Valve Timing Control (VTC) solenoid.
13. Remove the chain tensioners and the upper front covers.
14. Remove the front timing chain cover.
15. Installation is the reverse of the removal procedure. Make sure all mating surfaces are clean prior to installation. Apply a suitable sealant to the proper locations on the timing chain covers.

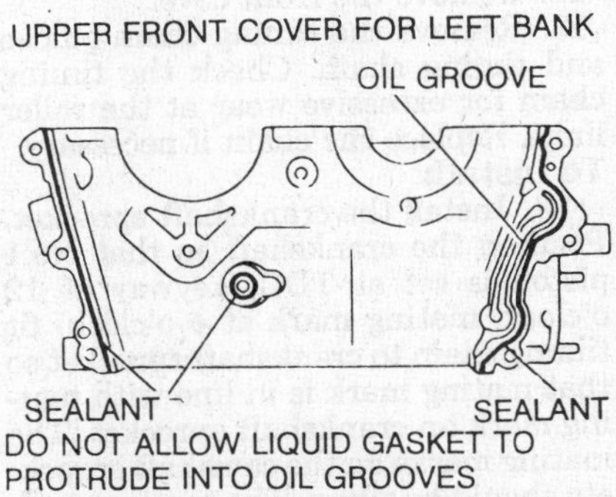

Upper front left cover sealant application areas—Q45

16. Tighten the cover bolts to 4.6–6.1 ft. lbs. (6.3–8.3 Nm) and the crankshaft pulley bolt to 260–275 ft. lbs. (353–373 Nm).

Front Cover Oil Seal

REPLACEMENT

G20

1. Raise and support the vehicle safely. Remove the engine under cover, right wheel and engine side cover.
2. Remove the drive belts.
3. Remove the crankshaft pulley using a suitable puller.
4. Pry the front seal out using a pry bar taking care not to damage the front cover.
5. Install a new seal lubricated with engine oil using a seal driver.
6. Install the crankshaft pulley, drive belts, engine covers and right wheel.
7. Start engine and check for leaks.

Q45

1. Disconnect the negative battery cable.
2. Raise and safely support the vehicle.
3. Remove the engine splash shield.
4. Remove the cooling fan and the engine gusset.
5. Remove the necessary accessory drive belts.
6. Remove the lower rear plate in order to remove the crankshaft pulley bolt.

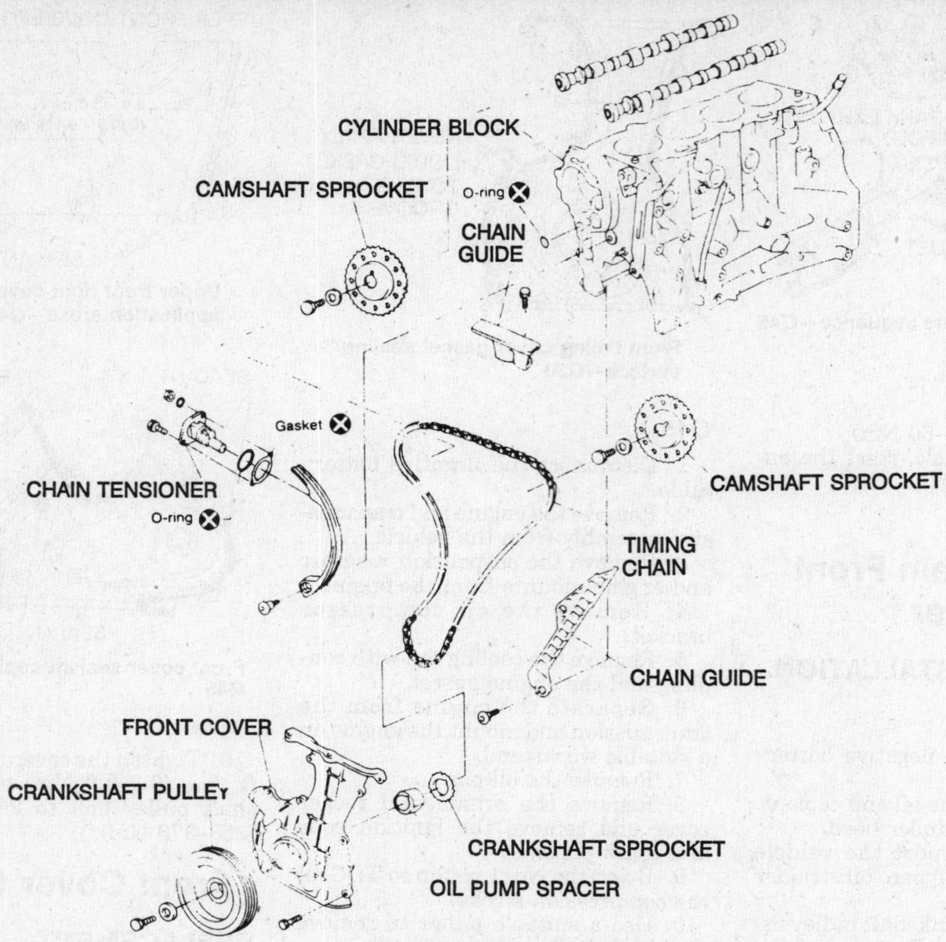

Timing chain assembly—G20

7. Remove the crankshaft pulley bolt and the crankshaft pulley.
8. Use a suitable tool to remove the front cover oil seal.
9. Installation is the reverse of the removal procedure. Lubricate the seal lip prior to installation. Tighten the crank pulley bolt to 260–275 ft. lbs. (353–373 Nm).

Timing Chain and Sprockets

REMOVAL & INSTALLATION

G20

1. Relieve the fuel system pressure and Disconnect the negative battery cable.
2. Drain the coolant from the radiator and engine block. Remove the radiator.
3. Raise and support the vehicle safely. Remove the right front wheel and engine side cover and lower the vehicle.
4. Remove the drive belts, water pump pulley, alternator and power steering pump.
5. Label and remove the vacuum hoses, fuel hoses and wire harness connectors.
6. Remove the cylinder head.
7. Raise and support the vehicle safely.
8. Remove the oil pan.
9. Remove the crankshaft pulley using a suitable puller.
10. Remove the engine front mount.
11. Remove the front cover.
12. Remove the timing chain guides and timing chain. Check the timing chain for excessive wear at the roller links. Replace the chain if necessary.

To install:

13. Install the crankshaft sprocket. Position the crankshaft so that No.1 piston is set at TDC (keyway at 12 o'clock, mating mark at 4 o'clock) fit timing chain to crankshaft sprocket so that mating mark is in line with mating mark on crankshaft sprocket. The mating marks on the camshaft sprockets should be silver. The mating mark on the crankshaft sprocket should be gold.
14. Install the timing chain and timing chain guides.
15. Install front engine mount.
16. Install the crankshaft pulley and set No.1 piston at TDC on the compression stroke.
17. Install the oil strainer, baffle plate and oil pan.
18. Install the cylinder head, camshafts, oil tubes and baffles. Position the left camshaft key at 12 o'clock and the right camshaft key at 10 o'clock.
19. Install the camshaft sprockets by lining up the mating marks on the timing chain with the mating marks on the camshaft sprockets. Tighten the camshaft bolts to 101–116 ft. lbs. (137–157 Nm).
20. Install the timing chain guide and distributor. Ensure rotor is at 5 o'clock position.
21. Install the chain tensioner. Press the cam stopper down and the press-in sleeve untill the hook can be engaged on the pin. When tensioner is bolted in position the hook will release automatically. Ensure the arrow on the outside faces the front of the engine.

22. Install all other components in reverse order of removal.

Q45

1. Disconnect the negative battery cable.
2. Remove the engine and transmission assembly from the vehicle.
3. Remove the suspension member and engine mounts from the engine.
4. Remove the air compressor bracket.
5. Remove the cooling fan with coupling and the engine gusset.
6. Separate the engine from the transmission and mount the engine on a suitable workstand.
7. Remove the oil pan.
8. Remove the ornamental rocker cover and remove the ignition coils and spark plugs.
9. Bring the No. 1 piston to TDC on the compression stroke.
10. Use a suitable puller to remove the crankshaft pulley.
11. Remove the rocker cover.
12. Remove the crank angle sensor and the VTC (Valve Timing Control) solenoid.
13. Remove the chain tensioners and the upper front covers.
14. Remove the front timing chain cover.

NOTE: The timing chain will not be disengaged or dislocated from the crankshaft sprocket unless the front cover is removed. The cast portion of the front cover is located on the lower side of the crankshaft sprocket so the timing chain is not disengaged from the sprocket.

15. Remove the VTC assembly and the camshaft sprocket.
16. Remove the oil pump chain and the timing chains.

NOTE: do not attempt to disassemble the VTC assembly since they are difficult to reassemble accurately in the field. If it should be disassembled, the VTC assembly must be replaced with a new one.

17. Use a suitable tool to remove the crankshaft sprocket.

To install:

18. Make sure all mating surfaces are clean before installation.
19. Install the VTC assembly and the exhaust cam sprocket on the right bank.
20. Make sure the camshafts are still correctly positioned and the piston in the No. 1 cylinder is still at TDC.
21. Install the timing chain on the right bank, aligning the mating marks on the chain with those on the crankshaft and camshaft sprockets.

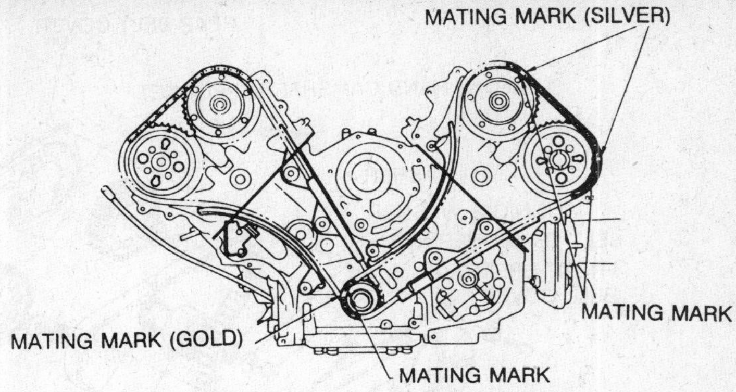

Left bank timing chain alignment—Q45

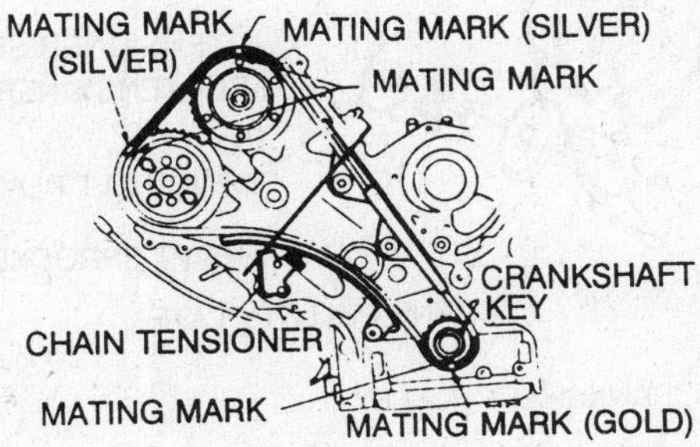

Right bank timing chain alignment—Q45

22. Install the chain tensioner on the right bank.
23. Turn the crankshaft approximately 120 degrees clockwise from the point where the No. 1 piston is at TDC on the compression stroke. At this point, the valves on the left bank still remain unlifted.
24. Correctly position the camshafts and rocker arms for the left cylinder head. Properly lubricate the rocker arms and camshafts prior to installation. Install the VTC assembly and the exhaust cam sprocket.
25. Install the timing chain on the left bank, aligning the mating marks on the chain with those on the crankshaft and camshaft sprockets.
26. Install the oil pump chain and sprockets.
27. Install the oil pump chain guides. Place a 0.04 in. (1.0mm) feeler gauge between the upper chain guide and chain before assembling the chain guides. The force applied to the chain is equivalent to the upper chain guide weight.
28. Apply suitable sealer and install the front covers.
29. Install the chain tensioner for the left bank.
30. Apply suitable sealer to the rubber plugs and install them on the cylinder head.
31. Install the crank angle sensor, VTC solenoid, rocker cover and crank pulley.
32. Installation of the remaining components is the reverse of the removal procedure.

Timing Belt Front Cover

REMOVAL & INSTALLATION

M30

1. Disconnect the negative battery cable.
2. Raise and support the front of the vehicle safely.
3. Remove the engine undercovers.
4. Drain the cooling system.
5. Remove the right front wheel.
6. Remove the engine side cover.
7. Remove the alternator, power steering and air conditioning compressor drive belts from the engine. When removing the power steering drive belt, loosen the idler pulley from the right side wheel housing.
8. Remove the upper radiator and water inlet hoses; remove the water pump pulley.
9. Remove the idler bracket of the compressor drive belt.
10. Remove the crankshaft pulley with a suitable puller.
11. Remove the upper and lower timing belt covers and gaskets.

To install:

12. Install the upper and lower timing belt covers with new gaskets.
13. Install the crankshaft pulley. Torque the pulley bolt to 90–98 ft. lbs. (123–132 Nm).
14. Install the compressor drive belt idler bracket.
15. Install the water pump pulley and torque the nuts to 12–15 ft. lbs. (16–21 Nm); install the upper radiator and water inlet hoses.
16. Install the drive belts.
17. Install the engine side cover.
18. Mount the front right wheel.
19. Install the engine undercovers.
20. Lower the vehicle.
21. Fill the cooling system and connect the negative battery cable.

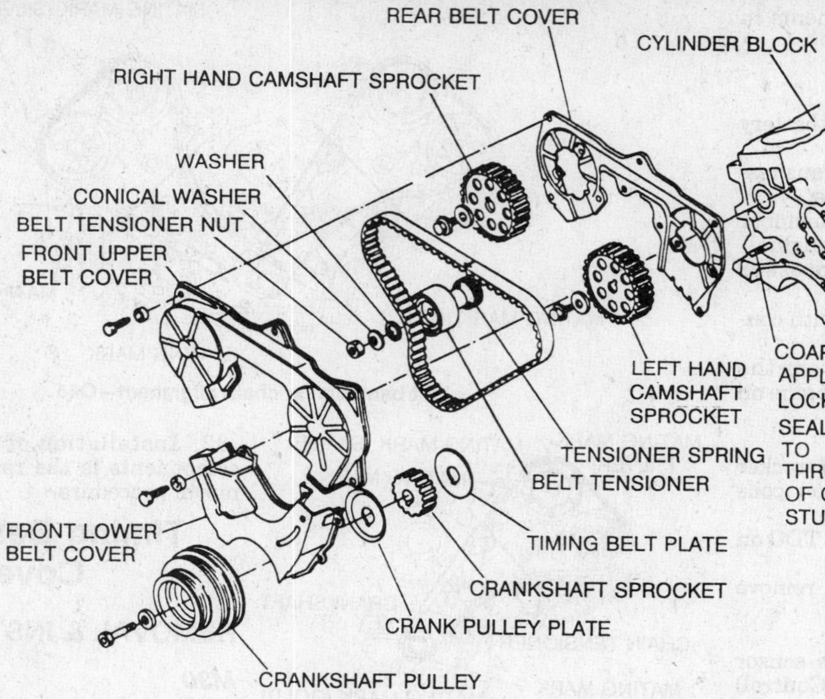

Timing belt installation—M30

OIL SEAL REPLACEMENT

M30

1. Disconnect the negative battery cable.
2. Remove the timing belt.
3. Remove the crankshaft sprocket.
4. Remove the oil pan and oil pump.
5. Using a suitable tool, pry the oil seal from the front cover.

NOTE: When removing the oil seal, be careful not the gouge or scratch the seal bore or crankshaft surface.

6. Wipe the seal bore with a clean rag.
7. Lubricate the lip of the new seal with clean engine oil.
8. Install the seal into the front cover with a suitable seal installer.
9. Install the oil pump and oil pan.
10. Install the crankshaft sprocket.
11. Install the timing belt.
12. Connect the negative battery cable.

Timing Belt and Tensioner

ADJUSTMENT

M30

1. Disconnect the negative battery cable.
2. Remove timing belt front covers.

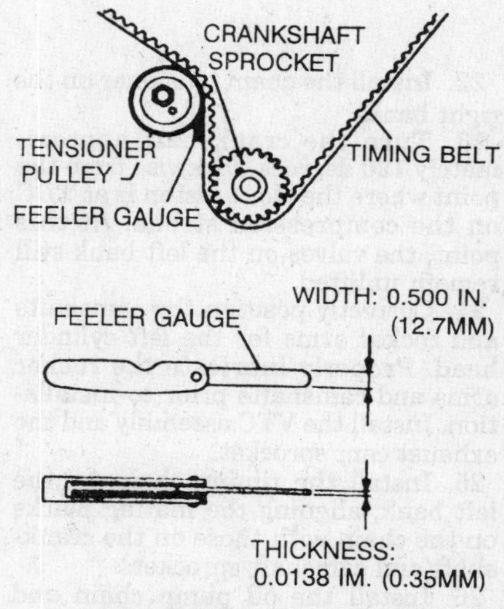

Setting timing belt tension—M30

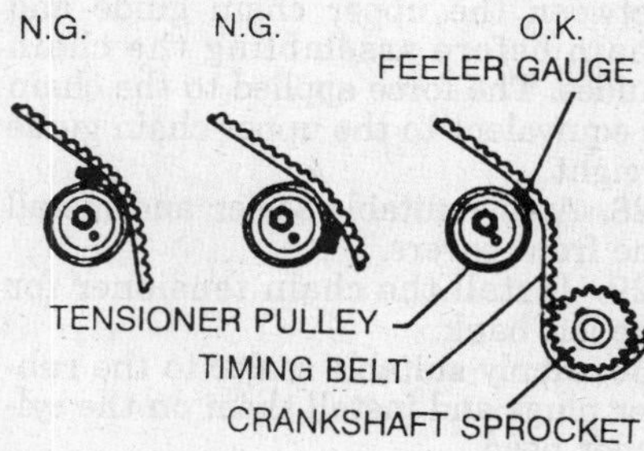

Proper feeler gauge position for setting timing belt tension—M30

3. Set engine to TDC No. 1 cylinder on its compression stroke.
4. Loosen the tensioner locknut, keeping the tension steady with the hexagonal wrench.
5. Turn tensioner 70–80 degrees clockwise with the hexagonal wrench. Temporarily tighten locknut.
6. Turn crankshaft clockwise at least 2 times, then slowly set the engine to TDC No. 1 cylinder on its compression stroke.
7. Push the middle of the timing belt between the right hand camshaft sprocket and tensioner pulley with a force of 22 lbs. (98 N).
8. Loosen the tensioner locknut, keeping the tensioner steady with the hexagonal wrench.
9. Set a feeler gauge 0.0138 in. (0.35mm) thick and 0.500 in. (12.7mm) wide between the timing belt and the tensioner pulley.
10. Turn the crankshaft clockwise and until the feeler gauge is on the tensioner pulley behind the timing belt.
11. Tighten the tensioner locknut, keeping tensioner steady with the hexagonal wrench.
12. Turn the crankshaft clockwise to remove the feeler gauge.
13. Turn the crankshaft clockwise at least 2 times, then slowly set the engine to TDC No. 1 on its compression stroke.
14. Install the timing belt covers.
15. Connect negative battery cable.

REMOVAL & INSTALLATION

M30

1. Disconnect the negative battery cable.
2. Raise and support the front of the vehicle safely.
3. Remove the engine undercovers.
4. Drain the cooling system.
5. Remove the front right side wheel.
6. Remove the engine side cover.
7. Remove the alternator, power steering and air conditioning compressor drive belts from the engine. When removing the power steering drive belt, loosen the idler pulley from the right side wheel housing.
8. Remove the upper radiator and water inlet hoses; remove the water pump pulley.
9. Remove the idler bracket of the compressor drive belt.
10. Remove the crankshaft pulley with a suitable puller.
11. Remove the upper and lower timing belt covers and gaskets.
12. Rotate the engine with a socket wrench on the crankshaft pulley bolt to align the punch mark on the left hand camshaft pulley with the mark on the upper rear timing belt cover; align the punchmark on the crankshaft with the notch on the oil pump housing; temporarily install the crankshaft pulley bolt to allow for crankshaft rotation.
13. Use a hex wrench to turn the belt tensioner clockwise and tighten the tensioner locknut just enough to hold the tensioner in position. Then, remove the timing belt.

To install:

14. Before installing the timing belt confirm that No. 1 cylinder is at TDC of the compression stroke. Install tensioner and tensioner spring. If stud is removed apply locking sealant to threads before installing.
15. Swing tensioner fully clockwise with hexagon wrench and temporarily tighten locknut.
16. Point the arrow on the timing belt toward the front belt cover. Align the white lines on the timing belt with the punch marks on all 3 pulleys.

NOTE: There are 133 total timing belt teeth. If timing belt is installed correctly there will be 40 teeth between left hand and right hand camshaft sprocket timing marks. There will be 43 teeth between left hand camshaft sprocket and crankshaft sprocket timing marks.

17. Loosen tensioner locknut, keeping tensioner steady with a hexagon wrench.
18. Swing tensioner 70–80 degrees clockwise with hexagon wrench and temporarily tighten locknut.
19. Turn crankshaft clockwise 2–3 times, then slowly set No. 1 cylinder at TDC of the compression stroke.
20. Push middle of timing belt between right hand camshaft sprocket and tensioner pulley with a force of 22 lbs.
21. Loosen tensioner locknut, keeping tensioner steady with a hexagon wrench.
22. Insert a 0.138 in. (0.35mm) thick and 0.5 in. (12.7mm) wide feeler gauge between the bottom of tensioner pulley and timing belt. Turn crankshaft clockwise and position gauge completely between tensioner pulley and timing belt. The timing belt will move about 2.5 teeth.
23. Tighten tensioner locknut, keeping tensioner steady with a hexagon wrench.
24. Turn crankshaft clockwise or counterclockwise and remove the gauge.
25. Rotate the engine 3 times, then set No. 1, to TDC, on its compression stroke.
26. Install the upper and lower timing belt covers with new gaskets.
27. Install the crankshaft pulley. Torque the pulley bolt to 90–98 ft. lbs. (123–132 Nm).
28. Install the compressor drive belt idler bracket.
29. Install the water pump pulley and torque the nuts to 12–15 ft. lbs. (16–21 Nm). Install the upper radiator and water inlet hoses.
30. Install the drive belts.
31. Install the engine side cover.
32. Mount the front right wheel.
33. Install the engine undercovers.
34. Lower the vehicle.
35. Fill the cooling system and connect the negative battery cable.

Timing Sprockets

REMOVAL & INSTALLATION

M30

1. Disconnect the negative battery cable.
2. Set the No. 1 piston to TDC of the compression stroke.
3. Remove the timing belt covers.
4. Remove the timing belt.
5. Using a suitable spanner wrench and a socket wrench, remove the camshaft pulley bolt and washer. Remove the front plate, O-ring and spring from the right (intake) camshaft to gain access to the sprocket bolt. The left camshaft sprocket is held in place by plate and 4 bolts.
6. Using a suitable puller, remove the crankshaft gear and timing belt plates from the crankshaft. Be careful not to gouge or scratch the surface of the crankshaft when removing the gear.
7. Inspect the timing gear teeth for wear and replace, as necessary.

To install:

8. Install the crankshaft gear with new Woodruff® keys.
9. Install the camshaft sprockets. Torque the sprocket bolts to 58–65 ft. lbs. (78–88 Nm); 90–98 ft. lbs. (123–132 Nm) for right (intake) and 10–14 ft. lbs. (14–19 Nm) for the left (exhaust).

NOTE: The right hand and left hand camshaft pulleys are different. Install them in their correct positions. The right hand pulley has an R3 identification mark and the left hand pulley has an L3.

10. Install the timing belt.
11. Install the timing belt covers.
12. Connect the negative battery cable.

Camshaft

REMOVAL & INSTALLATION

G20

1. Disconnect the negative battery cable. Remove the rocker cover and oil separator.
2. Rotate the crankshaft until the No.1 piston is at TDC on the compression stroke. Then rotate the crankshaft until the mating marks on the camshaft sprockets line up with the mating marks on the timing chain.
3. Remove the timing chain tensioner.
4. Remove the distributor.
5. Remove the timing chain guide.
6. Remove the camshaft sprockets. Use a wrench to hold the camshaft while loosening the sprocket bolt.
7. Loosen the camshaft bracket bolts in the opposite order of the torquing sequence.
8. Remove the camshaft.

To install:

9. Clean the left hand camshaft end bracket and coat the mating surface with liquid gasket. Install the camshafts, camshaft brackets, oil tubes and baffle plate. Ensure the left camshaft key is at 12 o'clock and the right camshaft key is at 10 o'clock.
10. The procedure for tightening camshaft bolts must be followed exactly to prevent camshaft damage. Tighten bolts as follows:
    a. Tighten right camshaft bolts 9 and 10 (in that order) to 1.5 ft. lbs. (2 Nm) then tighten bolts 1–8 (in that order) to the same specification.
    b. Tighten left camshaft bolts 11 and 12 (in that order) to 1.5 ft. lbs.

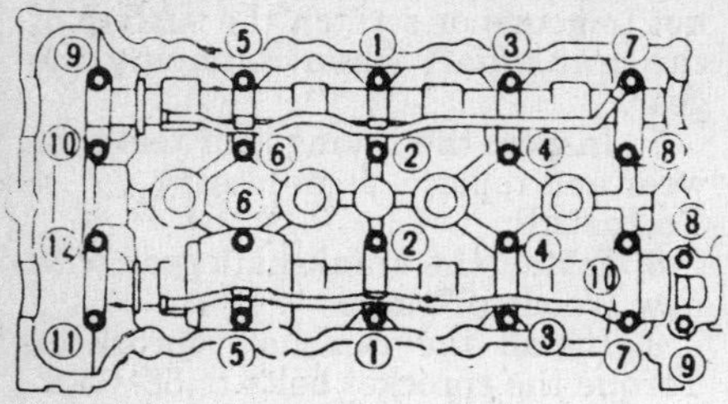

Camshaft bracket bolt torque sequence G20

(2 Nm) then tighten bolts 1–10 (in that order) to the same specification.

c. Tighten all bolts in sequence to 4.5 ft. lbs. (6 Nm).

d. Tighten all bolts in sequence to 6.5–8.5 ft. lbs. (9–12 Nm) for type A, B and C bolts, and 13–19 ft. lbs. (18–25 Nm) for type D bolts.

11. Line up the mating marks on the timing chain and camshaft sprockets and install the sprockets. Tighten sprocket bolts to 101–116 ft. lbs. (137–157 Nm).

12. Install the timing chain guide, distributor (ensure that rotor head is at 5 o'clock position) and chain tensioner.

13. Clean the rocker cover and mating surfaces and apply a continious bead of liquid gasket to the mating surface.

14. Install the rocker cover and oil separator. Tighten the rocker cover bolts as follows:

a. Tighten nuts 1, 10, 11, and 8 in that order to 3 ft. lbs. (4 Nm).

b. Tighten nuts 1–13 as indicated in the figure to 6–7 ft. lbs. (8–10 Nm).

M30

1. Disconnect the negative battery cable.
2. Drain the cooling system.
3. Remove the timing belt.
4. Remove the collector assembly.
5. Remove the intake manifold.
6. Remove the cylinder head.
7. Remove the rocker shafts with rocker arms. Bolts should be loosened in several steps in the proper sequence.
8. Remove hydraulic valve lifters and lifter guide. Hold hydraulic valve lifters with wire so they will not drop from lifter guide.
9. Using a dial gauge measure the camshaft endplay. If the camshaft endplay exceeds the limit (0.0012–0.0024 in.), select the thickness of a cam locate plate so the endplay is within specification. For example: if camshaft endplay measures 0.0031 in. (0.08mm) with shim 2 used, then change shim 2 to shim 3 so the camshaft endplay is 0.0020 in. (0.05mm).
10. Remove the camshaft front oil seal and slide camshaft out the front of the cylinder head assembly.

To install:

11. Install camshaft, locater plates, cylinder head rear cover and front oil seal. Set camshaft knock pin at 12 o'clock position. Install cylinder head with new gasket to engine.
12. Install valve lifter guide assembly. Assemble valve lifters in their original position. After installing them in the correct location remove the wire holding them in lifter guide.
13. Install rocker shafts in correct position with rocker arms. Tighten bolts, in 2–3 stages, to 13–16 ft. lbs. (18–22 Nm). Before tightening, be sure to set camshaft lobe at the position where lobe is not lifted or the valve closed. Set each cylinder 1 at a time or follow the procedure below. The cylinder head, intake manifold, collector and timing belt must be installed:

a. Set No. 1 piston at TDC of the compression stroke and tighten rocker shaft bolts for Nos. 2, 4 and 6 cylinders.

b. Set No. 4 piston at TDC of the compression stroke and tighten rocker shaft bolts for Nos. 1, 3 and 5 cylinders.

c. Torque specification for the rocker shaft retaining bolts is 13–16 ft. lbs. (18–22 Nm).

14. Fill the cooling system to the proper level.
15. Connect the negative battery cable.

Q45

1. Disconnect the negative battery cable.
2. Remove the engine and transmission assembly from the vehicle.
3. Remove the suspension member and engine mounts from the engine.
4. Remove the air compressor bracket.
5. Remove the cooling fan with coupling and the engine gusset.
6. Separate the engine from the transmission and mount the engine on a suitable workstand.
7. Remove the oil pan.
8. Remove the ornamental rocker cover and remove the ignition coils and spark plugs.
9. Bring the No. 1 piston to TDC on the compression stroke.
10. Use a suitable puller to remove the crankshaft pulley.
11. Remove the rocker cover.
12. Remove the crank angle sensor and the Valve Timing Control (VTC) solenoid.
13. Remove the chain tensioners and the upper front covers.
14. Remove the front timing chain cover.

NOTE: The timing chain will not be disengaged or dislocated from the crankshaft sprocket unless the front cover is removed. The cast portion of the front cover is located on the lower side of the crankshaft sprocket so the timing chain is not disengaged from the sprocket.

15. Remove the VTC assembly and the camshaft sprocket.
16. Remove the oil pump chain and the timing chains.

NOTE: do not attempt to disassemble the VTC assembly since they are difficult to reassemble accurately in the field. If it should be disassembled, the VTC assembly must be replaced with a new one.

17. Remove the camshaft brackets and the camshafts. Mark the parts so they can be reinstalled in their original positions.
18. Remove the rocker arms. Be sure to identify each rocker arm so it can be reinstalled in it's original position.

To install:

19. Make sure all mating surfaces are clean before installation.
20. Install the rocker arms, camshafts and camshaft brackets on the right bank. Properly lubricate the rocker arms and camshafts prior to installation. Tighten the camshaft bracket bolts to 9–10 ft. lbs. (12–14 Nm) in the proper sequence.
21. Install the VTC assembly and the exhaust cam sprocket on the right bank.
22. Make sure the camshafts are still correctly positioned and the piston in the No. 1 cylinder is still at TDC.
23. Install the timing chain on the right bank, aligning the mating marks on the chain with those on the crankshaft and camshaft sprockets.
24. Install the chain tensioner on the right bank.
25. Turn the crankshaft approximately 120 degrees clockwise from the point where the No. 1 piston is at TDC on the compression stroke. At this point, the valves on the left bank still remain unlifted.
26. Correctly position the camshafts and rocker arms for the left cylinder head. Properly lubricate the rocker arms and camshafts prior to installation. Tighten the camshaft bracket bolts to 9–10 ft. lbs. (12–14 Nm) in the proper sequence. Install the VTC assembly and the exhaust cam sprocket.
27. Install the timing chain on the left bank, aligning the mating marks on the chain with those on the crankshaft and camshaft sprockets.
28. Install the oil pump chain and sprockets.

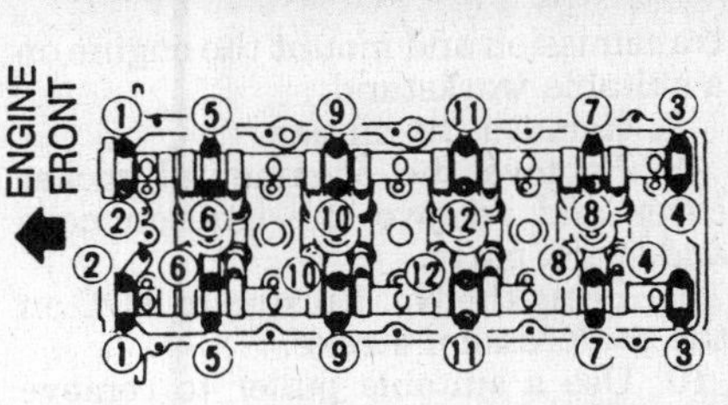

Camshaft bracket torque sequence—Q45

29. Install the oil pump chain guides. Place a 0.04 in. (1.0mm) feeler gauge between the upper chain guide and chain before assembling the chain guides. The force applied to the chain is equivalent to the upper chain guide weight.

30. Apply suitable sealer and install the front covers.

31. Install the chain tensioner for the left bank.

32. Apply suitable sealer to the rubber plugs and install them on the cylinder head.

33. Install the crank angle sensor, VTC solenoid, rocker cover and crank pulley.

34. Installation of the remaining components is the reverse of the removal procedure.

Piston and Connecting Rod

POSITIONING

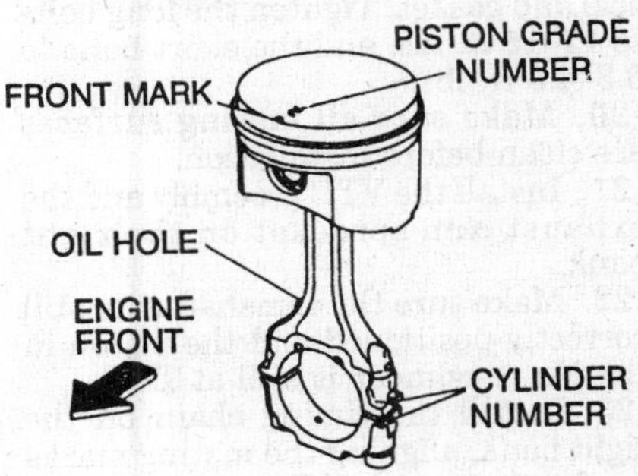

ENGINE LUBRICATION

Oil Pan

REMOVAL & INSTALLATION

G20

1. Raise and support the vehicle safely. Remove the engine under cover and drain the oil.

2. Remove the steel oil pan bolts in the proper sequence. Remove the steel

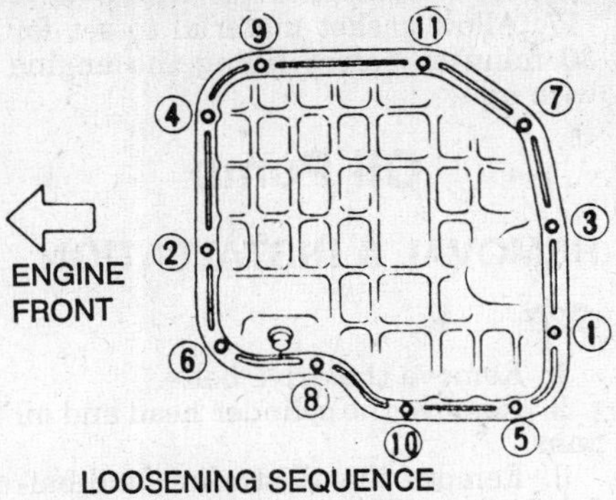

Steel oil pan bolt removal sequence—G20

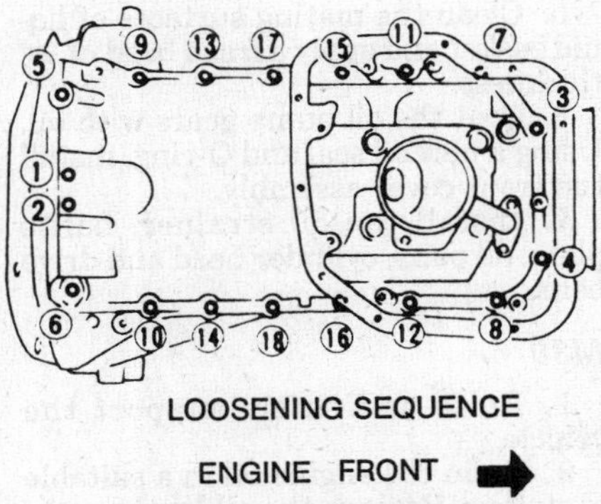

Aluminum oil pan bolt removal sequence—G20

oil pan. Insert tool KV10111100 between steel oil pan and aluminum oil pan to pry apart.

3. Remove the oil baffle bolts and oil baffle. Remove the front tube.

4. Set a suitable jack under the transaxle and raise the engine with and engine hoist.

5. If equipped with an automatic transaxle, remove the transaxle shift control cable.

6. Remove the compressor gussets, the rear cover plate and all aluminum oil pan bolts. Loosen aluminum oil pan bolts in the proper sequence.

7. Remove the two engine to transaxle bolts and refit the them into vacant vacant holes at the bottom of the oil pan. Remove the aluminum oil pan. Use tool KV10111100 to pry oil pan from block. Remove the engine to transaxle bolts.

To install:

8. Clean the oil pan rail of all liquid gasket and apply a new bead of 1/8" thickness to the oil pan rail.

9. Install the aluminum oil pan and torque bolts 1–16 to 12–14 ft. lbs. (16–19 Nm) and bolts 17–18 to 5–6 ft. lbs. (6–8 Nm) in the opposite order of removal.

10. Install the two engine to transaxle bolts, rear cover plate, compressor gussets, automatic transmission shift

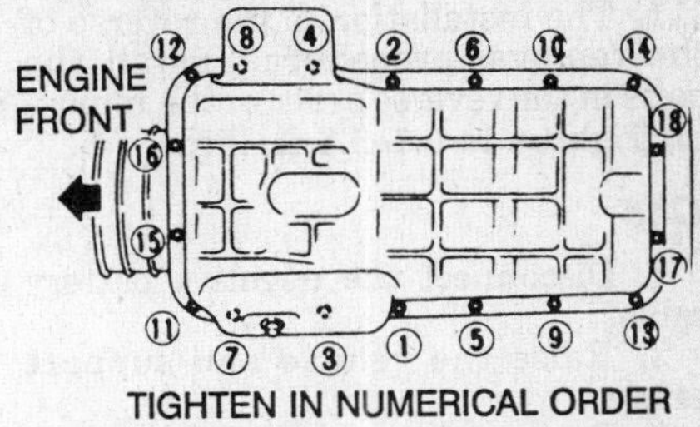

Oil pan bolt loosening sequence—M30

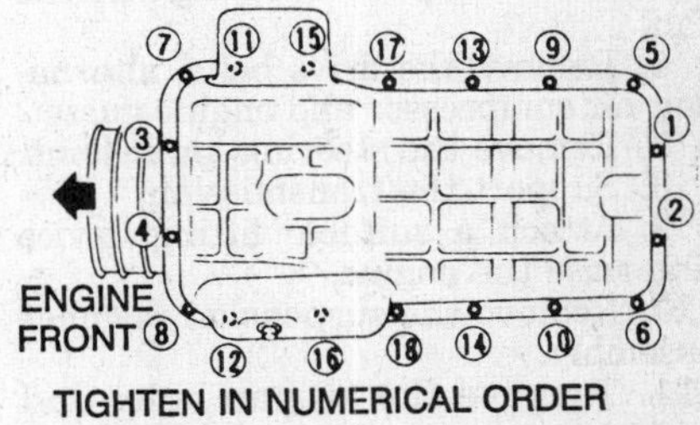

Oil pan bolt tightening sequence—M30

control cable (if equipped), center member, front tube and baffle plate.

11. Clean the oil pan rail of all liquid gasket and apply a new bead of 1/8" thickness to the oil pan rail.

12. Install the steel oil pan and install bolts untill snug. Tighten bolts in the reverse order of REMOVAL & wait 30 minutes before refilling crankcase with oil.

M30

1. Raise and safely support the vehicle.

2. Drain the engine oil in a suitable container. Remove the oil level gauge.

3. Remove the air duct.

4. Disconnect the air conditioning and brake booster vacuum hoses from their mounting brackets.

5. Remove the upper radiator mounting bolts and the automatic transmission oil cooler line mounting bolts.

6. Remove the oil pan mounting bolts, loosen them in numerical order. remove the oil pan using the proper tool to remove.

7. Loosen the front exhaust tube mounting bolts and remove the front stabilizer bar mounting brackets. Remove the right side stabilizer mounting bolt.

8. Loosen the left side stabilizer mounting bolt. Position a suitable transmission jacking device under the transmission case.

9. Remove the engine mounting bolts. Slowly, raise the transmission jack and remove the oil pan.

10. Thoroughly clean the mounting surfaces. Apply the proper sealant to the oil pump gasket, the rear oil seal retainer and the oil pan mounting surface.

11. The installation is the reverse of the removal procedure. Install the bolts in the reverse order of the removal. Tighten to 5.1–5.8 ft. lbs.

Q45

1. Disconnect the negative battery cable.
2. Raise the vehicle and support safely.
3. Remove the engine undercover.
4. Drain the engine oil
5. Remove the fan coupling with the fan.
6. Remove the drive belts, alternator, air compressor and engine gusset.
7. Remove the steering lower joint.
8. Support the transmission.
9. Attach a suitable lifting device and raise the engine.
10. Remove the suspension member assembly.
11. Remove the oil pan bolts and nuts.
12. Remove the oil pan from the engine block. Be careful not to damage the mating surface on the engine block.

To install:

13. Remove all gasket material from mating surfaces on the block and oil pan.
14. Apply a continuous bead of liquid gasket to the mating surface on the oil pan. Ensure that the bead is 0.138–0.177 in. (3.5–4.5mm) wide.
15. Install the oil pan. Install attaching bolts and nuts in sequence.
16. Complete the installation of the oil pan by reversing the removal procedure.

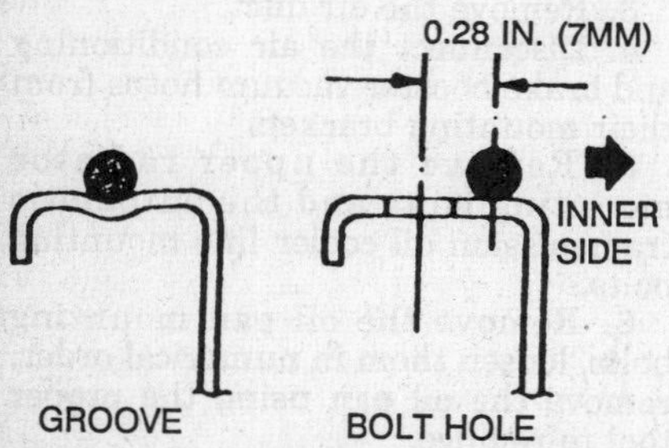

Oil pan liquid sealant bead—all engines

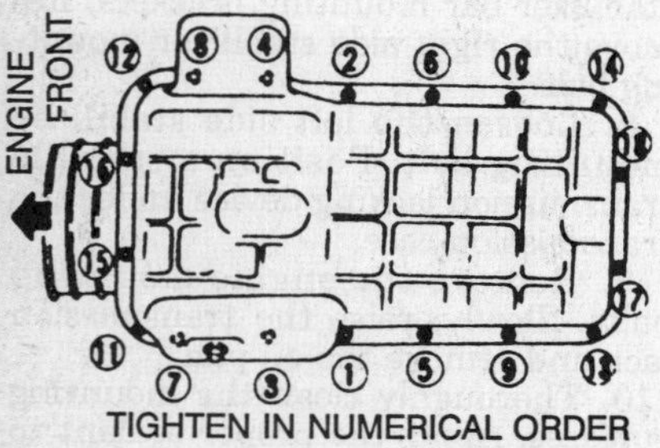

Oil pan installation torque sequence—Q45

17. Allow gasket material to set for 30 minutes before filling the engine with oil.

Oil Pump

REMOVAL & INSTALLATION

G20

1. Remove the drive belts.
2. Remove the cylinder head and oil pans.
3. Remove the oil strainer and baffle plate.
4. Remove the front cover assembly.

To install:

5. Clean the mating surfaces of liquid gasket and apply a fresh bead of 1/8" thickness.
6. Coat the oil pump gears with oil. Using a new oil seal and O-ring, install the front cover assembly.
7. Install the oil strainer, baffle plate, oil pans, cylinder head and drive belts.

M30

1. Raise and safely support the vehicle.
2. Drain the engine oil in a suitable container. Remove the oil level gauge.
3. Remove the oil pan.
4. Remove the oil pump mounting bolts and lift out the oil pump.
5. Always replace with a new oil seal and gasket. Apply oil to the inner and outer gears when installing.
6. The installation is the reverse of the removal procedure. Tighten the long mounting bolt to 9–12 ft. lbs. and the short bolts to 4.3–5.1 ft. lbs.

Q45

1. Disconnect the negative battery cable.
2. Remove the engine and transmission assembly from the vehicle.
3. Remove the suspension member and engine mounts from the engine.
4. Remove the air compressor bracket.
5. Remove the cooling fan with coupling and the engine gusset.
6. Separate the engine from the transmission and mount the engine on a suitable workstand.
7. Remove the oil pan.
8. Remove the ornamental rocker cover and remove the ignition coils and spark plugs.
9. Bring the No. 1 piston to TDC on the compression stroke.
10. Use a suitable puller to remove the crankshaft pulley.
11. Remove the rocker cover.
12. Remove the crank angle sensor and the Valve Timing Control (VTC) solenoid.
13. Remove the chain tensioners and the upper front covers.
14. Remove the front timing chain cover.

NOTE: The timing chain will not be disengaged or dislocated from the crankshaft sprocket unless the front cover is removed. The cast portion of the front cover is located on the lower side of the crankshaft sprocket so the timing chain is not disengaged from the sprocket.

15. Remove the VTC assembly and the camshaft sprocket.
16. Remove the oil pump chain and the timing chains.
17. Remove the mounting bolts and lift out the oil pump.

To install:

18. Thoroughly clean the mounting surfaces. Apply engine oil to the gears.
19. Install the oil pump with a new seal and gasket. Tighten the long bolts to 12–15 ft. lbs. and the short bolts to 3.3–4.3 ft. lbs.
20. Make sure all mating surfaces are clean before installation.
21. Install the VTC assembly and the exhaust cam sprocket on the right bank.
22. Make sure the camshafts are still correctly positioned and the piston in the No. 1 cylinder is still at TDC.
23. Install the timing chain on the right bank, aligning the mating marks on the chain with those on the crankshaft and camshaft sprockets.
24. Install the chain tensioner on the right bank.
25. Turn the crankshaft approximately 120 degrees clockwise from the point where the No. 1 piston is at TDC on the compression stroke. At this point, the valves on the left bank still remain unlifted.
26. Correctly position the camshafts and rocker arms for the left cylinder head. Properly lubricate the rocker arms and camshafts prior to installation. Install the VTC assembly and the exhaust cam sprocket.
27. Install the timing chain on the left bank, aligning the mating marks on the chain with those on the crankshaft and camshaft sprockets.

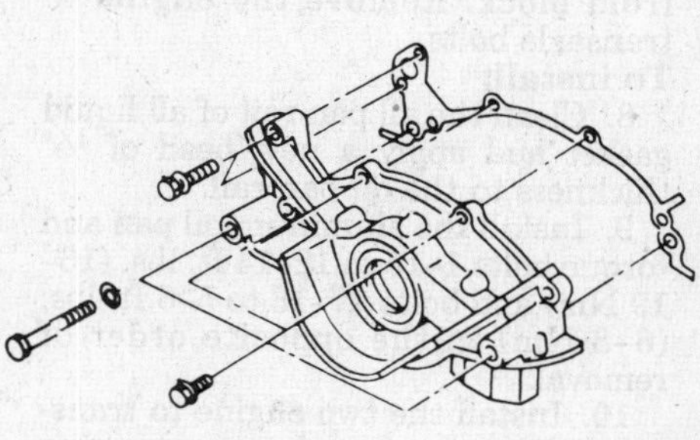

Oil pump assembly—M30

28. Install the oil pump chain and sprockets.
29. Install the oil pump chain guides. Place a 0.04 in. (1.0mm) feeler gauge between the upper chain guide and chain before assembling the chain guides. The force applied to the chain is equivalent to the upper chain guide weight.
30. Apply suitable sealer and install the front covers.
31. Install the chain tensioner for the left bank.
32. Apply suitable sealer to the rubber plugs and install them on the cylinder head.
33. Install the crank angle sensor, VTC solenoid, rocker cover and crank pulley.
34. Installation of the remaining components is the reverse of the removal procedure.

Rear Main Bearing Oil Seal

The rear main oil seal is a solid type seal located in the rear oil seal retainer at the rear of the engine.

REMOVAL & INSTALLATION

1. Raise and safely support the vehicle. Remove the transmission.
2. Remove the flywheel or drive plate.
3. Remove the rear oil seal retainer from the block.
4. Using a suitable prying tool, remove the oil seal from the retainer.
5. Thoroughly scrape the surface of the retainer to remove any traces of the existing sealant or gasket material.
6. Wipe the seal bore with a clean rag.
7. Apply clean engine oil to the new oil seal and carefully install it into the retainer using the proper seal installation tool.
8. Install the rear oil seal retainer into the engine, along with a new gasket.
9. Install the driveplate and transmission. Lower the vehicle.

ENGINE COOLING

Radiator

REMOVAL & INSTALLATION

G20

1. Disconnect the negative battery cable. Drain the coolant system, remove the upper radiator hose and reservoir tank.
2. Remove the lower radiator hose and transmission cooler lines.
3. Unplug the radiator fan motor connector and remove the radiator fan.
4. Remove all radiator attaching bolts and remove the radiator.

To install:

5. Lower the radiator into position. Take care not to damage the radiator fins as this will effect cooling efficiency.
6. Install all attaching bolts and tighten securely.
7. Install the radiator fan and reconnect the radiator fan motor connector.
8. Install the radiator upper and lower hoses, and the reservior tank.
9. Fill the cooling system, start the engine and allow it to reach normal operating temperature. Bleed the cooling system and check for leaks.

M30 and Q45

1. Disconnect the negative battery cable.
2. Drain the coolant.
3. On the Q45, remove the plastic cover over the radiator. Remove the upper hose and coolant reserve tank hose from the radiator.
4. Unbolt the shroud and move it backward in order to remove the fan and coupling. Remove the fan to water pump bolts and remove the fan, coupling, water pump pulley and shroud.
5. Raise the vehicle and support safely. Remove the lower hose from the radiator.
6. Disconnect and plug the automatic transmission cooler hoses. Disconnect the coolant thermo switch. Lower the vehicle.
7. Remove the mounting brackets or unbolt the radiator from the support and carefully lift out of the engine compartment.
8. Remove the cooling fans from the radiator.

To install:

9. Lower the radiator into position.
10. Install the mounting brackets or bolts.
11. Raise the vehicle and support safely. Connect the automatic transmission cooler lines and the thermo switch connector.
12. Connect the lower hose. Lower the vehicle.
13. Install the shroud, pulley, coupling and fan. Torque the water pump pulley nuts to 7 ft. lbs. (10 Nm). Adjust the belt.
14. Connect the upper hose and coolant reserve tank hose.
15. On the M30, open the air release plug. Fill the cooling system and check for leaks.
16. Connect the negative battery cable, run the vehicle until the thermostat opens, fill the radiator completely and check the automatic transmission fluid level. Recheck for coolant leaks.
17. Once the vehicle has cooled, recheck the coolant level.

Heater Core

REMOVAL & INSTALLATION

G20

It should be possible to remove the heater core without removing the dashboard.

1. Disconnect the negative battery cable.
2. With the temperature control lever set to the **HOT** position, drain the cooling system.
3. Disconnect the heater hoses a the drivers side of the heater unit.
4. Remove the glove compartment and the front panel from the center console.
5. Remove the radio and heater/air conditioner controls to remove the lower portion of the center console.
6. Disconnect the output vent ducts and remove the heater unit.
7. Disassemble the housing to remove the heater core.

To install:

8. Install the heater core and assemble the heater unit housing. Use new gaskets and seals as required, and check for smooth movement of the doors and linkage.
9. Install the heater unit and attach the ducts. Take care not to damage the gasket between the heater and cooling units.
10. Install the lower center console, the radio, and the heater controls. Before completing the assembl, connect the battery and adjust the door motor linkage.
11. Install the glove compartment and console panel.
12. Connect the heater hoses and refill the cooling system with the temperature control set at **HOT**. Bleed the cooling system.

M30

1. Disconnect the negative battery cable and allow 10 minutes to elapse before entering the vehicle.
2. Drain the coolant.
3. Disconnect the heater hoses from the heater core tubes and plug them.
4. Remove the steering column covers.
5. Remove the front pillar garnish and lower instrument covers.

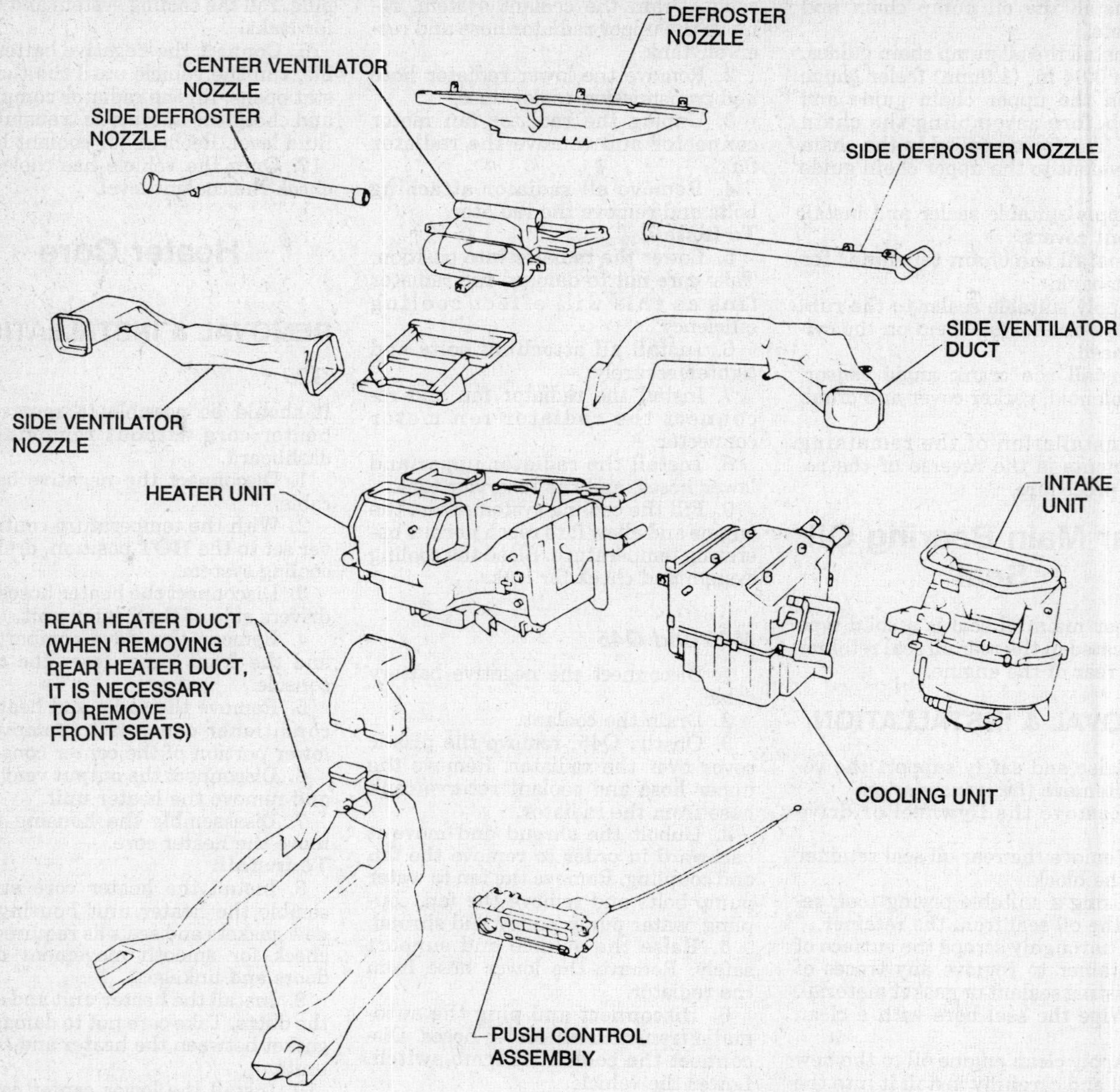

Heater and air conditioner assembly—G20

6. Remove the cluster lid and instrument cluster.
7. Remove the radio bezel, radio and climate control switch assembly.
8. Remove the glove box.
9. Remove the instrument reinforcement and the shift lever cover.
10. Remove the console assembly.
11. Remove the defroster grille and sensors.
12. Remove the hood lock cable bracket and rear heater ducts.
13. Remove the fuse block and disconnect the Super Multiple Junction (SMJ).
14. Remove the steering column mounting bolts and lower the column.
15. Remove the caps that cover the instrument panel securing screws, remove the screws and remove the instrument panel assembly.
16. Remove the air distribution ducts from the heater unit.
17. Disconnect all wires and cables that connect to the unit.
18. Remove the mounting bolts and nuts and remove the heater unit from the vehicle.
19. Disassemble and remove the heater core from the unit.

To install:

20. Clean the inside of the unit out, install the heater core and assemble the unit.
21. Install the unit to the vehicle and connect all wires and cables. Install the air distribution ducts.
22. Install the instrument panel assembly and snap the screw caps in place.
23. Raise and secure the steering column.
24. Connect the SMJ and install the fuse block.
25. Install the rear heater ducts and hood lock cable bracket.
26. Install the defroster grille and sensors.
27. Install the console assembly and shift lever cover.

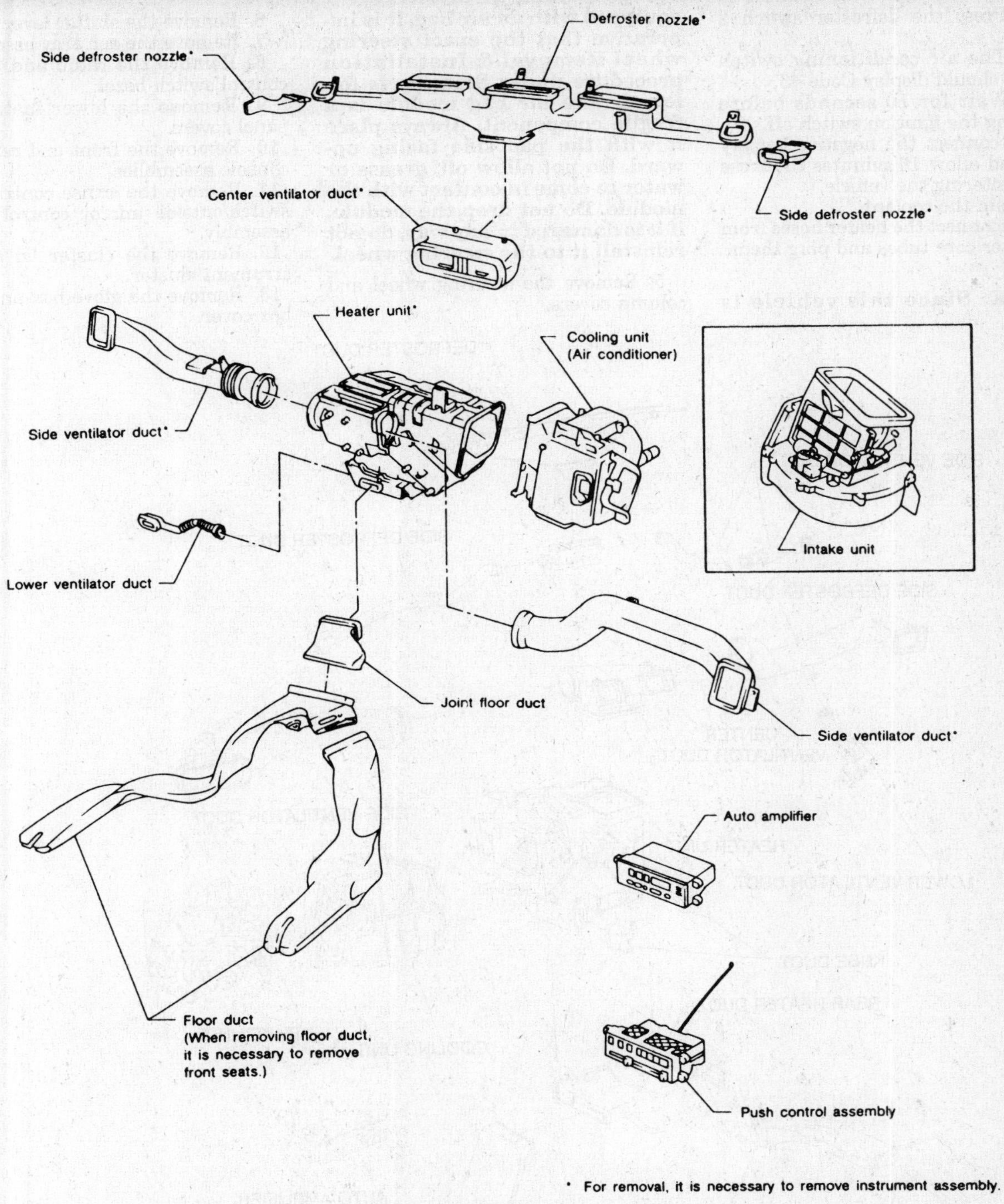

Instrument panel assembly – Q45

28. Install the instrument reinforcement and glove box.

29. Install the climate control switch assembly, radio and bezel.

30. Install the instrument cluster and lid.

31. Install the lower instrument panel covers and pillar garnish.

32. Install the steering column covers.

33. Connect the heater core tubes to the heater core tubes.

34. Open the air release plug. Fill the cooling system and check for leaks.

35. Connect the negative battery cable, run the vehicle until the thermostat opens, fill the radiator completely and check the automatic transmission fluid level. Recheck for coolant leaks.

36. Once the vehicle has cooled, recheck the coolant level.

Q45

1. In order to open the hot water valve, perform the following:
   a. Turn the ignition switch to the **ON** position.
   b. Within 10 seconds, press the **OFF** switch on the climate control switch assembly for at least 5 seconds.
   c. Press the temp-hotter switch 3 times.

d. Press the defroster switch 2 times.

e. The air conditioning switch panel should display Code 43.

f. Wait for 10 seconds before turning the ignition switch off.

2. Disconnect the negative battery cable and allow 10 minutes to elapse before entering the vehicle.

3. Drain the coolant.

4. Disconnect the heater hoses from the heater core tubes and plug them.

NOTE: Since this vehicle is equipped with an air bag, it is imperative that the exact steering wheel Removal & Installation procedure under Steering is followed. The air bag module is a fragile component. Always place it with the pad side facing upward. Do not allow oil, grease or water to come in contact with the module. Do not drop the module; if it is damaged in any way, do not reinstall it to the steering wheel.

5. Remove the steering wheel and column covers.

6. Remove the shifter lever bezel.

7. Remove the ash tray assembly.

8. Remove the radio and climate control switch bezel.

9. Remove the lower instrument panel covers.

10. Remove the front and rear floor console assemblies.

11. Remove the cruise control main switch/outside mirror control switch assembly.

12. Remove the cluster lid and instrument cluster.

13. Remove the glove box and glove box cover.

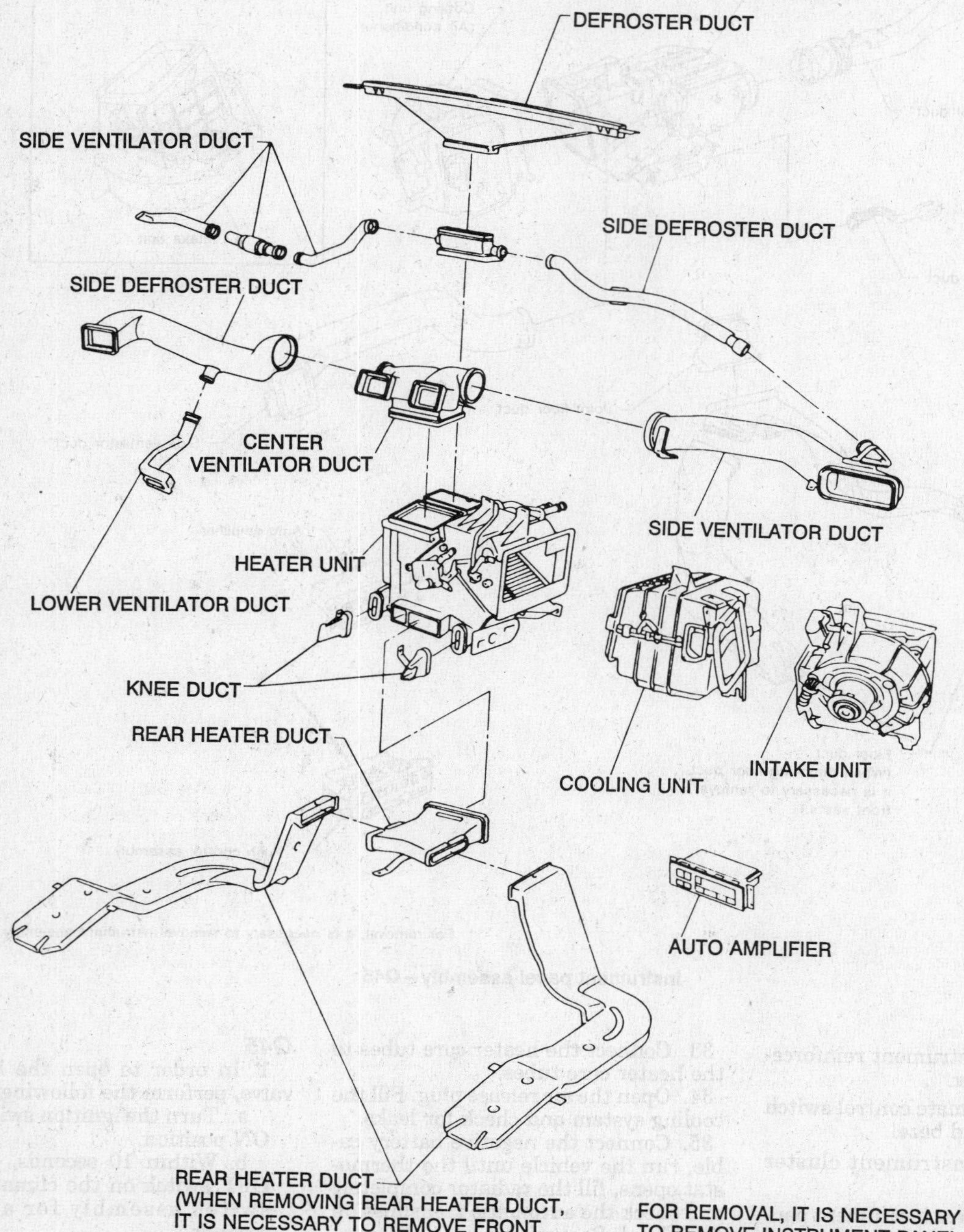

Heater and air conditioner assembly—Q45

14. Remove the cover on the lower right side of the instrument panel.
15. Remove the defroster grille.
16. Remove the radio and climate control switch assemblies.
17. Remove the remaining mounting screws remove the instrument panel assembly.
18. Remove the air distribution ducts from the heater unit.
19. Disconnect all wires and cables that connect to the unit.
20. Remove the mounting bolts and nuts and remove the heater unit from the vehicle.
21. Disassemble and remove the heater core from the unit.

To install:

22. Clean the inside of the unit out, install the heater core and assemble the unit. Connect the heater hoses.
23. Install the unit to the vehicle and connect all wires and cables. Install the air distribution ducts.
24. Install the instrument panel assembly. Install the cover on the lower right side.
25. Install the radio and climate control switch assemblies.
26. Install the defroster grille.
27. Install the glove box cover and glove box assembly.
28. Install the instrument cluster and cluster lid.
29. Install the cruise control main switch/outside mirror control switch assembly.
30. Install the console assemblies.
31. Install the lower instrument panel covers.
32. Install the radio and climate control switch bezel.
33. Install the ash tray assembly.
34. Install the shifter lever bezel.
35. Install the steering wheel and column covers.
36. Fill the cooling system and check for leaks.
37. Connect the negative battery cable, run the vehicle until the thermostat opens, fill the radiator completely and check the automatic transmission fluid level. Recheck for coolant leaks.
38. Once the vehicle has cooled, recheck the coolant level.

Water Pump

REMOVAL & INSTALLATION

G20

1. Drain the coolant from the radiator and engine block. The drain plug in the engine block is located at the left front of the cylinder block.
2. Remove the drive belts.
3. Loosen the water pump attaching bolts and remove the water pump. Take care not to drip coolant on the drive belts.
4. Clean all mating surfaces and place a 2–3mm bead of liquid gasket on the water pump mating surface.
5. Install water pump and tighten bolts to 12–15 ft. lbs. (16–21 Nm).
6. Using a radiator cap tester, or equivalent, check the system for leaks.
7. Refill with coolant and bleed the system of air.

M30

1. Disconnect the negative battery cable.
2. Drain the coolant from the radiator and from the drain plugs on both sides of the cylinder block.
3. Remove the timing belt covers.

NOTE: Use the proper precautions to avoid getting coolant on the timing belt.

4. Note the positioning of the clamp and disconnect the hose from the water pump.
5. Remove the water pump mounting bolts and remove the pump from the engine.

To install:

6. Thoroughly clean and dry the mating surfaces, bolts and bolt holes.
7. Apply liquid gasket to the water pump and install to the engine. Torque the bolts to 14 ft. lbs. (19 Nm).
8. Connect the hose and install the clamp in the same position as when it was removed to provide adequate clearance between it and the timing belt cover.
9. Open the air release plug. Fill the cooling system and check for leaks using a pressure tester before continuing.
10. Install the timing belt covers and all related parts.
11. Connect the negative battery cable, run the vehicle until the thermostat opens and fill the radiator completely. Recheck for coolant leaks.
12. Once the vehicle has cooled, recheck the coolant level.

Q45

1. Disconnect the negative battery cable.
2. Drain the coolant from the radiator and from the drain cocks on both sides of the cylinder block.
3. Unbolt the shroud and move it backward in order to remove the fan and coupling. Remove the fan to water pump bolts and remove the fan, coupling, water pump pulley and shroud.
4. Remove all necessary accessories to gain access to the water pump.
5. Note the positioning of the clamp and disconnect the hose from the water pump.
6. Remove the water pump mounting bolts and remove the pump from the engine.

To install:

7. Thoroughly clean and dry the mating surfaces, bolts and bolt holes.
8. Apply liquid gasket to the water pump and install to the engine. Torque the bolts to 14 ft. lbs. (19 Nm).
9. Connect the hose and install the clamp in the same position as when it was removed. Fill the cooling system and check for leaks using a pressure tester before continuing.
10. Install all removed accessories.
11. Install the shroud, pulley, cou-

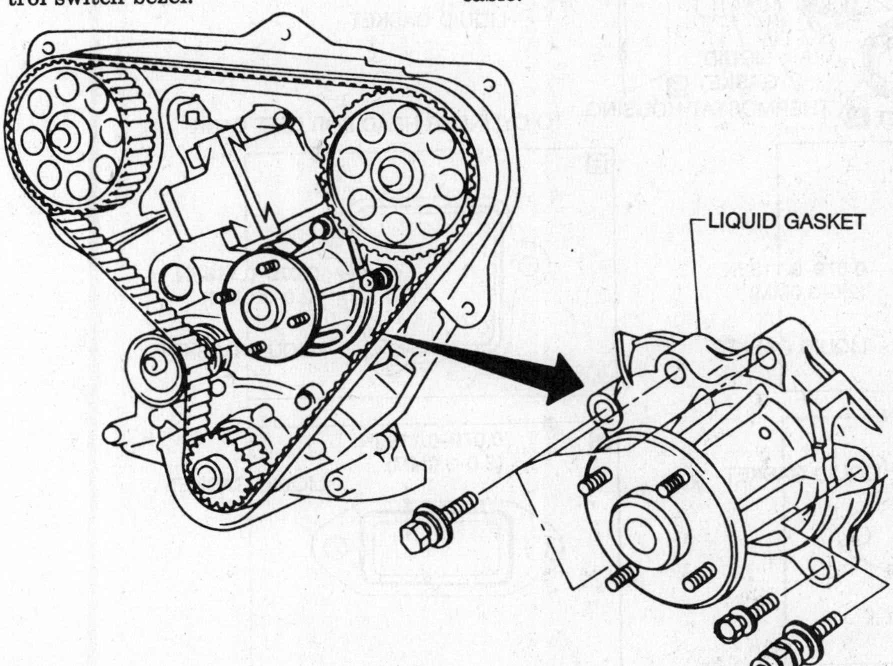

Water pump location—M30

pling and fan. Torque the water pump pulley nuts to 7 ft. lbs. (10 Nm). Adjust all belts.

12. Connect the negative battery cable, run the vehicle until the thermostat opens and fill the radiator completely. Recheck for coolant leaks.

13. Once the vehicle has cooled, recheck the coolant level.

Thermostat

REMOVAL & INSTALLATION

G20

1. Drain the engine coolant.
2. Remove the lower radiator hose.
3. Remove the water inlet, then remove the thermostat.
4. Install the new thermostat with the air bleeder or jiggle valve facing upward.
5. Clean all mating surfaces and apply a 2–3mm bead of liquid gasket to the water inlet.
6. Install the water pump inlet and tighten bolts to 5–6 ft. lbs. (6–8 Nm).
7. Install the lower radiator hose, refill and bleed the coolant system and check for leaks.

M30

1. Disconnect the negative battery cable. Drain the cooling system to below thermostat level.
2. Disconnect the upper radiator hose from the thermostat housing.
3. Remove the thermostat housing and thermostat.

To install:

4. Thoroughly clean and dry the mating surfaces, bolts and bolt holes.
5. Install the thermostat with the **UPR** mark and arrow at the top.
6. Apply liquid gasket to the thermostat housing. Install the housing and torque the bolts to 14 ft. lbs. (19 Nm).
7. Open the air release plug and fill the cooling system.
8. Connect the negative battery cable, run the vehicle until the thermostat opens and fill the radiator completely. Recheck for coolant leaks.
9. Once the vehicle has cooled, recheck the coolant level.

Q45

1. Disconnect the negative battery cable. Drain the cooling system to below thermostat level.
2. Remove the front ornament cover.
3. Disconnect the upper hose from the coolant inlet.
4. Remove the inlet and thermostat.

To install:

5. Thoroughly clean and dry the mating surfaces, bolts and bolt holes.
6. Install the thermostat with the jiggle valve at the top.

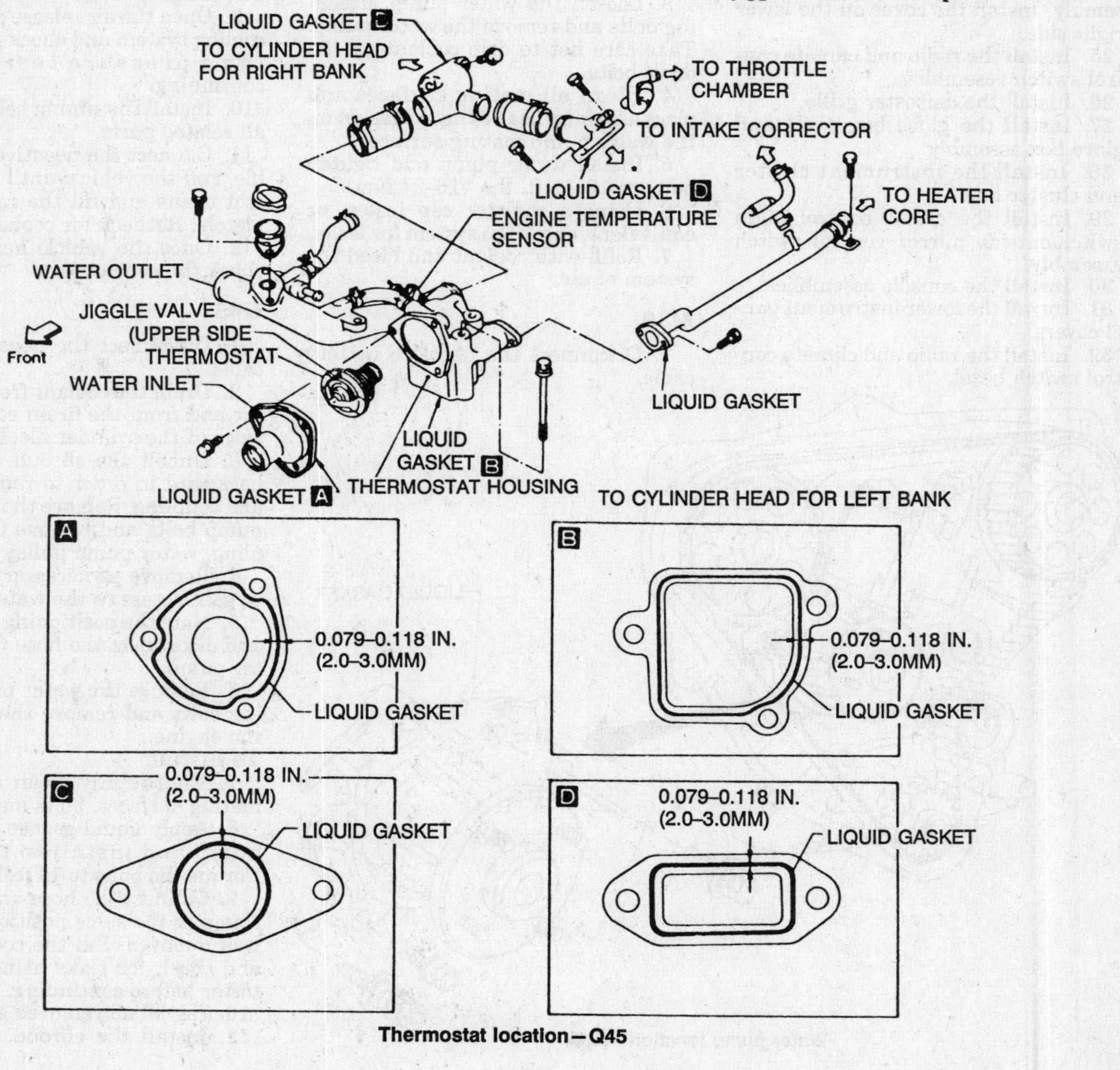

Thermostat location—Q45

7. Apply liquid gasket to the inlet. Install and torque the bolts to 14 ft. lbs. (19 Nm).
8. Fill the cooling system.
9. Connect the negative battery cable, run the vehicle until the thermostat opens and fill the radiator completely. Recheck for coolant leaks.
10. Once the vehicle has cooled, recheck the coolant level.

Cooling System Bleeding

G20

1. Set the heater temperature control lever to **MAX** hot position. Remove the radiator cap, air relief plug (located at the thermostat housing) and the air bleeder cap (located near the heater core).
2. Refill the reservoir bottle to the **MAX** line. Reinstall the the air relief plug when coolant spills from the hole. Reinstall the air bleeder cap.
3. Install a steel wire between the negative pressure valve and the seat of the radiator cap. Install the cap and warm the engine to normal operating temperature.
4. Run the engine at 2500 rpm for 10 seconds and return to idle. Repeat this 2–3 times. Turn the engine off and allow car to cool.
5. Remove the radiator cap and check the coolant level. If necessary refill the radiator with coolant up to the filler neck. Repeat Steps 9 and 10 several times.
6. Remove the radiator cap and remove the steel wire. Install the cap and warm the engine and check for the sound of coolant flow with engine running from idle to 4000 rpm. If a sound is heard, bleed air from the cooling system as follows:
   a. Cool engine and remove the air bleeder cap on the heater inlet hose.
   b. Attach a suitable transparent hose at the air bleeder pipe and put the opposite end of the hose into the coolant reservoir.
   c. Install the radiator cap with the steel wire inserted and check for proper connection of all coolant related hoses.
   d. Start the engine and check for bubbles in the reservoir tank.
   e. Set the heater control lever to **MAX** cool and run the engine up to 2300 rpm until the bubbles disappear in the hose.
   f. After bubbles disappear, set the heater control lever to **MAX** hot and listen for coolant system sound. If sound is heard, perform Steps A–E again.
   g. After all air has been bled from the system, remove the steel wire from the radiator cap, remove the transparent hose, install the air bleeder cap and check the coolant reservoir to ensure it is full.

M30

The M30 is equipped with a air release plug in line with a coolant hose on the right side of the ornamental collector cover. When filling the coolant system open the plug to bleed air from the system.

Q45

The Q45 uses a thermostat which is equipped with a jiggle valve. This valve bleeds air as the system is being filled, thus the cooling system requires no further bleeding.

ENGINE ELECTRICAL

NOTE: Disconnecting the negative battery cable on some vehicles may interfere with the functions of the on board computer systems and may require the computer to undergo a relearning process, once the negative battery cable is reconnected.

Distributor

REMOVAL

G20 and M30

1. Disconnect the negative battery cable.
2. Remove the splash shield (if equipped). Disconnect the distributor connectors.
3. Unscrew the distributor cap hold-down screws and lift off the distributor cap with all ignition wires still connected.
4. Matchmark the rotor to the distributor housing and the distributor housing to the engine.

NOTE: do not crank the engine during this procedure. If the engine is cranked, the matchmark must be disregarded.

5. Remove the hold-down bolt.
6. Remove the distributor from the engine.

INSTALLATION

Timing Not Disturbed

1. Install a new distributor housing O-ring.
2. Install the distributor in the engine so the rotor is aligned with the matchmark on the housing and the housing is aligned with the matchmark on the engine. Make sure the distributor is fully seated and the distributor gear is fully engaged.
3. Install and snug the hold-down bolt.
4. Connect the distributor pickup lead wires.
5. Install the distributor cap and tighten the screws. Install the splash shield.
6. Connect the negative battery cable.
7. Adjust the ignition timing and tighten the hold-down bolt.

Timing Disturbed

1. Install a new distributor housing O-ring.
2. Position the engine so the No. 1 piston is at TDC of its compression stroke and the mark on the vibration damper is aligned with **0** on the timing indicator.
3. Install the distributor in the engine so the rotor is aligned with the position of the No. 1 ignition wire on the distributor cap (4–5 o'clock position on the G20). Make sure the distributor is fully seated and that the distributor shaft is fully engaged.

NOTE: There are distributor cap runners inside the cap on 3.0L engine. Make sure the rotor is pointing to where the No. 1 runner originates inside the cap.

4. Install and snug the hold-down bolt.
5. Connect the distributor pickup lead wires.
6. Install the distributor cap and tighten the screws. Install the splash shield, if equipped.
7. Connect the negative battery cable.
8. Adjust the ignition timing and tighten the hold-down bolt.

Distributorless Ignition

REMOVAL & INSTALLATION

Q45

POWER TRANSISTOR UNIT

1. Disconnect the negative battery cable.
2. Remove the air intake duct, if necessary.
3. Disconnect the connector.
4. Remove the bolts that attach the unit to the ornamental rocker cover.
5. Remove the unit from the engine.
6. The installation is the reverse of the removal procedure.

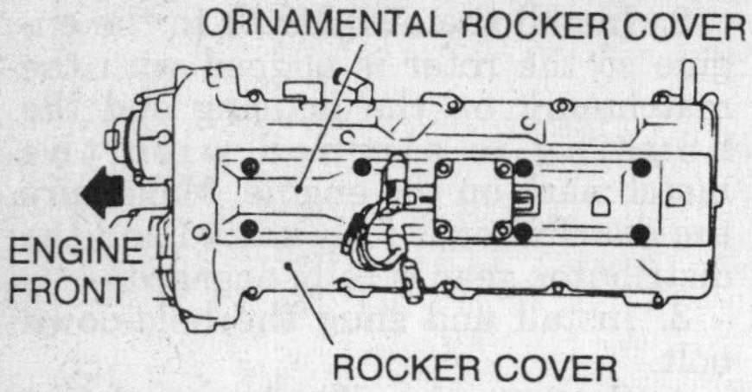

Ornamental rocker cover—4.5L engine

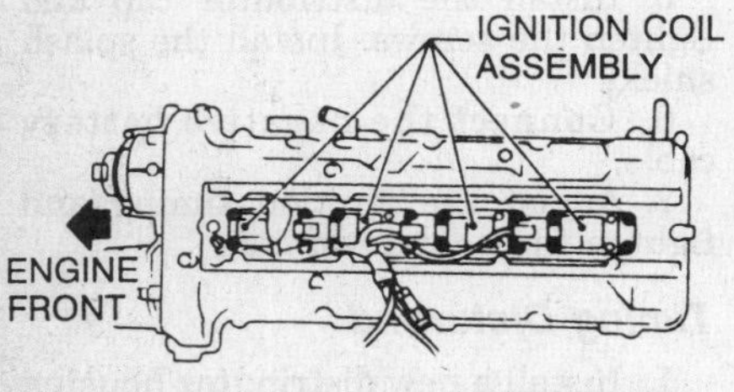

Ignition coil assembly—4.5L engine

IGNITION COIL

1. Disconnect the negative battery cable.
2. Remove the air intake duct, if necessary.
3. Disconnect the power transistor unit connector.
4. Remove the ornamental rocker cover.
5. Remove the ignition coil bracket mounting bolts and pull out the bracket with the ignition coils.
6. Separate the coil from the bracket and remove from the engine.
7. The installation is the reverse of the removal procedure.

CRANK ANGLE SENSOR

1. Disconnect the negative battery cable.
2. Remove the air intake duct.
3. Matchmark the position of the crankshaft sensor assembly to the head.
4. Disconnect the connector, remove the mounting bolts and remove the crank angle sensor from the engine.
5. The installation is the reverse of the removal procedure.
6. Check the ignition timing and adjust, if necessary.

Ignition Timing

ADJUSTMENT

1. Start the engine, set the parking brake and run the engine until at normal operating temperature. Keep all lights and accessories off.
2. Connect a timing light to the No. 1 cylinder spark plug wire.
3. Use the Nissan Consult System Checking tool in the Data Monitor mode to check engine rpm. Adjust, if necessary.
4. Aim the timing light at the timing scale.
5. On the G20, run the engine at 2000 rpm for 2 minuites and race engine 2–3 times under no load. Return engine to idle. Turn engine OFF and disconnect the throttle sensor harness connector. Start engine and race at 2000–3000 rpm 2–3 times. Check igntion timing with a timing light. Specification is 13–17° BTDC (manual transmission) and 13–17° BTDC in **N** (automatic transmission). Adjust timing as necessary by loosening distributor holddown clamp and rotating distributor.
6. On the M30 and Q45, run the engine at 2000 rpm for 2 minuites and race engine 2–3 times under no load. Return engine to idle. Check igntion timing with a timing light. Specification is 13–17° BTDC. Adjust timing as necessary by loosening distributor holddown clamp and rotating distributor.

Alternator

PRECAUTIONS

Several precautions must be observed with alternator equipped vehicles to avoid damage to the unit.

- If the battery is removed for any reason, make sure it is reconnected with the correct polarity. Reversing the battery connections may result in damage to the one-way rectifiers.
- When utilizing a booster battery as a starting aid, always connect the positive to positive terminals and the negative terminal from the booster battery to a good engine ground on the vehicle being started.
- Never use a fast charger as a booster to start vehicles.
- Disconnect the battery cables when charging the battery with a fast charger.
- Never attempt to polarize the alternator.
- do not use test lamps of more than 12 volts when checking diode continuity.
- do not short across or ground any of the alternator terminals.
- The polarity of the battery, alternator and regulator must be matched and considered before making any electrical connections within the system.
- Never separate the alternator on an open circuit. Make sure all connections within the circuit are clean and tight.
- Disconnect the battery ground terminal when performing any service on electrical components.
- Disconnect the battery if arc welding is to be done on the vehicle.

BELT TENSION ADJUSTMENT

G20

1. Disconnect the negative battery cable.
2. Loosen the nut that secures the T-bolt to the slotted adjustment bracket.
3. Turn the adjustment bolt until the belt deflects approximately 0.28–0.31 in. (alternator) and 0.16–0.20 in. (power steering) at its longest expanse.
4. Tighten the T-bolt nut to 12–16 ft. lbs. (16–22 Nm).
5. Connect the negative battery cable.

M30

1. Disconnect the negative battery cable.
2. Loosen the nut that secures the T-bolt to the slotted adjustment bracket.
3. Turn the adjustment bolt until the belt deflects approximately 0.3 in. at its longest expanse.
4. Tighten the T-bolt nut to 11 ft. lbs. (15 Nm).
5. Connect the negative battery cable.

Q45

1. Disconnect the negative battery cable.
2. Loosen the nut that secures the T-bolt to the alternator belt idler pulley.
3. Turn the adjustment bolt until the belt deflects approximately 0.3 in. at its longest expanse.
4. Tighten the T-bolt nut to 24 ft. lbs. (32 Nm).
5. Connect the negative battery cable.

REMOVAL & INSTALLATION

G20 and M30

1. Disconnect the negative battery cable.
2. Loosen the alternator belt and remove from the pulley.
3. Remove the adjusting bracket.
4. Disconnect the harness connector and cable from the rear of the alternator.

NOTE: The front mounting bolt cannot be removed separately because of insufficient clearance between the alternator and engine coolant inlet tube.

5. Remove the rear mounting bolt loosen the front mounting bolt.

6. Remove the alternator with the front mounting bolt.

To install:

7. The installation is the reverse of the removal procedure. Torque the mounting bolts to 15 ft. lbs. (20 Nm).
8. Adjust the belt so it deflects approximately 0.3 in. at its longest expanse.
9. Connect the negative battery cable and check the alternator for proper operation.

Q45

1. Disconnect the negative battery cable.
2. Remove the radiator shroud and cooling fan.
3. Drain a sufficient amount of coolant and remove the upper radiator hose.
4. Remove the upper alternator bracket and the air conditioner pipe mounting bracket.
5. Remove the idler pulley and belt.
6. Remove the 2 power steering cooler pipe mounting screws.
7. Remove the mounting through bolt.
8. Pull the alternator toward the radiator and remove the harness heat shield.
9. Disconnect the wires from the rear of the alternator and remove from the vehicle.

To install:

10. Position the alternator and connect the wires. Install the heat shield.
11. Install the mounting through bolt loosely.
12. Install the 2 power steering cooler pipe mounting screws.
13. Install the idler pulley and belt.
14. Install the air conditioner pipe mounting bracket and upper alternator bracket. Tighten the through bolt.
15. Adjust the belt so it deflects approximately 0.3 in. at its widest expanse.
16. Install the upper radiator hose and refill the cooling system.
17. Install the cooling fan and radiator shroud.
18. Connect the negative battery cable and check the alternator for proper operation.

Starter

REMOVAL & INSTALLATION

1. Disconnect the negative battery cable.
2. Raise the vehicle and support safely.
3. Remove the engine undercover.
4. Remove exhaust components, as required, in order to gain access to the starter.
5. Remove the starter mounting bolts and remove the starter.
6. The installation is the reverse of the removal procedure. Torque the mounting bolts to 25 ft. lbs. (34 Nm).
7. Connect the negative battery cable and check the starter for proper operation.

EMISSION CONTROLS

Please refer to "Emission Controls" in the Unit Repair section for system maintenance procedures. Due to the complex nature of modern electronic engine control systems, comprehensive diagnosis and testing procedures fall outside the confines of this repair manual. For complete information on diagnosis, testing and repair procedures concerning all modern engine and emission control systems, please refer to "Chilton's Guide to Fuel Injection and Electronic Engine Controls".

FUEL SYSTEM

Fuel System Service Precautions

Safety is the most important factor when performing not only fuel system maintenance but any type of maintenance. Failure to conduct maintenance and repairs in a safe manner may result in serious personal injury or death. Maintenance and testing of the vehicle's fuel system components can be accomplished safely and effectively by adhering to the following rules and guidelines.

- To avoid the possibility of fire and personal injury, always disconnect the negative battery cable unless the repair or test procedure requires that battery voltage be applied.
- Always relieve the fuel system pressure prior to disconnecting any fuel system component (injector, fuel rail, pressure regulator, etc.), fitting or fuel line connection. Exercise extreme caution whenever relieving fuel system pressure to avoid exposing skin, face and eyes to fuel spray. Please be advised that fuel under pressure may penetrate the skin or any part of the body that it contacts.
- Always place a shop towel or cloth around the fitting or connection prior to loosening to absorb any excess fuel due to spillage. Ensure that all fuel spillage (should it occur) is quickly removed from engine surfaces. Ensure that all fuel soaked cloths or towels are deposited into a suitable waste container.
- Always keep a dry chemical (Class B) fire extinguisher near the work area.
- do not allow fuel spray or fuel vapors to come into contact with a spark or open flame.
- Always use a backup wrench when loosening and tightening fuel line connection fittings. This will prevent unnecessary stress and torsion to fuel line piping. Always follow the proper torque specifications.
- Always replace worn fuel fitting O-rings with new. Do not substitute fuel hose or equivalent where fuel pipe is installed.

RELIEVING FUEL SYSTEM PRESSURE

1. Disable the fuel system either by pulling the fuel pump fuse, located in the interior fuse box, or by disconnecting the fuel pump relay or module located in the trunk.
2. Start the engine and run until it stalls.
3. Crank the engine 2–3 more times to ensure that all pressure is relieved.
4. Disconnect the negative battery cable. Install or reconnect the fuse, relay or module.
5. Erase the created code using a Nissan Consult Tester, or equivalent, when servicing is finished.

Fuel Tank

REMOVAL & INSTALLATION

1. Relieve the fuel system pressure.
2. Disconnect the negative battery cable. Raise and support the rear of the vehicle safely.
3. Using the proper equipment, drain the fuel tank.
4. Remove the fuel tank and filler neck protective plates. Remove the filler neck to quarter panel attaching bolts. Disconect the ventilation pipes and remove the fuel filler assembly.
5. Disconnect the wiring harness connector for the fuel pump/sending unit assembly.
6. Place a suitable jack under the center of the tank and apply slight pressure. Remove the tank retaining bolts.
7. Lower the tank and disconnect the fuel hoses from the pump/sending

unit assembly and plug them. Some vehicles are equipped with an inspection cover under the rear seat. Remove this cover to gain access to the fuel hoses prior to lowering the fuel tank.
8. Remove the fuel tank from the vehicle.

To install:

9. If the pump/sending unit was removed from the fuel tank, use a new O-ring and install the assembly on the fuel tank.
10. Install the fuel tank. Torque the retaining bolts to 24 ft. lbs. (33 Nm).
11. Connect all fuel lines and harness connections. Connect the filler neck and overflow tube.
12. Install the protective plates.
13. Lower the vehicle. Install the bolts that attach the filer neck to the quarter panel.
14. Connect the negative battery cable, start the engine and check for leaks.

Fuel Filter

REMOVAL & INSTALLATION

CAUTION

do not use conventional fuel filters, hoses or clamps when servicing this fuel system. They are not compatible with the high pressures of the injection system and could fail, causing personal injury. Use only components specifically designed for fuel injection.

1. Relieve the fuel system pressure.
2. Disconnect the negative battery cable.
3. Disconnect the fuel hoses from the fuel filter, located in the right side of the engine compartment.
4. Remove the filter mounting screws and remove from the vehicle.
5. Inspect all hoses and clamps for damage of any type. Replace parts, as required.
6. The installation is the reverse of the removal.

Fuel Pump

The fuel pump for all Infiniti models is located inside the fuel tank at the fuel gauge sender unit.

PRESSURE TESTING

1. Relieve the fuel system pressure.
2. Disconnect the fuel hose between the fuel filter and the fuel tube leading to the engine.
3. Install an appropriate fuel pressure gauge between the filter and tube.
4. Start the engine and check for fuel leaks.
5. Observe the fuel pressure. The specification is 34 psi at idle and 43 psi when the fuel pressure regulator vacuum hose is pinched off.
6. Stop the engine, disconnect the vacuum hose to the pressure regulator and plug it.
7. Connect a hand-held vacuum pump to the regulator.
8. Start the engine and observe the fuel pressure as the vacuum is varied. The fuel pressure should decrease as the vacuum is increased.

REMOVAL & INSTALLATION

G20

1. Release the fuel system pressure.
2. Remove the inspection hole cover located beneth the rear seat.
3. Disconnect the connectors and fuel tubes.
4. Remove the fuel gauge locking ring using tool SST-X38879, or equivalent.
5. Remove the fuel gauge assembly and disconnect the tubes and connector.
6. Remove the fuel pump by sliding it out on an angle.

To install:

7. Use a new O-ring on the fuel gauge assembly locking ring.
8. Install the new fuel pump and attach all fuel lines and connectors.
9. Using tool SST-X38879, or equivalent, tighten the locking ring to 22–26 ft. lbs. (30–35 Nm).
10. Install the inspection cover and test fuel system pressure at the injectors.

M30 and Q45

1. Relieve the fuel system pressure.
2. Disconnect the negative battery cable.
3. Remove the fuel tank.
4. Disconnect the wiring harness. Remove the fuel tank sender unit attaching bolts. Remove the fuel tank sender and discard the O-ring.
5. Remove the fuel pump from the sender unit.

To install:

6. Install the new fuel pump on the sender unit assembly.
7. Using a new O-ring, install the sender unit in the fuel tank. Tighten the bolts to 2 ft. lbs. (2–3 Nm).
8. Connect the wiring harness and install the fuel tank. Tighten the fuel tank attaching strap bolts to 20–27 ft. lbs. (26–36 Nm).
9. Connect the negative battery cable, start the engine and check for leaks.

Fuel Injection

IDLE SPEED ADJUSTMENT

The idle speed is controlled by the ECCS control unit. Adjustment is not required.

IDLE MIXTURE ADJUSTMENT

The idle mixture is controlled by the ECCS control unit. Adjustment is not required.

Fuel Injector

REMOVAL & INSTALLATION

G20

1. Disconnect the negative battery cable. Relieve fuel system pressure.
2. Disconnect injector harness connectors.
3. Disconnect vacuum hose from pressure regulator.
4. Disconnect fuel hoses from fuel tube assembly.
5. Remove injectors with fuel tube assembly. Loosen bolts in reverse order of torquing sequence.
6. To remove injector, push out of the fuel tube assembly.

NOTE: do not remove injector by pinching connector.

To install:

7. Replace or clean injector as necessary.
8. Install injector on fuel tube as-

Fuel tube torquing sequence – G20

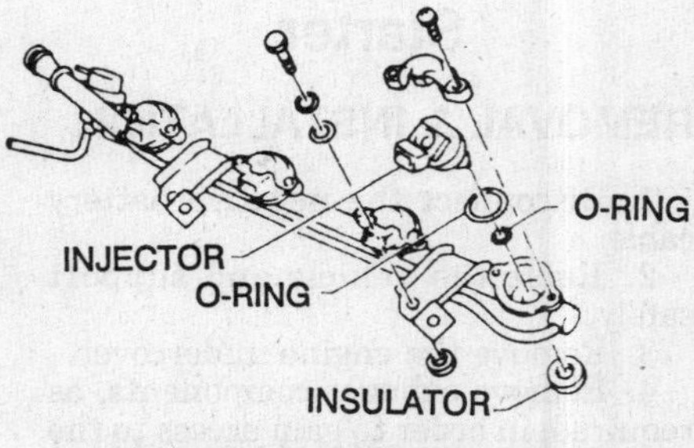

Fuel tube components – G20

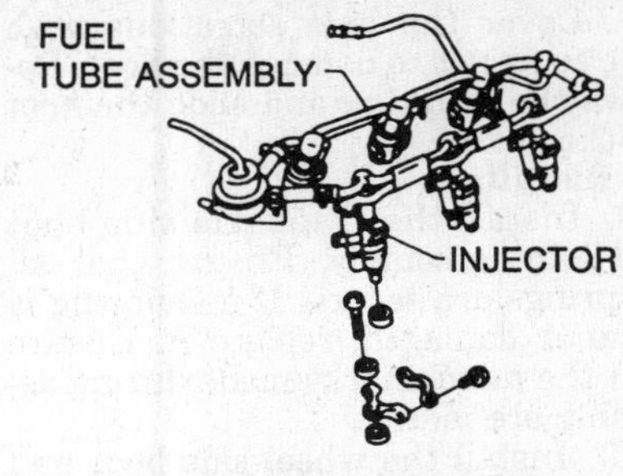

Fuel tube assembly–M30

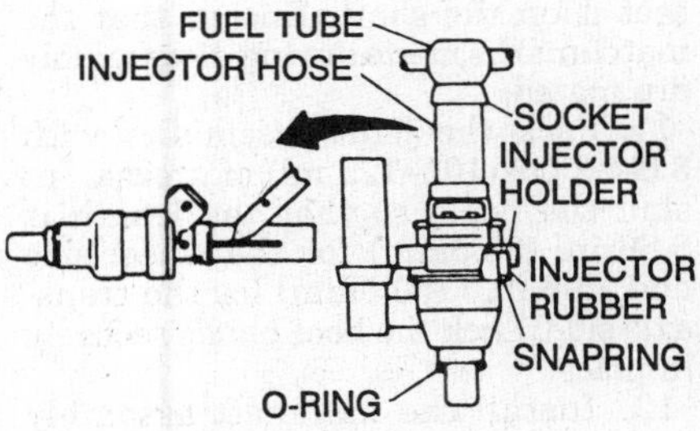

Injector assembly–M30

sembly using a new O-ring and insulator. Lubricate O-rings with silicone oil.

9. Install injectors with fuel tube assembly onto intake manifold. Tighten fuel tube assembly bolts in sequence to 7–8 ft. lbs. (9–10 Nm), and then retighten to 15–20 ft. lbs. (21–26 Nm).

10. Install fuel hoses, lubricating them with silicone oil.

11. Connect the injector harness connector, start the engine and check for leaks.

M30

1. Relieve the fuel system pressure. Disconnect the negative battery cable.
2. Disconnect the cruise control and throttle cables from the throttle body.
3. Remove the intake manifold collector.
4. Disconnect the vacuum hose from the fuel pressure regulator.
5. Disconnect and plug the fuel hoses.
6. Disconnect all injector harness connectors.
7. Disconnect the fuel temperature sensor connector.
8. Remove the injector fuel fuel tube assembly retaining bolts and remove the assembly from the engine.
9. Remove the injector(s) and short fuel hose(s) from the fuel tube. Do not reuse the rubber hose(s).

To install:

10. Wet the inside of the new rubber hose(s) with fuel.
11. Push the end of the rubber hose with hose sockets into the injector tail piece and fuel tube end as far as they

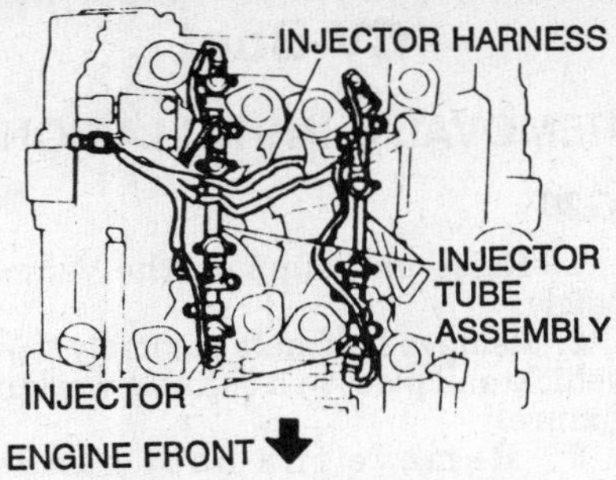

Fuel tube assembly–Q45

will go. Clamps are not used at these connections.

12. Install the injector fuel tube assembly.
13. Connect the fuel temperature sensor connector.
14. Connect all injector harness connectors.
15. Connect the fuel hoses and the regulator vacuum hose.
16. Install the intake manifold collector.
17. Connect the cruise control and throttle cables to the throttle body.
18. Connect the negative battery cable and check for leaks.

Q45

1. Relieve the fuel system pressure. Disconnect the negative battery cable.
2. Drain the coolant.
3. Remove the EGR control valve.
4. Remove the intake manifold collector.
5. Disconnect the harness connector(s) from the fuel injector(s).
6. Remove the injector(s) from the injector tube assembly. Do not reuse the O-ring(s).

To install:

7. Using new O-ring(s), install the injector(s) to the injector tube.
8. Connect the harness connector(s).
9. Install the intake manifold collector.
10. Install the EGR control valve.
11. Fill the cooling system.
12. Connect the negative battery cable and check for leaks.

DRIVE AXLE

NOTE: Final tightening of any suspension component must be performed with the suspension unladen with the tires on the ground.

Halfshaft

REMOVAL & INSTALLATION

G20

1. Raise and support the vehicle safely. Remove the wheel bearing lock nut.
2. Remove the brake caliper assembly and rotor. Using a piece of wire, position the caliper so that it is not supported by the brake line.
3. Separate the tie-rod from the ball joint.
4. Separate the kingpin from the knuckle.
5. Remove the halfshaft from the wheel hub/knuckle by lightly tapping it with a wood drift. Take care not to damage the CV boots.
6. Remove the halfshaft from the transaxle by prying outward with a suitable tool at the transaxle case.
7. On automatic transaxle models, remove the left halfshaft by tapping it out with a drift from the right side of the transaxle case. Take care not to damage the pinion mate shaft and side gear.

To install:

8. Drive a new oil seal into the transaxle. Set tool KV38106800, or equivalent, along the inner circumference of the oil seal.
9. Insert the halfshaft into the transaxle. Ensure that the serrations are aligned. Remove the tool.
10. Push the halfshaft inward and install the circular clip in the groove of the side gear. After inserting the clip, pull outward on the flange of the slide joint to ensure the clip is properly meshed with the side gear. If if pulls out, the clip was not installed properly.
11. Install the halfshaft into the wheel hub/knuckle. Tighten the upper knuckle nut to 72–87 ft. lbs. (98–118 Nm) and wheel bearing lock nut to 174–231 ft. lbs. (235–314 Nm).
12. Using a dial indicator, check wheel bearing axial end play. Specification calls for 0.0020 in. (0.05mm) or less.

M30

1. Raise the vehicle and support safely.
2. Remove the 6 bolts and nuts attaching the outer CV-joint to the companion flange.
3. Remove the inner CV-joint from the differential carrier by prying with a suitable tool.

To install:

4. Install the inner CV-joint into the differential carrier.
5. Connect the outer CV-joint to the companion flange with the 6 bolts and

nuts. Tighten to 20–27 ft. lbs. (27–37 Nm).

6. Lower the vehicle.

Q45

1. Raise the vehicle and support safely.
2. Remove the rear wheel.
3. Remove the differential side flange bolts and nuts and separate shaft.
4. Remove the cotter pin, adjusting cap, insulator, wheel bearing locknut and washer from halfshaft.
5. Remove the halfshaft by lightly tapping it with a copper hammer.
6. Remove the halfshaft assembly from the vehicle.

To install:

7. Insert halfshaft into wheel hub and install washer and wheel bearing locknut. Temporarily tighten the locknut.
8. Connect the halfshaft with the differential side flange. Install the nuts and bolts and tighten to 25–33 ft. lbs. (34–44 Nm).
9. Tighten the wheel bearing locknut to 152–203 ft. lbs. (206–275 Nm). Install the insulator, adjusting cap and a new cotter pin.
10. Install the rear wheel.
11. Lower the vehicle.

CV-Boot

REMOVAL & INSTALLATION

G20

1. Raise and support the vehicle safely.
2. Remove halfshaft assembly from vehicle and place in a suitable working fixture.
3. Remove the boot bands. Matchmark the transaxle side slide joint housing and the inner race before separating the joint assembly.
4. Remove the snapring and disassemble the slide joint housing.
5. Matchmark the inner race and halfshaft. Remove the snapring, then remove the ball cage, inner race and balls as a unit.
6. Cover the axle serrations with tape so as not to damage the boot. Remove the snapring and slide the boot off the shaft.
7. Install the wheel bearing locknut on the wheel side joint assembly. Matchmark the halfshaft and joint assembly. Using a suitable puller, separate the joint assembly.

NOTE: The wheel side joint assembly cannot be disassembled.

8. Cover the axle serrations with tape so as not to damage the boot. Remove the snapring and slide the boot off the shaft.

To install:

9. Install the transaxle side boot and joint assembly. Ensure that all snaprings are secure. If a snapring is loose or damaged, replace it. Ensure that the matchmarks made during assembly are mated.
10. Install the wheel side boot and joint by setting the joint assembly on the halfshaft. Lightly tap the joint to seat it on the shaft. Ensure that the matchmarks made during disassembly are mated.
11. Pack the joint assemblies with 3.6-4.2 Oz (105–125 ml) of grease. Install the boots so that the length is 3.86 in. (98.5mm) for the wheel side and 3.96 in. (100.5mm) for the transaxle side. Lock the boot bands securely in place.
12. Install the halfshaft assembly and lower the vehicle.

M30

1. Raise the vehicle and support safely.
2. Remove the halfshaft from the vehicle and place in a vise.

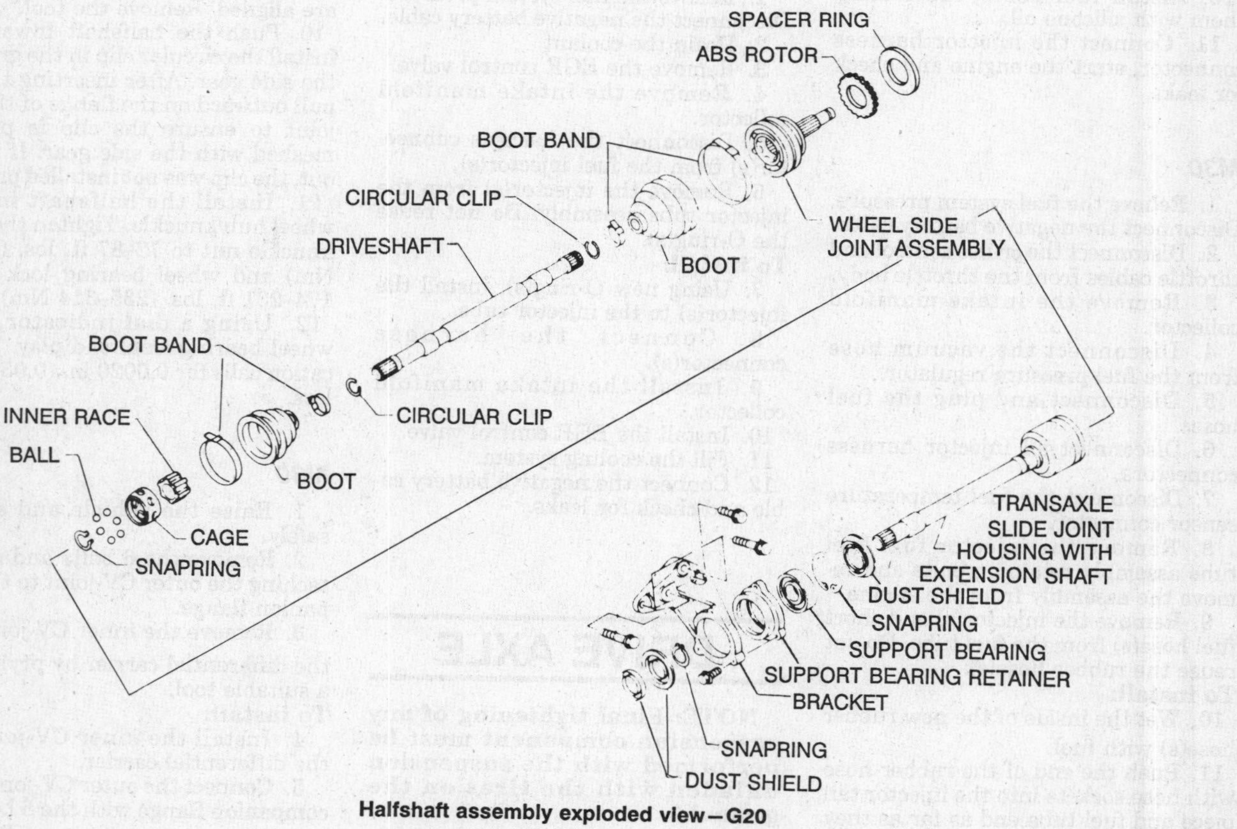

Halfshaft assembly exploded view—G20

3. Remove the plug seal from the slide joint housing by lightly tapping around the slide joint housing.
4. Remove the boot bands.
5. Put matchmarks on the slide joint housing, halfshaft and spider assembly before separating the joint assembly.
6. Remove the snapring on the halfshaft and remove spider assembly.
7. Remove the CV-joint housing.
8. Remove the boot from the shaft.

NOTE: Cover the shaft splines with tape to protect the boot.

To install:
9. Install boot onto shaft.
10. Install CV-joint housing onto shaft.
11. Install spider assembly onto shaft observing matchmarks made on disassembly. Ensure that the spider assembly chamfer faces the shaft.
12. Install snapring onto shaft.
13. With the CV-joint housing held vertically in the vise, install the coil spring, spring cap and new plug seal.

NOTE: The CV-joint housing is held vertically to prevent the coil spring from tilting or falling over.

14. Pack the halfshaft with the 6.52–6.88 oz. (185–195 g) of grease.
15. Set the boot so it does not swell or deform when installed.
16. Install a new large boot band and lock in place.
17. Install a new small boot band and lock in place.
18. Install halfshaft assembly in vehicle.
19. Lower vehicle.

Q45

1. Raise the vehicle and support safely.
2. Remove the halfshaft assembly from the vehicle and place in a vise.
3. Remove the boot bands on both inner and outer joints.
4. Put matchmarks on the slide joint housing and inner race before separating the joint assembly.
5. Remove large snapring retaining slide joint and remove slide joint from halfshaft.
6. Put matchmarks on the inner race and the halfshaft.
7. Remove small snapring and remove the ball cage, inner race and balls as a unit.
8. Remove the boot.
9. Before separating the joint assembly on the wheel side, put matchmarks on the halfshaft and joint assembly.

NOTE: The joint on the wheel side cannot be disassembled.

10. Separate the joint assembly from the halfshaft using a slide hammer or equivalent.
11. Remove the boot.

To install:
12. Apply tape to the halfshaft splines to prevent damage to the boots.
13. Install a new small boot band and a new boot on the wheel side of the halfshaft.
14. Set the joint assembly onto the halfshaft and seat the joint by lightly tapping it. Ensure that the matchmarks are aligned when assembling.
15. Pack the halfshaft with 6.00–6.70 oz. (170–190 g).
16. Set boot so it does not swell or deform when installed in the vehicle.
17. Lock new larger and smaller boot band securely with a suitable tool.
18. Install a new small boot band and a new boot on the differential side of the halfshaft.
19. Install the ball cage, inner race and balls as a unit. Ensure that the matchmarks are aligned when assembling.
20. Install a new large snapring.
21. Pack the halfshaft with 6.35–7.05 oz. (180–200 g) of grease.
22. Install slide joint housing and install a new small snapring.
23. Set the boot so it does not swell or deform when installed in the vehicle.
24. Lock the new larger and smaller boot bands securely with a suitable tool.
25. Install the halfshaft assembly in the vehicle.
26. Install the rear wheel.
27. Lower the vehicle.

Driveshaft and U-Joints

REMOVAL & INSTALLATION

M30 and Q45

1. Raise the vehicle and support safely.
2. Put matchmarks on the flanges and separate driveshaft from the differential carrier.
3. Remove driveshaft from the transmission and plug the rear opening of the extension housing.
4. Remove the bolts attaching the center bearing bracket.
5. Remove driveshafts from vehicle.
6. Inspect driveshaft runout. Runout should not exceed 0.024 in. (0.6mm).

To install:
7. Temporarily install the differential companion flange to the flange yoke. Observe the alignment marks made during removal.
8. Turn the driveshaft until alignment marks face straight upward. Securely fasten the driveshaft so the lower side wall of the concave flange yoke touches the lower side wall of the convex companion flange.
9. Remove the plug in the rear extension housing and install the driveshaft into the transmission.
10. Install the bolts attaching the center bearing bracket.

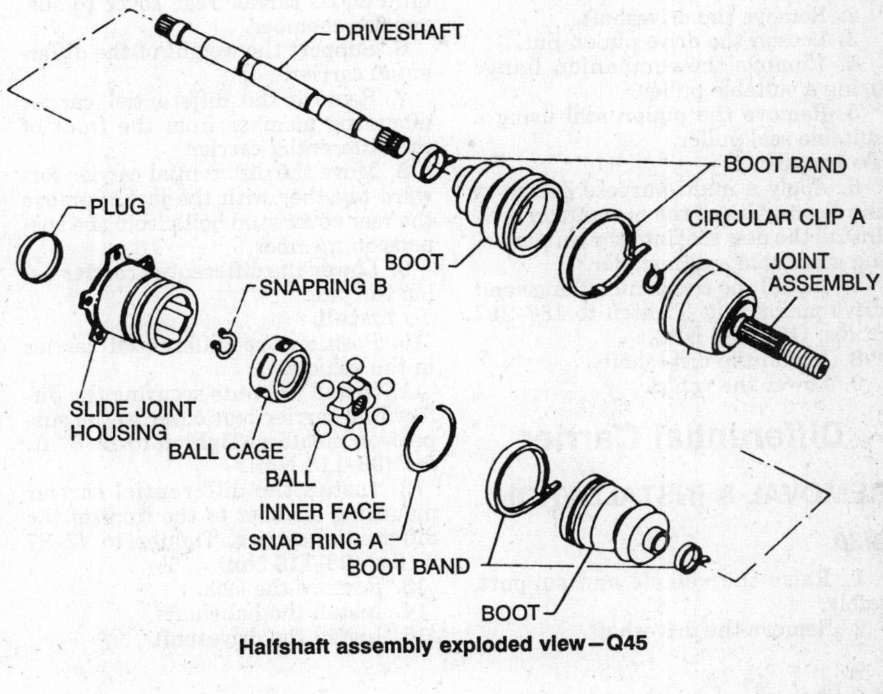

Halfshaft assembly exploded view—Q45

Front Wheel Hub and Knuckle

REMOVAL & INSTALLATION

G20

1. Raise and support the vehicle safely. Remove the wheel bearing lock nut.
2. Remove the brake caliper assembly and rotor. Using a piece of wire, position the caliper so that it is not supported by the brake line.
3. Separate the tie-rod from the ball joint.
4. Separate the kingpin from the knuckle.
5. Remove the halfshaft from the wheel hub/knuckle.
6. Remove lower ball joint nut and using tool HT2520000, or equivalent. Separate the knuckle from the transverse link and remove the wheel hub/knuckle from the vehicle.

To install:

7. Install the wheel hub/knuckle on the vehicle and tighten the lower ball joint nut to 52–64 ft. lbs. (71–86 Nm).
8. Install the halfshaft into the wheel hub/knuckle and tighten the kingpin nut to 72–87 ft. lbs. (98–118 Nm).
9. Install the tie-rod, wheel bearing lock nut and brake assembly.

Pinion Seal

REMOVAL & INSTALLATION

M30 and Q45

1. Raise the vehicle and support safely.
2. Remove the driveshaft.
3. Loosen the drive pinion nut.
4. Remove the companion flange using a suitable puller.
5. Remove the pinion seal using a suitable seal puller.

To install:

6. Apply a multi-purpose grease to the sealing lips of the new pinion seal. Install the new seal into the carrier using a suitable seal installer.
7. Install the companion flange and drive pinion nut. Tighten to 137–217 ft. lbs. (186–294 Nm).
8. Install the driveshaft.
9. Lower the vehicle.

Differential Carrier

REMOVAL & INSTALLATION

M30

1. Raise the vehicle and support safely.
2. Remove the driveshaft.

NOTE: Plug rear opening in transmission extension housing.

3. Remove the halfshafts.
4. Support the weight of the differential carrier.
5. Remove the nuts and bolts securing the differential carrier to the suspension member.
6. Remove the bolts and nuts securing the differential mounting insulator to the body.
7. Move the differential carrier toward the rear of the vehicle with the jack.
8. Lower the differential carrier using the jack.

To install:

9. Position the differential carrier in the vehicle.
10. Install bolts and nuts securing the differential mounting insulator to the body. Tighten bolts to 22–29 ft. lbs. (29–39 Nm). Tighten the nuts to 43–58 ft. lbs. (59–78 Nm).
11. Install the differential carrier to the suspension member. Tighten the nuts to 43–65 ft. lbs. (59–88 Nm).
12. Remove the jack.
13. Install the halfshafts.
14. Install the driveshaft.
15. Install the exhaust tube.
16. Lower the vehicle.

Q45

1. Raise the vehicle and support safely.
2. Remove the exhaust tube.
3. Remove the driveshaft.

NOTE: Plug rear opening in transmission extension housing.

4. Remove the halfshafts.
5. Remove the nuts securing the differential carrier rear cover to suspension member.
6. Support the weight of the differential carrier.
7. Remove the differential carrier mounting member from the front of the differential carrier.
8. Move the differential carrier forward together with the jack. Remove the rear cover stud bolts from the suspension member.
9. Lower the differential carrier using the jack.

To install:

10. Position the differential carrier in the vehicle.
11. Install the nuts securing the differential carrier rear cover to the suspension member. Tighten to 72–87 ft. lbs. (98–118 Nm).
12. Install the differential carrier mounting member to the front of the differential carrier. Tighten to 72–87 ft. lbs. (98–118 Nm).
13. Remove the jack.
14. Install the halfshafts.
15. Install the driveshaft.
16. Install the exhaust tube.
17. Lower the vehicle.

MANUAL TRANSAXLE

Transaxle Assembly

REMOVAL & INSTALLATION

G20

1. Disconnect the negative battery cable and disconnect the air duct.
2. Disconnect the clutch control cable and speedometer cable from the transaxle.
3. Disconnect the back-up light switch, neutral switch and ground harness connectors.
4. Remove the starter, shift control rod and support rod from the transaxle.
5. Drain the gear oil from the transaxle and remove the exhaust front tube.
6. Remove the halfshafts.
7. Support the engine with a suitable jack under the oil pan.
8. Remove the rear and left engine mounts
9. Raise the jack and remove the lower transaxle housing bolts. Lower the jack and remove the upper housing bolts. Keep the bolts in order as they are different lengths and must be returned to the same position.
10. Lower the transaxle.

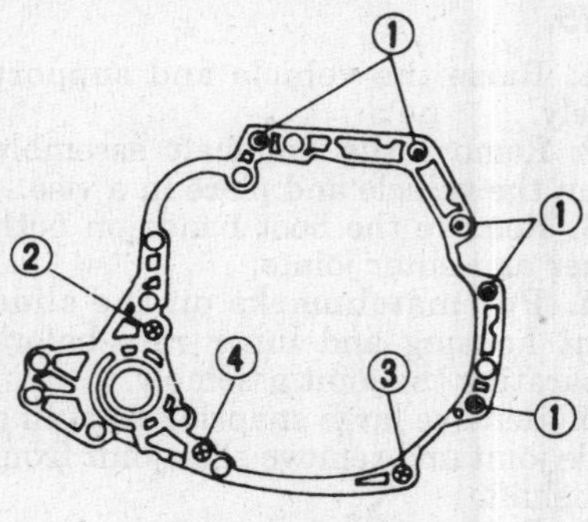

1. 51–59 ft. lbs. (70–79 Nm) – 2.17 in. (55mm) length
2. 51–59 ft. lbs. (70–79 Nm) – 2.56 in. (65mm) length
3. 22–30 ft. lbs. (30–40 Nm) – 1.38 in. (35mm) length
4. 22–30 ft. lbs. (30–40 Nm) – 1.77 in. (45mm) length

Transaxle mounting bolt torque specifications

To install:

11. Raise the transaxle into place and install the attaching bolts. Tighten bolts to the torque specified.
12. Install the rear and left engine mounts.
13. Install the driveshafts.
14. Install the shift control rods, support rod and starter on the transaxle.
15. Connect the back-up light switch, neutral switch and ground harness connectors.
16. Connect the clutch control cable and speedometer cable from the transaxle.
17. Connect the air duct and install the negative battery cable. Road test the vehicle.

CLUTCH

Clutch Assembly

REMOVAL & INSTALLATION

1. Disconnect the negative battery cable.
2. Raise and support the vehicle safely.
3. Remove the transaxle.
4. Insert tool KV30101000, or equivalent, into the clutch disc hub and loosen pressure plate bolts in 2–3 steps in sequence.
5. Remove the pressure plate and clutch disc as an assembly.
6. Remove the release bearing by pulling the bearing retainers outward from the transaxle case.
7. Inspect the clutch disc for surface wear. Measure from the friction surface to the top of the rivets. Wear limit is 0.012 in. (0.3mm). Replace clutch disc as necessary.
8. Inspect the contact surface of the flywheel for burns or discoloration. Check flywheel runnout. Maximum runnout is 0.0059 in. (0.15mm)
9. Using tools ST20050100 and ST20050010, or equivalent, check pressure plate diaphragm springs. Measure from the pressure plate/flywheel mating surface to the top of the diaphragm spring. Height should be 1.201–1.280 in. (30.5–32.5mm). Replace pressure plate as necessary.
10. Inspect the release bearing for damage. Spin the bearing to see that it rolls freely.

To install:

11. Lightly lubricate the transaxle input shaft, input shaft collar, clutch lever assembly and the clutch release bearing with a lithium based grease.

NOTE: Keep clutch disc and all clutch components clean during installation. Do not allow grease to contact the clutch disc.

12. Insert tool KV30101000, or equivalent, into the clutch disc hub. Install the clutch disc and pressure plate on the tool and tighten the pressure plate bolts to 16–22 ft. lbs. (22–29 Nm) in 2–3 steps using the sequence provided. Remove the tool.
13. Install release bearing in the transaxle. Ensure that the bearing retainer clips are fully engaged.
14. Install the transaxle.
15. Adjust clutch pedal height and free-play.
16. Lower vehicle, connect negative battery cable and road test vehicle.

PEDAL HEIGHT/FREE-PLAY ADJUSTMENT

1. Adjust pedal height with pedal stopper or automatic speed control device (ASCD) cancel switch. Pedal height specification is 6.28–6.67 in. (159.5–169.5mm) from the top of the pedal to the floot well (when measured at a 90 degree angle to the top of the pedal).
2. Adjust the withdrawl lever play on the top of the transaxle, by pushing the withdrawl lever until resistance is felt and then adjusting the nut. Turn the adjusting nut 2.5–3.5 turns back and then tighten the lock nut. Withdrawl lever play should be 0.0098–0.138 in. (2.5–3.5mm). Tighten the locknut to 2–3 ft. lbs. (3–4 Nm).
3. As a final check, measure pedal free travel at the center of the pedal pad. Pedal free travel should be 0.425–0.594 in. (10.8–15.1mm).
4. On U.S. models only, adjust the clearance between the stopper rubber and the threaded end of the clutch interlock switch while depressing the clutch pedal fully. Adjust the clearance to 0.039–0.004 in. (0.1–1.0mm).

Clutch Cable

REMOVAL & INSTALLATION

1. Raise and support the vehicle safely.
2. Loosen the lock nut and adjusting nut on the clutch cable at the withdrawl lever and disconnect the cable.
3. Disconnect the cable from the clutch pedal under the dash.
4. Remove any clips or ties holding the cable to the chassis and remove the cable.

To install:

5. Install the cable using the original routing.
6. Connect the cable at the clutch pedal and withdrawl lever.
7. Adjust the pedal height and free-play.
8. Lower the vehicle and road test.

AUTOMATIC TRANSMISSION

For further information on transmissions/transaxles, please refer to "Chilton's Guide to Transmission Repair".

Transmission Assembly

REMOVAL & INSTALLATION

M30 and Q45

1. Disconnect the negative battery cable.
2. Raise the vehicle and support safely.
3. Remove the exhaust tube.
4. Remove the fluid charging line.
5. Remove the oil cooler line.
6. Plug fluid charging and oil cooler fittings after removing lines.
7. Remove the control linkage from the selector lever.
8. Disconnect the neutral safety switch and solenoid harness connectors.
9. Disconnect the speedometer cable.
10. Remove the driveshaft. Insert plug into rear seal opening to prevent loss of fluid.
11. Remove the starter motor.
12. Support the transmission safely.
13. Remove the gusset securing the transmission to the engine. Remove the bolts attaching the transmission to the engine.

NOTE: The bolts securing the transmission to the engine are of differing lengths. Note the length of the bolts as they are removed.

14. Remove the bolts securing the torque converter to the flexplate.
15. Support the engine safely. Avoid jacking directly under the oil pan drain plug.
16. Remove the transmission from the vehicle.

To install:

17. Position the transmission in the vehicle and install the torque converter-to-flexplate bolts. Tighten to 33–43 ft. lbs. (44–59 Nm).

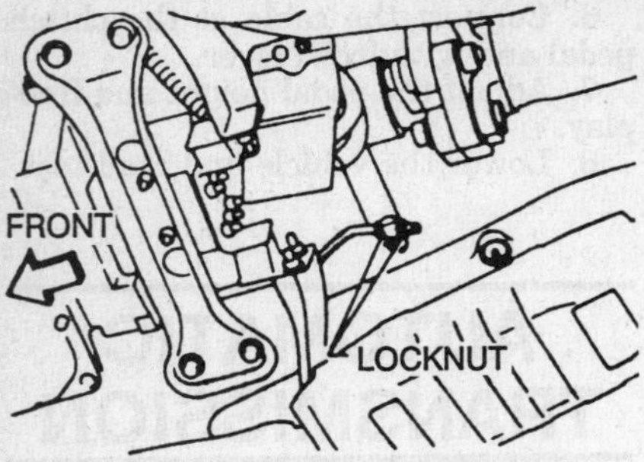

Manual control linkage locknut

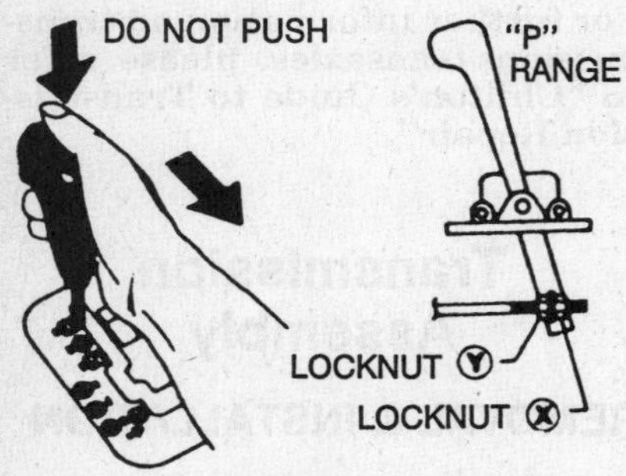

Piston positioning—Q45

18. Secure the transmission to the engine. Torque the:
60mm bolts to 29–36 ft. lbs. (39–49 Nm)
50mm bolts to 29–36 ft. lbs. (39–49 Nm)
45mm bolts to 29–36 ft. lbs. (39–49 Nm)
25mm bolts to 22–29 ft. lbs. (29–39 Nm)
20mm gusset bolts to 22–29 ft. lbs. (29–39 Nm)
19. Install the starter motor.
20. Install the driveshaft.
21. Connect the speedometer cable.
22. Connect the neutral safety switch and solenoid harness connectors.
23. Install the control linkage to the selector lever.
24. Install the fluid charging and oil cooler lines.
25. Connect the exhaust tube.
26. Lower the vehicle.
27. Connect negative battery cable.

SHIFT LINKAGE ADJUSTMENT

1. Place the selector lever in **P** range.
2. Loosen the locknuts.
3. Without pushing the button, pull selector lever toward **R** and tighten locknut X until it touches trunnion.
4. Back off locknut X one turn and tighten locknut Y to 8–11 ft. lbs. (11–15 Nm).

THROTTLE LINKAGE ADJUSTMENT

Adjust throttle wire endplay to 0.04–0.12 in. (1–3mm). Tighten locknut to 6–7 ft. lbs. (8–10 Nm).

AUTOMATIC TRANSAXLE

For further information on transmissions/transaxles, please refer to "Chilton's Guide to Transmission Repair".

Transaxle Assembly

REMOVAL & INSTALLATION

G20

1. Disconnect the negative battery cable and the air duct.
2. Raise and support the vehicle safely. Disconnect the transaxle solenoid harness and inhibitor switch harness connector. Disconnect the throttle wire at the engine side.
3. Drain the transaxle fluid.
4. Disconnect the control cable and transaxle coolant lines.
5. Remove the halfshafts, exhaust front tube and starter.
6. Remove the rear plate cover and the bolts securing the torque converter to the drive plate. Rotate the crankshaft to gain access to the bolts.
7. Support the engine with a suitable stand and use a suitable jack to support the transaxle.
8. Remove the transaxle mounting bolts and the transaxle-to-engine bolts. Lower the transaxle.

To install:

9. Place a straightedge across the bellhousing of the transaxle and measure the distance to the mounting bosses on the torque converter. The distance should be 0.626 in. (15.9mm). If not, the torque converter is not installed correctly.
10. Check the drive plate runout with a dial indicator. Maximum allowable runout is 0.008 in. (0.2mm).
11. Raise the transaxle into position and install the torque converter bolts. Tighten bolts to 33–43 ft. lbs. (44–59 Nm). Rotate the crankshaft to gain access to the bolts.
12. Install the halfshafts, exhaust front tube and starter.
13. Connect the control cable and transaxle coolant lines.
14. Connect the transaxle solenoid harness and inhibitor switch harness connector. Disconnect the throttle wire at the engine side.
15. Fill the transaxle with lubricant, install the negative battery cable and road test the vehicle.

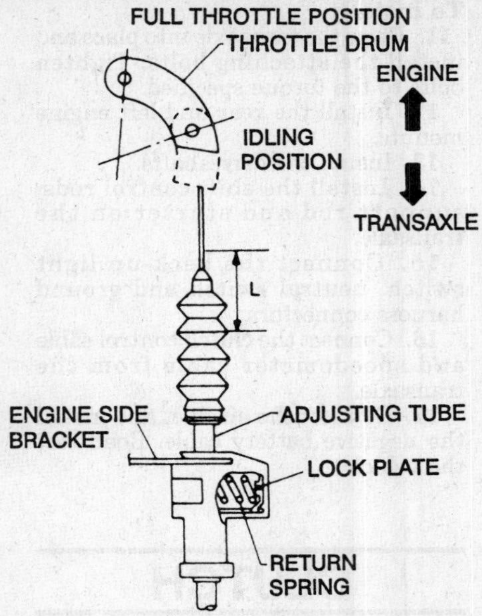

Throttle wire adjustment—G20

CONTROL CABLE ADJUSTMENT

The control cable is adjusted by loosening the locknut on the manual shaft (located at the top of the transaxle) and sliding the cable. After adjustment, move the selector lever from Park to Low range and make sure that the selector lever moves smoothly without making a sliding noise.

THROTTLE WIRE ADJUSTMENT

1. Turn ignition switch **OFF**.
2. Move adjusting tube toward the transaxle side while pressing the lock plate. Then return the lock plate to lock the adjusting tube.
3. Put a mark on the throttle wire to use as a reference while measuring.
4. Move the throttle drum from the idling position to the full throttle position quickly. The adjusting tube should move tn the direction of the engine side depressing the lock plate.
5. Ensure that the throttle wire stroke is 1.54–1.69 in. (39–43mm). If the throttle drum is too far toward the transaxle, kickdown range will greatly increase. If the throttle drum is too far toward the engine, kickdown will not occur.

6. After properly adjusting the throttle wire, ensure the parting line is as straight as possible.

FRONT SUSPENSION

Shock Absorbers

REMOVAL & INSTALLATION

G20

1. Raise and support the vehicle safely. Remove the shock absorber fixing bolt at the lower suspension member, and the three nuts inside the engine compartment. Do not remove the piston rod lock nut.
2. Remove the shock absorber assembly and place in a suitable holding device.
3. Using a prybar to hold the spring, loosen the piston rod lock nut.
4. Compress the spring with a spring compressor so that the shock mounting insulator can be turned by hand.
5. Remove the piston rod lock nut. Remove the shock absorber.

To install:

6. Inspect all components carefully for damage or wear. Replace as necessary.
7. Install shock absorber and tighten rod lock nut to 13–17 ft. lbs. (18–24 Nm).
8. Install the shock absorber assembly in the vehicle. Ensure the bend in the lower shock bracket faces rearward on the left side and forward on the right side of the vehicle.
9. Install the upper spring seat with the cutout facing the inside of the vehicle.
10. Tighten the upper shock mounting bolts to 31–40 ft. lbs. (42–54 Nm) and the lower through-bolt to 82–93 ft. lbs. (112–126 Nm). Final tightening must take place with the suspension loaded (vehicle at normal ride height).

Q45

1. Remove the upper shock absorber mounting insulator bolts.
2. Raise and safely support the vehicle.
3. Remove the lower shock mounting bolt and lift out the shock assembly.
4. Lower the vehicle.
5. The installation is the reverse of the removal procedure. Keep the following torques in mind:
   a. Tighten the upper mounting bolts to 30–35 ft. lbs.
   b. Tighten the piston rod locknut to 13–17 ft. lbs.
   c. Tighten the lower mounting bolt to 80–94 ft. lbs.

MacPherson Strut

REMOVAL & INSTALLATION

M30

1. Disconnect the negative battery cable.
2. Disconnect the sub-harness connector strut actuator mounting bolt.
3. Remove the strut assembly mounting nut. Do not remove the piston rod locknut on the vehicle.
4. Raise and safely support the vehicle. Remove the tension rod nuts and the strut-to-steering knuckle mounting bolts. Make sure the brake hose is not twisted.
5. Remove the strut assembly.
6. The installation is the reverse of the removal procedure. Keep the following torques in mind:

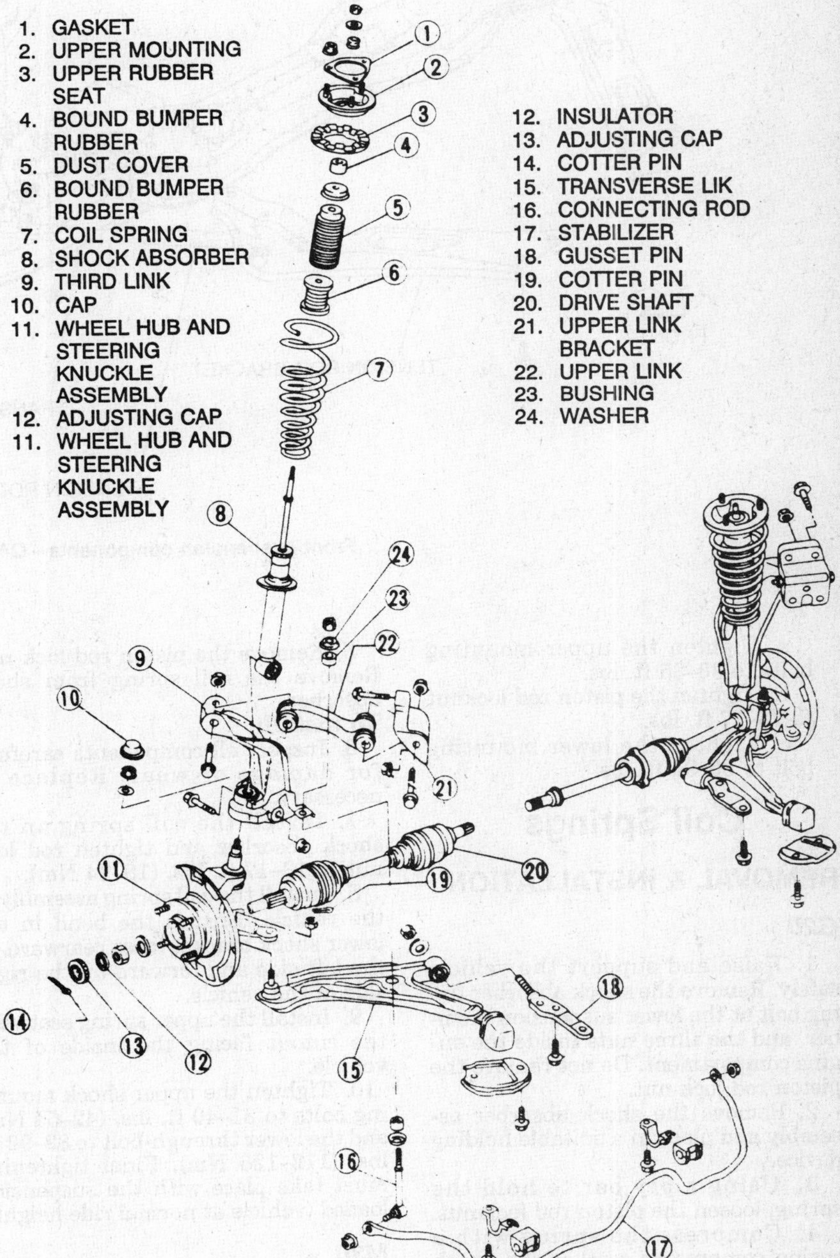

Front suspension components—G20

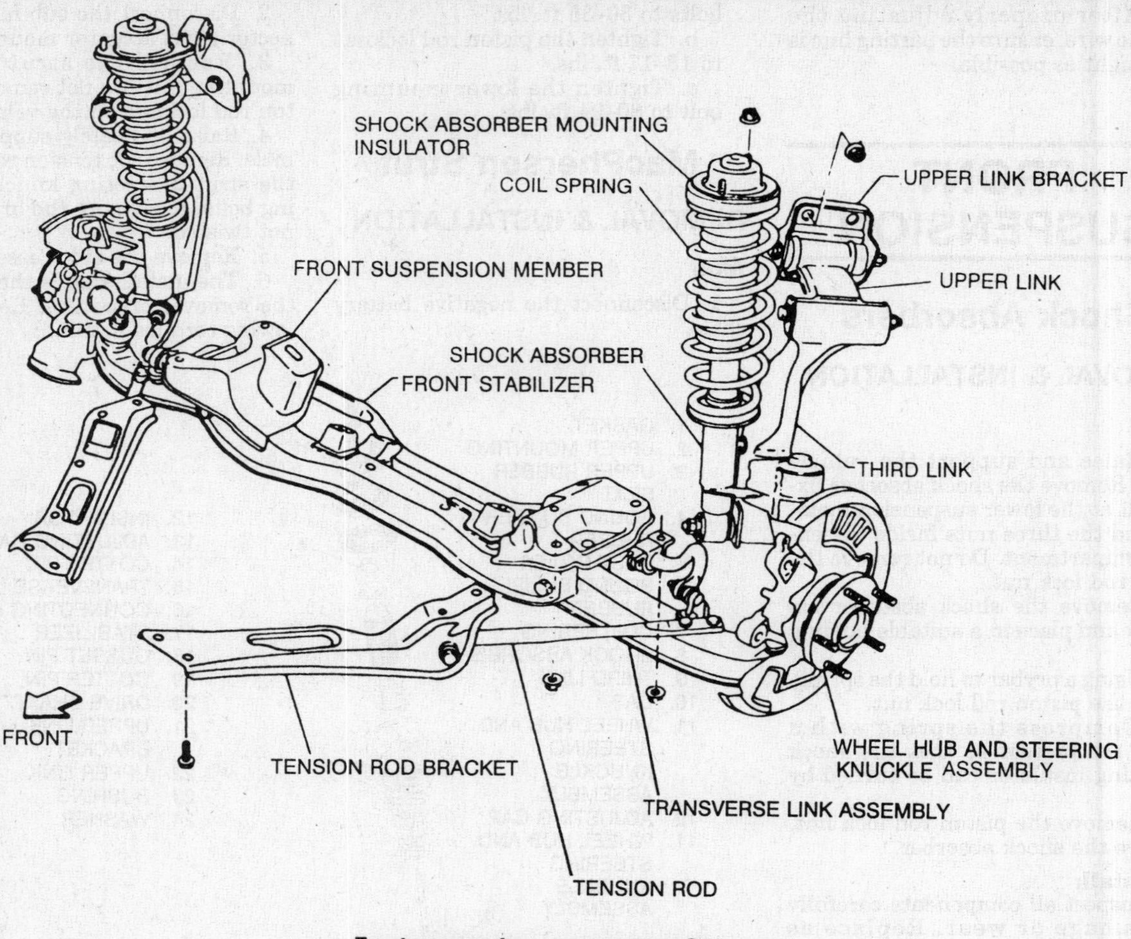

Front suspension components—Q45

a. Tighten the upper mounting bolts to 30–35 ft. lbs.

b. Tighten the piston rod locknut to 13–17 ft. lbs.

c. Tighten the lower mounting bolt to 80–94 ft. lbs.

Coil Springs

REMOVAL & INSTALLATION

G20

1. Raise and support the vehicle safely. Remove the shock absorber fixing bolt at the lower suspension member, and the three nuts inside the engine compartment. Do not remove the piston rod lock nut.
2. Remove the shock absorber assembly and place in a suitable holding device.
3. Using a pry bar to hold the spring, loosen the piston rod lock nut.
4. Compress the spring with a spring compressor so that the shock mounting insulator can be turned by hand.
5. Remove the piston rod lock nut. Remove the coil spring from shock absorber.

To install:

6. Inspect all components carefully for damage or wear. Replace as necessary.
7. Install the coil spring on the shock absorber and tighten rod lock nut to 13–17 ft. lbs. (18–24 Nm).
8. Install the coil spring assembly in the vehicle. Ensure the bend in the lower shock bracket faces rearward on the left side and forward on the right side of the vehicle.
9. Install the upper spring seat with the cutout facing the inside of the vehicle.
10. Tighten the upper shock mounting bolts to 31–40 ft. lbs. (42–54 Nm) and the lower through-bolt to 82–93 ft. lbs. (112–126 Nm). Final tightening must take place with the suspension loaded (vehicle at normal ride height).

M30

1. Disconnect the negative battery cable.
2. Disconnect the sub-harness connector strut actuator mounting bolt.
3. Remove the strut assembly mounting nut. Do not remove the piston rod locknut on the vehicle.
4. Raise and safely support the vehicle. Remove the tension rod nuts and the steering knuckle mounting bolts. Make sure the brake hose is not twisted.
5. Secure the strut assembly in suitable holding fixture and loosen the piston rod locknut. Do not remove it.
6. Compress the spring with the proper tool so the shock absorber mounting insulator can be turned by hand.
7. Remove the piston rod locknut and the coil spring assembly.
8. Remove the gland packing with the proper tool. Retract the piston, by pushing it down until it bottoms.
9. Slowly, remove the piston rod from the cylinder together with the piston guide.

To install:

10. Inspect the rubber parts for deterioration.

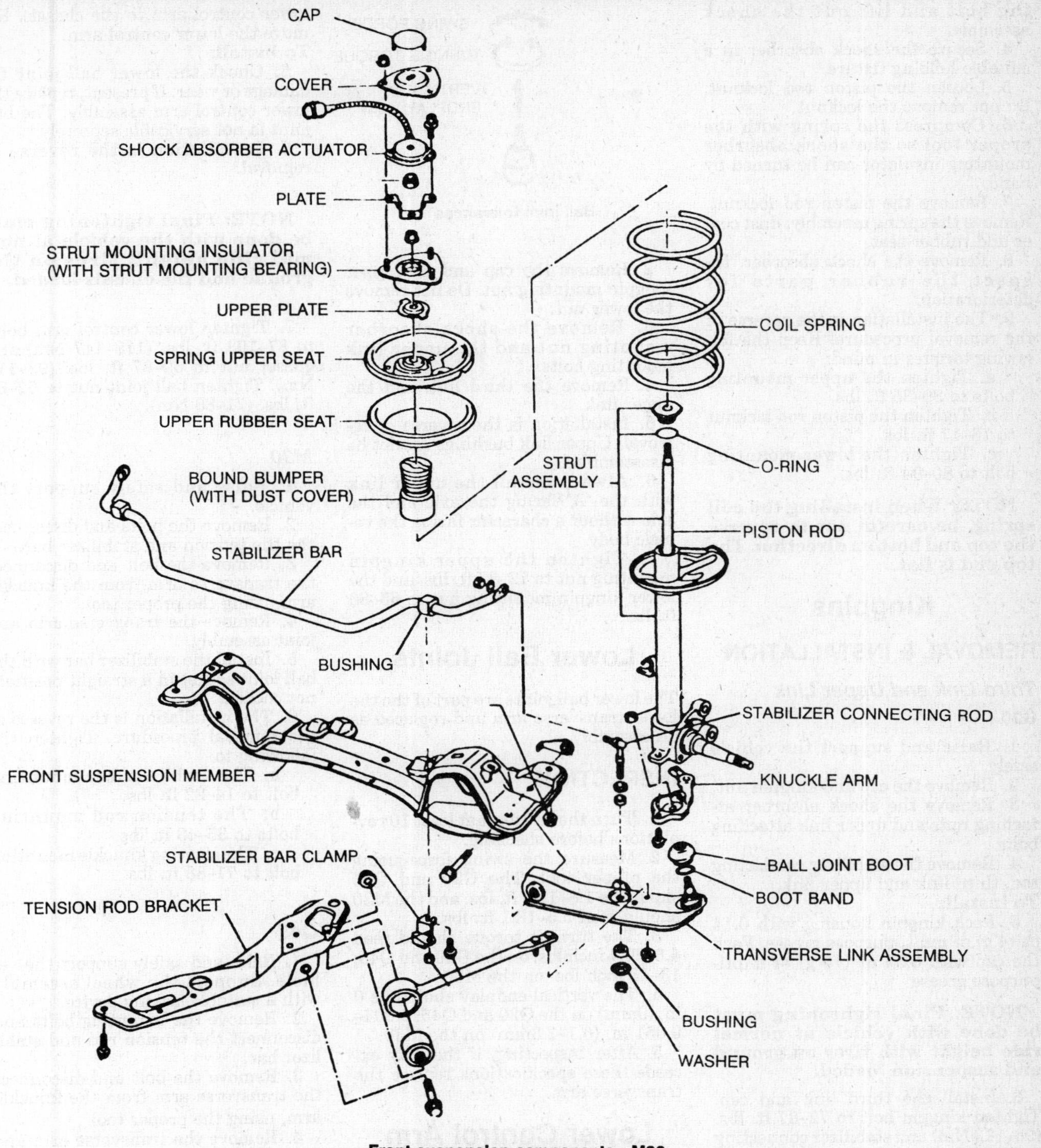

Front suspension components—M30

11. Lubricate the sealing lip of the gland packing.
12. Install the gland packing while covering the piston rod with tape so not to damage the oil sealing lip.
13. Tighten the gland packing to 51–94 ft. lbs. without the special tool.
14. The installation is the reverse of the removal procedure. The flat portion of the spring goes in the top position.
15. Install the spring seat with it's cutout facing the outer side of the vehicle. Tighten the following:
    a. The upper cover mounting bolts to 22–29 ft. lbs.
    b. The upper piston rod locknut to 43–58 ft. lbs.
    c. The lower strut-to-knuckle arm mounting bolts to 53–72 ft. lbs.

Q45

1. Remove the upper shock absorber mounting insulator bolts.
2. Raise and safely support the vehicle.
3. Remove the lower shock mount-

ing bolt and lift out the shock assembly.

4. Secure the shock absorber in a suitable holding fixture.

5. Loosen the piston rod locknut. Do not remove the locknut.

6. Compress the spring with the proper tool so the shock absorber mounting insulator can be turned by hand.

7. Remove the piston rod locknut. Remove the spring assembly, dust cover and rubber seat.

8. Remove the shock absorber. Inspect the rubber parts for deterioration.

9. The installation is the reverse of the removal procedure. Keep the following torques in mind:

a. Tighten the upper mounting bolts to 30–35 ft. lbs.

b. Tighten the piston rod locknut to 13–17 ft. lbs.

c. Tighten the lower mounting bolt to 80–94 ft. lbs.

NOTE: When installing the coil spring, be careful not to reverse the top and bottom direction. The top end is flat.

Kingpins

REMOVAL & INSTALLATION

Third Link and Upper Link

G20

1. Raise and support the vehicle safely.

2. Remove the cap and kingpin nut.

3. Remove the shock absorber attaching nuts and upper link attaching bolts.

4. Remove the stabilizer connecting rod, third link and upper link.

To install:

5. Pack kingpin housing with 0.14 oz (4 g) of multi-purpose grease. Pack the cap with 0.35 oz (10 g) of multi-purpose grease.

NOTE: Final tightening must be done with vehicle at normal ride height with tires on ground and suspension loaded.

6. Install the third link and cap. Tighten kingpin bolt to 72–87 ft. lbs. (98–118 Nm) and stabilizer connecting rod bolt to 12–16 ft. lbs. (16–22 Nm).

7. Install the upper link. Tighten upper link-to-third link through bolt to 82–93 ft. lbs. (112–126 Nm) and upper link-to-bracket bolt to 65–90 ft. lbs. (88–123 Nm).

Q45

1. Raise and safely support the vehicle. Support the wheel assembly with a suitable jacking device.

Ball joint tolerances

2. Remove the cap and the upper kingpin mounting nut. Do not remove the lower nut.

3. Remove the shock absorber mounting nut and the upper link mounting bolts.

4. Remove the third link and the upper link.

5. Installation is the reverse of removal. Upper link bushings cannot be disassembled.

6. Always install the upper link with the 'A' facing the axle and the side without a character facing the vehicle body.

7. Tighten the upper kingpin mounting nut to 72–87 ft. lbs. and the lower kingpin mounting nut to 65–80 ft. lbs.

Lower Ball Joints

The lower ball joints are part of the the lower transverse arm and replaced as an assembly.

INSPECTION

1. Turn the ball joint at least 10 revolutions before checking.

2. Measure the swing force using the proper tool. The G20 and Q45 should be 1.8–11.9 ft. lbs. and the M30 should read 5.5–18.1 ft. lbs.

3. The turning torque should read 4.3–30.4 inch lbs. on the G20 and Q45; 13–43 inch lbs. on the M30.

4. The vertical endplay should be 0 in. (0mm) on the G20 and Q45; 0.004–0.051 in. (0.1–1.3mm) on the M30.

5. After inspecting, if the play exceeds these specifications replace the transverse arm.

Lower Control Arm

REMOVAL & INSTALLATION

G20

1. Raise and support the vehicle safely.

2. Remove the stabilizer.

3. Support the steering knuckle with a suitable jack and remove the lower ball joint nut.

4. Remove the bolts attaching the lower control arm to the chassis. Remove the lower control arm.

To install:

5. Check the lower ball joint for damage or wear. If present, replace the lower control arm assembly. The ball joint is not servicable seprately.

6. Installation is the reverse of removal.

NOTE: Final tightening must be done with the vehicle at normal ride height, tires on the ground and the chassis loaded.

7. Tighten lower control arm bolts to 87–108 ft. lbs. (118–147 Nm) and gusset nut to 69–87 ft. lbs. (93–118 Nm). Tighten ball joint nut to 52–64 ft. lbs. (71–86 Nm).

M30

1. Raise and safely support the vehicle.

2. Remove the bolts and disconnect the the tension and stabilizer bar.

3. Remove the bolt and disconnect the transverse arm from the knuckle arm, using the proper tool.

4. Remove the transverse arm and joint assembly.

5. Install the stabilizer bar with the ball joint socket in a straight position, not cocked.

6. The installation is the reverse of the removal procedure. Tighten the following to:

a. The stabilizer bar mounting bolt to 14–22 ft. lbs.

b. The tension rod mounting bolts to 35–43 ft. lbs.

c. The steering knuckle mounting bolt to 71–88 ft. lbs.

Q45

1. Raise and safely support the vehicle. Support the wheel assembly with a suitable jacking device.

2. Remove the mounting bolts and disconnect the tension rod and stabilizer bar.

3. Remove the bolt and disconnect the transverse arm from the knuckle arm, using the proper tool.

4. Remove the transverse arm and joint assembly.

5. The installation is the reverse of the removal procedure. The final tightening must be at curb weight with the tires on the ground.

6. Tighten the following to:

a. The stabilizer bar mounting bolt to 14–22 ft. lbs.

b. The tension rod mounting bolts to 72–87 ft. lbs.

c. The steering knuckle mounting bolt to 65–80 ft. lbs.

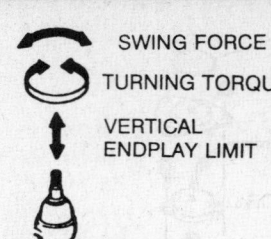

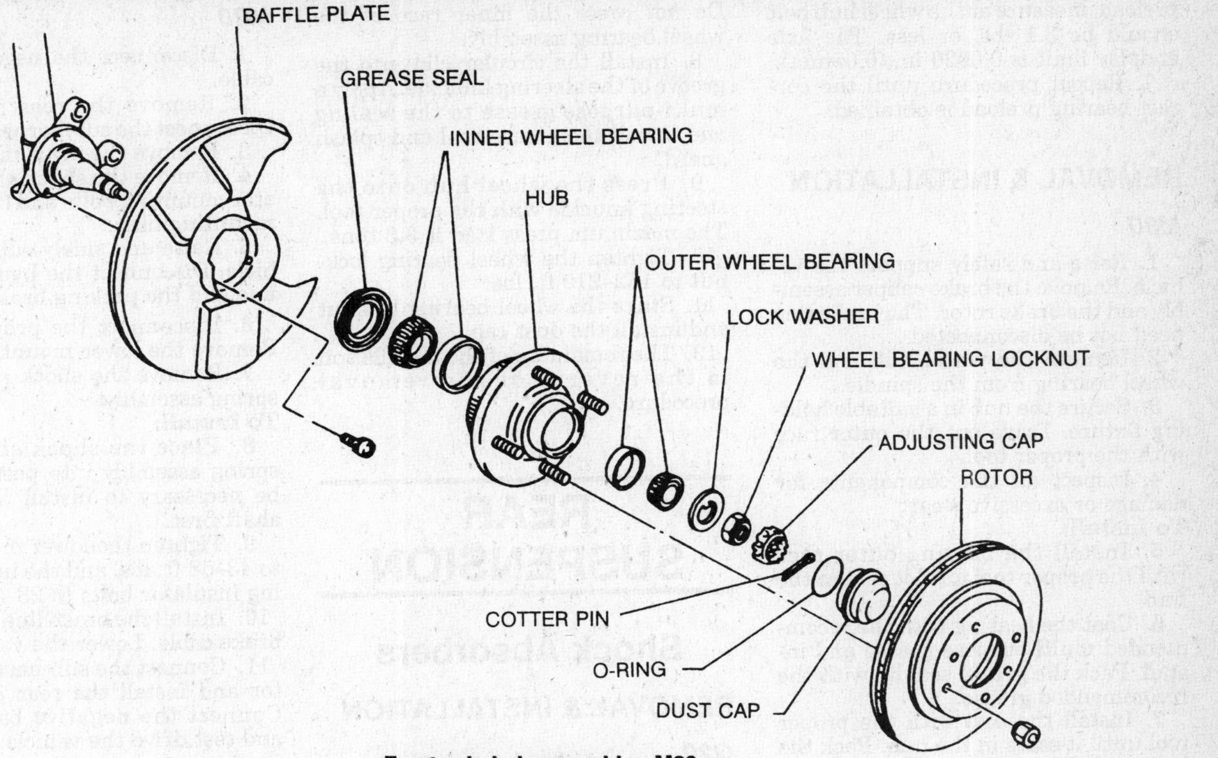

Front axle hub assembly—M30

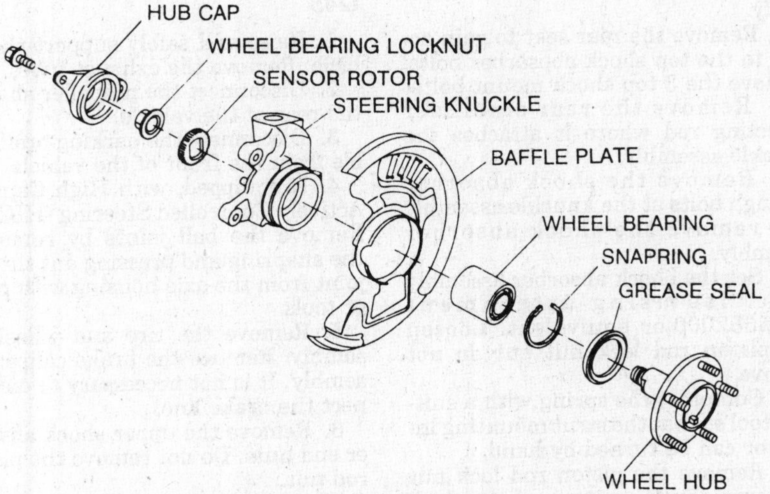

Front axle hub assembly—Q45

Stabilizer Bar

REMOVAL & INSTALLATION

1. Raise and support the vehicle safely.
2. Using a backup wrench to support the connecting rod, remove the stabilizer to connecting rod bolt.
3. Remove the stabilizer bracket bolts. Remove the stabilizer.

To install:

4. Install the stabilizer with the paint mark to the right of the bracket when viewed from the front of the vehcile. Install the bracket with the elongated hole toward the rear of the vehicle. Tighten the bracket bolts to 29–36 ft. lbs. (39–49 Nm). On the M30 tighten bolts to 40–47 ft. lbs. (54–64 Nm).

NOTE: Final tightening must be done with the vehicle at normal ride height, tires on the ground and the chassis loaded.

5. Ensure that the ball socket on the connecting rod is straight, then attach the connecting rod and tighten the bolt to 30–38 ft. lbs. (41–51 Nm). Use a backup wrench to keep the connecting rod straight.

Front Wheel Bearings

PRELOAD ADJUSTMENT

M30

1. Thoroughly clean all parts to prevent dirt entry.
2. Apply the recommended multipurpose grease to the following components:
   a. The rubbing surface of the spindle.
   b. The contact surface between the lock washer and the outer wheel bearing.
   c. The inside of the dust cap.
   d. The grease seal lip.
3. Tighten the wheel bearing lock to 25–29 ft. lbs. Turn the wheel hub several times in both directions to seat the wheel bearing correctly.
4. Again, tighten the wheel bearing to the specified torque. Turn back the wheel bearing locknut 90 degrees.
5. Install the adjusting cap and the locknut. Do not turn the nut back for cotter pin insertion. Align the cotter pin by re-tightening the nut within 15 degrees.
6. Measure the wheel bearing preload and the axle endplay limit with the proper tool. The wheel bearing

preload, measure at the wheel hub bolt should be 3.1 lbs. or less. The axle endplay limit is 0.0020 in. (0.05mm).

7. Repeat procedure until the correct bearing preload is obtained.

REMOVAL & INSTALLATION

M30

1. Raise and safely support the vehicle. Remove the brake caliper assembly and the brake rotor. The brake line need not be disconnected.
2. Remove the wheel hub and the wheel bearing from the spindle.
3. Secure the hub in a suitable holding fixture. Drive out the outer race with the proper tool.
4. Inspect all the components for damage or excessive wear.

To install:

5. Install the bearing outer race with the proper tool until it seat in the hub.
6. Coat the bearing with the recommended multi-purpose grease and install. Pack the grease seal lip with the recommended grease.
7. Install the seal with the proper tool until it seats in the hub. Pack the hub and dust cap with the recommended grease.
8. Adjust the wheel bearing preload.
9. The remainder of the installation is the reverse of the removal procedure.

Q45

1. Raise and safely support the vehicle. Remove the brake caliper assembly and the brake rotor. The brake line need not be disconnected.
2. Disconnect the tie rod and transverse arm from the steering knuckle assembly with the proper tool.

NOTE: The steering knuckle is made from aluminum alloy, Be careful not to hit the knuckle.

3. Remove the kingpin lower nut and the steering knuckle assembly. Secure the steering knuckle in a suitable holding fixture.
4. Remove the dust cap and the wheel bearing locknut. Remove the wheel hub with the proper tool.
5. Remove the circular clip and press out the bearing assembly from the steering knuckle with the proper tools.
6. Drive out the wheel bearing inner race from the wheel hub and remove the grease seal with the suitable tools.

To install:

7. Press a new wheel bearing assembly into the steering knuckle from outside the of the steering knuckle. The maximum press load is 3.9 tons. Do not press the inner race of the wheel bearing assembly.
8. Install the circular clip into the groove of the steering knuckle. Apply a multi-purpose grease to the sealing and install the grease seal and splash guard.
9. Press the wheel hub onto the steering knuckle with the proper tool. The maximum press load is 3.3 tons.
10. Tighten the wheel bearing locknut to 152–210 ft. lbs.
11. Stake the wheel bearing locknut and install the dust cap.
12. The remainder of the installation is the reverse of the removal procedure.

REAR SUSPENSION

Shock Absorbers

REMOVAL & INSTALLATION

G20

1. Raise and support the vehicle safely.
2. Remove the rear seat to gain access to the top shock abosorber bolts. Remove the 3 top shock mount bolts.
3. Remove the rear stabilizer conecting rod where it attaches the knuckle assembly.
4. Remove the shock absorber through bolts at the knuckle assembly and remove the shock absorber assembly.
5. Set the shock absorber assembly in a vise using attachment ST25652000,or equivalent. Loosen the piston rod lock nut but do not remove.
6. Compress the spring with a suitable tool so that the strut mounting insulator can be turned by hand.
7. Remove the piston rod lock nut and spring with compressor attached.

To install:

8. Replace the bound rubber bumpers. Install the coil spring on the shock absorber and tighten the piston rod locknut to 43–58 ft. lbs. (59–78 Nm). Gradually release the spring compressor. When the coil spring is located correctly, there should be 2 identification color codes on the lower side.
9. Installation is the reverse of removal.
10. Tighten the shock assembly upper attaching bolts to 31–40 ft. lbs. (42–54 Nm); lower attaching bolts to 72–87 ft. lbs. (98–118 Nm) and the stabilizer bar connecting rod bolts to 30–35 ft. lbs. (41–47 Nm)

M30

1. Disconnect the negative battery cable.
2. Remove the rear parcel shelf. Disconnect the sub-harness connector.
3. Remove the strut mounting cap.
4. Remove the shock absorber actuator mounting bolts and the upper end mounting nuts.
5. Raise and safely support the vehicle. Disconnect the hydraulic brake line and the parking brake cable.
6. Disconnect the propeller shaft. Remove the lower mounting bolt.
7. Remove the shock absorber and spring assembly.

To install:

8. Place the shock absorber and spring assembly into postion. It may be necessary to install the propeller shaft first.
9. Tighten the lower mounting bolt to 43–58 ft. lbs. and the upper mounting insulator bolts to 23–31 ft. lbs.
10. Install the brake line and parking brake cable. Lower the vehicle.
11. Connect the sub-harness connector and install the rear parcel shelf. Connect the negative battery cable and test drive the vehicle.

Q45

1. Raise and safely support the vehicle. Remove the exhaust tube.
2. Disconnect the propeller shaft at the rear of the vehicle.
3. Disconnect the parking brake cable from the front of the vehicle.
4. If equipped, with High Capacity Actively Controlled Steering (HICAS). Remove the ball joints by removing the snapring and pressing out the ball joint from the axle housing with proper tools.
5. Remove the tire and wheel assembly. Remove the brake caliper assembly. It is not neccessary to disconnect the brake line.
6. Remove the upper shock absorber end nuts. Do not remove the piston rod nut.
7. Remove the rear suspension mounting nuts. Draw out the rear axle and rear suspension assembly.
8. Remove the shock absorber upper and lower mounting nuts. Do not remove the piston rod nut.
9. Remove the shock absorber and spring assembly.

To install:

10. Install the shock absorber and spring assembly. It may be necessary to install the rear axle and suspension assembly prior to installing the shock assembly.
11. Tighten the lower shock mounting bolt to 57–72 ft. lbs. and the upper spring seat mounting bolts to 12–14 ft. lbs.

1. GASKET
2. STRUT MOUNTING INSULATOR
3. UPPER SPRING SEAT
4. DUST COVER
5. COIL SPRING
6. BOUND BUMPER
7. STRUT ASSEMBLY
8. CONNECTING ROD
9. KNUCKLE ASSEMBLY
10. BAFFLE PLATE
11. WHEEL HUB BEARING
12. COTTER PIN
13. CAP
14. PARALLEL LINK
15. MOUNTING BRACKET
16. BUSHING
17. CLAMP
18. STABILIZER BAR
19. RADIUS ROD

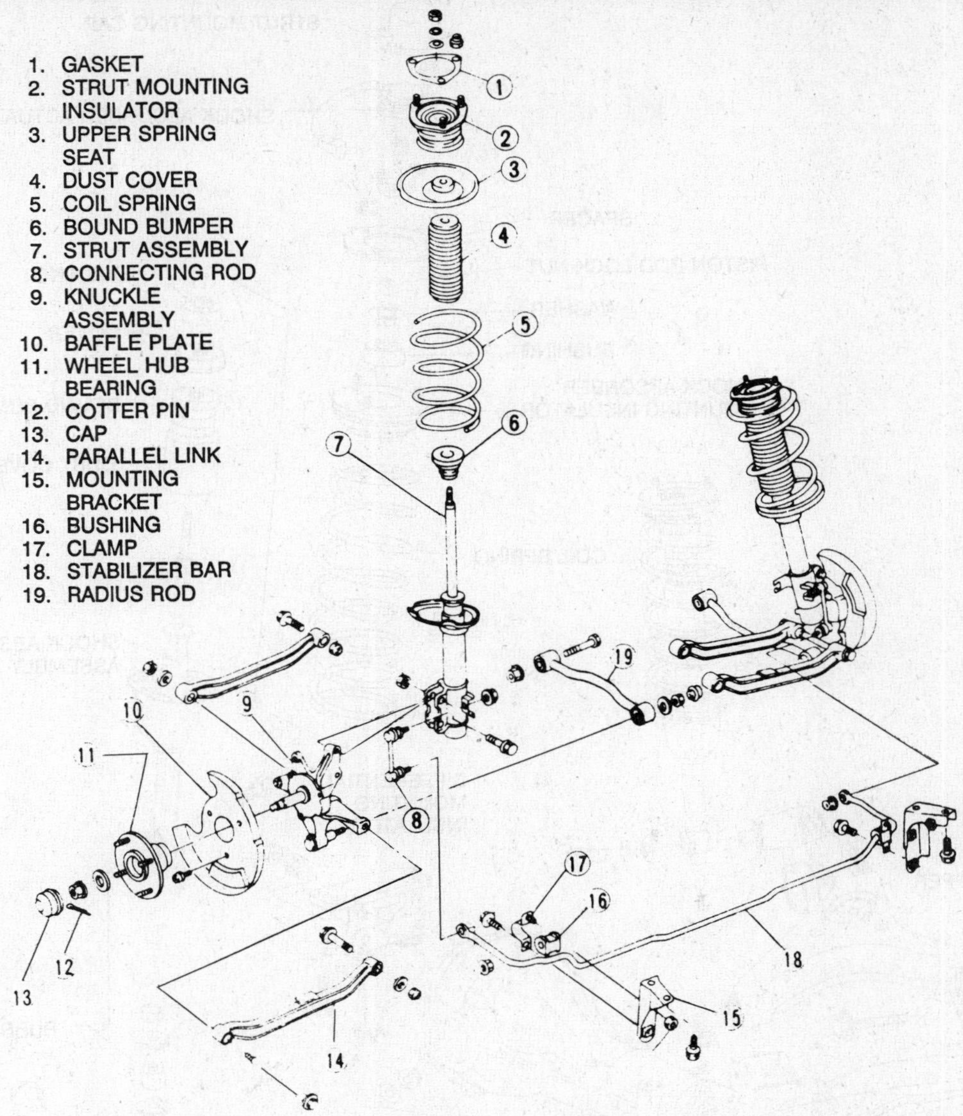

Rear suspension assembly—G20

Install the brake caliper, wheel and tire, parking brake cable, propeller shaft and exhaust tube.

13. Lower the vehicle and test drive.

Coil Springs

REMOVAL & INSTALLATION

G20

1. Raise and support the vehicle safely.
2. Remove the shock absorber assembly.
3. Set the shock absorber assembly in a vise using attachment ST25652000, or equivalent. Loosen the piston rod lock nut but do not remove.
4. Compress the spring with a suitable tool so that the strut mounting insulator can be turned by hand.
5. Remove the piston rod lock nut and spring with compressor attached.

To install:

6. Replace the bound rubber bumpers. Install the coil spring on the shock absorber and tighten the piston rod locknut to 43–58 ft. lbs. (59–78 Nm). Gradually release the spring compressor. When the coil spring is located correctly, there should be 2 identification color codes on the lower side.
7. Install the shock absorber assembly. Lower the vehicle.
8. Final tightening of all rubber parts should take place with the tires on the ground and the chassis at normal ride height.

M30

1. Raise and support the vehicle safely.
2. Remove the rear parcel shelf and disconnect the shock absorber actuator wiring connector.
3. Remove the strut mounting cap, spacer and insulator bolts.
4. Remove the lower mounting bolt. Remove the shock absorber assembly.
5. Place the shock absorber assembly in an appropriate holding fixture and loosen the piston lock nut.
6. Compress the coil spring with a

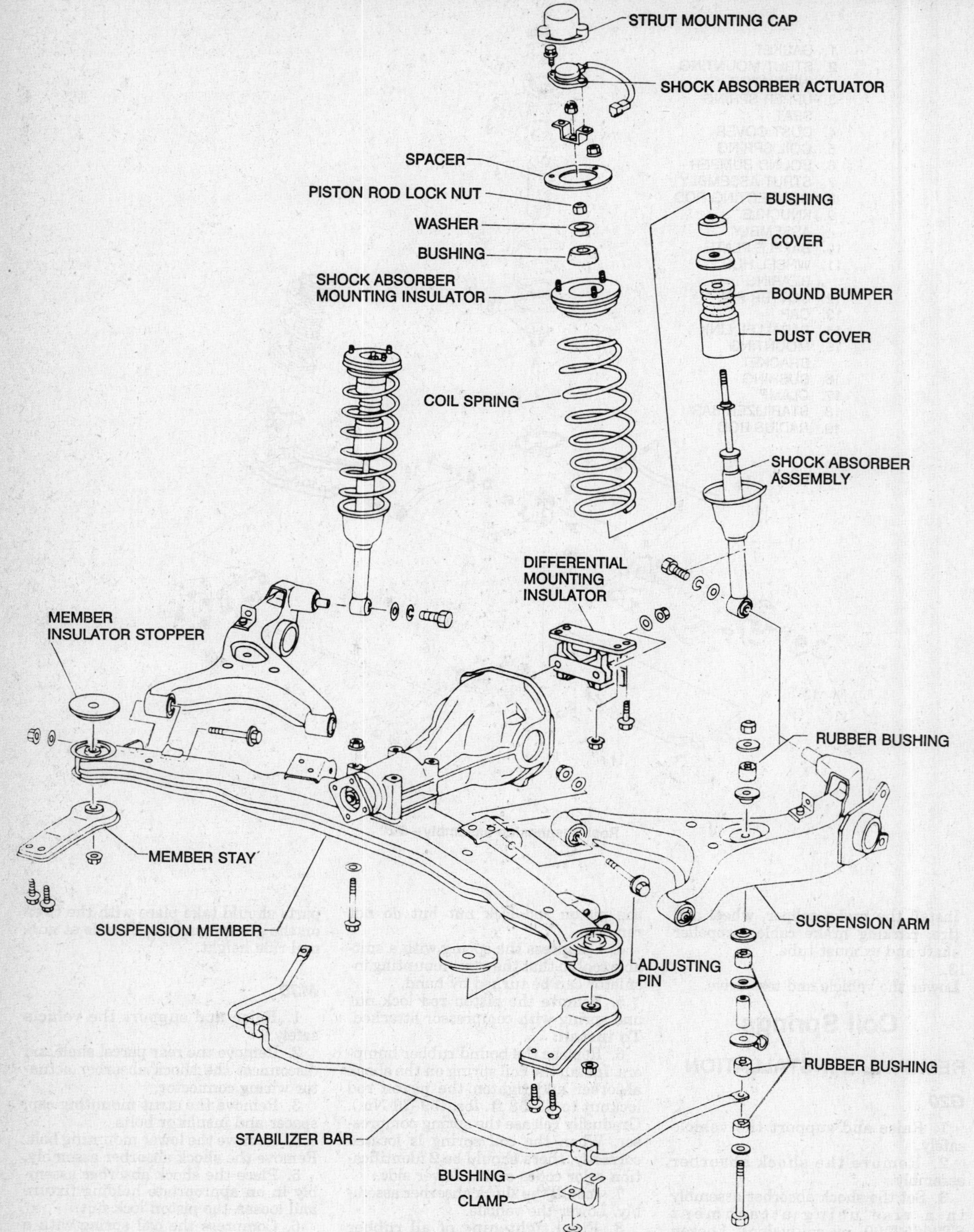
STRUT MOUNTING CAP
SHOCK ABSORBER ACTUATOR
SPACER
PISTON ROD LOCK NUT
WASHER
BUSHING
SHOCK ABSORBER MOUNTING INSULATOR
COIL SPRING
BUSHING
COVER
BOUND BUMPER
DUST COVER
SHOCK ABSORBER ASSEMBLY
DIFFERENTIAL MOUNTING INSULATOR
MEMBER INSULATOR STOPPER
RUBBER BUSHING
MEMBER STAY
SUSPENSION ARM
SUSPENSION MEMBER
ADJUSTING PIN
RUBBER BUSHING
STABILIZER BAR
BUSHING
CLAMP

Rear suspension assembly – M30

spring compressor, then remove the lock nut. Remove the coil spring with the compressor still attached.

To install:

7. Install the new coil spring with the spring compressor attached.
8. Assemble the shock absorber components and tighten the piston rod lock nut.
9. Installation is the reverse of removal.
10. Tighten the lower shock mount bolt to 43–58 ft. lbs. (59–78 Nm); shock insulator bolts to 23–31 ft. lbs. (31–42 Nm).
11. Final tightening of all rubber parts should take place with the tires on the ground and the chassis at normal ride height.

Q45

1. Raise and support the vehicle safely.
2. Remove the upper shock assembly attaching nuts. It may be necessary to remove the rear parcel shelf to gain access to the nuts.
3. Remove the lower shock assembly mounting bolt at the axle assembly.
4. Place the shock absorber assembly in a suitable holding fixture and loosen the piston rod lock nut.
5. Using a spring compressor, compress the coil spring, then remove the piston rod lock nut. Remove the coil spring with the compressor attached.

To install:

6. Install the spring compressor, if removed, and assemble the shock absorber assembly. Remove the spring compressor.
7. Installation is the reverse of removal.
8. Tighten the piston rod lock nut to 13–17 ft. lbs. (18–24 Nm); upper shock mount nuts to 12–14 ft. lbs. (16–19 Nm); lower shock mount bolts to 57–72 ft. lbs. (77–98 Nm).
9. Final tightening of all rubber parts should take place with the tires on the ground and the chassis at normal ride height.

Rear Control Arm

REMOVAL & INSTALLATION

G20

1. Raise and support the vehicle safely.
2. Remove the brake caliper assembly and rotor.
3. Support the suspension under the knuckle assembly with a floor jack.
4. Remove the parallel link attaching bolts. Remove the parallel link.
5. Installation is the reverse of removal.
6. Tighten all parallel link bolts to 80–94 ft. lbs. (108–127 Nm).

M30

1. Raise and support the vehicle. Remove the axle shaft assembly.
2. Remove the stabilizer bar bolt and disconnect the parking brake cable.
3. Disconnect the lower shock absorber bolt.
4. Matchmark the suspension arm to the pin and remove the pin.
5. Remove the lower suspension arm.
6. The installation is the reverse of the removal procedure. Adjust the rear alignment after installing the lower arm.
7. When installing, tighten the suspension arm pin to 72–87 ft. lbs. after installing the wheels and placing the vehicle on the ground under the unladen condition.

Q45

1. Raise and safely support the vehicle. Remove the exhaust tube.
2. Disconnect the propeller shaft at the rear of the vehicle.
3. Disconnect the parking brake cable from the front of the vehicle.
4. If equipped, with High Capacity Actively Controlled Steering (HICAS). Remove the ball joints by removing the snapring and pressing out the ball joint from the axle housing with proper tools.
5. Remove the tire and wheel assembly. Remove the brake caliper assembly. It is not neccessary to disconnect the brake line.
6. Remove the upper shock absorber end nuts. Do not remove the piston rod nut.
7. Remove the rear suspension mounting nuts. Draw out the rear axle and rear suspension assembly.
8. The installation is the reverse of the removal procedure. Tighten the lower arm adjusting pin bolts to 57–72 ft. lbs.

Rear Wheel Bearings

REMOVAL & INSTALLATION

G20

1. Raise and support the vehicle safely.
2. Remove the rear wheel and tire assembly.
3. Remove the rear wheel hub cap, cotter pin, lock nut, washer and wheel hub bearing.

NOTE: The wheel bearing is integral with the hub and cannot be serviced separately.

4. Installation is the reverse of removal. Tighten the wheel bearing lock nut to 137–188 ft. lbs. (186–255 Nm).

Rear Axle Assembly

REMOVAL & INSTALLATION

G20

1. Raise and support the vehicle safely.
2. Remove the rear wheel, disc brake caliper and rotor assembly.
3. Remove the rear parcel shelf to gain access to the upper shock mount and remove the shock absorber assembly.
4. Remove the stabilizer bar from the connecting rods. Remove the radius rods and parallel links from the rear hub assembly only.

To install:

5. Attach the parallel links and radius rods to the rear hub assembly. Tighten the parallel link bolts to 80–94 ft. lbs. (108–127 Nm) and the radius rod bolts to 65–80 ft. lbs. (88–108 Nm).
6. Install the stabilizer bar. Tighten the stabilizer bar-to-connecting rod nuts to 30–35 ft. lbs. (41–47 Nm).
7. Install the shock absorber assembly.
8. Install the rear disc brake rotor and caliper. Install the rear wheel and tire.

STEERING

Steering Wheel

CAUTION

On vehicles equipped with an air bag, the negative battery cable must be disconnected, before working on the system. Failure to do so may result in deployment of the air bag and possible personal injury.

REMOVAL & INSTALLATION

Except Air Bag

1. Disconnect the negative battery cable.
2. Ensure that the steering wheel and front tires are positioned in the straight ahead position.
3. Using an appropriate tool, pry the horn pad off the steering wheel.
4. Remove the steering wheel lock nut.
5. Using an appropriate puller, remove the steering wheel.

To install:

6. Apply multi-purpose grease to the entire surface of the trun signal

cancel pin and the horn contact clip ring.

7. Install the steering wheel and tighten the lock nut to 22–29 ft. lbs. (29–39 Nm).

8. Install the horn pad. Reconnect the negative battery cable.

With Air Bag

1. Make sure the wheels are pointing straight ahead. Disconnect the negative battery cable and allow 10 minutes to elapse.

2. Remove the lower lid from the steering column and disconnect the air bag module connector.

NOTE: The air bag module is a fragile component. Always place it with the pad side facing upward. Do not allow oil, grease or water to come in cantact with the module. Do not drop the module; if it is damaged in any way, do not reinstall it to the steering wheel.

3. Remove the side access lids, remove the left and right T50H Torx® bolts and discard them. These bolts are specially coated and should not be reused.

4. Carefully remove the air bag module and place in a safe location with the pad side facing upward.

5. Disengage the spiral cable and disconnect the horn connector. Remove the steering wheel hold-down nut.

6. Using an appropriate puller, remove the steering wheel.

7. Attach the spiral cable to the stopper.

8. Remove the steering column covers.

9. Disconnect the connector, remove the 4 mounting screws and remove the spiral cable.

10. Disconnect all combination switch connectors from underneath the steering column.

11. Remove the screws that fasten the combination switch to the steering column and remove from the vehicle.

To install:

12. Feed the wires down the column and install the combination switch to the steering column.

13. Connect the spiral cable connectors and install to the column. Disengage the stopper by pulling the 2 pin guides on the spiral cable unit.

14. Pull the spiral cable through the steering wheel opening and install the steering wheel, setting the pin guides.

15. Connect the horn connector and engage the spiral cable with the pawls in the steering wheel.

16. Install the hold-down nut and torque to 25 ft. lbs. (34 Nm).

17. Carefully position the air bag module. Install new Torx® bolts and torque to 15 ft. lbs. (20 Nm). Connect the air bag module connector.

18. Install the 3 access lids and the column covers.

19. Connect the negative battery cable.

20. Using the Nissan Consult System Checking tool, conduct self-diagnosis to ensure the system is operating properly.

21. If the Consult tool is not available, perform the following:

a. From the passanger seat, turn the ignition switch to the **ON** position.

b. Observe the **AIR BAG** warning light on the instrument cluster.

c. The warning light should illuminate for about 7 seconds, then go out.

d. If the warning light illuminates in any sequence except the above, perform the proper diagnostics before continuing.

Steering Column

REMOVAL & INSTALLATION

1. Disconnect the negative battery cable.

2. Remove the steering wheel.

3. Remove the upper and lower steering column covers.

4. Disconnect the steering column multi-function switch connectors.

5. Remove the bolts attaching the lower steering column bracket to the firewall.

6. Remove the set bolt from the steering column lower joint.

7. Remove the bolts attaching the steering column to the instrument panel.

8. Remove the steering column from the vehicle.

To install:

9. Position the steering column in the vehicle.

NOTE: Ensure that the lower steering column engages in the lower joint before the steering column is permanently attached.

10. Install and finger-tighten the bolts attaching the steering column to the instrument panel.

11. Tighten steering column-to-lower joint set bolt.

NOTE: When attaching the coupling, ensure that set bolt faces the cutout portion in the splines of the lower steering column shaft.

12. Install the bolts attaching the lower steering column bracket to the firewall. Tighten to 5.8–7.2 ft. lbs. (24–29 Nm).

13. Tighten the steering column-to-instrument panel bolts to 9–13 ft. lbs. (13–18 Nm).

14. Install the steering wheel.

15. Install the negative battery cable.

Power Steering Rack

ADJUSTMENT

1. With rack assembly removed from the vehicle and installed in a vise, set the rack to the neutral position without fluid in the gear.

2. Coat the adjusting screw with locking sealant and install the screw.

3. Lightly tighten the locknut.

4. Tighten the adjusting screw to 43–52 ft. lbs. (4.9–5.9 Nm).

5. Loosen the adjusting screw and retighten to 1.7 inch lbs. (0.2 Nm).

6. Move rack over its entire stroke several times.

7. Measure pinion rotating torque with the range of 180 degrees from the neutral position. Stop the gear at the point of maximum torque.

8. Loosen the adjusting screw, the retighten to 43 inch lbs. (4.9 Nm).

9. Loosen the adjusting screw 70–110 degrees.

10. Prevent the adjusting screw from turning and tighten the locknut to specified torque.

11. Measure the pinion rotating torque. Within ±100 from the neutral position: Average rotating torque should be 6.9–11.3 inch lbs. (0.8–1.3 Nm). Maximum torque deviation is 3.5 inch lbs. (0.4 Nm).

12. Check rack sliding force as follows:

a. Install the steering gear in the vehicle., but do not connect the tie rod-to-knuckle arm.

b. Connect all piping and fill with steering fluid.

c. Start the engine and bleed the air completely.

d. Disconnect the steering column lower joint from the gear.

e. Keep the engine at idle and make sure steering fluid has reached normal operating temperature.

f. While pulling tie rod slowly in the ±0.453 in. (±11.5mm) range from the neutral position, make sure the rack sliding force is within specification. On G20 and M30, the average rack sliding force should be 53–64 lbs. (235–284 N). On Q45, the average rack sliding force should be 37–51 lbs. (167–226 N).

g. Check sliding force outside the above range. Maximum allowable sliding force is not more than 9 lbs. (39 N) above the normal value.

REMOVAL & INSTALLATION

1. Disconnect the negative battery cable.
2. Raise the vehicle and support safely.
3. Remove the front wheels.
4. Disconnect the outer tie rods from the steering knuckle.
5. Remove the set screw from the lower steering column universal joint. Disconnect the shaft from the joint.
6. Remove the power steering fluid lines from the rack assembly.
7. Remove the bolts attaching the power steering rack assembly to the body.
8. Remove the rack assembly from the vehicle.
9. Complete the installation of the power steering rack assembly by reversing the removal procedure. Pay close attention to the following:
   a. Tighten the rack mounting bolts to 62–80 ft. lbs. (84–108 Nm).
   b. Initially, tighten the tie rod-to-steering knuckle nuts bolts to 22–29 ft. lbs. (29–39 Nm). Tighten the nut further to expose first pin hole and install a new cotter pin.
   c. Tighten the pinion shaft-to-universal joint set screw to 17–22 ft. lbs. (24–29 Nm).
   d. On G20 and M30 models, tighten low pressure power steering lines to 20–29 ft. lbs. (27–39 Nm). Tighten high pressure lines to 11–18 ft. lbs. (15–25 Nm).
   e. On Q45 models, tighten low pressure power steering lines to 27–30 ft. lbs. (36–40 Nm). Tighten high pressure lines to 22–26 ft. lbs. (30–35 Nm).

Power Steering Pump

REMOVAL & INSTALLATION

1. Disconnect the negative battery cable.
2. Remove the power steering belt pump drive belt.
3. Disconnect the power steering fluid lines.
4. Remove the power steering pump mounting bolts and remove the pump.
5. Complete the installation of the power steering pump by reversing the removal procedure.

BELT ADJUSTMENT

1. Loosen the power steering pump tension locknut.
2. On the G20 and M30, using a suitable belt tension gauge, adjust the belt tension to 0.55–0.63 in. (14–16mm) with a force of 22 lbs. (98 N) applied at the midpoint of the belt run between the crankshaft and power steering pump pulleys.
3. On the Q45, using a suitable belt tension gauge, adjust the belt tension to 0.35–0.39 in. (9–10mm) with a force of 22 lbs. (98 N) applied at the midpoint of the belt run between the crankshaft and power steering pump pulleys for vehicles without Super HICAS. Vehicles with Super HICAS, adjust the tension to 0.28–0.31 in. (7–8mm) with a force of 22 lbs. (98 N) applied at the midpoint of the belt run between the crankshaft and power steering pump pulleys.
4. Tighten the power steering pump tension locknut.

SYSTEM BLEEDING

1. Raise and support the vehicle safely.
2. Ensure that the reservoir is full.
3. Quickly turn the wheels from side to side lightly touching the steering stops.
4. Repeat steps 2 and 3 until the fluid level no longer decreases in the reservoir.
5. Start the engine.
6. Air in the system may cause one or all of the following:
   a. Air bubbles to appear in the reservoir.
   b. Generation of a clicking noise in the oil pump.
   c. Excessive buzzing in the oil pump.

Tie Rod Ends

REMOVAL & INSTALLATION

1. Raise the vehicle and support safely.
2. Remove the front wheel.
3. Matchmark the position of tie rod end locknut on the threaded section of the tie rod.
4. Loosen the tie rod end locknut.
5. Remove the cotter pin and tie rod end nut.
6. Separate the tie rod end from the steering knuckle using a suitable tool.
7. Remove the tie rod end from the tie rod.

To install:

8. Install the new tie rod end on the tie rod.
9. Install the tie rod end on the steering knuckle. Initially, tighten the tie rod-to-steering knuckle nuts bolts to 22–29 ft. lbs. (29–39 Nm). Tighten the nut further to expose first pin hole and install a new cotter pin.
10. Adjust the toe-in to the matchmark made on the threaded section of the tie rod. Tighten the locknut.
11. Install the front wheel.
12. Lower the vehicle.
13. Check alignment to verify proper toe-in setting.

BRAKES

For all brake system repair and service procedures not detailed below, please refer to "Brakes" in the Unit Repair section.

Master Cylinder

REMOVAL & INSTALLATION

NOTE: Prevent brake fluid from coming in contact with painted surfaces. Clean up any spills immediately.

1. Loosen the brake line flare nuts and remove brake lines from master cylinder fittings.
2. Remove the master cylinder mounting nuts.
3. Remove the master cylinder.

To install:

4. Bench bleed the master cylinder.
5. Install the master cylinder in the vehicle. Tighten bolts to 6–8 ft. lbs. (8–11 Nm).
6. Connect the brake lines to the master cylinder and finger-tighten the flare nuts.
7. Bleed the air from the brake lines. Tighten the flare nuts.

Proportioning Valve

The proportioning valve is integral to the master cylinder and cannot be serviced or removed separately.

Power Brake Booster

REMOVAL & INSTALLATION

1. Remove the master cylinder.
2. Remove the clevis pin connecting the brake pedal to the booster input rod.
3. Remove the brake pedal bracket to booster mounting nuts.
4. Remove the brake booster.
5. Complete the installation of the brake booster by reversing the removal procedure. Tighten the brake booster nuts to 9–12 ft. lbs. (13–16 Nm).

Brake Caliper

REMOVAL & INSTALLATION

NOTE: Prevent brake fluid from coming in contact with

painted surfaces. Clean up any spills immediately.

1. Raise the vehicle and support safely.
2. Remove the wheel.
3. Loosen the brake hose connecting bolt.
4. Remove the bolts connecting the caliper to the torque member.
5. Slide the caliper out from the rotor and remove the pad, shim and shim cover.
6. Remove the brake hose connecting bolt from the caliper.
7. Remove the caliper from the vehicle.
8. Complete the installation of the brake caliper by reversing the removal procedure. Tighten the caliper bolts to 16–23 ft. lbs. (22–31 Nm). Bleed the air from the system.

Disc Brake Pads

REMOVAL & INSTALLATION

1. Remove the cap from the master cylinder reservoir and extract a small amount of brake fluid from the reservoir.
2. Raise the vehicle and support safely.
3. Remove the wheel.
4. On Q45 models, if servicing the right front brake, disconnect the sensor harness by pushing the connector pin and pulling the connector. Remove the bracket from the cylinder body. If servicing the right rear brake, remove the sensor harness by pushing it toward the pad, turning it counterclockwise and removing it.
5. Remove the lower pin bolt.
6. Pivot the caliper body upward and remove pad retainers, inner and outer shims and pads.

To install:

7. Place the old pad in place over the caliper cylinders. Use a C-clamp to compress the cylinder pistons to allow for the added thickness of the new pads.
8. Install the new pads and install caliper on rotor. Install pin bolts.
9. Install connector harness, if removed. Pump brakes to seat pads and then refill master cylinder.

Brake Rotor

REMOVAL & INSTALLATION

1. Raise the vehicle and support safely.
2. Remove the wheel.
3. Remove the caliper from the torque member and support using a length of mechanics wire.
4. Remove the bolts attaching the torque member and remove.
5. Remove the rotor from the hub assembly.
6. Complete the installation of the brake rotor by reversing the removal procedure.

Parking Brake Cable

ADJUSTMENT

Shoe Clearance (except G20)

1. Remove adjuster hole plug and turn the adjuster wheel down until the brake is locked.

NOTE: Ensure that the parking brake control lever is completely released.

2. Return the adjuster wheel 7–8 notches on the M30 and 5–6 latches on the Q45.
3. Install the adjuster hole plug and ensure that there is not drag between the shoes and the brake drum when rotating the wheel.

Parking Brake Cable

1. On the M30 and Q45, adjust shoe clearance before adjusting the parking brake cable.
2. Loosen the adjuster locknut.
3. Rotate the adjuster to adjust cable.
4. Tighten locknut.
5. On G20 models, pull the lever control handle with 44 lbs. (196 N) of force. The parking brake should be set in 7–8 notches.
6. On M30 models, pull the lever control handle with 44 lbs. (196 N) of force. The parking brake should be set in 8–9 notches.
7. On Q45 models, depress the parking brake pedal with 44 lbs. (194 N) of force. The parking brake should be set within the pedal stroke of 3.54–4.13 in. (90–105mm).

REMOVAL & INSTALLATION

G20

1. Remove the center console and disconnect the warning lamp connector.
2. Remove the retaining bolts and slacken off the adjusting nut.
3. Remove the cable mounting bracket and lock spring at the rear caliper.
4. Remove the brake cable.
5. Installation is the reverse of removal.
6. Tighten cable holddown bracket bolts to 2–3 ft. lbs. (3–4 Nm). Adjust the brake cable.

M30

FRONT CABLE

1. It is necessary to cut the carpet directly behind the parking brake handle in order to access the front cable.
2. Remove the bolts attaching the parking brake handle and disconnect the parking brake front cable.
3. Remove the bolts attaching the bracket at the point where the cable passes through the floor.
4. Raise the vehicle and support safely.
5. Pull the cable through the hole in the floor and disconnect from the rear cable.
6. Connect the new cable and feed into the hole.
7. Connect the cable to the parking brake lever and install the floor bracket and parking brake handle bolts.
8. Adjust cable.

REAR CABLE

1. Raise the vehicle and support safely.
2. Remove rear wheels.
3. Remove the bolts attaching the rear disc brake caliper mounting bracket. Remove bracket with caliper attached. Support using a length of mechanics wire.
4. Remove 2 bolts attaching rear brake rotor and remove rotor.
5. Disconnect parking brake cable end from toggle lever. Remove cable from backing plate mounting.
6. Remove cable brackets.
7. Disconnect rear cable from equalizer.
8. Complete the installation of the rear parking brake cable by reversing the removal procedure. Pay close attention to the following:
    a. Install the rear parking brake cable into the backing plate by tapping the flanged section of the cable cover with a hammer and punch.
    b. Check the shoe clearance adjustment before adjusting the cable.
    c. Adjust parking brake cable.

Q45

FRONT CABLE

NOTE: It is possible to remove the front parking brake cable without removing the pedal assembly.

1. Raise the vehicle and support safely.
2. Disconnect the front cable from the equalizer.
3. Lower the vehicle.
4. Remove the clip retaining the cable end to the pedal assembly.
5. Remove the center console.
6. Remove the cable brackets and

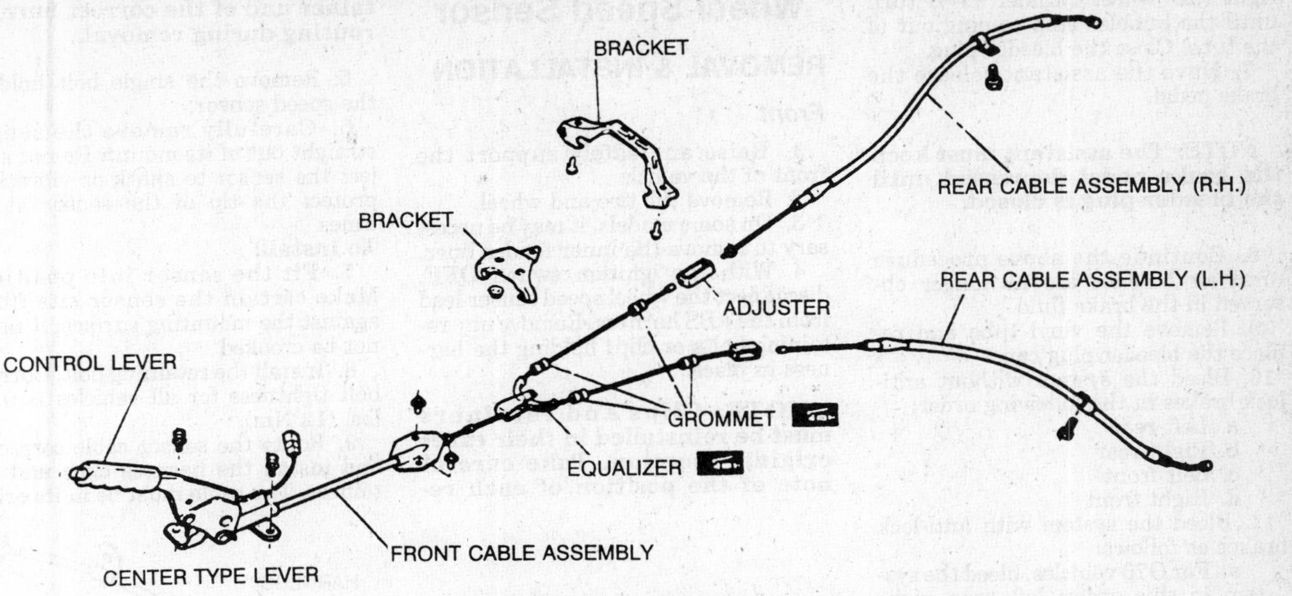

M30 parking brake cable assembly

remove the cable from the floor grommet.

7. Complete the installation of the front parking brake cable by reversing the removal procedure.

REAR CABLE

1. Raise the vehicle and support safely.
2. Remove rear wheels.
3. Remove the bolts attaching the rear disc brake caliper mounting bracket. Remove bracket with caliper attached. Support using a length of mechanics wire.
4. Remove 2 bolts attaching rear brake rotor and remove rotor.
5. Disconnect parking brake cable end from toggle lever. Remove cable from backing plate mounting.
6. Remove cable brackets.
7. Disconnect rear cable from equalizer.
8. Complete the installation of the rear parking brake cable by reversing the removal procedure. Pay close attention to the following:
   a. Install the rear parking brake cable into the backing plate by tapping the flanged section of the cable cover with a hammer and punch.
   b. Check the shoe clearance adjustment before adjusting the cable.
   c. Adjust parking brake cable.

Brake System Bleeding

1. Fill the brake master cylinder reservoir with brake fluid.
2. Raise and safely support the vehicle.
3. Connect a length of vinyl hose to the bleeder plug.
4. Submerge one end of the vinyl hose in a container filled with brake fluid. Connect the other end of the vinyl hose to the wheel cylinder bleeder plug.
5. Have an assistant slowly depress the brake pedal and hold it.
6. Open the bleeder plug of the

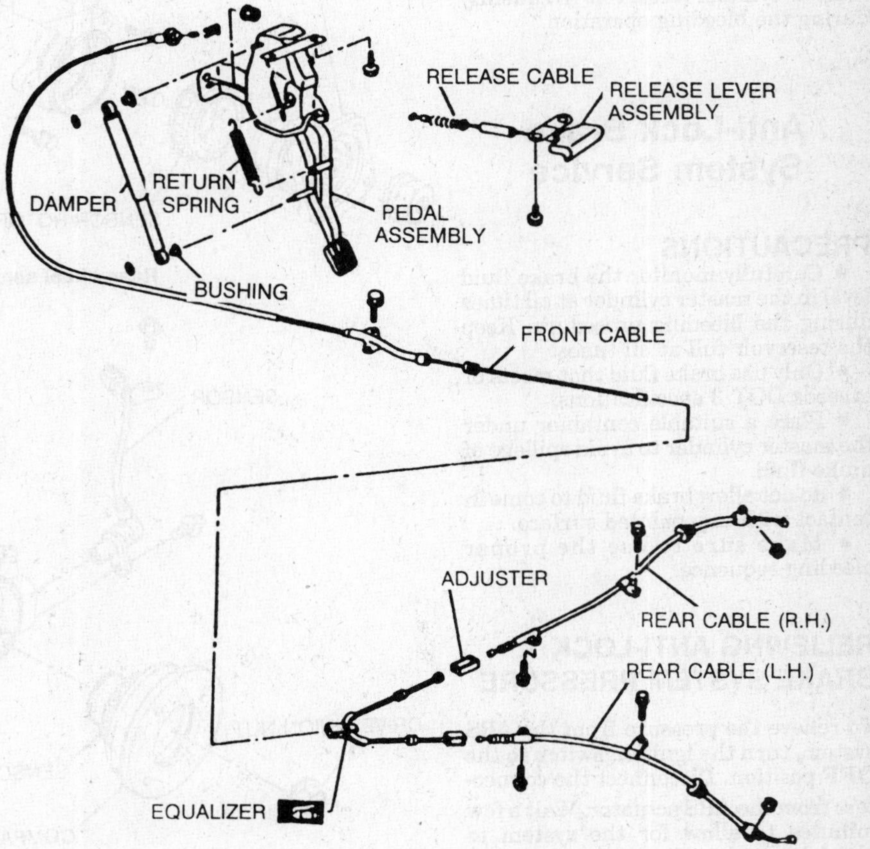

Q45 parking brake cable assembly

right rear wheel cylinder ⅓–½ turn until the bubbles stop coming out of the tube. Close the bleeder plug.
7. Have the assistant release the brake pedal.

NOTE: The assistant must keep the brake pedal depressed until the bleeder plug is closed.

8. Continue the above procedure until air bubbles are no longer observed in the brake fluid.
9. Remove the vinyl tube and replace the bleeder plug cap.
10. Bleed the system without anti-lock brakes in the following order:
   a. Left rear
   b. Right rear
   c. Left front
   d. Right front
11. Bleed the system with anti-lock brakes as follows:
   a. For G20 vehicles, bleed the system in this order: left rear, right front, right rear and left front.
   b. M30 and Q45 vehicles should be bled in this order: left rear, right rear, left front, right front, front side bleeder on ABS actuator and rear side bleeder on the ABS actuator.
12. Check the brake fluid level in the master cylinder reservoir frequently during the bleeding operation.

Anti-Lock Brake System Service

PRECAUTIONS

- Carefully monitor the brake fluid level in the master cylinder at all times during the bleeding procedure. Keep the reservoir full at all times.
- Only use brake fluid that meets or exceeds DOT 3 specifications.
- Place a suitable container under the master cylinder to avoid spillage of brake fluid.
- do not allow brake fluid to come in contact with any painted surface.
- Make sure to use the proper bleeding sequence.

RELIEVING ANTI-LOCK BRAKE SYSTEM PRESSURE

To relieve the pressure from the ABS system, turn the ignition switch to the **OFF** position. Disconnect the connectors from the ABS actuator. Wait a few minutes to allow for the system to bleed down, then disconnect the negative battery cable.

Wheel Speed Sensor

REMOVAL & INSTALLATION

Front

1. Raise and safely support the front of the vehicle.
2. Remove the tire and wheel.
3. On some models, it may be necessary to remove the inner fender liner.
4. With the ignition switch **OFF**, disconnect the wheel speed sensor lead from the ABS harness. Remove any retaining bolts or clips holding the harness in place.

NOTE: Clips and retainers must be reinstalled in their exact original location. Take careful note of the position of each retainer and of the correct harness routing during removal.

5. Remove the single bolt holding the speed sensor.
6. Carefully remove the sensor straight out of its mount. Do not subject the sensor to shock or vibration; protect the tip of the sensor at all times.

To install:

7. Fit the sensor into position. Make certain the sensor sits flush against the mounting surface; it must not be crooked.
8. Install the retaining bolt. Correct bolt tightness for all vehicles is 9 ft. lbs. (12 Nm).
9. Route the sensor cable correctly and install the harness clips and retainers. The cable must be in its origi-

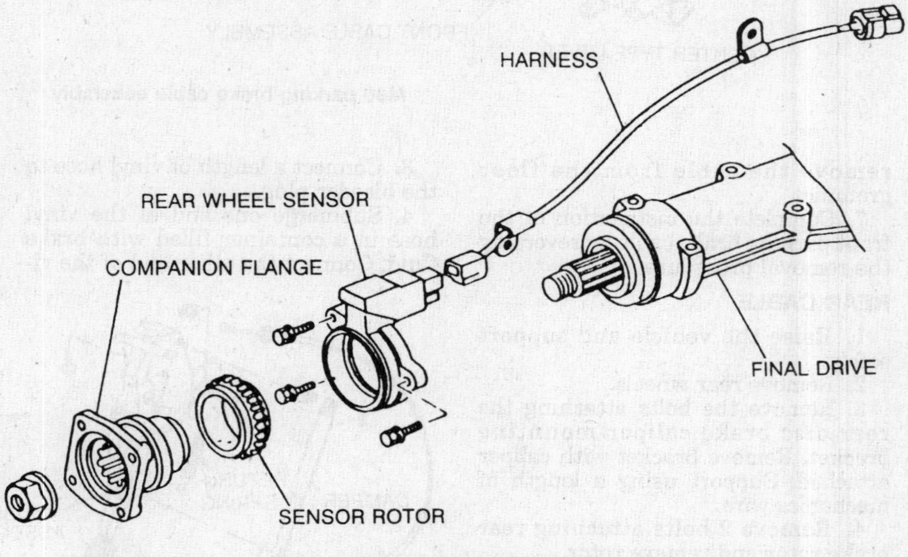

Rear wheel sensor assembly—M30

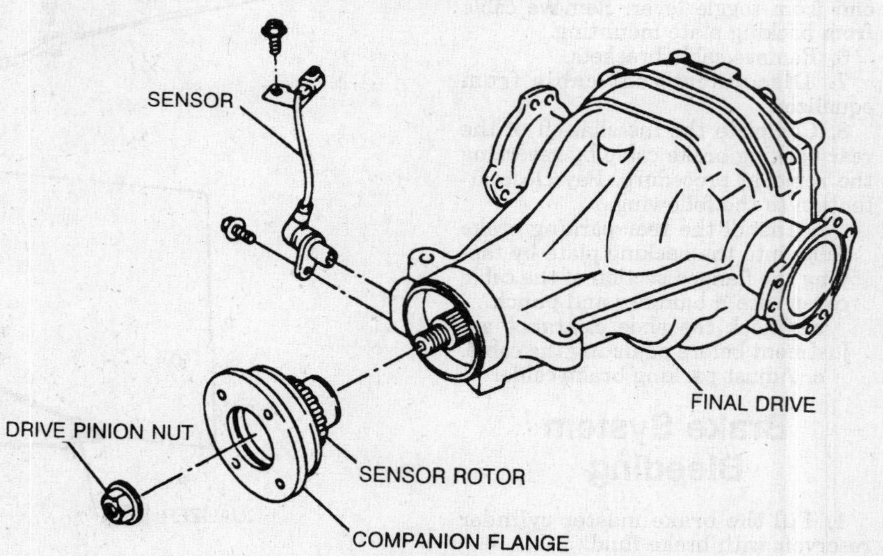

Rear wheel sensor assembly—Q45

nal position and completely clear of moving components.

10. Connect the sensor cable to the ABS harness.
11. Install the inner fender liner if it was removed.
12. Install the wheel and tire.
13. Lower the vehicle to the ground.

Rear

1. Raise and safely support the rear of the vehicle.
2. For G20, remove the tire and wheel.
3. Disconnect the wheel speed sensor lead from the ABS harness. Remove any retaining bolts or clips holding the harness in place.

NOTE: Clips and retainers must be reinstalled in their exact original location. Take careful note of the position of each retainer and of the correct harness routing during removal.

4. Remove the single bolt holding the speed sensor.
5. Carefully remove the sensor straight out of its mount. Do not subject the sensor to shock or vibration; protect the tip of the sensor at all times.

To install:

6. Fit the sensor into position. Make certain the sensor sits flush against the mounting surface; it must not be crooked.
7. Install the retaining bolt. Tighten the bolt to 9 ft. lbs. (12 Nm).
8. Route the sensor cable correctly and install the harness clips and retainers. The cable must be in its original position and completely clear of moving components.
9. Connect the sensor cable to the ABS harness.
10. On the G20, install the wheel and tire.
11. Lower the vehicle to the ground.

Hydraulic Actuator

REMOVAL & INSTALLATION

1. Disconnect the negative battery cable.
2. Drain the brake fluid from the system. Use a syringe or similar tool to empty the master cylinder reservoir. Connect a plastic tube to each brake bleeder. Proceed as if bleeding the brakes; pump each line clear of fluid by operating the brake pedal.
3. Disconnect the wiring connectors at the actuator.
4. Apply dots of colored paint to identify each actuator brake line and its correct port. Using tool GG 94310000 or its equivalent, carefully disconnect each brake line from the actuator.
5. Remove the retaining nuts holding the actuator. Make certain the brake lines are out of the way, then remove the actuator from the engine compartment.

To install:

6. Install the actuator into the engine compartment.
7. Connect the brake lines to the actuator temporarily, finger tight only. Make certain each is in the correct location.
8. Tighten the actuator mounting bolts to 12 ft. lbs. (16 Nm.)
9. Tighten the actuator brake line fittings to 11 ft. lbs. (15 Nm).
10. Connect the wiring connectors to the actuator.
11. Refill the system with DOT 3 brake fluid from unopened containers. Since the system was drained, a substantial amount may be required.
12. Bleed the system at all 4 wheels and at the ABS actuator if required. Each line may require repeated bleeding to eliminate all air within.
13. Connect the negative battery cable.

CHASSIS ELECTRICAL

CAUTION

It is possible for the air bag to inflate for 10 minutes after the battery has been disconnected. Therefore, disconnect the negative battery cable and wait 10 minutes before working on the system. Failure to do so may result in deployment of the air bag and possible personal injury.

Air Bag

DISARMING

Before servicing any component, turn the ignition switch **OFF**, disconnect the negative battery cable and wait for at least 10 minutes. This will disarm the air bag.

Heater Blower Motor

REMOVAL & INSTALLATION

1. Disconnect the negative battery cable. Remove the lower right side instrument panel cover.
2. Remove the screws that attach the blower housing to the intake unit.
3. Remove the housing and remove the blower motor from the housing.
4. The installation is the reverse of the removal procedure.
5. Connect the negative battery cable and check the climate control system for proper operation.

Windshield Wiper Motor

REMOVAL & INSTALLATION

1. Disconnect the negative battery cable.
2. Disconnect the leads at the motor.
3. Remove the motor mounting bolts.
4. Pull the motor out and remove the wiper motor linkage attaching nut.
5. Remove the motor from the firewall.
6. The installation is the reverse of the removal procedure.
7. Connect the negative battery cable and check all windshield wiper and washer functions for proper operation.

Instrument Cluster

REMOVAL & INSTALLATION

Except Q45

1. Disconnect the negative battery cable.
2. Remove the steering column covers.
3. Remove the screws that fasten the cluster lid to the instrument panel and remove the lid.
4. Remove the screws that fasten the instrument cluster to the instrument panel, pull the cluster out, disconnect all connectors and remove the cluster.
5. Disassemble the cluster, as required.
6. The installation is the reverse of the removal procedure.
7. Connect the negative battery cable and check all gauges for proper operation.

Q45

1. Disconnect the negative battery cable.
2. Remove the steering column covers.
3. Remove the gear shifter bezel from the console.
4. Remove the ash tray assembly.
5. Remove the screws that fasten the radio and climate control switch bezel to the instrument panel. Pull the bezel down and out, disconnect the rear window defogger switch and remove the bezel.
6. Remove the cruise control main

switch/outside mirror control switch assembly.

7. Remove the screws that fasten the cluster lid to the instrument panel and remove the lid.

8. Remove the screws that attach the instrument cluster to the instrument panel, pull the cluster out, disconnect all connectors and remove the cluster.

9. Disassemble the cluster, as required.

To install:

10. Assemble the cluster, connect all connectors and install to the instrument panel.

11. Connect the negative battery cable and check all gauges for proper operation. If everything is operating properly, disconnect the negative battery cable and proceed.

12. Install the cluster lid and cruise control main switch/outside mirror control switch assembly.

13. Install the radio and climate control switch bezel to the instrument panel.

14. Install the ash tray and gear shifter bezel.

15. Install the steering column covers.

16. Connect the negative battery cable and check all gauges for proper operation.

Radio

REMOVAL & INSTALLATION

1. Disconnect the negative battery cable.

2. Remove the trim panel surrounding the radio and heater control panel.

3. Remove the radio attaching screws.

4. Disconnect the radio wiring harness and remove the radio.

5. Installation is the reverse of removal.

6. Before installing the radio, ensure that the fuse located at the rear of the radio is good.

Combination Switch

REMOVAL & INSTALLATION

NOTE: On vehicles equipped with an airbag, it is imperative that the exact steering wheel Removal & Installation procedure is followed. The air bag module is a fragile component. Always place it with the pad side facing upward. Do not allow oil, grease or water to come in contact with the module. Do not drop the module; if it is damaged in any way, do not reinstall it to the steering wheel.

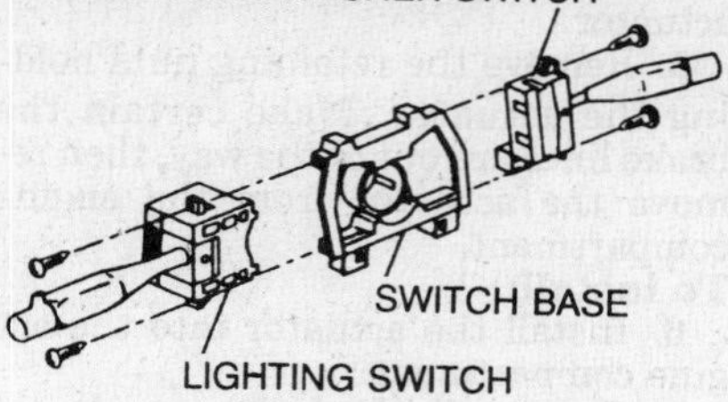

Combination switch assembly—G20 and M30

Except Q45

On the G20 and M30, the combination switch assembly is made up of 3 units— the windsheild wiper/washer switch is located to the right side of the steering column, the combination switch base is attached to the steering column, and the headlight, dimmer and turn signal switch is located to the right side of the steering column. The 2 switches can be removed without removing the switch base.

1. Disconnect the negative battery cable.

2. Remove the steering column covers. Disconnect the connector.

3. Remove the combination switch mounting screws and remove the switch from the switch base.

4. The installation is the reverse of the removal procedure.

5. Connect the negative battery cable and check all functions of the combination switch for proper operation.

Q45

On the Q45, the combination switch assembly cannot be disassembled and refers to all column-mounted stalk switches as a single unit.

Before replacing the switch, unplug the existing switch and plug in the replacement one to make sure all functions operate properly.

1. Make sure the wheels are pointing straight ahead. Disconnect the negative battery cable and allow 10 minutes to elapse.

2. Remove the steering wheel.

3. Disconnect all combination switch connectors from underneath the steering column.

4. Remove the screws that fasten the combination switch to the steering column and remove from the vehicle.

5. The installation is the reverse of the removal procedure.

6. Connect the negative battery cable and check all functions of the combination switch for proper operation.

Combination Switch Base

REMOVAL & INSTALLATION

NOTE: On vehicles equipped with airbags, it is imperative that the exact steering wheel Removal & Installation procedure under Steering is followed. The air bag module is a fragile component. Always place it with the pad side facing upward. Do not allow oil, grease or water to come in contact with the module. Do not drop the module; if it is damaged in any way, do not reinstall it to the steering wheel.

G20

1. Make sure the wheels are point-

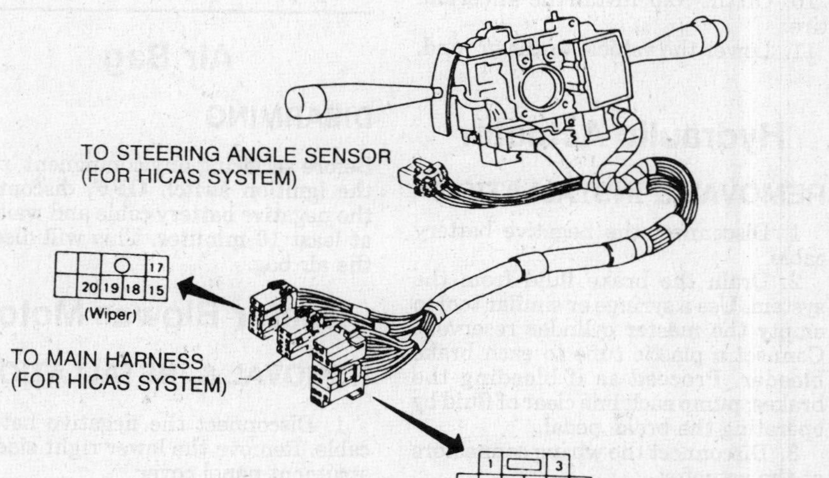

Combination switch assembly—Q45

ing straight ahead. Disconnect the negative battery cable and allow 10 minutes to elapse.

2. Remove the steering wheel.
3. Disconnect all combination and windshield wiper switch connectors and remove the switches from the switch base.
4. Insert a suitable tool between the combination switch base and the steering column. Lift the base and pull outward to remove.
5. The installation is the reverse of the removal procedure.
6. Connect the negative battery cable and check all functions of the combination and windshield wiper switches for proper operation.

M30

1. Make sure the wheels are pointing straight ahead. Disconnect the negative battery cable and allow 10 minutes to elapse.
2. Remove the steering wheel.
3. Disconnect all combination and windshield wiper switch connectors and remove the switches from the switch base.
4. To remove the combination switch base, remove the base attaching screw and turn after pushing it.
5. The installation is the reverse of the removal procedure.
6. Connect the negative battery cable and check all functions of the combination and windshield wiper switches for proper operation.

Ignition Lock/Switch

REMOVAL & INSTALLATION

NOTE: On vehicles equipped with airbags, it is imperative that the exact steering wheel Removal & Installation procedure under Steering is followed. The air bag module is a fragile component. Always place it with the pad side facing upward. Do not allow oil, grease or water to come in contact with the module. Do not drop the module; if it is damaged in any way, do not reinstall it to the steering wheel.

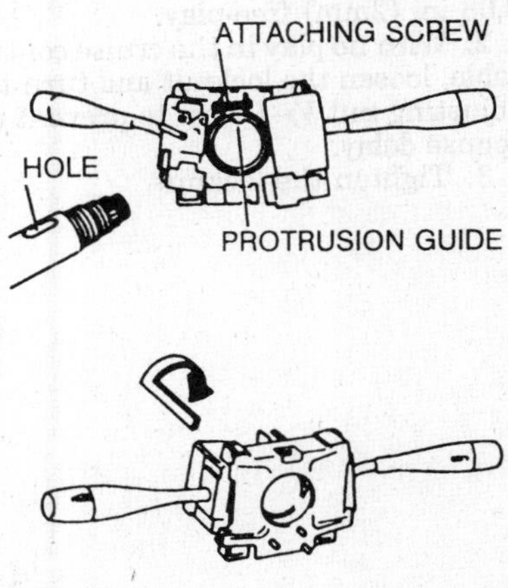

Combination switch base removal—M30

1. Make sure the wheels are pointing straight ahead. Disconnect the negative battery cable and allow 10 minutes to elapse.
2. Remove the steering wheel and combination switch or switch base.
3. Disconnect the ignition switch wiring.
4. Lower the steering column.
5. Using a hacksaw blade, cut a groove into the heads of the special self-shearing screws and remove the screws.
6. Remove the assembly from the column.

To install:

7. With the key inserted in the switch, install the assembly onto the column with new self-shearing screws. Tighten the screws gradually, testing the key for binding often. Tighten the screws until the heads shear off.
8. Raise and secure the steering column.
9. Install the combination switch or switch base, and steering wheel.
10. Connect the negative battery cable and check the ignition switch for proper operation in all positions.

Stoplight Switch

ADJUSTMENT

1. Measure the free height of the brake pedal at its bottom edge. The specification is 6.0 in. (152mm) for the manual transmission and 6.0–7.0 in. (152–177mm) for the automatic transaxle on the G20; 8.0 in. (205mm) for M30 and 7.4 in. (190mm) for Q45. Adjust by loosening the booster input rod locknut and turning the input rod.
2. The stoplight switch is to the right of the cruise control cancel switch on the bracket above the brake pedal. Adjust both switches during this procedure.
3. Measure the clearance between the threaded end of the switches and the pedal stopper. The specification is 0.025 in. (0.06mm) for both.
4. Adjust be loosening the locknut and turning each switch.
5. Check the pedal free-play. The specification is 0.08 in. (2.0mm); 0.04–0.12 in. (1–3mm) for the G20.
6. Make sure the brake lights illuminate when the pedal is depressed and they go out when the pedal is released.
7. Also, make sure the cruise control cancels when the brake pedal is depressed.

REMOVAL & INSTALLATION

1. Disconnect the negative battery cable.
2. Remove the locknut.
3. Remove the switch from the bracket above the brake pedal.
4. The installation is the reverse of the removal procedure.
5. Adjust the switch.
6. Connect the negative battery cable and check the switch for proper operation.

Neutral Safety Switch

ADJUSTMENT

1. Disconnect the negative battery cable.
2. Raise the vehicle and support safely.
3. Disconnect the manual control linkage from the manual shift shaft.
4. Set the manual shift shaft in the **N** detent.
5. Loosen the neutral safety switch mounting bolts.
6. Align the switch with the shift shaft by inserting a suitable pin in the alignment holes.
7. Tighten the switch mounting bolts and connect the control linkage.
8. Make sure the vehicle does not start in any gear except **P** or **N** and does start in both **P** and **N**.

REMOVAL & INSTALLATION

1. Disconnect the negative battery cable.
2. Raise the vehicle and support safely.
3. Disconnect the wires to the switch.
4. Remove the switch mounting screws.
5. Remove the switch from the transmission.
6. The installation is the reverse of the removal procedure.
7. Adjust the switch.
8. Make sure the vehicle does not start in any gear except **P** or **N** and does start in both **P** and **N**.

Clutch Switch

ADJUSTMENT

1. Ensure that the clutch pedal height and free travel adjustments are within specification.
2. With the clutch pedal fully depressed, measure the clearance between the rubber stopper and the threaded end of the clutch switch. Clearance should be 0.004–0.039 in. (0.1–1.0mm).

3. If the clearance is not within specificaiton, adjust the switch by loosening the locknut and adjusting the switch.

REMOVAL & INSTALLATION

1. Disconnect the negative battery cable.
2. Disconnect the wiring harness from the clutch switch.
3. Loosen the locknut and remove the clutch switch.
4. Installation is the reverse of removal.
5. Adjust the switch.

Fuses, Circuit Breakers and Relays

LOCATION

A fuse, fusible link and relay box is located in the engine compartment, near the battery. Release the latch and remove the protective covering to access the desired component. The radio has its own fuse behind it. Other various relays are located throughout the vehicle.

There is a second fuse box located behind an access door to the left of the steering column on the instrument panel. The circuit breaker for the power door locks and seats is located near this fuse box. The circuit breaker for the power windows and the sunroof is located behind the left side kick panel.

Computers

LOCATION

G20

ABS Acutator—is located at the right front of the engine compartment.

ABS Control Unit—is located at the right side kick panel.

ASCD Actuator—is located at the rear center of the engine compartment.

ASCD Control Unit—is located under the dash to the left of the steering column.

Automatic Seat Belt Control Unit—is located under the console.

Combination Flasher Unit—is located under the dash to the right of the steering column.

Diagnostic Connector—is located at the fuse block.

ECCS Control Unit—is located under the front of the console.

Electronic Control Unit (ECU)—see Electronic Concentrated Control System (ECCS) Unit.

Shift Lock Control Unit (automatic transmission)—is located under the left side of the dash.

Theft Warning Control Unit—is located under the dash to the right of the steering column.

M30

Sonar Suspension Control Unit—located to the left of the rear seat, behind the trim panel

Automatic Transmission Control Unit—located on the left side kick panel.

Time Control Unit—located on the left side kick panel.

Shift Lock Control Unit—located under the left side of the instrument panel.

Automatic Speed Control Device (ASCD) Control Unit—located to the left of the rear seat, behind the trim panel.

Theft Warning Control Unit—located on the left side kick panel.

Anti-Lock Brake System (ABS) Control Unit—located in the top front of the trunk.

Electronic Concentrated Control System (ECCS) Unit (for engine control)—located on the right side kick panel.

Air Bag Control Unit—located in the rear of the console.

Q45

Fuel Pump Control Module—located at the top front of the trunk.

High Captivity Actively Controlled Steering (HICAS) Control Unit (for 4WS)—located at the top front of the trunk.

Automatic Transmission Control Unit—located on the left side kick panel.

Time Control Unit—located on the left side kick panel.

Shift Lock Control Unit—located on the left side kick panel.

Automatic Drive Positioner Control Unit—located to the left of the steering column.

Automatic Speed Control Device (ASCD) Control Unit—located to the left of the console.

Power Steering Control Unit—located to the left of the console.

Theft Warning Control Unit—located behind the glove box.

Anti-Lock Brake System (ABS) Control Unit—located behind the glove box.

Electronic Concentrated Control System (ECCS) Unit (for engine control)—located on the right side kick panel.

Right Side Power Seat Control Unit—located under the right front seat.

Left Side Power Seat Control Unit—located under the left front seat.

Air Bag Control Unit—located in the rear of the console.

Flasher

LOCATION

The combination flasher unit is located under the instrument panel to the right of the steering column on the G20 and M30, and near the interior fuse box on the Q45.

Cruise Control

ADJUSTMENT

1. Adjust the throttle cable so it has 0.08 in. (2mm) free-play.
2. With no play in the cruise control cable, loosen the locknut and turn the adjusting nut ½–1 turn to prevent response delay.
3. Tighten the locknut.

Isuzu 10

I-Mark, Impulse, Stylus — All Models

SERIAL NUMBER IDENTIFICATION

Vehicle Identification Plate

The vehicle identification number is embossed on a plate, that is attached to the top left corner of the instrument panel. The number is visible through the windshield from the outside of the vehicle. The 8th digit of the number, indicates the engine model and the 10th digit represents the model year; example is J for 1988, K for 1989, L for 1990, M for 1991 and N for 1992.

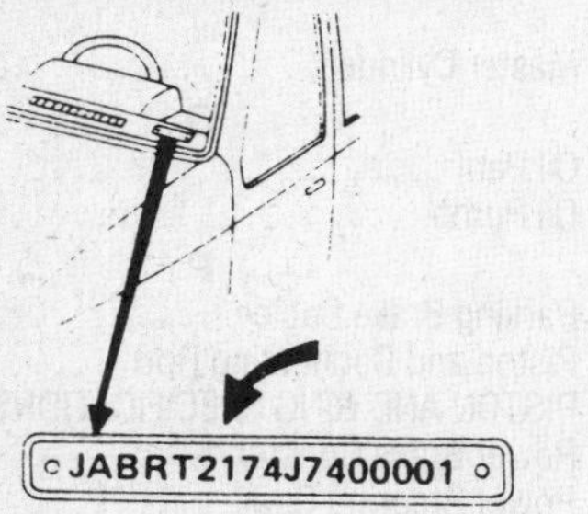

Vehicle identification plate location

Engine Number

The Impulse equipped with the 4ZC1-T and 4ZD1 engine, has the number stamped on the left rear corner of the engine block, near the engine to transaxle mounting. The Impulse and Stylus equipped with the 4XE1-W, 4XE1-WT and 4XE1-V engine and all I-Marks, have the number stamped on the flange near the transaxle mounting, toward the front of the vehicle.

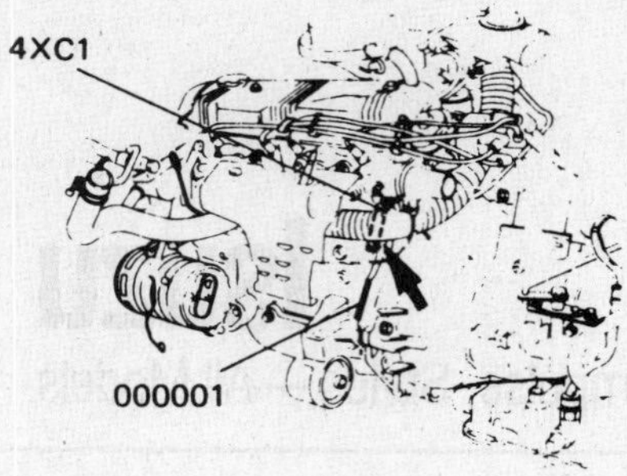

4XC1-T engine serial number location

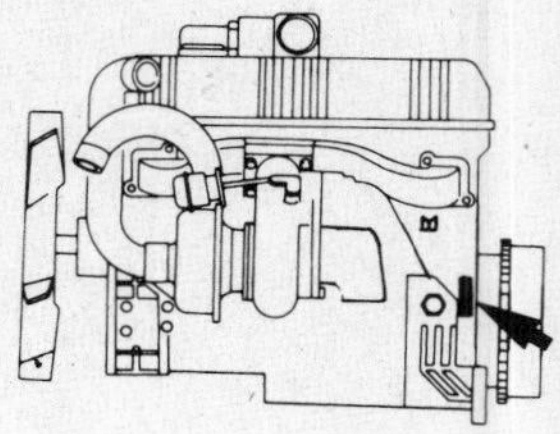

4ZC1-T and 4ZD1 engine serial number location

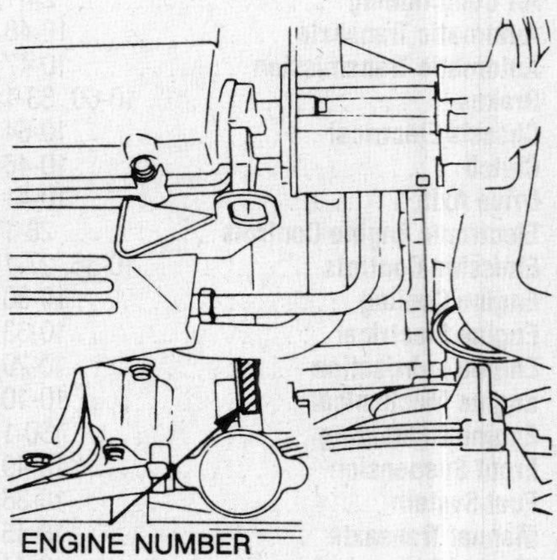

4XE1 engine serial number location

Vehicle Identification Number

The Vehicle Identification Number (VIN) appears on the vehicle identification plate and on the driver's door post pillar.

SPECIFICATIONS

ENGINE IDENTIFICATION

Year	Model	Engine Displacement cu. in. (cc/liter)	Engine Series Identification	No. of Cylinders	Engine Type
1988	I-Mark	90 (1471/1.5)	4XC1-U	4	OHC
	I-Mark (Turbo)	90 (1471/1.5)	4XC1-T	4	Turbo OHC
	Impulse (Turbo)	121.7 (1994/2.0)	4ZC1-T	4	Turbo OHC
	Impulse	138 (2254/2.3)	4ZD1	4	OHC
1989	I-Mark	90 (1471/1.5)	4XC1-U	4	OHC
	I-Mark (Turbo)	90 (1471/1.5)	4XC1-T	4	Turbo OHC
	I-Mark (DOHC)	92 (1588/1.6)	4XE1	4	DOHC
	Impulse (Turbo)	121.7 (1994/2.0)	4ZC1-T	4	Turbo OHC
	Impulse	138 (2254/2.3)	4ZD1	4	OHC
1990	Impulse	92 (1588/1.6)	4XE1-W	4	DOHC
1991-92	Impulse	92 (1588/1.6)	4XE1-W	4	DOHC
	Impulse (Turbo)	92 (1588/1.6)	4XE1-WT	4	Turbo DOHC
	Stylus	92 (1588/1.6)	4XE1-W	4	SOHC
	Stylus	92 (1588/1.6)	4XE1-V	4	DOHC

OHC—Overhead Camshaft DOHC—Double Overhead Camshaft SOHC—Single Overhead Camshaft

GENERAL ENGINE SPECIFICATIONS

Year	Model	Engine Displacement cu. in. (cc)	Fuel System Type	Net Horsepower @ rpm	Net Torque @ rpm (ft. lbs.)	Bore × Stroke (in.)	Com-pression Ratio	Oil Pressure @ rpm
1988	I-Mark	90 (1471)	2 bbl	70 @ 5400	87 @ 3400	3.03 × 3.11	9.6:1	49 @ 5200
	I-Mark (Turbo)	90 (1471)	EFI	110 @ 5400	120 @ 3400	3.03 × 3.11	8.0:1	49 @ 5200
	Impulse (Turbo)	121.7 (1994)	EFI	140 @ 5400	166 @ 3000	3.46 × 3.29	7.9:1	57 @ 1400
	Impulse	138 (2254)	EFI	110 @ 5000	127 @ 3000	3.52 × 3.54	8.6:1	57 @ 1400
1989	I-Mark	90 (1471)	2 bbl	70 @ 5400	87 @ 3400	3.03 × 3.11	9.6:1	49 @ 5200
	I-Mark (Turbo)	90 (1471)	EFI	110 @ 5400	120 @ 3400	3.03 × 3.11	8.0:1	49 @ 5200
	I-Mark (DOHC)	92 (1588)	EFI	125 @ 6800	138 @ 5400	3.15 × 3.11	9.8:1	49 @ 5200
	Impulse (Turbo)	121.7 (1994)	EFI	140 @ 5400	166 @ 3000	3.46 × 3.29	7.9:1	57 @ 1400
	Impulse	138 (2254)	EFI	110 @ 5000	127 @ 3000	3.52 × 3.54	8.6:1	57 @ 1400
1990	Impulse	92 (1588)	EFI	130 @ 6600	102 @ 4600	3.15 × 3.11	9.8:1	51–80 @ 3000
1991-92	Impulse	92 (1588)	EFI	130 @ 6600	102 @ 4600	3.15 × 3.11	9.8:1	51–80 @ 3000
	Impulse (Turbo)	92 (1588)	EFI	160 @ 6600	150 @ 4800	3.15 × 3.11	8.5:1	51–80 @ 3000
	Stylus	92 (1588)	EFI	95 @ 5800	97 @ 3400	3.15 × 3.11	9.1:1	51–80 @ 3000
	Stylus	92 (1588)	EFI	130 @ 6800	102 @ 4600	3.15 × 3.11	9.8:1	51–80 @ 3000

EFI—Electronic Fuel Injection

ENGINE TUNE-UP SPECIFICATIONS

Year	Model	Engine Displacement cu. in. (cc)	Spark Plugs Type	Spark Plugs Gap (in.)	Ignition Timing (deg.) MT	Ignition Timing (deg.) AT	Com-pression Pressure (psi)	Fuel Pump (psi)	Idle Speed (rpm) MT	Idle Speed (rpm) AT	Valve Clearance In.	Valve Clearance Ex.
1988	I-Mark	90 (1471)	BPR6ES11	0.040	3B	3B	177.8	3.8–4.7	750	1000	0.006	0.010
	I-Mark (Turbo)	90 (1471)	BPR6ES11	0.040	15B	NA	171.0	28.4 ①	950	NA	0.006	0.010
	Impulse (Turbo)	121.7 (1994)	BPR6ES11	0.040	12B	12B	178.0	35.6 ①	900 ②	900 ②	0.006	0.010
	Impulse	138 (2254)	BPR6ES11	0.040	12B	12B	178.0	35.6 ①	900 ②	900 ②	0.008	0.008
1989	I-Mark	90 (1471)	BPR6ES11	0.040	3B	3B	177.8	3.8–4.7	750	1000	0.006	0.010
	I-Mark (Turbo)	90 (1471)	BPR6ES11	0.040	15B	NA	171.0	28.4 ①	950	NA	0.006	0.010
	I-Mark (DOHC)	92 (1588)	BPR6ES11	0.040	16B	—	171.0	28.4 ①	950	—	Hyd.	Hyd.
	Impulse (Turbo)	121.7 (1994)	BPR6ES11	0.040	12B	12B	178.0	35.6 ①	900 ②	900 ②	0.006	0.010
	Impulse	138 (2254)	BPR6ES11	0.040	12B	12B	178.0	35.6 ①	900 ②	900 ②	0.008	0.008

ENGINE TUNE-UP SPECIFICATIONS

Year	Model	Engine Displacement cu. in. (cc)	Spark Plugs Type	Spark Plugs Gap (in.)	Ignition Timing (deg.) MT	Ignition Timing (deg.) AT	Compression Pressure (psi)	Fuel Pump (psi)	Idle Speed (rpm) MT	Idle Speed (rpm) AT	Valve Clearance In.	Valve Clearance Ex.
1990	Impulse	92 (1588)	BKR6E11	0.041	10B	10B	185	35–38	850	850	0.006	0.010
1991	Impulse	92 (1588)	BKR6E11	0.041	10B	10B	185	35–38	850	850	0.006	0.010
	Impulse (Turbo)	92 (1588)	BKR6E	0.030	10B	10B	185	35–38	900	900	0.006	0.010
	Stylus ③	92 (1588)	BPR6ES11	0.041	10B	10B	185	35–38	850	940	0.006	0.010
	Stylus ④	92 (1588)	BKR6E11	0.041	10B	10B	185	35–38	850	850	0.006	0.010
1992	SEE UNDERHOOD SPECIFICATIONS STICKER											

① At 900 rpm with vacuum hose of the pressure regulator connected
② ±50 rpm
③ SOHC
④ DOHC

FIRING ORDERS

NOTE: To avoid confusion, always replace spark plug wires one at a time.

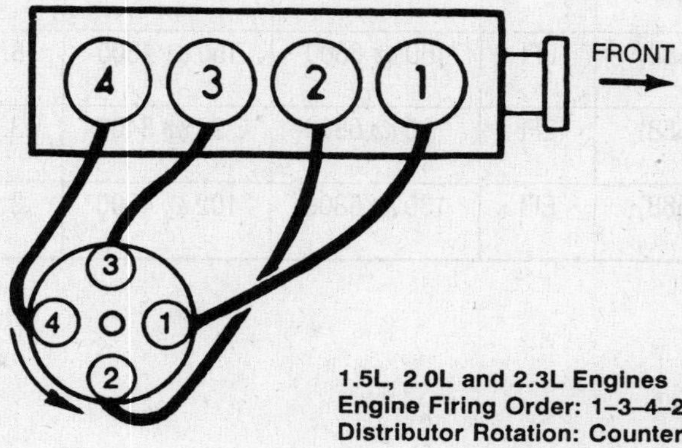

1.5L, 2.0L and 2.3L Engines
Engine Firing Order: 1–3–4–2
Distributor Rotation: Counterclockwise

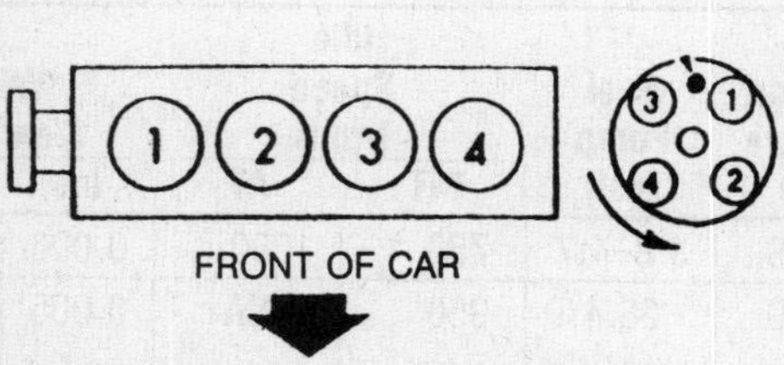

1.6L (4XE1) SOHC and DOHC Engines
Engine Firing Order: 1–3–4–2
Distributor Rotation: Counterclockwise

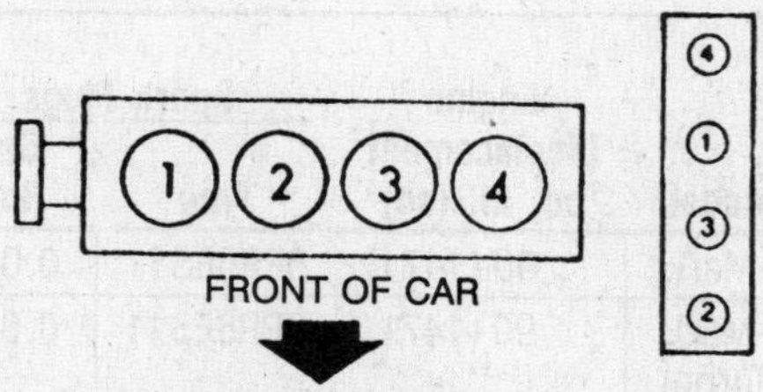

1.6L (4XE1) DOHC Turbo Engine
Engine Firing Order: 1–3–4–2
Distributorless Ignition System

CAPACITIES

Year	Model	Engine Displacement cu. in. (cc)	Engine Crankcase (qts.) with Filter	Engine Crankcase (qts.) without Filter	Transmission (pts.) 4-Spd	Transmission (pts.) 5-Spd	Transmission (pts.) Auto.	Drive Axle (pts.)	Fuel Tank (gal.)	Cooling System (qts.)
1988	I-Mark	90 (1471)	3.4 ③	3.0	—	5.0	13.6	NA	11.1	6.8
	I-Mark (Turbo)	90 (1471)	3.4 ③	3.0	—	5.0	NA	NA	11.1	7.5

CAPACITIES

Year	Model	Engine Displacement cu. in. (cc)	Engine Crankcase (qts.) with Filter	Engine Crankcase (qts.) without Filter	Transmission (pts.) 4-Spd	Transmission (pts.) 5-Spd	Transmission (pts.) Auto.	Drive Axle (pts.)	Fuel Tank (gal.)	Cooling System (qts.)
1988	Impulse (Turbo)	121.7 (1994)	3.8 ①	3.4	—	3.3	13.7	3.2	15.1	9.5
	Impulse	138 (2254)	3.8 ①	3.4	—	3.3	13.7	3.2	15.1	9.3
1989	I-Mark	90 (1471)	3.4 ③	3.0	—	5.0	13.6	NA	11.1	6.8
	I-Mark (Turbo)	90 (1471)	3.4 ③	3.0	—	5.0	NA	NA	11.1	7.5
	I-Mark (DOHC)	92 (1588)	3.2	3.0	—	5.0	—	—	11.1	6.8
	Impulse (Turbo)	121.7 (1994)	3.8 ①	3.4	—	3.3	13.7	3.2	15.1	9.5
	Impulse	138 (2254)	3.8 ①	3.4	—	3.3	13.7	3.2	15.1	9.3
1990	Impulse	92 (1588)	4.6	4.0	—	4.0	14.0	—	12.4	②
1991-92	Impulse	92 (1588)	4.6	4.0	—	4.0	14.0	⑤	12.4	②
	Impulse Turbo	92 (1588)	4.6	4.0	—	4.0	14.0	⑤	12.4	7.9
	Stylus ⑥	92 (1588)	3.2	3.6	—	4.0	14.0	—	12.4	②
	Stylus ⑦	92 (1588)	4.6	4.0	—	4.0	14.0	—	12.4	7.3

NA—Not Available
① The original fill is 5.0 qts.
② Manual transaxle—7.3 pts.
Automatic transaxle—7.8 pts.
③ Original fill 3.7 qts.

CAMSHAFT SPECIFICATIONS

All measurements given in inches.

Year	Engine Displacement cu. in. (cc)	Journal Diameter 1	Journal Diameter 2	Journal Diameter 3	Journal Diameter 4	Journal Diameter 5	Lobe Lift In.	Lobe Lift Ex.	Bearing Clearance	Camshaft End Play
1988	90 (1471)	1.021–1.022	1.021–1.022	1.021–1.022	1.021–1.022	1.021–1.022	1.426	1.426	0.0024–0.0044	0.0039–0.0071
	138 (2254)	1.339	1.339	1.339	1.339	1.339	1.451	1.451	0.0033–0.0051	0.0002–0.0059
	121.7 (1994)	1.339	1.339	1.339	1.339	1.339	1.451	1.451	0.0033–0.0051	0.0002–0.0059
1989	90 (1471)	1.021–1.022	1.021–1.022	1.021–1.022	1.021–1.022	1.021–1.022	1.426	1.426	0.0024–0.0044	0.0039–0.0071
	92 (1588)	1.021–1.022	1.021–1.022	1.021–1.022	1.021–1.022	1.021–1.022	1.536	1.536	0.0024–0.0044	0.0020–0.0060
	138 (2254)	1.339	1.339	1.339	1.339	1.339	1.451	1.451	0.0033–0.0051	0.0002–0.0059
	121.7 (1994)	1.339	1.339	1.339	1.339	1.339	1.451	1.451	0.0033–0.0051	0.0002–0.0059
1990	92 (1588)	1.021–1.022	1.021–1.022	1.021–1.022	1.021–1.022	1.021–1.022	1.503	1.503	0.0011–0.0031	0.0020–0.0060

CAMSHAFT SPECIFICATIONS

All measurements given in inches.

Year	Engine Displacement cu. in. (cc)	Journal Diameter 1	2	3	4	5	Lobe Lift In.	Ex.	Bearing Clearance	Camshaft End Play
1991-92	92 (1588) SOHC	1.021–1.022	1.021–1.022	1.021–1.022	1.021–1.022	1.021–1.022	1.426	1.426	0.0024–0.0044	0.0020–0.0060
	92 (1588) DOHC	1.021–1.022	1.021–1.022	1.021–1.022	1.021–1.022	1.021–1.022	1.503	1.503	0.0011–0.0031	0.0020–0.0060
	92 (1588) Turbo	1.021–1.022	1.021–1.022	1.021–1.022	1.021–1.022	1.021–1.022	1.525	1.525	0.0011–0.0031	0.0020–0.0060

SOHC—Single Overhead Camshaft
DOHC—Double Overhead Camshaft

CRANKSHAFT AND CONNECTING ROD SPECIFICATIONS

All measurements are given in inches.

Year	Engine Displacement cu. in. (cc)	Crankshaft Main Brg. Journal Dia.	Main Brg. Oil Clearance	Shaft End-play	Thrust on No.	Connecting Rod Journal Diameter	Oil Clearance	Side Clearance
1988	90 (1471)	1.8865–1.8873	0.0008–0.0020	0.0024–0.0095	2	1.5720–1.5726	0.0010–0.0023	0.0079–0.0138
	138 (2254)	2.2032–2.2038	0.0009–0.0020	0.0024–0.0099	3	1.9276–1.9282	0.0012–0.0024	0.0078–0.0130
	121.7 (1994)	2.2032–2.2038	0.0009–0.0020	0.0024–0.0099	3	1.9276–1.9282	0.0012–0.0024	0.0078–0.0130
1989	90 (1471)	1.8865–1.8873	0.0008–0.0020	0.0024–0.0095	2	1.5720–1.5726	0.0010–0.0023	0.0079–0.0138
	92 (1588)	1.8861–1.9171	0.0008–0.0020	0.0024–0.0095	2	1.5718–1.5728	0.0013–0.0024	0.0079–0.0138
	138 (2254)	2.2032–2.2038	0.0009–0.0020	0.0024–0.0099	3	1.9276–1.9282	0.0012–0.0024	0.0078–0.0130
	121.7 (1994)	2.0232–2.2038	0.0009–0.0020	0.0024–0.0099	3	1.9276–1.9282	0.0012–0.0024	0.0078–0.0130
1990	92 (1588)	2.0439–2.0448	0.0008–0.0020	0.0024–0.0095	2	1.5724–1.5728	0.0010–0.0023	0.0079–0.0138
1991-92	92 (1588) SOHC	2.0243	0.0008–0.0020	0.0024–0.0095	2	1.5526	0.0010–0.0023	0.0079–0.0138
	92 (1588) DOHC & Turbo	2.0439–2.0448	0.0008–0.0020	0.0024–0.0095	2	1.5724–1.5728	0.0010–0.0023	0.0079–0.0138

SOHC—Single Overhead Camshaft
DOHC—Double Overhead Camshaft

VALVE SPECIFICATIONS

Year	Engine Displacement cu. in. (cc)	Seat Angle (deg.)	Face Angle (deg.)	Spring Test Pressure (lbs.)	Spring Installed Height (in.)	Stem-to-Guide Clearance (in.) Intake	Exhaust	Stem Diameter (in.) Intake	Exhaust
1988	90 (1471)	45	45	47 @ 1.57	1.57	0.0009–0.0022	0.0012–0.0025	0.274–0.275	0.2740–0.2744
	138 (2254)	45	45	55.3 @ 1.62	1.62	0.0009–0.0022	0.0015–0.0031	0.315	0.3150
	121.7 (1994)	45	45	55.3 @ 1.62	1.62	0.0009–0.0022	0.0015–0.0031	0.315	0.3150

VALVE SPECIFICATIONS

Year	Engine Displacement cu. in. (cc)	Seat Angle (deg.)	Face Angle (deg.)	Spring Test Pressure (lbs.)	Spring Installed Height (in.)	Stem-to-Guide Clearance (in.)		Stem Diameter (in.)	
						Intake	Exhaust	Intake	Exhaust
1989	90 (1471)	45	45	47 @ 1.57	1.57	0.0009–0.0022	0.0012–0.0025	0.274–0.275	0.2740–0.2744
	92 (1588)	45	45	52 @ 1.52	1.52	0.0009–0.0022	0.0018–0.0025	0.234–0.235	0.2340–0.2350
	138 (2254)	45	45	55.3 @ 1.62	1.62	0.0009–0.0022	0.0015–0.0031	0.315	0.3150
	121.7 (1994)	45	45	55.3 @ 1.62	1.62	0.0009–0.0022	0.0015–0.0031	0.315	0.3150
1990	92 (1588)	45	45	44.1 @ 1.504	1.504	0.0009–0.0022	0.0012–0.0025	0.2348–0.2356	0.2346–0.2352
1991–92	92 (1588) SOHC	45	45	①	1.504	0.0009–0.0022	0.0012–0.0025	0.2348–0.2356	0.2346–0.2352
	92 (1588) DOHC	45	45	44.1 @ 1.504	1.504	0.0009–0.0022	0.0012–0.0025	0.2348–0.2356	0.2346–0.2352
	92 (1588) Turbo	45	45	44.1 @ 1.504	1.504	0.0009–0.0022	0.0012–0.0025	0.2348–0.2356	0.2346–0.2352

SOHC—Single Overhead Camshaft
DOHC—Double Overhead Camshaft
① 44 @ 1.504—Intake
55 @ 1.504—Exhaust

PISTON AND RING SPECIFICATIONS

All measurements are given in inches.

Year	Engine Displacement cu. in. (cc)	Piston Clearance	Ring Gap			Ring Side Clearance		
			Top Compression	Bottom Compression	Oil Control	Top Compression	Bottom Compression	Oil Control
1988	90 (1471)	0.0011–0.0019	0.0098–0.0138	NA	0.0039–0.0236	0.0010–0.0025	NA	NA
	90 (1471) Turbo	0.0011–0.0019	0.0106–0.0153	0.0098–0.0145	0.0039–0.0236	0.0010–0.0026	0.0008 0.0024	NA
	138 (2254)	0.0016–0.0024	0.0120–0.0180	0.0100–0.0160	0.0080–0.0280	0.0010–0.0024	0.0010–0.0024	NA
	121.7 (1994)	0.0008 0.0016	0.0120–0.0180	0.0100–0.0160	0.0080–0.0280	0.0010–0.0024	0.0010–0.0024	NA
1989	90 (1471)	0.0011–0.0019	0.0098–0.0138	NA	0.0039–0.0236	0.0010–0.0025	NA	NA
	90 (1471) Turbo	0.0011–0.0019	0.0106–0.0153	0.0098–0.0145	0.0039–0.0236	0.0010–0.0026	0.0008–0.0024	NA
	92 (1588)	0.0019–0.0027	0.0110–0.0157	0.0177–0.0236	0.0039–0.0236	0.0018–0.0032	0.0008 0.0024	NA
	138 (2254)	0.0016–0.0024	0.0120–0.0180	0.0100–0.0160	0.0080–0.0280	0.0010–0.0024	0.0010–0.0024	NA
	121.7 (1994)	0.0008 0.0016	0.0120–0.0180	0.0100–0.0160	0.0080–0.0280	0.0010–0.0024	0.0010–0.0024	NA
1990	92 (1588)	0.0019–0.0027	0.0110–0.0157	0.0177–0.0236	0.0039–0.0236	0.0018–0.0032	0.0008–0.0024	NA

PISTON AND RING SPECIFICATIONS

All measurements are given in inches.

Year	Engine Displacement cu. in. (cc)	Piston Clearance	Ring Gap Top Compression	Ring Gap Bottom Compression	Ring Gap Oil Control	Ring Side Clearance Top Compression	Ring Side Clearance Bottom Compression	Ring Side Clearance Oil Control
1991-92	92 (1588) SOHC	0.0011–0.0019	0.0110–0.0157	0.0177–0.0236	0.0039–0.0236	0.0018–0.0032	0.0008–0.0024	NA
	92 (1588) DOHC & Turbo	0.0019–0.0027	0.0110–0.0157	0.0177–0.0236	0.0039–0.0236	0.0018–0.0032	0.0008–0.0024	NA

NA—Not Available
SOHC—Single Overhead Camshaft
DOHC—Double Overhead Camshaft

TORQUE SPECIFICATIONS

All readings in ft. lbs.

Year	Engine Displacement cu. in. (cc)	Cylinder Head Bolts	Main Bearing Bolts	Rod Bearing Bolts	Crankshaft Pulley Bolts	Flywheel Bolts	Manifold Intake	Manifold Exhaust	Spark Plugs
1988	90 (1471)	③	68	25	108	22④	17	17	11–14
	138 (2254)	①	72	43	87	43	13–18	16	11–14
	121.7 (1994)	①	72	43	87	40	13–18	16	11–14
1989	90 (1471)	③	68	25	108	22④	17	17	11–14
	92 (1588)	③	65	⑤	123	⑥	17	30	11–14
	138 (2254)	①	72	43	87	43	13–18	16	11–14
	121.7 (1994)	①	72	43	87	40	13–18	16	11–14
1990	92 (1588)	③	65②	⑤	109	⑥	17	30	11–14
1991-92	92 (1588) SOHC	③	65②	⑤	109	⑥	17	30	11–14
	92 (1588) DOHC & Turbo	③	65②	⑤	109	⑥	17	30	11–14

SOHC—Single Overhead Camshaft
DOHC—Double Overhead Camshaft
① 1st step—62 ft. lbs.
2nd step—72 ft. lbs.
② See text for proper sequence
③ 1st step—29 ft. lbs.
2nd step—58 ft. lbs.
④ Turn the bolt an additional 45 degrees
⑤ 1st step—11 ft. lbs.
2nd step—turn an additional 45–60 degrees
⑥ 1st step—22 ft. lbs.
2nd step—turn an additional 45–60 degrees

BRAKE SPECIFICATIONS

All measurements in inches unless noted.

Year	Model	Lug Nut Torque (ft. lbs.)	Master Cylinder Bore	Brake Disc		Standard Brake Drum Diameter	Minimum Lining Thickness	
				Minimum Thickness	Maximum Runout		Front	Rear
1988	I-Mark	65 ①	0.810	0.378	0.0059	7.09	0.039	0.039
	I-Mark (Turbo)	65 ①	0.875	0.378	0.0059	7.09	0.039	0.039
	Impulse	87	0.875	0.654	0.0051	NA	0.120	0.120
	Impulse (Turbo)	87	0.875	0.654	0.0051	NA	0.120	0.120
1989	I-Mark	65 ①	0.810	0.378	0.0059	7.09	0.039	0.039
	I-Mark (Turbo)	65 ①	0.875	0.378	0.0059	7.09	0.039	0.039
	I-Mark (DOHC)	65 ①	0.875	0.378	0.0059	7.09	0.039	0.039
	Impulse	87	0.875	0.654	0.0051	NA	0.120	0.120
	Impulse (Turbo)	87	0.875	0.654	0.0051	NA	0.120	0.120
1990	Impulse	87	0.875	②	0.0059	—	0.039	0.039
1991-92	Impulse	87	0.875	0.866 ③	0.0059	—	0.039	0.039
	Stylus	87	0.875	0.866 ③	0.0059		0.039	0.039

NA—Not Available
DOHC—Double Overhead Camshaft
① Aluminum wheels—86 ft. lbs.
② Front—0.81
Rear—0.29
③ Rear—0.299

WHEEL ALIGNMENT

Year	Model	Caster		Camber		Toe-in (in.)	Steering Axis Inclination (deg.)
		Range (deg.)	Preferred Setting (deg.)	Range (deg.)	Preferred Setting (deg.)		
1988	I-Mark	$1\frac{3}{4}$P–$2\frac{3}{4}$P	$2\frac{1}{4}$P	$\frac{11}{16}$P–$1\frac{5}{16}$P	1P	0	$11\frac{13}{16}$
	I-Mark (Turbo)	$1\frac{3}{4}$P–$2\frac{3}{4}$P	$2\frac{1}{4}$P	$\frac{1}{2}$P–$1\frac{1}{2}$P	1P	0	$12\frac{1}{8}$
	Impulse	$1\frac{3}{4}$P–$2\frac{3}{4}$P	$2\frac{1}{4}$P	$\frac{1}{2}$N–$\frac{1}{2}$P	0	0	8
	Impulse (Turbo)	$4\frac{3}{4}$P–$5\frac{1}{4}$P	5P	$\frac{1}{2}$N–$\frac{1}{2}$P	0	0	8
1989	I-Mark	$1\frac{3}{4}$P–$2\frac{3}{4}$P	$2\frac{1}{4}$P	$\frac{11}{16}$P–$1\frac{5}{16}$P	1P	0	$11\frac{13}{16}$
	I-Mark (Turbo)	$1\frac{3}{4}$P–$2\frac{3}{4}$P	$2\frac{1}{4}$P	$\frac{1}{2}$P–$1\frac{1}{2}$P	1P	0	$12\frac{1}{8}$
	I-Mark (DOHC)	$1\frac{3}{4}$P–$2\frac{3}{4}$P	$2\frac{1}{4}$P	$\frac{1}{2}$P–$1\frac{1}{2}$P	1P	0	$12\frac{1}{8}$
	Impulse	$1\frac{3}{4}$P–$2\frac{3}{4}$P	$2\frac{1}{4}$P	$\frac{1}{2}$N–$\frac{1}{2}$P	0	0	8
	Impulse (Turbo)	$4\frac{3}{4}$P–$5\frac{1}{4}$P	5P	$\frac{1}{2}$N–$\frac{1}{2}$P	0	0	8
1990	Impulse	2P–4P	3P	$1\frac{1}{4}$N–$\frac{1}{4}$P	$\frac{1}{2}$N	0	NA
1991-92	Impulse	2P–4P	3P	$1\frac{1}{4}$N–$\frac{1}{4}$P	$\frac{1}{2}$N	$\frac{1}{32}$–$\frac{3}{32}$	NA
	Stylus	2P–4P	3P	$1\frac{1}{4}$N–$\frac{1}{4}$P	$\frac{1}{2}$N	$\frac{1}{32}$–$\frac{3}{32}$	NA

NOTE: Caster angle is pre-set and cannot be serviced
NA—Not Available
DOHC—Double Overhead Camshaft
N Negative
P Positive

ENGINE MECHANICAL

NOTE: Disconnecting the negative battery cable on some vehicles may interfere with the functions of the on board computer systems and may require the computer to undergo a relearning process, once the negative battery cable is reconnected.

Engine Assembly

REMOVAL & INSTALLATION

I-Mark

EXCEPT 4XE1 ENGINE

1. Relieve the fuel pressure and disconnect the battery cables. Scribe the position of the hood on the hood brackets and remove the hood.
2. Drain the oil from the transaxle case and engine oil pan into a suitable drain pan.
3. Drain the cooling system and if equipped with air conditioning, discharge the refrigerant from the air conditioning system.
4. Remove the air cleaner assembly and the throttle cable at the carburetor, on non-turbocharged model.
5. Remove, plug and tag the fuel pump inlet and return hoses on non-turbocharged model.
6. Remove the power steering hoses, air conditioning hose assembly, radiator and heater hoses and automatic transaxle cooler hoses, if equipped.
7. Remove the brake booster hose, clutch cable and select/shift cables, if equipped, speedometer cable and high tension cable.
8. Remove and tag all necessary electrical wires and remove the distributor from the engine.
9. Remove the battery and battery tray. Remove all the necessary wiring connectors from their respective sensor. Be sure to tag all connectors.
10. Remove and tag all necessary vacuum lines and remove the engine control cable.
11. Remove the halfshafts as follows:
    a. Raise and safely support the front of the vehicle. Remove the front wheels.
    b. Disconnect the tie rod end ball joints from the steering knuckles using special remover tool J-21687-02 or equivalent.
    c. Disconnect the lower arm end ball joints from the lower arm. Loosen but do not remove the nuts attaching the strut to the body.
    d. Using a suitable tool, pull out the halfshafts, being careful not to damage the oil seals.
12. Remove the front exhaust pipe. On turbocharged engine remove the exhaust pipe at wastegate manifold and vacuum line at the wastegate control valve. Attach a suitable engine chain hoist to the engine hanger located on the top of the engine.
13. Raise the engine just enough to remove the weight from the engine mounts and disconnect the mounts from the vehicle.
14. Remove the torque rod from the side of the body.
15. Making sure all hoses and wires are aside, carefully and slowly raise the engine and transaxle assembly out of the vehicle. Place the assembly on a suitable engine stand or equivalent.
16. Remove the torque rod with bracket from the engine. Remove the main wire harness assembly from the engine.
17. Place a suitable transmission jack or equivalent under the transaxle and separate the transaxle assembly from the engine.

To install:

18. Installation is the reverse of the removal procedure. Make all necessary adjustments and refill the engine, transaxle, cooling system, power steering reservoir and recharge the air conditioning system with the proper lubricants and to the specified amount.
19. When installing the engine pay close attention to the following torque specifications.
    a. Transaxle-to-engine mounting bolts—56 ft. lbs. (77 Nm).
    b. Torque rod bracket-to-frame bolt—40 ft. lbs. (56 Nm).
    c. Torque rod-to-bracket bolt—56 ft. lbs. (77 Nm).
    d. Engine mount bracket-to-block—28 ft. lbs. (39 Nm).
    e. Front and rear engine mounting through bolts and nuts—60 ft. lbs. (84 Nm). Always use new bolts and nuts.
    f. Right hand mounting rubber on body side—30 ft. lbs. (41 Nm). Right hand mounting rubber on engine side—45 ft. lbs. (62 Nm).
    g. Torque rod to body frame—42 ft. lbs. (58 Nm).
    h. Strut-to-body nuts—40 ft. lbs. (56 Nm).
    i. Tie rod end-to-steering knuckle ball joints—29 ft. lbs. (40 Nm).
20. After the installation is completed, road test the vehicle and then recheck the fluid levels and check for any leaks.

4XE1 ENGINE

1. Relieve the fuel pressure and disconnect the battery cables.
2. Remove the battery. Scribe lines on the inside of the engine hood and remove the engine hood.
3. Drain the engine coolant and the transaxle fluid into suitable drain pans.
4. Remove the air intake duct from the common chamber and the breather hose from the rear of the valve cover.
5. Remove the fast idle vacuum hose from the air intake duct and the pulse air hose from the reed valve side.
6. Remove the 3 air cleaner fixing bolts and remove the air cleaner and air cleaner bracket from the engine.
7. Remove and tag the following vacuum lines:
    a. The EGR vacuum hoses.
    b. Canister hose from the throttle valve side of the vacuum switching valve.
    c. Remove the MAP sensor hose from the common chamber side.
    d. Remove the pulse air hose from the common chamber side.
    e. Remove the canister hoses from the common chamber side.
    f. Remove the master vacuum hose from the master VAC tube at the common chamber side.
8. Disconnect the following wiring harness:
    a. On the electronic control gas injection harness, disconnect the 2 ECM ground connectors from the bracket located on top of the common chamber. Disconnect the 2 green and black multi-pin connectors on the top of the common chamber.
    b. Remove the high tension cable from the ignition coil. Remove the 2 primary connections from the coil. Disconnect the condenser plug. Remove the ignition coil with bracket.
    c. Disconnect 2 engine harness multi-pin plugs, near the battery tray. Remove the engine harness ground cable from the driver's inner fender. Disconnect the connector from the slow blow fuse.
    d. Disconnect the connectors from the EGR temperature sensor, throttle position sensor, MAT sensor, oxygen sensor and remove the engine ground from the valve cover.
9. Working from the left side of the engine, remove the clutch cable by loosening the 2 adjusting nuts.
10. Remove the 2 transaxle shift cables by disconnecting the cotter pin and removing the clip from the shift cable bracket.
11. Disconnect the speedometer cable from the transaxle.
12. Disconnect the heater hoses from the engine. Disconnect the upper radiator hose from the engine.
13. Disconnect the accelerator cable from the throttle body and at the common chamber.

14. Disconnect the fuel feed hose at the fuel filter outlet and disconnect the return fuel line near the fuel filter.
15. Working from the front of the engine, remove the coolant reservoir.
16. Remove the 2 bolts from the power steering pump bracket, without disconnecting the feed lines and position the pump aside from the engine.
17. Disconnect the lower radiator hose from the radiator.
18. Remove the air conditioner compressor bolts and without disconnecting the air conditioning lines, secure the compressor away from the engine.
19. Remove the cooling fan and shroud as one assembly.
20. Raise and support the vehicle safely. Remove both front wheels.
21. Loosen the strut tower nuts on both sides but do not remove.
22. Remove the stabilizer bar. Disconnect the control arms on both sides from the body.
23. Remove the front air deflector, if equipped.
24. Remove the tension rod fixing nut from both sides and remove the rod out of the bracket.
25. Remove the passenger side halfshaft fixing bracket from the engine block and pull the halfshaft out of the transaxle. Pull the halfshaft on the drivers side out of the transaxle.
26. Remove the exhaust pipe from the exhaust manifold.
27. Remove the engine by using a suitable engine hoist. Remove the passenger side engine mount from the body.
28. Disconnect the torque rod at the firewall. Remove both lower engine mounting bolts.
29. Remove the engine and transaxle out of the vehicle together as an assembly.

To install:

30. Using a suitable lifting device, install the engine into the vehicle and install the transaxle and engine mounting bolts.
31. Tighten the engine mount bolts to the following specifications:
   a. Lower engine mount-to-block bolts – 28 ft. lbs. (39 Nm).
   b. Lower engine mount through bolts and nuts – 60 ft. lbs. (84 Nm).
   c. Front and rear engine mount bracket-to-engine bolts – 28 ft. lbs. (39 Nm).
   d. Passenger side engine mount bracket-to-bracket bolts – 29 ft. lbs. (41 Nm).
   e. Passenger side mount-to-bracket bolts – 29 ft. lbs. (41 Nm).
   f. Passenger side mount-to-body bolt – 60 ft. lbs. (84 Nm).
   g. Torque rod through bolts – 40 ft. lbs. (56 Nm).
31. Install all hoses, electrical connectors and components to the original location.
32. Fill the engine and transaxle with the proper type and quantity of oil.
33. Fill the cooling system with the proper type and quantity of coolant.
34. Start the engine and check for leaks.

1988–89 Impulse

1. Relieve the fuel pressure, remove the battery cables, the battery clamp and the battery.
2. On 1988 vehicles, disconnect the headlight cover motor harness.
3. Scribe the location of the hood on the hood hinges and remove the hood.
4. Raise and safely support the vehicle.
5. Remove the engine splash shield and drain the cooling system.
6. Remove the air duct from the air cleaner to the throttle valve on non-turbocharged vehicles or the air ducts from the air cleaner to the turbocharger, turbocharger to intercooler and intercooler to throttle body on turbocharged vehicles.
7. Remove the radiator hoses and disconnect the overflow hose.
8. Remove the nuts attaching the fan blade and the fan pulley and remove the radiator stay and radiator mounting nuts.
9. Remove the radiator with the fan.
10. Remove the air cleaner intake duct on non-turbocharged vehicles or the intercooler and air duct on turbocharged vehicles.
11. Disconnect the wiring and cooler pipes on the condenser and remove the condenser with the receiver dryer and fan.
12. Disconnect the starter wires.
13. Disconnect the following hoses:
   a. VSV (vacuum switch valve) to common chamber.
   b. Fast idle solenoid to common chamber.
   c. Throttle valve to 3 way.
   d. Fast idle solenoid to pipe.
   e. 2 from the VSV to the throttle valve on turbocharged vehicles.
   f. Canister to common chamber.
   g. Common chamber to brake master cylinder.
   h. Injector blower duct to pipe.
   i. Common chamber to automatic cruise actuator.
14. Disconnect and plug the power steering hoses.
15. On turbocharged vehicles, disconnect and plug the oil cooler hoses at the oil filter adapter.
16. If equipped with automatic transmission, disconnect and plug the transmission cooler lines at the radiator.
17. Disconnect the fuel lines and remove the injector blower duct.
18. Disconnect and plug the air conditioning compressor hoses.
19. Remove the clip clamping the battery cables and disconnect the engine wiring cables at the connectors inside the relay box.
20. Disconnect the knock sensor wiring, the dropping resister wiring and the crank angle sensor wiring.
21. Disconnect the ground cables at the engine and remove the right side engine mount nut.
22. Disconnect the alternator wires, the oxygen sensor wires and the transmission switch wires.
23. Disconnect the accelerator control cable and the heater hoses.
24. On turbocharged vehicles, remove the air switching valve with hoses, disconnect the turbocharger water hose, remove the turbocharger heat protector and disconnect the control cable.
25. Remove the engine left side mount nut.
26. If equipped with a manual transmission, remove the gear shift knob, the front console assembly and the gear shift lever boot.
27. Remove the heat protector at the exhaust pipe flange, disconnect the exhaust pipe at the flange and the front exhaust pipe bracket from the transmission bracket.
28. Remove the clutch slave cylinder and disconnect the shift linkage at the link joint on vehicles with manual transmission.
29. Disconnect the speedometer cable.
30. Remove the driveshafts and install a plug to the transmission rear cover.
31. Attach a suitable lifting device and lift the engine slightly. Remove the rear engine mounting bracket bolts.
32. Check to make sure all parts have been removed or disconnected from the engine.
33. Raise the engine toward the front of the vehicle and remove the engine and transmission assembly from the vehicle.
34. Disconnect the transmission from the engine.

To install:

35. Attach the transmission to the engine.
36. Position the engine and transmission assembly in the vehicle and install the engine mount bolts.
37. Install the driveshaft and connect the shift linkage.
38. Connect the speedometer cable and on manual transmission vehicles, install the clutch slave cylinder.
39. Connect the exhaust pipe and the bracket and install the heat protector.

40. If equipped with a manual transmission, install the gear shift lever boot, the console assembly and the gear shift knob.
41. On turbocharged vehicles, connect the control cable, install the turbocharger heat protector, connect the turbocharger water hose and install the air switching valve with the hoses.
42. Connect the heater hoses and the accelerator control cable.
43. Connect the air conditioning compressor hoses and install the injector blower duct.
44. Connect the fuel lines.
45. If equipped with an automatic transmission, connect the transmission cooler lines at the radiator.
46. Connect the oil cooler hoses at the oil filter adapter on turbocharged vehicles.
47. Connect the power steering hoses.
48. Install the condenser with the receiver dryer and fan.
49. Connect the wiring and cooler pipes on the condenser.
50. On turbocharged vehicles, install the intercooler and air duct. On all others, install the air cleaner intake duct.
51. Install the radiator with the fan blade, the radiator mounting nuts and radiator stay, the fan blade and pulley attaching nuts and the radiator and overflow hoses.
52. Connect all wiring and vacuum lines.
53. Install the air cleaner to throttle valve duct on non-turbocharged vehicles and the air cleaner to turbocharger, turbocharger to intercooler and intercooler to throttle valve ducts on turbocharged vehicles.
54. Install the engine splash shield.
55. Install the hood, aligning the marks that were made during the removal procedure.
56. Install the battery.
57. Fill the cooling system with the proper type and quantity of coolant. Fill the crankcase with the proper type of oil to the required level.
58. Connect the negative battery cable, start the engine and check for leaks.
59. Fill the power steering reservoir with fluid and bleed the system.
60. Evacuate and charge the air conditioning system.

1990–92 Impulse and 1991-92 Stylus

1. Relieve the fuel system pressure, disconnect the battery cables and remove the battery and battery tray.
2. Mark the position of the hood on the hood hinges and remove the hood.
3. Raise and safely support the vehicle.
4. Drain the coolant from the radiator and the oil from the transaxle. Remove the left and right undercovers.
5. Disconnect the accelerator cable from the throttle valve and the common chamber.
6. Disconnect the breather hose from the intake air duct side and remove the intake air duct from the throttle valve.
7. Remove the intercooler and breather hose, Turbo only.
8. Remove the air cleaner cover and element and the air cleaner body.
9. Disconnect the MAP sensor hose from the MAP sensor side, the vacuum hose from the vacuum booster side and the 2 canister vacuum hoses from the common chamber and throttle valve.
10. Remove the bracket which supports the 2 canister pipes and the MAP sensor pipe from the common chamber.
11. Disconnect the following electrical connectors and tag them for reassembly:
   a. 2 cable harness connectors near the left front strut tower.
   b. Ignition wire from the ignition coil side.
   c. Bonding cable connector from the terminal at the thermostat housing flange.
   d. Both primary connections from the ignition coil.
   e. Ground cable from the driver's side inner fender.
   f. Positive cable terminal in fuse box.
   g. Cooling fan harness connector.
   h. Both harness connectors at the front of the battery.
   i. Oxygen sensor harness connector.
   j. Ground cable terminals from the right side of the common chamber.
   k. Bonding cable terminal from the right side of the induction control assembly.
   l. If equipped with an automatic transaxle, the 4 connectors of the automatic transaxle control system on the automatic transaxle assembly.
   m. Disconnect the oil cooler pipes, turbo only.
12. Remove the ignition coil with the battery bracket.
13. If equipped with a manual transaxle, loosen both adjusting nuts and remove the clutch cable and disconnect the cotter pin and clip from the shift cable bracket.
14. If equipped with an automatic transaxle, disconnect the cotter pin and joint from the shift cable lever.
15. Disconnect the heater hoses, radiator hoses, fuel lines and speedometer cable.
16. Remove the coolant reservoir tank.
17. Remove the power steering pump and air conditioning compressors and position them aside without disconnecting their lines.
18. If equipped with an automatic transaxle, disconnect the oil cooler lines from the radiator.
19. Remove the cooling fan and the shroud.
20. Remove the front wheel and tire assemblies.
21. Loosen but do not remove, the strut tower nuts.
22. Disconnect the lower ball joints and tie rod ends and remove the halfshafts.
23. Remove the engine splash shield and disconnect the exhaust pipe.
24. Attach a suitable engine hoist to the engine/transaxle assembly and lift the engine slightly to take weight off of the mounts.
25. Disconnect the torque rod from the center beam.
26. Remove the center bolt of the rear side engine mounting after removing the damper weight.
27. Remove the left side engine mount.
28. Remove the cruise control pump and VSV assembly and remove the right side engine mount.
29. If equipped with a manual transaxle, disconnect the axle shaft center bearing support bracket from the cylinder bracket.
30. Remove the engine/transaxle assembly.
31. Separate the engine from the transaxle.

To install:

32. Connect the transaxle to the engine and position the assembly in the vehicle.
33. Connect the halfshaft center bearing support bracket to the cylinder bracket.
34. Install the right side engine mount. Tighten the mount attaching bolt to 89 ft. lbs. (121 Nm), the mount attaching nut to 37 ft. lbs. (50 Nm) and the through bolt to 51 ft. lbs. (69 Nm).
35. Install the left side engine mount and tighten the through bolt to 51 ft. lbs. (69 Nm).
36. Install the rear side engine mount. Tighten the through bolt to 76 ft. lbs. (103 Nm) and the damper weight to 37 ft. lbs. (50 Nm).
37. Install the torque rod. Tighten the center beam side bolt to 51 ft. lbs. (69 Nm) and the engine side bolt to 95 ft. lbs. (128 Nm).
38. Connect the exhaust pipe and install the engine splash shield.
39. Install the halfshafts and connect the tie rods and lower ball joints. Tighten the lower ball joint nuts to 48

ft. lbs. (66 Nm) and the tie rod nuts to 29 ft. lbs. (39 Nm).

40. Tighten the strut tower nuts to 51 ft. lbs. (69 Nm) and install the front wheel and tire assemblies.

41. Install the cooling fan and shroud. On automatic transaxle vehicles, connect the transaxle cooling lines.

42. Install the power steering pump and air conditioning compressor, install and adjust the drive belts.

43. Install the coolant reservoir tank.

44. Connect the radiator and heater hoses, the fuel lines and the speedometer cable. Install the intercooler and hoses. Connect the oil cooler lines, Turbo only.

45. Connect the transaxle shift cable and on manual transaxle vehicles, connect the clutch cable.

46. Install the ignition coil with the battery bracket.

47. Connect all electrical connectors.

48. Install the bracket that supports the 2 canister pipes and the MAP sensor pipe to the common chamber.

49. Connect all vacuum hoses.

50. Install the air cleaner body and the element and cover.

51. Install the intake air duct and connect the breather hose to the intake air side.

52. Connect the accelerator cable and install the hood, aligning the marks that were made during the removal procedure.

53. Install the battery tray and the battery.

54. Fill the cooling system with the proper type and quantity of coolant. Fill the crankcase with the proper type of oil to the required level.

55. Connect the battery cables, start the engine and check for leaks. Adjust the clutch cable.

Cylinder Head

REMOVAL & INSTALLATION

I-Mark

4XC1 AND 4XC1-TURBO ENGINES

1. Relieve the fuel system pressure, disconnect the negative battery cable and drain the cooling system into a suitable drain pan.

2. Remove the air cleaner assembly and disconnect the flex hose along with the oxygen sensor at the exhaust manifold.

3. Disconnect the exhaust pipe bracket at the block and the exhaust pipe at the manifold. On turbocharged model remove exhaust pipe at wastegate manifold, and remove vacuum line for turbocharger control.

4. Disconnect the spark plug wires and remove the thermostat housing.

5. Rotate the engine until the engine is at TDC on the compression stroke of the No. 1 cylinder. Remove distributor cap and mark the distributor rotor to housing position and housing to cylinder head. Remove the distributor hold-down bolt and remove the distributor.

6. Remove the vacuum advance hoses and the ground cable at the cylinder head.

7. Disconnect the fuel lines at the fuel pump and at the carburetor and remove the secondary hoses and throttle cable on non-turbocharged model.

8. Remove engine wiring harness assembly from fuel injectors and fuel line from fuel injector pipe on turbocharged model.

9. Disconnect the vacuum switching valve electrical connector and the heater hoses.

10. Remove the alternator, power steering pump and air conditioning adjusting bolts, brackets and drive belts. Remove and tag all necessary electrical and vacuum lines.

11. Support the engine using a suitable vertical hoist. Remove the right hand motor mount and the bracket at the front cover.

12. Remove the crankshaft bolt and remove the boss and the crank pulley.

13. Remove the timing cover and be sure the mark on the cam pulley is aligned with the upper surface of the cylinder head. Also the dowel pin on the camshaft should be positioned at the top.

14. Disconnect the PCV hoses and remove the valve cover. If the cover sticks to the head, carefully strike the valve cover with a soft mallet.

15. Insert a hex wrench into the tension pulley hexagonal hole. Loosen the timing belt tension by rotating the tension pulley clockwise and remove the timing belt.

16. Remove the fuel pump, on non-turbocharged model, and disconnect the intake manifold coolant hoses.

17. Remove the cylinder head bolts, remove the bolts from both ends at the same time, working toward the middle and remove the cylinder head. Remove the manifolds from the cylinder head.

To install:

18. Installation is the reverse of the removal procedure. Make sure all gasket mating surfaces are clean prior to installation.

19. Check the cylinder head for flatness before installing, using a suitable straight-edge and feeler gauge. If the head is warped more than 0.008 in. (0.2mm) it must be resurfaced. If the head is warped more than 0.016 in. (0.4mm) it must be replaced.

20. Clean the gasket mating surfaces being careful not to damage the aluminum cylinder head. Using new seals and gaskets, apply oil to the head bolt threads and torque the head bolts. The head bolts are tightened, in 2 steps, first to 29 ft. lbs. (40 Nm), in sequence, and finally to 58 ft. lbs. (80 Nm), in sequence.

21. Properly adjust the timing belt tension and adjust the valve clearance.

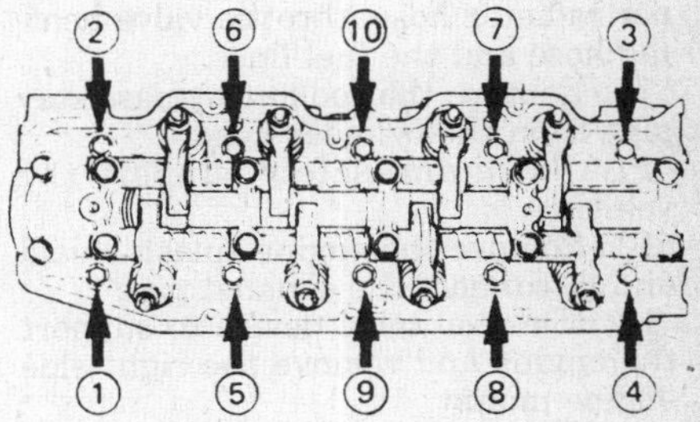

Cylinder head loosening sequence—4XC1 and 4XC1-Turbo engines

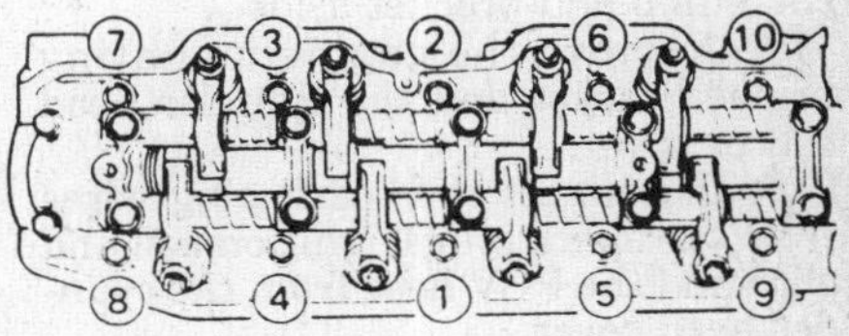

Cylinder head torque sequence—4XC1 and 4XC1-Turbo engines

4XE1 ENGINE

1. Relieve the fuel system pressure and disconnect the negative battery cable.

2. Mark the position of the hood on the hinge brackets and remove the hood.

3. Drain the cooling system.

4. Remove the accelerator cable from the throttle valve and common chamber.

5. Disconnect the breather hose from the intake air duct side.

6. Disconnect the MAP sensor hose from the MAP sensor side, the vacuum booster hose from the vacuum booster and the 2 canister hoses from the pipes on the common chamber.

7. Remove the bracket that supports the 2 canister pipes and MAP sensor pipe from the common chamber.

8. Disconnect the following electrical connectors and tag them for reassembly:

   a. Oxygen sensor.

   b. Ignition wire from the coil side.

   c. Bonding cable connector from the terminal at the thermostat housing flange.

   d. Temperature sensor and thermometer unit connector at the thermostat housing.

e. Ground cable terminals from the right side of the common chamber.

f. Bonding cable terminal from the right side of the induction control assembly.

g. The 2 cable harness connectors near the left front strut tower.

9. Disconnect the heater hose, upper radiator hose, throttle valve heating hose and the fuel lines.

10. Remove the coolant by-pass stay pipe from the cylinder head.

11. Raise and safely support the vehicle.

12. Remove the engine splash shield and disconnect the exhaust pipe.

13. Use a suitable device to support the engine and remove the right side engine mount.

14. Remove the coolant recovery tank with the bracket.

15. Remove the accessory drive belts.

16. Remove the 2 bolts from the power steering pump bracket, without disconnecting the feed lines and position the pump and bracket aside.

17. Remove the engine mounting bridge bracket and the cylinder head center cover.

18. Disconnect the spark plug wires from the spark plugs, ignition coil and clips and the PCV hose from the cylinder head cover.

19. Remove the upper timing belt cover, the power steering pump bracket and the cylinder head cover.

20. Remove the tension adjusting hole cover and align the camshaft pulley timing marks even with the top edge of the cylinder head.

21. Loosen the tension pulley lock bolt and turn the tension pulley clockwise to loosen the timing belt. Remove the timing belt from the camshaft pulleys.

22. Remove the cylinder head bolts, starting at each end of the cylinder head and working toward the center.

23. Raise the cylinder head and remove the coolant hose from the oil cooler.

24. Remove the cylinder head assembly.

To install:

25. Take note of the following points.

26. Make sure all mating surfaces are clean prior to installation.

27. Check the cylinder head for flatness before installing. If the head is warped more than 0.008 in. (0.2mm), it must be resurfaced. If the head is warped more than 0.016 in. (0.4mm), it must be replaced.

28. Use a new gasket and align the cylinder head on the block dowel pins. Apply engine oil to the threads and seating faces of the cylinder head bolts.

29. Tighten the bolts, in 2 steps, following the proper sequence. First, tighten the bolts to 29 ft. lbs. (40 Nm), in sequence, and finally to 59 ft. lbs. (79 Nm), in sequence.

30. Apply a 0.08–0.12 in. (2–3mm) width bead of sealant to the arched area of the No. 1 and No. 5 camshaft bearing caps. Tighten the cylinder head cover bolts to 27 inch lbs. (3 Nm) in the proper sequence.

31. Properly adjust the timing belt after installation.

32. Be sure to torque the following components to specification:

Upper timing cover – 7 ft. lbs. (10 Nm).

Engine mounting bridge bracket – 30 ft. lbs. (40 Nm).

Right side engine mount through bolt – 51 ft. lbs. (69 Nm).

Right side engine mount nut – 37 ft. lbs. (50 Nm).

Right side engine mount bolt – 89 ft. lbs. (121 Nm).

Exhaust pipe nuts – 42 ft. lbs. (57 Nm).

Impulse and Stylus

4ZD1, 4ZC1-T AND 4XE1 (SOHC) ENGINES

1. Relieve the fuel system pressure, disconnect the negative battery cable and drain the cooling system.

2. Rotate the engine until the engine is at TDC on the compression stroke of the No. 1 cylinder. Remove the distributor cap and mark the distributor rotor to housing position and housing to cylinder head position. Remove the distributor hold-down bolt and remove the distributor.

3. Disconnect the radiator inlet and outlet hoses and remove the radiator.

4. Remove the alternator and the air conditioner drive belts. Remove the engine cooling fan.

5. Remove the crankshaft pulley center bolt and remove the pulley and hub assembly.

6. Remove the air pump belt and move the air pump aside. If equipped, remove the air conditioning compressor and lay it to one side. Remove the compressor mounting bracket.

7. Remove the water pump pulley. Remove the top section of the front cover and the water pump.

8. Remove the lower section of the front cover.

9. Remove the tension spring. Loosen the top bolt of the tension pulley and draw the tension pulley fully to the water pump side.

10. Remove the timing belt.

11. Remove cam cover.

12. Sequentially, loosen and remove the rocker arm shaft tightening nuts from the outermost one and remove the rocker arm shaft with the bracket as an assembly.

13. Raise and safely support the vehicle and disconnect the exhaust pipe at the exhaust manifold. On turbocharged vehicles, disconnect the exhaust pipe from the wastegate manifold and remove the control cable for the turbocharger.

14. Lower the vehicle and disconnect all necessary lines, hoses and electrical connectors and tag them for reassembly.

15. Disconnect the accelerator linkage. On turbocharged vehicles remove the engine wiring harness assembly from the fuel injectors and fuel line from fuel injector pipe.

16. Remove the cylinder head bolts beginning with the outer bolts and working in towards the center on both sides.

17. Remove the cylinder head.

To install:

18. Make sure all mating surfaces are clean prior to installation.

19. Check the cylinder head surface for flatness before installing. If the head is warped more than 0.008 in. (0.2mm), it must be resurfaced. If the head is warped more than 0.016 in. (0.4mm), it must be replaced.

20. Install a new cylinder head gasket and the cylinder head, aligning them on the dowels on the cylinder block.

21. Apply a thin coat of engine oil to the cylinder head bolt threads and install the bolts. Tighten them, in 2 steps, in sequence. First tighten to 58 ft. lbs. (80 Nm), in sequence, and then to 72 ft. lbs. (100 Nm), in sequence.

22. Properly adjust the timing belt tension.

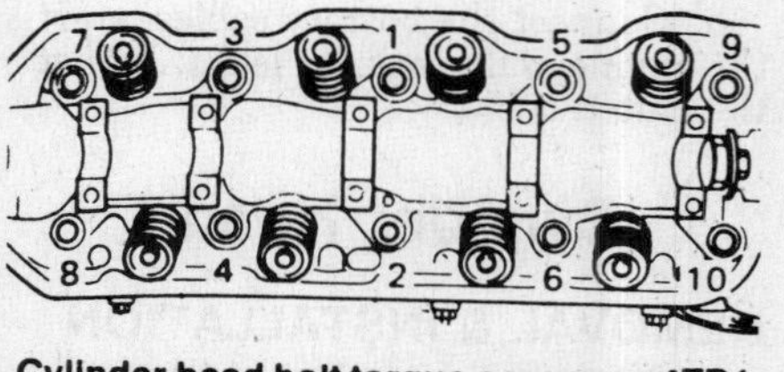

Cylinder head bolt torque sequence 4ZD1 and 4ZC1-T engines

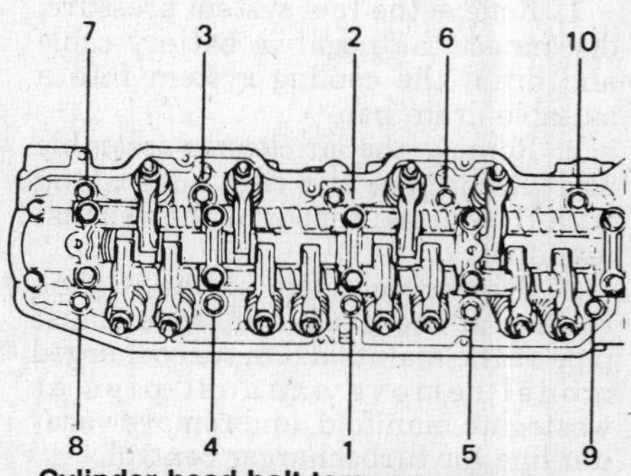

Cylinder head bolt torque sequence – 4XE1 (SOHC) engine

4XE1 (DOHC) ENGINE

1. Relieve the fuel system pressure and disconnect the negative battery cable.
2. Mark the position of the hood on the hinge brackets and remove the hood.
3. Drain the cooling system.
4. Remove the accelerator cable from the throttle valve and common chamber.
5. Disconnect the breather hose from the intake air duct side.
6. Disconnect the MAP sensor hose from the MAP sensor side, the vacuum booster hose from the vacuum booster and the 2 canister hoses from the pipes on the common chamber.
7. Remove the bracket that supports the 2 canister pipes and MAP sensor pipe from the common chamber.
8. Disconnect the following electrical connectors and tag them for reassembly:
   a. Oxygen sensor.
   b. Ignition wire from the coil side.
   c. Bonding cable connector from the terminal at the thermostat housing flange.
   d. Temperature sensor and thermometer unit connector at the thermostat housing.
   e. Ground cable terminals from the right side of the common chamber.
   f. Bonding cable terminal from the right side of the induction control assembly.
   g. The 2 cable harness connectors near the left front strut tower.
9. Disconnect the heater hose, upper radiator hose, throttle valve heating hose and the fuel lines.
10. Remove the coolant by-pass stay pipe from the cylinder head.
11. Raise and safely support the vehicle.
12. Remove the engine splash shield and disconnect the exhaust pipe.
13. Use a suitable device to support the engine and remove the right side engine mount.
14. Remove the coolant recovery tank with the bracket.
15. Remove the accessory drive belts.
16. Remove the 2 bolts from the power steering pump bracket, without disconnecting the feed lines and position the pump and bracket aside.
17. Remove the engine mounting bridge bracket and the cylinder head center cover.
18. Disconnect the spark plug wires from the spark plugs, ignition coil and clips and the PCV hose from the cylinder head cover.
19. Remove the upper timing belt cover, the power steering pump bracket and the cylinder head cover.
20. Remove the tension adjusting hole cover and align the camshaft pulley timing marks even with the top edge of the cylinder head.
21. Loosen the tension pulley lock bolt and turn the tension pulley clockwise to loosen the timing belt. Remove the timing belt from the camshaft pulleys.
22. Remove the cylinder head bolts, starting at each end of the cylinder head and working toward the center.
23. Raise the cylinder head and remove the coolant hose from the oil cooler.
24. Remove the cylinder head assembly.

To install:

25. Make sure all mating surfaces are clean prior to installation.
26. Check the cylinder head for flatness before installing. If the head is warped more than 0.008 in. (0.2mm), it must be resurfaced. If the head is warped more than 0.016 in. (0.4mm), it must be replaced.
27. Use a new gasket and align the cylinder head on the block dowel pins. Apply engine oil to the threads and seating faces of the cylinder head bolts.
28. Tighten the bolts, in 2 steps, in sequence. First, tighten the bolts to 29 ft. lbs. (40 Nm), in sequence, and finally to 59 ft. lbs. (79 Nm), in sequence.
29. Apply a 0.08–0.12 in. (2–3mm) width bead of sealant to the arched area of the No. 1 and No. 5 camshaft bearing caps. Tighten the cylinder head cover bolts to 53 inch lbs. (6 Nm) in the proper sequence.
30. Properly adjust the timing belt after installation.
31. Be sure to torque the following components to specification:

   Upper timing cover – 7 ft. lbs. (10 Nm).

   Engine mounting bridge bracket – 30 ft. lbs. (40 Nm).

   Right side engine mount through bolt – 51 ft. lbs. (69 Nm).

   Right side engine mount nut – 37 ft. lbs. (50 Nm).

   Right side engine mount bolt – 89 ft. lbs. (121 Nm).

   Exhaust pipe nuts – 42 ft. lbs. (57 Nm).

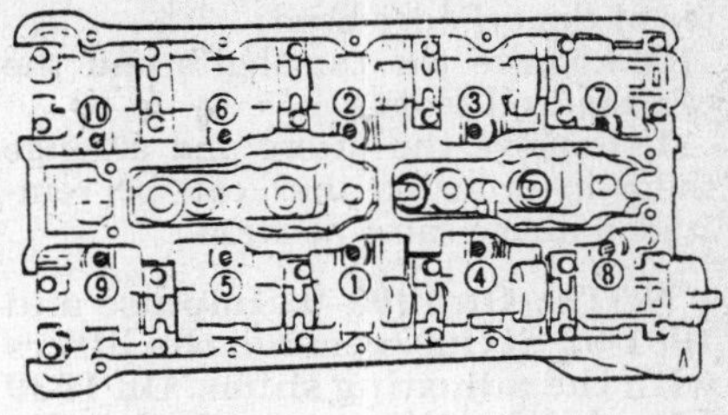

Cylinder head bolt torque sequence – 4XE1 engine

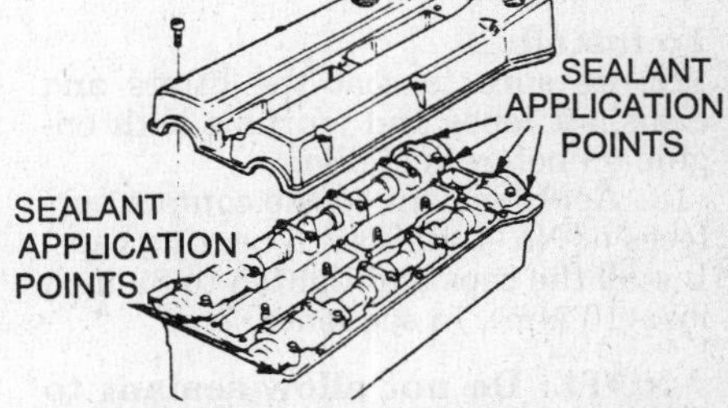

Installing the cylinder head cover – 4XE1 engine

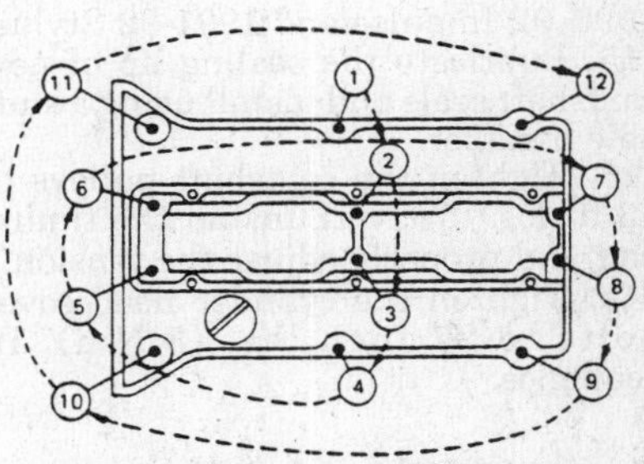

Cylinder head bolt torque sequence – 4XE1 engine

Valve Lifters

REMOVAL & INSTALLATION

4XE1 Engine

All other engines do not use valve lifters. The camshaft contacts the rocker arm which drives the valve.

1. Disconnect the negative battery cable.
2. Disconnect the PCV hoses and the center cover.
3. Disconnect the spark plug wires and remove the upper timing cover.
4. Remove the cylinder head cover.
5. Bring the No. 1 cylinder to TDC of the compression stroke. The timing marks on the camshafts should be even with the top edge of the cylinder head.
6. Loosen the camshaft pulley bolts and loosen the timing belt tension pulley bolt ½ turn. Loosen the timing belt by rotating the tension pulley and remove the belt from the camshaft pulleys.

NOTE: Be careful not to rotate the engine.

7. Remove the camshaft pulleys.
8. Remove the distributor cap and mark the position of the distributor rotor in relation to the distributor housing and the distributor housing in relation to the cylinder head. Remove the distributor.
9. Remove the camshaft bearing cap bolts, beginning with the end caps and working gradually toward the cen-

ter of the cylinder head.

10. Remove the camshafts and the camshaft oil seals.

11. Remove the lifters and arrange them in order so they can be reinstalled in the same position.

NOTE: On 1990–92 Impulse and 1991-92 Stylus, remove the lifters with the adjusting shims. On 1989 I-Mark, the lifters are hydraulic and should be submerged in engine oil when removed. This will prevent air from mixing with the oil in the lifter and causing noise when reinstalled.

To install:

12. Be sure to coat the lifters and camshaft lobes and journals with engine oil before installing.

13. Apply sealant to the contact surfaces of No. 1 and No. 5 bearing caps. Install the caps and tighten to 89 inch lbs. (10 Nm), in sequence.

NOTE: Do not allow sealant to contact the bearing surfaces of the bearing cap.

14. Adjust the valve clearance on 1990–92 Impulse and 1991-92 Stylus.

15. Lubricate the sealing lip of new camshaft seals and install using a suitable installer.

16. Tighten the camshaft pulleys to 43 ft. lbs. (59 Nm), install the timing belt and properly adjust the tension.

17. Tighten the cylinder head cover bolts to 27 inch lbs. (3 Nm), in sequence.

Valve Lash

ADJUSTMENT

Valve lash adjustments must be made when the engine is cold.

Except 4XE1 (DOHC) Engine

1. Disconnect the negative battery cable.

2. Remove the cylinder head cover.

3. Before proceeding further, check the rocker shaft bolts for looseness and tighten to 16 ft. lbs. (22 Nm), as necessary.

4. Remove the distributor cap. Turn the crankshaft 1 full turn in the normal direction of rotation and align the distributor rotor with the No. 1 cylinder spark plug wire on the distributor cap and align the notched line on the crank pulley with the **0** mark on the timing gear case cover.

5. The following valves can be adjusted: Intake–No. 1 and No. 2. Exhaust–No. 1 and No. 3.

6. Measure the clearance between the rocker arm and the valve stem using a suitable feeler gauge. Adjust by

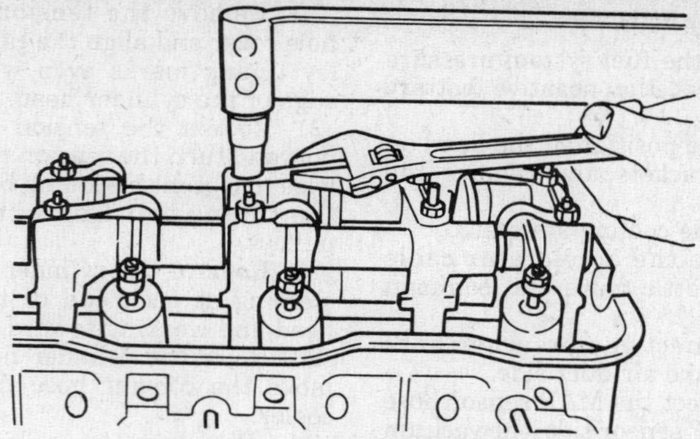

Checking rocker shaft for tightness

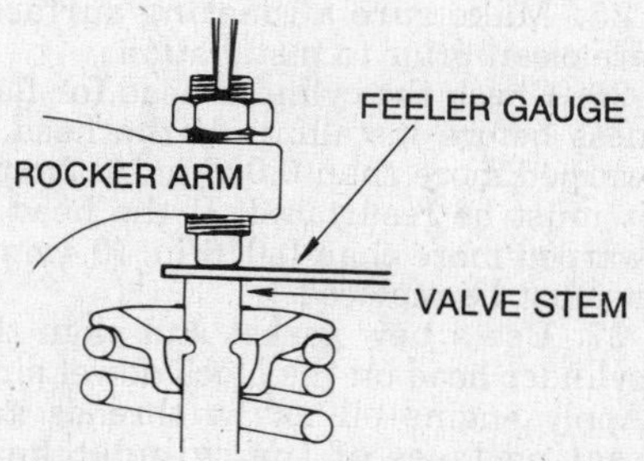

Adjusting the valve clearance

loosening the locknut and turning the adjustment stud until a slight drag is felt on the feeler gauge.

7. Adjust the valve clearance to the following specifications:

Except 4ZD1: Intake–0.006 in. (0.15mm), Exhaust–0.010 in. (0.25mm).

4ZD1 engine: Intake–0.008 in. (0.20mm), Exhaust–0.008 in. (0.20mm).

8. After adjustment is completed, rotate the engine 1 revolution until the notched line on the crank pulley is aligned with the **0** mark on the timing gear case cover and the distributor rotor is aligned with the No. 4 spark plug wire on the distributor cap.

9. The remaining valves can now be adjusted.

10. After all valves have been adjusted, install the remaining components in the reverse of their removal procedure.

4XE1 (DOHC) Engine

NOTE: The 4XE1 engine in 1989 I-Mark uses hydraulic lifters which require no adjustment.

1. Disconnect the negative battery cable.

2. Remove the cylinder head cover and remove the spark plugs.

3. Check the torque of the camshaft bearing cap bolts and tighten to 89 inch lbs. (10 Nm), as necessary.

4. Rotate the engine and align the notched line on the crank pulley with the **0** mark on the timing cover. The lifters on the No. 1 cylinder should have play and the lifters on the No. 4 cylinder should not. If not, rotate the engine 1 more revolution and align the marks again. The No. 1 cylinder is on TDC of the compression stroke.

5. The following valves can be measured for valve clearance: Intake–No. 1 and No. 2. Exhaust–No. 1 and No. 3.

6. Measure the clearance using a suitable feeler gauge and record the measurements of those that are out of adjustment.

7. Rotate the engine 1 revolution until the No. 4 cylinder is at TDC and measure the clearance of the remaining valves. Record the measurements of those that are out of adjustment.

8. Adjust the valve clearance by replacing the adjuster shim as follows:

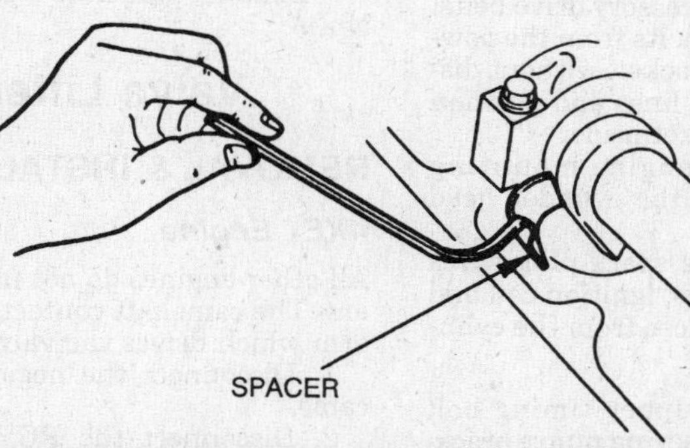

Installing valve adjustment spacer tool–Impulse with 4XE1 engine

a. Turn the camshaft and use the cam lift to press down the lifter.

b. Place spacer J-38413-2 or J-38413-3 or equivalent, on the upper circumference of the lifter from outside the cylinder head.

c. Release the cam lift by turning the camshaft and hold the lifter down with the spacer. Remove the adjuster shim to the spark plug side using a suitable tool.

NOTE: Before pressing down the lifter, turn the cutaway of the lifter toward the position at which the adjuster shim is easily removed.

d. Measure the thickness of the adjuster shim using a suitable micrometer. Calculate the needed thickness of the replacement shim by adding the thickness of the original shim to the required clearance. Select the available adjuster shim with a thickness as close as possible to the calculated values.

NOTE: Adjuster shims are available in 19 sizes in thickness increments of 0.05mm, ranging from 2.55–3.45mm.

e. Place the selected adjuster shim on the lifter and press down the lifter by turning the camshaft and removing the spacer.

9. After valve adjustment is completed, install the remaining components in the reverse order of their removal.

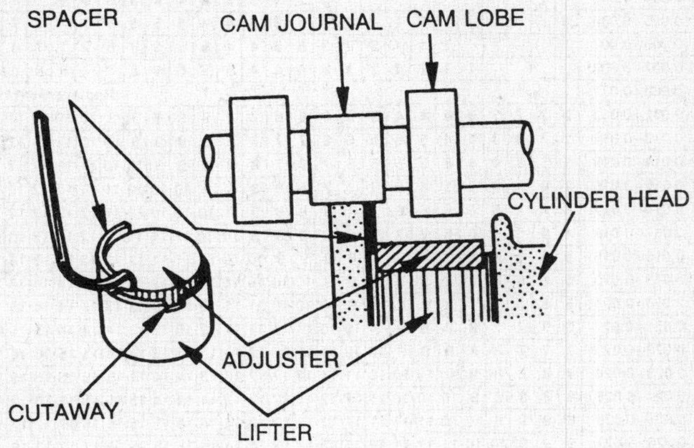

Valve adjuster shim removal—Impulse with 4XE1 engine

Measured clearance		Original Adjuster (Shim) Thickness (mm)																																																
mm	inch	2.52	2.54	2.56	2.58	2.60	2.62	2.64	2.66	2.68	2.70	2.72	2.74	2.76	2.78	2.80	2.82	2.84	2.86	2.88	2.90	2.92	2.94	2.96	2.98	3.00	3.02	3.04	3.06	3.08	3.10	3.12	3.14	3.16	3.18	3.20	3.22	3.24	3.26	3.28	3.30	3.32	3.34	3.36	3.38	3.40	3.42	3.44	3.46	3.48
0.000~0.025	0.000~0.001									1	1	2	2	2	3	3	4	4	4	5	5	6	6	6	7	7	8	8	8	9	9	10	10	10	11	11	12	12	12	13	13	14	14	14	15	15	16	16	16	17
0.026~0.050	0.001~0.002							1	1	1	2	2	3	3	3	4	4	5	5	5	6	6	7	7	7	8	8	9	9	9	10	10	11	11	11	12	12	13	13	13	14	14	15	15	15	16	16	17	17	17
0.051~0.075	0.002~0.003						1	1	1	2	2	3	3	3	4	4	5	5	5	6	6	7	7	7	8	8	9	9	9	10	10	11	11	11	12	12	13	13	13	14	14	15	15	15	16	16	17	17	17	18
0.076~0.100	0.003~0.004					1	1	2	2	2	3	3	4	4	4	5	5	6	6	6	7	7	8	8	8	9	9	10	10	10	11	11	12	12	12	13	13	14	14	14	15	15	16	16	16	17	17	18	18	18
0.101~0.200	0.004~0.008	Replacement not to be required																																																
0.201~0.225	0.008~0.009	2	2	2	3	3	4	4	4	5	5	6	6	6	7	7	8	8	8	9	9	10	10	10	11	11	12	12	12	13	13	14	14	14	15	15	16	16	16	17	17	18	18	18	19	19				
0.226~0.250	0.009~0.010	2	3	3	3	4	4	5	5	5	6	6	7	7	7	8	8	9	9	9	10	10	11	11	11	12	12	13	13	13	14	14	15	15	15	16	16	17	17	17	18	18	19	19	19					
0.251~0.275	0.010~0.011	3	3	3	4	4	5	5	5	6	6	7	7	7	8	8	9	9	9	10	10	11	11	11	12	12	13	13	13	14	14	15	15	15	16	16	17	17	17	18	18	19	19	19						
0.276~0.300	0.011~0.012	3	4	4	4	5	5	6	6	6	7	7	8	8	8	9	9	10	10	10	11	11	12	12	12	13	13	14	14	14	15	15	16	16	16	17	17	18	18	18	19	19								
0.301~0.325	0.012~0.013	4	4	4	5	5	6	6	6	7	7	8	8	8	9	9	10	10	10	11	11	12	12	12	13	13	14	14	14	15	15	16	16	16	17	17	18	18	18	19	19									
0.326~0.350	0.013~0.014	4	5	5	5	6	6	7	7	7	8	8	9	9	10	10	11	11	11	11	12	12	13	13	14	14	14	15	15	15	16	16	17	17	17	18	18	19	19	19										
0.351~0.375	0.014~0.015	5	5	5	6	6	7	7	7	8	8	9	9	9	10	10	11	11	11	12	12	13	13	13	14	14	15	15	15	16	16	17	17	17	18	18	19	19	19											
0.376~0.400	0.015~0.016	5	6	6	6	7	7	8	8	8	9	9	10	10	10	11	11	12	12	12	13	13	14	14	14	15	15	16	16	16	17	17	18	18	18	19	19													
0.401~0.425	0.016~0.017	6	6	6	7	7	8	8	8	9	9	10	10	10	11	11	12	12	12	13	13	14	14	14	15	15	16	16	16	17	17	18	18	18	19	19														
0.426~0.450	0.017~0.018	6	7	7	7	8	8	9	9	9	10	10	11	11	11	12	12	13	13	13	14	14	14	15	15	16	16	17	17	17	18	18	19	19	19															
0.451~0.475	0.018~0.019	7	7	7	8	8	9	9	9	10	10	11	11	12	12	12	13	13	13	14	14	15	15	15	16	16	17	17	17	18	18	19	19	19																
0.476~0.500	0.019~0.020	7	8	8	8	9	9	10	10	10	11	11	12	12	12	13	13	14	14	14	15	15	16	16	16	17	17	18	18	18	19	19																		
0.501~0.525	0.020~0.021	8	8	8	9	9	10	10	10	11	11	12	12	12	13	13	14	14	14	15	15	16	16	16	17	17	18	18	18	19	19																			
0.526~0.550	0.021~0.022	8	9	9	9	10	10	11	11	11	12	12	13	13	13	14	14	15	15	15	16	16	17	17	17	18	18	19	19	19																				
0.551~0.575	0.022~0.023	9	9	9	10	10	11	11	11	12	12	13	13	13	14	14	15	15	15	16	16	17	17	17	18	18	19	19	19																					
0.576~0.600	0.023~0.024	9	10	10	10	11	11	12	12	12	13	13	14	14	14	15	15	16	16	16	17	17	18	18	18	19	19																							
0.601~0.625	0.024~0.025	10	10	10	11	11	12	12	12	13	13	14	14	14	15	15	16	16	16	17	17	18	18	18	19	19																								
0.626~0.650	0.025~0.026	10	11	11	11	12	12	13	13	13	14	14	15	15	15	16	16	17	17	17	18	18	19	19	19																									
0.651~0.675	0.026~0.027	11	11	11	12	12	13	13	13	14	14	15	15	15	16	16	17	17	17	18	18	19	19	19																										
0.676~0.700	0.027~0.028	11	12	12	12	13	13	14	14	14	15	15	16	16	16	17	17	18	18	18	19	19																												
0.701~0.725	0.028~0.029	12	12	12	13	13	14	14	14	15	15	16	16	16	17	17	18	18	18	19	19																													
0.726~0.750	0.029~0.030	12	13	13	13	14	14	15	15	15	16	16	17	17	17	18	18	19	19	19																														
0.751~0.775	0.030~0.031	13	13	13	14	14	15	15	15	16	16	17	17	17	18	18	19	19	19																															
0.776~0.800	0.031~0.032	13	14	14	14	15	15	16	16	16	17	17	18	18	18	19	19																																	
0.801~0.825	0.032~0.033	14	14	14	15	15	16	16	16	17	17	18	18	18	19	19																																		
0.826~0.850	0.033~0.034	14	15	15	15	16	16	17	17	17	18	18	19	19	19																																			
0.851~0.875	0.034~0.035	15	15	15	16	16	17	17	17	18	18	19	19	19																																				
0.876~0.900	0.035~0.036	15	16	16	16	17	17	18	18	18	19	19																																						
0.901~0.925	0.036~0.037	16	16	16	17	17	18	18	18	19	19																																							
0.926~0.950	0.0365~0.0374	16	17	17	17	18	18	19	19	19																																								
0.951~0.975	0.037~0.038	17	17	17	18	18	19	19	19																																									
0.976~1.000	0.038~0.039	17	18	18	18	19	19																																											
1.001~1.025	0.039~0.040	18	18	18	19	19																																												
1.026~1.050	0.040~0.041	18	19	19	19																																													
1.051~1.075	0.041~0.042	19	19	19																																														
1.076~1.100	0.042~0.043	19																																																

Thickness of available adjuster (Shim)

NO in Chart	Thickness (mm)	NO in Chart	Thickness (mm)
1	2.55	11	3.05
2	2.60	12	3.10
3	2.65	13	3.15
4	2.70	14	3.20
5	2.75	15	3.25
6	2.80	16	3.30
7	2.85	17	3.35
8	2.90	18	3.40
9	2.95	19	3.45
10	3.00		

Intake adjuster shim chart—4XE1 (DOHC) engine

Measured clearance		Original Adjuster (Shim) Thickness (mm)																																																
mm	inch	2.52	2.54	2.56	2.58	2.60	2.62	2.64	2.66	2.68	2.70	2.72	2.74	2.76	2.78	2.80	2.82	2.84	2.86	2.88	2.90	2.92	2.94	2.96	2.98	3.00	3.02	3.04	3.06	3.08	3.10	3.12	3.14	3.16	3.18	3.20	3.22	3.24	3.26	3.28	3.30	3.32	3.34	3.36	3.38	3.40	3.42	3.44	3.46	3.48
0.000~0.025	0.000~0.001															1	1	2	2	2	3	3	4	4	4	5	5	6	6	6	7	7	8	8	8	9	9	10	10	10	11	11	12	12	12	13	13	14	14	14	15
0.026~0.050	0.001~0.002												1	1	1	2	2	3	3	3	4	4	5	5	5	6	6	7	7	7	8	8	9	9	9	10	10	11	11	11	12	12	13	13	13	14	14	14	15	15
0.051~0.075	0.002~0.003											1	1	1	2	2	3	3	3	4	4	5	5	5	6	6	7	7	7	8	8	9	9	9	10	10	11	11	11	12	12	13	13	13	14	14	15	15	15	16
0.076~0.100	0.003~0.004										1	1	2	2	2	3	3	4	4	4	5	5	6	6	6	7	7	8	8	8	9	9	10	10	10	11	11	12	12	12	13	13	14	14	14	15	15	16	16	16
0.101~0.125	0.004~0.005									1	1	2	2	2	3	3	4	4	4	5	5	6	6	6	7	7	8	8	8	9	9	10	10	10	11	11	12	12	12	13	13	14	14	14	15	15	16	16	16	17
0.126~0.150	0.005~0.006							1	1	1	2	2	3	3	3	4	4	5	5	5	6	6	7	7	7	8	8	9	9	9	10	10	11	11	11	12	12	13	13	13	14	14	15	15	15	16	16	17	17	17
0.151~0.175	0.006~0.007						1	1	1	2	2	3	3	3	4	4	5	5	5	6	6	7	7	7	8	8	9	9	9	10	10	11	11	11	12	12	13	13	13	14	14	15	15	15	16	16	17	17	17	18
0.176~0.200	0.007~0.008					1	1	2	2	2	3	3	4	4	4	5	5	6	6	6	7	7	8	8	8	9	9	10	10	10	11	11	12	12	12	13	13	14	14	14	15	15	16	16	16	17	17	18	18	18
0.201~0.300	0.008~0.012	Replacement not to be required																																																
0.301~0.325	0.012~0.013	2	2	2	3	3	4	4	4	5	5	6	6	6	7	7	8	8	8	9	9	10	10	10	11	11	12	12	12	13	13	14	14	14	15	15	16	16	16	17	17	18	18	18	19	19				
0.326~0.350	0.013~0.014	2	3	3	3	4	4	5	5	5	6	6	7	7	7	8	8	9	9	9	10	10	11	11	11	12	12	13	13	13	14	14	15	15	15	16	16	17	17	17	18	18	19	19	19					
0.351~0.375	0.014~0.015	3	3	3	4	4	5	5	5	6	6	7	7	7	8	8	9	9	9	10	10	11	11	11	12	12	13	13	13	14	14	15	15	15	16	16	17	17	17	18	18	19	19	19						
0.376~0.400	0.015~0.016	3	4	4	4	5	5	6	6	6	7	7	8	8	8	9	9	10	10	10	11	11	12	12	12	13	13	14	14	14	15	15	16	16	16	17	17	18	18	18	19	19								
0.401~0.425	0.016~0.017	4	4	4	5	5	6	6	6	7	7	8	8	8	9	9	10	10	10	11	11	12	12	12	13	13	14	14	14	15	15	16	16	16	17	17	18	18	18	19	19									
0.426~0.450	0.017~0.018	4	5	5	5	6	6	7	7	7	8	8	9	9	9	10	10	11	11	11	12	12	13	13	13	14	14	15	15	15	16	16	17	17	17	18	18	19	19	19										
0.451~0.475	0.018~0.019	5	5	5	6	6	7	7	7	8	8	9	9	9	10	10	11	11	11	12	12	13	13	13	14	14	15	15	15	16	16	17	17	17	18	18	19	19	19											
0.476~0.500	0.019~0.020	5	6	6	6	7	7	8	8	8	9	9	10	10	10	11	11	12	12	12	13	13	14	14	14	15	15	16	16	16	17	17	18	18	18	19	19													
0.501~0.525	0.020~0.021	6	6	6	7	7	8	8	8	9	9	10	10	10	11	11	12	12	12	13	13	14	14	14	15	15	16	16	16	17	17	18	18	18	19	19														
0.526~0.550	0.021~0.022	6	7	7	7	8	8	9	9	9	10	10	11	11	11	12	12	13	13	13	14	14	14	15	15	16	16	17	17	17	18	18	19	19	19															
0.551~0.575	0.022~0.023	7	7	7	8	8	9	9	9	10	10	11	11	12	12	12	13	13	13	14	14	15	15	15	16	16	17	17	17	18	18	19	19	19																
0.576~0.600	0.023~0.024	7	8	8	8	9	9	10	10	10	11	11	12	12	12	13	13	14	14	14	15	15	16	16	16	17	17	18	18	18	19	19																		
0.601~0.625	0.024~0.025	8	8	8	9	9	10	10	10	11	11	12	12	12	13	13	14	14	14	15	15	16	16	16	17	17	18	18	18	19	19																			
0.626~0.650	0.025~0.026	8	9	9	9	10	10	11	11	11	12	12	13	13	13	14	14	15	15	15	16	16	17	17	17	18	18	19	19	19																				
0.651~0.675	0.026~0.027	9	9	9	10	10	11	11	11	12	12	13	13	13	14	14	15	15	15	16	16	17	17	17	18	18	19	19	19																					
0.676~0.700	0.027~0.028	9	10	10	10	11	11	12	12	12	13	13	14	14	14	15	15	16	16	16	17	17	18	18	18	19	19																							
0.701~0.725	0.028~0.029	10	10	10	11	11	12	12	12	13	13	14	14	14	15	15	16	16	16	17	17	18	18	18	19	19																								
0.726~0.750	0.029~0.030	10	11	11	11	12	12	13	13	13	14	14	15	15	15	16	16	17	17	17	18	18	19	19	19																									
0.751~0.775	0.030~0.031	11	11	11	12	12	13	13	13	14	14	15	15	15	16	16	17	17	17	18	18	19	19	19																										
0.776~0.800	0.031~0.032	11	12	12	12	13	13	14	14	14	15	15	16	16	16	17	17	18	18	18	19	19																												
0.801~0.825	0.032~0.033	12	12	12	13	13	14	14	14	15	15	16	16	16	17	17	18	18	18	19	19																													
0.826~0.850	0.033~0.034	12	13	13	13	14	14	15	15	15	16	16	17	17	17	18	18	19	19	19																														
0.851~0.875	0.034~0.035	13	13	13	14	14	15	15	15	16	16	17	17	17	18	18	19	19	19																															
0.876~0.900	0.035~0.036	13	14	14	14	15	15	16	16	16	17	17	18	18	18	19	19																																	
0.901~0.925	0.036~0.037	14	14	14	15	15	16	16	16	17	17	18	18	18	19	19																																		
0.926~0.950	0.0365~0.0374	14	15	15	15	16	16	17	17	17	18	18	19	19	19																																			
0.951~0.975	0.037~0.038	15	15	15	16	16	17	17	17	18	18	19	19	19																																				
0.976~1.000	0.038~0.039	15	16	16	16	17	17	18	18	18	19	19																																						
1.001~1.025	0.039~0.040	16	16	16	17	17	18	18	18	19	19																																							
1.026~1.050	0.040~0.041	16	17	17	17	18	18	19	19	19																																								
1.051~1.075	0.041~0.042	17	17	17	18	18	19	19	19																																									
1.076~1.100	0.042~0.043	17	18	18	18	19	19																																											
1.101~1.125	0.043~0.044	18	18	18	19	19																																												
1.126~1.150	0.044~0.045	18	19	19	19																																													
1.151~1.175	0.045~0.046	19	19	19																																														
1.176~1.200	0.046~0.047	19																																																

Thickness of available adjuster (Shim)

NO in Chart	Thickness (mm)	NO in Chart	Thickness (mm)
1	2.55	11	3.05
2	2.60	12	3.10
3	2.65	13	3.15
4	2.70	14	3.20
5	2.75	15	3.25
6	2.80	16	3.30
7	2.85	17	3.35
8	2.90	18	3.40
9	2.95	19	3.45
10	3.00		

Note; Thickness mark is printed on the surface to be contacted with tappet.

Exhaust adjuster shim chart—4XE1 (DOHC) engine

Rocker Arms/Shafts

REMOVAL & INSTALLATION

1. Disconnect the negative battery cable.
2. Remove the cylinder head cover and the timing belt cover.
3. Bring the No. 1 cylinder to TDC on the compression stroke.
4. On I-Mark, loosen the bolts attaching the timing belt tension pulley. Turn the tension pulley clockwise to loosen the timing belt and tighten the attaching bolts to hold the tension pulley in place.
5. On Impulse, remove the tension spring and loosen the bolt. Draw the timing cover tension pulley fully to the water pump side to loosen the timing belt.
6. Loosen and remove the rocker arm shaft tightening bolts starting with the outermost one and working toward the center of the cylinder head.

To install:

7. Remove the rocker arm shafts.
8. Inspect the rocker arms and shafts for wear and replace, as necessary. When disassembling, label the components and keep them in order so they may be reinstalled in their original positions. Apply engine oil to the rocker arms and shafts when reassembling.
9. Installation is the reverse of the removal procedure.
10. On I-Mark, remove any oil from the contact surfaces on the No. 1 and No. 5 rocker brackets and apply sealant to the contact surfaces before installation.
11. On Impulse, remove any oil from the contact surface of the No. 1 rocker arm bracket and apply sealant to the contact surface before installation.
12. Tighten the rocker arm shaft bolts on I-Mark to 16 ft. lbs. (22 Nm) in the proper sequence.
13. On Impulse, tighten the rocker arm shaft bolts to 16 ft. lbs. (22 Nm) with the exception of the 2 small bolts at the front of the No. 1 bearing cap which are tightened to 5 ft. lbs. (8 Nm).
14. Properly adjust the timing belt tension.

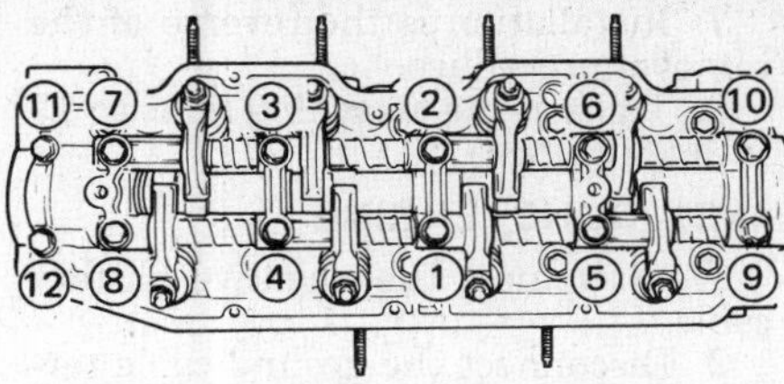

Rocker arm shaft bolt torque sequence—4XC1-U and 4XC1-T engines

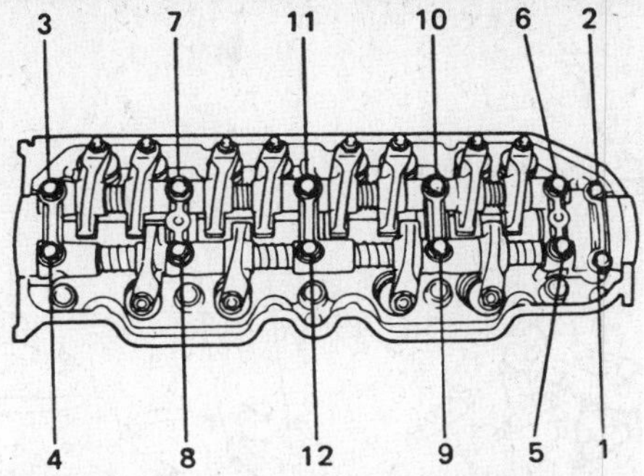

Rocker arm shaft bolt torque sequence—4XE1 (SOHC) engine

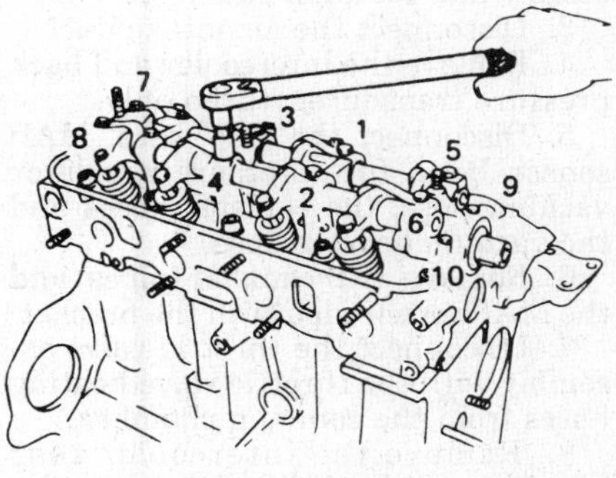

Rocker arm shaft bolt torque sequence—4ZD1 and 4ZC1-T engines

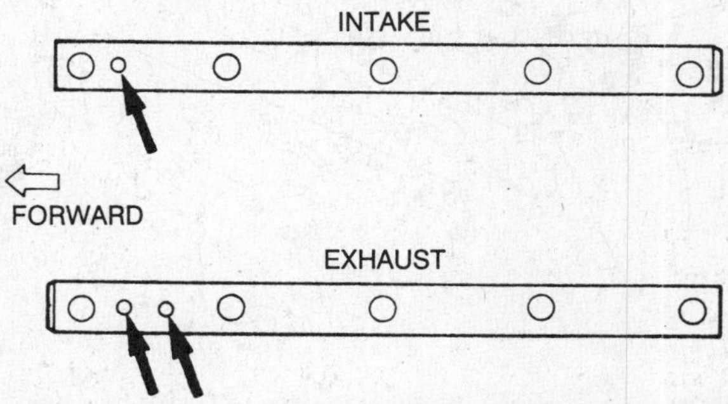

Rocker arm shaft positioning—4XC1, 4XC1-Turbo and 4XE1 (SOHC) engines

Intake Manifold

REMOVAL & INSTALLATION

I-Mark

4XC1-U ENGINE

1. Disconnect the negative battery cable and drain the coolant into a suitable drain pan.
2. Remove the bolt securing the alternator adjusting plate to the engine.
3. Disconnect and tag all hoses attached to the air cleaner assembly and remove the air cleaner.
4. Disconnect the air inlet temperature switch wiring connector. Disconnect and tag all the hoses, electrical connectors and control cable attached to the carburetor.
5. If equipped with air conditioning, disconnect the fast idle control vacuum hose, the pressure tank control valve hose, the distributor/3-way connector hose and the vacuum switching valve wiring connector.
6. Remove the carburetor attaching bolts, which are located underneath the intake manifold. Remove the carburetor and the EFE heater.
7. At the intake manifold, remove the PCV hose, the water bypass hose, the two heater hoses, the EGR valve canister hose, the distributor vacuum advance hose and the ground wires.
8. Disconnect the thermometer unit switch wiring connector, if equipped.
9. Remove the intake manifold attaching nuts and bolts and remove the intake manifold.

To install:

10. Clean the sealing surfaces of the intake manifold and the cylinder head.
11. Use a straight-edge and a feeler gauge to check the surfaces containing the cylinder head for excessive warpage. The inlet manifold must be replaced if the warpage is in excess of 0.0157 in. (0.4mm).
12. To install, use new gaskets and reverse the removal procedure. Torque the intake manifold to 17 ft. lbs. (24 Nm).
13. Adjust the engine control cable and the alternator belt tension.
14. Refill the engine with coolant, run the engine and check for leaks. Make all necessary adjustments and road test the vehicle, be sure to check for vacuum leaks around the intake manifold sealing surfaces.

4XC1-T ENGINE

1. Disconnect the negative battery cable.
2. Remove pressure regulator and oil separator.
3. Disconnect vacuum line from VSV and remove vacuum switching valve from bracket.
4. Remove oil separator/VSV bracket and hanger as an assembly.
5. Remove throttle valve assembly and engine harness assembly, mark or tag harness connections, if necessary.
6. Remove Idle Air Control Valve (IACV), Relief Valve and Map sensor.
7. Disconnect vacuum line from Back Pressure transducer and unclip transducer from hold-down bracket.
8. Remove EGR valve and adaptor plate.
9. Remove fuel injectors and fuel pipe connected to rail as one unit and position out of way then remove Intake Manifold with common chamber
10. To install, use new gaskets and reverse the removal procedures. Torque the intake manifold to 17 ft. lbs. (24 Nm). Start the engine and check for vacuum leaks.

4XE1 ENGINE

1. Disconnect negative battery cable and drain cooling system. Remove the air cleaner duct hose from the common chamber.
2. Disconnect and tag the following vacuum lines and wiring harness:
   a. Fast idle vacuum hose from the air duct hose.
   b. The MAP sensor vacuum hose from the common chamber.
   c. Disconnect the TPC valve vacuum hose from the common chamber.
   d. Disconnect the canister vacuum hose from the common chamber.

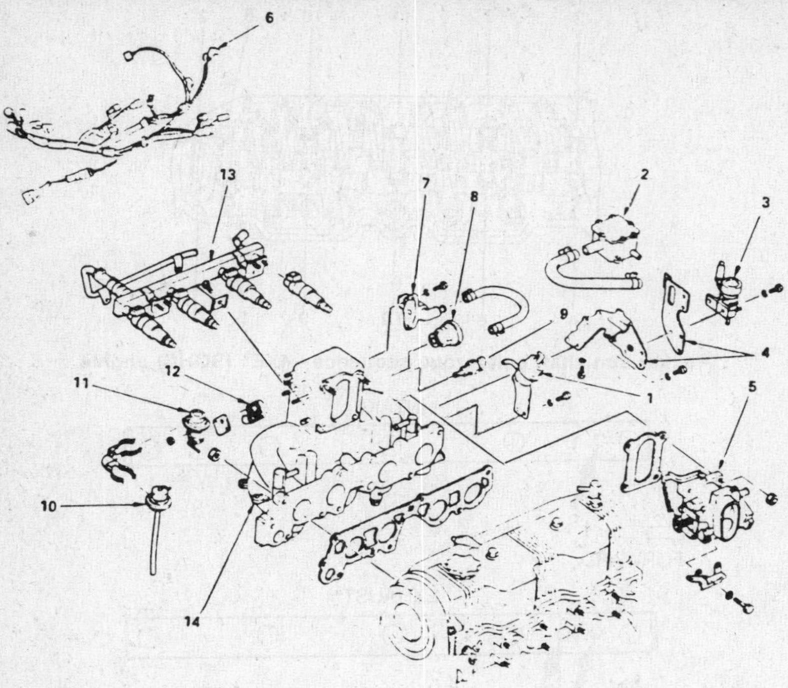

1. Pressure regulator
2. Oil separator
3. VSV
4. Bracket and hanger
5. Throttle valve assembly
6. Engine harness assembly
7. Idle air control valve
8. Relief valve
9. Map sensor
10. Back pressure transducer
11. EGR valve
12. Adaptor
13. Fuel injector with pipe
14. Intake manifold

Intake manifold assembly—I-Mark with 4XC1-T engine

e. Electronic control gas injection harness, disconnect the 2 ECM ground connectors from the bracket located on top of the common chamber. Disconnect the 2 green and black multi-pin connectors on the top of the common chamber.

f. Remove the MAT sensor connector.

3. Disconnect the accelerator cable from the throttle body and at the common chamber. Disconnect the PCV valve hose from the valve cover.
4. Disconnect the master vacuum hose from the master vacuum tube at the common chamber side.
5. Disconnect the induction valve vacuum hose from the common chamber. Disconnect the EGR valve and the throttle valve.
6. Remove the fuel hose clips from the common chamber. Remove the bolt and nuts retaining the common (intake) chamber to the engine. Remove the common (intake) chamber and gasket from the engine.

To install:

7. Clean the gasket mating surfaces and install the manifold.
8. Be sure to use a new gasket and tighten the common chamber retaining bolts and nuts to 17 ft. lbs. (24 Nm).
9. Reconnect all wiring, hoses and cables to the original location.
10. Start the engine and check for leaks.

Impulse and Stylus

EXCEPT 4XE1 (DOHC) ENGINE

1. Disconnect the negative battery cable.
2. Drain the cooling system.
3. Remove the air cleaner.
4. Disconnect the linkage to the throttle valve.
5. Tag and disconnect all wires and hoses.
6. Remove the 8 manifold attaching bolts and remove the manifold and common chamber as an assembly.
7. Installation is the reverse of the removal procedure.
8. Tighten the attaching bolts to 17 ft. lbs. (24 Nm).

4XE1 (DOHC) ENGINE

1. Disconnect the negative battery cable.
2. Disconnect the ground cable terminals from the common chamber, the 2 cable harness connectors from the throttle valve assembly and the MAT sensor harness connector and accelerator cable from the throttle valve assembly and common chamber.
3. Disconnect the air intake duct.
4. Remove the intercooler and back pressure transducer, turbo only.
5. Disconnect the PCV hose, MAP sensor hose, fuel pressure regulator vacuum hose, the canister hoses and the vacuum booster hose.
6. Remove the canister pipes and the MAP sensor pipe with the bracket.
7. Disconnect the throttle valve assembly with the throttle valve heating hoses from the common chamber.
8. Remove the intercooler rear bracket and breather hoses, turbo only.
9. Loosen the EGR pipe and valve retainers and clip.
10. Disconnect the induction control vacuum hose and the alternator harness clip.
11. Disconnect the common chamber bracket and the engine hanger from the common chamber and remove the common chamber.
12. Disconnect the VSV cable harness connector and the oil cooler pipe bracket from the bosses beneath the intake manifold.
13. Remove the EGR transducer and the VSV bracket.
14. Disconnect the fuel lines and remove the fuel injector harness and the fuel rail with the fuel injectors.
15. Remove the intake manifold.

To install:

16. Clean the gasket mating surfaces and install the manifold.
17. Be sure to use a new gasket and tighten the manifold retaining bolts and nuts to 17 ft. lbs. (24 Nm).
18. Reconnect all wiring, hoses and cables to the original location.
19. Start the engine and check for leaks.

Exhaust Manifold

REMOVAL & INSTALLATION

I-Mark

4XC1-U ENGINE

1. Disconnect the negative battery cable.
2. Disconnect the oxygen sensor connector.

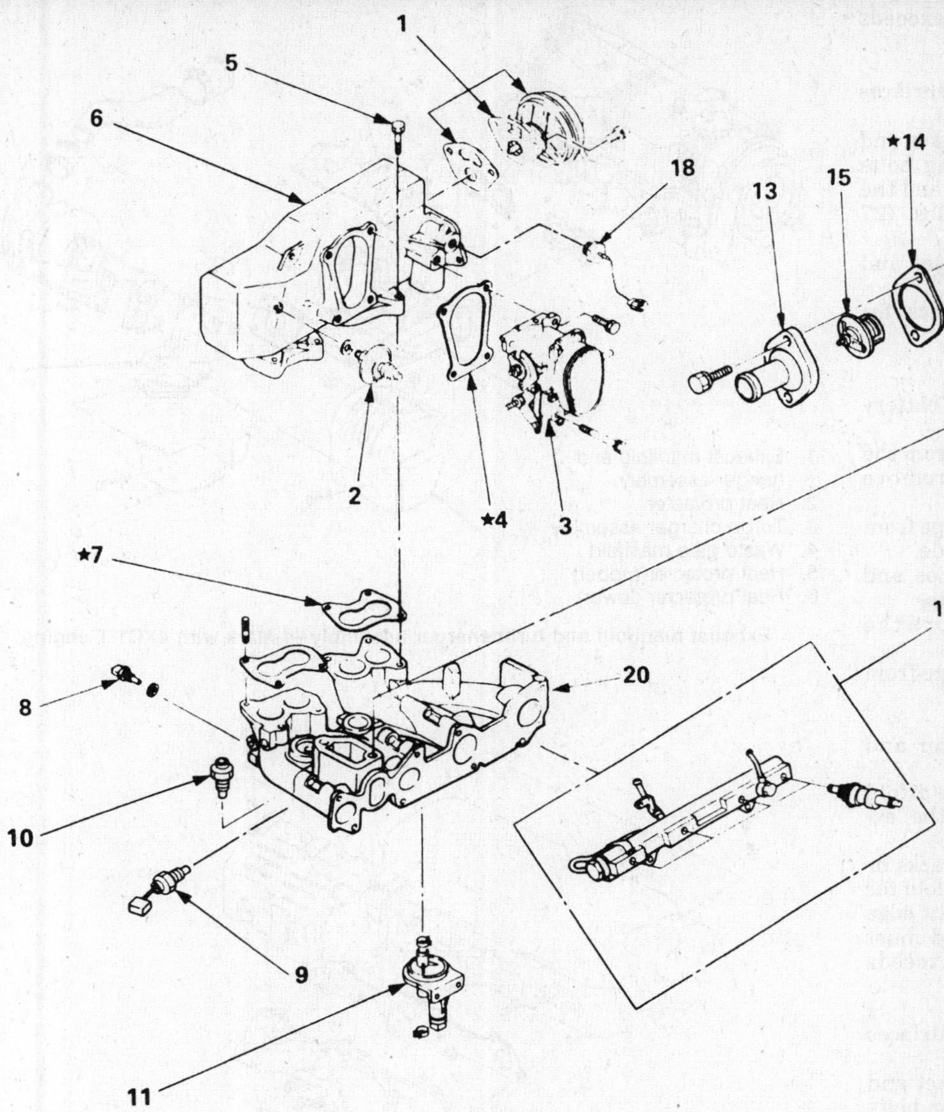

1. EGR valve
2. Dashpot (automatic transmission only)
3. Throttle valve assembly
4. Throttle valve gasket
5. Common bolt
6. Common chamber
7. Common chamber gasket
8. Water temperature sensor
9. Water temperature unit
10. Water temperature unit
11. Air regulator
12. Thermal valve
13. Water outlet pipe
14. Water outlet pipe gasket
15. Thermostat
16. Thermostat housing
17. Thermostat housing gasket
18. EGR temperature sensor (California only)
19. Fuel injector with pipe
20. Intake manifold

Intake manifold and common chamber—4ZD1 and 4ZC1-Turbo engines

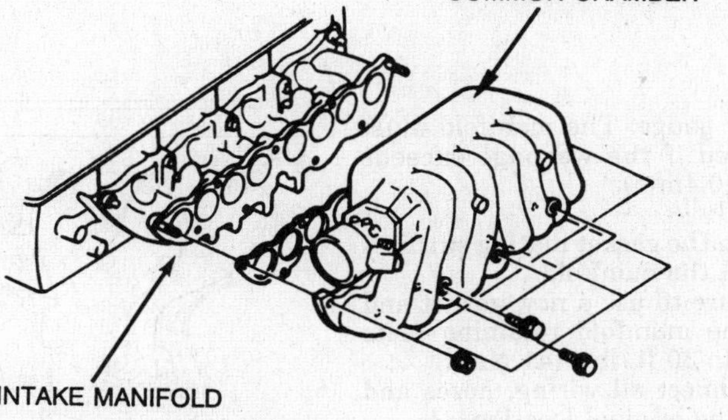

Intake manifold and common chamber—4XE1 engine

3. Remove the air cleaner assembly.
4. Remove the manifold hot air cover.
5. Raise and safely support the vehicle.
6. Disconnect the exhaust pipe from the exhaust manifold.
7. Lower the vehicle.
8. Remove the exhaust manifold mounting nuts and remove the exhaust manifold.
9. Check the manifold for cracks or other damage. Check the manifold for flatness using a suitable straight-edge and feeler gauge. The manifold must

be replaced if the warpage exceeds 0.016 in. (0.4mm).

To install:

10. Clean the gasket mating surfaces and install the manifold.
11. Be sure to use a new gasket and tighten the manifold retaining bolts and nuts to 17 ft. lbs. (24 Nm) and the exhaust pipe bolts to 20 ft. lbs. (27 Nm).
12. Reconnect all wiring, hoses and cables to the original location.
13. Start the engine and check for leaks.

4XC1-T ENGINE

1. Disconnect the negative battery cable.
2. Remove the heat shields from the turbocharger assembly and remove the manifold heat protector.
3. Disconnect the vacuum pipe from the wastegate and position aside.
4. Disconnect the water lines and the oil return and delivery lines.
5. Raise and safely support the vehicle.
6. Disconnect the exhaust pipe from the wastegate manifold.
7. Lower the vehicle.
8. Remove the turbocharger and wastegate as an assembly.
9. Remove the exhaust manifold mounting nuts and remove the exhaust manifold.
10. Check the manifold for cracks or other damage. Check the manifold for flatness using a suitable straight edge and feeler gauge. The manifold must be replaced if the warpage exceeds 0.016 in. (0.4mm).

To install:

11. Clean the gasket mating surfaces and install the manifold.
12. Be sure to use a new gasket and tighten the manifold retaining bolts and nuts to 17 ft. lbs. (24 Nm). Torque the exhaust pipe nuts to 25 ft. lbs. (34 Nm).
13. Reconnect all wiring, hoses and cables to the original location.
14. Start the engine and check for leaks.

4XE1 ENGINE

1. Disconnect the negative battery cable.
2. Raise and safely support the vehicle.
3. Disconnect the oxygen sensor connector.
4. Remove the pulse air bracket with the pipe and remove the heat protector and the EGR pipe.
5. Disconnect the exhaust pipe from the exhaust manifold.
6. Remove the mounting nuts and the exhaust manifold.
7. Check the manifold for cracks or other damage. Check the manifold for flatness using a suitable straightedge and feeler gauge. The manifold must be replaced if the warpage exceeds 0.016 in. (0.4mm).

To install:

8. Clean the gasket mating surfaces and install the manifold.
9. Be sure to use a new gasket and tighten the manifold retaining bolts and nuts to 30 ft. lbs. (41 Nm).
10. Reconnect all wiring, hoses and cables to the original location.
11. Start the engine and check for leaks.

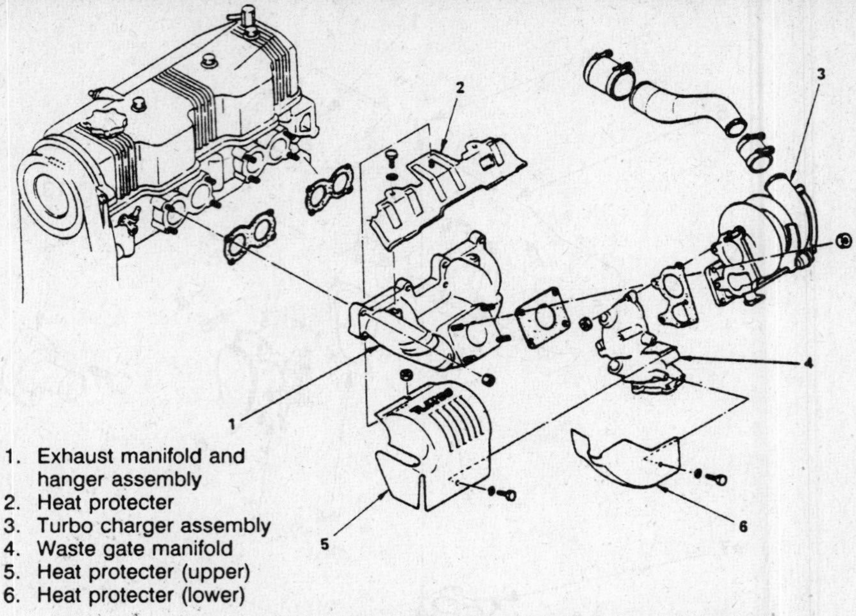

1. Exhaust manifold and hanger assembly
2. Heat protecter
3. Turbo charger assembly
4. Waste gate manifold
5. Heat protecter (upper)
6. Heat protecter (lower)

Exhaust manifold and turbocharger assembly – I-Mark with 4XC1-T engine

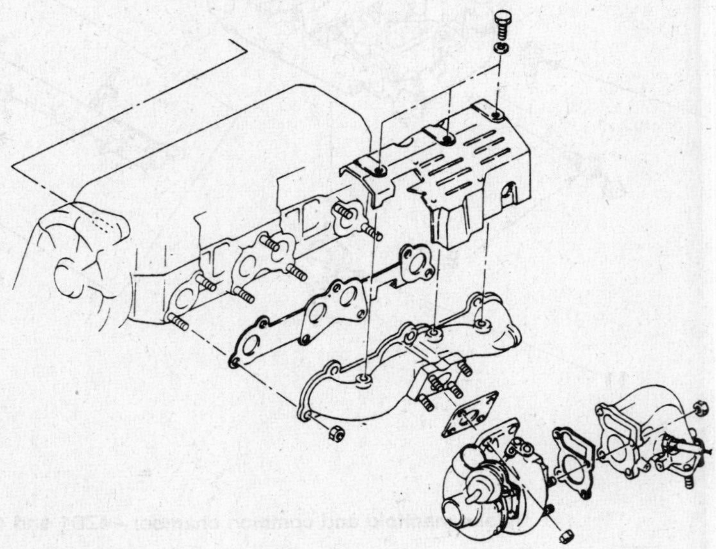

Exhaust manifold and turbocharger assembly – Impulse with 4ZC1-T

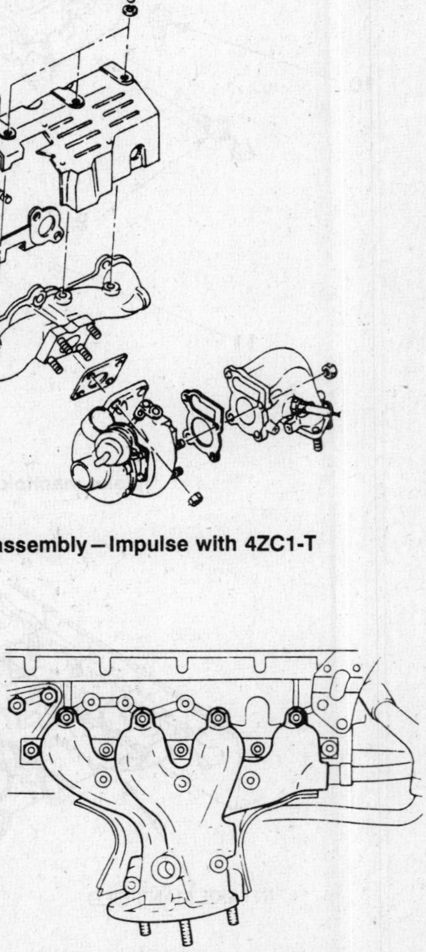

Exhaust manifold – 4XE1 engine

Impulse and Stylus

4ZD1 ENGINE

1. Disconnect the negative battery cable.
2. Disconnect the oxygen sensor connector and EGR pipe.
3. Remove the heat protector.
4. Raise and safely support the vehicle.
5. Disconnect the exhaust pipe from the exhaust manifold.
6. Lower the vehicle.
7. Remove the exhaust manifold nuts and the exhaust manifold.
8. Check the manifold for cracks or other damage. Check the manifold for flatness using a suitable straightedge and feeler gauge. The manifold must be replaced if the warpage exceeds 0.016 in. (0.4mm).

To install:

9. Clean the gasket mating surfaces and install the manifold.
10. Be sure to use a new gasket and tighten the manifold retaining bolts and nuts to 17 ft. lbs. (24 Nm).
11. Reconnect all wiring, hoses and cables to the original location.
12. Start the engine and check for leaks.

4ZC1-T ENGINE

1. Disconnect the negative battery cable.
2. Remove the heat protector.
3. Disconnect the hoses from the air cleaner and intercooler at the turbocharger.
4. Disconnect the oxygen sensor connector, the control cable and the vacuum hose from the turbocharger.
5. Disconnect the oil delivery pipe.
6. Raise and safely support the vehicle.
7. Disconnect the oil return hose, the water lines and the exhaust pipe from the turbocharger.
8. Lower the vehicle.
9. Remove the turbocharger from the exhaust manifold.
10. Remove the mounting nuts and the exhaust manifold.
11. Check the manifold for cracks or other damage. Check the manifold for flatness using a suitable straightedge and feeler gauge. The manifold must be replaced if the warpage exceeds 0.016 in. (0.4mm).

To install:

12. Clean the gasket mating surfaces and install the manifold.
13. Be sure to use a new gasket and tighten the manifold retaining bolts and nuts to 17 ft. lbs. (24 Nm).
14. Reconnect all wiring, hoses and cables to the original location.
15. Start the engine and check for leaks.

4XE1 ENGINE

1. Disconnect the negative battery cable.
2. Disconnect the oxygen sensor connector and the EGR pipe.
3. Remove the heat protector.
4. Raise and safely support the vehicle.
5. Disconnect the exhaust pipe from the exhaust manifold.
6. Lower the vehicle.
7. Remove the exhaust manifold nuts and the exhaust manifold.

To install:

8. Check the manifold for cracks or other damage. Check the manifold for flatness using a suitable straight-edge and feeler gauge. The manifold must be replaced if the warpage exceeds 0.016 in. (0.4mm).
9. Installation is the reverse of the removal procedure. Make sure all gasket mating surfaces are clean prior to installation.
10. Tighten the manifold mounting nuts to 30 ft. lbs. (39 Nm). 17 ft. lbs. (23 Nm) for the SOHC engine.

4XE1-T ENGINE

1. Disconnect the negative battery cable.
2. Drain the engine coolant.
3. Remove the intercooler, left and right undercovers, front exhaust pipe from the manifold, intake ducts from turbocharger and remove the power steering pump and move out of the way. Do not disconnect the lines.
4. Remove the bracket from the

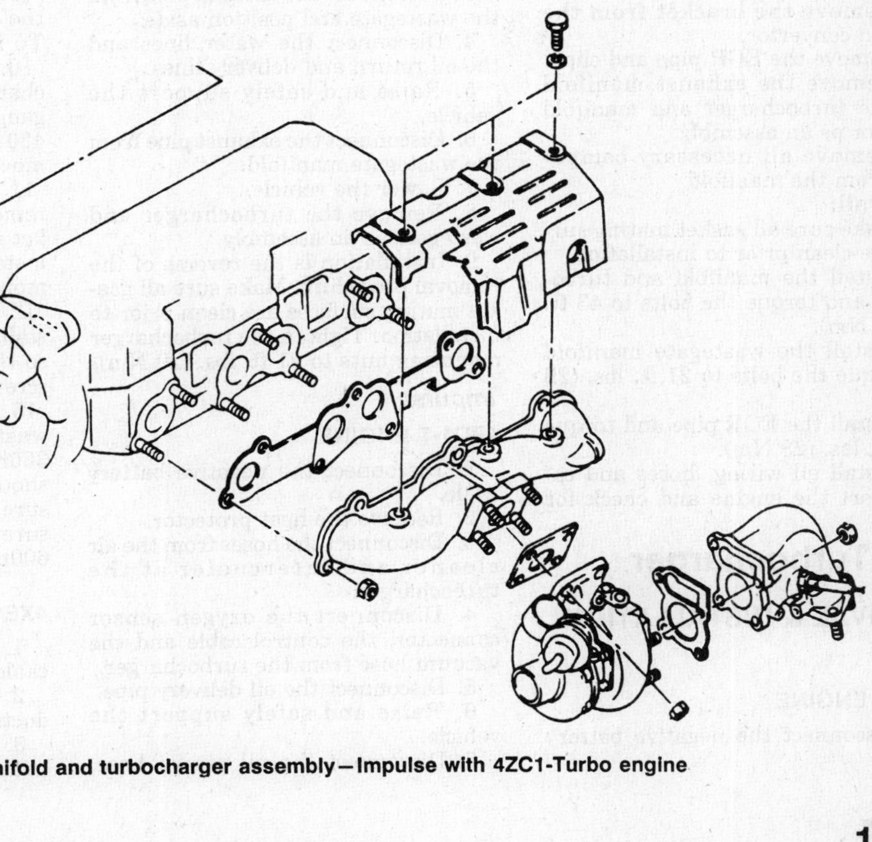

Exhaust manifold and turbocharger assembly—Impulse with 4ZC1-Turbo engine

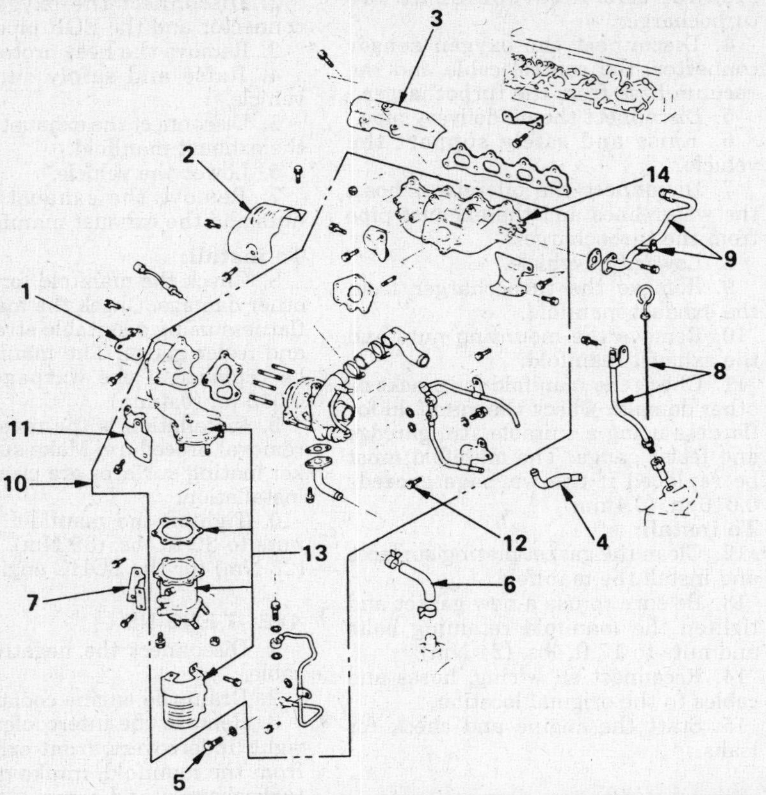

1. Bracket from wastegate manifold
2. Heat protector from turbocharger
3. Upper heat protector from exhaust manifold
4. Turbocharger coolant hose from coolant pipe
5. Turbocharger oil feed pipe joint bolt from cylinder block
6. Turbocharger oil return hose from turbocharger
7. Bracket from manifold convertor
8. Oil level gauge and guide tube from cylinder block
9. EGR pipe and clip
10. Exhaust manifold with turbocharger and manifold convertor
11. Heat protector from wastegate manifold
12. Coolant pipe bracket fixing bolts from exhaust manifold
13. Wastegate manifold with manifold convertor from turbocharger
14. Exhaust manifold

Exhaust manifold and turbocharger assembly—4XE1-Turbo engine

wastegate and heat protector from the turbocharger.

5. Disconnect the turbo coolant and oil lines.

6. Remove the bracket from the manifold convertor.

7. Remove the EGR pipe and clip.

8. Remove the exhaust manifold with the turbocharger and manifold convertor as an assembly.

9. Remove all necessary components from the manifold.

To install:

10. Make sure all gasket mating surfaces are clean prior to installation.

11. Install the manifold and turbocharger and torque the bolts to 43 ft. lbs. (59 Nm).

12. Install the wastegate manifold and torque the bolts to 21 ft. lbs. (28 Nm).

13. Install the EGR pipe and torque to 21 ft. lbs. (28 Nm).

14. Install all wiring, hoses and cables. Start the engine and check for leaks.

Turbocharger

REMOVAL & INSTALLATION

I-Mark

4XC1-T ENGINE

1. Disconnect the negative battery cable.

2. Remove the heat shields from the turbocharger assembly and remove the manifold heat protector.

3. Disconnect the vacuum pipe from the wastegate and position aside.

4. Disconnect the water lines and the oil return and delivery lines.

5. Raise and safely support the vehicle.

6. Disconnect the exhaust pipe from the wastegate manifold.

7. Lower the vehicle.

8. Remove the turbocharger and wastegate as an assembly.

9. Installation is the reverse of the removal procedure. Make sure all gasket mating surfaces are clean prior to installation. Tighten the turbocharger mounting nuts to 21 ft. lbs. (29 Nm).

Impulse

4ZC1-T ENGINE

1. Disconnect the negative battery cable.

2. Remove the heat protector.

3. Disconnect the hoses from the air cleaner and intercooler at the turbocharger.

4. Disconnect the oxygen sensor connector, the control cable and the vacuum hose from the turbocharger.

5. Disconnect the oil delivery pipe.

6. Raise and safely support the vehicle.

7. Disconnect the oil return hose, the water lines and the exhaust pipe from the turbocharger.

8. Lower the vehicle.

9. Remove the turbocharger from the exhaust manifold.

To install:

10. Before reinstalling the turbocharger, connect a suitable pressure gauge and make sure the pressure is 430–466mm Hg with the control rod moved 2mm.

11. Installation is the reverse of the removal procedure. Make sure all gasket mating surfaces are clean prior to installation. Tighten the turbocharger mounting nuts to 21 ft. lbs. (29 Nm).

12. Adjust the control cable after installation by turning the adjust nuts so the clearance between the stopper screw and lever is 0.02 in. (0.5mm).

13. Install the pressure gauge to the wastegate and apply approximately 360mm Hg to the wastegate. The rod should begin to move when this pressure is applied. Do not allow the pressure applied to the wastegate to exceed 600mm Hg.

4XE1 ENGINE

1. Disconnect the negative battery cable.

2. Remove the air cleaner and air duct.

3. Disconnect the intake and exhaust hoses from the turbocharger.

4. Disconnect the oil lines from the turbocharger.

5. Remove the turbocharger-to-exhaust manifold nuts, the turbocharger assembly-to-exhaust pipe nuts and the turbocharger assembly.

To install:

6. Clean the gasket mounting surfaces. Refill the turbocharger with clean engine oil.

7. Using a new gasket, install the turbocharger-to-exhaust manifold nuts to 16–23 ft. lbs. (21–33 Nm) and the turbocharger assembly-to-exhaust pipe nuts to 16–23 ft. lbs. (21–33 Nm).

Timing Belt Front Cover

REMOVAL & INSTALLATION

4XC1-U and 4XC1-T Engines

1. Disconnect the negative battery cable.

2. Place a wooden block on a suitable floor jack and position the jack under the oil pan. Raise the engine slightly.

3. Remove the rear side torque rod.

4. Remove the right side engine mount, then remove the body side bracket and engine side bracket.

5. Remove the necessary accessory drive belts.

6. Remove the 4 crank pulley bolts and the crank pulley.

7. Remove the 6 timing belt cover bolts and the timing belt cover.

To install:

8. Tighten the timing belt cover bolts and the crank pulley bolts to 7 ft. lbs. (10 Nm).

9. Tighten the body side engine mount bolts to 29 ft. lbs. (41 Nm) and the engine side engine mount bolts to 44 ft. lbs. (62 Nm). Tighten the torque rod bolt to 42 ft. lbs. (58 Nm).

4ZD1 and 4ZC1-T Engines

1. Disconnect the negative battery cable.

2. Drain the cooling system.

3. Disconnect the radiator hoses and remove the radiator.

4. Remove the necessary accessory drive belts and remove the engine cooling fan and pulley.

5. Remove the crank pulley bolt and the crank pulley.

6. Remove the timing belt cover.

7. Installation is the reverse of the removal procedure. Tighten the timing cover bolts to 5 ft. lbs. (8 Nm) and the crank pulley bolt to 86 ft. lbs. (120 Nm).

4XE1 Engines

1. Disconnect the negative battery cable.

2. On 1989 I-Mark, disconnect the clip securing the high pressure air conditioning line to the strut tower.

3. Use a suitable device to support the engine and remove the right side engine mount.

4. Remove the necessary accessory drive belts.

5. Remove the right side engine mounting bridge bracket and the torque rod.

6. Remove the crank pulley bolt and the crank pulley.

7. On 1990–92 Impulse and 1991-92 Stylus, remove the front exhaust pipe, the stud from the transaxle housing, the stiffener and the flywheel dust cover.

8. Remove the upper and lower timing belt covers.

To install:

9. Install the front cover and gaskets.

10. Tighten the timing belt cover bolts to 89 inch lbs. (10 Nm). Tighten the crank pulley bolt to 123 ft. lbs. (170 Nm) on 1989 I-Mark or to 109 ft. lbs. (147 Nm) on 1990–92 Impulse and 1991-92 Stylus.

OIL SEAL REPLACEMENT

1. Disconnect the negative battery cable.

2. Remove the timing belt cover and the timing belt.

3. Use a suitable tool to remove the crankshaft timing sprocket.

4. Remove the oil seal using a suitable removal tool.

5. Lubricate the lip of a new oil seal and install it using a suitable installation tool.

6. Install the remaining components in the reverse order of their removal.

Timing Belt and Tensioner

ADJUSTMENT

4XC1-U, 4XC1-T and 4XE1 (SOHC) Engines

1. Disconnect the negative battery cable.

2. Remove the timing belt cover.

3. Loosen the tension pulley bolt.

4. Insert an Allen wrench into the tension pulley hexagonal hole. Hold the pulley stationary and temporarily tighten the bolt.

5. Turn the crankshaft 2 complete revolutions in the reverse direction of normal rotation (counterclockwise) and align the crankshaft timing sprocket groove with the mark on the oil pump.

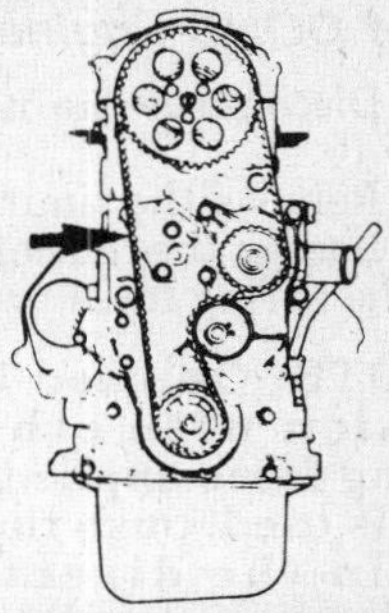

Timing belt tension measuring position—4XC1-U and 4XC1-T engines

6. Loosen the tension pulley bolt and apply tension to the belt.

7. Insert the Allen wrench into the tension pulley hexagonal hole. Hold the pulley stationary and tighten the bolt to 37 ft. lbs. (51 Nm).

8. Move the crankshaft back to about 50 degrees BTDC and then turn the crankshaft 2 complete revolutions in the reverse direction of normal rotation and align the crank timing sprocket groove with the mark on the oil pump.

9. Use a suitable belt tension gauge to check the timing belt tension at the required measuring position. The tension should be 39.6–48.4 lbs. (18–22 kg).

4ZD1 and 4ZC1-T Engines

1. Disconnect the negative battery cable.

2. Remove the timing belt cover.

3. Loosen bolt (B) to allow the tension spring to tighten the belt.

4. Temporarily tighten bolt (B).

5. Temporarily attach the crank pulley.

6. Turn the crankshaft 2 revolutions in the opposite direction of normal rotation to bring the oil seal retainer mark in line with the crank pulley mark.

7. Loosen bolt (B) and tighten the belt with the tension pulley.

8. Tighten bolt (B) to 13 ft. lbs. (19 Nm).

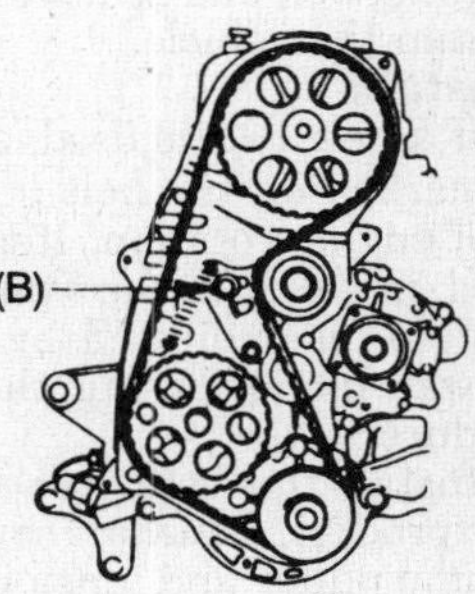

Timing belt tensioner bolt location—4ZD1 and 4ZC1-T engines

4XE1 DOHC Engine

1. Disconnect the negative battery cable.
2. Remove the timing belt covers.
3. Loosen the tensioner bolt and apply the spring force to the belt.

NOTE: On used belts, do not tension with other than the spring force applied. When a new belt is used, push the tension pulley in the direction of belt tension.

4. Tighten the tensioner bolt to 31 ft. lbs. (42 Nm).
5. Turn the crankshaft 2 complete revolutions and align the crankshaft and camshaft sprocket timing marks correctly.
6. Turn the crankshaft 60 degrees and measure the deflection of the belt. Deflection is measured with a down force of 22 lbs. (10 kg) applied to the timing belt at a point between the camshaft sprockets. The deflection should be 0.28–0.33 in. (7–8.5mm) for a new belt and 0.35–0.41 in. (9–10.5mm) for a used belt.

REMOVAL & INSTALLATION

4XC1-U, 4XC1-T and 4XE1 (SOHC) Engines

1. Disconnect the negative battery cable.
2. Remove the timing belt cover.
3. Bring the piston in No. 4 cylinder to TDC on the compression stroke. The crankshaft sprocket timing mark should be aligned with the triangular mark on the oil pump housing and the notch on the camshaft sprocket should be aligned with the left upper corner of the cylinder head and the dowel pin in the up position.
4. Remove the crank pulley bolt and the crank pulley, being careful not to disturb the position of the crankshaft.
5. Loosen the bolts retaining the tension pulley. Use a suitable Allen wrench to turn the tension pulley clockwise and relieve the tension on the timing belt.
6. Mark the direction of rotation on the timing belt and remove the timing belt from the vehicle.

To install:

7. If a new belt is used, set the letters marked on the belt in the direction of engine rotation. If the old belt is used, install it in the same direction as before, as indicated by the mark that was made during the removal procedure.
8. Install the belt over the crankshaft sprocket, camshaft sprocket, water pump pulley and tension pulley, in that order.

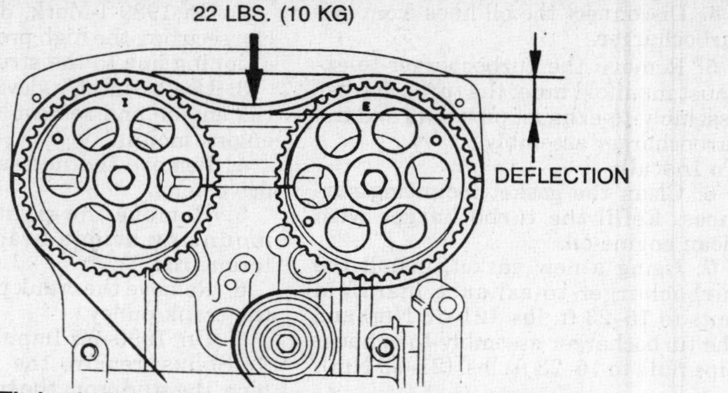

Timing belt deflection measuring position—4XE1 engine

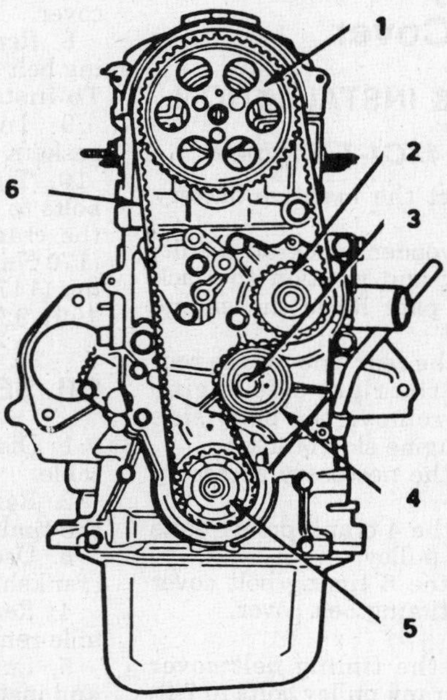

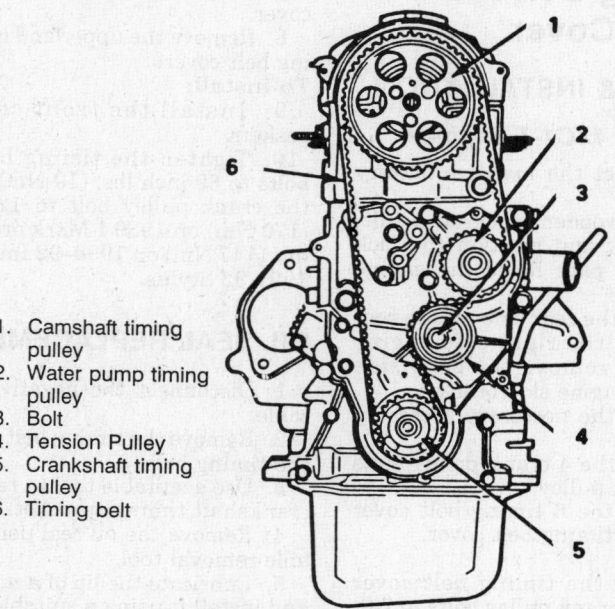

1. Camshaft timing pulley
2. Water pump timing pulley
3. Bolt
4. Tension Pulley
5. Crankshaft timing pulley
6. Timing belt

Timing belt installed—4XC1-U and 4XC1-T engines

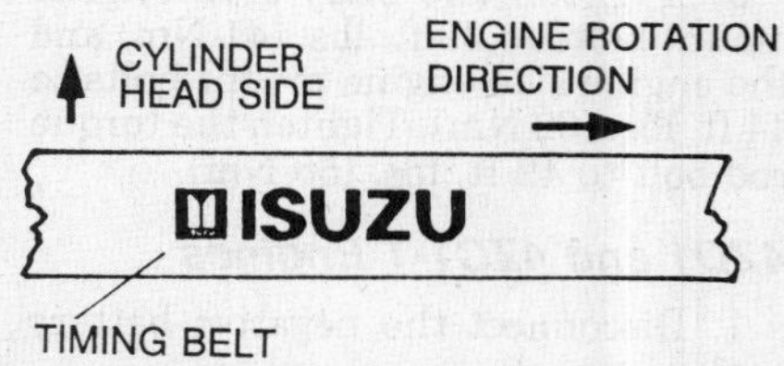

New timing belt positioning—4XC1-U and 4XC1-T engines

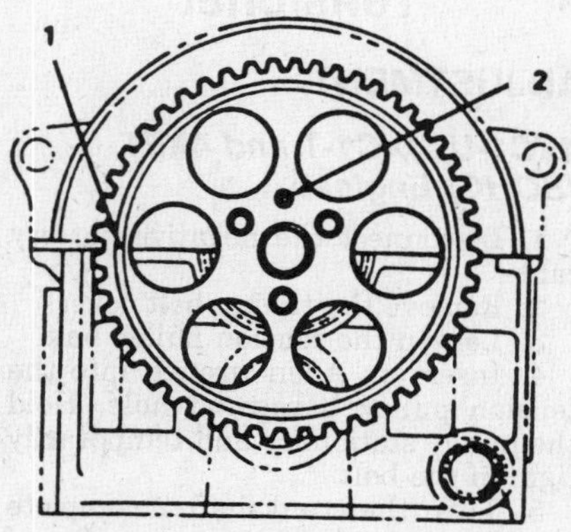

1 ALIGNMENT MARKS 2 DOWEL

Camshaft sprocket alignment—4XC1-U and 4XC1-T engines

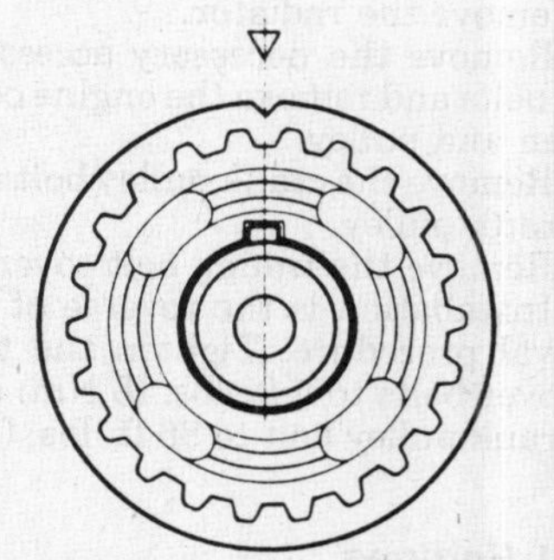

Crankshaft sprocket alignment—4XC1-U and 4XC1-T engines

NOTE: There must be no slack in the belt after it has been installed. The teeth of the belt and the teeth of the pulley must be in perfect alignment.

9. Properly tension the timing belt.
10. Install the crankshaft pulley hub with the taper face to the belt. Tighten the crank pulley bolt to 108 ft. lbs. (150 Nm).
11. Install the remainder of the components in the reverse order of their removal.

4ZD1 and 4ZC1-T Engines

1. Disconnect the negative battery cable.
2. Remove the timing belt cover.
3. Bring the piston in No. 4 cylinder to TDC on the compression stroke. The mark on the crankshaft timing sprocket should be aligned with the mark on the front seal retainer and the mark on the camshaft sprocket should be aligned with the mark on the front plate.
4. Remove the tension spring, loosen bolt (B) and draw the tension pulley fully to the water pump side. Tighten bolt (B).
5. Mark the rotational direction of the timing belt and remove the timing belt from the vehicle.

To install:

6. Make sure the timing marks are still aligned.
7. Install the timing belt, laying it over the crank sprocket, oil pump pulley, camshaft sprocket and tension pulley, in that order. If the old belt is being used, make sure it is installed in the rotational direction that was marked during the removal procedure.
8. Properly tension the timing belt.
9. Install the remainder of the components in the reverse order of their removal.

4XE1 (DOHC) Engine

1. Disconnect the negative battery cable.
2. Remove the timing belt cover.
3. Rotate the crankshaft to align the timing marks. The mark on the crankshaft timing sprocket should be aligned with the triangular mark on the oil pump and the keyway should be at the top of the crankshaft, towards the cylinder head. The marks on the camshaft sprockets should be directly across from one another, aligned with the top edge of the cylinder head.
4. Loosen the tension pulley attaching bolt ½ turn. Insert a suitable hex wrench into the tension pulley hexagonal hole and loosen the timing belt by rotating the tension pulley.
5. Mark the rotational direction of the timing belt and remove it from the vehicle.

To install:

6. Make sure the timing marks are still in alignment.
7. Lock the camshaft sprockets in position by inserting 6mm bolts through the camshaft sprockets and into the cylinder heads.
8. Install the timing belt. A new belt is installed correctly if the lettering can be read while viewing it from the passenger side fender. If the old belt is being used, it must be installed in the same direction as was marked during the removal procedure. The belt must be installed in the following order:
   a. Crankshaft timing sprocket.
   b. Water pump pulley.
   c. Idler pulley.
   d. Exhaust camshaft sprocket.
   e. Intake camshaft sprocket.
   f. Tensioner pulley.

NOTE: There must be no slack in the belt after it has been installed. The teeth of the belt and the teeth of the sprocket must be in perfect alignment.

9. Properly tension the belt.
10. Install the remainder of the components in the reverse order of their removal.

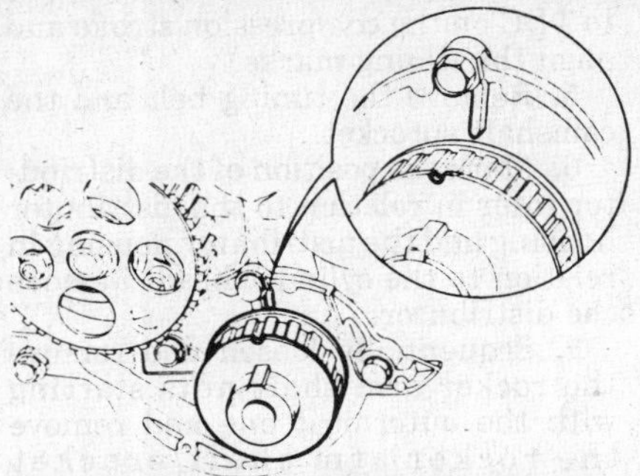

Aligning the crankshaft sprocket timing mark – 4ZD1 and 4ZC1-T engines

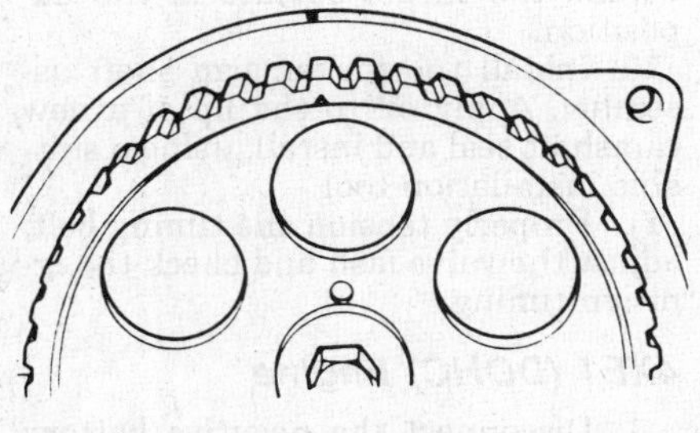

Aligning the camshaft sprocket timing marks – 4ZD1 and 4ZC1-T engines

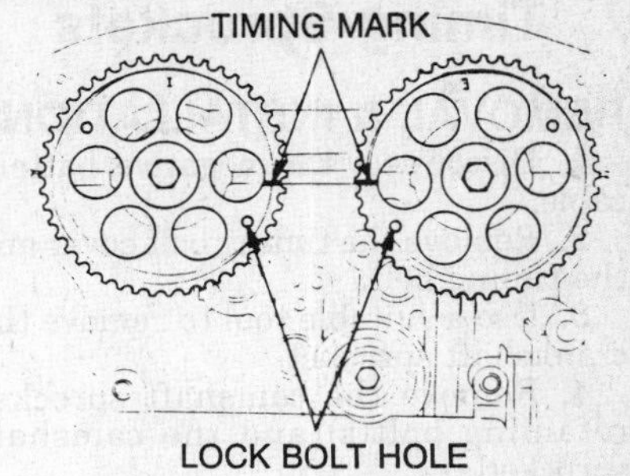

Aligning the camshaft sprocket timing marks – 4XE1 engine

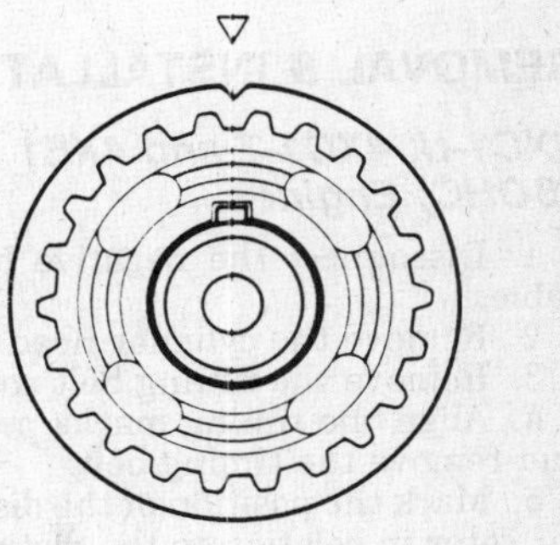

Aligning the crankshaft sprocket timing marks – 4XE1 engine

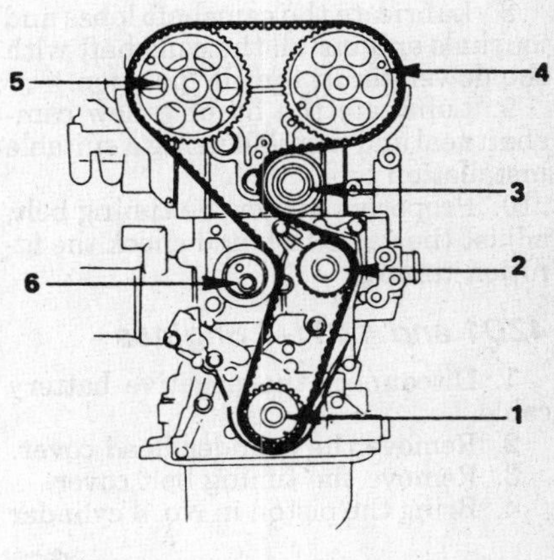

1. Crankshaft timing sprocket
2. Water pump pulley
3. Idler pulley
4. Exhaust camshaft sprocket
5. Intake camshaft sprocket
6. Tensioner pulley.

Timing belt installed – 4XE1 engine

Timing Sprockets

REMOVAL & INSTALLATION

1. Disconnect the negative battery cable.
2. Remove the timing belt cover and the timing belt.
3. Use a suitable tool to remove the crankshaft sprocket.
4. Remove the camshaft sprocket retaining bolt(s) and the camshaft sprocket(s).
5. Installation is the reverse of the removal procedure. Tighten the camshaft sprocket retaining bolt(s) to 7 ft. lbs. (10 Nm) on 4XC1-U and 4XC1-T engines and 43 ft. lbs. (59 Nm) on 4ZD1, 4ZC1-T and 4XE1 engines.

Camshaft

REMOVAL & INSTALLATION

4XC1-U, 4XC1-T and 4XE1 (SOHC) Engines

1. Disconnect the negative battery cable.
2. Remove the cylinder head cover.
3. Remove the timing belt cover.
4. Align the timing marks properly and remove the timing belt.
5. Mark the position of the distributor rotor in relation to the distributor housing and the distributor housing in relation to the cylinder head. Remove the distributor.
6. Remove the camshaft sprocket.
7. Remove the rocker arm shafts and remove the camshaft and seal.

To install:

8. Lubricate the camshaft lobes and journals and install the camshaft with the dowel pin in the UP position.
9. Lubricate the lip of a new camshaft seal and install it using a suitable installation tool.
10. Properly tension the timing belt, adjust the valve lash and check the ignition timing.

4ZD1 and 4ZC1-T Engines

1. Disconnect the negative battery cable.
2. Remove the cylinder head cover.
3. Remove the timing belt cover.
4. Bring the piston in No. 4 cylinder to TDC on the compression stroke and align the timing marks.
5. Remove the timing belt and the camshaft sprocket.
6. Mark the position of the distributor rotor in relation to the distributor housing and the distributor housing in relation to the cylinder head. Remove the distributor.
7. Sequentially loosen and remove the rocker arm shaft nuts starting with the outermost one and remove the rocker arm shaft/bracket assembly.
8. Remove the camshaft.

To install:

9. Lubricate the lobes and journals of the camshaft before installing. Install the camshaft with the mark just behind the thrust surface in the UP position.
10. Install the rocker arm shaft assembly. Apply oil to the lip of a new camshaft seal and install, using a suitable installation tool.
11. Properly tension the timing belt, adjust the valve lash and check the ignition timing.

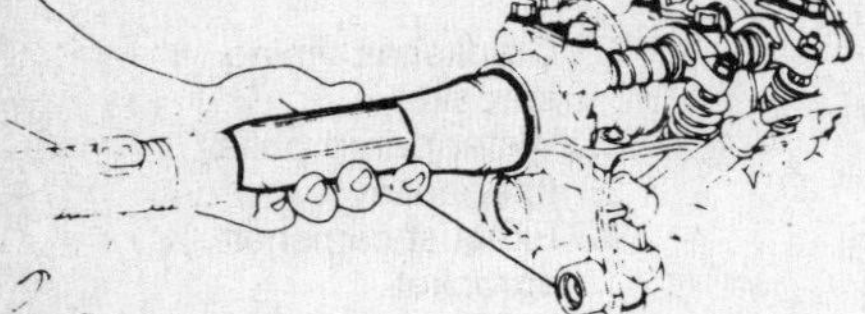

Installing the camshaft oil seal—4XC1-U and 4XC1-T engines

4XE1 (DOHC) Engine

1. Disconnect the negative battery cable.
2. Remove the cylinder head cover.
3. Bring the piston in the No. 1 cylinder to TDC on the compression stroke.
4. Remove the timing belt cover.
5. Loosen the camshaft sprocket bolts and loosen the timing belt tension.
6. Remove the camshaft sprockets.
7. Mark the position of the distributor rotor in relation to the distributor housing and the distributor housing in relation to the cylinder head. Remove the distributor.
8. Remove the camshaft bearing caps working from the outside caps toward the center of the cylinder head.
9. Remove the camshafts and camshaft oil seals.

To install:

10. Lubricate the lobes and journals of the camshafts thoroughly before installation. Install the camshafts with the dowel pins in the UP position.
11. Remove any oil from the contact surfaces of the No. 1 and No. 5 bearing caps and the cylinder head. Apply sealant to the contact surfaces. Do not get sealant on the bearing surface.
12. Install the bearing caps and torque in sequence to 89 inch lbs. (10 Nm). Adjust the valve clearance using the appropriate shims. Valve adjustment is not required on the 1989 I-Mark with the 4XE1 engine.
13. Lubricate the sealing lip of the new camshaft seals and install them using a suitable installation tool.
14. Properly tension the timing belt and check the ignition timing.

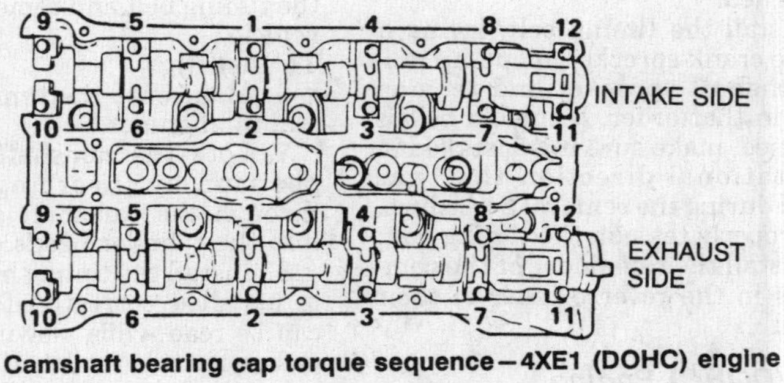

Camshaft bearing cap torque sequence—4XE1 (DOHC) engine

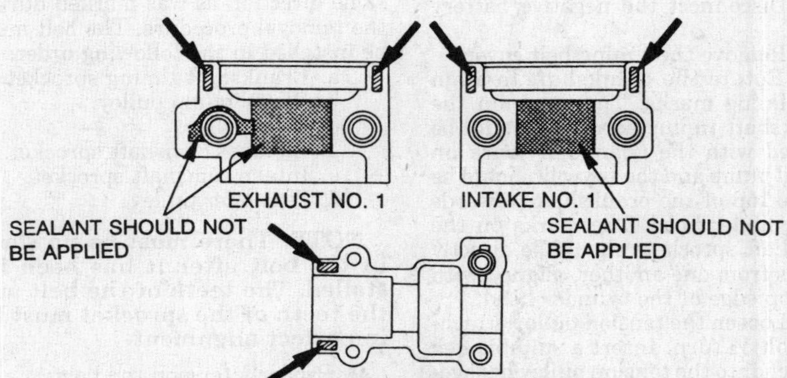

Sealant application points—4XE1 engine

Piston and Connecting Rod

Positioning

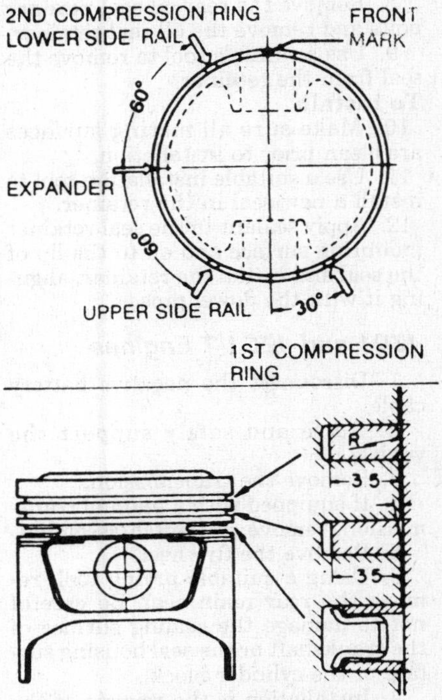

Piston ring positioning—all engines

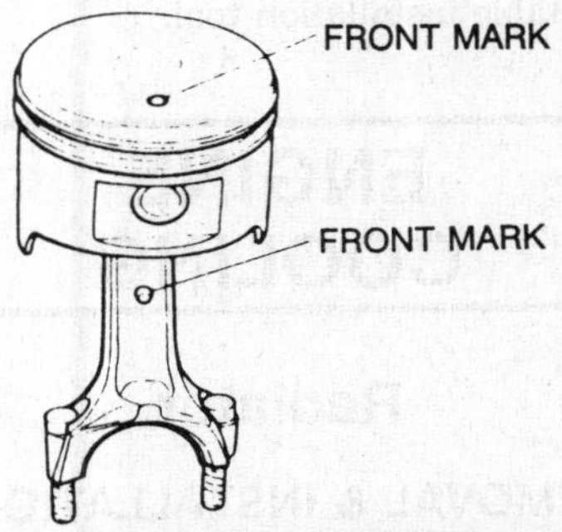

Piston and connecting rod alignment: marks to front of engine—4XC1-U and 4XC1-T engines

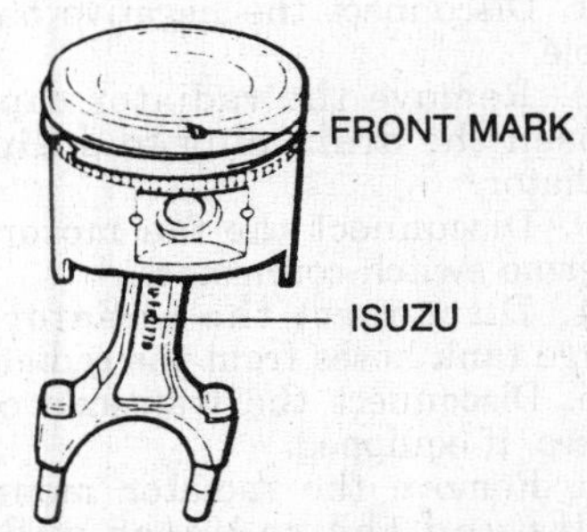

Piston and connecting rod alignment marks to front of engine—G200Z, 4ZD1 and 4ZC1-T engines

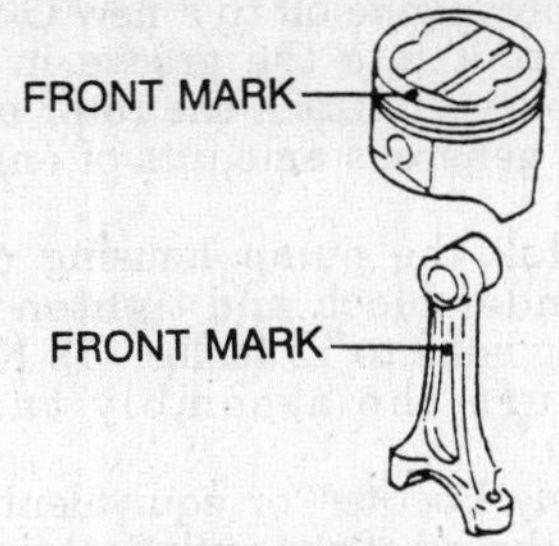

Piston and connecting rod alignment marks to front of engine—4XE1 engine

ENGINE LUBRICATION

Oil Pan

REMOVAL & INSTALLATION

1. Disconnect the negative battery cable.
2. Raise and safely support the vehicle.
3. Drain the engine oil and replace the drain plug.
4. Remove the torque rod and stiffener on 1990–92 Impulse.
5. Remove the engine splash shield.
6. Remove the exhaust header pipe, if necessary.
7. Remove the flywheel dust shield.
8. Remove the oil pan retaining bolts.
9. Raise the engine, if necessary, to allow sufficient room to remove the oil pan and remove the pan from the vehicle.

To install:

10. Make sure all mating surfaces are clean before installation.
11. On all except the 4XE1 engine, apply sealant to the indicated contact surfaces before installing the oil pan gasket and oil pan. On 4XE1 engines, apply a 0.18 in. (4.5mm) width bead of sealant to the contact surfaces of the oil pan. The oil pan must be installed within 30 minutes after sealant application.
12. Tighten the oil pan bolts to 7.2 ft. lbs. (10 Nm) on 4XC1-U, 4XC1-T and 4XE1 engines, 13 ft. lbs. (18 Nm) on 4ZD1 and 4ZC1-T engines.

Oil Pump

REMOVAL & INSTALLATION

4XC1-U, 4XC1-T and 4XE1 (SOHC) Engines

1. Disconnect the negative battery cable, raise and safely support the vehicle and drain the engine oil.
2. Place a wooden block on a suitable floor jack and place the jack under the oil pan. Slightly raise the engine.
3. Remove the rear side torque rod while the engine is slightly lifted.
4. Remove the right side engine mount, then remove the body side bracket and the engine side bracket.
5. Remove the necessary accessory drive belts.
6. Remove the 4 crank pulley bolts with the engine slightly lifted.
7. Remove the 6 timing cover bolts and remove the timing belt cover.
8. Make sure the crankshaft timing mark on the crankshaft pulley hub is aligned with the top dead center mark and the notch on the camshaft pulley

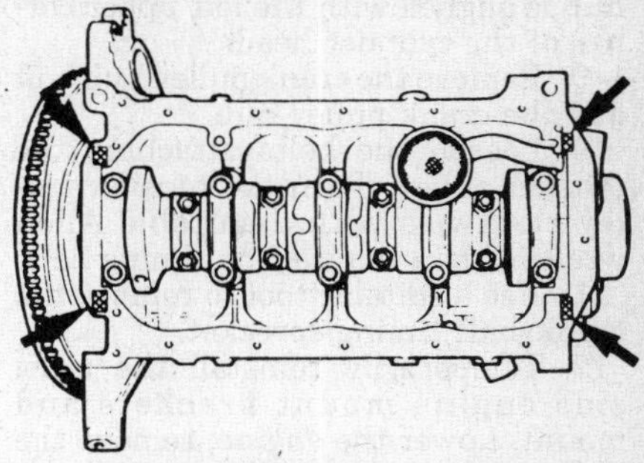

Sealant application points—4XC1-U and 4XC1-T engines

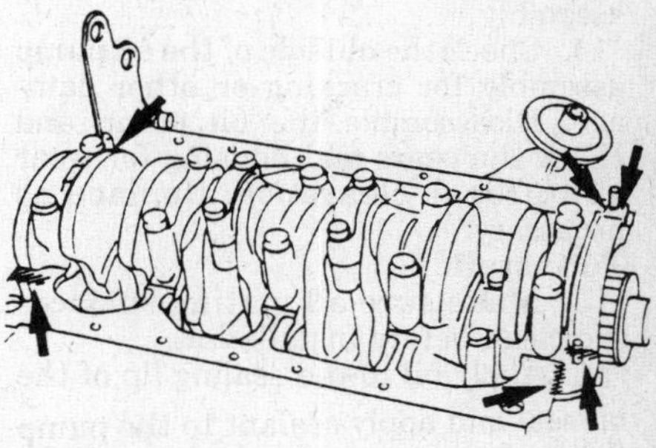

Sealant application points—4ZD1 and 4ZC1-T engines

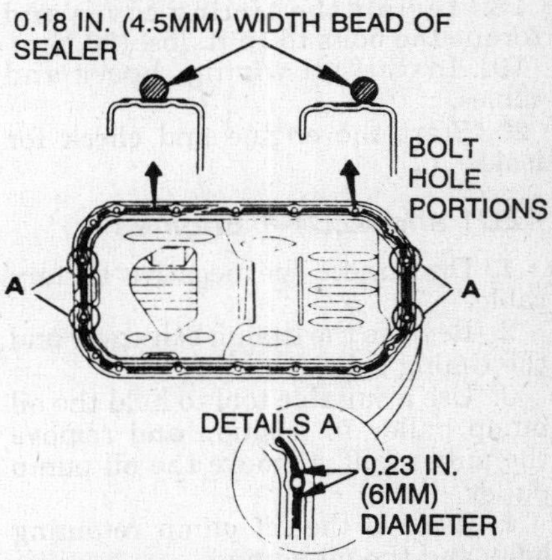

Oil pan sealer application—4XE1 engine

hub is aligned with the left upper corner of the cylinder head.

9. Remove the crank pulley hub bolt and the crank pulley hub.

10. Loosen the bolts attaching the tension pulley. Turn the tension pulley clockwise with a suitable Allen wrench, then remove the timing belt.

11. Use a suitable tool to remove the crankshaft timing sprocket.

12. Temporarily reinstall the right side engine mount brackets and mount. Lower the engine, remove the jack and remove the oil pan.

13. Remove the oil pump retaining bolts and remove the oil pump assembly.

14. Check the outside of the oil pump assembly for cracking or other damage. Disassemble the oil pump and check the gears and housing for wear and proper clearance. Replace as necessary.

To install:

15. Make sure all mating surfaces are clean before installation.

16. Apply oil to the sealing lip of the oil seal and apply sealant to the pump fitting face.

17. Install the pump and tighten the retaining bolts to 7.2 ft. lbs. (10 Nm).

18. Install the timing cover and torque the bolts to 15 ft. lbs. (20 Nm).

19. Install all wiring, hoses and cables.

20. Start the engine and check for leaks.

4ZD1 and 4ZC1-T Engines

1. Disconnect the negative battery cable.

2. Remove the timing belt cover and the timing belt.

3. Use a suitable tool to hold the oil pump pulley in position and remove the pulley bolt. Remove the oil pump pulley.

4. Remove the oil pump retaining bolts and the oil pump.

To install:

5. Clean the gasket mating surfaces and use new gaskets.

6. Apply a generous amount of engine oil to the rotor and install it with the chamfered side toward the cylinder block.

7. Apply engine oil to a new O-ring and insert it into the groove in the pump housing. Attach the rotor after applying generous amounts of engine oil.

8. Install the pump housing onto the cylinder block and tighten the mounting bolts to 13 ft. lbs. (19 Nm). Make sure the assembly turns smoothly.

9. Apply Loctite® or equivalent, to the first thread of the pulley retaining nut, install the pulley and tighten the nut to 55 ft. lbs. (77 Nm).

4XE1 (DOHC) Engine

1. Disconnect the negative battery cable, raise and safely support the vehicle and drain the engine oil.

2. Remove the timing belt cover and the timing belt.

3. Remove the crankshaft timing sprocket using a suitable tool.

4. Remove the oil pan.

5. Remove the oil pump bolts and the oil pump.

To install:

6. Check the outside of the oil pump assembly for cracking or other damage. Disassemble the oil pump and check the gears and housing for wear and proper clearance. Replace as necessary.

7. Installation is the reverse of the removal procedure. Make sure all mating surfaces are clean prior to installation.

8. Apply sealant to the oil pump fitting face, being careful not to get sealant on the oil ports. Apply engine oil to the oil seal lip and install the pump. Tighten the pump mounting bolts to 17 ft. lbs. (24 Nm).

Rear Main Bearing Oil Seal

REMOVAL & INSTALLATION

Except 4ZD1 and 4ZC1-T Engines

1. Disconnect the negative battery cable.

2. Raise and safely support the vehicle.

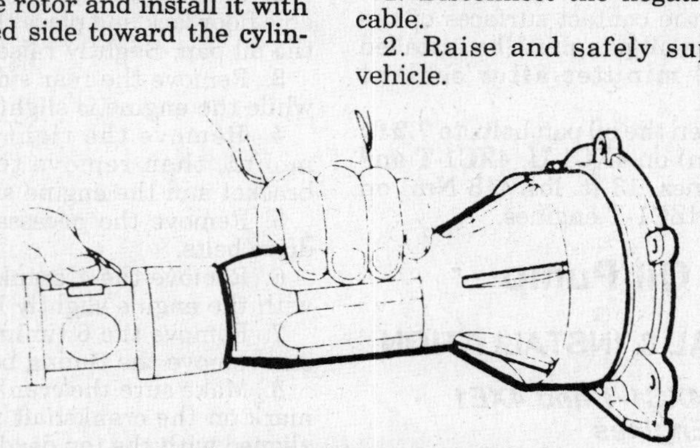

Installing rear main bearing oil seal

3. Drain the engine oil and replace the drain plug.

4. Remove the transaxle assembly.

5. Remove the clutch assembly on manual transmission vehicles.

6. Remove the flywheel.

7. Remove the oil pan.

8. Remove the rear oil seal retainer bolts and remove the oil seal retainer.

9. Use a suitable tool to remove the seal from the retainer.

To install:

10. Make sure all mating surfaces are clean prior to installation.

11. Use a suitable installation tool to install a new seal in the retainer.

12. Apply sealant to the seal retainer mounting surface and oil to the lip of the seal and install the retainer, aligning it with the dowel pins.

4ZD1 and 4ZC1-T Engines

1. Disconnect the negative battery cable.

2. Raise and safely support the vehicle.

3. Remove the transmission.

4. If equipped with a manual transmission, remove the clutch assembly.

5. Remove the flywheel.

6. Using a suitable prying tool, remove the rear main seal. Be careful not to damage the sealing surface of the crankshaft or the seal housing surface of the cylinder block.

7. Installation is the reverse of the removal procedure. Apply engine oil to the lip of the seal and install, using a suitable installation tool.

ENGINE COOLING

Radiator

REMOVAL & INSTALLATION

I-Mark, 1990–92 Impulse and 1991-92 Stylus

1. Disconnect the negative battery cable.

2. Remove the radiator cap and loosen the drain plug to drain the radiator.

3. Disconnect the fan motor and thermo switch connectors.

4. Disconnect the radiator and surge tank hoses from the radiator.

5. Disconnect the transaxle cooler hoses, if equipped.

6. Remove the radiator mounting bolts and the radiator and fan assembly.

7. Separate the cooling fan from the radiator.

8. Installation is the reverse of the removal procedure. Fill the radiator with enough water and anti-freeze to provide the required cooling, freezing and corrosion protection.

1988–89 Impulse

1. Disconnect the negative battery cable.
2. Remove the radiator cap and loosen the drain plug to drain the radiator.
3. Disconnect the radiator and surge tank hoses from the radiator. Disconnect the turbocharger water hose on vehicles equipped with the 4ZC1-T engine.
4. Remove the fan shroud and the stay.
5. Remove the radiator mounting bolts and the radiator.
6. Installation is the reverse of the removal procedure. Fill the radiator with enough water and anti-freeze to provide the required cooling, freezing and corrosion protection.

Heater Core

REMOVAL & INSTALLATION

Without Air Conditioning

I-MARK

1. Disconnect the negative battery cable.
2. Drain the cooling system.
3. Disconnect and plug the heater hoses at the heater core.
4. Remove the retaining clips holding the lower heater unit case and remove the lower case.
5. Remove the heater core assembly.
6. Installation is the reverse of the removal procedure. Fill the cooling system with the proper type and quantity of coolant.

IMPULSE

1988–89

1. Disconnect the negative battery cable.
2. Drain the cooling system.
3. Remove the instrument panel as follows:
    a. Remove the steering wheel and the gearshift knob.
    b. Remove the front console and the radio.
    c. Remove the lower dash panel, hood release lever and knee pad.
    d. Remove the instrument cluster hood and the steering column covers.
    e. Remove the instrument cluster and switch assembly.
    f. Remove the instrument panel front cover and the front pillar trim cover.
    g. Remove the glove box.
    h. Remove the instrument panel grille and the instrument panel cover assembly.
4. Disconnect and plug the heater hoses at the heater core and disconnect the electrical connectors from the heater unit.
5. Remove the heater unit mounting bolts and the heater unit.
6. Disassemble the heater unit and remove the heater core.
7. Installation is the reverse of the removal procedure. Fill the cooling system with the proper type and quantity of coolant.

1990–92 IMPULSE AND 1991-92 STYLUS

1. Disconnect the negative battery cable.
2. Drain the engine coolant into a suitable container.
3. Disconnect the heater hoses. Be careful not to damage the core by pulling on the hose to remove. Cut the hoses if they will not come off easily.
4. Remove the instrument panel.
5. Disconnect the resistor assembly.
6. Remove the duct (without air conditioning).
7. Remove the center vent duct and heater unit.
8. Disassemble the heater unit by removing the duct, mode control case, core assembly and heater core.

To install:

9. Install the heater core, making sure all the seals are in place.
10. Assemble the heater unit, making sure all the seals are in place.
11. Install the center vent duct and heater unit.
12. Install the duct (without air conditioning).
13. Connect the resistor assembly.
14. Install the instrument panel.
15. Connect the heater hoses. Be careful not to damage the core by pulling on the hose to remove. Cut the hoses if they will not come off easily.
16. Refill the engine coolant.
17. Connect the negative battery cable and check for leaks.

With Air Conditioning

I-MARK

1. Disconnect the negative battery cable.
2. Drain the cooling system.
3. Disconnect and plug the heater hoses at the heater core.
4. Remove the retaining clips holding the lower heater unit case and remove the lower case.
5. Remove the heater core assembly.
6. Installation is the reverse of the removal procedure. Fill the cooling system with the proper type and quantity of coolant.

1988–89 IMPULSE

1. Disconnect the negative battery cable.
2. Drain the cooling system and properly discharge the air conditioning system.
3. Remove the instrument panel as follows:
    a. Remove the steering wheel and the gearshift knob.
    b. Remove the front console and the radio.
    c. Remove the lower dash panel, hood release lever and knee pad.
    d. Remove the instrument cluster hood and the steering column covers.
    e. Remove the instrument cluster and switch assembly.
    f. Remove the instrument panel front cover and the front pillar trim cover.
    g. Remove the glove box.
    h. Remove the instrument panel grill and the instrument panel cover assembly.
4. Disconnect and plug the heater hoses at the heater core and disconnect the electrical connectors from the heater unit.
5. Disconnect and plug the air conditioning lines and disconnect the electrical connectors at the evaporator.
6. Remove the foot air duct, if equipped.
7. Remove the evaporator mounting nuts and the evaporator.
8. Remove the heater unit mounting bolts and the heater unit.
9. Disassemble the heater unit and remove the heater core.
10. Installation is the reverse of the removal procedure. Fill the cooling system with the proper type and quantity of coolant. Evacuate and recharge the air conditioning system.

1990–92 IMPULSE AND 1991-92 STYLUS

1. Disconnect the negative battery cable.
2. Drain the cooling system and properly discharge the air conditioning system.
3. Remove the instrument panel as follows:
    a. Pull the switch bezel out and disconnect the switch connectors.
    b. Pull the cigarette lighter bezel out and disconnect the electrical connectors, then remove the bezel.
    c. Disconnect the engine hood opener cable.
    d. Remove the knee pad assembly.
    e. Remove the 2 hinge pins from inside the glove box and remove the glove box.

f. Remove the front console bracket.
g. Remove the instrument cluster hood and the instrument cluster.
h. Remove the front hole covers and the front cover.
i. Remove the instrument panel assembly.
4. Disconnect and plug the heater hoses at the heater core.
5. Disconnect the resistor connector.
6. Disconnect and plug the air conditioning lines at the evaporator.
7. Disconnect the hose and the electrical connectors at the evaporator.
8. Remove the 3 mounting nuts and the evaporator.
9. Remove the center ventilator duct.
10. Remove the heater unit.
11. Disassemble the heater unit and remove the heater core.

To install:

12. Assemble the heater unit.
13. Install the heater unit into the vehicle.
14. Install the center ventilator duct.
15. Install the 3 mounting nuts and the evaporator.
16. Connect the hose and the electrical connectors at the evaporator.
17. Connect and plug the air conditioning lines at the evaporator.
18. Connect the resistor connector.
19. Connect the heater hoses at the heater core.
20. Install the instrument panel as follows:
a. Install the instrument panel assembly.
b. Install the front hole covers and the front cover.
c. Install the instrument cluster and hood.
d. Install the front console bracket.
e. Install the 2 hinge pins to the inside of the glove box.
f. Install the knee pad assembly.
g. Connect the engine hood opener cable.
h. Push the cigarette lighter and bezel.
i. Install the switch bezel.
21. Refill the cooling system and properly evacuate and recharge the air conditioning system.
22. Connect the negative battery cable.

Water Pump

REMOVAL & INSTALLATION

I-Mark

1. Disconnect the negative battery cable.
2. Drain the cooling system.
3. Place a wooden block over a suitable floor jack and jack up the oil pan slightly.
4. Remove the rear side torque rod.
5. Remove the right side engine mount, then remove the body side and engine side brackets.
6. Remove all necessary drive belts.
7. Remove the crank pulley and the timing cover.
8. Turn the crank pulley hub bolt to bring the No. 1 cylinder to TDC.
9. Remove the crank pulley hub bolt and the pulley hub, being careful not to disturb the position of the crank.
10. Loosen the tension pulley bolts, turn the tension pulley clockwise and remove the timing belt.
11. Remove the tension pulley and spring and remove the water pump.

To install:

12. Make sure all gasket mating surfaces are clean prior to installation. Tighten the water pump bolts to 17 ft. lbs. (24 Nm).
13. Properly adjust the timing belt tension and tighten the tension pulley bolt to 37 ft. lbs. (51 Nm). Tighten the crank pulley hub bolt to 108 ft. lbs. (150 Nm).
14. Fill the cooling system with the proper type and quantity of coolant.

Impulse

1988–89

1. Disconnect the negative battery cable.
2. Drain the cooling system.
3. Remove the necessary drive belts and the cooling fan and pulley.
4. Remove the water pump.
5. Installation is the reverse of the removal procedure. Make sure all gasket mating surfaces are clean prior to installation. Tighten the water pump mounting bolts to 17 ft. lbs. (24 Nm).

Impulse and Stylus

1990–92

1. Disconnect the negative battery cable.
2. Drain the cooling system.
3. Place a wooden block on a suitable floor jack and support the engine under the oil pan.
4. Remove the right side engine mount.
5. Remove the necessary drive belts.
6. Remove the crank pulley and the timing cover.
7. Bring the No. 1 cylinder to TDC.
8. Loosen the tension pulley lock bolt, turn the tension pulley clockwise and remove the timing belt.
9. Remove the water pump.

To install:

10. Make sure all gasket mating surfaces are clean prior to installation. Tighten the water pump mounting bolts to 17 ft. lbs. (24 Nm).
11. Properly adjust the timing belt tension and tighten the tension pulley bolt to 31 ft. lbs. (42 Nm). Tighten the crank pulley bolt to 109 ft. lbs. (147 Nm).
12. Fill the cooling system with the proper type and quantity of coolant.

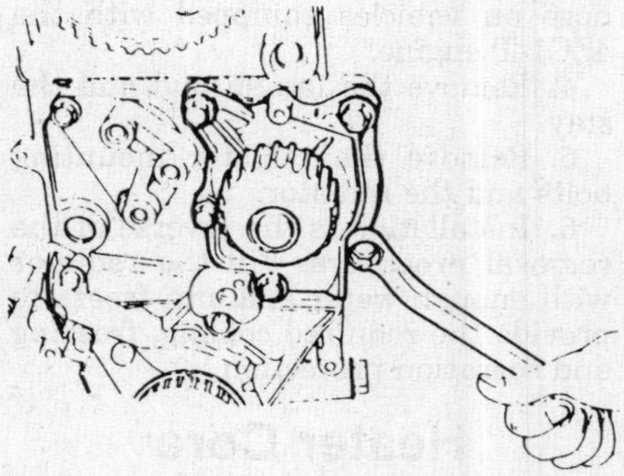

Water pump installation—I-Mark

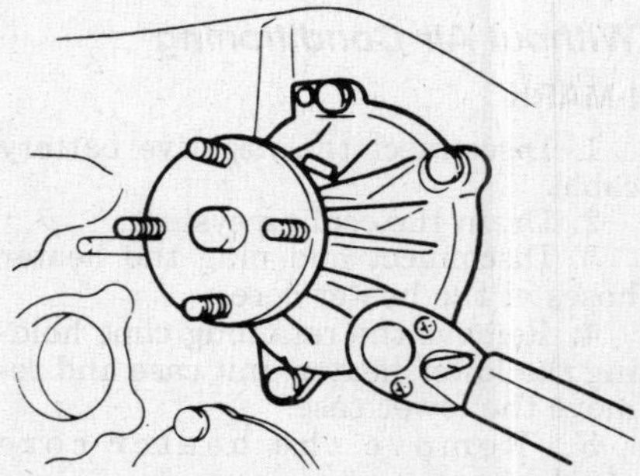

Water pump—4ZD1 and 4ZC1-Turbo

Thermostat

REMOVAL & INSTALLATION

1. Disconnect the negative battery cable.

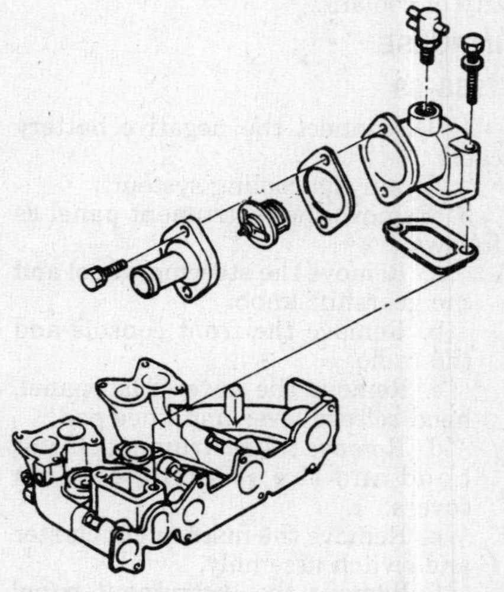

Exploded view of thermostat and housing 1988–89 Impulse

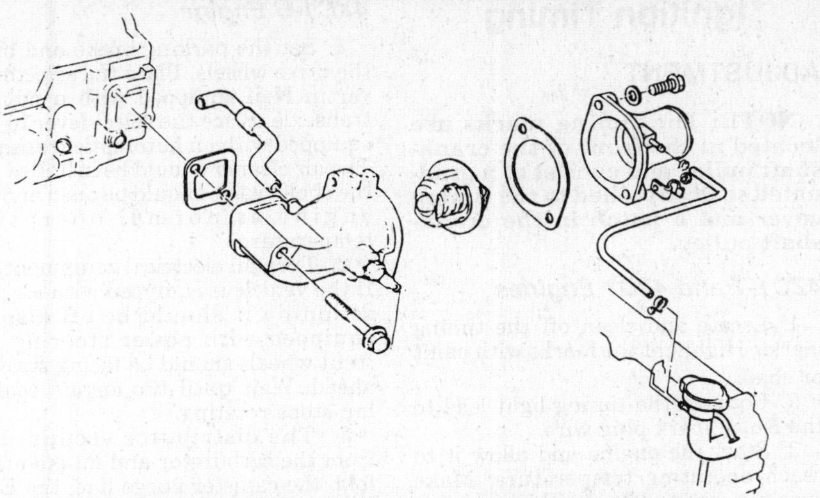

Exploded view of thermostat and housing—I-Mark except 4XE1 engine

2. Drain the cooling system.
3. Disconnect the radiator hose from the water outlet.
4. Remove the water outlet from the thermostat housing and remove the thermostat.
5. Installation is the reverse of the removal procedure. Make sure all gasket mating surfaces are clean prior to installation. Install the thermostat with the jiggle valve toward the water outlet and the spring toward the thermostat housing.

Cooling System Bleeding

1. Make sure the engine and radiator are cold before proceeding.
2. Remove the radiator cap and the coolant reserve tank cap.
3. Check that the radiator is full to the base of the filler neck and the coolant reserve tank is filled to a level between the **MAX** and **MIN** lines. Add coolant, as necessary.
4. Block the drive wheels and apply the parking brake. Place automatic transmissions/transaxles in **P** and manual transmissions/transaxles in neutral.
5. Run the engine, with the radiator cap removed, until the upper radiator hose is hot. With the engine idling, add coolant, as necessary, to the radiator until it is full. Install the radiator cap.
6. Allow the engine to cool down to outside air temperature and check the coolant level in the coolant reservoir. The coolant level should be at the **MAX** mark. Add coolant, as necessary.

ENGINE ELECTRICAL

NOTE: Disconnecting the negative battery cable on some vehicles may interfere with the functions of the on board computer systems and may require the computer to undergo a relearning process, once the negative battery cable is reconnected.

Distributor

REMOVAL

NOTE: The 4XE1 DOHC Turbo does not have a distributor. The engine is equipped with a Direct Ignition System (DIS). The camshaft angle sensor is removed in the same manner as the conventional distributor.

1. Turn the engine over and bring the No. 1 piston up to TDC on the compression stroke.
2. Disconnect the negative battery cable.
3. Remove the distributor cap and disconnect the electrical connectors from the distributor.
4. Mark the position of the distributor rotor in relation to the distributor housing and the distributor housing in relation to the cylinder head or block.
5. Remove the distributor hold-down bolt and remove the distributor.

INSTALLATION

Timing Not Disturbed

1. Install the distributor, aligning the marks that were made during the removal procedure.
2. Install the distributor hold-down bolt and tighten temporarily.
3. Connect the distributor electrical connectors and install the distributor cap.
4. Connect the negative battery cable, start the engine, check the ignition timing and adjust, as necessary. Tighten the distributor hold-down bolt.

Timing Disturbed

1. Disconnect the spark plug wire and remove the spark plug from the No. 1 cylinder.
2. Place a finger over the spark plug hole and rotate the crankshaft until compression is felt.
3. Align the notched line on the crankshaft pulley with the **0** mark on the timing scale of the timing cover.
4. Install the distributor so the distributor rotor points to the No. 1 spark plug wire tower of the distributor cap.
5. Install the distributor hold-down bolt and tighten temporarily.
6. Connect the distributor electrical connectors and install the distributor cap.

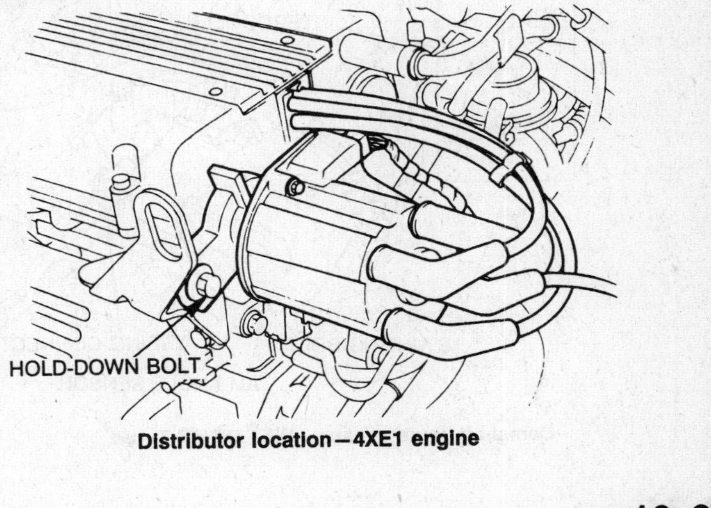

Distributor location—4XE1 engine

7. Install the No. 1 spark plug and connect the No. 1 spark plug wire.
8. Connect the negative battery cable, start the engine and adjust the ignition timing. Tighten the distributor hold-down bolt.

Camshaft Sensor

REMOVAL

1. Rotate the engine and bring up the No. 1 cylinder to top dead center of its compression stroke.

NOTE: To bring the engine to TDC of the No. 1 compression stroke, remove the spark plug for the No. cylinder. With the engine cool, turn the crankshaft over until compression is forced out of the spark plug hole. Watch the crankshaft damper while feeling for compression. When compression is felt, align the mark on the crankshaft damper with the 0° mark on the timing cover.

2. Disconnect the negative battery cable.
3. Remove the intercooler and disconnect the sensor electrical connector.
4. Mark the cylinder block and camshaft angle sensor before removing. Remove the mounting bolt and camshaft angle sensor.

INSTALLATION

1. Install the sensor to its original position and install the mounting bolt.
2. Connect the wiring connectors.
3. Install the intercooler and connect the battery cable.
4. Start the engine and adjust the timing.

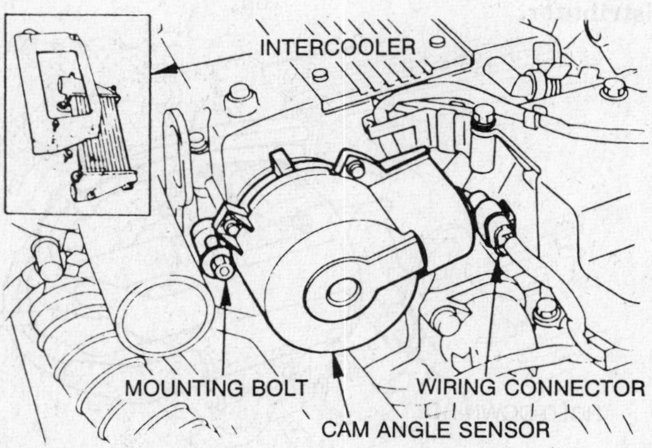

Camshaft angle sensor—4XE1 DOHC Turbo

Ignition Timing

ADJUSTMENT

NOTE: The timing marks are located at the front of the crankshaft pulley and consist of a graduated scale attached to the timing cover and a notch in the crankshaft pulley.

4ZC1-T and 4ZD1 Engines

1. Locate and clean off the timing marks. Highlight the marks with paint or chalk.
2. Connect the timing light lead to the No. 1 spark plug wire.
3. Start the engine and allow it to reach operating temperature. Make sure the engine idle speed is correct.
4. Aim the timing light at the marks and check the position of the crankshaft pulley notch on the timing scale. If necessary, adjust the timing by loosening the distributor hold-down bolt and turning the distributor to the correct specification.
5. After timing is set to specifications, tighten the distributor hold-down bolt and remove the timing light connections.

Timing indicator—I-Mark

4XC1-U Engine

1. Set the parking brake and block the drive wheels. Place the selector lever in **N** if equipped with a manual transaxle. Place the select lever in **P** if equipped with an automatic transaxle. The air cleaner should be installed and the choke valve should be open and the engine at normal operating temperature.
2. Turn all electrical equipment off. If the vehicle is equipped with air conditioning it should be off also. If equipped with power steering the front wheels should be facing straight ahead. Wait until the engine cooling fan stops rotating.
3. The distributor vacuum line from the carburetor and intake manifold, the canister purge line, the EGR vacuum line and the ITC valve vacuum line should be disconnected and plugged.
4. Connect the timing light lead to the No. 1 spark plug wire. Loosen the distributor hold-down bolt (slightly).
5. Align the notched line on the crankshaft pulley with the mark on the timing cover using the timing light.
6. While aligning the notched line on the crankshaft pulley, advance or retard the timing, as necessary, by turning the distributor clockwise or counterclockwise.
7. After the timing has been set to specifications, tighten the distributor hold-down bolt and re-check the timing.

NOTE: Be sure the distributor body does not move together with the hold-down bolt.

8. Reconnect all vacuum lines and remove all test equipment, except for the tachometer.
9. Re-check the idle speed and adjust, as necessary.

4XC1-T and 1989 I-Mark 4XE1 Engines

1. Check the ECM for trouble codes. If codes are found, repair or replace the problem sensor or circuit, as necessary.
2. Check all the vacuum lines for the proper routing.
3. Set the parking brake and block the drive wheels. Place the selector lever in **N**. The engine should be at normal operating temperature.
4. Turn all electrical equipment off. If equipped with air conditioning it should be off also. If equipped with power steering the front wheels should be facing straight ahead. Wait until the engine cooling fan stops rotating.
5. Connect the timing light lead to the No. 1 spark plug wire. Loosen the

distributor hold-down bolt (slightly).

6. Align the notched line on the crankshaft pulley with the mark on the timing cover using the timing light.

7. While aligning the notched line on the crankshaft pulley, advance or retard the timing, as necessary, by turning the distributor clockwise or counterclockwise.

8. After the timing has been set to specifications, tighten the distributor hold-down bolt and re-check the timing.

NOTE: Be sure the distributor body does not move together with the hold-down bolt. The ignition timing is controlled by the ECM according to the engine operating conditions. Therefore, there are no external vacuum lines to the distributor.

9. Remove all test equipment, except for the tachometer. Re-check the idle speed and adjust, as necessary.

1990–92 Impulse and 1991-92 Stylus

4XE1 Engine

NOTE: The 4XE1 Turbo engine is equipped with a Direct Ignition System. Timing can be adjusted by moving the cam angle sensor. The sensor is mounted in the same location as a conventional distributor for non-Turbo 4XE1 engines.

1. Apply the parking brake and start the engine.
2. Place the transaxle in **N**.
3. Make sure the CHECK ENGINE light is not on.
4. Locate the ALDL connector under the right hand side of the instrument panel. Connect the terminals **1** and **3** on both ends of the ALDL connector with a jumper wire.
5. Connect the timing light lead to the No. 1 spark plug wire.
6. Using the timing light, check that the center of fluctuation of a white notched line on the crankshaft pulley against the scale on the timing cover is between 9–11 degrees BTDC. If the fluctuation is too large, open the throttle valve a little to increase the engine speed to 1500–2000 rpm. The fluctuation will be reduced so the timing can be confirmed.
7. If the timing is incorrect, loosen the hold-down bolt on the distributor or cam angle sensor and turn the assembly clockwise or counterclockwise to adjust the timing. After adjustment, tighten the bolt to 17 ft. lbs. (24 Nm).

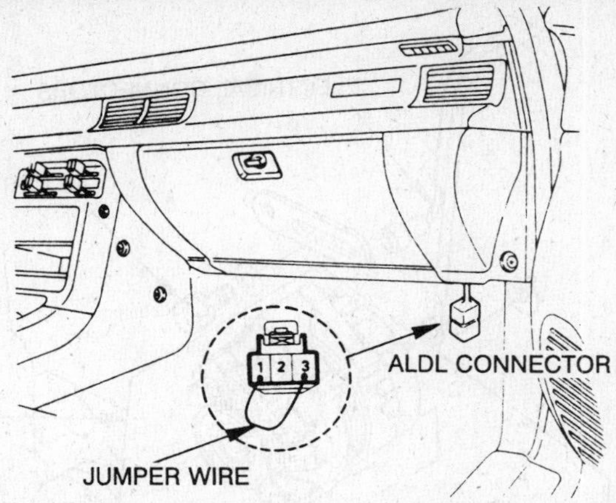

ALDL connector location – 1990–92 Impulse and 1991–92 Stylus

NOTE: When tightening the distributor or cam angle sensor hold-down bolt, make sure the assembly body does not rotate together with the mounting bolt.

8. Turn the ignition switch **OFF** and remove the jumper wire from the ALDL connector.

Alternator

PRECAUTIONS

Several precautions must be observed with alternator equipped vehicles to avoid damage to the unit.

- If the battery is removed for any reason, make sure it is reconnected with the correct polarity. Reversing the battery connections may result in damage to the one-way rectifiers.
- When utilizing a booster battery as a starting aid, always connect the positive to positive terminals and the negative terminal from the booster battery to a good engine ground on the vehicle being started.
- Never use a fast charger as a booster to start vehicles.
- Disconnect the battery cables when charging the battery with a fast charger.
- Never attempt to polarize the alternator.
- Do not use test lamps of more than 12 volts when checking diode continuity.
- Do not short across or ground any of the alternator terminals.
- The polarity of the battery, alternator and regulator must be matched and considered before making any electrical connections within the system.
- Never separate the alternator on an open circuit. Make sure all connections within the circuit are clean and tight.
- Disconnect the battery ground terminal when performing any service on electrical components.
- Disconnect the battery if arc welding is to be done on the vehicle.

BELT TENSION ADJUSTMENT

I-Mark, 1990–92 Impulse and 1991-92 Stylus

1. Check the belt tension between the pulleys using a suitable belt tension gauge. The tension should be 70–110 lbs. (95–149 N).
2. If the tension is incorrect, loosen the alternator pivot and adjusting bolts. Pry on the alternator until the proper belt tension is obtained.

1988–89 Impulse

1. Check the belt tension by applying approximately 22 lbs. (29 N) of finger pressure to the belt between the pulleys and measuring the belt deflection. The belt should deflect 0.16–0.2 in. (4–5mm).
2. If the tension is incorrect, loosen the alternator pivot and adjusting bolts. Pry on the alternator until the proper belt deflection is obtained.

REMOVAL & INSTALLATION

I-Mark and 1988–89 Impulse

1. Disconnect the negative battery cable.
2. Disconnect the electrical connections at the back of the alternator and tag the wires for reassembly.
3. Loosen the pivot and adjusting bolts and remove the alternator belt.
4. Remove the mounting bolts and the alternator.
5. Installation is the reverse of the removal procedure. Adjust the belt tension.

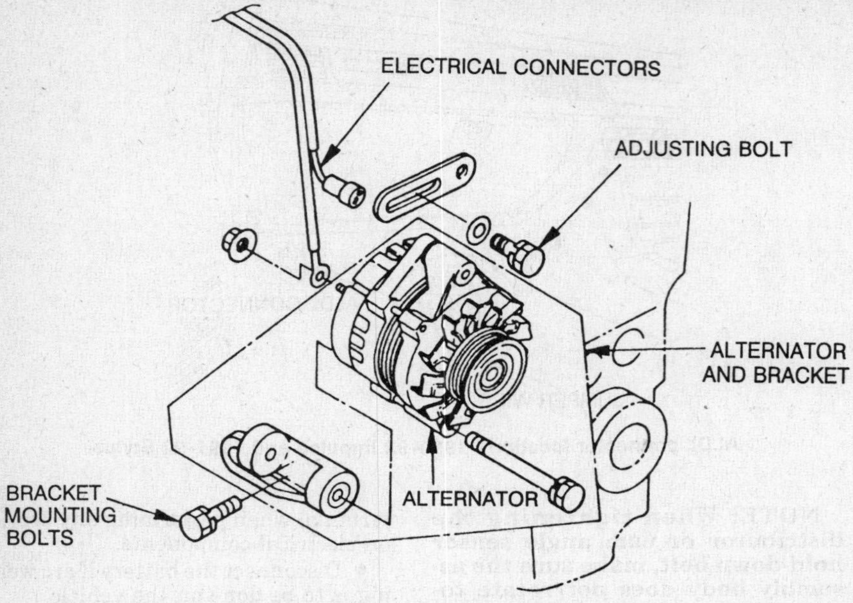

Alternator Installation—1990–92 Impulse and 1991–92 Stylus

1990–1992 Stylus and Impulse

1. **Disconnect the negative battery cable.**
2. Remove the right tie rod end, lower ball joint and driveshaft (4WD only).
3. Remove the adjuster plate bolt.
4. Disconnect the electrical connector.
5. Remove the alternator bracket bolts, alternator and bracket.

To install:

6. Install the alternator bracket bolts, alternator and bracket.
7. Connect the electrical connector.
8. Install the adjuster plate bolt.
9. Install the right tie rod end, lower ball joint and driveshaft (4WD only).
10. Connect the negative battery cable.

Starter

REMOVAL & INSTALLATION

1. Disconnect the negative battery cable.
2. Disconnect the electrical connectors at the starter.
3. Remove the starter mounting nuts or bolts and remove the starter.
4. Installation is the reverse of the removal procedure.

EMISSION CONTROLS

Please refer to "Emission Controls" in the Unit Repair section for system maintenance procedures. Due to the complex nature of modern electronic engine control systems, comprehensive diagnosis and testing procedures fall outside the confines of this repair manual. For complete information on diagnosis, testing and repair procedures concerning all modern engine and emission control systems, please refer to "Chilton's Guide to Fuel Injection and Electronic Engine Controls".

Emission Warning Lamps

Isuzu vehicles do not employ emission warning lamps that would indicate scheduled component replacement. A CHECK ENGINE lamp is used to indicate component malfunction. When the CHECK ENGINE lamp comes on, the vehicle should be checked as soon as possible, using the proper diagnostic procedure, to locate the malfunction.

FUEL SYSTEM

Fuel System Service Precautions

Safety is the most important factor when performing not only fuel system maintenance but any type of maintenance. Failure to conduct maintenance and repairs in a safe manner may result in serious personal injury or death. Maintenance and testing of the vehicle's fuel system components can be accomplished safely and effectively by adhering to the following rules and guidelines.

- To avoid the possibility of fire and personal injury, always disconnect the negative battery cable unless the repair or test procedure requires that battery voltage be applied.
- Always relieve the fuel system pressure prior to disconnecting any fuel system component (injector, fuel rail, pressure regulator, etc.), fitting or fuel line connection. Exercise extreme caution whenever relieving fuel system pressure to avoid exposing skin, face and eyes to fuel spray. Please be advised that fuel under pressure may penetrate the skin or any part of the body that it contacts.
- Always place a shop towel or cloth around the fitting or connection prior to loosening to absorb any excess fuel due to spillage. Ensure that all fuel spillage (should it occur) is quickly removed from engine surfaces. Ensure that all fuel soaked cloths or towels are deposited into a suitable waste container.
- Always keep a dry chemical (Class B) fire extinguisher near the work area.
- Do not allow fuel spray or fuel vapors to come into contact with a spark or open flame.
- Always use a backup wrench when loosening and tightening fuel line connection fittings. This will prevent unnecessary stress and torsion to fuel line piping. Always follow the proper torque specifications.
- Always replace worn fuel fitting O-rings with new. Do not substitute fuel hose or equivalent where fuel pipe is installed.

RELIEVING FUEL SYSTEM PRESSURE

1. Remove the fuel pump fuse from the fuse block or disconnect the harness connector at the tank.
2. Start the engine. It should run and then stall when the fuel in the lines is exhausted. When the engine

stops, crank the starter for about 3 seconds to make sure all pressure in the fuel lines is released.

3. Install the fuel pump fuse after repair is made.

Fuel Tank

REMOVAL & INSTALLATION

I-Mark

1. Disconnect the negative battery cable. Relieve the fuel pressure.
2. Drain the fuel using an approved pump and container.
3. Label and disconnect the fuel filler, breather, tank-to-evaporator and tank-to-return pipe hoses.
4. Disconnect the tank wiring.
5. Place a floor jack under the tank, remove the retaining bolts and lower the tank.

To install:

6. Raise the tank with the jack and install the retaining bolts. Torque the retainer to 15 ft. lbs. (20 Nm).
7. Connect the wiring and fuel hoses.
8. Connect the breather hose and fuel filler hose.
9. Connect the battery cable, fill the tank with fuel and check for leaks.

1988–89 Impulse

1. Disconnect the negative battery cable. Relieve the fuel system pressure.
2. Drain the fuel tank using an approved pump and container.
3. Disconnect the tank harness connectors, fuel filler assembly and remove the hoses guard.
4. Position a floor jack under the tank and remove the tank straps.
5. Lower the tank and disconnect any wiring and hoses.

To install:

6. Raise the tank and connect any wiring and hoses.
7. Install the tank straps and retainers. Torque the retainer to 15 ft. lbs. (20 Nm).
8. Install the fuel filler guard.
9. Connect the tank harness connectors.
10. Refill the tank and check for leaks.
11. Connect the battery cable.

1990–92 Impulse and 1991–92 Stylus

1. Disconnect the negative battery cable. Relieve the fuel system pressure.
2. Drain the tank with an approved pump and container.
3. Remove the rear and center exhaust pipes.
4. Remove the third driveshaft, fuel filler and air breather hose.
5. Disconnect the feed, return and evaporative hoses.
6. Disconnect the parking brake cable bracket and return spring.
7. Disconnect the tank harness connectors.
8. Place a floor jack under the tank and remove the tank retainers. Lower the tank and disconnect any wiring or hoses.

To install:

9. Raise the tank into position and install the tank retainers. Torque the retainers to 15 ft. lbs. (20 Nm).
10. Connect the tank harnesses and hoses.
11. Connect the parking brake cable bracket and return spring.
12. Connect the feed, return and evaporative hoses.
13. Install the third driveshaft, fuel filler and air breather hoses.
14. Install the rear and center exhaust pipes.
15. Refill the tank and check for leaks.
16. Connect the battery cable.

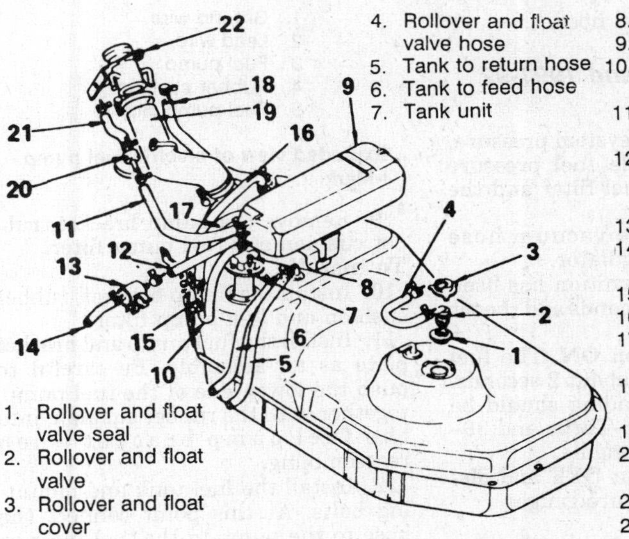

1. Rollover and float valve seal
2. Rollover and float valve
3. Rollover and float cover
4. Rollover and float valve hose
5. Tank to return hose
6. Tank to feed hose
7. Tank unit
8. Tank unit screws
9. Tank vapor separator
10. Separator to 3 way hose
11. Separator to 2 way valve hose
12. Separator to evaporator valve hose
13. Evaporator valve
14. Pipe to evaporator valve hose
15. 3 way joint
16. 3 way to tank hose
17. 3 way to rollover and float valve hose
18. Filler neck to tank hose
19. Fuel filler hose
20. 2 way valve, filler neck to separator
21. 2 way valve hose
22. Fuel filler neck

Fuel tank—1988–89 Impulse

1. Fuel filler hose
2. Breather hose
3. Evaporator to tank hose
4. Return to tank hose
5. Fuel pipe to tank hose
6. Wiring connector
7. Fuel tank

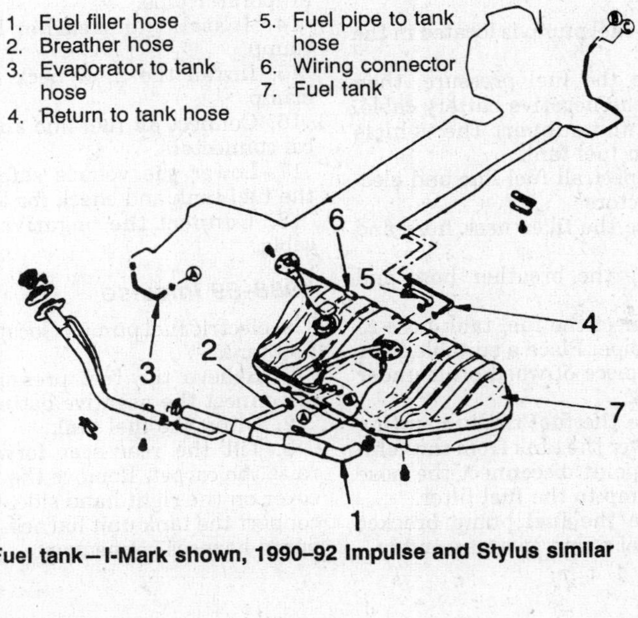

Fuel tank—I-Mark shown, 1990–92 Impulse and Stylus similar

Fuel Filter

REMOVAL & INSTALLATION

1. Disconnect the negative battery cable.
2. Relieve the fuel system pressure and remove the fuel filler cap.
3. On 1990–92 Impulse and 1991-92 Stylus, disconnect the engine harness connector and remove the air duct with the air cleaner cover.

4. Disconnect and plug the fuel lines at the fuel filter.
5. On Impulse, loosen the filter clamp bolt.
6. Remove the fuel filter.
7. Installation is the reverse of the removal procedure. Make sure the filter is installed in the proper direction of fuel flow.
8. Start the engine and check for leaks.

Mechanical Fuel Pump

PRESSURE TESTING

1. Disconnect the negative battery cable.
2. Remove the fuel inlet hose from the carburetor.
3. Connect a fuel pump pressure gauge between the inlet hose and the carburetor, using a tee fitting. Reconnect the negative battery cable.
4. Start the engine and run it at 750 rpm. Note the reading on the pressure gauge.
5. Stop the engine and disconnect the negative battery cable. Remove the pressure gauge and tee fitting.
6. Place the open end of the fuel line into a graduated container. Reconnect the negative battery cable.
7. Crank the engine for 30 seconds. Note the amount of gasoline delivered to the container, then safely dispose of the fuel.
8. Connect the fuel inlet line to the carburetor. Compare the results of the these tests to the specifications below:
   Fuel pressure – 3.84–4.70 psi.
   Fuel delivery – 0.428 qts. in 30 seconds.
9. If the fuel pump fails these tests, it should be replaced.

REMOVAL & INSTALLATION

1. Disconnect the fuel delivery and return hoses from the fuel pump.
2. Remove the bolts, fuel pump and heat insulator assembly.
3. After removing the fuel pump, cover the mounting face of the cylinder head to prevent oil discharge.
4. To install, reverse the removal procedures. Clean all mating surfaces prior to installation. Replace the heat insulator assembly.

Electric Fuel Pump

PRESSURE TESTING

I-Mark and 1988–89 Impulse

1. Relieve the fuel system pressure.
2. Disconnect the fuel hose between the pressure regulator and fuel distributor pipe.
3. Connect a suitable fuel pressure gauge across the pressure regulator and fuel distributor pipe correctly.

NOTE: On Impulse engines disconnect the Vacuum Switching Valve (VSV) harness at the connector.

4. Start the engine and measure the fuel pressure under the following conditions:
   a. Vacuum hose of the pressure regulator disconnected (Intake manifold side end of hose must be plugged). The pressure should read 35.6 psi. on I-Mark and 42.6 psi. on Impulse.
   b. Vacuum hose of the pressure regulator connected, at idling speed of 900 rpm. The pressure should read 28.4 psi. on I-Mark and 35.6 psi. on Impulse.
5. Remove the fuel pressure gauge and reconnect the fuel hose.

1990–92 Impulse and 1991-92 Stylus

1. Relieve the fuel system pressure.
2. Install a suitable fuel pressure gauge between the fuel filter and the fuel distributor pipe.
3. Disconnect the vacuum hose from the pressure regulator.
4. Make sure the ignition has been **OFF** for at least 10 seconds and the air conditioning is off.
5. Turn the ignition **ON**. The fuel pump should run for about 2 seconds. The fuel pressure reading should be 35–38 psi for the Non-Turbo and 39–47 psi for the Turbo engine.
6. Turn the ignition **OFF** and disconnect the fuel pressure gauge.

REMOVAL & INSTALLATION

I-Mark

The electric fuel pump is located in the fuel tank.

1. Relieve the fuel pressure, then disconnect the negative battery cable.
2. Raise and support the vehicle safely. Drain fuel tank.
3. Disconnect all fuel line and electrical connectors.
4. Remove the filler neck hose and clamp.
5. Remove the breather hose and clamp.
6. Disconnect the fuel tank hose to evaporator pipe. Place a suitable floor jack with a piece of wood on it under the fuel tank.
7. Remove the fuel tank mounting bolts and lower the tank from the vehicle. At this point disconnect the hose from the pump to the fuel filter.
8. Remove the fuel pump bracket plate and fuel pump as an assembly.

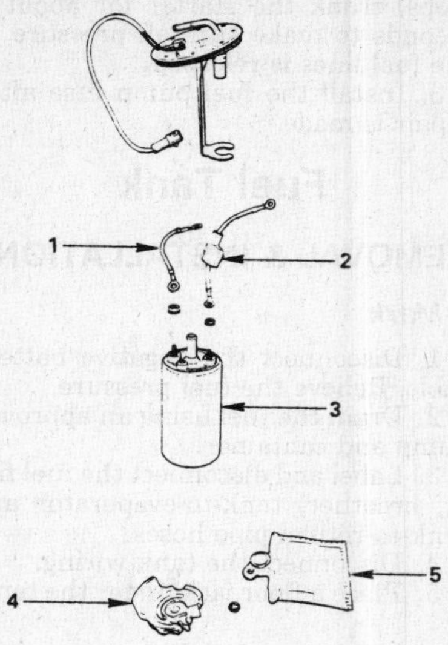

1. Ground wire
2. Lead wire
3. Fuel pump
4. Rubber cushion
5. Fuel pump filter

Exploded view of electric fuel pump – I-Mark

9. Remove the pump bracket, rubber cushion and fuel pump filter.

To install:

10. Install the pump bracket, rubber cushion and fuel pump filter.
11. Install the fuel pump and bracket plate as an assembly. Be careful to push the lower side of the fuel pump, together with the rubber cushion, into the fuel pump bracket when reassembling.
12. Install the fuel tank and mounting bolts. At this point connect the hose to the pump to the fuel filter as the tank in being raised into position.
13. Connect the fuel tank hose to evaporator pipe.
14. Install the breather hose and clamp.
15. Install the filler neck hose and clamp.
16. Connect all fuel line and electrical connectors.
17. Lower the vehicle safely. Refill the fuel tank and check for leaks.
18. Connect the negative battery cable.

1988–89 Impulse

The electric fuel pump is located in the fuel tank.

1. Relieve the fuel pressure, then disconnect the negative battery cable.
2. Drain the fuel tank.
3. Tilt the rear seat forward and raise the carpet. Remove the rear side cover on the right hand side, then disconnect the tank unit harness and fuel pump harness connectors.

4. Disconnect the fuel lines.
5. Remove the cap assembly, rubber fuel filler receiver and filler guard.
6. Support the tank and remove the tank strap holding bolts.
7. Remove the tank assembly from the vehicle.
8. Remove the fuel pump bracket plate and fuel pump as an assembly.
9. Remove the pump bracket, rubber cushion and fuel pump sock or filter.

To install:

10. Install the pump bracket, rubber cushion and fuel pump sock or filter.
11. Install the fuel pump and bracket plate as an assembly. Be careful, to push the lower side of the fuel pump, together with the rubber cushion, into the fuel pump bracket when reassembling.
12. Install the tank assembly into the vehicle.
13. Support the tank and install the tank strap holding bolts.
14. Install the cap assembly, rubber fuel filler receiver and filler guard.
15. Connect the fuel lines.
16. Connect the tank unit harness and fuel pump harness connectors.
17. Refill the fuel tank and check for leaks.
18. Connect the negative battery cable.

1990–92 Impulse and Stylus

The electric fuel pump is located in the fuel tank.

1. Relieve the fuel system pressure, then disconnect the negative battery cable.
2. Loosen the fuel filler cap and drain the fuel system.
3. Remove the fuel filler and air breather hoses and disconnect the fuel lines.
4. Remove the parking brake cable brackets and the parking brake return spring.
5. Disconnect the fuel gauge and fuel pump harness connectors.
6. Support the fuel tank and remove the mounting bolts. Lower the tank onto the exhaust pipe.
7. Peel off the harness fixing tape. Remove the fuel pump assembly attaching screws and pull the fuel pump assembly out of the tank.

To install:

8. Install the fuel pump assembly into of the tank and tighten the retaining screws.
9. Raise the tank and install the retaining bolts. Torque the bolts to 15 ft. lbs. (20 Nm).
10. Connect the fuel gauge and fuel pump harness connectors.
11. Install the parking brake cable brackets and the parking brake return spring.
12. Install the fuel filler and air breather hoses.
13. Refill the tank, check for leaks and install the fuel filler cap.
14. Connect the negative battery cable.

Carburetor

REMOVAL & INSTALLATION

1. Disconnect the negative battery terminal from the battery.
2. Remove the air cleaner.
3. Disconnect the harness connector and hoses.
4. Remove the accelerator cable from the carburetor.
5. Remove the bolts securing the carburetor to the intake manifold. Remove the carburetor and place a cover over the intake manifold.

To install:

6. Install the carburetor and torque the fixing bolts to 7.2 ft. lbs. (9.5 Nm).
7. Start the engine and check for leaks.

IDLE SPEED ADJUSTMENT

NOTE: Before adjusting the idle speed, be sure to check the ignition timing. If the ignition timing is off, adjust it to the proper specification before adjusting the idle speed.

1. Set the parking brake and block the drive wheels.
2. Place manual transaxles in neutral or automatic transaxles in **P** and start the engine.
3. Let the engine run until it reaches normal operating temperature. Leave the air cleaner installed and disconnect the distributor vacuum line from the carburetor, canister purge line, the EGR vacuum line and the Inlet Air Temperature Compensator (ITC) valve. Plug and mark all lines.
4. With the proper tachometer installed, the air conditioning and all electrical equipment turned off, check the idle speed and adjust, as necessary, by turning the throttle adjustment screw.
5. The idle speed should be 750 rpm for manual transaxle or 1000 rpm for automatic transaxle.
6. Disconnect and plug the vacuum advance hoses from the intake manifold and the carburetor. Maintain the engine speed between 1900–2100 rpm and set the dashpot to contact the dashpot shaft and with the adjust screw of the throttle lever.
7. If equipped with air conditioning, turn it **ON** and set the temperature control level to **MAX COLD**. Set the blower to its highest position.
8. Using the Fast Idle Control Device (FICD) adjusting screw, located at the tip of the carburetor lever, set the fast idle speed. It should be 850 rpm for manual transaxle or 980 rpm for automatic transaxle.

IDLE MIXTURE ADJUSTMENT

1. Set the parking brake and block the drive wheels.
2. Place manual transaxle in neutral or automatic transaxle in **P** and start the engine.
3. Remove the carburetor assembly.
4. Remove the idle mixture adjusting screw plug as follows:
   a. Cover the primary and second-

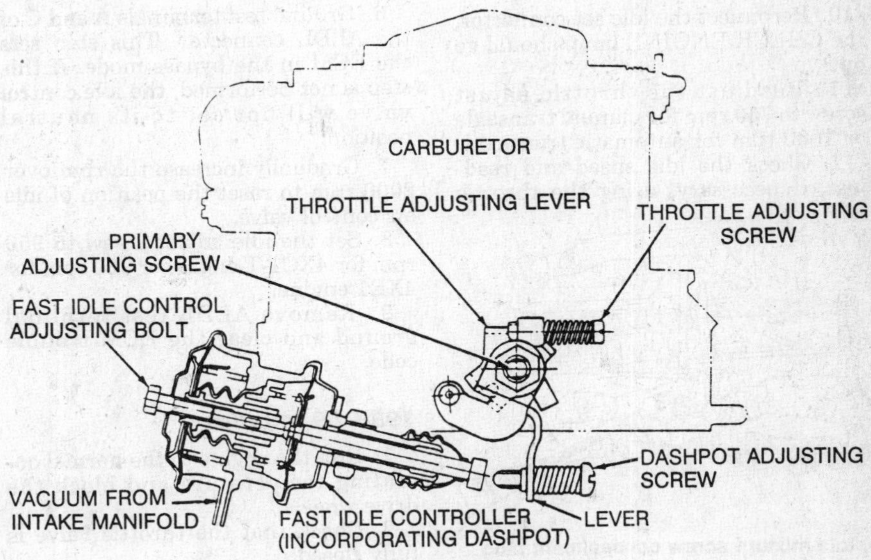

Carburetor adjusting screw locations

ary bores and other ports to prevent metal chips from entering the carburetor.

b. Mark the center of the plug with a punch.

c. Drill a hole in the plug to a depth less than 0.39 in. (10mm) and then thread a screw into the hole. Pull on the screw to remove the plug.

d. Remove the chips from the surroundings of the mixture screw with compressed air. Remove and inspect the idle mixture adjusting screw. Replace the screw if it is deformed or damaged from the drilling procedure.

e. Reinstall the idle mixture screw. Lightly seat the screw, then back it out 3 turns from fully closed for manual transaxle or 2 turns from fully closed on automatic transaxle.

NOTE: Do not overtighten the idle mixture screw.

5. Reinstall the carburetor and the air cleaner.
6. Adjust the idle speed.
7. Disconnect the idle set connector, located in the harness to right of the carburetor assembly. The CHECK ENGINE lamp will come on.
8. Adjust the throttle adjust screw to 750 rpm for manual transaxle or 1000 rpm for automatic transaxle.
9. Connect the positive lead of a suitable dwell meter to the duty monitor lead, located in the harness to the right of the carburetor assembly. Connect the negative lead of the dwell meter to ground. Place the meter dial on the 4 cylinder scale or 6 cylinder scale. Turn the idle mixture screw until the dwell meter reads 45 degrees on the 4 cylinder scale or 30 degrees on the 6 cylinder scale.
10. Reconnect the idle set connector, the CHECK ENGINE lamp should go out.
11. Readjust the throttle adjust screw to 750 rpm for manual transaxle or 1000 rpm for automatic transaxle.
12. Check the idle speed and readjust, if necessary, using the throttle adjust screw. Readjust the dashpot, if necessary.
13. If equipped with air conditioning, turn it **ON** and set the temperature control lever to **MAX COLD**. Set the blower to it's highest position. Use the Fast Idle Control Device (FICD) adjusting screw at the tip of the carburetor throttle lever to set the fast idle speed to 850 rpm for manual transaxle or 980 rpm for automatic transaxle.

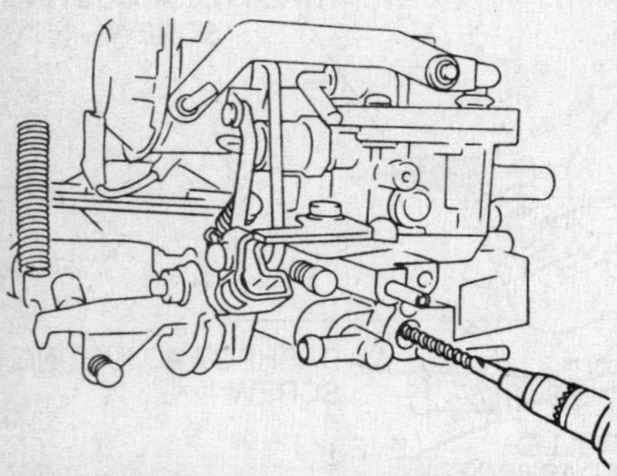

Idle mixture screw concealment plug removal

SERVICE ADJUSTMENTS

For all carburetor service adjustment procedures and specifications, please refer to "Carburetor Service" in the Unit Repair section.

Fuel Injection

IDLE SPEED ADJUSTMENT

I-Mark

NOTE: Before adjusting the idle speed, be sure to check the ignition timing. If the ignition timing is off, adjust it to the proper specification before adjusting the idle speed.

1. Set the parking brake.
2. Block the front wheels.
3. Place the select lever in neutral.
4. Make the idling speed adjustment with the engine at normal operating temperature, with the air conditioning turned **OFF** and front wheels facing straight ahead.

NOTE: All electrical equipment (lights, rear defogger, heater, etc.) should be turned OFF.

5. Make sure the CHECK ENGINE light is not on.
6. Ground test terminals **A** and **C** of the ALDL connector. This step sets the ECM in the bypass mode. If this step is not performed, the idle control valve will not set to its neutral position.
7. Gradually increase the rpm over 2000 rpm to reset the position of idle air control valve.
8. Set the idle adjust screw to 950 rpm for 4XC1-T engine or 900 rpm for 4XE1 engine.
9. Remove ALDL test terminal ground and clear the ECM trouble code.

1988–89 Impulse

1. Run the engine to the normal operating temperature and block the drive wheels.
2. Check that the throttle valve is fully closed.
3. Set the manual transmission in neutral or automatic transmission in **P** position.
4. With the air conditioner turned **OFF** and the harness of the pressure regulator Variable Switching Valve (VSV) disconnected, adjust the idle adjustment screw to 850–950 rpm.

1990–92 Impulse and 1991-92 Stylus

Idle speed is controlled by the Idle Air Control (IAC) valve and the ECM. Idle adjustment is only required when a new IAC valve is installed.

1. Set the parking brake and block the drive wheels.
2. Place the transaxle in neutral.
3. The air conditioning must be off and the front wheels facing straight ahead. All electrical equipment should be turned off.
4. Ground the test terminals of the ALDL connector.
5. Start the engine and increase rpm slightly to reset the pintle position of the IAC valve. A throttle opening of 10 percent or more should correspond to an engine speed of 2000 rpm or more.
6. Continue to run the engine until it reaches normal operating temperature. Check the ignition timing and adjust as necessary.
7. Set the idle adjust screw to 950 rpm.
8. Remove the test terminal ground and clear the ECM trouble code.

IDLE MIXTURE ADJUSTMENT

Idle mixture is controlled electronically by the fuel injection system. No adjustments are necessary.

Fuel Injector

REMOVAL & INSTALLATION

1. Disconnect the negative battery cable. Remove the air cleaner and duct.
2. Relieve the fuel system pressure.
3. Remove the intercooler assembly, Turbo only.
4. Disconnect the throttle cable, hoses and electrical connectors from the throttle body and remove the throttle body.
5. Disconnect the vacuum line at the pressure regulator and the electrical connectors at the fuel injectors.
6. Disconnect the fuel lines at the fuel rail.
7. Remove the fuel rail mounting bolts and remove the fuel rail and injectors.

To install:

8. Replace all O-rings with new

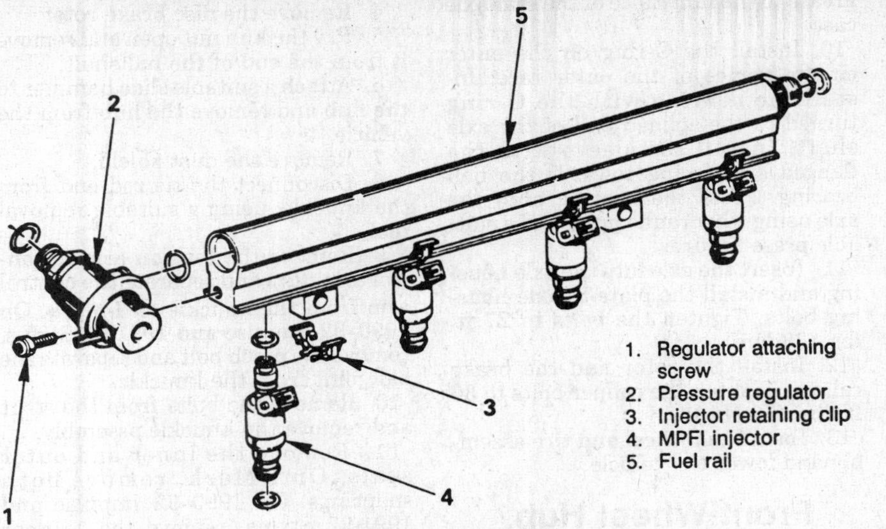

Fuel rail and injector

ones. Lubricate O-rings with engine oil during installation.

9. Install the fuel rail and mounting bolts. Torque the bolts to 10 ft. lbs. (13 Nm).
10. Connect the fuel lines at the fuel rail.
11. Connect the vacuum line at the pressure regulator and the electrical connectors at the fuel injectors.
12. Connect the throttle cable, hoses and electrical connectors to the throttle body.
13. Install the intercooler assembly, Turbo only.
14. Install the air cleaner. Connect the negative battery cable.

DRIVE AXLE

Halfshaft

REMOVAL & INSTALLATION

1. Raise and safely support the vehicle. Drain the oil from the transaxle and replace the drain plug.
2. Remove the front wheel and tire assembly.
3. Remove the hub nut.
4. Disconnect the tie rod end from the steering knuckle using a suitable removal tool.
5. Loosen the pinch bolt and disconnect the lower control arm from the knuckle.
6. Pull out the hub and knuckle assembly from the halfshaft carefully. When pulling the hub assembly, pull just enough to push the shaft off. If necessary, strike the end of the halfshaft with a suitable plastic hammer.
7. To detach the snapring, fitted on the spline of the inboard CV-joint, from the differential side gear, pry out the inboard joint using a suitable tool.

NOTE: Never pull out the halfshaft, to prevent damaging the CV-joint. To prevent damage to the CV-boots, be careful not to bring them into contact with other parts when removing the halfshaft assembly.

To install:

8. Tighten the ball joint pinch bolt to 48 ft. lbs. (65 Nm). Tighten the tie rod end nut to 29 ft. lbs. (39 Nm).
9. Replace the locking nub nut with a new one.
10. Apply grease to the hub nut fitting surfaces and the shaft threads. Tighten the nut to 137 ft. lbs. (186 Nm) and stake the nut.

CV-Boot

REMOVAL & INSTALLATION

1. Raise and safely support the vehicle.
2. Remove the halfshaft assembly.
3. Mount the halfshaft in a suitable Vice.

NOTE: The outer CV-joint cannot be disassembled. The inboard CV-joint must be removed to replace either CV-joint boot. The 1990–92 Impulse, Stylus and 1989 I-Mark equipped with automatic transaxle, on the right halfshaft only, a tripod inboard CV-joint is used. All other inboard CV-joints are the double offset type. A different removal and installation procedure is used for each.

4. Remove the double offset CV-joint as follows:
   a. Mark the position of the case in relation to the shaft using paint or ink. Do not use a punch.
   b. Remove the big end band clip.
   c. Remove the circular clip and remove the outer case.
   d. Turn the ball guide on an angle and move it to the center shaft side. Remove the balls.
   e. Remove the snapring. Mark the position of the ball retainer in relation to the shaft and remove the ball retainer.
   f. Remove the ball guide and small end band clip. Remove the boot.
   g. If the outer boot is to be replaced, remove the band clips and remove the outer boot.
5. Remove the Tripod CV-joint as follows:
   a. Mark the position of the case in relation to the shaft using paint or ink. Do not use a punch.
   b. Remove the big end band clip.
   c. Remove the outer case.
   d. Remove the snapring and pull the Tripod from the shaft. It may be necessary to use a brass drift to dislodge the Tripod from the splines.
   e. Remove the small band clip and boot.

To install:

6. When assembling, apply special grease to ½ of the space within the double offset joint and tripod joint. Make sure the band clips are securely tightened and the boot is not twisted.
7. Install the tripod CV-joint, boots and clips.
8. Install the halfshaft and lower the vehicle. Torque the halfshaft nut to 137 ft. lbs. (186 Nm) and stake the nut or install a new cotter pin.

Driveshaft and U-Joints

REMOVAL & INSTALLATION

1988-89 Impulse

1. Raise and safely support the vehicle.
2. Mark the position of the rear driveshaft in relation to both the center bearing flange and the differential flange.
3. Remove the rear driveshaft mounting nuts and bolts and remove the rear driveshaft.
4. Remove the center bearing bracket bolts and remove the center bearing bracket.
5. Remove the front driveshaft. Install a plug or equivalent in the rear of the transmission to prevent fluid loss.
6. Installation is the reverse of the

removal procedure. Torque the center bearing bracket bolts to 13 ft. lbs. (19 Nm). Torque the driveshaft-to-flange bolts to 24 ft. lbs. (33 Nm).

1990–92 Impulse with AWD

1. Raise and safely support the vehicle.
2. Apply alignment marks on the flange at the center bearing.
3. Remove the bolts at the rear axle side and center bearing.
4. Remove the shaft from the vehicle.
5. Installation is the reverse of the removal procedure. Align the marks and install the bolts. Torque the center bearing bolts to 14 ft. lbs. (19 Nm) and the rear bolts to 26 ft. lbs. (35 Nm).

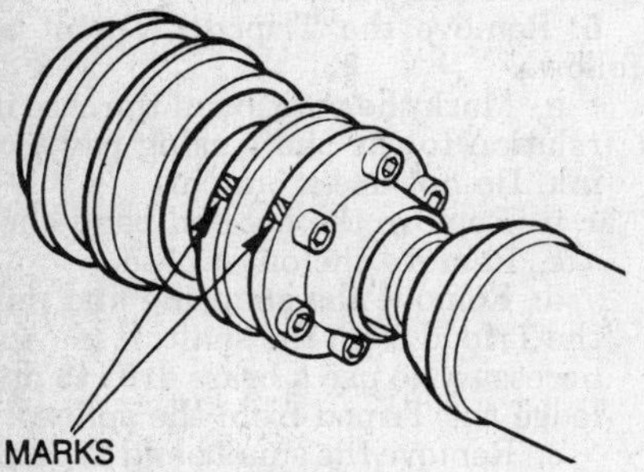

Halfshaft alignment marks — All Wheel Drive Impulse

Rear Axle Shaft, Bearing and Seal

REMOVAL & INSTALLATION

1. Raise and safely support the vehicle.
2. Remove the rear wheel and tire assembly.
3. Remove the 2 caliper attaching bolts and remove the caliper from the rotor. Support the caliper with a length of wire, do not let it hang from the brake hose.
4. Remove the disc brake rotor.
5. Working through the access hole, remove the plate-to-axle housing bolts.
6. Attach a suitable slide hammer to the axle flange and remove the axle from the axle housing.
7. Use a suitable tool to remove the oil seal from the axle housing.
8. Remove the axle bearing by applying pressure on the shaft with a hydraulic press, using a suitable press fixture.

To install:

9. Use a suitable tool to install the oil seal in the axle housing. Apply approximately 1.4 ozs. of wheel bearing grease to the inner face of the rear axle case.
10. Install the O-ring on the outer circumference of the outer race. Install the bearing with the O-ring turned to the splined end of the axle shaft. Install the sleeve with the flanged side facing towards the ball bearing. Press the bearing onto the axle using a hydraulic press and a suitable press fixture.
11. Insert the axle into the axle housing and install the plate-to-axle housing bolts. Tighten the bolts to 27 ft. lbs. (38 Nm).
12. Install the rotor and the brake caliper. Tighten the caliper bolts to 36 ft. lbs. (50 Nm).
13. Install the wheel and tire assembly and lower the vehicle.

Front Wheel Hub, Knuckle and Bearings

REMOVAL & INSTALLATION

NOTE: Do not remove the hub from the knuckle unless it is absolutely necessary.

1. Raise and safely support the vehicle.
2. Remove the front wheel and tire assembly.
3. Remove the brake caliper and support it with a length of wire. Do not let the caliper hang from the brake hose.

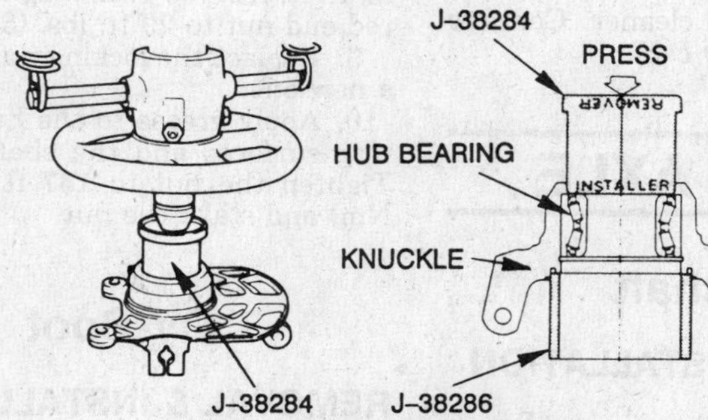

Removing from hub bearings

4. Remove the disc brake rotor.
5. Pry the hub nut open and remove it from the end of the halfshaft.
6. Attach a suitable slide hammer to the hub and remove the hub from the vehicle.
7. Remove the dust shield.
8. Disconnect the tie rod end from the knuckle, using a suitable removal tool.
9. Remove the tension arm-to-control arm nuts and separate the control arm from the knuckle on I-Mark. On 1990–92 Impulse and 1991-92 Stylus, remove the pinch bolt and separate the ball joint from the knuckle.
10. Remove the bolts from the strut and remove the knuckle assembly.
11. Remove the inner and outer seals. On I-Mark, remove both snaprings. On 1990–92 Impulse and 1991-92 Stylus, remove the 1 inner snapring.
12. On I-Mark, mount the knuckle in a hydraulic press and press out the inside inner race and bearing, using a suitable press fixture. Mount the hub in the press and, using a suitable fixture, press out the outside inner race.
13. On 1990–92 Impulse and 1991-92 Stylus, mount the knuckle in a hydraulic press and press out the bearing, using a suitable fixture.

To install:

14. On I-Mark, install the outer snapring and mount the knuckle in a hydraulic press. Using a suitable fixture, press in a new hub bearing and

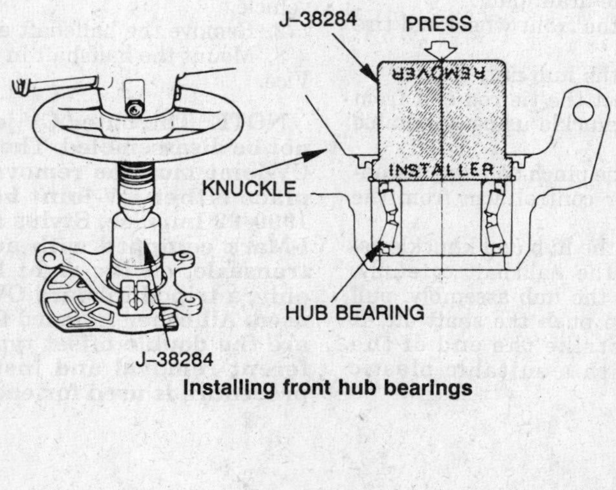

Installing front hub bearings

inner and outer races. Do not remove the inner race fixed cover which is attached to the inside of the bearing inner race. When forcing the hub bearing onto the knuckle, apply pressure to the bearing outer race.

15. On 1990–92 Impulse and 1991-92 Stylus, mount the knuckle in a hydraulic press and, using a suitable fixture, press in a new bearing.

16. Install the inner snapring and the inner and outer seals. Install the dust cover.

17. Mount the knuckle in a hydraulic press and press the hub onto the knuckle, using a suitable fixture. On I-Mark, the bearing inner race mounting cover must be removed before the hub can be forced onto the knuckle. Push the mounting cover down and away from the hub to remove it.

18. Install the hub and knuckle assembly onto the halfshaft and tighten the hub nut temporarily.

19. Install the strut bolts and tighten to 87 ft. lbs. (120 Nm) on I-Mark and 115 ft. lbs. (156 Nm) on 1990–92 Impulse and 1991-92 Stylus.

20. On I-Mark, connect the control arm to the tension arm and install the nuts. Tighten the nuts to 79 ft. lbs. (110 Nm). Always replace a used self-locking nut with a new one.

21. On 1990–92 Impulse and 1991-92 Stylus, connect the ball joint to the knuckle and tighten the pinch bolt to 48 ft. lbs. (65 Nm).

22. Connect the tie rod end to the knuckle and tighten the nut to 29 ft. lbs. (39 Nm).

23. Remove the hub nut and apply grease to the halfshaft's thread. Install the hub nut and tighten it to 137 ft. lbs. (186 Nm). Stake the nut after installation.

24. Install the brake rotor and the caliper.

25. Install the wheel and tire assembly and lower the vehicle.

Pinion Seal

REMOVAL & INSTALLATION

1. Raise and safely support the vehicle.
2. Remove the driveshaft and the differential flange.
3. Use a suitable tool to pry out the pinion seal.
4. Installation is the reverse of the removal procedure. Lubricate the lip of the seal after installation.
5. Tighten the flange nut to 130–202 ft. lbs. (180–280 Nm).

Differential Carrier

REMOVAL & INSTALLATION

Impulse with 4ZD1 and 4ZC1-T Engines

1. Raise and safely support the vehicle.
2. Drain the axle housing and replace the drain plug.
3. Remove the driveshaft.
4. Remove the axle shafts.
5. Remove the differential carrier mounting bolts and remove the differential carrier.
6. Installation is the reverse of the removal procedure. Tighten the carrier mounting bolts to 18 ft. lbs. (25 Nm). Fill the axle housing with the proper type and quantity of oil.

Axle Housing

REMOVAL & INSTALLATION

Except 1990–92 Impulse with AWD

1. Raise and safely support the vehicle.
2. Remove the rear wheel and tire assemblies.
3. Remove the driveshaft.
4. Remove the calipers and support them with a length of wire. Do not let the calipers hang by the brake hoses. Disconnect the parking brake cables.
5. Remove the exhaust pipe and muffler, if necessary.
6. Disconnect the sway bar from the axle housing.
7. Support the axle housing with a suitable jack.
8. Remove the panhard rod.
9. Disconnect the shock absorbers from the axle housing.
10. Lower the axle housing slowly, enough to remove the coil springs.
11. Disconnect the control arms from the axle housing and lower the axle housing from the vehicle.

To install:

12. Connect the control arms to the axle housing.
13. Raise the axle housing slowly, enough to install the coil springs.
14. Connect the shock absorbers to the axle housing.
15. Install the panhard rod.
16. Connect the sway bar to the axle housing. Torque to 58 ft. lbs. (78 Nm).
17. Install the exhaust pipe and muffler, if removed.
18. Install the calipers. Connect the parking brake cables.
19. Install the driveshaft. Torque the nut to 137 ft. lbs. (186 Nm).
20. Install the rear wheel and tire assemblies. Torque the lug nuts to 100 ft. lbs. (136 Nm).
21. Lower the vehicle.

1990–92 Impulse with AWD

1. Drain the rear differential fluid.

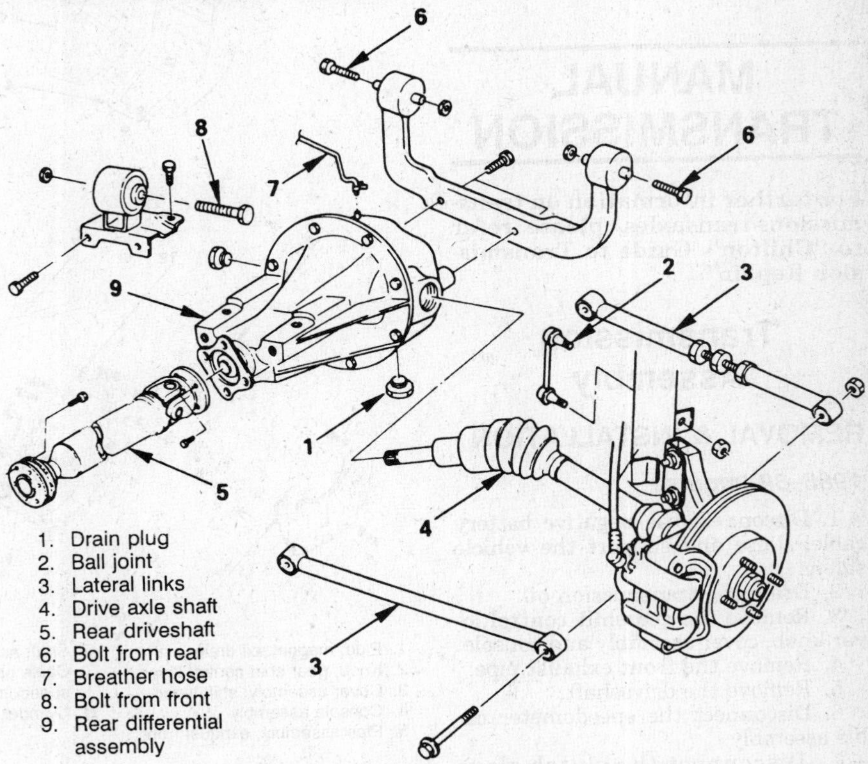

1. Drain plug
2. Ball joint
3. Lateral links
4. Drive axle shaft
5. Rear driveshaft
6. Bolt from rear
7. Breather hose
8. Bolt from front
9. Rear differential assembly

Rear differential assembly—All Wheel Drive Impulse

2. Disconnect the ball joint for both side of the stabilizer bar.
3. Disconnect the lateral links at the wheel side.

NOTE: Never pull out the driveshaft. Damage to the boots and joints may result.

4. Pry the inboard joint against the differential housing to dislodge the snapring inside the differential assembly using a suitable prybar. Remove the rear axle shafts from the vehicle.
5. Remove the rear driveshaft.
6. Support the differential assembly using a suitable jack. Remove the 2 bolts in the rear mounting and disconnect the breather hose.
7. Remove the bolt from the front mounting and remove the assembly from the vehicle.

To install:

8. Install the assembly into the vehicle and install the bolt to the front mounting.
9. Install the 2 bolts in the rear mounting and connect the breather hose.
10. Install the rear driveshaft.
11. Install the rear axle shafts to the vehicle.
12. Connect the lateral links at the wheel side.
13. Connect the ball joint for both side of the stabilizer bar.
14. Refill the rear differential with fluid.

MANUAL TRANSMISSION

For further information on transmissions/transaxles, please refer to "Chilton's Guide to Transmission Repair".

Transmission Assembly

REMOVAL & INSTALLATION

1988–89 Impulse

1. Disconnect the negative battery cable. Raise and support the vehicle safely.
2. Drain the transmission oil.
3. Remove the gearshift control lever knob, cover assembly and console.
4. Remove the front exhaust pipe.
5. Remove the driveshaft.
6. Disconnect the speedometer cable assembly.
7. Disconnect the clutch slave cylinder.
8. Remove the flywheel dust shield.
9. Position a jack under the transmission case, remove the engine rear mounting nuts, lower the transmission case slightly, then remove the bolts attaching the quadrant box cover to the transmission case.
10. Disconnect all electrical harness connectors.
11. Remove the control box assembly.
12. Remove the transmission-to-engine retaining bolts.

NOTE: The starter assembly is mounted in position with the bolts that are used for installing the transmission assembly to the engine. It may be necessary to move the starter assembly forward to prevent it from falling out when the bolts are removed.

To install:

13. Position the transmission assembly with the speedometer cable fitting face turned downward and slide the assembly forward, guiding the gear shaft into the pilot bearing.
14. Install and tighten the quadrant box cover to the transmission case bolts with a gasket fitted in position between the quadrant box cover and the transmission case, then install the engine rear mount.
15. When reconnecting the driveshaft, install the bolts from the extension shaft side and the nuts and washers on the driveshaft side and tighten to 26 ft. lbs. (36 Nm).
16. After tightening the rear engine mounting nuts to 14 ft. lbs. (19 Nm), raise the tab of the washers to prevent the nuts from loosening.
17. Install and tighten the drain plug to 29 ft. lbs. (40 Nm). Remove the filler plug and fill the transmission with the proper oil. The capacity is approximately 3.28 pints. Tighten the filler plug to 29 ft. lbs. (40 Nm).

LINKAGE ADJUSTMENT

The shift lever is mounted on top of the transmission extension housing. All linkage is inside the transmission and requires no adjustment.

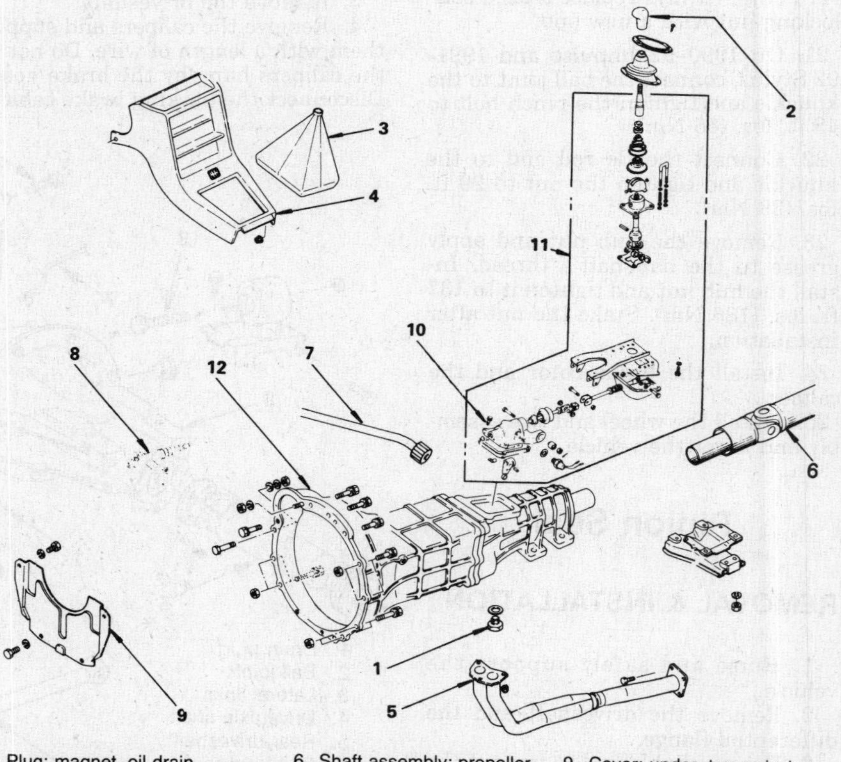

1. Plug; magnet, oil drain
2. Knob; gear shift control lever
3. Cover assembly; shift lever
4. Console assembly
5. Pipe assembly; exhaust front
6. Shaft assembly; propeller
7. Cable assembly; speedometer
8. Cylinder assembly; slave
9. Cover; under, transmission case
10. Bolts; quadrant cover to case
11. Box assembly; control
12. Transmission assembly

Manual transmission removal—1988–89 Impulse

MANUAL TRANSAXLE

For further information on transmissions/transaxles, please refer to "Chilton's Guide to Transmission Repair".

Transaxle Assembly

REMOVAL & INSTALLATION

I-Mark

1. Disconnect the negative battery cable.
2. Raise and safely support the vehicle. Remove the front wheel and tire assemblies.
3. Drain the oil from the transaxle and replace the drain plug.
4. Mark the position of the hood on the hood hinge brackets and remove the hood. Remove the air duct.
5. Disconnect the negative battery cable and the electrical connectors from the transaxle.
6. Disconnect the speedometer, clutch and shift cables from the transaxle.
7. Disconnect the right control arm end at the knuckle. Remove the left tension rod with the bracket. Disconnect both tie rod ends at the knuckle using a suitable removal tool.
8. Use a suitable tool to pry out the halfshafts. Be careful not to damage the transaxle oil seals when pulling out the halfshafts.
9. Remove the flywheel dust cover.
10. Attach a suitable hoist to the engine and support the transaxle with a suitable jack. Remove the motor mount bolts.
11. Remove the bolts of the center beam and lower the engine, tilting it away from the engine support fixture.
12. Remove the mounting bolts of the clutch housing to the engine. Remove the transaxle assembly from the vehicle.

NOTE: On turbocharged vehicles, removal of the transaxle assembly is possible only after the engine foot of the engine mounting has been removed.

To install:

13. Install the transaxle into the vehicle.
14. Torque the clutch housing-to-engine mounting bolts and the center beam mounting bolts to 56 ft. lbs. (77 Nm). Tighten the flywheel dust cover bolts to 4 ft. lbs. (6 Nm).
15. Tighten the tie rod nuts to 42 ft. lbs. (58 Nm). Tighten the tension rod mounting bolts on the body to 48 ft. lbs. (67 Nm) and on the control arm to 80 ft. lbs. (110 Nm).
16. Refill the transaxle with the proper type and quantity of oil and adjust the clutch cable.

1990–92 Impulse and 1991-92 Stylus

1. Disconnect the battery cables and remove the battery and battery tray.
2. Mark the position of the hood on the hinge brackets and remove the hood.
3. Raise and safely support the vehicle. Drain the oil from the transaxle and replace the drain plug.
4. Remove the air duct and air cleaner assembly.
5. Tag and disconnect the electrical connectors and disconnect the ground cable and engine wiring harness clamp from the transaxle side.
6. Remove the ignition coil ground wire and ignition wire and remove the battery bracket and ignition coil.
7. Disconnect the clutch cable and the shift and select cables from the transaxle. Disconnect the speedometer cable.
8. Remove the front wheel and tire assemblies.
9. Remove the engine splash shield.
10. Use a suitable tool to disconnect the tie rod ends from the steering knuckles.
11. Remove the pinch bolt and disconnect both ball joints at the steering knuckle.
12. Remove the remaining bolts attaching the center bearing support bracket to the engine. Disengage both halfshafts from the transaxle side.
13. Attach a suitable hoist to the engine. Support the transaxle using a suitable jack.
14. Remove the front exhaust pipe and the torque rod and bracket.
15. Remove the left transaxle mount. Remove the center beam with the rear transaxle mount.
16. Remove the engine stiffener and the flywheel dust cover.
17. Remove the transaxle-to-engine attaching bolts and lower the transaxle from the vehicle.

To install:

18. Raise the transaxle into position and install the transaxle-to-engine attaching bolts. Tighten the bolts to 69 ft. lbs. (94 Nm).
19. Install the flywheel dust shield and tighten the mounting bolts to 53 inch lbs. (6 Nm).
20. Install the engine stiffener and tighten the bolts to 28 ft. lbs. (38 Nm).
21. Install the center beam with the rear transaxle mount. Tighten the center beam-to-body bolts to 37 ft. lbs. (50 Nm), the center beam-to-crossmember bolts to 56 ft. lbs. (76 Nm) and the rear mount-to-transaxle bolt to 76 ft. lbs. (103 Nm).
22. Install the left transaxle mount. Tighten the left transaxle mount-to-transaxle bolt to 35 ft. lbs. (48 Nm) and the left transaxle mount-to-body bolt to 51 ft. lbs. (69 Nm).
23. Install the torque rod stud to the transaxle and tighten to 71 ft. lbs. (96 Nm). Install the torque rod to the center beam and the torque rod bracket to the engine. Tighten the bracket-to-engine bolt to 28 ft. lbs. (38 Nm), the torque rod-to-center beam bolt to 51 ft. lbs. (69 Nm) and the torque rod-to-transaxle bolt to 95 ft. lbs. (129 Nm).
24. Install the front exhaust pipe and tighten the pipe-to-manifold bolts to 49 ft. lbs. (67 Nm).
25. Install the halfshafts. Tighten the ball joint nuts to 48 ft. lbs. (65 Nm). Tighten the tie rod end nuts to 29 ft. lbs. (39 Nm).
26. Install the remainder of the components in the reverse order of their removal. Fill the transaxle with the proper type and quantity of oil. Adjust the clutch cable and shift linkage.

1990–92 Impulse with AWD

1. Disconnect the negative battery cable.
2. Remove the battery and battery bracket.
3. Remove the intercooler, air duct and air cleaner.
4. Disconnect all electrical harnesses and cables from the transaxle.
5. Remove the transfer case.
6. Disconnect the torque rod and bracket to the lower control arm.
7. Remove the left mounting rubber.
8. Place a transmission jack under the transaxle and raise the engine with an engine holding fixture.
9. Remove the transaxle-to-engine bolts.
10. Lower the transaxle case from the vehicle.

To install:

11. Raise the transaxle case into the vehicle.
12. Install the transaxle-to-engine bolts.
13. Install the left mounting rubber and remove the transmission jack.
14. Connect the torque rod and bracket to the lower control arm.
15. Install the transfer case.
16. Connect all electrical harnesses and cables to the transaxle.
17. Install the intercooler, air duct and air cleaner.
18. Install the battery and battery bracket.
19. Connect the negative battery cable and check operation.

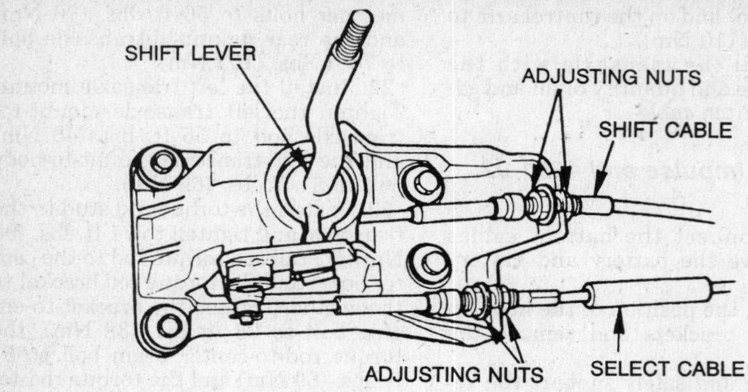

Shift linkage adjustment—1990–92 Impulse and 1991–92 Stylus

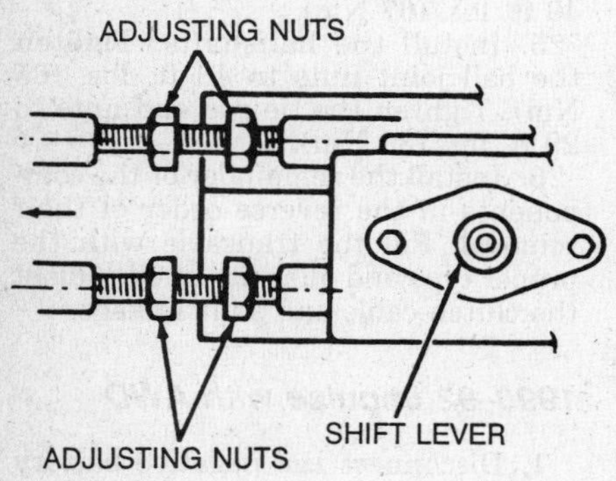

Shift linkage adjustment—I-Mark

LINKAGE ADJUSTMENT

1. Place the transaxle in neutral. Turn the adjusting nuts on the driver's side of the shift housing, until the gearshift lever is at right angle to the pivot case, as viewed from the side of the gear control.
2. After the adjustment tighten the 2 adjusting nuts securely.
3. Turn the adjusting nuts on the passenger's side of the shift housing, until the gearshift lever is at right angle to the pivot case as viewed from the rear of the gear control.
4. After the adjustment tighten the 2 adjusting nuts securely.

CLUTCH

Clutch Assembly

REMOVAL & INSTALLATION

1. Disconnect the negative battery cable.
2. Raise and safely support the vehicle.
3. Remove the transmission/transaxle assembly.
4. Mark the position of the pressure plate on the flywheel and remove the pressure plate and the clutch disc.
5. Remove the release bearing from the clutch fork.
6. Inspect the flywheel for scoring or heat cracks and resurface or replace, as necessary.
7. Inspect the pressure plate and clutch disc for wear or scoring and replace, as necessary.
8. Inspect the release bearing for wear or binding and replace, as necessary.

To install:

9. Install the flywheel if it was removed. Make sure the crankshaft flange and the mating surface of the flywheel are clean. Tighten the flywheel bolts in a criss-cross pattern.
10. Tighten the bolts on 4XC1-U, 4XC1-T and 4XE1 engines in 2 steps, first to 22 ft. lbs. (29 Nm) and then repeat the pattern turning the bolts another 45 degrees each. Tighten the bolts on 4ZD1 engines to 43 ft. lbs. (60 Nm). Tighten the bolts on 4ZC1-T engines to 40 ft. lbs. (55 Nm).
11. Lubricate the pilot bearing with a suitable grease. Install a suitable clutch alignment tool.
12. Install the clutch disc and the pressure plate on the flywheel. If the old pressure plate is used, make sure it is installed in the position that was marked during the removal procedure. Tighten the pressure plate bolts in a criss-cross pattern in 2–3 steps to avoid warping the pressure plate. The final bolt torque should be 13 ft. lbs. (18 Nm).
13. Remove the clutch alignment tool.
14. Apply suitable grease to the sliding surface of the release bearing and the contact surface of the clutch fork. Install the release bearing.
15. Install the remaining components in the reverse order of their removal. Adjust the clutch cable or pedal height, as required.

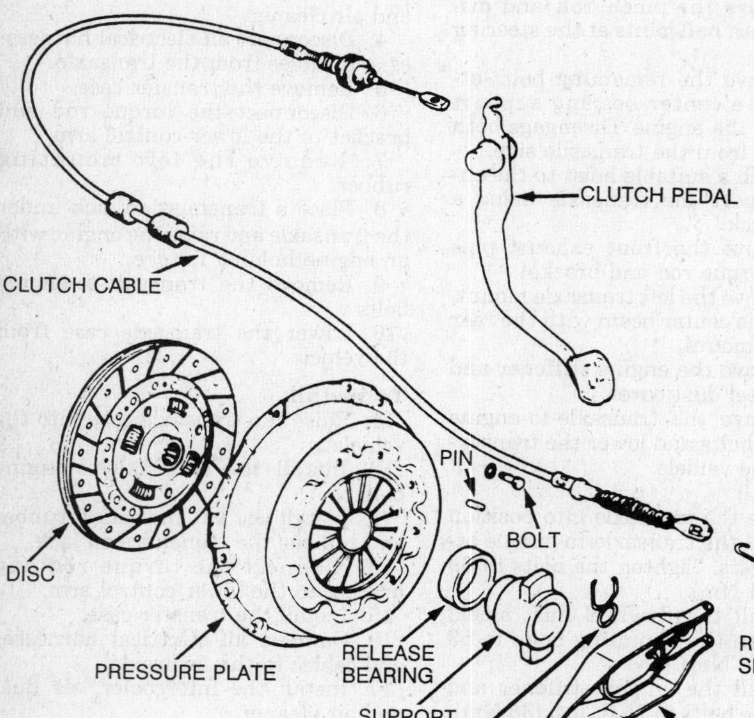

Exploded view of clutch components—except 1988–89 Impulse

PEDAL HEIGHT/FREE-PLAY ADJUSTMENT

1988–89 Impulse

1. Loosen the clutch switch locknut.
2. Turn the clutch switch until the clearance between the clutch switch and the pedal is 0.020–0.059 in. (0.5–1.5mm).
3. Tighten the locknut.

NOTE: After adjustment, make sure the pushrod is in contact with the piston in the master cylinder. The clutch pedal free-play and pedal stroke are self-adjusted.

Clutch Cable

ADJUSTMENT

1. Pull the clutch cable to the rear until the adjusting nut turns freely.
2. Turn the adjusting nut either clockwise or counterclockwise to adjust the cable length.
3. On 1990–92 Impulse and 1991-92 Stylus, repeat Step 2 so the play between the clutch release arm and the clutch cable is 0.04–0.12 in. (1–3mm).
4. The pedal free-play should be 0.39–0.79 in. (10–20mm) on SOHC vehicles or 0.19–0.59 in. (15–25mm) on DOHC vehicles.

REMOVAL & INSTALLATION

1. Disconnect the negative battery cable.
2. Raise and support safely the front of the vehicle.
3. Disconnect the clutch cable from the clutch housing and transaxle. Remove all necessary snaprings and locknuts to free the clutch cable.
4. Remove the clutch cable from the clutch pedal and pull the cable out from the engine compartment.
5. Installation is the reverse of the removal procedure. Adjust the clutch cable.

Clutch Master Cylinder

REMOVAL & INSTALLATION

1. Disconnect the negative battery cable.
2. Disconnect clutch pedal return spring and joint pin from clutch pedal.
3. Disconnect and plug the hydraulic line and the hose at the clutch master cylinder.
4. Remove 2 hold-down nuts from inside of passenger compartment.
5. Remove clutch master cylinder from engine compartment.
6. Installation is the reverse of the removal procedure. Bleed the hydraulic system.

Clutch Slave Cylinder

REMOVAL & INSTALLATION

1. Disconnect the negative battery cable.
2. Raise and safely support the vehicle.
3. Disconnect and plug the hydraulic line at the slave cylinder.
4. Remove the slave cylinder mounting bolts and disconnect the pushrod from the clutch fork. Remove the slave cylinder.
5. Installation is the reverse of the removal procedure. Bleed the hydraulic system.

Hydraulic Clutch System Bleeding

1. Fill the clutch fluid reservoir with brake fluid and keep it filled during the bleeding operation.
2. Remove the bleeder rubber cap and connect a clear plastic length of hose to the bleeder. Submerge the other end of the hose in a transparent container half filled with brake fluid.
3. Pump the clutch pedal several times and hold the pedal depressed.
4. With the pedal depressed, loosen the bleeder ½ turn to release brake fluid with air and tighten it immediately.
5. Repeat Steps 3 and 4 until air bubbles disappear completely from the fluid being forced out.
6. After bleeding is completed, check for pedal free-play and clutch disengagement, then check the brake fluid level in the reservoir.

AUTOMATIC TRANSMISSION

For further information on transmissions/transaxles, please refer to "Chilton's Guide to Transmission Repair".

Transmission Assembly

REMOVAL & INSTALLATION

1. Disconnect the negative battery cable.
2. Raise and safely support the vehicle. Drain the transmission and replace the drain plug.
3. Disconnect the throttle cable at the engine side. Remove the transmission dipstick.
4. Disconnect the electrical connectors and remove the starter.
5. Mark the position of the driveshaft on the differential flange and remove the driveshaft.
6. Disconnect the link rod at the shift lever side. Disconnect the speedometer cable.
7. Remove the exhaust header pipe.
8. Disconnect the transmission cooler lines from the transmission. Secure the lines close to the body to prevent damage during transmission removal.
9. Remove the flywheel dust cover and the engine splash shield.
10. Remove the 6 bolts attaching the converter to the flywheel. Rotate the engine at the crankshaft pulley to gain access to all of the bolts.
11. Support the transmission using a suitable jack. Support the rear of the engine to hold it in position when the transmission is removed.
12. Remove the rear transmission mount and remove the transmission-to-engine mounting bolts.
13. Remove the transmission, taking care not to let the torque converter slip out of the transmission.

To install:

14. Install the transmission and torque the transmission-to-engine mounting bolts to 47 ft. lbs. (65 Nm) and the converter attaching bolts to 13 ft. lbs. (19 Nm).
15. Install the driveshaft, aligning the marks that were made during the removal procedure. Tighten the bolts to 26 ft. lbs. (36 Nm).
16. Tighten the drain plug to 15 ft. lbs. (21 Nm). Adjust the link rod and the throttle cable.
17. Lower the vehicle and fill the transmission with approximately 4.2 qts. of the proper fluid. Start the engine and add fluid, as necessary, to bring the fluid to the proper level. Avoid racing the engine.

SHIFT LINKAGE ADJUSTMENT

1. Remove the adjustment nut attaching the select lever link to the control lever.
2. Move the manual valve lever forward to stop, then return to the **N** position (3rd stop).
3. Depress the manual valve lever toward the **R** position lightly and tighten the adjust nut.
4. Check that the control lever moves smoothly and that the position indicator works correctly.

THROTTLE LINKAGE ADJUSTMENT

1. Check that the throttle valve is held closed completely.
2. Adjust the setting of the adjustment nut, as necessary, so the clearance A between the inner cable stopper and the end of the rubber boot on the outer cable is adjusted to 0.032–0.059 in. (0.8–1.5mm).
3. Open the throttle valve fully and check that the inner cable stroke B is within the range of 1.30–1.34 in. (33–34mm).

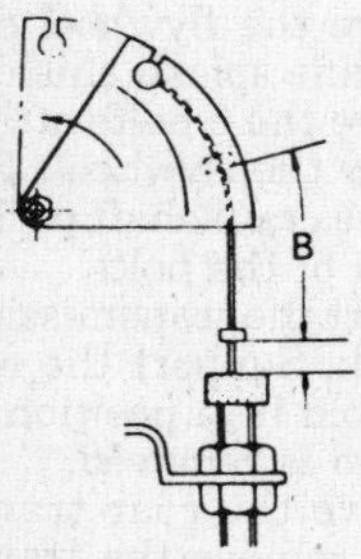

Throttle cable adjustment – 1988–89 Impulse

AUTOMATIC TRANSAXLE

For further information on transmissions/transaxles, please refer to "Chilton's Guide to Transmission Repair".

Transaxle Assembly

REMOVAL & INSTALLATION

I-Mark

1. Disconnect the negative battery cable.
2. Mark the position of the hood on the hinge brackets and remove the hood.
3. Raise and safely support the vehicle.
4. Drain the transaxle and replace the drain plug.
5. Tag and disconnect the electrical connectors from the transaxle. Disconnect the negative battery cable from the transaxle.
6. Disconnect the speedometer and the control cable from the transaxle.
7. Disconnect and plug the transaxle cooler lines at the transaxle. Disconnect the vacuum diaphragm hose from the transaxle.
8. Remove the air duct.
9. Remove the front wheel and tire assemblies.
10. Disconnect the right control arm end at the knuckle and remove the left tension rod with the bracket. Use a suitable removal tool to disconnect the tie rod ends at the steering knuckles.
11. Use a suitable tool to pry out the halfshafts. Be careful not to damage the transaxle seals when pulling out the halfshafts.
12. Attach a suitable hoist to the engine and support the transaxle with a suitable jack.
13. Disconnect the engine mounts and remove the center support. Remove the 3 upper transaxle-to-engine mounting bolts.
14. Remove the flywheel dust cover. Remove the 4 converter-to-flywheel mounting bolts.
15. Tilt the engine and lower the transaxle from the vehicle.

To install:

16. Raise the transaxle into position and install the transaxle-to-engine mounting bolts. Tighten the bolts to 56 ft. lbs. (77 Nm).
17. Install the converter-to-flywheel bolts and tighten to 30 ft. lbs. (41 Nm). Install the flywheel dust shield.
18. Install the center support and tighten the bolts to 56 ft. lbs. (77 Nm).
19. Install both lower engine mounts and tighten to 61 ft. lbs. (84 Nm).
20. Install the halfshafts into the transaxle. Connect the tie rod ends to the knuckles and tighten the nuts to 42 ft. lbs. (58 Nm).
21. Install the tension rod. Tighten the tension rod-to-body mounting bolts to 48 ft. lbs. (67 Nm) and the tension rod-to-knuckle nuts to 80 ft. lbs. (110 Nm). Always replace removed self-locking nuts with new ones.
22. Install the remaining components in the reverse order of their removal. Fill the transaxle with the proper type and quantity of fluid and adjust the shift control cable.

1990–92 Impulse and 1991-92 Stylus

1. Disconnect the battery cables and remove the battery and battery tray.
2. Mark the position of the hood on the hinge brackets and remove the hood.
3. Remove the air duct and air cleaner assembly.
4. Tag and disconnect the electrical connectors from the transaxle. Disconnect the ground cable and the engine wiring harness clamp from the transaxle side.
5. Disconnect the breather hose from the transaxle side.
6. Disconnect the ignition coil ground wire and ignition wire and remove the battery bracket and ignition coil.
7. Disconnect the shift cable from the transaxle side.
8. Raise and safely support the vehicle.
9. Disconnect the speedometer cable.
10. Remove the front wheel and tire assemblies.
11. Remove the engine splash shield. Disconnect the transaxle cooler lines from the radiator side.
12. Remove the pinch bolt and disconnect both ball joints at the steering knuckle.
13. Use a suitable tool to disconnect the tie rod ends from the steering knuckle.
14. Use a suitable tool to disengage both halfshafts from the transaxle side.
15. Attach a suitable hoist to the engine and support the transaxle with a suitable jack.
16. Remove the exhaust header pipe and the torque rod and bracket.
17. Remove the left transaxle mount and the center beam with the rear transaxle mount.
18. Remove the engine stiffener and the flywheel dust shield.
19. Remove the flywheel-to-converter attaching bolts.
20. Remove the transaxle-to-engine bolts and lower the transaxle from the vehicle.

To install:

21. Raise the transaxle into position and install the transaxle-to-engine bolts. Tighten the bolts to 69 ft. lbs. (94 Nm).
22. Install the flywheel-to-converter bolts and tighten to 33 ft. lbs. (44 Nm). Install the flywheel dust shield and tighten the mounting bolts to 53 inch lbs. (6 Nm).
23. Install the engine stiffener and tighten the bolts to 28 ft. lbs. (38 Nm).
24. Install the center beam with the rear transaxle mount. Tighten the center beam-to-body bolts to 37 ft. lbs. (50 Nm), the center beam-to-crossmember bolts to 56 ft. lbs. (76 Nm) and the rear mount-to-transaxle bolt to 76 ft. lbs. (103 Nm).
25. Install the left transaxle mount. Tighten the left transaxle mount-to-transaxle bolt to 35 ft. lbs. (48 Nm) and the left transaxle mount-to-body bolt to 51 ft. lbs. (69 Nm).
26. Install the torque rod stud to the transaxle and tighten to 71 ft. lbs. (96 Nm). Install the torque rod to the center beam and the torque rod bracket to the engine. Tighten the bracket-to-engine bolt to 28 ft. lbs. (38 Nm), the torque rod-to-center beam bolt to 51 ft. lbs. (69 Nm) and the torque rod-to-transaxle bolt to 95 ft. lbs. (129 Nm).
27. Install the front exhaust pipe and

tighten the pipe-to-manifold bolts to 49 ft. lbs. (67 Nm).

28. Install the halfshafts. Tighten the ball joint nuts to 48 ft. lbs. (65 Nm). Tighten the tie rod end nuts to 29 ft. lbs. (39 Nm).

29. Install the remainder of the components in the reverse order of their removal. Fill the transaxle with the correct quantity of Dexron® II ATF. Adjust the shift linkage.

SHIFT CABLE ADJUSTMENT

I-Mark

1. Loosen locknuts (1) and (2).
2. Place the link assembly (3) on the transaxle in the **N** position.
3. Move the shift lever to the **N** range.
4. Rotate the link assembly (3) clockwise.
5. The shift lever detent pin (4) must be pressing against the **N** wall (5).
6. Temporarily tighten nut (1).
7. Tighten nut (2).

Impulse and Stylus

1. Loosen the 2 adjusting nuts at the selector lever.
2. Shift the transaxle into the **N** detent.
3. Place the shift lever into the **N** position.
4. Hand tighten the front adjusting nut fully against the selector lever boss and then back it off approximately ½ turn.
5. Tighten the rear adjusting nut fully against the selector lever boss to 19 ft. lbs. (26 Nm).

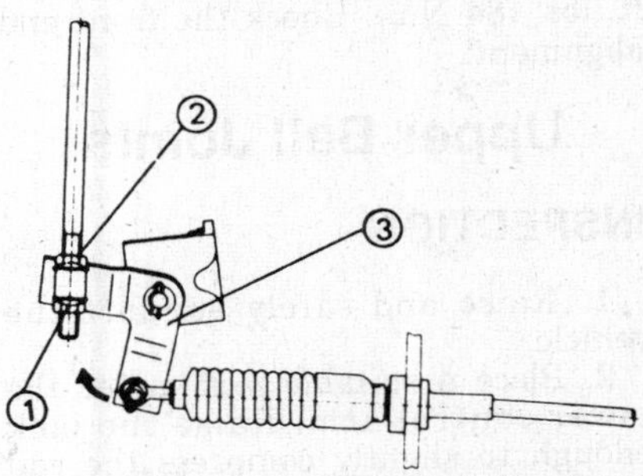

Shift link and cable end—I-Mark

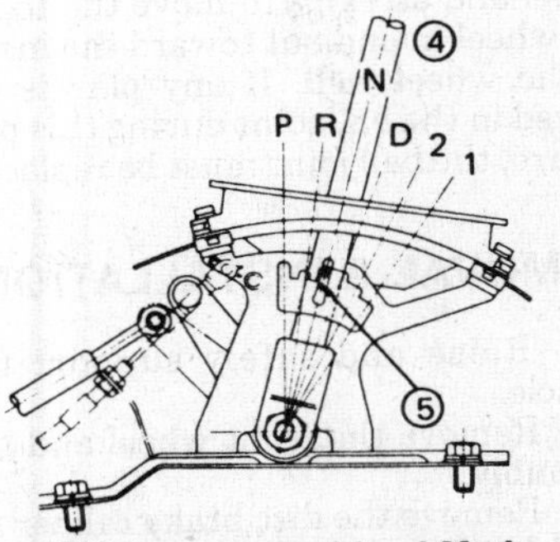

Shift lever assembly—I-Mark

TRANSFER CASE

Transfer Case Assembly

REMOVAL & INSTALLATION

All Wheel Drive Impulse

1. Disconnect the negative battery cable.
2. Drain the transaxle and transfer fluid.
3. Remove the steering shaft protector, intermediate shaft pinch bolt and steering shaft boot.
4. Raise the vehicle and support safely.
5. Remove the front wheels.
6. Remove the left halfshaft nut and both tie rod ends.
7. Disconnect the lower ball joints, remove all under covers, front exhaust pipe and rear driveshaft.
8. Remove the steering hoses and place a drain pan under to catch the fluid.
9. Support the engine with an engine hoist.
10. Remove the torque rod bolt at the center beam.
11. Remove the rear rubber dmounting bolt, center beam and crossmember with the steering unit and stabilizer.
12. Disconnect the FWD halfshafts.
13. Disconnect the speedometer cable and support the transfer case with a jack.
14. Remove the transfer-to-transaxle attaching bolts and remove the assembly.

To install:

15. Install the assembly and attaching bolts. Torque the bolts to 25 ft. lbs. (34 Nm).
16. Connect the speedometer cable and support the transfer case with a jack.
17. Connect the FWD halfshafts.
18. Install the rear rubber mounting bolt, center beam and crossmember with the steering unit and stabilizer.
19. Install the torque rod bolt at the center beam.
20. Install the steering hoses, all undercovers, front exhaust pipe and rear driveshaft.
21. Connect the lower ball joints and install the left halfshaft nut. Torque the nut to 137 ft. lbs. (186 Nm). Reconnect the tie rod ends.
22. Install the front wheels and lower the vehicle.
23. Install the steering shaft protector, intermediate shaft pinch bolt and steering shaft boot.
24. Refill the transaxle and transfer case with fluid.
25. Connect the battery cable and align the front end.

FRONT SUSPENSION

Shock Absorbers

REMOVAL & INSTALLATION

1. Raise and safely support the vehicle.
2. Remove the front wheel and tire assemblies.
3. Remove the shock absorber caps and remove the upper shock mounting nuts.
4. Remove the lower shock mounting bolts and remove the shock absorbers.
5. Installation is the reverse of the removal procedure. Tighten the lower shock mounting bolts to 40 ft. lbs. (56 Nm).

MacPherson Strut

REMOVAL & INSTALLATION

1. Raise and safely support the vehicle.
2. Remove the front wheel and tire assemblies.
3. Remove the brake hose clip at the strut bracket.
4. Disconnect and plug the brake hose at the caliper. Pull the brake hose back through the opening in the strut bracket.
5. Remove the strut-to-knuckle bolts.
6. Remove the strut mounting nuts and remove the strut.

To install:

7. Install the strut assembly and torque the strut mounting nuts to 40 ft. lbs. (56 Nm) on I-Mark or 50 ft. lbs. (68 Nm) on 1990–92 Impulse and 1991-92 Stylus.
8. Torque the strut-to-knuckle bolts to 86 ft. lbs. (120 Nm) on I-Mark or 115 ft. lbs. (156 Nm) on Impulse and Stylus.
9. Bleed the brake system.

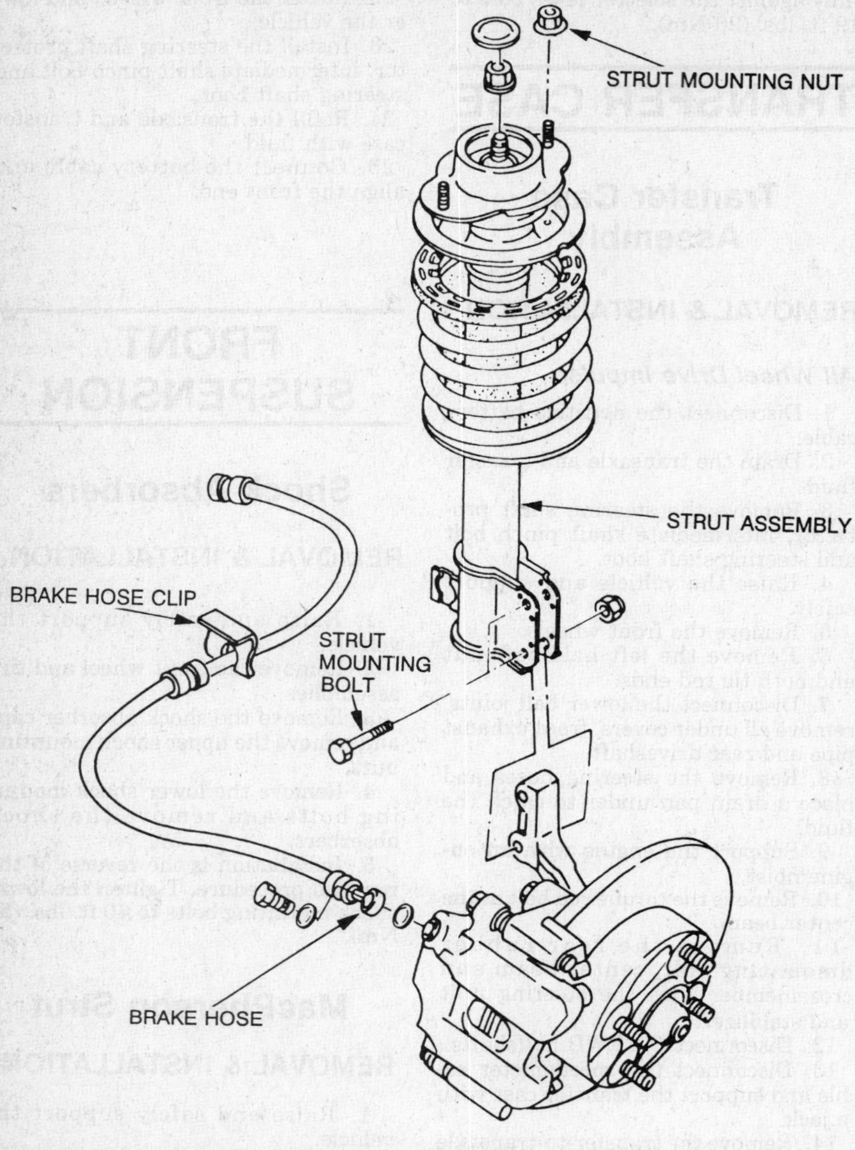

MacPherson strut assembly—I-Mark, 1990–92 Impulse and 1991–92 Stylus similar

Coil Springs

REMOVAL & INSTALLATION

1. Raise and safely support the vehicle.
2. Remove the front wheel and tire assembly.
3. Remove the disc brake caliper assembly. Support the caliper with a length of wire. Do not let the caliper hang from the brake hose.
4. Mark the position of the nuts on the front of the strut bar for reassembly, these nuts control the caster setting, and remove the strut bar.
5. Remove the sway bar brackets on both sides of the vehicle. Remove the bolt, tube and grommets holding the sway bar to the lower control arm. Move the sway bar aside.
6. Use a suitable tool to disconnect the tie rod end from the steering knuckle. Turn the steering wheel to move the tie rod end aside.
7. Place a suitable hydraulic floor jack under the lower control arm and apply slight upward pressure.

NOTE: Install a safety chain through 1 coil at the top of the spring and attach it to the upper control arm to prevent the spring from coming out unexpectedly.

8. Install a spring compressor tool J-36567 or equivalent, and slightly compress the spring.
9. Remove the upper and lower ball joint nuts. Using a suitable removal tool, disconnect the upper and lower ball joints from the steering knuckle. Remove the steering knuckle from the vehicle.
10. Slowly lower the lower control arm until the spring is free of the lower control arm. Slowly and evenly release the spring compressor until the spring is fully extended. Remove the safety chain and the coil spring.

To install:

11. Make sure the lower end of the coil spring is aligned with the notch in the lower control arm.
12. Install new self-locking nuts on the upper and lower ball joint studs and tighten the upper ball joint nut to 39 ft. lbs. (54 Nm) and the lower ball joint nut to 58 ft. lbs. (80 Nm).
13. Install a new self-locking nut on the sway bar end link and tighten to 19 ft. lbs. (26 Nm). Tighten the sway bar bracket bolt to 14 ft. lbs. (19 Nm).
14. Install the strut bar to the chassis and tighten the locknuts. Install the strut bar-to-lower arm bolts, aligning the marks that were made during the removal procedure. Tighten the small bolt to 47 ft. lbs. (65 Nm) and the large bolt to 114 ft. lbs. (158 Nm).
15. Tighten the tie rod end nut to 60 ft. lbs. (84 Nm). Check the front end alignment.

Upper Ball Joints

INSPECTION

1. Raise and safely support the vehicle.
2. Place a suitable jack under the lower control arm. Raise the jack enough to slightly compress the coil spring.
3. Have an assistant grasp the front wheel and attempt to move the top of the wheel in and out toward the inside of the wheel well. If any play is observed in the ball joint during this procedure, the ball joint must be replaced.

REMOVAL & INSTALLATION

1. Raise and safely support the vehicle.
2. Remove the front wheel and tire assembly.
3. Remove the disc brake caliper assembly. Support the caliper with a

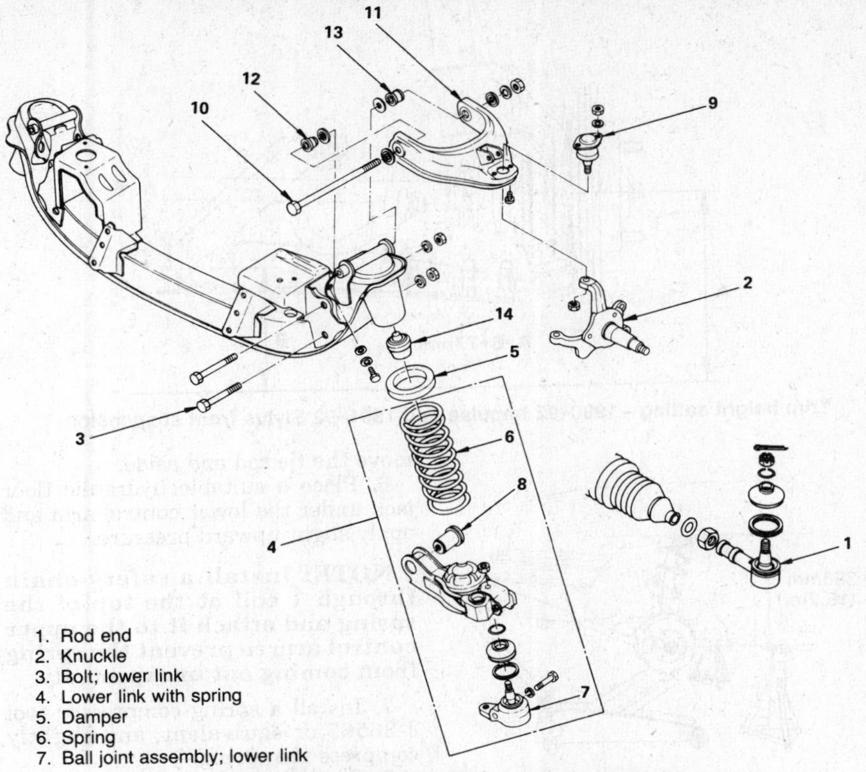

Front suspension assembly—1988–89 Impulse

length of wire. Do not let the caliper hang from the brake hose.

4. Place a suitable hydraulic floor jack under the lower control arm and apply slight upward pressure.

NOTE: Install a safety chain through 1 coil at the bottom of the spring and attach it to the lower control arm to prevent the spring from coming out unexpectedly.

5. Install a spring compressor tool J-36567 or equivalent, and slightly compress the spring.

6. Remove the upper ball joint nut and use a suitable tool to disconnect the upper ball joint from the steering knuckle.

7. Remove the upper ball joint from the upper control arm.

8. Installation is the reverse of the removal procedure. Install the ball joint with the cutaway portion turned outward. Tighten the ball joint mounting nuts to 40 ft. lbs. (56 Nm). Install a new self-locking nut on the upper ball joint stud and tighten to 39 ft. lbs. (54 Nm).

Lower Ball Joints

INSPECTION

1. Raise and safely support the vehicle.

2. On 1988–89 Impulse, place a suitable jack under the lower control arm. Raise the jack enough to slightly compress the coil spring.

3. Have an assistant grasp the front wheel and attempt to move the bottom of the wheel in and out toward the inside of the wheel well. If any play is observed in the ball joint during this procedure, the ball joint must be replaced.

REMOVAL & INSTALLATION

Except 1988–89 Impulse

1. Raise and safely support the vehicle.

2. Remove the front wheel and tire assembly.

3. Remove the pinch bolt and nut from the steering knuckle.

4. Remove the nuts and bolts retaining the lower ball joint to the lower control arm and remove the lower ball joint.

5. Installation is the reverse of the removal procedure. Tighten the bolts retaining the lower ball joint to the lower control arm to 80 ft. lbs. (110 Nm) on I-Mark and 115 ft. lbs. (156 Nm) on 1990–92 Impulse and 1991–92 Stylus. Tighten the pinch bolt to 51 ft. lbs. (70 Nm) on I-Mark and 48 ft. lbs. (65 Nm) on 1990–92 Impulse.

1988–89 Impulse

1. Raise and safely support the vehicle.

2. Remove the front wheel and tire assembly.

3. Mark the position of the nuts on the front of the strut bar for reassembly, these nuts control the caster setting, and remove the strut bar.

4. Remove the sway bar brackets on both sides of the vehicle. Remove the bolt, tube and grommets holding the sway bar to the lower control arm. Move the sway bar aside.

5. Use a suitable tool to disconnect the tie rod end from the steering knuckle. Turn the steering wheel to move the tie rod end aside.

6. Place a suitable hydraulic floor jack under the lower control arm and apply slight upward pressure.

NOTE: Install a safety chain through 1 coil at the top of the spring and attach it to the upper control arm to prevent the spring from coming out unexpectedly.

7. Install a spring compressor tool J-36567 or equivalent, and slightly compress the spring.

8. Remove the lower ball joint nut. Using a suitable removal tool, disconnect the lower ball joint from the steering knuckle.

9. Slowly lower the lower control arm until the ball joint stud is clear of the knuckle. Remove the lower ball joint mounting bolts and remove the lower ball joint.

To install:

10. Install the joint and torque the lower ball joint mounting bolt (A) to 76 ft. lbs. (105 Nm) and bolt (B) to 47 ft. lbs. (65 Nm). Tighten the lower ball joint nut to 58 ft. lbs. (80 Nm).

11. Install a new self-locking nut on the sway bar end link and tighten to 19 ft. lbs. (26 Nm). Tighten the sway bar bracket bolt to 14 ft. lbs. (19 Nm).

12. Install the strut bar to the chassis and tighten the locknuts. Install the strut bar-to-lower arm bolts, aligning the marks that were made during the removal procedure. Tighten the

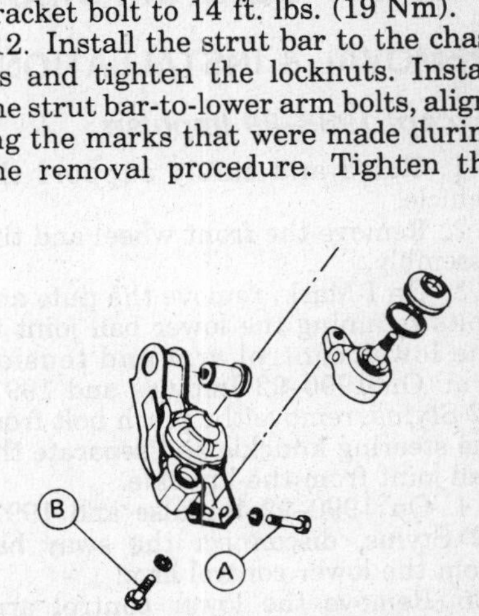

Lower ball joint mounting bolt installation—1988–89 Impulse

small bolt to 47 ft. lbs. (65 Nm) and the large bolt to 114 ft. lbs. (158 Nm).

13. Tighten the tie rod end nut to 60 ft. lbs. (84 Nm). Check the front end alignment.

Upper Control Arms

REMOVAL & INSTALLATION

1. Raise and safely support the vehicle.
2. Remove the front wheel and tire assembly.
3. Remove the disc brake caliper assembly. Support the caliper with a length of wire. Do not let the caliper hang from the brake hose.
4. Place a suitable hydraulic floor jack under the lower control arm and apply slight upward pressure.

NOTE: Install a safety chain through 1 coil at the bottom of the spring and attach it to the lower control arm to prevent the spring from coming out unexpectedly.

5. Install a spring compressor tool J-36567 or equivalent, and slightly compress the spring.
6. Remove the upper ball joint nut and use a suitable tool to disconnect the upper ball joint from the steering knuckle. Remove the upper control arm bolt and remove the upper control arm.
7. Installation is the reverse of the removal procedure. Install the upper control arm bolt but do not tighten it until after completion of all other installation work, with the wheels on the ground and, if possible, 2 assistants sitting in the front seats to simulate actual vehicle load. Tighten the bolt to 47 ft. lbs. (65 Nm).

Lower Control Arms

REMOVAL & INSTALLATION

Except 1988–89 Impulse

1. Raise and safely support the vehicle.
2. Remove the front wheel and tire assembly.
3. On I-Mark, remove the nuts and bolts retaining the lower ball joint to the lower control arm and tension arm. On 1990–92 Impulse and 1991-92 Stylus, remove the pinch bolt from the steering knuckle and separate the ball joint from the knuckle.
4. On 1990–92 Impulse and 1991-92 Stylus, disconnect the sway bar from the lower control arm.
5. Remove the lower control arm bolt(s) and remove the lower control arm.

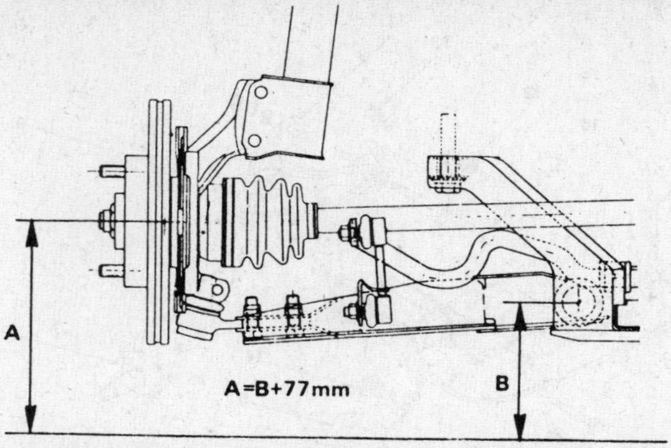

Trim height setting—1990–92 Impulse and 1991–92 Stylus front suspension

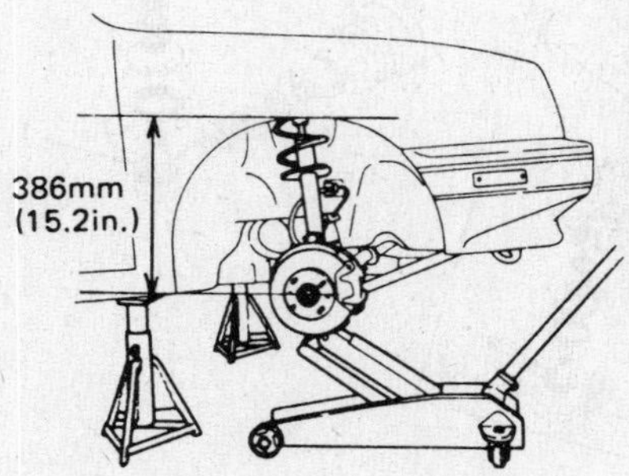

Trim height setting—I-Mark front suspension

To install:

6. Before tightening the lower control arm bolt(s), the trim height must be set.
7. Raise the lower control arm with a suitable jack and set the trim height to 15.2 in. (386mm) on I-Mark and 3 in. (77mm) from the center of the front lower control arm bushing to the center of the wheel hub on 1990–92 Impulse and 1991-92 Stylus.
8. Tighten the lower control arm bolt to 41 ft. lbs. (56 Nm) on I-Mark. On 1990–92 Impulse and 1991-92 Stylus, the torque is 50 ft. lbs. (68 Nm) on the rear bolt and 94 ft. lbs. (128 Nm) on the front bolt.

1988–89 Impulse

1. Raise and safely support the vehicle.
2. Remove the front wheel and tire assembly.
3. Mark the position of the nuts on the front of the strut bar for reassembly (these nuts control the caster setting) and remove the strut bar.
4. Remove the sway bar brackets on both sides of the vehicle. Remove the bolt, tube and grommets holding the sway bar to the lower control arm. Move the sway bar aside.
5. Use a suitable tool to disconnect the tie rod end from the steering knuckle. Turn the steering wheel to move the tie rod end aside.
6. Place a suitable hydraulic floor jack under the lower control arm and apply slight upward pressure.

NOTE: Install a safety chain through 1 coil at the top of the spring and attach it to the upper control arm to prevent the spring from coming out unexpectedly.

7. Install a spring compressor tool J-36567 or equivalent, and slightly compress the spring.
8. Remove the lower ball joint nut. Using a suitable removal tool, disconnect the lower ball joint from the steering knuckle.
9. Slowly lower the lower control arm. Remove the lower control arm bolt and remove the lower control arm.

To install:

10. Install the lower control arm bolt but do not tighten it until all components are reassembled and the vehicle is on the ground, preferably with a driver and passenger in the front seats to simulate average load. Tighten the bolt to 68 ft. lbs. (94 Nm).
11. Tighten the lower ball joint nut to 58 ft. lbs. (80 Nm).
12. Install a new self-locking nut on the sway bar end link and tighten to 19 ft. lbs. (26 Nm). Tighten the sway bar bracket bolt to 14 ft. lbs. (19 Nm).
13. Install the strut bar to the chassis and tighten the locknuts. Install the strut bar-to-lower arm bolts, aligning the marks that were made during the removal procedure. Tighten the small bolt to 47 ft. lbs. (65 Nm) and the large bolt to 114 ft. lbs. (158 Nm).
14. Tighten the tie rod end nut to 60 ft. lbs. (84 Nm). Check the front end alignment.

Sway Bar

REMOVAL & INSTALLATION

I-Mark

1. Raise and safely support the vehicle.

2. Remove the sway bar from the tension rod.

To install:

3. Install the sway bar when the vehicle is at no load.
4. Install the mounting rubbers and brackets to the tension rod. Align the front mounting rubbers with the rear side of the tension rod paint mark. Align the rear mounting rubbers rear end with the end of the sway bar.
5. Install the sway bar to the mounting rubbers and brackets on the right side of the vehicle. Loosely close and temporarily tighten the right hand side mounting rubbers and brackets.
6. Push the sway bar toward the vehicle center to install it to the left rear mounting rubber and bracket.
7. Install the sway bar to the left front mounting rubber and bracket.
8. Tighten the nuts to 6 ft. lbs. (8 Nm).
9. Lower the vehicle.

1988–89 Impulse

1. Raise and safely support the vehicle.
2. Remove the engine splash shield.
3. Remove the sway bar link assemblies from the ends of the sway bar and the lower control arms.
4. Remove the sway bar-to-body brackets and remove the sway bar.
5. Installation is the reverse of the removal procedure. Tighten the sway bar link nuts to 19 ft. lbs. (26 Nm) and the bracket bolts to 14 ft. lbs. (19 Nm).

1990–92 Impulse and 1991-92 Stylus

1. Raise and safely support the vehicle.
2. Remove the exhaust header pipe.
3. Disconnect and plug the power steering lines at the steering rack.
4. Remove the steering shaft boot nuts from inside the vehicle. Remove the steering shaft bolt.
5. Remove the pinch bolts and disconnect the lower ball joints from the steering knuckles.
6. Using a suitable tool, disconnect the tie rod ends from the steering knuckle.
7. Place a suitable jack under the engine and raise it to support the engine.
8. Disconnect the engine torque rod at the center beam.
9. Disconnect the rear engine mount. Remove 2 bolts at the center beam and 4 bolts at the crossmember. Lower the crossmember assembly with the steering rack carefully.
10. Remove the steering rack, then remove the sway bar from the crossmember.

NOTE: The sway bar cannot be removed unless the crossmember assembly is removed, however the sway bar bushings can be replaced without removing the crossmember.

11. Installation is the reverse of the removal procedure. Tighten the 4 crossmember bolts to 137 ft. lbs. (186 Nm) and the 2 center beam bolts to 37 ft. lbs. (50 Nm). Tighten the lower ball joint pinch bolts to 48 ft. lbs. (65 Nm) and the steering shaft bolt to 30 ft. lbs. (40 Nm). Tighten the tie rod nuts to 40 ft. lbs. (54 Nm).

Front Wheel Bearings

ADJUSTMENT

1. Raise and safely support the vehicle. Remove the front wheel and tire assembly.
2. Remove the dust cap and the cotter pin. Make sure the wheel bearings are clean and adequately greased before proceeding further.
3. Tighten the hub nut to 22 ft. lbs. (30 Nm) and turn the hub 2–3 turns in fore and aft directions to set the bearings.
4. Loosen the hub nut just enough, so it can be turned by hand. Hand tighten the hub nut using a suitable socket wrench and check that the hub has no play in it.
5. Attach a suitable pull scale to one of the hub studs. Adjust the tightness of the hub nut, so the hub begins to rotate when the pull scale is pulled forward with a force of 1.1–3.3 lbs. (1.3–4.2 N).
6. Install the cotter pin. If the cotter pin holes are not aligned, further turn the nut the minimum amount necessary in the direction of tightening.
7. Replace the dust cap, front wheel and tire assembly and lower the vehicle.

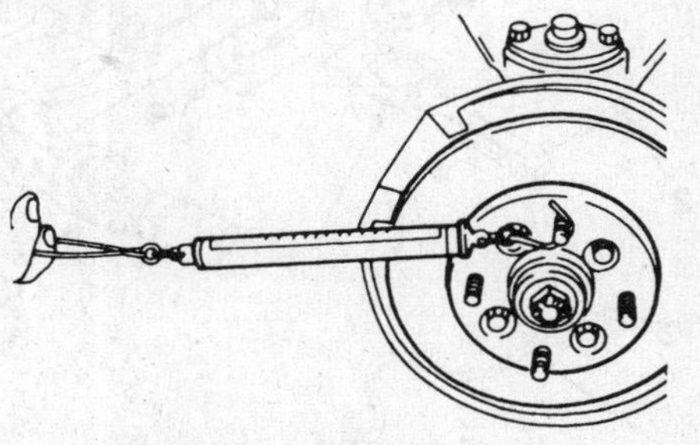

Front wheel bearing adjustment procedure – 1988–89 Impulse

REMOVAL & INSTALLATION

1. Raise and safely support the vehicle.
2. Remove the front wheel and tire assembly.
3. Remove the disc brake caliper and the caliper mounting bracket. Support the caliper with a length of wire, do not let it hang from the brake hose.
4. Remove the dust cap and the cotter pin. Remove the castle nut, washer and outer wheel bearing assembly. Remove the rotor/hub assembly with the inner wheel bearing and seal.
5. Use a suitable tool to remove the wheel seal. Remove the inner wheel bearing assembly.
6. Clean the wheel bearings, washer, castle nut and the inside of the hub thoroughly with a suitable cleaning solvent and allow to air dry.
7. Inspect the wheel bearing rollers, cages and races for pitting, cracking or other wear and replace, as necessary. Bearings and bearing races must always be replaced as a unit. If the bearings or bearing races are worn or damaged, drive the races out of the hub using suitable removal tools.

To install:

8. If the bearing races were removed, drive the new races into the hub using suitable installation tools. Make sure the inside of the hub is clean before installation.
9. Pack the wheel bearings using a suitable high temperature wheel bearing grease. If a bearing packer is not available, the bearings may be packed by hand, but care must be taken that grease thoroughly penetrates behind the bearing rollers.
10. Place a quantity of wheel bearing grease inside the hub between the races and coat the races with grease. Install the inner wheel bearing.
11. Use a suitable tool to install a new wheel seal.
12. Install the rotor/hub assembly onto the spindle and install the outer wheel bearing, washer and the castle nut. Adjust the wheel bearings, install a new cotter pin and install the dust cap.
13. Install the disc brake caliper mounting bracket and the disc brake caliper.
14. Install the front wheel and tire assembly and lower the vehicle.

REAR SUSPENSION

Shock Absorbers

REMOVAL & INSTALLATION

1. Raise and safely support the vehicle.

2. Support the rear axle using a suitable jack.
3. Remove the shock tower cover or trim plate cover, as necessary.
4. Remove the upper shock mounting nuts, washers and grommets.
5. Remove the lower shock mounting bolts, washers and grommets and remove the shock absorbers.
6. Installation is the reverse of the removal procedure. Tighten the lower shock mounting bolts to 30 ft. lbs. (41 Nm) on I-Mark and 20 ft. lbs. (28 Nm) on 1988–89 Impulse.

MacPherson Strut

REMOVAL & INSTALLATION

1. Raise and safely support the vehicle.
2. Remove the rear wheel and tire assembly. Open the rear hatch and remove the strut tower cover. Loosen the strut mount.
3. Remove the brake line clip at the strut. Disconnect and plug the brake hose at the strut. Disconnect the speed sensor cable if equipped with ABS brakes.
4. Disconnect the sway bar end. Remove the strut-to-knuckle bolts.
5. Open the rear hatch and remove the strut tower cover. Remove the strut mounting nuts and remove the strut.
6. Installation is the reverse of the removal procedure. Tighten the strut mounting nuts to 51 ft. lbs. (69 Nm) and the strut-to-knuckle bolts to 116 ft. lbs. (157 Nm).

Coil Springs

REMOVAL & INSTALLATION

1. Raise and safely support the vehicle.
2. Support the rear axle with a suitable jack.
3. Remove the lower shock mounting bolts and disconnect the shocks from the axle assembly.
4. Slowly lower the rear axle with the jack until the coil springs can be removed.

NOTE: Do not stress the brake hoses when lowering the rear axle.

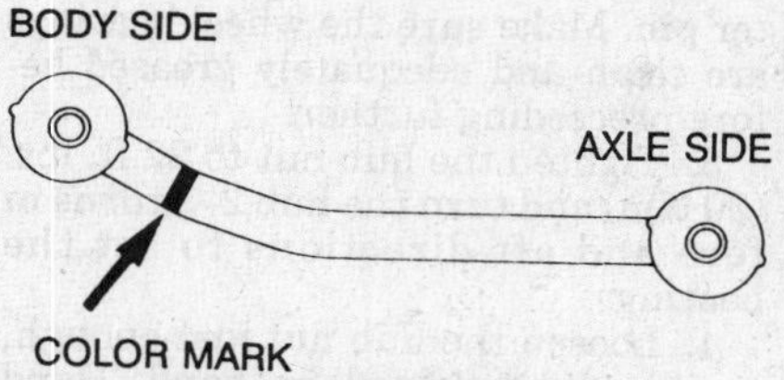

Upper control arm identification mark—1988–89 Impulse

5. Installation is the reverse of the removal procedure. Tighten the lower shock mounting bolts to 30 ft. lbs. (41 Nm) on I-Mark and 20 ft. lbs. (28 Nm) on 1988–89 Impulse.

Rear Control Arms

REMOVAL & INSTALLATION

1988–89 Impulse

UPPER ARM

1. Raise and safely support the vehicle.
2. Remove the rear wheel and tire assembly.
3. Remove the upper arm mounting bolts and remove the upper control arm.
4. Installation is the reverse of the removal procedure. The upper arms are color coded to indicate whether they are to be installed on the right hand or left hand side. Upper arms for right side installation are red and upper arms for left side installation are blue.
5. Tighten the upper control arm mounting bolts to 98 ft. lbs. (136 Nm).

LOWER ARM

1. Raise and safely support the vehicle.
2. Remove the rear wheel and tire assembly.

1. Lower control arm
2. Lower control arm-to-body bolt
3. Lower control arm-to-rear axle bolt
4. Upper control arm
5. Upper control arm-to-body bolt
6. Upper control arm-to-rear axle bolt
7. Bushing
8. Panhard rod
9. Cover
10. Nut and spring washer
11. Bolt
12. Nut and spring washer
13. Lower spring insulator
14. Spring
15. Upper spring insulator
16. Sway bar
17. Shackle bolts
18. Sway bar-to-rear axle bolts
19. Washer and grommet
20. Shock absorber
21. Washer and grommet
22. Nuts
23. Washer and grommet
24. Shock absorber
25. Washer and grommet
26. Nuts
27. Bushing and washer
28. Spring washer, cover and bushing
29. Bolt
30. Bushing and washer
31. Spring washer, cover and bushing
32. Bolt

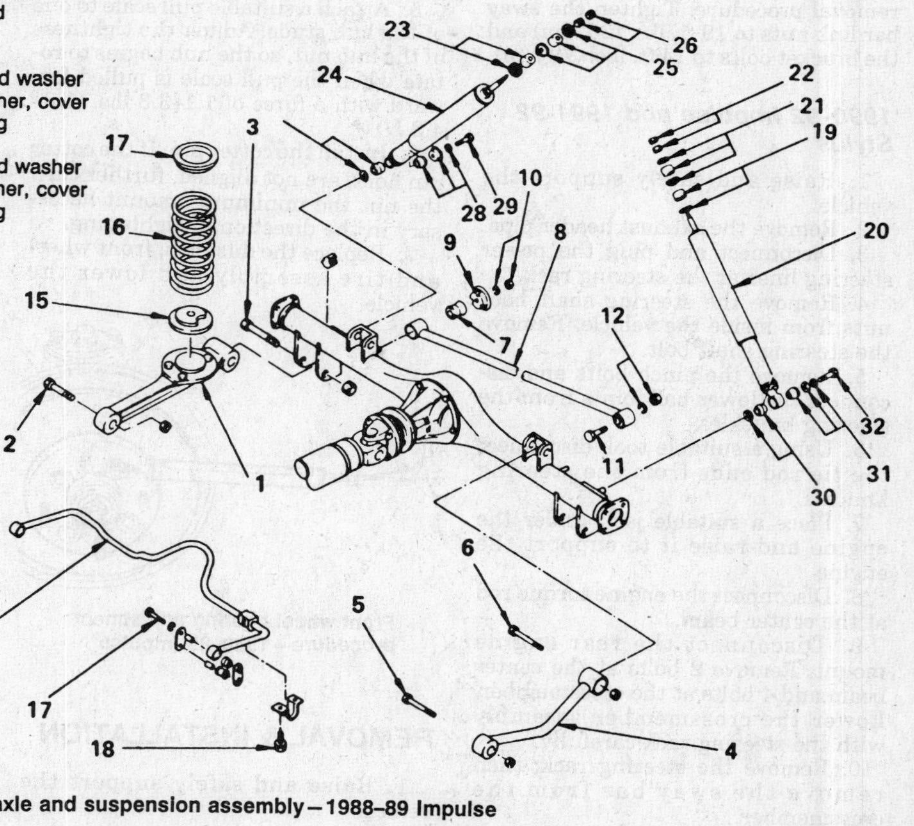

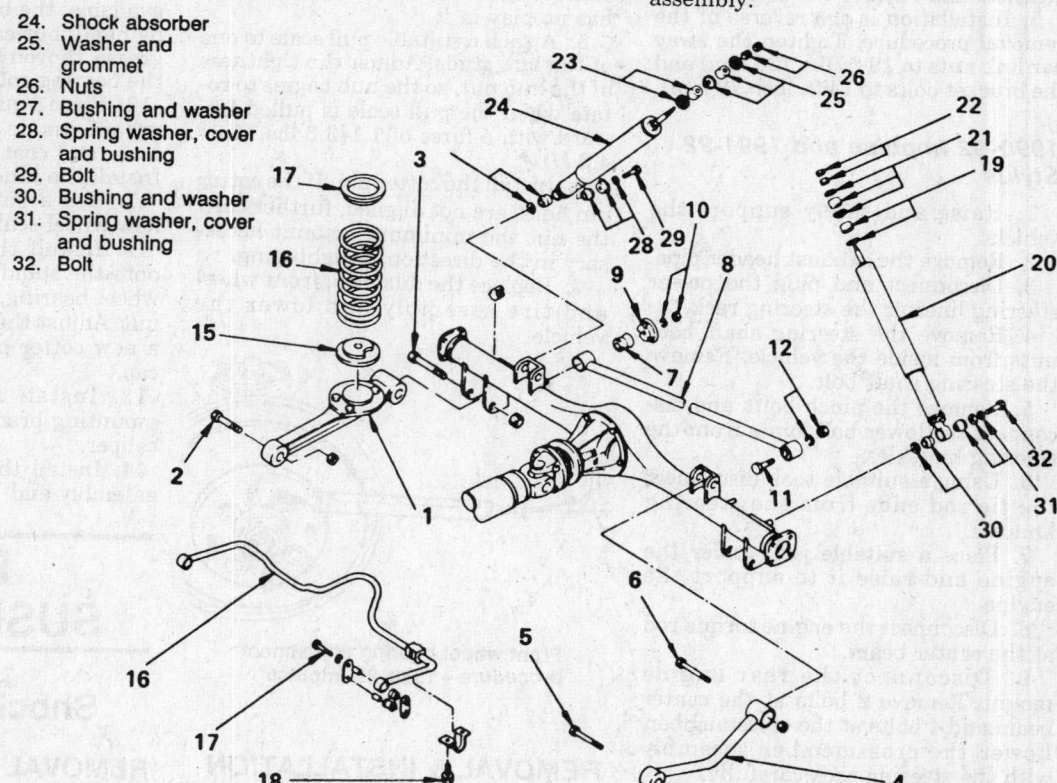

Rear axle and suspension assembly—1988–89 Impulse

3. Support the rear axle assembly with a suitable jack.
4. Remove the lower shock mounting bolts and disconnect the shock absorbers from the axle assembly.
5. Slowly lower the rear axle with the jack until the coil springs can be removed.

NOTE: Do not stress the brake hoses when lowering the rear axle.

6. Remove the lower control arm mounting bolts and remove the lower control arm.
7. Installation is the reverse of the removal procedure. Tighten the lower control arm mounting bolts to 118 ft. lbs. (163 Nm) and the lower shock mounting bolts to 20 ft. lbs. (28 Nm).

1990–92 Impulse and 1991-92 Stylus

TRAILING ARM

1. Raise and safely support the vehicle.
2. Remove the rear wheel and tire assembly.
3. Remove the trailing arm mounting bolts and remove the trailing arm.
4. Installation is the reverse of the removal procedure. Set the trim height before tightening the trailing arm mounting bolts. The trim height is set as follows:
   a. Raise the rear suspension assembly using a suitable jack until the center of the hub is 1.3 in. (33mm) above the center line of the body side lateral arm bushing.
   b. Tighten the trailing arm mounting bolts to 94 ft. lbs. (127 Nm).

LATERAL ARM

NOTE: Do not loosen the turn buckle of the trailing lateral arm unless it is absolutely necessary.

1. Raise and safely support the vehicle.
2. Remove the rear wheel and tire assembly.
3. Remove the lateral arm mounting bolts and remove the lateral arm.

NOTE: To remove the lateral arm on the left hand side, loosen the crossmember bolts and push the crossmember down as far as possible. This will prevent interference with the fuel tank and create space to pull the bolt free. Place a suitable jack under the crossmember for safety when dropping down the crossmember.

4. Installation is the reverse of the removal procedure. Tighten the crossmember mounting bolts to 94 ft. lbs. (127 Nm). Set the trim height before tightening the lateral arm mounting bolts. The trim height is set as follows:
   a. Raise the rear suspension assembly using a suitable jack until the center of the hub is 1.3 in. (33mm) above the center line of the body side lateral arm bushing.
   b. Tighten the lateral arm mounting bolts to 94 ft. lbs. (127 Nm).
5. Check the rear wheel alignment.

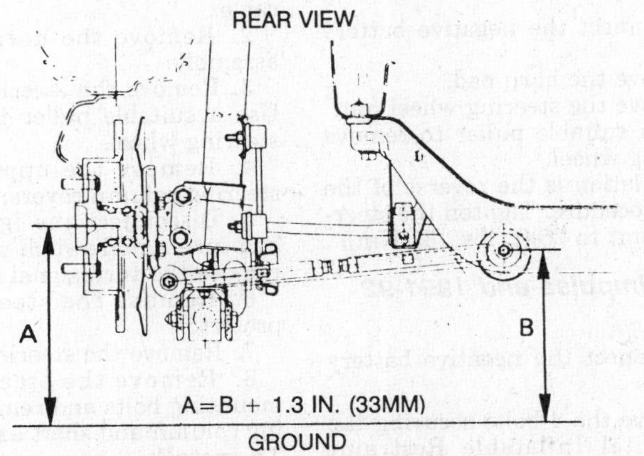

Trim height setting—1990–92 Impulse rear suspension

Rear Wheel Bearings

REMOVAL & INSTALLATION

I-Mark

1. Raise and safely support the vehicle.
2. Remove the rear wheel and tire assembly.
3. Remove the dust cap and the cotter pin. Remove the castle nut, washer and outer wheel bearing assembly. Remove the drum/hub assembly with the inner wheel bearing and seal.
4. Use a suitable tool to remove the wheel seal. Remove the inner wheel bearing assembly.
5. Clean the wheel bearings, washer, castle nut and the inside of the hub thoroughly with a suitable cleaning solvent and allow to air dry.
6. Inspect the wheel bearing rollers, cages and races for pitting, cracking or other wear and replace, as necessary. Bearings and bearing races must always be replaced as a unit. If the bearings or bearing races are worn or damaged, drive the races out of the hub using suitable removal tools.

To install:

7. If the bearing races were removed, drive the new races into the hub using suitable installation tools. Make sure the inside of the hub is clean before installation.
8. Pack the wheel bearings using a suitable high temperature wheel bearing grease. If a bearing packer is not available, the bearings may be packed by hand, but care must be taken that grease thoroughly penetrates behind the bearing rollers.
9. Place a quantity of wheel bearing grease inside the hub between the races and coat the races with grease. Install the inner wheel bearing.
10. Use a suitable tool to install a new wheel seal.
11. Install the drum/hub assembly onto the spindle and install the outer wheel bearing, washer and the castle nut. Adjust the wheel bearings, install a new cotter pin and install the dust cap.
12. Install the rear wheel and tire assembly and lower the vehicle.

1990–92 Impulse and 1991-92 Stylus

NOTE: The hub, hub bearing and the spindle are a single unit and cannot be disassembled. Replace the entire hub unit assembly if the hub axial play is excessive or abnormal noise occurs.

1. Raise and safely support the vehicle.
2. Remove the rear wheel and tire assembly.
3. Remove the disc brake caliper and caliper mounting bracket. Remove the disc brake rotor.
4. Remove the hub unit assembly mounting bolts and remove the hub assembly.
5. Installation is the reverse of the removal procedure. Tighten the hub unit assembly mounting bolts to 49 ft. lbs. (66 Nm).

ADJUSTMENT

I-Mark

1. Raise and safely support the vehicle.

2. Remove the rear wheel and tire assembly.
3. Remove the dust cap and the cotter pin.
4. Make sure the wheel bearings are adequately greased before proceeding further.
5. Tighten the nut to 22 ft. lbs. (30 Nm). Rotate the hub 2–3 times.
6. Loosen the nut completely.
7. Tighten the nut fully by hand.
8. Install the cotter pin.

NOTE: If the cotter pin holes are not aligned, tighten the nut just enough to align the holes.

Rear Axle Assembly

REMOVAL & INSTALLATION

I-Mark

1. Raise and safely support the vehicle.
2. Remove the rear wheel and tire assemblies.
3. Remove the brake drums.
4. Disconnect the parking brake cable from the parking brake levers and the brake backing plates. Remove the parking brake cable mounting bolts.
5. Remove the brake hose clip. Disconnect and plug the brake hose.
6. Remove the sway bar.
7. Support the rear axle with a suitable jack.
8. Remove the lower shock mounting bolts and disconnect the shocks from the rear axle.
9. Slowly lower the rear axle until the coil springs can be removed.
10. Remove the rear axle mounting bolts and remove the rear axle assembly.

To install:

11. Set the trim height before tightening the rear axle mounting bolts. Raise the axle with a suitable jack until the trim height is 15 in. (381mm) between the center of the wheel hub to the top of the wheel well.
12. Tighten the rear axle mounting bolts to 72 ft. lbs. (100 Nm).
13. Tighten the lower shock mounting bolts to 30 ft. lbs. (41 Nm) and the sway bar mounting bolts to 20 ft. lbs. (28 Nm).

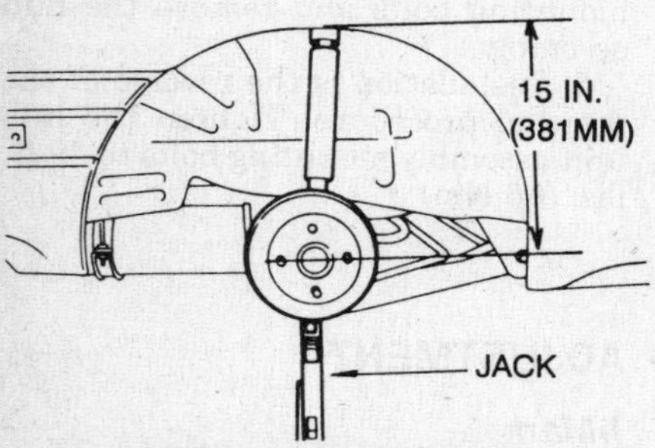

Trim height setting – I-Mark rear suspension

Rear Knuckle

REMOVAL & INSTALLATION

1. Raise and safely support the vehicle. Disconnect the negative battery cable.
2. Disconnect the ABS speed sensor, if equipped with ABS.
3. Remove the brake caliper and rotor.
4. Remove the hub nut, if equipped with All Wheel Drive.
5. Remove the hub assembly, retained by 4 bolts.
6. Remove the backing plate.
7. Disconnect the lateral link, trailing link and remove the knuckle from the vehicle.
8. Installation is the reverse of the removal procedure. Torque the knuckle-to-strut bolts to 116 ft. lbs. (157 Nm), trailing link-to-knuckle bolts to 94 ft. lbs. (127 Nm), lateral link-to-knuckle bolts to 116 ft. lbs. (157 Nm), hub-to-knuckle bolts to 49 ft. lbs. (66 Nm) and 4WD hub nut to 137 ft. lbs. (186 Nm).

STEERING

Steering Wheel

CAUTION

On vehicles equipped with an air bag, the negative battery cable must be disconnected, before working on the system. Failure to do so may result in deployment of the air bag and possible personal injury.

REMOVAL & INSTALLATION

Except 1990–92 Impulse and 1991-92 Stylus

1. Disconnect the negative battery cable.
2. Remove the horn pad.
3. Remove the steering wheel nut.
4. Use a suitable puller to remove the steering wheel.
5. Installation is the reverse of the removal procedure. Tighten the steering wheel nut to 25 ft. lbs. (35 Nm).

1990–92 Impulse and 1991-92 Stylus

1. Disconnect the negative battery cable.
2. Remove the 4 bolts securing the Supplemental Inflatable Restraint (SIR) module to the steering wheel.
3. Disconnect the module connector and remove the SIR module.

CAUTION

The SIR module should always be carried with the urethane cover away from the body. It should always be laid on a flat surface with the urethane side up. This is necessary because a free space is provided to allow the air cushion to expand in the unlikely event of accidental deployment. Otherwise, personal injury may result.

4. Disconnect the horn connector and remove the steering wheel nut.
5. Use a suitable puller to remove the steering wheel.

To install:

6. Mount the steering wheel on the column and tighten the steering wheel nut to 25 ft. lbs. (35 Nm).

NOTE: Be careful not to damage the harness section of the coil when installing the steering wheel. At this time, the cancel cam for the turn signal switch must be closely inserted into a boss hole in the steering wheel.

7. Support the SIR module and carefully connect the module connector. Pass the lead wire through the tabs on the plastic cover (wire protector) of the inflator to prevent the lead wire from being pinched.
8. Install the module retaining bolts and tighten them to 3.6 ft. lbs. (5 Nm).
9. Connect the negative battery cable. Turn the ignition switch to **ON** while watching the warning light. The light should flash 7–9 times and then go off. If the light does not operate correctly, there is a problem in the SIR system.

Steering Column

REMOVAL & INSTALLATION

I-Mark

1. Disconnect the negative battery cable.
2. Remove the horn button pad assembly.
3. Remove the steering wheel nut. Use a suitable puller to remove the steering wheel.
4. Remove the upper and lower steering column covers.
5. Disconnect the ignition switch and turn signal switch connectors and remove the turn signal switch.
6. Remove the steering column protector.
7. Remove the steering joint bolt.
8. Remove the steering column mounting bolts and remove the steering column and shaft assembly.

To install:

9. Align the serrated notches of the

stub shaft and the intermediate shaft. Tighten the bolt to 19 ft. lbs. (26 Nm).

10. Align the serrated notches of the steering shaft and the intermediate shaft. Partially tighten the steering shaft side joint part before tightening the steering column upper and lower brackets.

11. Pull the steering column rearward as the upper bracket bolt is tightened to 11 ft. lbs. (15 Nm). Tighten the lower bracket bolt to 11 ft. lbs. (15 Nm). Finish tightening the steering shaft side joint bolt to 19 ft. lbs. (26 Nm).

12. The remaining components are installed in the reverse order of their removal. Tighten the steering wheel nut to 25 ft. lbs. (35 Nm).

Impulse

1988–89

1. Disconnect the negative battery cable.
2. Remove the screw from the lower side of the horn button pad assembly. Slide the horn button pad towards the front of the vehicle and disconnect the electrical connector.
3. Remove the front console and the radio assembly.
4. Remove the lower dash panel and the hood lock release lever.
5. Remove the left knee pad and the steering column cover.
6. Remove the 2nd shaft-to-universal joint bolt and the steering stay instrument panel nut.
7. Disconnect the electrical connectors from the steering column.
8. Remove the steering column bracket nuts and bolts and remove the steering column and shaft assembly.

To install:

9. Install the steering column and bracket nuts and bolts.
10. Tighten the column-to-instrument panel nut to 13 ft. lbs. (18 Nm) and the steering column support bracket bolt to 4.3 ft. lbs. (6 Nm). Tighten the 2nd shaft-to-universal joint bolt to 18 ft. lbs. (25 Nm). Tighten the left knee pad bolts to 29 ft. lbs. (40 Nm) and the steering column nut to 25 ft. lbs. (35 Nm).

1990–92 Impulse and 1991–92 Stylus

NOTE: The wheels of the vehicle must be in the straight ahead position and the steering column in the LOCK position before disconnecting the steering column or steering shaft from the steering gear. Failure to do so will cause the coil assembly to become uncentered which will cause damage to the coil assembly.

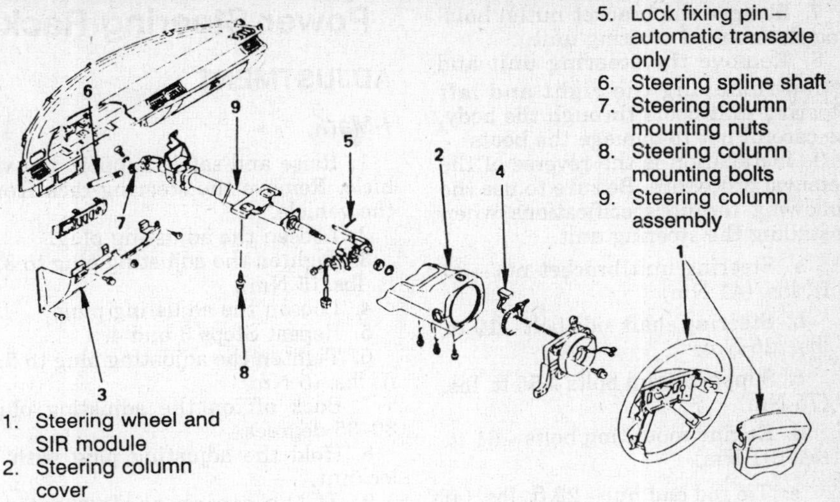

1. Steering wheel and SIR module
2. Steering column cover
3. Knee pad
4. Electrical connector
5. Lock fixing pin—automatic transaxle only
6. Steering spline shaft
7. Steering column mounting nuts
8. Steering column mounting bolts
9. Steering column assembly

Steering column assembly—1990–92 Impulse and 1991–92 Stylus

CAUTION

The SIR module should always be carried with the urethane cover away from the body. It should always be laid on a flat surface with the urethane side up. This is necessary because a free space is provided to allow the air cushion to expand in the unlikely event of accidental deployment. Otherwise, personal injury may result.

1. Disconnect the negative battery cable.
2. Remove the steering wheel according to the proper procedure.
3. Remove the steering column cover.
4. Remove the knee pad.
5. Disconnect the electrical connectors from the steering column.
6. If equipped with an automatic transaxle, remove the steering lock fixing pin.
7. Remove the steering shaft bolt from the steering unit side. Loosen the bolt on the steering column side, then remove the spline shaft from the steering unit side.
8. Remove the column mounting nuts and bolts and remove the column assembly.

To install:

9. Install the steering column and bracket nuts and bolts.
10. Tighten the column mounting nuts and bolts to 11 ft. lbs. (15 Nm). Tighten the steering spline shaft bolt to 30 ft. lbs. (40 Nm).

Manual Steering Rack

ADJUSTMENT

I-Mark

1. Raise and safely support the vehicle. Remove the steering rack from the vehicle.
2. Remove the adjusting plug and apply grease to the sliding surfaces of the plunger, spring seat and the inside of the adjusting plug. Replace the adjusting plug and tighten to 3.6 ft. lbs. (5 Nm).
3. Loosen the adjusting plug.
4. Repeat Steps 2 and 3.
5. Back down the adjusting plug by a maximum of 25 degrees. Hold the adjusting plug with a locknut. If the pinion shaft starting torque does not equal 0.4–0.9 ft. lbs. (0.6–1.2 Nm), the adjusting plug must be readjusted.
6. Tighten the adjusting plug locknut to 49 ft. lbs. (68 Nm).

REMOVAL & INSTALLATION

I-Mark

1. Disconnect the negative battery cable. Mark the position of the hood in relation to the hinge brackets and remove the hood.
2. Raise and safely support the vehicle. Remove both front wheel and tire assemblies.
3. Use a suitable tool to remove the tie rod ends from the steering knuckles.
4. Using a suitable engine hoist, slightly raise the engine. Support the lower part of the engine with a suitable jack.
5. Remove the engine mounting bolts.
6. Remove the front exhaust pipe hanger mounting rubber nut. Detach the mounting rubber from the beam. Remove the beam. Remove the intermediate shaft mounting bolt (steering shaft side bolt).

7. Remove the bracket nut(s) holding the manual steering unit.
8. Remove the steering unit and when removing the right and left steering unit boots through the body, be careful not to damage the boots.
9. Installation is the reverse of the removal procedure. Be sure to use the following torque specifications when installing the steering unit.

a. Steering unit bracket nut—30 ft. lbs. (41 Nm)

b. Steering shaft side bolt—19 ft. lbs. (26 Nm)

c. Support beam bolts—56 ft. lbs. (75 Nm)

d. Engine mounting bolts—61 ft. lbs. (81 Nm)

e. Tie rod end nut—29 ft. lbs. (40 Nm)

Stylus

1. Disconnect the negative battery cable.
2. Remove the steering column protector and spline shaft pinch bolt.
3. Remove the steering shaft boot nuts.
4. Mark the engine hood bolts and remove the hood with the help of an assistant.
5. Raise the vehicle and support safely. Remove the front wheels.
6. Disconnect the exhaust pipe.
7. Using a tie rod remover J-21687-02 or equivalent, remove the tie rod ends from the knuckle.
8. Using a ball joint remover tool, disconnect the ball joints from the knuckles.
9. Remove the crossmember assembly and manual steering assembly.

To install:

10. Install the steering assembly and torque the nuts to 51 ft. lbs. (69 Nm).
11. Install the crossmember and torque the bolts to 137 ft. lbs. (186 Nm).
12. Install the knuckle to the ball joints and torque the nuts to 37 ft. lbs. (50 Nm).
13. Install the tie rod ends and torque the nuts to 40 ft. lbs. (54 Nm).
14. Install the exhaust pipe and front wheels.
15. With an assistant, install the engine hood.
16. Install the steering shaft boot and nuts and the spline shaft pinch bolt. Torque the bolt to 30 ft. lbs. (40 Nm).
17. Install the front wheels and lower the vehicle.
18. Connect the battery cable and have the front end aligned.

Power Steering Rack

ADJUSTMENT

I-Mark

1. Raise and safely support the vehicle. Remove the steering rack from the vehicle.
2. Loosen the adjusting plug.
3. Tighten the adjusting plug to 3.6 ft. lbs. (5 Nm).
4. Loosen the adjusting plug.
5. Repeat Steps 3 and 4.
6. Tighten the adjusting plug to 3.6 ft. lbs. (5 Nm).
7. Back off on the adjusting plug 30–35 degrees.
8. Hold the adjusting plug with a locknut.
9. If the pinion shaft starting torque is different from 0.43–1.01 ft. lbs. (0.6–1.4 Nm), the adjusting plug must be readjusted. Tighten the adjusting plug locknut to 49 ft. lbs. (68 Nm).

1988–89 Impulse

1. Raise and safely support the vehicle. Remove the steering rack from the vehicle.
2. Loosen the adjustment plug, then tighten it to 5.42 ft. lbs. (7.5 Nm).
3. Back off the adjustment plug 30–35 degrees.
4. Make sure the rack shaft operates smoothly. Tighten the locknut to 50 ft. lbs. (70 Nm).

1990–92 Impulse and 1991-92 Stylus

1. Raise and safely support the vehicle. Remove the steering rack from the vehicle.
2. Loosen the adjustment plug.
3. Tighten the plug to 43.4 inch lbs. (4.9 Nm), loosen the plug and then tighten it again to 43.4 inch lbs. (4.9 Nm).
4. Back off the plug 26 degrees, then tighten the locknut.
5. Check the pinion shaft preload. It should be 5.3–14.1 inch lbs. (0.6–1.6 Nm). If the preload is not within specification, loosen the locknut and readjust the plug.
6. Tighten the locknut to 49 ft. lbs. (66 Nm).

REMOVAL & INSTALLATION

I-Mark

1. Disconnect the negative battery cable. Mark the position of the hood on the hinge brackets and remove the hood.
2. Raise and safely support the vehicle. Remove the front wheel and tire assemblies.
3. Use a suitable tool to disconnect the tie rod ends from the steering knuckle.
4. Attach a suitable hoist to the engine and raise the engine slightly. Support the lower part of the engine with a suitable jack. Remove the engine mounting bolts.
5. Remove the exhaust pipe hanger rubber mounting nut. Separate the mounting rubber and the beam. Remove the beam.
6. Remove the steering column protector and the steering shaft bolt. Remove the intermediate shaft mounting bolt and the intermediate shaft.
7. Disconnect the power steering lines from the rack.
8. Remove the steering rack mounting bolts and remove the rack from the vehicle.

To install:

9. Make sure the serrated notches on the intermediate shaft are aligned. Tighten the steering shaft bolt to 19 ft. lbs. (26 Nm).
10. Tighten the beam mounting bolts to 56 ft. lbs. (77 Nm). Tighten the engine mounting bolts to 61 ft. lbs. (84 Nm). Tighten the tie rod nuts to 29 ft. lbs. (40 Nm).
11. Fill the reservoir with the proper type of fluid and bleed the air from the system.

1988–89 Impulse

1. Raise and safely support the vehicle. Remove the front wheel and tire assemblies.
2. Remove the front disc brake calipers and support them with a length of wire. Do not let them hang from the brake hoses.
3. Remove the hub and rotor assembly and the brake dust shield.
4. Use a suitable tool to disconnect the tie rod ends from the steering knuckles.
5. Disconnect the steering shaft from the steering column.
6. Disconnect the power steering lines from the rack.
7. Remove the rack mounting bolts and remove the rack assembly.

To install:

8. After tightening the rack mounting bolts, lock each bolt securely by bending the washer at 2 opposed points.
9. Fill the reservoir with the proper type of fluid and bleed the air from the system.

1990–92 Impulse and 1991-92 Stylus

1. Disconnect the negative battery cable. Mark the position of the hood on the hinge brackets and remove the hood.
2. Remove the column protector

1. Disc brake; front
2. Hub and rotor; front brake
3. Rod end assembly; outer
4. Shaft; steering, 2nd
5. Pipe assembly; return
6. Pipe assembly; feed
7. Bolt; bracket to crossmember
8. Washer; spring, bracket to crossmember
9. Bracket; steering unit to crossmember

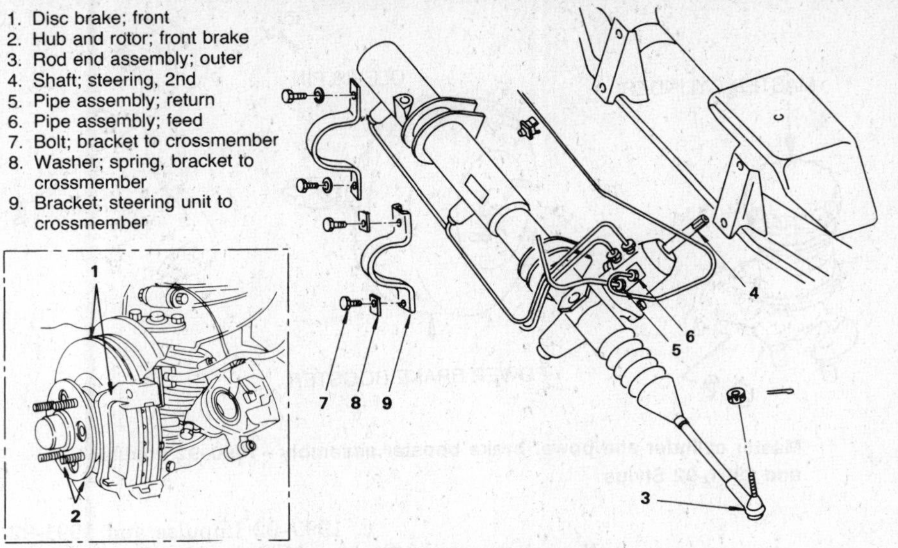

Power steering rack installation—1988–89 Impulse

and remove the spline shaft pinch bolts. Remove the steering shaft boot nuts.

3. Raise and safely support the vehicle. Remove the front wheel and tire assemblies.
4. Attach a suitable hoist to the engine and support it.
5. Remove the exhaust header pipe. Place a suitable drain pan below the steering rack and disconnect the power steering lines.
6. Use a suitable tool to disconnect the tie rod ends from the steering knuckle.
7. Disconnect the ball joints from the steering knuckle and remove the crossmember assembly. Separate the steering rack from the crossmember.

To install:

8. Install the steering rack and crossmember and mounting bolts. Connect the ball joints to the steering knuckle.
9. Connect the tie rod ends to the steering knuckle.
10. Install the exhaust header pipe. Connect the power steering lines.
11. Remove the engine hoist.
12. Install the front wheels and lower the vehicle.
13. Install the column protector, spline shaft pinch bolts and steering shaft boot nuts.
14. Connect the negative battery cable. With an assistant, install the engine hood.
15. Tighten the following components to specification:
    a. Steering rack-to-chassis bolts—37 ft. lbs. (50 Nm).
    b. Crossmember assembly—137 ft. lbs. (186 Nm).
    c. Front beam-to-chassis bolts—37 ft. lbs. (50 Nm).
    d. Ball joint—48 ft. lbs. (65 Nm).
    e. Tie rod ends—40 ft. lbs. (54 Nm).
    f. Exhaust pipe—49 ft. lbs. (67 Nm).
    g. Spline shaft pinch bolt—30 ft. lbs. (40 Nm).
16. Fill the reservoir with the proper type of fluid and bleed the air from the system. Adjust the front end alignment.

Power Steering Pump

REMOVAL & INSTALLATION

1. Disconnect the negative battery cable.
2. Remove the engine splash shield, if so equipped.
3. Disconnect the pressure and return lines from the pump.
4. Remove the drive belt from the pump pulley.
5. On 1988–89 Impulse, remove the pump pulley and then remove the pump and bracket assembly. On all others, remove the pump and bracket assembly and then remove the pulley and bracket from the pump.
6. Installation is the reverse of the removal procedure. Adjust the drive belt, fill the reservoir with the proper type of fluid and bleed the air from the system.

Belt Adjustment

Except 1988–89 Impulse

1. Loosen the pump adjusting bolt and the pivot bolt. Use a suitable tool to force the pump away from the engine until the correct belt tension is reached.
2. On all except 4XE1 engine equipped vehicles, exert a force of approximately 22 lbs. to the belt at a point midway between the pump pulley and the crank pulley. The belt tension is correct if the belt deflects 0.2–0.4 in. when this force is applied.
3. On 4XE1 engine equipped vehicles, use a suitable belt tension gauge and set the belt tension to 120–150 lbs. for a used belt and 130–160 lbs. for a new belt.

1988–89 Impulse

1. Move the idler pulley to adjust the belt tension.
2. The belt tension is correct if the belt deflects 0.4 in. when a force of approximately 22 lbs. is applied midway between the pulleys.

SYSTEM BLEEDING

1. Turn the wheels to the extreme left.
2. With the engine stopped, add power steering fluid to the **MIN** mark on the fluid indicator.
3. Start the engine and run it for 15 seconds at fast idle.
4. Stop the engine, recheck the fluid level and refill to the **MIN** mark.
5. Start the engine and turn the wheels from side to side 3 times.
6. Stop the engine and check the fluid level.

NOTE: If air bubbles are still present in the fluid, the procedure must be repeated.

Tie Rod Ends

REMOVAL & INSTALLATION

1. Raise and safely support the vehicle. Remove the front wheel and tire assembly.
2. Remove the nut from the tie rod end stud. Using a suitable removal tool, separate the tie rod from the steering knuckle.
3. Disconnect the retaining wire from the inner boot and pull back the boot.
4. Using a suitable tool, straighten the staked part of the locking washer between the tie rod and the rack.
5. Remove the tie rod from the rack.
6. Installation is the reverse of the removal procedure. Tighten the tie rod end nut to 30 ft. lbs. (40 Nm) on I-Mark, 60 ft. lbs. (84 Nm) on 1988–89 Impulse and 40 ft. lbs. (54 Nm) on 1990–92 Impulse and 1991-92 Stylus.
7. Adjust the toe setting of the front end alignment.

BRAKES

For all brake system repair and service procedures not detailed below, please refer to "Brakes" in the Unit Repair section.

Master Cylinder

REMOVAL & INSTALLATION

1. Disconnect the negative battery cable.
2. Disconnect the electrical connector from the master cylinder.
3. Disconnect and plug the brake lines at the master cylinder.
4. Remove the master cylinder mounting nuts and remove the master cylinder.
5. Installation is the reverse of the removal procedure. Tighten the mounting nuts to 9 ft. lbs. (13 Nm). Bleed the master cylinder.

Proportioning Valve

REMOVAL & INSTALLATION

Except 1988–89 Impulse

There are 2 proportioning valves attached to the master cylinder.

1. Disconnect the negative battery cable.
2. Disconnect and plug the brake lines at the master cylinder.
3. Remove the proportioning valves from the master cylinder.
4. Installation is the reverse of the removal procedure. Tighten the proportioning valves to 30 ft. lbs. (40 Nm). Bleed the brake system.

1988–89 Impulse

1. Disconnect the negative battery cable.
2. Disconnect and plug the brake lines at the proportioning valve.
3. Remove the proportioning valve.
4. Installation is the reverse of the removal procedure. Bleed the brake system.

Power Brake Booster

REMOVAL & INSTALLATION

1. Disconnect the negative battery cable.
2. Remove the air cleaner duct on Impulse vehicles.
3. Remove the master cylinder assembly.
4. Disconnect the vacuum hose from the power brake booster.

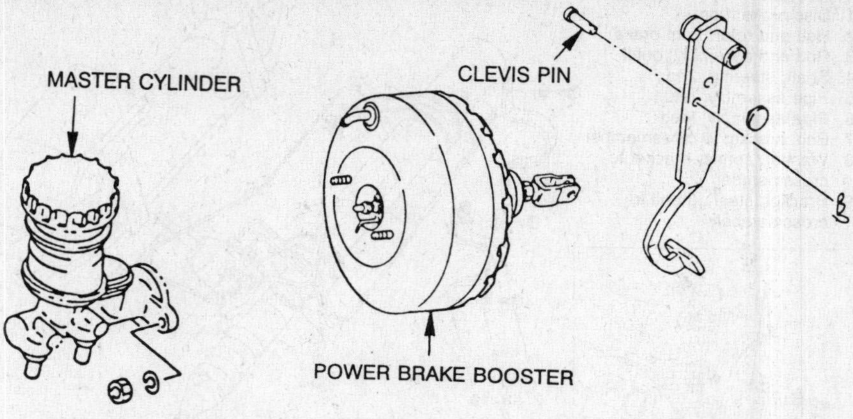

Master cylinder and power brake booster assembly – 1990–92 Impulse and 1991–92 Stylus

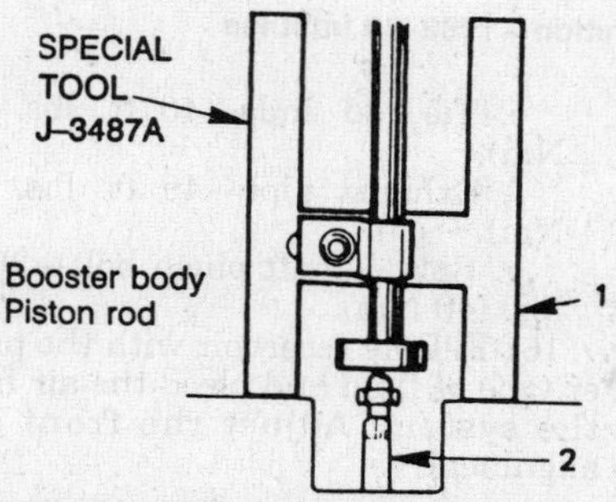

1. Booster body
2. Piston rod

Adjusting pushrod lengh using special tool – 1990–92 Impulse and 1991–92 Stylus

5. Remove the clevis pin from the brake pedal.
6. Remove the booster attaching nuts and remove the booster.

To install:

7. Install the power brake booster and tighten the booster attaching nuts to 13 ft. lbs. (18 Nm) for all, except 1988–89 Impulse or 9 ft. lbs. (13 Nm) for 1988–89 Impulse.
8. If a different booster is installed than was removed, or if the booster has been rebuilt, the pushrod must be adjusted before the master cylinder is installed. Adjust as follows:
   a. 1988–89 Impulse – Measure the distance from the flange face of the booster to the end of the pushrod. It should be 0.709–0.717 in. (18.0–18.2mm). Adjust, if necessary, by turning the locknut at the end of the pushrod.
   b. All other vehicles – Place pushrod gauge J-34873–A or equivalent, on the master cylinder and lower the pin until it's tip slightly touches the piston. Turn the pushrod gauge upside down and set it on the power brake booster. Adjust the pushrod length until the pushrod lightly touches the pin head.
   c. 1990–92 Impulse and 1991-92 Stylus – Measure the distance from the flange face of the booster to the end of the pushrod using a pushrod gauge J-34873–A. Turn the locknut at the end of the pushrod.

Brake Caliper

REMOVAL & INSTALLATION

Front

1. Raise and safely support the vehicle.
2. Remove the front wheel and tire assemblies.
3. Disconnect and plug the brake hose at the caliper.
4. Remove the caliper slide pin(s) and remove the caliper.
5. Installation is the reverse of the removal procedure. Tighten the slide pin(s) to 36 ft. lbs. (50 Nm) for all, except on 1988–89 Impulse or 27 ft. lbs. (37 Nm) for 1988–89 Impulse. Bleed the brake system.

Rear

1. Raise and safely support the vehicle.
2. Remove the rear wheel and tire assemblies.
3. Disconnect and plug the brake hose at the caliper.

NOTE: On 1990–92 Impulse and 1991-92 Stylus, the banjo bolt retaining the brake hose on the right hand side has left hand thread.

4. On 1990–92 Impulse and 1991-92 Stylus, disconnect the rear parking brake cable from the front cable, remove the brake cable from the cable support bracket and disconnect the brake cable from the brake lever.
5. Remove the lower slide pin and remove the caliper assembly.
6. Installation is the reverse of the

removal procedure. Tighten the slide pin to 14 ft. lbs. (20 Nm) on 1988–89 Impulse or 32 ft. lbs. (43 Nm) on 1990–92 Impulse. Bleed the brake system. On 1990–92 Impulse and 1991-92 Stylus, adjust the parking brake.

Disc Brake Pads

REMOVAL & INSTALLATION

1. Remove ½ brake fluid from the master cylinder.
2. Raise and safely support the vehicle.
3. Remove the wheel and tire assemblies.
4. Remove the brake caliper without disconnecting the brake line. Support the caliper with a length of wire. Do not let the caliper hang from the brake hose.
5. Remove the brake pads and shims. Inspect the brake rotor and machine or replace, as necessary. Check the minimum thickness specification when machining.

To install:

6. Use a suitable tool to push the caliper piston into it's bore. On 1990–92 Impulse and 1991-92 Stylus rear calipers, push the piston in by rotating it clockwise until it stops, then set the piston, aligning the uneven section on the piston surface with the caliper center.
7. Apply a thin coat of grease to the rear face of the brake pad and install the shim. Install the brake pads. On 1990–92 Impulse and 1991-92 Stylus rear calipers, the brake pad is provided with a pinion. The automatic adjuster becomes inoperative when the pinion is not placed correctly into the indentation of the piston.
8. Install the calipers. Install the wheel and tire assemblies and lower the vehicle.
9. Apply the brakes several times to seat the pads. Check the fluid in the master cylinder and add, as necessary.

Brake Rotor

REMOVAL & INSTALLATION

Except Front – 1988–89 Impulse

1. Raise and safely support the vehicle.
2. Remove the wheel and tire assembly.
3. Remove the caliper and the caliper bracket.
4. Remove the brake rotor.
5. Installation is the reverse of the removal procedure.

Front – 1988–89 Impulse

The rotor is part of the wheel hub on 1988–89 Impulse and must be removed as a unit.

1. Raise and safely support the vehicle.
2. Remove the front wheel and tire assembly.
3. Remove the caliper and the caliper bracket.
4. Remove the dust cap and cotter pin. Remove the castle nut, washer and the outer wheel bearing.
5. Remove the rotor/hub assembly with the inner wheel bearing.
6. If the rotor is machined while removed, the wheel bearings must be removed and the hub thoroughly cleaned before installation.
7. Installation is the reverse of the removal procedure. Properly adjust the wheel bearings and install a new cotter pin.

Brake Drums

REMOVAL & INSTALLATION

1. Raise and safely support the vehicle.
2. Remove the rear wheel and tire assembly.
3. Remove the dust cap and cotter pin. Remove the castle nut, washer and outer wheel bearing.
4. Remove the brake drum/hub assembly with the inner wheel bearing.
5. Installation is the reverse of the removal procedure. Properly adjust the wheel bearings and install a new cotter pin.

Brake Shoes

REMOVAL & INSTALLATION

1. Raise and safely support the vehicle.
2. Remove the rear wheel and tire assemblies.
3. Remove the brake drums.
4. Remove the brake return springs.
5. Remove the leading shoe holding pin and spring and the leading shoe.
6. Remove the self adjuster and the adjuster lever.
7. Remove the trailing shoe holding pin and spring.
8. Disconnect the parking brake cable from the trailing shoe and remove the trailing shoe. Remove the parking brake lever from the trailing shoe.

To install:

9. Apply a thin coat of suitable high temperature grease to the shoe contact pads on the brake backing plate prior to installation.
10. Check the brake drum for scoring or other wear and machine or replace, as necessary. Check the maximum brake drum diameter specification when machining. If the drum is machined, the wheel bearings must be removed and the hub thoroughly cleaned before reinstalling.
11. Adjust the brake shoes.

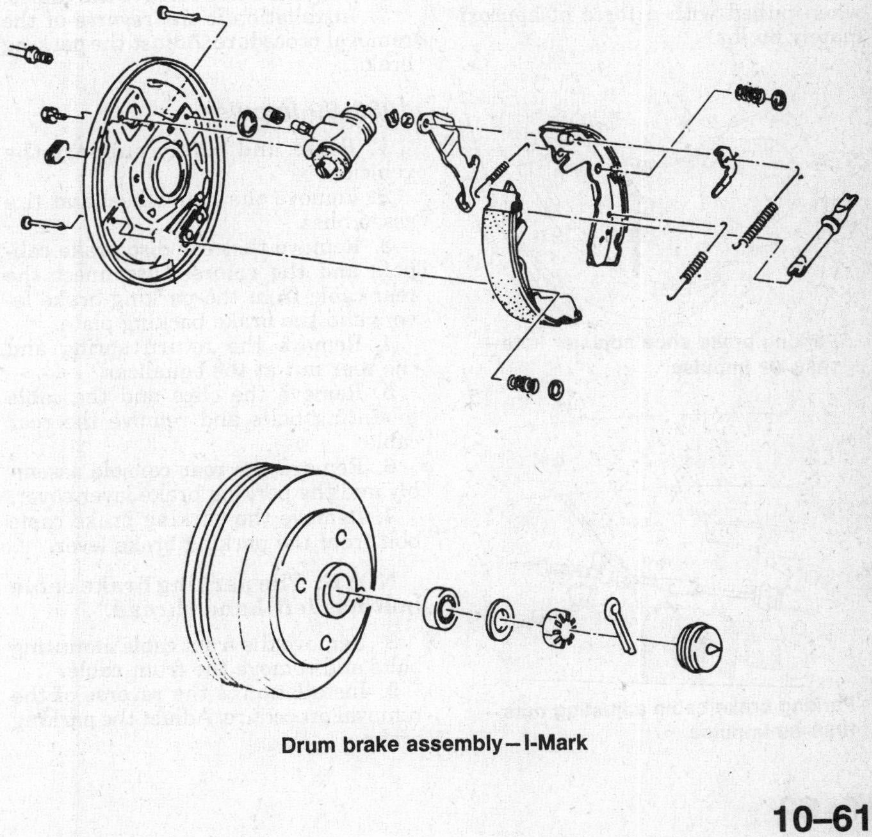

Drum brake assembly – I-Mark

Wheel Cylinder

REMOVAL & INSTALLATION

1. Raise and safely support the vehicle.
2. Remove the rear wheel and tire assembly.
3. Remove the brake drum and the brake shoes.
4. Disconnect and plug the brake line at the wheel cylinder.
5. Remove the wheel cylinder attaching bolts and the wheel cylinder.
6. Installation is the reverse of the removal procedure. Bleed the brake system.

Parking Brake Cable

ADJUSTMENT

1988–89 Impulse vehicles are equipped with parking brake shoes which are located inside the disc brake rotor. These shoes are adjusted by turning the adjuster until shoe contact can be felt, then backing off 6 notches.

1. Release the parking brake lever.
2. Adjust the brake shoes on I-Mark and 1988–89 Impulse.
3. On I-Mark, adjust the parking brake by turning the turnbuckle until the parking brake lever stroke is 7–9 notches when pulled with a force of approximately 66 lbs.
4. On 1988–89 Impulse, turn the nuts at the equalizer until the parking brake lever stroke is 11–13 notches when pulled with a force of approximately 66 lbs.

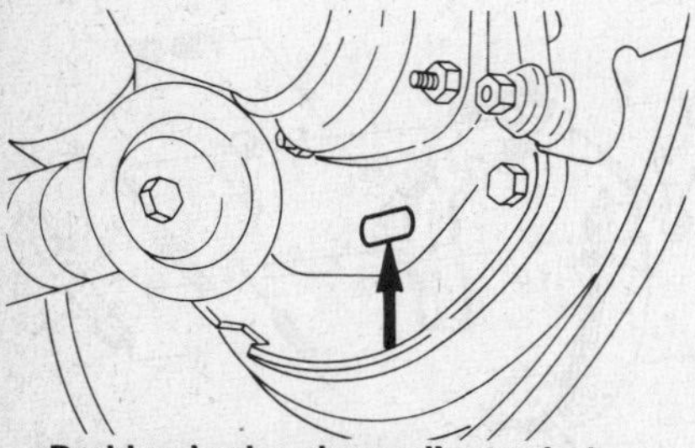

Parking brake shoe adjuster hole—1988–89 Impulse

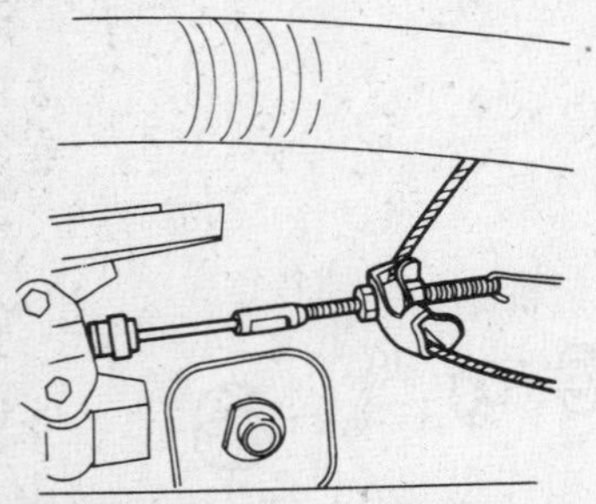

Parking brake cable adjusting nuts—1988–89 Impulse

NOTE: The parking brake shoes on 1988–89 Impulse must be broken down periodically or after replacement of the shoes or rotor in order to ensure effective operation. Drive the vehicle at about 30 mph on a safe, dry and level surface. Pull the brake lever up with a force of approximately 20 lbs. with the brake release button depressed. Drive the vehicle approximately ¼ mile with the parking brake partly applied. Repeat this operation 2–3 times.

5. On 1990–92 Impulse and 1991-92 Stylus, adjust the cable by turning the adjust nuts until the parking brake lever stroke is 7–8 notches when pulled with a force of approximately 66 lbs.

REMOVAL & INSTALLATION

I-Mark

1. Raise and safely support the vehicle.
2. Remove the rear wheel and tire assemblies.
3. Remove the brake drums.
4. Disconnect the rear cables from the parking brake levers and the brake backing plates.
5. Disconnect the front parking brake cable from the parking brake lever. Disconnect the front cable from the rear cables and remove the tension spring from the rear axle.
6. Remove the cable-to-body mounting bolts and remove the cables.
7. Installation is the reverse of the removal procedure. Adjust the parking brake.

1988–89 Impulse

1. Raise and safely support the vehicle.
2. Remove the rear wheel and tire assemblies.
3. Remove the rear disc brake calipers and the rotors. Disconnect the rear cable from the parking brake levers and the brake backing plate.
4. Remove the return spring and the rear nut at the equalizer.
5. Remove the clips and the cable mounting bolts and remove the rear cable.
6. Remove the rear console assembly and the parking brake lever cover.
7. Remove the parking brake cable bolt from the parking brake lever.

NOTE: The parking brake cable bolt has left hand thread.

8. Remove the front cable mounting bolts and remove the front cable.
9. Installation is the reverse of the removal procedure. Adjust the parking brake.

1990–92 Impulse and 1991-92 Stylus

1. Raise and safely support the vehicle.
2. Remove the rear wheel and tire assemblies.
3. Loosen the adjusting nuts at the left side rear cable.
4. Remove the right side rear cable from the bracket and disconnect it from the front cable.
5. Disconnect the rear cables from the parking brake levers at the disc brake assemblies.
6. Remove the cable mounting bolts and remove the rear cables.
7. Remove the console box.
8. Disconnect the front parking brake cable from the parking brake lever, remove the mounting nuts and remove the front cable.
9. Installation is the reverse of the removal procedure. Adjust the parking brake.

Brake System Bleeding

1. Set the parking brake and start the engine.

NOTE: The vacuum booster will be damaged if the bleeding operation is performed with the engine off.

2. Remove the master cylinder reservoir cap and fill the reservoir with brake fluid. Keep the reservoir at least half full during the bleeding operation.
3. If the master cylinder is replaced or overhauled, first bleed the air from the master cylinder and then from each caliper or wheel cylinder. Bleed the master cylinder as follows:
   a. Disconnect the left front wheel brake line from the master cylinder.
   b. Have an assistant depress the brake pedal slowly once and hold it depressed.
   c. Seal the delivery port of the master cylinder where the line was disconnected with a finger, then release the brake pedal slowly.
   d. Release the finger from the delivery port when the brake pedal returns completely.
   e. Repeat Steps c–e until the brake fluid comes out of the delivery port during Step c.
   f. Reconnect the brake line to the master cylinder.
   g. Have an assistant depress the brake pedal slowly once and hold it depressed.
   h. Loosen the front wheel brake line at the master cylinder.
   i. Retighten the brake line, then release the brake pedal slowly.

j. Repeat Steps g–i until no air comes out from the port when the brake line is loosened.

k. Bleed the air from the right front wheel brake line connection by repeating Steps a–j.

4. Bleed the air from each wheel in the following order: Left front caliper, Right rear caliper or wheel cylinder, Right front caliper, Left rear caliper or wheel cylinder. Bleed the air as follows:

a. Place the proper size box wrench over the bleeder screw.

b. Cover the bleeder screw with a transparent tube and submerge the free end of the tube in a transparent container containing brake fluid.

c. Have an assistant pump the brake pedal 3 times, then hold it depressed.

d. Remove the air along with the brake fluid by loosening the bleeder screw.

e. Retighten the bleeder screw, then release the brake pedal slowly.

f. Repeat Steps c–e until the air is completely removed. It may be necessary to repeat the bleeding procedure 10 or more times for front wheels and 15 or more times for rear wheels.

g. Go to the next wheel in sequence after each wheel is bled.

5. Depress the brake pedal to check if sponginess is felt after the air has been removed from all wheel cylinders and calipers. If the pedal feels spongy, the entire bleeding procedure must be repeated.

6. After the bleeding operation is completed on each individual wheel, check the level of brake fluid in the reservoir and replenish up to the **MAX** level, if necessary.

7. Install the master cylinder reservoir cap and stop the engine.

Anti-lock Brake System Service

Some diagnostic procedures require the installation of a pinout box tool J-35592 in order to prevent damage to the 35-pin EBCM connector. The pinout box should be installed prior to probing any circuit with a digital multi-meter.

PINOUT BOX INSTALLATION

1. Ensure that the ignition switch is in the **OFF** position when removing the 35-pin EBCM connector.

2. Disconnect the connector by depressing the locking plate and rotating the connector toward the front of the vehicle.

3. Inspect the 35-pin connector for damage. Install the pinout box tool J-35592 on the connector.

4. Proceed with self-diagnostic test.

SYSTEM SELF-DIAGNOSIS

The ABS is equipped with a self diagnostic capability which is used to isolate ABS failures.

Electrical failures in the ABS are detected by the ECBM, located under the passenger seat, and result in the Anti-Lock warning light illuminating. The Anti-Lock warning light is intended to inform the driver that a condition exists which results in the ABS being disabled. The Anti-Lock warning light is connected to a Light Emitting Diode (LED) on the EBCM. The LED assists the individual servicing the system by flashing trouble codes which pinpoint the defective component.

NOTE: If more than one failure in the system is detected, the first failure to occur will be flashed in code on the LED. Once this failure has been corrected, the next failure code will be will be flashed.

The EBCM (Electronic Brake Control Module) enters diagnostic mode any time a trouble code is set; the Anti-Lock light illuminates. To read the ABS trouble code, count the number of times the LED on the EBCM turns ON and OFF. Trouble codes are erased from the EBCM memory when the ignition key is turned OFF.

SERVICE PRECAUTIONS

- Do not use rubber hoses or other parts not specifically specified for the ABS system. When using repair kits, replace all parts included in the kit. Partial or incorrect repair may lead to functional problems.
- Lubricate rubber parts with clean, fresh brake fluid to ease assembly. Do not use lubricated shop air to clean parts; damage to rubber components may result.
- Use only brake fluid from an unopened container. Use of suspect or contaminated brake fluid can reduce system performance and/or durability.
- A clean repair area is essential. Perform repairs after components have been thoroughly cleaned; use only denatured alcohol to clean components. Do not allow components to come into contact with any substance containing mineral oil; this includes used shop rags.
- The ABS ECU is a microprocessor similar to other computer units in the vehicle. Insure that the ignition switch is OFF before removing or installing controller harnesses. Avoid static electricity discharge at or near the controller.
- Never disconnect any electrical connection with the ignition switch ON unless instructed to do so in a test.
- Avoid touching module connector pins.
- Leave new components and modules in the shipping package until ready to install them.
- To avoid static discharge, always touch a vehicle ground after sliding across a vehicle seat or walking across carpeted or vinyl floors.
- Never allow welding cables to lie on, near or across any vehicle electrical wiring.
- Do not allow extension cords for power tools or droplights to lie on, near or across any vehicle electrical wiring.
- If welding is to be performed on the vehicle using an electric arc welder, the EBCM and valve block connectors should be disconnected.
- Hydraulic units of the Anti-Lock Brake System are not separately serviceable and must be replaced as assemblies. Do not disassemble any component which is designated as non-serviceable.

RELIEVING ANTI-LOCK BRAKE SYSTEM PRESSURE

1. Set the parking brake and start the engine.

NOTE: The vacuum booster will be damaged if the bleeding operation is performed with the engine off.

2. Remove the master cylinder reservoir cap and fill the reservoir with brake fluid. Keep the reservoir at least half full during the bleeding operation.

3. If the master cylinder is replaced or overhauled, first bleed the air from the master cylinder and then from each caliper or wheel cylinder. Bleed the master cylinder as follows:

a. Disconnect the left front wheel brake line from the master cylinder.

b. Have an assistant depress the brake pedal slowly once and hold it depressed.

c. Seal the delivery port of the master cylinder where the line was disconnected with a finger, then release the brake pedal slowly.

d. Release the finger from the delivery port when the brake pedal returns completely.

e. Repeat Steps C–E until the brake fluid comes out of the delivery port during Step C.

f. Reconnect the brake line to the master cylinder.

g. Have an assistant depress the

brake pedal slowly once and hold it depressed.

h. Loosen the front wheel brake line at the master cylinder.

i. Retighten the brake line, then release the brake pedal slowly.

j. Repeat Steps G–I until no air comes out from the port when the brake line is loosened.

k. Bleed the air from the right front wheel brake line connection by repeating Steps A–J.

4. Bleed the air from each wheel in the following order: Left front caliper, Right rear caliper or wheel cylinder, Right front caliper, Left rear caliper or wheel cylinder. Bleed the air as follows:

a. Place the proper size box wrench over the bleeder screw.

b. Cover the bleeder screw with a transparent tube and submerge the free end of the tube in a transparent container containing brake fluid.

c. Have an assistant pump the brake pedal 3 times, then hold it depressed.

d. Remove the air along with the brake fluid by loosening the bleeder screw.

e. Retighten the bleeder screw, then release the brake pedal slowly.

f. Repeat Steps C–E until the air is completely removed. It may be necessary to repeat the bleeding procedure 10 or more times for front wheels and 15 or more times for rear wheels.

g. Go to the next wheel in sequence after each wheel is bled.

5. Depress the brake pedal to check if sponginess is felt after the air has been removed from all wheel cylinders and calipers. If the pedal feels spongy, the entire bleeding procedure must be repeated.

6. After the bleeding operation is completed on each individual wheel, check the level of brake fluid in the reservoir and replenish up to the **MAX** level, if necessary.

7. Install the master cylinder reservoir cap and stop the engine.

Hydraulic Unit

REMOVAL & INSTALLATION

1. Disconnect negative battery cable.
2. Raise and support vehicle safely.
3. Remove under cover to gain access to hydraulic unit.
4. Remove tire if needed and remove inner fender liner.
5. Disconnect harness connectors from hydraulic unit.
6. Remove radiator reservoir tank.
7. Remove brake lines using a flare nut wrench. Cap or tape brake line ends to prevent entry of foreign matter.
8. Disconnect hydraulic motor ground cable.
9. Remove bracket attaching bolt, hydraulic unit attaching nut, bracket and hydraulic unit.

To install:

10. Install the hydraulic unit and torque the hydraulic unit attaching nut, bracket attaching bolt and ground cable bolt to 17 ft. lbs. (22 Nm). Torque brake line to 9 ft. lbs. (12 Nm).
11. Bleed the brake system.

NOTE: Replace all components included in repair kits used to service this system. Lubricate rubber parts with clean, fresh brake fluid to ease assembly. Do not use lubricated shop air to clean parts, as damage to rubber components may result. Always bleed the braking system after repairing or replacing hydraulic components.

Electric Brake Control Unit (EBCM)

REMOVAL & INSTALLATION

The EBCM is located under the passenger side seat on the Impulse.

1. Disconnect the negative battery cable.
2. It may be necessary to move the passenger seat out of the way to gain access to the EBCM. If so, remove the seat attaching bolts and move the seat.
3. Remove the EBCM attaching bolts.
4. Remove the EBCM wiring harness connector. Remove the EBCM.
5. Installation is the reverse of removal. Tighten EBCM attaching bolts to 62 inch lbs. (7.0 Nm).

G-Sensor

REMOVAL & INSTALLATION

1. Disconnect negative battery cable.
2. Remove center console.
3. Remove G-Sensor wiring harness connector.
4. Remove G-Sensor attaching bolt. Remove the G-Sensor.
5. Place G-Sensor on a known level surface and check continuity between terminals. If no continuity, replace the G-Sensor.
6. Incline the G-Sensor to a 30 degree angle and retest for continuity. If continuity exists, replace the G-Sensor.

NOTE: Ensure that G-Sensor is installed in the correct direction.

7. Installation is the reverse of removal. Torque attaching bolts to 53 inch lbs. (5.4 Nm).

Speed Sensor

REMOVAL & INSTALLATION

1. Disconnect negative battery cable.
2. Raise and support vehicle safely.
3. Remove wheel assembly. Remove inner fender liner.
4. Disconnect speed sensor wire connector.
5. Remove sensor cable attaching bolts and screws.
6. Remove sensor attaching bolts. Remove sensor.

To install:

7. Inspect the speed sensor for damage.

a. Check speed sensor pole piece for dirt and remove.

b. Check the pole piece for damage and replace, if necessary.

c. Check for continuity while flexing the sensor cable. Replace sensor cable, if a short or open is found.

d. Check the sensor rotor for damage including tooth chipping. Replace the driveshaft assembly, if sensor rotor is damaged.

8. Install the speed sensor taking care not to damage the pole piece. Tighten the attaching bolt to 62 inch lbs. (6.8 Nm).
9. Check the clearance between the speed sensor pole piece and the rotor. Clearance should be 0.0079–0.0315 in. (0.20–0.80mm).
10. Install the sensor cable fixing bolt, tighten to 13 ft. lbs. (17 Nm) and screw to 9 ft. lbs. (12 Nm).
11. Ensure the white line marked on the cable is not twisted. If so, loosen connections and straighten cable.
12. Reconnect sensor wire connector, install inner liner, install wheel, lower vehicle and reconnect negative battery cable.

CHASSIS ELECTRICAL

Air Bag

DISARMING

CAUTION

On vehicles equipped with an air bag, the negative battery cable must be disconnected, before working on the system. Failure to do so may result in deployment of the air bag and possible personal injury.

Always wear gloves and safety glasses when handling a deployed SIR module and wash your hands with mild soap and water afterwards.

The SIR module should alway6s be carreied with the urethane cover away from the body and should always be laid on a flat survae with the urethane side up. This is necessary because a free space is provided to allow the air cushion to expand in the unlikey event of an accidental deployment.

Heater Blower Motor

REMOVAL & INSTALLATION

1. Disconnect the negative battery cable.
2. Disconnect the wire connector from the blower motor.
3. Disconnect the rubber hose on vehicles equipped with air conditioning.
4. Remove the retaining clip and the blower motor.
5. Installation is the reverse of the removal procedure.

Windshield Wiper Motor

REMOVAL & INSTALLATION

Front

1. Disconnect the negative battery cable.
2. Remove the wiper arm cap, nut and the wiper arm and blade.
3. Remove the cowl cover, if equipped.
4. Disconnect the wiper motor from the wiper linkage.
5. Disconnect the electrical connectors.
6. Remove the mounting bolts and the wiper motor.
7. Installation is the reverse of the removal procedure. Be sure to apply grease to the crank arm ball joint.

Rear

1. Disconnect the negative battery cable.
2. Remove the cover, nut and rear wiper arm and blade.
3. Remove the rear hatch trim panel, if equipped.
4. Disconnect the electrical connector at the wiper motor.
5. On I-Mark, remove the wiper motor mounting bolts, disconnect the wiper motor from the wiper linkage and remove the wiper motor.
6. On Impulse, remove the nut, cap washer and seal. Remove the wiper motor mounting bolts and remove the wiper motor.
7. Installation is the reverse of the removal procedure.

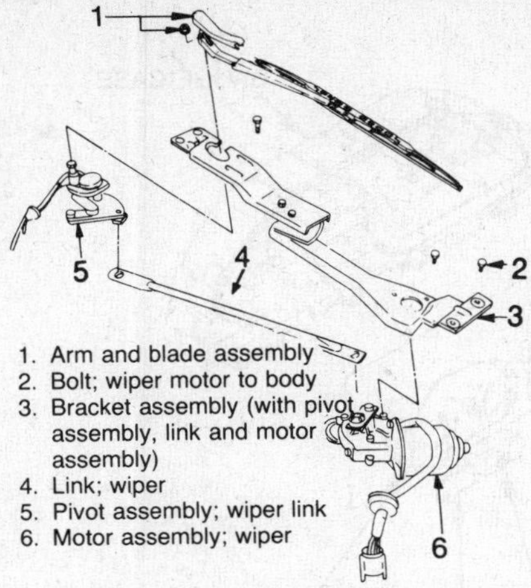

Front wiper motor and linkage— 1988–89 Impulse

Windshield Wiper Switch

REMOVAL & INSTALLATION

I-Mark, 1990–92 Impulse and 1991-92 Stylus

1. Disconnect the negative battery cable.
2. Remove the instrument cluster hood attaching screws. Disconnect the electrical connectors so the hood can be removed.
3. Remove the instrument cluster, if necessary.
4. Remove the windshield wiper switch.
5. Installation is the reverse of the removal procedure.

1988–89 Impulse

1. Disconnect the negative battery cable.
2. Remove the instrument cluster and switch assembly.
3. Remove the screws attaching the switch and the instrument cluster lower hood.
4. Remove the switch harness clip located at the lower rear of the instrument cluster.
5. Remove the lower hood and switch together and separate them.
6. Installation is the reverse of the removal procedure.

Instrument Cluster

REMOVAL & INSTALLATION

I-Mark

1. Disconnect the negative battery cable.
2. Remove the instrument cluster hood attaching screws and the instrument cluster hood.
3. Disconnect the electrical connectors at the instrument cluster.
4. Remove the instrument cluster attaching screws and the instrument cluster.
5. Remove the trip meter reset knob and the instrument cluster screen.
6. Remove the gauge surround and the buzzer, if equipped.
7. Remove the sockets and bulbs.
8. Remove the gauges.
9. Installation is the reverse of the removal procedure.

Impulse and Stylus

1988–89

1. Disconnect the negative battery cable.
2. Remove the steering wheel.
3. Remove the front console and radio assembly.
4. Remove the lower instrument panel cover and hood lock release lever.
5. Remove the left knee pad and the steering column covers.
6. Remove the instrument cluster hood.
7. Remove the nuts and bolts retaining the instrument cluster and pull the cluster outward to gain access to the speedometer cable and electrical connectors.
8. Disconnect the speedometer cable and the electrical connectors and remove the instrument cluster and switch assembly.
9. Remove the screws attaching the switch assemblies and the instrument cluster lower hood. Remove the switch harness clip at the lower rear of the instrument cluster and remove the switches and lower hood from the instrument cluster.
10. Remove the trip meter reset knob. Remove the instrument cluster screen by unlocking the hooks. Remove the gauge surround by unlocking the hooks.
11. Remove the speedometer by removing the retaining nuts and screws.

To install:

12. Install the speedometer.
13. Install the trip meter reset knob and cluster screen.
14. Install the screws attaching the switch assemblies and the instrument cluster lower hood.
15. Connect the speedometer cable and the electrical connectors.
16. Install the nuts and bolts retaining the instrument cluster.
17. Install the instrument cluster hood.
18. Install the left knee pad and the steering column covers.
19. Install the lower instrument panel cover and hood lock release lever.

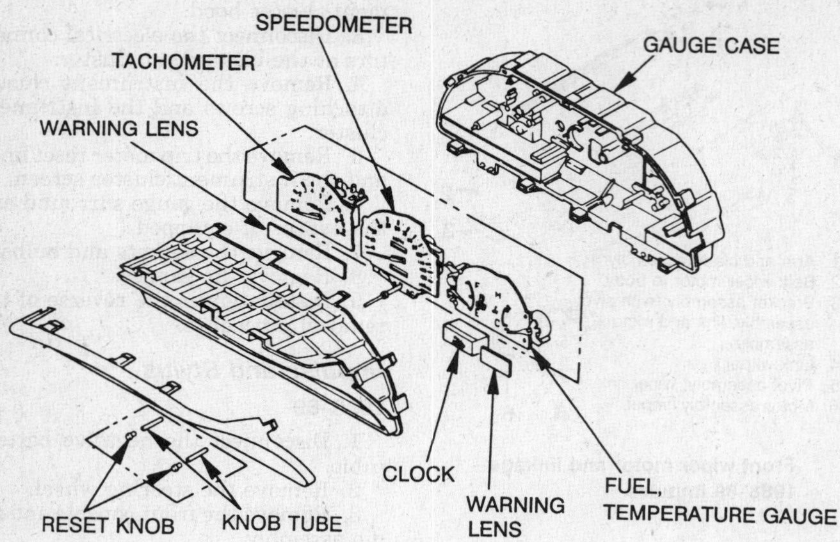

Instrument cluster assembly—1990–92 Impulse and 1991–92 Stylus

20. Install the front console and radio assembly.
21. Install the steering wheel.
22. Connect the negative battery cable.

1990–92

1. Disconnect the negative battery cable.
2. Remove the instrument cluster hood screw hole covers and remove the hood attaching screws.
3. Disconnect the lighting and windshield wiper switch connectors and remove the instrument cluster hood.
4. Disconnect the speedometer cable and electrical connectors from the instrument cluster.
5. Remove the instrument cluster.
6. Remove the trip meter and clock reset knob and the gauge screen.
7. Remove the instrument cluster bezel and remove the speedometer from the gauge case.

To install:

8. Install the instrument cluster bezel and speedometer to the gauge case.
9. Install the trip meter and clock reset knob and the gauge screen.
10. Install the instrument cluster.
11. Connect the speedometer cable and electrical connectors to the instrument cluster. Install the cluster hood.
12. Connect the lighting and windshield wiper switch connectors.
13. Install the instrument cluster hood screw hole covers.
14. Connect the negative battery cable.

Radio

REMOVAL & INSTALLATION

I-Mark

1. Disconnect the negative battery cable. Remove the screws retaining the radio cover (front console) and remove the cover.
2. Remove the radio and bracket.
3. Disconnect the electrical connector, speaker connectors and the antenna cable.
4. To install, reverse the removal procedures.

Impulse

1. Disconnect the negative battery cable.
2. Remove the floor console on 1990–92 vehicles.
3. Remove the front console panel.
4. Remove the radio bracket mounting screws. Remove the radio mounting screws. Disconnect the electrical leads and the antenna.
5. Installation is the reverse of the removal procedures.

Concealed Headlights

MANUAL OPERATION

1. Raise the hood and disconnect the electrical connector at the headlight cover motor.
2. Turn the manual operation knob located on the headlight cover motor to open or close the headlight covers.

Headlight Switch

REMOVAL & INSTALLATION

I-Mark, 1990–92 Impulse and 1991-92 Stylus

1. Disconnect the negative battery cable.

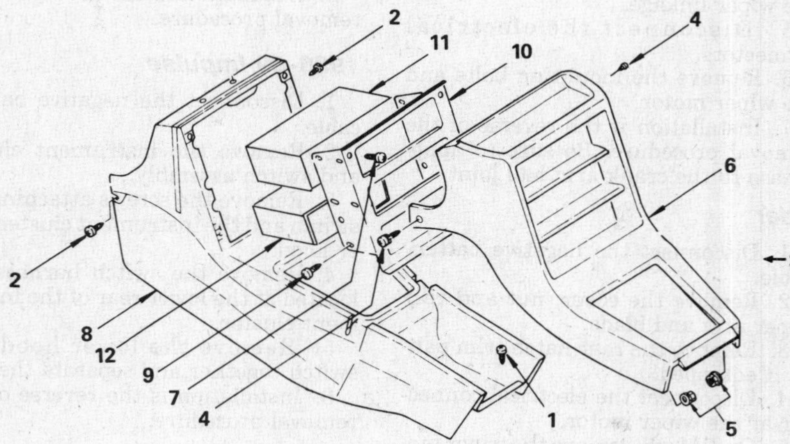

1. Screw; front console to body
2. Screws; front console to bracket
3. Console assembly
4. Screws; front console pad to front console
5. Flange nuts; front console pad to front console
6. Front console pad
7. Screws; bezel to front console
8. Front console
9. Screws; bezel to radio
10. Bezel
11. Cassette deck with FM/AM radio or FM/AM radio
12. Graphic equalizer or cassette deck

Radio installation—Impulse

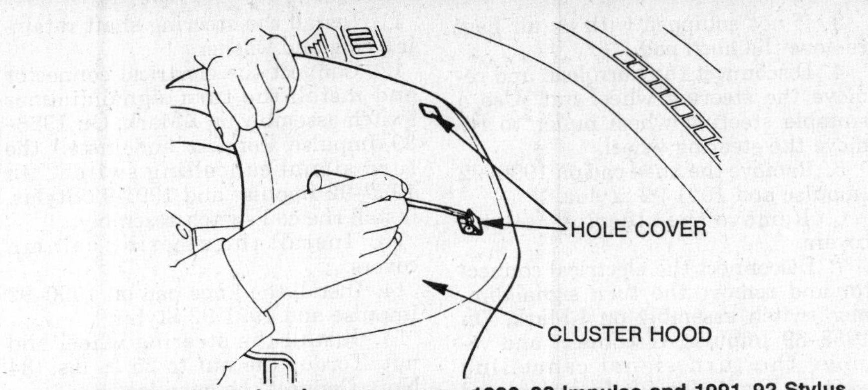

Removing the cluster hood screw covers—1990–92 Impulse and 1991–92 Stylus

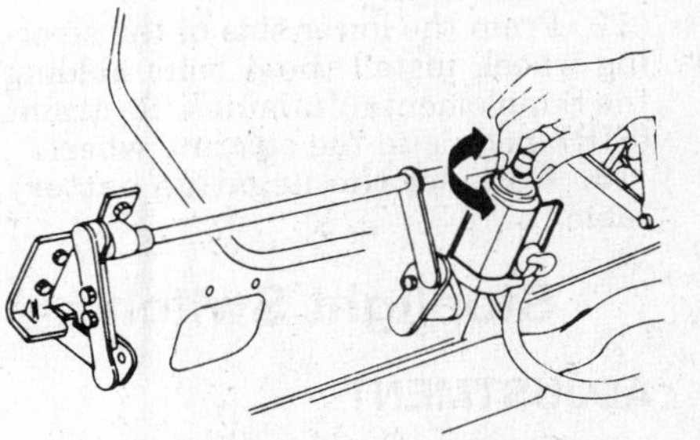

Headlight cover manual operation—Impulse

2. Remove the instrument cluster hood attaching screws. Disconnect the electrical connectors so the hood can be removed.
3. Remove the instrument cluster, if necessary.
4. Remove the headlight switch.
5. Installation is the reverse of the removal procedure.

1988–89 Impulse

1. Disconnect the negative battery cable.
2. Remove the instrument cluster and switch assembly.
3. Remove the screws attaching the switch and the instrument cluster lower hood.
4. Remove the switch harness clip located at the lower rear of the instrument cluster.
5. Remove the lower hood and switch together and separate them.
6. Installation is the reverse of the removal procedure.

Dimmer Switch

REMOVAL & INSTALLATION

I-Mark

1. Disconnect the negative battery cable.
2. Remove the steering wheel.
3. Remove the upper and lower steering column covers.
4. Disconnect the switch electrical connector from the harness.
5. Remove the switch attaching screws and remove the switch.

1988–89 Impulse

1. Disconnect the negative battery cable.
2. Remove the instrument cluster and switch assembly.
3. Remove the screws attaching the switch and the instrument cluster lower hood.
4. Remove the switch harness clip located at the lower rear of the instrument cluster.
5. Remove the lower hood and switch together and separate them.
6. Installation is the reverse of the removal procedure.

1990–92 Impulse and 1991-92 Stylus

1. Disconnect the negative battery cable.
2. Place the front wheels in the straight ahead position and the ignition key in the **LOCK** position.
3. From the lower side of the steering wheel, remove the 4 bolts holding the Supplemental Inflatable Restraint (SIR) module to the steering wheel. Disconnect the module connector and remove the SIR module.

CAUTION

The SIR module should always be carried with the urethane cover away from the body and should always be laid on a flat surface with the urethane side up. This is necessary because a free space is provided to allow the air cushion to expand in the unlikely event of accidental deployment. Otherwise, personal injury may result.

4. Disconnect the horn lead and remove the steering wheel nut. Use a suitable steering wheel puller to remove the steering wheel.
5. Remove the knee pad and the steering column cover.
6. Disconnect the electrical connector from the coil/switch assembly.
7. Remove the retaining screws and the coil/switch assembly.
8. Separate the coil from the switch.

To install:

9. Install the switch, trim panels and connect the electrical connectors.

NOTE: Whenever the coil/switch assembly is be replaced, the vehicle's front wheels must be straight ahead. Failure to do so will cause the coil assembly to be removed without being centered. Installing an uncentered coil assembly can cause damage to the coil assembly.

10. Install the steering wheel and torque the nut to 25 ft. lbs. (34 Nm).
11. Install and connect the horn pad or SIR module.
12. Connect the battery cable and check operation.

Turn Signal Switch

REMOVAL & INSTALLATION

I-Mark

1. Disconnect the negative battery cable.
2. Remove the steering wheel.
3. Remove the upper and lower steering column covers.
4. Disconnect the switch electrical connector from the harness.
5. Remove the switch attaching screws and remove the switch.

1988–89 Impulse

1. Disconnect the negative battery cable.
2. Remove the instrument cluster and switch assembly.
3. Remove the screws attaching the switch and the instrument cluster lower hood.
4. Remove the switch harness clip located at the lower rear of the instrument cluster.
5. Remove the lower hood and switch together and separate them.
6. Installation is the reverse of the removal procedure.

1990–92 Impulse and 1991-92 Stylus

1. Disconnect the negative battery cable.
2. Place the front wheels in the straight ahead position and the ignition key in the **LOCK** position.
3. From the lower side of the steering wheel, remove the 4 bolts holding the Supplemental Inflatable Restraint (SIR) module to the steering wheel. Disconnect the module connector and remove the SIR module.

CAUTION

The SIR module should always be carried with the urethane cover away from the body and should always be laid on a flat surface with the urethane side up. This is necessary because a free space is provided to allow the air cushion to expand in the unlikely event of accidental deployment. Otherwise, personal injury may result.

4. Disconnect the horn lead and remove the steering wheel nut. Use a suitable steering wheel puller to remove the steering wheel.
5. Remove the knee pad and the steering column cover.
6. Disconnect the electrical connector from the coil/switch assembly.
7. Remove the retaining screws and the coil/switch assembly.
8. Separate the coil from the switch.

To install:

9. Install the switch, trim panels and connect the electrical connectors.

NOTE: Whenever the coil/switch assembly is be replaced, the vehicle's front wheels must be straight ahead. Failure to do so will cause the coil assembly to be removed without being centered. Installing an uncentered coil assembly can cause damage to the coil assembly.

10. Install the steering wheel and torque the nut to 25 ft. lbs. (34 Nm).
11. Install and connect the horn pad or SIR module.
12. Connect the battery cable and check operation.

Ignition Lock/Switch

REMOVAL & INSTALLATION

1. Disconnect the negative battery cable.
2. If equipped with an air bag, perform the following procedure:
   a. Place the front wheels in the straight ahead position and the ignition key in the **LOCK** position.
   b. From the lower side of the steering wheel, remove the 4 bolts holding the Supplemental Inflatable Restraint (SIR) module to the steering wheel. Disconnect the module connector and remove the SIR module.

CAUTION

The SIR module should always be carried with the urethane cover away from the body and should always be laid on a flat surface with the urethane side up. This is necessary because a free space is provided to allow the air cushion to expand in the unlikely event of accidental deployment. Otherwise, personal injury may result.

3. If not equipped with an air bag, remove the horn pad.
4. Disconnect the horn lead and remove the steering wheel nut. Use a suitable steering wheel puller to remove the steering wheel.
5. Remove the knee pad on 1990–92 Impulse and 1991-92 Stylus.
6. Remove the steering column covers.
7. Disconnect the electrical connector and remove the turn signal/dimmer switch assembly on I-Mark. On 1988–89 Impulse, disconnect and remove the turn signal cancelling switch. On 1990–92 Impulse and 1991-92 Stylus, remove the coil/switch assembly.
8. Remove the steering shaft retaining ring and washer.
9. Disconnect the ignition switch electrical connector, remove the steering lock retaining screws and remove the lock/switch assembly.

NOTE: Whenever the coil/switch assembly is be replaced, the vehicle's front wheels must be straight ahead. Failure to do so will cause the coil assembly to be removed without being centered. Installing an uncentered coil assembly can cause damage to the coil assembly.

To install:

10. Install and connect the ignition switch electrical connector.
11. Install the steering shaft retaining ring and washer.
12. Connect the electrical connector and install the turn signal/dimmer switch assembly on I-Mark. On 1988–89 Impulse, connect and install the turn signal cancelling switch. On 1990–92 Impulse and 1991-92 Stylus, install the coil/switch assembly.
13. Install the steering column covers.
14. Install the knee pad on 1990–92 Impulse and 1991-92 Stylus.
15. Install the steering wheel and nut. Torque the nut to 25 ft. lbs. (34 Nm). Connect the horn lead.
16. If not equipped with an air bag, install the horn pad.
17. From the lower side of the steering wheel, install the 4 bolts holding the Supplemental Inflatable Restraint (SIR) module to the steering wheel.
18. Connect the negative battery cable.

Stoplight Switch

ADJUSTMENT

I-Mark, 1990–92 Impulse and 1991-92 Stylus

1. Loosen the locknut.
2. Place the tip of the switch so it rests gently against the rubber stopper on the pedal arm.

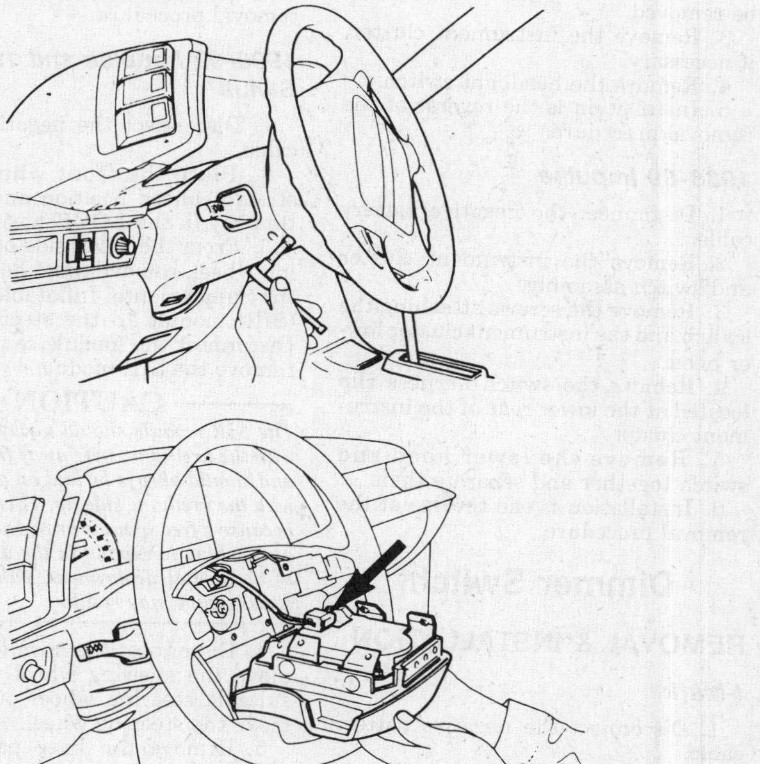

Supplemental Inflatable Restraint (SIR) module removal

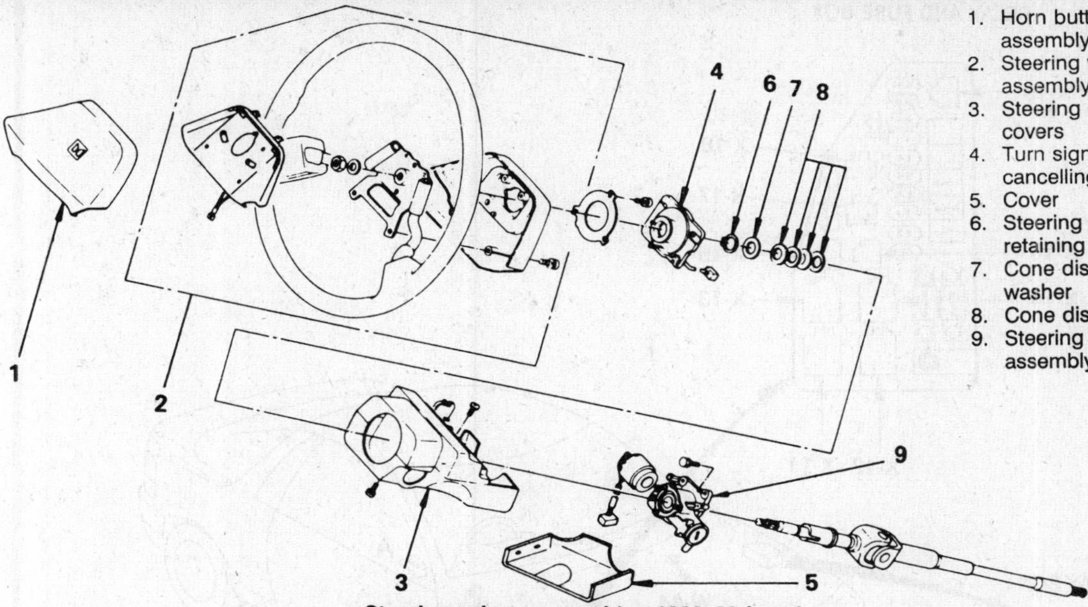

1. Horn button pad assembly
2. Steering wheel assembly
3. Steering column covers
4. Turn signal cancelling switch
5. Cover
6. Steering shaft retaining ring
7. Cone disc plain washer
8. Cone disc spring
9. Steering lock assembly

Steering column assembly—1988–89 Impulse

3. Rotate the stoplight switch until the switch housing contacts lightly with the pedal stopper.

NOTE: Do not attempt to force the brake pushrod into position during the stoplight switch adjustment procedure.

4. Tighten the locknut.

1988–89 Impulse

1. Adjust the brake pedal height to 5.630 in. (143mm) with the brake pedal pushrod.
2. Adjust the stoplight switch so the clearance between the switch and the brake pedal is 0.0039 in. (0.1mm).
3. On all except 1988 vehicles equipped with 4ZC1-T engine and automatic transmission and all 1989 vehicles, turn the stoplight switch ½ turn and lock into position, so the pedal height is reduced from 5.630 in. (143mm) to 5.52 in. (140mm).

REMOVAL & INSTALLATION

1. Disconnect the negative battery cable.
2. Disconnect the electrical connector from the switch.
3. Loosen the locknuts and remove the switch.
4. Installation is the reverse of the removal procedure. Adjust the switch according to the proper procedure.

Clutch Switch

REMOVAL & INSTALLATION

1. Disconnect the negative battery cable.
2. Disconnect the electrical connector at the switch.
3. Remove the switch retaining screws and the switch.
4. Installation is the reverse of the removal procedure.

Neutral Safety Switch

REMOVAL & INSTALLATION

1. Raise and safely support the vehicle.
2. Disconnect the electrical connector to the switch.
3. Remove the switch from the transmission.
4. Installation is the reverse of the removal procedure.

NOTE: This adjustment is necessary only if the engine will start with the shift selector in any range except N or P.

5. Loosen the neutral start switch bolt and set the shift selector into the **N** range.
6. Align the groove and the neutral basic line.
7. Hold it in position and torque the bolt to 9 ft. lbs. (12 Nm).

Fuses

LOCATION

The fuse box is located in the lower left hand corner of the dashboard on all I-Mark. On 1988–89 Impulse, the the fuse box is located on the left kick panel. On 1990–92 Impulse and 1991-92 Stylus, fuses are located in both the junction block located on the left kick panel and in the relay and fuse box located in the left side of the engine compartment.

All I-Marks except those equipped with the 4XC1-U engine have the air conditioning compressor fuse located behind the left side of the dashboard. The 1988–89 Impulse has the blower fan fuse located under the blower fan case.

Circuit Breakers

LOCATION

All 1988–89 Impulse have a 20A circuit breaker for the power windows and a 30A circuit breaker for the Automatic seat belts located in the fuse box. The 1990–92 Impulse and 1991-92 Stylus have 2 circuit breaker locations on the junction box, with only 1 being used for the power windows and sunroof.

Relays

LOCATION

I-Mark

FICD Relay—located in right rear corner of engine compartment.

EFE Heater Relay—located in right rear corner of engine compartment.

Automatic Choke Relay—located in right rear corner of engine compartment.

Fog Lamp Relay—located in left front corner of engine compartment.

Cooling Fan Relay—located in

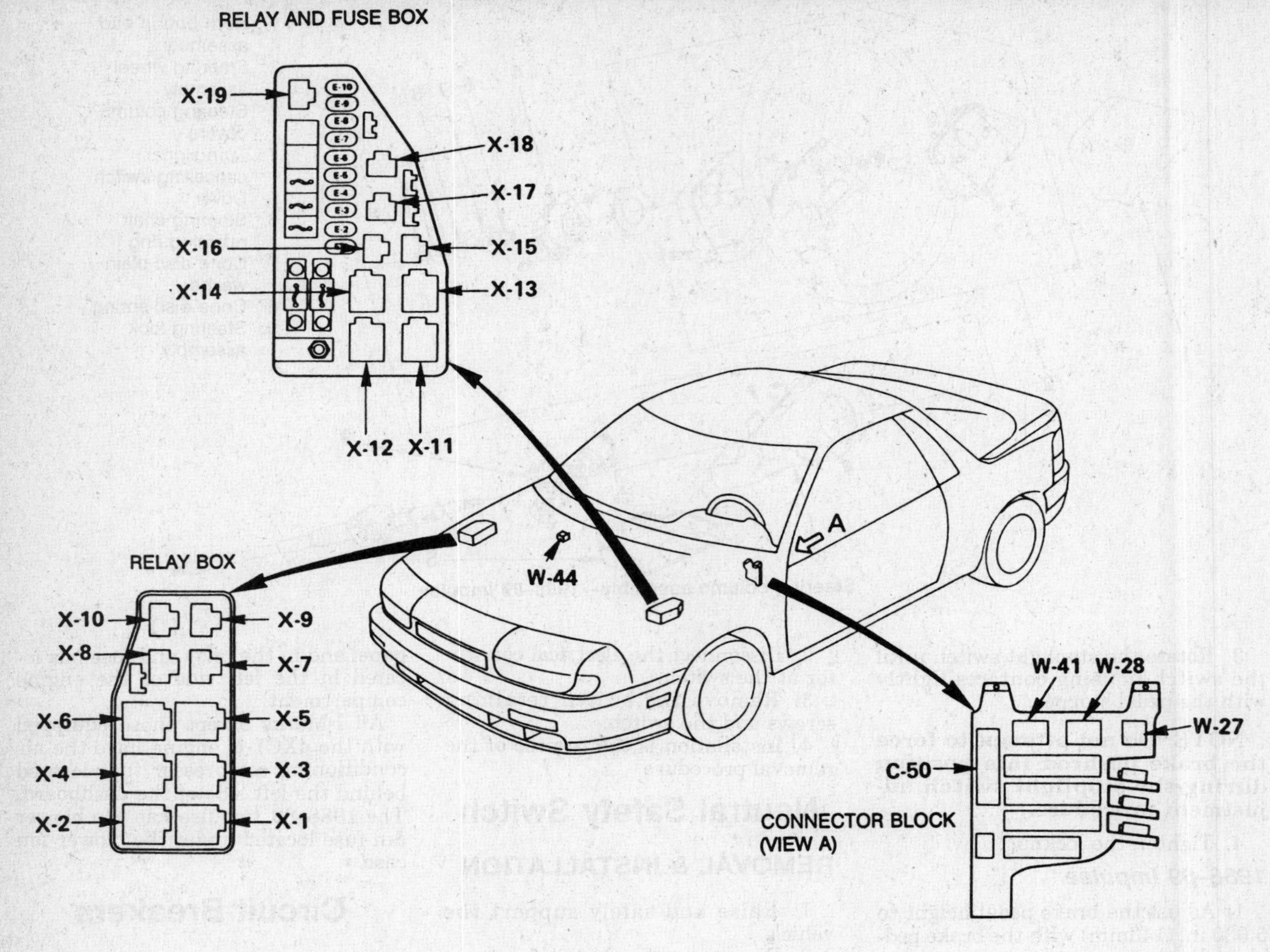

RELAY LIST

No. / Usage / Model	X-1	X-2	X-3	X-4	X-5	X-6	X-7	X-8	X-9	X-10
Usage	Headlight cover-1	—	Headlight cover-2	Neutral (A/T)	—	—	Fog light	Horn	Thermo. SW	A/C compressor
2WD	o	—	o	o	—	—	o	o	o	o
4WD	o	—	o	—	—	—	o	o	o	o

No. / Usage / Model	X-11	X-12	X-13	X-14	X-15	X-16	X-17	X-18	X-19
Usage	Lighting	Restart	—	Anti-theft	Radiator fan	Condenser fan	Main	Heater & A/C	Fuel pump
2WD	o	o	—	—	o	o	o	o	o
4WD	o	o	—	o	o	o	o	o	o

No. / Usage / Model	W-27	W-28	W-41	W-44	C-50
Usage	Flasher unit	RR defogger	Power window	Max. high	Rise up & intermittent
2WD	o	o	o	o	o
4WD	o	o	o	o	o

Relay location—1990–92 Impulse

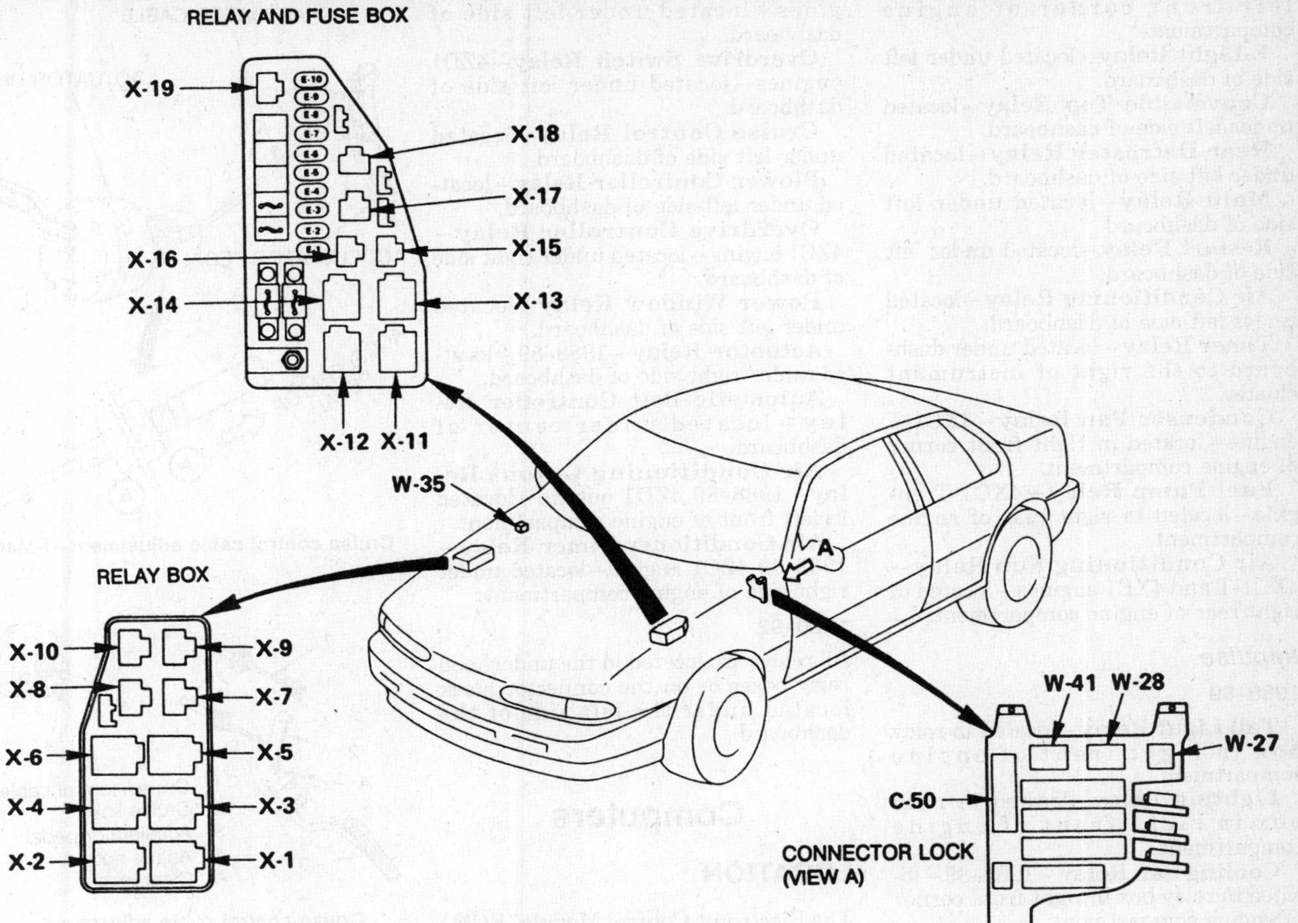

RELAY LIST

No. / Usage / Model	X-1	X-2	X-3	X-4	X-5	X-6	X-7	X-8	X-9	X-10
Usage	–	–	–	Neutral (A/T)	–	–	Fog light	Horn	Thermo. SW	A/C compressor
SOHC	–	–	–	o	–	–	o	o	o	o
DOHC	–	–	–	–	–	–	o	o	o	o

No. / Usage / Model	X-11	X-12	X-13	X-14	X-15	X-16	X-17	X-18	X-19
Usage	Lighting	Restart	–	–	Radiator fan	Condenser fan	Main	Heater & A/C	Fuel pump
SOHC	o	o	–	–	o	o	o	o	o
DOHC	o	o	–	–	o	o	o	o	o

No. / Usage / Model	W-27	W-28	W-35	W-41	C-50
Usage	Flasher unit	RR defogger	Upshift indecator	Power window	Rise up & intermittent
SOHC	o	o	o	–	–
DOHC	o	o	–	o	o

Relay location—1991–92 Stylus

left front corner of engine compartment.

E-Light Relay—located under left side of dashboard.

Convertible Top Relay—located under left side of dashboard.

Rear Defroster Relay—located under left side of dashboard.

Main Relay—located under left side of dashboard.

Restart Relay—located under left side of dashboard.

Air Conditioning Relay—located under left side of dashboard.

Timer Relay—located under dashboard to the right of instrument cluster.

Condenser Fan Relay—4XC1-T engine—located in right front corner of engine compartment.

Fuel Pump Relay—4XC1-T engine—located in right rear of engine compartment.

Air Conditioning Sub Relay—4XC1-T and 4XE1 engines—located in right rear of engine compartment.

Impulse

1988–89

Tail Light Relay—located in relay box in right front of engine compartment.

Lighting Relay—located in relay box in right front of engine compartment.

Cooling Fan Relay—1988–89—located in relay box in right front corner of engine compartment.

Air Conditioning and Charge Relay—located in relay box in right front of engine compartment.

Dimmer Relay—located in relay box in right front of engine compartment.

Horn Relay—located in right front of engine compartment.

Wiper Relays—located in right front of engine compartment.

Air Conditioning Relay—1988–89—located in right front of engine compartment.

Fast Idle Relay—located in left front of engine compartment.

Air Relay—4ZC1 engine—located in left front of engine compartment.

Fuel Pump Relay—located in right rear of engine compartment.

Main Relay—located in right rear of engine compartment.

Starter Relay—located in right rear of engine compartment.

Blower Relay—located in right rear of engine compartment.

Door Lock Relay—located under left side of dashboard.

Buzzer Relay—located under left side of dashboard.

Air Conditioning Blower—located under right side of dashboard.

Overdrive Relay—4ZD1 engines—located under left side of dashboard.

Overdrive Switch Relay—4ZD1 engines—located under left side of dashboard.

Cruise Control Relay—located under left side of dashboard.

Blower Controller Relay—located under left side of dashboard.

Overdrive Controller Relay—4ZC1 engine—located under right side of dashboard.

Power Window Relay—located under left side of dashboard.

Actuator Relay—1988–89—located under right side of dashboard.

Automatic Belt Controller Relay—located under center of dashboard.

Air Conditioning Cutout Relay—1988–89 4ZD1 engine—located in left front of engine compartment.

Air Conditioner Timer Relay—1988–89 4ZC1 engine—located under right side of engine compartment.

1990–92

All relays are located in the underhood relay boxes or on the connector block located under the left side of the dashboard.

Computers

LOCATION

The Electronic Control Module (ECM) is located under the right side of the dashboard on I-Mark and under the left side of the dashboard on Impulse.

Flashers

LOCATION

Flashers are located under the left side of the dashboard.

Cruise Control

ADJUSTMENT

I-Mark

1. Position the accelerator pedal so the engine runs at it's normal idling speed. Hold the accelerator pedal in this position.
2. Loosen locknuts **A** and **B**.
3. Pull the outer cable toward the accelerator pedal. There must be no play in the inner cable at this time.
4. Move locknut **A** until it makes contact with the bracket and then tighten.
5. Tighten locknut **B** to lock the actuator side cable in position.

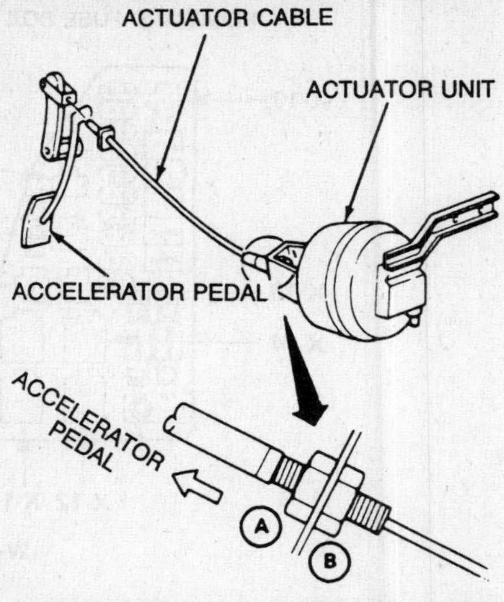

Cruise control cable adjustment—I-Mark

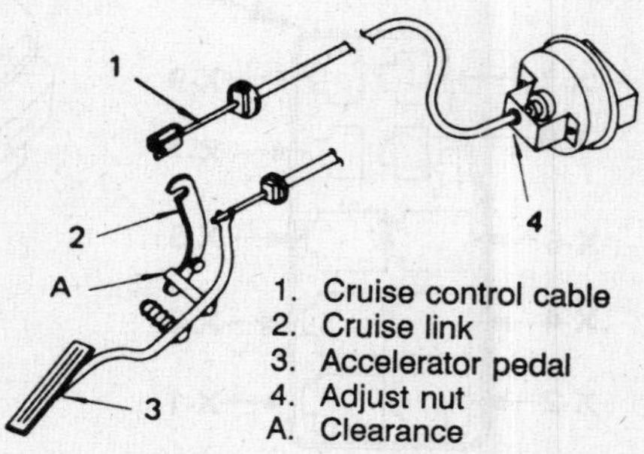

Cruise control cable adjustment—1988–89 Impulse

Impulse and Stylus

1988–89

1. Loosen the adjust nut.
2. Adjust the cable with the "U" fitting connected to the cruise link, so the clearance **A** is adjusted to zero and the accelerator pedal is not being pushed by the cruise link.
3. Tighten the adjust nut.

1990–92

1. Position the accelerator pedal so the engine runs at normal idling speed. Hold the accelerator pedal in this position.
2. Loosen the actuator attaching bolts.
3. Relocate the actuator so a clearance between the leading end of the slide plate on the actuator and guide pin is within 0.079 in. (2.0mm).
4. Tighten the actuator attaching bolts.

NOTE: When tightening the bolts, be careful not to tilt the actuator, which would cause an excessive force to be applied to the cable.

Lexus

ES250, LS400

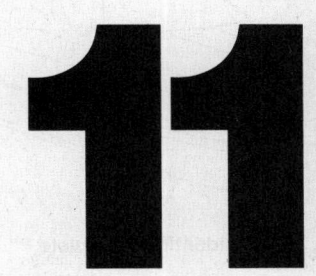

11

SERIAL NUMBER IDENTIFICATION

Vehicle identification plate

Vehicle Identification Plate

The vehicle identification plate is located at the top of the left instrument panel.

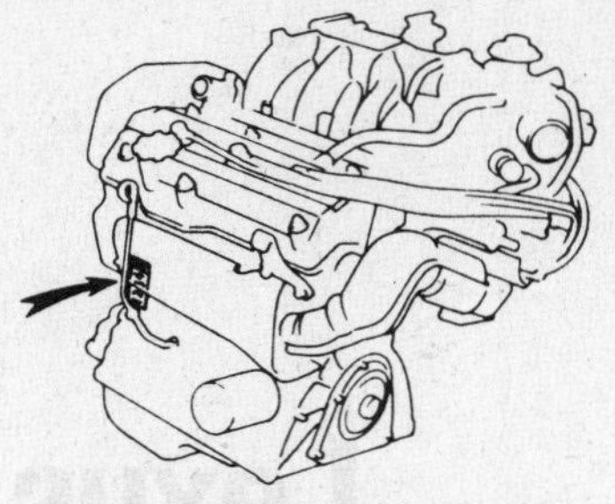

Engine serial number – ES250

Engine Number

The engine identification on the side of the engine block on the ES250 and the top of the engine block on the LS400.

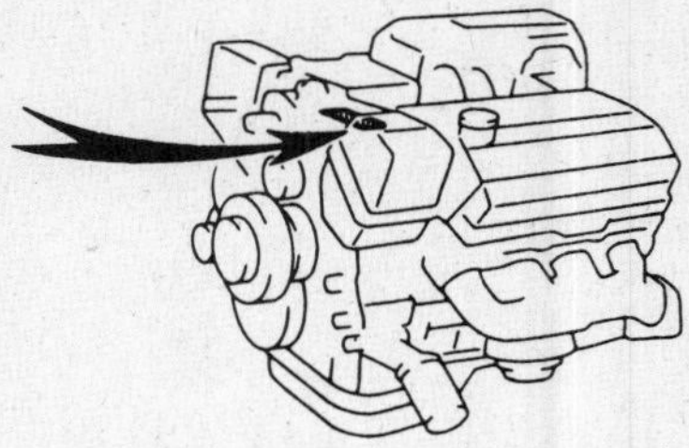

Engine serial number – LS400

Vehicle Identification Number

The vehicle identification number is also located on the left front door.

SPECIFICATIONS

ENGINE IDENTIFICATION

Year	Model	Engine Displacement cu. in. (cc/liter)	Engine Series Identification	No. of Cylinders	Engine Type
1990	ES250	153 (2508/2.5)	2V2-FE	6	DOHC
	LS400	242.1 (3969/4.0)	1UZ-FE	8	DOHC
1991-92	ES250	153 (2508/2.5)	2V2-FE	6	DOHC
	LS400	242.1 (3969/4.0)	1UZ-FE	8	DOHC

GENERAL ENGINE SPECIFICATIONS

Year	Model	Engine Displacement cu. in. (cc)	Fuel System Type	Net Horsepower @ rpm	Net Torque @ rpm (ft. lbs.)	Bore × Stroke (in.)	Compression Ratio	Oil Pressure @ rpm
1990	ES250	153 (2508)	EFI	156 @ 5600	160 @ 4400	3.44 × 2.74	9.0:1	43–78 @ 3000
	LS400	242.1 (3969)	EFI	250 @ 5600	260 @ 4400	3.44 × 3.25	10.0:1	36–71 @ 3000
1991-92	ES250	153 (2508)	EFI	156 @ 5600	160 @ 4400	3.44 × 2.74	9.0:1	43–78 @ 3000
	LS400	242.1 (3969)	EFI	250 @ 5600	260 @ 4400	3.44 × 3.25	10.0:1	36–71 @ 3000

GASOLINE ENGINE TUNE-UP SPECIFICATIONS

Year	Model	Engine Displacement cu. in. (cc)	Spark Plugs Type	Spark Plugs Gap (in.)	Ignition Timing (deg.) MT	Ignition Timing (deg.) AT	Compression Pressure (psi)	Fuel Pump (psi)	Idle Speed (rpm) MT	Idle Speed (rpm) AT	Valve Clearance In.	Valve Clearance Ex.
1990	ES250	153 (2508)	①	0.043	10B	10B	142	38–44	600–700	650–750	③	③
	LS400	242.1 (3969)	②	0.043	8–12B	8–12B	142	38–44	600–700	600–700	④	④

GASOLINE ENGINE TUNE-UP SPECIFICATIONS

Year	Model	Engine Displacement cu. in. (cc)	Spark Plugs Type	Spark Plugs Gap (in.)	Ignition Timing (deg.) MT	Ignition Timing (deg.) AT	Com-pression Pressure (psi)	Fuel Pump (psi)	Idle Speed (rpm) MT	Idle Speed (rpm) AT	Valve Clearance In.	Valve Clearance Ex.
1991	ES250	153 (2508)	①	0.043	10B	10B	142	38–44	600–700	650–750	③	③
	LS400	242.1 (3969)	②	0.043	8–12B	8–12B	142	38–44	600–700	600–700	④	④
1992	SEE UNDERHOOD SPECIFICATIONS STICKER											

① BCPR6EP11
② BKR6EP11
③ Intake—0.005 in. cold
Exhaust—0.015 in. cold
④ Intake—0.006–0.010 in. cold
Exhaust—0.010–0.014 in. cold

FIRING ORDERS

NOTE: To avoid confusion, always replace spark plug wires one at a time.

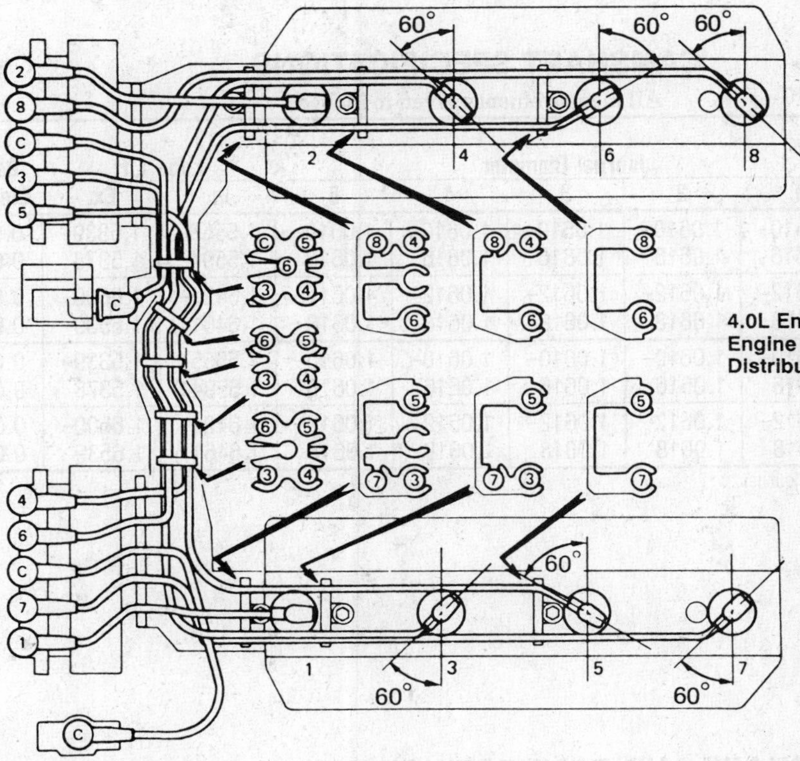

4.0L Engine
Engine Firing Order: 1-8-4-3-6-5-7-2
Distributorless Ignition System

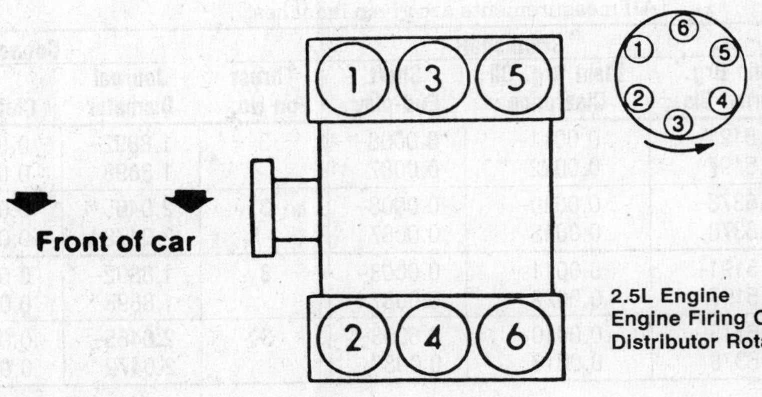

2.5L Engine
Engine Firing Order: 1-2-3-4-5-6
Distributor Rotation: Counterclockwise

CAPACITIES

Year	Model	Engine Displacement cu. in. (cc)	Engine Crankcase (qts.) with Filter	without Filter	Transmission (pts.) 4-Spd	5-Spd	Auto.	Drive Axle (pts.)	Fuel Tank (gal.)	Cooling System (qts.)
1990	ES250	153 (2508)	4.1	3.9	—	4.4	2.6	1.1	15.9	①
	LS400	242.1 (3969)	5.3	5	—	—	2.1	1.3	22.5	11.2
1991-92	ES250	153 (2508)	4.1	3.9	—	4.4	2.6	1.1	15.9	①
	LS400	242.1 (3969)	5.3	5	—	—	2.1	1.3	22.5	11.2

① MT—10.0 qts.
AT—9.9 qts.

CAMSHAFT SPECIFICATIONS

All measurements given in inches.

Year	Engine Displacement cu. in. (cc)	Journal Diameter 1	2	3	4	5	Lobe Lift In.	Ex.	Bearing Clearance	Camshaft End Play
1990	153 (2508)	1.0610–1.0616	1.0610–1.0616	1.0610–1.0616	1.0610–1.0616	1.0610–1.0616	1.5555–1.5594	1.5339–1.5378	0.0014–0.0028	0.0012–0.0031
	242.1 (3969) ①	1.0612–1.0618	1.0612–1.0618	1.0612–1.0618	1.0612–1.0618	1.0612–1.0618	1.6421–1.6461	1.6500–1.6539	0.0012–0.0026	0.0016–0.0035
1991-92	153 (2508)	1.0610–1.0616	1.0610–1.0616	1.0610–1.0616	1.0610–1.0616	1.0610–1.0616	1.5555–1.5594	1.5339–1.5378	0.0014–0.0028	0.0012–0.0031
	242.1 (3969) ①	1.0612–1.0618	1.0612–1.0618	1.0612–1.0618	1.0612–1.0618	1.0612–1.0618	1.6421–1.6461	1.6500–1.6539	0.0012–0.0026	0.0016–0.0035

① The exhaust camshaft thrust portion of journal diameter is 0.9433–0.9439 in.

CRANKSHAFT AND CONNECTING ROD SPECIFICATIONS

All measurements are given in inches.

Year	Engine Displacement cu. in. (cc)	Crankshaft Main Brg. Journal Dia.	Main Brg. Oil Clearance	Shaft End-play	Thrust on No.	Connecting Rod Journal Diameter	Oil Clearance	Side Clearance
19901	153 (2508)	2.5191–2.5197	0.0011–0.0022	0.0008–0.0087	3	1.8892–1.8898	0.0011–0.0026	0.0059–0.0130
	242.1 (3969)	2.6373–2.6378	0.0010–0.0018	0.0008–0.0087	3	2.0465–2.0472	0.0011–0.0021	0.0063–0.0114
1991-92	153 (2508)	2.5191–2.5197	0.0011–0.0022	0.0008–0.0087	3	1.8892–1.8898	0.0011–0.0026	0.0059–0.0130
	242.1 (3969)	2.6373–2.6378	0.0010–0.0018	0.0008–0.0087	3	2.0465–2.0472	0.0011–0.0021	0.0063–0.0114

VALVE SPECIFICATIONS

Year	Engine Displacement cu. in. (cc)	Seat Angle (deg.)	Face Angle (deg.)	Spring Test Pressure (lbs.)	Spring Installed Height (in.)	Stem-to-Guide Clearance (in.)		Stem Diameter (in.)	
						Intake	Exhaust	Intake	Exhaust
1990	153 (2508)	NA	44.5	41.0–47.2	1.677	0.0010–0.0024	0.0012–0.0026	0.2350–0.2356	0.2348–0.2354
	242.1 (3969)	NA	44.5	41.9–46.3	1.717	0.0010–0.0024	0.0012–0.0026	0.2350–0.2356	0.2348–0.2354
1991-92	153 (2508)	NA	44.5	41.0–47.2	1.677	0.0010–0.0024	0.0012–0.0026	0.2350–0.2356	0.2348–0.2354
	242.1 (3969)	NA	44.5	41.9–46.3	1.717	0.0010–0.0024	0.0012–0.0026	0.2350–0.2356	0.2348–0.2354

NA—Not Available

PISTON AND RING SPECIFICATIONS

All measurements are given in inches.

Year	Engine Displacement cu. in. (cc)	Piston Clearance	Ring Gap			Ring Side Clearance		
			Top Compression	Bottom Compression	Oil Control	Top Compression	Bottom Compression	Oil Control
1990	153 (2508)	0.0018–0.0026	0.0118–0.0213	0.0138–0.0244	0.0079–0.0224	0.0004–0.0031	0.0012–0.0028	—
	242.1 (3969)	0.0008–0.0016	0.0098–0.0177	0.0138–0.0236	0.0059–0.0197	0.0008–0.0024	0.0006–0.0022	—
1991-92	153 (2508)	0.0018–0.0026	0.0118–0.0213	0.0138–0.0244	0.0079–0.0224	0.0004–0.0031	0.0012–0.0028	—
	242.1 (3969)	0.0008–0.0016	0.0098–0.0177	0.0138–0.0236	0.0059–0.0197	0.0008–0.0024	0.0006–0.0022	—

TORQUE SPECIFICATIONS

All readings in ft. lbs.

Year	Engine Displacement cu. in. (cc)	Cylinder Head Bolts	Main Bearing Bolts	Rod Bearing Bolts	Crankshaft Pulley Bolts	Flywheel Bolts	Manifold		Spark Plugs
							Intake	Exhaust	
1990	153 (2508)	①	②	③	181	61	13	29	13
	242.1 (3969)	④	⑤	③	181	72	13	29	13
1991-92	153 (2508)	①	②	③	181	61	13	29	13
	242.1 (3969)	④	⑤	③	181	72	13	29	13

① Tighten in 3 steps:
1—tighten to 25 ft. lbs.
2—turn 90 degrees
3—turn 90 degrees

② Tighten in 2 steps:
1—tighten to 45 ft. lbs.
2—turn 90 degrees

③ Tighten in 2 steps:
1—tighten to 18 ft. lbs.
2—turn 90 degrees

④ Tighten in 2 steps:
1—tighten to 29 ft. lbs.
2—turn 90 degrees

⑤ Tighten in 2 steps:
1—tighten to 20 ft. lbs.
2—turn 90 degrees

BRAKE SPECIFICATIONS

All measurements in inches unless noted.

Year	Model	Lug Nut Torque (ft. lbs.)	Master Cylinder Bore	Brake Disc Minimum Thickness	Brake Disc Maximum Runout	Standard Brake Drum Diameter	Minimum Lining Thickness Front	Minimum Lining Thickness Rear
1990	ES250	76	NA	0.945 ②	0.0028 ③	NA	0.039	0.039
	LS400	76	NA	0.906 ①	0.0020	NA	0.039	0.039
1991-92	ES250	76	NA	0.945 ②	0.0028 ③	NA	0.039	0.039
	LS400	76	NA	0.906 ①	0.0020	NA	0.039	0.039

NA—Not available
① Rear—0.591 in.
② Rear—0.354 in.
③ Rear—0.0059 in.

WHEEL ALIGNMENT

Year	Model	Caster Range (deg.)	Caster Preferred Setting (deg.)	Camber Range (deg.)	Camber Preferred Setting (deg.)	Toe-in (in.)	Steering Axis Inclination (deg.)
1990	ES250 Front	11/12P–2 5/12P	1 2/3P	1/4N–1 1/4P	1/2P	1/16N–3/4	22
	Rear	—	—	1 5/12N–1/4P	2/3N	13/32–15/16	—
	LS400 Front	8 1/2P–10P ②	9 1/4P	2/3N–1 5/6P ①	1/12P	0–13/16	32 2/3 ③
	Rear	—	—	3/4N–3/4P	0 ④	0–13/16 ⑤	—
1991-92	ES250 Front	11/12P–2 5/12P	1 2/3P	1/4N–1 1/4P	1/2P	1/16N–3/4	22
	Rear	—	—	1 5/12N–1/4P	2/3N	13/32–15/16	—
	LS400 Front	8 1/2P–10P ②	9 1/4P	2/3N–1 5/6P ①	1/12P	0–13/16	32 2/3 ③
	Rear	—	—	3/4N–3/4P	0 ④	0–13/16 ⑤	—

① W/air suspension 5/6N–2/3P—Range
1/12N—Preferred
② W/air suspension 9 1/4P–10 7/12P—Range
9 5/6P—Preferred
③ W/air suspension 32 1/2 degree
④ W/air suspension 1 1/2N–0—Range
3/4N—Preferred
⑤ W/air suspension 11/32–7/8 in.

ENGINE MECHANICAL

NOTE: Disconnecting the negative battery cable on some vehicles may interfere with the functions of the on board computer systems and may require the computer to undergo a relearning process, once the negative battery cable is reconnected.

Engine Assembly

REMOVAL & INSTALLATION

ES250

1. Disconnect the negative battery cable. Disconnect the positive battery cable.
2. Remove the hood assembly. Remove the battery from the vehicle and disconnect the ground cable.
3. Remove the engine undercovers and drain the cooling system.
4. Raise and safely support the vehicle. Drain the engine oil in a suitable container. Lower the vehicle.
5. Remove the suspension upper brace.
6. Disconnect the igniter connector, noise filter connector and the high tension electrical cord.
7. Remove the ignition coil, igniter and bracket assembly.
8. Remove the radiator, alternator, alternator belt and the adjusting bar.
9. Remove the radiator reservoir tank and disconnect the accelerator cable from the throttle body.
10. Disconnect the throttle cable from the throttle body, on vehicles equipped with automatic transaxle.
11. Remove the cruise control actuator.
12. Disconnect the air flow meter connector, ISC valve air hose and the vacuum pipe air hose.
13. Disconnect the air cleaner hose, air cleaner cap, hoses and the air flow meter. Remove the mounting bolts and the air cleaner assembly.
14. Disconnect the following:
    a. Check connector
    b. Ground straps from the left side fender apron
    c. Connectors from the relay box
    d. Engine compartment wire connector.
15. Disconnect the following hoses:
    a. Brake booster vacuum hose to the air intake chamber.
    b. Air conditioning control valve vacuum hose.
    c. Charcoal canister vacuum hose.
16. Disconnect the ground strap from the transaxle.
17. Disconnect the heater hoses and

the fuel line hoses. Use a suitable container to catch any excess fuel.
18. Remove the starter, if equipped with manual transaxle.
19. Remove the clutch release cylinder and tube clamp, do not disconnect the tube, if equipped with manual transaxle.
20. Disconnect the speedometer cable and the transaxle control cable(s).
21. Remove the engine undercover and glove compartment box.
22. Disconnect the following connectors:
a. 3 engine and ECT Electronic Control Unit (ECU) connectors.
b. Circuit opening relay connector.
c. Cowl wire connector.
d. Instrument panel wire connector.
23. Remove the engine wire from the cowl panel.
24. Raise and safely support the vehicle. Remove the bolts and the suspension lower crossmember.
25. Disconnect the front exhaust pipe.
26. Disconnect the air conditioning wire connectors and remove the mounting bolts. Position and support the compressor aside.
27. Remove the halfshafts.
28. Remove the power steering pump and support. Do not disconnect the hoses.
29. Remove the mounting bolts and the engine cross member. Support the engine with a suitable jacking device.
30. Remove the front, center and rear engine mounting insulators and brackets.
31. Remove the power steering reservoir tank, without disconnecting the hoses. Remove the ground strap.
32. Remove the mounting brackets on both sides of the engine assembly. Lower the vehicle, keeping the engine supported.
33. Attach a suitable engine hoist to the engine hangers. Disconnect the clamps of the power steering oil cooler lines.
34. Remove the mounting insulators on both sides of the engine. Lift the engine and transaxle out of the vehicle.
35. Remove the starter, if equipped with automatic transaxle.
36. Separate the engine from the transaxle.

To install:

37. Assemble the engine to the transaxle. Install the starter, if equipped with an automatic transaxle. Tighten the mounting bolts to 29 ft. lbs. (40 Nm).
38. Lower the engine and the transaxle into the vehicle with a suitable engine hoist. Tilt the transaxle downward to clear the left side mounting bracket.
39. Align the right side and the left side mounting with the body bracket. Attach the right side mounting insulator to the mounting bracket and install the bolts.
40. Install the left side mounting bracket to the transaxle case and tighten the mounting bolts to 38 ft. lbs. (52 Nm).
41. Attach the left side mounting insulator to the mounting bracket. Tighten the mounting bolts to 38 ft. lbs. (52 Nm) and the through bolt to 64 ft. lbs. (87 Nm).
42. Tighten the nuts and bolts of the right side mounting insulator. Tighten the bolts to 47 ft. lbs. (64 Nm). Tighten the bracket nuts to 38 ft. lbs. (52 Nm) and the body nuts to 65 ft. lbs. (88 Nm).
43. Remove the engine hoist from the engine and connect the power steering cooler line clamp.
44. Connect the right side engine mounting brackets. Tighten the bolts to 38–48 ft. lbs. (52-65 Nm) and the nut 38 ft. lbs. (52 Nm).
45. Connect the left side engine mounting bracket. Tighten the bolt to 14 ft. lbs. (19 Nm) and the nuts to 38 ft. lbs. (52 Nm).
46. Install the automatic transaxle mounting bracket, if equipped. Tighten the 12mm nut to 15 ft. lbs. (20 Nm) and the 14mm nut to 38 ft. lbs. (52 Nm).
47. Install the power steering reservoir tank and connect the ground strap.
48. Install the front engine mounting bracket and insulator. Tighten the bolts to 57 ft. lbs. (77 Nm).
49. Install the center mounting bracket and insulator. Tighten the mounting bolts to 38 ft. lbs. (52 Nm).
50. Install the rear engine mounting bracket and insulator. Tighten the mounting bolts to 57 ft. lbs. (77 Nm).
51. Install the engine mounting cross member and tighten the bolts to 29 ft. lbs. (40 Nm).
52. Install and tighten the bolts holding the insulators to the cross member. Tighten the bolts to 54 ft. lbs. (73 Nm). Tighten the mounting insulator through bolts to 64 ft. lbs. (87 Nm).
53. Install the power steering pump and the halfshafts.
54. Install the air conditioning compressor and tighten the mounting bolts. Connect the electrical connection.
55. Raise and safely support the vehicle. Install the front exhaust pipe . Tighten the manifold nuts to 46 ft. lbs. (62 Nm) and the converter nuts 32 ft. lbs. (43 Nm).
56. Install the suspension cross member and tighten the bolts and nuts to 153 ft. lbs. (207 Nm). Lower the vehicle.
57. Push in the engine wire through the cowl panel.
58. Connect the following connectors:
a. 3 engine and ECT Electronic Control Unit (ECU) connectors
b. Circuit opening relay connector
c. Cowl wire connector
d. Instrument panel wire connector
59. Install the glove compartment box and the engine undercover.
60. Connect the transaxle control cables and the speedometer cable.
61. Install the clutch release cylinder and tube clamp, if equipped with manual transaxle.
62. Install the starter, if equipped with manual transaxle.
63. Connect the heater hoses and fuel line hoses. Connect the ground strap to the transaxle case.
64. Connect the following hoses:
a. Brake booster vacuum hose to the air intake chamber
b. Air conditioning control valve vacuum hose
c. Charcoal canister vacuum hose
65. Connect the following:
a. Check connector
b. Ground straps from the left side fender apron
c. Connectors from the relay box
d. Engine compartment wire connector
66. Install the air cleaner case and tighten the mounting bolt. Connect the air cleaner hose, air cleaner cap, hoses and the air flow meter.
67. Install the cruise control actuator.
68. Install the throttle control cable and adjust, if equipped with automatic transaxle.
69. Install the accelerator cable and adjust. Replace the radiator reservoir tank.
70. Install the alternator belt adjusting bar, the alternator and belt.
71. Install the radiator assembly.
72. Replace the ignition coil, igniter and bracket assembly. Connect the igniter connector, noise filter connector and the high tension electrical cord.
73. Install the suspension upper brace and tighten the mounting bolts to 47 ft. lbs.
74. Install the battery and refill the cooling system. Fill the engine crankcase to the proper oil level.
75. Install the engine undercovers and the hood assembly.
76. Connect the battery cables. Start the engine and check for leaks. Check the ignition timing. Adjust, if necessary. Recheck the cooling system and the oil level.

LS400

1. Disconnect the negative battery

cable and the positive battery cable. Remove the hood assembly.

2. Remove the dust covers and the air duct above the radiator assembly. Drain the cooling system.

3. Remove the battery from the vehicle. Raise and safely support the vehicle.

4. Remove the engine undercover and drain the engine oil, in a suitable container. Lower the vehicle.

5. Disconnect the radiator upper hose from the water inlet. Loosen the nuts holding the fluid coupling to the fan bracket.

6. Loosen the drive belt tension by turning the belt tensioner counterclockwise. Remove the drive belt.

7. Remove the radiator assembly.

8. Disconnect the air flow meter connector, the mounting bolts and the the air cleaner hose. Remove the air cleaner, the air flow meter and hose assembly.

9. Remove the igniter cover and disconnect the igniter connectors.

10. Remove the bolts, nut and the throttle body cover. Disconnect the accelerator and cruise control actuator cables.

11. Disconnect the air hose from the ISC valve and the power steering air control valve.

12. Disconnect the air connector pipe from the throttle body and remove the air connector pipe. Remove the bolt and connector pipe bracket.

13. Disconnect the air hose from the air intake chamber. Remove the power steering pump mounting bolts and nut. Position the pump aside.

14. Disconnect the coolant level sensor connector and remove the radiator reservoir tank. Remove the mounting bolts and reservoir tank bracket.

15. Disconnect the following hoses:

a. Heater and bypass hoses

b. Fuel hoses (plug the open end and catch the fuel in a suitable container)

c. Vacuum hose from the brake booster on the air intake chamber

d. Air conditioning control valve vacuum hoses

e. EVAP and BVSV vacuum hoses

16. Remove the relay box cover. Disconnect the connector and ground cables from the engine compartment relay box. Remove the ground straps from under the fender aprons.

17. Remove the cruise control actuator cover.

18. Remove the instrument panel undercover and the lower the trim panel the ECU for the engine and transmission.

19. Disconnect the glove box door, the glove box light and remove the glove box assembly.

20. Disconnect the ABS ECU and the heater air duct.

21. Disconnect the following connectors:

a. The 3 engine and Electronic Controlled Transmission (ECT) ECU connectors

b. Circuit opening relay connector

c. Cowl wire connector

d. Instrument panel wire connector

22. Remove the mounting bolts and pull out the engine wire from the cowl panel.

23. Raise and safely support the vehicle.

24. Remove the mounting bolts and disconnect the power steering oil cooler pipe from the oil pan.

25. Remove the mounting bolts and the steering damper.

26. Disconnect the grommet from the floor and the sub-oxygen sensor from the exhaust pipe. Disconnect the 2 sub-oxygen sensors.

27. Remove the sub-oxygen sensor covers and the exhaust pipe. Remove the exhaust pipe support brackets.

28. Remove the catalytic converters and the exhaust pipe heat insulator.

29. Remove the center floor crossmember braces. Remove the driveshaft.

30. Disconnect the shift control rod from the shift lever. Lower the vehicle.

31. Attach a suitable engine hoist to the engine hangers and support the engine.

32. Remove the nuts holding the engine mounting insulators to the front suspension crossmember.

33. Remove the rear engine mounting member. Disconnect the ground strap.

34. Lift out the engine with the transmission attached. Place the engine assembly on a suitable holding fixture. Separate the engine from the transmission.

To install:

35. Connect the engine to the transmission. Attach a suitable engine hoist to the engine hangers.

36. Lower the engine assembly into the vehicle. Insert the stud bolts of the front engine mounting brackets into the stud bolt holes of the front suspension crossmember.

37. Install the rear engine mounting member and tighten the bolts to 19 ft. lbs. (26 Nm) and the nuts to 10 ft. lbs. (14 Nm). Install the ground strap.

38. Remove the engine hoist. Raise and safely support the vehicle.

39. Install the nuts holding the engine mounting brackets to the front suspension crossmember. Tighten the nuts to 43 ft. lbs. (58 Nm).

40. Connect the transmission control rod to the shift lever. Install the propeller shaft.

41. Install the center floor crossmember braces. Tighten the bolts to 9 ft. lbs. (11 Nm).

42. Install the exhaust pipe heat insulator. Replace the catalytic converters and tighten the bolts to 46 ft. lbs. (65 Nm).

43. Install the front exhaust pipe and the sub-oxygen sensor covers. Tighten the bolts to 32 ft. lbs. (44 Nm). Install the sub-oxygen sensors to the exhaust pipe and tighten to 33 ft. lbs. (45 Nm).

44. Install the steering damper and tighten the mounting bolts to 20 ft. lbs. (27 Nm). Connect the engine wire to the wire bracket on the front suspension crossmember.

45. Install the power steering oil cooler pipe to the engine oil pan. Lower the vehicle.

46. Push in the engine wire through the cowl panel and install the wire retainer.

47. Connect the following connectors:

a. 3 engine and ECT ECU connectors

b. Circuit opening relay connector

c. Cowl wire connector

d. Instrument panel wire connector

48. Install the heater duct and the glove compartment.

49. Install the right side lower instrument panel pad and the engine and ECT electronic control units. Replace the right side instrument panel undercover.

50. Install the cruise control actuator. Connect the connectors and ground cables to the relay box.

51. Install the upper cover to the relay box. Connect the 2 ground straps to the underside of the fender aprons.

52. Connect the following hose:

a. The heater bypass water hoses

b. Fuel hoses

c. Vacuum hose to the brake booster on the air intake chamber

d. Vacuum hose to the EVAP BVSV

53. Install the air conditioning compressor. Tighten the bolts to 36 ft. lbs. (49 Nm) and the nut to 22 ft. lbs. (30 Nm). Connect the electrical connectors.

54. Install the radiator reservoir tank bracket, the reservoir and connect the coolant level sensor connector.

55. Install the power steering pump and tighten the mounting bolts to 29 ft. lbs. (40 Nm) and the nuts to 32 ft. lbs. (44 Nm). Connect the air hose to the air intake manifold.

56. Connect the air connector pipe to the throttle body. Connect the air hose

to the ISC valve and the power steering control valve.

57. Connect the accelerator cable and the cruise control actuator cable to the throttle body. Install the throttle body cover.

58. Connect the igniter connectors and install the igniter cover.

59. Connect the air cleaner hose to the intake air connector pipe.

60. Install the air cleaner, the air flow meter and hose assembly. Connect the air flow meter connector.

61. Install the radiator.

62. Temporarily install the fan pulley, the fan and the fluid coupling assembly. Install the drive belt by turning the belt tensioner counterclockwise.

63. Tighten the bolts holding the fluid coupling to the fan bracket to 16 ft. lbs. (22 Nm).

64. Install the battery. Replace the air ducts and dust covers.

65. Refill the cooling system and the crankcase to the proper levels.

66. Install the engine undercover and hood assembly.

67. Connect the battery cables. Start the engine and check for leaks. Check the timing.

68. Recheck the fluid levels.

Engine Mounts

REMOVAL & INSTALLATION

ES250

FRONT

1. Raise and safely support the vehicle.
2. Support the engine with a suitable jacking device.
3. Remove the nut and the through bolt.
4. Remove the insulator. Remove the mounting bolts and the bracket, if necessary.
5. The installation is the reverse of the removal procedure. Tighten the bolts to 57 ft. lbs. (77 Nm).

CENTER

1. Raise and safely support the vehicle.
2. Support the engine with a suitable jacking device.
3. Remove the nut and the through bolt.
4. Remove the insulator. Remove the mounting bolts and the bracket, if necessary.
5. The installation is the reverse of the removal procedure. Tighten the bolts to 38 ft. lbs. (52 Nm).

REAR

1. Raise and safely support the vehicle.
2. Support the engine with a suitable jacking device.
3. Remove the nut and the through bolt.
4. Remove the insulator. Remove the mounting bolts and the bracket, if necessary.
5. The installation is the reverse of the removal procedure. Tighten the bolts to 57 ft. lbs. (77 Nm). Remove the jacking device and lower the vehicle.

LS400

1. Raise and safely support the vehicle.
2. Support the engine with a suitable jacking device.
3. Remove the mounting nuts and the mounting insulator. Remove the mounting bracket, if necessary.
4. The installation is the reverse of the removal procedure. Tighten the bolts to 19 ft. lbs. (26 Nm) and the nuts to 10 ft. lbs. (14 Nm). Remove the jacking device and lower the vehicle.

Cylinder Head

REMOVAL & INSTALLATION

ES250

1. Disconnect the negative battery cable. Drain the cooling system.
2. Remove the suspension upper brace. Disconnect the throttle cable from the throttle body, if equipped with an automatic transaxle.
3. Disconnect the accelerator cable from the throttle body. Remove the cruise control actuator and the vacuum pump.
4. Disconnect the ISC hose, the vacuum pipe hose and the air cleaner hose.
5. Raise and safely support the vehicle. Remove the right side engine undercover.
6. Remove the suspension lower crossmember and the front exhaust pipe. Lower the vehicle.
7. Remove the alternator, the ISC valve and the throttle body.
8. Remove the EGR pipe, the EGR valve and the vacuum modulator.
9. Disconnect the (4) BVSV vacuum hoses, the fuel pressure VSV hose and the air conditioning control valve vacuum hose.
10. Remove the distributor and the exhaust crossover pipe.
11. Disconnect the cold start injector connector and remove the cold start injector tube.
12. Disconnect the following hoses:
   a. PCV hose
   b. Vacuum sensing hose
   c. Fuel pressure VSV hose
   d. Air conditioning control valve vacuum hose

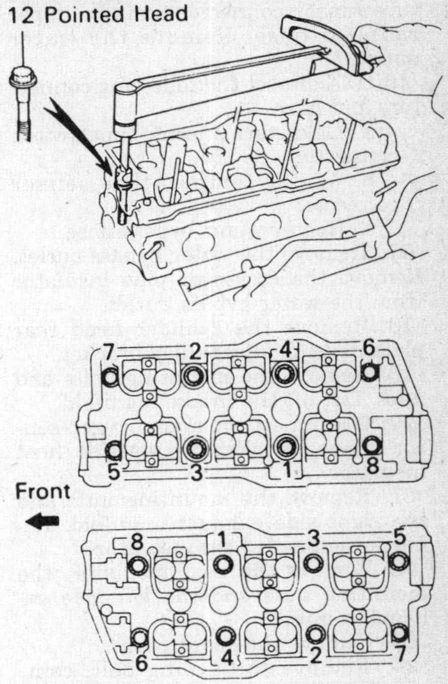

Cylinder head tightening sequence – ES250

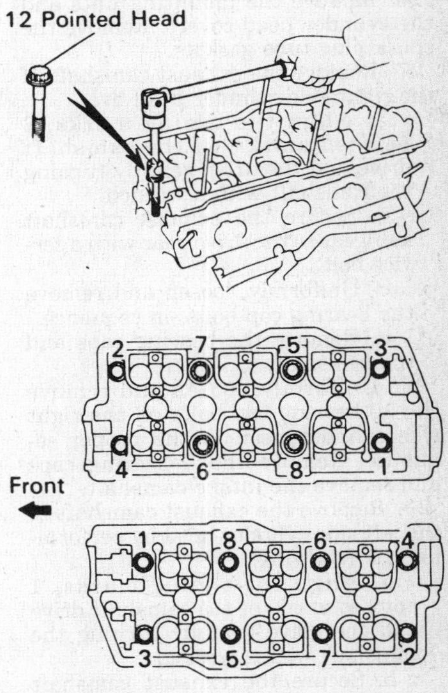

Cylinder head bolt removal sequence – ES250

   e. EGR gas temperature sensor connector – California only

13. Remove the mounting bolts and the brackets from the intake chamber. Remove the air intake chamber.
14. Remove the delivery pipes and the injectors.
15. Disconnect the water tempera-

ture sensor connector and the upper radiator hose. Remove the water outlet.

16. Disconnect the following connectors and hose:

a. Cold start injector time switch connector

b. Water temperature sensor connector

c. Heater water bypass hose

17. Remove the water bypass outlet. Remove the crossover pipe insulator from the water bypass outlet.

18. Remove the cylinder head rear plate and the idler pulley bracket.

19. Remove the mounting bolts and nuts. Lift off the intake manifold.

20. Disconnect the main oxygen sensor connector. Remove outside heat insulator.

21. Remove the mounting nuts and the right side exhaust manifold. Remove the inside heat insulator.

22. Remove the heat insulator, the mounting nuts and the left side exhaust manifold.

23. Remove the spark plugs.

24. Remove the timing belt, camshaft pulleys and the No. 2 idler pulley.

25. Remove the No. 3 timing cover.

26. Remove the mounting nuts and the cylinder head covers. Remove the spark plug tube gaskets.

27. Remove the exhaust camshaft of the right side cylinder head by:

a. Align the timing marks, 2 pointed marks, of the camshaft drive and the drive gear by turning the camshaft with a wrench.

b. Secure the exhaust camshaft sub-gear to the drive gear with a service bolt.

c. Uniformly, loosen and remove the bearing cap bolts, in sequence.

d. Remove the bearing caps and the camshaft.

28. Uniformly, loosen and remove the 10 bearing cap bolts on the right side cylinder head in the proper sequence. Remove the 5 bearing caps and remove the intake camshaft.

29. Remove the exhaust camshaft of the left side cylinder head by performing the following:

a. Align the timing marks, 1 pointed mark, of the camshaft drive and the drive gear by turning the camshaft with a wrench.

b. Secure the exhaust camshaft sub-gear to the drive gear with a service bolt.

c. Uniformly, loosen and remove the 8 bearing cap bolts in the proper sequence.

d. Remove the 4 bearing caps and the exhaust camshaft.

30. Uniformly, loosen and remove the bearing cap bolts on the right side cylinder head, in sequence. Remove

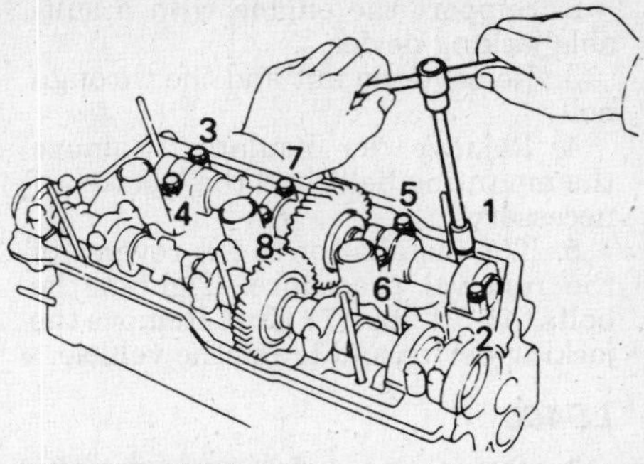

Right hand exhaust camshaft bolt removal sequence – ES250

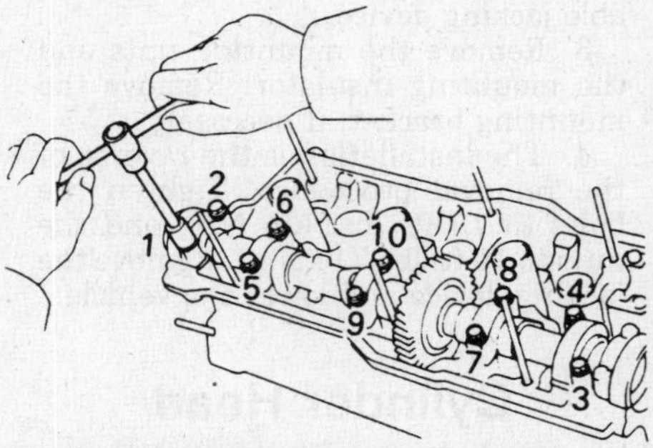

Right hand intake camshaft bolt removal sequence – ES250

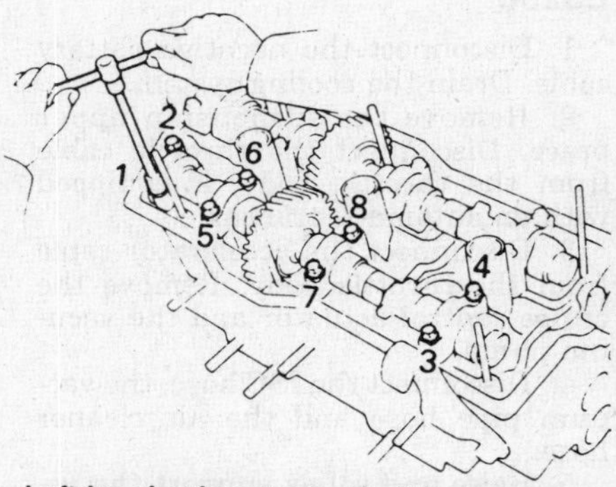

Left hand exhaust camshaft bolt removal sequence – ES250

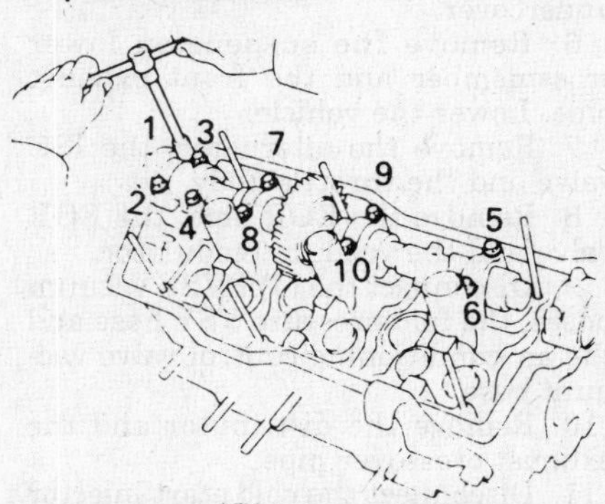

Left hand intake camshaft bolt removal sequence – ES250

the bearing caps and remove the intake camshaft.

31. Remove the recessed head bolts with the proper tool.

32. Uniformly, loosen and remove the cylinder head bolts in the proper sequence. Lift the cylinder head from the dowels on the cylinder block and place on wooden blocks.

NOTE: Be careful not to damage the contact surfaces of the cylinder head and cylinder block.

To install:

33. Place the cylinder head gasket in position on the cylinder block.

34. Install the cylinder head onto the cylinder block, aligning the dowels.

35. Install the head bolts and tighten, in sequences, using 3 steps:

a. Apply a light coat of engine oil on the threads of the bolts and install. Uniformly, tighten the cylinder head bolts in several passes, in sequence, to 25 ft. lbs. (34 Nm).

b. Mark the front of the cylinder head bolt with paint. Retighten the cylinder head bolts 90 degrees in the proper sequence.

c. Retighten the cylinder head bolts by an additional 90 degrees.

36. Apply a light coat of engine oil on the recessed head bolts. Install the head bolts and tighten to 13 ft. lbs. (18 Nm).

37. Install the left side engine hanger and tighten to 27 ft. lbs. (37 Nm).

NOTE: Since the thrust clearance of the camshaft is small, the camshaft must be held level while it is being installed. If the camshaft is not level, the portion of the cylinder head receiving the shaft thrust may crack or be damaged, causing the camshaft to seize or break.

38. Install the intake camshaft of the right side cylinder head by:

a. Apply a suitable multi-purpose grease to the thrust portion of the camshaft.

b. Apply seal packing to the No. 1 bearing cap. Install the bearing caps.

c. Apply a light coat of oil on the threads of the bearing cap bolts.

d. Install and uniformly tighten the bearing cap bolts to 12 ft. lbs. (16 Nm).

39. Install the exhaust camshaft of the right side cylinder head by:

a. Apply a suitable multi-purpose grease to the thrust portion of the camshaft.

b. Align the timing marks, 2 pointed marks, of the camshaft and the drive gears.

c. Place the camshaft on the cylinder head and install the bearing caps.

d. Apply a light coat of oil on the threads of the bearing cap bolts.

e. Install and uniformly tighten the bearing cap bolts to 12 ft. lbs. (16 Nm). Remove the service bolt.

40. Install the intake camshaft on the left side cylinder head by:

a. Apply a suitable multi-purpose grease to the thrust portion of the camshaft.

b. Place the intake camshaft at a 90 degree angle of the timing mark on the cylinder head, 1 pointed mark.

c. Apply seal packing to the No. 1 bearing cap. Install the bearing caps.

d. Apply a light coat of oil on the threads of the bearing cap bolts.

e. Install and uniformly tighten the bearing cap bolts to 12 ft. lbs. (16 Nm).

41. Install the exhaust camshaft on the left side cylinder head by:

a. Apply MP grease to the thrust portion of the camshaft.

b. Align the timing marks, 1 pointed mark, of the camshaft and the drive gears.

c. Place the camshaft on the cylinder head and install the bearing caps.

d. Apply a light coat of oil on the threads of the bearing cap bolts.

e. Install and uniformly tighten the bearing cap bolts to 12 ft. lbs. (16 Nm). Remove the service bolt.

42. Turn the camshaft and position the cam lobe upward. Check and adjust the valve clearance.

43. Apply a suitable multi-purpose grease to the the new camshaft oil seals and install with the proper tool.

44. Install the spark plug tube gaskets. Install the proper seal packing to the cylinder heads.

45. Install the cylinder head cover gasket and install with the seal washers and nuts. Tighten to 52 inch lbs.

46. Install the No. 3 timing belt cover and tighten the bolts 65 inch lbs.

47. Install the No. 2 idler pulley, the camshaft timing pulleys and the timing belt.

48. Install the spark plugs and tighten to 13 ft. lbs. (18 Nm).

49. Install the right side heat insulator and the right side exhaust manifold with a new gasket. Tighten the nuts to 29 ft. lbs. (40 Nm).

50. Install the outside heat insulator and connect the oxygen sensor connector.

51. Install the left side exhaust manifold with a new gasket and tighten the nuts to 29 ft. lbs. (40 Nm). Install the heat insulator.

52. Install the intake manifold with new gaskets. Tighten the nuts and bolts to 13 ft. lbs. (18 Nm).

53. Install the No. 2 idler pulley bracket and tighten to 13 ft. lbs. (18 Nm). Replace the rear cylinder head plate.

54. Install the crossover pipe heat insulator. Replace the water bypass outlet and tighten the mounting bolts to 14 ft. lbs. (19 Nm).

55. Connect the heater water bypass hose, cold start injector time switch connector and the water temperature switch.

56. Connect the upper radiator hose, the water temperature sensor connector and install the water outlet with a new gasket. Tighten the bolts to 73 inch lbs.

57. Install the fuel injectors and the delivery pipes.

58. Install the air intake chamber with new gaskets and tighten the mounting bolts to 32 ft. lbs. (44 Nm).

59. Install the intake chamber brackets.

60. Connect the following hoses:

a. PCV hose

b. Vacuum sensing hose

c. EGR gas temperature sensor connector (California only)

d. Air conditioning control valve air hose

61. Install the cold start injector tube and connect the cold start injector connector.

62. Install the exhaust crossover pipe with new gaskets and tighten the mounting bolts to 25 ft. lbs. (34 Nm) and the nuts to 29 ft. lbs. (40 Nm).

63. Install the distributor.

64. Install the vacuum pipe and connect the BVSV vacuum hoses and the air conditioning control valve vacuum hose.

65. Install the EGR valve and vacuum modulator. Tighten the mounting bolts to 13 ft. lbs. (18 Nm).Install the vacuum pipe hoses.

66. Install the EGR pipe with a new gasket and tighten the mounting bolts to 13 ft. lbs. (18 Nm) and the nut to 58 ft. lbs. (79 Nm).

67. Install the throttle body and ISC valve.

68. Raise and support the vehicle. Install the front exhaust pipe and the right side engine undercover. Lower the vehicle.

69. Install the air cleaner hose and cruise control actuator.

70. Install the accelerator cable and connect the throttle cable, if equipped with an automatic transaxle.

71. Install the suspension upper brace and fill the cooling system.

72. Connect the negative battery cable. Run the engine and check for leaks.

73. Adjust the timing. Recheck the fluid levels.

LS400

1. Disconnect the negative battery cable. Drain the cooling system.

2. Remove the camshaft timing pulleys.

3. Disconnect the accelerator cable, the throttle control cable, if equipped with automatic transmission and the cruise control actuator cable.

4. Remove the high tension cord cover and the right side ignition coil.

5. Remove the water inlet housing mounting bolts and disconnect the water bypass hose from the ISC valve.

6. Remove the water inlet and inlet housing assemblies. Remove the O-ring from the water inlet housing.

7. Remove the EGR pipe.

8. Disconnect the following:

a. VSV connector

b. Vacuum pipe hose

c. EGR water bypass pipe

d. Fuel pressure VSV

9. Disconnect the EGR vacuum hoses and remove the EGR VSV.

10. Disconnect the following hoses:

a. Water bypass pipe hose from the ISC valve.

b. Water bypass joint hose.

c. Vacuum pipe hoses.

11. Disconnect the EGR gas temperature sensor (California only). Remove the EGR valve adapter.

12. Disconnect the following:

a. Fuel pressure regulator vacuum hose.

b. Air intake chamber vacuum hose.

c. Vacuum hose from the EVAP BVSV.

13. Remove the mounting bolts, hoses and the vacuum pipe.

14. Remove the ISC valve.

15. Remove the throttle body sensor connectors and the water bypass pipe from the rear water bypass joint.

16. Remove the mounting bolts/nuts and disconnect the PCV valve. Remove the throttle body and gasket.

17. Disconnect the accelerator cable bracket and the brake booster vacuum union and hose.

18. Disconnect the cold start injector connector and the cold start injector tube from the right side delivery pipe.

19. Disconnect the check connector from the intake chamber and remove the mounting nuts and bolts.

20. Remove the air intake chamber and the cold start injector, tube and wire assembly.

21. Disconnect the engine wire from the intake manifold and from the right side cylinder head. Disconnect the heater hoses.

22. Remove the delivery pipes and the fuel injectors. Remove the mounting bolts and nuts. Lift up the intake manifold.

23. Remove the front and rear water bypass joint.

24. Raise and safely support the vehicle. Remove the front exhaust pipe

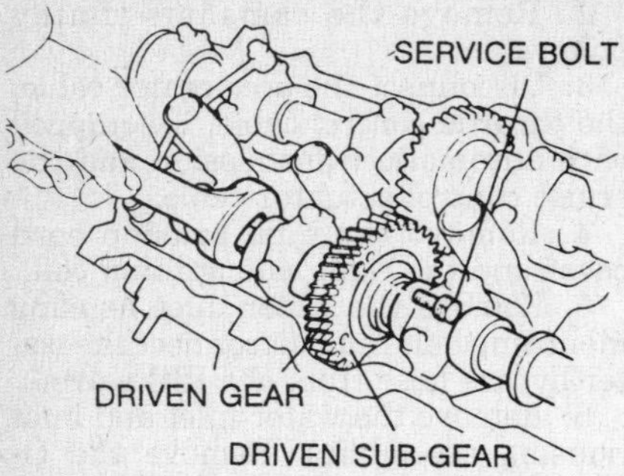

Securing the camshaft gears—LS400

and the main catalytic converters. Lower the vehicle.

25. Disconnect the right side oxygen sensor. Remove the mounting bolts and nuts and remove the right side exhaust manifold.

26. Remove the oil dipstick and guide. Disconnect the left side oxygen sensor.

27. Remove the mounting bolts and nuts and remove the left side exhaust manifold.

28. Remove the 2 engine hangers and the wire brackets from the right side cylinder head.

29. Remove the bolts, washers and the cylinder head cover. Remove the semi-circular plugs, if necessary.

30. Remove the exhaust camshaft of the right side cylinder head by:

a. Position the service bolt hole of the drive sub-gear to the upright position. Secure the camshaft sub-gear to drive gear with a service bolt.

b. Set the timing mark, 1 pointed mark, of the camshaft drive gear at approximately 10 degrees, by turning the camshaft with the proper tool.

c. Alternately, loosen and remove the bearing cap bolts holding the intake camshaft side of the oil feed pipe to the cylinder head.

d. Uniformly, loosen and remove the bearing cap bolts in sequence.

e. Remove the oil feed pipe and the bearing caps. Remove the camshaft.

31. Remove the intake camshaft from the right side cylinder head by:

a. Set the timing mark, 1 pointed mark, of the camshaft drive gear at approximately 45 degrees, by turning the camshaft with the proper tool.

c. Uniformly, loosen and remove the bearing cap bolts in the proper sequence.

d. Remove the bearing caps, oil seal and the intake camshaft.

32. Remove the exhaust camshaft of the left side cylinder head by:

a. Position the service bolt hole of the drive sub-gear to the upright position. Secure the camshaft sub-gear to drive gear with a service bolt.

NOTE: When removing the camshaft, make sure the torsional spring force of the sub-gear has been eliminated.

b. Set the timing mark, 2 pointed marks, of the camshaft drive gear at approximately 15 degrees, by turning the camshaft with the proper tool.

c. Alternately, loosen and remove the bearing cap bolts holding the intake camshaft side of the oil feed pipe to the cylinder head.

d. Uniformly, loosen and remove the bearing cap bolts in the proper sequence.

e. Remove the oil feed pipe and the bearing caps. Remove the camshaft.

33. Remove the intake camshaft from the left side cylinder head by:

a. Set the timing mark, 1 pointed mark, of the camshaft drive gear at approximately 60 degrees, by turning the camshaft with the proper tool.

c. Uniformly, loosen and remove the bearing cap bolts in the proper sequence.

d. Remove the bearing caps, oil seal and the intake camshaft.

34. Disconnect the ground straps and clamp of the engine wire from the rear of the cylinder heads.

35. Uniformly, loosen the head bolts of 1 side of the cylinder head then on the other side. Remove the head bolts and washers.

36. Lift out the cylinder head from the dowels on the cylinder block and place on a suitable holding fixture. Remove the gasket and clean the mounting surface.

To install:

37. Place new cylinder gasket into position on the cylinder block. Install the cylinder head.

38. The cylinder head bolts are tightened in 2 steps:

a. Apply a light coat of engine oil on the threads of the bolts. Temporarily, install the washers and bolts. Uniformly, tighten the head bolts one 1 side of the cylinder head in the proper sequence then the other side. Tighten the bolts to 29 ft. lbs. (40 Nm).

b. Mark the front the cylinder head bolt with paint. Retighten the cylinder head bolts in the proper se-

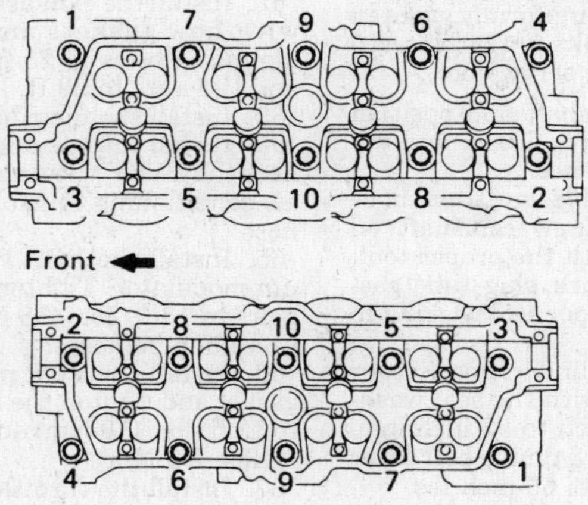

Cylinder head bolt removal sequence—LS400

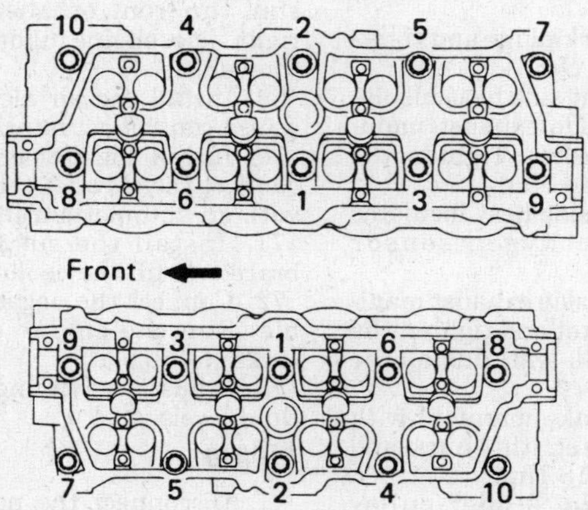

Cylinder head tightening sequence—LS400

quence 90 degrees. Check that the painted mark is at a 90 degree angle to the front.

39. Connect the engine wire to the cylinder head(s), tighten the clamps.

40. Install the circular plugs on the cylinder head with the cup side facing forward.

41. Remove any old packing and apply new seal packing to the bearing caps.

42. Install the bearing cap on the right side cylinder head, marked I1, in position with the arrow mark facing the rear. Install the bearing cap on the left side cylinder head, marked I6, in position with the arrow mark facing the front.

43. Apply a light coat of oil on the threads of the cap bolts. Install the nearing cap bolts with new washers and tighten to
12 ft. lbs. (16 Nm).

44. Install the right side cylinder head intake camshaft by:

a. Apply MP grease to the thrust portion of the camshaft.

b. Place the intake camshaft at a 45 degree angle of the timing mark, 1 pointed mark, on the cylinder head.

c. Remove any old packing and apply new seal packing to the bearing cap marked I6 and install the front bearing cap, marked I6 with the arrow facing rearward.

d. Align the arrows at the front and rear of the cylinder head with the bearing cap.

e. Install the remaining bearing caps in the proper sequence with the arrow mark facing rearward. Install the oil feed pipe and the mounting bolts.

f. Uniformly, tighten the bearing cap bolts in the proper sequence to 12 ft. lbs. (16 Nm).

45. Install the right side cylinder head exhaust camshaft by:

a. Set the timing mark, 1 pointed mark, of the camshaft drive gear at a 10 degree angle by turning the intake camshaft with the proper tool.

b. Apply MP grease to the thrust portion of the camshaft.

c. Align the timing marks, 1 pointed mark, of the camshaft drive and driven gears.

d. Place the exhaust camshaft ion the cylinder head. Install the rear bearing cap with the arrow mark facing rearward.

e. Align the arrow marks at the front and rear of the cylinder head with the mark on the bearing cap. Apply a light coat of oil on the threads of the bearing cap bolts.

f. Uniformly, tighten the bearing cap bolts in the proper sequence to 12 ft. lbs.

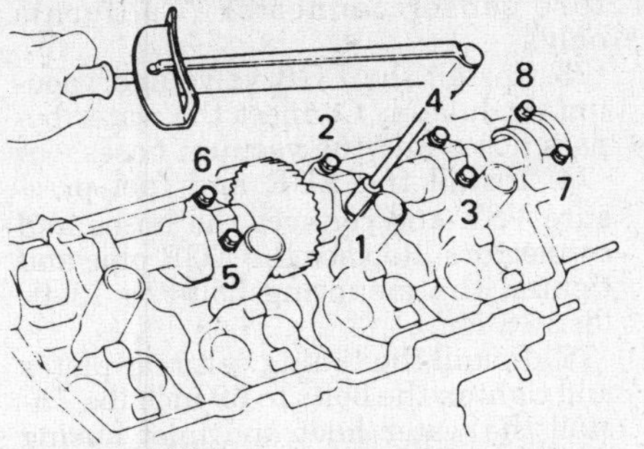

Right hand intake camshaft bearing cap bolt tightening sequence—LS400

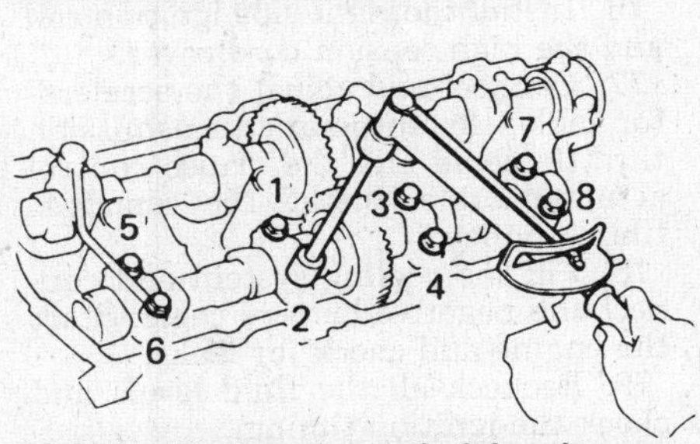

Right hand exhaust camshaft bearing cap bolt tightening sequence—LS400

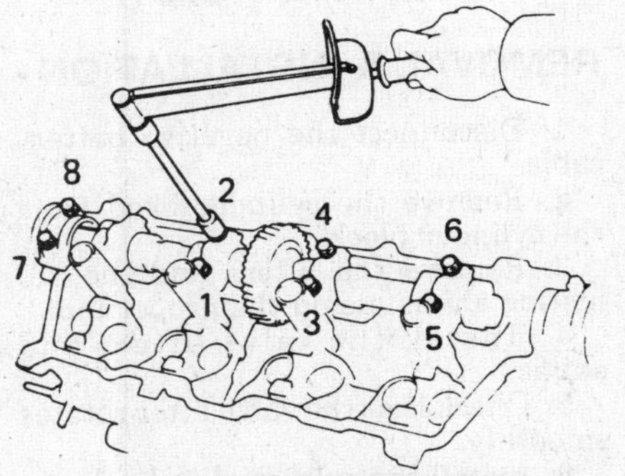

Left hand intake camshaft bearing cap bolt tightening sequence—LS400

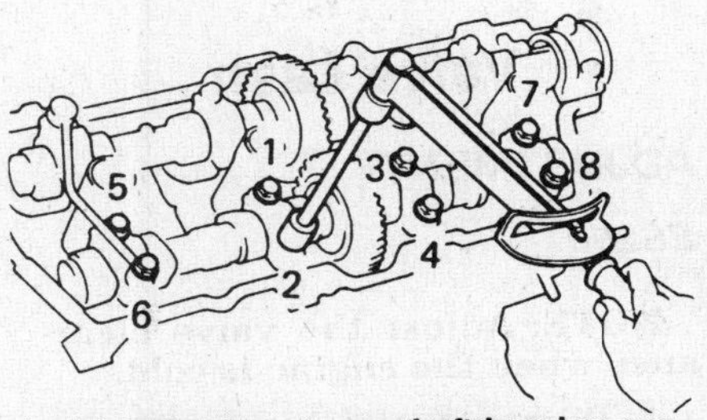

Left hand exhaust camshaft bearing cap bolt tightening sequence—LS400

g. Bring the service bolt installed upward by turning the camshaft with the proper tool. Remove the service bolt.

46. Install the left side cylinder head intake camshaft by:

a. Apply MP grease to the thrust portion of the camshaft.

b. Place the intake camshaft at a 60 degree angle of the timing mark (1 pointed mark) on the cylinder head.

c. Remove any old packing and apply new seal packing to the bearing cap marked I6 and install the front bearing cap, marked I1 with the arrow facing rearward.

d. Align the arrows at the front and rear of the cylinder head with the bearing cap. Apply a light coat of oil on the threads of the bearing cap bolts.

e. Install the remaining bearing caps in the proper sequence with the arrow mark facing rearward. Install the oil feed pipe and the mounting bolts.

f. Uniformly, tighten the bearing cap bolts in the proper sequence to 12 ft. lbs. (16 Nm).

47. Install the left side cylinder head exhaust camshaft by:

a. Set the timing mark, 2 dot marks, of the camshaft drive gear at a 15 degree angle by turning the intake camshaft with the proper tool.

b. Apply MP grease to the thrust portion of the camshaft.

c. Align the timing marks, 2 dot marks, of the camshaft drive and driven gears.

d. Place the exhaust camshaft ion the cylinder head. Install the rear bearing cap with the arrow mark facing rearward.

e. Align the arrow marks at the front and rear of the cylinder head with the mark on the bearing cap. Apply a light coat of oil on the threads of the bearing cap bolts.

f. Uniformly, tighten the bearing cap bolts in the proper sequence to 12 ft. lbs. (16 Nm).

g. Bring the service bolt installed upward by turning the camshaft with the proper tool. Remove the service bolt.

48. Install the camshaft oil seals with the proper tool. Install the semi-circular plugs with the proper seal packing.

49. Install the cylinder head covers with the proper seal packing and gasket. Tighten the mounting bolts to 52 inch lbs.

50. Install the engine wire bracket and hangers. Tighten the hanger bolts to 27 ft. lbs. (37 Nm).

51. Install the right side exhaust manifold with a new gasket and tighten the mounting bolts to 29 ft. lbs. (40 Nm). Connect the right side oxygen sensor connector.

52. Install the right side exhaust manifold with a new gasket and tighten the mounting bolts to 29 ft. lbs. (40 Nm). Connect the right side oxygen sensor connector.

53. Install the left side exhaust manifold with a new gasket and tighten the

mounting bolts to 29 ft. lbs. (40 Nm). Connect the left side oxygen sensor connector.

54. Install the oil dipstick and guide. Raise and safely support the vehicle.

55. Install the catalytic converters and front exhaust pipe. Lower the vehicle.

56. Install the front and rear water bypass joints. Tighten the mounting bolts to 13 ft. lbs. (18 Nm).

57. Install the intake manifold, using new gaskets. Tighten the mounting nuts and bolts to 13 ft. lbs. (18 Nm).

58. Install the delivery pipes and fuel injectors. Install the fuel return pipe with new gaskets. Tighten the union bolt to 26 ft. lbs. (35 Nm).

59. Connect the fuel hoses and the injector connectors. Connect the engine wire to the delivery pipes.

60. Connect the connectors on the left side delivery pipe, the water temperature sensor connector, cold start injector time switch connector and the water temperature sender gauge connector.

61. Connect the heater hoses and engine wire bracket. Install the engine wire to the bracket.

62. Install the cold start injector, tube and wire assembly. Tighten the mounting bolts to 69 inch lbs.

63. Install the air intake chamber with new gaskets and tighten the mounting bolts to 13 ft. lbs. (18 Nm).

64. Connect the cold start injector tube to the right side delivery pipe and tighten the union bolt to 11 ft. lbs. (15 Nm).

65. Connect the cold start injector connector. Install the accelerator cable bracket.

66. Install the brake booster union and connect the vacuum hose. Tighten the union bolt to 22 ft. lbs. (30 Nm).

67. Connect the water bypass hose to the throttle body and the PCV hose to the cylinder head cover.

68. Install the throttle body, using a new gasket. Tighten the mounting bolts to 13 ft. lbs. (18 Nm).

69. Install the water bypass pipe and connect the sensor connectors. Install the ISC valve and tighten the mounting bolts to 13 ft. lbs. (18 Nm). Connect the water bypass hose.

70. Install the vacuum pipe and the following hoses:

a. Fuel pressure regulator vacuum hose.

b. Vacuum hose to the upper port of the EVAP BVSV.

c. Air intake chamber vacuum hose.

d. Throttle body vacuum hoses.

71. Install the EGR valve adapter with a new gasket. Tighten the mounting bolts to 13 ft. lbs. (18 Nm).

72. Connect the EGR gas temperature sensor connector (California only).

73. Install the EGR valve and vacuum modulator. Connect the water bypass hoses and the vacuum hoses.

74. Install the EGR and fuel pressure VSV and connect the hoses and connectors. Replace the EGR pipe and tighten the mounting bolts to 13 ft. lbs. (18 Nm).

75. Install the timing belt rear plates and tighten the bolts to 69 inch lbs. Install the water inlet and inlet hosing and tighten the bolts to 13 ft. lbs. (18 Nm).

76. Install the right side ignition coil and the high tension cord cover.

77. Connect and adjust the accelerator cable, the automatic transmission throttle cable and the cruise control actuator cable. Install the camshaft timing pulley.

78. Fill the cooling system and connect the negative battery cable. Start the engine and check for leaks.

79. Recheck all the fluid levels and check the ignition timing.

Valve Lifters

REMOVAL & INSTALLATION

1. Disconnect the negative battery cable.

2. Remove the cylinder head from the cylinder block.

3. Remove the lifters and the adjusting shims, using the proper tool.

4. Install the valve lifters and shims.

5. Check that the valve lifter rotates smoothly.

6. Install the cylinder head.

7. Connect the negative battery cable.

Valve Lash

ADJUSTMENT

ES250

NOTE: Adjust the valve clearance when the engine is cold.

1. Disconnect the negative battery cable.

2. Remove the air intake chamber. Remove the cylinder head covers.

3. Turn the crankshaft pulley and align it's groove with the timing mark 0 of the No. 1 timing cover.

4. Check that the valve lifters on the No. 1 intake and exhaust are tight.

5. Measure the clearance between the valve lifter and the camshaft. Record the measurements on valves No. 1, 2, 3 and 6.

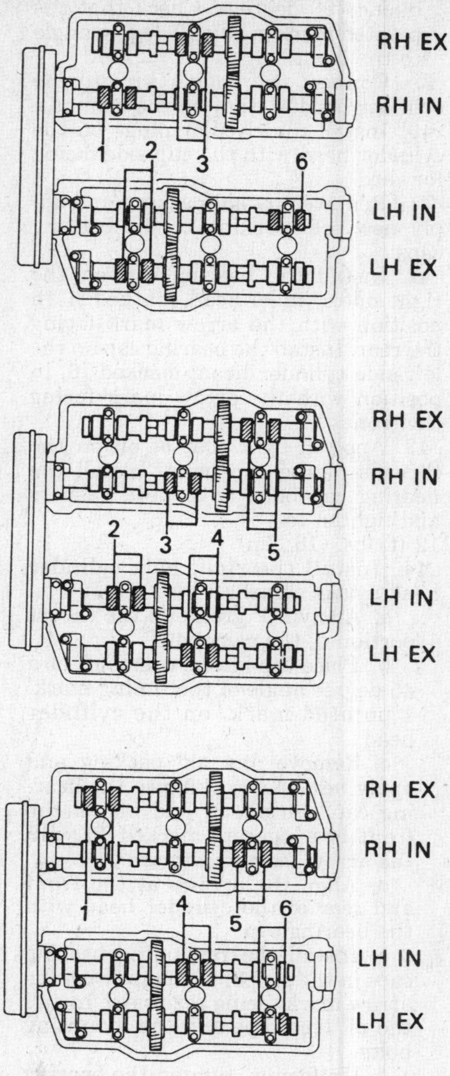

Valve adjusting sequence—ES250

a. The intake valve clearance cold is 0.005-0.009 in. (0.13-0.23mm).

b. The exhaust valve clearance cold is 0.011-0.015 in. (0.27-0.37mm).

6. Turn the crankshaft $^2/_3$ of a revolution and check the clearance on valves No. 2, 3, 4 and 5 and record.

7. Turn the crankshaft another $^2/_3$ of a revolution and check valves; 1, 4, 5 and 6 and record.

8. Remove the adjusting shim and turn the crankshaft to position the cam lobe of the camshaft on the adjusting valve upward. Press down the valve lifter with the proper tool and place the proper tool between the camshaft and the valve lifter. Remove the tool.

9. Remove the adjusting shim with the proper tool.

10. Install the specified valve shim on the valve lifter with the proper tool.

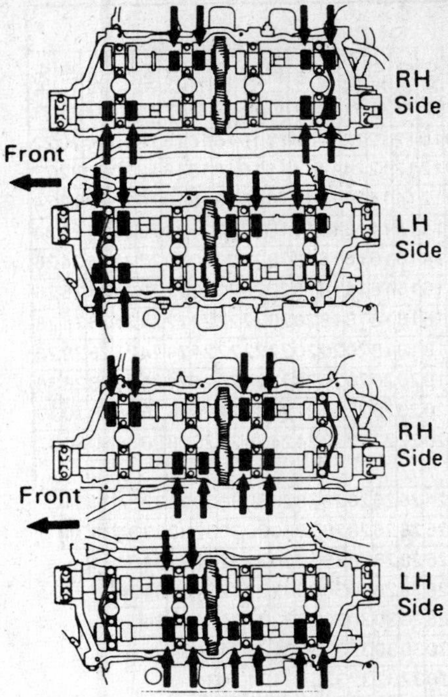

Valve adjusting sequence—LS400

11. Recheck the valve clearance.
12. Install the cylinder head covers and intake chamber.
13. Connect the negative battery cable.

LS400

1. Disconnect the negative battery cable.
2. Remove the No. 3 timing belt covers.
3. Disconnect the spark plug wires and remove the cylinder head covers.
4. Turn the crankshaft pulley and align it's groove with the timing mark 0 of the No. 1 timing cover. Check that the timing marks of the camshaft timing pulleys and timing belt rear plates are aligned. If not, turn the crankshaft 1 revolution (360 degrees) and align the mark.
5. Measure the clearance between the valve lifter and the camshaft on the valves in step 1 and record.
   a. The intake valve clearance cold is 0.006-0.010 in. (0.15-0.25mm).
   b. The exhaust valve clearance cold is 0.010-0.014 in. (0.25-0.35mm).
6. Turn the crankshaft 1 full revolution (360 degrees) and align the mark.
7. Measure the clearance between the valve lifter and the camshaft on the valves in step 2 and record.
8. Remove the adjusting shim and turn the crankshaft to position the cam lobe of the camshaft on the adjusting valve upward. Press down the valve lifter with the proper tool and place the proper tool between the camshaft and the valve lifter. Remove the tool.
9. Remove the adjusting shim with the proper tool.
10. Install the specified valve shim on the valve lifter with the proper tool.
11. Recheck the valve clearance. Install the cylinder head covers.
12. Connect the spark plug wires and install the No. 3 timing belt covers.
13. Connect the negative battery cable.

Intake Manifold

REMOVAL & INSTALLATION

ES250

1. Disconnect the negative battery cable. Drain the cooling system.
2. Remove the suspension upper brace. Disconnect the throttle cable from the throttle body, if equipped with an automatic transaxle.
3. Disconnect the accelerator cable from the throttle body. Remove the cruise control actuator and the vacuum pump.
4. Disconnect the ISC hose, the vacuum pipe hose and the air cleaner hose.
5. Raise and safely support the vehicle. Remove the right side engine undercover.
6. Remove the suspension lower crossmember and the front exhaust pipe. Lower the vehicle.
7. Remove the alternator, the ISC valve and the throttle body.
8. Remove the EGR pipe, the EGR valve and the vacuum modulator.
9. Disconnect the (4) BVSV vacuum hoses, the fuel pressure VSV hose and the air conditioning control valve vacuum hose.
10. Remove the distributor and the exhaust crossover pipe.
11. Disconnect the cold start injector connector and remove the cold start injector tube.
12. Disconnect the following hoses:
   a. PCV hose
   b. Vacuum sensing hose
   c. Fuel pressure VSV hose
   d. Air conditioning control valve vacuum hose
   e. EGR gas temperature sensor connector—California only
13. Remove the mounting bolts and the brackets from the intake chamber. Remove the air intake chamber.
14. Remove the delivery pipes and the injectors.
15. Disconnect the water temperature sensor connector and the upper radiator hose. Remove the water outlet.
16. Disconnect the following connectors and hose:
   a. Cold start injector time switch connector
   b. Water temperature sensor connector
   c. Heater water bypass hose
17. Remove the water bypass outlet. Remove the crossover pipe insulator from the water bypass outlet.
18. Remove the cylinder head rear plate and the idler pulley bracket.
19. Remove the mounting bolts and nuts. Lift off the intake manifold.

To install:

20. Install the intake manifold with new gaskets. Tighten the nuts and bolts to 13 ft. lbs. (18 Nm).
21. Install the No. 2 idler pulley bracket and tighten to 13 ft. lbs. (18 Nm). Replace the rear cylinder head plate.
22. Install the crossover pipe heat insulator. Replace the water bypass outlet and tighten the mounting bolts to 14 ft. lbs. (19 Nm).
23. Connect the heater water bypass hose, cold start injector time switch connector and the water temperature switch.
24. Connect the upper radiator hose, the water temperature sensor connector and install the water outlet with a new gasket. Tighten the bolts to 73 inch lbs.
25. Install the fuel injectors and the delivery pipes.
26. Install the air intake chamber with new gaskets and tighten the mounting bolts to 32 ft. lbs. (44 Nm).
27. Install the intake chamber brackets.
28. Connect the following hoses:
   a. PCV hose
   b. Vacuum sensing hose
   c. EGR gas temperature sensor connector (California only)
   d. Air conditioning control valve air hose
29. Install the cold start injector tube and connect the cold start injector connector.
30. Install the exhaust crossover pipe with new gaskets and tighten the mounting bolts to 25 ft. lbs. (34 Nm) and the nuts to 29 ft. lbs. (40 Nm).
31. Install the distributor.
32. Install the vacuum pipe and connect the BVSV vacuum hoses and the air conditioning control valve vacuum hose.
33. Install the EGR valve and vacuum modulator. Tighten the mounting bolts to 13 ft. lbs. (18 Nm). Install the vacuum pipe hoses.
34. Install the EGR pipe with a new gasket and tighten the mounting bolts to 13 ft. lbs. (18 Nm) and the nut to 58 ft. lbs. (79 Nm).
35. Install the throttle body and ISC valve.

Measured clearance (mm)	Installed Shim thickness (mm)																																												
	2.500	2.525	2.550	2.575	2.600	2.620	2.640	2.650	2.660	2.680	2.700	2.720	2.740	2.750	2.760	2.780	2.800	2.820	2.840	2.850	2.860	2.880	2.900	2.920	2.940	2.950	2.960	2.980	3.000	3.020	3.040	3.050	3.060	3.080	3.100	3.120	3.140	3.150	3.160	3.180	3.200	3.225	3.250	3.275	3.300
0.000 - 0.025														02	02	02	02	02	04	04	04	04	06	06	08	08	08	08	10	10	12	12	12	12	14	14	16	16	16	16	18	18	20	20	22
0.026 - 0.050												02	02	02	02	02	02	04	04	04	06	06	06	08	08	08	10	10	10	12	12	12	14	14	14	16	16	16	18	18	18	20	20	22	22
0.051 - 0.075											02	02	02	02	02	02	04	04	06	06	06	06	08	08	10	10	10	10	12	12	14	14	14	14	16	16	18	18	18	18	20	20	22	22	24
0.076 - 0.100										02	02	02	02	02	04	04	04	06	06	06	08	08	08	10	10	10	12	12	12	14	14	14	16	16	16	18	18	18	20	20	20	22	22	24	24
0.101 - 0.125								02	02	02	02	02	04	04	04	04	06	06	08	08	08	08	10	10	12	12	12	12	14	14	16	16	16	16	18	18	20	20	20	20	22	22	24	24	26
0.126 - 0.150						02	02	02	02	02	02	04	04	04	06	06	06	08	08	08	10	10	10	12	12	12	14	14	14	16	16	16	18	18	18	20	20	20	22	22	22	24	24	26	26
0.151 - 0.175					02	02	02	02	02	02	04	04	06	06	06	06	08	08	10	10	10	10	12	12	14	14	14	14	16	16	18	18	18	18	20	20	22	22	22	22	24	24	26	26	28
0.176 - 0.200				02	02	02	02	02	04	04	04	06	06	06	08	08	08	10	10	10	12	12	12	14	14	14	16	16	16	18	18	18	20	20	20	22	22	22	24	24	24	26	26	28	28
0.201 - 0.225			02	02	02	02	04	04	04	04	06	06	08	08	08	08	10	10	12	12	12	12	14	14	16	16	16	16	18	18	20	20	20	20	22	22	24	24	24	24	26	26	28	28	30
0.226 - 0.250		02	02	02	02	04	04	04	06	06	06	08	08	08	10	10	10	12	12	12	14	14	14	16	16	16	18	18	18	20	20	20	22	22	22	24	24	24	26	26	26	28	28	30	30
0.251 - 0.269		02	02	02	04	04	06	06	06	06	08	08	10	10	10	10	12	12	14	14	14	14	16	16	18	18	18	18	20	20	22	22	22	22	24	24	26	26	26	26	28	28	30	30	32
0.270 - 0.370																																													
0.371 - 0.375	04	06	06	08	08	08	10	10	10	12	12	12	14	14	14	16	16	16	18	18	18	20	20	20	22	22	22	24	24	24	26	26	26	28	28	28	30	30	30	32	32	34	34	34	
0.376 - 0.400	04	06	06	08	08	10	10	10	12	12	12	14	14	14	16	16	16	18	18	18	20	20	20	22	22	22	24	24	24	26	26	26	28	28	28	30	30	30	32	32	32	34	34	34	
0.401 - 0.425	06	06	08	08	10	10	12	12	12	12	14	14	16	16	16	16	18	18	20	20	20	20	22	22	24	24	24	24	26	26	28	28	28	28	30	30	32	32	32	32	34	34	34		
0.426 - 0.450	06	08	08	10	10	12	12	12	14	14	14	16	16	16	18	18	18	20	20	20	22	22	22	24	24	24	26	26	26	28	28	28	30	30	30	32	32	32	34	34	34	34			
0.451 - 0.475	08	08	10	10	12	12	14	14	14	14	16	16	18	18	18	18	20	20	22	22	22	22	24	24	26	26	26	26	28	28	30	30	30	30	32	32	34	34	34	34	34				
0.476 - 0.500	08	10	10	12	12	14	14	14	16	16	16	18	18	18	20	20	20	22	22	22	24	24	24	26	26	26	28	28	28	30	30	30	32	32	32	34	34	34	34	34					
0.501 - 0.525	10	10	12	12	14	14	16	16	16	16	18	18	20	20	20	20	22	22	24	24	24	24	26	26	28	28	28	28	30	30	32	32	32	32	34	34	34	34	34						
0.526 - 0.550	10	12	12	14	14	16	16	16	18	18	18	20	20	20	22	22	22	24	24	24	26	26	26	28	28	28	30	30	30	32	32	32	34	34	34	34	34								
0.551 - 0.575	12	12	14	14	16	16	18	18	18	18	20	20	22	22	22	22	24	24	26	26	26	26	28	28	30	30	30	30	32	32	34	34	34	34	34										
0.576 - 0.600	12	14	14	16	16	18	18	18	20	20	20	22	22	22	24	24	24	26	26	26	28	28	28	30	30	30	32	32	32	34	34	34	34	34											
0.601 - 0.625	14	14	16	16	18	18	20	20	20	20	22	22	24	24	24	24	26	26	28	28	28	28	30	30	32	32	32	32	34	34	34	34	34												
0.626 - 0.650	14	16	16	18	18	20	20	20	22	22	22	24	24	24	26	26	26	28	28	28	30	30	30	32	32	32	34	34	34	34	34														
0.651 - 0.675	16	16	18	18	20	20	22	22	22	22	24	24	26	26	26	26	28	28	30	30	30	30	32	32	34	34	34	34	34																
0.676 - 0.700	16	18	18	20	20	22	22	22	24	24	24	26	26	26	28	28	28	30	30	30	32	32	32	34	34	34	34	34																	
0.701 - 0.725	18	18	20	20	22	22	24	24	24	24	26	26	28	28	28	28	30	30	32	32	32	32	34	34	34	34	34																		
0.726 - 0.750	18	20	20	22	22	24	24	24	26	26	26	28	28	28	30	30	30	32	32	32	34	34	34	34	34																				
0.751 - 0.775	20	20	22	22	24	24	26	26	26	28	28	30	30	30	30	32	32	34	34	34	34	34	34																						
0.776 - 0.800	20	22	22	24	24	26	26	26	28	28	28	30	30	30	32	32	32	34	34	34	34	34																							
0.801 - 0.825	22	22	24	24	26	26	28	28	28	28	30	30	32	32	32	32	34	34	34	34	34																								
0.826 - 0.850	22	24	24	26	26	28	28	28	30	30	30	32	32	32	34	34	34	34	34																										
0.851 - 0.875	24	24	26	26	28	28	30	30	30	30	32	32	34	34	34	34	34																												
0.876 - 0.900	24	26	26	28	28	30	30	30	32	32	32	34	34	34	34	34																													
0.901 - 0.925	26	26	28	28	30	30	32	32	32	32	34	34	34	34	34																														
0.926 - 0.950	26	28	28	30	30	32	32	32	34	34	34	34	34																																
0.951 - 0.975	28	28	30	30	32	32	34	34	34	34	34																																		
0.976 - 1.000	28	30	30	32	32	34	34	34	34	34																																			
1.001 - 1.025	30	30	32	32	34	34	34	34	34																																				
1.026 - 1.050	30	32	32	34	34	34	34																																						
1.051 - 1.075	32	32	34	34	34																																								
1.076 - 1.100	32	34	34	34																																									
1.101 - 1.125	34	34	34																																										
1.126 - 1.150	34	34																																											
1.151 - 1.170	34																																												

New shim thicknesses mm (in.)

Shim No.	Thickness	Shim No.	Thickness
02	2.500 (0.0984)	20	2.950 (0.1161)
04	2.550 (0.1004)	22	3.000 (0.1181)
06	2.600 (0.1024)	24	3.050 (0.1201)
08	2.650 (0.1043)	26	3.100 (0.1220)
10	2.700 (0.1063)	28	3.150 (0.1240)
12	2.750 (0.1083)	30	3.200 (0.1260)
14	2.800 (0.1102)	32	3.250 (0.1280)
16	2.850 (0.1122)	34	3.300 (0.1299)
18	2.900 (0.1142)		

Exhaust valve adjusting shim selection using chart—LS400

Measured clearance (mm)	Installed shim thickness (mm)																																												
	2.500	2.525	2.550	2.575	2.600	2.620	2.640	2.650	2.660	2.680	2.700	2.720	2.740	2.750	2.760	2.780	2.800	2.820	2.840	2.850	2.860	2.880	2.900	2.920	2.940	2.950	2.960	2.980	3.000	3.020	3.040	3.050	3.060	3.080	3.100	3.120	3.140	3.150	3.160	3.180	3.200	3.225	3.250	3.275	3.300
0.000 - 0.025						02	02	02	02	02	04	04	04	06	06	06	08	08	08	10	10	10	12	12	12	14	14	14	16	16	16	18	18	18	20	20	20	22	22	22	24	24	26	26	28
0.026 - 0.050					02	02	02	02	02	04	04	06	06	06	06	08	08	10	10	10	10	12	12	14	14	14	14	16	16	18	18	18	18	20	20	22	22	22	22	24	24	26	26	28	28
0.051 - 0.075				02	02	02	02	04	04	04	06	06	06	08	08	08	10	10	10	12	12	12	14	14	14	16	16	16	18	18	18	20	20	20	22	22	22	24	24	24	26	26	28	28	30
0.076 - 0.100			02	02	02	04	04	04	04	06	06	08	08	08	08	10	10	12	12	12	12	14	14	16	16	16	16	18	18	20	20	20	20	22	22	24	24	24	24	26	26	28	28	30	30
0.101 - 0.125		02	02	02	04	04	04	06	06	06	08	08	08	10	10	10	12	12	12	14	14	14	16	16	16	18	18	18	20	20	20	22	22	22	24	24	24	26	26	26	28	28	30	30	32
0.126 - 0.129		02	02	04	04	04	06	06	06	08	08	08	10	10	10	12	12	12	14	14	14	16	16	16	18	18	18	20	20	20	22	22	22	24	24	24	26	26	26	28	28	28	30	30	32
0.130 - 0.230																																													
0.231 - 0.250	04	06	06	08	08	10	10	10	10	12	12	14	14	14	14	16	16	18	18	18	18	20	20	22	22	22	22	24	24	26	26	26	26	28	28	30	30	30	30	32	32	34	34	34	
0.251 - 0.275	06	06	08	08	10	10	10	12	12	12	14	14	14	16	16	16	18	18	18	20	20	20	22	22	22	24	24	24	26	26	26	28	28	28	30	30	30	32	32	32	34	34	34	34	
0.276 - 0.300	06	08	08	10	10	12	12	12	12	14	14	16	16	16	16	18	18	20	20	20	20	22	22	24	24	24	24	26	26	28	28	28	28	30	30	32	32	32	32	34	34	34	34		
0.301 - 0.325	08	08	10	10	12	12	12	14	14	14	16	16	16	18	18	18	20	20	20	22	22	22	24	24	24	26	26	26	28	28	28	30	30	30	32	32	32	34	34	34	34	34			
0.326 - 0.350	08	10	10	12	12	14	14	14	14	16	16	18	18	18	18	20	20	22	22	22	22	24	24	26	26	26	26	28	28	30	30	30	30	32	32	34	34	34	34	34	34				
0.351 - 0.375	10	10	12	12	14	14	14	16	16	16	18	18	18	20	20	20	22	22	22	24	24	24	26	26	26	28	28	28	30	30	30	32	32	32	34	34	34	34	34						
0.376 - 0.400	10	12	12	14	14	16	16	16	16	18	18	20	20	20	20	22	22	24	24	24	24	26	26	28	28	28	28	30	30	32	32	32	32	34	34	34	34	34							
0.401 - 0.425	12	12	14	14	16	16	16	18	18	18	20	20	20	22	22	22	24	24	24	26	26	26	28	28	28	30	30	30	32	32	32	34	34	34	34	34									
0.426 - 0.450	12	14	14	16	16	18	18	18	18	20	20	22	22	22	22	24	24	26	26	26	26	28	28	30	30	30	30	32	32	34	34	34	34	34	34										
0.451 - 0.475	14	14	16	16	18	18	18	20	20	20	22	22	22	24	24	24	26	26	26	28	28	28	30	30	30	32	32	32	34	34	34	34	34												
0.476 - 0.500	14	16	16	18	18	20	20	20	20	22	22	24	24	24	24	26	26	28	28	28	28	30	30	32	32	32	32	34	34	34	34	34													
0.501 - 0.525	16	16	18	18	20	20	20	22	22	22	24	24	24	26	26	26	28	28	28	30	30	30	32	32	32	34	34	34	34	34															
0.526 - 0.550	16	18	18	20	20	22	22	22	22	24	24	26	26	26	26	28	28	30	30	30	30	32	32	34	34	34	34	34	34																
0.551 - 0.575	18	18	20	20	22	22	22	24	24	24	26	26	26	28	28	28	30	30	30	32	32	32	34	34	34	34	34																		
0.576 - 0.600	18	20	20	22	22	24	24	24	24	26	26	28	28	28	28	30	30	32	32	32	32	34	34	34	34	34																			
0.601 - 0.625	20	20	22	22	24	24	24	26	26	26	28	28	28	30	30	30	32	32	32	34	34	34	34	34																					
0.626 - 0.650	20	22	22	24	24	26	26	26	26	28	28	30	30	30	30	32	32	34	34	34	34	34	34																						
0.651 - 0.675	22	22	24	24	26	26	26	28	28	28	30	30	30	32	32	32	34	34	34	34	34																								
0.676 - 0.700	22	24	24	26	26	28	28	28	28	30	30	32	32	32	32	34	34	34	34	34																									
0.701 - 0.725	24	24	26	26	28	28	28	30	30	30	32	32	32	34	34	34	34	34																											
0.726 - 0.750	24	26	26	28	28	30	30	30	30	32	32	34	34	34	34	34																													
0.751 - 0.775	26	26	28	28	30	30	30	32	32	32	34	34	34	34	34																														
0.776 - 0.800	26	28	28	30	30	32	32	32	32	34	34	34	34	34																															
0.801 - 0.825	28	28	30	30	32	32	32	34	34	34	34	34																																	
0.826 - 0.850	28	30	30	32	32	32	34	34	34	34	34																																		
0.851 - 0.875	30	30	32	32	34	34	34	34	34																																				
0.876 - 0.900	30	32	32	34	34	34	34	34																																					
0.901 - 0.925	32	32	34	34	34	34																																							
0.926 - 0.950	32	34	34	34	34																																								
0.951 - 0.975	34	34	34	34																																									
0.976 - 1.000	34	34	34																																										
1.001 - 1.025	34	34																																											
1.026 - 1.030	34																																												

New shim thicknesses mm (in.)

Shim No.	Thickness	Shim No.	Thickness
02	2.500 (0.0984)	20	2.950 (0.1161)
04	2.550 (0.1004)	22	3.000 (0.1181)
06	2.600 (0.1024)	24	3.050 (0.1201)
08	2.650 (0.1043)	26	3.100 (0.1220)
10	2.700 (0.1063)	28	3.150 (0.1240)
12	2.750 (0.1083)	30	3.200 (0.1260)
14	2.800 (0.1102)	32	3.250 (0.1280)
16	2.850 (0.1122)	34	3.300 (0.1299)
18	2.900 (0.1142)		

Intake valve adjusting shim selection using chart – LS400

Measured clearance mm (in.) \ Installed shim thickness mm (in.)	2.50 (0.0984)	2.52 (0.0992)	2.54 (0.1000)	2.55 (0.1004)	2.56 (0.1008)	2.58 (0.1016)	2.60 (0.1024)	2.62 (0.1031)	2.64 (0.1039)	2.65 (0.1043)	2.66 (0.1047)	2.67 (0.1051)	2.68 (0.1055)	2.69 (0.1059)	2.70 (0.1063)	2.71 (0.1067)	2.72 (0.1071)	2.73 (0.1075)	2.74 (0.1079)	2.75 (0.1083)	2.76 (0.1087)	2.77 (0.1091)	2.78 (0.1094)	2.79 (0.1098)	2.80 (0.1102)	2.81 (0.1106)	2.82 (0.1110)	2.83 (0.1114)	2.84 (0.1118)	2.85 (0.1122)	2.86 (0.1126)	2.87 (0.1130)	2.88 (0.1134)	2.89 (0.1138)	2.90 (0.1142)	2.91 (0.1146)	2.92 (0.1150)	2.93 (0.1154)	2.94 (0.1157)	2.95 (0.1161)	2.96 (0.1165)	2.97 (0.1169)	2.98 (0.1173)	2.99 (0.1177)	3.00 (0.1181)	3.01 (0.1185)	3.02 (0.1189)	3.03 (0.1193)	3.04 (0.1197)	3.05 (0.1201)	3.06 (0.1205)	3.08 (0.1213)	3.10 (0.1220)	3.12 (0.1228)	3.14 (0.1236)	3.15 (0.1240)	3.16 (0.1244)	3.18 (0.1252)	3.20 (0.1260)	3.22 (0.1268)	3.24 (0.1276)	3.25 (0.1280)	3.26 (0.1283)	3.28 (0.1291)	3.30 (0.1299)
0.000 — 0.020 (0.0000 — 0.0008)																		01	01	01	01	01	01	01	01	01	63	63	63	63	63	06	06	06	06	06	66	66	66	66	66	13	13	13	13	13	18	18	18	18	18	23	23	28	28	28	28	33	33	38	38	38	38	43	43
0.021 — 0.040 (0.0008 — 0.0016)																01	01	01	01	01	01	01	01	01	63	63	63	63	63	06	06	06	06	06	66	66	66	66	66	13	13	13	13	13	18	18	18	18	18	23	23	23	28	28	28	33	33	33	38	38	38	43	43	43	48
0.041 — 0.060 (0.0016 — 0.0024)														01	01	01	01	01	01	01	01	01	63	63	63	63	63	06	06	06	06	06	66	66	66	66	66	13	13	13	13	13	18	18	18	18	18	23	23	23	23	28	28	28	33	33	33	38	38	38	43	43	43	48	48
0.061 — 0.080 (0.0024 — 0.0031)												01	01	01	01	01	01	01	01	01	63	63	63	63	63	06	06	06	06	06	66	66	66	66	66	13	13	13	13	13	18	18	18	18	18	23	23	23	23	23	28	28	28	33	33	33	38	38	38	43	43	43	48	48	48
0.081 — 0.100 (0.0032 — 0.0039)										01	01	01	01	01	01	01	01	01	63	63	63	63	63	06	06	06	06	06	66	66	66	66	66	13	13	13	13	13	18	18	18	18	18	23	23	23	23	23	28	28	28	28	33	33	38	38	38	38	43	43	48	48	48	48	51
0.101 — 0.120 (0.0040 — 0.0047)									01	01	01	01	01	01	01	01	63	63	63	63	63	06	06	06	06	06	66	66	66	66	66	13	13	13	13	13	18	18	18	18	18	23	23	23	23	23	28	28	28	28	28	33	33	38	38	38	38	43	43	48	48	48	48	51	51
0.121 — 0.140 (0.0048 — 0.0055)								01	01	01	01	01	01	01	63	63	63	63	63	06	06	06	06	06	66	66	66	66	66	13	13	13	13	13	18	18	18	18	18	23	23	23	23	23	28	28	28	28	28	33	33	33	38	38	38	43	43	43	48	48	48	51	51	51	77
0.141 — 0.160 (0.0056 — 0.0063)							01	01	01	01	01	01	63	63	63	63	63	06	06	06	06	06	66	66	66	66	66	13	13	13	13	13	18	18	18	18	18	23	23	23	23	23	28	28	28	28	28	33	33	33	33	38	38	38	43	43	43	48	48	48	51	51	51	77	77
0.161 — 0.180 (0.0063 — 0.0071)						01	01	01	01	01	63	63	63	63	63	06	06	06	06	06	66	66	66	66	66	13	13	13	13	13	18	18	18	18	18	23	23	23	23	23	28	28	28	28	28	33	33	33	33	33	38	38	38	43	43	43	48	48	48	51	51	51	77	77	77
0.181 — 0.200 (0.0071 — 0.0079)				01	01	01	01	01	63	63	63	63	63	06	06	06	06	06	66	66	66	66	66	13	13	13	13	13	18	18	18	18	18	23	23	23	23	23	28	28	28	28	28	33	33	33	33	33	38	38	38	38	43	43	48	48	48	48	51	51	77	77	77	77	56
0.201 — 0.220 (0.0079 — 0.0087)			01	01	01	01	01	63	63	63	63	06	06	06	06	06	66	66	66	66	66	13	13	13	13	13	18	18	18	18	18	23	23	23	23	23	28	28	28	28	28	33	33	33	33	33	38	38	38	38	38	43	43	48	48	48	48	51	51	77	77	77	77	56	56
0.221 — 0.240 (0.0087 — 0.0094)		01	01	01	01	01	63	63	63	06	06	06	06	06	66	66	66	66	66	13	13	13	13	13	18	18	18	18	18	23	23	23	23	23	28	28	28	28	28	33	33	33	33	33	38	38	38	38	38	43	43	43	48	48	48	51	51	51	77	77	77	56	56	56	80
0.241 — 0.249 (0.0095 — 0.0098)		01	01	01	01	01	63	63	06	06	06	06	06	66	66	66	66	13	13	13	13	13	18	18	18	18	18	23	23	23	23	23	28	28	28	28	28	33	33	33	33	33	38	38	38	38	38	43	43	43	43	48	48	48	51	51	51	77	77	77	56	56	56	80	80
0.250 — 0.350 (0.0098 — 0.0138)																																																																		
0.351 — 0.360 (0.0138 — 0.0142)	63	06	06	06	06	66	66	13	13	13	13	18	18	18	18	18	23	23	23	23	23	28	28	28	28	28	33	33	33	33	33	38	38	38	38	38	43	43	43	43	43	48	48	48	48	48	51	51	51	51	51	77	77	56	56	56	56	80	80	61	61	61	61	61	
0.361 — 0.380 (0.0142 — 0.0150)	63	06	06	06	66	66	66	13	13	13	18	18	18	18	18	23	23	23	23	23	28	28	28	28	28	33	33	33	33	33	38	38	38	38	38	43	43	43	43	43	48	48	48	48	48	51	51	51	51	51	77	77	77	56	56	56	80	80	80	61	61	61	61	61	
0.381 — 0.400 (0.0150 — 0.0157)	06	06	66	66	66	66	13	13	18	18	18	18	18	23	23	23	23	23	28	28	28	28	28	33	33	33	33	33	38	38	38	38	38	43	43	43	43	43	48	48	48	48	48	51	51	51	51	51	77	77	77	77	56	56	80	80	80	80	61	61	61	61	61		
0.401 — 0.420 (0.0158 — 0.0165)	06	66	66	66	66	13	13	18	18	18	18	23	23	23	23	23	28	28	28	28	28	33	33	33	33	33	38	38	38	38	38	43	43	43	43	43	48	48	48	48	48	51	51	51	51	51	77	77	77	77	77	56	56	80	80	80	80	61	61	61	61				
0.421 — 0.440 (0.0166 — 0.0173)	66	66	66	13	13	13	18	18	18	23	23	23	23	23	28	28	28	28	28	33	33	33	33	33	38	38	38	38	38	43	43	43	43	43	48	48	48	48	48	51	51	51	51	51	77	77	77	77	77	56	56	56	80	80	80	61	61	61	61	61					
0.441 — 0.460 (0.0174 — 0.0181)	66	66	13	13	13	18	18	18	23	23	23	23	28	28	28	28	28	33	33	33	33	33	38	38	38	38	38	43	43	43	43	43	48	48	48	48	48	51	51	51	51	51	77	77	77	77	77	56	56	56	56	80	80	80	61	61	61	61	61						
0.461 — 0.480 (0.0181 — 0.0189)	66	13	13	13	18	18	18	23	23	23	28	28	28	28	28	33	33	33	33	33	38	38	38	38	38	43	43	43	43	43	48	48	48	48	48	51	51	51	51	51	77	77	77	77	77	56	56	56	56	56	80	80	80	61	61	61	61	61							
0.481 — 0.500 (0.0189 — 0.0197)	13	13	18	18	18	18	23	23	28	28	28	28	28	33	33	33	33	33	38	38	38	38	38	43	43	43	43	43	48	48	48	48	48	51	51	51	51	51	77	77	77	77	77	56	56	56	56	56	80	80	80	80	61	61	61	61	61								
0.501 — 0.520 (0.0197 — 0.0205)	13	18	18	18	18	23	23	28	28	28	28	33	33	33	33	33	38	38	38	38	38	43	43	43	43	43	48	48	48	48	48	51	51	51	51	51	77	77	77	77	77	56	56	56	56	56	80	80	80	80	80	61	61	61	61										
0.521 — 0.540 (0.0205 — 0.0213)	18	18	18	23	23	23	28	28	28	33	33	33	33	33	38	38	38	38	38	43	43	43	43	43	48	48	48	48	48	51	51	51	51	51	77	77	77	77	77	56	56	56	56	56	80	80	80	80	80	61	61	61	61	61											
0.541 — 0.560 (0.0213 — 0.0220)	18	18	23	23	23	28	28	28	33	33	33	33	38	38	38	38	38	43	43	43	43	43	48	48	48	48	48	51	51	51	51	51	77	77	77	77	77	56	56	56	56	56	80	80	80	80	80	61	61	61	61	61	61												
0.561 — 0.580 (0.0221 — 0.0228)	18	23	23	23	28	28	28	33	33	33	38	38	38	38	38	43	43	43	43	43	48	48	48	48	48	51	51	51	51	51	77	77	77	77	77	56	56	56	56	56	80	80	80	80	80	61	61	61	61	61	61	61													
0.581 — 0.600 (0.0229 — 0.0236)	23	23	28	28	28	28	33	33	38	38	38	38	38	43	43	43	43	43	48	48	48	48	48	51	51	51	51	51	77	77	77	77	77	56	56	56	56	56	80	80	80	80	80	61	61	61	61	61	61	61	61														
0.601 — 0.620 (0.0237 — 0.0244)	23	28	28	28	28	33	33	38	38	38	38	43	43	43	43	43	48	48	48	48	48	51	51	51	51	51	77	77	77	77	77	56	56	56	56	56	80	80	80	80	80	61	61	61	61	61	61	61	61																
0.621 — 0.640 (0.0244 — 0.0252)	28	28	28	33	33	33	38	38	38	43	43	43	43	43	48	48	48	48	48	51	51	51	51	51	77	77	77	77	77	56	56	56	56	56	80	80	80	80	80	61	61	61	61	61	61	61	61																		
0.641 — 0.660 (0.0252 — 0.0260)	28	28	33	33	33	38	38	38	43	43	43	43	48	48	48	48	48	51	51	51	51	51	77	77	77	77	77	56	56	56	56	56	80	80	80	80	80	61	61	61	61	61	61	61	61																				
0.661 — 0.680 (0.0260 — 0.0268)	28	33	33	33	38	38	38	43	43	43	48	48	48	48	48	51	51	51	51	51	77	77	77	77	77	56	56	56	56	56	80	80	80	80	80	61	61	61	61	61	61	61	61																						
0.681 — 0.700 (0.0268 — 0.0276)	33	33	38	38	38	38	43	43	48	48	48	48	48	51	51	51	51	51	77	77	77	77	77	56	56	56	56	56	80	80	80	80	80	61	61	61	61	61	61	61	61																								
0.701 — 0.720 (0.0276 — 0.0283)	33	38	38	38	38	43	43	48	48	48	48	51	51	51	51	51	77	77	77	77	77	56	56	56	56	56	80	80	80	80	80	61	61	61	61	61	61	61	61																										
0.721 — 0.740 (0.0284 — 0.0291)	38	38	38	43	43	43	48	48	48	51	51	51	51	51	77	77	77	77	77	56	56	56	56	56	80	80	80	80	80	61	61	61	61	61	61	61	61																												
0.741 — 0.760 (0.0292 — 0.0299)	38	38	43	43	43	48	48	48	51	51	51	51	77	77	77	77	77	56	56	56	56	56	80	80	80	80	80	61	61	61	61	61	61	61	61																														
0.761 — 0.780 (0.0300 — 0.0307)	38	43	43	43	48	48	48	51	51	51	77	77	77	77	77	56	56	56	56	56	80	80	80	80	80	61	61	61	61	61	61	61	61																																
0.781 — 0.800 (0.0307 — 0.0315)	43	43	48	48	48	48	51	51	77	77	77	77	77	56	56	56	56	56	80	80	80	80	80	61	61	61	61	61	61	61	61																																		
0.801 — 0.820 (0.0315 — 0.0323)	43	48	48	48	48	51	51	77	77	77	77	56	56	56	56	56	80	80	80	80	80	61	61	61	61	61	61	61	61																																				
0.821 — 0.840 (0.0323 — 0.0331)	48	48	48	51	51	51	77	77	77	56	56	56	56	56	80	80	80	80	80	61	61	61	61	61	61	61	61																																						
0.841 — 0.860 (0.0331 — 0.0339)	48	48	51	51	51	77	77	77	56	56	56	56	80	80	80	80	80	61	61	61	61	61	61	61	61																																								
0.861 — 0.880 (0.0339 — 0.0346)	48	51	51	51	77	77	77	56	56	56	80	80	80	80	80	61	61	61	61	61	61	61	61																																										
0.881 — 0.900 (0.0347 — 0.0354)	51	51	77	77	77	77	56	56	80	80	80	80	80	61	61	61	61	61	61	61	61																																												
0.901 — 0.920 (0.0355 — 0.0362)	51	77	77	77	77	56	56	80	80	80	80	61	61	61	61	61	61	61	61																																														
0.921 — 0.940 (0.0363 — 0.0370)	77	77	77	56	56	56	80	80	80	80	61	61	61	61	61	61	61																																																
0.941 — 0.960 (0.0370 — 0.0378)	77	77	56	56	56	80	80	80	61	61	61	61	61	61	61																																																		
0.961 — 0.980 (0.0378 — 0.0386)	77	56	56	56	80	80	80	61	61	61	61	61	61																																																				
0.981 — 1.000 (0.0386 — 0.0394)	56	56	80	80	80	80	61	61	61	61	61																																																						
1.001 — 1.020 (0.0394 — 0.0402)	56	80	80	80	80	61	61	61	61																																																									
1.021 — 1.040 (0.0402 — 0.0409)	80	80	80	61	61	61	61	61																																																										
1.041 — 1.060 (0.0410 — 0.0417)	80	80	61	61	61	61	61																																																												
1.061 — 1.080 (0.0418 — 0.0425)	80	61	61	61	61	61																																																													
1.081 — 1.100 (0.0426 — 0.0433)	61	61	61	61	61																																																														
1.101 — 1.120 (0.0433 — 0.0441)	61	61	61																																																																
1.121 — 1.140 (0.0441 — 0.0449)	61	61																																																																	
1.141 — 1.150 (0.0449 — 0.0453)	61																																																																		

New shim thickness mm (in.)

Shim No.	Thickness	Shim No.	Thickness
01	2.50 (0.0984)	38	2.95 (0.1161)
63	2.55 (0.1004)	43	3.00 (0.1181)
06	2.60 (0.1024)	48	3.05 (0.1201)
66	2.65 (0.1043)	51	3.10 (0.1220)
13	2.70 (0.1063)	77	3.15 (0.1240)
18	2.75 (0.1083)	56	3.20 (0.1260)
23	2.80 (0.1102)	80	3.25 (0.1280)
28	2.85 (0.1122)	61	3.30 (0.1299)
33	2.90 (0.1142)		

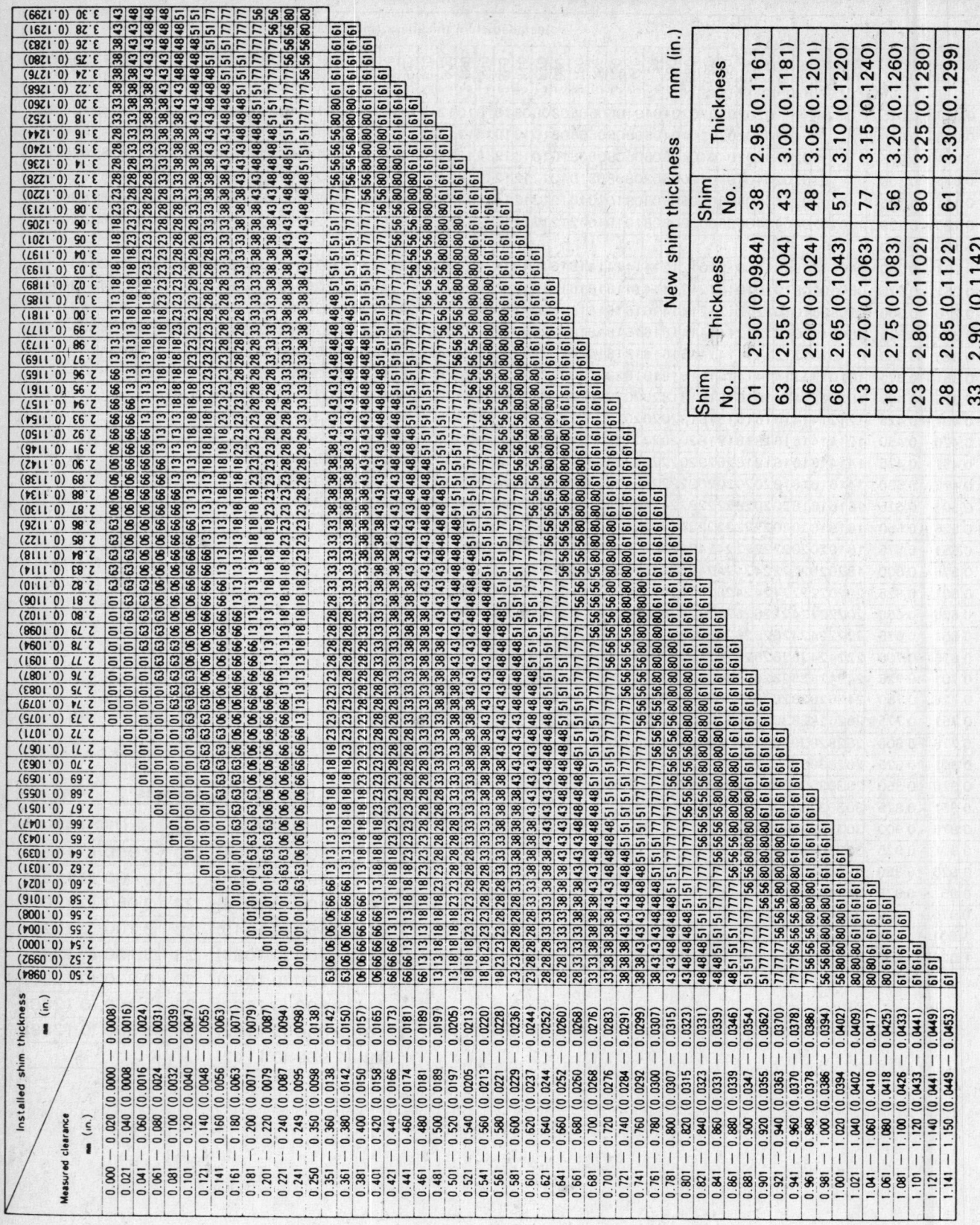

Exhaust valve adjusting shim selection using chart – ES250

Measured clearance mm (in.) \ Installed shim thickness mm (in.)	2.50 (0.0984)	2.52 (0.0992)	2.54 (0.1000)	2.55 (0.1004)	2.56 (0.1008)	2.58 (0.1016)	2.60 (0.1024)	2.62 (0.1031)	2.64 (0.1039)	2.65 (0.1043)	2.66 (0.1047)	2.67 (0.1051)	2.68 (0.1055)	2.69 (0.1059)	2.70 (0.1063)	2.71 (0.1067)	2.72 (0.1071)	2.73 (0.1075)	2.74 (0.1079)	2.75 (0.1083)	2.76 (0.1087)	2.77 (0.1091)	2.78 (0.1094)	2.79 (0.1098)	2.80 (0.1102)	2.81 (0.1106)	2.82 (0.1110)	2.83 (0.1114)	2.84 (0.1118)	2.85 (0.1122)	2.86 (0.1126)	2.87 (0.1130)	2.88 (0.1134)	2.89 (0.1138)	2.90 (0.1142)	2.91 (0.1146)	2.92 (0.1150)	2.93 (0.1154)	2.94 (0.1157)	2.95 (0.1161)	2.96 (0.1165)	2.97 (0.1169)	2.98 (0.1173)	2.99 (0.1177)	3.00 (0.1181)	3.01 (0.1185)	3.02 (0.1189)	3.03 (0.1193)	3.04 (0.1197)	3.05 (0.1201)	3.06 (0.1205)	3.08 (0.1213)	3.10 (0.1220)	3.12 (0.1228)	3.14 (0.1236)	3.15 (0.1240)	3.16 (0.1244)	3.18 (0.1252)	3.20 (0.1260)	3.22 (0.1268)	3.24 (0.1276)	3.25 (0.1280)	3.26 (0.1283)	3.28 (0.1291)	3.30 (0.1299)
0.000 — 0.020 (0.0000 — 0.0008)									01	01	01	01	01	01	01	01	63	63	63	63	63	06	06	06	06	06	66	66	66	66	66	13	13	13	13	13	18	18	18	18	18	23	23	23	23	23	28	28	28	28	28	33	33	38	38	38	38	43	43	48	48	48	48	51	51
0.021 — 0.040 (0.0008 — 0.0016)								01	01	01	01	01	01	01	63	63	63	63	63	06	06	06	06	06	66	66	66	66	66	13	13	13	13	13	18	18	18	18	18	23	23	23	23	23	28	28	28	28	33	33	33	33	38	38	38	43	43	43	48	48	48	51	51	51	77
0.041 — 0.060 (0.0016 — 0.0024)							01	01	01	01	01	01	63	63	63	63	63	06	06	06	06	06	66	66	66	66	66	13	13	13	13	13	18	18	18	18	18	23	23	23	23	23	28	28	28	28	28	33	33	33	33	38	38	38	43	43	43	48	48	48	51	51	51	77	77
0.061 — 0.080 (0.0024 — 0.0031)						01	01	01	01	01	63	63	63	63	63	06	06	06	06	06	66	66	66	66	66	13	13	13	13	13	18	18	18	18	18	23	23	23	23	23	28	28	28	28	28	33	33	33	33	33	38	38	38	43	43	43	48	48	48	51	51	51	77	77	77
0.081 — 0.100 (0.0032 — 0.0039)				01	01	01	01	01	63	63	63	63	63	06	06	06	06	06	66	66	66	66	66	13	13	13	13	13	18	18	18	18	18	23	23	23	23	23	28	28	28	28	28	33	33	33	33	33	38	38	38	38	43	43	48	48	48	48	51	51	77	77	77	77	56
0.101 — 0.120 (0.0040 — 0.0047)			01	01	01	01	01	63	63	63	63	06	06	06	06	06	66	66	66	66	66	13	13	13	13	13	18	18	18	18	18	23	23	23	23	23	28	28	28	28	28	33	33	33	33	33	38	38	38	38	38	43	43	48	48	48	48	51	51	77	77	77	77	56	56
0.121 — 0.140 (0.0048 — 0.0055)		01	01	01	01	01	63	63	63	06	06	06	06	06	66	66	66	66	66	13	13	13	13	13	18	18	18	18	18	23	23	23	23	23	28	28	28	28	28	33	33	33	33	33	38	38	38	38	38	43	43	43	48	48	48	51	51	51	77	77	77	56	56	56	80
0.141 — 0.149 (0.0056 — 0.0059)		01	01	01	01	63	63	63	06	06	06	06	66	66	66	66	66	13	13	13	13	13	18	18	18	18	18	23	23	23	23	23	28	28	28	28	28	33	33	33	33	33	38	38	38	38	38	43	43	43	43	48	48	48	51	51	51	77	77	77	56	56	56	80	80
0.150 — 0.250 (0.0059 — 0.0098)																																																																	
0.251 — 0.260 (0.0099 — 0.0102)	63	06	06	06	06	66	66	13	13	13	13	18	18	18	18	18	23	23	23	23	23	28	28	28	28	28	33	33	33	33	33	38	38	38	38	38	43	43	43	43	43	48	48	48	48	48	51	51	51	51	51	77	77	56	56	56	56	80	80	61	61	61	61	61	
0.261 — 0.281 (0.0103 — 0.0110)	63	06	06	06	66	66	66	13	13	13	18	18	18	18	18	23	23	23	23	23	28	28	28	28	28	33	33	33	33	33	38	38	38	38	38	43	43	43	43	43	48	48	48	48	48	51	51	51	51	51	77	77	77	56	56	56	80	80	80	61	61	61	61	61	
0.281 — 0.300 (0.0111 — 0.0118)	06	06	66	66	66	66	13	13	18	18	18	18	18	23	23	23	23	23	28	28	28	28	28	33	33	33	33	33	38	38	38	38	38	43	43	43	43	43	48	48	48	48	48	51	51	51	51	51	77	77	77	77	56	56	80	80	80	80	61	61	61	61	61		
0.301 — 0.320 (0.0119 — 0.0126)	06	66	66	66	66	13	13	18	18	18	18	23	23	23	23	23	28	28	28	28	28	33	33	33	33	33	38	38	38	38	38	43	43	43	43	43	48	48	48	48	48	51	51	51	51	51	77	77	77	77	77	56	56	80	80	80	80	61	61	61	61				
0.321 — 0.340 (0.0126 — 0.0134)	66	66	66	13	13	13	18	18	18	23	23	23	23	23	28	28	28	28	28	33	33	33	33	33	38	38	38	38	38	43	43	43	43	43	48	48	48	48	48	51	51	51	51	51	77	77	77	77	77	56	56	56	80	80	80	61	61	61	61	61					
0.341 — 0.360 (0.0134 — 0.0142)	66	66	13	13	13	18	18	18	23	23	23	23	28	28	28	28	28	33	33	33	33	33	38	38	38	38	38	43	43	43	43	43	48	48	48	48	48	51	51	51	51	51	77	77	77	77	77	56	56	56	56	80	80	80	61	61	61	61	61						
0.361 — 0.380 (0.0142 — 0.0150)	66	13	13	13	18	18	18	23	23	23	28	28	28	28	28	33	33	33	33	33	38	38	38	38	38	43	43	43	43	43	48	48	48	48	48	51	51	51	51	51	77	77	77	77	77	56	56	56	56	56	80	80	80	61	61	61	61	61							
0.381 — 0.400 (0.0150 — 0.0157)	13	13	18	18	18	18	23	23	28	28	28	28	28	33	33	33	33	33	38	38	38	38	38	43	43	43	43	43	48	48	48	48	48	51	51	51	51	51	77	77	77	77	77	56	56	56	56	56	80	80	80	80	61	61	61	61	61								
0.401 — 0.420 (0.0158 — 0.0165)	13	18	18	18	18	23	23	28	28	28	28	33	33	33	33	33	38	38	38	38	38	43	43	43	43	43	48	48	48	48	48	51	51	51	51	51	77	77	77	77	77	56	56	56	56	56	80	80	80	80	80	61	61	61	61										
0.421 — 0.440 (0.0166 — 0.0173)	18	18	18	23	23	23	28	28	28	33	33	33	33	33	38	38	38	38	38	43	43	43	43	43	48	48	48	48	48	51	51	51	51	51	77	77	77	77	77	56	56	56	56	56	80	80	80	80	80	61	61	61	61	61											
0.441 — 0.460 (0.0174 — 0.0181)	18	18	23	23	23	28	28	28	33	33	33	33	38	38	38	38	38	43	43	43	43	43	48	48	48	48	48	51	51	51	51	51	77	77	77	77	77	56	56	56	56	56	80	80	80	80	80	61	61	61	61	61	61												
0.461 — 0.480 (0.0181 — 0.0189)	18	23	23	23	28	28	28	33	33	33	38	38	38	38	38	43	43	43	43	43	48	48	48	48	48	51	51	51	51	51	77	77	77	77	77	56	56	56	56	56	80	80	80	80	80	61	61	61	61	61	61	61													
0.481 — 0.500 (0.0189 — 0.0197)	23	23	28	28	28	28	33	33	38	38	38	38	38	43	43	43	43	43	48	48	48	48	48	51	51	51	51	51	77	77	77	77	77	56	56	56	56	56	80	80	80	80	80	61	61	61	61	61	61	61	61														
0.501 — 0.520 (0.0197 — 0.0205)	23	28	28	28	28	33	33	38	38	38	38	43	43	43	43	43	48	48	48	48	48	51	51	51	51	51	77	77	77	77	77	56	56	56	56	56	80	80	80	80	80	61	61	61	61	61	61	61	61																
0.521 — 0.540 (0.0205 — 0.0213)	28	28	28	33	33	33	38	38	38	43	43	43	43	43	48	48	48	48	48	51	51	51	51	51	77	77	77	77	77	56	56	56	56	56	80	80	80	80	80	61	61	61	61	61	61	61	61																		
0.541 — 0.560 (0.0213 — 0.0220)	28	28	33	33	33	38	38	38	43	43	43	43	48	48	48	48	48	51	51	51	51	51	77	77	77	77	77	56	56	56	56	56	80	80	80	80	80	61	61	61	61	61	61	61	61																				
0.561 — 0.580 (0.0221 — 0.0228)	28	33	33	33	38	38	38	43	43	43	48	48	48	48	48	51	51	51	51	51	77	77	77	77	77	56	56	56	56	56	80	80	80	80	80	61	61	61	61	61	61	61	61																						
0.581 — 0.600 (0.0229 — 0.0236)	33	33	38	38	38	38	43	43	48	48	48	48	48	51	51	51	51	51	77	77	77	77	77	56	56	56	56	56	80	80	80	80	80	61	61	61	61	61	61	61	61																								
0.601 — 0.620 (0.0237 — 0.0244)	33	38	38	38	38	43	43	48	48	48	48	51	51	51	51	51	77	77	77	77	77	56	56	56	56	56	80	80	80	80	80	61	61	61	61	61	61	61	61																										
0.621 — 0.640 (0.0244 — 0.0252)	38	38	38	43	43	43	48	48	48	51	51	51	51	51	77	77	77	77	77	56	56	56	56	56	80	80	80	80	80	61	61	61	61	61	61	61	61																												
0.641 — 0.660 (0.0252 — 0.0260)	38	38	43	43	43	48	48	48	51	51	51	51	77	77	77	77	77	56	56	56	56	56	80	80	80	80	80	61	61	61	61	61	61	61	61																														
0.661 — 0.680 (0.0260 — 0.0268)	38	43	43	43	48	48	48	51	51	51	77	77	77	77	77	56	56	56	56	56	80	80	80	80	80	61	61	61	61	61	61	61	61																																
0.681 — 0.700 (0.0268 — 0.0276)	43	43	48	48	48	48	51	51	77	77	77	77	77	56	56	56	56	56	80	80	80	80	80	61	61	61	61	61	61	61	61																																		
0.701 — 0.720 (0.0276 — 0.0283)	43	48	48	48	48	51	51	77	77	77	77	56	56	56	56	56	80	80	80	80	80	61	61	61	61	61	61	61	61																																				
0.721 — 0.740 (0.0284 — 0.0291)	48	48	48	51	51	51	77	77	77	56	56	56	56	56	80	80	80	80	80	61	61	61	61	61	61	61	61																																						
0.741 — 0.760 (0.0292 — 0.0299)	48	48	51	51	51	77	77	77	56	56	56	56	80	80	80	80	80	61	61	61	61	61	61	61	61																																								
0.761 — 0.780 (0.0300 — 0.0307)	48	51	51	51	77	77	77	56	56	56	80	80	80	80	80	61	61	61	61	61	61	61	61																																										
0.781 — 0.800 (0.0307 — 0.0315)	51	51	77	77	77	77	56	56	80	80	80	80	80	61	61	61	61	61	61	61	61																																												
0.801 — 0.820 (0.0315 — 0.0323)	51	77	77	77	77	56	56	80	80	80	80	61	61	61	61	61	61	61	61																																														
0.821 — 0.840 (0.0323 — 0.0331)	77	77	77	56	56	56	80	80	80	61	61	61	61	61	61	61	61																																																
0.841 — 0.860 (0.0331 — 0.0339)	77	77	56	56	56	80	80	80	61	61	61	61	61	61	61																																																		
0.861 — 0.880 (0.0339 — 0.0346)	77	56	56	56	80	80	80	61	61	61	61	61	61																																																				
0.881 — 0.900 (0.0347 — 0.0354)	56	56	80	80	80	80	61	61	61	61	61																																																						
0.901 — 0.920 (0.0355 — 0.0362)	56	80	80	80	80	61	61	61	61																																																								
0.921 — 0.940 (0.0363 — 0.0370)	80	80	80	61	61	61	61	61																																																									
0.941 — 0.960 (0.0370 — 0.0378)	80	80	61	61	61	61	61																																																											
0.961 — 0.980 (0.0378 — 0.0386)	80	61	61	61	61	61																																																												
0.981 — 1.000 (0.0386 — 0.0394)	61	61	61	61	61																																																													
1.001 — 1.020 (0.0394 — 0.0402)	61	61	61																																																															
1.021 — 1.040 (0.0402 — 0.0409)	61	61																																																																
1.041 — 1.050 (0.0410 — 0.0413)	61																																																																		

New shim thickness mm (in.)

Shim No.	Thickness	Shim No.	Thickness
01	2.50 (0.0984)	38	2.95 (0.1161)
63	2.55 (0.1004)	43	3.00 (0.1181)
06	2.60 (0.1024)	48	3.05 (0.1201)
66	2.65 (0.1043)	51	3.10 (0.1220)
13	2.70 (0.1063)	77	3.15 (0.1240)
18	2.75 (0.1083)	56	3.20 (0.1260)
23	2.80 (0.1102)	80	3.25 (0.1280)
28	2.85 (0.1122)	61	3.30 (0.1299)
33	2.90 (0.1142)		

Intake valve adjusting shim selection using chart—ES250

36. Raise and support the vehicle. Install the front exhaust pipe and the right side engine undercover. Lower the vehicle.
37. Install the air cleaner hose and cruise control actuator.
38. Install the accelerator cable and connect the throttle cable, if equipped with automatic transaxle.
39. Install the suspension upper brace and fill the cooling system.
40. Connect the negative battery cable. Run the engine and check for leaks.
41. Adjust the timing. Recheck the fluid levels.

LS400

1. Disconnect the negative battery cable. Drain the cooling system.
2. Remove the camshaft timing pulleys.
3. Disconnect the accelerator cable, the throttle control cable, if equipped with automatic transmission and the cruise control actuator cable.
4. Remove the high tension cord cover and the right side ignition coil.
5. Remove the water inlet housing mounting bolts and disconnect the water bypass hose from the ISC valve.
6. Remove the water inlet and inlet housing assemblies. Remove the O-ring from the water inlet housing.
7. Remove the EGR pipe.
8. Disconnect the following:
   a. VSV connector
   b. Vacuum pipe hose
   c. EGR water bypass pipe
   d. Fuel pressure VSV
9. Disconnect the EGR vacuum hoses and remove the EGR VSV.
10. Disconnect the following hoses:
    a. Water bypass pipe hose from the ISC valve.
    b. Water bypass joint hose.
    c. Vacuum pipe hoses.
11. Disconnect the EGR gas temperature sensor, California only. Remove the EGR valve adapter.
12. Disconnect the following:
    a. Fuel pressure regulator vacuum hose.
    b. Air intake chamber vacuum hose.
    c. Vacuum hose from the EVAP BVSV.
13. Remove the mounting bolts, hoses and the vacuum pipe.
14. Remove the ISC valve.
15. Remove the throttle body sensor connectors and the water bypass pipe from the rear water bypass joint.
16. Remove the mounting bolts/nuts and disconnect the PCV valve. Remove the throttle body and gasket.
17. Disconnect the accelerator cable bracket and the brake booster vacuum union and hose.
18. Disconnect the cold start injector connector and the cold start injector tube from the right side delivery pipe.
19. Disconnect the check connector from the intake chamber and remove the mounting nuts and bolts.
20. Remove the air intake chamber and the cold start injector, tube and wire assembly.
21. Disconnect the engine wire from the intake manifold and from the right side cylinder head. Disconnect the heater hoses.
22. Remove the delivery pipes and the fuel injectors. Remove the mounting bolts and nuts. Lift up the intake manifold.

To install:

23. Install the intake manifold, using new gaskets. Tighten the mounting nuts and bolts to 13 ft. lbs. (18 Nm).
24. Install the delivery pipes and fuel injectors. Install the fuel return pipe with new gaskets. Tighten the union bolt to 26 ft. lbs. (35 Nm).
25. Connect the fuel hoses and the injector connectors. Connect the engine wire to the delivery pipes.
26. Connect the connectors on the left side delivery pipe, the water temperature sensor connector, cold start injector time switch connector and the water temperature sender gauge connector.
27. Connect the heater hoses and engine wire bracket. Install the engine wire to the bracket.
28. Install the cold start injector, tube and wire assembly. Tighten the mounting bolts to 69 inch lbs.
29. Install the air intake chamber with new gaskets and tighten the mounting bolts to 13 ft. lbs. (18 Nm).
30. Connect the cold start injector tube to the right side delivery pipe and tighten the union bolt to 11 ft. lbs. (15 Nm).
31. Connect the cold start injector connector. Install the accelerator cable bracket.
32. Install the brake booster union and connect the vacuum hose. Tighten the union bolt to 22 ft. lbs. (30 Nm).
33. Connect the water bypass hose to the throttle body and the PCV hose to the cylinder head cover.
34. Install the throttle body, using a new gasket. Tighten the mounting bolts to 13 ft. lbs. (18 Nm).
35. Install the water bypass pipe and connect the sensor connectors. Install the ISC valve and tighten the mounting bolts to 13 ft. lbs. (18 Nm). Connect the water bypass hose.
36. Install the vacuum pipe and the following hoses:
    a. Fuel pressure regulator vacuum hose.
    b. Vacuum hose to the upper port of the EVAP BVSV.
    c. Air intake chamber vacuum hose.
    d. Throttle body vacuum hoses.
37. Install the EGR valve adapter with a new gasket. Tighten the mounting bolts to 13 ft. lbs. (18 Nm).
38. Connect the EGR gas temperature sensor connector (California only).
39. Install the EGR valve and vacuum modulator. Connect the water bypass hoses and the vacuum hoses.
40. Install the EGR and fuel pressure VSV and connect the hoses and connectors. Replace the EGR pipe and tighten the mounting bolts to 13 ft. lbs. (18 Nm).
41. Install the timing belt rear plates and tighten the bolts to 69 inch lbs. Install the water inlet and inlet hosing and tighten the bolts to 13 ft. lbs. (18 Nm).
42. Install the right side ignition coil and the high tension cord cover.
43. Connect and adjust the accelerator cable, the automatic transmission throttle cable and the cruise control actuator cable. Install the camshaft timing pulley.
44. Fill the cooling system and connect the negative battery cable. Start the engine and check for leaks.
45. Recheck all the fluid levels and check the ignition timing.

Exhaust Manifold

REMOVAL & INSTALLATION

ES250

1. Disconnect the negative battery cable. Drain the cooling system.
2. Remove the suspension upper brace. Disconnect the throttle cable from the throttle body, if equipped with an automatic transaxle.
3. Disconnect the accelerator cable from the throttle body. Remove the cruise control actuator and the vacuum pump.
4. Disconnect the ISC hose, the vacuum pipe hose and the air cleaner hose.
5. Raise and safely support the vehicle. Remove the right side engine undercover.
6. Remove the suspension lower crossmember and the front exhaust pipe. Lower the vehicle.
7. Remove the alternator, the ISC valve and the throttle body.
8. Remove the EGR pipe, the EGR valve and the vacuum modulator.
9. Disconnect the (4) BVSV vacuum hoses, the fuel pressure VSV hose and the air conditioning control valve vacuum hose.
10. Remove the distributor and the exhaust crossover pipe.
11. Disconnect the cold start injector connector and remove the cold start injector tube.
12. Disconnect the following hoses:

a. PCV hose
b. Vacuum sensing hose
c. Fuel pressure VSV hose
d. Air conditioning control valve vacuum hose
e. EGR gas temperature sensor connector – California only

13. Remove the mounting bolts and the brackets from the intake chamber. Remove the air intake chamber.
14. Remove the delivery pipes and the injectors.
15. Disconnect the water temperature sensor connector and the upper radiator hose. Remove the water outlet.
16. Disconnect the following connectors and hose:

a. Cold start injector time switch connector
b. Water temperature sensor connector
c. Heater water bypass hose

17. Remove the water bypass outlet. Remove the crossover pipe insulator from the water bypass outlet.
18. Remove the cylinder head rear plate and the idler pulley bracket.
19. Remove the mounting bolts and nuts. Lift off the intake manifold.
20. Disconnect the main oxygen sensor connector. Remove outside heat insulator.
21. Remove the mounting nuts and the right side exhaust manifold. Remove the inside heat insulator. It may be necessary to remove the cylinder on the right side to provide ample clearance.
22. Remove the heat insulator, the mounting nuts and the left side exhaust manifold.

To install:

23. Install the right side heat insulator and the right side exhaust manifold with a new gasket. Tighten the nuts to 29 ft. lbs. (40 Nm).
24. Install the outside heat insulator and connect the oxygen sensor connector.
25. Install the left side exhaust manifold with a new gasket and tighten the nuts to 29 ft. lbs. (40 Nm). Install the heat insulator.
26. Install the intake manifold with new gaskets. Tighten the nuts and bolts to 13 ft. lbs. (18 Nm).
27. Install the No. 2 idler pulley bracket and tighten to 13 ft. lbs. (18 Nm). Replace the rear cylinder head plate.
28. Install the crossover pipe heat insulator. Replace the water bypass outlet and tighten the mounting bolts to 14 ft. lbs. (19 Nm).
29. Connect the heater water bypass hose, cold start injector time switch connector and the water temperature switch.
30. Connect the upper radiator hose, the water temperature sensor connector and install the water outlet with a new gasket. Tighten the bolts to 73 inch lbs.
31. Install the fuel injectors and the delivery pipes.
32. Install the air intake chamber with new gaskets and tighten the mounting bolts to 32 ft. lbs. (44 Nm).
33. Install the intake chamber brackets.
34. Connect the following hoses:

a. PCV hose
b. Vacuum sensing hose
c. EGR gas temperature sensor connector (California only)
d. Air conditioning control valve air hose

35. Install the cold start injector tube and connect the cold start injector connector.
36. Install the exhaust crossover pipe with new gaskets and tighten the mounting bolts to 25 ft. lbs. (34 Nm) and the nuts to 29 ft. lbs. (40 Nm).
37. Install the distributor.
38. Install the vacuum pipe and connect the BVSV vacuum hoses and the air conditioning control valve vacuum hose.
39. Install the EGR valve and vacuum modulator. Tighten the mounting bolts to 13 ft. lbs. (18 Nm). Install the vacuum pipe hoses.
40. Install the EGR pipe with a new gasket and tighten the mounting bolts to 13 ft. lbs. (18 Nm) and the nut to 58 ft. lbs. (79 Nm).
41. Install the throttle body and ISC valve.
42. Raise and support the vehicle. Install the front exhaust pipe and the right side engine undercover. Lower the vehicle.
43. Install the air cleaner hose and cruise control actuator.
44. Install the accelerator cable and connect the throttle cable, if equipped with an automatic transaxle.
45. Install the suspension upper brace and fill the cooling system.
46. Connect the negative battery cable. Run the engine and check for leaks.
47. Adjust the timing. Recheck the fluid levels.

LS400

1. Disconnect the negative battery cable. Drain the cooling system.
2. Remove the camshaft timing pulleys.
3. Disconnect the accelerator cable, the throttle control cable, if equipped with automatic transaxle and the cruise control actuator cable.
4. Remove the high tension cord cover and the right side ignition coil.
5. Remove the water inlet housing mounting bolts and disconnect the water bypass hose from the ISC valve.
6. Remove the water inlet and inlet housing assemblies. Remove the O-ring from the water inlet housing.
7. Remove the EGR pipe.
8. Disconnect the following:

a. VSV connector
b. Vacuum pipe hose
c. EGR water bypass pipe
d. Fuel pressure VSV

9. Disconnect the EGR vacuum hoses and remove the EGR VSV.
10. Disconnect the following hoses:

a. Water bypass pipe hose from the ISC valve.
b. Water bypass joint hose.
c. Vacuum pipe hoses.

11. Disconnect the EGR gas temperature sensor, California only. Remove the EGR valve adapter.
12. Disconnect the following:

a. Fuel pressure regulator vacuum hose.
b. Air intake chamber vacuum hose.
c. Vacuum hose from the EVAP BVSV.

13. Remove the mounting bolts, hoses and the vacuum pipe.
14. Remove the ISC valve.
15. Remove the throttle body sensor connectors and the water bypass pipe from the rear water bypass joint.
16. Remove the mounting bolts/nuts and disconnect the PCV valve. Remove the throttle body and gasket.
17. Disconnect the accelerator cable bracket and the brake booster vacuum union and hose.
18. Disconnect the cold start injector connector and the cold start injector tube from the right side delivery pipe.
19. Disconnect the check connector from the intake chamber and remove the mounting nuts and bolts.
20. Remove the air intake chamber and the cold start injector, tube and wire assembly.
21. Disconnect the engine wire from the intake manifold and from the right side cylinder head. Disconnect the heater hoses.
22. Remove the delivery pipes and the fuel injectors. Remove the mounting bolts and nuts. Lift up the intake manifold.
23. Remove the front and rear water bypass joint.
24. Raise and safely support the vehicle. Remove the front exhaust pipe and the main catalytic converters. Lower the vehicle.
25. Disconnect the right side oxygen sensor. Remove the mounting bolts and nuts and remove the right side exhaust manifold.
26. Remove the oil dipstick and guide. Disconnect the left side oxygen sensor.
27. Remove the mounting bolts and nuts and remove the left side exhaust manifold.

To install:

28. Install the right side exhaust manifold with a new gasket and tighten the mounting bolts to 29 ft. lbs. (40 Nm). Connect the right side oxygen sensor connector.
29. Install the left side exhaust manifold with a new gasket and tighten the mounting bolts to 29 ft. lbs. (40 Nm). Connect the left side oxygen sensor connector.
30. Install the oil dipstick and guide. Raise and safely support the vehicle.
31. Install the catalytic converters and front exhaust pipe. Lower the vehicle.
32. Install the front and rear water bypass joints. Tighten the mounting bolts to 13 ft. lbs. (18 Nm).
33. Install the intake manifold, using new gaskets. Tighten the mounting nuts and bolts to 13 ft. lbs. (18 Nm).
34. Install the delivery pipes and fuel injectors. Install the fuel return pipe with new gaskets. Tighten the union bolt to 26 ft. lbs. (35 Nm).
35. Connect the fuel hoses and the injector connectors. Connect the engine wire to the delivery pipes.
36. Connect the connectors on the left side delivery pipe, the water temperature sensor connector, cold start injector time switch connector and the water temperature sender gauge connector.
37. Connect the heater hoses and engine wire bracket. Install the engine wire to the bracket.
38. Install the cold start injector, tube and wire assembly. Tighten the mounting bolts to 69 inch lbs.
39. Install the air intake chamber with new gaskets and tighten the mounting bolts to 13 ft. lbs. (18 Nm).
40. Connect the cold start injector tube to the right side delivery pipe and tighten the union bolt to 11 ft. lbs. (15 Nm).
41. Connect the cold start injector connector. Install the accelerator cable bracket.
42. Install the brake booster union and connect the vacuum hose. Tighten the union bolt to 22 ft. lbs. (30 Nm).
43. Connect the water bypass hose to the throttle body and the PCV hose to the cylinder head cover.
44. Install the throttle body, using a new gasket. Tighten the mounting bolts to 13 ft. lbs. (18 Nm).
45. Install the water bypass pipe and connect the sensor connectors. Install the ISC valve and tighten the mounting bolts to 13 ft. lbs. (18 Nm). Connect the water bypass hose.
46. Install the vacuum pipe and the following hoses:
    a. Fuel pressure regulator vacuum hose.
    b. Vacuum hose to the upper port of the EVAP BVSV.
    c. Air intake chamber vacuum hose.
    d. Throttle body vacuum hoses.
47. Install the EGR valve adapter with a new gasket. Tighten the mounting bolts to 13 ft. lbs. (18 Nm).
48. Connect the EGR gas temperature sensor connector (California only).
49. Install the EGR valve and vacuum modulator. Connect the water bypass hoses and the vacuum hoses.
50. Install the EGR and fuel pressure VSV and connect the hoses and connectors. Replace the EGR pipe and tighten the mounting bolts to 13 ft. lbs. (18 Nm).
51. Install the timing belt rear plates and tighten the bolts to 69 inch lbs. Install the water inlet and inlet hosing and tighten the bolts to 13 ft. lbs. (18 Nm).
52. Install the right side ignition coil and the high tension cord cover.
53. Connect and adjust the accelerator cable, the automatic transmission throttle cable and the cruise control actuator cable. Install the camshaft timing pulley.
54. Fill the cooling system and connect the negative battery cable. Start the engine and check for leaks.
55. Recheck all the fluid levels and check the ignition timing.

Timing Belt Front Cover

REMOVAL & INSTALLATION

ES250

1. Disconnect the negative battery cable. Remove the suspension upper brace.
2. Remove the cruise control actuator.
3. Remove the power steering oil reservoir tank. Do not disconnect the hoses.
4. Raise and safely support the vehicle. Remove the right front tire and wheel assembly. Lower the vehicle.
5. Remove the alternator and power steering belts. Remove the right side fender apron seal.
6. Remove the right side engine support brackets. Raise the engine enough to remove the weight from the engine mounting on the right side, using a suitable jacking device.
7. Disconnect the power steering oil cooler lines and remove the right side engine mount.
8. Remove the air intake chamber and remove the spark plugs.
9. Remove the No. 2 timing belt cover and the right side engine mount bracket.
10. Turn the crankshaft pulley and align it's groove with the timing mark 0 of the No. 1 timing cover. Check that the timing marks of the camshaft timing pulleys and the No. 3 timing belt cover are aligned. If not, turn the crankshaft 1 full revolution (360 degrees).
11. Remove the timing belt tensioner and dust boot. Using the proper tool, loosen the tension between the left side and right side timing pulleys by slightly turning the right side camshaft timing pulley clockwise.
12. Remove the timing belt from the camshaft pulleys.
13. Remove the bolt, timing pulley and lock pin with the proper tool. Remove the 2 timing pulleys.
14. Remove the bolt and the No. 2 idler pulley.
15. Remove the crankshaft pulley bolt and the pulley, using the proper tool.
16. Remove the No. 1 timing cover.

To install:

17. Install the No. 1 timing belt cover and gasket.
18. Install the crankshaft pulley by aligning the pulley set key with the key groove of the pulley. Install the bolt with the proper tool and tighten to 181 ft. lbs. (245 Nm).
19. Install the No. 2 idler pulley and tighten the bolt to 29 ft. lbs. (40 Nm).
20. Install the left side camshaft timing pulley with the flange side facing outward. Align the knock pin hole of the camshaft with the pin groove of the timing pulley. Tighten the pulley bolt to 80 ft. lbs. (108 Nm).
21. Set the No. 1 cylinder to TDC compression by:
    a. Turn the crankshaft pulley and align it's groove with the 0 timing mark and the No. 1 timing cover.
    b. Turn the right side camshaft and align the knock pin hole of the camshaft with the timing mark of the No. 3 timing belt cover.
    c. Turn the left side camshaft and align the timing marks of the camshaft pulley with the timing mark of the No. 3 timing belt cover.
22. Install the timing belt to the left side camshaft timing pulley by:
    a. Check that the installation mark on the timing belt matches the end of the No. 1 timing cover. If not aligned, shift the meshing of the timing belt and the crankshaft pulley until they align.
    b. Using the proper tool, slightly, turn the left side camshaft timing pulley clockwise. Align the installation mark on the timing belt with the timing mark of the camshaft pulley and hang the timing belt on the left side camshaft pulley.
    c. Using the proper tool, align the timing marks of the left side cam-

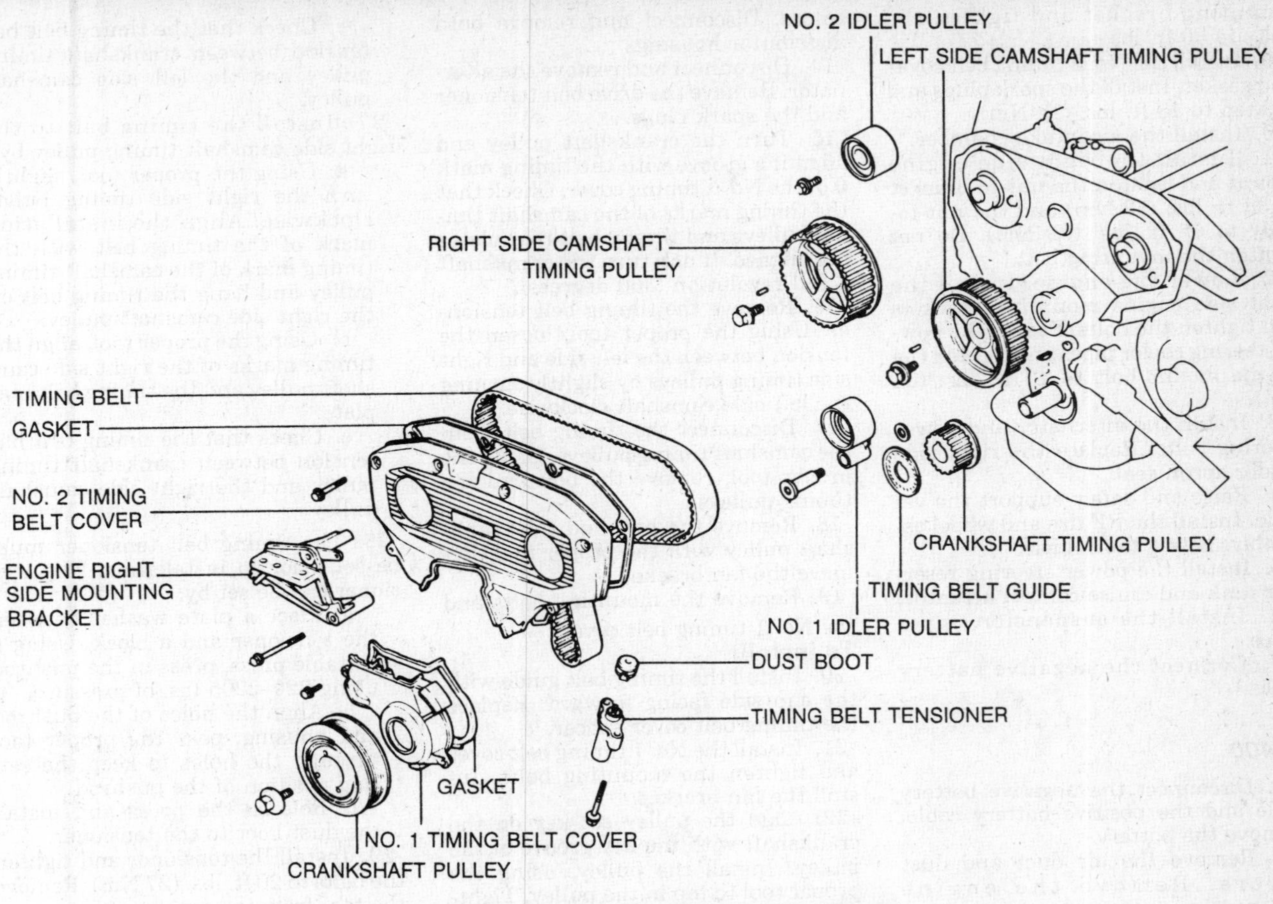

Exploded view of the timing belt assembly—ES250

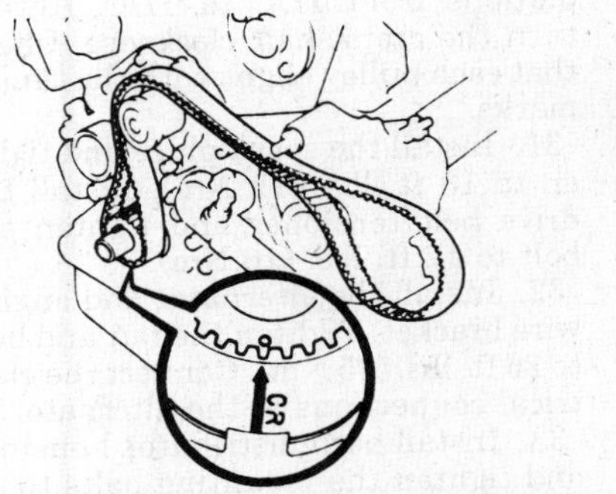

Aligning the timing belt—ES250

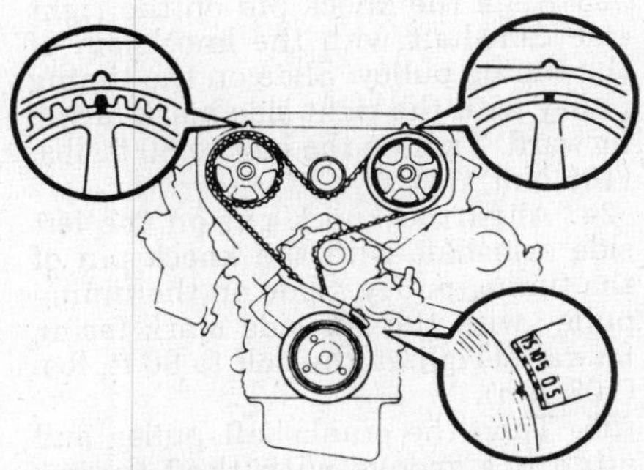

Aligning the timing belt and timing belt marks—ES250

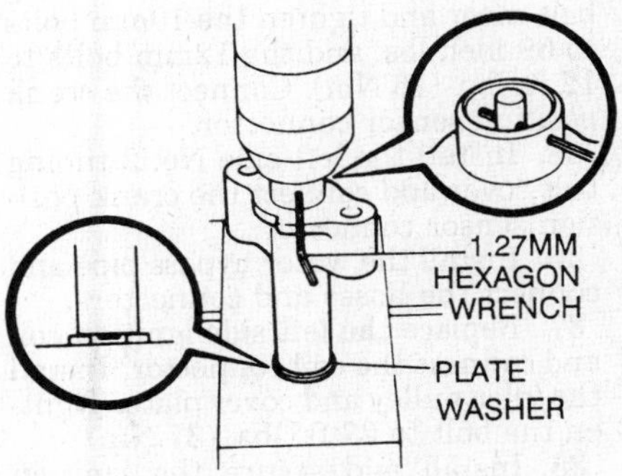

Setting the timing belt tensioner—ES250

shaft pulley and the No. 3 timing belt cover.

d. Check that the timing belt has tension between the crankshaft timing and the left side camshaft timing pulleys.

23. Install the timing belt to the right side camshaft timing pulley by:

a. Align the timing mark on the timing belt with the timing mark of the right side camshaft timing pulley.

b. Hang the timing belt on the right side camshaft timing pulley with the flange side facing inward.

c. Align the timing marks of the right side camshaft timing pulley and the No. 3 timing belt cover.

24. Slide the right side camshaft timing pulley on the camshaft. Align the knock pin hole of the camshaft with the knock pin groove of the pulley and install the knock pin. Tighten the bolt to 55 ft. lbs. (74 Nm).

25. The timing belt tensioner must be set prior to installation. The tensioner can be set by:

a. Place a plate washer between the tensioner and a block. Using a suitable press, press in the pushrod using 220–2205 lbs. of pressure.

b. Align the holes of the pushrod and housing, pass the proper tool through the holes to keep the setting position of the pushrod.

c. Release the press and install the dust boot to the tensioner.

26. Install the tensioner and tighten the bolts to 20 ft. lbs. (27 Nm). Remove the tool from the tensioner.

27. Turn the crankshaft pulley 2 revolutions from TDC to TDC. Always turn the crankshaft clockwise. Check that each pulley aligns with the timing marks.

28. Install the right side engine

mounting bracket and tighten the bolts to 30 ft. lbs.

29. Install the No. 2 timing belt cover and gasket. Install the spark plugs and tighten to 13 ft. lbs. (18 Nm).

30. Install the air intake chamber.

31. Install the right side engine mount and tighten the nut-to-bracket to 38 ft. lbs. (52 Nm) and the nut-to-body to 65 ft. lbs. (88 Nm). Do not tighten the mounting bolt.

32. Lower the engine. Install the right side engine mounting brackets and tighten the bolts. Fasten the power steering cooler pipes and tighten the engine mount bolt to 47 ft. lbs. (64 Nm).

33. Install the alternator and power steering belts. Replace the right side fender apron seal.

34. Raise and safely support the vehicle. Install the RF tire and wheel assembly. Lower the vehicle.

35. Install the power steering reservoir tank and cruise control actuator.

36. Install the suspension upper brace.

37. Connect the negative battery cable.

LS400

1. Disconnect the negative battery cable and the positive battery cable. Remove the battery.

2. Remove the air duct and dust covers. Remove the engine undercover.

3. Drain the cooling system. Remove the drive belt, fan, fluid coupling and fan pulley.

4. Remove the radiator, air cleaner and throttle body cover. Remove the air intake connector pipe.

5. Remove the air conditioning compressor and power steering pump. Do not disconnect the hoses.

6. Remove the upper high tension cord cover and the right side engine wire cover.

7. Disconnect the PCV hose and remove the left side engine wire cover. Remove the right side No. 3 timing cover.

8. Disconnect and tag the vacuum hoses and remove the left side engine wire cover. Disconnect the spark plug wires.

9. Remove the bolt, cover plate and idler pulley.

10. Disconnect the crank position sensor connector and remove the right side No. 2 timing belt cover.

11. Disconnect and remove the ignition coil. Disconnect the hoses and wires from the water bypass pipe. Remove the water bypass pipe.

12. Disconnect the crank position sensor connector and remove the No. 2 timing belt cover.

13. Remove the distributor caps and rotors. Disconnect and remove both distributor housings.

14. Disconnect and remove the alternator. Remove the drive belt tensioner and the spark plugs.

15. Turn the crankshaft pulley and align it's groove with the timing mark **0** of the No. 1 timing cover. Check that the timing marks of the camshaft timing pulleys and timing belt rear plates are aligned. If not, turn the crankshaft 1 full revolution (360 degrees).

16. Remove the timing belt tensioner. Using the proper tool, loosen the tension between the left side and right side timing pulleys by slightly turning the left side camshaft clockwise.

17. Disconnect the timing belt from the camshaft timing pulleys. Using the proper tool, remove the bolt and the timing pulleys.

18. Remove the bolt and the crankshaft pulley with the proper tool. Remove the fan bracket.

19. Remove the mounting bolts and the No. 1 timing belt cover.

To install:

20. Install the timing belt guide with the cup side facing forward. Replace the timing belt cover spacer.

21. Install the No. 1 timing belt cover and tighten the mounting bolts. Install the fan bracket.

22. Align the pulley set key on the crankshaft with the key groove of the pulley. Install the pulley, using the proper tool to tap in the pulley. Tighten the pulley bolt to 181 ft. lbs. (245 Nm).

23 Align the knock pin on the right side camshaft with the knock pin of the timing pulley. Slide on the timing pulley with the right side mark facing forward. Tighten the bolt to 80 ft. lbs. (108 Nm).

24. Align the knock pin on the left side camshaft with the knock pin of the timing pulley. Slide on the timing pulley with the left side mark facing forward. Tighten the bolt to 80 ft. lbs. (108 Nm).

25. Turn the crankshaft pulley and align it's groove with the**0** timing mark on the No. 1 timing belt cover. Using the proper tool, turn the crankshaft timing pulley and align the timing marks of the camshaft timing pulley and the timing belt rear plate.

26. Install the timing belt to the left side camshaft timing pulley by:

a. Using the proper tool, slightly turn the left side timing pulley clockwise. Align the installation mark of the timing belt with the timing mark of the camshaft timing pulley and hang the timing belt on the left side camshaft pulley.

b. Using the proper tool, align the timing marks of the left side camshaft pulley and the timing belt rear plate.

c. Check that the timing belt has tension between crankshaft timing pulley and the left side camshaft pulley.

27. Install the timing belt to the right side camshaft timing pulley by:

a. Using the proper tool, slightly turn the right side timing pulley clockwise. Align the installation mark of the timing belt with the timing mark of the camshaft timing pulley and hang the timing belt on the right side camshaft pulley.

b. Using the proper tool, align the timing marks of the right side camshaft pulley and the timing belt rear plate.

c. Check that the timing belt has tension between crankshaft timing pulley and the right side camshaft pulley.

28. The timing belt tensioner must be set prior to installation. The tensioner can be set by:

a. Place a plate washer between the tensioner and a block. Using a suitable press, press in the pushrod using 220–2205 lbs. of pressure.

b. Align the holes of the pushrod and housing, pass the proper tool through the holes to keep the setting position of the pushrod.

c. Release the press and install the dust boot to the tensioner.

29. Install the tensioner and tighten the bolts to 20 ft. lbs. (27 Nm). Remove the tool from the tensioner.

30. Turn the crankshaft pulley 2 revolutions from TDC to TDC. Always turn the crankshaft clockwise. Check that each pulley aligns with the timing marks.

31. Install the spark plugs and tighten to 13 ft. lbs. (18 Nm). Install the drive belt tensioner and tighten the bolt to 12 ft. lbs. (16 Nm).

32. Install the alternator and engine wire bracket. Tighten the nut and bolt to 26 ft. lbs. (35 Nm). Connect the electrical connections at the alternator.

33. Install both distributor housings and tighten the mounting bolts to 13 ft. lbs. (18 Nm). Replace the distributor rotors and caps.

34. Install the right side No. 2 timing belt cover and tighten the 10mm bolts to 69 inch lbs. and the 12mm bolts to 12 ft. lbs. (16 Nm). Connect the crank position sensor connector.

35. Install the left side No. 2 timing belt cover and connect the crank position sensor connector.

36. Install the water bypass pipe and connect the hoses and connectors.

37. Replace the left side ignition coil and connect the coil connector. Install the idler pulley and cover plate. Tighten the bolt to 27 ft. lbs. (37 Nm).

38. Install and secure the ignition wires. Install the right side No. 3 timing belt cover.

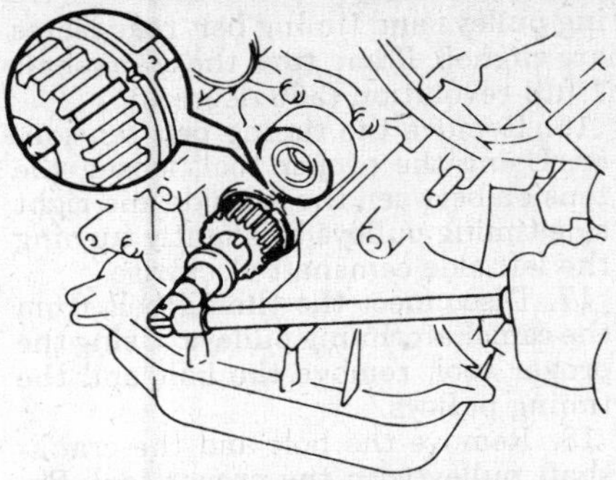

Aligning the timing mark of the crankshaft timing pulley and the oil pump body – LS400

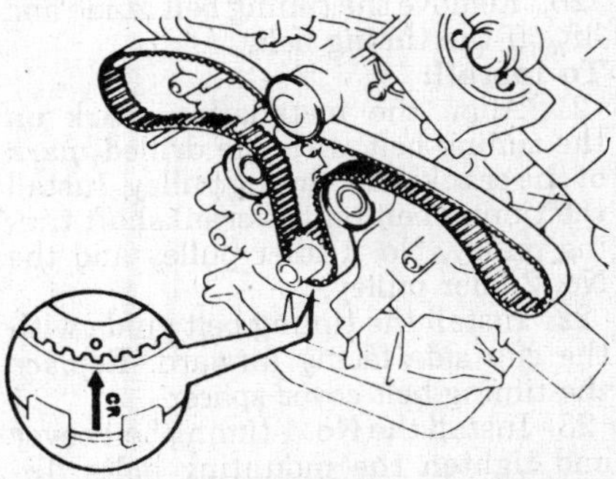

Align the installation mark on the timing belt with the drilled mark of the crankshaft timing pulley (if the belt is marked) – LS400

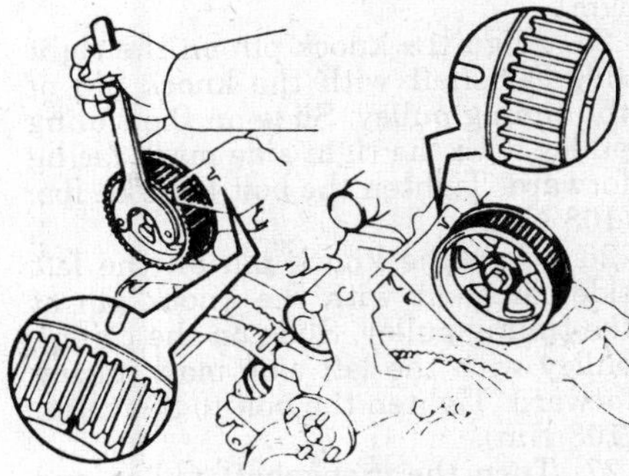

Aligning the timing marks of the camshaft timing pulley and timing belt rear plate – LS400

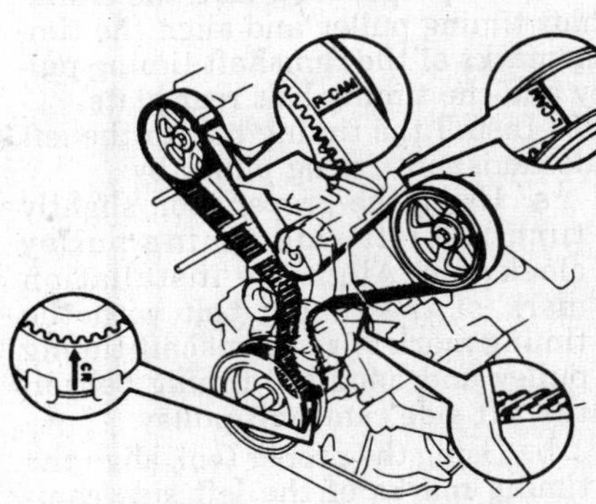

Checking the timing belt installation marks on a reinstalled used timing belt – LS400

39. Install the left side No. 3 timing belt cover and connect the vacuum hose and connectors. Install the right side engine wire cover.
40. Install the left side engine wire cover and connect the vacuum hoses.
41. Install the upper high tension cord covers. Fit the front side claw groove of the upper cover to claw of the lower cover.
42. Install the power steering pump and the air conditioning compressor.
43. Install the throttle body cover and the air cleaner.
44. Install the radiator, fan pulley, fan coupling, fan and drive belt.
45. Install the engine undercover and replace the battery.
46. Install the air ducts and dust covers. Connect the battery cables.
47. Refill the cooling system. Check the ignition timing.

OIL SEAL REPLACEMENT

The front cover oil seal replacement can and should be done when the front timing belt cover has been removed.
1. With the timing cover containing the oil seal removed, use a suitable seal removal tool and remove the front oil seal from the cover.
2. To install, apply a light coat of oil to the crankshaft and the lip of the new oil seal.
3. Using a suitable seal driver tool, drive in the new oil seal into the cover until the end of the seal sits squarely with the cover.
4. Reinstall the cover on the vehicle.

Timing Belt and Tensioner

REMOVAL & INSTALLATION

ES250

1. Disconnect the negative battery cable. Remove the suspension upper brace.
2. Remove the cruise control actuator.
3. Remove the power steering oil reservoir tank. Do not disconnect the hoses.
4. Raise and safely support the vehicle. Remove the right front tire and wheel assembly. Lower the vehicle.
5. Remove the alternator and power steering belts. Remove the right side fender apron seal.
6. Remove the right side engine support brackets. Raise the engine enough to remove the weight from the engine mounting on the right side, using a suitable jacking device.
7. Disconnect the power steering oil cooler lines and remove the right side engine mount.
8. Remove the air intake chamber and remove the spark plugs.
9. Remove the No. 2 timing belt cover and the right side engine mount bracket.
10. Turn the crankshaft pulley and align it's groove with the timing mark **0** of the No. 1 timing cover. Check that the timing marks of the camshaft timing pulleys and the No. 3 timing belt cover are aligned. If not, turn the crankshaft 1 full revolution (360 degrees).
11. Remove the timing belt tensioner and dust boot. Using the proper tool, loosen the tension between the left side and right side timing pulleys by slightly turning the right side camshaft timing pulley clockwise.
12. Remove the timing belt from the camshaft pulleys.
13. Remove the bolt, timing pulley and lock pin with the proper tool. Remove the 2 timing pulleys.
14. Remove the bolt and the No. 2 idler pulley.
15. Remove the crankshaft pulley bolt and the pulley, using the proper tool.
16. Remove the No. 1 timing cover.
17. Remove the timing belt guide and lift off the timing belt.

To install:

18. Align the installation mark on the timing belt with the drilled mark of the crankshaft timing pulley. Install the timing belt on the crankshaft timing pulley, No. 1 idler pulley and the No. 2 idler pulley.
19. Install the timing belt guide with the cup side facing outward.
20. Install the No. 1 timing belt cover and gasket.
21. Install the crankshaft pulley by aligning the pulley set key with the key groove of the pulley. Install the bolt with the proper tool and tighten to 181 ft. lbs. (245 Nm).
22. Install the No. 2 idler pulley and tighten the bolt to 29 ft. lbs. (40 Nm).
23. Install the left side camshaft timing pulley with the flange side facing outward. Align the knock pin hole of the camshaft with the pin groove of the timing pulley. Tighten the pulley bolt to 80 ft. lbs. (108 Nm).
24. Set the No. 1 cylinder to TDC compression by:
   a. Turn the crankshaft pulley and align it's groove with the 0 timing mark and the No. 1 timing cover.
   b. Turn the right side camshaft and align the knock pin hole of the camshaft with the timing mark of the No. 3 timing belt cover.
   c. Turn the left side camshaft and align the timing marks of the camshaft pulley with the timing mark of the No. 3 timing belt cover.
25. Install the timing belt to the left side camshaft timing pulley by:

a. Check that the installation mark on the timing belt matches the end of the No. 1 timing cover. If not aligned, shift the meshing of the timing belt and the crankshaft pulley until they align.

b. Using the proper tool, slightly, turn the left side camshaft timing pulley clockwise. Align the installation mark on the timing belt with the timing mark of the camshaft pulley and hang the timing belt on the left side camshaft pulley.

c. Using the proper tool, align the timing marks of the left side camshaft pulley and the No. 3 timing belt cover.

d. Check that the timing belt has tension between the crankshaft timing and the left side camshaft timing pulleys.

26. Install the timing belt to the right side camshaft timing pulley by:

a. Align the timing mark on the timing belt with the timing mark of the right side camshaft timing pulley.

b. Hang the timing belt on the right side camshaft timing pulley with the flange side facing inward.

c. Align the timing marks of the right side camshaft timing pulley and the No. 3 timing belt cover.

27. Slide the right side camshaft timing pulley on the camshaft. Align the knock pin hole of the camshaft with the knock pin groove of the pulley and install the knock pin. Tighten the bolt to 55 ft. lbs. (74 Nm).

28. The timing belt tensioner must be set prior to installation. The tensioner can be set by:

a. Place a plate washer between the tensioner and a block. Using a suitable press, press in the pushrod using 220–2205 lbs. of pressure.

b. Align the holes of the pushrod and housing, pass the proper tool through the holes to keep the setting position of the pushrod.

c. Release the press and install the dust boot to the tensioner.

29. Install the tensioner and tighten the bolts to 20 ft. lbs. (27 Nm). Remove the tool from the tensioner.

30. Turn the crankshaft pulley 2 revolutions from TDC to TDC. Always turn the crankshaft clockwise. Check that each pulley aligns with the timing marks.

31. Install the right side engine mounting bracket and tighten the bolts to 30 ft. lbs. (41 Nm).

32. Install the No. 2 timing belt cover and gasket. Install the spark plugs and tighten to 13 ft. lbs. (18 Nm).

33. Install the air intake chamber.

34. Install the right side engine mount and tighten the nut-to-bracket to 38 ft. lbs. (52 Nm) and the nut-to-body to 65 ft. lbs. (88 Nm). Do not tighten the mounting bolt.

35. Lower the engine. Install the right side engine mounting brackets and tighten the bolts. Fasten the power steering cooler pipes and tighten the engine mount bolt to 47 ft. lbs. (64 Nm).

36. Install the alternator and power steering belts. Replace the right side fender apron seal.

37. Raise and safely support the vehicle. Install the tire and wheel assembly. Lower the vehicle.

38. Install the power steering reservoir tank and cruise control actuator.

39. Install the suspension upper brace.

40. Connect the negative battery cable.

LS400

1. Disconnect the negative battery cable and the positive battery cable. Remove the battery.

2. Remove the air duct and dust covers. Remove the engine undercover.

3. Drain the cooling system. Remove the drive belt, fan, fluid coupling and fan pulley.

4. Remove the radiator, air cleaner and throttle body cover. Remove the air intake connector pipe.

5. Remove the air conditioning compressor and power steering pump. Do not disconnect the hoses.

6. Remove the upper high tension cord cover and the right side engine wire cover.

7. Disconnect the PCV hose and remove the left side engine wire cover. Remove the right side No. 3 timing cover.

8. Disconnect and tag the vacuum hoses and remove the left side engine wire cover. Disconnect the spark plug wires.

9. Remove the bolt, cover plate and idler pulley.

10. Disconnect the crank position sensor connector and remove the right side No. 2 timing belt cover.

11. Disconnect and remove the ignition coil. Disconnect the hoses and wires from the water bypass pipe. Remove the water bypass pipe.

12. Disconnect the crank position sensor connector and remove the No. 2 timing belt cover.

13. Remove the distributor caps and rotors. Disconnect and remove both distributor housings.

14. Disconnect and remove the alternator. Remove the drive belt tensioner and the spark plugs.

15. Turn the crankshaft pulley and align it's groove with the timing mark 0 of the No. 1 timing cover. Check that the timing marks of the camshaft timing pulleys and timing belt rear plates are aligned. If not, turn the crankshaft 1 full revolution (360 degrees).

16. Remove the timing belt tensioner. Using the proper tool, loosen the tension between the left side and right side timing pulleys by slightly turning the left side camshaft clockwise.

17. Disconnect the timing belt from the camshaft timing pulleys. Using the proper tool, remove the bolt and the timing pulleys.

18. Remove the bolt and the crankshaft pulley with the proper tool. Remove the fan bracket.

19. Remove the mounting bolts and the No. 1 timing belt cover.

20. Remove the timing belt guide and lift off the timing belt.

To install:

21. Align the installation mark on the timing belt with the drilled mark of the crankshaft timing pulley. Install the timing belt on the crankshaft timing pulley, No. 1 idler pulley and the No. 2 idler pulley.

22. Install the timing belt guide with the cup side facing forward. Replace the timing belt cover spacer.

23. Install the No. 1 timing belt cover and tighten the mounting bolts. Install the fan bracket.

24. Align the pulley set key on the crankshaft with the key groove of the pulley. Install the pulley, using the proper tool to tap in the pulley. Tighten the pulley bolt to 181 ft. lbs. (245 Nm).

25. Align the knock pin on the right side camshaft with the knock pin of the timing pulley. Slide on the timing pulley with the right side mark facing forward. Tighten the bolt to 80 ft. lbs. (108 Nm).

26. Align the knock pin on the left side camshaft with the knock pin of the timing pulley. Slide on the timing pulley with the left side mark facing forward. Tighten the bolt to 80 ft. lbs. (108 Nm).

27. Turn the crankshaft pulley and align it's groove with the0 timing mark on the No. 1 timing belt cover. Using the proper tool, turn the crankshaft timing pulley and align the timing marks of the camshaft timing pulley and the timing belt rear plate.

28. Install the timing belt to the left side camshaft timing pulley by:

a. Using the proper tool, slightly turn the left side timing pulley clockwise. Align the installation mark of the timing belt with the timing mark of the camshaft timing pulley and hang the timing belt on the left side camshaft pulley.

b. Using the proper tool, align the timing marks of the left side camshaft pulley and the timing belt rear plate.

c. Check that the timing belt has tension between crankshaft timing pulley and the left side camshaft pulley.

29. Install the timing belt to the right side camshaft timing pulley by:

a. Using the proper tool, slightly turn the right side timing pulley clockwise. Align the installation mark of the timing belt with the timing mark of the camshaft timing pulley and hang the timing belt on the right side camshaft pulley.

b. Using the proper tool, align the timing marks of the right side camshaft pulley and the timing belt rear plate.

c. Check that the timing belt has tension between crankshaft timing pulley and the right side camshaft pulley.

30. The timing belt tensioner must be set prior to installation. The tensioner can be set by:

a. Place a plate washer between the tensioner and a block. Using a suitable press, press in the pushrod using 220–2205 lbs. of pressure.

b. Align the holes of the pushrod and housing, pass the proper tool through the holes to keep the setting position of the pushrod.

c. Release the press and install the dust boot to the tensioner.

31. Install the tensioner and tighten the bolts to 20 ft. lbs. (27 Nm). Remove the tool from the tensioner.

32. Turn the crankshaft pulley 2 revolutions from TDC to TDC. Always turn the crankshaft clockwise. Check that each pulley aligns with the timing marks.

33. Install the spark plugs and tighten to 13 ft. lbs. (18 Nm). Install the drive belt tensioner and tighten the bolt to 12 ft. lbs. (16 Nm).

34. Install the alternator and engine wire bracket. Tighten the nut and bolt to 26 ft. lbs. (35 Nm). Connect the electrical connections at the alternator.

35. Install both distributor housings and tighten the mounting bolts to 13 ft. lbs. (18 Nm). Replace the distributor rotors and caps.

36. Install the right side No. 2 timing belt cover and tighten the 10mm bolts to 69 inch lbs. and the 12mm bolts to 12 ft. lbs. (16 Nm). Connect the crank position sensor connector.

37. Install the left side No. 2 timing belt cover and connect the crank position sensor connector.

38. Install the water bypass pipe and connect the hoses and connectors.

39. Replace the left side ignition coil and connect the coil connector. Install the idler pulley and cover plate. Tighten the bolt to 27 ft. lbs. (37 Nm).

40. Install and secure the ignition wires. Install the right side No. 3 timing belt cover.

41. Install the left side No. 3 timing belt cover and connect the vacuum hose and connectors. Install the right side engine wire cover.

42. Install the left side engine wire cover and connect the vacuum hoses.

43. Install the upper high tension cord covers. Fit the front side claw groove of the upper cover to claw of the lower cover.

44. Install the power steering pump and the air conditioning compressor.

45. Install the throttle body cover and the air cleaner.

46. Install the radiator, fan pulley, fan coupling, fan and drive belt.

47. Install the engine undercover and replace the battery.

48. Install the air ducts and dust covers. Connect the battery cables.

49. Refill the cooling system. Check the ignition timing.

Timing Sprockets

REMOVAL & INSTALLATION

ES250

1. Disconnect the negative battery cable.
2. Remove the timing belt assembly.
3. Remove the idler pulley with the proper tool.
4. Remove the crankshaft pulley.

NOTE: If the pulley cannot be removed by hand, carefully pry off the pulley with a suitable pry bar or puller.

To install:

5. Align the crankshaft timing pulley set key with the groove on the timing pulley. Slide on the crankshaft pulley with the flange side facing inward.
6. Install the No. 1 idler pulley, using the proper adhesive on the threads of the mounting bolt end. Install the bolt and washer with the proper tool. Tighten the bolt to 25 ft. lbs. (34 Nm). Check that the pulley bracket moves smoothly.
7. Align the installation mark on the timing belt with the drilled mark on the crankshaft pulley.
8. Install the timing belt assembly.
9. Connect the negative battery cable.

LS400

1. Disconnect the negative battery cable.
2. Remove the timing belt.
3. Remove the pulley bolt and the No. 2 idler pulley. Using the proper tool, remove the bolt and No. 1 idler pulley.
4. Remove the crankshaft timing pulley with the proper tool.

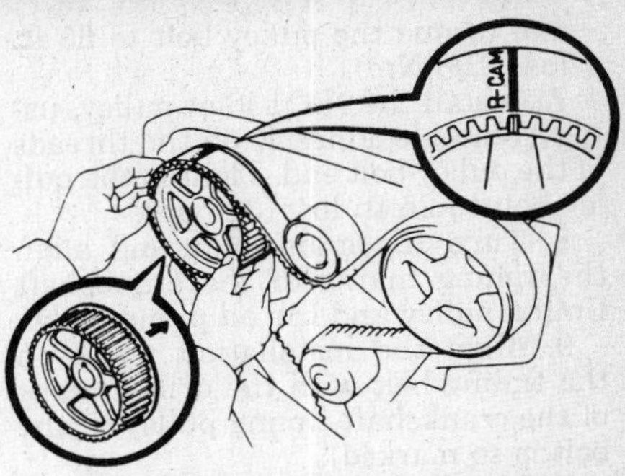

Installing the right hand camshaft pulley – ES250

To install:

5. When installing the right hand camshaft timing pulley, use the following procedure:

a. Align the knock pin on the camshaft with the knock pin groove of the timing pulley.

b. Slide the timing pulley, facing the **RH** mark forward.

c. Holding the camshaft gear still and torque the pulley bolt to 80 ft. lbs. (108 Nm).

6. When installing the right hand camshaft timing pulley, use the following procedure:

a. Align the knock pin on the camshaft with the knock pin groove of the timing pulley.

b. Slide the timing pulley, facing the **LH** mark forward.

c. Holding the camshaft gear still

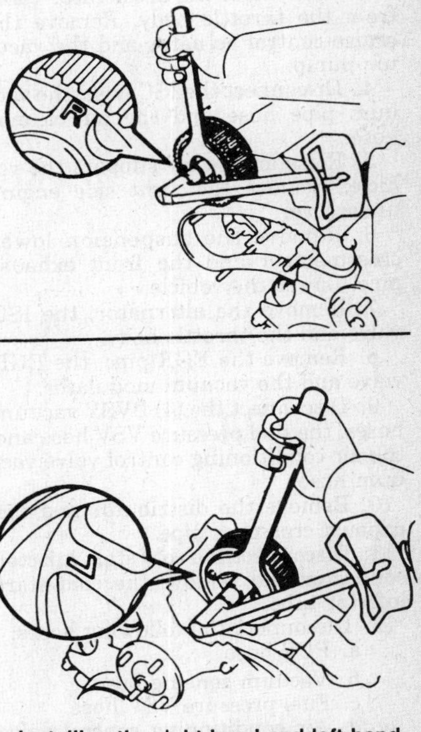

Installing the right hand and left hand camshaft pulley's – LS400

and torque the pulley bolt to 80 ft. lbs. (108 Nm).

7. Install the No. 1 idler pulley, using a suitable adhesive on the threads of the pulley bolt end. Tighten the pulley bolt to 25 ft. lbs. (34 Nm).

8. Turn the crankshaft and align the timing marks of the crankshaft timing pulley and the oil pump body.

9. Align the installation mark on the timing belt with the drilled mark of the crankshaft timing pulley (if the belt is so marked).

10. Install the timing belt assembly and connect the negative battery cable

Camshaft

REMOVAL & INSTALLATION

ES250

1. Disconnect the negative battery cable.

NOTE: Since the thrust clearance of the camshaft is small, the camshaft must be held level while it is being removed. If the camshaft is not kept level, the portion of the camshaft head receiving the shaft thrust may crack or be damaged, causing the camshaft to seize or break.

2. Remove the suspension upper brace. Disconnect the throttle cable from the throttle body, if equipped with an automatic transaxle.

3. Disconnect the accelerator cable from the throttle body. Remove the cruise control actuator and the vacuum pump.

4. Disconnect the ISC hose, the vacuum pipe hose and the air cleaner hose.

5. Raise and safely support the vehicle. Remove the right side engine undercover.

6. Remove the suspension lower crossmember and the front exhaust pipe. Lower the vehicle.

7. Remove the alternator, the ISC valve and the throttle body.

8. Remove the EGR pipe, the EGR valve and the vacuum modulator.

9. Disconnect the (4) BVSV vacuum hoses, the fuel pressure VSV hose and the air conditioning control valve vacuum hose.

10. Remove the distributor and the exhaust crossover pipe.

11. Disconnect the cold start injector connector and remove the cold start injector tube.

12. Disconnect the following hoses:
   a. PCV hose
   b. Vacuum sensing hose
   c. Fuel pressure VSV hose
   d. Air conditioning control valve vacuum hose
   e. EGR gas temperature sensor connector—California only

13. Remove the mounting bolts and the brackets from the intake chamber. Remove the air intake chamber.

14. Remove the delivery pipes and the injectors.

15. Disconnect the water temperature sensor connector and the upper radiator hose. Remove the water outlet.

16. Disconnect the following connectors and hose:
   a. Cold start injector time switch connector
   b. Water temperature sensor connector
   c. Heater water bypass hose

17. Remove the water bypass outlet. Remove the crossover pipe insulator from the water bypass outlet.

18. Remove the cylinder head rear plate and the idler pulley bracket.

19. Remove the mounting bolts and nuts. Lift off the intake manifold.

20. Disconnect the main oxygen sensor connector. Remove outside heat insulator.

21. Remove the mounting nuts and the right side exhaust manifold. Remove the inside heat insulator.

22. Remove the heat insulator, the mounting nuts and the left side exhaust manifold.

23. Remove the spark plugs.

24. Remove the timing belt, camshaft pulleys and the No. 2 idler pulley.

25. Remove the No. 3 timing cover.

26. Remove the mounting nuts and the cylinder head covers. Remove the spark plug tube gaskets.

27. Remove the exhaust camshaft of the right side cylinder head by:
   a. Align the timing marks, 2 pointed marks, of the camshaft drive and the drive gear by turning the camshaft with a wrench.
   b. Secure the exhaust camshaft sub-gear to the drive gear with a service bolt.
   c. Uniformly, loosen and remove the 8 bearing cap bolts in the proper sequence.
   d. Remove the 4 bearing caps and the camshaft.

28. Uniformly, loosen and remove the 10 bearing cap bolts on the right side cylinder head, in sequence. Remove the 5 bearing caps and remove the intake camshaft.

29. Remove the exhaust camshaft of the left side cylinder head by performing the following:
   a. Align the timing marks, 1 pointed mark, of the camshaft drive and the drive gear by turning the camshaft with a wrench.
   b. Secure the exhaust camshaft sub-gear to the drive gear with a service bolt.
   c. Uniformly, loosen and remove the 8 bearing cap bolts in the proper sequence.
   d. Remove the 4 bearing caps and the exhaust camshaft.

30. Uniformly, loosen and remove the 10 bearing cap bolts on the right side cylinder head, in sequence. Remove the 5 bearing caps and remove the intake camshaft.

To install:

31. Install the intake camshaft of the right side cylinder head by:
   a. Apply a suitable multi-purpose grease to the thrust portion of the camshaft.
   b. Apply seal packing to the No. 1 bearing cap. Install the bearing caps.
   c. Apply a light coat of oil on the threads of the bearing cap bolts.
   d. Install and uniformly tighten the bearing cap bolts to 12 ft. lbs. (16 Nm).

32. Install the exhaust camshaft of the right side cylinder head by:
   a. Apply a suitable multi-purpose grease to the thrust portion of the camshaft.
   b. Align the timing marks (2 pointed marks) of the camshaft and the drive gears.
   c. Place the camshaft on the cylinder head and install the bearing caps.
   d. Apply a light coat of oil on the threads of the bearing cap bolts.
   e. Install and uniformly tighten the bearing cap bolts to 12 ft. lbs. (16 Nm). Remove the service bolt.

33. Install the intake camshaft on the left side cylinder head by:
   a. Apply MP grease to the thrust portion of the camshaft.
   b. Place the intake camshaft at a 90 degree angle of the timing mark on the cylinder head, 1 pointed mark.
   c. Apply seal packing to the No. 1 bearing cap. Install the bearing caps.
   d. Apply a light coat of oil on the threads of the bearing cap bolts.
   e. Install and uniformly tighten the bearing cap bolts to 12 ft. lbs. (16 Nm).

34. Install the exhaust camshaft on the left side cylinder head by:
   a. Apply a suitable multi-purpose grease to the thrust portion of the camshaft.
   b. Align the timing marks, 1 pointed mark, of the camshaft and the drive gears.
   c. Place the camshaft on the cylinder head and install the bearing caps.
   d. Apply a light coat of oil on the threads of the bearing cap bolts.
   e. Install and uniformly tighten

the bearing cap bolts to 12 ft. lbs. (16 Nm). Remove the service bolt.

35. Turn the camshaft and position the cam lobe upward. Check and adjust the valve clearance.

36. Apply a suitable multi-purpose grease to the the new camshaft oil seals and install with the proper tool.

37. Install the spark plug tube gaskets. Install the proper seal packing to the cylinder heads.

38. Install the cylinder head cover gasket and install with the seal washers and nuts. Tighten to 52 inch lbs.

39. Install the No. 3 timing belt cover and tighten the bolts 65 inch lbs.

40. Install the No. 2 idler pulley, the camshaft timing pulleys and the timing belt.

41. Install the spark plugs and tighten to 13 ft. lbs. (18 Nm).

42. Install the right side heat insulator and the right side exhaust manifold with a new gasket. Tighten the nuts to 29 ft. lbs. (40 Nm).

43. Install the outside heat insulator and connect the oxygen sensor connector.

44. Install the left side exhaust manifold with a new gasket and tighten the nuts to 29 ft. lbs. (40 Nm). Install the heat insulator.

45. Install the intake manifold with new gaskets. Tighten the nuts and bolts to 13 ft. lbs. (18 Nm).

46. Install the No. 2 idler pulley bracket and tighten to 13 ft. lbs. (18 Nm). Replace the rear cylinder head plate.

47. Install the crossover pipe heat insulator. Replace the water bypass outlet and tighten the mounting bolts to 14 ft. lbs. (19 Nm).

48. Connect the heater water bypass hose, cold start injector time switch connector and the water temperature switch.

49. Connect the upper radiator hose, the water temperature sensor connector and install the water outlet with a new gasket. Tighten the bolts to 73 inch lbs.

50. Install the fuel injectors and the delivery pipes.

51. Install the air intake chamber with new gaskets and tighten the mounting bolts to 32 ft. lbs. (44 Nm).

52. Install the intake chamber brackets.

53. Connect the following hoses:

a. PCV hose

b. Vacuum sensing hose

c. EGR gas temperature sensor connector, California only

d. Air conditioning control valve air hose

54. Install the cold start injector tube and connect the cold start injector connector.

55. Install the exhaust crossover pipe with new gaskets and tighten the mounting bolts to 25 ft. lbs. (34 Nm) and the nuts to 29 ft. lbs. (40 Nm).

56. Install the distributor.

57. Install the vacuum pipe and connect the BVSV vacuum hoses and the air conditioning control valve vacuum hose.

58. Install the EGR valve and vacuum modulator. Tighten the mounting bolts to 13 ft. lbs. (18 Nm). Install the vacuum pipe hoses.

59. Install the EGR pipe with a new gasket and tighten the mounting bolts to 13 ft. lbs. (18 Nm) and the nut to 58 ft. lbs. (77 Nm).

60. Install the throttle body and ISC valve.

61. Raise and support the vehicle. Install the front exhaust pipe and the right side engine undercover. Lower the vehicle.

62. Install the air cleaner hose and cruise control actuator.

63. Install the accelerator cable and connect the throttle cable, if equipped with an automatic transaxle.

64. Install the suspension upper brace and fill the cooling system.

65. Connect the negative battery cable. Run the engine and check for leaks.

66. Adjust the timing. Recheck the fluid levels.

LS400

1. Disconnect the negative battery cable. Drain the cooling system.

2. Remove the camshaft timing pulleys.

3. Disconnect the accelerator cable, the throttle control cable, if equipped with automatic transmission and the cruise control actuator cable.

4. Remove the high tension cord cover and the right side ignition coil.

5. Remove the water inlet housing mounting bolts and disconnect the water bypass hose from the ISC valve.

6. Remove the water inlet and inlet housing assemblies. Remove the O-ring from the water inlet housing.

7. Remove the EGR pipe.

8. Disconnect the following:

a. VSV connector

b. Vacuum pipe hose

c. EGR water bypass pipe

d. Fuel pressure VSV

9. Disconnect the EGR vacuum hoses and remove the EGR VSV.

10. Disconnect the following hoses:

a. Water bypass pipe hose from the ISC valve.

b. Water bypass joint hose.

c. Vacuum pipe hoses.

11. Disconnect the EGR gas temperature sensor, California only. Remove the EGR valve adapter.

12. Disconnect the following:

a. Fuel pressure regulator vacuum hose.

b. Air intake chamber vacuum hose.

c. Vacuum hose from the EVAP BVSV.

13. Remove the mounting bolts, hoses and the vacuum pipe.

14. Remove the ISC valve.

15. Remove the throttle body sensor connectors and the water bypass pipe from the rear water bypass joint.

16. Remove the mounting bolts/nuts and disconnect the PCV valve. Remove the throttle body and gasket.

17. Disconnect the accelerator cable bracket and the brake booster vacuum union and hose.

18. Disconnect the cold start injector connector and the cold start injector tube from the right side delivery pipe.

19. Disconnect the check connector from the intake chamber and remove the mounting nuts and bolts.

20. Remove the air intake chamber and the cold start injector, tube and wire assembly.

21. Disconnect the engine wire from the intake manifold and from the right side cylinder head. Disconnect the heater hoses.

22. Remove the delivery pipes and the fuel injectors. Remove the mounting bolts and nuts. Lift up the intake manifold.

23. Remove the front and rear water bypass joint.

24. Raise and safely support the vehicle. Remove the front exhaust pipe and the main catalytic converters. Lower the vehicle.

25. Disconnect the right side oxygen sensor. Remove the mounting bolts and nuts and remove the right side exhaust manifold.

26. Remove the oil dipstick and guide. Disconnect the left side oxygen sensor.

27. Remove the mounting bolts and nuts and remove the left side exhaust manifold.

28. Remove the 2 engine hangers and the wire brackets from the right side cylinder head.

29. Remove the bolts, washers and the cylinder head cover. Remove the semi-circular plugs, if necessary.

30. Remove the exhaust camshaft of the right side cylinder head by:

a. Position the service bolt hole of the drive sub-gear to the upright position. Secure the camshaft sub-gear to drive gear with a service bolt.

b. Set the timing mark, 1 pointed mark, of the camshaft drive gear at approximately 10 degrees, by turning the camshaft with the proper tool.

c. Alternately, loosen and remove the bearing cap bolts holding the intake camshaft side of the oil feed pipe to the cylinder head.

d. Uniformly, loosen and remove the bearing cap bolts, in sequence.

e. Remove the oil feed pipe and the bearing caps. Remove the camshaft.

31. Remove the intake camshaft from the right side cylinder head by:

a. Set the timing mark, 1 pointed mark, of the camshaft drive gear at approximately 45 degrees, by turning the camshaft with the proper tool.

c. Uniformly, loosen and remove the bearing cap bolts in the proper sequence.

d. Remove the bearing caps, oil seal and the intake camshaft.

32. Remove the exhaust camshaft of the left side cylinder head by:

a. Position the service bolt hole of the drive sub-gear to the upright position. Secure the camshaft sub-gear to drive gear with a service bolt.

NOTE: When removing the camshaft, make sure the torsional spring force of the sub-gear has been eliminated.

b. Set the timing mark, 2 pointed marks, of the camshaft drive gear at approximately 15 degrees, by turning the camshaft with the proper tool.

c. Alternately, loosen and remove the bearing cap bolts holding the intake camshaft side of the oil feed pipe to the cylinder head.

d. Uniformly, loosen and remove the bearing cap bolts in the proper sequence.

e. Remove the oil feed pipe and the bearing caps. Remove the camshaft.

33. Remove the intake camshaft from the left side cylinder head by:

a. Set the timing mark, 1 pointed mark, of the camshaft drive gear at approximately 60 degrees, by turning the camshaft with the proper tool.

c. Uniformly, loosen and remove the bearing cap bolts, in sequence.

d. Remove the bearing caps, oil seal and the intake camshaft.

To install:

34. Remove any old packing and apply new seal packing to the bearing caps.

35. Install the bearing cap on the right side cylinder head, marked I1, in position with the arrow mark facing the rear. Install the bearing cap on the left side cylinder head, marked I6, in position with the arrow mark facing the front.

36. Apply a light coat of oil on the threads of the cap bolts. Install the nearing cap bolts with new washers and tighten to 12 ft. lbs. (16 Nm).

37. Install the right side cylinder head intake camshaft by:

a. Apply MP grease to the thrust portion of the camshaft.

b. Place the intake camshaft at a 45 degree angle of the timing mark (1 pointed mark) on the cylinder head.

c. Remove any old packing and apply new seal packing to the bearing cap marked I6 and install the front bearing cap, marked I6 with the arrow facing rearward.

d. Align the arrows at the front and rear of the cylinder head with the bearing cap.

e. Install the remaining bearing caps in the proper sequence with the arrow mark facing rearward. Install the oil feed pipe and the mounting bolts.

f. Uniformly, tighten the bearing cap bolts in the proper sequence to 12 ft. lbs. (16 Nm).

38. Install the right side cylinder head exhaust camshaft by:

a. Set the timing mark, 1 pointed mark, of the camshaft drive gear at a 10 degree angle by turning the intake camshaft with the proper tool.

b. Apply MP grease to the thrust portion of the camshaft.

c. Align the timing marks, 1 pointed mark, of the camshaft drive and driven gears.

d. Place the exhaust camshaft in the cylinder head. Install the rear bearing cap with the arrow mark facing rearward.

e. Align the arrow marks at the front and rear of the cylinder head with the mark on the bearing cap. Apply a light coat of oil on the threads of the bearing cap bolts.

f. Uniformly, tighten the bearing cap bolts in the proper sequence to 12 ft. lbs. (16 Nm).

g. Bring the service bolt installed upward by turning the camshaft with the proper tool. Remove the service bolt.

39. Install the left side cylinder head intake camshaft by:

a. Apply MP grease to the thrust portion of the camshaft.

b. Place the intake camshaft at a 60 degree angle of the timing mark, 1 pointed mark, on the cylinder head.

c. Remove any old packing and apply new seal packing to the bearing cap marked I6 and install the front bearing cap, marked I1 with the arrow facing rearward.

d. Align the arrows at the front and rear of the cylinder head with the bearing cap. Apply a light coat of oil on the threads of the bearing cap bolts.

e. Install the remaining bearing caps in the proper sequence with the arrow mark facing rearward. Install the oil feed pipe and the mounting bolts.

f. Uniformly, tighten the bearing cap bolts in the proper sequence to 12 ft. lbs. (16 Nm).

40. Install the left side cylinder head exhaust camshaft by:

a. Set the timing mark, 2 dot marks, of the camshaft drive gear at a 15 degree angle by turning the intake camshaft with the proper tool.

b. Apply MP grease to the thrust portion of the camshaft.

c. Align the timing marks, 2 dot marks, of the camshaft drive and driven gears.

d. Place the exhaust camshaft ion the cylinder head. Install the rear bearing cap with the arrow mark facing rearward.

e. Align the arrow marks at the front and rear of the cylinder head with the mark on the bearing cap. Apply a light coat of oil on the threads of the bearing cap bolts.

f. Uniformly, tighten the bearing cap bolts in the proper sequence to 12 ft. lbs. (16 Nm).

g. Bring the service bolt installed upward by turning the camshaft with the proper tool. Remove the service bolt.

41. Install the camshaft oil seals with the proper tool. Install the semi-circular plugs with the proper seal packing.

42. Install the cylinder head covers with the proper seal packing and gasket. Tighten the mounting bolts to 52 inch lbs.

43. Install the engine wire bracket and hangers. Tighten the hanger bolts to 27 ft. lbs. (37 Nm).

44. Install the right side exhaust manifold with a new gasket and tighten the mounting bolts to 29 ft. lbs. (40 Nm). Connect the right side oxygen sensor connector.

45. Install the right side exhaust manifold with a new gasket and tighten the mounting bolts to 29 ft. lbs. (40 Nm). Connect the right side oxygen sensor connector.

46. Install the left side exhaust manifold with a new gasket and tighten the mounting bolts to 29 ft. lbs. (40 Nm). Connect the left side oxygen sensor connector.

47. Install the oil dipstick and guide. Raise and safely support the vehicle.

48. Install the catalytic converters and front exhaust pipe. Lower the vehicle.

49. Install the front and rear water bypass joints. Tighten the mounting bolts to 13 ft. lbs. (18 Nm).

50. Install the intake manifold, using new gaskets. Tighten the mounting nuts and bolts to 13 ft. lbs. (18 Nm).

51. Install the delivery pipes and fuel injectors. Install the fuel return pipe

with new gaskets. Tighten the union bolt to 26 ft. lbs. (36 Nm).

52. Connect the fuel hoses and the injector connectors. Connect the engine wire to the delivery pipes.

53. Connect the connectors on the left side delivery pipe, the water temperature sensor connector, cold start injector time switch connector and the water temperature sender gauge connector.

54. Connect the heater hoses and engine wire bracket. Install the engine wire to the bracket.

55. Install the cold start injector, tube and wire assembly. Tighten the mounting bolts to 69 inch lbs.

56. Install the air intake chamber with new gaskets and tighten the mounting bolts to 13 ft. lbs. (18 Nm).

57. Connect the cold start injector tube to the right side delivery pipe and tighten the union bolt to 11 ft. lbs. (15 Nm).

58. Connect the cold start injector connector. Install the accelerator cable bracket.

59. Install the brake booster union and connect the vacuum hose. Tighten the union bolt to 22 ft. lbs. (30 Nm).

60. Connect the water bypass hose to the throttle body and the PCV hose to the cylinder head cover.

61. Install the throttle body, using a new gasket. Tighten the mounting bolts to 13 ft. lbs. (18 Nm).

62. Install the water bypass pipe and connect the sensor connectors. Install the ISC valve and tighten the mounting bolts to 13 ft. lbs. (18 Nm). Connect the water bypass hose.

63. Install the vacuum pipe and the following hoses:

a. Fuel pressure regulator vacuum hose.

b. Vacuum hose to the upper port of the EVAP BVSV.

c. Air intake chamber vacuum hose.

d. Throttle body vacuum hoses.

64. Install the EGR valve adapter with a new gasket. Tighten the mounting bolts to 13 ft. lbs. (18 Nm).

65. Connect the EGR gas temperature sensor connector (California only).

66. Install the EGR valve and vacuum modulator. Connect the water bypass hoses and the vacuum hoses.

67. Install the EGR and fuel pressure VSV and connect the hoses and connectors. Replace the EGR pipe and tighten the mounting bolts to 13 ft. lbs. (18 Nm).

68. Install the timing belt rear plates and tighten the bolts to 69 inch lbs. Install the water inlet and inlet housing and tighten the bolts to 13 ft. lbs. (18 Nm).

69. Install the right side ignition coil and the high tension cord cover.

70. Connect and adjust the accelerator cable, the automatic transmission throttle cable and the cruise control actuator cable. Install the camshaft timing pulley.

71. Fill the cooling system and connect the negative battery cable. Start the engine and check for leaks.

72. Recheck all the fluid levels and check the ignition timing.

Piston and Connecting Rod

Positioning

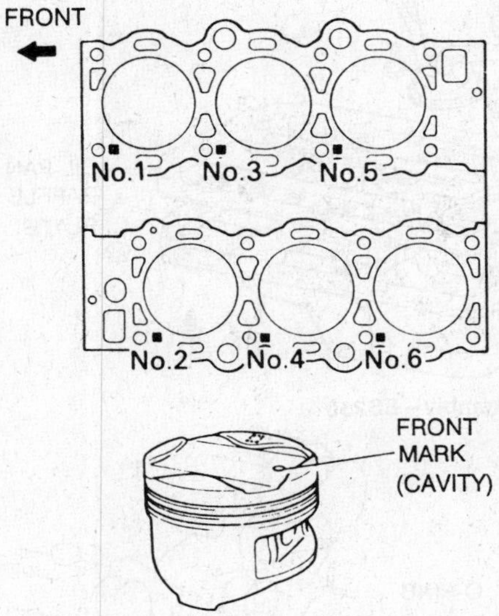

Piston installation location—ES250

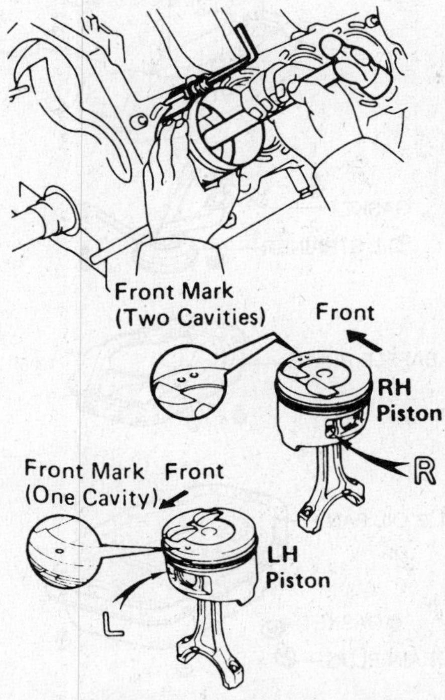

Piston installation location—LS400

ENGINE LUBRICATION

Oil Pan

REMOVAL & INSTALLATION

ES250

1. Disconnect the negative battery cable. Remove the hood assembly.

2. Raise and safely support the vehicle. Remove the engine undercovers.

3. Drain the engine oil, using a suitable container. Remove the suspension lower crossmember.

4. Disconnect the front exhaust pipe and remove the center engine support mount. Remove the front engine mount and bracket.

5. Remove the stiffener plate. Remove the oil dipstick.

6. Remove the mounting bolts and the oil pan assembly. Remove any old packing or sealer from the mounting surfaces.

7. The installation is the reverse of the removal procedure. Tighten the oil pan mounting bolts to 52 inch lbs. Tighten the stiffener plate mounting bolts to 27 ft. lbs. (37 Nm).

LS400

1. Disconnect the negative battery cable.

2. Raise and safely support the vehicle. Remove the engine undercover.

3. Drain the engine oil, using a suitable container.

4. Remove the mounting bolts and drop down the oil pan. Remove any old packing or sealer from the mounting surfaces.

5. The installation is the reverse of the removal procedure. Tighten the oil pan mounting bolts to 69 inch lbs.

Oil Pump

REMOVAL & INSTALLATION

ES250

1. Disconnect the negative battery cable. Remove the oil pan.

2. Remove the oil strainer and gasket.

2. Raise the engine using a suitable chain hoist. Remove the timing belt and pulleys.

3. Remove the alternator and the air conditioning compressor and bracket. Do not disconnect the refrigerant lines.

4. Remove the oil pump from the engine.

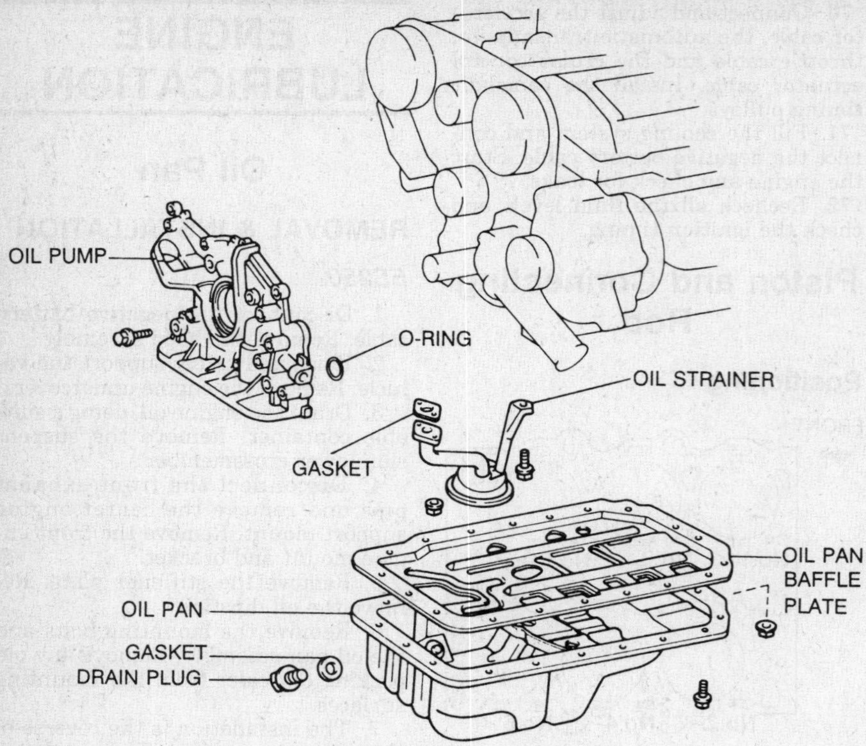

Oil pump assembly – ES250

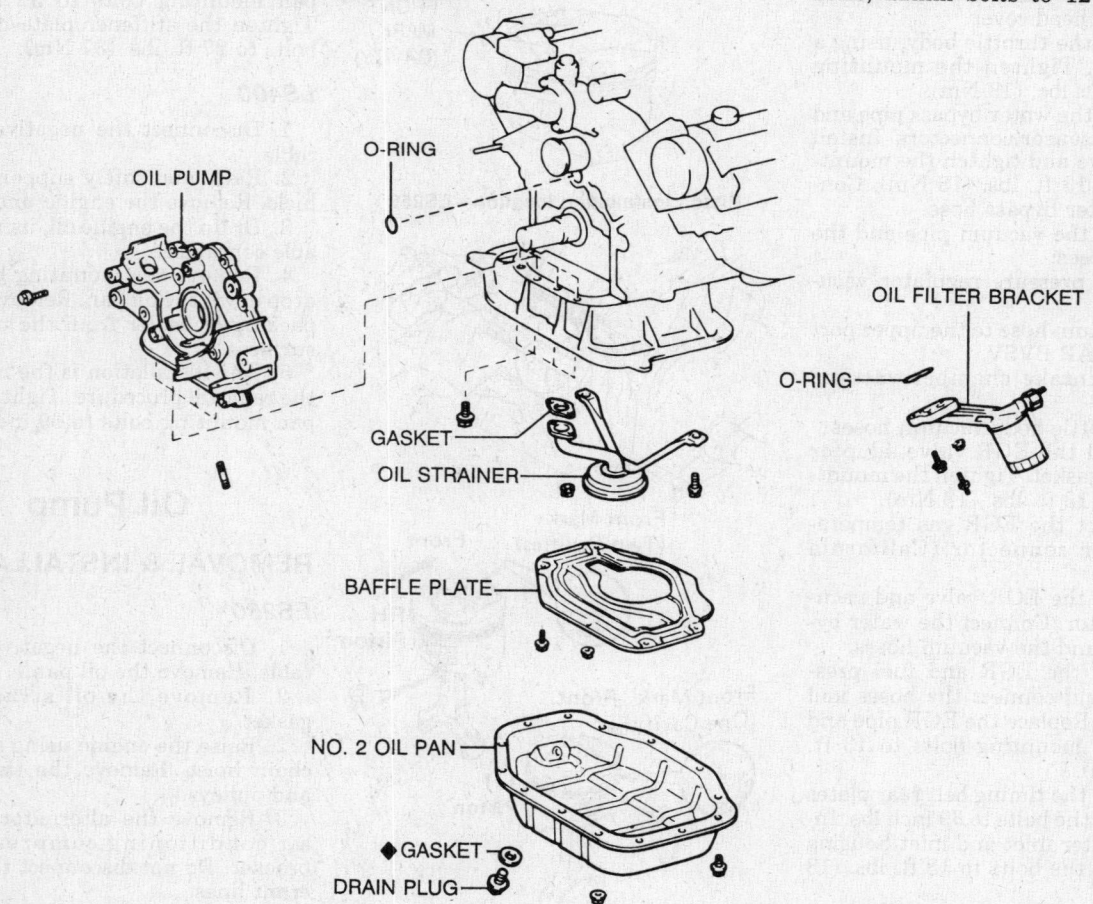

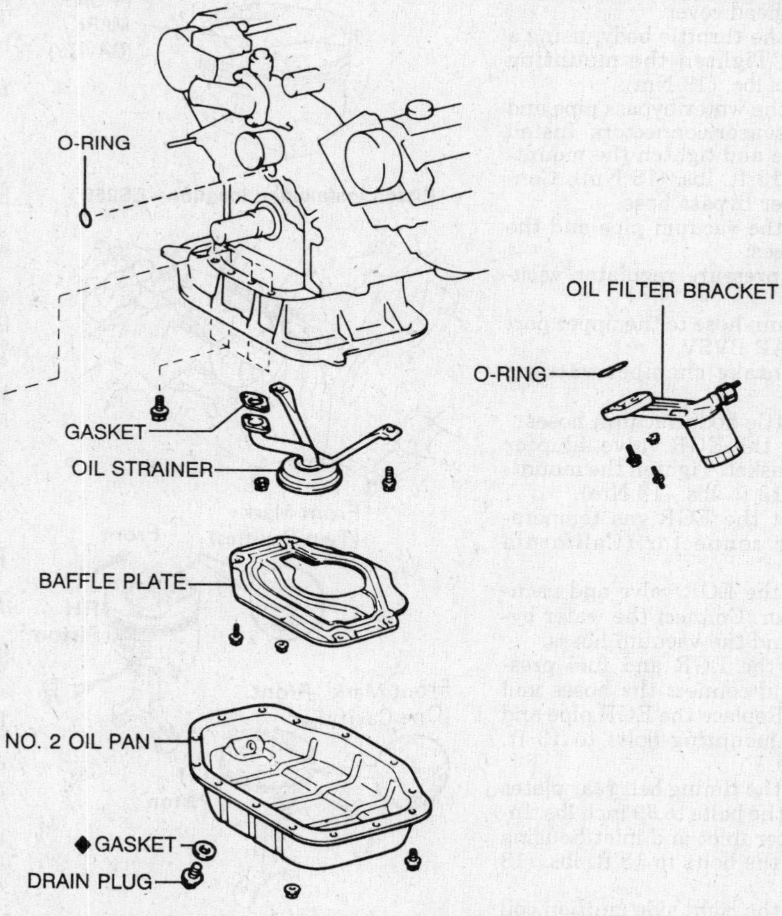

Exploded view of the oil pan – LS400

5. Installation is the reverse of the removal procedure.

LS400

1. Disconnect the negative battery cable. Raise and safely support the vehicle.
2. Remove the engine undercover.
3. Drain the oil, using a suitable container. Remove the oil dipstick.
4. Remove the mounting bolts and pull down the oil pan. Remove the baffle plate and the oil strainer.
5. Remove the timing belt, No. 1 and No. 2 idler and crankshaft pulleys. Remove the bolt and pickup sensor.
6. Remove the stud bolts and mounting bolts. Remove the oil pump with the proper tool to pry away from the cylinder block. Clean and remove any packing from the mounting surfaces.

To install:

7. Install a new O-ring and align the oil drive rotor groove with the pump body mark.
8. Install the pump to the crankshaft with the spine teeth of the drive gear engaged with the large teeth of the crankshaft. Tighten the mounting bolts; 12mm bolts to 12 ft. lbs. (16

Nm) and 14mm bolts to 22 ft. lbs. (30 Nm).

9. Install the pickup sensor with the bolt and tighten to 56 ft. lbs. (76 Nm). Replace the stud bolt.

10. Install the crankshaft timing pulley, No. 1 and No. 2 idler pulley and the timing belt. Install the oil strainer and tighten to 69 inch lbs.

11. Install the baffle plate and the oil pan. Tighten the mounting bolts to 69 inch lbs.

12. Install the engine undercover. Lower the vehicle.

13. Install the dipstick and refill the crankcase.

CHECKING

1. Measure the body clearance between the driven motor and the pump body with the proper tool. The standard body clearance is 0.0039–0.0069 in. The maximum clearance is 0.0118 in. If the body clearance is greater then maximum replace the rotors as a set.

2. Measure the rotor tip clearance between the drive and the driven gears with the proper tool. The standard clearance is 0.0043–0.0118 in. The maximum clearance is 0.0138 in.

3. Measure the side clearance between the rotors, using the proper tool. The standard side clearance is 0.0012–0.0035 in. The maximum side clearance is 0.0059 in. If the side clearance is greater than maximum, replace the rotors as set. If necessary, replace the oil pump assembly.

Rear Main Bearing Oil Seal

REMOVAL & INSTALLATION

1. Disconnect the negative battery cable. Raise and safely support the vehicle.

2. Remove the transmission.

3. Remove the clutch cover assembly and flywheel, if equipped with manual transaxle. Remove the drive plate, if equipped with an automatic transmission or transaxle.

4. Remove the oil seal retaining plate, complete with the oil seal.

5. Using a suitable tool pry the old seal from the retaining plate. Be careful not to damage the plate. The seal is solid type seal.

6. Install the new seal, carefully, by using a block of wood to drift it into place. Do not damage the seal as a leak will result.

7. Lubricate the lips of the seal with multipurpose grease. Installation is the reverse of removal.

ENGINE COOLING

NOTE: On the LS400, a design change has been made to the cylinder head coolant core plug and gasket. The newly designed plug has eliminated the sealant on the flat surface of the plug. There was also some molybdenum disulfide coating added to the plug gasket. Torque the new coolant core plug to 58 ft. lbs. (79 Nm). Always replace the screw plugs and gaskets as a set and only with the new style parts.

Radiator

REMOVAL & INSTALLATION

ES250

1. Disconnect both battery cables and remove the battery. Drain the cooling system.

2. Remove the ignition coil, igniter and bracket assembly.

3. Disconnect the radiator reservoir hose and the radiator hoses.

4. Remove the engine undercover and disconnect the cooling fan connectors.

5. Disconnect and plug the oil cooler lines, if equipped with an automatic transaxle.

6. Remove the mounting bolts, the supports and the radiator with the cooling fans attached.

7. Disconnect the cooling fans from the radiator.

8. The installation is the reverse of the removal procedure.

LS400

1. Disconnect the negative battery cable. Drain the cooling system.

2. Remove the air intake duct and disconnect and plug the automatic transmission lines.

3. Disconnect the cooling fan motor connector.

4. Disconnect the water hose from the coolant reservoir and remove both radiator hoses.

5. Remove the radiator supports and the radiator.

6. Remove both fan shrouds.

7. The installation is the reverse of the removal procedure.

Electric Cooling Fan

TESTING

1. Disconnect the fan motor connector.

2. Connect a suitable jumper wire between the battery and the fan motor connector.

3. If the fan does not run, replace the motor.

REMOVAL & INSTALLATION

ES250

1. Disconnect both battery cables and remove the battery. Drain the cooling system.

2. Remove the ignition coil, igniter and bracket assembly.

3. Disconnect the radiator reservoir hose and the radiator hoses.

4. Remove the engine undercover and disconnect the cooling fan connectors.

5. Disconnect and plug the oil cooler lines, if equipped with an automatic transaxle.

6. Remove the mounting bolts, the supports and the radiator with the cooling fans attached.

7. Disconnect the cooling fans from the radiator.

8. The installation is the reverse of the removal procedure.

LS400

1. Disconnect the negative battery cable. Remove the air cleaner cover on the right side.

2. Remove the clearance light and disconnect the electrical connector.

3. Remove the mounting bolts and nut. Remove the headlight together with the fog light.

4. Remove the parking light and disconnect the wire connector. Remove the engine undercover.

5. Disconnect the tube from the wind guide. Remove the mounting screws and lift off the guide.

6. Disconnect the mounting bolts. Remove the bumper and bumper retainer.

7. Remove the bumper reinforcement and both horn assemblies.

8. Disconnect the fan motor connectors.

9. Remove the mounting bolts and disconnect the wire from the mounting brackets. Remove the cooling fan(s).

To install:

10. Install the cooling fan(s) and tighten the mounting bolts. Connect the wire to the mounting brackets.

11. Connect the fan motor connector. Install the horn assemblies.

12. Replace the bumper reinforcement and replace the bumper and retainer.

13. The remainder of the installation is the reverse of the removal procedure.

Heater Core

REMOVAL & INSTALLATION

ES250

1. Disconnect the negative battery cable. Drain the cooling system.
2. Remove the console, if equipped, by removing the shift knob (manual), wiring connector and console attaching screws.
3. Remove the carpeting from the tunnel.
4. If necessary, remove the cigarette lighter and ash tray.
5. Remove the package tray, if access to the heater core is difficult.
6. Remove the bottom cover/intake assembly screws and withdraw the assembly.
7. Remove the cover from the water valve.
8. Remove the water valve.
9. Remove the hose clamps and remove the hoses from the core.
10. Remove the heater core.
11. Installation is the reverse of the removal procedure. Fill the cooling system to the proper level. Operate the heater and check for leaks.

LS400

1. Disconnect the negative battery cable. Drain the cooling system.
2. Properly discharge the air conditioning system.
3. Disconnect the mounting nuts and hoses. Remove the heater valve.
4. Remove the cooling and blower unit.
5. Disconnect the inlet and outlet water hoses. Remove the mounting nuts and the insulator retainer.
6. Remove the instrument cluster and the radio assembly with the air conditioning control attached.
7. Remove the undercover.

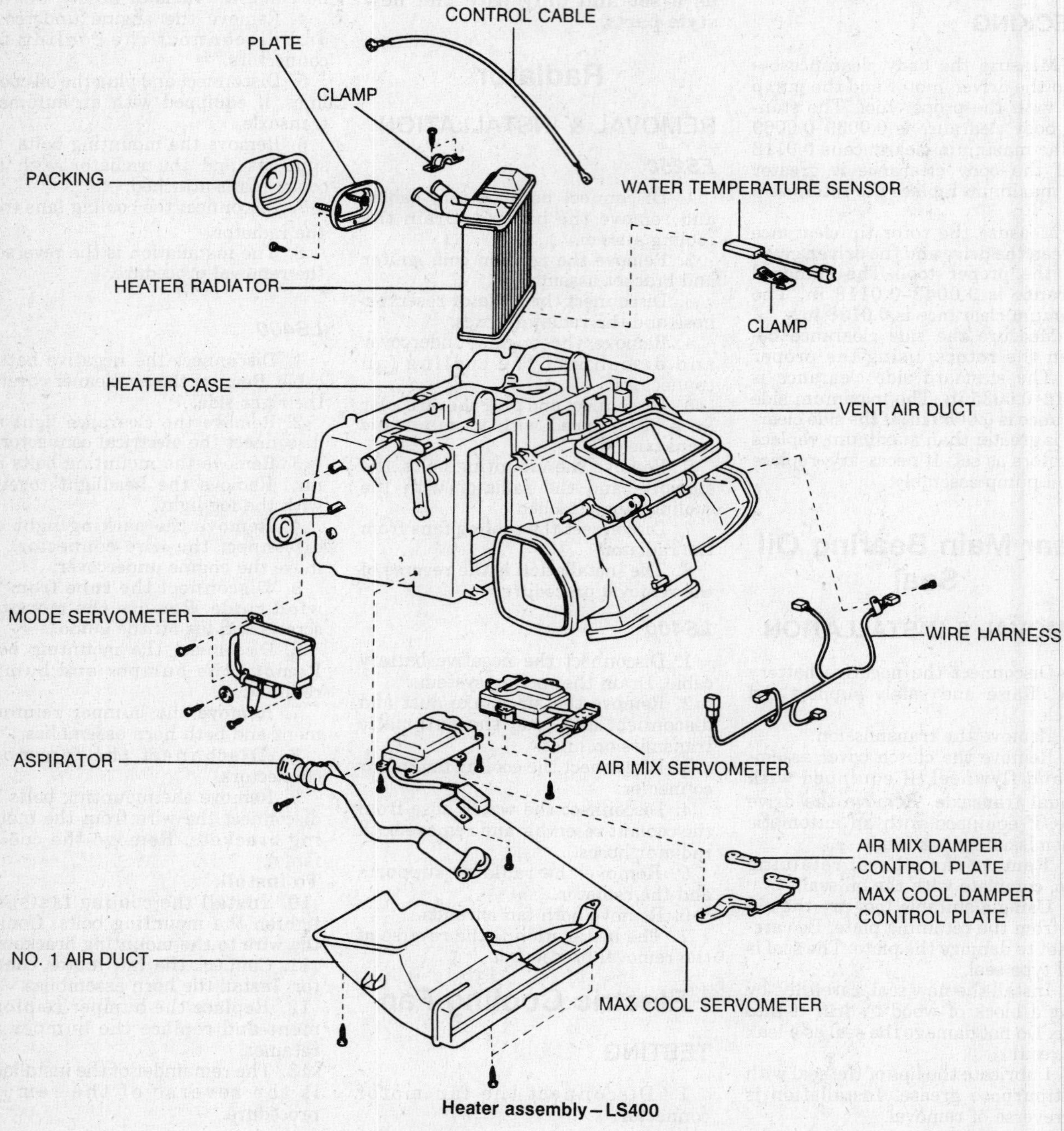

Heater assembly – LS400

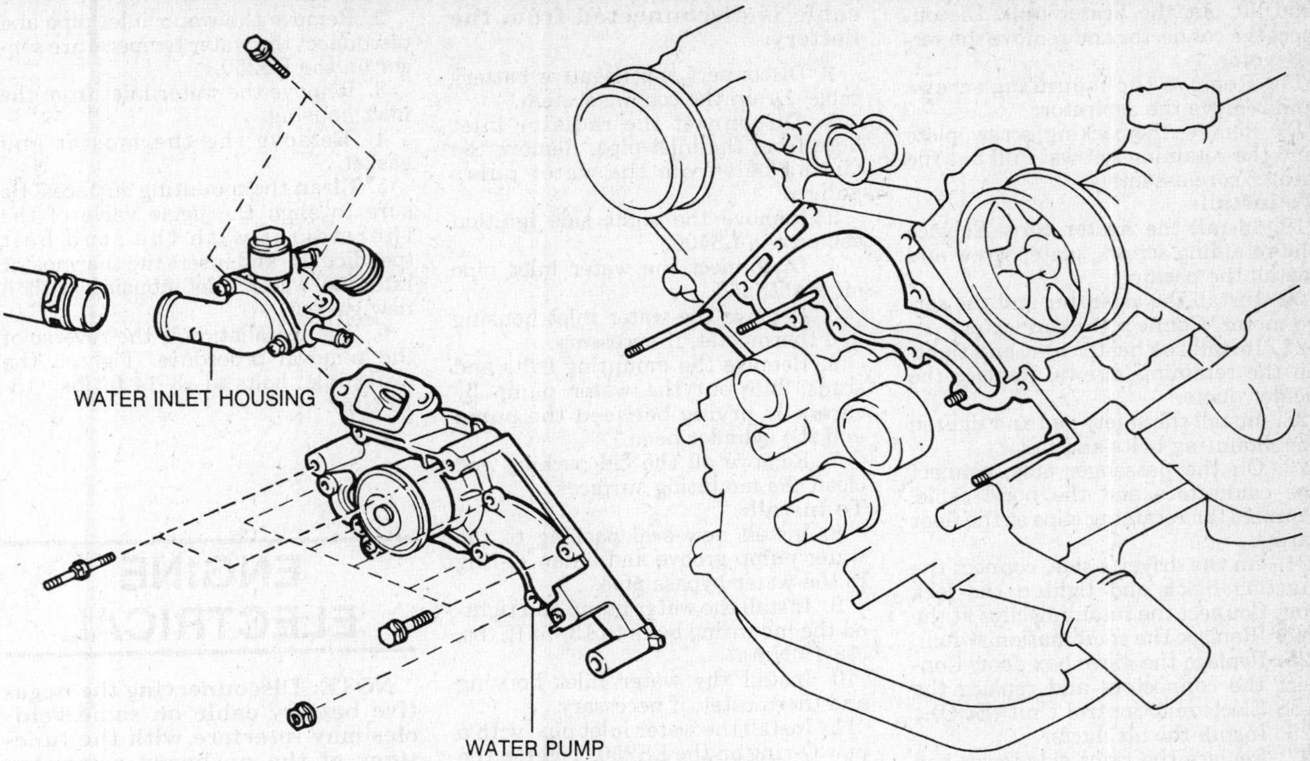

Water pump assembly—LS400

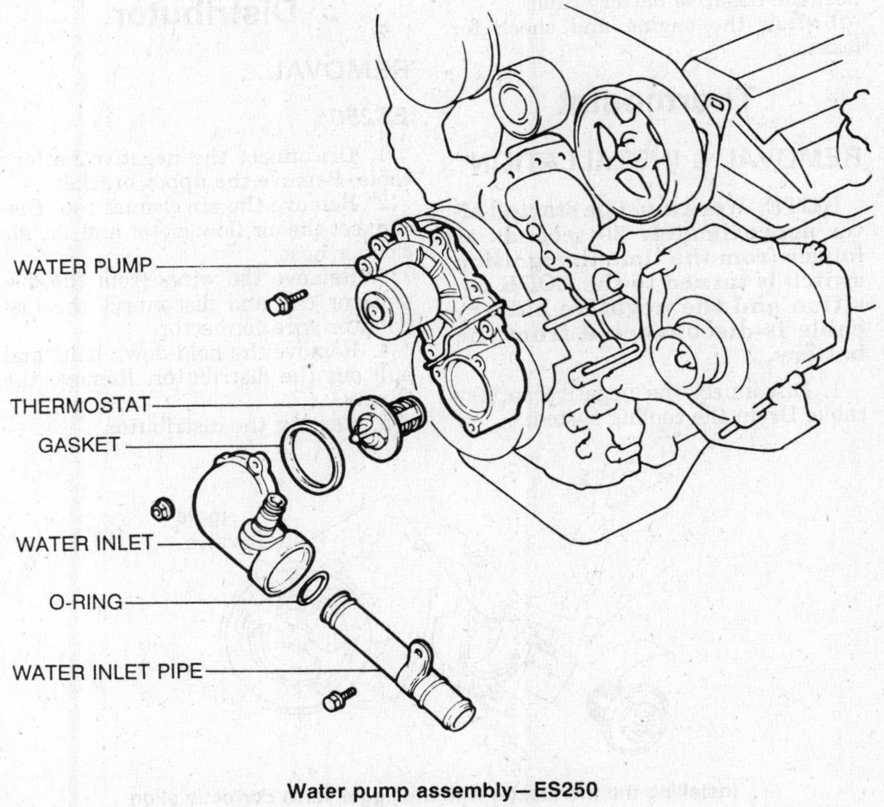

Water pump assembly—ES250

8. Remove the glove box by performing the following:

a. Remove the glove box compartment panel and disconnect the left side check arm from the door.

b. Remove the retaining clips. Insert the proper tool between the upper side of the compartment and the safety pad, pry out the compartment to remove.

c. Disconnect the connector from the glove box compartment.

9. Disconnect the mounting bolts and remove the right side lower pad. remove the connectors from the pad.

10. Remove the glove box door. Disconnect the connectors and remove the ABS Electronic Control Unit (ECU).

11. Remove the air ducts.

12. On the driver's side, loosen the lock bolt and disconnect the junction block. Disconnect the retaining clips at the floor. Remove the combination switch.

13. On the passenger side, disconnect the connectors and the bond cable. Disconnect the retaining clips at the floor carpet.

14. Remove the bolts and nut from the safety pad and lift out the pad.

15. Remove the heater ducts.

16. Remove the mounting screws

and lift out the heater unit. Disconnect the connector and remove the servo motor.

17. Remove the mounting screws and remove the aspirator.

18. Remove the packing, screw, plate and the retaining screws. Pull out the heater core assembly.

To install:

19. Install the heater core. Replace the retaining screws, plate, screw and install the packing.

20. Install the aspirator and the servo motor. Connect the connector.

21. Install the heater unit and tighten the retaining screws. Replace the heater ducts.

22. Install the safety pad and tighten the mounting bolts and nut.

23. On the passenger side, connect the connectors and the bond cable. Connect the retaining clips at the floor carpet.

24. On the driver's side, connect the junction block and tighten the lock bolt. Connect the retaining clips at the floor. Replace the combination switch.

25. Replace the glove box door. Connect the connectors and replace the ABS Electronic Control Unit (ECU).

26. Install the air ducts.

27. Replace the right side lower pad and tighten the mounting bolts. Replace the connectors to the pad.

28. Replace the glove box by performing the following:

a. Connect the connector to the glove box compartment.

b. Install the glove box and replace the retaining clips.

c. Replace the glove box compartment panel and connect the left side check arm from the door.

29. Install the undercover.

30. Replace the instrument cluster and the radio assembly with the air conditioning control attached.

31. Connect the inlet and outlet water hoses. Replace the mounting nuts and the insulator retainer.

32. Replace the cooling and blower unit.

33. Install the heater valve. Connect the mounting nuts and hoses.

34. Properly, evacuate and recharge the air conditioning system.

35. Refill the cooling system and connect the negative battery cable. Run the engine and check for leaks.

Water Pump

REMOVAL & INSTALLATION

NOTE: Work must be started after approximately 20 seconds or longer from the time the ignition switch is turned to the LOCK position and the negative battery cable is disconnected from the battery.

1. Disconnect the negative battery cable. Drain the cooling system.

2. Disconnect the radiator inlet hose from the inlet pipe. Remove the timing belt from the water pump pulley.

3. Remove the right side ignition coil on the LS400.

4. Disconnect the water inlet pipe on the ES250.

5. Remove the water inlet housing and thermostat, if necessary.

6. Remove the mounting bolts and studs. Lift out the water pump by carefully prying between the pump and the cylinder head.

7. Remove all the old packing and clean the mounting surfaces.

To install:

8. Install new seal packing to the water pump groove and a new O-ring to the water bypass pipe.

9. Install the water pump and tighten the mounting bolts to 13–14 ft. lbs. (18-19 Nm).

10. Install the water inlet housing and thermostat, if necessary.

11. Install the water inlet pipe with a new O-ring on the ES250. Tighten the mounting bolt to 14 ft. lbs. (19 Nm).

12. Replace the ignition coil on the LS400.

13. Install the timing belt and connect the inlet hose to the inlet pipe.

14. Refill the cooling system and connect the negative battery cable.

15. Run the engine and check for leaks.

Thermostat

REMOVAL & INSTALLATION

NOTE: Work must be started after approximately 20 seconds or longer from the time the ignition switch is turned to the LOCK position and the negative battery cable is disconnected from the battery.

1. Disconnect the negative battery cable. Drain the cooling system.

2. Remove the water inlet pipe and disconnect the water temperature sensor on the ES250.

3. Remove the water inlet from the inlet housing.

4. Remove the thermostat and gasket.

5. Clean the mounting surfaces. Be sure to align the jiggle valve of the thermostat with the stud bolt (pendicular) and insert the thermostat into the water inlet housing with a new gasket.

6. The installation is the reverse of the removal procedure. Tighten the water inlet bolts to 13–14 ft. lbs. (18-19 Nm).

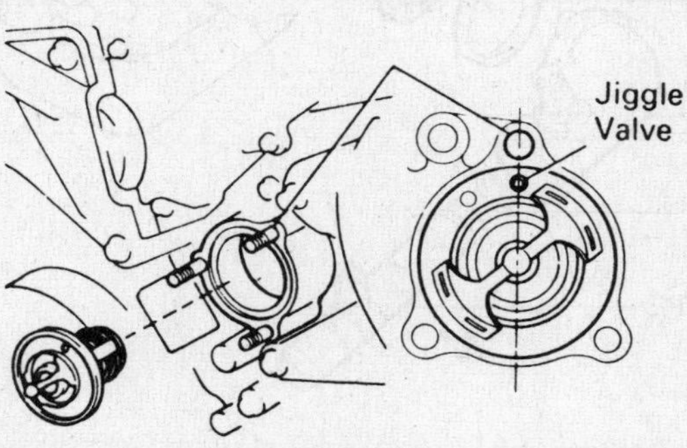

Installing the thermostat with the jiggle valve correctly align

ENGINE ELECTRICAL

NOTE: Disconnecting the negative battery cable on some vehicles may interfere with the functions of the on board computer systems and may require the computer to undergo a relearning process, once the negative battery cable is reconnected.

Distributor

REMOVAL

ES250

1. Disconnect the negative battery cable. Remove the upper bracket.

2. Remove the air cleaner top. Disconnect the air flow meter and the air cleaner hose.

3. Remove the wires from the distributor cap and disconnect the distributor wire connector.

4. Remove the hold-down bolts and pull out the distributor. Remove the O-ring.

5. Remove the distributor.

LS400

RIGHT

1. Disconnect the negative battery cable. Remove the air duct assembly.
2. Disconnect the air flow meter connector and air hose.
3. Remove the air flow meter assembly and the throttle body cover.
4. Disconnect the ISC and power steering idle-up air hose. Remove the No. 1 air hose.
5. Remove the high tension cable upper cover and the right side engine wire cover.
6. Disconnect the mounting bolts and remove the No. 3 timing belt cover.
7. Disconnect the sensor connector. Remove the mounting bolts, the wire cover and take off the No. 2 timing belt cover.
8. Disconnect the electrical wires from the distributor cap.
9. Remove the distributor cap and the rotor.
10. Remove the mounting bolts and lift out the distributor housing.

LEFT

1. Disconnect the negative battery cable. Drain the cooling system and remove the engine wire cover.
2. Remove the No. 2 junction block cover and the No. 3 timing belt cover.
3. Disconnect the water inlet housing hose and the reservoir tank hose.
4. Remove the mounting bolts and the water pipe from the No. 2 timing belt cover.
5. Disconnect the sensor connector, the connector boot and remove the No. 2 timing belt cover.
6. Disconnect the electrical connections from the cap and the housing. Remove the distributor cap and the rotor.
7. Remove the mounting screws and lift out the distributor housing.

INSTALLATION

Timing Not Disturbed

ES250

1. Install a new O-ring in the housing. Lubricate the O-ring with engine oil.
2. Align the cutout marks of the coupling and the housing.
3. Insert the distributor, aligning the line of the distributor housing with the cutout of the distributor attachment bearing cap. Tighten the left side hold-down bolts.
4. Install the rotor and the distributor cap. Connect the cables to the distributor cap. Align the spline of the distributor cap with the spline groove of the holder.
5. Connect the electrical connections to the distributor. Replace the air cleaner cap, air flow meter and the air cleaner hose.
6. Install the upper bracket and connect the negative battery cable.
7. Adjust the timing.

LS400 – RIGHT

1. Install the distributor housing. Replace the rotor and the distributor cap.
2. Connect the ignition cables to the distributor cap.
3. Connect the sensor connector. Install the No. 2 timing belt cover and boots.
4. Install the No. 3 timing belt cover and engine wire cover. Tighten the mounting bolts.
5. Install the electrical connection upper cover and the upper throttle body cover.
6. Replace the No. 1 air hose, the PS idle-up air hose and the ISC air hose.
7. Install the air flow meter assembly and connect the air hose and the meter connector.
8. Connect the negative battery cable. Adjust the timing.

LS400 – LEFT

1. Install the distributor housing and tighten the mounting bolts.
2. Install the rotor and replace the distributor cap.
3. Connect the ignition cables to the distributor cap.
4. Connect the sensor connector. Install the No. 2 timing belt cover and boots.
5. Connect the water inlet housing hose and the reservoir tank hose.
4. Replace the water pipe to the No. 2 timing belt cover.
6. Install the No. 3 timing belt cover and replace the No. 2 junction block cover.
7. Install the engine wire cover and tighten the mounting bolts.
8. Refill the cooling system and connect the negative battery cable. Adjust the timing.

Timing Disturbed

1. Turn the crankshaft pulley and the groove on the pulley with the timing mark **0** of the No. 1 timing belt cover.
2. Check that the timing marks of the camshaft timing pulleys and the No. 3 timing belt cover are aligned. If not, turn the crankshaft 1 revolution (360 degrees).
3. Position the slit of the intake camshaft (right side cylinder head) in the proper position.
4. Install the distributor following the proper procedure.

Ignition Timing

ADJUSTMENT

1. Allow the engine to reach normal operating temperature.
2. Connect a tachometer to terminal IG (-) of the check connector.

NOTE: Never allow the tachometer test probe to touch ground as it could result in damage to the igniter and or ignition coil. As some tachometers are not compatible with this ignition system, it is recommended to confirm the compatibility of the unit before use.

3. Check the idle speed, 650–750 rpm on the ES250 or 600–700 rpm on the LS400.
4. Connect the proper jumper wire to terminals TE1 and E1 of the check connector.
5. Connect the timing light to No. 1 cylinder on the ES250 and No. 6 cylinder on the LS400.
6. Start the engine and check the timing with the transmission in **N** position.
7. The ignition timing should be 10 degrees BTDC at idle on the ES250 or 8–12 degrees BTDC at idle on the LS400.
8. If not within specifications, loosen the hold-down bolts and adjust the timing by turning the distributor.
9. Tighten the hold-down bolts and recheck the timing and idle speed, adjust as necessary. Remove the jumper wires and test equipment.

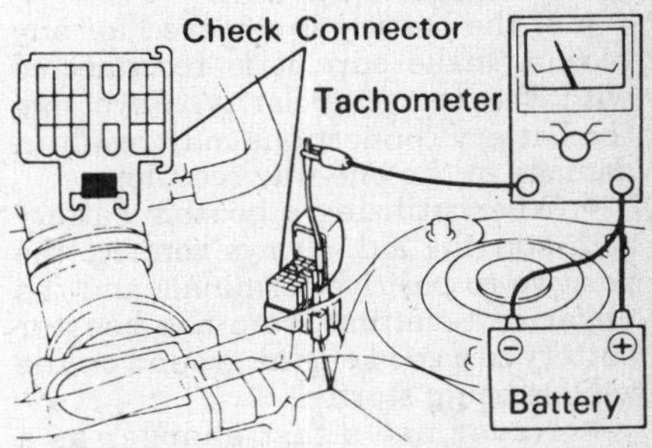

Connecting the tachometer into the check connector – ES250

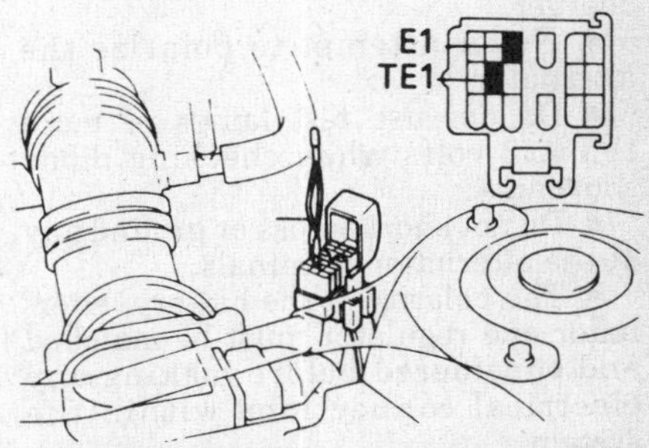

Installing the jumper wire into the TE1 and E1 terminals of the check connector – ES250

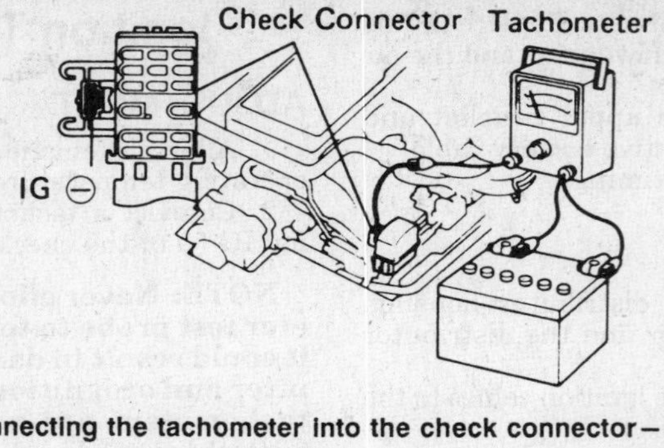

Connecting the tachometer into the check connector—LS400

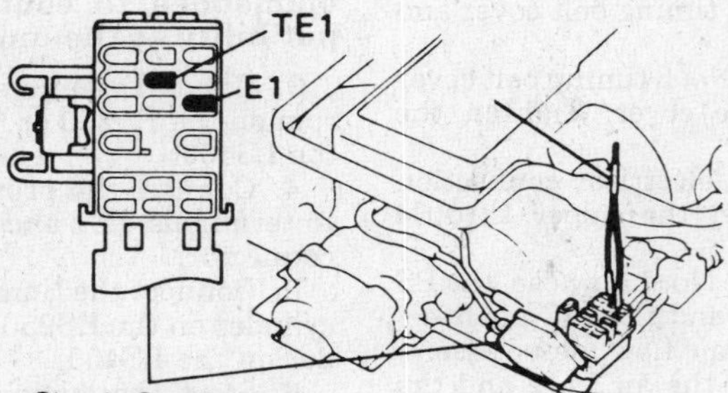

Installing the jumper wire into the TE1 and E1 terminals of the check connector—LS400

Alternator

PRECAUTIONS

Several precautions must be observed with alternator equipped vehicles to avoid damage to the unit.

- If the battery is removed for any reason, make sure it is reconnected with the correct polarity. Reversing the battery connections may result in damage to the one-way rectifiers.
- When utilizing a booster battery as a starting aid, always connect the positive to positive terminals and the negative terminal from the booster battery to a good engine ground on the vehicle being started.
- Never use a fast charger as a booster to start vehicles.
- Disconnect the battery cables when charging the battery with a fast charger.
- Never attempt to polarize the alternator.
- Do not use test lamps of more than 12 volts when checking diode continuity.
- Do not short across or ground any of the alternator terminals.
- The polarity of the battery, alternator and regulator must be matched and considered before making any electrical connections within the system.
- Never separate the alternator on an open circuit. Make sure all connections within the circuit are clean and tight.
- Disconnect the battery ground terminal when performing any service on electrical components.
- Disconnect the battery if arc welding is to be done on the vehicle.

BELT TENSION ADJUSTMENT

A belt tensioner is used to maintain the proper amount of pressure on the drive belt on the LS400. On the ES250:

1. Disconnect the negative battery cable.
2. Loosen the adjusting lock bolt and the pivot bolt.
3. Move the alternator to adjust the belt tension to 5 lbs. on a new belt or 20 lbs. on a used belt.
4. Tighten the adjusting and pivot bolts.
5. Connect the negative battery cable.

REMOVAL & INSTALLATION

ES250

1. Disconnect the negative battery cable.

NOTE: Work must be started after approximately 20 seconds or longer from the time the ignition switch is turned to the LOCK position and the negative battery cable is disconnected.

2. Remove the mounting bolts and disconnect the electrical connections at the alternator.
3. Disconnect the wiring harness from the mounting clip.
4. Remove the drive belt.
5. Remove the adjusting lock bolt, the pivot bolt and remove the alternator.

To install:

6. Mount the alternator on the alternator bracket with the pivot bolt and the adjusting bolt. Do not tighten.
7. Install the drive belt and adjust the belt tension.
8. Connect the electrical connections at the alternator and the wiring harness to the mounting clip.
9. Replace the mounting bolts and tighten to 64 ft. lbs. (87 Nm).
10. Connect the negative battery cable.

LS400

1. Disconnect the negative battery cable. Turn the ignition switch to the **LOCK** position.
2. Remove the drive belt and the engine undercover.
3. Disconnect the electrical connections at the alternator.
4. Remove the through bolt and nut. Remove the alternator.
5. The installation is the reverse of the removal procedure.

NOTE:On the LS400, there has been a problem on occasion with the alternator connector not being fully inserted into its socket. Should this condition exist, the vehicle may experience diminished alternator output which in turn will effect the battery charging system. When installing the connector to the alternator socket, be sure to reconnect the connector to the socket until a click is heard. Fasten the rubber boot and check to see that the rubber boot is fastened properly.

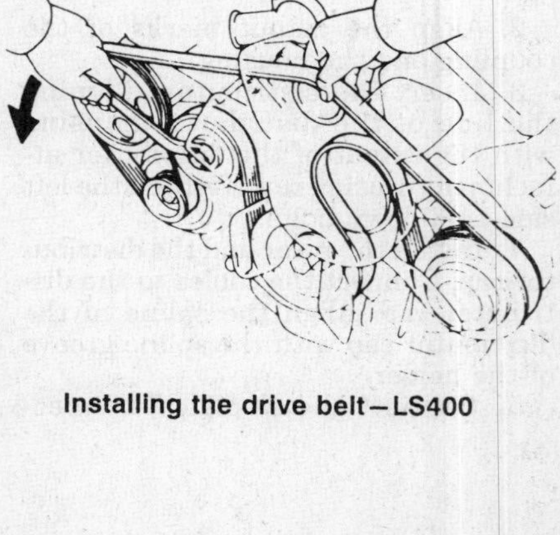

Installing the drive belt—LS400

Starter

REMOVAL & INSTALLATION

ES250

1. Disconnect both battery cables. Remove the battery and battery tray.

NOTE: Work must be started after approximately 20 seconds or longer from the time the ignition switch is turned to the LOCK position and the negative battery cable is disconnected.

2. Disconnect the noise filter connector, the igniter connector, the coil wire and the ground strap.
3. Disconnect the harness clamp and remove the igniter bracket.
4. Disconnect the electrical connections at the starter.
5. Remove the mounting bolts and the starter motor.

To install:

6. Install the starter and tighten the mounting bolts to 29 ft. lbs. (40 Nm).
7. Connect the electrical connections at the starter.
8. Install the igniter bracket and connect the harness clamp.
9. Connect noise filter connector, igniter connector, the coil wire and the ground strap.
10. Install the battery tray and the battery.
11. Connect the battery cables.

LS400

1. Disconnect the negative battery cable.

NOTE: Work must be started after approximately 20 seconds or longer from the time the ignition switch is turned to the LOCK position and the negative battery cable is disconnected.

2. Remove the intake chamber and the intake manifold.
3. Disconnect the electrical connections at the starter.
4. Remove the mounting bolts and the starter motor.

To install:

5. Install the starter and tighten the mounting bolts to 29 ft. lbs. (40 Nm).
6. Connect the electrical connections at the starter.
7. Install the intake manifold and the intake manifold.
8. Connect the negative battery cable.

EMISSION CONTROLS

Please refer to "Emission Controls" in the Unit Repair section for system maintenance procedures. Due to the complex nature of modern electronic engine control systems, comprehensive diagnosis and testing procedures fall outside the confines of this repair manual. For complete information on diagnosis, testing and repair procedures concerning all modern engine and emission control systems, please refer to "Chilton's Guide to Fuel Injection and Electronic Engine Controls".

FUEL SYSTEM

Fuel System Service Precautions

Safety is the most important factor when performing not only fuel system maintenance but any type of maintenance. Failure to conduct maintenance and repairs in a safe manner may result in serious personal injury or death. Maintenance and testing of the vehicle's fuel system components can be accomplished safely and effectively by adhering to the following rules and guidelines.

- To avoid the possibility of fire and personal injury, always disconnect the negative battery cable unless the repair or test procedure requires that battery voltage be applied.
- Always relieve the fuel system pressure prior to disconnecting any fuel system component (injector, fuel rail, pressure regulator, etc.), fitting or fuel line connection. Exercise extreme caution whenever relieving fuel system pressure to avoid exposing skin, face and eyes to fuel spray. Please be advised that fuel under pressure may penetrate the skin or any part of the body that it contacts.
- Always place a shop towel or cloth around the fitting or connection prior to loosening to absorb any excess fuel due to spillage. Ensure that all fuel spillage (should it occur) is quickly removed from engine surfaces. Ensure that all fuel soaked cloths or towels are deposited into a suitable waste container.
- Always keep a dry chemical (Class B) fire extinguisher near the work area.
- Do not allow fuel spray or fuel vapors to come into contact with a spark or open flame.
- Always use a backup wrench when loosening and tightening fuel line connection fittings. This will prevent unnecessary stress and torsion to fuel line piping. Always follow the proper torque specifications.
- Always replace worn fuel fitting O-rings with new. Do not substitute fuel hose or equivalent where fuel pipe is installed.

RELIEVING FUEL SYSTEM PRESSURE

1. Be sure the engine is cold.
2. Relieve the fuel pressure by slowly loosening the connection at the pressure regulator.
3. Be sure to place a rag under the pressure regulator to prevent the fuel from spilling on the engine.
4. Tighten the connection.

Fuel Tank

REMOVAL & INSTALLATION

NOTE: Before removing fuel system parts, clean them with a spray-type engine cleaner. Follow the instructions on the cleaner. Do not soak fuel system parts in liquid cleaning solvent.

CAUTION

The fuel injection system is under pressure. Release pressure slowly and contain spillage. Observe no smoking/no open flame precautions. Have a Class B-C (dry powder) fire extinguisher within arm's reach at all times.

ES250

1. Relieve the fuel system pressure. Remove the filler cap.
2. Using a siphon or pump, drain the fuel from the tank and store it in a proper metal container with a tight cap.
3. Disconnect the fuel pump and sending unit wiring at the connector.
4. Raise the vehicle and safely support it.
5. Loosen the clamp and remove the filler neck and overflow pipe from the tank.
6. Remove the supply hose from the tank. Wrap a rag around the fitting to collect escaping fuel. Disconnect the breather hose from the tank, again using a rag to control spillage.
7. Cover or plug the end of each dis-

connected line to keep dirt out and fuel in.

8. Support the fuel tank with a floor jack or transmission jack. Use a broad piece of wood to distribute the load. Be careful not to deform the bottom of the tank.

9. Remove the fuel tank protectors.

10. Remove the fuel tank support strap bolts.

11. Swing the straps away from the tank and lower the jack. Balance the tank. The tank is bulky and may have some fuel left in it. If its balance changes suddenly, the tank may fall.

12. Remove the fuel filler pipe extension, the breather pipe assembly and the sending unit assembly. Keep these items in a clean, protected area away from the car.

13. While the tank is out and disassembled, inspect it for any signs of rust, leakage or metal damage. If any problem is found, replace the tank. Clean the inside of the tank with water and a light detergent and rinse the tank thoroughly several times.

14. Inspect all of the lines, hoses and fittings for any sign of corrosion, wear or damage to the surfaces. Check the pump outlet hose and the filter for restrictions.

15. When reassembling, always replace the sealing gaskets with new ones. Also replace any rubber parts showing any sign of deterioration.

To install:

16. Connect the breather pipe assembly and the filler pipe extension.

NOTE: Tighten the breather pipe screw to 17 inch lbs. (2.0 Nm) and all other attaching screws to 35 inch lbs. (3.9 Nm).

17. Place the fuel tank on the jack and elevate it into place within the car. Attach the straps and install the strap bolts, tightening them to 29 ft. lbs. (40 Nm).

18. Install the fuel tank protectors.

19. Connect the breather hose to the tank pipe, the return hose to the tank pipe and the supply hose to its tank pipe. tighten the supply hose fitting to 21 ft. lbs. (28 Nm).

20. Connect the filler neck and overflow pipe to the tank. Make sure the clamps are properly seated and secure.

21. Lower the vehicle to the ground.

22. Connect the pump and sending unit electrical connectors to the harness.

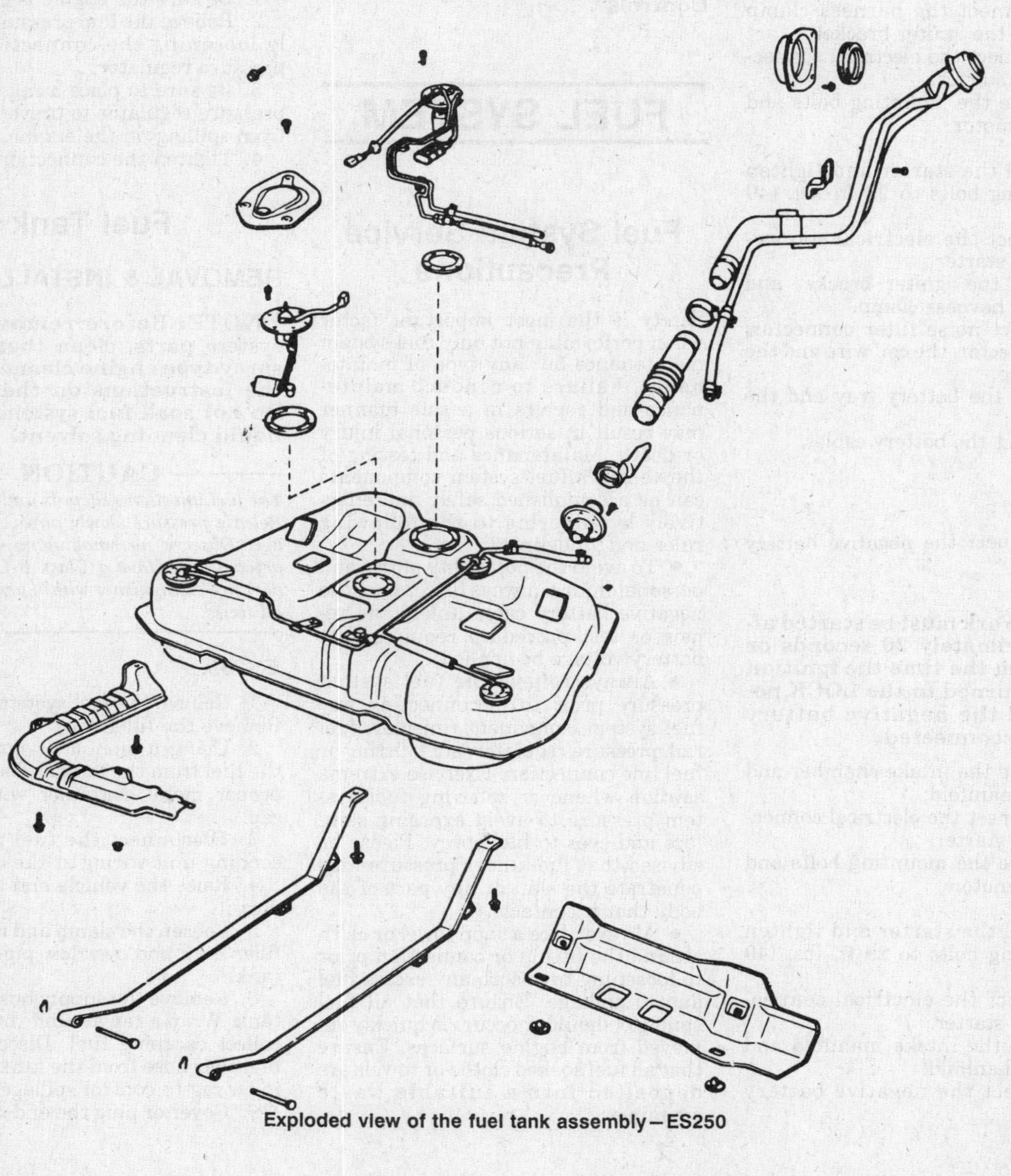

Exploded view of the fuel tank assembly – ES250

23. Using a funnel, pour the fuel that was drained from its container into the fuel filler.
24. Install the fuel filler cap.
25. Start the engine and check carefully for any sign of leakage around the tank and lines.

LS400

1. Relieve the fuel system pressure. Remove the filler cap.
2. Using a siphon or pump, drain the fuel from the tank and store it in a proper metal container with a tight cap.
3. Disconnect the fuel pump and sending unit wiring at the connectors.
4. Raise the vehicle and safely support it.
5. Loosen the clamp and remove the filler neck and overflow pipe from the tank.
6. Remove the supply hose from the tank. Wrap a rag around the fitting to collect escaping fuel. Disconnect the breather hose from the tank, again using a rag to control spillage.
7. Cover or plug the end of each disconnected line to keep dirt out and fuel in.
8. Support the fuel tank with a floor jack or transmission jack. Use a broad piece of wood to distribute the load. Be careful not to deform the bottom of the tank.
9. Remove the No. 1 and No. 2 fuel tank cover cases.
10. Remove the fuel tank retaining nuts and bolts.
11. Lower the jack and balance the tank with by hand. The tank is bulky and may have some fuel left in it. If its balance changes suddenly, the tank may fall.
12. Remove the fuel filler pipe extension, the breather pipe assembly and the sending unit assembly. Keep these items in a clean, protected area away from the car.
13. While the tank is out and disassembled, inspect it for any signs of rust, leakage or metal damage. If any problem is found, replace the tank. Clean the inside of the tank with water and a light detergent and rinse the tank thoroughly several times.
14. Inspect all of the lines, hoses and fittings for any sign of corrosion, wear or damage to the surfaces. Check the pump outlet hose and the filter for restrictions.
15. When reassembling, always replace the sealing gaskets and O-rings with new ones. Also replace any rubber parts showing any sign of deterioration.

To install:

16. Connect the breather pipe assembly and the filler pipe extension.

NOTE: Tighten the breather pipe screw to 26 in. lbs. (2.9 Nm) and all other attaching screws to 35 in. lbs. (3.9 Nm).

17. Place the fuel tank on the jack and elevate it into place within the car. Attach the retaining nuts and bolts, tightening them to 18 ft. lbs. (25 Nm).
18. Install the No. 1 and No. 2 fuel tank cover cases.
19. Connect the breather hose to the tank pipe, the return hose to the tank pipe and the supply hose to its tank pipe. Torque the inlet tank pipe screws to 26 in. lbs. (2.9 Nm). Install the fuel main tank tube and torque the bolt to 22 ft. lbs. (29 Nm). Install the No. 2 fuel return tank tube and torque the bolt to 18 ft. lbs. (25 Nm).
20. Connect the filler neck and overflow pipe to the tank. Make sure the clamps are properly seated and secure.
21. Lower the vehicle to the ground.
22. Connect the pump and sending unit electrical connectors to the harness.
23. Using a funnel, pour the fuel that was drained from its container into the fuel filler.
24. Install the fuel filler cap.
25. Start the engine and check carefully for any sign of leakage around the tank and lines.

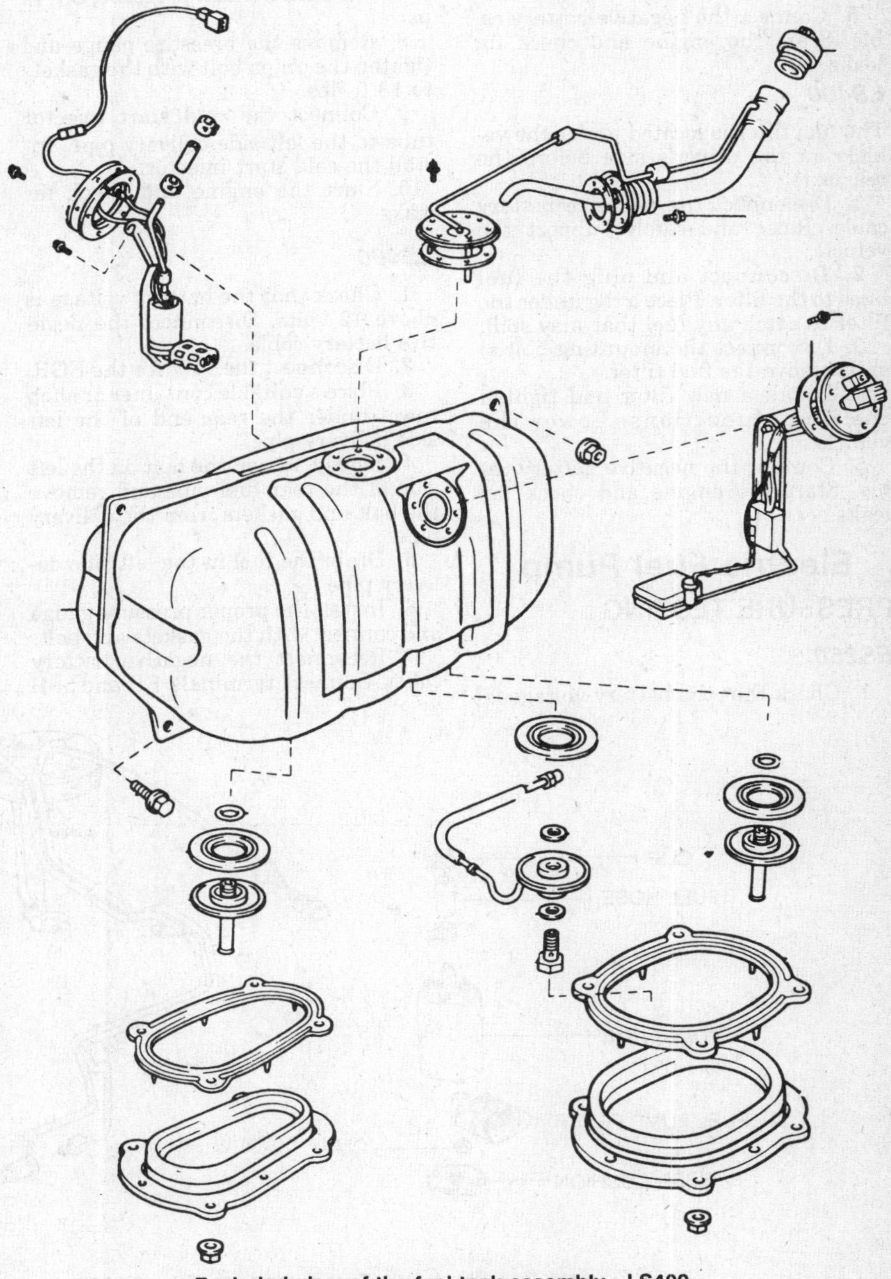

Exploded view of the fuel tank assembly—LS400

Fuel Filter

REMOVAL & INSTALLATION

ES250

The fuel filter is located under the hood, on the driver's side, by the fender well.

1. Disconnect the negative battery cable.
2. Disconnect and plug the fuel lines to the filter. Place a rag under the filter to catch any fuel that may spill.
3. Disconnect the mounting bolt(s) and remove the fuel filter.
4. Install a new filter and tighten the line connections.
5. Connect the negative battery cable. Start the engine and check for leaks.

LS400

The fuel filter is located under the vehicle on the driver's side before the rear axle.

1. Disconnect the negative battery cable. Raise and safely support the vehicle.
2. Disconnect and plug the fuel lines to the filter. Place a rag under the filter to catch any fuel that may spill.
3. Disconnect the mounting bolt(s) and remove the fuel filter.
4. Install a new filter and tighten the line connections. Lower the vehicle.
5. Connect the negative battery cable. Start the engine and check for leaks.

Electric Fuel Pump

PRESSURE TESTING

ES250

1. Check that the battery voltage is above 12 volts. Disconnect the negative battery cable.
2. Disconnect the cold start injector.
3. Place a suitable container or shop towel under the cold start injector connector.
4. Remove the union bolt and gaskets. Disconnect the cold start injector tube from the left side delivery pipe.
5. Install the proper pressure gauge to the left side delivery pipe and tighten the washers and bolt to 13 ft. lbs. (18 Nm).
6. Reconnect the negative battery cable and turn the ignition switch to the **ON** position.
7. Measure the fuel pressure; 38–44 psi.
8. Remove the pressure gauge and tighten the union bolt with the gaskets to 13 ft. lbs.
9. Connect the cold start injector tube to the left side delivery pipe. Install the cold start injector.
10. Start the engine and check for leaks.

LS400

1. Check that the battery voltage is above 12 volts. Disconnect the negative battery cable.
2. Disconnect the VSV for the EGR.
3. Place a suitable container or shop towel under the rear end of the left side delivery pipe.
4. Slowly, loosen the bolt on the left side of the rear fuel pipe and remove the bolt and gaskets from the delivery pipe.
5. Drain the fuel in the left side delivery pipe.
6. Install the proper pressure gauge and connect with the gaskets and bolt.
7. Reconnect the negative battery cable. Connect terminals FP and +B of the check connector with the proper tool.
8. Turn the ignition switch to the **ON** position.
9. Measure the fuel pressure; 38–44 psi.
10. Remove the pressure gauge. Install the bolt and gasket to the delivery pipe.
11. Connect the VSV for the EGR. Start the engine and check for leaks.

REMOVAL & INSTALLATION

ES250

1. Disconnect the negative battery cable. Raise and safely support the vehicle.
2. Drain the fuel from the fuel tank. Disconnect and clamp the fuel lines.
3. Remove both fuel tank protector shields. Support the tank with a suitable jacking device.
4. Remove the mounting bolts and take down both tank band straps. Lower the tank and disconnect the electrical connections.
5. Remove the fuel tank.
6. Remove the fuel pump bracket. Disconnect the fuel line and electrical connector.
7. Remove the mounting nuts and the pump assembly.

To install:

8. Install the fuel pump to the mounting bracket. Connect the mounting nuts, fuel line and electrical connectors.
9. Raise the fuel tank and connect the electrical connections.
10. Install the tank band straps and tighten the strap bolts to 29 ft. lbs.
11. Install both fuel tank protector shields. Connect the fuel lines.

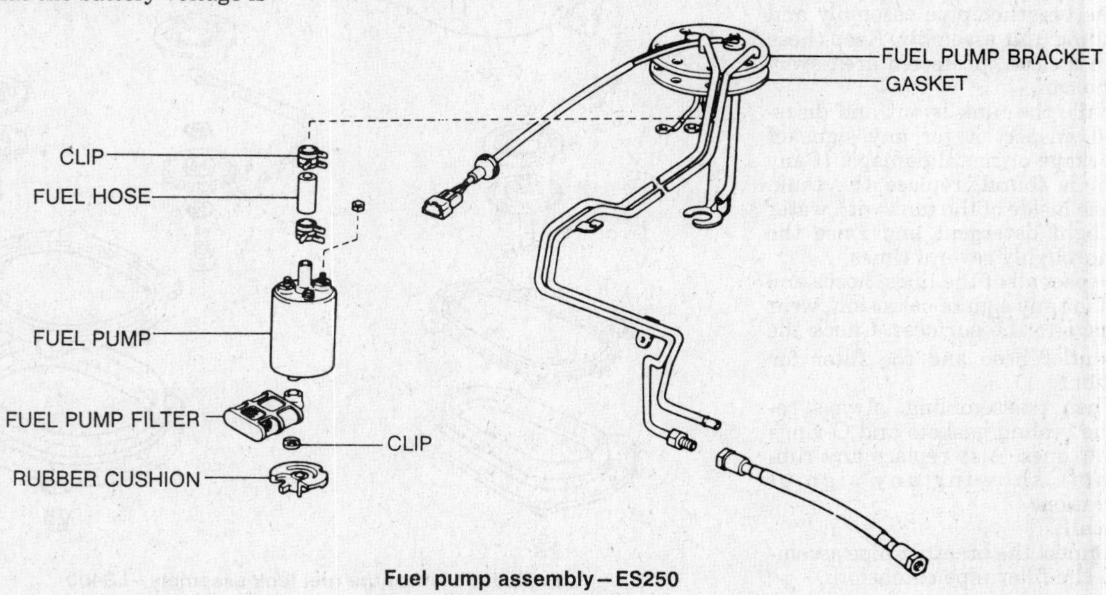

Fuel pump assembly—ES250

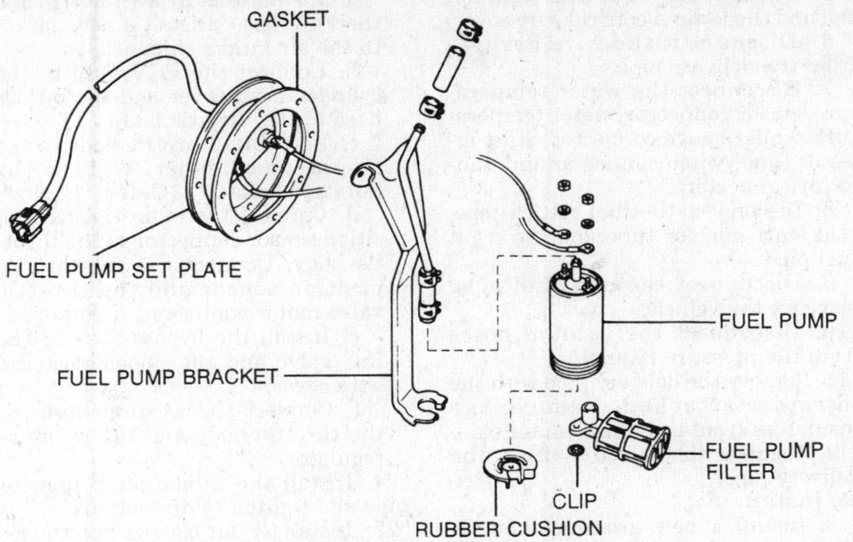

Fuel pump assembly—LS400

12. Remove the jacking device. Lower the vehicle.
13. Connect the negative battery cable. Fill the fuel tank.

LS400

1. Disconnect the negative battery cable. Drain the fuel from the fuel tank.
2. Remove the front trim panel from inside the trunk.
3. Remove the seat back assembly by performing the following:
   a. Disconnect the hook and remove the clips, using the proper tool.
   b. Disconnect the wire harness clamp and connectors.
   c. Disconnect the side hooks and remove the mounting bolts and seat back assembly.
4. Remove the partition panel plug. Disconnect the fuel pump connector.
5. Remove the mounting bolts and the fuel pump set plate. Move the clips aside and disconnect and plug the fuel hose.
6. Remove the mounting bolts and the fuel pump assembly.
7. The installation is the reverse of the removal procedure. Tighten the fuel pump mounting bolts to 48 inch lbs. and the fuel pump set plate bolts to 26 inch lbs.
8. Connect the negative battery cable.

Fuel Injection

IDLE SPEED ADJUSTMENT

The idle speed adjustment is controlled by the Electronic Control Unit (ECU).

IDLE MIXTURE ADJUSTMENT

The idle mixture adjustment is controlled by the Electronic Control Unit (ECU).

Fuel Injector

REMOVAL & INSTALLATION

ES250

1. Disconnect the negative battery cable. Drain the cooling system.
2. Disconnect the throttle cable from the throttle body and bracket on vehicles equipped with automatic transaxle.
3. Disconnect the accelerator and bracket from the throttle body and air intake chamber.
4. Remove the air cleaner cap, the air flow meter and the air cleaner hose.
5. Disconnect and tag the following hoses:
   a. PCV hoses
   b. Vacuum sensing hose
   c. Water bypass hose
   d. Fuel pressure VSV hose
   e. Emission control vacuum hoses
   f. ISC connector
   g. Throttle position sensor connector
   h. EGR gas temperature sensor for California only.
6. Remove the right side mounting bracket.
7. Disconnect the cold start injector and the cold start injector tube.
8. Disconnect and tag the following:
   a. Brake booster vacuum hose
   b. Power steering vacuum and air hoses
   c. Cruise control vacuum hose
   d. Ground strap connector
   e. Wire harness and clamp
9. Disconnect the EGR pipe. Disconnect the engine hanger and the air intake chamber stay from the air intake chamber.
10. Remove the air intake chamber. Disconnect and tag the cold start injector connector, water temperature sensor connector and the 6 injector connectors.
11. Disconnect the wire harness from the left side delivery pipe.
12. Disconnect the fuel inlet and return hoses. Remove the fuel line.
13. Remove the 2 bolts and the left side delivery pipe with the 3 injectors. Remove the 3 bolts and the right side delivery pipe with the fuel pipe and the injectors attached.
14. Pull out the 6 injectors from the delivery pipe. Remove the 6 insulators and the 4 spacers from the intake manifold.

To install:

15. Install a new grommet and O-ring to the injector. Apply a light coat of gasoline to the new O-ring.
16. Install the injectors to the delivery pipes, while turning to left and right.
17. Place the insulators and the spacers in the proper position on the intake manifold.
18. Install the 3 injectors together with the right side delivery pipe and the fuel pipe in the proper position on the intake manifold.
19. Install the 3 injectors together with the left side delivery pipe in the proper position on the intake manifold.

NOTE: Make sure the injectors rotate smoothly. If the injectors do not rotate smoothly, the probable cause is the incorrect installation of the O-rings. Replace the O-rings.

20. Position the injector connector upward and install the mounting bolts. Tighten to 9 ft. lbs. (12 Nm)
21. Install the No. 2 fuel pipe with new gaskets. Tighten the mounting bolts to 24 ft. lbs. (33 Nm).
22. Connect the fuel inlet and return hoses. Connect the wire harness clamps to the left side delivery pipe.
23. Connect the 6 injector connectors, the cold start injector connector and the water temperature connector.
24. Install the air intake chamber, using a new gasket. Tighten the mounting bolts to 32 ft. lbs. (44 Nm).
25. Connect the EGR pipe and tighten to 58 ft. lbs. (79 Nm). Connect the wire harness clamp.

26. Connect the air intake chamber stay and tighten the bolts to 27 ft. lbs. (37 Nm).
27. Install the engine hanger and tighten the mounting bolts to 27 ft. lbs. (37 Nm).
28. Connect the following:
a. Brake booster vacuum hose
b. Power steering vacuum and air hoses
c. Cruise control vacuum hose
d. Ground strap connector
e. Wire harness and clamp
29. Connect the cold start injector and the cold start injector tube.
30. Replace the right side mounting stay and tighten the bolts to 38 ft. lbs. (52 Nm).
31. Connect the following hoses:
a. PCV hoses
b. Vacuum sensing hose
c. Water bypass hose
d. Fuel pressure VSV hose
e. Emission control vacuum hoses
f. ISC connector
g. Throttle position sensor connector
h. EGR gas temperature sensor for California only.
32. Replace the air cleaner cap, the air flow meter and the air cleaner hose.
33. Connect the accelerator and bracket to the throttle body and air intake chamber.
34. Connect the throttle cable and bracket, if equipped with automatic transaxle.
35. Refill the cooling system. Connect the negative battery cable.

LS400

RIGHT

1. Disconnect the negative battery cable. Remove the air cleaner with the air flow meter.
2. Remove the throttle body by performing the following:
a. Disconnect the vacuum hoses from the throttle body and the pressure regulator.
b. Remove the upper electrical wire cover and the bypass pipe from the ISC valve.
c. Disconnect the throttle valve motor connector, the sub-throttle position sensor connector and the main throttle position sensor, if equipped.
d. Remove the nuts and bolts. Separate the throttle body from the intake chamber.
e. Remove the PCV hose from the cylinder head cover and the bypass pipe from the throttle body.
f. Remove the throttle body.
3. Remove the left side and right side engine wire covers.
4. Remove the right side and left side No. 3 timing belt covers.
5. Remove the right side ignition coil and the lower electrical wire cover.
6. Disconnect the wire harness from the delivery pipe.
7. Disconnect the water temperature sensor connector, water temperature sender gauge connector, start injector time switch connector and 4 injector connectors.
8. Disconnect the fuel return pipe, cold start injector tube and the front fuel pipe.
9. Disconnect the rear fuel pipe, leave on the vehicle.
10. Disconnect the vacuum hoses from the pressure regulator.
11 Remove the delivery pipe with the injectors attached. Remove the insulators from the intake manifold.
12. Remove the injectors from the delivery pipe.

To install:

13. Install a new grommet and O-ring to the injector. Apply a light coat of gasoline to the new O-ring.
14. Install the injectors to the delivery pipes, while turning to left and right. Make sure the injectors rotate smoothly.
15. Place the insulators and the spacers in the proper position on the intake manifold.
16. Install the injectors together with the delivery pipe in the proper position on the intake manifold.

NOTE: Make sure the injectors rotate smoothly. If the injectors do not rotate smoothly, the probable cause is the incorrect installation of the O-rings. Replace the O-rings.

17. Temporarily, install the rear fuel pipe and the front fuel pipe. Tighten the delivery pipe first, then the front and rear fuel pipes. Tighten the delivery pipe to 13 ft. lbs. (18 Nm) and the front and rear fuel pipes to 22 ft. lbs. (30 Nm).
18. Connect the cold start injector tube, tighten to 11 ft. lbs. (15 Nm). Install the fuel return pipe and tighten the union bolt to 22 ft. lbs. (30 Nm).
19. Connect the fuel injector connectors, the start injector time switch connector, the water temperature sensor connector and the water temperature sender gauge connector.
20. Install the wire harness to the delivery pipe and tighten the mounting bolts to 74 inch lbs.
21. Install the lower electrical wire cover and the right side ignition coil.
22. Install the right side and left side No. 3 timing covers.
23. Install the throttle body by performing the following:
a. Install the bypass hoses and the PCV hose. Install a new gasket to the air intake chamber.
b. Connect the PCV hose to the cylinder head cover and the bypass hose to the throttle body.
c. Install the throttle body to the air intake chamber. Tighten the nuts and bolts to 13 ft. lbs. (18 Nm).
d. Connect the main throttle position sensor connector to the throttle body. Connect the sub-throttle position sensor and the throttle valve motor connector, if equipped.
e. Install the bypass hose to the ISC valve and the upper electrical wire cover.
f. Connect the vacuum hoses to the throttle body and the pressure regulator.
24. Install the intake air connector pipe and tighten to 45 inch lbs.
25. Install the air cleaner and the air flow meter. Refill the engine coolant.
26. Connect the negative battery cable. Run the engine and check for leaks.

LEFT

1. Disconnect the negative battery cable. Drain the cooling system.
2. Disconnect the PCV hose. Remove the upper electrical wire cover and the left side wire cover.
3. Disconnect the following hoses:
a. Vacuum hoses from the air pipe
b. Vacuum hose from the BVSV
c. Vacuum hose from the VSV for the EGR
d. EGR vacuum modulator hose
4. Remove the EGR vacuum modulator with the mounting bracket. Disconnect the check connector from the bracket.
5. Disconnect the connectors and remove the mounting bolts from the 2 VSV's. Remove the VSV's.
6. Remove the water bypass pipes.
7. Disconnect the engine wire from the delivery pipe.
8. Disconnect the following connectors:
a. Distributor connector
b. Engine speed sensor connector
c. EGR gas temperature sensor connector, California only
d. Injector connectors
9. Remove the bolt, pulsation damper and disconnect the fuel inlet hose from the delivery pipe.
10. Disconnect the front fuel pipe and disconnect the rear fuel pipe.
11. Remove the delivery pipe with the injectors attached. Remove the insulators from the intake manifold.
12. Remove the injectors from the delivery pipe.

To install:

13. Install a new grommet and O-ring to the injector. Apply a light coat of gasoline to the new O-ring.
14. Install the injectors to the delivery pipes, while turning to left and right. Make sure the injectors rotate smoothly.
15. Place the insulators and the spacers in the proper position on the intake manifold.
16. Install the injectors together with the delivery pipe in the proper position on the intake manifold.

NOTE: Make sure the injectors rotate smoothly. If the injectors do not rotate smoothly, the probable cause is the incorrect installation of the O-rings. Replace the O-rings.

17. Temporarily, install the rear fuel pipe and the front fuel pipe. Tighten the delivery pipe first, then the front and rear fuel pipes. Tighten the delivery pipe to 13 ft. lbs. (18 Nm) and the front and rear fuel pipes to 22 ft. lbs. (30 Nm).
18. Connect the cold start injector tube, tighten to 11 ft. lbs. (15 Nm) Install the fuel return pipe and tighten the union bolt to 22 ft. lbs. (30 Nm).
19. Connect the fuel inlet hose to the delivery pipe and replace the pulsation damper and bolt. Tighten the damper to 22 ft. lbs. (30 Nm) and the bolt to 69 inch lbs.
20. Connect the following connectors:
    a. Distributor connector
    b. Engine speed sensor connector
    c. EGR gas temperature sensor connector, California only
    d. Injector connectors
21. Connect the engine wire to the delivery pipe and tighten the mounting bolts to 74 inch lbs.
22. Install the water bypass pipes.
23. Install the VSV's and tighten the bolts to 13 ft. lbs. (18 Nm). Connect the connectors.
24. Connect the check connector to the bracket. Replace the EGR vacuum modulator with the mounting bracket.
25. Connect the following hoses:
    a. Vacuum hoses from the air pipe
    b. Vacuum hose from the BVSV
    c. Vacuum hose from the VSV for the EGR
    d. EGR vacuum modulator hose
26. Connect the PCV hose. Replace the upper electrical wire cover and the left side wire cover.
27. Refill the cooling system. Connect the negative battery cable.
28. Run the engine and check for leaks.

DRIVE AXLE

Halfshaft

REMOVAL & INSTALLATION

ES250

FRONT

1. Raise and safely support the vehicle. Remove the front tire and wheel assembly. Remove the cotter pin and locknut cap. Loosen the bearing locknut while an assistant applies the brake.
2. Disconnect the steering knuckle from the lower ball joint with the proper tool.
3. Disconnect the tie rod end from the steering knuckle with the proper tool.
4. Place matchmarks on the halfshaft and center halfshaft. Using the proper tool, loosen the mounting bolts.

NOTE: Do not remove the bolts. Finger tighten them so the halfshaft does not fall.

5. Disconnect the halfshaft from the axle hub. Remove the left side halfshaft and the joint cover gasket.
6. Remove the bearing lock bolt and the snapring with the proper tool. Pull out the left side halfshaft with the center halfshaft.

NOTE: If the halfshaft cannot be pulled out, tap out the driveshaft with the proper tool.

7. Push the side gear shaft to the differential, in order to replace. Measure and note the distance between the transaxle case and the side gear shaft. Remove the side gear shaft with the proper tool.
8. If necessary, replace the side gear gear shaft oil seal.

To install:

9. Install the left side gear shaft, using a suitable tool to tap in the driveshaft until it makes contact with the pinion shaft. Ensure a new snapring is positioned securely in the groove of the side gear shaft.
10. Check that the side gear will not come out by hand. Push the side gear shaft to the differential and measure the distance between side gear shaft and the transaxle case. Make sure the distance is the same measurement taken before removing the side gear shaft.
11. Pack the side gear shaft with a suitable grease.
12. Align the matchmarks on the side gear shaft and the halfshaft. Install the left side halfshaft and finger tighten the bolts.
13. Install the right side halfshaft with the center halfshaft to the transaxle through the bearing bracket. Install the snapring.
14. Install the outboard joint side of the halfshaft to the axle hub. Temporarily, connect the steering knuckle to the lower ball joint.
15. Connect the tie rod end to the steering knuckle and tighten the bolt to 36 ft. lbs. (49 Nm). Tighten the lower ball joint mounting bolts to 83 ft. lbs. (113 Nm).
16. Tighten the hexagon bolts to 48 ft. lbs. (65 Nm).
17. Replace the front tire and wheel assembly. Lower the vehicle.

LS400

REAR

1. Raise and safely support the vehicle. Remove the rear tire and wheel assembly.
2. Remove the tail pipe O-rings and suspend the tail pipe, using a piece of wire.
3. Disconnect the height control sensor, if equipped.
4. Place matchmarks on the driveshaft and the side gear shaft. Remove the hexagon bolts and washers with the proper tool.
5. Hold the inboard joint side of the halfshaft so the outboard joint side does not bend too much. Tap the end of the halfshaft with the proper tool and disengage the axle hub.
6. Remove the halfshaft.

To install:

7. Insert the outboard joint side of the halfshaft and align the matchmarks on the side gear shaft and the halfshaft.
8. Install the hexagon bolts and tighten to 61 ft. lbs.
9. Connect the height control sensor, if equipped. Replace the O-rings supporting the tail pipe.
10. Install the bearing locknut and tighten to 253 ft. lbs. Replace the rear tire and wheel assembly.
11. Lower the vehicle.

CV-Boot

REMOVAL & INSTALLATION

ES250

FRONT

1. Raise and safely support the vehicle. Remove the front tire and wheel assembly.
2. Disconnect the center halfshaft from the right side halfshaft, using the proper tool. Remove the inboard joint clamp.

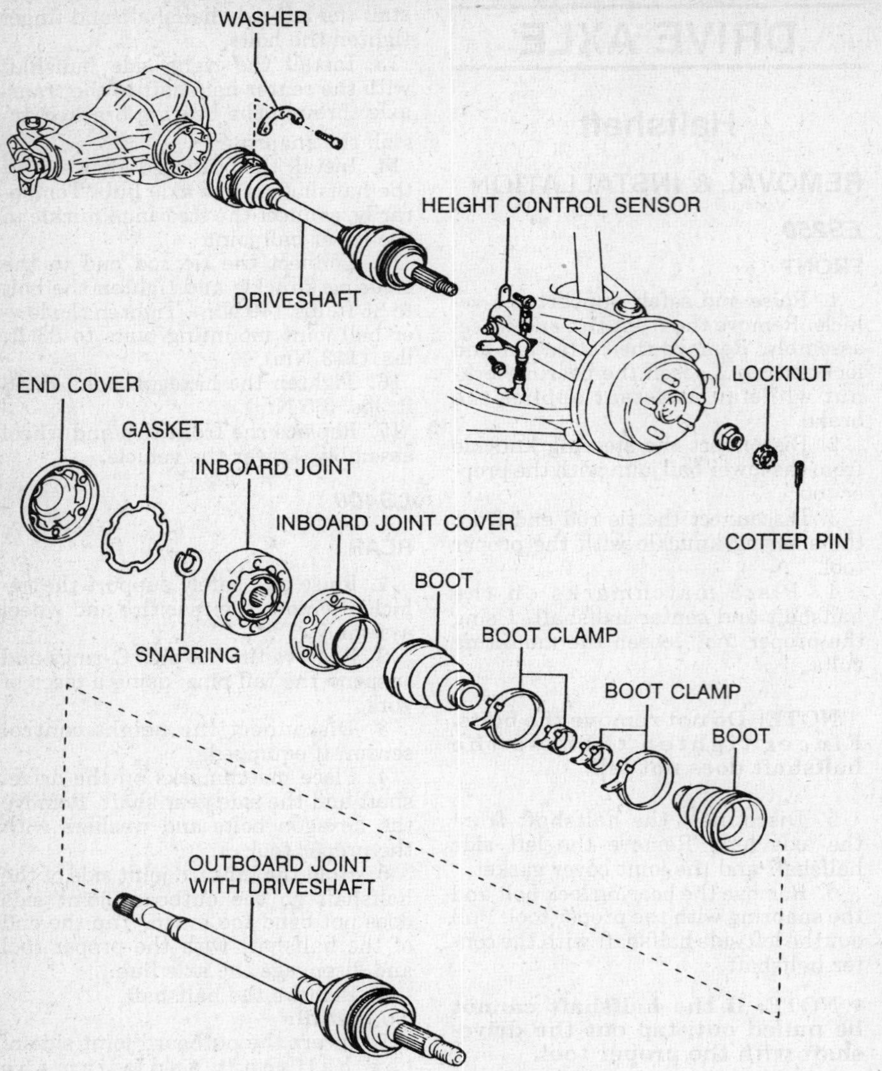

Rear axle shaft assembly – LS400

3. Place matchmarks on the inboard joint and the halfshaft. Remove the snapring with the proper tool.
4. Remove the inboard joint from the halfshaft, using the proper tool to press out the joint.
5. Remove the inboard joint from the inboard joint cove. When lifting the inboard joint, hold on to the inner race and outer race.
6. Remove the inboard joint boot and the outboard joint boot.
7. Reverse the removal procedure for installation. Pack the boot with a suitable grease prior to installation.

LS400

REAR

1. Raise and safely support the vehicle. Remove the rear tire and wheel assembly.
2. Remove the tail pipe O-rings and suspend the tail pipe, using a piece of wire.
3. Disconnect the height control sensor, if equipped.
4. Remove the halfshaft.
5. Secure the halfshaft in a suitable holding fixture. Tap out the end cover with the proper tools.
6. Remove the boot clamps from the inboard and outboard joint boots.
7. Place matchmarks on the inboard joint and halfshaft. Remove the snapring , using the proper pliers.
8. Press out the inboard joint from the halfshaft with the proper tools. Secure the inboard joint in a suitable holding fixture.
9. Tap out the inboard joint cover. Remove both the inboard and outboard
joint boots.
10. The installation is the reverse of the removal procedure. Pack the boots and the end cover with a suitable grease prior to installation.

Driveshaft and U-Joints

REMOVAL & INSTALLATION

LS400

1. Raise and safely support the vehicle.
2. Remove the front exhaust pipe and the heat insulator.
3. Remove the front and rear center floor crossmember braces.
4. Using the proper tool, loosen the adjusting nut on the driveshaft. Place matchmarks on the transmission flange and the flexible coupling.
5. Remove the bolts inserted from the transmission side.

NOTE: The bolts inserted from the driveshaft side should not be removed.

6. Place matchmarks on the differential flange and the flexible coupling.
7. Remove the bolts inserted from the differential side. Separate the flexible couplings from the transmission and the differential.
8. Remove the center support bearing set bolts and the adjusting washers, if equipped.

NOTE: When removing the set bolts, support the center support bearing so the transmission and intermediate shaft and the driveshaft and differential remain in straight line.

9. Push the rear driveshaft forward to compress the driveshaft and pull out the driveshaft from the centering pin of the differential.
10. Remove the driveshaft by pulling out toward the rear of the vehicle.

To install:

11. Apply a suitable grease to the flexible coupling centering bushings.
12. Insert the propeller shaft from the rear of the vehicle and connect the transmission and differential.
13. Temporarily, install the center support bearing set bolts with the adjusting washers, if equipped.
14. Align the matchmarks and connect the propeller shaft to the transmission. Insert the bolts from the transmission side and tighten to 58 ft. lbs. (79 Nm).
15. Align the matchmarks and install the driveshaft to the differential. Insert the bolts from the differential side and tighten to 58 ft. lbs. (79 Nm).
16. Tighten the center bearing support bolts to 27 ft. lbs. (37 Nm). Tighten the adjusting nut with the proper tool.

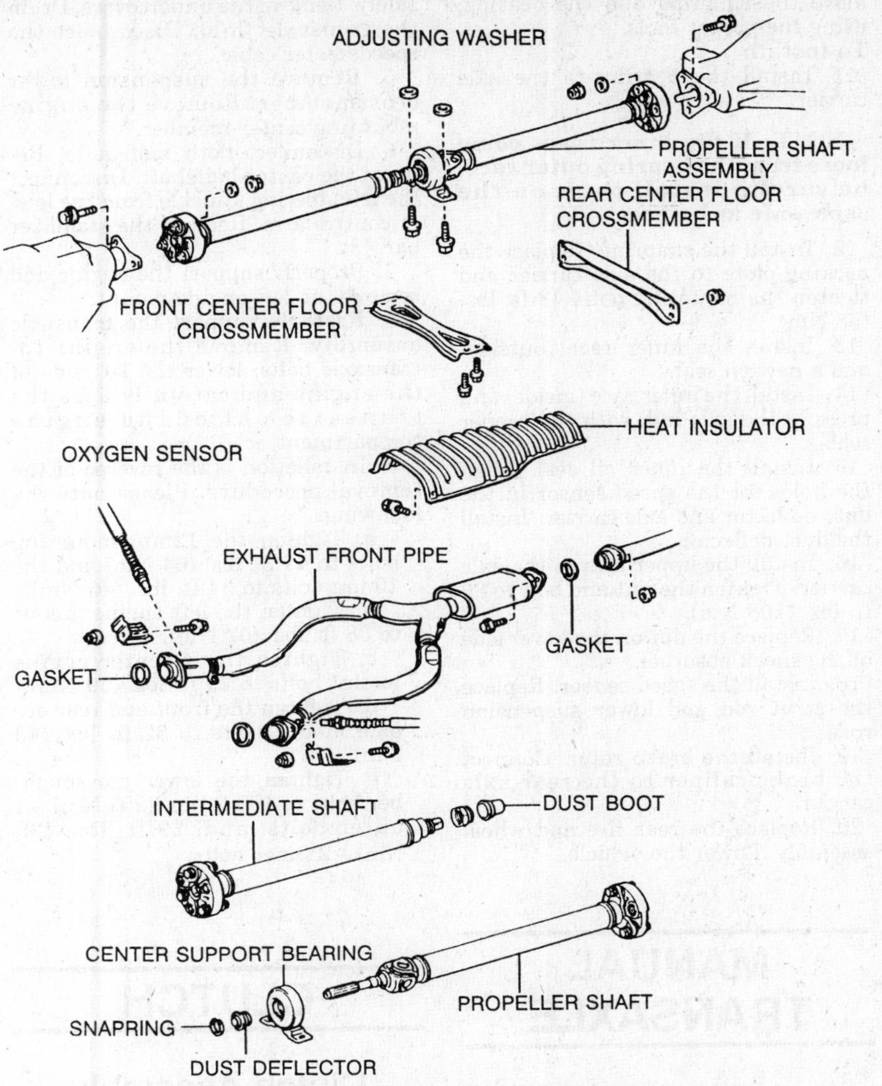

Exploded view of the driveshaft assembly—LS400

17. Install the front and rear crossmember braces and tighten to 9 ft. lbs. (12 Nm).
18. Install the heat insulator and the front exhaust pipe.
19. Lower the vehicle.

Front Wheel Hub, Knuckle and Bearings

REMOVAL & INSTALLATION

ES250

1. Raise and safely support the vehicle. Remove the front tire and wheel assembly.
2. Remove the brake caliper and support, using a piece of wire. Place match marks on the rotor disc and the axle hub. Remove the rotor.
3. Disconnect the ABS speed sensor.
4. Disconnect the lower ball joint from the steering knuckle with the proper tool. Disconnect the tie rod end from the steering knuckle.
5. Disconnect the steering knuckle from the shock absorber.
6. Remove the steering knuckle with the axle hub from the halfshaft.
7. Remove the dust protector and pry out the inner oil seal with the proper tool.
8. Remove the snapring from the steering knuckle and separate the dust cover from the steering knuckle.
9. Push out the axle hub and remove the inner race (inside) from the bearing, using the proper tool.
10. Remove the sensor control rotor from the axle hub. Remove the bearing inner race (outside) from the axle hub.
11. Pry out the oil seal with the proper tool and tap out the bearing.

To install:

12. Install the sensor control rotor to the axle hub and press in a new bearing into the steering knuckle.
13. Install the outer oil seal and the disc brake dust cover.
14. Apply a suitable multi-purpose grease between the oil seal lip, the oil seal and the bearing. Install the axle hub into the steering knuckle, using the proper tool.
15. Install the snapring into the steering knuckle.
16. Install the oil seal into the steering knuckle with a suitable seal installer tool. Apply a suitable multi-purpose grease to the oil seal lip.
17. Tap the dust deflector into the steering knuckle
18. Install the steering knuckle with the axle hub to the halfshaft. Tighten the lower mounting nuts to 224 ft. lbs. (304 Nm).
19. Connect the tie rod end to the steering knuckle and tighten the bolts to 36 ft. lbs. (49 Nm).
20. Tighten the lower ball joint mounting bolts to 83 ft. lbs. (113 Nm). Connect the ABS speed sensor.
21. Align the matchmarks and install the brake rotor. Replace the brake caliper and tighten the mounting bolts to 79 ft. lbs. (107 Nm).
22. Tighten the bearing locknut to 137 ft. lbs. (186 Nm). Replace the tire and wheel assembly.
23. Lower the vehicle.

LS400

1. If equipped with air suspension, move the height control switch in the trunk area to the **OFF** position.
2. Raise and safely support the vehicle. Remove the front tire and wheel assembly.
3. Disconnect the brake caliper from the steering knuckle and support with a piece of wire.
4. Place matchmarks on the disc brake rotor and the axle hub. Remove the brake rotor.
5. Remove the speed sensor from the steering knuckle. Loosen the axle shaft nut.
6. Remove the steering knuckle from the lower ball joint, using the proper tool. Remove the steering knuckle from the upper ball joint.
7. Remove the steering knuckle with the axle hub.
8. Remove the nut and the speed sensor rotor. Using the proper tool, remove the axle hub from the steering knuckle.
9. Remove the outside inner race from the axle and the oil seal from the steering knuckle, using the proper tools.

10. Remove the snapring and remove the bearing from the steering knuckle.

To install:

11. Using the proper tool, install the bearing to the steering knuckle. Replace the snapring.
12. Install the inner race (outside) and press in a new oil seal until it is flush with the end surface of the steering knuckle.
13. Install the brake dust cover to the steering knuckle and using the proper tools, press the axle hub to the steering knuckle.
14. Install the speed sensor. Install the steering knuckle to the lower ball joint and temporarily, tighten the bolts.
15. Install the steering knuckle to the upper arm and tighten the nut to 48 ft. lbs. (65 Nm).
16. Align the matchmarks and install disc rotor to the axle hub. Install the brake caliper and tighten the mounting bolts top 87 ft. lbs. (118 Nm).
17. Tighten the axle shaft nut to 147 ft. lbs. (199 Nm). Install the speed sensor to the steering knuckle.
18. Replace the front tire and wheel assembly. Lower the vehicle.
19. If equipped with air suspension, turn the height control switch to the **ON** position.

Rear Axle Shaft Bearing and Seal

REMOVAL & INSTALLATION

LS400

1. If equipped with air suspension, move the height control switch in the trunk area to the **OFF** position.
2. Raise and safely support the vehicle. Remove the rear tire and wheel assembly.
3. Disconnect the brake caliper from the rear axle carrier and support with a piece of wire.
4. Place matchmarks on the disc brake rotor and the axle hub. Remove the brake rotor.
5. Remove the speed sensor. Remove the strut rod and lower suspension rods.
6. Remove the nut on the lower side of the shock absorber. Do not remove the bolt.
7. Remove the upper arm set bolts and the bolt on the lower side of the shock absorber. Remove the axle with the arm.
8. Remove the upper arm and the dust deflector, using the proper tools. Remove the inner oil seal.
9. Remove the axle hub and the backing plate. Remove the inner race (outside) from the axle hub.
10. Pry out the outer oil seal. Remove the snapring and the bearing, using the proper tools.

To install:

11. Install the bearing to the axle carrier.

NOTE: If the inner races come loose from the bearing outer race, be sure to install them on the same side as before.

12. Install the snapring. Replace the backing plate to the axle carrier and tighten the mounting bolts 43 ft. lbs. (58 Nm).
13. Install the inner race (outside) and a new oil seal.
14. Install the inner race (inside) and press in the axle hub with the proper tools.
15. Install the inner oil seal. Align the holes for the speed sensor in the dust deflector and axle carrier. Install the dust deflector.
16. Install the upper arm to the axle carrier. Tighten the nut and bolt to 80 ft. lbs. (108 Nm).
17. Replace the nut on the lower side of the shock absorber.
18. Install the speed sensor. Replace the strut rod and lower suspension rods.
19. Install the brake rotor. Connect the brake caliper to the rear axle carrier.
20. Replace the rear tire and wheel assembly. Lower the vehicle.

MANUAL TRANSAXLE

For further information on transmissions/transaxles, please refer to "Chilton's Guide to Transmission Repair".

Transaxle Assembly

REMOVAL & INSTALLATION

ES250

1. Disconnect the negative battery cable. Remove the clutch release cylinder and tube clamp. Remove the clutch tube bracket.
2. Disconnect the control cables. Disconnect the back-up light switch electrical connector. Remove the ground strap.
3. Remove the starter assembly. Remove the transaxle upper mounting bolts.
4. Raise and support the vehicle safely. Remove the undercovers. Drain the transaxle fluid. Disconnect the speedometer cable.
5. Remove the suspension lower crossmember. Remove the engine mounting center member.
6. Disconnect both halfshafts. Remove the center halfshaft. Disconnect the left steering knuckle from the lower control arm. Remove the stabilizer bar.
7. Properly support the engine and remove the left engine mount.
8. Properly support the transaxle assembly. Remove the engine-to-transaxle bolts, lower the left side of the engine and carefully ease the transaxle out of the engine compartment.
9. Installation is the reverse of the removal procedure. Please note the following:
   a. Tighten the 12mm mounting bolts to 47 ft. lbs. (64 Nm) and the 10mm bolts to 34 ft. lbs. (46 Nm).
   b. Tighten the left engine mount to 38 ft. lbs. (52 Nm).
   c. Tighten the 4 center engine mount bolts to 29 ft. lbs. (39 Nm).
   d. Tighten the front and rear engine mount bolts to 32 ft. lbs. (43 Nm).
   e. Tighten the lower crossmember bolts to 153 ft. lbs. (207 Nm) – 4 outer bolts; and, 29 ft. lbs. (39 Nm) – 2 inner bolts.

CLUTCH

Clutch Assembly

REMOVAL & INSTALLATION

ES250

1. Disconnect the negative battery cable. Remove the transaxle assembly from the vehicle.
2. Place matchmarks on the flywheel and clutch cover. Remove the clutch pressure plate retaining bolts. Remove the pressure plate assembly.
3. Remove the clutch disc.
4. Installation is the reverse of the removal procedure.
5. Tighten the pressure plate mounting bolts to 14 ft. lbs. (19 Nm).

FREE-PLAY ADJUSTMENT

ES250

1. Adjust the clearance between the master cylinder piston and the pushrod to specification by loosening the pushrod locknut and rotating the

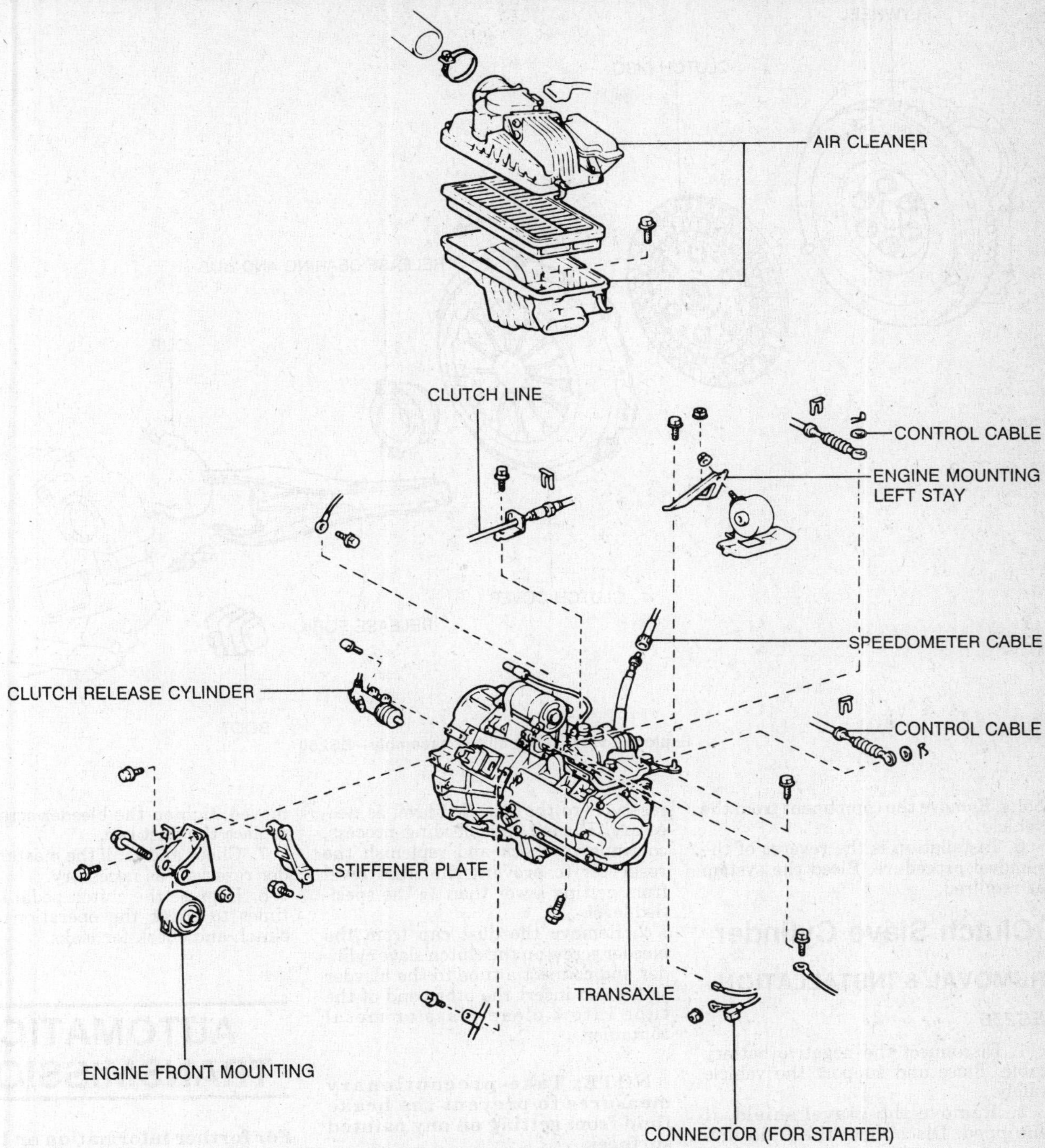

Manual transaxle assembly – ES250

pushrod while depressing the clutch pedal lightly.

2. Tighten the locknut when finished with the adjustment.

3. Adjust the release cylinder free-play by loosening the release cylinder pushrod locknut and rotating the pushrod until proper specification is obtained.

4. Measure the clutch pedal free-play after performing the adjustments. If it fails to fall within specification, repeat the procedure.

Clutch Master Cylinder

REMOVAL & INSTALLATION

ES250

1. Disconnect the negative battery cable.

2. Remove the ABS control relay on vehicles so equipped.

3. Remove the pushrod clevis pin and clip.

NOTE: On some vehicles it will be necessary to remove the under dash panel in order to gain access to the pushrod clevis pin.

5. Disconnect the fluid line. Remove the clutch master cylinder retaining

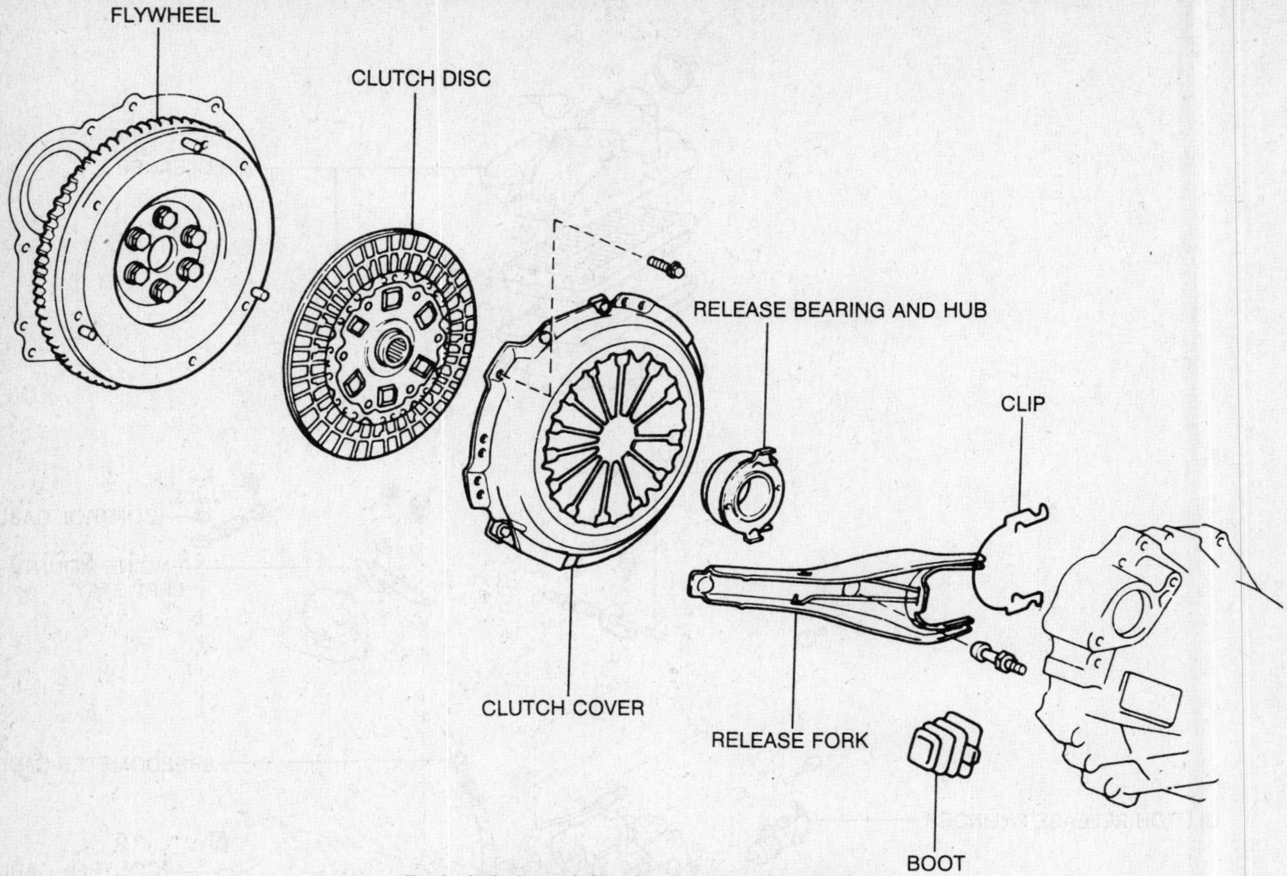

Exploded view of the clutch assembly—ES250

bolts. Remove the component from the vehicle.

6. Installation is the reverse of the removal procedure. Bleed the system as required.

Clutch Slave Cylinder

REMOVAL & INSTALLATION

ES250

1. Disconnect the negative battery cable. Raise and support the vehicle safely.
2. Remove the gravel shield, if equipped. Disconnect the fluid line from the assembly.
3. Remove the slave cylinder retaining bolts. Remove the clutch slave cylinder from the vehicle.
4. Installation is the reverse of the removal procedure. Bleed the system as required.

Hydraulic Clutch System Bleeding

ES250

1. Check and fill the clutch fluid reservoir to the specified level as necessary. During the bleeding process, continue to check and replenish the reservoir to prevent the fluid level from getting lower than ½ the specified level.
2. Remove the dust cap from the bleeder screw on the clutch slave cylinder and connect a tube to the bleeder screw and insert the other end of the tube into a clean glass or metal container.

NOTE: Take precautionary measures to prevent the brake fluid from getting on any painted surfaces.

3. Pump the clutch pedal several times, hold it down and loosen the bleeder screw slowly.
4. Tighten the bleeder screw and release the clutch pedal gradually. Repeat this operation until air bubbles disappear from the brake fluid being expelled out through the bleeder screw.
5. Repeat until all evidence of air bubbles completely disappears from the fluid being pumped out of the tube.
6. When the air is completely removed, tighten the bleeder screw and replace the dust cap.
7. Check and refill the master cylinder reservoir as necessary.
8. Depress the clutch pedal several times to check the operation of the clutch and check for leaks.

AUTOMATIC TRANSMISSION

For further information on transmissions/transaxles, please refer to "Chilton's Guide to Transmission Repair".

Transmission Assembly

REMOVAL & INSTALLATION

LS400

1. Disconnect the negative battery cable. Remove the air cleaner assem-

bly. Disconnect the transmission throttle cable.

2. Raise and support the vehicle safely. Drain the transmission fluid. Remove the driveshaft along with the center bearing.

3. Remove the exhaust pipe together with the catalytic converter. Disconnect the manual shift linkage. Remove the speedometer cable.

4. Disconnect the oil cooler lines. As necessary, remove the transmission oil filler tube. As required, remove the starter assembly. Remove the speedometer cable.

5. Remove both stiffener plates and the catalytic converter cover from the transmission housing and cylinder block.

6. Support the engine and transmission using the proper jacking device. Remove the rear crossmember.

7. Remove the torque converter cover. Remove the torque converter-to-engine retaining bolts.

8. Remove the bolts retaining the transmission to the engine. Carefully remove the transmission from the vehicle.

9. Installation is the reverse of the removal procedure. Tighten the transmission housing bolts to 47 ft. lbs. (64 Nm). Tighten the torque converter bolts to 20 ft. lbs. (27 Nm).

SHIFT LINKAGE ADJUSTMENT

1. Loosen the nut on the shift linkage. Push the selector lever all the way to the rear of the vehicle.

2. Return the lever 2 notches to the **N** shift position.

3. While holding the selector lever slightly toward the **R** shift position, tighten the connecting rod nut.

THROTTLE CABLE ADJUSTMENT

1. Remove the air cleaner.

2. Confirm that the accelerator linkage opens the throttle fully. Adjust the linkage as necessary.

3. Peel the rubber dust boot back from the throttle cable.

4. Loosen the adjustment nuts on the throttle cable bracket (cylinder head cover) just enough to allow cable housing movement.

5. Have an assistant depress the accelerator pedal fully.

6. Adjust the cable housing so the distance between its end and the cable stop collar is 0.04 in.

7. Tighten the adjustment nuts. Make sure the adjustment hasn't changed. Install the dust boot and the air cleaner.

NOTE: When reinstalling the transmission assembly after repair or replacement, it is critical that the correct torque converter mounting bolts are used. The use of a torque converter set bolt longer than specified will deform the front cover of the torque converter. This deformation can internally damage the torque converter lock-up disc. Material from the lock-up clutch may contaminate the transmission and valve body requiring additional repairs. This damage will not be immediately detected by a technician.

AUTOMATIC TRANSAXLE

For further information on transmissions/transaxles, please refer to "Chilton's Guide to Transmission Repair".

Transaxle Assembly

REMOVAL & INSTALLATION

ES250

1. Disconnect the negative battery cable. Remove the air flow meter and the air cleaner assembly.

2. Disconnect the transaxle wire connector. Disconnect the neutral safety switch electrical connector.

3. Disconnect the transaxle ground strap. Disconnect the throttle cable from the throttle linkage.

4. Remove the transaxle case protector. Disconnect the speedometer cable. Disconnect the control cable.

5. Disconnect the oil cooler hoses. Remove the upper starter retaining bolts, as required remove the starter assembly. Remove the upper transaxle housing bolts. Remove the engine rear mount insulator bracket set bolt.

6. Raise and support the vehicle safely. Drain the transaxle fluid.

7. Remove the left front fender apron seal. Disconnect both halfshafts.

8. Remove the suspension lower crossmember assembly. Remove the center halfshaft.

9. Remove the engine mounting center crossmember. Remove the stabilizer bar. Remove the left steering knuckle from the lower control arm.

10. Remove the torque converter cover. Remove the torque converter retaining bolts.

11. Properly support the engine and transaxle assembly. Remove the rear engine mounting bolts. Remove the remaining transaxle-to-engine retaining bolts.

12. Carefully remove the transaxle assembly from the vehicle.

13. Installation is the reverse of the removal procedure. Tighten the 12mm transaxle housing bolts to 47 ft. lbs. (64 Nm); tighten the 10mm bolts to 34 ft. lbs. (46 Nm). Tighten the rear engine mount set bolts to 38 ft. lbs. (52 Nm). Tighten the torque converter mounting bolts to 20 ft. lbs. (27 Nm).

SHIFT CABLE ADJUSTMENT

ES250

1. Loosen the swivel nut on the selector lever.

2. Push the lever fully toward the right side of the vehicle.

3. Return the lever 2 notches to the **N** position.

4. Set the shift lever in the **N** position.

5. While holding the selector lever slightly toward the **R** shift position tighten the swivel nut to 48 inch lbs. (5.4 Nm).

THROTTLE CABLE ADJUSTMENT

ES250

1. Remove the air cleaner.

2. Confirm that the accelerator linkage opens the throttle fully. Adjust the linkage as necessary.

3. Peel the rubber dust boot back from the throttle cable.

4. Loosen the adjustment nuts on

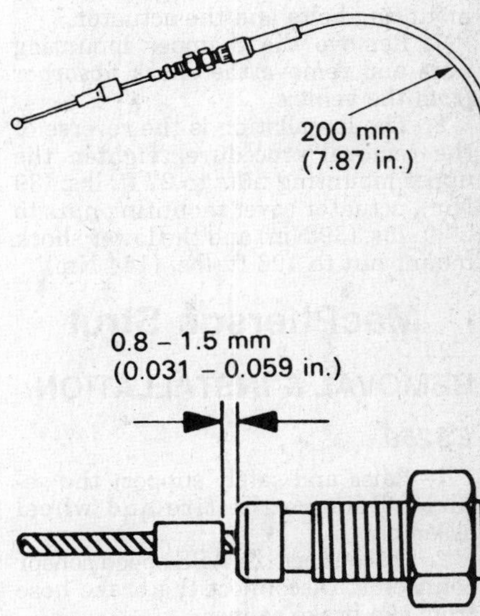

Throttle cable adjustment—ES250

the throttle cable bracket (cylinder head cover) just enough to allow cable housing movement.

5. Depress the accelerator pedal fully.

6. Adjust the cable housing so the distance between its end and the cable stop collar is 0.04 in.

7. Tighten the adjustment nuts. Make sure the adjustment hasn't changed. Install the dust boot and the air cleaner.

FRONT SUSPENSION

Pneumatic Cylinder

REMOVAL & INSTALLATION

LS400

1. Move the height control switch, located in the trunk area to the **OFF** position.

2. Raise and safely support the vehicle. Remove the steering knuckle from the upper ball joint with the proper tool.

3. Support the steering knuckle using a piece of wire. Disconnect speed sensor wire connector.

4. Remove the height control sensor link from the shock absorber lower bracket.

5. Disconnect the shock absorber from the lower mounting bracket. Remove the grommet and disconnect the air tube from the shock absorber.

6. Remove the mounting bolts and the actuator cover. Remove the mounting bolts and the actuator.

7. Remove the 3 upper mounting nuts and remove the shock absorber from the vehicle.

8. The installation is the reverse of the removal procedure. Tighten the upper mounting nuts to 27 ft. lbs. (39 Nm), actuator cover mounting nuts to 27 ft. lbs. (39 Nm) and the lower shock mount nut to 106 ft. lbs. (144 Nm).

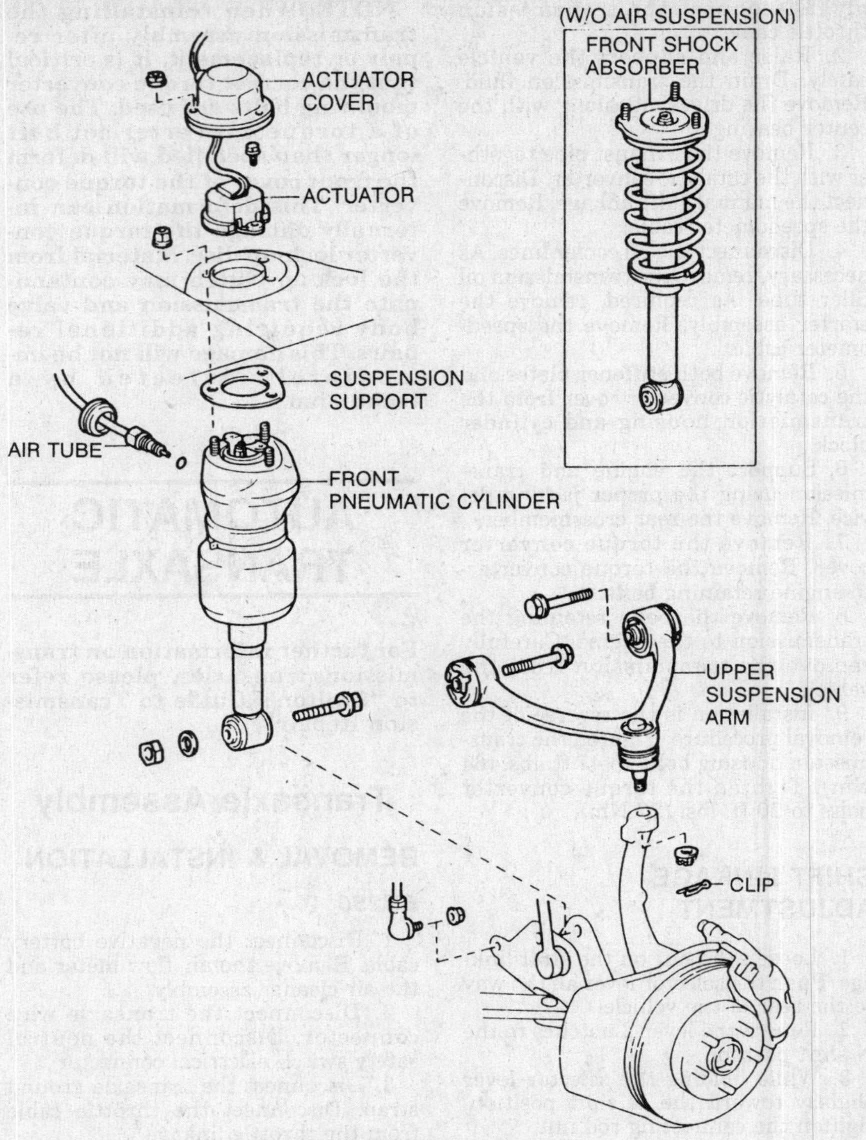

Front suspension assembly—LS400

MacPherson Strut

REMOVAL & INSTALLATION

ES250

1. Raise and safely support the vehicle. Remove the tire and wheel assembly.

2. Disconnect the ABS speed sensor connector. Disconnect the brake hose from the brake caliper.

3. Disconnect the steering knuckle and the strut assembly from the lower mount.

4. Remove the upper mounting nuts from the top suspension mount and remove the strut assembly.

5. The installation is the reverse of the removal procedure. Tighten the upper strut mount nuts to 47 ft. lbs. (64 Nm) and the lower strut mount nuts to 224 ft. lbs. (304 Nm).

NOTE: On the ES250, the front strut bar cushion retainers have been changed to a new resin formed casting cover to improve the front suspension noise during suspension compression and rebound.

LS400

1. Raise and safely support the vehicle. Remove the tire and wheel assembly.

2. Remove the steering knuckle from the upper ball joint with the proper tool. Support the steering knuckle using a piece of wire.

3. Disconnect the strut assembly from the lower strut bracket. Remove the plug from the upper strut mount.

4. Loosen the nut on the middle of the strut mount support. Do not remove it.

5. Remove the other 3 mounting nuts and remove the strut assembly with the coil spring from the vehicle.

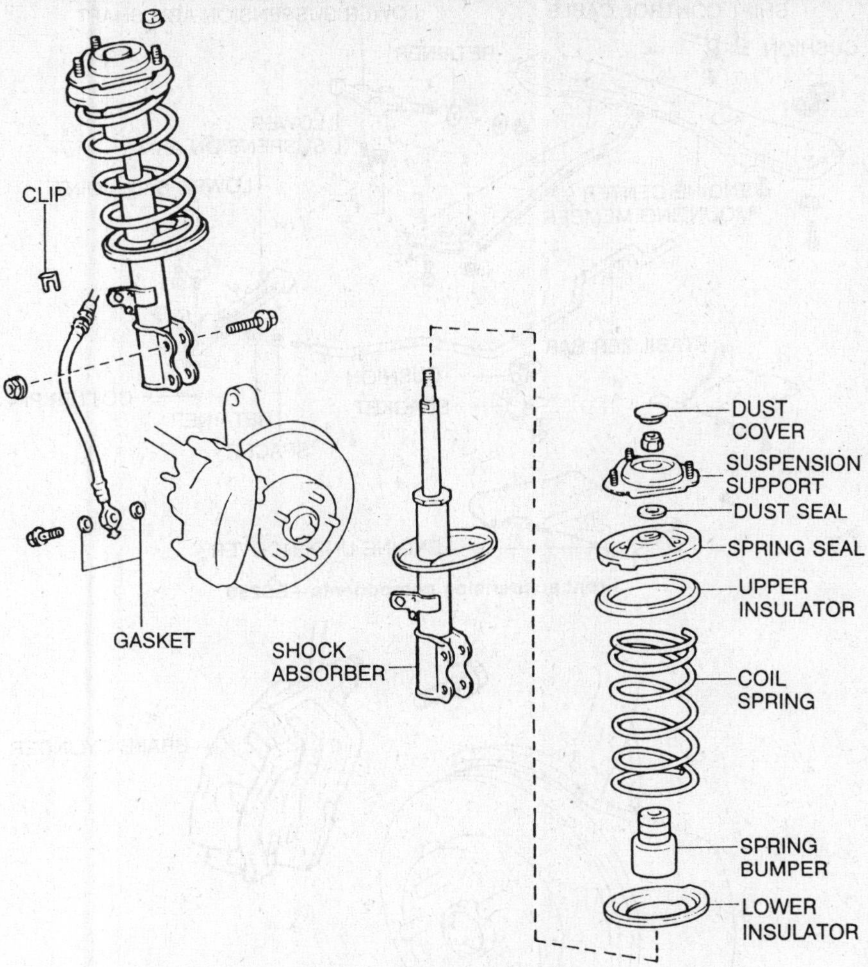

Front strut assembly—ES250

6. The installation is the reverse of the removal procedure. Tighten the upper strut mount nuts to 27 ft. lbs. (37 Nm) and the lower strut mount nut to 106 ft. lbs. (144 Nm).

Strut Bar

REMOVAL & INSTALLATION

LS400

1. Disconnect the negative battery cable. Raise and support the vehicle safely.
2. Remove the steering knuckle with the axle hub.
3. Remove the strut assembly or the pneumatic cylinder, depending on what the vehicle is equipped with.
4. Remove the strut assembly lower bracket.
5. Place matchmarks on the screw part and nut of the strut bar.
6. Remove the nut and washer from the front side of the strut bar.
7. Remove the 2 nuts and remove the strut bar from the lower arm.
8. Remove the 3 nuts and remove the strut bar cushion and strut bar.

To install:

9. Installation is the reverse order of the removal procedure.Be sure to align the match marks on the strut bar when installing it.
10. Use the following torque specifications during installation:
   a. Torque the 2 strut bar bolts to 59 ft. lbs. (80 Nm).
   b. Torque the 2 strut bar nuts to 121 ft. lbs. (164 Nm).
   c. Torque the through bolt of the strut assembly bracket to 83 ft. lbs. (113 Nm) and the other bolt to 43 ft. lbs. (59 Nm).
   d. Lower the vehicle and after stabilizing the suspension, torque the front strut bar nut to 87 ft. lbs. (118 Nm).

Stabilizer Bar

REMOVAL & INSTALLATION

ES250

1. Disconnect the negative battery cable. Raise and support the vehicle safely.
2. Remove the lower suspension crossmember.
3. Remove the nuts and retainers holding the stabilizer bar to the lower suspension arms.
4. Remove the stabilizer bar brackets.
5. Remove the the 2 engine under covers. Remove the control cable clamp bolts from the engine center mounting member, if so equipped. Remove the 10 bolts and engine center mounting member.
6. Pull off the stabilizer bar from the lower suspension arms. Remove the retainers and spacers from the stabilizer bar.

To install:

7. Installation is the reverse order of the removal procedure.
8. Use the following torque specifications during installation:
   a. Torque the stabilizer bar bracket cushion bolts to 94 ft. lbs. (127 Nm).
   b. Torque the 2 bolts on the right hand end and the 2 bolts on the left hand end of the engine center mounting member to 29 ft. lbs. (39 Nm) and the other retaining bolts to 32 ft. lbs. (43 Nm).
   c. Torque the suspension lower crossmember bolts to 112 ft. lbs. (152 Nm).
   d. Torque the stabilizer bar mounting nuts 156 ft. lbs. (212 Nm).

LS400

1. Disconnect the negative battery cable. Raise and support the vehicle safely.
2. Remove the steering knuckle with the axle hub.
3. Remove the strut assembly or the pneumatic cylinder, depending on what the vehicle is equipped with.
4. Remove the strut assembly lower bracket.
5. Place matchmarks on the screw part and nut of the strut bar.
6. Remove the nut and washer from the front side of the strut bar.
7. Remove the 2 nuts and remove the strut bar from the lower arm.
8. Remove the 3 nuts and remove the strut bar cushion and strut bar.
9. Remove the right and left stabilizer bar bushings.
10. Remove the strut bar brackets with the stabilizer bar. Remove the stabilizer bar and inspect the stabilizer bar link as follows.
   a. Flip the ball joint stud back and forth 5 times.
   b. Using a torque gauge, turn the stud continuously 1 turn per 2-4 seconds and take the torque reading on the 5th turn.

c. Turning torque should be 0.4-13 in. lbs.

To install:

11. Installation is the reverse order of the removal procedure. Be sure to install the stabilizer bar through the right and left strut bar brackets. Install the strut bar brackets. Insert the 2 strut bar bolts before hand into the holes in the lower arm. Torque the bolts to 53 ft. lbs. (72 Nm).

12. Use the following torque specifications during installation:

a. Torque the 2 strut bar bolts to 59 ft. lbs. (80 Nm).

b. Torque the 2 strut bar nuts to 121 ft. lbs. (164 Nm).

c. Torque the through bolt of the strut assembly bracket to 83 ft. lbs. (113 Nm) and the other bolt to 43 ft. lbs. (59 Nm).

d. Lower the vehicle and after stabilizing the suspension, torque the front strut bar nut to 87 ft. lbs. (118 Nm).

e. Torque the stabilizer bar links to 70 ft. lbs. (95 Nm).

f. Torque the stabilizer bar bushings to 21 ft. lbs. (28 Nm).

Lower Ball Joints

INSPECTION

1. Raise and safely support the vehicle. Place a wooden block under the tire and wheel assembly.

2. Lower the jacking device until there is about ½ the load on the front spring.

3. Make sure the front wheels are in a straight forward position and block the wheel with wheel chocks.

4. Move the lower arm up and down and check that the ball joint has no excessive play. The ball joint vertical play limit is 0 in. (0mm) on the ES250 or 0.012 in. (0.3mm) on the LS400.

REMOVAL & INSTALLATION

ES250

1. Raise and safely support the vehicle. Remove the tire and wheel assembly.

2. Loosen the nut holding the stabilizer bar to the lower suspension arm.

3. Loosen the nut holding the lower suspension arm to the lower suspension arm shaft.

4. Disconnect the lower ball joint from the lower suspension arm with the proper tool.

5. Remove the mounting bolts and using a suitable prybar, push down the lower suspension arm, the remove the ball joint.

6. The installation is the reverse of the removal procedure. Tighten the

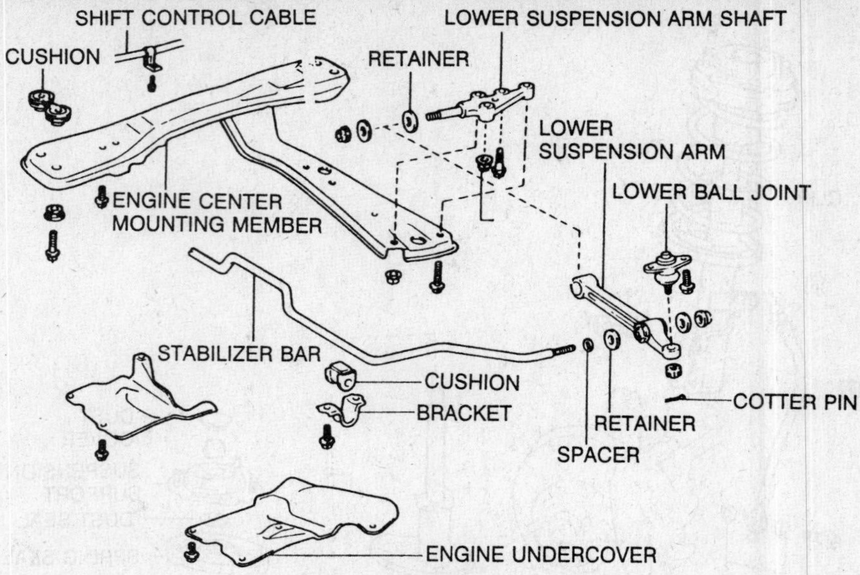

Front suspension components—ES250

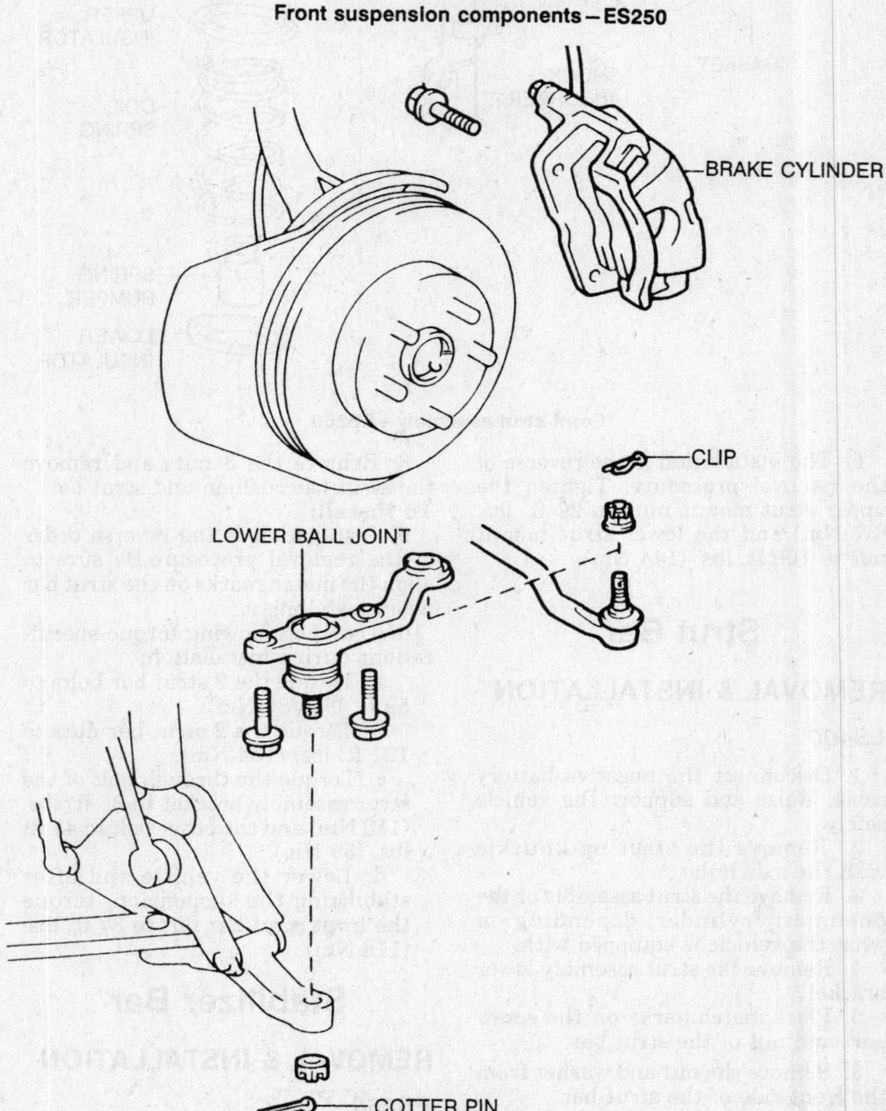

Lower ball joint components—LS400

ball joint mounting bolts to 83 ft. lbs. (113 Nm)., the castle nut to 90 ft. lbs. (12 Nm), the stabilizer bar—lower suspension arm nuts to 156 ft. lbs. (212 Nm) and the lower arm shaft—lower suspension arm nuts to 156 ft. lbs. (212 Nm).

7. Check the front wheel alignment.

LS400

1. If equipped with air suspension, move the height control switch, located in the trunk area to the **OFF** position.
2. Raise and safely support the vehicle. Remove the tire and wheel assembly.
3. Disconnect the brake caliper and support, using a piece of wire.
4. Loosen the lower mounting bolts. Do not remove.
5. Disconnect the tie rod end from the steering arm with the proper tool. Remove the bolts and disconnect the lower ball joint from the steering knuckle.
6. Remove the nut and disconnect the lower ball joint from the lower arm with the proper tool.
7. The installation is the reverse of the removal procedure. Tighten the ball joint castle nut to 112 ft. lbs. (152 Nm), the tie rod end nut to 43 ft. lbs. (58 Nm) and the lower ball joint bolts to 83 ft. lbs. (113 Nm).

Upper Control Arms

REMOVAL & INSTALLATION

LS400

1. Raise and safely support the vehicle. Remove the tire and wheel assembly.
2. Remove the shock absorber or pneumatic cylinder, if equipped.
3. Remove the mounting bolts and the upper suspension arm.
4. Install the upper suspension arm and tighten the mounting bolts to 83 ft. lbs. (113 Nm).
5. Reverse the remainder of the procedures for installation.

Lower Control Arms

REMOVAL & INSTALLATION

ES250

1. Raise and safely support the vehicle.
2. Remove the nut holding the stabilizer to the lower suspension arm. Remove the nut holding the lower suspension arm to the lower suspension arm shaft.
3. Remove the mounting bolts and disconnect the lower ball joint from the steering knuckle with the proper tool.
4. Remove the suspension lower crossmember. Remove the mounting bolt and the lower arm with the shaft.
5. Remove the nut and disconnect the lower ball joint from the lower arm with the proper tool.
6. The installation is the reverse of the removal procedure. Tighten the lower arm shaft-to-body bolts to 153 ft. lbs. (207 Nm), ball joint-to-steering knuckle bolts to 83 ft. lbs. (113 Nm) and the lower suspension arm mounting nuts to 156 ft. lbs. (212 Nm).

LS400

1. Raise and safely support the vehicle. Remove the tire and wheel assembly.
2. Remove the shock absorber or pneumatic cylinder, if equipped.
3. Disconnect the tie rod end from the steering knuckle with the proper tool. Remove the lower shock bracket.
4. Remove the nuts and disconnect the lower strut bar from the lower arm.
5. Place matchmarks on the camber adjusting cam. Remove the nut, the adjusting cam and the lower arm with the lower ball joint.

NOTE: To help remove the adjusting cam, fully pull the steering wheel toward the lower arm being removed.

To install:

6. Insert the camber adjusting cam from the rear side of the vehicle and temporarily tighten the nut. Put 2 strut bar bolts into the holes of the lower arm beforehand.
7. Connect the strut bar to the lower arm and tighten the bolts to 121 ft. lbs. (164 Nm). Install the lower shock bracket.
8. Connect the tie rod to the steering arm and tighten the nut to 43 ft. lbs. (58 Nm). Install a new clip.
9. Install the shock absorber or pneumatic cylinder, if equipped.
10. Install the steering knuckle with the axle hub. Replace the tire and wheel assembly.
11. Lower the vehicle and stabilize the suspension.
12. Support the lower arm with a suitable jacking device and remove the tire and wheel assembly.
13. Align the matchmarks and tighten the nut to 185 ft. lbs. (250 Nm). Replace the tire and wheel assembly. Remove the jacking device.
14. Check the front end alignment.

Sway Bar

REMOVAL & INSTALLATION

ES250

1. Raise and safely support the vehicle. Remove the suspension lower crossmember.
2. Remove the mounting nuts and retainers holding the suspension bar to the lower suspension arms.
3. Remove the stabilizer bar brackets.
4. Remove the 2 engine undercovers and the automatic transaxle control cables from the engine center mounting member.
5. Remove the bolts and lower the engine center mounting member. Pull the stabilizer bar from the suspension arms.
6. Remove the retainers and the spacers from the stabilizer bar.
7. The installation is the reverse of the removal procedure. Tighten the stabilizer bar brackets to 94 ft. lbs. (127 Nm), the engine center mounting member to 29–32 ft. lbs. (40-44 Nm), the lower crossmember to 153 ft. lbs. (207 Nm) and the stabilizer bar mounting nuts to 156 ft. lbs. (212 Nm).
8. Check the front wheel alignment.

Front Wheel Bearings

REMOVAL & INSTALLATION

LS400

1. Raise and safely support the vehicle. Remove the tire and wheel assembly.
2. Remove the steering knuckle with the axle hub. Remove the nut and the speed sensor rotor.
3. Remove the 4 bolts and shift the brake dust cover towards the hub side (outside). Remove the axle shaft from the steering knuckle with the proper tool.
4. Using the proper tool, remove the inner race (outside) from the axle shaft. Pry out the oil seal from the steering knuckle with the proper tool.
5. Remove the snapring and press out the bearing from the steering knuckle with the proper tools.

To install:

6. Using the proper tool, press the new bearing into the steering knuckle. Install the snapring, using suitable snapring pliers.
7. Install the inner race (outside) and press in the new oil seal until it is flush with the end surface of the steering knuckle.
8. Install the brake dust cover to the steering knuckle. Press the axle

hub to the steering knuckle with the proper tool.

9. Install the speed sensor rotor and the steering knuckle.

10. Replace the tire and wheel assembly. Lower the vehicle.

REAR SUSPENSION

Pneumatic Cylinder

REMOVAL & INSTALLATION

LS400

1. Remove the rear seat cushion and seat back. Remove the rear scuff plates and the roof side inner trim panel and the speaker panel.
2. Remove the trunk trim panel. Move the height control switch, located in the trunk area to the **OFF** position.
3. Raise and safely support the vehicle. Remove the tire and wheel assembly.
4. Disconnect the stabilizer links from the stabilizer bar.
5. Disconnect and support the brake caliper, using a piece of wire. Do not disconnect the brake line.
6. Disconnect the height control sensor link from the suspension arm. Remove the nut on the lower side of the shock absorber. Do not remove the bolt.
7. Support the rear axle assembly with a suitable jacking device. Remove the grommet and disconnect the air tube from the shock absorber.
8. Remove the mounting bolts and the actuator cover. Remove the mounting bolts and the actuator.
9. Remove the upper mounting nuts and lower the rear axle assembly. Remove the bolt on the lower side of the shock absorber.
10. Remove the shock absorber.

To install:

11. Install the shock to the vehicle and tighten the upper mounting nuts to 43 ft. lbs. (58 Nm).
12. Install the actuator and replace the actuator cover. Tighten the mounting nuts to 13 ft. lbs. (18 Nm).
13. Install new O-rings and connect the air line to the shock absorber. Tighten the fitting to 13 ft. lbs. (18 Nm).
14. Install the shock to the rear axle carrier. Insert the bolt from the vehicle's rear and temporarily tighten the nut.
15. Connect the height control sensor link to suspension arm and tighten to 48 inch lbs.

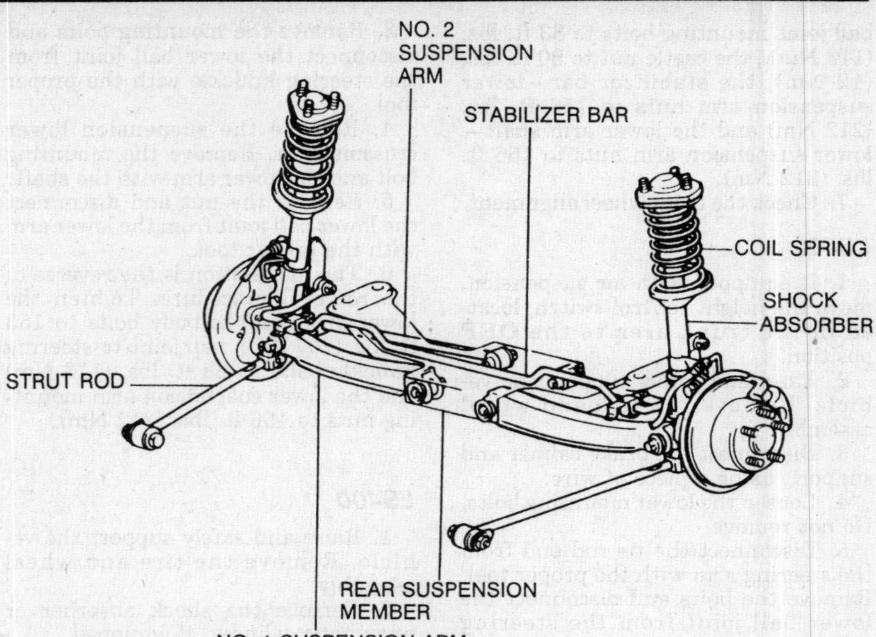

Rear suspension – ES250

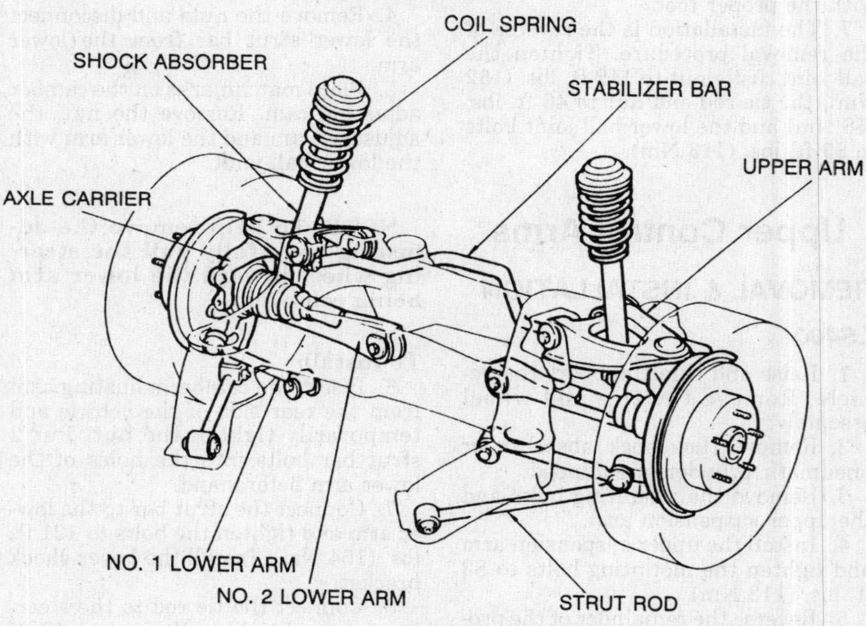

Rear suspension – LS400

16. Install the rear brake caliper to the rear axle carrier and tighten the mounting bolts to 77 ft. lbs. (104 Nm).
17. Connect the stabilizer links and tighten to 26 ft. lbs. (35 Nm).
18. Stabilize the suspension by:
    a. Install the tire and wheel assembly and lower the vehicle.
    b. Move the height control switch to the **ON** position. Start the engine an fill the pneumatic cylinder with air.
    c. Bounce the vehicle up and down several times to stabilize the suspension.
19. Raise and safely support the vehicle. Remove the tire and wheel assembly.
20. Support the rear axle carrier with a suitable jacking device. Tighten the lower shock bolt to 101 ft. lbs. (137 Nm).
21. Replace the tire and wheel assembly. Lower the vehicle.
22. Install the rear seat cushion and seat back. Replace the rear scuff

plates, the roof side inner trim panel and the speaker panel.
23. Install the trunk trim panel. Check the rear wheel alignment.

MacPherson Strut

REMOVAL & INSTALLATION

LS250

1. Raise and safely support the vehicle. Remove the tire and wheel assembly.
2. Disconnect and plug the brake line from the strut assembly.
3. Disconnect the stabilizer link from the strut assembly.
4. Remove the rear seat back and package tray trim. Remove the dust cover from the upper suspension support. Loosen the nut but do not remove.
5. Remove the strut mounting bolts and disconnect the strut assembly. Remove the upper mounting bolts and remove the strut assembly.
6. The installation is the reverse of the removal procedure. Tighten the upper mounting bolts to 29 ft. lbs., lower strut-to-axle bolts to 166 ft. lbs., upper center mounting nut to 36 ft. lbs. and the stabilizer link top 47 ft. lbs.

LS400

1. Remove the rear seat cushion and seat back. Remove the rear scuff plates and the roof side inner trim panel and the speaker panel.
2. Raise and safely support the vehicle. Remove the tire and wheel assembly.
3. Remove the rear halfshaft and disconnect the stabilizer links.
4. Disconnect and support the brake caliper. Do not disconnect the brake line.
5. Remove the nut on the lower side of the strut. Do not remove the bolt.
6. Support the rear axle assembly with a suitable jacking device. Remove the 3 nuts and the strut cap.
7. Loosen the nut in the middle of the suspension support. Do not remove it.
8. Remove the other 3 mounting bolts. Lower the rear axle assembly and remove the bolt on the lower side of the strut assembly.
9. Remove the strut assembly with the coil spring.

To install:

10. Install the strut assembly with the coil spring to the vehicle and tighten the nuts to 47 ft. lbs. (64 Nm). Tighten the nut in the middle of the suspension support to 20 ft. lbs. (27 Nm).
11. Install the strut to the rear axle carrier. Install the bolt from the rear of the vehicle and temporarily tighten the nut.
12. Install the brake caliper and tighten the mounting bolts to 77 ft. lbs. (104 Nm). Connect the stabilizer links and tighten to 26 ft. lbs. (35 Nm).
13. Install the rear halfshaft. Replace the tire and wheel assembly. Lower the vehicle.
14. Bounce the vehicle up and down to stabilize the suspension.
15. Raise and safely support the vehicle. Remove the tire and wheel assembly.
16. Support the rear axle assembly with a suitable jacking device. Tighten the bolt to 101 ft. lbs. (137 Nm).
17. Install the tire and wheel assembly. Lower the vehicle.
18. Replace the rear scuff plates and the roof side inner trim panel and the speaker panel.
19. Replace the rear seat cushion and seat back.
20. Check the rear wheel alignment.

Rear Control Arms

REMOVAL & INSTALLATION

LS400

UPPER

1. Raise and safely support the vehicle. Remove the tire and wheel assembly.
2. Remove the rear axle carrier with the upper arm assembly.
3. Install the axle carrier in a suitable holding fixture.
4. Disconnect the upper control arm from the axle carrier.
5. The installation is the reverse of the removal procedure. Tighten the upper arm-to-axle bolt to 80 ft. lbs. (108 Nm).

Rear Wheel Bearings

REMOVAL & INSTALLATION

ES250

1. Raise and safely support the vehicle. Remove the tire and wheel assembly.
2. Remove the rear brakes and rotor assembly.
3. Remove the axle hub mounting bolts and remove the axle hub with the parking brake assembly.
4. Using the proper tool, unstake the locknut and remove.
5. Push the rear axle shaft off the axle hub with the proper tool. Remove the bearing inner race (inside) with the proper tool.
6. Using the proper tool, pull off the

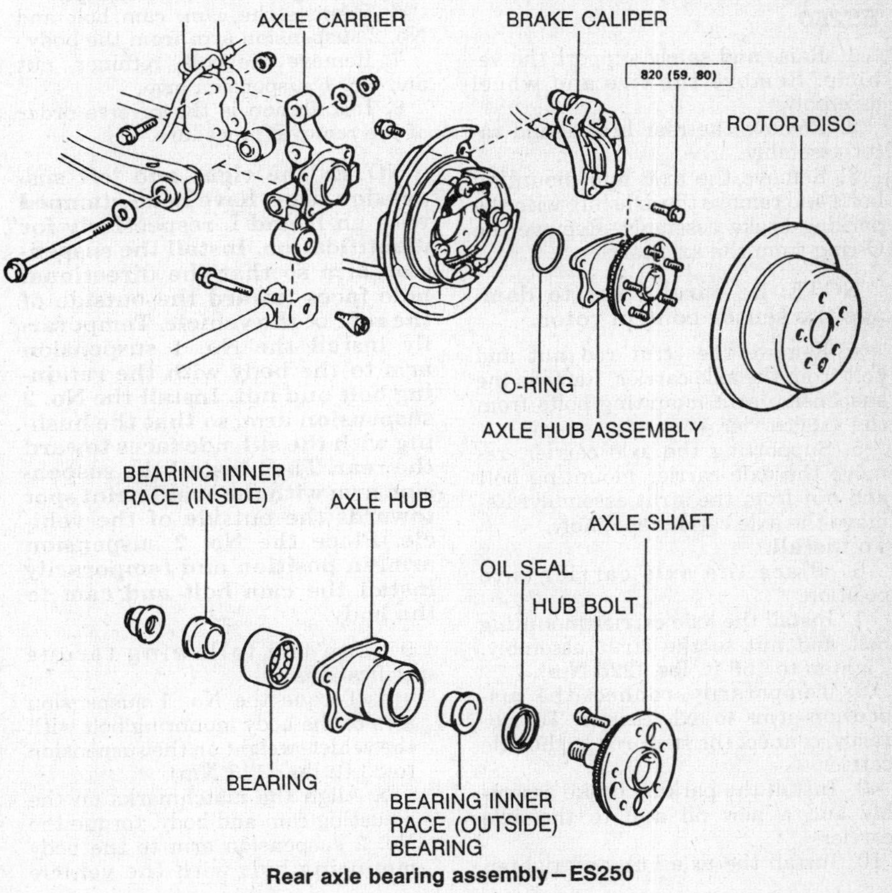

Rear axle bearing assembly—ES250

bearing inner race (outside) from the axle shaft. Remove the oil seal.

7. Install the inner race (outside) of the bearing to be removed, using the proper tool, press out the bearing.

NOTE: Always replace the bearing as an assembly.

To install:

8. Apply a suitable multi-purpose grease to the outer race of a new bearing. Press the bearing into the axle hub with the proper tool.

9. Install a new bearing inner race (outside) and drive in a new oil seal with the proper tool. Apply a suitable multi-purpose grease to the seal lip.

10. Install a new bearing inner race (inside) and press the inner races onto the axle shaft, using the proper tool. Install and tighten the locknut to 90 ft. lbs. (122 Nm). Stake the nut.

11. Install the parking brake assembly and a new oil seal to the axle carrier.

12. Install the axle hub and tighten the bolts to 59 ft. lbs. (80 Nm).

13. Install the tire and wheel assembly. Lower the vehicle.

Rear Axle Assembly

REMOVAL & INSTALLATION

ES250

1. Raise and safely support the vehicle. Remove the tire and wheel assembly.

2. Remove the rear brakes and rotor assembly.

3. Remove the axle hub mounting bolts and remove the axle hub with the parking brake assembly. Remove the O-ring from the axle carrier.

NOTE: Be careful not to damage the sensor control rotor.

4. Remove the strut rod nut and bolt from the axle carrier. Remove the suspension arm mounting bolts from the axle carrier.

5. Supporting the axle carrier, remove the axle carrier mounting bolt and nut from the strut assembly. Remove the axle carrier assembly.

To install:

6. Place the axle carrier into position.

7. Install the axle carrier mounting bolt and nut to the strut assembly. Tighten to 166 ft. lbs. (225 Nm).

8. Temporarily, connect the suspension arms to axle carrier. Temporarily, connect the strut rod to the axle carrier.

9. Install the parking brake assembly and a new oil seal to the axle carrier.

10. Install the axle hub and tighten the mounting bolts to 59 ft. lbs. (80 Nm).

11. Install the rotor and disc brake assembly. Stabilize the suspension.

12. Tighten the axle carrier-to strut rod mounting bolts to 83 ft. lbs. (113 Nm) and the suspension arm-to-axle carrier bolts to 134 ft. lbs. (182 NM). Tighten with the vehicle weight on the suspension.

13. Check the rear wheel alignment.

Suspension Arms

REMOVAL & INSTALLATION

ES250

1. Disconnect the negative battery cable. Raise and support the vehicle safely.

2. Disconnect the ABS speed sensor wire clamp from the No. 1 suspension arm.

3. Remove the bolt and nut holding the strut rod to the axle carrier and remove the strut rod from the carrier.

4. Place a match mark on the adjusting cam and body of the No. 1 and No. 2 suspension arms. Remove the volt and nut holding the No. 1 and No. 2 suspension arms to the axle carrier.

5. Remove the fuel tank protector by removing the 2 bolts and clips.

6. Remove the cam, cam bolt and No. 2 suspension arm from the body.

7. Remove the bolt, retainer, nut and No. 1 suspension arm.

8. Installation is the reverse order of the removal procedure.

NOTE: The right and left suspension arms have been stamped with an R and L respectively for identification. Install the suspension arm so that the directional hole faces toward the outside of the rear of the vehicle. Temporarily install the No. 1 suspension arm to the body with the retaining bolt and nut. Install the No. 2 suspension arm, so that the bushing with the slit side faces toward the rear. Then install the suspension arm with the small paint spot towards the outside of the vehicle. Place the No. 2 suspension arm, in position and temporarily install the cam bolt and cam to the body.

9. Use the following torque specifications:

a. Torque the No. 1 suspension arm to the body mounting bolt with the vehicle weight on the suspension to 83 ft. lbs. (113 Nm).

b. Align the matchmarks on the adjusting cam and body, torque the No. 2 suspension arm to the body mounting bolt with the vehicle weight on the suspension to 83 ft. lbs. (113 Nm).

c. Torque the No. 1 and No. 2 suspension arms to the axle carrier mounting bolt and nut with the vehicle weight on the suspension to 134 ft. lbs. (181 Nm).

d. Torque the strut rod to the axle carrier mounting bolt with the vehicle weight on the suspension to 83 ft. lbs. (113 Nm).

e. Reconnect the negative battery cable and Check the rear wheel alignment.

Strut Rod

REMOVAL & INSTALLATION

ES250

1. Disconnect the negative battery cable. Raise and support the vehicle safely.

2. Disconnect the ABS speed sensor wire clamp from the No. 1 suspension arm.

3. Remove the bolt and nut holding the strut rod to the axle carrier and remove the strut rod from the carrier. If necessary, remove the rear floor heat upper insulator on the right hand side.

4. Installation is the reverse order of the removal procedure. Position the strut rod to the body and temporarily install the nut and bolt. Be sure the lip of the nut is resting on the flange of the bracket.

5. Torque the strut rod to the body mounting bolt with the vehicle weight on the suspension to 83 ft. lbs. (113 Nm).

6. Torque the strut rod to the axle carrier mounting bolt with the vehicle weight on the suspension to 83 ft. lbs. (113 Nm).

7. Reconnect the negative battery cable and Check the rear wheel alignment.

Stabilizer Bar

REMOVAL & INSTALLATION

ES250

1. Disconnect the negative battery cable. Raise and support the vehicle safely.

2. Remove the tail pipe assembly.

3. Remove the stabilizer links. If the ball joint stud turns together with the nut, use a hexagon wrench to hold the stud.

4. Disconnect the ABS speed sensor wire clamp. Remove the 4 bolts, stabilizer brackets and cushions.

5. Using a suitable jack and a block of wood, support the fuel tank. Remove the 2 fuel tank band installation bolts.

6. Lower the fuel tank approximately 1.57 in. (40 mm) and then remove the stabilizer bar from the body.

7. Rotate the ball joint stud in all directions, If the movement is not smooth and free, replace the stabilizer link.

To install:

8. Installation is the reverse order of the removal installation. Use the following torque specifications:

a. After placing the fuel tank in its proper position, torque the tank strap retaining bolts to 29 ft. lbs. (39 Nm).

b. Torque the stabilizer bar cushions retaining bolts to 14 ft. lbs. (19 Nm).

c. Torque the stabilizer link to 47 ft. lbs. (64 Nm).

d. Torque the tail pipe clamp to 29 ft. lbs. (39 Nm).

Lower Suspension Arm and Strut Rod

REMOVAL & INSTALLATION

LS400

1. Disconnect the negative battery cable. If the vehicle is equipped with air suspension. move the height control ON/OFF switch (located in the luggage compartment) to the **OFF** position.

2. Raise and support the vehicle safely and remove the rear wheel assembly.

3. Disconnect the strut rod from the rear axle carrier. Remove the strut rod.

4. Disconnect the height control sensor link if so equipped, from the No. 1 suspension arm.

5. Place matchmarks on the adjusting cam and body. Remove the adjusting cam. Remove the nut on the axle carrier side of the No. 1 lower suspension arm.

6. Using a tie rod removal tool, remove the No. 1 suspension arm.

7. Disconnect the stabilizer bar link from the No. 2 suspension arm.

8. Place matchmarks on the adjusting cam and body. Remove the adjusting cam. Remove the No. 2 lower suspension arm. Remove the arm cover.

9. Inspect the No. 1 lower suspension arm ball joint as follows.

a. Flip the ball joint stud back and forth 5 times.

b. Using a torque gauge, turn the stud continuously 1 turn per 2–4 seconds and take the torque reading on the 5th turn.

c. Turning torque should be 7.4–30 inch lbs.

d. If not within specifications, replace the No. 1 suspension arm.

To install:

10. Temporarily install the No. 2 and No. 1 lower suspension arms. Install the bolt and nut into the No. 2 suspension arm, connected the No. 2 lower suspension arm to the axle carrier and torque the bolts to 121 ft. lbs. (164 Nm). Install a new nut to the No. 1 suspension arm ball joint and torque it to 43 ft. lbs. (59 Nm).

11. Connect the height control sensor link to the No. 1 lower suspension arm with a new nut. Torque it to 48 inch lbs.

12. Temporarily install the strut rod.

13. Install the rear wheel assemblies and lower the vehicle.

14. Move the height control switch back to the **ON** position, if so equipped. Bounce the vehicle up and down several times to stabilize the suspension.

15. Raise and safely support the vehicle. Remove the rear wheel assemblies.

16. Support the axle carrier with a suitable jack. Torque the bolt and nut of the strut rod to 136 ft. lbs. (184 Nm). Torque the nut on the body side of the No. 1 lower suspension arm to 136 ft. lbs. (184 Nm). Torque the nut on the body side of the No. 2 lower suspension arm to 136 ft. lbs. (184 Nm).

17. Install the rear wheel assemblies, lower the vehicle and torque the rear wheel retaining nuts to 76 ft. lbs. (103 Nm).

18. Reconnect the negative battery cable. Check the rear wheel alignment.

STEERING

Steering Wheel

CAUTION

On vehicles equipped with an air bag, the negative battery cable must be disconnected, before working on the system. Failure to do so may result in deployment of the air bag and possible personal injury. Work must be started after approximately 20 seconds or longer from the time the ignition switch is turned to the LOCK position and the negative battery cable is disconnected from the battery. If the wiring connector of the airbag system is disconnected with the ignition switch at ON or ACC, diagnostic coded will be recorded.

REMOVAL & INSTALLATION

1. Disconnect the negative battery cable. Position the front wheels in a straight ahead position.

2. Loosen the screw until the groove along the screw circumference catches on the screw case.

3. Pull the wheel pad away from the steering wheel and disconnect the air bag connector.

NOTE: When removing the wheel pad, take care not to pull the air bag harness connector. When storing the wheel pad, keep the upper surface of the pad facing upward.

4. Disconnect the wire connector and remove the set nut. Place matchmarks on the steering wheel and main shaft.

5. Remove the steering with a suitable puller.

To install:

6. Check that the front wheels are facing straight ahead. Center the spiral cable. When centering the spiral cable be sure to use the following procedure:

a. Check that the front wheels are pointing straight ahead.

b. Turn the spiral cable counterclockwise by hand until it becomes harder to turn the cable. The spiral cable will rotate approximately 2½ turns to either the left or right of the center.

c. Then rotate the spiral cable clockwise approximately 2½ turns to align the red mark.

7. Align the matchmarks and install the steering wheel. Tighten the set nut to 26 ft. lbs. (35 Nm).

8. Connect the connector.

9. Install the steering wheel pad after confirming that the circumference groove of the screws is caught on the screw case.

NOTE: Make sure that the wheel pad is installed to the specified torque. If the wheel pad has been dropped, or there are cracks, dents or other defects in the case or connector, replace the wheel pad with a new one. When installing the wheel pad, be sure that the wires and connectors do not interfere with other parts and are not pinched between other parts.

10. Tighten the screws to 65 inch lbs. Connect the negative battery cable.

Steering Column

REMOVAL & INSTALLATION

ES250

1. Disconnect the negative battery cable. Remove the steering wheel.

2. Using a centering punch, mark the center of the tapered-head bolts.

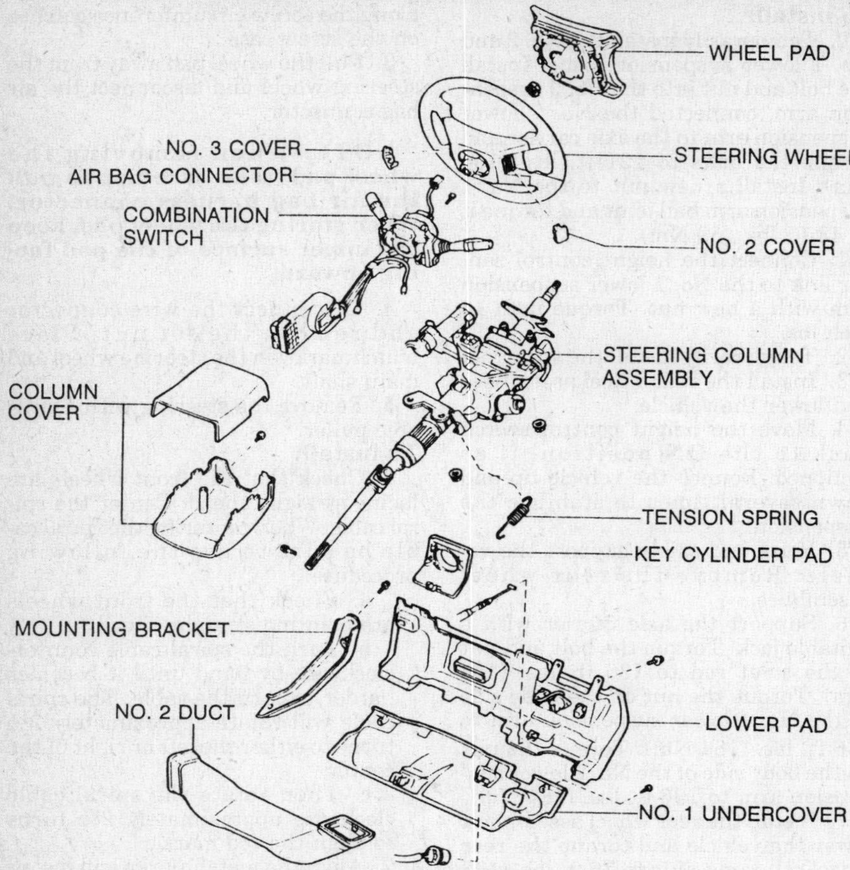

Steering column assembly—LS400

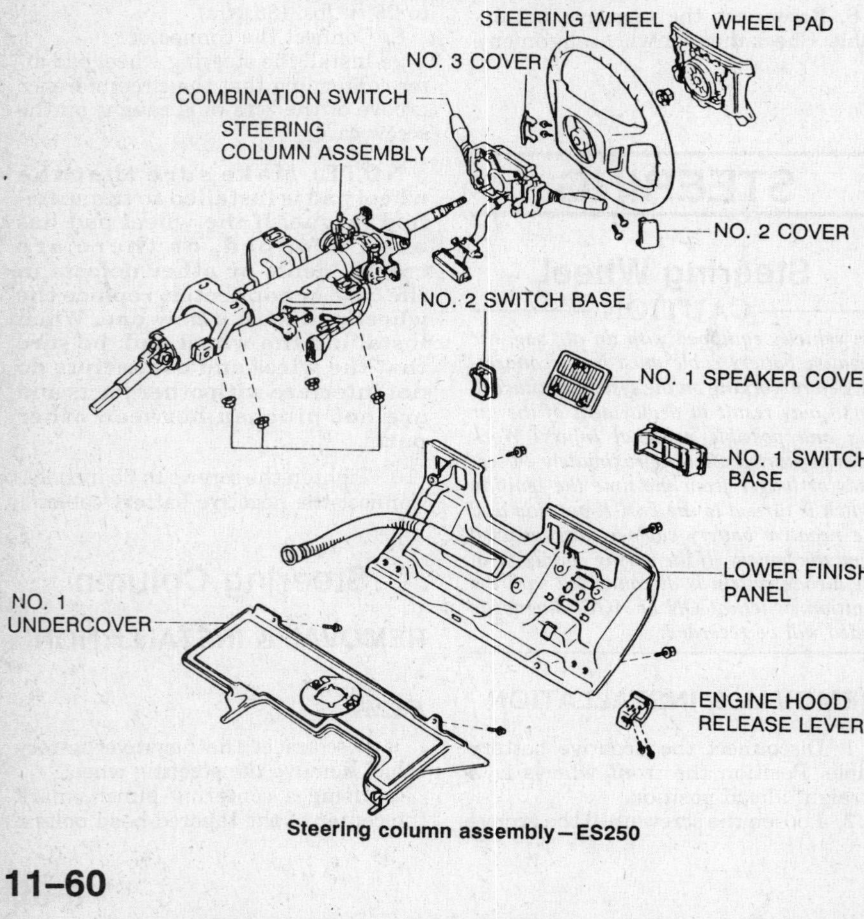

Steering column assembly—ES250

3. Drill into the tapered-head bolts, using a 0.12–0.16 in. (0.3–0.4mm) drill. Remove the tapered-head bolts.
4. Use a screw extractor and remove the bolts and separate the upper bracket and column tube. Remove the thrust stopper set bolts and disconnect the snapring.
5. Remove the main shaft. Using the proper tool, remove the snapring and thrust stopper.
6. The installation is the reverse of the removal procedure. Tighten the thrust stopper bolts to 9 ft. lbs. (12 Nm).

LS400

1. Disconnect the negative battery cable. Remove the steering wheel.
2. Disconnect the ignition key light. Remove the No. 2 intermediate shaft.
3. Remove the lower dust cover and gasket. Remove the tilt position sensor.
4. Remove the turn signal bracket, the lower shield protectors and connector bracket.
5. Using a centering punch, mark the center of the tapered-head bolts.
6. Drill into the tapered-head bolts, using a 0.16–0.20 in. (0.4–0.5mm) drill. Remove the tapered-head bolts.
7. Use a screw extractor and remove the bolts and separate the upper bracket and the break-away bracket.
8. Remove the nuts, spacers, springs, bushings and bond cable. Using the proper tool, remove the 2, No. 2 tilt steering shafts.
9. Remove the tilt steering gear assembly. Disconnect the nut and telescopic lever serration attachment.
10. Install a double nut on the lock bolt and remove the lock bolt. Remove the steering column tube upper stop bolt.
11. Remove the snapring from the mainshaft. Disconnect the break away bracket from the main shaft.
12. Remove the No. 1 and No. 2 telescopic spring seats and compression ring from the lower tube.
13. Remove the steering shaft thrust stopper from the break-away bracket. Remove the lock wedges from the break-away bracket.
14. Remove the upper tube with the main shaft and the tilt housing support.
15. Remove the main shaft using the proper tool to compress the mainshaft spring and remove the snapring.
16. Remove the main shaft from the upper tube. Remove the spring, thrust collar and bearing.

To install:

17. Install the spring, thrust collar and spring to the main shaft. Insert the main shaft in the upper tube.
18. Using the proper tool, compress

the main shaft spring and install the snapring.

19. Install the tilt steering upper housing support. Tighten to 9 ft. lbs. (12 Nm).

20. Insert the upper tube and the mainshaft into the lower tube. Press in the 2 tilt steering shafts.

21. Replace the lock wedges to the break-away bracket. Install the steering shaft thrust stopper to the break-away bracket.

22. Install the No. 1 and No. 2 telescopic spring seats and compression spring. Install the break-away bracket to the main shaft.

23. Install the snapring to the main shaft and the steering column tube stopper bolt. Tighten the bolt to 14 ft. lbs. (19 Nm).

24. Pull the steering column away from the break-away bracket and facing the body installation surface of the break-away bracket upward, temporarily install the lock bolt.

25. Install a double nut on the lock bolt. Tighten to 12 ft. lbs. (16 Nm), loosen once, then tighten again to 65 inch lbs.

26. Install the compression spring and ball, washer, lever, collar and bolt. Tighten the bolt to 19 ft. lbs. (26 Nm), with the depression of the lever matching the depression of the ball.

27. Rotate the telescopic lever until it touches the break-away bracket. Install the serration attachment so the alignment marks on the telescopic lever align with the serration attachment. Tighten to 9 ft. lbs. (12 Nm).

28. Install the tilt steering gear assembly with the motor by:

a. Install the bushings and support stopper bolts.

b. Press in the 2, No. 2 tilt steering shafts, using the proper tools.

c. Install the bushing, spring and spacer to each stopper bolt.

d. Install the bond cable and the locknuts. Tighten to 26 inch lbs.

29. Install the upper bracket and tighten the tapered-head bolts until the bolt heads break off.

30. Install the connector bracket and tighten to 43 inch lbs. Replace both protective covers.

31. Install the turn signal bracket so the upper surface is parallel with the upper surface of the capsule of the break-away bracket.

32. Install the tilt position sensor. Install the lower dust cover, using the proper tool to press in the dust seal.

33. Install the No. 2 intermediate shaft and tighten to 26 ft. lbs. (35 Nm). Replace the ignition key cylinder lamp.

34. Check that there is no axial play when the lever is turned fully counterclockwise.

35. Check that the main shaft moves smoothly when the lever is turned clockwise

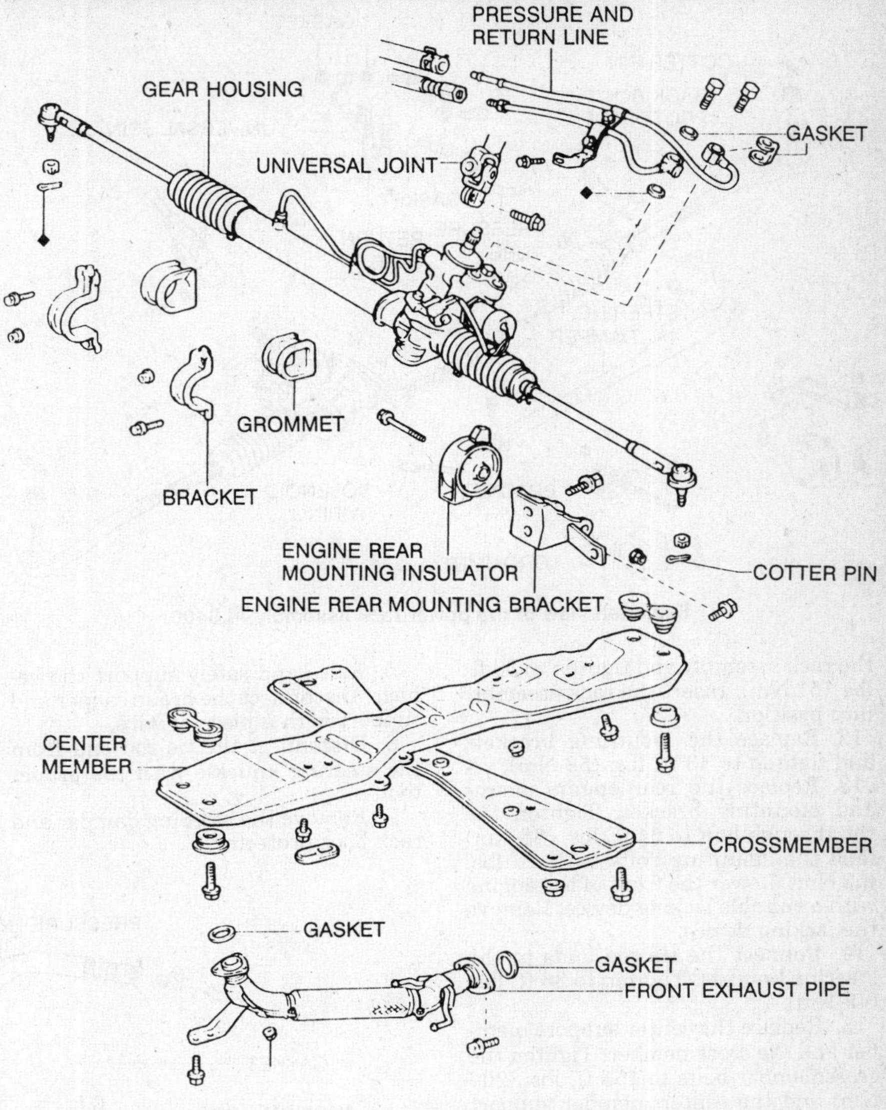

Exploded view of the power rack assembly—ES250

Power Steering Rack

REMOVAL & INSTALLATION

ES250

1. Place the front wheels in a straight ahead position, secure the steering wheel with a suitable device to prevent the wheel from turning.

2. Place matchmarks in the universal joint and control valve shaft. Disconnect the joint.

3. Disconnect and plug the hydraulic lines to rack assembly.

4. Raise and safely support the vehicle. Disconnect the front exhaust pipe.

5. Remove the center support member and the crossmember.

6. Matchmark and disconnect the tie rod ends from the steering knuckle with the proper tool.

7. Remove the rear engine mount and mounting bracket. Raise the front of the engine with a suitable jacking device.

NOTE: Do not over-tilt the engine

8. Remove the mounting brackets and slide the gear housing to the right side to put the tie rod end in the body panel.

9. Pull the gear housing out through the opening in the left side lower side of the vehicle body.

NOTE: Do not damage the turn pressure tube and the transmission control cables.

10. Secure the gear housing in a suitable holding fixture. Remove the return and pressure tubes.

To install:

11. Connect the hydraulic tubes to

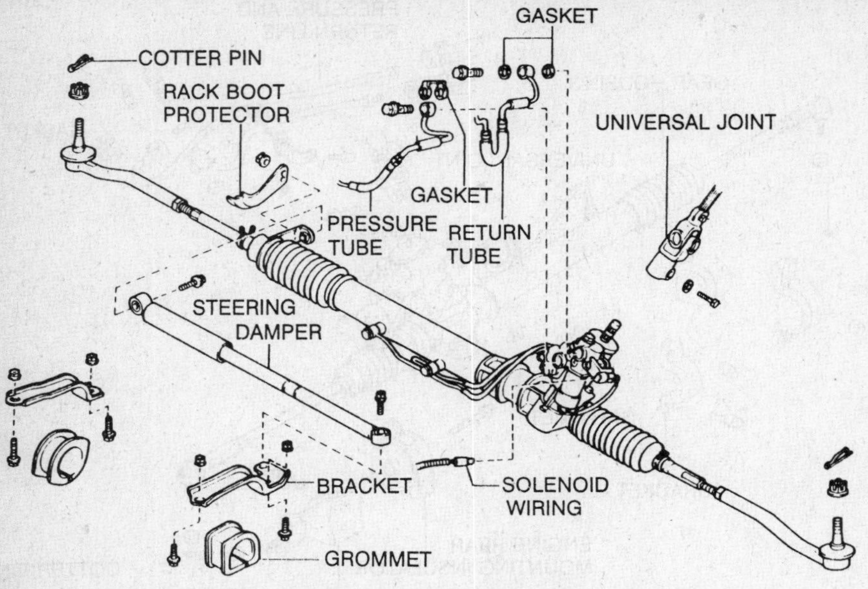

Exploded view of the power rack assembly—LS400

the rack assembly and tighten to 38 ft. lbs. (51 Nm). Insert the rack assembly into position.

12. Replace the mounting brackets and tighten to 43 ft. lbs. (58 Nm).

13. Replace the rear engine mount and mounting bracket. Tighten the the through bolt to 64 ft. lbs. (85 Nm) (and the mounting bolts to 38 ft. lbs. (52 Nm). Lower the front of the engine with a suitable jacking device. Remove the jacking device.

14. Connect the tie rod ends to the steering knuckle. Tighten to 36 ft. lbs. (49 Nm).

15. Replace the center support member and the crossmember. Tighten the crossmember bolts to 153 ft. lbs. (207 Nm) and the center member support bolts to 29 ft. lbs. (39 Nm).

16. Install the front exhaust pipe and lower the vehicle.

17. Connect the hydraulic lines to the rack assembly.

18. Align the matchmarks on the universal joint and the control valve shaft and connect. Tighten the connecting bolt to 26 ft. lbs. (35 Nm).

19. Check the steering wheel center point and the toe-in. Refill the power steering reservoir.

LS400

1. Disconnect the negative battery cable.

2. Place the front wheels in a straight ahead position, secure the steering wheel with a suitable device to prevent the wheel from turning.

3. Place matchmarks in the universal joint and control valve shaft. Disconnect the joint.

4. Disconnect and plug the hydraulic lines to rack assembly.

5. Raise and safely support the vehicle. Disconnect the brake caliper and support with a piece of wire.

6. Disconnect the tie rod end from the steering knuckle with the proper tool.

7. Remove the steering damper and rack boot protector.

8. Disconnect the solenoid wiring and remove the mounting grommets and brackets. Remove the rack assembly.

9. The installation is the reverse of the removal procedure. Follow the following torque specifications:

a. Joint connecting bolt—26 ft. lbs. (35 Nm).

b. Tie rod end nuts—43 ft. lbs. (58 Nm).

c. Steering damper bolts—20 ft. lbs. (27 Nm).

d. Mounting bracket bolts—56 ft. lbs. (76 Nm).

Power Steering Pump

REMOVAL & INSTALLATION

ES250

1. Disconnect the negative battery cable. Disconnect the hydraulic lines at the pump assembly.

2. Raise and safely support the vehicle. Disconnect the right side tie rod end with the proper tool.

3. Remove the lower crossmember and the front fender apron seal.

4. Loosen the adjusting and through bolt and push the power steering pump forward. Remove the drive belt.

5. Remove the adjusting bolt and

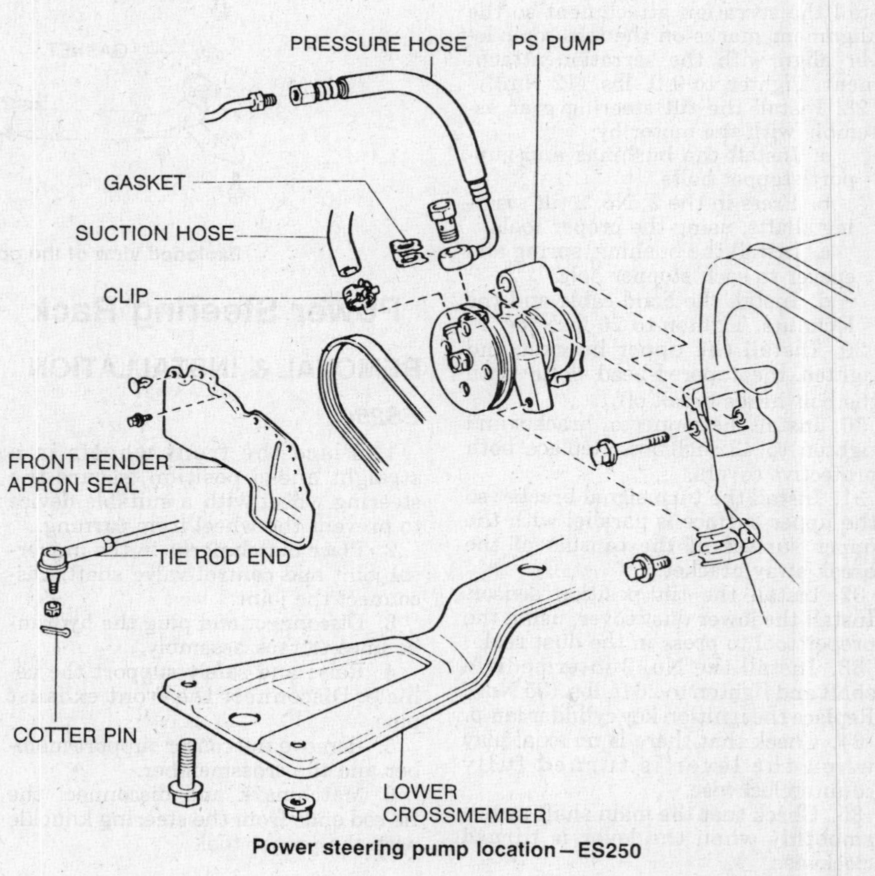

Power steering pump location—ES250

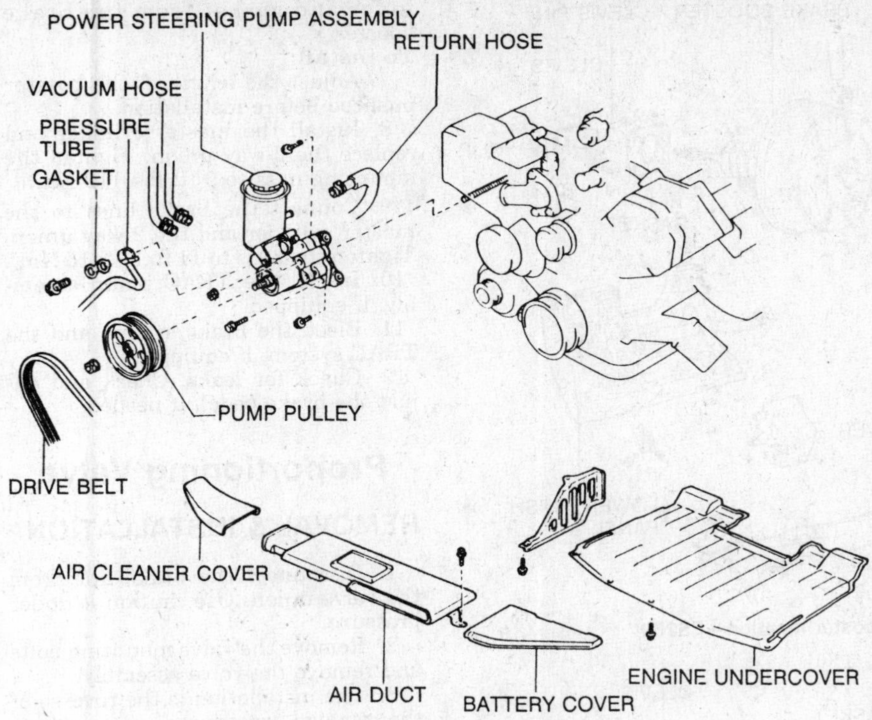

Power steering pump location—LS400

through bolt, then remove the power steering pump.

6. Remove the pressure hose from the pump.

7. The installation is the reverse of the removal procedure. Follow the following torque specifications:

a. Power steering pump mounting bolts—29 ft. lbs. (39 Nm).

b. Lower crossmember bolts—153 ft. lbs. (207 Nm).

c. Tie rod end nuts—36 ft. lbs. (49 Nm).

d. Pressure line fitting—27–33 ft. lbs. (37-45 Nm).

LS400

1. Disconnect the negative battery cable.

2. Remove the air cleaner cover, air duct and battery cover.

3. Turn the drive belt tensioner counterclockwise and remove the drive belt. Remove the pump pulley with the proper tool.

4. Raise and safely support the vehicle. Remove the engine undercover.

5. Disconnect the hydraulic and vacuum lines at the pump assembly.

6. Remove the pump mounting bolts and the pump assembly.

7. The installation is the reverse of the removal procedure. Follow the following torque specifications:

a. Pump mounting bolts—29 ft. lbs. (40 Nm).

b. Pump pulley bolt—32 ft. lbs. (44 Nm).

c. Pressure line fitting—36 ft. lbs. (49 Nm).

8. Bleed the power steering system.

NOTE: Note the curvature of the No. 2 and No. 3 power steering return tubes and the rear left hand engine cover have been modified. The new curvature is a 50 degree bend. Previous and new parts are interchangeable as a set only.

BELT ADJUSTMENT

On the LS400, a belt tensioner is used. There is no need for belt adjustment.

ES250

1. Disconnect the negative battery cable.

2. Loosen the power steering pump adjusting bolt and the through bolt.

3. Using the proper tool, move the pump assembly to attain the proper belt tension.

4. Tighten the pump mounting bolts.

5. Connect the negative battery cable.

SYSTEM BLEEDING

1. Check that the fluid level in the reservoir tank is at the maximum level.

2. Start the engine and turn the steering wheel from lock to lock until the air bubbles are removed from the fluid.

3. Stop the engine and measure the fluid level.

4. Make sure the rise of the fluid is not over 0.020 in.

Tie Rod Ends

REMOVAL & INSTALLATION

1. Raise and safely support the vehicle.

2. Place matchmarks on the threads of the tie rod end and the rack end.

3. On LS400, disconnect the brake caliper and suspend with a piece of wire. Do not disconnect the lines.

4. Remove the cotter pin and nut. Disconnect the tie rod end from the steering knuckle with the proper tool.

5. Unscrew the tie rod end from the rack end.

To install:

6. Install the tie rod end to the rack end, counting the same number of threads as were removed.

7. Connect the tie rod end to the steering knuckle. Tighten the nut to 41–43 ft. lbs.

8. Connect the caliper, if removed.

9. Lower the vehicle. Check the toe-in.

BRAKES

For all brake system repair and service procedures not detailed below, please refer to "Brakes" in the Unit Repair section.

Master Cylinder

REMOVAL & INSTALLATION

ES250

1. Disconnect the negative battery cable. Remove the charcoal canister.

2. Remove the fluid from the reservoir, using a suitable syringe.

3. Disconnect and plug the hydraulic lines to the master cylinder.

4. Disconnect the level warning switch connector. Remove the mounting nuts and the union and clamp.

5. Remove the master cylinder from the booster and remove the gasket.

To install:

6. Adjust the length of the booster pushrod before installing the master cylinder.

7. Install the master cylinder with a new gasket. Replace the union, clamp

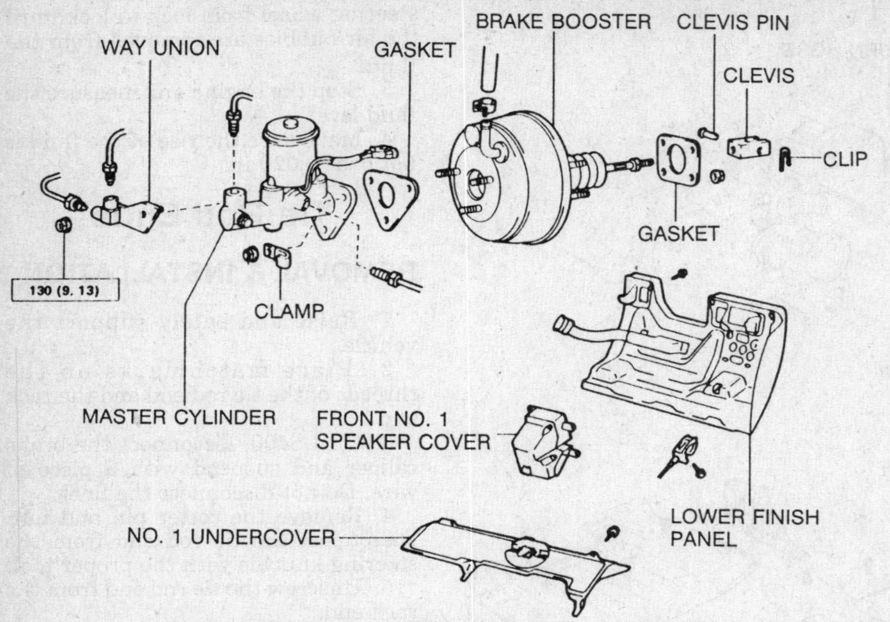

Brake master cylinder and booster location—ES250

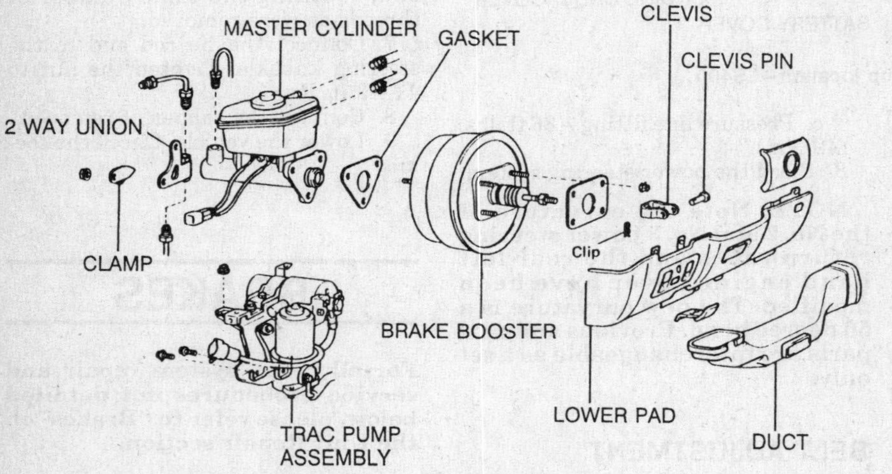

Brake master cylinder and booster location—LS400

and tighten the mounting nuts to 9 ft. lbs. (12 Nm).

8. Connect the brake lines with the proper tool and tighten the union nuts to 11 ft. lbs. (15 Nm).

9. Connect the level warning switch connector. Fill the reservoir with brake fluid.

10. Bleed the brake system and check for leaks.

LS400

1. Disconnect the negative battery cable.

2. If equipped with a Traction Control System (TRAC), perform the following procedure:

a. Remove the air cleaner.

b. Connect a vinyl tube from a container to the bleeder plug of the TRAC actuator, then loosen the bleeder plug with the ignition in the **OFF** position.

c. Tighten the plug when the fluid stops flowing out.

NOTE: The fluid is under high pressure and could spray out with great force, use caution.

3. Remove the fluid from the reservoir, using a suitable syringe.

4. If equipped with a TRAC, disconnect the hydraulic lines, mounting bolts and connector. Remove the TRAC pump assembly.

5. Disconnect the brake lines from the master cylinder.

6. Remove the mounting nuts and 2-way union. Remove the master cylinder and gasket from the brake booster.

To install:

7. Adjust the length of the booster pushrod before installation.

8. Install the master cylinder and replace the 2-way union. Tighten the mounting nuts to 9 ft. lbs. (12 Nm).

9. Connect the brake lines to the master cylinder and the 2-way union. Tighten the nuts to 11 ft. lbs. (15 Nm).

10. Install the TRAC pump assembly, if equipped.

11. Bleed the brake system and the TRAC system, if equipped.

12. Check for leaks. Check and adjust the brake pedal, if needed.

Proportioning Valve

REMOVAL & INSTALLATION

1. Disconnect the brake line from the valve union. Use caution if under pressure.

2. Remove the valve mounting bolts and remove the valve assembly.

3. The installation is the reverse of the removal procedure.

4. Bleed the brake system. Check for leaks.

Power Brake Booster

REMOVAL & INSTALLATION

ES250

1. Disconnect the negative battery cable. Remove the master cylinder and the vacuum hose.

2. Remove the No. 1 undercover and the lower finish panel. Disconnect the pedal return spring and the theft deterrent horn.

3. Remove the clip and clevis pin from the operating rod.

4. Remove the pedal bracket mounting bolt, the steering support nuts and the break-away bracket nuts. Lower the steering column.

5. Remove the mounting nuts and pull out the booster and gasket.

To install:

6. Adjust the length of the pushrod by:

a. Install the gasket on the master cylinder. Place the proper tool on the gasket and lower the pin of the tool until it's tip slightly touches the pin.

b. Turn the tool upside down and set it on the booster.

c. Measure the distance pushrod, the clearance is 0 in. (0mm). Adjust the booster pushrod length until the pushrod slightly touches the pin head.

7. Install the booster and gasket.

Replace the clevis pin to the operating rod.

8. Install and tighten the mounting nuts to 9 ft. lbs. (12 Nm).

9. Insert the clevis pin to the clevis and brake pedal. Install the clip to the clevis pin.

10. Lift up the steering column and install the steering support nuts and break-away nuts. Tighten to 19 ft. lbs. (26 Nm).

11. Install the pedal bracket mounting bolt and tighten to 13 ft. lbs. (18 Nm). Replace the pedal return spring.

12. Install the master cylinder and connect the vacuum hose to the brake booster.

13. Fill the reservoir with brake fluid and bleed the brake system. Check for leaks.

14. Check and adjust the brake pedal. Tighten the clevis locknut to 19 ft. lbs. (26 Nm).

15. Install the lower finish panel and No. 1 undercover. Connect the negative battery cable.

LS400

1. Disconnect the negative battery cable.

2. Remove the TRAC pump assembly, if equipped.

3. Remove the master cylinder and the vacuum hose from the booster.

4. Remove the No. 1 undercover and the lower finish panel. Disconnect the pedal return spring and the theft deterrent horn.

5. Remove the clip and clevis pin from the operating rod.

6. Remove the brake line between the TRAC accumulator and the actuator from the clamp.

7. Remove the booster mounting nuts and lift out the booster.

To install:

8. Adjust the length of the pushrod by:

a. Install the gasket on the master cylinder. Place the proper tool on the gasket and lower the pin of the tool until it's tip slightly touches the pin.

b. Turn the tool upside down and set it on the booster.

c. Measure the distance pushrod, the clearance is 0 in. (0mm). Adjust the booster pushrod length until the pushrod slightly touches the pin head.

9. Install the brake booster and gasket. Replace the clevis on the operating rod.

10. Install and tighten the mounting nuts to 9 ft. lbs. (12 Nm).

11. Insert the clevis pin into the clevis and brake pedal and install the clip on the clevis pin.

12. If equipped with a TRAC, install the brake tube between the TRAC accumulator and actuator to the No. 2 clamp.

13. Install the No. 1 undercover and lower finish panel.

14. Install the master cylinder and connect the vacuum hose to the booster.

15. Replace the TRAC pump, if equipped.

16. Fill the brake reservoir with brake fluid and bleed the brake system.

17. Check and adjust the brake pedal. Check for leaks.

18. Connect the negative battery cable.

Brake Caliper

REMOVAL & INSTALLATION

Front

1. Raise and safely support the vehicle. Remove the front tire and wheel assembly.

2. Disconnect and plug the brake line at the caliper.

3. Remove the mounting bolts and the caliper assembly.

4. Remove the brake pads, shims, springs and indicators from the caliper.

5. The installation is the reverse of the removal procedure.

6. Tighten the mounting bolts to 29 ft. lbs. (40 Nm) on the ES250 or 25 ft. lbs. (34 Nm) on the LS400.

7. Bleed the brake system.

Rear

1. Raise and safely support the vehicle. Remove the rear tire and wheel assembly.

2. Disconnect and plug the brake line at the caliper.

3. Remove the mounting bolts and the caliper assembly.

4. Remove the brake pads, shims, springs and indicators from the caliper.

5. The installation is the reverse of the removal procedure.

6. Tighten the mounting bolts to 29 ft. lbs. (40 Nm) on the ES250 or 25 ft. lbs. (34 Nm) on the LS400.

7. Bleed the brake system.

Disc Brake Pads

REMOVAL & INSTALLATION

Front

1. Raise and safely support the vehicle. Remove the front tire and wheel assembly.

3. Remove the mounting bolts and lift off the caliper assembly. Support the caliper. Do not disconnect the brake line.

4. Remove the brake pads, shims, springs and indicators from the caliper.

To install:

5. Install the pad support plates and the pad wear indicator plate on the inside pad.

6. Apply disc brake grease to both sides of the anti-squeal shims to each pad.

7. Install the inside pad with the wear indicator facing upward. Install the outside pad.

8. Install the anti-squeal springs. Press in the caliper piston with the proper tool and install the caliper. Tighten the mounting bolts to 29 ft. lbs. (40 Nm) on the ES250 or 25 ft. lbs. (34 Nm) on the LS400.

9. Install the front tire and wheel assembly. Lower the vehicle.

Rear

1. Raise and safely support the vehicle. Remove the rear tire and wheel assembly.

3. Remove the mounting bolts and lift off the caliper assembly. Support the caliper. Do not disconnect the brake line.

4. Remove the brake pads, shims, springs and indicators from the caliper.

To install:

5. Install the pad support plates and the pad wear indicator plate on the inside pad.

6. Apply disc brake grease to both sides of the anti-squeal shims to each pad.

7. Install the inside pad with the wear indicator facing upward. Install the outside pad.

8. Install the anti-squeal springs. Press in the caliper piston with the proper tool and install the caliper. Tighten the mounting bolts to 29 ft. lbs. (40 Nm) on the ES250 or 25 ft. lbs. (34 Nm) on the LS400.

9. Install the rear tire and wheel assembly. Lower the vehicle.

Brake Rotor

REMOVAL & INSTALLATION

Front

1. Raise and safely support the vehicle. Remove the front tire and wheel assembly.

2. Remove the mounting bolts and lift off the caliper assembly. Support the caliper. Do not disconnect the brake line.

3. Remove the torque plate from the steering knuckle, if equipped.

4. Remove the hub bolts and pull the brake rotor.

5. The installation is the reverse of

the removal procedure. Tighten the torque plate bolts to 79 ft. lbs. (107 Nm) on the ES250 and 87 ft. lbs. (118 Nm) on the LS400.

Rear

1. Raise and safely support the vehicle. Remove the rear and wheel assembly.
2. Remove the 2 retaining bolts from the rear disc brake caliper. Suspend the caliper with a suitable piece of wire so as not to stretch the brake hose.
3. Place matchmarks on the rotor disc and the rear axle shaft. Remove the 2 rotor retaining screws from the rotor and remove the rotor.

NOTE:If the rotor disc cannot be removed easily, return the shoe adjuster until the wheel turns freely.

4. The installation is the reverse of the removal procedure. Tighten the caliper retaining bolts to 34 ft. lbs. (47 Nm) on the ES250 and 77 ft. lbs. (104 Nm) on the LS400.

Parking Brake Shoes

REMOVAL & INSTALLATION

1. Raise and safely support the vehicle. Remove the rear and wheel assembly. On the ES250 models, remove the axle carrier mounting bolt and nut.
2. Remove the 2 retaining bolts from the rear disc brake caliper. Suspend the caliper with a suitable piece of wire so as not to stretch the brake hose.
3. Place matchmarks on the rotor disc and the rear axle shaft. Remove the 2 rotor retaining screws from the rotor and remove the rotor.

NOTE:If the rotor disc cannot be removed easily, return the shoe adjuster until the wheel turns freely.

4. Remove the parking brake shoe return springs.
5. Slide out the front shoe and remove the shoe adjuster. Remove the shoe strut with spring. Disconnect the tension spring and remove the front shoe.
6. Slide out the rear shoe. Disconnect the tension spring from the rear shoe. Disconnect the parking brake cable from the parking brake shoe lever. Remove the shoe hold down spring cups, springs and pin. Remove the rear shoe.
7. Installation is the reverse order of the removal procedure. On the ES250 models, torque the disc brake caliper bolts to 34 ft. lbs. (47 Nm). Torque the axle carrier bolt and nut of the upper side to 166 ft. lbs. (226 Nm).
8. On the LS400 models, torque the disc brake caliper bolts to 77 ft. lbs.

Parking Brake Cable

ADJUSTMENT

ES250

1. Pull the parking brake lever all the way up and count the number of clicks, should be 5–8 clicks.
2. Before adjusting the parking brake, make sure the rear parking brake shoe clearance is adjusted. Adjust by:
   a. Raise and safely support the vehicle. Remove the rear tire and wheel assembly.
   b. Remove the hole plug.
   c. Turn the adjuster and expand the shoes until the rotor locks.
   d. Return the adjuster 8 notches.
   e. Install the hole plug. Replace the tire and wheel assembly.
   f. Lower the vehicle.
3. Remove the console box.
4. Loosen the locknut and turn the adjusting nut until the lever travel is correct.
5. Tighten the locknut to 48 inch lbs. Install the console box.

LS400

1. Depress the parking brake all the way and count the number of clicks, should be 5–7 clicks.
2. Before adjusting the parking brake, make sure the rear parking brake shoe clearance is adjusted. Adjust by:
   a. Raise and safely support the vehicle. Remove the rear tire and wheel assembly.
   b. Remove the hole plug.
   c. Turn the adjuster and expand the shoes until the rotor locks.
   d. Return the adjuster 8 notches.
   e. Install the hole plug. Replace the tire and wheel assembly.
   f. Lower the vehicle.
3. Raise and safely support the vehicle.
4. Loosen the lock adjuster locknut and adjuster until the parking brake pedal travel is correct. Tighten the locknut.
5. The installation is the reverse of the removal procedure.

Brake System Bleeding

1. Fill the reservoir to the maximum level with brake fluid.
2. To bleed the master cylinder:
   a. Disconnect the brake lines from the master cylinder.
   b. Depress the brake pedal and hold it.
   c. Block off the outer holes with fingers and release the brake pedal. Repeat 2–3 times.
3. To bleed the wheels:
   a. Start bleeding the brakes from the farthest point.
   b. Connect a vinyl tube to the brake cylinder bleeder plug and insert the other end of the tube in a ½ full container of brake fluid.
   c. Press on the brake pedal and loosen the bleeder plug until brake fluid comes out.
   d. Repeat until there is no more air bubbles in the fluid. Tighten the bleeder screw.
   e. Repeat the procedure for each wheel.
4. To bleed TRAC Control System:
   a. Remove the air cleaner, then temporarily reinstall it so the engine can be started.
   b. Connect a vinyl tube to the bleeder plug of the TRAC actuator, then loosen the bleeder plug.
   c. Start the engine, then operate the TRAC pump motor until all the air has been bled out of the fluid.
   d. Tighten the bleeder screw and stop the engine. Install the air cleaner.

Anti-Lock Brake System Service

The ABS system controls the hydraulic pressure of all 4 wheels during sudden braking and braking on slippery road surfaces, preventing the wheels from locking.

ABS Actuator

REMOVAL & INSTALLATION

ES250

1. Disconnect the negative battery cable. Remove brake fluid using a proper syringe.
2. Remove the actuator cover and disconnect the connectors from the actuator.
3. Remove the cover bracket and stud bolt.
4. Disconnect the brake tubes from the actuator, mounting nuts, wave washers and washers.
5. Remove the actuator from the actuator bracket.

To install:

6. Install the actuator to the actuator bracket. Replace the washers, wave washers and tighten the mounting nuts to 48 inch lbs.

7. Connect the brake tubes to the actuator with the proper tool and tighten to 11 ft. lbs. (15 Nm).
8. Connect the connectors to the actuator. Install the stud and bracket.
9. Install the actuator cover.
10. Fill the brake reservoir to the proper level and bleed the brake system.
11. Connect the negative battery cable. Check for leaks.

LS400

1. Disconnect the negative battery cable. Remove brake fluid using a proper syringe.
2. Disconnect the dust cover, air cleaner and the air duct.
3. Disconnect the brake line from the ABS, actuator with the proper tool. Disconnect the brake lines from the TRAC actuator, if equipped.
4. Remove the mounting bolts from the ABS actuator or TRAC actuator, if equipped.
5. Disconnect the connectors and remove the actuator.

To install:

6. Install the ABS actuator or TRAC actuator, if equipped. Tighten the mounting bolts to 9 ft. lbs. (12 Nm). Connect the actuator connectors.
7. Using the proper tool, connect the brake lines to the actuator and tighten to 11 ft. lbs. (15 Nm).
8. Connect the dust cover, air cleaner and the air duct.
9. Bleed the brake system and connect the negative battery cable.
10. Check for leaks.

Front Speed Sensor

REMOVAL & INSTALLATION

ES250

1. Disconnect the negative battery cable.
2. Raise and safely support the vehicle. Remove the tire and wheel assembly.
3. Disconnect the speed sensor connector.
4. Remove the front hub and steering knuckle assembly.
5. Remove the nut and the speed sensor rotor. Do not scratch the serrations of the speed sensor rotor.
6. The installation is the reverse of the removal procedure.

LS400

1. Disconnect the negative battery cable.
2. Raise and safely support the vehicle. Remove the tire and wheel assembly.
3. Disconnect the speed sensor connector.
4. Remove the front axle hub from the steering knuckle. Remove the sensor control rotor from the axle hub with the proper tool.
5. The installation is the reverse of the removal procedure.

Rear Speed Sensor

REMOVAL & INSTALLATION

ES250

1. Disconnect the negative battery.
2. Remove the rear seat cushion and rear side seat back cushion.
3. Disconnect the sensor connector and pull out the sensor wire harness. Remove the 2 clamp bolts holding the sensor wire harness to the body and suspension arm.
4. Remove the axle carrier mounting bolt and nut of the upper side. Remove the 2 bolts and remove the disc brake caliper assembly, Be sure to suspend the disc brake caliper with a piece of safety wire.
5. Remove the rotor disc.
6. Remove the 4 axle hub mounting bolts. Remove the rear axle hub. Remove the backing plate with the parking brake assembly and O-ring.
7. Remove the speed sensor bolts and remove the speed sensor from the backing plate.
8. Installation is the reverse order of the removal procedure. Torque speed sensor retaining bolts to 69 in. lbs. Install the rear hub retaining bolts and 59 ft. lbs. (80 Nm). Torque the brake caliper retaining bolts to 34 ft. lbs. (47 Nm). Torque the axle carrier mounting bolt and nut of the upper side to 166 ft. lbs. (226 Nm).

LS400

1. Disconnect the negative battery cable.
2. Raise and safely support the vehicle. Remove the tire and wheel assembly.
3. Disconnect the speed sensor connector.
4. Remove the rear axle hub and backing plate.
5. Remove the mounting bolts and the speed sensor from the backing plate.
6. The installation is the reverse of the removal procedure.

CHASSIS ELECTRICAL

Air Bag

DISARMING

On vehicles equipped with an air bag, the negative battery cable must be disconnected, before working on the system. Failure to do so may result in deployment of the air bag and possible personal injury. Work must be started after approximately 20 seconds or longer from the time the ignition switch is turned to the **LOCK** position and the negative battery cable is dis-

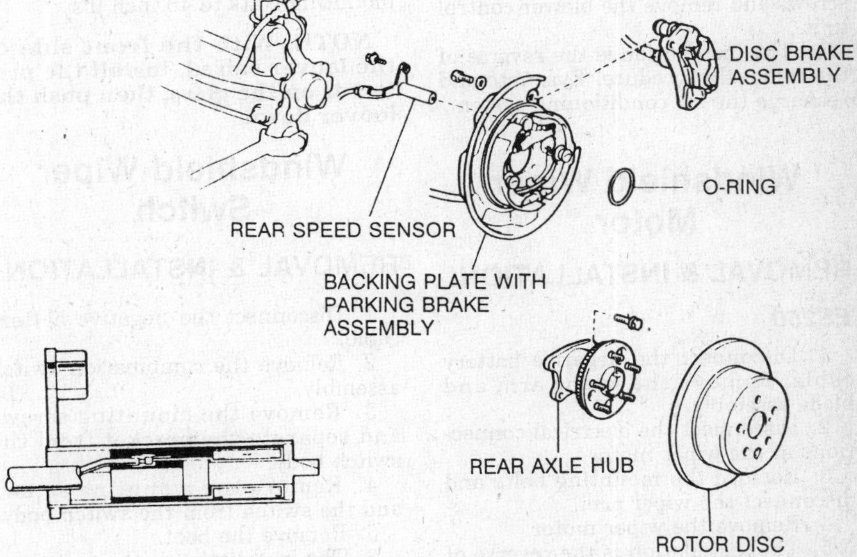

Exploded view of the ABS rear wheel speed sensor

connected from On vehicles equipped with an air bag, the negative battery cable must be disconnected, before working on the system.

Heater Blower Motor

REMOVAL & INSTALLATION

ES250

1. Disconnect the negative battery cable.
2. Disconnect the electrical connection at the blower motor.
3. Remove the mounting bolts and the blower motor.
4. The installation is the reverse of the removal procedure.

LS400

1. Disconnect the negative battery cable. Remove the right side instrument panel cover.
2. Properly discharge the air conditioning system. Remove the cruise control actuator.
3. Remove the bolt and disconnect the actuator tubes.
4. Disconnect both the liquid and the suction tubes for the air conditioner.
5. Remove the cover plate to the control assembly.

NOTE: Cap the open tubes to keep moisture out of the system.

6. Disconnect the drain hose clamp. Remove the mirror control electronic unit.
7. Disconnect the electrical connections and the mounting bracket from the blower control.
8. Disconnect the vehicle side wire harness from the blower unit.
9. Remove the mounting nuts and screws and remove the blower control unit.
10. The installation is the reverse of the removal procedure. Evacuate and recharge the air conditioning system.

Windshield Wiper Motor

REMOVAL & INSTALLATION

ES250

1. Disconnect the negative battery cable. Remove the wiper arm and blade assembly.
2. Disconnect the electrical connections at the wiper motor.
3. Remove the mounting bolts and disconnect the wiper arm.
4. Remove the wiper motor.
5. The installation is the reverse of the removal procedure.

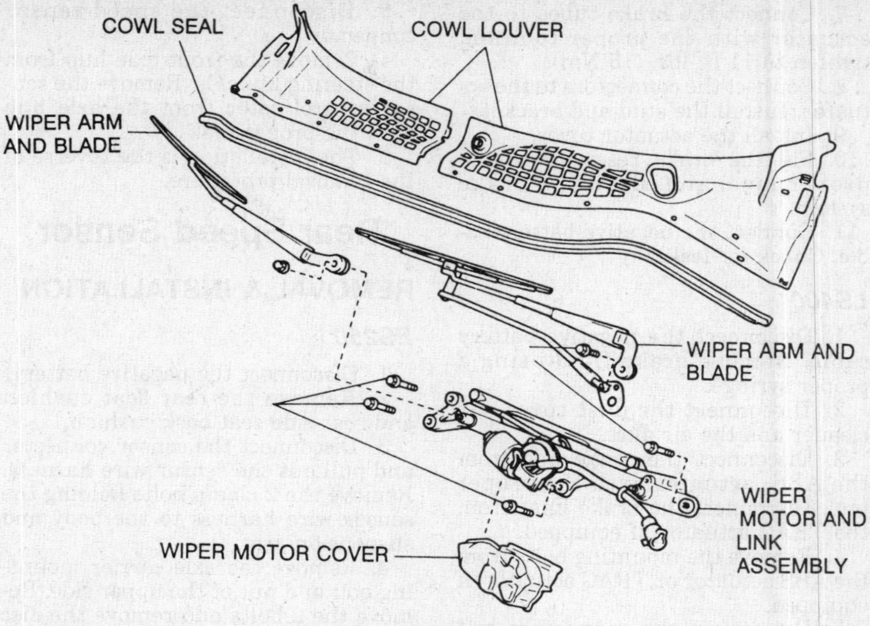

Wiper motor location—LS400

LS400

1. Disconnect the negative battery cable. Remove the wiper arm and blade assembly.
2. Remove the hood to cowl seal and remove the cowl louver.

NOTE: Raise the front side of the cowl louver up to remove the louver.

3. Remove the mounting bolts and disconnect the electrical connection.
4. Raise the front side of the wiper motor and link assembly up. Remove the wiper motor and link assembly.
5. Remove the wiper motor cover.
6. The installation is the reverse of the removal procedure. Tighten the mounting bolts to 48 inch lbs.

NOTE: With the front side of the louver raised, install the protector on the glass, then push the louver down.

Windshield Wiper Switch

REMOVAL & INSTALLATION

1. Disconnect the negative battery cable.
2. Remove the combination switch assembly.
3. Remove the mounting screws and separate the bracket from the switch body.
4. Remove the mounting screws and the switch from the switch body.
5. Remove the boot.
6. The installation is the reverse of the removal procedure.

Instrument Cluster

REMOVAL & INSTALLATION

ES250

1. Disconnect the negative battery cable.

NOTE: Work must be started after approximately 20 seconds or longer from the time the ignition switch is turned to the LOCK position and the negative battery cable is disconnected.

2. Remove the steering wheel.
3. Pry out the clips and pull the front pillar molding upward and remove.
4. Remove the left lower dash panel. Disconnect the hood release lever and remove the panel cover.
5. Pry out the switch bases and the speaker panel with the proper tool. Remove the mounting screws and lower the lower column panel.
6. Remove the mounting screws and the cluster panel. Disconnect the electrical connections.
7. Remove the steering column cover.
8. Disconnect the mounting screws, the speedometer cable and the electrical connections. Remove the instrument cluster and the speedometer.
9. The installation is the reverse of the removal procedure.

LS400

1. Disconnect the negative battery cable.

NOTE: Work must be started after approximately 20 seconds or longer from the time the ignition switch is turned to the LOCK position and the negative battery cable is disconnected.

2. Remove the steering wheel. Remove the left and right front pillar moldings.
3. Remove the steering column cover and the upper console panel, with the proper tool, to prevent damaging the cover.
4. Remove the front ash receptacle and disconnect the electrical connection.
5. Pry out the front side of the lower console cover and remove by sliding the cover forward.
6. Remove the mounting screws and pry out the lower console box. Remove the cup holder.
7. Pry out the rear end panel and disconnect the wire connector. Remove the console box.
8. Remove the left lower trim panel. Remove the hood release lever and disconnect the cable from the lever.
9. Remove the mounting bolts and the left trim pad. Disconnect the wire connections and hose from the pad.
10. Remove the key cylinder pad and disconnect the park brake lever. Remove the outer mirror switch assembly and disconnect the electrical connection.
11. Remove the mounting screws and carefully pry out the instrument cluster. Remove the speedometer from the cluster.
12. The installation is the reverse of the removal procedure.

NOTE: A rattle or squeak noise from the glove box door arear may be apparent when the glove box door is closes and/or the vehicle is driven over rough roads. This condition may be caused by the movement of the glove box door check arm against the glove box door and/or the check arm through hole in the dash.

Speedometer

REMOVAL & INSTALLATION

1. Disconnect the negative battery cable.
2. Remove the combination meter from the instrument cluster.
3. Remove the speedometer from the combination meter.
4. Installation is the reverse order of the removal procedure.

Combination Switch

The combination switch incorporates the headlight switch, turn signal switch, dimmer switch and the windshield wiper switch.

REMOVAL & INSTALLATION

1. Disconnect the negative battery cable.
2. On the ES250 models, remove the instrument panel No. 1 under cover subassembly.Remove the instrument panel lower finish panel and cluster finish panel.
3. On the LS400 models, remove the under cover, lower pad, key cylinder pad. The No. 3 finish panel mounting bracket. The No. 2 heater to register duct.

NOTE: Position the front wheels in a straight ahead position. Work must be started after approximately 20 seconds or longer from the time the ignition switch is turned to the LOCK position and the negative battery cable is disconnected from the battery. If the wiring connector of the airbag system is disconnected with the ignition switch at ON or ACC, diagnostic coded will be recorded.

4. Remove the steering wheel center pad.

NOTE: When removing the wheel pad, take care not to pull the air bag harness connector. When storing the wheel pad, keep the upper surface of the pad facing upward. Since the air bag con-

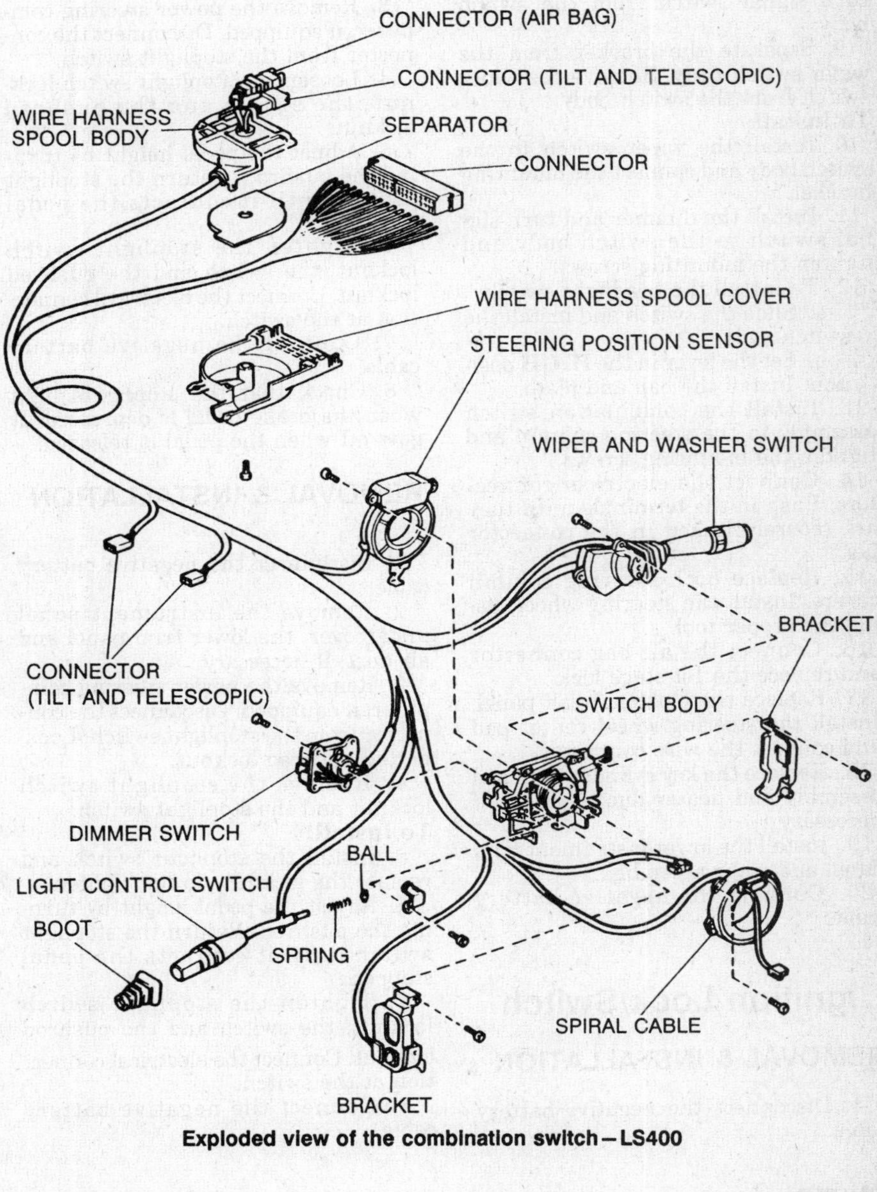

Exploded view of the combination switch – LS400

nector has a 2-stage lock, remove the 1st stage lock and disconnect the connector.

5. Remove the steering wheel assembly.
6. Remove the steering column cover. Remove the 4 combination switch retaining switch. Disconnect the connectors and remove the combination switch assembly from the steering column.
7. To remove the headlight switch:
   a. Remove the mounting screws and separate the bracket from the switch body.
   b. Remove the screws and the ball set plate from the switch body.
   c. Remove the ball and slide out the switch from the switch body with the spring.
   d. Remove the boot.
8. Loosen the mounting screws and remove the dimmer switch and the turn signal switch from the switch body.
9. Separate the bracket from the wiper switch body. Remove the wiper switch from the switch body.

To install:

10. Install the wiper switch to the switch body and connect the mounting bracket.
11. Install the dimmer and turn signal switch to the switch body and tighten the mounting screws.
12. To install the headlight switch:
   a. Slide the switch and install the switch body.
   b. Set the lever in the **HIGH** position. Install the ball and plate.
13. Install the combination switch assembly to the steering column and tighten the mounting screws.
14. Connect the electrical connectors. Push in the terminals until they are securely locked in the connector lug.
15. Replace both steering column covers. Install the steering wheel, using the proper tool.
16. Connect the air bag connector and replace the 1st stage lock.
17. Replace the cluster finish panel. Install the steering wheel center pad and connect the wire connector.
18. Replace the key cylinder trim pad assembly and heater duct register, if necessary.
19. Install the lower instrument trim panel and cover assembly.
20. Connect the negative battery cable.

Ignition Lock/Switch

REMOVAL & INSTALLATION

1. Disconnect the negative battery cable.
2. Remove the lower trim panel and pad assembly, if necessary.
3. Remove the key cylinder trim panel and pad assembly. Disconnect the key cylinder lamp assembly.
4. Remove the trim panel mounting bracket and the heater duct register, if necessary.
5. Remove the mounting screws and disconnect the electrical connections. Remove the ignition switch.
6. The installation is the reverse of the removal procedure.

Stoplight Switch

ADJUSTMENT

1. Disconnect the negative battery cable.
2. Remove the instrument panel undercover, the lower trim panel and air duct, if necessary.
3. Remove the power steering computer, if equipped. Disconnect the connector from the stoplight switch.
4. Loosen the stoplight switch locknut, the switch and the pushrod locknut.
5. Adjust the pedal height by turning the pushrod. Return the stoplight switch until it contacts the pedal stopper.
6. Tighten the stoplight switch locknut, the switch and the pushrod locknut. Connect the electrical connection at the switch.
7. Connect the negative battery cable.
8. Check that the stoplights light when the brake pedal is depressed. It goes off when the pedal is released.

REMOVAL & INSTALLATION

1. Disconnect the negative battery cable.
2. Remove the instrument panel undercover, the lower trim panel and air duct, if necessary.
3. Remove the power steering computer, if equipped. Disconnect the connector from the stoplight switch. Loosen the pushrod locknut.
4. Remove the stoplight switch locknut and the stoplight switch.

To install:

5. Install the stoplight switch and replace the locknut, do not tighten.
6. Adjust the pedal height by turning the pushrod. Return the stoplight switch until it contacts the pedal stopper.
7. Tighten the stoplight switch locknut, the switch and the pushrod locknut. Connect the electrical connection at the switch.
8. Connect the negative battery cable.
9. Check that the stoplights light when the brake pedal is depressed and off when the pedal is released.

Clutch Switch

ADJUSTMENT

ES250

1. Disconnect the negative battery cable.
2. Remove the lower dash trim panel and disconnect the air duct.
3. Loosen the locknut and disconnect the connection at the clutch switch. Turn the clutch switch until the pedal height is 7.52–7.91 in. from stop.
4. Tighten the locknut and connect the electrical connection at the switch.
5. Connect the air duct and replace the lower dash panel.
6. Connect the negative battery cable.

Neutral Safety Switch

ADJUSTMENT

1. Disconnect the negative battery cable. Raise and safely support the vehicle.
2. Loosen the neutral start switch bolt and place the shift lever in the **N** position.
3. Align the groove and the neutral basic line.
4. Hold in position and tighten the bolt to 9 ft. lbs.
5. Lower the vehicle and connect the negative battery cable.

Fuses, Circuit Breakers and Relays

LOCATION

There is a fuse block, relay center and circuit breaker center located on the left lower portion of the instrument panel. There is a fuse block and relay center located on the right lower portion of the instrument panel. There is a fuse block and relay center located in the engine compartment on the driver's side.

Flashers

LOCATION

The turn signal and hazard flashers are located on the left lower section of the instrument panel.

12

Mazda

323, 626, 929, MX6, Miata, Protege, RX-7—All Models

SERIAL NUMBER IDENTIFICATION

Vehicle Identification Plate

The serial number is on a plate located on the driver's side windshield pillar and is visible through the glass.

A Vehicle Identification Number (VIN) plate, bearing the serial number and other data, is attached to the cowl.

Engine Number

The engine number is located on a plate which is attached to the engine block, just behind or below the distributor or on a machined pad at the right front side of the engine block.

The engine number consists of an identification number followed by a 6-digit production number.

The 626, MX-6 and RX-7 engine serial number is located on the rear of the alternator bracket, stamped on the engine block. The 323 and Protege engine serial number is stamped on the left side of the engine block, just below the cylinder head. The 929 engine serial number is stamped on the cylinder block, just below the distributor. The Miata engine serial number is stamped on the cylinder block, right in front of the intake manifold.

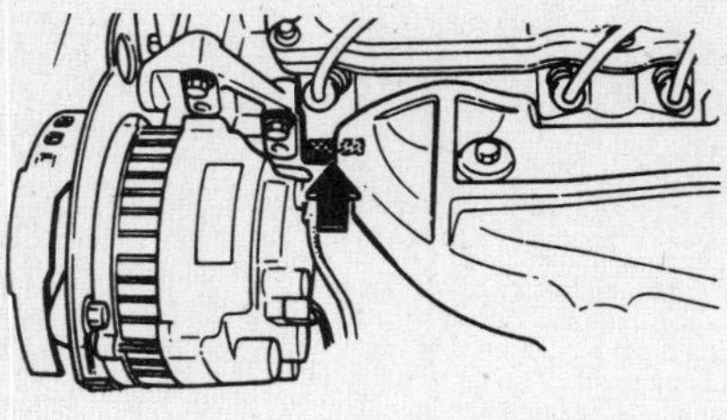

Engine number location

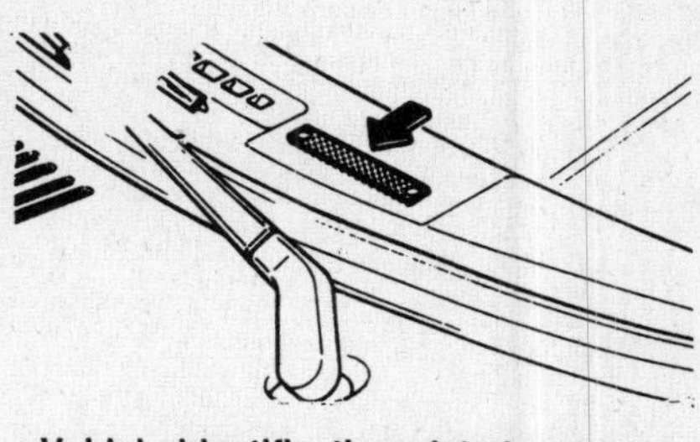

Vehicle identification plate location

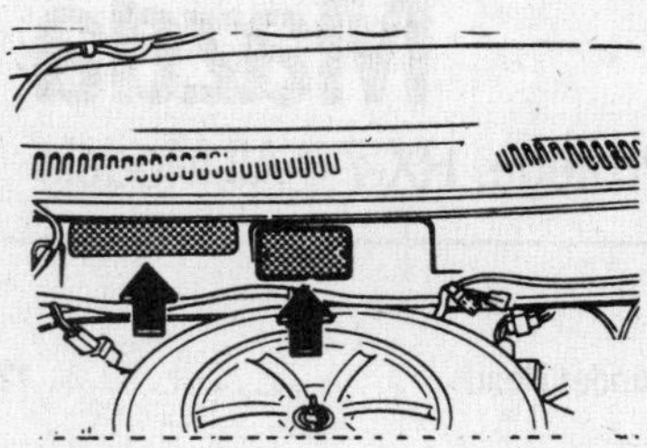

Chassis number (left) and model plate (right) location

Vehicle Identification Label

In addition to the serial numbers, other important vehicle information and specifications can be found on the underhood emission sticker, vacuum hose routing diagram, tire pressure label and motor vehicle safety certification label.

Transmission/Transaxle Number

The transmission/transaxle model and serial number are either stamped on a plate that is bolted to the transmission case or stamped directly on the case. The location varies from model-to-model.

SPECIFICATIONS

ENGINE IDENTIFICATION

Year	Model	Engine Displacement cu. in. (cc/liter)	Engine Series Identification	No. of Cylinders	Engine Type
1988	323	97.4 (1597/1.6)	B6	4	SOHC
	323	97.4 (1597/1.6)	B6	4	DOHC-Turbo
	626	133.2 (2184/2.1)	F2	4	SOHC
	626	133.2 (2184/2.1)	F2	4	SOHC-Turbo
	MX6 GT	133.2 (2184/2.1)	F2	4	SOHC
	MX6	133.2 (2184/2.1)	F2	4	SOHC-Turbo
	RX7	80.0 (1308/1.3)	13B	—	Rotary
	RX7	80.0 (1308/1.3)	13B	—	Rotary-Turbo
	929	180.2 (2954/2.9)	JE	6	DOHC
1989	323	97.4 (1597/1.6)	B6	4	SOHC
	323	97.4 (1597/1.6)	B6	4	DOHC-Turbo
	626	133.2 (2184/2.1)	F2	4	SOHC
	626 GT	133.2 (2184/2.1)	F2	4	SOHC-Turbo
	MX6	133.2 (2184/2.1)	F2	4	SOHC

ENGINE IDENTIFICATION

Year	Model	Engine Displacement cu. in. (cc/liter)	Engine Series Identification	No. of Cylinders	Engine Type
1989	MX6 GT	133.2 (2184/2.1)	F2	4	SOHC-Turbo
	RX7	80.0 (1308/1.3)	13B	—	Rotary
	RX7	80.0 (1308/1.3)	13B	—	Rotary-Turbo
	929	180.2 (2954/2.9)	JE	6	SOHC
1990	323	97.4 (1597/1.6)	B6E	4	SOHC
	Protege	112.2 (1839/1.8)	BPE	4	SOHC
	Protege LX	112.2 (1839/1.8)	BPD	4	DOHC
	626	133.2 (2184/2.1)	F2	4	SOHC
	626 GT	133.2 (2184/2.1)	F2	4	SOHC-Turbo
	MX6	133.2 (2184/2.1)	F2	4	SOHC
	MX6 GT	133.2 (2184/2.1)	F2	4	SOHC-Turbo
	929	180.2 (2954/3.0)	JE	6	SOHC
	929	180.2 (2954/3.0)	JE	6	DOHC
	RX7	80.0 (1308/1.3)	13B	—	Rotary
	RX7	80.0 (1308/1.3)	13B	—	Rotary-Turbo
	Miata	97.4 (1597/1.6)	B6ZE	4	DOHC
1991-92	323	97.4 (1597/1.6)	B6E	4	SOHC
	Protege	112.5 (1839/1.8)	BPE	4	SOHC
	Protege LX	112.5 (1839/1.8)	BPD	4	DOHC
	626	133.3 (2184/2.2)	F2	4	SOHC
	626 GT	133.3 (2184/2.2)	F2	4	SOHC-Turbo
	MX6	133 (2184/2.2)	F2	4	SOHC
	MX6 GT	133 (2184/2.2)	F2	4	SOHC-Turbo
	929	180 (2954/3.0)	JE	6	SOHC
	929S	180 (2954/3.0)	JE	6	DOHC
	RX7	80 (1308/1.3)	13BRE	—	Rotary
	RX7	80 (1308/1.3)	13BRE	—	Rotary-Turbo
	Miata	97.4 (1597/1.6)	B6ZE	4	DOHC

SOHC—Single Overhead Cam
DOHC—Double Overhead Cam

GENERAL ENGINE SPECIFICATIONS

Year	Model	Engine Displacement cu. in. (cc)	Fuel System Type	Net Horsepower @ rpm	Net Torque @ rpm (ft. lbs.)	Bore × Stroke (in.)	Compression Ratio	Oil Pressure @ rpm
1988	323	97.4 (1597)	EFI	82 @ 5000	92 @ 2500	3.07 × 3.29	9.3:1	50–64
	323 Turbo	97.4 (1597)	EFI	132 @ 6000	136 @ 3000	3.07 × 3.29	7.9:1	50–64
	626	133.2 (2184)	EFI	97 @ 5000	120 @ 2500	3.39 × 3.70	8.6:1	43–57
	626 Turbo	133.2 (2184)	EFI	125 @ 5000	155 @ 3500	3.39 × 3.70	7.8:1	43–57
	MX6	133.2 (2184)	EFI	97 @ 5000	120 @ 2500	3.39 × 3.70	8.6:1	43–57
	MX6 Turbo	133.2 (2184)	EFI	125 @ 5000	155 @ 3500	3.39 × 3.70	7.8:1	43–57
	929	180.2 (2954)	EFI	158 @ 5500	170 @ 4000	3.54 × 3.05	8.5:1	50–64

GENERAL ENGINE SPECIFICATIONS

Year	Model	Engine Displacement cu. in. (cc)	Fuel System Type	Net Horsepower @ rpm	Net Torque @ rpm (ft. lbs.)	Bore × Stroke (in.)	Com-pression Ratio	Oil Pressure @ rpm
1989	323	97.4 (1597)	EFI	82 @ 5000	92 @ 2500	3.07 × 3.29	9.3:1	50–64
	323 Turbo	97.4 (1597)	EFI	132 @ 6000	136 @ 3000	3.07 × 3.29	7.9:1	50–64
	626	133.2 (2184)	EFI	110 @ 4700	130 @ 3000	3.39 × 3.70	8.6:1	43–57
	626 GT	133.2 (2184)	EFI	145 @ 4300	190 @ 3500	3.39 × 3.70	7.8:1	43–57
	MX6	133.2 (2184)	EFI	110 @ 4700	130 @ 3000	3.39 × 3.70	7.8:1	43–57
	MX6 GT	133.2 (2184)	EFI	145 @ 4300	110 @ 3500	3.39 × 3.70	8.6:1	43–57
	929	180.2 (2954)	EFI	158 @ 5500	170 @ 4000	3.54 × 3.05	8.5:1	41–61
1990	323	97.4 (1597)	EFI	82 @ 5000	92 @ 2500	3.07 × 3.29	9.3:1	50–64
	Protege	112.2 (1839)	EFI	103 @ 5500	111 @ 4000	3.27 × 3.35	8.9:1	50–64
	Protege LX	112.2 (1839)	EFI	125 @ 6500	114 @ 4500	3.27 × 3.35	9.0:1	50–64
	626	133.2 (2184)	EFI	110 @ 4700	130 @ 3000	3.39 × 3.70	8.6:1	43–57
	626 GT	133.2 (2184)	EFI	145 @ 4300	190 @ 3500	3.39 × 3.70	7.8:1	43–57
	MX6	133.2 (2184)	EFI	110 @ 4700	130 @ 3000	3.39 × 3.70	8.6:1	43–57
	MX6 GT	133.2 (2184)	EFI	145 @ 4300	190 @ 3500	3.39 × 3.70	7.8:1	43–57
	929	180.2 (2954)	EFI	158 @ 5500	170 @ 4000	3.54 × 3.05	8.5:1	53–75
	929 S	180.2 (2954)	EFI	190 @ 5600	191 @ 4500	3.54 × 3.05	8.5:1	53–75
	Miata	97.4 (1597)	EFI	116 @ 6500	100 @ 5500	3.10 × 3.30	9.4:1	43–57
1991-92	323	97.4 (1597)	EFI	82 @ 5000	92 @ 2500	3.07 × 3.29	9.3:1	50–64
	Protege	112.2 (1839)	EFI	103 @ 5500	111 @ 4000	3.27 × 3.35	8.9:1	50–64
	Protege LX	112.2 (1839)	EFI	125 @ 6500	114 @ 4500	3.27 × 3.35	9.0:1	50–64
	626	133.2 (2184)	EFI	110 @ 4700	130 @ 3000	3.39 × 3.70	8.6:1	43–57
	626 GT	133.2 (2184)	EFI	145 @ 4300	190 @ 3500	3.39 × 3.70	7.8:1	43–57
	MX6	133.2 (2184)	EFI	110 @ 4700	130 @ 3000	3.39 × 3.70	8.6:1	43–57
	MX6 GT	133.2 (2184)	EFI	145 @ 4300	190 @ 3500	3.39 × 3.70	7.8:1	43–57
	929	180.2 (2954)	EFI	158 @ 5500	170 @ 4000	3.50 × 3.90	8.5:1	53–75
	929 S	180.2 (2954)	EFI	190 @ 5600	191 @ 4500	3.50 × 3.90	8.9:1	53–75
	Miata	97.4 (1597)	EFI	116 @ 6500 ①	100 @ 5500 ②	3.10 × 3.30	9.4:1 ③	43–57

EFI—Electronic Fuel Injection
① Automatic—105 @ 6000
② Automatic—100 @ 4000
③ Automatic—9.0:1

GENERAL ENGINE SPECIFICATIONS—ROTARY ENGINE

Year	Model	Engine Displacement cu. in. (cc)	Fuel System Type	Net Horsepower @ rpm	Net Torque @ rpm (ft. lbs.)	Rotor Displacement (cu. in.)	Com-pression Ratio	Oil Pressure @ 3000 rpm
1988	RX7	80.0 (1308)	EFI	146 @ 6500	138 @ 3500	40	9.4:1	64–78
	RX7 Turbo	80.0 (1308)	EFI	182 @ 6500	183 @ 3500	40	8.5:1	64–78
1989	RX7	80.0 (1308)	EFI	146 @ 6500	138 @ 3500	40	8.5:1	64–78
	RX7 Turbo	80.0 (1308)	EFI	182 @ 6500	183 @ 3500	40	9.4:1	64–78
1990	RX7	80.0 (1308)	EFI	160 @ 7000	140 @ 4000	40	9.7:1	64–78
	RX7 Turbo	80.0 (1308)	EFI	200 @ 6500	196 @ 3500	40	9.0:1	64–78

GENERAL ENGINE SPECIFICATIONS—ROTARY ENGINE

Year	Model	Engine Displacement cu. in. (cc)	Fuel System Type	Net Horsepower @ rpm	Net Torque @ rpm (ft. lbs.)	Rotor Displacement (cu. in.)	Compression Ratio	Oil Pressure @ 3000 rpm
1991-92	RX7	80.0 (1308)	EFI	160 @ 7000	140 @ 4000	40	9.7:1	64-78
	RX7 Turbo	80.0 (1308)	EFI	200 @ 6500	196 @ 3500	40	9.0:1	64-78

EFI—Electronic Fuel Injection

GASOLINE ENGINE TUNE-UP SPECIFICATIONS

Year	Model	Engine Displacement cu. in. (cc)	Spark Plugs Type	Spark Plugs Gap (in.)	Ignition Timing (deg.) MT	Ignition Timing (deg.) AT	Compression Pressure (psi)	Fuel Pump (psi)	Idle Speed (rpm) MT	Idle Speed (rpm) AT	Valve Clearance In.	Valve Clearance Ex.
1988	323	97.4 (1597)	BPR5ES-11	0.040	2B	2B	135-192	38-40	850	850	Hyd.	Hyd.
	323 Turbo	97.4 (1597)	DCPR6E-11	0.040	12B	12B	109-156	24-31	850	850	Hyd.	Hyd.
	626	133.2 (2194)	ZFR5A-11	0.040	6B	6B	114-162	34-40	750	750	Hyd.	Hyd.
	626 Turbo	133.2 (2184)	ZFR5A-11	0.040	9B	9B	98-139	34-40	750	750	Hyd.	Hyd.
	MX6	133.2 (2184)	ZFR5A-11	0.040	6B	6B	114-162	34-40	750	750	Hyd.	Hyd.
	MX6 Turbo	133.2 (2184)	ZFR5A-11	0.040	9B	9B	98-139	34-40	750	750	Hyd.	Hyd.
	929	180.2 (2954)	ZFR5A-11	0.040	15B	15B	114-164	38-46	650	650	Hyd.	Hyd.
1989	323	97.4 (1597)	BPR5ES-11	0.041	2B	2B	135-192	38-40	650	650	Hyd.	Hyd.
	323 Turbo	97.4 (1597)	BCPR-6ES11	0.041	12B	12B	109-156	24-31	850	850	Hyd.	Hyd.
	626	133.2 (2194)	ZFR5A-11	0.041	6B	6B	114-162	34-40	750	750	Hyd.	Hyd.
	626 GT	133.2 (2184)	ZFR5A-11	0.041	9B	9B	98-139	34-40	750	750	Hyd.	Hyd.
	MX6	133.2 (2184)	ZFR5A-11	0.041	6B	6B	114-162	34-40	750	750	Hyd.	Hyd.
	MX6 GT	133.2 (2184)	ZFR5A-11	0.041	9B	9B	98-139	34-40	750	750	Hyd.	Hyd.
	929	180.2 (2954)	ZFR5A-11	0.041	15B	15B	114-164	38-46	650	650	Hyd.	Hyd.
1990	323	97.4 (1597)	BPR5ES-11	0.041	7B	7B	135-192	38-46	750	750	Hyd.	Hyd.
	Protege	112.2 (1839)	BKR5E-11	0.041	5B	5B	121-173	38-46	750	750	Hyd.	Hyd.
	Protege LX	112.2 (1839)	BKR5E-11	0.041	10B	10B	128-182	38-46	750	750	Hyd.	Hyd.
	626	133.2 (2184)	ZFR5A-11	0.041	6B	6B	114-162	38-40	750	750	Hyd.	Hyd.
	626 GT	133.2 (2184)	ZFR5A-11	0.041	9B	9B	98-139	34-40	750	750	Hyd.	Hyd.
	MX6	133.2 (2184)	ZFR5A-11	0.041	6B	6B	114-162	34-40	750	750	Hyd.	Hyd.
	MX6 GT	133.2 (2184)	ZFR5A-11	0.041	9B	9B	98-139	34-40	750	750	Hyd.	Hyd.
	929	180.2 (2954)	ZFR5A-11	0.041	15B	15B	114-164	38-46	650	650	Hyd.	Hyd.
	929 S	180.2 (2954)	ZFR5A-11	0.041	8B	8B	125-179	31-38	700	700	Hyd.	Hyd.
	Miata	97.4 (1597)	BKR5E-11	0.041	10B	8B	135-192	31-38	850	850	Hyd.	Hyd.
1991-92	323	97.4 (1597)	BPR5ES-11	0.041	7B	7B	135-192	38-46	750	750	Hyd.	Hyd.
	Protege	112.2 (1839)	BKR5E-11	0.041	5B	5B	121-173	38-46	750	750	Hyd.	Hyd.
	Protege LX	112.2 (1839)	BKR5E-11	0.041	10B	10B	128-182	38-46	750	750	Hyd.	Hyd.
	626	133.2 (2184)	ZFR5A-11	0.041	6B	6B	114-162	38-40	750	750	Hyd.	Hyd.
	626 Turbo	133.2 (2184)	ZFR5A-11	0.041	9B	9B	98-139	34-40	750	750	Hyd.	Hyd.
	MX6	133.2 (2184)	ZFR5A-11	0.041	6B	6B	114-162	34-40	750	750	Hyd.	Hyd.

GASOLINE ENGINE TUNE-UP SPECIFICATIONS

Year	Model	Engine Displacement cu. in. (cc)	Spark Plugs Type	Spark Plugs Gap (in.)	Ignition Timing (deg.) MT	Ignition Timing (deg.) AT	Compression Pressure (psi)	Fuel Pump (psi)	Idle Speed (rpm) MT	Idle Speed (rpm) AT	Valve Clearance In.	Valve Clearance Ex.
1991-92	MX6 Turbo	133.2 (2184)	ZFR5A-11	0.041	9B	9B	98–139	34–40	750	750	Hyd.	Hyd.
	929	180.2 (2954)	ZFR5A-11	0.041	15B	15B	114–164	38–46	650	650	Hyd.	Hyd.
	929 S	180.2 (2954)	ZFR5A-11	0.041	8B	8B	125–179	31–38	700	700	Hyd.	Hyd.
	Miata	97.4 (1597)	BKR5E-11	0.041	10B	8B	135–192	31–38	850	850	Hyd.	Hyd.
1992	SEE UNDERHOOD SPECIFICATIONS STICKER											

B—Before top dead center
Hyd.—Hydraulic

TUNE-UP SPECIFICATIONS—ROTARY ENGINE

Year	Engine Displacement cu. in. (cc)	Spark Plugs Type	Spark Plugs Gap (in.)	Distributor	Ignition Timing (degrees) Leading	Ignition Timing (degrees) Trailing	Idle Speed (rpm) MT	Idle Speed (rpm) AT①
1988	80.0 (1308)	SD10A② SD11③	0.080	Electronic	5A	20A	750	750
1989	80.0 (1308)	SD10A② SD11A③	0.080	Electronic	5A	20A	750	750
	80.0 (1308)	SD10A② SD11A③	0.080	Electronic	5A	20A	750	750
1990	80.0 (1308)	BUR7EQ② BUR9EQ③	0.043–0.067	Electronic	5A	20A	750	750
1991	80.0 (1308)	BUR7EQ② BUR9EQ③	0.043–0.067	Electronic	5A	20A	750	750
1992	REFER TO UNDERHOOD SPECIFICATION STICKER							

A—After top dead center
① Transmission in Drive
② Leading
③ Trailing

FIRING ORDERS

NOTE: To avoid confusion, always replace spark plug wires one at a time.

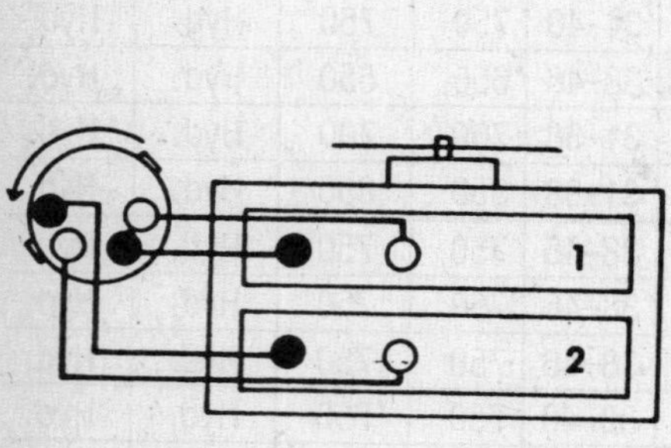

1.3L Rotary Engine
Distributorless Ignition System

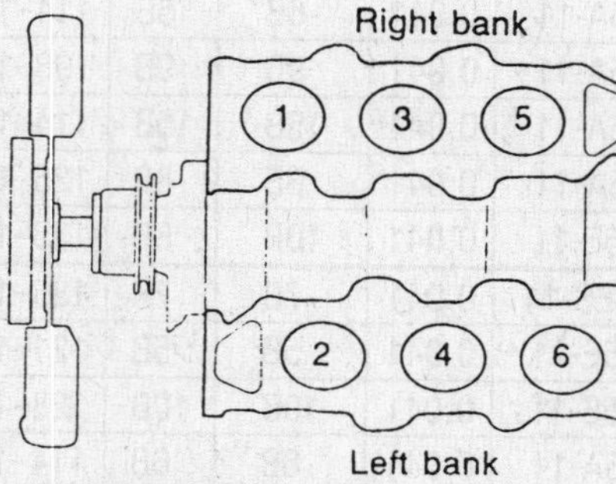

2.9L Engine
Engine Firing Order: 1–2–3–4–5–6
Distributorless Ignition

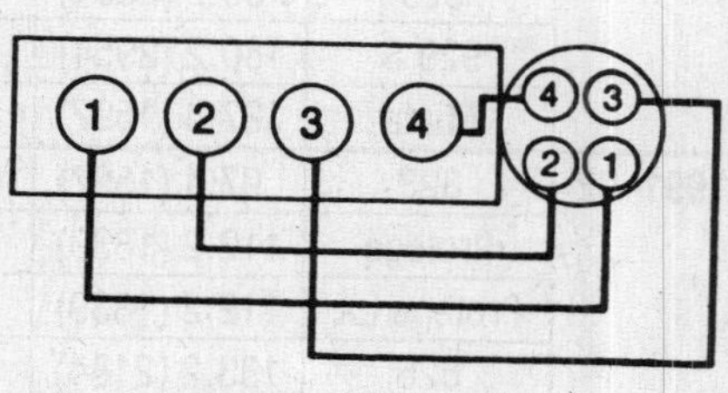

1.6L, 1.8L, 2.2L Engines
Engine Firing Order: 1–3–4–2
Distributor Rotation: Counterclockwise

CAPACITIES

Year	Model	Engine Displacement cu. in. (cc)	Engine Crankcase (qts.) with Filter	Engine Crankcase (qts.) without Filter	Transmission (pts.) 4-spd	Transmission (pts.) 5-Spd	Transmission (pts.) Auto.	Drive Axle (pts.)	Fuel Tank (gal.)	Cooling System (qts.)
1988	323	97.4 (1597)	3.6①	3.2①	—	6.8②	13.2	2.8	12.7	5.3③
	626	133.2 (2184)	4.9	4.1	—	7.2⑤	13.2	NA	15.9④	7.9
	MX6	133.2 (2184)	4.9	4.1	—	7.2⑤	13.2	NA	15.9④	7.9
	RX7	80.0 (1308)	6.1	4.7	—	5.2	15.8	2.8	16.6	9.2⑥
	929	180.2 (2954)	5.7	4.8	—	5.2	15.4	2.8	18.5	9.9
1989	323	97.4 (1597)	3.6①	3.2①	—	7.6⑦	13.4	—	12.7	5.3⑧
	626	133.2 (2184)	4.9	4.1	—	7.0⑤	14.4	—	15.9④	7.9
	MX6	133.2 (2184)	4.9	4.1	—	7.0⑤	14.4	—	15.9④	7.9
	RX7	80.0 (1308)	6.1	4.7	—	5.2	15.8	1.3	16.6	9.2⑥
	929	180.2 (2954)	5.7	9.8	—	5.2	15.4	—	18.5	9.9④
1990	323	97.4 (1597)	3.6①	3.2①	—	7.6	13.4	—	12.7	5.3⑧
	Protege	112.2 (1839)	3.8	3.4	—	7.1	12.2	—	13.2⑨	5.3⑧
	626	133.2 (2184)	4.9	4.1	—	7.0⑤	14.4	—	15.9④	7.9
	MX6	133.2 (2184)	4.9	4.1	—	7.0⑤	14.4	—	15.9④	7.9
	RX7	80.0 (1308)	6.1	4.7	—	5.2	15.8	1.3	16.6	9.2⑥
	929	180.2 (2954)	5.7	9.8	—	5.2	15.4	—	18.5	9.9④
	Miata	97.4 (1597)	4.0	3.5	—	4.2	—	0.7	11.9	6.3
1991-92	323	97.4 (1597)	3.6①	3.2①	—	7.6	13.4	—	12.7	5.3⑧
	Protege	112.2 (1839)	3.8	3.4	—	7.1	12.2	—	13.2⑨	5.3⑧
	626	133.2 (2184)	4.9	4.1	—	7.0⑤	14.4	—	15.9④	7.9
	MX6	133.2 (2184)	4.9	4.1	—	7.0⑤	14.4	—	15.9④	7.9
	RX7	80.0 (1308)	6.1	4.7	—	5.2	15.8	1.3	16.6	9.2⑥
	929	180.2 (2954)	5.7	9.8	—	5.2	15.4	—	18.5	9.9④
	Miata	97.4 (1597)	4.0	3.5	—	4.2	4.2	0.7	11.9	6.3

① DOHC engine
3.8 with filter
3.4 without filter
② DOHC engine—7.2
③ DOHC engine and all automatic transaxle—6.3
④ 4 wheel steering—15.0
⑤ Turbocharged engine—7.8
⑥ Non-turbocharged engine—7.7
⑦ With 4WD
Transaxle capacity—7.6 pts.
Transfer carrier capacity—1 pt.
⑧ Automatic—6.3
⑨ Sedan, 14.5 gal.

CRANKSHAFT AND CONNECTING ROD SPECIFICATIONS

All measurements are given in inches.

Year	Engine Displacement cu. in. (cc)	Crankshaft Main Brg. Journal Dia.	Crankshaft Main Brg. Oil Clearance	Crankshaft Shaft End-play	Crankshaft Thrust on No.	Connecting Rod Journal Diameter	Connecting Rod Oil Clearance	Connecting Rod Side Clearance
1988	97.4 (1597)	1.9961–1.9668	0.0009–0.0017 ②	0.0031–0.0111	4	1.7693–1.7699	0.0011–0.0027	0.0043–0.0103
	133.2 (2184)	2.3597–2.3604	0.0010–0.0017 ①	0.0031–0.0071	3	2.0055–2.0061	0.0011–0.0026	0.0040–0.0100
	180.2 (2954)	2.4385–2.4392	0.0010–0.0015	0.0031–0.0111	NA	2.0842–2.0848	0.0009–0.0025	0.0070–0.0130

CRANKSHAFT AND CONNECTING ROD SPECIFICATIONS

All measurements are given in inches.

Year	Engine Displacement cu. in. (cc)	Crankshaft: Main Brg. Journal Dia.	Crankshaft: Main Brg. Oil Clearance	Crankshaft: Shaft End-play	Crankshaft: Thrust on No.	Connecting Rod: Journal Diameter	Connecting Rod: Oil Clearance	Connecting Rod: Side Clearance
1989	97.4 (1597)	1.9961–1.9668	0.0009–0.0017 ②	0.0031–0.0111	4	1.8898–1.8904	0.0011–0.0027	0.0043–0.0103
	133.2 (2184)	2.3597–2.3604	0.0010–0.0017 ①	0.0031–0.0071	3	2.1261–2.1266	0.0011–0.0026	0.0120
	180.2 (2954)	2.4385–2.4392	0.0010–0.0015	0.0031–0.0111	NA	2.2047–2.2053	0.0009–0.0025	0.0070–0.0130
1990	97.4 (1597)	1.9961–1.9668	0.0007–0.0014	0.0031–0.0111	4	1.7693–1.7699	0.0011–0.0027	0.0043–0.0103
	112.2 (1839)	1.9961–1.9668	0.0007–0.0014	0.0031–0.0111	4	1.7693–1.7699	0.0011–0.0027	0.0043–0.0103
	133.2 (1839)	2.3597–2.3604	0.0010–0.0017 ①	0.0031–0.0071	3	2.1261–2.1266	0.0011–0.0026	0.0120
	180.2 (2954)	2.4385–2.4392	0.0010–0.0015	0.0031–0.0111	NA	2.2047–2.2053	0.0009–0.0025	0.0070–0.0130
1991-92	97.4 (1597)	1.9961–1.9668	0.0007–0.0014	0.0031–0.0111	4	1.7693–1.7699	0.0011–0.0027	0.0043–0.0103
	112.2 (1839)	1.9961–1.9668	0.0007–0.0014	0.0031–0.0111	4	1.7693–1.7699	0.0011–0.0027	0.0043–0.0103
	133.2 (1839)	2.3597–2.3604	0.0010–0.0017 ①	0.0031–0.0071	3	2.1261–2.1266	0.0011–0.0026	0.0120
	180.2 (2954)	2.4385–2.4392	0.0010–0.0015	0.0031–0.0111	NA	2.2047–2.2053	0.0009–0.0025	0.0070–0.0130

NA—Not Available
① Bearing No. 3—0.0012–0.0019
② DOHC engine—0.0010–0.0017

VALVE SPECIFICATIONS—PISTON ENGINE

Year	Engine Displacement cu. in. (cc)	Seat Angle (deg.)	Face Angle (deg.)	Spring Squareness limit	Spring Free Length: Outer	Spring Free Length: Inner	Stem-to-Guide Clearance: Intake	Stem-to-Guide Clearance: Exhaust	Stem Diameter: Intake	Stem Diameter: Exhaust
1988	97.4 (1597)	45	45	0.0590	1.7200	1.7200	0.0010–0.0024	0.0011–0.0026	0.2744–0.2750	0.2744–0.2750
	97.4 (1597) ①	45	45	NA	1.8580	1.8580	0.0010–0.0024	0.0012–0.0026	0.2350–0.2360	0.2350–0.2360
	133.2 (2184)	45	45	0.0670	1.9490	1.9490	0.0010–0.0024	0.0012–0.0026	0.2740–0.2750	0.2740–0.2750
	180.2 (2954)	45	45	②	③	③	0.0010–0.0024	0.0012–0.0026	0.2740–0.2750	0.2760–0.2770

VALVE SPECIFICATIONS—PISTON ENGINE

Year	Engine Displacement cu. in. (cc)	Seat Angle (deg.)	Face Angle (deg.)	Spring Squareness limit	Spring Free Length Outer	Spring Free Length Inner	Stem-to-Guide Clearance Intake	Stem-to-Guide Clearance Exhaust	Stem Diameter Intake	Stem Diameter Exhaust
1989	97.4 (1597)	45	45	0.0590	1.7200	1.7200	0.0010–0.0024	0.0011–0.0026	0.2740–0.2750	0.2740–0.2750
	97.4 (1597) ①	45	45	0.0620	1.8580	1.8580	0.0010–0.0024	0.0012–0.0026	0.2350–0.2356	0.2348–0.2354
	133.2 (2184)	45	45	0.0670	1.9490	1.9490	0.0010–0.0024	0.0012–0.0026	0.2744–0.2750	0.2742–0.2748
	180.2 (2954)	45	45	②	④	③	0.0010–0.0024	0.0012–0.0026	0.2744–0.2750	0.3159–0.3165
1990	97.4 (1597)	45	45	0.0600	1.7200	1.7200	0.0010–0.0024	0.0012–0.0026	0.2744–0.2750	0.2742–0.2748
	97.4 (1597) ①	45	45	0.0660	0.8900 ⑥	1.8900 ⑥	0.0010–0.0024	0.0012–0.0026	0.2350–0.2356	0.2348–0.2354
	112.2 (1839)	45	45	0.0600 ⑤	0.8150	1.7170	0.0010–0.0024	0.0012–0.0026	0.2350–0.2356	0.2348–0.2354
	112.2 (1839) ①	45	45	0.0640	0.8210	1.8210	0.0010–0.0024	0.0012–0.0026	0.2350–0.2356	0.2348–0.2354
	133.2 (2184)	45	45	0.0670	1.9490	1.9840	0.0010–0.0024	0.0012–0.0026	0.2744–0.2750	0.2742–0.2748
	180.2 (2954)	45	45	④	⑤	⑤	0.0010–0.0024	0.0012–0.0026	0.2744–0.2750	0.3159–0.3165
1991–92	97.4 (1597)	45	45	0.0600	1.7200	1.7200	0.0010–0.0024	0.0012–0.0026	0.2744–0.2750	0.2742–0.2748
	97.4 (1597) ①	45	45	0.0660	0.8900 ⑥	1.8900 ⑥	0.0010–0.0024	0.0012–0.0026	0.2350–0.2356	0.2348–0.2354
	112.2 (1839)	45	45	0.0600 ⑤	0.8150	1.7170	0.0010–0.0024	0.0012–0.0026	0.2350–0.2356	0.2348–0.2354
	112.2 (1839) ①	45	45	0.0640	0.8210	1.8210	0.0010–0.0024	0.0012–0.0026	0.2350–0.2356	0.2348–0.2354
	133.2 (2184)	45	45	0.0670	1.9490	1.9840	0.0010–0.0024	0.0012–0.0026	0.2744–0.2750	0.2742–0.2748
	180.2 (2954)	45	45	④	⑤	⑤	0.0010–0.0024	0.0012–0.0026	0.2744–0.2750	0.3159–0.3165

① DOHC engine
② Intake
 Outer—0.0700
 Inner—0.0640
 Exhaust
 Outer—0.0830
 Inner—0.0740
③ Intake
 Outer—2.0000
 Inner—1.9835
 Exhaust
 Outer—2.3820
 Inner—2.1300
④ Intake
 Outer—2.0000
 Inner—1.8350
 Exhaust
 Outer—2.2950
 Inner—2.0910
⑤ Intake, 0.0630
⑥ Exhaust, 1.9020

PISTON AND RING SPECIFICATIONS

All measurements are given in inches.

Year	Engine Displacement cu. in. (cc)	Piston Clearance	Ring Gap Top Compression	Ring Gap Bottom Compression	Ring Gap Oil Control	Ring Side Clearance Top Compression	Ring Side Clearance Bottom Compression	Ring Side Clearance Oil Control
1988	97.4 (1597)	0.0010–0.0026	0.0079–0.0157	0.0590–0.0120	0.0080–0.0280	0.0012–0.0026	0.0012–0.0026	—
	133.2 (2184)	0.0014–0.0030	0.0080–0.0138	0.0060–0.0120	0.0080–0.0276 ①	0.0012–0.0028	0.0012–0.0028	—
	180.2 (2954)	0.0019–0.0026	0.0080–0.0138	0.0060–0.0120	0.0080–0.0280	0.0012–0.0028	0.0012–0.0028	—
1989	97.4 (1597)	0.0010–0.0026	0.0079–0.0157	0.0060–0.0120	0.0080–0.0280	0.0012–0.0026	0.0012–0.0026	—
	133.2 (2184)	0.0014–0.0030	0.0080–0.0138	0.0060–0.0120	0.0080–0.0276 ①	0.0012–0.0028	0.0012–0.0028	—
	180.2 (2954)	0.0019–0.0026	0.0080–0.0140	0.0060–0.0120	0.0080–0.0280	0.0010–0.0030	0.0010–0.0030	—
1990	97.4 (1597)	0.0015–0.0020	0.0060–0.0120	0.0060–0.0120	0.0080–0.0280	0.0012–0.0026 ②	0.0012–0.0026 ②	—
	112.2 (1839)	0.0015–0.0020	0.0060–0.0120	0.0060–0.0120	0.0080–0.0280	0.0012–0.0026	0.0012–0.0026	—
	133.2 (2184)	0.0014–0.0030	0.0080–0.0138	0.0060–0.0120	0.0080–0.0276	0.0012–0.0028	0.0012–0.0028	—
	180.2 (2954)	0.0019–0.0026	0.0080–0.0140	0.0060–0.0120	0.0080–0.0280	0.0010–0.0030	0.0010–0.0030	—
1991-92	97.4 (1597)	0.0015–0.0020	0.0060–0.0120	0.0060–0.0120	0.0080–0.0280	0.0012–0.0026 ②	0.0012–0.0026 ②	—
	112.2 (1839)	0.0015–0.0020	0.0060–0.0120	0.0060–0.0120	0.0080–0.0280	0.0012–0.0026	0.0012–0.0026	—
	133.2 (2184)	0.0014–0.0030	0.0080–0.0138	0.0060–0.0120	0.0080–0.0276	0.0012–0.0028	0.0012–0.0028	—
	180.2 (2954)	0.0019–0.0026	0.0080–0.0140	0.0060–0.0120	0.0080–0.0280	0.0010–0.0030	0.0010–0.0030	—

① Turbocharged engine—0.012–0.0354
② DOHC engine—0.0012–0.0028

TORQUE SPECIFICATIONS—PISTON ENGINES

All readings in ft. lbs.

Year	Engine Displacement cu. in. (cc)	Cylinder Head Bolts	Main Bearing Bolts	Rod Bearing Bolts	Crankshaft Pulley Bolts	Flywheel Bolts	Manifold Intake	Manifold Exhaust	Spark Plugs
1988	97.4 (1597)	56-60	40-43	35-38 ①	36-45	71-76	14-19	12-17 ②	11-17
	133.2 (2184)	59-64	61-65	48-51	36-45	71-76	14-22	14-22	11-17
	180.2 (2954)	③	④	⑤	116-123	76-81	14-19	16-21	11-17
1989	97.4 (1597)	56-60	40-43	35-38 ①	9-13	71-76	14-19	12-17 ②	11-17
	133.2 (2184)	59-64	61-65	48-51	36-45	71-76	14-22	14-22	11-17
	180.2 (2954)	③	④	⑤	7-11	76-81	14-19	16-21	11-17

TORQUE SPECIFICATIONS—PISTON ENGINES

All readings in ft. lbs.

Year	Engine Displacement cu. in. (cc)	Cylinder Head Bolts	Main Bearing Bolts	Rod Bearing Bolts	Crankshaft Pulley Bolts	Flywheel Bolts	Manifold Intake	Manifold Exhaust	Spark Plugs
1990	97.4 (1597)	56–60	40–43	35–38	9–13	71–76	14–19	12–17 ②	11–17
	112.2 (1839)	56–60	40–43	36–38	9–13	71–76	14–19	12–17 ⑥	11–17
	133.2 (2184)	59–64	61–65	48–51	36–45	71–76	14–22	14–22	11–17
	180.2 (2954)	③	④	⑤	7–11	76–81	14–19	16–21	11–17
1991-92	97.4 (1597)	56–60	40–43	35–38	9–13	71–76	14–19	12–17 ②	11–17
	112.2 (1839)	56–60	40–43	36–38	9–13	71–76	14–19	12–17 ⑥	11–17
	133.2 (2184)	59–64	61–65	48–51	36–45	71–76	14–22	14–22	11–17
	180.2 (2954)	③	④	⑤	7–11	76–81	14–19	16–21	11–17

① DOHC Engine—48–51
② DOHC Engine—29–42
③ 14 ft. lbs.
Paint a mark on each bolt head. Using the mark as a reference, tighten the bolts 90 degrees in the proper sequence. Retighten each bolt another 90 degrees in the proper sequence.
④ 14 ft. lbs.
Paint a mark on each bolt head. Using the mark as a reference, tighten the bolts 90 degrees. Tighten the bolts again 45 degrees.
⑤ 22 ft. lbs.
Paint a mark on each bolt head. Using the mark as a reference, tighten the bolts 90 degrees.
⑥ DOHC Engine—28–34

TORQUE SPECIFICATIONS—ROTARY ENGINE

All readings in ft. lbs.

Engine Displacement cu. in. (cc)	Front Cover	Bearing Housing	Rear Stationary Gear	Eccentric Shaft Pulley Bolt	Flywheel-to-Eccentric Shaft Nut	Manifolds Intake	Manifolds Exhaust	Oil Pan	Tension Bolts
80.0 (1308)	12–17	12–17	12–17	80–98	290–360	14–19	23–34	6–8	23–29

BRAKE SPECIFICATIONS

All measurements in inches unless noted.

Year	Model	Lug Nut Torque (ft. lbs.)	Master Cylinder Bore	Brake Disc Minimum Thickness	Brake Disc Maximum Runout	Standard Brake Drum Diameter	Minimum Lining Thickness Front	Minimum Lining Thickness Rear
1988	323	65–87	0.875	0.390 ③	0.003	7.870	0.120	0.040
	626	65–87	0.875	0.940 ①	0.004	9.000	0.039	0.031
	MX6	65–87	0.875	0.940 ①	0.004	9.000	0.039	0.031
	929	65–87	0.875	0.790 ①	0.004	—	0.039	0.031
	RX7	65–87	0.875	0.790 ②	0.004	—	0.035 ②	0.031
1989	323	65–87	0.875	0.390 ④	0.004	7.870	0.039	0.039 ③
	626	65–87	0.875	0.790 ④	0.004	9.000	0.039	0.040
	MX6	65–87	0.875	0.790 ④	0.004	9.000	0.039	0.040
	929	65–87	0.875	0.790 ④	0.004	—	0.039	0.031
	RX7	65–87	0.875	0.790 ②	0.004	—	0.035 ②	0.031

BRAKE SPECIFICATIONS

All measurements in inches unless noted.

Year	Model	Lug Nut Torque (ft. lbs.)	Master Cylinder Bore	Brake Disc Minimum Thickness	Brake Disc Maximum Runout	Standard Brake Drum Diameter	Minimum Lining Thickness Front	Minimum Lining Thickness Rear
1990	323	65–87	0.875	0.790 ⑥	0.004	9.000	0.080	0.040
	Protege	65–87	0.875	0.790 ⑥	0.004	9.000	0.080	0.040
	626	65–87	0.875	0.870 ③	0.004	9.000	0.080	0.040
	MX6	65–87	0.875	0.870 ③	0.004	9.000	0.080	0.040
	929	65–87	0.875	0.790 ⑦	0.004	6.690	0.080	0.080 ④
	RX7	65–87	0.875	0.790 ⑧	0.004	—	0.080 ②	0.040
	Miata	65–87	0.875	0.630 ⑥	0.004	—	0.040	0.040
1991-92	323	65–87	0.875	0.790 ⑥	0.004	9.000	0.080	0.040
	Protege	65–87	0.875	0.790 ⑥	0.004	9.000	0.080	0.040
	626	65–87	0.875	0.870 ③	0.004	9.000	0.080	0.040
	MX6	65–87	0.875	0.870 ③	0.004	9.000	0.080	0.040
	929	65–87	0.875	0.790 ⑦	0.004	6.690	0.080	0.080 ④
	RX7	65–87	0.875	0.790 ⑧	0.004	—	0.080 ②	0.040
	Miata	65–87	0.875	0.630 ⑥	0.004	—	0.040	0.040

NOTE: Minimum lining thickness is as recommended by the manufacturer. Due to variations in state inspection regulations, the minimum allowable thickness may be different than specified.

① Rear disc—0.350 in.
② Rear disc
  14 in. wheels—0.310 in.
  15 in. wheels—0.710 in.
③ Rear disc—0.310 in.
④ Rear disc—0.390 in.
⑤ Rear disc—14 in. wheel—0.040 in.
  15 in. wheel—0.790 in.
⑥ Rear disc—0.280 in.
⑦ Rear disc—0.630 in.
⑧ Rear disc—0.310 solid, 0.710 vented
⑨ Parking brake shoe—0.040 in.

WHEEL ALIGNMENT

Year	Model	Caster Range (deg.)	Caster Preferred Setting (deg.)	Camber Range (deg.)	Camber Preferred Setting (deg.)	Toe-in (in.)	Steering Axis Inclination (deg.)
1988	323 (2WD)	13/16P–2 5/16P	1 9/16P	5/16P–1 5/16P	13/16P	3/64–13/64	12 3/8
	323 (4WD)	1 1/16P–2 9/16P	1 13/16P	9/16P–1 9/16P	1/16P	3/64–13/64	12
	626 ①	5/16P–1 5/16P	1 3/16P	7/16N–1 1/16P	5/16P	0–1/4	12 13/16
	MX6 ①	5/16P–1 15/16P	1 3/16P	7/16N–1 1/16P	5/16P	0–1/4	12 13/16
	929	3 3/4P–5 1/4P	4 1/2P	1/4P–1 3/4P	1P	0–1/4	12 11/16
	RX-7	3 15/16P–5 7/16P	4 5/8P	3/16N–13/16P	5/16P	0–1/4	13 3/4
1989	323 (2WD)	13/16P–2 5/16P	1 9/16P	5/16P–1 5/16P	13/16P	3/64–13/36	12 3/8
	323 (4WD)	1 1/16P–2 9/16P	1 13/16P	9/16P–1 9/16P	1 1/16P	3/64–13/64	12
	626 ①	5/16P–1 5/16P	1 3/16P	7/16N–1 1/16P	5/16P	0–1/4	12 13/16
	MX6 ②	5/16P–1 15/16P	1 3/16P	7/16N–1 1/16P	5/16P	0–1/4	12 13/16
	929	3 3/4P–5 1/4P	4 1/2P	1/4P–1 3/4P	1P	0–1/4	12 11/16
	RX-7	3 15/16P–5 7/16P	4 5/8P	3/16N–13/16P	5/16P	0–1/4	13 3/4

WHEEL ALIGNMENT

Year	Model	Caster Range (deg.)	Caster Preferred Setting (deg.)	Camber Range (deg.)	Camber Preferred Setting (deg.)	Toe-in (in.)	Steering Axis Inclination (deg.)
1990	323 (2WD)	1 15/16P–2 13/16P	2 1/6P	13/16N–1 1/16P ③	1/16N ④	1/32–7/32	12 7/16
	Protege	1 5/16P–2 13/16P	2 1/6P	13/16N–1 1/16P ③	1/16N ④	1/32–7/32	12 7/16
	626 ①	1 5/16P–1 15/16P	1 3/16P	7/16N–1 1/16P	5/16P	0–1/4	12 13/16
	MX6 ①	5/16P–1 15/16P	1 3/16P	7/16N–1 1/16P	5/16P	0–1/4	12 13/16
	929	3 3/4P–5 1/4P	4 1/2P	1/4P–1 3/4P	1P	0–1/4	12 11/16
	RX-7	3 15/16P–5 7/16P	4 5/8P	3/16N–1 3/16P	5/16P	0–1/4	13 3/4
	Miata	3 3/4P–5 1/4P	4 1/2P	3/8N–1 1/8P ⑤	3/8P ⑥	0–1/4	11 5/16
1991-92	323 (2WD)	1 15/16P–2 13/16P	2 1/6P	13/16N–1 1/16P ③	1/16N ④	1/32–7/32	12 7/16
	Protege	1 5/16P–2 13/16P	2 1/6P	13/16N–1 1/16P ③	1/16N ④	1/32–7/32	12 7/16
	626 ①	1 5/16P–1 15/16P	1 3/16P	7/16N–1 1/16P	5/16P	0–1/4	12 13/16
	MX6 ①	5/16P–1 15/16P	1 3/16P	7/16N–1 1/16P	5/16P	0–1/4	12 13/16
	929	3 3/4P–5 1/4P	4 1/2P	1/4P–1 3/4P	1P	0–1/4	12 11/16
	RX-7	3 15/16P–5 7/16P	4 5/8P	3/16N–1 3/16P	5/16P	0–1/4	13 3/4
	Miata	3 3/4P–5 1/4P	4 1/2P	3/8N–1 1/8P ⑤	3/8P ⑥	0–1/4	11 5/16

N—Negative
P—Positive
① Rear Alignment:
2 Wheel Steering
Camber—1 1/4N–1/4P
Toe In—1/4N–1/4P
4 Wheel Steering
Camber—3/4N–3/4P
Toe In—0–1/2P
② Rear—7/16N–9/16P
③ Rear—1 1/16N–7/16P
④ Rear—5/16P
⑤ Rear—1 1/4N–1/4N
⑥ Rear—3/4N

ECCENTRIC SHAFT SPECIFICATIONS—ROTARY ENGINE

All measurements are given in inches.

Engine Type	Journal Diameter Main Bearing	Journal Diameter Rotor Bearing	Oil Clearance Main Bearing	Oil Clearance Rotor Bearing	Eccentric Shaft End-play Normal	Eccentric Shaft End-play Limit	Minimum Shaft Run-out
13B	1.6918–1.6923	2.9122–2.9128	0.0016–0.0031	0.0016–0.0031	0.0016–0.0028	0.0035	0.0047

ROTOR AND HOUSING SPECIFICATIONS—ROTARY ENGINE

All measurements are given in inches.

Engine Type	Rotor Side ① Clearance	Rotor Width	Housings Front and Rear Distortion Limit	Housings Front and Rear Wear Limit	Housings Rotor Width	Housings Rotor Distortion Limit	Housings Intermediate Distortion Limit	Housings Intermediate Wear Width
13B	0.0047–0.0083	3.1420–3.1440	0.0016	0.0039	3.1485–3.1500	0.0024	0.0016	0.0039

① New

SEAL CLEARANCES—ROTARY ENGINE

All measurements are given in inches.

Engine Type	Apex Seals: To Side Housing Normal	Apex Seals: To Side Housing Limit	Apex Seals: To Rotor Groove Normal	Apex Seals: To Rotor Groove Limit	Side Seal: To Rotor Groove Normal	Side Seal: To Rotor Groove Limit	Side Seal: To Corner Seal Normal	Side Seal: To Corner Seal Limit
13B	0.0051–0.0075	—	0.0024–0.0040	0.0059	0.0011–0.0031	0.0039	0.0020–0.0059	0.0160

PISTON ENGINE MECHANICAL

NOTE: Disconnecting the negative battery cable on some vehicles may interfere with the functions of the on board computer systems and may require the computer to undergo a relearning process, once the negative battery cable is reconnected.

Engine Assembly

REMOVAL & INSTALLATION

323 and Protege

FRONT WHEEL DRIVE

1. The engine and transaxle are lifted as a unit from the vehicle. To relieve the fuel system pressure:
   a. Raise the rear seat and locate the fuel pump connector. If the connector is not accessible under the seat, locate the main fuel pump relay to the right of the ECU behind the center console.
   b. With the engine running at idle, disconnect the power to the fuel pump and allow the engine to stall. Turn the ignition switch **OFF**.
   c. Under the hood, wrap a rag around the fuel filter outlet fitting and loosen the clamp to remove the hose.
2. Disconnect and remove the battery and the battery box. Raise and safely support the vehicle.
3. Remove the splash pan under the vehicle and drain the engine and transaxle oil and the coolant. Dispose of these fluids properly.
4. Remove the air cleaner assembly, including the air flow meter and all of the ducting. Remove the oil dip stick.
5. Remove the radiator hoses and remove the radiator with the fan and shroud as a single assembly. On vehicles with automatic transaxle, don't forget to disconnect the cooler hoses at the bottom of the radiator.
6. Disconnect the throttle and cruise control cables, if equipped, and the speedometer cable.
7. Label and disconnect the vacuum hoses and wiring.
8. Disconnect the fuel supply and return hoses and the heater hoses.
9. Disconnect the exhaust pipe from the manifold. If equipped with secondary air, disconnect the pipe.
10. Without disconnecting the hydraulic hoses, remove the power steering pump and hang it from the body with wire. Make sure the hoses are not kinked or strained and that the engine will come out past the pump.
11. Without disconnecting the coolant hoses, remove the air conditioning compressor and hang it from the body with wire. Make sure the hoses are not kinked or strained and that the engine will come out past the compressor.
12. Disconnect the clutch cable and shift linkage. On vehicles with hydraulic clutch, remove the slave cylinder from the transaxle without disconnecting the hydraulic line.
13. Remove the nuts and press the tie rod ends out of the steering knuckles.
14. Remove the pinch bolts from the steering knuckle and pry the control arm down to slip lower ball joint out of the knuckle.
15. Carefully pry the inner CV-joints out of the transaxle and support the halfshafts so they do not hang by the outer CV-joints.
16. Attach a chain to the lifting eyes on the engine and lift the engine slightly to take up the slack. Check to make sure all wiring and hoses are disconnected.
17. Remove the engine mount nuts and carefully lift the engine and transaxle out of the vehicle as a unit. The rubber mounts should come out with the engine. It may be easier to remove the front–to–rear engine mount member under the vehicle.
18. With the transaxle properly supported, remove the starter. On vehicles with automatic transaxle, remove the flywheel–to–torque converter bolts. Remove all stiffener brackets and other mounts to separate the transaxle from the engine.

To install:

19. Carefully fit the engine and transaxle together and attach the front and rear mounts and brackets. Torque the transaxle–to–engine bolts, front motor mount bolts and the stiffener bracket bolts to 37 ft. lbs. (50 Nm).
20. On vehicles with automatic transaxle, install the flywheel–to–torque converter bolts and torque to 36 ft. lbs. (49 Nm). Install the starter and torque the bolts to 18 ft. lbs. (25 Nm).
21. Carefully install the engine/transaxle and align the mounts to install the long bolts. Install the mount member and all other mounting bolts before torquing any of them.
22. Torque the mount member bolts to 66 ft. lbs. (89 Nm).
23. Torque the front and then the rear mount nuts to 38 ft. lbs. (52 Nm).
24. Install the right side mount and torque the nuts to 76 ft. lbs. (103 Nm). Insert the long bolt through the rubber mount and torque to 69 ft. lbs. (93 Nm). Install the dynamic balancer over the mount nuts and torque to 59 ft. lbs. (80 Nm).
25. Install the left side mount and start all the nuts and bolts. Torque the mount–to–body bolts to 45 ft. lbs. (61 Nm) and the mount–to–transaxle nuts to 69 ft. lbs. (93 Nm).
26. Grease the halfshaft splines and press them into the transaxle. Make sure they are firmly held in place by the circlip..
27. Install the lower ball joint and torque the clamping bolt to 43 ft. lbs. (59 Nm). Install the tie rod ends and torque the nut to 42 ft. lbs. (57 Nm), then tighten as required to install a new cotter pin.
28. Attach the stabilizer bar and tighten the nuts so there is ¾ in. (19mm) of thread showing.
29. Reconnect the clutch and/or shift linkage.
30. Install a new gasket and connect the exhaust pipe to the manifold. Use new self-locking nuts and torque to 34 ft. lbs. (46 Nm).
31. Connect the wiring and all water, fuel and vacuum hoses.

32. Install the air conditioner compressor and power steering pump, if equipped.
33. Install the radiator and connect the hoses.
34. Install the air cleaner and air flow meter assembly and all the ducting. Connect and adjust the throttle cable as required.
35. Refill all the fluids and install the battery. Make sure to bleed the cooling system and check the ignition timing.

FOUR WHEEL DRIVE

1. The engine and transaxle are lifted as a unit from the vehicle. To relieve the fuel system pressure:
   a. Raise the rear seat and locate the fuel pump connector. If the connector is not accessible under the seat, locate the main fuel pump relay to the right of the ECU behind the center console.
   b. With the engine running at idle, disconnect the power to the fuel pump and allow the engine to stall. Turn the ignition switch **OFF**.
   c. Under the hood, wrap a rag around the fuel filter outlet fitting and loosen the clamp to remove the hose.
2. Disconnect and remove the battery and the battery box. Raise and safely support the vehicle.
3. Remove the splash pan under the vehicle and drain the engine and transaxle oil and the coolant. Dispose of these fluids properly.
4. Remove the air cleaner assembly, including the air flow meter and all of the ducting. Remove the oil dip stick.
5. Remove the radiator hoses and remove the radiator with the fan and shroud as a single assembly. On vehicles with automatic transaxle, don't forget to disconnect the cooler hoses at the bottom of the radiator.
6. Disconnect the throttle and cruise control cables, if equipped, and the speedometer cable.
7. Label and disconnect the vacuum hoses and wiring.
8. Disconnect the fuel supply and return hoses and the heater hoses.
9. Disconnect the exhaust pipe from the manifold. If equipped with secondary air, disconnect the pipe.
10. Without disconnecting the hydraulic hoses, remove the power steering pump and hang it from the body with wire. Make sure the hoses are not kinked or strained and that the engine will come out past the pump.
11. Without disconnecting the coolant hoses, remove the air conditioning compressor and hang it from the body with wire. Make sure the hoses are not kinked or strained and that the engine will come out past the compressor.
12. Disconnect the clutch cable and shift linkage. On vehicles with hydraulic clutch, remove the slave cylinder from the transaxle without disconnecting the hydraulic line.
13. Remove the nuts and press the tie rod ends out of the steering knuckles.
14. Remove the pinch bolts from the steering knuckle and pry the control arm down to slip lower ball joint out of the knuckle.
15. Carefully pry the left inner CV-joint out of the transaxle and support the halfshaft so it does not hang by the outer CV-joint.
16. Remove the bolts from the right side halfshaft joint and carefully pry the intermediate shaft out of the transaxle. Support the shafts so they do not hang by the outer CV-joint.
17. Matchmark the drive shaft to the front and rear flanges and remove the drive shaft from the vehcile. Keep all the spacers and bushings in order and tag them for reassembly. Each spacer and bushing must be returned to its original position.
18. Remove the exhaust manifold. Remove the water inlet pipe and metal by-pass pipe.
19. Attach a chain to the lifting eyes on the engine and lift the engine slightly to take up the slack. Check to make sure all wiring and hoses are disconnected.
20. Remove the engine mount nuts and carefully lift the engine and transaxle out of the vehicle as a unit. The rubber mounts should come out with the engine. It may be easier to remove the front–to–rear engine mount member under the vehicle.
21. With the transaxle properly supported, remove the starter. On vehicles with automatic transaxle, remove the flywheel–to–torque converter bolts.
22. To remove the center differential lock motor:
   a. Remove the set bolt and lock sensor switch.
   b. At the end of the motor, remove the plug and use a small screwdriver to turn the shift rod ½ turn clockwise. This will disengage the mechanism.
   c. Remove the bolts and remove the differential lock motor.
23. Remove any stiffener brackets and other mounts to separate the transaxle from the engine.

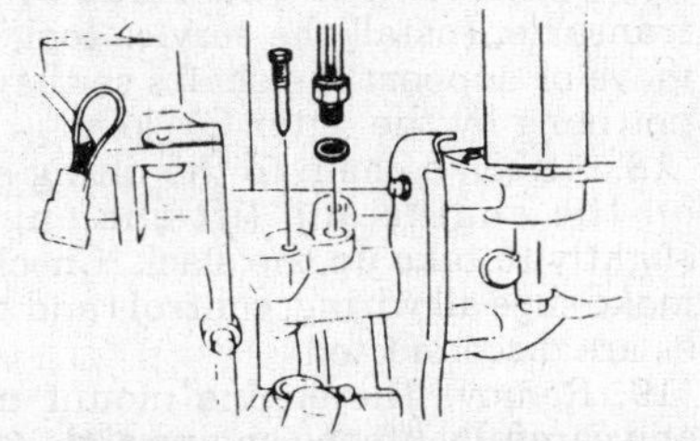
Remove the set bolt and sensor switch from the differential lock motor on 4 wheel drive 323 and Protege

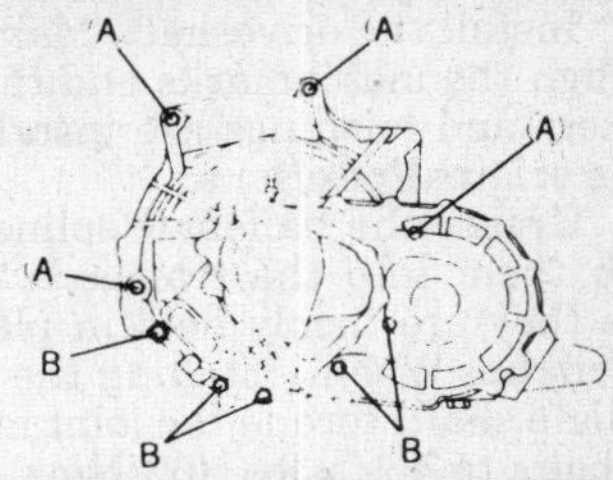

Manual 323 and Protege with 4 wheel drive, bolt torques are different

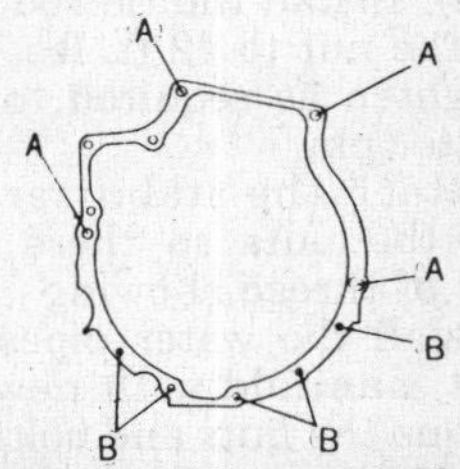

Automatic 323 and Protege with 4 wheel drive, bolt torques are different

To install:

24. Carefully fit the engine and transaxle together and torque bolts **A** to 66 ft. lbs. (89 Nm) and bolts **B** to 38 ft. lbs. (52 Nm).
25. On automatic transaxle, install the flywheel–to–torque converter bolts and torque to 36 ft. lbs. (49 Nm).
26. Attach the front mount and and torque the bolts to 38 ft. lbs. (52 Nm). Install the rear bracket and torque the bolts to 38 ft. lbs. (52 Nm).
27. Install the starter and torque the bolts to 18 ft. lbs. (25 Nm).
28. Before installing the locking differential motor, make sure the flat side of the shift rod is facing up and install a new O-ring. Install the motor and use a screwdriver to turn the shift rod ½ turn counterclockwise to engage the linkage. Torque the mounting bolts, set bolt and sensor to 22 ft. lbs. (29 Nm).
29. Carefully install the engine/transaxle and align the mounts to install the long bolts. Install the mount member and all other mounting bolts before torquing any of them.
30. Torque the mount member bolts to 66 ft. lbs. (89 Nm).
31. Torque the front and then the rear mount nuts to 38 ft. lbs. (52 Nm).
32. Install the right side mount and torque the nuts to 76 ft. lbs. (103 Nm). Insert the long bolt through the rubber mount and torque to 69 ft. lbs. (93 Nm). Install the dynamic balancer over the mount nuts and torque to 59 ft. lbs. (80 Nm).
33. Install the left side mount and start all the nuts and bolts. Torque the mount–to–body bolts to 45 ft. lbs. (61 Nm) and the mount–to–transaxle nuts to 69 ft. lbs. (93 Nm).

34. Install the driveshaft. Make sure to align the matchmarks and that the spacers and bushings are installed in their original positions.
35. Grease the halfshaft splines and push them into the transaxle. Make sure they are firmly held in place by the circlip. When installing the intermediate shaft, torque the joint mounting bolts to 46 ft. lbs. (62 Nm).
36. Install the lower ball joint and torque the clamping bolt to 43 ft. lbs. (59 Nm). Install the tie rod ends and torque the nut to 42 ft. lbs. (57 Nm), then tighten as required to install a new cotter pin.
37. Attach the stabilizer bar and tighten the nuts so there is ¾ in. (19mm) of thread showing.
38. Install the water pipes and the exhaust manifold with new gaskets and torque the nuts and bolts to 18 ft. lbs. (24 Nm).
39. Install a new gasket and connect the exhaust pipe to the manifold. Use new self-locking nuts and torque to 34 ft. lbs. (46 Nm).
40. Reconnect the clutch and/or shift linkage.
41. Connect the wiring and all water, fuel and vacuum hoses.
42. Install the air conditioner compressor and power steering pump, if equipped.
43. Install the radiator and connect the hoses.
44. Install the air cleaner and air flow meter assembly and all the ducting. Connect and adjust the throttle cable as required.
45. Refill all the fluids and install the battery. Make sure to bleed the cooling system and check the ignition timing.

626 and MX-6

1. The engine and transaxle are lifted as a unit from the vehicle. To relieve the fuel system pressure:
   a. Locate the fuel pump relay on the fuse panel under the left side of the dashboard.
   b. With the engine running at idle, remove the relay and allow the engine to stall. Turn the ignition switch **OFF**.
   c. Under the hood, wrap a rag around the fuel filter outlet fitting and loosen the clamp to remove the hose.
2. Disconnect and remove the battery and the battery box. Raise and safely support the vehicle.
3. Remove the splash shields at the front of the inner fenders and drain the engine and transaxle and the cooling system. Dispose of these fluids properly.
4. Remove the air cleaner assembly, including the air flow meter and all of the ducting. On turbocharged models, remove the ducting and cover the turbo outlet to protect the turbine.
5. Remove the radiator hoses and remove the radiator with the fan and shroud as a single assembly. On vehicles with automatic transaxle, don't forget to disconnect the cooler hoses at the bottom of the radiator.
6. Disconnect the throttle and cruise control cables, if equipped, and the speedometer cable.
7. Label and disconnect the vacuum hoses and wiring.
8. Disconnect the fuel supply and return hoses and the heater hoses. Plug the fuel hoses to prevent leakage.
9. Disconnect the exhaust pipe from the manifold. If equipped with secondary air, disconnect the pipe.
10. Without disconnecting the hydraulic hoses, remove the power steering pump and hang it from the body with wire. Make sure the hoses are not kinked or strained and that the engine will come out past the pump.
11. Without disconnecting the coolant hoses, remove the air conditioning compressor and hang it from the body with wire. Make sure the hoses are not kinked or strained and that the engine will come out past the compressor.
12. On vehicles with manual transaxle, remove the clutch slave cylinder without disconnecting the hydraulic line. Disconnect the shift linkage.
13. On vehicles with automatic transaxle, disconnect the shift cable.
14. Remove the nuts and press the tie rod ends out of the steering knuckles.
15. Remove the pinch bolts from the steering knuckle and pry the control arm down to slip lower ball joint out of the knuckle.

NOTE: When the halfshafts are removed from the transaxle, the differential side gears must be held in place with service tool 49 G027 003 (manual) or 40 G030 455 (automatic), or equivalent.

16. Carefully pry the left inner CV-joint out of the transaxle and support the halfshaft so it does not hang by the outer CV-joint. Install the service tool.
17. Remove the bolts from the right side halfshaft joint and carefully pry the intermediate shaft out of the transaxle. Install the service tool. Remove or support the shafts so they do not hang by the outer CV-joint.
18. Attach a chain to the lifting eyes on the engine and lift the engine slightly to take up the slack. Check to make sure all wiring, controls and hoses are disconnected.
19. Remove the engine mount nuts and carefully lift the engine and transaxle out of the vehicle as a unit. The rubber mounts should be unbolt from the body and come out with the engine.
20. With the transaxle properly supported, remove the starter. On vehicles with automatic transaxle, remove the flywheel–to–torque converter bolts.
21. Remove any stiffener brackets and other mounts to separate the transaxle from the engine.

To install:

22. Carefully fit the engine and transaxle together and torque the bolts to 66 ft. lbs. (89 Nm).
23. On automatic transaxle, install the flywheel–to–torque converter bolts and torque to 45 ft. lbs. (61 Nm).
24. On manual transaxle, when installing the clutch cover, seal the inner corners with silicone sealant.
25. Attach the stiffener brackets and torque the bolts to 38 ft. lbs. (52 Nm).
26. Attach the front mount to the engine and torque the bolts to 38 ft. lbs. (52 Nm) on non-turbo models or to 68 ft. lbs. (92 Nm) on turbo-charged models.
27. Install the starter and torque the bolts to 18 ft. lbs. (25 Nm).
28. Carefully install the engine/transaxle with just the front and rear mounts attached. When the unit is in place, install the left and right mounts. Start all nuts and bolts first, then torque the following:
   Left mount–to–transaxle nuts—66 ft. lbs. (89 Nm).
   Left mount–to–body bolts—40 ft. lbs. (54 Nm).
   Right mount–to–engine nuts—76 ft. lbs. (103 Nm).
   Right mount long bolt—69 ft. lbs. (93 Nm).
   Front mount–to–body nut—69 ft. lbs. (93 Nm).
   Rear mount long bolt—86 ft. lbs. (117 Nm).
29. Use new gaskets and attach the exhaust pipe. Torque the pipe–to–manifold or turbocharger nuts to 36 ft. lbs. (49 Nm) and the pipe–to–catalytic converter nuts to 69 ft. lbs. (94 Nm).
30. Connect the manual shift linkage and install the clutch cylinder. Torque the shift rod bolt to 17 ft. lbs. (22 Nm) and the extension bar bolt to 34 ft. lbs. (46 Nm).
31. Grease the halfshaft splines and push them into the transaxle. Make sure they are firmly held in place by the circlip. When installing the intermediate shaft, torque the joint mounting bolts to 46 ft. lbs. (62 Nm).
32. Install the lower ball joint and torque the clamping bolt to 40 ft. lbs. (54 Nm). Install the tie rod ends and torque the nut to 42 ft. lbs. (57 Nm), then tighten as required to install a new cotter pin.
33. Attach the stabilizer bar and tighten the nuts so there is 0.80 in.

(20mm) of thread showing at the top of the long mounting bolt.

34. On automatic transaxle, connect and adjust the shift cable as required.

35. Connect the wiring and all water, fuel and vacuum hoses.

36. Install the air conditioner compressor and power steering pump, if equipped.

37. Install the radiator and connect the hoses.

38. Install the air cleaner and air flow meter assembly and all the ducting. Connect and adjust the throttle cable as required.

39. Refill all the fluids and install the battery. Make sure to bleed the cooling system and check the ignition timing.

929

1. The transmission must be removed from below before removing the engine. Matchmark and remove the hood.

2. To relieve fuel system pressure:

a. Locate the fuel pump connector in the trunk or the pump relay behind the driver's kick panel.

b. With the engine running at idle, disconnect power to the fuel pump.

c. After the engine stalls, turn the ignition switch **OFF**. Wrap a rag around the fitting and loosen the fuel filter inlet or outlet hose clamp.

3. Remove the battery. Remove the fresh air duct and the air cleaner/air flow meter assembly.

4. Disconnect the accelerator cable from the throttle body.

5. Remove the cooling fan pulley bolts, the cooling fan and cowling.

6. Remove the drive belts, spark plug wires and spark plugs.

7. Disconnect the evaporative canister and brake hoses.

8. Disconnect and plug the fuel hoses at the fuel rails.

9. Label and disconnect all engine wiring and vacuum hoses. Remove the heater hoses.

10. Working from under the vehicle, remove the engine undercover.

11. Remove the upper and lower radiator hoses.

12. If equipped with an automatic transmission, disconnect the automatic transmission fluid hoses from the radiator.

13. Disconnect the radiator wiring harness and remove the radiator.

14. Remove the alternator and the alternator strap.

15. If equipped, disconnect the air conditioning compressor and bracket from it's mounting and tie the unit off to side of the vehicle with the lines still connected.

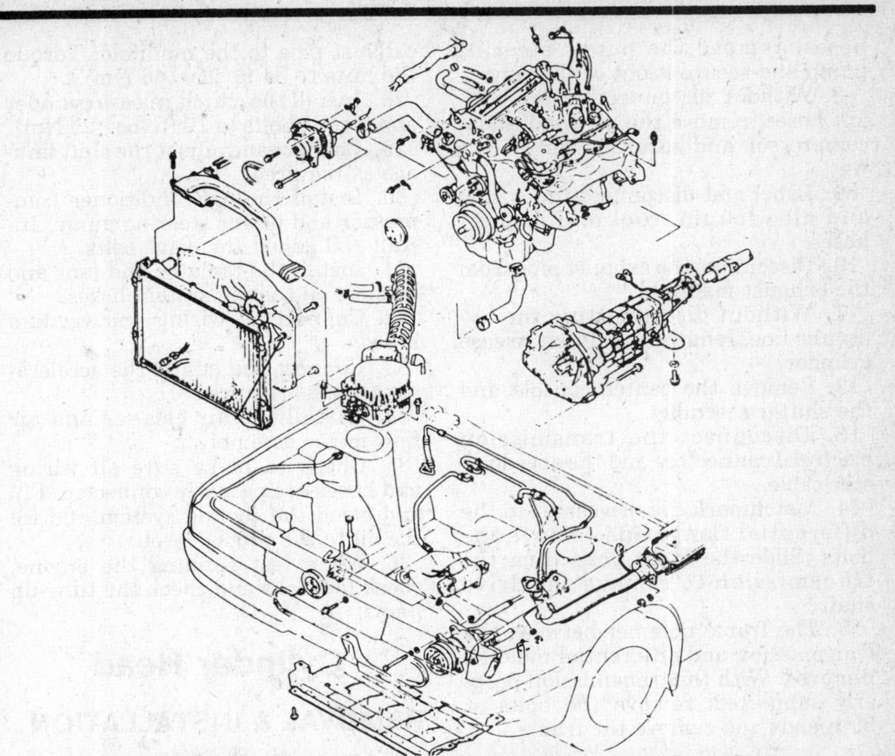

The transmission must be removed to remove the engine from the 929

16. Without disconnecting the hydraulic hoses, remove the power steering pump and secure it out of the way. On the S model with the DOHC engine, the pump pulley must be removed to access the pump bolts. Tool No. 49 W023 585A or equivalent can be used to hold the pulley while removing the center nut.

17. Remove the exhaust pipe(s) from the exhaust manifold and the catalytic converter.

18. Connect a lifting strap to the engine lifting bracket and attach a hoist to the sling and tension the hoist.

19. Remove the transmission.

20. Remove the engine mounting nuts and lift the engine from the vehicle.

To install:

21. Carefully lower the engine into the vehicle and install the mounting nuts. Do not torque them yet.

22. Install the transmission. When the engine and transmission are securely bolted together, torque the engine mount–to–body nuts to 36 ft. lbs. (49 Nm).

23. Install the exhaust pipe(s) with a new gasket and torque the nuts to 36 ft. lbs. (49 Nm).

24. Install the steering pump, air conditioner compressor and the alternator. If the pump pulley was removed, torque the nut to 43 ft. lbs. (59 Nm).

25. Install the radiator and connect all coolant hoses. Install the fan and shroud and adjust the drive belts.

26. Connect the wiring and all vacuum and fuel system hoses.

27. Install the air cleaner and air flow meter assemblies and connect the throttle cable and shift linkage.

28. Install the proper oils and coolant and bleed the cooling system. Make sure to check for leaks when test running the engine.

Miata

1. The engine and transmission are lifted from the vehicle as a unit. Matchmark and remove the hood.

2. To relieve fuel system pressure:

a. Remove the steering column under cover and locate the fuel pump relay.b. With the engine running at idle, remove the relay to disconnect power to the fuel pump.

c. After the engine stalls, turn the ignition switch **OFF**. Wrap a rag around the fitting and loosen the fuel filter inlet or outlet hose clamp.

3. Disconnect the battery (in the trunk). Remove the fresh air duct and the air cleaner/air flow meter assembly.

4. Disconnect the accelerator cable from the throttle body.

5. Raise and safely support the vehicle and remove the under cover. Drain the engine and transmission oil and the coolant.

6. Remove the radiator, shroud and cooling fans as an assembly.

7. Remove the accessory drive belts. Without disconnecting the hydraulic

hoses, remove the power steering pump and secure it out of the way.

8. Without disconnecting the coolant hoses, remove the air conditioner compressor and secure it out of the way.

9. Label and disconnect the wiring and all vacuum, fuel and coolant hoses.

10. Disconnect the exhaust pipe from the exhaust manifold.

11. Without disconnecting the hydraulic line, remove the clutch release cylinder.

12. Remove the center console and the shifter assembly.

13. Disconnect the transmission electrical connectors and the speedometer cable.

14. Matchmark the driveshaft to the differential flange and remove the bolts. Slide the input yoke from the transmission to remove the drive shaft.

15. The frame member between the transmission and differential must be removed. With the transmission properly supported, remove the bolts at both ends and remove the frame.

NOTE: Do not remove the spacers from the frame. If they are removed, the entire frame must be replaced as a unit.

16. Remove the transmission-to-frame bolts.

17. Install lifting equipment onto the engine and make sure all hoses, wires and cables are disconnected.

18. Remove the engine mount nuts and lift the engine and transmission as an assembly from the vehicle.

19. Remove the starter.

20. On vehicles with automatic transmission, remove the torque converter–to–flywheel bolts.

21. Remove the bolts to separate the transmission from the engine.

To install:

22. Assemble the engine to the transmission and torque the bolts to 66 ft. lbs. (89 Nm). On vehicles with automatic transmission, install the torque converter–to–flywheel bolts and torque to 40 ft. lbs. (54 Nm).

23. Install the starter and torque the bolts to 38 ft. lbs. (52 Nm).

24. Carefully install the engine and transmission assembly into the vehicle. Start but do not tighten the mount nuts.

25. Install the transmission–to–differential frame and torque the bolts to 91 ft. lbs. (124 Nm). Torque the engine mount nuts to 58 ft. lbs. (78 Nm).

26. Connect the transmission wiring and speedometer cable and install the driveshaft. Torque the driveshaft bolts to 22 ft. lbs. (30 Nm).

27. Use a new gasket and attach the exhaust pipe to the manifold. Torque the nuts to 34 ft. lbs. (46 Nm).

28. Install the clutch release cylinder and torque bolts to 19 ft. lbs. (25 Nm).

29. Connect and adjust the shift linkage as required.

30. Install the air conditioner compressor and power steering pump. Install and adjust the drive belts.

31. Install the radiator and fans and connect all cooling system hoses.

32. Connect all wiring and vacuum hoses.

33. Connect and adjust the accelerator cable as required.

34. Install the air cleaner and air flow meter assembly.

35. Check to make sure all wiring and hoses are properly connected. Fill and bleed the cooling system and fill the oil to the proper level.

36. When test running the engine, check for leaks and check the tune-up items.

Cylinder Head

REMOVAL & INSTALLATION

323 and Protege

1. Relieve the fuel system pressure. Be carefull not to run the engine long enough to warm it up.

2. Disconnect the negative battery cable and remove the engine under cover.

3. Remove the air cleaner and air flow meter assembly.

4. Remove the spark plugs and the distributor assembly.

5. Remove all of the air ducting, including the air bypass valve, if equipped. On turbocharged engines, cover the turbocharger inlet to keep dirt out.

6. Drain the cooling system and disconnect the radiator and heater hoses. On turbocharged engines, remove the radiator and fans as an assembly.

7. Disconnect the exhaust pipe from the engine and remove the exhaust manifold. On turbocharged engines, remove the manifold and the turbocharger as an assembly.

8. On DOHC engines, remove the coolant bypass pipe.

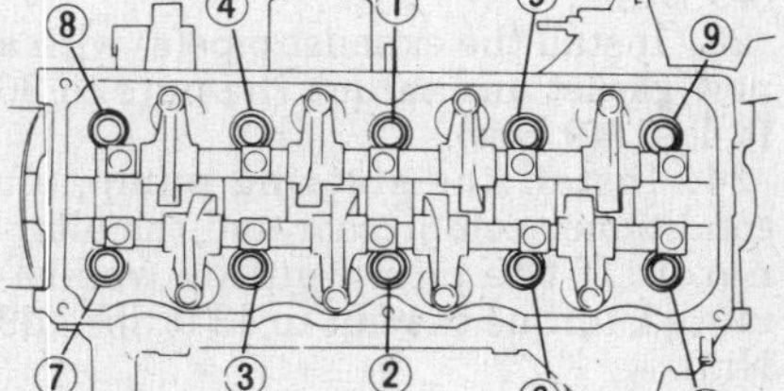

Cylinder head bolt torque sequence—323 and Protege with SOHC engine

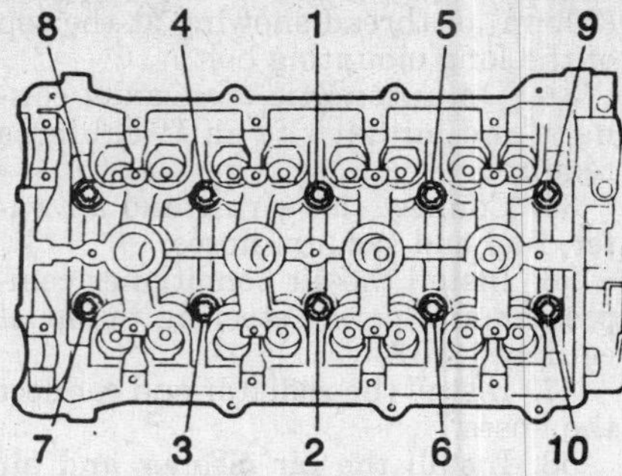

Cylinder head bolt torque sequence—323, Protege and Miata with DOHC

9. Disconnect the accelerator cable.

10. Label and disconnect all required electrical connections, vacuum hoses and fuel line couplings.

11. Remove the intake manifold bracket and the intake manifold.

12. Remove the cylinder head cover bolts and the cylinder head cover.

13. Remove the timing belt cover assembly and turn the engine to TDC No. 1 cylinder. Make sure the timing marks on the crankshaft and camshaft sprockets are properly aligned and mark the direction of rotation of the belt.

14. Loosen the timing belt tensioner and remove the belt.

15. When everything is disconnected, loosen the cylinder head bolts in the reverse of the tightening sequence. Remove the bolts and lift the head off the engine.

To install:

16. Thoroughly, clean the cylinder head and the block contact surfaces. Examine the head gasket and check the cylinder head for warping and cracks. The maximum allowable contact distortion is 0.006 in. (0.15mm).

17. Clean the cylinder head bolts and the threads in the block. Make sure the bolts turn freely in the block.

18. Lay the new gasket on the surface of the block.

NOTE: Turbocharged and non-turbocharged engines use different cylinder head gaskets. To ensure proper sealing and compression, make sure the proper gasket is being installed.

19. Set the cylinder head carefully on the gasket.

20. Lubricate the bolt threads and seat surfaces with clean engine oil and install them. Torque the bolts in 3 steps to 56–60 ft. lbs. (75–81 Nm) in the proper sequence.

21. Make sure the engine is not at TDC on any piston and align the camshaft sprocket timing marks. Set the engine to TDC of No. 1 cylinder, install the timing belt and set the tension. Carefully rotate the crankshaft 2

turns to make sure the timing marks still line up.

22. Apply a thin bead of sealant to the cylinder head cover and install the new gasket. Install the cover and torque the cover nuts to 78 inch lbs. (9 Nm).

23. Use new gaskets and install the manifolds. Torque the intake manifold nuts to 19 ft. lbs. (25 Nm). Torque the SOHC engine exhaust manifold bolts to 17 ft. lbs. (23 Nm). On DOHC engines, torque the exhaust manifold nuts to 34 ft. lbs. (46 Nm).

24. Use a new gasket to connect the exhaust pipe and torque the nuts or bolts 34 ft. lbs. (46 Nm).

25. If removed, install the radiator and connect all cooling system hoses.

26. Install the distributor and spark plugs.

27. Connect all vacuum and fuel system hoses and connect all wiring.

28. Connect the accelerator cable and adjust as required.

29. Complete the remainder of the installation. Change the oil and fill and bleed the cooling system. Run the engine to check for leaks.

626 and MX-6

1. Relieve the fuel system pressure. Be carefull not to run the engine long enough to warm it up.

2. Disconnect the negative battery cable and drain the cooling system.

3. Disconnect the appropriate wiring and remove the spark plugs and the distributor.

4. Disconnect the accelerator cable and if equipped, the cruise control cable.

5. Remove the air intake pipe and disconnect the fuel hose. Cover the fuel hose to catch leakage.

6. Remove the upper radiator hose, water bypass hose, heater hose, oil cooler hose (turbo only) and brake vacuum hose.

7. Remove the 3-way and EGR solenoid valve assemblies.

8. Label and disconnect the wiring and vacuum hoses.

9. Remove the vacuum chamber and exhaust manifold shield.

10. Remove the EGR pipe, turbo oil pipes, if equipped, and exhaust pipe.

11. Remove the exhaust manifold. On turbo-charged engines, remove the manifold and turbocharger as an assembly.

12. Remove the intake manifold bracket and the intake manifold.

13. Loosen the air conditioning compressor and bracket and position the compressor aside. Do not disconnect the refrigerant lines.

14. Remove the upper timing belt cover.

15. To remove the timing belt, perform the following:

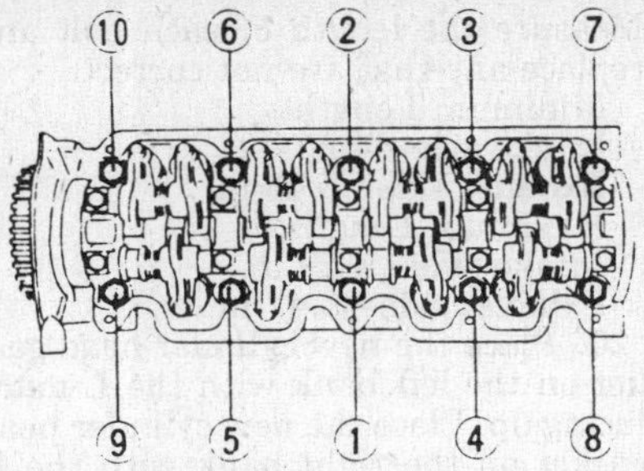

Cylinder head bolt torque sequence for 626 and MX-6; loosen in reverse sequence

a. Rotate the crankshaft so the **1** on the camshaft pulley is aligned with the timing mark on the front housing.

b. When timing marks are aligned, loosen the timing belt tensioner lock bolt. Pull the tensioner as far out as it will go and temporarily tighten the lock bolt.

c. Lift the timing belt from the camshaft pulley and position it aside.

16. Remove the cylinder head cover and gasket.

17. Loosen the cylinder head bolts in the reverse of the tightening sequence, and remove the cylinder head and head gasket.

To install:

18. Thoroughly clean the cylinder head and cylinder block contact surfaces. Check the cylinder head for warping and cracks. The maximum allowable contact distortion is 0.006 in. (0.15mm). Inspect the cylinder head bolts for damaged threads and make sure they turn freely in the threads in the block.

19. Lay the new gasket on the surface of the block.

NOTE: Turbocharged and non-turbocharged engines use different cylinder head gaskets. To ensure proper sealing and compression, make sure the proper type gasket is being installed.

20. Set the cylinder head carefully on the gasket.

21. Lubricate the bolt threads and seat surfaces with clean engine oil and install them. Torque the bolts in 3 steps to 59–64 ft. lbs. (80–86 Nm) in the proper sequence.

22. Apply sealant to the 4 corners of the cylinder head and install the cover with a new gasket. Torque the cover nuts to 69 inch lbs. (8 Nm).

23. Make sure the camshaft pulley and front housing timing marks are aligned and install the timing belt. Set the tension and carefully rotate the crankshaft 2 turns to make sure the timing marks still line up.

24. Use a new gasket and install the intake manifold. Torque the nuts and bolts to 22 ft. lbs. (30 Nm). Install the bracket.

25. Use a new gasket and install the exhaust manifold. Torque the nuts to 36 ft. lbs. (49 Nm). On turbo-charged engines, connect the oil lines.

26. Connect the exhaust pipe with a new gasket and torque the nuts to 34 ft. lbs. (46 Nm). Install the bracket, if equipped.

27. Connect all the coolant, vacuum and fuel system hoses.

28. Install the distributor and spark plugs and connect all the wiring

29. Connect the accelerator cable and adjust as required.

30. Complete the remainder of the installation. Change the oil and fill and bleed the cooling system. Run the engine to check for leaks.

929

SOHC ENGINE

1. Relieve the fuel system pressure. Be carefull not to run the engine long enough to warm it up.

2. Disconnect the negative battery cable and drain the cooling system.

3. Remove the air intake ducting.

4. Disconnect the appropriate wiring and remove the spark plugs and the distributor.

5. Disconnect the accelerator cable and if equipped, the cruise control cable.

6. Label and disconnect the wiring, fuel lines and vacuum hoses. Remove the upper radiator hose and disconnect the heater hoses as required.

7. Remove the fan shroud and the fan from the front of the engine.

8. Mark the intake tubes so they can be returned to their original position. It may be easier to remove the throttle body before removing the tubes and plenum chamber.

9. Remove the intake manifold. Loosen the bolts ½ turn at a time in the reverse order of the torque sequence. This is important to prevent warping.

10. Remove the cylinder head covers.

11. Disconnect the exhaust pipes and EGR tube and remove the exhaust manifolds.

12. Remove the accessory drive belts and the timing belt cover and crankshaft pulley.

13. Position the engine at TDC on No. 1 cylinder so all the timing sprocket matchmarks are aligned. Mark the direction of rotation of the belt, remove the belt tensioner and the timing belt.

14. Remove the seal plate at the front of the heads.

15. Loosen the cylinder head bolts ½ turn at a time in the reverse of the torque sequence.

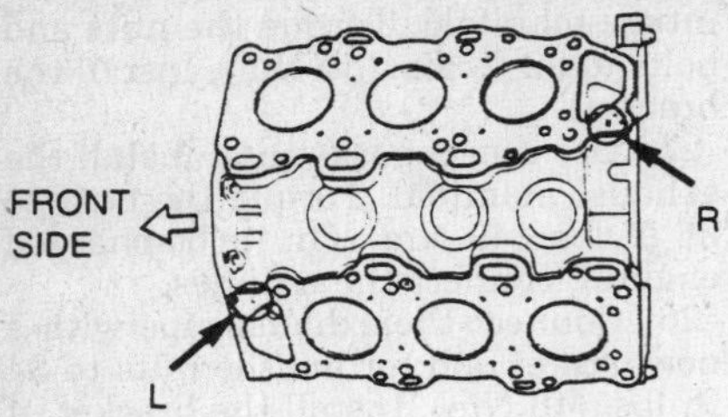

Cylinder head gasket installation—929

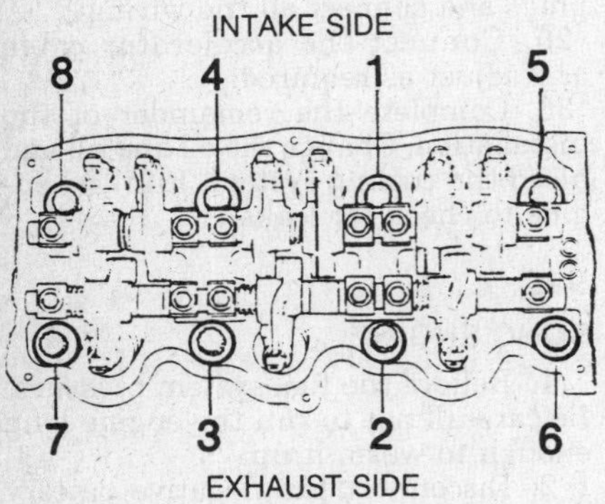

Cylinder head bolt torque sequence—929

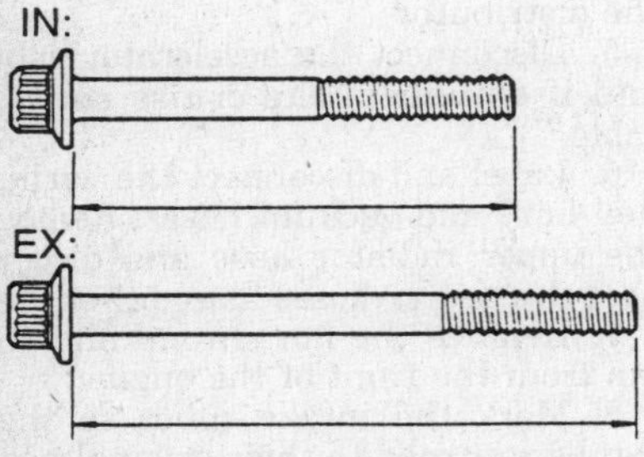

Measure the cylinder head bolts, replace any that are out of spec

16. Remove the bolts and lift the heads off the engine.

To install:

17. Thoroughly clean the cylinder head and cylinder block contact surfaces. Check the cylinder head for warping and cracks. The maximum allowable contact distortion is 0.004 in. (0.10mm). Make sure the bolt threads in the block are clean and that the bolts turn freely.

18. Check the oil control orifice plug projection at the cylinder block. Projection should be 0.209–0.224 in. (5.3–5.7mm). Apply clean engine oil to a new O-ring, position it to the control plug and set the plugs in the block.

19. The cylinder head bolts are designed to stretch when torqued. If they are stretched beyond a specific length, they will become brittle and weak. Measure the length of each bolt and replace any that are not correct.

Minimum Length:
Intake side—4.25 in.
Exhaust side—5.43 in.
Maximum Length:
Intake side—4.29 in.
Exhaust side—5.47 in.

20. Place the new cylinder head gasket on the left bank with the **L** mark facing up. Place the new cylinder head gasket on the right bank with the **R** mark facing up.

21. Carefully place the cylinder heads onto the block. Tighten the head bolts in the following manner:

a. Coat the threads and the seating faces of the head bolts with clean engine oil.

b. Torque the bolts, in sequence, to 14 ft. lbs.

c. Place a paint mark on the head of each bolt.

d. Using this mark as a reference, tighten the bolts, in sequence, an additional 90 degrees in 1 smooth motion.

e. Repeat Step d.

22. Install the seal plate.

23. Install the timing belt and set the tensioner. Turn the crankshaft 2 full turns to make sure the timing marks still line up.

24. Install the cylinder head covers with new gaskets and seal washers and torque the bolts to 39 in. lbs. (4.4 Nm). Do not over torque. Install the timing belt cover.

25. Install the intake manifold with new gaskets. Torque the manifold bolts in the proper sequence in 2–3 steps to 19 ft. lbs. (25 Nm).

26. If the intake plenum was removed, make sure to install a new O-ring at the drain hole. Install the intake tubes with new O-rings and torque the nuts to 19 ft. lbs. (25 Nm). If the throttle body was removed, install it with a new gasket and torque the nuts to 19 ft. lbs. (25 Nm).

27. Install the exhaust manifolds with new gaskets and torque the nuts to 21 ft. lbs. (28 Nm). Install the exhaust pipes with new gaskets and torque the nuts to 34 ft. lbs. (46 Nm).

28. Complete the installation of the remaining components in reverse of the removal procedure. Fill and bleed the cooling system and change the oil. Make all the required tune-up adjustments.

DOHC ENGINE

1. Relieve the fuel system pressure. Be careful not to run the engine long enough to warm it up.

2. Disconnect the negative battery cable and drain the cooling system.

3. Remove the intake manifold cover and the air intake ducting. It may be easier to also remove the air cleaner and air flow meter assembly.

4. Disconnect the appropriate wiring and remove the spark plugs and the distributor.

5. Disconnect the accelerator cable.

6. Label and disconnect the wiring, fuel lines and vacuum hoses. Remove the upper radiator hose and disconnect the heater hoses as required.

7. Remove the fan shroud and the fan from the front of the engine.

8. Remove the upper portion of the intake manifold, with the throttle body and by-pass air control valve.

9. Remove the intake manifold. Loosen the bolts ½ turn at a time in the reverse order of the torque sequence. This is important to prevent warping.

10. Disconnect the exhaust pipes and EGR tube and remove the exhaust manifolds.

11. Remove the accessory drive belts and the timing belt cover and crankshaft pulley.

12. Position the engine at TDC on No. 1 cylinder so all the timing sprocket matchmarks are aligned. Mark the direction of rotation of the belt, remove the belt tensioner and the timing belt.

13. Remove both spark plug hall covers and cylinder head covers. Hold the camshaft from turning with the wrench flat on the shaft and remove the drive sprockets.

14. Note how the camshaft bearing caps are numbered and indexed. Make sure they are assembled in the same loaction. Remove the front caps on all 4 camshafts.

15. Loosen the camshaft bearing cap bolts ½ turn at a time in the reverse to the torque sequence. Remove the caps and lift out the camshafts and rocker arms. Make sure the rocker arms are returned to the same position.

16. Loosen the cylinder head bolts ½ turn at a time in the reverse of the torque sequence.

17. Remove the bolts and lift the heads off the engine.

To install:

18. Thoroughly clean the cylinder head and cylinder block contact surfaces. Check the cylinder head for warping and cracks. The maximum allowable contact distortion is 0.006 in. (0.15mm). Make sure the bolt threads in the block are clean and that the bolts turn freely.

19. Check the oil control orifice plug projection at the cylinder block. Projection should be 0.209–0.224 in. (5.3–5.7mm). Apply clean engine oil to a new O-ring, position it to the control plug and set the plugs in the block.

20. The cylinder head bolts are designed to stretch when torqued. If they

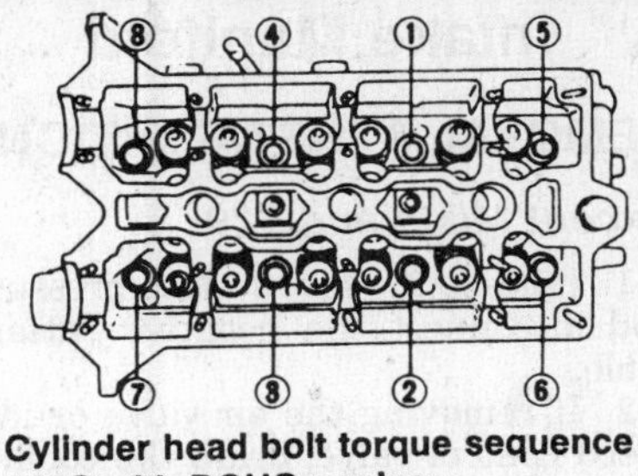

Cylinder head bolt torque sequence—929S with DOHC engine

are stretched beyond a specific length, they will become brittle and weak. Measure the length of each bolt and replace any that are not correct.

Minimum Length: 4.25 in. (108mm)
Maximum Length:4.29 in. (109mm)

21. Place the new cylinder head gasket on the left bank with the **L** mark facing up. Place the new cylinder head gasket on the right bank with the **R** mark facing up.
22. Carefully place the cylinder heads onto the block. Tighten the head bolts in the following manner:
   a. Coat the threads and the seating faces of the head bolts with clean engine oil.
   b. Torque the bolts, in sequence, to 14 ft. lbs.
   c. Place a paint mark on the head of each bolt.
   d. Using this mark as a reference, tighten the bolts, in sequence, an additional 90 degrees in 1 smooth motion.
   e. Repeat Step d.
23. Install the rocker arms and camshafts, making sure the bearing caps are returned to their original position.

NOTE: Camshaft installation requires a different bearing cap torque sequence for left and right bank cylinder heads.

24. With the camshafts installed, install the drive sprockets. Torque the right bank sprocket bolts to 59 ft. lbs. (80 Nm) and the left bank sprocket bolts to 19 ft. lbs. (25 Nm).
25. Apply silicone sealant to the front bearing caps and install the cylinder head covers with new gaskets. Torque the nuts to 39 in. lbs. (4.4 Nm).
26. Remove the old gasket or sealant from the cylinder head covers and apply a thin bead of silicone sealant when installing the spark plug hall covers.
27. Turn the camshafts and the crankshaft to align the timing marks and install the timing belt and tensioner. Install the timing belt cover.
28. Install and adjust the accessory drive belts.
29. Use new gaskets and install the exhaust manifolds. Torque the nuts to 21 ft. lbs. (28 Nm). Use new gaskets to connect the exhaust pipes and torque the nuts to 36 ft. lbs. (49 Nm).
30. Use new gaskets to install the intake manifold. Make sure to use the proper lock and flat washers and torque the nuts in 3 steps in the proper sequence to 19 ft. lbs. (25 Nm).
31. Make sure the fuel injection components are properly installed and connected and install the upper intake manifold assembly with a new gasket. Torque the bolts to 19 ft. lbs. (25 Nm).
32. Complete the installation and adjustment of the remaining components. Fill and bleed the cooling system.
33. When test running the engine, check for leaks and check the ignition timing adjustment.

Miata

1. Relieve the fuel system pressure. Be carefull not to run the engine long enough to warm it up.
2. Disconnect the battery negative cable and drain engine cooling system.
3. Remove the air cleaner and air flow meter assembly and the inlet air ducting. Disconnect the accelerator cable.
3. Disconnect the radiator and heater hoses. If desired, remove the radiator and fans as an assembly by removing the 4 radiator bolts.
4. Disconnect the heater hoses.
5. Label and disconnect all required electrical connections, vacuum hoses and fuel line couplings.
6. Remove the ignition coil pack assembly with the spark plug wires.
7. Remove the exhaust manifold from the cylinder head and secure it aside or disconnect it from the exhaust pipe.
8. Remove the cylinder head cover and the timing belt front cover.
9. Turn the engine to TDC No. 1 cylinder. Make sure the timing marks on the crankshaft and camshaft sprockets are properly aligned and mark the direction of rotation of the belt. Loosen the tensioner and remove the timing belt.
9. Remove the intake manifold bracket.
10. When everything is disconnected, loosen the cylinder head bolts in the reverse of the tightening sequence. Remove the bolts and lift the head off the engine.

To install:

11. Thoroughly, clean the cylinder head and the block contact surfaces. Examine the head gasket and check the cylinder head for warping and cracks. The maximum allowable contact distortion is 0.006 in. (0.15mm).
12. Clean the cylinder head bolts and the threads in the block. Make sure the bolts turn freely in the block.
13. Lay the new gasket on the surface of the block and set the cylinder head carefully on the gasket.
14. Lubricate the bolt threads and seat surfaces with clean engine oil and install them. Torque the bolts in 3 steps to 56–60 ft. lbs. (75–81 Nm) in the proper sequence. The sequence is the same as the DOHC Protege engine.
15. Make sure the engine is not at TDC on any piston and align the camshaft sprocket timing marks. Set the engine to TDC of No. 1 cylinder, install the timing belt and set the tension. Carefully rotate the crankshaft 2 turns to make sure the timing marks still line up. Install the front cover.
16. Apply a thin bead of sealant to the cylinder head cover and install the new gasket. Install the cover and torque the cover nuts to 78 inch lbs. (9 Nm).
17. Use new gaskets to install the exhaust manifold and torque the nuts to 34 ft. lbs. (46 Nm). If the pipe was disconnected, install a new gasket and torque the nuts to 34 ft. lbs. (45 Nm).
18. If removed, install the radiator and connect all cooling system hoses.
19. Install the ignition coil pack and spark plugs.
20. Connect all vacuum and fuel system hoses and connect all wiring.
21. Connect the accelerator cable and adjust as required.
22. Complete the remainder of the installation. Change the oil and fill and bleed the cooling system. Run the engine to check for leaks.

Valve Lifters

The lifter used in all engines is actually a Hydraulic Lash Adjuster (HLA), mounted either in the rocker arm or directly over the valve. They cannot be adjusted or repaired and should be removed only if obviously worn or damaged. If the HLA is removed from the rocker arm, a new O-ring must be fitted for installation.

REMOVAL & INSTALLATION

SOHC Engine

1. Remove the rocker arm assembly.
2. Inspect the HLA mounted in the rocker arm. If there is no obvious damage or excess play, the HLA need not be removed.

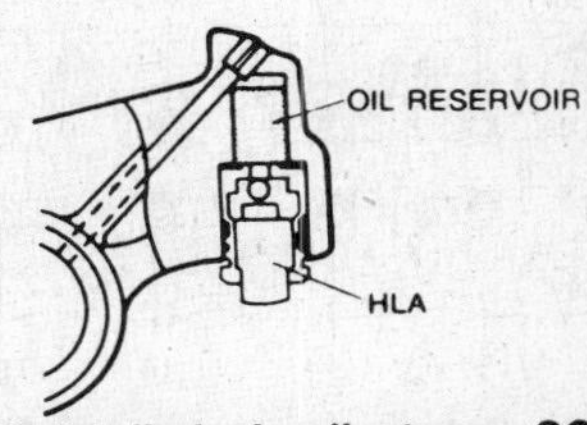

Hydraulic lash adjuster on SOHC engines

3. To remove the HLA, pull it out of the rocker arm. Use pliers if necessary, but this will probably damage the part.
4. When installing the HLA, coat a new O-ring with oil and install it onto the HLA. Fill the oil reservoir in the rocker arm and carefully push the HLA into place.

DOHC Engine

EXCEPT 929S

1. Remove the camshaft assembly.
2. Mark each HLA so it can be returned to the same bore. In necessary, use a magnet to lift it out.
3. Inspect the HLA for excess wear or damage. Hold the part between thumb and finger and try to collapse it. If it moves, it must be replaced.
4. Before installing the HLA, lubricate the part and the bore in the head with engine oil. Make sure it moves freely in the bore.

929S

1. Remove the camshafts and rocker arms.
2. The HLA should only be removed if necessary. If it cannot be removed with fingers, use a rag to protect it and pull it out with pliers.
3. Inspect the HLA for excess wear or damage. Hold the part between thumb and finger and try to collapse it. If it moves even after being filled with oil, it must be replaced.
4. Before installation, submerge the HLA in engine oil and insert a pin into the hole in the plunger. This will allow the HLA to fill with oil.
5. Remove any excess oil from the bore in the head and insert the HLA.

Rocker Arms/Shafts

REMOVAL & INSTALLATION

323 and Protege SOHC Engine

1. Disconnect the negative battery cable.
2. Disconnect the spark plug wires and the breather hoses and remove the cylinder head cover.
3. Loosen the rocker arm assembly bolts 1 turn at a time in the reverse order of the torque sequence. This is important to prevent warping. Remove the rocker arm assembly with the bolts.

To install:

4. Set the assembly in place and install the bolts finger tight. Make sure the rocker arms and spacers are properly positioned. Turn the bolts 1 turn at a time in the proper sequence till the assembly is seated in the head.
5. Torque the bolts in 3 steps in the proper sequence to 21 ft. lbs. (28 Nm).
6. Install the cover and connect the wires and hoses.

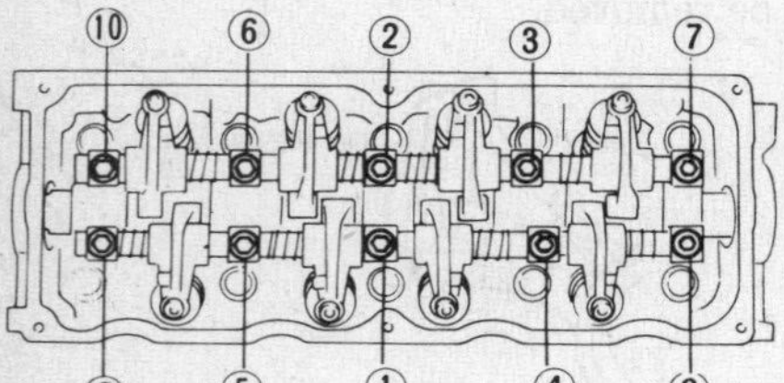

Rocker shaft torque sequence – 323, Protege, 626 and MX-6

626, MX-6 and 929 SOHC

1. Disconnect the negative battery cable.
2. Disconnect the spark plug wires and the breather hoses and remove the cylinder head cover.
3. Remove the front timing belt cover and set the engine at TDC of No. 1 cylinder.
4. Loosen the timing belt tension and remove the belt from the camshaft sprocket.
5. The rocker arm shafts and camshaft bearing caps are held by the same bolts. Loosen the bolts 1 turn at a time in the reverse order of the torque sequence. This is important to prevent bending the camshaft and rocker arms. Remove the rocker arm assembly with the bolts.

To install:

6. If the camshaft was removed, make sure it is installed with the No. 1 cylinder lobes pointing up.
7. Set the assembly in place and install the bolts finger tight. Make sure the rocker arms and spacers are not caught between the shaft and the cap. Turn the bolts in 1 turn at a time in the proper sequence till the assembly is seated in the head.
8. Torque the bolts in 3 steps in the proper sequence to 20 ft. lbs. (26 Nm).
9. Install the camshaft sprocket and torque the bolt to 48 ft. lbs. (65 Nm).
10. Align the timing marks, install the belt and adjust the tension. Rotate the crankshaft 2 turns and check that the timing marks still align.
11. Install the covers and connect the wires and hoses.

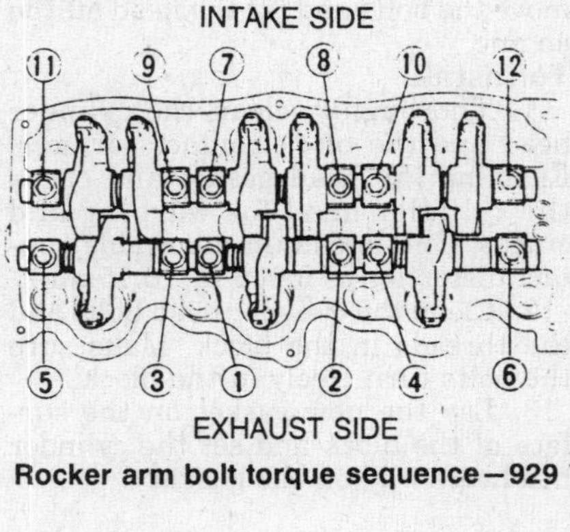

Rocker arm bolt torque sequence – 929

Intake Manifold

REMOVAL & INSTALLATION

Except Miata and 929

1. Release the fuel system pressure and disconnect the negative battery cable.
2. If removing the air valve or idle speed control valve, drain the cooling system.
3. At the throttle body, remove the air inlet duct and disconnect the accelerator cable.
4. Label and disconnect all hoses and wiring from the manifold and the throttle body. If equipped, disconnect the EGR pipe.
5. The manifold can be considered 3 separate units; the throttle body, the upper chamber and the intake runners. To remove the upper chamber and throttle body as an assembly, some of the bolts go through the bottom of the intake runner casting and are reached from underneath. Remove the nuts and bolts and remove the upper chamber assembly.
6. If the entire intake manifold assembly is to be removed, disconnect the fuel injector wiring and hoses and remove the injectors and delivery pipe as an assembly.
7. Remove the support bracket and remove the manifold.

To install:

7. Carefully clean the gasket mating surfaces. Visually inspect the intake manifold and upper chamber and, if equipped, make sure the shutter valves move freely. There may be a

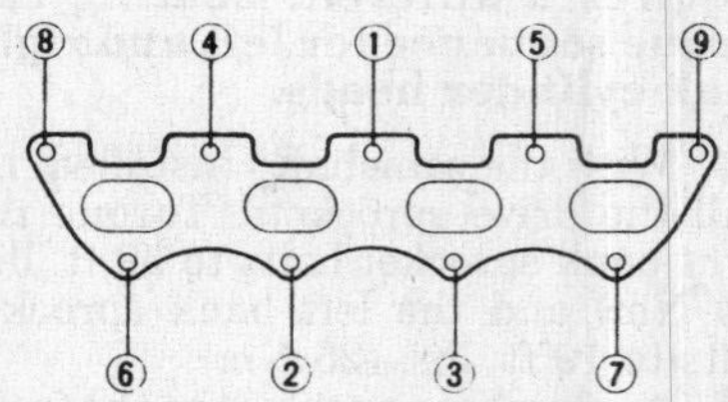

Intake manifold–to–cylinder head torque sequence – 323 and Protege with BP series engine

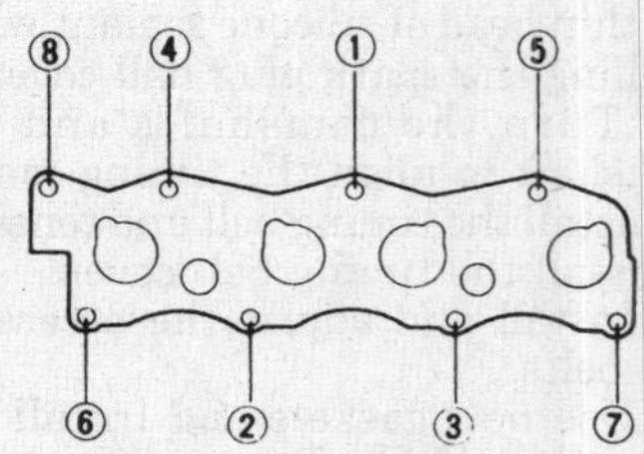

Intake manifold–to–cylinder head torque sequence – 323 and Protege with B6 series engine

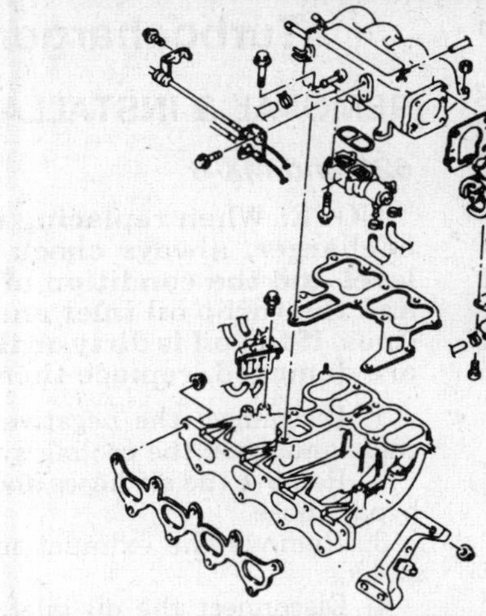

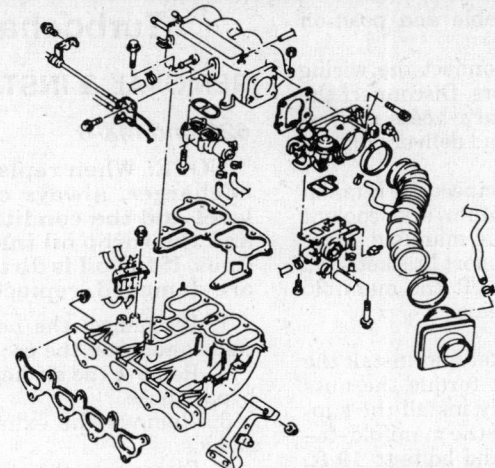

Intake manifold assembly on 323 BP series engine; other engines similar

thin sealer coating on the throttle plate or bore; do not remove it.

8. Use new gaskets when installing the manifold and torque the manifold–to–cylinder head nuts or bolts to 19 ft. lbs. (25 Nm). On 323 and Protege, be sure to use the proper torque sequence.

9. If the upper chamber or throttle body were removed, use new gaskets to install them and torque the nuts and bolts to 19 ft. lbs. (25 Nm).

10. When installing the idle speed control valve and the air valve, use new O-rings and torque the bolts to 40 in. lbs. (4.5 Nm).

11. Install the remaining components in the reverse order of removal.

12. Refill and bleed the cooling system and run the engine to check for leaks.

929 SOHC

1. Release the fuel system pressure and disconnect the negative battery cable.

2. If removing the complete intake manifold, drain the coolant. If removing only the upper chamber and throttle body, disconnect and plug the small coolant hoses from the idle speed control valve and the air valve.

3. Disconnect the air inlet duct from the air cleaner and remove it from the throttle body. Disconnect the throttle cable.

4. Remove the front cover from the manifold and label and disconnect the wiring and vacuum hoses from the TICS and purge air control solenoid valves.

5. Label and disconnect the wiring and vacuum hoses at the throttle body and the EGR valve.

6. Remove the wiring bracket from the right side and lay it aside on the valve cover.

7. There are 6 tubes connecting the upper and lower sections of the manifold which are not interchangable. Mark the tubes so they can be returned to their original position, remove the nuts from the upper end and pull the tubes off the manifold.

8. If all wiring and hoses are disconnected, the upper chamber can be lifted off by removing the 2 nuts from the top.

9. To remove the intake manifold, disconnect the fuel hoses at the pressure regulator and disconnect the wiring for the fuel injectors.

10. Drain the cooling system and disconnect the hoses and sensor wires at the front of the manifold.

11. Loosen the intake manifold nuts 1 turn at a time in the reverse order of the torque sequence. This is important to prevent warping the manifold. Remove the manifold and gaskets.

To install:

12. Remove all gasket material from the seating surface on the manifold and the engine. Be careful to not let the old gasket material fall into the intake ports. Carefully inspect the manifold for cracks or damage. Make sure the manifold and cylinder head mating surfaces are clean and make good contact.

13. Place the new intake manifold gaskets onto the cylinder heads and lower the manifold over the gaskets. Install the washers with the white paint marks facing up. Install the nuts and torque them in the proper sequence in 2 stages to 19 ft. lbs. (25).

14. Connect the coolant hoses and sensor wires and the fuel injector hoses and wires.

15. Before installing the upper chamber, be sure to replace the O-ring for the drain hole at the rear stud. Install the upper chamber and torque the nuts to 19 ft. lbs. (25 Nm).

16. Use new gaskets and O-rings and

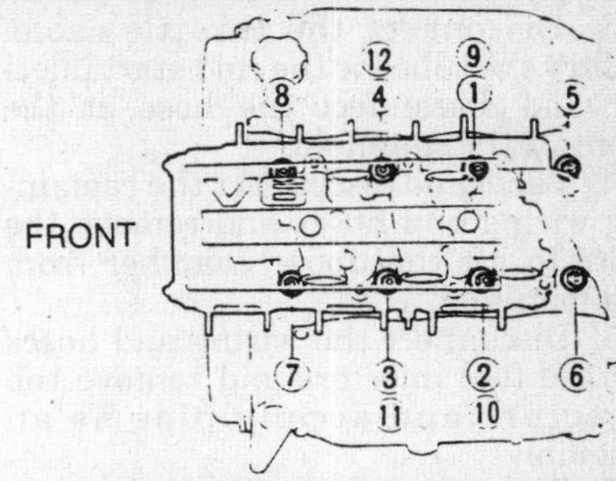

Intake manifold torque sequence—929 with SOHC

install the connecting tubes in their original positions. If necessary, lightly lubricate the O-rings with silicone spray when pushing the tubes into the lower section. Torque the nuts to 19 ft. lbs. (25 Nm).

17. Install the wiring bracket on the right side and begin connecting the coolant and vacuum hoses.

18. Connect the wiring, install the front cover and fill the cooling system.

929S DOHC

1. The manifold consists of an upper and lower section, with the fuel injectors on the lower section. Release the fuel system pressure, disconnect the negative battery cable and drain the coolant.

2. Remove the cover on top of the manifold, the air duct above the radiator and the air inlet duct from the throttle body.

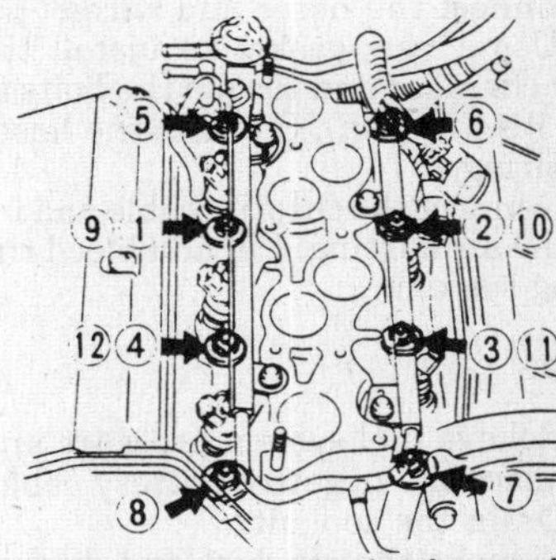

Torque sequence for intake manifold nuts—929S with DOHC

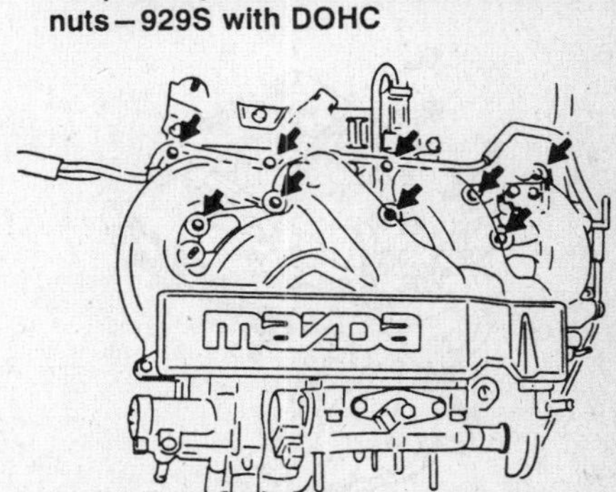

Upper manifold chamber bolt locations—929S

3. Disconnect the throttle cable. Follow the tube for the cold start injector and disconnect the hose at the front of the manifold.

4. Label and disconnect the remaining wiring and hoses and remove the bolts to lift the upper chamber from the manifold.

5. Disconnect the wiring and hoses for the fuel injectors and remove the injectors and supply pipe as an assembly.

6. Disconnect the coolant hoses and sensor wiring. Disconnect the EGR hoses and sensor wiring.

7. Loosen the manifold nuts 1 turn at a time in the reverse of the torque sequence. This is important to prevent warping. Remove the nuts and lift the manifold off the engine.

To install:

8. Remove all gasket material from the seating surface on the manifold and the engine. Be careful to not let the old gasket material fall into the intake ports. Carefully inspect the manifold for cracks or damage. Make sure the manifold and cylinder head mating surfaces are clean and make good contact.

9. Place the new intake manifold gaskets onto the cylinder heads and lower the manifold over the gaskets. Install the washers with the white paint marks facing up. Install the nuts and torque them in the proper sequence in 2 stages to 19 ft. lbs. (25).

10. Connect the coolant hoses and sensor wires. Install the fuel injectors and connect the hoses and wires.

11. Use a new gasket to install the upper chamber and torque the bolts to 19 ft. lbs. (25 Nm). Connect the hoses and wiring.

12. Connect the throttle cable and install the air ducting. Fill and bleed the cooling system.

Miata

1. Relieve fuel system pressure and disconnect the negative battery cable.

2. Drain the coolant.

3. Remove the air duct and disconnect the throttle cable and position sensor wiring.

4. Label and disconnect the wiring from the fuel injectors. Disconnect the fuel supply and return hoses and remove the injectors and delivery pipe as an assembly.

5. Label and disconnect the remaining wiring and vacuum and coolant hoses from the intake manifold.

6. Remove the support brackets and remove the bolts to lift the manifold assembly off the engine.

To install:

7. Use a new gasket and install the manifold but do not torque the nuts and bolts yet. Loosely install the support brackets, torque the manifold–to–cylinder head nuts and bolts to 19 ft. lbs. (25 Nm), then torque the support bracket fasteners to the same value.

8. Install the fuel injectors and connect the wiring and hoses.

9. Connect the remaining vacuum and water hoses and connect and adjust the throttle cable as required.

10. Connect the water hoses and fill the cooling system. Install the air ducts and run the engine to check for leaks.

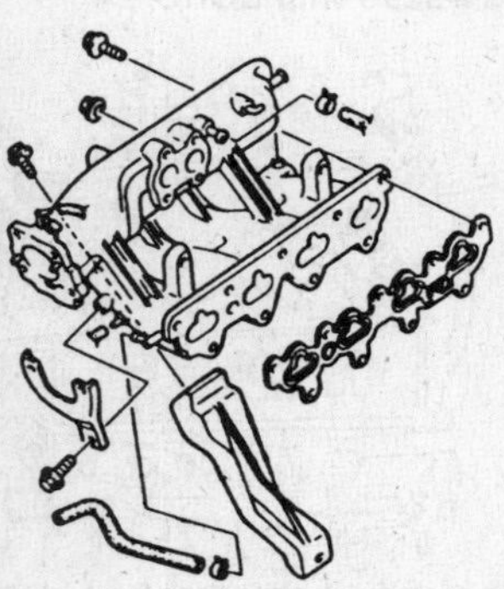

Miata intake manifold assembly removal

Exhaust Manifold

REMOVAL & INSTALLATION

Except 929

1. Disconnect the negative battery cable. Remove the heat shield cover, if equipped. Remove the attaching nuts from the exhaust pipe at the manifold.

2. If equipped with a turbocharger, remove the air duct and turbocharger-to-manifold mounting bolts.

3. Remove the exhaust manifold nuts or bolts and remove the manifold from the engine.

4. Installation is the reverse of the removal procedure. Use new gaskets and torque the nuts or bolts to 34 ft. lbs. (46 Nm).

929

1. Remove the air ducting as required and remove the heat shields.

2. On the 929 with the SOHC engine, there is an exhaust system crossover pipe that runs behind the engine. Remove the nuts or bolts and remove the pipe.

3. Disconnect the exhaust pipes from the manifolds, remove the nuts and remove the manifolds from the engine.

4. Installation is the reverse of removal. Use new gaskets and torque the nuts and bolts to 22 ft. lbs. (30 Nm).

Turbocharger

REMOVAL & INSTALLATION

626 and MX-6

NOTE: When replacing the turbocharger, always check the oil level and the condition of the oil and the turbo oil inlet and outlet lines. If the oil is dirty or the lines are damaged, replace them.

1. Disconnect the negative battery cable and drain the cooling system.

2. Remove the air hoses and the air bypass hose.

3. Remove the exhaust manifold shields.

4. Disconnect the oil inlet and return pipes from the turbocharger and plug the ends.

5. Disconnect and plug the water hoses from the water pipe.

6. Disconnect the EGR pipe from the exhaust manifold.

7. Remove the oxygen sensor.

8. Disconnect the front pipe from the turbocharger and set the gasket aside. Remove the bolt from the turbocharger joint pipe.

9. Support the turbocharger and remove the exhaust manifold nuts. Remove the turbo and the manifold as an assembly. Cover the turbocharger ports with a clean rag or masking tape to prevent the entry of foreign matter.

NOTE: Do not drop the turbocharger or carry it around by the actuating rod. When laying the unit down, do so with the turbine shaft in the horizontal position. Do not bend the actuator mounting or rod.

To install:

10. Use a wire wheel or brush to remove all the sealant and gasket material from the turbocharger and exhaust manifold mating surfaces. Use all new gaskets.

11. Pour about 1.2 oz.(25cc) of clean engine oil into the opening for the turbo oil line.

12. Attach the turbo to the exhaust manifold and torque the nuts to 29 ft. lbs. (39 Nm). Attach the assembly to the engine using new gaskets and torque the nuts to 34 ft. lbs. (46 Nm).

13. Torque the turbocharger joint pipe bolt to 46 ft. lbs. (62 Nm) and the turbocharger bracket bolts to 30 ft. lbs. (41 Nm).

14. Install the front pipe and the oxygen sensor.

15. Connect the EGR pipe to the exhaust manifold and the water hose to the inlet pipe.

16. Connect the oil inlet and return pipes to the turbo connections. Install the exhaust manifold insulators.

17. Install the air bypass valve and air hoses.
18. Fill the cooling system to the proper level and connect the negative battery cable.
19. Disconnect the ignition coil and crank the engine for 20 seconds to fill the oil line. Reconnect the coil and start the engine. Run the engine at idle for 20 seconds, then stop the engine.
20. Disconnect the negative battery cable. Depress and hold the brake pedal for 5 seconds to clear the malfunction code from the control unit.

323 and Protege

NOTE: When replacing the turbocharger, always check the oil level and the condition of the oil and the turbo oil inlet and outlet lines. If the oil is dirty or the lines are damaged, replace them.

1. Disconnect the negative battery cable. Remove the engine under cover and drain the cooling system.
2. Remove the radiator and fan as an assembly.
3. Remove the exhaust manifold shields.
4. Remove the turbocharger air inlet and outlet ducting and disconnect the exhaust pipe from the turbocharger.
5. Disconnect the turbocharger coolant and oil lines.
6. Support the turbocharger and remove the nuts and bolts from the exhaust manifold. Remove the turbocharger and the exhaust manifold as an assembly.7. Remove the nuts and lift the turbocharger from the exhaust manifold studs. Cover the turbocharger ports to prevent the entry of foreign matter.

To install:

8. Use a wire wheel or brush to remove all the sealant and gasket material from the turbocharger and exhaust manifold mating surfaces. Use all new gaskets.
9. Pour about 1.2 oz.(25cc) of clean engine oil into the opening for the turbo oil line.
10. Attach the turbo to the exhaust manifold and torque the nuts to 29 ft. lbs. (39 Nm). Attach the assembly to the engine using new gaskets and torque the nuts to 34 ft. lbs. (46 Nm).
11. Make sure all fittings are clean and connect the oil and coolant lines.
12. Connect the exhaust pipe and torque the nuts to 35 ft. lbs. (47 Nm). Be sure to use a new gasket and lockwashers.
13. Install the air ducting and heat shields.
14. Install the radiator, connect the coolant hoses and refill the system.
15. Disconnect the ignition coil and crank the engine for 20 seconds to fill the oil line. Reconnect the coil and start the engine. Run the engine at idle for at least 30 seconds.

Timing Belt Cover, Belt and Tensioner

REMOVAL & INSTALLATION

323 and Protege

SOHC ENGINE

1. Disconnect the negative battery cable. Remove the engine side cover. Remove the air conditioning belt and the power steering pump belt.
2. Remove the alternator drive belt and alternator. Remove the water pump pulley. Remove the upper and the lower timing belt covers.
3. Turn the crankshaft so the 2 timing marks on the camshaft sprocket align with the marks on the cylinder head and the cylinder head cover. This will be TDC on No. 1 cylinder.
4. Remove the crankshaft pulley mounting bolts and the pulley along with the baffle plate.
5. Remove the belt tensioner lock bolt, the tensioner wheel and the spring.
6. Remove the timing belt and mark an arrow in the direction of rotation on the timing belt.

To install:

7. Inspect the belt for oil soaking or excessive wear, peeling, cracking or hardening. Inspect the tensioner for free and smooth rotation and replace it if it does not turn smoothly. Inspect the sprockets and replace all damaged components, as necessary.
8. Be sure the timing marks on the sprockets are properly aligned.

NOTE: If the crankshaft or camshaft must be turned to align the marks and either one stops before turning a full revolution, this means a valve is contacting a piston. Do not force it. Turn both shafts as required to align the timing marks.

9. Install the timing belt tensioner and spring. Temporarily tighten the bolt with the spring fully extended.
10. Install the timing belt. If using the old timing belt, be sure it is reinstalled in the same direction of rotation. Also make sure there is no oil, grease or dirt on the timing belt.
11. Loosen the tensioner lock bolt to set the tension, then torque the bolt to 19 ft. lbs. (25 Nm).
12. Turn the crankshaft twice in the direction of rotation to check alignment of the timing marks. If they are not aligned, remove the timing belt tensioner and timing belt and repeat Steps 8–12.

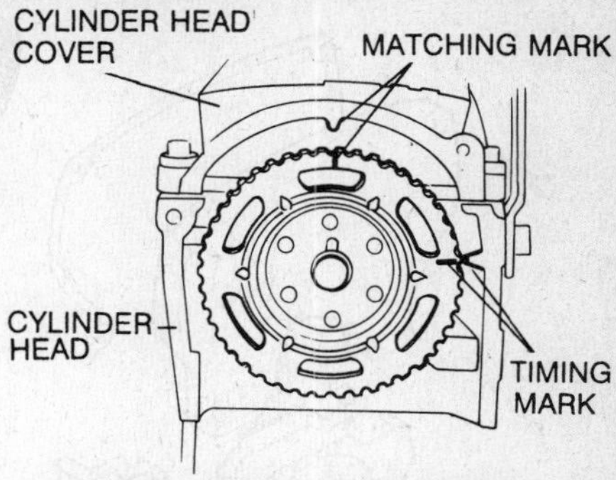

Timing mark alignment—323 and Protege with SOHC engine

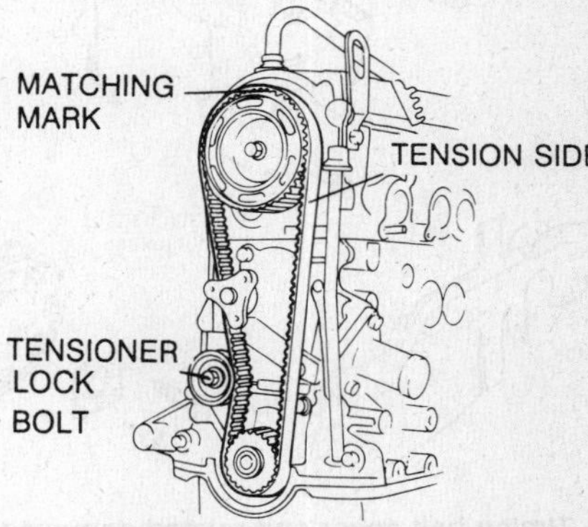

Timing belt and tensioner—323 and Protege with SOHC engine

13. Check the timing belt tension. The timing belt deflection should be 0.35–0.51 in. (9–13mm) at 22 lbs. (10 kg) of pressure.
14. Complete the installation by reversing the removal procedure.

DOHC ENGINE

1. Disconnect the negative battery cable and remove the engine side cover.
2. Remove the accessory drive belts, the water pump pulley and crankshaft pulley. If necessary, remove the right side engine mount nuts and lower the engine to remove the pulleys.
3. Remove the timing cover assembly retaining bolts. Remove the upper, middle and lower timing covers from their mountings. Remove the baffle plate from behind the crankshaft pulley.
4. Turn the crankshaft to position the engine at TDC of No. 1 cylinder. The exhaust and intake camshaft sprocket marks will be aligned with the marks on the seal plate and with each other. The crankshaft mark will align with the pointer above the sprocket.
5. Remove the timing belt tensioner and spring and slide the timing belt from the sprockets. If the old belt is to be installed again, mark the direction of rotation.

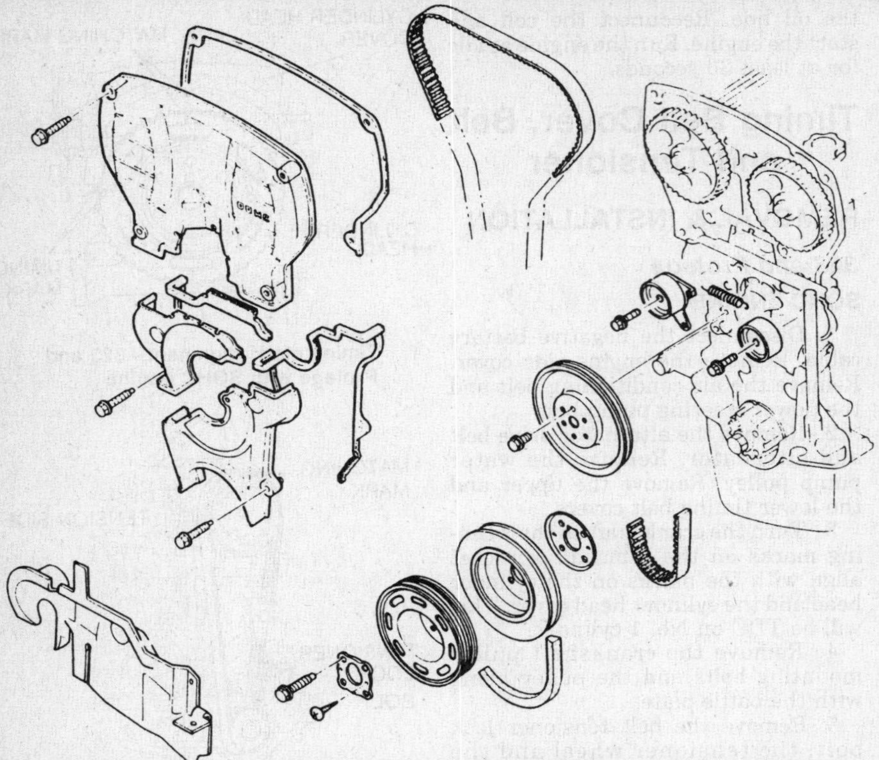

Timing belt cover and related components – 323 and Protege with DOHC engine

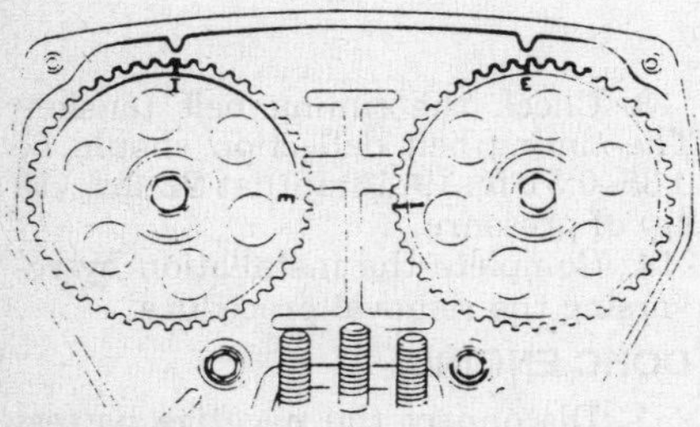

Camshaft sprocket timing mark alignment – 323 and Protege with DOHC engine

To install:

6. Align the timing marks on the crankshaft and camshaft sprockets with the marks on the engine.

NOTE: If the crankshaft or camshaft must be turned to align the marks and either one stops before turning a full revolution, this means a valve is contacting a piston. Do not force it. Turn both shafts as required to align the timing marks.

7. Install the spring and tensioner with the bolt. Move the tensioner until the spring is fully extended and temporarily tighten the tensioner bolt to hold it in place.

8. Install the timing belt by keeping the right side of the belt as tight as possible. Old belts must be installed in the original direction of rotation.

9. Loosen the tensioner lock bolt to apply tension to the belt, then torque the tensioner lock bolt to 38 ft. lbs. (51 Nm).

10. Turn the crankshaft twice in the normal direction of rotation and make sure all the timing marks are aligned.

11. Measure the tension between the camshaft sprockets. The deflection should be between 0.33–0.45 in. (8.4–11.5mm) between the pulleys. If not correct, repeat Steps 6–10 or replace the tensioner spring to adjust the tension to specification.

12. Installation of the remaining components is reverse of removal. Make sure the curved surface of the crankshaft pulley baffle plate faces outward. Torque the crankshaft pulley bolts to 13 ft. lbs. (17 Nm) and the engine mount bracket to 63 ft. lbs. (85 Nm). Adjust the drive belt tension and check the ignition timing.

626 and MX-6

1. Disconnect the negative battery cable and remove the spark plug wires and spark plugs.

2. Remove the accessory drive belts and the timing belt cover.

3. Unbolt the crankshaft pulley and remove the upper and lower timing belt covers. Remove the crankshaft pulley baffle plate and make a note of how it is installed. The curved side should be facing out.

4. Turn the crankshaft until the 1

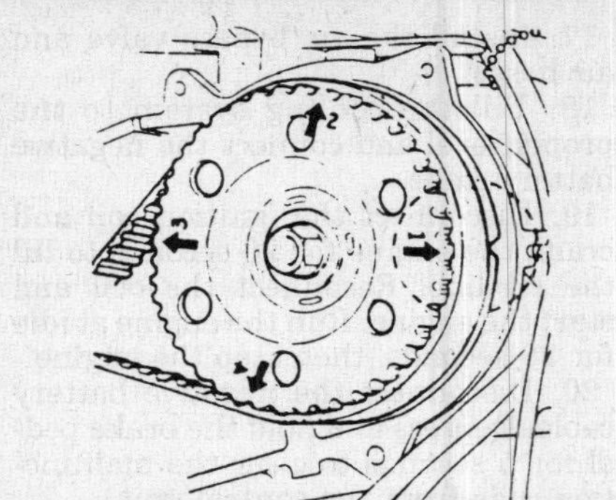

Camshaft pulley and front housing timing mark alignment – 626 and MX-6

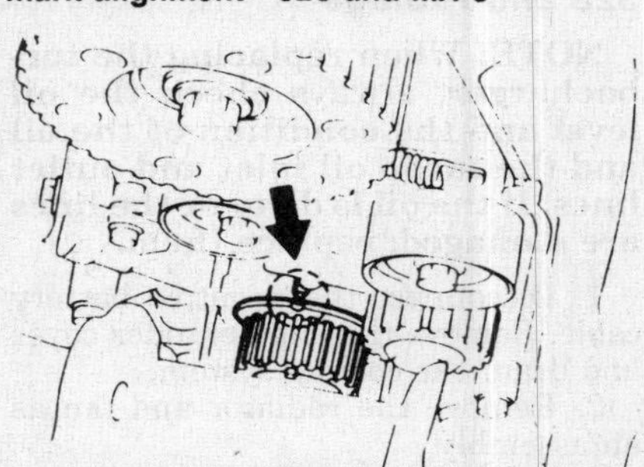

Timing belt pulley and front housing alignment marks – 626 and MX-6

mark on the camshaft is aligned with the mark on top of the front housing. Unbolt and remove the tensioner and the tensioner spring.

5. Remove the timing belt. If the timing belt is to be reused, mark the direction of rotation.

To install:

6. Inspect the belt for oil soaking or excessive wear, peeling, cracking or hardening. Inspect the tensioner for free and smooth rotation and replace it if it does not turn smoothly. Inspect the sprockets and replace all damaged components, as necessary.

7. Align the timing mark on the timing belt pulley with the matchmark on the lower front housing. Recheck the camshaft pulley and and front housing alignment marks. Turn the camshaft as necessary to re-align the marks.

NOTE: If the crankshaft or camshaft must be turned to align the marks and either one stops before turning a full revolution, this means a valve is contacting a piston. Do not force it. Turn both shafts as required to align the timing marks.

8. Install the spring and tensioner with the bolt. Move the tensioner until the spring is fully extended and temporarily tighten the tensioner bolt to hold it in place.

9. Install the timing belt. Make sure there is no slack at the side of the water pump and idler pulleys. Old belts

should be installed in the original direction of rotation.

10. Loosen the lock bolt to release the tensioner and turn the crankshaft twice in the direction of engine rotation to align all the timing marks.

11. Make sure all timing marks are aligned correctly. If not, remove the timing belt tensioner and timing belt and repeat Steps 7–10 until all the timing marks are correctly aligned. Torque the tensioner lock bolt to 38 ft. lbs. (52 Nm). Check the timing belt deflection. Correct deflection is 0.30–0.33 in. (7.6–8.4mm). If the deflection is not correct, loosen the tensioner lock bolt and adjust the tension by repeating Steps 10 and 11 or replace the tensioner spring.

12. Install the crankshaft pulley baffle plate so the curved side faces outward. Torque the pulley screws to 13 ft. lbs. (18 Nm).

13. Complete the installation of the remaining components in reverse of the removal procedure. Connect the negative battery cable, check and/or adjust the ignition timing and the idle speed.

929 and 929S

1. Disconnect the negative battery cable and remove the spark plugs and inlet air ducting. It may be easier to also remove the air cleaner and air flow meter assembly.

2. Drain the cooling system and remove the upper radiator hose, by-pass hose, radiator fan and shroud. On 929S, remove the distributor.

3. Remove the accessory drive belts and the idler pulley(s). If necessary, remove the power steering pump and/or air conditioner compressor and move it aside without disconnecting the hoses.

4. Remove the crankshaft pulley. On 929S, the torue on the center bolt of the crankshaft pulley is very high. A special tool is available that fits into the holes on the pulley to keep the crankshaft from turning.

5. Remove the timing belt covers and turn the engine to align the timing marks on the crankshaft and camshaft sprockets. This will put the engine at TDC on No. 1 cylinder.

6. If the old belt is to be installed again, mark the direction of rotation. Remove the upper idler pulley and slide the belt off the sprockets. Remove the automatic tensioner.

To install:

NOTE: Prior to installation of the timing belt, the automatic tensioner must be loaded. For this operation, a 1 ton press (or larger) is required.

7. To load the tensioner, place a flat washer on the bottom of the tensioner body to prevent damage to the body. Make sure the hole in the body and the hole in the tensioner rod line up with each other.

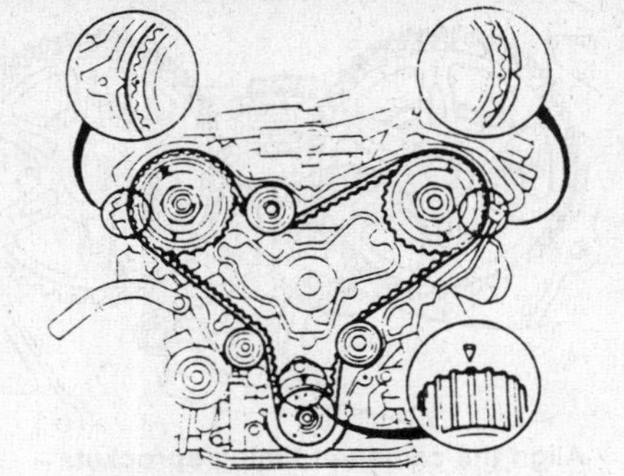

Timing belt alignment marks – 929

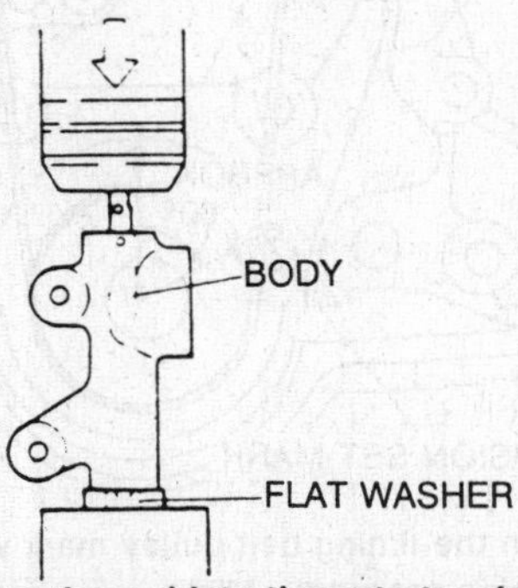

Pressing the rod into the auto tensioner body – 929 and 929S

8. Press the rod into the tensioner body using an arbor press or vise. Do not use more than 2000 lbs. of pressure. If more force is required, the tensioner is faulty and must be replaced.

9. Once the rod is fully pressed into the body, insert a bent pin or small Allen wrench through the body to hold the rod in place.

10. Remove the unit from the press and install it onto the engine. Torque the mounting bolt to 19 ft. lbs. (25 Nm) and leave the pin in place.

NOTE: If the crankshaft or camshaft must be turned to align the marks and either one stops before turning a full revolution, this means a valve is contacting a piston. Do not force it. Turn both shafts as required to align the timing marks.

11. Make sure all the timing marks are aligned properly and that the belt is clean and in good condition. With the upper idler pulley removed, slide the timing belt onto each sprocket and pulley in the correct sequence. Install the upper idler pulley and torque the mounting bolt to 38 ft. lbs. (51 Nm).

12. Rotate the crankshaft twice in the normal direction of rotation to make sure all the timing marks align correctly. If not, remove the belt and repeat Step 11.

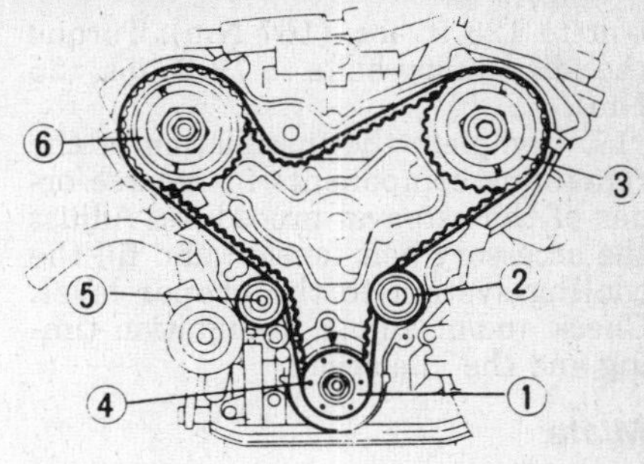

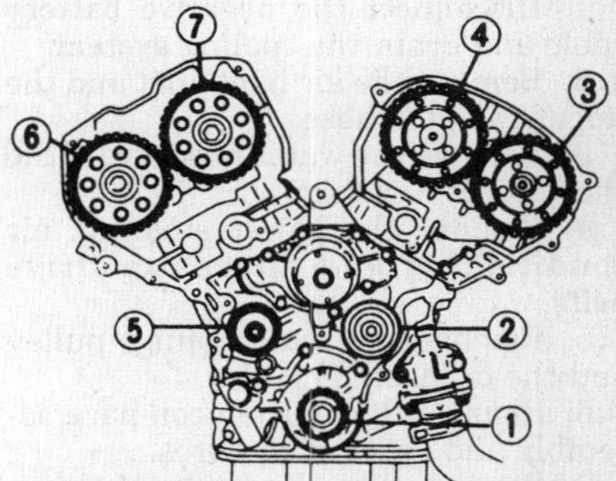

Install timing belt onto pulleys and sprockets in the correct order; No. 5 is the automatic tensioner pulley

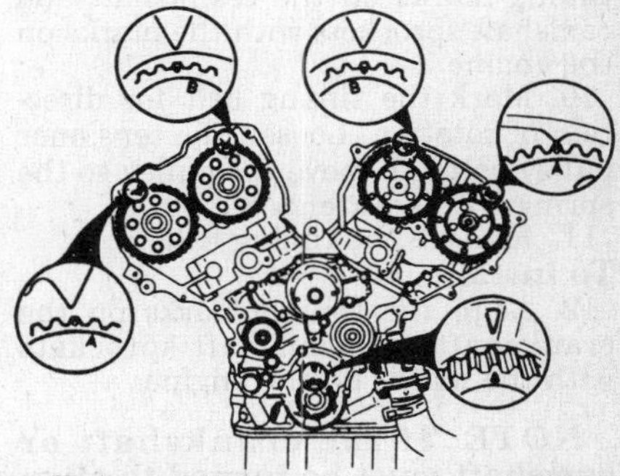

Timing belt alignment marks – 929S

13. Remove the pin from the automatic tensioner. Turn the crankshaft twice in the normal direction of rotation and check timing mark alignment again.

14. Check the timing belt deflection between the tensioner pulley and the right bank camshaft sprocket. The deflection should be 0.20–0.28 in. (5–7mm) on 929, 0.24–0.31 in. (6–8mm) on 929S, with 22 lbs. (10 kg) of pressure. Excessive belt deflection is caused by tensioner failure or a stretched timing belt.

15. When the belt is properly installed and tension is set, clean the inside of the covers and install the covers and gaskets. The timing belt must be kept free of oil and dirt.

16. Install the accessories and drive belts. On 929, torque the crankshaft pulley bolts to 11 ft. lbs. (15 Nm). On 929S, torque the center crankshaft

bolt to 123 ft. lbs. (167 Nm). Torque the idler pulley bolts to 19 ft. lbs. (25 Nm).

17. Complete the installation of the remaining components in reverse order of the removal procedure. Adjust the accessory belt tension and fill the cooling system to the proper level. Check and/or adjust the ignition timing and the idle speed.

Miata

1. Disconnect the negative battery cable and drain the cooling system.
2. Remove the air inlet duct and the upper radiator hose.
3. Remove the water hoses from the thermostat housing.
4. Remove the power steering, air conditioning and alternator drive belts.
5. Remove the water pump pulley and the crankshaft pulley.
6. Remove the ignition coil pack assembly and spark plug wires.
7. Remove the cylinder head cover.
8. Remove the upper, middle and lower timing belt covers.
9. Turn the crankshaft to align the timing marks on the crankshaft and camshaft sprockets with the marks on the engine.
10. Mark the timing belt for direction of rotation. Loosen the tensioner pulley bolt and move the pulley so the spring is fully extended.
11. Remove the timing belt.

To install:

12. Align the timing marks on the crankshaft and camshaft sprockets with the marks on the engine.

NOTE: If the crankshaft or camshaft must be turned to align the marks and either one stops before turning a full revolution, this means a valve is contacting a piston. Do not force it. Turn both shafts as required to align the timing marks.

13. Install the spring and tensioner with the bolt. Move the tensioner until the spring is fully extended and temporarily tighten the tensioner bolt to hold it in place.
14. Install the timing belt by keeping the right side of the belt as tight as possible. Old belts must be installed in the original direction of rotation.
15. Loosen the tensioner lock bolt to apply tension to the belt, then torque the tensioner lock bolt to 38 ft. lbs. (51 Nm).
16. Turn the crankshaft twice in the normal direction of rotation and make sure all the timing marks are aligned.
17. Measure the tension between the camshaft sprockets. The deflection should be between 0.33–0.45 in. (8.4–11.5mm) between the pulleys. If not

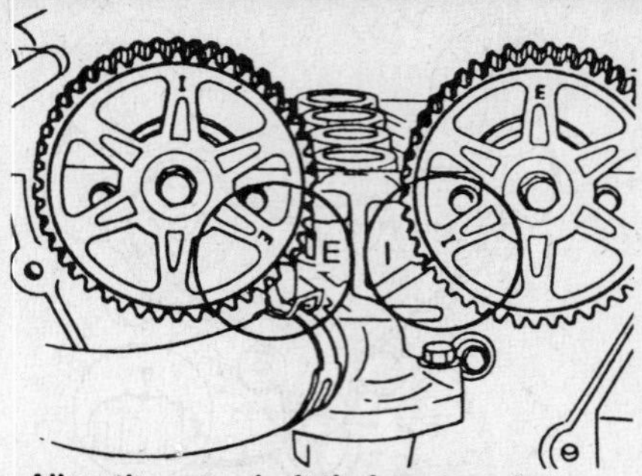

Align the camshaft timing sprockets – Miata

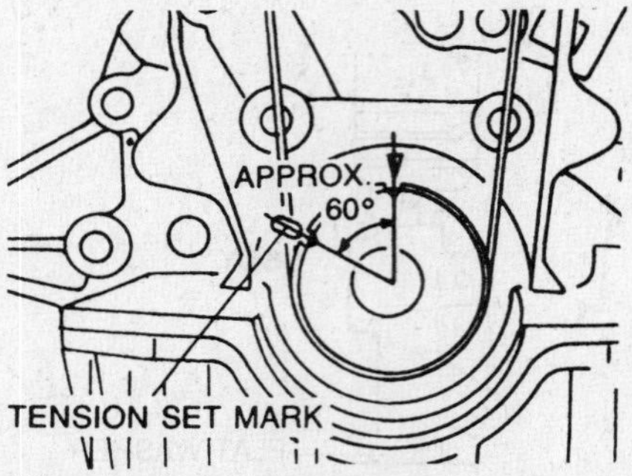

Align the timing belt pulley mark with the tension set mark – Miata

correct, repeat Steps 6–10 or replace the tensioner spring to adjust the tension to specification.

18. Installation of the remaining components is reverse of removal. Make sure the curved surface of the crankshaft pulley baffle plate faces outward. Torque the crankshaft pulley bolts to 13 ft. lbs. (17 Nm) and the engine mount bracket to 63 ft. lbs. (85 Nm). Adjust the drive belt tension and check the ignition timing.
19. Before installing the cylinder head cover, apply silicone sealer to the head at both ends of the intake camshaft and the front end of the exhaust camshaft.
20. Install the timing belt covers and the cylinder head cover.
21. Install the ignition coil pack assembly and the spark plug wires.
22. Install the crankshaft pulley and torque the bolts to 13 ft. lbs. (17 Nm).
23. Install the water pump pulley and torque the bolts to 8 ft. lbs. (11 Nm).
24. Install the accessory drive belts and adjust the tension.
25. Install the coolant hoses and the air ducts. Connect the battery negative cable and refill cooling system.

Front Oil Seal

REMOVAL & INSTALLATION

929S, 323, Protege and Miata

1. Remove the timing belt covers and the timing belt.
2. Remove the crankshaft sprocket bolt. A special tool is available that fits onto the sprocket to hold the crankshaft while loosening the center bolt.
3. Use a steering wheel puller to remove the sprocket.
4. Pry out the old seal, being careful to not damage the crankshaft. It may be necessary to cut the seal lip with a razor knife.

To install:

5. Lubricate the crankshaft and the seal and carefully press the seal into the oil pump. Use an appropriate seal installation tool and make sure the seal fits flush with the pump body.
6. When installing the sprocket, fit the tapered end of the Woodruff key towards the oil pump body.
7. When installing the crankshaft center bolt, use the tool to keep the crankshaft from turning. Torque the bolt to 87 ft. lbs. (118 Nm) on SOHC engines, 123 ft. lbs. (167 Nm) on DOHC engines.
8. Install the timing belt and check the ignition timing.

626, MX6 and 929

On these vehicles, the oil pump must be removed and disassembled to change the front oil seal.

Camshaft

REMOVAL & INSTALLATION

323, Protege and 929

SOHC ENGINE

NOTE: Since the camshaft is withdrawn from the front of the engine, this procedure assumes either the engine or the cylinder head has been removed from the vehicle.

1. Remove the cylinder head from the engine and remove the distributor.
2. Hold the camshaft with an open end wrench at the front bearing and loosen the bolt to remove the camshaft sprocket.
3. Carefully loosen the rocker arm shaft bolts 1 turn at a time in the reverse of the torque sequence to prevent bending the shafts. Remove the rocker arm assembly.
4. Remove the camshaft thrust plate from the rear bearing tower. Carefully slide the camshaft out without damaging the bearing surfaces in the head.

To install:

5. If a new seal is being installed, lubricate the seal and carefully press it into place with a suitable installation tool until the seal is flush with the head.
6. Liberally lubricate the camshaft

bearing surfaces in the head and carefully insert the camshaft into place.

7. Install the thrust plate. On the B6 series engine and on the 929, torque the retaining bolt to 95 in. lbs. (11 Nm). On the BP series engine, the thrust plate is held in place by the rocker arm assembly.

8. Install the rocker arm assembly and torque the bolts carefully in the proper sequence.

9. Turn the camshaft and install the drive sprocket so the dowel pin is up. Install and torque the bolt to 45 ft. lbs. (61 Nm) on 323 and Protege, 59 ft. lbs. (80 Nm) on 929.

10. Install the cylinder head.

323, Protege and Miata

DOHC ENGINE

1. Disconnect the negative battery cable and remove the spark plugs.

2. Remove the timing cover and the timing belt.

3. Hold the camshafts with an open end wrench and loosen the drive sprocket bolt to remove the sprockets.

4. The camshaft bearing caps must be returned to their original position. If they do not already have a position number and an arrow pointing towards the front of the engine, mark the caps for installation.

5. Carefully loosen the bearing cap bolts 1 turn at a time in the reverse of the torque sequence to avoid damage to the camshafts or bearings. Lift the camshafts out of the head. If the hydraulic lash adjusters are to be removed, mark them with a felt pen so they can be returned to the same bore.

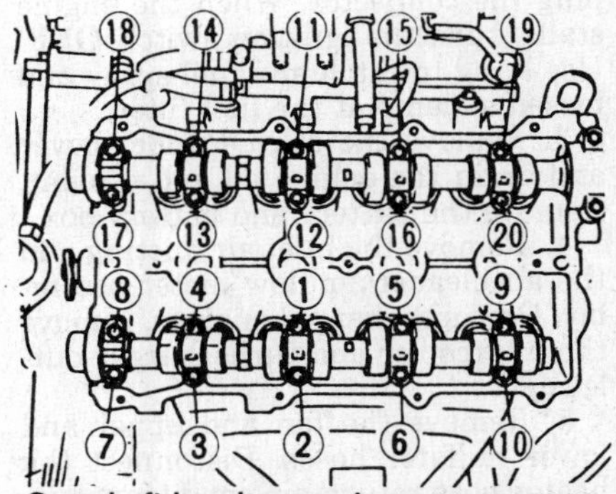

Camshaft bearing cap torque sequence on 4 cylinder DOHC engine

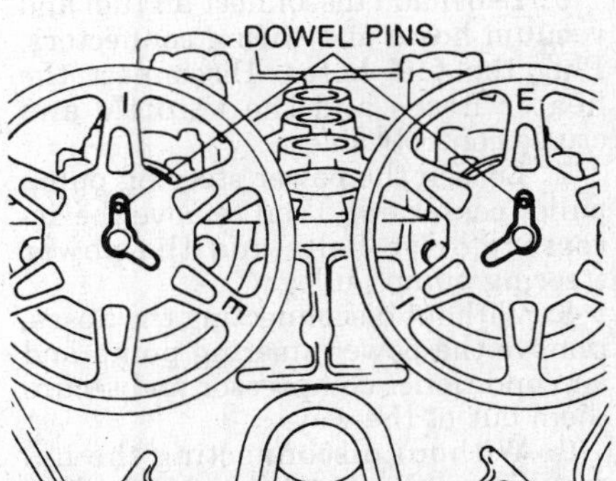

Installing the camshaft sprockets on 4 cylinder DOHC engine

To install:

6. Make sure the crankshaft is at TDC of No. 1 cylinder.

7. With the lash adjusters installed, lay the camshafts in place with the sprocket dowel pins at the top. Apply silicone sealant where the front bearing caps contact the cylinder head and install the bearing caps. Make sure the caps are in their original position, then start the bolts.

8. Tighten the bolts 1 turn at a time in sequence to draw the camshafts down against the valve springs evenly. When the caps are seated, torque the bolts in the proper sequence in 2 steps to 10 ft. lbs. (14 Nm). Do not over torque.

9. Lubricate and install the oil seals. The seals must be flush with the edge of the bearing cap.

10. When the sprockets are properly installed, the intake camshaft will have the **I** at the top and the exhaust camshaft will have the **E** at the top. The camshafts are properly positioned to install the timing belt. Hold the camshaft with a wrench and torque the sprocket bolt to 45 ft. lbs. (61 Nm).

11. Install the timing belt and set the tension. Rotate the crankshaft 2 turns to make sure the timing is correct.

12. Install the covers and complete the installation of the remaining components.

626 and MX6

1. Disconnect the negative battery cable and remove the timing belt cover, cylinder head cover and the timing belt.

2. To remove the sprocket, insert a pin through one of the sprocket holes and through the hole in the rear timing belt housing to hold the camshaft. Remove the bolt and sprocket and the rear timing belt housing.

3. The rocker arm shafts and camshaft bearing caps are held by the same bolts. Loosen the bolts 1 turn at a time in the reverse order of the torque sequence. This is important to prevent bending the camshaft and rocker arms. Remove the rocker arm assembly with the bolts.

4. If necessary, mark the camshaft bearing caps so they can be returned to their original position and lift the camshaft out.

To install:

5. Lubricate and install the camshaft with the No. 1 cylinder lobes pointing up. Set the bearing caps into their original positions.

6. Set the rocker arm assembly in place and install the bolts finger tight. Make sure the rocker arms and spacers are not caught between the shaft and the cap. Turn the bolts in 1 turn at a time in the proper sequence till the assembly is seated in the head.

7. Torque the bolts in 3 steps in the proper sequence to 20 ft. lbs. (26 Nm).

8. Install a new seal into the rear timing belt housing and install the housing with a new gasket. Install the camshaft sprocket and torque the bolt to 48 ft. lbs. (65 Nm).

9. Align the timing marks, install the belt and adjust the tension. Rotate the crankshaft 2 turns and check that the timing marks still align.

10. Install the remaining components and check ignition timing.

929S

DOHC ENGINE

The camshafts are identified by a number found between the center lobes. Make sure the correct number is on the camshaft before installation or serious engine damage could result from incorrect valve timing.

Right bank intake camshaft – No. 421
Right bank exhaust camshaft – No. 441
Left bank intake camshaft – No. 431
Left bank exhaust camshaft – No. 451

1. Remove the accessory drive belts and the timing belt cover and crankshaft pulley.

2. Position the engine at TDC on No. 1 cylinder so all the timing sprocket matchmarks are aligned. Mark the direction of rotation of the belt, remove the belt tensioner and the timing belt.

3. Remove both spark plug hall covers and cylinder head covers. Hold the camshaft from turning with the wrench flat on the shaft and remove the drive sprockets.

4. Note how the camshaft bearing caps are numbered and indexed. Make sure they are assembled in the same loaction. Remove the front caps on all 4 camshafts.

5. Loosen the camshaft bearing cap bolts ½ turn at a time in the reverse of the torque sequence. Remove the caps and lift out the camshafts and rocker arms. Make sure the rocker arms are returned to the same position.

To install:

6. Lubricate and install the camshaft seal onto the end of the camshaft.

7. Lubricate and install the rocker arms and camshafts, making sure everything is returned to its original position.

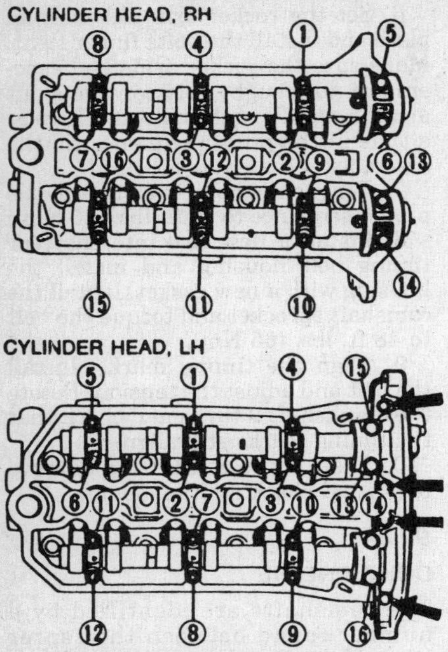

Camshaft bearing cap bolt torque sequence on 929S is different for left and right bank

8. The right and left bank bearing cap bolts are torqued in a different sequence. Torque the bolts in the proper sequence in 2 steps to 24 ft. lbs. (32 Nm). Torque the front cap bolts (arrows) to 8 ft. lbs. (11 Nm).

9. With the camshafts installed, install the drive sprockets. Torque the right bank sprocket bolts to 59 ft. lbs. (80 Nm) and the left bank sprocket bolts to 19 ft. lbs. (25 Nm).

10. Make sure the timing marks are properly aligned and install and adjust the timing belt. Turn the crankshaft 2 turns and check that the timing marks are properly aligned.

11. Apply silicone sealant to the front bearing caps and install the cylinder head covers with new gaskets. Torque the nuts to 39 in. lbs. (4.4 Nm). Use sealant when installing the spark plug hall covers.

12. Install the remaining components and check the ignition timing.

Piston and Connecting Rod

Positioning

The "F" marks (arrow) face the front of the engine

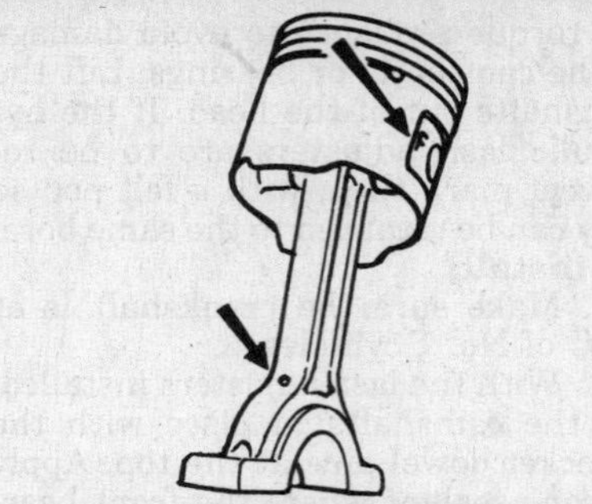

Piston and connecting rod positioning

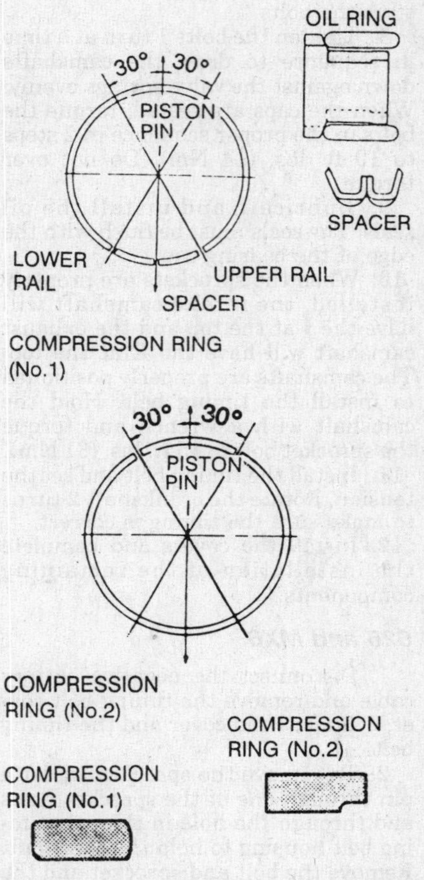

Ring positioning—323 and Protege

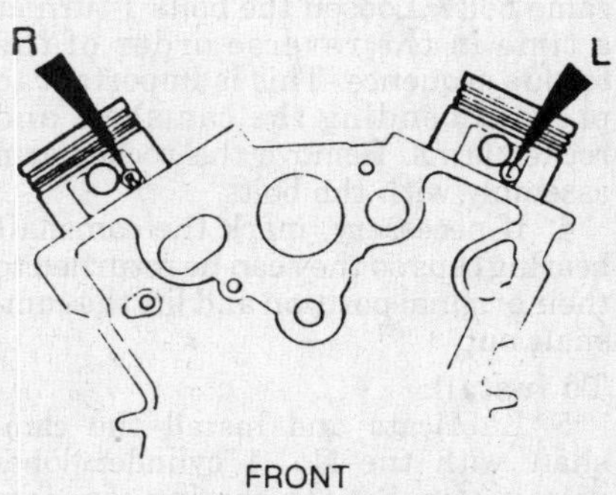

Piston positioning—929

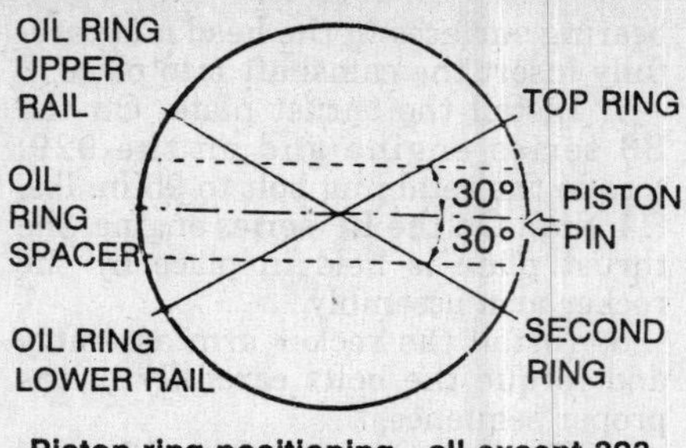

Piston ring positioning—all except 323 and Protege

ROTARY ENGINE MECHANICAL

Engine Assembly

REMOVAL & INSTALLATION

RX-7

1. The engine is lifted out of the vehicle. Scribe matchmarks where the hinges meet the hood and remove the hood.

2. To relieve the fuel system pressure, locate the fuel pump connector under the carpet in the luggage compartment. Start the engine, then unplug the connector. When the engine stalls, turn the ignition switch **OFF**. Use a rag to catch any fuel spray and loosen a clamp at the fuel filter.

3. Remove the engine under cover and drain the engine oil and coolant. Remove the battery and battery box.

4. Remove the inlet air ducting and the air cleaner/air flow meter assembly. On turbocharged engines, remove the intercooler and turbocharger outlet duct.

5. Remove the fan and upper and lower radiator hoses. Disconnect the heater hose return pipe and transmission coolant hose, if equipped, and remove the radiator and shroud.

6. Label and disconnect all fuel and vacuum hoses and wiring connectors. Plug the fuel hoses. Disconnect the heater hoses and the throttle and cruise control cables.

7. Loosen the power steering pump pulley center nut, then remove the accessory drive belts and the power steering pump pulley.

8. Without disconnecting the hoses, remove the power steering pump and air conditioner compressor and secure them out of the way.

9. Without disconnecting the hydraulic hose, remove the clutch slave cylinder and secure it out of the way.

10. Disconnect the oil cooler pipes at

the engine and cover the fittings to protect the cooler.

11. Remove the starter and the exhaust system heat shields. Remove the front exhaust pipe or, on turbocharged engines, the 2 catalytic converters

12. On automatic transmission, remove the cover and remove the torque converter-to-flywheel bolts.

13. Make sure all wiring and hoses are disconnected and attach the lifting equipment. Remove the engine-to-transmission bolts and the engine mount bolts. Pull the engine away from the transmission and lift it out of the vehicle.

To install:

13. Carefully lower the engine into the vehicle and mate it to the transmission. Torque the engine-to-transmission bolts and the engine mount nuts to 35 ft. lbs. (47 Nm).

14. On automatic transmission, install the torque converter-to-flywheel bolts and torque to 35 ft. lbs. (47 Nm). Install the cover.

15. Install the starter and torque the bolts to 34 ft. lbs. (46 Nm). Connect the wiring. Use new gaskets to install the exhaust pipe or catalytic converters and torque the nuts to 59 ft. lbs. (80 Nm). Install the shields.

16. Use new sealing washers to connect the rear oil cooler pipe and torque the banjo bolt to 50 ft. lbs. (68 Nm). Torque the front oil cooler fitting to 40 ft. lbs. (54 Nm). Install the clutch cylinder.

17. Install the air conditioner compressor and power steering pump and pulley. Torque the pulley bolt to 36 ft. lbs. (49 Nm). Install the drive belts and adjust the tension.

18. Connect all hoses and wiring. Connect the accelerator cable and adjust as required.

19. When installing the cooling fan and radiator shroud, adjust the shroud so there is about $^{15}/_{16}$ in. (24mm) of fan clearance at the top.

20. Connect all vacuum hoses and install the air ducting and battery. Fill all fluids to the proper level and run the engine to check for leaks and adjustments.

DISASSEMBLY

NOTE: Because of the design of the rotary engine, it is not practical to attempt component removal and installation. Procedures described here are for complete engine disassembly and assembly.

1. Remove the clutch disc and cover and mount the engine on a work stand.

2. Label the vacuum and wiring connections and remove the harness as an assembly.

3. Remove the intake and exhaust systems. If equipped, cap the turbocharger inlets and outlets to protect the turbine wheels.

4. Remove the fuel and oil injectors and cap the nozzles and inlet tubes to keep them clean.

5. Remove the water pump.

6. Remove the oil metering pump and cover the openings to protect the internal parts.

7. When all accessories and mounts have been removed, invert the engine to remove the oil pan, oil strainer and gasket.

8. Identify the front and rear rotor housings with paint or a felt tip pen. These are common parts and must be identified to be assembled in their respective locations.

9. Turn the engine right side up and remove the eccentric shaft pulley and

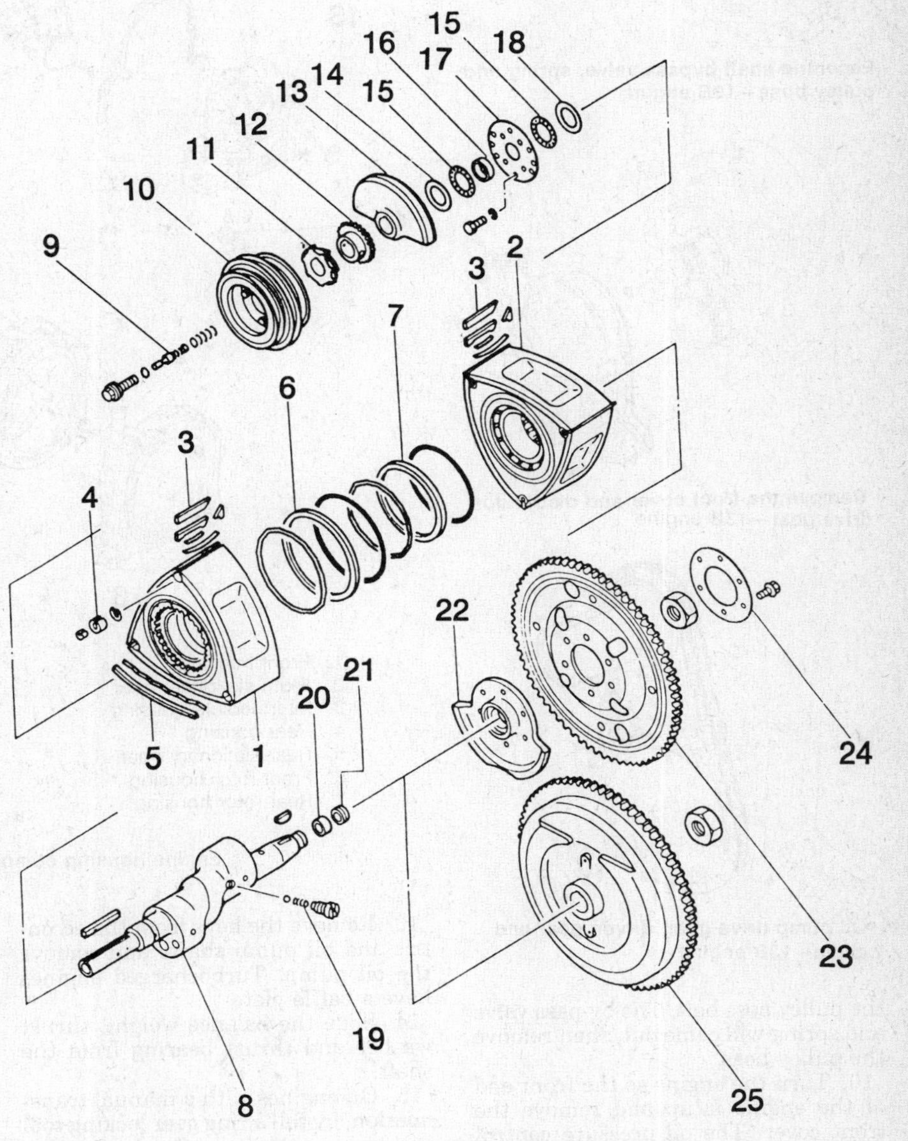

1. Front rotor
2. Rear rotor
3. Apex seal
4. Corner seal
5. Side seal
6. Outer oil seal
7. Inner oil seal
8. Eccentric shaft
9. Oil by-pass valve
10. Eccentric shaft pulley
11. Distributor drive gear
12. Oil pump drive sprocket
13. Balance weight
14. Thrust washer
15. Thrust needle bearing
16. Spacer
17. Plate
18. Thrust plate
19. Oil jet valve
20. Pilot bearing (M/T)
21. Oil seal (M/T)
22. Counter weight (A/T)
23. Drive plate (A/T)
24. Back plate (A/T)
25. Flywheel (M/T)

Internal components of the 13B rotary engine

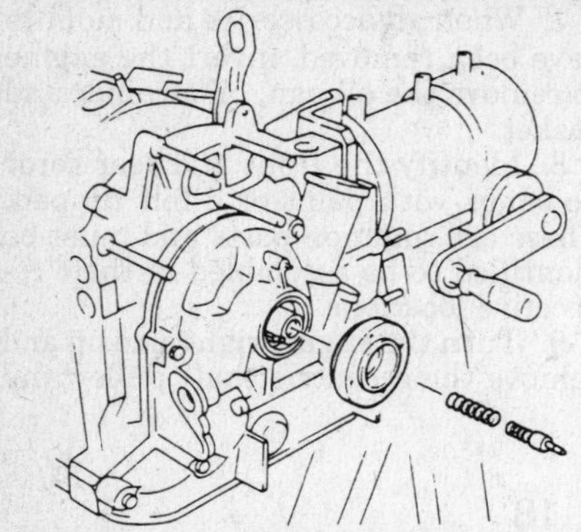

Eccentric shaft bypass valve, spring and pulley boss—13B engine

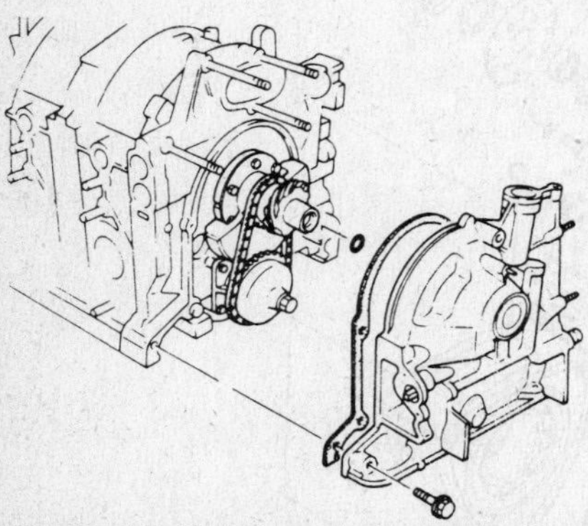

Remove the front cover and distributor drive gear—13B engine

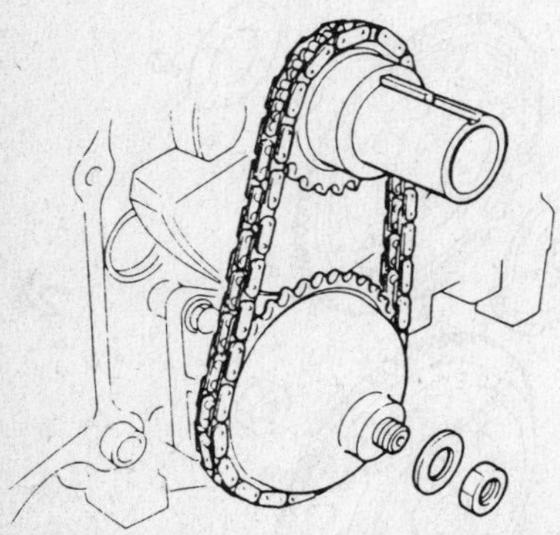

Oil pump drive gear, driven gear and chain—13B engine

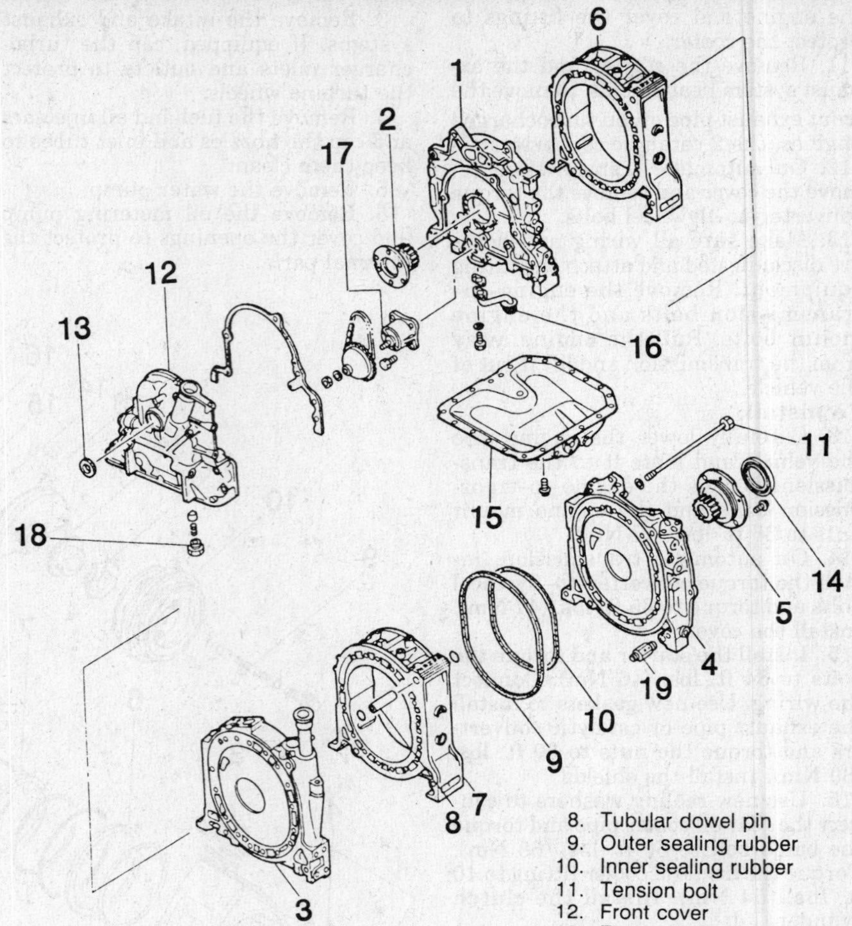

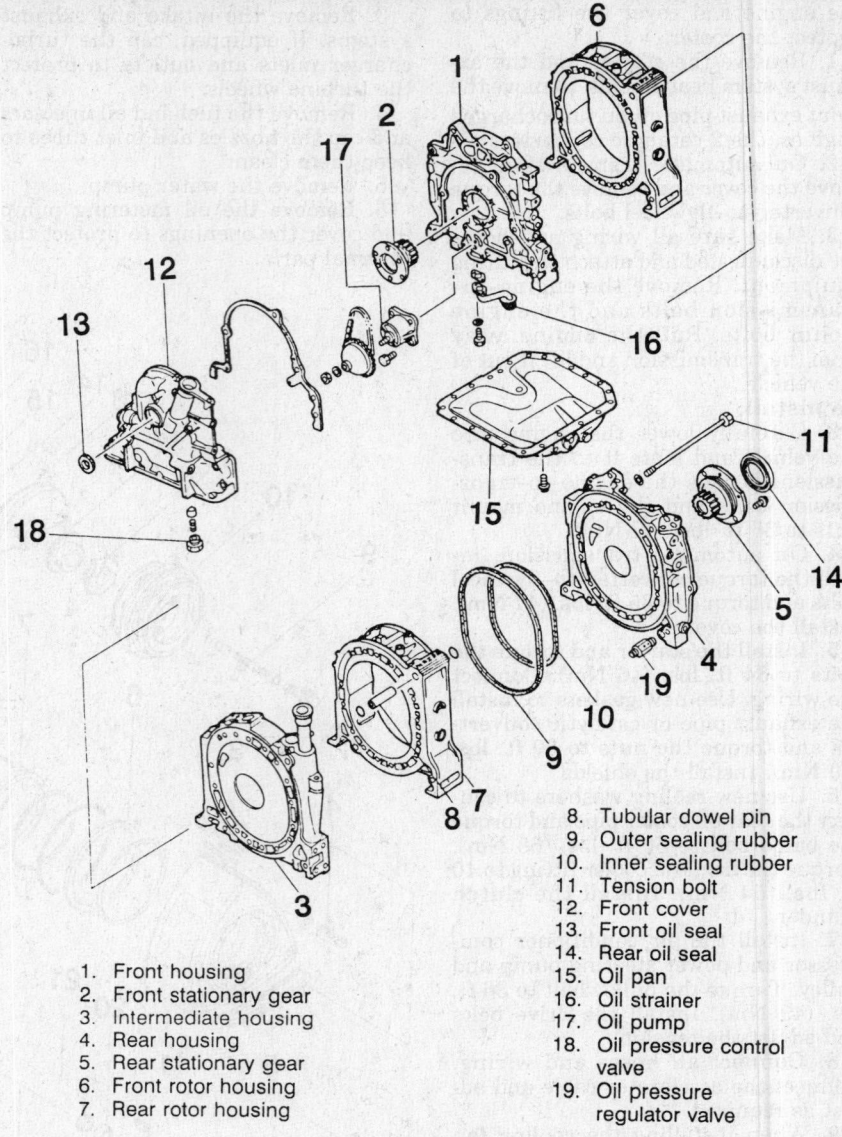

1. Front housing
2. Front stationary gear
3. Intermediate housing
4. Rear housing
5. Rear stationary gear
6. Front rotor housing
7. Rear rotor housing
8. Tubular dowel pin
9. Outer sealing rubber
10. Inner sealing rubber
11. Tension bolt
12. Front cover
13. Front oil seal
14. Rear oil seal
15. Oil pan
16. Oil strainer
17. Oil pump
18. Oil pressure control valve
19. Oil pressure regulator valve

Engine housing components—13B engine

the pulley boss bolt. The by-pass valve and spring will come out, then remove the pulley boss.

10. Turn the engine so the front end of the engine is up and remove the front cover. The oil pressure control valve is next to the oil pan gasket surface.

11. Remove the oil slinger and distributor drive gear from the shaft. Unbolt and remove the chain adjuster.

12. Remove the locknut and washer from the oil pump driven sprocket. Slide the oil pump drive sprocket and driven sprocket, together with the drive chain off the eccentric shaft and oil pump simultaneously.

13. Remove the keys from the eccentric and oil pump shafts and remove the oil pump. Turbocharged engines have a baffle plate.

14. Slide the balance weight, thrust washer and thrust bearing from the shaft.

15. On engines with a manual transmission, install a ring gear locking tool and remove the large flywheel-to-eccentric shaft nut. Use a puller to remove the flywheel

16. On engines with an automatic transmission, remove the 6 bolts and remove the drive plate. Use the special tool No. 49 1881 055, or equivalent, to hold the counter weight and remove the large nut. Use a puller to remove the counter weight.

NOTE: On engines equipped with an automatic transmission drive plate, do not hold the drive plate to hold the eccentric shaft when removing or installing the large nut. The torque on the nut is over 300 ft. lbs. (400 Nm) and the drive plate is not designed to withstand that much torque.

17. Working at the rear of the engine, loosen the long tension bolts 1 turn at a time in the reverse order of the assembly torque sequence to prevent distortion. Since they are not all the same, mark tension bolts to replace them in their original holes during reassembly.

18. Lift the rear housing off the shaft. Remove any seals that are stuck to the rotor sliding surface of the rear

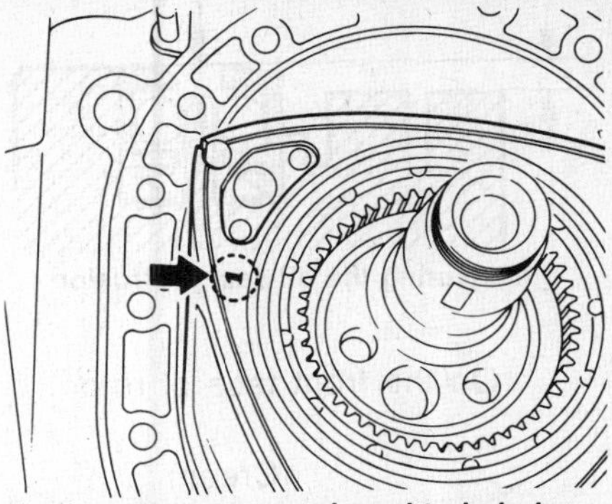

Seal grooves are numbered to help keep seals in order for inspection

housing and reinstall them in their original locations.

19. Remove all the corner seal assemblies and the side seal assemblies from the rear side of the rotor. Mazda has a special tray which holds all the seals and keeps them segregated to prevent mistakes during reassembly. Each seal groove is marked with numbers near the grooves on the rotor face to prevent confusion.

20. Remove the tubular dowels from the rear rotor housing using the appropriate puller.

21. Lift the rear rotor housing away from the rear rotor, being careful not to drop the apex seals on the rear rotor. Remove the O-ring from the upper dowel hole.

22. Remove each apex seal, side piece and spring from the rear rotor and segregate them.

23. Remove the rear rotor from the eccentric shaft and place it upside down on a clean rag. The coated surfaces on the rotor must be protected from damage, do not place the rotor on a hard surface.

24. Remove each seal and spring from the other side of the rotor and segregate them. If some of the seals fall off the rotor, be careful not to change the original position of each seal. Identify the bottom of each apex seal with a felt tip pen.

25. When removing the oil seals from the rotors, do not exert heavy pressure at only one place on the seal or it could be deformed. Replace the O-rings in the oil seal when the engine is overhauled.

26. Hold the intermediate housing down and remove the dowels from it using an appropriate puller.

27. Position the eccentric shaft as required and lift the intermediate housing off. If any rotor seals stick to it, put them back into their original position.

28. Remove the front rotor housing and remove the seals and segregate them.

29. Lift the eccentric shaft out, then remove the front rotor.

COMPONENT INSPECTION & REPLACEMENT

Front, Intermediate and Rear Housings

1. Check the housing for signs of gas or water leakage.

2. Remove the sealing compound from the housing surface with a cloth or brush soaked in solvent or thinner.

3. Remove the carbon deposits from the front housing with extra fine emery cloth.

4. Check for distortion by placing a straight-edge on the surface of the housing. Measure the clearance between the straight-edge and the housing with a feeler gauge. If the clearance is greater than 0.0016 in. (0.04mm) at any point, replace the housing.

5. Use a dial indicator to check for wear on the rotor contact surfaces of the housing. If the wear is greater than 0.004 in. (0.10mm), replace the housing.

NOTE: The wear at either end of the minor axis is greater than at any other point on the housing. However, this is normal and should not be cause for concern.

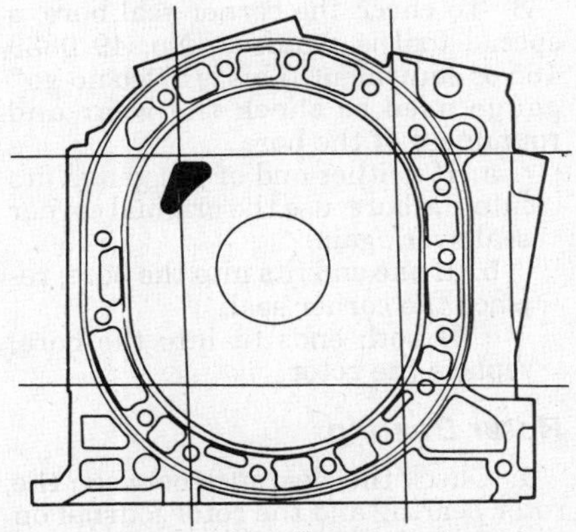

Measure housing distortion along the lines

Front Stationary Gear and Main Bearing

1. Examine the teeth of the stationary gear for wear or damage.

2. Be sure the main bearing shows no signs of excessive wear, scoring or flaking.

3. Check the main bearing to eccentric journal clearance by measuring the shaft journal diameter and the bearing inside diameter with a vernier caliper. The standard clearance is 0.0016–0.0031 in. (0.04–0.78mm).

4. Remove the securing bolts, if used, and press the stationary gear and main bearing assembly out of the housing.

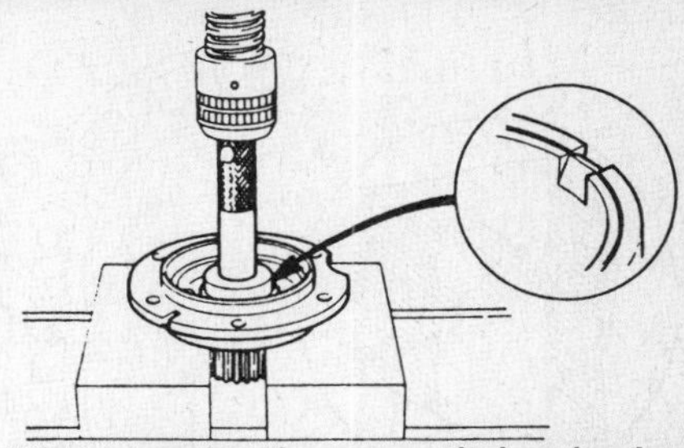

When pressing the new main bearing into the stationary gear, make sure the lug in the bearing aligns with the space on the gear

5. Press the main bearing from the stationary gear.

6. Press a new main bearing into the stationary gear, making sure to align the locating lug.

7. Align the slot in the stationary gear flange with the dowel pin in the housing and press the gear into place. Install the securing bolts, if required.

Rear Stationary Gear and Main Bearing

Inspect the rear stationary gear and main bearing in a similar manner to the front. In addition, examine the O-ring, which is located in the stationary gear, for signs of wear or damage. To replace the stationary gear, use the following procedure.

1. Remove the rear stationary gear securing bolts.

2. Drive the stationary gear from the rear housing with a brass drift.

3. Apply a light coating of grease to a new O-ring and fit it into the groove on the stationary gear.

4. Apply sealer to the flange of the stationary gear.

5. Install the stationary gear on the housing so the slot on its flange aligns with the pin on the rear housing. On later engines align the bearing lug with the housing slot. Use care not to damage the O-ring during installation.

6. Tighten the stationary gear bolts, evenly, in several stages, to 17 ft. lbs. (23 Nm).

Rotor Housings

1. Examine the inner margin of both housings for signs of gas or water leakage.

2. Wipe the inner surface of each housing with a clean cloth to remove the carbon deposits.

3. Clean all of the rust deposits out of the cooling passages of each rotor housing.

4. Remove the old sealer using the proper removal solvent.

5. Examine the chromium plated inner surfaces for scoring, flaking or other signs of damage. If any are present, the housing must be replaced.

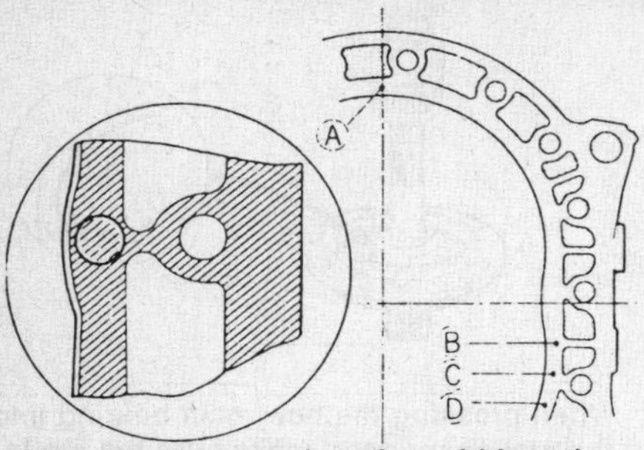

Measure the rotor housing width at the indicated points

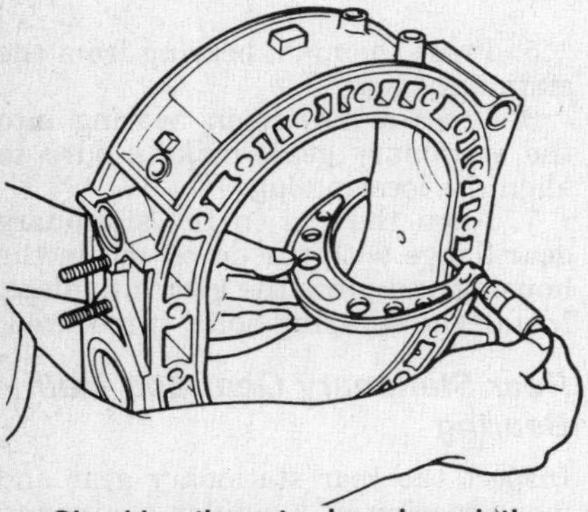

Checking the rotor housing width

6. Check the rotor housings for distortion by placing a straight-edge on the axes. If distortion exceeds 0.002 in. (0.05mm), replace the rotor housing.

7. Check the widths of both rotor housings, at a minimum of 8 points near the trochoid surfaces of each housing, using a vernier caliper or micrometer.

8. If the difference between the maximum and minimum values obtained is greater than 0.0024 in. (0.06mm), replace the housing. A housing in this condition will be prone to gas and coolant leakage.

Rotors

1. Check the rotor for signs of blow-by around the side and corner seal areas.

2. The color of the carbon deposits on the rotor should be brown, just as in a piston engine. Usually the carbon deposits on the leading side of the rotor are brown, while those on the trailing side tend toward black, as viewed from the direction of rotation.

3. Remove the carbon from the rotor with a plastic brush or extra fine emery paper. Be careful not to remove the surface coating, especially around the seal grooves.

4. Wash the rotor in solvent and blow it dry with compressed air.

5. Examine the internal gear for cracks or damaged teeth. If the internal gear is damaged, the rotor and gear must be replaced as a single assembly.

6. With the oil seal removed, check

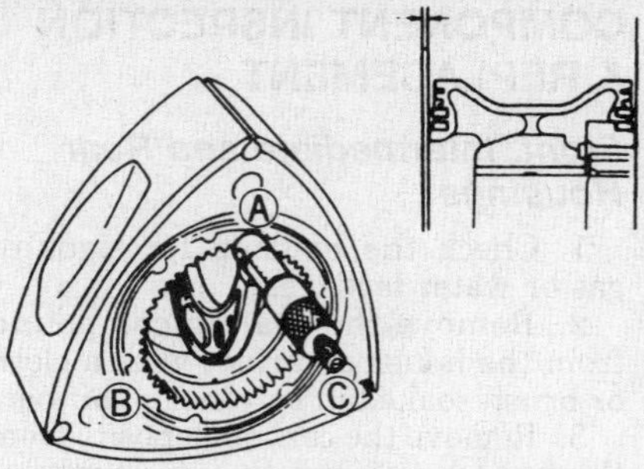

Measure the rotor width at the points indicated and use a straight edge to measure seal land protrusion

the land protrusions by placing a straight-edge over the lands. Measure the gap between the straight-edge and the land with a feeler gauge in a number of places. Standard clearance is 0.0047–0.0083 in. (0.12–0.21mm), minimum cleanance is 0.0039 in. (0.10mm).

7. Measure the rotor width with a vernier caliper or micrometer. Compare the rotor width against the width of the rotor housing which was measured above. Replace the rotor if the difference between both measurements is less than 0.0039 in. (0.10mm).

8. To check the corner seal bore, a special tool is required; No. 49 0839 165 or equivalent. This is a "go no-go" gauge used to check the wear and roundness of the bore.

a. If neither end of the gauge fits into the bore, use the original corner seal over again.

b. If one end fits into the bore, replace the corner seal.

c. If both ends fit into the bore, replace the rotor.

Rotor Bearing

1. Check the clearance between the rotor bearing and the rotor journal on the eccentric shaft. Measure the inner diameter of the rotor bearing and the outer diameter of the journal. The wear limit is 0.0039 in. (0.10mm). Re-

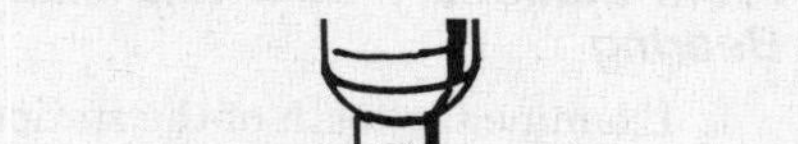

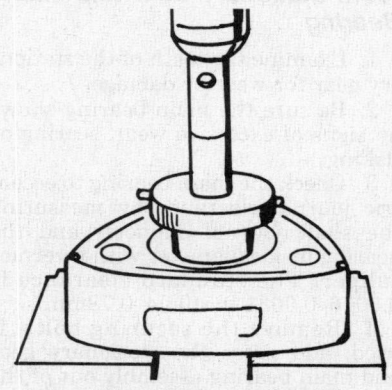

Installing a new rotor bearing

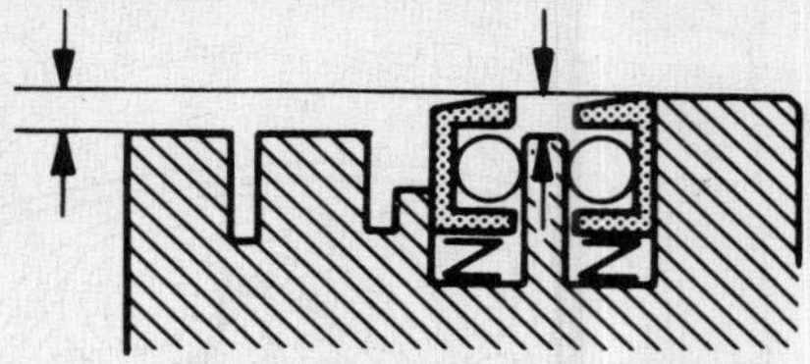

Measuring the oil seal protrusion

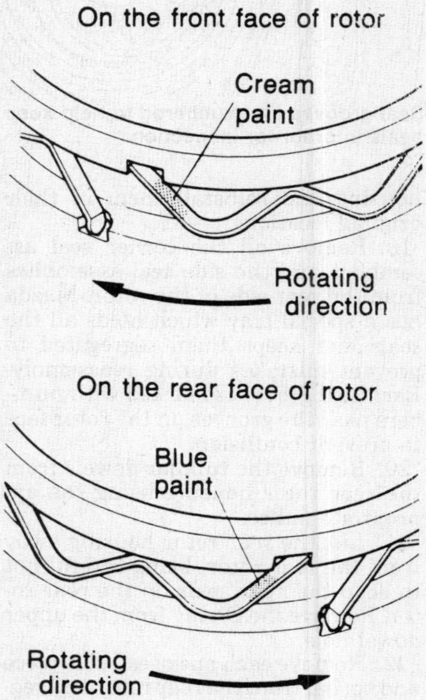

The oil seal springs are identified by paint marks: cream color on the front face of each rotor, blue on the rear face

place the bearing if it exceeds specification.

2. The bearing must be pressed out of the rotor past internal gear, being careful not to damage the gear.

3. Place the rotor on the press support with internal gear faced upward. Press the new rotor bearing into the rotor so the bearing lug is aligned with the slot of the rotor bore. Press the new bearing until it is flush with the rotor boss.

Oil Seal

1. Measure the oil seal while it is mounted in the rotor. The width should be no more than 0.020 in. (0.5mm). The protrusion above the inner portion of the rotor should be no less than 0.020 in. (0.5mm).

2. To replace the seal, pry the old seal out by inserting the tool into the slots on the rotor. Be careful not to damage the rotor.

3. Make sure the spring moves freely in the groove. Fit both oil seal springs into their respective grooves

and arrange them so the gaps are opposite each other on the rotor. Make sure the ends fit into the stopper holes in the groove.

NOTE: The springs are color coded. The cream colored spring goes into the front face of each rotor. The blue colored spring goes into the rear face of each rotor. Do not mix these up.

4. Install the O-ring into the seal. Coat the oil seal groove and the oil seal with engine oil.
5. Be sure the white mark is on the bottom side of each seal when it is installed. Gently press the oil seal into the groove. The old seal can be used as an installation tool, but use only finger pressure.
6. After installation, press the seal slowly and make sure it moves freely in and out of the groove against the spring.

Apex Seals

1 Wash the seals and the springs in cleaning solution. Be careful not to damage the finish. Check the apex seals for cracks or wear.
2. Measure the combined height of the upper and lower seals at 2 points. The standard height is 0.315 in. (8mm), minimum is 0.256 in. (6.5mm). Replace the short seal spring if the combined height is less than 0.295 in. (7.5mm).
3. Put the 2 seals together, top to top, and measure the gap between them to check for warping. If the ends of the seals touch and there is a gap in the middle, replace the seals. If there is a gap at the ends and the middle touches, the seals are still good. Do this with all 3 seals.
4. Using a feeler gauge, check the side clearance between the apex seal and the groove in the rotor. Install the seal into its proper groove and insert the gauge until its tip contacts the bottom of the groove.
   a. Non-turbocharged apex seal-to-rotor standard clearance:

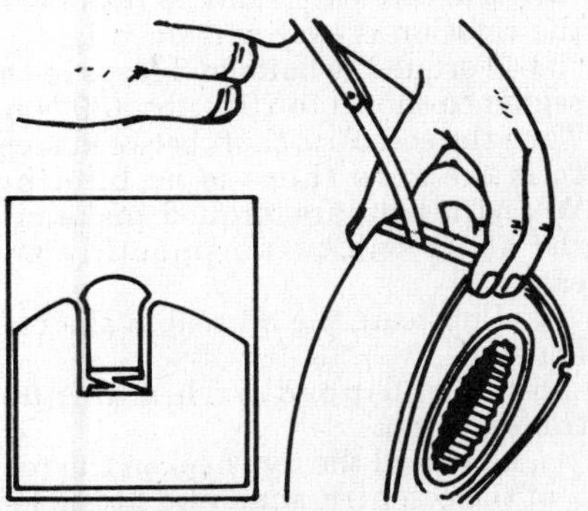

Check the gap between the apex seal and groove with a feeler gauge

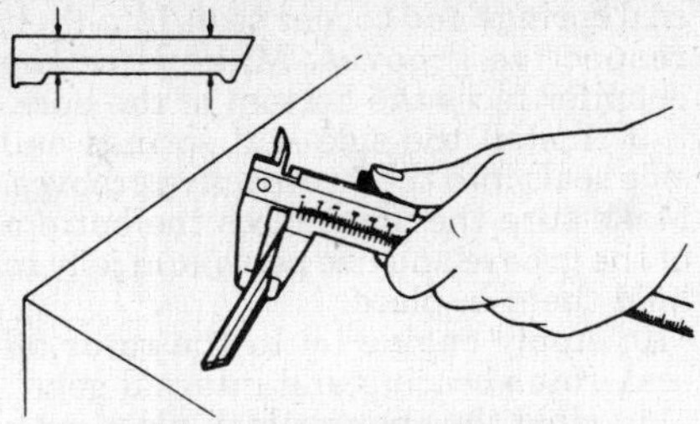

Measure the combined height of the apex seals

0.0024–0.0040 in. (0.062–0.102mm).
   b. Turbocharged apex seal-to-rotor standard clearance: 0.0020–0.0040 in. (0.051–0.101mm).
   c. Maximum clearance, both engines: 0.0059 in. (0.15mm)
5. With the spring on a flat surface, measure the free height with a vernier caliper. The long spring should be 0.181 in. (4.6mm) high, the short spring should be 0.067 in. (1.7mm). See Step 3.

Side Seals

1. Remove the carbon deposits from the side seals and their springs. Check the side seals for cracks and wear.
2. Install the side seals and springs and make sure the seal moves freely against the spring in the groove. The seal should protrude from the groove a minimum of 0.020 in. (0.5mm).
3. Check the clearance between the side seals and their grooves with a feeler gauge. Standard clearance is 0.0011–0.0031 in. (0.028–0.078mm). Maximum clearance is 0.0039 in. (0.10mm).
4. Check the clearance between the side seals and the corner seals with all installed in the rotor. The side seal touches the corner seal with its curved end. Measure the gap at the flat end. The standard clearance is 0.0020–0.0059 in. (0.05–0.15mm). The maximum clearance is 0.016 in. (0.40mm).
5. If the side seal must be replaced, adjust the side seal-corner seal standard clearance by lapping the flat (non-finished) end of the side seal.

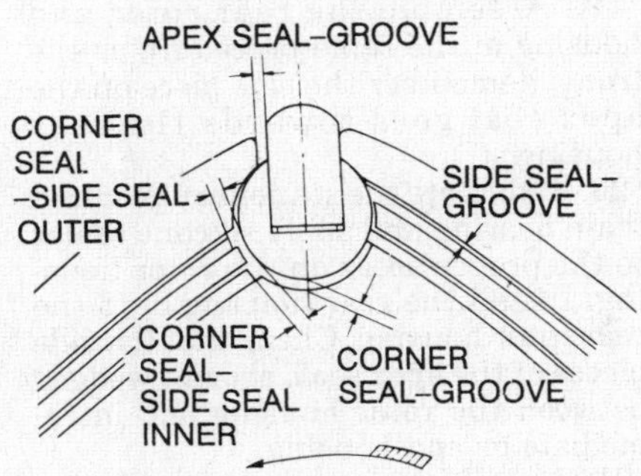

Check the clearance of the seals at the points indicated

Corner Seals

1. Clean the carbon deposits and carefully examine the corner seals, springs and soft seals.
2. Install the corner seal assemblies and make sure each seal moves freely against the spring in the groove.
3. Corner seal clearance is measured by measuring seal protrusion above the rotor. The protrusion should be a minimum of 0.020 in. (0.5mm).

Eccentric Shaft

1. Wash the eccentric shaft in solvent and blow out the oil passages with compressed air. Check the shaft for wear, cracks or other signs of damage; remove the oil jet and check the spring and ball for free movement.
2. To make sure the shaft is straight, check eccentric shaft run-out. Rotate the shaft slowly and note the dial indicator reading. If run-out is greater than specification, replace the eccentric shaft.
3. Check the operation of the roller pilot bearing for smoothness by inserting a mainshaft into the bearing and rotating it. Examine the bearing for signs of wear or damage. Replace the bearings if necessary, using the special removal and installation tools 49 0823 073 and 49 0823 072, or equivalent.

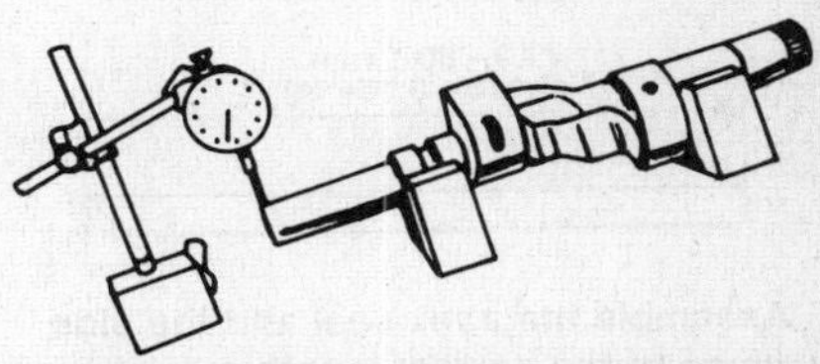

Position the dial indicator and rotate the shaft to measure the runout

Needle Bearing and Thrust Plate

1. Inspect the needle thrust bearing in the rotor housing end plate for wear and damage.
2. Inspect the bearing housing and the thrust plate for wear and damage.

Oil Pump Drive Chain and Sprocket

1. Lay the chain on a flat surface and check the entire length for broken links.
2. Check the oil pump drive and driven sprockets for damaged teeth. Replace as necessary.

ASSEMBLY

1. Replace all oil control seals, gaskets and O-rings with new parts. Place the front rotor on a rubber pad or cloth with the front end facing up.

2. Install the rotor oil seals as described earlier.

3. Mount the front housing into the engine stand and install the thrust plate with the chamfer facing the housing. Install the bearing and plate and torque the bolts to 17 ft. lbs. (23 Nm).

4. Turn the housing so the inner face is up. The inner and outer housing seals are painted on one edge. Use petrolium jelly to hold the seals in place and install them so the painted edges face each other, white edge in, blue edge out. Make sure the seals are not twisted and position the seams at 12 o'clock and 3 o'clock on the housing.

5. Lubricate the housing contact surfaces, stationary gear and main bearing with engine oil. Do not oil the housing seals just installed.

6. Assemble the apex seal side piece to the apex seal using an anaerobic sealing compound such as Loctite 312 or equivalent. Make sure the total length of the assembled part is 3.146–3.154 in (79.7–80.1mm). Wipe away any excess sealer.

7. Install the upper and lower apex seals without the springs into their respective grooves so each side piece is at the rear of the rotor. The springs will be installed later.

8. Install the new soft seals, corner

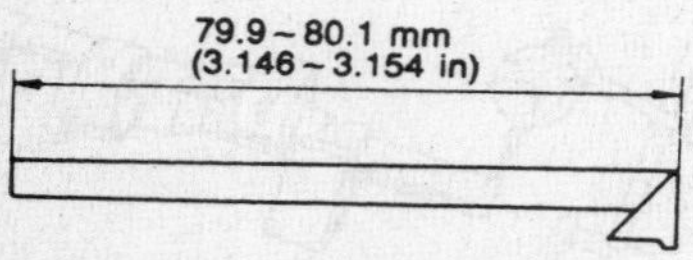

Assemble the apex seal and the side piece to the correct length

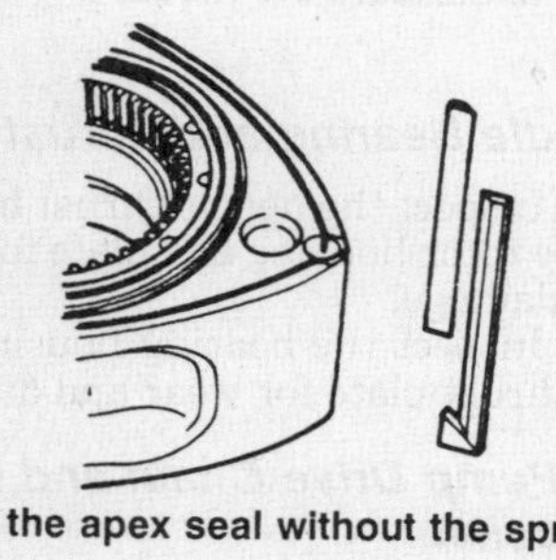

Install the apex seal without the springs

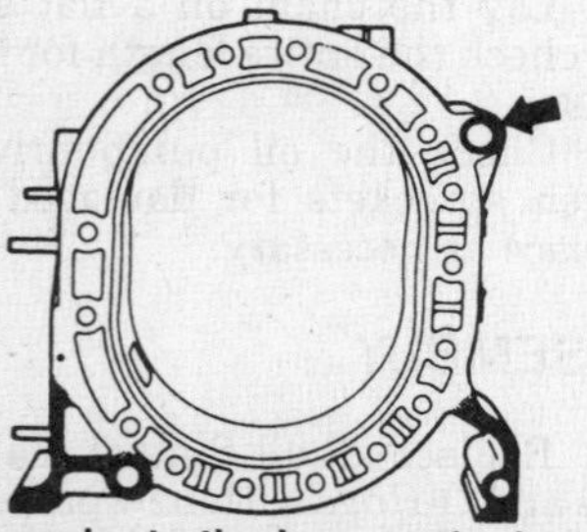

Apply sealer to the legs on the bottom and install a new O-ring

seal springs and corner seals into their respective grooves. Make sure the chamfer faces the bottom of the bore.

9. Install the side seal springs and side seals into their respective grooves. Make sure the paint faces the bottom of the groove and use petrolium jelly to hold them in place.

10. Apply engine oil to the rotor oil seal, rotor bearing and internal gear.

11. Hold the apex seals in place with a rubber band or used O-ring and place the rotor on the front housing. Mesh the internal and stationary gears so one of the rotor apexes is at 6 o'clock or 12 o'clock. Be careful not to set the rotor in the housing seal.

12. Make sure the pilot bearing in the eccentric shaft is properly installed and greased. Oil the shaft and rotor journal and carefully install the shaft through the rotor and housing.

13. Use petrolium jelly to hold the new O-ring in place on the housing and apply a silicone sealer to the 2 outer "legs" on the bottom of the housing. Lubricate the walls of the housing and carefully slip it over the rotor.

14. Lubricate the dowels with the engine oil and insert them through the front rotor housing holes and into the front housing.

15. Remove the rubber band or O-ring from the installed rotor and slide the apex seal springs into place, small spring first.

16. Install the corner seal, side seal assemblies and housing seals onto the rear side of the rotor and housing using the same procedure as before. Make sure the painted edge faces the side of the seal groove.

17. Lubricate the rotor between the inner and outer oil seals and lubricate the intermediate housing. Be carefull not to oil the housing seal rubber groove. Install a new O-ring and apply sealer to the housing legs as before.

18. Turn the eccentric shaft so the rear rotor journal faces the intake and exhaust port side. Lift the shaft about 1 in. (25mm) and install the intermediate housing over the shaft and onto the front housing. Be careful not to lift the shaft too far.

19. Assemble the rear rotor and housing in the same procedure as the front. Remember the side piece on the apex seal goes towards the rear housing.

20. Lubricate the stationary gear and main bearing and apply silicone sealer to the proper areas on the rotor housing. Install the rear housing onto the rear rotor housing. Check that the side pieces of the apex seals are not wedged between the rotor housing and intermediate or end housing.

21. Install a new washer on each tension bolt, and lubricate the threads and sealing washer of each bolt with

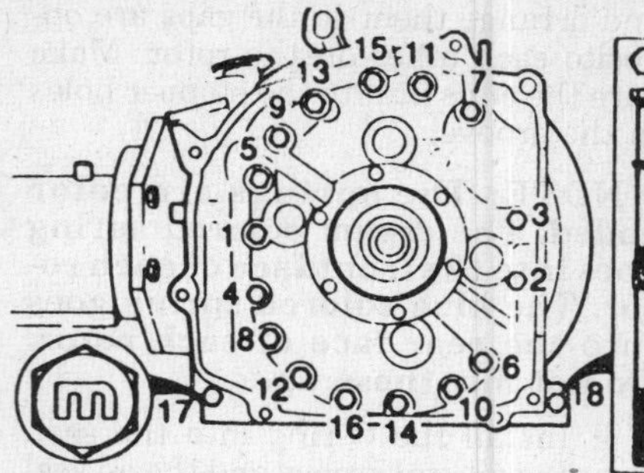

Assembly torque sequence; make sure the marked bolt is in position 17, the bolt with the tube in position 18

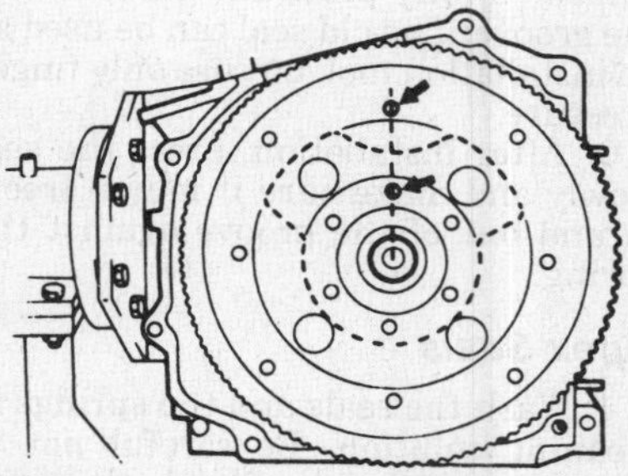

Install the automatic transmission drive plate as shown relative to the counter weight

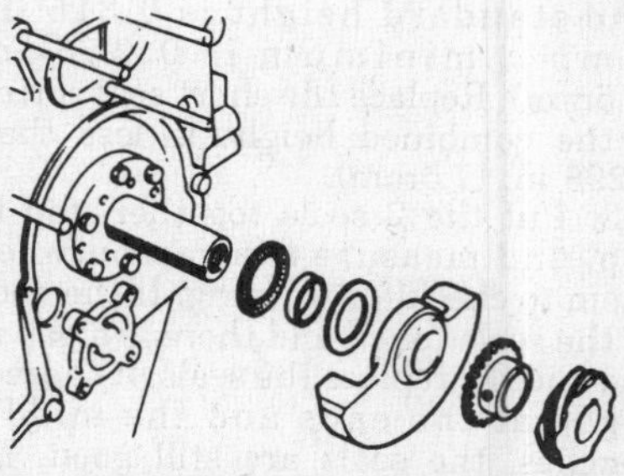

Front thrust bearing and balance weight installation

engine oil. Start the bolts in the holes, making sure to install them into their original position. On later engines, install the correct bolts into positions 17 and 18.

22. Tighten the bolts finger tight and turn the eccentric shaft to make sure the rotation is easy and smooth.

23. Torque the bolts in 3 steps in the sequence shown to 29 ft. lbs. (39 Nm). Turn the eccentric shaft between steps to make sure there is no binding. When the bolts are torqued, make sure the shaft still turns smoothly and easily.

24. Lubricate the oil seal in the rear housing.

25. If equipped with a manual transmission:

a. Install the flywheel on the rear of the eccentric shaft and fit the key into place.

b. Apply thread locking compound to the eccentric shaft and sealer to the contact surface of the nut.

c. Install the flywheel locknut. Hold the flywheel securely and tighten the nut to 362 ft. lbs. (491 Nm).

26. If equipped with an automatic transmission:

a. Install the key and counterweight.

b. Apply thread locking compound to the eccentric shaft threads and sealer to the contact surface of the nut.

c. Install and torque the nut to 362 ft. lbs. (491 Nm).

NOTE: On engines equipped with an automatic transmission, do not hold the drive plate to hold the eccentric shaft when removing or installing the large nut. The torque on the nut is over 300 ft. lbs. (400 Nm) and the drive plate is not designed to withstand that much torque.

d. Install the driveplate on the counterweight as shown and torque the nuts to 45 ft. lbs. (61 Nm).

27. Turn the engine so the front faces up. Temporarily install the end play spacer, needle bearing, thrust plate, balance weight, oil pump drive sprocket and distributor drive gear onto the front of the eccentric shaft. Install the pulley and bolt and torque to 98 ft. lbs. (132 Nm). Lubricate the needle bearing.

28. Use a dial indicator to check the eccentric shaft end play. Standard end play is 0.0016–0.0028 in. (0.040–0.070mm), maximum is 0.0035 in. (0.009mm). Different thickness spacers are available from Mazda, replace the spacer as required.

29. After checking or setting the end play, remove the pulley and install the oil pump. Torque the bolts to 87 in. lbs. (9.8 Nm). Install the key, sprockets and drive chain and the lock washer and nut. Torque the nut to 34 ft. lbs. (46 Nm) and bend the lock washer tab.

30. If removed, install the oil pressure control valve and torque to 36 ft. lbs. (49 Nm). Install a new front oil seal and O-ring and gasket and install the front cover to the engine. Torque the bolts to 17 ft. lbs. (23 Nm).

31. Install the pulley with the old lock bolt and tighten it by hand. Without pushing the eccentric shaft, remove the bolt and measure the shaft depth inside the pulley. If there is more than 0.096 in. (2.44mm) between the end of the shaft and the face of the pulley, the thrust bearing is caught behind the spacer and must be installed correctly.

32. When the front cover is properly installed and the shaft depth in the pulley is correct, install the by-pass valve and spring into the end of the eccentric shaft and install a new O-ring onto a new lock bolt. Apply a sealer to the bolt flange, install the bolt and torque it to 98 ft. lbs. (132 Nm).

33. Turn the engine upside down and trim the excess front cover gasket. Install a new oil strainer gasket and the strainer and torque the bolts to 7 ft. lbs. (10 Nm).

34. Make sure the oil pan gasket surface is clean and apply a bead of silicone sealer. The sealer should be no more than ¼ in. wide and should be inside the bolt holes. Install the gasket, apply a second bead to the gasket and install the oil pan. Start the engine mount bolt and torque the oil pan bolts to 8 ft. lns. (11 Nm) in 2 steps.

35. Install the engine mount and torque the bolts to 69 ft. lbs. (93 Nm).

36. The remaining accessory parts can be installed in the reverse order of removal. Observe the following torque values:

Metering oil pump – 8 ft. lbs. (11 Nm)

Oil pipe and injectors – 15 ft. lbs. (20 Nm) with new gaskets

Water pump assembly – 20 ft. lbs. (26 Nm) in a crisscross pattern, remember to shim studs that don't contact the gasket

Exhaust manifold – 34 ft. lbs.

Intake manifold and bracket and plenum chamber – 19 ft. lbs.

Fuel injectors and delivery pipe – 19 ft. lbs. (26 Nm)

Turbocharger-to-exhaust manifold – 40 ft. lbs. (54 Nm)

Manual transmission clutch plate – 20 ft.lbs. (26 Nm)

Intake Manifold

REMOVAL & INSTALLATION

1. Relieve the fuel system pressure and disconnect the negative battery cable.
2. Drain the cooling system.
3. Remove the air inlet ducting and disconnect the throttle cable.
4. Label and disconnect the wiring and vacuum hoses. Remove the manifold support bracket.
5. The manifold assembly is basically 3 sections; the plenum chamber and throttle body, the upper or extension manifold, and the intake manifold which bolts to the engine. Remove the extension manifold and plenum chamber together as an assembly, since this contains all the working parts of the air intake system.
6. If the intake manifold is to be left on the engine, cover the openings.

To install:

7. Installation is the reverse of removal. Use new gaskets and torque all nuts and bolts to 19 ft. lbs. (26 Nm).
8. Connect and adjust the throttle cable as required.
9. Fill and bleed the cooling system and check idle speed adjustment.

Exhaust Manifold

REMOVAL & INSTALLATION

1. Disconnect the negative battery cable.
2. Remove the intake manifold assembly.
3. Remove the exhaust shields.
4. Disconnect the oxygen sensor connector and route the wiring so it will be readily removed with the exhaust manifold.
5. Disconnect the exhaust pipe from the manifold or turbocharger.
6. On turbocharged engines, remove the turbocharger and waste gate assembly and cover the air, water and oil openings to keet them clean.
7. Remove the nuts to remove the exhaust manifold from the engine. Cover the exhaust ports to keep dirt out of the engine.

To install:

8. Installation is the reverse of removal. Use a new gasket and torque the manifold-to-engine nuts to 34 ft. lbs. (46 Nm).
9. On turbocharged engines:

a. Use a new gasket and lock washers, install the turbocharger and torque the nuts to 40 ft. lbs. (54 Nm).

b. Use a new gasket and torque the oil return pipe bolts to 19 ft. lbs. (25 Nm).

c. Use a new gasket and torque the catalytic converter-to-turbocharger nuts to 40 ft. lbs. (54 Nm).

10. Use a new gasket and connect the exhaust pipe to the manifold. Torque the nuts to 59 ft. lbs. (80 Nm).
11. Install the heat shields, connect the oxygen sensor wire and install the intake manifold assembly.

Turbocharger

REMOVAL & INSTALLATION

1. Disconnect the negative battery cable. Drain the cooling system.
2. Disconnect the air hoses from the air pump and remove the air pump from the engine.
3. Remove the air ducting leading to and from the turbocharger.
4. Disconnect the connector from the air control valve and remove the valve from the engine.
5. Disconnect the split air pipe from

the engine and remove the pipe along with the gasket.

6. Disconnect the water hose and the water pipe from the engine and remove them.

7. Disconnect the supply and return oil pipes from the turbo and cover the openings.

8. Remove the shields and disconnect the front catalytic converter from the exhaust manifold. Remove the gasket and discard it.

9. Unstake the retainer tabs from the retainer plate with a small prying tool. Remove the nuts and washers that secure the turbocharger to the exhaust manifold studs and remove the turbocharger from the engine. Cover all the turbo openings to prevent the entry of dirt and foreign matter. Remove the turbocharger gasket and discard it.

To install:

10. Thoroughly clean the exhaust manifold and turbocharger contact surfaces and check the exhaust manifold for warping with a metal straightedge.

11. Install a new turbocharger gasket onto the exhaust manifold and carefully guide the turbo over the mounting studs and onto the gasket. Install the nuts and new lock washers. Torque the nuts to 40 ft. lbs. (54 Nm). Once the nuts are torqued, crimp the tabs on the nut retaining plate to prevent the nuts from loosening. Remove the protective covers from the turbo openings.

12. Connect the catalytic converter to the exhaust manifold with a new gasket. Torque the nuts to 40 ft. lbs. (54 Nm).

13. Install the remaining components in reverse of the removal procedure. Use new gaskets:

a. Oil pipes.
b. Water pipe and water hose.
c. Air pump and air hoses.
d. Split air pipe.
e. Air control valve.
f. Air ducting.
g. Heat shields.

14. Fill the cooling system and connect the negative battery cable. Start the engine and check for leaks. Make all the necessary adjustments.

ENGINE LUBRICATION

Oil Pan

REMOVAL & INSTALLATION

323, Protege, 626 and MX6

1. Disconnect the negative battery cable and raise and safely support the vehicle.

2. Remove the engine under cover, if equipped, and drain the oil.

3. Remove the exhaust pipe from the exhaust manifold and from the catalytic converter.

4. If equipped, disconnect the wiring for the temperature and level sensors.

5. Except on the BP series engines, remove the engine–to–transaxle brackets and the clutch cover as required.

6. Remove the bolts and remove the oil pan. It may be necessary to pry the pan away from the engine; be careful not to damage the gasket surfaces.

7. Above the oil pan is a main bearing support plate that helps stabilize the engine block. This must also be removed along with the oil pick-up strainer and a new gasket installed.

To install:

8. When cleaning the sealing surfaces of the pan, block and bearing support, make sure to also clean the old sealer off the bolts. Failure to clean the bolts could cause cracking at the block.

9. Apply silicone sealer to the new rubber end gaskets and press them into place on the engine.

10. Apply a bead of silicone sealer to the main bearing support and the oil pan, making sure the bead is inside the bolt holes. Install the bearing support and pick-up strainer with a new gasket.

11. Install the pan and bolts and torque the bolts in 2 steps to:

B6 engine – 8 ft. lbs. (11 Nm)

BP engine – 8 ft. lbs. (11 Nm) on pan bolts, 38 ft. lbs. (52 Nm) on pan–to–transaxle bolts

626 and MX6 – 9 ft. lbs. (12 Nm)

929 and 929S

1. Disconnect the negative battery cable and raise and safely support the vehicle.

2. Remove the engine under cover and drain the oil.

3. Remove the bolts and carefully pry the oil pan from the engine. Be careful not to damage the gasker surfaces.

To install:

4. Clean the old sealer from the pan, engine and pan bolts. Apply a bead of silicone sealer to the pan, making sure it is inside the bolt holes.

5. Install the pan and torque the bolts to 7 ft. lbs. (10 Nm). Do not overtorque.

Miata

1. The engine must be raised and the cross member must be lowered. Disconnect the negative battery cable and raise and safely support the vehicle.

2. Remove the engine under cover and drain the oil. Remove the dip stick and tube.

3. Remove both lower engine mount nuts and disconnect the steering shaft joint at the steering rack.

4. Attach lifting equipment and lift the engine slightly.

5. Support the crossmember with a jack and remove the 2 nuts and 2 bolts on each side. Lower the crossmember slightly.

6. Remove the oil pan bolts and lower the pan enough to remove the pickup tube bolts. If prying the pan is necessary, pry at the pan–to–transmission face. Be careful not to damage the block sealing surfaces.

7. Raise the engine and lower the crossmember enough to gain about 4 in. (100mm) of clearance between the pan and the steering rack. Remove the oil pan and carefully pry the baffle off the block.

To install:

8. Make sure all the old sealer is cleaned off the pan, baffle, block and bolts. Make sure all surfaces are clean and dry.

9. Apply a bead of silicone sealer to the baffle, making sure the bead is inside the bolt holes. Install the baffle with a few of the pan bolts and allow the sealer to set up long enough to hold the baffle in place when the bolts are removed.

10. Apply silicone sealer to each end of the block and press the rubber end gaskets into place. Make sure the notches point out.

11. Install the oil pick-up tube with a new gasket and torque the bolts to 8 ft. lbs. (11 Nm).

12. Apply a bead of silicone sealer to the oil pan, making sure the bead is inside the bolt holes. Fit the pan into place and torque the pan–to–engine bolts to 8 ft. lbs. (11 Nm). Start at the center and work out. Torque the pan–to–transmission bolts to 66 ft. lbs. (89 Nm).

13. Raise the crossmember and set the engine into place. Torque the crossmember nuts to 87 ft. lbs. (118 Nm) and the bolts to 61 ft. lbs. (83 Nm).

14. Torque the engine mount nuts to 58 ft. lbs. (78 Nm).

15. Connect the steering rack and install the dip stick. Fill the engine with the correct amount of oil and run it to check for leaks.

RX7

1. The engine must be lifted slightly. Disconnect the negative battery cable and raise and safely support the vehicle.

2. Remove the under cover and drain the oil. Disconnect the oil level and temperature sensor wires, if equipped.
3. Attach engine lifting equipment and remove the fan from the front of the engine.
4. Remove the engine mount nuts and raise the engine slightly. Remove the right side mount from the vehicle.
5. Remove the oil pan bolts and carefully pry the pan off the engine. Be careful not to damage the sealing surfaces.

To install:

6. Clean the old sealer off the engine, pan, and bolts.
7. The pan can be installed with or without a gasket. In either case, apply a bead of silicone sealer to the gasket and/or the pan. Make sure the bead is inside the bolt holes.
8. Install the pan and torque the bolts to 8 ft. lbs. (11 Nm).
9. Install the engine mount and torque the bolt to 69 ft. lbs. (93 Nm).
10. Lower the engine into place and torque the mounting nuts to 34 ft. lbs. (46 Nm). Torque fan nuts to 9 ft. lbs. (12 Nm).

Oil Pump

REMOVAL & INSTALLATION

Except RX7

1. Raise and safely support the vehicle and disconnect the negative battery cable.
2. Remove the oil pan and oil pick-up tube.
3. Remove the timing belt cover and timing belt. The oil pump is behind the crankshaft sprocket.
4. Remove the crankshaft sprocket and unbolt and remove the oil pump.
5. Installation is the reverse of removal. Be sure to use new gaskets and O-rings. Torque 8mm bolts to 19 ft. lbs. (25 Nm), 10mm bolts to 38 ft. lbs. (52 Nm).

RX7

1. This engine uses 2 oil pumps; a metering pump bolted to the front cover that feeds the oil injectors and a main pressure pump behind the front cover. To remove the main pump, the front cover and oil pan must be removed.
2. Disconnect the negative battery cable and raise and safely support the vehicle.
3. Remove the engine under cover and drain the oil.
4. Remove the accessory drive belts and the engine cooling fan.
5. Without disconnecting the hoses, remove the power steering pump and the air conditioner compressor and secure them aside. Remove the pump/compressor bracket.
6. Disconnect the oil lines and remove the oil metering pump. Cap the pump openings and the lines to protect the injectors and pump.
7. Raise the engine as required and remove the oil pan and pick-up tube.
8. Remove the eccentric shaft pulley bolt, by-pass valve and spring and the pulley.
9. Remove the crank angle sensor and turn the eccentric shaft so the balance weight is at the bottom. Remove the front cover.
10. Remove the distributor drive gear. Bend the lock tab away from the nut and remove the nut and washer from the front of the oil pump.
11. Slide the pump sprocket and drive sprocket with the chain off the engine. Remove the bolts to remove the pump.

To install:

12. Remove the balance weight and make sure the thrust bearing and spacer are properly positioned. Pull out on the eccentric shaft when sliding the balance weight back on so the thrust bearing does not fall behind the spacer.
13. Install the oil pump and torque the bolts to 7 ft. lbs. (10 Nm).
13. Install the sprockets and chain. When installing the distributor drive gear, the chamfer faces the front cover.
14. It is recommended that the front cover oil seal be replaced. Press the old seal out, lubricate the new seal and press it in with a seal tool or socket.
15. Make sure the thrust bearing is not caught behind the spacer and install the oil pump. Torque the bolts to 87 in. lbs. (9.8 Nm). Install the key, sprockets and drive chain and the lock washer and nut. Torque the nut to 34 ft. lbs. (46 Nm) and bend the lock washer tab.
16. If removed, install the oil pressure control valve into the front cover and torque to 36 ft. lbs. (49 Nm). Install a new O-ring and gasket and install the front cover to the engine. Torque the bolts to 17 ft. lbs. (23 Nm).
17. Temporarily install the pulley with the old lock bolt and tighten it by hand. Without pushing the eccentric shaft, remove the bolt and measure the shaft depth inside the pulley. If there is more than 0.096 in. (2.44mm) between the end of the shaft and the face of the pulley, the thrust bearing is caught behind the spacer and must be installed correctly.
18. A new pulley lock bolt is recommended. Install the by-pass valve and spring into the end of the eccentric shaft and install a new O-ring onto the lock bolt. Apply a sealer to the bolt flange, install the bolt and torque it to 98 ft. lbs. (132 Nm).

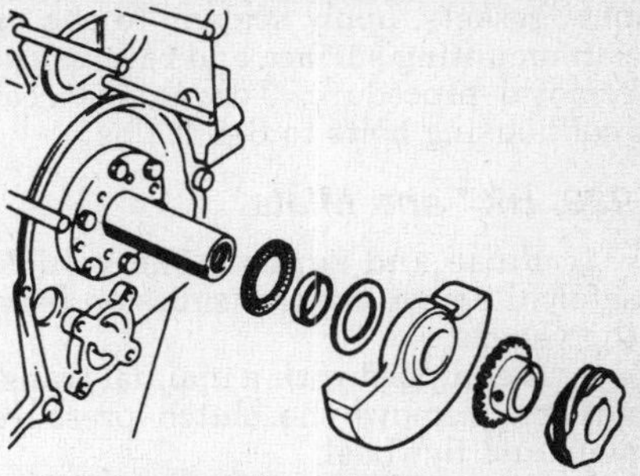

Make sure the thrust bearing does not fall behind the end play spacer—RX7

19. At the oil pan sealing surface, trim the excess front cover gasket. Install a new oil strainer gasket and the strainer and torque the bolts to 7 ft. lbs. (10 Nm).
20. Install the oil pan and engine mount and lower the engine onto the mounts.
21. The remaining accessory parts can be installed in the reverse order of removal. Observe the following torque values:

Metering oil pump—8 ft. lbs. (11 Nm)

Oil pipe fittings—15 ft. lbs. (20 Nm) with new gaskets

Water pump assembly—20 ft. lbs. (26 Nm) in a crisscross pattern, remember to shim studs that don't contact the gasket

Fan nuts—8 ft. lbs. (12 Nm).

Rear Main Bearing Oil Seal

REMOVAL & INSTALLATION

323, Protege, 626 AND MX-6

1. Disconnect the negative battery cable. Raise and support the vehicle safely. Remove the transaxle.
2. If equipped with a manual transaxle, remove the pressure plate, the clutch disc and the flywheel. If equipped with an automatic transaxle, remove the driveplate from the crankshaft.
3. Remove the rear oil pan-to-seal housing bolts.
4. Remove the rear main seal housing bolts and the housing from the engine.
5. Remove the oil seal from the rear main housing.
6. Clean the gasket mounting surfaces.
7. To install the new seal, coat the seal and the housing with oil. Press the seal into place until it is flush with the housing.
8. To complete the installation, use

new gaskets, apply sealant to the oil pan mounting surface and reverse the removal procedure. Torque the rear seal housing bolts to 6–8 ft. lbs.

929, RX7 and Miata

1. Raise and support the vehicle safely. Remove the transmission from the vehicle.
2. If equipped with a manual transmission, remove the clutch pressure plate and flywheel.
3. If equipped with automatic transmission, remove the drive plate assembly.
4. Drain the engine oil.
5. Carefully pry the old seal out of the housing. Do not damage the housing.
6. Apply clean engine oil to the new seal and press it into place until it is flush with the housing. The old seal can be used as an installation tool.
7. Install the flywheel and transmission.

ENGINE COOLING

Radiator

REMOVAL & INSTALLATION

Except RX7 and 929

1. Disconnect the negative battery cable and drain the coolant.
2. Disconnect the wiring to the electric fans.
3. Remove the upper and lower radiator hoses. If equipped with automatic transaxle, disconnect the cooling lines and plug the hoses.
4. Remove the bolts and lift the radiator and fans out as an assembly.
5. Installation is the reverse of removal.

929 and 929S

1. Disconnect the negative battery cable and drain the coolant.
2. Disconnect the vacuum lines and remove the air inlet duct.
3. Remove the upper and lower radiator hoses. If equipped with an automatic transmission, disconnect and plug the cooling hoses.
4. Remove the fasteners for the fan and shroud and lift them out together. It may be easier to remove the shroud and move it back towards the engine to access the nuts holding the fan/clutch assembly.
5. Remove the bolts and lift the radiator out of the vehicle.
6. Installation is the reverse of removal. Make sure the fan turns without contacting the shroud.

RX7

1. Remove the battery and battery box and drain the coolant.
2. Remove the fan and the air inlet duct.
3. Remove the radiator hoses and disconnect the heater hose. If equipped with an automatic transmission, disconnect the plug the cooling lines.
4. Disconnect the coolant level sensor wiring, remove the bolts and lift the radiator and shroud out together.
5. Installation is the reverse of removal. Make sure the fan turns without contacting the shroud.

Heater Core

REMOVAL & INSTALLATION

CAUTION

On models equipped with an air bag system, the air bag must be disarmed and removed before removing the steering wheel. Accidental deployment of the air bag could cause severe personal injury.

1. If the vehicle is equipped with an air bag, properly disarm it.
2. On all except 929, the dashboard must be removed to remove the heater core assembly. Disconnect the negative battery cable and drain the coolant.
3. Disconnect the heater hoses at the engine firewall.
4. Remove the shift lever knob and the center console.
5. Remove the steering wheel and upper and lower column covers.
6. Remove the gauge cluster hood and remove the gauge panel. Don't forget to disconnect the speedometer cable.
7. Remove the heater ducts and the glove compartment assembly.
8. Remove the trim pieces as required and remove the heater controls and radio.
9. Loosen the steering column bolts as required and lower the column.
10. Pry out the defroster grille or the center caps between the grille sections and remove the screws or bolts.
11. With all wires and control cables disconnected, remove the bolts and remove the dashboard from the vehicle.
12. Remove the screws and remove the heater unit from the firewall.
13. Remove the screws to split the case and remove the heater core from the housing. Take notice of how the air control doors fit together.

To install:

14. Install the heater core and air control doors and fit the halves of the housing together.
15. Install the housing and connect the ducts.
16. Install the dashboard and connect the wiring and control cables.
17. Install the gauge panel and connect the wiring.
18. Assemble the steering column and install the steering wheel.
19. Fill the cooling system and test the heater.

Water Pump

REMOVAL & INSTALLATION

Except RX-7

1. Disconnect the negative battery cable and drain the cooling system.
2. Remove the timing belt covers and the timing belt.
3. On 626 and MX6, remove the idler pulley.
4. On the BP series engines, remove the idler pulley and tensioner pulley and unbolt the lower inlet pipe.
5. Remove the bolts and remove the water pump. Take note of the position of the rubber seals on the 626 and MX6 pump.

To install:

6. On 626 and MX6, secure the rubber gaskets to the pump with silicone sealer.
7. Make sure the sealing surfaces are clean. Use a new gasket and install the pump to the block. Torque the nuts or bolts to 19 ft. lbs. (25 Nm).
8. On BP series engines, use a new gasket and attach the inlet pipe to the pump. Torque the bolts to 19 ft. lbs.
9. Install the timing belt, covers and accessory drive belts. Fill the cooling system and check for leaks.
10. When installing the fan on 929, make sure the fan turns freely without contacting the shroud.

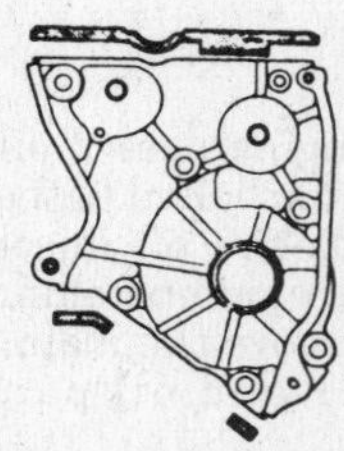

Replace the rubber gaskets on the water pump—626 and MX6

RX7

1. Disconnect the negative battery cable and drain the cooling system.
2. Turn the eccentric shaft so the pulley mark lines up with the indicator pin.

3. Remove the fan, accessory drive belts, air pump, alternator and water pump pulley. Remove the belt drive pulley from the eccentric shaft.

4. Disconnect the coolant hoses and the sensor wiring from the water pump.

5. Since the nuts that hold the housing onto the engine must be removed to remove the pump from the housing, the housing must also be removed to replace the gasket behind it. Remove the nuts and slide the water pump assembly off the studs.

6. Remove the bolts and remove the pump from the housing.

To install:

7. Make sure the housing sealing surface is clean of old gasket material. Use a new gasket and assemble the pump to the housing. Torque the bolts to 19 ft. lbs. (25 Nm).

8. Make sure the sealing surfaces are clean and install the pump housing with a new gasket.

9. Connect the hoses and sensor wiring and fill the system to check for leaks.

10. If there are no leaks, install the remaining parts and adjust the belt tensions. Torque the drive pulley–to–eccentric shaft bolts to 8 ft. lbs. (12 Nm).

11. When installing the fan, make sure it turns freely without contacting the shroud.

Thermostat

REMOVAL & INSTALLATION

RX-7

1. Disconnect the negative battery cable. Drain the engine coolant. Remove the upper hose from thermostat housing.

2. Disconnect the water thermo switch connector. Remove the thermostat housing and the thermostat.

3. Installation is the reverse of removal. Always install a new mounting gasket. When installing the thermostat, make sure the jiggle pin is facing up. Align the marks on the upper hose with the cover marks during installation.

Except RX-7

1. Disconnect the negative battery cable. Drain the cooling system.

2. Disconnect the coolant hose and the coolant temperature switch lead (except 929) from the thermostat housing.

3. Remove the thermostat housing bolts, housing, and thermostat. Clean all old gasket material off the housing and engine.

4. When installing the new thermostat, the spring goes towards the engine and the jiggle pin goes on top.

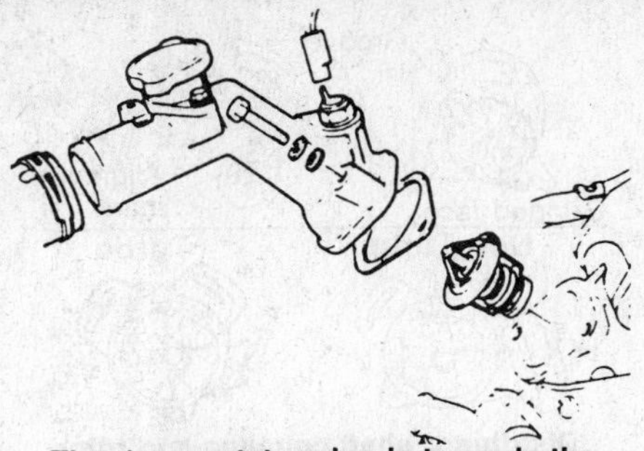

The thermostat spring is towards the engine, the jiggle pin at the top – RX7 shown

5. On 929, the mark on the housing is at the top.

6. Install the thermostat with a new gasket and torque the housing bolts to 19 ft. lbs. (25 Nm).

7. Connect the wiring and hose and fill the cooling system to check for leaks.

ENGINE ELECTRICAL

Distributor

REMOVAL

323 and Protege

1. Disconnect the negative battery cable to prevent accidental cranking once the distributor is removed.

2. Disconnect and label the vacuum hose(s) and the electrical connector from the distributor.

3. Turn the crankshaft pulley until the No. 1 piston is at TDC mark, pulley mark aligned with **T** mark on the timing belt cover.

4. Remove the distributor hold-down bolt and withdraw the distributor from the cylinder head.

5. Remove the O-ring seal and discard it.

626 and MX-6

1. Disconnect the negative battery cable to prevent accidental cranking once the distributor is removed. Remove the distributor cap without disconnecting the wires and move it aside.

2. On the non-turbocharged engine, disconnect the vacuum hoses and wiring. On the turbocharged engine, disconnect the electrical coupler.

3. Turn the crankshaft pulley until the No. 1 piston is at TDC.

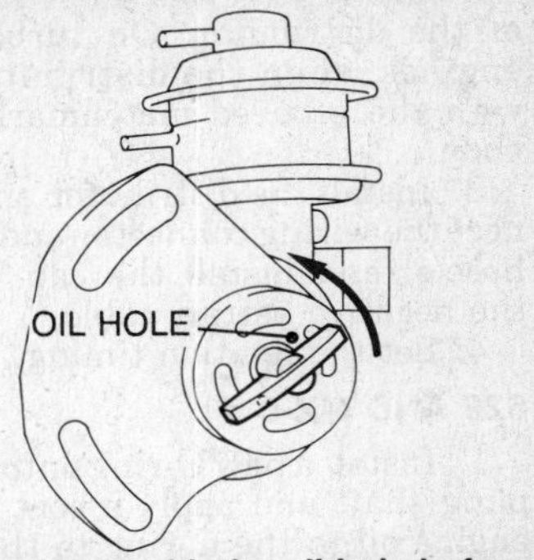

Align the blade with the oil hole before installing the distributor – 323 without turbocharger

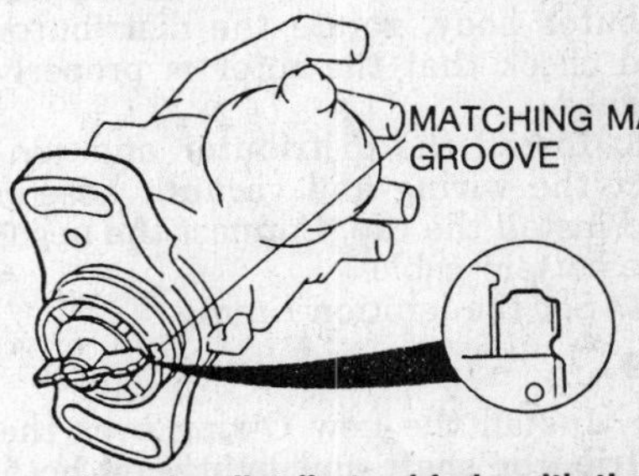

Align the distributor blade with the groove in the housing – 323 with turbocharger

4. Loosen the lock bolts and remove the distributor.

5. Remove the O-ring from the coupling shaft and discard it.

929

1. Disconnect the negative battery cable to prevent accidental cranking once the distributor is removed. Remove the distributor cap without disconnecting the wires and move it aside.

2. Disconnect the distributor electrical connector.

3. Turn the crankshaft pulley until the No. 1 piston is at TDC, yellow mark on pulley aligned with **T** mark on the timing scale.

4. Loosen the lock bolt and remove the distributor.

5. Remove the O-ring from the distributor shaft and discard it.

INSTALLATION

Timing Not Disturbed

323 AND PROTEGE

1. Coat the new O-ring seal with a light film of clean engine oil and fit it into the cylinder head opening.

NOTE: Make sure the No. 1 piston is at TDC before installing the distributor.

2. On non-turbocharged engines, turn the distributor blade so it aligns with the small oil holes in the bottom

of the distributor. On turbocharged engines, align the distributor blade with the grooved matchmark on the body.

3. Install the distributor and reconnect the wiring connector and vacuum hose(s) and install the cap. Connect the negative battery cable.

4. Set the ignition timing.

626 AND MX-6

1. Install a new O-ring onto the coupling shaft and apply a coat of clean engine oil to the O-ring to the driven gear.

2. Align the shaft coupling blade with the alignment marks on the distributor body, rotate the distributor and check that the rotor is properly aligned.

3. Install the distributor and connect the wiring and vacuum hose(s) and install the cap. Connect the negative battery cable.

4. Set the ignition timing.

929

1. Install the new O-ring over the distributor shaft and lightly oil both the O-ring and the driven gear with clean engine oil.

2. On SOHC engines, align the match marks on the distributor housing with the driven gear.

3. Install the distributor, connect the wiring and install the cap. Connect the negative battery cable.

4. Set the ignition timing. Torque the lock bolt to 14–18 ft. lbs. once the timing is set.

Timing Disturbed

323 AND PROTEGE

1. Rotate the the crankshaft to position the No. 1 piston on TDC of its compression stroke.

2. Coat the new O-ring seal with a light film of clean engine oil and fit it into the cylinder head opening.

NOTE: Make sure the No. 1 piston is at TDC before installing the distributor.

3. On non-turbocharged engines, turn the distributor blade so it aligns with the small oil holes in the bottom of the distributor. On turbocharged engines, align the distributor blade with the grooved matchmark on the body.

4. Install the distributor and connect the wiring and vacuum hose(s) and install the cap. Connect the negative battery cable.

5. Set the ignition timing.

626 AND MX-6

1. Rotate the the crankshaft to position the No. 1 piston on TDC of its compression stroke.

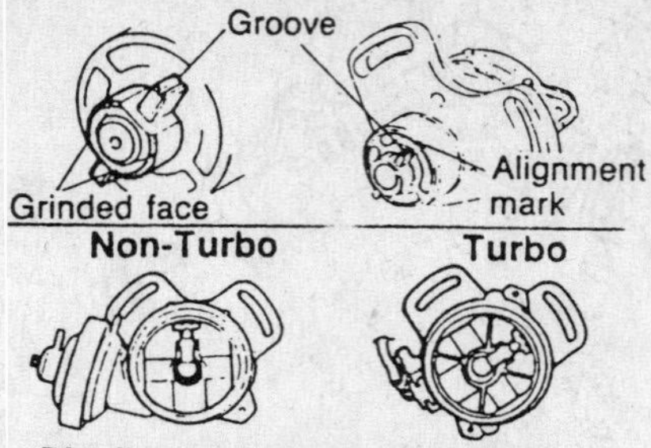

Distributor shaft coupling and rotor alignment – 626 and MX-6

2. Install a new O-ring onto the coupling shaft and apply a coat of clean engine oil to the O-ring to the driven gear.

3. Align the shaft coupling blade with the alignment marks on the distributor body, rotate the distributor and check that the rotor is properly aligned.

4. Install the distributor and connect the wiring and vacuum hose(s) and install the cap. Connect the negative battery cable.

5. Set the ignition timing.

929

1. Rotate the the crankshaft to position the No. 1 piston on TDC of its compression stroke.

2. Install the new O-ring over the distributor shaft and lightly oil both the O-ring and the driven gear with clean engine oil.

3. On SOHC engines, align the match marks on the distributor housing with the driven gear.

4. Install the distributor, connect the wiring and install the cap. Connect the negative battery cable.

5. Set the ignition timing. Torque the lock bolt to 14–18 ft. lbs. once the timing is set.

Distributorless Ignition

REMOVAL & INSTALLATION

Crank Angle Sensor

RX-7

1. Disconnect the negative battery cable.

2. Turn the eccentric shaft to position the pulley to the leading timing mark on the pulley, usually painted yellow.

3. Disconnect the crank angle sensor electrical connector and remove the crank angle sensor locknut.

4. Slowly pull up on the crank angle sensor and remove it from the vehicle.

To install:

5. Align the matching marks on the crank angle sensor housing and the driven gear.

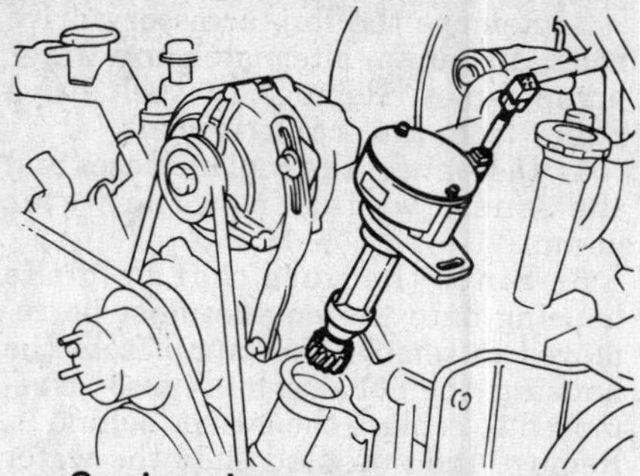

Crank angle sensor removal and installation – RX-7

6. Make sure the eccentric shaft pulley is set to the leading timing mark.

7. Install the crank angle sensor and the locknut.

8. Connect the battery negative cable and check the ignition timing. Tighten locknut to 70–90 inch lbs.

Ignition Coil

MIATA

1. Mark the position of the spark plug wires in the coil pack.

2. Disconnect the spark plug wires and the ignition coil electrical connector.

3. Remove the 3 ignition coil-to-engine bolts and the ignition coil assembly from the vehicle.

4. Reverse procedure to install. Torque ignition mounting bolts to 19 ft. lbs. (25 Nm).

Alternator

PRECAUTIONS

- Before disconnecting the battery, obtain the radio theft protection code.
- Do not reverse the battery connections or the alternator and various control computers will be damaged.
- Do not use high voltage testers when testing the rectifiers.
- Do not disconnect the battery or the alternator while the engine is running. Do not start the engine with the alternator disconnected.
- Do not ground or cross any alternator connections while the engine is running.
- Remember that terminal **B** on the alternator is always the battery connection.
- Disconnect the battery cables when charging the battery with a fast charger.
- Disconnect the battery ground terminal when performing any service on electrical components.
- Disconnect the battery and control computers if electric welding is to be done anywhere on the vehicle.

BELT TENSION ADJUSTMENT

Check tension by using a suitable belt tension gauge, installed midway between the crankshaft or eccentric shaft and alternator pulleys. Used belts should deflect about 0.55–67 in. (14–17mm) for rotary engines or 0.31–0.35 in. (8–9mm) for piston engines. New belts should deflect slightly less.

1. To adjust the alternator drive belt, loosen the alternator mounting bolt and adjusting bar bolt. Move the alternator to obtain the correct belt tension and tighten the bolts. Run the engine for about 5 minutes and recheck the belt tension.
2. To adjust the power steering pump drive belt, loosen the mounting bolt and locknut. Turn the adjusting bolt, if equipped or manually move the pump until the correct belt tension is obtained. Tighten the locknut, run the engine for about 5 minutes and recheck the belt tension.
3. If equipped with air conditioning, loosen the locknut on the idler pulley and turn the adjusting bolt until the correct belt tension is obtained. Tighten the locknut, run the engine for about 5 minutes and recheck the belt tension.

REMOVAL & INSTALLATION

1. Disconnect the negative battery cable. Label and disconnect all electrical leads from the alternator.

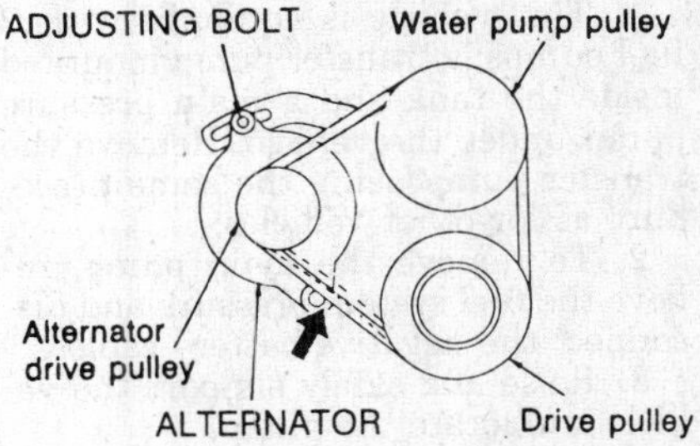

Alternator belt adjustment – piston engines

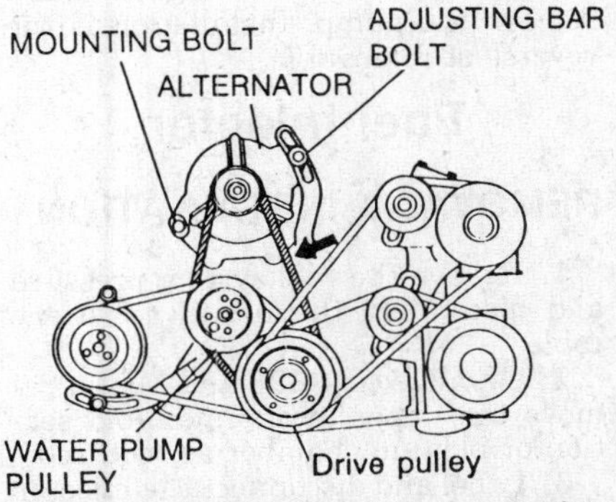

Alternator belt adjustment – rotary engines

2. Remove the alternator adjusting link bolt and remove the belt from the alternator. Do not remove the adjusting link.
3. Remove the pivot bolt and remove the alternator. On some vehicles the alternator is removed from under the vehicle.
4. Installation is the reverse of removal. Adjust the drive belt tension.
5. Connect the negative battery cable. Run the engine for about 5 minutes and recheck the belt tension.

Starter

REMOVAL & INSTALLATION

1. Disconnect the negative battery cable from the battery.
2. On 626, MX6, 929 and Miata, raise and safely support the vehicle and remove the engine under cover, if equipped.
3. Label and disconnect the wiring.
4. Remove the starter and manifold brackets as required, remove the bolts and remove the starter.
5. When installing, torque the mounting bolts to 34 ft. lbs. (41 Nm).

EMISSION CONTROLS

Please refer to "Emission Controls" in the Unit Repair section for system maintenance procedures. Due to the complex nature of modern electronic engine control systems, comprehensive diagnosis and testing procedures fall outside the confines of this repair manual. For complete information on diagnosis, testing and repair procedures concerning all modern engine and emission control systems, please refer to "Chilton's Guide to Fuel Injection and Electronic Engine Controls".

FUEL SYSTEM

Fuel System Service Precautions

Safety is the most important factor when performing not only fuel system maintenance but any type of maintenance. Failure to conduct maintenance and repairs in a safe manner may result in serious personal injury or death. Maintenance and testing of the vehicle's fuel system components can be accomplished safely and effectively by adhering to the following rules and guidelines.

- To reduce the possibility of fire and personal injury, always disconnect the negative battery cable unless the repair or test procedure requires that battery voltage be applied.
- Always relieve the fuel system pressure prior to disconnecting any fuel system component (injector, fuel rail, pressure regulator, etc.), fitting or fuel line connection. Exercise extreme caution whenever relieving fuel system pressure to avoid exposing skin, face and eyes to fuel spray. Please be advised that fuel under pressure may penetrate the skin or any part of the body that it contacts.
- Always place a shop towel or cloth around the fitting or connection prior to loosening to absorb any excess fuel due to spillage. Ensure that all fuel spillage (should it occur) is quickly removed from engine surfaces. Ensure that all fuel soaked cloths or towels are deposited into a suitable waste container.
- Always keep a dry chemical (Class B) fire extinguisher near the work area.
- Do not allow fuel spray or fuel vapors to come into contact with a spark or open flame.
- Always use a backup wrench when loosening and tightening fuel line connection fittings. This will prevent unnecessary stress and torsion to fuel line piping. Always follow the proper torque specifications.
- Always replace any removed fuel fitting gaskets and O-rings with new ones. Do not substitute fuel hose or equivalent where fuel pipe is installed.

RELIEVING FUEL SYSTEM PRESSURE

1. Locate the fuel pump connector or relay:

On 323 and Protege, raise the rear seat and locate the fuel pump connector. If the connector is not accessable under the seat, locate the main fuel pump relay to the right of the engine computer behind the center console.

On 626 and MX6, locate the fuel pump relay on the fuse panel under the left side of the dashboard.

On 929, locate the fuel pump connector in the trunk or the pump relay behind the driver's kick panel.

On RX7, locate the fuel pump connector under the carpet in the luggage compartment.

On Miata, remove the steering column under cover and locate the fuel pump relay.

2. If the engine is cold, do not let it run long enough to warm up. With the engine running at idle, disconnect the power to the fuel pump and allow the engine to stall.

3. Turn the ignition switch **OFF**.

4. Under the hood, wrap a rag around the fuel filter outlet fitting and loosen the clamp to remove the hose.

Fuel Filter

REMOVAL & INSTALLATION

EXCEPT RX-7

1. Properly relieve the fuel system pressure. Disconnect the negative battery cable. Remove the screw and wire retainer bracket.

2. Remove the inlet and outlet fuel lines.

3. Remove the fuel filter. On some vehicles, it may be necessary to remove the bracket with the fuel filter.

4. Installation is the reverse of the removal procedure. Install the filter in the proper direction. If the fuel filter is equipped with union bolt fittings, use new metal crush gaskets. Start the engine and check for leaks.

RX-7

1. Relieve the fuel system pressure. Disconnect the negative battery cable. Raise and support the vehicle safely.

2. Loosen the clips at both ends of the filter and place a collection pan under it to catch excess fuel.

3. Disconnect the fuel filter lines and remove the filter from its retainer.

4. Install the new filter, paying close attention to the direction of the filter in relation to the direction of the fuel flow.

5. Turn the ignition switch to **ON** to activate the fuel pump and check the fuel filter connections for leaks.

Fuel Tank

REMOVAL & INSTALLATION

Except 4 Wheel Drive

1. Relieve the fuel system pressure and drain the tank.

2. Remove the fuel pump and label and disconnect all hoses from the tank.

3. Raise and safely support the vehicle and disconnect the filler tube hose.

4. Loosen the strap retaining nuts or bolts. The tank may be stuck to the body, carefully pry the tank away and allow it to rest on the straps.

5. Remove the strap nuts or bolts and lower the tank out of the vehicle.

6. Installation is the reverse of removal. Use a new gasket when installing the pump.

323 with 4 Wheel Drive

1. Relieve the fuel system pressure and drain the tank.

2. Remove the tank gauge unit and the transfer pump unit from the tank.

3. Raise and safely support the vehicle. Remove the rear muffler and exhaust pipe.

4. Matchmark the flanges so they can be returned to their original position and remove the drive shaft.

5. Remove the main fuel pump and drain the tank into a proper container.

6. Disconnect any remaining hoses and wires and disconnect the fuel filler tube.

7. Loosen the strap retaining nuts or bolts. The tank may be stuck to the body, carefully pry the tank away and allow it to rest on the straps.

8. Remove the strap nuts or bolts and lower the tank out of the vehicle.

To install:

9. Install the tank and connect the filler tube and hoses.

10. Use new gaskets and install the gauge unit and the transfer pump. Connect the wiring and hoses.

11. Install the main fuel pump and connect the wiring and hoses.

12. Install the drive shaft and torque the flange nuts to 22 ft. lbs. (30 Nm). Make sure to install the same center bearing shims and torque the nuts to 38 ft. lbs. (52 Nm).

13. Use new gaskets and install the exhaust pipe and muffler. Torque the nuts to 34 ft. lbs. (46 Nm).

Electric Fuel Pump

PRESSURE TESTING

1. Relieve fuel system pressure and disconnect the filter outlet hose.

2. Install a 0–100 psi gauge with a tee fitting and reconnect the hose.

3. Locate the yellow 2 pin test connector.

On early 323 and all 626, the connector is near the windshield wiper motor.

On 929 the test connector is near the air flow meter connector.

On RX7, the test connector is near the oil filler cap.

On later 323 and Miata, use the indicated **F/P** terminals of the large diagnostic connector.

4. Jumper the connector terminals and turn the ignition switch **ON** to run the fuel pump. Check for proper system pressure as listed in the specification chart.

5. To test maximum pump pressure, stop the pump and disconnect the negative battery cable. Relieve the system pressure and disconnect the fuel hose from the tee fitting.

6. Connect the gauge directly to the fuel supply hose. Make sure the gauge is securely connected to the hose.

7. Run the pump again only as long as necessary. Pump pressure should be 64–85 psi. and should hold at least 20 psi. for 5 minutes after the pump is turned off.

REMOVAL & INSTALLATION

Except 4 Wheel Drive

1. The pump is mounted inside the fuel tank. Relieve the fuel system pressure and disconnect the negative battery cable.

2. Depending on the vehicle, lift the mat in the luggage compartment or remove the rear seat to locate the fuel pump cover plate. Unplug the connector and remove the plate.

3. Disconnect the fuel feed and return hoses. Wrap a clean rag around the fuel lines to catch any fuel spray, then plug the lines to prevent leakage.

4. Remove the mounting screws and lift the fuel pump and gauge assembly from the fuel tank.

5. Installation is the reverse of the removal procedure. Be careful not to allow any dirt or other foreign material to contaminate the fuel tank while the unit is removed. Use a new cover plate gasket.

323 with 4 Wheel Drive

1. This vehicle is equipped with 2 fuel pumps; a transfer pump mounted inside the tank and a main pressure pump under the vehicle. Remove the transfer pump using the same procedure as for other vehicles.

2. To remove the main pump, relieve the fuel system pressure and disconnect the negative battery cable.

3. Raise and safely support the vehicle and locate the pump.

4. Disconnect the wiring, then disconnect the hoses and plug them to prevent fuel leakage.

5. Remove the mounting bolts and remove the pump. Installation is the reverse of removal.

Fuel Injector

REMOVAL & INSTALLATION

1. Relieve the fuel system pressure and disconnect the negative battery cable.

2. On all except 323 and Miata, remove the upper intake manifold section or plenum chamber as required.

3. Label and disconnect the injector wiring.

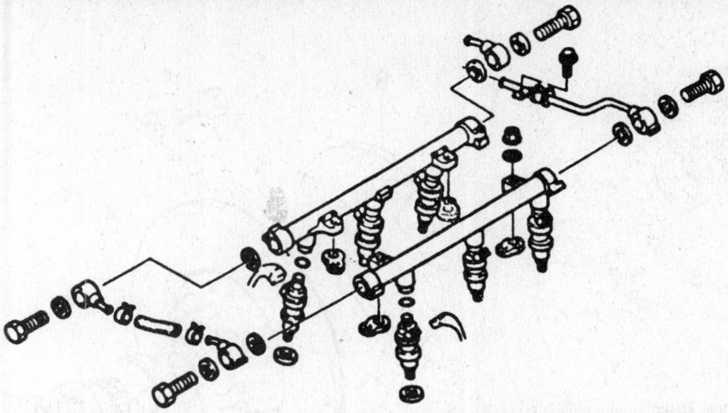

Fuel injectors and delivery pipe are removed as an assembly—929 shown

4. Remove the nuts to remove the injectors and fuel delivery pipe as an assembly. The injectors are fitted with O-rings at top and bottom and spacers between the injector and manifold.

5. Installation is the reverse of removal. Use new O-rings and be careful to install the spacers and wiring correctly.

DRIVE AXLE

Halfshaft

REMOVAL & INSTALLATION

323, Protege, 626 and MX6

NOTE: When loosening or tightening an axle nut or bolt, make sure the vehicle is on all 4 wheels. Axle nut torque is high enough that attempting to loosen it may cause the vehicle to fall.

1. Remove the wheel hub cap as required and pry back the lock tab on the axle bolt or nut. Loosen the bolt.

2. Raise and support the vehicle safely. Drain the transaxle fluid and remove the front wheels and the splash shield.

3. Disconnect the stabilizer bar control link from the lower arm. Note that some of the link is exposed beyond the locknut.

4. Disconnect the tie rod end from the steering knuckle.

5. On all except 1988–89 323, remove the pinch bolt and pry the control arm down to remove the ball joint from the steering knuckle.

6. To remove the left side halfshaft, carefully insert a prybar between the halfshaft and the transaxle case. Lightly tap on the end of the bar to disengage the circlip inside the transaxle. Be careful not to damage the oil seal.

7. To remove the right side halfshaft, insert a prybar between the halfshaft and the joint shaft and separate them. Unbolt the joint shaft bearing bracket and remove the joint shaft from the transaxle in the same maner.

8. Remove the halfshaft locknut or bolt and pull the front hub outward and to the rear, disconnecting the halfshaft from the front hub. If it is stuck in the hub, use a bearing puller to push the halfshaft out of the splined hub.

9. Install differential side gear holder 49–G030–455, or equivalent, to hold the gears in place and prevent misalignment.

To install:

10. Check the condition of the axle seal, replace if necessary. Install a new circlip, lubricate the seal and press the halfshaft or joint shaft into the transaxle.

11. If the joint shaft was removed, torque the bracket bolts to 46 ft. lbs. (62 Nm).

12. Make sure the splines are clean and insert the halfshaft into the hub. Start but do not torque the locknut or bolt.

13. Connect the lower ball joint to the control arm and torque the pinch bolt to 40 ft. lbs. (54 Nm).

14. Connect the tie rod end and torque the nut to 33 ft. lbs. (44 Nm), then tighten as required to install a new cotter pin.

15. Connect the stabilizer bar linkage and adjust the length of the exposed threads on the link, then tighten the locknut.

1988–89 323—0.43 in. (10.8mm)
1990–92 323—0.75 in. (19mm)
626 and MX6—0.79 in. (20.1mm)

16. With the wheel installed and the vehicle on the ground, torque the axle nut or bolt to 235 ft. lbs. (319 Nm). Bend the tab on the lock washer or stake the nut in place. Refill the transaxle fluid.

Driveshaft and U-Joints

REMOVAL & INSTALLATION

RX-7, 929 and Miata

1. Raise and support the vehicle safely. Matchmark the flanges on the driveshaft and pinion so they may be installed in their original position.

2. Remove exhaust components as required. On 929, remove the center bearing bolts.

NOTE: On 929, take note of the spacers on the center bearing bushings. These must be installed in the same place. On RX7, note the balance washers on the rear flange bolts. These must be installed in the same place.

3. Remove the nuts that attach the driveshaft to the companion flange of the rear axle. Lower the back end of the driveshaft and slide the front end out of the transmission.

4. Plug the hole in the transmission with the main shaft turning holder tool 49–0259–440 or 49–S120–440, or equivalent, to prevent fluid from leaking.

5. Installation is the reverse of removal. Torque the flange bolts to 43 ft. lbs. (59 Nm) on 929 and RX7, 22 ft. lbs. (30 Nm) on Miata. On 929, torque the center bearing bolts to 38 ft. lbs. (52 Nm).

6. On RX-7, if the driveshaft was replaced and unusual noise and vibration is noticed, correct the problem with balance washers positioned on various places on the companion flange.

7. If noise and vibration is exhibited on the 929, the problem may be corrected by using different size bolts and spacers on the center bearing support.

323 and Protege With 4WD

1. Raise and support the vehicle safely.

2. Matchmark the front and rear driveshaft flanges for assembly reference. Stuff a rag in the double offset joint to prevent damage to the boot by the propeller shaft. Remove the front and rear flange nuts.

3. Remove the center bearing bolts.

NOTE: Take note of the spacers on the center bearing bushings. These must be installed in the same place.

4. Lower and remove the driveshaft from the vehicle.

5. Installation is the reverse of the removal procedure. Torque the rear flange bolts to 38 ft. lbs. (52 Nm) and the front flange bolts to 22 ft. lbs. (30 Nm). Check that the front and rear shafts are aligned. If not, adjust the height of the center bearing support

with shims. Both shims must be the same thickness.

Front Wheel Hub, Knuckle and Bearing

REMOVAL & INSTALLATION

323 and Protege

NOTE: When loosening or tightening axle nut or bolt, make sure the vehicle is on the ground. Axle nut torque is high enough that attempting to loosen it may cause the vehicle to fall off the support.

1. Loosen the axle nut and the lug nuts. Raise the vehicle and support it safely. Remove the tire and wheel.
2. Without disconnecting the hydraulic hose, remove the brake caliper and hang it from the body. Do not let it hang by the hose. On 1990–92 models, remove the brake disc.
3. Remove the cotter pin and nut and separate the tie rod end from the steering knuckle.
4. Remove the pinch bolt and push the control arm down to separate the ball joint from the knuckle.
5. Remove the 2 bolts and separate the knuckle from the strut.
6. Be careful not to distort the back plate. If damaged or removed, a new one must be pressed onto the knuckle.
7. To remove the hub and bearings on 1988–89 models:
   a. Matchmark the brake disc to the hub and use the special tool 49 G030 725 or equivalent to press the hub out of the knuckle.
   b. Press the hub out of the disc and press the bearing out of the hub.
   c. Remove the seal and press the bearing and spacer out of the knuckle.
8. To remove the hub and bearing on 1990–92 models:
   a. Remove the seal and properly support the assembly to press the hub out from the back.
   b. If the inner bearing race stays on the hub, use a chisel to move it far enough to grab it with a bearing puller.
   c. Remove the snapring and press the bearing out of the knuckle.

To install:

9. Carefully inspect all parts for wear or damage, replace as necessary. Always insall a new seal and bearing.
10. To install the new bearing on 1988–89 model:
    a. Press the outer race into the knuckle.
    b. The bearing pre-load must be adjusted. Insert the bearing and spacer and install the special tool 49 B001 727, or equivalant, and torque

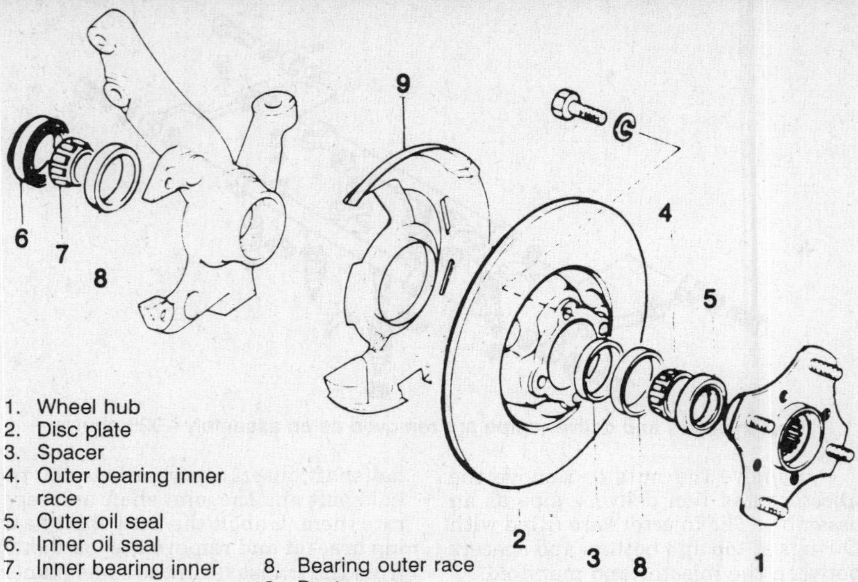

1. Wheel hub
2. Disc plate
3. Spacer
4. Outer bearing inner race
5. Outer oil seal
6. Inner oil seal
7. Inner bearing inner race
8. Bearing outer race
9. Dust cover

Front bearing and hub assembly – 1988–89 323

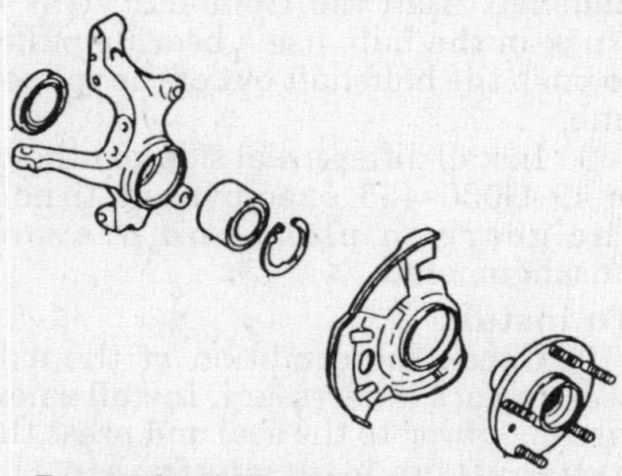

Front wheel bearing and knuckle assembly – 626, MX6, 1990–92 323 and Protege: rear assembly on 323 with 4 wheel drive similar

the nut to 145 ft. lbs. (196 Nm) in 3 steps.
    c. Attach a spring scale to the caliper mounting hole and measure the force required to turn the knuckle on the tool. It should be 0.53–2.55 lbs. (2.4–11.4 Nm) for vehicles with a 13 in. wheel; 0.48–2.35 lbs. (2.2–10.6 Nm) for vehicles with a 14 in. wheel.
    d. To adjust the pre-load, change the spacer as required. Spacers are avaliable in 12 sizes in 0.0015 in. (.04mm) increments. Changing the spacer 1 size increment will change the pre-load by about 1.7–3.5 in. lbs. (0.2–0.4 Nm).
11. To install the new bearing on 1990–92 model:
    a. Press the bearing into the knuckle. Make sure to press only on the outer race.
    b. Install the snapring, support the inner race and press the hub into the bearing. Failure to properly support the bearing races for pressing will ruin the bearing.
12. Install a new seal, fit the knuckle onto the strut and install the bolts. Torque the bolts to 85 ft. lbs. (115 Nm).
13. Fit the lower ball joint into place, install the clamp bolt and torque it to 40 ft. lbs. (58 Nm).
14. Connect the tie rod end, torque the nut to 42 ft. lbs. (57 Nm) and tighten as required to install a new cotter pin.
15. Install the brake disc and caliper and torque the caliper bolts to 36 ft. lbs. (49 Nm).
16. Install the washer and a new axle nut. With all 4 wheels on the ground, torque the nut to 174 ft. lbs. (235 Nm) on 1988–89 models, 235 ft. lbs. (318 Nm) on 1990–92 models. Stake the nut into place.
17. Check and adjust the wheel alignment.

626 and MX6

NOTE: When loosening or tightening axle nut or bolt, make sure the vehicle is on the ground. Axle nut torque is high enough that attempting to loosen it may cause the vehicle to fall off the support.

1. Loosen the axle nut and the lug nuts. Raise the vehicle and support it safely. Remove the tire and wheel.
2. Without disconnecting the hydraulic hose, remove the brake caliper and hang it from the body. Do not let it

hang by the hose. Remove the brake disc. If equipped with ABS, remove the speed sensor.

3. Remove the cotter pin and nut and separate the tie rod end from the steering knuckle.

4. Before removing the nut and locknut to remove the stabilizer bar link from the control arm, measure how much thread protrudes above the locknut. The link must be assembled to the same dimension.

5. Remove the pinch bolt and push the control arm down to separate the ball joint from the knuckle.

6. Remove the 2 bolts and separate the knuckle from the strut.

7. Be careful not to distort the back plate. If damaged or removed, a new one must be pressed onto the knuckle.

8. To remove the hub and bearing:

a. Remove the seal and properly support the assembly by inserting long bolts through the knuckle mounting holes. Lay the bolts on the press table and allow the knuckle to hang.

b. Press the hub out from the back on the knuckle.

c. If the inner bearing race stays on the hub, use a chisel to move it far enough to grab it with a bearing puller.

d. Remove the snapring and press the bearing out of the knuckle.

To install:

9. Carefully inspect all parts for wear or damage, replace as necessary. Always insall a new seal and bearing.

10. To install the new bearing:

a. Press the knuckle onto the new bearing. Make sure to support the bearing only on the outer race.

b. Install the snapring, support the inner race and press the hub into the bearing. Failure to properly support the bearing races for pressing will ruin the bearing.

11. Install a new seal, fit the knuckle onto the strut and install the bolts. Torque the bolts to 85 ft. lbs. (115 Nm).

12. Fit the lower ball joint into place, install the clamp bolt and torque it to 40 ft. lbs. (58 Nm).

13. Connect the tie rod end, torque the nut to 42 ft. lbs. (57 Nm) and tighten as required to install a new cotter pin.

14. Install the brake disc and caliper and torque the caliper bolts to 72 ft. lbs. (98 Nm). Install the speed sensor.

15. Assemble the stabilizer bar link, making sure to properly adjust the length. Tighten the locknuts together.

16. Install the washer and a new axle nut. With all 4 wheels on the ground, torque the nut to 235 ft. lbs. (318 Nm). Stake the nut into place.

17. Check and adjust the wheel alignment.

MANUAL TRANSMISSION

For further information on transmissions/transaxles, please refer to "Chilton's Guide to Transmission Repair".

Transmission Assembly

REMOVAL & INSTALLATION

RX-7

1. Disconnect the negative battery cable. Raise the vehicle and support safely. Drain the transmission.

2. Remove the shifter knob and boot and remove the gearshift lever.

3. Unbolt the clutch release cylinder from the transmission but do not disconnect the hydraulic line. Route the assembly aside.

4. Disconnect the exhaust pipe at the manifold and at the muffler and remove the pipe and the catalytic converter. Remove the heat shield covers.

5. Matchmark the driveshaft flanges and the position of the balance washers. Remove the driveshaft and disconnect the speedometer cable and wiring.

6. Disconnect the electrical wiring and remove the starter. Remove the clutch housing inspection cover.

7. Properly support the transmission with a transmission jack. Disconnect the rear mount and remove the mount crossmember.

8. Remove the transmission case-to-engine bolts. Slide the transmission rearward until the main shaft clears the clutch disc and carefully lower the transmission from the vehicle.

To install:

9. Lightly lubricate the main shaft spline and release bearing fork contact points with molybdenum grease. Install the release bearing.

10. Install the rear mount and crossmember onto the transmission. Torque the mount-to-transmission nuts to 17 ft. lbs. (23 Nm) but do not torque the single center nut yet.

11. Fit the transmission into place, making sure the main shaft fits easily into the clutch disc spline. Push the transmission up against the engine and start all the bolts before tightening any of them. Torque the bolts to 38 ft. lbs. (52 Nm).

12. Bolt the crossmember to the body and torque the bolts to 34 ft. lbs. (46 Nm). Torque the center crossmember mount nut to 59 ft. lbs. (80 Nm).

13. Install the starter and connect the wiring. Install the clutch housing inspection plate.

14. Connect the speedometer cable and wiring and install the exhaust heat shields.

15. Install the driveshaft, making sure to align the matchmarks and balance washers correctly.

16. When installing the exhaust pipes, use new gaskets and torque the flange nuts to 34 ft. lbs. (46 Nm).

17. Install the clutch release cylinder and shift lever and refill the transmission with the correct oil.

Miata

1. Disconnect the negative battery cable. Raise the vehicle and support safely. Drain the transmission.

2. Remove the shifter knob and the center console and remove the gearshift lever.

3. Remove the engine under cover and disconnect the exhaust pipe from the manifold. Remove the entire exhaust system as an assembly.

4. Matchmark the driveshaft flange at the rear and remove the driveshaft.

5. Without disconnecting the hydraulic hose, remove the clutch release cylinder and set it aside. Disconnect the wiring and remove the starter.

6. Disconnect the speedometer cable and remove the wiring from the frame member.

7. Support the transmission with a jack. To remove the frame member, refer to the differential removal and installation procedure.

8. Remove the bolts from the clutch housing and slide the transmission back away from the engine. Lower the transmission from the vehicle.

To install:

9. Lightly lubricate the main shaft spline and the release bearing fork contact points with molybdenum grease and install the fork. Place a wood block on a floor jack and use it to tilt the engine up in front.

10. Carefully guide the transmission into place, making sure the main shaft spline fits properly into the clutch disc. Start all the transmission-to-engine bolts, then torque them to 66 ft. lbs. (89 Nm).

11. Raise the transmission into place and install the transmission-to-differential frame and all the bolts, leaving all the bolts loose. Torque the frame-to-transmission bolts, then the frame-to-differential bolts to 91 ft. lbs. (124 Nm). Install the bracket and torque the bracket-to-transmission bolts to 40 ft. lbs. (54 Nm), the frame bolts to 91 ft. lbs. (124 Nm).

12. Install the driveshaft and torque the nuts to 22 ft. lbs. (30 Nm).

13. Install the starter and connect the speedometer cable and wiring.

14. Install the clutch release cylinder.
15. Use a new gasket and install the exhaust system. Torque the nuts to 34 ft. lbs. (46 Nm). Install the engine under cover.
16. If the transmission was disassembled, pour about 3 oz. of oil into the shift lever mounting hole before installing the lever.

MANUAL TRANSAXLE

For further information on transmissions/transaxles, please refer to "Chilton's Guide to Transmission Repair".

Transaxle Assembly

REMOVAL & INSTALLATION

323 and Protege

WITHOUT 4WD

1. Raise and safely support the vehicle and remove the front wheels. Remove the battery and battery box and the air cleaner and ducting.
2. Remove the splash shield and drain the transaxle oil.
3. Without disconncting the hydraulic hose, remove the clutch release cylinder. Disconnect the speedometer cable and the wiring from the transaxle.
4. Disconnect the wiring and remove the starter.
5. Disconnect the shift linkage and the extension bar.
6. Disconnect the exhaust pipe from the manifold and the catalytic converter and remove the pipe.
7. Disconnect the tie rod ends and lower ball joints and remove the halfhsafts.
8. Install the necessary lifting equipment and support the engine from above. Remove the lower mounting frame and support the transaxle from below with a jack.
9. Remove the front and left rear mounts and allow the engine/transaxle to tilt towards the left.
10. Remove the bolts and slide the transaxle away from the engine to lower it out of the vehicle.

To install:

11. Lightly lubricate the main shaft spline and the release bearing fork contact points with molybdenum grease and install the release bearing.
12. Carefully guide the transaxle into place, making sure the main shaft spline fits properly into the clutch disc. Start all the transaxle-to-engine bolts, then torque them to 66 ft. lbs. (89 Nm).
13. Install the left rear mount but do not torque the bolts yet.
14. Install the halfshafts, making sure the inner joint is firmly seated into place. Torque the extension shaft bracket bolts to 46 ft. lbs. (62 Nm).
15. Assemble the suspension. Torque the lower ball joint pinch bolt and the tie rod end nuts to 43 ft. lbs. (59 Nm). If equipped with a stabilizer bar, adjust the link with ¾ in. (19mm) of thread showing above the locknut.
16. Install the front mount to the transaxle and torque the bolts to 38 ft. lbs. (52 Nm).
17. Install the lower mounting frame and torque the frame-to-body nuts and bolts to 66 ft. lbs. (89 Nm). Torque the mount-to-frame nuts and bolts to 38 ft. lbs. (52 Nm).
18. Connect the shift linkage and control rod.
19. Install the starter and connect the wiring. Install the clutch release cylinder.
20. connect the speedometer cable and transaxle wiring.
21. Install the splash shields and wheels.
22. Install the air cleaner and battery and fill the transaxle with oil.

WITH 4WD

1. Remove the battery and the air cleaner assembly and the ducting. Raise and support the vehicle safely and remove the under cover. Drain the transaxle and transfer case oil and remove the front wheels.
2. Disconnect the speedometer cable in the center. Remove the clutch release cylinder bolt, clip and the clutch release cylinder.
3. Disconnect the wiring from the transaxle and the starter.
4. Disconnect the shift linkage cables from the transaxle by removing the pins and cable retaining clips. Route the cables aside.
5. Matchmark the flanges and remove the driveshaft.
6. Remove the exhaust pipe and catalytic converter and the heat shield.
7. Disconnect the tie rod ends, the lower ball joints and the stabilizer rod links.
8. Remove the halfshafts and the right side extension shaft. Install special tool No. 40 B027 001 or equivalent into the differential to prevent the gears from becomming misaligned inside the case.
9. To remove the differential lock motor, remove the sensor switch, insert a small screwdriver into the hole and turn the rod ½ turn counterclockwise. Remove the bolts and remove the motor.
10. Remove the starter and bracket.
11. Install the necessary lifting equipment and support the engine from above. Remove the crossmember, the front mount and the mount member.
12. Remove the stiffener plate and support the transaxle with a jack.
13. Remove the bolts and separate the engine and transaxle. Carefully lower the unit out of the vehicle.

To install:

14. Lightly lubricate the main shaft spline and the release bearing fork contact points with molybdenum grease and install the release bearing.
15. Carefully guide the transaxle into place, making sure the main shaft spline fits properly into the clutch disc. Start all the transaxle-to-engine bolts, then torque the large ones to 59 ft. lbs. (80 Nm), the smaller ones to 38 ft. lbs. (52 Nm). Install the stiffener plate.
16. Install the engine mount member and all the mount bolts before tightening any of them. Torque the member-to-body bolts to 66 ft. lbs. (89 Nm) and the mount-to-frame nuts and bolts to 38 ft. lbs. (52 Nm). When all the mounts are connected, remove the lifting equipment.
17. Install the center differential lock motor and the starter and connect the wiring.
18. Install the halfshafts, making sure the inner joint is firmly seated into place. Torque the extension shaft bracket bolts to 46 ft. lbs. (62 Nm).
19. Install the exhaust pipe with new gaskets, torque the nuts to 34 ft. lbs. (46 Nm).
20. Install the crossmember, torque the bolts to 86 ft. lbs. (117 Nm).
21. Install the driveshaft, making sure to align the marks. Torque the nuts to 22 ft. lbs. (30 Nm).
23. Assemble the suspension. Torque the lower ball joint pinch bolt and the tie rod end nuts to 43 ft. lbs. (59 Nm). If equipped with a stabilizer bar, adjust the link with ¾ in. (19mm) of thread showing above the locknut.
24. Connect the shift linkage.
25. Connect the speedometer cable and all the wiring and install the clutch release cylinder.
26. Fill the transaxle and transfer case with oil and install the under cover.
27. Install the remaining components and adjust the shift linkage as required.

626 and MX-6

1. Remove the battery and battery carrier.
2. Remove the air ducting and the air cleaner and air flow meter assembly.

3. Unplug the wiring and remove the fuse block.
4. Disconnect the speedometer cable and the transaxle grounds.
5. Raise and support the vehicle safely and remove the front wheels and splash shield. Drain the transaxle oil.
6. Remove the clutch release cylinder and disconnect the tie rod ends.
7. Remove the stabilizer control links. Remove the nuts and bolts from the lower control arm ball joints and pull the lower control arms downward to separate them from the steering knuckles. Be careful not to damage the ball joint dust boots.
8. Insert a small prybar between the left driveshaft and the transaxle case and tap the end of the lever to uncouple the driveshaft from the differential side gear. Pull the front hub forward and separate the driveshaft from the transaxle. Remove the left joint shaft bracket. Separate the right driveshaft and joint shaft in the same manner as the left.

NOTE: Do not insert the lever too deeply between the shaft and the case or the oil seal lip could be damaged. To avoid damage to the oil seal, hold the CV-joint at the differential and pull the driveshaft straight out.

9. Once both drive and joint shafts are removed, install differential side gear holders 49–G030–455 (turbo), 49–G027–003 or their equivalents, in the differential side gears to hold them in place and prevent misalignment.
10. Remove the engine–to–transaxle gusset plates and under cover. Remove the extension bar and the control rod. Remove the manifold bracket and the starter.
11. Suspend the engine from the engine hanger with a suitable lifting device or engine support fixture.
12. Remove the front and left engine mounts and bracket. Disconnect the rubber hanger from the crossmember, then remove the crossmember and left side lower control arm as an assembly.
13. Lean the engine towards the transaxle and support the transaxle with a jack. Remove the transaxle-to-engine bolts and slide the transaxle back and out from under the vehicle.

To install:

14. Lightly lubricate the main shaft spline and the release bearing fork contact points with molybdenum grease and install the release bearing.
15. Carefully guide the transaxle into place, making sure the main shaft spline fits properly into the clutch disc. Start all the transaxle–to–engine bolts, then torque them to 86 ft. lbs. (117 Nm).
16. Install the front mount and torque nuts and bolts to 66 ft. lbs. (89 Nm).
17. Install the left mount crossmember and torque the mount–to–transaxle bolts to 38 ft. lbs. (52 Nm). Torque the crossmember bolts to 40 ft. lbs. (54 Nm), the nuts to 69 ft. lbs. (93 Nm).
18. Install the starter and bracket and torque the bolts to 38 ft. lbs. (52 Nm). Connect the wiring
19. Connect the shift linkage and the extension bar. Install the transaxle–to–engine gusset plates and torque the bolts to 38 ft. lbs. (52 Nm).
20. Install the halfshafts. When assembling the suspension pieces, torque the ball joint pinch bolts and the tie rod ends to 40 ft. lbs. (54 Nm). Install new cotter pins.
21. Assemble the stabilizer bar links and adjust them so that 0.8 in. (20mm) of thread protrudes above the locknut.
22. Install the clutch release cylinder and the splash shields.
23. Connect all wiring and the speedometer cable. Install the air flow meter and air cleaner assembly and the fuse box.
24. Fill the transaxle with the proper amount of oil and install the air ducting and battery.

CLUTCH

Clutch Assembly

REMOVAL & INSTALLATION

1. Disconnect the negative battery cable. Remove the transmission or transaxle.
2. Remove the clutch cover, if equipped. Remove the pressure plate bolts, the pressure plate and clutch disc from the flywheel.
3. Remove the flywheel only if the flywheel surface is damaged or there is trouble in removing the pilot bearing.

On the RX-7, use the flywheel box wrench tool 49–0820–035, or equivalent, to remove the flywheel nut and the flywheel. On all others, remove the bolts and the flywheel.

4. From the clutch housing, unhook the return spring from the throw out bearing and remove the bearing.
5. Remove the bolt holding the release fork and release lever together. Pull the release lever and remove the key and the release fork until the retaining spring frees itself from the ball stud.
6. When installing the flywheel onto a rotary engine, torque the flywheel nut to 289–362 ft. lbs. with no more than a 3 foot extension on the wrench.
7. When installing the pressure plate assembly, tighten the bolts 2 turns at a time to avoid distorting the plate. Torque the bolts to 20 ft. lbs. (26 Nm).

PEDAL HEIGHT ADJUSTMENT

1. Do not remove the floor mat.
2. Loosen the locknut on the stopper bolt or switch.
3. Turn the adjusting bolt until the clearance between the upper surface of the pedal pad and the firewall is within specification.
4. After adjustment, tighten the locknut.
5. Pedal height is as follows:
   RX-7 – 8.46–8.86 in.
   626 and MX-6 – 8.52–8.72 in.
   323 and Protege – 9.02–9.22 in.
   929 – 8.46–8.66 in.
   Miata – 6.89–7.28 in.

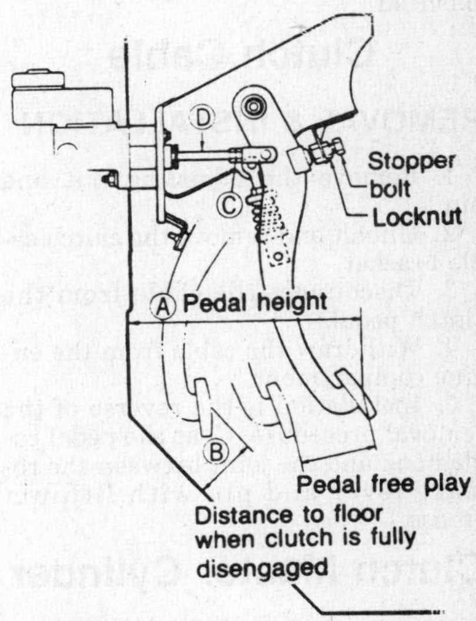

Pedal height adjustment on hydraulic clutch

PEDAL FREE-PLAY ADJUSTMENT

Cable Clutch

323 AND PROTEGE

1. Depress the clutch pedal 7 times and straighten the clutch cable in the cable bracket.
2. At the transaxle, depress the release lever by hand and pull the slack out of the cable. Measure the gap at **A**. Turn the adjusting nut until there is about 0.080 in. (2mm) clearance between the cable pin and the lever.
3. After adjustment, ensure that when the clutch is disengaged, the distance between the floor and the upper

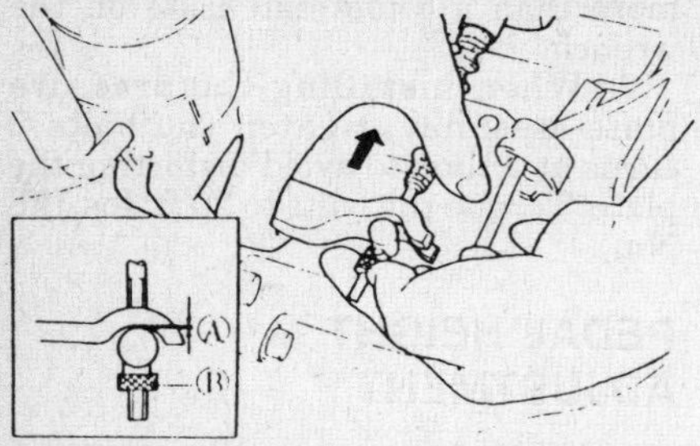

Clutch cable adjustment on 323; measure free play at point A

center of the pedal is about 3.3 in. (85mm) with the floor mat installed.
4. Recheck the pedal height and adjust, if necessary.

Hydraulic Clutch

1. Loosen the locknut on the clutch master cylinder pushrod.
2. Turn the pushrod to obtain 0.02–0.12 in. (0.5–3.0mm) free-play between the pedal and the pushrod.
3. Tighten the locknut on the pushrod.

Clutch Cable

REMOVAL & INSTALLATION

1. Remove the adjusting nut and pin.
2. Unbolt and remove the clutch cable bracket.
3. Disconnect the cable from the clutch pedal.
4. Withdraw the cable from the engine compartment.
5. Installation is the reverse of the removal procedure. Coat the pedal cable hook and the joint between the release lever and pin with lithium grease.

Clutch Master Cylinder

REMOVAL & INSTALLATION

1. If equipped, remove the ABS relay box located forward of the brake power booster on the driver's side. Disconnect the negative battery cable. Disconnect the hydraulic line at the master cylinder using a suitable line wrench.
2. Remove the blower duct, as required.
3. Remove the 2 master cylinder nuts and the clutch master cylinder.
4. Installation is the reverse of the removal procedure. Use new gaskets. Bleed the hydraulic system.

Clutch Slave Cylinder

REMOVAL & INSTALLATION

1. Disconnect the negative battery cable. Remove the air cleaner. Unscrew the hydraulic line at the body mounting bracket.
2. As required, raise and support the vehicle safely to gain access to the clutch slave cylinder bolts.
3. Unhook the release fork return spring from the cylinder.
4. Remove the clutch cylinder bolts and the clutch cylinder.
5. Installation is the reverse of removal. Bleed the hydraulic system.

Hydraulic Clutch System Bleeding

1. Remove the rubber cap from the bleeder screw on the release cylinder.
2. Place a bleeder tube over the end of the bleeder screw.
3. Submerge the other end of the tube in a jar half filled with hydraulic brake fluid.
4. Slowly pump the clutch pedal fully and allow it to return slowly, several times.
5. While pressing the clutch pedal to the floor, loosen the bleeder screw until the fluid starts to run out. Then close the bleeder screw. Keep repeating this Step, while watching the hydraulic fluid in the jar. As soon as the air bubbles disappear, close the bleeder screw.
6. During the bleeding procedure the reservoir must be kept at least ¾ full.

AUTOMATIC TRANSMISSION

For further information on transmissions/transaxles, please refer to "Chilton's Guide to Transmission Repair".

Transmission Assembly

REMOVAL & INSTALLATION

RX-7

1. Disconnect the negative battery cable. Raise the vehicle and support safely.
2. Disconnect the exhaust pipe at the manifold and at the muffler and remove the pipe and the catalytic converter. Remove the heat shield covers.
3. Matchmark the driveshaft flanges and the position of the balance washers. Remove the driveshaft and cap the extension housing to prevent the fluid from leaking from the housing.
4. Disconnect the speedometer cable and wiring.
5. Disconnect the vacuum and oil pipes and plug the ends to prevent leakage.
6. Disconnect the electrical wiring and remove the starter.
7. Disconnect the shift rod from the transmission.
8. Remove the dip stick gauge and filler pipe.
9. Remove the inspection cover at the bottom of the torque converter housing and remove the torque converter–to–flywheel bolts.
10. Properly support the transmission with a transmission jack. Disconnect the rear mount and remove the mount crossmember.
11. Remove the transmission case-to-engine bolts. Slide the transmission rearward and carefully lower it from the vehicle. Do not let the torque converter fall out of the transmission.

To install:

12. Support the torque converter so it does not fall out of the transmission. Install the rear mount and crossmember onto the transmission. Torque the mount–to–transmission nuts to 17 ft. lbs. (23 Nm) but do not torque the single center nut yet.
13. Install the top cover onto the front of the housing and install the vacuum pipe and wiring bracket.
14. Fit the transmission into place and start all the bolts before tightening any of them. Torque the bolts to 38 ft. lbs. (52 Nm).
15. Use new gaskets and connect the oil pipes, torque to 26 ft. lbs. (35 Nm).
16. Bolt the crossmember to the body and torque the bolts to 34 ft. lbs. (46 Nm). Remove the jack and torque the center crossmember mount nut to 59 ft. lbs. (80 Nm).
17. Install the torque converter–to–fly wheel bolts and torque to 25 ft. lbs. (34 Nm). Install the cover.
18. Install the starter and connect all the wiring and vacuum line.
19. Connect the speedometer cable and install the filler tube and dip stick.
20. Install the driveshaft, making sure to align the matchmarks and balance washers correctly.
20. Install the heat shields and the exhaust system. Use new gaskets and torque the flange nuts to 34 ft. lbs. (46 Nm).
21. Connect and adjust the shift rod as required.
22. Fill the transmission with the correct amount of fluid and run the engine to warm the transmission. Watch for leaks.
23. Move the selector through all gears and check for proper fluid level.

929

1. Disconnect the negative battery cable. Remove the transmission oil dipstick.
2. Raise and support the vehicle safely. Disconnect the shift rod.
3. Remove the front exhaust pipe and catalytic converter. Remove the heat shield.
4. Matchmark the flanges and remove the driveshaft. Make sure the spacers on the center bearing are returned to their original position. Cap the extension housing to prevent fluid leakage.
5. Disconnect all the wiring and the speedometer cable.
6. Remove the starter and the inspection cover from the bottom of the torque converter housing.
7. Disconnect the oil pipes and the vacuum pipe. Plug the oil pipe to prevent leakage.
8. Support the engine from above as required. Place a transmission jack under the transmission and remove the crossmember assembly.
9. Remove the torque converter–to–flywheel bolts.
10. Remove the transmission-to-engine bolts and the transmission from the vehicle. Do not let the torque converter fall out.

To install:

11. Properly support the torque converter so it does not fall out. Install the rear mount and crossmember onto the transmission and torque the bolts to 21 ft. lbs. (28 Nm).
12. Fit the transmission into place and start all the bolts before tightening any of them. Torque the transmission–to–engine bolts to 38 ft. lbs. (52 Nm).
13. Install the torque converter–to–flywheel bolts and torque to 25 ft. lbs. (34 Nm). Install the inspection cover.
14. Bolt the crossmember to the body and torque the bolts to 42 ft. lbs. (57 Nm).
15. Connect the oil coolant pipes and make sure the clamps do not interfere with any other parts.
16. Install the starter and connect all wiring and vacuum lines.
17. Install the driveshaft and torque the bolts to 43 ft. lbs. (59 Nm). Make sure the center bearing spacers are properly installed and torque the bolts to 38 ft. lbs. (52 Nm).
18. Use new gaskets and install the exhaust system. Torque the flange nuts to 34 ft. lbs. (46 Nm).
19. Connect the shift rod adn install the dip stick tube.
20. With everything installed, fill the transmission with the proper amount of fluid and run the engine to warm the transmission. Move the selector through all gears and check the fluid level.

AUTOMATIC TRANSAXLE

For further information on transmissions/transaxles, please refer to "Chilton's Guide to Transmission Repair".

Transaxle Assembly

REMOVAL & INSTALLATION

323 and Protege

1. Raise and safely support the vehicle and remove the front wheels. Remove the battery and battery box and the air cleaner and ducting.
2. Remove the splash shield and drain the transaxle oil.
3. Disconnect the speedometer cable, throttle cable, shift cable and the wiring from the transaxle.
4. Disconnect the wiring and remove the starter.
5. On 4WD models, matchmark the flanges and remove the driveshaft.
6. Disconnect the exhaust pipe from the manifold and the catalytic converter and remove the pipe.
7. Disconnect the tie rod ends and lower ball joints and remove the halfhsafts. Use special tool 49 G030 455 or equivalent to hold the differential side gears in place when the halfshafts are removed.
8. On 4WD models, to remove the differential lock motor, remove the sensor switch. Insert a small screwdriver into the hole and turn the rod ½ turn counterclockwise. Remove the bolts and remove the motor.
9. If equipped, remove the torque converter–to–flywheel nuts.
10. Disconnect the oil cooler hoses and plug them to prevent leakage.
11. Install the necessary lifting equipment and support the engine from above. Remove the lower mounting frame and support the transaxle from below with a jack.
12. Remove the front and left rear mounts and allow the engine/transaxle to tilt towards the left.
13. Remove the bolts and slide the transaxle away from the engine to lower it out of the vehicle. Do not let the torque converter fall out.

To install:

14. Make sure the torque converter is properly placed and carefully guide the transaxle into place. Start all the transaxle–to–engine bolts, then torque them to 59 ft. lbs. (80 Nm).
15. Install the left rear mount but do not torque the bolts yet.
16. Install the torque converter–to–flywheel nuts and torque to 25 ft. lbs. (34 Nm).
17. On 4WD models, install the differential lock motor.
18. Install the halfshafts, making sure the inner joint is firmly seated into place. Torque the extension shaft bracket bolts to 46 ft. lbs. (62 Nm).
19. Assemble the suspension. Torque the lower ball joint pinch bolt and the tie rod end nuts to 43 ft. lbs. (59 Nm). If equipped with a stabilizer bar, adjust the link with ¾ in. (19mm) of thread showing above the locknut.
20. Install the front mount to the transaxle and torque the bolts to 38 ft. lbs. (52 Nm).
21. Install the lower mounting frame and torque the frame–to–body nuts and bolts to 66 ft. lbs. (89 Nm). Torque the mount–to–frame nuts and bolts to 38 ft. lbs. (52 Nm).
22. Connect the cooler hoses, making sure the clamp does not interfere with other parts.
23. Connect the shift cable, speedometer cable, throttle cable and the wiring.
24. Install the starter and connect the wiring.
25. On 4WD models, install the driveshaft, making sure to align the matchmarks. Torque the nuts to 22 ft. lbs. (30 Nm).
26. Install the splash shields and wheels.
27. Install the air cleaner and battery.

626 and MX-6

1. Remove the battery and battery carrier.
2. Remove the air ducting and the air cleaner and air flow meter assembly.
3. Unplug the wiring and remove the fuse block.
4. Disconnect the speedometer cable and the transaxle grounds.
5. Raise and support the vehicle safely and remove the front wheels and splash shield. Drain the transaxle fluid.
6. Disconnect the fluid cooler hoses.
7. Remove the stabilizer control links and disconnect the tie rod ends. Remove the nuts and bolts from the lower control arm ball joints and pull the lower control arms downward to separate them from the steering knuckles. Be careful not to damage the ball joint dust boots.
8. Insert a small prybar between the left driveshaft and the transaxle case and tap the end of the lever to uncouple the driveshaft from the differential side gear. Pull the front hub forward and separate the driveshaft from the transaxle. Remove the left joint shaft bracket. Separate the right

driveshaft and joint shaft in the same manner as the left.

NOTE: Do not insert the lever too deeply between the shaft and the case or the oil seal lip could be damaged. To avoid damage to the oil seal, hold the CV-joint at the differential and pull the driveshaft straight out.

9. Once both drive and joint shafts are removed, install differential side gear holders 49–G030–455 (turbo), 49–G027–003 or their equivalents, in the differential side gears to hold them in place and prevent misalignment.
10. Remove the engine–to–transaxle gusset plates and under cover. Remove the extension bar and the control rod. Remove the manifold bracket and the starter.
11. Remove the torque converter–to–flywheel bolts.
12. Suspend the engine from the engine hanger with a suitable lifting device or engine support fixture.
13. Remove the front and left engine mounts and bracket. Disconnect the rubber hanger from the crossmember, then remove the crossmember and left side lower control arm as an assembly.
14. Lean the engine towards the transaxle and support the transaxle with a jack. Remove the transaxle-to-engine bolts and slide the transaxle back and out from under the vehicle. Do not let the torque converter fall out.

To install:

15. Carefully guide the transaxle into place, making sure the torque converter fits properly into place. Start all the transaxle–to–engine bolts, then torque them to 86 ft. lbs. (117 Nm).
16. Install the front mount and torque nuts and bolts to 66 ft. lbs. (89 Nm).
17. Install the left mount crossmember and torque the mount–to–transaxle bolts to 38 ft. lbs. (52 Nm). Torque the crossmember bolts to 40 ft. lbs. (54 Nm), the nuts to 69 ft. lbs. (93 Nm).
18. Install the torque converter–to–flywheel bolts and torque to 45 ft. lbs. (61 Nm). Connect the oil cooler hoses.
19. Install the starter and bracket and torque the bolts to 38 ft. lbs. (52 Nm). Connect the wiring
20. Connect the shift cable. Install the transaxle–to–engine gusset plates and torque the bolts to 38 ft. lbs. (52 Nm).
21. Install the halfshafts. When assembling the suspension pieces, torque the ball joint pinch bolts and the tie rod ends to 40 ft. lbs. (54 Nm). Install new cotter pins.
22. Assemble the stabilizer bar links and adjust them so that 0.8 in. (20mm) of thread protrudes above the locknut. Install the splash shields.
23. Connect all wiring and the speedometer cable. Install the air flow meter and air cleaner assembly and the fuse box.
24. Fill the transaxle with the proper amount of oil and install the air ducting and battery.

FRONT SUSPENSION

Shock Absorbers

REMOVAL & INSTALLATION

Miata

1. Raise adn safely support the vehicle and remove the front wheel. Remove the upper stabilizer bar link bolt.
2. Remove the cotter pin and nut and disconnect the upper ball joint.
3. Remove the upper mount nuts and push the control arm down. Remove the lower mount bolt and lift the shock absorber/spring unit out.
4. Installation is the reverse of removal. torque the lower mount bolt to 69 ft. lbs. (93 Nm), the upper mount nuts to 27 ft. lbs. (36 Nm), the lower ball joint nut to 45 ft. lb. (61 Nm) and install a new cotter pin.

MacPherson Strut

REMOVAL & INSTALLATION

Except Miata

1. Raise and support the vehicle safely. Remove the tire and wheel assembly.
2. Remove the brake line clip or bolt from the strut assembly. If equipped, remove the ABS harness bracket.
3. Unfasten the nuts which secure the upper strut mount to the top of the wheel arch. If equipped, remove the Adjustable Shock Aborbers (ASA) rubber cap, electrical connectors and actuator assembly from its mounting before removing the strut retaining nuts.
4. Unfasten the 2 bolts that secure the lower end of the strut to the steering knuckle arm.
5. Remove the strut assembly from the vehicle.
6. Installation is the reverse of the removal. Torque the upper strut mount nuts to 27 ft. lbs. (36 Nm) on RX7, 30 ft. lbs. (40 Nm) on front wheel drive models, 46 ft. lbs. (63 Nm) on 929. Torque the lower mount bolts to 86 ft. lbs. (117 Nm).

Upper Control Arm

REMOVAL & INSTALLATION

Miata

1. Raise and safely support the vehicle and remove the front wheel.
2. Remove the engine undercover and remove the lower shock absorber mount bolt.
3. Remove the cotter pin and separate the upper ball joint from the steering knuckle.
4. Remove the through bolt and remove the arm.
5. Installation is the reverse of removal. Torque the long bolt to 102 ft. lbs. (138 Nm) and the ball joint nut to 46 ft. lbs (62 Nm). Install a new cotter pin.

Lower Control Arms

REMOVAL & INSTALLATION

323, Protege, 626 and MX-6

1. Raise and support the vehicle safely. Remove the wheels and the splash shield.
2. Before removing stabilizer link through bolt, measure how much of the bolt protrudes beyond the locknut.
3. Remove the control arm inner bushing bolts.
4. Remove the pinch bolt and separate the ball joint from the steering knuckle. Remove the lower control arm.
5. Installation is the reverse of the removal procedure. Torque the bushing bolts with the vehicle resting on the wheels to 78 ft. lbs. (106 Nm). Adjust the stabilizer link through bolt so the correct amount of thread shows above the locknut.

RX-7

1. Raise the vehicle and support it safely. Remove front wheel and splash shield.
2. Disconnect the front stabilizer link from the control arm.
3. Remove the pinch bolt, then separate the lower ball joint from the steering knuckle.
4. Remove the front control arm mounting bolt.
5. Remove the control arm bushing bracket bolts and lower the control arm from the vehicle.
6. Installation is the reverse of removal. Torque the pinch bolt to 37 ft. lbs (50 Nm) and the bushing bolts to 60 ft. lbs. (81 Nm).

929

1. Raise and support the vehicle safely. Remove the tire and wheel assembly.
2. Before removing stabilizer link through bolt, measure how much of the bolt protrudes beyond the locknut.
3. Disconnect the tie rod end from the steering knuckle arm.
4. Remove the compression rod bolts.
5. Remove the inner bushing bolt and the lower control arm from the vehicle.
6. Installation is the reverse of the removal procedure. Torque the bushing bolt to 69 ft. lbs. (93 Nm), the compression rod and basll joint nuts to 91 ft. lbs. (124 Nm) and the tie rod end nut to 55 ft. lbs. (75 Nm).
7. When installing the stabilizer link bolt, adjust the through bolt so the correct amount of thread shows above the locknut.

Stabilizer Bar

REMOVAL & INSTALLATION

1. Raise and safely support the vehicle and remove the front wheels.
2. Before removing stabilizer link-to-control arm through bolt, measure how much of the bolt protrudes beyond the locknut.
3. Before removing the bushings, mark the position of the bushing on the bar.
4. Remove the through bolts and the bushings and remove the bar from the vehicle.
5. Installation is the reverse of removal. On RX7, torque the bushing bolts to 20 ft. lbs. (54 Nm). On all others, torque the through bolt locknut to 27 ft. lbs. (36 Nm) and the bushing bolts to 40 ft. lbs. (54 Nm).

Front Wheel Bearings

REMOVAL & INSTALLATION

RX7

1. This vehicle uses a standard tapered roller bearing and grease seal. Raise and safely support the vehicle and remove the front wheel.
2. Without disconnecting the hydraulic hose, remove the brake caliper and hang it from the body. Remove the disc.
3. Remove the grease cup, cotter pin and set cover. Remove the axle nut and washer and slide the hub and bearings off the spindle.
4. Clean the grease off the bearings. If the bearing shows excess wear or damage, pry out the old seal and use a press or a soft drift pin to remove the inner bearing races from the hub.

To install:

5. Press the bearing races into the hub.
6. Pack both bearings with clean axle grease and install the inner bearing into the hub. Carefully install the new seal and lubricate the lip with a small amount of grease.
7. Install the hub, bearing, thrust washer and nut onto the spindle. To set the bearing pre-load:
   a. Torque the nut to 22 ft. lbs. (29 Nm), then loosen it again.
   b. Use a pull scale hooked to one of the wheel studs and tighten the bearing nut until 2.2 lbs. (9.8 N) is required to turn the hub.
   c. Install the set cover and a new cotter pin and measure the pull again.
8. Install the grease cap. If equipped with ABS, install the speed sensor and measure the gap between the sensor and rotor: it should be 0.016–0.039 in. (0.4–1.0mm).
9. Install the disc and caliper and torque the caliper bolts to 72 ft. lbs. (98 Nm).

929 and Miata

1. Raise and safely support the vehicle and remove the front wheels.
2. Remove the hub cap and have and assistant hold the brake pedal. Carefully loosen and remove the spindle nut. The torque is fairly high, be careful the vehicle does not fall.
3. Remove the brake disc. Without disconnecting the hydraulic hose, remove the brake caliper and hang it from the body. Do not let it hang by the hose. Remove the ABS speed sensor.
4. Remove the spindle nut and slide the hub off the spindle. On Miata, the hub and bearing are a single assembly and must be replaced as a unit if the bearing is faulty.
5. On 929, remove the seal and snapring and use a press to remove the front wheel bearing from the hub. Once removed, the bearing must be replaced.

To install:

6. On 929, press a new bearing into the hub and install a new seal. Be sure to press only on the outer race or the bearing will be ruined. Install a new seal.
7. Install the hub with a new axle nut. Do not torque the nut yet.
8. Install the brake disc and caliper. Torque the caliper bolts to 86 ft. lbs. (117 Nm) on 929, or to 51 ft. lbs. (69 Nm) on Miata.
9. Have an assistant hold the brake and torque the axle nut to 130 ft. lbs. (177 Nm) on 929, or to 159 ft. lbs. (216 Nm) on Miata. Install the ABS speed sensor.

INSPECTION

1. Raise and support the vehicle safely. Remove the tire and wheel assembly.
2. Remove the caliper assembly and position it aside.
3. Position a dial indicator gauge against the dust cap. Push and pull the disc in and out, the axial direction.
4. Measure the endplay of the wheel bearing. On 929 and Miata, endplay should not exceed 0.002 in. (0.05mm). On RX7, there should be no bearing endplay.
5. If endplay is greater than specification, the RX7 bearing can be repacked and adjusted. On 929 and Miata, replace the wheel bearing.

REAR SUSPENSION

MacPherson Strut

REMOVAL & INSTALLATION

Except Miata

1. As required, remove the side trim panels from the inside of the trunk or the rear seat and trim.
2. If equipped with Asjustable Shock Absorber (ASA) system, disconnect the wiring and remove the cap. Loosen and remove the top mounting nuts from the strut mounting block assembly.
3. Raise and safely support the vehicle and remove the rear wheels. The suspension will droop when the weight lifts off the wheels.
4. On 1988–89 323 with 2WD, remove the long bolt connecting the trailing arm to the bottom of the strut.
5. Unclip the brake line or wiring retainers as required and unbolt the bottom strut mount. Lift the shock absorber and spring out as an assembly.
6. Installation is the reverse of removal. Torque the following:
   a. 323 and 626 lower strut mount bolts – 86 ft. lbs (117 Nm).
   b. 929 and RX7 lower strut mount bolt – 69 ft. lbs. (93 Nm).
   c. 323 trailing arm bolt – 50 ft. lbs. (68 Nm).
   d. 323, 929 and RX7 upper strut mount nuts – 22 ft. lbs.(29 Nm).
   e. 626 upper strut mount nuts – 46 ft. lbs. (63 Nm).

Miata

1. Raise and safely support the vehicle and remove the rear wheels.
2. On the left side, remove the fuel filler pipe protector panel.
3. Remove the bolt from the lower stabilizer bar connecting link.
4. Remove the upper mount nuts and lower mount bolt and lift the spring and shock absorber out as an assembly.
5. Installation is the reverse of removal. Torque the upper mount nuts to 27 ft. lbs. (36 Nm), the lower mount bolt to 69 ft. lbs. (93 Nm) and the stabilizer link bolt to 40 ft. lbs. (54 Nm).

Rear Control Arms

REMOVAL & INSTALLATION

323, Protege, 626 and MX6

1. Raise and safely support the vehicle and remove the wheels.
2. Before disconnecting the stabilizer bar link, measure the length of the threads protruding above the locknut. Remove the nuts and through bolt to disconnect the stabilizer bar.
3. Remove the nuts and bolts as required and remove the control arms.
4. Installation is the reverse of removal. Make sure to properly adjust the stabilizer bar before tightening the locknuts and check rear wheel alignment.5. Torque the following:
   a. Inner control arm through bolt–70 ft. lbs. (95 Nm).
   b. 323 outer control arm through bolt–55 ft. lbs. (75 Nm).
   c. 626 outer control arm through bolt–86 ft. lbs. (117 Nm).

929

1. Raise and safely support the vehicle and remove the rear wheels.
2. Before disconnecting the stabilizer bar link, measure the length of the threads protruding above the locknut. Disconnect the stabilizer bar from the upper control arm.
3. Remove the cotter pins and nuts and use a ball joint press to disconnect the upper and lower control arms from the hub support.
4. Remove the bolts to remove the front and rear control arms from the sub-frame. The ball joint dust boots can be replaced but the joints themselves are part of the control arm and cannot be replaced separately.
5. Installation is the reverse of removal. Torque the ball joint nuts to 55 ft. lbs. (75 Nm) and tighten as required to install a new cotter pin.
6. Torque the upper arm and lower front arm–to–sub-frame bolts to 55 ft. lbs. (75 Nm). After aligning the rear wheels, torque the lower rear arm eccentric bolt to 86 ft. lbs. (117 Nm).

Miata

1. Raise and safely support the vehicle and remove the rear wheels.
2. Disconnect the stabilizer bar link and the lower shock absorber mounting bolt from the lower arm.
3. Remove the nuts and eccentric bolts and the outer through bolt and remove the lower control arm. The bushings can be replaced separately.

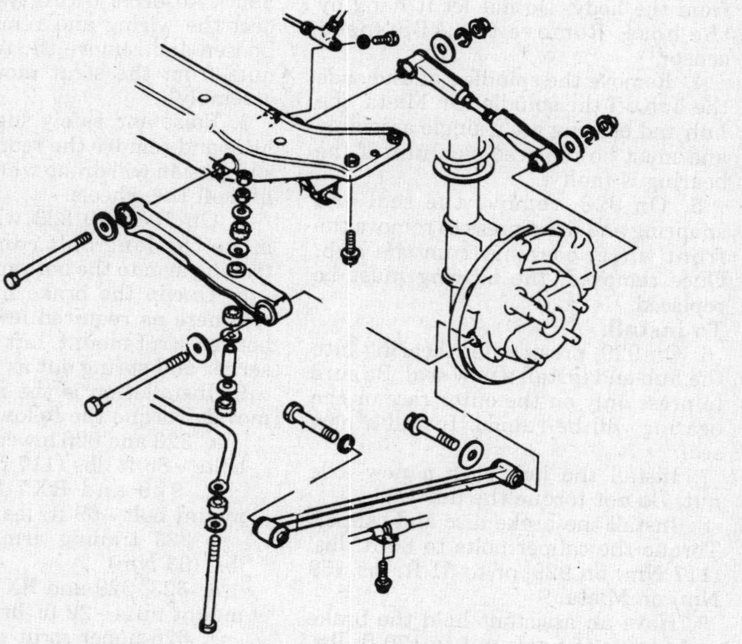

Rear suspension on 626, 323 similar

4. To remove the upper control arm, either re-install the lower control arm or remove the brake caliper, halfshaft and hub carrier. Do not allow the carrier to be supported only by the halfshaft.
5. Installation is the reverse of removal. Torque the upper arm through bolts ot 49 ft. lbs. (67 Nm). Torque the lower arm outer through bolt to 55 ft. lbs. (75 Nm). After aligning the rear wheels, torque the eccentric bolts to 70 ft. lbs. (95 Nm).

Rear Wheel Bearings

REMOVAL & INSTALLATION

323 and Protege with 2WD

1988–89

1. Raise and safely support the vehicle and remove the rear wheels.
2. If equipped with disc brakes, without disconnecting the hydraulic hose, remove the caliper and hang it from the body. Do not let it hang by the hose.
3. Unstake the axle nut and remove the nut. Remove the brake disc or drum with the bearings inside. If the drum is difficult to remove, push up on the parking brake operating lever.
4. Clean and inspect the tappered roller bearings. If they are worn or damaged, remove the inner race from the disc or drum with a drift pin or press.
5. To remove the spindle, remove the back plate. On drum brakes, hang the plate from the spring without disconnecting the hudraulic hose.
6. Loosen the spindle–to–shock absorber bolts, remove the long lower link bolt, then remove the shock absorber bolts and remove the spindle.

To install:

7. Pack the bearings with clean axle grease and install the inner bearing and race with a new seal.
8. Install the spindle and start all the bolts before tightening any of them. Torque the spindle–to–shock absorber bolts, then the lower link bolt to 86 ft. lbs. (117 Nm).
9. Install the brake back plate and torque the bolts to 49 ft. lbs. (67 Nm).
10. Install the caliper if equipped and install the disc or drum with a new axle nut.
11. To adjust the bearing pre-load, torque the nut to 22 ft. lbs. (9.8 Nm), then loosen it again. Hook a pull scale to one of the wheel studs and tighten the nut until it takes 1.9 lbs. (8.5 N) of pull to turn the wheel. Stake the nut in place.

1990–92

1. Raise and safely support the ve-

hicle and remove the rear wheels and the hub cap.

2. Have an assistant hold the brake or set the parking brake and loosen the axle nut.

3. Without disconnecting the hydraulic hose, remove the brake caliper and hang it from the body. Do not let it hang by the hose. Remove the brake disc.

4. Remove the axle nut and remove the hub. The bearing is part of the hub and cannot be replaced separately.

5. To remove the spindle, remove the back plate and remove the 4 suspension bolts.

To install:

6. Install the spindle and start all the bolts. Torque all spindle–to–suspension bolts to 93 ft. lbs. (117 Nm).

7. Install the hub with a new axle nut, do not toque the nut yet.

8. Install the back plate, disc and caliper and torque the caliper bolts to 44 ft. lbs. (60 Nm).

9. Hold the brake and torque the axle nut to 174 ft. lbs. (235 Nm). Stake the nut in place.

626 and MX6

1. Raise and safely support the vehicle and remove the rear wheels.

2. Hold the brake to remove the center axle nut. If equipped with drum brakes, remove the drum.

3. Without disconnecting the hudraulic hose, remove the disc brake caliper and hang it from the body. Do not let it hang by the hose. Slide the disc and hub off the spindle.

4. To remove the spindle, remove the back plate and ABS speed sensor. if equipped.

5. Remove the suspension bolts as required and remove the spindle.

6. The bearing is pressed into the hub and retained with a snapring. Remove the old seal and snapring and press the old bearing out.

To install:

7. Press a new bearing into the hub, install the snapring and install a new seal.

8. Install the spindle to the suspension and start all bolts before tightening any of them. Torque all the spindle–to–suspension bolts to 86 ft. lbs. (117 Nm).

9. Install the back plate and brakes. Torque the caliper bolts to 49 ft. lbs. (67 Nm).

10. Install a new spindle nut and torque it to 130 ft. lbs. (177 Nm). Stake the nut into place.

ADJUSTMENT

1988–89 323 and Protege

1. Raise and support the vehicle safely. Remove the tire and wheel assembly.

2. Remove the dust cap and torque the locknut to 18–21 ft. lbs.

3. Turn the wheel assembly to seat the bearing properly. Loosen the locknut until it can be turned by hand.

4. Hook a spring scale to a wheel lug stud in order to measure the oil seal drag. Pull the spring scale squarely.

5. Measure the oil seal drag when the wheel hub starts to turn and record the measurement.

6. Add the oil seal drag value to the standard bearing preload of 0.6–1.9 lbs. Turn the locknut slowly until the standard bearing preload is obtained.

1990–92 323 and Protege 626 and MX-6

1. Raise and support the vehicle safely. Remove the tire and wheel assembly.

2. Remove and properly support the caliper assembly.

3. Position a dial indicator gauge against the dust cap. Push and pull the disc brake rotor or brake drum in and out in the axial direction and measure the endplay of the wheel bearing.

4. Endplay should be 0.0079 in. Correct by replacing the wheel bearing.

STEERING

Steering Wheel

CAUTION

If equipped with an air bag, the vehicle battery and the system's own back-up battery must be disconnected before removing the steering wheel. Failure to do so may result in deployment of the air bag and possible personal injury.

REMOVAL & INSTALLATION

Without Air Bag

1. Disconnect the negative battery cable. Remove the horn pad button assembly. If equipped with a 4 spoke steering wheel, pull the center cap toward the wheel top.

2. Make matchmarks on the steering wheel and steering shaft. Never strike the steering shaft with a hammer, as damage to the column may result.

3. Remove the wheel using a suitable puller.

4. Installation is the reverse of removal. Torque the steering wheel nut to 36 ft. lbs. (49 Nm).

With Air Bag

1. Disarm the air bag.

2. At the back of the steering wheel hub, remove the nuts that hold the air bag assembly and remove the air bag. Place it in a safe place, pad side up.

3. Matchmark the wheel to the shaft and remove the nut. Use a puller to remove the wheel.

4. When installing, the clockspring must be reset.

a. Make sure the front wheels are straight ahead.

b. Turn the clockspring all the way to the right.

c. Turn the clockspring back about 2 ¾ turns and aligh the marks.

d. Connect the wiring and install the steering wheel.

5. Torque the steering wheel nut to 36 ft. lbs. (49 Nm). Install the air bag unit.

Steering Column

REMOVAL & INSTALLATION

1. Disconnect the negative battery cable and remove the steering wheel.

2. Remove the upper and lower column covers and the combination switch.

3. Remove the dash board panels and air ducts as required.

4. Disconnect the universal joint where the column passes through the floor.

5. Disconnect the automatic transmission interlock cable from the ignition switch and unbolt the column from the vehicle.

6. Installation is the reverse of removal.

Steering Rack

REMOVAL & INSTALLATION

Without 4WD

1. Disconnect the negative battery cable. Raise and safely support the vehicle and remove the front wheels and under cover.

2. On RX7, remove the stabilizer bar. On 929, remove the plate under the steering rack. On 626 with power steering, disconnect the manual transaxle shift linkage and extension rod. On vehicles with 4WD, remove the battery and the battery tray.

3. Disconnect the tie rod ends.

4. Remove the steering column universal joint bolt. Matchmark the joint to the pinion shaft.

5. On vehicles with power steering, disconnect the hydraulic lines and drain the fluid into a container.

6. Remove the bracket nuts or bolts

and remove the steering rack from the vehicle.

To install:

7. When fitting the rack into place, make sure the pinion shaft and universal joint matchmarks are correctly aligned. Fit the shaft into the joint and start the steering rack bracket bolts.
8. On 929, torque the bracket bolts to 61 ft. lbs. (83 Nm). On all others, torque the nuts or bolts to 33 ft. lbs. (44 Nm).
9. Install the universal joint bolt and torque it to 20 ft. lbs (26 Nm). Make sure the joint has at least ¼ in. (6mm) of clearance and can turn freely.
10. Connect the tie rod ends and torque the nuts to 33 ft. lbs. (44 Nm), then tighten as required to install a new cotter pin.
11. Connect the power steering hydraulic lines and refill the system with the correct fluid.

With 4WD

1. Remove the battery and battery tray. Raise and safely support the vehicle and remove the front wheels and front crossmember.
2. Disconnect the tie rod ends and remove the bolt from the steering column universal joint. Matchmark the universal joint to the pinion shaft.
3. Disconnect the hydraulic lines and drain the fluid into a container.
4. To remove the steering rack, the front–to–rear engine mount member must be removed:
   a. Use the proper lifting equipment and support the engine from above.
   b. Disconnect the front and rear engine mounts from the mount member and remove the mount member.
   c. Remove the rear mount from the engine.
5. Remove the exhaust pipe and catalytic converter.
6. Matchmark the flanges and remove the driveshaft.
7. Lower the engine gradually until the lower left rack mounting bolt can be reached. Do not lower too far or the halfshaft joints will be damaged.
8. Remove the mount bolts and move the rack to the left to remove the steering rack.

To install:

9. When fitting the rack into place, make sure the pinion shaft and universal joint matchmarks are correctly aligned. Fit the shaft into the joint and start the steering rack bracket bolts. Torque the nuts and bolts to 38 ft. lbs. (52 Nm).
10. Use new gaskets and install the exhaust pipe. Torque the flange nuts to 34 ft. lbs. (46 Nm).
11. Install the driveshaft and torque the flange nuts to 22 ft. lbs. (30 Nm).
12. Install the engine mount and mount member. Torque the mount member nuts to 66 ft. lbs. (89 Nm), the mount–to–engine nuts to 38 ft. lbs. (52 Nm).
13. Connect the fluid lines. Install the universal joint bolt and torque it to 20 ft. lbs. (27 Nm).
14. Connect the tie rod ends and torque the nuts to 42 ft. lbs. (57 Nm) and tighten as required to install a new cotter pin.
15. Install the front crossmember and torque the bolts to 86 ft. lbs. (117 Nm).
16. Fill the pump with fluid, bleed and test the system.

626 and MX6

1. Remove the battery and battery tray. Raise and safely support the vehicle and remove the front wheels and front crossmember.
2. Disconnect the tie rod ends and remove the bolt from the steering column universal joint. Matchmark the universal joint to the pinion shaft.
3. Disconnect the hydraulic lines and drain the fluid into a container.
4. Remove the exhaust pipe and catalytic converter.
5. Disconnect the front–to–rear steering angle transfer shaft.
6. Remove the left engine mount.
7. Remove the nuts and bolts from the sub frames and allow the frame members to hang down.
8. Remove the stabilizer bar.
9. Remove the mounting bolts and remove the steering rack.

To install:

10. When fitting the rack into place, make sure the pinion shaft and universal joint matchmarks are correctly aligned. Fit the shaft into the joint and start the steering rack bracket bolts. Torque the nuts and bolts to 38 ft. lbs. (52 Nm).
11. Install the stabilizer bar and adjust the links.
12. Attach the sub frames and torque the nuts to 69 ft. lbs. (93 Nm), the bolts to 40 ft. lbs. (54 Nm).
13. Install the engine mount.
14. Connect the front–to–rear steering shaft.
15. Use new gaskets and install the exhaust pipe and catalytic converter.
16. Connect the hydraulic lines and install the front crossmember.
17. Connect the tie rod ends. Torque the nuts to 33 ft. lbs. (44 Nm) and tighten as required to install a new cotter pin.
18. With all parts installed, fill and bleed the system and adjust the rear steering angle transfer shaft. This is escential.

Power Steering Pump

REMOVAL & INSTALLATION

1. Disconnect the negative battery cable. Disconnect and plug the fluid hoses from the pump.
2. Remove all necessary drive belts. Remove the alternator and or the air conditioning compressor, if necessary.
3. Loosen the pump belt adjusting bolt, slide the pump aside and remove the belt. On some vehicles, it may be necessary to remove the pump pulley before removing the pump. On 626, MX-6 and 929 use tool 49-W023-585 or equivalent, to hold the pulley stationary while the pulley lock bolt is removed.
4. Support the pump, remove the mounting bolts and lift out the pump.
5. Installation is the reverse of removal.

BELT ADJUSTMENT

Adjust belt to give approximately a ½ in. (13mm) deflection with 22 lbs. (10 kg) force at the midpoint of its longest stretch.

SYSTEM BLEEDING

1. Check the fluid level. Add fluid, as required.
2. Turn the steering wheel full cycle, in both directions, 5 times with the engine **OFF**.
3. Recheck the fluid level again and add, as required.
4. Start the engine and allow to warm up at idle. Turn the steering wheel full cycle, in both directions, 5 times with the engine running.
5. Top off the fluid reservoir.

Tie Rod Ends

REMOVAL & INSTALLATION

1. Raise and support the vehicle safely.
2. Place alignment marks on the tie rod end, adjusting nut and shaft. Disconnect the tie rod end from the center link and knuckle arm, using the proper removal tool.
3. Remove the tie rod end from the vehicle.
4. Install the tie rod end to the center link and knuckle arm. Be sure to use new cotter pins.

BRAKES

Master Cylinder

REMOVAL & INSTALLATION

NOTE: If equipped with a fluid reservoir located separately from the master cylinder, remove the lines which run between the 2 and plug the lines to prevent leakage. If equipped with Anti-Lock brakes, relieve the system pressure.

1. Disconnect the negative battery cable. Disconnect the fluid level sensor, if equipped.
2. Using a suitable wrench, disconnect and plug the brake fluid lines from the master cylinder; if equipped with ABS, "banjo" type fittings are used. Collect all the excess fluid in a small container.
3. Remove the proportioning bypass valve attaching bolts and valve, if equipped.
4. Remove the master cylinder-to-power brake unit bolts and the master cylinder from the vehicle. Remove the clutch pipe holder, if equipped.
5. Installation is the reverse of removal. Bleed the brake system. If equipped with ABS, use new crush washers on the banjo fittings.

Proportioning Valve

REMOVAL & INSTALLATION

1. If equipped with Anti-Lock brakes, relieve the system pressure. Disconnect the negative battery cable. Disconnect and plug the brake lines at the valve assembly. The 323, 626 and the MX-6 use a dual proportioning valve.
2. Remove the valve bolts and the valve.
3. Installation is the reverse of the removal procedure. Bleed the system. Inspect the fluid lines for leakage.

Power Brake Booster

REMOVAL & INSTALLATION

1. Disconnect the negative battery cable. Remove the blower air duct, if equipped. Disconnect the vacuum hose from the power booster assembly.
2. On some vehicles, it may be possible to remove the power brake booster without disconnecting the brake fluid lines from the master cylinder; if so, remove the cylinder and position it aside. If equipped with Anti-Lock brakes, relieve the system pressure.
3. As required, disconnect the master cylinder fluid lines and remove it from the vehicle.
4. Remove the cotter pin and disconnect the clevis pin from the booster yoke at the brake pedal.
5. Remove the booster nuts and the booster from the vehicle.
6. Installation is the reverse of removal. Use a new gasket. Bleed the system.

Brake Caliper

REMOVAL & INSTALLATION

1. Raise and safely support the vehicle. If equipped with Anti-Lock brakes, relieve the system pressure.
2. Remove the tire and wheel assembly.
3. Disconnect and plug the brake line.
4. Disconnect the brake cable, if equipped.
5. Remove the caliper pin bolts that the sliding caliper moves on and lift the caliper off the carrier.
6. Installation is the reverse of removal. Clean and lightly grease the pin bolts and tighten them to the correct torque or the caliper may not slide properly.

   323 and Protege – 36 ft. lbs. (49 Nm).

   626, MX6 and RX7 – 30 ft. lbs. (41 Nm).

   929 – 17 ft. lbs. (29 Nm).

   Miata – 65 ft. lbs. (88 Nm).

Disc Brake Pads

REMOVAL & INSTALLATION

Front

323, PROTEGE AND RX7 WITH 4 PISTON CALIPER

1. Raise and safely support the vehicle and remove the front wheels.
2. Use large pliers or a C-clamp and squeeze the pads away from the disc.
3. Remove the retaining pin clips and the pins and pull the pads out of the caliper. Note the placement of the shims.
4. Clean the caliper with a brush and press the piston all the way into the bore. Do not use compressed air to clean brake parts.
5. Install the new pads and shims. Install the pins and clips and pump the brake pedal to adjust the brakes.

626, 929 AND MIATA

1. Raise and safely support the vehicle. Remove the front wheels.
2. Remove both caliper pin bolts and pull the caliper off the disc. Tie the caliper up to prevent putting tension and stress on the brake hose.
3. Remove the outer pad by using a suitable prying tool to release the clip. Then, remove the inner pad.
4. Remove about ½ the brake fluid from the reservoir in the master cylinder. Use a C-clamp or special holding tool 49–0221–600C or equivalent with an old pad to depress the caliper piston back into the caliper.
5. Fit the new pads into the carrier and install the caliper. Make sure the caliper pin bolts are clean and lubricated and torque to the proper specification.

Rear

1. Raise and safely support the vehicle and remove the rear wheels.
2. Disconnect the parking brake cable.
3. Remove the caliper pin bolts and lift the caliper off the carrier. Hang the caliper from the spring with wire to avoid hanging it by the hydraulic hose.
4. Remove the pads and shims. Note how the shims are arranged.
5. The caliper piston must be returned to the bottom of its adjustment.
   a. On 323 and Miata, insert an Allen wrench through the adjustment hole in the back of the caliper. Turn the wrench counterclockwise until the piston in fully retracted.
   b. On other vehicles, use special tool 49 FA18 602 or equivalent to screw the piston into the bore.
   c. The 929 uses shoe type parking brakes that are separate from the disc brake system. Press the piston into the bore with a C-clamp and an old pad.
6. Install the new pads with the shims in their proper order.
7. Make sure the caliper pin bolts are clean and lightly greased and install the caliper. Torque the pin bolts to the correct specification and connect the parking brake cable.
8. On 323 and Miata, use the Allen wrench to adjust the piston outward until the pads contact the disc, then turn the wrench counterclockwise about ⅓ turn.
9. Pump the brake pedal to adjust the brakes.

Brake Rotor

REMOVAL & INSTALLATION

1. Raise and safely support the vehicle. Remove the wheel.
2. Remove the caliper.
3. Remove the rotor screws and the rotor from the vehicle. On rear brakes,

it may be necessary to loosen the parking brake.

4. Installation is the reverse of removal.

Brake Drums

REMOVAL & INSTALLATION

1. Raise and safely support the vehicle and remove the wheel.
2. On 626, remove the center axle nut. The torque is quite high, make sure the vehicle is properly supported.
3. The drum should slide off the wheel studs. If necessary, loosen the parking brake adjustment.
4. After installing the drum, adjust the parking brake. On 626, install a new axle nut and torque to 130 ft. lbs. (177 Nm).

Brake Shoes

REMOVAL & INSTALLATION

Except 929

1. Raise the vehicle and support safely. Remove the wheels and the brake drum. Clean the dirt from the brake components with a dry brush.
2. To ease removal of the leading shoe and installation of the return spring later, push up on the quadrant to release the self adjuster.
3. Use brake pliers to remove the return springs. Then, use needle nose pliers to remove the holding pins from the backing plate.
4. Push the bottoms of the shoes outward in order to release them from the anchors and then unhook them at the wheel cylinder. Remove the leading shoe first. Both rear wheels should be done if either side shows excessive wear.

To install:

5. Lightly lubricate with brake grease the shoe and cylinder contact points, shoe anchor points and backing plate projections.
6. Install the shoes to the back plate, then install the springs and adjuster.
7. Install the drum and apply the brakes several times to take up the adjustment before the vehicle is driven.

929

1. The parking brake on this vehicle is a drum brake inside the disc brake rotor. Raise and safely support the vehicle and remove the wheel and caliper.
2. Remove the screws to remove the disc. If necessary, insert a tool through the hole in the disc and back down the adjuster wheel.
3. Disconnect the parking brake cable.
4. Remove the return springs, then the clips and pull the shoes, adjuster and operating lever off as an assembly.
5. Installation is the reverse of removal. Screw the adjuster all the way in and make sure the threads face the front shoe on the left, rear shoe on the right.
6. To adjust, turn the adjuster until the disc will not turn, then back it down so there is no drag.

Wheel Cylinder

REMOVAL & INSTALLATION

1. Raise and support the vehicle safely. Remove the tire and wheel assembly.
2. Remove the brake drum and brake shoes.
3. Disconnect and plug the brake lines.
4. Remove the stud nuts and bolt attaching the wheel cylinder to the backing plate and remove the wheel cylinder.
5. Installation is the reverse of removal. Be sure to bleed the system after installation.

Brake System Bleeding

Without Anti-Lock Brakes

1. Clean all dirt from around the master cylinder reservoir cap.
2. If a bleeder tank is used, follow the manufacturer's instruction.
3. Remove the filler cap and fill the master cylinder reservoir to the lower edge of the filler neck.
4. Clean off the bleeder connections at all of the wheel cylinders and disc brake calipers. Attach a bleeder hose and fixture to the right wheel cylinder bleeder screw and place the end of the hose in a suitable glass jar, submerged in clean brake fluid.
5. Open the bleeder valve ½–¾ of a turn. Have an assistant depress the brake pedal and allow it to return slowly. Continue this pumping action, stopping with each up and down motion, to force any air from the system.
6. When bubbles cease to appear at the end of the bleeder hose, close the bleeder valve and remove the hose. Check the level of the brake fluid in the master cylinder reservoir and add fluid, if necessary.
7. After the bleeding operation at each caliper or wheel cylinder has been completed, fill the master cylinder reservoir and replace the filler cap.

Anti-Lock Brake System Service

PRECAUTION

- Properly relieve the system pressure and disable the system before loosening any hydraulic fitting.
- Before preforming any arc welding on the vehicle, disconnect the Electronic Brake Control Module.
- When placing the vehicle in a paint drying oven, do not expose the Electronic Brake Control Module to temperatures in excess of 185°F (85°C) for longer than 2 hrs.
- Never disconnect or connect Electronic Brake Control Module or its components connectors with the ignition switch **ON**.
- Never disassemble any component of the Anti-Lock Brake System (ABS) which is designated non-serviceable; the component must be replaced as a unit.

RELIEVING ANTI-LOCK BRAKE SYSTEM PRESSURE

CAUTION

The Anti-Lock Brake System pressure must be properly relieved and the system disabled before loosening any hydraulic fitting. The system is capable of generating fluid pressure high enough to penetrate skin, which could be fatal.

1. With the ignition switch **OFF** and the negative battery cable disconnected, pump the brake pedal a minimum of 25 times using approximately 50 lbs. of pedal force. When a noticeable change in pedal feel occurs, the accumulator is discharged.
2. When a definite increase in pedal effort is felt, stroke the pedal a few additional times.
3. If the job requires power while the hydraulic system is open, disconnect the wiring to the ABS hydraulic unit before connecting the battery cable.

Hydraulic Unit

REMOVAL & INSTALLATION

1. Disconnect the negative battery cable and depressurize the Anti-Lock Brake system pressure.
2. Disconnect the wiring from the hydraulic unit.
3. Disconnect and plug the brake lines and hoses from the hydraulic unit. Cap the openings.
4. Remove the hydraulic unit attaching bolts and the hydraulic unit from the vehicle.

5. Installation is the reverse of removal. Remove the caps only when ready to connect the hydraulic fittings. Bleed the system.

CHASSIS ELECTRICAL

Air Bag

PRECAUTIONS

- An air bag is an explosive device. Handle with extreme caution.
- Always disconnect the battery the air bag connector before removing the steering wheel or beginning work on the air bag system.
- Air bag components must not be repaired or opened. Always use new parts, including the wiring harness.
- Always place a removed air bag unit with the horn pad facing up. Put it in a safe place where it will not be disturbed.
- The air bag unit must not be exposed to grease, fluids, or cleaning agents.
- The air bag unit must not be exposed to temperatures above 194°F (90°C) at any time. Even the heat of a soldering iron can damage or ignite the charge.
- Storage and transport of air bags is subject to rules governing explosive devices and should be done only in the original package.
- Failure to follow proper safety precautions may result in personal injury through accidental firing of the air bag, or through failure of the air bag in an accident.

DISARMING

1. Disconnect the negative battery cable.
2. On RX7, remove the panel under the left side of the dashboard.
3. On Miata, remove the bottom steering column cover.
4. Locate and unplug the blue and orange connector. Protect this connector from accidental electrical contact, including static electricity.
5. The air bag is now disarmed and the battery can be connected to perform electrical testing on other systems.
6. When ready to reconnect the air bag, disconnect the battery, connect the air bag connector, install the cover or panel and make sure no one is in the vehicle when connecting the battery.

Heater Blower Motor

REMOVAL & INSTALLATION

1. Disconnect the negative battery cable.
2. Remove the dash undercover which is located on the passenger side of the vehicle, if equipped. Remove the glove box. Disconnect the multi-connector to the blower motor.
3. Remove the stay of the steel plate provided in the upper part of the glove box. Remove the right side defroster hose for clearance, if necessary.
4. Remove the air duct in between the blower unit and the heater unit. If equipped with sliding heater controls, move the control to the **HOT** position and disconnect the control wire, if necessary.
5. On 323 and Protege, remove the mounting screws and remove the air selector wire harness connector. Remove the screws and the blower motor.
6. Installation is the reverse of removal.

Windshield Wiper Motor

REMOVAL & INSTALLATION

1. Disconnect the negative battery cable. Remove the wiper arms.
2. Remove the cowl plate screws, move the cowl plate up at the front and disconnect the washer hose. Remove the cowl plate.
3. Disconnect the wires from the wiper motor.
4. Unbolt and remove the motor.
5. Installation is the reverse of removal. Check the system for proper operation.

Instrument Cluster

REMOVAL & INSTALLATION

Except RX7 and Miata

1. Disconnect the negative battery cable.
2. Remove the screws and remove the gauge assembly hood. Disconnect the switches as required.
3. Remove the screws and pull the gauge assembly out until the electrical connectors are visible. Disconnect the speedometer cable and all the connectors from the rear of the gauge panel and remove it.
4. Installation is the reverse of removal.

RX7

1. Disconnect the negative battery cable and the air bag connector.
2. Remove the steering wheel.
3. Remove the 3 screws in front of the cluster glass and the 2 screws under the cluster switch panel and pull the switch panel out far enough to disconnect the switches.
4. Remove the 4 instrument cluster screws and pull the cluster out far enough to disconnect the wiring and speedometer cable.
5. Installation is the reverse of removal. Torque the steering wheel nut to 36 ft. lbs. (49 Nm).

Miata

1. Disconnect the negative battery cable and the air bag connector.
2. Remove the screws from the top inside and the bottom outside of the gauge hood and pull the hood off the clips.
3. Remove the screws and pull the gauge assembly out far enough to disconnect the speedometer cable and wiring.
4. Installation is the reverse of removal.

Radio

REMOVAL & INSTALLATION

323 and Protege

1. If the vehicle is equipped with a theft protected radio, obtain the reset code. Disconnect the negative battery cable.
2. On 1988–89 model, remove the ash tray and remove the 2 screws at the bottom.
3. Carefully pry the panel trim out of the dashboard. Disconnect the lighter wire, if equipped.
4. Remove the screws and pull the radio out of the dashboard far enough to disconnect the wiring.
5. Installation is the reverse of removal.

626 and MX6

1. If the vehicle is equipped with a theft protected radio, obtain the reset code. Disconnect the negative battery cable.
2. Remove the ash tray and the storage box under it and remove the screws to remove the face panel.
3. Remove the screws and pull the radio out far enough to disconnect the wiring.
4. Installation is the reverse of removal.

929

1. If the vehicle is equipped with a theft protected radio, obtain the reset code. Disconnect the negative battery cable.
2. On the trim panel above the ra-

dio, pull out the center plug on the push clips and remove the clips. Remove the plastic nut and remove the panel.

3. Reach behind the radio and remove the 2 nuts.
4. Pull the radio out far enough to disconnect the wiring.
5. Installation is the reverse of removal.

RX7

1. If the vehicle is equipped with a theft protected radio, obtain the reset code. Disconnect the negative battery cable.
2. Remove the cover around the shifter boot.
3. Carefully pry out the center dashboard vent.
4. Remove the ash tray and remove the screws to remove the center console facing. Disconnect the lighter wiring.
5. Remove the screws to remove the radio. The headrest speaker amplifier is next to the speaker behind the driver's seat. The woofer amplifier is behind the glove compartment.

Miata

1. If the vehicle is equipped with a theft protected radio, obtain the reset code. Disconnect the negative battery cable.
2. Remove the ash tray and the shifter knob and boot.
3. Remove the center console.
4. Pass a piece of string through the center dashboard vent, hold onto the panel and pull the vents out of the dashboard.
5. Remove the screws to remove the console face and remove the screws to remove the radio.

Concealed Headlights

MANUAL OPERATION

RX-7 and Miata

Next to each headlight is the retractor motor that automatically raises and lower the light. If the motor fails, the light can be raised or lowered manually by turning the knob on the top of each motor. If the light is positioned up or down manually but moves to the wrong position when the lights or ignition switch are turned **ON**, unplug the motor connector and leave the lights up until repairs can be made.

Combination Switch

REMOVAL & INSTALLATION

Without Air Bag

1. Disconnect the negative battery cable.
2. Remove the steering wheel.
3. Remove the steering column covers.
4. Disconnect the electrical connectors.
5. Remove the stop ring from the shaft.
6. Remove the switch retaining screws. Remove the combination switch from its mounting.
7. Installation is the reverse of the removal procedure. On all vehicles except 323, once the combination switch is in place and the covers are installed, set the front wheels straight and align the combination switch and steering angle sensor marks.

With Air Bag

1. Disconnect the negative battery cable and disarm the air bag system.
2. Remove the steering wheel and the column covers.
3. Disconnect the wiring and remove the screws to remove the combination switch.
4. Installation of the switch is the reverse of removal. To install the steering wheel, the clockspring must be reset.
   a. Make sure the front wheels are straight ahead.
   b. Turn the clockspring all the way to the right. Don't force it.
   c. Turn the clockspring back about 2 ¾ turns and aligh the marks.
   d. Connect the wiring and install the steering wheel.

Ignition Lock/Switch

REMOVAL & INSTALLATION

Except RX-7

1. Disconnect the negative battery cable.
2. Remove the steering wheel.
3. Remove the steering column covers.
4. Disconnect the electrical connectors.
5. Remove the stop ring from the shaft.
6. Remove the switch screws and the combination switch.
7. Remove the instrument frame brace. Disconnect the switch wires.
8. Use a chisel to make slots in the lock screws. Remove the screws.
9. Installation is the reverse of the removal procedure. Install the new screws until the head twists off. Make sure the lock operates properly while tightening the new locking screws.

RX-7

1. Disconnect the negative battery cable.
2. Remove the steering wheel.
3. Remove the steering column covers.
4. Remove the air duct and disconnect the combination switch.
5. Remove the combination switch assembly.
6. Place a protector under the steering lock assembly to protect the steering shaft from the shock of the hammer blows.
7. Using a chisel, make grooves in the heads of the lock installation screws and remove the screws from the column jacket.
8. To install, use new screws and tighten the switch screws, until their heads break off. Make sure the lock operates properly while tightening the new locking screws.
9. To complete the installation, reverse the removal procedure.

Stoplight Switch

REMOVAL & INSTALLATION

1. Disconnect the negative battery cable.
2. Disconnect the stoplight switch wire connector from the switch.
3. Remove the switch retainer and outer washer from the pedal pin. Slide the stoplight switch out of the brake pedal bracket and remove the switch.
4. Installation is the reverse of removal.

Clutch Switch

REMOVAL & INSTALLATION

1. Disconnect the negative battery cable.
2. Disconnect the switch wiring.
3. Loosen the locknut and remove the switch.
4. Installation is the reverse of removal. Adjust the switch.
5. Adjust the cruise control switch to set maximum pedal height to be even with the brake pedal.

Fuses, Circuit Breakers and Relays

LOCATION

Fuse Panel

MIATA

The main fuse block is located on the right side of engine compartment. A second fuse block is located to the left of the steering column, just above the clutch pedal.

RX-7

The main fuse block is located in the

engine compartment, near the radiator support. The fuse panel and joint box are located behind the drivers side left kick panel.

323 AND PROTEGE

The fuse blocks are located in the engine compartment, near the battery.

626 AND MX-6

The main fuse block is located in the engine compartment, near the battery.

929

The main fuse block is located in the engine compartment, near the battery. The fuse panel is located on the drivers side left kick panel, near the brake pedal.

Relays

RX-7

Cooling Fan Relay—located in front of engine near radiator.

Power Antenna Relay—located in rear of vehicle, near antenna.

Dimmer Relay—located in front of engine near radiator.

323 and Protege

Horn Relay—located in engine compartment, on left fender.

Headlight Relay—located in engine compartment, on left fender.

Windshield Wiper Relay—located in engine compartment, on left fender.

929

Various relays can be found in the relay box, located in the engine compartment near the battery.

MIATA

Horn Relay—located behind the instrument panel, on the driver's side.

Cooling Fan Relay—located in the engine compartment, on the right fender.

Headlight Relay—located in the engine compartment, on the left fender.

626 AND MX-6

Various relays can be found in the relay boxes, located in the engine compartment on the firewall and under the instrument panel on the drivers side.

Computers

LOCATION

RX-7

Central Processing Unit (CPU)—located behind the drivers side kick panel, near the fuse block.

Anti-Lock Brake Control Unit—located under the right hand side of the instrument panel.

323 and Protege

Cruise Control Unit—located under the left side of the instrument, near the instrument panel relay box.

Engine RPM Control Unit—located in the center of the instrument panel, near the EGI control unit.

626 AND MX-6

Anti-Lock Brake Control Unit—located under the driver's front seat.

EGI Electronic Control Unit—located below the center of the instrument panel.

Electronic Control Unit (ECU)—located in the engine compartment, on the right fender.

929

Anti-Lock Brake Control Unit—located behind the passenger's side front kick panel.

Engine Control Unit—located in the engine compartment, on the right fender.

Miata

Cruise Control Unit—located behind the driver's side kick panel, near the instrument panel.

Flashers

LOCATION

RX-7

The turn signal/hazard flasher unit is located in the left kick panel, on the bottom of the CPU.

323 and Protege

The turn signal/hazard flasher unit is located on the instrument panel relay box.

626 and MX-6

The turn signal/hazard flasher unit is located below the left side side of the instrument panel.

929

The turn signal/hazard flasher unit is located in the relay box, under the instrument panel on the steering column.

Miata

The turn signal/hazard flasher unit is located behind the center of the instrument panel.

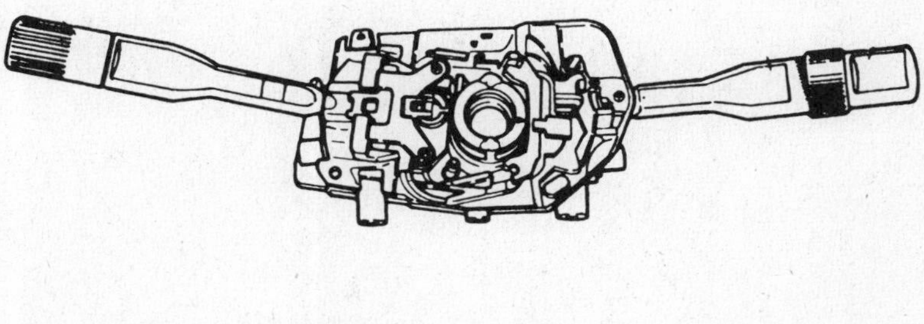

Typical combination switch mounting 626

USA WITHOUT CRUISE CONTROL SYSTEM

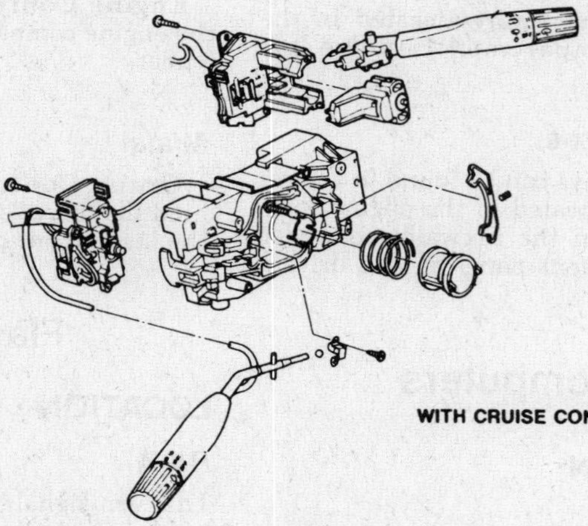

WITH CRUISE CONTROL SYSTEM

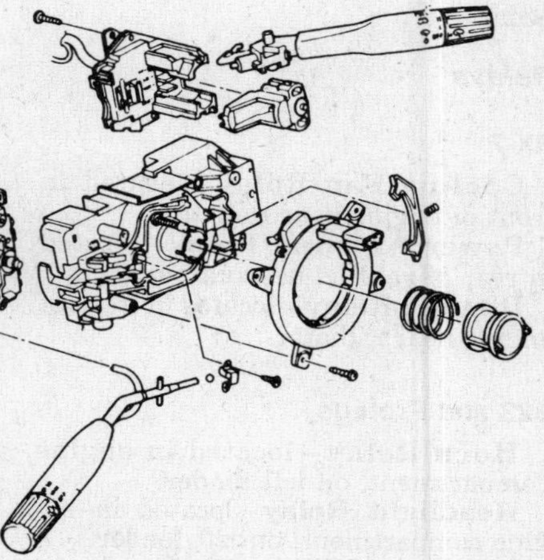

Typical combination switch mounting-323

13

Mercedes-Benz

190, 260, 300, 420, 560—All Models

SERIAL NUMBER IDENTIFICATION

Vehicle Identification Plate

The Vehicle Identification Plate is located in the left window post and consists of a 17 digit number.

Model Designation

Mercedes-Benz uses model numbers instead of names. Throughout this section, model names such as 107, 124, 126, 129 and 201 will appear.

Model 107 – 560SL
Model 124 – 260E, 300D, 300E, 300CE, 300TE, 500E
Model 126 – 300SE, 300SEL, 350SD, 350SDL, 420SEL, 560SEC, 560SEL
Model 129 – 300SL, 500SL
Model 201 – 190D, 190E

C–coupe
D–diesel
E–fuel injected
L–long
S–sports model
SL–super light (sports car)
T–station wagon

Engine Number

The Engine Number is located in rear of the engine block and consists of a 10 digit number.

Transmission Number

Mercedes-Benz vehicles for the U.S. market have been equipped with either a 5 speed manual transmission or with a fully automatic 4 speed unit.

Serial numbers on the manual transmission are located on a pad on the side cover of the transmission (left side).

Automatic transmission serial numbers are located on a metal plate which is attached to the driver's side of the transmission.

Example model 300 E: WDB E A 30 D X L B 123456

WDB E A 30 D X L B 123456

Manufacturer

Model

C = 126, D = 201, E = 124, F = 129

Attribute code

A = Gasoline, B = Diesel, D = 4MATIC

Model designation

124.030

Restraint system

D = Seat belts + SRS with driver airbag
E = Seat belts + SRS with driver and front passenger airbag

Check digit

Model year

L = 1990, M = 1991

Manufacturing plant

A – E = Sindelfingen
F – H = Bremen

Chassis end number

Example of the Vehicle Identification Number

SPECIFICATIONS

ENGINE IDENTIFICATION

Year	Model	Engine Displacement cu. in. (cc/liter)	Engine Series Identification	No. of Cylinders	Engine Type
1988	190E	140 (2299/2.3)	M102	4	102.985
	190E	159 (2599/2.6)	M103	6	103.942
	190D	152 (2497/2.5)	OM602	5	602.911
	260E	159 (2599/2.6)	M103	6	103.940
	300E	181 (2962/3.0)	M103	6	103.983
	300CE	181 (2962/3.0)	M103	6	103.983
	300TE	181 (2962/3.0)	M103	6	103.983
	300SE	181 (2962/3.0)	M103	6	103.981
	300SEL	181 (2962/3.0)	M103	6	103.981
	420SEL	256 (4196/4.2)	M116	8	116.965
	560SEL	338 (5547/5.6)	M117	8	117.967
	560SEC	338 (5547/5.6)	M117	8	117.968
	560SL	338 (5547/5.6)	M117	8	117.968
1989	190E	159 (2599/2.6)	M103	6	103.942
	190D	152 (2497/2.5)	OM602	5	602.911
	260E	159 (2599/2.6)	M103	6	103.940
	300E	181 (2962/3.0)	M103	6	103.983
	300CE	181 (2962/3.0)	M103	6	103.983
	300TE	181 (2962/3.0)	M103	6	103.983
	300SE	181 (2962/3.0)	M103	6	103.981
	300SEL	181 (2962/3.0)	M103	6	103.981
	420SEL	256 (4196/4.2)	M116	8	116.965
	560SEL	338 (5547/5.6)	M117	8	117.967
	560SEC	338 (5547/5.6)	M117	8	117.968
	560SL	338 (5547/5.6)	M117	8	117.968
1990	190E	159 (2599/2.6)	M103	6	103.942
	300E	181 (2962/3.0)	M103	6	103.983
	300CE	181 (2962/3.0)	M103	6	104.980
	300SE	181 (2962/3.0)	M103	6	103.981
	300SL	181 (2962/3.0)	M104	6	104.981
	300TE	181 (2962/3.0)	M103	6	103.983
	300SEL	181 (2962/3.0)	M103	6	103.981
	420SEL	256 (4196/4.2)	M116	8	116.965
	500SL	304 (4973/5.0)	M119	8	119.960
	560SEC	338 (5547/5.6)	M117	8	117.968
	560SEL	338 (5547/5.6)	M117	8	117.968

ENGINE IDENTIFICATION

Year	Model	Engine Displacement cu. in. (cc/liter)	Engine Series Identification	No. of Cylinders	Engine Type
1991-92	190E	140 (2298/2.3)	M102	4	102.985
	190E	158 (2597/2.6)	M103	6	103.942
	300D	152 (2497/2.5)	OM602	5	602.962
	300E	158 (2597/2.6)	M103	6	103.940
	300E	180 (2960/3.0)	M103	6	103.983
	300CE	180 (2960/3.0)	M104	6	104.980
	300TE	180 (2960/3.0)	M103	6	103.983
	300SE	180 (2960/3.0)	M103	6	103.981
	300SEL	180 (2960/3.0)	M103	6	103.981
	300SL	180 (2960/3.0)	M104	6	104.981
	350SD	210 (3449/3.5)	OM602	6	603.970
	350SDL	210 (3449/3.5)	OM602	6	603.970
	420SEL	256 (4196/4.2)	M116	8	116.965
	500E	303 (4972/5.0)	M119	8	119.960
	500SL	303 (4972/5.0)	M119	8	119.960
	560SEC	338 (5547/5.6)	M117	8	117.968
	560SEL	338 (5547/5.6)	M117	8	117.968

GENERAL ENGINE SPECIFICATIONS

Year	Model	Engine Displacement cu. in. (cc)	Fuel System Type	Net Horsepower @ rpm	Net Torque @ rpm (ft. lbs.)	Bore × Stroke (in.)	Com-pression Ratio	Oil Pressure @ 2000 rpm
1988	190D	152 (2497)	DFI	93 @ 4600	122 @ 2400	3.43 × 3.31	22:1	55
	190E	140 (2299)	CIS-E	130 @ 5100	146 @ 3500	3.76 × 3.16	9.0:1	55
	190E	159 (2599)	CIS-E	158 @ 5800	162 @ 4600	3.26 × 3.16	9.2:1	55
	260E	159 (2599)	CIS-E	158 @ 5800	162 @ 4600	3.26 × 3.16	9.2:1	55
	300E	181 (2962)	CIS-E	177 @ 5700	188 @ 4400	3.48 × 3.16	9.2:1	55
	300CE	181 (2962)	CIS-E	177 @ 5700	188 @ 4400	3.48 × 3.16	9.2:1	55
	300SE	181 (2962)	CIS-E	177 @ 5700	188 @ 4400	3.48 × 3.16	9.2:1	55
	300SEL	181 (2962)	CIS-E	177 @ 5700	188 @ 4400	3.48 × 3.16	9.2:1	55
	300TE	181 (2962)	CIS-E	177 @ 5700	188 @ 4400	3.48 × 3.16	9.2:1	55
	420SEL	256 (4196)	CIS-E	201 @ 5200	228 @ 3600	3.62 × 3.11	9.0:1	55
	560SEC	338 (5547)	CIS-E	238 @ 4800	287 @ 3500	3.80 × 3.73	9.0:1	55
	560SEL	338 (5547)	CIS-E	238 @ 4800	287 @ 3500	3.80 × 3.73	9.0:1	55
	560SL	338 (5547)	CIS-E	227 @ 4750	279 @ 3250	3.80 × 3.73	9.0:1	55
1989	190D	152 (2497)	DFI	93 @ 4600	122 @ 2400	3.43 × 3.31	22:1	55
	190E	159 (2599)	CIS-E	158 @ 5800	162 @ 4600	3.26 × 3.16	9.2:1	55
	260E	159 (2599)	CIS-E	158 @ 5800	162 @ 4600	3.26 × 3.16	9.2:1	55
	300E	181 (2962)	CIS-E	177 @ 5700	188 @ 4400	3.48 × 3.16	9.2:1	55
	300CE	181 (2962)	CIS-E	177 @ 5700	188 @ 4400	3.48 × 3.16	9.2:1	55
	300SE	181 (2962)	CIS-E	177 @ 5700	188 @ 4400	3.48 × 3.16	9.2:1	55
	300SEL	181 (2962)	CIS-E	177 @ 5700	188 @ 4400	3.48 × 3.16	9.2:1	55

GENERAL ENGINE SPECIFICATIONS

Year	Model	Engine Displacement cu. in. (cc)	Fuel System Type	Net Horsepower @ rpm	Net Torque @ rpm (ft. lbs.)	Bore × Stroke (in.)	Com-pression Ratio	Oil Pressure @ 2000 rpm
1989	300TE	181 (2962)	CIS-E	177 @ 5700	188 @ 4400	3.48 × 3.16	9.2:1	55
	420SEL	256 (4196)	CIS-E	201 @ 5200	228 @ 3600	3.62 × 3.11	9.0:1	55
	560SEC	338 (5547)	CIS-E	238 @ 4800	287 @ 3500	3.80 × 3.73	9.0:1	55
	560SEL	338 (5547)	CIS-E	238 @ 4800	287 @ 3500	3.80 × 3.73	9.0:1	55
	560SL	338 (5547)	CIS-E	227 @ 4750	279 @ 3250	3.80 × 3.73	9.0:1	55
1990	190E	159 (2599)	CIS-E	158 @ 5800	162 @ 4600	3.26 × 3.16	9.2:1	55
	300E	181 (2962)	CIS-E	177 @ 5700	188 @ 4400	3.48 × 3.16	9.2:1	55
	300CE	181 (2962)	CIS-E	177 @ 5700	188 @ 4400	3.48 × 3.16	9.2:1	55
	300SE	181 (2962)	CIS-E	177 @ 5700	188 @ 4400	3.48 × 3.16	9.2:1	55
	300SL	181 (2962)	CIS-E	228 @ 6300	201 @ 4600	3.48 × 3.16	10.0:1	55
	300TE	181 (2962)	CIS-E	177 @ 5700	188 @ 4400	3.48 × 3.16	9.2:1	55
	300SEL	181 (2962)	CIS-E	177 @ 5700	188 @ 4400	3.48 × 3.16	9.2:1	55
	420SEL	256 (4196)	CIS-E	201 @ 5200	228 @ 3600	3.62 × 3.11	9.0:1	55
	500SL	304 (4973)	CIS-E	322 @ 5500	332 @ 4000	3.80 × 3.35	10.0:1	55
	560SEC	338 (5547)	CIS-E	238 @ 4800	287 @ 3500	3.80 × 3.73	9.0:1	55
	560SEL	338 (5547)	CIS-E	238 @ 4800	287 @ 3500	3.80 × 3.73	9.0:1	55
1991-92	190E	140 (2298)	CIS-E	130 @ 5100	146 @ 3500	3.76 × 3.16	9.0:1	55
	190E	158 (2597)	CIS-E	158 @ 5800	162 @ 4600	3.26 × 3.16	9.2:1	55
	300D	152 (2497)	DFI	121 @ 4600	165 @ 2400	3.43 × 3.31	22.0:1	55
	300E	158 (2597)	CIS-E	158 @ 5800	162 @ 4600	3.26 × 3.16	9.2:1	55
	300E	180 (2960)	CIS-E	177 @ 5700	188 @ 4400	3.48 × 3.16	9.2:1	55
	300CE	180 (2960)	CIS-E	217 @ 6400	195 @ 4600	3.48 × 3.16	10.0:1	55
	300TE	180 (2960)	CIS-E	177 @ 5700	188 @ 4400	3.48 × 3.16	9.2:1	55
	300SE	180 (2960)	CIS-E	177 @ 5700	188 @ 4400	3.48 × 3.16	9.2:1	55
	300SEL	180 (2960)	CIS-E	177 @ 5700	188 @ 4400	3.48 × 3.16	9.2:1	55
	300SL	180 (2960)	CIS-E	228 @ 6300	201 @ 4600	3.48 × 3.16	10.0:1	55
	350SD	210 (3499)	DFI	134 @ 4000	229 @ 2000	3.50 × 3.60	22.0:1	55
	350SDL	210 (3499)	DFI	134 @ 4000	229 @ 2000	3.50 × 3.60	22.0:1	55
	420SEL	256 (4196)	CIS-E	201 @ 5200	228 @ 3600	3.62 × 3.10	9.0:1	55
	500E	303 (4972)	CIS-E	322 @ 5700	355 @ 3900	3.80 × 3.35	10.0:1	55
	500SL	303 (4972)	CIS-E	322 @ 5700	332 @ 4000	3.80 × 3.35	10.0:1	55
	560SEC	338 (5547)	CIS-E	238 @ 4800	287 @ 3500	3.80 × 3.73	9.0:1	55
	560SEL	338 (5547)	CIS-E	238 @ 4800	287 @ 3500	3.80 × 3.73	9.0:1	55

CIS-E—Electronic Continuous Injection System
DFI—Diesel Fuel Injection

GASOLINE ENGINE TUNE-UP SPECIFICATIONS

Year	Model	Engine Displacement cu. in. (cc)	Spark Plugs Type	Spark Plugs Gap (in.)	Ignition Timing (deg.) MT	Ignition Timing (deg.) AT	Compression Pressure (psi)	Fuel Pump (psi)	Idle Speed (rpm) MT	Idle Speed (rpm) AT	Valve Clearance In.	Valve Clearance Ex.
1988	190E	140 (2299)	S9YC	.032	10B	10B	125	29–58	750	750	Hyd.	Hyd.
	190E	159 (2599)	S12YC	.032	9B	9B	125	91–94	700	700	Hyd.	Hyd.
	260E	159 (2599)	S12YC	.032	9B	9B	125	91–94	700	700	Hyd.	Hyd.
	300E	181 (2962)	S9YC	.032	TDC	TDC	125	91–94	650	650	Hyd.	Hyd.
	300CE	181 (2962)	S9YC	.032	—	TDC	125	91–94	—	650	Hyd.	Hyd.
	300SE	181 (2962)	S9YC	.032	—	TDC	125	91–94	—	650	Hyd.	Hyd.
	300SEL	181 (2962)	S9YC	.032	—	TDC	125	91–94	—	650	Hyd.	Hyd.
	300TE	181 (2962)	S9YC	.032	—	TDC	125	91–94	—	650	Hyd.	Hyd.
	420SEL	256 (4196)	N9YC	.032	—	TDC	125	91–94	—	650	Hyd.	Hyd.
	560SEC	338 (5547)	N9YC	.032	—	TDC	125	91–94	—	650	Hyd.	Hyd.
	560SEL	338 (5547)	N9YC	.032	—	TDC	125	91–94	—	650	Hyd.	Hyd.
	560SL	338 (5547)	N9YC	.032	—	TDC	125	91–94	—	650	Hyd.	Hyd.
1989	190E	159 (2599)	S12YC	.032	9B	9B	125	91–94	700	700	Hyd.	Hyd.
	260E	159 (2599)	S12YC	.032	9B	9B	125	91–94	700	700	Hyd.	Hyd.
	300E	181 (2962)	S9YC	.032	TDC	TDC	125	91–94	650	650	Hyd.	Hyd.
	300CE	181 (2962)	S9YC	.032	—	TDC	125	91–94	—	650	Hyd.	Hyd.
	300SE	181 (2962)	S9YC	.032	—	TDC	125	91–94	—	650	Hyd.	Hyd.
	300SEL	181 (2962)	S9YC	.032	—	TDC	125	91–94	—	650	Hyd.	Hyd.
	300TE	181 (2962)	S9YC	.032	—	TDC	125	91–94	—	650	Hyd.	Hyd.
	420SEL	256 (4196)	N9YC	.032	—	TDC	125	91–94	—	650	Hyd.	Hyd.
	560SEC	338 (5547)	N9YC	.032	—	TDC	125	91–94	—	650	Hyd.	Hyd.
	560SEL	338 (5547)	N9YC	.032	—	TDC	125	91–94	—	650	Hyd.	Hyd.
	560SL	338 (5547)	N9YC	.032	—	TDC	125	91–94	—	650	Hyd.	Hyd.
1990	190E	159 (2599)	S12YC	.032	9B	9B	125	91–94	700	700	Hyd.	Hyd.
	300E	181 (2962)	S9YC	.032	—	TDC	174	91–94	—	650	Hyd.	Hyd.
	300CE	181 (2962)	S9YC	.032	—	TDC	174	91–94	—	650	Hyd.	Hyd.
	300SE	181 (2962)	S9YC	.032	—	TDC	174	91–94	—	650	Hyd.	Hyd.
	300SL	181 (2962)	C12YCC	.032	TDC	TDC	174	91–94	—	—	Hyd.	Hyd.
	300TE	181 (2962)	S9YC	.032	—	TDC	174	91–94	—	650	Hyd.	Hyd.
	300SEL	181 (2962)	S9YC	.032	—	TDC	174	91–94	—	650	Hyd.	Hyd.
	420SEL	256 (4196)	N9YC	.032	—	TDC	125	91–94	—	650	Hyd.	Hyd.
	500SL	304 (4973)	C10YCC	.032	—	TDC	125	91–94	—	700	Hyd.	Hyd.
	560SEC	338 (5547)	N9YC	.032	—	TDC	125	91–94	—	650	Hyd.	Hyd.
	560SEL	338 (5547)	N9YC	.032	—	TDC	125	91–94	—	650	Hyd.	Hyd.

GASOLINE ENGINE TUNE-UP SPECIFICATIONS

Year	Model	Engine Displacement cu. in. (cc)	Spark Plugs Type	Spark Plugs Gap (in.)	Ignition Timing (deg.) MT	Ignition Timing (deg.) AT	Compression Pressure (psi)	Fuel Pump (psi)	Idle Speed (rpm) MT	Idle Speed (rpm) AT	Valve Clearance In.	Valve Clearance Ex.
1991-92	190E	140 (2298)	S12YC	.032	5B	5B	174	77-80	750	750	Hyd.	Hyd.
	190E	158 (2597)	S12YC	.032	7-11B	7-11B	174	77-80	700	700	Hyd.	Hyd.
	300E	158 (2597)	S12YC	.032	—	7-11B	174	77-80	—	650	Hyd.	Hyd.
	300E	180 (2960)	S12YC	.032	—	6-11B	174	77-80	—	650	Hyd.	Hyd.
	300CE	180 (2960)	S12YC	.032	—	NA	174	90-93	—	650	Hyd.	Hyd.
	300TE	180 (2960)	S12YC	.032	—	6-11B	174	77-80	—	650	Hyd.	Hyd.
	300SE	180 (2960)	S12YC	.032	—	6-11B	174	77-80	—	650	Hyd.	Hyd.
	300SEL	180 (2960)	S12YC	.032	—	6-11B	174	77-80	—	650	Hyd.	Hyd.
	300SL	180 (2960)	C12YC	.032	NA	NA	174	90-93	700	650	Hyd.	Hyd.
	420SEL	256 (4196)	N9YC	.032	—	NA	124	90-93	—	650	Hyd.	Hyd.
	500E	303 (4972)	C12YC	.032	—	NA	174	90-93	—	650	Hyd.	Hyd.
	500SL	303 (4972)	C12YC	.032	—	NA	174	90-93	—	650	Hyd.	Hyd.
	560SEC	338 (5547)	N9YC	.032	—	NA	125	90-93	—	650	Hyd.	Hyd.
	560SEL	338 (5547)	N9YC	.032	—	TDC	124	90-93	—	650	Hyd.	Hyd.

NA—Not available

DIESEL ENGINE TUNE-UP SPECIFICATIONS

Year	Engine Displacement cu. in. (cc)	Valve Clearance Intake Type	Valve Clearance Exhaust (in.)	Intake Valve Opens (deg.)	Injection Pump Setting (deg.)	Injection Nozzle Pressure (psi) New	Injection Nozzle Pressure (psi) Used	Idle Speed (rpm)	Cranking Compression Pressure (psi)
1988	152 (2497) 190D	Hyd.	Hyd.	12A	15A	1564-2103	1740	660-700	284-327
1989	152 (2497) 190D	Hyd.	Hyd.	12A	15A	1564-2103	1740	660-700	284-327
1990	152 (2497) 190D	Hyd.	Hyd.	12A	15A	1564-2103	1740	660-700	284-327
1991-92	152 (2497) 300D	Hyd.	Hyd.	12A	15A	1564-2103	1740	660-700	261-464
	210 (3449) 350SDL	Hyd.	Hyd.	12A	15A	1564-2103	1740	610-650	261-464

FIRING ORDERS

NOTE: To avoid confusion, always replace spark plug wires one at a time.

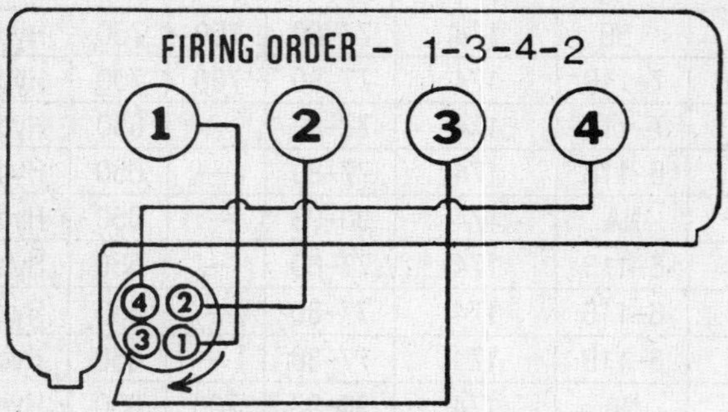

4 Cylinder Gasoline Engine
Engine Firing Order: 1–3–4–2
Distributor Rotation: Clockwise

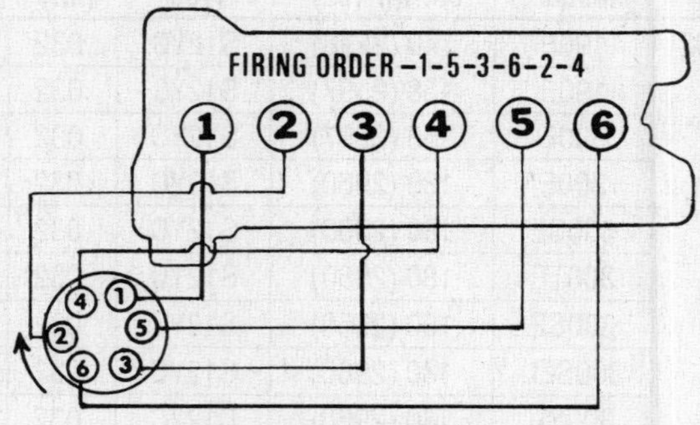

6 Cylinder Gasoline Engine
Engine Firing Order: 1–5–3–6–2–4
Distributor Rotation: Clockwise

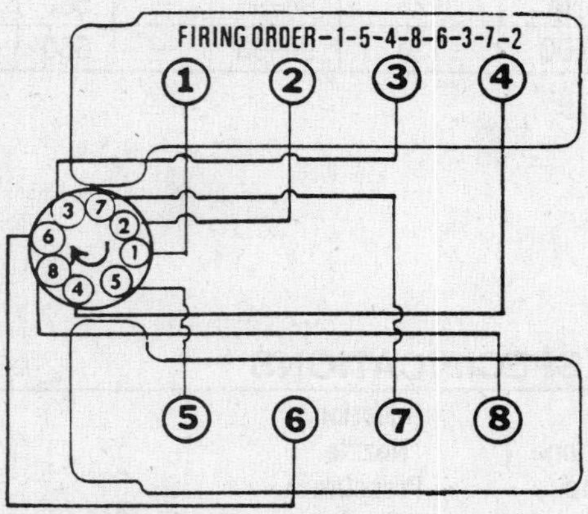

8 Cylinder, Except 5.0L Gasoline Engine
Engine Firing Order: 1–5–4–8–6–3–7–2
Distributor Rotation: Clockwise

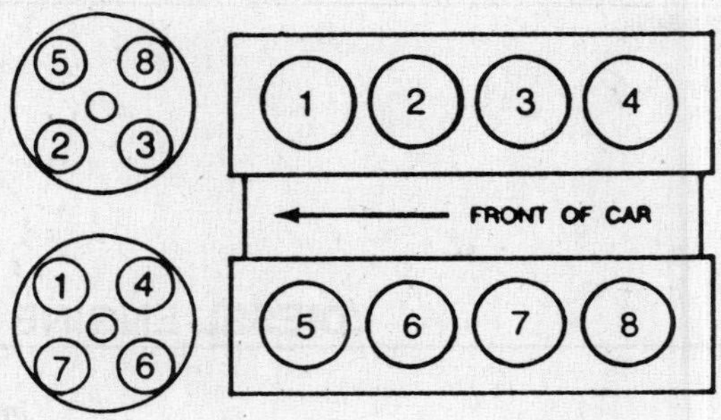

5.0L DOHC Gasoline Engine
Engine Firing Order: 1–5–4–8–6–3–7–2
Distributor Rotation: Clockwise

CAPACITIES

Year	Model	Engine Displacement cu. in. (cc)	Engine Crankcase (qts.) with Filter	Engine Crankcase (qts.) without Filter	Transmission (pts.) 4-Spd	Transmission (pts.) 5-Spd	Transmission (pts.) Auto.	Drive Axle (pts.)	Fuel Tank (gal.)	Cooling System (qts.)
1988	190D	152 (2497)	8.0	7.5	—	—	11.6	1.5	14.5	8.5
	190E	140 (2299)	4.8	4.3	—	3.2	12.7	2.2	14.5	9.0
	190E	159 (2599)	6.4	5.9	—	3.2	12.7	2.3	14.5	9.5
	260E	159 (2599)	6.4	5.9	—	3.2	13.1	2.3	18.5	9.5
	300E	181 (2962)	6.4	5.9	—	3.2	13.1	2.3	18.5	8.5
	300CE	181 (2962)	6.4	5.9	—	—	13.1	2.3	18.5	10.6
	300SE	181 (2962)	6.4	5.9	—	—	13.1	2.3	23.8	10.6
	300SEL	181 (2962)	6.4	5.9	—	—	13.1	2.3	23.8	10.6
	300TE	181 (2962)	6.4	5.9	—	—	13.1	2.3	19.0	10.6
	420SEL	256 (4196)	8.5	8.0	—	—	16.2	2.7	23.8	13.8
	560SEC	338 (5547)	8.5	8.0	—	—	16.2	2.7	23.8	13.8
	560SEL	338 (5547)	8.5	8.0	—	—	16.2	2.7	23.8	13.8

CAPACITIES

Year	Model	Engine Displacement cu. in. (cc)	Engine Crankcase (qts.) with Filter	without Filter	Transmission (pts.) 4-Spd	5-Spd	Auto.	Drive Axle (pts.)	Fuel Tank (gal.)	Cooling System (qts.)
1988	560SL	338 (5547)	8.5	8.0	—	—	16.2	2.7	22.4	13.8
1989	190D	152 (2497)	8.0	7.5	—	—	11.6	1.5	14.5	8.5
	190E	159 (2599)	6.4	5.9	—	3.2	12.7	2.3	14.5	9.5
	260E	159 (2599)	6.4	5.9	—	3.2	13.1	2.3	18.5	9.5
	300E	181 (2962)	6.4	5.9	—	3.2	13.1	2.3	18.5	8.5
	300CE	181 (2962)	6.4	5.9	—	—	13.1	2.3	18.5	10.6
	300SE	181 (2962)	6.4	5.9	—	—	13.1	2.3	23.8	10.6
	300SEL	181 (2962)	6.4	5.9	—	—	13.1	2.3	23.8	10.6
	300TE	181 (2962)	6.4	5.9	—	—	13.1	2.3	19.0	10.6
	420SEL	256 (4196)	8.5	8.0	—	—	16.2	2.7	23.8	13.8
	560SEC	338 (5547)	8.5	8.0	—	—	16.2	2.7	23.8	13.8
	560SEL	338 (5547)	8.5	8.0	—	—	16.2	2.7	23.8	13.8
	560SL	338 (5547)	8.5	8.0	—	—	16.2	2.7	22.4	13.8
1990	190E	159 (2599)	6.4	5.9	—	3.2	12.7	2.3	14.5	9.5
	300E	181 (2962)	6.4	5.9	—	—	13.1	2.3	18.5	8.5
	300CE	181 (2962)	6.4	5.9	—	—	13.1	2.3	18.5	10.6
	300SE	181 (2962)	6.4	5.9	—	—	13.1	2.3	23.8	10.6
	300SL	181 (2962)	8.0	7.5	—	3.4	12.6	2.8	21.1	12.2
	300TE	181 (2962)	6.4	5.9	—	—	13.1	2.3	19.0	10.6
	300SEL	181 (2962)	6.4	5.9	—	—	13.1	2.3	23.8	10.6
	420SEL	256 (4196)	8.5	8.0	—	—	16.2	2.7	23.8	13.8
	500SL	304 (4973)	8.5	8.0	—	—	16.2	2.8	21.1	15.9
	560SEC	338 (5547)	8.5	8.0	—	—	16.2	2.7	23.8	13.8
	560SEL	338 (5547)	8.5	8.0	—	—	16.2	2.7	23.8	13.8
1991-92	190E	140 (2298)	5.3	4.8	—	3.2	12.7	2.3	14.5	9.5
	190E	158 (2597)	6.4	5.9	—	3.2	12.7	2.3	18.5	10.0
	300D	152 (2497)	8.0	7.5	—	—	13.1	2.3	18.5	11.7
	300E	158 (2597)	6.9	6.4	—	—	13.1	2.3	18.5	8.5
	300E	180 (2960)	6.9	6.4	—	—	13.1	2.3	18.5	9.5
	300CE	180 (2960)	8.0	7.5	—	—	13.1	2.3	18.5	9.5
	300TE	180 (2960)	6.9	6.4	—	—	13.1	2.3 ①	19.0	9.5
	300SE	180 (2960)	6.9	6.4	—	—	13.1	2.3	23.8	8.5
	300SEL	180 (2960)	6.9	6.4	—	—	13.1	2.3	23.8	11.7
	300SL	180 (2960)	8.0	7.5	—	3.2	13.1	2.3	21.0	11.7
	350SD	210 (3449)	8.0	7.5	—	—	13.1	2.7	21.6	12.7
	350SDL	210 (3449)	8.0	7.5	—	—	13.1	2.7	21.6	12.7
	420SEL	256 (4196)	8.5	8.0	—	—	16.2	2.7	23.8	13.8
	500E	303 (4972)	8.5	8.0	—	—	16.2	2.7	23.8	14.3
	500SL	303 (4972)	8.5	8.0	—	—	16.2	2.3	21.1	14.3
	560SEC	338 (5547)	8.5	8.0	—	—	16.2	2.7	23.8	13.8
	560SEL	338 (5547)	8.5	8.0	—	—	16.2	2.7	23.8	13.8

① 4 Matic: 4.2
4 Matic with ASD: 11.6
Hydropneumatic Suspension: 9.9

CAMSHAFT SPECIFICATIONS

All measurements given in inches.

Year	Engine Displacement cu. in. (cc)	Journal Diameter 1	2	3	4	5	Lobe Lift In.	Ex.	Bearing Clearance	Camshaft End Play
1988	140 (2299)	1.260	1.260	1.260	1.260	1.260	NA	NA	NA	NA
	152 (2497)	1.260	1.260	1.260	1.260	1.260	NA	NA	NA	NA
	159 (2599)	1.378	1.831	1.831	1.831	1.831	0.394	0.413	0.002	0.004
	181 (2962)	1.378	1.831	1.831	1.831	1.831	0.394	0.413	0.002	0.004
	256 (4196)	1.377	1.936	1.936	1.944	1.944	NA	NA	0.0016	0.004
	338 (5547)	1.377	1.936	1.936	1.944	1.944	NA	NA	0.0016	0.004
1989	152 (2497)	1.260	1.260	1.260	1.260	1.260	NA	NA	NA	NA
	159 (2599)	1.378	1.831	1.831	1.831	1.831	0.394	0.413	0.002	0.004
	181 (2962)	1.378	1.831	1.831	1.831	1.831	0.394	0.413	0.002	0.004
	256 (4196)	1.377	1.936	1.936	1.944	1.944	NA	NA	0.0016	0.004
	338 (5547)	1.377	1.936	1.936	1.944	1.944	NA	NA	0.0016	0.004
1990	159 (2599)	1.378	1.831	1.831	1.831	1.831	0.394	0.413	0.002	0.004
	181 (2962)	1.378	1.831	1.831	1.831	1.831	0.394	0.413	0.002	0.004
	256 (4196)	1.377	1.936	1.936	1.944	1.944	NA	NA	0.0016	0.004
	338 (5547)	1.377	1.936	1.936	1.944	1.944	NA	NA	0.0016	0.004
	304 (4973)	1.377	1.936	1.936	1.944	1.944	NA	NA	0.0016	0.004
1991-92	140 (2298)	1.260	1.260	1.260	1.260	1.260	NA	NA	0.002	0.003
	152 (2497)	1.218	1.218	1.218	1.218	1.218	NA	NA	0.002	NA
	158 (2597)	1.218	1.218	1.218	1.218	1.218	NA	NA	0.002	0.004
	180 (2960)	1.218	1.218	1.218	1.218	1.218	NA	NA	0.002	0.004
	180 (2960) ①	1.179	1.179	1.179	1.179	1.179	NA	NA	0.002	0.002–0.008
	210 (3449)	1.218	1.218	1.218	1.218	1.218	NA	NA	0.002	NA
	256 (4196)	1.378	1.937	1.937	1.945	1.945	NA	NA	0.002	0.003–0.006
	303 (4972)	1.100	1.100	1.100	1.100	1.100	NA	NA	0.002	0.002
	338 (5547)	1.378	1.937	1.937	1.945	1.945	NA	NA	0.002	0.003–0.006

NA—Not Available
① Dual Overhead Cam

CRANKSHAFT AND CONNECTING ROD SPECIFICATIONS

All measurements are given in inches.

Year	Engine Displacement cu. in. (cc)	Crankshaft Main Brg. Journal Dia.	Main Brg. Oil Clearance	Shaft End-play	Thrust on No.	Connecting Rod Journal Diameter	Oil Clearance	Side Clearance
1988	140 (2299)	2.281–2.282	0.001–0.003	0.004–0.010	①	1.887–1.888	0.001–0.003	NA
	152 (2497)	2.281–2.282	0.001–0.003	0.004–0.010	①	1.887–1.888	0.001–0.003	NA
	159 (2599)	2.360–2.361	NA	NA	①	2.031–2.032	0.001–0.002	NA
	181 (2962)	2.360–2.361	NA	NA	①	2.031–2.032	0.001–0.002	NA
	256 (4196)	2.517–2.518	0.002–0.003	0.004–0.009	①	1.887–1.888	0.004–0.009	0.009–0.014
	338 (5547)	2.517–2.518	0.002–0.003	0.004–0.009	①	1.887–1.888	0.004–0.009	0.009–0.014

CRANKSHAFT AND CONNECTING ROD SPECIFICATIONS

All measurements are given in inches.

Year	Engine Displacement cu. in. (cc)	Crankshaft Main Brg. Journal Dia.	Crankshaft Main Brg. Oil Clearance	Crankshaft Shaft End-play	Crankshaft Thrust on No.	Connecting Rod Journal Diameter	Connecting Rod Oil Clearance	Connecting Rod Side Clearance
1989	152 (2497)	2.281–2.282	0.001–0.003	0.004–0.010	①	1.887–1.888	0.001–0.003	NA
	159 (2599)	2.360–2.361	NA	NA	①	2.031–2.032	0.001–0.002	NA
	181 (2962)	2.360–2.361	NA	NA	①	2.031–2.032	0.001–0.002	NA
	256 (4196)	2.517–2.518	0.002–0.003	0.004–0.009	①	1.887–1.888	0.004–0.009	0.009–0.014
	338 (5547)	2.517–2.518	0.002–0.003	0.004–0.009	①	1.887–1.888	0.004–0.009	0.009–0.014
1990	159 (2599)	2.360–2.361	NA	NA	①	2.031–2.032	0.001–0.002	NA
	181 (2962)	2.360–2.361	NA	NA	①	2.031–2.032	0.001–0.002	NA
	256 (4196)	2.517–2.518	0.002–0.003	0.004–0.009	①	1.887–1.888	0.004–0.009	0.009–0.014
	338 (5547)	2.517–2.518	0.002–0.003	0.004–0.009	①	1.887–1.888	0.004–0.009	0.009–0.014
	304 (4973)	2.517–2.518	0.002–0.003	0.004–0.009	①	1.887–1.888	0.004–0.009	0.009–0.014
1991-92	140 (2298)	2.281–2.282	0.001–0.002	0.004–0.010	3	1.887–1.888	0.001–0.002	NA
	152 (2497)	2.281–2.282	0.001–0.002	0.004–0.010	4	1.887–1.888	0.001–0.002	NA
	158 (2597)	2.360–2.361	NA	NA	①	2.031–2.032	0.001–0.002	NA
	180 (2960)	2.360–2.361	NA	NA	①	2.031–2.032	0.001–0.002	NA
	180 (2960)	2.360–2.361	NA	NA	①	2.031–2.032	0.001–0.002	NA
	210 (3449)	2.281–2.282	0.001–0.002	0.004–0.010	5	1.887–1.888	0.001–0.002	NA
	256 (4196)	2.517–2.518	0.001–0.002	0.004–0.004	3	2.045–2.046	0.001–0.002	NA
	303 (4972)	2.517–2.518	0.001–0.002	0.004–0.009	3	2.045–2.046	0.001–0.002	NA
	338 (5547)	2.517–2.518	0.001–0.002	0.004–0.009	3	2.045–2.046	0.001–0.002	NA

NA Not available

① Center main on 5 main bearing engines; rear main on 7 main bearing engines; 3rd from front on 300D (5 cylinder)

VALVE SPECIFICATIONS

Year	Engine Displacement cu. in. (cc)	Seat Angle (deg.)	Face Angle (deg.)	Spring Test Pressure (lbs.)	Spring Installed Height (in.)	Stem-to-Guide Clearance (in.)		Stem Diameter (in.)	
						Intake	Exhaust	Intake	Exhaust
1988	140 (2299)	45 ①	45 ①	191	1.929	0.004	0.004	0.314	0.353
	152 (2497)	45 ①	45 ①	169	2.000	0.004	0.004	0.314	0.353
	159 (2599)	45	45	NA	NA	0.004	0.004	0.314	0.353
	183 (2996)	45	45	191	1.929	0.004	0.004	0.274	0.313
	181 (2962)	45	45	NA	NA	0.004	0.004	0.314	0.353
	256 (4196)	45	45	194	1.200	0.004	0.004	0.353	0.352
	338 (5547)	45	45	194	1.200	0.004	0.004	0.353	0.352
1989	152 (2497)	45 ①	45 ①	169	2.000	0.004	0.004	0.314	0.353
	159 (2599)	45	45	NA	NA	0.004	0.004	0.314	0.353
	181 (2962)	45	45	NA	NA	0.004	0.004	0.314	0.352
	256 (4196)	45	45	194	1.200	0.004	0.004	0.353	0.352
	338 (5547)	45	45	194	1.200	0.004	0.004	0.353	0.352
1990	159 (2599)	45	45	NA	NA	0.004	0.004	0.314	0.353
	181 (2962)	45	45	NA	NA	0.004	0.004	0.314	0.352
	256 (4196)	45	45	194	1.200	0.004	0.004	0.353	0.352
	338 (5547)	45	45	194	1.200	0.004	0.004	0.353	0.352
	304 (4973)	45	45	194	1.200	0.004	0.004	0.315	0.354
1991-92	140 (2298)	45 ①	45 ①	191	1.929	0.004	0.004	0.314	0.353
	152 (2497)	45 ①	45 ①	158	2.000	0.004	0.004	0.314	0.353
	158 (2597)	45	45	NA	NA	0.004	0.004	0.352	0.353
	180 (2960)	45	45	NA	NA	0.004	0.004	0.314	0.353
	180 (2960)	45	45	NA	NA	0.004	0.004	0.314	0.353
	210 (2449)	45	45 ①	158	1.062	0.004	0.004	0.314	0.352
	256 (4196)	45	45	194	1.200	0.004	0.004	0.353	0.352
	303 (4972)	45	45	194	1.200	0.004	0.004	0.315	0.354
	338 (5547)	45	45	194	1.200	0.004	0.004	0.353	0.352

NA Not available
① Plus 15 minutes

PISTON AND RING SPECIFICATIONS

All measurements are given in inches.

Year	Engine Displacement cu. in. (cc)	Piston Clearance	Ring Gap			Ring Side Clearance		
			Top Compression	Bottom Compression	Oil Control	Top Compression	Bottom Compression	Oil Control
1988	140 (2299)	0.001–0.002	0.008–0.016	0.008–0.016	0.008–0.016	0.004–0.005	0.002–0.003	0.001–0.003
	152 (2497)	0.001–0.002	0.008–0.016	0.008–0.016	0.008–0.016	0.004–0.005	0.003–0.004	0.001–0.002
	159 (2599)	0.001–0.002	0.008–0.016	0.008–0.016	0.010–0.016	0.004–0.005	0.003–0.004	0.001–0.002
	181 (2962)	0.001–0.002	0.008–0.016	0.008–0.016	0.010–0.016	0.004–0.005	0.003–0.004	0.001–0.002
	256 (4196)	0.001	0.008–0.016	0.008–0.016	0.010–0.016	0.004–0.005	0.003–0.004	0.001–0.002
	338 (5547)	0.001	0.010–0.017	0.010–0.017	0.010–0.016	0.004–0.005	0.001–0.002	0.001–0.002

PISTON AND RING SPECIFICATIONS

All measurements are given in inches.

Year	Engine Displacement cu. in. (cc)	Piston Clearance	Ring Gap			Ring Side Clearance		
			Top Compression	Bottom Compression	Oil Control	Top Compression	Bottom Compression	Oil Control
1989	152 (2497)	0.001–0.002	0.008–0.016	0.008–0.016	0.010–0.016	0.004–0.005	0.003–0.004	0.001–0.002
	159 (2599)	0.001–0.002	0.008–0.016	0.008–0.016	0.010–0.016	0.004–0.005	0.003–0.004	0.001–0.002
	181 (2962)	0.001–0.002	0.008–0.016	0.008–0.016	0.010–0.016	0.004–0.005	0.003–0.004	0.001–0.002
	256 (4196)	0.001	0.008–0.016	0.008–0.016	0.010–0.016	0.004–0.005	0.003–0.004	0.001–0.002
	338 (5547)	0.001	0.010–0.017	0.010–0.017	0.010–0.016	0.004–0.005	0.001–0.002	0.001–0.002
1990	159 (2599)	0.001–0.002	0.008–0.016	0.008–0.016	0.010–0.016	0.004–0.005	0.003–0.004	0.001–0.002
	181 (2962)	0.001–0.002	0.008–0.016	0.008–0.016	0.010–0.016	0.004–0.005	0.003–0.004	0.001–0.002
	256 (4196)	0.001	0.008–0.016	0.008–0.016	0.010–0.016	0.004–0.005	0.003–0.004	0.001–0.002
	338 (5547)	0.001	0.010–0.017	0.010–0.017	0.010–0.016	0.004–0.005	0.001–0.002	0.001–0.002
	304 (4973)	0.001	0.010–0.017	0.010–0.017	0.010–0.016	0.004–0.005	0.001–0.002	0.001–0.002
1991-92	140 (2298)	0.001	0.012–0.022	0.012–0.022	0.010–0.020	0.003–0.006	0.001–0.004	0.002–0.004
	152 (2497)	0.001	0.010–0.040	0.008–0.040	0.008–0.040	0.003–0.008	0.002–0.006	0.001–0.004
	158 (2597)	0.001–0.002	0.008–0.016	0.008–0.016	0.010–0.016	0.003–0.006	0.001–0.004	0.002–0.004
	180 (2960)	0.001–0.002	0.008–0.016	0.008–0.016	0.010–0.016	0.003–0.006	0.001–0.004	0.002–0.004
	180 (2960)	0.001–0.002	0.008–0.016	0.008–0.016	0.010–0.016	0.003–0.006	0.001–0.004	0.002–0.004
	210 (3449)	0.001	0.010–0.040	0.008–0.040	0.008–0.040	0.003–0.008	0.002–0.006	0.001–0.004
	256 (4196)	0.003	0.012–0.040	0.014–0.031	0.010–0.031	0.002–0.006	0.002–0.003	0.001–0.003
	303 (4972)	0.003	0.014–0.040	0.014–0.031	0.010–0.031	0.002–0.006	0.002–0.003	0.001–0.003
	338 (5547)	0.003	0.014–0.040	0.014–0.031	0.010–0.031	0.002–0.006	0.002–0.003	0.001–0.003

TORQUE SPECIFICATIONS

All readings in ft. lbs.

Year	Engine Displacement cu. in. (cc)	Cylinder Head Bolts	Main Bearing Bolts	Rod Bearing Bolts	Crankshaft Pulley Bolts	Flywheel Bolts	Manifold Intake	Manifold Exhaust	Spark Plugs
1988	140 (2299)	51 ⑦	65	⑥	145	25	NA	NA	15
	152 (2497)	④ ①	40	⑤	218	25	18	NA	—
	159 (2599)	70 ⑩	65	22	217	22	NA	NA	15
	181 (2962)	65 ⑨	65	33	218	⑥	NA	18	15
	256 (4196)	44 ⑧	③	33	289	⑥	NA	NA	15
	338 (5547)	44 ⑧	②	33	187	⑥	NA	20	15
1989	152 (2497)	④ ①	40	⑤	218	25	18	NA	—
	159 (2599)	70 ⑩	65	22	217	22	NA	NA	15
	181 (2962)	65 ⑨	65	33	218	⑥	NA	18	15
	256 (4196)	44 ⑧	③	33	289	⑥	NA	NA	15
	338 (5547)	44 ⑧	②	33	187	⑥	NA	20	15
1990	159 (2599)	70 ⑩	65	22	217	22	NA	NA	15
	181 (2962)	65 ⑨	65	33	218	⑥	NA	18	15
	256 (4196)	44 ⑧	③	33	289	⑥	NA	NA	15
	338 (5547)	44 ⑧	②	33	187	⑥	NA	20	15
	304 (4973)	75 ⑪	②	33	187	⑥	NA	20	15
1991-92	140 (2298)	⑭	⑬	⑥	150	⑫	NA	NA	15
	152 (2497)	⑯	⑬	⑥	236	⑫	18	18	—
	158 (2597)	⑮	⑬	⑥	218	⑤	NA	NA	15
	180 (2960)	⑮	⑬	⑥	218	⑤	NA	NA	15
	180 (2960)	⑮	⑬	⑥	218	⑤	NA	NA	15
	210 (3449)	⑯	⑬	⑥	236	⑤	18	18	—
	256 (4196)	⑰	③	⑱	218	⑤	NA	NA	15
	303 (4972)	⑭	③	⑱	295	⑤	NA	NA	15
	338 (5547)	⑰	③	⑱	218	⑤	NA	NA	15

NA Not available

① See text

② M 10 bolts—37 ft. lbs.
M 12 bolts—72 ft. lbs.

③ M 10 bolts—43 ft. lbs.
M 12 bolts—58 ft. lbs.

④ M 10 bolts:
1st step—18 ft. lbs.
2nd step—29 ft. lbs., setting time 10 minutes
3rd step—90 degrees torquing angle
4th step—29 degrees torquing angle
M 8 bolts—18 ft. lbs.

⑤ 1st step—22–25 ft. lbs.
2nd step—90–100 degrees torquing angle

⑥ 1st step—22–25 ft. lbs.
2nd step—90–100 degrees torquing angle

⑦ M 12 bolts:
1st step—29 ft. lbs.
2nd step—51 ft. lbs., setting time 10 minutes
3rd step—90 degrees torquing angle
4th step—90 degrees torquing angle
M 8 bolts—18 ft. lbs.

⑧ 1st step—22 ft. lbs.
2nd step—44 ft. lbs., setting time 10 minutes
3rd step—Loosen bolts and retighten to 44 ft. lbs.

⑨ 1st step—29 ft. lbs.
2nd step—51 ft. lbs., setting time 10 minutes
3rd step—90 degrees torquing angle
4th step—90 degrees torquing angle

⑩ 1st step—70 ft. lbs.
2nd step—90 degrees torquing angle
3rd step—90 degrees torquing angle

⑪ 1st step—75 ft. lbs.
2nd step—90 degrees torquing angle
3rd step—90 degrees torquing angle

⑫ Automatic transmission:
1st step—22 ft. lbs.
2nd step—90 degrees torquing angle
Manual transmission:
1st step—37 ft. lbs.
2nd step—90 degrees torquing angle

⑬ 1st step—40 ft. lbs.
2nd step—90 degrees torquing angle

⑭ 1st step—40 ft. lbs.
2nd step—90 degrees torquing angle
3rd step—90 degrees torquing angle

⑮ 1st step—52 ft. lbs.
2nd step—90 degrees torquing angle
3rd step—90 degrees torquing angle

⑯ Allen head bolts
1st step—11 ft. lbs.
2nd step—26 ft. lbs.
3rd step—90 degrees torquing angle and wait 10 minutes
4th step—90 degrees torquing angle

⑰ 1st step—22 ft. lbs.
2nd step—44 ft. lbs.
3rd step—warm engine to 176°F (80°C) and retorque to 44 ft. lbs.

⑱ 1st step—30–40 ft. lbs.
2nd step—90–100 degrees torquing angle

BRAKE SPECIFICATIONS

All measurements in inches unless noted.

Year	Model	Lug Nut Torque (ft. lbs.)	Master Cylinder Bore	Brake Disc Minimum Thickness	Brake Disc Maximum Runout	Standard Brake Drum Diameter	Minimum Lining Thickness Front	Minimum Lining Thickness Rear
1988	190D	75	④	0.35⑥	0.005⑦	—	0.08	0.08
	190E	75	④	0.35⑥	0.005⑦	—	0.08	0.08
	260E	75	②	①⑤	0.005⑦	—	0.08	0.08
	300E	75	②	①⑤	0.005⑦	—	0.08	0.08
	300CE	75	②	①⑤	0.005⑦	—	0.08	0.08
	300SE	75	②	③⑤	0.005	—	0.08	0.08
	300SEL	75	②	③⑤	0.005	—	0.08	0.08
	300TE	75	②	①⑤	0.005⑦	—	0.08	0.08
	420SEL	75	②	③⑤	0.005	—	0.08	0.08
	560SEC	75	②	③⑤	0.005	—	0.08	0.08
	560SEL	75	②	③⑤	0.005	—	0.08	0.08
	560SL	75	②	③⑤	0.005	—	0.08	0.08
1989	190D	75	④	0.35⑥	0.005⑦	—	0.08	0.08
	190E	75	④	0.35⑥	0.005⑦	—	0.08	0.08
	260E	75	②	①⑤	0.005⑦	—	0.08	0.08
	300E	75	②	①⑤	0.005⑦	—	0.08	0.08
	300CE	75	②	①⑤	0.005⑦	—	0.08	0.08
	300SE	75	②	③⑤	0.005	—	0.08	0.08
	300SEL	75	②	③⑤	0.005	—	0.08	0.08
	300TE	75	②	①⑤	0.005⑦	—	0.08	0.08
	420SEL	75	②	③⑤	0.005	—	0.08	0.08
	560SEC	75	②	③⑤	0.005	—	0.08	0.08
	560SEL	75	②	③⑤	0.005	—	0.08	0.08
	560SL	75	②	③⑤	0.005	—	0.08	0.08
1990	190E	75	④	0.35⑥	0.005⑦	—	0.08	0.08
	300E	75	②	①⑤	0.005⑦	—	0.08	0.08
	300CE	75	②	①⑤	0.005⑦	—	0.08	0.08
	300SE	75	②	③⑤	0.005	—	0.08	0.08
	300SL	75	②	⑧	0.005	—	0.08	0.08
	300TE	75	②	①⑤	0.005⑦	—	0.08	0.08
	300SEL	75	②	③⑤	0.005	—	0.08	0.08
	420SEL	75	②	③⑤	0.005	—	0.08	0.08
	500SL	75	②	⑧	0.005	—	0.08	0.08
	560SEC	75	②	③⑤	0.005	—	0.08	0.08
	560SEL	75	②	③⑤	0.005	—	0.08	0.08
1991-92	190E	75	④	0.35⑥	0.005⑦	—	0.08	0.08
	190E	75	④	0.35⑥	0.005⑦	—	0.08	0.08
	300D	75	②	①⑤	0.005	—	0.08	0.08
	300E	75	②	①⑤	0.005	—	0.08	0.08
	300E	75	②	①⑤	0.005	—	0.08	0.08
	300CE	75	②	①⑤	0.005	—	0.08	0.08
	300TE	75	②	①⑤	0.005	—	0.08	0.08

BRAKE SPECIFICATIONS

All measurements in inches unless noted.

Year	Model	Lug Nut Torque (ft. lbs.)	Master Cylinder Bore	Brake Disc Minimum Thickness	Brake Disc Maximum Runout	Standard Brake Drum Diameter	Minimum Lining Thickness Front	Minimum Lining Thickness Rear
1991-92	300SE	75	②	③ ⑤	0.005	—	0.08	0.08
	300SEL	75	②	③ ⑤	0.005	—	0.08	0.08
	300SL	75	②	⑧	0.005	—	0.08	0.08
	300SD	75	②	③ ⑤	0.005	—	0.08	0.08
	300SDL	75	②	③ ⑤	0.005	—	0.08	0.08
	420SEL	75	②	③ ⑤	0.005	—	0.08	0.08
	500E	75	②	③ ⑤	0.005	—	0.08	0.08
	500SL	75	②	⑧	0.005	—	0.08	0.08
	560SEC	75	②	③ ⑤	0.005	—	0.08	0.08
	560SEL	75	②	③ ⑤	0.005	—	0.08	0.08

① Caliper with 57mm piston diameter—0.44 in.
Caliper with 60mm piston diameter—0.42 in.
② Pushrod circuit—15/16 in.
Floating circuit—3/4 in.
③ Caliper with 57mm piston diameter—0.81 in.
Caliper with 60mm piston diameter—0.79 in.
④ Pushrod circuit—7/8 in.
Floating circuit—11/16 in.
⑤ Rear disc—0.33 in.
⑥ Rear disc—0.30 in.
⑦ Rear disc—0.006 in.
⑧ Front disc—1.10 in.
Rear disc—0.35 in.

WHEEL ALIGNMENT

Year	Model	Caster Range (deg.)	Caster Preferred Setting (deg.)	Camber Range (deg.)	Camber Preferred Setting (deg.)	Toe-in (in.)	Steering Axis Inclination (deg.)
1988	190D	10-11	10 1/2	1/2N-0	3/16N	3/32	NA
	190E	10-11	10 1/2	1/2N-0	3/16N	3/32	NA
	260E	9 11/16-10 11/16	10 3/16	5/16N-3/16P	0	3/16	NA
	300E	9 11/16-10 11/16	10 3/16	5/16N-3/16P	0	3/16	NA
	300CE	9 11/16-10 11/16	10 3/16	5/16N-3/16P	0	3/16	NA
	300SE	10-11	10 1/2	1/2N-0	3/16N	3/32	NA
	300SEL	10-11	10 1/2	1/2N-0	3/16N	3/32	NA
	300TE	9 11/16-10 11/16	10 3/16	5/16N-3/16P	0	3/16	NA
	420SEL	10-11	10 1/2	1/2N-0	3/16N	3/32	NA
	560SEC	10-11	10 1/2	1/2N-0	3/16N	3/32	NA
	560SEL	10-11	10 1/2	1/2N-0	3/16N	3/32	NA
	560SL	10-11	10 1/2	1/2N-0	3/16N	3/32	NA
1989	190D	10-11	10 1/2	1/2N-0	3/16N	3/32	NA
	190E	10-11	10 1/2	1/2N-0	3/16N	3/32	NA
	260E	9 11/16-10 11/16	10 3/16	5/16N-3/16P	0	3/16	NA
	300E	9 11/16-10 11/16	10 3/16	5/16N-3/16P	0	3/16	NA
	300CE	9 11/16-10 11/16	10 3/16	5/16N-3/16P	0	3/16	NA
	300SE	10-11	10 1/2	1/2N-0	3/16N	3/32	NA
	300SEL	10-11	10 1/2	1/2N-0	3/16N	3/32	NA
	300TE	9 11/16-10 11/16	10 3/16	5/16N-3/16P	0	3/16	NA
	420SEL	10-11	10 1/2	1/2N-0	3/16N	3/32	NA
	560SEC	10-11	10 1/2	1/2N-0	3/16N	3/32	NA
	560SEL	10-11	10 1/2	1/2N-0	3/16N	3/32	NA
	560SL	10-11	10 1/2	1/2N-0	3/16N	3/32	NA

WHEEL ALIGNMENT

Year	Model	Caster Range (deg.)	Caster Preferred Setting (deg.)	Camber Range (deg.)	Camber Preferred Setting (deg.)	Toe-in (in.)	Steering Axis Inclination (deg.)
1990	190E	10–11	10 1/2	1/2N–0	3/16N	3/32	NA
	300E	9 11/16–10 11/16	10 3/16	5/16N–3/16P	0	3/16	NA
	300CE	9 11/16–10 11/16	10 3/16	5/16N–3/16P	0	3/16	NA
	300SE	10–11	10 1/2	1/2N–0	3/16N	3/32	NA
	300SL	10–11	10 1/2	3/4N–1/4N	1/2N	1/4	NA
	300SEL	10–11	10 1/2	1/2N–0	3/16N	3/32	NA
	300TE	9 11/16–10 11/16	10 3/16	5/16N–3/16P	0	3/16	NA
	420SEL	10–11	10 1/2	1/2N–0	3/16N	3/32	NA
	500SL	10–11	10 1/2	3/4N–1/4N	1/2N	1/4	NA
	560SEC	10–11	10 1/2	1/2N–0	3/16N	3/32	NA
	560SEL	10–11	10 1/2	1/2N–0	3/16N	3/32	NA
1991–92	190E-2.3	9 11/16–10 11/16	10 3/16	1/2N–0	5/16N	3/32	NA
	190E-2.6	10–11	10 1/2	11/16N–3/16P	5/16N	3/32	NA
	300D	9 3/16–10 3/16	9 11/16	5/16N–3/16P	0	3/32	NA
	300E	9 3/16–10 3/16	9 11/16	5/16N–3/16P	0	3/32	NA
	300E	9 3/16–10 3/16	9 11/16	5/16N–3/16P	0	3/32	NA
	300CE	9 3/16–10 3/16	9 11/16	5/16N–3/16P	0	3/32	NA
	300TE	9 3/16–10 3/16	9 11/16	5/16N–3/16P	0	3/32	NA
	300SE	9 1/4–10 1/4	9 3/4	3/16N–3/16P	0	7/64	NA
	300SEL	9 1/4–10 1/4	9 3/4	3/16N–3/16P	0	7/64	NA
	300SL	10–11	10 1/2	3/4N–1/4p	1/2N	1/4	NA
	350SD	9 1/4–10 1/4	9 3/4	3/16N–3/16P	0	7/64	NA
	350SDL	9 1/4–10 1/4	9 3/4	3/16N–3/16P	0	7/64	NA
	420SEL	9 1/4–10 1/4	9 3/4	3/16N–3/16P	0	7/64	NA
	500E	9 3/16–10 3/16	9 11/16	5/16N–3/16P	0	3/32	NA
	500SL	10–11	10 1/2	3/4N–1/4P	1/2N	1/4	NA
	560SEC	9 1/4–10 1/4	9 3/4	3/16N–3/16P	0	7/64	NA
	560SEL	9 1/4–10 1/4	9 3/4	3/16N–3/16P	0	7/64	NA

GASOLINE ENGINE MECHANICAL

NOTE: Disconnecting the negative battery cable on some vehicles may interfere with the functions of the on board computer systems and may require the computer to undergo a relearning process, once the negative battery cable is reconnected.

Engine Assembly

NOTE: Care should be taken when working on Mercedes-Benz engines, since there are many aluminum parts which can be damaged, if carelessly handled.

REMOVAL & INSTALLATION

NOTE: In all cases, Mercedes-Benz engines and transmissions are removed as a unit.

CAUTION

Air conditioner lines should not be indiscriminately disconnected without taking proper precautions. It is best to swing the compressor aside while still connected to its hoses. Never do any welding around the compressor-heat may cause an explosion. Also, the refrigerant, while inert at normal room temperature, breaks down under high temperature into hydrogen fluoride and phosgene (among other products), which are highly poisonous.

Except V8 Engines

1. Remove the hood, drain the cooling system and disconnect the battery. While not strictly necessary, it is better to remove the battery completely to prevent breakage by the engine as it is lifted out.
2. Remove the fan shroud and radiator.
3. Disconnect all heater hoses and oil cooler lines. Plug all openings to keep out dirt.
4. Remove the air cleaner and all fuel, vacuum and oil hoses (e.g., power

steering and power brakes). Plug all openings to keep out dirt.

5. Remove the viscous coupling and fan.

6. Disconnect the accelerator linkage.

7. Disconnect all ground straps and electrical connections; it is a good idea to tag each wire for easy reassembly.

8. Detach the gearshift linkage and the exhaust pipes from the manifolds.

9. Loosen the steering relay arm and move it aside, along with the center steering rod and hydraulic steering damper.

10. The hydraulic engine shock absorber should be removed.

11. Remove the hydraulic line from the clutch housing and the oil line connectors from the automatic transmission.

12. Unbolt the clutch slave cylinder from the bell housing after removing the return spring.

13. Remove the exhaust pipe bracket from the transmission. Support bellhousing or place a cable sling under the oil pan, to support the engine.

14. Mark the position of the rear engine support and unbolt the 2 outer bolts, remove the top bolt at the transmission and pull the support out.

15. Disconnect the speedometer cable and the front driveshaft U-joint. Push the driveshaft back and wire it aside.

16. Unbolt the engine mounts on both sides and, on 4 cylinder engines, the front limit stop.

17. Unbolt the power steering fluid reservoir and swing it aside; then, using a chain hoist and cable, lift the engine and transmission upward and outward. An angle of about 45 degrees will allow the vehicle to be pushed backward while the engine is coming up.

To install:

18. Using a chain hoist and cable, position the engine and transmission into the vehicle. An angle of about 45 degrees will allow the vehicle to be pushed backward while the engine is going in. Install the power steering reservoir.

19. Install the engine mounts on both sides and, on 4 cylinder engines, the front limit stop. Torque the bolts to 25 ft. lbs. (34 Nm).

20. Connect the speedometer cable and the front driveshaft U-joint.

21. Install the top bolt at the transmission and support.

22. Install the exhaust pipe bracket to the transmission.

23. Install the clutch slave cylinder to the bell housing and connect the turn spring.

24. Install the hydraulic line to the clutch housing and the oil line connectors to the automatic transmission.

25. Install the hydraulic engine shock absorber and torque to 30 ft. lbs. (41 Nm).

26. Tighten the steering relay arm, along with the center steering rod and hydraulic steering damper.

27. Connect the gearshift linkage and the exhaust pipes to the manifolds. Torque the manifold bolts to 20 ft. lbs. (25 Nm).

28. Connect all ground straps and electrical connections.

29. Connect the accelerator linkage.

30. Install the viscous coupling and fan.

31. Install the all fuel, vacuum and oil hoses (e.g., power steering and power brakes). Plug all openings to keep out dirt. Install the air cleaner.

32. Connect all heater hoses and oil cooler lines.

33. Install the fan shroud and radiator.

34. Install the hood, refill the cooling system and connect the battery.

35. Bleed the hydraulic clutch, power steering, power brakes and fuel system.

V8 Engines

EXCEPT 5.0L DOHC

NOTE: Removal of a V8 engine equipped with air conditioning, may require discharging the air conditioning system. Use caution; always wear safety glasses, Freon is lethal.

1. Remove the hood.

2. Drain the cooling system.

3. Remove the radiator and fan shroud.

4. Remove the cable plug from the temperature switch.

5. Remove the battery, battery frame and air filter.

6. Drain the power steering reservoir and windshield washer reservoir.

7. Disconnect and plug the high pressure and return lines on the power steering pump.

8. Detach the fuel lines from the fuel filter, pressure regulator, and pressure sensor.

9. If equipped, loosen the line to the supply and anti-freeze tanks. If equipped, disconnect the lines to the hydro-pneumatic suspension.

10. Disconnect the cables from the ignition coil and transistor ignition switchbox.

11. Disconnect the brake vacuum lines.

12. Detach the cable connections for the following:

a. Venturi control unit
b. Temperature sensor
c. Distributor
d. Temperature switch
e. Cold start valve

13. Remove the regulating shaft by pushing it in the direction of the firewall.

14. Disconnect the thrust and pullrods.

15. Disconnect the heater lines.

16. Detach the lines to the oil pressure and temperature gauges.

17. Remove the ground strap from the vehicle.

18. Detach the cables from the alternator, terminal bridge, and battery. Remove the battery.

19. Position a lifting sling on the engine and take up the slack in the chain.

20. Remove the left side engine mount and loosen the hex nut on the right side mount.

21. Remove the exhaust system. Remove the connecting rod chain on the rear level control valve and loosen the torsion bar slightly. Raise the vehicle slightly at the rear and remove the exhaust system in the rearward direction.

22. Disconnect the hand brake cable.

23. Remove the shield plate from the transmission tunnel.

24. Place a block of wood between the transmission and cross-yoke so the engine will not sag when the rear mount is removed.

25. Loosen the driveshaft intermediate bearing and the driveshaft slide.

26. Support the transmission.

27. Mark the installation of the crossmember and remove the crossmember. Remove the rear engine carrier with the engine mount.

28. Unbolt the front U-joint flange on the transmission and push it back. Do not loosen the clamp nut on the intermediate bearing. Support the driveshaft.

29. Disconnect the speedometer shaft, shift rod, control pressure rod, regulating linkage, for automatic transmissions, kickdown switch cable, starter lockout switch cable, and the cable for the back-up light switch.

30. Remove the front engine mounting bolt and remove the engine at approximately a 45 degree angle.

31. Installation is the reverse of removal. Lower the engine until it is behind the front axle carrier. Support the transmission and lower the engine into its compartment. While lowering the engine, install the right-hand shock mount.

32. Fill the engine with all required fluids and start the engine. Check for leaks.

To install:

33. Lower the engine until it is behind the front axle carrier. Support the transmission. While lowering the engine, install the right-hand shock mount.

34. Install the front engine mounting bolt and torque to 40 ft. lbs. (54 Nm).

35. Connect the speedometer shaft, shift rod, control pressure rod, regulating linkage, for automatic transmissions, kickdown switch cable, starter lockout switch cable, and the cable for the back-up light switch.
36. Install the front U-joint flange on the transmission.
37. Raise the transmission into place. Install the rear engine carrier, engine mount and crossmember. Torque the bolts to 35 ft. lbs. (47 Nm).
38. Tighten the driveshaft intermediate bearing and the driveshaft slide.
39. Install the shield plate to the transmission tunnel.
40. Connect the hand brake cable.
41. Install the exhaust system and torque the nuts to 25 ft. lbs. (34 Nm). Install the connecting rod chain on the rear level control valve and tighten the torsion bar slightly.
42. Install the left side engine mount and tighten the hex nut on the right side mount to 15 ft. lbs. (20 Nm).
43. Connect the cables to the alternator and terminal bridge. Install the battery.
44. Install the ground strap to the vehicle. Do not connect the battery terminals at this time.
45. Connect the lines to the oil pressure and temperature gauges.
46. Connect the heater lines.
47. Connect the thrust and pullrods.
48. Install the regulating shaft.
49. Connect the cable connections for the following:
 a. Venturi control unit
 b. Temperature sensor
 c. Distributor
 d. Temperature switch
 e. Cold start valve
50. Connect the brake vacuum lines.
51. Connect the cables to the ignition coil and transistor ignition switchbox.
52. Tighten the line to the supply and anti-freeze tanks. If equipped, connect the lines to the hydro-pneumatic suspension.
53. Connect the fuel lines to the fuel filter, pressure regulator, and pressure sensor.
54. Connect and plug the high pressure and return lines on the power steering pump.
55. Refill the power steering reservoir and windshield washer reservoir.
56. Install the battery, battery frame and air filter.
57. Install the cable plug to the temperature switch.
58. Install the radiator and fan shroud.
59. Refill the cooling system and check all engine fluids.
60. Install the hood with the help of an assistant, connect the battery cable, start the engine and check for leaks.

5.0L DOHC ENGINE

1. Disconnect the negative battery cable.
2. Mark the hood fasteners. With the help of an assistant, remove the engine hood.
3. Remove the air filter, engine compartment bottom panel and air conditioning guard plate to the condenser.
4. Drain the engine coolant and remove the radiator.
5. Disconnect the heater hoses at the firewall.
6. Remove the viscous fan coupling and fan.
7. Remove the air conditioning belt, if so equipped.
8. Label and disconnect all engine wiring, vacuum lines, throttle linkage, power steering and oil cooler lines.
9. Remove the wheelhouse assembly.
10. Completely remove the exhaust system.
11. Remove the starter motor shield and air conditioning compressor.
12. Mark and disconnect the driveshaft from the transmission.
13. Label and disconnect all transmission wiring and control cables.
14. Support the transmission using a suitable jack. Disconnect the transmission mount and remove the crossmember.
15. Install engine lifting device.
16. Disconnect the engine mounts.
17. Remove the guard plate at the engine firewall.
18. Raise the engine very carefully and disconnect any connected components.

To install:

19. With the help of an assistant, lower the engine into the vehicle on a 45 degree angle.
20. Install the guard plate at the engine firewall.
21. Connect the engine mounts and torque to 25 ft. lbs. (34 Nm).
22. Support the transmission using a suitable jack. Connect the transmission mount and install the crossmember. Torque the bolts to 30 ft. lbs. (41 Nm).
23. Connect all transmission wiring and control cables.
24. Connect the driveshaft to the transmission. Use new nuts and torque to 15 ft. lbs. (20 Nm).
25. Install the starter motor shield and air conditioning compressor.
26. Install the exhaust system and torque the manifold nuts to 20 ft. lbs. (25 Nm).
27. Install the wheelhouse assembly.
28. Connect all engine wiring, vacuum lines, throttle linkage, power steering and oil cooler lines.
29. Install the air conditioning belt, if so equipped.
30. Install the viscous fan coupling and fan.
31. Connect the heater hoses at the firewall.
32. Install the radiator and refill the engine coolant.
33. Install the air filter, engine compartment bottom panel and air conditioning guard plate to the condenser.
34. With the help of an assistant, install the engine hood.
35. Connect the negative battery cable, refill all fluids, start the engine and check for leaks.

Engine Mounts

REMOVAL & INSTALLATION

CAUTION

The engine has to be raised to take the tension off the engine mount. Make sure the engine safely supported before attempting to remove the mount. Failure to secure the engine may result in personal injury.

1. Disconnect the negative battery cable.
2. Remove the air scoops, fan cowl rings, engine compartment bottom panels or any vehicle undercovers that are in the way.
3. Place a suitable jack and a piece of wood under the engine. Do not jack under the oil pan.
4. Raise the engine far enough to take the tension off the engine mount. Use blocks of wood between the frame and engine in case the jack slips.
5. Remove the engine mount retaining bolts and engine mount.

To install:

6. Install the mount and bolts. Be careful not dislodge the jack while servicing the mount.
7. Torque the bolts to 18 ft. lbs. (25 Nm).
8. Lower the jack and connect the battery cable.

Cylinder Head

REMOVAL & INSTALLATION

4 Cylinder Engine

EXCEPT 190E-16 (DOHC)

This is fairly straight forward but some caution must be observed to ensure that the valve timing is not disturbed.

NOTE: The cylinder head should be removed cold.

1. Disconnect the negative battery cable. Drain the radiator and remove

all hoses and wires. Tag all wires to ensure easy reassembly.

2. The cylinder head cover on the 190E is removed with the spark plug cables and distributor cap still attached to it.

3. The rockers and their supports must be removed together.

4. Mark the chain, sprocket and cam for ease of assembly.

5. Using a suitable puller, remove the camshaft sprocket.

6. Remove the sprocket and chain and wire it aside.

7. Make sure the chain is securely wired so it will not slide into the engine.

8. Unbolt the manifolds and exhaust header pipe and push them aside.

9. Loosen the cylinder head hold-down bolts in the reverse order of that shown in torque diagrams for each model. It is good practice to loosen each bolt a little at a time, working around the head, until all are free. This prevents unequal stresses in the metal.

10. Reach into the engine compartment and gradually work the head loose from each end by rocking it. Never, under any circumstances, use a prybar between the head and block to pry, as the head will be scarred badly and may be ruined.

To install:

11. Clean the gasket mating surfaces and check for warpage.

12. With the help of an assistant, install the cylinder head onto the engine.

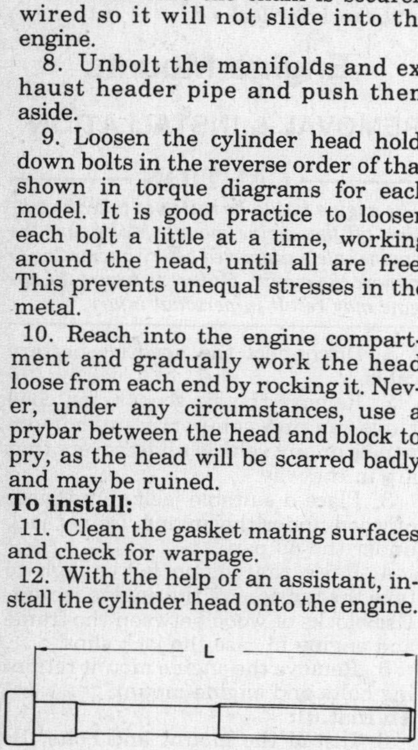

Cylinder head bolt stretch is measured at "L"

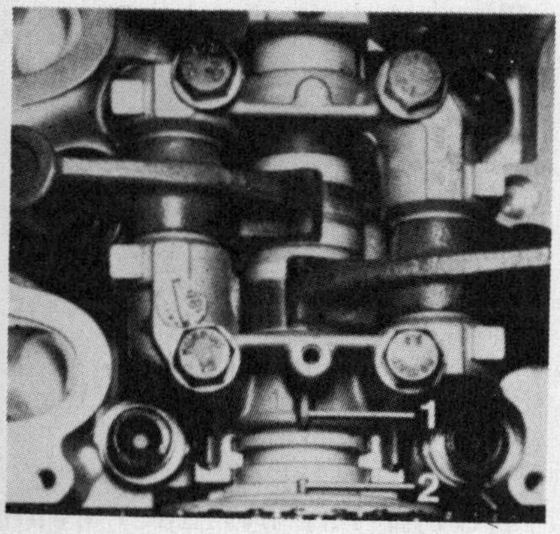

The timing marks on the camshaft bearing cap (1) and the camshaft (2) should be in alignment when the No. 1 cylinder is at TDC

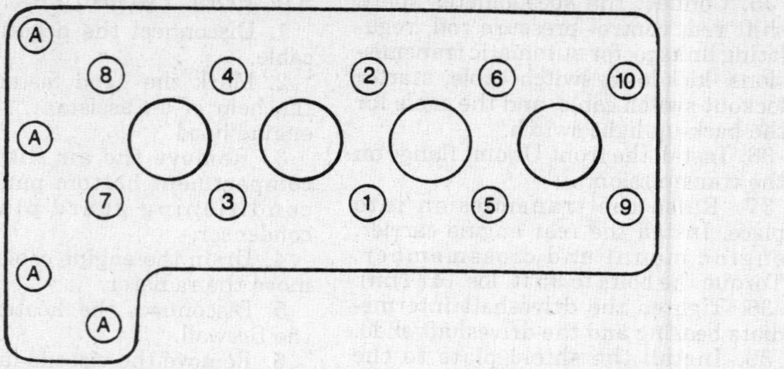

Cylinder head torque sequence—4 cylinder engine, (A) bolts to 18 ft. lbs. (22 Nm)

Torque the head in 4 steps, in sequence. The 1st step to 18 ft. lbs. (25 Nm), 2nd step to 29 ft. lbs. (39 Nm) plus allow to set for 10 minutes, 3rd step to 90 degrees torquing angle and 4th step to 90 degrees torquing angle.

13. Install the manifolds and exhaust header pipe. Torque the pipe to 20 ft. lbs. (27 Nm).

14. Install the sprocket and timing chain. Align the timing marks.

15. Install the the rockers and their supports.

16. Install the cylinder head cover and spark plug wires.

17. Reconnect all wiring, hoses and cables to the cylinder head.

18. Connect the negative battery cable. Refill the radiator.

190E-16 (DOHC)

NOTE: The cylinder head should be removed cold.

1. Disconnect the negative battery cable. Drain the engine coolant and disconnect the radiator and heater hoses at the cylinder head.

2. Tag and disconnect the intake air temperature sensor lead and the crankcase ventilation hose where they connect to the air cleaner.

3. Remove the 2 air cleaner mounting nuts. Lift the housing at the rear until it releases from its holding studs, slide it backwards slightly and remove it from the air flow sensor.

4. Loosen the oil dipstick mounting bracket screw and pull the dipstick from the crankcase.

5. Loosen the screw in the center of the serpentine belt tensioner ¼–½ turn, loosen the tensioning nut by turning it counterclockwise and remove the belt.

6. Remove the exhaust manifold.

7. Remove the ignition cable cover-to-cylinder head cover screws and the cover. Label and disconnect the cables and position them aside.

8. Loosen the clamp on the rear heater hose and pull it off the water outlet.

9. Disconnect the breather line at the cylinder head cover.

10. Remove the 6 mounting nuts and lift up the cylinder head cover.

NOTE: If the cylinder head cover sticks to the head, do not use a hammer to loosen it as it may crack. Try to break the seal by pushing at both corners on 1 side or the other by hand.

11. Set the No. 1 piston to TDC, of the compression stroke, by turning the crankshaft in the directions of normal engine rotation. When the 2 punch marks in the camshaft sprockets are aligned, the engine will be at TDC.

12. Place matchmarks on the camshaft sprocket and the timing chain.

13. Remove the alternator air duct. Label and disconnect the electrical leads. Pull the harness through the component compartment wall and position it aside. Remove the alternator.

14. Using a 32mm wrench, unscrew the chain tensioner.

15. Label and disconnect the two pump lines and remove the pump from the exhaust side camshaft. Remove the 3 screws and pull the pump flange from the end of the camshaft.

16. Remove the 2 mounting nuts and lift the chain slide from the front of the engine.

NOTE: If equipped, remove the sheet metal bracket that is attached to the 2 front cylinder head bolts and the 2 eyes at the timing chain housing cover.

17. Remove the 4 mounting screws from each camshaft sprocket. Knock the camshaft back slightly with a rubber mallet; be careful. Remove the front bearing caps and pull off the 2 sprockets.

NOTE: Secure the timing chain so it will not slip into the crankcase.

18. Loosen the water bypass hose clamp. Remove the mounting screws

and pull out the water inlet/thermostat housing.

19. Use a Allen wrench to remove the 2 return pipe mounting screws and pull it from the cylinder head. If the pipe sticks, rotate it clockwise slightly and force it out with a suitable prybar.

20. Unbolt the intake manifold and push it aside.

21. Loosen the cylinder head bolts in the reverse order of the tightening sequence. Loosen each bolt, a little at a time, working around the head until all are free; this will prevent unequal stress on the aluminum head.

22. Reach into the engine compartment and gradually work the head loose from the cylinder block. Never use a prybar to pry the head free.

To install:

23. Clean the gasket mating surfaces and check for warpage. With an assistant, install the cylinder head onto the engine. Please note the following:

a. Measure the cylinder head bolts prior to installation. A new bolt is 110mm long from the bottom of bolt head to the end of the bolt. If it has stretched to more than 113mm, replace it.

b. Tighten the cylinder head bolts, in steps, in sequence.

c. Use new O-rings and a new flange gasket when installing the return pipe.

d. Install the intake camshaft sprocket first; right side when facing the vehicle. Be certain that the matchmarks are aligned.

e. Make sure the 2 punch marks in the sprockets are in alignment.

24. Install the remaining components. Refill the engine fluids, start the engine and check for leaks.

6 Cylinder SOHC

NOTE: The cylinder head on the 190E with 2.6L engine, 260E, 300E, 300SE, 300TE, and 300SEL should be removed cold, with the camshaft, intake and exhaust manifolds attached.

1. Disconnect the negative battery cable. Remove the engine undercovers from below.

2. Drain the engine coolant. Drain the engine oil.

3. Remove the air filter.

4. Remove the distributor cap mounting bolts. Unbolt the cylinder head cover and remove it with the ignition wires and distributor cap still attached.

NOTE: Distributor cap removal will require a 5mm T-shaped Allen wrench about 80mm in length.

5. Loosen the 3 Allen screws and lift off the distributor rotor.

6. Using a 6mm Allen wrench, unscrew the distributor driver and remove it. Carefully, pry off the protective cover.

7. Remove the mounting screws for the cylinder head front cover and carefully knock the cover off with a rubber mallet.

8. Rotate the crankshaft so the No. 1 cylinder is set at TDC of the compression stroke.

9. Unscrew the timing chain tensioner plug and remove the compression spring.

10. Use a 17mm Allen-head socket and unscrew the tensioner threaded ring.

11. Insert an M8 screw into the tensioner bore, tilt it slightly and ease the tensioner from the bore. If the tensioner is difficult to remove, loosen the socket head screw above the tensioner bore slightly; this should facilitate removal.

12. Matchmark the camshaft sprocket to the camshaft by putting a dab of paint next to the hole in the sprocket with the dowel pin.

13. Matchmark the camshaft sprocket to the timing chain.

14. Remove the mounting screws and pull off the camshaft sprocket. Secure the timing chain so it will not slip into the crankcase.

15. Remove the slide rail bolt with an impact puller.

16. Unscrew the oil dipstick guide tube bracket and pull out the dipstick and tube.

17. Unscrew the upper intake manifold mounting bolt. Loosen the lower bolt.

18. Loosen the hose clamp and remove the coolant hose at the water pump.

19. Unscrew the exhaust pipe at both flanges.

20. Disconnect the automatic transmission dipstick tube at the cylinder head and position it aside.

21. Tag and disconnect all wiring, electrical leads and vacuum hoses connected to or in the way of the cylinder head.

22. Disconnect the fuel feed and return lines, plug them and position them aside.

23. Disconnect the accelerator pedal Bowden cable.

24. Loosen the cylinder head bolts in the reverse order of the tightening sequence. Loosen each bolt, a little at a time, working around the head until all bolts are free; this will prevent unequal stress on the aluminum head.

25. Reach into the engine compartment and gradually work the head loose from the cylinder block. Never use a prybar to pry the head free.

To install:

26. Position a new cylinder head gasket on the cylinder block.

27. Connect the water pump coolant hose to the head and position the head on the block. There are 2 dowel pins for locating purposes.

28. Measure the length of the cylinder head bolts from the underside of the bolt head to the end of the bolt. If the length exceeds 108.4mm, the bolts

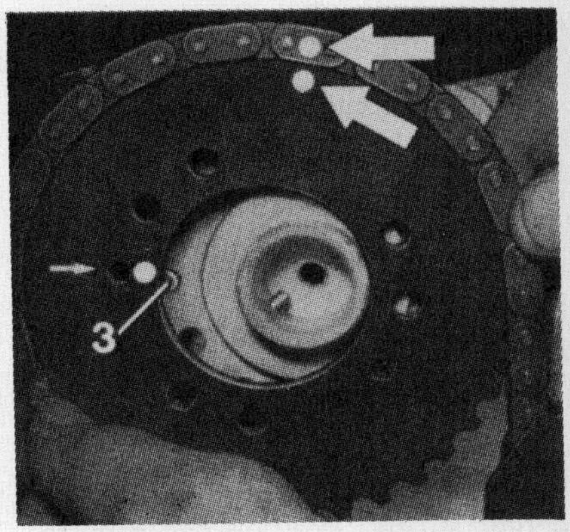

Matchmark the camshaft sprocket on the 260E, 300E, 300CE, 300SEL and 300TE

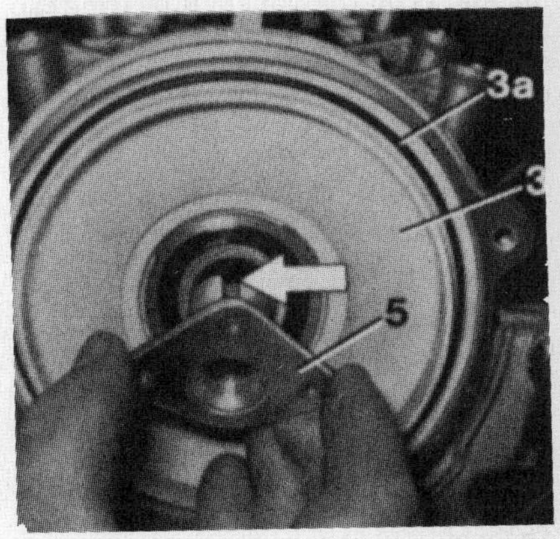

The groove (arrow) on the driver (5) must engage the pin in the camshaft—260E, 300E, 300CE, 300SE, 300SEL and 300TE

Removing the chain tensioner on the 260E, 300E, 300CE, 300SE, 300SEL and 300TE

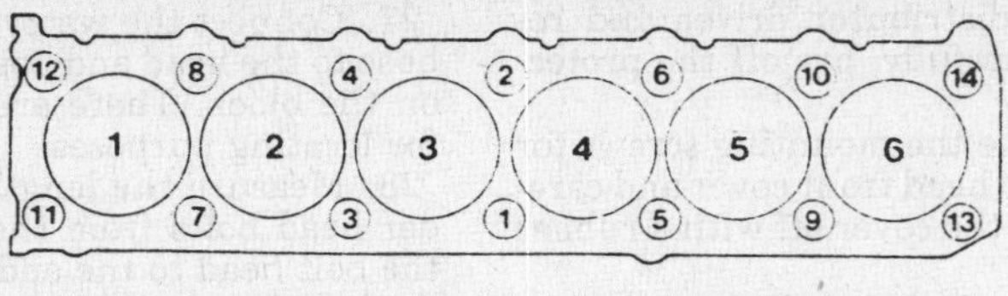

Cylinder head bolt torque sequence—6 cylinder engines

must be replaced with new stretch bolts.

29. Install the cylinder head bolts and tighten them a little at a time, in sequence.

30. Install the camshaft sprocket and tighten the bolts to 8 ft. lbs. (11 Nm). Be sure the dowel pin is in the hole marked previously and that the matchmarks on the timing chain and sprocket are aligned.

31. Slide the chain tensioner housing into the bore. Screw in the threaded ring and tighten it to 22 ft. lbs. (30 Nm). Install the thrust bolt with the detent spring. Position the compression spring and a new seal. Tighten the plug to 37 ft. lbs. (50 Nm).

32. Check the alignment of the timing marks on the camshaft bearing cap and the camshaft. When they are aligned, the engine should be at TDC of the compression stroke.

33. Install a new elastic gasket into the groove of the timing chain housing cover and mount the front cover. Tighten the 2 lower screws first. Torque all screws to 15.5 ft. lbs. (21 Nm).

34. Install the protective cover with a new seal. Install the distributor driver so the groove engages the pin on the camshaft. Tighten the screw to 15.5 ft. lbs. (21 Nm).

35. Installation of the remaining components is in the reverse order of removal.

NOTE: When refilling the coolant system on the 300E, 300CE, 300SE, 300TE and 300SEL, always remove the hex head plug on the left side of the cylinder head and fill the hole with coolant until it overflows. Install the plug and fill the coolant system. When filling the coolant system on the 190E with 2.6L engine and 260E, open the vent screw approximately 2 turns, start the engine and run at idle.

6 Cylinder DOHC

NOTE: The cylinder head on the 1990–92 300CE and 300SL should be removed cold, with the camshafts, intake and exhaust manifolds attached.

1. Disconnect the negative battery cable. Remove the engine undercovers from below.

2. Drain the engine coolant. Drain the engine oil.

3. Remove the air filter.

4. Remove the distributor cap mounting bolts. Unbolt the cylinder head cover and remove it with the ignition wires and distributor cap still attached.

5. Remove the camshaft adjuster armature for the variable valve timing.

6. Remove the upper guide track and secure the camshafts from movement.

7. Disconnect the cable from the reference value transmitter at the right wheel well (300SL).

8. Label and disconnect all electrical wiring, hoses and cables from the cylinder head assembly.

9. Disconnect the exhaust pipe from the manifold.

10. Label and remove the air injection line, dipstick guide, crankcase breather and fuel lines from the fuel distributor.

11. Remove the cylinder head bolts in the opposite sequence then torquing.

12. Tap the head with a plastic hammer. With an assistant or suitable hoist, remove the cylinder head making sure all components are disconnected.

To install:

13. Clean the gasket mating surfaces and check for warpage. Match old gasket to new for perfect match.

14. Turn the No. 1 piston to TDC.

15. Install the gasket. With a suitable hoist, lower the cylinder head onto the engine and oil and loosely install the head bolts.

16. Turn the camshafts such that the lower edges of the holes in the camshaft flange is level with the top of the cylinder head. Secure the 5mm pins.

17. Torque the cylinder head bolts in 3 steps, in sequence. The 1st step to 52 ft. lbs. (70 Nm), 2nd step to 90 degrees angle of rotation and 3rd step to 90 degrees of rotation.

18. Install the timing chain and remove the pins from the camshaft flange.

19. Install all wiring, hoses and cables.

20. Install the exhaust pipe and torque to 20 ft. lbs. (25 Nm).

21. Install the remaining components. Refill the engine fluids, connect

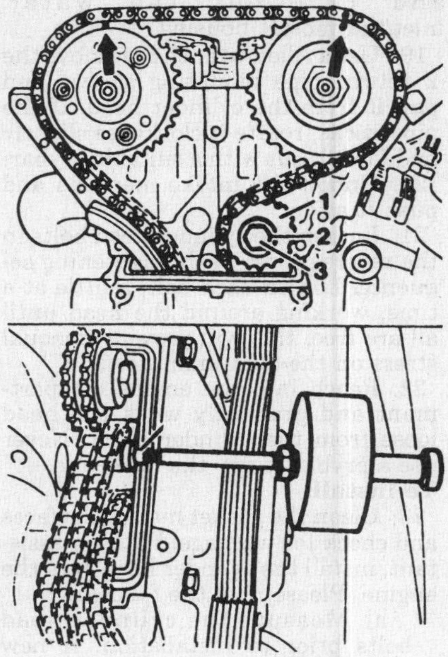

1. Idler gear
2. Washer
3. Idler bolt

Camshaft sprocket alignment—3.0L DOHC engine

the battery cable, start the engine and check for leaks.

V8 Engines

EXCEPT 5.0L DOHC

NOTE: Before removing the cylinder head from a V8, obtain the 4 special tools necessary to torque the head bolts; without them it will be impossible. Do not confuse the left and right side head gaskets. The left side has 2 attaching holes in the timing chain cover, the right side has only 1 hole.

Cylinder heads can only be removed with the engine cold.

1. Drain the cooling system.

2. Remove the battery.

3. Remove the air cleaner. Remove the fan and fan shroud.

4. Pull the cable plug from the temperature sensor.

5. Detach the vacuum hose from the venturi control unit.

6. Remove the following electrical connections:

   a. Injection valves.
   b. Distributor
   c. Venturi control unit
   d. Temperature sensor and temperature switch
   e. Starting valve
   f. Temperature switch for the auxiliary fan.

7. Loosen the ring line on the fuel distributor.

8. Loosen the screws on the injection valves and pressure regulator or

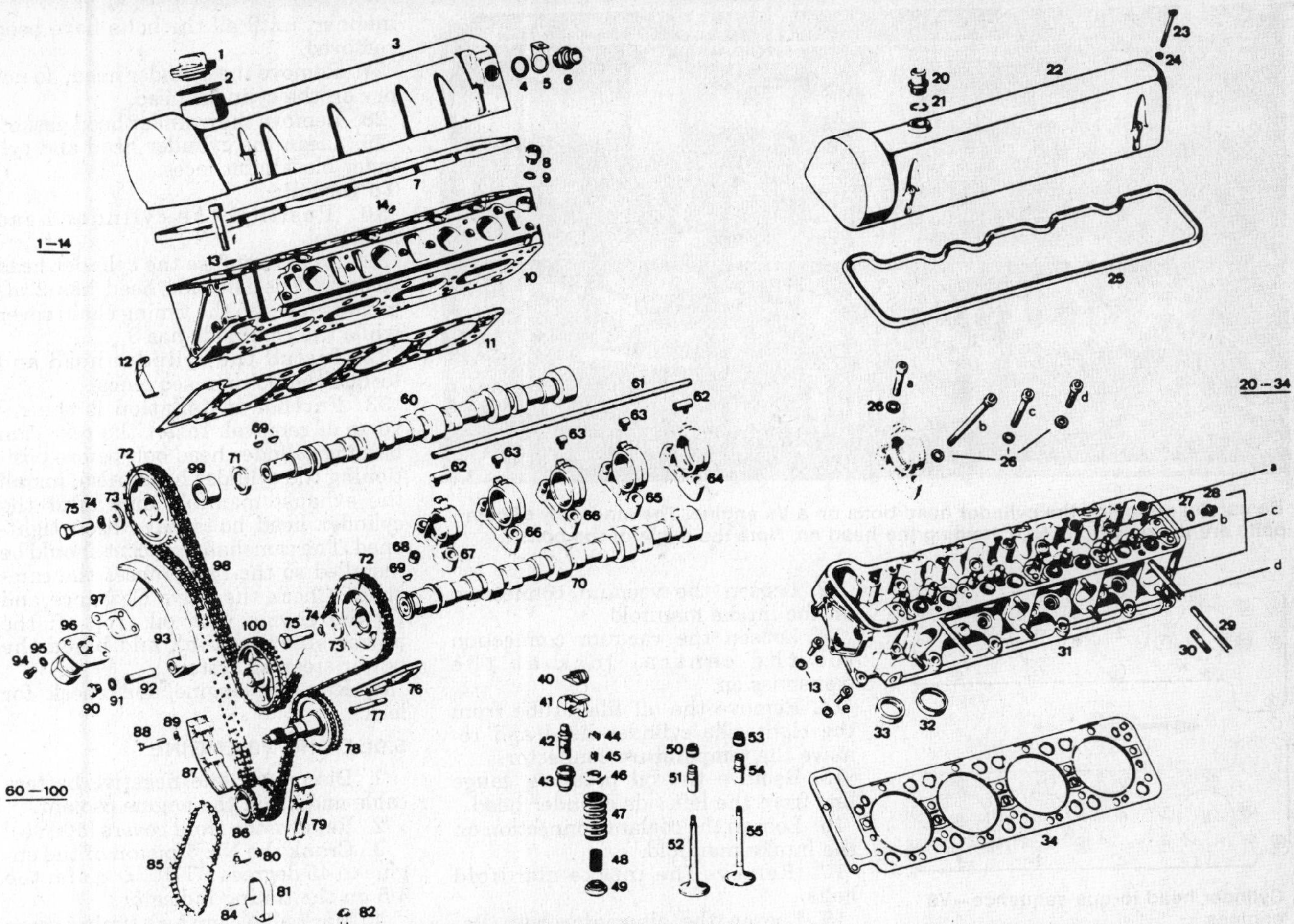

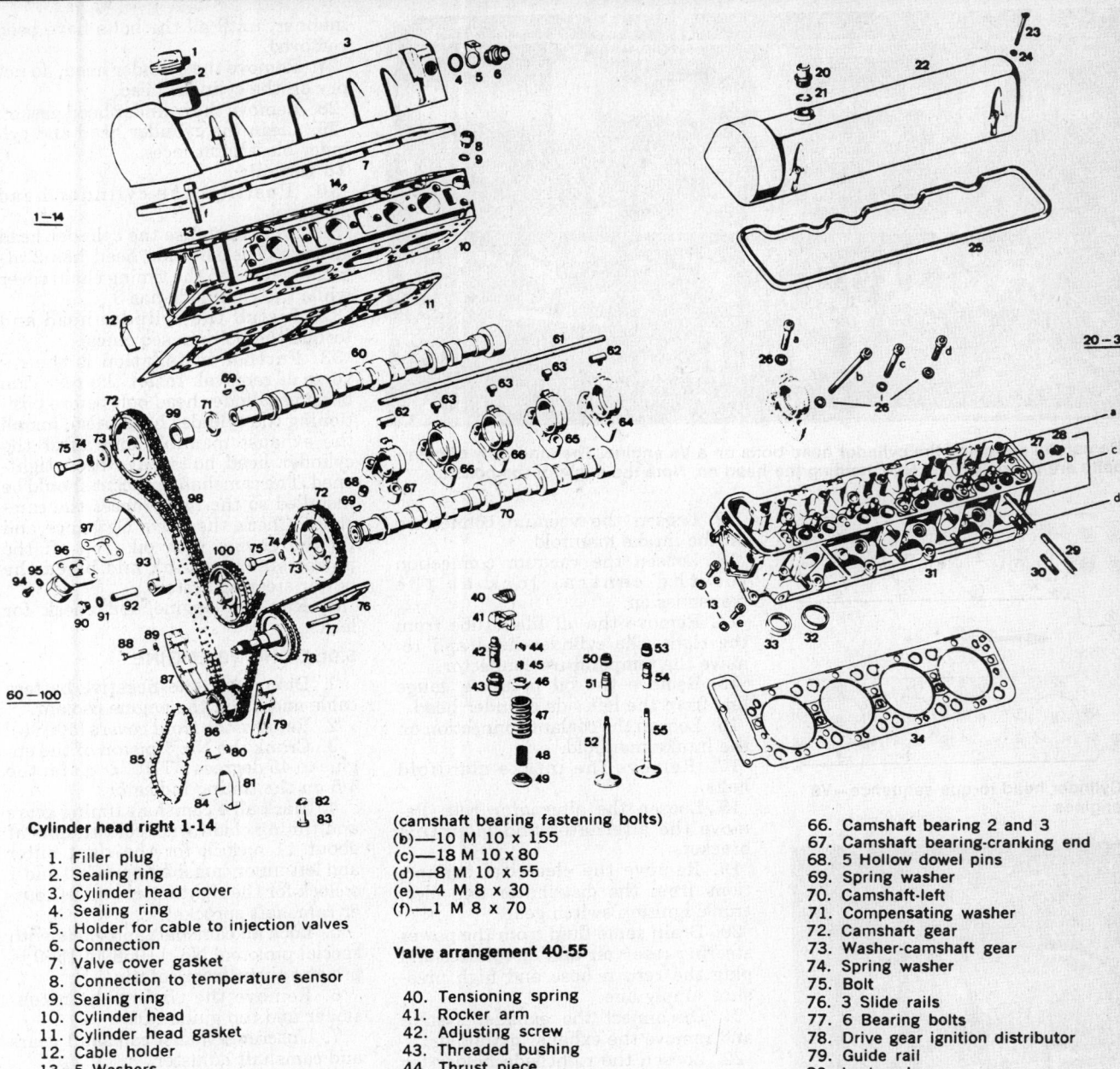

Cylinder head right 1-14

1. Filler plug
2. Sealing ring
3. Cylinder head cover
4. Sealing ring
5. Holder for cable to injection valves
6. Connection
7. Valve cover gasket
8. Connection to temperature sensor
9. Sealing ring
10. Cylinder head
11. Cylinder head gasket
12. Cable holder
13. 5 Washers
14. Hollow dowel pins

Cylinder head left 20-34

20. Connection
21. Sealing ring
22. Cylinder head cover
23. 8 Screws
24. 8 Sealing rings
25. Cylinder head cover gasket
26. 36 Washers
27. Sealing ring
28. Screw connection oil pressure gauge
29. 3 Studs
30. 13 Studs
31. Cylinder head
32. Valve seat ring—intake
33. Valve seat ring—exhaust
34. Cylinder head gasket

Cylinder head bolts

(a)—10 M 10 x 50chrauben)
(camshaft bearing fastening bolts)
(b)—10 M 10 x 155
(c)—18 M 10 x 80
(d)—8 M 10 x 55
(e)—4 M 8 x 30
(f)—1 M 8 x 70

Valve arrangement 40-55

40. Tensioning spring
41. Rocker arm
42. Adjusting screw
43. Threaded bushing
44. Thrust piece
45. Valve cone piece
46. Valve spring retainer
47. Outer valve spring
48. Inner valve spring
49. Rotator
50. Intake valve seal
51. Exhaust valve guide
52. Intake valve
53. Exhaust valve seal
54. Exhaust valve guide
55. Exhaust valve

Engine timing 60-100

60. Camshaft-right
61. Oil pipe (external lubrication)
62. Connecting piece
63. Connecting piece

Oil pipe to camshaft bearing

64. Camshaft bearing-flywheel end
65. Camshaft bearing 4
66. Camshaft bearing 2 and 3
67. Camshaft bearing-cranking end
68. 5 Hollow dowel pins
69. Spring washer
70. Camshaft-left
71. Compensating washer
72. Camshaft gear
73. Washer-camshaft gear
74. Spring washer
75. Bolt
76. 3 Slide rails
77. 6 Bearing bolts
78. Drive gear ignition distributor
79. Guide rail
80. Lockwasher
81. Spring—chain tensioner, oil pump
82. Washer
83. Screw
84. Clamp
85. Single roller chain (oil pump drive)
86. Crankshaft gear
87. Slide rail
88. 4 Screws
89. 4 Spring washers
90. Plug
91. Sealing ring
92. Bearing bolt
93. Tensioning lever
94. 2 Bolts
95. 2 Spring washers
96. Chain tensioner
97. Gasket
98. Double roller chain
99. Spacer ring
100. Idler gear

Exploded view of the V8 cylinder head, except 5.0L engine

Be careful removing the cylinder head bolts on a V8 engine. The inner row of cam bolts are the only bolts NOT holding the head on. Note the angle of the bolts

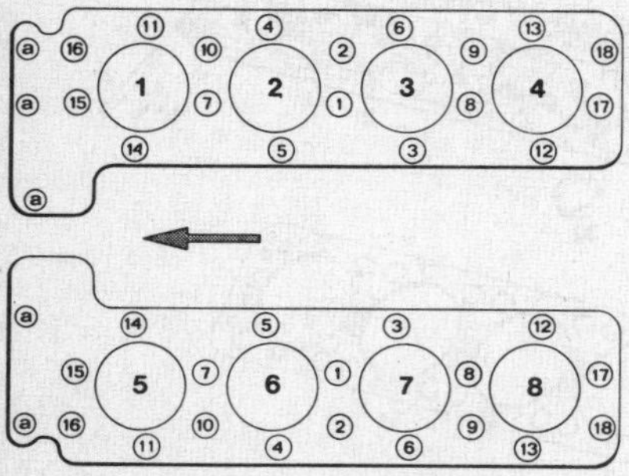

Cylinder head torque sequence—V8 engines

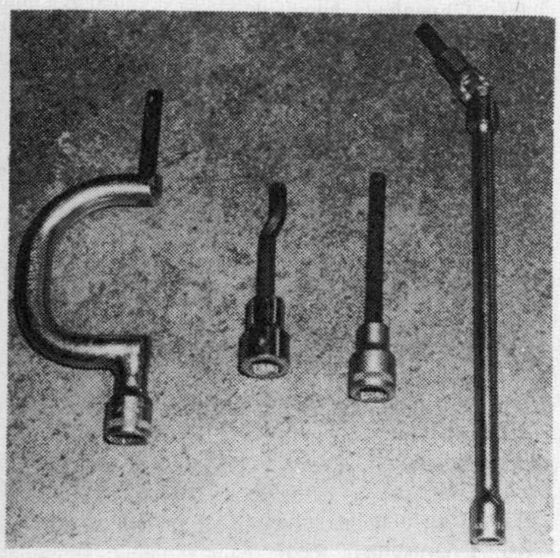

Without these tools, it is practically impossible to service the V8 cylinder head.

mixture regulator. Remove the ring line with the injection valves and pressure regulator.

9. Plug the holes for the injection valves in the cylinder head.

10. Remove the regulating shaft by disconnecting the pull rod and the thrust rod.

11. Remove the ignition cable plug.

12. Loosen the vacuum connection on the intake manifold.

13. Loosen the vacuum connection for the central lock at the transmission.

14. Remove the oil filler tube from the right side cylinder head and remove the temperature connector.

15. Remove the oil pressure gauge line from the left side cylinder head.

16. Loosen the coolant connection on the intake manifold.

17. Remove the intake manifold bolts.

18. Loosen the alternator belt. Remove the alternator and mounting bracket.

19. Remove the electrical connections from the distributor and electronic ignition switch gear.

20. Drain some fluid from the power steering reservoir and disconnect and plug the return hose and high pressure supply line.

21. Disconnect the exhaust system and remove the exhaust manifolds.

22. Loosen the right side holder for the engine damper.

23. Remove the right side chain tensioner.

24. Matchmark the camshaft, camshaft sprocket and chain. Remove the camshaft sprocket and chain after removing the cylinder head cover. Be sure to hang the chain and sprocket to prevent it from falling into the timing chain case.

25. Remove the upper slide rail. Remove the distributor and remove the inner slide rail on the left cylinder head. Remove the rail after the camshaft sprocket.

26. Unscrew the cylinder head bolts; this should be done with a cold engine. Unscrew the bolts in the reverse order of the torque sequences. Unscrew all the bolts, a little at a time, in the same manner, until all the bolts have been removed.

27. Remove the cylinder head; do not pry on the cylinder head.

28. Remove the cylinder head gasket.

29. Clean the cylinder head and cylinder block joint faces.

To install:

30. Position the cylinder head gasket.

31. Do not confuse the cylinder head gaskets. The left side head has 2 attaching holes in the timing chain cover while the right side has 3.

32. Install the cylinder head and torque the bolts in sequence.

33. Further installation is the reverse of removal. Insert the rear cam bearing cylinder head bolt before positioning the cylinder head. Also, install the exhaust manifold only after the cylinder head bolts have been tightened. The camshaft sprocket should be installed so the flange faces the camshaft. Check the valve clearance and fill the engine with oil. Top off the power steering tank and bleed the power steering system.

34. Run the engine and check for leaks.

5.0L DOHC V8 ENGINE

1. Disconnect the negative battery cable and drain the engine coolant.

2. Remove the front covers (500SL).

3. Crank the No. 1 piston of the engine to 45 degrees BTDC. Look for the 4/5 on the timing indicator.

4. Mark all 4 camshaft timing gears and timing chain with colored dots at about 11 o:clock for the right outer and left inner camshaft sprocket and 1 o:clock for the right inner and left outer camshaft sprocket.

5. Lock all camshaft sprockets with special pins tool No. 119589001500 to prevent them from rotating.

6. Remove the timing chain tensioner and top guide rails.

7. Unscrew exhaust camshaft gears and camshaft adjuster.

8. Remove the engine cover and intake manifold.

9. Label and disconnect all electrical wiring, hoses and cables from the manifolds and cylinder head(s).

10. If equipped with an automatic transmission, remove the dipstick guide tube.

11. Disconnect the exhaust system from the head.

12. Remove the cylinder head cover very carefully.

NOTE: Make sure the engine is cold before removing cylinder head bolts.

13. Loosen the cylinder head bolts in the opposite sequence as torquing. A special Torx® like socket is needed to remove the bolts.

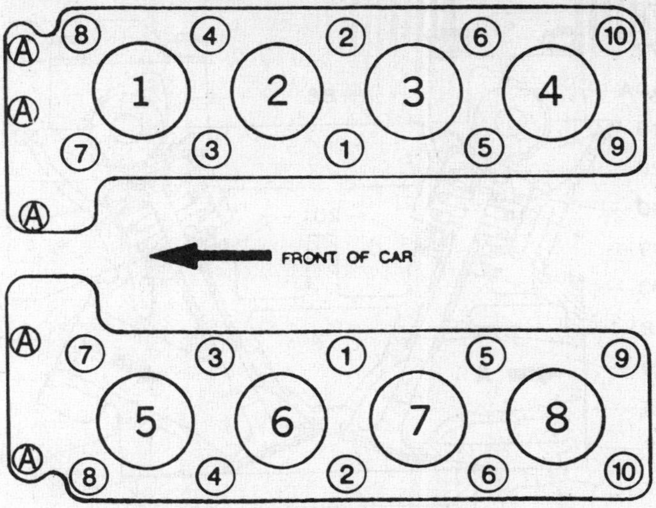

Cylinder head bolt torque sequence—5.0L DOHC engine, (A) bolts to 15 ft. lbs. (20 Nm)

14. Install a suitable lifting device onto the cylinder head and remove.

To install:

15. Clean the gasket mating surfaces and check for warpage.
16. Check the head bolts for stretch. The maximum length is 6.40 in. (162.70mm). Replace if excessive.
17. Install the cylinder head gasket and head. Torque the bolts in sequence, in steps. The 1st step to 40 ft. lbs. (55 Nm), 2nd step to 90 degrees angle of rotation and 3rd step to 90 degrees angle of rotation after waiting 10 minutes. Torque the M8 bolts near the timing sprockets to 18 ft. lbs. (25 Nm).
18. Connect the exhaust pipe and torque to 20 ft. lbs. (25 Nm).
19. Install the remaining components.
20. Connect all wiring, hoses and cables.
21. Refill the engine with fluids, connect the battery cable, start the engine and check for leaks.

Valve Lifters

REMOVAL & INSTALLATION

DOHC Engines

All engines are equipped with hydraulic valve lifters. They can be checked by placing the camshaft on the base circle and use a non-metal tool to push the lifter down. If it is easier to press one lifter in comparison to the others, this one must be replaced.

Remove the cylinder head cover, camshaft and pull the lifter out with a magnetic tool.

All Others

Temporarily removed valve lifters must be reinstalled in their original locations. When replacing worn rocker arms, the camshaft must also be replace. If the rocker arm or hydraulic lifter is replaced, check the base setting.

Remove the rocker arm and unscrew the valve lifter with a 24mm socket.

Valve Lash

CHECKING BASE SETTING

The base setting is the clearance between the upper edge of the cylindrical part of the plunger and the lower edge of the retaining cap (dimension A) when the cam lobe is vertical.

NOTE: A dial indicator with an extension and a measuring thrust piece (MBNA *100 589 16 63 00), 0.187 in. thick are necessary to perform this adjustment.

1. Turn the cam lobe to a vertical position, relative to the rocker arm.
2. Attach a dial indicator and tip extension and insert the extension through the bore in the rocker arm onto the head plunger. Preload the dial indicator by 0.08 in. and zero the instrument.
3. Depress the valve with a valve spring compressor. The lift on the dial indicator should be 0.028–0.075 in.
4. If the lift is excessive, the base setting can be changed by installing a new thrust piece.
5. Remove the dial indicator.
6. Remove the rocker arm.
7. Remove the thrust piece and insert the measuring disc.
8. Install the rocker arm and repeat Steps 1–3.
9. Select a thrust piece according to the table. If the measured valve was 0–0.002 in. and the 0.2146 in. thrust piece will not give the proper base setting, use the 0.2283 in. thrust piece.
10. Remove the dial indicator and the rocker arm. Install the selected thrust piece.
11. Reinstall the rocker arm and dial indicator and repeat Steps 1–3.

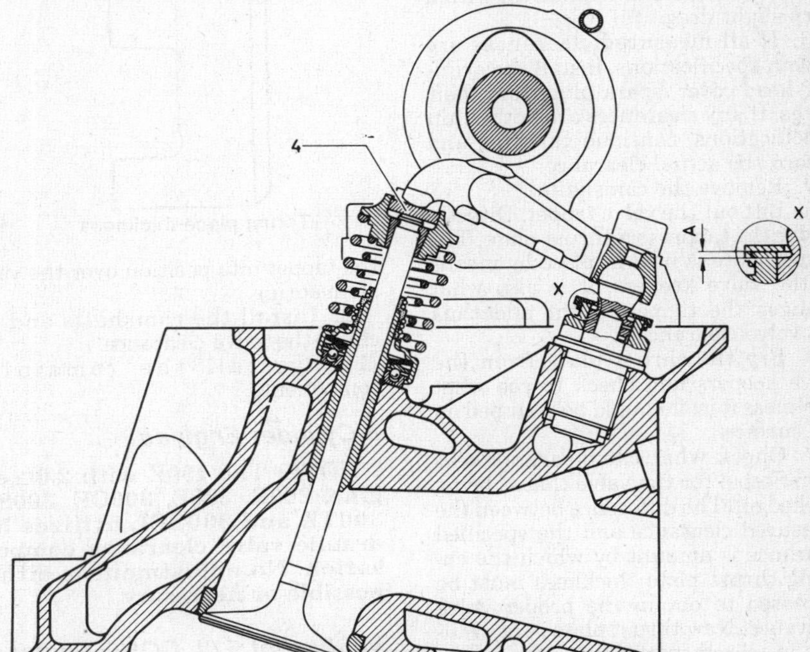

Cutaway of valve train showing hydraulic valve lifter. Dimension "A" is base setting clearance.

ADJUSTMENT

4 Cylinder Engine

190 SERIES

The 190E (SOHC) utilizes hydraulic valve clearance compensation. No adjustment is either possible or necessary. The 190E-16 (DOHC) uses mechanical lash adjusters, adjustable by means of tappets and thrust washers.

Valve clearance on the 190E-16 is measured between the cam base circle and the cup-type valve tappet.

1. Tag and disconnect the spark plug wires and position them aside.
2. Remove the spark plugs and the cylinder head cover.
3. Note the position of the intake and exhaust valves. Viewed from the front of the vehicle, the exhaust valves are on the left and the intake valves are on the right.
4. Using a wrench on the crankshaft pulley bolt, rotate the crankshaft until the heel of the camshaft lobe is in the position.

NOTE: Do not rotate the engine using the camshaft sprocket bolt. The strain will distort the timing chain tensioner rail. Always rotate the engine in the direction of normal rotation only.

5. To measure the valve clearance, insert a feeler gauge of the specified thickness between the heel of the camshaft lobe and the top of the valve tappet. Clearance is correct if the blade can be inserted and withdrawn with a very slight drag.
6. If all measured clearances are within specifications, install the cylinder head cover, spark plugs and their wires. If any clearances are not within specifications, continue checking and record the actual clearance.
7. Remove the camshafts.
8. Lift out the valve tappet. Directly under the tappet is a thrust plate. The thrust plate is held in place by means of the valve keepers; it is also what changes the tappet height and thus the valve clearance.
9. Pry the thrust plate from the valve keepers and check to see what thickness it is; it should be stamped on the surface.
10. Check what the clearance was from Step 6 for the valve that is being worked on. The difference between the measured clearance and the specified clearance is amount by which the existing thrust plate thickness must be increased to obtain the proper valve clearance. New thrust plates are available in increments of 0.05mm.
11. When the proper thickness of the new thrust plate has been determined, press it into the valve keepers and drop

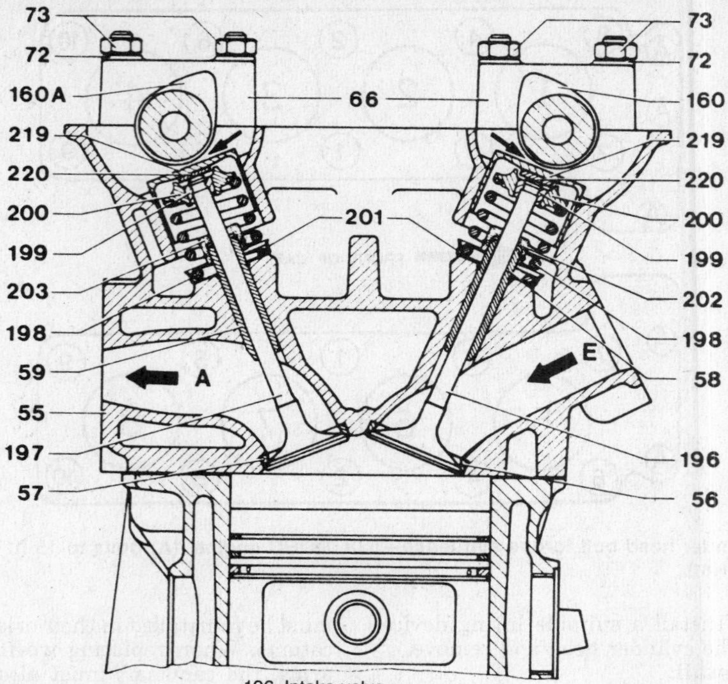

55 Cylinder head
56 Valve seat ring, intake
57 Valve seat ring, exhaust
58 Valve guide, intake
59 Valve guide, exhaust
160E Camshaft, intake
160A Camshaft, exhaust
196 Intake valve
197 Exhaust valve
198 Valve spring
199 Valve keeper
200 Valve spring retainer
201 Thrust ring
202 Valve stem seal, intake valve
203 Valve stem seal, exhaust valve
219 Valve tappet
220 Thrust plate
E Intake
A Exhaust

On the 190E-16 DOHC, measure valve clearance between the valve tappet and the heal of the camshaft lobe (small arrow)

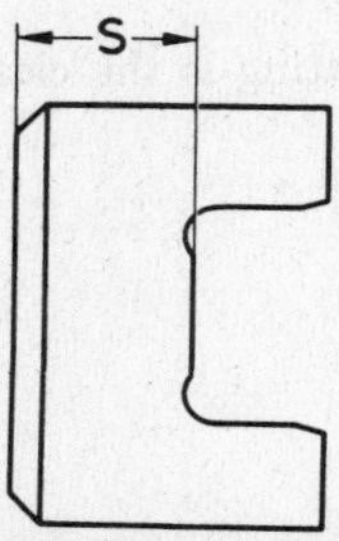

Thrust piece thickness

the tappet into position over the valve stem/spring.

12. Install the camshafts and recheck the valve clearance.
13. Install the remaining components.

6 Cylinder Engines

NOTE: The 190E with 2.6L engine, 260E, 300E, 300CE, 300SE, 300TE and 300SEL utilizes hydraulic valve clearance compensation. No adjustment is either possible or necessary.

V8, Except 5.0L DOHC Engines

NOTE: V8 engines use hydraulic valve lifters and require no periodic adjustment.

A valve adjusting wrench (crow's foot) is required to accurately measure torque on all models

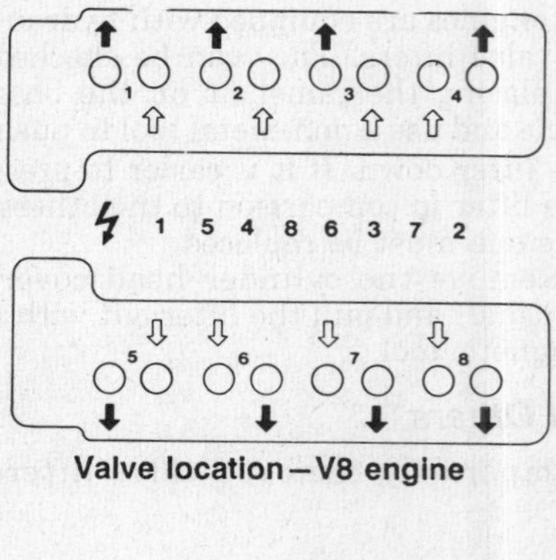

Valve location – V8 engine

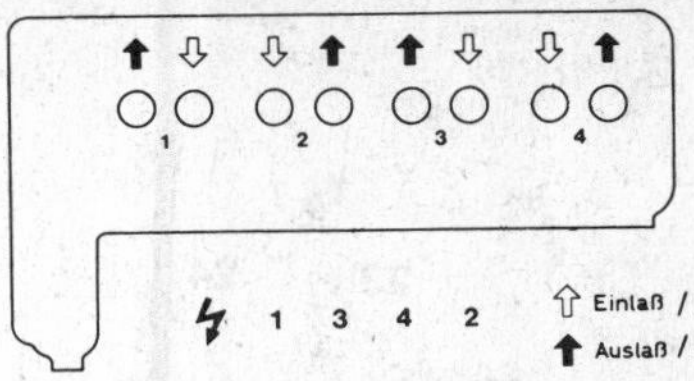

Valve location—SOHC 4 cylinder engines

The valve clearance is measured between the sliding surface of the rocker arm and the heel of the camshaft lobe. The highest point of the camshaft lobe should be at a 90 degree angle to the sliding surface of the rocker arm.

1. Disconnect the negative battery cable. Loosen the venting line and remove the regulating linkage. Remove the valve cover.
2. Disconnect the cable from the ignition coil.
3. Identify all of the valves as intake or exhaust.
4. Begin with the No. 1 cylinder, crank the engine with the starter to position the heel of the camshaft approximately over the sliding surface of the rocker arm.
5. Rotate the crankshaft by means of a socket wrench on the crankshaft pulley bolt until the heel of the camshaft lobe is perpendicular to the sliding surface of the rocker arm.

NOTE: Do not rotate the engine using the camshaft sprocket bolt. The strain will distort the timing chain tensioner rail. Always rotate the engine in the direction of normal rotation only.

6. Some models have holes in the vibration damper plate to assist in crankshaft rotation. In this case, a small prybar can be used to carefully rotate the crankshaft.
7. To measure the valve clearance, insert a feeler blade of the specified thickness between the heel of the camshaft lobe and the sliding surface of the rocker arm. The clearance is correct if the blade can be inserted and withdrawn with a very slight drag.
8. If adjustment is necessary, it can be done by turning the ball pin head at the hex collar. If the clearance is too small, increase it by turning the ball pin head in. If the clearance is too large, turn the ball pin head out.

NOTE: If the adjuster turns very easily or it the proper clearance can't be obtained, check the torque on the adjuster using a special adapter (crow's foot).

9. When the ball pin head is turned, the adjusting torque should be 14–29 ft. lbs. (19–39 Nm). If the torque is lower, either the adjusting screw, the threaded bolt or both will have to be replaced. If the valve clearance is too small and the ball pin head cannot be screwed in far enough to correct it, a thinner pressure piece should be installed in the spring retainer. To replace the pressure piece, the rocker arm must be removed.
10. Install the regulating linkage, valve cover gasket, and valve cover. Be sure the gasket is seated properly.
11. Connect the cable to the coil and the venting line. Run the engine and check for leaks at the valve cover.

Rocker Arms

REMOVAL & INSTALLATION

Except 190E, 260E, 300E, 300CE, 300SE, 300TE, 300SEL, 300SL and 500SL

Before removing the rocker arm(s), be sure they are identified as to their position relative to the camshaft lobe. They should be installed in the same place as they were before disassembly.

Be very careful removing the thrust pieces; they can easily fall into the engine.

1. Disconnect the negative battery cable. Remove the rocker arm cover or covers.
2. Force the clamping spring from the notch in the top of the rocker arm. Slide it in an outward direction across the ball socket or the rocker arm.

NOTE: Turn the engine over each time to relieve any load from the rocker arm.

3. On V8 models, the clamping spring must be forced from the adjusting screw with a small prybar.
4. Force the valve down to remove load from the rocker arm.

NOTE: Do not depress the spring too far. When the piston is up as it should be, the valve will hit the piston. As the spring goes down, the thrust piece will fall into the engine.

5. Lift the rocker arm from the ball pin and remove the rocker arm.

To install:

6. Force the rocker arm down until the rocker arm and its ball socket can be installed in the top of the pin.
7. Install the rocker arms.
8. Slide the clamping spring across the ball socket of the rocker arm until it rests in the notch of the rocker arm.
9. On V8 models, engage the clamping spring into the recess of the adjusting screw.
10. Check and, if necessary, adjust the valve clearance.
11. After completion of the adjustment, check to be sure the clamping springs are correctly seated.
12. Install the rocker arm cover and connect any hoses or lines that were disconnected.
13. Run the engine and check for leaks at the rocker arm cover.

190E, 260E, 300E, 300CE, 300SE, 300TE, 300SEL, 300SL and 500SL

NOTE: The 190E-16, 3.0L DOHC and 5.0L DOHC utilize double overhead camshafts acting directly on valve tappets. There are no rocker arms on this engine.

Rocker arms on this engine are individually mounted on rocker arm shafts that fit into either side of the camshaft bearing brackets.

1. Disconnect the negative battery cable. Remove the cylinder head cover. The cover is removed with the spark plug wires and distributor cap still connected.
2. Tag each rocker arm and shaft so they are identified as to their position relative to the camshaft. They should always be install in the same place as they were before disassembly.
3. The rocker arm shaft is held axially and rotationally by a bearing bracket fastening bolt. Remove the bolt on the side of the bearing bracket that allows access to the exposed end of the rocker shaft.
4. Thread a bolt (M8) into the end of the rocker arm shaft and slowly ease the shaft from the bearing bracket.

NOTE: Support the rocker arm/lifter assembly while removing the shaft so it will not drop onto the cylinder head.

Carefully, forcing the valve down with a small prybar will remove the load on the hydraulic valve tappet and ease the removal of the shaft. Do not depress the spring too far. When the piston is up as it should be, the valve will hit the piston. As the spring goes down the thrust piece will fall into the engine.

5. Replace the bearing bracket bolt and tighten it to 11 ft. lbs. (15 Nm) until ready to replace the rocker shaft.

To install:

6. Position the rocker arm between the 2 bearing brackets and slide the shaft into place.

NOTE: The circular groove on the end of the rocker shaft must align with the mounting bolt shank to ensure proper positioning.

To remove the rocker shaft on the 190E SOHC, thread a bolt into the hole (D). On installation, the dished groove (arrow) must always line up with the mounting bolt shank

7. Replace the bearing bracket mounting bolt.
8. Repeat Steps 3–7 for all remaining rocker arm/shaft assemblies. Turn the engine over each time to relieve any load from the rocker arm.
9. Replace the cylinder head cover.

Intake Manifold

REMOVAL & INSTALLATION

190E With 2.3L Engine

1. Disconnect the negative battery cable. Remove mixture control unit with air guide housing.
2. Disconnect fuel lines.
3. Remove holder for starter cable.
4. Remove electric lines and vacuum lines.
5. Remove supporting holder for intake manifold.
6. Remove engine suspension eye.
7. Remove fastening nuts and bolt.
8. Remove intake manifold.
9. Clean and test flange surfaces with straight-edge, machine on surface plate, if required.

To install:

10. Use new gasket and reverse removal procedure. Check idle.

Replace the rubber connecting pieces on the intake manifold, anytime the manifold is removed

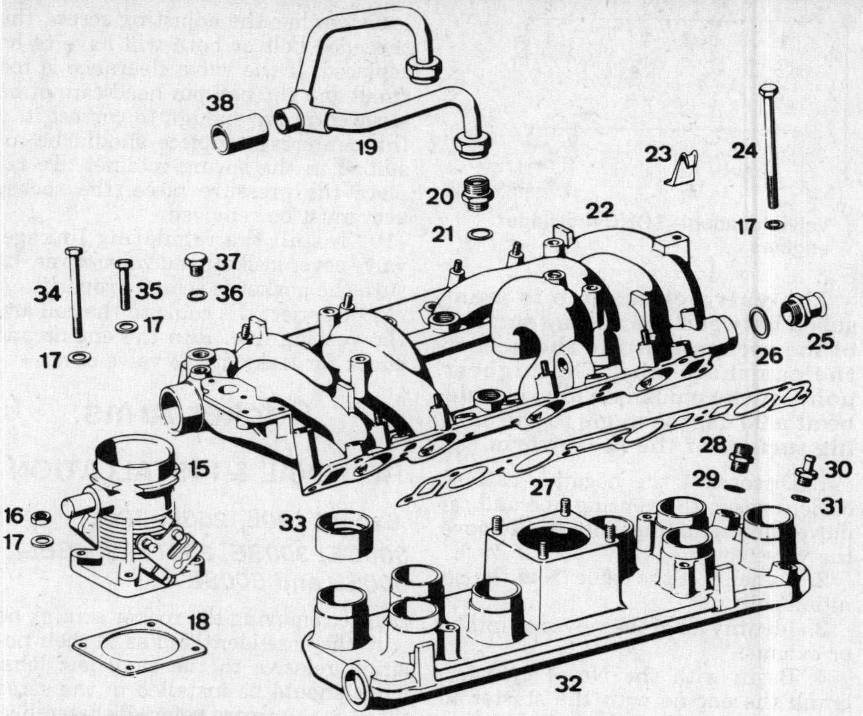

15. Valve connection
16. Nut
17. Washer
18. Gasket
19. Idle speed air line
20. Screw connection
21. Sealing ring
22. Upper intake manifold
23. Holder
24. Hex bolt
25. Connection
26. Sealing ring
27. Gasket
28. Screw connection
29. Sealing ring
30. Screw connection
31. Sealing ring
32. Bottom intake manifold
33. Rubber connecting piece
35. Hex bolt
34. Hex bolt
36. Sealing ring
37. Plug
38. Hose

V8 intake manifold

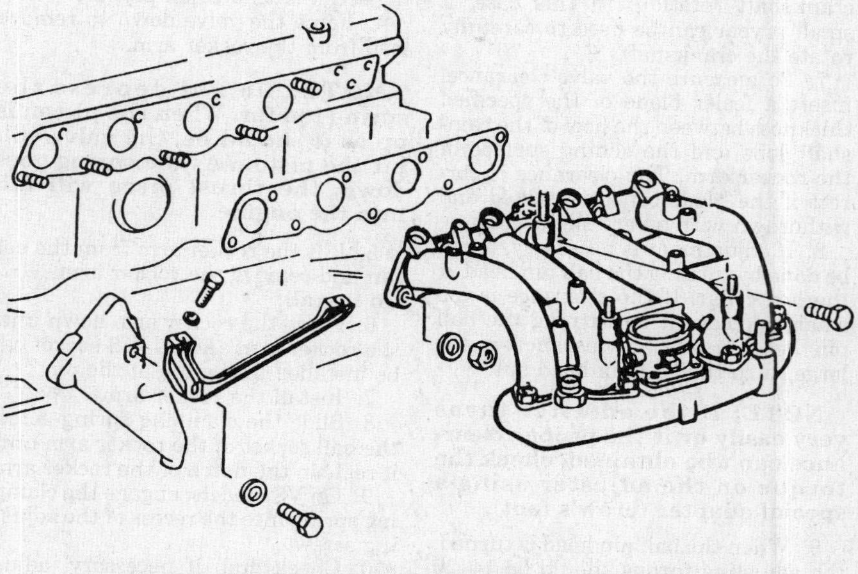
Intake manifold servicing – 190E with 2.3L or 2.3-16 engine

6 Cylinder Engines

1. Disconnect the negative battery cable and drain the engine coolant.
2. Label and disconnect all vacuum, electrical and cable connectors from the intake manifold.
3. Disconnect the line support rail.
4. Remove the intake manifold mounting bolts.
5. Remove the engine compartment rear lining, if so equipped.
6. Remove the guide tube for the dipstick.
7. Remove the mounting bolts for both the manifold and supports.
8. Carefully remove the manifold and components.

To install:

9. Clean the gasket mating surfaces and check for warpage.
10. Using a new gasket that matches the old one, install the manifold and mounting hardware.
11. Evenly torque the manifold in 2 steps.
12. Reconnect all wiring, hoses and cables. Check for leakage.
13. Check and adjust the throttle linkage and idle speed.
14. Refill the engine fluids, connect the battery cable, start the engine and check for leaks.

V8 Engine

1. Disconnect the negative battery cable. Partially drain the coolant.
2. Remove the air cleaner and engine cover.
3. Disconnect the regulating linkage and remove the longitudinal regulating shaft.
4. Label and pull off all cable plug connections.
5. Disconnect and plug the fuel lines on the pressure regulator and starting valve.
6. Unscrew the nuts on the injection valves and set the injection valves aside.
7. Remove the 16 attaching bolts from the intake manifold.
8. Loosen the hose clip on the thermostat housing hose and disconnect the hose.
9. Remove the intake manifold. If a portion of the manifold must be replaced, disassembly the intake manifold. Replace the rubber connections during reassembly.
10. Intake manifold installation is the reverse of removal. Replace all seals and gaskets. Adjust the linkage and idle speed.

Exhaust Manifold

REMOVAL & INSTALLATION

4 and 6 Cylinder Engines

1. Disconnect the negative battery cable.
2. Disconnect the exhaust pipe from the manifold.
3. Remove the exhaust support from the transmission, if so equipped.
4. Disconnect the air injection tube from the manifold, if so equipped.
5. Remove the exhaust manifold retaining bolts and manifold. Be careful with the manifold coupler tube between the 2 halves. Replace if damaged.

To install:

6. Install the manifold and torque the bolts in an even pattern. Use a new gasket.
7. Install the exhaust pipe with new self-locking nuts. Torque the pipe to 25 ft. lbs. (34 Nm).
8. Install the remaining components and connect the battery cable.
9. Start the engine and check for leaks.

V8 Engine

1. Disconnect the negative battery cable. Unbolt the exhaust pipes from the manifolds.
2. Disconnect the rubber mounting ring from the exhaust system.
3. Loosen the shield plate on the exhaust manifold.
4. When removing the left side exhaust manifold, remove the shield plate for the engine mount together with the engine damper.
5. Unbolt the manifold from the engine.
6. Pull the manifold off of the mounting.
7. Installation is the reverse of removal. Torque the manifold bolts to 18 ft. lbs. (25 Nm).

Timing Chain Front Cover

REMOVAL & INSTALLATION

4 and 6 Cylinder Engines

1. Disconnect the negative battery cable and drain the engine coolant.
2. Remove the engine cover, air cleaner and intake scoop.
3. Remove the viscous fan clutch, fan belt and mounting pulleys.
4. Disconnect all wiring and hoses from the front of the engine.
5. Remove the upper timing cover and mark the timing chain and sprockets.
6. Loosen and rotate the timing chain tensioning device.
7. Remove the power steering pump, air pump and fan bearing support.
8. Remove the TDC sensor, alternator and bracket.
9. Remove the timing covers, top and bottom.

To install:

10. Install the cover and torque the M6 bolts to 7 ft. lbs. (10 Nm). Torque the M8 bolts to 15 ft. lbs. (21 Nm).
11. Install the water pump and torque to 15 ft. lbs. (21 Nm).
12. Install the remaining components in reverse order.
13. Refill the engine with coolant, start the engine and check for leaks.

V8 Engines

1. Disconnect the negative battery cable and drain the radiator. Remove the air cleaner and valve cover.
2. Remove the viscous fan coupling and loosen the serpentine belt.
3. Remove the front top cover, serpentine belt and tensioner.
4. Set the engine No. 1 piston to TDC with the crankshaft pointer at the 4/5 mark on the pulley.
5. Remove the engine compartment panel.
6. Remove the left exhaust pipe and cover the starter mounting opening.
7. Lock the camshafts in place using a holding pin 119589001500 or equivalent to prevent the engine from rotating.
8. Mark all 4 camshafts sprockets and chain with colored dots.
9. Remove the armature for the variable cam adjuster. Take off as an assembly and do not disassemble the unit.
10. Remove the alternator, bracket and timing chain tensioner.
11. Remove the top chain guide rails.
12. Remove the radiator, hoses and fan carrier.
13. Remove the thermostat housing and serpentine belt pulleys.
14. Remove the air pump and bracket. Remove the water pump cover and pump.

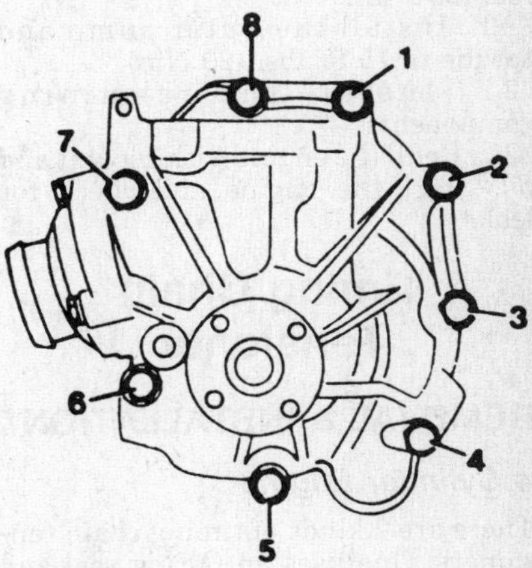

1. Bolt M8 × 60
2. Bolt M8 × 60
3. Bolt M8 × 60
4. Bolt M8 × 85
5. Bolt M8 × 60
6. Bolt M8 × 90
7. Bolt M8 × 35
8. Bolt M8 × 65

Water pump bolts—5.0L DOHC engine

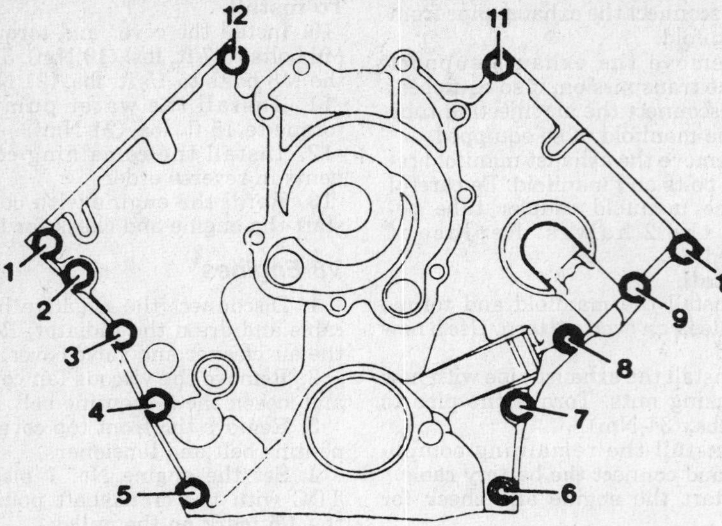

1. Bolt M8 × 50
2. Bolt M8 × 75
3. Bolt M8 × 70
4. Bolt M8 × 150
5. Bolt M8 × 50
6. Hex socket bolt M8 × 80
7. Bolt M8 × 30
8. Bolt M8 × 60
9. Hex socket bolt M8 × 50
10. Bolt M8 × 50
11. Bolt M8 × 50
12. Bolt M8 × 50

Front timing cover bolts—5.0L DOHC engine, other V8 engines similar

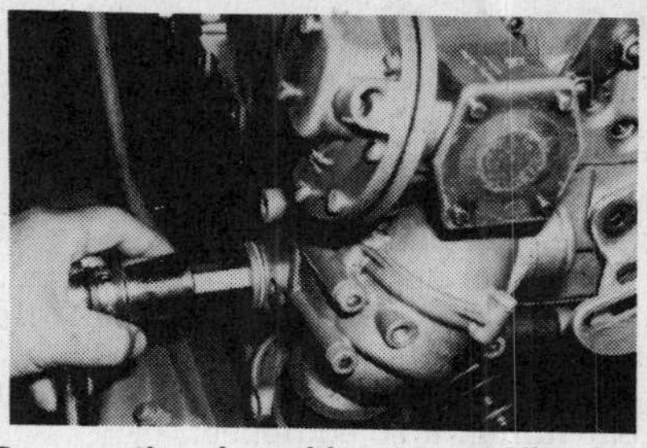

Remove the plug with a 17mm Allen key

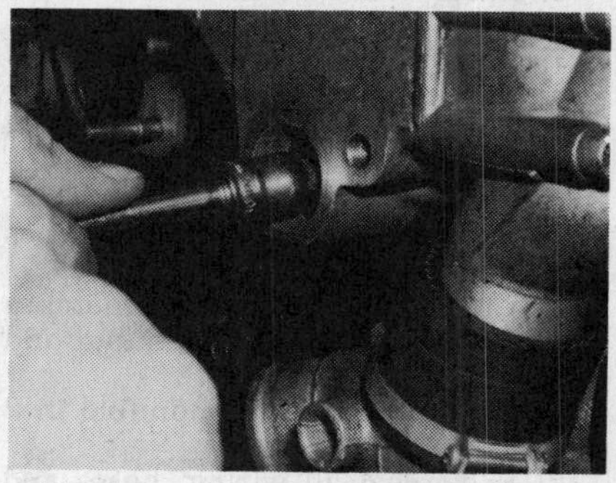

Tighten the tensioner until it "clicks"

15. Remove the air conditioning compressor and move out of the way. Secure to the side of the vehicle with wire. Remove the compressor bracket.
16. Remove the TDC sensor.
17. Remove the oil pump and dipstick tube. Remove the top section of oil sump.
18. Remove the front cover bolts and from cover with the coil pump chain.

To install:

19. Install the front cover, new gasket and bolts. Torque the bolts to 15 ft. lbs. (20 Nm).
20. Install the water pump and torque to 15 ft. lbs. (20 Nm).
21. Install the remaining components.
22. Refill the engine with coolant and oil. Start the engine and check for leaks.

Timing Chain Tensioner

REMOVAL & INSTALLATION

4 Cylinder Engine

There are 2 kinds of timing chain tensioners. One uses an O-ring seal and the other a flat gasket. Do not install a flat gasket on a tensioner meant to be use with an O-ring.

Chain tensioners should be replaced as a unit if defective.

1. Disconnect the negative battery cable. Drain the coolant. If equipped with an air conditioner, disconnect the compressor and mounting bracket and lay it aside; do not disconnect the refrigerant lines.
2. Remove the thermostat housing.
3. Loosen and remove the chain tensioner; be careful of loose O-rings. On the 190, remove the tensioner cap nut and the tension spring; the tensioner body can be unscrewed with an Allen wrench.
4. Check the O-rings or gasket and replace, if necessary.
5. To fill the chain tensioner, place the tensioner, pressure bolt down, in a container of SAE 10 engine oil, at least up to the flat flange. Using a drill press, depress the pressure bolt slowly, about 7–10 times; be sure this is done slowly and uniformly.
6. Install the chain tensioner. Tighten the bolts evenly. Tighten the cap nut on the 190 to 51 ft. lbs. (70 Nm).

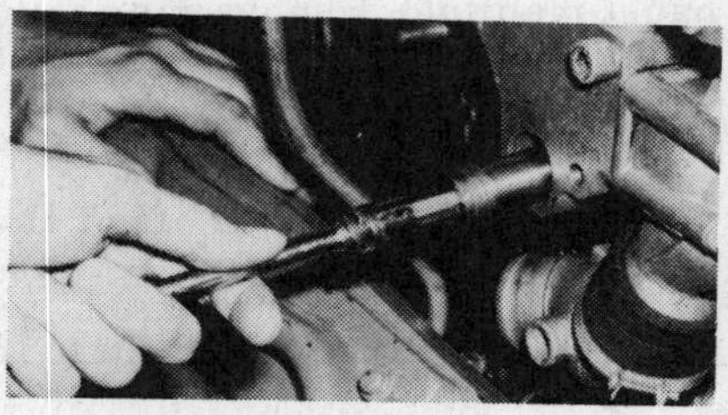

Remove the chain tensioner with a 10mm Allen key

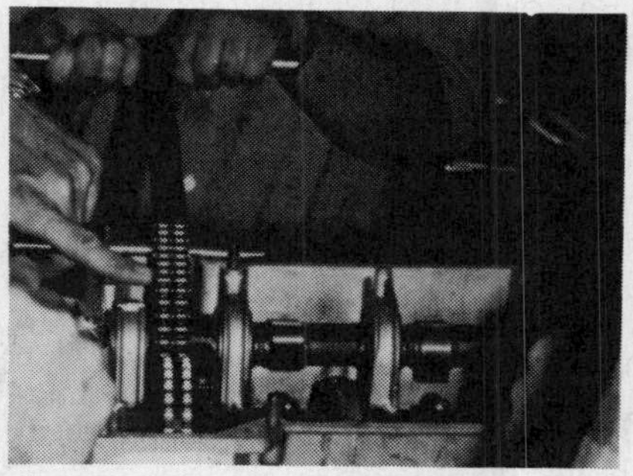

Crank the engine by hand until the new chain has come all the way through the engine. Be sure to keep tension on chain

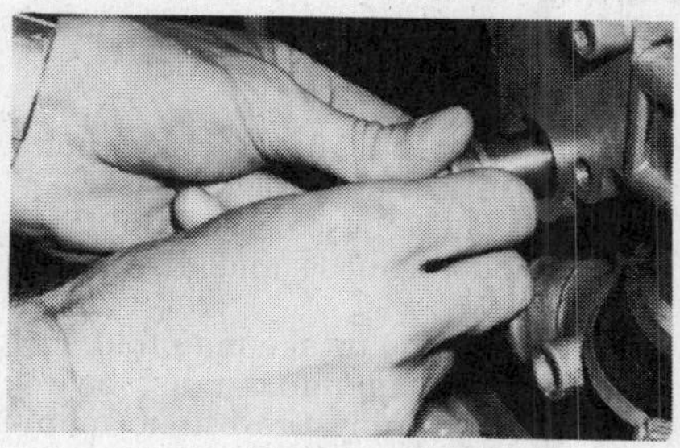

Install the chain tensioner

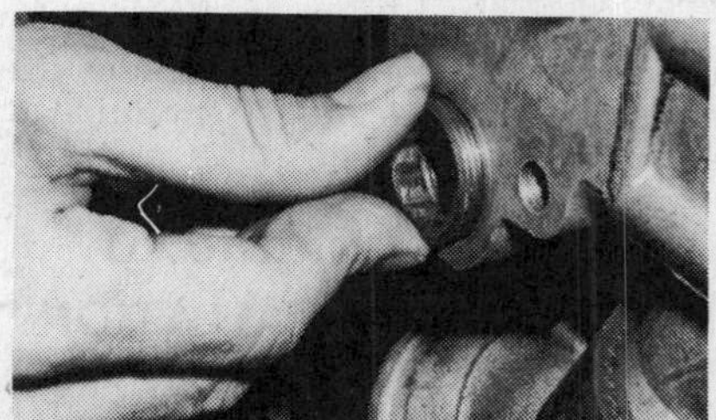

Remove the threaded plug

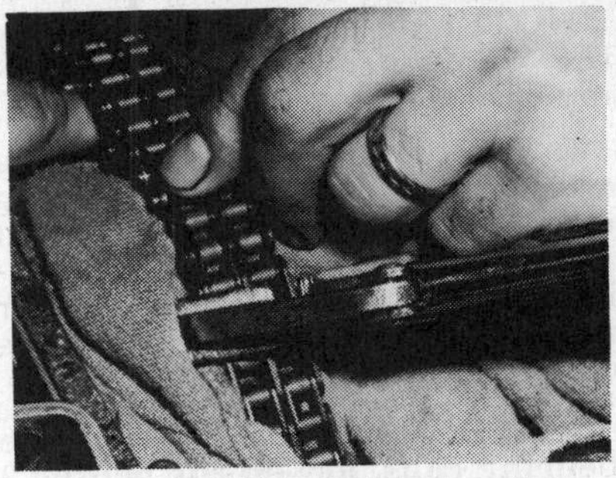

Clamp the chain again, cover the opening and remove the old chain from the master link. Connect both ends of the new chain

The inside bolt (arrow) on the V8 chain tensioner can only be reached by inserting a long, straight Allen key under the exhaust manifold

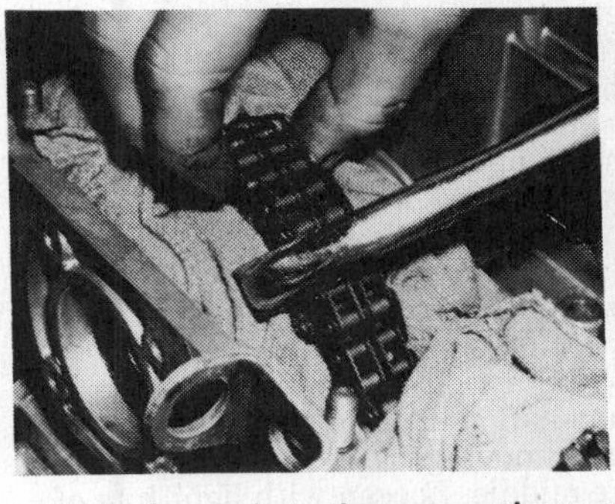

Clamp the chain to the gear and cover the opening with rags

6 Cylinder and V8 Engines

EXCEPT 5.0L DOHC ENGINE

The chain tensioner is connected to the engine oil circuit. Bleeding occurs once oil pressure has been established and the tensioner is filling with oil.

A venting hole has been installed in the tensioner to prevent oil foaming. If there is a lot of timing chain noise, use

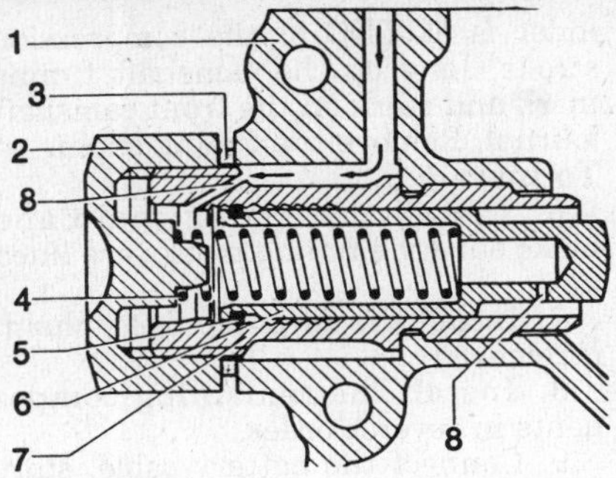

1. Crankcase
2. Cap nut
3. Seal ring
4. Compression
5. Detent spring
6. Thrust pin
7. Chain tensioner housing
8. Supply hole 1.1 mm dia.
9. Orifice 1.2 mm dia.

Cross section of the timing chain tensioner—190E (190D similar)

this type of tensioner, which is identified by a white paint dot on the cap.

Service procedures for tensioners and rails on the different V8's are similar. Arrangement and shape and size of parts however, is slightly different.

1. Disconnect the negative battery cable. On California models, disconnect the line from the tensioner.
2. Remove the bolts and the tensioner; the inside bolts will probably require a long, straight 6mm Allen key to bypass the exhaust manifold. It is a tight fit.
3. Place the tensioner vertically in a container of engine oil. Operate the pressure bolt to fill the tensioner. After filling, it should permit compression very slowly under considerable force. If not, replace the tensioner with a new unit.
4. Install the tensioner and tighten the bolts evenly.

5.0L DOHC ENGINE

1. Disconnect the negative battery cable.
2. Set the No. 1 piston on TDC with the arrow pointer near the pulley at the 4/5 mark.
3. Remove the right hand cylinder head cover.
4. Secure the right hand exhaust camshaft with a pin tool 119589001500 or equivalent so the camshaft does not rotate.
5. Remove the serpentine belt from the alternator and remove the alternator.
6. Remove the timing chain tensioner from the right side of the engine block.

To install:

7. Install the tensioner and torque the bolts to 15 ft. lbs. (20 Nm).
8. Install the alternator and cylinder head cover.
9. Connect the battery cable, start the engine and check for abnormal noise or vibration.

Timing Chain

REPLACEMENT

Unbroken

An endless timing chain is used on production engines but a split chain with a connecting link is used for service. The endless chain can be separated with a "chain breaker." Only 1 master link (connecting link) should be used on a chain.

1. Disconnect the negative battery cable. Remove the spark plugs.
2. Remove the valve cover(s).
3. Clamp the chain to the camshaft gear and cover the opening of the timing chain case with rags. On 6 cylinder and V8 engines, remove the rocker arms from the right side camshaft.
4. Separate the chain with a chain breaker.

To install:

5. Attach a new timing chain to the old chain with a master link.
6. Using a socket wrench on the crankshaft, slowly rotate the engine in the direction of normal rotation. Simultaneously, pull the old chain through until the master link is uppermost on the camshaft sprocket; be sure to keep tension on the chain throughout this procedure.
7. Disconnect the old timing chain and connect the ends of the new chain with the master link. Insert the new connecting link from the rear so the lock washers can be seen from the front.

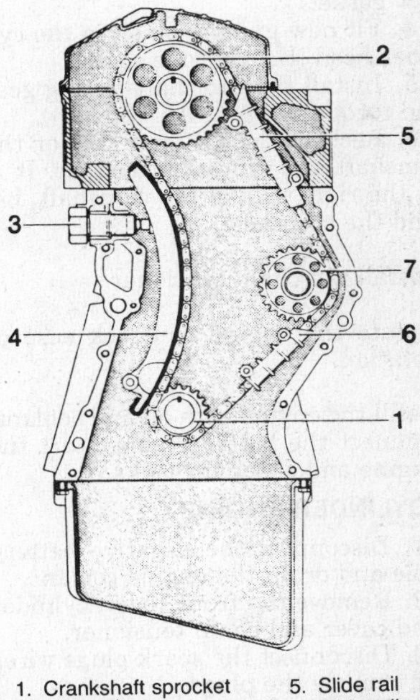

1. Crankshaft sprocket
2. Camshaft sprocket
3. Chain tensioner
4. Tensioning rail
5. Slide rail
6. Slide rail
7. Idler gear

Timing chain assembly—190E SOHC

8. Rotate the engine until the timing marks align. Check the valve timing. Once the new chain is assembled, rotate the engine, by hand, through a least 1 complete revolution to be sure everything is Ok.

Broken

4 CYLINDER ENGINE

1. Disconnect the negative battery cable and drain the engine coolant.
2. Remove engine accessories and timing chain covers.
3. Remove the cylinder head assembly.
4. Remove the tensioner blade, guiding blade and idler gear shaft with timing chain.
5. Using a puller if needed, remove the crankshaft gear.

To install:

6. Install the chain guiding blade with the open side forward. Located above the crankshaft to the right.
7. Drive the crankshaft gear with the straight pin in alignment.
8. Rivet the open roller chain with a connector link and riveting tool 000589584300 or equivalent.
9. Install a new idler gear shaft together with the chain and clip. Torque the bolt to 4 ft. lbs. (5 Nm).
10. Put the chain around the crankshaft gear and fit into the tensioner blade.
11. Fit the timing chain cover and install the TDC sensor and vibration damper.
12. Install the thermostat housing.
13. Install the cylinder head with a new gasket.
14. Fit new guiding blade in the cylinder head. Use silicone sealer.
15. Install the camshaft timing gear and torque to 59 ft. lbs. (80 Nm).
16. Recheck the TDC marking on the camshaft and vibration damper. It is on the right side of the camshaft, behind the sprocket.
17. Install the chain tensioner.
18. Rotate the engine to check ease of rotation.
19. Refill the engine with oil and coolant. Connect the battery cable, start the engine and check for leaks.

6 CYLINDER ENGINE

1. Disconnect the negative battery cable and drain the engine coolant.
2. Remove the front cover, cylinder head cover and chain tensioner.
3. Disconnect the spark plugs wires and remove the plugs.
4. The oil and oil pump may have to be removed to disconnect the timing chain from the crankshaft pulley.
5. Turn the engine so the No. 1 cylinder is at TDC of the compression stroke. Line up the camshaft timing mark and mark on the front camshaft journal. Remove the timing chain.

To install:

6. Install the timing chain and make sure the timing marks are lined up.
7. Install the oil pump drive chain, if removed.
8. Install the remaining components in reverse order.
9. Connect the battery cable, start the engine and check for leaks.

Camshaft

REMOVAL & INSTALLATION

190E

NOTE: On the 190E it is always a good idea to replace the rocker arms and shafts whenever the camshaft is replaced.

1. Disconnect the negative battery cable. Remove the valve cover.
2. Remove the chain tensioner.
3. Remove the rocker arms and shafts.
4. Set the crankshaft at TDC for the No. 1 piston and make sure the timing marks on the camshaft are in alignment.
5. Using a 24mm open-end wrench, hold the rear of the camshaft, flats are provided, loosen and remove the camshaft bolt. Carefully, slide the gear and chain off the shaft and wire them securely so they won't slip into the case.

Loosen the camshaft bolts

On the 190E SOHC, the mark (arrow) on the camshaft collar (B) must always be aligned

NOTE: Be careful not to lose the Woodruff® key while removing the gear.

6. The camshaft is secured on the cylinder head by means of the bearing caps. Remove them and keep them in order. Each cap is marked by a number punched into its side; this number must match the number cast into the cylinder head.

NOTE: When removing the bearing caps, loosen the center 2 first and move on to the outer ones.

7. Remove the camshaft.

To install:

8. Always make sure the No. 1 cylinder is at TDC and all timing marks are aligned.
9. Tighten the bearing caps to 15 ft. lbs. (21 Nm). The camshaft gear retaining bolt should be tightened to 58 ft. lbs. (80 Nm).

NOTE: Be certain not to forget the Woodruff® key.

10. Install the chain tensioners and rocker arm cover.
11. Install the remaining components. Connect the battery cable, start the engine and check for leaks.

6 Cylinder Engine

1. Disconnect the negative battery cable.
2. Set the No. 1 piston to TDC at the front crankshaft pulley.
3. Remove the front engine covers and rocker arm cover.
4. Remove the timing chain tensioner.
5. Mark the position of the camshaft gear(s) relative to the camshaft using a colored marker next to the locating pin on the cam gear. The camshaft timing will change if the camshaft sprocket is turned.
6. Remove the oil pipe from the camshaft journals, noting the spray direction.
7. Starting at the ends, loosen the camshaft journal bolts one turn at a time, moving from the outer ends to the middle. Loosen with light hits of a plastic hammer. Be careful not to allow the valve lifters to drop out of the rocker arms. The DOHC engine does not have rocker arms.
8. Remove the ball sockets and remove the camshaft(s).

To install:

9. Oil the camshaft journals and lobes. Install the camshaft and align timing marks.
10. Make sure all the valve lifters are in place. Install the rocker arms, camshaft journals and bolts. Torque the journals evenly to 15 ft. lbs. (21 Nm).
11. Make sure the timing marks are

at TDC of the No. 1 cylinder. Install the timing chain and tensioners. The camshaft timing marks are at 12 o:clock.

12. Install the rocker arm cover and remaining components.

13. Connect the battery cable, start the engine and check for leaks.

V8 Engines

Experience shows that the right side camshaft is always the 1st one to require replacement. When the V8 camshaft is removed, keep the pedestals with the camshaft. In particular, make sure the 2 left side rear cam pedestals are not swapped. The result will be no oil pressure. Always replace the oil gallery pipe with the camshaft.

CAUTION

The camshafts are extremely sensitive to fracturing and must not be tensioned when removing and installing the bearing covers. When the camshafts are moved on the DOHC engines, the other camshafts must be secured with the locking pins to prevent turning.

1. Disconnect the negative battery cable. Remove the valve cover.

2. Remove the tensioning springs and rocker arms.

3. Using a wrench on the crankshaft pulley, crank the engine around until the No. 1 piston is at TDC on the compression stroke. The 4/5 mark on the crankshaft pulley for the DOHC engine. Using some stiff wire, hang the camshaft gear so the chain will not slip off the gears.

4. Remove the camshaft gear.

5. Unbolt the camshaft, camshaft bearing pedestals and the oil pipe. Note the angle of the bolts that do not hold the head to the block.

6. Install the bearing pedestals and camshaft. On the left side camshaft, the outer bolt on the rear bearing must be inserted prior to installing the bearings or it will not clear the power brake until. Tighten the bolts from the inside out. Torque camshaft bearing cap bolts to 37 ft. lbs. (50 Nm). When finished tightening, the camshaft should rotate freely.

7. Check the oil pipes for obstructions and replace, if necessary.

8. When install the oil pipes, also check the 3 inner connecting pipes.

9. Install the compensating washer so the keyway below the notch slides over the Woodruff® key of the camshaft.

10. Install the rocker arms and tensioning springs.

11. Adjust the valve clearance and check the valve timing.

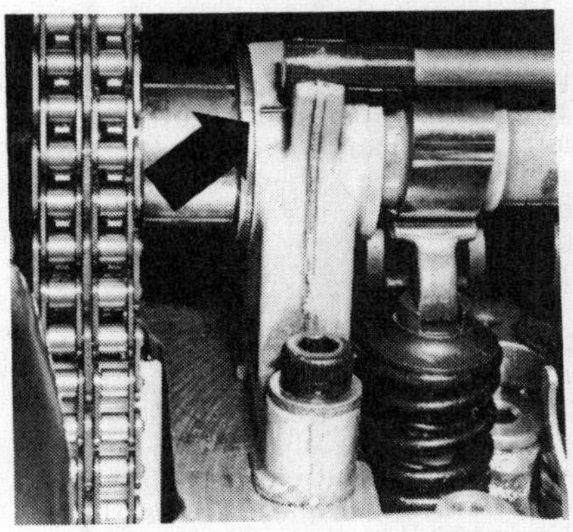

Note that the timing marks on the right-hand cam do not exactly align. This is because the timing chain travels farther on the right side than on the left

Valve Timing

Checking valve timing is too inaccurate at the standard tappet clearance; therefore timing values are given for an assumed tappet clearance of 0.4mm. The engines are not measured at 0.4mm but rather at 2mm.

1. To check the timing, remove the rocker arm cover and spark plugs. Remove the tensioning springs. On the 6 cylinder engine, install the testing thrust pieces. Eliminate all valve clearance.

2. Install a degree wheel.

NOTE: If the degree wheel is attached to the camshaft as shown, values read from it must be doubled.

3. A pointer must be made from a bent section of $^{3}/_{16}$ in. brazing rod or coat hanger wire and attached to the engine.

4. With a 22mm wrench on the crankshaft pulley, turn the engine, in the direction of rotation, until the TDC mark on the vibration damper registers with the pointer and the distributor rotor points to the No. 1 cylinder mark on the housing. The camshaft timing marks should align at this point.

NOTE: Due to the design of the chain tensioner on V8 engines, the right side of the chain travels slightly farther than the left side. This means the right side cam will be almost 7 degrees retarded compared to the left side and both marks will not simultaneously align.

5. Turn the loosened degree wheel until the pointer aligns with the 0 degree (OT) mark and tighten it in this position.

6. Continue turning the crankshaft in the direction of rotation until the camshaft lobe of the associated valve is vertical, e.g., point away from the rocker arm surface. To take up tappet clearance, insert a feeler gauge, thick enough to raise the valve slightly from its seat, between the rocker arm cone and the pressure piece.

7. Attach the indicator to the cylinder head so the feeler rests against the valve spring retainer of the No. 1 cylinder intake valve. Preload the indicator at least 0.008 in. and set to 0, make sure the feeler is exactly perpendicular on the valve spring retainer. It may be necessary to bleed down the chain tensioner at this time to facilitate readings.

8. Turn the crankshaft, in the normal direction of rotation, using a wrench on the crankshaft pulley, until the indicator reads 0.016 in. less than 0 reading.

9. Note the reading of the degree wheel at this time, remembering to double the reading, if the wheel is mounted to the camshaft sprocket.

10. Turn the crankshaft until the valve is closing and the indicator again reads 0.016 in. less than 0 reading. Make sure, at this time, that preload has remained constant, note the read-

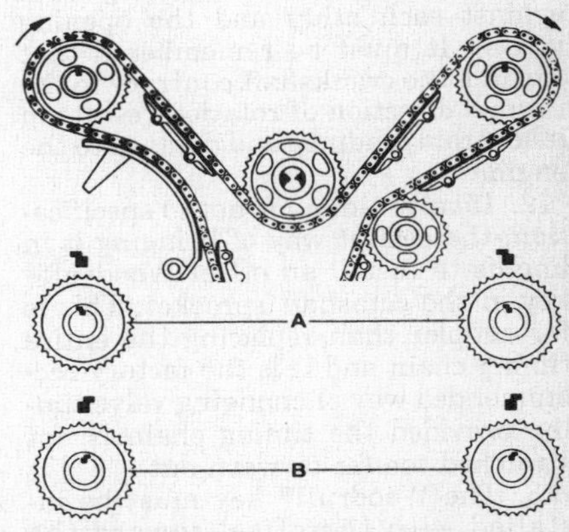

With installation position "A" opening begins earlier
With installation position "B" opening begins later

Offset Woodruff® keys—V8 engine

The V8 timing marks on the left-hand camshaft

Check the oil pipes (arrow) – V8 engine

ing of the degree wheel. The difference between the 2 degree wheel reading is the timing angle, number of degrees the valve is open, for that valve.

11. The other valves may be checked in the same manner, comparing them against each other and the opening values. It must be remembered that turning the crankshaft contrary to the normal direction of rotation results in inaccurate readings and damage to the engine.

12. If valve timing is not to specification, the easiest way of bringing it in line, is to install an offset Woodruff® key in the camshaft sprocket. This is far simpler than replacing the entire timing chain and it is the factory-recommended way of changing valve timing provided the timing chain is not stretched too far or worn out.

13. The Woodruff® key must be installed with the offset toward the right, in the normal direction of rotation, to effect advanced valve opening; toward the left to retard.

14. Advancing the intake valve opening too much can result in piston and/or valve damage, the valve will hit the piston. To check the clearance between the valve head and the piston, the crankshaft must be positioned at 5 degrees ATDC, on intake stroke. The procedure is essentially the same as for measuring valve timing.

15. As before, the dial indicator is set to 0 after being preloaded, then the valve is depressed until it touches the top of the piston. As the normal valve head-to-piston clearance is approximately 0.035 in., the dial indicator must be preloaded at least 0.042 in. so there will be enough movement for the feeler.

16. If the clearance is much less than 0.035 in., the cylinder head must be removed and checked for carbon deposits. If none exist, the valve seat must be cut deeper into the head. Always set the ignition timing after installing an offset key.

Piston and Connecting Rod

POSITIONING

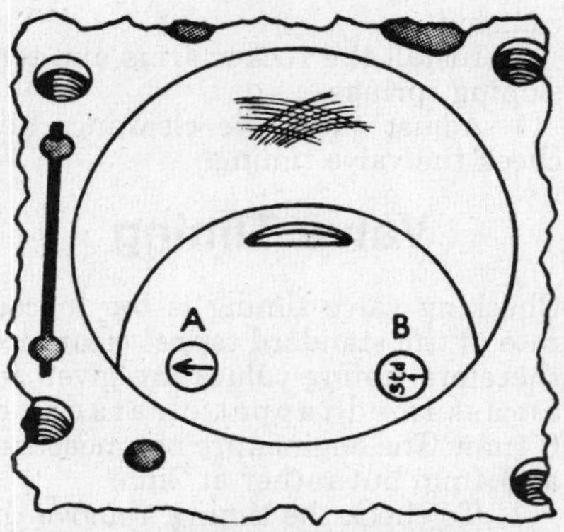

Pistons normally are marked with an arrow (a) indicating front and a weight or size marking (b).

DIESEL ENGINE MECHANICAL

NOTE: Disconnecting the negative battery cable may interfere with the functions of the on board computer systems and may require the computer to undergo a relearning process, once the negative battery cable is reconnected.

Engine Assembly

REMOVAL & INSTALLATION

NOTE: In all cases, Mercedes-Benz engines and transmissions are removed as a unit.

CAUTION

Air conditioner lines should not be indiscriminately disconnected without taking proper precautions. It is best to swing the compressor aside while still connected to its hoses. Never do any welding around the compressor-heat may cause an explosion. Also, the refrigerant, while inert at normal room temperature, breaks down under high temperature into hydrogen fluoride and phosgene (among other products), which are highly poisonous.

1. Remove the hood, drain the cooling system and disconnect the battery. While not strictly necessary, it is better to remove the battery completely to prevent breakage by the engine as it is lifted out.
2. Remove the fan shroud, radiator and disconnect all heater hoses and oil cooler lines.
3. Remove the air cleaner and all fuel, vacuum and oil hoses, e.g., power steering and power brakes.
4. Plug all openings to keep out dirt.
5. Remove the viscous coupling and fan.
6. Disconnect the accelerator linkage.
7. Disconnect all ground straps and electrical connections. It is a good idle to tag each wire for easy reassembly.
8. Detach the gearshift linkage and the exhaust pipes from the manifolds.
9. Loosen the steering relay arm and pull it aside, along with the center steering rod and hydraulic steering damper.
10. The hydraulic engine shock absorber should be removed.
11. Remove the hydraulic line from the clutch housing and the oil line connectors from the automatic transmission.
12. Unbolt the clutch slave cylinder from the bell housing after removing the return spring.
13. Remove the exhaust pipe bracket from the transmission. Support the bellhousing or place a cable sling under the oil pan, to support the engine. On turbocharged models, disconnect the exhaust pipes at the turbocharger.
14. Mark the position of the rear engine support and unbolt the 2 outer bolts, then remove the top bolt at the transmission and pull the support out.
15. Disconnect the speedometer cable and the front driveshaft U-joint. Push the driveshaft back and wire it aside.
16. Unbolt the engine mounts on both sides.
17. Unbolt the power steering fluid reservoir and swing it aside; then, using a chain hoist and cable, lift the engine and transmission upward and outward. An angle of about 45 degrees will allow the vehicle to be pushed backward while the engine is coming up.

To install:

18. With an assistant, install the engine and transmission into the vehicle using a chain hoist and cable. Lower the engine and transmission downward at an angle of about 45 degrees.
19. Connect the engine mounts on both sides and torque the bolts to 30 ft. lbs. (40 Nm).
20. Connect the speedometer cable and the front driveshaft U-joint.
21. Install the rear engine support and bolt the 2 outer bolts.

22. Install the exhaust pipe bracket to the transmission. On turbocharged models, connect the exhaust pipes at the turbocharger. Torque the pipe bolts to 25 ft. lbs. (34 Nm).
23. Install the clutch slave cylinder to the bell housing and install the return spring.
24. Install the hydraulic line to the clutch housing and the oil line connectors to the automatic transmission.
25. The hydraulic engine shock absorber should be installed.
26. Tighten the steering relay arm to 15 ft. lbs. (20 Nm). Connect the steering damper to the steering linkage.
27. Attach the gearshift linkage and the exhaust pipes to the manifolds. Torque the manifold bolts to 25 ft. lbs. (34 Nm).
28. Connect all ground straps and electrical connections.
29. Connect the accelerator linkage.
30. Install the viscous coupling and fan.
31. Install all fuel, vacuum and oil hoses, e.g., power steering, power brakes and air cleaner.
32. Install the radiator, fan shroud and connect all heater hoses and oil cooler lines.
33. Refill the engine coolant and install the hood.
34. Install the battery and connect.
35. Bleed the hydraulic clutch, power steering, power brakes and fuel system.

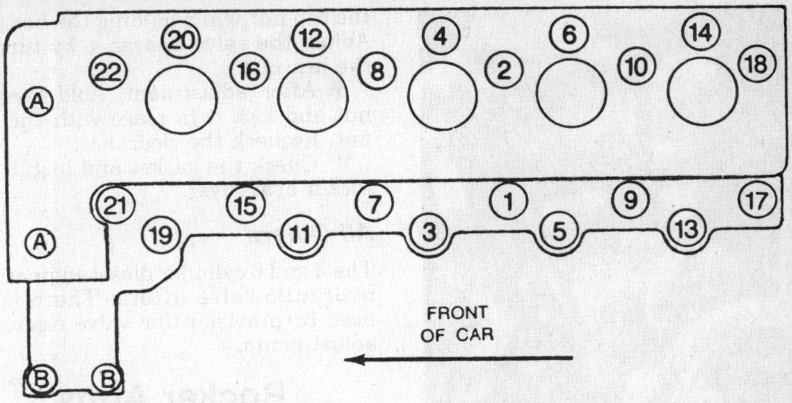

Cylinder head bolt torque sequence—5 cylinder diesel engine

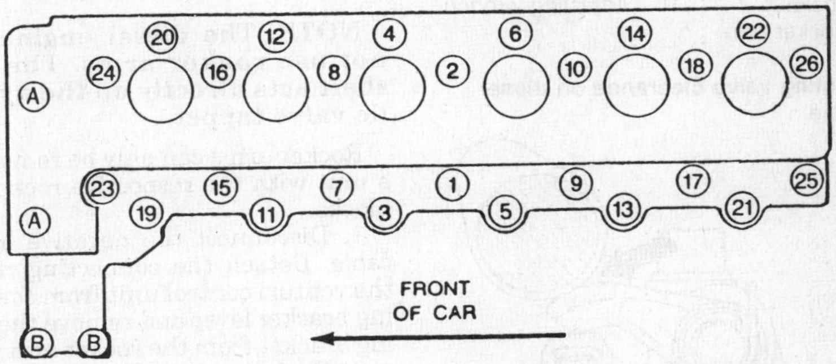

Cylinder head bolt torque sequence—6 cylinder diesel engine

Cylinder Head

REMOVAL & INSTALLATION

Use care to ensure that the valve timing is not disturbed.

1. Disconnect the negative battery cable. Drain the radiator and remove all hoses and wires. Tag all wires to ensure easy reassembly.
2. Remove the camshaft cover and associated throttle linkage.
3. Remove the camshaft sprocket nut.
4. Remove the rockers and their supports must be removed together.
5. Mark the chain, sprocket and cam for ease of assembly.
6. Using a suitable puller, remove the camshaft sprocket.
7. Remove the sprocket and chain and wire it aside.

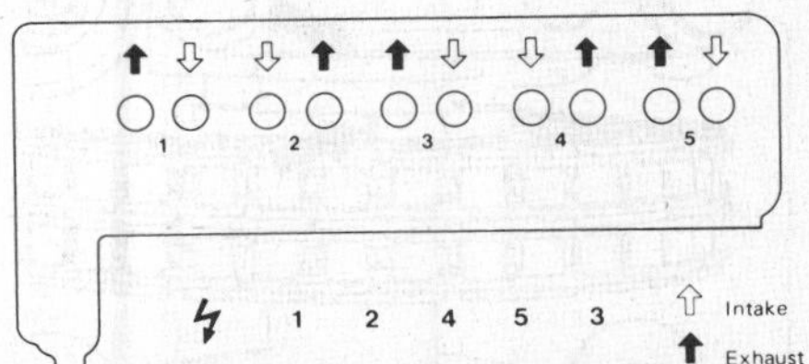

Vale arrangement—5 cylinder diesel engine

NOTE: Make sure the chain is securely wired so it will not slide into the engine.

8. Unbolt the manifolds and exhaust header pipe and push them aside.
9. Loosen the cylinder head hold-down bolts, in the reverse order of the torque sequence. It is good practice to loosen each bolt, a little at a time, working around the head, until all are free.
10. Reach into the engine compartment and gradually work the head loose from each end by rocking it. Never use a prybar between the head and block to pry, as the head will be scarred badly and may be ruined.

NOTE: All diesel engines utilize cylinder head stretch-bolts. These bolts undergo a permanent stretch each time they are tightened. When a maximum length is reached, they must be discarded and replace with new bolts. When tightening the head bolts on these engines, it is imperative that the steps listed under "Torque Specifications" are followed exactly.

To install:

11. Clean the gasket mating surfaces. Install the cylinder head using an approved hoist.
12. Torque the cylinder head bolts in sequence, in 4 steps. The 1st step to 11 ft. lbs. (15 Nm), 2nd step to 26 ft. lbs. (35 Nm), 3rd step to 90 degrees torquing angle and wait 10 minutes and the 4th step to 90 degrees torquing angle.
13. Install the manifolds and exhaust header pipe. Torque the pipe to 24 ft. lbs. (34 Nm).
14. Install the sprocket and chain.
15. Install the rockers and their supports. Torque to 15 ft. lbs. (20 Nm).
16. Install the camshaft sprocket nut.
17. Install the camshaft cover and associated throttle linkage.
18. Connect the negative battery cable. Refill the radiator and install all hoses and wires.

Valve Lash

ADJUSTMENT

190D

1. Disconnect the negative battery cable. Remove the valve cover and note the position of the intake and exhaust valves.
2. Turn the engine with a socket and breaker bar on the crankshaft pul-

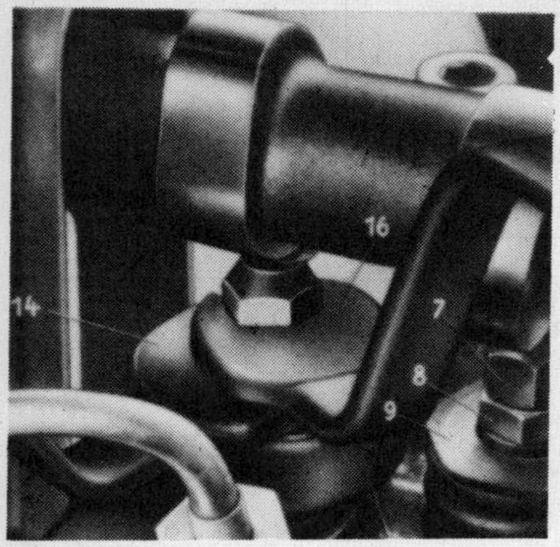

7. Capnut
8. Locknut
9. Rocker arm
14. Holding wrench
16. Adjusting wrench

Adjusting valve clearance on diesel engine

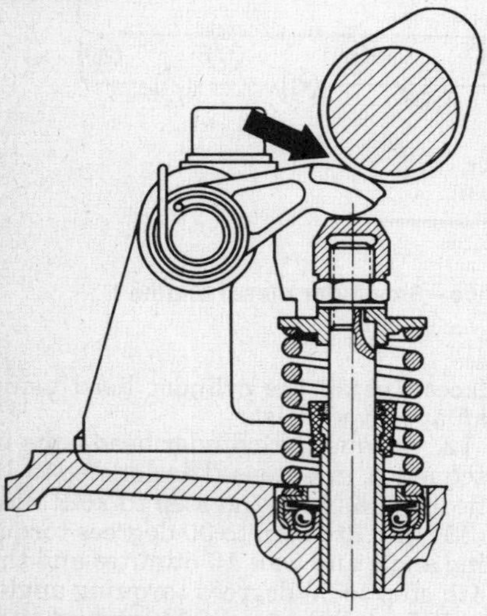

Measure valve clearance on diesel engines at arrow

ley or by using a remote starter, hooked to the battery terminal and the large, uppermost starter solenoid terminal. Due to the extremely high compression pressures, it will be considerably easier to use a remote starter. If a remote starter is not available, the engine can be bumped into position with the normal starter.

NOTE: Do not turn the engine backwards or use the camshaft sprocket bolt to rotate the engine.

3. Measure the valve clearance when the heel of the camshaft lobe is directly over the sliding surface of the rocker arm. The lobe of the camshaft should be vertical to the surface of the rocker arm. The clearance is correct when the specified feeler gauge can be pulled through with a very slight drag.

4. To adjust the clearance, loosen the cap nut while holding the hex nut. Adjust the valve clearance by turning the hex nut.

5. After adjustment, hold the cap nut and lock it in place with the hex nut. Recheck the clearance.

6. Check the gasket and install the rocker arm cover.

All Others

The 5 and 6 cylinder diesel engines use hydraulic valve lifters. There is no need or provision for valve clearance adjustments.

Rocker Arms

REMOVAL & INSTALLATION

NOTE: The diesel engine does not use rocker arms. The camshaft acts directly on the hydraulic valve tappet.

Rocker arms can only be removed as a unit with the respective rocker arm blocks.

1. Disconnect the negative battery cable. Detach the connecting rod for the venturi control unit from the bearing bracket lever and remove the bearing bracket from the rocker arm cover.

2. Remove the air vent line from the rocker arm cover and the rocker arm cover.

3. Remove the stretch-bolts from the rocker arm blocks and the blocks with the rocker arms. Turn the crankshaft in each case so the camshaft does not put any load on the rocker arms.

NOTE: Turn the crankshaft with a socket wrench on the crankshaft pulley bolt. Do not rotate the engine by turning the camshaft sprocket.

4. Before installing the rocker arms, check the sliding surfaces of the ball cup and rocker arms. Replace any defective parts.

To install:

5. Assembly the rocker arm blocks and insert new stretch-bolts.

6. Tighten the stretch-bolts. In each case, position the camshaft so there is no load on the rocker arms.

7. Check to be sure the tension clamps have engaged with the notches of the rocker arm blocks.

8. Reinstall the rocker arm cover, air vent line and bearing bracket for the reverse lever. Attach the connecting rod for the venturi control unit to the reversing lever.

9. Make sure during acceleration, the control cable can move freely without binding.

10. Start the engine and check the rocker arm cover for leaks.

Turbocharger

REMOVAL & INSTALLATION

1. Disconnect the negative battery cable. Remove the air filter.

2. Disconnect the electrical cable from the temperature switch.

3. Loosen the lower hose clamp on the air duct that connects the air filter with the compressor housing.

4. Remove the vacuum line and crankcase breather pipe.

5. Remove the air filter and air intake duct.

6. Disconnect the oil line at the turbocharger.

7. Remove the air filter mounting bracket.

8. Disconnect the turbocharger at the exhaust flange.

9. Disconnect and remove the pipe bracket on the automatic transmission.

10. Push the exhaust pipe rearward.

11. Remove the mounting bracket at the intermediate flange.

12. Unbolt and remove the turbocharger.

13. Remove the intermediate flange and oil return line at the turbocharger.

To install:

14. Before installing the turbocharger, install the oil return line and intermediate flange.

15. Install the flange gasket between the turbocharger and exhaust mani-

1. Mounting bracket
2. Intermediate flange
3. Turbocharger

Remove the mounting nuts (arrow) to remove the turbocharger

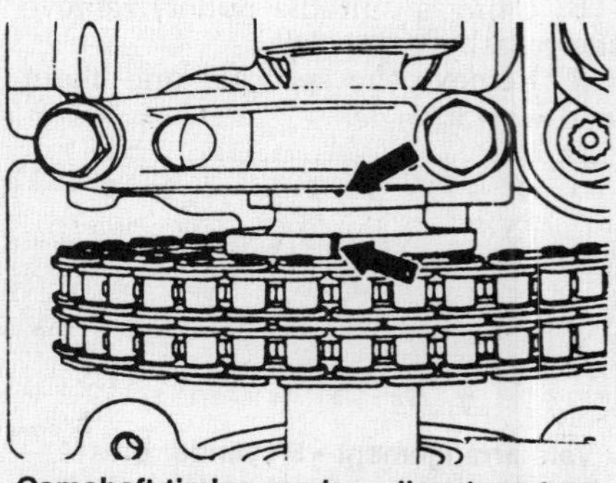

Camshaft timing marks – diesel engines

fold with the reinforcing bead toward the exhaust manifold. Use only heat proof nuts and bolts.

16. Fill a new turbocharger with ¼ pint of engine oil through the engine oil supply bore before operating.

Timing Chain Tensioner

FILLING

1. Place chain tensioner with thrust pin, in downward direction in, engine oil SAE 10 up to above collar on hex head.
2. Slowly press thrust pin 7–10 times up to stop, by means of a press or an upright drill press.
3. Upon filling, the chain tensioner should permit compressing very slowly only, uniformly and at considerable force.
4. To prevent peak pressures of chain tensioner against tensioning rail, a modified valve disk will be installed in chain tensioner.

REMOVAL & INSTALLATION

There are 2 kinds of timing chain tensioners. One uses an O-ring seal and the other a flat gasket. Do not install a flat gasket on a tensioner meant to be use with an O-ring.

Chain tensioners should be replaced as a unit, if defective.

1. Disconnect the negative battery cable. Drain the coolant. If equipped with air conditioning, disconnect the compressor, mounting bracket and lay it aside. Do not disconnect the refrigerant lines. Drain the coolant from the block.
2. Remove the thermostat housing.
3. Loosen and remove the chain tensioner; be careful of loose O-rings. On the 190, remove the tensioner cap nut and the tension spring. The tensioner body can be unscrewed with an Allen wrench.

To install:

4. Check the O-rings or gasket and replace, if necessary.
5. To fill the chain tensioner, place the tensioner, pressure bolt down, in a container of SAE 10 engine oil, at least up to the flat flange. Using a drill press, depress the pressure bolt slowly, about 7–10 times. Be sure this is done slowly and uniformly.
6. Install the chain tensioner. Tighten the bolts evenly. Tighten the cap nut on the 190 to 51 ft. lbs. (70 Nm).

Timing Chain

REPLACEMENT

1. Disconnect the negative battery cable. Remove cylinder head cover.
2. Remove injection nozzles.
3. Remove chain tensioner.
4. Remove fan and fan cover.
5. Connect new timing chain with connecting link to old timing chain.
6. Slowly, rotate crankshaft in rotating direction of engine, while simultaneously pulling up the old timing chain until the connecting link comes to rest against uppermost point of camshaft timing gear.

NOTE: Timing chain should remain in mesh while rotating camshaft and crankshaft timing gears.

7. Take off old timing chain and connect ends of new timing chain with connecting link. For this purpose, secure chain ends with wire on camshaft timing gear.

To install:

NOTE: Use only a rivet-type connecting link. Do not use connecting link that use a retaining spring.

8. Insert connecting link from the rear into timing chain.
9. Put separately enclosed outer flange of connecting link, with punched in IWIS identification, into pressing-on tool. The outer flange is held magnetically.
10. Place pressing-on tool on connecting link and press on flange up to stop, while holding pressing-on tool on vertical level.
11. Rearrange plunger, of assembly tool, so the notch is pointing forward.
12. Hold assembly tool on handle and rivet chain bolts individually. Tightening torque of spindle approximately 22–26 ft. lbs. (30–35 Nm).
13. Check chain bolt rivet and rivet again, if required.
14. Install chain tensioner.
15. Rotate crankshaft and check adjusting mark at TDC position of engine.

NOTE: If the adjusting mark is wrong, check timing of camshaft and begin of delivery of injection pump.

16. Install cylinder head cover and tighten to 7.5 ft. lbs. (10 Nm).
17. Install fan and fan cover.

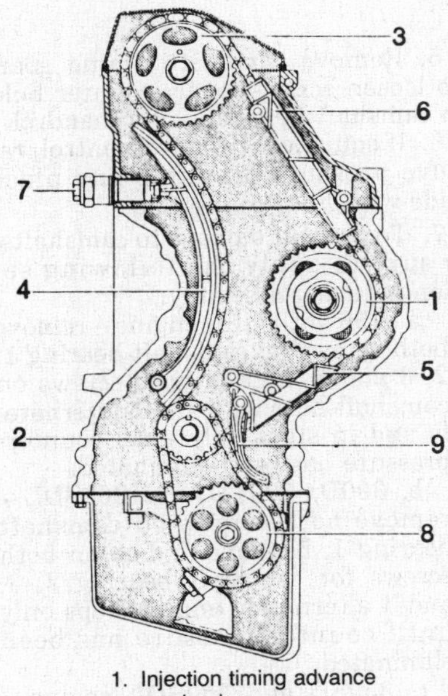

1. Injection timing advance mechanism
2. Crankshaft sprocket
3. Camshaft sprocket
4. Tensioning rail
5. Slide rail
6. Slide rail
7. Chain tensioner
8. Oil pump drive gear
9. Tensioning lever, chain, oil pump drive

Timing chain assembly—190D

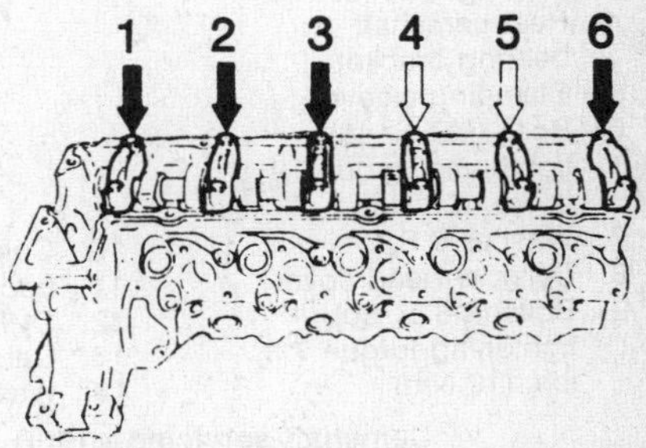

Camshaft bearing cap servicing sequence—190D with 2.5L or 2.5L Turbo engine

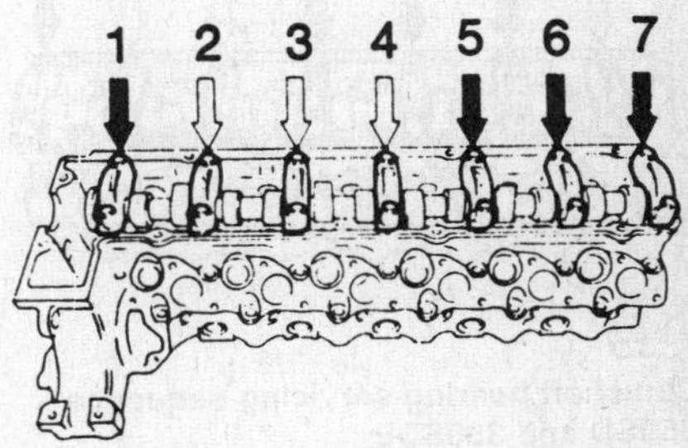

Camshaft bearing cap servicing sequence—300D and 300SDL

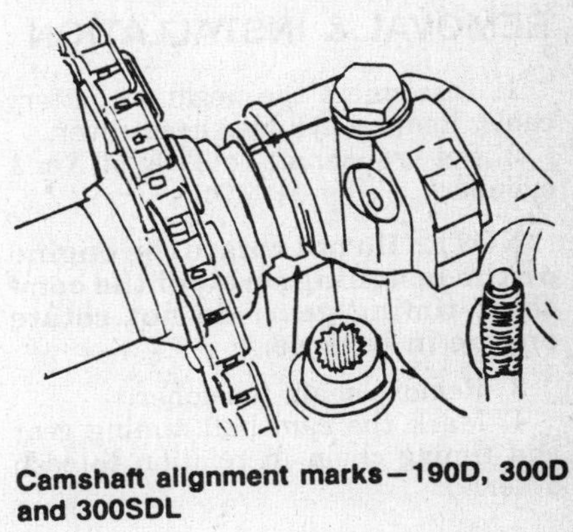

Camshaft alignment marks—190D, 300D and 300SDL

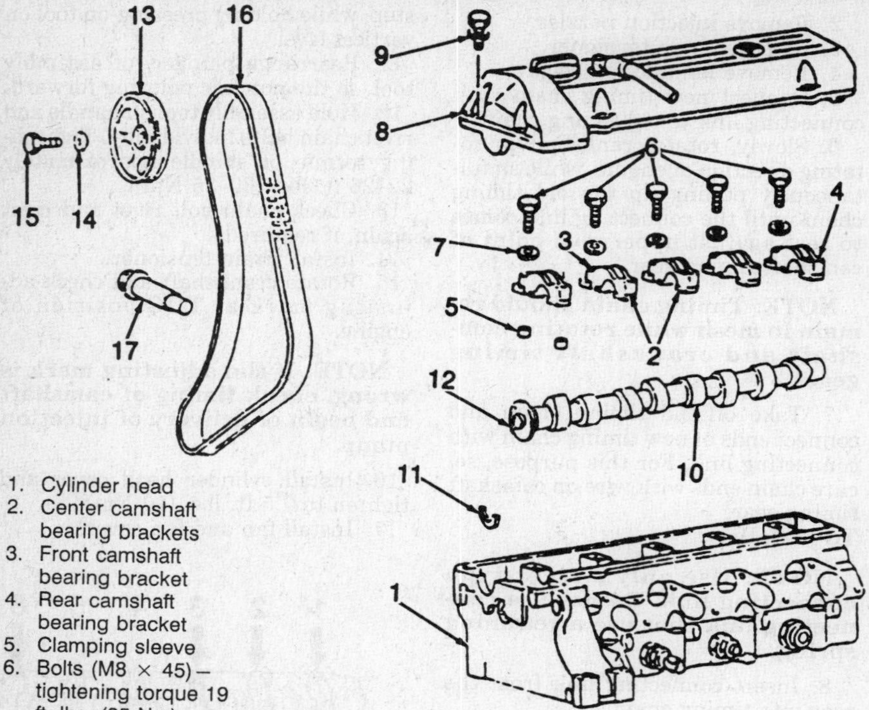

1. Cylinder head
2. Center camshaft bearing brackets
3. Front camshaft bearing bracket
4. Rear camshaft bearing bracket
5. Clamping sleeve
6. Bolts (M8 × 45)—tightening torque 19 ft. lbs. (25 Nm)
7. Washer B 8,4
8. Cylinder head cover
9. Bolts (M6 × 30)—tightening torque 7 ft. lbs. (10 Nm)
10. Camshaft
11. Lock washer
12. Cylinder pin
13. Camshaft sprocket
14. Washer B 10
15. Bolt (M10 × 50)—tightening torque 48 ft. lbs. (65 Nm)
16. Timing chain
17. Chain tensioner

Camshaft servicing—190D, 300D and 300SDL, 350SD similar

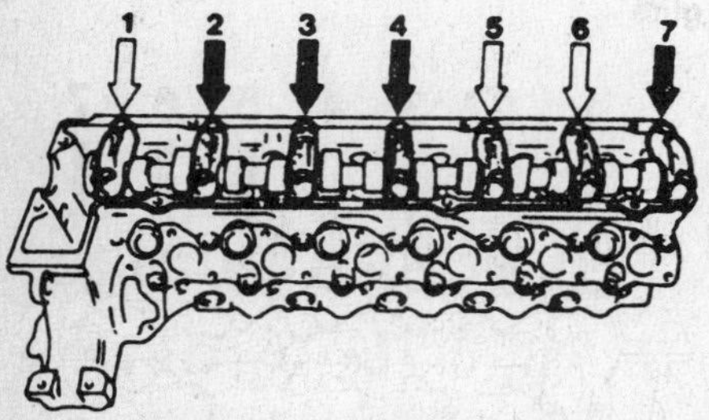

Camshaft bearing servicing sequence—350SD and 350SDL

Camshaft

REMOVAL & INSTALLATION

1. Disconnect the negative battery cable. Remove cylinder head cover.
2. Set crankshaft to TDC of No. 1 cylinder.

NOTE: Do not rotate the engine on the fastening screw of the camshaft timing gear; do not rotate engine in reverse.

3. Remove chain tensioner.
4. Mark the camshaft timing gear and timing chain in relation to each other.
5. Remove camshaft timing gear. To loosen screws, apply counter hold on camshaft by means of a mandrel.
6. If equipped with level control, remove pressure oil pump and place aside with lines connected.
7. To prevent damage to camshafts, be sure to apply the following sequence during assembly:
   a. 190 with 2.5L engine—remove both screws on camshaft bearing 1, 2, 3 and 6. Loosen both screws on camshaft bearing 4 and 5 alternately and in steps only until counterpressure has been eliminated.
   b. 300D, 300TD and 300SDL—remove both screws on camshaft bearing 1, 5, 6 and 7. Loosen both screws for camshaft bearing 2, 3, and 4 alternately and in steps only until counter-pressure has been eliminated.
   c. 350SD and 350SDL engine—remove both bolts on camshaft bearings 1, 5 and 6 (dark arrows). Loosen both bolts for camshaft bearings 2, 3, 4 and 7 in one turn increments.
8. Remove camshaft in upward direction.
9. Remove the camshaft alignment circlip and check for misaligned condition.
10. Pull out valve tappet by means of solenoid lifter tool 102 589 03 40 00 or equivalent.
11. Check valve tappet for condition, visual checkup, and renew, if required.

To install:

NOTE: Install valve tappets only at the same spot where they were installed. If a new camshaft has been installed or if the cylinder head has been machined, check camshaft for easy operation.

12. Insert circlip for axial locating into cylinder head.
13. Lubricate camshaft and place into cylinder head, without valve tappet.
14. Tighten camshaft bearing caps uniformly to 18.5 ft. lbs. (25 Nm). Pay attention to identification of bearing caps.
15. When checking for easy operation, the camshaft can be rotated by means of a hex socket screw M10 × 30, which is screwed in through camshaft timing gear instead of fastening screw. If the camshaft can be rotated with effort only, proceed as follows:
    a. Loosen camshaft bearing caps individually. Turn camshaft, if required.
    b. Repeat, until tight bearing point has been found.
    c. Check camshaft for runout.
16. Lubricate valve tappets and insert. Pay attention to sequence.
17. Lubricate camshaft and place into cylinder head so the TDC mark is vertical.
18. Install camshaft bearing caps.

NOTE: Screw in camshaft bearing screws 4 and 5 on 190D with 2.5L engine or 2, 3, 4 and 7 on 300D, 300TD, 300SDL, 350SD and 350SDL engine, opposite the loosening sequence. The remaining camshaft bearing caps can be installed at will. Pay attention to tightening torques of 18 ft. lbs. (25 Nm).

19. Mount camshaft timing gear. Pay attention to color marks. Tighten fastening screw for camshaft timing gear to 48 ft. lbs. (65 Nm). For this purpose, apply counter hold to camshaft timing gear by means of a steel pin or suitable tool.
20. Install chain tensioner.
21. If equipped with level control, mount pressure oil pump and driver.
22. Check engine for TDC marks.
23. Mount cylinder head cover.
24. Run engine, check for leaks.

ENGINE LUBRICATION

Oil Pump

REMOVAL & INSTALLATION

190E With 2.3L Engine

1. Disconnect the negative battery cable. Remove timing housing cover.
2. Remove screw and oil suction pipe with oil strainer, as well as oil pump cover.
3. Remove oil pump gear wheels from timing housing cover.
4. Check driving sleeve for damage or drive surface dents. Replace driving sleeve, if required.

NOTE: If driving sleeve cannot be pulled from crankshaft manually, remove crankshaft timing gear together with driving sleeve.

5. Carefully, clean separating surfaces on timing housing cover and oil pump cover.

To install:

6. Lubricate oil pump gear wheels and insert into timing housing cover.

NOTE: Renew oil pump gear wheels only in pairs. For this reason, they are supplied as a spare part in a set, rotor set, only.

7. Renew sealing ring on connection of oil pump cover.
8. Mount oil pump cover on timing housing cover. Position oil suction pipe with oil strainer and with new gasket on oil pump cover. Screw in fastening screws and tighten to 7.5 ft. lbs. (10 Nm).

NOTE: Pay attention to correct installation of gasket between flange of oil suction pipe and oil pump cover.

9. Check oil pump for easy operation.
10. Slip driving sleeve on crankshaft.
11. Install timing housing cover.
12. Check for leaks with engine running.

6 Cylinder Engine

1. Drain the oil and remove the oil pan and gasket.
2. Remove the oil sump.
3. Remove the pump sprocket and bolt. Remove the chain and sprocket as an assembly.
4. Remove the oil pump mounting bolts and remove the pump.
5. Install a new gasket, oil pump and torque the mounting bolts to 18 ft. lbs. (25 Nm).

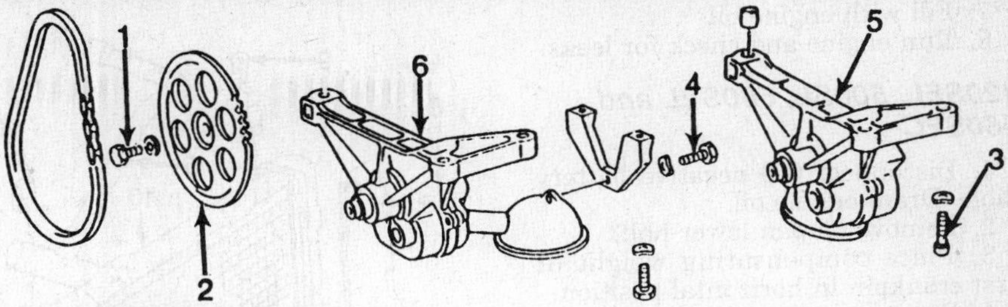

1. Screw (Tightening torque 19 ft. lbs. (25 Nm)
2. Oil pump sprocket
3. Screw (Tightening torque 19 ft. lbs. (25 Nm)
4. Screw (engines 602 and 603 only) (Tightening torque 7 ft. lbs. (10 Nm)
5. Oil pump (engine 601)
6. Oil pump (engine 602 and 603)

Oil pump servicing – 190D, 300D and 300SDL

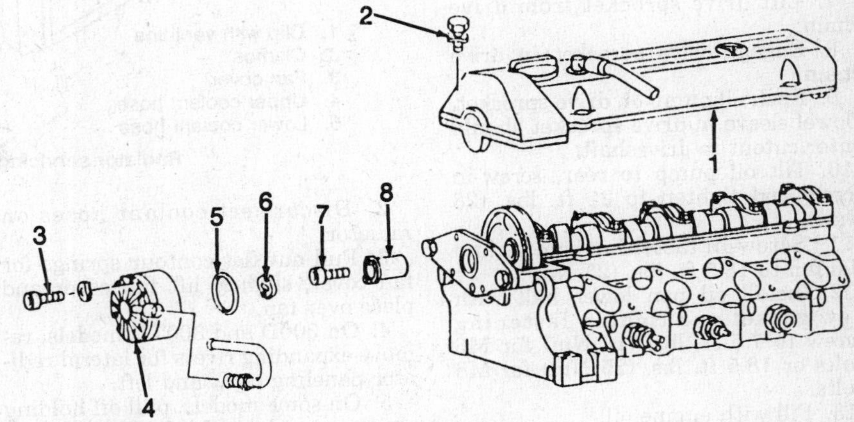

1. Cylinder head cover
2. Screw (Tightening torque 7 ft. lbs. (10 Nm))
3. Screw
4. Hydraulic oil pump
5. O-ring
6. Driven plate
7. Hex, head socket screw
8. Drive sleeve

Hydraulic oil pump servicing – 190D, 300D and 300SDL

6. Install the pump sprocket and torque to 24 ft. lbs. (32 Nm).
7. Install the oil pan and fill with oil.
8. Start the engine and check for leaks.

190D and 300D, 300TD, 300SDL, 350SD and 350SDL

1. Disconnect the negative battery cable. Remove oil pan.
2. Remove screw from sprocket and remove sprocket from drive shaft.
3. Remove screws and remove oil pump.
4. On 190D with 2.5L engine and 300D, 300TD and 300SDL, additional screw on intake manifold holder.

To install:

5. Position oil pump and torque screw to 18.5 ft. lbs. (25 Nm).
6. On 190D with 2.5L engine and 300D, 300TD and 300SDL, the additional screw is torqued to 7.5 ft. lbs. (10 Nm).
7. Engage sprocket in chain and mount on driveshaft.

NOTE: Mount sprocket so the rise points toward oil pump and the trochoid shape corresponds with that on oil pump shaft.

8. Install oil pan.
9. Run engine, check for leaks.

560SL

1. Disconnect the negative battery cable. Drain engine oil.
2. Remove oil pan.
3. Place compensating weight of first crankpin in horizontal position.
4. Remove fastening screw, tilt oil pump forward, remove drive chain from drive sprocket and remove oil pump.

To install:

5. Tilt oil pump forward and place drive chain on drive sprocket, screw-in fastening screw and tighten to 18.5 ft. lbs. (25 Nm).
6. Install oil pan with new gasket and tighten fastening screws to 7.5 ft. lbs. (10 Nm) for M6 bolts or 18.5 ft. lbs. (25 Nm) for M8 bolts.

7. Fill with engine oil.
8. Run engine and check for leaks.

420SEL, 500SL, 560SEL and 560SEC

1. Disconnect the negative battery cable. Drain engine oil.
2. Remove oil pan lower half.
3. Place compensating weight of first crankpin in horizontal position.
4. Remove screws.
5. Loosen screw on drive sprocket, tilt oil pump toward rear and remove screw.
6. Push drive sprocket away from oil pump by means of a suitable tool and remove oil pump.

To install:

7. Lift drive sprocket from drive chain.
8. Engage drive sprocket in drive chain.
9. Push oil pump on drive sprocket. Dowel sleeve in drive sprocket should enter cutout in driveshaft.
10. Tilt oil pump to rear, screw-in screw and tighten to 21 ft. lbs. (28 Nm).
11. Screw-in fastening screws and tighten to 18.5 ft. lbs. (25 Nm).
12. Install oil pan lower half with new gasket and tighten fastening screw to 7.5 ft. lbs. (10 Nm) for M6 bolts or 18.5 ft. lbs. (25 Nm) for M8 bolts.
13. Fill with engine oil.
14. Run engine and check for leaks.

ENGINE COOLING

Radiator

REMOVAL & INSTALLATION

190D, 300D, 300TD, 300SDL, 350SD and 350SDL

NOTE: After refilling the cooling system, the system may have to be bleed. Remove a cooling sensor or equivalent at the highest point in the engine's cooling system. Fill the radiator until coolant spills out of the hole. Apply thread sealing tape to the component and install. Finish filling the radiator to the proper level.

1. Disconnect the negative battery cable. If equipped with an automatic transmission, pinch oil lines from or to transmission with special tool 000 589 40 37 00 or equivalent, displacing coil spring slightly laterally and removing from radiator for this purpose.

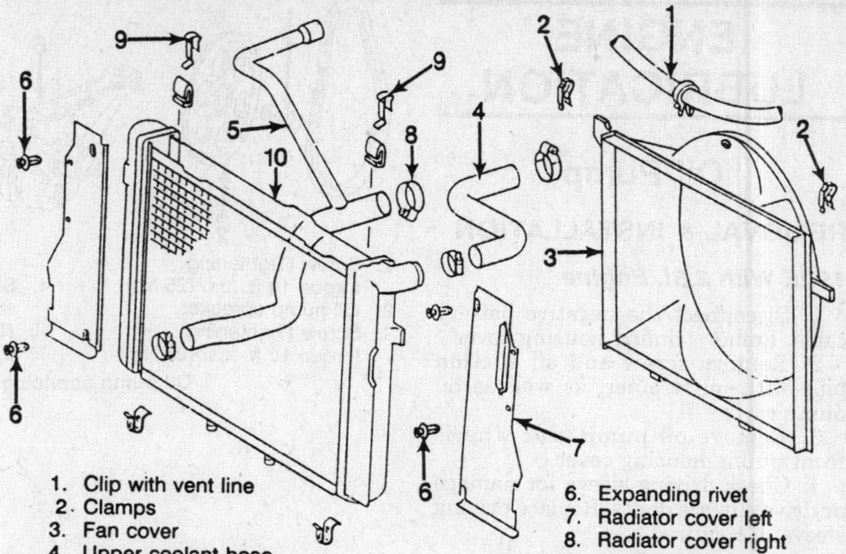

Radiator servicing—190D, 300D and 300SDL

2. Disconnect coolant hoses on radiator.
3. Pull out flat contour springs for fan cover, slightly lift fan cover and place over fan.
4. On 300D and 300TD models, remove expanding rivets for lateral radiator paneling right and left.
5. On some models, pull off holding clamps at right and left below.
6. Pull out flat contour springs for radiator and lift out radiator.

To install:

7. Reverse the removal procedure. Take note that the fastening mounts of the radiator are correctly introduced into rubber grommets of lower holders and the holders of the fan cover into holding lugs on radiator.
8. Fill with coolant, pressure test cooling system with tester and check for leaks.

EXCEPT 190D, 300D, 300TD, 300SDL, 350SD and 350SDL

NOTE: After refilling the cooling system, the system may have to be bleed. Remove a cooling sensor or equivalent at the highest point in the engine's cooling system. Fill the radiator until coolant spills out of the hole. Apply thread sealing tape to the component and install. Finish filling the radiator to the proper level.

1. Disconnect the negative battery cable. Remove the radiator cap.
2. Unscrew the radiator drain plug and drain the coolant from the radiator. If all of the coolant in the system is to be drained, move the heater controls to **WARM** and open the drain cocks on the engine block.
3. If equipped with an oil cooler, drain the oil from the cooler.
4. If equipped, loosen the radiator shell.
5. Loosen the hose clips on the top and bottom radiator hoses and remove the hoses from the connections on the radiator.
6. Unscrew and plug the bottom line on the oil cooler.
7. If equipped with an automatic transmission, unscrew and plug the lines on the transmission cooler.
8. Disconnect the right and left side rubber loops and pull the radiator up and from the body.

To install:

9. Inspect and replace any hoses which have become hardened or spongy.
10. Install the radiator shell and radiator, if the shell was removed, from the top and connect the top and bottom hoses to the radiator.
11. Bolt the shell to the radiator.
12. Attach the rubber loops or position the retaining spring, as applicable.
13. Position the hose clips on the top and bottom hoses.
14. Attach the lines to the oil cooler.
15. If equipped with an automatic transmission, connect the lines to the transmission cooler.
16. Move the heater levers to the **WARM** position and slowly add coolant, allowing air to escape.
17. Check the oil level and fill if necessary. Run the engine for about 1 minute at idle with the filler neck open.
18. Add coolant to the specified level. Install the radiator cap and turn it un-

til it seats in the 2nd notch. Run the engine and check for leaks.

Water Pump

REMOVAL & INSTALLATION

190E With 2.3L Engine

NOTE: After refilling the cooling system, the system may have to be bleed. Remove a cooling sensor or equivalent at the highest point in the engine's cooling system. Fill the radiator until coolant spills out of the hole. Apply thread sealing tape to the component and install. Finish filling the radiator to the proper level.

1. Disconnect the negative battery cable. Drain coolant.
2. Remove air cleaner.
3. Remove radiator.
4. Loosen hose clamps and disconnect heater return line and coolant hose from coolant pump.
5. Remove fan.
6. Remove pulley of water pump.
7. Remove hex socket screws, slacken water pump V-belt and remove.
8. Pull cable from magnetic body, remove fastening screws and remove magnetic body.
9. Remove screws and place alternator with front holder aside.
10. Loosen lower hose clamp of bypass line, unscrew fastening screws and remove water pump.
11. Carefully clean sealing surfaces on water pump housing and timing housing cover.

To install:

12. Apply gasket adhesive to gasket and pump.
13. Install pump and torque bolts to 7.5 ft. lbs. (10 Nm).
14. Complete installation by reversing removal procedure. Fill coolant system and check for leaks.

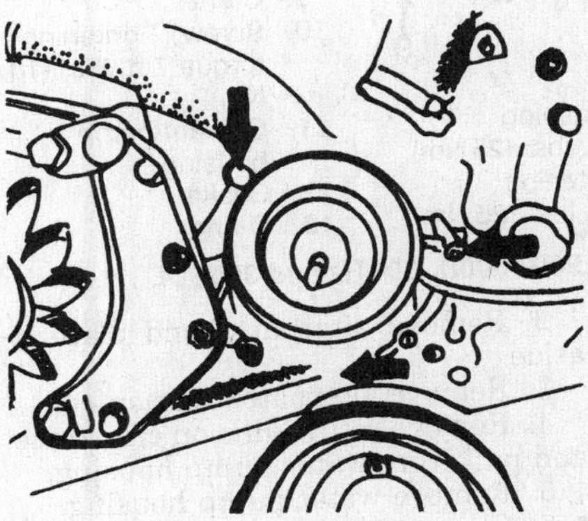

Water pump servicing—190E with 2.3L or 2.3L-16 engine

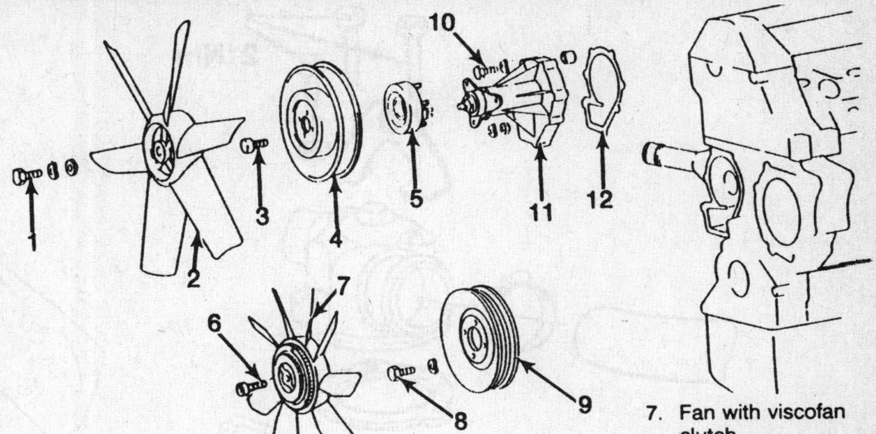

1. Hex head screw (engine 601) (Tightening torque 19 ft. lbs. (25 Nm)
2. Fan
3. Hex head socket screw (Tightening torque 7 ft. lbs. (10 Nm)
4. Pulley
5. Magnet, body
6. Hex head socket screw (engines 602,603) (Tightening torque 33 ft. lbs. (45 Nm)
7. Fan with viscofan clutch
8. Hex head screw (Tightening torque 7 ft. lbs. (10 Nm)
9. Pulley
10. Hex head screw (Tightening torque 7 ft. lbs. (10 Nm)
11. Coolant pump
12. Gasket

Water pump servicing—190D, 300D, 300TD, 300SDL and 350SDL

190D, 300D, 300TD, 300SDL and 350SDL Engine

NOTE: After refilling the cooling system, the system may have to be bleed. Remove a cooling sensor or equivalent at the highest point in the engine's cooling system. Fill the radiator until coolant spills out of the hole. Apply thread sealing tape to the component and install. Finish filling the radiator to the proper level.

1. Disconnect the negative battery cable. Remove fan and remove together with fan cover, if necessary.
2. On 190D with 2.5L engine and 300D, 300TD, 300SDL and 350SDL, loosen fan cover and place on fan. Remove viscous fan clutch using prybar element 103 589 01 09 00 or equivalent, torque wrench 001 589 72 21 00 or equivalent, and counter holder 603 589 00 40 00 or equivalent, for this purpose.
3. Remove fastening screws and pulley.
4. Remove hex nuts and magnet body.
5. The magnet carrier is glued to the water pump housing and should not be pulled off.
6. Remove water pump housing.
7. Clean sealing surfaces.

To install:

8. Insert water pump with a new gasket and tighten combination screws to 7.5 ft. lbs. (10 Nm).
9. Mount water pump with a new gasket and tighten combination screws to 7.5 ft. lbs. (10 Nm).
10. Mount magnet body and plug on cable.
11. Mount pulley and tighten fastening screws to 7.5 ft. lbs. (10 Nm).
12. Complete installation by reversing removal procedure.

6 Cylinder Engines

1. Disconnect the negative battery cable and drain the cooling system.
2. Remove the air scoop hose and viscous fan clutch.
3. Loosen and remove the serpentine belts.
4. Remove the water pump drive pulley.
5. Remove the bottom engine cover, if so equipped.
6. Do not disconnect refrigerant hoses. Remove the air conditioning compressor and lay aside.
7. Remove all water and heater hoses from the pump.
8. Remove the tandom pump and move to the side. Remove the thermostat housing bolts and housing.
9. Remove the pump mounting bolts and slide the pump up and out of the engine compartment.

To install:

10. If removed, install the thermostat so that the ball valve is at the highest point. Install a new seal.
11. Install a new water pump seal and torque the retaining bolts to 18 ft. lbs. (25 Nm).
12. Install the remaining components. Torque the water pump pulley bolts to 7 ft. lbs. (10 Nm) and compressor bolts to 18 ft. lbs. (25 Nm).

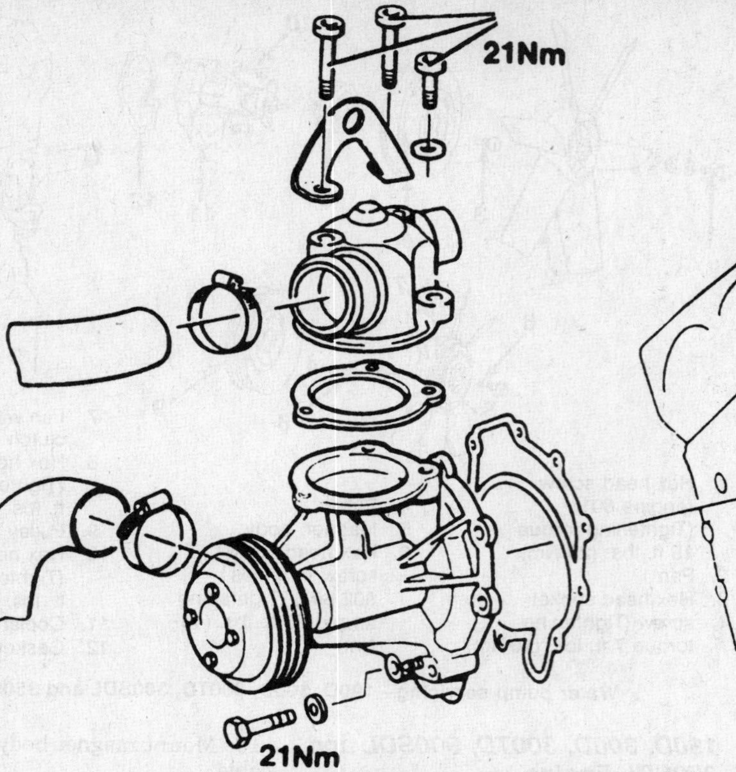

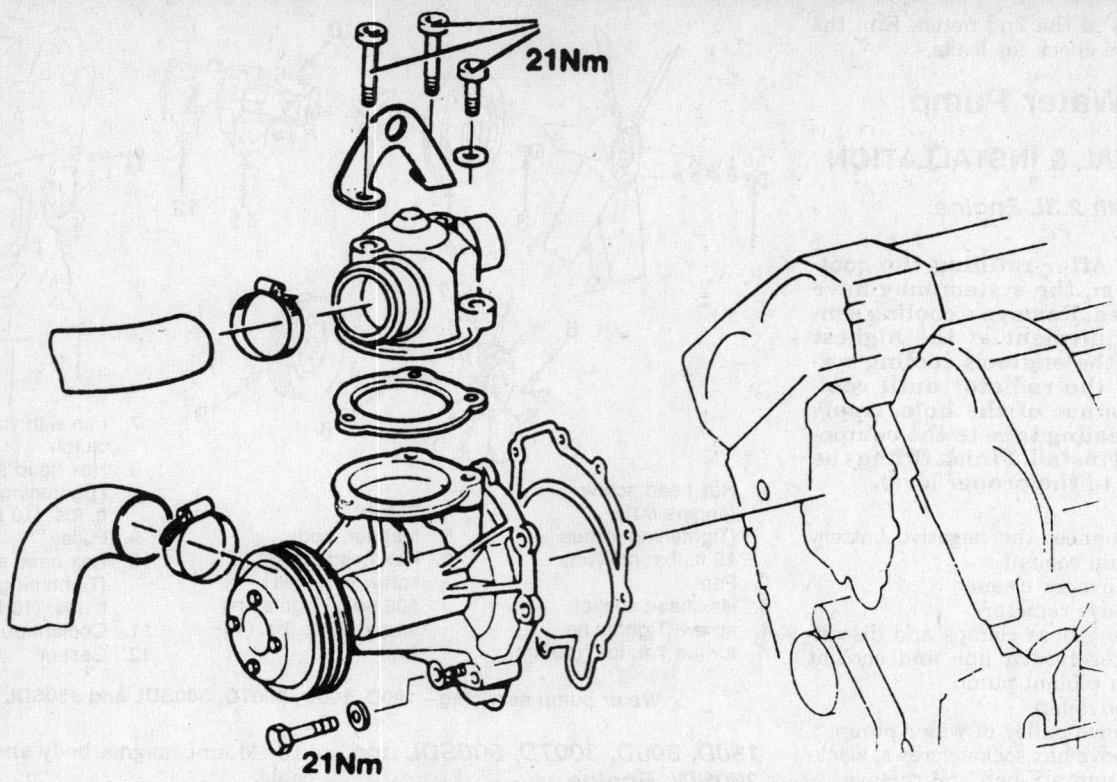

Water pump assembly – 5.0L DOHC, other V8 engines similar

8 Cylinder Engines

1. Disconnect the negative battery cable. Drain the water from the radiator and block.
2. Remove the air cleaner.
3. Loosen and remove the drive belt.
4. Disconnect the upper water hose from the radiator and thermostat housing.
5. Remove the fan and coupling.
6. Remove the hose from the intake (top) connection of the water pump.
7. Set the engine at TDC. Matchmark the distributor and engine and remove the distributor. Crank the engine with a socket wrench on the crankshaft pulley bolt or with a small prybar inserted in the balancer. Crank in the normal direction of rotation only.
8. Turn the balancer so the recesses provide access to the mounting bolts. Remove the mounting bolts. Rotate the engine in the normal direction of rotation only.
9. Remove the water pump.

To install:

10. Clean the mounting surfaces of the water pump and block.
11. Installation is the reverse of removal. Always use a new gasket. Set the engine at TDC and install the distributor rotor points to the notch on the distributor housing. Fill the cooling system and check and adjust the ignition timing.

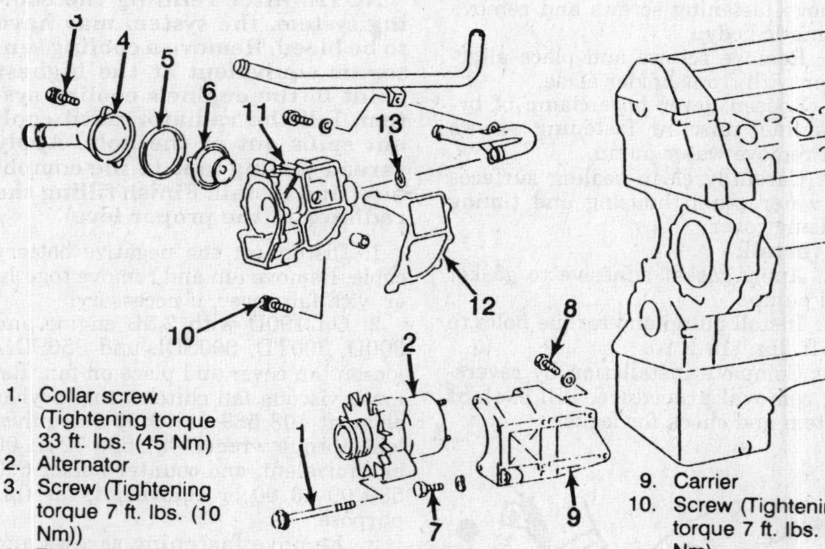

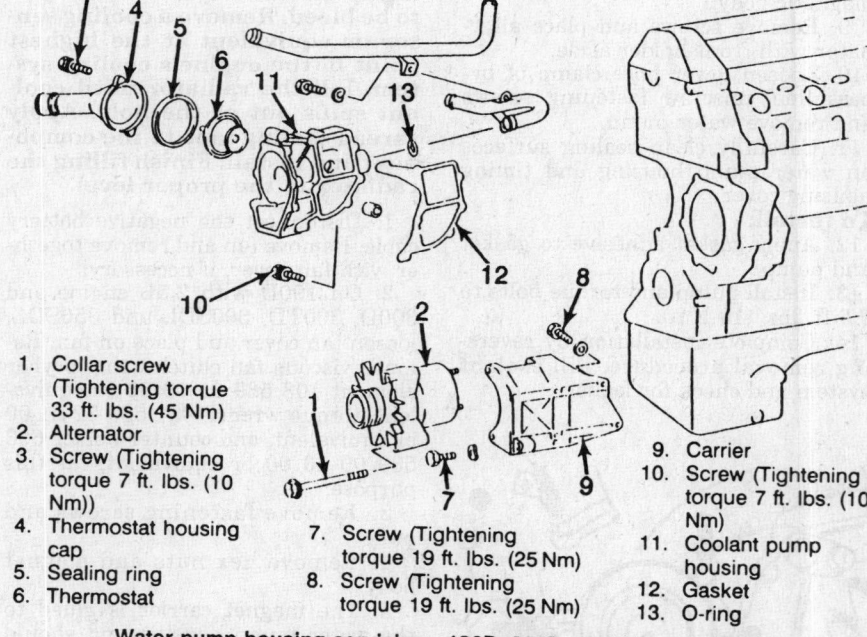

1. Collar screw (Tightening torque 33 ft. lbs. (45 Nm)
2. Alternator
3. Screw (Tightening torque 7 ft. lbs. (10 Nm))
4. Thermostat housing cap
5. Sealing ring
6. Thermostat
7. Screw (Tightening torque 19 ft. lbs. (25 Nm)
8. Screw (Tightening torque 19 ft. lbs. (25 Nm)
9. Carrier
10. Screw (Tightening torque 7 ft. lbs. (10 Nm)
11. Coolant pump housing
12. Gasket
13. O-ring

Water pump housing servicing – 190D, 300D, 300TD and 300SDL

Water Pump Housing

REMOVAL & INSTALLATION

190D, 300D, 300TD, 300SDL, 350SD and 350SDL

1. Disconnect negative terminal on battery.
2. Remove alternator and place it aside.
3. Remove alternator carrier.
4. Remove return line on crankcase and pull from water pump housing.
5. Remove water pump housing.
6. Clean sealing surfaces.

To install:

7. Renew O-ring on return line.

NOTE: Keep O-ring free of grease. For better assembly, immerse O-ring into coolant.

8. Plug coolant pump housing on return line and screw with a gasket to crankcase, tighten to 7.5 ft. lbs. (10 Nm).
9. Screw return line to crankcase.
10. Mount alternator carrier and tighten screws to 18.5 ft. lbs. (25 Nm).
11. Mount alternator and tighten screw to 33 ft. lbs. (45 Nm). Connect negative battery terminal.

Thermostat

REMOVAL & INSTALLATION

Except V8 Engines

The thermostat housing is a light metal casting attached directly to the cylinder head, except on the 190D where it is attached to the side of the water pump housing; and the 190E with the 2.6L engine, 260E, 300E, 300CE, 300SE, 300TE, 300SEL and 300SL where it is under a plastic cover atop the water pump.

1. Disconnect the negative battery cable. Open the radiator cap and depressurize the system.
2. Open the radiator drain cock and partially drain the coolant. Drain enough coolant to bring the coolant level below the level of the thermostat housing.
3. Remove the thermostat housing cover bolts and cover.
4. Note the installation position of the thermostat and remove it.

To install:

5. Installation is the reverse of removal. Be sure the thermostat is positioned with the ball valve at the highest point and the bolts are tightened evenly against the seal. On the 190D, the recess in the thermostat casing should be located above the lug in the thermostat housing. On the 190E with 2.6L engine, 260E, 300E, 300CE, 300SE, 300TE, 300SEL and 300SL, the ball valve must be at its highest point in the housing to allow complete venting of gas bubbles.
6. Refill the cooling system and check for leaks.

NOTE: When refilling the coolant system on the 300E, 300CE, 300SE, 300TE, 300SL and 300SEL, always remove the hex-head plug on the left side of the cylinder head and fill the hole with coolant until it overflows. Install the plug and fill the coolant system. When filling the coolant system on the 190E with 2.6L and the 260E, open the vent screw on top of the thermostat housing approximately 2 turns, start the engine run the engine at idle.

Aligning the thermostat on the 190D

V8 Engines

1. Drain the coolant from the radiator and block.
2. Remove the air cleaner.
3. Disconnect the battery and remove the alternator.
4. Unscrew the housing cover on the side of the water pump and remove the thermostat.
5. If a new thermostat is to be installed, always install a new sealing ring.

To install:

6. Install the thermostat so the ball valve is at the highest point in the housing.
7. Be sure to tighten the screws on the housing cover evenly to prevent leaks. Refill the cooling system and check of leaks.

ENGINE ELECTRICAL

NOTE: Disconnecting the negative battery cable on some vehicles may interfere with the functions of the on board computer systems and may require the computer to undergo a relearning process, once the negative battery cable is reconnected.

Distributor

REMOVAL & INSTALLATION

NOTE: The distributor on the 190E with 2.6L engine, 260E, 300E, 300CE, 300SE, 300TE, 300SL and 300SEL is part of the cylinder head. With the exception of the cap and rotor, it is not readily removable.

The removal and installation procedures for all distributors on Mercedes-Benz vehicles are basically similar. Certain minor differences may exist from model-to-model.

1. The distributor is usually located on the front side or front of the engine.
2. Disconnect the negative battery cable. Remove the dust, cover, distributor cap, cable plug connections and vacuum line.
3. Rotating the engine in the normal direction, crank it until the distributor rotor points to the mark on the rim of the distributor housing. This indicates the No. 1 cylinder. Check the crankshaft pulley for the 0 mark.
4. The engine can be cranked with a socket wrench on the balancer bolt or with a prybar inserted in the balancer.
5. Matchmark the distributor body and the engine so the distributor can be returned to its original position.
6. Remove the distributor hold-down bolt and withdraw the distributor from the engine.

NOTE: Do not crank the engine while the distributor is removed.

7. Installation is the reverse of removal. Insert the distributor so the matchmarks on the distributor and engine are aligned.
8. Tighten the clamp bolt and check the dwell angle and ignition timing.

Ignition Timing

ADJUSTMENT

Before attempting to set the timing, read the "Ignition Timing Specifications" chart carefully and determine at what speed the timing should be set and whether the vacuum should be connected or disconnected.

NOTE: It is a good idea to paint the appropriate timing mark with dayglow or white paint to make it quickly and easily visible.

On engines with transistorized coil ignition, the timing light may or may not work depending on the construction of the light.

Gasoline Engines

NOTE: All gasoline engines utilize the new EZL electronic ignition system. Although service checking of ignition timing is possible, no adjustment is either possible or necessary.

1. Raise the hood and connect a tachometer.
2. Connect a timing light.
3. Run the engine at the specified speed and read the firing point on the balancing plate or vibration damper while shining the light on it.

NOTE: The balancer on some engines has 2 timing scales. If in doubt as to which scale to use, rotate the crankshaft, in the direc-

tion of rotation only, until the distributor rotor is aligned with the notch on the distributor housing (No. 1 cylinder). In this position, the timing pointer should be at TDC on the proper timing scale.

4. Adjust the ignition timing by loosening the distributor clamp bolt and rotating the distributor. To advance the timing, rotate the distributor in the opposite direction of normal rotation. To retard the timing, rotate the distributor in the direction of normal rotation.

5. Once the timing has been adjusted, recheck the timing once more to be sure it has not been disturbed.

6. Remove the timing light and tachometer and connect any wires that were removed.

Diesel Engines

The diesel uses no distributor, so requires no ignition timing adjustment.

Alternator

PRECAUTIONS

Several precautions must be observed with alternator equipped vehicles to avoid damage to the unit.

- If the battery is removed for any reason, make sure it is reconnected with the correct polarity. Reversing the battery connections may result in damage to the one-way rectifiers.
- When utilizing a booster battery as a starting aid, always connect the positive to positive terminals and the negative terminal from the booster battery to a good engine ground on the vehicle being started.
- Never use a fast charger as a booster to start vehicles.
- Disconnect the battery cables when charging the battery with a fast charger.

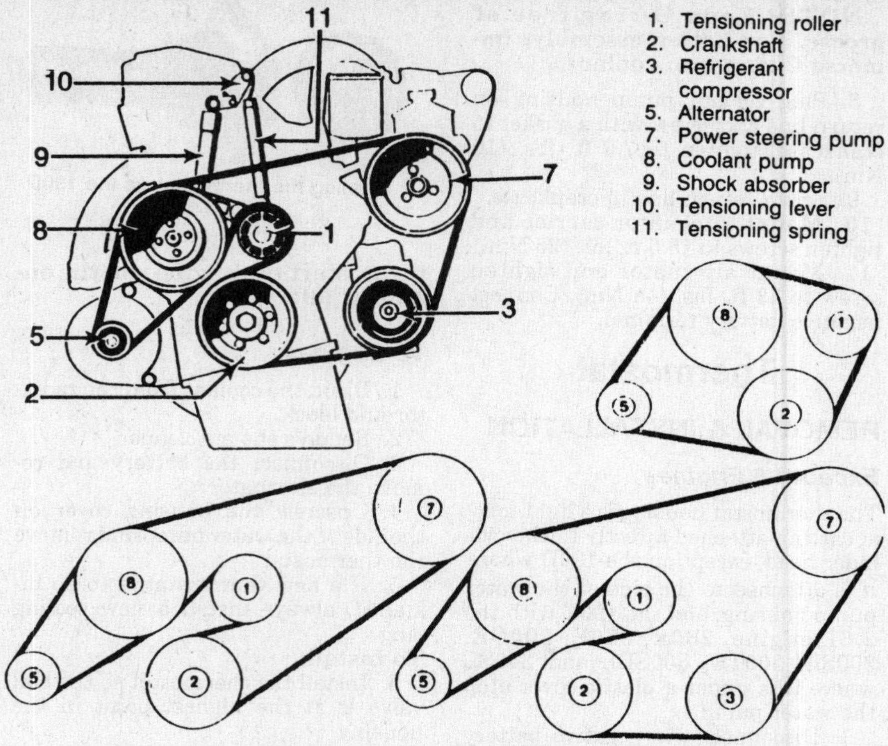

1. Tensioning roller
2. Crankshaft
3. Refrigerant compressor
5. Alternator
7. Power steering pump
8. Coolant pump
9. Shock absorber
10. Tensioning lever
11. Tensioning spring

Poly V-belt and accessories – 190D, 300D and 300SDL

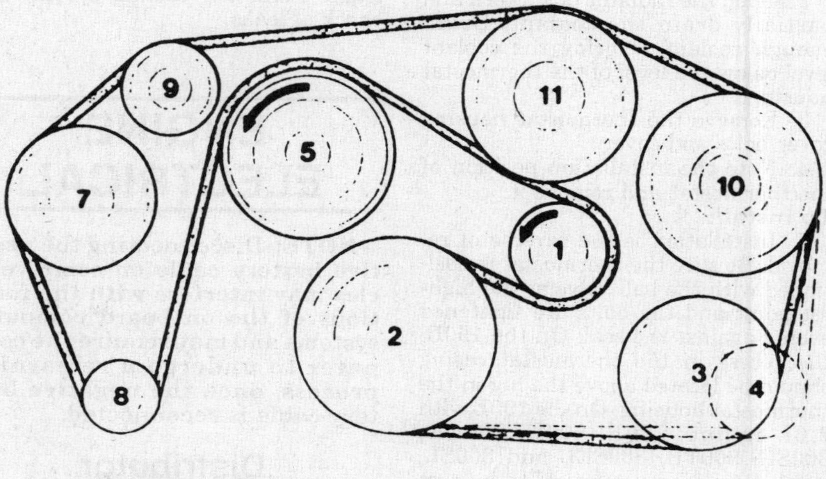

Belt length: 2445mm

1. Tensioner pulley
2. Crankshaft
3. A/C compressor
4. Lower guide pulley
5. Fan
7. Air pump
8. Alternator
9. Upper guide pulley
10. Steering pump
11. Coolant pump

Poly V-belt routing – 3.0L DOHC engine

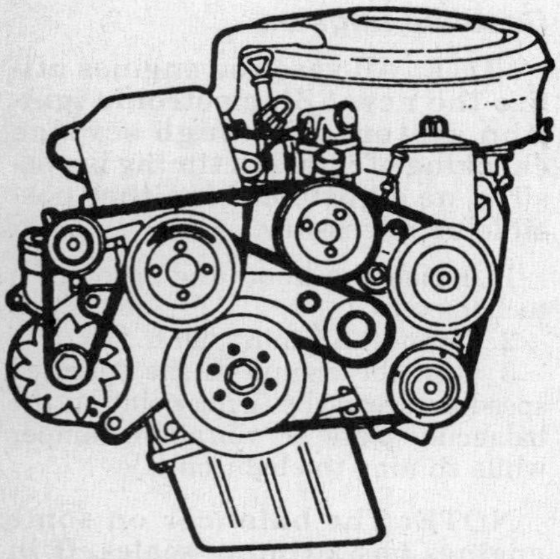

Poly V-belt routing – 2.6L and 3.0L engines

- Never attempt to polarize the alternator.
- Do not use test lights of more than 12 volts when checking diode continuity.
- Do not short across or ground any of the alternator terminals.
- The polarity of the battery, alternator and regulator must be matched and considered before making any electrical connections within the system.
- Never separate the alternator on an open circuit. Make sure all connections within the circuit are clean and tight.

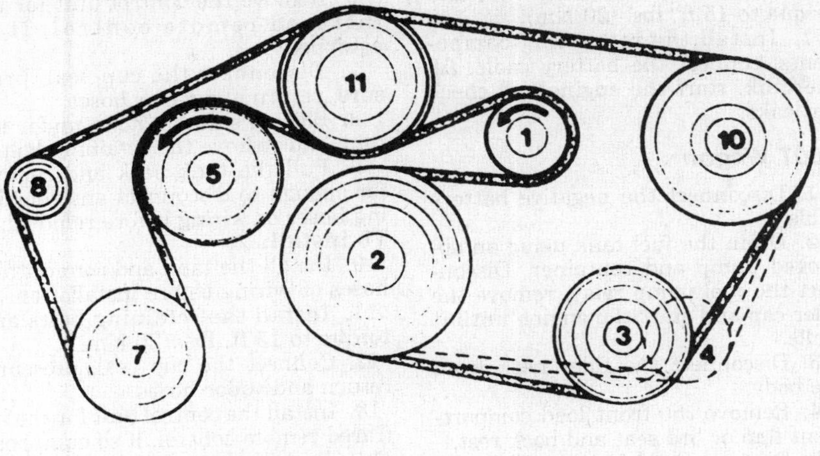

1. Tensioner pulley
2. Crankshaft
3. A/C compressor
4. Guide pulley
5. Fan
7. Air pump
8. Alternator
10. Steering pump
11. Coolant pump

Poly V-belt routing—5.0L DOHC engine

- Disconnect the battery ground terminal when performing any service on electrical components.
- Disconnect the battery if arc welding is to be done on the vehicle.

BELT TENSION ADJUSTMENT

All alternator dive belts should be tensioned to approximately ½ in. deflection under thumb pressure at the middle of the longest span.

NOTE: All models utilize a single V-belt with automatic tensioning. No adjustment is necessary.

REMOVAL & INSTALLATION

Viewing the engine from the front, the alternator is located on either side, usually down low. Because of the location, it is sometimes easier to remove the alternator from under the vehicle. The following is a general procedure for all models.

1. Disconnect the negative battery cable. Locate the alternator and disconnect and identify all wires.
2. Loosen the adjusting (pivot) bolt or the adjusting mechanism and swing the alternator in toward the engine.
3. Remove the drive belt from the alternator pulley.
4. The alternator can now be removed from its mounting bracket or the bracket and alternator can be removed from the engine.
5. Installation is the reverse of removal.
6. Tighten all of the drive belts that were loosened.

Starter

REMOVAL & INSTALLATION

260E, 300E, 300CE and 300TE

1. Disconnect negative battery terminal.
2. Remove complete air cleaner.
3. Remove holder on intake manifold.
4. Remove engine compartment enclosure.
5. Disconnect electric wires for oil level and oil pressure sensor.
6. Remove starter.

To install:

7. Reverse removal procedure.

All Other Models

1. Remove all wires from the starter and tag them for location.
2. Disconnect the battery cable.
3. Some models use a starter shroud to protect the unit. Remove this if necessary. Unbolt the starter from the bell housing and remove the ground cable.
4. Remove the starter from under the vehicle.
5. Installation is the reverse of removal. Torque the bolts to 15 ft. lbs. (20 Nm). Be sure to replace all wires and washers in their original location.

EMISSION CONTROLS

Please refer to "Emission Controls" in the Unit Repair section for system maintenance procedures. Due to the complex nature of modern electronic engine control systems, comprehensive diagnosis and testing procedures fall outside the confines of this repair manual. For complete information on diagnosis, testing and repair procedures concerning all modern engine and emission control systems, please refer to "Chilton's Guide to Fuel Injection and Electronic Engine Controls".

Emission Warning Lights

RESETTING

560 Series

1. The instrument cluster must be partially removed on certain models. Using a steel wire with a small hook on the end, slip the wire between the right side of the cluster and the dashboard. Turn the hook to engage the cluster and the dashboard. Turn the hook to engage the cluster and gently pull the edge of the cluster from the retaining clips.
2. Remove the oxygen sensor bulb at the extreme lower corner of the cluster. Press the cluster back into position. No reset switch is provided.

GASOLINE FUEL SYSTEM

Fuel System Service Precautions

Safety is the most important factor when performing not only fuel system maintenance but any type of maintenance. Failure to conduct maintenance and repairs in a safe manner may result in serious personal injury or death. Maintenance and testing of the vehicle's fuel system components can be accomplished safely and effectively by adhering to the following rules and guidelines.

- To avoid the possibility of fire and

personal injury, always disconnect the negative battery cable unless the repair or test procedure requires that battery voltage be applied.

• Always relieve the fuel system pressure prior to disconnecting any fuel system component (injector, fuel rail, pressure regulator, etc.), fitting or fuel line connection. Exercise extreme caution whenever relieving fuel system pressure to avoid exposing skin, face and eyes to fuel spray. Please be advised that fuel under pressure may penetrate the skin or any part of the body that it contacts.

• Always place a shop towel or cloth around the fitting or connection prior to loosening to absorb any excess fuel due to spillage. Ensure that all fuel spillage (should it occur) is quickly removed from engine surfaces. Ensure that all fuel soaked cloths or towels are deposited into a suitable waste container.

• Always keep a dry chemical (Class B) fire extinguisher near the work area.

• Do not allow fuel spray or fuel vapors to come into contact with a spark or open flame.

• Always use a backup wrench when loosening and tightening fuel line connection fittings. This will prevent unnecessary stress and torsion to fuel line piping. Always follow the proper torque specifications.

• Always replace worn fuel fitting O-rings with new. Do not substitute fuel hose or equivalent where fuel pipe is installed.

Fuel Tank

REMOVAL & INSTALLATION

190, 300, 420 and 560 Series Sedans

1. Disconnect the negative battery cable.
2. Drain the fuel tank using an approved pump and container. Disconnect the fuel pump relay, remove the filler cap and start the engine until it stalls.
3. Disconnect the filler neck from the body and fuel gauge sender.
4. Remove the tank covering.
5. Disconnect all visible pressure, return and vapor lines.
6. Support the tank using an approved hoist.
7. Remove the tank retaining nuts and lower the tank far enough to disconnect any connected hoses or wiring.

To install:

8. Empty the remaining fuel from the tank.
9. Install the tank and align the cushions.
10. Install the retaining nuts and torque to 15 ft. lbs. (20 Nm).
11. Install the remaining components, connect the battery cable, fill the tank, start the engine and check for leaks.

300T Wagon

1. Disconnect the negative battery cable.
2. Drain the fuel tank using an approved pump and container. Disconnect the fuel pump relay, remove the filler cap and start the engine until it stalls.
3. Disconnect the filler neck from the body.
4. Remove the front load compartment flap or 3rd seat and back rest.
5. Disconnect the breather line and fuel gauge sender wire.
6. Disconnect any accessible hose and wiring before removal.
7. Remove the rear portion of the exhaust system.
8. Remove the splashs and bracket level pipe .
9. Remove the front retaining bolts. Push the exhaust and level pipe to the left.
10. Install a suitable hoist to the tank and remove the remaining tank retainers. Lower the tank far enough to disconnect any hoses or wiring.

To install:

11. Install a suitable hoist to the tank and install tank and retainers. Connect any hoses or wiring.
12. Install the front retaining bolts and torque to 15 ft. lbs. (20 Nm).
13. Install the splashs and bracket level pipe .
14. Install the rear portion of the exhaust system.
15. Connect any accessible hose and wiring.
16. Connect the breather line and fuel gauge sender wire.
17. Install the front load compartment flap or 3rd seat and back rest.
18. Connect the filler neck to the body.
19. Refill the fuel tank.
20. Connect the negative battery cable and check for leaks.

300 and 500SL Series

1. Disconnect the negative battery cable.
2. Drain the fuel tank using an approved pump and container. Disconnect the fuel pump relay, remove the filler cap and start the engine until it stalls.
3. Disconnect the filler neck from the body.
4. Remove the convertible partition wall.
5. Remove the cover from the filler neck.
6. Remove the control unit for the infrared remote control, if so equipped.
7. Disconnect the cup seal, pressure, return and vapor hoses.
8. Place a suitable hoist under the tank and remove the retaining bolts.
9. Pull the tank back and remove far enough to disconnect any remaining hoses or wiring before removal.

To install:

10. Install the tank and connect any hoses or wiring before installation.
11. Install the retaining bolts and torque to 15 ft. lbs. (20 Nm).
12. Connect the cup seal, pressure, return and vapor hoses.
13. Install the control unit for the infrared remote control, if so equipped.
14. Install the cover to the filler neck.
15. Install the convertible partition wall.
16. Connect the filler neck to the body.
17. Refill the fuel tank and check for leaks.
18. Connect the negative battery cable.

Fuel Filter

REMOVAL & INSTALLATION

Two types of filters are used, depending on the vehicle. Both are located between the rear axle and the fuel tank.

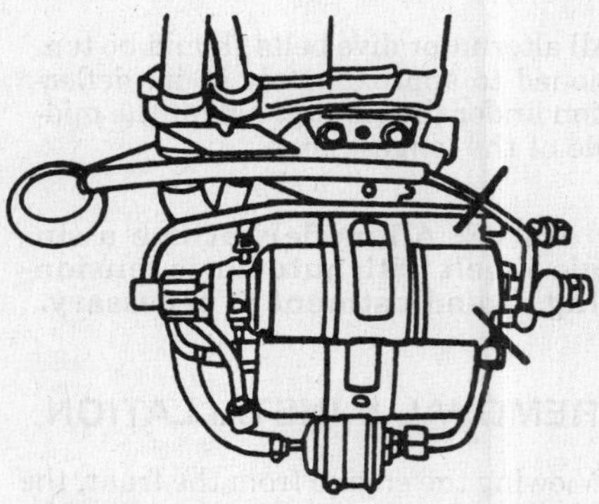

Fuel pump/filter assembly—190E-16 shown, others similar

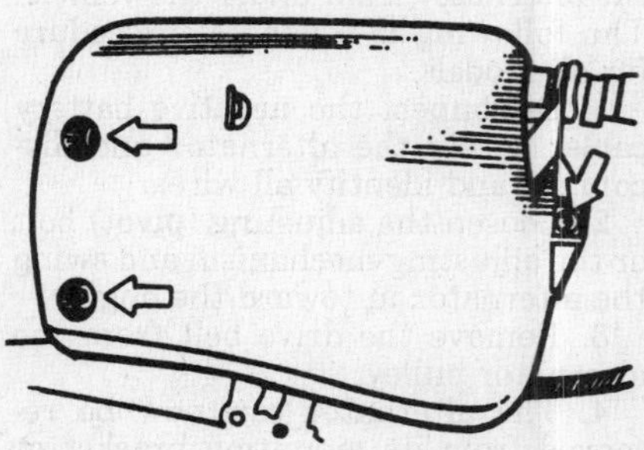

Fuel pump package cover—190 series, 260E, 300E, 300CE, 300SEL and 300TE

1. Unscrew the cover box.
2. Remove the gas cap, fuel pump cover and pressure hoses.
3. Loosen the screws and remove the filter. Remove the connecting plug from the old filter and install it on a new filter using a new gasket.

To install:

4. Install a new filter in the direction of flow.
5. Replace the attaching screws.
6. Install the pressure hoses, pump cover and gas cap.
7. Install the fuel filter in the holder by positioning it in the center of the transparent holder. Be sure the plastic sleeve between the fuel filter and fuel pump is installed. Galvanic corrosion may occur in cases of direct contact between these components.
8. Replace the cover box and check for proper sealing.

Electric Fuel Pump

NOTE: Do not confuse the electric fuel pump with the injection pump.

All Mercedes-Benz fuel injected engines are equipped with electric fuel pumps. The electric fuel pump is located under the rear floor panel. The fuel return line was also eliminated and a check ball installed in its place. The fuel pump uses a replaceable check valve on the outside of the pump which can be replaced separately. Some later model vehicles use 2 fuel pumps in the same location.

Two types of fuel pumps have been used. One, the large pump, has been replaced with a new small design which has a bypass system to prevent vapor lock.

TESTING DELIVERY VOLUME

1. Disconnect the inlet hose at the pressure regulator. Connect a fuel hose to the fuel pressure regulator inlet hose.
2. Place the hose in an approved measuring container.
3. Turn the ignition key **ON** and measure the volume. The minimum value is 1.06 quart (1 liter) in 40 seconds.
4. Replace the fuel filter and test again. If still low, replace the fuel pump.

PRESSURE TEST

1. Disconnect the fuel line at the fuel distributor and connect a pressure gauge 103589002100 or equivalent to the hose.
2. Connect an adapter 102589066300 or equivalent to the plug fitting at the fuel distributor.
3. Jump the fuel pump relay at the manifold and check the pressure. If not within specifications, check the fuel filter and pump pressure at the pump.
4. Disconnect the pressure line at the fuel pump.
5. Connect a pressure gauge and adapter to the fuel pump outlet at the pump. The pressure should be 29–58 psi (2–4 bar).
6. Reconnect all fuel hoses and check for leaks.

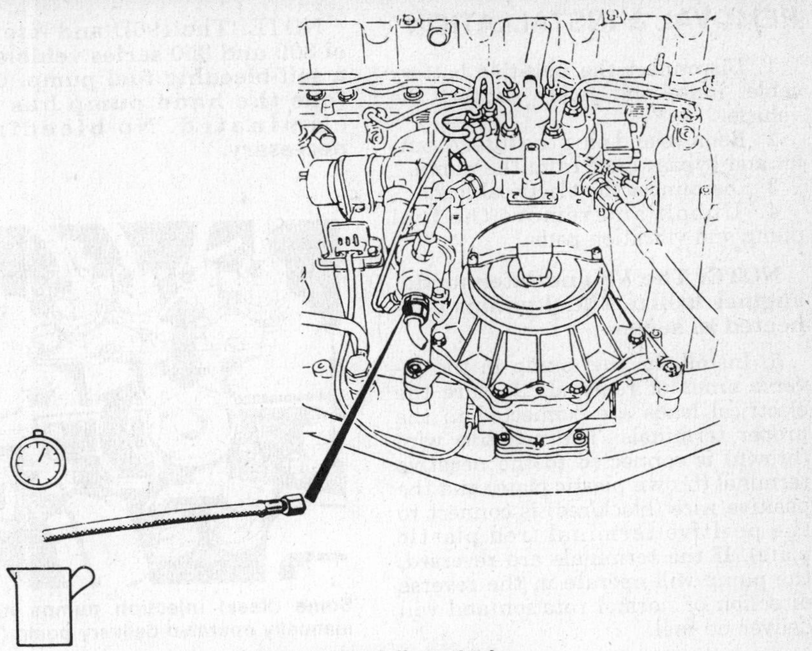

Fuel pump delivery test

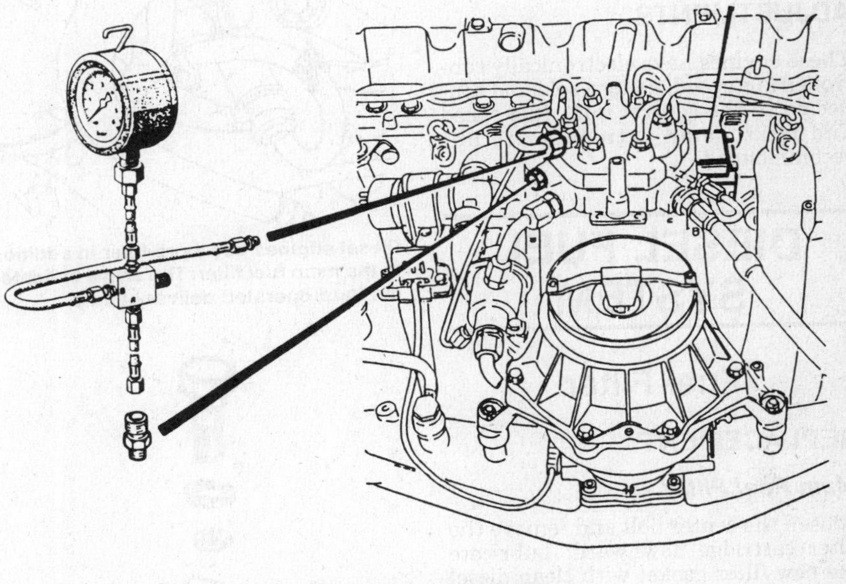

Fuel pump pressure test

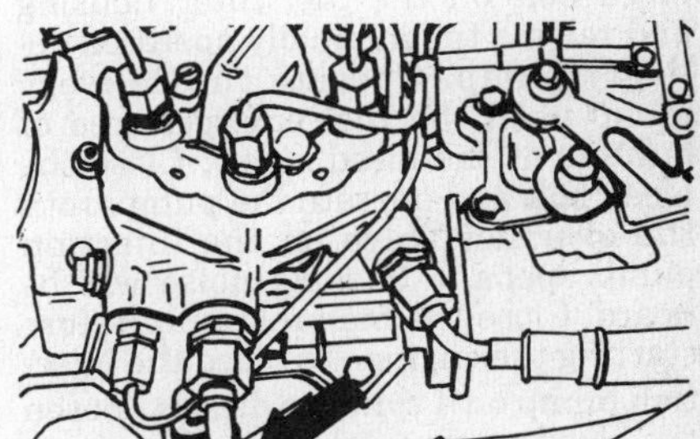

Testing fuel pump pressure

REMOVAL & INSTALLATION

1. Disconnect the negative battery cable. Raise and safely support the vehicle.
2. Remove and plug the intake, outlet and bypass lines from the pump.
3. Disconnect the electrical leads.
4. Unbolt and remove the fuel pump and vibration pads.

NOTE: The V8 and later model engines utilize 2 fuel pumps connected in series.

5. Install the fuel pump in the reverse order of removal. Be sure the electrical leads are connected to the proper terminals. The negative wire (brown) is connected to the negative terminal (brown plastic plate) and the positive wire (black/red) is connect to the positive terminal (red plastic plate). If the terminals are reversed, the pump will operate in the reverse direction of normal rotation and will deliver no fuel.

Fuel Injection

ADJUSTMENTS

These engines have electronically controlled idle speed, using a solenoid connected to the control unit. Idle speed and mixture adjustments are not recommended.

DIESEL FUEL SYSTEM

Fuel Filter

REPLACEMENT

Main Fuel Filter

Loosen the center bolt and remove the filter cartridge downward. Lubricate the new filter gasket with clean diesel fuel and install a new filter cartridge.

To bleed the fuel filter: Loosen the bleed bolt on the fuel filter housing and release the manually operated delivery pump. Operate the delivery pump until the fuel emerges free of bubbles at the bleed screw. Close the bleed bolt and operate the pump until the overflow valve on the injection pump opens, a buzzing noise will be heard. Close the manual pump before starting the engine. To bleed the injection pump on 4 cylinder diesels, loosen the bleed screw on the injection pump and keep pumping the hand pump until fuel emerges free of bubbles.

NOTE: The 190D and late model 300 and 350 series vehicles use a self-bleeding fuel pump, therefore the hand pump has been eliminated. No bleeding is necessary.

Some diesel injection pumps have a manually operated delivery pump (1)

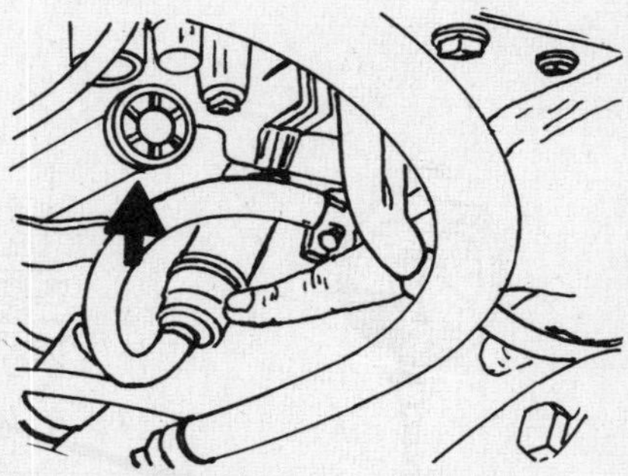

Diesel engines use a pre-filter in addition to the main fuel filter. The arrow indicates the hard operated delivery pump

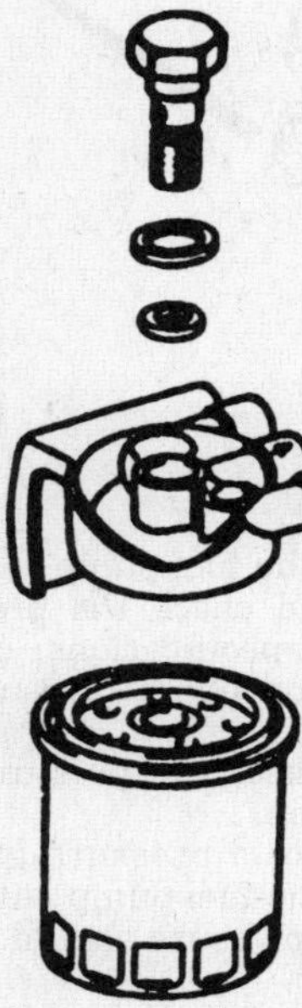

Diesel fuel filter

Diesel Pre-Filter

Diesel engines use a pre-filter in addition to the main fuel filter, since even the most minute particle of dirt will clog the injection system. The pre-filter is located in the line just before it enters the injection pump at the fuel pump. It is transparent plastic.

To replace it, simply unscrew the clamps on each end and remove the old filter. Install a new filter and bleed the system.

Diesel Injection Pump

REMOVAL & INSTALLATION

1. Disconnect the negative battery cable.
2. Remove the cylinder head cover.
3. Position the crankshaft pulley at 15 degrees after TDC. Fix the camshaft gear and injection pump in place with a cable strap, or equivalent.
4. Remove the timing chain tensioner.
5. Remove the camshaft gear bolt and gear. Make sure the crankshaft is held in place during this procedure.
6. Remove the injection pump

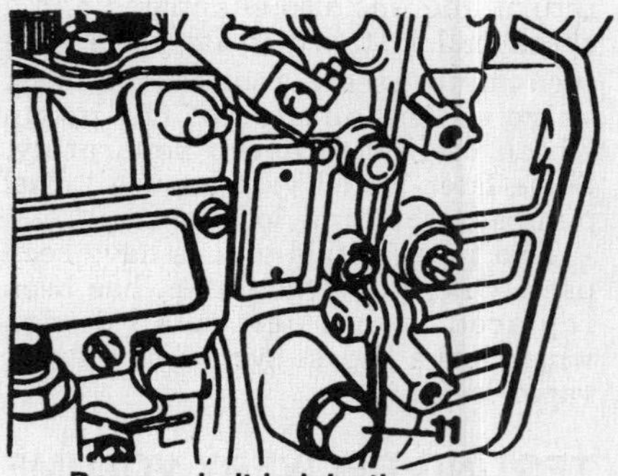

Remove bolt in injection pump

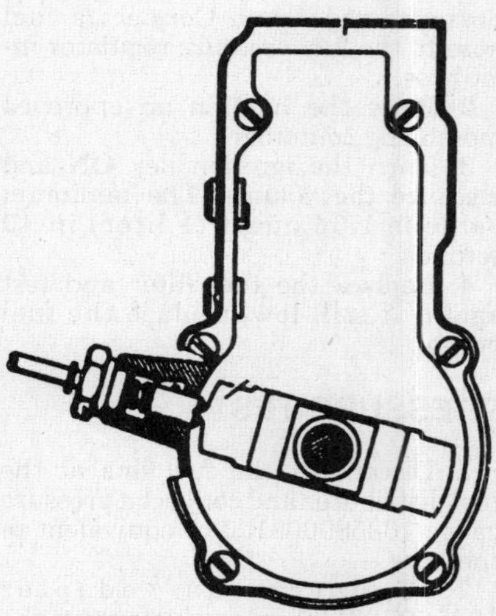

Injection pump timing mark

sprocket bolt, holding the crankshaft in place. The bolt is a left hand thread.

7. Remove the locking pin and remove the mounting bracket.

8. Press the timing chain out of the way with strips of metal and universal pliers to withdraw the chain.

9. Remove the injection pump lines and retainers. Carefully lift the pump away from the engine.

To install:

10. Remove the locking bolt from the side of the injection pump. The pump will be damaged when the engine is started if this procedure is not followed.

11. Install the pump and reconnect the injector lines.

12. Reconnect the timing chain and install components.

13. Test start of injector (injector pump timing) delivery with a digital tester.

IDLE SPEED ADJUSTMENT

All engines are equipped with an electronic idle speed control system. The rpm sensor picks up the engine speed and transmits it in the form of AC voltage to the EDS control unit. The EDS control unit processes the rpm signal and performs a nominal-actual value comparison. The idle speed is held constant by the magnetic actuator independent of engine load. Idle speed should not need adjustment during normal operation of the vehicle. Perform idle speed adjustment only if the injection pump has been overhauled.

1. Connect the digital tester to the pulse generator at the crankshaft pulley.

2. Check to see if the accelerator cable is operating properly. Adjust the cable from inside the vehicle so that the lever is resting on the throttle stop.

3. Start the engine and allow to warm to operating temperature.

4. Disconnect the 2 pin connector at the idle speed actuator.

5. Loosen the idle speed bolt lock nut and adjust to the proper idle specification.

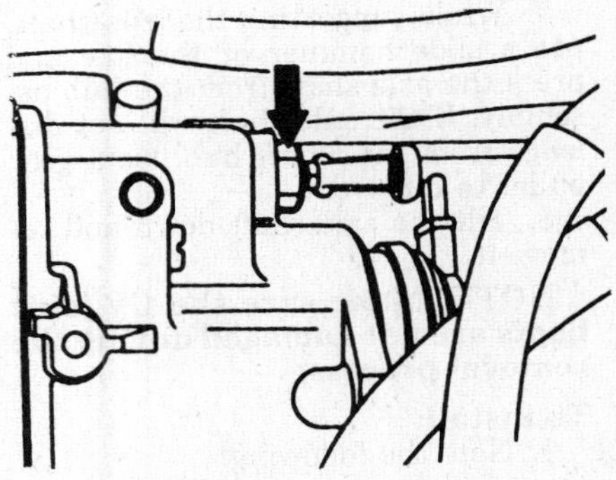

Idle speed adjustment bolt

6. Reconnect all connectors and remove test equipment.

Diesel Injection Timing

ADJUSTMENT

1. Remove the screw plug from the side of the injection pump governor housing. Collect the lost oil.

2. Install a position sensor into the hole and connect the tester to the vehicle battery and position sensor.

3. Turn the crankshaft by hand until lamps A and B on the tester light simultaneously. Take the reading on the crankshaft pulley.

4. Loosen the injection pump and adjust until the timing pointer is at 15 degrees after TDC and both lamps are ON.

5. Install the oil and plug. Torque the plug to 22 ft. lbs. (30 Nm).

6. Recheck the oil level.

Fuel Injector

REMOVAL & INSTALLATION

CAUTION

The fuel injector is under extreme pressure. Always wear eye protection while servicing the diesel injection system. Failure to do so may cause personal injury.

1. Disconnect the negative battery cable.

2. Place shop rags around the injector. Using a flare nut and backup wrench, loosen the metal pipe slightly until the pressure is released. Remove the pipe and hoses.

3. Remove the injector using special tool 001589650900 or equivalent.

4. Install a new nozzle shim and install the injector.

5. Connect the injector pipe and torque to 15 ft. lbs. (20 Nm).

6. Replace the oil leak hose if damaged.

DRIVE AXLE

These vehicles use either 2 or 3 piece driveshaft to connect the transmission to a hypoid independent rear axle. All models covered in this book use independent rear suspension with open or enclosed driveshaft to the rear wheels.

Driveshaft and U-Joints

REMOVAL & INSTALLATION

NOTE: Steps 1–3 apply to 4 cylinder and V8 engines. Matchmark all driveshaft connections prior to removal.

1. Raise and support the vehicle safely.

2. Fold the torsion bar down after disconnecting the level control linkage, if equipped.

3. Remove the exhaust system.

4. Remove the heat shield from the frame.

5. Support the transmission and remove the rear engine mount crossmember.

6. Without sliding the rubber sleeve back, loosen the clamp nut approximately 2 turns, the rubber sleeve will slide along.

NOTE: On 3 piece driveshafts, only the front clamp nut need be loosened.

7. Unscrew the U-joint mounting flange from the U-joint plate.

8. Bend back the locktabs and remove the driveshaft-to-rear axle pinion yoke bolts.

9. Remove the intermediate bearing(s)-to-frame bolts, push the driveshaft together slightly and remove it from the vehicle.

10. Try not to separate the driveshafts. If necessary, matchmark all components so they can be reassembled in the same order.

To install:

11. Always use new self-locking nuts.

12. After the driveshaft is installed, rock the vehicle back and forth, several times, to settle the driveshaft.

13. Make sure neither intermediate shaft is binding against either intermediate bearing and the clearance between the intermediate bearing and the driveshaft is the same at both ends.

Rear Axle Shafts

NOTE: The rubber covered joints are filled with special oil. If they are disassembled for any reason, they must be refilled with special oil.

REMOVAL & INSTALLATION

Except 190D, 190E, 260E, 300D, 300TD, 300E, 300CE, 300SE, 300TE, 300SEL, 300SDL, 420SEL, 560SEC and 560SEL

WITHOUT TORQUE COMPENSATOR (TORSION BAR)

1. Raise and safely support the vehicle. Remove the wheel and center axle hold-down bolt (in hub).

2. Remove the brake caliper and suspend it from a hook.

3. Drain the differential oil and support the differential housing.
4. Unbolt the rubber mount from the chassis and the differential housing and remove the differential housing cover to expose the ring and pinion gears.
5. Press the shaft from the axle flange. If necessary, loosen the shock absorber.
6. Using a prybar, remove the axle lock ring inside the differential case.
7. Pull the axle from the housing by pulling the splined end from the side gears, with the spacer.

NOTE: Axle shafts are stamped R and L for right and left units. Always use new lock rings.

To install:

8. Install the axle shaft and seat the bearing. Torque the axle nut to 148–175 ft. lbs. (200–240 Nm).
9. Fill the rear axle with oil. New radial seal rings are used on all models. Lubricate the outside diameter of rubber covered radial sealing rings with hypoid gear lubricant prior to installation.

NOTE: Check endplay of the lock ring in the groove. If necessary, install a thicker lock ring or spacer to eliminate all endplay, while still allowing the lock ring to rotate. Do not allow the joints in the axle shaft to hang free or the joint bearing may be damaged and leak.

WITH TORQUE COMPENSATOR (TORSION BAR)

1. Raise the vehicle and support safely. Drain the oil from the rear axle.
2. Disconnect and plug the brake lines.
3. Loosen the connecting rod and unscrew the torsion bar bearing bracket. Lower the exhaust system slightly and remove the torsion bar.
4. Loosen the shock absorber.
5. Remove the bolt which attaches the rear axle shaft to the rear axle shaft flange.
6. Disconnect the brake cable control. Remove the bracket from the wheel carrier, remove the rubber sleeve and push back the cover.
7. Force the rear axle shaft from the flange with a suitable tool.
8. Support the rear axle.
9. Remove the rubber mount.
10. Clean the axle housing and remove the cover fan from the housing.

NOTE: The axle shafts are the floating type and can be compressed in the constant velocity joints.

11. Remove the locking ring from the end of the axle shafts which engage the side gears in the differential.

Removing the lock-ring (26) from the axle shaft with pliers (1) or a screwdriver

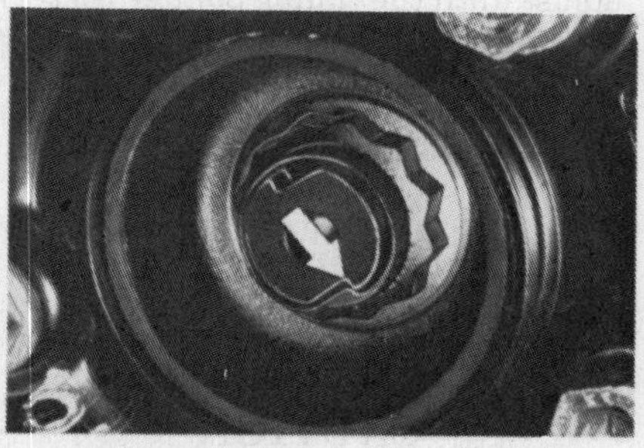

Lock the collar nut on the 190 at the crush flange (arrow)

Axle shaft markings (r)

12. Disengage the axle shaft from the side gear and remove the axle shaft together with the spacer.

NOTE: Do not hang the outer constant velocity joint in a free position, unsupported, as the shaft may be damaged and the constant velocity joint housing may leak.

To install:

13. Install the axle shaft and torque the shaft-to-hub bolt to 148–175 ft. lbs. (200–240 Nm).
14. If either axle shaft is replace, be sure the proper replacement shaft is installed. Axle shafts are marked L and R for left and right.
15. Check the endplay between the lock ring on the axle shaft and the side gear. There should be no noticeable endplay but the lock ring should be able to turn in the groove.
16. Be sure to bleed the brakes and fill the rear axle with the proper quantity and type of lubricant. New radial seal rings are used on all models. Lubricate the outside diameter of rubber covered radial seal rings with hypoid gear lubricant prior to installation.

190D, 190E, 260E, 300D, 300TD, 300E, 300CE, 300SE, 300SL, 300TE, 300SEL, 300SDL, 350SD, 420SEL, 500SL, 500E, 560SEC and 560SEL

1. Loosen but do not remove, the axle shaft collar nut.
2. Raise and safely support the vehicle.
3. Disconnect the axle shaft from the hub assembly. On the 190, make sure while loosening the locking screws, the bit is seated properly in the multi-tooth profile of the screws.

NOTE: Make sure the socket bit is inserted straight into the self-locking screws at the inner CV-joint.

4. Remove the self-locking screws that attach the inner CV-joint to the connecting flange on the differential. Always loosen the screws in a crosswise manner.

NOTE: Make sure the end cover on the inner CV-joint is not damaged when separated from the connecting flange.

5. When removing the right axle shaft, remove the wiring guide rail for the 300SL and 500SL models. On vehicles equipped with ASD or level control, disconnect the control rod from the level control unit.
6. Raise the differential with a suitable jack.
7. While supporting the axle shaft, use a slide hammer or the like and press the axle shaft from the hub assembly. If the axle shaft will not dislodge from the wheel hub use a gear puller to remove.
8. Tilt the axle shaft down and remove it.

NOTE: Make sure the CV-joint boots are not damaged during the removal process.

To install:

9. Note the following:
 a. Always clean the connecting flanges before installation.

b. Always use new self-locking screws. Lubricate the screw threads and contact faces with oil before installing. Use anti-seize compound on the hub splines.

c. 190, 260E, 300E, 300CE and 300TE – tighten the screws to 51 ft. lbs. (70 Nm).

d. 300D, 300SE, 300SL, 300SEL, 300TD, 300SDL, 500SL, 420SEL, 560SEC and 560SEL – tighten the screws to 51 ft. lbs. (70 Nm) for M10 bolts or 100 ft. lbs. (135 Nm) for M12 × 1.5 bolts.

e. All other models – tighten the screws to 90–105 ft. lbs. (125–145 Nm). Always tighten the screws in a criss-cross pattern.

f. Torque the axle shaft collar nut to 148–175 ft. lbs. (200–240 Nm) on the 300SL and 500SL. Replace the washer.

g. Tighten the axle shaft collar nut to 203–230 ft. lbs. (280–320 Nm) on the 190 or 22 ft. lbs. (30 Nm) on the others. On the 190, lock the collar nut at the crush flange.

Rear Axle Seal

REMOVAL & INSTALLATION

Independent Suspension

1. Raise the vehicle and support safely.
2. Drain the differential oil and remove the axle shaft.
3. Remove the end cover from the differential. Remove the locking ring from the inner axle shaft, inside the carrier assembly.
4. Use 2 suitable prybars to dislodge the inner axle shaft from the differential assembly.
5. Pry out the axle seal using a suitable prybar.

To install:

6. Install the seal using a seal installer.
7. Install the inner axle shaft and new locking ring.
8. Install the differential cover and refill with oil.
9. Install the remaining components.

Front Axle Shafts

REMOVAL & INSTALLATION

4-Matic

1. Raise the vehicle and safely support. Remove the front wheel.
2. Remove the axle hub nut and washer and engine compartment cover.
3. Remove the self-locking bolts from the inside CV-joint.
4. Turn the steering wheel to the right. Guide the front axle shaft upwards behind the connecting flange and remove by pushing the vehicle down at the same time. Be careful not to over extend the inner CV-joint.

To install:

4. Install the axle shaft and connect with the inner joint and wheel hub.
5. Torque the hub nut and washer to 200–250 ft. lbs. (280–320 Nm) and the self-locking bolts to 51 ft. lbs. (70 Nm).

Front Axle Bearing and Seal

REMOVAL & INSTALLATION

4-Matic

1. Raise the vehicle and safely support.
2. Remove the front axle shaft, brake hose support, caliper and brake disc.
3. Remove the axle shaft flange using a hub puller tool 201589006100 or equivalent with the thrust piece on the axle flange.
4. Remove the snapring from the knuckle.
5. Heat the bearing area of the knuckle and draw the bearing out of the knuckle with a removal tool 201589044300 or equivalent.
6. Using a puller, remove the bearing inner race.

To install:

7. Install the bearing using a installation tool 201589044300 or equivalent until it contacts the inside stop.
8. Install the snapring and axle shaft flange with an installation tool 201589044300 or equivalent.
9. Install the remaining components.
10. Torque the hub nut and washer to 200–250 ft. lbs. (280–320 Nm) and the self-locking bolts to 51 ft. lbs. (70 Nm).

MANUAL TRANSMISSION

For further information on transmissions/transaxles, please refer to "Chilton's Guide to Transmission Repair".

Transmission Assembly

REMOVAL & INSTALLATION

With Engine

The transmission should only be removed with the engine as a unit, since the transmission-to-clutch housing bolts can only be reached from the inside. Once the engine/transmission unit has been removed from the vehicle, the transmission and bellhousing must be separated from the engine, as follows:

See the "Engine" section to remove the engine/transmission.

1. After removing the engine/transmission unit, unbolt the bellhousing from the engine. The bolts which hold the transmission to the bellhousing cannot be reached except from inside the bellhousing.
2. Remove the starter from its mounting position and pull the transmission and bellhousing from the engine.
3. The bolts which secure the bellhousing to the transmission are now visible and can be removed to separate the bellhousing and transmission.

To install:

4. Connect the engine, bellhousing and transmission, after coating the splines of the mainshaft with grease.
5. Install the starter.
6. Further installation is the reverse of removal.

Without Engine

190D, 190E, 260E, 300E, 300CE, 300SE, 300TE AND 300SEL

1. Disconnect the battery.
2. Cover the insulation mat in the engine compartment to prevent damage.
3. If equipped with an auxiliary heater, be sure the water hose is aside.
4. Support the transmission.
5. Unbolt the engine mounts at the rear transmission cover.
6. Unbolt the engine carrier on the floor frame.
7. Unscrew the exhaust holder at the transmission. Note the number and positioning of all washers.
8. Unscrew the clamping strap and remove the exhaust pipe holder.
9. Remove the intermediate bearing shield plate.
10. Loosen the clamp nut on the driveshaft.
11. Loosen but do not remove, the intermediate bearing bolts.
12. Unbolt the driveshaft on the transmission so the companion plate remains with the driveshaft.
13. Carefully push the driveshaft as far to the rear as permitted.

NOTE: On the 190E, the fitted sleeves on the universal flange must be loosened before separating the flange from the companion plate. This will require a cylindrical mandrel.

14. Disconnect the exhaust system at the rear suspension and suspend it with wire.
15. Loosen and remove the input shaft for the tachometer.
16. Loosen and remove the tachometer drive shaft on the rear transmission case cover. Unclip the clip for the tachometer drive shaft from its holder.
17. Unscrew the holder for the line to the clutch housing. Unscrew the clutch slave cylinder and move it toward the rear until the pushrod is clear of the housing.
18. Push off the clip locks and remove the shift rods from the intermediate levers on the shift bracket. Note the position of the disc springs.

NOTE: When the shift rods are disconnected, do not move the shift lever into reverse or the back-up light switch could be damaged.

19. Unbolt the starter and remove it.
20. Remove all transmission-to-intermediate flange screw. Remove the upper 2 last.
21. Rotate the transmission approximately 45 degrees to the left, slide it from the clutch plate and remove it downward.

NOTE: Make sure the input shaft has cleared the clutch plate before tilting the transmission.

To install:

22. Lightly grease the centering lug and splines on the transmission input shaft.

NOTE: Position the clutch slave cylinder and line above the transmission before beginning installation.

23. Move the transmission into the clutch so one gear step engages. Rotate the mainshaft back and forth until the splines on the input shaft and clutch plate are aligned.
24. Move the transmission all the way in and tighten the transmission-to-intermediate flange screws.
25. Install the starter.
26. Install the clutch slave cylinder with the proper plastic shims.
27. Installation of the remaining components is in the reverse order of removal. Please note the following:
   a. After installing the driveshaft, roll the vehicle back and forth and tighten the intermediate bearing free of tension.
   b. Tighten the driveshaft clamp nut to 22–29 ft. lbs. (30–40 Nm).
   c. Make sure of the proper positioning of all washers, spacers and shims.

CLUTCH

Clutch Assembly

REMOVAL & INSTALLATION

1. To remove the clutch, first remove the transmission.
2. Loosen the clutch pressure plate hold-down bolts, evenly, 1–1½ turns at a time, until tension is relieved. Never remove 1 bolt at a time, as damage to the pressure plate is possible.
3. Examine the flywheel surface for blue heat marks, scoring or cracks. If the flywheel is to be machined, always machine both sides.

To install:

4. Coat the splines with high temperature grease and place the clutch disc against the flywheel, centering it with a clutch pilot shaft.
5. Tighten the pressure plate hold-down bolts evenly, 1–1½ turns at a time, until tight, then remove the pilot shaft. Torque the bolts to 18 ft. lbs. (25 Nm) in the same rotation as the 1–1½ turns.

NOTE: Most clutch plates have the flywheel side marked Kupplungsseite. Do not assume the pressure springs always face the transmission.

6. Apply grease to the release fork and release bearing slide before installing the transmission.
7. Install the transmission and remaining components.
8. Check the freedom of movement of the clutch by selecting reverse gear with the engine running. The transmission should shift into reverse with little or no gear noise.

ADJUSTMENT

Without Brake Bleeding Device

1. Loosen locknut on adjusting screw on master cylinder.
2. Turn adjusting screw in such a manner that pushrod will travel the idle path **B** up to piston first when pedal is actuated. If a line mark is shown on head of adjusting screw, make sure during inspection or during adjustment that this line mark is pointing toward the rear.

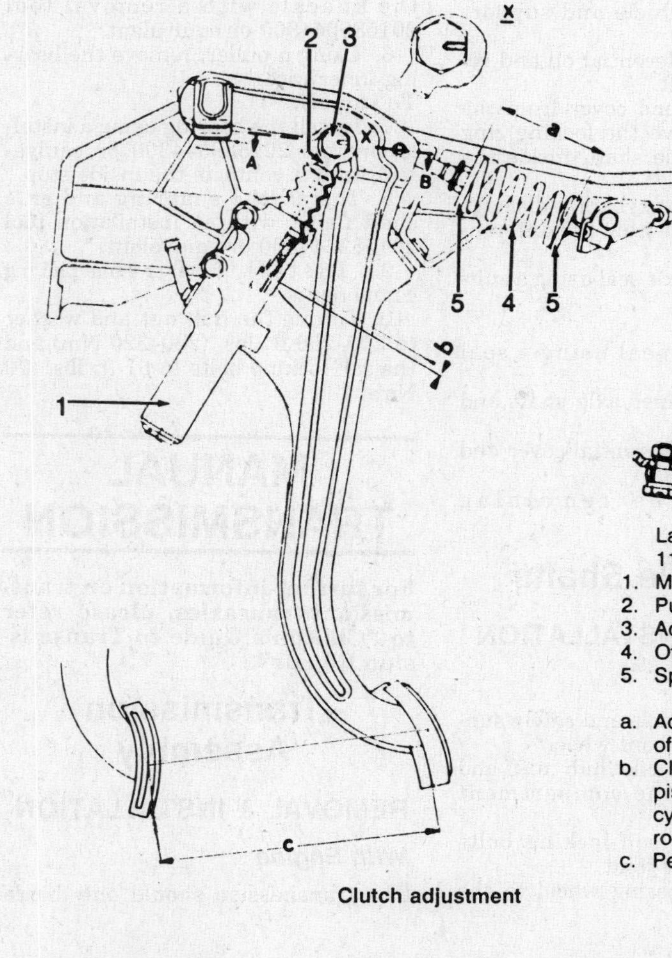

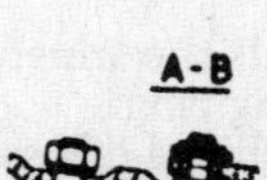

Layout model 107, 114, 115 and 116
1. Master cylinder
2. Push rod
3. Adjusting screw
4. Off-center spring
5. Spring retainer

a. Adjusting dimension of off-center spring
b. Clearance between piston in master cylinder and push rod
c. Pedal travel (lash)

Clutch adjustment

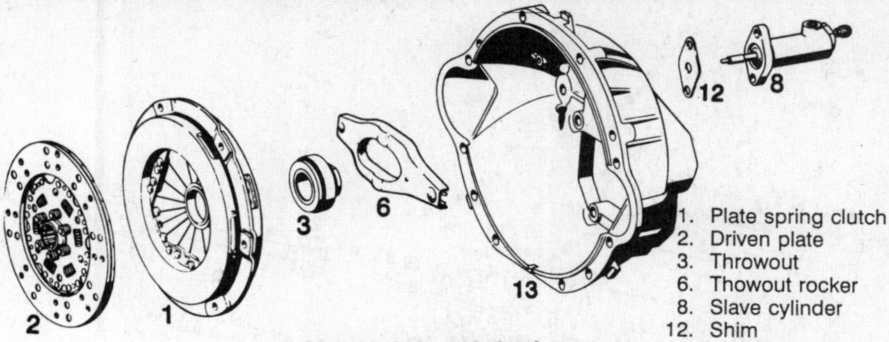

Exploded view of clutch

With Brake Bleeding Device

1. Draw brake fluid from expansion tank up to connection for clutch actuation.
2. Pull connecting hose to master cylinder from expansion tank.
3. Open venting screw on clutch sleeve cylinder and evacuate clutch system by stepping repeatedly on clutch pedal.
4. Insert plastic hose of approximately 1 meter in length, 8mm in diameter, into connecting hose and immerse other end into a container filled with water.
5. Remove the left instrument panel under cover.
6. Clamp sheet metal approximately 1.5mm thick between upper pedal stop and rubber buffer.
7. Fill brake bleeding device with air and set working pressure to 0.5 bar gauge pressure.
8. Adjust brake bleeding device depending on make, in such a manner, that air is blown from clutch system and bubbles will rise in water tank.

NOTE: Place water tank into left hand leg room to gain advantage.

9. Turn adjusting screw on clutch pedal only until airflow is interrupted and no more bubbles are rising in water tank. Tighten locknut on adjusting screw.
10. Remove sheet metal at upper pedal stop. Bubbles should rise again in water tank. Set brake bleeding device to 0.
11. Attach connecting hose to expansion tank.
12. Vent clutch actuation.
13. Check clutch actuation for function with engine running.

Clutch Master Cylinder

REMOVAL & INSTALLATION

Except 300SL and 500SL

1. Disconnect the negative battery cable. Remove cover under instrument panel at left.
2. Remove floor mat at left.
3. To prevent contamination inside vehicle, draw fluid from respective chamber of combination clutch and expansion tank.
4. Unscrew line on master cylinder.
5. Pull off connecting hose on combination brake and clutch expansion tank.
6. Loosen piston rod for brake unit (brake booster) on brake pedal.
7. Pull cable plug from stop light switch.
8. Unscrew nuts for attaching pedal carrier to fire wall.
9. Move pedal assembly to the rear until screw plate of pedal carrier is free from threaded bolt of brake unit (brake booster) and holder at top on water tank.
10. Remove pedal assembly in downward direction, while paying attention to connecting hose for master cylinder and remove master cylinder.

To install:

11. Install any fallen rubber mounts.
12. Install the master cylinder and torque the retaining bolts to 15 ft. lbs. (20 Nm).
13. Install the remaining components. Bleed the hydraulic system.

300 and 500SL

1. Disconnect the negative battery cable.
2. Remove the instrument panel undercover.
3. Siphon the fluid from the master cylinder.
4. Remove the hydraulic line from the cylinder.
5. Remove the pedal carrier, return spring and slacken the over-center spring. Disengage the clutch and remove the circlip and pushrod, spring and spring seat.
6. Remove the master cylinder retaining nuts and master cylinder.

To install:

7. Install the master cylinder and torque the nuts to 15 ft. lbs. (20 Nm).
8. Fit the pushrod and spring seat so that the stiffening rib on the push rod points upwards and the larger radius on the spring seat is on the left hand side.
9. Install the remaining components and bleed the system.

Clutch Slave Cylinder

REMOVAL & INSTALLATION

1. Detach and plug the pressure line from the slave cylinder.
2. Remove the screws from the slave cylinder.
3. Remove the slave cylinder, pushrod and spacer.

To install:

4. Place the grooved side of the spacer in contact with the housing and hold it in position.
5. Install the slave cylinder and pushrod into the housing; be sure the dust cap is properly seated.
6. Install the attaching screws.
7. Connect the pressure line to the slave cylinder.
8. Bleed the slave cylinder.

Bleeding the Slave Cylinder

The same principle is used as in bleeding the brakes.

1. Check the brake fluid level in the compensating tank and fill to maximum level.
2. Put a hose on the bleeder screw of the right front caliper and open the bleeder screw.
3. Have a helper depress the brake pedal until the hose is full and there are no air bubbles. Be sure the bleeder screw is closed each time the pedal is released.
4. Put the free end of the hose on the bleeder screw of the slave cylinder and open the bleeder screw.
5. Keep stepping on the brake pedal. Close the bleeder screw on the caliper and release the brake pedal. Open the bleeder screw and repeat the process until no air bubbles show up at the mouth of the inlet line of the compensating tank. Between operations, check and, if necessary, refill the compensating tank.
6. Close the bleeder screws on the caliper and slave cylinder and remove the hose.
7. Check the clutch operation and the fluid level.

AUTOMATIC TRANSMISSION

For further information on trans-

missions/transaxles, please refer to "Chilton's Guide to Transmission Repair".

Transmission Assembly

REMOVAL & INSTALLATION

722.3 (W4A040) Transmission

EXCEPT 190D, 190E, 260E, AND 4-MATIC

1. Disconnect negative battery terminal.
2. Remove holder for oil filler pipe on cylinder head.
3. Disengage engine longitudinal regulating shaft.
4. Force off ball socket.
5. Disconnect control wire for control pressure.
6. Pull out lock and loosen control wire.
7. Raise and safely support the vehicle.
8. Remove cross yoke center place.
9. Remove drain plug on oil pan and drain oil.
10. Remove drain plug on torque converter and drain oil.
11. Remove cover plate.
12. Remove screws for driving plate torque converter.
13. Place a fitting wooden block between engine oil pan and cross yoke.
14. Loosen exhaust system on plug connection and remove.
15. Remove crossbeam together with rear engine mount.
16. Remove cable strap and cable on kickdown solenoid valve. Remove fastening screw for impulse transmitter and pull out impulse transmitter.

NOTE: Disconnect tachometer shaft, if equipped with a mechanical tachometer.

17. Remove exhaust support.
18. Remove exhaust shielding plate.
19. Loosen propeller shaft clamping nut and contract propeller shaft, as much as possible.
20. Remove plug for starter lock out switch.

NOTE: Starter lock out switch plug is secured by a lock, white plastic ring. Prior to pushing off plug, turn lock in upward direction. Carefully, push off plug at cable outlet and tongue, by means of 2 suitable tools.

21. Pull off plug.
22. Disconnect control rod on range selector lever.
23. Remove holder and pull off vacuum line.

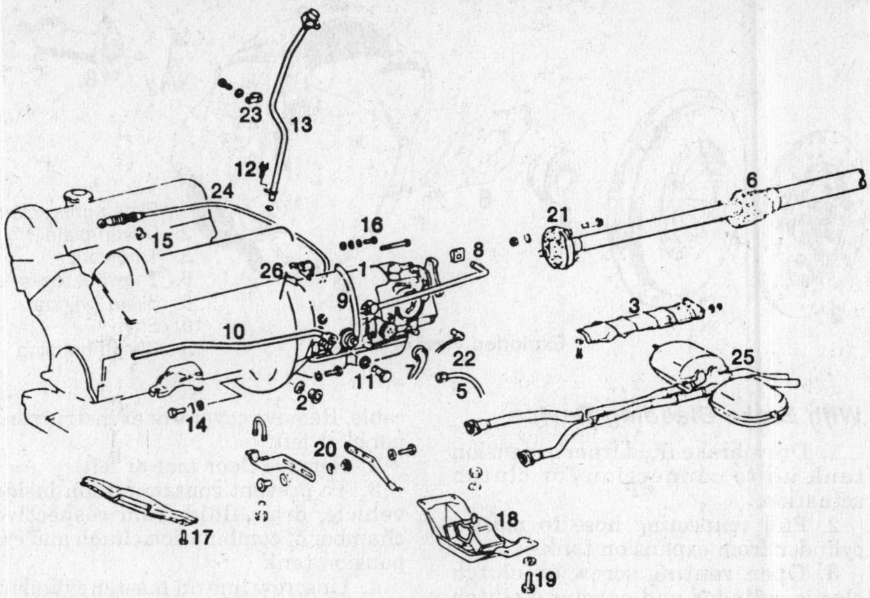

1. Transmission
2. Oil drain plug front, torque converter (Tightening torque 10 ft. lbs. (14 Nm)
3. Shielding plate
4. Cross member-center piece
5. Cable for kickdown solenoid valve
6. Clamping nut of propeller shaft (Tightening torque 22 ft. lbs. (30 Nm)
7. Plug starter lockout switch
8. Shiftrod and clips
9. Vacuum line
10. Oil lines to oil cooler
11. Hollow screw and sealing rings
12. Fastening screw
13. Oil filling pipe
14. Screws for fastening converter (Tightening torque 31 ft. lbs. (42 Nm))
15. Fuse, comtrol pressure cable
16. Fastening screws
17. Self-locking hex head screws
18. Cross member with rear engine mount
19. Fastening screws
20. Exhaust support
21. Unscrew companion plate on flexible flange
22. Impulse sensor, speedometer
23. Holder, oil filling pipe
24. Control pressure cable
25. Exhaust system
26. Cover position sensor EZL

Automatic transmission and related components—300D, 300E, 300CE, 300TE, 300SE, 300SEL, 300TD, 420SEL, 560SL, 560SEL and 560SEC

24. Remove oil cooler feed line.
25. Remove oil cooler return line.
26. Remove fastening screw for oil filler pipe and push oil filler pipe in upward direction.
27. Remove all fastening screw except for 2 lateral screws.
28. Slightly, lift transmission with mount 116 589 06 62 00 or equivalent, for pit lift.
29. Remove lateral screws.
30. Push transmission to rear as far as propeller shaft permits and lower carefully.

To install:

31. Reverse removal procedure taking attention to the following:
    a. Replace sealing rings for forward and return flow lines.
    b. Torque propeller shaft clamping nuts to 22 ft. lbs. (30 Nm).
    c. Torque driveplate screw to 31 ft. lbs. (42 Nm).
    d. Screw in drain plug on oil pan and on torque converter and torque to 10.5 ft. lbs. (14 Nm).
    e. Replace self-locking screws on cross yoke center piece and torque to 33 ft. lbs. (45 Nm).
    f. Adjust cable for control pressure.

722.4 (W4A020) Transmission

190D, 190E AND 260E

NOTE: Attach a 300mm square sheetmetal panel to unit compartment wall to protect insulating mat during all jobs where the transmission is lowered at the rear. Disconnect exhaust assembly at rear mounting bracket and fasten by means of a wire approximately 50cm lower. If equipped with an auxiliary heater, make sure the water hose is not damaged when lowering transmission. On 123 models, disconnect engine throttle control.

1. Disconnect negative cable on battery.

2. Remove holder for oil filling pipe on cylinder head and holder on valve cover.

3. If equipped with a fuel injection engine:

a. Force off ball socket.

b. Disconnect cable control for control pressure. Pull out lock and remove cable control.

c. Force plastic ball socket apart by means of a prybar and pull holding bracket from slotted lever.

4. Force off ball socket. Compress holding clips and disengage cable control for control pressure.

5. On 190D and 300D and 300TD, force off ball socket, remove holding clips and disengage control pressure cable control.

6. Raise and safely support the vehicle.

7. Remove drain plug on oil pan as well as torque converter and drain oil.

8. Install drain plug with new seals and tighten to 10.5 ft. lbs. (14 Nm).

9. Remove screw for driven plate torque converter.

10. Remove crossmember with rear engine mount.

11. Remove exhaust support. Remove companion plate with articulated flange-transmission.

12. Disconnect exhaust system on rear suspension.

13. Remove shielding plate.

14. Remove propeller shaft clamping nut and run together propeller shaft, as much as possible.

15. Pull off cable on kickdown solenoid valve.

16. Remove speedometer shaft. If equipped with an electronic speedometer, remove impulse transmitter.

17. Disconnect control rod on floor shift.

18. Remove fastening clip for speedometer shaft.

19. Swivel locking bracket in upward direction and pull plug from starter lockout switch.

20. Pull vacuum line form vacuum control unit.

21. Remove socket screw from oil filler pipe and pull out oil filler pipe in upward direction.

22. Remove oil cooler lines and fastening clamps.

23. Remove all fastening screws on transmission-to-engine except the 2 screws at left and right.

24. On 190D, insert holding device for torque converter into vent grille cutout and screw in stud until it is entering the socket of oil drain plug.

25. Slightly lift transmission with mounting for pit lift.

26. Remove remaining screws.

27. Slide transmission, to the extent propeller shaft permits, to the rear and carefully lower.

To install:

28. Install new sealing rings on oil cooler line.

29. Torque driven plate-to-converter screws to 31 ft. lbs. (42 Nm).

30. Install the transmission using a transmission jack.

31. Install remaining bolts and torque to 20 ft. lbs. (27 Nm).

23. Install the torque converter bolts.

24. Install oil cooler lines and fastening clamps.

25. Install the vacuum line form vacuum control unit.

26. Install fastening clip for speedometer shaft.

27. Connect control rod on floor shift.

28. Connect the cable on kickdown solenoid valve.

29. Install the shielding plate.

30. Connect the exhaust system on rear suspension.

31. Install the exhaust support.

32. Install the crossmember with rear engine mount.

33. Connect the cable control for control pressure. Pushing in the lock.

34. Lower the vehicle and connect negative cable on battery. Refill the transmission with fluid and road test.

4-MATIC

1. Disconnect the negative battery cable and drain the transmission fluid from the pan and torque converter.

2. Cover the insulating cover with suitable sheet metal to prevent damage.

3. Release the pressure from the hydraulic system by switching over the lever on the service valve near the firewall. To the test position.

4. Remove the oil filler pipe at the cylinder head and remove the ball socket from the throttle control cable.

5. Remove the complete exhaust system with side support and shielding plates.

6. Disconnect the oil lines from the distributor and central lock connector.

7. Raise the transmission with a jack and remove the rear crossmember.

8. Remove the driveshafts and center bearing using a large box wrench and special tool 001589662100 or equivalent.

9. Remove the bellhousing cover plate and torque converter bolts(6).

10. Disconnect the shift control linkage, wiring harnesses and speedometer.

11. Disconnect the oil filler tube.

12. Remove the transmission-to-engine bolts and lower the transmission. Pull the transmission/transfer case backwards and lower the transmission carefully, making sure all components are disconnected.

To install:

13. If renewing transmission, disconnect the transfer case from the transmission.

14. Raise the transmission/transfer into place and install retaining bolts. Torque the bolts to 40 ft. lbs. (55 Nm).

15. Connect the oil filler tube.

16. Connect the shift control linkage, wiring harnesses and speedometer.

17. Install the bellhousing cover plate and torque converter bolts(6). Apply Loctite® to the bolts and torque the bolts to 29 ft. lbs. (40 Nm).

18. Install the driveshafts and center bearing using a large box wrench and special tool 001589662100 or equivalent.

19. Install the rear crossmember.

20. Connect the oil lines to the distributor and central lock connector.

21. Install the complete exhaust system with side support and shielding plates.

22. Install the oil filler pipe at the cylinder head and the ball socket from the throttle control cable.

23. Remove sheet metal from the insulating cover.

24. Connect the negative battery cable and refill the transmission fluid.

SELECTOR ROD LINKAGE ADJUSTMENT

NOTE: Before performing this adjustment on any vehicle, be sure it is resting on its wheels. No part of the vehicle may be raised for this adjustment.

Column Mounted Linkage

W4A 040

1. Loosen the counter nut on the rear selector rod while holding both recesses of the front selector rod with an open end wrench.

2. Disconnect the selector rod from the selector lever.

3. Set the selector lever on the transmission and on the column to **N**.

4. Adjust the selector rod until the bearing pin is aligned with the bearing bushing in the selector lever.

5. Connect the rear selector lever to the selector rod and secure it with the lock. Be sure the clearance of the selector lever in **D** and **S** is equal.

6. Tighten the locknut on the rear selector rod while holding the front selector rod as in Step 1.

Floor Mounted Linkage

NOTE: The vehicle must be standing with the weight normally distributed on all 4 wheels.

1. Disconnect the selector rod from the selector lever.

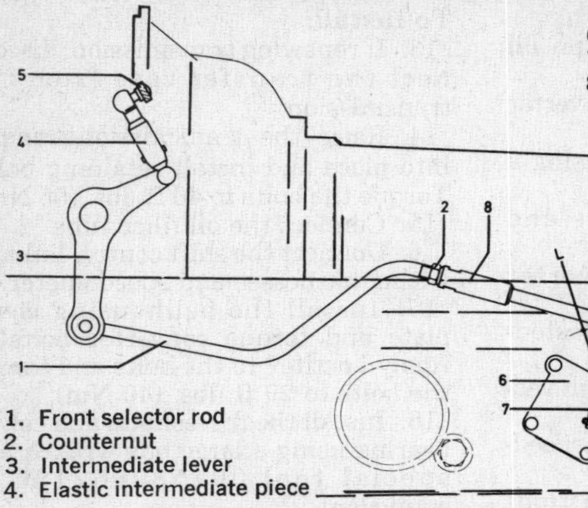

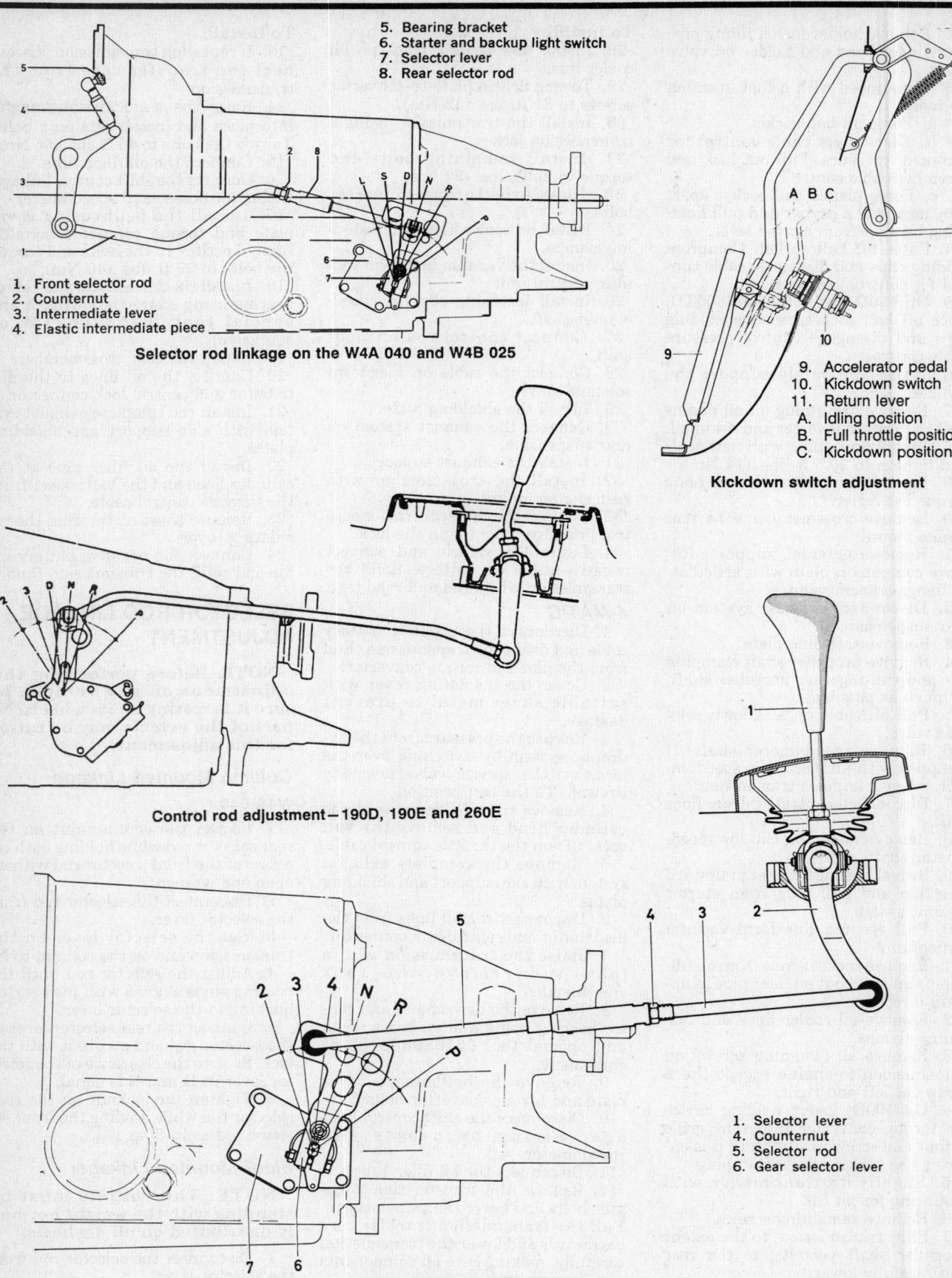

1. Front selector rod
2. Counternut
3. Intermediate lever
4. Elastic intermediate piece
5. Bearing bracket
6. Starter and backup light switch
7. Selector lever
8. Rear selector rod

Selector rod linkage on the W4A 040 and W4B 025

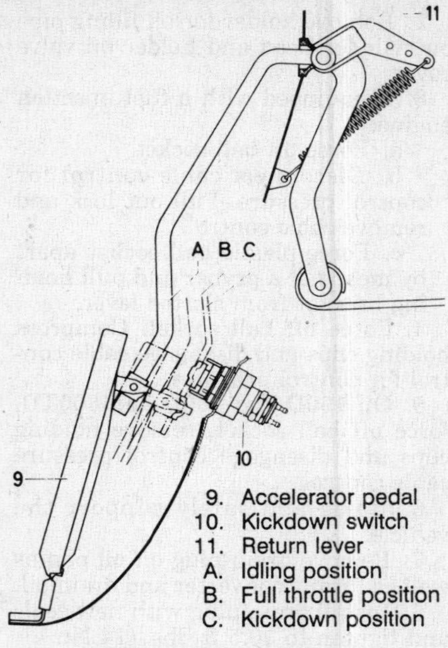

9. Accelerator pedal
10. Kickdown switch
11. Return lever
A. Idling position
B. Full throttle position
C. Kickdown position

Kickdown switch adjustment

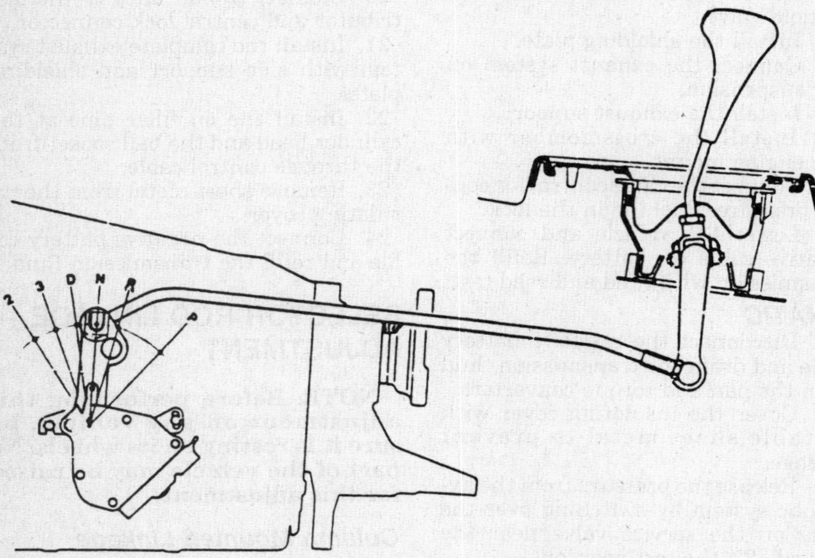

Control rod adjustment—190D, 190E and 260E

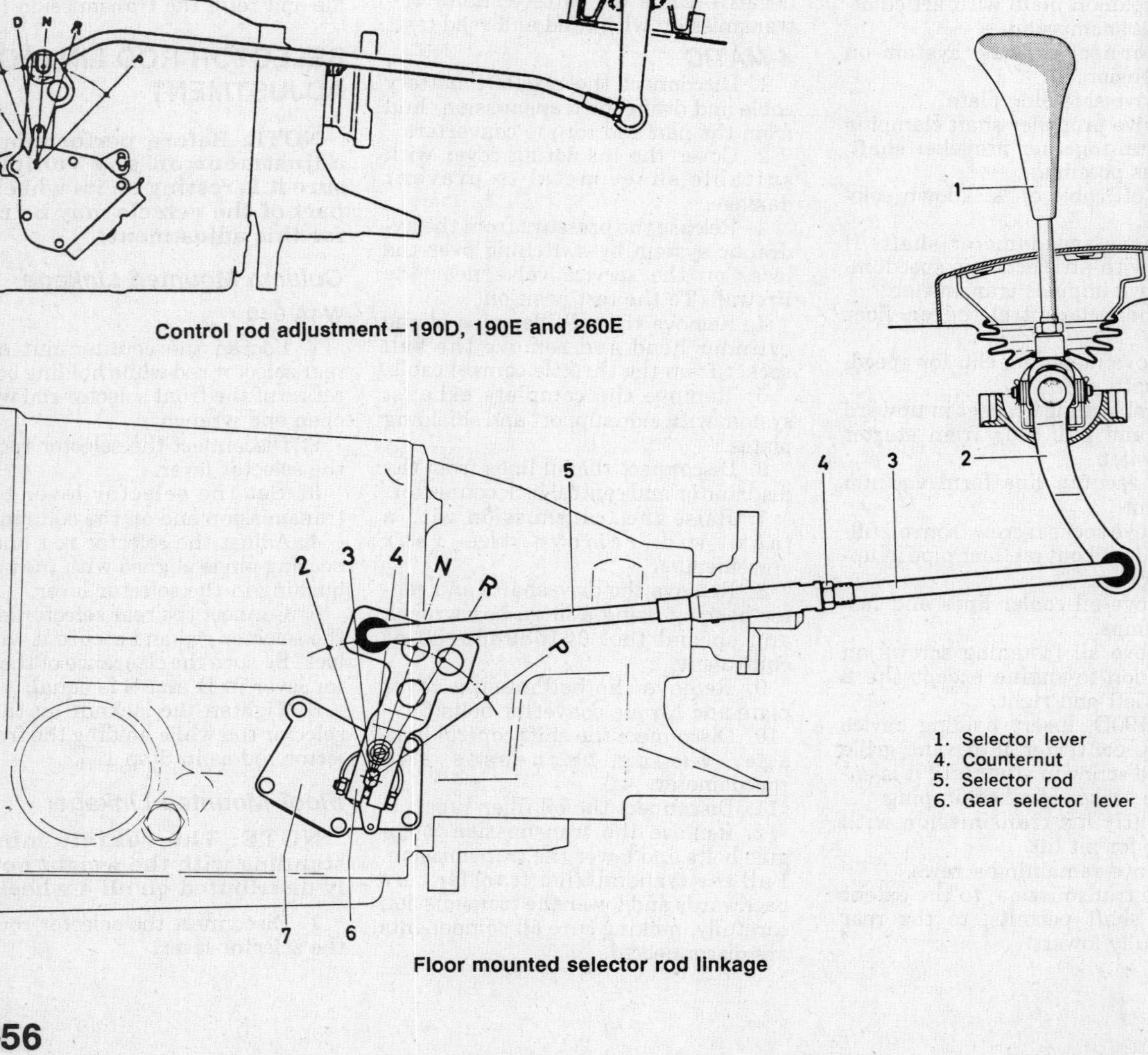

1. Selector lever
4. Counternut
5. Selector rod
6. Gear selector lever

Floor mounted selector rod linkage

2. Set the selector lever in **N** and make sure there is approximately 1mm clearance between the selector lever and the **N** stop of the selector gate.

3. Adjust the length of the selector rod so it can be attached free of tension.

4. Retighten the counter nut.

KICKDOWN SWITCH ADJUSTMENT

1. The kickdown position of the solenoid valve is controlled by the accelerator pedal.

2. Push the accelerator pedal against the kickdown limit stop. In this position the throttle lever should rest against the full load stop of the venturi control unit.

3. Adjustments are made by loosening the clamping screw on the return lever on the accelerator pedal shaft and turning the shaft. Tighten the clamping screw again.

FRONT SUSPENSION

Shock Absorbers

REMOVAL & INSTALLATION

560SL

1. Raise and safely support the vehicle.

2. When removing the shock absorbers, note the position of all mounting hardware.

3. Raise the hood and locate the upper shock absorber mount.

4. Support the lower control arm.

5. Unbolt the mount for the shock absorber at the top. Remove the coolant expansion tank to allow access to the right front shock absorber.

6. Remove the nuts which secure the shock absorber to the lower control arm.

7. Push the shock absorber piston rod in, install the stirrup, and remove the shock absorber.

8. Remove the stirrup, since this must be install on replacement shock absorber.

9. Installation is the reverse of removal. Always use new bushing when installing replacement shock absorber.

Except 190D, 190E, 260E, 300E, 300CE, 300SE, 300TE, 300SEL and 350SD

1. Raise and safely support the vehicle. Support the lower control arm.

2. Loosen the nuts on the upper shock absorber mount. Remove the plate and ring.

3. Place the shock absorber vertical to the lower control arm and remove the lower mounting bolts.

4. Remove the shock absorber; be sure to disconnect and plug the pressure line on models with level control.

5. Installation is the reverse or removal. On Bilstein shocks, do not confuse the upper and lower plates.

Damper Strut

REMOVAL & INSTALLATION

190D, 190E, 260E, 300E, 300CE, 300SE, 300TE, 300SEL, 350SD and 500E

1. Raise and safely support the vehicle. Remove the wheel.

2. Using a spring compressor, compress the spring until any load is removed from the lower control arm.

NOTE: When using a spring compressor, be sure a least 7½ coils are engaged before applying tension.

3. Support the lower control arm. Loosen the retaining bolt for the upper end of the damper strut by holding the inner piston rod with an Allen wrench and unscrew the nut. Never use an impact wrench on the retaining nut. Disconnect the hydraulic line from the ADS actuator, if so equipped.

CAUTION

Never unscrew the nut with the axle half at full rebound the spring may fly out with considerable force, causing personal injury.

4. Unbolt the 2 screws and 1 nut and disconnect the lower damper strut from the steering knuckle.

5. Remove the strut down and forward. Be sure to disconnect and plug the pressure line on models with level control. Secure the steering knuckle in position so it won't tilt.

To install:

6. Note the following:

   a. When attaching the lower end of the damper strut to the steering knuckle, first position all 3 screws; next tighten the 2 lower screws to 72 ft. lbs. (100 Nm); finally, tighten the nut on the upper clamping connection screw to 54 ft. lbs. (75 Nm).

   b. Tighten the retaining nut on the upper end of the damper strut to 44 ft. lbs. (60 Nm).

Coil Springs

REMOVAL & INSTALLATION

190D, 190E, 260E, 300E, 300CE, 300SE, 300TE, 300SEL and 500E

1. Raise and safely support the vehicle. Remove the wheel.

2. Remove the engine compartment lining under the vehicle, if equipped.

3. Install a spring compressor so at least 7½ coils are engaged.

4. Support the lower control arm and loosen the retaining nut at the upper end of the damper strut.

CAUTION

Never loosen the damper strut retaining nut unless the wheels are on the ground, the control arm is supported or the springs have been removed; personal injury may result.

5. Lower the control arm slightly and remove the spring toward the front.

To install:

6. On installation, position the spring between the control arm and the upper mount so when the control arm is raised, the end of the lower coil will be seated in the impression in the control arm.

7. Raise the control arm until the spring is held securely.

8. Using a new nut, tighten the upper end of the damper strut to 44 ft. lbs. (60 Nm).

9. Slowly ease the tension on the spring compressor until the spring is seated properly and remove the compressor.

10. Installation of the remaining components is in the reverse order of removal.

560SL

NOTE: Be extremely careful when attempting to remove the front springs as they are compressed and under considerable load.

1. Raise and safely support the vehicle. Remove the wheels.

2. Remove the front shock absorber and disconnect the sway bar.

3. Punchmark the position of the eccentric adjusters and loosen the hex bolts.

4. Support the lower control arm.

5. Knock out the eccentric pins and gradually lower the arm until spring tension is relieved.

6. The spring can now be removed.

NOTE: Check caster and camber after installing a new spring.

7. Installation is the reverse of removal.

8. For ease of installation, tape the rubber mounts to the springs.

9. If the eccentric adjusters were not match marked, install the eccentric bolts.

All Other Models

EXCEPT 300SL AND 500SL

1. Raise and safely support the vehicle. Support the lower control arm.
2. Remove the wheel. Unbolt the upper shock absorber mount.
3. Install a spring compressor and compress the spring.
4. Remove the front spring with the lower mount.
5. Installation is the reverse of removal. Tighten the upper shock absorber suspension.

300SL AND 500SL

CAUTION

The spring is under extreme tension. Do not remove the spring unless an approved spring compressor is installed and is secure. If this caution is not followed, severe personal injury may result. Use Mercedes-Benz spring compressor 124589063100 or equivalent.

1. Raise the vehicle and support safely.
2. Install an approved spring compressor with the upper tensioning plate in the center of the spring and the lower tensioning plate at the bottom.
3. Turn the compressor with the tensioning cylinder far enough for about 8 windings to be covered.
4. Push the compressor cylinder through both plates and lock in the top by turning 90 degrees.
5. Compress the front spring and remove the top strut mounting only if the spring can not be removed at this time. Disconnect the ADS sensor hydraulic pipe if so equipped.
6. Place a jack under the lower control arm. Lower the front axle after disconnecting the top damper strut mounting.
7. Remove the spring with the rubber mount towards the front.
8. Place the spring in a vise and release the tension.

To install:

9. Clean the spring contact surfaces of dirt or contamination.
10. Compress the spring in the area of 8 spring windings and fit the rubber mount onto the front spring with a screwing movement.
11. Install the spring so that the winding end is located in the embossed surface of the control arm. Raise the control arm and install a new damper strut nut and washer. Torque the nut to 44 ft. lbs. (60 Nm).
12. Release the tension on the spring and remove the spring compressor.
13. Connect the ADS sensor pipe, if so equipped.
14. Lower the vehicle and check ride height and headlight setting.

Steering Knuckle and Ball Joints

INSPECTION

1. Raise and safely support the vehicle. Check the steering knuckles or ball joints, by raising the front spring plate. This unloads the front suspension to allow the maximum play to be observed.
2. Late model ball joints need to be replaced only if dried out with plainly visible wear and/or play.

REMOVAL & INSTALLATION

190D, 190E, 260E, 300E, 300CE, 300SE, 300TE, 300SEL and 500E

CAUTION

The spring is under extreme tension. Do not remove the spring unless an approved spring compressor is installed and is secure. If this caution is not followed, severe personal injury may result. Use Mercedes-Benz spring compressor 124589063100 or equivalent.

1. Raise and safely support the vehicle. Remove the wheel.
2. Install a spring compressor on the spring.
3. Remove the brake caliper and wire it aside; be careful not to damage the brake line.
4. Remove the brake disc and wheel hub.

NOTE: If equipped equipped with ABS, remove the speed sensor.

5. Unscrew the 3 socket-head bolts and remove the brake backing plate from the steering knuckle.
6. Tighten the spring compressor until all tension and/or lead has been removed from the lower control arm.
7. Disconnect the steering knuckle arm from the steering knuckle; this is the arm attached to the tie rod.

CAUTION

There must be no tension on the lower control arm; otherwise personal injury may result.

8. Unscrew the 3 bolts and disconnect the lower end of the damper strut from the steering knuckle.
9. Remove the hex-head clamp nut at the supporting joint, lower ball joint.
10. Remove the steering knuckle.
11. Installation is in the reverse order of removal. Tighten the supporting joint clamp nut to 70 ft. lbs. (125 Nm).

560SL

CAUTION

The spring is under extreme tension. Do not remove the spring unless an approved spring compressor is installed and is secure. If this caution is not followed, severe personal injury may result. Use Mercedes-Benz spring compressor 124589063100 or equivalent.

1. This should only be done with the front shock absorber installed. If, however, the front shock absorber has been removed, the lower control arm should be supported and the spring should be clamped with a spring tensioner. In this case, the hex nut on the guide joint should not be loosened without the spring tensioner installed.
2. Raise and safely support the vehicle.
3. Remove the wheel.
4. Remove the brake caliper.
5. Unbolt the steering relay lever from the steering knuckle. For safety, install spring clamps on the front springs.
6. Remove the hex nuts from the upper and lower ball joints.
7. Remove the ball joints from the steering knuckle with the aid of a puller.
8. Remove the steering knuckle.
9. Installation is the reverse of removal. Be sure the seats for the pins of the ball joints are free of grease.
10. Bleed the brakes.

350SD, 350SDL, 420SEL, 500SL, 560SEC and 560SEL

CAUTION

The spring is under extreme tension. Do not remove the spring unless an approved spring compressor is installed and is secure. If this caution is not followed, severe personal injury may result. Use Mercedes-Benz spring compressor 124589063100 or equivalent.

1. Raise and safely support the vehicle. For safety, install an approved coil spring compressor on the front spring. Support the lower control arms.
2. Remove the wheel.
3. Remove the steering knuckle arm from the steering knuckle.
4. Remove and suspend the brake caliper.
5. Remove the front wheel hub.

NOTE: If equipped equipped with ABS, disconnect the speed sensor.

6. Loosen the brake hose holder on the cover plate.
7. Loosen the guide joint nut and remove the joint from the steering knuckle.
8. Loosen the nut on the support joint.
9. Swivel the steering knuckle outward and force the ball joint from the lower control arm.
10. Remove the steering knuckle.
11. If necessary, remove the cover plate from the steering knuckle.

To install:

12. Install the steering knuckle and the cover plate.
13. Install the ball joint into the lower control arm.
14. Torque the nut on the support joint to 30 ft. lbs. (40 Nm).
15. Install the brake hose holder on the cover plate.
16. Install the front wheel hub.
17. Install the brake caliper.
18. Install the wheel.
19. Lower the vehicle.

Upper Control Arm

NOTE: The 190D, 190E, 260E, 300E, 300CE, 300SE, 300TE, 300SL, 300SEL, 500E and 500SL models have no upper control arm.

REMOVAL & INSTALLATION

Except 560SL

1. Raise and safely support the vehicle.
2. Remove the wheel.
3. Loosen the nut on the guide joint.
4. Remove the guide joint from the steering knuckle.
5. Secure the steering knuckle with a hook on the upper control arm stop to prevent it from tilting.
6. Loosen the clamp screw and separate the upper control arm from the torsion bar.
7. Loosen the upper control arm bearing at the front and remove the upper control arm.
8. Installation is the reverse of removal. Use new self-locking nuts and check the front wheel alignment.

560SL

1. The front shock absorbers should remain installed. Never loosen the hex nuts of the ball joints with the shock absorber removed, unless a spring clamp is installed.
2. Raise and safely support the vehicle. Remove the wheel.
3. Support the front end.
4. Remove the steering arm from the steering knuckle.
5. Separate the brake line and brake hose from each other and plug the openings.
6. Support the lower control arm and unscrew the nuts from the ball joints.
7. Remove the ball joints from the steering knuckle.
8. Loosen the bolts on the upper control arm and remove the upper control arm.
9. Installation is the reverse of removal.

NOTE: Mount the front hex bolt from the rear in a forward direction and the rear hex bolt from the front in a rearward direction.

10. Bleed the brakes.

Lower Control Arm

REMOVAL & INSTALLATION

Except 190D, 190E, 260E, 300E, 300CE, 300SE, 300SL, 300TE, 300SEL, 500SL and 560SL

The lower control arm is the same as the front axle half. For safety install a spring compressor on the coil spring.

1. Raise and safely support the vehicle. Remove the wheels.
2. Remove the front shock absorber. Loosen the top mount first.
3. Remove the front springs.
4. Separate and plug the brake lines.
5. Remove the track rod from the steering knuckle arm.
6. Matchmark the position of the eccentric bolts on the bearing of the lower control arm in relation to the crossmember.
7. Remove the shield from the cross yoke.
8. Support the front axle half.
9. Loosen the eccentric bolt on the front and rear bearing of the lower control arm and knock them out.
10. Remove the bolt from the cross yoke bearing.
11. Loosen the screw at the opposite end of the cross yoke bearing.
12. Pull the cross yoke bearing down slightly.
13. Loosen the support of the upper control arm on the torsion bar. Remove the clamp screw from the clamp.
14. Remove the upper control arm bearing on the front end.
15. Remove the front axle half.

To install:

16. Install the front axle half. Tighten the eccentric bolts on the lower control arm bearing to 35 ft. lbs. (47 Nm). with the vehicle resting on the wheels. Check the front end alignment after complete reassembly.
17. Install the upper control arm bearing on the front end.
18. Tighten the support of the upper control arm on the torsion bar to 25 ft. lbs. (34 Nm). Install the clamp screw to the clamp.
19. Tighten the screw at the opposite end of the cross yoke bearing.
20. Install the bolt to the cross yoke bearing.
21. Torque the eccentric bolt on the front and rear bearing of the lower control arm to 35 ft. lbs. (47 Nm).
22. Install the shield from the cross yoke.
23. Install the track rod to the steering knuckle arm.
24. Reconnect the brake lines.
25. Install the front springs.
26. Install the front shock absorber.
27. Install the front wheel and lower the vehicle.

190D, 190E, 260E, 300E, 300CE, 300SE, 300TE, 300SEL, 350SD and 500E

1. Remove the engine compartment lining at the bottom of the vehicle, if equipped.
2. Raise and safely support the vehicle. Remove the wheel.
3. Support the lower control arm and disconnect the torsion bar bearing at the control arm.
4. Remove the spring.
5. Disconnect the tie rod at the steering knuckle and press out the ball joint with the proper tool.
6. Remove the brake caliper and position it aside; do not damage the brake line.
7. Remove the brake disc/wheel hub assembly.
8. Disconnect the lower end of the damper strut from the steering knuckle and remove the knuckle.
9. Mark the position of the inner eccentric pins, relative to the frame, on the bearing of the control arm.
10. Unscrew and remove the pins.
11. Remove the lower control arm.

To install:

12. Note the following:
    a. Tighten the eccentric bolts on the inner arm to 130 ft. lbs. (180 Nm).
    b. To facilitate torsion bar installation, raise the opposite side of the lower control arm.
    c. Tighten the clamp nut on the tie rod ball joint to 25 ft. lbs. (35 Nm).
    d. When installing the rear torsion bar bushing, on the 300E, 300CE, 300SE, 300TE and 300SEL, the flats on the cone must be vertical.

560SL

1. Since the front shock absorber acts as a deflection stop for the front wheels, the lower shock absorber attaching point should not be loosened unless the vehicle is resting on the wheels or unless the lower control arm is supported.
2. Raise and safely support the vehicle.
3. Support the lower control arm.
4. Loosen the lower shock absorber attachment.
5. Unscrew the steering arm from the steering knuckle.
6. Separate the brake line and brake hose and plug the openings.
7. Remove the front spring.
8. Unscrew the hex nuts on the ball joints.
9. Remove the lower ball joint and remove the lower control arm.

To install:

10. Install the lower ball joint and lower control arm. Torque the ball joint nut to 25 ft. lbs. (34 Nm).
11. Install the front spring.
12. Separate the brake line and brake hose and plug the openings.
13. Install the steering arm to the steering knuckle.
14. Torque the lower shock absorber attachment to 20 ft. lbs. (27 Nm).
15. Lower the vehicle. Bleed the brakes and check the front end alignment.

300SL and 500SL

1. Raise and safely support the vehicle.
2. Remove the lower engine compartment cover.
3. Remove the sway bar nuts and retaining bracket.
4. Using an approved spring compressor, remove the front spring.
5. Mark the control arm mounting and eccentric bolts.
6. Remove the lower control arm mounting bolts.
7. Remove the steering knuckle from the control arm by removing the clamp bolt.

To install:

8. Check the control arm support and mounting for damage.
9. Install the control arm and torque the steering knuckle to 90 ft. lbs. (125 Nm). Fill the separating slot with sealing compound to prevent contamination.
10. Insert the front eccentric bolt from the back to the front. The rear eccentric bolt from the front to the rear. Do not tighten until the vehicle weight is on the suspension.
11. Install the sway bar and torque the nuts to 15 ft. lbs. (20 Nm).
12. Install the front spring and lower the vehicle.
13. Torque the eccentric bolts to 88 ft. lbs. (120 Nm) with the vehicle weight on the suspension.
14. Adjust the front wheel alignment.

Sway Bar

REMOVAL & INSTALLATION

1. Raise the vehicle and support safely. Remove the front wheels and under covers.
2. Loosen but do not remove the sway bar-to-control arm mountings.
3. Remove the sway bar-to-frame mountings and the sway bar.
4. Install the sway bar and torque the mountings to 30 ft. lbs. (40 Nm).

Front Wheel Bearings

ADJUSTMENT

1. Tighten the clamp nut until the hub can just be turned.
2. Slacken the clamp nut and seat the bearings on the spindle by rapping the spindle sharply with a hammer.
3. Attach a dial indicator, with the pointer indexed, onto the wheel hub.
4. Check the endplay of the hub by pushing and pulling on the flange. The endplay should be approximately 0.0004–0.0008 in.
5. Make an additional check by rotating the washer between the inner race of the outer bearing and the clamp nut. It should be able to be turned by hand.
6. Check the position of the suppressor pin in the wheel spindle and the contact spring in the dust cap.
7. Pack the dust cap with 20–25 grams of wheel bearing grease and install the cap.
8. Install the brake caliper and bleed the brakes.

REMOVAL & INSTALLATION

If the wheel bearing play is being checked for correct setting only, it is not necessary to remove the caliper. It is only necessary to remove the brake pads.

1. Raise the vehicle and support safely. Remove the brake caliper.
2. Pull the cap from the hub with a pair of channel-lock pliers. Remove the radio suppression spring, if equipped.
3. Loosen the socket screw of the clamp nut on the wheel spindle. Remove the clamp nuts and washer.
4. Remove the front wheel hub and brake disc.
5. Remove the inner race with the roller cage of the outer bearing.
6. Using a brass or aluminum drift, carefully tap the outer race of the inner bearing until it can be removed with the inner race, bearing cage and seal.
7. In the same manner, tap the outer race of the bearing off the hub.
8. Separate the front hub from the brake disc.

To install:

9. To assemble, press the outer races into the front wheel hub.
10. Pack the bearing cage with bearing grease and insert the inner race with the bearing into the wheel hub.
11. Coat the sealing ring with sealant and press it into the hub.
12. Pack the front wheel hub with 45–55 grams of wheel bearing grease. The races of the tapered bearing should be well packed and also apply grease to the front faces of the rollers. Pack the front bearings with the specified amount of grease. Too much grease will cause overheating of the lubricant and it may lose its lubricity. Too little grease will not lubricate properly.
13. Coat the contact surface of the sealing ring on the wheel spindle with Molykote paste or equivalent.
14. Press the wheel hub onto the wheel spindle.
15. Install the inner race and cage of the outer bearing.
16. Install the steel washer and the clamp nut.

REAR SUSPENSION

Shock Absorbers

REMOVAL & INSTALLATION

190D, 190E, 260E, 300E, 300CE, 300SE, 300TE, 300SEL, 350SD, 500E and 560SL

1. Raise and safely support the vehicle.
2. From inside the trunk (sedans), remove the rubber cap, locknut and hex nut from the upper mount of the shock absorber.
3. Unbolt the mounting for the rear shock absorber at the bottom and remove the shock absorber. Be sure to disconnect and plug the pressure line on the 190E-16.
4. Installation is the reverse of removal. Torque the shock bolts to 25 ft. lbs. (34 Nm).

All Others Models

1. Remove the rear seat and backrest.

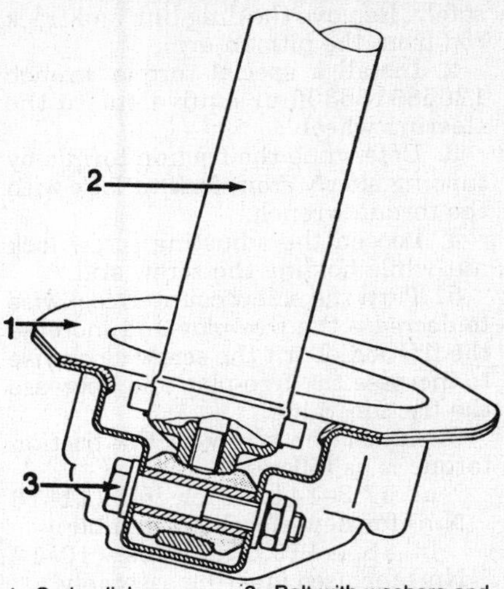

1. Spring link
2. Shock absorber
3. Bolt with washers and self-locking nut

Rear shock absorber lower mount—190D and 190E

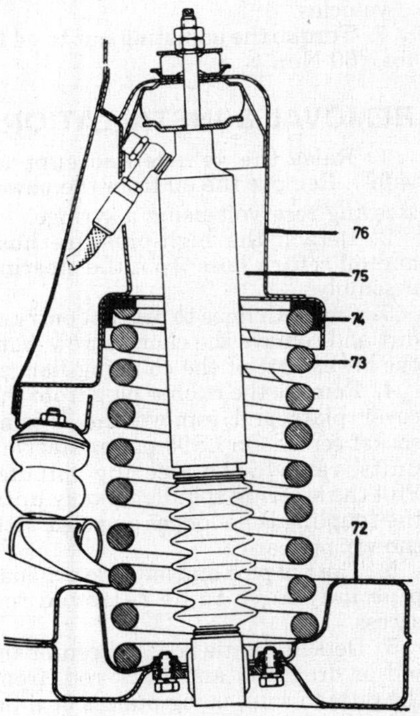

72. Semi-trailing arm
73. Rear spring
74. Rubber mounting
75. Shock absorber or spring strut
76. Dome on frame floor

Rear spring—except 190D, 190E, 260E, 300E, 300CE, 300TE and 560SL

2. Remove the cover from the rear wall.
3. Raise and safely support the vehicle. Support the trailing arm.
4. Loosen the nuts on the upper mount. Remove the washer and rubber ring.
5. Loosen the lower mount and remove the shock absorber downward.
6. Installation is the reverse of removal. Tighten the upper mounting nut to 15 ft. lbs. (20 Nm) to the end of the threads.

Springs

REMOVAL & INSTALLATION

190D, 190E, 260E, 300E, 300CE, 300SE, 300TE, 300SL, 300SEL, 500E and 500SL

1. Raise and safely support the vehicle. Remove the wheel.
2. Disconnect the holding clamps for the spring link cover and remove the cover. Disconnect the ADS hydraulic pipe, if so equipped.
3. Install a spring compressor and compress the spring until the spring link is free of all load.
4. Disconnect the lower end of the shock absorber.
5. Increase the tension on the spring compressor and remove the spring.

To install:

6. Note the following:
   a. Position the spring so the end of the lower coil is seated in the impression of the spring seat and the upper coil seats properly in the rubber mount in the frame floor.
   b. Do not release tension on the spring compressor until the lower end of the shock absorber is connected and tightened to 47 ft. lbs. (65 Nm).

560SL

1. Raise and safely support the vehicle.
2. Remove the rear shock absorber.
3. Raise the control arm to a horizontal position. Install a spring compressor to aid in this operation.
4. Carefully, lower the control arm until it contacts the stop on the rear axle support.
5. Remove the spring and spring compressor with great care.
6. Installation is the reverse of removal. For ease of installation, attach the rubber seats to the springs with masking tape.

All Others

1. Raise and safely support the vehicle. Support the trailing arm.
2. Remove the rear shock absorber.
3. Be sure the upper shock absorber attachment is released first.
4. Compress the spring with a spring compressor.
5. Remove the rear spring with the rubber mount.
6. Installation is the reverse or removal. When installing the shock absorber, tighten the lower mount first.

Lower Control Arm

REMOVAL & INSTALLATION

1. Raise the vehicle and safely support. Remove the wheels.
2. Remove the control arm cover, if so equipped.
3. Support the lower control arm using a suitable jack.
4. Disconnect the roll-over switch (300SL and 500SL) and any other pipe or wiring harness. Remove the serrated bolts from the frame transverse link and reinforcement plate.
5. Remove the rear spring using a suitable spring compressor.
6. Disconnect the shock absorber, lower ball joint and sway bar.
7. Remove the inner bushing bolt and lower control arm.

To install:

8. Install the lower control arm and loosely install the bolt and nut.
9. Connect the knuckle and torque the ball joint nut to 15 ft. lbs. (20 Nm).
10. Install the spring using a suitable spring compressor.
11. Install the spring link and sway bar. Torque the bolts to 44 ft. lbs. (60 Nm).
12. Install the remaining components and lower the vehicle.
13. Torque the inner bushing bolt to 51 ft. lbs. (70 Nm) with the vehicle weight resting on the suspension.

STEERING

Steering Wheel

CAUTION

Some vehicles are equipped with a Supplemental Restraint System (SRS). Improper maintenance, including incorrect removal and installation of related components, can lead to personal injury caused by unintentional activation of the Airbag.

REMOVAL & INSTALLATION

Without SRS

1. Disconnect the negative battery cable. On 560SL, pry the 3-pointed star trademark from the center padding. On all other models, remove the padded plate. Pull at one corner near the wheel spokes.
2. Unscrew the hex nut from the steering shaft and remove the spring washer and the steering wheel.

NOTE: All models use an Allen screw in place of the hex nut. The Allen screw must be replaced, if removed.

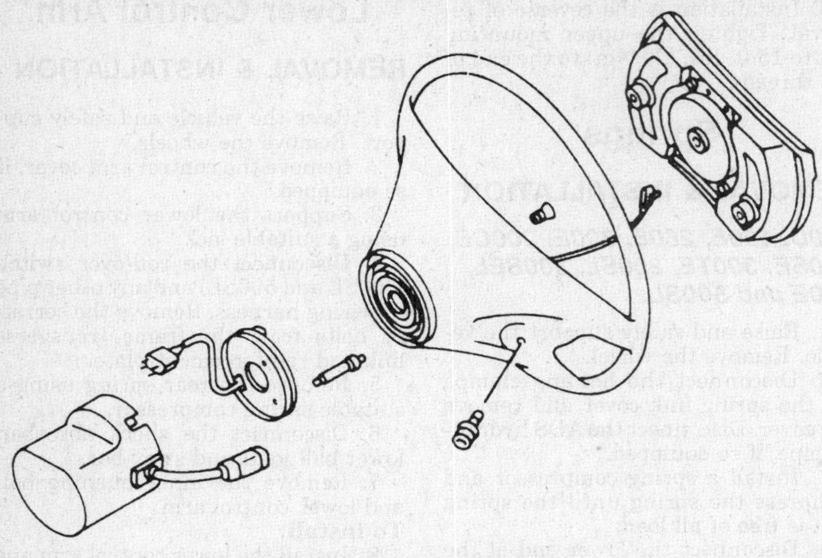

Steering wheel with air bag

3. Installation is the reverse of removal. Be sure the alignment mark on the steering shaft is pointing upward and be sure the slightly curved spoke of the steering wheel is down. Torque the steering shaft nut to 25 ft. lbs. (34 Nm).

With SRS

1. Disconnect the negative battery cable.
2. From behind the steering wheel, remove the air bag retaining screws. Pull the air bag out far enough to disconnect the electrical connector.
3. Set the steering wheel in a horizontal position and lock the wheel in place.
4. Remove the countersunk nut from the steering spindle.
5. Matchmark the steering shaft and wheel. Remove the steering wheel with a suitable steering wheel puller.

To install:

6. Install the steering wheel and torque the shaft nut to 60 ft. lbs. (80 Nm).
7. Install the air bag and connect the battery cable.

Steering Column

REMOVAL & INSTALLATION

CAUTION

Some vehicles are equipped with a Supplemental Restraint System (SRS). Improper maintenance, including incorrect removal and installation of related components, can lead to personal injury caused by unintentional activation of the Airbag.

1. Disconnect the negative battery cable.
2. Remove the instrument panel under covers and steering wheel.
3. Remove the instrument cluster.
4. Matchmark the steering shaft to the power steering gear and remove the locking bolt.
5. Remove the cover plate and combination switch.
6. Remove the ignition switch and protective bushing.
7. Remove the stiffening strut and disconnect the electrical connection.
8. Remove the 2 bolts holding the steering column jacket to the crossmember.
9. Remove the lower retaining bolts. Disconnect the column from the floor and remove the steering column from the vehicle. Do not hammer on the shaft or steering tube.

To install:

10. Before installation: set the steering spindle in a straight ahead position, carefully slide the steering column jacket into the steering clutch and make sure the steering clutch contacts the sealing ring and the screw is installed through the groove in the corrugated pipe flange.
11. Install the steering column and set the sealing sleeve into the firewall.
12. Install the column retaining bolts and torque to 15 ft. lbs. (20 Nm).
13. Install the remaining components. Torque the steering shaft-to-steering gear bolt to 18 ft. lbs. (25 Nm).
14. Connect the battery cable and check operation before driving the vehicle.

Power Steering Gear

ADJUSTMENT

1. Raise the vehicle and support safely. Remove the drag link and track rod from the pitman arm.
2. Install a special torque wrench 126589136300 or equivalent to the steering wheel.
3. Determine the friction torque by turning slowly from lock to lock with the torque wrench.
4. Loosen the adjusting screw lock nut while holding the screw still.
5. Turn the screw counterclockwise to decrease the free-play and increase the friction. Turn the screw clockwise to increase the free-play and decrease the friction value.
6. Adjust the screw so the friction torque is as follows:
   a. 97.3–141.5 inch lbs. (11–16 Nm) for new steering assembly.
   b. 88.4–115.0 inch lbs. (10–13 Nm) for used steering assembly.
   c. 97.3–123.8 inch lbs. (11–14 Nm) for vehicles with air bag.
   d. 44.2–62.0 inch lbs. (5–7 Nm) for SD, SDL, SE, SEC and SEL vehicles.
7. Torque the adjusting nut to 44 ft. lbs. (60 Nm).

REMOVAL & INSTALLATION

1. Raise the vehicle and support safely. Remove the oil from the power steering reservoir using a syringe.
2. Detach the high-pressure hose and oil return hose from the steering assembly.
3. Cap both lines to prevent entry of dirt and remove the clamp screw from the lower part of the coupling flange.
4. Remove the rubber plug from the cover plate and remove the U-joint socket screw. On LS90 power steering units, remove the steering spindle. Pull the steering spindle up only until the coupling is no longer engaged with the worm gear.
5. The tail pipe and left side exhaust pipe may have to be removed for access.
6. Detach the tie rod and center tie rod or drag link and track rod, from the pitman arm, using pullers or a tie rod splitter.
7. Remove the hex-head bolts that hold the gearbox to the frame, press the worm shaft stub from the steering coupling and remove the gearbox from under the vehicle.

To install:

8. First install the pitman arm, if removed, aligning the matchmarks. Tighten the pitman arm nut to 110 ft. lbs. and install the cotter pin. Use new self-locking nuts to attach the gear to the frame.
9. Remove the screw plug from the steering box. Turn the worm shaft until the center of the power piston is directly below the bore in the housing. Check dimension (a) which can be al-

tered by changing the position of the pitman arm on its shaft.
10. Center the steering wheel.
11. Press the worm shaft stub into the steering shaft coupling, making sure not to damage the serrations.

NOTE: Install assembly pin as for manual steering.

12. Install and tighten the hex-head screws that hold the gearbox to the chassis, install and tighten the coupling clamp screw.
13. Install the plug in the gearbox, using a new gasket; attach the tie rods to the pitman arm and make sure the steering knuckle arms rest against their stops at full left and right lock.
14. Check toe-in and correct if necessary. Remove the dust covers from the fluid lines, reconnect the high and low pressure lines.
15. Fill the reservoir and connect a hose between the bleed screw on the steering and the reservoir.
16. Open the bleed screw and, with engine running, bleed the system and top up.

Power Steering Pump

REMOVAL & INSTALLATION

1. Disconnect the negative battery cable. Remove the supply tank nut.
2. Remove the spring and damping plate.
3. Drain the oil from the tank with a syringe.
4. Loosen and remove the expanding and return hoses from the pump. Plug all connections and pump openings.
5. If necessary for clearance, loosen the radiator shell. Loosen the mounting bolts and move the pump toward the engine by using the toothed wheel. Remove the belt. Remove the pulley and the pump.
6. Loosen the plate nut and the support bolt.
7. Push the pump toward the engine and remove the belts from the pulley.
8. Unscrew the mounting bolts and remove the pump and carrier.

To install:

9. Transfer the pulley to the new pump. Install the pump and mounting bolts. Torque the bolts to 18 ft. lbs. (25 Nm).
10. Install the drive belt and adjust the tensioner.
11. Install the expanding and return hoses to the pump. Torque the fittings to 29–34 ft. lbs. (40–45 Nm).
12. Refill the oil to the tank.
13. Install the spring and damping plate.
14. Connect the negative battery cable.

SYSTEM BLEEDING

1. The reservoir should be full of power steering fluid.
2. Raise the vehicle and support it safely.
3. Turn the steering wheel fully to the right and left until no air bubbles appear in the fluid. Maintain the reservoir level.
4. Lower the vehicle and with the engine idling, turn the wheels fully to the right and left. Stop the engine.
5. Repeat the procedure until no air bubbles pass through the tube.
8. Refill the reservoir as needed and check that no further bubbles are present in the fluid. An abrupt rise in the fluid level after stopping the engine is a sign of incomplete bleeding. This will cause noise from the pump or control valve.

Tie Rod Ends

REMOVAL & INSTALLATION

1. Mark the tie rod and adjusting sleeve with tape.
2. Remove the self-locking nuts from the rod joint.
3. Using a tie rod separator 129589106300 or equivalent, remove the tie rod end from the steering arm.
4. Loosen the locking bolt and turn the tie rod from the adjusting sleeve.
5. Check the tie rod end for excessive backlash and damaged boots, replace if necessary.

To install:

6. Install tie rod end to the marked location. Torque the self-locking nut and adjusting sleeve nut to 34 ft. lbs. (50 Nm). Torque the collar band bolt to 15 ft. lbs. (20 Nm).
7. Have the front end aligned.

BRAKES

For all brake system repair and service procedures not detailed below, please refer to "Brakes" in the Unit Repair section.

Master Cylinder

REMOVAL & INSTALLATION

1. Disconnect the negative battery cable. To remove the master cylinder, first open a bleed screw at one front and one rear wheel.
2. Pump the pedal to empty the reservoir completely. Make sure both reservoirs are completely drained.
3. Disconnect the switch connectors using a small prybar. Disconnect the brake lines at the master cylinder. Plug the ends with bleed screw caps or the equivalent.
4. Unbolt the master cylinder from the power brake unit and remove; do not loose the O-ring in the flange groove of the master cylinder.
5. Installation is the reverse of removal. Be sure to replace the O-ring between the master cylinder and the power brake unit, since this must be absolutely tight. Torque the nuts to 12–15 ft. lbs. Be sure both chambers are completely filled with brake fluid and bleed the brakes.

Power Brake Booster

REMOVAL & INSTALLATION

300SDL, 300SE, 300SEL, 350SD, 420SEL, 560SEC and 560SEL

1. Disconnect the negative battery cable. If equipped with a manual transmission, remove hose to master cylinder.
2. Remove master cylinder.
3. Loosen vacuum line to brake unit.
4. Remove cover under dash panel on driver's side.
5. Remove lock and remove collar bolt to release pushrod.
6. Remove nuts for fastening booster unit to front end and remove booster unit.

NOTE: Care must be taken when handling brake unit. It is made of plastic and may break.

To install:

7. Position brake unit to front end and install attaching nuts. Torque to 11 ft. lbs. (15 Nm).
8. Install collar bolt and lock to pushrod and brake pedal.
9. Install cover under dash panel.
10. Connect vacuum line to brake unit and torque nut to 22 ft. lbs. (30 Nm).
11. Install master cylinder.
12. If equipped with a manual transmission, connect hose to master cylinder on expansion tank.

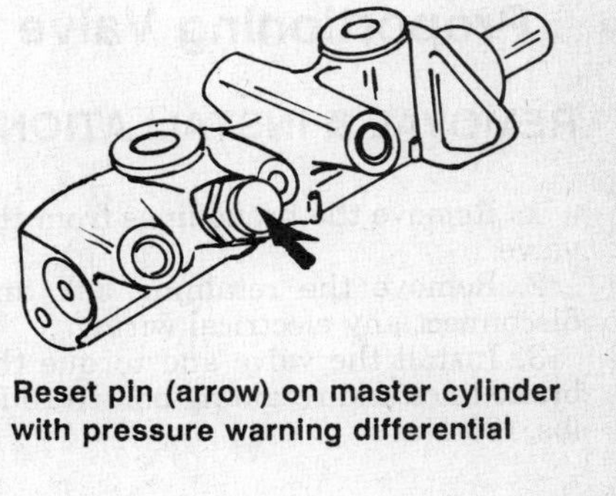

Reset pin (arrow) on master cylinder with pressure warning differential

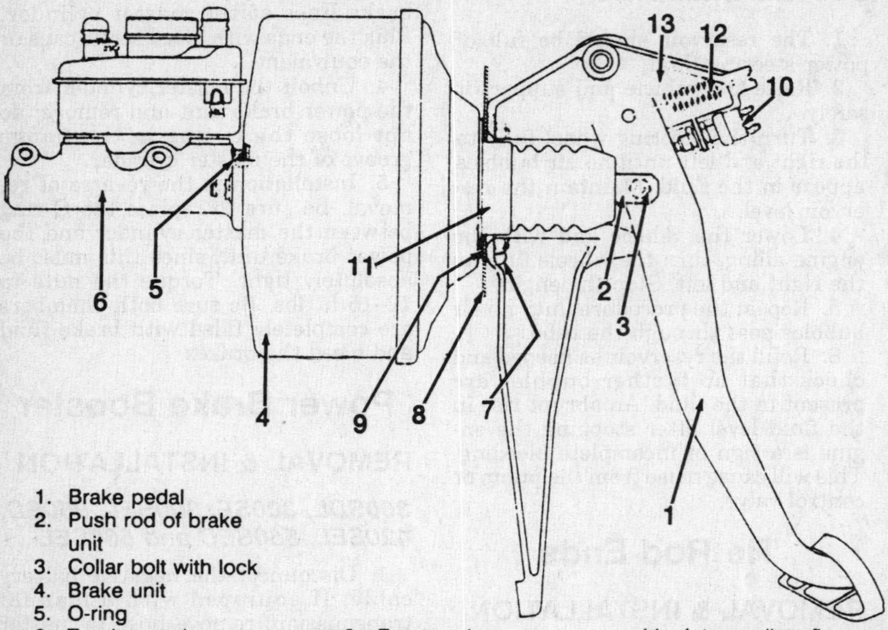

1. Brake pedal
2. Push rod of brake unit
3. Collar bolt with lock
4. Brake unit
5. O-ring
6. Tandem-main cylinder
7. Hex nut
8. Front end
9. Gasket
10. Stop lamp switch
11. Intermediate flange
12. Restoring spring
13. Carrier

Power braking system—300SDL, 300SE, 300SEL, 420SEL, 560SEC and 560SEL

190D, 190E, 260E, 300E, 300CE, 300SL, 500E and 500SL

1. Disconnect the negative battery cable.
2. Remove the master cylinder and instrument panel cover.
3. Disconnect the vacuum line and disconnect the pedal return spring.
4. Remove the lock pin from the brake booster push rod.
5. Remove the booster retaining nuts and remove the brake booster.

To install:

6. Install the booster and torque the nuts and brake lines to 11 ft. lbs. (15 Nm).
7. Connect the vacuum line and install the master cylinder.
8. Bleed the master cylinder.

Proportioning Valve

REMOVAL & INSTALLATION

1. Remove the brake lines from the valve.
2. Remove the retaining bolt and disconnect any electrical wiring.
3. Install the valve and torque the brake lines and retaining bolt to 15 ft. lbs. (20 Nm).

Disc Brake Pads

REMOVAL & INSTALLATION

190D, 190E, 260E, 300E, 300CE, 300TE and 500E

FRONT AXLE

1. Raise and safely support the vehicle.
2. Remove the front wheel assemblies.
3. Lift the 2 holding lugs located laterally on the cover of the plug connection by means of a suitable tool and open the cover. Do not use force. Remove the cable of clip sensor from the plug connection on the floating caliper. Do not pull on the cable.
4. Remove the lower caliper bolt while applying counter hold to the sliding bolt.
5. Swing the cylinder housing upward and engage with a suitable hook to the wheel housing. Remove both brake pads from the brake carrier.
6. Pull the clip sensor from the back plate of the lining.
7. Draw some brake fluid from the expansion tank.
8. Push the piston back with resetting device 000 589 52 43 00 or equivalent.

To install:

9. Place the new brake pads into the brake carrier. Make sure the spring clamp is located in parallel with the upper edge of the lining.
10. Complete the installation by reversing the removal procedure. Torque the caliper bolt to 26 ft. lbs. (35 Nm).
11. Install the wheel assemblies and lower the vehicle. Check the brake fluid lever and replenish, if necessary.

NOTE: Prior to moving the vehicle, apply the brake pedal several times to adjust the brake pads to the brake disc.

REAR AXLE

1. Raise and safely support the vehicle.
2. Remove the rear wheel assemblies.
3. Knock the holding pin from the fixed caliper by means of a punch. Remove the cross spring.
4. Push the brake pads from the fixed caliper by means of a pushing lever.

NOTE: If the brake pads are rusted, use a puller for removal of the stuck brake pads.

5. Clean the guide surface for the brake pad in the fixed caliper with a brake caliper brush.
6. Draw some brake fluid from the expansion tank.
7. Push both the pistons back with a resetting device.

To install:

8. Place the new brake pads into the brake carrier.
9. Complete the installation by reversing the removal procedure.
10. Install the wheel assemblies and lower the vehicle. Check the brake fluid lever and replenish, if necessary.

NOTE: Prior to moving the vehicle, apply the brake pedal several times to adjust the brake pads to the brake disc.

300D, 300TD 300SE, 300SL, 300SEL, 350SD, 420SEL, 500SL, 560SEC, 560SEL and 560SL

FIXED CALIPER

1. Raise and safely support the vehicle.
2. Remove the wheel assemblies.
3. On fixed calipers with a brake lining wear indicator, pull the cables of the clip sensors from the plug connection on the fixed caliper.
4. On Teves® fixed caliper, knock the holding pin from the caliper by means of a punch.
5. On Bendix® and Girling® fixed calipers, pull both locking eyes from the holding pins and remove the holding pins.
6. Pull the clip sensor from the backing plate or brake lining. Simulta-

neously, remove the cross spring or spring holding lining.

NOTE: Renew the clip sensor only, when the insulating layer of the contact pin is rubbed through or in the event of damage on a part of the sensor, including the line insulation.

7. Force the brake pads from the fixed caliper by means of the forcing lever.

NOTE: If the brake pad is rusted, use a puller for removal of the stuck brake pads.

8. Clean the guide for the brake pad in the fixed caliper with a brake caliper brush.
9. Draw a slight amount of brake fluid from the expansion tank.
10. Push both the pistons back with a resetting device.

To install:

11. Place the new brake pads into the brake carrier.
12. Complete the installation by reversing the removal procedure.
13. Install the wheel assemblies and lower the vehicle. Check the brake fluid lever and replenish, if necessary.

NOTE: Prior to moving the vehicle, apply the brake pedal several times to adjust the brake pads to the brake disc.

FLOATING CALIPER

1. Raise and safely support the vehicle.
2. Remove the wheel assemblies.
3. Lift the 2 holding lugs located laterally on the cover of the plug connection by means of a suitable tool and open the cover. Do not use force. Remove the cable of clip sensor from the plug connection on the floating caliper; do not pull on the cable.
4. Remove the upper caliper bolt while applying counter hold to the sliding bolt.
5. Fold the cylinder housing in a downward direction and attach to the torsion bar by means of a suitable hook. Remove both brake pads from the brake carrier.
6. Pull the clip sensor from the lining backup plate.

NOTE: The wear indicator on the floating caliper is the inside of the brake pad only. Renew the clip sensor only, when the insulating layer of the contact pin is rubbed through or in the event of damage on a part of the sensor, including the line insulation.

7. Clean the contact surface of the brake pads in the brake carrier.
8. Draw a slight amount of brake fluid from the expansion tank.
9. Push the piston back with resetting device.

To install:

10. Place the new brake pads into the brake carrier; make sure the spring clamp is located in parallel with the upper edge of the lining.
11. Complete the installation by reversing the removal procedure. Torque the caliper bolt to 26 ft. lbs. (35 Nm).
12. Install the wheel assemblies and lower the vehicle. Check the brake fluid lever and replenish, if necessary.

NOTE: Prior to moving the vehicle, apply the brake pedal several times to adjust the brake pads to the brake disc.

Brake Shoes

REMOVAL & INSTALLATION

1. Raise and safely support the vehicle.
2. Remove the rear wheel assemblies.
3. Remove the caliper bolts and the floating caliper from the wheel carrier.
4. Hang the floating caliper, including the brake hose, on the rear spring by means of a suitable hook.
5. Turn the rear axle shaft flange in such a manner that 1 tapped hole faces the spring. Compress the spring slightly with installer, turn the tool by approximately 90 degrees, disconnect the spring from the covering ring and remove it.
6. Remove the spring on the other brake shoe.
7. Disconnect the return spring with the remover and installer from the brake shoes.
8. Pull both brake shoes apart until they can be removed over the rear axle flange.
9. Disconnect the return spring from the brake shoes and remove the adjusting device.
10. Push the bolt from the expanding lock and remove the expanding lock from the brake cable control.
11. Installation is the reverse of removal procedures. Torque the caliper bolts to 38 ft. lbs. (50 Nm).

Brake Rotor

REMOVAL & INSTALLATION

1. Raise the vehicle and support safely. Remove the wheel.
2. Remove the brake caliper.
3. Remove the rotor retaining screws, if so equipped and remove the rotor.
4. Install the rotor and torque the retaining screw to 11 ft. lbs. (15 Nm).
5. Install the caliper and wheel. Torque the wheel lugs to 75 ft. lbs. (100 Nm).

Bleeding Brake System

Without Anti-skid Control (ASR)

1. Carefully clean all dirt from around the master cylinder filler cap.
2. If a bleeder tank is used, follow the manufacturer's instructions.
3. Remove the filler cap and refill the master cylinder to the lower edge of the filler neck.
4. Clean off the bleeder connections at all of the wheel cylinders or disc brake calipers. Attach the bleeder hose and fixture to the right rear caliper bleeder screw and place the end of the tube in a glass jar, submerged in brake fluid.
5. Open the bleeder valve ½–¾ turn. Have an assistant depress the brake pedal and allow it to return slowly. Continue this pumping action to force any air out of the system.
6. When bubbles cease to appear at the end of the bleeder hose, close the bleeder valve and remove the hose. Check the level of the brake fluid in the master cylinder and add fluid, if necessary.
7. After the bleeding operation at each caliper has been completed, refill the master cylinder reservoir and replace the filler plug.

NOTE: Never reuse brake fluid which has been removed from the lines through the bleeding process because it contains air bubbles and dirt. Do not allow the master cylinder to run dry. The entire system will have to be rebled if this happens.

With Anti-skid Control (ASR)

1. Turn the ignition switch **OFF**. Remove the master cylinder cap.
2. Drain the pressure storage tank through the bleed screw labeled **SP** at the ABS/ASR unit.
3. Install a brake bleeding device according to the manufacturers instructions.
4. Start the engine and bleed screw **SP** until clean brake fluid is free of bubbles.
5. Turn the engine **OFF**. Bleed all 4 calipers starting with the furthest from the master cylinder.
6. Fill the reservoir and install the cap.
7. Check brake operation before moving the vehicle.

Anti-Lock Brake System Service

PRECAUTIONS

- When welding with an electric welding unit, unplug the electric control unit.
- During paint jobs, the electronic control unit may be exposed to a maximum of 203°F (95°C) for up to 2 hours or 185°F (85°C) if more time is needed.
- When removing the rear axle centerpiece, make sure the correct toothed wheel with the correct ratio for the wheel speed sensor is installed. If a wheel with the wrong number of teeth is installed, this fault will not show up when checking the system with the ABS tester. The stopping distance, however, will be increased during controlled braking.
- If work was done to non-ABS brake components, a simple operational test will be sufficient. This means that after driving about 5 mph, the yellow warning light on the instrument panel should go out if the ABS system is intact.
- If ABS components have been replaced, the entire system should be checked using the appropriate Bosch tester in combination with brake test bench or an adaptor in combination with a multimeter.

RELIEVING ANTI-LOCK BRAKE SYSTEM PRESSURE

Open the bleeder screw **SP** on the hydraulic ABS/ASR unit about 1 turn. Allow the entire contents to drain into a collection bottle.

Anti-Lock Brake (ABS) Hydraulic Unit

REMOVAL & INSTALLATION

300SE, 300SEL, 350SD, 420SEL, 560SEC and 560SEL

1. With ignition switch **OFF**, disconnect battery negative terminal.
2. Disconnect brake lines from hydraulic unit and seal open lines with blind plugs.

NOTE: Do not loosen sealed center bolt and 2 socket screws.

3. Remove cover fastening screw and remove cover.
4. Disconnect grounding strap from pump motor.
5. Disconnect stress relief and remove plug.

NOTE: Two relays for pump motor or for solenoid valves can be replaced.

6. Remove mounting nuts and hydraulic unit.

To install:

7. Mount hydraulic unit on mounting bracket and attach 12 terminal plug and attach stress relief.
8. Install hydraulic unit cover and screw.
9. Connect brake lines to hydraulic unit. Torque line nuts to 10 ft. lbs. (14 Nm).

NOTE: Do not interchange brake lines.

10. Connect ground terminal of battery.

190, 300D, 300E, 300CE, 300TE, 300SL, and 500SL

1. Disconnect the negative battery cable.
2. Bleed the ABS system and remove the ABS/ASR protective cover.
3. Disconnect all electrical and hydraulic lines from the hydraulic unit. Use flarenut wrenches to remove the lines. Plug all lines to prevent contamination.
4. Remove the retaining bolts and remove the unit with the mounting brackets.

To install:

5. Install the unit and torque the bolts to 15 ft. lbs. (20 Nm).
6. Connect the lines and torque to 11 ft. lbs. (15 Nm).
7. Use the following symbols for reference to the ABS hydraulic unit:

vl – front left caliper
vr – front right caliper
hl – rear left caliper
hr – rear right caliper
H – master cylinder rear axle
V – master cylinder front axle
BA – precharging pump for ASR
P – pump
E – input pressure storage tank
SP – pressure storage tank bleed screw

7. Connect all electrical wiring.
8. Bleed the ABS/ASR system.

Rear Speed Sensor

REMOVAL & INSTALLATION

The speed sensors are located at the rear drive axle assembly. The drive axle assembly has to lowered to remove the sensor. Remove the retaining screw and pull the sensor out of the housing.

CHASSIS ELECTRICAL

CAUTION

On vehicles equipped with an air bag, the negative battery cable must be disconnected, before working on the system. Failure to do so may result in deployment of the air bag and possible personal injury.

Air Bag

DISARMING

CAUTION

The air bag units must be fitted when the battery is disconnected and negative terminal of the test connector disconnected. If the work is interrupted for any reason, the air bag must be returned to the storage package. Do not leave the components unsupervised.

1. Switch the ignition key to the **0** position.
2. Disconnect the negative battery terminal. Cover the terminal so there is no chance of an accidental connection completing the electrical circuit and arming the SRS system.
3. Remove the foot mat in the passenger side footwell and unscrew the foot rest.
4. Locate and disconnect the red test connector for the air bag connection.

Heater Blower Motor

REMOVAL & INSTALLATION

260E, 300D, 300E, 300SE, 300TE, 300SEL, 300TD, 500E

1. Disconnect the negative battery cable. Remove the cover from under the right side of the instrument panel.
2. Disconnect the plug from the blower motor.
3. Unscrew the contact plate screw, lift the contact plate and disconnect both wires to the series resistor.
4. Loosen the blower motor flange screws and lift out the blower motor.
5. Installation is in the reverse order or removal.

560SL

1. Disconnect the negative battery cable. Working in the engine compartment, unscrew the 8 mounting screws and remove the panel which covers the blower motor.
2. Disconnect the plug from the series resistor at the firewall.
3. Remove the mounting bolts and remove the series resistor.

4. Unscrew the 4 blower motor retaining nuts and lift out the motor.
5. Installation is in the reverse order of removal. Be sure the rubber sealing strip is not damaged.

350SD, 420SEL, 560SEC and 560SEL

1. Disconnect the negative battery cable. Remove the cover from under the right side of the instrument panel.
2. Remove the cover for the blower motor and disconnect the 2-prong plug.
3. Remove the blower motor flange bolts and the blower motor.
4. Installation is in the reverse order of removal.

190D and 190E

1. Disconnect the negative battery cable. Open the hood to a 90 degree position and remove the wiper arms.
2. Disconnect the retaining clips for the air intake cover at the firewall.
3. Remove the rubber sealing strip from the cover and remove the retaining screw. Slide the cover from the lower windshield trim strip and remove it.
4. Disconnect the vacuum line from the heater valve.
5. Remove the heater cover retaining screws.
6. Pull up the rubber sealing strip from the engine side of the defroster plenum (firewall), unscrew the retaining screws and pull up and out on the blower motor cover.
7. Loosen the cable straps on the connecting cable and disconnect the plug.
8. Unscrew the mounting bolts and remove the blower motor.
9. Installation is in the reverse order of removal.

300SL and 500SL

1. Disconnect the negative battery cable.
2. Remove the wiper motor and transmission.
3. Remove the right and left water drain and center partition wall. The wall has 2 rivets that have to be removed.
4. Disconnect any vacuum hoses and heater housing.
5. Remove the blower controller and cover.
6. Unclip the retaining strap and flat blade connector at the motor.
7. Detach and remove the motor from the housing.
8. Installation is in the reverse order of removal. Make sure the motor is correctly inserted into the mounting bracket before reassembly.

Recess slot in the instrument cluster

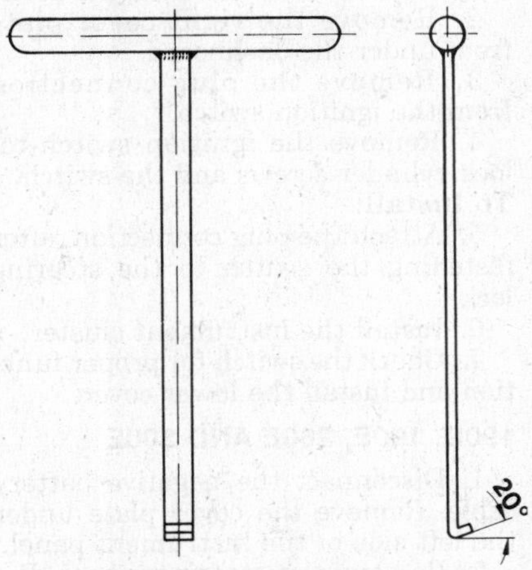

Fabricated tool for removing the instrument cluster

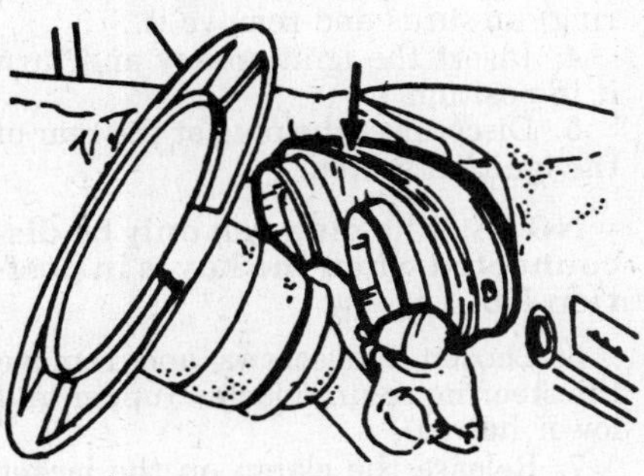

Instrument panel removal showing rubber retaining strip

Instrument Cluster

REMOVAL & INSTALLATION

190D, 190E, 350SD, 420SEL, 560SEC, 560SEL and 560SL

1. Disconnect the negative battery cable. Remove the cover under the left side of the instrument panel.
2. Disconnect the defroster ducting which runs behind the instrument cluster.
3. Unscrew the speedometer cable from below and push the cluster out far enough to disconnect all connections on the back of the instrument cluster.
4. Remove the 5 clips which secure the instrument cluster and remove it.
5. Installation is in the reverse order of removal.

260E, 300D, 300E, 300CE, 300SE, 300TE, 300SEL, 500E

1. Disconnect the negative battery cable. Remove the instrument panel undercover.
2. Remove the speedometer cable from the slips on the panel under the instrument cluster. This will allow the cable come out when the cluster is removed.
3. Fabricate a removal tool (hook) and insert it between the padding and the cluster at the top of the left side of the cluster. Rotate the tool and pull out until it rests against the detent.
4. Pull the instrument cluster out evenly on both sides. Be sure the speedometer cable slides into the recess on the brake pedal cover.
5. Reach behind the cluster, unscrew the speedometer cable. Label and disconnect all electrical leads.
6. Remove the instrument cluster.

To install:

7. When installing, position the cluster and reconnect the speedometer cable and all electrical leads.
8. Push the cluster backwards into its recess.

NOTE: Be certain the speedometer cable slides back into the footwell area without buckling behind the cluster.

300D and 300TD

1. Disconnect the negative battery cable.
2. Remove the instrument cluster slightly by hand. Don't pull on the edge of the glass.
3. A removal hook can be fabricated and inserted between the instrument cluster and the dashboard.
4. Guide the removal hook up to the right to the recess and pull the instrument cluster out.
5. Pull it out as far as possible and disconnect the speedometer cable, electrical connections and oil pressure line.

To install:

6. Reconnect the electrical connections, oil pressure line and speedometer cable. To avoid speedometer cable noise, guide it into the largest radius possible.
7. Push the instrument cluster firmly into the dashboard.

300SL and 500SL

1. Disconnect the negative battery cable.
2. Remove the steering wheel.
3. Insert a pulling hook 126589033300 or equivalent. There are 3 slots to insert the hook tool. One

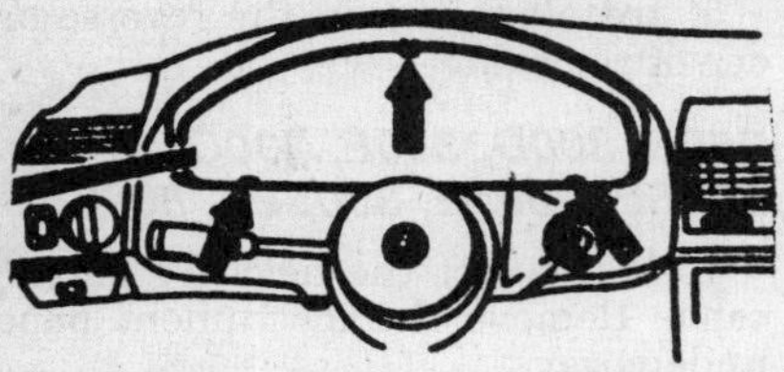

Instrument cluster removal—300SL and 500SL shown, others similar

is top center and 2 others are on the bottom between the steering column and edge of the cluster. Carefully insert the tool with the curvature pointing down at the side between the cluster and instrument panel and turn 90 degrees. Pull the hook out until it catches in the recess in the cluster.

4. After the spring clips release, pull the cluster out far enough to disconnect the wiring and speedometer cable.

5. Installation is the reverse of the removal procedure.

Speedometer

After removing the instrument cluster, disassembly the cluster and remove the speedometer assembly from the cluster.

Combination Switch

REMOVAL & INSTALLATION

190D, 190E, 260E, 300D, 300E, 300CE, 300SE, 300TE, 300SEL, 300TD and 560SL

1. Disconnect the negative battery cable. Remove the rubber sleeve on the switch and unscrew the retaining screws.

2. Pull the switch out slightly, loosen the screws for the cable connection of the twin carbon contacts and pull out the cable.

3. Remove the cover under the left side of the instrument panel.

4. Disconnect the plug and remove the switch.

5. Installation is in the reverse order of removal.

300SL, 350SD, 420SEL, 500SL, 560SEC and 560SEL

1. Disconnect the negative battery cable. Remove the steering wheel.

2. Remove the cover under the left side of the instrument panel.

3. Unscrew the switch retaining screws.

4. Disconnect the 14-prong plug under the instrument panel.

5. Remove the switch.

6. Installation is in the reverse order of removal.

Ignition Switch

REMOVAL & INSTALLATION

With Ignition Switch In Dashboard

EXCEPT 190D, 190E, 260E, 300E, 300CE, 300SE, 300TE, 300SEL AND 500E

1. Disconnect the negative battery cable. Remove the instrument cluster.

2. Remove the right cover plate from under the dashboard.

3. Remove the plug connection from the ignition switch.

4. Remove the ignition switch-to-lock cylinder screws and the switch.

To install:

5. Attach the plug connection, after fastening the switch to the steering lock.

6. Install the instrument cluster.

7. Check the switch for proper function and install the lower cover.

190D, 190E, 260E AND 300E

1. Disconnect the negative battery cable. Remove the cover plate under the left side of the instrument panel.

2. Remove the steering wheel. Remove the instrument cluster.

3. Pry the cylinder rosette (trim ring) upwards and remove it.

4. Insert the ignition key and turn it to position 1.

5. Disconnect the plug at the rear of the ignition switch.

NOTE: The plug can only be disconnected when the key is in position No. 1.

6. Loosen the screws and remove the steering column jacket (upper and lower halves).

7. Release the clamp on the jacket tube. Press in the lock-pin in position 1 and pull the steering lock out slightly from the jacket tube holder.

8. Pull off the ignition key at the right bottom section, slightly to the rear. Swivel the steering lock so the lock cylinder clears its hole in the instrument panel.

9. Unscrew the retaining screws and remove the ignition switch from the back of the steering lock.

10. Installation is in the reverse order of removal. Remember to reconnect the switch to the steering lock.

Lock Cylinder

REMOVAL & INSTALLATION

Key Can Be Removed In Position No. 1

1. Disconnect the negative battery cable. Turn the key to position **1** and remove the key.

2. Pry the cover sleeve from the lock cylinder with a small prybar.

3. Using a bent paper clip, hook onto the cover sleeve and remove the sleeve. Do not remove the rosette in the dashboard.

4. Insert the paper clip between the rosette and the steering lock and push in the lock pin. Remove the lock cylinder slightly with the key.

5. Insert the paper clip into the locking hole and pull the lock cylinder completely out.

6. Installation is the reverse or removal. Turn the lock cylinder to position **1** and insert it into the steering lock, make sure the lock pin engages. Push the cover sleeve into position **1**.

7. Make sure the cylinder operates properly.

Key Can Not Be Removed in Position No. 1

EXCEPT 190D, 190E, 260E, 300E, 300CE, 300SE, 300TE, 300SEL, 500E

Because of legal requirements, the lock was changed from the previous version, so the key can only be removed in position **0**.

1. Disconnect the negative battery cable. Turn the key to position **1**.

2. Lift the cover sleeve to the edge of the key and turn the key to position **0**.

3. Remove the key and cover sleeve.

4. Insert the key into the lock cylinder and turn to position **1** (90 degrees to the right), push in the lock pin and remove the lock cylinder.

To install:

5. Turn the lock cylinder to position **1** and insert the lock cylinder, make sure the locking pin engages.

6. Turn the key to position **0** and remove the key.

7. Place the cover sleeve on the steering lock, insert and turn the key and push in the cover sleeve in position **1**.

8. Check the locking cylinder for proper function.

190D, 190E, 260E, 300E, 300CE, 300SE, 300TE, 300SEL AND 500SL

1. Disconnect the negative battery cable. Pry the cylinder rosette (trim ring) upwards and remove it.

2. Insert the ignition key and turn it to position **1**.

3. Using a bent paper clip, insert each end into the holes on either side of the lock cylinder. Press the clip ends inward; the pressure will unlock the cylinder from the steering lock.

4. Grasp the key and with pressure still on the paper clip, pull the ignition

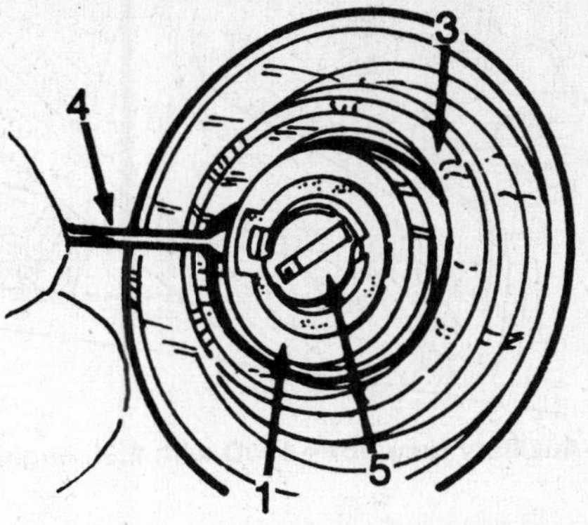

1. Steering lock 4. Steel wire (paper clip)
3. Rosette 5. Locking cylinder

Ignition lock cylinder removal from the instrument panel (both types)

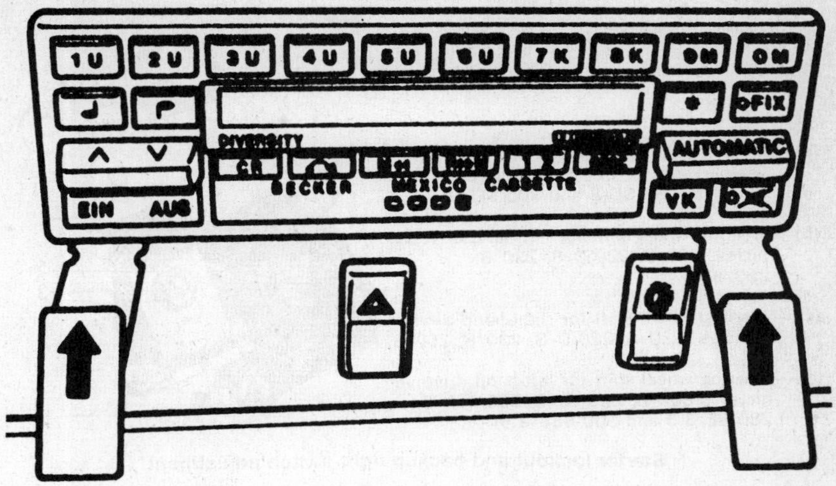

Radio removal with special tool

key/lock cylinder assembly from the steering lock.

5. Remove the paper clip, turn the key to position **0** and remove it. Slide the lock cylinder from the cover.

To install:

6. Insert the lock cylinder just enough so the ridge on the cylinder body engages the groove in the steering lock.

7. Slide the cover onto the lock cylinder so the detent is on the left side.

8. Insert the ignition key, turn it to position **1**, push the lock cylinder and its cover into the steering lock.

NOTE: When the ignition key is in position 1 and is aligned with the mark on the cover, the detent on the cover is also aligned with the ridge on the steering lock. This is the only manner in which the lock cylinder/cover can be installed in the steering lock.

9. Check that the lock cylinder functions properly, if so, install the rosette.

Radio

REMOVAL & INSTALLATION

190E and 190D

1. Disconnect the negative battery cable.
2. Remove the 2 screws holding in the ashtray.
3. Disconnect the wiring and remove the radio.
4. Installation is the reverse of removal.

260E, 300E, 300D, 300CE, 300TE, 300SE, 300SEL, 350SD, 420SEL, 500E, 560SEC, 560SL and 560SEL,

1. Disconnect the negative battery cable. Remove the ashtray.
2. Press down on the clamping levers underneath the radio.
3. Slide the radio out of the bracket and disconnect the wiring harness.
4. Installation is the reverse of removal.

300SL, 500SL and 560SL

1. Disconnect the negative battery cable.
2. Remove the wooden bezel and ashtray (560SL only). Insert special tool 129589000500 or equivalent under the radio cover up to the marks. Pull out and push in with the tool to dislodge the radio.
3. Pull the radio from the instrument panel and disconnect the wiring harness.
4. Installation is the reverse of removal.

Headlight Switch

REMOVAL & INSTALLATION

1. Disconnect the negative battery cable.
2. Remove the knob and instrument panel undercover.
3. Remove the nut and turn the switch to the left and withdraw.
4. Disconnect the wiring.

To install:

5. Install the switch and turn to the right.
6. Install the nut, connect the wiring and install the remaining components.

Stoplight Switch

ADJUSTMENT

1. Press down on the brake pedal.
2. Install the switch and turn until the locating lug engages.
3. Release the brake pedal. The actuation travel is automatically adjusted.

REMOVAL & INSTALLATION

1. Disconnect the negative battery cable. Disconnect electrical connector from stoplight switch. Remove the instrument panel undercover.
2. Press on the locating lug of the switch. Turn and pull the switch out to obtain the maximum travel.
3. Remove stoplight switch.

To install:

4. Install and adjust the switch.
5. Connect the electrical connectors and battery cable.

Starter Lockout and Back-Up Light Switch

ADJUSTMENT

1. Disconnect the selector rod and move the shift selector, on the transmission, to position **N**.
2. Tighten the clamping screw prior to making adjustments.
3. Loosen the adjusting screw and insert the locating pin through the driver into the locating hole in the shift housing.
4. Tighten the adjusting screw and remove the locating pin.
5. Move the selector lever to position **N** and connect the selector rod so there is no tension.
6. Make sure the engine cannot be started in **N** or **P**.

1. Selector range lever
2. Washer
3. Adjusting screw
4. Shaft
5. Locating pin
6. Clamping screw

(a)—Column shift for left-hand and right-hand drive vehicles 220/8, 220 D/8, 230/8, 280 S/8, 280 SE/8 and 300 SEL/8.

(b)—Steering wheel shift for left-hand drive vehicles (220/8, 220 D/8, 230/8, 250/8)

(c)—Steering wheel shift for right-hand drive vehicles (220/8, 220 D/8, 230/8, 250/8)

(d)—Steering wheel shift for left-hand drive vehicles (280S/8, 280 SE/8, 300 SEL/8, 280 SE/3.5 and 300 SEL/3.5)

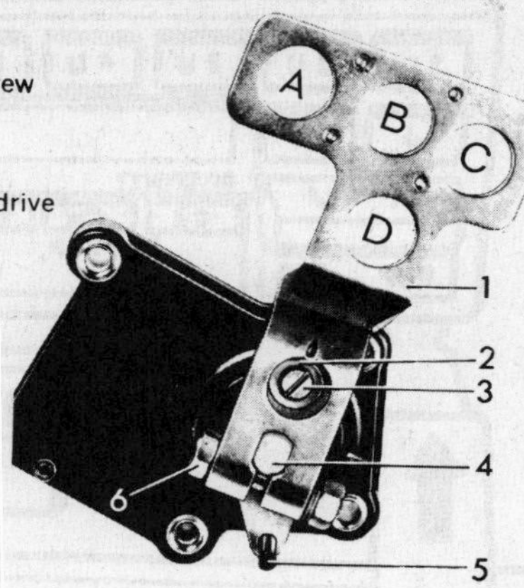

Starter lockout and backup light switch adjustment

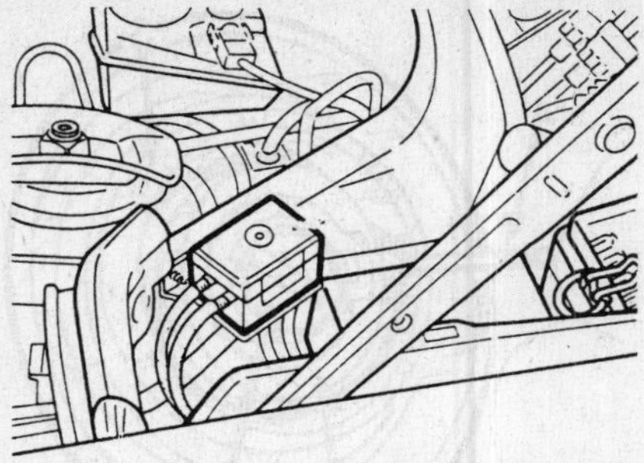

Auxiliary fuse box—190D with 2.5L engine

Fuses

A listing of the protected equipment and the amperage of the fuse is printed in the lid of the fuse box. Spare fuses and a tool for removing and installing fuses are contained in the vehicle's tool kit.

LOCATION

560SL

The fuse and relay box is located at the passenger's side kickpanel.

300D and 300TD

The fuse box is located in the engine compartment on the driver's side, next the brake master cylinder. Some models have separate fuse boxes or inline fuses for additional equipment. The radio is usually fused with a separate inline glass fuse behind the radio and the ignition is unfused.

190D, 190E, 260E, 300E, 300CE, 300SE, 300TE, 300SEL, 300SD, 420SEL, 560SEC, 560SEL and 560SEC

The fuse box is located in the engine compartment, on the driver's side, next to the brake master cylinder. Some models may have separate fuse boxes or inline fuses in the engine compartment for additional equipment. The radio is usually fused with a separate inline glass fuse behind the radio and the ignition is unfused. The fuse box also contains various relays.

An auxiliary heater fuse holder is located in the relay block in section (K) for the 126 models.

300SL and 500SL

The main fuse block is located next to the power brake booster. The 1st auxiliary fuse block (F19) is located behind the power brake booster. The 2nd auxiliary fuse block (F20) is located in the trunk.

Circuit Breakers

LOCATION

107 Models

An overvoltage protection relay is located in the number 5 section of the fuse/relay block, right kickpanel.

124 Models

5, 7 and 9 pole circuit breakers are located in a block behind the battery (K1, K1/1 and K1/2).

126 Models

5 pole circuit breaker is located in the (K1) section, left side firewall.

201 Models

5 pole circuit breaker is located in the (R4) section of the fuse/relay block.

Mitsubishi

Cordia, Diamante, Eclipse, Galant, Mirage, Precis, Sigma, Starion, Tredia, 3000-GT—All Models

14

SERIAL NUMBER IDENTIFICATION

Vehicle Identification Plate

The Vehicle Identification Number (VIN) is mounted at the left side of the dash, visible through the windsheild, on all models except the Precis. The Precis VIN is located at the center of the firewall in the engine compartment.

All models except the Precis use a standard 17 digit VIN code. The eighth digit identifies the engine:

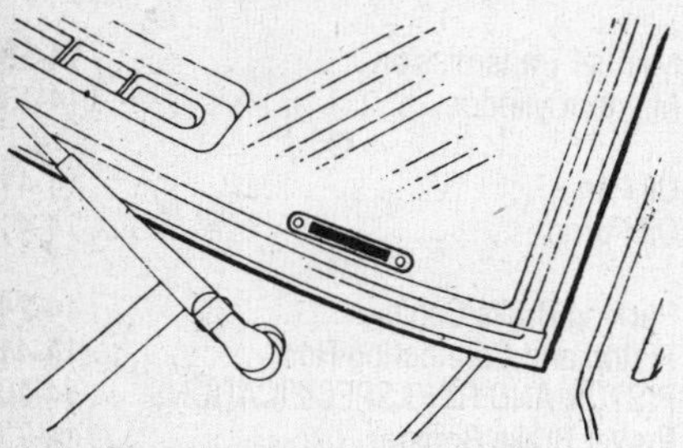

VIN location—except Precis

A—1.5L (92 cu. in.) SOHC
B—3.0L (181.4 cu. in.) DOHC
C—3.0L (181.4 cu. in.) DOHC Turbo
R—2.0L (122 cu. in.) DOHC
T—1.8L (107 cu. in.) SOHC
U—2.0L (122 cu. in.) DOHC Turbo
V—2.0L (122 cu. in.) SOHC
Y—1.6L (98 cu. in.) DOHC

The tenth digit identifies the model year:

I—1988
J—1989
K—1990
L—1991
M—1992

A vehicle information code plate is riveted onto the front of the right side wheel house or onto the firewall, depending on vehicle. The plate shows vehicle code, engine vehicle, transaxle vehicle and body color code.

Chassis Number

A chassis number plate is located on the top center of the firewall in the engine compartment.

Engine Number

The engine vehicle and serial numbers in all cases are stamped on the block near the front of the engine. In most cases, they are located on the right side.

Transaxle Number

The transaxle identification number is located below the engine number on the vehicle information code plate.

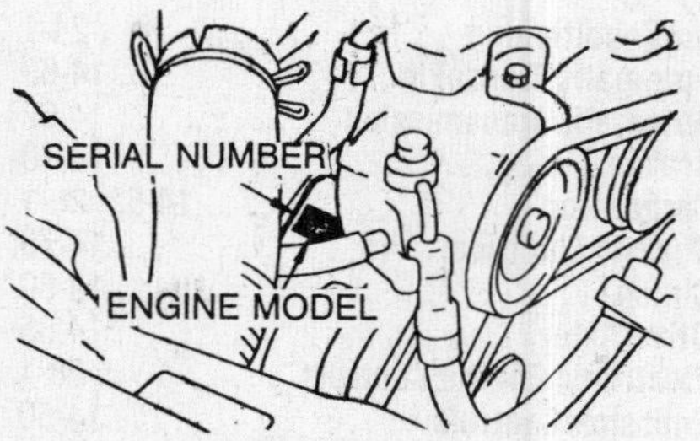

Engine number location

SPECIFICATIONS

ENGINE IDENTIFICATION

Year	Model	Engine Displacement cu. in. (cc/liter)	Engine Series Identification	No. of Cylinders	Engine Type
1988	Cordia	109.5 (1795/1.8)	G62B	4	SOHC
	Cordia	121.9 (1997/2.0)	G63B	4	SOHC
	Tredia	109.5 (1795/1.8)	G62B	4	SOHC
	Tredia	121.9 (1997/2.0)	G63B	4	SOHC
	Starion	155.9 (2555/2.5)	G54B	4	SOHC
	Mirage	89.6 (1468/1.5)	G15B	4	SOHC
	Mirage	97.4 (1597/1.6)	G32B	4	SOHC
	Galant	143.4 (2350/2.3)	G64B	4	SOHC
	Galant	181.4 (2972/3.0)	6G72	6	SOHC
	Precis	89.6 (1468/1.5)	G15B	4	SOHC
1989	Starion	155.9 (2555/2.5)	G64B	4	SOHC
	Mirage	89.6 (1468/1.5)	4G15	4	SOHC
	Mirage	97.4 (1597/1.6)	4G61	4	DOHC
	Galant	122 (1997/2.0)	4G63	4	SOHC & DOHC
	Sigma	181.4 (2972/3.0)	6G72	6	SOHC
	Precis	89.6 (1468/1.5)	G15B	4	SOHC
1990	Mirage	89.6 (1468/1.5)	4G15	4	SOHC
	Mirage	97.3 (1595/1.6)	4G61	4	DOHC
	Galant	122 (1997/2.0)	4G63	4	SOHC & DOHC
	Sigma	181.4 (2972/3.0)	6G72	6	SOHC
	Precis	89.6 (1468/1.5)	G15B	4	SOHC

ENGINE IDENTIFICATION

Year	Model	Engine Displacement cu. in. (cc/liter)	Engine Series Identification	No. of Cylinders	Engine Type
1990	Eclipse	107 (1755/1.8)	4G37	4	SOHC
	Eclipse	122 (1997/2.0)	4G63	4	DOHC
1991-92	Mirage	89.6 (1468/1.5)	4G15	4	SOHC
	Mirage	97.3 (1595/1.6)	4G61	4	DOHC
	Galant	122 (1997/2.0)	4G63	4	SOHC & DOHC
	Precis	89.6 (1468/1.5)	G15B	4	SOHC
	Eclipse	107 (1755/1.8)	4G37	4	SOHC
	Eclipse	122 (1997/2.0)	4G63	4	DOHC
	3000GT	181.4 (2972/3.0)	6G72	6	DOHC
	Diamante	181.4 (2972/3.0)	6G72	6	SOHC & DOHC

SOHC—Single Overhead Camshaft
DOHC—Double Overhead Camshaft

GENERAL ENGINE SPECIFICATIONS

Year	Model	Engine Displacement cu. in. (cc)	Fuel System Type	Net Horsepower @ rpm	Net Torque @ rpm (ft. lbs.)	Bore × Stroke (in.)	Com-pression Ratio	Oil Pressure @ rpm
1988	Cordia	109.5 (1795)	ECI	116 @ 5500	129 @ 3000	3.17 × 3.46	7.5:1	63 ①
	Cordia	121.9 (1997)	Carb.	88 @ 5000	108 @ 3500	3.35 × 3.46	8.5:1	63 ①
	Tredia	109.5 (1795)	ECI	116 @ 5500	129 @ 3000	3.17 × 3.46	7.5:1	63 ①
	Tredia	121.9 (1997)	Carb.	88 @ 5000	108 @ 3500	3.35 × 3.46	8.5:1	63 ①
	Starion	155.9 (2555)	ECI	145 @ 5000	185 @ 2500	3.59 × 3.86	7.0:1	63 ①
	Mirage	89.6 (1468)	Carb.	68 @ 5000	82 @ 3500	2.97 × 3.23	9.4:1	63 ①
	Mirage	97.4 (1597)	ECI	102 @ 5500	122 @ 3000	3.03 × 3.39	7.6:1	63 ①
	Galant	143.4 (2350)	MPI	110 @ 4500	138 @ 3500	3.41 × 3.94	8.5:1	63 ①
	Galant	181.1 (2972)	MPI	142 @ 5000	168 @ 2500	3.59 × 2.99	8.9:1	63 ①
	Precis	89.6 (1468)	Carb.	68 @ 5500	82 @ 3500	2.97 × 3.23	9.4:1	63 ①
1989	Starion	155.9 (2555)	ECI	145 @ 5000	185 @ 2500	3.59 × 3.86	7.0:1	63 ①
	Mirage	86.6 (1468)	Carb.	68 @ 5000	82 @ 3500	2.97 × 3.23	9.4:1	63 ①
	Mirage	97.4 (1597)	ECI	102 @ 5500	122 @ 3000	3.03 × 3.39	7.6:1	63 ①
	Galant	122 (1997) ②	MPI	120 @ 5000	116 @ 4500	3.35 × 3.46	8.5:1	11.4 @ 750
	Galant	122 (1997) ③	MPI	135 @ 6000	125 @ 5000	3.35 × 3.46	9.0:1	11.4 @ 750
	Sigma	181.1 (2972)	MPI	142 @ 5000	168 @ 2500	3.59 × 2.99	8.9:1	63 ①
	Precis	89.6 (1468)	Carb.	68 @ 5500	82 @ 3500	2.97 × 3.23	9.4:1	63 ①
1990	Mirage	89.6 (1468)	MPI	81 @ 5500	91 @ 3000	2.97 × 3.23	9.4:1	11.4 @ 750
	Mirage	97.3 (1595)	MPI	113 @ 6500	99 @ 5000	3.24 × 2.95	9.2:1	11.4 @ 750
	Galant	122 (1997) ②	MPI	120 @ 5000	116 @ 4500	3.35 × 3.46	8.5:1	11.4 @ 750
	Galant	122 (1997) ③	MPI	135 @ 6000	125 @ 5000	3.35 × 3.46	9.0:1	11.4 @ 750
	Sigma	181.4 (2972)	MPI	142 @ 5000	168 @ 2500	3.59 × 2.99	8.9:1	11.4 @ 750
	Precis	89.6 (1468)	MPI	81 @ 5500	91 @ 3000	2.97 × 3.23	9.4:1	11.4 @ 750
	Eclipse	107 (1755)	MPI	92 @ 5000	105 @ 3500	3.17 × 3.39	9.0:1	11.4 @ 750
	Eclipse	122 (1997)	MPI	135 @ 6000	125 @ 3000	3.35 × 3.46	9.0:1	11.4 @ 750
	Eclipse	122 (1997)	MPI ⑥	190 @ 6000 ⑤	203 @ 3000	3.35 × 3.46	7.8:1	11.4 @ 750

GENERAL ENGINE SPECIFICATIONS

Year	Model	Engine Displacement cu. in. (cc)	Fuel System Type	Net Horsepower @ rpm	Net Torque @ rpm (ft. lbs.)	Bore × Stroke (in.)	Com-pression Ratio	Oil Pressure @ rpm
1991–92	Mirage	89.6 (1468)	MPI	81 @ 5500	91 @ 3000	2.97 × 3.23	9.4:1	11.4 @ 750
	Mirage	97.3 (1595)	MPI	113 @ 6500	99 @ 5000	3.24 × 2.95	9.2:1	11.4 @ 750
	Galant	122 (1997) ②	MPI	120 @ 5000	116 @ 4500	3.35 × 3.46	8.5:1	11.4 @ 750
	Galant	122 (1997) ③	MPI	135 @ 6000	125 @ 5000	3.35 × 3.46	9.0:1	11.4 @ 750
	Precis	89.6 (1468)	MPI	81 @ 5500	91 @ 3000	2.97 × 3.23	9.4:1	11.4 @ 750
	Eclipse	107 (1755)	MPI	92 @ 5000	105 @ 3500	3.17 × 3.39	9.0:1	11.4 @ 750
	Eclipse	122 (1997)	MPI	135 @ 6000	125 @ 3000	3.35 × 3.46	9.0:1	11.4 @ 750
	Eclipse	122 (1997)	MPI ④	190 @ 6000 ⑤	203 @ 3000	3.35 × 3.46	7.8:1	11.4 @ 750
	3000GT	181.4 (2972)	MPI ③	222 @ 6000	201 @ 4500	3.59 × 2.99	10.0:1	11.4 @ 750
	3000GT	181.4 (2972)	MPI ③④	300 @ 6000	307 @ 2500	3.59 × 2.99	8.0:1	11.4 @ 750
	Diamante	181.4 (2972)	MPI ②	175 @ 5500	185 @ 5500	3.59 × 2.99	10.0:1	11.4 @ 750
	Diamante	181.4 (2972)	MPI ③	202 @ 6000	199 @ 3000	3.59 × 2.99	10.0:1	11.4 @ 750

MPI—Multi-Point Injection
ECI—Electronic Controlled Injection
Carb.—Carbureted
① Relief valve opening pressure
② Single overhead camshaft
③ Double overhead camshaft
④ Turbocharged
⑤ GSX model-195

TUNE-UP SPECIFICATIONS

Year	Model	Engine Displacement cu. in. (cc)	Spark Plugs Type	Spark Plugs Gap (in.)	Ignition Timing (deg.) MT	Ignition Timing (deg.) AT	Com-pression Pressure (psi)	Fuel Pump (psi)	Idle Speed (rpm) MT	Idle Speed (rpm) AT	Valve Clearance ① In.	Valve Clearance ① Ex.
1988	Cordia	109.5 (1795)	BUR7EZ-11	0.035–0.039	5B	5B	170 ③	35–47	750	750	0.006	0.010
	Cordia	121.9 (1997)	BPR6ES-11	0.035–0.039	5B	5B	170 ③	2.4–3.4	700 ②	750 ②	Hyd.	Hyd.
	Tredia	109.5 (1795)	BUR7EZ-11	0.035–0.039	5B	5B	170 ③	35–47	750	750	0.006	0.010
	Tredia	121.9 (1997)	BPR6ES-11	0.035–0.039	5B	5B	170 ③	2.4–3.4	700 ②	750 ②	Hyd.	Hyd.
	Starion	155.9 (2555)	BP6ES-11	0.039–0.043	10B	10B	170 ③	35–47	750	850	0.006	0.010
	Mirage	89.6 (1468)	W20EP-10	0.039–0.043	3B	3B	170 ③	—	700	750	0.006	0.010
	Mirage	97.4 (1597)	BUR7EZ-11	0.035–0.039	8B	8B	170 ③	36	700	—	0.006	0.010
	Galant	143.4 (2350)	BPR6ES-11	0.039–0.043	—	5B	170 ③	36	—	750	Hyd.	Hyd.
	Galant	181.4 (2972)	PGR5A-11	0.039–0.043	5B	5B	170 ③	38	700	700	Hyd.	Hyd.
	Precis	89.6 (1468)	W20EP-10	0.039–0.043	3B	3B	164 ③	—	700	750	0.006	0.010
1989	Starion	155.9 (2555)	BP6ES-11	0.039–0.043	10B	10B	170 ③	35–47	750	850	0.006	0.010
	Mirage	89.6 (1468)	W20EP-10	0.039–0.043	3B	3B	170 ③	—	700	750	0.006	0.010
	Mirage	97.4 (1597)	BUR7EZ-11	0.035–0.039	8B	8B	170 ③	36	700	—	0.006	0.010
	Galant	122 (1997)	BPR6ES-11	0.039–0.043	5	5B	125	47–50	750	750	Hyd.	Hyd.
	Galant	122 (1997)	BPR6ES-11	0.039–0.043	5	5B	137	47–50	750	750	Hyd.	Hyd.
	Sigma	181.4 (2972)	PGR5A-11	0.039–0.043	5B	5B	170 ③	38	700	700	Hyd.	Hyd.
	Precis	89.6 (1468)	W20EP-10	0.039–0.043	5B	5B	164 ③	2.8–3.6	800	800	0.006	0.010

TUNE-UP SPECIFICATIONS

Year	Model	Engine Displacement cu. in. (cc)	Spark Plugs Type	Spark Plugs Gap (in.)	Ignition Timing (deg.) MT	Ignition Timing (deg.) AT	Compression Pressure (psi)	Fuel Pump (psi)	Idle Speed (rpm) MT	Idle Speed (rpm) AT	Valve Clearance ① In.	Valve Clearance ① Ex.
1990	Eclipse	107 (1755)	BPR6ES-11	0.039–0.043	5B	5B	131	47–50	700	700	Hyd.	Hyd.
	Eclipse	122 (1997)	BPR6ES	0.028–0.031	5B	5B	137	47–50	700	700	Hyd.	Hyd.
	Eclipse	122 (1997)	BPR6ES-11	0.039–0.043	5B	5B	114	36–38	700	700	Hyd.	Hyd.
	Mirage	89.6 (1468)	BPR6ES-11	0.040–0.043	5B	5B	137	47–50	750	750	0.006	0.010
	Mirage	97.3 (1595)	BPR6ES	0.028–0.031	8B	8B	170 ③	47–50	750	750	Hyd.	Hyd.
	Galant	122 (1997)	BPR6ES-11	0.039–0.043	5B	5B	125	47–50	750	750	Hyd.	Hyd.
	Galant	122 (1997)	BPR6ES-11	0.039–0.043	5	5B	137	47–50	750	750	Hyd.	Hyd.
	Sigma	181.4 (2972)	PGR5A-11	0.039–0.043	—	5B	119	47–53	—	700	Hyd.	Hyd.
	Precis	89.6 (1468)	W20EP-10	0.039–0.043	5B	5B	164 ③	—	800	800	0.006	0.010
1991	Eclipse	107 (1755)	BPR6ES-11	0.039–0.043	5B	5B	131	47–50	700	700	Hyd.	Hyd.
	Eclipse	122 (1997)	BPR6ES	0.028–0.031	5B	5B	137	47–50	700	700	Hyd.	Hyd.
	Eclipse	122 (1997)	BPR6ES-11	0.039–0.043	5B	5B	114	36–38	700	700	Hyd.	Hyd.
	Mirage	89.6 (1468)	BPR6ES-11	0.040–0.043	5B	5B	137	47–50	750	750	0.006	0.010
	Mirage	97.3 (1595)	BPR6ES	0.028–0.031	8B	8B	170 ③	47–50	750	750	Hyd.	Hyd.
	Galant	122 (1997)	BPR6ES-11	0.039–0.043	5B	5B	125	47–50	750	750	Hyd.	Hyd.
	Galant	122 (1997)	BPR6ES-11	0.039–0.043	5B	5B	137	47–50	750	750	Hyd.	Hyd.
	Precis	89.6 (1468)	W20EP-10	0.039–0.043	5B	5B	164 ③	—	800	800	0.006	0.010
	3000GT	181.4 (2972)	PFR6J-11	0.039–0.043	5B	5B	139 ③	47–50	750	750	Hyd.	Hyd.
	3000GT	181.4 (2972)	PFR6J-11	0.039–0.043	5B	5B	115	43–45	750	750	Hyd.	Hyd.
1992	SEE UNDERHOOD SPECIFICATIONS STICKER											

① Jet Valve—0.010
② With air conditioning
Manual transaxle—750
Automatic transaxle—850
③ Standard Valve—136 Limit

FIRING ORDER

NOTE: To avoid confusion, always replace spark plug wires one at a time.

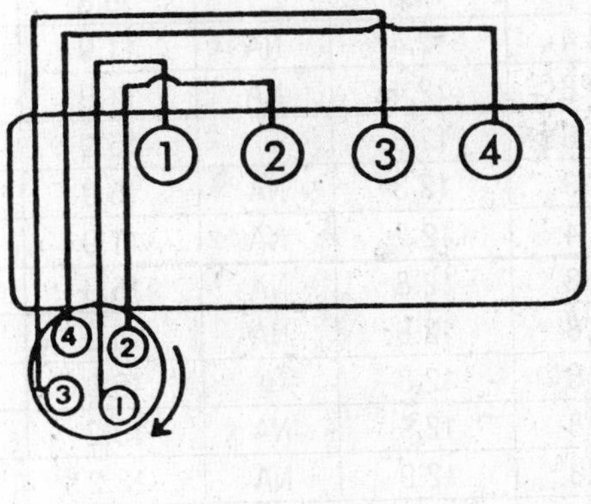

1.5L, 1.6L, 1.8L, 2.0L, 2.3L and 2.5L
Engine Firing Order: 1–3–4–2
Distributor Rotation: Clockwise

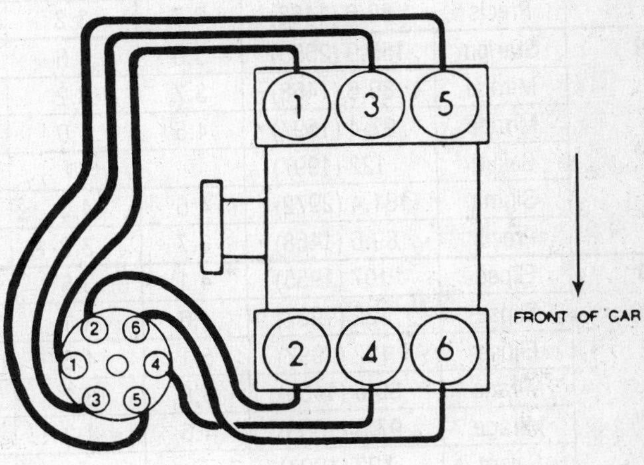

3.0L SOHC Engine
Engine Firing Order: 1–2–3–4–5–6
Distributor Rotation: Counterclockwise

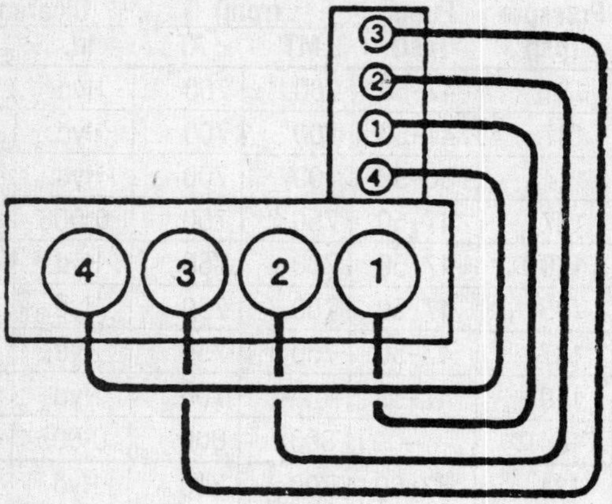

2.0L DOHC Engine
Engine Firing Order: 1–3–4–2
Distributorless Ignition System

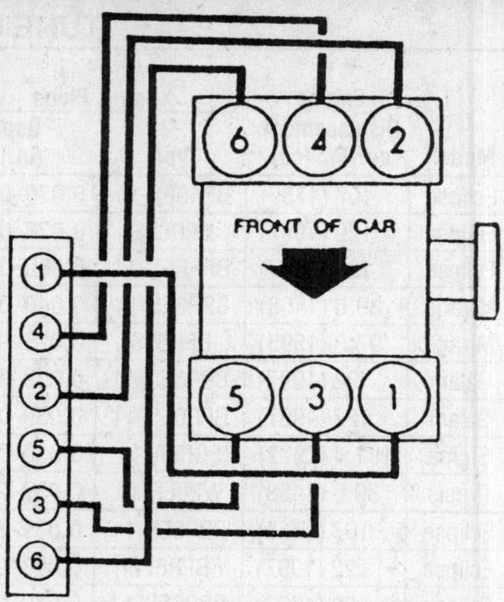

3.0L DOHC Engine
Engine Firing Order: 1–2–3–4–5–6
Distributorless Ignition System

CAPACITIES

Year	Model	Engine Displacement cu. in. (cc)	Engine Crankcase (qts.) with Filter	without Filter	Transmission (pts) 4-Spd	5-Spd	Auto.	Drive Axle (pts.)	Fuel Tank (gal.)	Cooling System (qts.)
1988	Cordia	109.5 (1795)	4.5	4.0	4.4	4.4	12.2	NA	13.2	7.4
	Cordia	121.9 (1997)	4.5	4.0	4.4	4.4	12.2	NA	13.2	7.4
	Tredia	109.5 (1795)	4.5	4.0	4.4	4.4	12.2	NA	13.2	7.4
	Tredia	121.9 (1997)	4.5	4.0	4.4	4.4	12.2	NA	13.2	7.4
	Starion	155.9 (2555)	5.0	4.5	4.8	—	14.8	2.7	19.8	9.7
	Mirage	89.6 (1468)	3.7	3.2	—	4.4	12.4	NA	11.9	5.3
	Mirage	97.4 (1597)	4.5	4.0	—	4.8	12.4	NA	11.9	5.3
	Galant	143.4 (2350)	4.5	4.0	—	—	12.4	NA	15.9	7.4
	Galant	181.4 (2972)	4.5	4.0	—	5.3	12.3	NA	15.9	9.7
	Precis	89.6 (1468)	3.7	3.2	—	4.4	—	NA	11.9	5.3
1989	Starion	155.9 (2555)	5.0	4.5	4.8	—	14.8	2.7	19.8	9.7
	Mirage	89.6 (1468)	3.7	3.2	—	4.4	12.4	NA	11.9	5.3
	Mirage	97.4 (1597)	4.5	4.0	—	4.8	12.4	NA	11.9	5.3
	Galant	122 (1997)	①	①	—	3.8 ②	12.8	④	15.9 ③	7.6
	Sigma	181.4 (2972)	4.5	4.0	—	5.3	12.3	NA	15.9	9.7
	Precis	89.6 (1468)	3.7	3.2	—	4.4	12.2	NA	11.9	5.3
1990	Elipse	107 (1955)	4.1	3.6	—	3.8	12.8	NA	15.9	6.6
	Elipse	122 (1997)	4.6	4.1	—	3.8	12.8	NA	15.9	7.6
	Elipse	122 (1997)	5.1	4.6	—	4.6 ②	12.8	④	15.9	7.6
	Mirage	89.6 (1468)	3.6	3.2	—	3.8	12.9	NA	13.2	5.3
	Mirage	97.3 (1595)	4.6	4.2	—	3.8	12.9	NA	13.2	5.3
	Galant	122 (1997)	①	①	—	3.8 ②	12.8	④	15.9 ③	7.6
	Sigma	181.4 (2972)	4.5	4.0	—	—	12.2	NA	15.9	9.7
	Precis	89.6 (1468)	3.7	3.2	—	4.4	12.2	NA	11.9	5.3

CAPACITIES

Year	Model	Engine Displacement cu. in. (cc)	Engine Crankcase (qts.) with Filter	Engine Crankcase (qts.) without Filter	Transmission (pts) 4-Spd	Transmission (pts) 5-Spd	Transmission (pts) Auto.	Drive Axle (pts.)	Fuel Tank (gal.)	Cooling System (qts.)
1991-92	Elipse	107 (1955)	4.1	3.6	—	3.8	12.8	NA	15.9	6.6
	Elipse	122 (1997)	4.6	4.1	—	3.8	12.8	NA	15.9	7.6
	Elipse	122 (1997)	5.1	4.6	—	4.6 ②	12.8	④	15.9	7.6
	Mirage	89.6 (1468)	3.6	3.2	—	3.8	12.9	NA	13.2	5.3
	Mirage	97.3 (1595)	4.6	4.2	—	3.8	12.9	NA	13.2	5.3
	Galant	122 (1997)	①	①	—	3.8 ②	12.8	④	15.9 ③	7.6
	Precis	89.6 (1468)	3.7	3.2	—	4.4	12.2	NA	11.9	5.3
	3000GT	181.4 (2972)	4.7	4.2 ⑤	—	4.8	15.8	⑥	19.8	8.5

① SOHC
Without filter—3.6 qts.
With filter—4.1 qts.
DOHC
Without filter—4.1 qts.
With filter—4.6 qts.

② 2WD; 4.8 on 4WD
③ 2WD; 16.4 on 4WD
④ 4WD model—transfer—1.26; Rear axle—1.48
⑤ Turbo 5.2 with filter
4.7 without filter
⑥ AWD model—transfer .58; rear axle 2.32

CRANKSHAFT AND CONNECTING ROD SPECIFICATIONS

All measurements are given in inches.

Year	Engine Displacement cu. in. (cc)	Crankshaft Main Brg. Journal Dia.	Crankshaft Main Brg. Oil Clearance	Crankshaft Shaft End-play	Crankshaft Thrust on No.	Connecting Rod Journal Diameter	Connecting Rod Oil Clearance	Connecting Rod Side Clearance
1988	109.5 (1795)	2.244	0.0008–0.0020	0.0020–0.0071	3	1.772	0.0008–0.0020	0.004–0.010
	121.9 (1997)	2.244	0.0008–0.0020	0.0020–0.0071	3	1.772	0.0008–0.0020	0.004–0.010
	155.9 (2555)	2.362	0.0008–0.0020	0.0020–0.0071	3	2.087	0.0008–0.0024	0.004–0.010
	89.6 (1468)	1.889	0.0008–0.0020	0.0020–0.0071	3	1.653	0.0004–0.0024	0.004–0.010
	97.4 (1597)	2.244	0.0008–0.0020	0.0020–0.0071	3	1.772	0.0004–0.0024	0.004–0.010
	143.4 (2350)	2.244	0.0008–0.0020	0.0020–0.0071	3	2.087	0.0008–0.0024	0.004–0.010
	181.4 (2972)	2.362	0.0008–0.0019	0.0020–0.0098	3	1.969	0.0006–0.0018	0.008–0.016
1989	121.9 (1997)	2.244	0.0008–0.0020	0.0020–0.0071	3	1.772	0.0008–0.0020	0.004–0.010
	155.9 (2555)	2.362	0.0008–0.0020	0.0020–0.0071	3	2.087	0.0008–0.0024	0.004–0.010
	89.6 (1468)	1.889	0.0008–0.0020	0.0020–0.0071	3	1.653	0.0004–0.0024	0.004–0.010
	97.4 (1597)	2.244	0.0008–0.0020	0.0020–0.0071	3	1.772	0.0008–0.0020	0.004–0.010
	181.4 (2972)	2.362	0.0008–0.0019	0.0020–0.0981	3	1.969	0.0006–0.0018	0.008–0.016
	181.4 (2972)	2.362	0.0008–0.0019	0.0020–0.0098	3	1.969	0.0006–0.0018	0.008–0.016

CRANKSHAFT AND CONNECTING ROD SPECIFICATIONS

All measurements are given in inches.

Year	Engine Displacement cu. in. (cc)	Crankshaft Main Brg. Journal Dia.	Main Brg. Oil Clearance	Shaft End-play	Thrust on No.	Connecting Rod Journal Diameter	Oil Clearance	Side Clearance
1990	107 (1755)	2.244	0.0008–0.0020	0.0020–0.0070	3	1.772	0.0008–0.0020	0.004–0.010
	122 (1997)	2.244	0.0008–0.0020	0.0020–0.0070	3	1.772	0.0008–0.0010	0.004–0.010
	121.9 (1997)	2.244	0.0008–0.0020	0.0020–0.0070	3	1.772	0.0008–0.0020	0.004–0.010
	89.6 (1468)	1.889	0.0008–0.0018	0.0020–0.0071	3	1.653	0.0006–0.0017	0.004–0.010
	97.3 (1595)	2.244	0.0008–0.0020	0.0020–0.0071	3	1.772	0.0008–0.0020	0.004–0.010
	181.4 (2972)	2.358	0.0008–0.0019	0.0020–0.0098	3	1.965	0.0006–0.0018	0.004–0.010
1991–92	107 (1755)	2.244	0.0008–0.0020	0.0020–0.0070	3	1.772	0.0008–0.0020	0.004–0.010
	122 (1997)	2.244	0.0008–0.0020	0.0020–0.0070	3	1.772	0.0008–0.0010	0.004–0.010
	121.9 (1997)	2.244	0.0008–0.0020	0.0020–0.0070	3	1.772	0.0008–0.0020	0.004–0.010
	89.6 (1468)	1.889	0.0008–0.0018	0.0020–0.0071	3	1.653	0.0006–0.0017	0.004–0.010
	97.3 (1595)	2.244	0.0008–0.0020	0.0020–0.0071	3	1.772	0.0008–0.0020	0.004–0.010
	181.4 (2972)	2.358	0.0007–0.0017	0.0020–0.0098	3	1.965	0.0006–0.0018	0.004–0.010

VALVE SPECIFICATIONS

Year	Engine Displacement cu. in. (cc)	Seat Angle (deg.)	Face Angle (deg.)	Spring Test Pressure (lbs. @ in.)	Spring Installed Height (in.)	Stem-to-Guide Clearance (in.) Intake	Exhaust	Stem Diameter (in.) Intake	Exhaust
1988	109.5 (1795)	45	45	62 @ 1.591	1.591	0.0010–0.0022	0.0020–0.0035	0.315	0.315
	121.9 (1997)	45	45	72 @ 1.591	1.591	0.0010–0.0022	0.0020–0.0035	0.315	0.315
	155.9 (2555)	45	45	72 @ 1.591	1.591	0.0012–0.0024	0.0020–0.0035	0.315	0.315
	89.6 (1468)	45	45	53 @ 1.469	1.417	0.0008–0.0020	0.0020–0.0035	0.315	0.315
	97.4 (1597)	45	45	62 @ 1.469	1.469	0.0012–0.0024	0.0020–0.0035	0.315	0.315
	143.4 (2350)	45	45	72 @ 1.591	1.591	0.0012–0.0024	0.0020–0.0035	0.322	0.315
	181.4 (2972)	44	45	74 @ 1.591	1.591	0.0012–0.0024	0.0020–0.0035	0.314	0.313

VALVE SPECIFICATIONS

Year	Engine Displacement cu. in. (cc)	Seat Angle (deg.)	Face Angle (deg.)	Spring Test Pressure (lbs. @ in.)	Spring Installed Height (in.)	Stem-to-Guide Clearance (in.)		Stem Diameter (in.)	
						Intake	Exhaust	Intake	Exhaust
1989	121.9 (1997)	44	45	72 @ 1.591	1.591	0.0012–0.0024	0.0020–0.0035	0.315	0.315
	155.9 (2555)	45	45	72 @ 1.591	1.591	0.0012–0.0024	0.0020–0.0035	0.315	0.315
	89.6 (1468)	45	45	53 @ 1.469	1.417	0.0008–0.0020	0.0020–0.0035	0.315	0.315
	97.4 (1597)	45	45	62 @ 1.469	1.469	0.0012–0.0024	0.0020–0.0035	0.315	0.315
	143.4 (2350)	45	45	72 @ 1.591	1.591	0.0012–0.0024	0.0020–0.0035	0.322	0.315
	181.4 (2972)	44	45	74 @ 1.591	1.591	0.0012–0.0024	0.0020–0.0035	0.314	0.313
1990	107 (1755)	44	45	62 @ 1.469	1.469	0.0012–0.0024	0.0020–0.0035	0.315	0.315
	122 (1997)	44	45	66 @ 1.575	1.575	0.0008–0.0019	0.0020–0.0033	0.259	0.258
	121.9 (1997)	44	45	72 @ 1.591	1.591	0.0012–0.0024	0.0020–0.0035	0.315	0.315
	89.6 (1468)	44	45	53 @ 1.469	1.417	0.0008–0.0020	0.0020–0.0035	0.260	0.260
	97.3 (1595)	44	45	62 @ 1.469	1.469	0.0008–0.0019	0.0020–0.0033	0.259	0.258
	181.4 (2972)	44	45	74 @ 1.591	1.591	0.0012–0.0024	0.0020–0.0035	0.314	0.314
1991-92	107 (1755)	44	45	62 @ 1.469	1.469	0.0012–0.0024	0.0020–0.0035	0.315	0.315
	122 (1997)	44	45	66 @ 1.575	1.575	0.0008–0.0019	0.0020–0.0033	0.259	0.258
	121.9 (1997)	44	45	72 @ 1.591	1.591	0.0012–0.0024	0.0020–0.0035	0.315	0.315
	89.6 (1468)	44	45	53 @ 1.469	1.417	0.0008–0.0020	0.0020–0.0035	0.260	0.260
	97.3 (1595)	44	45	62 @ 1.469	1.469	0.0008–0.0019	0.0020–0.0033	0.259	0.258
	181.4 (2972)	44	45	62 @ 1.492	1.492	0.0008–0.0020	0.0020–0.0035	0.260	0.260

PISTON AND RING SPECIFICATIONS

All measurements are given in inches.

Year	Engine Displacement cu. in. (cc)	Piston Clearance	Ring Gap			Ring Side Clearance		
			Top Compression	Bottom Compression	Oil Control	Top Compression	Bottom Compression	Oil Control
1988	109.5 (1795)	0.0008–0.0016	0.0100–0.0180	0.0080–0.0160	0.0080–0.0200	0.002–0.004	0.001–0.002	—
	121.9 (1997)	0.0008–0.0016	0.0100–0.0180	0.0080–0.0160	0.0080–0.0200	0.002–0.004	0.001–0.002	—
	155.9 (2555)	0.0008–0.0016	0.0120–0.0200	0.0100–0.0160	0.0120–0.0310	0.002–0.004	0.001–0.002	—
	89.6 (1468)	0.0008–0.0016	0.0080–0.0160	0.0080–0.0160	0.0080–0.0280	0.0012–0.0028	0.0008–0.0024	—
	97.4 (1597)	0.0008–0.0016	0.0080–0.0160	0.0080–0.0160	0.0080–0.0280	0.0012–0.0028	0.0008–0.0024	—
	143.4 (2350)	0.0008–0.0016	0.0100–0.0180	0.0080–0.0160	0.0080–0.0280	0.0012–0.0028	0.0008–0.0024	—
	181.4 (2972)	0.0008–0.0016	0.0118–0.0177	0.0098–0.0157	0.0079–0.0276	0.0012–0.0035	0.0008–0.0024	—
1989	121.9 (1997)	0.0004–0.0012	0.0098–0.0157	0.0079–0.0138	0.0079–0.0276	0.0012–0.0028	0.0008–0.0024	—
	155.9 (2555)	0.0008–0.0016	0.0120–0.0200	0.0100–0.0160	0.0120–0.0310	0.002–0.004	0.001–0.002	—
	89.6 (1468)	0.0008–0.0016	0.0080–0.0160	0.0080–0.0160	0.0080–0.0280	0.0012–0.0028	0.0008–0.0024	—
	97.4 (1597)	0.0008–0.0016	0.0080–0.0160	0.0080–0.0160	0.0080–0.0280	0.0012–0.0028	0.0008–0.0024	—
	143.4 (2350)	0.0008–0.0016	0.0100–0.0180	0.0080–0.0160	0.0080–0.0280	0.0012–0.0028	0.0008–0.0024	—
1990	107 (1755)	0.0004–0.0012	0.0118–0.0177	0.0079–0.0138	0.0079–0.0276	0.0018–0.0033	0.0008–0.0024	—
	122 (1997)	0.0008–0.0016	0.0098–0.0177	0.0138–0.0197	0.0079–0.0276	0.0012–0.0028	0.0012–0.0028	—
	121.9 (1997)	0.0004–0.0012	0.0098–0.0157	0.0079–0.0138	0.0079–0.0276	0.0012–0.0028	0.0008–0.0024	—
	89.6 (1468)	0.0008–0.0016	0.0079–0.0138	0.0079–0.0138	0.0079–0.0276	0.0012–0.0028	0.0008–0.0024	—
	97.3 (1595)	0.0008–0.0016	0.0098–0.0157	0.0138–0.0197	0.0079–0.0276	0.0012–0.0028	0.0012–0.0028	—
	181.4 (2972)	0.0008–0.0016	0.0118–0.0177	0.0098–0.0157	0.0118–0.0354	0.0012–0.0035	0.0008–0.0024	—
1991-92	107 (1755)	0.0004–0.0012	0.0118–0.0177	0.0079–0.0138	0.0079–0.0276	0.0018–0.0033	0.0008–0.0024	—
	122 (1997)	0.0008–0.0016	0.0098–0.0177	0.0138–0.0197	0.0079–0.0276	0.0012–0.0028	0.0012–0.0028	—
	121.9 (1997)	0.0004–0.0012	0.0098–0.0157	0.0079–0.0138	0.0079–0.0276	0.0012–0.0028	0.0008–0.0024	—
	89.6 (1468)	0.0008–0.0016	0.0079–0.0138	0.0079–0.0138	0.0079–0.0276	0.0012–0.0028	0.0008–0.0024	—
	97.3 (1595)	0.0008–0.0016	0.0098–0.0157	0.0138–0.0197	0.0079–0.0276	0.0012–0.0028	0.0012–0.0028	—
	181.4 (2972)	0.0012–0.0020	0.0118–0.0177	0.0177–0.0236	0.0079–0.0236	0.0012–0.0028	0.0008–0.0024	—

TORQUE SPECIFICATIONS

All readings in ft. lbs.

Year	Engine Displacement cu. in. (cc)	Cylinder Head Bolts	Main Bearing Bolts	Rod Bearing Bolts	Crankshaft Pulley Bolts	Flywheel Bolts	Manifold Intake	Manifold Exhaust	Spark Plugs
1988	109.5 (1795)	73–79	38	37	80–94	94–101	11–14	11–14	14–22
	121.9 (1997)	73–79	38	37	80–94	94–101	11–14	11–14	14–22
	155.9 (2555)	73–79	55–61	33	80–94	94–101	11–14	11–14	14–22
	89.6 (1468)	58–61	38	24	51–72	94–101	11–14	11–14	14–22
	97.4 (1597)	58–61	38	24	80–93	94–101	11–14	11–14	14–22
	143.4 (2350)	73–79	38	33	80–94	94–101	11–14	11–14	14–22
	181.4 (2972)	73–79	55–61	38	109–115	53–55	11–14	11–16	14–22
1989	155.9 (2555)	73–79	55–61	33	80–94	94–101	11–14	11–14	14–22
	89.6 (1468)	58–61	38	24	51–72	94–101	11–14	11–14	14–22
	97.4 (1597)	58–61	38	24	80–93	94–101	11–14	11–14	14–22
	122 (1997) ①	65–72	36–40	36–38	80–94	94–101	11–14	11–14	14–22
	122 (1997) ②	65–72	47–51	36–38	80–94	94–101	④	18–22	14–22
	181.4 (2972)	73–79	55–61	38	109–115	53–55	11–14	11–16	14–22
1990	107 (1755)	51–54	37–39	24–25	80–94	94–101	11–14	11–14	14–22
	122 (1997)	65–72	47–51	36–38	80–94	94–101	④	18–22	14–22
	155.9 (2555)	73–79	55–61	33	80–94	94–101	11–14	11–14	14–22
	89.6 (1468)	51–54	36–40	③	51–72	94–101	13–18	11–14	14–22
	97.3 (1595)	65–72	47–51	36–38	80–93	94–101	②	18–22	14–22
	122 (1997) ①	65–72	36–40	36–38	80–94	94–101	11–14	11–14	14–22
	122 (1997) ②	65–72	47–51	36–38	80–94	94–101	④	18–22	14–22
	181.4 (2972)	66–72	55–61	38	130–137	53–55	11–14	11–16	14–22
1991-92	107 (1755)	51–54	37–39	24–25	80–94	94–101	11–14	11–14	14–22
	122 (1997)	65–72	47–51	36–38	80–94	94–101	④	18–22	14–22
	89.6 (1468)	51–54	36–40	③	51–72	94–101	13–18	11–14	14–22
	97.3 (1595)	65–72	47–51	36–38	80–93	94–101	②	18–22	14–22
	122 (1997) ①	65–72	36–40	36–38	80–94	94–101	11–14	11–14	14–22
	122 (1997) ②	65–72	47–51	36–38	80–94	94–101	④	18–22	14–22
	181.4 (2972)	76–83	58	38	130–137	55	11–14	33	14–22
	181.4 (2972) ⑤	87–94	58	38	130–137	55	9–11	22 ⑥	18

① Single overhead camshaft
② Double overhead camshaft
③ Tighten to 14.5 ft. lbs.; 2nd back off; 3rd tighten to 14.5; 4th tighten additional ¼ turn
④ Torque bolts to 11–14 ft. lbs., torque nuts to 22–30 ft. lbs.
⑤ Turbo
⑥ See text

BRAKE SPECIFICATIONS

All measurements in inches unless noted.

Year	Model	Lug Nut Torque (ft. lbs.)	Master Cylinder Bore	Brake Disc Minimum Thickness	Brake Disc Maximum Runout	Maximum Brake Drum Diameter	Minimum Lining Thickness Front	Minimum Lining Thickness Rear
1988	Cordia	50–57 ②	0.87	—	0.650	8.000	0.040	0.040
	Tredia	50–57 ②	0.87	—	0.650	8.000	0.040	0.040
	Starion	50–57 ②	0.94	—	0.880	—	0.040	—
	Mirage	50–57 ②	0.81 ④	—	0.450 ③	7.100	0.040	0.040
	Galant	43–52 ②	0.94	—	0.650	—	0.040	0.040
	Precis	51–58 ①	0.81	—	0.450	—	0.040	0.040

BRAKE SPECIFICATIONS

All measurements in inches unless noted.

Year	Model	Lug Nut Torque (ft. lbs.)	Master Cylinder Bore	Brake Disc Minimum Thickness	Brake Disc Maximum Runout	Maximum Brake Drum Diameter	Minimum Lining Thickness Front	Minimum Lining Thickness Rear
1989	Starion	50–57②	0.94	—	0.880	—	0.040	—
	Mirage	50–57②	0.81④	—	0.450③	7.100	0.040	0.040
	Galant	65–80②	⑨	0.882	0.003	8.100⑦	0.080	0.080
	Precis	69–78⑩	0.81	0.670	0.006	7.200	0.040	0.040
	Sigma	43–52②	0.94	—	0.004	—	0.079	0.039
1990	Eclipse	87–101②	0.87⑪	0.882	0.003	⑦	0.080	0.080
	Mirage⑤	65–80	0.81	0.449	0.006	7.200	0.080	0.080
	Mirage⑥	65–80	0.87	0.882	0.006	⑦	0.080	0.080
	Galant	65–80	⑨	0.882	0.003	8.100⑦	0.080	0.080
	Precis	69–78⑩	0.81	0.670	0.006	7.200	0.040	0.040
	Sigma	65–79	0.94	0.882	0.004	⑧	0.079	0.079
1991–92	Eclipse	87–101②	0.87⑪	0.882	0.003	⑦	0.080	0.080
	Mirage⑤	65–80	0.81	0.449	0.006	7.200	0.080	0.080
	Mirage⑥	65–80	0.87	0.882	0.006	⑦	0.080	0.080
	Galant	65–80	⑨	0.882	0.003	8.100⑦	0.080	0.080
	Precis	69–80	0.81	0.670	0.006	7.200	0.040	0.040
	3000GT	87–101	1.0625⑪	⑫	0.0031	—	0.080	0.080

① With aluminum wheels—57–72
② With aluminum wheels—66–81
③ Turbo 0.650
④ Turbo 0.87
⑤ 1500 Engine
⑥ 1600 Engine
⑦ Rear disc minimum thickness—0.331
⑧ Rear disc minimum thickness—0.646
⑨ 2WD w/o ABS—0.87
2WD w/ ABS—0.94
4WD w/o ABS—0.94
4WD w/ ABS—1.00
⑩ Turbo 0.94
⑪ Without ABS 1.0
⑫ Front 0.880 FWD 1.12 AWD
Rear 0.650 FWD 0.720 AWD

WHEEL ALIGNMENT

Year	Model	Caster Range (deg.)	Caster Preferred Setting (deg.)	Camber Range (deg.)	Camber Preferred Setting (deg.)	Toe-in (in.)	Steering Axis Inclination (deg.)
1988	Cordia	5/16–1 5/16P	13/16P	1/16N–5/16P	7/16P⑤	1/8N–1/8P	7 1/16
	Tredia	5/16–1 5/16P	13/16P	1/16N–15/16P	7/16P⑤	1/8N–1/8P	7 1/16
	Starion	5 5/16–6 5/16P	5 13/16P	1N–0P	1/2N	13/64P–13/64P	—
	Mirage	1/2–1 1/2P⑥	1P	1/2N–1/2P	0P①	1/8N–1/8P	5 3/4
	Galant	5/32–1 5/32P	21/32P	0–1P	1/2P⑧	1/8N–1/8P	6 5/8
	Precis	1/2–1/8P	3/16P	0–1P	1/2P⑦	1/16P–5/32P	12 11/16
1989	Starion	5 5/16–6 5/16P	5 13/16P	1N–0P	1/2N	13/64P–13/64P	—
	Mirage	1/2–1 1/2P⑥	1P	1/2N–1/2P	0P①	1/8N–1/8P	5 3/4
	Galant	1 1/2–2 1/2P	2P	3/16N–13/16P	5/16P⑧	1/8N–1/8P	—
	Precis	1/2–1 1/8P	13/16P	0–1P	1/2P⑦	1/16P–5/32P	12 11/16
	Sigma	3 1/16–1 3/16P	11/16P	0–1P⑨	1/2P⑩	1/8N–1/8P	NA

WHEEL ALIGNMENT

Year	Model	Caster Range (deg.)	Caster Preferred Setting (deg.)	Camber Range (deg.)	Camber Preferred Setting (deg.)	Toe-in (in.)	Steering Axis Inclination (deg.)
1990	Eclipse (11)	1 27/32–2 27/32P	2 11/32P	1/4N–3/4P	1/4P (13)	1/8N–1/8P	14 3/32
	Eclipse (12)	1 29/32–2 29/32P	2 13/32P	13/32N–19/32P	3/32P (13)	1/8N–1/8P	14 3/32
	Mirage	1 11/16–2 11/16P	2 3/16P	1/2N–1/2P	0P (1)	1/8N–1/8P	—
	Galant	1 1/2–2 1/2P	2P	3/16N–13/16P	5/16P (8)	1/8N–1/8P	—
	Precis	1/2–1 1/8P	13/16P	0–1P	1/2P (7)	1/16P–5/32P	12 11/16
	Sigma	3/16–1 3/16P	11/16P	0–1P (9)	1/2P (10)	1/8N–1/8P	—
1991-92	Eclipse (11)	1 27/32–2 27/32P	2 11/32P	1/4N–3/4P	1/4P (13)	1/8N–1/8P	14 3/32
	Eclipse (12)	1 29/32–2 29/32P	2 13/32P	13/32N–19/32P	3/32P (13)	1/8N–1/8P	14 3/32
	Mirage	1 11/16–2 11/16P	2 3/16P	1/2N–1/2P	0P (1)	1/8N–1/8P	—
	Galant	1 1/2–2 1/2P	2P	3/16N–13/16P	5/16P (8)	1/8N–1/8P	—
	Precis	1/2–1 1/8P	13/16P	0–1P	1/2P (7)	1/16P–5/32P	12 11/16
	3000GT	3 1/2–4 1/2P	4P	1/2N–1/2P (14)	(14)	1/8N–1/8P (15)	—

N Negative
P Positive
(1) Rear—11/16N
(2) Rear—5/16P
(3) Rear—9/16N
(4) Rear—21/32N
(5) Rear—11/16P
(6) With power steering—11/16–2 3/16
(7) Rear—5/8N
(8) Rear—3/4N
(9) Rear—1N–1/4N
(10) Rear—3/4N
(11) 1.8L engine
(12) 2.0L engine
(13) Rear—3/4P
(14) Rear:
FWD 0° ± 1/2°
AWD 3/16° ± 1/2°
(15) Rear: 0

ENGINE MECHANICAL

NOTE: Disconnecting the negative battery cable on some vehicles may interfere with the functions of the on board computer systems and may require the computer to undergo a relearning process, once the negative battery cable is reconnected.

Engine Assembly

REMOVAL & INSTALLATION

NOTE: All engine and transaxle assemblies are removed as a unit.

Cordia and Tredia

NOTE: Use care when disconnecting the refrigerant lines. Escaping refrigerant will freeze any surface it contacts, including skin and eyes.

1. Matchmark and then unbolt and remove the hood. Disconnect both battery connectors and remove the battery.
2. Drain the engine coolant and transaxle fluid. Disconnect the heater hoses from the engine, as required.
3. Disconnect the power steering fluid return hose at the reservoir and drain the fluid into a clean container.
4. Drain and remove the radiator overflow and windshield washer tanks.
5. Disconnect the upper and lower radiator hoses at both ends and remove them. If equipped with air conditioning, disconnect the hoses as close as possible to the condenser unit in front of the radiator and cap the openings securely. Then, remove the radiator or radiator and condensor assemble.
6. Remove the battery tray. Disconnect both heater hoses from the side of the engine.
7. Remove the air cleaner. On turbocharged vehicles, disconnect the turbocharger intake hose.
8. Disconnect the brake booster vacuum hose. On turbocharged vehicles, disconnect the oil cooler hoses at the engine. If equipped with an automatic transaxle, disconnect the cooler lines.
9. If equipped with a manual transaxle, disconnect the clutch cable. If equipped with an automatic transaxle, disconnect the shift control cable. Disconnect the speedometer cable from the transaxle.
10. Disconnect the accelerator cable at the side of the engine. Disconnect the engine ground strap at the right front fender. If equipped with an air conditioner, disconnect the hoses at the compressor and cap all openings securely.
11. Disconnect the power steering hoses at the side of the pump. Cap all openings. Disconnect the coil low and high tension wires. Label the low tension wires for reassembly to the proper terminals. Disconnect the battery negative cable from the engine.
12. Label and disconnect the alternator connectors. Disconnect the oil pressure sending unit wire.
13. Remove the vacuum unit and solenoid valve screws and disconnect the electrical connector; move the unit aside.
14. Disconnect both smaller vacuum hoses from the purge control valve, remove the screw and move the unit aside.
15. Loosen the clamps and disconnect the vacuum hoses going to the evaporative emissions canister.
16. Disconnect the fuel return hose from the carburetor or injection mixer. Disconnect the fuel supply hose at the fuel filter.
17. Raise the vehicle and support it safely. Then, disconnect the exhaust pipe at the manifold. Fasten the exhaust pipe with wire to keep it from falling.
18. If equipped with a manual transaxle, disconnect the shift control rod and extension and remove them.
19. Disconnect the left and right side

strut bars and stabilizer bars where they connect to the lower control arms. Then, remove the bolts fastening the control arms on both sides to the rearward crossmember.

20. Disconnect the lower arm ball joint at the steering knuckle on both sides. Then, disconnect the strut bar and stabilizer bar at the lower control arm. Using a prybar inserted between the transaxle case and driveshaft, carefully pry the halfshaft out of the transaxle on each side. Plug the openings in the transaxle to prevent dirt from getting in.

21. Carefully lower all parts to the crossmember. Discard the retaining clips for the halfshafts. They must be replaced.

22. Attach a cable securely supported by a lift and pulley arrangement to each engine lifting point. Put tension on all the cables to support the engine securely.

23. Remove the nut from the left side engine mount insulator. Remove the 4 front roll bracket bolts located on the side of the front crossmember.

24. Remove the bolt from the rear roll insulator. Remove the nuts attaching the left engine mount insulator to the fender.

25. From inside the right fender shield, detach the protective cap and then remove the transaxle insulator bracket bolts. Remove the bolts connecting the transaxle mount insulator.

26. Remove the bolts to the shift control selector. Remove the wiring connector going to the transaxle. Disconnect vacuum hoses.

27. Remove the transaxle insulator bracket. Increase the tension on the lifting cables so the engine weight is supported entirely by the cables and none of the weight is on the mounts. Remove the bolts passing through the insulators of the rear roll stop and left bracket.

28. Make sure all items are disconnected from the engine/transaxle assembly. Press downward on the transaxle to guide the assembly and lift it carefully out of the vehicle.

To install:

29. When installing the engine, be careful to ensure that engine compartment wiring and hoses do not catch on engine wiring and hoses do not catch on engine parts. Torque bolts as follows:

Left engine mount insulator nut (large) – 43–58 ft. lbs.

Left engine mount insulator nuts (small) – 22–29 ft. lbs.

Left engine mount bracket-to-engine bolts/nuts – 36–47 ft. lbs.

Transaxle mount insulator nut – 43–58 ft. lbs.

Transaxle insulator bracket-to-fender shield bolts – manual 40–43 ft. lbs.

Transaxle insulator bracket-to-fender shield bolts – automatic 22–29 ft. lbs.

Transaxle mounting bracket-to-automatic transaxle nuts – 43–58 ft. lbs.

Transaxle mounting bracket bolts – 22–29 ft. lbs.

Rear roll stop insulator nut – 22–29 ft. lbs.

Rear roll stop-to-rear crossmember nuts – 43–58 ft. lbs.

Rear roll stop bracket-to-rear roll stop stay bolt – 22–29 ft. lbs.

Front roll stop insulator nut – 36–47 ft. lbs.

Front roll bracket-to-front crossmember nuts – 29–36 ft. lbs.

30. After the engine is securely mounted, replenish all fluids. Operate the engine checking carefully for leaks. Check all gauges for proper readings. Adjust clutch and shift linkage. Adjust the accelerator cable. Recharge the air conditioner.

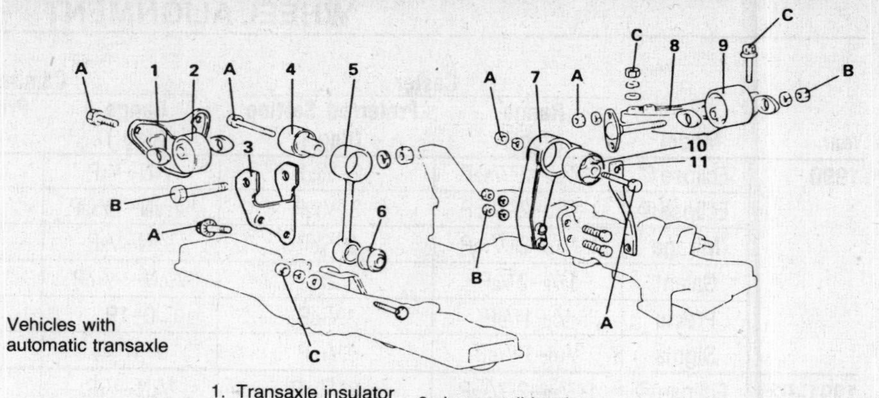

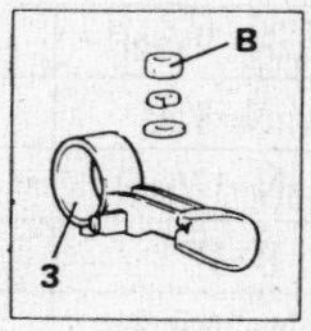

1. Transaxle insulator bracket
2. Transaxle mount insulator
3. Transaxle mount bracket
4. Upper roll insulator
5. Front roll rod
6. Lower roll insulator
7. Rear roll stopper bracket
8. Left mount bracket
9. Left mount insulator
10. Rear roll insulator
11. Rear roll stopper stay

	Nm	ft.lbs.
A	30–40	22–29
B	60–80	43–58
C	50–65	36–47

Cordia, Tredia engine/transaxle mounting

Starion

NOTE: Use care when disconnecting the refrigerant lines. Escaping refrigerant will freeze any surface it contacts, including skin and eyes.

1. Matchmark and then unbolt and remove the hood. Drain the cooling system. Disconnect the accelerator cable at the injection mixer. Disconnect both battery cables.

2. Disconnect both heater hoses at the block and the brake booster vacuum hose at the intake manifold.

3. Disconnect the fuel hoses at the injection system.

4. Disconnect the high tension wire at the center of the distributor. Disconnect the temperature sensor wire. Disconnect the intake manifold ground cable connector. First label the wires for reassembly and then disconnect the starter motor wiring harness.

5. Disconnect the power steering pump hoses. Remove the power steering pump.

6. Unbolt and disconnect both engine oil cooler hoses at the oil filter adapter. Plug all openings to prevent the entry of dirt and leakage of oil.

7. Label and then disconnect the alternator wiring. Disconnect the engine ground cable. Disconnect both plugs for the electronic injection wiring harness.

8. Disconnect the vacuum hose from the boost sensor, located on the firewall.

9. Remove the rear catalytic converter.

10. Unscrew and disconnect the speedometer cable at the transmission. Disconnect the wiring for the oil pressure gauge sending unit on the block.

11. If equipped with an automatic transmission, disconnect both oil cooler hoses and plug all openings. On all vehicles, disconnect the back-up light switch harness at the plug located under the transaxle.

12. Remove the propeller shaft.

13. Remove the clutch slave cylinder.

14. Put the gearshift in **N**. Unbolt the gearshift lever assembly.

15. Securely support the engine using a suitable engine crane. Support the transmission with a jack. Then, remove the front and rear engine mounting nuts and bolts, and the rear crossmember. Raise the assembly slightly and remove all the front support brackets and insulators. Then, gradually lower the transmission jack

and pull the engine and transaxle assembly out by raising the front of the engine so the transmission will clear the firewall.

To install:

16. When reassembling mounts, make sure all holes are properly aligned and that mounts are not distorted. On both front insulators, make sure the locating boss and hole in the insulator are in alignment.
17. Install the engine and transmission, clutch slave cylinder and propeller shaft.
18. Reconnect the oil cooler lines, back-up light switch and speedometer cable.
19. Install the catalytic converter, boost sensor vacuum hose and engine wiring. Install the power steering pump and other accessories. Reconnect the ignition and fuel system and any other component disconnected for removal.
20. Fill the engine with oil, the radiator with coolant and start the engine. Check the timing and idle speed. Allow the engine to reach operating temperature and check for leaks.

Mirage

NOTE: Use care when disconnecting the refrigerant lines. Escaping refrigerant will freeze any surface it contacts, including skin and eyes.

1. Remove the hood. Remove the air cleaner assembly. Disconnect both battery cables and then remove the batter. Unbolt and remove the battery tray.
2. Disconnect the electrical connectors for the backup lights and engine harness, located near the battery tray. If equipped with a 5 speed, disconnect the select control valve connector. Disconnect both alternator harness connectors and the oil pressure sending unit.
3. Label and then disconnect the automatic transaxle oil cooler hoses. Avoid spilling oil and cap the openings.
4. Drain the cooling system. Then disconnect and remove the upper and lower hoses and remove them. Remove the radiator.
5. Label and then disconnect all low tension wires and the one high tension wire going to the coil from the distributor. Disconnect the engine ground.
6. Disconnect the brake booster vacuum hose at the intake manifold. Disconnect the cap power steering lines, as required. Remove pump as needed.
7. Disconnect the fuel supply, return and vapor hoses at the side of the engine.
8. If equipped with a turbocharger, the 3 electrical connectors must be disconnected – 1 for the idle speed control system and 2 for the injection system. All are located on the injection mixer.
9. Disconnect the heater hoses from the side of the engine. Disconnect the accelerator cable at the side of the injection mixer and the block.
10. If equipped with a manual transaxle, disconnect the clutch control cable. If equipped with an automatic transaxle, disconnect the shift control cable from the transaxle.
11. Unscrew and disconnect the speedometer cable at the transaxle.
12. Raise and securely support the vehicle. Remove the drain plug and drain the transaxle fluid. Disconnect the exhaust pipe at the manifold. Then, suspend the pipe securely with wire.
13. If equipped with a manual transaxle, remove the shift control rod and extension rod.
14. Disconnect the stabilizer bar at both lower control arms. Remove the bolts that attach the lower control arms to the body on either side. Support the arms from the body.
15. If equipped with a turbocharger, disconnect and remove the oil cooler tube from the side of the engine.
16. Disconnect the halfshaft at the transaxle on both sides. Then, seal off the openings in the transaxle. Make sure to replace the circlips holding the driveshafts in the transaxle. Support the halfshafts from the body.
17. Attach a crane-type lift, via chains or cables, to both the engine lifting hooks. Put just a little tension on the cables. Then, remove the nut and bolt from the front roll stopper; unbolt the brace from the top of the engine damper.
18. Separate the rear roll stopper from the crossmember. Remove the attaching nut from the left mount insulator bolt, but do not remove the bolt.
19. Raise the engine just enough that the crane is supporting its weight. Check that everything is disconnected from the engine.
20. Remove the blind cover from the inside of the right fender inner shield. Then, remove the blind cover from the inside of the right front fender inner shied. Remove the transaxle bracket bolts.
21. Remove the left mount insulator bolt. Then, press downward on the transaxle while lifting the engine/transaxle assembly to guide it up and out of the vehicle.

To install:

22. During installation, first install all nuts and bolts with the weight of the engine carried by the crane. Tighten just slightly. Then, allow the weight of the engine to sit on the mounts and torque parts as follows.
23. Install the engine and transaxle assembly and securely tighten all mountings. Install the halfshafts.
24. Install the turbocharger oil cooler lines, stabilizer bars, exhaust system, coolant, vacuum and fuel hoses, and the accelerator cable. Reconnect all engine electrical leads.
25. Install the shift rod or cable on the transaxle. On manual transaxle, install the clutch control cable.
26. Fill the engine with oil and the radiator with coolant. Start the engine and check the timing and idle speed. Bring the engine to operating temperature and check for leaks.

Galant With 4 Cylinder Engine

NOTE: Use care when disconnecting the refrigerant lines. Escaping refrigerant will freeze any surface it contacts, including skin and eyes.

1. Matchmark the position of the hood hinges on the hood and remove it. Drain the engine coolant. Remove the drain plug and drain the transaxle fluid.
2. Remove the air cleaner. Disconnect the battery cables, remove the battery and then unbolt and remove the battery tray.
3. Carefully label and then disconnect each of 3 connectors of the engine wiring harness.
4. Disconnect the ground wire at the right side wheel house. Drain the power steering fluid by disconnecting a hose at the low point of the system. Plug all openings.
5. If equipped with a electronically controlled suspension, remove the compressor and reserve tank.
6. Disconnect the transaxle control cable by moving the adjusting nut on the transaxle end of the cable and pulling the cable out of the fitting on the transaxle. Keep the rearward nut from moving to preserve the adjustment.
7. Disconnect the alternator connectors and the oil pressure sending unit. Disconnect the high tension cable at the coil.
8. Disconnect the engine ground cable at the firewall. Disconnect the brake booster vacuum hose.
9. Label and then disconnect the eight connectors for the sensors of the electronic fuel injection at the injection mixing body. Disconnect the accelerator and speed control cables nearby.
10. Disconnect the fuel supply and return hoses at the electronic injection mixing body.
11. Mark and then disconnect the transaxle oil cooler hoses. Cap the openings to keep oil in and dirt out.

Disconnect both radiator hoses and both heater hoses at the engine.
12. Unscrew and then disconnect the speedometer cable at the transaxle. Disconnect the power steering hoses at the pump and cap all openings.
13. Raise and securely support the vehicle. Disconnect the exhaust pipe at the exhaust manifold. Then, hang the pipe to the body with wire.
14. Disconnect the steering knuckles from the lower arm ball joints. Disconnect the tie rods from the steering knuckle. Remove the halfshafts from the transaxle. Make sure to cap all openings to keep fluid in and dirt out. Discard the retaining clips that hold the shafts in the transaxle and replace them.
15. Attach a lifting crane to both engine lifting hooks and put very slight tension on the chains or cables. Then remove the nut, but not the bolt, coupling the engine mount bracket to the body. Remove the upper installation nuts of the front and rear roll stopper brackets.
16. Detach the protective cap from inside the right fender shield and then remove the transaxle bracket bolts. Then, remove the bolts connecting the transaxle mount insulator to the bracket and remove the bracket.
17. Increase the tension on the engine lifting mechanism until the weight of the engine is borne by the lift instead of its mounts. Remove the bolts of the rear roll stopper bracket, the engine mount bracket and the front roll stopper bracket. Confirm that all cable, wires and linkages are disconnected.
18. Tilting the transaxle side downward, carefully lift the engine/transaxle assemble out of the vehicle.

To install:
19. Make sure nothing gets pinched or bent when installing the engine. When installing rubber insulators, make sure they are not twisted.
20. Install the engine and transaxle assembly and tighten mounts securely. Install the halfshafts and reassemble the front suspension.
21. Install the exhaust system, speedometer cable, power steering pump, transaxle cooler lines, all hoses (vacuum, fuel and coolant), accelerator and cruise control cables, and the engine electrical harness.
22. Install the transaxle control cable and all other items removed from the engine.
23. Fill the engine with oil and the radiator with coolant. Start the engine and check the timing and idle speed. Bring the engine to operating temperature and check for leaks.

Galant and Sigma With V6 Engine

NOTE: Use care when disconnecting the refrigerant lines. Escaping refrigerant will freeze any surface it contacts, including skin and eyes.

1. Matchmark the position of the hood hinges on the hood and remove it. Drain the engine coolant. Remove the drain plug and drain the transaxle fluid.
2. Remove the air cleaner. Disconnect the battery cables, remove the battery and then unbolt and remove the battery tray.
3. Carefully label and then disconnect each of the connectors of the engine wiring harness.
4. Remove the accelerator cable bracket screws. Remove the accelerator cable and cruise control cable from the throttle body linkage.
5. Disconnect and plug the high pressure fuel lines and fuel return lines from the engine.
6. Open the cover of the fusible link box and disconnect the alternator wiring. Disconnect and tag all necessary wiring including the pulse generator connector, if equipped with an automatic transaxle.
7. If equipped with an automatic transaxle, disconnect the transaxle control cables. Remove the control cable bracket bolts and disconnect the transaxle control cables from the transaxle. Disconnect and plug the oil cooler lines.
8. If equipped with electronic controlled suspension, remove the air compressor.
9. If equipped with a manual transaxle, remove the clutch release cylinder bolts. Remove the clutch oil tube bracket and secure the clutch release cylinder and the clutch oil tube assembly on the chassis side with some wire.
10. Disconnect the speedometer. Disconnect and plug the heater hoses. Remove all the radiator hoses and remove the radiator.
11. Disconnect and plug the air conditioner lines from the compressor. Release the engine block section clamp and disconnect the discharge flexible hose. Remove the clip from the power steering oil pump section and disconnect the suction flexible hose.
12. Remove the engine under cover panels. Remove the stabilizer nut and tie rod end cotter pin. Using the special tie rod tool, disconnect the tie rod end from the steering knuckle. Using the suitable tool, disconnect the lower arm ball joint from the knuckle. Loosen the nut but do not remove it.
13. Remove the left side halfshaft (after removing the circlip, cotter pin and halfshaft nut) by inserting a suitable tool between the bearing bracket and the halfshaft and then pry the halfshaft from the bearing bracket. Be sure to pull the halfshaft out from the bearing bracket as an assembly with the hub knuckle and other parts. Suspend the removed halfshaft with a wire to prevent the joint section from bending sharply.
14. Remove the right side halfshaft circlip and locknut, then remove the halfshaft from the hub by using the special hub removal tool. Once the hub is remove, use a suitable tool and pull the halfshaft out from the transaxle.
15. Remove the rubber hangers from the exhaust system. Disconnect the oxygen sensor connector. Remove the bolts or nuts from the front exhaust pipe and remove the pipe from the engine. Be sure to suspend the disconnected exhaust pipe with a wire to keep it from bending sharply.
16. Remove the distributor cap and connect a suitable engine lifting device to the proper locations on the engine and raise the lifting device just enough to take the slack out of the cable.
17. Remove the front roll stopper bracket mount bolt, engine damper and rear roll stopper bracket mount bolt. Remove the bolts and nuts that fasten the engine mount bracket to the body.
18. To remove the transaxle mount bracket, first remove the 4 plugs from the fender shield. Remove the transaxle mount bracket bolts with care to prevent them from falling into the fender shield.
19. Check that all cables, hoses, electrical harness connections and wires are disconnected from the side of the engine. Slowly remove the engine and transaxle assembly upwards from the engine compartment with the engine lift. Once the engine and transaxle assembly has cleared the vehicle, place it on a suitable engine stand.

To install:
20. Note the following during installation:
   a. When installing the halfshaft nut, install the washer and wheel bearing nut in the proper direction. If the position of the cotter pin holes does not match, tighten the nut up to 188 ft. lbs. and then install the cotter pin in the first matching holes and bend it securely.
   b. After the installation is complete, check all the remove and disconnect components for proper operation and adjust or repair as necessary.

21. Install the engine and transmission assembly and tighten the mounting bolts securely. Install the roll stopper brackets and engine dampeners. Install the exhaust system and oxygen sensor.

22. Install the halfshafts and reassemble the front suspension. Install the air conditioning compressor, power steering pump, speedometer cable, heater hoses and radiator.

23. Install the clutch control cables and the clutch oil system. Install the air compressor on electronic control suspension. Install the engine compartment wiring harness. Reconnect the fuel lines and accelerator cable.

24. Install the remainder of the components removed. Fill the engine with oil and the radiator with coolant. Start the engine and check the timing and idle speed. Bring the engine to normal operating temperature and check for leaks.

Precis

NOTE: Use care when disconnecting the refrigerant lines. Escaping refrigerant will freeze any surface it contacts, including skin and eyes.

1. Remove the air cleaner assembly by disconnecting all hoses, unbolting it and removing it. Disconnect both battery cables and then remove the battery. Unbolt and remove the battery tray.

2. Disconnect the electrical connectors for the back-up lights and engine harness, located near the battery tray. If equipped with a 5 speed, disconnect the select control valve connector. Disconnect both alternator harness connectors and the oil pressure sending unit.

3. First label and then disconnect the transaxle oil cooler hoses. Avoid spilling oil and cap the openings.

4. Drain the cooling system through the cock on the bottom of the radiator and the plug in the block. Then disconnect and remove the upper and lower hoses and remove them. Remove the radiator.

5. Label and then disconnect all low tension wires and the 1 high tension wire going to the coil from the distributor. Disconnect the engine ground.

6. Disconnect the brake booster vacuum hose at the intake manifold.

7. Disconnect the fuel supply, return, and vapor hoses at the side of the engine. Avoid spilling fuel.

8. Disconnect the heater hoses from the side of the engine.

9. If equipped with a manual transaxle, disconnect the clutch control cable. If equipped with an automatic transaxle, disconnect the shifter control cable from the transaxle.

10. Unscrew and disconnect the speedometer cable at the transaxle.

11. Raise and securely support the vehicle. Remove the drain plug and drain the transaxle fluid. Disconnect the exhaust pipe at the manifold. Then, suspend the pipe securely with wire.

12. If equipped with a manual transaxle, remove the shift control rod and extension rod.

13. Disconnect the stabilizer bar at both lower control arms. Remove the bolts that attach the lower control arms to the body on either side. Support the arms from the body.

14. Disconnect the halfshafts at the transaxle on both sides. Then, seal off the openings in the transaxle. Make sure to replace the circlips holding the halfshafts in the transaxle. Support the halfshafts from the body.

15. Attach a crane-type lift, via chains or cables, to both the engine lifting hooks. Put just a little tension on the cables. Then, remove the nut and bolt from the front roll stopper; unbolt the brace from the top of the engine damper.

16. Separate the rear roll stopper from the crossmember. Remove the attaching nut from the left mount insulator bolt, but do not remove the bolt.

17. Raise the engine just enough that the crane is supporting its weight. Check that everything is disconnected from the engine.

18. Remove the blind cover from the inside of the right fender inner shield. Remove the transaxle bracket bolts.

19. Remove the left mount insulator bolt. Then, press downward on the transaxle while lifting the engine/transaxle assembly to guide it up and out of the vehicle.

To install:

20. Installation is the reverse of removal. First install all nuts and bolts with the weight of the engine carried by the crane. Tighten just slightly. Then, allow the weight of the engine to sit on the mounts, and torque parts as follows:

Left mount large insulator nut—65–80 ft. lbs.

Left mount small insulator nut—22–29 ft. lbs.

Left mount bracket-to-engine nuts/bolts—36–47 ft. lbs.

Transaxle mount insulator nut—65–80 ft. lbs.

Transaxle insulator bracket-to-side frame bolts—22–29 ft. lbs.

Transaxle bracket assembly-to-automatic transaxle nuts—65–80 ft. lbs.

Transaxle mount bracket-to-manual transaxle bolts—40–43 ft. lbs.

Rear roll insulator nut—33–43 ft. lbs.

Rear roll stopper bracket-to-crossmember assembly bolts—22–29 ft. lbs.

Front roll insulator nut—33–43 ft. lbs.

Front roll stopper bracket-to-crossmember assembly bolts—33–40 ft. lbs.

Lower roll insulator roll-to-damper bracket bolt—22–29 ft. lbs.

Center crossmember body—43–58 ft. lbs.

21. Finally, replenish all fluids. Adjust the transaxle and accelerator linkages. Start the engine and check for leaks as well as proper gauge operation. Replace the hood and have the air conditioner recharged.

Eclipse

1. Relieve fuel system pressure.

2. Disconnect the negative battery cable.

3. Matchmark hood and hinges and remove hood assembly.

4. Drain the engine coolant and remove the radiator assembly.

5. Remove the transaxle.

6. Disconnect and tag for assembly reference the connections for the accelerator cable, heater hoses, brake vacuum hose, connection for vacuum hoses, high pressure fuel line, fuel return line, oxygen sensor connection, coolant temperature gauge connection, coolant temperature sensor connector, connection for thermo switch sensor, if equipped with automatic transaxle, the connection for the idle speed control, the motor position sensor connector, the throttle position sensor connector, the EGR temperature sensor connection (California vehicles), the fuel injector connectors, the power transistor connector, the ignition coil connector, the condensor and noise filter connector, the distributor and control harness, the connections for the alternator and oil pressure switch wires.

7. Remove the air conditioner drive belt and then the air conditioning compressor. Leave the hoses attached. Do not discharge the system. Wire the compressor aside.

8. Remove the power steering pump and wire back aside.

9. Remove the exhaust manifold to head pipe nuts. Discard the gasket.

10. Attach a hoist to the engine and take up the engine weight. Remove the engine mount bracket. Remove any torque control brackets (roll stoppers). Note that some engine mount pieces have arrows on them for proper assembly. Double check that all cables, hoses, harness connectors, etc., are disconnected from the engine. Lift the engine slowly from the engine compartment.

To install:

11. Installation is the reverse of the removal procedure. Install the engine and torque control brackets.

12. Install the exhaust pipe, power steering pump and air conditioning compressor.

13. Checking the tags installed at removal, reconnect all electrical and vacuum connections.

14. Install the transaxle.
15. Refill with engine oil, coolant, install the hood and adjust the drive belts.
16. Adjust the accelerator cable as required.

3000 GT

1. Relieve fuel system pressure.
2. Disconnect the negative battery cable.
3. Matchmark the hood and hinges and remove the hood assembly. Remove the air cleaner assembly and all adjoining air intake duct work.
4. Drain the engine coolant and remove the radiator assembly and intercooler.
5. Remove the transaxle.
6. Disconnect and tag for assembly reference the connections for the accelerator cable, heater hoses, brake vacuum hose, connection for vacuum hoses, high pressure fuel line, fuel return line, oxygen sensor connection, coolant temperature gauge connection, coolant temperature sensor connector, connection for thermo switch sensor, if equipped with automatic transaxle, the connection for the idle speed control, the motor position sensor connector, the throttle position sensor connector, the EGR temperature sensor connection (California vehicles), the fuel injector connectors, the power transistor connector, the ignition coil connector, the condenser and noise filter connector, the distributor and control harness, the connections for the alternator and oil pressure switch wires.
7. Remove the air conditioner drive belt and the air conditioning compressor. Leave the hoses attached. Do not discharge the system. Wire the compressor aside.
8. Remove the power steering pump and wire aside.
9. Remove the exhaust manifold to head pipe nuts. Discard the gasket.
10. Attach a hoist to the engine and take up the engine weight. Remove the engine mount bracket. Remove any torque control brackets (roll stoppers). Note that some engine mount pieces have arrows on them for proper assembly. Double check that all cables, hoses, harness connectors, etc., are disconnected from the engine. Lift the engine slowly from the engine compartment.

To install:

11. Install the engine and secure all control brackets.
12. Install the exhaust pipe, power steering pump and air conditioning compressor.
13. Checking the tags installed at removal, reconnect all electrical and vacuum connections.
14. Install the transaxle.
15. Install the radiator assembly and intercooler.
16. Install the air cleaner assembly.
17. Fill the engine with the proper amount of engine oil. Connect the negative battery cable.
18. Refill the cooling system. Start the engine, allow it to reach normal operating temperature. Check for leaks.
19. Check the ignition timing and adjust if necessary.
20. Install the hood.
21. Road test the vehicle and check all functions for proper operation.

Engine Mounts

REMOVAL & INSTALLATION

Except 3000 GT

1. Disconnect the negative battery cable.
2. Raise and safely support the engine so it is not resting on the engine mount. One suggested way is a block of wood between a floor jack and the oil pan. Use care not to bend or damage any components.
3. Remove the engine mount bracket and body connection through bolt. Take note of the position of the arrow on the oval shaped stopper plate. This is important.
4. Remove the engine bracket. Some engines may use an additional small strap that should be removed.
5. Remove the stopper plate.

To install:

6. Installation is the reverse of the removal procedure. Note the arrows on the stopper plates and make sure they are installed properly. Torque the engine mount-to-body bolts as well as the engine mount-to-engine nuts to 36–47 ft. lbs. Torque the stopper through bolt to 33–43 ft. lbs.
7. Lower engine. Reconnect the negative battery cable.

3000 GT

1. Disconnect the negative battery cable. Remove the air cleaner and all necessary air duct work.
2. Raise and safely support the engine so it is not resting on the engine mount. One suggested way is a block of wood between a floor jack and the oil pan. Use care not to bend or damage any components.
3. Remove the engine mount bracket and body connection through bolt. Take note of the position of the arrow on the oval shaped mounting stopper plate. This is important.
4. Remove the engine mounting bracket and stopper plate.
5. Lower mounts (roll stoppers) are removed by removing the through bolt, then the frame bolts. On 3000-GT, the condenser and fan assembly and front catalytic converter must first be removed to gain access to the front mount.

To install:

6. The installation is the reverse of the removal procedure. Note the arrows on the stopper plates and make sure they are installed properly.
7. The front lower mount through bolt nut should not be tightened until the full weight of the engine is on the mount. Torque specifications are as follows:

Upper mount to engine nuts and bolts – 72–87 ft. lbs. (100–120 Nm)

Upper mount through bolt nut – 45–60 ft. lbs. (33–43 Nm)

Lower mount through bolt nut – 36–43 ft. lbs. (47–60 Nm).

Cylinder Head

REMOVAL & INSTALLATION

4-Cylinder Except 1.5L, 1.6L Engines and Eclipse

1. Turn the engine until the No. 1 piston is at TDC on the compression stroke. Disconnect the negative battery cable. Remove the air cleaner assembly.
2. Drain the engine coolant. Remove the upper radiator hose and disconnect the heater hoses.
3. Disconnect the fuel lines, wiring harnesses, distributor vacuum lines, spark plug wires (from plugs), purge valves, accelerator linkage and water temperature unit wire.
4. Remove the distributor and (if necessary) the fuel pump from the cylinder head.
5. Remove the nuts connecting the exhaust pipe to the manifold or turbocharger. Lower the exhaust pipe.
6. Remove the turbocharger and/or exhaust manifold.
7. Remove the intake manifold assembly.
8. On 1.8L, 2.0L and 2.3L engines:
   a. Remove the upper, outer front cover. Align the timing mark on the cylinder head with the mark on the camshaft sprocket, engine should already be on the No. 1 piston TDC of the compression stroke.
   b. Matchmark the timing belt with the timing mark on the camshaft sprocket using a felt tip marker.
   c. Remove the sprocket and insert a 2 in. piece of rubber or other material between the camshaft sprocket and sprocket holder on the lower front cover, to hold the sprocket and belt so the valve timing will not be changed.

d. Remove the timing belt upper under cover and the rocker arm cover.

9. On 2.5L engines:

a. Remove the rocker arm cover.

b. Position the camshaft sprocket dowel pin at the 12 o'clock position with the timing mark TDC at the front of the timing case cover; engine should already be on the No. 1 piston TDC of the compression stroke.

c. Match the timing chain with the timing mark on the camshaft sprocket. Take a soft piece of wire and secure the chain and sprocket together at the timing mark and opposite side.

d. Remove the camshaft sprocket bolt, gear and sprocket from the camshaft.

10. Except on the Starion, a special hex head wrench will be needed. Mitsubishi part MD998051-01 or equivalent. Loosen and remove the cylinder head bolts in 2–3 stages to avoid cylinder head warpage.

11. Remove the cylinder head from the engine.

12. Clean the cylinder head and block mating surfaces and install a new cylinder head gasket.

To install:

13. Position the cylinder head on the engine block, engage the dowel pins front and rear and install the cylinder head bolts.

14. The bolts must be torqued to cold specification, which is 65–72 ft. lbs. in 2 equal stages. Torque the bolts in order to 32.5–36 ft. lbs. then, repeat the operation torquing them to the full torque. Note that on the Starion, the front head bolts, attaching the head only to the timing cover, are torqued to 11–15 ft. lbs. Torque them to about 7 ft. lbs. the first time around.

15. Install the timing belt upper undercover, 1.8L and 2.0L engines.

16. Locate the camshaft in original position. Pull the camshaft sprocket and belt or chain upward and install on the camshaft.

NOTE: If the dowel pin and the dowel pin hole does not align between the sprocket and the spacer or camshaft, move the camshaft by bumping either of both projections provided at the rear of No. 2 cylinder exhaust cam of the camshaft, with a light hammer or other tool, until the hole and pin align. Be certain the crankshaft does not turn.

17. Install the camshaft sprocket bolt and the distributor gear and tighten.

18. Install the timing belt upper front cover and spark plug cable support.

19. Apply sealant to the intake manifold gasket on both sides. Position the gasket and install the intake manifold. Tighten the nuts to specifications. Be sure no sealant enters the jet air passages, when equipped.

20. Install the exhaust manifold gaskets and the manifold assembly. Tighten the nuts to specifications.

21. Connect the exhaust pipe to the exhaust manifold and install the fuel pump. Install the purge valve.

22. Install the water temperature gauge wire, heater hoses and the upper radiator hose.

23. Connect the fuel lines, accelerator linkage, vacuum hoses and the spark plug wires.

24. Fill the cooling system and connect the battery ground cable. Install the distributor.

25. Temporarily adjust the valve clearance to the cold engine specifications.

26. Install the gasket on the rocker arm cover and temporarily install the cover on the engine.

27. Start the engine and bring it to normal operating temperature. Stop the engine and remove the rocker arm cover.

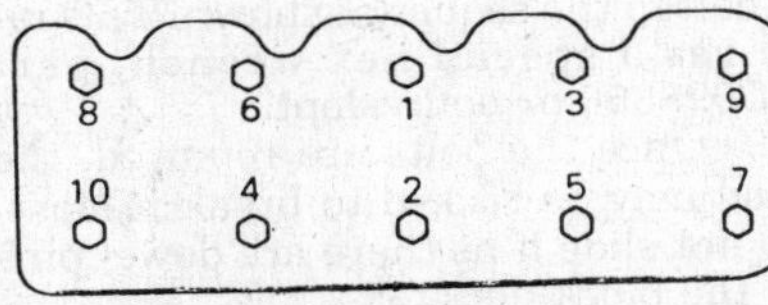

Engine head bolt torque sequence for 1.8L, 2.0L and 2.3L engines

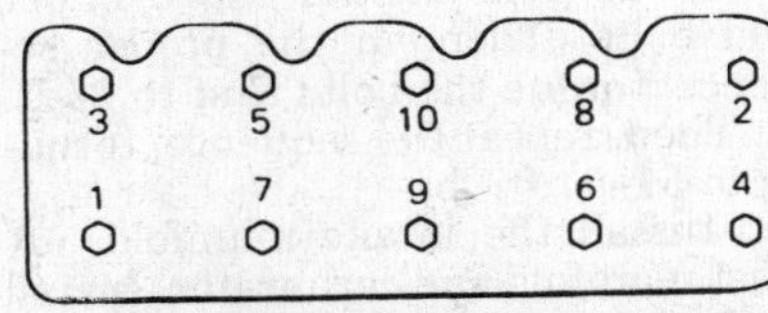

Engine head bolt torque removal sequence for 1.8L, 2.0L and 2.3L engines

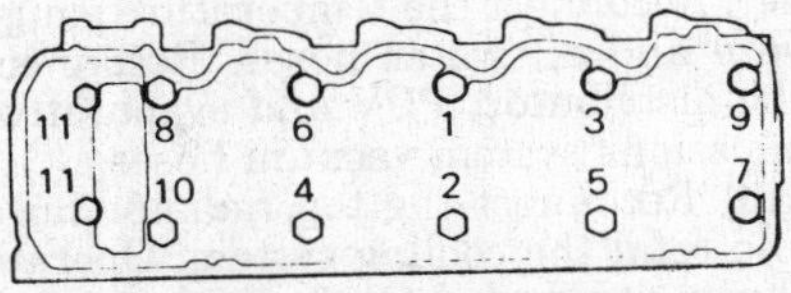

Engine head bolt torque sequence for the 2.5L engine

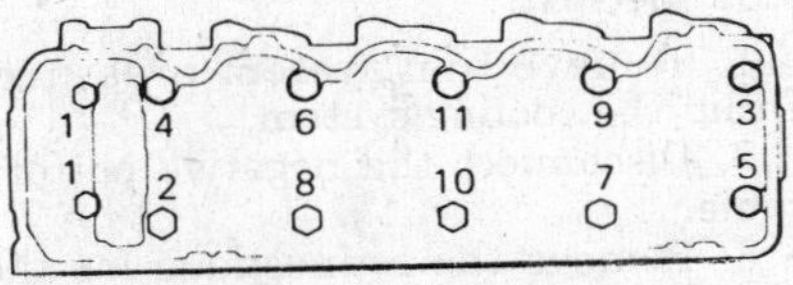

Engine head bolt removal sequence for the 2.5L engine

28. Adjust the valves to hot engine specifications.

29. Reinstall the rocker arm cover and tighten securely.

30. Install the air cleaner, hoses, purge valve hose, and any other removed unit.

1.5L Engine

1. Disconnect the negative battery cable. Drain the cooling system and then disconnect the upper radiator hose. Remove the PCV hose that runs between the air cleaner and the rocker cover.

2. Remove the air cleaner. Disconnect the fuel lines. Label and disconnect any vacuum lines running to the cylinder head, manifold, or carburetor from other parts of the engine compartment. Disconnect the heater hoses going to the head.

3. Label and disconnect the spark plug wires. Remove the rocker cover. Turn the crankshaft over until the TDC timing marks align and both No. 1 cylinder valves are closed; both rockers are off the cams. Then, remove the distributor.

4. Remove the carburetor, intake manifold and the exhaust manifold.

5. Remove the timing belt cover. Note the location of the camshaft sprocket timing mark. Loosen both timing belt tensioner bolts and lever it toward the water pump as far as it will go. Retighten the adjusting bolt to hold the tensioner in this position. Pull the timing belt off the camshaft sprocket but leave it engaged with the other sprockets.

6. Using a hex-type wrench, part MD998360, loosen the head bolts in the proper sequence. When all have been loosened, remove them. Then, pull the head off the engine block, rocking it slightly to break it loose, if necessary.

7. Remove the gasket. If pieces of the gasket adhere to the head or block deck, scrape them off carefully, using a scraper that will not scratch the surfaces. Make sure none of the pieces gets into the engine.

To install:

8. Install a new head gasket (without any sealer) and then position the head on the block deck. Install all the bolts finger-tight. Torque them, in sequence, first to 25 ft. lbs., then, to 58–61 ft. lbs.

9. Install the timing belt on the camshaft tensioner and rotate the camshaft sprocket backward so the belt is tight on what is normally the tension side. Make sure all the timing marks are now aligned.

10. Loosen the timing belt tensioner adjusting bolt and allow spring tension to tension the belt. Make sure all timing marks are still aligned. If not, the

belt is out of time and must be shifted with the tensioner shifted back toward the water pump and locked there. Torque the adjusting bolt (on the right side and working through a slot) to 15–18 ft. lbs. After the tensioner adjusting bolt is torqued, torque the hinged bolt located on the opposite side. Don't torque the bolt first or the tension on the belt will be too great.

11. Turn the crankshaft 1 full turn in the normal direction of rotation. Loosen first the tensioner pivot bolt and the adjusting bolt. Torque them exactly as before—adjusting bolt (working in the slot) first. This extra step is necessary to ensure the timing belt is properly seated before final tension is adjusted.

12. Install the intake and exhaust manifolds, carburetor, distributor, air cleaner, rocker cover, timing belt cover, and all hoses in reverse of the above procedures. Refill the cooling system. Operate the engine and check for leaks.

1.6L Engine

1. Disconnect the negative battery cable. Drain engine coolant and then disconnect the upper radiator hose at the thermostat. Remove PCV and canister purge hoses.

2. Remove the air cleaner. Disconnect the fuel line. Disconnect vacuum hoses at the distributor and canister purge control valve.

3. Label and then disconnect the spark plug wires. Remove the rocker cover. Turn the crankshaft over until the TDC timing marks align and both No. 1 cylinder valves are closed (both rockers are off the cams). Then, remove the distributor.

4. Disconnect the heater hose at the intake manifold. Disconnect the water hose leading from the cylinder head and carburetor water jacket.

5. Disconnect the temperature gauge sending unit wire at the head. Remove the fuel pump.

6. Remove the exhaust manifold. Turn the crankshaft until No. 1 piston is at TDC of its compression stroke. Align the timing mark on the upper cover at the rear of the timing belt with the mark on the camshaft sprocket to do this. Use some sort of marker to mark the relationship between the timing belt and the mark on the cam sprocket.

7. Remove the sprocket-to-camshaft bolt, hold the sprocket in position and work to keep the belt from slipping off. Then, rest the sprocket on the sprocket holder provided on the lower front cover. If necessary, slip a short piece of used timing belt or other thin, flexible object between the holder and the sprocket to keep tension and avoid losing belt timing. Be sure not to turn the crankshaft throughout this work.

8. Remove the bolts from the timing belt rear cover and remove the cover.

9. Remove the cylinder head bolts. Loosen in three stages, going from bolt to bolt in the sequence shown. This requires a special hex wrench, part MD998360 or equivalent.

10. Once the bolts are removed, the head may be rocked to break it loose. Do not slide it as there are dowel pins on the block deck.

To install:

11. Install a new head gasket without sealant, and install the head over the dowel pins. Install the cylinder head bolts. Then, in the proper sequence, torque the bolts first to 25 ft. lbs. Then, repeat the sequence, torquing to 51–54 ft. lbs.

12. Install the intake manifold, exhaust manifold and carburetor. Install the fuel pump and reconnect all fuel lines.

13. Install the distributor. Reconnect the plug wires in the proper firing order. Reconnect the temperature gauge wire and all water hoses. Reconnect the distributor, PCV and evaporative emissions system vacuum hoses.

14. Reconnect the top radiator hose and refill the cooling system. Operate the engine and check for leaks.

Eclipse

1.8L ENGINE

1. Relieve fuel system pressure. Drain the cooling system.

2. Disconnect the negative battery cable.

3. Remove the connections for the air intake hose and the breather hose.

4. Remove the connection for the accelerator cable. There will be 2 cables, if equipped with cruise control.

5. Remove the connection for the high pressure fuel line.

6. Remove the upper radiator hose, the water breather hose, the water bypass hose and the heater hose.

7. Remove the connector for the PCV hose.

8. Remove the spark plug cables.

9. Remove the fuel return line.

10. Remove the vacuum line for the brake booster.

11. Remove the electrical connections for the oxygen sensor, engine coolant temperature gauge unit and the water temperature sensor.

12. Remove the electrical connections for the ISC motor, throttle position sensor, distributor, MPS, fuel injectors, EGR temperature sensor (California vehicles), power transistor, condenser and ground cable.

13. Remove the engine control wiring harness.

14. Remove the clamp that holds the power steering pressure hose to the engine bracket.

15. Place a jack and wood block under the oil pan and carefully lift just enough to take the weight off the engine bracket and remove the bracket.

16. Remove the rocker cover, gasket and half-round seal.

17. Remove the timing belt front upper cover.

18. Remove the camshaft sprocket. First rotate the crankshaft clockwise until the timing marks on the sprocket align. Remove the sprocket with the timing belt attached and place on the timing belt front lower cover. Remove the timing belt rear upper cover.

19. Remove the exhaust pipe self-locking nuts and separate the exhaust pipe from the exhaust manifold. Discard the gasket.

20. Loosen the cylinder head bolts according to sequence in 2–3 steps and lift off the cylinder head assembly.

To install:

21. Installation is the reverse of the removal process. Check the cylinder head for cracks, damage or engine coolant leakage. Remove scale, sealing compound and carbon. Clean oil passages thoroughly. Check the head for flatness. End to end, the head should be within 0.002 in. normally with 0.008 in. the maximum allowed out of true. The total thickness allowed to be removed from the head and block is 0.008 in. maximum.

22. Place a new head gasket on the cylinder block with any identification marks facing upward.

23. Carefully install the cylinder head on the block. Tighten the bolts in sequence and torque in 3 steps to 51–54 ft. lbs.

24. Complete installation by reversing the removal procedure.

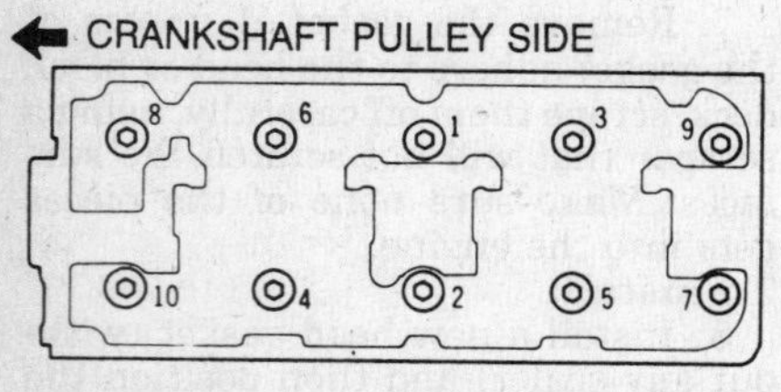

Engine head bolt torque sequence for 1.5L and 1.6L engines

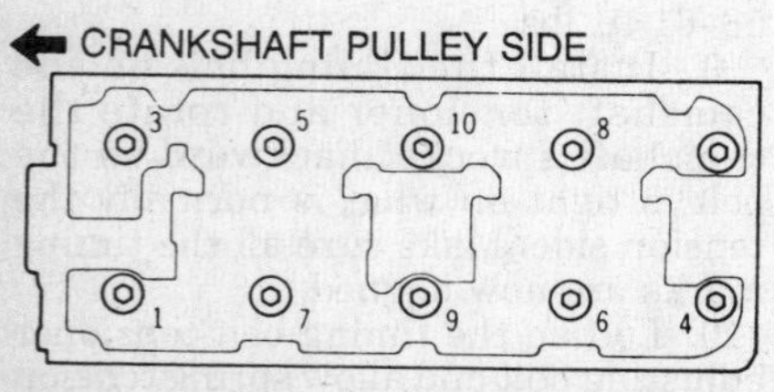

Engine head bolt removal sequence for 1.5L and 1.6L engines

2.0L ENGINE

1. Relieve fuel system pressure. Drain the cooling system.
2. Disconnect the negative battery cable.
3. Remove the connection for the accelerator cable. There will be 2 cables, if equipped with cruise control.
4. Remove the electrical connections for the oxygen sensor, engine coolant temperature sensor, the engine coolant temperature gauge unit and the engine coolant temperature switch on vehicles with air conditioning.
5. Remove the electrical connections for the ISC motor, throttle position sensor, crankshaft angle sensor, fuel injectors, ignition coil, power transistor, noise filter, knock sensor on turbocharged engines, EGR temperature sensor (California vehicles) and ground cable.
6. Remove the engine control wiring harness.
7. Remove the upper radiator hose and the overflow tube.
8. Remove the connections for the air intake hose on turbocharged models, and the breather hose. Remove the large bellows-type air intake hose.
9. Remove the connection for the high pressure fuel line.
10. Remove the small vacuum hoses.
11. Remove the heater hose and water bypass hose.
12. Remove the PCV hose.
13. If turbocharged, remove the vacuum hoses, water line, eyebolt connection for the oil line for the turbo.
14. Remove the fuel return hose.
15. Remove the brake booster hose.
16. Remove the timing belt.
17. Remove the rocker cover and the half-round seal.
18. On non-turbo vehicles, remove the exhaust pipe self-locking nuts and separate the exhaust pipe from the exhaust manifold. Discard the gasket.
19. On turbocharged engines, remove the sheetmetal heat protector.
20. Loosen the cylinder head bolts according to sequence in 2–3 cycles and lift off the cylinder head assembly.

To install:

21. Installation is the reverse of the removal process. Check the cylinder head for cracks, damage or engine coolant leakage. Remove scale, sealing compound and carbon. Clean oil passages thoroughly. Check the head for flatness. End to end, the head should be within 0.002 in. normally with 0.008 in. the maximum allowed out of true. The total thickness allowed to be removed from the head and block is 0.008 in. maximum.
22. Place a new head gasket on the cylinder block with any identification marks facing upward.
23. Carefully install the cylinder head on the block. Tighten the bolts in sequence and torque in 3 steps to 65–72 ft. lbs.
24. Complete installation by reversing the removal procedure.

3.0L SOHC Engine

1. Bring the No. 1 cylinder up to **TDC** of its compression stroke. Disconnect the negative battery cable. Drain the engine coolant into a suitable drain pan. Remove the air intake plenum.
2. Remove the upper radiator hose. Disconnect the engine control wiring harness.
3. Disconnect and plug the fuel lines, be sure to release the fuel pressure in the fuel system first. Remove the intake manifold and manifold gasket.
4. Disconnect and tag the spark plug wires. Disconnect the oxygen sensor connector.
5. Remove the self-locking nuts from the exhaust pipe and then suspend the exhaust pipe with a piece of wire.
6. Remove the distributor assembly and the air intake plenum stay. Remove any bolts that attach hoses or pipes to the cylinder heads.
7. Remove the heat protector bolts, then remove the heat protector along with the exhaust manifolds.
8. Remove the oil level gauge guide and the exhaust manifold gaskets.
9. Remove the timing belt and camshaft sprocket. Use the camshaft special tool MB990767 or equivalent, to hold the camshaft sprocket still so as to break loose the camshaft sprocket bolt.
10. Remove the timing belt rear cover. Remove the small retaining bolt from the side of the cylinder heads.
11. Remove the rocker arm cover and rocker arm cover gasket.
12. Remove the cylinder head bolts, in sequence, in 2–3 cycles, and remove the cylinder heads and gaskets from the engine.
13. Installation is the reverse order of the removal procedure, except for the following:
    a. Using a suitable torque wrench, torque the head bolts, in sequence, using 2–3 steps, to a final torque of 73–79 ft. lbs. (hot) or 65–72 ft. lbs. (cold).
    b. Be sure to coat all O-rings with clean engine oil and apply a suitable sealant to the cut-out ends of the rocker arm covers.
    c. Be sure to use special tool MB990767 or equivalent, to hold the camshaft sprocket still so as to torque the camshaft sprocket bolt to 58–72 ft. lbs.

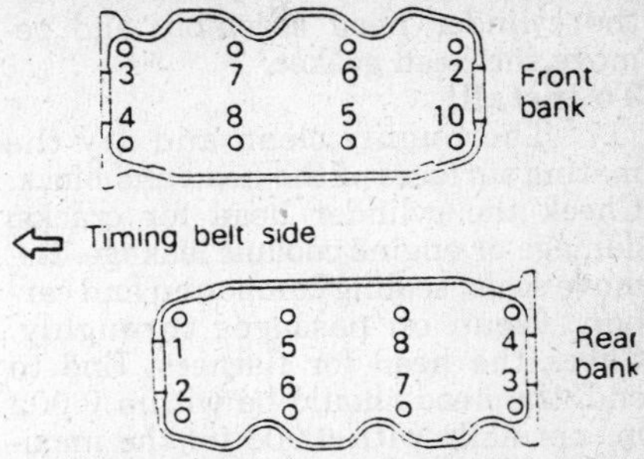

Cylinder head bolt removal sequence—3.0L engine

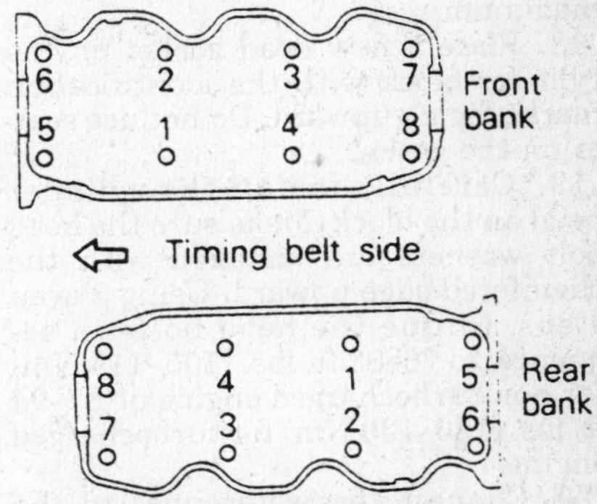

Cylinder head bolt torque sequence—3.0L engine

3.0L DOHC Engine

1. Reieve fuel system pressure. Disconnect the negative battery cable.
2. Drain the cooling system.
3. Remove the air intake hoses.
4. Remove air intake plenum and intake manifold.
5. Remove the turbocharger if equipped, and exhaust manifold.
6. Remove the timing belt.
7. Remove the triple pipe assembly across the top of the engine.
8. Remove the breather hose.
9. Remove the spark plug cable center cover and remove the spark plug cables.
10. When removing the valve cover, note that bolts for the front head are black and bolts for the rear head are green. Also, all bolts are 10mm long except the 1 closest to the sprockets on the rear head which is 20mm long.
11. To remove the intake camshaft sprocket, hold the camshaft with a wrench on the hexagon near the end of the camshaft and remove the bolt.
12. Remove the center rear timing belt cover.
13. Remove the ignition coil.
14. Disconnect all water hoses from the thermostat housing and remove the housing.
15. Disconnect the water inlet from the front head.
16. Loosen the cylinder head mounting bolts in 3 steps, starting from the outside and working inward. Lift off

the cylinder head assembly and remove the head gasket.

To install:

17. Thoroughly clean and dry the mating surfaces of the head and block. Check the cylinder head for cracks, damage or engine coolant leakage. Remove scale, sealing compound and carbon. Clean oil passages throughly. Check the head for flatness. End to end, the head should be within 0.002 in. normally with 0.008 in. the maximum allowed out of true. The total thickness allowed to be removed from the head and block is 0.008 in. maximum.
18. Place a new head gasket on the cylinder block with the identification marks facing upward. Do not use sealer on the gasket.
19. Carefully install the cylinder head on the block. Make sure the head bolt washers are installed with the chamfered edge upward. Using 3 even steps, torque the head bolts in sequence, to 76–83 ft. lbs. (105–115 Nm) for non-turbocharged engine or 87–94 ft. lbs. (120–130 Nm) for turbocharged engine.
20. Connect the water inlet to the front head.
21. Replace the gaskets and install the thermostat housing and connect the hoses.
22. Install the ignition coil and center rear timing belt cover.
23. Using the same procedure as in removal, install the intake camshaft sprocket. Torque the retaining bolt to 60–70 ft. lbs. (81–95 Nm).
24. Apply sealer to the lower edges of the half-round portions of the belt-side of the new gasket and install the valve cover. Make sure green bolts are installed on the rear head and black bolts are installed on the front head. Also, make sure the longest bolt is installed in its proper location closest to the sprockets on the rear head. Tighten the bolts in the proper sequence to 26 inch lbs. Then retighten bolts 1–6 to 36 inch lbs.
25. Connect the spark plug cables and install the center cover.
26. Install the breather hose.
27. Install the triple pipe assembly across the top of the engine.
28. Install the timing belt and all related items.
29. Using all new gaskets, install the intake manifold, air intake plenum, turbocharger and exhaust manifold, following the proper torque sequences.
30. Install the air intake hoses.
31. Change the engine oil.
32. Fill the system with coolant.
34. Connect the negative battery cable, run the vehicle until the thermostat opens, fill the radiator completely.
35. Check and adjust the idle speed and ignition timing.
36. Once the vehicle has cooled, recheck the coolant level.

Valve Lash

ADJUSTMENT

Valve lash must be adjusted on all engines not equipped with automatic lash adjusters. Some engines have an unusual third valve of very small size called a jet valve. The jet valve must be adjusted, whether the engine uses automatic lash adjusters for the normal intake and exhaust valves or not. Thus, on some engines, there are 3 valves per cylinder that must be adjusted.

1. Run the engine until operating temperature is reached.
2. Turn **OFF** the engine and block the wheels.
3. Remove all necessary components in order to gain access to the rocker cover.
4. Remove the spark plugs from the cylinder head for easy operations.
   a. On Starion, remove the air intake pipe and remove the rocker cover.
   b. On Cordia and Tredia, disconnect the oxygen sensor connecting joint. Remove the engine bracket mounting, be sure to place a block of wood on the oil pan and raise it into place for the duration of the operation. Remove the upper front timing belt cover, remove the air cleaner assembly (2.0L engine) and the air intake pipe (1.8L engine) and remove the rocker cover.
   c. On all other vehicles, remove the air cleaner or air intake pipe assembly and remove the rocker cover.
5. Turn each cylinder head bolt in the sequence back just until it is loose. Torque the cylinder head bolts in the proper sequence to specification.
6. Position the engine at **TDC** with No. 1 cylinder at the firing position. Turn the engine by using a wrench on the bolt in the front of the crankshaft until the **0** degree timing mark on the timing cover aligns with the notch in the front pulley. On some vehicles turn the crankshaft clockwise until the notch on the pulley is aligned with the **T** mark on the timing belt lower cover.
7. Observe the valve rockers for No. 1 cylinder. If both are in identical positions with the valves up, the engine is in the right position. If not, rotate the engine exactly 360 degrees until the **0** degree timing mark is again aligned. Each jet valve is associated with an intake valve that is on the same rocker lever. In this position, adjust all the valves marked **A**, including associated jet valves which are located on the rockers on the intake side only.
8. To adjust the appropriate jet valves, first loosen the regular (larger) intake valve adjusting stud by loosening the locknut and backing the stud off 2 turns. Note that this particular step is not required on engines that have automatic lash adjusters.
9. Loosen the jet valve (smaller) adjusting stud locknut, back the stud out slightly and insert the feeler gauge between the jet valve and stud. Make

Adjusting valve clearance

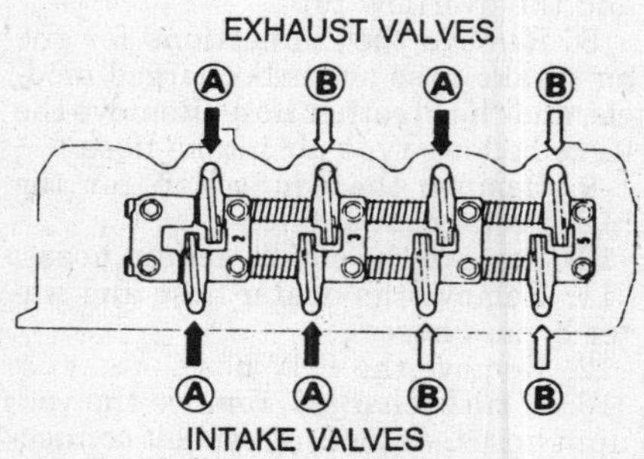

"A" and "B" valve adjusting positions

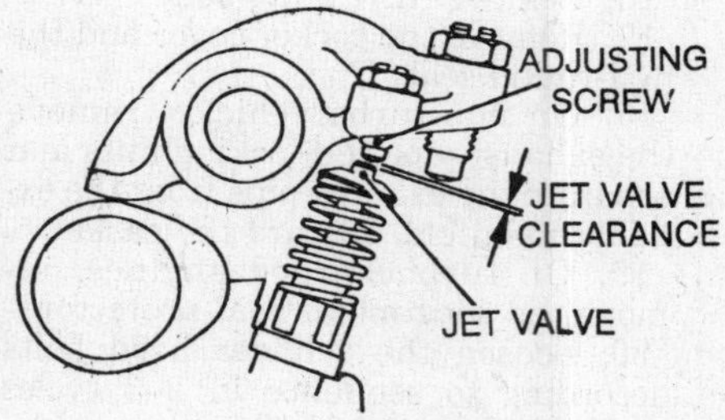

Jet valve adjusting

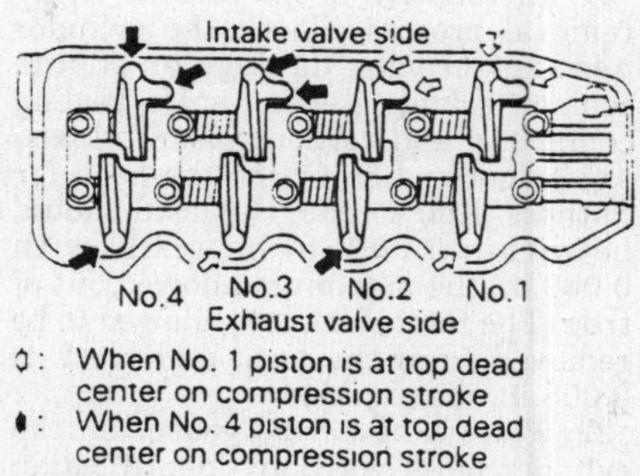

Typical valve adjustment sequence

sure the gauge lies flat on the top of the jet valve. Be careful not to twist the gauge or otherwise depress the jet valve spring, rotate the jet valve adjusting stud back in until it just touches the gauge. Tighten the locknut. Make sure the gauge still slides very easily between the stud and jet valve and that they both are still just touching the gauge.

NOTE: The clearances must not be too tight.

10. Repeat the entire procedure for the other jet valves associated with rockers labeled **A**.
11. On engines without automatic lash adjusters, repeat the procedure for the intake valves labeled **A**.
12. Repeat the basic adjustment procedure for exhaust valves labeled **A** on engines without automatic lash adjusters.
13. Turn the engine exactly 360 degrees, until the timing marks are again aligned at **O** degrees BTDC.
14. On engines with automatic lash adjusters, after the jet valves and rockers on the intake side and labeled **B** are adjusted, the valve adjustment procedure is completed. If equipped without automatic lash adjusters, adjust the regular intake and exhaust valves labeled **B**.
15. Reinstall the cam cover. Run the engine to check for oil leaks.

JET VALVE ADJUSTMENT

NOTE: An incorrect jet valve clearance would affect the emission levels and could also cause engine troubles, so the jet valve clearance must be correctly adjusted. Adjust the jet valve clearance before adjusting the intake valve clearance. Furthermore, the cylinder head bolts should be retightened before making this adjustment. The jet valve clearance should be adjusted with the adjusting screw on the intake valve side fully loosened.

1. Start the engine and let it run at idle until it reaches normal operating temperature.
2. Remove all spark plugs from the cylinder head for easy operation.
3. On Starion—remove the air intake pipe and remove the rocker cover.
4. On Cordia and Tredia—disconnect the oxygen sensor connecting joint. Remove the engine bracket mounting, be sure to place a block of wood on the oil pan and raise into to place for the duration of the operation. Remove the upper front timing belt cover, remove the air cleaner assembly (2.0L engine) and the air intake pipe (1.8L engine) and remove the rocker arm.
5. On all other vehicles—remove the air cleaner or air intake pipe assembly and remove the rocker cover.
6. Put the engine at TDC with No. 1 cylinder at the firing position. Turn the engine by using a wrench on the bolt in the front of the crankshaft until the **0** degree timing mark on the timing cover aligns with the notch in the front pulley; on some vehicles turn the crankshaft clockwise until the notch on the pulley is aligned with the **T** mark on the timing belt lower cover. This will bring both No. 1 and No. 4 cylinder pistons to **TDC**.

NOTE: Never turn the crankshaft counterclockwise.

7. Move the rocker arms on the No. 1 and No. 4 cylinders up and down by hand to determine if the piston in that cylinder is at TDC on the compression stroke. If the intake and exhaust rocker arms do not move, the piston in that cylinder is not at **TDC** on the compression stroke.
8. Measure the jet valve clearance at point **A**.

NOTE: Measure the valve clearance when the No. 1 cylinder or the No. 4 cylinder pistons are at TDC on the compression stroke. Then give the crankshaft 1 clockwise turn to bring the other cylinder piston to TDC on compression stroke.

9. If the jet valve clearance is not 0.010 in. (hot) and 0.007 in. (cold), loosen the rocker arm locknut of the intake valve and loosen the adjusting screw at least 2 turns or more.
10. Loosen the jet valve locknut and adjust the clearance using a feeler gauge while turning the adjusting screw.

NOTE: The jet valve spring has a small tension and the adjustment is somewhat delicate. Be careful not to push in the jet valve by turning the adjusting screw in too much.

11. Tighten the adjusting screw until it touches the feeler gauge. Turn the locknut to secure it, while holding the rocker arm adjusting screw with a suitable tool to keep it from turning.
12. Check the intake and exhaust valve clearance, if it is not within specifications, adjust the valves as follows:
    a. Loosen the locknut on the adjusting screw for the valve. Turn the adjusting screw counterclockwise and insert the proper size feeler gauge between the valve stem and the adjusting screw.
    b. Tighten the adjusting screw until it touches the feeler gauge. Turn the locknut to secure it, while holding the rocker arm adjusting screw with a suitable tool to keep it from turning.
13. Turn the engine by using a wrench on the bolt in the front of the crankshaft 360 degrees until the **0** degree timing mark on the timing cover aligns with the notch in the front pulley; on some vehicles, turn the crankshaft clockwise until the notch on the pulley is aligned with the **T** mark on the timing belt lower cover.
14. Repeat Steps 9 through 13 on the other valves (marked **B**) for clearance adjustment.
15. Reinstall all the remove components in the reverse order of the removal procedure.

Rocker Arms/Shafts

REMOVAL & INSTALLATION

4 Cylinder Engines Except 1.5L and 1.6L Engines

NOTE: If equipped with hydraulic lash adjusters, 8 special holders, tool MD998443, or equivalent, are required to retain the hydraulic lash adjusters when disassembling the valve train.

1. Disconnect the negative battery cable.
2. Remove the rocker cover and, on Cordia, Tredia and Galant the upper timing belt cover.
3. Loosen the camshaft sprocket bolt until it can be turned by hand.
4. Turn the engine over until the camshaft sprocket timing mark aligns with the timing mark on the cylinder head on the Cordia, Tredia and Galant. On the Starion, the timing mark on the sprocket ends up on the extreme right of the sprocket bolt as viewed from the front. In both cases the TDC mark on the front crankshaft pulley must align with the timing scale on the front cover.
5. Remove the camshaft sprocket bolt and without allowing tension on the timing chain or belt to be lost, place the sprocket in the sprocket holder of the front cover or lower timing belt cover. Make sure the crankshaft is not turned throughout the work.
6. On 1988 Cordia, Tredia and Galant with hydraulic lash adjusters, put the special clips on the 8 hydraulic adjusters at the outer ends of all 8 rocker arms. Note that these clips go over the lash adjusters that actuate the large intake valves, not on the small adjusting screw for the smaller jet valves.
7. Loosen but do not remove the camshaft bearing cap bolts. After all bolts have been loosened, remove them and then, holding the ends so the

assembly stays together, remove the rocker shaft assembly from the cylinder head. Note that the rear most cam bearing cap is not associated with the rocker shafts on the Starion and need not be removed.

8. Keep all parts in original order. Assemble the parts of the rocker assembly as follows:

a. Cordia, Tredia and Galant install left and right side rocker shafts into the front bearing cap. Notches in the ends of the shaft must be upward.

b. Install the bolts for the front cap to retain the shafts in place. Note that the left rocker shaft is longer than the right rocker shaft.

c. Install the wave washer onto the left rocker shaft with the bulge forward.

d. Coat the inner surfaces of the rockers and the upper bearing surfaces of the bearing caps with clean engine oil and assemble rockers, springs and the remaining bearing caps in the order in which removed. The intake rockers are the only ones with the jet valve actuators.

NOTE: The rockers are labeled for cylinders 1–3 and 2–4 because the direction the jet valve actuator faces changes. Use bolts to hold the caps in place after each is assembled.

To install:

9. When the assembly is complete, install it onto the head and start all bolts into the head and tighten finger-tight.

10. On the Starion install the right and left rocker shafts into the front bearing cap. Note that shafts can be identified by the fact that the rear end of the left side shaft has a notch. Align the mating marks of the front of the rocker shaft with that on the front bearing cap. Insert the front bolts.

11. Install the waved washers on both sides with the bulge in the washers facing forward.

12. Install the rockers, shafts, caps and bolts in their original positions, using the bolts to hold each cap in place after it is installed.

13. Oil the inner surfaces of the rockers and the upper bearing surfaces of the caps with clean engine oil prior to assembly. Note that the valve actuating ends of the rockers must face outward and that only the intake side rockers have the jet valve actuator.

14. When the assembly is complete, install it onto the head and start all the bolts into the threads, tightening them finger-tight.

15. Torque the attaching bolts for the rocprocket goes into the hole in the front of the cam.

16. Install the bolt. Torque it to

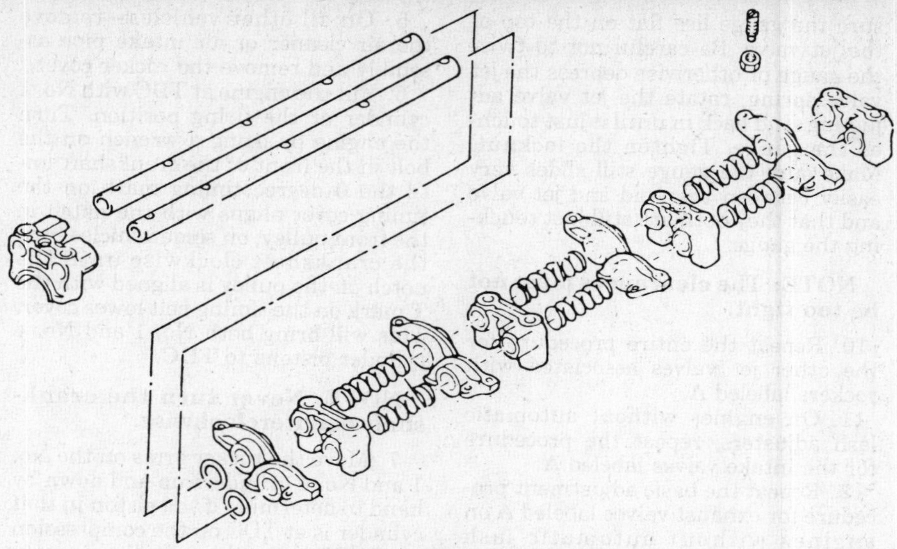

Typical rocker arm and shaft assembly

37–43 ft. lbs. on the Starion or 59–72 ft. lbs. on the Cordia, Tredia and Galant.

17. Adjust the valves.

18. Apply sealant to the top surface oker assembly 14–15 ft. lbs. going from the center outward.

19. Without removing tension from the timing chain or belt, lift the sprocket out of the holder and position it against the front of the cam. Make sure the locating tang on the sf the semicircular seals in the head and then install the valve cover. Install the upper timing belt cover on the Cordia, Tredia and Galant.

1.5L Engine

1. Disconnect the negative battery cable. Remove the PCV hose running from the rocker cover and the air cleaner. Remove the air cleaner.

2. Remove the upper timing belt cover. Remove the rocker cover.

3. Loosen the rocker shaft bolts, but do not remove them. After all bolts are loosened remove the rocker shaft, rocker arms and springs as an assembly.

4. Be sure to keep all parts in original order, for reinstallation.

To install:

5. Assemble all the parts, noting the differences between intake and exhaust parts. The intake rocker shaft is much longer; the intake rocker shaft springs are over 3 in. long, while those for the exhaust side are less than 2 in. long; intake rockers have the extra adjusting screw for the jet valve; rockers are labeled 1–3 and 2–4 for the cylinder with which they are associated. Torque the rocker shaft bolts to 15–19 ft. lbs.

6. Adjust the valve clearances. This step may be omitted only if all parts are being reused. Install the rocker cover with a new gasket, torquing the bolts to 1–1.5 ft. lbs. Install the air cleaner and PCV valve. Remember that there is no timing belt cover in place and keep fingers clear. Run the engine at idle speed until it is hot. Then remove the valve cover again and adjust the valve clearances with the engine hot. Finally, replace the rocker cover and timing belt cover, air cleaner and PCV valve.

1.6L Engine

1. Disconnect the negative battery cable. Remove the air cleaner assembly. Label and then disconnect the spark plug high tension wires. Remove the upper front timing belt cover.

2. Turn the crankshaft until No. 1 piston is at TDC of its compression stroke. Align the timing mark on the upper cover at the rear of the timing belt with the mark on the camshaft sprocket; mark the relationship between the timing belt and the mark on the cam sprocket.

3. Remove the sprocket-to-camshaft bolt; hold the sprocket in position to keep the belt from slipping off. Then, rest the sprocket on the sprocket holder provided on the lower front cover. If necessary, slip a short piece of used timing belt or other thin, flexible object between the holder and the sprocket to keep tension and avoid losing belt timing. Be sure not to turn the crankshaft throughout this work.

4. Remove the upper cover located behind the timing belt. Remove the rocker cover. Loosen the camshaft bearing cap bolts without pulling them out of the caps and remove the caps, rockers and shafts as an assembly.

5. Be sure to keep all parts in original order, for reinstallation.

To install:

6. Lubricate all wear surfaces with clean engine oil and then insert both rocker shafts into the front bearing cap with the cuts at the top/front of the caps at the tops. Note that the longer shaft goes on the left side (facing the crankshaft pulley). Note that the intake rockers only have the jet valve actuators and that the waved washers are installed behind the last set of rockers with the bulge at the center of the washer facing the crankshaft pulley. After each cap goes on and the holes are aligned, install the bolts to keep it in place. Note that if the camshaft front oil seal has been damaged it must be replaced.

7. Lubricate the wear surfaces of the cam bearing caps and then install them. Torque the bolts to 14–15 ft. lbs.

8. Install the timing belt rear cover. Pull the camshaft sprocket upward and install it to the camshaft. Turn the camshaft slightly if necessary to make the dowel pin fit into the hole in the sprocket. Make sure the mating mark made when these parts were disassembled are still aligned so the camshaft will be in time. Make corrections as necessary. Install the sprocket bolt and torque it to 44–57 ft. lbs.

9. Install the timing belt upper cover and spark plug high tension wire supports. Adjust the valve clearances. Apply sealant to the top of the front bearing cap and rear of the head where the rocker cover seals and then install the rocker cover. Install new gaskets and install the rocker cover, torquing the bolts to 4–5 ft. lbs. Reconnect the spark plug wires and install the air cleaner and PCV and evaporative emissions hoses. Run the engine at idle speed until it is hot. Then, remove the rocker cover and again set the valves with the engine hot.

3.0L SOHC Engine

1. Disconnect the negative battery cable. Remove the rocker arm covers.
2. To remove the rocker arm assembly, remove the camshaft bearing cap retaining bolts and remove the rocker arm assembly from the cylinder head. Install tool MD998443-01 or equivalent, over the auto lash adjuster to keep it from falling out.
3. When disassembling the rocker arm shaft assembly, be sure to remove one rocker arm and spring at a time and keep all the parts in their original order.
4. Check the rocker arms and rocker arm shafts for any cracks, distortion, wear or heat damage and replace as necessary.

To install:

5. Reassemble the rocker arm shaft assemblies in the reverse order of disassembly. If the auto lash adjuster should fall out during disassembly, reinstall it from under the rocker arm, using caution so as not to spill the diesel fuel inside the adjuster. Install tool MD998443-01 or equivalent, over the auto lash adjuster to keep it from falling out.
6. Apply a small amount of a suitable sealant on the 4 corners of the cylinder head just in front of the bearing caps at the end of the rocker arm shaft.

NOTE: Be sure the sealant does not swell out onto the cam journal surface of the cylinder head. If it swells out, immediately wipe it off before it can dry.

7. Attach the rocker arm shaft assemblies so the arrow mark on the bearing cap faces in the same direction as the arrow on the cylinder head.

NOTE: The arrow marks face each other on the rocker arm shaft assemblies. Since bearing caps number 1 and 4 look alike, check the number stamped on the cap. Be sure to coat the inside of the bearing cap and rocker arm with clean engine oil before assembling them.

8. Insert bearing cap No. 1 so the notch on the end of the shaft faces in the direction as shown in the illustration provided and insert the bolts. Be sure the oil groove faces downward and the oil port is located on the rocker shaft side **A**.
9. Install all the bearing cap bolts and torque them to 15 ft. lbs. Remove the auto lash adjuster special tool.
10. Apply a suitable sealant to the rocker arm covers and install the covers to the cylinder head. Reinstall any other removed components and reconnect the negative battery cable.

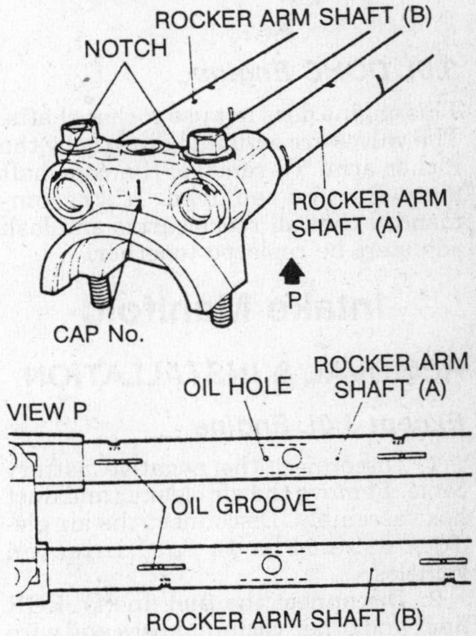

Installing the bearing cap No. 1 – 3.0L engine

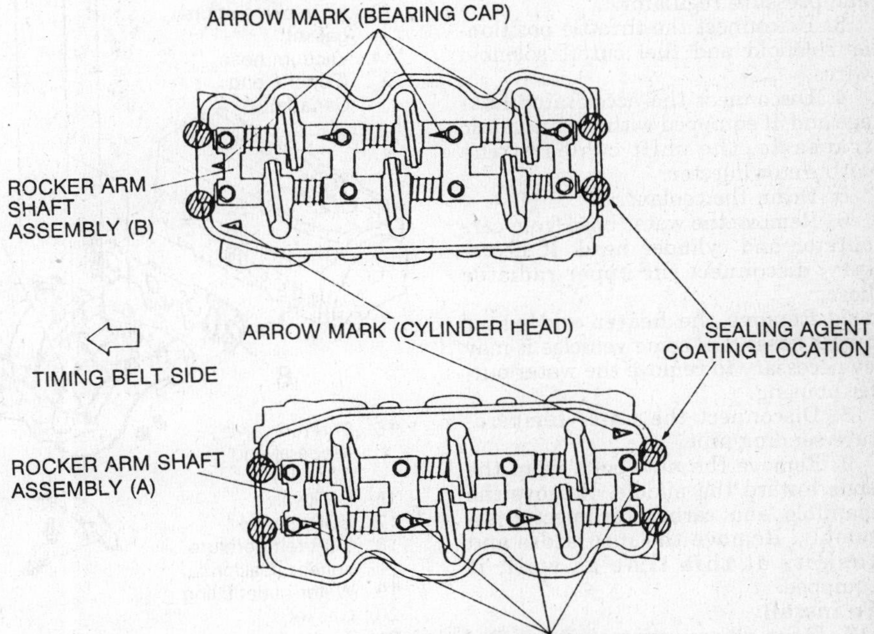

Rocker arm shaft assemblies installation – 3.0L engine

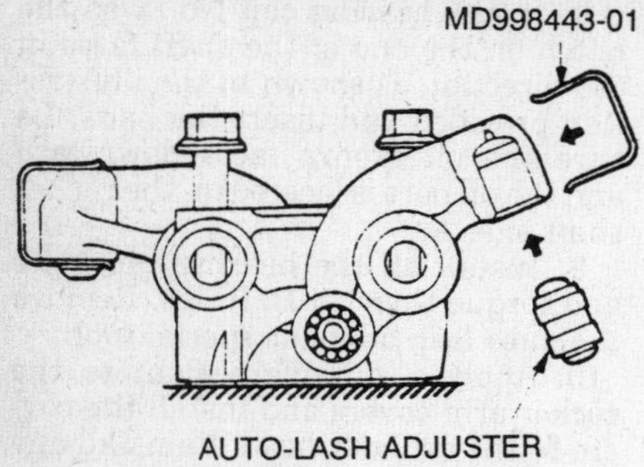

Auto lash adjuster installation – 3.0L engine

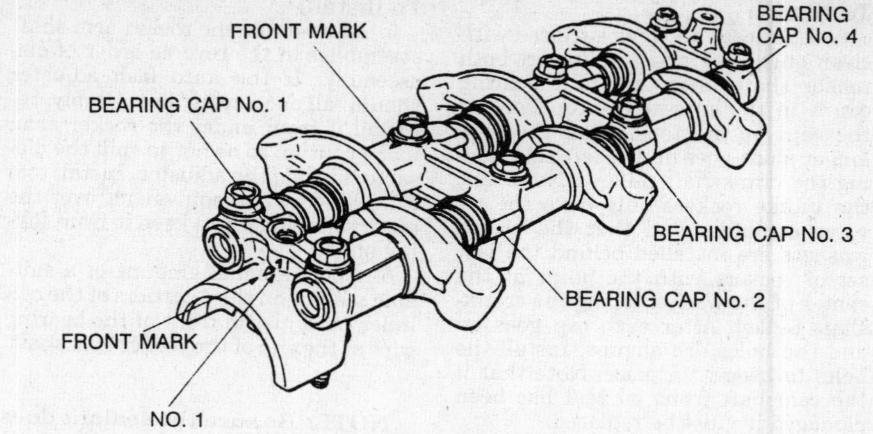

Rocker arm assembly for the 3.0L engine

3.0L DOHC Engine

This engine does not use rocker shafts. The valves are actuated directly by the rocker arm. To remove, the camshaft must first be removed. It's recommended that all rocker arms and lash adjusters be replaced together.

Intake Manifold

REMOVAL & INSTALLATION

Except 3.0L Engine

1. Disconnect the negative battery cable. Remove the air cleaner and duct hose assembly. Disconnect the air plenum assembly on fuel injected vehicles.
2. Disconnect the fuel line(s), EGR lines and other vacuum hoses and wire harness connectors. On the fuel injected vehicles, it is necessary to remove the fuel delivery pipe, fuel injectors and pressure regulator.
3. Disconnect the throttle positioner solenoid and fuel cutoff solenoid wires.
4. Disconnect the accelerator linkage and, if equipped with an automatic transaxle, the shift cables at the carburetor/injector.
5. Drain the coolant.
6. Remove the water hose from carburetor and cylinder head. If necessary, disconnect the upper radiator hose.
7. Remove the heater and water outlet hoses. On some vehicles it may be necessary to remove the water outlet housing.
8. Disconnect the water temperature sending unit.
9. Remove the nuts/bolts from the ends toward the middle. Remove the manifold and carburetor/injector assembly. Remove the insulators and gaskets at this time as well, if equipped.

To install:

10. Clean all mounting surfaces. Before reinstalling the manifold, coat both sides with gasket sealer. Install

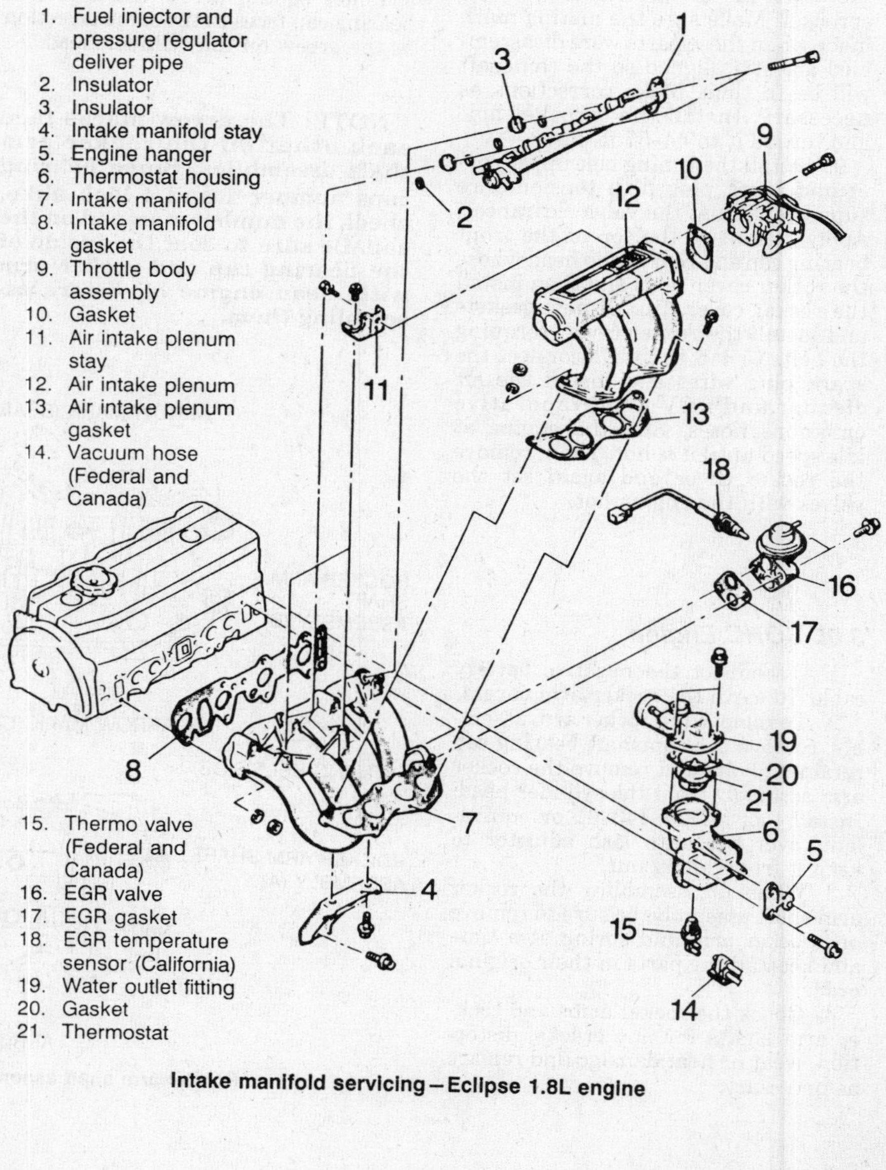

Intake manifold servicing – Eclipse 1.8L engine

nuts/bolts starting from the center toward the ends. Torque nuts/bolts to the proper torque, in the proper sequence.

11. Install the water temperature sending unit and heater and water outlet hoses. Fill the system with coolant.

12. Install the accelerator linkage, all fuel lines and all wiring previously removed.

13. Install the air plenum and all other components removed. Start the engine and check for leaks.

3.0L Engine

1. Relieve the fuel system pressure.
2. Disconnect battery negative cable and drain the cooling system.
3. Remove the air intake hose(s).
4. Disconnect the throttle control cables from the throttle body.
5. Matchmark and disconnect the vacuum hoses including the brake booster hose.
6. Disconnect all harness connectors.
7. Disconnect EGR components on California vehicles.
8. Remove the plenum retaining bracket.
9. Remove the plenum retaining nuts and bolts and remove the air intake plenum. Discard the gasket.
10. Disconnect the high pressure and return fuel hoses.
11. Matchmark and disconnect the vacuum hoses.
12. Disconnect the harness connector.
13. Remove the fuel rail with the injectors attached.
14. On SOHC engines, disconnect the water hoses. On DOHC engines, remove the timing belt upper cover.
15. Remove the intake manifold mounting nuts; turbocharged engines have cone disc springs under some of the nuts which should be removed. Remove the intake manifold and discard the gaskets.

To install:

16. Check all items for cracks, clogging and warpage. Maximum warpage is 0.008 in. (0.2mm). Replace all questionable parts.
17. Thoroughly clean and dry the mating surfaces of the heads, intake manifold and air intake plenum.
18. Install new intake manifold gaskets to the heads with the adhesive side facing up.
19. Place the manifold on the heads install the cone disc springs and/or the lock washers.
20. Lubricate the studs lightly with oil, then install the nuts following this procedure:
    a. Tighten the nuts on the front bank to 26–43 inch lbs. (3–5 Nm).
    b. Tighten the nuts on the rear bank to 9–11 ft. lbs. (12–15 Nm).
    c. Tighten the nuts on the front bank to 9–11 ft. lbs. (12–15 Nm).
    d. Repeat Steps B and C.
    e. On non-turbocharged engines only, tighten the nuts to a final torque of 13–14 ft. lbs. (18–19 Nm).
21. On SOHC engines, connect the water hoses. On DOHC engines, install the timing belt upper cover.
22. Install the fuel rail assembly.
23. Connect the harness connector and vacuum hoses.
24. Replace the O-ring and connect the fuel hoses.
25. Install a new intake air plenum gasket and install the plenum. Tighten the retaining nuts and bolts evenly and gradually to 13 ft. lbs. (18 Nm).
26. Install the retaining bracket.
27. Connect EGR components on California vehicles.

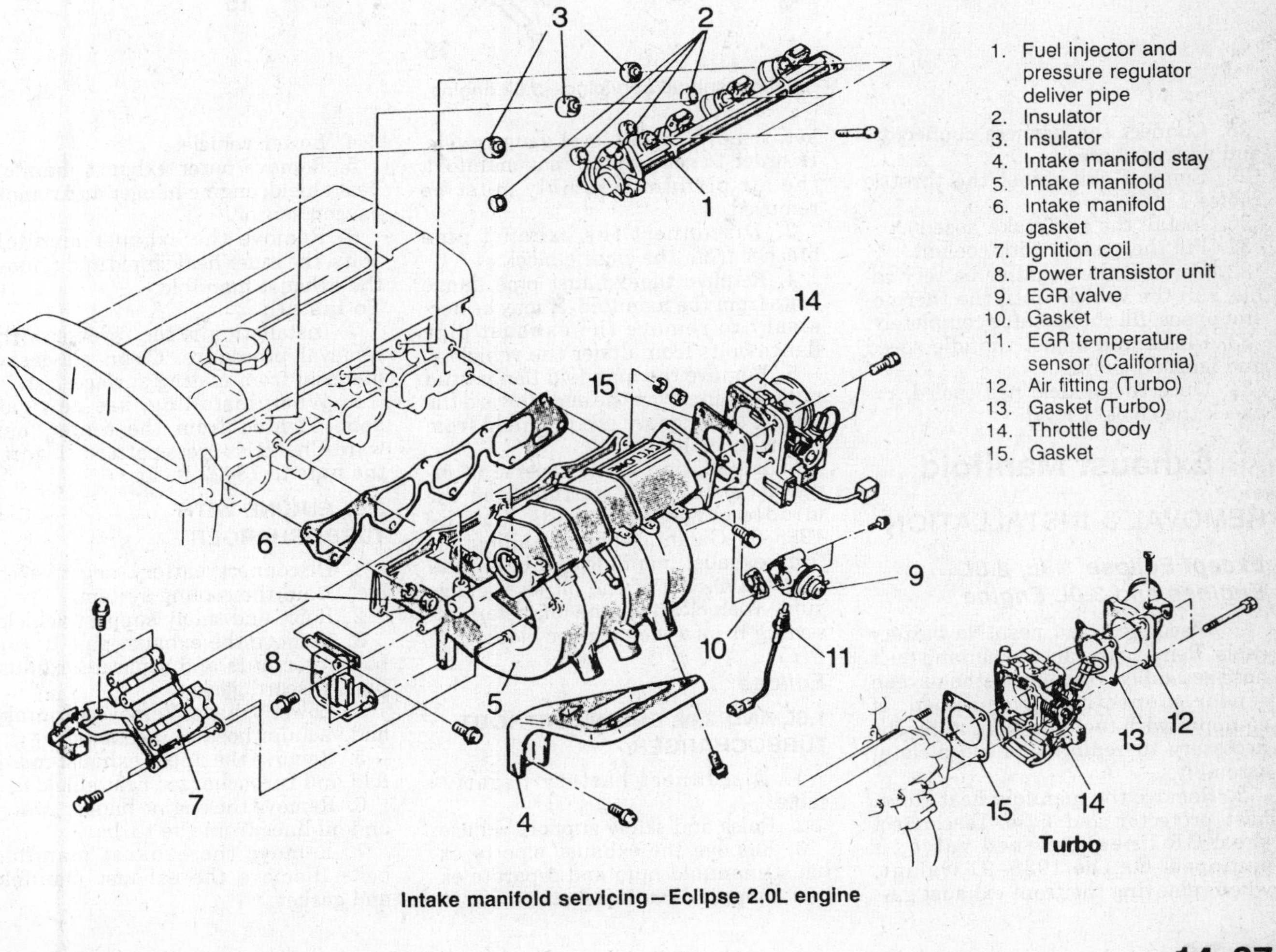

1. Fuel injector and pressure regulator deliver pipe
2. Insulator
3. Insulator
4. Intake manifold stay
5. Intake manifold
6. Intake manifold gasket
7. Ignition coil
8. Power transistor unit
9. EGR valve
10. Gasket
11. EGR temperature sensor (California)
12. Air fitting (Turbo)
13. Gasket (Turbo)
14. Throttle body
15. Gasket

Intake manifold servicing—Eclipse 2.0L engine

1. Air hose connection
2. Accelerator cable
3. Throttle body assembly
4. Throttle body gasket
5. Air pipe
6. Vacuum hose
7. Brake booster vacuum hose
8. Harness connector
9. Clutch booster vacuum hose
10. EGR temperature sensor (California)
11. EGR valve
12. EGR valve gasket
13. EGR pipe installaiton bolts
14. EGR pipe gasket
15. Intake plenum stay
16. Intake plenum bolts
17. Intake plenum nuts
18. Intake plenum
19. Intake plunum gasket

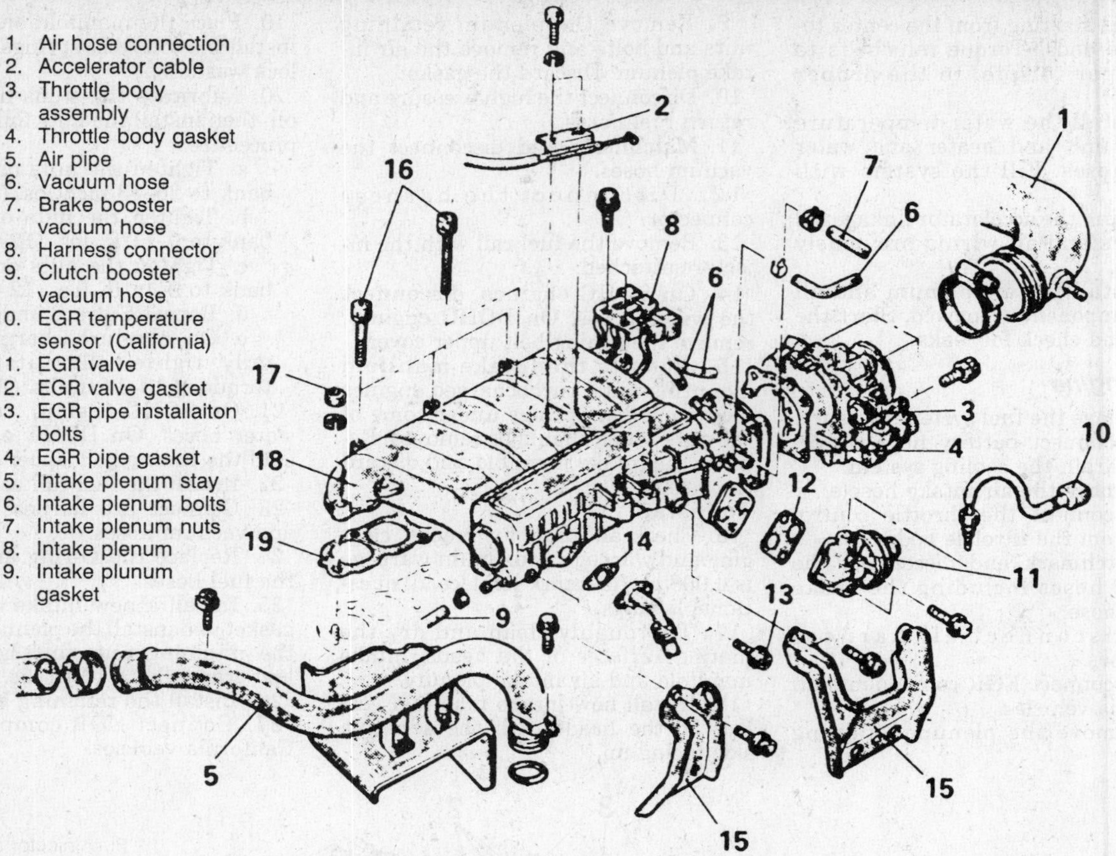

Intake manifold servicing—3.0L engine

28. Connect the harness connectors and vacuum hoses.
29. Connect and adjust the throttle cables.
30. Install the air intake hose(s).
31. Fill the system with coolant.
32. Connect the negative battery cable, run the vehicle until the thermostat opens, fill the radiator completely.
33. Check and adjust the idle speed and ignition timing.
34. Once the vehicle has cooled, recheck the coolant level.

Exhaust Manifold

REMOVAL & INSTALLATION

Except Eclipse 1.8L, 2.0L Engines and 3.0L Engine

1. Disconnect the negative battery cable. Remove the air cleaner and duct hose assembly. Disconnect the oxygen sensor connection, if equipped. If equipped with fuel injection, it may be necessary to remove the air plenum assembly.
2. Remove the manifold heat stove, heat protector and hose. Disconnect the EGR lines and reed valve, if equipped. On the 1988–91 Galant, when removing the front exhaust gasket, remove the oil level gauge guide. In order to remove the rear manifold, the air plenum assembly must be removed.
3. Disconnect the exhaust pipe bracket from the engine block.
4. Remove the exhaust pipe flange bolts from the manifold. It may be necessary to remove the exhaust pipe flange bolts from under the vehicle.
5. Remove the manifold flange stud nuts starting from the ends toward the middle and remove the manifold from the cylinder head.
6. Installation is the reverse of removal. Install nuts starting from the middle toward the ends. On the 1988–91 Galant, when installing the front exhaust manifold and be sure to coat the O-ring of the oil level gauge guide with clean engine oil, before inserting it into the cylinder block.

Eclipse

1.8L AND 2.0L ENGINE WITHOUT TURBOCHARGER

1. Disconnect battery negative cable.
2. Raise and safely support vehicle.
3. Remove the exhaust pipe to exhaust manifold nuts and separate exhaust pipe. Discard gasket.
4. Lower vehicle.
5. Remove outer exhaust manifold heat shield, engine hanger and remove oxygen sensor.
6. Remove the exhaust manifold bolts, the inner heat shield and remove the exhaust manifold.

To install:

7. Installation is the reverse of the removal procedure. Clean all gasket material from mating surfaces.
8. When installing, use new gaskets. Tighten from the center, outwards in a criss-cross pattern. Tighten the nuts to 18–22 ft. lbs.

2.0L ENGINE WITH TURBOCHARGER

1. Disconnect battery negative cable. Drain the cooling system.
2. Raise and safely support vehicle.
3. Remove the exhaust pipe to turbocharger nuts and separate exhaust pipe. Discard gasket.
4. Lower vehicle. Remove air intake and vacuum hose connections.
5. Remove the upper exhaust manifold and turbocharger heat shields.
6. Remove the engine hanger, water and oil lines from the turbo.
7. Remove the exhaust manifold bolts. Remove the exhaust manifold and gasket.

To install:

8. Installation is the reverse of removal. Clean all gasket material from mating surfaces.
9. When installing, use new gaskets. Tighten from the center, outwards in a criss-cross pattern. Tighten the nuts to 18–22 ft. lbs.

3.0L Engine

WITHOUT TURBOCHARGER

1. Disconnect battery negative cable.
2. Raise the vehicle and support safely.
3. Remove the exhaust pipe to exhaust manifold nuts and separate exhaust pipe. Discard gasket.
4. Lower vehicle.
5. Remove electric cooling fan assembly if necessary. If removing the front manifold on 3.0L engine, remove the dipstick tube. If removing the front manifold from 3.0L DOHC engine, remove the alternator.
6. Disconnect necessary EGR components.
7. Remove outer exhaust manifold heat shield, engine hanger and remove the oxygen sensor.
8. Remove the exhaust manifold mounting bolts, the inner heat shield and remove the exhaust manifold.

To install:

8. Clean all gasket material from the mating surfaces and check the manifold for damage.
9. Install a new gasket and install the manifold. Tighten the nuts to in a criss-cross pattern to:

   SOHC engines – 11–14 ft. lbs. (15–20 Nm).

   DOHC engine – 33 ft. lbs. (45 Nm).
10. Install the heat shields.
11. Connect EGR components.
12. Install the electric cooling fan assembly, dipstick tube or alternator.
13. Install a new gasket and connect the exhaust pipe.
14. Connect the negative battery cable and check for exhaust leaks.

WITH TURBOCHARGER

1. Disconnect the negative battery cable.
2. Drain the engine coolant.
3. Disconnect the exhaust pipe from the turbocharger and remove the turbocharger assembly.
4. Remove the heat shield.
5. Remove the mounting nuts and remove the exhaust manifold. Note that cone disc springs are installed at all lower mounting points.

To install:

6. Clean all gasket material from the mating surfaces and check the manifold for damage.
7. Install new gaskets and install the manifold. Make sure all cone disc springs are in their original locations with the grooved side facing the nut. Tighten the manifold nuts using the following procedure:
   a. Tighten all but the outer 2 nuts to 22 ft. lbs. (30 Nm).
   b. Tighten the outer 2 nuts to 34–38 ft. lbs. (47–53 Nm).
   c. Loosen the outer 2 nuts, then torque them to 22 ft. lbs. (30 Nm).
8. Install the heat shield.
9. Install the turbocharger assembly.
10. Fill the cooling system.
11. Connect the negative battery cable and check for exhaust leaks.

Turbocharger

REMOVAL & INSTALLATION

Except 2.0L and 3.0L Engines

1. Disconnect the negative battery cable. Remove the air cleaner and turbocharger inlet ducting. Unbolt the turbocharger discharge hose going to the injection mixer.
2. Disconnect the oxygen sensor at the catalytic converter to protect it.
3. Unbolt and remove the large heat shield that covers the top of the turbocharger.
4. Remove the nuts fastening the turbo to the catalytic converter. On Cordia, Tredia and Mirage, disconnect the oil return pipe at the oil pan. On Starion, disconnect the oil return hose at the oil return pipe and the timing chain cover.
5. Disconnect the oil supply line at the turbo and at the oil filter bracket.
6. Disconnect the turbocharger from the exhaust manifold and remove the unit from the vehicle.
7. Installation is the reverse of the removal procedure. Replace all gaskets, as required. Pour clean engine oil into the oil supply fitting before connecting the oil supply pipe.

2.0L Engine

1. Disconnect negative battery cable.
2. Drain engine oil, cooling system and remove radiator. If equipped with air conditioning, remove the condensor fan assembly with the radiator.
3. Disconnect the oxygen sensor connector and remove sensor. Pull out oil dipstick and tube.
4. Remove the air intake bellows hose, the wastegate vacuum hose, the connections for the air outlet hose and the upper and lower heat shields. Unbolt the power steering pump and bracket assembly and leaving the hoses connected, wire it back aside.
5. Remove the self-locking exhaust manifold nuts, the triangular engine hanger bracket, the eyebolt and gaskets that connect the oil feed line to the turbo center section and the cooling water lines. The water line under the turbo has a threaded connection.
6. Remove the exhaust pipe nuts and gasket and lift off the exhaust manifold. Discard the gasket. Remove both through bolts and 2 nuts that hold the exhaust manifold to the turbo.
7. Remove both cap screws from the oil return line, under the turbo. Discard the gasket. Separate the turbo from the exhaust manifold. both water pipes and oil feed line can still be attached.

To install:

8. Visually check the turbine wheel (hot side) and compressor wheel (cold side) for cracking or other damage. Check whether the turbine wheel and the compressor wheel can be easily turned by hand. Check for oil leakage. Check whether or not the wastegate valve remains open. If any problem is found, replace the part.

NOTE: Many turbocharger failures are due to oil supply problems. Heat soak after hot shutdown can cause the engine oil in the turbocharger and oil lines to 'coke.' Often the oil feed lines will become partially or completely blocked with hardened particles of carbon, blocking oil flow. Always check the oil feed pipe and oil return line for clogging. Clean these tubes well. Always use new gaskets above and below the oil feed eyebolt fitting. Use care that no particles of dirt of old gasket enter the oil passage hole and that no portion of the new gasket blocks the passage.

9. The wastegate can be checked with a pressure tester. Apply approximately 9 psi to the actuator and make sure the rod moves. Do not apply more than 10.3 psi or the diaphragm in the wastegate may be damaged. Do not attempt to adjust the wastegate valve.
10. Installation is the reverse of the removal process. Note that the oil feed line should be primed with clean engine oil. None of the self-locking nuts should be reused. Replace all locking nuts. Before installing the threaded connection for the water inlet pipe, apply light oil to the inner surface of the pipe flange.
11. Fill the engine crankcase, cooling system and reconnect the battery negative cable.

3.0L Engine

RIGHT SIDE (FRONT) TURBOCHARGER

1. Disconnect the negative battery cable.
2. Remove the radiator.
3. Remove the right side transaxle bracket.
4. Remove the front exhaust pipe.
5. Carefully matchmark, diagram or photograph all air intake hoses and pipes along the front of the engine. It is imperative that all of these pieces are installed in the exact same positions when assembling. Remove the hoses and pipes and keep covered in a clean area.
6. Remove the alternator.
7. Remove the oil dipstick tube.
8. Remove the turbocharger heat protector.
9. Remove the water feed pipes.
10. Remove the oxygen sensor.
11. Remove the oil return line.
12. Remove the exhaust extension fitting and bracket.
13. Remove all air conditioning components preventing removal of the turbocharger.
14. Remove the oil feed tube.
15. Remove the turbocharger to exhaust manifold bolts and remove the turbocharger assembly.

To install:

16. Visually check the turbine wheel (hot side) and compressor wheel (cold side) for cracking or other damage. Check whether the turbine wheel and the compressor wheel can be easily turned by hand. Check for oil leakage. Check whether or not the wastegate valve remains open. If any problem is found, replace the part.
17. Clean all mating surfaces. Pour clean engine oil through the oil pipe feed hole in the turbocharger.
18. Install a new gasket and ring a install the turbocharger to the manifold. Torque the bolts to 40–47 ft. lbs. (55–65 Nm).
19. Replace the eye-bolt rings and install the oil feed pipe.
20. Install the removed air conditioning components.
21. Install the exhaust extension fitting and bracket with a new gasket. Torque the nuts to 40–47 ft. lbs. (55–65 Nm).
22. Install the oil return line with new gaskets.
23. Install the oxygen sensor.
24. Replace the eye-bolt rings and install the water feed pipes.
25. Install the turbocharger heat protector.
26. Install the dipstick tube.
27. Install the alternator.
28. Install all air intake hoses and pipes along the front of the engine. Make sure all are in their proper positions.
29. Install a new gasket and connect the front exhaust pipe.
30. Install the right side transaxle bracket.
31. Install the radiator.
32. Fill the system with coolant.
33. Connect the negative battery cable and check for exhaust leaks.

LEFT SIDE (REAR) TURBOCHARGER

1. Remove the battery.
2. Drain the coolant.
3. Remove the front exhaust pipe.
4. Disconnect the accelerator cable from the throttle body.
5. Remove the intake air hose, the air pipe across the top of the engine and its heat shield.
6. Remove the clutch booster vacuum hose and disconnect the accelerator cable from the pedal.
7. Remove the air intake hoses coming from the air cleaner box.
8. Remove the oxygen sensor and the turbocharger heat protector.
9. Remove the EGR pipe if equipped.
10. Remove the oil feed pipe.
11. Remove the EGR valve if equipped.
12. Remove the water feed pipes.
13. Remove the exhaust extension fitting and bracket.
14. Remove the inner heat protector.
15. Remove the oil return tube.
16. Remove the turbocharger to exhaust manifold nuts and remove the turbocharger assembly.

To install:

17. Visually check the turbine wheel (hot side) and compressor wheel (cold side) for cracking or other damage. Check whether the turbine wheel and the compressor wheel can be easily turned by hand. Check for oil leakage. Check whether or not the wastegate valve remains open. If any problem is found, replace the part.
18. Clean all mating surfaces. Pour clean engine oil through the oil pipe feed hole in the turbocharger.
19. Install a new gasket and ring a install the turbocharger to the manifold. Torque the nuts to 40–47 ft. lbs. (55–65 Nm).
20. Install the oil return line with new gaskets.
21. Install the inner heat protector.
22. Install the exhaust extension fitting and bracket with a new gasket. Torque the nuts to 40–47 ft. lbs. (55–65 Nm).
23. Replace the eye-bolt rings and install the water feed pipes.
24. Install the EGR valve if equipped.
25. Replace the eye-bolt rings and install the oil feed pipe.
26. Install the EGR pipe if equipped.
27. Install the turbocharger heat protector and oxygen sensor.
28. Install the air intake hoses coming from the air cleaner box. Make sure the triangular aligning marks are engaged.
29. Connect the accelerator cable to from the pedal and install the clutch booster vacuum hose.
30. Install the heat shield, the air pipe across the top of the engine and the air intake hose.
31. Connect the accelerator cable to the throttle body.
32. Install a new gasket and connect the front exhaust pipe.
33. Fill the system with coolant.
34. Install the battery.
35. Connect the negative battery cable and check for exhaust leaks.

Timing Chain Front Cover

REMOVAL & INSTALLATION

2.5L Engine

1. Disconnect the negative battery cable. Unbolt the clutch fan. Unbolt the fan shroud and then remove the shroud and clutch fan together. Remove the pulley and belt.
2. Remove the crankshaft bolt. With a puller, remove the crankshaft pulley.
3. Remove the rocker cover. Then, remove both front bolts from the cylinder head; these screw into and seal the top of the timing cover.
4. Remove the oil pan bolts (front and side) that screw into the timing cover.
5. Drain the cooling system and remove the coolant hose leading to the water pump. Remove the alternator or other accessories that are in the way of the timing cover.
6. Unbolt and remove the timing cover.
7. Clean all the gasket surfaces. If the oil pan gasket was damaged in removing the front cover, carefully cut the oil pan gasket off flush with the front of the block on both sides and remove the cut off pieces of gasket.
8. Carefully pry the old oil seal out of the cover without scratching the bore into which the seal fits. Then, install a new seal with an installer such as parts MD998376-01 and MB990938-01.

To install:

9. Install new gaskets to the cylinder block. If necessary, cut an exact replacement for the oil pan gasket section that was removed from a new pan gasket. Insert this piece of gasket onto the front of the pan in the exact posi-

tion of the old piece and use liquid sealer on the joint between both sections of gasket on both sides.

10. Install the chain cover and bolts. Torque bolts to 9–10.5 ft. lbs.

11. Lightly coat the outside diameter of the crankshaft pulley boss with clean engine oil. Install the pulley onto the crankshaft and use the crankshaft bolt to force the pulley all the way on. Torque the crankshaft bolt to 80–94 ft. lbs.

12. Install the front head bolts (2) and torque to 11–15 ft. lbs. Install the oil pan bolts and torque them to 4.5–5.5 ft. lbs.

13. Install all hoses and accessories and refill the cooling system.

Timing Chain and Sprockets

REMOVAL & INSTALLATION

2.5L Engine

1. Disconnect the negative battery cable.
2. Remove the rocker cover.
3. Put the engine on TDC No. 1 cylinder firing position by turning the crankshaft until the TDC timing marks align and both front valves are fully closed (rockers off the cams).
4. Remove the timing cover.
5. Unbolt and remove the 3 silent shaft chain guides.
6. Unscrew and remove the oil pump drive sprocket bolt and the left silent shaft sprocket bolt.
7. When the bolts are removed, pull these 2 sprockets off and then disengage the chain from the crankshaft sprocket. Note that both sprockets are identical, but that the oil pump drive sprocket is installed with the concave side toward the engine while the left silent shaft sprocket has the concave side out.
8. Remove the sprockets and the chain. Remove the crankshaft sprocket for the silent shaft chain.
9. The timing chain tensioner maintains constant spring pressure on the chain. Fasten the follower plunger so it will not be forced out of the body of the oil pump. Securely run wire around the follower and the left side of the oil pump.
10. Remove the camshaft sprocket bolt and pull the sprocket off the camshaft. Separate the chain from the sprockets and remove it. Pull the sprocket for the camshaft timing chain that is on the crankshaft off, keeping it and the silent shaft drive sprockets in order for correct installation.
11. Inspect the tensioner follower and replace it if the follower shows a deep grooving where the chain was ridden against it. To replace it, remove the wire holding it in place and allow the spring to gradually push it out of the oil pump body.
12. Replace the rubber seal that goes in the oil pump body and the spring behind it when replacing the tensioner follower. Make sure the thinner part of the follower faces downward. Wire the new follower in place just as the old one was.
13. If the timing chain right and left guides show heavy grooving, they should be replaced by unbolting them. Sprockets should be replaced if the teeth are deformed from wear or there are any obvious cracks.

To install:

14. To install first install the crankshaft sprocket for the camshaft timing chain onto the crankshaft. Install the sprocket so the teeth are on the crankshaft or inner end of the sprocket. Engage the camshaft timing chain with the camshaft sprocket so the chrome plated link straddles the timing mark on the front of the sprocket.
15. Wrap the lower end of the chain around the crankshaft sprocket so the chrome link straddles the timing mark and make sure the chain rides inside the chain guides on both sides.
16. Rest the camshaft sprocket on the sprocket holder and get the camshaft bolt. Engage the cam shaft sprocket with the front of the camshaft so the prong on the camshaft flange fits into the hole in the sprocket, install the bolt and torque to 37–43 ft. lbs. Remove the wire holding the tensioner follower.
17. Install the crankshaft silent shaft chain sprocket, facing so the teeth are on the outer end of the sprocket.
18. Assemble the oil pump drive gear and left silent shaft sprockets to the left silent shaft chain with the chrome plated links straddling the timing marks on each. Make sure the concave

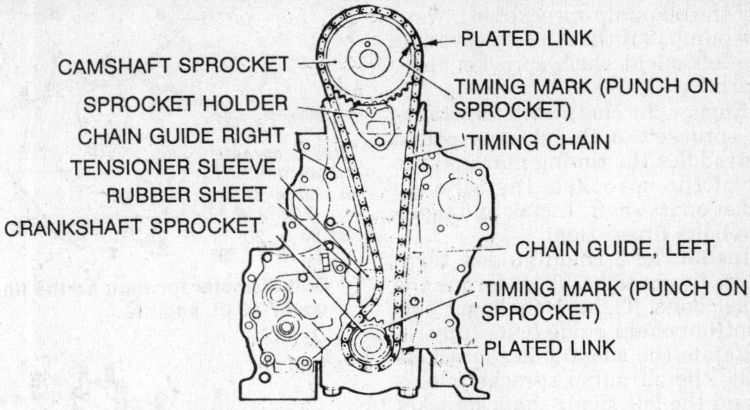

2.5L camshaft drive timing marks

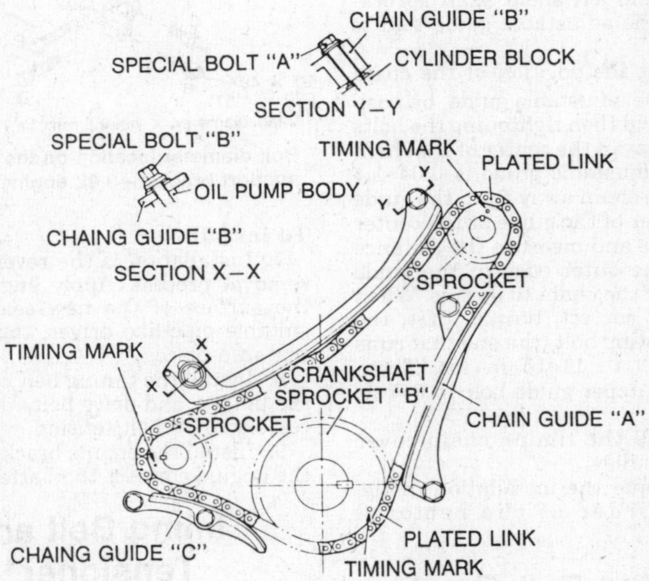

2.5L silent shaft chain timing marks

side of the oil pump sprocket is toward the oil pump, but that the concave side of the left silent shaft sprocket faces outward.

19. Engage the chain with the crankshaft sprocket so the chrome plated link straddles the timing mark on the front of the sprocket. Install each sprocket on its shaft. Install and tighten the bolts finger-tight.

20. Install the 3 chain guides, turning bolts finger-tight. Then torque the sprocket bolts. Tighten the right side and bottom chain guide bolts fully.

21. Rotate the silent shaft sprockets slightly, the oil pump sprocket clockwise and the left silent shaft sprocket counter clockwise so the slack in the chain all goes to the span between the oil pump and left silent shaft sprockets, near the adjustable guide that is still loose.

22. Adjust the position of the chain guide **B**, the adjustable guide, by positioning it and then tightening the bolts until the play in the center of the chain near the adjustable guide is 0.04–1.4 in. Pull the chain away from the guide at the center of the guide at the center of the guide and measure the distance between the outer edge of the guide and edge of the chain to do this. When the play is correct, torque, first, the guide adjusting bolt (the one that runs in the slot) to 11–15 ft. lbs. Then, torque the upper guide bolt to 6–7 ft. lbs.

23. Install the timing chain cover and front pulley.

24. Continue the installation in the reverse order of the removal procedure.

Timing Belt Front Cover

REMOVAL & INSTALLATION

1. Disconnect the negative battery cable.

2. Raise and safely support vehicle. Remove the under panel.

3. Place a wooden block between a jack and the oil pan. Slightly raise the engine and remove the engine mount bracket.

4. Remove all accessory drive belts, tension pulley bracket, water pump pulley, crankshaft compressor pulley and crankshaft pulley. The crankshaft pulley may be difficult to remove since the crankshaft will tend to turn when the center bolt is loosened. Use an old drive belt, wrap it around the pulley and draw it tight to hold the pulley.

5. Remove the upper and lower timing belt covers.

6. If removal of the front crank seal is necessary, pry the seal from the case cover.

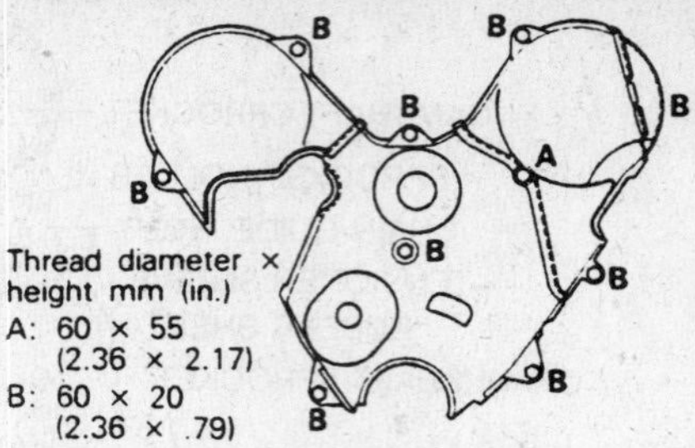

Bolt diameter location on the timing cover—3.0L engine

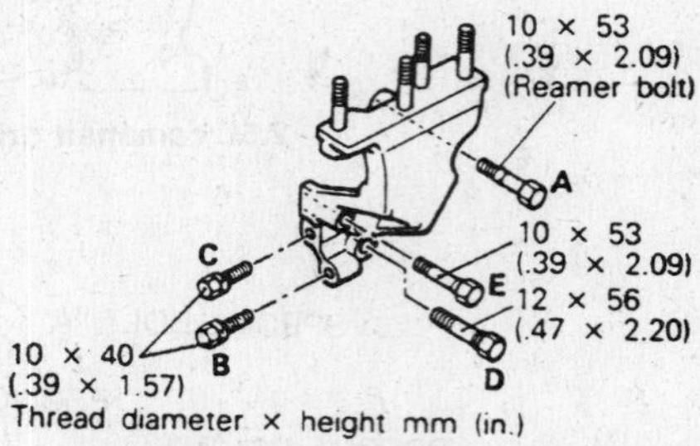

Bolt diameter location on the engine support bracket—3.0L engine

To install:

7. Installation is the reverse of the removal process. Apply engine oil to the surface of the new seal. With a suitable pipe-like driver, tap the new seal into place.

8. Install the timing belt covers and the pulleys and drive belts. Adjust accessory drive belt tension.

9. Install the engine brackets, lower the engine connect the battery cable.

Timing Belt and Tensioner

REMOVAL & INSTALLATION

4 Cylinder Except 1.5L, 1.6L Engines and Eclipse

NOTE: Timing belt and sprocket removal procedures are combined because the procedures are interrelated. Belts are kept in place to permit sprocket bolts to be loosened. If only replacing the belt(s), skip the steps related to removing or replacing the sprockets, unless it is noted that a sprocket must be removed to gain access to a belt related part.

1. Disconnect the negative battery cable. Remove the timing belt cover. Rotate the engine until the timing marks on the camshaft sprocket and cylinder head or rear belt cover and the crankshaft sprocket and front cover are perfectly aligned.

2. Loosen the timing belt tensioner adjusting bolt and the mounting bolt, shift the tensioner as far as it will go toward the left or water pump side (so belt tension is lost) and then retighten the adjusting bolt. If the belt is to be reused, draw an arrow on it in the direction of rotation. Remove the belt. Hold the tensioner in position to remove the tensioner adjusting bolt. Slowly release tension and remove the bolt the tension, spring and spacer.

3. Remove the bolt and the camshaft sprocket.

4. If replacing the inner timing belt, which drives the oil pump and right silent shaft, or need to remove the sprockets, proceed as follows; otherwise, proceed with Step 5:

a. Remove the crankshaft front sprocket bolt and remove the front crankshaft sprocket and flange. Remove the plug from the left side of the block. Insert a suitable tool about 0.3 in. in diameter and about 2.5 in. long or longer into the hole to keep the left silent shaft in position.

b. Remove the oil pump sprocket retaining nut and remove the nut and the sprocket. Loosen the right silent shaft sprocket bolt until it can be turn by hand.

c. Then, remove the inner tensioner bolts and remove the tensioner. Remove the inner timing belt. Then, remove the large crankshaft sprocket from the crankshaft and the right silent shaft bolt, sprocket and spacer.

5. Inspect all components as required. Replace defective parts as needed.

To install:

6. Install the larger crankshaft sprocket onto the crankshaft with the flatter or flanged side forward and the boss which is there to extend the sprocket forward from the front of the crankshaft at the rear. Align the timing mark on the sprocket with the mark on the front case. Apply a light coating of engine oil to the inner surface of the right silent shaft spacer and install the spacer. The chamfer must face inward, toward the engine. Then, install the right silent shaft sprocket and bolt and tighten the bolt finger-tight. Align the timing mark on this sprocket also with the timing mark on the front case.

7. Install the inner belt over the sprockets so the timing marks are in alignment and the upper side is under slight tension. Then, install the inner belt tensioner with the center of the pulley on the left side of the bolt and the flange of the pulley facing the front of the engine. Lift the tensioner until there is tension on the inner belt's upper length. Hold the tensioner in exactly this position and tighten the tensioner bolt. Make sure the turning of the bolt does not alter the position of the tensioner, or belt tension will be

excessive. Then, tighten the right silent shaft retaining flange bolt to 25–28 ft. lbs.

8. Check to make sure the timing marks effected by this belt are in alignment. Shift the position of the belt's teeth and retension, if necessary. Depressing the belt's upper span should enable it to depress about 0.2–0.3 in. Adjust the tension again to product this amount of deflection, if necessary.

9. Torque the right silent shaft bolt to 25–28 ft. lbs. Then, install the flange and crankshaft sprocket onto the crankshaft. The concave (inner) side of the flange must face to the rear so as to fit the curved front of the inner crankshaft sprocket. The flat side of the outer crankshaft sprocket must face the flange, to the rear. Finally, install the washer and bolt to the front of the crankshaft and torque it to 80–94 ft. lbs.

10. Install the camshaft sprocket to the camshaft and torque the bolt to 58–72 ft. lbs.

11. Install the spacer and main timing belt tensioner, installing the bolts finger-tight. Install the spring between the locking tang on the right side of the tensioner and the tang on the right side of the water pump, just above the tensioner. This will force the tensioner to turn counterclockwise on the pivot bolt. Push the tensioner all the way toward the water pump and lock it by tightening the adjusting bolt.

12. Check alignment of all timing marks: the mark on the camshaft sprocket must align with the mark on the head; the mark on the crankshaft sprocket must align with that on the front case; and the mark on the oil pump sprocket must align with that on the front case.

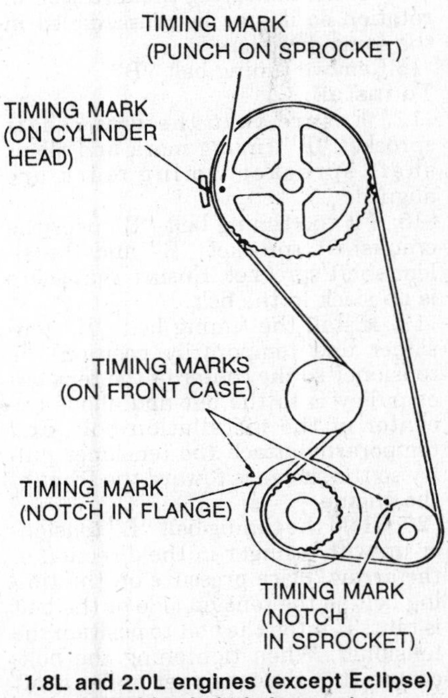

1.8L and 2.0L engines (except Eclipse) camshaft drive belt timing marks

13. Install the timing belt. The belt should be fitted over the sprockets in order: first the crankshaft, the oil pump and the camshaft sprocket. The (right) side of the belt which is normally straight must be straight during installation so the timing marks will remain aligned when the belt is actually tensioned. Remove the suitable tool installed to keep the silent shaft in position and replace the plug. Making sure there is no tension on the pivot bolt, loosen the tensioner adjusting bolt so the spring applies tension to the belt. Make sure the belt remains completely engaged with the teeth on the camshaft sprocket and that all timing marks remain aligned. Correct if necessary. Tighten the adjusting bolt. Finally, tighten the pivot bolt. Make sure to tighten the bolts in that order or tension will not be correct. Recheck alignment of the timing marks.

14. Turn the engine one full turn clockwise only. Loosen the tensioner pivot bolt and then the adjusting bolt. Allow the tensioner spring to again position the tensioner without interference from bolt friction. Tighten the adjusting bolt. Tighten the pivot bolt. Try to pry the belt outward by hand. The distance between the back of the belt and seal line will be about 0.55 in., if the tension is correct.

15. Continue the installation in the reverse order of the removal procedure.

1.5L Engine

1. Disconnect the negative battery cable. Remove the cooling fan, spacer, water pump pulley and belt. Remove the timing belt cover.

2. Turn the crankshaft until the timing marks on the camshaft sprocket and cylinder head are aligned. Loosen the tensioning bolt, it runs in the slotted portion of the tensioner, and the pivot bolt on the timing belt tensioner and lever the tensioner as far as it will go toward the water pump. Tighten the adjusting bolt. Mark the timing belt with an arrow showing direction of rotation, if reusing it.

3. Pull the timing belt off the camshaft sprocket. Remove the camshaft sprocket.

4. Remove the crankshaft pulley. Then, remove the timing belt.

5. Remove the crankshaft sprocket bolts and remove the crankshaft sprocket and flange, noting the direction of installation for each. Remove the timing belt tensioner.

6. Inspect all components, as required. Replace defective parts as needed.

To install:

7. To install, first reinstall the flange and crankshaft sprocket. The flange must go on first with the chamfered area outward. The sprocket is installed with the boss forward and the studs for the fan belt pulley outward. Install and torque the crankshaft sprocket bolt to 51–72 ft. lbs. Install the camshaft sprocket and bolt, torque it to 47–54 ft. lbs.

8. Align the timing marks of the camshaft sprocket. Check that the crankshaft timing marks are still in alignment, the locating pin on the front of the crankshaft sprocket is aligned with a mark on the front case.

9. To install the tensioner assembly, mount the tensioner, spring and spacer with the bottom end of the spring free. Then, install the bolts and tighten the adjusting bolt slightly with the tensioner moved as far as possible away from the water pump. Install the free end of the spring into the locating tang on the front case. Position the belt over the crankshaft sprocket and then over the camshaft sprocket. Make sure the belt is straight on the right side, where there's no tensioner. Slip the back of the belt over the tensioner wheel. Turn the camshaft sprocket in the opposite of its normal direction of rotation until the straight side of the belt is tight and make sure the timing marks align. If not, shift the belt one tooth at a time in the appropriate direction until this occurs.

10. Install the crankshaft pulley, making sure the pin on the crankshaft sprocket fits through the hole in the rear surface of the pulley. Install the bolts and torque to specification.

11. Loosen the tensioner bolts so the tensioner works, without the interference of any friction, under spring pressure. Make sure the belt follows the curve of the camshaft pulley so the teeth are engaged all the way around. Correct the path of the belt, if necessary. Torque the tensioner adjusting bolt to 15–18 ft. lbs. Torque the tensioner pivot bolt to the same figure. Bolts must be torqued in that order or tension won't be correct.

12. Turn the crankshaft 1 turn clockwise until timing marks again align to seat the belt. Loosen both tensioner bolts and let the tensioner position itself under spring tension as before. Finally, torque the bolts in the proper order exactly as before. Check belt tension by pulling the belt toward the water pump side of the tensioner wheel. The belt should move toward the pump until the teeth are about ¼ of the way across the head of the tensioner adjusting bolt. Retension the belt, if necessary.

13. Install the timing belt covers and remaining cooling system parts in the

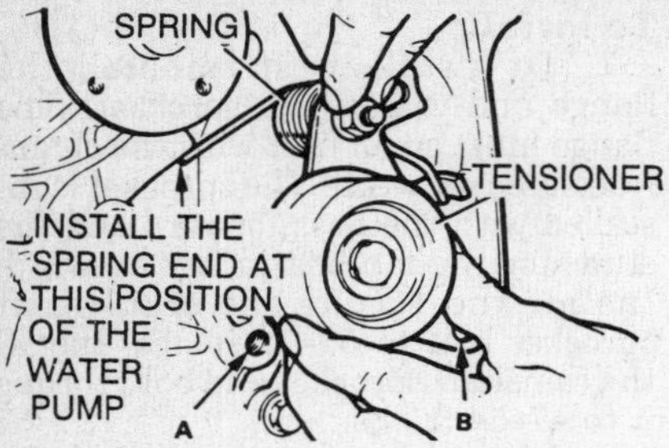

Installing belt tensioner spring—1.6L engine

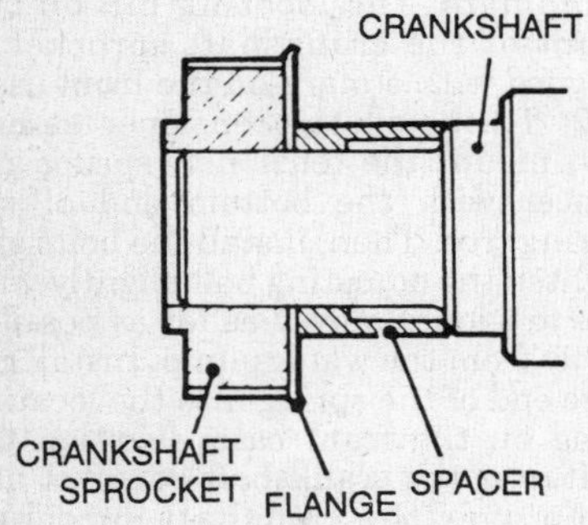

Installation of the crankshaft sprocket—1.6L engine

reverse of the removal procedure.

1.6L Engine

1. Disconnect the negative battery cable. Remove the crankshaft pulley. Remove the upper and lower timing belt covers. Rotate the crankshaft until all timing marks are aligned. There is a pin on the crankshaft sprocket which serves as the timing mark. It aligns with a pin protruding from the block behind the sprocket.
2. Remove the crankshaft sprocket bolt and loosen the other sprocket bolts.
3. Loosen the tensioner and adjusting bolts, shift the tensioner all the way to the left and retighten the adjusting bolt. Mark the timing belt with an arrow in the direction of rotation if it may be reused. Remove the timing belt.
4. Remove the camshaft sprocket, crankshaft sprocket and flange. If necessary, the crankshaft sprocket may be pulled off with a puller such as Mitsubishi part MD998311.
5. Remove the tensioner.
6. Inspect all components, as required. Replace defective parts as required.
7. Install the spacer, flange and crankshaft sprocket. The spacer is installed with the larger opening to the rear, so it fits tightly over the crankshaft at the front. Then install the flange with the slightly concave side backward. Finally, install the sprocket with the flat side rearward and boss forward. Install the sprocket bolt and torque to 80–93 ft. lbs. Make sure the sprocket and block timing marks are still aligned. Also check the timing marks for the oil pump drive sprocket and make sure they are aligned.
8. Apply a thin coating of clean engine oil to the outer circumference of the camshaft spacer and install the spacer onto the camshaft. Install the camshaft sprocket and bolt to 44–57 ft. lbs. Make sure the timing marks are aligned. Then install the crankshaft pulley so the engine can be turned. The bolts may be finger-tight.
9. Install the tensioner by first installing the spring, then the tensioner itself and then by installing and tightening the nut (finger-tight) used for adjusting the tensioner. Make sure the bent end of the spring goes to the right. Rock the tensioner, as necessary, until the pivot hole and bolt hole in the block align and install the pivot bolt. The spring must be installed so the bent end will work against the tab on the tensioner and the straight end works against the tab on the water pump body. Engage the ends of the spring with the tabs. Push the tensioner as far as it will go toward the water pump and then tighten the adjusting nut.
10. Install the timing belt, first over the crankshaft sprocket and then onto the oil pump sprocket. With the right side straight, engage the belt with the camshaft sprocket. Then, loosen the tensioner adjusting nut so the tensioner will tension the belt.
11. Push the tensioner slightly toward the adjusting nut so the belt teeth will be forced to mesh with the sprocket teeth. Make sure all teeth have meshed. With the tensioner under spring tension only, tighten the adjusting nut and then the pivot bolt. Check to make sure all timing marks are in alignment and make corrections, if necessary.
12. Turn the crankshaft one full turn in the normal direction of rotation, until all timing marks again align. Turn the engine smoothly and do not allow it to turn backwards. Don't grab the belt to test tension during this procedure. Loosen the tensioner adjusting nut and mounting bolt, again allowing it to adjust under spring pressure along. Torque the tensioner adjusting nut to 16–21 ft. lbs. Finally, tighten the bolt.
13. Test the tension on the belt by grasping the right edge of the rear timing belt cover and pull the center of the belt span toward it. With reasonable pressure, the belt should move to within just under ½ in. (0.47 in.) from the seal line. Repeat Steps 11 and 12 if the tension isn't correct.
14. Remove the crankshaft pulley. Install the timing belt lower front cover. Then install the upper cover.

Eclipse

1.8L ENGINE

1. Disconnect the negative battery terminal. Remove the under cover.
2. Remove the power steering pressure hose clamp.
3. Support the bottom of the engine and remove the engine mount bracket.
4. Remove the power steering belt.
5. If equipped with air conditioning, remove the air conditioning belt.
6. Remove the alternator belt.
7. Remove the water pump pulley.
8. Remove the crankshaft pulley, damper pulley and adapter.
9. Remove the timing belt front upper and lower covers and gaskets.
10. Remove the crankshaft sprocket bolt access cover and remove the crankshaft sprocket bolt and washer.
11. To remove the timing belt, turn the crankshaft clockwise and align the timing marks.

NOTE: The crankshaft must always be turned clockwise.

12. If the timing belt is to be reused, make a mark on the back of the timing belt to indicate the direction of rotation so it may be reassembled in the same direction.
13. Remove the timing belt tensioner, tensioner spacer and tensioner spring and remove the timing belt.
14. To remove timing belt "B", remove the timing belt "B" tensioner.
15. If the timing belt "B" is to be reused, make a mark on the back of the timing belt indicating the direction of rotation so it may be reassembled in the same direction.
16.Remove timing belt "B".

To install:

17. Ensure that the crankshaft sprocket "B" timing mark and silent shaft sprocket timing mark are aligned.
18. Fit the timing belt "B" over the crankshaft sprocket "B" and the silent shaft sprocket. Ensure that there is no slack in the belt.
19. Install the timing belt "B" tensioner and temporarily position the tensioner so the center of the tensioner pulley is to the left and above the center of the installation bolt, and temporarily attach the tensioner pulley so the flange is toward the front of the engine.
20. Hold the timing belt "B" tensioner up with a finger in the direction of the arrow, place pressure on the timing belt so the tension side of the belt is taut. Tighten the bolt to position the tensioner. When tightening the bolt, ensure that the tensioner pulley shaft does not rotate with the bolt. Allowing

it to rotate with the bolt can cause excessive tension on the belt. Belt deflection should be 0.20–0.28 in. (5–7mm).

21. Install the timing belt tensioner spring, spacer and belt tensioner. Place the upper end of the tensioner spring against the water pump body. Move the tensioner fully toward the water pump and temporarily tighten the bolt.

22. Ensure that the timing marks of the camshaft sprocket, the crankshaft sprocket and the oil pump sprocket are all aligned. When aligning the timing mark of the oil pump sprocket, remove the plug of the cylinder block; then insert the shaft of a prybar with a shaft diameter of 0.31 in. (8mm) into the plug hole and be sure the prybar's shaft can be inserted at least 2.4 in. (60mm). Do not remove the prybar until the timing belt is completely attached. If the prybar's shaft can be inserted only to a depth of about 0.79–0.98 in. (20–25mm) because it contacts the silent shaft, turn the sprocket by one rotation and align the timing mark once again; then, check again to be sure the prybar's shaft can be inserted at least 2.4 in. (60mm).

23. Install the timing belt. While making sure tension side of the belt is not slackened, install the timing belt onto the crankshaft sprocket, oil pump sprocket and camshaft sprocket in that order. If reusing the timing belt, be sure to install the timing belt in the direction of the marked arrow.

24. Loosen the belt tensioner nut. This will apply pressure to the belt.

25. Be sure each sprocket's timing mark is aligned.

26. Turn the crankshaft clockwise by 2 teeth of the camshaft sprocket. This is to apply the proper amount of tension on the timing belt.

27. Apply force on the tensioner toward turning direction, such that no portion of the belt raises out of the camshaft sprocket, place the belt on the camshaft sprocket such that the belt sprocket teeth are fully engaged.

28. Tighten the tensioner installation bolt and tensioner spacer in that order. Be sure to tighten the bolt first.

29. Check to see that a 0.40 in. (12mm) clearance between the outside of the belt and the cover by grasping the tension side, between the camshaft sprocket and the oil pump sprocket, of the center part of the timing belt between thumb and finger.

30. Install the crankshaft sprocket bolt and washer and torque to 80–94 ft. lbs. (110–130mm). Install the access cover.

31. Install the timing belt front upper and lower covers and gaskets.

32. Install the crankshaft pulley, adapter and damper pulley. Torque bolts to 11–13 ft. lbs. (15–18mm).

33. Install the water pump pulley. Torque to 6–7 ft. lbs. (8–10mm).

34. Install all drive belts and adjust.

35. Install the engine mount.

36. Install the power steering pressure hose clamp. Install the under cover. Connect the negative battery cable.

2.0L ENGINE

1. Disconnect the negative battery terminal. Remove the under cover.

2. Remove the power steering pressure hose clamp.

3. Support the bottom of the engine and remove the engine mount bracket.

4. Remove the alternator belt.

5. Remove the power steering belt. Remove the tensioner pulley bracket.

6. If equipped with air conditioning, remove the air conditioning belt.

7. Remove the water pump pulley.

8. Remove the crankshaft pulley.

9. Remove the timing belt front upper and lower covers and gaskets.

10. Remove the center cover, breather hose, PCV hose, spark plug cables and rocker cover with the semi-circular gasket.

11. Remove the plug rubber.

12. Turn the crankshaft clockwise and align the timing marks. Remove the auto tensioner.

NOTE: The crankshaft must always be turned clockwise.

13. If the timing belt is to be reused, make a mark on the back of the timing belt to indicate the direction of rotation so it may be reassembled in the same direction.

14. Remove the timing belt tensioner pulley and arm and remove the timing belt.

15. To remove timing belt "B", remove the timing belt "B" tensioner.

16. If the timing belt "B" is to be reused, make a mark on the back of the timing belt indicating the direction of rotation so it may be reassembled in the same direction.

17. Remove timing belt "B".

To install:

18. Ensure that the crankshaft sprocket "B" timing mark and silent shaft sprocket timing mark are aligned.

19. Fit the timing belt "B" over the crankshaft sprocket "B" and the silent shaft sprocket. Ensure that there is no slack in the belt.

20. Install the timing belt "B" tensioner and temporarily position the tensioner so the center of the tensioner pulley is to the left and above the center of the installation bolt, and temporarily attach the tensioner pulley so the flange is toward the front of the engine.

21. Hold the timing belt "B" tensioner up with a finger in the direction of the arrow, place pressure on the timing belt so the tension side of the belt is taut. Tighten the bolt to position the tensioner. When tightening the bolt, ensure that the tensioner pulley shaft does not rotate with the bolt. Allowing it to rotate with the bolt can cause excessive tension on the belt. Belt deflection should be 0.20–0.28 in. (5–7mm).

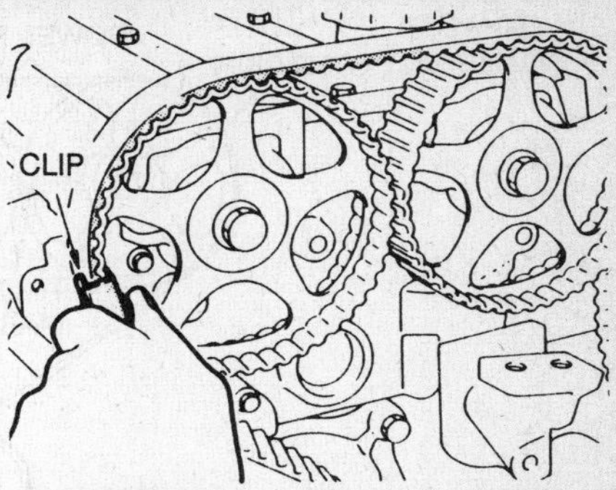

Timing belt place holding clamps during installation—Eclipse 2.0L engine

22. To install the auto tensioner, the tensioner should be reset.

a. Keep the auto tensioner level and clamp it in a vice with soft jaws. If the plug at the bottom of the tensioner protrudes, install flat washers to prevent the plug from contacting the vise.

b. Push in the rod little by little with the vise until the set hole in the rod is aligned with the hole in the cylinder.

c. Install a 0.055 in. (1.4mm) wire into the set holes.

d. Unclamp the tensioner from the vise.

23. Install the auto tensioner and tighten bolts to 14–20 ft. lbs. (20–27mm).

24. Install the tensioner pulley onto the arm. Position the pinholes in the tensioner pulley shaft to the left of the center bolt. Tighten the center bolt finger-tight. Do not remove the wire from the auto tensioner.

25. Ensure that the timing marks of the camshaft sprocket, the crankshaft and the oil pump sprockets are aligned. When aligning the timing mark of the oil pump sprocket, remove the plug of the cylinder block; then insert the shaft of a prybar with a shaft diameter of 0.31 in. (8mm) into the plug hole and be sure the prybar's shaft can be inserted at least 2.4 in. (60mm). Do not remove the prybar until the timing belt is completely attached. If the prybar's shaft can be inserted only to a depth of about 0.79–0.98 in. (20–25mm) because it contacts the silent shaft, turn the sprocket by one rotation and align the timing mark once again; then check again to be sure that the prybar's shaft

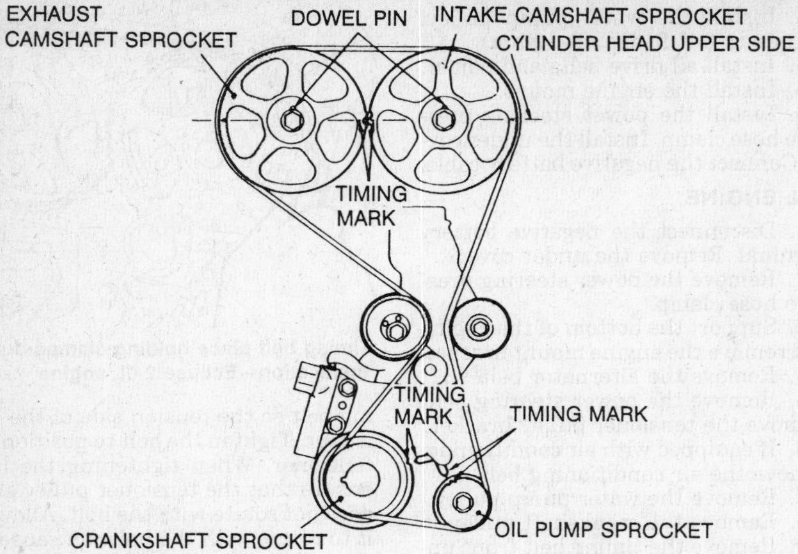

Timing belt alignment marks—Eclipse 2.0L engine

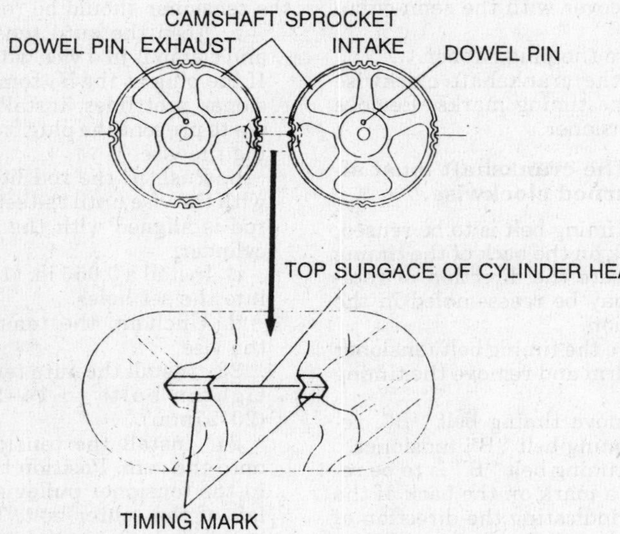

Camshaft sprocket alignment marks—Eclipse 2.0L engine

can be inserted at least 2.4 in. (60mm).

26. Install the timing belt over the intake side camshaft sprocket and fix a clip to the hold the belt.

27. Install the timing belt over the exhaust side sprocket, aligning the timing marks with the cylinder head top surface using 2 wrenches and fix a clip to hold the belt.

28. Install the timing belt over the idler pulley, the oil pump sprocket, the crankshaft sprocket and the tension pulley in that order. Remove the clips.

29. Lift up the tensioner pulley in the direction of the arrow and tighten the center bolt.

30. Check to see that all timing marks are aligned. Remove the prybar and install the plug.

31. Turn the crankshaft ¼ turn counterclockwise. Then, turn it clockwise until the timing marks are aligned again.

32. To adjust the timing belt tension:

a. Loosen the center bolt, and then attach special tool MD998752 and a torque wrench capable of measurement within a range of 0–2.5 ft. lbs. (0–3 Nm). Apply a torque of 1.88–2.03 ft. lbs. (2.6–2.8 Nm).

b. Holding the tensioner pulley with the special tool and the torque wrench, tighten the center bolt to 31–40 ft. lbs. (43–55 Nm).

c. Install special tool MD998738 into the engine left support bracket until its end makes contact with the tensioner arm. At that point, screw the special tool in some more and then remove the set wire attached to the auto tensioner. Remove the special tool.

d. Rotate the crankshaft 2 complete turns clockwise and leave it as is for about 15 minutes. Then, measure the auto tensioner protrusion; distance between the tensioner arm and the auto tensioner body. Measurement should be 0.15–0.18 in. (3.8–4.5mm).

e. If the clearance between the tensioner arm and the auto tensioner body cannot be measured, screw in the special tool MD998738 until it contacts the tensioner arm. From that point of contact, further the screw in the special tool, screwing it in until the pushrod of the auto tensioner body is caused to move backward and the tensioner arm contacts the auto tensioner body. Be sure the amount the special tool has been screwed in, when the pushrod moves backward, is 2½–3 turns.

33. Install the plug rubber.

34. Install the rocker cover with gasket, spark plug cables, PCV hose, breather hose and center cover.

35. Install the timing belt front upper and lower covers and gaskets.

36. Install the crankshaft pulley and torque to 14–22 ft. lbs. (20–30mm).

37. Install the water pump pulley. Torque to 6–7 ft. lbs. (8–10mm).

38. Install all drive belts and adjust.

39. Install the engine mount.

40. Install the power steering pressure hose clamp. Install the under cover. Connect the negative battery cable.

3.0L SOHC Engine

1. If possible, position the engine so the No. 1 cylinder is at TDC of its compression stroke. Disconnect the negative battery cable. Remove the timing covers from the engine.

2. If the same timing belt will be reused, mark the direction of the timing belt's rotation for installation in the same direction. Make sure the engine is positioned so the No. 1 cylinder is at the TDC of its compression stroke and the sprockets' timing marks are aligned with the engine's timing mark indicators.

3. Loosen the timing belt tensioner bolt and remove the belt. If the tensioner is not being removed, position it as far away from the center of the engine as possible and tighten the bolt.

4. If the tensioner is being removed, paint the outside of the spring to ensure that it is not installed backwards.

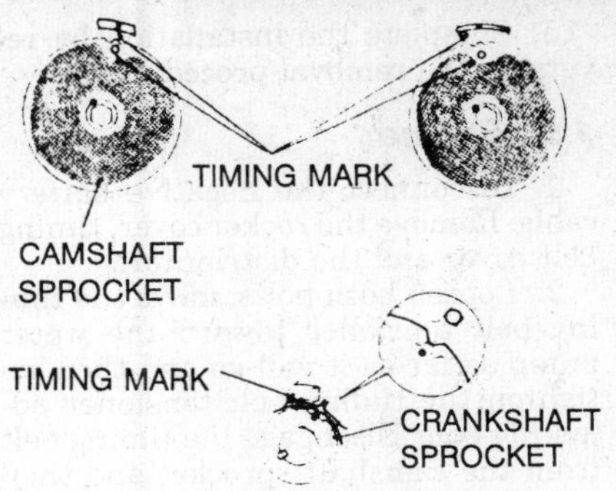

Aligning the timing marks—3.0L SOHC engine

Unbolt the tensioner and remove it along with the spring.

To install:

5. Install the tensioner, if removed, and hook the upper end of the spring to the water pump pin and the lower end to the tensioner in exactly the same position as originally installed. If not already done, position both camshafts so the marks align with those on the rear. Rotate the crankshaft so the timing mark aligns with the mark on the oil pump.

6. Install the timing belt on the crankshaft sprocket and while keeping the belt tight on the tension side, install the belt on the front camshaft sprocket.

7. Install the belt on the water pump pulley, then the rear camshaft sprocket and the tensioner.

8. Rotate the front camshaft counterclockwise to tension the belt between the front camshaft and the crankshaft. If the timing marks became misaligned, repeat the procedure.

9. Install the crankshaft sprocket flange.

10. Loosen the tensioner bolt and allow the spring to apply tension to the belt.

11. Turn the crankshaft 2 full turns in the clockwise direction until the timing marks align again. Now that the belt is properly tensioned, torque the tensioner lock bolt to 21 ft. lbs. (29 Nm). Measure the belt tension between the rear camshaft sprocket and the crankshaft with belt tension gauge. The specification is 46–68 lbs. (210–310 N).

12. Install the timing belt covers and all related parts.

13. Connect the negative battery cable and road test the vehicle.

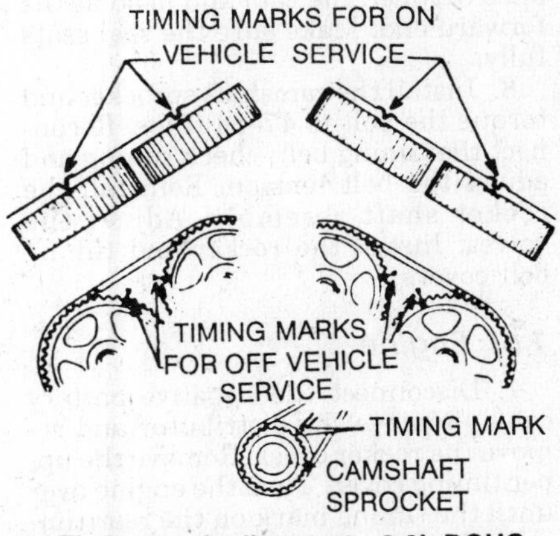

Timing mark alignment—3.0L DOHC engine

3.0L DOHC Engine

1. If possible, position the engine so the No. 1 cylinder is at TDC of its compression stroke. Disconnect the negative battery cable. Remove the timing covers from the engine.

2. If the same timing belt will be reused, mark the direction of the timing belt's rotation for installation in the same direction. Make sure the engine is positioned so the No. 1 cylinder is at the TDC of its compression stroke and the sprockets' timing marks are aligned with the engine's timing mark indicators on the valve covers or head.

3. Loosen the timing belt tensioner bolt and remove the belt.

4. Remove the tensioner assembly.

To install:

5. If the auto tensioner rod is fully extended, reset it as follows:

   a. Clamp the tensioner in a soft-jaw vice in level position.

   b. Slowly push the rod in with the vice until the set hole in the rod is aligned with the hole in the cylinder.

   c. Insert a stiff wire into the set holes to retain the position.

   d. Remove the assembly from the vice.

6. Leave the retaining wire in the tension and install to the engine. Torque the retaining bolts to 17 ft. lbs. (24 Nm).

7. If the timing marks of the camshaft sprockets and crankshaft sprocket are not aligned at this point, proceed as follows:

NOTE: Keep fingers out from in between the camshaft sprockets. The sprockets may move unexpectedly because of valve spring pressure and could pinch fingers.

   a. Align the mark on the crankshaft sprocket with the mark on the front case. Then move the sprocket 3 teeth clockwise to lower the piston so the valve can't touch the piston when the camshafts are being moved.

   b. Turn each camshaft sprocket 1 at a time to align the timing marks with the mark on the valve cover or head. If the intake and exhaust valves of the same cylinder are opened simultaneously, they could interfere with each other. Therefore, if any resistance is felt, turn the other camshaft to move the valve.

   c. Align the timing mark of the crankshaft sprocket, then continue 1 tooth farther in the counterclockwise direction to facilitate belt installation.

8. Using 4 spring loaded paper clips to hold the belt on the cam sprockets, install the belt to the sprockets in the following order:

1st—exhuast camshaft sprocket for the front head
2nd—intake camshaft sprocket for the front head
3rd—water pump pulley
4th—intake camshaft sprocket for the rear head
5th—exhuast camshaft sprocket for the rear head
6th—idler pulley
7th—crankshaft sprocket
8th—tensioner pulley

9. Turn the tensioner pulley so its pin holes are located above the center bolt. Then press the tensioner pulley against the timing belt and simultaneously tighten the center bolt.

10. Make certain that all timing marks are still aligned. If so, remove the 4 clips.

11. Turn the crankshaft ¼ turn counterclockwise, then turn it clockwise until all timing marks are aligned.

12. Loosen the center bolt on the tensioner pulley. Using tool MD998767 or equivalent and a torque wrench, apply a torque of 7.2 ft. lbs. (10 Nm). Tighten the tensioner bolt; make sure the tensioner doesn't rotate with the bolt.

13. Remove the set wire attached to the auto tensioner, if the wire was not previously removed.

14. Rotate the crankshaft 2 complete turns clockwise and let it sit for approximately 5 minutes. Then, make sure the set pin can easily be inserted and removed from the hole in the tensioner.

15. Measure the auto tensioner protrusion (the distance between the tensioner arm and auto tensioner body) to ensure that it is within 0.15–0.18 in. (3.8–4.5mm). If out of specification, repeat Step 1–4 until the specified value is obtained.

16. Install the timing belt covers and all related items.

17. Connect the negative battery cable.

Camshaft

REMOVAL & INSTALLATION

1.8L, 2.0L (Except Eclipse) and 2.3L Engines

1. Disconnect the negative battery

cable. Remove the distributor. Remove the rocker cover, disconnect the camshaft sprocket and remove the rocker arm shaft and cam bearing assembly. The camshaft may then be lifted off the top of the cylinder head.

2. Check and replace defective components, as required.

3. Thoroughly lubricate the camshaft bearing journals, the bearing saddles in the cylinder head and the inner surfaces of the caps with clean engine oil. Then continue the installation in the reverse order of the removal procedure.

Eclipse

1.8L ENGINE

1. Relieve the fuel system pressure.
2. Disconnect the battery negative cable.
3. Remove the distributor.
4. Remove the rocker cover, timing belt cover and timing belt.
5. Remove the camshaft sprocket and oil seal.
6. Loosen both rocker arms assembly uniformly and remove.
7. Remove the camshaft rear cover, rear cover gasket, thrust plate and camshaft thrust case. Remove the camshaft.
8. After the camshaft has been removed, check the following:
   a. Check the camshaft journals for wear or damage.
   b. Check the fuel pump drive eccentric cam and distributor drive gear tooth surfaces.
   c. Check the cam lobes for damage. Also, check the cylinder head oil holes for clogging.

To install:

9. Lubricate the camshaft with heavy engine oil and slide it into the head.
10. Insert the camshaft thrust case in cylinder head with the threaded hole facing upward and align the threaded hole with the bolt hole in the cylinder head. Install and firmly tighten the attaching bolt.
11. Check the camshaft endplay between the thrust case and camshaft. The camshaft endplay should be 0.0020–0.0080 in. (0.5–0.20mm). If the endplay is not within specification, replace the camshaft thrust bearing.
12. When installing the oil seal, coat the external surface with engine oil. Position the seal on the camshaft end and drive into place using tool MD998306 or equivalent.
13. Complete installation by reversing the removal procedure.

2.0L ENGINE

1. Relieve the fuel system pressure.
2. Disconnect battery negative cable.
3. Remove the accelerator cable connection.
4. Remove the timing belt cover and timing belt.
5. Remove the center cover, breather and PCV hoses and spark plug cables.
6. Remove the rocker cover, semi-circular packing, throttle body stay, crankshaft angle sensor, both camshaft sprockets and oil seals.
7. Loosen the bearing cap bolts in 2–3 steps. Label and remove both camshaft bearing caps.

NOTE: If the bearing caps are difficult to remove, use a plastic hammer to gently tap the rear part of the camshaft.

8. Remove the intake and exhaust camshafts.
9. After the camshaft has been removed, check the following:
   a. Check the camshaft journals for wear or damage.
   b. Check the cam lobes for damage. Also, check the cylinder head oil holes for clogging.

To install:

10. To install, lubricate the camshafts with heavy engine oil and position the camshafts on the cylinder head.

NOTE: Do not confuse the intake camshaft with the exhaust camshaft. The intake camshaft has a split on its rear end for driving the crank angle sensor.

11. Make sure the dowel pin on both camshaft sprocket ends are located on the top.
12. Install the bearing caps. Tighten the caps, in sequence, in 2–3 steps. No. 2 and 5 caps are of the same shape. Check the markings on the caps to identify the cap number and intake/exhaust symbol. Only **L** (intake) or **R** (exhaust) is stamped on No. 1 bearing cap. Also, make sure the rocker arm is correctly mounted on the lash adjuster and the valve stem end.
13. Apply a coating of engine oil to the oil seal. Using tool MD998307 or equivalent, press-fit the seal into the cylinder head.
14. Align the punch mark on the crank angle sensor housing with the notch in the plate. With the dowel pin on the sprocket side of the intake camshaft at top, install the crank angle sensor on the cylinder head.

NOTE: The crank angle sensor can be installed with the punch mark positioned opposite the notch; however, that position will result in incorrect fuel injection and ignition timing.

15. Complete the installation by reversing the removal procedure.

1.5L Engine

1. Disconnect the negative battery cable. Remove the rocker cover, timing belt cover and the distributor.
2. Loosen both bolts, move the timing belt tensioner toward the water pump as far as it will go and then retighten the timing belt tensioner adjusting bolt. Disengage the timing belt from the camshaft sprocket and then unbolt and remove the sprocket. The timing belt may be left engaged with the crankshaft sprocket and tensioner.
3. Remove the rocker shaft assembly. Remove the small, square cover that sits directly behind the camshaft on the transaxle side of the head. Remove the camshaft thrust case tightening bolt that sits on the top of the head right near that cover.
4. Very carefully slide the entire camshaft out of the head through the hold in the camshaft side of the head, being sure the cam lobes do not strike the bearing bores in the head.

To install:

5. Check and replace defective components as required.
6. Lubricate all journal and thrust surfaces with clean engine oil and then insert the camshaft into the engine, again keeping the cam lobes from touching the bearing bores. Make sure the camshaft goes in with the threaded hole in the top of the thrust case straight upward and align the bolt hole in the thrust case and the cylinder head surface once the camshaft is all the way inside the head. Install the thrust case bolt and tighten firmly. Finally, install the rear cover with a new gasket and install and tighten the 4 bolts.
7. Coat the external surface of the front oil seal with engine oil. With a special installer part MD998306-01 or equivalent, drive the reusable or new front camshaft oil seal into the clearance between the cam and head at the forward end. Make sure the seal seats fully.
8. Install the camshaft sprocket and torque the bolt to 47–54 ft lbs. Reconnect the timing belt, check timing and adjust the belt tension. Reinstall the rocker shaft assembly. Adjust the valves. Install the rocker and timing belt covers.

1.6L Engine

1. Disconnect the negative battery cable. Remove the distributor and remove the rocker cover. Remove the upper timing cover. Turn the engine over until the timing mark on the rear timing belt cover aligns with the mark on the camshaft sprocket. It's a good idea

to mark the timing belt itself to align with the marks on the sprocket and rear timing belt cover to make Precise reassembly easier. Remove the camshaft sprocket from the camshaft and remove the rocker arms and shafts assembly.

2. Pull the camshaft front oil seal off the front of the camshaft. Remove the camshaft.

To install:

3. Check and replace defective components, as required.

4. Thoroughly lubricate the camshaft bearing journals, the bearing saddles in the cylinder head and the inner surfaces of the caps with clean engine oil.

5. Install the camshaft onto the cylinder head, being careful not to damage any of the camshaft journals. Install the rocker arm and shaft assembly to the head, torque the bolts to 14–15 ft. lbs.

6. Coat the outside diameter of the front end of the camshaft with clean engine oil. Then, with a special tool such as MD998354-01 or equivalent, tap a new front seal in, using a hammer. Install the rear timing belt cover.

7. Turn the camshaft so the dowel pin on the front aligns with the hole in the sprocket. If necessary to turn the cam, exert force on either of both projections behind the No. 2 cylinder exhaust valve cam. Reconnect the camshaft drive sprocket by lifting it off the rest and installing it to the camshaft with the dowel pin going through the hole in the sprocket. Torque the sprocket bolt to 44–57 ft. lbs.

8. Install the remaining parts and adjust the valves.

2.5L Engine

1. Disconnect the negative battery cable. Remove the distributor. Remove the rocker cover and rocker shaft assembly. Remove also the rear bearing cap bolts and the cap.

2. Remove the camshaft from the head.

To install:

3. Check and replace defective components as required.

4. Thoroughly lubricate the camshaft bearing journals, the bearing saddles in the cylinder head and the inner surfaces of the caps with clean engine oil. Install the camshaft onto the cylinder head, being careful not to damage any of the camshaft journals. Apply a sealer to the outside diameter of the circular seal for the rear bearing and install it in the head with one side directly in contact with the rear of the camshaft. The packing will end up under the rearmost portion of the rear bearing cap. Then, install and torque the rocker shaft/bearing cap assembly. Include the rear bearing cap, using the same torque. Refit the cam sprocket and chain to the camshaft.

5. Also inspect the semicircular seal that goes in the front of the timing chain cover and seal the top with an adhesive such as 3M Super Weatherstrip Adhesive 801k or equivalent.

6. Adjust the valve clearances. Install the rocker cover and all other parts removed earlier. Start the engine and idle it until after the temperature gauge indicates normal operating temperature. Remove the rocker cover again and adjust the valves with the engine hot.

3.0L SOHC Engine

1. Disconnect the negative battery cable. Remove the valve covers and timing belt.

2. Install auto lash adjuster retainer tools MD998443 or equivalent, on the rocker arms.

3. If removing the right side (front) camshaft on 3.0L engine, remove the distributor extension.

4. Remove the camshaft bearing caps but do not remove the bolts from the caps.

5. Remove the rocker arms, rocker shafts and bearing caps, as an assembly.

6. Remove the camshaft from the cylinder head.

7. Inspect the bearing journals on the camshaft, cylinder head, and bearing caps.

To install:

8. Lubricate the camshaft journals and camshaft with clean engine oil and install the camshaft in the cylinder head.

9. Align the camshaft bearing caps with the arrow mark depending on cylinder numbers and install in numerical order.

10. Apply sealer at the ends of the bearing caps and install the assembly.

11. Torque the bearing cap bolts in the following sequence: No. 3, No. 2, No. 1 and No. 4 to 85 inch lbs. (10 Nm).

12. Repeat the sequence increasing the torque to 15 ft. lbs. (20 Nm).

13. Install the distributor extension if it was removed.

14. Install the timing belt, valve cover and all related parts.

15. Connect the negative battery cable and check for leaks.

3.0L DOHC Engine

1. Relieve the fuel system pressure.

2. Disconnect battery negative cable.

3. Remove the timing belt cover and timing belt.

4. Remove the center cover, breather and PCV hoses, and spark plug cables.

5. Remove the rocker cover, semicircular packing, throttle body stay, crankshaft angle sensor, both camshaft sprockets, and oil seals.

6. Remove the crank angle sensor and adaptor.

7. Loosen the bearing cap bolts in 2–3 steps. Label and remove all camshaft bearing caps.

NOTE: If the bearing caps are difficult to remove, use a plastic hammer to gently tap the rear part of the camshaft.

8. Remove the intake and exhaust camshafts.

9. Check the camshaft journals for wear or damage. Check the cam lobes for damage. Also, check the cylinder head oil holes for clogging.

To install:

10. To install, lubricate the camshafts with heavy engine oil and posi-

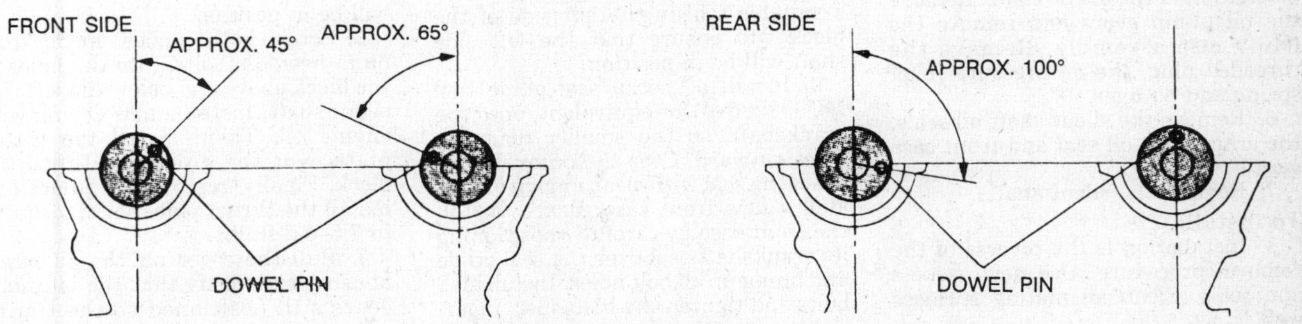

Camshaft alignment—3.0L DOHC engine

tion the camshafts on the cylinder head.

NOTE: Do not confuse the intake camshaft with the exhaust camshaft. The intake camshaft has a V stamped on the hexagon and the exhaust camshaft has a C.

11. Make sure the dowel pin on both camshaft sprocket ends are located as shown.
12. Install the bearing caps. Tighten the caps in sequence and in 2 or 3 steps. Caps 2, 3 and 4 have a front mark. Install with the mark aligned with the front mark on the cylinder head. Intake caps have **I** stamped on the cap and exhaust caps have **E**. Also, make sure the rocker arm is correctly mounted on the lash adjuster and the valve stem end. Torque the retaining bolts to 15 ft. lbs. (20 Nm).
13. Apply a coating of engine oil to the oil seals and install.
14. Install the timing belt, valve cover and all related parts.
15. Connect the negative battery cable and check for leaks.

Intermediate Shaft

REMOVAL & INSTALLATION

Eclipse

1.8L ENGINE

1. Disconnect the negative battery cable.
2. Remove the oil filter, oil pressure switch, oil gauge sending unit and oil filter bracket and gasket.
3. Drain engine oil. Remove engine oil pan, oil screen and gasket.
4. Remove the front engine cover which is also the oil pump cover. Different length bolts are used. Take note of their locations. If the cover sticks to the block, look for a special slot provided and pry with a suitable tool. Discard the shaft seal and gasket.
5. Remove the oil pump driven gear flange bolt. When loosening this bolt, first insert a suitable tool approximately 3/8 in. diameter into the plug hole on the left side of the cylinder block to hold the silent shaft. Remove the oil pump gears and remove the front case assembly. Remove the threaded plug, the oil pressure relief spring and plunger.
6. Remove the silent shaft oil seals, the crankshaft oil seal and front case gasket.
7. Remove the silent shafts.

To install:

8. Installation is the reverse of the removal procedure. Use new gaskets and seals. Clean all mating surfaces well.
9. Use care to get the proper length bolt in the correct location on the timing cover as well as the oil pump cover.
10. Refill with engine oil. Install new filter. Check for leaks.

Silent Shafts

REMOVAL & INSTALLATION

4 Cylinder Engines Except 2.5L Engine

NOTE: A special oil seal guide MD998285 or equivalent, is needed to complete this operation.

1. Disconnect the negative battery cable. Remove the timing belt covers, timing belts and sprockets.
2. Drain the oil and remove the oil filter. Then, remove the oil pan and gasket. Remove the oil pick-up and gasket.
3. Remove the oil pressure relief plunger plug and gasket, and then remove the spring and plunger from the oil filter bracket. Remove the 4 bracket bolts, the oil filter mount and gasket.
4. Remove the cap and gasket that cover the oil pump driven gear shaft. This is located on the right side of the front case at the front of the engine, just above the protruding silent shaft.
5. Using a long socket, remove the retaining bolt from the oil pump driven gear, behind the plug removed earlier.
6. Remove the front case bolts, the case and the gasket. Slide the silent shafts from the block, noting their installation angles.

To install:

7. Inspect the silent shaft bearing journals for signs of excessive wear or seizure. If there are signs of critical wear problems, the bushings should also be inspected. The bushings may be replaced by pulling them out and pressing new ones in, using special tools. This is done with the crankshaft removed, since it normally is required only at time of major engine overhaul.
8. Lubricate the silent shaft bearing journals with clean engine oil and install the shafts into the block. Insert the shafts into their original position, a suitable tool in the left side of the block will ensure that the left side shaft will be in position.
9. Install a special seal guide tool MD998285-01 or equivalent, onto the crankshaft, so the smaller diameter faces outward. Coat the outer diameter of the seal with clean engine oil. Install a new front case gasket. Install the front case by carefully positioning its crankshaft seal over the seal guide and lining up all bolt holes. Install all 8 bolts and tighten the bolts just finger-tight.
10. Install the oil filter bracket gasket, the bracket and 4 bolts; torque the front case bolts to 15–19 ft. lbs. and the oil filter bracket bolts to 11–15 ft. lbs.
11. Install the remaining parts in reverse of the removal procedure.

2.5L Engine

NOTE: Two long, 8mm bolts are needed to pull out the silent shaft thrust plates and 2 guides, made by cutting the heads off 6mm bolts about 2 in. long.

1. Disconnect the negative battery cable. Remove the timing cover, chains and sprockets. Before removing both sprocket bolts, put a wrench on the flange bolt which attaches the upper oil pump gear to the center of the right side silent shaft and turn it just enough to break it loose.
2. Screw 8mm bolts into the bolt holes in the thrust plate and turn them evenly to pull the thrust plate out of the block. Then, remove the left silent shaft.
3. Remove the oil pump bolts. Then, pull the oil pump and gasket straight off the front of the block. The right side silent shaft will come out with the pump. Be careful to support the pump and shaft in such a way that the rear shaft bearing will not be damaged. Remove the bolt from the center of the oil pump driven (upper) gear. Separate the silent shaft and key from the oil pump driven gear by sliding it out. Remove the oil pump gasket.

To install:

4. Inspect the silent shaft bearing journals for signs of excessive wear or seizure. If there are signs of critical wear problems, the bushings should also be inspected. The bushings may be replaced by pulling them out and pressing new ones in , using special tools. This is done with the crankshaft removed, since it normally is required only at time of major engine overhaul.
5. Lubricate the left silent shaft bearing journals with clean engine oil and install the shaft into the block. Insert the shaft into their original position, a suitable tool in the left side of the block will ensure that the shaft will be in position.
6. Screw both guides, made from 6mm headless bolts, into the holes in the block above and below the left side silent shaft. Install a new O-ring with engine oil. Then, install the thrust plate over the guides and into the block. Finally, remove the guides and install the thrust plate bolts, torquing to 7.5–8.5 ft. lbs.
7. Pull the cover off the oil pump housing and verify that the oil pump gears still positioned so the timing marks are aligned. Install the cover over the guide pins and pour about 0.6

cu. in. of clean engine oil into the oil pump outlet, which is at top right of the pump cover. Install the oil pump gasket to the rear of the pump, it may be necessary to use grease to hold it in position. Then, position the pump in its installed direction and engage the key of the right silent shaft to the slot in the oil pump driven gear. Slide the shaft all the way into the pump driven gear and then install the bolt and torque to 44–50 ft. lbs. Lubricate the right side silent shaft bearing journals with clean engine oil and then insert the shaft into the block and install the oil pump. Install the oil pump bolts and torque to 7.5–8.5 ft. lbs.

8. Install the sprockets, timing chains, tensioners and front cover.

Piston and Connecting Rod

POSITIONING

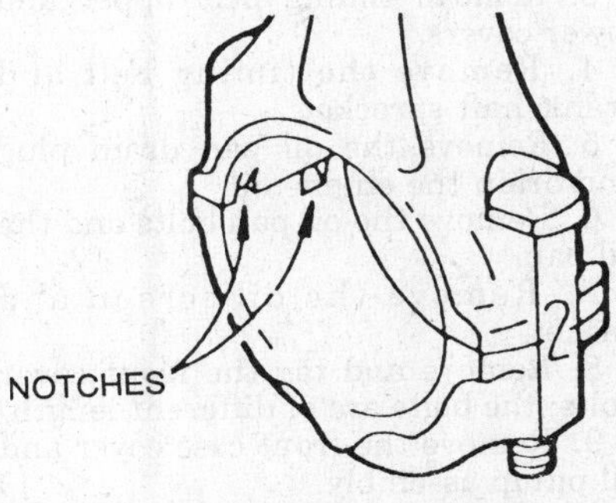

Connecting rod cap installation

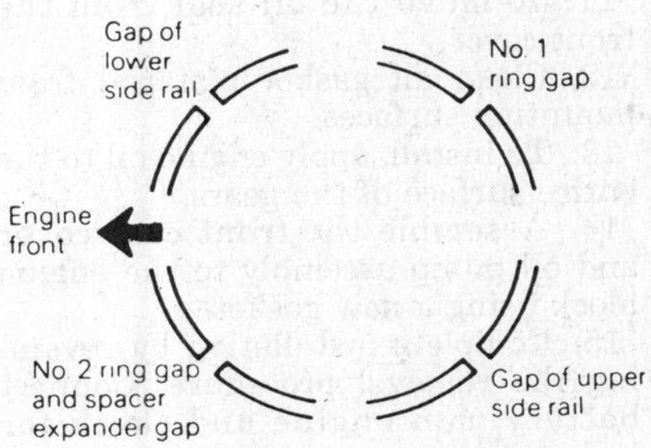

Piston ring positioning

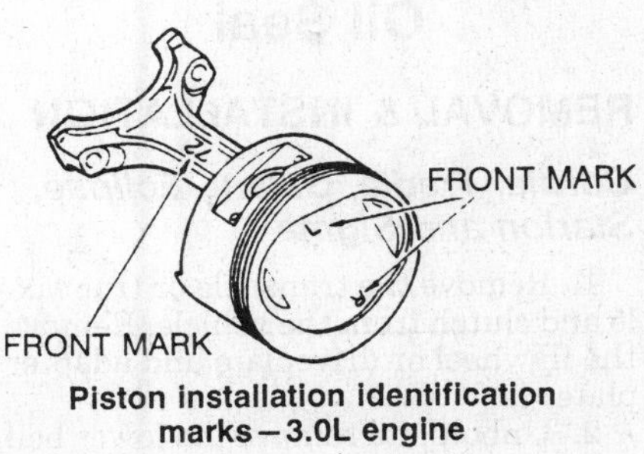

Piston installation identification marks – 3.0L engine

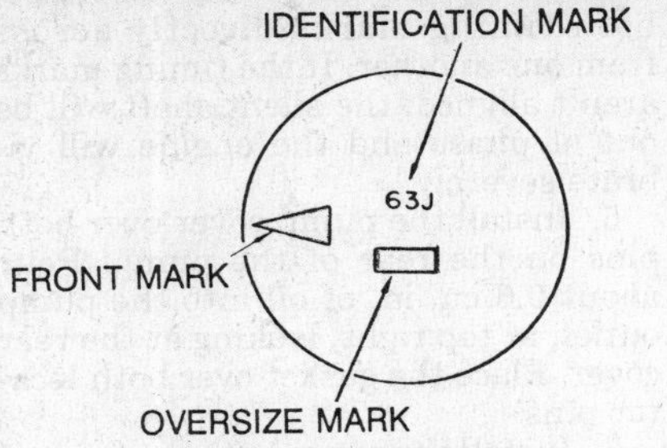

Piston installation

ENGINE LUBRICATION

Oil Pan

REMOVAL & INSTALLATION

1. The oil pan must be pulled downward as much as 6 in. to clear the oil pickup. In nearly all applications, this requires that the engine mounts be disconnected and the engine raised to clear a crossmember under the shallower section of the pan. First, survey the area under the engine to determine whether or not there is clearance, in case, the engine can be left in place.

2. Disconnect the negative battery cable. Drain the oil pan into a suitable container. Disconnect all those hoses and wires that would prevent the engine from being lifted the required distance for removal of the pan. If necessary, remove the starter, transaxle mounts, bell housing and oil filter.

3. Raise and safely support the vehicle and remove the oil pan bolts. Hook a lift to the hooks on the cylinder head and support the engine.

4. Remove the through bolts from the engine mounts. Raise the engine far enough to gain clearance, as necessary. Remove the oil pan from the vehicle.

To install:

5. Installation is the reverse of the removal procedure.

6. On the Mirage 1.6L engine, coat the 4 seams on the gasket surface for the block with a liquid sealer. These are the joints between the front cover and block on the front and the rear oil seal case and the block at the rear.

7. If equipped with gasketless pans, use liquid sealer MD997110 or equivalent. Cut the end of the tube off at the smallest diameter and run a bead of sealer around the entire groove in the oil pan. It should be about 0.16 in. thick. Run the head in back of the bolt holes. The pan should be installed within 15 minutes of the time the sealer is applied. Position the pan, install the bolts and tighten finger-tight.

8. Torque the bolts alternately and in several stages.

Oil Pump

REMOVAL & INSTALLATION

1.8L, 2.0L (Except Eclipse) and 2.3L Engines

1. Disconnect the negative battery cable. Remove the timing belt cover, timing belts and sprockets. Drain the oil pan.

2. The front oil pan bolts screw into the front case, onto which the oil pump is mounted using a seal. Remove the oil pan.

3. Remove the oil filter. Remove the oil screen and gasket. Remove the oil relief plunger plug and gasket. Then, remove the relief spring and plunger from the oil filter bracket.

4. Remove the 4 oil filter bracket bolts and remove the bracket.

5. Remove the cap from the oil pump area of the front case. This is slightly to the right and above the silent shaft on the driver's side of the vehicle. Remove the plug from the left side of the block (near the front case) and insert a suitable tool at least 2.4 in. long to retain the position of the silent shaft.

6. Remove the retaining bolt for the left silent shaft retaining bolts. Use a deep well socket. Remove the bolts and the front case from the front of the block.

7. Remove the oil pump cover bolts from the rear of the front case and remove the oil pump cover. Remove the gears from the front case.

To install:

8. Install the oil pump cover to the front case and torque the 5 bolts to 11–13 ft. lbs.

9. Install a special oil seal guide tool MD998285-01, to the front of the crankshaft, with the smaller diameter facing outward. Install a new front case gasket to the block. Install the front case and install and tighten the 8 bolts slightly. Remove the seal guide.

10. Install the oil pump gear and left silent shaft retaining bolt and torque to 25–28 ft. lbs.

11. Install the oil filter bracket and gasket. Tighten all the front case bolts to 15–19 ft. lbs. and those going through the oil filter bracket to 11–15 ft. lbs. Install the cap that covers the oil pump shaft.

12. Coat the oil pressure relief plunger with clean engine oil and insert it into the bore, followed by the spring. Install the plug and gasket and torque the plug to 29–36 ft. lbs.

13. Install the oil screen and gasket.
14. Install the oil pan in reverse of removal. Install the sprockets, tensioners and timing belts and tension them to specification. Install the timing covers and engine accessories. Make sure to refill the oil pan with the full capacity of clean engine oil. Idle the engine and make sure oil pressure builds up within a reasonable length of time.

1.5L and 1.6L Engines

NOTE: On the 1.5L engine, the front case must be removed to gain access to the oil pump. On the Mirage with the 1.6L engine, the oil pump is bolted to the front of the front case. If necessary to leave the front case in place, remove the timing belt covers and belts to gain access to the pump. If doing a complete overhaul, follow the procedure in order to replace the oil pan gasket and other parts.

1. Disconnect the negative battery cable. Remove the timing belt cover and timing belt.
2. Drain the oil and then remove the oil pan. Unbolt the oil pick-up and screen from the front case and remove it.
3. Remove the front cover with the oil pump assembled to it. On the 1.5L engine, pull the cover straight off to avoid damaging the crankshaft seal.
4. Put the cover on a clean bench. Remove the oil pump relief valve plug and gasket, spring and plunger.
5. Remove the attaching nut and then remove the oil pump sprocket on the Mirage with the 1.6L engine. On the 1.5L engine, turn the cover over. Then, remove the bolts and remove the oil pump cover from the case.
6. Installation is the reverse of the removal procedure. Repair or replace defective components as required.

2.5L Engine

1. Disconnect the negative battery cable.
2. Remove the front cover. Remove the bolt for the right side silent shaft, just above the oil pump drive sprocket. Remove the silent shaft drive chain and timing chain tensioner. Do not remove the timing chain. Leave the crankshaft sprocket that drives the silent shaft chain in place.
3. Remove the oil pump relief valve plug, spring and plunger. Remove the oil pump bolts and the pump assembly and gasket. Remove the cover from the rear of the oil pump.

To install:

4. Oil the oil pump gears and the inner walls of the pump housing thoroughly with engine oil. Install the gears into the oil pump housing with both timing marks directly across from one another. If the timing marks aren't aligned, the silent shaft will be out of phase and the engine will vibrate severely.
5. Install the pump cover over both pins on the rear of the pump. Pour about 0.6 cu. in. of oil into the pump outlet, at top right, looking at the rear cover. Place the gasket over both locator pins.
6. Install the pump onto the front of the block, engaging the keyway slot in the upper oil pump gear with the key on the right silent shaft and fitting the locating pins into the holes in the front of the block. Install the oil pump bolts and torque in several stages and alternately to 7.5–8.5 ft. lbs. Install the bolt that attaches the right silent shaft to the upper oil pump gear.
7. Remove the securing wire from the timing chain tensioner. Reinstall the oil pump relief valve spring, plunger and cap and torque the cap to 22–32 ft. lbs.
8. Install the timing chains and sprockets. When the timing chain for the silent shafts is installed, torque the bolt that attaches the right silent shaft gear to 44–50 ft. lbs.
9. Install the front cover. Make sure the engine oil pan is full to the correct level. Start the engine, idling it and making sure oil pressure is built up within a reasonable length of time. Check for leaks and repair, as necessary.

3.0L Engine

1. Bring the No. 1 piston to TDC of its compression stroke. Disconnect the negative battery cable and drain the engine oil into a suitable container.
2. Remove the timing belt covers, timing belt and sprockets.
3. Connect a suitable engine lift to the engine and take up the slack in the chain.
4. Remove the crankshaft sprocket. Remove the front 2 transaxle mounts.
5. Remove the oil pressure switch, oil filter, oil filter bracket and bracket gasket.
6. Remove the oil pan, oil pan gasket, oil pump screen and screen gasket.
7. Remove the oil pump pressure relief valve assembly. Remove the crankshaft front oil seal.
8. Remove the oil pump case, oil pump gasket, oil pump cover, oil pump outer rotor and oil pump inner rotor.
9. Installation is the reverse order of the removal procedure. Be sure to check all moving parts and replace, as necessary. Use new gaskets and O-rings when necessary, replace the front crankshaft seal with a new one. Apply a suitable sealant to the oil pan gasket and coat the threads of the oil pressure switch with a suitable sealant. Torque the oil pump bolts to 10 ft. lbs.

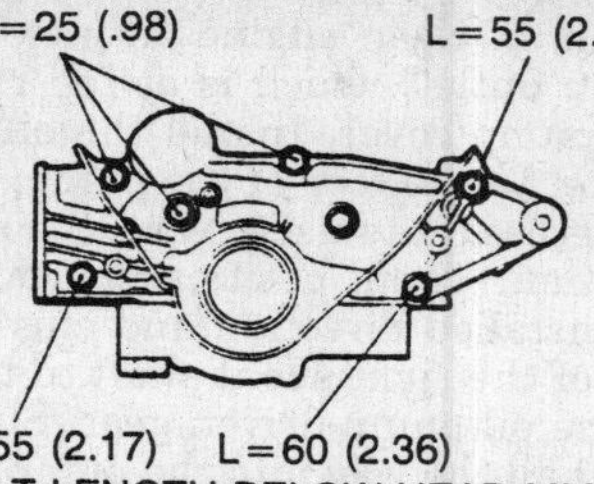

Bolt diameter location on the oil pump assembly—3.0L engine

Eclipse

1.8L AND 2.0L ENGINES

1. Disconnect the negative battery cable.
2. Remove the front engine mount bracket and accessory drive belts.
3. Remove timing belt upper and lower covers.
4. Remove the timing belt and crankshaft sprocket.
5. Remove the oil pan drain plug and drain the engine oil.
6. Remove the oil pan bolts and the oil pan.
7. Remove the oil screen and gasket.
8. Remove and tag the front cover bolts; the bolts are of different length.
9. Remove the front case cover and oil pump assembly.
10. Check the oil pump housing and gears for cracks, wear and other damage.
11. Remove the oil seal from the front cover.
12. Clean all gasket material from mounting surfaces.
13. To install, apply engine oil to the entire surface of the gears.
14. Assemble the front case cover and oil pump assembly to the engine block using a new gasket.
15. Complete installation by reversing the removal procedure. Connect battery, run engine and check for leaks.

Rear Main Bearing Oil Seal

REMOVAL & INSTALLATION

Cordia, Tredia, Galant, Eclipse, Starion and Sigma

1. Remove the transaxle or transaxle and clutch from the vehicle. Remove the flywheel or driveplate and adapter plate.
2. Unbolt and remove the lower bell

housing cover from the rear of the engine. Remove the rear plate from the upper portion of the rear of the block.

3. The lower surface of the oil seal case seals against the oil pan gasket or sealer at the rear. On engines with a gasket, carefully separate the gasket from the bottom of the seal case with a moderately sharp instrument. Loosen the oil pan bolts slightly at the rear to make it easier to separate both surfaces. If the gasket is damaged, the oil pan will have to be removed and the gasket replaced. If using sealant, unbolt and lower the oil pan and clean both surfaces, apply new sealer and reinstall the oil pan after Step 7 is completed.

4. Remove the oil seal case bolts and pull it straight off the rear of the crankshaft. Remove the case gasket.

5. Remove the seal retainer or oil separator from the case, and then pry out the seal. Inspect the sealing surface at the rear of the crankshaft. If a deep groove is worn into the surface, the crankshaft will have to be replaced. Lubricate the sealing surface with clean engine oil.

To install:

6. Using a seal installer such as MD998376–01 and MD990938–01, install the new seal into the bore of rear oil seal case in such a way that the flat side of the seal will face outward when the case is installed on the engine. The inside of the seal must be flush with the inside surface of the seal case.

7. Install the retainer or oil separator directly over the seal with the small hole located directly at the bottom. Then, install a new gasket onto the block surface and install the seal case to the rear of the block. Retorque pan bolts, as necessary. Refill the oil pan if necessary.

8. Install the rear plate and bell housing cover. Install the flywheel or driveplate and the transaxle in reverse of the removal procedure.

Mirage and Precis

1. Remove the transaxle or manual transaxle and clutch from the vehicle. Remove the flywheel or driveplate and adapter plate.

2. Unbolt and remove the rear plate from the rear of the block. On the Mirage with the 1.6L engine, use a moderately sharp instrument and separate the rear portion of the oil pan gasket from the lower surface of the rear main seal case on the back of the block. Loosen the oil pan bolts slightly at the rear to make it easier to separate both surfaces. If the gasket is damaged, drain the oil pan and remove it. On the 1.5L engine, drain the oil pan and remove it, as the sealing surfaces must be cleaned and new sealer applied all around.

3. Unbolt the oil seal case and then pull it straight back and off the crankshaft. Remove the case gasket. Pry the old seal out of the case.

To install:

4. Inspect the sealing surface at the rear of the crankshaft. If a deep groove is worn into the surface, the crankshaft will have to be replaced. Press a new seal into the case with a special seal installing tool such as MD998011, or equivalent. The seal must be pressed in square until it bottoms in the case.

5. Oil the crankshaft sealing surfaces and the lips of the new seal. On the Mirage with the 1.6L engine spread a liquid sealer thoroughly around those areas which butt up against the block and oil pan gasket at the bottom surface and on the front at both sides. Then, install the seal, gasket and seal case straight over the crankshaft sealing surface. Install and tighten the 5 case bolts.

6. On the 1.5L engine, install sealer and reinstall the oil pan. On the Mirage with the 1.6L engine reinstall the pan with a new gasket, if necessary, or retorque pan bolts, as necessary.

7. Reinstall the transaxle. Make sure the engine oil pan is refilled with clean engine oil, if necessary.

3000-GT

1. Disconnect the negative battery cable.

2. Remove the transaxle from the vehicle.

3. Remove the flywheel/ring gear assembly.

4. If the crankshaft rear oil seal case is leaking, remove it. Otherwise, just remove the oil seal. Some engines have a separator that should also be removed.

To install:

5. Install the separator. Lubricate the inner diameter of the new seal with clean engine oil.

6. Install the oil seal in the crankshaft rear oil seal case using tool MD998376 or equivalent. Press the seal all the way in without tilting it. Force the oil separator into the oil seal case so the oil hole in the separator is at 7 o'clock position.

7. Install the seal case with a new gasket.

8. Install the flywheel and transaxle.

9. Connect the negative battery cable and check for leaks.

ENGINE COOLING

Radiator

REMOVAL & INSTALLATION

Cordia, Tredia and Mirage

1. Disconnect the negative battery cable. Disconnect the electrical connector for the fan motor. Drain the coolant into a clean container.

2. Disconnect the upper and lower radiator hoses and the overflow tank at the radiator. If equipped with an automatic transaxle, disconnect both hoses for the cooler at the lower tank and plug all openings. Then, remove both bolts from the rear of the radiator, lift the unit out of the bushings at the front crossmember and remove it.

3. Remove the fan and electric motor from the radiator, transferring it to a new unit, if necessary.

4. Install the unit in reverse order, making sure the prongs on the lower tank fit securely into the bushings on the crossmember. Refill the radiator with clean coolant and run the engine until the thermostat opens. Refill the radiator with coolant as necessary, install the cap and then fill the overflow tank. If the vehicle has an automatic transaxle, check the fluid level and if necessary refill.

Galant and Sigma

1. Drain the engine coolant into a suitable container, remove the radiator cap while draining the coolant.

2. Remove the battery, coolant overflow tube and coolant reserve tank bracket and tank.

3. Remove the upper radiator hose. Disconnect the electrical fan motor connection, thermosensor connections on the condenser fan and the radiator fan.

4. Remove the oil cooler lines from the radiator, if equipped with an automatic transaxle.

5. Remove the radiator fan motor, lower radiator hose and radiator bracket.

6. Remove any remaining radiator retaining bolts and remove the radiator.

7. Installation is the reverse order of the removal procedure. Replace the radiator bushings, if necessary.

Starion

1. Disconnect and remove the battery. Drain the coolant into a clean container.

2. Remove both bolts on either side of the radiator and remove the upper and lower fan shrouds. Disconnect the upper and lower hoses at the radiator. Disconnect the overflow tank hose at the filler cap opening.
3. Then, remove the 4 radiator bolts, 2 on either side and remove the radiator.
4. Install the unit in reverse order. Refill the radiator with clean coolant and run the engine until the thermostat opens. Refill the radiator with coolant as necessary, install the cap and then fill the overflow tank.

Precis

1. Disconnect the negative battery cable. Remove the splash shield from under the vehicle.
2. Drain the radiator.
3. Remove the fan shroud and disconnect the fan motor wiring harness.
4. Disconnect the radiator hoses and, if equipped, the automatic transaxle cooler hoses.
5. Disconnect the expansion tank hose.
6. Remove the radiator bolts and lift out the radiator and fan assembly.
7. Installation is the reverse of removal.

Eclipse

1. Disconnect the negative battery cable.
2. Drain the cooling system.
3. Disconnect the overflow tube. Some vehicles may also require removal of the overflow tank.
4. Disconnect upper and lower radiator hoses.
5. Disconnect electrical connectors for cooling fan and air conditioning condensor fan, if equipped.
6. Disconnect thermo sensor wires.
7. Disconnect and plug automatic transaxle cooler lines, if used.
8. Remove the upper radiator mounts and lift out the radiator/fan assembly.

To install:

9. At installation, use proper mix of coolant suitable for use in engines with aluminum components.
10. Reconnect the negative battery cable.
11. Check automatic transaxle fluid level and refill as necessary.

3000-GT

1. Disconnect the negative battery cable.
2. Drain the cooling system.
3. Disconnect the overflow tube. Some vehicles may also require removal of the overflow tank.
4. Disconnect upper and lower radiator hoses.
5. Disconnect electrical connectors for cooling fan and air conditioning condenser fan, if equipped. Remove the fan assembly if necessary.
6. Disconnect thermo sensor wires.
7. Disconnect and plug automatic transaxle cooler lines, if used.
8. Remove the upper radiator mounts and lift out the radiator/fan assembly.
9. Service the lower mounts as required.

To install:

10. Install the radiator and fan assembly, if removed as an assembly.
11. Connect the automatic transaxle cooler lines, if disconnected.
12. Connect the thermo wires.
13. Install the fan if removed separately.
14. Install the radiator hoses.
15. Install the overflow tube and reservoir, if removed.
16. Fill the system with coolant.
17. Connect the negative battery cable, run the vehicle until the thermostat opens, fill the radiator completely and check the automatic transaxle fluid level, if equipped.
18. Once the vehicle has cooled, recheck the coolant level.

Heater Core

REMOVAL & INSTALLATION

Cordia, Tredia, Mirage, Galant and Sigma

1. Disconnect the battery ground cable.
2. Set the heater control lever to **WARM**.
3. Drain the cooling system.
4. Remove the instrument panel.
5. Remove the duct from between the heater unit and the blower case.
6. Disconnect the coolant hoses at the heater case.
7. Unbolt and remove the heater case.
8. Remove the hose and pipe clamps and remove the water valve, if equipped.
9. Remove the core from the case.

To install:

10. Set the mixing damper to the closed position and, with the damper in that position, install the rod so the water valve is fully closed.
11. Place the damper lever in the **VENT** position and adjust the linkage so the **FOOT/DEF** damper opens to the **DEF** side and the **VENT** damper is level with the separator.
12. Install the hoses. They are marked for flow direction.
13. The remainder of assembly is the reverse of disassembly.

Starion

1. Drain the cooling system and disconnect the negative battery cable.
2. Place the heater control in the **WARM** position.
3. Disconnect the heater hose at the engine firewall.
4. Remove the instrument panel and the center console.
5. Remove the center ventilator duct, defroster duct and lap duct.
6. Remove the center reinforcement and heater control assembly.
7. Remove the heater assembly bolts and the heater assembly.
8. Remove the heater core.
9. Install in reverse order.

Eclipse

1. Disconnect the negative battery cable.
2. Drain the cooling system.
3. Remove the floor console by first removing the plugs, then the screws retaining the side covers and the small cover piece in front of the shifter. Remove the shifter knob, manual transaxle, and the cup holder. Remove both small pieces of upholstery to gain access to retainer screws. Disconnect the 2 electrical connectors at the front of the console. Remove the shoulder harness guide plates and remove the console assembly.
4. Remove the instrument panel assembly. Use the following procedure:
   a. Locate the rectangular plugs in the knee protector on either side of the steering column. Pry these plugs out, remove the screws. Remove the screws from the hood lock release lever and remove the knee protector.
   b. Remove the upper and lower column covers.
   c. Remove the narrow panel covering the instrument cluster cover screws, and take out the cover.
   d. Remove the radio panel and take out the radio.
   e. Remove the center air outlet assembly by reaching through the grill and pushing the side clips out with a small flat-tip tool while carefully prying the outlet free.
   f. Pull the heater control knobs off and remove the heater control panel assembly.
   g. Open the glove box, remove the plugs from the sides and remove the glove box assembly.
   h. Remove the instrument gauge cluster and the speedometer adapter by disconnecting the speedometer cable at the transaxle, pulling the cable sightly towards the vehicle interior, then giving a slight twist on the adapter to release it.
   i. Remove the left and right speaker covers from the top of the instrument panel.

j. Remove the center plate below the heater controls.

k. Remove the heater control assembly installation screws.

l. Remove the lower air ducts.

m. Drop the steering column by removing the bolts.

n. Remove the instrument panel screws, bolts and the instrument panel assembly.

5. Remove both stamped steel reinforcement pieces.

6. Remove the lower duct work from the heater box.

7. Remove the upper center duct.

8. Vehicles without air conditioning will have a square duct in place of the evaporator. Remove this duct, if not equipped with air conditioning. If equipped with air conditioning, remove the evaporator assembly after properly discharging the air conditioning system. Disconnect and cap the refrigerant lines at the evaporator. Remove the wiring harness connectors and the electronic control unit. Remove the drain hose and lift out the evaporator unit.

9. With the evaporator removed, take out the heater unit. To prevent bolts from falling inside the blower assembly, set the inside/outside air-selection damper to the position that permits outside air introduction.

10. Remove the cover plate around the heater tubes and remove the core fastener clips. Pull the heater core from the heater box, being careful not to damage the fins or tank ends.

To install:

11. Installation is the reverse of the removal procedure. Install the heater core to the heater box. Install the clips and cover.

12. Install the evaporator and the automatic transaxle ELC box.

13. Install the heater box and connect the duct work.

14. Connect all wires and control cables.

15. Install the instrument panel assembly and the console.

16. Connect the battery negative cable, start engine and bleed cooling system. Recharge the air conditioning system.

3000-GT

1. Disconnect the negative battery cable.

NOTE: If equipped with an air bag, be sure to disarm it before entering the vehicle.

2. Drain the coolant and disconnect the heater hoses from the core tubes.

3. To remove the console, perform the following:

a. Remove the cup holder and console plug.

b. Remove the rear console.

c. Remove the radio bezels and radio.

d. Remove the switch bezel.

e. Remove the side covers and front console garnish.

f. If equipped with a manual transaxle, remove the shifter knob.

g. Remove the mounting screws and remove the console assembly.

4. Remove the hood lock release handle from the instrument panel.

5. Remove the interior and dash lights rheostat and switch bezel to its right.

6. Remove the driver's knee protector. Remove the steering column covers.

7. Remove the glove box and cover.

8. Remove the center air outlet assembly.

9. Remove the climate control switch assembly.

10. Remove the instrument cluster bezel and cluster.

11. If equipped with front speakers, remove them. If not, remove the plug in their place.

12. Disconnect the wiring harnesses on the right side of the instrument panel.

13. Remove the steering shaft support bolts and lower the steering column.

14. Remove the instrument panel mounting hardware and remove the instrument panel from the vehicle.

15. Remove the center reinforcement.

16. Remove the foot warmer ducts and lap duct.

17. If equipped with air conditioning, remove the evaporator case mounting bolt and nut to allow clearance for heater unit removal.

18. Remove the center duct above the heater unit.

19. Remove the heater unit and disassemble on a workbench. Remove the heater core from the heater case.

To install:

20. Thoroughly clean and dry the inside of the case and install the heater core and all related parts.

21. Install the heater unit to the vehicle and install the mounting screws.

22. Install the center duct above the unit.

23. Secure the evaporator case with the bolt and nut.

24. Install the lap duct and foot warmer ducts.

25. Install the center reinforcement.

26. Install the instrument panel by reversing its removal procedure.

27. Install the hood lock release cable handle.

28. Install the console.

29. Fill the cooling system.

30. Connect the negative battery cable and check the entire climate control system for proper operation and leaks.

Water Pump

REMOVAL & INSTALLATION

Cordia, Tredia, Mirage, Precis

1. Disconnect the negative battery cable. Loosen the 4 bolts attaching the water pump pulley to the pulley flange. Loosen the alternator bolts, slide the alternator toward the engine and remove the belt. Drain the radiator.

2. Remove the 4 bolts attaching the water pump pulley to the pump flange and remove the pulley. Remove the timing belt covers and timing belt tensioner.

3. Remove the 5 water pump bolts. Remove the pump and gasket, disconnecting the outlet at the water pipe (don't lose the O-ring).

4. Clean gasket surfaces and coat a new gasket with sealer. Then, position the gasket on the front of the block with all bolt holes aligned. Replace the O-ring for the outlet water pipe.

To install:

5. Install the pump over a new gasket, connecting the outlet water pipe.

6. Install the remaining parts in reverse order. Final tightening of the water pump pulley bolts is done most easily after the V-belt has been installed and tensioned some what. Recheck tension after the pulley bolts are tightened. Close the radiator drain and refill the system. Run the engine until the thermostat opens and then add coolant until the level stabilizes before replacing the radiator cap. Check for leaks.

Galant and Sigma

1. Bring the No. 1 piston to TDC of its compression stroke. Disconnect the negative battery cable and drain the engine oil and coolant into a suitable containers.

2. Remove the timing belt covers, timing belt and sprockets and crankshaft sprocket.

3. Remove the water pump bolts. Remove the pump and gasket, disconnecting the outlet at the water pipe (don't lose the O-ring).

To install:

4. Clean gasket surfaces and coat a new gasket with sealer. Then, position the gasket on the front of the block with all bolt holes aligned. Replace the O-ring for the outlet water pipe.

5. Install the pump over a new gasket, connecting the outlet water pipe.

6. Complete the installation by reversing the order of the removal procedure. Torque the water pump bolts to 14–19 ft. lbs.

Starion

1. Disconnect the negative battery cable. Drain the radiator.
2. Loosen the 4 nuts attaching the clutch fan to the water pump studs; then, loosen the adjusting and mounting bolts for the alternator, rock it toward the engine and remove the belt. Remove the 4 bolts for the fan shrouds from the rear of the radiator and remove the upper and lower shrouds. Loosen the nuts and remove them together with the lockwashers. Remove the fan clutch unit, storing the fan clutch in its normal altitude to keep the fluid from migrating to the wrong portions of the unit. Remove the pulley from the studs.
3. Disconnect the lower radiator hose at the pump by loosening the clamp and pulling the hose off. Then, remove the mounting bolts from the pump and then remove the pump and gasket from the front of the block.

To install:

4. Clean both gasket surfaces thoroughly and coat both sides of a new gasket and both gasket surfaces with sealer.
5. Install the gasket onto the block and then position the water pump over the gasket. Install the bolts in the proper positions.
6. Install the remaining parts in reverse order.
7. Refill the radiator with clean antifreeze and water mixed 50/50. Run the engine until the thermostat opens, refill the radiator as necessary, install the cap and check for leaks.

Eclipse

1. Disconnect the negative battery cable.
2. Drain cooling system.
3. Remove engine under cover.
4. Remove the timing belt.
5. Remove the alternator bracket.
6. Remove the water pump, gasket and O-ring where the water inlet pipe joins the pump.

To install:

7. Installation is the reverse of the removal process. Clean both gasket surfaces of the water pump and block. Install the new O-ring into the groove on the front end of the water inlet pipe. Do not apply oils or grease to the O-ring. Wet with water only. Install the gasket and pump assembly and tighten the bolts. Note the marks on the bolt heads. Those marked **4** should be torqued to 9–11 ft. lbs. Those bolts marked **7** should be torqued from 14–20 ft. lbs.
8. Reinstall the alternator bracket.
9. Install the timing covers, drive pulleys and belts and the undercover.
10. Refill with coolant.
11. Reconnect the negative battery cable, start engine and bleed cooling system.

3000-GT

1. Disconnect the negative battery cable.
2. Drain the cooling system.
3. Remove the engine undercover.
4. Remove the timing belt.
5. Disconnect the hoses from the pump, if equipped. Remove the alternator bracket.
6. Remove the water pump, gasket and O-ring where the water inlet pipe joins the pump.

To install:

7. Thoroughly clean and dry both gasket surfaces of the water pump and block.
8. Install a new O-ring into the groove on the front end of the water inlet pipe. Do not apply oils or grease to the O-ring. Wet with water only.
9. Install the gasket and pump assembly and tighten the bolts. Note the marks on the bolt heads. Those marked **4** should be torqued to 9–11 ft. lbs. Those bolts marked **7** should be torqued from 14–20 ft. lbs.
10. Connect the hoses to the pump.
11. Reinstall the timing belt and related parts.
12. Install the engine undercover.
13. Fill the system with coolant.
14. Connect the negative battery cable, run the vehicle until the thermostat opens and fill the radiator completely.
15. Once the vehicle has cooled, recheck the coolant level.

Thermostat

REMOVAL & INSTALLATION

1. Disconnect the negative battery cable. Drain the coolant below the level of the thermostat.
2. Remove both retaining bolts and lift the thermostat housing off the intake manifold with the hose still attached. If careful, it is not necessary to remove the upper radiator hose.
3. Lift the thermostat out of the manifold.
4. Installation is the reverse of the removal procedure.

Cooling System Bleeding

All vehicles are equipped with a self-bleeding thermostat. Slowly fill the cooling system in the conventional manner; air will vent through the jiggle valve in the thermostat. Run the vehicle until the thermostat has opened and continue filling the radiator. Recheck the coolant level after the vehicle has cooled.

ENGINE ELECTRICAL

NOTE: Disconnecting the negative battery cable on some vehicles may interfere with the functions of the on board computer systems and may require the computer to undergo a relearning process, once the negative battery cable is reconnected.

Distributor

REMOVAL

1. Rotate the engine until the No. 1 piston is at TDC of the compression stroke. Disconnect the negative battery cable. Remove all necessary components in order to gain access to the distributor assembly.
2. Remove the distributor cap with the spark plug wires attached and position it out of the way. Disconnect the distributor wiring connector and vacuum hoses. Be sure to tag all the wires and vacuum lines for easy installation.
3. Remove the distributor base retaining nut. Remove the distributor.
4. Before installation check that the No. 1 piston is at TDC of the compression stroke, then align the marks on the bottom of the distributor housing, just above the drive gear, with the punch mark on the distributor drive gear.

INSTALLATION

Timing Not Disturbed

1. To install, double check that the crankshaft mark and timing mark are still aligned and that the engine has not been turned. Align the mating mark on the distributor housing described above with the mating mark (punch) on the distributor driven gear.
2. Install the distributor assembly while aligning the mating mark on the distributor attaching flange with the center of the hold-down stud.
3. Connect all wiring and reinstall distributor cap and seal.
4. Connect the negative battery cable. Start engine and set timing. Then tighten distributor mounting nut.

Timing Disturbed

1. Remove the spark plug from No. 1 cylinder and position a compression gauge or a thumb over the spark plug hole.

2. Slowly crank the engine until compression pressure starts to build.
3. Continue cranking the engine so the timing marks align with the TDC mark.
4. Install the distributor assembly while aligning the mating mark on the distributor attaching flange with the center of the hold-down stud.
5. Connect all wiring and reinstall distributor cap and seal.
6. Connect the negative battery cable. Start engine and set timing. Then tighten distributor mounting nut.

NOTE: Some engines may be sensitive to the routing of the distributor sensor wires. If routed near the high-voltage coil wire or the spark plug wires, the electromagnetic field surrounding the high voltage wires could generate an occasional disruption of the ignition system operation.

Distributorless Ignition

REMOVAL & INSTALLATION

Crank Angle Sensor

1. Disconnect the battery negative cable.
2. The crank sensor is driven off the back of the intake camshaft. To remove, turn the crankshaft by hand so the No. 1 cylinder piston is at TDC.
3. Disconnect the multi-wire connector.
4. Remove the retainer bolts and lift the sensor from the cylinder head.

To install:
5. At installation, align the punch mark on the crank angle sensor housing with the notch in the plate, then install the crank sensor. Make sure the flat drive tang registers into the slot in the camshaft.
6. Reconnect the multi-wire connector.
7. Reconnect the battery negative cable and check the timing.

Ignition Coil

1. Disconnect the battery negative cable.
2. The ignition coil is mounted on the front of the intake manifold. Remove and tag the spark plug cables.
3. Remove the mounting bolts and remove from engine.

To install:
4. Install coil to manifold. Install bolt and tighten to 15–19 ft. lbs.
5. Install spark plug cables in the correct locations.
6. Reconnect the battery negative cable.

Power Transistor

1. Disconnect the battery negative cable.
2. The power transistor is mounted on the front of the intake manifold. Remove and retaining screw and disconnect the wires to remove.

To install:
3. Install power transistor to manifold. Install screw and tighten.
4. Reconnect the battery negative cable.
5. Install the distributor to the cylinder head while aligning the mark on the base attaching flange with center of the hold-down stud. Install the retaining nut, distributor cap and vacuum hoses. Start the engine and adjust the ignition timing.

Ignition Timing

ADJUSTMENT

1. Start the engine and allow it to reach normal operating temperature.
2. Apply the hand brake and position the gear selector in neutral, if equipped with a manual transaxle, or **P**, if equipped with an automatic transaxle. Turn **OFF** all accessories and stop the engine.
3. Install a tachometer and check the idle speed. Adjust as necessary.
4. Ground the appropriate ignition timing terminals prior to timing check. By grounding this terminal, it prohibits the engine computer from attempting to adjust the ignition timing durning servicing.
5. Disconnect and plug the vacuum advance lines at the distributor, if equipped. On the Cordia and Tredia 1.8L engine, disconnect the boost sensor connector, located in the engine compartment.
6. Stop the engine and connect the timing light according to manufacturers instructions.
7. Start the engine and allow it to idle. Point the timing light at the mark on the front cover and read the basic timing by noting the position of the groove in the front pulley in relation to the timing mark or scale on the front cover. If the timing is incorrect, loosen the distributor (crank angle sensor) mounting bolt. Turn the distributor (crank angle sensor) slightly to adjust timing.
8. When the reading is correct, tighten the distributor (crank angle sensor) mounting bolt and turn the engine **OFF**. Remove the jumper wire from the ignition timing adjusting terminal and install the water-proof cover.
9. Start the engine and check the actual timing (the timing without the terminal grounded). This reading should be 5 degrees more than the basic timing. This value may increase according to altitude. As long as the basic timing is correct, the engine is timed correctly. Also, actual timing may fluctuate because of slight variation accomplished by the ECU. The basic timing, though, should remain steady.
10. Turn the engine OFF and disconnect the timing apparatus and tachometer.

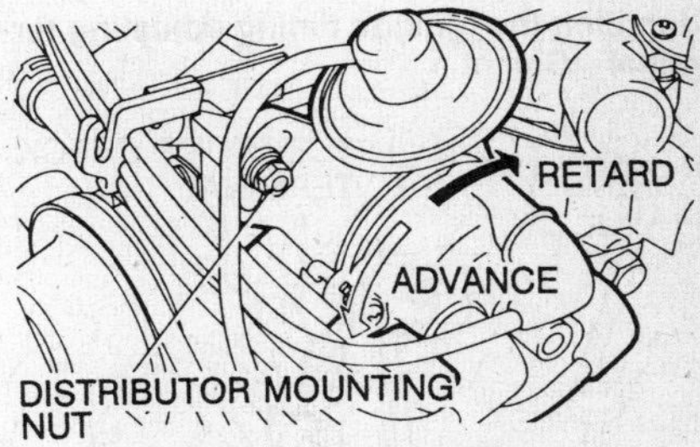

Adjusting ignition timing

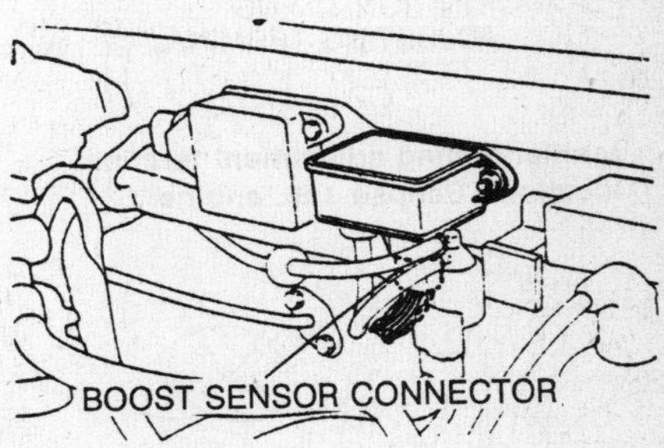

Boost sensor connector location – Cordia/Tredia

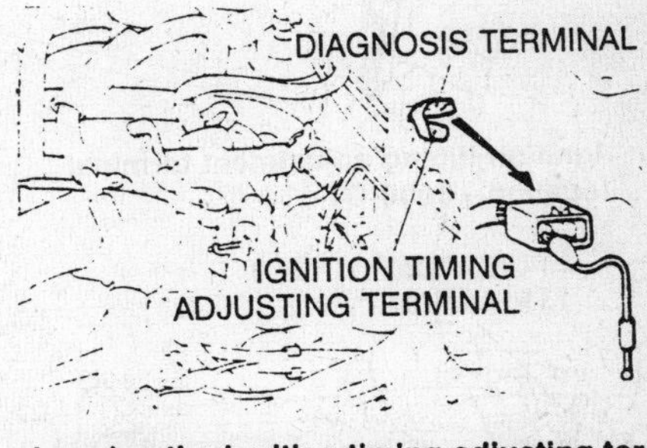

Jumping the ignition timing adjusting terminal – Mirage

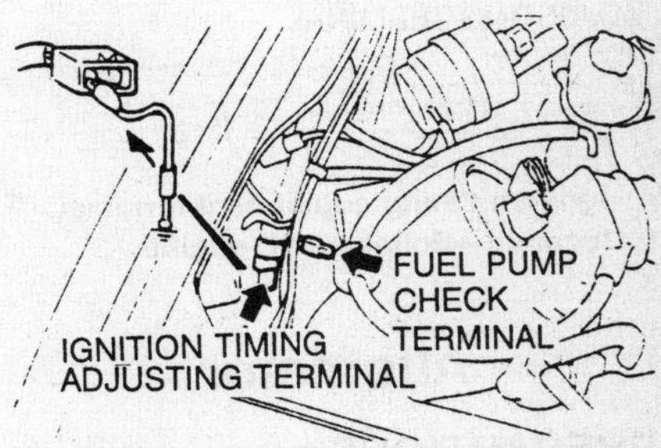

Jumping the ignition timing adjusting terminal – Starion

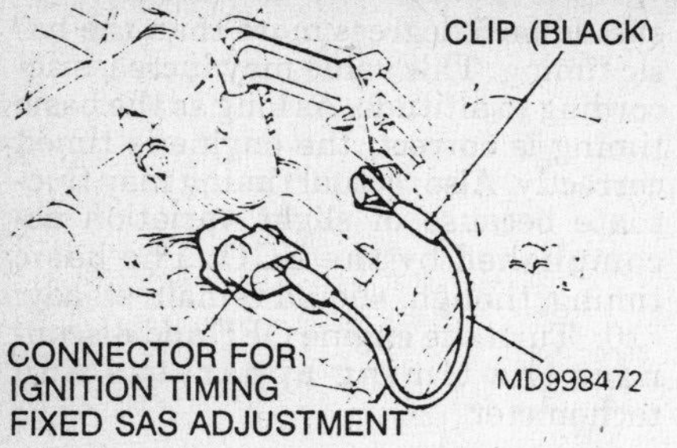

Jumping the ignition timing adjusting terminal – Galant

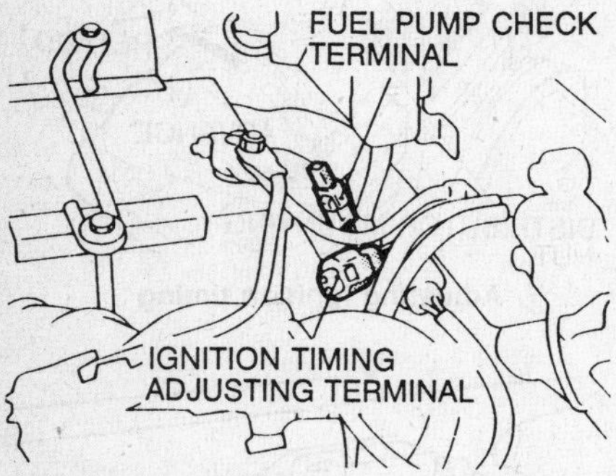

Ignition timing adjustment terminal location – Eclipse 1.8L engine

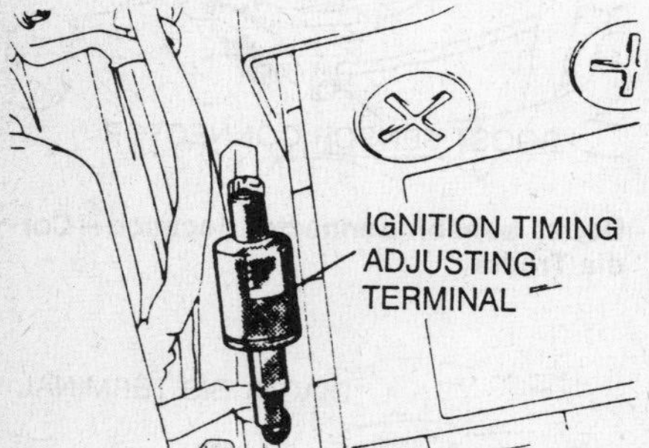

Ignition timing adjustment terminal location – 3000-GT

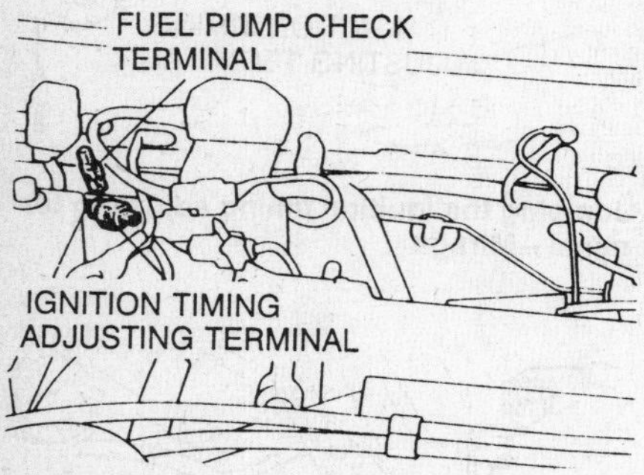

Ignition timing adjustment terminal location – Eclipse 2.0L engine

Alternator

PRECAUTIONS

In order to prevent damage to the alternator observe the following precautions:

- Reversing the battery connections will result in damage to the diodes.
- Booster cables should be connected from positive to positive and the negative cable from the booster battery connected to a good ground on the engine of the vehicle with the dead battery.
- Never use a fast charger as a booster to start the vehicle.
- When servicing the battery with a fast charger always disconnect the battery cables.
- Never attempt to polarize an alternator.
- Avoid long soldering times when replacing diodes or transistors. Prolonged heat is damaging to alternators.
- Do not use test lamps of more than 12 volts for checking diode continuity.
- Do not short across or ground any of the alternator terminals.
- The polarity of the battery, alternator and regulator must be matched and considered before making any electrical connections within the system.
- Never operate the alternator on an open circuit. Make sure all connections within a circuit are clean and tight.
- Disconnect the negative (or both) battery terminals when performing any service on the electrical system.
- Disconnect the negative battery cable if arc welding is to be done on any part of the vehicle.

BELT TENSION ADJUSTMENT

Except Sigma and Eclipse

1. Check the drive belt(s) for cracking, fraying or any other deterioration. Replace the drive belt if suspect.
2. Loosen the alternator pivot nut.
3. Loosen the lock bolt of the belt tension adjuster.
4. Using the adjustment bolt, adjust the belt tension to specification. Belt tension is proper when the belt can be deflected at midpoint $^{9}/_{32}$–$^{11}/_{32}$ in.
5. Tighten the lock bolt.
6. Tighten the alternator pivot nut.

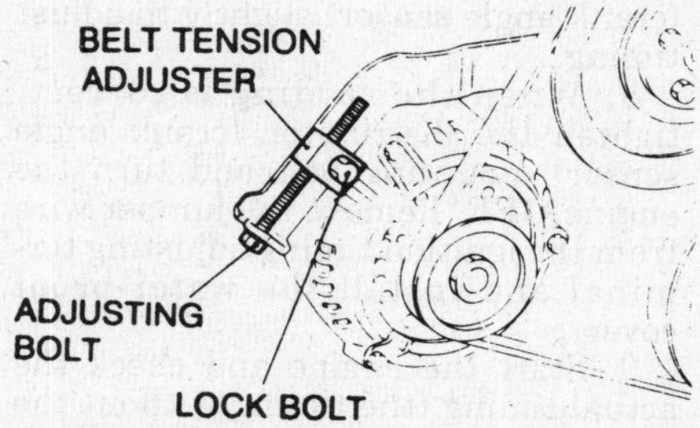

Alternator belt adjustment – except Sigma

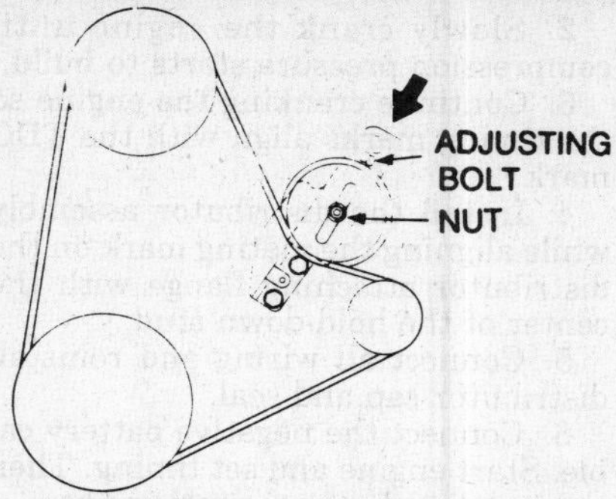

Alternator belt adjustment – Sigma

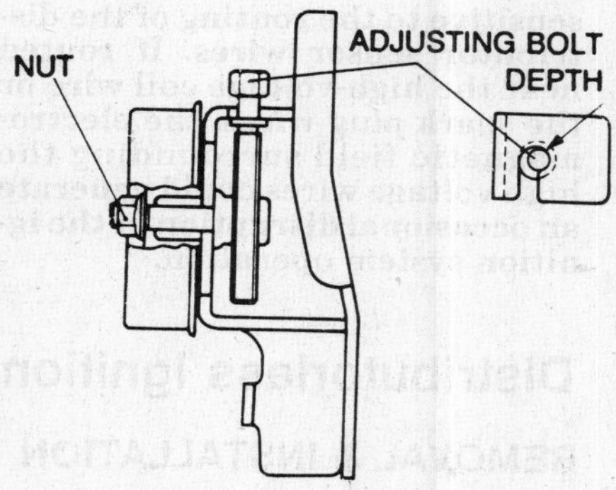

Tension pulley adjustment bolt

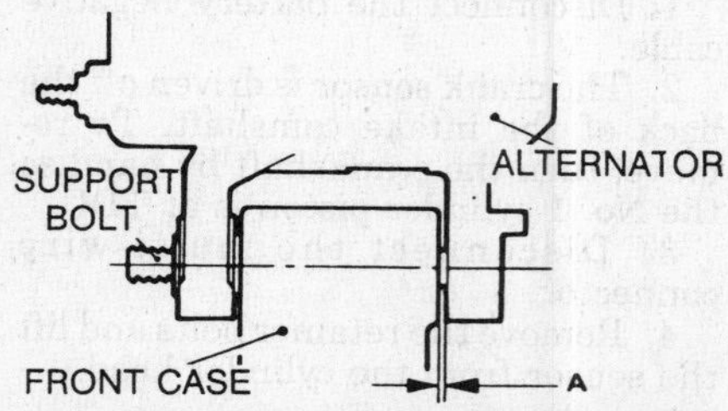

Gap "A" shows where you should measure the clearance for installation of alternator mounting shims

Sigma

1. Check the drive belt(s) for cracking, fraying or any other deterioration. Replace the drive belt if suspect.
2. To increase the belt tension, loosen the nut ⅛ turn, turn the left threaded adjusting bolt clockwise and displace the tension pulley slightly.

NOTE: Put the adjusting bolt into the recess at the far depth of the elongated hole on the tension bracket.

3. Tighten the nut to 28–43 ft. lbs. (39–60 Nm).

Eclipse

1. Place a straight-edge along the belt between 2 pulleys.
2. Measure the deflection with a force of about 22 lbs. applied midway between both pulleys. Deflection should be:

a. 1.8L engine – 0.315–0.433 in. (8.0–11.0mm)

b. 2.0L engine – 0.354–0.453 in. (9.0–11.5mm)

3. Belt tension can also be checked with a tension gauge. Measure between any 2 pulleys. The value should be 55–110 lbs. (250–500 N).

4. Several different alternator belt adjustment methods are used. If using a special bracket with a tension screw, loosen the locknut first. Then turn the screw clockwise to loosen the tension and counterclockwise to increase the tension.

REMOVAL & INSTALLATION

Except Eclipse, Galant and Sigma

1. Disconnect the negative battery cable.
2. Remove all necessary components in order to gain access to the alternator assembly.
3. On some vehicles, it may be necessary to remove the air conditioning compressor and position it aside in order to gain access to the alternator.
4. On the 1988 Mirage, it will be necessary to remove the air conditioning condenser fan motor and the power steering pump with bracket. On the 1988 Tredia and Cordia, it will be necessary to remove the power steering pump.

NOTE: After removing the engine bracket mounting, be sure to place a block of wood on the oil pan and raise the engine into to place for the duration of the operation.

5. Disconnect the alternator electrical. Note or label the wires so they can be reinstalled correctly.
6. Remove the top mounting bolt. Loosen the lower mounting nut. Slide the alternator over in its attaching bracket and remove the fan belt.
7. Remove the lower mounting nut and bolt. Remove the alternator from the vehicle.
8. Installation is the reverse of the removal procedure. Replace shims, as required, in their respective places. Adjust the drive belt, as required.

Galant and Sigma

1. Disconnect the negative battery cable.
2. Remove the front engine mounting bracket.
3. Remove the power steering pressure hose nut.
4. Remove the air conditioner low pressure line bolt.
5. Remove the drive belt tensioner with bracket.
6. Disconnect high tension spark plug wires 2, 4 and 6.
7. Remove the distributor cap and timing belt cover cap.
8. Disconnect the alternator electrical connections and remove the alternator.

To install:

9. Install the alternator and tighten all attaching bolts securely. Install the distributor cap and timing belt cover cap. Install spark plug wires 2, 4 and 6.
10. Install the drive belt tensioner and bracket, air conditioner low pressure line bolt, power steering pressure hose nut and the engine mounting bracket.
11. Adjust the alternator belt to specifications.

Eclipse

1.8L ENGINE

1. Disconnect the negative battery cable.
2. If equipped with air conditioning, remove condensor electric fan motor and shroud assembly. Then remove air conditioner compressor drive belt.
3. Remove alternator and water pump belts.
4. Remove both water pump pulleys.
5. Remove the alternator top brace, then disconnect the alternator wiring.
6. Remove alternator.
7. At installation, adjust drive belts, reconnect battery negative cable.

2.0L ENGINE

1. Disconnect the negative battery cable.

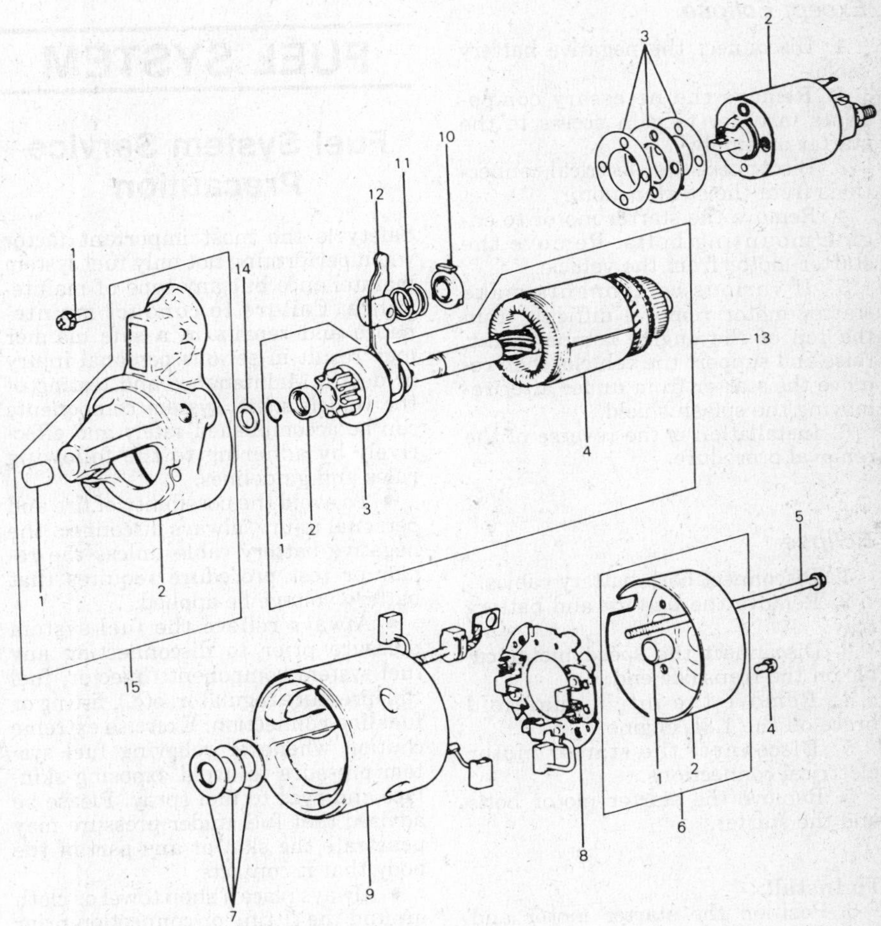

1. Screw (2)
2. Magnetic switch
3. Packing
4. Screw (2)
5. Through bolt (2)
6. Rear bracket assembly
7. Washer
8. Brush holder
9. Yoke assembly
10. Spring retainer
11. Lever spring
12. Lever
13. Armature assembly
14. Washer
15. Front bracket assembly

NOTE
Numbers show order of disassembly.
For reassembly, reverse order of disassembly.

Typical starter motor components

2. Remove the left cover panel from under the vehicle.
3. If equipped with air conditioning, remove condensor electric fan motor and shroud assembly.
4. Remove alternator and water pump belts.
5. Remove both water pump pulleys.
6. Remove the alternator top brace and disconnect the alternator wiring.
7. Remove alternator.
8. At installation, adjust belt tension, reinstall cover panel and reconnect battery negative cable.

Starter

REMOVAL & INSTALLATION

Except Eclipse

1. Disconnect the negative battery cable.
2. Remove the necessary components in order to gain access to the starter assembly.
3. Disconnect the electrical connections from the starter motor.
4. Remove the starter motor to engine mounting bolts. Remove the starter motor from the vehicle.
5. If various components make starter motor removal difficult from the top of the engine compartment, raise and support the vehicle, then remove the starter from under after removing the splash shield.
6. Installation is the reverse of the removal procedure.

Eclipse

1. Disconnect both battery cables.
2. Remove the battery and battery tray.
3. Disconnect the speedometer cable on the transaxle end.
4. Remove the intake manifold brace on the 1.8L engine.
5. Disconnect the starter motor electrical connections.
5. Remove the starter motor bolts and the starter.

To install:

6. Position the starter motor and install the bolts.
7. Reconnect the electrical connections.
8. Install the intake manifold brace on the 1.8L engine and reconnect the speedometer cable.
9. Install the battery tray and battery.
10. Reconnect battery cables and check starter motor operation.

EMISSION CONTROLS

Please refer to "Emission Controls" in the Unit Repair section for system maintenance procedures. Due to the complex nature of modern electronic engine control systems, comprehensive diagnosis and testing procedures fall outside the confines of this repair manual. For complete information on diagnosis, testing and repair procedures concerning all modern engine and emission control systems, please refer to "Chilton's Guide to Fuel Injection and Electronic Engine Controls".

FUEL SYSTEM

Fuel System Service Precaution

Safety is the most important factor when performing not only fuel system maintenance but any type of maintenance. Failure to conduct maintenance and repairs in a safe manner may result in serious personal injury or death. Maintenance and testing of the vehicle's fuel system components can be accomplished safely and effectively by adhering to the following rules and guidelines.

- To avoid the possibility of fire and personal injury, always disconnect the negative battery cable unless the repair or test procedure requires that battery voltage be applied.
- Always relieve the fuel system pressure prior to disconnecting any fuel system component (injector, fuel rail, pressure regulator, etc.), fitting or fuel line connection. Exercise extreme caution whenever relieving fuel system pressure to avoid exposing skin, face and eyes to fuel spray. Please be advised that fuel under pressure may penetrate the skin or any part of the body that it contacts.
- Always place a shop towel or cloth around the fitting or connection prior to loosening to absorb any excess fuel due to spillage. Ensure that all fuel spillage (should it occur) is quickly removed from engine surfaces. Ensure that all fuel soaked cloths or towels are deposited into a suitable waste container.
- Always keep a dry chemical (Class B) fire extinguisher near the work area.
- Do not allow fuel spray or fuel vapors to come into contact with a spark or open flame.
- Always use a backup wrench when loosening and tightening fuel line connection fittings. This will prevent unnecessary stress and torsion to fuel line piping. Always follow the proper torque specifications.
- Always replace worn fuel fitting O-rings with new. Do not substitute fuel hose or equivalent, where fuel pipe is installed.

RELIEVING FUEL SYSTEM PRESSURE

Fuel Injected

1. Disconnect the fuel pump harness connector at the fuel tank side.
2. Start the engine and after it stops by itself, turn the ignition switch to the **OFF** position.
3. Disconnect the negative battery terminal. Reconnect the fuel pump harness connector and then reconnect the negative battery terminal.

Fuel Tank

REMOVAL & INSTALLATION

1. Relieve fuel system pressure.
2. Disconnect the negative battery cable.
3. Raise the vehicle and support safely.
4. Drain the fuel from the fuel tank into an approved container.
5. Disconnect the return hose, high pressure hose and all other hoses and connectors connected to the pump/sending unit.
6. Disconnect the filler and vent hoses. Place a suitable support under the tank and remove the retaining nuts.
7. Lower the tank from the vehicle.

To install:

8. Install the fuel tank and all related items to the vehicle. Secure all tank retaining nuts.
9. Connect the negative battery cable and check the entire system for proper operation and leaks.

Fuel Filter

REMOVAL & INSTALLATION

1. On carbureted vehicles, remove the inlet and outlet fuel lines from the filter connections after loosening the fuel line clamps. Remove the old filter. Install the new filter in the reverse order.
2. On fuel injected and turbocharged vehicles, the under hood filter is replaced after first reducing fuel line

pressure. Hold the side filter nut securely and remove the mounts. Disconnect the lines and remove the filter.

3. On the 1988–91 Galant vehicles, remove the air cleaner assembly and the compressor for the electronic controlled suspension, if equipped.

4. Disconnect the fuel lines and remove the fuel filter mounting bolt and then remove the fuel filter assembly.

5. Install the new fuel filter in the reverse order of the removal procedure.

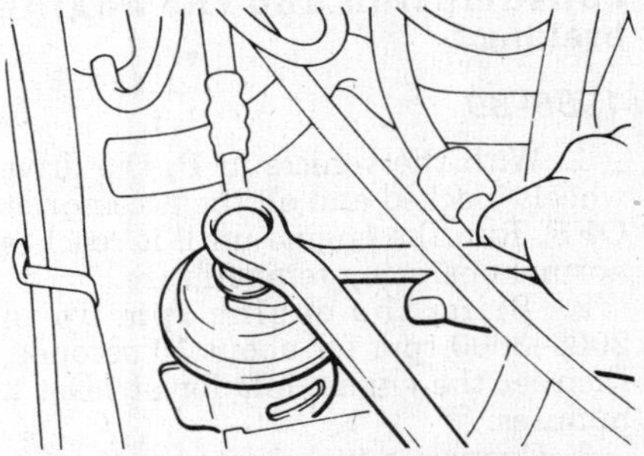

Fuel filter location—turbocharged and fuel injected engines

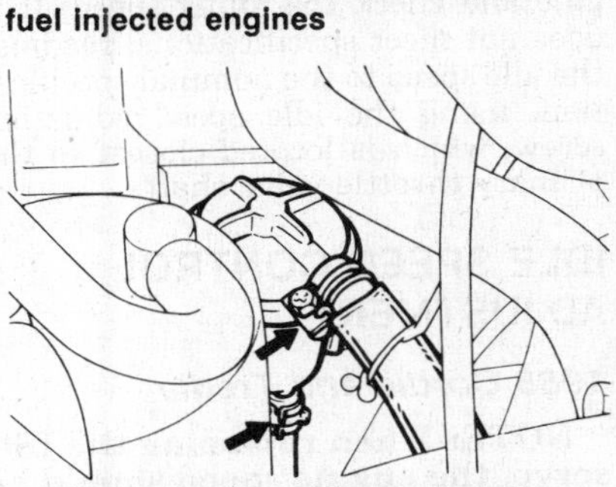

Fuel filter with carbureted engines

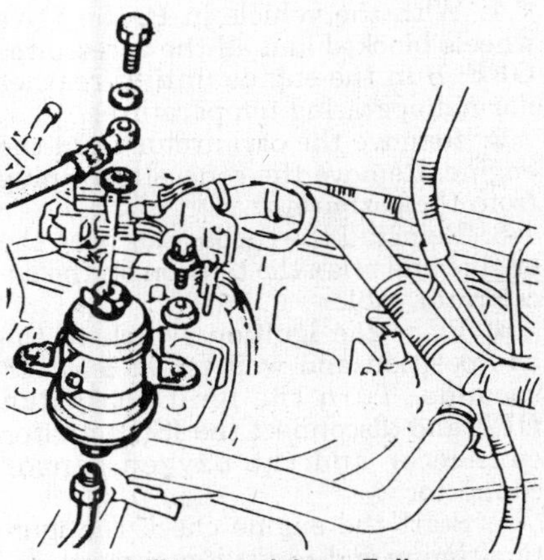

Fuel filter servicing—Eclipse

Mechanical Fuel pump

PRESSURE TESTING

1. Disconnect the fuel line from the carburetor. Attach a pressure tester to the end of the line.

2. Crank the engine. If fuel pump pressure is not within specification replace the fuel pump.

REMOVAL & INSTALLATION

1. Disconnect the negative battery cable. Remove the distributor cap to check the direction of the rotor, and then turn the engine over until the pointer near the front pulley is at Top Dead Center and the rotor points to the ignition wire for No. 1 cylinder, indicating that No. 1 is at firing position. Disconnect the negative battery cable.

2. Disconnect the fuel lines by using a pair of pliers to shift clamps away from the nipples on the pump and then pulling the lines off with a twisting motion. Note the locations at which lines connect.

3. Remove both bolts from the head, and then remove the pump, spacer and gasket(s) from the head. While pulling the pump off the head, catch the pushrod which is located just behind the pump.

To install:

4. Inspect the pump as follows: There is a small breather hole in the area of the pump above the diaphragm which vents the pump's upper chamber. Leakage of fuel or oil here indicates that the pump's diaphragm or oil seal is leaking and that the unit should be replaced. Also, inspect the end of the pushrod and the wear surface where the pushrod engages with the pump operating lever. Replace the pushrod or pump if there is obvious wear. If the camshaft end of the pushrod is badly worn, remove the cam cover and inspect the camshaft eccentric which operates the fuel pump for excessive wear.

5. Clean the gasket surfaces of the insulator, pump and cylinder head. Insert both bolts through the pump's base. Slide a new gasket, the insulator, and a second new gasket into position over both bolts. Turn the pump so its mounting surface faces the cylinder head.

6. Locate the pump pushrod against the cupped surface of the operating lever and angle it upward in the position it was in during removal. Hold the pushrod at that angle and insert it into the bore in the head. Once the pushrod is in the cylinder head bore, release it and move the pump toward the head following the installation angle of the pushrod. Start both bolts into the bores in the head and tighten them finger-tight.

7. Tighten the bolts alternately and evenly. Inspect the hoses for cracks (even hairline cracks can leak) and replace if necessary. Then, reconnect the fuel hoses. Make sure the hoses are installed all the way onto the nipples and then work the clamps into position. Make sure the clamps are located well past the bulged portion of the nipples but do not sit at the extreme inner ends of the hoses. Replace the distributor cap. Start the engine and check for leaks.

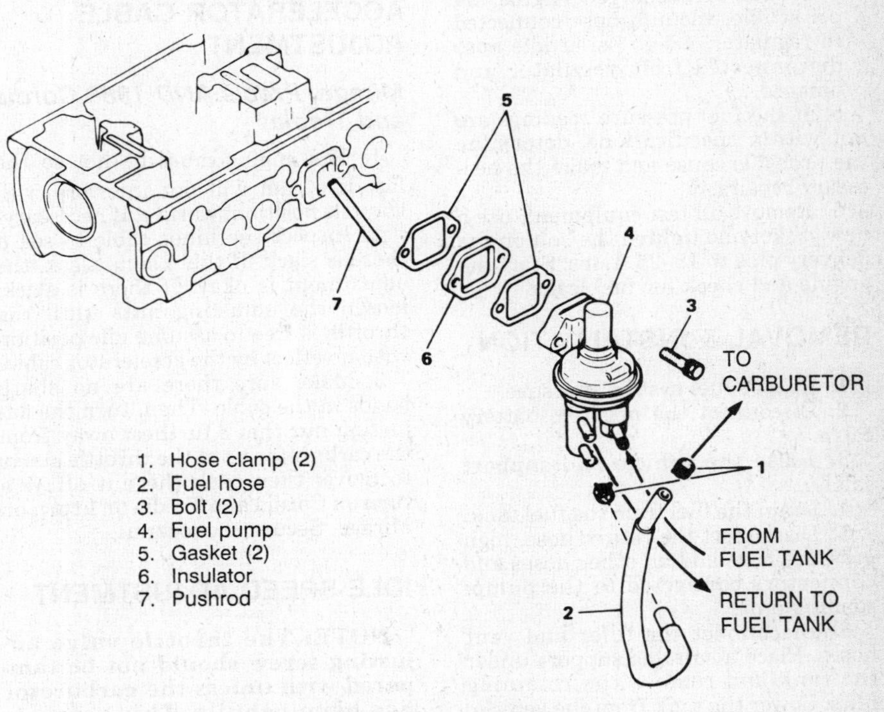

Mechanical fuel pump

Electric Fuel Pump

PRESSURE TESTING

1. Relieve the fuel pressure as follows:
   a. Disconnect the fuel pump harness connector at the fuel tank side.
   b. Start the engine and after it stops by itself, turn the ignition switch to the **OFF** position.
   c. Disconnect the negative battery terminal. Reconnect the fuel pump harness connector and then reconnect the negative battery terminal.
2. Install a suitable fuel pressure gauge to the fuel delivery pipe, be sure to tighten the bolt at 18–25 ft. lbs.
3. Apply voltage to the terminal for the fuel pump drive and activate the fuel pump; then, with fuel pressure thus applied, check the there is no fuel leakage from the pressure gauge or the special tool connection pipe.
4. Disconnect the vacuum hose from the pressure regulator and plug the hose end. Measure the fuel pressure during idling.
   a. All engines except 2.0L turbocharged engine: 38 psi at idle, vacuum hose connected to regulator, 47–50 psi at idle hose disconnected from regulator and plugged.
   b. 2.0L turbocharged engine: 27 psi at idle, vacuum hose connected to regulator, 36–38 psi at idle hose disconnected from regulator and plugged.
   c. 3.0L turbocharged engine: 34 psi at idle, vacuum hose connected to regulator, 43–45 psi at idle hose disconnected from regulator and plugged.
5. If the fuel pressure readings are not within specifications, determine the probable cause and make the necessary repairs.
6. Remove all test equipment, use a new gasket and tighten the bolt on the delivery pipe to 18–25 ft. lbs. Start the engine and check for fuel leaks.

REMOVAL & INSTALLATION

1. Relieve fuel system pressure.
2. Disconnect the negative battery cable.
3. Raise the vehicle and support safely.
4. Drain the fuel from the fuel tank.
5. Disconnect the return hose, high pressure hose and all other hoses and connectors connected to the pump/ sending unit.
6. Disconnect the filler and vent hoses. Place a suitable support under the tank and remove the retaining nuts. Lower the tank from the vehicle.
7. Remove the fuel pump cover (if equipped), retaining nuts and fuel pump/ sending unit assembly from the tank.

To install:

8. Install the replacement pump using a new gasket. Be certain the pump is installed in the same location, facing the same direction as before.
9. Install the fuel tank and all related items to the vehicle. Secure all tank retaining nuts.
10. Connect the negative battery cable and check the entire system for proper operation and leaks.

Carburetor

REMOVAL & INSTALLATION

1. Disconnect the negative battery cable. Remove the solenoid valve wiring.
2. Disconnect the air cleaner breather hose, air duct and vacuum tube.
3. Remove the air cleaner.
4. Remove the air cleaner case.
5. Disconnect the accelerator and shift cables at the carburetor.
6. Disconnect the purge valve hose; remove the vacuum compensator and fuel lines.
7. Drain the coolant.
8. Remove the water hose between the carburetor and the cylinder head.
9. Remove the carburetor.
10. Installation is the reverse of removal.

ACCELERATOR CABLE ADJUSTMENT

Mirage, Precis AND 1988 Cordia and Tredia

1. The engine must be hot so the fast idle cam will not interfere with throttle position; warm it if necessary.
2. Inspect the inner cable to see if there is slack. If there is no slack, the adjustment is okay. If there is slack, loosen the adjusting nuts until the throttle is free to assume idle position with no effect by the accelerator cable.
3. Make sure there are no sharp bends in the cable. Then, turn the adjusting nut that's furthest away from the carburetor until the throttle starts to move; then back the nut off ½ a turn on Cordia and Tredia or 1 turn on Mirage. Secure the locknut.

IDLE SPEED ADJUSTMENT

NOTE: The throttle valve adjusting screw should not be tampered with unless the carburetor has been rebuilt. This screw is preset and determines the relationship between the throttle valve and the free lever, and has been accurately set at the factory. If this setting is disturbed, the throttle opener adjustment and or dashpot adjustment cannot be done accurately. Also the improper setting (throttle valve opening) will increase the exhaust gas temperature and deceleration, which in turn will reduce the life of the catalyst greatly and deteriorate the exhaust gas cleaning performance. It will also effect the fuel consumption and the engine braking.

1988–89

1. With the vehicle in **P**, the drive wheels blocked and all the accessories **OFF**. Run the engine until it reaches normal operating temperature.
2. Bring the engine rpm up to 2000–3000 rpm for about 10 seconds, then let the engine idle for at least 2 minutes.
3. Connect a tachometer to the engine and check the idling speed. If it does not meet specifications, readjust the idle speed to the nominal specification, using the idle speed adjusting screw, which is located closest to the primary throttle valve shaft.

IDLE SPEED CONTROL ADJUSTMENT

1988 Cordia and Tredia

NOTE: When replacing the ISC servo, the engine speed should be adjusted.

1. With the vehicle in **P**, the drive wheels blocked and all the accessories **OFF**. Run the engine until it reaches normal operating temperature.
2. Remove the carburetor from the engine. Remove the concealment plug from the carburetor.
3. Reinstall the carburetor onto the engine and relax the tension on the accelerator cable.
4. Place the ignition switch to the one position and wait for at least 18 seconds. Turn the ignition switch **OFF** and disconnect the ISC actuator connector and the oxygen sensor connector.
5. Start the engine check the ignition timing and adjust if necessary. Increase the engine speed between 2000–3000 rpm, 2–3 times and let the engine idle for 30 seconds.
6. Adjust the mixture adjusting screw for a concentration of 0.1–0.3 percent. Adjust the engine rpm to the specified speed by using the ISC adjustment screw.
7. Turn the idle mixture screw, secondary air supply screw, until the en-

gine reaches its highest rpm. Turn the screw ⅔ of turn in the reverse direction from that point.

8. Race the engine 2–3 times. Check to be sure the CO and engine rpm are still adjusted to specifications. If they are not, readjust as necessary.

9. Adjust the tension of the accelerator cable. The cable should have enough play so as not to interfere with idle switch.

10. Reconnect the ISC actuator connector and the oxygen sensor connector.

Precis

1. Make sure the gear selector is in **N** or **P**. Run the engine at fast idle until the cooling system reaches 185°F or more, not to exceed 205°F. Then, race the engine at 2000–3000 rpm for more than 5 seconds. Release the throttle. All accessories including the electric cooling fan must be off.

2. Idle the engine for a full 2 minutes. Connect a tachometer between the (−) terminal of the coil and a good ground while the engine is idling. If the idle speed is not to the specification, adjust the idle speed screw (which is located closest to the primary throttle valve shaft) to obtain the proper idle speed specifications.

3. The idle speed should be as follows:

   a. Manual Transaxle – 600–800 rpm

   b. Automatic Transaxle – 650–850 rpm

   c. Canada – 820–880 rpm

IDLE-UP SPEED ADJUSTMENT

1988 Cordia, Tredia, Mirage and 1988–89 Precis

WITHOUT AIR CONDITIONING

NOTE: Adjustment condition – lights, electric cooling fan and all accessories are OFF and transaxle is in N.

1. Make sure the curb idle speed is within specifications, adjust if necessary.

2. By using the auxiliary lead wire, activate the idle up solenoid valve. Apply the intake manifold vacuum to the throttle opener and activate the throttle opener.

3. Open the throttle slightly, to engine speed of about 2000 rpm, and then slowly close it.

4. Adjust the engine speed to the specifications with the idle-up adjusting screw.

5. After repeating Step 3, check the engine speed.

6. Remove the auxiliary lead wire used in Step 2 and reconnect the idle-up solenoid valve wiring.

WITH AIR CONDITIONING

1. With the vehicle in **P**, the drive wheels blocked and all the accessories **OFF**. Run the engine until it reaches normal operating temperature.

2. Disconnect the electric cooling fan connector. If equipped with power steering, set the tires in the straight-ahead position to prevent the pump from being loaded. Set the steering wheel in the stationary position.

3. Be sure the curb idle speed is within the specifications, adjust, if necessary.

4. With the air conditioner on, adjust the engine speed to the specified speed with the throttle opener setting screw, idle-up adjusting screw.

5. Reconnect the electric cooling fan connector and turn the air conditioning **ON** and **OFF** several times to check the operation of the throttle opener.

Fuel Injection

IDLE SPEED ADJUSTMENT

Except Eclipse and 3000-GT

1. Run the engine until normal operating temperature is reached. Make sure all lights and accessories are turned **OFF**.

2. Apply the parking brake and block the wheels. Position the gear selector in **N** and stop the engine.

3. Attach a tachometer and timing light. Start the engine and increase the engine speed to 2000–3000 rpm several times, return to idle and check the ignition timing, adjust if necessary.

4. Remove the rubber cap covering the idle speed adjuster switch, leaving the cable connector connected. The idle adjuster switch is located on the throttle linkage. Adjust the idle speed.

5. If the idle adjustment screw must be turned more than 1 turn during adjustment, disconnect the connector from the speed adjust switch and plug it into the dummy terminal on the injector base. Adjust to correct idle speed and reconnect to the idle switch. Remove the tachometer and timing light.

Eclipse

1.8L ENGINE

The electronic system controls the idle speed and adjustment of the idling speed is usually unnecessary. The idle speed may be checked using the following procedure:

1. Warm the engine to operating temperature, leave lights, electric cooling fan and accessories **OFF**. The transaxle should be in **N** or **P** for automatic transaxle. The steering wheel in a neutral position for vehicles with power steering.

2. Check the ignition timing and adjust, if necessary.

3. Connect a tachometer to the CRC filter connector. Use a paper clip for a tach adapter.

4. Run the engine for more than 5 seconds at 2000–3000 rpm. Allow the engine to idle for 2 minutes. Check the idle rpm. Curb idle should be 600–800 rpm.

5. If adjustment is required, slacken the accelerator cable.

6. Connect a digital voltmeter between terminal 19 throttle position sensor output voltage) of the engine control unit and terminal 24 (ground).

7. Set the ignition switch to **ON**, without starting the engine, and hold it in that position for 15 seconds or more. Turn the ignition switch **OFF**.

8. Disconnect the connectors of the idle speed control servo and lock the idle speed control plunger at the initial position. Back out the fixed Speed Adjusting Screw (SAS).

9. Start the engine and allow to idle. Basic idle speed should be 650–750 rpm. A new engine may idle a little lower. If the vehicle stalls or has a very low idle speed, suspect a deposit build-up on the throttle valve which must be cleaned.

10. If the idle speed is wrong, adjust with the idle speed control adjusting screw. Use a hexagon wrench if possible. Turn in the fixed SAS until the engine speed rises. Then back out the fixed SAS until the Touch Point where the engine speed does not fall any longer, is found. Back out the fixed SAS an additional ½ turn from the touch point.

11. Stop the engine. Turn the ignition switch to **ON** but do not start engine. Check that the output voltage from the throttle position sensor is 0.48–0.52 volts. If it is out of specification, adjust by loosening the throttle position sensor screws and rotating the throttle position sensor. Turning the throttle position sensor clockwise increases the output voltage. After adjustment, tighten screws firmly.

12. Turn the ignition switch **OFF**.

13. Adjust the free-play of the accelerator cable, reconnect the connectors of the idle speed control servo and remove the voltmeter.

14. Start the engine and check the curb idle. It should be 600–800 rpm.

15. Turn the ignition switch to **OFF**, disconnect the negative battery cable for more than 10 seconds and reconnect. This clears any trouble codes introduced during testing.

16. Restart the engine, allow to run for 5 minutes and check for good idle quality.

2.0L ENGINE

The electronic system controls the idle speed and adjustment of the idling speed is usually unnecessary. The idle speed may be checked using the following procedure:

1. Warm the engine to operating temperature, leave lights, electric cooling fan and accessories **OFF**. The transaxle should be in **N**. The steering wheel in a neutral position for vehicles with power steering.
2. Check the ignition timing and adjust, if necessary.
3. Connect a tachometer to the special terminal under the hood.
4. Run the engine for more than 5 seconds at 2000–3000 rpm. Allow the engine to idle for 2 minutes. Check the idle rpm. Curb idle should be 650–850 rpm.
5. If adjustment is required, disconnect the waterproof female connector used for ignition timing adjustment. Connect this terminal to ground using a jumper wire.
6. Locate the self-diagnosis terminal under the dashboard and connect terminal No. 10 to ground with a jumper wire.
7. Start the engine and allow to idle. Check that the basic idle speed is 650–850 rpm. If the idle speed deviates from this speed, check the following:
   a. A new engine will idle more slowly. Break-in should take approximately 300 miles.
   b. If the vehicle stalls or has a very low idle speed, suspect a deposit buildup on the throttle valve which must be cleaned.
   c. If the idle speed is high even though the speed adjusting screw is fully closed, check that the idle position switch, fixed speed adjusting screw, position has changed. if, adjust the idle position switch.
   d. If after all these checks the idle is still out of specification, it is probable that there is leakage resulting from deterioration of the Fast-Idle Air Valve (FIAV) and the throttle body will need to be replaced.
8. Turn the ignition switch **OFF** and stop the engine. Disconnect the jumper wire from the diagnosis connector, disconnect the jumper wire from the ignition timing connector and reconnect the waterproof connector. Disconnect the tachometer.
9. Restart the engine, allow to run for 5 minutes and check for good idle quality.

3000-GT

1. Warm the engine to operating temperature, leave lights, electric cooling fan and accessories **OFF**. The transaxle should be in **N**. The steering wheel in a neutral position for vehicles with power steering.
2. Check the ignition timing and adjust, if necessary.
3. Connect a tachometer to the special terminal under the hood.
4. Run the engine for more than 5 seconds at 2000–3000 rpm. Allow the engine to idle for 2 minutes. Check the idle rpm. Curb idle should be 750 ± 100 rpm.
5. If adjustment is required, disconnect the waterproof female connector used for ignition timing adjustment. Connect this terminal to ground using a jumper wire.
6. Locate the self-diagnosis terminal under the dashboard and connect terminal No. **10** to ground with a jumper wire.
7. Start the engine and allow to idle. Check that the basic idle speed is at specification. On 3000-GT, the tachometer reading will be ⅓ of the actual engine speed. Multiply the reading by 3 to figure the actual engine speed. If the idle speed deviates from this speed, check the following:
   a. A new engine will idle more slowly. Break-in should take approximately 300 miles.
   b. If the vehicle stalls or has a very low idle speed, suspect a deposit buildup on the throttle valve which must be cleaned.
   c. If the idle speed is high even though the speed adjusting screw is fully closed, check that the idle position switch (fixed speed adjusting screw) position has changed. If so, adjust the idle position switch.
   d. If after all these checks the idle is still out of specification, it is probable that there is leakage resulting from deterioration of the Fast-Idle Air Valve (FIAV) and the throttle body will need to be replaced.
8. Turn the ignition switch **OFF** and stop the engine. Disconnect the jumper wire from the diagnosis connector, disconnect the jumper wire from from the ignition timing connector and reconnect the waterproof connector. Disconnect the tachometer.
9. Restart the engine, allow to run for 5 minutes and check for good idle quality.

Fuel Injector

REMOVAL & INSTALLATION

Except 3000-GT

1. Relieve fuel system pressure.
2. Disconnect the negative battery cable.
3. Remove the high pressure line where it connects to the delivery pipe, or fuel rail. The O-ring must be replaced. Do not reuse.
4. Remove the return line from the pressure regulator. Also remove the vacuum line from the regulator.
5. Remove the fuel pressure regulator. The O-ring must be replaced. Do not reuse.
6. Remove the injector electrical connectors.
7. Remove the delivery pipe/injector assembly. Use care. Do not let the injectors drop.
8. Pull injectors from the delivery pipe for replacement.

To install:

9. Injectors can be checked for electrical resistance between the terminals. Resistance should be 13–16 ohms. If out of specification, replace injector.
10. Install the grommets on the injector first, then the O-ring. Apply light oil to lubricate the O-rings. Install the injector by pushing into the delivery pipe while turning back and forth. The injector should turn smoothly. If it does not, the O-ring may be trapped or dislodged. Remove the injector, check and insert injector again.
11. The O-ring for the pressure regulator and high pressure line should also be lubed with light oil or gasoline.
12. After assembly, the fuel pressure should be checked.

3000-GT

1. Relieve the fuel system pressure.
2. Disconnect the negative battery cable.
3. Drain the coolant.
4. Disconnect all components from the air intake plenum and remove the plenum from the intake manifold. Discard the gaskets.
5. Wrap the connection with a shop towel and disconnect the high pressure fuel line at the fuel rail.
6. Disconnect the fuel return hose and remove the O-ring.
7. Disconnect the vacuum hose from the fuel pressure regulator. Remove the fuel pressure regulator and O-ring.
8. Disconnect the connectors from each injector.
9. Remove the fuel pipe connecting the fuel rails. Remove the injector rail retaining bolts. Make sure the rubber mounting bushings do not get lost.
10. Lift the rail assemblies up and away from the engine.
11. Remove the injectors from the rail by pulling gently. Discard the lower insulator. Check the resistance through the injector. The specification for 3.0L turbocharged engine is 2–3 ohms at 70°F (20°C). The specification for the 3.0L fuel injected is 13–15 ohms at 70°F (20°C).

To install:

12. Install a new grommet and O-ring to the injector. Coat the O-ring with light oil.
13. Install the injector to the fuel rail.
14. Replace the seats in the intake manifold. Install the fuel rails and injectors to the manifold. Make sure the rubber bushings are in place before tightening the mounting bolts.
15. Tighten the retaining bolts to 72 inch lbs. (11 Nm). Install the fuel pipe with new gasket.
16. Connect the connectors to the injectors.
17. Replace the O-ring, lightly lubricate it and connect the fuel pressure regulator.
18. Connect the fuel return hose.
19. Replace the O-ring, lightly lubricate it and connect the high pressure fuel line.
20. Using new gaskets, install the intake plenum and all related items. Torque the plenum mounting bolts to 13 ft. lbs. (18 Nm).
21. Fill the cooling system.
22. Connect the negative battery cable and check the entire system for proper operation and leaks.

DRIVE AXLE

Halfshaft

REMOVAL & INSTALLATION

Except Center Bearing

1. Remove the hub center cap and remove the cotter pin, then loosen the driveshaft (axle) nut. Loosen the wheel lug nuts.
2. Raise and safely support the vehicle. Remove the front wheels. Remove the drive axle (hub) nut and remove the engine splash shield.
3. Using the tools required, remove the tie rod ends, stabilizer bar nut, the lower arm ball joint nut and lower am ball joint.
4. Disconnect the oxygen sensor connection, if necessary. Drain the transaxle fluid.
5. Remove any retaining circlips. Insert a suitable tool between the transaxle case, on the raised rib, and the halfshaft double offset joint case. Do not insert the tool too deeply or the seal may be damaged. Move the tool to the right to withdraw the left halfshaft; to the left to remove the right halfshaft.
6. Plug the transaxle case with a clean rag to prevent dirt from entering the case.
7. Use a puller driver mounted on the wheel studs to push the halfshaft from the front hub. Take care to prevent the spacer from falling out of place.

To install:

8. Insert the halfshaft into the hub first, then install the transaxle end. Torque the drive axle nut, if equipped, to 144–188 ft. lbs.

NOTE: Always use a new DOJ retaining ring every time the driveshaft is removed.

9. Install the center support bearing bracket bolts, if removed.
10. Install the lower ball joint and tie rod end on the steering knuckle.
11. Install the wheels, lower the vehicle and test drive.

Center Bearing

NOTE: If the vehicle is going to be rolled while the halfshafts are out of the vehicle, obtain 2 outer CV-joints or proper equivalent tools and install to the hubs. If the vehicle is rolled without the proper torque applied to the front wheel bearings, the bearings will no longer be usable.

1. Disconnect the negative battery cable.
2. Remove the cotter pin, halfshaft nut and washer. It is recommended that the halfshaft nut is removed while the vehicle is on the floor with the brakes applied.
3. Raise the vehicle and support safely. Remove the lower ball joint and the tie rod end from the steering knuckle.
4. On vehicles with an inner shaft, remove the center support bearing bracket bolts and washers.
5. On vehicles with an inner shaft, remove the halfshaft by setting up a puller on the outside wheel hub and pushing the halfshaft from the front hub. Then tap the joint case with a plastic hammer to remove the halfshaft shaft and inner shaft from the transaxle.
6. On vehicles without an inner shaft, remove the halfshaft by setting up a puller on the outside wheel hub and pushing the halfshaft from the front hub. After pressing the outer shaft, insert a prybar between the transaxle case and the halfshaft and pry the shaft from the transaxle. Do not pull on the shaft; doing so damages the inboard joint. Do not insert the prybar too far or the oil seal in the case may be damaged.

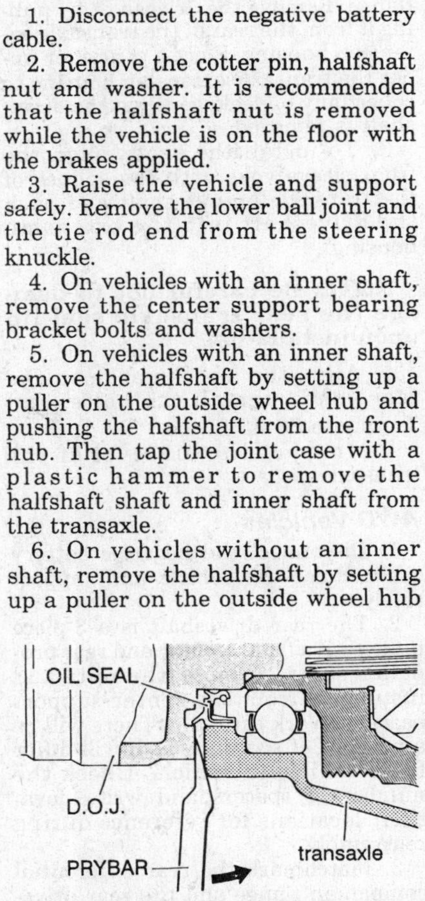

Half shaft removal—except Mirage with center bearing

To install:

7. Inspect the halfshaft boot for damage or deterioration. Check the ball joints and splines for wear.
8. Replace the circlips on the ends of the halfshafts.
9. Insert the halfshaft into the transaxle. Make sure it is fully seated.
10. Pull the strut assembly out and install the other end to the hub.
11. Install the center bearing bracket bolts and tighten to 33 ft. lbs. (45 Nm).
12. Install the washer so the chamfered edge faces outward. Install the nut and tighten temporarily.
13. Install the tie rod end and ball joint.
14. Install the wheel and lower the vehicle to the floor. Tighten the axle nut with the brakes applied. Tighten the nut to a maximum torque of 188 ft. lbs. (260 Nm) maximum. Install the cotter pin and bend it securely.

CV-Boot

These vehicles used several different types of joints. Engine size, transaxle type, whether the joint is an inboard or outboard joint, even which side of the vehicle is being serviced will make a difference in joint type. Proper identification is important when ordering parts. Be sure to properly identify the joint before attempting joint or boot replacement. Look for identification numbers at the big end of the boots and on the end of the metal retainer bands.

The 4 types of joints used are the Birfield Joint, (BJ), the Tripod Joint (TJ), the Double Offset Joint (DOJ) and the Rzeppa Joint (RJ). In addition, some left hand shafts will have a round dynamic damper installed on the shaft. Special grease is generally used with these joints and is often supplied with the replacement joint and/or boot. Do not use regular chassis grease.

In most cases, a specification is called out for the distance between the large and small boot bands. This is so the boot will not be installed either too loose or too tight which could cause early wear and cracking, allowing the grease to get out and water and dirt in, leading to early joint failure.

REMOVAL & INSTALLATION

Except Double Offset Joint

Although joint types vary, the basic procedures are the same, with the exception of the Double Offset Joint. The following is a general procedure which should apply to most applications.

1. Remove the halfshaft.
2. Remove the snapring next to the tripod joint spider from the halfshaft with snapring pliers and remove the spider from the shaft. Do not disassemble the spider and use care in handling.
3. Side cutter pliers can be used to cut the metal retaining bands.
4. If the boot is be reused, wrap vinyl tape around the spline part of the shaft so the boot will not be damaged when removed. Remove the dynamic damper, if used, and boots from the shaft.

To install:

5. Double check that the correct replacement parts are being installed. Wrap vinyl tape around the splines to protect the boot and install the boots and damper, if used, in the correct order.
6. Fill the inside of the boot with the specified grease. Often the grease supplied in the replacement parts kit is meant to be divided in half, with half being used to lube the joint and half being used inside the boot. Keep grease off the rubber part of the dynamic damper, if used.
7. Secure the boot bands with the halfshaft horizontal.

Double Offset Joint

1. Remove the halfshaft. The Double Offset Joint (DOJ) is bigger than other joints and in these applications, is only used as an inboard joint.
2. Side cutter pliers can be used to cut the metal retaining bands.
3. Locate and remove the large circlip at the base of the joint. Remove the outer race, the body of the joint.
4. Makematch marks on the shaft, DOJ inner race and cage. Remove the joint balls and the small snapring from the shaft. With a brass drift pin, tap lightly and evenly around the inner race to remove the race and then the inner cage from the shaft.
5. If the boot is to be reused, wipe the grease from the splines and wrap the splines in vinyl tape before sliding the boot from the shaft.

To install:

6. Be sure to tape the shaft splines before installing the boots. Fill the inside of the boot with the specified grease. Often the grease supplied in the replacement parts kit is meant to be divided in half, with half being used to lube the joint and half being used inside the boot.
7. Install the cage onto the halfshaft so the small diameter side of the cage is installed first. Align the matchmarks made at disassembly on the inner race and shaft. With a brass drift pin, tap lightly and evenly around the inner race to install the race until it comes into contact with the rib of the shaft. Apply the specified grease to the inner race and cage and fit them together aligning the matchmarks. Insert the balls into the cage.
8. Install the outer race, the body of the joint, after filling with the specified grease. The outer race should be filled with this grease.
9. Tighten the boot bands securely.

Driveshaft and U-Joints

REMOVAL & INSTALLATION

Except AWD

1. Raise and support the vehicle safely. Matchmark the rear flange yoke and the differential pinion flange.
2. Remove the bolts from the rear flange. Remove the driveshaft by pulling it from the rear of the transaxle extension housing. Place a container under the transaxle extension housing to collect any oil leakage when the driveshaft is removed.
3. To install the shaft, align the front sleeve yoke with the splines of the transaxle output shaft and push the driveshaft into the extension housing.

NOTE: Be careful not to damage the rear transaxle seal lip upon installation

4. Align the matchmarks on the rear yokes, install the bolts, and Torque to 36–43 ft. lbs.
5. Inspect the oil level of the transaxle.

AWD Vehicles

1. Disconnect the negative battery cable. Raise the vehicle and support safely.
2. The rear driveshaft is a 3-piece unit, with a front, center and rear propeller shaft. Remove the nuts and insulators from the center support bearing. Work carefully. There will be a number of spacers which will differ from vehicle to vehicle. Check the number of spacers and write down their locations for reference during reassembly.
3. Matchmark the rear differential companion flange and the rear driveshaft flange yoke. Remove the companion shaft bolts and remove the driveshaft, keeping it as straight as possible so as to ensure that the boot is not damaged or pinched. Use care to keep from damaging the oil seal in the output housing of the transfer case.

NOTE: Damage to the boot can be avoided and work will be easier if a piece of cloth or similar material is inserted in the boot.

4. Do not lower the rear of the vehicle or oil will flow from the transfer case. Cover the opening to keep dirt out.

To install:

5. Install the driveshaft to the vehicle and align the matchmarks at the rear yoke. Install the bolts and torque to 22–25 ft. lbs. (30–35 Nm) or 36–43 ft. lbs. (50–60 Nm) on 3000-GT.
6. Install the center support bearing with all spacers in place. Torque the retaining nuts to 22–25 ft. lbs. (30–35 Nm).
7. Check the fluid levels in the transfer case and rear differential case.

Front Axle Shafts, Bearing and Seal

REMOVAL & INSTALLATION

1. Remove the hub cotter pin, axle nut and washer.
2. Raise and safely support vehicle. Remove front wheels. Remove the brake caliper and hang by a wire.
3. Remove the ball joint from the lower arm and disconnect the tie rod end.
4. Remove the halfshaft.
5. Unbolt the lower end of the strut and remove the hub and steering knuckle assembly.
6. Set up a puller with the knuckle/hub in a vise and pull the hub from the knuckle. If the hub and knuckle are disassembled by hitting them with a hammer, the bearing will be damaged.
7. Once the hub and outer bearing inner race are removed with a puller, the bearing outer races can be removed by tapping out with a brass drift pin and a hammer.

To install:

8. Apply a thin coat of grease to the outside of the outer races and install into the hub with a bearing driver.
9. Apply multi-purpose grease to the bearings, inside surface of the hub and the lip of the grease seal. Place the outside bearing into the knuckle and install the seal with a driver.
10. The hub is assembled to the knuckle with a puller. Draw the parts together firmly to seat the bearings. Use a small torque wrench to check

the bearing turning torque. It should be 11 inch lbs. or less. Check that the bearings feel smooth when rotated. A dial indicator is used to check endplay which should be 0.008 in. or less.

11. Apply a thin coat of grease to the lip of the halfshaft side axle seal and drive into place until it contacts the inner bearing outer race.

12. Install the lower end of the strut and assemble the steering knuckle assembly after installing the halfshaft.

13. Install the disc brake caliper, hub and the wheel. Lower the vehicle and test drive.

Rear Axle Shafts, Bearing and Seal

REMOVAL & INSTALLATION

Except Starion

1. Disconnect the negative battery cable. Raise the vehicle and support safely.

2. Remove the bolts that attach the rear halfshaft to the companion flange.

3. Use a prybar to pry the inner shaft out of the differential case. Don't insert the prybar too far or the seal could be damage.

4. Remove the rear halfshaft from the vehicle.

5. If equipped with ABS, remove the rear wheel speed sensor.

6. Remove the caliper, pads and brake rotor.

7. Hold the axle shaft stationary and remove the axle shaft self-locking nut and washer.

8. Using a slide hammer, separate the axle shaft from the companion flange and remove.

9. Use a vice and gear puller tool to disassemble the axle shaft and companion flange assemblies.

To install:

10. Assemble the axle shaft and companion shaft assemblies using new parts as required.

11. Install the axle shaft to the housing and slide the axle shaft over it. Install the washer and new self-locking nut. Hold the axle shaft stationary and torque the nut to 116–159 ft. lbs. (160–220 Nm). Torque to 188–217 ft. lbs. (260–300 Nm) for turbocharged 3000-GT.

12. Install the brake rotor, pads and caliper.

13. Install the ABS rear wheel speed sensor.

14. Replace the circlip and install the rear halfshaft to the differential case. Make sure it snaps in place. Torque the companion flange bolts to 40–47 ft. lbs. (55–65 Nm).

15. Check the fluid level in the rear differential.

Starion

1. Raise and support the vehicle safely. Remove the tire and wheel.

2. Disconnect the parking brake cable and remove the disc brake pads and caliper. Leave the brake line connected to the caliper and use a piece of wire to suspend the caliper aside.

3. Remove the halfshaft mounting bolts. Remove the lower control arm-to-knuckle bolt.

4. Remove the strut assembly retaining bolts and remove the axle housing assembly.

5. Remove the axle shaft from the axle housing as follows:

   a. Remove the companion flange nut. Using a plastic hammer tap the axle shaft out of the axle housing.

   b. Remove the spacer, outer bearing, dust cover, companion flange, dust cover, oil seal, inner bearing, axle housing and dust cover.

To install:

6. Reassemble the axle shaft into the axle housing and install the axle housing by reversing the removal procedure. Torque the companion flange nut to 188–217 ft. lbs. Torque the strut bolts to 36–51 ft. lbs., the halfshaft mounting bolt/nuts to 40–47 ft. lbs. and the lower control arm bolt/nut to 51–58 ft. lbs.

Front Wheel Hub, Knuckle and Bearings

REMOVAL & INSTALLATION

1. Disconnect the negative battery cable.

2. Remove the cotter pin, halfshaft nut and washer. It is recommended that the halfshaft nut is removed while the vehicle is on the floor with the brakes applied.

3. Raise the vehicle and support safely. If equipped with ABS, remove the front wheel speed sensor. Remove the ball joint and tie rod end from the steering knuckle.

4. Remove the caliper and pads and suspend with a wire.

5. On vehicles with an inner shaft, remove the center support bearing bracket bolts and washers. Remove the halfshaft by setting up a puller on the outside wheel hub and pushing the halfshaft from the front hub. Then tap the joint case with a plastic hammer to remove the halfshaft shaft and inner shaft from the transaxle.

6. On vehicles without an inner shaft, remove the halfshaft by setting up a puller on the outside wheel hub and pushing the halfshaft from the front hub. After pressing the outer shaft, insert a prybar between the transaxle case and the halfshaft and pry the shaft from the transaxle.

7. On 3000-GT with AWD, the front hub/bearing assembly can be serviced at this point as a unit. If the knuckle is being removed, proceed. All others require knuckle removal.

8. Unbolt the lower end of the strut and remove the hub and steering knuckle assembly.

9. Set up a puller with the knuckle/hub in a vise and pull the hub from the knuckle. Do not use a hammer to accomplish this or the bearing will be damaged.

10. Once the hub and outer bearing inner race are removed with a puller, the bearing outer races can be removed by tapping out with a brass drift pin and a hammer.

To install:

11. Assemble the hub/knuckle assembly with pressing tools, using new parts as required.

12. Install the knuckle assembly to the vehicle and install the strut bolts.

13. On AWD 3000-GT, torque the front hub/bearing assembly nuts to 76 ft. lbs. (105 Nm).

14. Apply a thin coat of grease to the outside of the outer races and install into the hub with a bearing driver.

15. Apply multi-purpose grease to the bearings, inside surface of the hub and the lip of the grease seal. Place the outside bearing into the knuckle and install the seal with a driver.

16. The hub is assembled to the knuckle with a puller. Draw the parts together firmly to seat the bearings. Use a small torque wrench to check the bearing turning torque. It should be 16 inch lbs. or less. Check that the bearings feel smooth when rotated.

17. Apply a thin coat of grease to the lip of the halfshaft side axle seal and drive into place until it contacts the inner bearing outer race.

18. Replace the circlips on the ends of the halfshafts.

19. Insert the halfshaft into the transaxle. Make sure it is fully seated.

20. Pull the strut assembly out and install the other end to the hub.

21. Install the center bearing bracket bolts and tighten to 33 ft. lbs. (45 Nm).

22. Install the washer so the chamfered edge faces outward. Install the nut and tighten temporarily.

23. Install the tie rod end and ball joint.

24. Install the wheel and lower the vehicle to the floor. Tighten the axle nut with the brakes applied. Tighten the nut to a maximum torque of 188 ft. lbs. (260 Nm) maximum. Install the cotter pin and bend it securely.

Pinion Seal

REMOVAL & INSTALLATION

Front Differential

1. Disconnect the negative battery cable.
2. Remove the front halfshaft.
3. Using a suitable prying tool, pry the seal from the case.

To install:

4. Apply a thin coat of multi-purpose grease to the seal lip and the seal contact surface.
5. Install the new seal with a suitable driver.
6. Install the front halfshaft.

Rear Differential

1. Raise the vehicle and support safely.
2. Matchmark the rear driveshaft and companion flange and remove the shaft. Don't let it hang from the transaxle. Tie it up to the underbody.
3. Hold the companion flange stationary and remove the large self-locking nut in the center of the companion flange.
4. With a suitable puller, remove the flange. Pry the old seal out.

To install:

5. Apply a thin coat of multi-purpose grease to the seal lip and the companion flange seal contacting surface. Install the new seal with a suitable driver.
6. Install the companion flange. Install a new locknut and torque to 116–160 ft. lbs. (157–220). The rotation torque of the drive pinion should be about 2 inch lbs.
7. Install the rear driveshaft.

MANUAL TRANSMISSION

REMOVAL & INSTALLATION

1. Disconnect the negative battery cable. Remove the air cleaner. Remove the starter.
2. Remove the top transaxle bolts from the bell housing.
3. From inside the vehicle, raise the console assembly and remove the dust cover retaining plate at the shift lever.
4. Place the transaxle in the **N** position. Remove the control lever assembly.
5. Raise the vehicle and support it safely. Drain the transaxle. Disconnect the speedometer and the backup light switch.
6. Remove the driveshaft. Disconnect the exhaust pipe. Remove the clutch cable or slave cylinder and linkage.

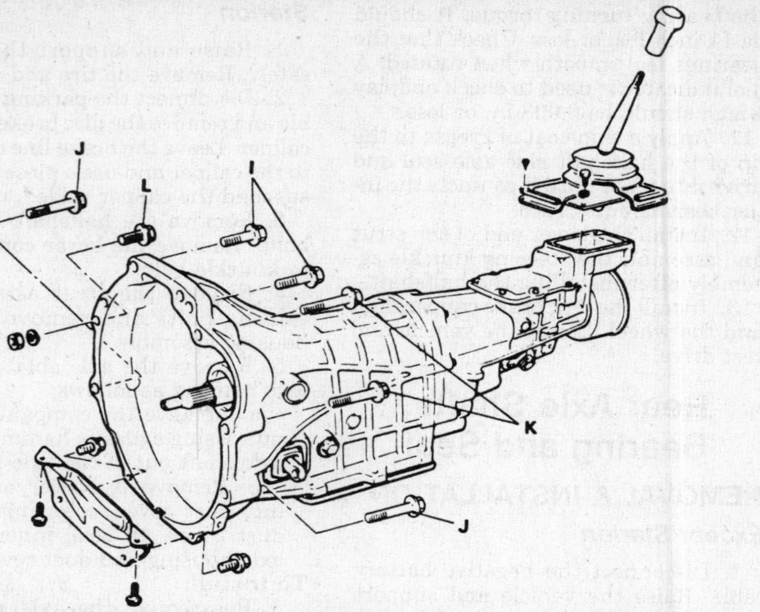

Torque labeled bolts as follows on the Starion transaxle: I and J—31–40 ft. lbs; K—16–23 ft. lbs.; L—14–20 ft. lbs.

7. Support the engine and transaxle and remove the engine rear support bracket. Drain the transaxle, as required.
8. Remove the bell housing cover and bolts, move the transaxle rearward and lower it carefully to the floor. Remove the transaxle from under the vehicle.

To install:

9. Lift the transmission into position and guide into position. Install the bell housing bolts.
10. Install the engine rear support bracket. Install the clutch assembly and reconnect the exhaust system. Install the speedometer cable and back-up light switch.
11. Ensure that the transmission is in **N** position and install the shift control lever assembly. Install the top transmission bolts. Install the console assembly inside the vehicle.
12. Install the starter. Fill the transmission with gear oil, lower the vehicle and test drive.

MANUAL TRANSAXLE

REMOVAL & INSTALLATION

Cordia and Tredia

1. Disconnect the negative battery cable. Remove the battery and battery tray. Remove the coolant and windshield reservoir tanks. Remove the air cleaner and housing.
2. Disconnect from the transaxle; the clutch cable or slave cylinder, speedometer cable, back-up light harness, starter motor and the upper bolts connecting the engine to the transaxle.
3. Raise and support the vehicle safely.
4. Remove the front wheels, Remove the engine splash shield.
5. Remove the shift rod and extension. It may be necessary to remove any heat shields.
6. Drain the transaxle fluid.
7. Remove the right and left halfshafts from the transaxle case.
8. Disconnect the range selector cable, if equipped. Remove the engine rear cover.
9. Support the weight of the engine from above. Remove the bell housing cover. Support the transaxle and remove the remaining lower bolts.
10. Remove the transaxle mount insulator bolt and the cover from inside the front fender shield. Remove the insulator bracket bolts and remove the bracket. Remove the transaxle mount bracket.
11. Remove, slide away from the engine, and lower the transaxle.

To install:

12. Install the transaxle and tighten the attaching bolts. Install the transaxle mounts, insulators and brackets.
13. Install the bellhousing cover, connect the range selector cable and install the engine rear cover.
14. Install the halfshafts and fill the

transaxle with fluid. Lower the vehicle.

15. Install the clutch cable or slave cylinder, speedometer cable, back-up light harness, starter motor and the upper bolts connecting the engine to the transaxle and all other parts removed.

16. Adjust the gearshift lever and range selector lever.

Galant and Sigma

1. Disconnect the battery cables and remove the battery and battery tray.

2. Drain the engine coolant and the transaxle fluid into suitable containers.

3. Disconnect the connections for the air flow sensor, purge control solenoid valve and air cleaner assembly.

4. Disconnect the air intake hose and breather hose and remove the air cleaner assembly. Raise and support the vehicle safely.

5. Disconnect the transaxle control cables. Disconnect the lower radiator hose and the water inlet pipe B. Disconnect all the connections on the water pipe assembly.

6. Disconnect the back-up light switch connector and the engine wiring harness connector.

7. Place a suitable transaxle jack under the transaxle to support. Remove the transaxle mount bracket cap and the mount bracket.

8. Remove the air compressor from the vehicles equipped with electronic controlled suspension.

9. Disconnect the clutch release cylinder and the clutch tube bracket. Disconnect the speedometer cable. Remove the starter assembly.

10. Remove the engine under cover panel. Disconnect the tie rod end from the steering knuckle. Disconnect the lower arm ball joint.

11. Disconnect the left side halfshaft and bearing bracket.

12. Remove the driveshaft nut from the right side shaft and remove the halfshaft, circlip, bolt, bearing bracket, shaft assembly and circlip.

13. Remove the transaxle stay (bracket). Remove the remaining transaxle retaining bolts, pull the transaxle clear from the engine and lower it away from the vehicle.

To install:

14. Lift the transaxle into position and install the attaching bolts. Install the transaxle bracket.

15. Install the driveshaft nut from the right side shaft and remove the halfshaft, circlip, bolt, bearing bracket, shaft assembly and circlip. Install the left side halfshaft and bearing bracket.

16. Install the tie rod end and lower

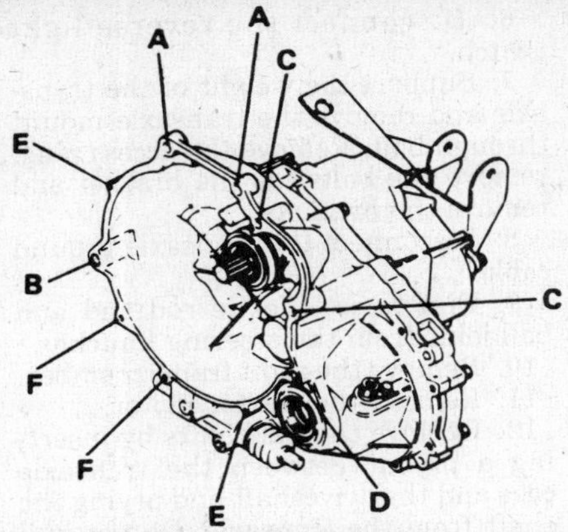

Torque Cordia/Tredia and Mirage transaxle bolts as follows: A – 31–40 ft. lbs.; B – 31–40 ft. lbs.; C – 16–23 ft. lbs.; D – 22–25 ft.

ball joint on the steering knuckle. Install the engine under cover.

17. Connect the clutch release cylinder and the clutch tube bracket. Connect the speedometer cable. Install the starter assembly.

18. Install the air compressor for the electronic controlled suspension (if equipped), the transaxle mount bracket and support, back-up light switch and wiring harness, transaxle control cables, radiator hoses and water pipe connections.

19. Install the air cleaner, fill the radiator with coolant, fill the transaxle with fluid, install the battery and adjust the gearshift lever and range selector lever.

Mirage and Precis

1. Remove the battery and battery tray. On turbocharged vehicles, remove the air cleaner housing assembly.

2. On 5 speed transaxles, disconnect the electrical connector for the selector control valve. On turbocharged vehicles, remove the actuator bolts, remove the actuator to shaft pin and then remove the actuator. Replace the collar with a new part.

3. Disconnect and remove the speedometer and clutch cables.

4. Disconnect the back-up lamp electrical connector. Remove the starter motor electrical harness.

5. Remove the 6 transaxle bolts accessible from the top side of the transaxle.

6. Unbolt and remove the starter motor.

7. Raise the vehicle and support it safely. Then, remove the splash shield from under the engine. Drain the transaxle fluid.

8. Disconnect the extension rod and the shift rod at the transaxle end and lower them.

9. Disconnect the stabilizer bar at the lower control arm.

10. Remove the halfshaft.

11. Support the transaxle from below with a floor jack or similar device. Make sure the support is widely enough spread that the transaxle pan will not be damaged. Then, remove the 5 attaching bolts and remove the bell housing cover.

12. Remove the lower bolts attaching the transaxle to the engine.

13. Remove the transaxle insulator mount bolt. Remove the cover from inside the right fender shield and remove the transaxle support bracket.

14. Remove the transaxle mount bracket.

15. Pull the assembly away from the engine and then lower it from the vehicle.

To install:

16. Install the transaxle and tighten all attaching bolts. Install all transaxle mounts.

17. Install the halfshafts, stabilizer bar, shift rods, splash sheild, starter, back-up light switch connector, speedometer and clutch cables.

18. Install the selector control valve connector on 5 speed transaxles.

19. Install all other parts removed. Fill the transaxle with fluid and adjust the clutch cable. Make sure the gearshift lever works correctly.

Eclipse

1. Disconnect the negative and positive battery cables and remove the battery.

2. Remove the Auto-Cruise Actuator and bracket underhood, on the passenger side inner fender wall.

3. Drain the transaxle oil. On AWD the transfer case also has a drain plug.

4. Remove the air intake hose.

5. Remove the cotter pin securing the select and shift cables and remove the cable ends from the transaxle.

6. Remove the connection for the clutch release cylinder and without disconnecting the hydraulic line, secure at the body side aside.

7. Disconnect the backup light switch and the speedometer cable.

8. Disconnect the starter electrical connections and remove the starter motor.

9. Remove the transaxle mount bracket.

10. Raise and safely support vehicle and remove the under cover.

11. Remove the cotter pin and disconnect the tie rod end from the steering knuckle.

12. Remove the self-locking nut and remove the lower arm ball joint.

13. Remove the halfshafts by inserting a prybar between the transaxle case and the driveshaft and prying the shaft from the transaxle. Do not pull on the driveshaft. Doing so damages the inboard joint. Use the prybar. Do

not insert the prybar so far that the oil seal in the case is damaged. On AWD, remove the right halfshaft. The left halfshaft can be removed by tapping with a plastic hammer. Remove the shaft with the hub and knuckle as an assembly. Don't tap on the center bearing or it will be damaged. Tie the shafts back aside. Note the circle clip on the end of the inboard shafts. These should not be reused.

14. On AWD, disconnect the front exhaust pipe.

15. On AWD, remove the transfer case by removing the attaching bolts, moving the transfer case to the left and lowering the front side. Remove it from the rear driveshaft. Be careful of the oil seal. Do not allow the prop shaft to hang; tie it up. Cover the transfer case openings to keep out dirt.

16. Remove the underpan from the transaxle bell housing. On AWD, also remove the crossmember and the triangular gusset.

17. Remove the transaxle lower coupling bolt. It is just above the halfshaft opening on 2WD or transfer case opening on AWD.

18. Remove the transaxle assembly. On turbocharged equipped vehicles, take care to prevent damaging the lower radiator hose with the transaxle housing. Wind tape around the lower hose and put tape on the transaxle housing. Support the transaxle assembly, move the transaxle to the right and lower it.

To install:

19. Installation is the reverse of the removal procedure, with the following points to watch. When installing the halfshafts, always use new circlips on the axle ends. Take care to get the inboard joint parts straight, not bent relative to the axle. Care must be taken to ensure that the oil seal lip of the transaxle is not damaged by the serrated part of the driveshaft.

20. When bolting up the starter, make sure the ground cable is securely fastened.

21. Make sure the vehicle is level when refilling the transaxle. Use Hypoid gear oil or equivalent, GL-4 or higher. Check transaxle and transfer case on AWD.

3000-GT

1. Remove the battery and battery tray. Raise the vehicle and support safely. Drain the transaxle oil and remove the transfer case if equipped.

2. Remove the left side splash shield.

3. Remove the air cleaner assembly and all adjoining duct work.

4. Disconnect the shifter control cables and speedometer connector.

5. Remove the clutch release cylinder.

6. Disconnect the reverse light switch.

7. Support the weight of the transaxle and remove the transaxle mount through bolt. Remove the access plug, remove the bolts for the bracket and remove the brackets.

8. Disconnect the transaxle ground cable.

9. Disconnect the tie rod end and ball joint from the steering knuckle.

10. Remove the right frame member.

11. Remove the starter motor.

12. Remove the halfshafts by inserting a prybar between the transaxle case and the driveshaft and prying the shaft from the transaxle. Do not pull on the driveshaft. Doing so damages the inboard joint. Use the prybar. Do not insert the prybar so far the oil seal in the case is damaged. On AWD, remove the right side shaft as just described. The left side shaft can be removed by tapping with a plastic hammer. Remove the shaft with the hub and knuckle as an assembly. Don't tap on the center bearing or it will be damaged. Tie the shafts aside. Note the circle clip on the end of the inboard shafts. These should not be reused.

13. Remove the transaxle brackets.

14. Remove the transaxle assembly. On turbocharged vehicles, take care to prevent damaging the lower radiator hose with the transaxle housing. Wind tape around the lower hose and put tape on the transaxle housing. Support the transaxle assembly using the proper jack, move the transaxle away from the engine and lower it.

To install:

15. Install the transaxle to the engine and install the mounting bolts.

16. When installing the halfshafts, use new circlips on the axle ends. Take care to get the inboard joint parts straight, not bent relative to the axle. Care must be taken to ensure that the oil seal lip of the transaxle is not damaged by the serrated part of the driveshaft.

17. Install the starter motor and cover.

18. Install the right side frame member.

19. Install the ball joint and tie rod to the steering knuckle.

20. Connect the transaxle ground cable.

21. Install the side mount brackets and install the access plug.

22. Connect the reverse light switch.

23. Install the clutch release cylinder.

24. Connect the shifter control cables and speedometer connector.

25. Install the transfer case on AWD vehicles.

26. Install the air cleaner assembly and all adjoining duct work.

27. Install the left side splash shield.

28. Install the battery tray and battery.

29. Make sure the vehicle is level when refilling the transaxle. Use Hypoid gear oil or equivalent, GL-4 or higher.

30. Connect the negative battery cable and check the transaxle and transfer case for proper operation. Make sure the reverse lamps come on when in reverse.

LINKAGE ADJUSTMENT

There are 2 cables, the select cable and the shift cable.

1. On the transaxle, put select lever in **N** and move the transaxle shift lever to put it in **4th** gear. Depress the clutch, if necessary, to shift.

2. Move the shift lever in the vehicle to the **4th** gear position until it contacts the stop.

3. Turn the adjuster turn buckle so the shift cable eye aligns with the eye in the gear shift lever. When installing the cable eye, make sure the flange side of the plastic bushing at the shift cable end is on the cotter pin side.

4. The cables should be adjusted so the clearance between the shift lever and both stoppers are equal when the shift lever is moved to 3rd and 4th gear. Move the shift lever to each position and check that the shifting is smooth.

CLUTCH

REMOVAL & INSTALLATION

1. Disconnect the negative battery cable.

2. Remove the transaxle assembly from the vehicle.

3. Remove the pressure plate attaching bolts. If the pressure plate is to be reused, loosen the bolts in succession, 1 or 2 turns at a time to prevent warping the the cover flange.

4. Remove the pressure plate release bearing assembly and the clutch disc. Do not use solvent to clean the bearing.

5. Inspect the condition of the clutch components and replace any worn parts.

To install:

6. Inspect the flywheel for heat damage or cracks. Use new bolts and replace if necessary.

7. Using the proper alignment tool, install the clutch disc to the flywheel. Install the pressure plate assembly and tighten the pressure plate bolts evenly to 11–16 ft. lbs. (15–22 Nm). Remove the alignment tool.

8. Apply a very light coat of high temperature grease to the clutch fork at the ball pivot and where the fork contacts the bearing. Also a little bit of grease can be applied to end of the release cylinder's pushrod and to the pushrod hole on the fork. Apply a light coat of grease on the transaxle input shaft splines.
9. Install a new clutch release bearing. Pack its inner surface with grease.
10. Install the transaxle assembly and check for proper clutch operation.

PEDAL HEIGHT/FREE-PLAY ADJUSTMENT

Except Eclipse and 3000-GT

1. Measure the clutch pedal clevis pin play at the pedal pad, it should be 6.9–7.1 in. for Cordia and Tredia or 7.4–7.6 in. for Starion.
2. Measure the clutch pedal height from the surface of the pad to the floor, it should be 0.04–0.12 in.
3. If adjustment is required, turn the clutch switch to adjust the pedal height then tighten the locknut.
4. To adjust the clevis pin play turn the pushrod and then tighten the locknut.
5. If adjustment can not be made there is probably either air in the system or the clutch cylinder or clutch disc is defective.
6. Bleed the air from the system as follows.
7. Loosen the bleeder screw at the clutch slave cylinder.
8. Push the clutch pedal down slowly while the bleeder screw is opened.
9. Hold the pedal down and tighten the bleeder screw.
10. Check the clutch master cylinder and refill with fluid, if necessary. Repeat the bleeding procedure several times until all air is dispelled from the system.

Eclipse

1. Measure the clutch pedal height from the face of the pedal pad to the firewall. If the pedal height is not within 6.70–6.89 in. (170–175mm), adjustment is necessary.
2. Measure the clutch pedal clevis pin play at the face of the pedal pad. If the clutch pedal clevis pin play is not within 0.04–0.12 in. (1–3mm), adjustment is necessary.
3. If the clutch pedal height or clevis pin play are not within the standard value, adjust as follows:
   a. For vehicles without cruise control, turn and adjust the bolt so the pedal height is the standard value and tighten the locknut.
   b. Vehicles with auto-cruise control system, disconnect the clutch switch connector and turn the switch to obtain the standard clutch pedal height. Then, lock with the locknut.
   c. Turn the pushrod to adjust the clutch pedal clevis pin play to agree with the standard value and secure the pushrod with the locknut.

NOTE: When adjusting the clutch pedal height or the clutch pedal clevis pin play, be careful not to push the pushrod toward the master cylinder.

   d. Check that when the clutch pedal is depressed all the way 5.9 in. (149mm), the interlock switch switches over from **ON** to **OFF**.

3000-GT

1. Measure the clutch pedal height from the face of the pedal pad to the firewall. If the pedal height is not within 6.93–7.17 in (176–182mm) adjustment is necessary.
2. Measure the clutch pedal clevis pin play at the face of the pedal pad. If the clutch pedal clevis pin play is not within 0.24–0.51 in. (6–13mm), adjustment is necessary.
3. If the clutch pedal height or clevis pin play are not within the standard value, adjust as follows:
   a. For vehicles without cruise control, turn and adjust the bolt so the pedal height is the standard value, then tighten the locknut.
   b. Vehicles with auto-cruise control system, disconnect the clutch switch connector and turn the switch to obtain the standard clutch pedal height. Then, lock with the locknut.
   c. Turn the pushrod to adjust the clutch pedal clevis pin play to agree with the standard value and secure the pushrod with the locknut.

NOTE: When adjusting the clutch pedal height or the clutch pedal clevis pin play, be careful not to push the pushrod toward the master cylinder.

   d. Check that when the clutch pedal is depressed all the way, the interlock switch switches over from **ON** to **OFF**.

Clutch Cable

ADJUSTMENT

1. Depress the clutch pedal by hand. The free-play distance (when tension is felt), should be 0.6–0.8 in.
2. If the free-play is too great or too little, turn the outer cable adjusting nut for adjustment.
3. After adjustment is made, depress the clutch pedal several times and recheck.

REMOVAL & INSTALLATION

1. Disconnect the negative battery cable.
2. Remove the cable retaining clamps.
3. Remove the cotter pin from the clutch actuating arm at the transaxle.
4. Rotate the adjusting wheel counterclockwise to loosen the cable and remove the cable from the vehicle.
5. The installation is the reverse of the removal procedure.
6. Lubricate all pivot points. Adjust the cable.

Clutch Master Cylinder

REMOVAL & INSTALLATION

1. Disconnect the negative battery cable. Loosen the bleeder screw and drain the clutch fluid; on the Eclipse, it may be necessary to remove the air filter for access. Disconnect the pushrod at the clutch pedal by removing the cotter pin.
2. Disconnect the hydraulic tube at the master cylinder. Remove the clutch master cylinder.
3. Installation is the reverse of removal. Use a new cotter pin. Bleed the system. Torque the bolts or nuts to 9 ft. lbs. (13 Nm). Torque the clutch tube connection at the master cylinder to 11 ft. lbs. (15 Nm).

Clutch Slave Cylinder

REMOVAL & INSTALLATION

1. Disconnect the negative battery cable. Loosen the bleeder screw and drain the fluid.
2. Disconnect the clutch tube or hose from the slave cylinder.
3. Unbolt and remove the slave cylinder.
4. Installation is the reverse of removal. Torque the clutch tube eye bolt to 14–18 ft. lbs.

Hydraulic Clutch System Bleeding

1. Fill the reservoir with brake fluid.
2. Loosen the bleed screw, have the clutch pedal pressed to the floor.
3. Tighten the bleed screw and release the clutch pedal.
4. Repeat the bleeding operation until the fluid is free of air bubbles.

NOTE: It is suggested to attach a hose to the bleeder and place the other end into a container at least ½ full of brake fluid during the bleeding operation. Do not allow the reservoir to run out of fluid during the bleeding operation.

AUTOMATIC TRANSMSISSION

For further information on transmissions/transaxles, please refer to "Chilton's Guide to Transmission Repair".

Transmission Assembly

REMOVAL & INSTALLATION

1. Disconnect the negative battery cable. Raise and support the vehicle safely. Loosen the oil pan screws, tap the oil pan at one corner to break it loose and then allow the fluid to drain out one side. Remove the pan and remove the remaining fluid.
2. Remove its attaching bolt and then remove the transaxle pan filler tube by pulling it upward and out of the transaxle case.
3. Remove both top transaxle attaching bolts from the converter housing.
4. Disconnect the starter wiring and remove the starter.
5. Disconnect the oil cooler hoses at the metal tubes near the engine block. Then, unbolt and remove the tubes and their mountings from the block.
6. Remove the 4 bolts and remove the converter housing cover. Remove the torque converter bolts.
7. Disconnect the speedometer cable. Disconnect the transaxle control rod and the connection lever at the cross shaft assembly.
8. Disconnect the transaxle ground cable. Remove the driveshaft.
9. Support the rear of the transaxle with a floor jack. Unbolt the transaxle rear support bracket by removing 2 bolts on either side. Then, unbolt the bracket from the transaxle.
10. Remove the remaining bolts from the area of the converter housing. Separate the transaxle from the engine and remove it.

To install:

11. Prior to installation, check the distance between the front of the bell housing and the torque converter driveplate bolts with a straightedge and ruler. The distance must be at least 1.38 in. Install the transmission pan.
12. Lift the transmission into position and install the attaching bolts. Install the transmission rear support and the driveshaft. Connect the transmission ground strap.
13. Install the speedometer cable, transmission control rod and connection lever. Install the oil cooler hoses and install the converter cover.
14. Install the starter. Fill the transmission with fluid and lower the vehicle.
15. Check that the transmission will start only in **N** and **P** positions and that the backup light lights in **R** position.

KICKDOWN BAND ADJUSTMENT

1. Remove the transaxle oil pan.
2. Loosen the band adjusting stem locknut and turn the stem outward. Then, turn the stem inward with a torque wrench until the required torque reaches 5–7 ft. lbs. Then, back the stem off exactly 2 turns.
3. Hold the adjustment and torque the locknut to 11–29 ft. lbs.

AUTOMATIC TRANSAXLE

For further information on transmissions/transaxles, please refer to "Chilton's Guide to Transmission Repair".

Transaxle Assembly

REMOVAL & INSTALLATION

Except Eclipse and 3000-GT

NOTE: The transaxle and converter must be removed and installed as an assembly.

1. Disconnect the negative battery cable. Remove the battery tray. Remove the coolant reservoir and windshield washer tank. Remove the air cleaner and housing. Where so equipped, disconnect also the pulse generator connector and solenoid valve connector.
2. Disconnect the throttle control cable at the carburetor and the manual control cable at the transaxle.
3. Disconnect from the transaxle; the neutral safety switch connector, fluid cooler hose and the 4 upper bolts connecting the engine to the transaxle.
4. Raise and support the vehicle safely.
5. Remove the front wheels. Remove the engine splash shield.
6. Drain the transaxle fluid.
7. Remove the right and left halfshafts from the transaxle case. Remove the strut bars and the stabilizer bar from the lower arms.
8. Disconnect the speedometer cable. Disconnect and plug the oil cooler hoses. Remove the starter motor.
9. Remove the lower cover from the converter housing. Remove the 3 bolts that connect the converter to the engine drive plate.

NOTE: Never support the full weight of the transaxle on the engine driveplate.

10. Turn and force the converter back and away from the engine driveplate.
11. Support the weight of the engine from above. Support the transaxle and remove the remaining bolts.
12. Remove the transaxle mount insulator bolt.
13. Remove and lower the transaxle and converter as an assembly.

To install:

14. Lift the transaxle into position and install attaching bolts. Install all insulators, brackets and mounts.
15. Install the coverter bolts and cover. Install the speedometer cable, starter and oil cooler hoses.
16. Install the halfshafts, strut bars, stabilizer bar and all other components removed or disconnected.
17. Fill the transaxle with fluid, install the front wheels and lower the vehicle.

Eclipse

1. Disconnect the battery cables and remove the battery.
2. On vehicles equipped with Autocruise, remove the control actuator and bracket.
3. Drain the transaxle fluid.
4. Remove the air cleaner assembly.
5. Remove the adjusting nut and disconnect the shift cable.
6. Disconnect and tag as required the electrical connectors for the solenoid, neutral safety switch (inhibitor switch), the pulse generator kickdown servo switch and oil temperature sensor.
7. Disconnect the speedometer cable and oil cooler lines.
8. Disconnect the wires to the starter motor and remove the starter.
9. Remove the upper transaxle to engine bolts.
10. Remove the transaxle bracket.
11. Raise and safely support vehicle and remove the sheet metal under guard.
12. Remove the tie rod ends and the ball joints from the steering knuckle.

13. Remove the halfshafts by inserting a prybar between the transaxle case and the driveshaft and prying the shaft from the transaxle. Do not pull on the driveshaft. Doing so damages the inboard joint. Use the prybar. Do not insert the prybar so far the oil seal in the case is damaged. Tie the halfshafts aside.
14. Remove the lower bell housing cover and remove the bolts holding the flexplate to the torque converter. These are special bolts. Do not loose. To remove, turn the engine crankshaft with a box wrench and bring the bolts into position one at a time. After removing the bolts, push the torque converter toward the transaxle so it doesn't stay on the engine side and allow oil to pour out the converter hub.
15. Remove the lower transaxle to engine bolts and remove the transaxle assembly.

To install:

16. Installation is the reverse of the removal process. After the torque converter has been mounted on the transaxle, install the transaxle assembly on the engine. If the torque converter is first mounted on the engine, a damaged oil seal in the transaxle could result. Tighten the driveplate bolts to 34–38 ft. lbs. Install the bell housing cover.
17. Install the halfshafts to the transaxle and connect the tie rods and ball joint connections.
18. Install the underguard and the bracket. Reconnect the cable controls, oil cooler lines and electrical connections.
19. Lower vehicle, refill with Dexron® or Dexron® II automatic transaxle fluid. Start engine and allow to idle for 2 minutes. Apply parking brake and move selector through each gear position , ending in **N**. Recheck fluid level and add, if necessary. Fluid level should be between the marks in the **HOT** range.

3000-GT

1. Remove the battery, battery tray and washer tank.
2. Remove the air cleaner assembly and adjoining duct work.
3. Disconnect the shifter control cable.
4. Disconnect and plug the oil cooler hoses.
5. Disconnect the inhibitor switch, kickdown servo switch, pulse generator, oil temperature sensor, shift control solenoid valve, and ground cable.
6. Disconnect the speedometer cable.
7. Raise the vehicle and support safely. Remove the undercovers.
8. Support the weight of the transaxle and remove the mount bracket. Remove the upper bellhousing bolts.
9. Disconnect the tie rod end and ball joint from the steering knuckle.
10. Remove the right frame member.
11. Remove the starter.
12. Remove the halfshafts by inserting a prybar between the transaxle case and the driveshaft and prying the shaft from the transaxle. Do not pull on the driveshaft. Doing so damages the inboard joint. Use the prybar. Do not insert the prybar so far the oil seal in the case is damaged. Tie the halfshafts aside.
13. Remove the remaining mounting brackets.
14. Remove the bellhousing cover plate.
15. Remove the special bolts holding the flexplate to the torque converter.
16. After removing the bolts, push the torque converter toward the transaxle so it doesn't stay on the engine side and allow oil to pour out the converter hub.
17. Remove the lower transaxle to engine bolts and remove the transaxle assembly.

To install:

18. After the torque converter has been mounted on the transaxle, install the transaxle assembly on the engine. Tighten the driveplate bolts to 34–38 ft. lbs. (46–53 Nm). Install the bell housing cover.
19. Install the mounting brackets.
20. Replace the circlips and install the halfshafts to the transaxle.
21. Install the starter and frame member.
22. Install the tie rods and ball joint to the steering arm.
23. Install the upper bellhousing bolts.
24. Install the transaxle mounting bracket.
25. Install the undercovers.
26. Connect the speedometer cable.
27. Connect the inhibitor switch, kickdown servo switch, pulse generator, oil temperature sensor, shift control solenoid valve, and ground cable.
28. Connect the oil cooler hoses.
29. Connect the shifter control cable.
30. Install the air cleaner assembly and adjoining duct work.
31. Install the washer tank, battery tray and battery.
32. Refill with Dexron II, Mopar ATF Plus type 7176, or equivalent automatic transaxle fluid.
33. Start the engine and allow to idle for 2 minutes. Apply parking brake and move selector through each gear position, ending in **N**. Recheck fluid level and add if necessary. Fluid level should be between the marks in the **HOT** range.

SHIFT LINKAGE ADJUSTMENT

1. The shifter cable adjustment is done at the neutral safety switch (inhibitor switch). Locate the switch on the transaxle and not the alignment hole in the arm and the body of the switch. Place the selector lever in **N**. Place the manual lever of the transaxle in **N**.
2. Align the holes on the switch.
3. If the cable needs to be adjusted, loosen the nut on the cable end and pull the cable end by hand until the alignment holes match. Tighten the nut. Check that the transaxle shifts and conforms to the positions of the selector lever.

THROTTLE LINKAGE ADJUSTMENT

1. Check that the throttle lever is in the curb idle position, with the engine **OFF** but at normal operating temperature.
2. At the lower cable bracket, raise the cone shaped cover to uncover a small fitting on the cable. By loosening the locknut and adjuster nut, make the distance between the fitting on the cable and the lower collar 0.020–0.060 in.
3. With the throttle in the wide open position, check that the cable does not bind.

NOTE: Not all vehicles use a throttle linkage. The throttle position sensor on some vehicles feeds an electric signal to the transaxle so no linkage adjustment is required. If the throttle position sensor itself needs to be adjusted, use the following procedure:

THROTTLE CABLE ADJUSTMENT

1. Run the engine to normal operating temperature and make sure the throttle lever on the carburetor is in the curb idle position.
2. Raise the cover on the throttle cable to expose the nipple.
3. Loosen the lower cable bolt.
4. Move the lower cable bracket until the distance between the nipple and the top of the cable end is 0.5mm.
5. Tighten the lower cable bracket bolt and check the adjustment by pulling the cable upward with the throttle plate in the wide open position. The cable should move freely.

DETENT CABLE ADJUSTMENT

1. Several special factory tools may be required for this operation. Locate the detent switch on the transaxle. Remove the road dirt from around it. Remove the snapring and pull out the kickdown servo switch.
2. To keep the piston in the transaxle from turning, the special tool has fingers or pawls that engage the slots in the piston to hold it while adjustment is made. Do not press in on the piston with the special tool or its equal. With the piston restrained from turning, loosen the locknut. Using the special tool, tighten to 7.2 ft. lbs. (10 Nm) and return or back off the adjustment 2 times. Then tighten to 3.6 ft. lbs. (5 Nm). Finally back off the adjustment 2–2¼ turns and making sure the piston does not turn, tighten the locknut.
3. Install a new O-ring into the groove in the switch, install the switch and fit the snapring in place.

TRANSFER CASE

Transfer Case

REMOVAL & INSTALLATION

1. Disconnect the battery negative cable.
2. Raise and safely support vehicle.
3. Disconnect the front exhaust pipe.
4. Unbolt the transfer case assembly and remove by sliding it off the rear driveshaft. Be careful not to damage the oil seal in the transfer case output housing. Do not let the rear driveshaft hang. Tie up with wire. Cover the opening in the transaxle to keep oil from dripping and to keep dirt out.

To install:

5. Installation is the reverse of the removal procedure. Use care when installing the rear driveshaft to the transfer case output shaft. Tighten the transfer case to transaxle bolts to 40–43 ft. lbs. On 3000-GT, tighten bolts to 64 ft. lbs. (88 Nm).
6. Install the exhaust pipe using a new gasket.
7. Check oil levels in transaxle and transfer case.

LINKAGE ADJUSTMENT

There are 2 cables, the select cable and the shift cable.

1. On the transaxle, put select lever in **N** and move the transfer case shift lever to **N**.
2. Turn the adjuster turn buckle so the shift cable eye aligns with the eye in the select lever. When installing the cable eye, make sure the flange side of the plastic bushing at the shift cable end is on the cotter pin side.
3. The cables should be adjusted so the clearance between the shift lever and both stoppers are equal when the shift lever is moved to 3rd and 4th gear. Move the shift and select levers to each position and check that the shifting is smooth.

FRONT SUSPENSION

MacPherson Strut

REMOVAL & INSTALLATION

Cordia and Tredia

1. Disconnect the negative battery cable. Raise and support the vehicle safely.
2. Remove the front wheel. Remove the brake line from the strut.
3. Disconnect the strut assembly from the steering knuckle by removing both bolts/nuts. Support the strut and remove both nuts and washers fastening it to the wheel well. Remove the strut.
4. Installation is the reverse of the removal procedure. When installing the strut, apply a non hardening sealer to the mating surfaces of the strut and knuckle arm.

Starion

1. Disconnect the negative battery cable. Raise and support the vehicle safely. Remove the tire and wheel. Remove the caliper. Remove the front hub with disc and dust cover.
2. Disconnect the stabilizer linkage and the lower. Remove the strut assembly, knuckle arm and strut insulator retaining bolts and remove the strut assembly from the wheelhouse.
3. Installation is the reverse of the removal procedure. When installing the strut, apply a non hardening sealer to the mating surfaces of the strut and knuckle arm.

Galant, Mirage and Sigma

1. Raise and support the vehicle safely. Remove the front wheels. Detach the brake hose bracket at the strut, if necessary.
2. If equipped with Electronic Control Suspension (ECS), remove the height sensor bolt. Remove the lower nuts and bolts attaching the strut to the steering knuckle. If equipped with ECS, ensure that the height sensor rod is detached from the lower control arm.
3. If equipped with ECS, disconnect the air line joint, then the connector, O-ring and air tube.
4. Remove the dust cover from the top of the strut on the wheel well. Support the strut from under. Install the socket wrench on the nut at the top of the strut and a box or open end wrench on the socket. Then, install the Allen wrench through the center of the socket, long part downward. Hold the Allen wrench in place, if necessary, by using a small diameter pipe as a cheater. Turn the socket to loosen the nut. Remove the nut and the lower and remove the strut.
5. To install, reverse the removal procedure. Torque the bolts attaching the bottom of the strut to the knuckle to 53–63 ft. lbs. on the Mirage or 65–76 ft. lbs. on the Galant and Sigma. The nut at the top of the strut must be torqued with the shaft of the shock held from turning with the Allen wrench, as during the loosening process. Since it is not usually possible to use a torque wrench on the flats of a socket, estimate the torque; it should be 36–43 ft. lbs.

Precis

1. Raise and support the vehicle safely. Remove the front wheels. Detach the brake hose bracket at the strut.
2. Remove the 4 nuts securing the strut to the fender well.
3. Unbolt the strut lower end from the knuckle.
4. Remove the strut from the vehicle.
5. Installation is the reverse of removal. Torque the strut to knuckle bolts to 55–65 ft. lbs.; the strut to fender well nuts to 7–11 ft. lbs.

Eclipse

1. Raise and safely support vehicle.
2. Remove the brake hose and tube bracket. Do not pry the brake hose and tube clamp away when removing it.
3. Remove the strut lower bolts. Support the lower arm. Use a piece of wire to suspend the knuckle to keep the weight off the brake hose.
4. Before removing the top bolts, make matchmarks on the body and the strut insulator for proper reassembly. If this plate is installed improperly, the wheel alignment will be wrong. Remove the strut upper bolts and pull the strut/spring from the vehicle.

To install:

5. Inspect the strut for signs of oil leakage or damage. Installation is the reverse of the removal process. Check that the top plate is properly installed or alignment will be affected. Torque the strut to knuckle bolts to 80–94 ft. lbs.

3000-GT

1. Disconnect the negative battery cable. Raise and safely support vehicle.
2. Remove the brake hose and tube bracket. Do not pry the brake hose and tube clamp away when removing it.
3. Support the lower arm and remove the strut to knuckle bolts. Use a piece of wire to suspend the knuckle to keep the weight off the brake hose.
4. If equipped with ECS, disconnect the ECS connector at the top of the strut.
5. Before removing the top bolts, make matchmarks on the body and the strut insulator for proper reassembly. If this plate is installed improperly, the wheel alignment will be wrong. Remove the strut upper bolts and remove the strut assembly from the vehicle.

To install:

6. Install the strut to the vehicle and install the top bolts.
7. Connect the ECS connector.
8. Install to the knuckle and install the bolts.
9. Install the brake hose bracket.
10. Perform a front end alignment.

Torsion Bars

REMOVAL & INSTALLATION

1. Raise and support the vehicle safely.
2. Remove the torsion bar locknut. Remove the torsion bar adjusting nut. Remove the seat holding nut.
3. Before removing the anchor bolt measure the protrusion through the assembly, this will aid in reinstallation of the assembly. Remove the anchor bolt that retains the torsion bar to its mounting on the frame.
4. Remove the nuts that retain the torsion bar to the control arm.
5. Remove the torsion bar from the vehicle.
6. Installation is the reverse of the removal procedure. Adjust the alignment and the torsion bar, as required.

Ball Joints

NOTE: The lower ball joints on some vehicles are not serviceable. If defective, the entire lower arm must be replaced. The ball joints cam be check using the procedure below.

INSPECTION

1. Wiggle the ball joint a few times to make sure it is free.
2. Double-nut the stud and use a torque wrench to measure how much torque is required to turn it. Starting torque should be:
   a. Mirage: 48 inch lbs. (5.5 Nm) or less.
   b. Galant, Sigma and Eclipse: 26–87 inch lbs. (3–10 Nm).
   c. 3000-GT: 86–191 inch lbs. (10–22 Nm).
3. If the stud has more resistance than specified, replace the lower arm assembly. If the resistance is less, it may still be reused unless it has excessive play.
4. A new grease boot can be installed using a large socket for a driver.

REMOVAL & INSTALLATION

Front Wheel Drive

1. Raise and support the vehicle safely. Remove the tire and wheel.
2. Disconnect the stabilizer bar and strut from the lower arm.
3. Remove the ball joint nut and separate the ball joint from the front knuckle. The ball joint stud must be pressed off; use special tool MB991113 or equivalent.
4. Remove the lower control arm by removing the bolt(s)/nut(s) attaching it to the crossmember.
5. Remove the dust cover from the ball joint. Remove the snapring.

To install:

6. Press the ball joint out of the lower control arm.
7. Press the new ball joint into place. Install a new snapring with snapring pliers.
8. Apply multipurpose grease to the lip and to the inside of the dust cover. Use a special tool (and hammer) such as MB990800-3-01, or equivalent, to drive a new dust cover. It must go in and make contact with the snapring.
9. Install the lower control arm to the crossmember, making sure there is not torque on it. Install and tighten the bolt and nut.

Rear Wheel Drive

1. Raise and support the vehicle safely. Remove the tire and wheel.
2. Remove the strut end from the steering knuckle.
3. Remove the ball joint-to-knuckle arm nut. A tool is necessary, which can be bolted to the holes in the knuckle arm and which will then press downward on the center of the ball stud tool MB990241-01 or equivalent.
4. Remove the ball joint to control arm nuts and bolts and remove the ball joint.
5. Install in reverse order. Torques in ft. lbs.: ball joint bolts 43–51 ft. lbs., strut to knuckle arm bolts 58–72 ft. lbs., ball stud nut 43–52 ft. lbs.

Lower Control Arm

REMOVAL & INSTALLATION

Cordia and Tredia

1. Raise and support the vehicle. Remove the tire and wheel.
2. Disconnect the stabilizer bar and strut bar from the lower control arm by removing the 1 attaching bolt for the stabilizer bar and both bolts for the strut bar.
3. Remove the ball stud nut and then press the ball joint stud out of the knuckle with a tool such as MB991113.
4. Remove the nut and bolt attaching the inner end of the stabilizer bar to the crossmember and pull the stabilizer bar and bushing out of the crossmember.
5. Installation is the reverse of removal. Install all parts and tighten nuts and bolts just snug. Then, complete tightening, torquing the lower arm to crossmember attaching nut/bolt to 87–108 ft. lbs. and the ball joint stud nut to 43–52 ft. lbs. Torque the strut rod to stabilizer bar bolt/nut to 43–50 ft. lbs.

Starion

1. Raise and support the vehicle safely. Remove the tire and wheel.
2. Disconnect the stabilizer bar where the link bolts to the control arm by removing the nut under the arm. Remove the nut and bolt attaching the strut bar to the control arm.
3. Disconnect the tie rod at the knuckle arm. Use a fork like tool such as MB990778-01, or equivalent, a standard type tool for pulling ball joint studs. First, loosen the stud nut until it is near the top of the threads and then hammer the tool between the ball joint of the tie rod end and the knuckle arm. When the ball stud comes loose, remove the nut and disconnect the stud.
4. Unbolt the strut from the knuckle arm.
5. Unbolt the inner end of the ball joint assembly to disconnect it from the outer end of the control arm.
6. Remove the nut, bolt and lockwasher and pull the inner end of the control arm out of the crossmember.
7. Installation is the reverse of removal. Torque the bolt fastening the control arm to the crossmember to 58–69 ft. lbs.; the bolts attaching the ball joint to the outer end to 43–51 ft. lbs.; the ball stud nut to 43–52 ft. lbs.; and the strut bolts to 58–72 ft. lbs.

Tighten the nut for the stabilizer bar link until 0.59–0.67 in. of thread shows below the bottom of the nut.

Mirage, Galant and Sigma

1. Raise and support the vehicle safely. Remove the tire and wheel.
2. On the Mirage, remove the under cover.
3. Disconnect the stabilizer bar from the lower arm. On the Galant, remove the nut at the top and remove the washer and bushing, keeping them in order. On the Mirage, remove the nut from under the control arm and take off the washer and spacer.
4. On the Galant with electronically controlled suspension, if removing the right arm, disconnect the height sensor rod from the lower arm. Loosen the ball joint stud nut and then press the stud out of the control arm, using a fork like tool MB990778–01, or equivalent, and hammer on the Galant; on the Mirage, remove the stud nut and press the tool off with a tool such as MB991113, or equivalent.
5. On the Galant, remove the nuts and bolts which retain the bushings to the crossmember at the front and which retain the bushing retainer to the crossmember at the rear and pull the arm out. On the Mirage, remove the bolts which retain the spacer at the rear and the nut and washers on the front of the lower arm shaft (at the front). Slide the arm forward, off the shaft and out of the busing.

To install:

6. Replace the dust cover on the ball joint. The new cover must be greased on the lip and inside with No. 2 EP Multi-purpose grease and pressed on with a tool such as MB990800 and a hammer until it is fully seated.
7. Installation is the reverse of removal.
8. On the Galant, make sure the nut on the stabilizer bar bolt is torqued to give 0.63–0.7 in. of thread exposed between the top of the nut and the end of the link.
9. On the Mirage, the nut must be torqued until the link shows 0.83–0.91 in. of threads below the bottom of the nut. Also on the Mirage, the washer for the lower arm must be installed as shown. The left side arm shaft has a left hand thread. Finally tighten the arm shaft to the lower arm with the weight of the vehicle with no passengers or luggage on the front suspension. On the Mirage, torque the nut for the lower arm shaft to 69–87 ft. lbs.; the bolts for the spacer at the rear to 43–58 ft. lbs. and the ball stud nut to 43–52 ft. lbs.
10. On the Galant, torque the nut for the nut/bolt to crossmember to 69–87 ft. lbs., the ball stud to 42–50 ft. lbs. and the bolt for retaining the rear bushing to the body to 58–72 ft. lbs.

Precis

1. Raise and support the vehicle safely. Remove the tire and wheel.
2. Remove the under cover.
3. Disconnect the stabilizer bar from the lower arm. Remove the nut from under the control arm and take off the washer and spacer.
4. Remove the ball joint stud nut and press the tool off with a tool such as MB991113, or equivalent.
5. Remove the bolts which retain the spacer at the rear and the nut and washers on the front of the lower arm shaft, at the front. Slide the arm forward, off the shaft and out of the bushing.
6. Replace the dust cover on the ball joint. The new cover must be greased on the lip and inside with No. 2 EP Multi-purpose grease and pressed on with a tool such as MB990800, or equivalent, and a hammer until it is fully seated.

To install:

7. Installation is the reverse of removal. The nut on the stabilizer bar bolt must be torqued until the link shows 21–23mm of threads below the bottom of the nut.
8. Ensure that the washer for the lower arm is installed correctly.
9. The left side arm shaft has a left hand thread.
10. Tighten the arm shaft to the lower arm with the weight of the vehicle with no passengers or luggage on the front suspension. Torque the nut for the lower arm shaft to 69–87 ft. lbs.; the bolts for the spacer at the rear to 43–58 ft. lbs.; the ball stud nut to 43–52 ft. lbs.

3000-GT

1. Disconnect the negative battery cable.
2. Raise the vehicle and support safely.
3. Remove the sway bar and links.
4. Disconnect the ball joint stud from the steering knuckle.
5. Remove the inner mounting frame-through bolt and nut.
6. Remove the rear mount bolts. Remove the clamp if equipped.
7. Remove the rear rod bushing if servicing.

To install:

8. Assemble the control arm and bushing.
9. Install the control arm to the vehicle and install the through bolt. Replace the nut and snug temporarily.
10. Install the rear mount clamp, bolts and replacement nuts. Torque the bolts to 70 ft. lbs. (95 Nm). The nut is torqued to 30 ft. lbs. (41 Nm).
11. Connect the ball joint stud to the knuckle. Install a new nut and torque to 43–52 ft. lbs. (60–72 Nm).
12. Install the sway bar and links.
13. Lower the vehicle to the floor for the final torquing of the frame mount through bolt.
14. Once the full weight of the vehicle is on the floor, torque the nuts to 75–90 ft. lbs. (102–122 Nm).
15. Connect the negative battery cable.

Sway Bar

REMOVAL & INSTALLATION

Except 3000-GT

1. Disconnect the negative battery cable.
2. Raise and safely support vehicle. Remove the front exhaust pipe if necessary.
3. Remove the tie rod end from the steering knuckle.
4. Remove the center crossmember rear bolts.
5. Remove the stabilizer link bolts. On the ball stud type, hold ball stud with a hex wrench and remove the self-locking nut with a box wrench.
6. Remove the stabilizer bar mounts and remove the bar from the vehicle.
7. The installation is the reverse of the removal procedure. Lubricate all rubber parts when installing. Note that the bar brackets are marked left and right.
8. Tighten link bolts with rubber bushings just until the bushings are squashed to the width of the washer.

3000-GT

1. Disconnect the negative battery cable.
2. Raise the vehicle and support safely.
3. Remove the front exhaust pipe and engine undercover.
4. Remove the left and right frame members.
5. On AWD vehicles with automatic transaxle, remove the transfer case bracket and transfer case.
6. Remove the sway bar link.
7. Remove the sway bar brackets and remove the sway bar from the vehicle.

To install:

8. Note that the bar brackets are marked left and right. Lubricate all rubber parts and install the bushings, the sway bar and brackets.
9. Install the sway bar link.
10. Install the transfer case and bracket.
11. Install the frame members.

12. Install the engine undercover and exhaust pipe.
13. Connect the negative battery cable.

Front Wheel Bearings

NOTE: This section pertains to rear wheel drive vehicles only.

ADJUSTMENT

1. Remove the wheel and dust cover. Remove the cotter pin and lock cap from the nut.
2. Torque the wheel bearing nut to 14.5 ft. lbs. and then loosen the nut. Retorque the nut to 3.6 ft. lbs. and install the lock cap and cotter pin.
3. Install the dust cover and the wheel.

REMOVAL & INSTALLATION

1. Raise and support the vehicle safely. Remove the tire and wheel. Remove the caliper.
2. Pry off the dust cap. Tap out and discard the cotter pin. Remove the locknut.
3. Being careful not to drop the outer bearing, pull off the brake disc and wheel hub.
4. Remove the grease inside the wheel hub.
5. Using a brass drift, carefully drive the outer bearing race out of the hub.
6. Remove the inner bearing seal and bearing.

To install:

7. Check the bearings for wear or damage and replace them, if necessary.
8. Coat the inner surface of the hub with grease.
9. Grease the outer surface of the bearing race and drift it into place in the hub.
10. Pack the inner and outer wheel bearings with grease. If the brake disc has been removed and/or replaced, tighten the retaining bolts to specification.
11. Install the inner bearing in the hub. Being careful not to distort it, install the oil seal with its lip facing the bearing. Drive the seal on until its outer edge is even with the edge of the hub.
12. Install the hub/disc assembly on the spindle, being careful not to damage the oil seal.
13. Install the outer bearing, washer and spindle nut. Adjust the bearing.

REAR SUSPENSION

MacPherson Strut

REMOVAL & INSTALLATION

Starion

1. Raise and safely support the vehicle on the frame rails. Position a floor jack under the lower control arm and raise it slightly.
2. Disconnect the rear brake hose from the strut assembly.
3. Disconnect the axle shaft from the wheel side flange.
4. Remove the strut assembly to axle housing bolts. Separate the strut assembly from the axle housing. Lower the floor jack and push down on the housing while opening the coupling with a small prybar.
5. Remove the upper strut nuts from under the side trim in rear hatch.
6. Remove the strut assembly.
7. Install in reverse order. Tighten the top nuts to 18–25 ft. lbs. and the lower to 36–51 ft. lbs.

Galant

1. Raise and safely support the vehicle. Remove the rear wheels.
2. Place a floor jack under the axle/arm assembly and raise it slightly. Remove the forward trim from the trunk and remove the cap and strut nuts and washers.
3. Remove the nut, pull the through bolt out where the strut connects with the axle/arm assembly and remove the strut assembly.
4. Installation is the reverse of removal. Torque the upper strut nuts to 33–40 ft. lbs. without Electronic Level Control (ELC) or 18–25 ft. lbs. with Electronic Level Control (ELC) and the lower through bolt to 72–87 ft. lbs. without Electronic Level Control (ELC) or 58–72 ft. lbs. with Electronic Level Control (ELC).

Eclipse

1. Remove the trim panel inside the trunk area for access to the top nuts.
2. Remove the top cap and nuts.
3. Raise and safely support vehicle.
4. Remove the brake tube bracket bolt, then remove the shock absorber lower bolt and remove the shock absorber/spring assembly from the vehicle.

To install:

5. Installation is the reverse of the removal procedure. It is recommended that the self-locking nuts used at the top and bottom mount not be reused, but replaced.

3000-GT

1. Disconnect the negative battery cable. Remove the trim panel inside the trunk or hatch area for access to the top mounting nuts.
2. Remove the top cap and mounting nuts. Disconnect the ECS connector if equipped.
3. Raise and safely support vehicle.
4. Remove the brake tube bracket bolt if necessary, then remove the strut lower mounting bolt.
5. Remove the rear strut assembly from the vehicle.
6. Installation is the reverse of the removal procedure. Do not tighten the lower mounting nut until the full weight of the vehicle is on the ground.

Rear Control Arm

REMOVAL & INSTALLATION

Except Starion and Eclipse

1. Raise and safely support the vehicle. Remove the rear wheels. Remove the rear brake assemblies. As required, remove the muffler.
2. Disconnect the parking brake cable from the suspension arm on both sides.
3. Raise the suspension arm on both sides just slightly and remove both lower shock absorber attaching bolts. Lower the jack and when it can be disengaged, remove the spring. Keep the spring in the position it was in when installed so it can be installed in the same direction.
4. Disconnect the brake hoses at the suspension arms. Support the rear suspension assembly while removing both bolts on either side and remove the assembly.
5. Installation is the reverse of removal. Lower shock mounting bolts are torqued to 47–58 ft. lbs. Suspension assembly-to-body bolts are torqued to 51–65 ft. lbs. on Cordia and Tredia or 36–51 ft. lbs. on Mirage.

Starion

1. Raise and support the vehicle safely.
2. Disconnect the parking brake from the control arm brackets. Disconnect the stabilizer bar.
3. Remove the nut and bolt connecting the lower control arm to the front support.
4. Matchmark the relationship between the crossmember and the eccentric bushing so alignment can be restored at assembly. Remove the nut and bolt connecting the lower control arm to the crossmember.

5. Remove the lower control arm from the vehicle.
6. Installation is the reverse of removal.

Eclipse

1. Raise and safely support vehicle.
2. Remove the locknut for the lower arm ball joint.
3. Remove the rear stabilizer bar link to the lower arm.
4. Remove the inboard lower arm pivot bolt and separate the arm from the vehicle.

To install:

5. Installation is the reverse of the removal procedure. It is recommended that the self-locking nuts not be reused, but replaced. Torque the inboard pivot bolt and nut to 65–80 ft. lbs., and the ball joint stud nut to 43–52 ft. lbs.

3000-GT

1. Disconnect the negative battery cable. On FWD 3000-GT, remove the rear strut assembly. Raise and safely support vehicle. Remove the brake line clamp bolt.
2. Remove the ball joint(s) from the rear trailing arm/steering knuckle.
3. If removing the lower arm, disconnect the sway bar link from the arm.
4. Matchmark and remove the inboard lower arm pivot bolt, if necessary, and remove the arm from the vehicle.

To install:

5. Installation is the reverse of the removal procedure. Replace all self-locking nuts. Do not torque the inboard pivot nuts until the full weight of the vehicle is on the ground.
6. Torque nuts to 101–116 ft. lbs. (140–160 Nm).
7. Perform a rear wheel alignment.

Rear Wheel Bearings

ADJUSTMENT

Mirage and Precis

After performing service to the wheel hub, tighten the hub nut to 108–145 ft. lbs (147–196 Nm). Check the hub end play. If not within specification, replace the hub nut or wheel bearing.

Eclipse, Galant (w/rear disc brakes) and 3000-GT

After performing service to the wheel hub, tighten the hub nut to 144–188 ft. lbs (200–260 Nm). Check the hub end play. If not within specification, replace the hub nut or wheel bearing.

Galant (w/rear drum brakes) and Sigma

After performing service to the wheel hub, tighten the hub nut to 14 ft. lbs. (20 Nm), then loosen the nut fully. Finally, retighten the nut to 7 ft. lbs. (10 Nm). Check the hub end play. If not within specification, replace the hub nut or wheel bearing.

REMOVAL & INSTALLATION

1. Raise and support the vehicle safely. Remove the rear wheels.
2. If equipped with ABS, carefully disconnect the speed sensor and move it to a safe location. Remove the dust cap, cotter pin and cap nut.
3. If equipped with drum brakes, remove the large center nut and the drum. If equipped with disc brakes, remove the brake caliper and hang it out of the way, then remove the rotor.
4. If equipped with ABS, the toothed rotor for the speed sensor is attached to the hub. This rotor must not be damaged in any way.
5. Remove the outer bearing from the hub and invert the unit to pty out the oil seal. Remove the inner bearing.

To install:

6. If replacing the bearings, use a punch to remove the bearing races from the hub. Clean everything throughly. Install new races with an appropriate driver.
7. Pack the wheel bearings with brearing grease and install. Apply a light coat of grease to the oil seal and install.
8. Install the hub on the spindle and tighten the center nut hand tight.
9. Adjust the bearing torque and check the endplay. Install the brake caliper, if removed. Install the wheel and lower the vehicle.

Rear Axle Assembly

REMOVAL & INSTALLATION

1. Raise the vehicle and support safely.
2. Remove the tire and wheel assembly.
3. If equipped with ABS, remove the bolts holding the speed sensor bracket to the trailing arm and remove the sensor assembly from the vehicle.

NOTE: The speed sensor has a pole piece projecting from it. This exposed tip must be protected from impact or scratches. Do not allow the pole piece to contact the toothed wheel during removal or installation.

4. If equipped with rear disc brakes, remove the caliper from the disc and remove the brake disc.
5. Remove the dust cap and bearing nut. Do not use an air gun to remove the nut.
6. Remove the outer wheel bearing.
7. Remove the drum and/or axle hub with the inner wheel bearing and the grease seal.
8. Remove the parking brake cable, brake hose, tube bracket and brake shoes with backing plate from the axle.
9. Remove the lateral rod mounting bolt and nut and secure the lateral rod to the axle beam with a piece of wire.
10. Using the proper equipment, slightly raise the torsion axle and arm assembly. Remove lower strut mounting bolt.
11. Remove the front trailing arm mount bolts and remove the rear axle assembly.

To install:

12. Install the rear axle assembly to the vehicle and install the strut mounting bolts. Install the front mount bolts and lateral rod bolts. Do not tighten these until the full weight of the vehicle is on the ground.
13. Install the backing plate, brake shoes, cable and hose.
14. To determine if the self-locking nut is reusable:
   a. Screw in the self-locking nut until about $\frac{1}{10}$ in. of the spindle is showing.
   b. Measure the torque required to turn the self-locking nut counterclockwise.
   c. The lowest allowable torque is 48 inch lbs. (5.5 Nm). If the measured torque is less than the specification, replace the nut.
15. Remove the drum and/or axle hub. Lubricate and install the outer wheel bearing to the spindle, if equipped. Torque the self-locking nut to 108–145 ft. lbs. (150–200 Nm).
16. Install the tounged washer and a new self-locking nut. Torque the nut to 144–188 ft. lbs. (200–260 Nm), align with the indentation in the spindle, and crimp.
17. Install the grease cap and brake parts.
18. Temporarily install the speed sensor to the knuckle; tighten the bolts only finger-tight.
19. Route the cable correctly and loosely install the clips and retainers. All clips must be in their original position and the sensor cable must not be twisted. Improper installation may cause cable damage and system failure.

NOTE: The wiring in the harness is easily damaged by twisting and flexing. Use the white stripe on the outer insulation to keep the sensor harness properly placed.

20. Use a brass or other non-magnetic feeler gauge to check the air gap between the tip of the pole piece and the toothed wheel. Correct gap is 0.012–0.035 in. (0.3–0.9mm). Tighten the 2 sensor bracket bolts to 10 ft. lbs. (14 Nm) with the sensor located so the gap is the same at several points on the toothed wheel. If the gap is incorrect, it is likely that the toothed wheel is worn or improperly installed.
21. Install the wheel.
22. Lower the vehicle so the full weight of the vehicle is on the floor.
23. Torque the front trailing arm bolt to 94–108 ft. lbs. (130–150 Nm).
24. Torque the lateral rod nut to 58–72 ft. lbs. (80–100 Nm).

STEERING

Steering Wheel

NOTE: If equipped with an air bag, be sure to disarm it before entering the vehicle.

REMOVAL & INSTALLATION

Except 3000-GT

1. Disconnect the negative battery cable. Pry off the steering wheel center foam pad or remove the screws from the back, depending on vehicle. Disconnect the electrical connector for the horn.
2. Remove the steering wheel retaining nut after marking the wheel and shaft position.
3. Using a steering wheel puller, remove the steering wheel.
4. Installation is the reverse of the removal procedure. Be sure the front wheels are in a straight ahead position.

3000-GT

1. Disconnect the negative battery cable.
2. Remove the air bag module mounting nut from behind the steering wheel.
3. Carefully disconnect the module connector.
4. Store the air bag module in a clean, dry place with the pad cover facing up.
5. Remove the steering wheel retaining nut. Matchmark the steering wheel to the shaft. Use a steering wheel puller to remove the wheel. Do not use a hammer or the collapsible mechanism in the column could be damaged.

To install:

6. Center the clock spring by aligning the **NEUTRAL** mark on the clock spring with the mating mark on the casing.
7. Line up the matchmarks and install the steering wheel. Torque the retaining nut to 29 ft. lbs. (40 Nm).

Steering Column

NOTE: If equipped with an air bag, be sure to disarm it before entering the vehicle.

REMOVAL & INSTALLATION

1. Disconnect the negative battery cable.
2. Remove the instrument panel undercover or knee protector.
3. Remove the trim clip, foot shower duct and lap shower duct.
4. Remove the steering wheel and column upper and lower cover. Disconnect the key interlock cable if equipped.
5. Disconnect all connector to column-mounted items.
6. Remove the band from the steering joint cover and remove the joint assembly and gear box pinch bolt.
7. Remove the screws that attach the rubber seal to the firewall.
8. Remove the lower and upper column mounting bolts.
9. Remove the steering column assembly.

To install:

10. Install the column so the splines are inserted around the rack input shaft. Install the pinch bolt.
11. Install the mounting bolts.
12. Install the rubber seal screws.
13. Connect the connectors and interlock cable.
14. Install the column covers.
15. Install the remaining interior pieces.
16. Connect the negative battery cable and check all column-mounted switches for proper operation.

Manual Steering Rack

ADJUSTMENT

1. Remove the rack and pinion assembly.
2. Mount rack in a vise and with a small torque wrench and an adapter to connect to the input shaft, position the rack at its center. Tighten the rack support cover, the bottom plug, to 11 ft. lbs. In the neutral position, rotate the shaft clockwise 1 turn in 4–6 seconds. Return the rack support cover 30–60 degrees and adjust the torque from 0–90 degrees to 5–11 inch lbs. and from 90–650 degrees to 2–9 inch lbs.
3. When adjusting, set to the high side of the specification. Make sure

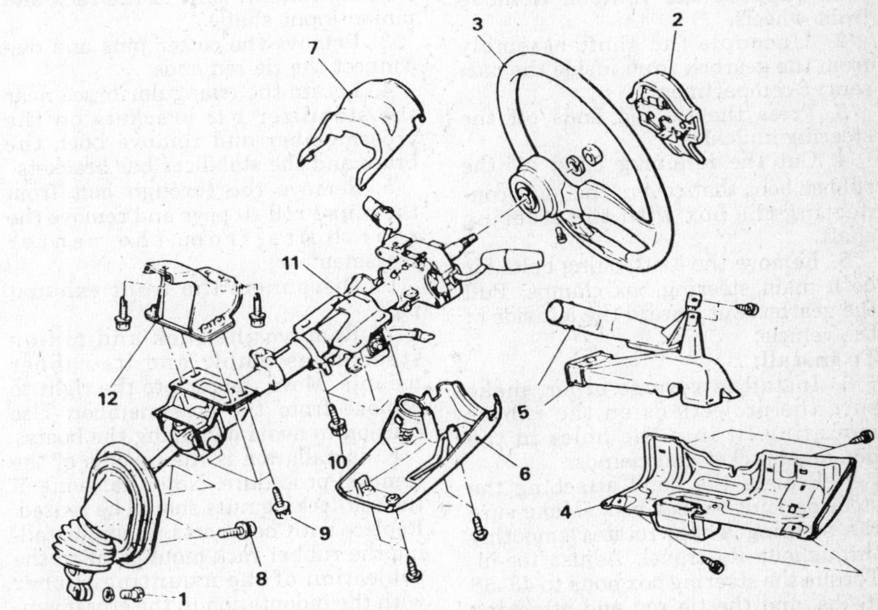

1. Joint assembly and gear box connecting bolt
2. Horn pad
3. Steering wheel
4. Instrument under cover
5. Foot shower duct and lap shower duct
6. Column cover lower
7. Column cover upper
8. Cover attaching bolt
9. Lower bracket installation bolt
10. Tilt bracket installation bolt
11. Steering column assembly
12. Column support

Steering wheel and column servicing—Eclipse

there is no ratcheting or catching when operating the rack. If the rack cannot be adjusted to specification, check the rack support cover components or replace. After adjusting, lock the rack support cover with the locking nut.

REMOVAL & INSTALLATION

Cordia and Tredia

1. Raise and support the vehicle safely. Remove the front wheels.
2. Remove the bolt connecting the steering shaft universal joint with the steering gear. Before removing the bolt, mark its location and be sure the wheels are pointed straight.
3. Remove the tie rod ends from the hub knuckles. Disconnect bolts located near the inner tie rods on the crossmember. Remove right side sub-member from the No. 2 crossmember. Remove the gearbox from the No. 2 crossmember. Pull the gear box out from the right side of the vehicle.
4. Installation is the reverse of removal. Observe the following torques:
   Gear box-to-No. 2 crossmember–43–58 ft. lbs.
   Tie rod-to-rack–58–72 ft. lbs.
   Tie rod end locknut–36–40 ft. lbs.
   Tie rod-to-knuckle–17–25 ft. lbs.

Mirage

1. Support the vehicle. Remove front wheels.
2. Uncouple the shaft assembly from the gearbox from inside the passenger compartment.
3. Press the tie rod ends off the steering knuckles.
4. Cut the retaining band off the rubber boot that covers the joint connecting the box with the steering shaft.
5. Remove the 4 attaching bolts for both main steering box clamps. Pull the gearbox out toward the left side of the vehicle.

To install:

6. Install in reverse order, make sure the projections on the rubber mounting fit into the holes in the housing bracket and clamps.
7. Replace the band attaching the steering joint rubber boot. Make sure the steering wheel rotates smoothly throughout its travel. Adjust toe-in. Torque the steering box bolts to 43–58 ft. lbs. and the tie rod end attaching nuts to 11–25 ft. lbs.

Precis

1. Raise and support the vehicle safely. Remove the bolt which secures the universal joint in the steering shaft to the gearbox; inside the vehicle where the steering linkage passes through the toe board.
2. Remove the cotter pin from the tie rod end ball stud and loosen the nut. Press the ball stud out of the steering knuckle with a vise like tool such as MB991113 or equivalent; then remove the nut. Do the same on the other side.
3. Cut the band off the steering joint rubber boot.
4. Remove both attaching bolts from the gearbox housing clamp on either side and pull the gearbox out the left side of the vehicle. Work slowly to keep the unit from being damaged.

To install:

5. Install the unit in reverse order.
6. There are rubber tabs on the inside and outside of the sleeve. The larger tab must go on the inside.
7. Always use a new band for the steering joint rubber boot. Adjust toe-in.
8. Use the following torques: bracket attaching bolts, 43–58 ft. lbs.; ball stud nut, 17 ft. lbs., then turn farther to align castellations with the cotter pin hole and install a new cotter pin.
9. Turn the steering wheel back and forth to test steering and support the vehicle safely.

Eclipse

1. Disconnect the negative battery cable. Raise and safely support vehicle.
2. Remove the bolt holding lower steering column joint to the rack and pinion input shaft.
3. Remove the cotter pins and disconnect the tie rod ends.
4. Locate the triangular brace near the stabilizer bar brackets on the crossmember and remove both the brace and the stabilizer bar brackets.
5. Remove the through bolt from the round roll stopper and remove the rear bolts from the center crossmember.
6. Disconnect the front exhaust pipe.
7. Remove the rack and pinion steering assembly and its rubber mounts. Move the rack to the right to remove from the crossmember. Use caution to avoid damaging the boots.
8. Installation is the reverse of the removal procedure. Note that none of the self-locking nuts should be reused. Replace with new parts. When installing the rubber rack mounts, align the projection of the mounting rubber with the indentation in the crossmember. Check the steering wheel position with the front wheels straight ahead. Align the front end, if necessary.

Power Steering Rack

ADJUSTMENT

1. Disconnect the negative battery cable.
2. Raise and support the vehicle safely.
3. Remove the steering rack assembly.
4. Secure the steering rack assembly in a vise. Do not clamp the vise jaws on the steering housing tubes. Clamp the vise jaws only on the housing cast metal.
5. Remove the steering gear housing end plug from the steering gear shaft bore using tool 6103 or equivalent.
6. Remove the preload adjustment cap locknut from the steering gear housing bore using tool 6097 or equivalent.
7. Loosen the preload adjustment cap. Retorque the preload adjustment cap to 45–50 inch lbs. (5–6 Nm), then back off the plug by turning it 45–50 degrees counterclockwise.
8. Secure the preload adjustment cap with a new locknut using tool 6097 or equivalent. Do not allow the adjustment cap to rotate when tightening the locknut.
9. Install the end plug using tool 6103 or equivalent. Complete installation by reversing the removal procedure.

REMOVAL & INSTALLATION

Cordia and Tredia

1. Raise and support the vehicle safely. Remove the bolt attaching the steering shaft universal joint to the gearbox.
2. Remove the cotter pin from the tie rod end ball stud and then loosen the nut. Press the ball stud out of the steering knuckle with a vise like tool such as MB991113 or equivalent. Remove the nut and pull the stud out of the knuckle.
3. Place a drain pan under the gearbox and then disconnect the pressure and return hose connectors with a flare nut wrench and allow the fluid to drain.
4. Disconnect the hose from the bottom of the fuel filter and plug it.
5. Remove the fuel line clips to permit the fuel line to move.
6. Remove the brace from the rear engine roll stop.
7. Remove the crossmember support bracket from the No. 2 crossmember, located on the right side of the vehicle.
8. Unbolt and remove both bolts in each gearbox clamp, working from the engine compartment side.
9. Pull the gearbox out the right side of the vehicle, working carefully to keep the unit from being damaged.

To install:

10. Install the gearbox from the right side and attach the clamps. Install the

crossmember support bracket and the brace from the rear engine roll stop.

11. Install the clips for the fuel lines and the hose to the bottom of the fuel filter.

12. Connect the pressure and return hoses to the gearbox and fill the system with fluid.

13. Assemble the steering knuckle and connect the steering shaft to the universal joint.

Starion

1. Raise and support the vehicle safely. Remove the clamp bolt which connects the steering box input shaft to the steering shaft.

2. Place a drain pan under, disconnect the pressure and return hoses at the gearbox.

3. Press the pitman arm off the gearbox with the special tool.

4. Remove the 4 attaching nuts and remove the steering box.

5. Install in reverse order, torquing the steering box mounting nuts/bolts to 25–29 ft. lbs. and the pitman arm shaft retaining nut to 25–33 ft. lbs. Fill the power steering pump with fluid, as required.

Mirage

1. Raise and support the vehicle safely. Remove the bolt which secures the universal joint in the steering shaft to the gearbox. It's just inside the vehicle where the steering linkage passes through the toė board.

2. Remove the cotter pin from the tie rod end ball stud and loosen the nut. Press the ball stud out of the steering knuckle with a vice like tool such as MB991113 or equivalent; then remove the nut. Do the same on the other side.

3. Cut the tab off the steering joint rubber boot. Place a drain pan under the steering box. Then, disconnect the pressure and return hoses at the gearbox.

4. Remove the stabilizer bar.

5. Remove the rear roll stopper-to-centermember bolt and move the rear roll stopper forward.

6. Remove both bolts in the clip on either side of the gearbox and remove the unit carefully out the left of the vehicle. Avoid damaging the rubber boots.

To install:

7. Install the gearbox from the left side of the vehicle and install the retaining clips. Install the roll stopper, centermember bolt and stabilizer bar.

8. Connect the pressure and return hoses and fill the system with fluid.

9. Assemble the steering knuckle and connect the steering shaft to the universal joint. Tighten the universal joint nut to 11–17 ft. lbs. (15–23 Nm). Install the wheels and lower the vehicle.

Galant and Sigma

1. Raise and support the vehicle safely. If equipped with electronically controlled suspension, remove the stabilizer bar.

2. Remove the cotter pin from the tie rod and ball stud and loosen the nut. Press the ball stud out of the steering knuckle with a vice like tool such as MB991113 or equivalent; then remove the nut. Do the same on the other side.

3. Drain the fluid from the system. Then, disconnect the pressure and return hoses at the gearbox.

4. Remove the bolt attaching the steering shaft universal joint to the gearbox. Disconnect the connector for the solenoid valve.

5. Remove the front bolt from the center crossmember. Remove the exhaust pipe hanger from the crossmember. Remove the front roll stopper bolt.

6. Disconnect the oxygen sensor connection. Disconnect the exhaust pipe at the front and lower it aside. Remove the stabilizer bar and bracket. Press the rear of the center crossmember downward.

7. Move the rack all the way to the right. Then, remove the bolts from the brackets. Tilt the gearbox downward and remove it toward the left. Avoid damaging the rubber boots.

To install:

8. Install the rack from the left side. Install the stabilizer bar and bracket.

9. Install the exhaust system and oxygen sensor. Install the steering shaft universal joint and tighten the bolt to 11–14 ft. lbs. (15–20 Nm). Connect the solenoid valve.

10. Assemble the steering knuckle, install the wheel and lower the vehicle.

Eclipse

1. Disconnect the negative battery cable.

2. Raise the vehicle and support it safely.

3. Drain the power steering fluid.

4. Remove the bolt holding lower steering column joint to the rack and pinion input shaft.

5. Disconnect the return and high pressure lines from the rack assembly.

3. Remove the cotter pins and disconnect the tie rod ends.

4. Locate the triangular brace near the stabilizer bar brackets on the crossmember and remove both the brace and the stabilizer bar brackets.

5. Remove the through bolt from the round roll stopper and remove the rear bolts from the center crossmember.

6. Disconnect the front exhaust pipe.

7. Remove the rack and pinion steering assembly and its rubber mounts. Move the rack to the right to remove from the crossmember. Use caution to avoid damaging the boots.

To install:

8. Installation is the reverse of the removal procedure. Note that none of the self-locking nuts should be reused. Replace with new parts. When installing the rubber rack mounts, align the projection of the mounting rubber with the indentation in the crossmember.

9. Refill the system and bleed out the air. Check the steering wheel position with the front wheels straight ahead. Align the front end, if necessary.

3000-GT

1. Disconnect the negative battery cable.

2. Disconnect the front exhaust pipe.

3. If equipped with AWD, remove the transfer case assembly.

4. Remove the bolt holding lower steering column joint to the rack and pinion input shaft.

5. Remove the cotter pins and disconnect the tie rod ends.

6. Remove the left and right frame members.

7. Remove the stabilizer bar bracket.

8. If equipped with 4 wheel steering, disconnect the lines going to the rear pump.

9. Remove the rack and pinion steering assembly and its rubber mounts. Move the rack to the right to remove from the crossmember. Use caution to avoid damaging the boots.

To install:

10. Install the rack and install the mounting bolts. When installing the rubber rack mounts, align the projection of the mounting rubber with the indentation in the crossmember. Install the pinch bolt.

11. Connect the lines going to the 4 wheel steering rear pump and to the rack itself.

12. Install the frame members and torque the bolts to 50 ft. lbs. (68 Nm).

13. Connect the tie rods and instal new cotter pins.

14. Install the transfer case and front exhaust pipe.

15. Refill the reservoir and bleed the system.

16. Peform a front end alignment.

Power Steering Pump

REMOVAL & INSTALLATION

Front

1. Disconnect the battery negative cable.
2. Remove the pressure switch connector from the side of the pump.
3. If the alternator is located under the oil pump, cover it with a shop towel to protect it from oil.
4. Disconnect the return fluid line. Remove the reservoir cap and allow the return line to drain the fluid from the reservoir. If the fluid is contaminated, disconnect the ignition high tension cable and crank the engine several times to drain the fluid from the gearbox.
5. Disconnect the pressure line.
6. Remove the pump drive belt and unbolt the pump from its bracket.

To install:

7. Install the pump, wrap the belt around the pulley and tighten the bolts.
8. Replace the O-rings and connect the pressure line. Connect the pressure line so the notch in the fitting aligns and contacts the pump's guide bracket.
9. Connect the return line.
10. Connect the pressure switch connector.
11. Adjust the belt tension and tighten the adjusting bolts.
12. Refill the reservoir and bleed the system.

Rear

1. Disconnect the negative battery cable. Raise the vehicle and support safely.
2. Drain the power steering fluid.
3. Remove the main muffler assembly.
4. Remove the rear shock absorber lower mounting bolts.
5. Remove the 2 small crossmember brackets.
6. Using the proper equipment, support the weight of the rear differential. Remove the large self-locking crossmember mounting nuts on the differential side.
7. Disconnect the pressure and suction hoses from the fittings on the pump.
8. Remove the pump retaining bolt and remove the pump from the rear differential assembly. Do not attempt to disassemble the pump; it is not serviceable.

To install:

9. Replace the O-ring and install the pump assembly to the differential. Make sure the housing is fully seated and the gear is fully engaged. Install the retaining bolt.
10. Replace the O-ring and connect the fluid lines to the pump.
11. Install the large self-locking crossmember mounting nuts on the differential side. Torque to 80–94 ft. lbs. (110–130 Nm). Remove the support equipment.
12. Install the 2 small crossmember brackets.
13. Install the shock mounting bolts.
14. Install the muffler assembly.
15. Refill the reservoir and bleed the system.

CAUTION

Extreme caution should be taken when testing the rear steering pump. Ensure that the vehicle is supported safely and that all components are torqued to specification prior be testing.

16. To check and see if the system is functioning:
    a. Raise the vehicle safely so all 4 wheels turn freely.
    b. Run the vehicle at 50 mph.
    c. Turn the steering wheel quickly to the left and right and make sure the rear wheels steer in the same direction as the front wheels.

BELT ADJUSTMENT

1. Press the belt in about the center between the power steering pump pulley and the pulley it shares, usually the water pump pulley. With reasonable pressure applied, about 22 lbs., the belt should deflect about ¼–⅜ in.
2. Adjustment can be made by loosening the 3 bolts that hold the pump. Place a suitable bar or lever between the body of the pump and gently pry to get the desired tension.
3. Retighten the 3 bolts and check again.

SYSTEM BLEEDING

Front

1. Raise the vehicle and support safely.
2. Manually turn the pump pulley a few times.
3. Turn the steering wheel all the way to the left and to the right 5 or 6 times.
4. Disconnect the ignition high tension cable and, while operating the starter motor intermittently, turn the steering wheel all the way to the left and right 5–6 times for 15–20 seconds. During bleeding, make sure the fluid in the reservoir never falls below the lower position of the filter. If bleeding is attempted with the engine running, the air will be absorbed in the fluid. Bleed only while cranking.
5. Connect ignition high tension cable, start engine and allow to idle.

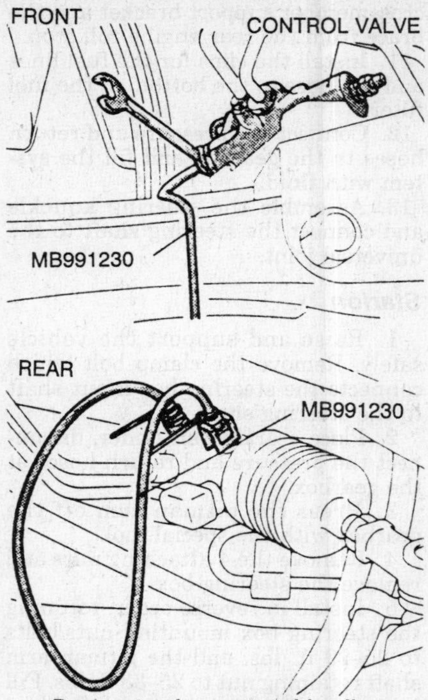

Power steering system bleeding

6. Turn the steering wheel left and right until there are no air bubbles in the reservoir. Confirm that the fluid is not milky and the level is up to the specified position on the gauge. Confirm that there is is very little change in the fluid level when the steering wheel is turned. If the fluid level changes more than 0.2 in. the air has not been completely bled. Repeat the process.

Rear

1. Bleed the front steering system.
2. Start the engine and let it idle.
3. Loosen the bleeder screw on the left side of the control valve and install special tool MB991230 to the bleeder.
4. Turn the steering wheel all the way to the left, then immediately turn it half way back. Confirm that air has discharged with the fluid.
5. Repeat Step 4 two or three times as required, to remove all air from the rear system. Stop the engine.
6. Loosen the power cylinder (rear steering gear) bleeder screw about ⅛ turn and install the same special tool with the rotation prevention metal fixtures to prevent the bleeder from opening more.
7. Start the engine and run to 50 mph to circulate the fluid.
8. Maintain a speed of 20 mph and turn the steering wheel back and forth. Air should be discharged through the tube of the special tool and into the oil reservoir.
9. Repeat until all air is removed from the power cylinder.

Tie Rod Ends

REMOVAL & INSTALLATION

1. Disconnect the battery negative cable.
2. Raise the vehicle and support safely.
3. Wire brush the threads on the tie rod shaft and lubricate with penetrating oil. Loosen the locknut.
4. Remove the cotter pin and nut and press the tie rod end from the steering knuckle.
5. Hold the tie rod shaft with locking pliers and turn the tie rod end off, counting the number of turns for installation.
6. The installation is the reverse of the removal procedure. Install the tie rod end the same number of turns that it took to remove the old one.
7. Perform a front end alignment.

BRAKES

For all brake system repair and service procedures not detailed below, please refer to "Brakes" in the Unit Repair section.

Master Cylinder

REMOVAL & INSTALLATION

1. Disconnect the negative battery cable. On the Starion, Mirage with turbocharger and Precis, the brake fluid reservoir is separate from the master cylinder. Disconnect the hoses at the master cylinder and plug them or drain the fluid into a container.
2. Disconnect the electrical connector for the fluid level sensor. Remove the proportioning valve bracket, if equipped.
3. Disconnect all the brake tubes. Remove the nuts and lock washers attaching the master cylinder to the booster. Remove the master cylinder from the vehicle.
4. Install in the reverse order of removal, torquing the attaching nuts to 6–9 ft. lbs. Refill the reservoirs with approved, new fluid and bleed the system thoroughly.

Proportioning Valve

REMOVAL & INSTALLATION

1. Disconnect battery negative cable.
2. On most vehicles the proportioning valve is located on the body under the master cylinder. Disconnect and tag for proper reassembly the brake lines, disconnect the electrical connector and unbolt the valve from the body.

To install:

3. Reverse the removal procedure to install. Use care not to cross thread any connections. Tighten flared brake lines to 9–12 ft. lbs. Add fluid and bleed brakes.

Power Brake Booster

REMOVAL & INSTALLATION

1. Disconnect the negative battery cable. Disconnect the vacuum supply line from the brake booster.
2. Remove the master cylinder. It may be possible to position the master cylinder aside rather then disconnect the fluid lines.
3. Disconnect the pushrod from the brake pedal.
4. Remove the mounting bolts from the firewall. Remove the power brake booster.
5. Installation is the reverse of the removal procedure. Tighten mounting bolts to 10 ft. lbs. (14 Nm). Bleed the brake system.

Brake Caliper

REMOVAL & INSTALLATION

Front

1. Disconnect the battery negative cable. Raise and safely support vehicle, remove appropriate wheel assembly.
2. Drain a portion of the brake fluid from the master cylinder reservoir.
3. Disconnect the front brake hose. Hold the nut on the brake hose side, loosen the flared brake line nut.
4. Remove the caliper lock pins and remove the caliper.

To install:

5. Reverse the removal procedure to install. Tighten the flared brake line nut to 11 ft. lbs (15 Nm). Tighten the caliper mounting bolts to 65 ft. lbs. (90 Nm). Make sure the brake hose is not twisted after installation. Refill brake fluid as required and bleed brakes.

Rear

1. Disconnect battery negative cable. Raise and safely support vehicle, remove appropriate wheel assembly.
2. Drain a portion of the brake fluid from the master cylinder reservoir.
3. Disconnect the parking brake cable and the rear brake hose. Hold the nut on the brake hose side, loosen the flared brake line nut.
4. Remove the rear caliper assembly.

To install:

5. Reverse the removal procedure to install. Tighten the flared brake line nut to 11 ft. lbs (15 Nm). Tighten the caliper mounting bolts to 36–43 ft. lbs. (50–60 Nm). Make sure the brake hose is not twisted after installation. Refill brake fluid as required and bleed brakes. Adjust parking brake if required.

Disc Brake Pads

REMOVAL & INSTALLATION

1. Disconnect battery negative cable.
2. Raise the vehicle and support safely.
3. Remove appropriate wheel assembly.
4. On the front of AWD 3000-GT, remove the pad retaining pins and pull the pads out of the caliper body.
5. On others, remove the caliper from its adaptor but do not allow the caliper to hang by the brake line. On some vehicles, the caliper can be flipped up by leaving the upper pin in place and using it as a pivot point. Take note of the clips, pins, anti-squeal shims and other parts for reference at assembly.
6. On vehicles with rear disc brakes, it may help to loosen the parking brake cable from inside the car and disconnect the parking brake end from the rear caliper.

To install:

7. Use a large C-clamp to compress the piston(s) back into the caliper bore. On rear disc brakes with the parking brake mechanism incorporated into the caliper, a special tool is needed to turn the piston back into the bore.
8. Install the pads and all other small parts. Note that rear disc pads on calipers with the parking brake mechanism incorporated into the caliper should have a projection on the back side of the shoe that fits into the rear caliper piston.
9. Install the caliper. Make sure the brake hose is not twisted after installation. Connect the parking brake cable if disconnected.

Brake Rotor

REMOVAL & INSTALLATION

Starion, Mirage and Precis

1. Loosen the large driveshaft nut while the vehicle is still on the ground with the brakes applied. Then raise and safely support vehicle. Remove appropriate wheel assembly.

2. Remove the axle end nut and lock washer.
3. Remove the caliper from its bracket. Do not allow the caliper to hang by the brake line. Remove the brake pads.
4. On front rotors:
a. Remove the ball joint from the lower control arm.
b. Use and puller to push the halfshaft through the rotor/hub assembly.
c. Remove the bolts and separate the rotor from the hub.

To install:

5. Assemble the rotor and hub. Tighten the nuts to 40 ft. lbs. (54 Nm).
6. Install the assembly to the vehicle.
7. Install the washer so the chamfered edge faces outward. Install the nut and tighten temporarily.
8. Install the ball joint.
9. Install the brake components.
10. Install the wheel and lower the vehicle to the floor. Tighten the axle nut with the brakes applied. Tighten the nut to a maximum torque of 188 ft. lbs. (260 Nm) maximum. Install the cotter pin and bend it securely.

3000-GT, Galant, Sigma and Eclipse

1. Raise and safely support vehicle, remove appropriate wheel assembly.
2. Disconnect the parking brake cable, if servicing the rear wheels. Remove the caliper from its bracket. Do not disconnect the brake line. Do not allow the caliper to hang by the brake line.
3. The rotor is held to the hub by 2 small bolts. Remove the bolts and pull off the rotor.
4. Installation is the reverse of the removal process.

Brake Drums

REMOVAL & INSTALLATION

1. Raise and safely support vehicle, remove appropriate wheel assembly.
2. Remove the grease cap and remove the center wheel bearing nut, if equipped.
3. Remove the outer bearing and brake drum assembly.

To install:

4. Installation is the reverse of the removal process. Check the drum inside diameter. The wear limit is 7.20 in. (182mm). Inspect the rear hub nut. Replace if unusable.

Brake Shoes

REMOVAL & INSTALLATION

1. Raise the vehicle and support safely. Remove appropriate wheel assembly.
2. Remove the brake drum. Remove the shoe to shoe spring.
3. Take note of the springs and clips for proper reassembly. Remove the shoe hold-down clips and remove the shoes.

To install:

4. Thoroughly clean and dry the backing plate. To prepare the backing plate, lubricate the bosses, anchor pin and parking brake actuating lever pivot surface lightly with lithium-based grease.
5. Remove, clean and dry all parts still on the old shoes. Lubricate the star wheel shaft threads with anti-sieze lubricant and transfer all parts to their proper locations on the new shoes.
6. Install shoes to the vehicle.
7. Connect the parking brake cable.
8. Adjust the star wheel.
9. To determine if the self-locking nut is reusable:
a. Screw in the self-locking nut until about $\frac{1}{10}$ in. of the spindle is showing.
b. Measure the torque required to turn the self-locking nut counterclockwise.
c. The lowest allowable torque is 48 inch lbs. (5.5 Nm). If the measured torque is less than the specification, replace the nut.
10. Remove any grease from the linings and install the drum to the spindle.
11. Lubricate and install the outer wheel bearing.
12. Torque the self-locking nut to 108–145 ft. lbs. (150–200 Nm).
13. Install the grease cap.

Wheel Cylinder

REMOVAL & INSTALLATION

1. Raise and support the rear of the vehicle. Remove the wheel and brake drum. Remove the brake shoes.
2. Place a container under the brake backing plate to catch the brake fluid that will run out of the wheel cylinder.
3. Disconnect the brake line and remove the cylinder mounting bolts. Remove the cylinder from the backing plate.
4. Installation is the reverse of the removal procedure. Tighten bolts to 9–13 ft. lbs. (21–18 Nm). Bleed the brake system.

Parking Brake Cable

ADJUSTMENT

Rear Drum Brakes

1. Make sure the parking brake cable is free and is not frozen or sticking. With the engine running, forcefully depress the brake pedal 5–6 times. Check the parking brake stroke. It should be 5–7 notches. If not, adjust using the following procedure.
2. Remove the floor console by prying out the coin holder, box tray and remote mirror switch or if not so equipped, the cover. Remove the small cover around the seat belt from the console side. The console is in 2 pieces. Remove the screws from the center section and remove the rear part of the console.
3. Loosen the locknut then loosen the adjusting to the end of the cable and free the parking brake cable. Repeat the procedure to pull the parking brake lever back with a force of about 44 lbs. until the lever stoke ceases to change. If the lever stroke does not change, the automatic adjustment mechanism is functioning normally and the clearance between the shoe and drum is correct.
4. Rotate the adjusting nut to adjust the parking brake stroke to the 5–7 notch setting. After making the adjustment check there is no looseness between the adjusting nut and the parking brake lever, then tighten the locknut.

NOTE: Do not adjust the parking brake too tight. If the number of notches is less than specification, the cable has been pulled too much and the automatic adjuster will fail. Use the 5–7 notch specification.

5. After adjusting the lever stroke, raise the vehicle. With the parking brake lever in the released position, turn the rear wheel to confirm that the rear brakes are not dragging.

Rear Disc Brakes

1. Make sure the parking brake cable is free and is not frozen or sticking. With the engine running, forcefully depress the brake pedal 5–6 times. Check the parking brake stroke. It should be 5–7 notches. If not, adjust using the following procedure.
2. Remove the floor console by prying out the coin holder, box tray, and remote mirror switch, or if not so equipped, the cover. Remove the small cover around the seat belt from the console side. The console is in 2 pieces. Remove the screws from the center section and remove the rear part of the console.
3. Loosen the locknut then loosen the adjusting to the end of the cable and free the parking brake cable. Repeat the procedure to pull the parking brake lever back with a force of about 44 lbs. until the lever stoke ceases to change. If the lever stroke does not

change, the automatic adjustment mechanism is functioning normally and the clearance between the shoe and drum is correct.

4. Check to be sure the distance between the stopper and the parking brake lever at the caliper side is 0.078 in. (2mm) or less. If the clearance between the parking lever, on the caliper, and the stopper exceeds 0.078 in., the probable causes are brake cable sticking, improper cable installation or a malfunction of the automatic adjuster in the caliper which will require disassembling the caliper.

5. Turn the adjusting nut to get the brake lever stroke to specification, 5–7 notches.

NOTE: Do not adjust the parking brake too tight. If the number of notches is less than specification, the cable has been pulled too much and the automatic adjuster will fail. Use the 5–7 notch specification.

6. After making the adjustment, check to be sure there is no play between the adjusting nut and the parking brake lever, then tighten the locknut. After adjusting the lever stroke, raise the vehicle. With the parking brake lever in the released position, turn the rear wheel to confirm that the rear brakes are not dragging.

REMOVAL & INSTALLATION

Front Wheel Drive

1. Disconnect the negative battery cable.

NOTE: If equipped with an air bag, be sure to disarm it before entering the vehicle.

2. Remove the floor console by prying out the coin holder, box tray and remote mirror switch, if equipped, or the cover. Remove the small cover around the seat belt from the console side. Remove the screws from the center section and remove the rear part of the console.
3. Remove the rear seat cushion.
4. Remove the center cable clamp and grommet.
5. Raise the vehicle and support safely.
6. At the rear wheel, remove the brake drum or disc and disconnect the cable end from the parking brake strut lever or actuator. If necessary, compress the retaining strips to remove the cable from the backing plate.
7. Unfasten any other frame retainers and remove the cables.

To install:

8. The installation is the reverse of the removal procedure.
9. Adjust the rear brakes and parking brake cables.
10. Connect the negative battery cable and check the rear wheels to confirm that the rear brakes are not dragging.
11. Check that the parking brake holds the vehicle on an incline.

Rear Wheel Drive

1. As required, remove the console and rear seat.
2. Raise and support the rear of the vehicle.
3. Disconnect all clevis pin connecting and the cable ends.
4. Pull the cable through the floor.
5. Install in reverse order.
6. Adjust the cable. Apply sealer to the edge of the grommet at the floor opening. Check the parking brake indicator, the light should come on when the brake is applied 1 notch.

Brake System Bleeding

System Dry

1. If the master cylinder is dry, disconnect the brake tube from the master cylinder.
2. With an assistant, have one person slowly depressing the brake pedal and holding it down. The other person should use a finger to close the outlet port of the master cylinder and then the first person should release the pedal. Repeat these steps 3–4 times. Keep the cylinder full of brake fluid. This operation bleeds the master cylinder.
3. Connect the brake tube to the master cylinder.
4. Start the engine. Using the bleeder screws are the calipers and wheel cylinders, bleed the brakes at the wheels in the following sequence: right rear, left front, left rear, right front.
5. Do not allow the master cylinder reservoir to run out of fluid.

Normal Service Bleeding

1. Press the brake pedal several times until resistance is felt.
2. With the brake pedal depressed, loosen the bleeder screw $^1/_3$–½ turn and then tighten it before the fluid pressure is gone.
3. Release the brake pedal. Repeat this procedure until there are no more air bubbles in the brake fluid.

Anti-Lock System Brake Service

PRECAUTIONS

- Certain components within the ABS system are not intended to be serviced or repaired individually. Only those components with REMOVAL & installation procedures should be serviced.
- Do not use rubber hoses or other parts not specifically specified for the ABS system. When using repair kits, replace all parts included in the kit. Partial or incorrect repair may lead to functional problems and require the replacement of components.
- Lubricate rubber parts with clean, fresh brake fluid to ease assembly. Do not use lubricated shop air to clean parts; damage to rubber components may result.
- Use only DOT 3 brake fluid from an unopened container.
- If any hydraulic component or line is removed or replaced, it may be necessary to bleed the entire system.
- A clean repair area is essential. Always clean the reservoir and cap thoroughly before removing the cap. The slightest amount of dirt in the fluid may plug an orifice and impair the system function. Perform repairs after components have been thoroughly cleaned; use only denatured alcohol to clean components. Do not allow ABS components to come into contact with any substance containing mineral oil; this includes used shop rags.
- The Anti-Lock control unit is a microprocessor similar to other computer units in the vehicle. Ensure that the ignition switch is **OFF** before removing or installing controller harnesses. Avoid static electricity discharge at or near the controller.
- If any arc welding is to be done on the vehicle, the ALCU connectors should be disconnected before welding operations begin.

FILLING THE SYSTEM

The brake fluid reservoir is part of the normal brake system and is filled or checked in the usual manner. Always clean the reservoir cap and surrounding area thoroughly before removing the cap. Fill the reservoir only to the **FULL** or **MAX** mark; do not overfill. Use only fresh DOT 3 brake fluid from unopened containers. Do not use any fluid containing a petroleum base. Do not use any fluid which has been exposed to water or moisture. Failure to use the correct fluid will affect system function and component life.

BLEEDING THE SYSTEM

Galant and Sigma

The complete brake system must be bled any time a line, hose or component is loosened or removed. Any air trapped within the lines can affect pedal feel and system function. Bleeding the system is performed in the usual manner with an assistant in the car to pump the brake pedal, but these systems require bleeding of several ports in addition to the usual wheel locations.. Make certain the fluid level in the reservoir is maintained at or near correct levels during bleeding operations.

Use of a box-end or brake bleeder wrench is recommended to avoid damage to bleeder port fittings; a socket and ratchet will be needed to bleed the delay valve on Sigma V6.

With the ignition **OFF**, depress the brake pedal several times until pedal feel changes to a noticeably stiffer resistance. Slowly pump the pedal a few more times; with the pedal depressed, loosen the bleeder screw ⅓–½ turn. Tighten the bleeder screw before the fluid pressure is gone. Release the pedal, pump again slowly, hold the pedal depressed and repeat the bleeding process until no air bubbles are seen in the fluid.

NOTE: The brake pedal will develop an unusually heavy or hard feel during bleeding of the rear brake on FWD vehicles. This results from fluid constriction within the delay valve and is not a sign of malfunction.

Correct bleeding order for Sigma V6 is: left rear wheel, left rear delay valve, right front wheel, outboard bleeder port on hydraulic unit, right rear wheel, right rear delay valve, left front wheel and inboard bleeder port on hydraulic unit.

Bleeding order for Galant FWD vehicles is: right rear wheel, right rear delay valve, left front wheel, inboard bleeder port on hydraulic unit, left rear wheel, left rear delay valve, right front wheel and outboard bleeder port on the hydraulic unit.

Bleed Galant AWD brakes in this order: right rear wheel, left front wheel, inboard bleeder port on the hydraulic unit, left rear wheel, right front wheel and outboard bleeder port on the hydraulic unit.

Eclipse and 3000 GT

LINES AND CALIPERS

The brake system must be bled any time a line, hose or component is loosened or removed. Any air trapped within the lines can affect pedal feel and system function. Bleeding the system is performed in the usual manner with an assistant in the car to pump the brake pedal. Make certain the fluid level in the reservoir is maintained at or near correct levels during bleeding operations.

With the ignition **OFF**, depress the brake pedal several times until pedal feel changes to a noticeably stiffer resistance. Slowly pump the pedal a few more times; with the pedal depressed, loosen the bleeder screw ⅓–½ turn. Tighten the bleeder screw before the fluid pressure is gone. Release the pedal, pump again slowly, hold the pedal depressed and repeat the bleeding process until no air bubbles are seen in the fluid.

If bleeding is necessary at all wheels, begin at the right rear, then the left front, left rear and right front wheels.

MASTER CYLINDER

If the master cylinder has been emptied of fluid, it must be bled separately from the rest of the system. Since the cylinder has no check valve, air can become trapped within it. To bleed the brake master cylinder after it has been drained:

1. Disconnect the 2 brake lines from the master cylinder. Plug the lines immediately. The brake fluid reservoir must be in place and connected to the master cylinder. Check the fluid level before beginning.
2. An assistant should slowly depress and hold the brake pedal.
3. With the pedal held down, use fingers to plug each outlet port on the master cylinder; release the brake pedal.
4. Repeat Steps 2 and 3 three or four times. The air will be bled from the cylinder.
5. Connect the brake lines to the master cylinder and tighten the fittings to 10 ft. lbs. (13.5 Nm).
6. Start the engine, allowing the system to pressurize and self-check. Shut the ignition **OFF** and bleed the brake lines at the wheels.

Hydraulic Unit

REMOVAL & INSTALLATION

Eclipse

1. Use a syringe or similar device to remove as much fluid as possible from the reservoir. Some fluid will be spilled from lines during removal of the hydraulic unit; protect adjacent painted surfaces.
2. On turbocharged engines, remove the center intercooler duct. Loosen the clamps and remove the bolts holding the duct to the air cleaner.
3. Disconnect the brake lines from the hydraulic unit. Correct reassembly is critical. Label or identify the lines before removal. Plug each line immediately after removal.
4. Remove the cover from the relay box. Disconnect the electrical harness to the hydraulic unit.
5. Disconnect the hydraulic unit ground strap from the chassis.
6. Remove the 3 nuts holding the hydraulic unit. Remove the unit upwards.

NOTE: The hydraulic unit is heavy; use care when removing it. The unit must remain in the upright position at all times and be protected from impact and shock.

7. Set the unit upright supported by blocks on the workbench. The hydraulic unit must not be tilted or turned upside down. No component of the hydraulic unit should be loosened or disassembled.
8. The bracket assemblies and relays may be removed if desired.

To install:

9. Install the relays and brackets if they were removed.
10. Install the hydraulic unit into the vehicle, keeping it upright at all times.
11. Install the retaining nuts and tighten them.
12. Connect the ground strap to the chassis bracket. Connect the hydraulic unit wiring harness.
13. Install the cover on the relay box.
14. Connect each brake line loosely to the correct port and double check the placement. Tighten each line to 10 ft. lbs. (13.5 Nm).
15. Fill the reservoir to the **MAX** line with brake fluid.
16. Bleed the master cylinder, then bleed the brake lines.
17. Install the intercooler air duct if it was removed.

3000 GT

1. Remove the splash shield from beneath the vehicle.
2. Use a syringe or similar device to remove as much fluid as possible from the reservoir. Some fluid will be spilled from lines during removal of the hydraulic unit; protect adjacent painted surfaces.
3. Lift the relay box with the harness attached and position it out of the way.
4. Remove the air intake duct.
5. Disconnect the brake lines from the hydraulic unit. Correct reassembly is critical. Label or identify the lines before removal. Plug each line immediately after removal. It will be necessary to hold the relay box out of the way to allow wrench access.
6. Disconnect the wiring harness connections at the hydraulic unit.
7. Disconnect the hydraulic unit ground strap from the chassis.
8. Remove the 3 bolts holding the hydraulic unit bracket. Remove the unit and the bracket.

NOTE: The hydraulic unit is heavy; use care when removing it. The unit must remain in the up-

right position at all times and be protected from impact and shock.

7. Set the unit upright supported by blocks on the workbench. The hydraulic unit must not be tilted or turned upside down. No component of the hydraulic unit should be loosened or disassembled.
8. Loosen the nut holding the bracket to the hydraulic unit and remove the bracket.
9. Disconnect the external ground wire from the bracket.

To install:

10. Install the bracket if was removed. Connect the ground wire to the bracket.
11. Install the hydraulic unit into the vehicle, keeping it upright at all times.
12. Install the retaining nuts and tighten them.
13. Connect the hydraulic unit wiring harness.
14. Connect each brake line loosely to the correct port and double check the placement. Tighten each line to 11 ft. lbs. (15 Nm).
15. Fill the reservoir to the MAX **line with brake fluid.**
16. Bleed the master cylinder, then bleed the brake lines.
17. Secure the relay box in position and install the air duct.
18. Install the splash shield.

Sigma and Galant

1. Use a syringe or similar device to remove as much fluid as possible from the reservoir. Some fluid will be spilled from lines during removal of the hydraulic unit; protect adjacent painted surfaces.
2. Remove the splash shield from the left front wheel house or fender area.
3. Remove the coolant reserve tank. On Galant, remove the coolant reservoir bracket.
4. Remove the dust shield from below the hydraulic unit.
5. Disconnect the brake hoses and lines from the hydraulic unit. Correct reassembly is critical. Label or identify the lines before removal. Plug each line and each port immediately after removal.
6. Remove the cover from the relay box. Disconnect the electrical harness to the hydraulic unit.
7. Remove the bolts holding the 3 mounting brackets to the vehicle; remove the unit downward and out of the vehicle.

NOTE: The hydraulic unit is heavy; use care when removing it. The unit must remain in the upright position at all times and be protected from impact and shock.

8. Set the unit upright supported by blocks on the workbench. The hydraulic unit must not be tilted or turned upside down. No component of the hydraulic unit should be loosened or disassembled.
9. The brackets and relays may be removed if desired.

To install:

10. Install the brackets and relays if they were removed. Tighten the bracket bolts to 16 ft. lbs. (22 Nm).
11. Install the hydraulic unit into the vehicle, keeping it upright at all times.
12. Install the retaining bolts holding the brackets to the vehicle. Tighten the bolts to 16 ft. lbs. (22 Nm).
13. Connect the hydraulic unit wiring harness.
14. Install the cover on the relay box.
15. Connect each brake line loosely to the correct port and double check the placement. Tighten each line to 10 ft. lbs. (13.5 Nm).
16. Fill the reservoir to the **MAX** line with brake fluid.
17. Bleed the brake system.
18. Install the dust shield and the coolant reserve tank with its bracket.
19. Install the fender splash shield.
20. Check ABS system function by turning the ignition **ON** and observing the dashboard warning lamp. Test drive the vehicle and confirm system operation.

Anti-Lock Control Unit

REMOVAL & INSTALLATION

1. Ensure that the ignition switch is **OFF** throughout the procedure.
2. For Eclipse and 3000 GT, remove the interior right rear quarter trim panel. Depending on the model, removal of the rear seat back and/or cushion may be required. For Sigma V6 and Galant, remove the left side luggage compartment trim panel.
3. Release the lock on the bottom of the connector; disconnect the multi-pin connector from the control unit. On Eclipse, access may be easier if the external ground is disconnected from the bracket.
4. Remove the retaining nuts and remove the control unit from its bracket. The bracket may be removed if desired.

To install:

5. Place the bracket in position if it was removed. Install the controller and tighten the retaining nuts.
6. Connect the ground wire to the bracket if it was removed. Insure a proper, tight connection. The ground must be connected before the multi-pin harness is connected.
7. Connect the multi-pin connector and secure the lock.
8. Install the rear quarter trim panel or the luggage compartment trim.

G-Sensor

REMOVAL & INSTALLATION

NOTE: The G-Sensor is found only on vehicles equipped with All Wheel Drive (AWD).

Eclipse

1. Insure that the ignition switch is **OFF** throughout the procedure.
2. Remove the rear seat cushion.
3. Disconnect the wiring harness to G–sensor.
4. Remove the retaining bolts and remove the sensor.
5. To install, position the sensor, tighten the retaining bolts and connect the harness.
6. Install the rear seat cushion.

3000 GT

1. Remove the rearmost console assembly.
2. Remove the front console assembly.
3. Disconnect the G–sensor wiring harness.
4. Remove the G–sensor from the bracket. Remove the bracket if desired.

To install:

5. Reinstall the bracket if it was removed. Tighten the bolts to 4 ft. lbs. (5 Nm.)
6. Install the G–sensor and connect the wiring harness.
7. Install the front and rear console assemblies.

Galant

1. For the front sensor, remove the console assembly.
2. For the rear sensor, remove the trunk floor mat.
3. Disconnect the G–sensor wiring harness.
4. Remove the cover from the rear sensor. Remove the sensor from the bracket. Remove the bracket if desired.

To install:

5. Reinstall the bracket if it was removed.
6. Install the G–sensor and connect the wiring harness. Tighten the retaining bolts to 8 ft. lbs. (11 Nm)
7. Install the cover on the rear G–sensor.
8. Install the console and/or the trunk floor mat or carpet.

Wheel Speed Sensors

CAUTION

Vehicles equipped with air bag systems will have wiring and system components in the fender or wheel well area. The ABS components must be correctly identified before beginning repairs. Improper work procedures may cause impaired function of the ABS and/or SRS systems

1. Raise and safely support the vehicle.
2. Remove the wheel and tire.
3. Remove the inner fender or splash shield.
4. Beginning at the sensor end, carefully disconnect or release each clip and retainer along the sensor wire. Take careful note of the exact position of each clip; they must be reinstalled in the identical position. Rear wheel sensor harnesses will be held by plastic wire ties; these may be cut away but must be replaced at reassembly.
5. Disconnect the sensor connector at the end of the harness.
6. Remove the 2 bolts holding the speed sensor bracket to the knuckle and remove the assembly from the vehicle.

NOTE: The speed sensor has a pole piece projecting from it. This exposed tip must be protected from impact or scratches. Do not allow the pole piece to contact the toothed wheel during removal or installation.

7. Remove the sensor from the bracket.

To install:

8. Assemble the sensor onto the bracket and tighten the bolt to 10 ft. lbs. (14 Nm). Note that the brackets are different for the left and right front wheels. Each bracket has identifying letters stamped on it.
9. Temporarily install the speed sensor to the knuckle; tighten the bolts only finger tight.
10. Route the cable correctly and loosely install the clips and retainers. All clips must be in their original position and the sensor cable must not be twisted. Improper installation may cause cable damage and system failure.

NOTE: The wiring in the harness is easily damaged by twisting and flexing. Use the white stripe on the outer insulation to keep the sensor harness properly placed.

11. Use a brass or other non-magnetic feeler gauge to check the air gap between the tip of the pole piece and the toothed wheel. Correct gap is 0.012–0.035 inch (0.3–0.9 mm). Tighten the 2 sensor bracket bolts to 10 ft. lbs. (14 Nm) with the sensor located so that the gap is the same at several points on the toothed wheel. If the gap is incorrect, it is likely that the toothed wheel is worn or improperly installed.
12. Tighten the screws and bolts for the cable retaining clips.
13. Install the inner fender or splash shield.
14. Install the wheel and tire. Lower the vehicle to the ground.

Front Toothed Wheel Rings

REMOVAL & INSTALLATION

1. Raise and safely support the vehicle.
2. Remove the wheel and tire.
3. Remove the wheel speed sensor and disconnect sufficient harness clips to allow the sensor and wiring to be moved out of the work area.

NOTE: The speed sensor has a pole piece projecting from it. This exposed tip must be protected from impact or scratches. Do not allow the pole piece to contact the toothed wheel during removal or installation.

4. Remove the front hub and knuckle assembly.
5. Remove the hub from the knuckle.
6. Support the hub in a vise with protected jaws. Remove the retaining bolts from the toothed wheel and remove the toothed wheel.

To install:

7. Fit the new toothed wheel onto the hub and tighten the retaining bolts to 7 ft. lbs. (10 Nm).
8. Assemble the hub to the knuckle
9. Install the hub and knuckle assembly to the vehicle.
10. Install the wheel speed sensor.
11. Install the wheel and tire.
12. Lower the vehicle to the ground.

Rear Toothed Wheel Rings

REMOVAL & INSTALLATION

Front Wheel Drive

1. Raise and safely support the vehicle.
2. Remove the wheel and tire.
3. Remove the wheel speed sensor and disconnect sufficient harness clips to allow the sensor and wiring to be moved out of the work area.

NOTE: The speed sensor has a pole piece projecting from it. This exposed tip must be protected from impact or scratches. Do not allow the pole piece to contact the toothed wheel during removal or installation.

4. Remove the hub assembly.
5. Support the hub in a vise with protected jaws. Remove the retaining bolts from the toothed wheel and remove the toothed wheel.

To install:

6. Fit the new toothed wheel onto the hub and tighten the retaining bolts to 7 ft. lbs. (10 Nm).
7. For all ABS equipped vehicles except Sigma V6, install the hub assembly to the vehicle. The center hub nut is not reusable. The new nut must be tightened to 144–188 ft. lbs. (200–260 Nm). After the nut is tightened, align the nut with the spindle indentation and crimp the nut in place.
8. For Sigma V6, assemble and install the hub, outer bearing, tongued washer and locknut. To set the wheel bearing end play:
   a. Tighten the locknut to 14 ft. lbs. (20 Nm).
   b. Rotate the hub 180 degrees or more counterclockwise, then return it to the original position. Repeat the rotation and return at least 3 more times. Temporarily fitting the brake disc will make the hub easier to turn.
   c. Loosen the locknut to 0 ft. lbs., then retighten it to 7 ft. lbs. (10 Nm).
   d. Again rotate the hub at least 180 degrees counterclockwise and return it to its original position.
   e. Reset the locknut to 7 ft. lbs. (10 Nm).
   f. Install the lock cap and cotter pin. If the cotter pin will not align with the holes in the lock cap, reposition the cap. If no alignment is possible, loosen the locknut by no more than 15 degrees.
   g. Once the cotter pin is in place, rotate the hub at least 180 degrees counterclockwise and return it to its original position.
9. Install the wheel speed sensor.
10. Install the wheel and tire.
11. Lower the vehicle to the ground.

All Wheel Drive

1. Raise and safely support the vehicle.
2. Remove the wheel and tire.
3. Disconnect the parking brake cable at the caliper.
4. Remove the speed sensor and its O-ring. Disconnect sufficient clamps and wire ties to allow the sensor to be moved well out of the work area.

NOTE: The speed sensor has a pole piece projecting from it. This

exposed tip must be protected from impact or scratches. Do not allow the pole piece to contact the toothed wheel during removal or installation.

5. Remove the brake caliper and brake disc.
6. Remove the 3 retaining nuts and bolts holding the outer end of the driveshaft to the companion flange. Swing the axle shaft away and support it with stiff wire. Do not overextend the joint in the axle; do not allow it to hang of its own weight.
7. Remove the retaining nut and washer on the back of the driveshaft. Use special tool MB 990767 or equivalent to counterhold the hub.
8. Remove the companion flange from the knuckle.
9. Using an axle puller which bolts to the wheel lugs, remove the axle shaft assembly.
10. Fit the shaft assembly in a press with the toothed wheel completely supported by a bearing plate such as special tool MB 990560 or its equivalent.
11. Press the toothed wheel off the axle shaft.

To install:

12. Press the new toothed wheel onto the shaft with the groove facing the axle shaft flange.
13. Install the axle shaft to the knuckle and fit the companion flange in place.
14. Install the lock washer and a new self-locking nut on the axle shaft. Tighten the nut to 116–159 ft. lbs. (160–220 Nm).
15. Swing the axle assembly into place and install the 3 nuts and bolts. Tighten each to 45 ft. lbs. (61 Nm).
16. Install the brake disc and caliper.
17. Install the wheel speed sensor. Always use a new O-ring.
18. Connect the parking brake cable to the caliper.
19. Install the wheel and tire; lower the vehicle to the ground.

CHASSIS ELECTRICAL

Air Bag

DISARMING

1. Position the front wheels in the straight ahead position and place the key in the **LOCK** position.
2. Disconnect the negative battery cable and insulate the cable end with high-quality electrical tape or similar non-conductive wrapping.
3. Wait at least 1 minute before working on the vehicle. The air bag system is designed to retain enough voltage to deploy for a short period of time even after the battery has been disconnected.
4. If necessary, enter the vehicle from the passenger side and turn the key to unlock the steering column.

Heater Blower Motor

REMOVAL & INSTALLATION

Cordia and Tredia

1. Disconnect the negative battery cable. Unscrew the 1 attaching bolt and remove the lower cover from under the right side of the instrument panel. Remove the mounting screws at the front and remove the glove box. Remove the cowl side trim.
2. Disconnect the air selector control wire. Disconnect the discharge duct at the blower.
3. Disconnect the electrical connector. Remove the 4 mounting bolts and remove the blower assembly.
4. Installation is the reverse of the removal procedure.

Starion

1. Disconnect the negative battery cable. Remove the under cover from the bottom of the dash panel, under the glove box. Then, open the glove box door and pull the glove box forward while pressing inward on both sides of the glove box. This will allow the door to drop.
2. Remove the screws attaching the glove box door hinge to the bottom of the dashboard and then remove the assembly.
3. Disconnect the fresh air/recirculated air change over cable from the blower housing. Disconnect the blower electrical connector.
4. Remove the 3 bolts and the blower assembly. The blower motor and fan may be removed from the case.
5. Installation is the reverse of removal.

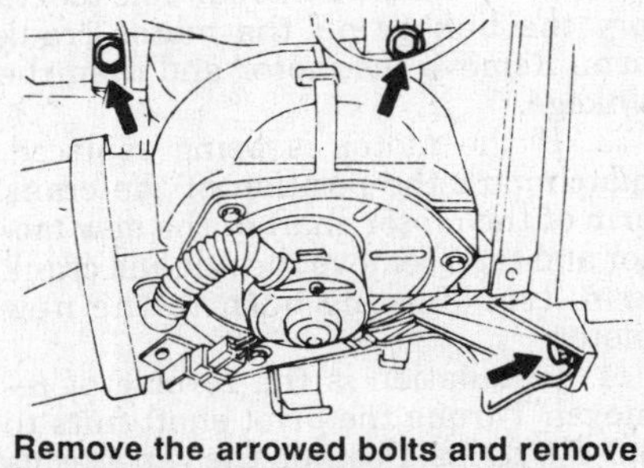

Remove the arrowed bolts and remove the blower motor (Cordia and Tredia)

Mirage

1. Disconnect the negative battery cable. Open the glove box door, release the hinges and remove the glove box.
2. Remove the 4 Phillips screws and the parcel tray.
3. Disconnect the recirculation/fresh air change over control wire. Disconnect the electrical connector and the duct from the blower assembly.
4. Remove the 3 bolts and the blower assembly. Remove the blower motor screws and the motor from the blower assembly.
5. Installation is the reverse of the removal procedure.

Galant and Sigma

1. Disconnect the negative battery cable. Remove the screw covers and both screws from the undercover, located at the bottom of the instrument panel on the right side. Remove the undercover.
2. Remove the installation screws and the instrument undercover, located under the steering column. Remove the under frame installation screws located under that cover and the passenger side under frame.
3. Remove both stops from the bottom of the glove box at the front. Remove the glove box installation screws and the glove box. Remove the duct leading into the blower unit, which is accessible through the glove box door.
4. Remove the 4 bolts and the blower assembly. Disconnect the electrical connector and the inside/outside air change over control vacuum hose before pulling the unit out.
5. Installation is the reverse of removal.

Precis

NOTE: To remove either the blower or core, the heater case must be removed.

1. Disconnect the battery ground.
2. Place the control in the **HOT** position.
3. Drain the cooling system.
4. Remove the heater hoses from the core tubes.
5. Remove the lower instrument panel section.
6. Remove the center console.
7. Disconnect the ducts at the heater case.
8. Disconnect the heater control cable at the case.
9. If equipped with air conditioning, discharge the system and remove the evaporator.
10. Unbolt and remove the heater case.

To install:

11. Install the heater case, heater control cable and ducts.
12. Install the center console, lower instrument panel and heater hoses.
13. Fill the cooling system and recharge the refrigerant. Adjust the control cable.

Eclipse

1. Disconnect battery negative cable.
2. Remove the right duct, if equipped.
3. Remove the molded hose from the blower assembly.
4. Remove the blower motor assembly.
5. Remove the packing seal.
6. Remove the fan retaining nut and fan in order to renew the motor.

To install:

7. Check that the blower motor shaft is not bent and the packing is in good condition. Clean all parts of dust, etc.
8. Install the blower motor then connect the motor terminals to battery voltage. Check that the blower motor operates smoothly. Reverse the polarity and check that the blower motor operates smoothly in the reverse direction.
9. Install duct, if removed. Connect battery negative cable.

3000-GT

1. Disconnect the negative battery cable.

NOTE: If equipped with an air bag, be sure to disarm it before entering the vehicle.

2. Remove the right side duct, as required. On 3000-GT, remove the instrument panel undercover.
3. Remove the cooling tube from the blower assembly.
4. Remove the blower motor assembly.
5. Remove the packing seal.
6. Remove the fan retaining nut and fan in order to renew the motor.

To install:

7. Check that the blower motor shaft is not bent and that the packing is in good condition. Clean all parts of dust, etc.
8. Assemble the motor and fan. Install the blower motor and connect the wiring.
9. Install the cooling tube.
10. Install the duct or undercover.
11. Connect the negative battery cable and check the entire climate control system for proper operation.

Windshield Wiper Motor

REMOVAL & INSTALLATION

Cordia and Tredia

1. Remove the wiper blade and arm assembly.
2. Remove the cover from the access hole or the deck panel, guide panel and garnish depending on mode.
3. Remove the wiper drive bolts at the arm pivots. On the Tredia, remove the washer nozzle.
4. Loosen the wiper motor bolts. Disconnect the wiper motor and linkage and remove.
5. Install in reverse order.

Starion

1. Remove the wiper arms. Remove the pivot shaft nuts and washers and push the pivot shafts into the area behind the cowl.
2. Remove the cover from the access hole for the wiper motor on the right side of the cowl, under the hood. Then, remove the motor bolts. Pull the motor into the best possible position for access and use a flat suitable tool to pry the linkage off the motor crank arm.
3. If the linkage is being replaced, it can be worked out of the cowl at this time. If the motor is being replaced, matchmark the position of the crank arm of the motor shaft of the new motor and then remove the nut and crank arm, transferring both to the new motor.
4. Installation is the reverse of removal. Make sure the wiper blades stop about 0.5 in. from the lower windshield molding. Torque the wiper arm attaching nuts to 7–12 ft. lbs.

Mirage and Precis

1. Remove the wiper arms. Remove the air inlet and cowl front center trim panels. Remove the 3 pivot shaft nuts and push the pivot shafts into the area under the cowl.
2. Remove the motor bolts. Pull the motor into the best possible position for access and use a flat suitable tool to pry the linkage off the motor crank arm. Remove the motor and then the linkage.
3. If the motor is being replaced, matchmark the position of the crank arm of the motor shaft of the new motor and then remove the nut and crank arm, transferring both to the new motor.
4. Installation is the reverse of removal. Torque the pivot shaft nuts to 4.3–5.8 ft. lbs. Position the wiper arms so the blades are about 0.6 in. above the lower windshield molding on the driver's side and 0.8 in. above it on the passenger's side. Torque the wiper arm nuts to 7.2–12 ft. lbs. Make sure the wiper motor is securely grounded.

Galant and Sigma

1. Remove the wiper arms. Remove the front deck and inlet trim.
2. Remove the 3 bolts for each pivot shaft and push the shafts into the area behind the panel. Disconnect the electrical connector. Remove the 3 motor mounting bolts and the motor and linkage as an assembly.
3. If the motor is being replaced, matchmark the relationship between linkage and motor, as it is critical. If the linkage only is being replaced, pry the connection off the end of the motor crank arm with a flat suitable tool.
4. Install in reverse order. Make sure the wiper arms sit in their original positions when in parked position. Make sure the wiper motor is securely grounded.

Eclipse

FRONT

1. Disconnect the negative battery cable.
2. Remove the windshield wiper arms by unscrewing the cap nuts and lifting the arms from the linkage posts.
3. Remove the front garnish panel.
4. Remove the air inlet trim pieces.
5. Remove the hole cover.
6. Remove the wiper motor by loosening the bolts, removing the motor assembly and disconnecting the linkage.

NOTE: Because the installation angle of the crank arm and the motor has been factory set, do not remove them unless necessary. If they must be removed, remove them only after marking their positions.

To install:

7. Install the windshield wiper motor and connect the linkage.
8. Reinstall all the trim pieces and connect the negative battery cable.
9. Reinstall the wiper blades. Note that the driver's side wiper arm should be marked **D** and the passenger's side wiper arm should be marked **A**. The identification marks should be located at the base of the arm, near the pivot. Install the arms so the blades are 1 in. from the garnish molding when parked. Wet down the windshield glass and test wiper motor at all speeds.

REAR

1. Disconnect the negative battery cable.
2. Remove the rear wiper arm by re-

moving the cover, unscrewing the nut and lifting the arm from the linkage post.

3. Remove the large interior trim panel. Use a plastic trim stick to unhook the trim clips of the liftgate trim.

4. If equipped with rear air spoiler, remove grommet.

5. Remove the rear wiper assembly; do not loose the grommet for the wiper post.

To install:

6. Install the motor and grommet. Mount the grommet so the arrow on the grommet is pointing upward.

7. Connect the negative battery cable.

8. Reinstall the wiper blade, wet down the rear glass and test the wiper motor.

9. If operation is satisfactory, fit the tabs on the upper part of the liftgate trim into the liftgate clips and secure the liftgate trim.

3000-GT

FRONT

1. Disconnect the negative battery cable.

2. Remove the windshield wiper arms by unscrewing the cap nuts and lifting the arms from the linkage posts.

3. Remove the access hole cover.

4. Remove the wiper motor mounting bolts.

5. Detatch the motor crank arm from the wiper linkage and remove the motor.

NOTE: The installation angle of the crank arm and motor has been factory set; do not remove them unless it is necessary to do so. If they must be removed, remove them only after marking their mounting positions.

To install:

6. Install the windshield wiper motor and connect the linkage.

7. Install the access hole cover.

8. Reinstall the wiper blades. Note that the driver's side wiper arm should be marked **D** and the passenger's side wiper arm should be marked **A**. The identification marks should be located at the base of the arm, near the pivot. Install the arms so the blades are parallel to the garnish molding when parked.

9. Connect the negative battery cable and check the wiper system for proper operation.

REAR

1. Disconnect the negative battery cable.

2. Remove the liftgate lower trim. Remove the clips that hold the trim by using the following procedure:

a. Remove the clip by pressing down on the center pin with a suitable blunt pointed tool. Press down a little more than $^1/_{16}$ in. (2mm). This releases the clip. Pull the clip outward to remove it.

b. Do not push the pin inward more than necessary because it may damage the grommet, or if pushed too far, the pin may fall in. Once the clips are removed, use a plastic trim stick to pry the trim cover loose.

3. Remove the rear spoiler, center brace and center brake light.

4. Lift the small cover, remove the retaining nut and remove the wiper arm and spacer.

5. Remove the mounting bolts and remove the wiper motor.

To install:

6. Install the motor and install the retaining bolts.

7. Install the spacer, wiper arm and retaining nut. The arm should be positioned so the upper tip points to the upper left corner of the rear window when parked. Connect the battery and check the operation of the motor before proceeding. If satisfactory, disconnect the cable and proceed.

8. Install the rear spoiler and related parts.

9. Install the interior trim piece.

10. Connect the negative battery cable and recheck the system for proper operation.

Windshield Wiper Switch

REMOVAL & INSTALLATION

Cordia, Tredia, Starion and Precis

1. Disconnect the negative battery cable.

2. Remove the steering column lower trim panel.

3. Remove the steering wheel and the steering column cable band.

4. Disconnect the electrical connections from the switch assembly.

5. Remove the switch screws and the switch assembly from the vehicle.

6. Installation is the reverse of the removal procedure.

Galant, Mirage and Sigma

1. Disconnect the negative battery cable.

2. If equipped with tilt wheel lower the steering wheel to its lowest position.

3. Remove the steering wheel, the steering column lower trim panel, the steering column upper trim panel and the steering column cable band.

4. Disconnect the electrical connections from the switch assembly.

5. Remove the switch screws and the switch assembly from the vehicle.

6. Installation is the reverse of the removal procedure.

Eclipse

1. Disconnect the negative battery cable.

2. Remove the horn pad by removing the screw from behind the steering wheel and pressing the pad upward.

3. Remove the steering wheel.

NOTE: Make mating marks on the steering wheel and the steering wheel shaft. Use a steering wheel puller to remove the steering wheel; do not hammer on the steering wheel to remove it or the collapsible mechanism may be damaged.

4. Locate the rectangular plugs in the knee protector on either side of the steering column. Pry these plugs out, remove the screws. Remove the screws from the hood lock release lever and remove the knee protector.

5. Remove the upper and lower column covers.

6. Remove the lap cooler ducts.

7. Remove the band retaining the switch wiring.

8. Remove the column switch.

To install:

9. Installation is the reverse of the removal procedure. Take care that no wires are pinched or out of place.

10. Torque the steering wheel-to-column nut to 25–33 ft. lbs.

11. Check the steering wheel position with the wheels straight ahead.

12. Connect battery negative cable.

3000-GT

1. Disconnect the negative battery cable.

NOTE: If equipped with an air bag, be sure to disarm it before entering the vehicle.

2. Remove the steering wheel:

a. Remove the air bag module mounting nut from behind the steering wheel.

b. Carefully disconnect the module connector.

c. Store the air bag module in a clean, dry place with the pad cover facing up.

d. Remove the steering wheel retaining nut and use a steering wheel puller to remove the wheel. Do not use a hammer or the collapsible mechanism in the column could be damaged.

3. Remove the hood lock release handle.

4. Remove the switches from the knee protector below the steering column, and remove the exposed retain-

ing screws. Then remove the knee protector.

5. Remove the column covers.
6. Remove necessary duct work and disconnect the windshield wiper switch connectors.
7. Remove the retaining screws and remove the windshield wiper switch assembly from the steering column.

To install:

8. Install the wiper switch to the steering column and connect the connectors.
9. Install any removed duct work.
10. Install the column covers.
11. Install the knee protector and switches.
12. Install the hood release handle.
13. Center the clock spring by aligning the **NEUTRAL** mark on the clock spring with the mating mark on the casing. Then install the steering wheel and torque the retaining nut to 29 ft. lbs. (40 Nm).
14. Connect the negative battery cable and check the windshield wiper and washer for proper operation.

Instrument Cluster

REMOVAL & INSTALLATION

Cordia and Tredia

1. Disconnect the negative battery cable.
2. Remove the screws at the top of the instrument cluster trim panel. Remove the trim panel.
3. Disconnect the speedometer cable from the back of the speedometer.
4. Remove the cluster screws and pull the cluster forward.
5. Disconnect the electrical connectors and remove the cluster. Install in reverse order.

Starion

1. Disconnect the negative battery cable.
2. Remove the meter trim hood screws. Pull out and down on the side of the hood.
3. Disconnect the plug connectors on both sides of the cluster.
4. Remove the cluster screws and nuts. Pull the lower sides of the cluster up and disconnect the speedometer cable.
5. Disconnect the plug connectors at the rear of the cluster and remove the cluster. Install in reverse order.

Mirage

1. Disconnect the negative battery cable. Remove both meter hood attaching screws, located at the bottom and tile the lower meter hood outward. Pull the hood downward to release the locking tangs at the top and remove it.
2. Remove the 4 meter assembly screws, 2 at top and 2 at the bottom, and pull the unit outward. Disconnect the speedometer cable and all connectors and remove the unit.
3. Installation is the reverse of removal.

Galant and Sigma

1. Disconnect the negative battery cable. Remove both meter hood screw covers located along the bottom of the hood using a suitable tool. Remove the screws and pull off the hood.
2. Remove the 4 meter assembly screws, 2 on each side, pull the assembly outward slightly and then disconnect the electrical connectors, speedometer cable and adapter. Remove the assembly.
3. Installation is the reverse of removal.

Eclipse

1. Disconnect the negative battery cable.
2. Remove the screw cover at the side of the bezel.
3. Remove the instrument cluster bezel.
4. Remove the instrument cluster.

NOTE: If the speedometer cable adapter must be serviced, disconnect the cable at the transaxle end. Pull the cable slightly toward the vehicle interior, release the lock by turning the adapter to the right or left, and then remove the adapter.

To install:

5. Installation is the reverse of the removal procedure. Use care not to damage the printed circuit board or any gauge components.
6. Connect battery negative cable.

3000-GT

1. Disconnect the negative battery cable.

NOTE: If equipped with an air bag, be sure to disarm it before entering the vehicle.

2. Remove the hood lock release handle and switches from the knee protector below the steering column. Then remove the exposed retaining screws and remove the knee protector.
3. Remove the screw cover at the side of the bezel.
4. Remove the instrument cluster bezel.
5. Remove the instrument cluster. Disassemble and remove gauges or the speedometer as required.

NOTE: If the speedometer cable adapter must be serviced, disconnect the cable at the transaxle end. Pull the cable slightly toward the vehicle interior, release the lock by turning the adapter to the right or left and remove the adapter.

6. The installation is the reverse of the removal procedure. Use care not to damage the printed circuit board or any gauge components.
7. Connect the negative battery cable and check all cluster-related items for proper operation.

Radio

REMOVAL & INSTALLATION

1. Disconnect battery negative cable.

NOTE: If equipped with an air bag, be sure to disarm it before entering the vehicle.

2. Remove the panel from around the radio. On some models the panel is retained with screws. On others, use a plastic trim tool to pry the lower part of the radio panel loose. Remove it from the center console.
3. Remove the radio or radio/tape player. Depending on the speaker installation, it may save time at installation to identify and tag all wires before they are disconnected.
4. Separate amplifiers and/or CD player can be removed by first removing the side cover of the console box.
5. Remove the mounting brackets from the radio.

To install:

6. The installation is the reverse of the removal procedure. Make all electrical and antenna connections before fastening the radio assembly in place.
7. Install the center panel.
8. Connect the negative battery cable and check the entire audio system for proper operation.

Concealed Headlights

MANUAL OPERATION

If the headlight covers will not raise electrically, remove the fusible link from the relay box, then remove the boot on the rear area of the pop-up motor and turn the manual knob clockwise until the cover is open. Perform this procedure on both the left and right sides.

Headlight Switch

REMOVAL & INSTALLATION

Except Eclipse, Galant, Mirage and Sigma

1. Disconnect the negative battery cable.

2. Remove the steering column lower trim panel.
3. Remove the steering wheel and the steering column cable band.
4. Disconnect the electrical connections from the switch assembly.
5. Remove the switch screws and the switch assembly from the vehicle.
6. Installation is the reverse of the removal procedure.

Galant, Mirage and Sigma

1. Disconnect the negative battery cable.
2. If equipped with tilt wheel, lower the steering wheel to its lowest position.
3. Remove the steering wheel and the steering column lower trim panel. Remove the steering column upper trim panel and the steering column cable band.
4. Disconnect the electrical connections from the switch assembly.
5. Remove the switch retaining screws and the switch assembly from the vehicle.
6. Installation is the reverse of the removal procedure.

Combination Switch

REMOVAL & INSTALLATION

Except Eclipse and 3000-GT

1. Disconnect the negative battery cable. Remove the steering wheel and have the tilt handle in the lowest position.
2. Remove the combination meter and column covers.
3. Remove the connectors from the column switch and the column switch from the column tube.

NOTE: Some vehicles may have the turn signal and hazard switches mounted on a base plate. Removal of the attaching screws will allow these switches to be removed without removal of the remaining switches.

4. Switch installation is the reverse of removal. Be sure the switch is centered in the column or self canceling will be affected.

Eclipse

1. Disconnect the negative battery cable.
2. Remove the horn pad by removing the screw from behind the steering wheel and then pressing the pad upward.
3. Remove the steering wheel.

NOTE: Make mating marks on the steering wheel and the steering wheel shaft. Use a steering wheel puller to remove the steering wheel. Do not hammer on the steering wheel to remove it or the collapsible mechanism may be damaged.

4. Locate the rectangular plugs in the knee protector on either side of the steering column. Pry these plugs out and remove the screws. Remove the screws from the hood lock release lever and the knee protector.
5. Remove the upper and lower column covers.
6. Remove the lap cooler ducts.
7. Remove the band retaining the switch wiring.
8. Remove the column switch.

To install:

9. At installation, take care that no wires are pinched or out of place.
10. Torque the steering wheel-to-column nut to 25–33 ft. lbs.
11. Check the steering wheel position with the wheels straight ahead.
12. Connect battery negative cable.

3000-GT

NOTE: The headlights, turn signals and dimmer switch are all built into 1 multi-function combination switch that is mounted on the left side of the steering column.

1. Disconnect the negative battery cable.

NOTE: If equipped with an air bag, be sure to disarm it before entering the vehicle.

2. Remove the steering wheel:
   a. Remove the air bag module mounting nut from behind the steering wheel.
   b. Carefully disconnect the module connector.
   c. Store the air bag module in a clean, dry place with the pad cover facing up.
   d. Remove the steering wheel retaining nut and use a steering wheel puller to remove the wheel. Do not use a hammer or the collapsible mechanism in the column could be damaged.
3. Remove the hood lock release handle.
4. Remove the switches from the knee protector below the steering column, and remove the exposed retaining screws. Then remove the knee protector.
5. Remove the column covers.
6. Remove necessary duct work and disconnect the combination switch connectors.
7. Remove the retaining screws and remove the combination switch assembly from the steering column.

To install:

8. Install the switch to the steering column and connect the connectors.
9. Install any removed duct work.
10. Install the column covers.
11. Install the knee protector and switches.
12. Install the hood release handle.
13. Center the clock spring by aligning the **NEUTRAL** mark on the clock spring with the mating mark on the casing. Then install the steering wheel and torque the retaining nut to 29 ft. lbs. (40 Nm).
14. Connect the negative battery cable and check all functions of the combination switch for proper operation.

Ignition Switch

REMOVAL & INSTALLATION

Except Eclipse and 3000-GT

1. Disconnect the negative battery cable. Cut a notch in the lock bracket bolt head with a hacksaw.
2. Remove the bolt and lock.
3. Remove the column cover and the ignition switch.
4. Install both lock and switch in reverse of removal.

NOTE: When installing lock, the bolt should be tightened until the head is crushed. When installing switch, install the switch bolt loosely and insert and work the key a few times to make sure everything checks out before tightening the bolt.

Eclipse

1. Disconnect the negative battery cable.
2. Remove the lower instrument panel knee protector.
3. Remove the lower steering column cover.
4. Remove the clip that holds the wiring against the steering column.
5. Unplug the ignition switch from the steering lock cylinder.
6. Insert the key into the steering lock cylinder and turn to the **ACC** position.
7. With a small pointed tool, push the lock pin of the steering lock cylinder inward and pull the lock out.

NOTE: Vehicles equipped with automatic transaxle safety-lock systems will have a key interlock cable installed in a slide lever on the side of the key lock.

To install:

8. Installation is the reverse of the removal process. Make sure the lock pin snaps into place. Install the ignition switch plug carefully and make sure no wires are pinched.
9. Reconnect the negative battery cable.

3000-GT

1. Disconnect the negative battery cable.

NOTE: If equipped with an air bag, be sure to disarm it before entering the vehicle.

2. Remove the steering wheel:
a. Remove the air bag module mounting nut from behind the steering wheel.
b. Carefully disconnect the module connector.
c. Store the air bag module in a clean, dry place with the pad cover facing up.
d. Remove the steering wheel retaining nut and use a steering wheel puller to remove the wheel. Do not use a hammer or the collapsible mechanism in the column could be damaged.
3. Remove the hood lock release handle.
4. Remove the switches from the knee protector below the steering column, and remove the exposed retaining screws. Then remove the knee protector.
5. Remove the column covers.
6. Remove necessary duct work and disconnect the windshield wiper and combination switch connectors.
7. Remove the retaining screws and remove the entire column switch/clock spring assembly from the steering column.
8. If damaged, remove the illumination ring, key reminder switch harness and ignition switch harness.
9. To remove the lock cylinder, insert the key and place in the **ACC** position. With a small pointed tool, push the lock pin of the steering lock cylinder inward and pull the lock out.

To install:
10. Install the lock cylinder; make sure the lock pin snaps into place.
11. Install any other removed items, making sure no wires are pinched.
12. Install the column switch/clock spring assembly to the steering column and connect the connectors.
13. Install any removed duct work.
14. Install the column covers.
15. Install the knee protector and switches.
16. Install the hood release handle.
17. Center the clock spring by aligning the **NEUTRAL** mark on the clock spring with the mating mark on the casing. Then install the steering wheel and torque the retaining nut to 29 ft. lbs. (40 Nm).
18. Connect the negative battery cable and check all functions of column-mounted switches and the ignition switch for proper operation.

Stoplight Switch

ADJUSTMENT

1. The stoplight switch works off the brake pedal lever. To adjust, disconnect the electrical connection and loosen the switch locknut.
2. Screw the switch inward until it contacts the stop on the brake pedal arm. Back out the switch ½–1 full turn. The gap between the switch plunger and the brake lever stop should be 0.020–0.040 in.
3. Tighten the locknut, connect the wires.
4. Check that the stoplights are not on unless the pedal is depressed.

REMOVAL & INSTALLATION

1. Disconnect the negative battery cable.
2. Locate the stoplight switch above the brake pedal lever.
3. Disconnect the wiring connectors from the switch and unscrew the switch.
4. Installation is the reverse of the removal process. Install the replacement switch and adjust to 0.020–0.040 in. clearance.
5. Reconnect the stoplight wires.
6. Reconnect the negative battery cable.

Clutch Switch

ADJUSTMENT

The clutch interlock switch is located at the top of the clutch pedal arm. Note that there may be 2 switches; one will be a cruise control cut-out switch.

1. Clutch interlock switch adjustment is made with the pedal depressed its full stock of 6 in.
2. Measure the gap between the switch plunger and the arm stop. The gap should be 0.140 in.
3. If adjustment is necessary, loosen the locknut and turn and adjust.
4. After completing the adjustment, check that the pedal free-play, measured at the face of the pedal pad, is 0.240–0.510 in. The distance between the pedal pad and the firewall when the clutch is disengaged should be 2.80 in. or more. If these dimensions are not right, the hydraulic clutch system will probably need to be bled.

REMOVAL & INSTALLATION

1. Disconnect the negative battery cable.
2. Locate the interlock switch above the clutch pedal lever.
3. Disconnect the wiring connectors from the switch and unscrew the switch.
4. Installation is the reverse of the removal procedure. Install the replacement switch and adjust to 0.140 in. clearance.
5. Reconnect the interlock wires.
6. Reconnect the negative battery cable.

Neutral Safety Switch

ADJUSTMENT

1. Locate the neutral safety switch on the top of the transaxle. Note that several different cable attaching methods have been used. The procedure here can be used as a general guide for all.
2. Place the selector lever in **N**.
3. Loosen both adjusting nuts to free up the cable and lever.
4. Place the safety switch manual control lever in **N**.
5. Note that one end of the safety switch manual control lever has a 12mm wide square end. There is also a 12mm wide tab on the switch body flange. Loosen both retaining bolts and turn the safety switch until these portions align. Tighten the bolts, making sure the switch doesn't move.
6. Loosen the adjuster nuts and gently pull the cable to remove any slack. Gently tighten adjusting nut until it just starts to contact the adjuster. Secure adjusting nut with its locknut then turn nut to lock.
7. Verify that the switch lever moves to positions corresponding to each position of the selector lever.
8. To test, apply the parking and service brake securely. Place the selector in **R**. Turn the ignition key to the **START** position. Slowly move the selector lever upward until it clicks and fits in the notch of the **P** range. If the starter motor operates when the lever makes a click the **P** position is correct. Slowly move the lever to the **N** range. Using the same procedure, if the starter operates when the selector fits in the **N** position, adjustment is correct. Also check that the vehicle doesn't begin to move and the lever doesn't stop between **P-R-N-D**.

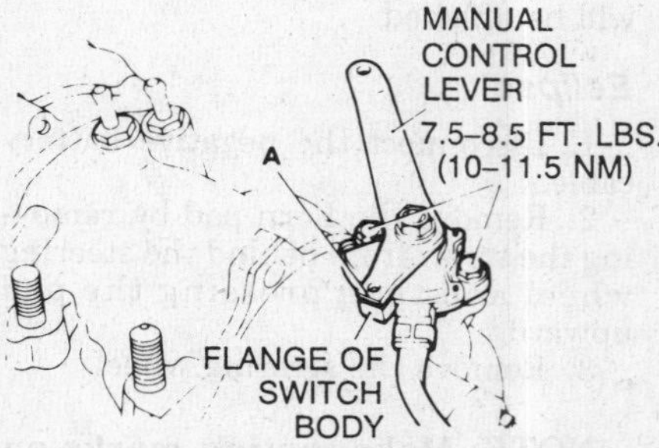

Neutral start switch adjustment. "A" denotes the small end of the lever

Fuses and Circuit Breakers

LOCATION

3000-GT

Fuses are located at the right front of the engine compartment, at the left side of the engine compartment and under the left side of the dash.

Fusible links (maxi-fuses) are located at the right front of the engine compartment.

Eclipse

Fuses are located at the right front of the engine compartment, the left rear of the engine compartment and under the left side of the dash.

Fusible links are located at the battery. Sub-fusible links are located at the right front of the engine compartment.

Galant

Fuses are located at the right front of the engine compartment and under the left side of the dash.

Fusible links are located at the battery and at the right side of the engine compartment.

Mirage

Fuses are located under the left side of the dash. Dedicated fuses for the air conditioner are located at the left front of the engine compartment. Other dedicated fuses are located at the right front of the engine compartment.

Fusible links are located at the right front of the engine compartment.

Precis

Fuses are located in the fuse box at the left kick panel and the relay box under the left side of the dash.

Fusible links are located in a holder next to the battery or in a holder at the left front of the engine compartment.

Sigma

Fuses are located under the left side of the dash and at the right side of the engine compartment.

Fusible links are located at the battery and at the right side of the engine compartment.

Starion

Fuses are located at the left side of the dash. Fusible links are located at the left front of the engine compartment near the battery.

Computers, Relays and Flashers

LOCATION

3000-GT

A/C Control Unit—is located under the center of the dash.

A/C Relay Box—is located at the left front of the engine compartment.

ABS Control Unit—is located next to the rear seat at the right quarter panel.

ABS Motor Relay—is located at the right front of the engine compartment.

ABS Relay— Relay—is located in the relay box at the right side of the engine compartment.

ABS Valve Relay—is located at the right front of the engine compartment.

Active Aero Control Unit—is located at the left quarter panel inside the trunk compartment.

Active Exhaust Control Unit—is located at the left rear of the trunk compartment.

Alternator Relay—is located in the relay box at the right side of the engine compartment.

Automatic transaxle Control Unit—at the front of the console under the dash.

Control Relay—is located at the right front of the console.

Defogger Relay—is located in the interior relay box under the left side of the dash.

Door Lock Relay—is located in the interior relay box under the left side of the dash.

ECS Control Unit—is located at the right quarter panel behind the right rear wheel.

Engine ECU—is located under the front of the console.

ETACS Unit—is located under the left side of the dash.

Fog Light Relay—is located in the relay box at the right side of the engine compartment.

Hazard Flasher/Turn Signal Unit—is located at the left kick panel.

Headlight Relay—is located in the relay box at the right side of the engine compartment.

Horn Relay—is located in the relay box at the right side of the engine compartment.

Interior Relay Box—is located under the left side of the dash.

Magnetic Clutch Relay—is located in the A/C relay box at the left front of the engine compartment.

Pop-Up Motor Relay—is located in the relay box at the right side of the engine compartment.

Power Window Relay—is located in the interior relay box under the left side of the dash.

Radiator Fan Motor Control Relay—is located in the A/C relay box at the left front of the engine compartment.

Radiator Fan Motor Relay—is located in the relay box at the right side of the engine compartment.

Rear Intermittent Wiper Relay—is located next to the rear seat at the left quarter panel.

Relay Box—is located at the right side of the engine compartment.

SRS Diagnosis Unit—is located under the console.

Starter Relay—is located in the relay box at the right side of the engine compartment.

Taillight Relay—is located in the relay box at the right side of the engine compartment.

Theft Alarm Starter Relay—is located at the left front of the console.

Turn Signal/Hazard Flasher Unit—is located at the left kick panel.

Eclipse

A/C Control Unit—is located under the right side of the dash near the A/C unit.

A/C Dedicated Fuse—is located at the left rear of the engine compartment.

ABS Control Unit—is located behind the rear seat.

ABS Motor Relay—is located the right side of the engine compartment.

ABS Power Relay—is located behind the rear seat.

ABS Valve Relay—is located the right side of the engine compartment.

Alternator Relay—is located at the relay box on the right side of the engine compartment.

Automatic Seat Belt Control Unit—is located at the left quarter panel behind the left seat.

Automatic Seat Belt Motor Relay—is located at the left quarter panel behind the left seat.

Automatic Transaxle Control Unit (2.0L engine)—is located at the front of the console under the dash.

Condenser Fan Motor Changeover Relay—is located at the relay box on the left side of the engine compartment.

Condenser Fan Motor Relay—is located at the relay box on the left side of the engine compartment.

Cruise Control Unit—is located under the left side of the dash.

Door Lock Control Unit—is located at the right kick panel.

Door Lock Relay—is located at the relay box under the left side of the dash.

Engine Control Unit (2.0L engine)—is located at the front of the console under the dash.

Fog Light Relay—is located at the relay box on the right side of the engine compartment.

Headlight Relay—is located at the relay box on the right side of the engine compartment.

Heater Relay—is located under the left side of the dash.

Intermittent Front Wiper Relay—is located inside the steering column.

Intermittent Rear Wiper Relay—is located at the left quarter panel behind the left seat.

Magnet Clutch Relay—is located at the left rear of the engine compartment.

Pop Up Relay—is located at the relay box on the right side of the engine compartment.

Power Window Relay—is located at the relay box on the right side of the engine compartment.

Radiator Fan Motor Relay—is located at the relay box on the right side of the engine compartment.

Starter Relay—is located at the relay box under the left side of the dash.

Taillight Relay—is located at the relay box on the right side of the engine compartment.

Theft Alarm Control Unit—is located at the right kick panel.

Theft Alarm Relay—is located at the relay box under the left side of the dash.

Theft Alarm Starter Relay (automatic transaxle)—is located under the left side of the dash.

Theft Alarm Starter Relay (manual transaxle)—is located under the front of the console.

Galant

A/C Compressor Control Unit—is located under the dash near the glove box.

A/C Compressor Relay—is located in the relay box at the right side of the engine comartment.

ABS Control Unit—is located at the left side of the trunk.

ABS Motor Relay—is located at the ABS hydraulic unit.

ABS Power Relay—is located at the left kick panel.

ABS Valve Relay—is located at the ABS hydraulic unit.

Air Conditioner Relay—is located in the relay box at the right side of the engine compartment.

Alternator Relay—is located in the relay box at the right side of the engine comartment.

Automatic Seat Belt Control Unit—is located under the rear of the console.

Automatic Seat Belt Motor Relay—is located at the left side door pillar near the floor.

Automatic Transaxle Control Unit—is located under the left side of the dash.

Cruise Control Unit—is located under the left side of the dash.

Defogger Relay—is located at the relay box under the left side of the dash.

Door Lock Control Unit—is located inside the ETACS control unit.

Door Lock Relay—is located at the relay box under the left side of the dash.

ECS Air Compressor Relay—is located at the right side of the engine compartment.

ECS Control Unit—is located at the right side of the trunk.

ECS Power Relay—is located at the left kick panel.

ECS Relay—is located in the relay box at the right side of the engine comartment.

ECS Return Pump Relay—is located at the right side of the engine compartment.

ECS Solenoid Valve Power Relay—is located at the relay box on the right side of the engine compartment.

Electronic Timer and Control System Control Unit—is located under the left side of the dash.

Engine Control Unit—is located under the right side of the dash.

EPS Control Unit—is located under the center of the dash at the front of the console.

Fog Light Relay—is located in the relay box at the right side of the engine comartment.

Headlight Relay—is located in the relay box at the right side of the engine comartment.

Heater Relay—is located under the left side of the dash.

MPI Control Relay—is located under the right side of the dash.

Power Window Relay—is located at the relay box under the left side of the dash.

Radiator Fan Motor Relay—is located in the relay box at the right side of the engine comartment.

Sunroof Relay—is located at the sunroof motor in the headliner.

Taillight Relay—is located in the relay box at the right side of the engine comartment.

Theft Alarm Headlight Relay—is located at the right front of the engine compartment.

Theft Alarm Horn Relay—is located at the left rear of the engine compartment.

Theft Alarm Starter Relay—is located at the relay box under the left side of the dash.

Wiper Relay—is located next to the wiper motor.

Mirage

A/C Compressor Control Unit—is located above the evaporator under the center of the dash.

A/C Compressor Relay 1—is located at the left side of the engine compartment.

A/C Compressor Relay 2—is located at the relay box on the right side of the engine compartment.

Alternator Relay—is located at the relay box on the right side of the engine compartment.

Automatic Seat Belt Control Unit—is located under the console near the parking brake lever.

Automatic Seat Belt Motor Relay—is located at the left side door pilar.

Automatic transaxle Control Unit—is located under the center of the dash.

Cruise Control Unit—is located under the left side of the dash.

Door Lock Control Unit—is located at the left kick panel.

Door Lock Relay—is located at the left kick panel.

Headlight Relay—is located at the relay box on the right side of the engine compartment.

Heater Relay—is located at the left kick panel.

MPI Control Unit (1.5L engine)—is located next to the blower motor under the dash.

MPI Control Unit (1.6L engine)—is located at the right kick panel.

Power Window Relay—is located at the relay box on the right side of the engine compartment.

Precis

A/C Relay—is located under the left side of the dash.

Automatic transaxle Control Unit (TACU)—is located under the left front seat.

Blower Relay—is located under the left side of the dash.

Condenser Fan Relay—is located at the left front of the engine compartment.

Daytime Running Light Unit (Canada)—is located at the right front of the engine compartment.

Electronic Control Unit (ECU)—is located under the dash on the left side.

Fuse Box Relay (red)—is located in the fusible link box, at the battery.

Fuse Box—is located in the left kick panel.

Hazard Flasher—is located in the relay box under the left side of the dash.

Headlight Relay—is located in the fusible link box, at the battery.

Main Fuse Box—is located on the side of the battery.

Main-Fusible Link Box—is located at the battery.

MPI Relay—is located under the left side of the dash.

Radiator Fan Motor Relay—is located at the left front of the engine compartment.

Relay Box—is located under the left side of the dash.

Starter Relay—is located in the relay box under the left side of the dash.

Sub-Fusible Link Box—is located at the left front of the engine compartment.

Taillight Relay—is located in the relay box under the left side of the dash.

Transaxle Control Unit (TCU)—is located under the right front seat.

Sigma

A/C Condenser Fan Motor Relay—is located at the left front of the engine compartment.

A/C Control Unit—is located under the center of the dash.

A/C Power Relay—is located under the center of the dash.

A/C Relay—is located at the right side of the engine compartment in the relay box.

ABS Motor Relay—is located at the hydraulic unit.

ABS Relay—is located at under the left side of the dash in the relay box.

ABS Valve Relay—is located at the hydraulic unit.

Alternator Relay—is located at the right front of the engine compartment.

Cruise Control Unit—is located under the left side of the dash, near the ETACS unit.

Defogger Relay—is located at under the left side of the dash in the relay box.

Door Lock Control Relay—is located at the left kick panel.

ECS Compressor Relay—is located at the right front of the engine compartment.

ECS Control Unit—is located at the right side of the trunk.

Electronic Control Unit—is located at the left quarter panel in the trunk.

Electronic Timer and Alarm Control System (ETACS) Control Unit—is located under the left side of the dash.

Engine Control Unit (ECU)—is located next to the blower motor.

Headlight Relay—is located at the right side of the engine compartment in the relay box.

Heater Relay—is located at under the left side of the dash in the relay box.

Magnetic Clutch Relay—is located at the right side of the engine compartment in the relay box.

MPI Control Relay—is located next to the blower motor.

Power Window Relay—is located at the right side of the engine compartment in the relay box.

Radiator Fan Relay—is located at the right side of the engine compartment in the relay box.

Start Inhibitor Relay—is located next to the battery.

Sunroof Relay—is located at the right kick panel.

Taillight Relay—is located at the right side of the engine compartment in the relay box.

Transaxle Control Unit—is located under the center of the dash.

Transaxle Oil Cooler Fan Motor Relay—is located at the right front of the engine compartment.

Turn Signal/Hazard Flasher Unit—is located at under the left side of the dash in the relay box.

Wiper Control Relay—is located at the right front of the engine compartment.

Starion

A/C Compressor Clutch Relay—is located on the relay block on the left front inner fender panel.

A/C Heater Blower Motor Relay—is located on the blower motor housing.

A/C Power Relay—is located on the evaporator housing.

Automatic A/C Control Unit—is located behind the automatic A/C panel.

Automatic Seat Belt Control Unit—is located in the center console, in the passenger's compartment.

Blower Motor High Speed Relay—is located under the right hand side of the instrument panel.

Blower Motor Relay—is located under the left hand side of the instrument panel, to the right of the steering column.

Blower Motor Starter Cut-Out Relay—is located under the right hand side of the instrument panel.

Condenser Fan Relay—is located on the left hand inner fender panel in the engine compartment.

Cooling Fan Relays—are located on the right front corner of the engine compartment.

ECI Control Relay—is located behind the right hand side kick panel.

ECI Control Unit—is located behind the right hand side kick panel.

ECU Control Relay—is located above the ECU on the right side kick panel.

Electronic Control Unit (ECU)—is located on the right hand side kick panel.

Electronic Time & Alarm Control System (ETACS) Unit—is located under the driver's seat.

Fog Light Relay—is located near the right headlight door.

Hazard Flasher—is located under the left hand side instrument panel.

Headlight Lighting Relay—is located on the front relay panel.

Left Headlight Pop-Up Relay—is located on the front relay panel.

Overdrive Relay—is located under the left side of the instrument panel.

Passing Control Relay—is located under the right side of the instrument panel near radio.

Passing Control Relay (w/Theft Alarm System)—is located on the left hand side of the firewall.

Pop-Up Relay—is located on the left hand front corner of the engine compartment.

Power Antenna Relay—is located in the front of the left hand side rear taillight.

Power Door Lock Relay—is located under the left hand side of the instrument panel.

Power Door Un-Lock Relay—is located under the left hand side of the instrument panel.

Power Window Relay—is located on the front relay panel, in the left hand front corner of the engine compartment.

Radiator Fan Relay—is located on the right hand front corner of the engine compartment.

Rear Brake Lock-Up Control Relay—is located under the left hand side of the instrument panel.

Rear Brake Lock-Up Control Unit—is located under the right hand rear luggage compartment panel.

Relay Panels—are located on the left hand side kick panel and the front of the engine compartment.

Speed Control Unit—is located behind the trim panel on the rear of the luggage compartment.

Starter Inhibitor Relay—is located on the left hand front corner in the engine compartment.

Taillight Lighting Relay—is located on the front relay panel.

Turn Signal Flasher—is located under the left hand side of the instrument panel.

Vacuum Pump Relay—is located on the right hand rear corner of the engine compartment.

Wiper Power Relay—is located on the front relay panel, on the left front inner fender panel.

Wiper Low & High Relay—is located on the front relay panel, on the left front inner fender panel.

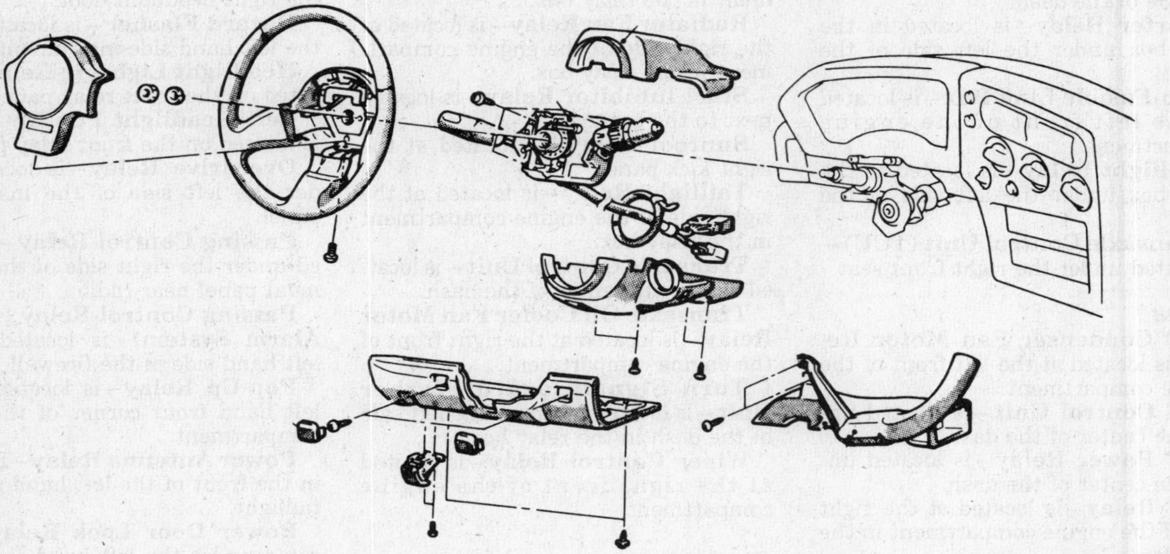

Combination switch – Eclipse

Nissan 15

200SX, 240SX, 300ZX, Maxima, Pulsar, Sentra, Stanza — All Models

SERIAL NUMBER IDENTIFICATION

Vehicle Identification Plate

The vehicle identification plate is attached to the hood ledge or the firewall. The VIN plate is mounted on the front of the left strut housing on the 1988 300ZX and on the radiator core on 1990–92 300ZX. The identification plate gives the vehicle type, model, engine displacement in cc, SAE horsepower rating, wheelbase, engine number and chassis number.

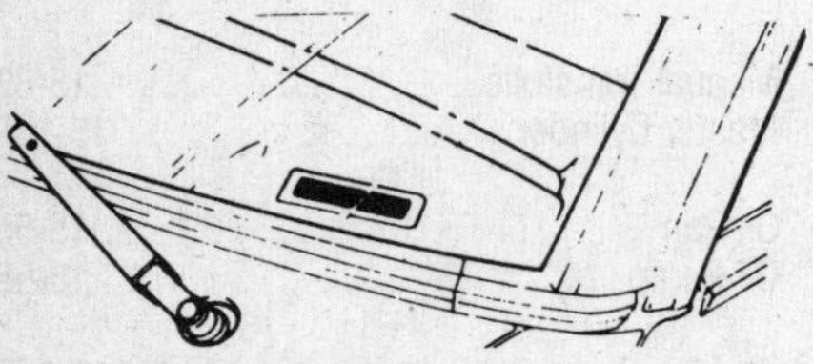

VIN location

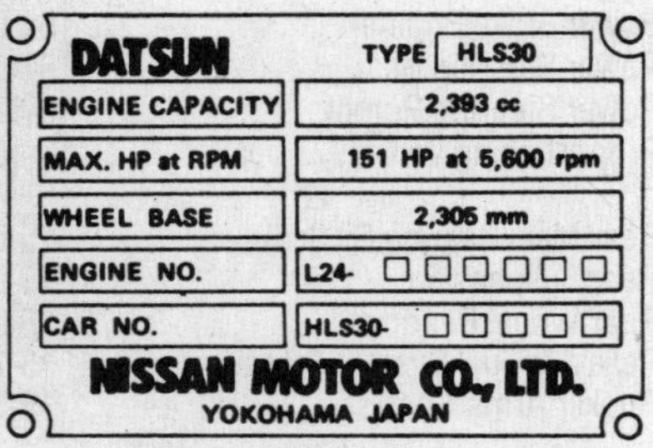

DATSUN
TYPE HLS30
ENGINE CAPACITY
2,393 cc
MAX. HP at RPM
151 HP at 5,600 rpm
WHEEL BASE
2,305 mm
ENGINE NO.
L24-
CAR NO.
HLS30-
NISSAN MOTOR CO., LTD.
YOKOHAMA JAPAN

Vehicle identification plate

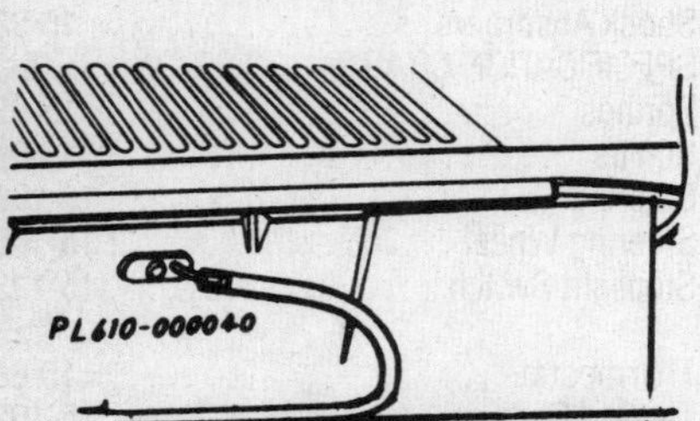

Chassis number location

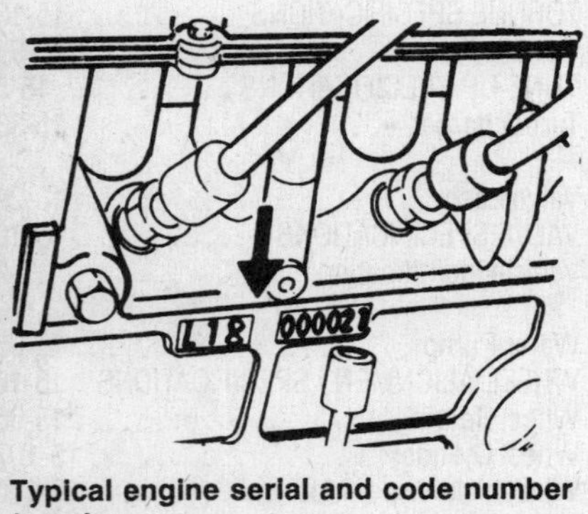

Typical engine serial and code number location

Engine Number

On most vehicles, the engine number is stamped on the right side top edge of the cylinder block. On the 200SX, the number is stamped on the left rear edge of the block, next to the bell housing, looking from driver side seat. On 240SX, the number is stamped on the block just below the valve cover looking from the driver's seat. On the 300ZX, the number is stamped on the right rear edge of the right cylinder bank, looking from driver side seat. On the Maxima, the number can be found on the driver's side edge of the front cylinder bank, looking from driver's seat. On the 1990–92 Stanza, the number is stamped on the cylinder block just below the valve cover looking down at the front of the engine. The engine serial number is preceded by the engine model code.

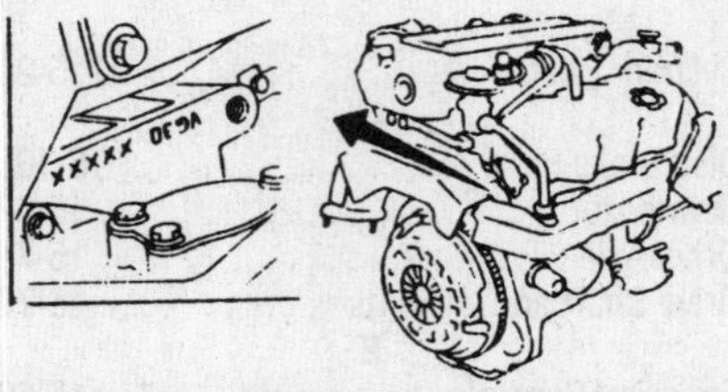

Typical engine serial and code number location

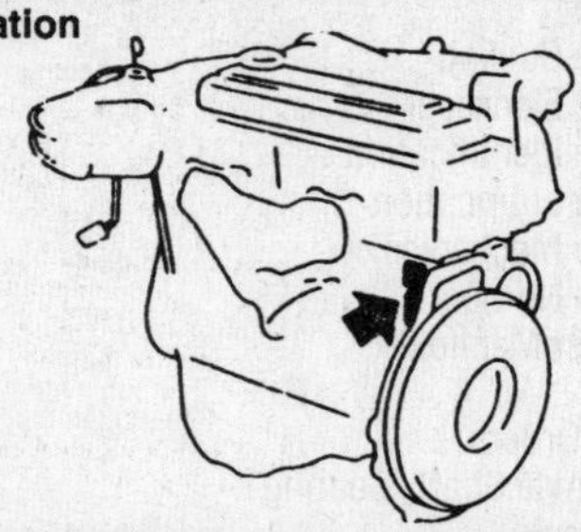

Engine indentification number location – most engines

Engine serial number location on 240SX and 1990–92 Stanza

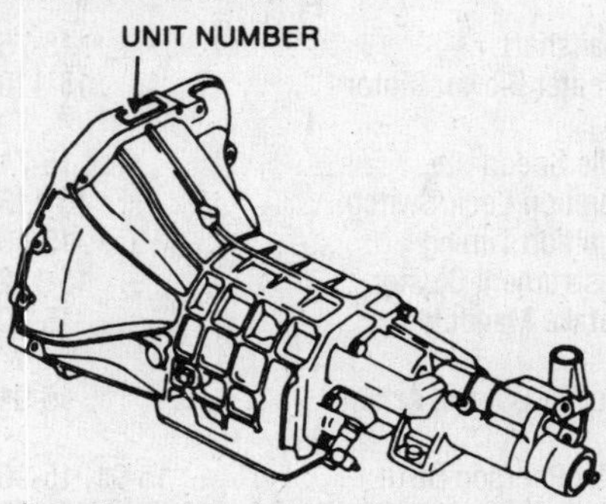

Location of the manual transmission serial number

Chassis Number

The chassis number is on the firewall under the hood on all models. On the 240SX, the chassis number plate is affixed to the firewall next to the wiper

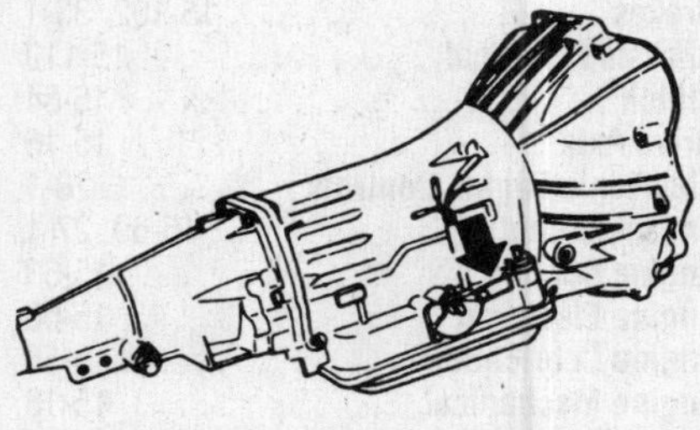

Location of the automatic transmission serial number

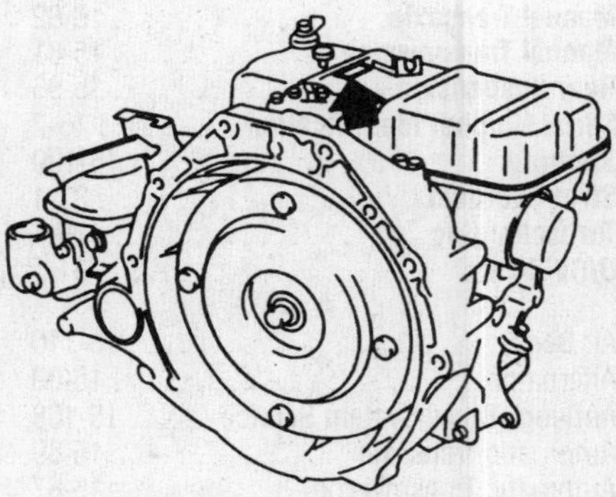

Transaxle serial number location automatic

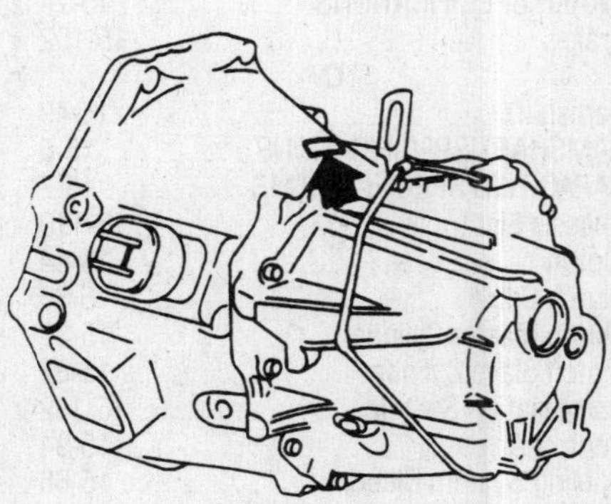

Transaxle serial number location manual

motor on the passenger's side of the engine compartment. All vehicles also have the chassis number (vehicle identification number) on a plate attached to the top of the instrument panel on the driver's side, visible through the windshield. The chassis serial number is preceded by the model designation. All models have an Emission Control information label affixed to the firewall or on the underside of the hood.

Transmission/ Transaxle Number

The transmission/transaxle identification number tag is attached to the upper area or side area of the unit.

SPECIFICATIONS

ENGINE IDENTIFICATION

Year	Model	Engine Displacement cu. in. (cc/liter)	Engine Series Identification	No. of Cylinders	Engine Type
1988	200SX	120.4 (1974/2.0)	CA20E	4	SOHC
		180.6 (2960/3.0)	VG30E	6	SOHC
	300ZX	180.6 (2960/3.0)	VG30E	6	SOHC
		180.6 (2960/3.0)	VG30ET (Turbo)	6	SOHC
	Maxima	180.6 (2690/3.0)	VG30E	6	SOHC
	Pulsar	97.4 (1597/1.6)	E16i	4	SOHC
		110.3 (1809/1.8)	CA18DE	4	DOHC
	Sentra	97.4 (1597/1.6)	E16i	4	SOHC
	Stanza	120.4 (1974/2.0)	CA20E	4	SOHC
1989	240SX	145.8 (2389/2.4)	KA24E	4	SOHC
	300ZX	180.6 (2960/3.0)	VG30E	6	SOHC
		180.6 (2960/3.0)	VG30ET (Turbo)	6	SOHC
	Maxima	180.6 (2690/3.0)	VG30E	6	SOHC
	Pulsar	97.5 (1597/1.6)	GA16i	4	SOHC
		110.3 (1809/1.8)	CA18DE	4	DOHC
	Sentra	97.5 (1597/1.6)	GA16i	4	SOHC
	Stanza	120.4 (1974/2.0)	CA20E	4	SOHC
1990	240SX	145.8 (2389/2.4)	KA24E	4	SOHC
	300ZX	180.6 (2960/3.0)	VG30DE	6	DOHC
		180.6 (2960/3.0)	VG30DETT (Twin Turbo)	6	DOHC
	Maxima	180.6 (2960/3.0)	VG30E	6	SOHC
	Pulsar	97.5 (1597/1.6)	GA16i	4	SOHC
	Sentra	97.5 (1597/1.6)	GA16i	4	SOHC
	Stanza	145.8 (2389/2.4)	KA24E	4	SOHC
1991-92	240SX	145.8 (2389/2.4)	KA24DE	4	DOHC
	300ZX	180.6 (2960/3.0)	VG30DE	6	DOHC
		180.6 (2960/3.0)	VG30DETT (Twin Turbo)	6	DOHC
	Maxima	180.6 (2960/3.0)	VG30E	6	SOHC
	Sentra	97.5 (1597/1.6)	GA16DE	4	DOHC
		122 (1998/2.0)	SR20DE	4	DOHC
	Stanza	145.8 (2389/2.4)	KA24E	4	SOHC

SOHC—Single overhead camshaft
DOHC—Double overhead camshaft

GENERAL ENGINE SPECIFICATIONS

Year	Model	Engine Displacement cu. in. (cc)	Fuel System Type	Net Horsepower @ rpm	Net Torque @ rpm (ft. lbs.)	Bore × Stroke (in.)	Com-pression Ratio	Oil Pressure @ rpm
1988	200SX	120.4 (1974)	EFI	99 @ 5200	116 @ 2800	3.33 × 3.46	8.5:1	60.5 @ 3200
		180.6 (2960)	EFI	165 @ 5200	168 @ 3600	3.43 × 3.27	9.0:1	59 @ 3200
	300ZX	180.6 (2960)	EFI	165 @ 5200	174 @ 3600	3.43 × 3.27	9.0:1	59 @ 3200
		180.6 (2690) ①	EFI	205 @ 5200	227 @ 3600	3.43 × 3.27	8.3:1	58.5 @ 3200
	Maxima	180.6 (2960)	EFI	157 @ 5200	168 @ 3600	3.43 × 3.27	9.0:1	59 @ 3200
	Pulsar	97.4 (1597)	EFI	70 @ 5000	94 @ 2800	2.99 × 3.46	9.4:1	64 @ 3200
		110.3 (1809)	EFI	125 @ 6400	115 @ 4800	3.27 × 3.29	10.0:1	67 @ 2000
	Sentra	97.4 (1597)	EFI	70 @ 5000	94 @ 2800	2.99 × 3.46	9.4:1	64 @ 3000
	Stanza	120.4 (1974)	EFI	97 @ 5200	114 @ 2800	3.33 × 3.46	8.5:1	58 @ 3000
1989	240SX	145.8 (2389)	EFI	140 @ 5600	152 @ 4400	3.50 × 3.78	9.1:1	65 @ 3000
	300ZX	180.6 (2960)	EFI	165 @ 5200	174 @ 4000	3.43 × 3.27	9.0:1	59 @ 3200
		180.6 (2960) ①	EFI	205 @ 5200	227 @ 3600	3.43 × 3.27	8.3:1	58 @ 3200
	Maxima	180.6 (2690)	EFI	160 @ 5200	182 @ 2800	3.43 × 3.27	9.0:1	59 @ 3200
	Pulsar	97.5 (1597)	EFI	90 @ 6000	96 @ 3200	2.99 × 3.46	9.4:1	64 @ 3000
		110.3 (1809)	EFI	96 @ 3200	115 @ 4800	3.27 × 3.29	9.5:1	67 @ 2000
	Sentra	97.5 (1597)	EFI	90 @ 6000	96 @ 3200	2.99 × 3.47	9.4:1	64 @ 3000
	Stanza	120.4 (1974)	EFI	94 @ 5400	114 @ 2800	3.33 × 3.47	8.5:1	61 @ 3200
1990	240SX	145.8 (2389)	EFI	140 @ 5600	152 @ 4400	3.50 × 3.78	8.6:1	60–70 @ 3000
	300ZX	180.6 (2960)	EFI	222 @ 6400	198 @ 4800	3.43 × 3.27	10.5:1	51–65 @ 3000
		180.6 (2960) ②	EFI	③	283 @ 3600	3.43 × 3.27	8.1:1	51–65 @ 3000
	Maxima	180.6 (2960)	EFI	160 @ 5200	181 @ 2800	3.43 × 3.27	9.0:1	53–65 @ 3200
	Pulsar	97.5 (1597)	EFI	90 @ 6000	96 @ 3200	2.99 × 3.47	9.4:1	57–71 @ 3000
	Sentra	97.5 (1597)	EFI	90 @ 6000	96 @ 3200	2.99 × 3.47	9.4:1	57–71 @ 3000
	Stanza	145.8 (2389)	EFI	138 @ 5600	148 @ 3200	3.50 × 3.78	8.6:1	60–70 @ 3000
1991–92	240SX	145.8 (2389)	EFI	155 @ 5600	160 @ 4400	3.50 × 3.78	8.6:1	60–70 @ 3000
	300ZX	180.6 (2960)	EFI	222 @ 6400	198 @ 4800	3.43 × 3.27	10.5:1	51–65 @ 3000
		180.6 (2960) ②	EFI	③	283 @ 3600	3.43 × 3.27	8.5:1	51–65 @ 3000
	Maxima	180.6 (2960)	EFI	160 @ 5200	181 @ 2800	3.43 × 3.27	9.0:1	53–65 @ 3200
	Sentra	97.5 (1597)	EFI	110 @ 6000	108 @ 4000	2.99 × 3.46	9.5:1	50–64 @ 3000
		122 (1998)	EFI	140 @ 6400	132 @ 4800	3.39 × 3.39	9.5:1	46–57 @ 3200
	Stanza	145.8 (2389)	EFI	138 @ 5600	148 @ 4400	3.50 × 3.78	8.6:1	60–70 @ 3000

EFI: Electronic Fuel Injection
① Turbo
② Twin Turbo
③ MT: 300 @ 6400
AT: 280 @ 6400

ENGINE TUNE-UP SPECIFICATIONS

Year	Model	Engine Displacement cu. in. (cc)	Spark Plugs Type	Spark Plugs Gap (in.)	Ignition Timing (deg.) ⑬ MT	Ignition Timing (deg.) ⑬ AT	Compression Pressure (psi)	Fuel Pump (psi)	Idle Speed (rpm) MT	Idle Speed (rpm) AT	Valve Clearance In.	Valve Clearance Ex.
1988	200SX	120.4 (1974)	②	0.039–0.043	15B	15B ④	171	36 ⑥	750	750 ④	Hyd.	Hyd.
		180.6 (2690)	BCPR6ES-11	0.039–0.043	20B	20B	173	37 ④ ⑦	700	700	Hyd.	Hyd.
	300ZX	180.6 (2960)	BCPR6ES-11	0.039–0.043	15B	20B	173	37 ⑦	700 ④	700 ④	Hyd.	Hyd.
		180.6 (2690) ①	BCPR6E-11	0.039–0.043	10B	15B	169	44 ⑦	700	650 ④	Hyd.	Hyd.
	Maxima	180.6 (2960)	BCPR6ES-11	0.039–0.043	15B	20B	173	30 ⑤	750	700 ④	Hyd.	Hyd.
	Pulsar	97.4 (1597)	BPR6ES-11	0.039–0.043	7B	7B	181	14 ⑤	800	700 ④	Hyd.	Hyd.
		110.3 (1809)	PFR6A-11	0.039–0.043	15B	15B	199	36 ⑧	800	700 ④	Hyd.	Hyd.
	Sentra	97.4 (1597)	BPR6ES-11	0.039–0.043	7B	7B	181	14 ⑨	800	700 ④	0.011	0.011
	Stanza	120.4 (1974)	②	0.039–0.043	15B	15B	171	43.4	750	700	Hyd.	Hyd.
1989	240SX	145.8 (2389)	ZFR5D-11	0.039–0.043	15B	15B	192	33 ⑧	750	750	Hyd.	Hyd.
	300ZX	180.6 (2690)	BCPR6ES-11	0.039–0.043	15B	20B	173	37 ⑦	700	700 ④⑩	Hyd.	Hyd.
		180.6 (2960) ①	BCPR6ES-11	0.039–0.043	10B	15B	169	44 ⑦	700	650	Hyd.	Hyd.
	Maxima	180.6 (2960)	BKR6ES-11	0.039–0.043	15B	15B	181	36 ⑧	750	700	Hyd.	Hyd.
	Pulsar	97.5 (1597)	BCPR5ES-11	0.039–0.043	7B	7B ⑪	181 ⑪	43 ⑤	800 ③	750 ③	Hyd.	Hyd.
		110.3 (1809)	PFR6A-11	⑫	15B	15B	199	36 ⑧	800	700 ④	Hyd.	Hyd.
	Sentra	97.5 (1597)	BCPR5ES-11	0.039–0.043	7B	7B	181	43 ⑤	800 ⑪	700 ④⑪	Hyd.	Hyd.
	Stanza	120.4 (1974)	BCPR5ES-11	0.039–0.043	15B	15B	171	37 ⑧	750	700 ④	Hyd.	Hyd.
1990	240SX	145.8 (2389)	ZFRSE-11	0.039–0.043	15B	15B	175	⑮	750	750	Hyd.	Hyd.
	300ZX	180.6 (2960)	PFR6B-11	0.039–0.043	15B	15B	186	⑮	770	750	Hyd.	Hyd.
		180.6 (2960) ①	PFR5B-11B	0.039–0.043	15B	15B	186	⑮	770	750	Hyd.	Hyd.
	Maxima	180.6 (2960)	BKR6ES-11	0.039–0.043	15B	15B	173	⑮	750	700	Hyd.	Hyd.
	Pulsar	97.5 (1597)	BCPRSES-11	0.039–0.043	7B ⑪	7B ⑪	181	⑮	800 ③	900 ③	Hyd.	Hyd.
	Sentra	97.5 (1597)	BCPRSES-11	0.039–0.043	7B ⑪	7B ⑪	181	⑮	800 ⑯	900 ⑯	Hyd.	Hyd.
	Stanza	145.8 (2389)	ZFRSF-11	0.039–0.043	15B	15B	175	⑮	700	700	Hyd.	Hyd.

ENGINE TUNE-UP SPECIFICATIONS

Year	Model	Engine Displacement cu. in. (cc)	Spark Plugs Type	Spark Plugs Gap (in.)	Ignition Timing (deg.) ⑬ MT	Ignition Timing (deg.) ⑬ AT	Compression Pressure (psi)	Fuel Pump (psi)	Idle Speed (rpm) MT	Idle Speed (rpm) AT	Valve Clearance In.	Valve Clearance Ex.
1991	240SX	145.8 (2359)	BKRSE-11	0.039–0.043	20B	20B	175	⑮	750	750	0.012–0.015	0.013–0.016
	300ZX	180.6 (2960)	PFR6B-11	0.039–0.043	15B	15B	186	⑮	700	770	Hyd.	Hyd.
		180.6 (2960) ①	PFR5B-11B	0.039–0.043	15B	15B	186	⑮	700	750	Hyd.	Hyd.
	Maxima	180.6 (2960)	BKRGES-11	0.039–0.043	15B	15B	173	⑮	750	700	Hyd.	Hyd.
	Sentra	97.5 (1597)	BKR5E	0.039–0.043	10B	10B	192	⑮	800	800	0.015	0.016
		122 (1998)	BKR6E	0.031–0.035	15B	15B	178	⑮	800	800	Hyd.	Hyd.
	Stanza	145.8 (2389)	ZFR5E-11	0.039–0.043	15B	15B	175	⑮	750	750	Hyd.	Hyd.
1992	SEE UNDERHOOD SPECIFICATIONS STICKER											

NOTE: The Underhood Specifications sticker often reflects tune-up specification changes made in production. Sticker figures must be used if they disagree with those in this chart

MT—Manual transmission
AT—Automatic transmission
NA—Not adjustable
B—Before Top Dead Center
Hyd.—Hydraulic valve lash adjusters
① Turbocharged model
② Intake side: BCPR6ES-11
Exhaust side: BCPRSES-11
③ Idle speed is computer controlled; not adjustable
④ In drive position
⑤ At idle speed
⑥ Fuel pressure is measured at idle speed between the fuel filter and injector body
⑦ The moment the gas pedal is fully depressed
⑧ Fuel pressure is measured at idle speed between the fuel filter and fuel pipe with the vacuum hose connected at the pressure regulator
⑨ 4WD model—36.6 psi at idle speed
⑩ 600 rpm at high altitudes
⑪ With throttle sensor harness connected
With throttle sensor harness disconnected—7° BTDC ±5°
⑫ Spark plug gap not adjustable
⑬ Ignition timing tolerance: ±2°
⑭ Idle speed tolerance: ±50 rpm
⑮ Fuel pressure is measured at idle speed between the fuel filter and fuel pipe (engine) side
36.3 psi—with pressure regulator vacuum hose connected
43.4 psi—with pressure regulator vacuum hose disconnected
⑯ Hot

FIRING ORDERS

NOTE: To avoid confusion, always replace spark plug wires one at a time.

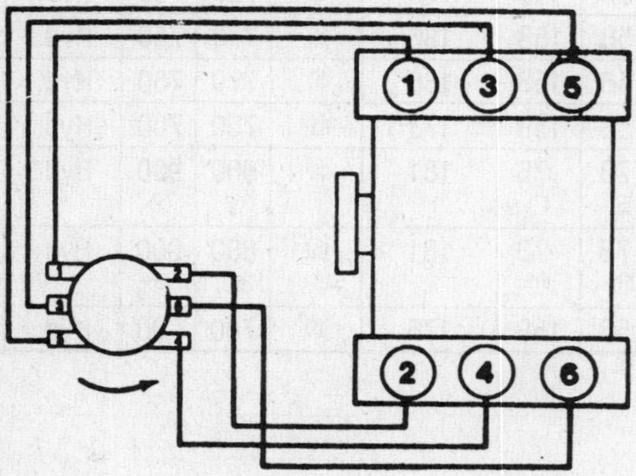

VG30E and VG30ET Series Engines
Engine Firing Order: 1-2-3-4-5-6
Distributor Rotation: Counterclockwise

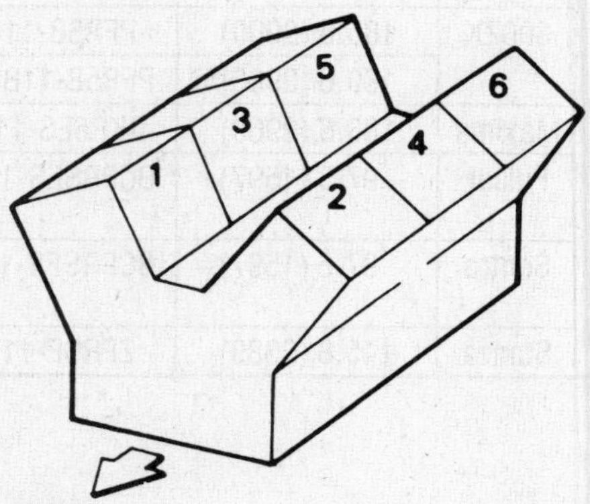

VG30DE and VG30DETT Series Engines
Engine Firing Order: 1-2-3-4-5-6
Distributorless Ignition System

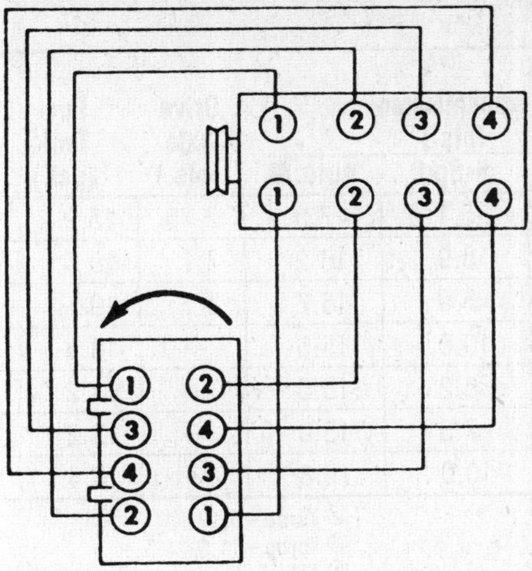

CA20E Engine
Engine Firing Order: 1–3–4–2
Distributor Rotation: Counterclockwise

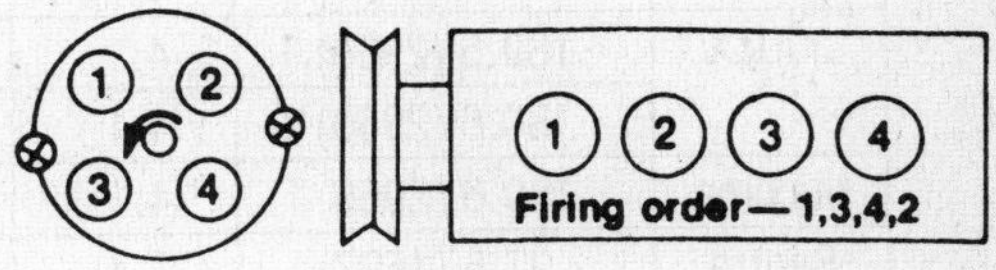

E16I, KA24E, SR20DE, GA16I and GA16DE Engines
Engine Firing Order: 1–3–4–2
Distributor Rotation: Counterclockwise

CAPACITIES

Year	Model	Engine Displacement cu. in. (cc)	Engine Crankcase (qts.) with Filter	Engine Crankcase (qts.) without Filter	Transmission (pts.) 4-Spd	Transmission (pts.) 5-Spd	Transmission (pts.) Auto. ■	Drive Axle (pts.)	Fuel Tank (gal.)	Cooling System (qts.)
1988	200SX	120.4 (1974)	3.9	3.4	—	4.25	14.8	①	14	9.1
		180.6 (2690)	4.5	4.0	—	4.25	14.8	2.75	14	9.6
	300ZX	180.6 (2960)	4.25	3.9	—	4.25	14.8	2.75	19	11.1 ②
	Maxima	180.6 (2960)	4.5	4.1	—	10.0	14.5	—	15.9	9.75
	Pulsar	97.4 (1597)	3.4	2.9	—	5.75	13.2	—	13.25	③
		110.3 (1809)	3.7	3.3	—	10.0	14.4	—	13.25	⑥
	Sentra	97.4 (1597)	3.4	3.0	5.7	5.9	13.2	⑦	13.25 ⑧	③
	Stanza	120.4 (1974)	3.75	3.25	—	10.0	14.4	⑦	15.9 ⑤	7.75 ⑨
1989	240SX	145.8 (2389)	3.75	3.4	—	5.1	17.5	2.75	15.9	7.1 ⑪
	300ZX	180.6 (2960)	4.25	3.8	—	4.25 ⑫	14.8	2.75	19	11.1 ⑬
	Maxima	180.6 (2960)	4.5	4.1	—	10.0	15.5	—	15.9	8.75
	Pulsar	97.4 (1597)	3.4	3.0	—	5.75	6.6	—	13.25	⑭
		110.3 (1809)	3.75	3.25	—	10.0	7.25	—	13.25	5.9
	Sentra	97.5 (1597)	3.4	3.0	5.75	5.9	13.25	⑦	13.25 ⑧	⑩
	Stanza	120.4 (1974)	3.75	3.25	—	10.0	7.25	—	15.9	7.75
1990	240SX	145.8 (2389)	3.75	3.4	—	5.1	17.5	2.75	15.9	7.1 ⑪
	300ZX	180.6 (2960)	4.4	3.9	—	5.9	16.2	3.1	19	10.6
		180.6 (2960) ③	4.4	3.9	—	5.9	16.2	3.1	19	10.6
	Maxima	180.6 (2960)	4.5	4.1	—	10.0	15.5	—	15.9	8.75
	Pulsar	97.5 (1597)	3.4	3.0	—	5.9	13.2	—	13.25	⑭
	Sentra	97.5 (1597)	3.4	3.0	5.75	5.9	13.2	⑦	13.25 ⑧	⑩
	Stanza	145.8 (2389)	3.75	3.25	—	10.0	15.8	—	16.4	7.9

CAPACITIES

Year	Model	Engine Displacement cu. in. (cc)	Engine Crankcase (qts.) with Filter	Engine Crankcase (qts.) without Filter	Transmission (pts.) 4-Spd	Transmission (pts.) 5-Spd	Transmission (pts.) Auto.■	Drive Axle (pts.)	Fuel Tank (gal.)	Cooling System (qts.)
1991-92	240SX	145.8 (2389)	3.75	3.4	—	5.1	17.5	2.75	15.9	7.1 ⑪
	300ZX	180.6 (2960)	4.1	3.7	—	5.9	16.2	3.1	19	10.6
		180.6 (2960) ③	4.1	3.7	—	5.9	16.2	3.1	19	10.6
	Maxima	180.6 (2960)	4.1	3.7	—	10.0	15.5	—	18.5	8.7
	Sentra	97.5 (1597)	3.4	3.0	—	6.2	15.0	—	13.2	5.5
		122 (1998)	3.7	3.4	—	7.5	15.0	—	13.2	6.2
	Stanza	145.8 (2389)	3.7	3.2	—	10.0	15.8	—	16.4	7.9

■ Figure is for drain and refill
— Not applicable
① Solid rear axle—2.1
IRS—2.75
② Turbo—11.5
③ Twin turbo engine
④ 4WD—12.4
⑤ 4WD—13.25
⑥ MT—5.9; AT—6.1
⑦ Rear differential carrier on 4WD—2.1
Transfer case on 4WD—2.2
⑧ 4WD—12.4
⑨ Station wagon with heater 7.1
and without heater 6.1
⑩ MT—5.75
2WD w/AT—5.75
4WD w/AT—6.25
⑪ Reservoir Capacity—.75
⑫ Turbo—5.1
⑬ Turbo—11.6
⑭ MT—5.75
AT—6.25

CAMSHAFT SPECIFICATIONS

All measurements given in inches.

Year	Engine Displacement cu. in. (cc)	Journal Diameter 1	Journal Diameter 2	Journal Diameter 3	Journal Diameter 4	Journal Diameter 5	Lobe Lift In.	Lobe Lift Ex.	Bearing Clearance	Camshaft End Play
1988	CA20E 120.4 (1974)	1.8085–1.8092	1.8085–1.8092	1.8085–1.8092	1.8085–1.8092	1.8077–1.8055	0.335	0.374	0.0040 ①	0.0028–0.0055
	V-Series 180.6 (2960)	1.8866–1.8874 ②	1.8472–1.8480	1.8472–1.8480	1.8472–1.8480	1.6701–1.6709	NA	NA	0.0018–0.0035	0.0012–0.0024
	E16i 97.4 (1597)	1.6515–1.6522	1.6498–1.6505	1.0515–1.6522	1.6498–1.6505	1.6515–1.6522	NA	NA	0.0014–0.0030	0.0059–0.0114
	CA18DE 110.3 (1809)	1.0998–1.1006	1.0998–1.1006	1.0998–1.1006	1.0998–1.1006	1.0998–1.1006	0.335	0.335	0.0018–0.0035	0.0028–0.0059
1989	CA20E 120.4 (1974)	1.8085–1.8092	1.8085–1.8092	1.8085–1.8092	1.8085–1.8092	1.8077–1.8055	0.335	0.374	0.0040 ①	0.0028–0.0055
	KA24E 145.8 (2389)	1.2967–1.2974	1.2967–1.2974	1.2967–1.2974	1.2967–1.2974	1.2967–1.2974	0.409	0.409	0.0018–0.0035	0.0028–0.0059
	V-Series 180.6 (2960)	1.8866–1.8874 ②	1.8472–1.8480	1.8472–1.8480	1.8472–1.8480	1.6701–1.6709	NA	NA	0.0018–0.0035	0.0012–0.0024
	GA16i 97.5 (1597)	1.6510–1.6518	1.6510–1.6518	1.6510–1.6518	1.6510–1.6518	1.6510–1.6518	NA	NA	0.0018–0.0035	0.0012–0.0059
	CA18DE 110.3 (1809)	1.0998–1.1006	1.0998–1.1006	1.0998–1.1006	1.0998–1.1006	1.0998–1.1006	0.335	0.335	0.0018–0.0035	0.0028–0.0059

CAMSHAFT SPECIFICATIONS

All measurements given in inches.

Year	Engine Displacement cu. in. (cc)	Journal Diameter 1	2	3	4	5	Lobe Lift In.	Lobe Lift Ex.	Bearing Clearance	Camshaft End Play
1990	GA16i 97.5 (1597)	1.6510–1.6518	1.6510–1.6518	1.6510–1.6518	1.6510–1.6518	1.6510–1.6518	NA	NA	0.0018–0.0035	0.0012–0.0051
	KA24E 145.8 (2389)	1.2967–1.2974	1.2967–1.2974	1.2967–1.2974	1.2967–1.2974	1.2967–1.2974	0.409	0.409	0.0018–0.0035	0.0028–0.0059
	VG30DE 180.6 (2960)	1.0998–1.1006	1.0998–1.1006	1.0998–1.1006	1.0998–1.1006	1.0998–1.1006	NA	NA	0.0018–0.0035	0.0018–0.0035
	VG30DETT 180.6 (2960)	1.0998–1.1006	1.0998–1.1006	1.0998–1.1006	1.0998–1.1006	1.0998–1.1006	NA	NA	0.0018–0.0035	0.0018–0.0035
	VG30E 180.6 (2960)	1.8866–1.8874 ②	1.8472–1.8480	1.8472–1.8480	1.8472–1.8480	1.6732–1.6742	NA	NA	0.0024–0.0041	0.0012–0.0024
1991–92	KA24DE 145.8 (2389)	1.0998–1.1006	0.9423–0.9431	0.9423–0.9431	0.9423–0.9431	0.9423–0.9431	NA	NA	0.0018–0.0035	0.0028–0.0059
	VG30DE 180.6 (2960)	1.0998–1.1006	1.0998–1.1006	1.0998–1.1006	1.0998–1.1006	1.0998–1.1006	NA	NA	0.0018–0.0034	0.0012–0.0031
	VG30DETT 180.6 (2960)	1.0998–1.1006	1.0998–1.1006	1.0998–1.1006	1.0998–1.1006	1.0998–1.1006	NA	NA	0.0018–0.0034	0.0012–0.0031
	VG30E 180.6 (2960)	1.8866–1.8874 ②	1.8472–1.8480	1.8472–1.8480	1.8472–1.8480	1.6701–1.6709	NA	NA	0.0018–0.0035	0.0012–0.0024
	GA16DE 97.5 (1597)	1.0998–1.1006	0.9423–0.9431	0.9423–0.9431	0.9423–0.9431	0.9423–0.9431	NA	NA	0.0018–0.0034	0.0045–0.0074
	SR20DE 122.0 (1998)	1.0998–1.1006	1.0998–1.1006	1.0998–1.1006	1.0998–1.1006	1.0998–1.1006	0.394	0.362	0.0018–0.0034	0.0022–0.0055
	KA24E 145.8 (2389)	1.2967–1.2974	1.2967–1.2974	1.2967–1.2974	1.2967–1.2974	1.2967–1.2974	0.409	0.409	0.0018–0.0035	0.0028–0.0059

NA Not available
① Clearance limit
② Front of engine, left hand camshaft only

CRANKSHAFT AND CONNECTING ROD SPECIFICATIONS

All measurements are given in inches.

Year	Engine Displacement cu. in. (cc)	Crankshaft Main Brg. Journal Dia.	Main Brg. Oil Clearance	Shaft End-play	Thrust on No.	Connecting Rod Journal Diameter	Oil Clearance	Side Clearance
1988	CA20E 120.4 (1974)	2.0847–2.0852	0.0016–0.0024	0.0120	3	1.7701–1.7706	0.0008–0.0024	0.0080–0.0120
	V-Series 180.6 (2960)	2.4790–2.4793	0.0011–0.0022	0.0020–0.0067	4	1.9760–1.9675	0.0006–0.0021	0.0079–0.0138
	E16i 97.4 (1597)	1.9661–1.9671	①	0.0020–0.0065	3	1.5733–1.5738	0.0004–0.0017	0.0040–0.0146
	CA18DE 110.3 (1809)	2.0847–2.0856	0.0008–0.0019	0.0020–0.0091	3	1.7698–1.7706	0.0007–0.0018	0.0079–0.0138

CRANKSHAFT AND CONNECTING ROD SPECIFICATIONS

All measurements are given in inches.

Year	Engine Displacement cu. in. (cc)	Crankshaft Main Brg. Journal Dia.	Main Brg. Oil Clearance	Shaft End-play	Thrust on No.	Connecting Rod Journal Diameter	Oil Clearance	Side Clearance
1989	CA20 120.4 (1974)	2.0847–2.0852	0.0008–0.0019	0.0020 0.0071	3	1.7701–1.7706	0.0004–0.0014	0.0080–0.0120
	KAE24 145.8 (2389)	2.3609–2.3612	0.0008–0.0019	0.0020–0.0071	3	1.9672–1.9675	0.0004–0.0017	0.0080–0.0160
	V-Series 180.6 (2960)	2.4790–2.4793	0.0011–0.0022	0.0020–0.0067	4	1.9667–1.9675	0.0006–0.0021	0.0079–0.0138
	GA16i 97.5 (1597)	1.9668–1.9671	0.0008–0.0017	0.0024–0.0071	3	1.5731–1.5738	0.0004–0.0014	0.0079–0.0185
	CA18DE 110.3 (1809)	2.0847–2.0856	0.0008–0.0019	0.0020–0.0071	3	1.7698–1.7706	0.0007–0.0018	0.0079–0.0138
1990	GA16i 97.5 (1597)	1.9668–1.9671	0.0008–0.0017	0.0024–0.0071	3	1.5731–1.5738	0.0004–0.0014	0.0079–0.0185
	KA24E 145.8 (2389)	2.3609–2.3612	0.0008–0.0019	0.0020–0.0071	3	1.7701–1.7706	0.0004–0.0014	0.0080–0.0120
	VG30DE 180.6 (2960)	2.4790–2.4793	0.0011–0.0022	0.0020–0.0071	4	1.9672–1.9675	0.0011–0.0019	0.0079–0.0138
	VG30DETT 180.6 (2960)	2.4790–2.4793	0.0011–0.0022	0.0020–0.0071	4	1.9672–1.9675	0.0011–0.0019	0.0079–0.0138
	VG30E 180.6 (2960)	2.4790–2.4793	0.0011–0.0022	0.0020–0.0067	4	1.9667–1.9675	0.0006–0.0021	0.0079–0.0138
1991-92	KA24DE 145.8 (2389)	2.3609–2.3612	2.3609–2.3612	2.3609–2.3612	3	1.9672–1.9675	0.0004–0.0014	0.0080–0.0160
	VG30DE 180.6 (2960)	2.4790–2.4793	2.4790–2.4793	2.4790–2.4793	4	1.9672–1.9675	0.0011–0.0019	0.0079–0.0138
	VG30DETT 180.6 (2960)	2.4790–2.4793	2.4790–2.4793	2.4790–2.4793	4	1.9672–1.9675	0.0011–0.0019	0.0079–0.0138
	VG30E 180.6 (2960)	2.4790–2.4793	2.4790–2.4793	2.4790–2.4793	4	1.9667–1.9675	0.0006–0.0021	0.0079–0.0138
	GA16DE 97.5 (1597)	1.9668–1.9671	1.9668–1.9671	1.9668–1.9671	3	1.5735–1.5738	0.0004–0.0014	0.0079–0.0185
	SR20DE 122 (1998)	2.1643–2.1646	2.1643–2.1646	2.1643–2.1646	3	1.8885–1.8887	0.0008–0.0018	0.0079–0.0138
	KA24E 145.8 (2389)	2.3609–2.3612	2.3609–2.3612	2.3609–2.3612	3	1.8491–1.9675	0.0004–0.0014	0.0080–0.0160

① No. 1, 3 & 5—0.0012–0.0022
No. 2 & 4—0.0012–0.0036

VALVE SPECIFICATIONS

Year	Engine Displacement cu. in. (cc)	Seat Angle (deg.)	Face Angle (deg.)	Spring Test Pressure (lbs.)	Spring Installed Height (in.)	Stem-to-Guide Clearance (in.) Intake	Exhaust	Stem Diameter (in.) Intake	Exhaust
1988	CA20E 120.4 (1974)	45°	45°30′	129.9 ③	1.959 ④	0.0008–0.0021	0.0016–0.0029	0.2742–0.2748	0.2734–0.2740
	V-Series 180.6 (2960)	45°	45°30′	118 @ 1.18 ②	1.575 ①	0.0008–0.0021	0.0016–0.0029	0.2742–0.2748	0.3136–0.3138
	E16i 97.4 (1597)	45°	45°30′	—	1.543	0.0008–0.0020	0.0018–0.0030	0.2744–0.2750	0.2734–0.2740

VALVE SPECIFICATIONS

Year	Engine Displacement cu. in. (cc)	Seat Angle (deg.)	Face Angle (deg.)	Spring Test Pressure (lbs.)	Spring Installed Height (in.)	Stem-to-Guide Clearance (in.)		Stem Diameter (in.)	
						Intake	Exhaust	Intake	Exhaust
1988	CA18DE 110.3 (1809)	45°	45°30′	⑤	⑥	0.0008–0.0021	0.0016–0.0029	0.2348–0.2354	0.2341–0.2346
1989	CAE20 120.4 (1974)	45°	45°30′	⑦	⑧	0.0008–0.0021	0.0016–0.0029	0.2742–0.2748	0.2734–0.2740
	KAE24 145.8 (2389)	45°	45°30′	⑨	⑩	0.0008–0.0021	0.0016–0.0028	0.2742–0.2748	0.3129–0.3134
	V-Series 180.6 (2960)	45°	45°30′	117.7 @ 1.181 ⑪	2.016 ⑤	0.0008–0.0021	0.0016–0.0029	0.2742–0.2748	0.3136–0.3138 ⑬
	GA16i 97.5 (1597)	45°	45°30′	⑫	1.634 ⑬	0.0008–0.0020	0.0008–0.0020	0.2348–0.2354	0.2582–0.2587
	CA18DE 110.3 (1809)	45°	45°30′	⑭	⑥	0.0008–0.0020	0.0008–0.0020	0.2348–0.2354	0.2582–0.2587
1990	GA16i 97.5 (1597)	45°	45°15′ 45°45′	⑫	⑬	0.0008–0.0020	0.0012–0.0022	0.2348–0.2354	0.2582–0.2587
	KA24E 145.8 (2389)	45°	45°	⑨	⑩	0.0008–0.0021	0.0016–0.0028	0.2742–0.2748	0.3129–0.3134
	VG30DE 180.6 (2960)	45°15′ 45°45′	45°	⑮	⑥	0.0008–0.0021	0.0016–0.0028	0.2348–0.2354	0.2341–0.2346
	VG30DETT 180.6 (2960)	45°15′ 45°45′	45°	⑮	⑥	0.0008–0.0021	0.0016–0.0028	0.2348–0.2354	0.2341–0.2346
	VG30E 180.6 (2960)	45°15′ 45°45′	45°	⑯	⑰	0.0008–0.0021	0.0016–0.0029	0.2742–0.2748	0.3136–0.3138
1991–92	KA24DE 145.8 (2389)	45°15′ 45°45′	45°	123 @ 1.024	⑱	0.0008–0.0021	0.0016–0.0029	0.2742–0.2748	0.2734–0.2740
	VG30DE 180.6 (2960)	45°15′ 45°45′	45°	120 @ 1.043	⑥	0.0008–0.0021	0.0016–0.0029	0.2348–0.2354	0.2341–0.2346
	VG30DETT 180.6 (2960)	45°15′ 45°45′	45°	120 @ 1.043	⑥	0.0008–0.0021	0.0016–0.0029	0.2348–0.2354	0.2341–0.2346
	VG30E 180.6 (2960)	45°15′ 45°45′	45°	⑲	⑰	0.0008–0.0021	0.0016–0.0029	0.2742–0.2748	0.3136–0.3138
	GA16DE 97.5 (1597)	45°15′ 45°45′	45°	NA	⑳	0.0008–0.0020	0.0016–0.0028	0.2152–0.2157	0.2144–0.2150
	SR20DE 122 (1998)	45°15′ 45°45′	45°	130 @ 1.181	㉑	0.0008–0.0021	0.0016–0.0029	0.2348–0.2354	0.2341–0.2346
	KA24E 145.8 (2389)	45°30′ 45°30′	45°	⑨	⑩	0.0008–0.0021	0.0016–0.0028	0.2742–0.2748	0.3129–0.3134

① Outer; Inner—1.378
② Outer; Inner—57 @ 0.98
③ Outer; Inner—56 @ 0.965
④ Outer; Inner—1.736
⑤ 0.650 in. @ 121 lbs. of load
⑥ 1.697 in. free height
⑦ Outer—129.9 @ 2.32
Inner—66.6 @ 1.19
⑧ Free height:
Outer—1.959
Inner—1.736
⑨ Intake:
Outer—135.8 @ 1.480
Inner—63.9 @ 1.283
Exhaust:
Outer—144 @ 1.343
Inner—73.9 @ 1.146
⑩ Free height:
Intake—outer; 2.261, inner; 2.100
Exhaust—outer; 1.343, inner; 1.887
⑪ Outer; Inner—57.3 @ 0.9840
Maxima; 300ZX—0.3128-0.3134
⑫ Intake—110.0 @ 1.331
Exhaust—122.6 @ 1.346
⑬ Free height:
Intake—2.071
Exhaust—2.154
⑭ 162 @ 2.9
⑮ 26.5 @ 1.043
⑯ 25 @ 0.984
⑰ Free height:
Outer—2.016
Inner—1.736
⑱ 1.756—Free height
⑲ Outer—117 lbs. @ 1.181
Inner—57 lbs. @ 0.984
⑳ Free height:
Intake—2.071
Exhaust—2.154
㉑ 1.9433—Free height

PISTON AND RING SPECIFICATIONS

All measurements are given in inches.

Year	Engine Displacement cu. in. (cc)	Piston Clearance	Ring Gap: Top Compression	Ring Gap: Bottom Compression	Ring Gap: Oil Control	Ring Side Clearance: Top Compression	Ring Side Clearance: Bottom Compression	Ring Side Clearance: Oil Control
1988	CA20E 120.4 (1974)	0.0010–0.0018	0.0098–0.0201	0.0059–0.0122	0.0079–0.0299	0.0016–0.0029	0.0012–0.0025	—
	V-Series 180.6 (1960)	0.0010–0.0018	0.0083–0.0173 ④	0.0071–0.0173	0.0079–0.0299	0.0016–0.0029	0.0012–0.0025	0.0006–0.0075
	E16i 97.4 (1597)	0.0009–0.0017	①	②	0.0079–0.0236	0.0016–0.0029	0.0012–0.0025	③
	CA18DE 110.3 (1809)	0.0006–0.0014	0.0087–0.0154	0.0075–0.0177	0.0079–0.0299	0.0016–0.0029	0.0012–0.0025	0.0010–0.0033
1989	CA20E 120.4 (1974)	0.0010–0.0018	0.0098–0.0201	0.0059–0.0122	0.0079–0.0299	0.0016–0.0029	0.0012–0.0025	—
	KA24E 145.8 (2389)	0.0008–0.0016	0.0110–0.0169	⑤	0.0079–0.0236	0.0016–0.0031	0.0012–0.0028	0.0026–0.0053
	V-Series 180.6 (2960)	0.0010–0.0018	0.0083–0.0173 ④	0.0071–0.0173	0.0079–0.0299	0.0016–0.0029	0.0012–0.0025	0.0006–0.0075
	GA16i 97.5 (1597)	0.0006–0.0014	0.0079–0.0138	0.0146–0.0205	0.0079–0.0236	0.0016–0.0031	0.0012–0.0028	
	CA18DE 110.3 (1809)	0.0010–0.0018	0.0098–0.0201	0.0059–0.0122	0.0079–0.0299	0.0016–0.0029	0.0012–0.0025	—
1990	GA16i 97.5 (1597)	0.0006–0.0014	0.0079–0.0138	0.0146–0.0205	0.0079–0.0236	0.0016–0.0031	0.0012–0.0028	—
	KA24E 145.8 (2389)	0.0008–0.0016	0.0110–0.0169	⑤	0.0079–0.0236	0.0016–0.0031	0.0012–0.0028	—
	VG30DE 180.6 (2960)	0.0006–0.0014	0.0083–0.0157	0.0197–0.0299	0.0079–0.0299	0.0016–0.0029	0.0012–0.0025	—
	VG30DETT 180.6 (2960)	0.0010–0.0018	0.0083–0.0157	0.0197–0.0299	0.0079–0.0299	0.0016–0.0029	0.0012–0.0025	—
	VG30E 180.6 (2960)	0.0006–0.0014	0.0083–0.0173	0.0071–0.0173	0.0079–0.0299	0.0016–0.0029	0.0012–0.0025	—
1991–92	KA24DE 145.8 (2389)	0.0008–0.0016	0.0110–0.0205	0.0177–0.0272	0.0079–0.0272	0.0016–0.0031	0.0012–0.0028	—
	VG30DE 180.6 (2960)	0.0006–0.0010	0.0083–0.0157	0.0197–0.0299	0.0079–0.0299	0.0016–0.0029	0.0012–0.0025	0.0006–0.0075
	VG30DETT 180.6 (2960)	0.0010–0.0018	0.0083–0.0157	0.0197–0.0299	0.0079–0.0299	0.0016–0.0029	0.0012–0.0025	0.0006–0.0075
	VG30E 180.6 (2960)	0.0006–0.0014	0.0083–0.0173	0.0071–0.0173	0.0079–0.0299	0.0016–0.0029	0.0012–0.0025	0.0006–0.0075
	GA16DE 97.5 (1597)	0.0006–0.0014	0.0079–0.0138	0.0146–0.0205	0.0079–0.0236	0.0016–0.0031	0.0012–0.0028	—
	SR20DE 122 (1998)	0.0004–0.0012	0.0079–0.0118	0.0138–0.0197	0.0079–0.0236	0.0018–0.0031	0.0012–0.0026	—
	KA24E 145.8 (2389)	0.0008–0.0016	0.0110–0.0205	⑤	0.0079–0.0272	0.0016–0.0031	0.0012–0.0028	—

① Type 1: 0.0055–0.0102
Type 2: 0.0079–0.0118

② Type 1: 0.0110–0.0146
Type 2: 0.0059–0.0098

③ Type 1: 0.0026–0.0055
Type 2: 0.0002–0.0069

④ Turbocharged engine
0.0083–0.0122

⑤ For rings punched with R or T—0.0177–0.0236
For rings punched with N—0.0217–0.0276

TORQUE SPECIFICATIONS

All readings in ft. lbs.

Year	Engine Displacement cu. in. (cc)	Cylinder Head Bolts	Main Bearing Bolts	Rod Bearing Bolts	Crankshaft Pulley Bolts	Flywheel Bolts	Manifold Intake	Manifold Exhaust	Spark Plugs
1988	CA20E 120.4 (1974)	③	33–40	24–27	90–98	72–80	14–19	14–22	14–22
	V-Series 180.6 (2960)	40-47 ①	67–74	④	90–98	72–80	②	13–16	14–22
	E16i 97.4 (1597)	⑤	36–43	23–27	80–94	58–65 ⑥	12–15	12–15	14–22
	CA18DE 110.3 (1809)	⑦	33–40	④	105–112	61–69	14–19	27–35	14–22
1989	CA20E 120.4 (1974)	③	33–40	24–27	90–98	72–80	14–19	14–22	14–22
	KA24E 145.8 (2389)	⑧	34–38	④	87–116	⑬	12–15	14–22	
	V-Series 180.6 (2960)	①	67–74	④	90–98	72–80 ⑨	②	13–16	14–22
	GA16i 97.5 (1597)	①	34–38	⑩	132–152	69–76	12–15	12–15	14–22
	CA18DE 110.3 (1809)	⑦	33–40	⑪	105–112	61–69	12–15	12–15	14–22
1990	GA16i 97.5 (1597)	①	34–38	⑩	98–112	⑮	12–15	12–15	14–22
	KA24E 145.8 (2389)	⑧	34–38	④	87–116	⑫	12–15	12–15	14–22
	VG30DE 180.6 (2960)	①	64–74	④	159–174	61–69	⑭	17–20	14–22
	VG30DETT 180.6 (2960)	①	64–74	④	159–174	61–69	⑭	20–23	14–22
	VG30E 180.6 (2960)	①	67–74	④	90–98	61–69 ⑬	⑭	13–16	14–22
1991–92	KA24DE 145.8 (2389)	①	34–38	④	105–112	⑫	12–14	27–35	14–22
	VG30DE 180.6 (2960)	①	64–74	④	159–174	61–69	⑭	17–20	14–22
	VG30DETT 180.6 (2960)	①	64–74	④	159–174	61–69	⑭	20–23	14–22
	VG30E 180.6 (2960)	①	67–74	④	90–98	61–69	⑭	13–16	14–22
	GA16DE 97.5 (1597)	①	34–38	⑩	98–112	⑮	12–15	16–21	14–22

TORQUE SPECIFICATIONS

All readings in ft. lbs.

Year	Engine Displacement cu. in. (cc)	Cylinder Head Bolts	Main Bearing Bolts	Rod Bearing Bolts	Crankshaft Pulley Bolts	Flywheel Bolts	Manifold Intake	Manifold Exhaust	Spark Plugs
1991-92	SR20DE 122 (1998)	①	54-61	⑪	105-112	61-69	13-15	27-35	14-22
	KA24E 145.8 (2389)	⑧	34-38	④	87-116	⑫	12-15	12-15	14-22

① See text
② Intake bolt: 12-14 ft. lbs.
Intake nut: 17-20 ft. lbs.
③ Tighten in 2 steps:
1st—22 ft. lbs.
2nd—58 ft. lbs.
Then loosen all bolts completely
Final torque is in 2 steps:
1st—22 ft. lbs.
2nd—54-61 ft. lbs.
(If angle torquing, tighten bolt 8 to 83-88 degrees and all other bolts to 75-80 degrees clockwise.)
NOTE: No. 8 bolt is the longest bolt.
④ Tighten in 2 steps:
1st—10-12 ft. lbs.
2nd—28-33 ft. lbs.
(If angle torquing, tighten bolts to 60-65 degrees clockwise.)
⑤ Tighten in 2 steps:
1st—22 ft. lbs.
2nd—51 ft. lbs.
Then loosen all bolts completely.
Final torque in 2 steps:
1st—22 ft. lbs.
2nd—51-54 ft. lbs.
⑥ A/T Drive Plate: 69-76 ft. lbs.
⑦ Tighten in 2 steps:
1st—22 ft. lbs.
2nd—76 ft. lbs.
Then loosen all bolts completely.
Final torque in 2 steps:
1st—22 ft. lbs.
2nd—76 ft. lbs.
(If angle torquing, tighten all bolts to 85-90 degrees clockwise.)
⑧ Tighten in 2 steps:
1st—22 ft. lbs.
2nd—58 ft. lbs.
Then loosen all bolts completely.
Final torque is in 2 steps:
1st—22 ft. lbs.
2nd—54-61 ft. lbs.
(If angle torquing in 2nd step, turn all bolts 80 to 85 degrees clockwise with an angle torque wrench.)
⑨ 300ZX Maxima—61 to 69
⑩ Tighten in 2 steps:
1st—10 to 12 ft. lbs.
2nd—17-21 ft. lbs.
(If angle torquing in 2nd step, turn all nuts 35-40 degrees with an angle torque wrench.)
⑪ Tighten in 2 steps:
1st—10-12 ft. lbs.
2nd—30-33 ft. lbs.
(If angle torquing in 2nd step, turn all nuts 60 to 65 degrees with an angle torque wrench.)
⑫ M/T flywheel—105-112
A/T driveplate—69-76
⑬ Flywheel (M/T) or driveplate (A/T)
⑭ Tighten intake nut in two steps:
1st—2.2-3.6 ft. lbs.
2nd—17-20 ft. lbs.
Tighten intake bolt in two steps:
1st—2.2-3.6 ft. lbs.
2nd—12-14 ft. lbs.
⑮ M/T flywheel—61-69
A/T driveplate—69-76

BRAKE SPECIFICATIONS

All measurements in inches unless noted.

Year	Model	Lug Nut Torque (ft. lbs.)	Master Cylinder Bore	Brake Disc Minimum Thickness	Brake Disc Maximum Runout	Standard Brake Drum Diameter	Minimum Lining Thickness Front	Minimum Lining Thickness Rear
1988	200SX	87-108	0.938	0.630 ①	0.0028 ②	—	0.080	0.080
	300ZX	72-87	0.938	0.787 ③	0.0028 ②	—	0.080	0.080
	Maxima	72-87	1.000	0.787 ①	0.0028 ②	—	0.079	0.079
	Pulsar	72-87	④	⑤	0.0028	8.000	0.079	0.059
	Sentra	72-87	⑧	⑤	0.0028	8.000 ⑥	0.079	0.059
	Stanza	72-87	④	0.787	0.0028	9.000 ⑦	0.079	0.059

BRAKE SPECIFICATIONS

All measurements in inches unless noted.

Year	Model	Lug Nut Torque (ft. lbs.)	Master Cylinder Bore	Brake Disc Minimum Thickness	Brake Disc Maximum Runout	Standard Brake Drum Diameter	Minimum Lining Thickness Front	Minimum Lining Thickness Rear
1989	240SX	72–87	0.875	0.709 ⑨	0.0028 ⑥	—	0.079	0.079
	300ZX	72–87	0.937	0.787 ⑯	0.0028 ⑥	—	0.079	0.079
	Maxima	72–87	1.000	0.787 ①	0.0028 ⑥	9.06	0.079	0.059
	Pulsar	72–87	⑩	⑪	0.0028	8.05	0.079	0.059
	Sentra	72–87	⑫	⑤	0.0028	8.05 ⑬	0.079	0.059
	Stanza	72–87	1.000	0.787	0.0028	9.06	0.079	0.059
1990	240SX	72–87	⑭	⑮	0.0028	—	0.079	0.059
	300ZX	72–87	⑯	⑰	0.0028	—	0.079	0.079
	300ZX Twin Turbo	72–87	—	⑱	0.0028	—	0.079	0.079
	Maxima	72–87	⑲	⑳	0.0028	9.000	0.079	㉑
	Pulsar	72–87	⑩	0.394	0.0028	8.000	0.079	0.059
	Sentra	72–87	⑫	㉒	0.0028	㉓	0.079	0.059
	Stanza	72–87	㉔	㉕	0.0028	9.000	0.079	㉑
1991-92	240SX	72–87	⑭	⑮	0.0028	—	0.079	0.059
	300ZX	72–87	⑯	⑰	0.0028	—	0.079	0.079
	300ZX Twin Turbo	72–87	—	⑱	0.0028	—	0.079	0.079
	Maxima	72–87	⑲	⑳	0.0028	9.000	0.079	㉑
	Sentra	72–87	⑫	㉒	0.0028	㉓	0.079	0.059
	Stanza	72–87	㉔	㉕	0.0028	9.000	0.079	㉑

NOTE: Minimum lining thickness is as recommended by the manufacturer. Due to variation in state inspection regulations, the minimum allowable thickness may be different than recommended.

—Not applicable

① Front disc on V6 models—0.787
Rear disc on all models—0.354

② Rear disc—0.0028

③ Front disc on Turbo—0.945
rear disc on all models—0.709

④ Pulsar
with CA16DE—1.000
with E16i—0.9380
with CA18DE—1.000
Sentra
with gasoline engine—0.938
Stanza
All except 2WD wagon—1.000
2WD wagon—0.9380

⑤ Pulsar
with CA16DE—0.630
with E16i—0.394
with CA18DE—0.630
Sentra
Gasoline engine except wagon 0.394
Gasoline engine wagon—0.630
4WD wagon—0.630

⑥ 4WD—9.000

⑦ Wagon—9.000

⑧ 2WD wagon—0.938
4WD wagon—1.000

⑨ Front; Rear—0.079

⑩ Pulsar
with CA18DE
large—1.000
small—0.812
With GA16i
large—0.937
Small—0.750

⑪ Pulsar
with CA18DE 0.630
with GA16i 0.394

⑫ Sentra 2WD and 4WD
large—1.000
small—0.812

⑬ 4WD—9.06

⑭ With ABS—0.937
without ABS—0.575

⑮ Front disc:
with ABS—0.709
without ABS—0.787
Rear disc: 0.315

⑯ With ABS—0.941
without ABS—0.937

⑰ Front disc—0.945
Rear disc—0.630

⑱ Front disc—1.102
Rear disc—0.630

⑲ GXE, SE (w/o ABS),
GXE (with ABS)—0.937
SE with ABS—1.000

⑳ Front disc—0.787
Rear disc—0.354

㉑ Rear drum—0.059
Rear disc—0.079

㉒ 2WD except wagon—0.394
2WD wagon and all 4WD—0.630

㉓ 2WD—8.000
4WD—9.000

㉔ With ABS—1.000
without ABS—0.937

㉕ Front—0.787
Rear—0.354

WHEEL ALIGNMENT

Year	Model	Caster Range (deg.)	Caster Preferred Setting (deg.)	Camber Range (deg.)	Camber Preferred Setting (deg.)	Toe-in (in.)	Steering Axis Inclination (deg.)
1988	200SX (Front)	2 3/4P–4 1/4P	—	3/8N–1 1/16P	—	1/64P–1/10P	12 3/4
	(Rear)	—	—	1 1/4N–1/4P	—	5/64P–0	—
	300ZX (Front)	5 13/16P–7 5/16P	—	9/16N–15/16P	—	1/32P–1/8P	13 11/16
	(Rear)	—	—	1 15/16N–7/16N	—	①	—
	Maxima (Front)	1 1/4P–2 3/4P	—	7/16N–1/16P	—	1/16P–1/4P	14 1/2
	(Rear)	—	—	1 3/16N–5/16P	—	3/32P–1/4P	—
	Pulsar (Front)	1 3/16P–2 11/16P	—	1 1/4N–1/4P	—	②	14 13/16
	(Rear)	—	—	2N–1/2N	—	①	—
	Sentra 2WD (Front—Cpe)	7/8P–2 3/8P	—	1 1/16N–7/16P	—	③	14 3/4
	(Rear Cpe)	—	—	1 15/16N–7/16N	—	1/32P–1/8P	—
	(Front exc Cpe)	3/4P–2 1/4P	—	15/16N–9/16P	—	③	14 1/2
	(Rear exc Cpe)	—	—	1 7/8N–3/8N	—	0–3/16P	—
	Sentra 4WD (Front)	1/8P–1 5/8P	—	7/8N–5/8P	—	④	13 15/16
	(Rear)	—	—	7/8N–5/8P	—	0–3/16P	—
	Stanza (Front)	1 1/4P–2 3/4P	—	7/16N–1 1/16P	—	1/32P–1/8P	14 5/8
	(Rear)	—	—	1 3/16N–5/16P	—	3/32P–1/4P	—
	Stanza Wagon (Front 2WD)	3/4P–2 1/4P	—	1/4N–1 1/4P	—	1/16P–9/64P	12
	(Rear 2WD)	—	—	1N–1P	—	⑤	—
	(Front 4WD)	9/16P–2 1/16P	—	1/2N–1 1/16P	—	1/64P–1/16P	11 3/4
	(Rear 4WD)	—	—	0–1 1/2P	—	⑥	—
1989	240SX (Front)	6P–7 1/2P	—	1 1/2N–0	—	0–3/16P	13 1/4
	(Rear)	—	—	2N–1/2N	—	1/16–3/32	—
	300ZX (Front)	5 13/16P	—	9/16N–15/16P	—	1/32–1/8	13 7/16
	(Rear)	—	—	1 15/16N–7/16N	—	1/16–3/32	—
	Maxima (Front)	1/2P–2P	—	1N–1/2P	—	1/32–1/8	14 3/8
	(Rear)	—	—	1 5/16N–3/16	—	1/32–1/8P	—
	Pulsar (Front)	1 3/16P–2 11/16P	—	1 1/4N–1/4P	—	⑦	14 13/16
	(Rear)	—	—	2N–1/2N	—	1/16–3/32	—
	Sentra 2WD (Front—Cpe)	7/8P–2 3/8P	—	1 1/16N–7/16P	—	1/32–1/16 ⑧	14 3/4
	(Rear Cpe)	—	—	1 15/16N–7/16N	—	1/32P–1/8P ⑨	—
	(Front exc Cpe)	3/4P–2 1/4P	—	15/16N–9/16P	—	1/32–1/16 ⑧	14 1/2
	(Rear exc Cpe)	—	—	1 7/8N–3/8N	—	0–3/16P	—
	Sentra 4WD (Front)	1/8P–1 5/8P	—	7/8N–5/8P	—	③	13 15/16
	(Rear)	—	—	7/8N–5/8P	—	0–3/16	—
	Stanza (Front)	1 5/16P–2 13/16P	—	7/16N–1 1/16P	—	1/16–5/32	14 5/8
	(Rear)	—	—	1 3/16N–5/16P	—	3/32–5/16 ⑤	—

WHEEL ALIGNMENT

Year	Model	Caster Range (deg.)	Caster Preferred Setting (deg.)	Camber Range (deg.)	Camber Preferred Setting (deg.)	Toe-in (in.)	Steering Axis Inclination (deg.)
1990	240SX (Front)	6P–7 1/2P	—	1 1/2N–0	—	1/32–3/32	13 1/4
	(Rear)	—	—	1 5/8N–5/8N	—	1/32–3/16	—
	300ZX (Front)	9P–10 1/2P	—	1 9/16N–1/16N	—	0–3/32	12 15/16
	(Rear)	—	—	1 5/8N–5/8N	—	1/64–3/16	—
	Maxima (Front)	1/2P–2P	—	1N–1/2P	—	1/32–1/8	14 3/8
	(Rear)	—	—	1 5/16N–3/16P	—	0–5/32P	—
	Pulsar (Front)	1 3/16P–2 11/16P	—	1 1/4N–1/4P	—	⑦	14 13/16
	(Rear)	—	—	2N–1/2N	—	①	—
	Sentra 2WD (Front—Cpe)	7/8P–2 3/8P	—	1 1/16N–7/16N	—	③	14 3/4
	(Rear Cpe)	—	—	2N–1/2N	—	①	—
	(Front exc Cpe)	3/4P–2 1/4P	—	15/16N–9/16P	—	⑧	14 1/2
	(Rear exc Cpe)	—	—	1 7/8N–3/8N	—	0–3/16	—
	Sentra 4WD (Front)	1/8P–1 5/8P	—	7/8N–5/8P	—	③	13 15/16
	(Rear)	—	—	7/8N–5/8P	—	0–3/16	—
	Stanza (Front)	5/8P–2 1/16P	—	1/2N–1P	—	1/16–1/8	14 1/2
	(Rear)	—	—	1 7/16N–3/16	—	0–5/16	—
1991-92	240SX (Front)	6P–7 1/2P	—	1 1/2N–0	—	1/32–3/32	13 1/4
	(Rear)	—	—	1 5/8N–5/8N	—	1/64–3/16	—
	300ZX (Front)	9P–10 1/2P	—	1 9/16N–1/16N	—	0–3/32	12 15/16
	(Rear)	—	—	1 5/8N–5/8N	—	1/64–3/16	—
	Maxima (Front)	1/2P–2P	—	1N–1/2P	—	1/32–1/8	14 3/8
	(Rear)	—	—	1 5/16N–3/16P	—	0–5/32P	—
	Sentra 2WD (Front—Cpe)	7/8P–2 3/8P	—	1 1/16N–7/16N	—	③	14 3/4
	(Rear Cpe)	—	—	2N–1/2N	—	①	—
	(Front exc Cpe)	3/4P–2 1/4P	—	15/16N–9/16P	—	⑧	14 1/2
	(Rear exc Cpe)	—	—	1 7/8N–3/8N	—	0–3/16	—
	Sentra 4WD (Front)	1/8P–1 5/8P	—	7/8N–5/8P	—	③	13 15/16
	(Rear)	—	—	7/8N–5/8P	—	0–3/16	—
	Stanza (Front)	5/8P–2 1/16P	—	1/2N–1P	—	1/16–1/8	14 1/2
	(Rear)	—	—	1 7/16N–3/16	—	0–5/16	—

N—Negative
P—Positive
① 1/16 Toe Out—3/32 Toe In
② 1/16 Toe Out—1/16 Toe In
③ 1/32 Toe Out—1/16 Toe In
④ 1/32 Toe Out—1/32 Toe In
⑤ 3/32 Toe Out—5/16 Toe In
⑥ 5/32 Toe Out—0
⑦ 1/16 Toe Out—1/16 Toe In
⑧ 1/32 Toe Out—1/16 Toe In
⑨ 1/32 Toe Out—1/8 Toe In

ENGINE MECHANICAL

NOTE: Disconnecting the negative battery cable on some vehicles may interfere with the functions of the on board computer systems and may require the computer to undergo a relearning process, once the negative battery cable is reconnected.

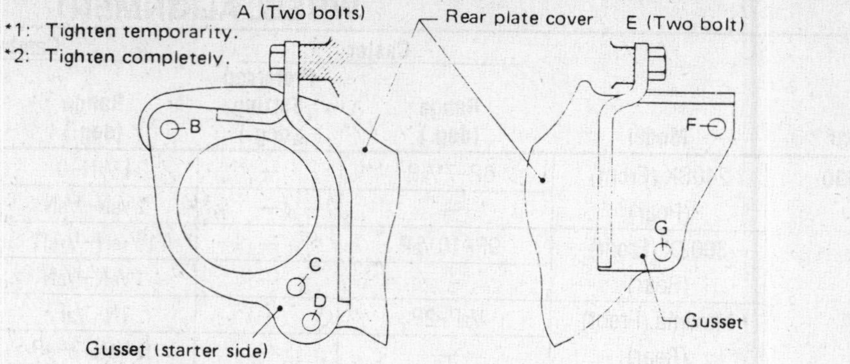

1st	2nd	3rd	4th	5th	6th
A*1	D*2	A*2	F*2 and G*2	E*2	B*2 and C*2

On 1988–89 300ZX torque the engine gussets in 6 stages to 22–29 ft. lbs.

Engine Assembly

REMOVAL & INSTALLATION

200SX, 240SX and 1988–89 300ZX

1. Mark the hood hinge relationship and remove the hood.
2. Release the fuel system pressure and disconnect the negative battery cable.
3. Drain the cooling system and transmission fluid.
4. Remove the radiator after disconnecting the automatic transmission coolant tubes, if equipped.
5. Remove the air cleaner.
6. Remove the fan and pulley.
7. Disconnect or remove following:
   Water temperature gauge wire
   Oil pressure sending unit wire
   Ignition distributor primary wire
   Starter motor connections
   Fuel hose
   Alternator leads
   Heater hoses
   Throttle and choke connections
   Engine ground cable and all wiring harnesses
   Any interfering engine accessory

CAUTION

On vehicles with air conditioning, it is necessary to remove the compressor and the condenser from their mounts. Do not attempt to disconnect any of the air conditioner hoses.

8. Disconnect the power brake booster hose from the engine.
9. Remove the clutch operating cylinder and return spring, if equipped.
10. Disconnect the speedometer cable from the transmission. Disconnect the back up light switch and any other wiring or attachments to the transmission.
11. Disconnect the column shift linkage. Remove the floor shift lever.
12. Raise and safely support the vehicle. Detach the exhaust pipe from the exhaust manifold. Remove the front section of the exhaust system. On 300ZX, remove the right side exhaust manifold and exhaust connecting tube section.
13. Mark the relationship of the flanges and disconnect the driveshaft.
14. Suppport the transmission with a jack. Remove the rear cross member, if required.
15. Attach a hoist to the lifting hooks on the engine at either end of the cylinder head. Support the engine with a suitable jack.
16. Unbolt the front engine mount brackets from the block. Tilt and remove the engine by lowering the jack under the transmission and raising the hoist.

To install:

17. Lower the engine into the vehicle and align the block with with the front mount brackets. On the 200SX (V6) and 300ZX, torque the engine gusset bolts in 6 stages to 22–29 ft. lbs. (29–39 Nm). When installing the engine on automatic transmission equipped 200SX (4 cyl.), adjust the rear mounting insulator to 0.451–0.569 in. (11.1–14.5mm). Torque the engine mount bolts to 33–43 ft. lbs. (44–59 Nm), 32–41 ft. lbs. (43–55 Nm) on 240SX and 33–44 ft. lbs. (45–60 Nm) on 300ZX.

NOTE: Never loosen the front engine mount insulator cover nuts on a 200SX (4 cyl.); if removed, the insulator will malfunction due to oil loss.

18. Install the rear cross member, if removed.
19. Connect the driveshaft. Make sure the driveshaft flanges are aligned properly.
20. On 300ZX, install the right exhaust manifold and connecting tube section. On the other models, install the front exhaust section and connect the exhaust pipe to the manifold.

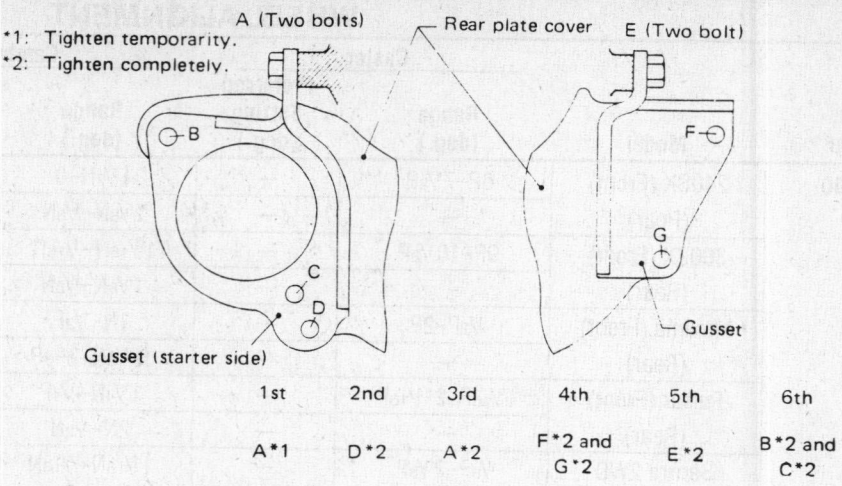

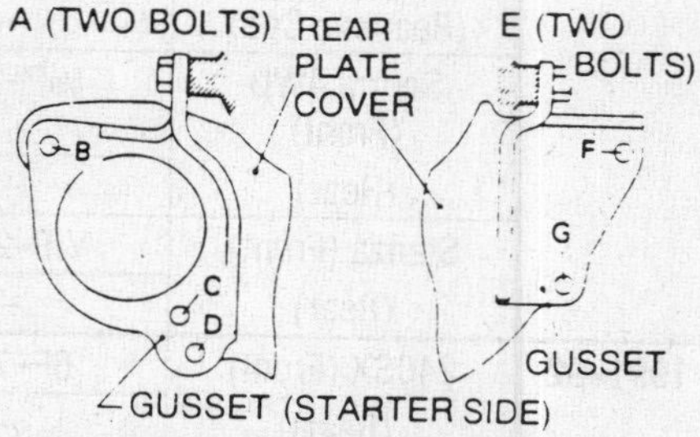

Torquing the 200SX (V6) engine gussets

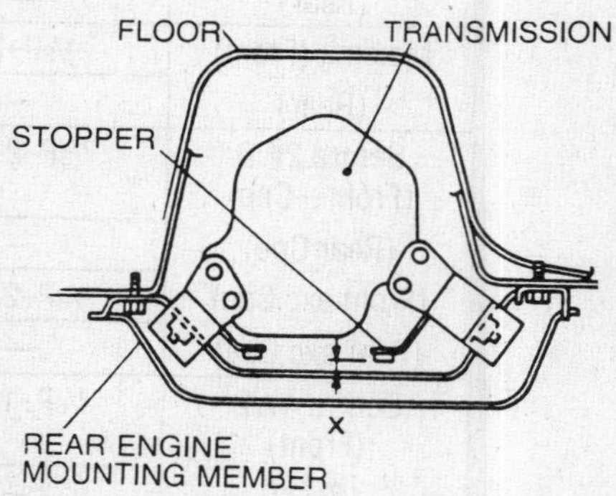

Adjust the rear mount stopper clearance (X) to 13mm ± 1.5mm – 200SX (4cyl.) with automatic transmission

21. Install the floor shift lever and connect the column shift linkage.
22. Connect the back-up light switch and any other wiring to the transmission. Connect the speedometer cable.
23. Install the clutch return spring and operating cylinder, if removed.
24. Connect the power brake booster hose to the engine.
25. Connect all engine hoses and electrical wires. Install any removed engine accessory.
26. Install the fan and pulley.

27. Install the air cleaner.
28. Install the radiator and connect the transmission cooling lines, if equipped.
29. Fill the transmission and cooling system to the proper levels.
30. Install the hood and connect the negative battery cable.
31. Make all the necessary engine adjustments. Road test the vehicle for proper operation.

1990–92 300ZX

WITH MANUAL TRANSMISSION

1. Mark the hood hinge relationship and remove the hood.
2. Release the fuel system pressure, disconnect the negative battery cable and raise and safely support the vehicle.
3. Remove the undercover.
4. Drain the coolant from both sides of the block and from the radiator.
5. Drain the oil pan.
6. Disconnect and label all engine vacuum hoses, fuel piping, harnesses and connectors.
7. Disconnect and remove the front exhaust tube sections.
8. Mark the relationship of the flanges and disconnect the driveshaft.
9. Remove the radiator.
10. Remove the drive belts.
11. Remove the cooling fan and coupling.
12. Remove the power steering pump, alternator, starter and clutch operating cylinder.
13. Discharge the air conditioning system and remove the compressor from the engine. Disconnect the air conditioning tube clamps.
14. Disconnect the steering column lower joint from the steering rack.
15. Remove the tension rod retaining bolts on both sides.
16. Loosen the transverse link bolts on both sides.
17. Support the rear suspension member using the proper equipment.
18. Install engine slingers to the block and connect a suitable lifting device to the slingers. Tension the lifting device slightly.
19. Remove the rear suspension member retaining bolts and center nut.
20. Remove the engine mount bracket bolts from both sides and slowly lower the transmission jack. Lift the engine from the vehicle.

To install:

21. Lower the engine into the vehicle and slowly raise the transmission jack. Install the engine mount bracket bolts. Torque the bolts to 30–38 ft. lbs. (40–42 Nm).
22. Install the rear suspension bolts and center nut. Torque the bolts to 38–48 ft. lbs. (51–65 Nm) and the center nut to 26–33 ft. lbs. (35–45 Nm).

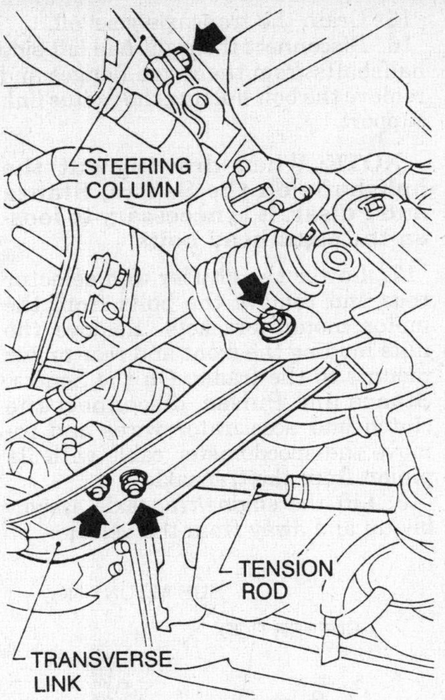

Steering column, tension rod and transverse link attachment points

23. Remove the jack and disconnect the engine hoist.
24. Torque the transverse link bolts to 80–94 ft. lbs. (108–127 Nm).
25. Install the tension rod retaining bolts and torque them to 80–94 ft. lbs. (108–127 Nm).
26. Connect the steering column lower joint to the steering rack. Torque the lower joint bolt to 17–22 ft. lbs. (24–29 Nm).
27. Connect the air conditioning tube clamps and mount the air conditioning compressor on the engine.
28. Install the clutch operating cylinder, starter, alternator and power steering pump.
29. Install the cooling fan and coupling.
30. Install the drive belts.
31. Install the radiator.
32. Install the driveshaft. Make sure the flanges are aligned properly. On non-turbo models, torque the flange bolts to 29–33 ft. lbs. (39–45 Nm) and 40–47 ft. lbs. (54–64 Nm) on turbocharged models.
33. Connect and install the front exhaust tube sections.
34. Connect the engine connectors, harnesses, fuel piping and vacuum hoses.
35. Install the undercover.
36. Fill the transmission and cooling system to the proper levels.
37. Install the hood and connect the negative battery cable.
38. Make all the necessary engine adjustments. Charge the air conditioning system.

WITH AUTOMATIC TRANSMISSION

1. Mark the hood hinge relationship and remove the hood.
2. Relieve the fuel system pressure, disconnect the negative battery cable and raise and support the vehicle safely.
3. Remove the undercover.
4. Drain the coolant from both sides of the block and from the radiator.
5. Drain the oil pan.
6. Disconnect and label all engine vacuum hoses, fuel piping, harnesses and connectors.
7. Disconnect and remove the front exhaust tube sections.
8. Mark the relationship of the flanges and disconnect the driveshaft.
9. Remove the radiator.
10. Remove the drive belts.
11. Remove the cooling fan and coupling.
12. Remove the power steering pump, alternator, starter and clutch operating cylinder.
13. Remove the transmission.
14. Connect an engine hoist to the engine lifting brackets and tension the hoist.
15. Remove the engine mount bracket bolts and slowly lift the engine from the vehicle.

To install:

16. Lower the engine into the vehicle and install the engine mount bracket bolts. Torque the bolts to 30–38 ft. lbs. (40–42 Nm).
17. Install the clutch operating cylinder, starter, alternator and power steering pump.
18. Install the cooling fan and coupling.
19. Install the drive belts.
20. Install the radiator.
21. Install the driveshaft. Make sure the flanges are aligned properly. On non-turbo models, torque the flange bolts to 29–33 ft. lbs. (39–45 Nm) and 40–47 ft. lbs. (54–64 Nm) on turbocharged models.
22. Connect and install the front exhaust tube sections.
23. Connect the engine connectors, harnesses, fuel piping and vacuum hoses.
24. Install the undercover.
25. Fill the transmission and cooling system to the proper levels.
26. Install the hood and connect the negative battery cable.
27. Make all the necessary engine adjustments. Charge the air conditioning system. Road test the vehicle for proper operation.

Maxima, Pulsar, Sentra and Stanza

It is recommended that the engine and transaxle be removed as a unit. If need be, the units may be separated after removal.

NOTE: On the 1989–92 Sentra (GA16i and GA16DE engines), the engine cannot be removed separately from the tranaxle. Remove the engine and the transaxle as a unit. If equipped with 4WD, remove the engine, transaxle and transfer case together.

1. Mark the hood hinge relationship and remove the hood.
2. Release the fuel system pressure, disconnect the negative battery cable and raise and support the vehicle safely.
3. Drain the cooling system and the oil pan.
4. Remove the air cleaner and disconnect the throttle cable.
5. Disconnect or remove the following:
   - Drive belts
   - Ignition wire from the coil to the distributor
   - Ignition coil ground wire and the engine ground cable
   - Block connector from the distributor
   - Fusible links
   - Engine harness connectors
   - Fuel and fuel return hoses
   - Upper and lower radiator hoses
   - Heater inlet and outlet hoses
   - Engine vacuum hoses
   - Carbon canister hoses and the air pump air cleaner hose
   - Any interfering engine accessory: power steering pump, air conditioning compressor or alternator
   - Driveshaft from transfer for 4WD vehicles. Make sure to matchmark flanges
6. Remove the air pump air cleaner.
7. Remove the carbon canister.
8. Remove the auxiliary fan, washer tank, grille and radiator (with fan assembly).
9. Remove the clutch cylinder from the clutch housing for manual transaxles.
10. Remove both buffer rods without altering the length of the rods. Disconnect the speedometer cable.
11. Remove the spring pins from the transaxle gear selector rods.
12. Install engine slingers to the block and connect a suitable lifting device to the slingers. Do not tension the lifting device at this point.
13. Disconnect the exhaust pipe at both the manifold connection and the clamp holding the pipe to the engine.
14. On the Sentra, Pulsar and 1988–89 Stanza, remove the lower ball joint.
15. Drain the transaxle gear oil.
16. Disconnect the right and left side halfshafts from their side flanges and remove the bolt holding the radius link support.

NOTE: When drawing out the halfshafts on the Sentra, Stanza and Pulsar, it is necessary to loosen the strut head bolts.

17. Lower the shifter and selector rods and remove the bolts from the motor mount brackets. Remove the nuts holding the front and rear motor mounts to the frame. On the Sentra, Stanza and Pulsar, disconnect the clutch and accelerator wires and remove the speedometer cable with its pinion from the transaxle.
18. Lift the engine/transaxle assembly up and away from the vehicle.

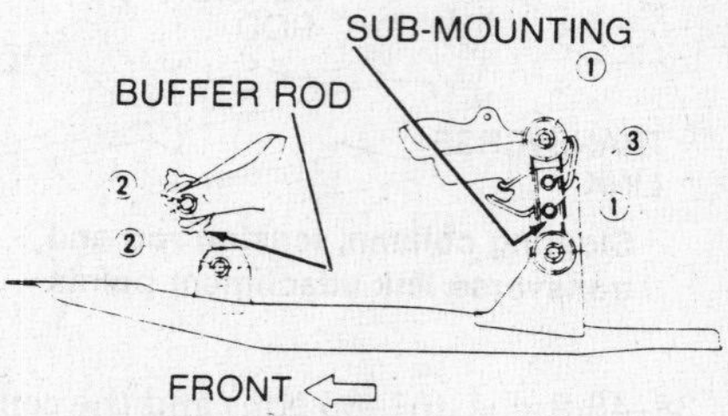

On Stanza Wagon (2WD), tighten the buffer rod and sub-mounting bolts in the order shown

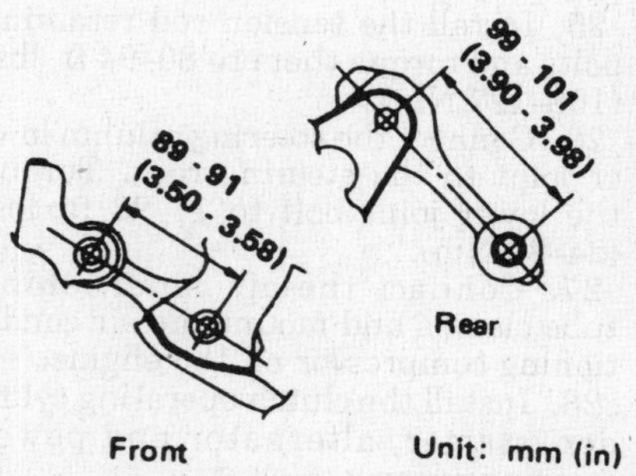

Front and rear buffer rod length adjustment on 1988–90 Pulsar (E16i and GA16i), Maxima and Sentra

To install:

19. Lower the engine transaxle assembly into the vehicle. When lowering the engine onto the frame, make sure to keep it as level as possible.
20. Check the clearance between the frame and clutch housing and make sure the engine mount bolts are seated in the groove of the mounting bracket.
21. After installing the motor mounts, adjust and install the buffer rods. On the 1988–90 Pulsar with E16i and GA16i engines, 1989–92 Maxima and 1989–92 Sentra: front should be 3.50–3.58 in. (89–91mm), and the rear, 3.90–3.98 in. (99–101mm).
22. On the Stanza Wagon (2WD), tighten the engine mount bolts first, then apply a load to the mounting insulators before tightening the buffer rod and sub-mounting bolts.
23. On the Sentra, Stanza and Pulsar, connect the clutch and accelerator wires and remove the speedometer cable with its pinion from the transaxle.
24. Raise the shifter and selector rods to their normal operating positions.
25. Connect the halfshafts.
26. On the Sentra, Pulsar and 1988–89 Stanza, connect the lower ball joint.
27. Connect the exhaust pipe to the manifold connection and the clamp holding the pipe to the engine.
28. Disconnect the lifting device and remove the engine slingers.
29. Insert the spring pins into the transaxle gear selector rods.
30. Connect the speedometer cable.
31. Mount the clutch cylinder onto the clutch housing.
32. Install the auxiliary fan, washer tank, grille and radiator (with fan assembly).
33. Install the carbon canister.
34. Install the air pump air cleaner.
35. Install or connect all hoses, belts, harnesses, connectors and components that were necessary to remove the engine.
36. Connect the throttle cable and install the air cleaner.
37. Fill the transaxle and cooling system to the proper levels.
38. Install the hood and connect the negative battery cable.
39. Make all the necessary engine adjustments. Charge the air conditioning system. Road test the vehicle for proper operation.

Cylinder Head

REMOVAL & INSTALLATION

NOTE: To prevent distortion or warping of the cylinder head, allow the engine to cool completely before removing the head bolts.

CA18DE Engine

1. Crank the engine until the No. 1 piston is at TDC of the compression stroke. Relieve the fuel system pressure and disconnect the negative battery cable . Drain the cooling system, remove the air cleaner assembly and raise and safely support the vehicle.
2. Loosen the alternator and remove all drive belts. Remove the alternator.
3. Disconnect the air duct at the throttle chamber.

4. Tag and disconnect all lines, electrical harnesses, hoses and wires which may interfere with cylinder head removal.
5. Remove the ornament cover.
6. Disconnect the oxygen sensor wire.
7. Remove the 2 exhaust heat shield covers.
8. Unbolt the exhaust manifold and wire the entire assembly aside.
9. Disconnect the EGR tube at the passage cover and then remove the passage cover and gasket.
10. Disconnect and remove the crank angle sensor from the upper front cover.

NOTE: Put an aligning mark on crank angle sensor and timing belt cover.

11. Remove the support stay from under the intake manifold assembly.
12. Unbolt the intake manifold and remove it along with the collector and throttle chamber.
13. Remove the fuel injectors as an assembly.
14. Remove the upper and lower front covers.

NOTE: Remove engine mount bracket but support engine under oil pan with wooden blocks.

15. Remove the timing belt and camshaft sprockets.

NOTE: When the timing belt has been removed, never rotate the crankshaft and camshaft separately because the valves will hit the top of the pistons.

16. Remove the camshaft cover.
17. Remove the breather separator.
18. Gradually loosen the cylinder head bolts in several stages, in the proper sequence.
19. Carefully remove the cylinder head from the block, pulling the head up evenly from both ends. If the head seems stuck, do not pry it off. Tap lightly around the lower perimeter of the head with a rubber mallet to help break the joint.

To install:

20. Thoroughly clean both the cylinder block and head mating surfaces. Avoid scratching either.
21. Lay the cylinder head gasket onto the block and lower the head onto the gasket.
22. When installing the bolts tighten the 2 center bolts temporarily to 15 ft. lbs. (20 Nm) and install the head bolts loosely. After the breather separator, camshaft cover, timing belt, camshaft sprockets and front cover have been installed, torque all the head bolts in the proper sequence as follows. Tighten all bolts to 22 ft. lbs. (29 Nm). Retighten all bolts to 76 ft. lbs. (103 Nm). Loosen all bolts completely and then re-tighten them once again to 22 ft. lbs. (29 Nm). Tighten all bolts to a final torque of 76 ft. lbs. (103 Nm) or 85–95 degrees if using an angle torque wrench.

LOOSENING ORDER

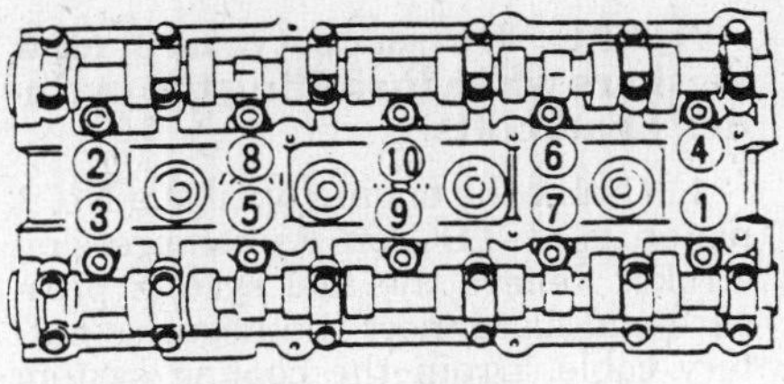

Cylinder head loosening sequence—CA18DE engine

Tightening order

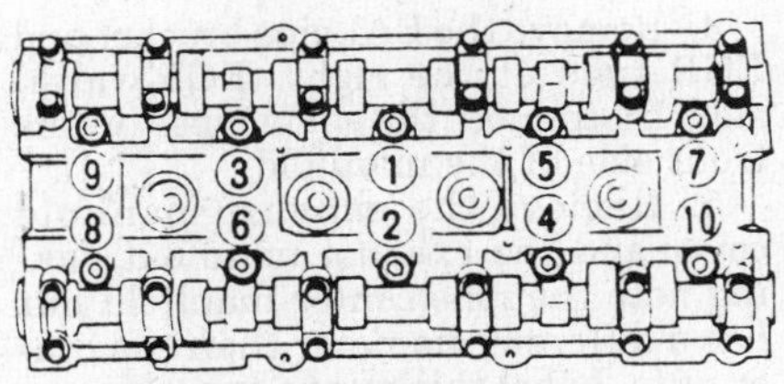

Cylinder head tightening sequence—CA18DE engine

NOTE: Newer models use cupped washers on the cylinder head bolts, always make sure the flat side of the washer is facing downward before tightening the cylinder head bolts.

23. Install the fuel injector assembly. Use new O-rings and insulators as required.
24. Install the intake manifold assembly and intake manifold stay.
25. Install the crank angle sensor. Make sure the sensor and upper front matchmarks are aligned properly.
26. Install the passage cover and gasket. Connect the EGR tube to the passage cover.
27. Install the exhaust manifold.
28. Install the exhaust manifold heat shield covers.
29. Install the ornament cover.
30. Connect all lines, electrical harnesses, hoses and wires.
31. Connect the air duct to the throttle chamber.
32. Install the alternator and drive belts.
33. Install the air cleaner assembly.
34. Fill the cooling system to the proper level and connect the negative battery cable.
35. Make all the necessary engine adjustments. Road test the vehicle for proper operation.

CA20E Engine

1. Relieve the fuel system pressure, disconnect the negative battery cable and drain the cooling system.
2. Remove the air intake pipe
3. Remove the cooling fan and radiator shroud.
4. Remove the alternator drive belt, power steering pump drive belt and the air conditioner compressor drive belt, if equipped.
5. Position the No. 1 cylinder at TDC of the compression stroke and remove the upper and lower timing belt covers.
6. Loosen the timing belt tensioner and return spring, then remove the timing belt.

NOTE: When the timing belt has been removed, do not rotate the crankshaft and the camshaft separately, because the valves will hit the tops of the pistons.

7. Remove the exhaust manifold.
8. Remove the camshaft pulley.
9. Remove the water pump pulley.
10. Remove the crankshaft pulley.
11. Remove the alternator adjusting bracket.
12. Remove the water pump.
13. Remove the oil pump.
14. Loosen the cylinder head bolts in sequence and in several steps.
15. Remove the cylinder head and manifolds as an assembly.

To install:

16. Clean the cylider head gasket surfaces.
17. Lay the cylinder head gasket onto the block and lower the head onto the gasket.
18. Install the cylinder head bolts.

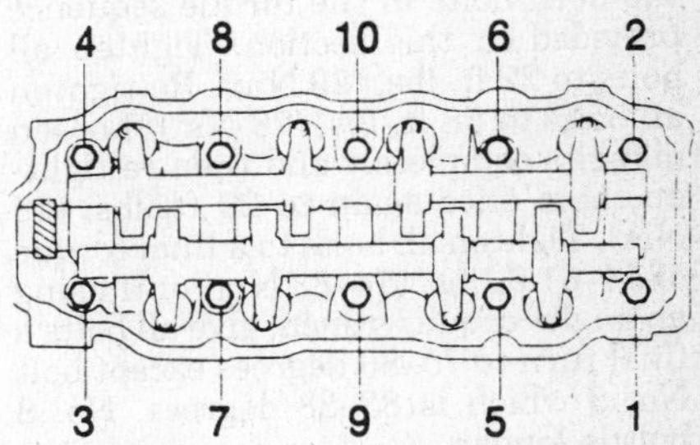

Cylinder head bolt loosening sequence—CA20E engine

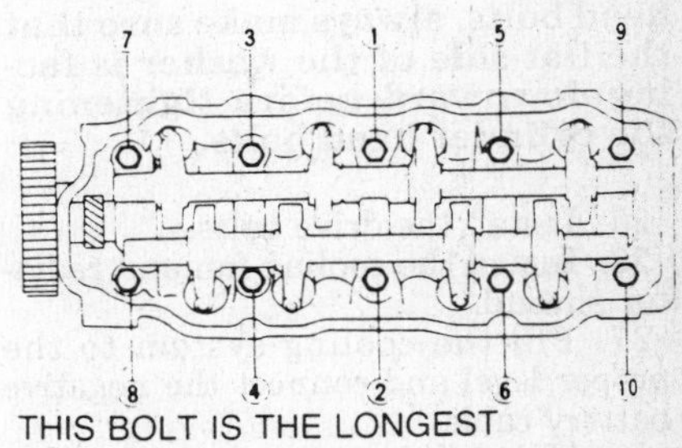

Cylinder head bolt tightening sequence—CA20E engine

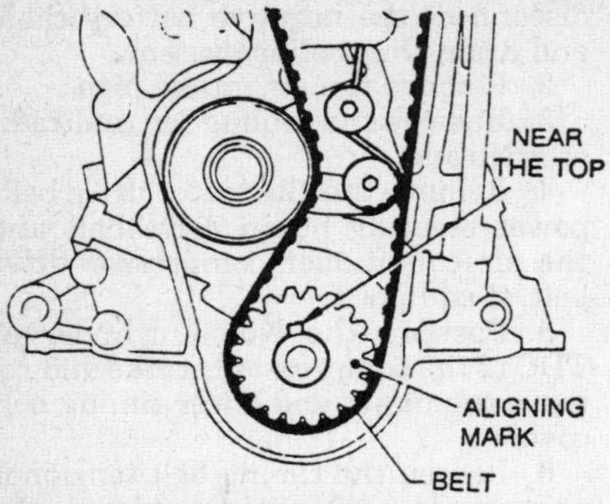

Make sure the crankshaft pulley key is near the top—C series engine

When installing the bolts, tighten the two center bolts temporarily to 15 ft. lbs. and install the head bolts loosely. They will be torqued after the timing belt and front cover are installed.
19. Install the oil pump.
20. Install the water pump.
21. Install the alternator adjusting bracket.
22. Install the crankshaft, water pump and camshaft pulleys.
23. Install the exhaust manifold.

NOTE: Before installing the timing belt, be certain the crankshaft pulley key is near the top and that the camshaft knock pin or sprocket aligning mark is at the top.

24. Install the timing belt and timing belt covers. After the timing belt and covers have been installed, torque all the head bolts in the torque sequence provided in this section. Tighten all bolts to 22 ft. lbs. (29 Nm). Re-tighten all bolts to 58 ft. lbs. (78 Nm). Loosen all bolts completely and then re-tighten them once again to 22 ft. lbs. (29 Nm). Tighten all bolts to a final torque of 54–61 ft. lbs. (74–83 Nm) or if using an angle torque wrench, give all bolts a final turn to 75–80 degrees except bolt No. 8 which is 83–88 degrees. No. 8 bolt is longer.

NOTE: Newer models use cupped washers on the cylinder head bolts, always make sure that the flat side of the washer is facing downward before tightening the cylinder head bolts.

25. Install the drive belts.
26. Install the cooling fan and radiator shroud.
27. Fill the cooling system to the proper level and connect the negative battery cable.
28. Make all the necessary engine adjustments. Road test the vehicle for proper operation.

E16i Engine

NOTE: Be sure to use new washers when installing the cylinder head bolts.

1. Crank the engine until the No. 1 piston is at TDC on its compression stroke. Relieve the fuel system pressure and disconnect the negative battery cable. Drain the cooling system and remove the air cleaner assembly.
2. Remove the alternator.
3. Remove the distributor, with all wires attached.
4. Remove the EAI pipe bracket and EGR tube at the right (EGR valve) side. Disconnect the same pipes on the front side of the manifold.
5. Remove the exhaust manifold cover and the exhaust manifold, taking note that the center manifold nut has a different diameter than the other nuts. Label this nut to ensure proper installation.
6. Remove the air conditioning compressor bracket and power steering pump bracket, if equipped.
7. Disconnect the carburetor throttle linkage, fuel line, and all vacuum and electrical connections.
8. Remove the intake manifold.
9. Remove water pump drive belt and pulley.
10. Remove crankshaft pulley.
11. Remove the rocker (valve) cover.
12. Remove upper and lower dust cover on the camshaft timing belt shroud.
13. Mark the relationship of the camshaft sprocket to the timing belt and the crankshaft sprocket to the timing belt with paint or a grease pencil. This will make setting everything up during reassembly much easier if the engine is disturbed during disassembly.
14. Remove the belt tensioner pulley.
15. Mark an arrow on the timing belt showing direction of engine rotation and slide the belt off the sprockets.
16. Loosen the head bolts in reverse of the tightening sequence and carefully remove the cylinder head from the block, pulling the head up evenly from both ends. If the head seems stuck, do not pry it off. Tap lightly around the lower perimeter of the head with a rubber mallet to help break the seal. Label all head bolts with tape, as they must go back in their original positions.

To install:

17. Thoroughly clean both the cylinder block and head mating surfaces. Avoid scratching either.
18. Turn the crankshaft and set the No. 1 cylinder at TDC on its compression stroke. This causes the crankshaft timing sprocket mark to be aligned with the cylinder block cover mark.
19. Align the camshaft sprocket mark with the cylinder head cover mark. This causes the valves for No. 1 cylinder to position at TDC on the compression stroke.
20. Place a new gasket on the cylinder block.

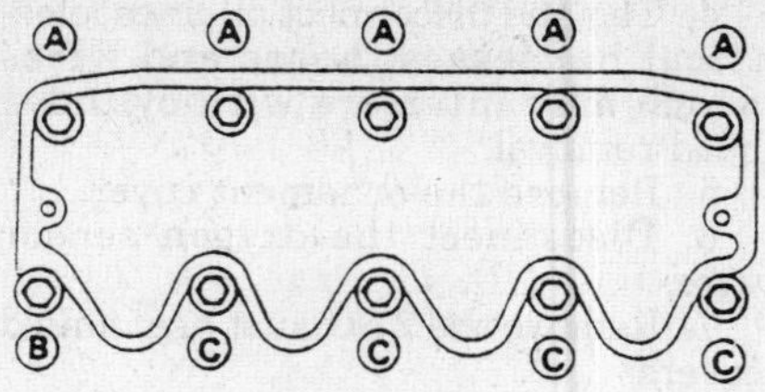

Cylinder head bolt location on E16i engines

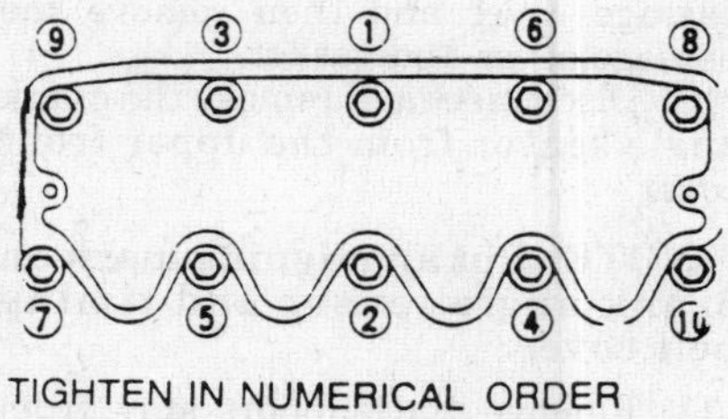

Cylinder head torque sequence—E series engine

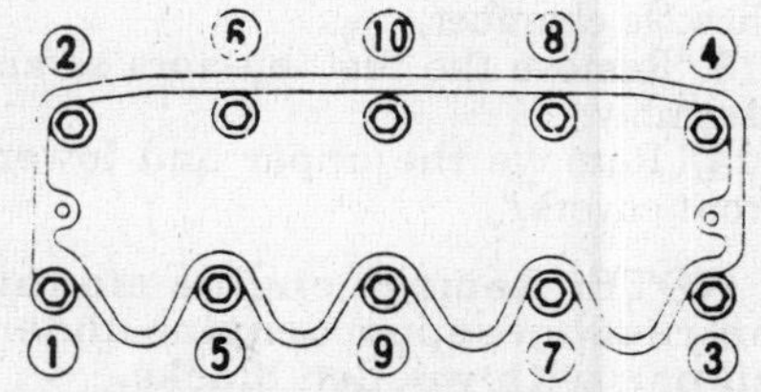

Loosen the cylinder head bolts, in stages in the order shown—E series engine

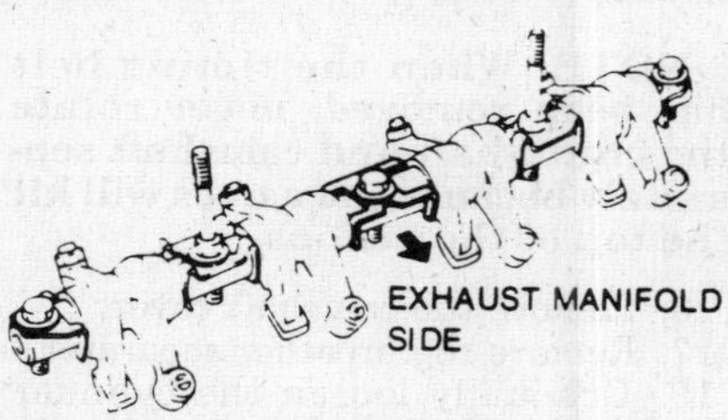

Make sure the cutout on the E series engine rocker shaft faces the exhaust manifold

NOTE: There are 3 different size head bolts used on the E16i engine. Bolt (A) is 3.74 in. (95mm), bolt (B) is 4.33 in. (110mm) and bolt (C) is 3.15 in. (80mm). Measure the length of each bolt prior to installation and make sure they are installed in their proper locations on the head.

21. Install the cylinder head on the block and tighten the bolts as follows:
a. Tighten all bolts to 22 ft. lbs. (29 Nm), then retighten them all to 51 ft. lbs. (69 Nm).

b. Loosen all bolts completely, and then retighten them again to 22 ft. lbs. (29 Nm).

c. Tighten all bolts to a final torque of 51–54 ft. lbs. (69–74 Nm); or if an angle wrench is used, turn each bolt until they have achieved the specified number of degrees—bolts 1, 3, 6, 8 and 9: 45–50 degrees; bolt 7: 55–60 degrees and bolts 2, 4, 5 and 10: 40–45 degrees.

22. Install the timing belt.

23. Install the upper and lower dust covers on the camshaft timing belt shroud.

24. Install the rocker arm cover.

25. Install the crankshaft pulley, water pump pulley and drive belt.

26. Install the intake manifold.

27. Connect the throttle linkage, fuel line, and all vacuum and electrical connections.

28. Install the air conditioning compressor bracket and the power steering pump bracket, if equipped.

29. Install the exhaust manifold and exhaust manifold cover. Make sure the center manifold nut, which has a different diameter, is installed in the proper location.

30. Connect the EAI exhaust pipes and tubing.

31. Install the distributor and connect the spark plug wiring.

32. Install the alternator and air cleaner.

33. Fill the cooling system to the proper level and connect the negative battery cable.

34. Make all the necessary engine adjustments. Road test the vehicle for proper operation.

GA16i Engine

1. Disconnect the negative battery cable, drain the cooling system and relieve the fuel system pressure.

2. Disconnect the exhaust tube from the exhaust manifold.

3. Remove the intake manifold support bracket.

4. Remove the air cleaner assembly.

5. Disconnect the center wire from the distributor cap.

6. Remove the rocker arm cover.

7. Remove the distributor.

8. Remove the spark plugs.

9. Set the No. 1 cylinder at TDC of the compression stroke by rotating the engine until the cut out machined in the rear of the camshaft is horizontally aligned with the cylinder head.

10. Hold the camshaft sprocket stationary with the proper tool and loosen the sprocket bolt. Place highly visible and accurate paint or chalk alignment marks on the camshaft sprocket and the timing chain, then slide the sprocket from the camshaft and lift the timing chain from the sprocket.

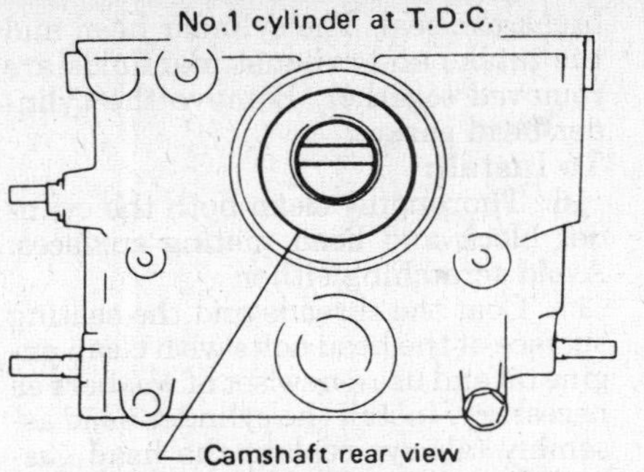

Whem camshaft is aligned as shown, the No. 1 piston is at TDC—GA16i engine

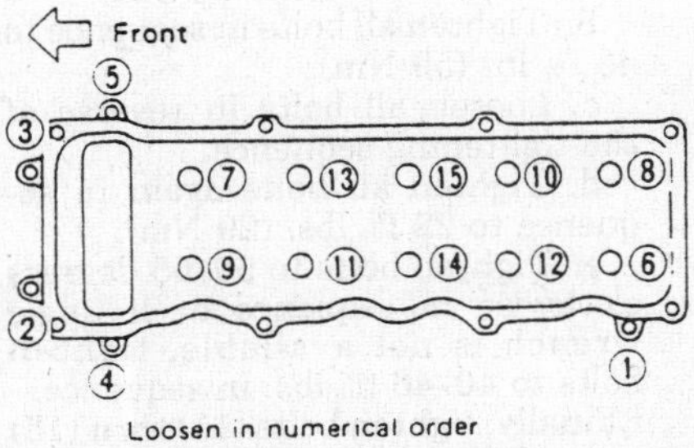

Loosen the cylinder head bolts in several stages in the order shown—GA16i and GA16DE engines

Remove the sprocket. The timing chain will not fall off the crankshaft sprocket unless the front cover is removed. This is due to the cast portion of the front cover located on the lower side of the crankshaft sprocket which acts a stopper mechanism. For this reason a chain stopper (wedge) is not required to remove the cylinder head.

11. Loosen the cylinder bolts in 2–3 stages to prevent warpage and cracking of the head. One of the cylinder head bolts is longer than the rest. Mark this bolt and make a note of its location.

12. Carefully remove the cylinder head from the block, pulling the head up evenly from both ends. If the head seems stuck, do not pry it off. Tap lightly around the lower perimeter of the head with a rubber mallet to help break the seal. The cylinder head and the intake and exhaust manifolds are removed together. Remove the cylinder head gasket.

To install:

13. Thoroughly clean both the cylinder block and head mating surfaces. Avoid scratching either.

14. Turn the crankshaft and set the No. 1 cylinder at TDC on its compression stroke. This is done by aligning the timing pointer with the appropriate timing mark on the pulley. To ensure that the No. 1 piston is at TDC, verify that the knock pin in the front of the camshaft is set at the top.

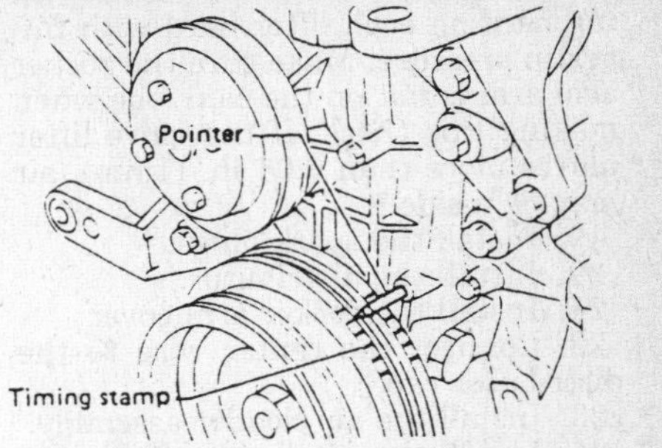

When the crankshaft pulley marks are aligned as shown, the No. 1 piston is at TDC

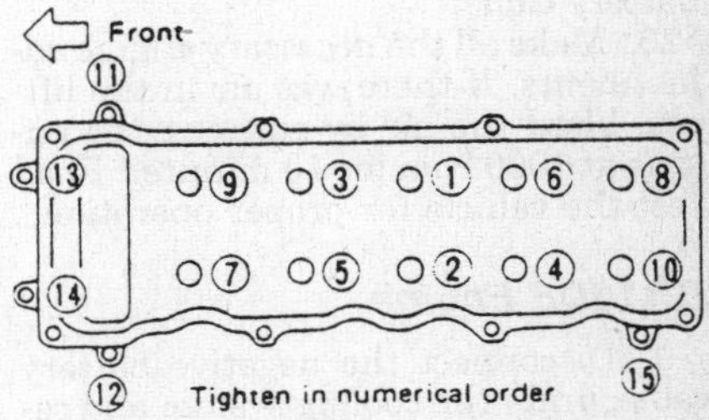

Cylinder head bolt tightening sequence—GA16i and GA16DE engines. Bolt (1) is the longest bolt

15. Place a new gasket on the block and lower the head onto the gasket.

NOTE: These engines use 2 different length cylinder head bolts. Bolt (1) is 5.24 in. (133mm) while bolts (2) thru (10) are 4.33 in. (110mm). Do not confuse the location of these bolts.

16. Coat the threads and the seating surface of the head bolts with clean engine oil and use a new set of washers. Install the cylinder head bolts in their proper locations and tighten as follows:

a. Tighten all the bolts in sequence to 22 ft. lbs. (30 Nm).

b. Tighten all bolts in sequence to 47 ft. lbs.(64 Nm).

c. Loosen all bolts in reverse of the tightening sequence.

d. Tighten all bolts again to 22 ft. lbs. (30 Nm).

e. If an angle torque wrench is not available, torque the bolts in sequence to 43–51 ft. lbs. (58–59 Nm). If using an angle torque wrench for this step, tighten bolt (1) 80–85 degrees clockwise and bolts (6) thru (10) 60–65 degrees clockwise.

f. Finally, tighten bolts (11) thru (15) to 4.6– 6.1 ft. lbs. (6.3–8.3 Nm).

17. Place the timing chain on the camshaft sprocket using the alignment marks. Slide the sprocket and timing chain onto the camshaft and install the center bolt.

18. At this point, check the hydraulic valve lifters for proper operation push-

ing hard on each lifter hard with fingertip pressure. Make sure the rocker arm arm is not on the cam lobe when making this check. If the valve lifter moves more than 0.04 in. (1mm), air may be inside it.

19. Install the spark plugs.
20. Install the distributor.
21. Install the rocker arm cover.
22. Connect the center wire to the distributor cap.
23. Install the air cleaner assembly.
24. Install the intake manifold support bracket.
25. Fill the cooling system to the proper level and connect the negative battery cable.
26. Make all the necessary engine adjustments. If there was air in the lifters, bleed the air by running the engine at 1000 rpm for 10 minutes. Road test the vehicle for proper operation.

GA16DE Engine

1. Disconnect the negative battery cable, drain the cooling system and relieve the fuel system pressure.
2. Remove all drive belts. Disconnect the exhaust tube from the exhaust manifold.
3. Remove the power steering bracket.
4. Remove the air duct to intake manifold collector.
5. Remove the front right side wheel, splash cover and front undercovers.
6. Remove the front exhaust pipe and engine front mounting bracket.
7. Remove the rocker arm cover.
8. Remove the distributor cap. Remove the spark plugs.
9. Set the No. 1 cylinder at TDC of the compression stroke.
10. Mark and remove the distributor assembly.
11. Remove the cam sprocket cover and gusset. Remove the water pump pulley. Remove the thermostat housing.
12. Remove the chain tensioner, chain guide. Loosen idler sprocket bolt.
13. Remove the camshaft sprocket bolts, camshaft sprocket, camshaft brackets and camshafts. Remove the idler sprocket bolt. These parts should be reassembled in their original position. Bolts should be loosen in 2 or 3 steps.
14. Loosen the cylinder bolts in 2–3 stages to prevent warpage and cracking of the head and note location of all head bolts.
15. Carefully remove the cylinder head from the block, pulling the head up evenly from both ends. If the head seems stuck, do not pry it off. Tap lightly around the lower perimeter of the head with a rubber mallet to help break the seal. The cylinder head and the intake and exhaust manifolds are removed together. Remove the cylinder head gasket.

To install:

16. Thoroughly clean both the cylinder block and head mating surfaces. Avoid scratching either.
17. Coat the threads and the seating surface of the head bolts with clean engine oil and use a new set of washers as necessary. Install the cylinder head assembly (always replace the head gasket). Install head bolts (with washers) in their proper locations and tighten as follows:
    a. Tighten all the bolts in sequence to 22 ft. lbs. (29 Nm).
    b. Tighten all bolts in sequence to 43 ft. lbs.(59 Nm).
    c. Loosen all bolts in reverse of the tightening sequence.
    d. Tighten all bolts again in sequence to 22 ft. lbs. (29 Nm).
    e. Tighten bolts to 50–55 degress clockwise in sequence or if angle wrench is not available, tighten bolts to 40–46 ft. lbs. in sequence.
    f. Finally, tighten bolts (11) thru (15) to 4.6–6.1 ft. lbs. (6.3–8.3 Nm).
18. Install the upper timing chain assembly.
19. Install all other components in the reverse order of the removal procedure. Refill and check all fluid levels. Road test the vehicle for proper operation.

KA24E Engine

NOTE: After completing this procedure, allow the rocker cover to cylinder head rubber plugs to dry for 30 minutes before starting the engine. This will allow the liquid gasket sealer to cure properly.

1. Release the fuel system pressure.
2. Disconnect the negative battery cable and drain the cooling system.
3. On 240SX, remove the power steering drive belt, power steering pump, idler pulley and power steering brackets.
4. Tag and disconnect all the vacuum hoses, water hoses, fuel tubes and wiring harnesses necessary to gain access to cylinder head.
5. Disconnect the air induction hose from the collector assembly.
6. Detach the accelerator bracket. If necessary mark the position and remove the accelerator cable wire end from the throttle drum.
7. Unbolt the intake manifold collector from the intake manifold.
8. Remove the intake manifold.
9. Unplug the exhaust gas sensor and remove the exhaust cover and exhaust pipe at exhaust manifold connection. Remove the exhaust manifold from the cylinder head.
10. Remove the rocker cover. If cover sticks to the cylinder head, tap it with a rubber hammer. Be careful not to strike the rocker arms when removing the rocker arm cover.

NOTE: After removing the rocker cover matchmark the timing chain with the camshaft sprocket with paint or equivalent.

11. Set No. 1 cylinder piston at TDC on its compression stroke. The No. 1 will be at TDC when the timing pointer is aligned with the red timing mark on the crankshaft pulley.
12. Loosen the camshaft sprocket bolt. Do not turn engine when removing the bolt.
13. Support the timing chain with the proper tool.
14. Remove the camshaft sprocket.
15. Remove the front cover-to-cylinder head retaining bolts.

NOTE: The cylinder head bolts should be loosened in 2-3 steps in the correct order to prevent head warpage or cracking.

16. Remove the cylinder head bolts in the correct sequence. Lift the cylinder head off the engine block. It may be necessary to tap the head lightly with a rubber mallet to loosen it.

To install:

17. Confirm that the No. 1 piston is at TDC on its compression stroke as follows: Align timing mark with the red (0 degree) mark on the crankshaft pulley. Make sure the distributor rotor head is set at No. 1 on the distributor cap. Confirm that the knock pin on the camshaft is set at the top position.
18. Install the cylinder head with a new gasket and torque the head bolts in numerical order using the following 5 step procedure:
    a. Torque all bolts to 22 ft. lbs. (29 Nm).
    b. Torque all bolts to 58 ft. lbs. (78 Nm).
    c. Loosen all bolts completely.
    d. Torque all bolts to 22 ft. lbs. (29 Nm).
    e. Torque all bolts to 54–61 ft. lbs. (74–83 Nm), or if an angle wrench is used, turn all bolts 80–85 degrees clockwise.

NOTE: Do not rotate crankshaft and camshaft separately, or valves will hit the tops of the pistons.

19. Remove the tool from the timing chain. Position the timing chain on the camshaft sprocket by aligning each matchmark. Install the camshaft sprocket to the camshaft.
20. Hold the camshaft sprocket stationary, and tighten the sprocket bolt

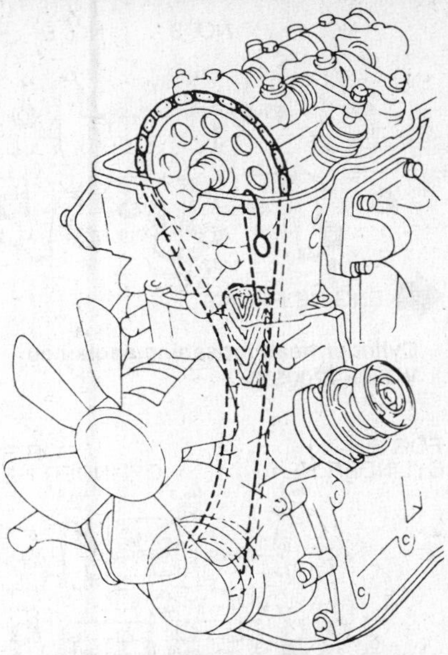
On KA24E engine, support the timing chain with a special tool when removing the cylinder head

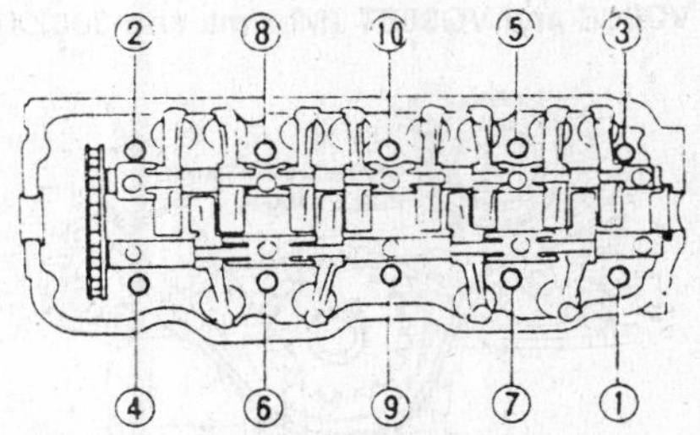

Cylinder head bolt loosening sequence – KA24E and KA24DE engines

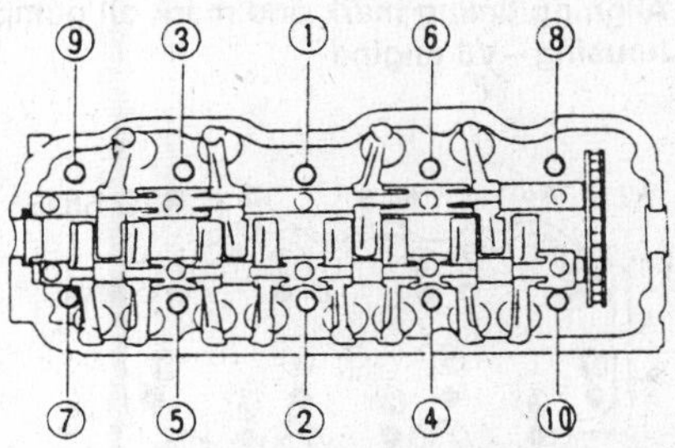

Cylinder head bolt tightening sequence – KA24E and KA24DE engines

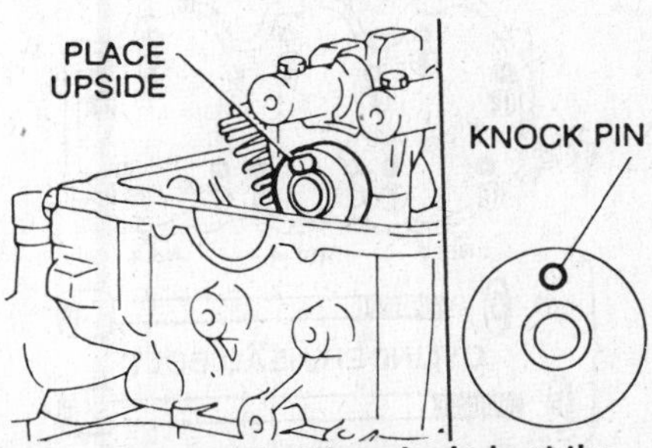

When the camshaft knock pin is at the top No. 1 piston is at TDC – KA24E engine

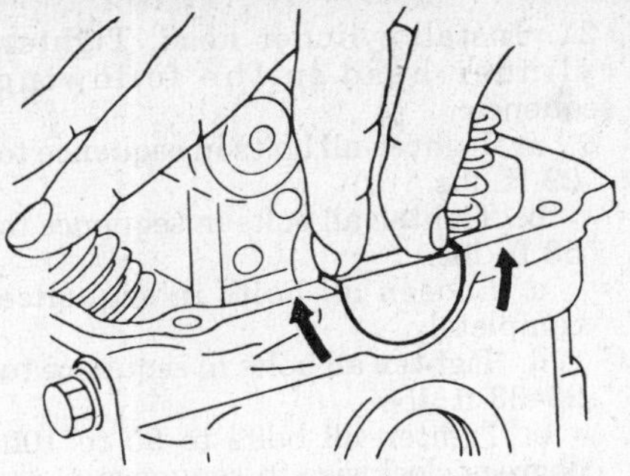
Rubber plug installation on KA24E engine

to 87–116 ft. lbs. (118–157 Nm). Install front cover-to-cylinder head retaining bolts. Torque the 6mm bolts to 5–6 ft. lbs. (7–8 Nm) and the 8mm bolts to 12–15 ft. lbs. (16–21 Nm).

21. Install the intake manifold and collector assembly with new gaskets.
22. Install the exhaust manifold with new gaskets.
23. Apply liquid gasket to the rubber plugs and install the rubber plugs in the correct location in the cylinder head. The seating surface of the rubber plugs must be clean and dry. The rubber plugs should be installed within 5 minutes of the sealant application. After the sealant is applied and the rubber plugs are in place, rock the plugs back and forth a few times to distribute the sealant evenly. Wipe the excess sealant from the cylinder head with a clean rag.
24. Install the rocker cover with new gasket.
25. Attach the accelerator bracket and cable if removed.
26. Connect all the vacuum hoses, water hoses, fuel tubes and electrical connections that were removed to gain access to cylinder head.
27. Reconnect the air induction hose to collector assembly.
28. Install the spark plugs and spark plug wires in the correct location.
29. On 240SX, install the power steering brackets, idler pulley, and power steering pump.
30. Install the drive belts.
31. Fill the cooling system and connect the negative battery cable.
32. Make all the necessary engine adjustments. Road test the vehicle for proper operation.

KA24DE Engine

1. Release the fuel system pressure.
2. Disconnect the negative battery cable and drain the cooling system. Drain the engine oil.
3. Remove all vacuum hoses, fuel lines, wires, electrical connections as necessary.
4. Remove the front exhaust pipe and A.I.V. pipe.
5. Remove the air duct, cooling fan with coupling and radiator shroud.
6. Remove the the fuel injector tube assembly with injectors.
7. Disconnect and mark spark plug wires. Remove the spark plugs.
8. Set No. 1 piston at TDC on compression stroke. Remove the rocker cover assembly.
9. Mark and remove the distributor assembly.
10. Remove the cam sprocket, brackets and camshafts. These parts should be reassembled in their original position. Bolts should be loosened in 2–3 steps.
11. Loosen cylinder head bolts in two or three steps in sequence.
12. Remove the cam sprocket cover. Remove the upper chain tensioner and upper chain guides.
13. Remove the upper timing chain and idler sprocket bolt. Lower timing chain will not disengaged from the crankshaft sprocket.
14. Remove the cylinder head with the intake manifold, collector and exhaust manifold assembly.

To install:

15. Check all components for wear. Replace as necessary. Clean all mating surfaces and replace the cylinder head gasket.
16. Install cylinder head. Tighten cylinder head in the following sequence:
    a. Tighten all bolts in sequence to 22 ft. lbs.
    b. Tighten all bolts in sequence to 59 ft. lbs.
    c. Loosen all bolts in sequence completely.
    d. Tighten all bolts in sequence to 18–25 ft. lbs.
    e. Tighten all bolts to 86 to 91 degrees clockwise, or if an angle wrench is not available, tighten all bolts to in sequence to 55–62 ft. lbs.
17. Install upper timing chain assemble in the correct position. Align all timing marks.
18. Install all other components in the reverse order of the removal procedure. Refill and check all fluid levels. Road test the vehicle for proper operation.

SR20DE Engine

1. Release the fuel pressure. Disconnect the negative battery cable.
2. Raise and safely support the vehicle. Remove the engine undercovers.
3. Remove the front right wheel and engine side cover.
4. Drain the cooling system. Remove the radiator assembly.
5. Remove the air duct to intake manifold.
6. Remove the drive belts and water pump pulley.

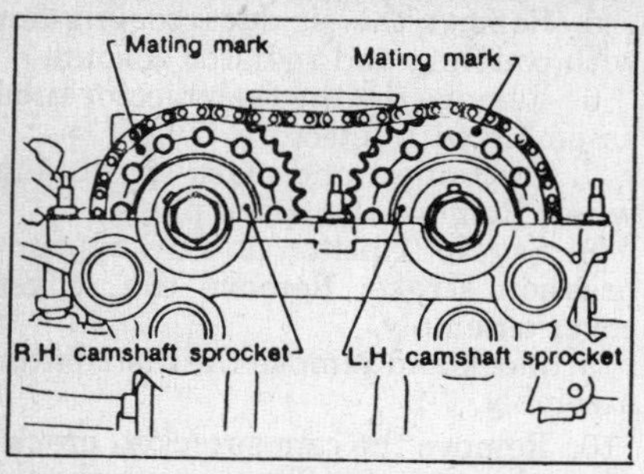

Cam sprocket correct position-SR20DE engine

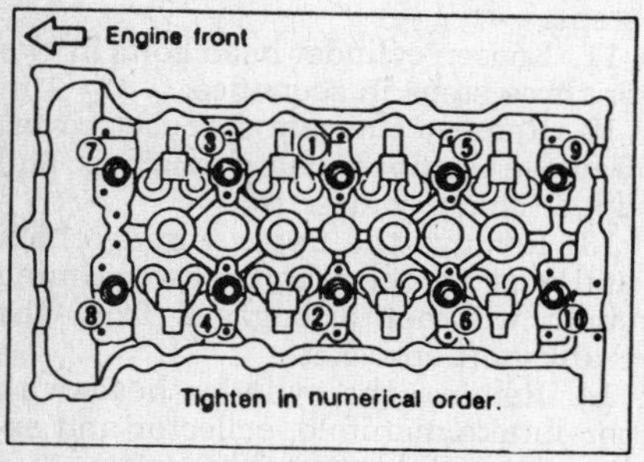

Cylinder head torque sequence – SR20DE engine

7. Remove the alternator and power steering pump.
8. Remove all vacuum hoses, fuel hoses, wires, electrical connections.
9. Remove all spark plugs.
10. Remove the A.I.V. valve and resonator.
11. Remove the rocker cover and oil separator.
12. Remove the intake manifold supports, oil filter bracket and power steering bracket.
13. Set No. 1 at TDC on the compression stroke. Rotate crankshaft until mating marks on camshaft sprockets are in the correct position.
14. Remove the timing chain tensioner.
15. Mark and remove the distributor assembly. Remove the timing chain guide and camshaft sprockets.
16. Remove the camshafts, camshaft brackets, oil tubes and baffle plate. Keep all parts in order for correct installation.
17. Remove the water hose from the cylinder block and water hose from the heater.
18. Remove the starter motor. Remove the water pipe bolt.
19. Remove the cylinder outside bolts. Remove the cylinder head bolts in 2 or 3 steps. Remove the cylinder head completely with manifolds attached.

To install:

20. Check all components for wear. Replace as necessary. Clean all mating surfaces and replace the cylinder head gasket.
21. Install cylinder head. Tighten cylinder head in the following sequence:
   a. Tighten all bolts in sequence to 29 ft. lbs.
   b. Tighten all bolts in sequence to 58 ft. lbs.
   c. Loosen all bolts in sequence completely.
   d. Tighten all bolts in sequence to 25–33 ft. lbs.
   e. Tighten all bolts to 90 to 100 degrees clockwise in sequence
   f. Tighten all bolts additional 90 to 100 degrees clockwise in sequence. Do not turn any bolt 180 to 200 degrees clockwise all at once.
22. Install all other components in the reverse order of the removal procedure. Refill and check all fluid levels. Road test the vehicle for proper operation.

VG30E and VG30ET Engines (200SX and 1988–89 300ZX)

NOTE: On all models, a special hex head wrench ST10120000 (J24239-01) or equivalent will be needed to remove and install the cylinder head bolts.

1. Disconnect the negative battery cable.
2. Relieve the fuel system pressure.
3. Remove the timing belt.

NOTE: Never rotate the crankshaft and camshaft separately after the timing belt has been removed or the valves will hit the tops of the pistons.

4. Set the No. 1 cylinder at TDC on its compression stroke.
5. Drain the coolant from the cylinder block.
6. Remove the collector cover and collector. Loosen the bolts starting from the ends and work towards the center. On the 200SX, remove the collector together with the throttle chamber, EGR valve and IAA unit.
7. Remove the intake manifold with fuel tube assembly. Loosen the intake manifold bolts starting from the front of the engine and proceed in criss-cross pattern.
8. Remove the power steering pump bracket.
9. Remove the exhaust collector bracket.
10. Disconnect the exhaust manifold balance and connecting tubes.
11. Remove the bolts securing the camshaft pulleys and rear timing cover.
12. Discharge the air conditioning system system and remove the compressor and compressor bracket. Remove the rocker covers.
13. Loosen the cylinder head bolts in

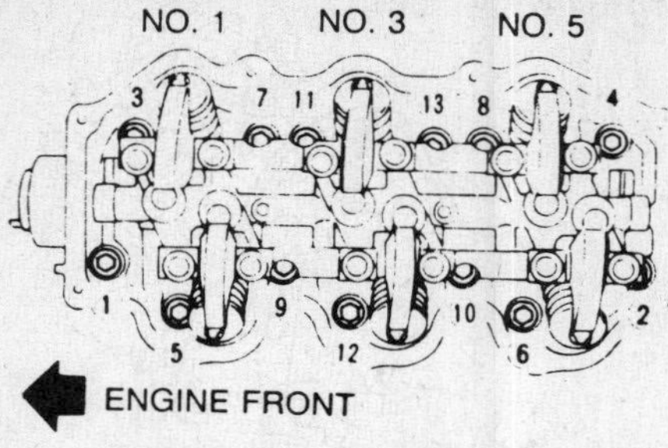

Cylinder head loosening sequence VG30E (200SX)

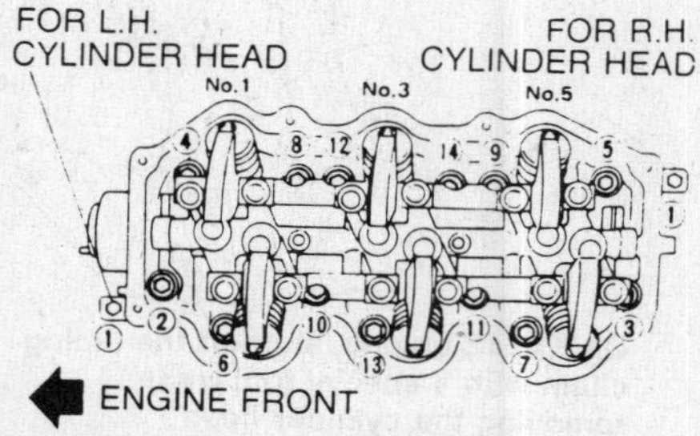

Cylinder head loosening sequence VG30E and VG30ET (Maxima and 300ZX)

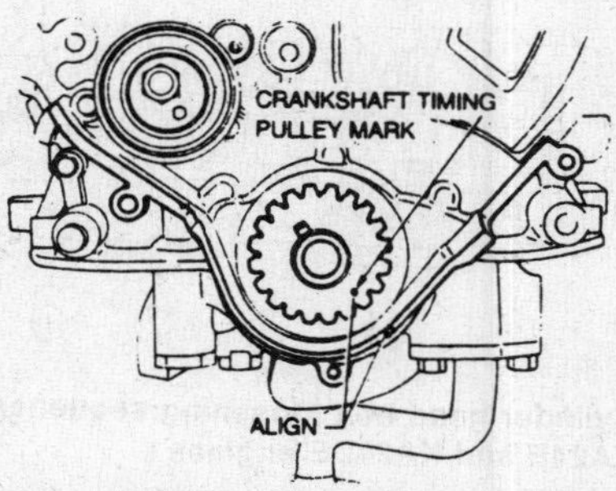

Aligning timing mark and mark oil pump housing – V6 engine

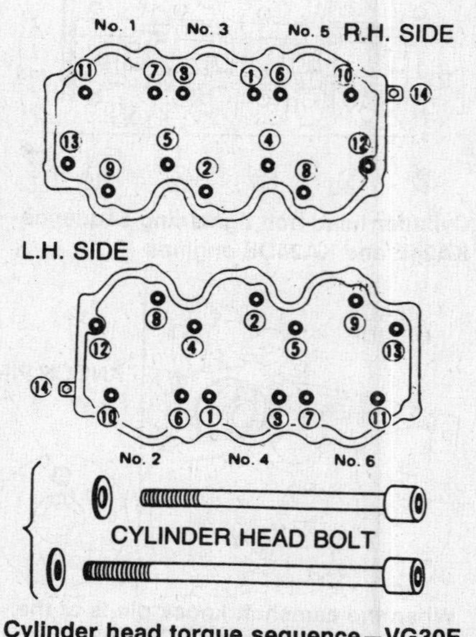

Cylinder head torque sequence – VG30E and VG30ET (Maxima and 300ZX)

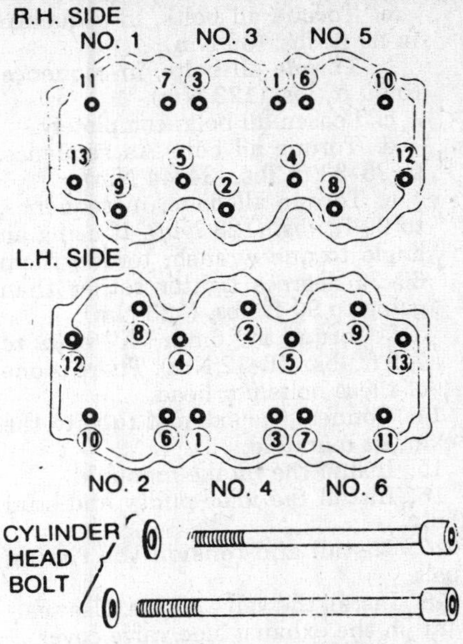

Cylinder head torque sequence – VG30E (200SX)

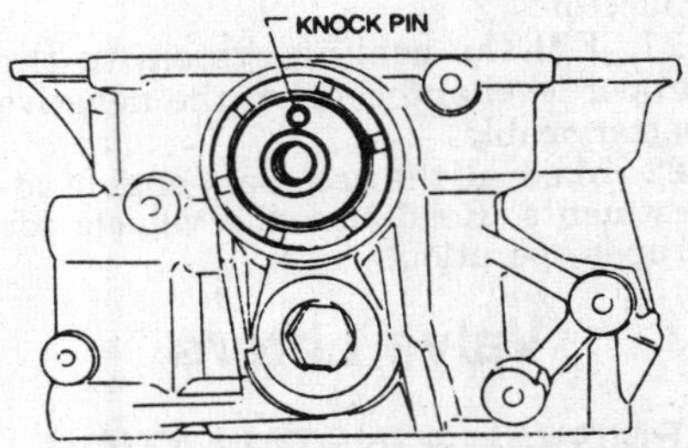

Knock pin of camshaft facing upward – V6 engine

the proper sequence. Remove the cylinder head with the exhaust manifolds attached. It may be necessary to tap the head lightly with a rubber mallet to loosen it.

To install:

14. Make sure the No. 1 cylinder is set at TDC on its compression stroke as follows:
   a. Align the crankshaft timing mark with the mark on the oil pump housing.
   b. The knock pin in the front end of the camshaft should be facing upward.

NOTE: Do not rotate crankshaft and camshaft separately because valves will hit the tops of the pistons.

15. Position the cylinder head and gasket on the block and tighten the cylinder head bolts as follows using the proper sequence:
   a. Tighten all bolts to 22 ft. lbs. (29 Nm).
   b. Tighten all bolts to 43 ft. lbs. (59 Nm).
   c. Loosen all bolts completely.
   d. Tighten all bolts to 22 ft. lbs. (29 Nm).
   e. Tighten all bolts to 40–47 ft. lbs. (54–64 Nm) or if using an angle wrench, turn all bolts 60–65 degrees clockwise.
16. Tighten the rear timing belt cover.
17. Install the camshaft pulley and tighten to 58–65 ft. lbs. (79–88 Nm).

NOTE: The right hand and left hand camshaft pulleys are different parts. Install them in the correct positions. The right hand pulley has an "R3" identification mark and the left hand pulley has an "L3".

18. Install the timing belt and adjust the tension.
19. Install the front upper and lower belt covers.
20. Install the rocker covers, compressor bracket and air conditioning compressor.
21. Install the intake manifold and fuel tube and tighten both the nuts and bolts as follows: first to 2–4 ft. lbs. (3–5 Nm), then to 17–20 ft. lbs. (24–27 Nm).
22. Connect the exhaust manifold balance and connecting tubes. Tighten the exhaust manifold connecting tube and tighten to 16–20 ft. lbs. (22–27 Nm).
23. Install the exhaust collector bracket.
24. Install the power steering pump bracket.
25. Install the intake manifold and fuel tube assembly. Make sure to tighten the bolts in 2–3 stages using the proper torque sequence.
26. Install the collector and collector cover. When installing the collector cover, always use a new gasket. On the 200SX and 1988–89 300ZX, tighten the throttle chamber-to-collector bolts in two stages; 6.5–8 ft. lbs. (9–11 Nm) and then to 13–16 ft. lbs. (18–22 Nm).
27. Install and tension the drive belts.
28. Fill the cooling system to the proper level and connect the negative battery cable.
29. Make all the necessary engine adjustments. Charge the air conditioning system.

VG30E Engine (Maxima)

NOTE: A special hex head wrench ST10120000 (J24239-01) or equivalent will be needed to remove and install the cylinder head bolts.

1. Relieve the fuel system pressure and disconnect the negative batttery cable.
2. Drain the cooling system. On 1989–92 Maxima, there are 2 cylinder block drain plugs. The left side drain plug is located beside the oil level gauge and the right side drain plug is located behind the right hand halfshaft boot.
3. Remove the timing belt.

NOTE: Do not rotate either the crankshaft or camshaft from this point onward, or the valves could be bent by hitting the tops of the pistons.

4. Disconnect and tag all vacuum and water hoses connected to the intake collector.
5. On 1989–92 Maxima, remove the distributor, ignition wires and disconnect the accelerator and cruise control (ASCD) cables from the intake manifold collector.
6. Remove the collector cover and the collector from the intake manifold. On 1989–92 Maxima, there are upper and lower collector covers. Disconnect and tag all harness connectors and vacuum lines to gain access to the cover retaining bolts on these models.
7. Remove the intake manifold and fuel tube assembly. Loosen the intake manifold bolts starting from the front of the engine and proceed in crisscross pattern towards the center.
8. Remove the exhaust collector bracket.
9. Remove the exhaust manifold covers.
10. Disconnect the exhaust manifold from the exhaust pipe.
11. Remove the camshaft pulleys and the rear timing cover securing bolts. Remove the rocker arm covers.
12. On 1989–92 Maxima, separate the air conditioning compressor and alternator from the their mounting brackets. Remove the mounting brackets. Do not disconnect the refrigerant lines from the compressor or serious injury will result.
13. Remove the cylinder head bolts in the correct sequence. Lift the cylinder head off the engine block with the exhaust manifolds attached. It may be necessary to tap the head lightly with a rubber mallet to loosen it.

To install:

14. Make sure the No. 1 cylinder is set at TDC on its compression stroke as follows:
   a. Align the crankshaft timing mark with the mark on the oil pump housing.
   b. The knock pin in the front end of the camshaft should be facing upward.

NOTE: Do not rotate crankshaft and camshaft separately because valves will hit piston head.

15. Install the cylinder head with a new gasket. Apply clean engine oil to

the threads and seats of the bolts and install the bolts with washers in the correct position. Note that bolts 4, 5, 12, and 13 are 4.95 in. (127mm) long. The other bolts are 4.13 in. (106mm) long.

16. Torque the bolts in the proper sequence as follows:

a. Torque all bolts, in sequence, to 22 ft. lbs. (29 Nm).

b. Torque all bolts, in sequence, to 43 ft. lbs. (58 Nm).

c. Loosen all bolts completely.

d. Torque all bolts, in sequence, to 22 ft. lbs. (29 Nm).

e. Torque all bolts, in sequence, to 40–47 ft. lbs. (54–64 Nm). If using an angle torque wrench, torque them 60–65 degrees tighter rather than going to 40–47 ft. lbs. (54–64 Nm).

17. On 1989–92 models, install the alternator and air conditioner compressor mounting brackets. Mount the compressor and alternator.

18. Install the rear timing cover bolts. Install the camshaft pulleys. Make sure the pulley marked R3 goes on the right and that marked L3 goes on the left. Align the timing marks if necessary and then install the timing belt and adjust the belt tension.

19. Connect the exhaust manifold to the exhaust pipe.

20. Install the exhaust manifold covers.

21. Install the exhaust collector bracket.

22. Install the intake manifold and fuel tube assembly.

23. Install the intake manifold collector cover.

24. On 1989–92 models, connect the accelerator and cruise control cables to the intake manifold and install the distributor and ignition wires.

25. Connect the vacuum and water hoses to the intake collector.

26. Install and tension the timing belt.

27. Fill the cooling system and connect the negative battery cable.

28. Make all the necessary engine adjustments. Road test the vehicle for proper operation.

VG30DE and VG30DETT Engines

1. Relieve the fuel system pressure and disconnect the negative battery cable.

2. Drain the cooling system.

3. Remove the intake manifold collector.

4. Remove the injector pipe assembly.

5. Remove the valve covers.

6. Remove the timing belt.

7. Remove the idler pulley and idler pulley stud bolt.

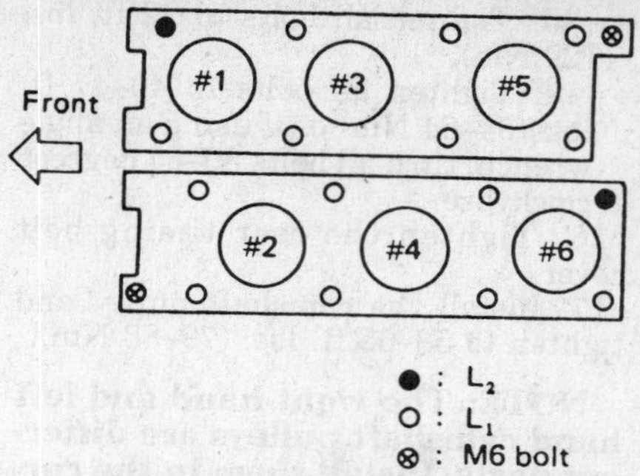

Torque the 6mm "X" bolts to 7–9 ft. lbs. (10–12 Nm)

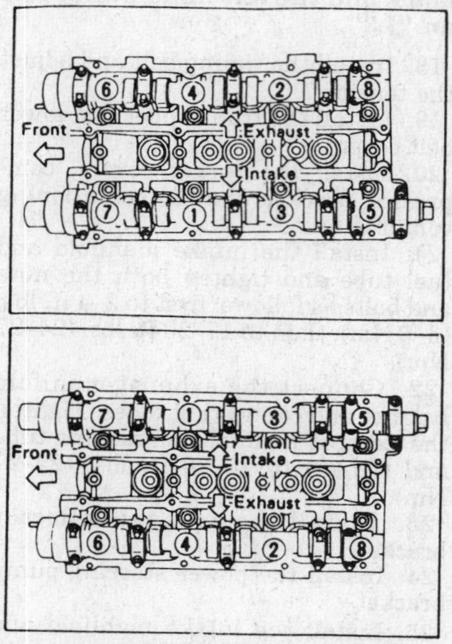

Cylinder head torque sequence—VG30DE and VG30DETT engines

8. Remove the intake manifold.

9. Disconnect the exhaust tube from the exhaust manifold.

10. Loosen the cylinder head bolts (in reverse order of installation sequence) in 2–3 stages. Lift the cylinder head off the engine block wih the exhaust manifolds attached It may be necessary to tap the head lightly with a rubber mallet to loosen it.

To install:

11. Make sure the No. 1 cylinder is set at TDC on its compression stroke as follows:

a. Align the crankshaft timing mark with the mark on the oil pump housing.

b. Align camshaft sprocket timing mark with the mark on the rear timing belt cover.

12. Install the cylinder head with a new gasket. Apply clean engine oil to the threads and seats of the bolts and install the bolts with washers in the correct position.

13. Torque the bolts in the proper sequence as follows:

a. Torque all bolts, in sequence, to 29 ft. lbs. (39 Nm).

b. Torque all bolts, in sequence, to 90 ft. lbs. (123 Nm).

c. Loosen all bolts completely.

d. Torque all bolts, in sequence, to 25–33 ft. lbs. (34–44 Nm).

e. Torque all bolts, in sequence, to 90 ft. lbs. (123 Nm). If using an angle torque wrench, torque them 60–70 degrees tighter rather than going to 90 ft. lbs. (123 Nm).

f. Torque the 6mm "X" bolts to 7–9 ft. lbs. (10–12 Nm). There is one of these bolts per head.

14. Connect the exhaust tube to the exhaust manifold.

15. Install the intake manifold.

16. Install the idler pulley and stud bolt.

17. Install and tension the timing belt.

18. Install the valve covers. Use sealant on the exhaust side valve cover.

19. Install the injector pipe assembly.

20. Install the intake manifold collector.

21. Fill the cooling system to the proper level and connect the negative battery cable.

22. Make all the necessary engine adjustments. Road test the vehicle for proper operation.

Valve Lifters

REMOVAL & INSTALLATION

1. Disconnect the negative battery cable.

2. Remove the cylinder head, if required.

3. Remove the rocker arms and shafts.

4. Withdraw the lifters from the head or from the bore in the rocker. Tag each lifter to the corresponding cylinder head opening or rocker. If the lifter is installed in the rocker, remove the snapring first. Be careful not to bend the snapring during removal.

NOTE: Do not lay the lifters on their sides because air will be allowed to enter the lifter. When storing lifters, set them straight up. To store lifters on their sides, they must be soaked in a bath of clean engine oil.

5. Install the lifters in their original locations. Use new lifter snaprings as needed. New lifters should be soaked in a bath of clean engine oil prior to installation to remove the air.

6. Install the rocker arms and shafts.

7. Install the cylinder head and leave the valve cover off.

8. Check the lifters for proper oper-

ation by pushing hard on each lifter with fingertip pressure. If the valve lifter moves more than 0.04 in. (1mm), air may be inside it. Make sure the rocker arm is not on the cam lobe when making this check. If there was air in the lifters, bleed the air by running the engine at 1000 rpm for 10 minutes.

Valve Lash

Hydraulic valve lifters are used on all engines except on those models listed below. Engines with hydraulic lifters do not require periodic valve adjustment, because the lifter automatically compensates for any required adjustment. Hydraulic valve lifters are best maintained through regular, scheduled engine oil and filter changes.

ADJUSTMENT

E16i Engine

1. Run the engine until it reaches normal operating temperature and shut if off.
2. Remove the rocker cover.
3. Bring the No. 1 piston at TDC on the compression stroke. There are at least two ways to do it; bump the engine over with the starter or turn it over by using a wrench on the front pulley attaching bolt. The easiest way to find TDC is to turn the engine over slowly with a wrench, after first removing No. 1 plug, until the piston is at the top of its stroke and the TDC timing mark on the crankshaft pulley is in alignment with the timing mark pointer. At this point, the valves for No. 1 should be closed.

NOTE: Make sure both valves are closed with the valve springs up as high as they will go. An easy way to find the compression stroke is to remove the distributor cap and see toward which spark plug lead the rotor is pointing. If the rotor points to the No. 1 spark plug lead, the No. 1 cylinder is on its compression stroke. When the rotor points to the No. 2 spark plug lead, the No. 2 cylinder is on its compression stroke etc.

4. With No. 1 piston at TDC of the compression stroke, use a feeler gauge and check the clearance on valves, check valves No. 1, 2, 3 and 6.
5. To adjust the clearance, loosen the locknut and turn the adjuster with a tool while holding the locknut. The correct size feeler gauge should pass with a slight drag between the rocker arm and the valve stem.
6. Turn the crankshaft one full revolution to position the No. 4 piston at TDC of the compression stroke. Adjust valves No. 4, 5, 7 and 8.
7. Replace the valve cover with a new cover gasket or sealing compound.

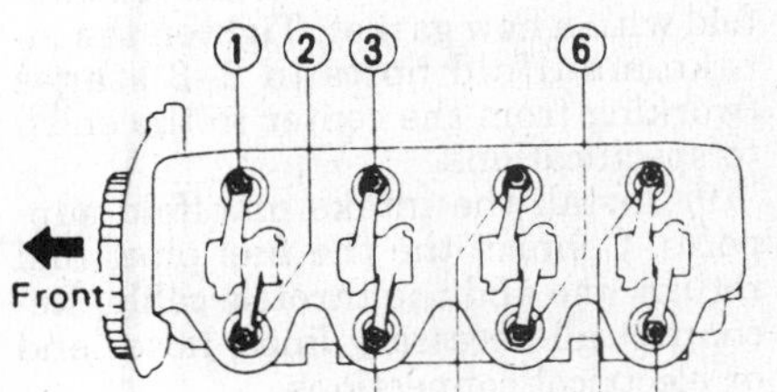

Valve adjustment sequence – E16i engine

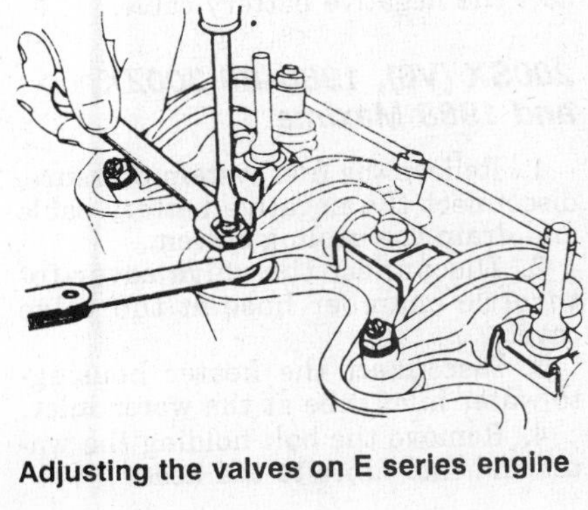

Adjusting the valves on E series engine

GA16DE Engine

1. Run the engine until it reaches normal operating temperature and shut if off.
2. Remove the rocker cover and all spark plugs.

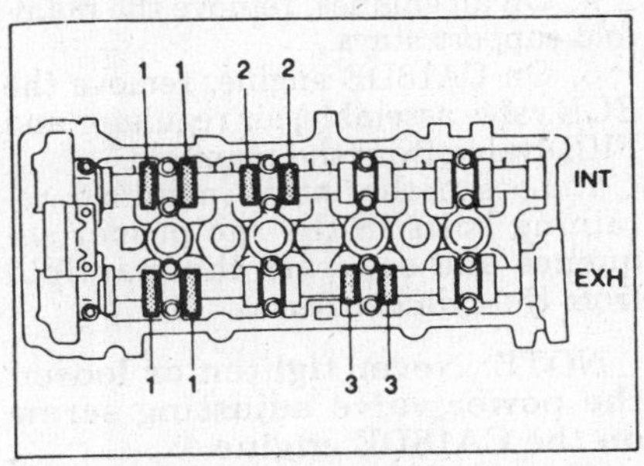

Valve adjustment step No. 1 – GA16DE and KA24DE engines

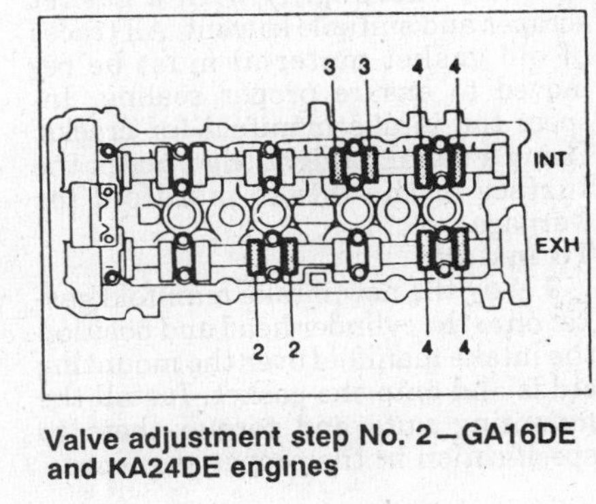

Valve adjustment step No. 2 – GA16DE and KA24DE engines

3. Set No. 1 cylinder at TDC on compression stroke. Align pointer with TDC mark on crankshaft pulley. Check that the valve lifters on No. 1 cylinder are loose and valve lifters on No. 4 are tight. If not turn crankshaft one revolution 360 degrees and align as above.
4. Check both No. 1 intake and both No. 1 exhaust valves, both No. 2 intake valves and both No. 3 exhaust valves. Using a feeler gauge, measure the clearance between valve lifter and camshaft. Record any valve clearance measurements which are out of specification. Intake Valve clearance (hot) is 0.008–0.019 in. and exhaust valve clearance (hot) is 0.012–0.023 in.
5. Turn crankshaft one revolution 360 degrees and align mark on crankshaft pulley with pointer. Check both No. 2 exhaust valves, both No. 3 intake valves, both No. 4 intake valves and both No. 4 exhaust valves. Using a feeler gauge, measure the clearance between valve lifter and camshaft. Record any valve clearance measurements which are out of specification. Intake valve clearance (hot) is 0.008–0.019 in. and exhaust valve clearance (hot) is 0.012–0.023 in.
6. If all valve clearances are within specification, install all related parts as necessary.
7. If adjustement is necessary, adjust valve clearance while engine is cold by removing adjusting shim. Determine replacement adjusting shim size using formula. Using a micrometer determine thickness of removed shim. Calculate thickness of new adjusting shim so valve clearance comes within specified valves. R = thickness of removed shim, N = thickness of new shim, M = measured valve clearance.

   INTAKE: N = R + (M - 0.0146 in.)
   EXHAUST: N = R + (M- 0.0157 in.)
8. Shims are available in 50 sizes (thickness is stamped on shim-this side always installled down), select new shims with thickness as close as possible to calculate valve.

KA24DE Engine

1. Run the engine until it reaches normal operating temperature and shut if off.
2. Remove the rocker cover and all spark plugs.
3. Set No. 1 cylinder at TDC on compression stroke. Align pointer with TDC mark on crankshaft pulley. Check that the valve lifters on No. 1 cylinder are loose and valve lifters on No. 4 are tight. If not turn crankshaft one revolution 360 degrees and align as above.
4. Check both No. 1 intake and both

No.1 exhaust valves, both No. 2 intake valves and both No. 3 exhaust valves. Using a feeler gauge, measure the clearance between valve lifter and camshaft. Record any valve clearance measurements which are out of specification. Intake Valve clearance (hot) is 0.012–0.015 in. and exhaust valve clearance (hot) is 0.013–0.016 in.

5. Turn crankshaft one revolution 360 degrees and align mark on crankshaft pulley with pointer. Check both No. 2 exhaust valves, both No. 3 intake valves, both No. 4 intake valves and both No. 4 exhaust valves. Using a feeler gauge, measure the clearance between valve lifter and camshaft. Record any valve clearance measurements which are out of specification. Intake valve clearance (hot) is 0.012–0.015 in. and exhaust valve clearance (hot) is 0.013–0.016 in.

6. If all valve clearances are within specification, install all related parts as necessary.

7. If adjustement is necessary, adjust valve clearance while engine is cold by removing adjusting shim. Determine replacement adjusting shim size using formula. Using a micrometer determine thickness of removed shim. Calculate thickness of new adjusting shim so valve clearance comes within specified valves. R = thickness of removed shim, N = thickness of new shim, M = measured valve clearance.

INTAKE: N = R + (M - 0.0138 in.)
EXHAUST: N = R + (M- 0.0146 in.)

8. Shims are available in 37 sizes (thickness is stamped on shim-this side always installled down), select new shims with thickness as close as possible to calculated value.

Rocker Arms/Shaft

REMOVAL & INSTALLATION

NOTE: All rocker shaft removal and installation procedures are given in "Camshaft, Removal and Installation".

Intake Manifold

REMOVAL & INSTALLATION

Pulsar and Sentra

1. Relieve the fuel system pressure, disconnect the negative battery cable and drain the cooling system.
2. Remove the air cleaner assembly.
3. Disconnect the throttle linkage, electrical connections, fuel and vacuum lines from the throttle body or throttle chamber.
4. The throttle body/throttle chamber can be removed from the manifold at this point or can be removed as an assembly with the intake manifold.

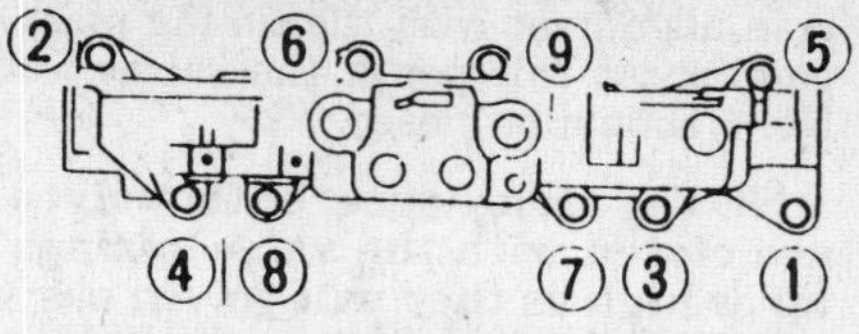

Intake manifold nut loosening sequence – GA16i, GA16DE and SR20DE engines

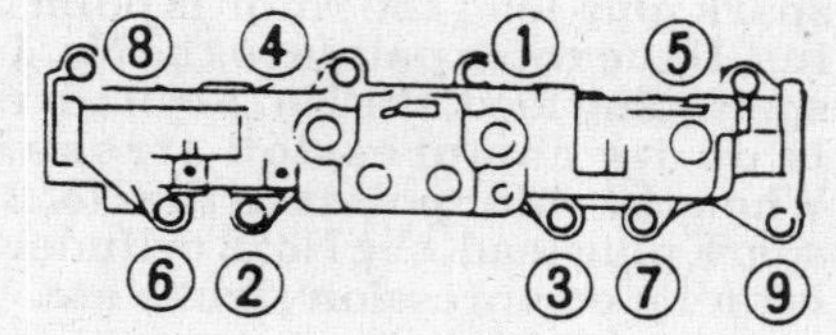

Intake manifold nut tightening sequence – GA16i, GA16DE and SR20DE engines

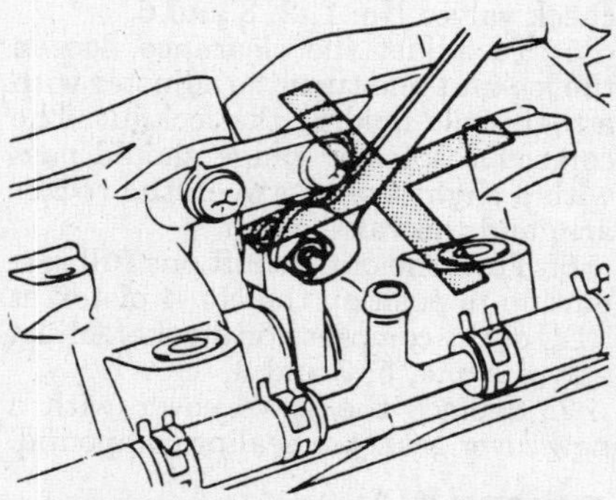
Never touch this bolt on the CA18DE engine

5. On all engines, remove the manifold support stays.
6. On CA18DE engine, remove the EGR valve assembly, air regulator and FICD valve from the manifold.
7. Loosen the intake manifold retaining bolts in the the proper sequence and separate the manifold from the cylinder head.

NOTE: Never tighten or loosen the power valve adjusting screw on the CA18DE engine.

8. Remove the intake manifold gasket and clean all the gasket contact surfaces thoroughly with a gasket scraper and suitable solvent. All traces of old gasket material must be removed to ensure proper sealing. Inspect the intake manifold for cracks. Using a metal straight edge, check the surface of the intake manifold for warpage.

To install:

9. Lay the new intake manifold gasket onto the cylinder head and position the intake manifold over the mounting studs and onto the gasket. Install the mounting nuts and torque them to specification in the proper sequence.
10. On CA18DE engine, install the EGR valve assembly, air regulator and FICD valve onto the manifold.
11. On all engines, install the manifold support stays.
12. If removed, install the throttle body or throttle chamber.
13. Connect the throttle linkage, electrical connections, fuel and vacuum lines.
14. Install the air cleaner.
15. Fill the cooling system to the proper level and connect the negative battery cable.
16. Road test the vehicle for proper operation.

200SX (4 Cylinder) and 1988–89 Stanza

1. Disconnnect the negative battery cable and drain the cooling system.
2. Remove the air cleaner hoses.
3. Remove the radiator hoses from the manifold.
4. Relieve the fuel pressure. Remove the throttle cable and disconnect the fuel pipe and the fuel return line. Plug the fuel pipe to prevent spilling fuel.
5. Remove all remaining wires, tubes and the EGR and PCV tubes from the rear of the intake manifold. Remove the manifold supports.
6. Unbolt and remove the intake manifold. Remove the manifold with the fuel injectors/injection body, EGR valve, fuel pipes and associated running gear still attached.
7. Remove the intake manifold gasket and clean the gasket surfaces.

To install:

8. Install the intake manifold manifold with a new gasket. Tighten the intake manifold bolts in 2–3 stages (working from the center to the ends) to specifications.
9. Install the intake manifold supports. Connect the the fuel pipe, fuel return line and the throttle cable. Reconnect all necessary lines, hoses and or electrical connections.
10. Connect the radiator hoses to the intake manifold. Connect the air cleaner hoses.
11. Fill the cooling system and connect the negative battery cable.

200SX (V6), 1988–89 300ZX and 1988 Maxima

1. Relieve the fuel system pressure, disconnect the negative battery cable and drain the cooling system.
2. Disconnect the valve cover-to-throttle chamber hose at the valve cover.
3. Disconnect the heater housing-to-water inlet tube at the water inlet.
4. Remove the bolt holding the water and fuel tubes to the head.

5. Remove the heater housing-to-thermostat housing tube.
6. Remove the intake collector cover and then remove the collector itself.
7. Disconnect the fuel line and remove the intake manifold bolts. Remove the intake manifold assembly, with the fuel tube assembly still attached, from the vehicle.

To install:

8. Install the intake manifold manifold with a new gasket. Tighten the intake manifold bolts in 2–3 stages in the proper sequence to specifications.
9. Connect the fuel line.
10. Install the intake manifold collector with a new gasket. Install the collector cover.
11. Connect the heater-to-thermostat housing tube.
12. Attach the water and fuel tubes to the cylinder head with the mounting bolt.
13. Connect the valve cover-to-throttle chamber hose.
14. Fill the cooling system to the proper level and connect the negative battery cable.
15. Road test the vehicle for proper operation. Make all the necessary engine adjustments.

1990–92 300ZX

1. Relieve the fuel system pressure, disconnect the negative battery cable and drain the cooling system.
2. Disconnect the air inlet hoses from both throttle chambers.
3. Disconnect the throttle cable from the accelerator drum located in the middle of the throttle chambers.
4. Disconnect the electrical connectors and vacuum lines from both throttle chambers.
5. Disconnect and tag the electrical wire connectors and vacuum lines from the intake manifold collector.
6. Unbolt and remove the intake manifold collector with the throttle chambers attached. Remove the collector gasket.
7. Disconnect the fuel supply and return lines from the injector assembly. Plug the lines to prevent leakage.
8. Remove the injector assembly from the intake manifold.
9. Remove the intake manifold and gaskets.

To install:

10. Install the intake manifold with a

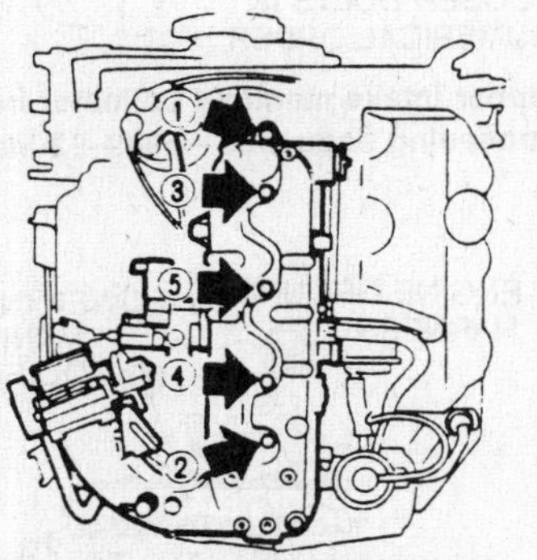

Intake manifold collector cover bolt removal sequence – 200SX (V6), 1988–89 300ZX and 1988 Maxima

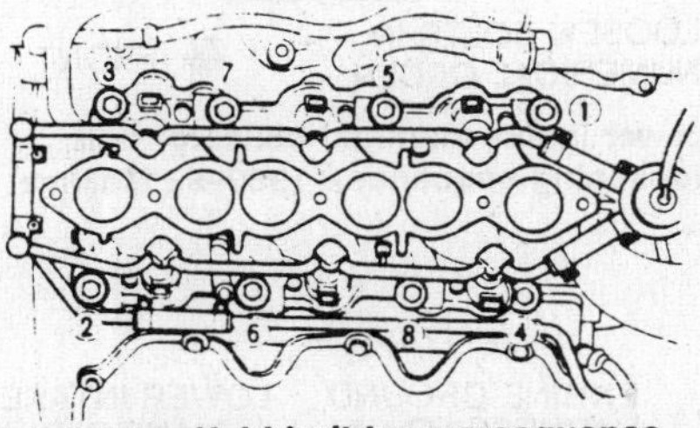

Intake manifold bolt torque sequence – 200SX (V6), 1988–89 300ZX and 1988 Maxima

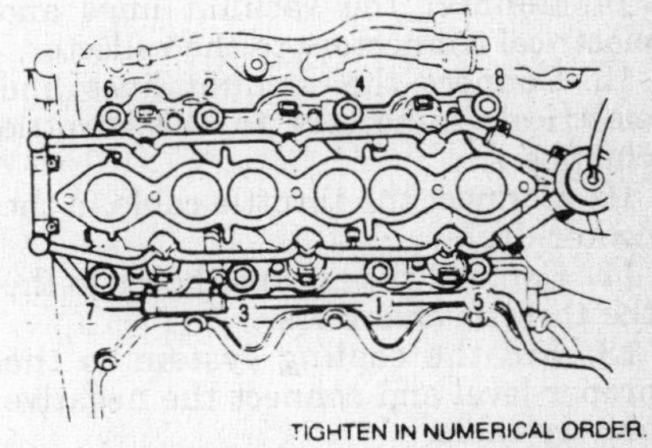

Intake manifold bolt removal sequence – 200SX (V6), 1988–89 300ZX and 1988 Maxima

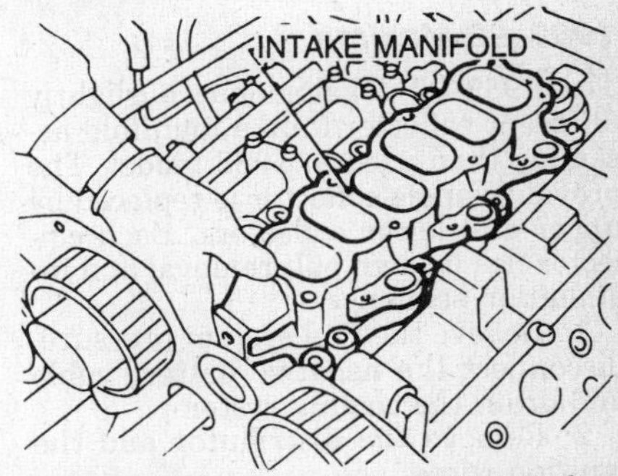

Intake manifold removal and installation – 1990–92 300ZX

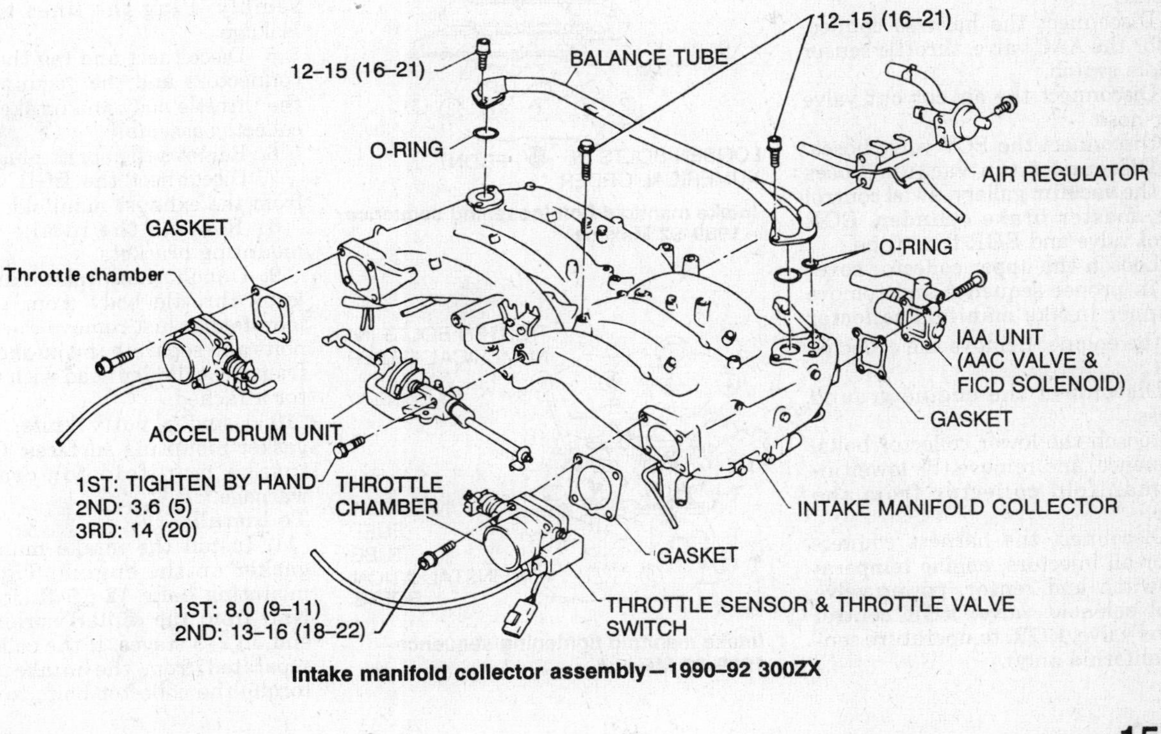

Intake manifold collector assembly – 1990–92 300ZX

new gasket. Tighten the intake manifold bolts to specification.

11. Install the fuel injectors with new insulators and O-rings.

12. Connect the injector supply and return lines.

13. Install the intake manifold collector with a new gasket. Torque the collector bolts to 12–15 ft. lbs. (16–21 Nm).

14. Connect the vacuum lines and electrical connectors to the collector.

15. Connect the vacuum lines and electrical connectors to the throttle chambers.

16. Connect the throttle cable to the center drum.

17. Connect the air inlet hoses to the the throttle chambers.

18. Fill the cooling system to the proper level and connect the negative battery cable.

19. Make all the necessary engine adjustments. Road test the vehicle for proper operation.

1989–92 Maxima

The 1989–92 Maxima has a slightly different collector/intake manifold assembly than used in 1988 model. The previous single collector is replaced by upper and lower collectors. Each collector has its own bolt removal and installation sequence.

1. Relieve the fuel system pressure, disconnect the negative battery cable and drain the cooling system.

2. Remove the distributor and the ignition wires.

3. Disconnect the Automatic Speed Control Device (ASCD) and accelerator wires from the intake manifold collector.

4. Disconnect the harness connectors for the AAC valve, throttle sensor and idle switch.

5. Disconnect the air cut out valve water hose.

6. Disconnect the PCV valve hoses.

7. Disconnect the vacuum hoses from the vacuum gallery, swirl control valve, master brake cylinder, EGR control valve and EGR flare tube.

8. Loosen the upper collector cover bolts in proper sequence and remove the upper intake manifold collector from the engine. Remove the collector gasket.

9. Disconnect the engine ground harness.

10. Loosen the lower collector bolts, in sequence, and remove the lower intake manifold collector from the engine.

11. Disconnect the harness connectors for all injectors, engine temperature switch and sensor, power valve control solenoid valve, EGR control soleniod valve, EGR. temperature sensor (California only).

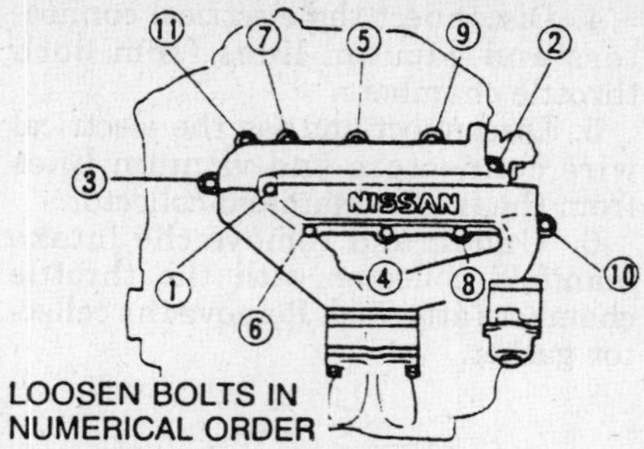

Upper intake manifold collector bolt loosening sequence – 1989–92 Maxima

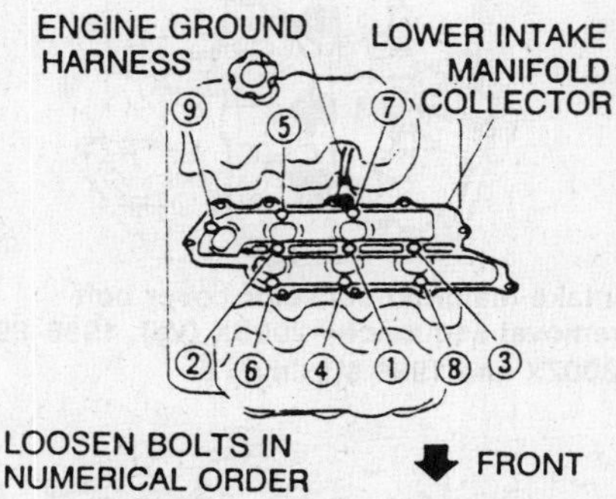

Lower intake manifold collector bolt loosening sequence – 1989–92 Maxima

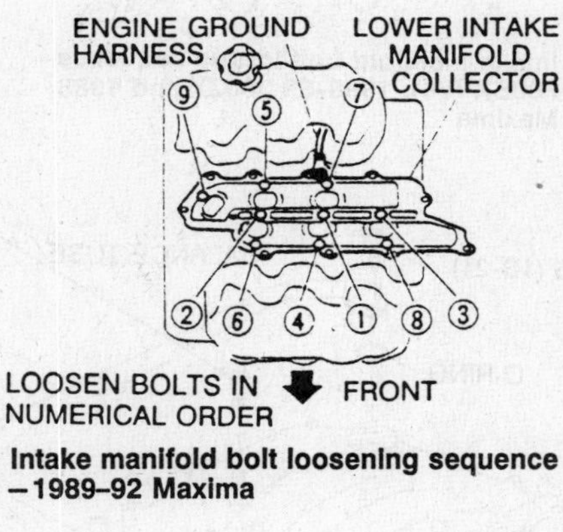

Intake manifold bolt loosening sequence – 1989–92 Maxima

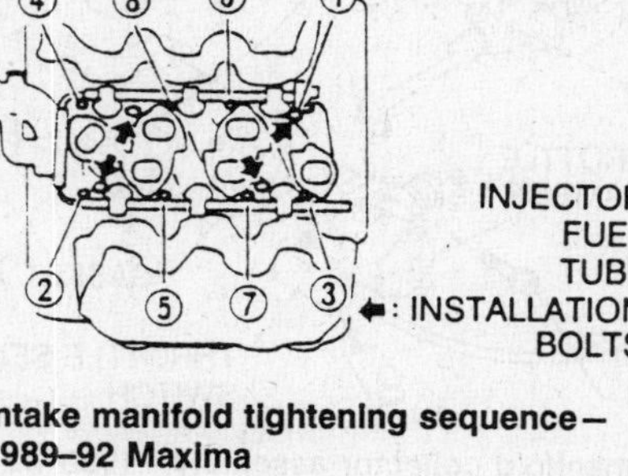

Intake manifold tightening sequence – 1989–92 Maxima

12. Disconnect the vacuum gallery hoses.

13. Disconnect the pressure regulator valve vacuum hose, heater hose, fuel feed and return hose.

14. Remove the intake manifold and fuel tube assembly. Loosen intake manifold bolts in numerical order.

To install:

15. Install the intake manifold and fuel tube assembly with a new gasket. Tighten the manifold bolts and nuts in 2–3 stages in sequence.

16. Connect the hoses and electrical wires to the intake manifold and fuel tube.

17. Install the upper and lower collector and collector cover with new gaskets. Tighten collector to intake manifold bolts in 2–3 stages by reversing the removal sequence.

18. Connect the vacuum lines, hoses, cables and brackets to the collector cover and collector assembly.

19. Install the distributor and ignition wires.

20. Fill the cooling system to the proper level and connect the negative battrey cable.

21. Make all the necessary engine adjustments. Road test the vehicle for proper operation.

240SX and 1990–92 Stanza

1. Relieve the fuel system pressure, disconnect the negative battery cable and drain the cooling system.

2. Remove the air duct between the air flow meter and the throttle body.

3. Disconnect the throttle cable.

4. Disconnect the fuel supply and return lines from the fuel injector assembly. Plug the lines to prevent leakage.

5. Disconnect and tag the electrical connectors and the vacuum hoses to the throttle body and intake manifold/collector assembly.

6. Remove the spark plug wires.

7. Disconnect the EGR valve tube from the exhaust manifold.

8. Remove the intake manifold mounting brackets.

9. Unbolt the intake manifold collector/throttle body from the intake manifold or just remove the mounting bolts and separate the intake manifold from the cylinder head with the collector attached.

10. Using a putty knife, clean the gasket mounting surfaces. Check the intake manifold for cracks and warpage.

To install:

11. Install the intake manifold and gasket on the engine. Tighten the mounting bolts 12–15 ft. lbs. (16–20 Nm) from the center working to the end, in 2–3 stages. If the collector was separated from the intake manifold, torque the collector bolts to 12–15 ft.

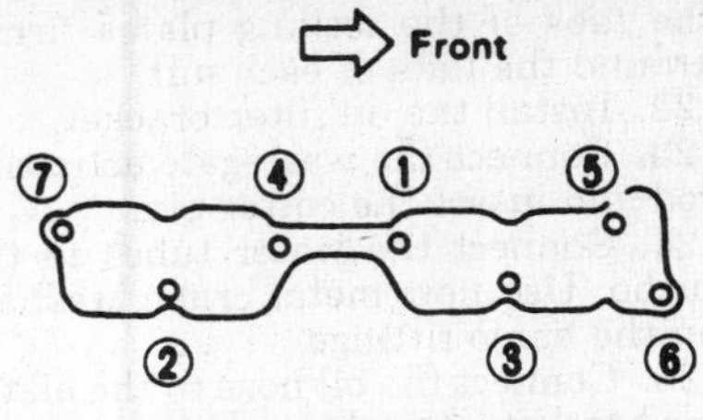

Intake manifold bolt torque sequence – 240SX and 1990–92 Stanza

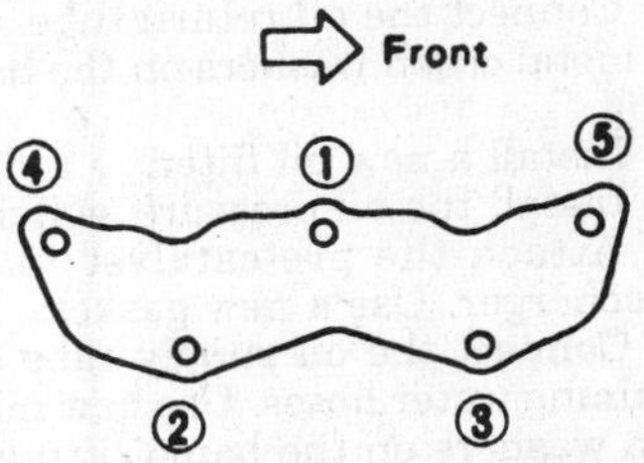

Intake manifold collector bolt torque sequence – 240SX and 1990–92 Stanza

lbs. (16–20 Nm) from the center working to the end.

12. Install intake manifold mounting brackets.
13. Connect the EGR valve tube to the exhaust manifold.
14. Install the spark plug wires.
15. Connect the electrical connectors and the vacuum hoses to the throttle body and intake manifold/collector assembly.
16. Connect the fuel line(s) to the fuel injector assembly.
17. Connect the throttle cable.
18. Connect the air duct between the air flow meter and the throttle body.
19. Fill the cooling system to the proper level and connect the negative battery cable.
20. Make all the necessary engine adjustments. Road test the vehicle for proper operation.

Exhaust Manifold

REMOVAL & INSTALLATION

NOTE: If any fuel system components must be removed, make to relieve the fuel system pressure first. If the engine is equipped with a turbocharger, it may be easier to remove the exhaust manifolds, with the turbo(s) atttached.

1. Disconnect the negative battery cable. Raise and support the vehicle safely.
2. Remove the undercover and dust covers, if equipped.
3. Remove the air cleaner or collector assembly, if necessary for access.
4. Remove the heat shield(s), if equipped.

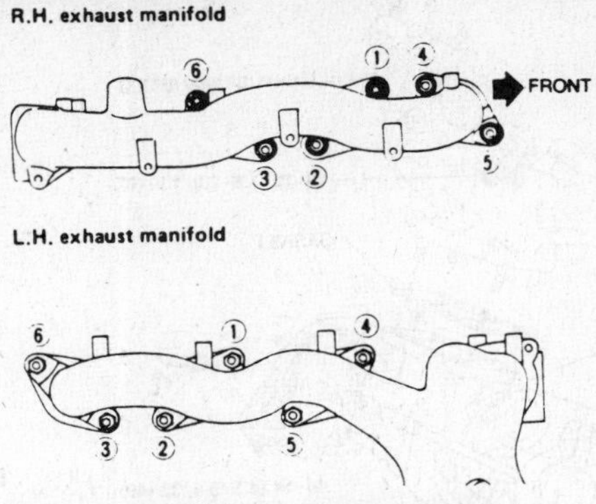

Exhaust manifold torque sequence – VG30E (200SX and 1988–89 Maxima)

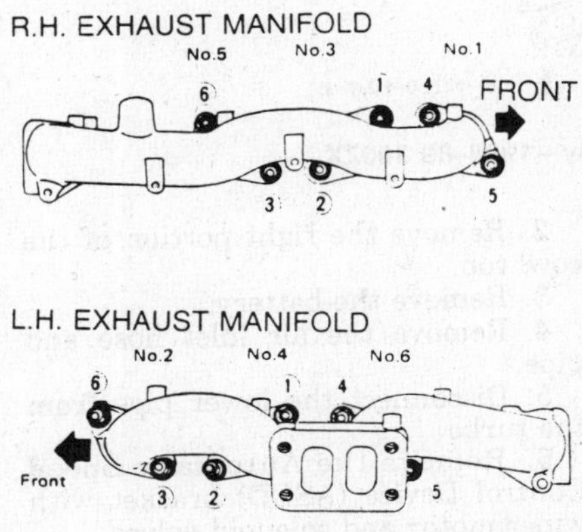

Exhaust manifold torque sequence – VG30E and VG30ET (1988–89 300ZX)

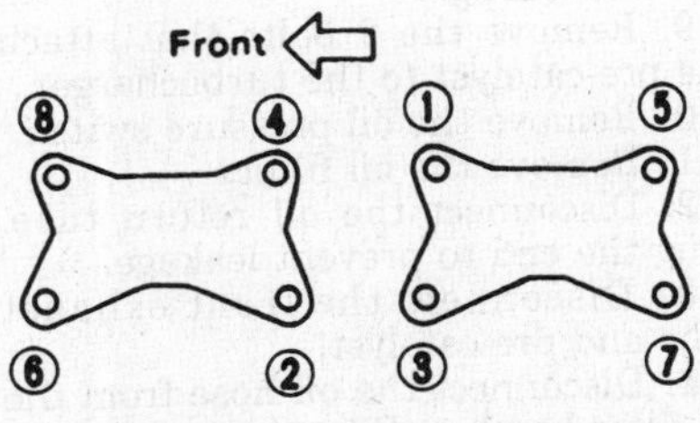

Exhaust manifold torque sequence – 240SX and 1990–92 Stanza

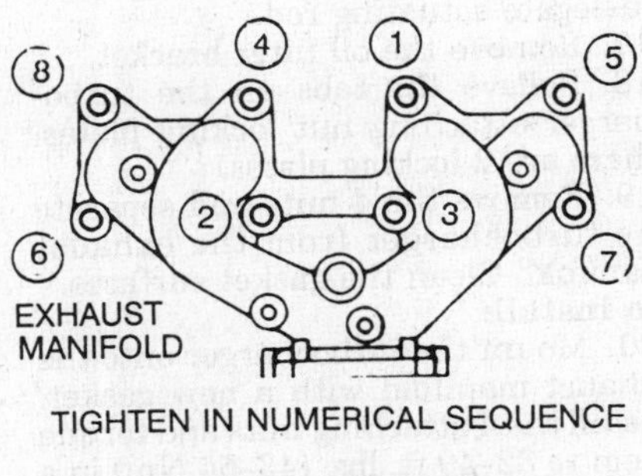

Exhaust manifold torque sequence – GA16i

5. Disconnect the exhaust pipe from the exhaust manifold or the turbo outlet.
6. Remove or disconnect the temperature sensors, oxygen sensors, air induction pipes, bracketry and other attachments from the manifold.
7. Disconnect the EAI and EGR tubes from their fittings if so equipped.
8. Loosen and remove the exhaust manifold attaching nuts and remove the manifold(s) from the block. Discard the exhaust manifold gaskets and replace with new.
9. Clean the gasket surfaces and check the manifold for cracks and warpage.

To install:

10. Install the exhaust manifold with a new gasket. Torque the manifold fasteners from the center outward in several stages to specifications.
11. Connect the EAI and EGR tubes to the connections on the manifold as necessary.
12. Install or connect the temperature sensors, oxygen sensors, air induction pipes, bracketry and other attachments to the manifold.
13. Connect the exhust pipe to the manifold or turbo outlet using a new gasket.
14. Install the heat shields.
15. Install the air cleaner or collector assembly.
16. Install the undercovers and dust covers.
17. Connect the negative battery cable.

Turbocharger

REMOVAL & INSTALLATION

NOTE: If the turbocharger is being replaced, always drain the crankcase and replace the oil and filter to ensure a clean oil supply. This is especially true in cases of complete turbo failure where there is the possibility of metal particles entering the engine's lubricating system and damaging the new turbocharger.

1988–89 300ZX Turbo

1. Disconnect the negative battery cable.
2. Discharge the air conditioning system and remove the compressor and compressor mounting bracket.
3. Disconnect the exhaust front tube.
4. Disconnect the center cable.
5. Remove the heat insulator for the brake master cylinder.
6. Disconnect the air duct and hoses.
7. Disconnect the exhaust manifold

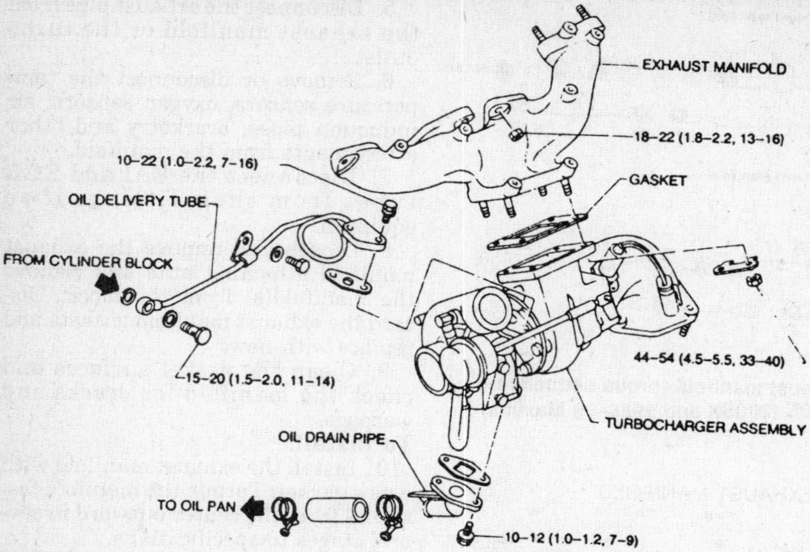

Turbocharger assembly – 1988–89 300ZX

connecting tube and remove the heat shield plate.

8. Disconnect the oil supply tube and return hose. Plug the ends of the hoses to prevent leakage.
9. Disconnect the water inlet line.
10. Remove the turbocharger from the exhaust manifold.

To install:

11. Add 25cc of clean engine oil to the turbocharger oil passages before installing.
12. Mount the turbocharger onto the manifold using new gaskets. Torque the turbocharger-to-manifold nuts to 33–40 ft. lbs. (45–54 Nm).
13. Connect the water inlet tube. Use new metal crush washers on the banjo fitting.
14. Connect the oil supply and return tubes. Use new metal crush washers on the banjo fittings.
15. Install the heat shield plate and connect the exhaust manifold tube.
16. Connect the air duct and hoses.
17. Install the brake master cylinder heat insulator.
18. Connect the center cable.
19. Connect the front exhaust tube.
20. Install the compressor bracket and mount the compressor.
21. Connect the negative battery cable and charge the air conditioning system.
22. Disconnect the coil and crank the engine for 20 seconds to ensure that oil reaches the center bearings. Connect the coil and start the engine, letting it idle for 30 seconds to ensure the proper operation of the turbocharger.

1990–92 300ZX Twin Turbo

RIGHT

1. Drain the cooling system and the oil pan.
2. Remove the right portion of the cowl top.
3. Remove the battery.
4. Remove the air inlet hose and pipe.
5. Disconnect the lower pipe from the turbo.
6. Remove the Automatic Speed Control Device (ASCD) bracket with wiper motor and solenoid valves.
7. Unplug the exhaust gas harness connector.
8. Disconnect the turbo water hoses and oil supply tube. Plug the ends to prevent leakage.
9. Remove the 2 bolts that attach the pre-catalyst to the turbocharger.
10. Remove the oil pressure switch.
11. Remove the oil filter.
12. Disconnect the oil return tube. Plug the end to prevent leakage.
13. Disconnect the front exhaust tube and pre-catalyst.
14. Disconnect the oil hose from the oil filter bracket. Plug the end to prevent leakage.
15. Disconnect the remaining water tubes from the turbocharger. Plug the ends to prevent leakage.
16. Remove the cotter pin from the wastegate actuating rod.
17. Remove the oil filter bracket.
18. Relieve the tabs on the turbocharger attaching nut locking plates. There are 2 locking plates.
19. Remove the 4 nuts and separate the turbocharger from the exhaust manifold. Clean the gasket surfaces.

To install:

20. Mount the turbocharger onto the exhaust manifold with a new gasket. Install the 4 attaching nuts and torque them to 32–40 ft. lbs. (43–54 Nm) in a criss-cross pattern.
21. Once the nuts are torqued, bend the tabs of the locking plates firmly around the flats of each nut.
22. Install the oil filter bracket.
23. Connect the wastegate actuating rod and insert the cotter pin.
24. Connect the water tubes to the turbo. Use new metal crush washers on the banjo fittings.
25. Connect the oil hose to the oil filter bracket.
26. Connect the front exhaust tube and pre-catalyst. Use new gaskets.
27. Connect the oil return tube. Use new metal crush washers on the banjo fitting.
28. Install a new oil filter.
29. Install the oil pressure switch.
30. Attach the pre-catalyst to the turbocharger. Use a new gasket.
31. Connect the oil supply tube and remaining water hoses. Use new metal crush washers on the banjo fittings.
32. Plug in the exhaust gas harness connector.
33. Mount the solenoid valves, wiper motor and Automatic Speed Control Device (ASCD) bracket.
34. Conect the lower pipe to the turbo.
35. Install the air inlet hose and pipe.
36. Install the battery.
37. Install the right portion of the top cowl.
38. Fill the crankcase and cooling system to the proper levels.
39. Start the engine and check for leaks.

LEFT

1. Drain the cooling system and the oil pan.
2. Remove the brake master cylinder and brake booster.
3. Remove the air inlet hose and pipe.
4. Disconnect the lower pipe from the turbocharger.
5. Disconnect the water tubes. Plug the the tube ends to prevent leakage.
6. Remove the 2 bolts that attach the pre-catalyst to the turbocharger.
7. Remove the front exhaust tube and pre-catalyst.
8. Disconnect the steering column lower joint from the steering rack.
9. Disconnect the oil return tube and remaining water tubes. Plug the tube ends to prevent leakage.
10. Disconnect the EGR tube and remove the wastegate valve actuator bracket.
11. Remove the exhaust manifold cover.
12. Remove the exhaust manifold attaching nuts. Remove the turbocharger and exhaust manifold together as one unit. Release the tabs on the attaching nut locking plates. There are 2 locking plates. Remove the 4 nuts and separate the turbocharger from the exhaust manifold. Clean the gasket surfaces.

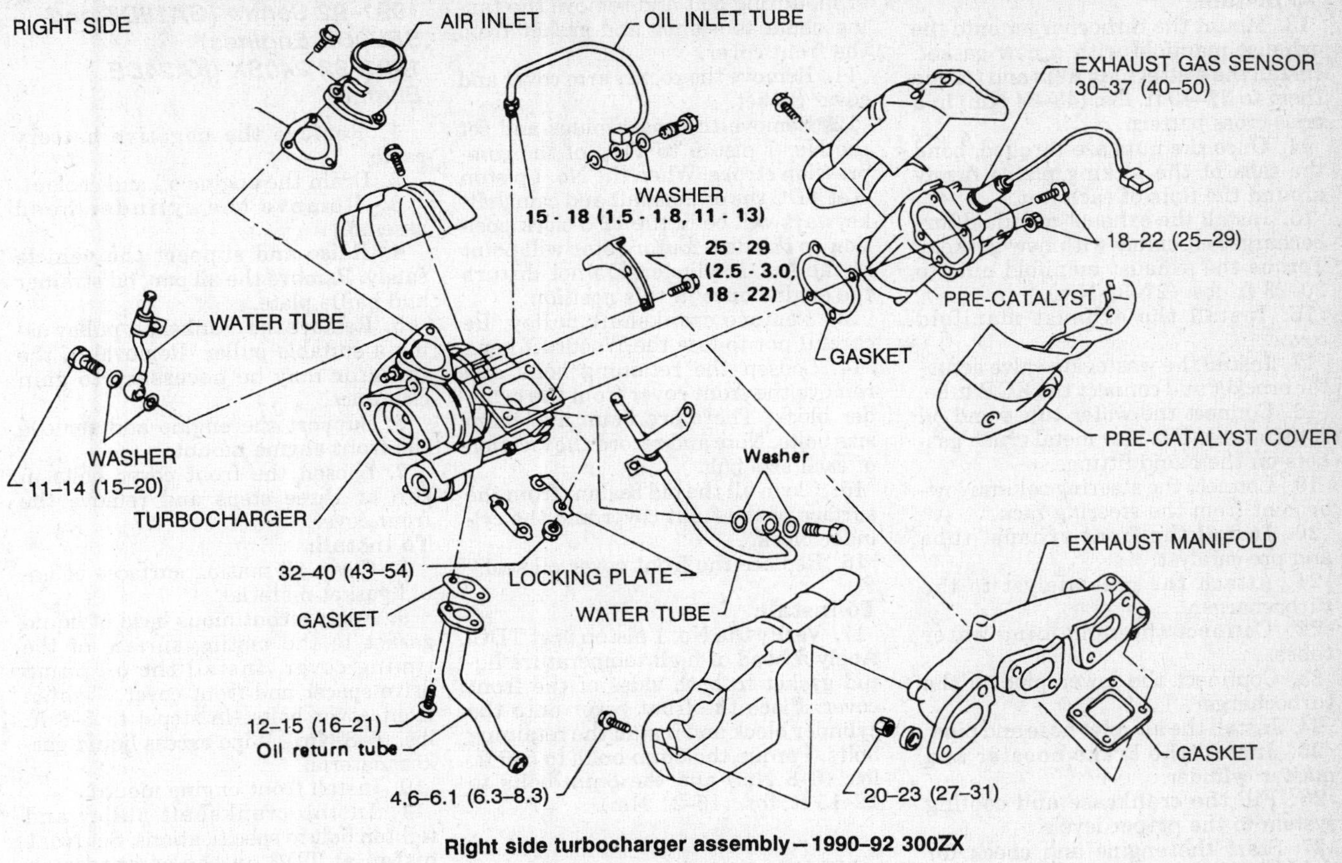

Right side turbocharger assembly—1990–92 300ZX

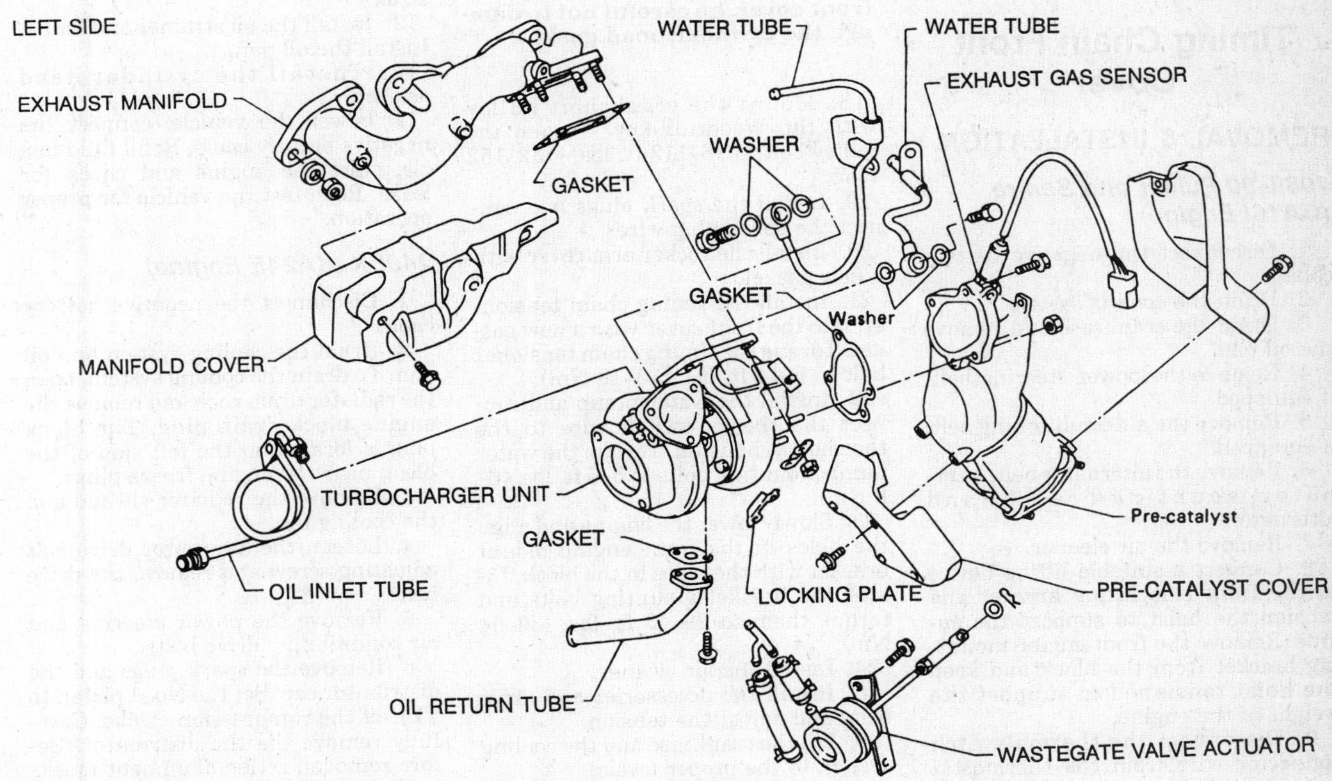

Left side turbocharger assembly—1990–92 300ZX

To install:

13. Mount the turbocharger onto the exhaust manifold with a new gasket. Install the 4 attaching nuts and torque them to 32–40 ft. lbs. (43–54 Nm) in a criss-cross pattern.
14. Once the nuts are torqued, bend the tabs of the locking plates firmly around the flats of each nut.
15. Install the exhaust manifold/turbocharger assembly with new gaskets. Torque the exhaust manifold nuts to 20–23 ft. lbs. (27–31 Nm).
16. Install the exhaust manifold cover.
17. Install the wastegate valve actuator bracket and connect the EGR tube.
18. Connect the water tubes and oil return tube. Use new metal crush gaskets on the banjo fittings.
19. Connect the steering column lower joint from the steering rack.
20. Install the front exhaust tube and pre-catalyst.
21. Attach the pre-catalyst to the turbocharger.
22. Connect the remaining water tubes.
23. Connnect the lower pipe to the turbocharger.
24. Install the air inlet hose and pipe.
25. Install the brake booster and master cylinder.
26. Fill the crankcase and cooling system to the proper levels.
27. Start the engine and check for leaks.

Timing Chain Front Cover

REMOVAL & INSTALLATION

1989–90 Pulsar and Sentra (GA16I Engine)

1. Disconnect the negative battery cable.
2. Drain the cooling system.
3. Drain the crankcase and remove the oil pan.
4. Remove the power steering belt, if equipped.
5. Remove the air conditioning belt, if equipped.
6. Remove the alternator belt, alternator mounting bracket and alternator.
7. Remove the air cleaner.
8. Connect a suitable lifting device to the front side lifting bracket and tension the hoist to support the engine. Remove the front engine mounting bracket from the block and keep the hoist tensioned to support the weight of the engine.
9. Disconnect the thermo switch connector wire from the thermostat housing and remove the water pump.
10. Loosen the timing chain tensioner mounting bolt and remove the timing chain tensioner and gasket from the front cover.
11. Remove the rocker arm cover and cover gasket.
12. Remove the spark plugs and set the No. 1 piston to TDC of the compression stroke. When the No. 1 piston is at TDC the crankshaft and camshaft keyways will be in the 12 o'clock position or the distributor rotor will point to the No. 1 cylinder. Do not disturb the engine once in this position.
13. Remove crankshaft pulley. Be careful not to lose the Woodruff key.
14. Loosen the retaining bolts and remove the front cover from the cylinder block. There are 6mm and 8mm size bolts. Note and record the location of each size bolt.
15. Clean all the old sealant from the surface of the front cover and the cylinder block.
16. Replace the front cover oil seal.

To install:

17. Verify the No. 1 piston is at TDC. Apply a bead of high temperature liquid gasket to both sides of the front cover. Place the front cover onto the cylinder block and install the retaining bolts. Torque the 6mm bolts to 5–6 ft. lbs. (6–8 Nm) and the 8mm bolts to 12–15 ft. lbs. (16–21 Nm).

NOTE: When installing the front cover, be careful not to damage the cylinder head gasket.

18. Mount the crankshaft pulley with the Woodruff key. Torque the pulley bolt to 98–112 ft. lbs. (132–152 Nm).
19. Install the spark plugs and connect the spark plug wires.
20. Install the rocker arm cover with a new gasket.
21. Install the timing chain tensioner onto the front cover with a new gasket. Torque the timing chain tensioner bolt to 9–14 ft. lbs. (13–19 Nm).
22. Install the water pump and connect the thermo-switch wire to the thermostat housing. Torque the water pump mounting bolts to 5–6 ft. lbs. (6–8 Nm).
23. Slowly lower the engine and align the holes in the front engine mount bracket with the holes in the block. Install the bracket mounting bolts and torque them to 29–40 ft. lbs. (39–54 Nm).
24. Install the air cleaner.
25. Install the accessories and drive belts and adjust the tension.
26. Fill the crankcase and the cooling system to the proper levels.
27. Connect the negative battery cable.

1991–92 Sentra (GA16DE and SR20DE Engines)
1991–92 240SX (KA24DE Engine)

1. Remove the negative battery cable.
2. Drain the engine oil and coolant.
3. Remove the cylinder head assembly.
4. Raise and support the vehicle safely. Remove the oil pan, oil strainer and baffle plate.
5. Remove the crankshaft pulley using a suitable puller. Removal of the radiator may be necessary to gain clearance.
6. Support the engine and remove the front engine mount.
7. Loosen the front cover bolts in two or three steps and remove the front cover.

To install:

8. Clean all mating surfaces of liquid gasket material.
9. Apply a continious bead of liquid gasket to the mating surface of the timing cover. Install the oil pump drive spacer and front cover. Tighten front cover bolts (in steps) to 5–6 ft. lbs. (6–8 Nm). Wipe excess liquid gasket material.
10. Install front engine mount.
11. Install crankshaft pulley and tighten bolt to specifications. Set No. 1 piston at TDC on the compression stroke.
12. Install the oil strainer and baffle. Install the oil pan.
13. Install the cylinder head assembly.
14. Lower the vehicle, connect the negative battery cable, Refill fluid levels, start the engine and check for leaks. Road test the vehicle for proper operation.

240SX (KA24E Engine)

1. Disconnect the negative battery cable.
2. Drain the cooling system and oil pan. To drain the cooling system, open the radiator drain cock and remove the engine block drain plug. The block plug is located on the left side of the block near the engine freeze plugs.
3. Remove the radiator shroud and the cooling fan.
4. Loosen the alternator drive belt adjusting screw and remove the drive belt.
5. Remove the power steering and air conditioning drive belts.
6. Remove the spark plugs and the distributor cap. Set the No. 1 piston to TDC of the compression stroke. Carefully remove the the distributor. Before removal, scribe alignment marks in the timing cover and flat portion of the oil pump/distributor drive spindle.

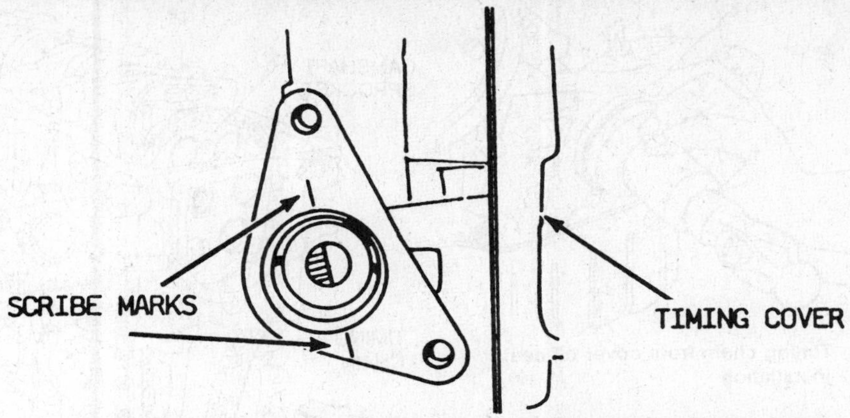

Aligning the timing cover and distributor/oil pump drive spindle – KA24E engine

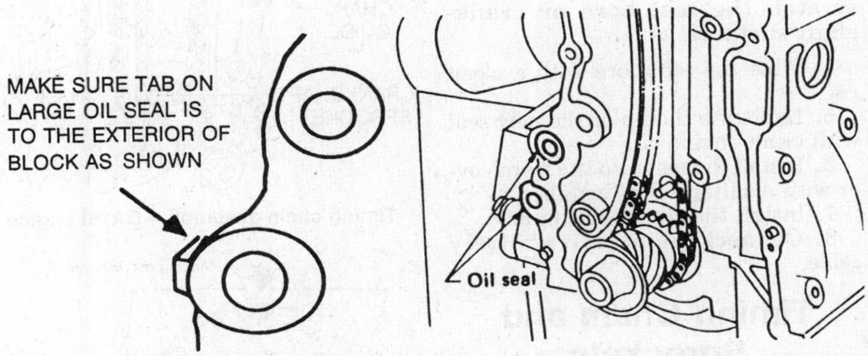

Cylinder block timing chain cover seals on KA24E engine. Make sure tab on larger seal is positioned as shown

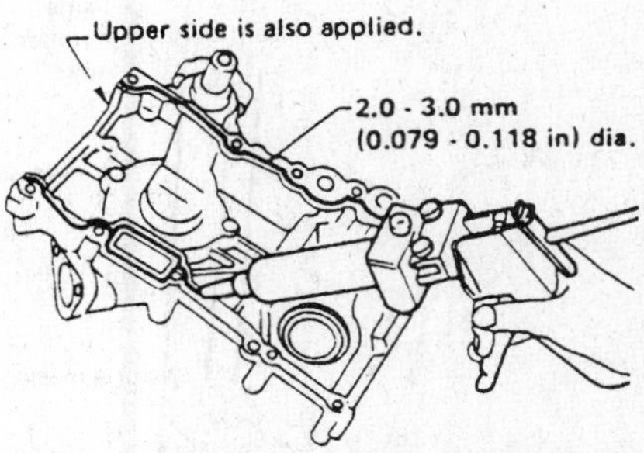

Applying sealant to front cover – KA24E engine

This alignment is critical and if not done properly, it could cause difficulty is aligning the distributor and setting the timing.

7. Remove the power steering pump, idler pulley and the power steering brackets.
8. Remove the air conditioning compressor idler pulley.
9. Remove the crankshaft pulley bolt and remove the crankshaft pulley with a 2 jawed puller.
10. Remove the oil pump attaching screws, and withdraw the pump and its drive spindle.
11. Remove the rocker arm cover.
12. Remove the oil pan.
13. Remove the bolts holding the front cover to the front of the cylinder block, the 4 bolts which retain the front of the oil pan to the bottom of the front cover, and the 4 bolts which are screwed down through the front of the cylinder head and into the top of the front cover. Carefully pry the front cover off the front of the engine. Clean all the old sealant from the surface of the front cover and the cylinder block.
14. Replace the crankshaft oil seal and the 2 timing chain cover oil seals in the block. These two seals should be installed in the block and not in the timing cover.

To install:

15. Verify the No. 1 piston is at TDC of the compression stroke. Apply a very thin bead of high temperature liquid gasket to both sides of the front cover and to where the cover mates with the cylinder head. Apply a light coating of grease to the crankshaft and timing cover oil seals and carefully bolt the front cover to the front of the engine.

NOTE: When installing the front cover, be careful not to damage the cylinder head gasket or to disturb the position of the oil seals in the block. Make sure the tab on the larger block oil seal is pointing to the exterior of the block.

16. Install new rubber plugs in the cylinder head.
17. Install the oil pan.
18. Install the rocker arm cover.
19. Before installing the oil pump, place the gasket over the shaft and make sure the mark on the drive spindle faces (aligned) with the oil pump hole. Install the oil pump and distributor driving spindle into the front cover with a new gasket.
20. Install the crankshaft pulley and bolt. Torque the pulley bolt to 87–116 ft. lbs. (118–157 Nm).
21. Install the distributor and the spark plugs.
22. Install the compressor idler pulley. Install power steering pump brackets, idler pulley and power steering pump. Install the drive belts and adjust the tension.
23. Install the radiator shroud and the cooling fan.
24. Refill the cooling system and crankcase to the proper levels.
25. Connect the negative battery cable.
26. Start the engine, check/set the ignition timing and check for engine leaks. Road test the vehicle for proper operation.

1990–92 Stanza (KA24E Engine)

1. Disconnect the negative battery cable.
2. Raise the front of the vehicle and support safely.
3. Remove the right front wheel.
4. Remove the dust cover and undercover.
5. Drain the oil pan.
6. Set the No. 1 piston at TDC of the compression stroke.
7. Remove the alternator and air conditioning compressor drive belts.
8. Remove the alternator and adjusting bar.
9. Remove the oil separator.
10. Remove the power steering pump pulley, pump stay and mounting bracket.
11. Discharge the air conditioning system and remove the compressor and mounting bracket.
12. Remove the crankshaft pulley and oil pump drive boss.
13. Remove the oil pan.
14. Remove the oil strainer mounting bolt.
15. Remove the bolts that attach the front cover to the head and the block.
16. Remove the rocker cover.
17. Support the engine with a suitable lifting device.
18. Unbolt the right side engine mount bracket from the block and lower the engine.

19. Remove the front cover.
20. Clean all the old sealant from the surface of the front cover and the cylinder block.
21. Replace the crankshaft oil seal and the 2 timing chain cover oil seals in the block. These two seals should be installed in the block and not in the timing cover.

To install:

22. Verify the No. 1 piston is at TDC. Apply a very thin bead of high temperature liquid gasket to both sides of the front cover and to where the cover mates with the cylinder head. Apply a light coating of grease to the crankshaft and timing cover oil seals and carefully mount the front cover to the front of the engine.

NOTE: When installing the front cover, be careful not to damage the cylinder head gasket or to disturb the position of the oil seals in the block. Make sure the tab on the larger block oil seal is pointing to the exterior of the block.

23. Install new rubber plugs in the cylinder head.
24. Raise the engine and install the right engine mount bracket bolts. Torque the bolts to 58–65 ft. lbs. (78–88 Nm).
25. Install the rocker arm cover.
26. Install the front cover bolts.
27. Install the oil strainer mounting bolt.
28. Install the oil pan.
29. Install the oil pump drive boss and the crankshaft pulley. Torque the pulley bolt to 87–116 ft. lbs. (118–157 Nm).
30. Install the air conditioning compressor bracket and mount the compressor.
31. Install the power steering bracket, pump stay and power steering pump.
32. Install the oil separator.
33. Install the dust cover and undercover.
34. Mount the right front wheel and lower the vehicle.
35. Fill the crankcase to the proper level and charge the air conditioning system.
36. Make all the necessary engine adjustments.

Front Cover Oil Seal

REPLACEMENT

1. Disconnect the negative battery cable.
2. Remove the crankshaft pulley.
3. Using a suitable tool, pry the oil seal from the front cover.

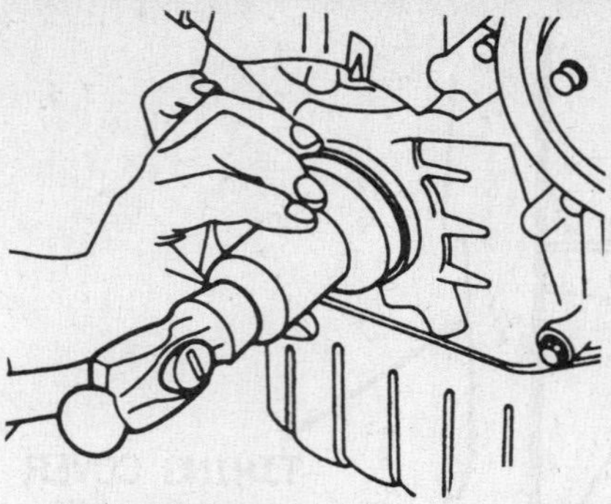

Timing chain front cover oil seal installation

NOTE: When removing the oil seal, be careful not the gouge or scratch the seal bore or crankshaft surfaces.

4. Wipe the seal bore with a clean rag.
5. Lubricate the lip of the new seal with clean engine oil.
6. Install the seal into the front cover with a suitable seal installer.
7. Install the crankshaft pulley.
8. Connect the negative battery cable.

Timing Chain and Sprockets

REMOVAL & INSTALLATION

1989–90 Pulsar and Sentra (GA16i Engine)

1. Disconnect the negative battery cable.
2. Set the No. 1 piston at TDC of the compression stroke.
3. Remove the front cover.
4. If necessary, define the timing marks with chalk or paint to ensure proper alignment.
5. Hold the camshaft sprocket stationary with a spanner wrench or similar tool and remove the camshaft sprocket bolt.
6. Remove the chain guides.
7. Remove the camshaft sprocket.
8. Remove the oil pump spacer.
9. Remove the crankshaft sprocket and timing chain.

To install:

10. Verify that the No. 1 piston is at TDC of the compression stroke. The crankshaft keyways should be at the 12 o'clock position.
11. Install the camshaft sprocket, bolt and washer. The alignment mark must face towards the front. When installing the washer, place the non-chamfered side of the washer towards the face of camshaft sprocket. Tighten the bolt just enough to hold the sprocket in place.

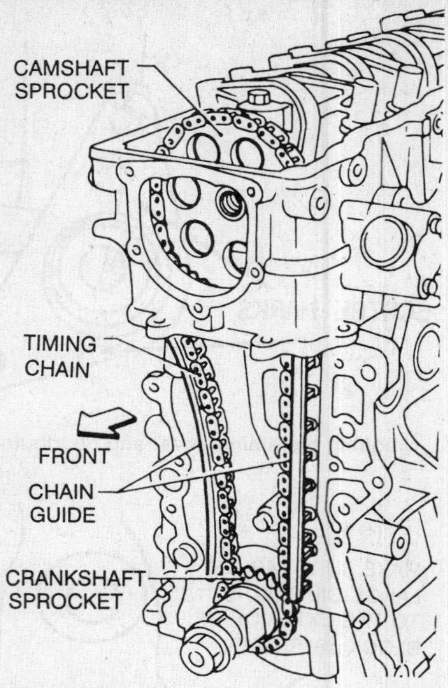

Timing chain assembly—GA16i engine

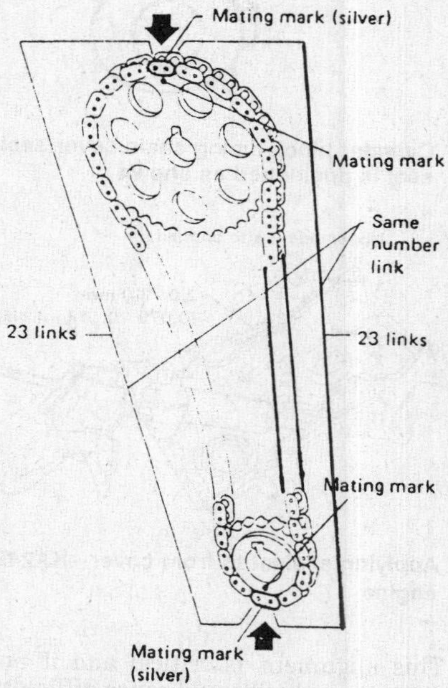

Timing chain and sprocket alignment marks—GA16i engine

12. Install the crankshaft sprocket making sure the alignment mark is facing the front.
13. Install the timing chain by aligning the silver links at the 12 o'clock and 6 o'clock positions on the chain with the timing marks on the crankshaft and camshaft sprockets. The number of links between the 2 silver links are the same for the left and the right sides of the chain, so either side

of the chain may be used to align the sprocket timing marks.

14. Torque the camshaft sprocket bolt to 72–94 ft. lbs. (98–128 Nm) once the chain is in place and aligned.

15. Install the chain guides and tenioner. Use a new tensioner gasket and torque the tensioner and chain guide bolts to 9–14 ft. lbs. (13–19 Nm). When installing the chain guide, move the guide in the direction that applies tension to the chain.

16. Install the front cover.

17. Connect the negative battery cable. Road test the vehicle for proper operation.

1991–92 Sentra (GA16DE Engine)

1. Disconnect the negative battery cable. Relieve the fuel pressure.
2. Remove the cylinder head assembly.
3. Remove the idle sprocket shaft from the rear side.
4. Remove the upper timing chain assembly.
5. Remove the center member.
6. Remove the oil pan assembly, oil strainer and crankshaft pulley.
7. Support engine and remove the engine front mounting bracket.
8. Remove the front cover. One retaining bolt for the front cover assembly is located on the water pump.
9. Remove the idler sprocket.
10. Remove the lower timing chain assembly, oil pump drive spacer, chain guide, crankshaft sprocket.

To install:

11. Confirm that No. 1 piston is set at TDC on compression stroke. Install the chain guide.
12. Install crankshaft sprocket and lower timing chain. Set timing chain by aligning its mating mark with the one on the crankshaft sprocket. Make sure sprocket's mating mark faces engine front. The number of links between the alignment marks are the same for the left and right side.
13. Install the front cover assembly.
14. Install engine front mounting.
15. Install oil strainer, oil pan assembly and crankshaft pulley.
16. Install center member.
17. Set idler sprocket by aligning the mating mark on the larger sprocket with the silver mating mark on the lower timing chain.
18. Install upper timing chain and set it by aligning the mating mark on the smaller sprocket with the silver mating marks on the upper timing chain. Make sure sprocket marks face engine front.
19. Install idler sprocket shaft.
20. Install the cylinder head assembly.
21. Install all remaining components in reverse order of removal.
22. Connect the negative battery cable. Refill all fluid levels. Road test the vehicle for proper operation.

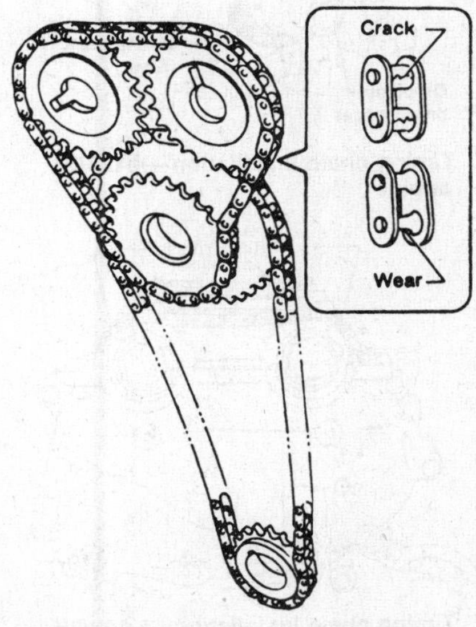

Timing chain assembly–GA16DE engine

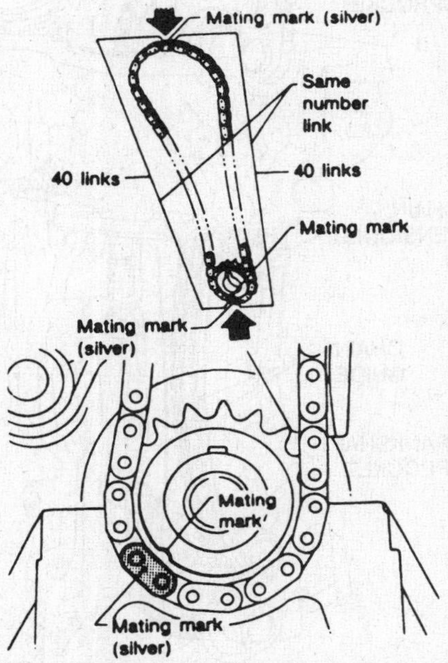

Timing chain installation–GA16DE engine

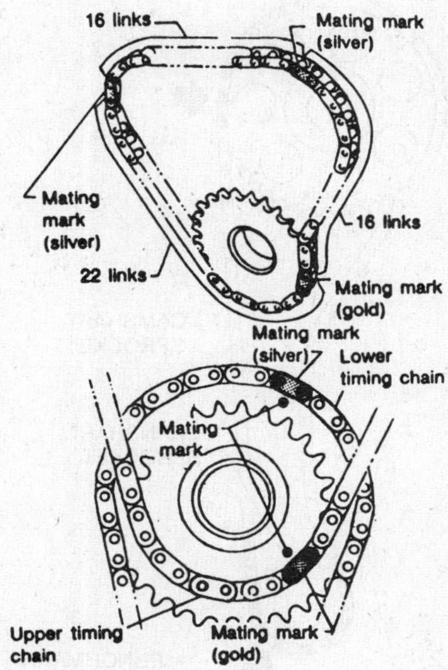

Timing chain installation–GA16DE engine

1991–92 Sentra (SR20DE Engine)

1. Relieve the fuel system pressure and remove the negative battery cable.
2. Drain the coolant from the radiator and engine block. Remove the radiator.
3. Remove the right front wheel and engine side cover.
4. Remove the drive belts, water pump pulley, alternator and power steering pump.
5. Label and remove the vacuum hoses, fuel hoses and wire harness connectors.
6. Remove the cylinder head.
7. Raise and support the vehicle safely.
8. Remove the oil pan.
9. Remove the crankshaft pulley using a suitable puller.
10. Remove the engine front mount.

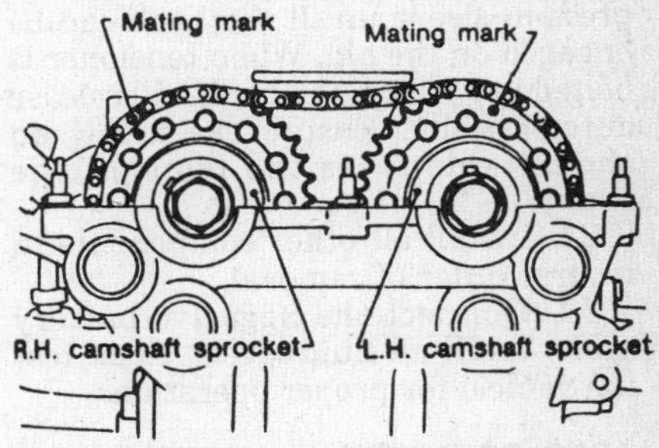

Timing chain installation–SR20DE engine

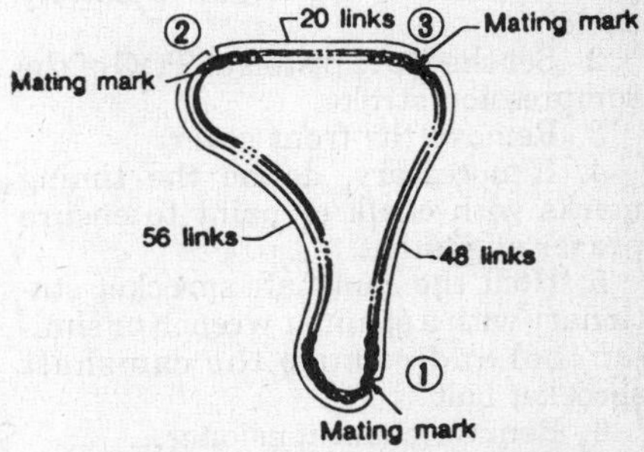

Timing chain installation–SR20DE engine

11. Remove the front cover.
12. Remove the timing chain guides and timing chain. Check the timing chain for excessive wear at the roller links. Replace the chain if necessary.

To install:

13. Install the crankshaft sprocket. Position the crankshaft so that No.1 piston is set at TDC (keyway at 12 o'clock, mating mark at 4 o'clock) fit timing chain to crankshaft sprocket so that mating mark is in line with mating mark on crankshaft sprocket. The mating marks on timing chain for the camshaft sprockets should be silver. The mating mark on the timing chain for the crankshaft sprocket should be gold.
14. Install the timing chain and timing chain guides.
15. Install front engine mount.
16. Install the crankshaft pulley and set No.1 piston at TDC on the compression stroke.
17. Install the oil strainer, baffle plate and oil pan.
18. Install the cylinder head, camshafts, oil tubes and baffles. Position the left camshaft key at 12 o'clock and the right camshaft key at 10 o'clock.
19. Install the camshaft sprockets by lining up the mating marks on the timing chain with the mating marks on the camshaft sprockets. Tighten the camshaft bolts to 101–116 ft. lbs. (137–157 Nm).
20. Install the timing chain guide and distributor. Ensure rotor is at 5 o'clock position.
21. Install the chain tensioner. Press the cam stopper down and the press-in sleeve untill the hook can be engaged on the pin. When tensioner is bolted in position the hook will release automatically. Ensure the arrow on the outside faces the front of the engine.
22. Install all other components in reverse order of removal.
23. Connect the negative battery cable. Refill all fluid levels. Road test the vehicle for proper operation.

1989–90 240SX and 1990–92 Stanza (KA24E Engine)

1. Disconnect the negative battery cable.
2. Set the No. 1 piston at TDC of the compression stroke.
3. Remove the front cover.
4. If necessary, define the timing marks with chalk or paint to ensure proper alignment.
5. Hold the camshaft sprocket stationary with a spanner wrench or similar tool and remove the camshaft sprocket bolt.
6. Remove chain tensioner.
7. Remove the chain guides.
8. Remove the timing chain.
9. Remove the sprocket oil slinger, oil pump drive gear and crankshaft gear.

To install:

10. Install the crankshaft sprocket, oil pump drive gear and oil slinger onto the end of the crankshaft. Make sure the crankshaft sprocket timing marks face toward the front.

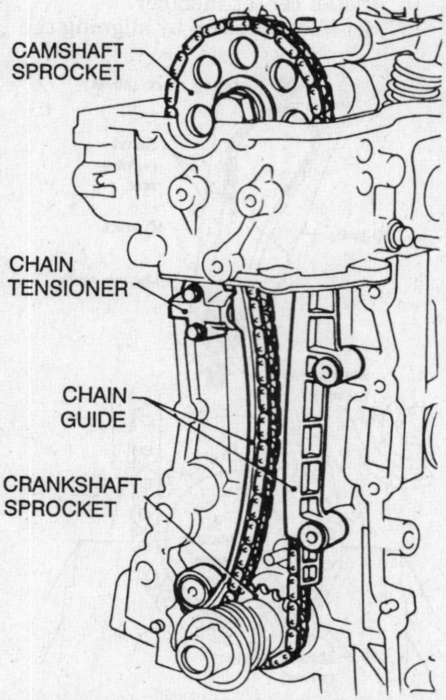

Timing chain assembly – KA24E engine

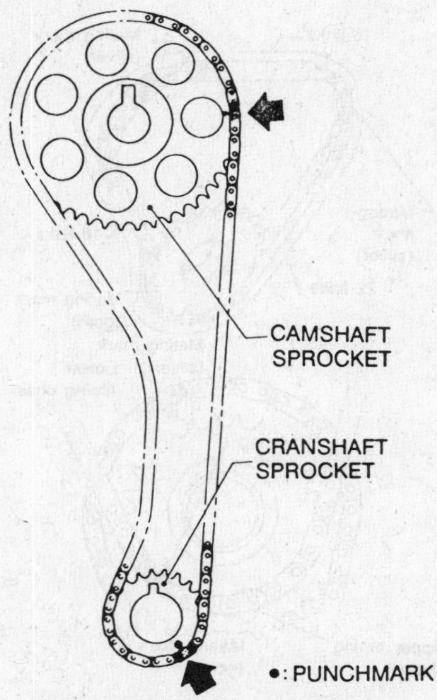

Timing chain and sprocket alignment marks – KA24E engine

11. Install the camshaft sprocket, bolt and washer. The alignment mark must face towards the front. Tighten the bolt just enough to hold the sprocket in place.
12. Verify that the No. 1 piston is at TDC of the compression stroke. The crankshaft keyways should be at the 12 o'clock position.
13. Install the timing chain by aligning the marks on the chain with the marks on the crankshaft and camshaft sprockets. Torque the camshaft sprocket bolt to 87–116 ft. lbs. (118–157 Nm) once the timing chain is in place and aligned.
14. Install the chain tensioner and chain guide.
15. Install the front cover.
16. Connect the negative battery cable.

1991–92 240SX (KA24DE Engine)

1. Release the fuel system pressure.

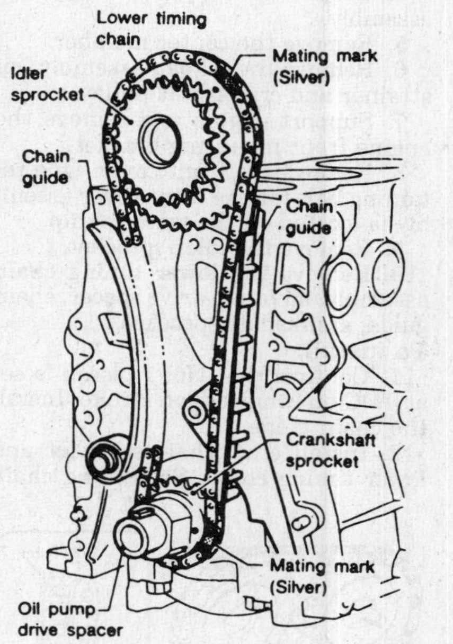

Timing chain installation – KA24DE engine

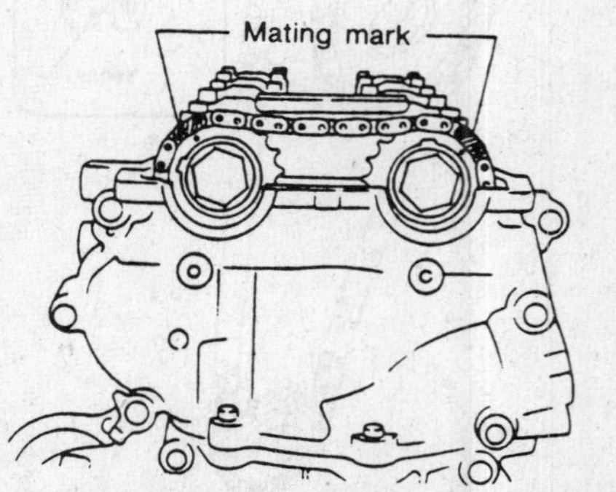

Timing chain installation – KA24DE engine

2. Disconnect the negative battery cable and drain the cooling system. Drain engine oil.
3. Remove the cylinder head assembly.
4. Remove the oil pan.
5. Remove the oil strainer, crankshaft pulley.
6. Remove the front cover assembly.
7. Remove the lower timing chain tensioner, tension arm, lower timing chain guide.
8. Remove the lower timing chain and idler sprocket.

To install:

9. Check all components for wear. Replace as necessary. Clean all mating surfaces and replace the cylinder head gasket.
10. Install crankshaft sprocket. Make sure that mating marks of crankshaft sprocket face front of the engine.
11. Rotate crankshaft so that No. 1 piston is set a TDC position.
12. Install idler sprocket and lower timing chain.
13. Install chain tension arm, chain guide and lower timing chain tensioner.
14. Install front cover assembly.
15. Install crankshaft pulley, oil strainer and oil pan.
16. Install the cylinder head assembly.
17. Install all remaining components in reverse order of removal.
18. Connect the negative battery cable. Refill all fluid levels. Road test the vehicle for proper operation.

Timing Belt Front Cover

REMOVAL & INSTALLATION

CA18DE Engine

1. Disconnect the negative battery cable.
2. Drain the cooling system.
3. Remove the upper radiator hose.
4. Remove the right side engine undercover.
5. Remove the power steering and air conditioning compressor drive belts.
6. Remove the water pump pulley.
7. Matchmark the crank angle sensor to the upper front cover and the remove it. Carefully position it aside.
8. Remove the upper front cover.
9. Align the timing marks on the camshaft pulley sprockets and then remove the crankshaft pulley.

NOTE: The crankshaft pulley may be reached by removing the side cover from inside the right hand wheel opening.

10. Remove the lower front cover.

To install:

11. Install the lower front cover with a new gasket.
12. Install the crankshaft pulley with its washer. Torque the pulley bolt to 105–112 ft. lbs. (145–152 Nm).
13. Install the crank angle sensor so the matchmarks made previously line up and tighten the bolts to 5.1–5.8 ft. lbs. (7–8 Nm).
14. Install the water pump pulley.
15. Install the power steering pump and air conditioning compressor drive belts. Adjust the belt tension.
16. Install the right side undercover.
17. Install the upper radiator hose.
18. Fill the cooling system to the proper level.
19. Connect the negative battery cable.

CA20E Engine

1. Disconnect the negative battery cable.
2. On 200SX: disconnect the air intake duct, remove the cooling fan and radiator shroud and remove the exhaust side spark plugs.
3. On Stanza: raise and support the front of the vehicle safely, remove the right front wheel, remove the exhaust side spark plugs and remove the dust cover and undercover.
4. Remove the alternator drive belt.
5. Remove the air conditioner compressor drive belt.
6. Remove the crankshaft pulley.
7. Remove the crankshaft damper.
8. Remove the water pump pulley.
9. Remove the upper and lower timing belt covers and gaskets. If the gaskets are in good condition after removal, they can be reused. If they are in way damaged or broken, replace them.

To install:

10. Install timing belt cover with new gaskets, as required. Torque the front cover bolts evenly to 2.2–3.6 ft. lbs. (3–5 Nm).
11. Install the water pump pulley. Torque the pulley bolts to 4–7 ft. lbs. (6–10 Nm).
12. Install the crankshaft damper. Torque the bolts to 90–98 ft. lbs. (123–132 Nm).
13. Install the crankshaft pulley. Torque the pulley bolts to 9–10 ft. lbs. 12–14 Nm).
14. Install the drive belts and adjust the drive belt tension.
15. On Stanza: install the undercover and dust cover, install the exhaust side spark plugs, mount the right front wheel and lower the vehicle.
16. On 200SX: install the exhaust side spark plugs, install the radiator shroud and cooling fan and connect the air intake duct.
17. Connect the negative battery cable.

E16i Engine

1. Disconnect the negative battery cable.
2. Drain the cooling system.
3. Remove the front right side splash cover.
4. Remove the front right side undercover.
5. Remove the air conditioning belt (if so equipped) and alternator belt.
6. Remove the alternator.
7. Remove the power steering belt, if equipped.
8. Remove the water pump pulley.
9. Remove the crankshaft pulley.

NOTE: The crankshaft pulley is accessible after removing the side cover from the right side wheel house.

10. Loosen and remove the 8 Torx head bolts securing the timing covers and remove the upper and lower covers.

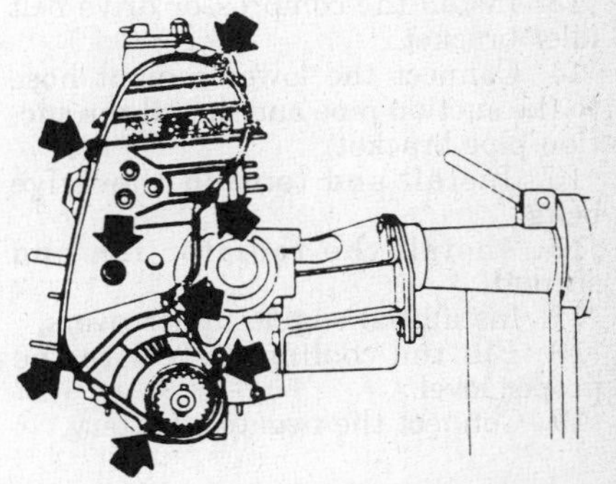

Front cover removal—E series engine

To install:

11. Install the upper and lower timing belt covers with new gaskets. Tighten the upper cover bolts to 4–5 ft. lbs. (7–8 Nm) and the lower cover bolts to 2–4 ft. lbs. (3–5 Nm).
12. Install the crankshaft pulley. Torque the pulley bolt to 80–94 ft. lbs. (108–127 Nm).
13. Install the water pump pulley. Torque the pulley bolts to 6–8 ft. lbs. (8–11 Nm).
14. Install the power steering belt.
15. Install the alternator.
16. Install the alternator and air conditioning belts.
17. Install the right side engine and splash covers.
18. Fill the cooling system to the proper level and adjust the drive belt tension.
19. Connect the negative battery cable.

VG30E and VG30ET Engines (1988–89 300ZX)

1. Disconnect the negative battery cable.

2. Drain the cooling system.
3. Remove the engine undercovers.
4. Remove the radiator shroud and fan.
5. Remove the power steering, alternator and air conditioning compressor drive belts.
6. Remove the suction pipe bracket and disconnect the lower coolant hose from the suction pipe.
7. Remove compressor drive belt idler bracket.
8. Set No. 1 cylinder at TDC of the compression stroke.
9. Remove the crankshaft pulley.
10. Remove the front upper and lower belt covers and gaskets.

To install:

11. Install the upper and lower timing belt covers with new gaskets. Torque the covers bolts to 2–4 ft. lbs. (3–5 Nm).
12. Install the crankshaft pulley. Torque the pully bolt to 90–98 ft. lbs. (123–132 Nm).
13. Install the compressor drive belt idler bracket.
14. Connect the lower coolant hose to the suction pipe and install the suction pipe bracket.
15. Install and tension the drive belts.
16. Install the radiator fan and shroud.
17. Install the engine undercovers.
18. Fill the cooling system to the proper level.
19. Connect the negative battery.

VG30DE and VG30DETT Engines (1990–92 300ZX)

1. Disconnect the negative battery cable.
2. Remove the engine undercover.
3. Drain the cooling system.
4. Remove the radiator.
5. Remove the drive belts.
6. Remove the cooling fan and cooling fan coupling.
7. Remove the crankshaft pulley bolt.
8. Remove the starter and lock the flywheel ring gear using a suitable locking device. This is done to prevent the crankshaft gear from turning during removal and installation.
9. Remove the crankshaft pulley using a suitable puller.
10. Remove the water inlet and outlet housings.
11. Remove the timing belt covers and gaskets.

To install:

12. Install the timing belt covers with new gaskets. Torque the cover bolts to 2–4 ft. lbs. (3–5 Nm).
13. Install the water inlet and outlet housings with new gaskets.
14. Install the crankshaft pulley. Torque the pulley bolt to 159–174 ft. lbs. (21–235 Nm).
15. Remove the flywheel locking device and install the starter.
16. Install the cooling fan and cooling fan coupling.
17. Install and tension the drive belts.
18. Install the radiator.
19. Fill the cooling system to the proper level.
20. Connect the negative battery cable.

VG30E Engine (1988–92 Maxima)

1. Disconnect the negative battery cable.
2. Raise and support the front of the vehicle safely.
3. Remove the engine undercovers.
4. Drain the cooling system.
5. Remove the right front wheel.
6. Remove the engine side cover.
7. On 1988 vehicles, remove the engine coolant reservoir tank, radiator hoses and the Automatic Speed Control Device (ASCD) actuator; remove the lower coolant hose support bracket and disconnect the lower hose from the suction pipe.
8. Remove the alternator, power steering and air conditioning compressor drive belts from the engine. When removing the power steering drive belt, loosen the idler pulley from the right side wheel housing.
9. On 1989–92 vehicles, remove the upper radiator and water inlet hoses; remove the water pump pulley.
10. Remove the idler bracket of the compressor drive belt.
11. Remove the crankshaft pulley with a suitable puller.
12. Remove the upper and lower timing belt covers and gaskets.

To install:

13. Install the upper and lower timing belt covers with new gaskets.
14. Install the crankshaft pulley. Torque the pulley bolt to 90–98 ft. lbs. (123–132 Nm).
15. Install the compressor drive belt idler bracket.
16. On 1989–92 vehicles, install the water pump pulley and torque the nuts to 12–15 ft. lbs. (16–21 Nm); install the upper radiator and water inlet hoses.
17. Install the drive belts.
18. On 1988 vehicles, connect the lower coolant hose to the suction pipe and install the hose support bracket; install the Automatic Speed Control Device (ASCD) actuator, radiator hoses and engine coolant reservoir tank.
19. Install the engine side cover.
20. Mount the front right wheel.
21. Install the engine undercovers.
22. Lower the vehicle.
23. Fill the cooling system and connect the negative battery cable.

OIL SEAL REPLACEMENT

Except 300ZX

1. Disconnect the negative battery cable.
2. Remove the crankshaft pulley.
3. Using a suitable tool, pry the oil seal from the front cover.

NOTE: When removing the oil seal, be careful not the gouge or scratch the seal bore or crankshaft surfaces.

4. Wipe the seal bore with a clean rag.
5. Lubricate the lip of the new seal with clean engine oil.
6. Install the seal into the front cover with a suitable seal installer.
7. Install the crankshaft pulley.
8. Connect the negative battery cable.

300ZX

1. Disconnect the negative battery cable.
2. Remove the timing belt.
3. Remove the crankshaft sprocket.
4. Remove the oil pan and oil pump.
5. Using a suitable tool, pry the oil seal from the front cover.

NOTE: When removing the oil seal, be careful not the gouge or scratch the seal bore or crankshaft surface.

6. Wipe the seal bore with a clean rag.
7. Lubricate the lip of the new seal with clean engine oil.
8. Install the seal into the front cover with a suitable seal installer.
9. Install the oil pump and oil pan.
10. Install the crankshaft sprocket.
11. Install the timing belt.
12. Connect the negative battery cable.

Timing Belt and Tensioner

REMOVAL & INSTALLATION

CA18DE Engine (1988–89 Pulsar)

1. Disconnect the negative battery cable.
2. Drain the cooling system.
3. Remove the upper radiator hose.
4. Remove the right side engine undercover.
5. Loosen the power steering pump and the air conditioning compressor and then remove the drive belts.
6. Remove the water pump pulley.

7. Matchmark the crank angle sensor to the upper front cover and the remove it. Carefully position it aside.
8. Remove the water pump pulley.
9. Position a floor jack under the engine and raise it just enough to support the engine.
10. Remove the upper engine mount bracket at the right side of the upper front cover.
11. Remove the upper front cover.
12. Align the timing marks on the camshaft pulley sprockets and then remove the crankshaft pulley.

NOTE: The crankshaft pulley may be reached by removing the side cover from inside the right hand wheel opening.

13. Remove the lower front cover.
14. Loosen the tensioner pulley nut to slacken the timing belt and then slide off the belt.

To install:

NOTE: Do not bend or twist the timing belt. Never rotate the crankshaft and camshaft separately with the timing belt removed. Make sure the timing belt is free of any oil, water or debris.

15. Install the crankshaft sprocket with the sprocket plates.
16. Before installing the timing belt, ensure that the No. 1 piston is at TDC of the compression stroke. All sprocket timing marks should be aligned with the marks on the case.

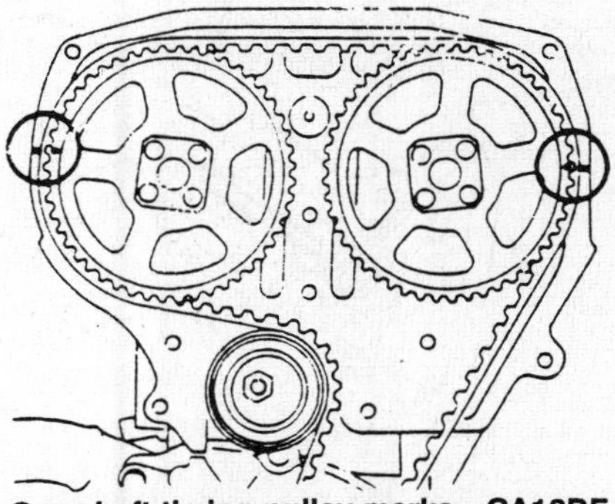
Camshaft timing pulley marks – CA18DE engine

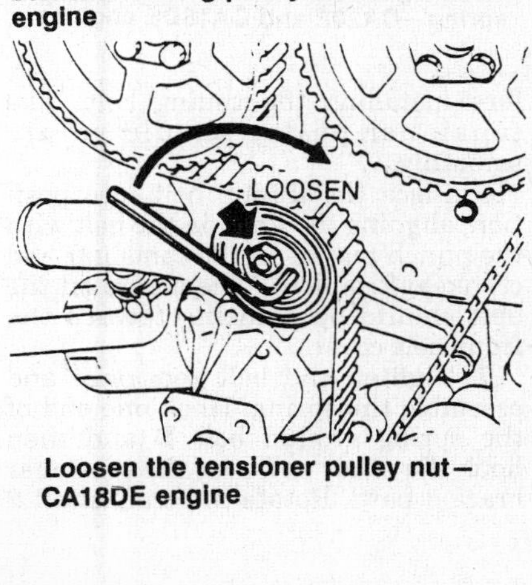

Loosen the tensioner pulley nut – CA18DE engine

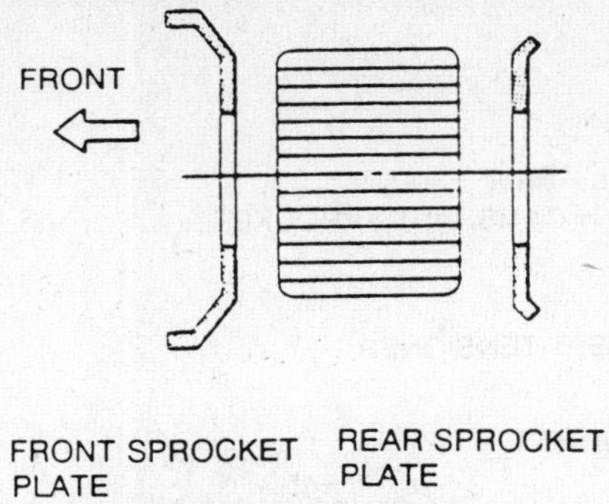

Crankshaft sprocket plate installation – CA18DE engine

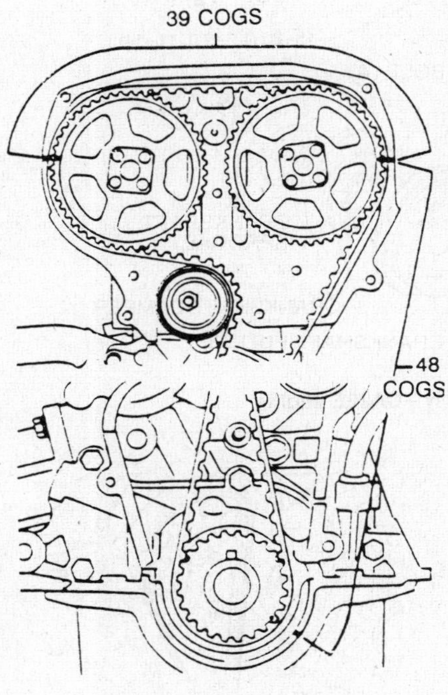

Timing belt timing mark alignment – CA18DE engine

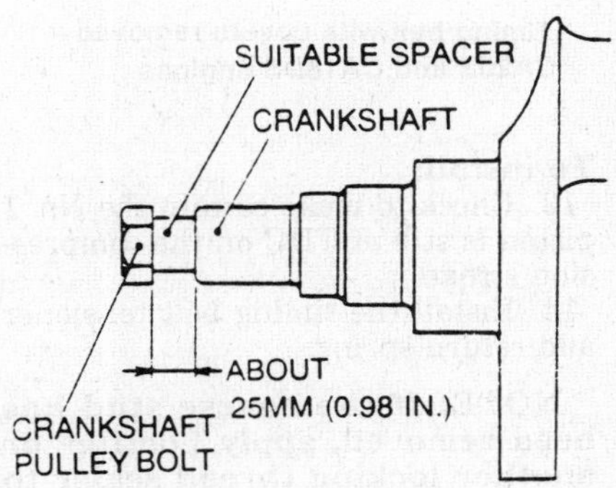

A spacer must be installed between the crankshaft and pulley bolt head before rotating the engine – CA18DE engine

NOTE: When the timing belt is on and in position, there should be 39 cogs between the timing mark on each of the camshaft sprocket and 48 cogs between the mark on the right camshaft sprocket and the mark on the crankshaft sprocket.

17. Loosen the timing belt tensioner pulley nut.
18. Temporarily install the crankshaft pulley bolt and then rotate the engine two complete revolutions.

NOTE: Fabricate and install a 0.98 in. (25mm) thick spacer between the end of the crankshaft and the head of the crankshaft pulley bolt to prevent bolt damage.

19. Tighten the tensioner pulley bolt to 16–22 ft. lbs. (22–29 Nm).
20. Install the upper and lower front covers with new gaskets.
21. Install the crankshaft pulley with its washer and tighten it to 105–112 ft. lbs. (145–152 Nm).
22. Install the engine mount bracket.
23. Install the water pump pulley. Install the crank angle sensor so the matchmarks made previously line up and tighten the bolts to 5.1–5.8 ft. lbs. (7–8 Nm).
24. Install the water pump pulley.
25. Installl the power steering pump and air conditioning compressor drive belts. Adjust the belt tension.
26. Install the right side undercover.
27. Install the upper radiator hose.
28. Fill the cooling system to the proper level.
29. Connect the negative battery cable.

CA20E Engine (1988 200SX and 1988–89 Stanza)

1. Disconnect the negative battery cable.
2. On 200SX, disconnect the air intake duct remove the cooling fan and radiator shroud; remove the exhaust side spark plugs.
3. On Stanza, raise and support the front of the vehicle safely. Remove the right front wheel. Remove the exhaust side spark plugs. Remove the dust cover and undercover.
4. Set the No. 1 piston at TDC of the compression stroke. The timing marks will all be aligned.
5. Remove the alternator drive belt.
6. Remove the air conditioner compressor drive belt.
7. Remove the crankshaft pulley.
8. Remove the crankshaft damper.
9. Remove the water pump pulley.
10. Remove the upper and lower timing belt covers and gaskets. If the gaskets are in good condition after remov-

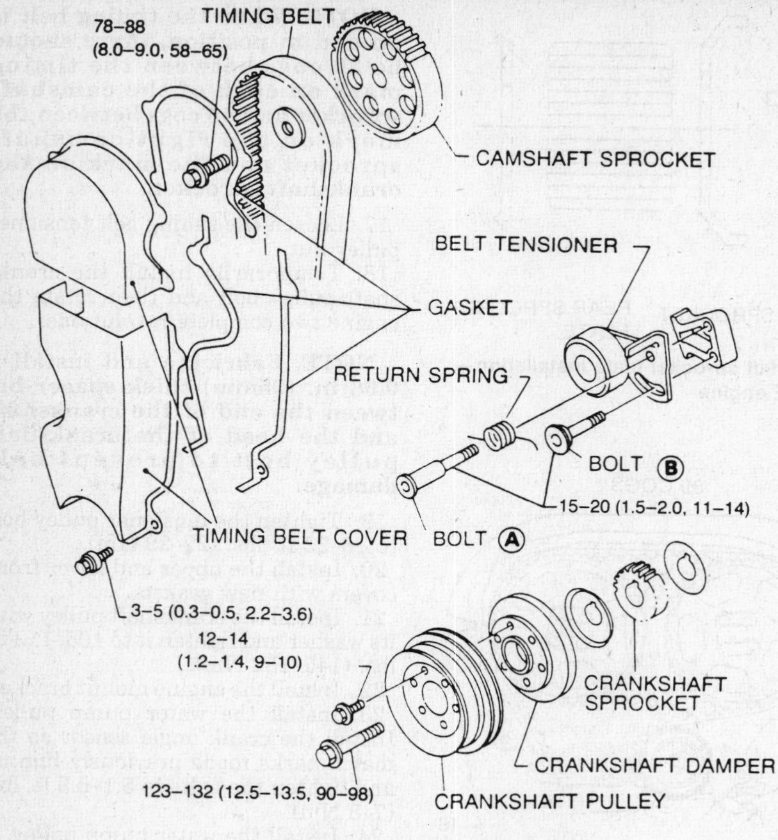

Timng belt assembly – CA20E engine

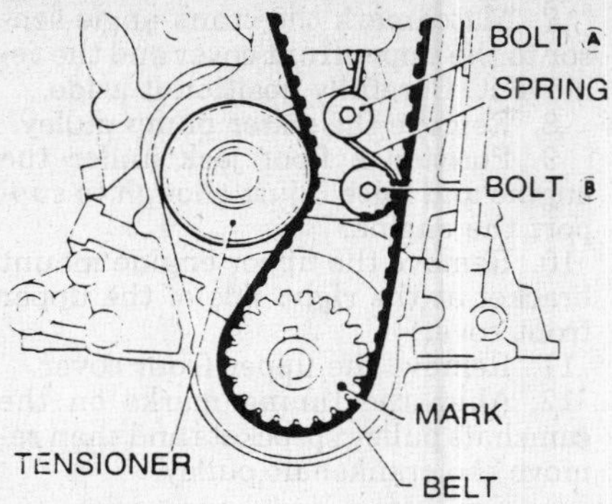

Setting the tensioner spring – CA20E and CA18DE engines

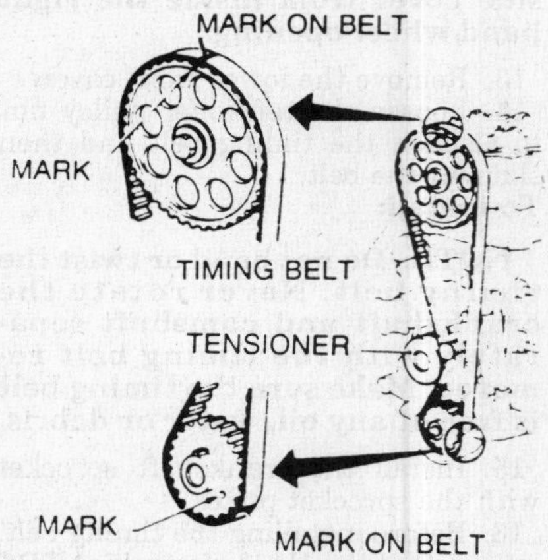

Timing belt installation – CA20E and CA18DE engines

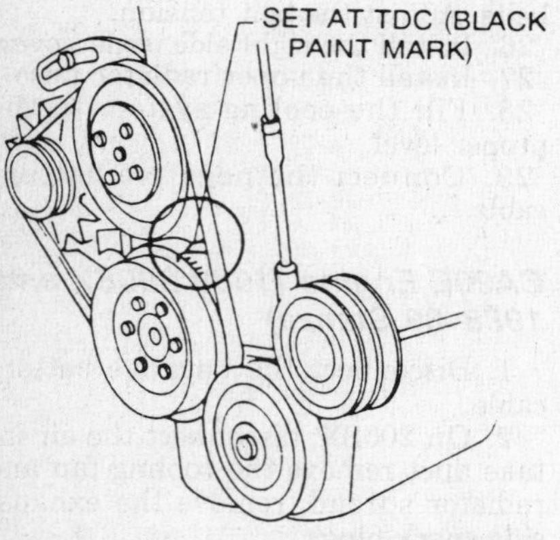

Setting No. 1 piston to TDC – CA20E engine

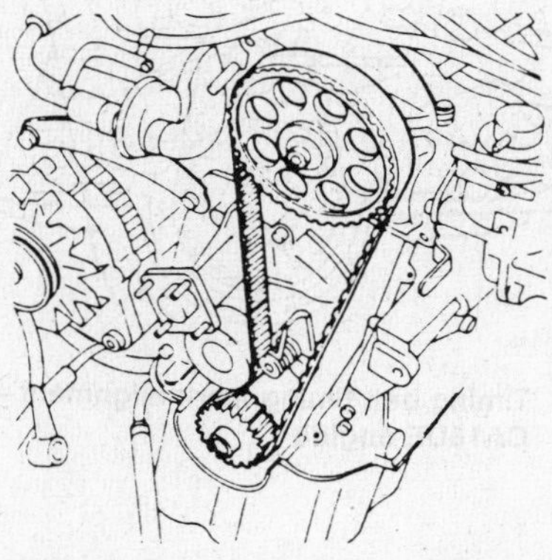

Timing belt with covers removed – CA20E and CA18DE engines

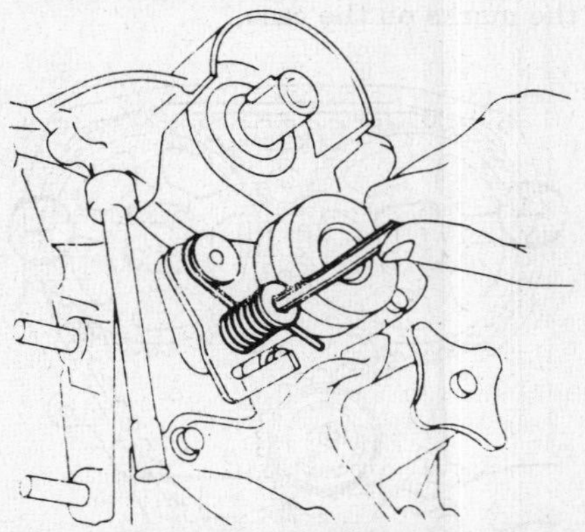

Installing the belt tensioner and return spring – CA20E and CA18DE engines

al, they can be reused; if they are damaged or broken, replace them.

11. Loosen the timing belt tensioner and return spring. Remove the timing belt.

12. Carefully inspect the condition of the timing belt. There should be no breaks or cracks anywhere on the belt. Be particularly careful when checking around the bottom of the cog teeth, where they the main belt; cracks often show up here first. Evidence of any wear or damage on the belt calls for replacement.

To install:

13. Check to make certain the No. 1 piston is still at TDC on the compression stroke.

14. Install the timing belt tensioner and return spring.

NOTE: If the coarse stud has been removed, apply Loctite® or another locking thread sealer to the stud threads prior to installation.

15. Make sure the tensioner mounting bolts are not securely tightened before installing the timing belt. The tensioner pulley should rotate smoothly.

16. Place the timing belt into position, aligning the lines on the belt with the punch marks on the camshaft and crankshaft pulleys. The arrow on the belt should be pointing toward the front belt covers.

17. Tighten the belt tensioner and assemble the spring. Hook one end of the spring around bolt **B** and then hook the other end over the tensioner bracket pawl. Rotate the crankshaft 2

complete revolutions clockwise, tighten bolt **B** and then bolt **A**.
18. Install timing belt cover with new gaskets, as required. Torque the front cover bolts evenly to 2.2–3.6 ft. lbs. (3–5 Nm).
19. Install the water pump pulley. Torque the pulley bolts to 4–7 ft. lbs. (6–10 Nm).
20. Install the crankshaft damper. Torque the bolts to 90–98 ft. lbs. (123–132 Nm).
21. Install the crankshaft pulley. Torque the pulley bolts to 9–10 ft. lbs. 12–14 Nm).
22. Install the drive belts and adjust the drive belt tension.
23. On Stanza, install the undercover and dust cover. Install the exhaust side spark plugs. Mount the right front wheel and lower the vehicle.
24. On 200SX, install the exhaust side spark plugs. Install the radiator shroud and cooling fan. Connect the air intake duct.
25. Connect the negative battery cable.

E16i Engine (1988 Pulsar and Sentra)

1. Disconnect the negative battery cable. Raise and safely support the vehicle.
2. Drain the cooling system.
3. Remove the front right side splash cover.
4. Remove the front right side undercover.
5. Remove the air conditioning belt and alternator belt.
6. Remove the power steering belt (if equipped).
7. Set the No. 1 piston to TDC of the compression stroke.
8. Remove the water pump pulley.
9. Remove the crankshaft pulley.

NOTE: The crankshaft pulley is accessible after removing the side cover from the right side wheel house.

10. Position a floor jack under the engine and raise it just enough to support the engine. Unbolt the right side engine mounting bracket from the block.
11. Loosen and remove the 8 Torx head bolts securing the timing covers and remove the upper and lower covers.
12. Mark the relationship of the camshaft sprocket to the timing belt and the crankshaft sprocket to the timing belt with paint or a grease pencil. This will make setting everything up during reassembly much easier if the engine is disturbed during disassembly.
13. Loosen the timing belt tensioner locknut and rotate the tensioner clockwise. Retighten the locknut.
14. Mark a rotational, direction arrow on the timing belt and then remove the belt.

NOTE: After removing the timing belt, do not rotate the crankshaft or camshaft separately or the valves will hit the pistons.

15. Remove the belt tensioner and its return spring.

To install:

16. Check that the timing marks on the camshaft sprocket and upper front cover and on the crankshaft sprocket and lower front cover are in alignment. This will ensure that the No. 1 piston is at TDC of its compression stroke.
17. Install the timing belt tensioner and return spring temporarily.
18. Rotate the tensioner about 70–80 degrees clockwise and then tighten the locknut.
19. Install the timing belt.
20. Loosen the tensioner locknut so the tensioner pushes on the timing belt and then turn the camshaft sprocket about 20 degrees clockwise (2 cogs).

NOTE: All spark plugs must be removed before turning the camshaft sprocket.

21. Prevent the the tensioner from spinning and tighten the locknut to 12–15 ft. lbs. (16–21 Nm).

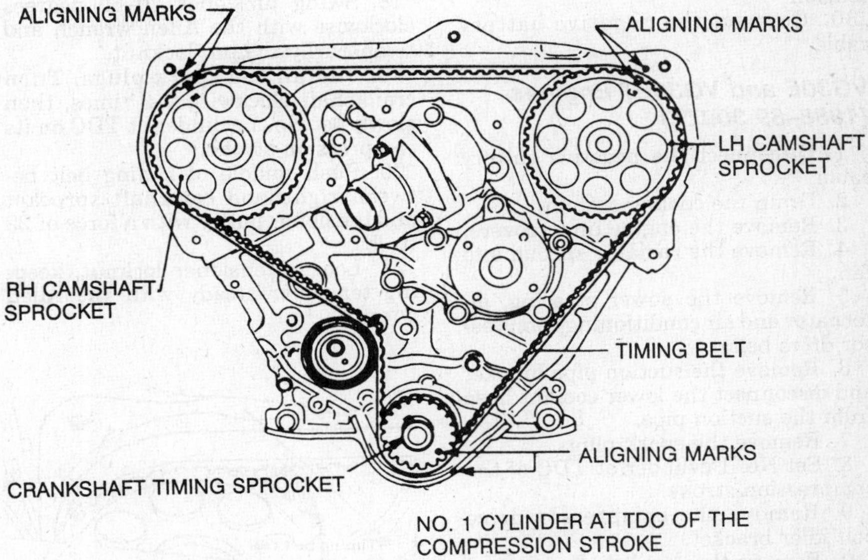

Timing belt installation and timing mark alignment—VG30E and VG30ET engines

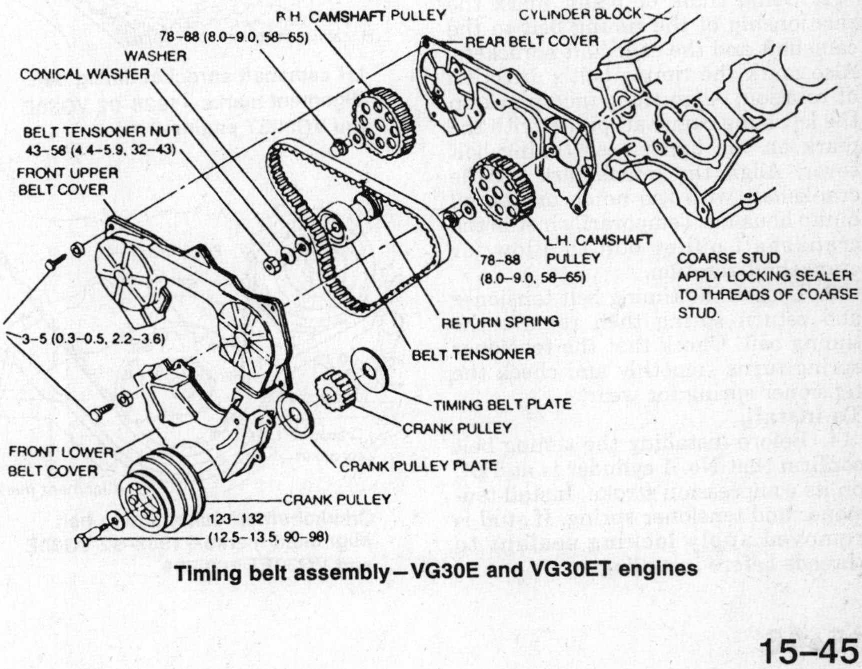

Timing belt assembly—VG30E and VG30ET engines

22. Install the upper and lower timing belt covers with new gaskets. Tighten the upper cover bolts to 4–5 ft. lbs. (7–8 Nm) and the lower cover bolts to 2–4 ft. lbs. (3–5 Nm).
23. Attach the right side engine mounting bracket.
24. Install the crankshaft pulley. Torque the pulley bolt to 80–94 ft. lbs. (108–127 Nm).
25. Install the water pump pulley. Torque the pulley bolts to 6–8 ft. lbs. (8–11 Nm).
26. Install the power steering belt.
27. Install the alternator and air conditioning belts.
28. Install the right side engine and splash covers.
29. Fill the cooling system to the proper level and adjust the drive belt tension.
30. Connect the negative battery cable.

VG30E and VG30ET Engines (1988–89 300ZX)

1. Disconnect the negative battery cable.
2. Drain the cooling system.
3. Remove the engine undercovers.
4. Remove the radiator shroud and fan.
5. Remove the power steering, alternator and air conditioning compressor drive belts.
6. Remove the suction pipe bracket and disconnect the lower coolant hose from the suction pipe.
7. Remove the spark plugs.
8. Set No. 1 cylinder at TDC of the compression stroke.
9. Remove the compressor drive belt idler bracket.
10. Remove the crankshaft pulley.
11. Remove the front upper and lower belt covers and gaskets.
12. Using chalk or paint, mark the relationship of the timing belt to the camshaft and the camshaft sprockets. Also mark the timing belt's direction of rotation. Align the punch mark on the left hand camshaft pulley with the mark on the upper rear timing belt cover. Align the punchmark on the crankshaft with the notch on the oil pump housing. Temporarily install the crankshaft pulley bolt to allow for crankshaft rotation.
13. Loosen the timing belt tensioner and return spring then remove the timing belt. Check that the tensioner spring turns smoothly and check the tensioner spring for wear.

To install:

14. Before installing the timing belt confirm that No. 1 cylinder is at TDC on its compression stroke. Install tensioner and tensioner spring. If stud is removed apply locking sealant to threads before installing.
15. Swing the tensioner fully clockwise with hexagon wrench and temporarily tighten locknut.
16. Point the arrow on the timing belt toward the front belt cover. Align the white lines on the timing belt with the punch marks on all 3 pulleys.

NOTE: There are 133 total timing belt teeth. If timing belt is installed correctly, there will be 40 teeth between left hand and right hand camshaft sprocket timing marks. There will be 43 teeth between left hand camshaft sprocket and crankshaft sprocket timing marks.

17. Loosen tensioner locknut, keeping tensioner steady with an Allen wrench.
18. Swing tensioner 70–80 degrees clockwise with the Allen wrench and temporarily tighten locknut.
19. Install the spark plugs. Turn crankshaft clockwise 2–3 times, then slowly set No. 1 cylinder at TDC on its compression stroke.
20. Push middle of timing belt between righthand camshaft sprocket and tensioner pulley with a force of 22 ft. lbs.
21. Loosen tensioner locknut, keeping tensioner steady with the Allen wrench.

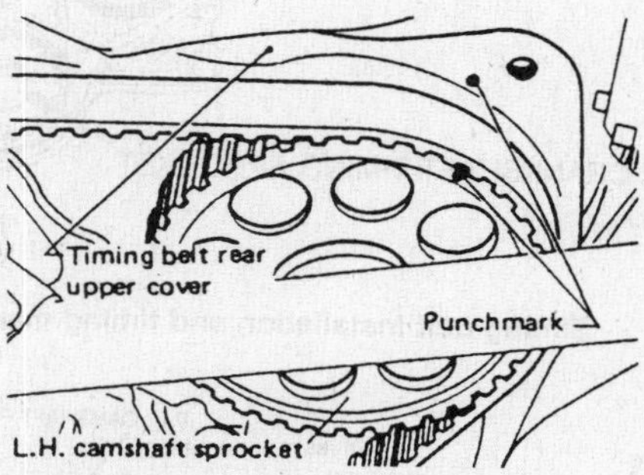

LH camshaft sprocket timing belt alignment marks – 1988–92 VG30E and VG30ET engines

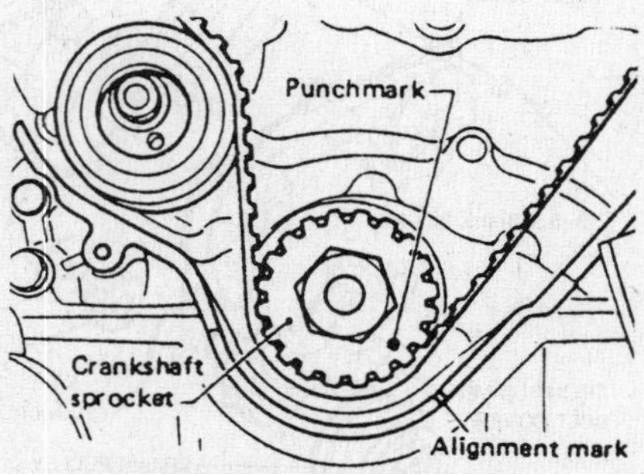

Crankshaft sprocket timing belt alignment marks – 1988–92 VG30E and VG30ET engines

22. Insert a 0.138 in. (0.35mm) thick and 0.5 in. (12.7mm) wide feeler gauge between the bottom of tensioner pulley and timing belt. Turn crankshaft clockwise and position gauge completely between tensioner pulley and timing belt. The timing belt will move about 2.5 teeth.
23. Tighten tensioner locknut, keeping tensioner steady with the Allen wrench.
24. Turn crankshaft clockwise or counterclockwise and remove the gauge.
25. Rotate the engine 3 times, then set No. 1 at TDC on its compression stroke.
26. Check timing belt deflection on 1988 vehicles only. Timing belt deflection is 13.0–14.5mm at 22 lbs. of pressure. If it is out of specified range, readjust the timing belt by repeatng Steps 14–25.
27. Install the upper and lower timing belt covers and complete the remainder of the installation in reverse of the removal procedure.

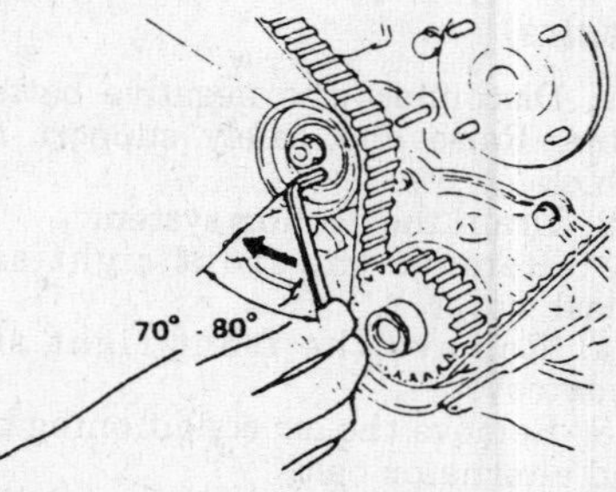

Swing the tensioner 70–80 degress clockwise

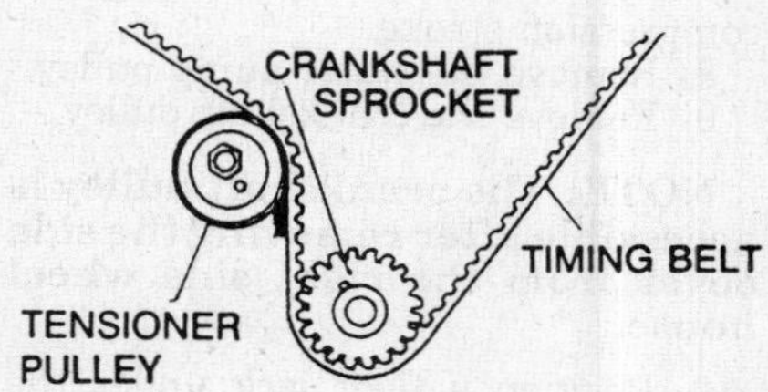

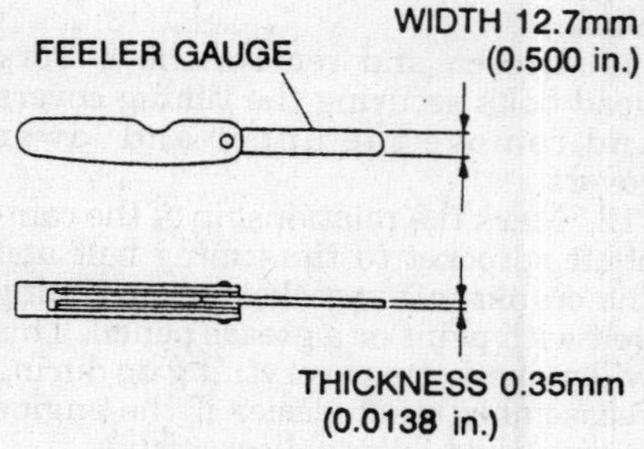

Checking timing belt adjustment with a feeler gauge

VG30DE and VG30DETT Engines (1990–92 300ZX)

1. Disconnect the negative battery cable.
2. Remove the engine undercover.
3. Drain the cooling system.
4. Remove the radiator.
5. Remove the drive belts.
6. Remove the cooling fan and cooling fan coupling.
7. Remove the crankshaft pulley bolt.
8. Remove the starter and lock the flywheel ring gear using a suitable locking device. This is done to prevent the crankshaft gear from turning during removal and installation.
9. Remove the crankshaft pulley using a suitable puller, then remove the locking device.
10. Remove the water inlet and outlet housings.
11. Remove the timing belt covers and gaskets.
12. Install a suitable 6mm stopper bolt in the tenioner arm of the auto tensioner so the length of the pusher does not change.
13. Set the No. 1 piston at TDC of the compression stroke.
14. Remove the auto-tensioner and the timing belt.

To install:

15. Check the auto-tensioner for oil leaks in the pusher rod and diaphragm. If oil is evident, replace the auto-tensioner assembly.
16. Verify that the No. 1 piston is at TDC of the compression stroke.
17. Align the timing marks on the camshaft and crankshaft sprockets with the timing marks on the rear timing belt cover and the oil pump housing.
18. Remove all the spark plugs.
19. With a feeler gauge, check the clearance between the tensioner arm and the pusher of the auto-tensioner. The clearance should be 0.16 in. (4mm) with a slight drag on the feeler gauge. If the clearance is not as specified, mount the tensioner in a vise and adjust the clearance. When the clearance is set, insert the stopper bolt into the tensioner arm to retain the adjustment.

NOTE: When adjusting the clearance, do not push the tensioner arm with the stopper bolt fitted, because damage to the threaded portion of the bolt will result.

20. Mount the auto-tensioner and tighten nuts and bolts by hand.
21. Install the timing belt. Ensure the timing sprockets are free of oil and water. Do not bend or twist the timing belt. Align the white lines on the belt

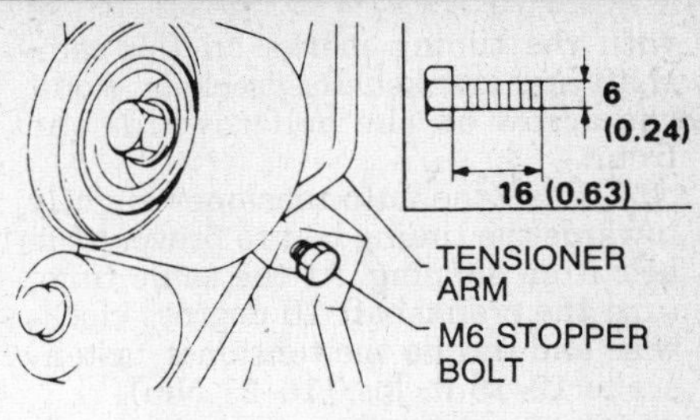

Installing suitable stopper bolt into auto-tensioner arm—VG30DE and VG30DETT engines

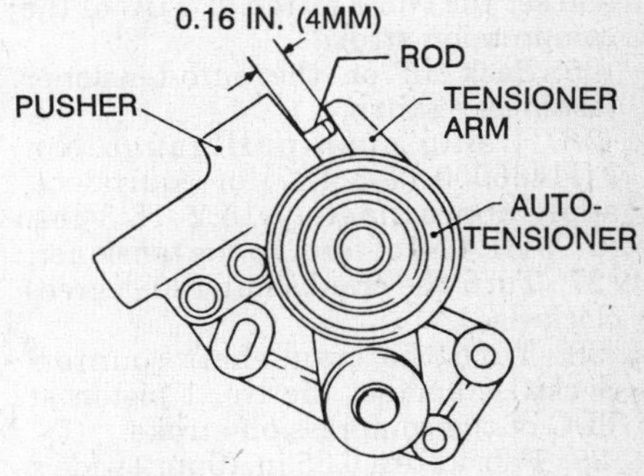

Checking tensioner arm and pusher clearance—VG30DE and VG30DETT engines

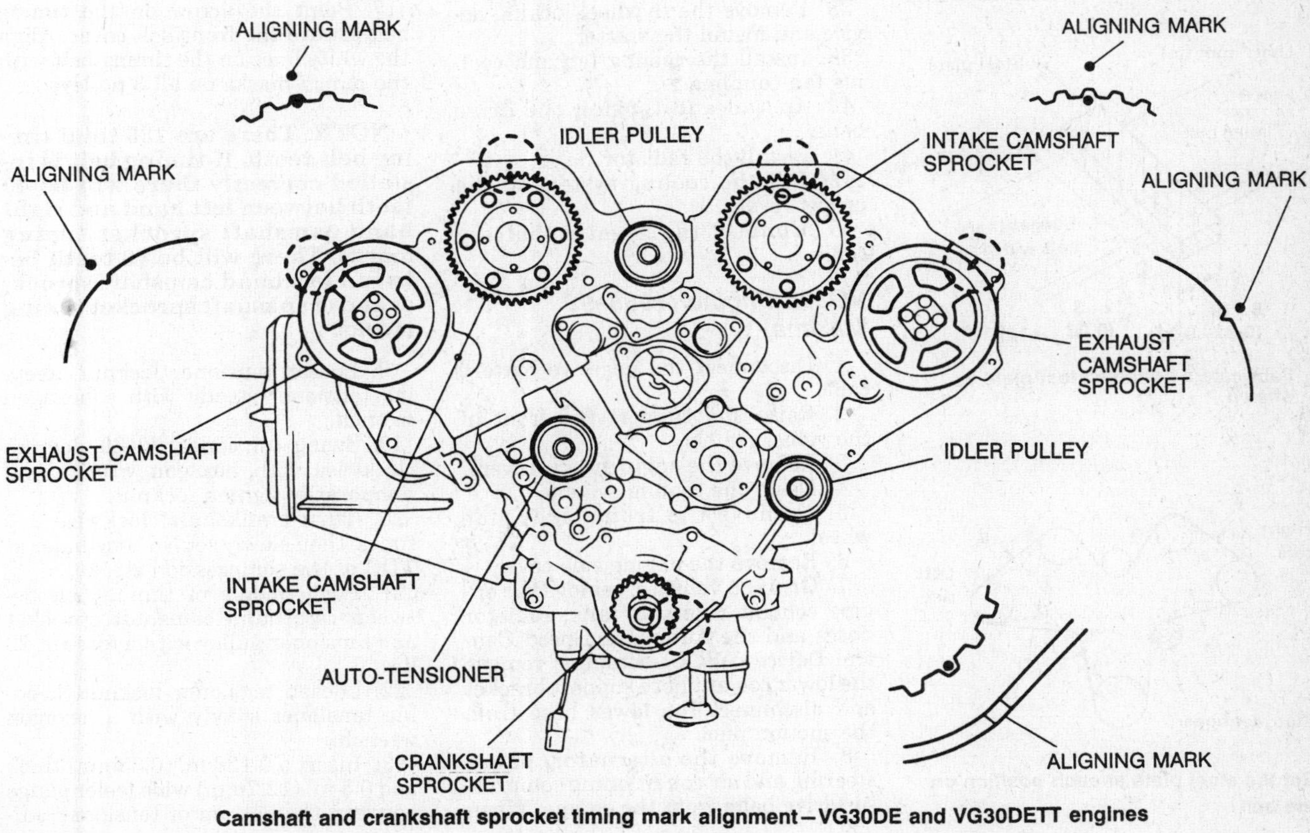

Camshaft and crankshaft sprocket timing mark alignment—VG30DE and VG30DETT engines

with the timing marks on the camshaft and crankshaft sprockets. Point the arrow on the belt towards the front.

22. Push the auto-tensioner slightly towards the timing belt to prevent the belt from slipping. At the same time, turn the crankshaft 10 degrees clockwise and torque the tensioner fasteners to 12–15 ft. lbs. (16–21 Nm).

NOTE: Do not push the tensioner too hard because it will create excessive tension on the belt.

23. Turn the crankshaft 120 degrees counterclockwise.

24. Turn the crankshaft clockwise and set the No. 1 piston at TDC of the compression stroke.

25. Back off on the auto-tensioner fasteners ½ turn.

26. Using push-pull gauge No. EG1486000 (J–38387) or equivalent, apply approximately 15.2–18.3 lbs. (67.7–81.4 N) of force to the tensioner.

27. Turn the crankshaft 120 degrees clockwise.

28. Turn the crankshaft counterclockwise and set the No. 1 piston at TDC of the compression stroke.

29. Fabricate a 0.35 in. (9mm) wide x 0.10 in. (2mm) deep steel plate. The length of the plate should be slightly longer than the width of the belt.

30. Set the steel plate at positions **A**, **B**, **C** and **D** of the timing belt mid-way between the pulleys as shown. Using the push-pull gauge or equivalent, apply approximately 11 lbs. (49 N) of force to the tensioner and check (and record) the belt deflection at each position with the steel plate in place. The timing belt deflection at each position should be 0.217–0.256 in. (5.5–6.5mm). Another means of determining the belt deflection is to add all deflection readings and divide them by 4. This average deflection should be 0.217–0.256 in. (5.5–6.5mm).

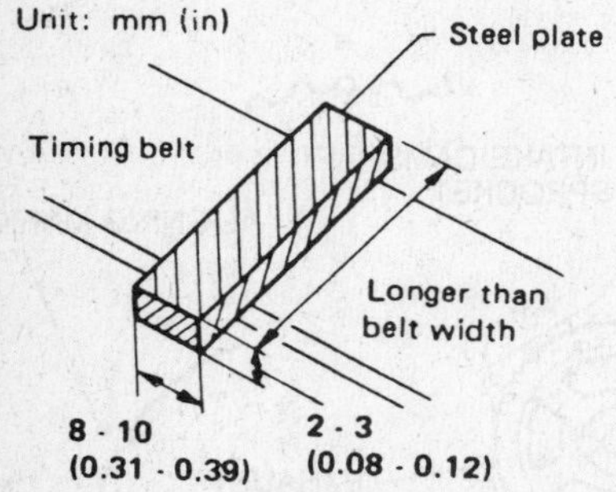

Fabricate a suitable steel plate as shown

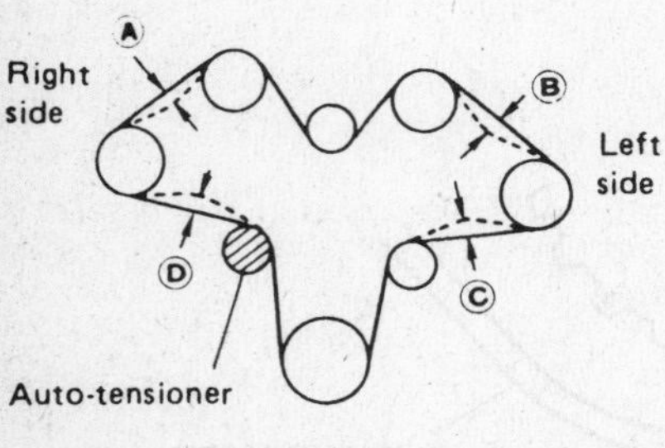

Set the steel plate at each position on the belt

31. If the belt deflection is not as specified, repeat Steps 22–30 until the belt deflection is correct.

32. Once the belt is properly tensioned, torque the auto-tensioner fasteners to 12–15 ft. lbs. (16–21 Nm).

33. Remove the stopper bolt from the tensioner and wait 5 minutes. After 5 minutes, check the clearance between the tensioner arm and the pusher of the auto-tensioner. The clearance should remain at 0.138–0.205 in. (3.5–5.2mm).

34. Make sure the belt is installed and aligned properly on each pulley and timing sprocket. There must be no slippage or misalignment.

35. Install the timing belt covers with new gaskets. Torque the covers bolts to 2–4 ft. lbs. (3–5 Nm).

36. Install the water inlet and outlet housings with new gaskets.

37. Install the crankshaft pulley. Torque the pulley bolt to 159–174 ft. lbs. (21–235 Nm).

38. Remove the flywheel locking device and install the starter.

39. Install the cooling fan and cooling fan coupling.

40. Install and tension the drive belts.

41. Install the radiator.

42. Fill the cooling system to the proper level.

43. Connect the negative battery cable.

VG30E Engine (1988–92 Maxima)

1. Disconnect the negative battery cable.
2. Raise and support the front of the vehicle safely.
3. Remove the engine undercovers.
4. Drain the cooling system.
5. Remove the front right side wheel.
6. Remove the engine side cover.
7. On 1988 vehicles, remove the engine coolant reservoir tank, radiator hoses and the Automatic Speed Control Device (ASCD) actuator; remove the lower coolant hose support bracket and disconnect the lower hose from the suction pipe.
8. Remove the alternator, power steering and air conditioning compressor drive belts from the engine. When removing the power steering drive belt, loosen the idler pulley from the right side wheel housing.
9. On 1989–92 vehicles, remove the upper radiator and water inlet hoses; remove the water pump pulley.
10. Remove the idler bracket of the compressor drive belt.
11. Remove the crankshaft pulley with a suitable puller.
12. Remove the upper and lower timing belt covers and gaskets.
13. Rotate the engine with a socket wrench on the crankshaft pulley bolt to align the punch mark on the left hand camshaft pulley with the mark on the upper rear timing belt cover; align the punchmark on the crankshaft with the notch on the oil pump housing; temporarily install the crankshaft pulley bolt to allow for crankshaft rotation.
14. Use a hex wrench to turn the belt tensioner clockwise and tighten the tensioner locknut just enough to hold the tensioner in position. Then, remove the timing belt.

To install:

15. Before installing the timing belt confirm that No. 1 cylinder is at TDC on its compression stroke. Install tensioner and tensioner spring. If stud is removed apply locking sealant to threads before installing.
16. Swing tensioner fully clockwise with hexagon wrench and temporarily tighten locknut.
17. Point the arrow on the timing belt toward the front belt cover. Align the white lines on the timing belt with the punch marks on all 3 pulleys.

NOTE: There are 133 total timing belt teeth. If timing belt is installed correctly there will be 40 teeth between left hand and right hand camshaft sprocket timing marks. There will be 43 teeth between left hand camshaft sprocket and crankshaft sprocket timing marks.

18. Loosen tensioner locknut, keeping tensioner steady with a hexagon wrench.
19. Swing tensioner 70–80 degrees clockwise with hexagon wrench and temporarily tighten locknut.
20. Turn crankshaft clockwise 2–3 times, then slowly set No. 1 cylinder at TDC of the compression stroke.
21. Push middle of timing belt between righthand camshaft sprocket and tensioner pulley with a force of 22 lbs.
22. Loosen tensioner locknut, keeping tensioner steady with a hexagon wrench.
23. Insert a 0.138 in. (0.35mm) thick and 0.5 in. (12.7mm) wide feeler gauge between the bottom of tensioner pul-

ley and timing belt. Turn crankshaft clockwise and position gauge completely between tensioner pulley and timing belt. The timing belt will move about 2.5 teeth.

24. Tighten tensioner locknut, keeping tensioner steady with a hexagon wrench.

25. Turn crankshaft clockwise or counterclockwise and remove the gauge.

26. Rotate the engine 3 times, then set No. 1 at TDC on its compression stroke.

27. Install the upper and lower timing belt covers with new gaskets.

28. Install the crankshaft pulley. Torque the pulley bolt to 90–98 ft. lbs. (123–132 Nm).

29. Install the compressor drive belt idler bracket.

30. On 1989–92 vehicles, install the water pump pulley and torque the nuts to 12–15 ft. lbs. (16–21 Nm). Install the upper radiator and water inlet hoses.

31. Install the drive belts.

32. On 1988 vehicles, connect the lower coolant hose to the suction pipe and install the hose support bracket; install the Automatic Speed Control Device (ASCD) actuator, radiator hoses and engine coolant reservoir tank.

33. Install the engine side cover.

34. Mount the front right wheel.

35. Install the engine undercovers.

36. Lower the vehicle.

37. Fill the cooling system and connect the negative battery cable.

Timing Sprockets

REMOVAL & INSTALLATION

1. Disconect the negative battery cable.

2. Set the No. 1 piston to TDC of the compression stroke.

3. Remove the timing belt covers.

4. Remove the timing belt.

5. Using a suitable spanner wrench and a socket wrench, remove the camshaft pulley bolt and washer.

   a. On CA18DE engine, the sprocket is held to the end of the cmashaft by a plate with 4 bolts.

   b. On E16i engine, the camshaft and jackshaft sprockets are held in place by a plate and 3 bolts. Pull the camshaft sprocket(s) from the camshaft(s). Be careful not to lose the Woodruff key.

   c. On 1990–92 300ZX, remove the front plate, O-ring and spring from the right (intake) camshaft to gain access to the sprocket bolt.

   d. On 1990–92 300ZX, the left camshaft sprocket is held in place by plate and 4 bolts.

6. Using a suitable puller, remove the crankshaft gear and timing belt plates from the crankshaft. Be careful not to gouge or scratch the surface of the crankshaft when removing the gear.

7. Inspect the timing gear teeth for wear and replace as necessary.

To install:

8. Install the crankshaft gear with new Woodruff keys.

9. Install the camshaft sprockets. Torque the sprocket bolts to 58–65 ft. lbs. (78–88 Nm) on CA20, VG30E and VG30ET engines; 90–98 ft. lbs. (123–132 Nm) for right (intake) and 10–14 ft. lbs. (14–19 Nm) for the left (exhaust) on 1990–92 300ZX; 10–14 ft. lbs. (14–19 Nm) on CA18DE engine; 7–9 ft. lbs. (9–12 Nm) on E16i engine.

NOTE: On VG30E and VG30ET engines, the right hand and left hand camshaft pulleys are different. Install them in their correct positions. The right hand pulley has an R3 identification mark and the left hand pulley has an L3.

10. Install the timing belt.

11. Install the timing belt covers.

12. Connect the negative battery cable.

Camshaft

REMOVAL & INSTALLATION

CA18DE Engine (1988–89 Pulsar)

1. Disconnect the negative battery cable and relieve the fuel system pressure.

2. Drain the cooling system and remove the air cleaner assembly.

3. Crank the engine until the No. 1 piston is at TDC on its compression stroke.

4. Remove the drive belts.

5. Remove the alternator.

6. Disconect the air duct at the throttle chamber.

7. Tag and disconnect all lines, hoses and wires which may interfere with removal of the cylinder.

8. Remove the ornament cover.

9. Disconnect the oxygen sensor wire.

10. Remove the 2 exhaust heat shield covers.

11. Unbolt the exhaust manifold and wire the entire assembly aside to gain removal clearance for the cylinder head.

12. Disconnect the EGR tube at the passage cover and then remove the passage cover and its gasket.

13. Disconnect and remove the crank angle sensor from the upper front cover.

NOTE: Put an aligning mark on crank angle sensor and timing belt cover.

14. Remove the support stay from under the intake manifold assembly.

15. Unbolt the intake manifold and remove it along with the collector and throttle chamber.

16. Disconnect and remove the fuel injectors as an assembly.

NOTE: Upper and lower front timing belt cover must be removed. Support engine under oil pan with floor jack or equivalent then remove the upper engine mount bracket at the right side of the front cover.

17. Remove the timing belt.

18. Remove the camshaft cover and remove the cylinder head.

19. Remove the breather separater.

20. While holding the camshaft sprockets, remove the 4 mounting bolts and then remove the sprockets themselves.

21. Remove the timing belt tensioner pulley. Remove the rear timing belt cover.

22. Loosen the camshaft bearing caps in several stages, in the correct order. Remove the bearing caps, but be sure to keep them in order.

23. Remove the front oil seals and then lift out the camshafts.

24. Check the camshaft runout, endplay, wear and journal clearance.

To install:

25. Position the camshafts in the cylinder head so the knock pin on each is on the outboard side.

NOTE: The exhaust side camshaft has splines to accept the crank angle sensor.

26. Position the camshaft bearing caps and finger-tighten them. Each cap has an ID mark (E1, E2, I1, I2 etc.) and a directional arrow stamped into its top surface.

27. Coat the new oil seal with engine oil (on the lip) and install it on each camshaft end.

28. Tighten the camshaft bearing cap bolts to 7–9 ft. lbs. (9–12 Nm) in the order shown.

29. Install the cylinder head and cover.

30. Install the rear timing cover and tighten the 4 bolts to 5–6 ft. lbs. (7–8 Nm).

31. Install the timing belt tensioner and tighten it to 16–22 ft. lbs. (22–29 Nm).

32. Install the camshaft sprockets and tighten the bolts to 10–14 ft. lbs. (14–19 Nm) while holding the camshaft in place.

33. Install the cylinder head and related components.

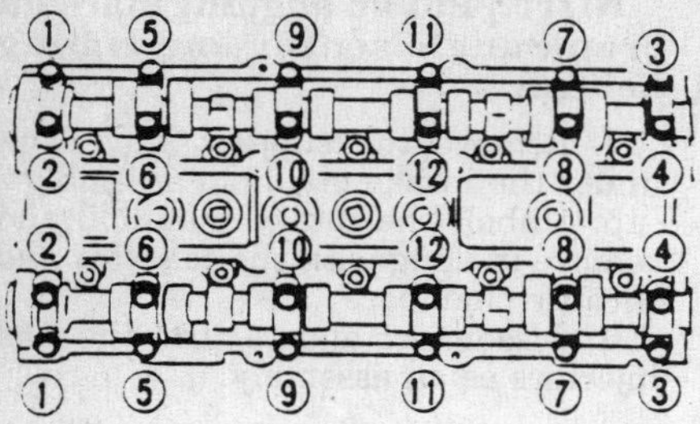

Loosen the camshaft bearing cap bolts in this order—CA18DE engine

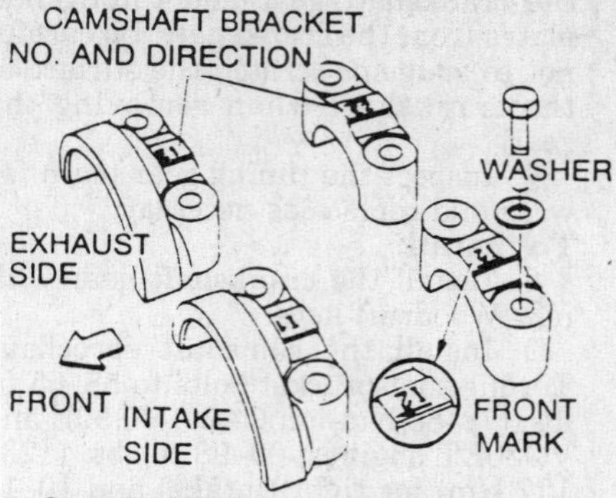

Camshaft bearing cap positioning—CA18DE engine

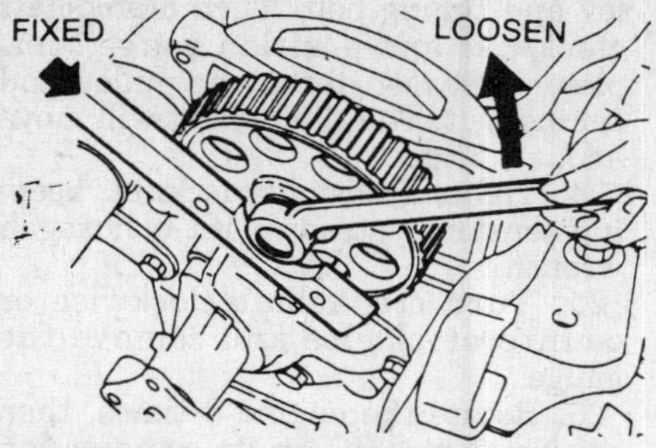

Loosening the camshaft sprocket—C series engine

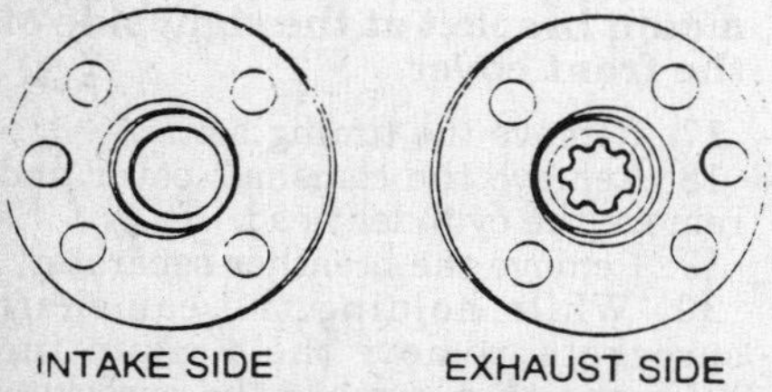

The exhaust side camshaft is splined—CA18DE engine

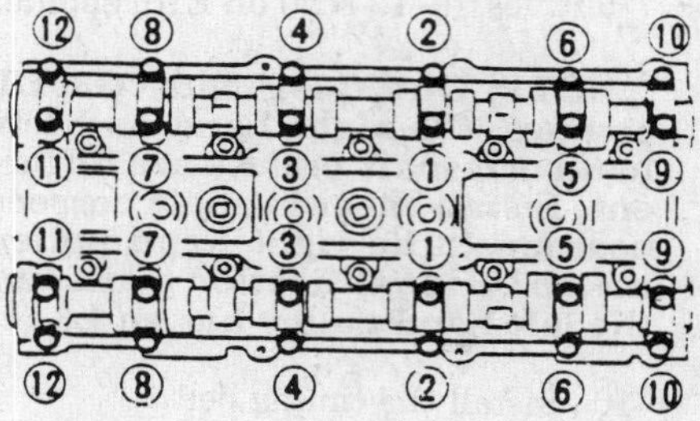

Tighten the camshaft bearing caps in this order—CA18DE engine

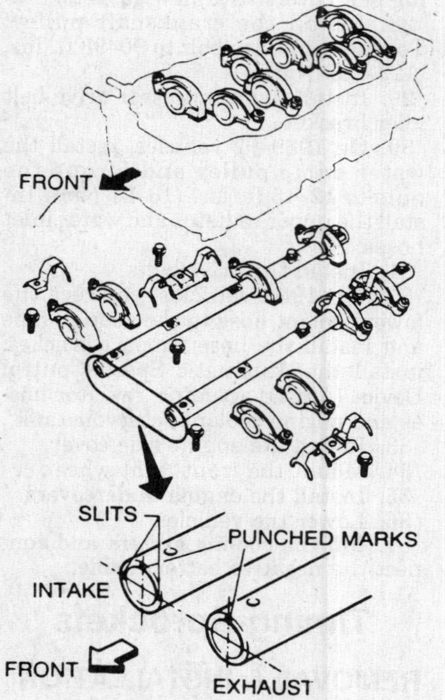

Rocker shaft assembly—C series engine

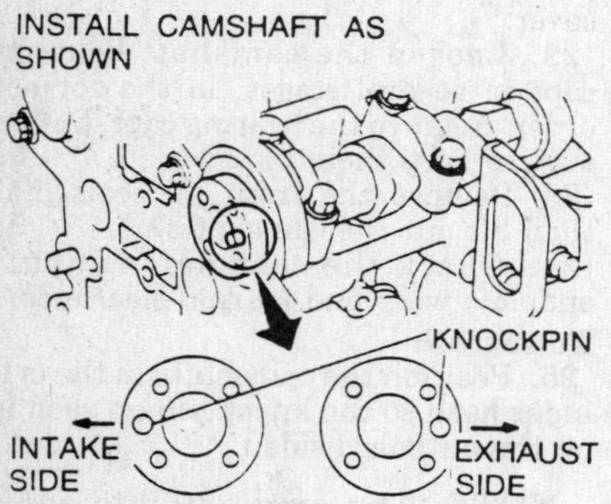

Install the camshaft as shown—CA18DE engine

Timing belt tensioner installation—CA18DE engine

34. Fill the cooling system to the proper level and connect the negative battery cable.

CA20E Engine (1988 200SX and 1988–89 Stanza)

1. Disconnect the negative battery cable and relieve the fuel system pressure.
2. Set the No. 1 piston to TDC of the compression stroke.
3. Remove the timing belt.
4. Remove the valve rocker cover.
5. Fully loosen all rocker arm adjusting screws (the valve adjusting screws). Loosen the rocker shaft mounting bolts in 2–3 stages and then remove the rocker shafts as an assembly. Keep all components in the correct order for reassembly.
6. Hold the camshaft pulley and remove the pulley mounting bolt. Remove the pulley. Remove the camshaft thrust plate.
7. Carefully pry the camshaft oil seal out of the front of the cylinder head.
8. Slide the camshaft out the front of the cylinder head, taking extreme care not to score any of the journals.

To install:

9. Coat the camshaft with clean engine oil.
10. Carefully slide the camshaft into the cylinder head, coat the end with oil and install a new oil seal. Install the camshaft thrust plate and wedge the camshaft with a small wooden block inserted between one of the cams and the cylinder head. Torque the thrust plate bolt to 58–65 ft. lbs. (78–88 Nm). Remove the wooden block.
11. Lubricate the rocker shafts lightly with clean engine oil and install them, with the rocker arms, into the head. Both shafts have punch marks on their leading edges, while the intake shaft is also marked with 2 slits on its leading edge.

NOTE: To prevent the rocker shaft springs from slipping out of the shaft, insert the bracket bolts into the shaft prior to installation.

12. Tighten the rocker shaft bolts gradually, in 2–3 stages to 13–16 ft. lbs. (18–22 Nm).
13. Install the camshaft pulley and then install the timing belt.
14. Adjust the valves as required and install the cylinder head cover.
15. Connect the negative battery cable.

E16i Engine (1988 Pulsar)

1. Disconnect the negative battery cable.
2. Remove the timing belt.
3. Remove the rocker shaft along with the rocker arms. Loosen the bolts gradually, in 2–3 stages.
4. Carefully slide the camshaft out the front of the cylinder head.
5. Check the camshaft runout, endplay, wear and journal clearance.

To install:

6. Slide the camshaft into the cylinder head carefully and then install a new oil seal. Coat the lip of the new seal with clean engine oil prior to installation.
7. Install the rear timing belt cover.
8. Set the camshaft so the knockpin faces upward and then install the camshaft sprocket so its timing mark aligns with the one on the rear timing cover.
9. Install the timing belt.
10. Coat the rocker shaft and the interior of the rocker arm with engine oil. Install them so the punch mark on the shaft faces forward and the oil holes in the shaft face down. The cutout in the center retainer on the shaft should face the exhaust manifold side of the engine.
11. Make sure the valve adjusting screws are loose and then tighten the shaft bolts to 13–15 ft. lbs. (18–21 Nm) in several stages, from the center out. The first and last mounting bolts should have a new bolt stopper installed.
12. Adjust the valves and connect the negative battery cable.

GA16i Engine (1989–90 Pulsar and 1989–90 Sentra)

1. Disconnect the negative battery cable.
2. Remove the timing chain.
3. Remove the cylinder head with manifolds attached.
4. Remove the intake and exhaust manifolds from the cylinder head. Loosen the bolts in 2–3 stages in the proper sequence.
5. Loosen the rocker arm shaft bolts in 2–3 stages and lift the rocker arm/shaft assembly from the cylinder head. The rocker arm shaft is marked with an **F** to indicate that it faces towards the front of the engine. Place a similar mark on the cylinder head for your own reference.
6. Loosen the thrust plate retaining bolt.
7. Withdraw the camshaft and the thrust plate from the front of the cylinder head. The thrust plate is located to the camshaft with a key. Retain this key.

To install:

8. Clean all cylinder head, intake

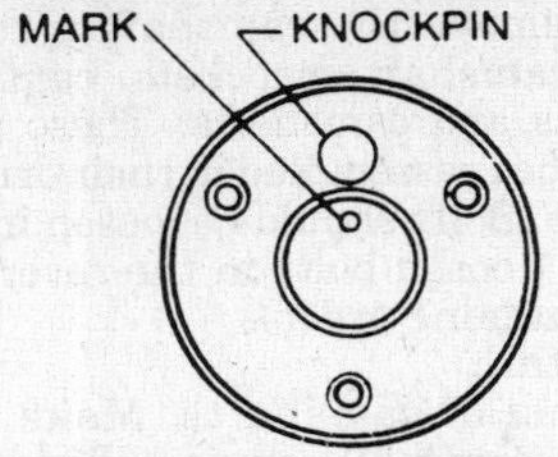

Camshaft positioning – E-Series engines

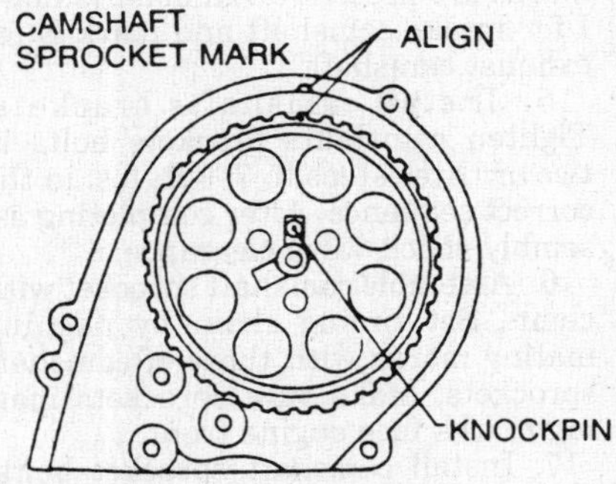

Camshaft sprocket alignment – E-Series engines

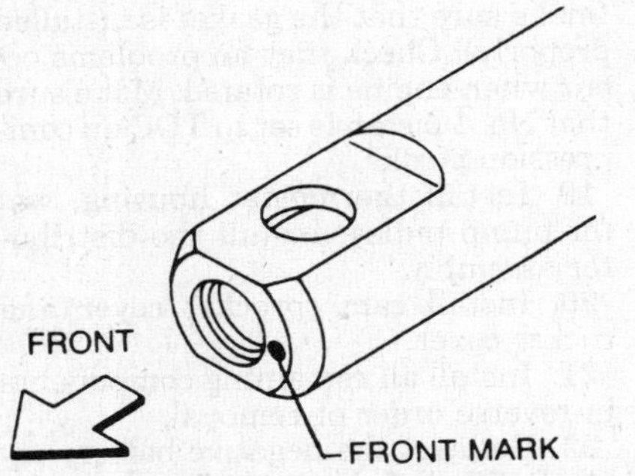

The punch mark on the rocker shaft should face forward – E-Series engines

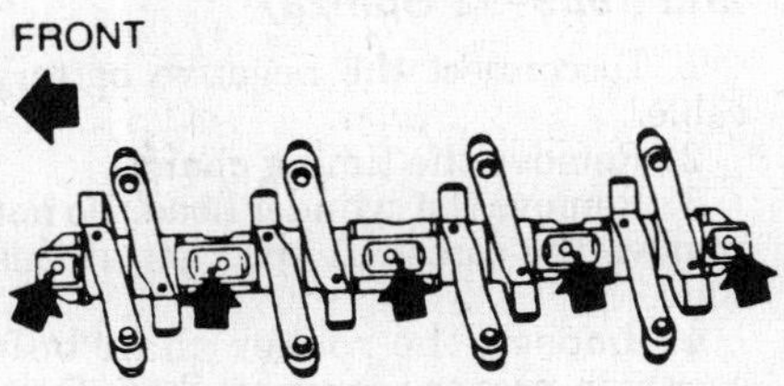

The oil holes must be facing down – E series engine

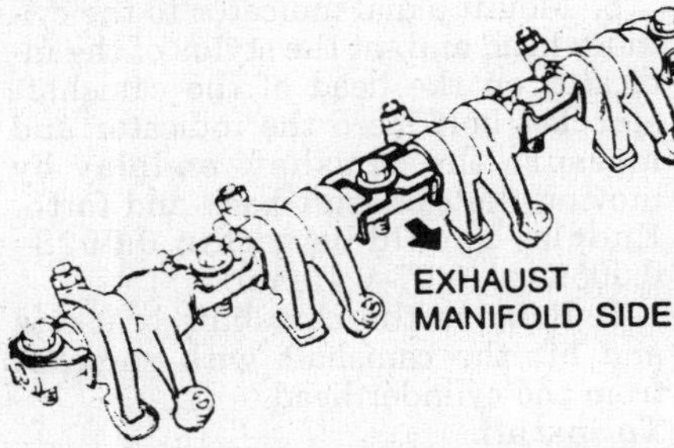

The center retainer cut-out should face the exhaust manifold – E series engine

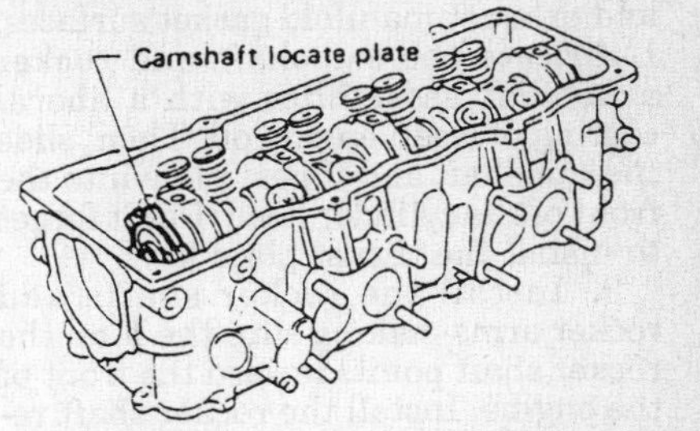

Rocker arm/shaft positioning – GA16i engine

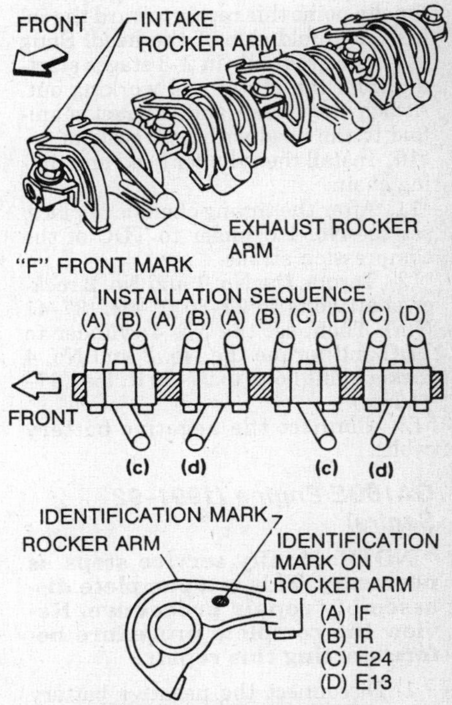

Rocker arm shaft identification – GA16i engines

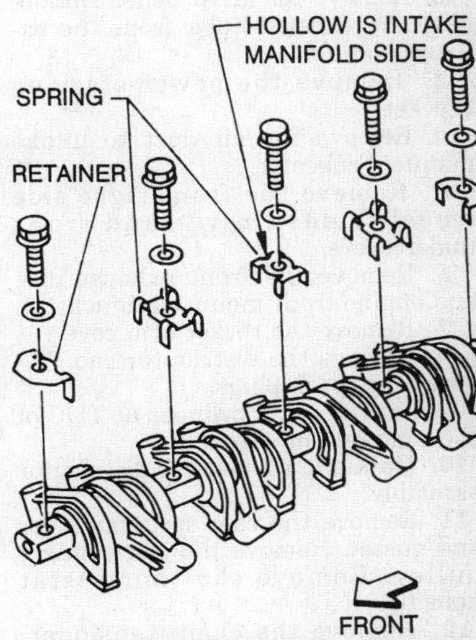

Rocker arm shaft bolt retainer positioning – GA16i engines

and exhaust manifold gasket surfaces. Lubricate the camshaft and rocker arm/shaft assemblies with a liberal coating of clean engine oil. Then, slide the camshaft and thrust plate into the front of the cylinder head. Don't forget to install the thrust plate key.

9. Install the rocker shafts and rocker arms making sure the **F** on the rocker shaft points toward the front of the engine. Install the rocker shaft retaining bolts, spring clips and washers. The center spring clip has a recess cut into one side. When installing the center clip point this recess toward the intake manifold side of the head. Snug the bolts gradually in 2–3 stages starting from the center and working out. Attach the intake and exhaust manifold to the head with new gaskets.

10. Install the cylinder head and timing chain.

11. After the timing chain is in place, set the No. 1 cylinder to TDC of the compression stroke.

12. Torque the No. 1 and No. 2 rocker shaft bolts to 27–30 ft. lbs. (37–41 Nm). Then, set the No. 4 cylinder to TDC and torque the No. 3 and No. 4 rocker shaft bolts to 27–30 ft. lbs. (37–41 Nm).

13. Connect the negative battery cable.

GA16DE Engine (1991–92 Sentra)

NOTE: Modify service steps as necessary. This is a complete disassembly repair procedure. Review the complete procedure before starting this repair.

1. Disconnect the negative battery cable, drain the cooling system and relieve the fuel system pressure.
2. Remove all drive belts. Disconnect the exhaust tube from the exhaust manifold.
3. Remove the power steering bracket.
4. Remove the air duct to intake manifold collector.
5. Remove the front right side wheel, splash cover and front undercovers.
6. Remove the front exhaust pipe and engine front mounting bracket.
7. Remove the rocker arm cover.
8. Remove the distributor cap. Remove the spark plugs.
9. Set the No. 1 cylinder at TDC of the compression stroke.
10. Mark and remove the distributor assembly.
11. Remove the cam sprocket cover and gusset. Remove the water pump pulley. Remove the thermostat housing.
12. Remove the chain tensioner, chain guide. Loosen idler sprocket bolt.
13. Remove the camshaft sprocket bolts, camshaft sprockets, camshaft brackets and camshafts. These parts should be reassembled in their original position. Bolts should be loosen in 2 or 3 steps (loosen bolts in the reverse of the tightening order).

To install:

14. Install camshafts. Make sure that the camshafts are installed in the correct position. Note identification marks are present on camshafts mark I for intake camshaft and mark E for exhaust camshaft.
15. Install camshafts brackets. Tighten camshafts brackets bolts in two or three steps to 7–9 ft. lbs. in the correct sequence. After completing assembly check valve clearance.
16. Assemble camshaft sprocket with chain. Set timing chain by aligning mating marks with those of camshaft sprockets. Make sure sprockets mating marks face engine front.
17. Install camshaft sprocket bolts. Install upper chain tensioner and chain guide.
18. Install lower chain tensioner (make sure that the gasket is installed properly). Check that no problems occur when engine is rotated. Make sure that No. 1 piston is set to TDC on compression stroke.
19. Install thermostat housing, water pump pulley. Install the distributor assembly.
20. Install cam sprocket cover and rocker cover.
21. Install all remaining components in reverse order of removal.
22. Connect the negative battery cable. Refill all fluid levels. Road test the vehicle for proper operation.

KA24E Engine (1989–90 240SX and 1989–92 Stanza)

1. Disconnect the negative battery cable.
2. Remove the timing chain.
3. Remove the cylinder head. Do not remove the camshaft sprocket at this time.
4. Loosen the rocker shaft bolt evenly in proper sequence. Start from the outside and work toward the center.
5. Mount a dial indicator to the cylinder head and set the stylus of the indicator on the head of the camshaft sprocket bolt. Zero the indicator and measure the camshaft endplay by moving the camshaft back and forth. Endplay should be within 0.0028–0.0059 in. (0.07–0.15mm).
6. Remove the camshaft brackets and lift the camshaft with sprocket from the cylinder head.

To install:

7. Clean all cylinder head, intake and exhaust manifold gasket surfaces. Lubricate the camshaft and rocker arm/shaft assemblies with a liberal coating of clean engine oil. Lay the camshaft and sprocket into the cylinder head so the knock pin is at the front of the head at the 12 o'clock postion. Install the camshaft brackets. The camshaft bracket directional arrows must face the toward the front of the engine.
8. Install the rocker shaft and rocker arms. Both intake and exhaust rocker shafts are stamped with an **F** mark. This mark must face the front of the engine during installation. Install the rocker arm bolts and spring clips so the cut outs are facing as shown. Torque the rocker arm bolts in the proper sequence to 27–30 ft. lbs. (37–41 Nm).
9. Install the timing chain.
10. Install the cylinder head. Use new rubber plugs when installing the cylinder head.
11. Connect the negative battery cable.

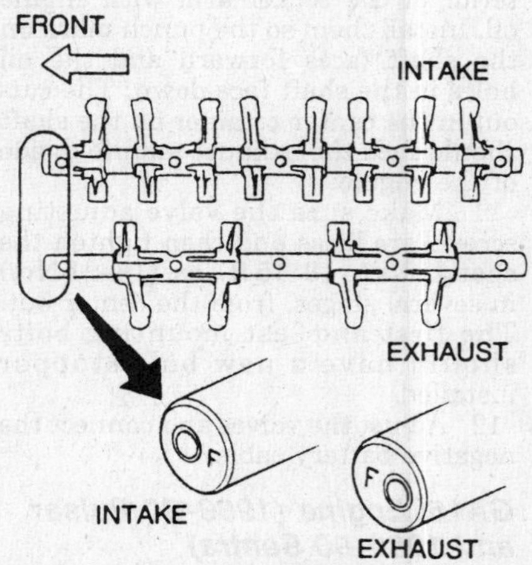

Rocker arm shaft positioning – KA24E engines

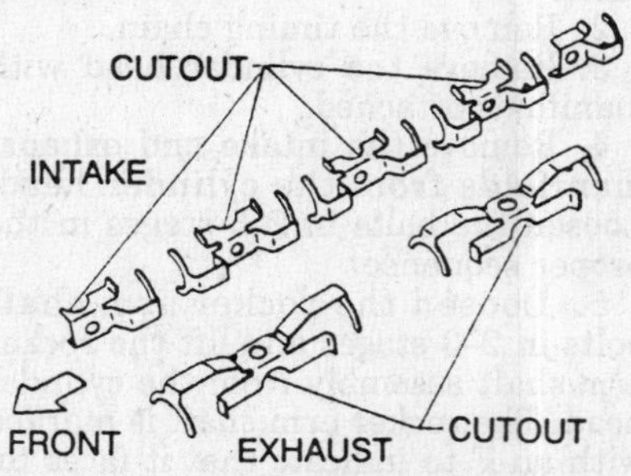

Spring clip installation – KA24E engines

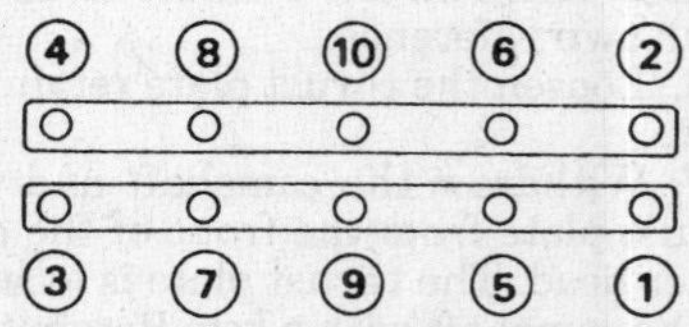

Rocker shaft bolt LOOSENING sequence – KA24E engines. Tighten in reverse of loosening sequence

KA24DE Engine (1991–92 240SX)

NOTE: Modify service steps as necessary. This is a complete disassembly repair procedure. Review the complete procedure before starting this repair.

1. Release the fuel system pressure.
2. Disconnect the negative battery cable and drain the cooling system. Drain the engine oil.
3. Remove all vacuum hoses, fuel lines, wires, electrical connections as necessary.
4. Remove the front exhaust pipe and A.I.V. pipe.
5. Remove the air duct, cooling fan with coupling and radiator shroud.

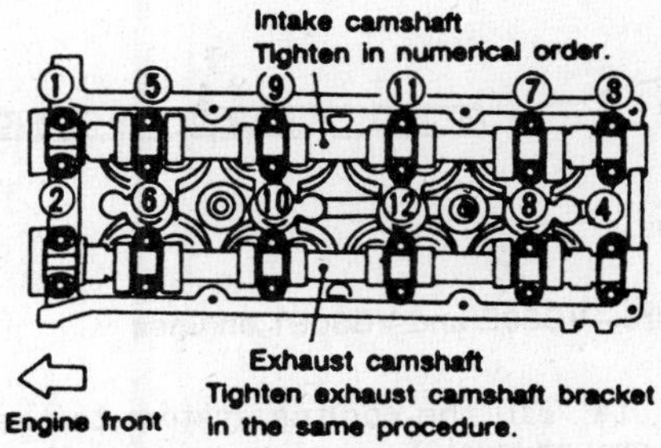

Camshaft bracket torque sequence – KA24DE engine

6. Remove the the fuel injector tube assembly with injectors.
7. Disconnect and mark spark plug wires. Remove the spark plugs.
8. Set No. 1 piston at TDC on compression stroke. Remove the rocker cover assembly.
9. Mark and remove the distributor assembly.
10. Remove the cam sprocket, brackets and camshafts. These parts should be reassembled in their original position. Bolts should be loosened in 2 or 3 steps (loosen all bolts in the reverse of the tightening order).

To install:

11. Install camshafts and camshafts brackets. Torque camshaft brackets in two or three steps in sequence. After completing assembly check valve clearance.
12. Install camshaft sprockets.
13. Install chain guide between both camshaft sprockets and distributor assembly.
14. Install all remaining components in reverse order of removal.
15. Connect the negative battery cable. Refill all fluid levels. Road test the vehicle for proper operation.

SR20DE Engine (1991–92 Sentra)

1. Disconnect the negative battery cable. Remove the rocker cover and oil separator.
2. Rotate the crankshaft until the No.1 piston is at TDC on the compression stroke. Then rotate the crankshaft until the mating marks on the camshaft sprockets line up with the mating marks on the timing chain.
3. Remove the timing chain tensioner.
4. Remove the distributor.
5. Remove the timing chain guide.
6. Remove the camshaft sprockets. Use a wrench to hold the camshaft while loosening the sprocket bolt.
7. Loosen the camshaft bracket bolts in the opposite order of the torquing sequence.
8. Remove the camshaft.

To install:

9. Clean the left hand camshaft end bracket and coat the mating surface with liquid gasket. Install the camshafts, camshaft brackets, oil tubes and baffle plate. Ensure the left camshaft key is at 12 o'clock and the right camshaft key is at 10 o'clock.
10. The procedure for tightening camshaft bolts must be followed exactly to prevent camshaft damage. Tighten bolts as follows:
    a. Tighten right camshaft bolts 9 and 10 (in that order) to 1.5 ft. lbs. (2 Nm) then tighten bolts 1–8 (in that order) to the same specification.
    b. Tighten left camshaft bolts 11 and 12 (in that order) to 1.5 ft. lbs. (2 Nm) then tighten bolts 1–10 (in that order) to the same specification.
    c. Tighten all bolts in sequence to 4.5 ft. lbs. (6 Nm).
    d. Tighten all bolts in sequence to 6.5–8.5 ft. lbs. (9–12 Nm) for type A, B and C bolts, and 13–19 ft. lbs. (18–25 Nm) for type D bolts.
11. Line up the mating marks on the timing chain and camshaft sprockets and install the sprockets. Tighten sprocket bolts to 101–116 ft. lbs. (137–157 Nm).
12. Install the timing chain guide, distributor (ensure that rotor head is at 5 o'clock position) and chain tensioner.
13. Clean the rocker cover and mating surfaces and apply a continious bead of liquid gasket to the mating surface.
14. Install the rocker cover and oil separator. Tighten the rocker cover bolts as follows:
    a. Tighten nuts 1, 10, 11, and 8 in that order to 3 ft. lbs. (4 Nm).
    b. Tighten nuts 1–13 as indicated in the figure to 6–7 ft. lbs. (8–10 Nm).
15. Connect the negative battery cable. Refill all fluid levels. Road test the vehicle for proper operation.

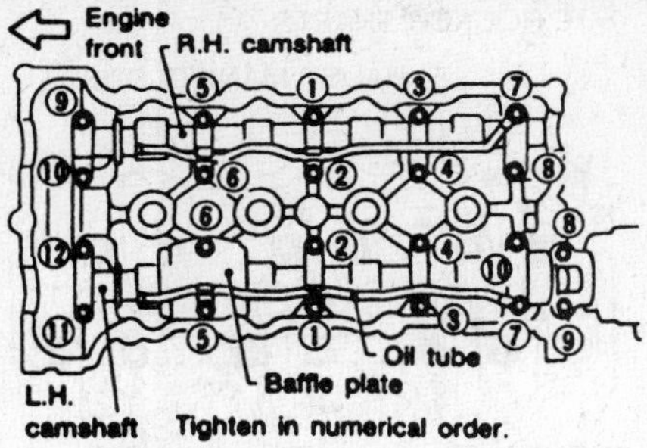

Camshaft bracket torque sequence – SR20DE engine

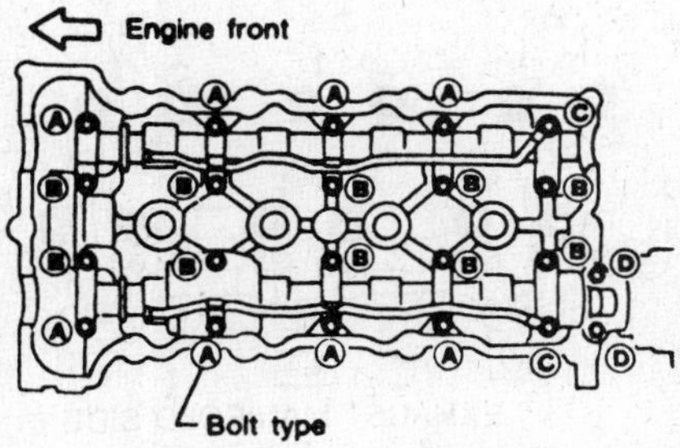

Camshaft bolt location – SR20DE engine

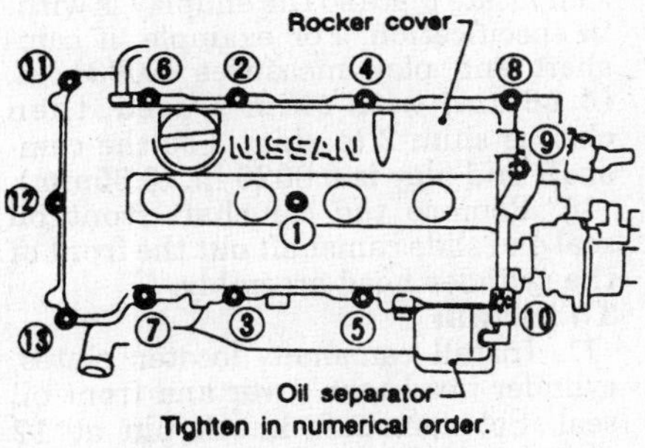

Rocker cover torque sequence – SR20DE engine

VG30E and VG30ET Engines (1988–89 300ZX and 1988–92 Maxima)

1. Disconnect the negative battery cable.
2. Drain the cooling system.
3. Remove the timing belt.
4. Remove the collector assembly.
5. Remove the intake manifold.
6. Remove the cylinder head.
7. Remove the rocker shafts with rocker arms. Bolts should be loosened in several steps in the proper sequence.
8. Remove hydraulic valve lifters and lifter guide. Hold hydraulic valve lifters with wire so they will not drop from lifter guide.
9. Using a dial gauge measure the camshaft endplay. If the camshaft endplay exceeds the limit (0.0012–0.0024 in.), select the thickness of a

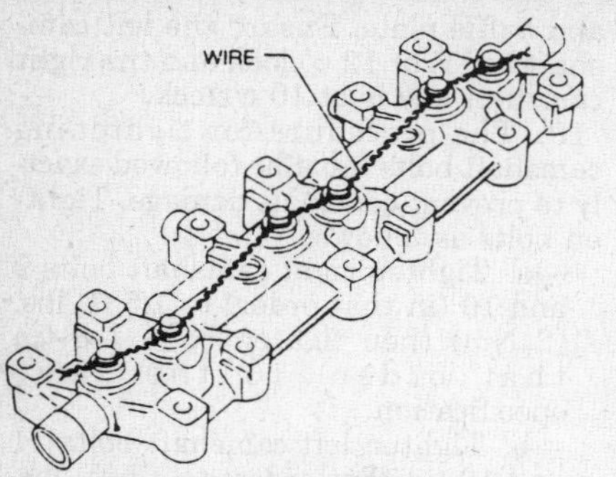

Holding the V6 valve lifters in place

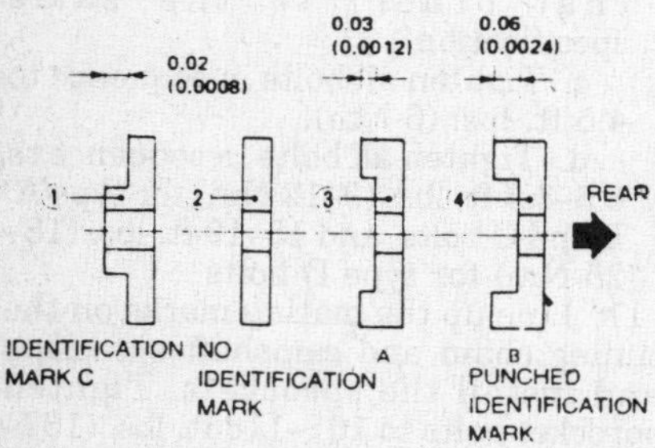

Select shim thickness so that camshaft thickness is within specifications

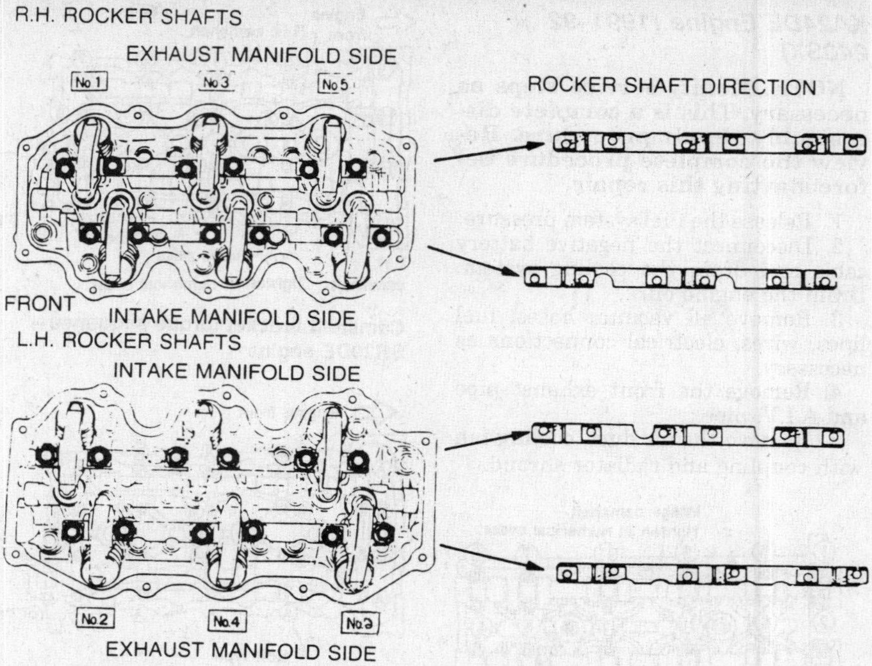

Rocker shaft/arm installation procedure – VG30E and VG30ET engines

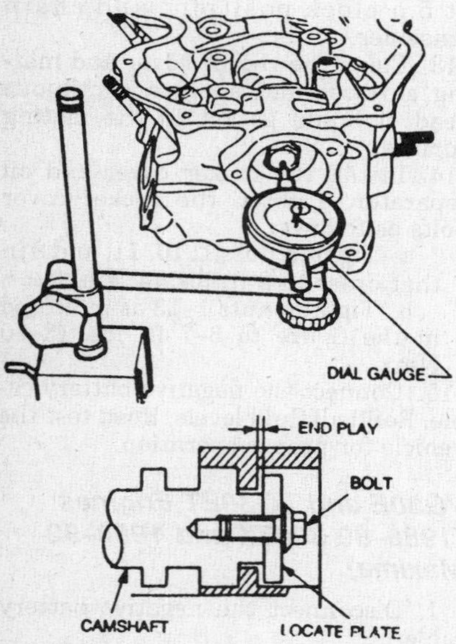

Using a dial indicator to measure camshaft endplay – V6 engine

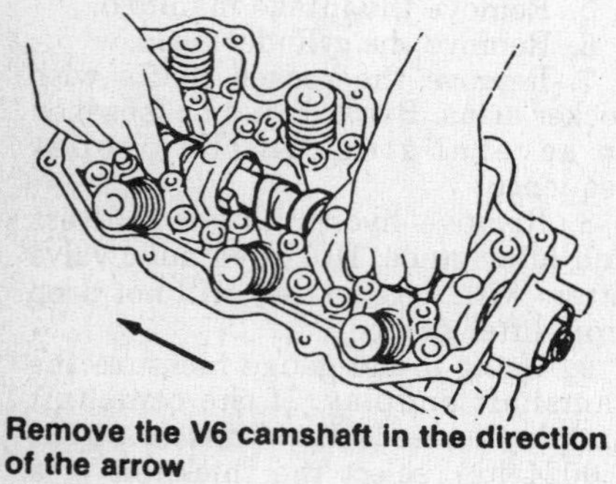
Remove the V6 camshaft in the direction of the arrow

cam locate plate so the endplay is within specification. For example, if camshaft end play measures 0.0031 in. (0.08mm) with shim 2 used, then change shim 2 to shim 3 so the camshaft end play is 0.0020 in. (0.05mm).

10. Remove the camshaft front oil seal and slide camshaft out the front of the cylinder head assembly.

To install:

11. Install camshaft, locater plates, cylinder head rear cover and front oil seal. Set camshaft knock pin at 12 o'clock position. Install cylinder head with new gasket to engine.

12. Install valve lifter guide assembly. Assemble valve lifters in their original position. After installing them in the correct location remove the wire holding them in lifter guide.

13. Install rocker shafts in correct position with rocker arms. Tighten bolts in 2–3 stages to 13–16 ft. lbs. (18–22 Nm). Before tightening, be sure to set camshaft lobe at the position where lobe is not lifted or the valve closed. Set each cylinder 1 at a time or follow the procedure below. The cylinder head, intake manifold, collector and timing belt must be installed:

a. Set No. 1 piston at TDC of the compression stroke and tighten rocker shaft bolts for No. 2, No. 4 and No. 6 cylinders.

b. Set No. 4 piston at TDC of the compression stroke and tighten rocker shaft bolts for No. 1, No. 3 and No. 5 cylinders.

c. Torque specification for the rocker shaft retaining bolts is 13–16 ft. lbs. (18–22 Nm).

14. Fill the cooling system to the proper level.

15. Connect the negative battery cable.

VG30DE and VG30DETT Engines (1990–92 300ZX)

1. Disconnect the negative battery cable.
2. Drain the cooling system.
3. Remove the the timing belt.
4. Remove the cylinder head with the exhaust manifold.
5. Separate the exhaust manifold from the cylinder head.
6. Remove the camshaft sprockets. Remove the front plate, O-ring and spring from the right (intake) camshaft to gain access to the sprocket bolt. The left camshaft sprocket is held in place by plate and 4 bolts.
7. Remove the rear timing belt cover.
8. Mount a dial indicator and set the stylus of the indicator on the end of the camshaft. Zero the indicator and measure the camshaft endplay by moving the camshaft back and forth. Endplay should be within 0.0012–0.0031 in. (0.03–0.08mm).
9. Remove the camshaft brackets. Loosen the bolts in the proper sequence gradually in 2–3 stages.
10. Gently pry the camshaft oil seals from the cylinder head.
11. Remove the timing control solenoid valves.
12. Remove the camshafts.

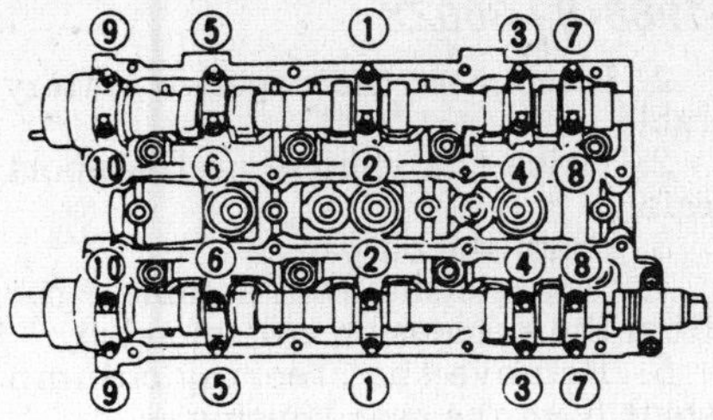

Camshaft bracket torque sequence—VG30DE and VG30DETT engines

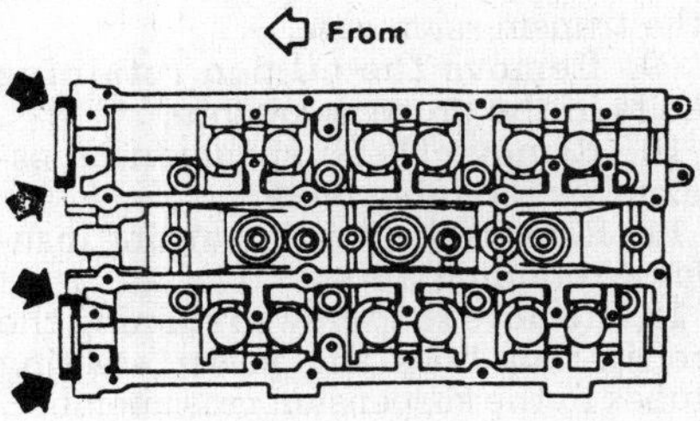

When installing front side camshaft bracket, apply liquid gasket as shown—VG30DE and VG30DETT engines

To install:

13. Install the camshafts so the knock pins are aligned properly. The exhaust side camshaft (left side) has a spline that accepts the crank angle sensor.
14. Install the timing control solenoid valves. Torque the bracket bolts to 12–18 ft. lbs. (16–25 Nm). Apply liquid gasket to the valve seating surface before installation.
15. Install the camshaft brackets. Torque the bracket bolts in sequence to 7–9 ft. lbs. (9–12 Nm). Tighten the bolts gradually in 2–3 stages. When installing the front camshaft brackets, apply liquid gasket to the bracket seating surface.
16. Coat the lips of the new camshaft seals with clean engine oil and install the seals into the cylinder head.
17. Install the rear timing belt covers. Torque the cover bolts to 5–6 ft. lbs. (6–8 Nm).
18. Install the camshaft sprockets. Torque the right side (intake) sprocket bolt 90–98 ft. lbs. (123–132 Nm) and the left side (exhaust) sprocket retainer bolts to 10–14 ft. lbs. (14–19 Nm). When tightening the sprocket fasteners, make sure to hold the camshafts stationary.
19. Mount the exhaust manifold to the head with new gaskets.
20. Install the cylinder head.
21. Install the timing belt.
22. Fill the cooling system to the proper level.
23. Connect the negative battery cable.

Piston and Connecting Rod

POSITIONING

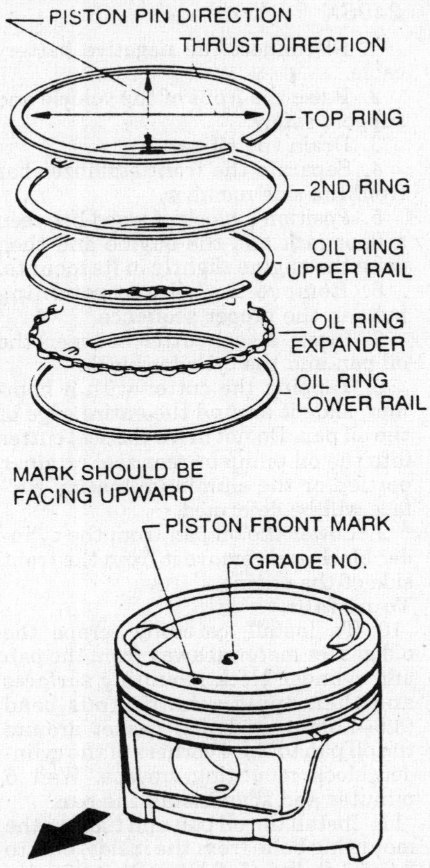

Piston ring identification and positioning—all engines

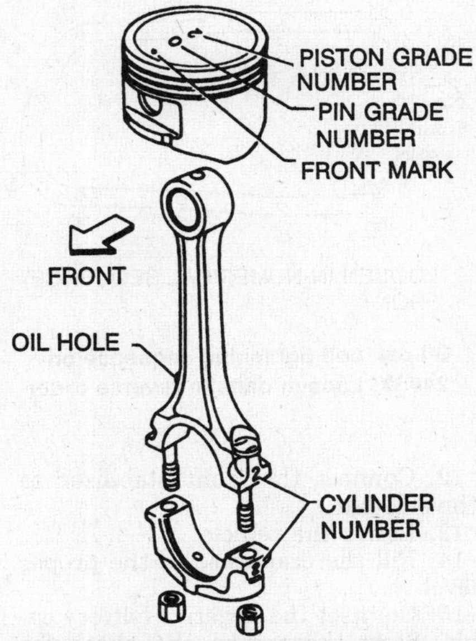

Piston and connection rod positioning—all engines

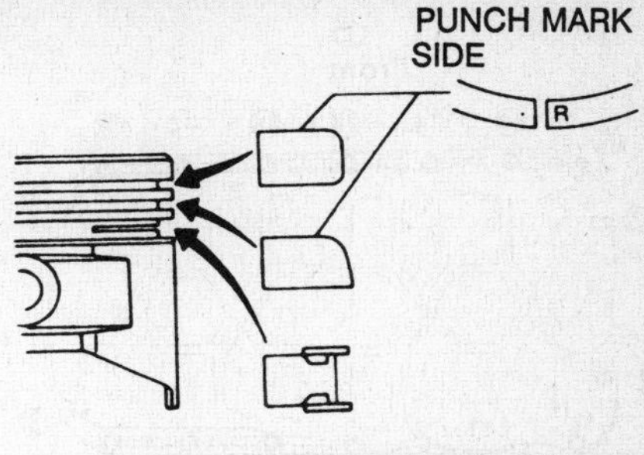

Piston ring installation—all engines

ENGINE LUBRICATION

Oil Pan

REMOVAL & INSTALLATION

200SX

1. Disconnect the negative battery cable.
2. Raise the front of the vehicle and support safely.
3. Drain the oil pan.
4. Remove the power steering bracket from the suspension crossmember.
5. Separate the stabilizer bar from the transverse link.
6. Separate the tension rod from the transverse link.
7. Remove the front engine mounting insulator nuts.
8. Lift the engine.
9. Loosen the oil pan bolts in the proper sequence.
10. Remove the suspension crossmember bolts and remove the screws that secure the power steering oil tubes to the crossmember.
11. Lower the suspension crossmember until there is sufficient clearance to remove the oil pan.
12. Insert a seal cutter between the oil pan and the cylinder block.
13. Tapping the seal cutter with a hammer, slide the cutting tool around the entire edge of the oil pan. Do not drive the seal cutter into the oil pump or rear seal retainer portion or the aluminum mating surface will be deformed.
14. Lower the oil pan from the cylinder block and remove it from the front side of the engine.

To install:

15. Carefully scrape the old gasket material away from the pan and cylinder block mounting surfaces.
16. First apply sealant to the oil pump gasket and rear oil seal retainer gasket surfaces. Then, apply a contin-

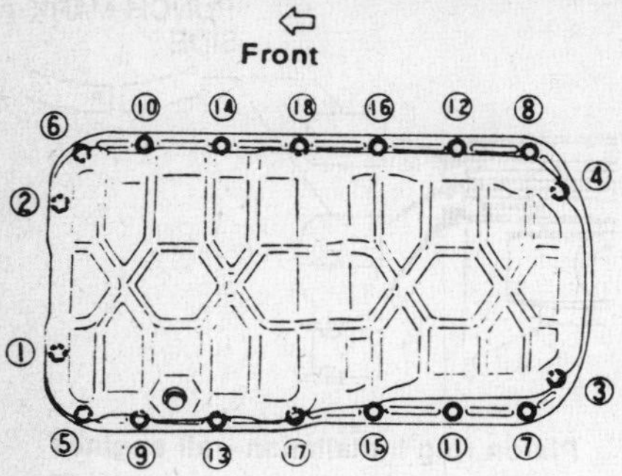

Oil pan bolt tightening sequence on 200SX. Loosen in reverse order

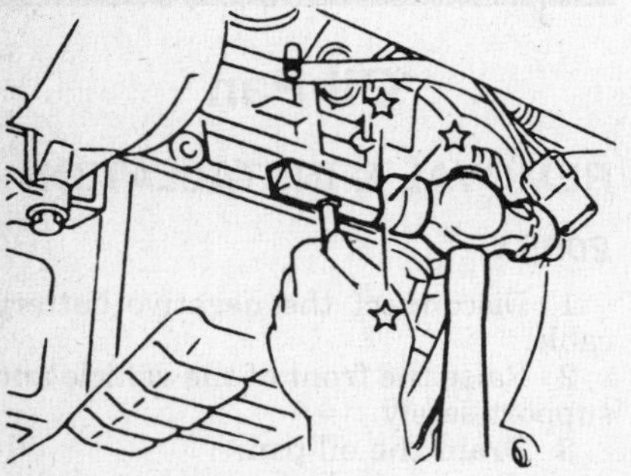

Using a seal cutter on the oil pan

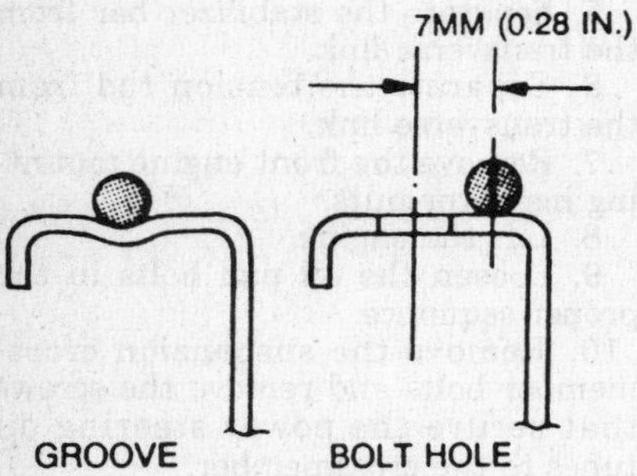

Apply sealer on the inside of the bolt holes

uous bead (3.5–4.5mm) of liquid gasket around the oil pan to the 4 corners of the cylinder block mounting surface. Wait 5 minutes and then install the pan. Tighten the oil pan bolts in sequence to 5–6 ft. lbs. (6–8 Nm).

17. Raise the crossmember from the lowered position. Attach the power steering tubes and install the crossmember bolts.
18. Install the front engine mounting insulator nuts.
19. Connect the tension rod and stabilizer bar to the transverse link.
20. Attach the power steering bracket to the crossmember.
21. Lower the vehicle.
22. Fill the crankcase to the proper level.
23. Connect the negative battery cable. Start the engine and check for leaks.

240SX

1. Disconnect the negative battery cable.
2. Raise the front of the vehicle and support safely.
3. Drain the oil pan.
4. Separate the front stabilizer bar from the side member.
5. Position a block of wood between a floor jack and the engine and then raise the engine slightly in its mounts.
6. Remove the oil pan retaining bolts in the proper sequence.
7. Insert a seal cutter between the oil pan and the cylinder block.
8. Tapping the cutter with a hammer, slide it around the entire edge of the oil pan. Do not drive the seal cutter into the oil pump or rear seal retainer portion or the aluminum mating surface will be deformed.
9. Lower the oil pan from the cylinder block and remove it from the front side of the engine.

To install:

10. To install, carefully scrape the old gasket material away from the pan and cylinder block mounting surfaces and then apply a continuous bead (3.5–4.5mm) of liquid gasket around the oil pan to the 4 corners of the cylinder block mounting surface. Wait 5 minutes and then install the pan.
11. Install the oil pan and tighten the mounting bolts from the inside, out, to 3.6–5.1 ft. lbs. (5–7 Nm). Wait 30 minutes before refilling the crankcase to allow for the sealant to cure properly.

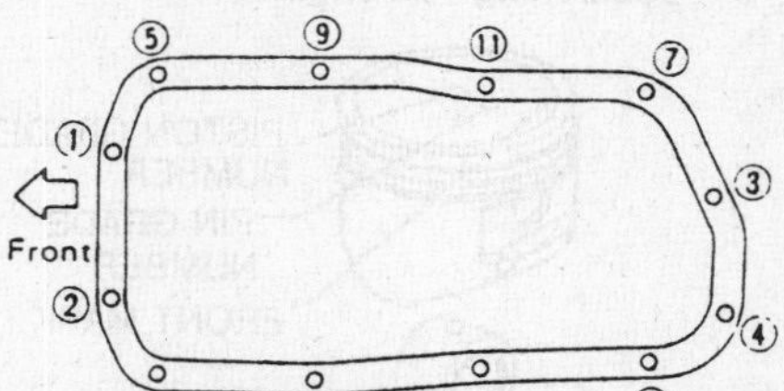

Oil pan bolt tightening sequence on 240SX. Loosen bolts in reverse order

12. Connect the front stabilizer to the side bar.
13. Lower the vehicle.
14. Fill the crankcase to the proper level.
15. Connect the negative battery cable. Start the engine and check for leaks.

1988–89 300ZX

1. Disconnect the negative battery cable.
2. Raise the vehicle and support safely.
3. Drain the oil pan.
4. Separate the front stabilizer bar from the suspension crossmember.
5. Remove the steering column shaft from the gear housing.
6. Separate the tension rod retaining nuts from the transverse link.
7. Raise and support the engine.
8. Remove the rear plate cover from the transmission case.
9. Remove the oil pan retaining bolts in the proper sequence.
10. Remove the suspension crossmember retaining bolts.
11. Remove the strut mounting insulator retaining nuts.
12. Remove the screws retaining the refrigerant lines and power steering tubes to the suspension crossmember.
13. Lower the suspension crossmember.
14. Insert a seal cutter between the oil pan and the cylinder block.
15. Tapping the cutter with a hammer, slide it around the entire edge of the oil pan. Do not drive the seal cutter into the oil pump or rear seal retainer portion or the aluminum mating surface will be deformed.
16. Lower the oil pan from the cylinder block and remove it from the front rear of the engine.

To install:

17. Carefully scrape the old gasket material away from the pan and cylinder block mounting surfaces and then apply a continuous bead (3.5–4.5mm) of liquid gasket around the oil pan and to the 4 corners of the cylinder block mounting surface. Wait 5 minutes and then install the pan.
18. Install tighten the pan mounting bolts from the inside, out, to 5–6 ft. lbs. (7–8 Nm). Wait 30 minutes before refilling the crankcase to allow for the sealant to cure properly.
19. Raise the crossmember from the lowered postion and attach the power steering and refrigerant lines.
20. Install and tighten the strut mounting insulator nuts.
21. Install the crossmember bolts.
22. Install the rear cover plate to the transmission case.
23. Lower the engine.
24. Connect the tension rod to the transverse link.
25. Connect the steering column shaft to the gear housing.
26. Connnect the front stabilizer bar to the crossmember.
27. Lower the vehicle.
28. Fill the crankcase to the proper level.
29. Connect the negative battery ca-

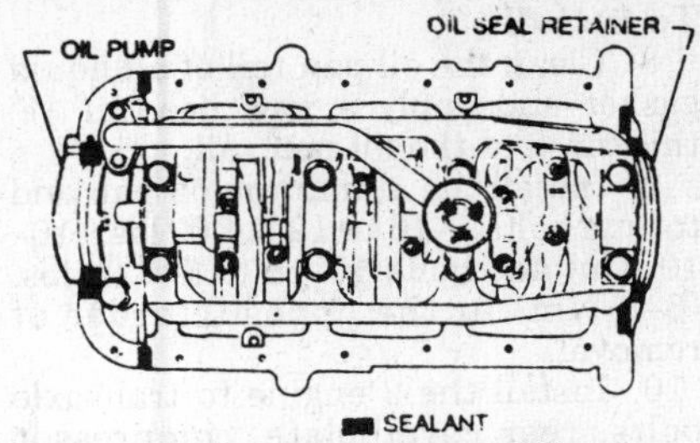

Apply sealant to these areas before installing the oil pan gasket—V6 engine

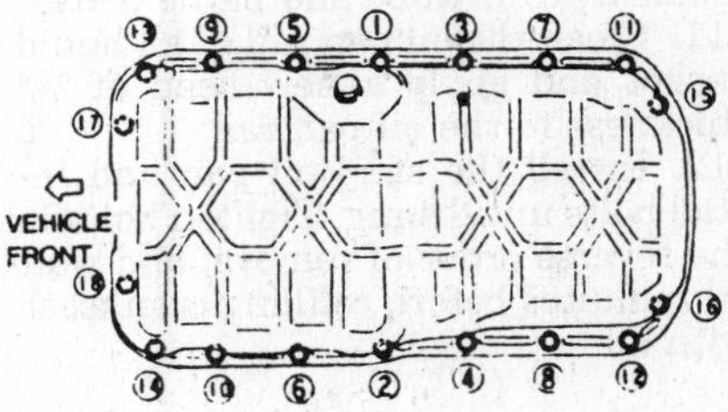

Oil pan bolt tightening sequence on Maxima and 1988–89 300ZX. Loosen bolts in reverse order

ble. Start the engine and check for leaks.

1990–92 300ZX

1. Disconnect the negative battery cable.
2. Raise the front of the vehicle and support safely.
3. Remove the engine undercover.
4. Drain the oil pan.
5. Remove the oil filter and bracket.
6. Remove the rear engine gussets from both sides.
7. Disconnect the air conditioning tube clamps.
8. Disconnect the lower steering column joint from the steering rack.
9. Remove the tension rod and transverse link bolts from both sides.
10. Support the suspension member with a suitable transmission jack. Install engine lifting slingers, connect a lifting device to the slingers and lift the engine.
11. Remove the suspension member bolts and lower the suspension member.
12. Remove the engine mounting bolts from both sides and slowly lower the transmission jack.
13. Remove the oil pan bolts in the proper sequence.
14. Insert a seal cutter between the oil pan and the cylinder block.
15. Tapping the cutter with a hammer, slide it around the entire edge of the oil pan. Do not drive the seal cutter into the oil pump or rear seal retainer portion or the aluminum mating surface will be deformed.
16. Lower the oil pan from the cylinder block and remove it.

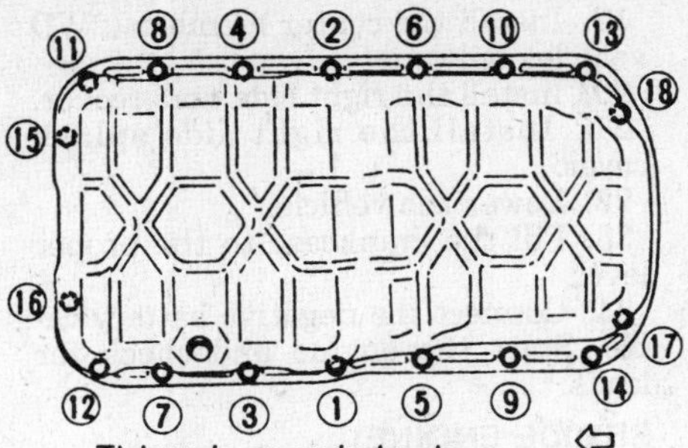

Oil pan bolt tightening sequence on 1990–92 300ZX. Loosen bolts in reverse order

To install:

17. Carefully scrape the old gasket material away from the pan and cylinder block mounting surfaces and then apply a continuous bead (3.5–4.5mm) of liquid gasket around the oil pan and to the 4 corners of the cylinder block mounting surface. Wait 5 minutes and then install the pan.
18. Install tighten the pan mounting bolts from the inside, out, to 4–6 ft. lbs. (6–8 Nm). Wait 30 minutes before refilling the crankcase to allow for the sealant to cure properly.
19. Slowly raise the transmission jack, then install the engine mounting bolts.
20. Raise the suspension member and install the mounting bolts.
21. Lower the engine, disconnect the lifting device and remove the lifting slingers.
22. Install the tension rod and transverse link bolts.
23. Connect the lower steering column joint to the steering rack.
24. Install the air conditioning tube clamps.
25. Install the rear engine gussets.
26. Install the oil filter bracket with a new oil filter.
27. Install the engine undercover.
28. Lower the vehicle.
29. Fill the crankcase to the proper level.
30. Connect the negative battery cable. Start the engine and check for leaks.

Maxima

1. Disconnect the negative battery cable.
2. Raise the front of the vehicle and support safely.
3. Drain the oil pan.
4. Remove the engine lower covers.
5. Using a suitable jack and block of wood, support the engine in the area of the crank pulley.
6. Remove the engine mounting insulator fasteners.
7. Remove the center crossmember.
8. Remove the oil pan bolts in the proper sequence.
9. Insert a seal cutter between the oil pan and the cylinder block.
10. Tapping the cutter with a hammer, slide it around the entire edge of the oil pan. Do not drive the seal cutter into the oil pump or rear seal retainer portion or the aluminum mating surface will be deformed.
11. Lower the oil pan from the cylinder block and remove it.

To install:

12. Carefully scrape the old gasket material away from the pan and cylinder block mounting surfaces and then apply a thin continuous bead of liquid gasket around the oil pan and to the 4 corners of the cylinder block mounting surface. Do the same to the oil pan gasket; both upper and lower surfaces. Wait 5 minutes and then install the pan. Wait 30 minutes before refilling the crankcase to allow the sealant to cure properly.
13. Install the oil pan and tighten the mounting bolts from the inside, out, to 5–6 ft. lbs. (7–8 Nm) in the proper sequence.
14. Install the center crossmember assembly.
15. Install the engine mount insulator fasteners.
16. Lower the engine.
17. Connnect the front exhaust pipe.
18. Install the engine lower covers.
19. Fill the crankcase to the proper level.
20. Connect the negative battery cable. Start the engine and check for leaks.

1988–89 Stanza

1. Disconect the negative battery cable.
2. Drain the oil pan.
3. Raise and support the front of the vehicle safely.
4. Remove the front exhaust tube section and the center crossmember.
5. Remove the oil pan bolts.
6. Insert a seal cutter between the oil pan and the cylinder block.
7. Tapping the cutter with a hammer, slide it around the entire edge of the oil pan. Do not drive the seal cutter into the oil pump or rear seal retainer portion or the aluminum mating surface will be deformed.
8. Lower the oil pan from the cylinder block and remove it.

To install:

9. Carefully scrape the old gasket material away from the pan and cylinder block mounting surfaces and then apply a thin continuous bead of liquid gasket around the oil pan and to the 4 corners of the cylinder block mounting surface. Do the same to the oil pan gas-

ket; both upper and lower surfaces. Wait 5 minutes and then install the pan. Wait 30 minutes before refilling the crankcase to allow the sealant to cure properly.

10. Install the oil pan and tighten the mounting bolts from the inside, out, to 4–5 ft. lbs. (5–7 Nm).
11. Install the center crossmember and front exhaust tube section.
12. Lower the vehicle.
13. Fill the crankcase to the proper level.
14. Connect the negative battery cable. Start the engine and check for leaks.

Pulsar, Sentra and 1990–92 Stanza

(EXCEPT SR20DE ENGINE)

1. Disconnect the negative battery cable.
2. Raise the vehicle and support safely.
3. Drain the oil pan.
4. Remove the right side splash cover.
5. Remove the right side undercover.
6. Remove the center member (2WD vehicles only).
7. Remove the forward section of the exhaust pipe.
8. Remove the front buffer rod and its bracket (1988 vehicles only).
9. Remove the engine gussets (1988 vehicles only).
10. Remove the oil pan bolts.
11. Insert a seal cutter between the oil pan and the cylinder block.
12. Tapping the cutter with a hammer, slide it around the entire edge of the oil pan. Do not drive the seal cutter into the oil pump or rear seal retainer portion or the aluminum mating surface will be deformed.
13. Lower the oil pan from the cylinder block and remove it.

To install:

14. Carefully scrape the old gasket material away from the pan and cylinder block mounting surfaces and then apply a thin continuous bead of liquid gasket around the oil pan and to the 4 corners of the cylinder block mounting surface. Do the same to the oil pan gasket; both upper and lower surfaces. Wait 5 minutes and then install the pan. Wait 30 minutes before refilling the crankcase to allow the sealant to cure properly.
15. Install the oil pan and tighten the mounting bolts from the inside, out, to 5–6 ft. lbs. (7–8 Nm).
16. Install the engine gussets (1988 vehicles only).
17. Install the front buffer rod and its bracket (1988 vehicles only).
18. Install the forward section of the exhaust pipe using new gaskets.
19. Install the center member (2WD vehicles only).
20. Install the right side undercover.
21. Install the right side splash cover.
22. Lower the vehicle.
23. Fill the crankcase to the proper level.
24. Connect the negative battery cable. Start the engine and check for leaks.

SR20DE ENGINE)

1. Raise and support the vehicle safely. Remove the engine under cover and drain the oil.
2. Remove the steel oil pan bolts in the proper sequence. Remove the steel oil pan. Insert tool KV10111100 or equivalent between steel oil pan and aluminum oil pan to pry apart.
3. Remove the oil baffle bolts and oil baffle. Remove the front tube.
4. Set a suitable jack under the transaxle and raise the engine with and engine hoist.
5. If equipped with an automatic transaxle, remove the transaxle shift control cable.
6. Remove the compressor gussets, the rear cover plate and all aluminum oil pan bolts. Loosen aluminum oil pan bolts in the proper sequence.
7. Remove the 2 engine-to-transaxle bolts and refit the them into vacant holes at the bottom of the oil pan. Remove the aluminum oil pan. Use tool KV10111100 or equivalent to pry oil pan from block. Remove the engine to transaxle bolts.

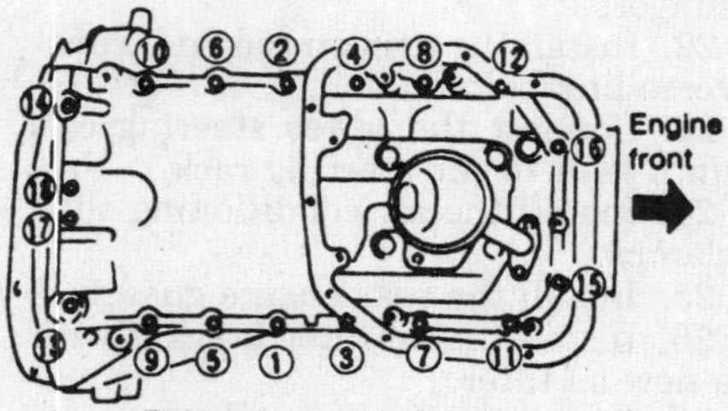

Tighten in numerical order.

Aluminum oil pan installation torque sequence—SR20DE engine

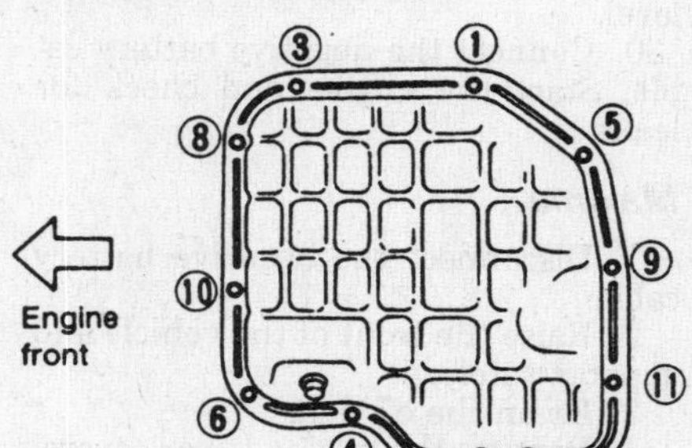

Tighten in numerical order.

Steel oil pan installation torque sequence—SR20DE engine

To install:

8. Clean the oil pan rail of all liquid gasket and apply a new bead of 1/8" thickness to the oil pan rail.
9. Install the aluminum oil pan and torque bolts 1–16 to 12–14 ft. lbs. (16–19 Nm) and bolts 17–18 to 5–6 ft. lbs. (6–8 Nm) in the opposite order of removal.
10. Install the 2 engine to transaxle bolts, rear cover plate, compressor gussets, automatic transmission shift control cable (if equipped), center member, front tube and baffle plate.
11. Clean the oil pan rail of all liquid gasket and apply a new bead of 1/8" thickness to the oil pan rail.
12. Install the steel oil pan and install bolts untill snug. Tighten bolts in the reverse order of removal and wait 30 minutes before refilling crankcase with oil.

Oil Pump

REMOVAL & INSTALLATION

CA18DE and CA20E Engines

1. Disconnect the negative battery cable.
2. Drain the oil pan.
3. Remove all accessory drive belts.
4. Remove the alternator.
5. Remove the timing belt covers.
6. Remove the timing belt.
7. On 200SX and the Stanza Wagon, unbolt the engine from its mounts and lift or jack the engine up from the unibody. On the Stanza (except Wagon) and Pulsar, remove the center member from the body.
8. Remove the oil pan.
9. Remove the oil pump assembly along with the oil strainer. Remove the O-ring from the oil pump body and replace it.
10. Replace the front seal.

To install:

11. If installing a new or rebuilt oil pump, first pack the pump full of petroleum jelly to prevent the pump from cavitating when the engine is started. Apply RTV sealer to the front oil seal end of the pan prior to installation.
12. Install the pump and torque the oil pump mounting bolts to 8–12 ft. lbs. (12–16 Nm). Make sure the oil pump body O-ring is properly seated.
13. Install the oil pan.
14. On the Stanza (except Wagon) and Pulsar, install the center member. On 200SX and the Stanza Wagon, lower and re-mount the engine.
15. Install the timing belt.
16. Install the timing belt covers.
17. Install the alternator.
18. Install and tension the drive belts.

19. Fill the crankcase to the proper level.
20. Connect the negative battery cable. Start the engine and check for leaks.

E16i Engine

1. Disconnect the negative battery cable.
2. Drain the oil pan.
3. Remove all accessory drive belts.
4. Remove the alternator.
5. Disconnect the oil pressure gauge harness.
6. Remove the oil filter.
7. Remove the oil pump and gasket.

To install:

8. If installing a new or rebuilt oil pump, first pack the pump full of petroleum jelly to prevent the pump from cavitating when the engine is started.
9. Mount the pump on the engine using a new gasket. Torque the pump mounting bolts to 7–9 ft. lbs. (10–12 Nm).
10. Install a new oil filter.
11. Connect the oil pressure gauge harness.
12. Install the alternator.
13. Install and tension the drive belts.
14. Fill the crankcase to the proper level.
15. Connect the negative battery cable. Start the engine and check for leaks.

GA16i Engine and 1990–92 Stanza Engine (KA24E)

The oil pump used on the GA16i engine consists of an inner and outer gear located in the front cover. Removal of the front cover is necessary to gain access to the oil pump.

1. Disconnect the negative battery cable.
2. Remove the front cover with the strainer tube.
3. Loosen the oil pump cover retaining screw and mounting bolts and separate the oil pump cover from the front cover.
4. Remove the oil pump inner and outer gears.

To install:

5. Thoroughly clean the oil pump cover mating surfaces and the gear cavity.
6. Install the outer gear into the cavity.
7. Install the inner gear so the grooved side is facing up (towards the oil pump cover). Make sure the gears mesh properly and pack the pump cavity with petroleum jelly.
8. Install the oil pump cover. On GA16i engines, torque the retaining screws to 2.2–3.6 ft. lbs. (3–5 Nm) and the bolts to 3.6–5.1 ft. lbs. (5–7 Nm).

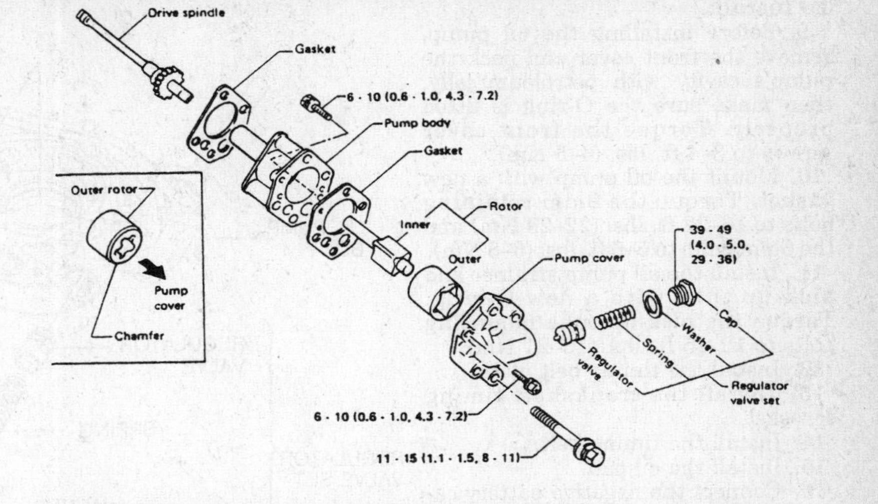

Align the punch mark on the drive spindle with the oil hole—240SX

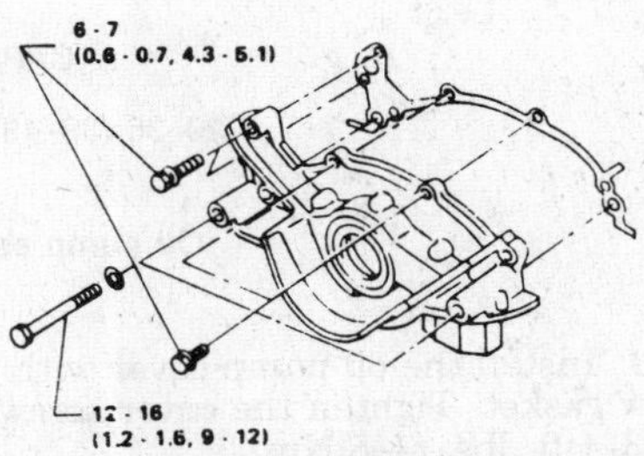

Oil pump installation—V6 engine

On KA24E engines, torque the cover screws to 2.2–3.6 ft. lbs. (3–5 Nm) and the bolts to 12–15 ft. lbs. (16–21 Nm).

9. Install the front cover with a new seal.
10. Connect the negative battery cable. Start the engine and check for leaks.

GA16DE, KA24DE and SR20DE Engines

1. Remove the drive belts.
2. Remove the cylinder head and oil pans.
3. Remove the oil strainer and baffle plate.
4. Remove the front cover assembly. Remove the oil pump.

To install:

5. Clean the mating surfaces of liquid gasket and apply a fresh bead of ⅛ in. thickness.
6. Coat the oil pump gears with oil. Using a new oil seal and O-ring, install the front cover assembly.
7. Install the oil strainer, baffle plate, oil pans, cylinder head and drive belts.

KA24E Engine (240SX)

1. Disconnect the negative battery cable.
2. Drain the oil pan.
3. Turn the crankshaft so No. 1 piston is at TDC on its compression stroke.
4. Remove the distributor cap and mark the position of the distributor rotor in relation to the distributor base with a piece of chalk.
5. Remove the splash shield.
6. Remove the oil pump body with the drive spindle assembly.

To install:

7. To install, fill the pump housing with engine oil, align the punch mark on the spindle with the hole in the pump. No. 1 piston should be at TDC on its compression stroke.
8. With a new gasket and seal placed over the drive spindle, install the oil pump and drive spindle assembly. Make sure the tip of the drive spindle fits into the distributor shaft notch securely. The distributor rotor should be pointing to the matchmark made earlier.
9. Install the splash shield.
10. Install the distributor cap.
11. Fill the crankcase to the proper level.
12. Connect the negative battery cable. Start the engine and check for leaks. Check the ignition timing.

VG30E, VG30ET, VG30DE and VG30DETT Engines

1. Disconnect the negative battery cable.
2. Remove the oil pan.
3. Remove the timing belt.
4. Remove the crankshaft timing sprocket using a suitable puller.
5. Remove the timing belt plate.
6. Remove the oil pump strainer and pick-up tube from the oil pump.
7. Remove the mounting bolts and remove the oil pump and gasket.
8. Replace the oil pump seal.

To install:

9. Before installing the oil pump, remove the front cover and pack the pump's cavity with petroleum jelly, then make sure the O-ring is fitted properly. Torque the front cover screws to 3–4 ft. lbs. (4–5 Nm).
10. Mount the oil pump with a new gasket. Torque the 8mm retaining bolts to 16–22 ft. lbs. (22–29 Nm) and the 6mm bolts to 5–6 ft. lbs. (6–8 Nm).
11. Install the oil pump strainer and pick-up tube with a new O-ring. Torque the pick-up tube mounting bolts to 12–15 ft. lbs. (16–21 Nm).
12. Install the timing belt plate.
13. Install the crankshaft timing sprocket.
14. Install the timing belt.
15. Install the oil pan.
16. Connect the negative battery cable. Start the engine and check for leaks.

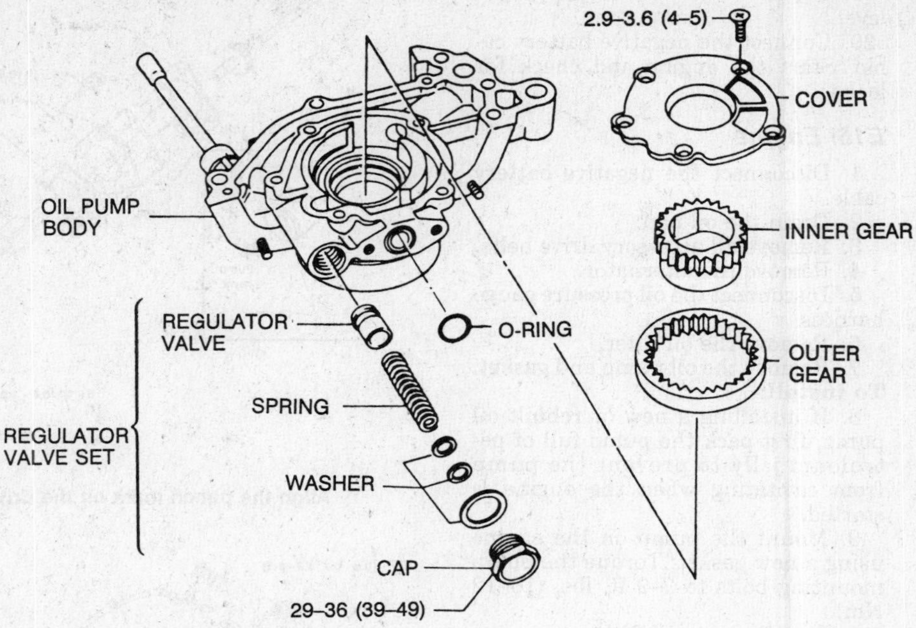

Oil pump assembly – C-Series engines

CHECKING

CA18DE, CA20E, VG30E, VG30ET, VG30DE and VG30DETT Engines

1. Remove the oil pump cover and gasket.
2. Disassemble the regulator valve components. Visually inspect all parts for wear and damage.
3. Make sure the regulator valve moves smoothly in the valve bore. Make sure the valve spring is sturdy.
4. Coat the regulator valve with clean engine oil and check that it falls into the valve bore by its own weight. Assemble the regulator valve components. Torque the valve cap to 29–36 ft. lbs. (39–49 Nm).
5. Inspect the oil pressure relief valve for movement, cracks and damage by pushing the ball in. If necessary, install a new valve by prying the old valve out and tapping the new valve in place.
6. Check the body-to-outer gear clearance. It should be 0.0043–0.0079 in. (0.11–0.20mm).
7. Check the inner gear-to-crescent clearance. It should be 0.0047–0.0091 in. (0.12–0.23mm).
8. Check the outer gear-to-crescent clearance. It should be 0.0083–0.0126 in. (0.21–0.32mm).
9. Check the body (housing)-to-inner gear clearance. It should be 0.0020–0.0035 in. (0.05–0.09mm).
10. Check the body (housing)-to-outer gear clearance. It should be 0.0020–0.0043 in. (0.05–0.11mm).
11. If any of the clearances exceed the specified limits, replace the gear set or the entire oil pump assembly.
12. Coat the inner and outer gears with clean engine oil prior to installation.
13. Install the oil pump cover with a new gasket. Tighten the cover screws to 3–4 ft. lbs. (4–5 Nm).

E16i and KA24E (240SX) Engines

NOTE: Do not disassemble the inner rotor and drive gear.

1. Disassemble the regulator valve components. Visually inspect all parts for wear and damage.
2. Make sure the regulator valve moves smoothly in the valve bore. Make sure the valve spring is sturdy.
3. Coat the regulator valve with clean engine oil and check that it falls into the valve bore by its own weight. Assemble the regulator valve components. Torque the valve cap to 29–36 ft. lbs. (39–49 Nm).
4. Inspect the oil pressure relief valve for movement, cracks and damage by pushing the ball in. If necessary, install a new valve by prying the old valve out and tapping the new valve in place.
5. Check the rotor tip clearance. It should be less than 0.0047 in. (0.12mm).
6. Check the outer rotor-to-body clearance. It should be 0.0059–0.0083 in. (0.15–0.21mm).
7. With the pump body gasket installed, check the side clearance. It should be 0.0020–0.0047 in. (0.05–0.12mm) for E16i engine; and 0.0016–0.0031 in. (0.04–0.08mm) on KA24E engines.
8. If any of the clearances exceed the specified limits, replace the gear set or the entire oil pump assembly.
9. Coat the inner and outer gears with clean engine oil prior to installation.

GA16i and KA24E (1990–92 Stanza) Engines

1. Remove the oil pump cover and gasket.
2. Disassemble the regulator valve components. Visually inspect all parts for wear and damage.
3. Make sure the regulator valve moves smoothly in the valve bore. Make sure the valve spring is sturdy.
4. Coat the regulator valve with clean engine oil and check that it falls into the valve bore by its own weight. Assemble the regulator valve components. Torque the valve cap to 29–36 ft. lbs. (39–49 Nm).
5. Inspect the oil pressure relief valve for movement, cracks and damage by pushing the ball in. If necessary, install a new valve by prying the old valve out and tapping the new valve in place.
6. Check the body-to-outer gear clearance. It should be 0.0043–0.0079 in. (0.11–0.20mm).
7. Check the inner gear-to-crescent clearance. It should be 0.0085–0.0129 in. (0.217–0.327mm).
8. Check the outer gear-to-crescent clearnce. It should be 0.0083–0.0126 in. (0.21–0.32mm).
9. Check the body-to-inner gear

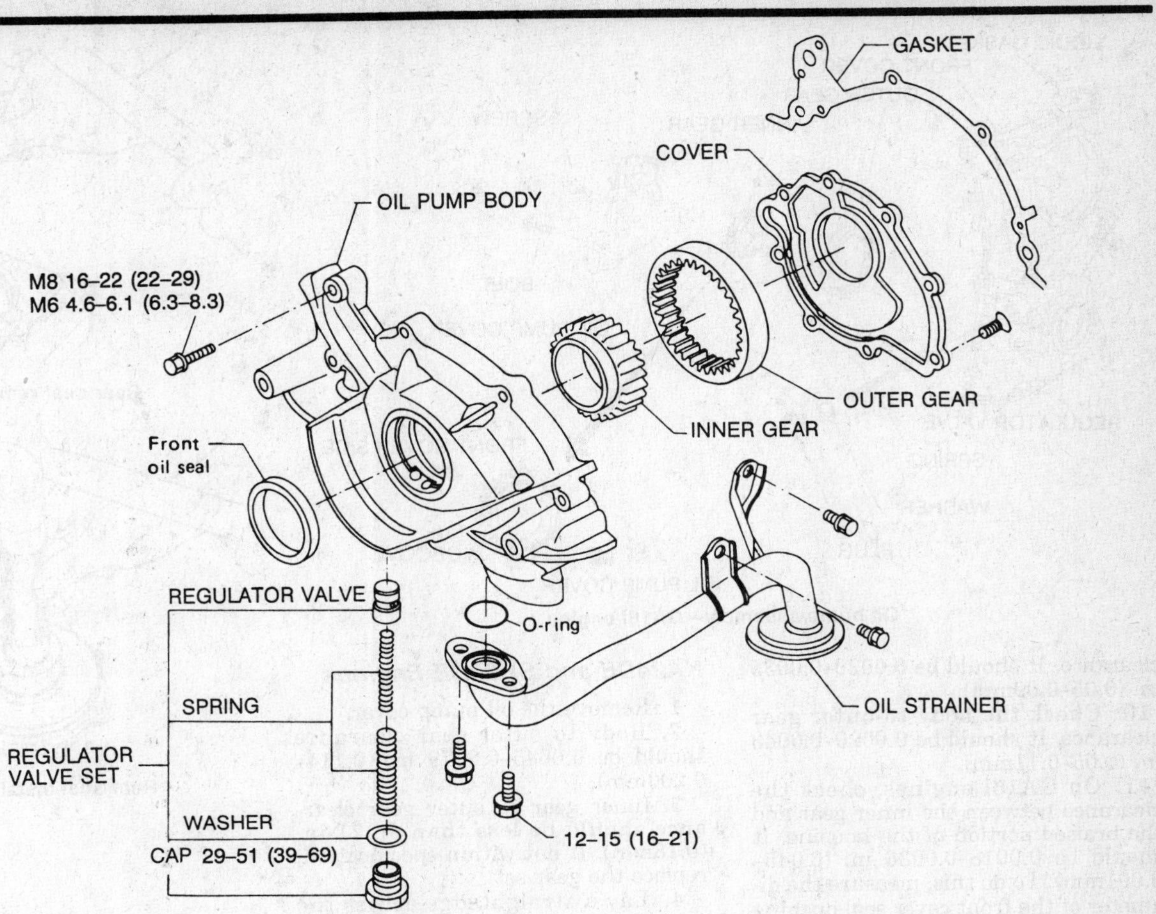

Oil pump assembly—V6 engines (typical)

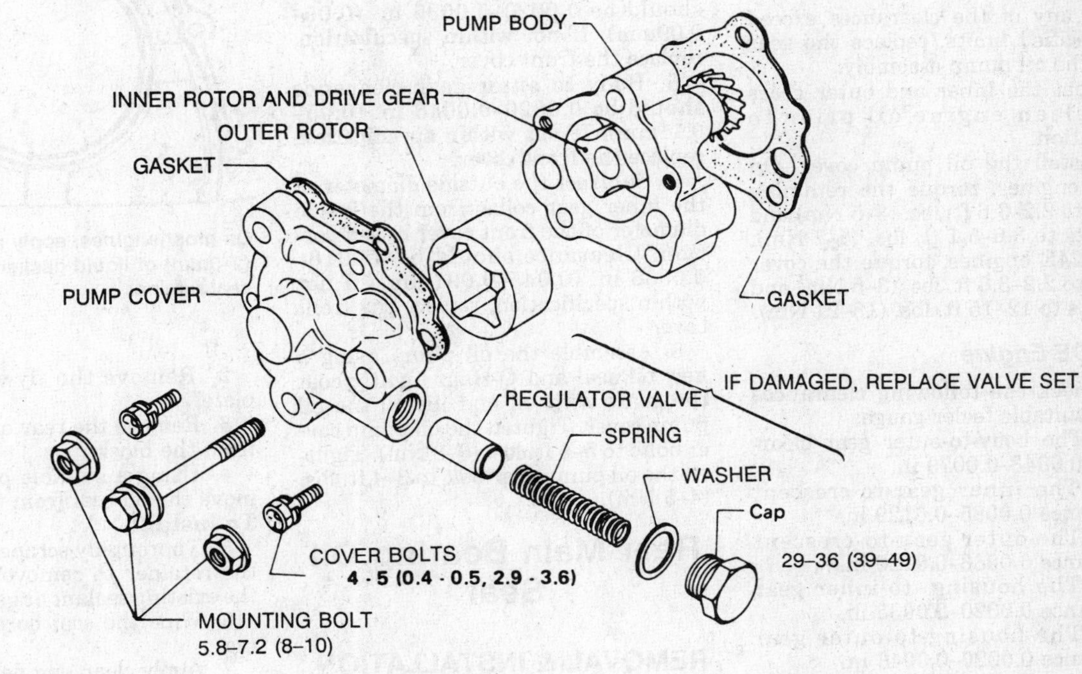

Oil pump assembly—E16i engine

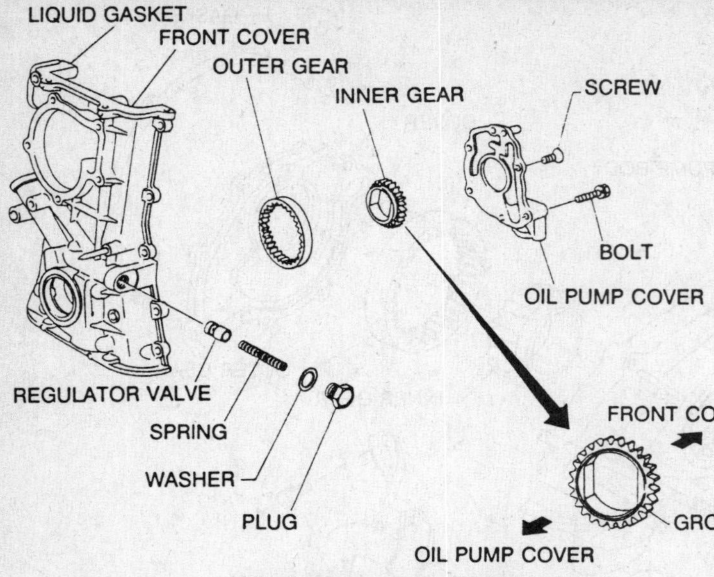

Oil pump assembly—GA16i engine

clearance. It should be 0.0020–0.0035 in. (0.05–0.09mm).

10. Check the body-to-outer gear clearance. It should be 0.0020–0.0043 in. (0.05–0.11mm).

11. On GA16i engines, check the clearance between the inner gear and the braised portion of the housing. It should be 0.0018–0.0036 in. (0.045–0.091mm). To do this, measure the diameter of the front cover seal opening with an inside micrometer, then measure the diameter of the inner gear race. Subtract the 2 readings to obtain the clearance.

12. If any of the clearances exceed the specified limits, replace the gear set or the oil pump assembly.

13. Coat the inner and outer gears with clean engine oil prior to installation.

14. Install the oil pump cover. On GA16i engines, torque the retaining screws to 2.2–3.6 ft. lbs. (3–5 Nm) and the bolts to 3.6–5.1 ft. lbs. (5–7 Nm). On KA24E engines, torque the cover screws to 2.2–3.6 ft. lbs. (3–5 Nm) and the bolts to 12–15 ft. lbs. (16–21 Nm).

GA16DE Engine

1. Check the following clearances with a suitable feeler gauge:
   a. The body-to-outer gear clearance 0.0043–0.0079 in.
   b. The inner gear-to-crescent clearance 0.0085–0.0129 in.
   c. The outer gear-to-crescent clearance 0.0083–0.0126 in.
   d. The housing -to-inner gear clearance 0.0020–0.0035 in.
   e. The housing-to-outer gear clearance 0.0020–0.0043 in.
2. If any clearance exceeds the limit, replace the gear set or the entire gear assembly.

KA24DE and SR20DE Engines

1. Remove the oil pump cover.
2. Body to outer gear clearance should be 0.0045–0.0079 in. (0.114–0.200mm).
3. Inner gear to outer gear clearance should be less than 0.071 in. (0.18mm). If not within specification, replace the gear set.
4. Lay a straightedge across the surface of the gears and check the gear to body clearance. If not within specification, replace the front cover.
5. Body to inner gear clearance should be 0.0020–0.0035 in. (0.05–0.09mm). If not within specification, replace the front cover.
6. Body to outer gear clearance should be 0.0020–0.0043 in. (0.05–0.11mm). If not within specification, replace the front cover.
7. Subtract the outside diameter of the inner gear collar from the inside diameter of the front cover raised portion. Clearance should be 0.0018–0.0036 in. (0.045–0.091mm). If not within specification, replace the front cover.
8. Assemble the oil pump using a new oil seal and O-ring. Lightly coat the gears with oil and install the oil pump cover. Tighten the oil pump cover bolts to 5–6 ft. lbs. (6–8 Nm). Tighten the oil pump gear bolt to 3–4 ft. lbs. (4–5 Nm).

Rear Main Bearing Oil Seal

REMOVAL & INSTALLATION

1. Remove the transmission or transaxle.

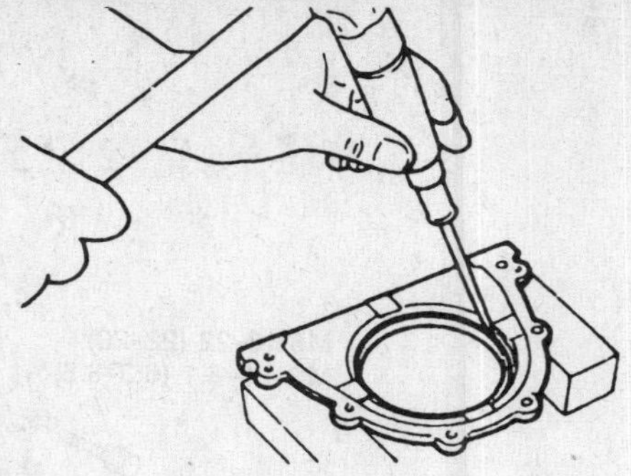
Rear seal removal

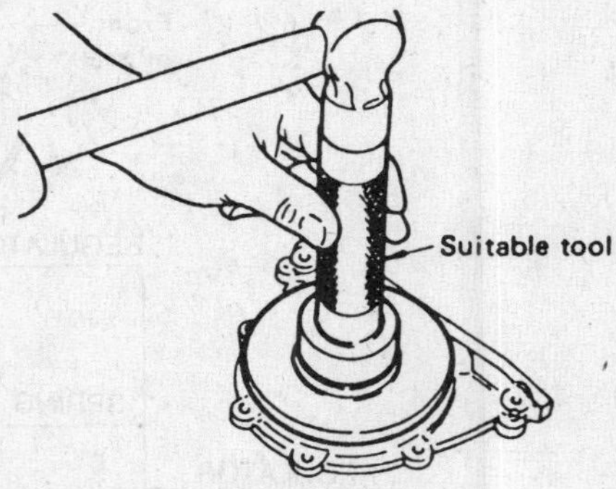

Rear seal installation

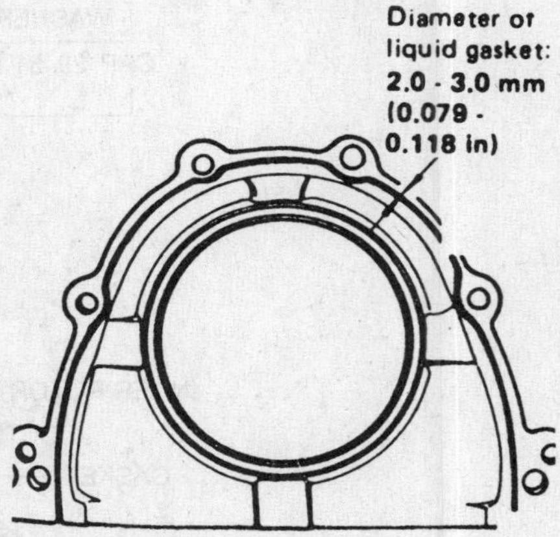

On most engines, apply a 0.08–0.12 in. (2–3mm) of liquid gasket to the rear oil seal retainer

2. Remove the flywheel or drive plate.
3. Remove the rear oil seal retainer from the block.
4. Using a suitable prying tool, remove the oil seal from the retainer.

To install:

5. Thoroughly scrape the surface of the retainer to remove any traces of the existing sealant or gasket material.
6. Wipe the seal bore with a clean rag.
7. Apply clean engine oil to the new oil seal and carefully install it into the retainer using the proper seal installation tool.

8. Install the rear oil seal retainer into the engine, along with a new gasket. Apply a 0.08–0.12 in. (2–3mm) of liquid gasket to the rear oil seal retainer prior to installation as necessary. Torque the bolts to 3–6 ft. lbs. (4–8 Nm).
9. Install the flywheel or driveplate.
10. Install the transmission or transaxle.

ENGINE COOLING

Radiator

REMOVAL & INSTALLATION

1. Disconnect the negative battery cable.
2. Drain the cooling system.
3. Remove the undercover, if equipped.
4. Disconnect the reservoir tank hose.
5. Disconnect all temperature switch connectors.
6. Remove the front bumper on the 1988 Maxima and 1988–89 300ZX.
7. Disconnect and plug the transmission or transaxle cooling lines from the bottom of the radiator, if equipped.
8. On rear wheel drive vehicles, remove the fan shroud and position the shroud over the fan and clear of the radiator. On front wheel drive vehicles, discharge the air conditioning system, then unbolt and remove the condenser and radiator fan assembly from the radiator if necessary.
9. Disconnect the upper and lower hoses from the radiator.
10. Remove the radiator retaining bolts or the upper supports.
11. Lift the radiator off the mounts and out of the vehicle.

To install:

12. Lower the radiator onto the mounts and bolt in place.
13. Install the lower shroud, if removed.
14. Connect the upper and lower radiator hoses.
15. On rear wheel drive vehicles, install the fan shroud. On front wheel drive vehicles, install the condenser and radiator fan assembly as necessary.
16. Connect the transaxle or transmission cooling lines, if removed.
17. On 1988 Maxima and 1988–89 300ZX, install the front bumper.
18. Plug in the temperature switch connectors.
19. Connect the reservoir tank hose.
20. Fill the cooling system to the proper level.
21. Connect the negative battery cable.
22. Start the engine and check for leaks.

Electric Cooling Fan

TESTING

Maxima

1. Warm up the engine to normal

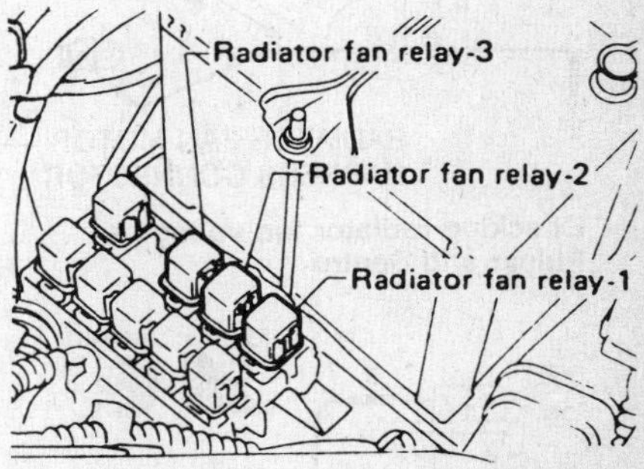

Radiator fan relay locations – Maxima

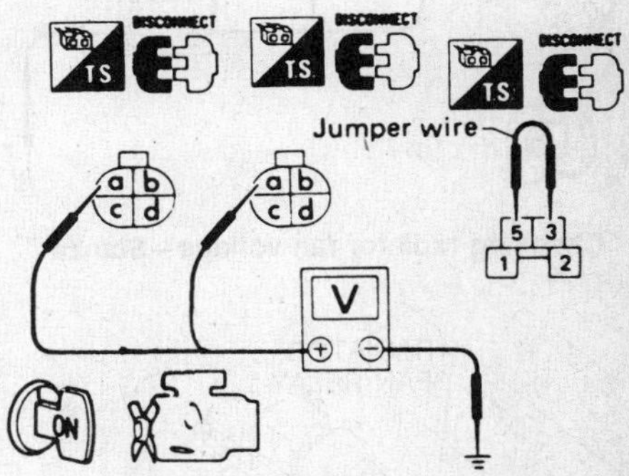

Checking radiator fan voltage – Maxima

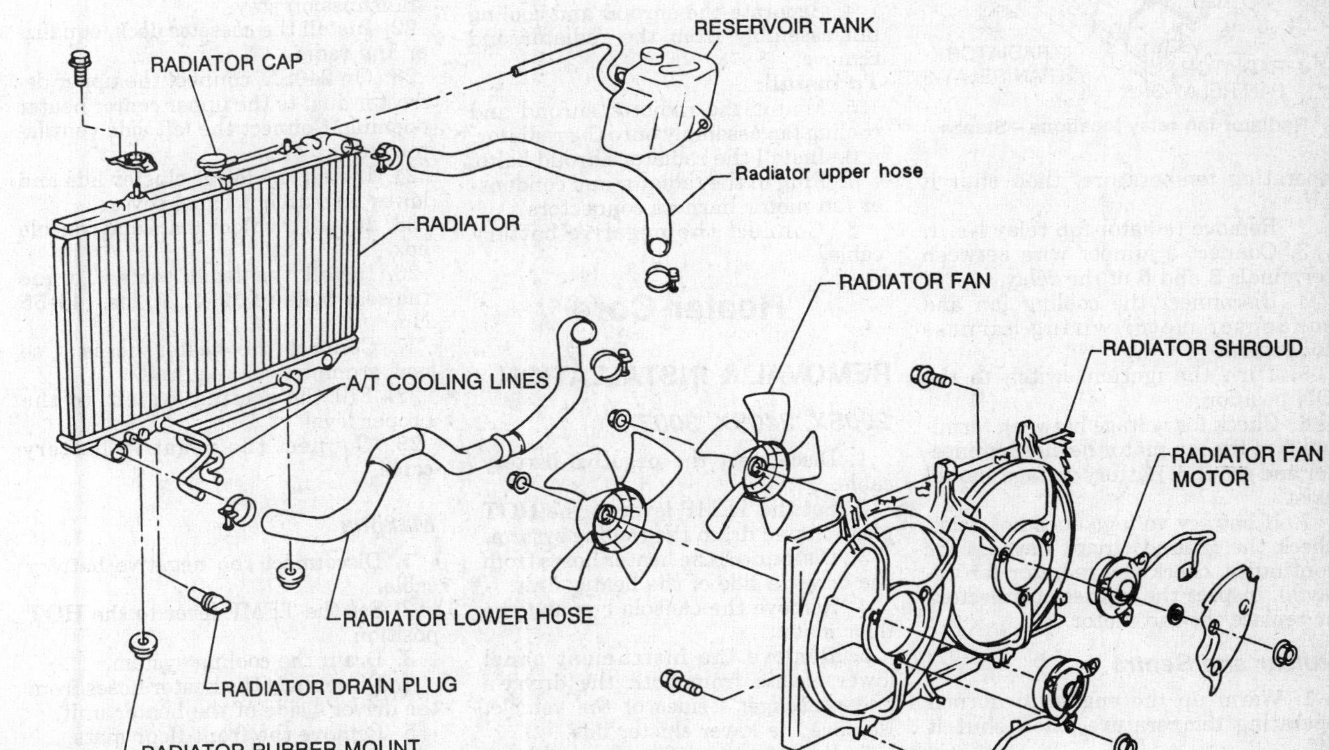

Typical radiator assembly

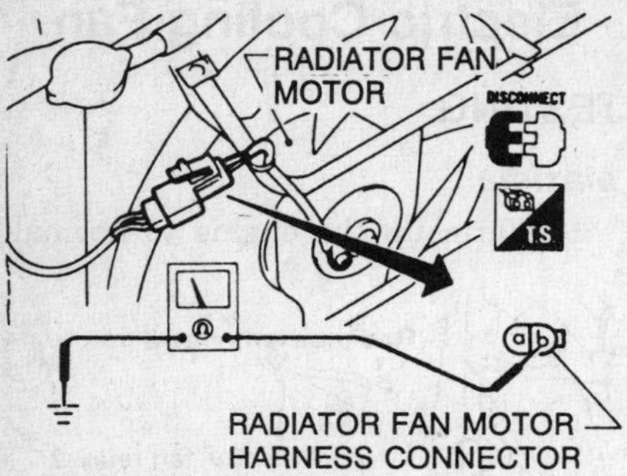

Checking radiator fan voltage – Pulsar and Sentra

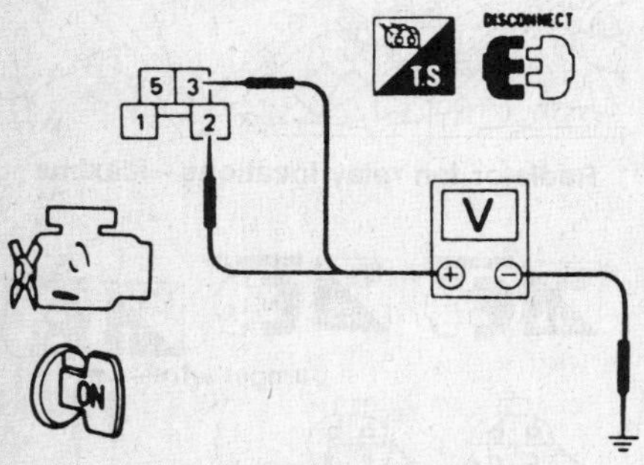

Checking radiator fan voltage – Stanza

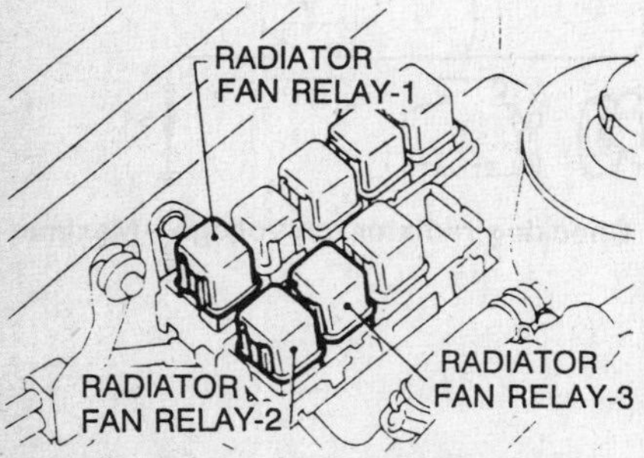

Radiator fan relay locations – Stanza

operating temperature, then shut it off.

2. Remove radiator fan relay No. 1.
3. Connect a jumper wire between terminals **3** and **5** of the relay.
4. Disconnect the cooling fan and condenser motor wiring harness connectors.
5. Turn the igntion switch to the **ON** position.
6. Check for voltage between terminal **A** of the fan motor harness connector and ground. Battery voltage should exist.
7. If battery voltage does not exist, check the ground circuit harness for continuity, check the radiator fan relay(s), inspect the harness connectors or replace the fan motor.

Pulsar and Sentra

1. Warm up the engine to normal operating temperature, then shut it off.
2. Disconnect the cooling fan motor wiring harness connector.
3. Turn the ignition switch to the **ON** position.
4. Check for voltage beteen terminal **A** of the fan motor harness connector and ground. Battery voltage should exist.
5. If battery voltage does not exist, check the ground circuit harness for continuity, check the radiator fan relays, or replace the fan motor.

Stanza

1. Warm up the engine to normal operating temperature, then shut it off.
2. Remove radiator fan relays No. 1 and 2.
3. Turn the ignition switch to the **ON** position.
4. Check for voltage beteen terminals **2** and **3** of the relays and ground. Battery voltage should exist.
5. If battery voltage does not exist, check the ground circuit harness for continuity, check the radiator fan relay(s), inspect the harness connectors or replace the fan motor.

REMOVAL & INSTALLATION

1. Disconnect the negative battery cable.
2. Unplug the condenser and radiator fan motor wiring harness connectors.
3. Remove the radiator shroud bolts.
4. Separate the shroud and cooling fan assembly from the radiator and remove.

To install:

5. Mount the radiator shroud and cooling fan assembly onto the radiator.
6. Install the radiator shroud bolts.
7. Plug in the radiator and condenser fan motor harness connectors.
8. Connect the negative battery cable.

Heater Core

REMOVAL & INSTALLATION

200SX, 240SX, 300ZX

1. Disconnect the negative battery cable.
2. Set the TEMP lever to the **HOT** position and drain the cooling system.
3. Disconnect the heater hoses from the driver's side of the heater unit.
4. Remove the console box and the floor mats.
5. Remove the instrument panel lower covers from both the driver's and passenger's sides of the vehicle. Remove the lower cluster lids.
6. Remove the left side ventilator duct. On 240SX, detach the defroster duct from the upper center heater unit opening.
7. Remove the radio, equalizer and stereo cassette deck as required.
8. Remove the instrument panel-to-transmission tunnel stay.
9. Remove the rear heater duct from the floor of the vehicle.
10. Remove the center ventilator duct.
11. Remove the left and right side ventilator ducts from the lower heater outlets.
12. Disconnect and label the wiring harness connections.
13. Separate the heating unit Remove the 2 screws at the bottom sides of the heater unit and the 1 screw at the top of the unit and remove the unit together with the heater control assembly.
14. Separate the heater case halves and slide the core from the case.

To install:

15. Install the heater core and assemble the heater case halves. Use new gaskets and seals as required.
16. Mount the heater unit/control assembly and install the upper and lower attaching screws.
17. Plug in the wiring harness connectors.
18. Connect the left and right side ducts to the lower heater outlets.
19. Connect the center ventilator duct.
20. Connect the rear heater duct.
21. Attach the instrument panel-to-transmission stay.
22. Install the cassette deck, equalizer and radio.
23. On 240SX, connect the upper defroster duct to the upper center heater opening. Connect the left side ventilator duct.
24. Install the lower cluster lids and lower instrument panel covers.
25. Install the floor mats and console box.
26. Install the front seats. Torque the seat bolts to 32–41 ft. lbs. (43–55 Nm).
27. Connect the heater hoses. Use new grommets as required.
28. Fill the cooling system to the proper level.
29. Connect the negative battery cable.

Maxima

1. Disconnect the negative battery cable.
2. Set the TEMP lever to the **HOT** position.
3. Drain the cooling system.
4. Disconnect the heater hoses from the driver's side of the heater unit.
5. Remove the front floor mats.
6. Remove the instrument panel

lower covers from both the driver's and passenger's sides of the vehicle.
7. Remove the left side ventilator duct.
8. Remove the instrument panel.
9. Remove the rear heater duct from the floor of the vehicle.
10. Disconnect the wiring harness connectors.
11. Separate the heating unit from the cooling unit. Remove the 2 screws at the bottom sides of the heater unit and the 1 screw from the top of the unit. Lift out the heater together with the heater control assembly.
12. Remove the center vent cover and heater control assembly, loosening the clips and screws.
13. Remove the screws securing the door shafts.
14. Remove the clips from the case and split the case. Remove the core.
15. Separate the heater case halves and slide the core from the case.

To install:

16. Install the heater core and assemble the heater case halves. Use new gaskets and seals as required.
17. Install the door shaft retaining screws.
18. Install the heater control assembly and center vent cover.
19. Mount the heater unit/control assembly and install the upper and lower attaching screws.
20. Plug in the wiring harness connectors.
21. Install the rear heater duct.
22. Install the instrument panel.
23. Install the left side ventilator duct.
24. Install the instrument panel lower covers.
25. Install the floor mats.
26. Connect the heater hoses. Use new grommets as required.
27. Fill the cooling system to the proper level.
28. Connect the negative battery cable.

Pulsar and Sentra

1. Disconnect the negative battery cable.
2. Set the TEMP lever to the maximum **HOT** position and drain the engine coolant.
3. Disconnect the heater hoses at the engine compartment.
4. Remove the instrument panel assembly.
5. Remove the heater control assembly.
6. If equipped with air conditioning, separate the heating unit from the cooling unit.
7. Remove the heater unit assembly.
8. Remove the case clips and split the case. Remove the core.

To install:

9. Install the heater core and assemble the heater case halves. Use new gaskets and seals as required. Always check the operation of the air mix door when re-attaching the heater case halves.
10. Mount the heater unit and connect it the cooling unit, if equipped.
11. Install the heater control assembly.
12. Install the instrument panel.
13. Connect the heater hoses. Use new grommets as required.
14. Fill and bleed the cooling system.
15. Connect the negative battery cable.

Stanza

1. Disconnect the negative battery cable.
2. Set the TEMP lever to the maximum HOT position and drain the engine coolant.
3. Disconnect the heater hoses at the engine compartment.
4. Remove the instrument panel assembly.
5. Remove the heater control assembly.
6. Remove pedal bracket mounting bolts, steering column mounting bolts, brake and clutch pedal cotter pins.
7. Move the pedal bracket and steering column to the left.
8. Disconnect the air mix door control cable and heater valve control lever, then remove the control lever.
9. Remove the core cover and remove the core.

To install:

10. Install the core and cover. Use new seals and gaskets as required.
11. Install the control and heater valve levers. Connect the air mix door control cable.
12. Move the steering column and brake pedal bracket to the right. Install the clutch and brake pedal cotter pins and steering column and brake pedal bolts.
13. Install the heater control assembly.
14. Install the instrument panel.
15. Connect the heater hoses to the core. Use new grommets as required.
16. Fill and bleed the cooling system.
17. Connect the negative battery cable.

Water Pump

REMOVAL & INSTALLATION

4 Cylinder Engine

1. Disconnect the negative battery cable.
2. Drain the coolant from the radiator and cylinder block.
3. Remove all the drive belts.
4. Unbolt the water pump pulley and the water pump attaching bolts.
5. Separate the water pump with the gasket, if installed, from the cylinder block.
6. Remove all gasket material or sealant from the water pump mating surfaces. All sealant must be removed from the groove in the water pump surface also.

To install:

7. Apply a continuous bead of high temperature liquid gasket to the water pump housing mating surface. The housing must be attached to the cylinder block within 5 minutes after the sealant is applied. After the pump housing is bolted to the block, wait at least 30 minutes for the sealant to cure before starting the engine.
8. Position the water pump (and gasket) onto the block and install the attaching bolts. Torque the small retaining bolts to about 5 ft. lbs. and large retaining bolts 12–14 ft. lbs.
9. Install the water pump pulley.
10. Install the drive belts and adjust the tension.

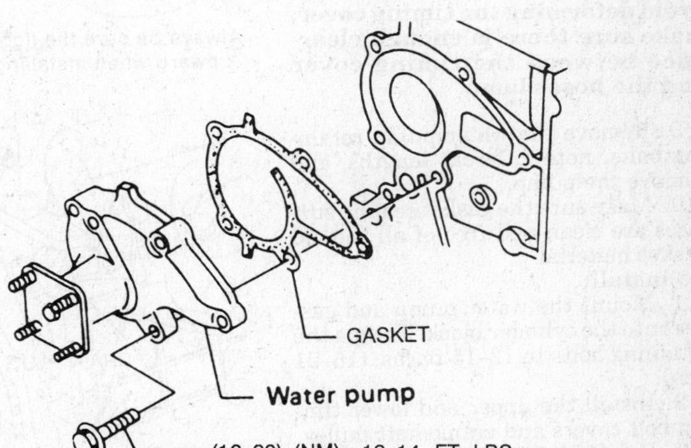

Water pump assembly—CA20E engine

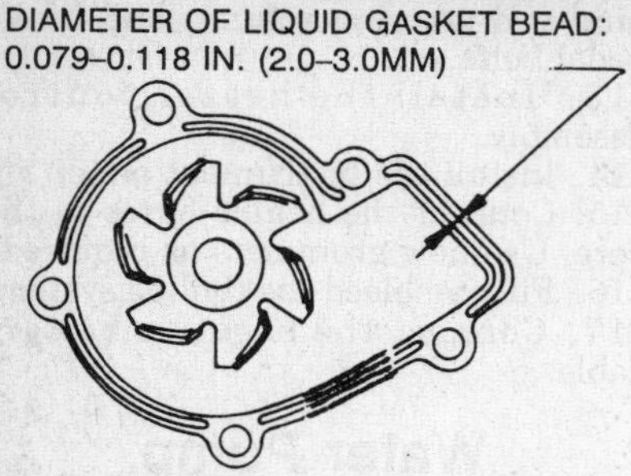

Apply a continuous bead oh high temperature sealant to the water pump housing mating surface

11. Fill the cooling system to the proper level.
12. Connect the negative battery cable.

6 Cylinder Engine

1. Disconnect the negative battery cable and drain the coolant from the radiator and the left side drain cocks on the cylinder block. On 1989–92 Maxima and 1990–92 300ZX, there are 2 cylinder block drain plugs; one on the right side of the cylinder block behind the right halfshaft boot and one on the left side of the block next to the oil level gauge.
2. On 1990–92 300ZX, remove the undercover and the radiator.
3. Remove the radiator shroud.
4. Remove the power steering, compressor and alternator drive belts.
5. Remove the cooling fan and coupling.
6. Disconnect the water pump hoses.
7. On 1990–92 300ZX, unbolt and remove the inlet and outlet pipes from the block.
8. Remove the water pump pulley, then the upper and lower timing covers.

NOTE: Be careful not to get coolant on the timing belt and to avoid deforming the timing cover, make sure there is enough clearance between the timing cover and the hose clamp.

9. Remove the water pump retaining bolts, note different lengths, and remove the pump.
10. Make sure the gasket sealing surfaces are clean and free of all the old gasket material.

To install:

11. Mount the water pump and gasket onto the cylinder block. Torque the retaining bolts to 12–15 ft. lbs. (16–21 Nm).
12. Install the upper and lower timing belt covers and crankshaft pulley.
13. On 1990–92 300ZX, install the inlet and outlet pipes and torque the nuts and bolts to 12–14 ft. lbs. (16–19 Nm).
14. Connect the water pump hoses.
15. Install the cooling fan and coupling.
16. Install and tension the drive belts.
17. Install the radiator shroud.
18. On 1990–92 300ZX, install the undercover and radiator.
19. Fill the cooling system and connect the negative battery cable.

Thermostat

REMOVAL & INSTALLATION

1. Disconnect the negative battery cable and drain the coolant from the radiator and the left side drain cocks on the cylinder block. On 1989–92 Maxima and 1990–92 300ZX, there are 2 cylinder block drain plugs; one on the right side of the cylinder block behind the right halfshaft boot and one on the left side of the block next to the oil level gauge.
2. On 1990–92 300ZX, remove the undercover.
3. On GA16i engines, disconnect the water temperature switch connector from the thermostat housing.
4. Remove the radiator hose from the water outlet side and remove the bolts securing the water outlet to the cylinder head.

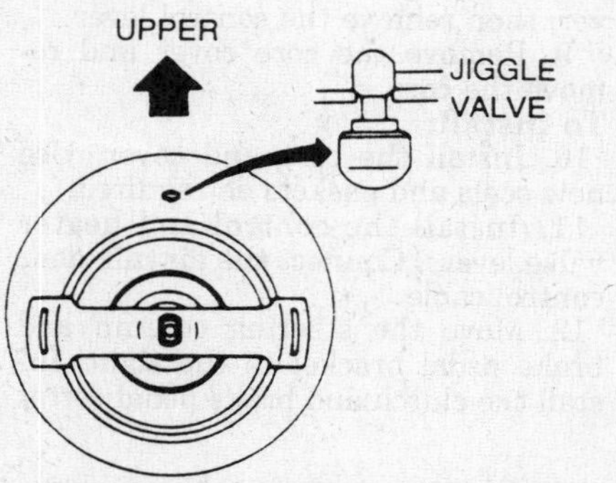

Always be sure the jiggle valve is facing upward when installing the thermostat

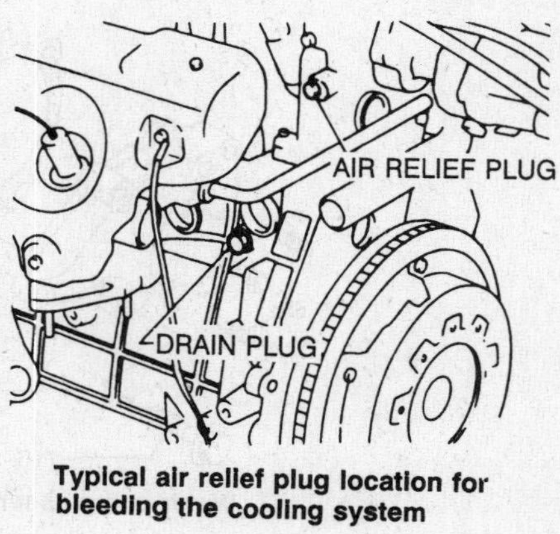

Typical air relief plug location for bleeding the cooling system

5. On 200SX (V6) and 300ZX, remove the radiator shroud, drive belts for 1990–92 300ZX, cooling fan and coupling and water inlet pipe.
6. Remove the thermostat and clean off the old gasket or sealant from the mating surfaces.

To install:

7. Install the thertmostat with a new gasket. When installing the thermostat, be sure to install a new gasket or sealant and be sure the air bleed hole in the thermostat is facing the left side or upward on the engine. The jiggle valve must always face up. Also make sure the new thermostat to be installed is equipped with a air bleed hole. Some thermostats have the word TOP stamped next to the jiggle valve. Again, the word TOP and the jiggle valve must be facing up.
8. On 200SX (V6) and 300ZX, install the water inlet pipe, cooling fan and coupling, drive belts for 1990–92 300ZX and radiator shroud.
9. Install the water outlet and upper radiator hose.
10. On GA16i engines, connect the water temperature switch connector to the thermostat housing.
11. On 1990–92 300ZX, install the undercover.
12. Fill the cooling system and connect the negative battery cable.

Cooling System Bleeding

1. Remove the radiator cap.
2. Fill the radiator and reservoir tank with the proper type of coolant. If equipped with an air relief plug, remove the plug and add coolant until it spills out the air relief opening. Install the plug.
3. Install and tighten the radiator cap.
4. Start the engine and allow the coolant to come up to operating temperature. On 4 cylinder engine, allow the electric cooling fan to come on at least once. Run the heater at full force and with the temperature lever in the HOT position. Be sure the heater control valve is functioning.
5. Shut the engine off and recheck the coolant level, refill as necessary.

ENGINE ELECTRICAL

NOTE: Disconnecting the negative battery cable on some vehicles may interfere with the functions of the on board computer

systems and may require the computer to undergo a relearning process, once the negative battery cable is reconnected.

Distributor

NOTE: The CA18DE (used in the 1988–89 Pulsar) and VG30DE and VG30DETT engines (used in the 1990–92 300ZX) do not use a conventional distributor and high tension wires. Instead these engines use small ignition coils fitted directly to each spark plug. The ECU controls the coils by means of a crank angle sensor and other engine parameter gathering equipment.

REMOVAL

1. Disconnect the negative battery cable.
2. Release the retaining clips and lift the distributor cap straight up. It will be easier to install the distributor if the wiring is not disconnected from the cap. If the wires must be removed from the cap, label the wires according to cylinder number to aid in installation and avoid confusion.
3. Disconnect the distributor wiring harness.
4. Disconnect and label the vacuum lines, if equipped.
5. Note the position of the rotor in relation to the base. Scribe a mark on the base of the distributor and on the engine block to facilitate reinstallation. Align the marks with the direction the rotor is pointing.
6. Remove the bolt(s) which hold the distributor to the engine.
7. Lift the distributor assembly from the engine.

NOTE: Once the distributor is removed, try not to disturb the position of the rotor.

INSTALLATION

Timing Not Disturbed

1. Insert the distributor shaft and assembly into the engine.
2. Align the distributor and engine matchmarks with the rotor. Make sure the vacuum advance diaphragm is pointed in the same direction as it was pointed originally. This will be done automatically if the marks on the engine and the distributor are lined up with the rotor. On 240SX, make sure the distributor driving spindle is properly aligned before inserting the distributor into the front cover.
3. Install the distributor hold-down bolt and clamp. Leave the screw loose enough so the distributor can be moved with moderate hand pressure.
4. Connect the vacuum lines, if equipped.
5. Connect the primary wire to the coil.
6. Install the distributor cap on the distributor housing. Secure the distributor cap with the spring clips.
7. Install the spark plug wires if removed. Make sure the wires are pressed all the way into the top of the distributor cap and firmly onto the spark plug.
8. Set the ignition timing.

Timing Disturbed

NOTE: If the crankshaft has been turned or the engine disturbed in any manner (i.e., disassembled and rebuilt) while the distributor was removed or if the marks were not drawn, it will be necessary to initially time the engine. Follow the procedure given below.

1. It is necessary to place the No. 1 cylinder in the firing position to correctly install the distributor. To locate this position, the ignition timing marks on the crankshaft front pulley are used.
2. Remove the No. 1 cylinder spark plug. Turn the crankshaft until the piston in the No. 1 cylinder is moving up on the compression stroke. This can be determined by placing a thumb over the spark plug hole and feeling the air being forced out of the cylinder. Stop turning the crankshaft when the timing marks are aligned. On 240SX, the driving spindle must be properly aligned to accept the distributor.
3. Oil the distributor housing lightly where the distributor mounts to the block.
4. Install the distributor so the rotor, which is mounted on the shaft, points toward the No. 1 spark plug terminal tower position when the cap is installed. Lay the cap on top of the distributor and make a mark on the side of the distributor housing just below the No. 1 spark plug terminal. Make sure the rotor points toward that mark when installing the distributor.
5. When the distributor shaft has reached the bottom of the hole, move the rotor back and forth slightly until the driving lug on the end of the shaft enters the slots cut in the end of the oil pump shaft and the distributor assembly slides down into place.
6. When the distributor is correctly installed, the reluctor teeth should be aligned with the pick-up coil. This can be accomplished by rotating the distributor body after it has been installed in the engine. Once again, line up the marks made before the distributor was removed.
7. Install the distributor hold-down bolt.

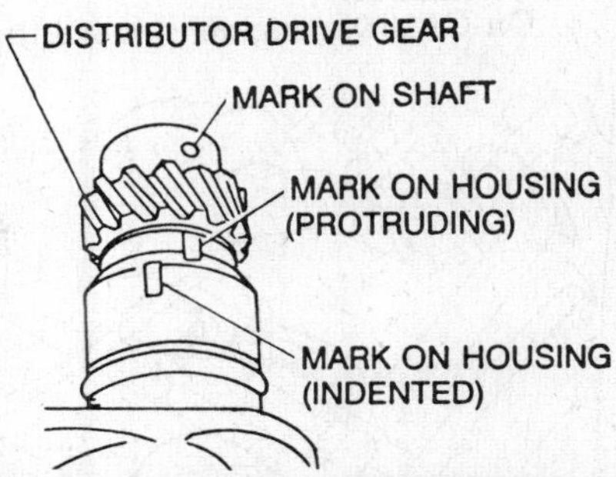

Distributor shaft and housing alignment marks–V-Series engines, except 1990–92 300ZX

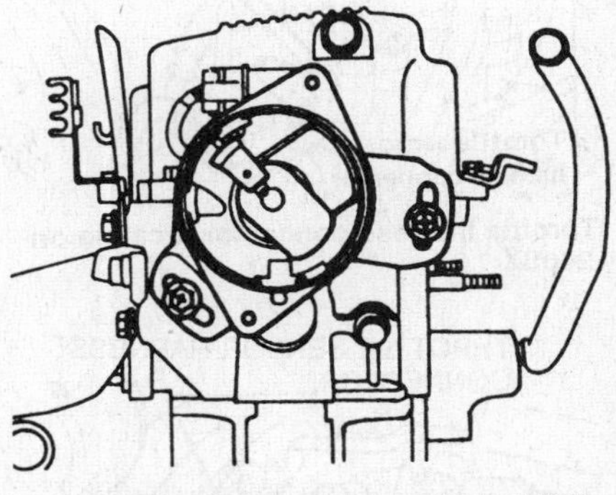

Distributor rotor at No.1 cylinder TDC postion–1990–92 Stanza

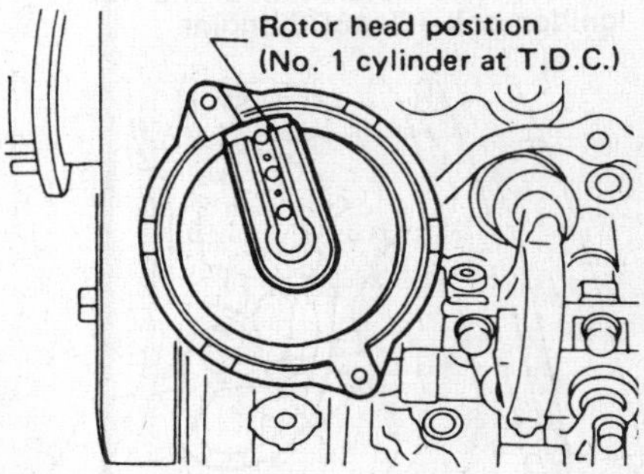

Distributor rotor at No.1 cylinder TDC postion–V series engines (except 1990–92 300ZX)

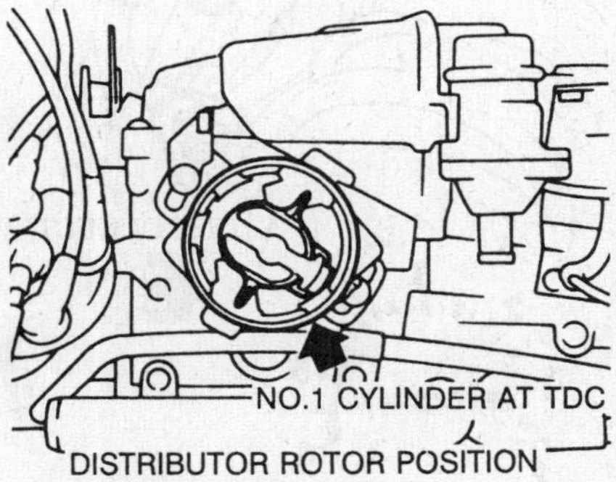

Distributor rotor at No.1 cylinder TDC postion–GA16i engines

8. Install the spark plug into the No. 1 spark plug hole and continue with the remainder of the distributor installation procedure.

Ignition Timing

NOTE: The 200SX (CA20E) and 1988–89 Stanza models use a dual electronic ignition. The firing order is 1–3–4–2 and the rotor is designed with a 135 degree offset to fire both spark plugs at the same time.

ADJUSTMENT

200SX, 240SX, 1988–89 300ZX, Maxima, 1988 Pulsar, 1988 Sentra and Stanza

1. Locate the timing marks on the crankshaft pulley and the front of the engine.
2. Clean off the timing marks.
3. Use chalk or white paint to color the mark on the crankshaft pulley and the mark on the scale which will indicate the correct timing when aligned with the notch on the crankshaft pulley.
4. Connect a tachometer to the engine.
5. Attach a timing light to the engine, according to the manufacturer's instructions.
6. Start the engine and allow to reach normal operating temperature.
7. Check that the idle speed is set to specifications. Adjust as necessary.
8. Aim the timing light and illuminate the timing marks. If the marks on the pulley and the engine are aligned when the light flashes, the timing is correct. Turn off the engine and remove the tachometer and the timing light. If the marks are not in alignment, proceed with the following steps.
9. On 240SX and 1990–92 Stanza, disconnect the throttle sensor harness connector.
10. Loosen the distributor lockbolt(s) just enough so the distributor can be turned with little effort.
11. Start the engine.
12. With the timing light aimed at pulley and the marks on the engine, turn the distributor in the direction of rotor rotation to retard the spark, and in the opposite direction of rotor rotation to advance the spark. Align the marks on the pulley and the engine with the flashes of the timing light. Tighten the hold-down bolt.
13. Disconnect the test equipment. On 240SX and 1990–92 Stanza, connect the throttle harness connector.

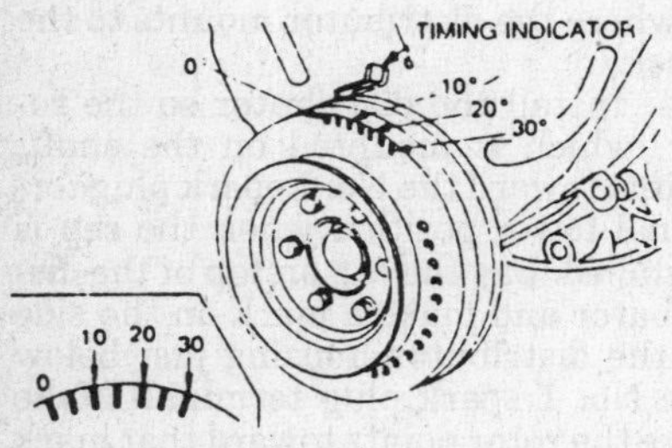

300ZX V6 timing marks

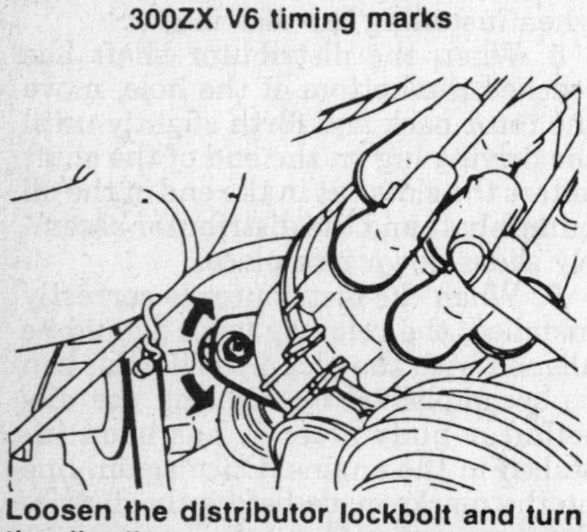
Loosen the distributor lockbolt and turn the distributor slightly to advance (upper arrow) or retard (lower arrow) the timing

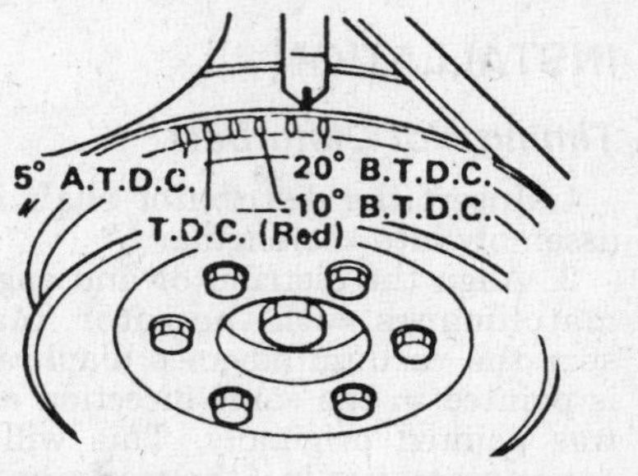

Ignition timing marks—240SX and 1990–92 Stanza

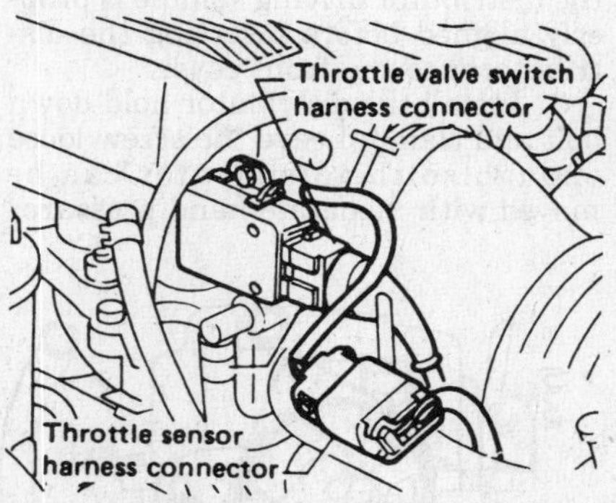

Throttle harness connector location on 240SX

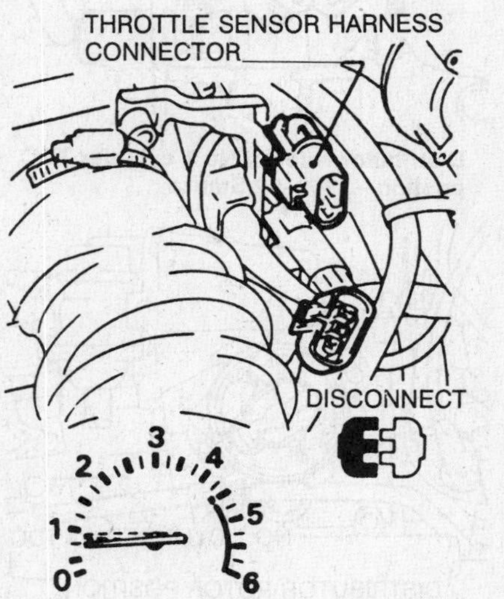

Throttle harness connector location on 1990–92 Stanza

1990–92 300ZX and 1988–89 Pulsar (CA18DE Engine)

NOTE: The CA18DE, VG30DE and VG30DETT engines do not utilize a conventional distributor and high tension wires. Instead they use small ignition coils fitted directly to each spark plug. The ECU controls the coils by means of a crank angle sensor from which it receives piston position and engine speed information. The ECU takes the information from the crank angle sensor and sends it to the power transistor which controls the engine timing.

1. Run the engine until it reaches normal operating temperature.
2. Check the idle speed and adjust as necessary.
3. On CA18DE engine, disconnect the air duct and both air hoses at the throttle chamber.
4. On CA18DE engine, remove the

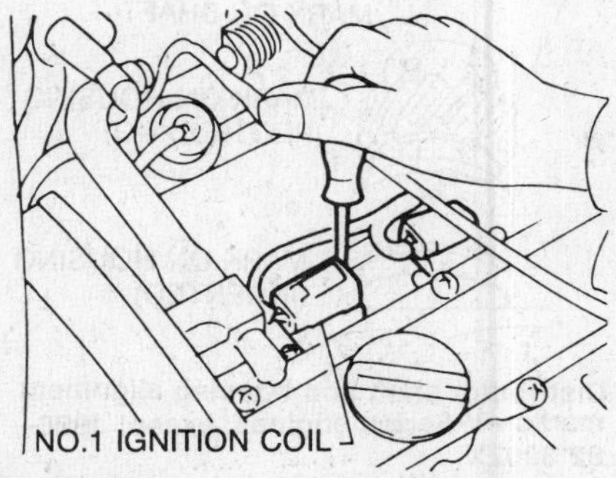

Removal and installation of the No. 1 ignition coil—CA18DE engine

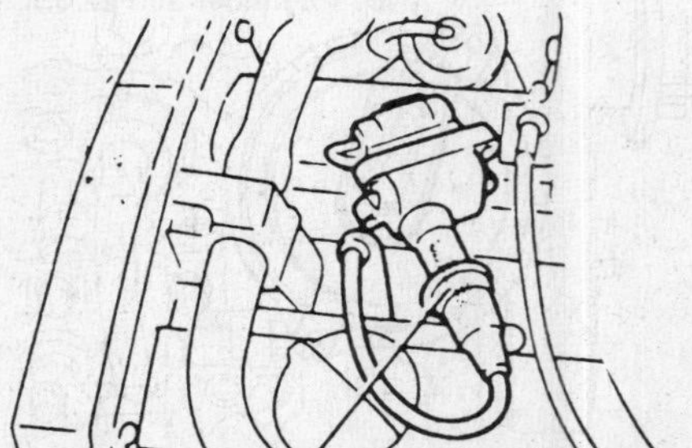

Timing light connection—CA18DE engine

ornament cover between the camshaft covers. The acceleration wire need not be removed to remove the ornament cover.

5. Remove the ignition coil at the No. 1 cylinder.

6. Connect the No. 1 ignition coil to the No. 1 spark plug with a suitable high tension wire.

7. Use an inductive pick-up type timing light and clamp it to the wire connected in Step 6.

8. Reconnect the air duct and hoses and then start the engine.

9. Check the ignition timing. If not to specifications, turn off the engine and loosen the 3 crank angle sensor mounting bolts slightly.

NOTE: The crank angle sensor can be found attached to the upper front cover.

10. Restart the engine and adjust the timing by turning the sensor body slightly until the timing is within specifications. Clockwise rotation retards the timing and counterclockwise rotation advances it.

1989–90 Pulsar (GA16i Engine) and 1989–92 Sentra (GA16i, CA16DE and SR20DE Engines)

1. Run the engine until the water temperature indicator points to the middle of the gauge.

2. Run the engine for 1–2 minutes with no load; all electrical accessories in the **OFF** position.

3. Connect a timing light to the engine and illuminate the timing marks. If the timing is not within specification, proceed to adjust.

4. To adjust the timing, stop the engine and disconnect the throttle sensor connector. Loosen the distributor hold-down bolt just enough to allow the distributor to be turned by hand.

5. Start the engine and race it 2–3 times with no load and then allow the engine to run at idle speed.

6. Adjust the ignition timing by rotating the distributor either clockwise or counterclockwise.

7. Tighten the distributor hold-down bolt and stop the engine.

8. Connect the throttle sensor connector and remove the timing light.

Alternator

PRECAUTIONS

The following precautions must be observed to prevent alternator and regulator damage:

- Be absolutely sure of correct polarity when installing a new battery or connecting a battery charger.
- Do not short across or ground any alternator or regulator terminals.
- Disconnect the battery ground cable before replacing any electrical unit.
- Never operate the alternator with any of the leads disconnected.
- When steam cleaning the engine, be careful not to subject the alternator to excessive heat or moisture.
- When charging the battery, remove it from the vehicle or disconnect the alternator output terminal.

BELT TENSION ADJUSTMENT

The correct belt tension for all alternators is about ¼–½ in. play on the longest span of the belt.

1. Loosen the alternator pivot and mounting bolts.

2. Pry the alternator toward or away from the engine until the tension is correct. Use a hammer handle or wooden prybar.

3. When the tension is correct, tighten the bolts and check the adjustment. Be careful not to over-tighten the belt, which will lead to alternator bearing failure.

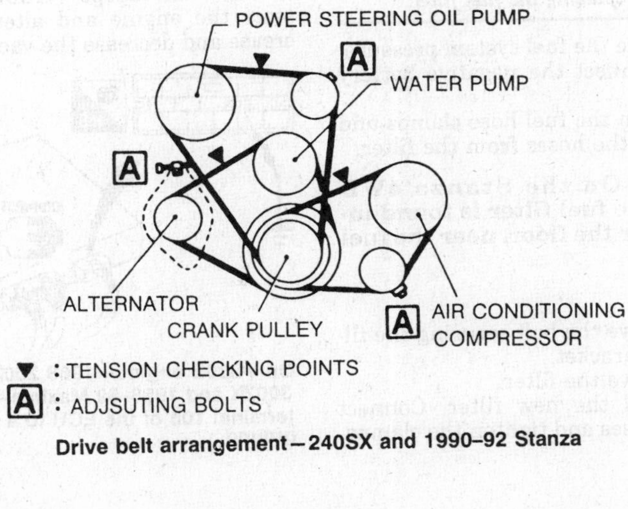

Drive belt arrangement – 240SX and 1990–92 Stanza

REMOVAL & INSTALLATION

1. Disconnect the negative battery cable.

2. Disconnect the 2 lead wires and harness connector from the alternator.

3. Loosen the drive belt adjusting bolt and remove the belt.

4. Unscrew the alternator attaching bolts and remove the alternator from the vehicle. On the 1988–89 300ZX, first remove the front stabilizer bar bolts and pull the stabilizer bar down. On 1990–92 300ZX, remove the lower radiator hose bracket and pull the hose upward to gain the clearance to remove the alternator.

5. Installation is in the reverse order of removal. Adjust the drive belt tension.

Starter

REMOVAL & INSTALLATION

1. Disconnect the negative battery cable.

2. Remove the starter heat shield (300ZX) and harness clamps, if equipped. On 1990–92 Stanza with automatic transaxle, remove the harness connectors from the harness connector bracket.

3. Disconnect and label the wires from the terminals on the solenoid.

4. Remove the 2 bolts which secure the starter to the flywheel housing and pull the starter forward and out.

5. On 1991–92 Sentra vehicles remove the starter motor from under vehicle on SR20DE engine. On GA16DE engine (manual transaxle) remove the starter motor from the transaxle side and from the engine side on automatic transaxle applications.

6. To install, reverse the removal procedure. Check the starter for proper operation.

EMISSION CONTROLS

Please refer to "Emission Controls" in the Unit Repair section for system maintenance procedures. Due to the complex nature of modern electronic engine control systems, comprehensive diagnosis and testing procedures fall outside the confines of this repair manual. For complete information on diagnosis, testing and repair procedures concerning all

modern engine and emission control systems, please refer to "Chilton's Guide to Fuel Injection and Electronic Engine Controls".

FUEL SYSTEM

Fuel System Service Precaution

Failure to conduct fuel system maintenance and repairs in a safe manner may result in serious personal injury. Maintenance and testing of the vehicle's fuel system components can be accomplished safely and effectively by adhering to the following rules and guidelines.

- To avoid the possibility of fire and personal injury, always disconnect the negative battery cable unless the repair or test procedure specifically requires that battery voltage be applied.
- Always relieve the fuel system pressure prior to disconnecting any fuel system component (injector, fuel rail, pressure regulator, etc.), fitting or fuel line connection. Exercise extreme caution whenever relieving fuel system pressure to avoid exposing skin, face and eyes to fuel spray. Be advised that fuel under pressure may penetrate the skin or any part of the body that it comes in contact with.
- Always place a shop towel or cloth around the fitting or connection prior to loosening to absorb any excess fuel due to spillage. Ensure that all fuel spillage (should it occur) is quickly removed from engine surfaces. Ensure that all fuel soaked cloths or towels are deposited into a suitable waste container.
- Always have a properly charged Class B dry chemical or CO2 fire extinguisher in the vicinity of the work area and always ensure work areas are adequately ventilated.
- Do not allow fuel spray or fuel vapors to come in contact with spark or open flame. Remember that smoking and fuel maintenance do not mix!
- Always use a backup wrench when loosening and tightening fuel line connection fittings. This will prevent unnecessary stress and torsion to fuel line piping. Always follow the proper torque specifications.
- Always replace worn fuel fitting O-rings with new ones. Do not substitute fuel hose or equivalent, where rigid fuel pipe is called for.
- Always use common sense.

RELIEVING FUEL SYSTEM PRESSURE

1. Remove the fuel pump fuse from the fuse block, fuel pump relay or disconnect the harness connector at the tank while engine is running.
2. It should run and then stall when the fuel in the lines is exhausted. When the engine stops, crank the starter for about 10 seconds to make sure all pressure in the fuel lines is released.
3. Install the fuel pump fuse, relay or harness connector after repair is made.

Fuel Tank

REMOVAL & INSTALLATION

1. Disconnect the negative battery cable.
2. Drain the fuel from the tank unit.
3. Remove the access plate from the trunk or rear seat area.
4. Disconnect all fuel lines and connections.
5. Raise and safely support the vehicle.
6. Remove the fuel tank protector if so equipped. Disconnect the fuel filler tube or filler hose at the fuel tank.
7. Remove the gas tank assembly strap retaining bolts and slowly lower the tank assembly down from the vehicle.
8. Installation is the reverse of the removal procedure. Replace all gas tank line hose clamps as necessary. Always torque gas tank assembly strap retaining bolts evenly.

Fuel Filter

REMOVAL & INSTALLATION

CAUTION

Make sure to relieve the fuel system pressure before replacing the fuel filter.

1. Relieve the fuel system pressure.
2. Disconnect the negative battery cable.
3. Loosen the fuel hose clamps and disconnect the hoses from the filter.

NOTE: On the Stanza 4WD Wagon, the fuel filter is found inline, under the floor, near the fuel pump.

4. Remove the bolt securing the filter to the bracket.
5. Remove the filter.
6. Install the new filter. Connect the fuel hoses and tighten the clamps.

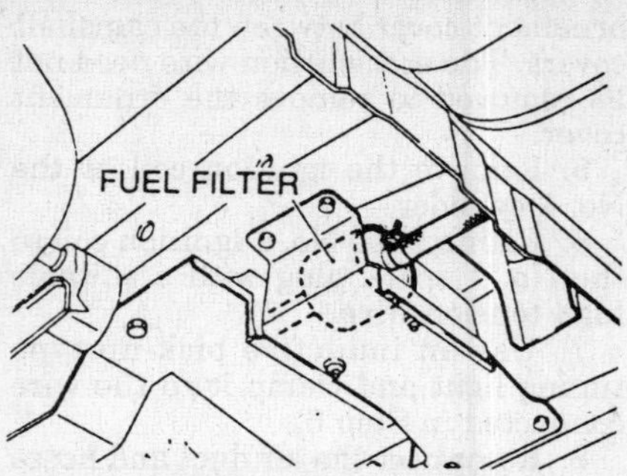

The fuel filter is found under the floor on the Stanza wagons (4WD)

8. Replace the fuel pump fuse, relay or connector.
9. Connect the negative battery cable. Start the engine and check for leaks.

Electric Fuel Pump

PRESSURE TESTING

Except Pulsar and Sentra With GA16i Engine

1. Relieve the fuel system pressure.
2. Remove the air duct, if required.
3. Connect a fuel pressure gauge between the fuel feed pipe and the fuel filter outlet.
4. Start the engine and read the fuel pressure. If the pressure is not as specified, replace the pump. If the pump output pressure is okay, go to Step 5 to check the pressure regulator.
5. Stop the engine and disconnect the fuel pressure regulator vacuum hose from the intake manifold.
6. Plug the intake manifold with a rubber cap.
7. On VG30E engines for 1988 200ZX, 1988–89 300ZX and 1988–89 Maxima, connect a jumper wire from terminal No. 108 of the ECU to a suitable body ground.
8. Connect a vacuum pump to the fuel pressure regulator.
9. On all except VG30E engines, start the engine and alternately increase and decrease the vacuum while

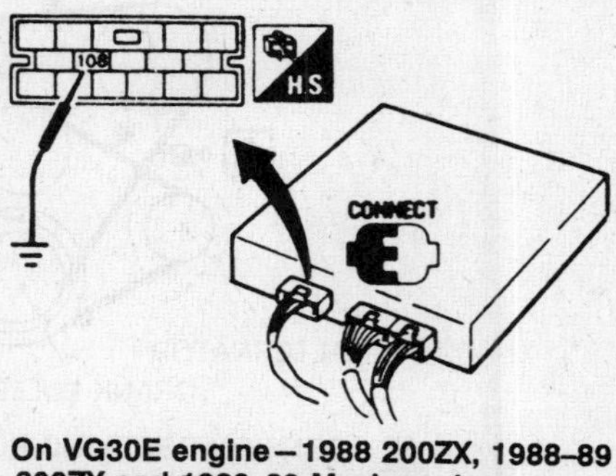

On VG30E engine – 1988 200ZX, 1988–89 300ZX and 1988–89 Maxima – jump terminal 108 of the ECU to a body ground

watching the gauge. On VG30E engine, turn the ignition switch to the **ON** position without starting the engine. Fuel pressure should decrease as the vacuum is increased. If the pressure is incorrect, replace the pressure regulator. After replacement of the regulator, repeat the pressure test. If still incorrect, check the fuel lines for kinks or blockage, and replace the pump as necessary.

Pulsar and Sentra With GA16I Engine

1. Relieve the fuel system pressure.
2. Disconnect the fuel inlet hose from the electro-injection unit.
3. Connect a pressure gauge to the electro-injection unit inlet opening and connect the fuel inlet hose to the gauge.
4. Start the engine and check the fuel line and gauge connections for fuel leaks.
5. Read the fuel pressure. If the pressure is not as specified, replace the pump.
6. Release the fuel system pressure and disconnect the gauge.
7. Connect the fuel inlet hose to the electro-injection unit.

REMOVAL & INSTALLATION

The fuel pump is located in the fuel tank on all models except the 2WD Stanza Wagon. On 2WD Stanza Wagon the fuel pump is located in line to the fuel tank. In tank fuel pumps are accessible either by lifting up the rear seat or through an opening in the trunk compartment.

1. Relieve the pressure from the fuel system, then disconnect the negative battery cable.
2. Open the trunk, remove the mat and flip up the fuel pump access plate in the trunk floor. If there is no access plate in the luggage compartment, check under the rear seats.
3. On 2WD Stanza Wagon, clamp the hose between the fuel tank and the fuel pump to prevent gas from spilling out of the tank. Disconnect and plug the fuel outlet hose and remove the pump from the mounting bracket.
4. Disconnect the inlet and outlet tubes from the fuel pump.
5. Unbolt and remove the pump from the top of the fuel tank. Discard the O-ring seal or gasket.

To install:

6. Install the pump with a new gasket or O-ring seal. Tighten the pump retaining bolts and connect the fuel hoses. Be sure to use new clamps and that all hoses are properly seated on the fuel pump and the fuel pump hoses.
7. Install the fuel pump access plate.
8. Connect the pump wiring harness.
9. Connect the negative battery cable.

Fuel Injection

IDLE SPEED ADJUSTMENT

Before adjusting the idle speed, visually check the following items first: air cleaner for clogging, hoses and ducts for leaks, EGR valve for proper operation, all electrical connectors, gaskets and the throttle valve and throttle valve switch operation.

1988–89 Stanza and 200SX (CA20E ENGINE)

1. Connect a tachometer and timing light to the engine.
2. Turn all electrical accessories and air conditioner to the **OFF** position.
3. Warm up engine to normal operating temperature.
4. Run the engine at 2000 rpm for about 2 minutes without load.
5. Race the engine 2–3 times and allow to idle.
6. Check the idle speed.
7. If the idle speed is not within specifications, disconnect the Auxiliary Air Control (AAC) valve and throttle valve switch harness connectors.
8. Adjust the idle speed by turning the idle speed adjusting screw.
9. Connect the AAC and throttle valve switch connectors.
10. Check the timing and adjust as necessary.
11. Stop the engine and remove the test equipment.

240SX and 1990–92 Stanza

1. Connect a tachometer and timing light to the engine.
2. Turn all electrical accessories and air conditioner to the **OFF** position.
3. Warm up engine to normal operating temperature.
4. Run the engine at 2000 rpm for about 2 minutes without load.
5. Race the engine 2–3 times and allow to idle.
6. Check the idle speed in the **N** position for both manual and automatic transmission models.
7. To adjust the idle speed, first disconnect the throttle sensor harness connector.
8. Adjust the idle speed by turning the idle speed adjusting screw.
9. Stop the engine. Connect the throttle sensor harness connector.
10. Remove the test equipmemt.

300ZX and 1988 Maxima

1. Connect a tachometer and timing light to the engine.
2. Turn all electrical accessories and air conditioner to the **OFF** position.
3. Warm up engine to normal operating temperature.
4. Run the engine at 2000 rpm for about 2 minutes without load.
5. On 1988–89 300ZX non-turbo and Maxima, disconnect harness connector at idle-up solenoid valve.
6. Race the engine 2–3 times and allow to idle.
7. Check the timing and adjust as necessary.
8. Check the idle speed. To adjust the idle speed on 1988–89 300ZX non-turbo and Maxima, turn idle speed adjusting screw. Connect idle-up solenoid on 300ZX and Maxima models.
9. To adjust idle speed on 1988–89 300ZX turbo and all 1990–92 300ZX models, stop engine and disconnect harness connector at Auxiliary Air Control (AAC) valve. Start engine and adjust idle speed to specifications. Stop engine and reconnect control valve. Start engine and ensure idle speed is correct.
10. Remove the test equipment.

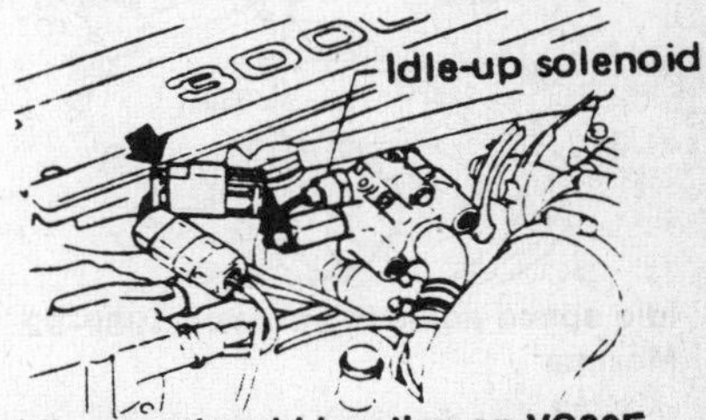

Idle-up solenoid location on VG30E engines

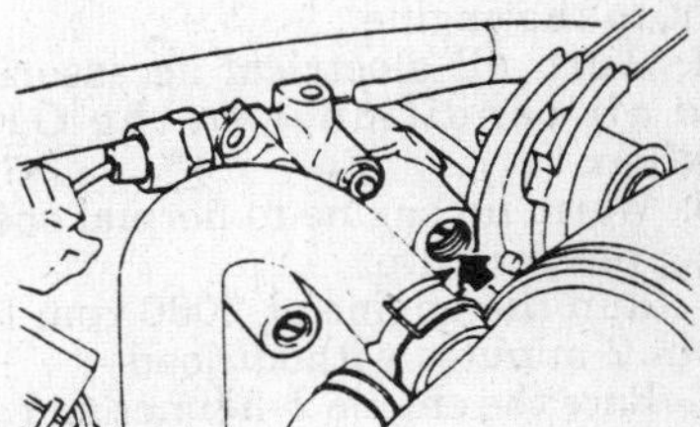

Idle speed adjustment screw location on VGE30 engines

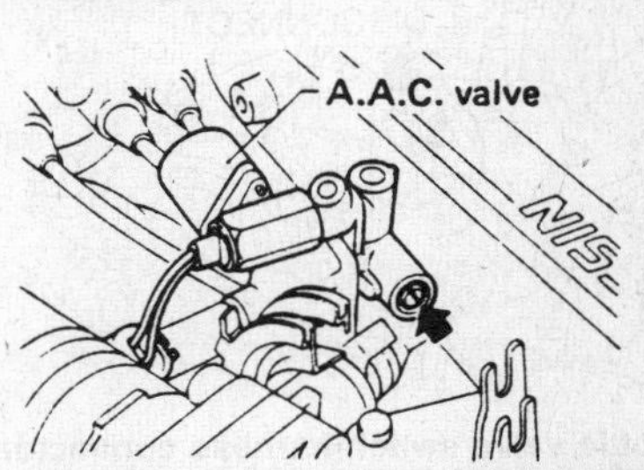

AAC valve location on 1988–89 300ZX turbo engines

1989–92 Maxima

1. Connect a tachometer and timing light to the engine.
2. Turn all electrical accessories and air conditioner to the **OFF** position.
3. Warm up engine to normal operating temperature.
4. Run the engine at 2000 rpm for about 2 minutes without load.
5. Race the engine 2–3 times and allow to idle for 1 minute.
6. Check the timing and adjust as necessary.
7. Check the idle speed in the **N** position for both manual and automatic transaxle models.
8. To adjust the idle speed, close the Auxiliary Air Control (AAC) valve by turning the diagnostic mode selector on the ECU fully clockwise.
9. Adjust the idle speed by turning the idle speed adjusting screw with transaxle in the **N** position.
10. Operate the AAC valve by turning the diagnostic mode selector on the ECU. fully conterclockwise.
11. Stop the engine and remove the test equipment.

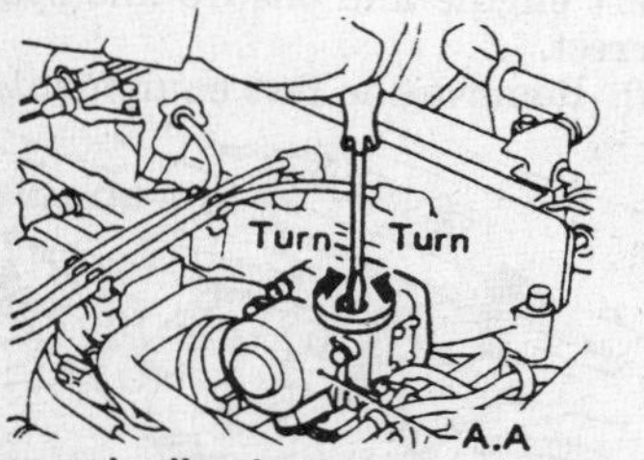

Idle speed adjusting screw – 1989–92 Maxima

Pulsar and Sentra

1. Connect a tachometer and timing light to the engine.
2. Turn all electrical accessories and air conditioner to the **OFF** position.
3. Warm up engine to normal operating temperature.
4. Run the engine at 2000 rpm for about 2 minutes without load.
5. Race the engine 2–3 times and allow to idle for 1 minute.

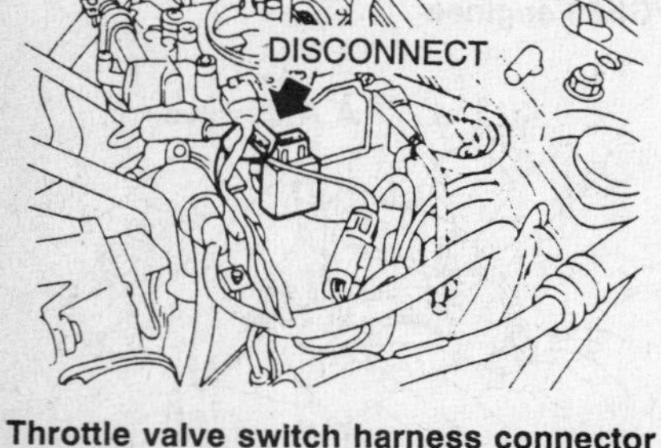

Throttle valve switch harness connector location – 1988 Pulsar and Sentra with E16i engine

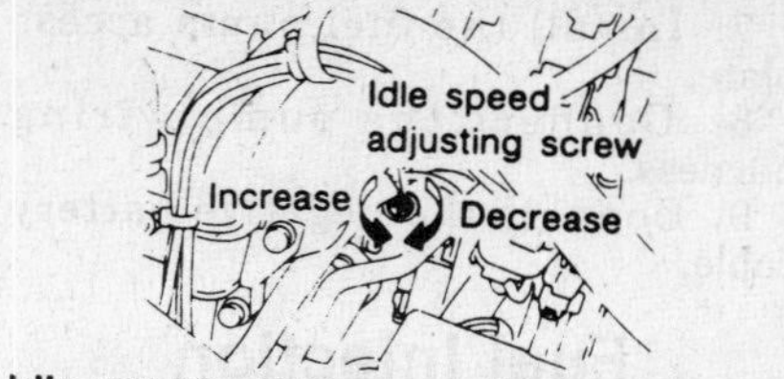

Idle speed adjustment – SR20De engine

6. Check the timing and adjust as necessary.
7. Check the idle speed and adjust as necessary.
8. Stop the engine and disconnect the throttle sensor connector.
9. Start the engine and adjust the idle speed by turning the throttle adjusting screw.
10. Stop the engine and reconnect the throttle valve switch connector.
11. Remove the test equipment.

IDLE MIXTURE ADJUSTMENT

1988–89 Stanza and 200SX (CA20E ENGINE)

1. Connect a tachometer and timing light to the engine.
2. Turn all electrical accessories and air conditioner to the **OFF** position.
3. Warm up engine to normal operating temperature.
4. Check the idle speed and ignition timing. Adjust as necessary.
5. Run the engine at 2000 rpm for about 2 minutes without any load. The green ECU inspection light should flash on and off at least 9 times in 10 seconds at 2000 rpm.
6. Race the engine 2–3 times and allow to idle.
7. Set the ECU to the No. 2 diagnosis mode and disconnect the throttle valve switch connector. The red and green lights on the ECU should flash together. If they do, then the idle mixture is correct and no further adjustment is required. If they don't, then continue with the remainder of the procedure.
8. Stop the engine and remove the air flow meter from the vehicle.
9. Drill a small hole in the seal plug which covers the variable resistor and remove the plug from the air flow meter.
10. Install the air flow meter.
11. Warm up engine to normal operating temperature.
12. Set the ECU to the No. 2 diagnosis mode, then adjust the idle mixture by turning the variable resistor until the red and green lights on the ECU flash together. If the mixture still can't be adjusted, replace the air flow meter.
13. Install a new seal plug and tap it into place with a suitable tool.
14. Connect the throttle valve switch connector and remove the test equipment.

200SX (VG30E ENGINE), 240SX, 300ZX, Maxima, and 1990–92 Stanza

1. Connect a tachometer and timing light to the engine.
2. Turn all electrical accessories and air conditioner to the **OFF** position.
3. Warm up engine to normal operating temperature.
4. Check the idle speed and ignition timing. Adjust as necessary.
5. Run the engine at 2000 rpm for about 2 minutes without any load. The green ECU inspection lamp should flash on and off at least 5 times in 10 seconds at 2000 rpm.
6. Race the engine 2–3 times and allow to idle.
7. Set the ECU to the No. 2 diagnosis mode by turning the diagnostic mode selector screw on ECU fully counterclockwise. Disconnect the throttle valve switch connector. The red and green lights on the ECU should flash together. If they do, then the idle mixture is correct and no further adjustment is required. If they don't, then continue with the remainder of the procedure.
8. Stop the engine and disconnect the engine temperature sensor harness connector from the sensor. Connect a 2.5 kilo-ohm resistor across the terminals of the engine temperature harness connector. The sensor is located on the cylinder head.
9. On 240SX and 1990–92 Stanza, disconnect the AIV hose and plug the AIV pipe. On 1990–92 300ZX, disconnect the AIV control solenoid valve harness connector.
10. Start the engine and run for 5 minutes, then race the engine 2–3 times and allow to idle.
11. Check the CO content and make sure the engine runs smoothly. The idle mixture on these vehicles is controlled by the ECU, and is not adjustable. However, the following components should be checked before identifying the ECU as the source of the problem.

Exhaust gas sensor(s)
Exhaust gas sensor harness
Fuel pressure regulator
Air flow meter
Fuel injectors
Engine temperature sensor

12. Stop the engine. Remove the resistor from the engine temperature switch harness connector and plug in the connector. Connect the AIV hose.
13. Remove the test equipment.

Pulsar and Sentra

1. Connect a tachometer and timing light to the engine.
2. Turn all electrical accessories and air conditioner to the **OFF** position.
3. Warm up engine to normal operating temperature.
4. Check the idle speed and ignition timing. Adjust as necessary.
5. Run the engine at 2000 rpm for about 2 minutes without any load. The green ECU inspection lamp should flash on and off at least 5 times in 10 seconds at 2000 rpm.
6. Race the engine 2–3 times and allow to idle.
7. Set the ECU to the No. 2 diagnosis mode. The red and green lights on the ECU should flash together. If they do, then the idle mixture is correct and no further adjustment is required. If they don't then continue with the remainder of the procedure.
8. Stop the engine.
9. On E16i and GA16i engines, remove the throttle body from the vehicle. On CA18DE engine, remove the air flow meter from the vehicle.
10. Drill a small hole in the seal plug which covers the variable resistor and remove the plug from the air flow meter or throttle body.
11. Install the throttle body or air flow meter.
12. Warm up engine to normal operating temperature.
13. Set the ECU to the No. 2 diagnosis mode, then adjust the idle mixture by turning the variable resistor until the red and green lights on the ECU flash together. Turning counterclockwise lowers the CO content and clockwise raises it. If the mixture still can't be adjusted, replace the air flow meter or the throttle body.
14. Install a new seal plug and tap it into place with a suitable tool.
15. Remove the test equipment.

Fuel Injector

REMOVAL & INSTALLATION

200SX (VG30E Engine), 300ZX and 1988 Maxima

1. Relieve the fuel system pressure.
2. Disconnect the negative battery cable.
3. Disconnect the hoses and electrical wiring from the intake collector. Label each hose and wire to ensure proper placement during installation.
4. On 200SX, remove the intake collector cover.
5. Remove the intake collector and gasket.
6. Remove the fuel tube retaining bolts.

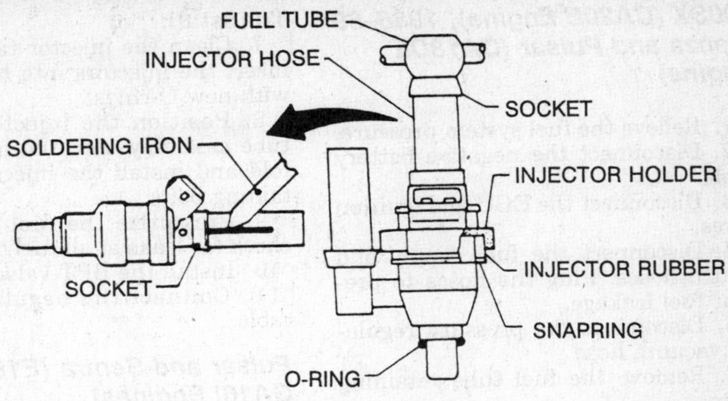

Removing the injector braided reinforcement hose

7. Remove the injector retaining bolts and remove the injector, fuel tubes and pressure regulator as an assembly.
8. Using a soldering iron or hot, sharp knife, slice the braided reinforcement hose from the socket end to the fuel tube end. Be careful not to allow the tool to contact the injector tail piece or the socket plastic connector. Pull the hose from the injector and repeat the procedure for the remaining injectors.

To install:

9. Clean the exterior of the injector tail piece and fuel tube end. Install new O-rings.
10. Wet the inside of the new fuel tube with clean fuel.
11. Push the end of the rubber hose and hose socket onto the injector tail piece and fuel tube end as far as it will go. Clamps are not required. Repeat the procedure for the remaining injectors.
12. Position and install the injector, fuel tube and pressure regulator assembly. Install the injector and fuel tube retaining bolts. Pressurize the fuel system and check for leaks at all fuel connections.
13. Install the intake collector.
14. On 200SX, install the intake collector cover and gasket. Position the gasket so the silicone rubber portion is facing down.
15. Connect all the hoses and electrical wiring to the intake collector.
16. Connect the negative battery cable.

1989–92 Maxima

1. Relieve the fuel system pressure.
2. Disconnect the negative battery cable.
3. Disconnect the automatic speed control device cable and accelerator cable from the intake manifold collector.
4. Disconnect the Auxiliary Air Control (AAC) valve, throttle sensor and idle switch connectors.
5. Disconnect the air cut valve water hose. Plug the end to prevent leakage.
6. Disconnect the PCV hoses.
7. Disconnect the vacuum gallery, power valve actuator, master brake cylinder and EGR control valve vacuum hoses.
8. Loosen and disconnect the EGR flare tube.
9. Remove the upper manifold collector from the engine.
10. Disconnect the engine ground harness from the lower intake collector manifold and remove the manifold from the engine.
11. Disconnect pressure regulator vacuum hose, fuel supply and return tubes and injector electrical connectors.
12. Remove the fuel injector tube assembly.
13. Withdraw the injectors from the fuel tube.

To install:

14. Insert the fuel injector(s) into the fuel tubes with new O-rings.
15. Install the injector and fuel tube assembly.
16. Connect the injector electrical connectors, fuel supply and return tubes and pressure regulator vacuum hose. Pressurize the fuel system and check for leaks at all fuel connections.
17. Install the lower intake collector manifold and connect the engine ground harness.
18. Install the upper collector manifold.
19. Connect and tighten the EGR flare tube.
20. Connect the vacuum and PCV hoses, air cut valve water hose and electrical connectors.
21. Connect the accelerator cable and automatic speed control device cable to the intake manifold collector. Adjust the cables.
22. Connect the negative battery cable.

200SX (CA20E Engine), 1988–89 Stanza and Pulsar (CA18DE Engine)

1. Relieve the fuel system pressure.
2. Disconnect the negative battery cable.
3. Disconnect the ECU and ignition wires.
4. Disconnect the fuel supply and return hoses. Plug the hoses to prevent fuel leakage.
5. Disconnect the pressure regulator vacuum hose.
6. Remove the fuel tube retaining bolts.
7. Remove the injector retaining bolts and remove the injector, fuel tubes and pressure regulator as an assembly. Be careful not to smack the injectors or bend the fuel tube.
8. Using a soldering iron or hot, sharp knife, slice the braided reinforcement hose from the socket end to the fuel tube end. Be careful not to allow the tool to contact the injector tail piece or the socket plastic connector. Pull the hose from the injector and repeat the procedure for the remaining injectors.

To install:

9. Clean the exterior of the injector tail piece and fuel tube end. Install new O-rings.
10. Wet the inside of the new fuel tube with clean fuel.
11. Push the end of the rubber hose and hose socket onto the injector tail piece and fuel tube end as far as it will go. Clamps are not required. Repeat the procedure for the remaining injectors.
12. Position and install the injector, fuel tube and pressure regulator assembly. Install the injector and fuel tube retaining bolts. Pressurize the fuel system and check for leaks at all fuel connections.
13. Connect the pressure regulator vacuum hose.
14. Connect the fuel supply and return hoses.
15. Connect the ECU and ignition wires.
16. Connect the negative battery cable.

240SX

1. Relieve the fuel system pressure.
2. Disconnect the negative battery cable.
3. Remove the BPT valve.
4. Remove the fuel tube retaining bolts.
5. Remove the fuel tube and injector assembly from the intake manifold.
6. Withdraw the injectors from the fuel tube.

To install:

7. Clean the injector tail piece and insert the injectors into the fuel tube with new O-rings.
8. Position the injector and fuel tube assembly onto the intake manifold and install the injector tube retaining bolts.
9. Pressurize the fuel system and check for leaks at all fuel connections.
10. Install the BPT valve.
11. Connect the negative battery cable.

Pulsar and Sentra (E16i and GA16i Engines)

1. Relieve the fuel system pressure.
2. Disconnect the negative battery cable.
3. Remove the injector cover plates.
4. Using the proper tool, carefully withdraw the fuel injector straight up from the throttle body. Be careful not to damage the injector terminals during removal.
5. Remove the injector upper and lower O-rings. Install a new lower O-ring.

To install:

6. Using a 13mm socket or suitable tool, carefully push the injector into the throttle body. Make sure the injector terminals are aligned properly. Be careful not to bend the injector terminals during installation.
7. Position the new upper injector O-ring and install it with a 19mm socket or other suitable tool.

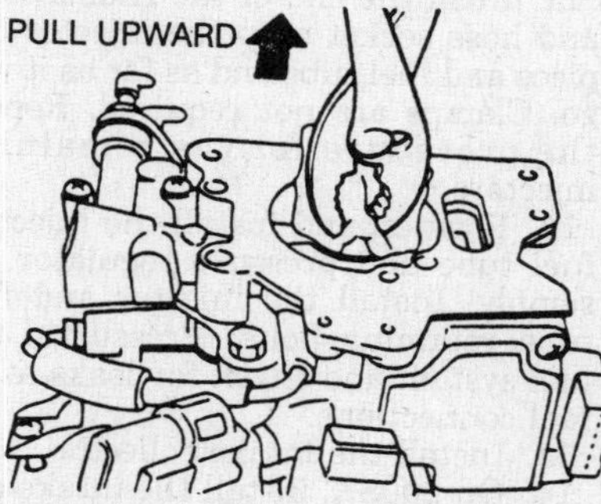

Injector removal on E16i and GA16i engines

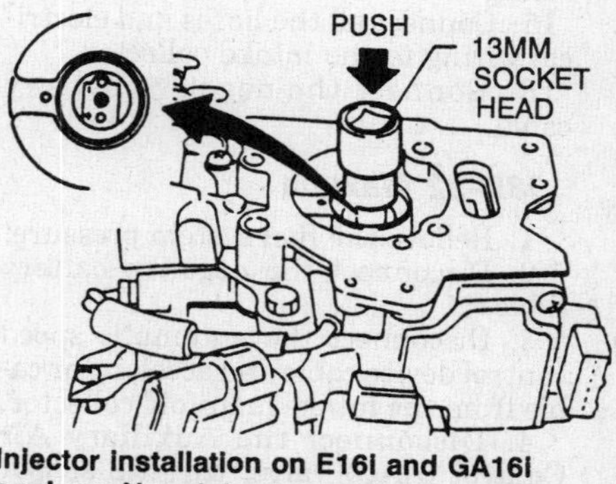

Injector installation on E16i and GA16i engines. Note injector terminal alignment

8. Install the lower (white) injector cover plate. Do not over tighten the cover screws.
9. Install the injector cover without the rubber boot. Make sure the 2 O-rings (large and small) properly seated in the cover. Make sure there is a good connection between the injector terminal and the injector cover terminal. When this connection is verified, install the cover boot.
10. Connect the negative battery cable.
11. Start the engine and check for leaks at all fuel connections.

1991–92 Sentra (GA16DE and SR20DE Engines)

1. Relieve the fuel system pressure. Disconnect the negative battery cable.
2. Disconnect the fuel injector wiring harness connectors and vacuum line from the fuel pressure regulator.
3. Disconnect the fuel hoses from the fuel tube assembly.
4. Remove the injectors with fuel tube assembly.
5. Installation is the reverse of the removal procedure. Install injectors with fuel tube assembly to intake manifold torque all reataining bolts in two steps to 15–20 ft. lbs. Check for fuel leaks after installation is complete.

1990–92 Stanza

1. Relieve the fuel system pressure.
2. Disconnect the negative battery cable.
3. Disconect the air duct.
4. Disconnect the supply and return hoses from the fuel tube. Plug the ends to prevent leakage.
5. Disconnect the vacuum line from the fuel pressure regulator.
6. Detach the accelerator cable bracket.
7. Disconnect the fuel injector wiring harness connectors.
8. Remove the fuel tube retaining bolts.
9. Pull the fuel tube and injector assembly from the intake manifold. Remove the injector assembly out from the No. 4 injector side.

To install:

10. Remove the O-rings and insulators and install new ones.
11. Install the injector and fuel tube assembly into the intake manifold.
12. Install the injector tube retaining bolts.
13. Connect the injector wiring harness conectors.
14. Attach the accelerator cable bracket.
15. Connect the pressure regulator vacuum line.
16. Connect the fuel supply and return hoses.
17. Connect the air duct.

18. Connect the negative battery cable.
19. Start the engine and check for leaks at all fuel connections.

DRIVE AXLE

Halfshaft

REMOVAL & INSTALLATION

Front Wheel Drive

This procedure applies to all 2WD drive vehicles and to the front halfshafts on 4WD vehicles. Removal and installation of the rear halfshafts on 4WD vehicles is described below.

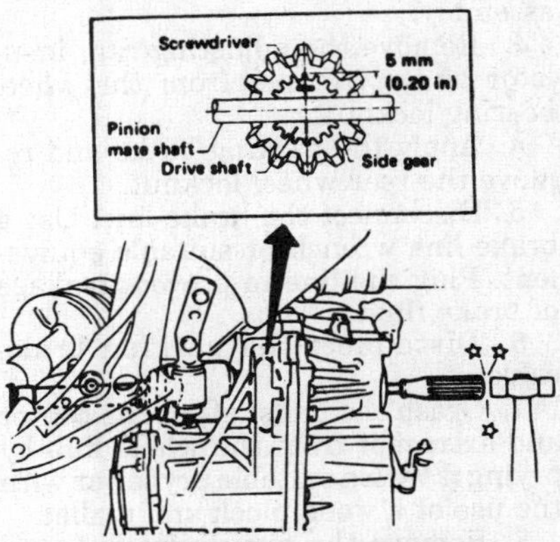

Left halfshaft removal on automatic transaxle vehicles—Maxima, Stanza and Stanza Wagon

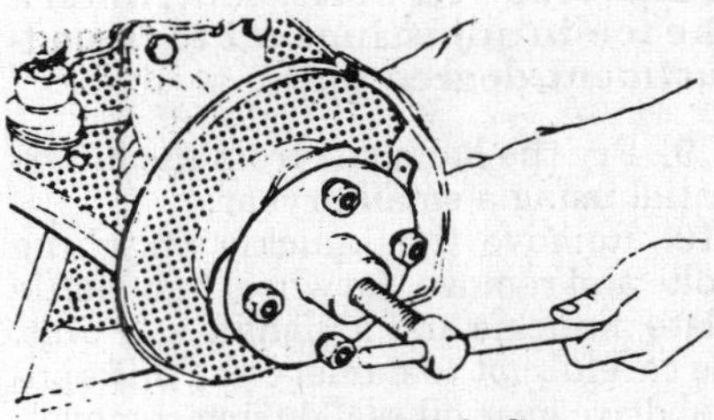

Removing halfshaft

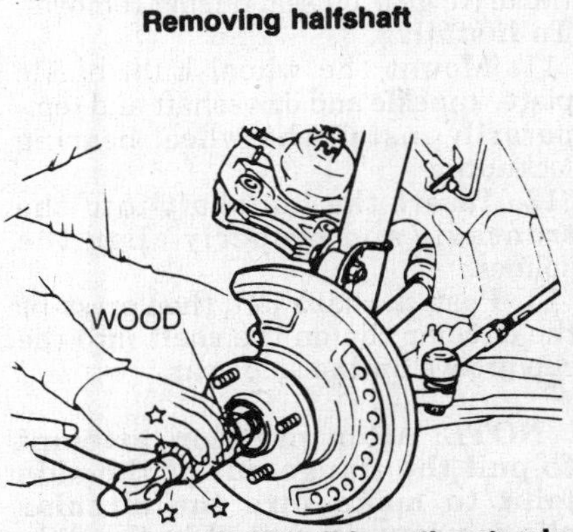

Separating the halfshaft from the steering knuckle

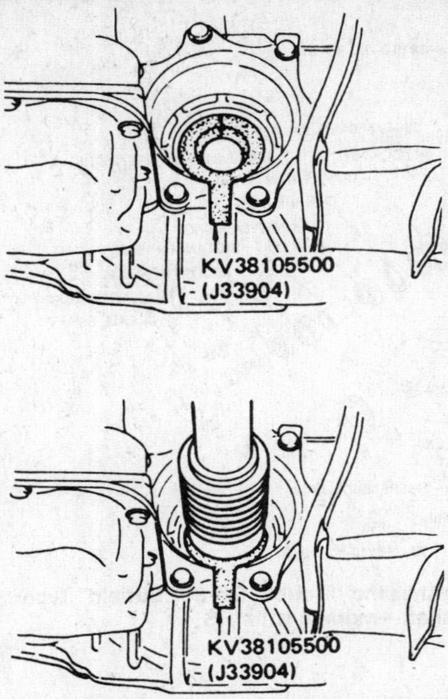

Halfshaft alignment tools used on front wheel drive vehicles

NOTE: Installation of the halfshafts will require a special tool for the spline alignment of the halfshaft end and the transaxle case. Do not perform this procedure without access to this tool or suitable equivalent. The tool is J-34296, J-34297 or J-33904 depending on the vehicle.

1. Raise the vehicle and support safely.
2. Remove the wheel and tire assembly.
3. Withdraw the cotter pin from the castellated nut on the wheel hub.
4. Depress the brake pedal and remove the wheel bearing locknut.
5. Remove the brake caliper assembly without disconnecting the brake line. Support the caliper with wire.
6. Separate the halfshaft from the steering knuckle by tapping it with a block of wood and a mallet.
7. Remove the tie rod ball joint. Remove the 3 mounting nuts for the lower ball joint and then pull it down.

NOTE: Always use a new nut when replacing the tie rod ball joint.

8. Using a suitable tool, reach through the engine crossmember and carefully tap the right side inner CV-joint out of the transaxle case.
9. Using a block of wood and a suitable jack, support the engine under the oil pan.
10. Remove the support bearing bracket and bearing retainer bolts from the engine and then withdraw the right halfshaft (except Pulsar with E16i and GA16i and Sentra).
11. On vehicles with manual transaxles, carefully insert a small prybar between the left CV-joint inner flange and the transaxle case mounting surface and pry the halfshaft out of the case. Withdraw the shaft from the steering knuckle and remove it.
12. On vehicles with automatic transaxles, insert a dowel through the right side halfshaft hole and use a small mallet to tap the left halfshaft out of the transaxle case. Withdraw the shaft from the steering knuckle and remove it.

NOTE: Be careful not to damage the pinion mating shaft and the side gear while tapping the left halfshaft out of the transaxle case.

To install:

13. When installing the shafts into the transaxle, use a new oil seal and then install an alignment tool along the inner circumference of the oil seal.
14. Insert the halfshaft into the transaxle, align the serrations and then remove the alignment tool.
15. Push the halfshaft, then press-fit the circular clip on the shaft into the clip groove on the side gear.

NOTE: After insertion, attempt to pull the flange out of the side joint to make sure the circular clip is properly seated in the side gear and will not come out.

16. Connect the tie rod end ball joint.
17. Insert the driveshaft into the steering knuckle.
18. Mount the brake caliper assembly.
19. Install the wheel bearing locknut. Torque the nut to 174–231 ft. lbs. (235–314 Nm) on Maxima and Stanza; 145–203 ft. lbs. (196–275 Nm) on 1988–89 Pulsar; 145–203 ft. lbs. (196–275 Nm) on Sentra and 1990 Pulsar. When tightening the nut, apply the brake pedal.
20. Install a new cotter pin into the wheel bearing locknut.
21. Mount the wheel and tire assembly.
22. Lower the vehicle.

Rear Wheel Drive

EXCEPT SENTRA (4WD) AND STANZA WAGON (4WD)

NOTE: When removing the rear halfshafts, cover the CV-boots with cloth to prevent damage.

1. Raise and support the rear of the vehicle safely.
2. Remove the rear wheel and tire assembly.
3. Remove the adjusting cap and

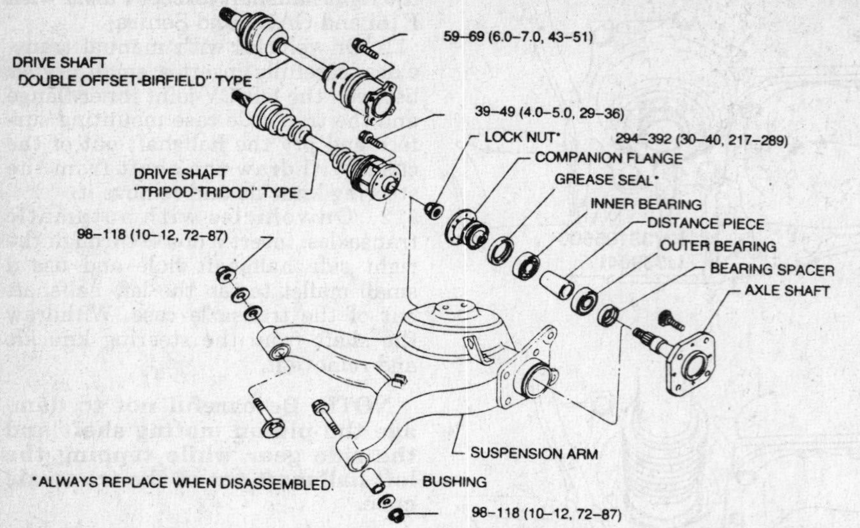

Exploded view of the rear axle shown with either the "Double Off-Set Birfield" type driveshaft or the "Tripod-Tripod" type driveshaft—models with IRS

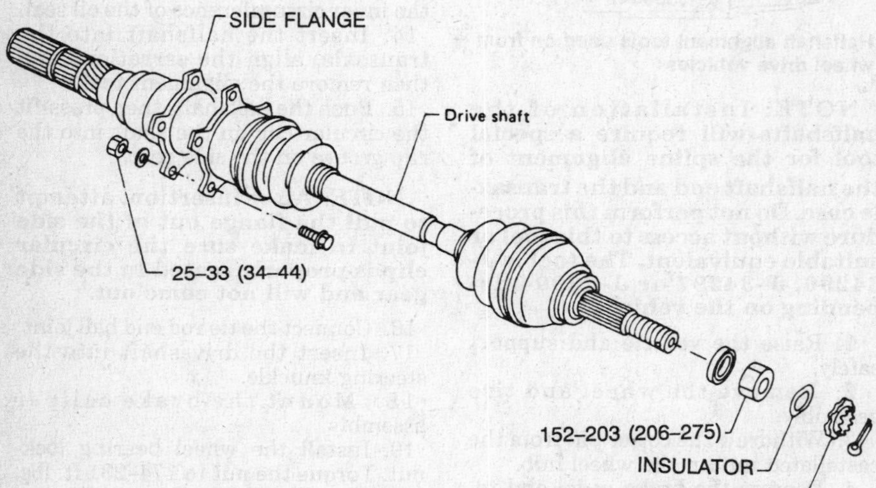

Typical rear halfshaft assembly on rear wheel drive vehicles (1990–92 300ZX shown)

cotter pin from the wheel bearing locknut.

4. Apply the parking brake and remove the rear wheel locknut.
5. On 200SX and 1988–89 300ZX, remove the spring seat stay.
6. Disconnect the halfshaft from the differential side by removing the flange bolts.
7. Grasp the halfshaft at the center and extract if from the wheel hub by prying it with a suitable prybar or with the use of a wood block and mallet.

NOTE: To protect the threads of the shaft, temporarily install the locknut when loosening the shaft from the wheel hub.

To install:

8. Insert the shaft into the wheel hub and temporarily install the locknut.

NOTE: Take care not to damage the oil seal or either end of the halfshaft during installation.

9. Connect the halfshaft to the differential and install the flange bolts. On 240SX and 300ZX, torque the flange bolts to 25–33 ft. lbs. (34–44 Nm). On 200SX with CA20E and VG30E engines, torque the flange bolts to 29–36 ft. lbs. (39–49 Nm).
10. On 200SX and 1988–89 300ZX, install the spring seat stay.
11. Apply the parking brake and tighten the locknut. Torque the locknut to 152–210 ft. lbs. (206–284 Nm) on 200SX and 1988–89 300ZX; 174–231 ft. lbs. on 240SX and 154–203 ft. lbs. (206–275 Nm) on 1990–92 300ZX.
12. Install a new locknut cotter pin and install the adjusting cap.
13. Mount the rear wheel and tire assembly.
14. Lower the vehicle.

Sentra (4WD) and Stanza Wagon (4WD)

This procedure applies to removal and installation of the rear halfshafts only.

NOTE: When removing the rear halfshafts, cover the CV-boots with cloth to prevent damage.

1. Raise and support the rear of the vehicle safely.
2. Remove the rear wheel and tire assembly.
3. Remove the adjusting cap, insulator and cotter pin from the wheel bearing locknut.
4. Apply the parking brake and remove the rear wheel locknut.
5. Disconnect the brake line. Use a brake line wrench or suitable equivalent. Plug the line to prevent leakage of brake fluid.
6. Disconnect the parking brake cable.
7. Grasp the halfshaft at the center and extract if from the wheel hub by prying it with a suitable prybar or with the use of a wood block and mallet.
8. Remove the transverse link and radius rod attaching bolts.

NOTE: Before removing the transverse rod bolts, matchmark the toe-in adjusting bolt to the adjustment degree plate.

9. Pry the halfshaft from the differential using a small prybar.
10. Remove the knuckle attaching bolts and remove the wheel hub, baffle plate, knuckle and halfshaft as a unit. Be careful not to damage the differential drive gear oil seal during removal.

To install:

11. Mount the wheel hub, baffle plate, knuckle and driveshaft and temporarily install the wheel bearing locknut.
12. Insert the halfshaft into the transaxle and properly align the splines.
13. Push the halfshaft, then press-fit the circular clip on the shaft into the clip groove on the side gear.

NOTE: After insertion, attempt to pull the flange out of the side joint to make sure the circular clip is properly seated in the side gear and will not come out.

14. Tighten the knuckle attaching bolts.
15. Install the transverse link and radius rod attaching (fixing) bolts. Make sure the toe-in bolt matchmarks are aligned properly.
16. Connect the parking brake cable and brake line.
17. Install the rear wheel bearing nut and adjust the rear wheel bearing pre-load.
18. Install adjusting cap, insulator and a new locknut cotter pin.
19. Mount the rear wheel and tire assembly.
20. Lower the vehicle.
21. Adjust the parking brake cable and bleed the brakes.

CV-Boot

REMOVAL & INSTALLATION

Transaxle Side

1. Remove the halfshaft and mount in a protected jaw vise.
2. Remove the boot bands.
3. Matchmark the slide joint housing and spider assembly to the halfshaft.
4. Remove the slide joint housing from the halfshaft.
5. Remove the spider snapring.
6. Remove the spider assembly from the halfshaft.
7. Cover the driveshaft splined end with tape to protect the CV-boot.
8. Remove the CV-boot.

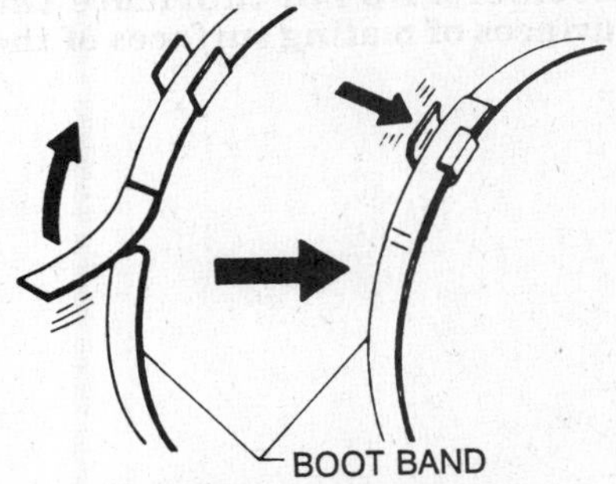

Installing the CV-boot bands

To install:

9. Install the CV-boot with a new boot band.
10. Install the spider assembly. Make sure the matchmarks are aligned properly.
11. Install a new spider snapring. Make sure the snapring seats evenly in the groove of the shaft.
12. Pack the CV-boot with grease.
13. Install the remaining boot bands. Tighten and crimp the bands using the proper tool.

Wheel Side

1. Remove the halfshaft and mount in a protected jaw vise.
2. Matchmark the joint assembly to the shaft.
3. Remove the joint assembly from the shaft using a suitable puller. Install the axle nut to prevent damage to the threads when removing the joint.
4. Remove the boot bands.
5. Cover the halfshaft splined end with tape to protect the CV-boot.
6. Remove the CV-boot.

To install:

7. Install the CV-boot with a new boot band.
8. Install the joint assembly by tapping lightly. Make sure the axle nut is installed to prevent damage to the threads. Make sure the matchmarks are aligned properly.
9. Pack the CV-boot with the proper grade and amount of grease.
10. Install the remaining boot bands. Tighten and crimp the bands using the proper tool.

Driveshaft and U-Joints

REMOVAL & INSTALLATION

200SX, 240SX and 300ZX

1. Release the hand brake.
2. Raise and safely support the vehicle. On 300ZX, remove the the front pipe and the heat shield plate.
3. Matchmark the flanges on the driveshaft and differential so the driveshaft can be reinstalled in its original orientation; this will help maintain drive line balance.
4. Unbolt the rear flange and the center bearing.
5. Withdraw the driveshaft from the transmission and pull the driveshaft down and back to remove.
6. Plug the transmission extension housing to prevent oil leakage.

To install:

7. Lubricate the sleeve yoke splines with clean engine oil prior to installation. Insert the driveshaft into the transmission and align the flange matchmarks.
8. Install the flange and the center bearing bolts.
9. On 200SX and 240SX, torque the center bearing support bracket bolts to 19–29 ft. lbs. (25–39 Nm). On 1990–92 300ZX, torque the center bearing bolts to 43–58 ft. lbs. (59–78 Nm).
10. On 200SX with CA20E and VG30E engines, torque the flange bolts to 29–33 ft. lbs. (39–44 Nm); 240SX and 1988–89 300ZX torque to 29–33 ft. lbs. (39–44 Nm). On 1990–92 300ZX turbo, torque the flange bolts to 47–54 ft. lbs. On 1990–92 300ZX non-turbo, torque the flange bolts to 29–33 ft. lbs. (39–44 Nm).
11. On 300ZX, install the the front pipe and the heat shield plate.

Sentra (4WD) and Stanza Wagon (4WD)

1. Raise and safely support the vehicle. Mark the relationship of the driveshaft flange to the differential flange.
2. Unbolt the center bearing bracket.

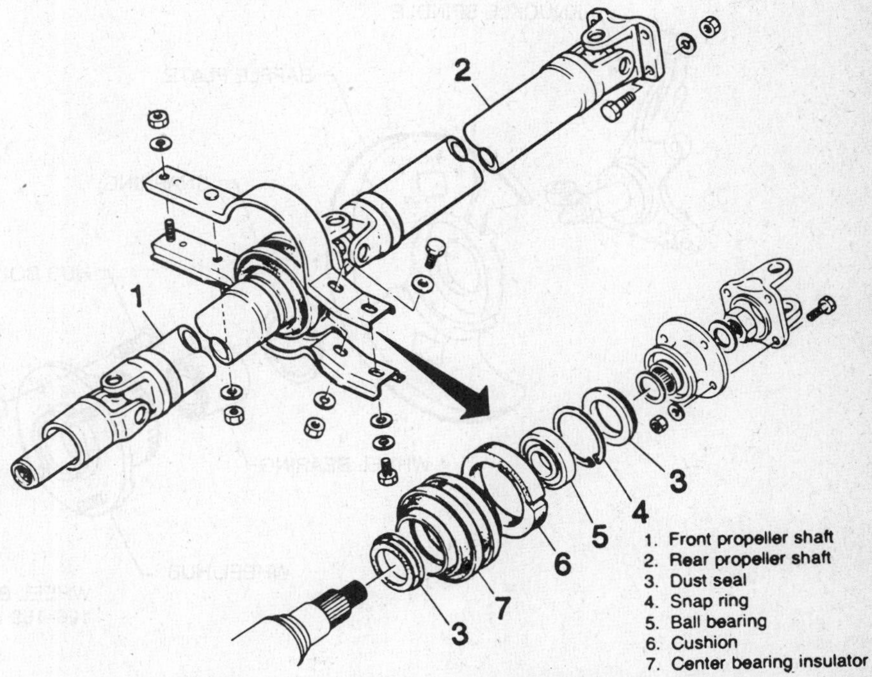

Two piece driveshaft with center bearing and three U-points

3. Unbolt the driveshaft flange from the differential flange.

4. Pull the driveshaft back under the rear axle. Plug the rear of the transmission to prevent oil or fluid loss.

5. To install, align the flange matchmarks made in Step 1. Torque the front and rear flange bolts to 25–33 ft. lbs. (34–44 Nm). On Sentra torque the center bracket bolts to 19–29 ft. lbs. (25–39 Nm). On Stanza Wagon, torque the center bearing bolts to 23–31 ft. lbs. (31–42 Nm).

Front Axle Shaft, Bearing and Seal

REMOVAL & INSTALLATION

200SX and 1988–89 300ZX

1. Raise and support the vehicle safely.

2. Remove the front wheels. Work off center hub cap by using thin tool. If necessary tap around it with a soft hammer while removing. Pry off cotter pin and take out adjusting cap. Apply the parking brake firmly and remove the wheel bearing nut. The nut will require a good deal of force to remove it.

3. Unbolt the caliper and move it aside. Do not disconnect the hose from the caliper. Do not allow the caliper to hang by the hose; support the caliper with a length of wire or rest it on a suspension member.

4. Remove the wheel hub, disc brake rotor and bearing from the spindle. During removal, capture the outer bearing to prevent it from hitting the ground.

5. To replace the bearing outer race, drive it out with a suitable brass drift and mallet.

To install:

6. Install the new bearing outer race using a suitable race installation tool.

7. Install a new oil seal so the words "BEARING SIDE" face the inner side of the hub. Coat the lip of the seal with multi-purpose grease.

8. Pack the bearings, hub, hub cap and hub cap O-ring with multi-purpose grease. If the hub cap O-ring is crimped, replace it.

9. Install the inner and outer bearings.

10. Install the wheel hub and rotor disc onto the spindle.

11. Coat the threaded portion of the spindle shaft and the contact surface between the lock washer and outer wheel bearing with multi-purpose grease.

12. Install the wheel bearing locknut and adjust the bearing pre-load. Use a new cotter pin.

13. Mount the brake caliper assembly.

14. Install the front wheels and lower the vehicle.

240SX

1. Raise and support the vehicle safely.

2. Remove the front wheels.

3. Work off center hub cap by using thin tool. If necessary tap around it with a soft hammer while removing. Pry off cotter pin and take out adjusting cap.

4. Apply the parking brake firmly and remove the wheel bearing nut. The nut will require a good deal of force to remove it.

5. Unbolt the caliper and move it aside. Do not disconnect the hose from the caliper. Do not allow the caliper to hang by the hose; support the caliper with a length of wire or rest it on a suspension member.

6. Pull the brake disc and wheel hub from the spindle.

7. Separate the tie rod and lower ball joints using the proper tool.

8. Place matchmarks on the strut lower bracket and camber adjusting pin for assembly reference. Remove the lower bracket bolts and nuts. Remove the wheel hub and knuckle assembly.

9. Remove the bearing retaining ring from the wheel hub.

10. Press the bearing assembly from the wheel hub. Apply pressure from the outside of the hub to remove the bearing.

To install:

11. Press the new bearing assembly into the hub from the inside.

NOTE: Do not press the on the inner race of the wheel bearing assembly. Do not lubricate the surfaces of mating surfaces of the

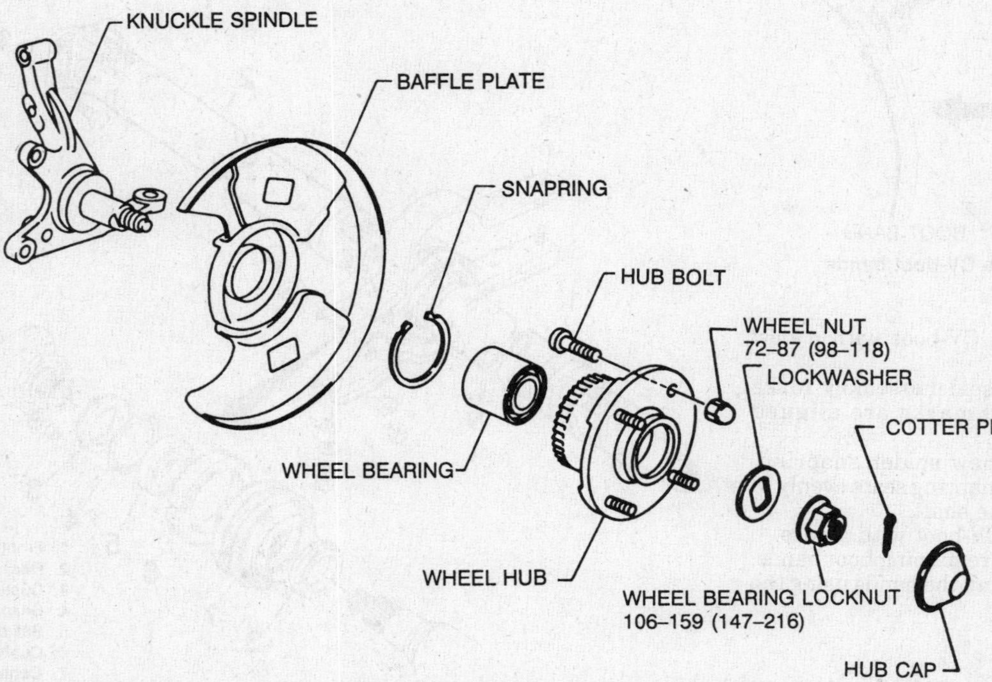

Front axle and wheel hub assembly—240SX

wheel bearing outer race and wheel with grease or oil. Be careful not to damage the grease seal.

12. Install the bearing retaining ring.
13. Coat the lip of the grease seal with multi-purpose grease.
14. Manuever the wheel hub and axle assembly onto the lower mounting bracket and install the bracket bolts and nuts. Make sure the matchmarks on the bracket and the camber adjusting pin are aligned properly.
15. Connect the lower and tie rod ball joints.
16. Push the brake disc and wheel hub onto the spindle.
17. Install the brake caliper assembly.
18. Apply the parking brake and torque the wheel bearing locknut to 108–159 ft. lbs. (147–216 Nm). Mount a dial indicator so the stylus of the dial rests on the face of the hub and check the wheel bearing axial endplay by attempting to rock the wheel hub in and out. The endplay should be 0.0012 in. or less.
19. Install a new locknut cotter pin. Install the bearing hub cap after packing it with multi-purpose grease.
20. Mount the the front wheels and lower the vehicle.

1990–92 300ZX

1. Raise and support the vehicle safely.
2. Remove the front wheels.
3. Unbolt the caliper and move it aside. Do not disconnect the hose from the caliper. Do not allow the caliper to hang by the hose; support the caliper with a length of wire or rest it on a suspension member.
4. Separate the tie rod and lower ball joints using the proper tool.

NOTE: The steering knuckle is made of an aluminum alloy. Be careful no to strike it when removing the ball joints.

5. Remove the kin pin lower nut and remove the steering knuckle assembly.
6. Remove the hub cap, wheel bearing locknut, sensor rotor (with ABS) or washer (without ABS).
7. Remove the wheel hub with a suitable drift.
8. Remove the wheel bearing retaining ring.
9. Press the wheel bearing from the knuckle.
10. Drive out the wheel bearing inner race to the outside of the wheel hub.
11. Remove the grease seal and splash guard (baffle plate).

To install:

12. From the outside of the knuckle, press the new wheel bearing assembly into the knuckle.

NOTE: Do not press the on the inner race of the wheel bearing assembly. Do not lubricate the surfaces of mating surfaces of the wheel bearing outer race and wheel with grease or oil. Be careful not to damage the grease seal.

13. Install the bearing retaining ring. Make sure it seats evenly in the groove of the knuckle.
14. Coat the lip of the grease seal with multi-purpose grease and install.
15. Install the splash guard.
16. Press the wheel hub into the steering knuckle.
17. Install the washer (without ABS), sensor rotor (with ABS) and wheel bearing locknut. Torque the locknut to 152–210 ft. lbs. (206–284 Nm). Stake the locknut tabs using a small cold chisel.
18. Place the hub cap onto the knuckle and tap it into place using a rubber or plastic mallet. Once the cap is seated lightly into the knuckle, install the cap retaining bolts and torque to 8–12 ft. lbs. (11–16 Nm).
19. Mount the steering knuckle assembly and tighten the lower king pin nut.
20. Connect the tie rod and lower ball joints using the proper tool.
21. Install the brake caliper assembly.
22. Prior to checking the bearing pre-load, spin the wheel hub at least 10 revolutions in both directions to seat the bearing. Check the wheel bearing preload and axial end play as follows:
    a. Pre-load—connect a spring scale of known calibration to a wheel hub bolt and measure the turning torque. If an NSK wheel bearing is used, the turning torque should be 1.3–8.4 lbs. (5.9–37.3 N). For NTN bearings, the turning torque should be 1.8–13.0 lbs. (7.8–57.9 N).
    b. Axial endplay—mount a dial indicator so the stylus of the dial rests on the face of the hub and check the wheel bearing axial endplay by attempting to rock the wheel hub in and out. The endplay should be 0.0020 in. (0.05mm) or less.
23. Mount the front wheels and lower the vehicle.

Rear Axle Shaft, Bearings and Seal

REMOVAL & INSTALLATION

200SX and 1988–89 300ZX

1. Block the front wheels.
2. Raise and support the vehicle safely. Remove the front wheels.
3. Apply the parking brake firmly. This helps hold the stub axle while removing the axle nut. Also, hold the stub axle at the outside while removing the nut from the axle shaft side. The nut will require a good deal of force to remove, so be sure to hold the stub axle firmly. Discard the axle nut and replace with new nut.
4. On vehicles with rear disc brakes, unbolt the caliper and move it aside. Do not disconnect the hose from the caliper. Do not allow the caliper to hang by the hose; support the caliper with a length of wire or rest it on a suspension member.
5. Remove the brake disc on vehicles with rear disc brakes. Remove the brake drum on vehicles with drum brakes.
6. Remove the stub axle with a slide hammer and an adapter. The outer wheel bearing will come off with the stub axle.
7. Unbolt and remove the companion flange from the lower arm.
8. Remove and discard the grease seal and inner bearing from the lower arm using a drift made for the purpose or a length of pipe of the proper diameter. The outer bearing can be removed from the stub axle with a puller. If the grease seal or the bearings are removed, new parts must be used on assembly.

To install:

9. Clean all the parts to be reused in solvent.
10. Sealed-type bearings are used. When the new bearings are installed, the sealed side must face out. Install the sealed side of the outer bearing facing the wheel, and the sealed side of the inner bearing facing the differential.
11. Press the outer bearing onto the stub axle.
12. The bearing housing is stamped with an A, C or no mark. Select a spacer (distance piece) on the stub axle that matches the letter stamped on the bearing housing except is there is no mark. Bearing housings with no mark always accept a B spacer.
13. Install the stub axle into the lower arm.
14. Install the new inner bearing into the lower arm with the stub axle in place. Install a new grease seal.
15. Install the companion flange onto the stub axle.
16. Install a new stub axle nut. Tighten to 152–210 ft. lbs. (206–284 Nm).
17. Install the brake disc or drum, and the caliper if removed.
18. Install the rear wheels and lower the vehicle.

240SX and 1990–92 300ZX

1. Block the front wheels.
2. Raise and support the rear of the vehicle and remove the rear wheels. Remove the cotter pin, adjusting cap and insulator.
3. Apply the parking brake firmly to hold the rear halfshaft while removing the axle nut. Hold the stub axle at the outside while removing the nut from the axle shaft side. The nut will require a good deal of force to remove.
4. Unbolt the caliper and move it aside. Do not disconnect the hose from the caliper. Do not allow the caliper to hang by the hose; support the caliper with a length of wire or rest it on a suspension member. Remove the brake disc.
5. Separate the halfhaft from the axle housing by lightly tapping it. Cover the driveshaft boots with a shop towel to prevent damage.
6. Unbolt and remove the axle housing from the vehicle. Remove the 4 bolts that hold the wheel bearing, flange and hub to the axle housing.
7. Press the wheel bearing from the axle hub. Mount the hub in a vise and remove the inner race using a bearing replacer/puller tool. Discard the inner race. If the grease seals are being replaced, replace them as a set.
8. Clean all parts in a suitable solvent. Check the wheel hub and axle housing for cracks, preferably using the dye penetrant method. Check the wheel bearing seating surface for roughness, seizure or other damage that may interfere with proper bearing function. Check the rubber bushing for wear.

To install:

9. Place the hub on a block of wood and seat the inner race using a suitable drift. Be careful not to damage the grease seals during installation of the inner race.
10. Press the bearing into the hub using a suitable drift.
11. Mount the axle housing. Torque the axle housing bolts to 58–72 ft. lbs. (78–98 Nm) on both the 240SX and 1990–92 300ZX.
12. Insert the halfshaft into the wheel hub. Lubricate the halfshaft splines prior to installation. Make sure the splines are aligned properly.
13. Install the caliper assembly.
14. Install the wheel bearing locknut. On 240SX, torque the nut to 174–231 ft. lbs. (235–314 Nm). On 1990–92 300ZX, torque the nut to 152–203 ft. lbs. (206–275 Nm). Install the insulator and fit adjusting cap. Install a new cotter pin.
15. On 1990–92 300ZX, check the axial endplay as follows before mounting the rear wheels: mount a dial indicator so the stylus of the dial rests on the face of the hub and check the wheel bearing axial endplay by attempting to rock the wheel hub in and out. The endplay should be 0.0020 in. (0.05mm) or less.
16. Mount the rear wheels and lower the vehicle.

Front Wheel Hub, Knuckle and Bearings

REMOVAL & INSTALLATION

Maxima, Pulsar, Sentra and Stanza

1. Raise and support the vehicle safely.
2. Remove the front wheels.
3. Remove the brake rotor.
4. Remove the cotter pin, adjusting cap and insulator.
5. Apply the parking brake firmly and remove the wheel bearing nut. The nut will require a good deal of force to remove it.
6. Unbolt the caliper and move it aside. Do not disconnect the hose from the caliper. Do not allow the caliper to hang by the hose; support the caliper with a length of wire or rest it on a suspension member.
7. Separate the tie rod end from the steering knuckle using the proper tool.
8. Disconnect the halfshaft from the transaxle using the proper tool or by tapping on it with a block of wood and a mallet.

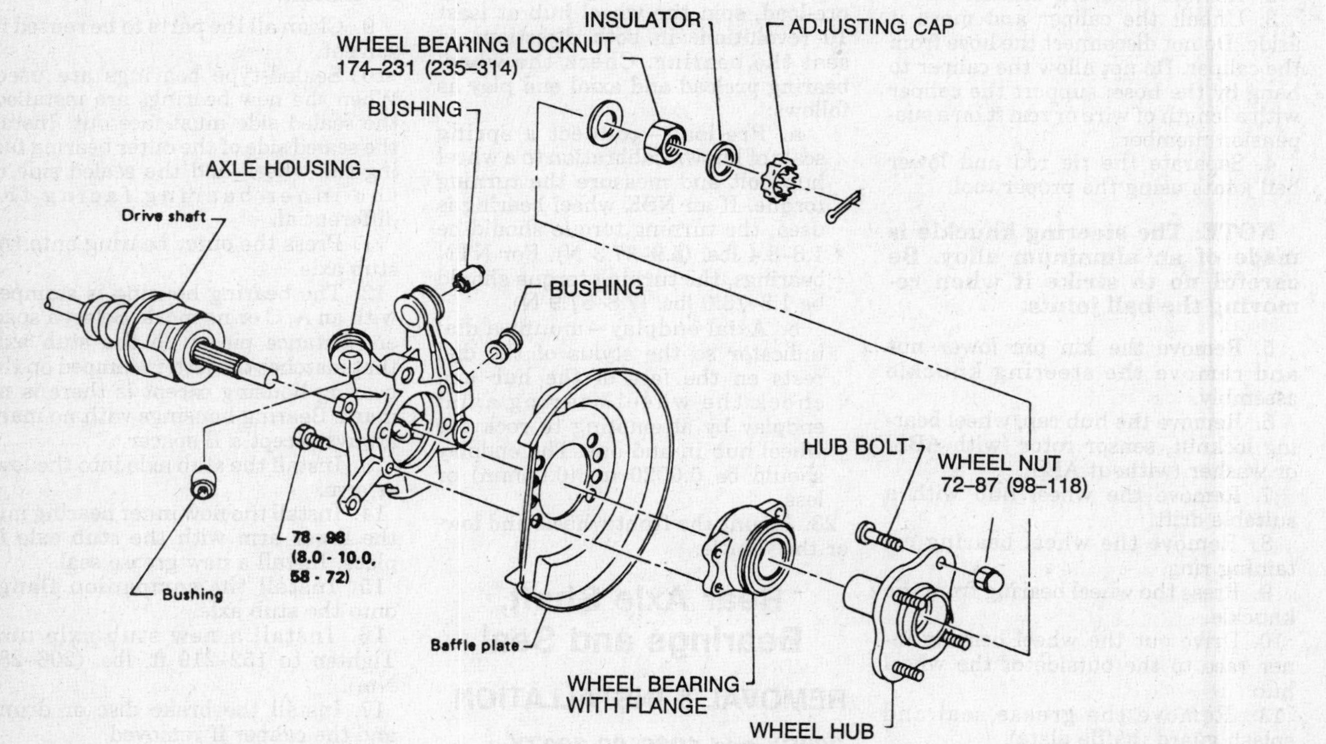

Rear axle housing assembly—240SX shown—1990–92 300ZX similar

NOTE: Cover the CV-boots with cloth to prevet damage when removing the halfshafts.

9. Remove the nuts and bolt that attach the knuckle to the strut. Make sure to place a visible matchmark on the adjusting pin and knuckle mounting bracket before removing these fasteners.
10. Remove the lower arm bolts.
11. On Pulsar and Sentra, loosen the lower ball joint nut and separate the knuckle from the lower ball joint stud using the proper tool.
12. Remove the knuckle and hub assembly.
13. Drive out the hub and outside inner race with a suitable tool.
14. Withdraw the outside inner race from the wheel hub.
15. On Maxima and Stanza, remove the outer and grease seals from the hub at this time, then press the outer race from the hub.
16. On Pulsar and Sentra, press the inside inner race from the hub. Set the race aside for use in removal of the wheel bearing.
17. Remove the wheel bearing retainer with the proper tool. On Maxima and Stanza, there are retainers on both sides of the hub. After both retainers are removed, the bearing can be pressed from the hub at this time.
18. On Pulsar and Sentra, place the inside inner race set aside in Step 16 on top of the wheel bearing and press the bearing out of the hub. Apply pressure to the inside of the knuckle to remove the bearing.
19. Clean all parts in a suitable solvent. Check the wheel hub and axle housing for cracks, preferrably using the dye penetrant method. Check the wheel bearing seating surface for roughness, seizure or other damage that may interfere with proper bearing function.

To install:

20. On Maxima and Stanza, install the inner bearing retainer.
21. Press the new bearing into the knuckle by applying pressure to the outside of the knuckle. Do not exceed 3.3 tons of pressure.

NOTE: Do not press the on the inner race of the wheel bearing assembly. Do not lubricate the surfaces of mating surfaces of the wheel bearing outer race and wheel with grease or oil. Be careful not to damage the grease seal.

22. Install the remaining bearing retainer. Make sure it seats evenly in the groove of the knuckle.
23. Coat the lip of the seal with multi-purpose grease. On Maxima and Stanza, install the inner and outer grease seals. Make sure the lip of the seal(s) faces the inside of the hub.
24. Press the hub into the knuckle. Do not exceed 3.3 tons of pressure.
25. Clamp the knuckle portion in a vise and apply a pre-load of 3.5–5.0 tons to the outside (wheel bolt side) of the bearing with a suitable press. Spin the knuckle several turns in both directions and make sure the bearing spins freely and does not bind.
26. Mount the knuckle and hub assembly.
27. On Pulsar and Sentra, connect the lower ball joint to the knuckle.
28. Install the lower arm bolts.
29. Install the knuckle-to-strut fasteners. Make sure the adjusting pin matchmarks are aligned properly.
30. Install the halfshafts.
31. Connect the tire rod end to the steering knuckle using the proper tool.
32. Install the brake caliper assembly.
33. Install the wheel bearing locknut. Torque the nut to 174–231 ft. lbs. (235–314 Nm) on Maxima and Stanza; 145–203 ft. lbs. (196–275 Nm) on 1988–89 Pulsar; 145–203 ft. lbs. (196–275 Nm) on Sentra and 1990 Pulsar. When tightening the nut, apply the brake pedal.
34. Install the insulator and adjusting cap. Install a new cotter pin into the wheel bearing locknut.
35. Check the axial endplay as follows: mount a dial indicator so the stylus of the dial rests on the face of the hub and check the wheel bearing axial endplay by attempting to rock the wheel hub in and out. The endplay should be 0.0020 in. (0.05mm) or less.
36. Mount the front wheels.
37. Lower the vehicle.

Differential Carrier

REMOVAL & INSTALLATION

1. Raise the rear of the vehicle and support safely. Drain the oil from the differential. Position a floor jack underneath the differential unit.
2. Disconnect the brake hydraulic lines and the parking brake cable. On 240SX, remove the brake caliper leaving the brake line connected. Plug the brake lines to prevent lakage.
3. Disconnect the sway bar from the control arms on either sides (not required on 1990–92 300ZX).
4. Remove the rear exhaust pipe.
5. Disconnect the driveshaft and the rear axle shafts.
6. Remove the rear shock absorbers from the control arms. On 1990–92 300ZX, remove the nuts that attach the differential rear cove to the suspension member.
7. Unbolt the differential unit from the chassis at the differential mounting insulator. On 1990–92 300ZX, remove the mounting member from the front of the final drive.
8. Lower the rear assembly out of the vehicle using the floor jack. It is best to have at least one other person helping to balance the assembly. After the final drive is removed, support the center suspension member to prevent damage to the insulators.
9. During installation, torque the rear cover-to-insulator nuts to 72–87 ft. lbs. (98–118 Nm); mounting insulator-to-chassis bolts to 22–29 ft. lbs. (30–39 Nm); strut nuts to 51–65 ft. lbs. (69–81 Nm); sway bar-to-control arm nuts to 12–15 ft. lbs. (16–21 Nm). On 240SX and 300ZX, torque the drive shaft flange bolts to 25–33 ft. lbs. (34–44 Nm); on 200SX with CA20E and VG30E engines, torque the flange bolts to 29–36 ft. lbs. (39–49 Nm).

MANUAL TRANSMISSION

For further information on transmissions/transaxles, please refer to "Chilton's Guide to Transmission Repair".

Transmission Assembly

REMOVAL & INSTALLATION

200SX, 240SX and 300ZX

1. Disconnect the negative battery cable.
2. Raise and support the vehicle safely.
3. On 1988–89 300ZX, remove the exhaust front pipe, catalytic converter and exhaust manifold conecting tube. On 1990–92 300ZX, remove the exhaust pipe section from the manifold and remove the support bracket from the transmission.
4. Unbolt the driveshaft at the rear and remove. If there is a center bearing, unbolt it from the crossmember. Seal the end of the transmission extension housing to prevent leakage.
5. Disconnect the speedometer drive cable from the transmission.
6. On 200SX, 1988–89 300ZX non-turbocharged and all 1990–92 300ZX, remove the shifter lever. On 1988–89 300ZX turbocharged, remove the shift knob and console boot finisher. On 240SX, disconnect the control rod front the shift lever.

NOTE: On the 1988–89 300ZX turbo, the shifter boot must not be removed from the shift lever.

7. Remove the clutch operating cylinder from the clutch housing.
8. Support the engine with a large wood block and a jack under the oil pan. Do not place the jack under the oil pan drain plug.
9. Unbolt the transmission from the crossmember. Support the transmission with a jack and remove the crossmember.
10. Lower the rear of the engine to allow clearance.
11. Unplug the back-up light, neutral and overdrive switch connectors.
12. Unbolt the transmission. Lower and remove it to the rear.

NOTE: The transmission bolts are different lengths. Tagging the transmission-to-engine bolts upon removal will facilitate proper tightening during installation.

To install:

13. Raise the transmission onto the engine and install the mounting bolts. Torque the bolts as follows:
   a. 200SX with 4 cylinder engine – tighten bolts (1) and (2) to 29–36 ft. lbs. (39–49 Nm) and bolt (3) to 22–29 ft. lbs. (29–39 Nm).
   b. 200SX (V6) and 1988–89 300ZX – tighten the long mounting bolts (65mm and 60mm) to 29–36 ft. lbs. (39–49 Nm). Tighten the short bolts (55mm and 25mm) to 22–29 ft. lbs. (29–39 Nm).
   c. 240SX – tighten bolts (1), (2) and (4) to 29–36 ft. lbs. (39–49 Nm) and bolt (3) to 22–29 ft. lbs. (29–39 Nm)
   d. 1990–92 300ZX – tighten bolts (1), (2) and (3) to 29–36 ft. lbs. (39–49 Nm). Tighten bolts (4) and (5) to 22–29 ft. lbs. (29–39 Nm).
14. Plug in the back-up light, neutral and overdrive switch connectors.
15. Install the crossmember.
16. Install the clutch operating cylinder.
17. Install the shifter lever, shift knob and console boot finisher or control rod.
18. Connect the speedometer drive cable.
19. Install the driveshaft. Torque the flange bolts to 29–33 ft. lbs. (34–44 Nm). On 1990–92 300ZX, torque the center bearing bracket nuts to 19–29 ft. lbs. (25–39 Nm).
20. On 1990–92 300ZX, connect the exhaust tube section to the manifolds and attach the support bracket to the transmission. On 1988–89 300ZX, install the exhaust front tube, catalytic converter and exhaust manifold conecting tube.

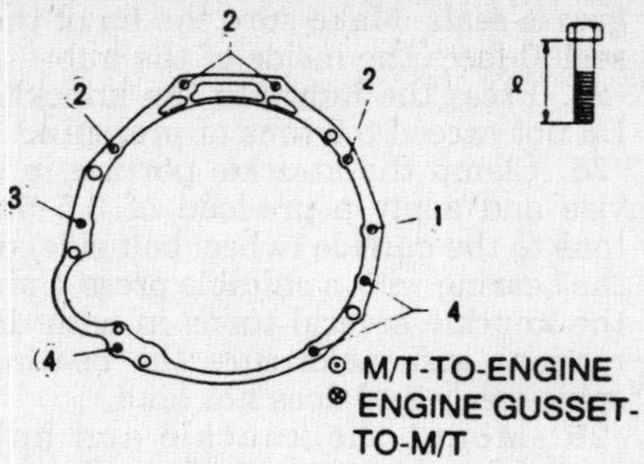

Transmission mounting bolt locations on 1988–89 300ZX and 200SX (V6); bolt (1) is 65mm, bolt (2) is 60mm, bolt (3) is 55mm and bolt 4 is 25mm

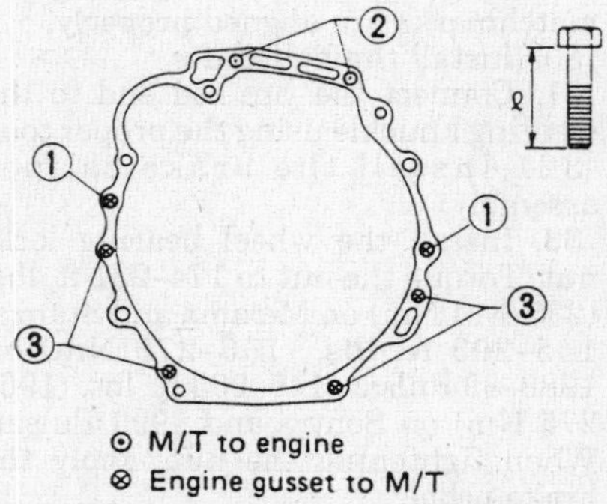

Transmission mounting bolt locations on 200SX (4 cylinder); bolt (1) is 75mm, bolt (2) is 65mm and bolt (3) is 25mm

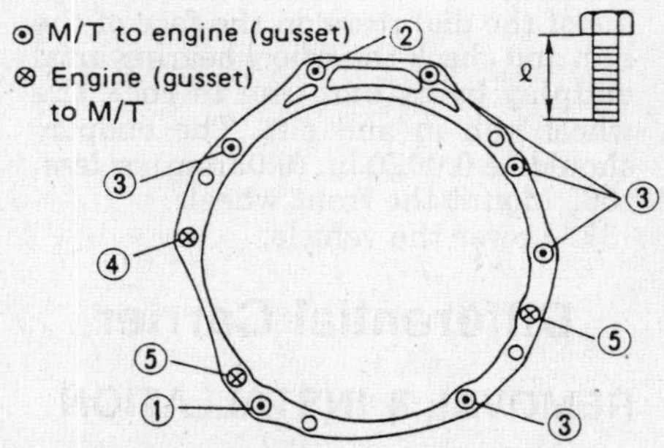

Transmission mounting bolt locations on 1990–92 300ZX; bolt (1) is 100mm, bolt (2) is 65mm, bolt (3) is 60mm, bolt (4) is 55mm and bolt (5) is 25mm

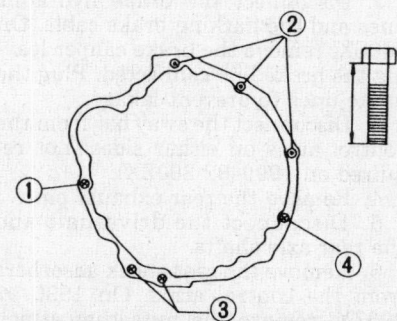

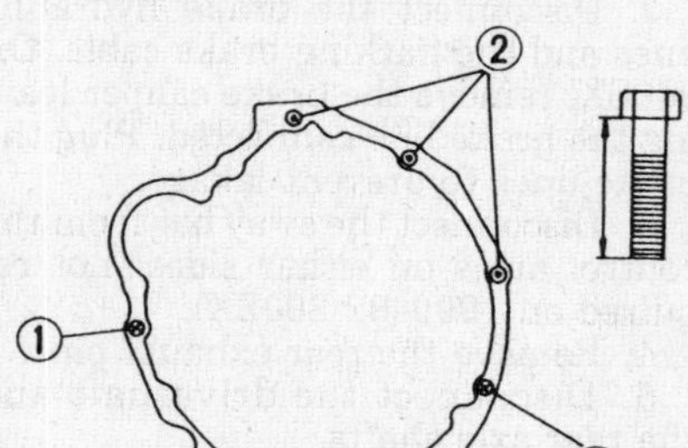

Transmission mounting bolt locations on 240SX; bolt (1) is 70mm, bolt (2) is 60mm, bolt (3) is 30mm and bolt (4) is 25mm

21. Lower the vehicle and connect the negative battery cable.

MANUAL TRANSAXLE

For further information on transmissions/transaxles, please refer to "Chilton's Guide to Transmission Repair".

Transaxle Assembly

REMOVAL & INSTALLATION

Except 1990–92 Stanza

1. Disconnect the negative battery cable.
2. Remove the battery and battery bracket.
3. Remove the air duct, air cleaner box and air flow meter.
4. Raise the front of the vehicle and support safely.
5. Drain the transaxle oil.
6. On Stanza Wagon (4WD) and Sentra (4WD) vehicles, remove the transfer case.
7. Withdraw the halfshafts from the transaxle. On Stanza Wagon (4WD), remove only the left halfshaft.

NOTE: When removing halfshafts, use care not to damage the lip of the oil seal. After shafts are removed, insert a steel bar or wooden dowel of suitable diameter to prevent the side gears from rotating and falling into the differential case.

8. On 1989–92 Maxima, remove the clutch operating cylinder from the transaxle.
9. Remove the wheel well protector(s).
10. Separate the control rod and support rod from the transaxle.
11. Remove the engine gusset securing bolt and the engine mounting.
12. Remove the clutch control cable from the operating lever.
13. Disconnect speedometer cable from the transaxle.
14. Disconnect the wires from the reverse (back-up), neutral and overdrive switches. On 1989–92 Maxima, disconnect the speed and position switch sensors from the transaxle also.
15. Support the engine by placing a jack under the oil pan, with a wooden block placed between the jack and pan for protection.
16. Support the transaxle with a hydraulic floor jack.

17. Remove the engine mounting securing bolts.

NOTE: Most of the transaxle mounting bolts are different lenghts. Tagging the bolts upon removal will facilitate proper tightening during installation.

18. Remove the bolts attaching the transaxle to the engine.
19. Using the hydraulic floor jack as a carrier, carefully lower the transaxle down and away from the engine.

To install:

20. Before installing, clean the mating surfaces on the engine rear plate and clutch housing. On Sentra (4WD) and Stanza Wagon (4WD), apply sealant KP510-00150 or equivalent.
21. Apply a light coat of a lithium-based grease to the spline parts of the clutch disc and the transaxle input shaft.
22. Raise the transaxle into place and bolt it to the engine. Install the engine mounts. Torque the tranasxle mounting bolts as follows:
    a. 1988 Maxima – tighten bolts (1), (2) and (3) to 32–43 ft. lbs. (43–58 Nm). Tighten bolt (4) to 22–30 ft. lbs. (30–40 Nm) and bolt (5) to 12–15 ft. lbs. (16–21 Nm).
    b. 1989–92 Maxima – tighten bolt (1) to 12–15 ft. lbs. (16–21 Nm), bolt (2) to 22–30 ft. lbs. (30–40 Nm), bolts (3) and (4) to 32–43 ft. lbs. (43–58 Nm). Torque the front and rear gusset bolts to 22–30 ft. lbs. (30–40 Nm).
    c. 1988 Pulsar/Sentra (E16S and E16i) – tighten bolts (1) and (3) to 12–15 ft. lbs. (16–22 Nm). Tighten bolts (2) and (4) to 14–22 ft. lbs. (20–29 Nm). Bolts (3) and (4) are found all Sentra models.
    d. 1988 Pulsar (CA18DE) – On CA18DE engines, tighten bolts (1) and (2) to to 32–43 ft. lbs. (43–58 Nm) and bolts (3) to 22–30 ft. lbs. (30–40 Nm).
    e. 1989 Pulsar (CA18DE) – tighten bolts (1) and (2) to 32–43 ft. lbs. (43–58 Nm) and bolt (3) to 22–30 ft. lbs. (30–40 Nm).
    f. Pulsar and 2WD Sentra – tighten all bolts to 12–15 ft. lbs. (16–21 Nm).
    g. Sentra 4WD – torque all the bolts to 22–30 ft. lbs.
    h. 1988–89 Stanza – tighten bolts (1), (2) and (3) to 32–43 ft. lbs. (39–49 Nm). Tighten bolt (4) to 22–30 ft. lbs. (30–40 Nm).
23. On 1989–92 Maxima, connect the speed and position switch sensor wires. Connect the reverse (back-up), neutral and overdrive switch wires.
24. Connect the speedometer cable to the transaxle.
25. Connect the clutch cable to the operating lever.

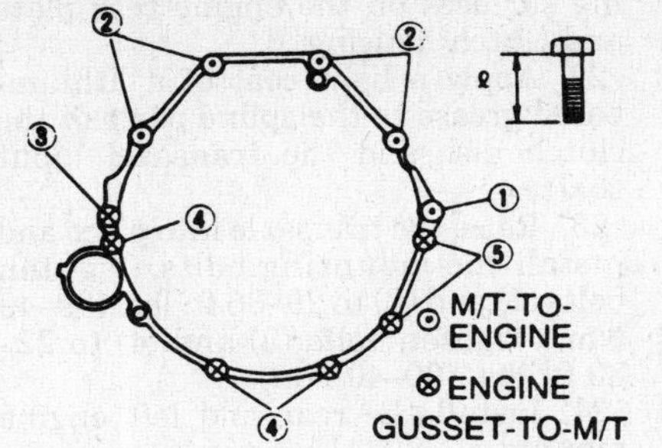

Transaxle mounting bolt locations on 1988 Maxima; bolt (1) is 65mm, bolt (2) is 55mm, bolt (3) is 60mm and bolts (4) and (5) are 25mm

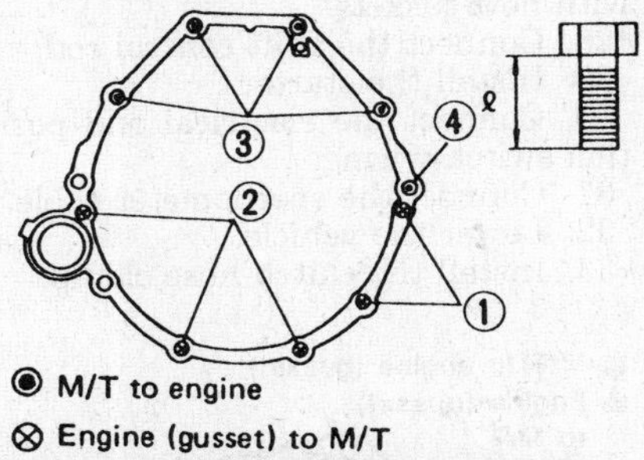

Transaxle mounting bolt locations on 1989–92 Maxima; bolts (1) and (2) are 25mm, bolt (3) is 55mm and bolt (4) is 65mm

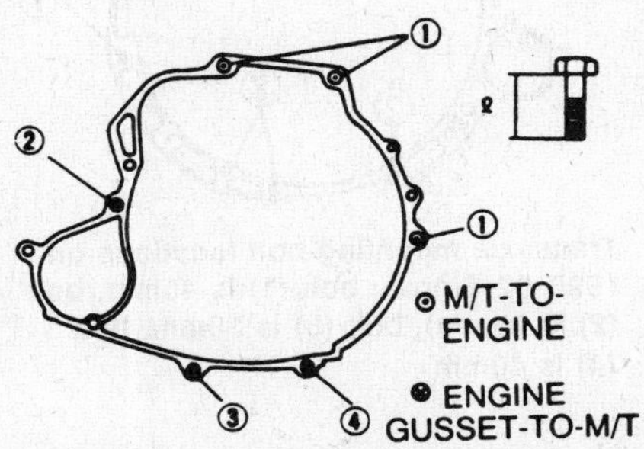

Transaxle mounting bolt locations on 1988 Pulsar/Sentra (E16S and E16i); bolt (1) is 70mm, bolt (2) is 40mm, bolt (3) is 25mm, bolt (4) is 20mm

26. Connect the control and support rods to the transaxle.
27. Install the wheel well protectors.
28. On 1989–92 Maxima, install the clutch operating cylinder.
29. Install the halfshafts.
30. On Stanza Wagon (4WD) and Sentra (4WD) vehicles, install the transfer case.
31. Lower the vehicle.
32. Install the air duct, air cleaner box and air flow meter.
33. Install the battery and battery bracket.
34. Connect the negative battery cable.
35. Remove the filler plug and fill the transaxle to the proper level with fluid that meets API GL-4 specifications.

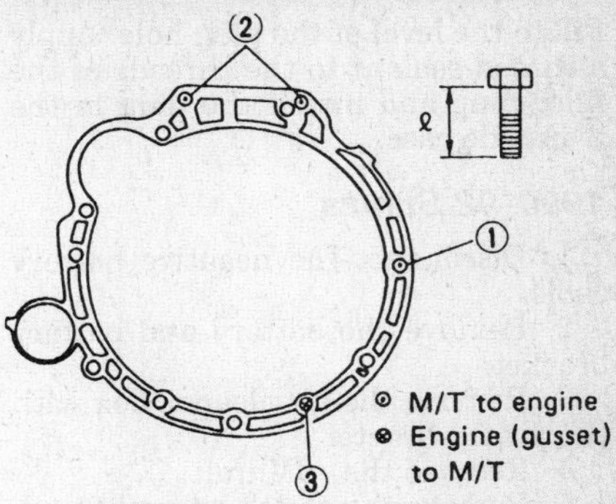

Transaxle mounting bolt locations on 1989 Pulsar (CA18DE engine); bolt (1) is 125mm, bolt (2) is 65mm, bolt (3) is 45mm. Bolt (1) has a nut

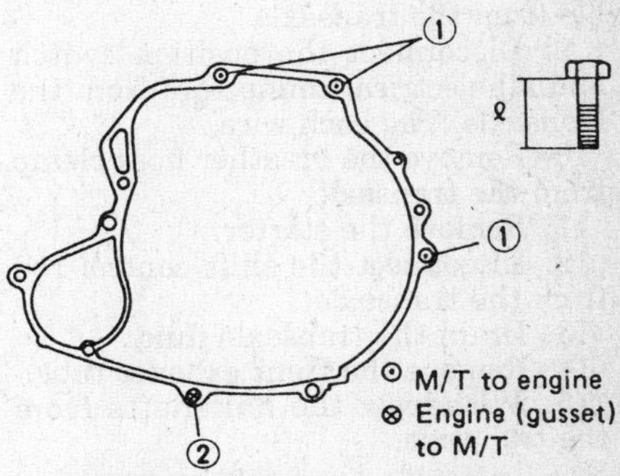

Transaxle mounting bolt locations on 1989–91 Pulsar and 2WD Sentra (GA16i engine); bolt (1) is 70mm and bolt (2) is 25mm

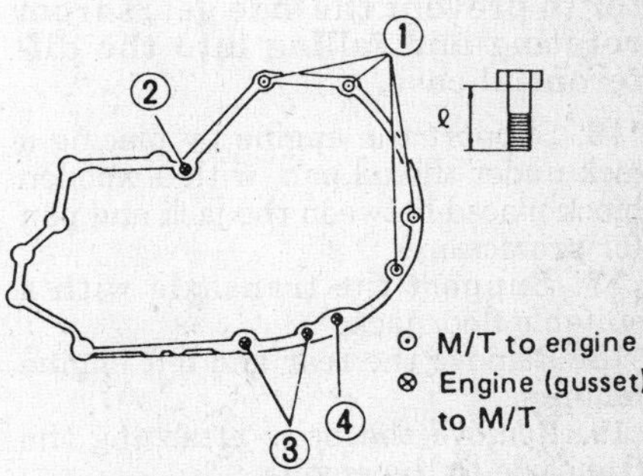

Transaxle mounting bolt locations on 1988–91 Sentra 4WD; bolt (1) is 70mm, bolt (2) is 40mm, bolt (3) is 20mm), bolt (4) is 55mm

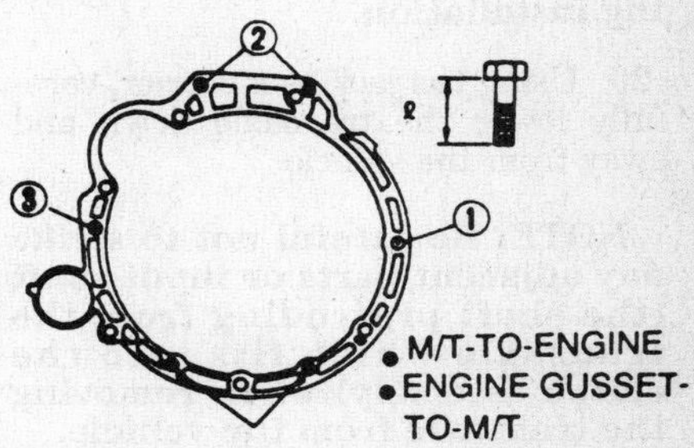

On 1988–89 Stanza; bolt (1) is 120mm, bolt (2) is 65mm, bolt (3) is 70mm, bolt (4) is 25mm. Bolt (1) has a nut

Fill to the level of the plug hole. Apply a thread sealant to the threads of the filler plug and install the plug in the transaxle case.

1990–92 Stanza

1. Disconnect the negative battery cable.
2. Remove the battery and battery bracket.
3. Remove the air cleaner box with the air flow meter.
4. Remove the AIV unit.
5. Remove the clutch operating cylinder from the transaxle.
6. Remove the clutch hose clamp.
7. Raise and support the vehicle safely.
8. Disconnect the speedometer cable from the transaxle.
9. Disconnect the position switch and all electrical connectors from the transaxle. Tag each wire.
10. Remove the breather hose clamp from the transaxle.
11. Remove the starter.
12. Disconnect the shift control rod from the transaxle.
13. Drain the transaxle fluid.
14. Remove the front exhaust tube.
15. Withdraw the halfshafts from the transaxle.

NOTE: When removing halfshafts, use care not to damage the lip of the oil seal. After shafts are removed, insert a steel bar or wooden dowel of suitable diameter to prevent the side gears from rotating and falling into the differential case.

16. Support the engine by placing a jack under the oil pan, with a wooden block placed between the jack and pan for protection.
17. Support the transaxle with a suitable floor jack.
18. Remove the rear and left engine mounts.
19. Remove the bolts attaching the transaxle to the engine.

NOTE: The transaxle mounting bolts are different lengths. Tagging the bolts upon removal will facilitate proper tightening during installation.

20. Using the jack as a carrier, carefully lower the transaxle down and away from the vehicle.

NOTE: Be careful not to strike any adjacent parts or input shaft (the shaft protruding from the transaxle which fits into the clutch assembly) when removing the transaxle from the vehicle.

To install:

21. Before installing, clean the mating surfaces on the engine rear plate and clutch housing.
22. Apply a light coat of a lithium-based grease to the spline parts of the clutch disc and the transaxle input shaft.
23. Raise the transaxle into place and install the mounting bolts. Tighten bolts (1) and (2) to 29–36 ft. lbs. (39–49 Nm). Tighten bolts (3) and (4) to 22–30 ft. lbs. (30–40 Nm).
24. Install the rear and left engine mounts.
25. Remove the transaxle and engine supports.
26. Install the halfshafts.
27. Install the front exhaust tube with new gaskets.
28. Connect the shift control rod.
29. Install the starter.
30. Connect the electrical and position switch wiring.
31. Connect the speedometer cable.
32. Lower the vehicle.
33. Install the clutch hose clamp.

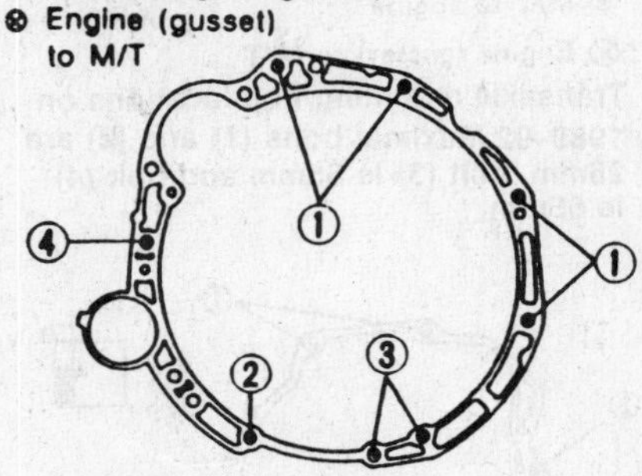

Transaxle mounting bolt locations on 1990–92 Stanza; bolt (1) is 45mm, bolt (2) is 25mm), bolt (3) is 30mm, bolt (4) is 40mm

34. Install the clutch operating cylinder.
35. Install the AIV unit.
36. Install the air cleaner box and air flow meter.
37. Install the battery bracket and battery.
38. Connect the negative battery cable.
39. Remove the filler plug and fill the transaxle to the proper level with fluid that meets API GL-4 specifications. Fill to the level of the plug hole. Apply a thread sealant to the threads of the filler plug and install the plug in the transaxle

CLUTCH

Clutch Assembly

REMOVAL & INSTALLATION

1. Remove the transmission or transaxle.
2. Insert a clutch aligning bar or similar tool all the way into the clutch disc hub. This must be done so as to support the weight of the clutch disc during removal.
3. Mark the clutch assembly-to-flywheel relationship with paint or a center punch so the clutch assembly can be assembled in the same position from which it is removed.
4. Loosen the pressure plate bolts in criss-cross fashion, a turn at a time to gradually relieve the spring pressure. Remove the bolts once the spring pressure is relieved.
5. Remove the pressure plate and

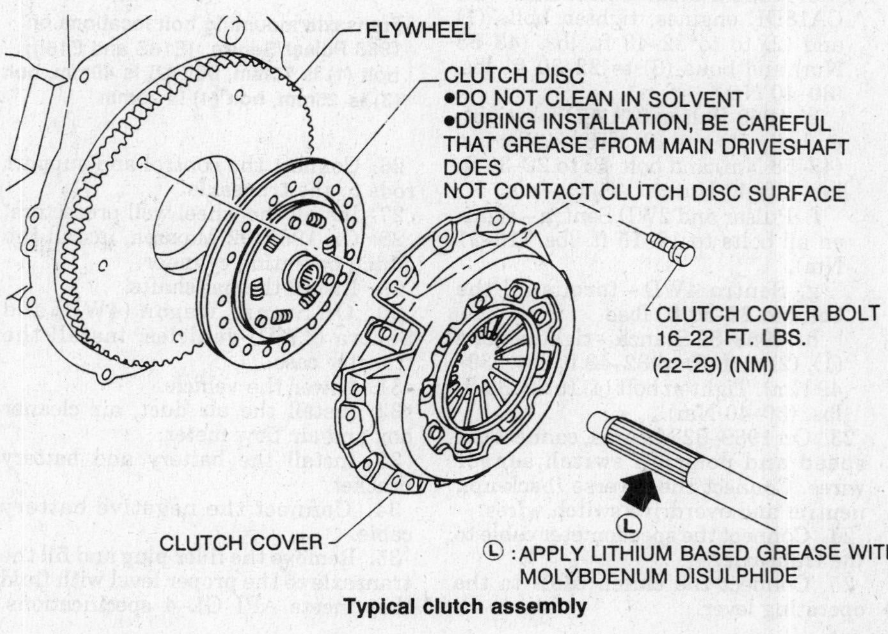

Typical clutch assembly

clutch disc. Inspect the pressure plate or scoring for roughness, and reface or replace as necessary. Slight roughness can be smoothed with a fine emery cloth. Inspect the clutch disc for worn or oily facings, loose rivets and broken or loose springs, and replace.

6. Remove the release mechanism. On Pulsar and Sentra, the clutch lever is removed by aligning the lever retaining pins with the clutch cavity, then driving out the pins with a suitable pin punch. Inspect the release sleeve and lever contact surfaces for wear, rust or any other damage. Replace if necessary.

7. Inspect the pressure plate for wear, scoring, etc., and reface or replace as necessary. Minor imperfections or discoloration may be removed with emery cloth.

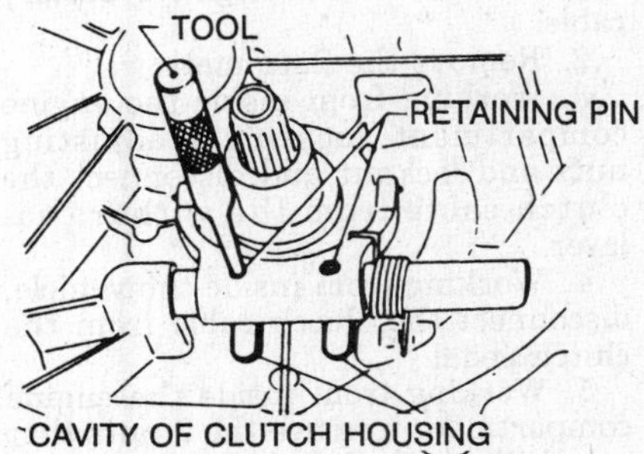

Clutch lever retaining pin removal—Pulsar and Sentra

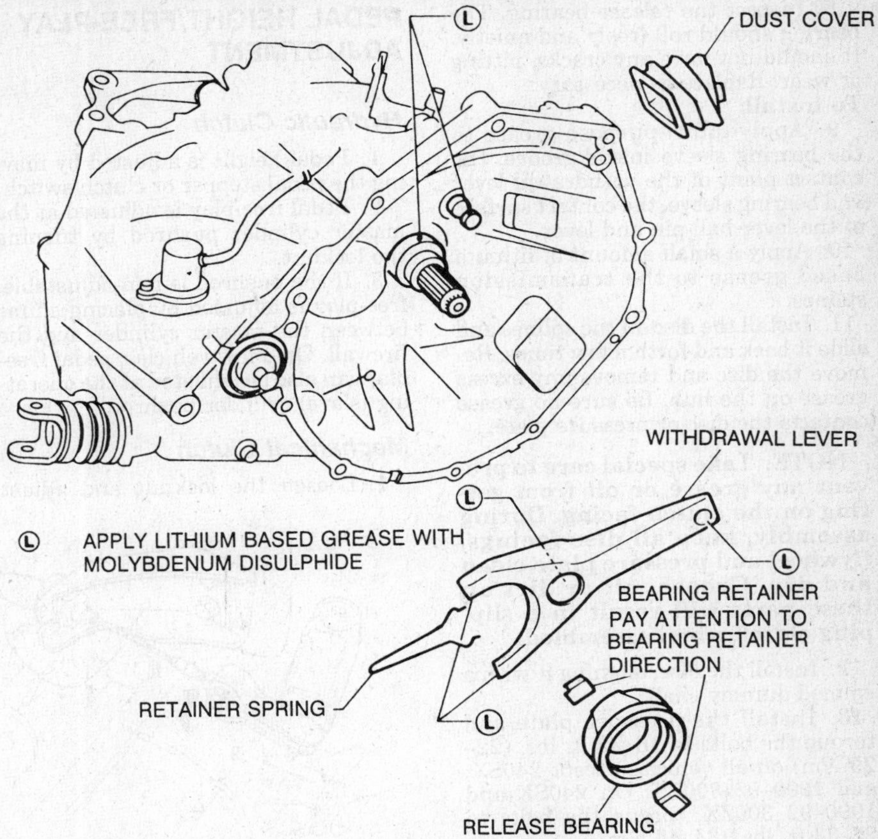

Clutch release mechanism—except Pulsar and Sentra

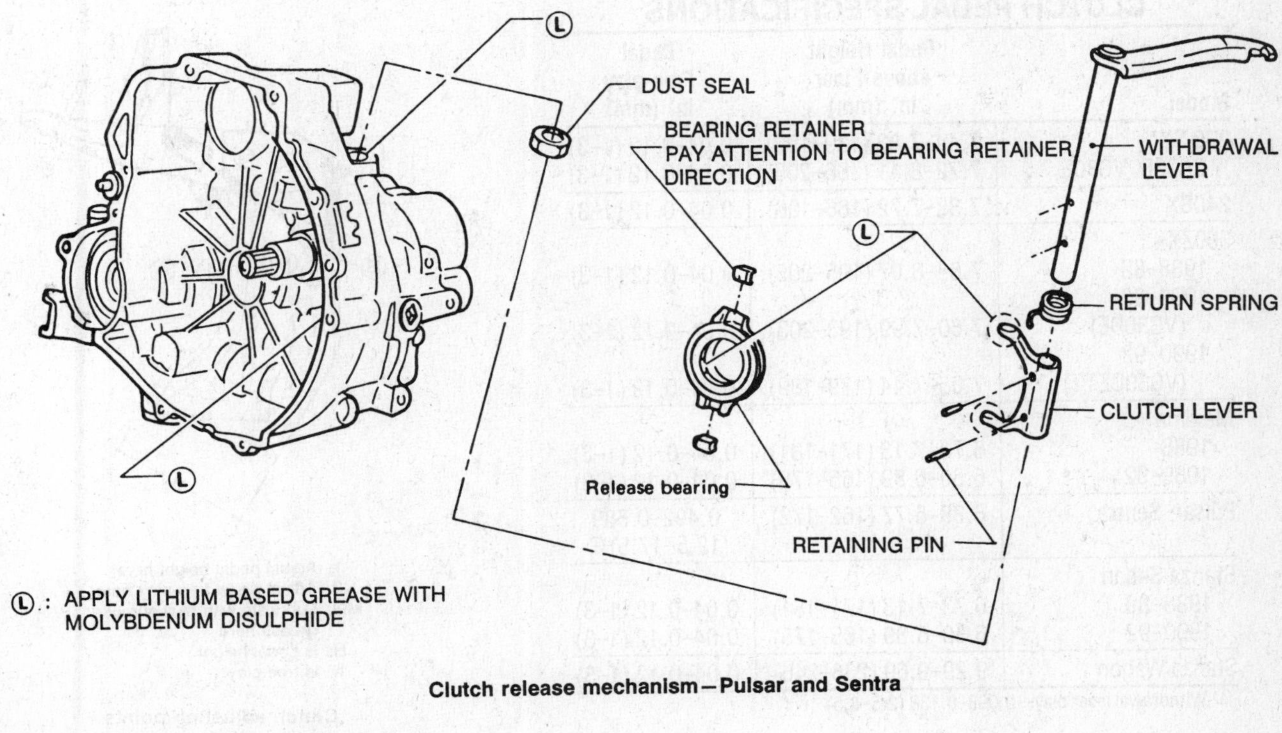

Clutch release mechanism—Pulsar and Sentra

8. Inspect the release bearing. The bearing should roll freely and quietly. It should not have any cracks, pitting or wear. Replace as necessary.

To install:

9. Apply multi-purpose grease to the bearing sleeve inside groove, the contact point of the withdrawal lever and bearing sleeve, the contact surface of the lever ball pin and lever.

10. Apply a small amount of lithium based grease to the transmission splines.

11. Install the disc on the splines and slide it back and forth a few times. Remove the disc and remove any excess grease on the hub. Be sure no grease contacts the disc or pressure plate.

NOTE: Take special care to prevent any grease or oil from getting on the clutch facing. During assembly, keep all disc facings, flywheel and pressure plate clean and dry. Grease, oil or dirt on these parts will result in a slipping clutch when assembled.

12. Install the disc, aligning it with a splined dummy shaft.

13. Install the pressure plate and torque the bolts to 16–22 ft. lbs. (22–29 Nm) on all vehicles except 240SX, and 1990–92 300ZX. On 240SX and 1990–92 300ZX, torque the bolts to 25–33 ft. lbs. (34–44 Nm).

14. Remove the dummy shaft.

15. Install the transmission or transaxle.

PEDAL HEIGHT/FREE-PLAY ADJUSTMENT

Hydraulic Clutch

1. Pedal height is adjusted by moving the pedal stopper or clutch switch.

2. Pedal free-play is adjusted at the master cylinder pushrod by turning the locknut.

3. If the pushrod is non-adjustable, free-play is adjusted by placing shims between the master cylinder and the firewall. On a few vehicles, pedal free-play can also be adjusted at the operating (slave) cylinder pushrod.

Mechanical Clutch

1. Loosen the locknut and adjust the pedal height by means of the pedal stopper. Tighten the locknut.

2. Push the withdrawal lever in by hand until resistance is felt. Adjust withdrawal lever play at the lever tip end with the locknuts. Withdrawal lever play should be 0.0198–0.138 in. (2.5–3.5mm).

3. Depress and release the clutch pedal several times and then recheck the withdrawal lever play again. Readjust if necessary.

4. Measure the pedal free travel at the center of the pedal pad.

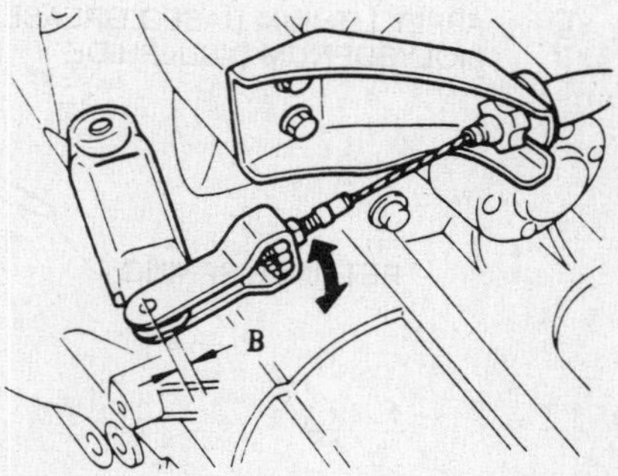

Clutch withdrawal lever adjustment on Pulsar and Sentra. Arrow shows locknut adjustment

Clutch Cable

REMOVAL & INSTALLATION

1. Disconnect the negative battery cable.

2. Remove the floor mats.

3. Working from inside the engine compartment, loosen the adjusting nuts and locknut and disconnect the clutch cable from the withdrawal lever.

4. Working from inside the vehicle, disconnect the clutch cable from the clutch pedal.

5. Working from inside the engine compartment, remove the 2 nuts that attach the end of the cable to the fire wall.

6. From inside the engine compartment, pull the clutch cable through the firewall and remove it.

CLUTCH PEDAL SPECIFICATIONS

Model	Pedal Height above Floor in. (mm)	Pedal Free-play in. (mm)
200 SX	7.44–7.83 (189–199)	0.04–0.12 (1–3)
CA20E, VG30E	7.72–8.11 (196–206)	0.04–0.12 (1–3)
240SX	7.32–7.72 (186–196)	0.04–0.12 (1–3)
300ZX		
1988–89	7.68–8.07 (195–205)	0.04–0.12 (1–3)
1990–92 (VG30DE)	7.60–7.99 (193–203)	0.04–0.12 (1–3)
1990–92 (VG30DETT)	7.05–7.44 (179–189)	0.04–0.12 (1–3)
Maxima		
1988	6.73–7.13 (171–181)	0.04–0.12 (1–3)
1989–92	6.50–6.89 (165–175)	0.04–0.12 (1–3)
Pulsar, Sentra	6.38–6.77 (162–172)	0.492–0.689 (12.5–17.5) ①
Stanza Sedan		
1988–89	6.73–7.13 (171–181)	0.04–0.12 (1–3)
1990–92	6.50–6.89 (165–175)	0.04–0.12 (1–3)
Stanza Wagon	9.29–9.69 (236–246)	0.04–0.12 (1–3)

① Withdrawal lever play—0.098–0.138 (2.5–3.5)

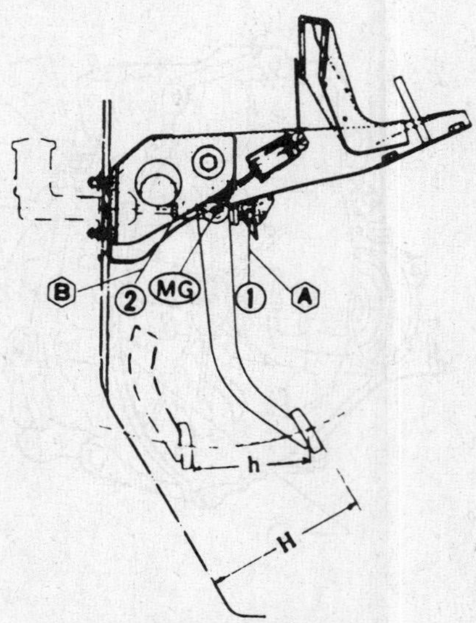

1. Adjust pedal height here
2. Adjust pedal free-play here
MG. Lubricate with multipurpose grease here
H. is pedal height
h. is free play

Clutch adjusting points

To install:

7. Route the clutch cable through the passenger compartment.
8. Position the cable end over the studs on the firewall and install the 2 mounting nuts. Torque the nuts to 6–8 ft. lbs. (9–11 Nm).
9. Connect the clutch cable to the clutch pedal.
10. Connect the clutch cable to the withdrawal lever.
11. Lubricate the pedal fulcrum pin and pivot points with lithium based grease.
12. Adjust the cable and the clutch switch.
13. Check the clutch for proper engagement.
14. Install the floor mats.
15. Connect the negative battery cable.

Clutch Master Cylinder

REMOVAL & INSTALLATION

1. Disconnect the negative battery cable.
2. Disconnect the clutch pedal arm from the pushrod.
3. Disconnect the clutch hydraulic line from the master cylinder. Plug the end of line to prevent leakage.
4. Remove the nuts attaching the master cylinder and remove the master cylinder and pushrod toward the engine compartment side.
5. Install the master cylinder in the reverse order of removal.
6. Bleed the clutch hydraulic system and make all necessary clutch adjustments.

Clutch Slave Cylinder

REMOVAL & INSTALLATION

1. Remove the slave cylinder attaching bolts and the pushrod from the shift fork.
2. Disconnect the flexible fluid hose from the slave cylinder and remove the unit form the vehicle. Plug the end of the hose.
3. Install the slave cylinder in the reverse order of removal and bleed the clutch hydraulic system.

Hydraulic Clutch System Bleeding

Bleeding is required to remove air trapped in the hydraulic system. This operation is necessary whenever the system has been leaking or opened for maintenance. The bleed screw is located on the clutch slave (operating) cylinder.

Some vehicles are also equipped with a clutch damper mechanism. The clutch damper mechanism is bled in exactly the same manner as the operating cylinder. It should be bled along with the operating cylinder.

1. Remove the bleed screw dust cap.
2. Attach a transparent vinyl tube to the bleed screw, immersing the free end in a clean container of clean brake fluid.
3. Fill the master cylinder with the proper fluid.
4. Open the bleed screw about ¾ turn.
5. Depress the clutch pedal quickly. Hold it down. Have an assistant tighten the bleed screw. Allow the pedal to return slowly.
6. Repeat Steps 2 and 5 until no more air bubbles are seen in the fluid container.
7. Remove the bleed tube. Replace the dust cap. Refill the master cylinder.
8. Bleed the clutch damper, if equipped.

AUTOMATIC TRANSMISSION

For further information on transmissions/transaxles, please refer to "Chilton's Guide to Transmission Repair".

Transmission Assembly

REMOVAL & INSTALLATION

200SX, 240SX and 300ZX

1. Disconnect the battery cable.
2. Remove the accelerator linkage.
3. Detach the shift linkage.
4. Disconnect the neutral safety switch and downshift solenoid wiring.
5. Raise and safely support the vehicle. Remove the drain plug and drain the torque converter. If there is no converter drain plug, drain the transmission. If there is no transmission drain plug, remove the pan to drain. Replace the pan to keep out dirt.
6. Remove the front exhaust pipe.
7. Remove the vacuum tube and speedometer cable.
8. Disconnect the fluid cooler and charging tubes. Plug the tube ends to prevent leakage.
9. Lower the driveshaft and remove the starter.
10. Support the transmission with a jack under the oil pan. Support the engine also.
11. Remove the rear crossmember.
12. Mark the relationship between the torque converter and the driveplate. Remove the bolts holding the converter to the driveplate through the access hole at the front, under the engine by rotating the crankshaft. Unbolt the transmission from the engine and remove it.

NOTE: The transmission bolts are different lengths. Tag each bolt according to location to ensure proper installation. This is particularly important on the 240SX and 1990–92 300ZX.

13. Check the driveplate runout with a dial indictator. Runout must be no more than 0.020 in.

To install:

14. If the torque converter was removed from the engine for any reason, after it is installed, the distance from the face of the converter to the edge of the converter housing must be checked prior to installing the transmission. This is done to ensure proper installation of the torque converter. On 200SX and 1988–89 300ZX, the dimension should be 1.38 in. (35mm) or more. On 240SX and 1990–92 300ZX (non-turbocharged), the dimension should be 1.02 in. (26mm) or more. On 1990–92 300ZX (turbocharged), the dimension should be 0.98 in. (25mm) or more.
15. Raise the transmission and bolt the driveplate to the converter and transmission to the engine. Torque the driveplate-to-torque converter and converter housing-to-engine bolts to 29–36 ft. lbs. (39–49 Nm) on all except 240SX and 1990–92 300ZX. On these vehicles, torque the transmission mounting bolts as follows: On 240SX, tighten bolts (1) and (2) to 29–36 ft. lbs. (39–49 Nm); tighten bolt (3) to 22–29 ft. lbs. (29–39 Nm); tighten the gusset-to-engine bolts to 22–29 ft. lbs. (29–39 Nm). On 1990–92 300ZX tighten bolts (1), (2), (3), (6) and (7) to 29–36 ft. lbs. (39–49 Nm). Tighten bolts (2) and (5) to 22–29 ft. lbs. (29–39 Nm). Tighten the engine gusset bolts to 22–29 ft. lbs. (29–39 Nm).

NOTE: After the converter is installed, rotate the crankshaft several times to make sure the transmission rotates freely and does not bind.

16. Install the rear crossmember.
17. Remove the engine and transmission supports.
18. Install the starter and connect the driveshaft. Torque the flange bolts to 29–33 ft. lbs. (34–44 Nm) on all except 1990–92 300ZX (turbo). On 1990–

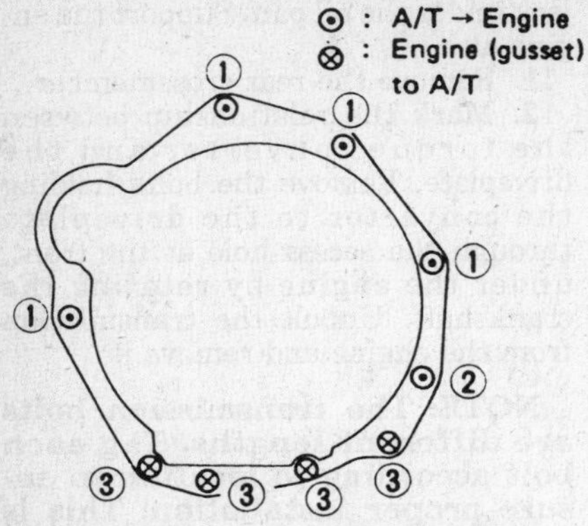

Transmission mounting bolt locations on 240SX; bolt (1) is 40mm, bolt (2) is 50mm, bolt (3) is 25mm and the gusset bolts are 20mm

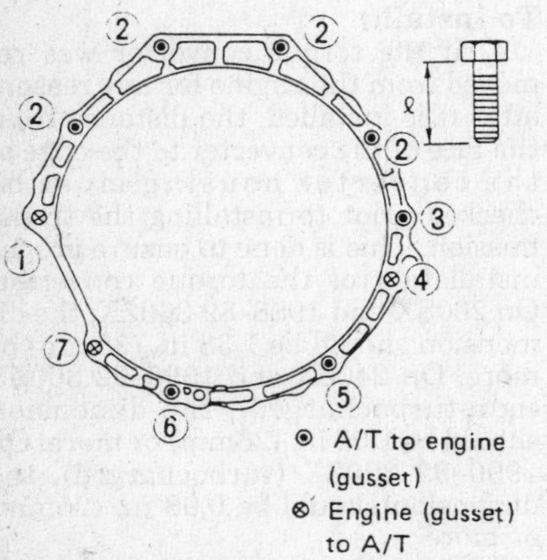

Transmission mounting bolt locations on 1990–92 300ZX (turbo and non-turbo)

92 300ZX (turbo), torque the flange bolts to 40–47 ft. lbs. (54–64 Nm).

19. Unplug, connect and tighten the fluid cooler tubes.
20. Connect the speedometer cable and the vacuum tube.
21. Connect the front exhaust pipe using new gaskets.
22. Connect the switch wiring to the transmission.
23. Connect the shift linkage.
24. Connect the negative battery cable, fill the transmission to the proper level and make any necessary adjustment.
25. Perform a road test and check the fluid level.

SHIFT LINKAGE ADJUSTMENT

200SX, 240SX and 300SX

If the detents cannot be felt or the pointer indicator is improperly aligned while shifting from the **P** range to

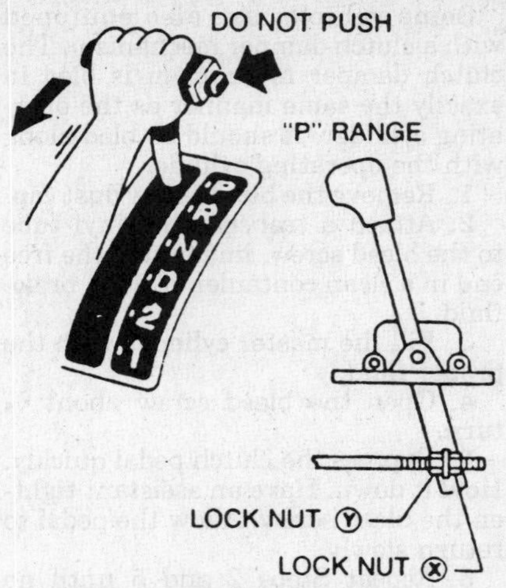

Manual control linkage adjustment—200SX, 240SX and 300ZX

range **1**, the linkage should be adjusted.

1. Place the shifter in the **P** position.
2. Loosen the locknuts.
3. Tighten the outer locknut **X** until it touches the trunnion, pulling the selector lever toward the **R** range side without pushing the button.
4. Back off the outer locknut **X** ¼–½ turns and then tighten the inner locknut **Y** to 5–11 ft. lbs. (8–15 Nm).
5. Move the selector lever from **P** to **1**. Make sure it moves smoothly.

NOTE: The 1988–92 300ZX has an automatic transmission interlock system. This interlock system prevents the transmission selector from being shifted from the P position unless the brake pedal is depressed.

KICKDOWN SWITCH ADJUSTMENT

When the accelerator pedal is depressed, a click can be heard just before the pedal bottoms out. If the click is not heard, loosen the locknut and extend the switch until the pedal lever makes contact with the switch and the switch clicks.

On 1990–92 300ZX, before adjusting the kickdown switch, make sure the accelerator cable is properly adjusted. Then, check the clearance between the stopper rubber and the threaded end of the switch with the accelerator cable fully depressed. The clearance should be 0.012–0.039 in. (0.3–1.0mm). If the clearance is not as

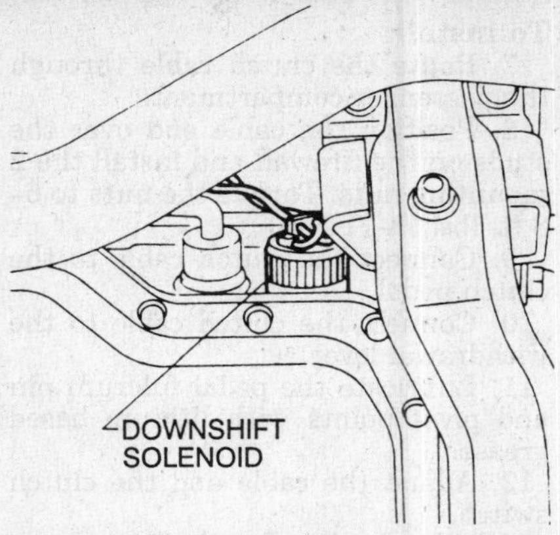

Downshift solenoid location—200SX and 1987–89 300ZX

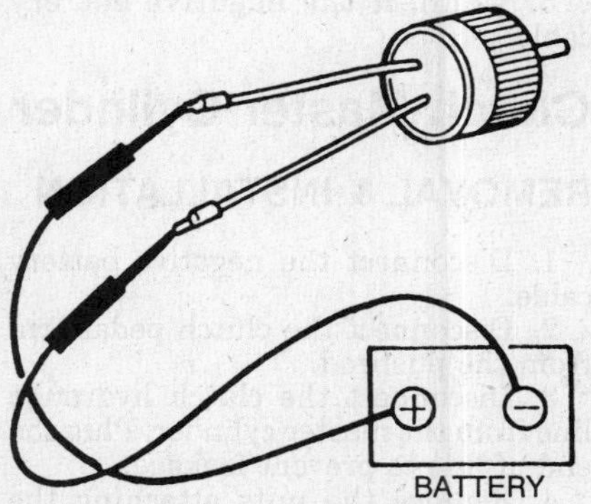

Check the downshift solenoid by applying battery voltage

specified, adjust by loosening the switch locknut and turning the switch in or out. Tighten the locknut and check the clearance again.

DOWNSHIFT SOLENOID CHECK

200SX and 1988–89 300ZX

The solenoid is controlled by a downshift switch on the accelerator linkage inside the vehicle. To test the switch and solenoid operation, preform the following:

1. Turn the ignition to the **ON** position.
2. Push the accelerator all the way down to actuate the switch.
3. The solenoid should "click" when actuated. The solenoid is screwed into the outside of the case. If there is no click, check the switch, wiring, and solenoid.
4. To remove the solenoid, first drain 2–3 pints of fluid, then unscrew the unit.
5. Apply battery voltage to the switch and listen for the click. If no click is audible, replace the switch or repair the wiring.

AUTOMATIC TRANSAXLE

For further information on transmissions/transaxles, please refer to "Chilton's Guide to Transmission Repair".

Transaxle Assembly

REMOVAL & INSTALLATION

1988 Maxima

NOTE: The engine/transaxle unit must be removed and installed as a unit. After removal, the transaxle may be separated from the engine.

1. Remove the transaxle/engine as an assembly.
2. Remove the transaxle-to-engine mounting bolts and then carefully draw out the rear plate.
3. Remove the bolts securing the torque converter to the driveplate.
4. Before removing the torque converter, use chalk or paint to matchmark at least 2 parts so they may be replaced in their original positions during installation. Remove the torque converter.
5. Check the driveplate runout with a dial indicator. Runout must be no more than 0.020 in.
6. If the torque converter was removed from the engine for any reason, after it is installed, the distance from the face of the converter to the edge of the converter housing must be checked prior to installing the transaxle. This is done to ensure proper installation of the torque converter. The dimension should be 0.709 in. (18mm) or more.
7. During installation of the transaxle/engine assembly, observe the following:
   a. When installing the torque converter to the driveplate, be certain the matchmarks made during removal are in alignment. Apply Loctite® or a similar sealing compound to the converter-to-driveplate bolts before installation.
   b. After the torque converter has been reinstalled, rotate the crankshaft a few times to ensure that the transaxle rotates freely, with no binding.
   c. Adjust the control cable and check the inhibitor switch.
   d. After installation of the engine/transaxle assembly into the vehicle, fill the transaxle and engine with the proper amounts of fluids, then road test the vehicle.

Pulsar, Sentra, Stanza and 1989–92 Maxima

1. Disconnect the negative battery cable.
2. Raise and support the vehicle safely.
3. Remove the left front tire.
4. Drain the transaxle fluid.
5. Remove the left side fender protector.
6. Remove the halfshafts.

NOTE: Be careful not to damage the oil seals when removing the halfshafts. After removing the halfshafts, install a suitable bar so the side gears will not rotate and fall into the differential case.

7. On Stanza Wagon, disconnect and remove the forward exhaust pipe.
8. Disconnect the speedometer cable.
9. Disconnect the throttle wire (cable) connection.
10. Remove the control cable rear end from the unit and remove the oil level gauge tube.
11. Place a suitable jack under the transaxle and engine. Do not place the jack under the oil pan drain plug. Support the engine with wooden blocks placed between the engine and the center member.
12. Disconnect the oil cooler and charging tubes. Plug the tube ends to prevent leakage.
13. Remove the engine motor mount securing bolts, as required.
14. Remove the starter motor and disconnect all electrical wires from the transaxle.
15. Loosen and remove all but 3 of the bolts holding the transaxle to the engine. Leave the 3 bolts in to support the weight of the transaxle while removing the converter bolts.
16. Remove the driveplate or dust covers.
17. Remove the bolts holding the torque torque converter to the driveplate. Rotate the crankshaft to gain access to each bolt. Before separating the torque converter, place chalk marks on 2 parts for alignment purposes during installation.

NOTE: The transaxle bolts are different lengths. Tag each bolt according to location to ensure proper installation.

18. Remove the 3 temporary bolts. Move the jack gradually until the transaxle can be lowered and removed from the vehicle through the left side wheel housing.
19. Check the driveplate runout with a dial indicator. Runout must be no more than 0.020 in.

To install:

20. If the torque converter was removed from the engine for any reason, after it is installed, the distance from the face of the converter to the edge of the converter housing must be checked prior to installing the transaxle. This is done to ensure proper installation of the torque converter. On Maxima, the distance should be 0.71 in. (18mm) or more. On Pulsar with RL3F01A transaxles, it should be 0.831 in. (21mm) or more. On Pulsar with RL4F02A transaxles, it should be 0.748 in. (19mm) or more. On Stanza, it should be 0.75 in. (19mm) or more.

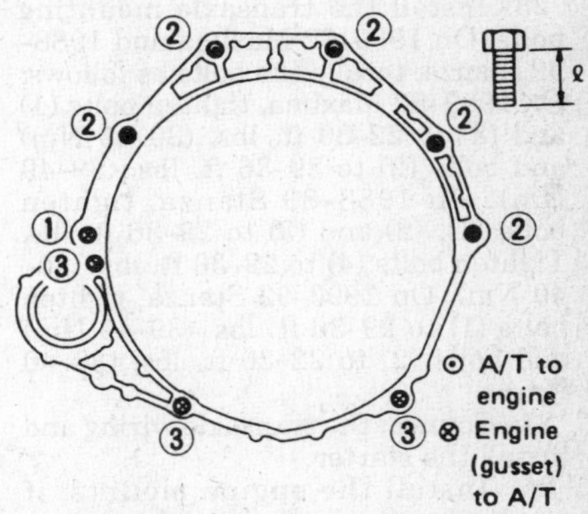

Transaxle mounting bolt locations on 1989–92 Maxima; bolt (1) is 60mm, bolt (2) is 45mm, bolt (3) is 25mm

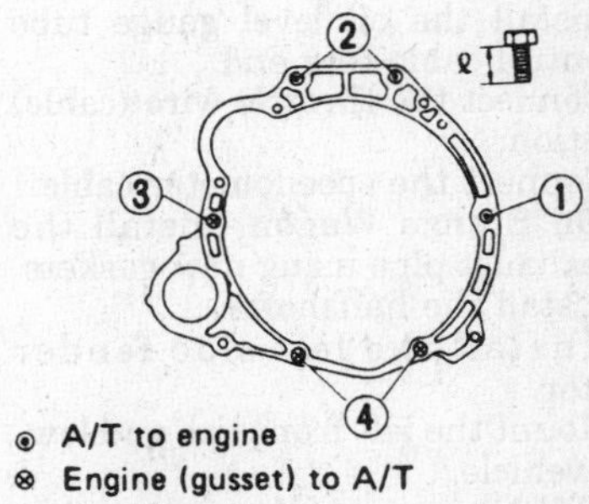

Transaxle mounting bolt locations on 1988–89 Stanza; bolt (1) is 85mm, bolt (2) is 50mm, bolt (3) is 70mm, bolt (4) is 25mm

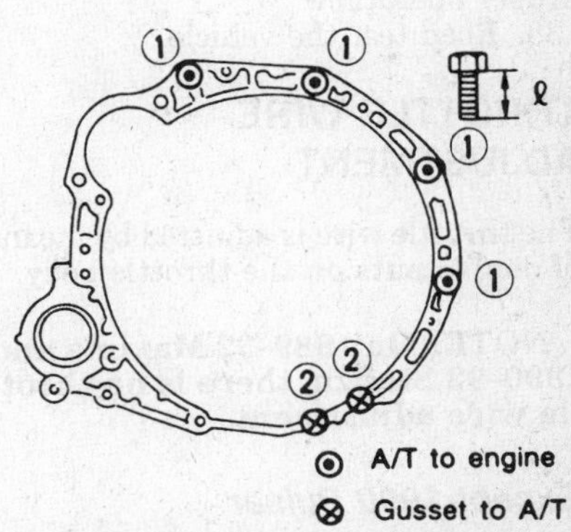

Transaxle mounting bolt locations on 1990–92 Stanza; bolt (1) is 45mm and bolt (2) is 20mm

21. Raise the transaxle onto the engine and install the torque coverter-to-driveplate bolts. Torque the bolts to specification. Install 3 bolts to support the transaxle while tighten the converter bolts.

NOTE: After the converter is installed, rotate the crankshaft several times to make sure the transaxle rotates freely and does not bind.

22. Install the driveplate or dust covers.
23. Install the transaxle mounting bolts. On 1989–92 Maxima and 1988–92 Stanza, torque the bolts as follows: On 1989–92 Maxima, tighten bolts (1) and (3) to 22–30 ft. lbs. (30–40 Nm) and bolts (2) to 29–36 ft. lbs. (39–49 Nm). On 1988–89 Stanza, tighten bolts (1), (2) and (3) to 29–36 ft. lbs. Tighten bolts (4) to 22–30 ft. lbs. (30–40 Nm). On 1990–92 Stanza, tighten bolts (1) to 29–36 ft. lbs. (39–49 Nm) and bolts (2) to 22–30 ft. lbs. (30–40 Nm)
24. Connect the transaxle wiring and install the starter.
25. Install the engine mounts, if removed.
26. Connect the oil cooler and charging tubes.
27. Remove the engine and transaxle supports.
28. Install the oil level gauge tube and control cable rear end.
29. Connect the throttle wire (cable) connection.
30. Connect the speedometer cable.
31. On Stanza Wagon, install the front exhaust pipe using new gaskets.
32. Install the halfshafts.
33. Install the left side fender protector.
34. Mount the left front tire and lower the vehicle.
35. Fill the transaxle and engine with the proper amounts of fluids.
36. Adjust the control cable and throttle wire.
37. Check the inhibitor switch for proper operation.
38. Road test the vehicle.

THROTTLE WIRE ADJUSTMENT

The throttle wire is adjusted by means of double nuts on the throttle body.

NOTE: On 1989-92 Maxima and 1990–92 Stanza, there is no throttle wire adjustment.

Except 1990 Pulsar

1. Loosen the adjusting nuts.
2. With the throttle fully opened, turn the threaded shaft inward as far as it will go and then tighten the first nut against the bracket.

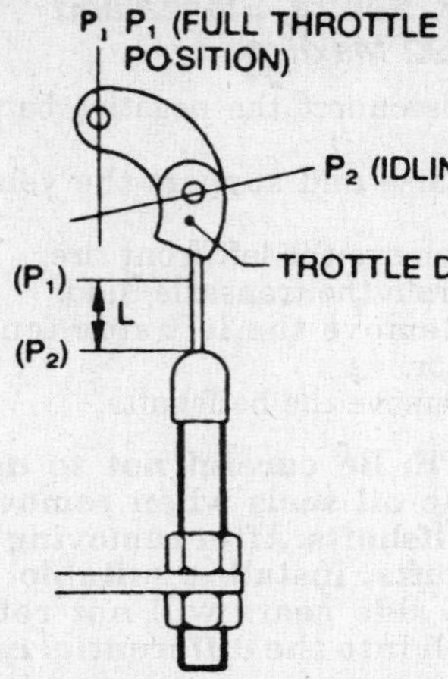

Throttle wire stroke – Sentra, 1988–89 Pulsar and 1988 Maxima

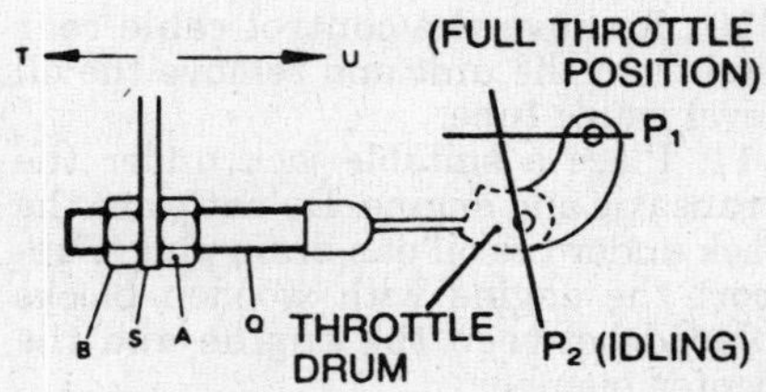

Throttle wire adjustment – Sentra, 1988–89 Pulsar and 1987–88 Maxima

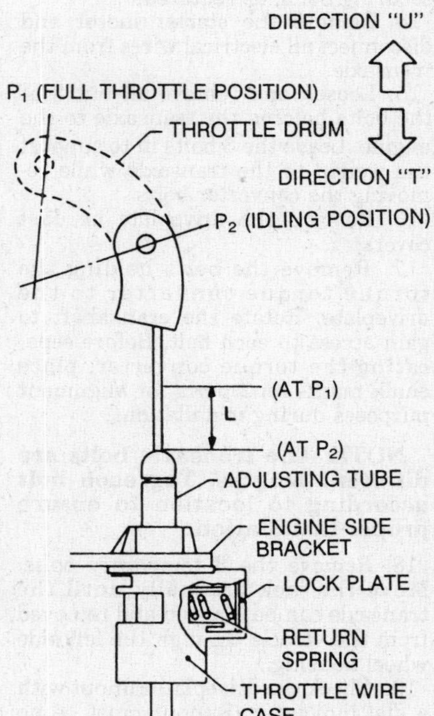

Throttle wire stroke adjustment – 1990 Pulsar

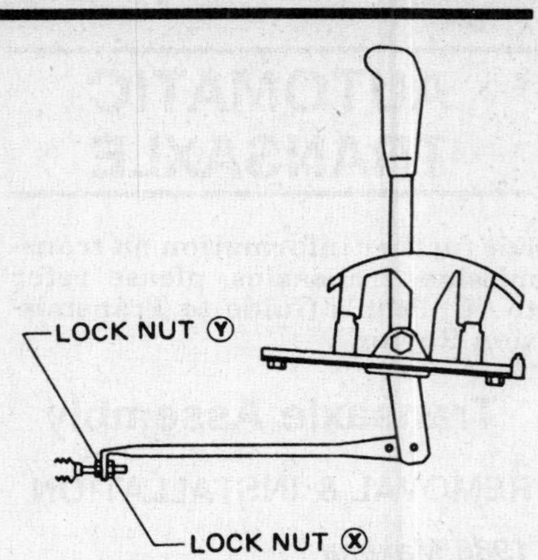

Automatic transaxle control cable adjustment – all vehicles

3. Back off the first nut ¾–1¼ turns on the 1988 Maxima; and 2¾–3¼ turns on the 1988–89 Stanza (including Wagon) and then tighten the second nut against the bracket. On 1988–89 Pulsar back off the nut 2¾–3¼ turns (RL4F02A transaxles) and 1–1½ turns (RL3F01A transaxles) and tighten the nut.
4. Tighten both double nuts to 5.8–7.2 ft. lbs. (8–10 Nm). The throttle drum should be held securely in the full open position.
5. On 1988–89 Stanza and and 1988 Maxima vehicles, check that the throttle wire stroke between full throttle and idling is 1.54–1.69 in. (39–43mm). On the 1988–90 Pulsar/Sentra it should be 1.079–1.236 in. (27.4–31.4mm).

1990 Pulsar

1. Remove the air cleaner cover.
2. While pressing on the lock plate, move the adjusting tube in the proper direction.
3. Return the lock plate to its original position.
4. Move the throttle drum from position P_1 to P_2 quickly.
5. Check that the throttle wire stroke **(L)** between full throttle and idling is 1.079–1.236 in. (27.4–31.4mm). Marking the throttle wire with paint dabs or a colored marker will help in measuring the throttle wire stroke.
6. Adjust the throttle wire stroke only if the throttle and accelerator wires are installed. After adjustment, make sure the parting line is straight.

CONTROL CABLE ADJUSTMENT

On all vehicles, move the selector from the **P** range through each gear to the **1** range. At each gear selection, the detent should be felt. If the detents cannot be felt or if the gear shift indicator pointer is not aligned properly, then the control, cable must be adjusted as follows:

1. Position the control lever (gear selector) in **P**.
2. Connect the control cable end to the lever in the transaxle unit and tighten the cable securing bolt.
3. Move the control lever from **P** to the **1** position. Be certain the lever works smoothly and quietly.
4. Position the lever in **P** once again. Make sure the lever locks into this position.
5. Loosen the cable adjusting locknuts.
6. While holding the select rod horizontal, tighten locknut **X** until it contacts the end of the rod. Then tighten locknut **Y**.
7. Move the control lever through all of its detents again and check for smooth and quiet operation.
8. Lubricate the spring washer at the end of the cable with multi-purpose grease.

TRANSFER CASE

Transfer Case Assembly

REMOVAL & INSTALLATION

Sentra and Stanza Wagon

1. Disconnect the negative battery cable.

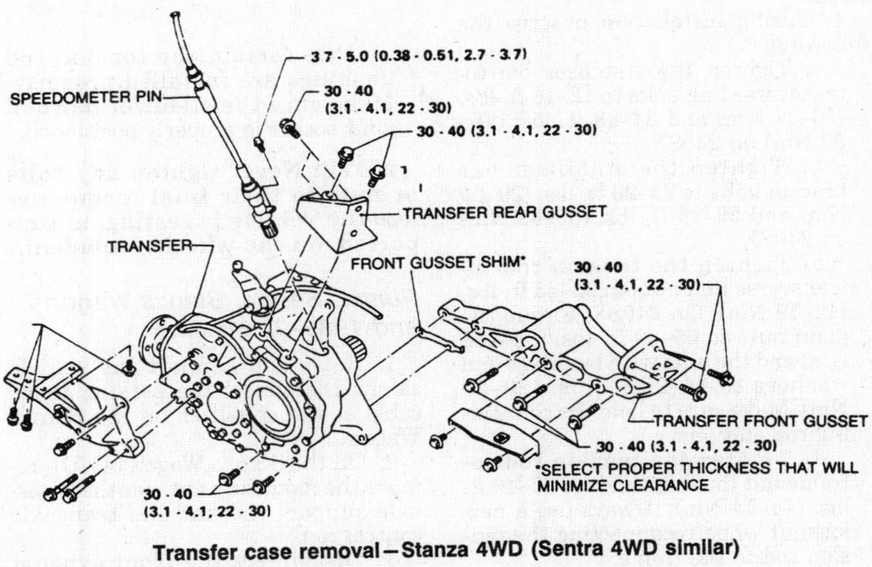

Transfer case removal—Stanza 4WD (Sentra 4WD similar)

2. Drain the gear oil from the transaxle and the transfer case.
3. Disconnect and remove the forward exhaust pipe.
4. Using chalk or paint, matchmark the flanges on the driveshaft and then unbolt the driveshaft from the transfer case.
5. On Sentra, unbolt and remove the transaxle support rod from the transfer case.
6. Unbolt and remove the transfer control actuator from the side of the transfer case as necessary.
7. Disconnect and remove the right side halfshaft.
8. Disconnect the speedometer gear from the transfer case.
9. Unbolt and remove the front, rear and side transfer case gussets (support members).
10. Use a hydraulic floor jack and a block of wood to support the transfer case, remove the transfer case-to-transaxle mounting bolts and then remove the case.

To install:

11. Lubricate the lips of the transfer side oil seal (in transaxle), adapter oil seal (in transfer) and driveshaft oil seal. Use a suitable multi-purpose grease.
12. Apply KP510-00150 or equivalent sealant to the ring gear oil seal seating surface prior to installation of the transfer case.
13. Raise and and mount the transfer. Tighten the transfer case-to-transaxle mounting bolts and the transfer case gusset mounting bolts to 22–30 ft. lbs. (30–40 Nm) on 1988 vehicles. On all other years, torque the transfer rear gusset bolts to 29–36 ft. lbs. (39–49 Nm).

NOTE: Be careful not to damage the transaxle oil seal when inserting thew splined portion of the transfer ring gear into the transaxle.

14. Install the front, rear and side transfer case gussets.
15. Connect the speedometer cable.
16. Install the right side halfshaft.
17. Connect the transfer control actuator to the side of the transfer case as necessary.
18. On Sentra, connect the transfer support rod to the transfer case.
19. Connect the driveshaft to the transfer case by aligning the matchmarks. Torque the driveshaft bolts to 25–33 ft. lbs. (34–44 Nm) on both Sentra and Stanza.

NOTE: When connecting the drive shaft, be careful not to damage the driveshaft and adapter oil seals.

20. Connect the forward exhaust pipe using new gaskets.
21. Fill the transfer case to the proper level with gear oil. The transfer case and the transaxle use different types and weights of lubricant.
22. Connect the negative battery cable.
23. Check the transfer case for proper operation.

FRONT SUSPENSION

MacPherson Strut

REMOVAL & INSTALLATION

1. Raise and support the vehicle safely.
2. Remove the front wheels.
3. Disconnect and plug the brake line if it interferes with removal of the strut.
4. Disconnect the tension rod and stabilizer bar from the transverse link.
5. Unbolt the steering arm from the lower end of the strut.
6. Support the bottom of the strut with a jack or equivalent. On 240SX, place matchmarks on the strut lower bracket and camber adjusting pin for assembly reference.
7. Open the hood and remove the nuts holding the top of the strut. On 300ZX and Maxima equipped with adjustable or sonar suspension shocks, disconnect the electrical lead from the actuating unit.
8. Lower the jack slowly and cautiously until the strut assembly can be removed.

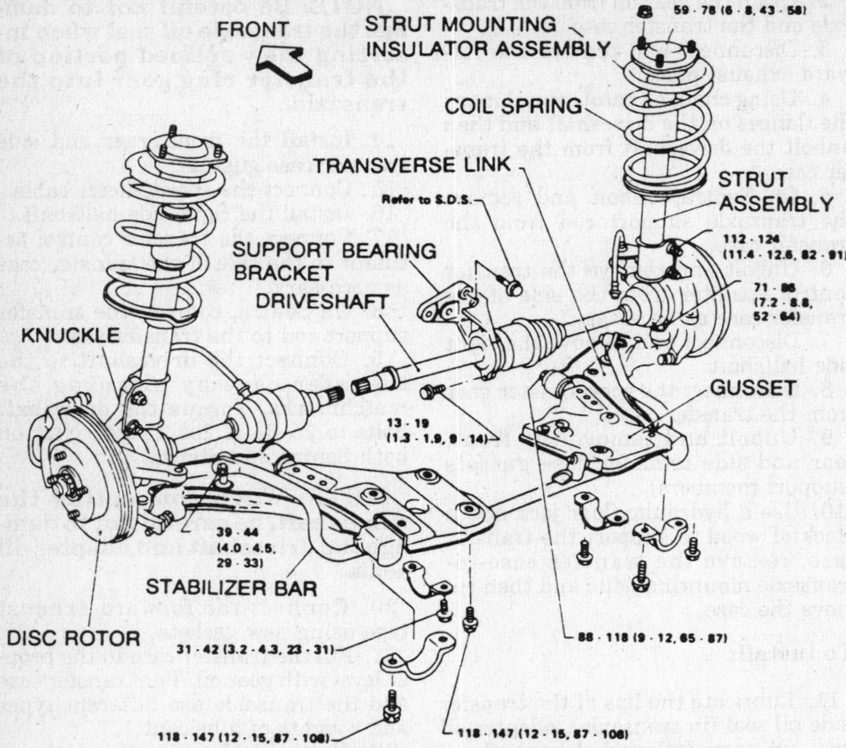

Strut-type front suspension—front wheel drive models

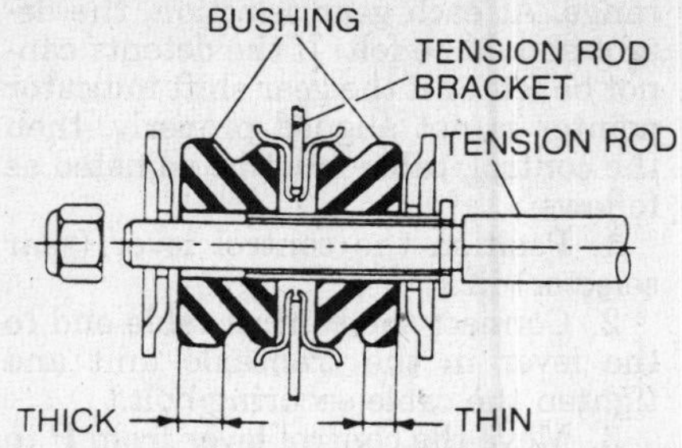

Tension rod bushing positioning—rear wheel drive models

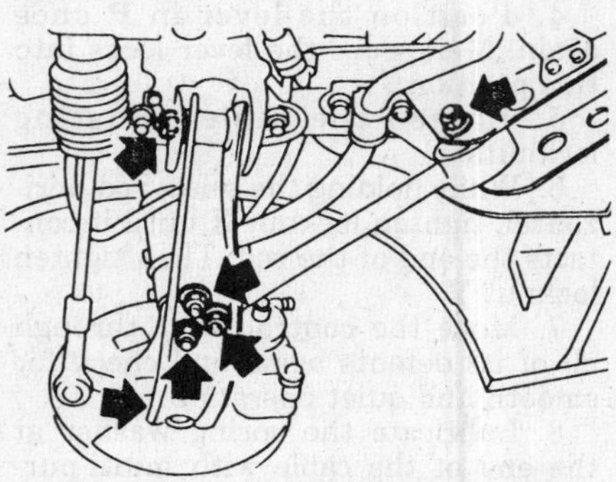
Tension rod and stabilizer bar ataching points—240SX

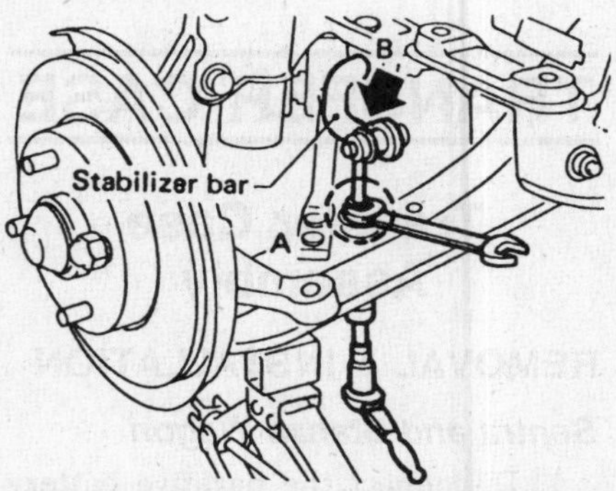

Hold the stabilizer connecting rod with a wrench when removing and installing the mounting nuts

9. During installation, observing the following:

a. The self locking nuts holding the top of the strut must be replaced.

c. On 240SX, make sure the matchmarks on the bracket and the camber adjusting pin are aligned properly.

d. On 1989–92 Maxima with sonar suspension, before installing the actuator ensure the output shaft on the inside of the actuating unit is aligned with the shock absorber control rod. If this is not done, the actuator will be damaged.

Tension Rod and Stabilizer Bar

REMOVAL & INSTALLATION

200SX, 240SX and 300SX

1. Raise and support the vehicle safely.
2. Remove the tension rod-to-frame lock nuts.
3. Remove the 2 mounting bolts at the transverse link, lower control arm, and then slide out the tension rod.
4. On 240SX, to remove the tension rod, remove the bolt and nut that holds the rod to the tension rod bracket (through the bushing), then swing the rod upward and remove the tranverse link bolts, nuts, bushings and washers. If the bushings are worn replace them.
5. Unbolt the stabilizer bar at each transverse link or connecting rod. On 240SX, engage the flats of stabilizer bar connecting rod with a wrench to keep the rod from moving when removing the nuts.
6. Remove the 4 stabilizer bar bracket bolts, and remove the stabilizer bar.
7. During installation observe the following:

a. Tighten the stabilizer bar-to-transverse link bolts to 12–16 ft. lbs. (16–22 Nm) and 34–38 ft. lbs. (46–52 Nm) on 240SX.

b. Tighten the stabilizer bar bracket bolts to 22–29 ft. lbs. (29–39 Nm) and 29–36 ft. lbs. (39–49 Nm) on 240SX.

c. Tighten the tension rod-to-transverse link nuts to 31–43 ft. lbs. (42–59 Nm). On 240SX, torque the plain nuts to 65–80 ft. lbs. (88–108 Nm) and the nuts with bushings and washers to 14–22 ft. lbs. (20–29 Nm). Make sure to hold the connecting rod stationary.

d. Tighten the tension rod-to-frame nut (bushing end) to 33–40 ft. lbs. (44–54 Nm). Always use a new locknut when reconnecting the tension rod to the frame.

e. Be certain the tension rod bushings are installed properly. Make sure the stabilizer bar ball joint socket is properly positioned.

NOTE: Never tighten any bolts or nuts to their final torque unless the vehicle is resting, unsupported, on the wheels (unladen).

Pulsar, Sentra, Stanza Wagon and 1989–92 Maxima

1. Raise and support the vehicle safely. Disconnect the parking brake cable at the equalizer on the Stanza Wagon.
2. On the Stanza Wagon (4WD), remove the mounting nuts for the transaxle support rod and the transaxle control rod.
3. Disconnect the front exhaust

pipe at the manifold and position it aside (not required on Maxima).

4. On the Stanza Wagon (4WD), matchmark the flanges and then separate the driveshaft from the transfer case.

5. Remove the stabilizer bar-to-transverse link (lower, control arm) mounting bolts. Engage the flats of stabilizer bar connecting rod with a wrench to keep the rod from moving when removing (and installing) the bolts.

6. Matchmark the stabilizer bar to the mounting clamps.

7. Remove the stabilizer bar mounting clamp bolts and then pull the bar out, around the link and exhaust pipe.

8. Installation is the reverse of the removal procedure. Never tighten the mounting bolts unless the vehicle is resting on the ground with normal weight upon the wheels. On Pulsar and Sentra, be sure the stabilizer bar ball joint socket is properly positioned.

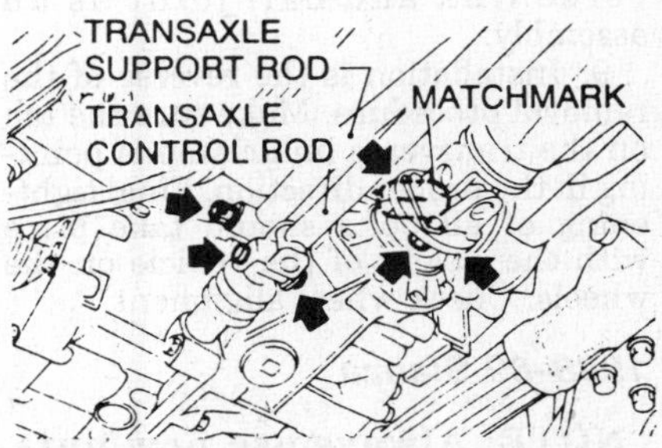

Removing the stabilizer bar on the Stanza wagon (4wd)

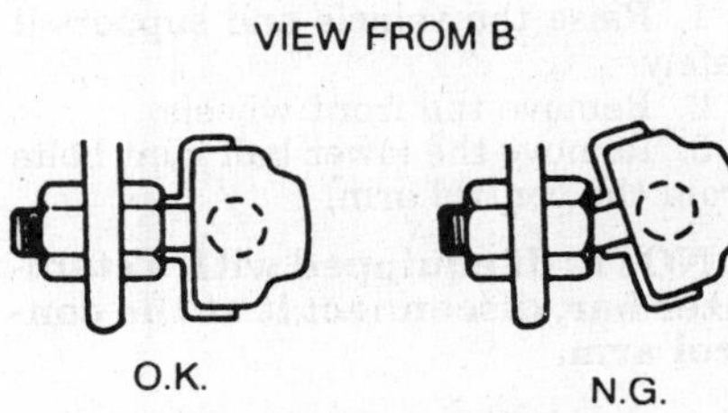

Ball joint socket positioning

Lower Ball Joints

INSPECTION

Dial Indicator Method

1. Raise and support the vehicle safely.

2. Clamp a dial indicator to the transverse link and place the tip of the dial on the lower edge of the brake caliper.

3. Zero the indicator.

4. Make sure the front wheels are straight ahead and the brake pedal is fully depressed.

5. Insert a long prybar between the transverse link and the inner rim of the wheel.

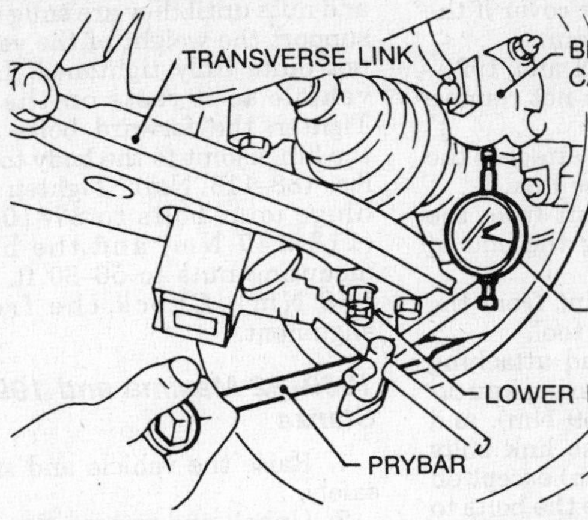

Measuring ball joint endplay with a dial indicator

6. Push down and release the prybar and observe the reading (deflection) on the dial indicator. Take several readings and use the maximum dial indicator deflection as the ball joint vertical endplay. Make sure to **0** the indicator after each reading. If the reading is not within specifications, replace the transverse link or the ball joint. Ball joint vertical endplay specifications are as follows:

200SX and 240SX – 0 in. (0mm)
300ZX
1988–89 – 0.098 in. (2.5mm) or less
1990–92 – 0 in. (0mm)
Maxima
1988 – 0.098 in. (2.5mm) or less
1989–92 – 0 in. (0mm)
Pulsar
1988 – 0.098 in. (2.5mm) or less
1989–90 – 0 in. (0mm)
Sentra
1988–92 – 0 in. (0mm)
Stanza
1988–89 – 0.004–0.039 in. (0.1–1.0mm)
1990–92 – 0 in. (0mm)
Stanza Wagon – 0.098 in. (2.5mm) or less

Visual Approximation Method

The lower ball joint should be replaced when play becomes excessive. An effective way to visually approximate ball joint verticle endplay without the use of a dial indicator is to preform the following:

1. Raise and safely support the vehicle until the wheel is clear of the ground. Do not place the jack under the ball joint; it must be unloaded.

2. Place a long prybar under the tire and move the wheel up and down. Keep one hand on top of the tire while doing this.

3. If ¼ in. or more of play exists at the top of the tire, the ball joint should be replaced. Be sure the wheel bearings are properly adjusted before making this measurement. A double check can be made; while the tire is being moved up and down, observe the ball joint. If play is seen, replace the ball joint.

REMOVAL & INSTALLATION

Rear Wheel Drive

On 200SX and 1988–89 300SX, there is a plugged hole in the bottom of the joint for installation of a grease fitting. The ball joint should be greased every 30,000 miles.

NOTE: The transverse link (lower control arm) must be removed and then the ball joint must be pressed out.

1. Raise and support the vehicle safely.

2. Remove the front wheels.

3. Separate the knuckle arm from the tie rod using the proper tool.

4. Separate the knuckle arm from the strut.

5. Remove the stabilizer bar and tension rod.

6. Remove the transverse link and knuckle arm.

7. Separate the knuckle arm from the ball joint with a suitable press.

8. Replace the transverse link/ball joint assembly.

9. Installation is the reverse of the removal procedure.

Front Wheel Drive

1. Raise and support the vehicle safely.

2. Remove the front wheels.

3. Remove the wheel bearing locknut.

4. Separate the tie rod end ball joint from the steering knuckle with a ball joint remover, being careful not to

damage the ball joint dust cover if the ball joint is to be used again.

5. On 1989–92 Maxima and 1990–92 Stanza, loosen, but do not remove the strut upper nuts.

6. Remove the nut that attaches the ball joint to the transverse link.

7. Separate the halfshaft from the knuckle by lightly taping the end of the shaft.

8. Separate the ball joint from the knuckle using the proper tool.

9. Tighten the ball stud attaching nut (from ball joint-to-steering knuckle) to 22–29 ft. lbs. (30–39 Nm), and the ball joint-to-transverse link bolts to 40–47 ft. lbs. (54–64 Nm) except on Stanza. On Stanza, torque the bolts to 56–80 ft. lbs. (76–108 Nm).

Lower Control Arm (Transverse Link)

REMOVAL & INSTALLATION

200SX, 240SX and 300ZX

1. Raise and support the vehicle safely.
2. Remove the front wheels.
3. Remove the cotter pin and castle nut from the side rod (steering arm) ball joint and separate the ball joint from the side rod using the proper tool.
4. Separate the steering knuckle arm from the MacPherson strut.
5. Remove the tension rod and stabilizer bar from the lower arm.
6. Remove the nuts or bolts connecting the lower control arm (transverse link) to the suspension crossmember.
7. Remove the lower control arm (transverse link) with the suspension ball joint and knuckle arm still attached.
8. When installing the control arm, temporarily tighten the nuts and/or bolts securing the control arm to the suspension crossmember. Tighten them fully only after the vehicle is sitting on its wheels. Lubricate the ball joints after assembly.

1988–89 Maxima

1. Raise and support the vehicle safely.
2. Remove the front wheels.
3. Remove the nut fastening the link between the stabilizer bar and the control arm to the control arm.
4. Remove the 3 nuts fastening the ball joint to the lower control arm.
5. Remove the 2 bolts attaching the front and rear hinge joints of the control arm to the body.
6. Remove the control arm.
7. Installation is the reverse of the removal procedure. Tighten all bolts and nuts until they are snug enough to support the weight of the vehicle, but not quite fully tightened. Lower the vehicle so it rests on the ground. Tighten the forward bolts attaching the hinge joint to the body to 65–87 ft. lbs. (88–118 Nm). Tighten the rear hinge joint bolts to 87–108 ft. lbs. (118–147 Nm) and the ball joint mounting nuts to 56–80 ft. lbs. (76–108 Nm). Check the front end alignment.

1989–92 Maxima and 1990–92 Stanza

1. Raise the vehicle and support it safely.
2. Unbolt and remove the stabilizer bar. The bar is removed by unfastening the clamp bolts and the bolts that hold the bar to the transverse link gusset plate. When removing the clamps, note the relationship between the clamp and paint mark on the bar.
3. Unbolt and remove the transverse link and gusset.
4. Inspect the transverse link, gusset and bushings for cracks, damage and deformation.
5. To install, bolt the transverse link and gusset into place. Lower the vehicle and torque the the bolts and nuts in the proper sequence as illustrated. Torque the nuts to 30–35 ft. lbs. (41–51 Nm) and the bolts to 87–108 ft. lbs. (118–147 Nm). The vehicle must at curb weight and the tires must be on the ground. After installation is complete, check the front end alignment.

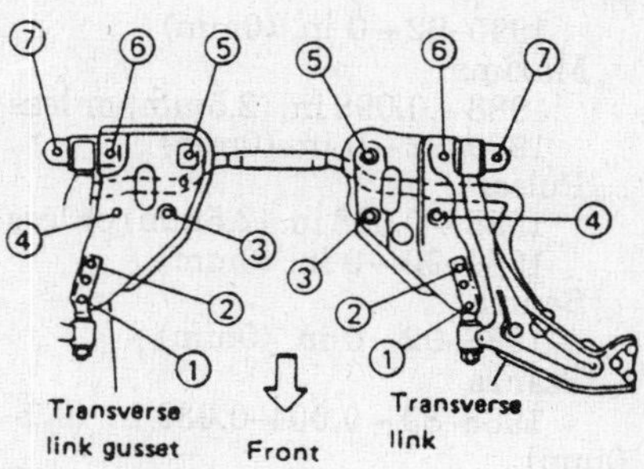

Transverse link and gusset bolt torque sequence – 1989–92 Maxima and 1990–92 Stanza

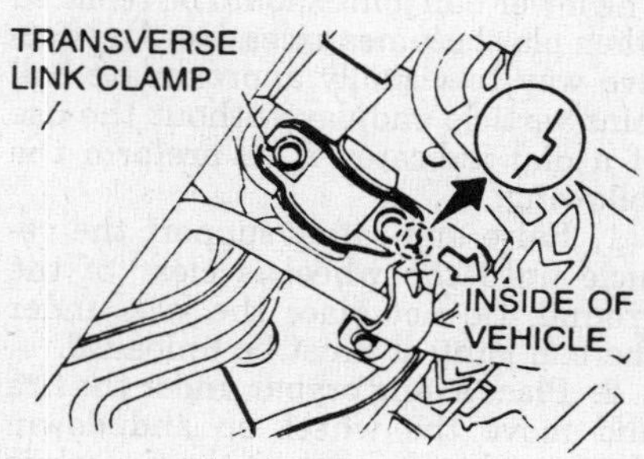

Transverse link clamp positioning – Pulsar and Sentra

Pulsar and Sentra

1. Raise the vehicle and support it safely.
2. Remove the front wheels.
3. Remove the wheel bearing locknut.
4. Remove the tie rod ball joint with a suitable puller.
5. Remove the lower strut-to-knuckle mounting bolts and separate the strut from the knuckle.
6. Separate the outer end of the halfshaft from the steering knuckle by carefully tapping it with a rubber mallet. Be sure to cover the CV-joints with a shop rag.
7. Using a suitable ball joint removal tool, separate the lower ball joint stud from the steering knuckle.
8. Unbolt and remove the transverse link and ball joint as an assembly.
9. Installation is the reverse of the removal procedure. Make sure the tab on the transverse link clamp is pointing in the proper direction. Final tightening of all bolts should take place with the weight of the vehicle on the wheels. Check wheel alignment.

1988–89 Stanza

NOTE: Always use new nuts when installing the ball joint to the control arm.

1. Raise the vehicle and support it safely.
2. Remove the front wheels.
3. Remove the lower ball joint bolts from the control arm.

NOTE: If equipped with a stabilizer bar, disconnect it at the control arm.

4. Remove the control arm-to-body bolts.
5. Remove the gusset.
6. Remove the control arm.
7. Installation is the reverse of the removal procedure using the following torque specifications: gusset-to-body bolts to 87–108 ft. lbs. (118–147 Nm); control arm securing nut to 87–108 ft. lbs. (118–147 Nm) and lower ball joint-to-control arm nuts to 56–80 ft. lbs. (76–108 Nm). When installing the link, tighten the nut securing the link spindle to the gusset. Final tightening should be made with the weight of the vehicle on the wheels.

NOTE: On the Stanza Wagon, make sure to torque the gusset bolts in the proper sequence.

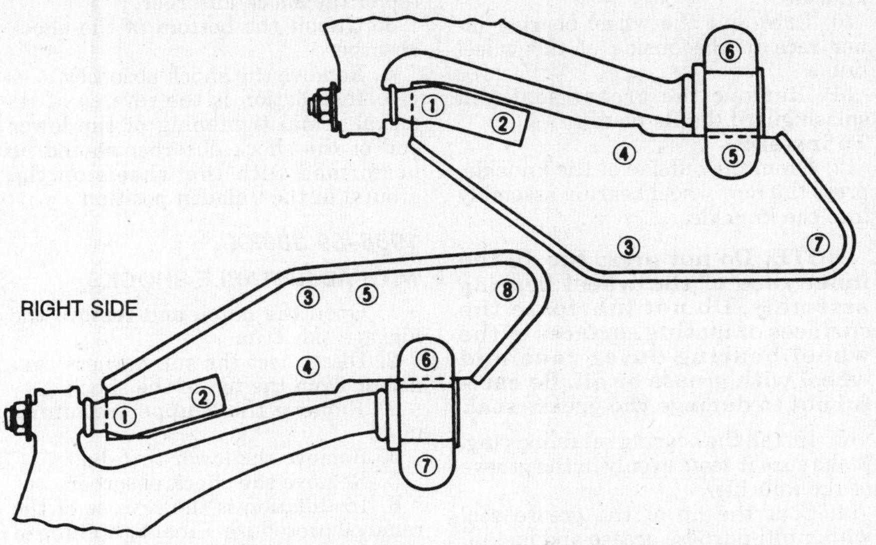

Transverse link gusset bolt torque sequence – Stanza wagon

Front Wheel Bearings

ADJUSTMENT

NOTE: For wheel bearing procedures on front wheel drive vehicles, please refer to the "Drive Axle" section.

200SX and 1988–89 300ZX

1. Raise and support the vehicle safely.
2. Remove the front wheels.
3. While rotating the brake disc, torque wheel bearing lock nut to 18–22 ft. lbs.
4. Loosen locknut approximately 60 degrees on all vehicles. Install adjusting cap and align groove of nut with hole in spindle. If alignment cannot be obtained, change position of adjusting cap. Also, if alignment cannot be obtained, loosen locknut slightly but not more than 15 degrees.
5. Install the front wheels and lower the vehicle.

240SX

There is no procedure for torquing the front wheel bearings due to the design of the bearing. Once the final torque is applied to the wheel bearing axle nut and the axial play is checked, no further adjustment is either necessary or possible.

Check the torque of the wheel bearing locknut. This value is 108–159 ft. lbs. (147–216 Nm). Then, mount a dial indicator to the face of the hub and check the axial play. It should not exceed 0.0012 in. (0.03mm). If the axial play is not as specified, replace the wheel bearing.

1990–92 300ZX

1. Raise the vehicle and suppport safely.
2. Remove the front wheels.
3. Prior to checking the bearing pre-load, spin the wheel hub at least 10 revolutions in both directions to seat the bearing.
4. To check the pre-load: connect a spring scale of known calibration to a wheel hub bolt and measure the turning torque. If an NSK wheel bearing is used, the turning torque should be 1.3–8.4 lbs. (5.9–37.3 N). For NTN bearings, the turning torque should be 1.8–13.0 lbs. (7.8–57.9 N).
5. To check the axial endplay: mount a dial indicator so the stylus of the dial rests on the face of the hub and check the wheel bearing axial endplay by attempting to rock the wheel hub in and out. The endplay should be 0.0020 in. (0.05mm) or less.
6. Mount the front wheels and lower the vehicle.

REMOVAL & INSTALLATION

200SX and 1988–89 300ZX

1. Raise and support the vehicle safely.
2. Remove the front wheels. Work off center hub cap by using thin tool. If necessary tap around it with a soft hammer while removing. Pry off cotter pin and take out adjusting cap. Apply the parking brake firmly and remove the wheel bearing nut. The nut will require a good deal of force to remove it.
3. Unbolt the caliper and move it aside. Do not disconnect the hose from the caliper. Do not allow the caliper to hang by the hose; support the caliper with a length of wire or rest it on a suspension member.
4. Remove the wheel hub, disc brake rotor and bearing from the spindle. During removal, capture the outer bearing to prevent it from hitting the ground.
5. To replace the bearing outer race, drive it out with a suitable brass drift and mallet.

To install:

6. Install the new bearing outer race using a suitable race installation tool.
7. Install a new oil seal so the words "BEARING SIDE" face the inner side of the hub. Coat the lip of the seal with multi-purpose grease.
8. Pack the bearings, hub, hub cap and hub cap O-ring with multi-purpose grease. If the hub cap O-ring is crimped, replace it.
9. Install the inner and outer bearings.
10. Install the wheel hub and rotor disc onto the spindle.
11. Coat the threaded portion of the spindle shaft and the contact surface between the lockwasher and outer wheel bearing with multi-purpose grease.
12. Install the wheel bearing locknut and adjust the bearing pre-load as described above. Use a new cotter pin.
13. Mount the brake caliper assembly.
14. Install the front wheels and lower the vehicle.

240SX

1. Raise and support the vehicle safely.
2. Remove the front wheels.
3. Work off center hub cap by using a suitable thin tool. If necessary tap around it with a soft hammer while removing. Pry off cotter pin and take out adjusting cap.
4. Apply the parking brake firmly and remove the wheel bearing nut. The nut will require a good deal of force to remove it.
5. Unbolt the caliper and move it aside. Do not disconnect the hose from the caliper. Do not allow the caliper to hang by the hose; support the caliper with a length of wire or rest it on a suspension member.
6. Pull the brake disc and wheel hub from the spindle.
7. Separate the tie rod and lower ball joints using the proper tool.
8. Place matchmarks on the strut lower bracket and camber adjusting

pin for assembly reference. Remove the lower bracket bolts and nuts. Remove the wheel hub and knuckle assembly.
9. Remove the bearing retaining ring from the wheel hub.
10. Press the bearing assembly from the wheel hub. Apply pressure from the outside of the hub to remove the bearing.
To install:
11. Press the new bearing assembly into the hub from the inside.

NOTE: Do not press the on the inner race of the wheel bearing assembly. Do not lubricate the surfaces of mating surfaces of the wheel bearing outer race and wheel with grease or oil. Be careful not to damage the grease seal.

12. Install the bearing retaining ring.
13. Coat the lip of the grease seal with multi-purpose grease.
14. Manuever the wheel hub and axle assembly onto the lower mounting bracket and install the bracket bolts and nuts. Make sure the matchmarks on the bracket and the camber adjusting pin are aligned properly.
15. Connect the lower and tie rod ball joints.
16. Push the brake disc and wheel hub onto the spindle.
17. Install the brake caliper assembly.
18. Check the wheel bearing preload.
19. Install a new locknut cotter pin. Install the bearing hub cap after packing it with multi-purpose grease.
20. Mount the the front wheels and lower the vehicle.

1990–92 300ZX

1. Raise and support the vehicle safely.
2. Remove the front wheels.
3. Unbolt the caliper and move it aside. Do not disconnect the hose from the caliper. Do not allow the caliper to hang by the hose; support the caliper with a length of wire or rest it on a suspension member.
4. Separate the tie rod and lower ball joints using the proper tool.

NOTE: The steering knuckle is made of an aluminum alloy. Be careful no to strike it when removing the ball joints.

5. Remove the kinpin lower nut and remove the steering knuckle assembly.
6. Remove the hub cap, wheel bearing locknut, sensor rotor (with ABS) or washer (without ABS).
7. Remove the wheel hub with a suitable drift.
8. Remove the wheel bearing retaining ring.
9. Press the wheel bearing from the knuckle.
10. Drive out the wheel bearing inner race to the ouside of the wheel hub.
11. Remove the grease seal and splash guard (baffle plate).
To install:
12. From the outside of the knuckle, press the new wheel bearing assembly into the knuckle.

NOTE: Do not press the on the inner race of the wheel bearing assembly. Do not lubricate the surfaces of mating surfaces of the wheel bearing outer race and wheel with grease or oil. Be careful not to damage the grease seal.

13. Install the bearing retaining ring. Make sure it seats evenly in the groove of the knuckle.
14. Coat the lip of the grease seal with multi-purpose grease and install.
15. Install the spash guard.
16. Press the wheel hub into the steering knuckle.
17. Install the washer (without ABS), sensor rotor (with ABS) and wheel bearing locknut. Torque the locknut to 152–210 ft. lbs. (206–284 Nm). Stake the locknut tabs using a small cold chisel.
18. Place the hub cap onto the knuckle and tap it into place using a rubber or plastic mallet. Once the cap is seated lightly into the knuckle, install the cap retaining bolts and torque to 8–12 ft. lbs. (11–16 Nm).
19. Mount the steering knuckle assembly and tighten the lower king pin nut.
20. Connect the tie rod and lower ball joints using the proper tool.
21. Install the brake caliper assembly.
22. Ajust the wheel bearing pre-load and axial endplay.
23. Mount the front wheels and lower the vehicle.

REAR SUSPENSION

Shock Absorbers

REMOVAL & INSTALLATION

200SX and Stanza Wagon (2WD)

1. Open the trunk and remove the cover panel, if necessary, to expose the shock mounts. Pry off the mount covers, if equipped.
2. Remove the 2 nuts holding the top of the shock absorber.
3. Unbolt the bottom of the shock absorber.
4. Remove the shock absorber.
5. Installation is the reverse of removal. Final tightening of the lower end of the shock absorber should be performed with the wheels on the ground in the unladen position.

1988–89 300ZX

WITH ADJUSTABLE SHOCKS

1. Open the hatch and remove the luggage side trim.
2. Disconnect the sub-harness connector from the top of the shock.
3. Remove the 2 upper retaining nuts.
4. Remove the lower thru-bolt.
5. Remove the shock absorber.
6. Installation is the reverse of the removal procedure. Final tightening of the shock absorber upper and lower end should be performed with the wheels on the ground in the unladen position. Torque the bottom bolt to 43–58 ft. lbs. (59–78 Nm) and top nuts to 23–31 ft. lbs. (31–42 Nm)

WITHOUT ADJUSTABLE SHOCKS

1. Open the hatch and remove the luggage side trim.
2. Remove the 2 upper retaining nuts.
3. Remove the lower thru-bolt.
4. Remove the shock absorber.
5. Installation is the reverse of the removal procedure. Final tightening of the shock absorber upper and lower end should be performed with the wheels on the ground in the unladen position. Torque the bottom bolt to 43–58 ft. lbs. (59–78 Nm) and the 2 top nuts to 23–31 ft. lbs. (31–42 Nm).

MacPherson Strut

REMOVAL & INSTALLATION

240SX and 1990–92 300ZX

1. Block the front wheels.
2. Raise and support the vehicle safely.

NOTE: The vehicle should be far enough off the ground so the rear spring does not support any weight.

3. Working inside the luggage compartment, turn and remove the caps above the strut mounts. Remove the strut mounting nuts.
4. Remove the mounting bolt for the strut at the lower arm (transverse link) and then lift out the strut.
5. Installation is in the reverse order of removal. Install the upper end first and secure with the nuts snugged

down but not fully tightened. Attach the lower end of the strut to the transverse link and the tighten the upper nuts to 12–14 ft. lbs. (16–19 Nm). Tighten the lower mounting bolt to 65–80 ft. lbs. (88–108 Nm).

Pulsar and Sentra (2WD)

1. Raise and support the rear of the vehicle safely.
2. Remove the rear wheels.
3. Disconnect the brake tube and parking brake cable.
4. If necessary, remove the brake assembly and wheel bearing.
5. Disconnect the parallel links and radius rod from the strut or knuckle.
6. Support the strut assembly.
7. Remove the strut upper end nuts and then remove the strut from the vehicle.
8. Installation is the reverse of the removal procedure. Tighten the radius rod-to-knuckle nuts to 43–61 ft. lbs. (59–83 Nm), the strut-to-knuckle and parallel link-to-knuckle bolts to 72–87 ft. lbs. (98–118 Nm) and the strut-to-body nuts to 18–22 ft. lbs. (25–29 Nm).

Sentra (4WD) and Stanza Wagon (4WD)

1. Block the front wheels.
2. Raise and support the vehicle safely.
3. Position a suitable floor jack under the transverse link on the side of the strut to be removed. Raise it just enough to support the strut.
4. Open the rear of the vehicle and remove the 3 nuts that attach the top of the strut to the body.
5. Remove the rear wheels.
6. Remove the brake line from its bracket and position it aside. Do not disconnect the brake line.
7. Remove the 2 lower strut-to-knuckle mounting bolts.
8. Carefully lower the floor jack and remove the strut.
9. Installation is the reverse order of removal. Final tightening of the strut mounting bolts should take place with the wheels on the ground and the vehicle unladen. Tighten the upper strut-to-body nuts to 33–40 ft. lbs. (45–60 Nm). Tighten the lower strut-to-knuckle bolts to 111–120 ft. lbs. (151–163 Nm).

1988 Maxima and 1988–89 Stanza

RIGHT STRUT

1. Unclip the rear brake line at the strut. Do not disconnect it.
2. Remove the radius rod mounting bolt, radius rod mounting bracket.
3. Remove the 2 parallel link mounting bolts.
4. Remove the rear seat and parcel shelf.
5. On Maxima with adjustable suspension, disconnect the sub-harness connector and the connector from the cap. Grasp the cap connector from both sides during removal to avoid damage. This connector is very sensitive.
6. Position a suitable floor jack under the strut and raise it just enough to support the strut.

NOTE: Do not support the strut at the parallel links or the radius rods.

7. Remove the 3 upper strut mounting nuts and then lift out the strut and rear axle assembly.
8. Installation is the reverse of the removal procedure. Tighten all bolts sufficiently to safely support the vehicle and then lower the vehicle to the ground so it rests on its own weight. Tighten the upper strut mounting nuts to 23–31 ft. lbs. (31–42 Nm); the radius rod bracket bolts to 43–58 ft. lbs. (59–78 Nm) and the parallel link mounting bolts to 65–87 ft. lbs. (88–118 Nm).

LEFT STRUT

1. Unclip the rear brake line at the strut. Do not disconnect it.
2. Remove the radius rod mounting bolt.
3. Remove the stabilizer bar connecting bracket.
4. Remove the suspension crossmember mounting nuts.
5. Remove the strut upper mounting nuts.
6. Remove the left suspension assembly and the crossmember.
7. Installation is the reverse of the removal procedure. Tighten all bolts sufficiently to safely support the vehicle and then lower the vehicle to the ground so it rests on its own weight. Tighten the upper strut mounting nuts to 23–31 ft. lbs. (31–42 Nm); the radius rod bracket bolts to 43–58 ft. lbs. (59–78 Nm) and the parallel link mounting bolts to 65–87 ft. lbs. (88–118 Nm).

1989–92 Maxima and 1990–92 Stanza

1. Unclip the rear brake line at the strut. Do not disconnect it.
2. Disconnect the parking brake at the equalizer.
3. Remove the parallel link mounting bolts, radius rod mounting bolts, stabilizer mounting bolts, stabilizer connecting brackets and parking brake cable mounting bracket bolts.
4. Remove the rear seat and parcel shelf.
5. Remove the 3 upper strut mounting nuts and then lift out the strut.
6. Installation is the reverse of the removal procedure. Tighten all bolts sufficiently to safely support the vehicle and then lower the vehicle to the ground so it rests on its own weight. Tighten the upper strut mounting nuts to 31–40 ft. lbs. (42–54 Nm); parallel link mounting bolts to 65–87 ft. lbs. (88–118 Nm); connecting rod bracket nuts to 30–35 ft. lbs. (41–47 Nm); stabilizer bar mounting bolts to 43–58 ft. lbs. (59–78 Nm) and radius rod mounting bolts to 65–87 ft. lbs. (88–118 Nm).

Coil Springs

CAUTION

Coil springs are under considerable tension and can exert enough force to cause bodily injury. Exercise extreme caution when working with them.

REMOVAL & INSTALLATION

200SX and 1988–89 300ZX

This suspension is similar to the IRS MacPherson strut type, except this type utilizes separate coil springs and shock absorbers, instead of strut units.

1. Compress the coil spring with a suitable spring compressor.
2. Raise the vehicle and support safely.
3. Compress the coil spring until it is of sufficient length to be removed. Remove the spring.
4. When installing the spring, be sure the upper and lower spring seat rubbers are not twisted and have not slipped off when installing the coil spring.

Stanza Wagon (2WD)

1. Raise the vehicle and support safely.
2. Remove the rear wheels.
3. Release the parking brake.
4. Remove the inner hub cap, the cotter pin and the wheel bearing locknut. Remove the brake drum.
5. Disconnect and plug the hydraulic brake line.
6. Disconnect the parking brake cable.
7. Remove the 4 brake backing plate mounting bolts and then slide the backing plate along with the inner wheel bearing off of the rear axle.
8. Disconnect the rear stabilizer bar.
9. Unbolt the anchor arm bracket and then remove the inner bushing bracket mounting bolts. Remove the torsion bar.
10. Installation is in the reverse or-

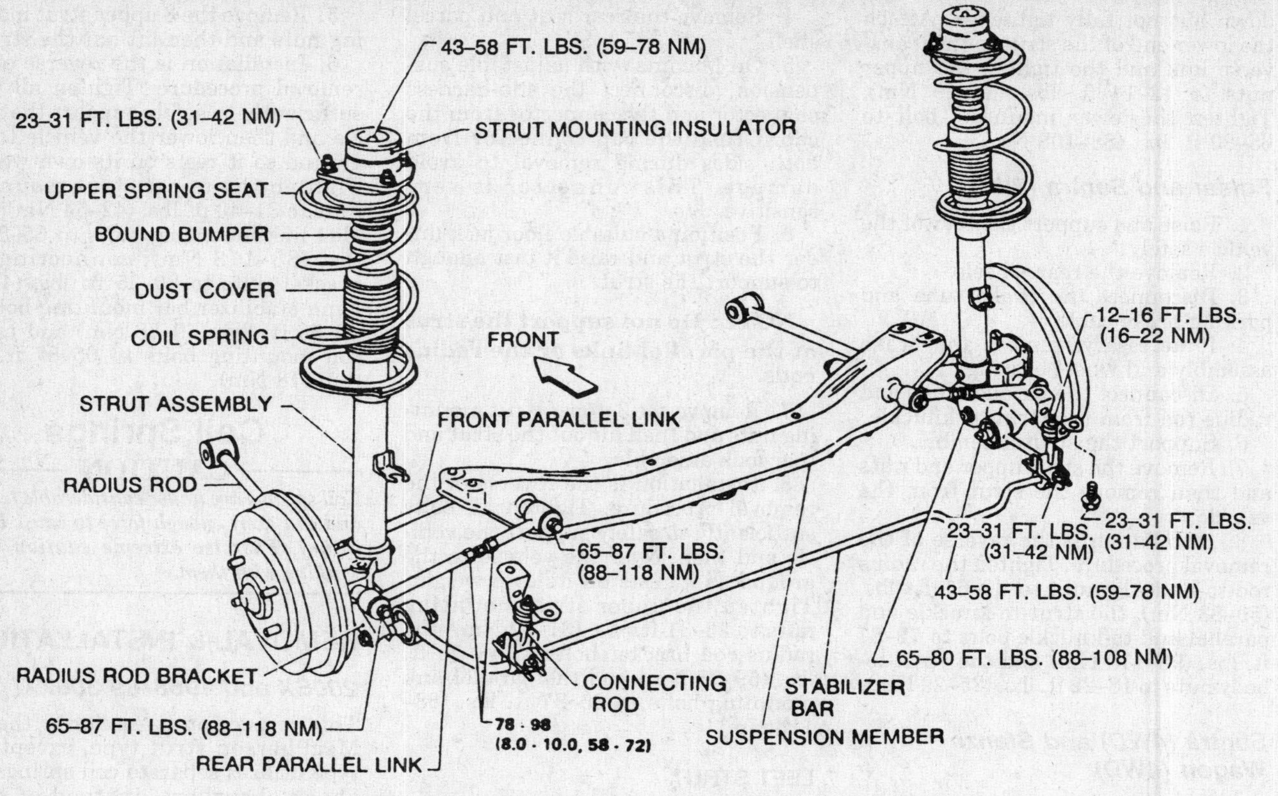

Typical MacPherson strut-type rear suspension – front wheel drive models

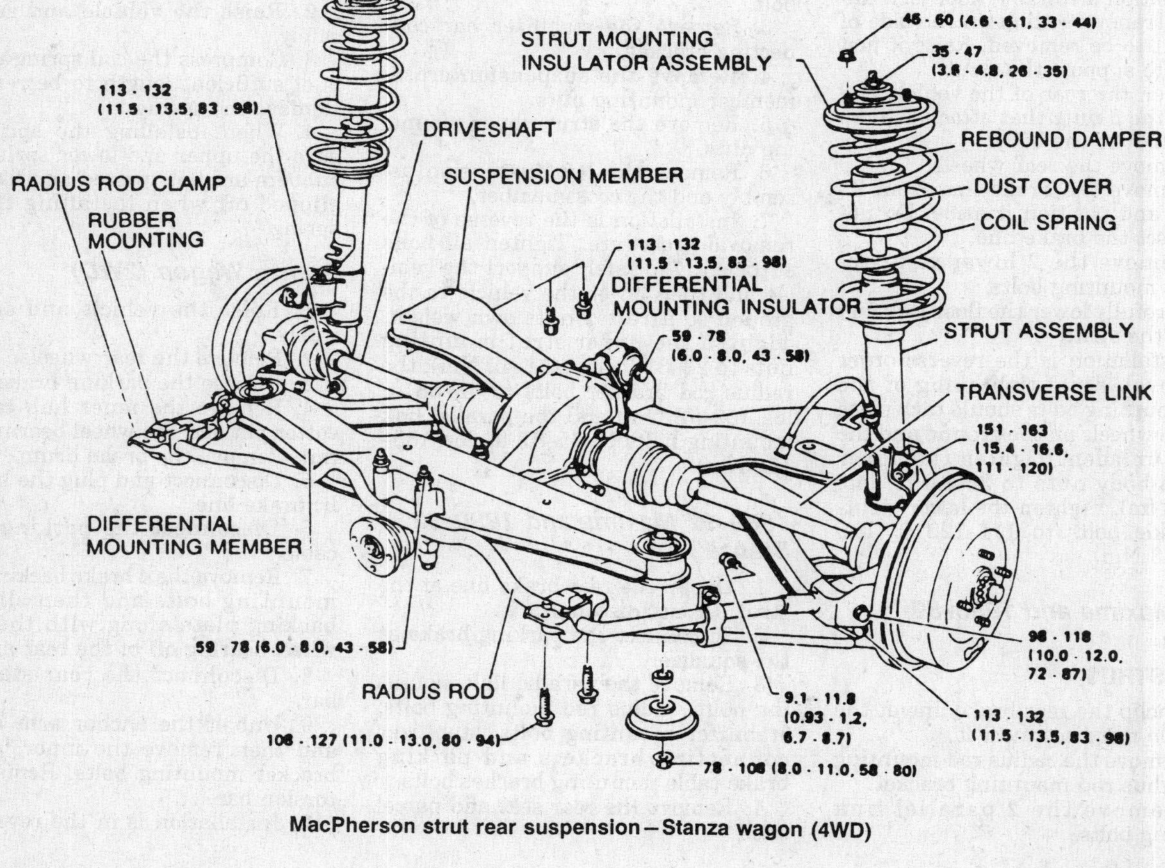

MacPherson strut rear suspension – Stanza wagon (4WD)

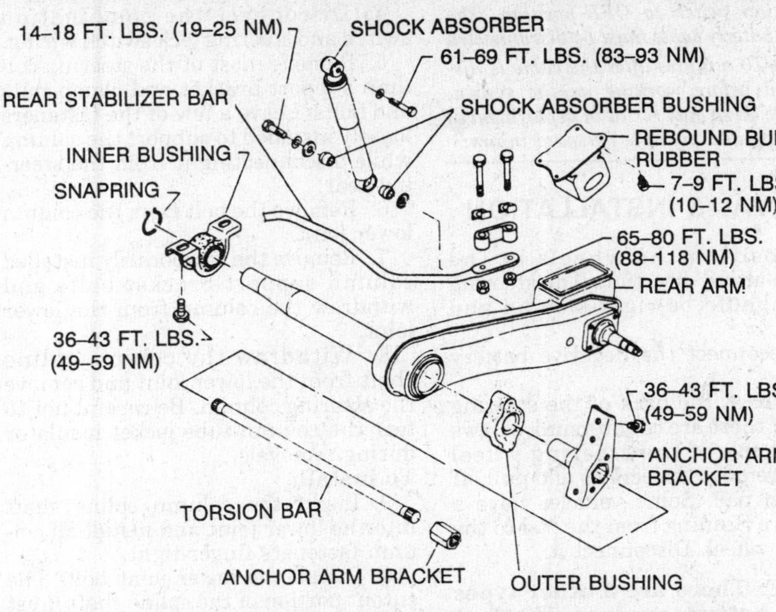

Typical torsion bar rear suspension – Stanza wagon (2wd)

der of removal. Tighten the inner bushing and anchor arm mounting bolts to 36–43 ft. lbs. (49–59 Nm). Tighten the stabilizer bar bolts to 65–80 ft. lbs. (88–108 Nm).

Rear Wheel Bearings

NOTE: For wheel bearing procedures on rear wheel drive models, please refer to "Rear Axle Shaft" in the Drive Axle section.

REMOVAL & INSTALLATION

Maxima, Pulsar, Sentra (2WD) and Stanza (2WD)

1. Raise and support the vehicle safely.
2. Remove the rear wheels.
3. On Maxima and 1990–92 Stanza, remove the brake caliper assembly and support it with wire. The brake hose need not be disconnected. Do not depress the brake pedal while the caliper is supported or the piston will pop out.
4. Work off center hub cap by using thin tool. If necessary tap around it with a soft hammer while removing.
5. Remove the cotter pin, take out adjusting cap and wheel bearing lock nut.
6. Remove drum or disc with bearing inside. On Maxima and 1990–92 Stanza, a disc rotor is used instead of a brake drum.

NOTE: On all Pulsar and Sentra vehicles, a circular clip holds inner wheel bearing in brake hub. On 1989–92 Maxima, Pulsar, Sentra and 1990–92 Stanza, the rear wheel bearing is a sealed unit which combines the bearing, inner and outer races and grease seal. This bearing is retained by a circlip.

7. Remove bearing from drum using long brass drift pin or an arbor press (1989–92 Maxima, Pulsar, Sentra and 1990–92 Stanza).
8. Pack the bearings.
9. Installation is the reverse of the removal procedire: During installation, observe the following:
   a. On 1989–92 Maxima, Pulsar, Sentra and 1990–92 Stanza, the bearing must be pressed into the brake drum or brake disc.
   b. Do not press the inner race of the bearing; do not coat the wheel bearing and outer hub mating surfaces with oil or grease and do not damage the grease seal.
   c. Adjust wheel bearings as outlined below.

Sentra (4WD) and Stanza (4WD)

1. Raise and support the vehicle safely.
2. Remove wheel bearing locknut while depressing brake pedal.
3. Disconnect brake hydraulic line and parking brake cable.
4. Separate halfshaft from knuckle by slightly tapping it with suitable tool. Cover axle boots with waste cloth so as not to damage them when removing halfshaft.
5. Remove all knuckle retaining bolts and nuts. Make a matchmark before removing adjusting pin.
6. Separate the hub from the knuckle using a suitable tool.
7. Drive out the inner (outside) race using a suitable press.
8. Remove the outer grease seal.
9. Drive the inner race (inside) from the hub. The inner grease seal will be removed with it.
10. Remove inner and outer circular clips.
11. Remove the bearings.
12. Drive out the outer race using a suitable tool.

To install:

13. Install the inner circlip in the knuckle groove.
14. Press in the new outer race from the outside of the knuckle.

NOTE: Do not apply grease the wheel bearing outer race and knuckle surfaces.

15. Pack the bearings and the grease seal lip with grease.
16. Install the outer circlip in the knuckle groove.
17. Install the inner races uisng the proper tool, then install the inner grease seal. Be careful not to damage the grease seal.
18. Press the hub into the knuckle.
19. Complete the installation of the remaining components in reverse of the removal procedure. Adjust the wheel bearings as described below.

ADJUSTMENT

1988 Maxima and 1988–89 Stanza 2WD

Before adjusting the rear wheel bearings on these vehicles apply multi-purpose grease to the following parts: threaded portion of the wheel spindle, mating surfaces of the lock washer and outer wheel bearing, inner hub cap and grease seal lip.

1. Tighten the wheel bearing nut to 18–25 ft. lbs. (25–34 Nm).
2. Turn the wheel several times in both directions to seat the bearing correctly.
3. Loosen the wheel bearing nut until there is no pre-load and then tighten it to 6.5–8.7 ft. lbs. (9–12 Nm). Turn the wheel several times again and then retighten it to the same torque again.
4. Install the adjusting cap and align any of its slots with the hole in the spindle.

NOTE: If necessary, loosen the locknut as much as 15 degrees in order to align the spindle hole with one in the adjusting cap.

5. Rotate the hub in both directions several times while measuring its starting torque and axial play. The axial play should be 0 in. (0mm). The

starting torque with grease seal should be 6.9 inch lbs. or less. When measured at wheel hub bolt, starting torque should be 3.1 lbs. (13.7 N).

6. Correctly measure the rotation from the starting force toward the tangential direction against the hub bolt. The above figures do not allow for any "dragging" resistance. When measuring starting torque, confirm that no "dragging" exists. No wheel bearing axial play can exist at all.

7. Spread the cotter pin and install the inner hub cap.

1989–92 Maxima, Pulsar, Sentra, 1990–92 Stanza 2WD and Stanza 4WD

Due to a bearing change on these models, there is no procedure for torquing the rear wheel bearings. Once the final torque is applied to the wheel bearing axle nut and the axial play is checked, no further adjustment is either necessary or possible.

Check the torque of the wheel bearing locknut. This value is 137–188 ft. lbs. on Maxima, Pulsar, Sentra 2WD and 1990–92 Stanza. On Sentra/Stanza 4WD, the torque value is 174–231 ft. lbs. Rotate the hub and make sure the bearing turn smoothly and quietly. Then, mount a dial indicator to the face of the hub and check the axial play. It should not exceed 0.0020 in. (0.05mm). If the axial play is not as specified, replace the necessary component.

Rear Axle Assembly

REMOVAL & INSTALLATION

1. Raise and support the vehicle safely.
2. Remove the rear wheels.
3. Disconnect the brake line and parking brake cable.
4. Work off center hub cap by using thin tool. If necessary tap around it with a soft hammer while removing.
5. Remove the cotter pin, take out adjusting cap and wheel bearing lock nut.
6. Remove drum and wheel hub assembly.
7. Unbolt and remove the knuckle/spindle assembly.
8. Installation is the reverse of the removal procedure. Adjust the rear wheel bearings and bleed the brakes.

STEERING

Steering Wheel

CAUTION

On vehicles equipped with an air bag, turn the ignition switch to OFF position. The negative battery cable must be disconnected and wait 10 minutes after the cable is disconnected before working on the system. Failure to do so may result in deployment of the air bag and possible personal injury.

REMOVAL & INSTALLATION

1. Position the wheels in the straight-ahead direction. The steering wheel should be right-side up and level.
2. Disconnect the negative battery cable.
3. Look at the back of the steering wheel. If there are countersunk screws in the back of the steering wheel spokes, remove the screws and pull off the horn pad. Some vehicles have a horn wire running from the pad to the steering wheel. Disconnect it.

NOTE: There are 3 other types of horn buttons or rings. The first simply pulls off. The second, which is usually a large, semi-triangular pad, must be pushed up, then pulled off. The third must be pushed in and turned clockwise.

4. Remove the rest of the horn switching mechanism, noting the relative location of the parts. Remove the mechanism only if it interferes with removal of the steering wheel.
5. Matchmark the top of the steering column shaft and the steering wheel flange.
6. Remove the attaching nut and remove the steering wheel with a puller.

NOTE: Do not strike the shaft with a hammer; which may cause the column to collapse.

To install:

7. Install the steering wheel by aligning the punch marks. Do not drive or hammer the wheel into place, or the collapsible steering column may collapse. Before installing the horn pad, apply multi-purpose grease to the surface of the cancel pin and horn contact slip ring.
8. Tighten the steering wheel nuts to 22–29 ft. lbs. (29–39 Nm).
9. Reinstall the horn button, pad, or ring.
10. Connect the negative battery cable.

Steering Column

REMOVAL & INSTALLATION

1. Disconnect the negative battery cable.
2. Remove the steering wheel.
3. Remove the steering column covers.
4. Disconnect the combination switch and steering lock switch wiring.
5. Remove most of the steering column support bracket and clamp nuts and bolts. Leave a few of the fasteners loosely installed to support the column while disconnecting it from the steering gear.
6. Remove the bolt from the column lower joint.
7. Remove the temporarily installed column support bracket bolts and withdraw the column from the lower joint.
8. Withdraw the column spline shaft from the lower joint and remove the steering column. Be careful not to tear the column tube jacket insulator during removal.

To install:

9. Insert the column spline shaft into the lower joint and install all column fasteners finger-tight.
10. Install the lower joint bolt. The cutout portion of the spline shaft must perfectly aligned with the bolt. Torque the bolt to 17–22 ft. lbs. (23–30 Nm). Tighten the steering bracket and clamp fasteners gradually. While tightening, make sure no stress is placed on the column.
11. Connect the combination switch and steering lock switch wiring.
12. Install the steering column covers.
13. Install the steering wheel.
14. Connect the negative battery cable.
15. After the installation is complete, turn the steering wheel from stop to stop and make sure it turns smoothly. The number of turns to the left and right stops must be equal.

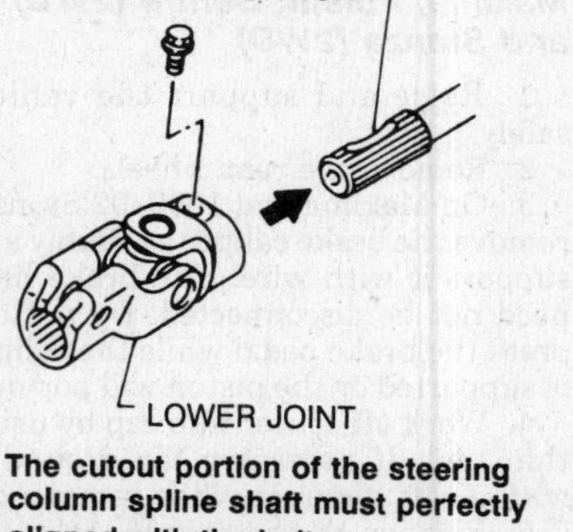

The cutout portion of the steering column spline shaft must perfectly aligned with the bolt

Manual Steering Rack and Pinion

REMOVAL & INSTALLATION

Sentra

1. Raise and support the vehicle safely and remove the wheels.

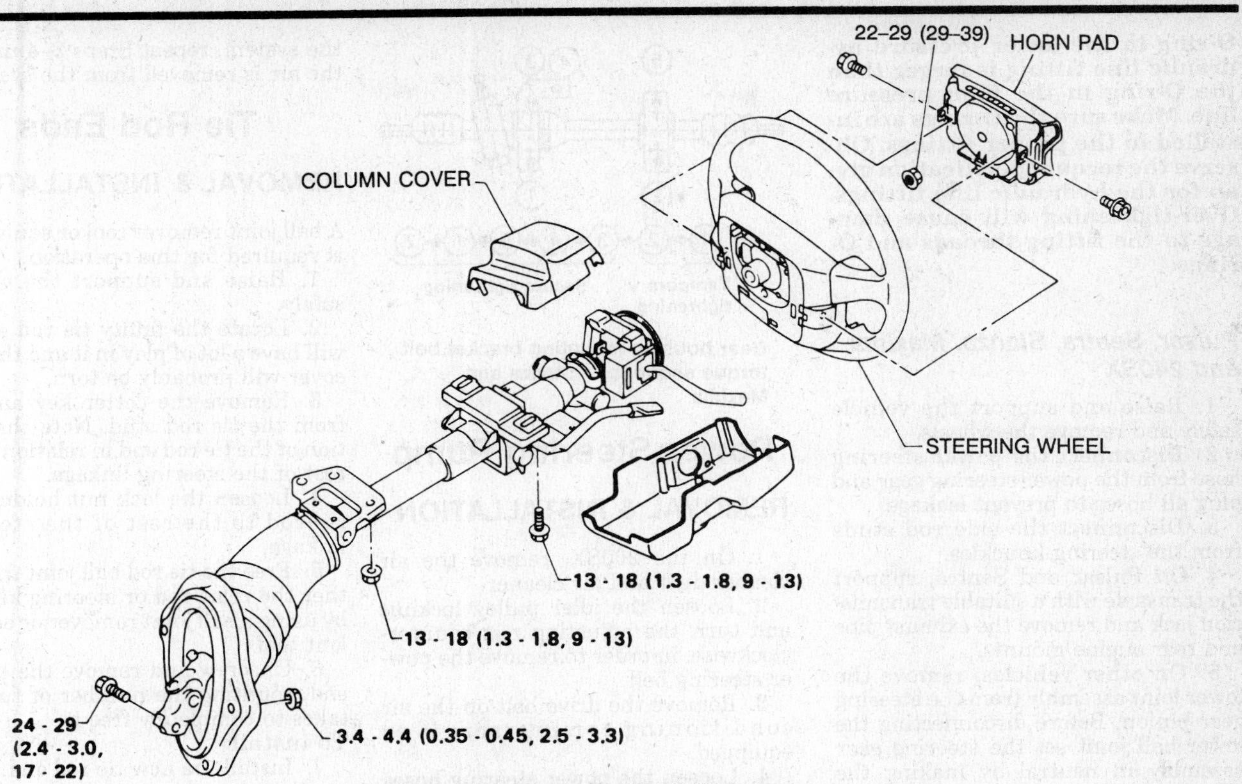

Steering column assembly—1990 Pulsar

2. Disconnect the tie rod from the steering knuckle and loosen the steering gear attaching bolts.

3. Remove the bolt securing the lower joint to the steering gear pinion and remove the lower joint from the pinion.

4. Remove the bolts holding the steering gear housing to the body, and remove the steering gear and linkage assembly from the vehicle.

5. Installation is the reverse order of the removal procedure. When fitting the lower U-joint, make sure the attaching bolt is aligned perfectly with the cut out in the splined end of the steering column shaft. Torque the steering gear mounting clamp bolts to 54–72 ft. lbs. (73–97 Nm). Torque the tie rod end nuts to 22–29 ft. lbs. (29–39 Nm).

Power Steering Rack and Pinion

REMOVAL & INSTALLATION

200SX

1. Raise and support the vehicle safely.

2. Remove the air cleaner and remove the bolt securing the U-joint to the worm shaft.

3. Disconnect the hoses from the power steering gear and plug the hoses to prevent leakage.

4. Remove the pitman arm from the sector shaft using a suitable tool and remove the steering gear mounting bolts.

5. Remove the exhaust pipe mounting nut.

6. Disconnect the control cable or linkage for the transmission and position it aside.

7. Remove the steering gear from the vehicle.

8. Installation is the reverse of the removal procedure. Torque the mounting clamp bolts to 29–36 ft. lbs. (39–49 Nm) and the tie rod end nuts to 40–72 ft. lbs. (54–98 Nm). Torque the worm shaft U-joint bolt to 17–22 ft. lbs. (24–29 Nm). Refill the power steering pump, start the engine and bleed the system.

300ZX

1. Block the rear wheels. Raise and support the vehicle safely.

2. Position an oil catch pan under the power steering gear, remove the hydraulic lines from the gear and drain the oil. Plug the lines to prevent leakage.

3. Loosen the steering column lower joint shaft bolt.

4. Before disconnecting the lower ball joint set the steering gear assembly in neutral by making the wheels straight. Loosen the bolt and disconnect the lower joint. Matchmark the pinion shaft to the pinion housing to record the neutral gear position.

5. Remove the tie rod end-to-knuckle arm cotter pins and castle nuts.

6. Separate the tie rods from the knuckle arms using a suitable puller.

7. Remove the steering gear housing-to-suspension crossmember bolts.

8. Position a floor jack under the engine and raise it just enough to support the engine. Loosen the engine mounting bolts and raise the engine about ½ in.

9. Remove the steering gear and linkage from the vehicle.

To install:

10. Installation is the reverse of the removal procedure observing the following:

a. Tighten the gear housing mouting bracket bolts to 29–36 ft. lbs. (39–49 Nm) on 1988–89 vehicles, and 65–80 ft. lbs. (88–108 Nm) on 1990–92 vehicles.

b. Torque the tie rod end nuts to 22–29 ft. lbs. (29–39 Nm).

c. On 1990–92 vehicles, torque the high pressure hydraulic line fitting to 22–26 ft. lbs. (36–40 Nm) and lower pressure fitting to 27–30 ft. lbs. (36–40 Nm).

d. When attaching the lower joint, set the left and right dust boots to equal deflection. Refill the power steering pump, start the engine and bleed the system.

NOTE: On 1990–92 vehicles, the

O-ring in the lower pressure hydraulic line fitting is larger than the O-ring in the high pressure line. Make sure the O-rings are installed in the proper fittings. Observe the torque specification given for the hydraulic line fittings. Over-tightening will cause damage to the fitting threads and O-rings.

Pulsar, Sentra, Stanza, Maxima and 240SX

1. Raise and support the vehicle safely and remove the wheels.
2. Disconnect the power steering hose from the power steering gear and plug all hoses to prevent leakage.
3. Disconnect the side rod studs from the steering knuckles.
4. On Pulsar and Sentra, support the transaxle with a suitable transmission jack and remove the exhaust pipe and rear engine mounts.
5. On other vehicles, remove the lower joint assembly from the steering gear pinion. Before disconnecting the lower ball joint set the steering gear assembly in neutral by making the wheels straight. Loosen the bolt and disconnect the lower joint. Matchmark the pinion shaft to the pinion housing to record the neutral gear position.
6. Remove the steering gear and linkage assembly from the vehicle.

To install:

7. Installation is the reverse of the removal procedure observing the following:
   a. Make sure the pinion shaft and pinion housing are aligned properly.
   b. Torque the high pressure hydraulic line fitting to 11–18 ft. lbs. (15–25 Nm) and lower pressure fitting to 20–29 ft. lbs. (27–39 Nm).
   c. When attaching the lower joint, set the left and right dust boots to equal deflection.
   d. On Maxima and Stanza, torque the gear housing mounting bracket bolts to 54–72 ft. lbs. (73–97 Nm) using the proper sequence.
8. Refill the power steering pump, start the engine and bleed the system. Refill the power steering pump, start the engine and bleed the system.

NOTE: On most vehicles, the O-ring in the lower pressure hydraulic line fitting is larger than the O-ring in the high pressure line. Make sure the O-rings are installed in the proper fittings. Observe the torque specification given for the hydraulic line fittings. Over-tightening will cause damage to the fitting threads and O-rings.

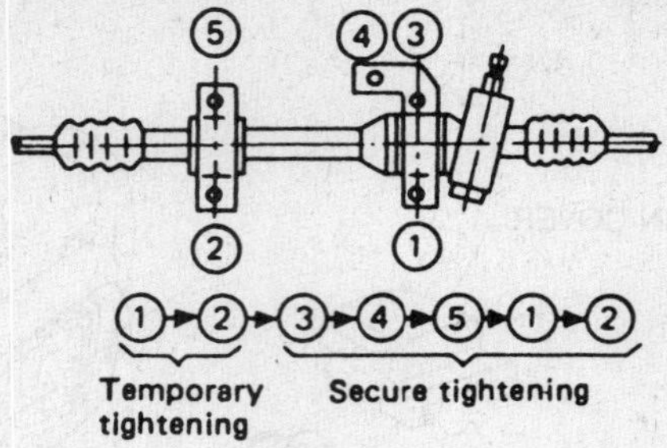

Gear housing mounting bracket bolt torque sequence—Stanza and Maxima

Power Steering Pump

REMOVAL & INSTALLATION

1. On the 200SX, remove the air cleaner duct and air cleaner.
2. Loosen the idler pulley locknut and turn the adjusting nut counterclockwise, in order to remove the power steering belt.
3. Remove the drive belt on the air conditioning compressor, if so equipped.
4. Loosen the power steering hoses at the pump and remove the bolts holding the power steering pump to the bracket.
5. Disconnect and plug the power steering hoses and remove the pump from the vehicle.
6. Installation is the reverse of the removal procedure. Fill and bleed the power steering system.

BELT ADJUSTMENT

1. Loosen the tension adjustment and mounting bolts.
2. Move the pump toward or away from the engine so the belt deflects ¼–½ in. midway between the idler pulley and the pump pulley under moderate thumb pressure.
3. Tighten the bolts and recheck the tension adjustment.

SYSTEM BLEEDING

1. Check the level in the power steering pump reservoir. Add fluid as necessary to the proper level.
2. Safely raise and support the vehicle until the wheels are just off the ground.
3. With the engine running, quickly turn the steering wheel all the way to the left and all the way to the right 10 times.
4. Stop the engine and check to see if any more fluid is required in the pump reservoir. Add fluid as necessary.
5. If all the air cannot be bled from the system, repeat Steps 3–4 until all the air is removed from the system.

Tie Rod Ends

REMOVAL & INSTALLATION

A ball joint remover tool or equivalent, is required for this operation.

1. Raise and support the vehicle safely.
2. Locate the faulty tie rod end. It will have a lot of play in it and the dust cover will probably be torn.
3. Remove the cotter key and nut from the tie rod stud. Note the position of the tie rod end in relation to the rest of the steering linkage.
4. Loosen the lock nut holding the tie rod to the rest of the steering linkage.
5. Free the tie rod ball joint from either the relay rod or steering knuckle by using a ball joint remover or equivalent tool.
6. Unscrew and remove the tie rod end, counting the number of turns it takes to completely free it.

To install:

7. Install the new tie rod end, turning it in exactly the same number of turns for removal. Make sure it is correctly positioned in relation to the rest of the steering linkage.
8. Fit the ball joint and nut. Torque the tie rod end to 22–36 ft. lbs. (29–39 Nm) on all vehicles except 200SX. On 200SX, torque the tie rod end nut to 40–72 ft. lbs. (54–98 Nm). Once the specified torque is reached, tighten further until the nut groove is aligned with the first pin hole. Install a new cotter pin.
9. Check and adjust the toe of the vehicle as needed.

BRAKES

For all brake system repair and service procedures not detailed below, please refer to "Brakes" in the Unit Repair section.

Master Cylinder

REMOVAL & INSTALLATION

1. Clean the outside of the cylinder thoroughly, particularly around the cap and fluid lines.
2. Disconnect the fluid lines and cap them to keep dirt out.
3. On vehicles with a fluid level gauge, disconnect the electrical connector.
4. Remove the clevis pin connecting

the pushrod to the brake pedal arm inside the vehicle.

5. Unbolt the master cylinder from the firewall and remove along with gasket. If the pushrod is not adjustable, there will be shims between the cylinder and the firewall. These shims, or the adjustable pushrod, are used to adjust brake pedal free-play.

6. Installation of the master cylinder is the reverse of the removal procedure. Bleed the brakes.

Proportioning Valve

The proportioning valve is incorporated into the master cylinder. Consequently, removal and installation procedures are limited to replacement of the master cylinder unit as a whole.

Power Brake Booster

REMOVAL & INSTALLATION

1. Remove the master cylinder.
2. Remove the vacuum hose at the power brake booster.
3. Remove the pushrod from the brake pedal.
4. From under the instrument panel, remove the cowl-to-booster nuts. Remove the brake booster.
5. Installation is in the reverse order of removal. Bleed the brake system.

Brake Calipers

REMOVAL & INSTALLATION

1. Raise the vehicle and support safely.
2. Remove the front or rear wheels.
3. Disconnect the brake line from the caliper. Remove the metal gaskets from the brake hose fitting and discard them.
4. Disconnect the parking brake cable.
5. Remove the brake pads if they interfere with caliper removal.
6. Remove the brake caliper mounting bolts.
7. Remove the brake caliper assembly.
8. Installation is the reverse of the removal procedure. Torque the caliper bolts to 40–72 ft. lbs. (54–98 Nm). Use new brake hose fitting gaskets. Bleed the brake system and adjust the parking brake cable.

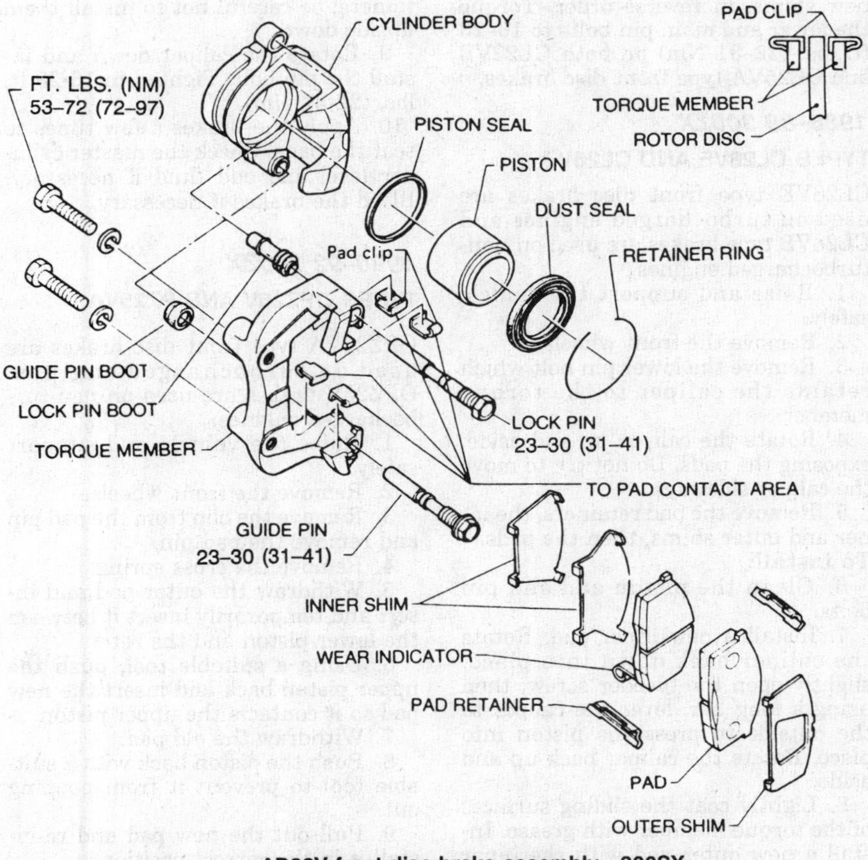

AD22V front disc brake assembly—200SX

Front Disc Brake Pads

REMOVAL & INSTALLATION

200SX and Maxima

TYPE AD22V

AD22V type front disc brakes are used on 200SX with CA20E engines.

1. Raise the vehicle and support safely.
2. Remove the front wheels.
3. Remove the lower caliper guide pin.
4. Rotate the brake caliper body upward.
5. Remove the brake pad retainer and the inner and outer pad shims.
6. Remove the brake pads.

NOTE: Do not depress the brake pedal when the caliper body is raised. The brake piston will be forced out of the caliper.

To install:

7. Clean the piston end of the caliper body and the pin bolt holes. Be careful not to get oil on the brake rotor.
8. Pull the caliper body to the outer side and install the inner brake pad.
9. Install the outer pad, shim and pad retainer.
10. Reposition the caliper body and then tighten the guide pin bolt to 23–30 ft. lbs. (31–41 Nm).
11. Apply the brakes a few times to seat the pads before driving out on the road.

TYPES CL25VB AND CL28VB

CL28VB type front disc brakes are used on 200SX with VG30E engines and 1988 Maxima. CL25VB type brakes are used on 1989–92 Maxima.

1. Raise the vehicle and support safely.
2. Remove the front wheels.
3. Remove the pin (lower) bolt from the caliper.
4. Swing the caliper body upward on the upper bolt.
5. Remove the pad retainers and inner and outer shims.

NOTE: Do not depress the brake pedal when the cylinder body is in the raised position or the piston will pop out of the cylinder.

To install:

6. Check the level of fluid in the master cylinder. If the fluid is near the maximum level, use a clean syringe to remove fluid until the level is down well below the lip of the reservoir.
7. Use a large C-clamp or piston expansion tool to press the caliper piston back into the caliper, to allow room for the installation of the thicker new pads.

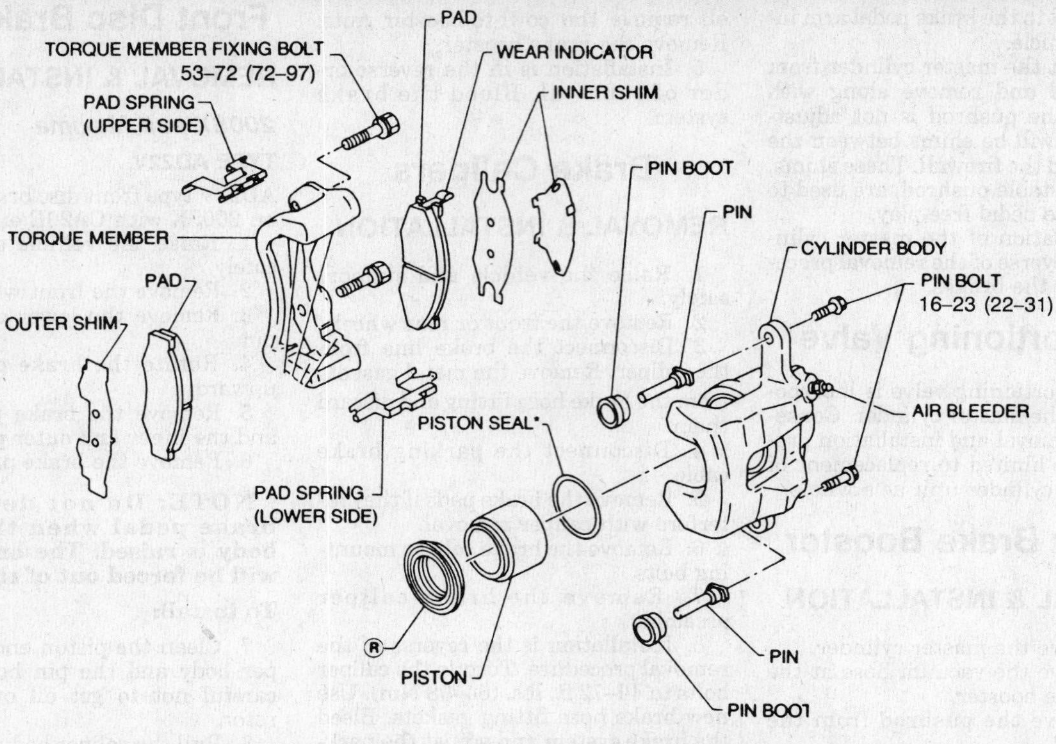

CL28VB front brake disc assembly – 200SX and 1988–89 300ZX (non-turbo)

8. Install the new pads, utilizing new shims, in reverse order. Torque the lower pin bolt to 16–23 ft. lbs. (22–31 Nm) on both types of brakes.

240SX

TYPES CL22VB AND CL25VA

CL22VB type front disc brakes are used on vehicles without ABS and CL25VA type brake are used on vehicles with ABS.

1. Raise and support the vehicle safely.
2. Remove the front wheels.
3. Remove the pin (lower) bolt from the caliper.
4. Swing the caliper body upward on the upper bolt.
5. Remove the pad retainers and inner and outer shims.

NOTE: Do not depress the brake pedal when the cylinder body is in the raised position or the piston will pop out of the cylinder.

6. Check the level of fluid in the master cylinder. If the fluid is near the maximum level, use a clean syringe to remove fluid until the level is down well below the lip of the reservoir.
7. Use a large C-clamp or piston expansion tool to press the caliper piston back into the caliper, to allow room for the installation of the thicker new pads.
8. Install the new pads, utilizing new shims, in reverse order. Torque the lower and main pin bolts to 16–23 ft. lbs. (22–31 Nm) on both CL22VB and CL25VA type front disc brakes.

1988–89 300ZX

TYPES CL28VE AND CL28VB

CL28VE type front disc brakes are used on turbocharged engines and CL28VB type brakes are used on non-turbocharged engines.

1. Raise and support the vehicle safely.
2. Remove the front wheels.
3. Remove the lower pin bolt which retains the caliper to the torque member.
4. Rotate the caliper up and aside, exposing the pads. Do not try to move the caliper sideways.
5. Remove the pad retainers, the inner and outer shims, then the pads.

To install:

6. Clean the piston end and pin bolts.
7. Install a new inner pad. Rotate the caliper back down into place, slightly open the bleeder screw, then using a long bar, lever the caliper to the outside to press the piston into place. Rotate the caliper back up and aside.
8. Lightly coat the sliding surfaces of the torque member with grease. Install a new outer pad with the inner and outer shims. Install the pad retainers; be careful not to install them upside down.
9. Rotate the caliper down and install the pin bolt. Tighten to 16-23 ft. lbs. (22-31 Nm).
10. Apply the brakes a few times to seat the pads. Check the master cylinder level and add fluid if necessary. Bleed the brakes if necessary.

1990–92 300ZX

TYPES OPZ25V AND PZ25VA

OPZ25VA type front disc brakes are used on turbocharged engines. OPZ25V brakes are used on non-turbocharged engines.

1. Raise the vehicle and support safely.
2. Remove the front wheels.
3. Remove the clip from the pad pin and remove the pad pin.
4. Remove the cross spring.
5. Withdraw the outer pad and insert and temporarily insert it between the lower piston and the rotor.
6. Using a suitable tool, push the upper piston back and insert the new pad so it contacts the upper piston.
7. Withdraw the old pad.
8. Push the piston back with a suitable tool to prevent it from popping out.
9. Pull out the new pad and re-install it in the correct position.

10. Repeat steps 5–9 for the inner pad.

11. Install the cross spring, pad pin and pad clip.

Pulsar, Sentra and Stanza

TYPES AD18B, AD18V, CL18B, CL25VA AND CL28VA

CL18B, AD18B and AD18V type front disc brakes are used on Pulsar and Sentra depending on engine application. CL28VA type brakes are used on 1988–89 Stanza and Stanza Wagon. The 1990–92 Stanza uses CL25VA front disc brakes on both ABS equipped and non-ABS vehicles.

1. Raise the vehicle and support safely.
2. Remove the front wheels.
3. Remove the bottom guide pin (Stanza and Sentra) or the lock pin (Pulsar) from the caliper and swing the caliper cylinder body upward.
4. Remove the brake pad retainers and the pads.

To install:

5. Install the brake pads and caliper assembly.
6. Install the wheels and lower the vehicle.
7. Apply the brakes a few times to seat the pads. Check the master cylinder and add fluid if necessary. Bleed the brakes, if necessary.

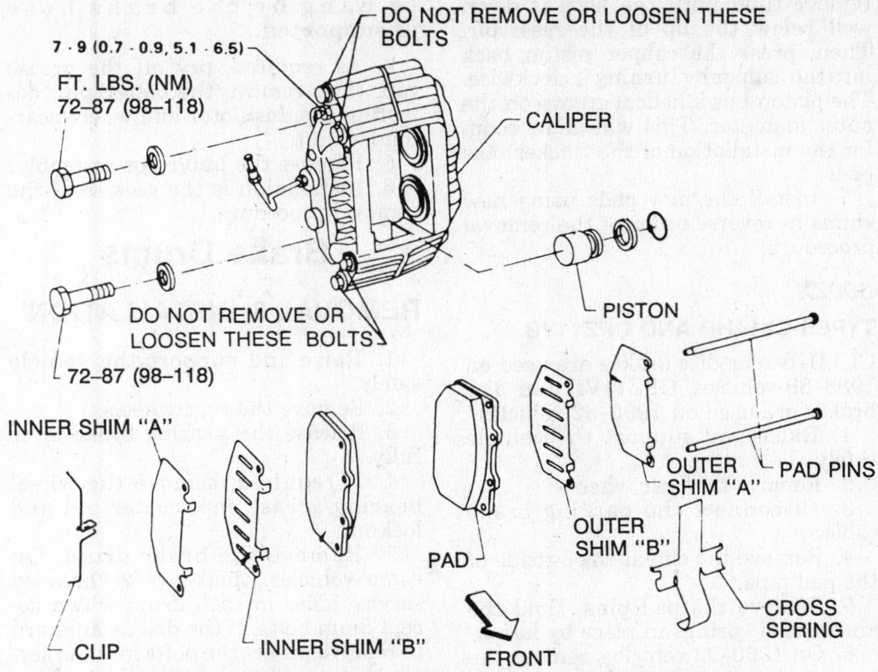

OPZ25V and OPZ25VA type front disc brake assembly – 1990–92 300ZX

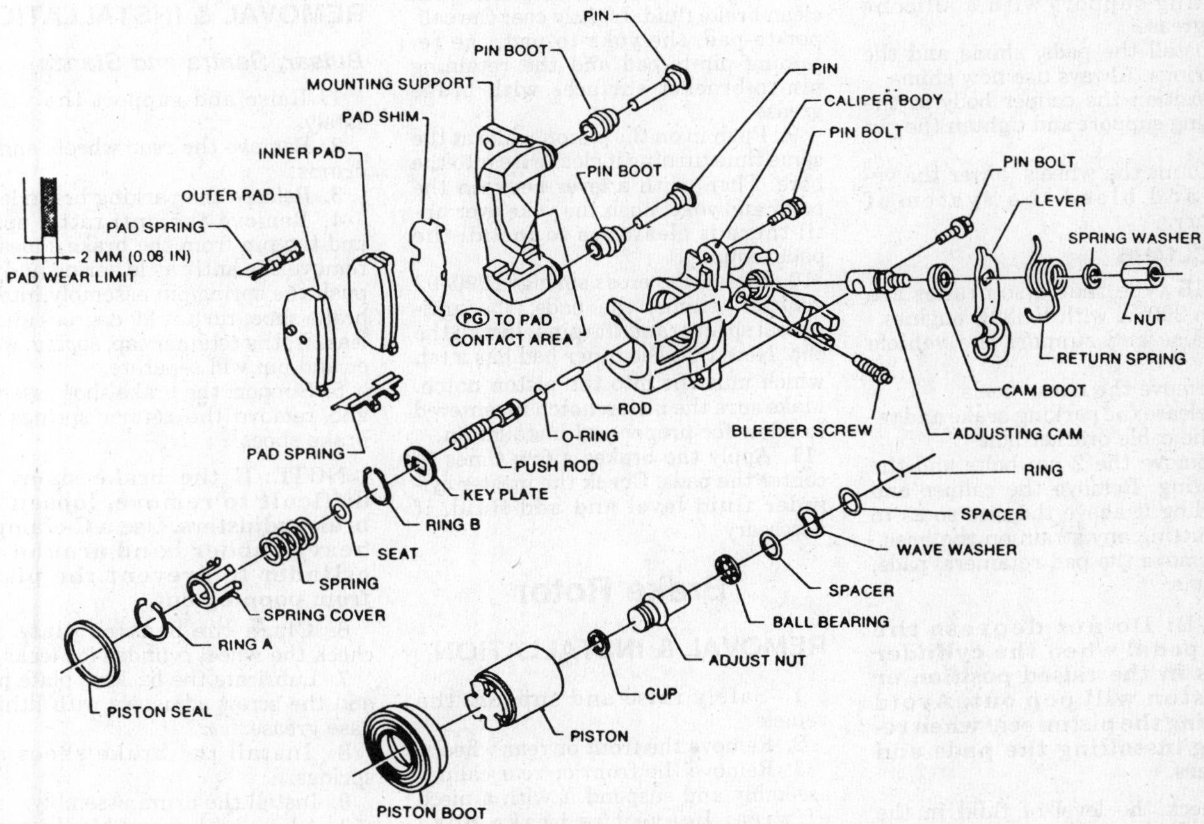

CL11H rear disc brake assembly

Rear Disc Brake Pads

REMOVAL & INSTALLATION

200SX, 240SX, Maxima and 1990–92 Stanza With ABS

TYPES CL11H AND CL9H

CL11H type rear disc brakes are used on 200SX with CA20E engines and on the 1988 Maxima. CL9H rear disc brake are used on 240SX and 1989–92 Maxima on both ABS equipped and non-ABS vehicles. CL9H rear disc brakes are also used on 1990–92 Stanza with ABS.

1. Raise and support the vehicle safely.
2. Remove the rear wheels.
3. Release the parking brake and remove the cable bracket bolt.
4. Remove the pin bolts and lift off the caliper body.
5. Pull out the pad springs and then remove the pads and shims.

To install:

6. Clean the piston end of the caliper body and the area around the pin holes. Be careful not to get oil on the rotor.
7. Using the proper tool, carefully turn the piston clockwise back into the caliper body. Take care not to damage the piston boot.
8. Coat the pad contact area on the mounting support with a silicone based grease.
9. Install the pads, shims and the pad springs. Always use new shims.
10. Position the caliper body in the mounting support and tighten the pin bolts.
11. Mount the wheels, lower the vehicle and bleed the system if necessary.

TYPE CL14HB

CL14HB type rear disc brakes are used on 200SX with VG30E engines.

1. Raise and support the vehicle safely.
2. Remove the rear wheels.
3. Release the parking brake and remove the cable bracket bolt.
4. Remove the 2 pin bolts and the lock spring. Remove the caliper and suspending it above the disc so as to avoid putting any strain on the hose.
5. Remove the pad retainers, pads, and shims.

NOTE: Do not depress the brake pedal when the cylinder body is in the raised position or the piston will pop out. Avoid damaging the piston seal when removing/installing the pads and retainers.

6. Check the level of fluid in the master cylinder. If the fluid is near the maximum level, use a clean syringe to remove fluid until the level is down well below the lip of the reservoir. Then, press the caliper piston back into the caliper by turning it clockwise. The piston has a helical groove on the outer diameter. This will allow room for the installation of the thicker new pads.
7. Install the new pads using new shims in reverse order of the removal procedure.

300ZX

TYPES CL14HB AND OPZ11VB

CL14HB rear disc brakes are used on 1988–89 vehicles. OPZ11VB rear disc brakes are used on 1990–92 vehicles.

1. Raise and support the vehicle safely.
2. Remove the rear wheels.
3. Disconnect the parking brake cable.
4. Remove the clip at the outside of the pad pins.
5. Remove the pad pins. Hold the anti-squeal springs in place by hand.
6. On 1990–92 vehicles, remove the cross spring.
7. Remove the pads.

NOTE: When the pads are removed, do not depress the brake or else the piston will pop out.

8. Clean the end of the piston with clean brake fluid. Lightly coat the caliper-to-pad, the yoke-to-pad, the retaining pin-to-pad and the retaining pin-to-bracket surfaces with brake grease.
9. Push in on the piston while at the same time turning it clockwise into the bore. Then, with a lever between the rotor and yoke, push the yoke over until there is clearance to install the pads, equally.
10. Install the cross spring (1990–92 vehicles) shims, the pads, the anti-squeal springs and the pins. Install the clip. Note that the inner pad has a tab which must fit into the piston notch. Make sure the piston notch is centered to allow for proper pad installation.
11. Apply the brakes a few times to center the pads. Check the master cylinder fluid level and add fluid, if necessary.

Brake Rotor

REMOVAL & INSTALLATION

1. Safely raise and support the vehicle.
2. Remove the front or rear wheels.
3. Remove the front or rear caliper assembly and suspend it with a piece of wire. Leave the brake hose connected.

NOTE: Do not allow the caliper to hang by the brake hose unsupported.

4. As required, pry off the grease cap, then remove the cotter pin, adjusting cap, insulator and wheel bearing locknut.
5. Remove the hub/rotor assembly.
6. Installation is the reverse of the removal procedure.

Brake Drums

REMOVAL & INSTALLATION

1. Raise and support the vehicle safely.
2. Remove the rear wheels.
3. Release the parking brake lever fully.
4. If required, remove the wheel bearing grease cap, cotter pin and locknut.
5. Remove the brake drum. On some vehicles, there are 2 threaded service holes in each drum which accept 8mm bolts. If the drums are hard to remove, insert the bolts into the service holes and screw them in to force the drum away from the axle.
6. Installation is the reverse of the removal procedure.

Brake Shoes

REMOVAL & INSTALLATION

Pulsar, Sentra and Stanza

1. Raise and support the vehicle safely.
2. Remove the rear wheels and the drums.
3. Release the parking brake lever.
4. Remove the anti-rattle spring and the pin from the brake shoes. To remove the anti-rattle spring and pin, push the spring/pin assembly into the brake shoe, turn it 90 degrees and release it; the retainer cap, spring, washer and pin will separate.
5. Support the brake shoe assembly and remove the return springs and brake shoes.

NOTE: If the brake shoes are difficult to remove, loosen the brake adjusters. Use a C-clamp or heavy rubber band around the cylinder to prevent the piston from popping out.

6. Clean the backing plate and check the wheel cylinder for leaks.
7. Lubricate the backing plate pads and the screw adjusters with lithium base grease.
8. Install the brake shoes and springs.
9. Install the drum assembly.
10. Adjust brakes and bleed the system if necessary.

Wheel Cylinder

REMOVAL & INSTALLATION

1. Raise and support the vehicle safely.
2. Remove the tire and wheel assembly.
3. Remove the brake drum and brake shoes.
4. Disconnect the hydraulic line from the wheel cylinder. Plug the line to prevent leakage.
5. Remove the wheel cylinder from the brake backing plate.
6. Remove the dust boot and take out the piston. Discard the piston cup. The dust boot can be reused although it is best to replace it.
7. Wash all of the components in clean brake fluid.
8. Inspect the piston and piston bore. Replace any components that are severely corroded, scored or worn. The piston and piston bore may be polished lightly with crocus cloth; move the cloth around the piston bore, not in and out.
9. Wash the wheel cylinder and piston in clean brake fluid.
10. Coat all new components to be installed with clean brake fluid.
11. Assemble the cylinder and install it on the backing plate. Connect the hydraulic line.
12. Install the brake shoes and brake drum.
13. Install the wheel and tire assembly.
14. Lower the vehicle.
15. Bleed the brake system.

Parking Brake Cable

ADJUSTMENT

1. Pull up the hand brake lever, counting the number of notches for full engagement. Full engagement should be:
   - 200SX: 7–8 notches
   - 240SX: 6–8 notches
   - 300ZX
     - 1988–89: 8–10 notches
     - 1990–92: 8–10 notches
   - Maxima
     - 1988: 11–13 notches
     - 1989–92: 8–11 notches
   - Pulsar: 7–11 notches
   - Sentra
     - 1988: 11–13 notches
     - 1989–92: 7–11 notches
   - Stanza sedan
     - 1988: 11–13 notches
     - 1989–92: 11–13 notches
   - Stanza Wagon
     - 2WD: 11–17 notches
     - 4WD: 8–9 notches
2. Release the parking brake.
3. Except on 200SX, adjust the lever stroke by loosening the locknut and tightening the adjusting nut to reduce the number of notches necessary for engagement. Tighten the locknut. The locknut and adjuster can be found inside the handbrake assembly, in the passenger compartment. Some vehicles just have an adjusting nut. Access to the locknut is gained by removing the parking brake console or through an access hole in the console itself. On 200SX, the lever stroke is adjusted by turning the equalizer under the vehicle.
4. Check the adjustment and repeat as necessary.
5. After adjustment, check to see that the rear brake levers, at the calipers, return to their full off positions when the lever is released, and that the rear cables are not slack when the lever is released.
6. To adjust the parking brake light, bend the light switch plate down so the light comes on when the lever is engaged 1–2 notches.

REMOVAL & INSTALLATION

Front Cable

1. Remove the parking brake console box.
2. Remove the heat insulator, if equipped.
3. Remove the front passenger seat, if required.
4. Disconnect the warning lamp switch plate connector.
5. Unbolt the lever from the floor.
6. Working from under the vehicle, remove the locknut, adjusting nut and equalizer.
7. Pull the front cable out through the compartment and remove it from the vehicle.

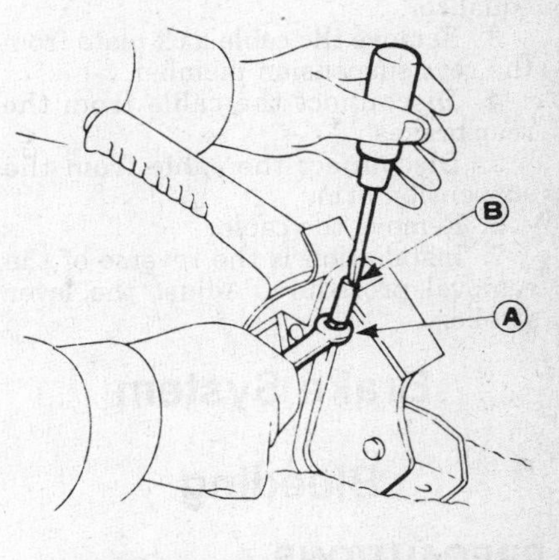

Typical parking brake adjustment

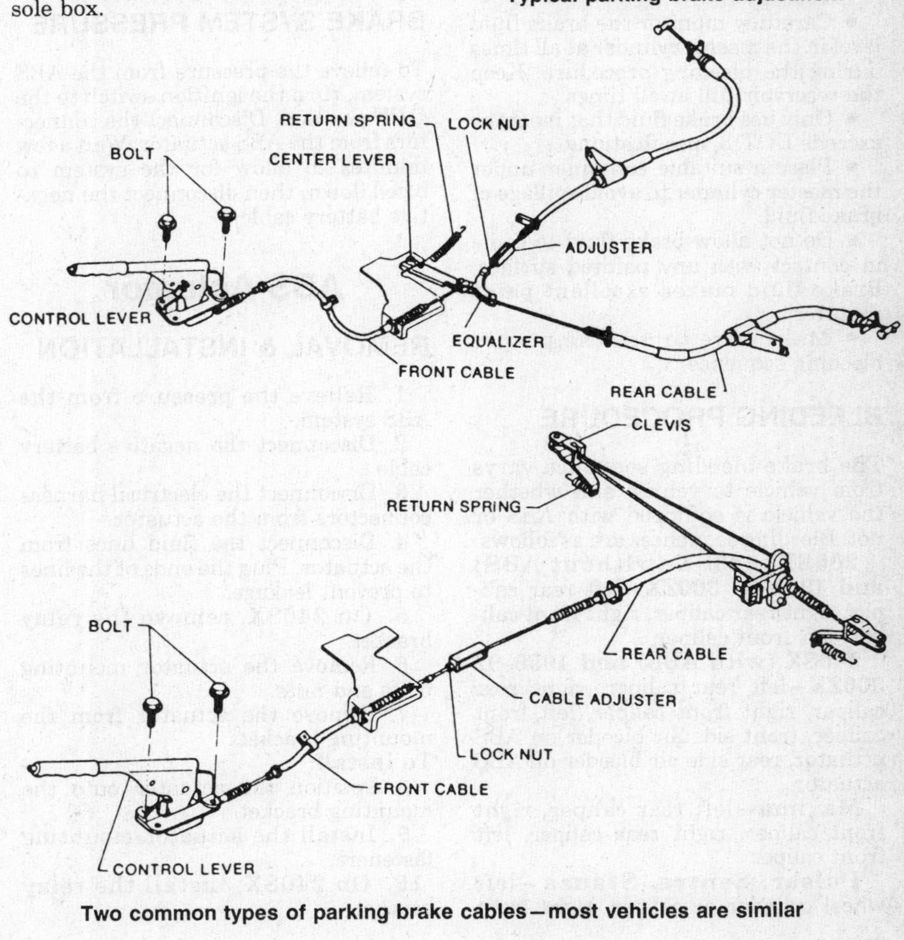

Two common types of parking brake cables—most vehicles are similar

NOTE: On some vehicles it may be necessary to separate the front cable from the lever by breaking the pin.

8. Installation is the reverse of the removal procedure. Adjust the lever stroke.

Rear Cable

1. Back off on the adjusting nut or equalizer to loosen the cable tension.
2. Working from underneath the vehicle, disconnect the cable at the equalizer.
3. Remove the cable lock plate from the rear suspension member.
4. Disconnect the cable from the rear brakes.
5. Disconnect the cable from the suspension arm.
6. Remove the cable.
7. Installation is the reverse of the removal procedure. Adjust the lever stroke.

Brake System

Bleeding

PRECAUTIONS

- Carefully monitor the brake fluid level in the master cylinder at all times during the bleeding procedure. Keep the reservoir full at all times.
- Only use brake fluid that meets or exceeds DOT 3 specifications.
- Place a suitable container under the master cylinder to avoid spillage of brake fluid.
- Do not allow brake fluid to come in contact with any painted surface. Brake fluid makes excellent paint remover.
- Make sure to use the proper bleeding sequence.

BLEEDING PROCEDURE

The brake bleeding sequence varys from vehicle to vehicle and whether the vehicle is equipped with ABS or not. Bleeding sequences are as follows:

200SX, 240SX (without ABS) and 1988–89 300ZX—left rear caliper, right rear caliper, right front caliper, left front caliper.

240SX (with ABS) and 1990–92 300ZX—left rear caliper, right rear caliper, right front caliper, left front caliper, front side air bleeder on ABS actuator, rear side air bleeder on ABS actuator

Maxima—left rear caliper, right front caliper, right rear caliper, left front caliper

Pulsar, Sentra, Stanza—left wheel cylinder or caliper, right front caliper, right rear wheel cylinder or caliper, left front caliper

To bleed the brakes, use the following procedure:

1. If equipped with ABS, turn the ignition switch to the **OFF** position and disconnect the connectors from the ABS actuator. Wait a few minutes to allow for the system to bleed down, then disconnect the negative battery cable.
2. Connect a transparent vinyl tube to the bleeder valve. Submerge the tube in a container half filled with clean brake fluid.
3. Fully depress the brake pedal several times.
4. With the brake pedal depressed, open the air bleeder valve to release the air.
5. Close the air bleeder valve.
6. Release the brake pedal slowly.
7. Repeat Steps 3–6 until clear fluid flows from the air bleeder valve.
8. Check the fluid level in the master cylinder reservoir and add as necessary.

Anti-Lock Brake System Service

RELIEVING ANTI-LOCK BRAKE SYSTEM PRESSURE

To relieve the pressure from the ABS system, turn the ignition switch to the **OFF** position. Disconnect the connectors from the ABS actuator. Wait a few minutes to allow for the system to bleed down, then disconnect the negative battery cable.

ABS Actuator

REMOVAL & INSTALLATION

1. Relieve the pressure from the ABS system.
2. Disconnect the negative battery cable.
3. Disconnect the electrical harness connectors from the actuator.
4. Disconnect the fluid lines from the actuator. Plug the ends of the lines to prevent leakage.
5. On 240SX, remove the relay bracket.
6. Remove the actuator mounting bolts and nuts.
7. Remove the actuator from the mounting bracket.

To install:

8. Position the actuator onto the mounting bracket.
9. Install the actuator mounting fasteners.
10. On 240SX, install the relay bracket.
11. Connect the fluid lines and the harness connectors.
12. Connect the negative battery cable.
13. Bleed the brake system.

ABS Front Wheel Sensor

REMOVAL & INSTALLATION

1. Raise and support the vehicle safely.
2. Remove the front wheels.
3. Disconnect the sensor harness connector.
4. Detach the sensor mounting brackets.
5. Unbolt the sensor from the rear of the steering knuckle.
6. Withdraw the sensor from the sensor rotor. Remove the sensor mounting brackets from the sensor wiring.

NOTE: During removal and installation, take care not to damage the sensor or the teeth of the rotor.

To install:

7. Transfer the mounting brackets to the new sensor. Insert the sensor through the opening in the the rear of the knuckle and engage the sensor with the rotor teeth.
8. Install the sensor mounting bolts. Check and adjust the sensor-to-rotor clearance as described below. Once the clearance is set, tighten the sensor mounting bolt(s) to 8–12 ft. lbs. (11–16 Nm) on 240SX (8–11 ft. lbs. on 1992 Sentra) and 300ZX and 13–17 ft. lbs. (18–24 Nm) on Maxima and Stanza.
9. Position and install the sensor mounting brackets. Make the sure the sensor wiring is routed properly.
10. Connect the sensor harness connector.
11. Mount the front wheels and lower the vehicle.

WHEEL SENSOR CLEARANCE ADJUSTMENT

1. Install the sensor.
2. Check the clearance between the edge of the sensor and rotor teeth using a feeler gauge. Clearances should be as follows:
   a. On 240SX, front wheel sensor clearance should be 0.0108–0.0295 in. (0.275–0.75mm).
   b. On 300ZX, front wheel sensor clearance should be 0.0087–0.0280 in. (0.22–0.71mm).
   c. On Maxima and Stanza, the clearance should be 0.008–0.039 in. (0.2–1.0mm).

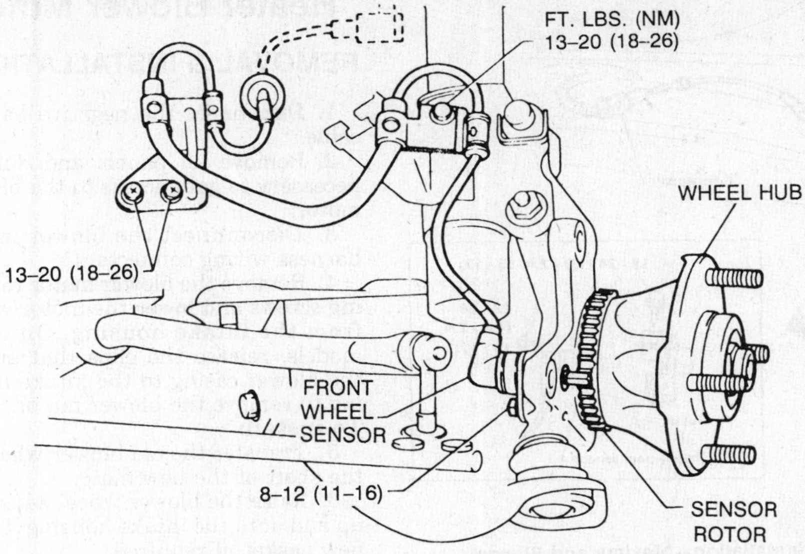

ABS front wheel sensor removal and installation – 240SX

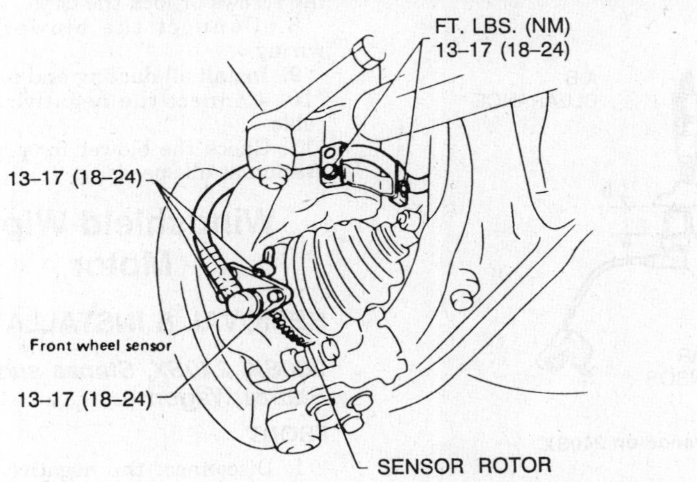

ABS front wheel sensor removal and installation – Maxima and Stanza

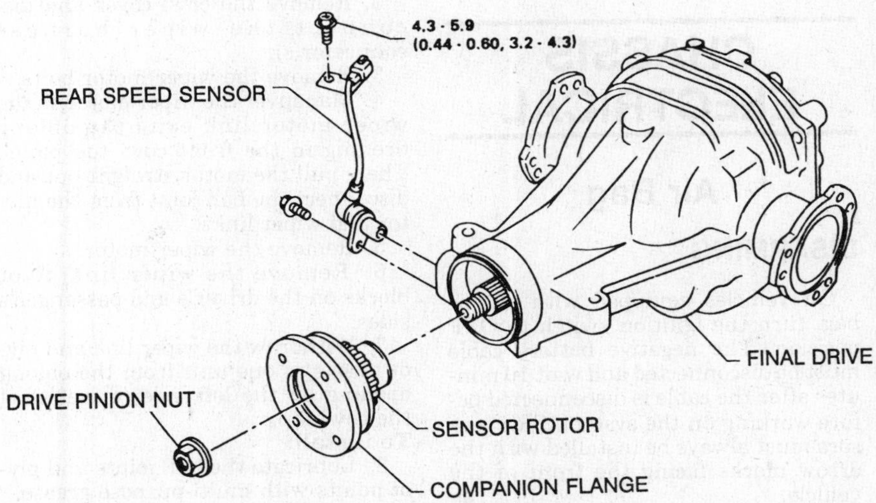

ABS rear wheel sensor removal and installation – 240SX

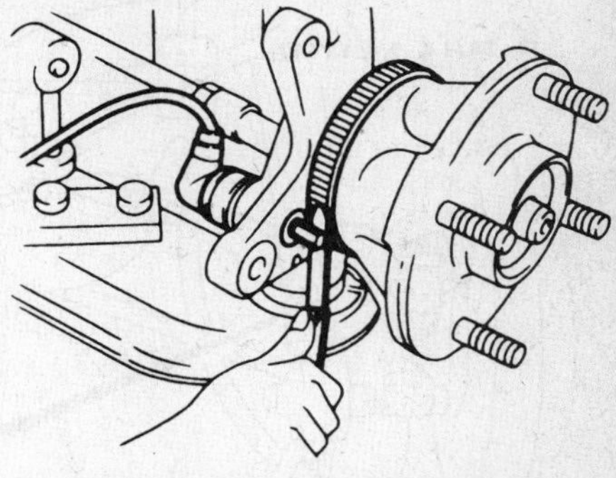
Checking front wheel sensor-to-rotor clearance

3. To adjust the clearance, loosen the sensor mounting bolt(s) and move the sensor back and forth until the clearance is as specified.

4. Once the clearance is set, tighten the sensor mounting bolt(s) to 8–12 ft. lbs. (11–16 Nm) on 240SX and 300ZX and 13–17 ft. lbs. (18–24 Nm) on Maxima and Stanza.

ABS Rear Wheel Sensor

REMOVAL & INSTALLATION

240SX, 1990–92 300ZX, Maxima and Stanza

1. Raise and support the vehicle safely.
2. Remove the rear wheels.
3. Disconnect the sensor harness connector.
4. Detach the sensor mounting brackets.
5. Remove the sensor mounting bolts.
6. On Maxima and Stanza, withdraw the sensor from the rear gusset. On 240SX, the sensor is located on the side of the differential carrier near the driveshaft companion flange. On 300ZX, there are 2 sensors on the side of the differential near each halfshaft.
7. Remove the sensor mounting brackets from the sensor wiring.

To install:

8. Transfer the mounting brackets to the new sensor.
9. Install the sensor. Check and adjust the sensor-to-rotor clearance as described below. Once the clearance is set, tighten the sensor mounting bolt(s) to to 13–20 ft. lbs. (18–26 Nm).
10. Install the sensor mounting brackets. Make the sure the sensor wiring is routed properly.
11. Connect the sensor harness connector.
12. Mount the rear wheels and lower the vehicle.

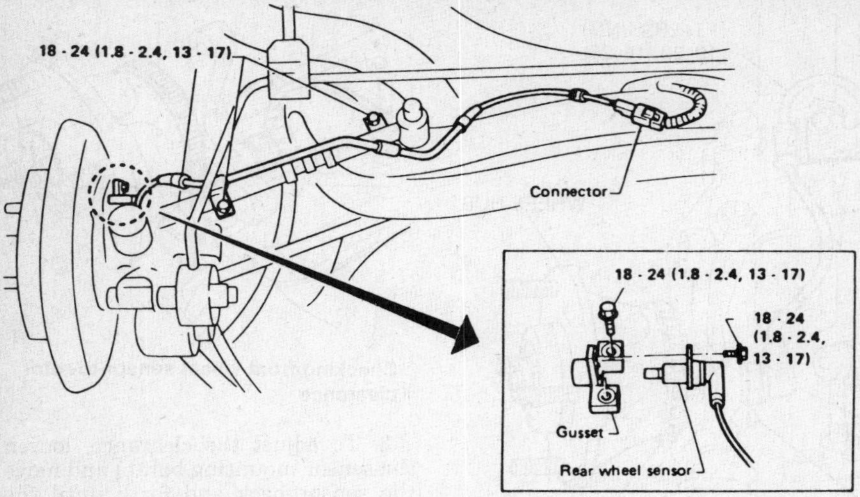

ABS rear wheel sensor removal and installation – Maxima and Stanza

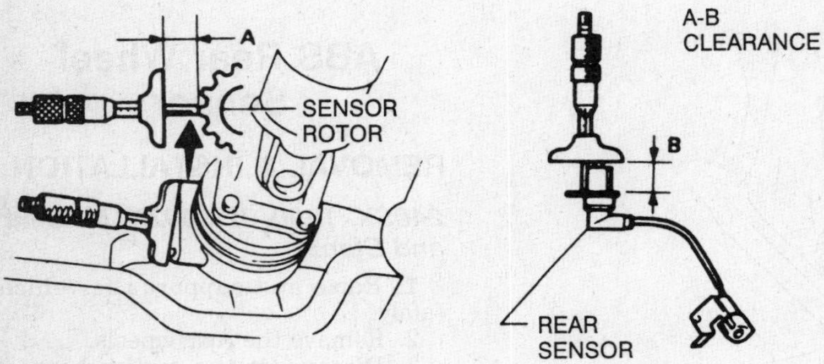

Checking rear wheel sensor-to-rotor clearance on 240SX

WHEEL SENSOR CLEARANCE ADJUSTMENT

1. Install the rear wheel sensor.
2. Check the clearance between the edge of the sensor and rotor teeth using a feeler gauge. Clearances should be as follows:
   a. On 240SX, rear wheel sensor clearance should 0.0138–0.0246 in. (0.035–0.625mm).
   b. On 300ZX, rear wheel sensor clearance should be 0.0024–0.0366 in. (0.06–0.93mm).
   c. On Maxima and Stanza, the clearance should be 0.008–0.039 in. (0.2–1.0mm).
3. To adjust the clearance, loosen the sensor mounting bolt(s) and move the sensor back and forth until the clearance is as specified.
4. Once the clearance is set, tighten the sensor mounting bolt(s) to to 13–20 ft. lbs. (18–26 Nm).

CHASSIS ELECTRICAL

Air Bag

DISARMING

On vehicles equipped with an air bag, turn the ignition switch to OFF position. The negative battery cable must be disconnected and wait 10 minutes after the cable is disconnected before working on the system. SRS sensors must always be installed with the arrow marks facing the front of the vehicle.

Heater Blower Motor

REMOVAL & INSTALLATION

1. Disconnect the negative battery cable.
2. Remove all panels and ducting necessary to gain access to the blower motor.
3. Disconnnect the blower motor harness wiring connectors.
4. Remove the blower motor retaining screws and lower the motor/wheel from the intake housing. On some models, release the clips that attach the blower casing to the intake housing to remove the blower motor.

To install:

5. Transfer the old blower wheel to the shaft of the new motor.
6. Raise the blower/wheel assembly up and onto the intake housing. Use a new gasket, if required.
7. Install the blower motor retaining screws or lock the clips.
8. Connect the blower motor wiring.
9. Install all ducting and panels.
10. Connect the negative battery cable.
11. Check the blower for proper operation at all speeds.

Windshield Wiper Motor

REMOVAL & INSTALLATION

200SX, 240SX, Stanza and Stanza Wagon

FRONT

1. Disconnect the negative battery cable and make sure the wiper switch is in the **OFF** position.
2. Remove the wiper arm.
3. Remove the cowl cover and disconnect the wiper harness connector(s).
3. Remove the wiper motor bolts.
4. Maneuver the wiper motor so the wiper motor link exits the oblong opening in the front cowl top panel. Then, pull the motor straight out and disconnect the ball joint from the motor and wiper links.
5. Remove the wiper motor.
6. Remove the wiper link pivot blocks on the driver's and passenger's sides.
7. Withdraw the wiper link and pivot blocks as one unit from the oblong opening on the left side (driver's) of the cowl top.

To install:

8. Lubricate the ball joints and pivot points with multi-purpose grease.

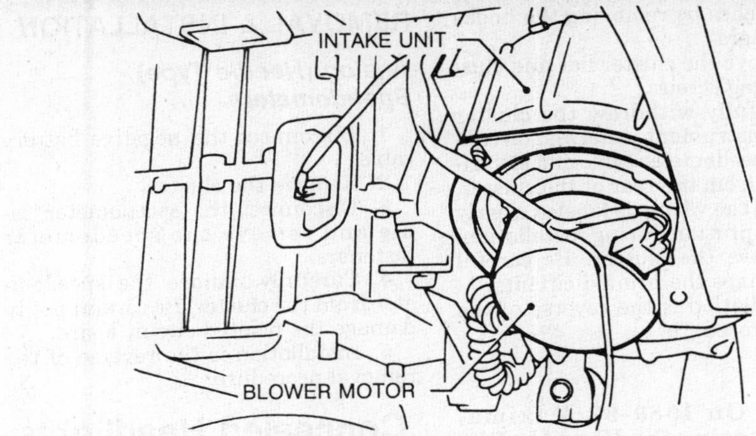

Typical blower motor mounting

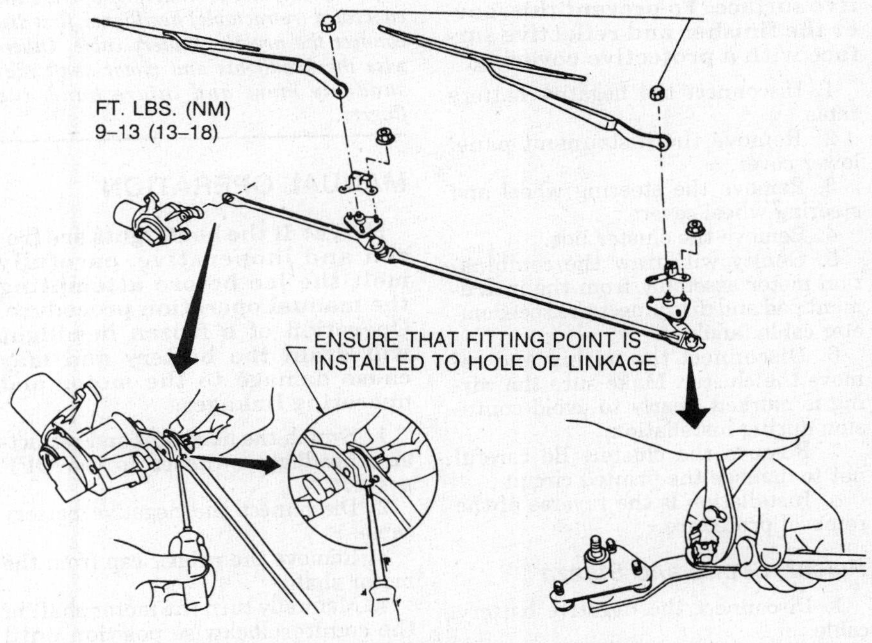

Wiper motor and arm assembly—Pulsar and Sentra

9. Position the wiper link and pivot block as one unit in the cowl top through the oblong hole.

10. Before installing the pivot blocks on the cowl top, hold the end of the motor side link at the hole in the front cowl top panel and insert the motor link ball pin into the wiper link hole.

11. Mount the wiper motor and install the bolts.

12. Connect the wiper motor wiring and install the cowl cover.

13. Attach the wiper arm. To reduce wiper arm looseness, prior to connecting the wiper arm, make sure the motor spline shaft and pivot area is completely free of debris and corrosion. Wire brush as necessary.

NOTE: On 200SX, one wiper arm is longer than the other. The driver's side arm is marked with a "D" and the passenger's side with an "A". Make sure they are installed on their respective sides.

14. Connnect the negative battery cable and check the wipers for proper operation.

REAR

1. Disconnect the negative battery cable.
2. Lift up the rear hatch.
3. Separate the rear wiper arm from the motor shaft.
4. Disconnect the wiper motor wiring harness connector.
5. Unbolt and remove the rear wiper motor from the hatch.
6. Installation is the reverse of the removal procedure. Check the wipers for proper operation.

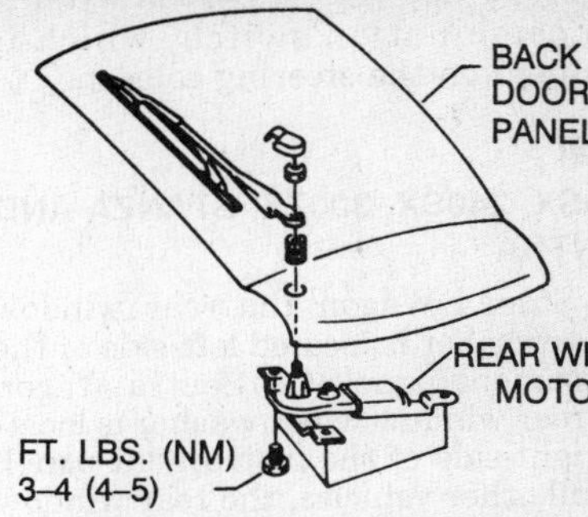

Rear wiper assembly—240SX

Pulsar and Sentra

The wiper motor is on the firewall under the hood. The operating linkage is on the firewall inside the vehicle.

1. Disconnect negative battery cable. Detach the motor wiring plug.
2. Working from inside the vehicle, remove the nut connecting the linkage to the wiper shaft.
3. Unbolt and remove the wiper motor from the firewall.
4. Installation is the reverse of the removal procedure. To reduce wiper arm looseness, prior to connecting the wiper arm, make sure the motor spline shaft and pivot area is completely free of debris and corrosion. Wire brush as necessary.

300ZX and Maxima

The wiper motor and operating linkage is on the firewall under the hood.

1. Disconnect the negative battery cable.
2. Lift the wiper arms. Remove the securing nuts and detach the arms.
3. Remove the air intake grille.
4. Remove the nuts holding the wiper pivots to the body.
5. Open the hood and unscrew the motor from the firewall.
6. Disconnect the wiring connector and remove the wiper motor with the linkage.
7. Installation is the reverse of the removal procedure. To reduce wiper arm looseness, prior to connecting the wiper arm, make sure the motor spline shaft and pivot area is completely free of debris and corrosion. Wire brush as necessary.

NOTE: If the wipers do not park correctly, adjust the position of the automatic stop cover on the wiper motor.

Windshield Wiper Switch

REMOVAL & INSTALLATION

Front

The windshield wiper switch is part of

the combination switch, which is mounted on the steering column.

Rear

200SX, 240SX, 300ZX, STANZA AND SENTRA

On Stanza Wagon the rear window wiper/washer is located left-side of the instrument panel. On Sentra Wagon the rear window wiper/washer is located right-side of the instrument panel. On all other vehicles, the rear window wiper switch is located on the right side of the instrument panel.

1. Remove the instrument cluster.
2. Remove the nut that attaches the combination switch to the dash.
3. Disconnect the electrical connectors from the rear of the switch, then remove it.
4. Installation is the reverse of the removal procedure.

Instrument Cluster

REMOVAL & INSTALLATION

200SX and 240SX

NOTE: On 240SX, when removing the Head-Up Display (HUD) finisher, be careful not to scratch the HUD's reflective surface. To prevent this, cover the finisher and reflective surface with a protective covering.

1. Disconnect the negative battery cable.
2. Remove the steering wheel and steering wheel covers, as required.
3. Remove the screws holding the cluster lid in place and remove the lid.
4. On 200SX, remove the 2 screws and 7 pawls to release the cluster. On 240SX, the cluster is held with 3 screws.
5. Carefully withdraw the cluster from the instrument panel and disconnect the speedometer cable (analog) and electrical wiring from the rear of the cluster. Make sure the wiring is labeled clearly to avoid confusion during installation.
6. Remove the cluster. Be careful not to damage the printed circuit.
7. Installation is the reverse of the removal procedure.

300ZX

1. Disconnect the negative battery cable.
2. Remove the steering wheel and steering wheel covers.
3. To remove the combination switch, first remove the combination switch lower mounting nut, then remove the switch.
4. Remove the left and right instrument switches by removing the hooks and fasteners.
5. Remove the cluster lids and cluster retaining screws.
6. Carefully withdraw the cluster from the instrument panel and disconnect the speedometer cable and electrical wiring from the rear of the cluster. Make sure the wiring is labeled clearly to avoid confusion during installation.
7. Remove the cluster. Be careful not to damage the printed circuit.
8. Installation is the reverse of the removal procedure.

Maxima

NOTE: On 1989–92 Maxima, when removing the Head-Up Display (HUD) finisher, be careful not to scratch the HUD's reflective surface. To prevent this, cover the finisher and reflective surface with a protective covering.

1. Disconnect the negative battery cable.
2. Remove the instrument panel lower cover.
3. Remove the steering wheel and steering wheel covers.
4. Remove the cluster lids.
5. Genlty withdraw the combination meter assembly from the instrument pad and disconnect the speedometer cable (analog).
6. Disconnect the wiring and remove the cluster. Make sure the wiring is marked clearly to avoid confusion during installation.
7. Remove the cluster. Be careful not to damage the printed circuit.
8. Installation is the reverse of the removal procedure.

Pulsar, Sentra and Stanza

1. Disconnect the negative battery cable.
2. Remove the steering wheel and the steering column covers.
3. Remove the instrument cluster lid by removing its screws.
4. Remove the instrument cluster screws.
5. Gently withdraw the cluster from the instrument pad and disconnect all wiring and speedometer cable. Make sure the wires are marked clearly to avoid confusion during installation. Be careful not to damage the printed circuit.
6. Remove the cluster.
7. Installation is the reverse of removal.

Speedometer

NOTE: If equipped with a digital speedometer, the entire cluster assembly must be replaced if the speedometer is faulty.

REMOVAL & INSTALLATION

Analog (Needle Type) Speedometers

1. Disconnect the negative battery cable.
2. Remove the cluster.
3. Disconnect the speedometer cable and remove the speedometer fasteners.
4. Carefully remove the speedometer from the cluster. Be careful not to damage the printed circuit board.
5. Installation is the reverse of the removal procedure.

Concealed Headlights

CAUTION

Before attempting to manually operate the concealed (retractable) headlights, first disconnect the negative battery cable. Otherwise the headlights and motor shaft may suddenly move and injure hand and fingers.

MANUAL OPERATION

NOTE: If the headlights are frozen and inoperative, carefully melt the ice before attempting the manual operation procedure. Operation of a frozen headlight will drain the battery and may cause damage to the motor and operating linkages.

1. Switch the headlight and retractable headlight switches to the **OFF** position.
2. Disconnect the negative battery cable.
3. Remove the rubber cap from the motor shaft.
4. Manually turn the motor shaft in the counterclockwise position until the headlights are in the desired position (open or closed).
5. Install the motor shaft cap.
6. Connect the negative battery cable.
7. Check the head lights for proper operation.

Combination Switch

REMOVAL & INSTALLATION

1. Disarm air bag if equipped. Disconnect the battery ground cable.
2. Remove the steering wheel. On 1989–92 Maxima with sonar suspension, remove the steering angle sensor from the steering column.
3. Remove the steering column covers.

NOTE: At this point, the individual switch assemblies can be removed without removing the

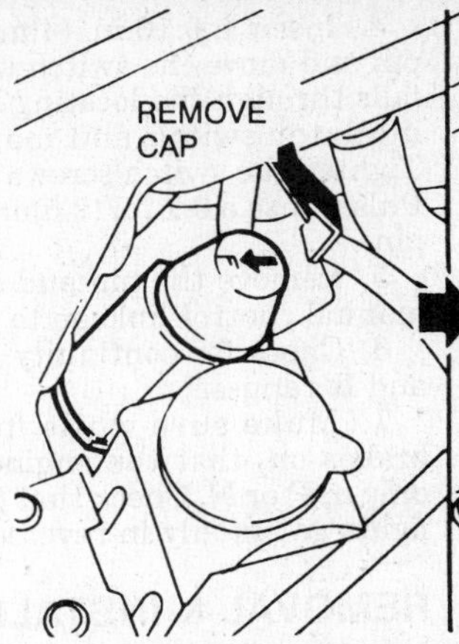

Manual operation of retractable headlights

combination switch base assembly. To service an individual switch/stalk assembly, disconnect the electrical lead and remove the 2 stalk-to-base mounting screws. If the switch base must be removed, proceed with the remainder of the removal procedure.

4. Disconnect the electrical plugs from the switch.
5. Remove the retaining screws, push down on the base of the switch with moderate pressure and twist the switch from the steering wheel shaft.
6. Installation is the reverse of the removal procedure. Check the switch functions for proper operation. Many vehicles have turn signal switches that have a tab which must fit into a hole in the steering shaft. This fit is necessary in order for the system to return the switch to the neutral position after the turn has been made. Be sure to align the tab and the hole when installing.

Ignition Lock/Switch

REMOVAL & INSTALLATION

The steering lock/ignition switch/warning buzzer switch assembly is attached to the steering column by special screws or bolts whose heads shear off on installation. The screws must be drilled out to remove the assembly or removed with an appropriate tool.

1. Disconnect the negative battery cable.
2. Remove the steering wheel, steering column covers and combination switch.
3. Disconnect the switch wiring.
4. Lower the steering column, as required.
5. Break the self-shear screws with a drill or other appropriate tool.
6. Remove the steering lock from the column.

To install:

7. Install the steering lock onto the

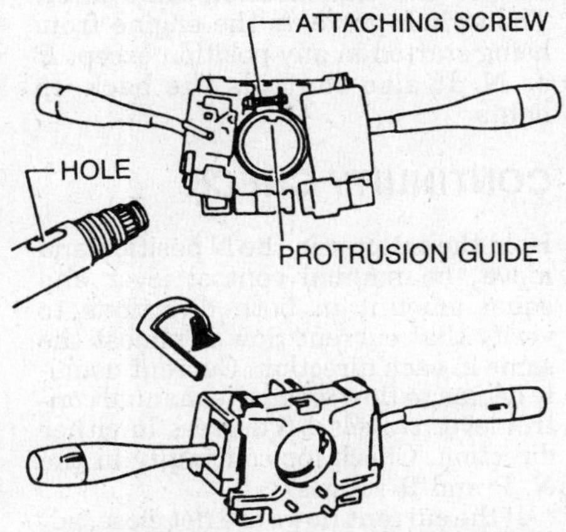

Combination switch removal, installation and alignment

column with new self-shear bolts or screws. Tighten the bolts or screws until the heads shear off.

8. Raise and secure the steering column.
9. Install the combination switch, steering column covers and steering wheel.
10. Connect the negative battery cable.

Stoplight Switch

ADJUSTMENT

1. Before adjustment, check the clearance between the pedal stopper and the threaded end of the stoplight switch. The clearance should be 0.012–0.039 in. (0.3–1.0mm) for all vehicles.
2. If the clearance is not as specified, adjust by loosening the switch locknut and moving the switch in and out as required until the clearance is as specified.
3. Tighten the locknut.
4. Depress the brake pedal and have an assistant verify that the brake lights illuminate.

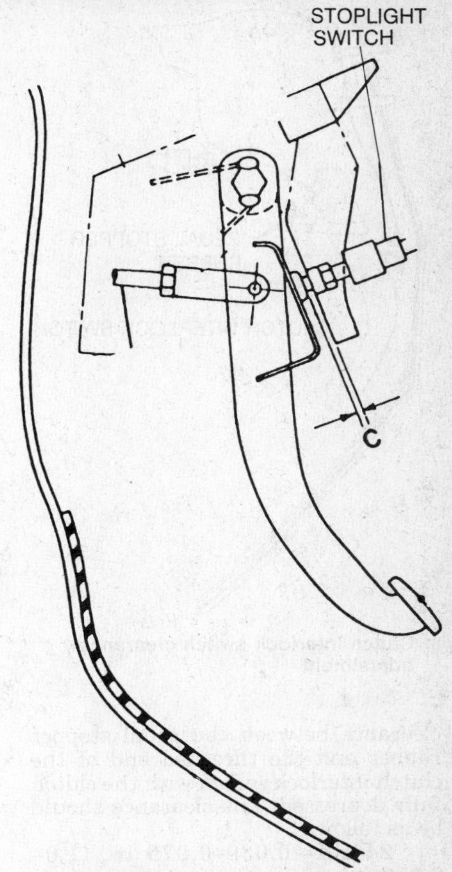

Stoplight switch clearance adjustment

NOTE: If equipped with Automatic Speed Control Device (ASCD), the ASCD cancel switch must be adjusted with the stoplight switch.

REMOVAL & INSTALLATION

1. Remove the floor mats.
2. Disconnect the multi-connector from the switch.
3. Note and record the amount of threads exposed on the switch.
4. Loosen the locknut and adjusting nuts and remove the switch from the mounting bracket.
5. Place the new switch in the mounting bracket.
6. Install and tighten the adjusting and locknuts so the same amount of threads is exposed as recorded in Step 3 or adjust the switch.
7. Connect the multi-connector and check that the brake lights illuminate when the brake pedal is depressed.
8. Replace the floor mats.

Clutch Switch

ADJUSTMENT

1. Before adjustment, check the

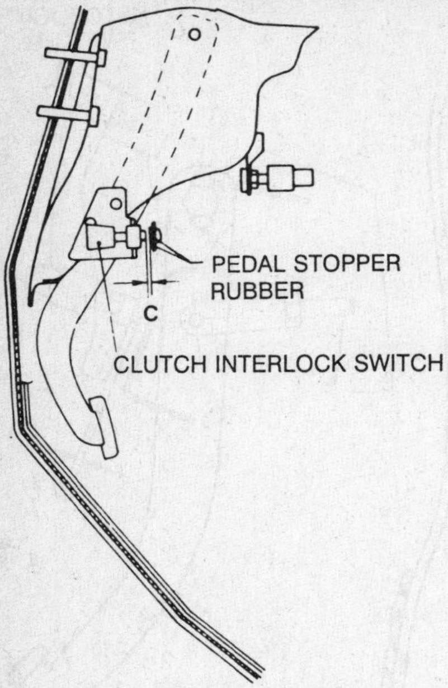

Clutch interlock switch clearance adjustment

clearance between the pedal stopper rubber and the threaded end of the clutch interlock switch with the clutch fully depressed. The clearance should be as follows:

240SX – 0.039–0.079 in. (1.0–2.0mm)

300ZX:

1988–89 – 0.059–0.138 in. (1.5–3.5mm)

1990–92 – 0.039–0.188 in. (1.0–3.0mm)

Maxima:

1988 – 0.039–0.079 in. (1.0–2.0mm)

1989–92 – 0.004–0.039 in. (0.1–1.0mm)

Pulsar – 0.004–0.039 in. (0.1–1.0mm)

Sentra – 0.004–0.039 in. (0.1–1.0mm)

Stanza:

1988 – 0.039–0.079 in. (1.0–2.0mm)

1989–92 – 0.039–0.079 in. (1.0–2.0mm)

Stanza Wagon – 0.059–0.138 in. (1.5–3.5mm)

2. If the clearance is not as specified, adjust by loosening the switch locknut and moving the switch in and out as required until the clearance is as specified.
3. Tighten the locknut and make sure the vehicle won't start with the transmission or transaxle in any gear except **N**.

REMOVAL & INSTALLATION

1. Remove the floor mats.
2. Disconnect the multi-connector from the switch.
3. Note and record the amount of threads exposed on the switch.
4. Loosen the locknut and adjusting nuts and remove the switch from the mounting bracket.
5. Place the new switch in the mounting bracket.
6. Install and tighten the adjusting and lock nuts so the same amount of threads is exposed as recorded in Step 3 or adjust the switch.
7. Connect the multi-connector.
8. Replace the floor mats.

Neutral Safety Switch

The switch unit is bolted to the left side of the transmission shift lever. The switch prevents the engine from being started in any position except **P** or **N**. It also controls the back-up lights.

CONTINUITY CHECK

Hold the selector in the **N** position and move the manual control lever and equal amount in both directions to verify that current flow is almost the same in each direction. Current usually begins to flow when the manual control lever travels 1.5 degrees in either direction. Check for continuity in the **N**, **P** and **R** ranges.

If the current flows are not close, adjust or replace the switch.

ADJUSTMENT

200SX and 1988–89 300ZX

1. Shift the manual valve to the **N** position. In this position, the valve should be vertical.
2. Remove the machine screw from the rotor and loosen the hex head attaching bolts.
3. Insert a 0.079 in. (2mm) pin into the machine screw hole and move the switch until the pin falls into the hole.
4. Remove the pin and tighten the bolts equally.
5. Check for continuity in the **N**, **P** and **R** ranges.
6. With the brakes on, ensure that the engine will start only in **P** or **N**. Check that the back-up lights go on only in reverse.

240SX, 1990–92 300ZX, Maxima, Pulsar, Sentra and Stanza

1. Disconnect the manual control linkage from the manual shaft.
2. Set the manual shaft to the **N** position.
3. Loosen the inhibitor switch mounting screws enough to allow for movement of the switch.
4. Insert a 0.16 in. (4mm) diameter pin and move the switch until the pin falls through the locating holes in the inhibitor switch and manual shaft. Tighten the switch screws equally. On Pulsar, use a 0.1 in. (2.5mm) diameter pin.
5. Remove the pin and connect the manual control linkage to the shaft.
6. Check for continuity in the **N**, **P** and **R** ranges.
7. Make sure while holding the brakes on, that the engine will start only in **P** or **N**. Check that the back-up lights go on only in reverse.

REMOVAL & INSTALLATION

1. On 1988 Stanza Wagon, remove the battery, battery bracket, air cleaner and air flow meter. On 1988 Maxima, remove the battery, air cleaner and the air flow meter/air damper/solenoid valve assembly.
2. Disconnect the manual control shaft from the control lever.
3. Disconnect the switch harness connector.
4. Remove the switch attaching bolts and screws.
5. Remove the switch.
6. Installation is the reverse of the removal procedure.
7. Adjust the switch and check for continuity.

Fuses, Circuit Breakers and Relays

LOCATION

200SX

Fusible Link Holder – left side engine compartment to rear of battery; off negative battery cable.

Fuse Panel – extreme right side of dash behind glove box.

Circuit Breaker – left side of dash above stoplight switch; on top of combination flasher unit.

Fan Motor Relay (CA20E engines) – left side engine compartment to rear of fusible link holder.

Fan Motor Relay (VG30E engines) – left side engine compartment.

A/C Relay (VG30E engines only) – left side engine compartment to rear of fusible link holder.

Fuel Pump Relay – above right license lamp.

Ignition Relay – extreme right side of dash above fuse block

240SX

Fusible Links – left side engine compartment behind battery; in fuse and relay box.

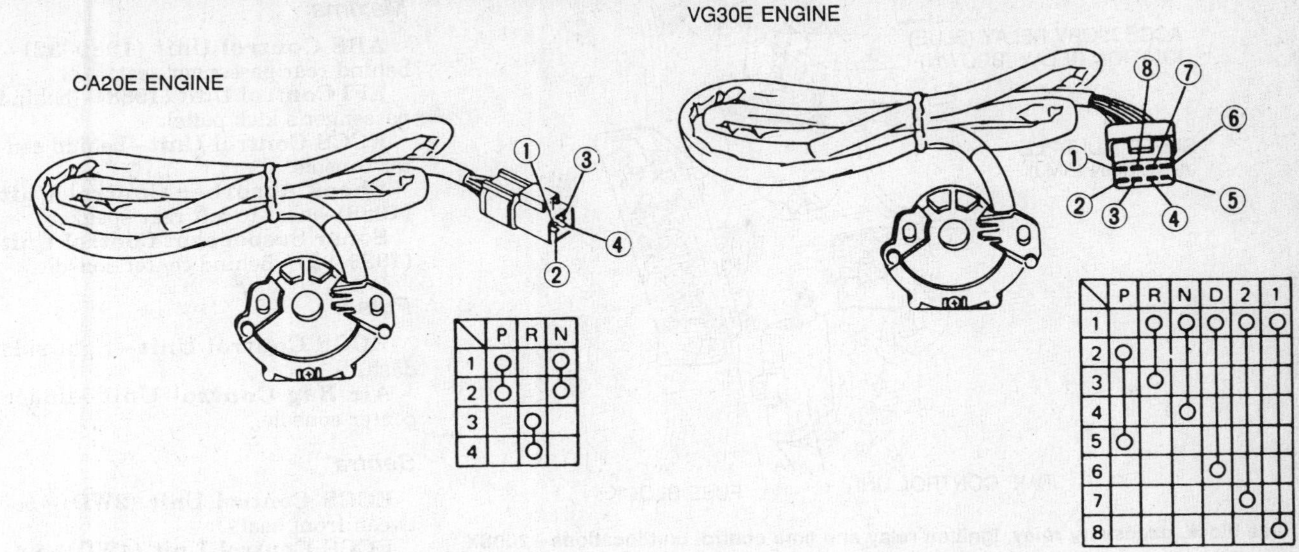

Checking neutral safety switch continuity—200SX and 1987–89 300ZX

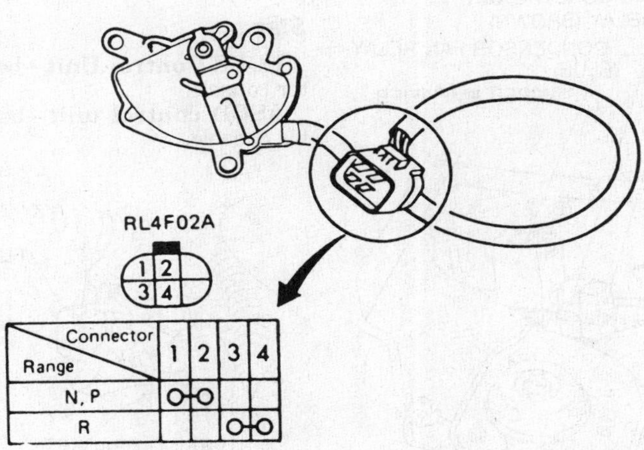

Checking neutral safety switch continuity—Pulsar, Sentra, 1988–89 Maxima and 1988–89 Stanza

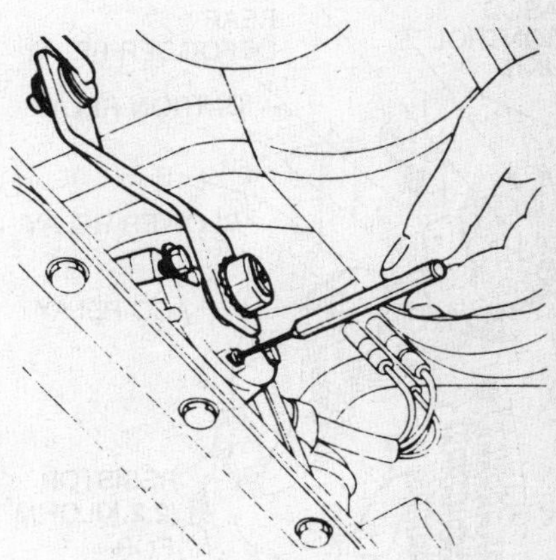

Neutral safety and back-up light switch adjustment—200SX and 1988–89 300ZX

Fuse Panels—(1) left side engine compartment behind battery and (2) under driver's side kick panel.

Fuel Pump Relay—left side engine compartment behind battery; in-fuse and relay box.

A/C Relay—right side engine compartment in front of radiator; in relay box.

ECCS Relay—left side engine compartment behind battery; in fuse and relay box.

Circuit Breaker—behind driver's kick panel.

300ZX

Fusible links (1988–89)—left side engine compartment attached to left fender.

Fusible links (1990–92)—in front of battery off negative battery cable.

Relay Box (1988–89)—left side engine compartment; attached to left fender.

Relay and Fuse Box (1990–92)—right side engine compartment

Maxima

Fusible Link Holder (1988)—right side engine compartment rear of battery.

Fusible Link Holder (1989–92)—right side engine compartment; in fuse and relay box.

Fuse panel—behind driver's kick panel.

Relay Box (1989–92)—right side engine compartment; in front of battery.

Pulsar

Fusible Link Holder—off negative battery cable.

Fuse Panel—behind driver's kick panel.

Sentra

Fusible Link Holder—off negative battery cable.

Fuse Panel—behind driver's kick panel.

Stanza

Fusible Link Holder—off negative battery cable.

Fuse Panel—behind driver's kick panel.

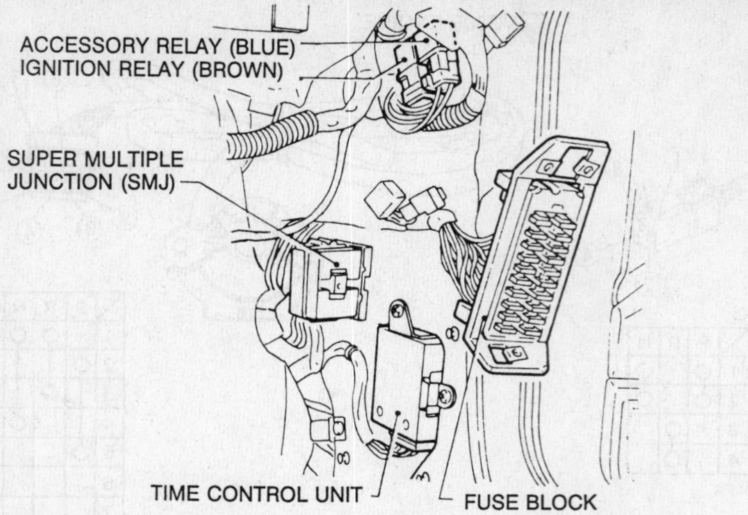

Fuse block, accessory relay, ignition relay and time control unit locations – 200SX

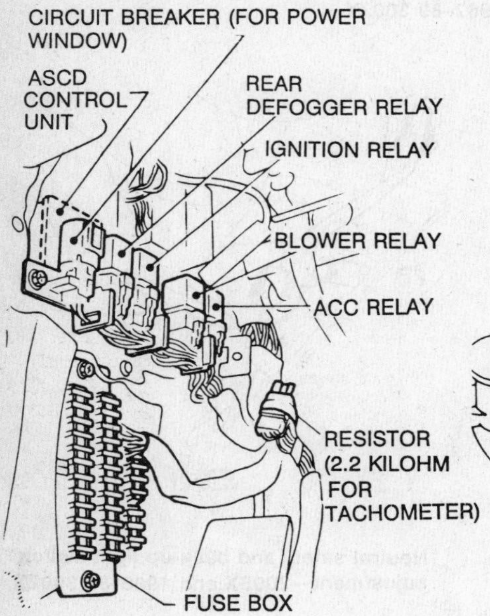

Relay box and fuse panel locations – 300ZX

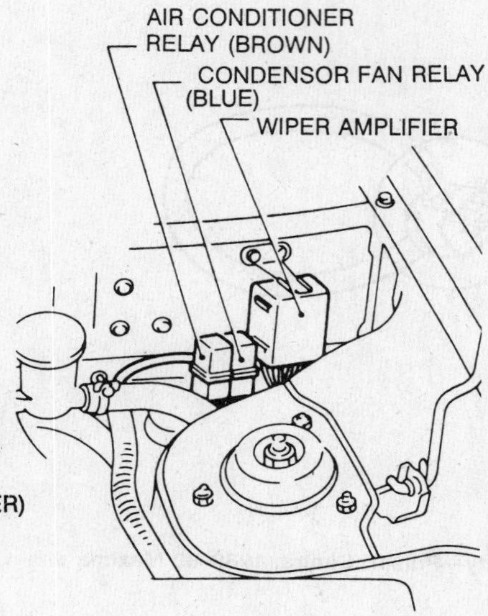

Accessory relay locations – Pulsar

Computers

LOCATION

200SX

ECCS Control Unit – under driver's side kick panel.

Time Control Unit – extreme right side of dash underneath glove box; below fuse block.

240SX

ECCS Control Unit – behind center console.

ASCD Control Unit – behind center console.

Automatic Transmission Control Unit – behind center console.

Timer Control Unit – behind driver's kick panel; left of fuse block.

300ZX

ECCS Control Unit (1988–89) – behind passenger's kick panel.

ECCS Control Unit – behind center console.

Timer Control Unit (1988–89) – behind passenger's kick panel.

Timer Control Unit (1990–92) – behind driver's side kick panel.

Maxima

ABS Control Unit (1989–92) – behind rear passenger seat.

EFI Control Unit (1988) – behind passenger's kick panel.

ECCS Control Unit – behind center console.

Shock Absorber Control Unit (1988) – next to left rear speaker.

Sonar Suspension Control Unit (1989–92) – behind center console.

Pulsar

ECCS Control Unit – right side dash.

Air Bag Control Unit – under center console.

Sentra

ECCS Control Unit (2WD) – between front seats.

ECCS Control Unit (4WD) – under left side of passenger seat.

Transfer Control Unit – extreme left side behind driver's kick panel.

Stanza

ECCS Control Unit – behind center console.

ASCD control unit – behind center console.

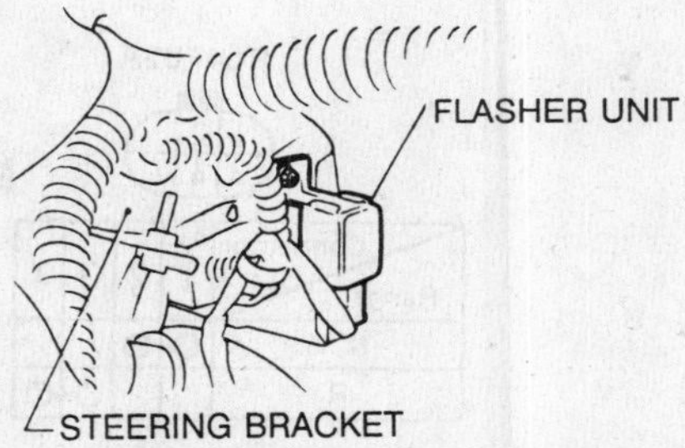

Flasher unit location – 300ZX

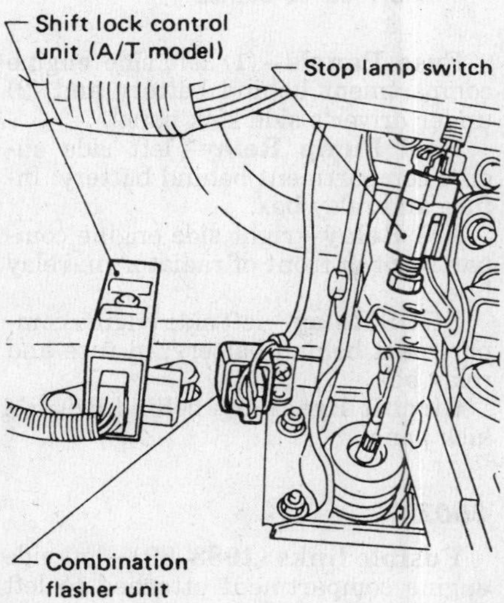

Flasher unit location – 1989–91 Sentra

16

Porsche

911, 924, 928, 944—All Models

SPECIFICATIONS

ENGINE IDENTIFICATION

Year	Model	Engine Displacement cu. in. (cc/liter)	Engine Series Identification	No. of Cylinders	Engine Type
1988	911	193 (3164/3.2)	930.21	6	①
	911 Turbo	201 (3299/3.3)	930.68	6	①
	924S	151 (2479/2.5)	M44/07 ⑤	4	SOHC
	928S4	302 (4957/5.0)	M28/43 ④	8	DOHC
	944	151 (2479/2.5)	M44/07 ⑤	4	SOHC
	944S	151 (2479/2.5)	M44/40 ⑥	4	DOHC
	944 Turbo	151 (2479/2.5)	M44/51	4	SOHC
1989	911	193 (3164/3.2)	930.25	6	①
	911 Turbo	201 (3299/3.3)	930.68	6	①
	911 Carrera 4	219 (3600/3.6)	M64/01	6	①
	928S4	302 (4957/5.0)	M28/41, 42	8	DOHC
	944	164 (2681/2.7)	M44/11, 12	4	SOHC
	944S2	181 (2969/3.0)	M44/41	4	DOHC
	944 Turbo	151 (2479/2.5)	M44/51	4	SOHC
1990	911 Carrera 2 ⑦	219 (3600/3.6)	M64/01	6	SOHC ①
	911 Carrera 4 ⑧	219 (3600/3.6)	M64/01	6	SOHC ①
	928S4	302 (4957/5.0)	M28/41, 42	8	DOHC
	944S2	183 (2990/3.0)	M44/41	4	DOHC

ENGINE IDENTIFICATION

Year	Model	Engine Displacement cu. in. (cc/liter)	Engine Series Identification	No. of Cylinders	Engine Type
1991-92	911 Carrera 2 ⑦	219 (3600/3.6)	M64/01	6	①
	911 Carrera 4 ⑧	219 (3600/3.6)	M64/01	6	①
	911 Turbo	201 (3299/3.3)	930.69	6	①
	928GT	302 (4957/5.0)	M28/47	8	DOHC
	928S4	302 (4957/5.0)	M28/42	8	DOHC
	944S2	183 (2990/3.0)	M44/41	4	DOHC

DOHC Dual Overhead Camshaft
SOHC Single Overhead Camshaft
① Air cooled, 6 cylinder, horizontally opposed, rear-mounted
② M28/20 with automatic transaxle
③ M44/04 with automatic transaxle
④ M28/44 with automatic transaxle
⑤ M44/08 with automatic transaxle
⑥ 16 valve engine
⑦ Includes Carrera 2 Coupe, Targa, Cabriolet
⑧ Includes Carrera 4 Coupe, Targa, Cabriolet

GENERAL ENGINE SPECIFICATIONS

Year	Model	Engine Displacement cu. in. (cc)	Fuel System Type	Net Horsepower @ rpm	Net Torque @ rpm (ft. lbs.)	Bore × Stroke (in.)	Compression Ratio	Oil Pressure @ rpm
1988	911	193 (2164)	DME	214 @ 5900	195 @ 4800	3.74 × 2.93	9.5:1	50 @ 5000
	911 Turbo	201 (3299)	KE	282 @ 5500	278 @ 4000	3.82 × 2.93	7.0:1	60 @ 5500
	924S	219 (3600)	DME	147 @ 5800	140 @ 3000	3.94 × 3.11	9.7:1	50–70 @ 5500
	928S4	302 (4957)	LH	316 @ 6000	317 @ 3000	3.94 × 3.11	10.0:1	70 @ 5500
	944	164 (2681)	DME	147 @ 5800	144 @ 3000	3.94 × 3.11	9.7:1	50–70 @ 5500
	944S	181 (2969)	DME	188 @ 6000	170 @ 4300	3.94 × 3.11	10.9:1	50–70 @ 5500
	944 Turbo	151 (2479)	DME	220 @ 5800	243 @ 3500	3.94 × 3.11	8.0:1	50–70 @ 5500
1989	911	193 (2164)	DME	214 @ 5900	195 @ 4500	3.74 × 2.93	9.5:1	66 @ 5000
	911 Turbo	201 (3299)	KE	282 @ 5500	288 @ 4000	3.82 × 2.93	7.0:1	66 @ 5500
	911 Carrera 4	219 (3600)	DME	247 @ 6100	247 @ 4800	3.94 × 3.01	11.3:1	74 @ 5000
	928S4	302 (4957)	LH	316 @ 6000	317 @ 3000	3.94 × 3.11	10.0:1	74 @ 4000
	944	164 (2681)	DME	162 @ 5800	166 @ 4200	4.09 × 3.11	10.9:1	52 @ 6000
	944S2	181 (2969)	DME	208 @ 5800	207 @ 4100	4.09 × 3.11	10.9:1	52 @ 6000
	944 Turbo	151 (2479)	DME	247 @ 6000	258 @ 4000	3.94 × 3.11	8.0:1	52 @ 6000
1990	911 Carrera 2	219 (3600)	DME	247 @ 6100	228 @ 4800	3.94 × 3.01	11.3:1	73 @ 5000
	911 Carrera 4	219 (3600)	DME	247 @ 6100	228 @ 4800	3.94 × 3.01	11.3:1	73 @ 5000
	928S4	302 (4957)	LH	326 @ 6200 ①	317 @ 4100	3.94 × 3.11	10.0:1	74 @ 4000
	944S2	183 (2990)	DME	208 @ 5800	207 @ 4100	4.09 × 3.46	10.9:1	58 @ 5000
1991-92	911 Carrera 2	219 (3600)	DME	247 @ 6100	228 @ 4800	3.94 × 3.01	11.3:1	73 @ 5000
	911 Carrera 4	219 (3600)	DME	247 @ 6100	228 @ 4800	3.94 × 3.01	11.3:1	73 @ 5000
	911 Turbo	201 (3299)	KE	315 @ 5750	332 @ 4500	3.89 × 2.93	7.0:1	66 @ 5500
	928S4	302 (4957)	LH	316 @ 6000	317 @ 3000	3.94 × 3.11	10.0:1	74 @ 4000
	928GT	302 (4957)	LH	326 @ 6200	317 @ 4100	3.94 × 3.11	10.0:1	74 @ 4000
	944S2	183 (2990)	DME	208 @ 5800	207 @ 4100	4.09 × 3.46	10.9:1	58 @ 5000

DME—Digital Motor Electronic Fuel Injection
KE—Bosch Electronic CIS Fuel Injection
LH—Bosch Air Flow Controlled Fuel Injection

① 326 w/Manual Transmission
316 w/Automatic Transmission

TUNE-UP SPECIFICATIONS

Year	Model	Engine Displacement cu. in. (cc)	Spark Plugs Type	Spark Plugs Gap (in.)	Ignition Timing (deg @ rpm)	Com-pression Pressure (psi)	Fuel Pump (psi)	Idle Speed (rpm)	Valve Clearance In.	Valve Clearance Ex.
1988	911	193 (2164)	WR7DC	0.028	26B @ 4000	①	34–40	780–820	0.004	0.004
	911 Turbo	201 (3299)	W3DP	0.028	26B @ 4000	①	34–40	850–950	0.004	0.004
	924S	219 (3600)	WR7DC	0.028	5B @ 840	①	34–40	800–880	Hyd.	Hyd.
	928S4	302 (4957)	WR7DC	0.028	10° ± 2 BTDC ②	①	56	800–880	Hyd.	Hyd.
	944	164 (2681)	WR7DC	0.028	10B @ 850	①	34–40	800–880	Hyd.	Hyd.
	944S	181 (2969)	WR7DC	0.028	10° ± 3 BTDC ③	①	NA	800–880	Hyd.	Hyd.
	944 Turbo	151 (2479)	WR6DS	0.028	5B @ 840	①	34–40	800–880	Hyd.	Hyd.
1989	911	193 (2164)	WR7DC	0.028	0° ± 3 BTDC ②	①	34–40	880 ± 20	0.004	0.004
	911 Turbo	201 (3299)	W3DPO	0.028	26B @ 4000	①	34–40	900 ± 50	0.004	0.004
	911 Carrera 4	219 (3600)	FR5DTC	0.031	0° ± 3	①	53–59	880 ± 40	0.004	0.004
	928S4	302 (4957)	WR7DC	0.028	10° ± 2 BTDC ②	①	56	675 ± 25	Hyd.	Hyd.
	944	164 (2681)	WR7DC	0.028	5° ± 3 BTDC ②	①	53–59	840 ± 40	Hyd.	Hyd.
	944S2	181 (2969)	WR5DC	0.028	10° ± 3	①	48 ⑤	840 ± 40	Hyd.	Hyd.
	944 Turbo	151 (2479)	WR7DC	0.028	5° ± 3 BTDC ②	①	34–40	840 ± 40 ③	Hyd.	Hyd.
1990	911 Carrera 2	219 (3600)	FR5DTC ④	0.031	0° ± 3°	①	48–50	880 ± 90 ②	0.004	0.004
	911 Carrera 4	219 (3600)	FR5DTC ④	0.031	0° ± 3°	①	48–50	880 + 40 ②	0.004	0.004
	928S4	302 (4957)	WR7DC ④	0.028	10° ± 2°BTDC ②	①	56	675 ± 25 ②	Hyd.	Hyd.
	944S2	183 (2990)	WR5DC ④	0.028	10° ± 3°	①	48 ⑤	840 ± 40 ②	Hyd.	Hyd.
1991	911 Carrera 2	219 (3600)	FR5DTC ④	0.032	0° ± 3°	①	45–52	880 ± 40 ②	0.004	0.004
	911 Carrera 4	219 (3600)	FR5DTC ④	0.032	0° ± 3°	①	45–52	880 + 40 ②	0.004	0.004
	911 Turbo	201 (3299)	W4DPO	0.024	26B @ 4000	①	34–40	900 ± 50	0.004	0.004
	928S4	302 (4957)	WR7DC ④	0.028	10° ± 2°BTDC ②	①	56	675 ± 25 ②	Hyd.	Hyd.
	928GT	302 (4957)	WR7DC	0.028	10° ± 2°BTDC ②	①	56	675 ± 25 ②	Hyd.	Hyd.
	944S2	183 (2990)	WR5DC ④	0.028	10° ± 3°	①	53–59	840 ± 40 ②	Hyd.	Hyd.
1992	REFER TO UNDERHOOD SPECIFICATIONS STICKER									

NOTE: The Underhood Specifications sticker often reflects tune-up specifications changes made in production. Sticker Figures must be used if they disagree with those in this chart

B Before Top Dead Center
BTDC Before Top Dead Center

① All cylinders should be within 22 psi of the highest reading
② Checking only, not adjustable
③ With idle stabilization system disconnected
④ Bosch number
⑤ At Idle

FIRING ORDERS

NOTE: To avoid confusion, always replace spark plug wires one at a time.

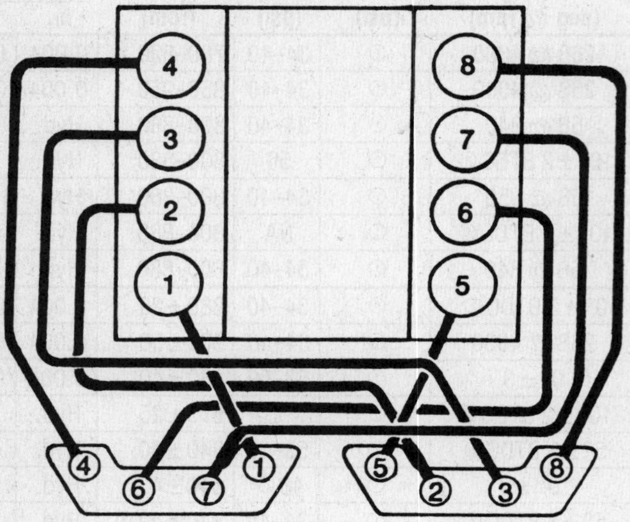

4957cc Engine
Engine Firing Order: 1–3–7–2–6–5–4–8
Distributorless Ignition System

FRONT
928 and 928S Models

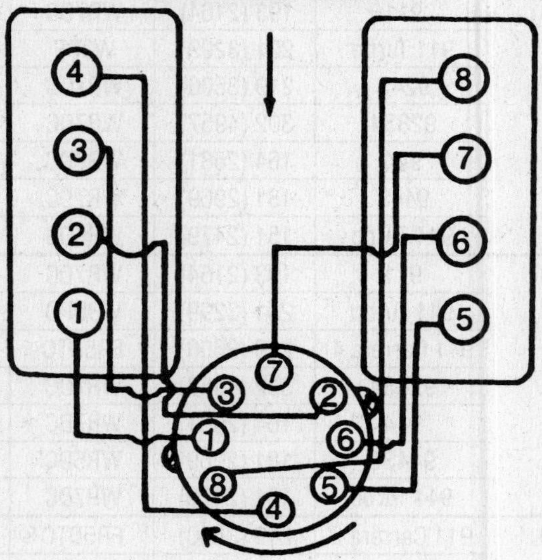

4957cc Engine
Engine Firing Order: 1–3–7–2–6–5–4–8
Distributor Rotation: Clockwise

FIRING ORDER: 1-6-2-4-3-5

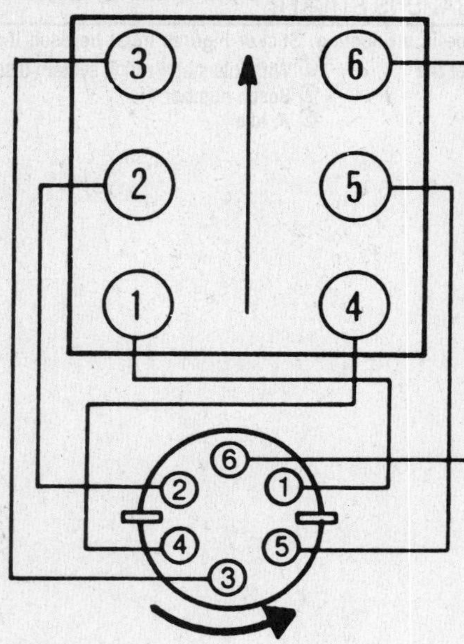

1364cc & 3299cc Engines
Engine Firing Order: 1–6–2–4–3–5
Distributor Rotation: Counterclockwise

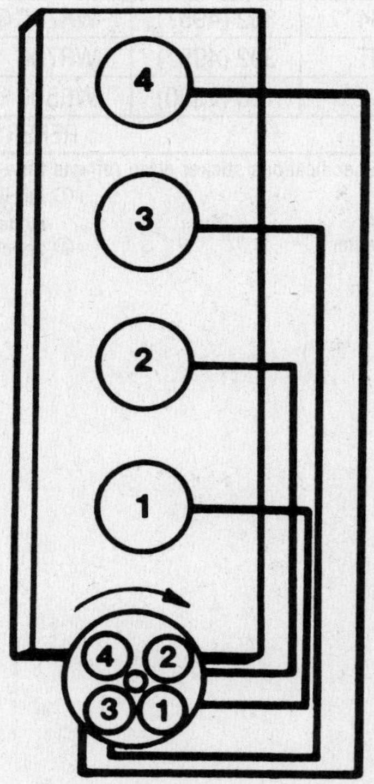

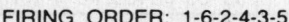

2479cc, 2681cc & 2969cc Engines
Engine Firing Order: 1–3–4–2
Distributor rotation: Clockwise

FIRING ORDERS

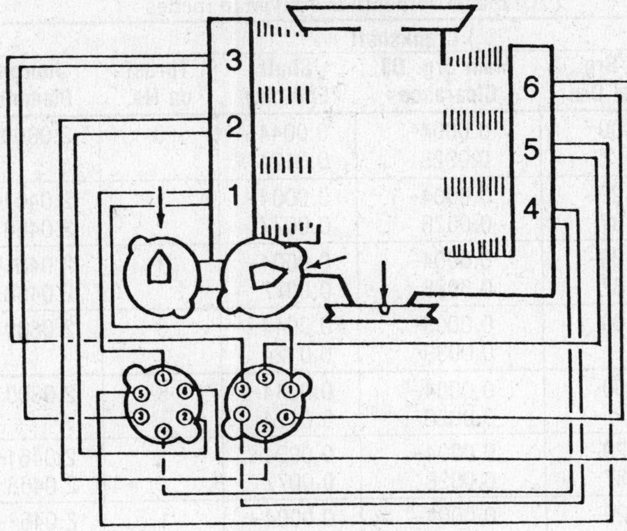

3600cc Engine
Engine Firing Order: 1-6-2-4-3-5
Distributor Rotation: Clockwise

CAPACITIES

Year	Model	Engine Displacement cu. in. (cc)	Engine Crankcase (qts.) with Filter	Engine Crankcase (qts.) without Filter	Transaxle (pts.) Manual	Transaxle (pts.) Auto.	Drive Axle (pts.)	Fuel Tank (gal.)	Cooling System (qts.)
1988	911	193 (3164)	10.6	—	6.6	—	—	22.5	—
	911 Turbo	201 (3299)	10.6	—	7.8	—	—	22.5	—
	924S	151 (2479)	6.4	—	5.5	6.0	—	17.4	9.0
	928S4	302 (4957)	8.5	8.0	8.0	17.0	—	22.7	17.0
	944	151 (2479)	6.4	—	5.5	6.0	—	21.1	8.5
	944S	151 (2479)	6.4	—	5.5	6.0	—	21.1	8.5
	944 Turbo	151 (2479)	6.9	—	5.5	—	—	21.1	9.0
1989	911	193 (2164)	14.0	—	6.0	—	—	22.5	—
	911 Turbo	201 (3299)	14.0	—	8.0	—	—	22.5	—
	911 Carrera 4	219 (3600)	12.0	—	8.0 ①	—	—	20.3	—
	928S4	302 (4957)	8.5	8.0	8.2	16.0	—	22.7	17.0
	944	164 (2681)	5.8	—	3.6	8.4	—	21.1	7.2
	944S2	181 (2969)	5.8	—	3.6	8.4	—	21.1	7.2
	944 Turbo	151 (2479)	7.2	—	4.1	—	—	21.1	9.0
1990	911 Carrera 2	219 (3600)	12.2	—	7.2	—	—	20.3	—
	911 Carrera 4	219 (3600)	12.2	—	7.2 ①	—	—	20.3	—
	928S4	302 (4957)	8.5	8.0	8.2	16.0	—	22.7	17.0
	944S2	183 (2990)	6.9	—	3.6	8.4	—	21.1	8.2
1991-92	911 Carrera 2	219 (3600)	12.2	—	7.5	19.0	—	20.3	—
	911 Carrera 4	219 (3600)	12.2	—	8.0 ①	19.0	—	20.3	—
	911 Turbo	201 (3299)	14.0	—	8.0	—	—	20.3	—
	928S4/GT	302 (4957)	8.0	—	9.6	19.7	—	22.7	17.0
	944S2	183 (2990)	6.9	—	4.3	—	—	21.1	8.2

① Front final drive 2.5 pts.

CRANKSHAFT AND CONNECTING ROD SPECIFICATIONS

All measurements are given in inches

Year	Engine Displacement cu. in. (cc)	Crankshaft Main Brg. Journal Dia.	Main Brg. Oil Clearance	Shaft End-play	Thrust on No.	Connecting Rod Journal Diameter	Oil Clearance	Side Clearance
1988	151 (2479)	2.8000	0.0004–0.0028	0.0044–0.0124	3	2.0800	0.0008–0.0028	—
	193 (3164)	2.2429–2.2437	0.0004–0.0028	0.0004–0.0077	1	2.0461–2.0468	0.0012–0.0035	0.0079–0.0158
	201 (3299)	2.2429–2.2437	0.0004–0.0028	0.0004–0.0077	1	2.0461–2.0468	0.0012–0.0035	0.0079–0.0158
	302 (4957)	2.8000	0.0008–0.0039	0.0044–0.0124	3	2.0800	0.0008–0.0028	—
1989	151 (2479)	2.8000	0.0004–0.0028	0.0044–0.0124	3	2.0800	0.0008–0.0028	—
	193 (3164)	2.2429–2.2437	0.0004–0.0028	0.0004–0.0077	1	2.0461–2.0468	0.0012–0.0035	0.0079–0.0158
	201 (3299)	2.2429–2.2437	0.0004–0.0028	0.0004–0.0077	1	2.0461–2.0468	0.0012–0.0035	0.0079–0.0158
	302 (4957)	2.8000	0.0008–0.0039	0.0044–0.0124	3	2.0800	0.0008–0.0028	—
1990	183 (2990)	2.7547–2.7555	0.0007–0.0038	0.0039–0.0157	3	2.0460–2.0468	0.0013–0.0036	—
	219 (3600)	2.3610–2.3618	0.0004–0.0028	0.0004–0.0077	1	2.1642–2.1649	0.0012–0.0035	0.0079–0.0158
	302 (4957)	2.7547 2.7555	0.0007–0.0038	0.0043–0.0122	3	2.0460 2.0468	0.0007–0.0027	—
1991-92	183 (2990)	2.7547–2.7555	0.0007–0.0038	0.0039–0.0157	3	2.0460–2.0468	0.0013–0.0036	—
	201 (3299)	2.2429–2.2437	0.0004–0.0028	0.0004–0.0077	1	2.0461–2.0468	0.0012–0.0035	0.0079–0.0158
	219 (3600)	2.3610–2.3618	0.0004–0.0028	0.0004–0.0077	1	2.1642–2.1649	0.0012–0.0035	0.0079–0.0158
	302 (4957)	2.7547 2.7555	0.0007–0.0038	0.0043–0.0122	3	2.0460 2.0468	0.0007–0.0027	—

VALVE SPECIFICATIONS

Year	Engine Displacement cu. in. (cc)	Seat Angle (deg.)	Face Angle (deg.)	Spring Test Pressure (lbs.)	Spring Installed Height (in.)	Stem-to-Guide Clearance (in.) Intake	Exhaust	Stem Diameter (in.) Intake	Exhaust
1988	193 (3164)	45	45	176.4 @ 1.21 ①	1.3779 ②	0.030–0.057	0.050–0.077 ④	0.353	0.352
	201 (3299)	45	45	176.4 @ 1.21 ①	1.3779 ②	0.030–0.057	0.050–0.077 ④	0.353	0.352
	302 (4957)	45	45	—	—	0.020	0.020 ④	0.352	0.352
	151 (2479)	45	45	—	—	0.032	0.032 ④	0.352	0.352
1989	193 (3164)	45	45	176.4 @ 1.21 ①	1.3779 ②	0.030–0.057	0.050–0.077 ④	0.353	0.352
	201 (3299)	45	45	176.4 @ 1.21 ①	1.3779 ②	0.030–0.057	0.050–0.077 ④	0.353	0.352
	302 (4957)	45	45	—	—	0.020	0.020 ④	0.352	0.352
	151 (2479)	45	45	—	—	0.032	0.032 ④	0.352	0.352

VALVE SPECIFICATIONS

Year	Engine Displacement cu. in. (cc)	Seat Angle (deg.)	Face Angle (deg.)	Spring Test Pressure (lbs.)	Spring Installed Height (in.)	Stem-to-Guide Clearance (in.)		Stem Diameter (in.)	
						Intake	Exhaust	Intake	Exhaust
1990	183 (2990)	45	45	—	1.6141 ③	0.001–0.001 ⑤	0.002–0.002 ⑤	0.353	0.352
	219 (3600)	45	45	—	—	0.001–0.002 ⑤	0.002–0.003 ⑤	0.352	0.351
	302 (4957)	45	45	—	1.6141	0.001–0.001 ⑤	0.002–0.002 ⑤	0.274	0.273
1991–92	183 (2990)	45	45	—	1.6141 ③	0.001–0.001 ⑤	0.002–0.002 ⑤	0.353	0.352
	201 (3299)	45	45	176.4 @ 1.21 ①	1.3779 ②	0.030–0.057	0.050–0.077 ④	0.353	0.352
	219 (3600)	45	45	—	1.3583 ⑥	0.001–0.002 ⑤	0.002–0.003 ⑤	0.352	0.351
	302 (4957)	45	45	—	1.6141	0.001–0.001 ⑤	0.002–0.002 ⑤	0.274	0.273

① 165.3 @ 1.25 for exhaust valve
② 1.3976 in. for exhaust
③ 1.5748 in. for exhaust
④ With the valve in the guide, with the valve stem flush with the end of the valve guide, measure the side play with a dial indicator on the side of the valve head.
⑤ Measure guide I.D., valve stem O.D., and subtract for clearance.
⑥ 1.3189 in. for exhaust

PISTON AND RING SPECIFICATIONS

All measurements are given in inches

Year	Engine Displacement cu. in. (cc)	Piston Clearance	Ring Gap			Ring Side Clearance		
			Top Compression	Bottom Compression	Oil Control	Top Compression	Bottom Compression	Oil Control
1988	193 (3164)	0.0060	0.0040–0.0080	0.0040–0.0080	0.0060–0.0120	0.0030–0.0040	0.0020–0.0030	0.0010–0.0020
	201 (3299)	0.0060	0.0040–0.0080	0.0040–0.0080	0.0060–0.0120	0.0030–0.0040	0.0020–0.0030	0.0010–0.0020
	302 (4957)	0.0009 0.0018 ①	0.0080–0.0180	0.0080–0.0180	0.0150–0.0550	0.0020–0.0030	0.0020–0.0030	0.0010–0.0050
	151 (2479)	0.0031	0.0080–0.0180	0.0080–0.0180	0.0150–0.0550	0.0020–0.0030	0.0020–0.0030	0.0010–0.0050
1989	193 (3164)	0.0060	0.0040–0.0080	0.0040–0.0080	0.0060–0.0120	0.0030–0.0040	0.0020–0.0030	0.0010–0.0020
	201 (3299)	0.0060	0.0040–0.0080	0.0040–0.0080	0.0060–0.0120	0.0030–0.0040	0.0020–0.0030	0.0010–0.0020
	302 (4957)	0.0009 0.0018 ①	0.0080–0.0180	0.0080–0.0180	0.0150–0.0550	0.0020–0.0030	0.0020–0.0030	0.0010–0.0050
	151 (2479)	0.0031	0.0080–0.0180	0.0080–0.0180	0.0150–0.0550	0.0020–0.0030	0.0020–0.0030	0.0010–0.0050
1990	183 (2990)	0.0003–0.0012 ①	0.0078–0.0157	0.0078–0.0157	0.0118–0.0236	0.0015–0.0028	0.0011–0.0024	0.0007–0.0021
	219 (3600)	0.0009–0.0016	0.0078–0.0157	0.0078–0.0157	0.0118–0.0236	0.0027–0.0040	0.0015–0.0028	0.0007–0.0020
	302 (4957)	0.0009–0.0018 ①	0.0078–0.0157	0.0078–0.0157	0.0078–0.0574	0.0023–0.0040	0.0015–0.0028	0.0005–0.0049

PISTON AND RING SPECIFICATIONS

All measurements are given in inches

Year	Engine Displacement cu. in. (cc)	Piston Clearance	Ring Gap			Ring Side Clearance		
			Top Compression	Bottom Compression	Oil Control	Top Compression	Bottom Compression	Oil Control
1991-92	183 (2990)	0.0003–0.0012 ①	0.0078–0.0157	0.0078–0.0157	0.0118–0.0236	0.0015–0.0028	0.0011–0.0024	0.0007–0.0021
	201 (3299)	0.0020 0.0030	0.0060–0.0120	0.0060–0.0120	0.0060–0.0120	0.0030–0.0040	0.0020–0.0030	0.0010–0.0020
	219 (3600)	0.0009–0.0016	0.0078–0.0157	0.0078–0.0157	0.0118–0.0236	0.0027–0.0040	0.0015–0.0028	0.0007–0.0020
	302 (4957)	0.0009–0.0018 ①	0.0078–0.0157	0.0078–0.0157	0.0078–0.0574	0.0023–0.0040	0.0015–0.0028	0.0005–0.0049

① Wear limit 0.0031 in.

TORQUE SPECIFICATIONS

All readings in ft. lbs.

Year	Engine Displacement cu. in. (cc)	Cylinder Head Bolts	Main Bearing Bolts	Rod Bearing Bolts	Crankshaft Pulley Bolts	Flywheel Bolts	Manifold		Spark Plugs
							Intake	Exhaust	
1988	151 (2479)	⑦ ⑥	⑤	55.3	155 ⑩	65	15	15	18–22
	193 (3164)	③ ④	25	①	58 ②	65	18	14–17	18–22
	201 (3299)	③ ④	25	①	58 ②	65	18	14–17	18–22
	302 (4957)	⑧	⑨	54	213	65	17	15	18–22
1989	151 (2479)	⑦ ⑥	⑤	55.3	155 ⑩	65	15	15	18–22
	193 (3164)	③ ④	25	①	58 ②	65	18	14–17	18–22
	201 (3299)	③ ④	25	①	58 ②	65	18	14–17	18–22
	302 (4957)	⑧	⑨	54	213	65	17	15	18–22
1990	183 (2990)	⑫	⑤	54	155 ⑫	65	14	15	18–22
	219 (3600)	③	⑪	⑧	173	66	15	15	18–22
	302 (4957)	⑬	⑭	55	215	65	11	14–17	18–22
1991-92	183 (2990)	⑮	⑯	54	155 ⑫	65	14	15	18–22
	201 (3299)	③ ④	25	①	58 ②	65	18	14–17	18–22
	219 (3600)	③	⑪	⑧	173	66	15	15	18–22
	302 (4957)	⑬	⑭	55	215	65	11	14–17	18–22

① Step 1—14 ft. lbs.
Step 2—Turn additional 90° ± 2°
② If equipped with air conditioning—123
③ Step 1—11 ft. lbs.
Step 2—Turn additional 90° ± 2°
④ Apply a thin coat of Optimoly HT
⑤ M12 bolts
Step 1—14 ft. lbs.
Step 2—29 ft. lbs.
Step 3—54 ft. lbs.
M10 bolts
Step 1—14 ft. lbs.
Step 2—36 ft. lbs.
M6 bolts
Step 1—6 ft. lbs.
M8 bolts
Step 1—14 ft. lbs.
⑥ Dip studs in engine oil
⑦ Tighten in 3 steps (in order each time)
1st—14 ft. lbs.
2nd—36 ft. lbs.
3rd—65 ft. lbs.
30 minutes later, loosen each bolt ¼ turn then repeat the tightening sequence.
⑧ Step 1—14 ft. lbs.
Step 2—turn additional 90°
Step 3—turn additional 90°
⑨ M12 bolts
Step 1—14 ft. lbs.
Step 2—29 ft. lbs.
Step 3—54 + 3.6 ft. lbs.
M10 bolts
Step 1—14 ft. lbs.
Step 2—36 + 3.6 ft. lbs.
⑩ Gear wheel to crankshaft
⑪ Crankcase Studs M10—29 ft. lbs.
Nuts or Bolts M8—17 ft. lbs.
⑫ Step 1—14 ft. lbs.
Step 2—36 ft. lbs.
Step 3—65 ft. lbs.
⑬ With hexbolts, see ⑨
With studs:
Step 1—14 ft. lbs.
Step 2—turn additional 90°
Step 3—turn additional 90°
Step 4—turn additional 90°
⑭ M12 bolts
Step 1—22 ft. lbs.
Step 2—40 ft. lbs.
Step 3—55 ft. lbs.
M10 bolts
Step 1—15 ft. lbs.
Step 2—37 ft. lbs.
⑮ Step 1—15 ft. lbs.
Step 2—turn additional 60°
Step 3—turn additional 90°
⑯ M12 bolts
Step 1—22 ft. lbs.
Step 2—turn additional 60°
M10 bolts
Step 1—15 ft. lbs.
Step 2—37 ft. lbs.
M8 bolts
Step 1—15 ft. lbs.
M6 bolts
Step 1—7.5 ft. lbs.

BRAKE SPECIFICATIONS

All measurements in inches unless noted

Year	Model	Lug Nut Torque (ft. lbs.)	Master Cylinder Bore	Brake Disc Minimum Thickness	Brake Disc Maximum Runout	Maximum Brake Drum Diameter	Minimum Lining Thickness Front	Minimum Lining Thickness Rear
1988	911	94	0.813	0.890	0.004	—	0.080	0.080
	911 Turbo	94	0.937	④	0.004	—	0.080	0.080
	924S	94	0.940	0.807 ⑤	0.004	—	0.080	0.080
	928S4	94	0.950	1.228	0.004	—	0.080	0.080
	944	94	0.940	0.807 ⑤	0.004	—	0.080	0.080
	944S	94	0.940	0.807 ⑤	0.004	—	0.080	0.080
	944 Turbo	94	0.940	0.807 ⑤	0.004	—	0.080	0.080
1989	911	94	0.810	0.940	0.004	—	0.080	0.080
	911 Turbo	94	0.940	1.260 ①	0.004	—	0.080	0.080
	911 Carrera 4	94	0.940	1.100 ②	0.004	—	0.080	0.080
	928S4	94	0.940	1.260 ③	0.004	—	0.080	0.080
	944	94	0.940	0.810 ④	0.004	—	0.080	0.080
	944S2	94	0.940	0.810 ④	0.004	—	0.080	0.080
	944 Turbo	94	0.940	1.260 ③	0.004	—	0.080	0.080
1990	911 Carrera 2	94	0.940	1.100 ②	0.004	—	0.080	0.080
	911 Carrera 4	94	0.940	1.100 ②	0.004	—	0.080	0.080
	928S4	94	0.940	1.260 ③	0.004	—	0.080	0.080
	944S2	94	0.940	0.810 ④	0.004	—	0.080	0.080
1991-92	911 Carrera 2	94	0.940	1.050 ⑦	0.004	—	0.080	0.080
	911 Carrera 4	94	0.940	1.050 ⑦	0.004	—	0.080	0.080
	911 Turbo	94	0.940	1.205 ⑥	0.004	—	0.080	0.080
	928S4/GT	94	⑧	1.205 ⑦	0.004	—	0.080	0.080
	944S2	94	⑧	1.050 ⑦	0.004	—	0.080	0.080

① Rear—1.100
② Rear—0.950
③ Rear—0.940
④ Front—1.205
Rear—0.790
⑤ Rear—0.788
⑥ Rear—1.050
⑦ Rear—0.890
⑧ Dual bore—0.940/0.813

WHEEL ALIGNMENT

Year	Model	Caster Range (deg.)	Caster Preferred Setting (deg.)	Camber Range (deg.)	Camber Preferred Setting (deg.)	Toe-in (in.)	Steering Axis Inclination (deg.)
1988	911	5 13/16P–6 5/16P	6 1/16P	3/16P	0	1/8	—
	911 Turbo	5 13/16P–6 5/16P	6 1/16P	3/16P	0	1/8	—
	924S	2 1/4P–3P	2 1/2P	9/16N–1/16N	5/16N	5/64	—
	928S4	3P–4P	3 1/2P	11/16N–5/16N	1/2N	5/32	—
	944	2 1/4P–3P	2 1/2P	9/16N–1/16N	5/16N	5/64	—
	944S	2 1/4P–3P	2 1/2P	9/16N–1/16N	5/16N	5/64	—
	944 Turbo	2 1/4P–3P	2 1/2P	9/16N–1/16N	5/16N	5/64	—

WHEEL ALIGNMENT

Year	Model	Caster Range (deg.)	Caster Preferred Setting (deg.)	Camber Range (deg.)	Camber Preferred Setting (deg.)	Toe-in (in.)	Steering Axis Inclination (deg.)
1989	911	5 13/16P–6 5/16P	6 1/16P	3/16P	0	1/8	—
	911 Turbo	5 13/16P–6 5/16P	6 1/16P	3/16P	0	1/8	—
	928S4	3P–4P	3 1/2P	11/16N–5/16N	1/2N	5/32	—
	944	2 1/4P–3P	2 1/2P	9/16N–1/16N	5/16N	5/64	—
	944S2	2 1/4P–3P	2 1/2P	9/16N–1/16N	5/16N	5/64	—
	944 Turbo	2 1/4P–3P	2 1/2P	9/16N–1/16N	5/16N	5/64	—
1990	911 Carrera 2	5 13/16P–6 5/16P	6 1/16P	3/16N–3/16P	0	11/64	—
	911 Carrera 4	3 13/16P–4 7/16P	4 3/16P	3/16N–3/16P	0	1/4	—
	928S4	3P–4P	3 1/2P	11/16N–5/16N	1/2N	1/8	—
	944S2	2 1/4P–3P	2 1/2P	9/16N–1/16N	5/16	5/64	—
1991–92	911 Carrera 2	4 3/16P–4 11/16P	4 7/16P	3/16N–3/16P	0	11/64	—
	911 Carrera 4	3 13/16P–4 7/16P	4 3/16P	3/16N–3/16P	0	11/64	—
	911 Turbo	4 5/32P–4 11/16N	4 7/16P	5/32N–5/32P	0	7/32	—
	928S4/GT	3P–4P	3 1/2P	11/16N–5/16N	1/2N	1/8	—
	944S2	2 1/4P–3P	2 1/2P	3/16P–3/16N	0	5/64	—

P—Positive
N—Negative

Saab 17

900, 9000—All Models

SERIAL NUMBER IDENTIFICATION

Vehicle Identification Plate

900 Series

The vehicle serial number is located in 2 places; it is stamped on a plate at the lower left corner of the windshield or is punched in the vehicle body under the left side of the rear seat cushion.

The vehicle serial number is located on the right side of the rear cross beam in the luggage compartment.

9000 Series

These vehicles have the chassis number plate located on the inner right fender panel and the left fire wall area of the engine compartment. The chassis number is also punched in the vehicle body, left of the right rear light, behind the panel in the luggage compartment.

Engine Number

The engine identification number is stamped on a plate which is secured to the upper portion of the engine directly forward of the fuel injection unit.

SPECIFICATIONS

ENGINE IDENTIFICATION

Year	Model	Engine Displacement cu. in. (cc/liter)	Engine Series Identification	No. of Cylinders	Engine Type
1988	900	121 (1985/2.0)	B201	4	SOHC 8-Valve
	900	121 (1985/2.0)	B202 (Turbo)	4	DOHC 16-Valve
	900	121 (1985/2.0)	B202	4	DOHC 16-Valve
	9000	121 (1985/2.0)	B202 (Turbo)	4	DOHC 16-Valve
1989	900	121 (1985/2.0)	B202 (Turbo)	4	DOHC 16-Valve
	900	121 (1985/2.0)	B202	4	DOHC 16-Valve
	9000	121 (1985/2.0)	B202 (Turbo)	4	DOHC 16-Valve
1990	900	121 (1985/2.0)	B202 (Turbo)	4	DOHC 16-Valve
	900	121 (1985/2.0)	B202	4	DOHC 16-Valve
	9000	121 (1985/2.0)	B202 (Turbo)	4	DOHC 16-Valve
1991-92	900	121 (1985/2.0)	B202 (Turbo)	4	DOHC 16-Valve
	900	129 (2119/2.1)	B212	4	DOHC 16-Valve
	9000	140 (2290/2.3)	B234 (Turbo)	4	DOHC 16-Valve
	9000	140 (2290/2.3)	B234	4	DOHC 16-Valve

SOHC—Single Overhead Camshaft
DOHC—Double Overhead Camshaft

GENERAL ENGINE SPECIFICATIONS

Year	Model	Engine Displacement cu. in. (cc)	Fuel System Type	Net Horsepower @ rpm	Net Torque @ rpm (ft. lbs.)	Bore × Stroke (in.)	Compression Ratio	Oil Pressure @ rpm
1988	900	121 (1985)	Fuel Injection	110 @ 5250	119 @ 3500	3.543 × 3.071	9.25:1	64-71 ①
	900S	121 (1985)	Fuel Injection	125 @ 5500	123 @ 3000	3.543 × 3.071	10.1:1	51-74 ①
	900 Turbo	121 (1985)	Fuel Injection	160 @ 5500 ②	188 @ 3000 ③	3.543 × 3.071	9.0:1	64-71 ①
	9000	121 (1985)	Fuel Injection	160 @ 5500	188 @ 3000	3.543 × 3.071	9.0:1	64-71 ①
	9000S	121 (1985)	Fuel Injection	125 @ 5500	125 @ 3000	3.543 × 3.071	10.0:1	51-74 ①

GENERAL ENGINE SPECIFICATIONS

Year	Model	Engine Displacement cu. in. (cc)	Fuel System Type	Net Horsepower @ rpm	Net Torque @ rpm (ft. lbs.)	Bore × Stroke (in.)	Com-pression Ratio	Oil Pressure @ rpm
1989	900	121 (1985)	Fuel Injection	125 @ 5500	123 @ 3000	3.543 × 3.071	10.1:1	51–74 ①
	900S	121 (1985)	Fuel Injection	125 @ 5500	123 @ 3000	3.543 × 3.071	10.1:1	51–74 ①
	900 Turbo	121 (1985)	Fuel Injection	160 @ 5500 ②	188 @ 3000 ③	3.543 × 3.071	9.0:1	64–71 ①
	9000	121 (1985)	Fuel Injection	160 @ 5500	188 @ 3000	3.543 × 3.071	9.0:1	64–71 ①
	9000S	121 (1985)	Fuel Injection	125 @ 5500	125 @ 3000	3.543 × 3.071	10.0:1	51–74 ①
1990	900	121 (1985)	Fuel Injection	125 @ 5500	123 @ 3000	3.543 × 3.071	10.1:1	51–74 ①
	900S	121 (1985)	Fuel Injection	125 @ 5500	123 @ 3000	3.543 × 3.071	10.1:1	51–74 ①
	900 Turbo	121 (1985)	Fuel Injection	160 @ 5500 ②	188 @ 3000 ③	3.543 × 3.071	9.0:1	64–71 ①
	9000	121 (1985)	Fuel Injection	160 @ 5500	188 @ 3000	3.543 × 3.071	9.0:1	64–71 ①
	9000S	121 (1985)	Fuel Injection	125 @ 5500	125 @ 3000	3.543 × 3.071	10.0:1	51–74 ①
1991-92	900 Turbo	121 (1985)	Fuel Injection	160 @ 5500 ④	188 @ 3000 ③	3.54 × 3.07	9.0:1	52–75 ①
	900	129 (2119)	Fuel Injection	140 @ 6000	133 @ 2900	3.66 × 3.07	10.1:1	52–75 ①
	9000	140 (2290)	Fuel Injection	150 @ 5500	157 @ 3800	3.54 × 3.54	10.0:1	52–75 ①
	9000 Turbo	140 (2290)	Fuel Injection	200 @ 5000	244 @ 2000 ⑤	3.54 × 3.54	8.5:1	52–75 ①

① At 2000 rpm
② SPG—165 @ 5500
③ SPG—195 @ 3000
④ SPG—175 @ 5500
⑤ 222 @ 2000 w/auto trans.

ENGINE TUNE-UP SPECIFICATIONS

Year	Model	Engine Displacement cu. in. (cc)	Spark Plugs Type	Spark Plugs Gap (in.)	Ignition Timing (deg.) MT	Ignition Timing (deg.) AT	Com-pression Pressure (psi)	Fuel Pump (psi)	Idle Speed (rpm) MT	Idle Speed (rpm) AT	Valve Clearance In.	Valve Clearance Ex.
1988	900 ⑧	121 (1985)	①	0.024–0.028	④	④	NA	⑦	875	875	0.008–0.010	0.016–0.018
	900	121 (1985)	③	0.024–0.028	⑥	⑥	NA	⑦	875	875	Hyd.	Hyd.
	900 Turbo	121 (1985)	②	0.024–0.028	⑤	⑤	NA	⑦	875	875	Hyd.	Hyd.
	9000	121 (1985)	②	0.024–0.028	⑤	⑤	NA	⑦	875	875	Hyd.	Hyd.
1989	900	121 (1985)	⑨	0.024–0.028	⑥	⑥	NA	⑦	875	875	Hyd.	Hyd.
	900 Turbo	121 (1985)	②	0.024–0.028	⑤	⑤	NA	⑦	875	875	Hyd.	Hyd.
	9000	121 (1985)	②	0.024–0.028	⑤	⑤	NA	⑦	875	875	Hyd.	Hyd.

ENGINE TUNE-UP SPECIFICATIONS

Year	Model	Engine Displacement cu. in. (cc)	Spark Plugs Type	Spark Plugs Gap (in.)	Ignition Timing (deg.) MT	Ignition Timing (deg.) AT	Compression Pressure (psi)	Fuel Pump (psi)	Idle Speed (rpm) MT	Idle Speed (rpm) AT	Valve Clearance In.	Valve Clearance Ex.
1990	900	121 (1985)	⑨	0.024–0.028	⑥	⑥	NA	⑦	875	875	Hyd.	Hyd.
	900 Turbo	121 (1985)	②	0.024–0.028	⑤	⑤	NA	⑦	875	875	Hyd.	Hyd.
	9000	121 (1985)	⑨	0.024–0.028	⑥	⑥	NA	⑦	875	875	Hyd.	Hyd.
	9000 Turbo	121 (1985)	⑩	0.024–0.028	⑤	⑤	NA	⑦	875	875	Hyd.	Hyd.
1991	900	129 (2119)	⑨	0.024–0.028	⑥	⑥	NA	⑦	875	875	Hyd.	Hyd.
	900 Turbo	121 (1985)	②	0.024–0.028	⑤	⑤	NA	⑦	875	875	Hyd.	Hyd.
	9000	140 (2290)	⑨	0.024–0.028	⑥	⑥	NA	⑦	875	875	Hyd.	Hyd.
	9000 Turbo	140 (2290)	⑩	0.024–0.028	⑤	⑤	NA	⑦	875	875	Hyd.	Hyd.
1992	SEE UNDERHOOD SPECIFICATION STICKER											

NA—Not Available
Hyd.—Hydraulic
① BP6ES, W7DC, N9Y, N9YC, BP7ES
② BCP7EV
③ BCP6ES, C9YC, F7DC
④ 20° @ 2000 rpm
18° @ 2000 rpm Canada with manual transmission
23° @ 2000 rpm Canada with automatic transmission
⑤ 16° BTDC @ 850 rpm
⑥ 14° @ 850 rpm
⑦ Fuel line pressure before the control pressure regulator is 66.9–69.7 (setting valve), and 48.5–54.0 psi (warm engine) after the control pressure regulator (located in fuel distributor).
⑧ SOHC—Single Overhead Camshaft
⑨ BCP5ES
⑩ BCPR7ES

FIRING ORDERS

NOTE: To avoid confusion, always replace spark plug wires one at a time.

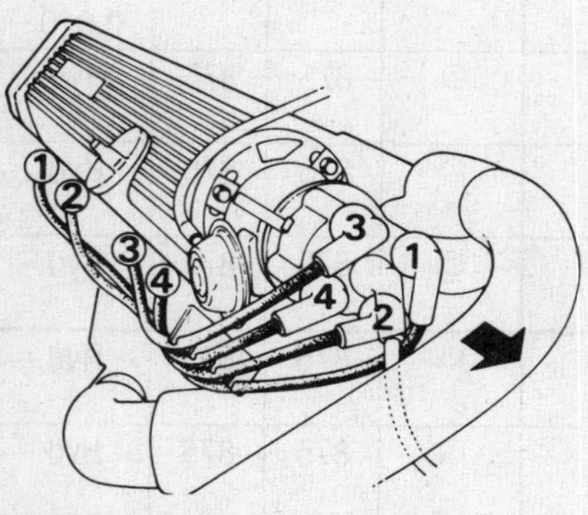

8 valve engine
Engine Firing Order: 1–3–4–2
Distributor Rotation: Counterclockwise

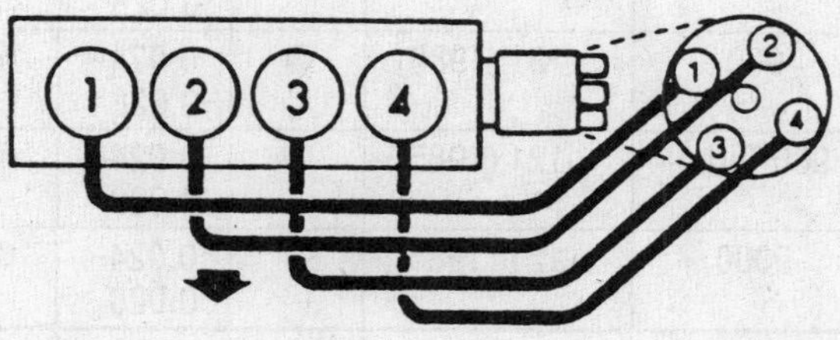

16 valve engine
Engine Firing Order: 1–3–4–2
Distributor Rotation: Counterclockwise
NOTE: Some 16 valve engines use a Direct Ignition (DI) system. This system does not use a distributor

CAPACITIES

Year	Model	Engine Displacement cu. in. (cc)	Engine Crankcase (qts.) with Filter	Engine Crankcase (qts.) without Filter	Transmission (pts.) 4-Spd	Transmission (pts.) 5-Spd	Transmission (pts.) Auto.	Drive Axle (pts.)	Fuel Tank (gal.)	Cooling System (qts.)
1988	900 ①	121 (1985)	4.0	3.5	5.2	6.4	17	2.6 ②	18.0	10.5
	900 Turbo	121 (1985)	4.5	4.0	5.2	6.4	17	2.6 ②	18.0	10.5
	900	121 (1985)	4.0	3.5	5.2	6.4	17	2.6 ②	18.0	10.5
	9000	121 (1985)	4.5	4.0	5.2	6.4	17	2.6 ②	18.0	10.5
1989	900 Turbo	121 (1985)	4.5	4.0	5.2	6.4	17	2.6 ②	18.0	10.5
	900	121 (1985)	4.0	3.5	5.2	6.4	17	2.6 ②	18.0	10.5
	9000	121 (1985)	4.5	4.0	5.2	6.4	17	2.6 ②	18.0	10.5
1990	900 Turbo	121 (1985)	4.5	4.0	5.2	6.4	17	2.6 ②	18.0	10.5
	900	121 (1985)	4.0	3.5	5.2	6.4	17	2.6 ②	18.0	10.5
	9000	121 (1985)	4.5	4.0	5.2	6.4	17	2.6 ②	18.0	10.5
1991-92	900	129 (2119)	4.0	3.5	5.2	6.4	17	2.6 ②	18.0	10.5
	900 Turbo	121 (1985)	4.5	4.0	5.2	6.4	17	2.6 ②	18.0	10.5
	900 Turbo	121 (1985)	4.5	4.0	5.2	6.4	17	2.6 ②	18.0	10.5
	9000	140 (2290)	4.5	4.0	5.2	6.4	17	2.6 ②	18.0	10.5
	9000 Turbo	140 (2290)	4.5	4.0	5.2	6.4	17	2.6 ②	18.0	10.5

① SOHC—Single Overhead Camshaft
② 3.0 for Borg Warner Type 37

CAMSHAFT SPECIFICATIONS

Year	Engine Displacement cu. in. (cc)	Journal Diameter 1	Journal Diameter 2	Journal Diameter 3	Journal Diameter 4	Journal Diameter 5	Lobe Lift In.	Lobe Lift Ex.	Bearing Clearance	Camshaft End Play
1988	121 (1985) ①	1.1394	1.1394	1.1394	1.1394	1.1394	0.425	0.433	NA	0.0031–0.0098
	121 (1985)	1.1387–1.1392	1.1387–1.1392	1.1387–1.1392	1.1387–1.1392	1.1387–1.1392	②	③	NA	0.0031–0.0138
1989	121 (1985) ③	1.1387–1.1392	1.1387–1.1392	1.1387–1.1392	1.1387–1.1392	1.1387–1.1392	③	③	NA	0.0031–0.0138
1990	121 (1985)	1.1387–1.1392	1.1387–1.1392	1.1387–1.1392	1.1387–1.1392	1.1387–1.1392	②	③	NA	0.0031–0.0138
1991-92	121 (1985)	1.1387–1.1392	1.1387–1.1392	1.1387–1.1392	1.1387–1.1392	1.1387–1.1392	②	③	NA	0.0031–0.0138
	129 (2119)	1.1387–1.1392	1.1387–1.1392	1.1387–1.1392	1.1387–1.1392	1.1387–1.1392	②	③	NA	0.0031–0.0138
	140 (2290)	1.1387–1.1392	1.1387–1.1392	1.1387–1.1392	1.1387–1.1392	1.1387–1.1392	0.340	0.340	NA	0.0031–0.0138

① SOHC—Single Overhead Camshaft
② Non-turbocharged—0.425
Turbocharged—0.3406/0.2618
③ Non-turbocharged—0.433
Turbocharged—0.3406

CRANKSHAFT AND CONNECTING ROD SPECIFICATIONS

All measurements are given in inches.

Year	Engine Displacement cu. in. (cc)	Crankshaft Main Brg. Journal Dia.	Main Brg. Oil Clearance	Shaft End-play	Thrust on No.	Connecting Rod Journal Diameter	Oil Clearance	Side Clearance
1988	121 (1985)	2.283–2.284	0.0008–0.0024	0.003–0.011	3	2.2047–2.2054	0.0010–0.0024	NA
1989	121 (1985)	2.283–2.284	0.0008–0.0024	0.003–0.011	3	2.2047–2.2054	0.0010–0.0024	NA
1990	121 (1985)	2.283–2.284	0.0008–0.0024	0.003–0.011	3	2.2047–2.2054	0.0010–0.0024	NA
1991-92	121 (1985)	2.283–2.284	0.0008–0.0024	0.003–0.011	3	2.2047–2.2054	0.0010–0.0024	NA
	129 (2119)	2.283–2.284	0.0008–0.0024	0.003–0.011	3	2.2047–2.2054	0.0010–0.0024	NA
	140 (2290)	2.283–2.284	0.0008–0.0024	0.003–0.011	3	2.2047–2.2054	0.0010–0.0024	NA

NA—Not available

VALVE SPECIFICATIONS

Year	Engine Displacement cu. in. (cc)	Seat Angle (deg.)	Face Angle (deg.)	Spring Test Pressure (lbs.)	Spring Installed Height (in.)	Stem-to-Guide Clearance (in.) Intake ①	Stem-to-Guide Clearance (in.) Exhaust ①	Stem Diameter (in.) Intake	Stem Diameter (in.) Exhaust
1988	121 (1985) ②	45	44.5	178–198 @ 1.16	1.56	0.020	0.020	0.3134–0.3139	0.3132–0.3142
	121 (1985)	45	44.5	133–145 @ 1.18	1.45	0.020	0.020	0.2740–0.2746	0.2738–0.2748
1989	121 (1985)	45	44.5	133–145 @ 1.18	1.45	0.020	0.020	0.2740–0.2746	0.2738–0.2748
1990	121 (1985)	45	44.5	133–145 @ 1.18	1.45	0.020	0.020	0.2740–0.2746	0.2738–0.2748
1991-92	121 (1985)	45	44.5	133–145 @ 1.18	1.45	0.020	0.020	0.2740–0.2746	0.2738–0.2748
	129 (2119)	45	44.5	133–145 @ 1.18	1.45	0.020	0.020	0.2740–0.2746	0.2738–0.2748
	140 (2290)	45	44.5	133–145 @ 1.18	1.45	0.020	0.020	0.2740–0.2746	0.2738–0.2748

① Measured on valve head; raised 0.12 in. (3 mm) above seat
② Single Overhead Camshaft

PISTON AND RING SPECIFICATIONS

All measurements are given in inches.

Year	Engine Displacement cu. in. (cc)	Piston Clearance	Ring Gap Top Compression	Ring Gap Bottom Compression	Ring Gap Oil Control	Ring Side Clearance Top Compression	Ring Side Clearance Bottom Compression	Ring Side Clearance Oil Control
1988	121 (1985) ①	0.0009–0.0020	0.014–0.022	0.012–0.018	0.015–0.055	0.002–0.003	0.002–0.003	NA
	121 (1985)	0.0009–0.0020	0.013–0.021	0.011–0.017	0.014–0.055	0.002–0.003	0.002–0.003	NA
1989	121 (1985)	0.0009–0.0020	0.013–0.021	0.011–0.017	0.014–0.055	0.002–0.003	0.002–0.003	NA

PISTON AND RING SPECIFICATIONS

All measurements are given in inches.

Year	Engine Displacement cu. in. (cc)	Piston Clearance	Ring Gap Top Compression	Ring Gap Bottom Compression	Ring Gap Oil Control	Ring Side Clearance Top Compression	Ring Side Clearance Bottom Compression	Ring Side Clearance Oil Control
1990	121 (1985)	0.0009–0.0020	0.013–0.021	0.011–0.017	0.014–0.055	0.002–0.003	0.002–0.003	NA
1991-92	121 (1985)	0.0009–0.0020	0.013–0.021	0.011–0.017	0.014–0.055	0.002–0.003	0.002–0.003	NA
	129 (2119)	0.0009–0.0020	0.013–0.021	0.011–0.017	0.014–0.055	0.002–0.003	0.002–0.003	NA
	140 (2290)	0.0004–0.0010	0.012–0.020	0.012–0.018	0.015–0.055	0.002–0.003	0.002–0.003	NA

① SOHC—Single Overhead Camshaft
NA—Not available

TORQUE SPECIFICATIONS

All readings in ft. lbs.

Year	Engine Displacement cu. in. (cc)	Cylinder Head Bolts	Main Bearing Bolts	Rod Bearing Bolts	Crankshaft Pulley Bolts	Flywheel Bolts	Manifold Intake	Manifold Exhaust	Spark Plugs
1988	121 (1985)	① ②	40	80	140	43	13	18	18–21
1989	121 (1985)	① ②	40	80	140	43	13	18	18–21
1990	121 (1985)	① ②	40	80	140	43	13	18	18–21
1991-92	121 (1985)	① ②	40	80	140	43	13	18	18–21
	129 (2119)	① ②	40	80	140	43	13	18	18–21
	140 (2290)	④	15③	15③	140	43	16	13	18–21

① 1st stage—43 ft. lbs.
2nd stage—72 ft. lbs.—run engine to warm. Allow 30 minutes cool time—Retighten to 72 ft. lbs.
3rd stage—Tighten each bolt another 1/4 (90 degrees) of a turn.

② Turbo 16-Valve
1st stage—45 ft. lbs.
2nd stage—67 ft. lbs. Engine to normal operating temp. Allow to cool for 30 minutes.
3rd stage—Tighten another 90 degrees turn (1/4 turn). Retorque after 1200 miles or after engine reaches normal operating temperature.

③ Plus 1/4 turn or 90 degrees

④ 1st stage—44 ft. lbs.
2nd stage—59 ft.lbs.
3rd stage—Tighten another 1/4 turn or 90 degrees.

BRAKE SPECIFICATIONS

All measurements in inches unless noted.

Year	Model	Lug Nut Torque (ft. lbs.)	Master Cylinder Bore	Brake Disc Minimum Thickness	Brake Disc Maximum Runout	Standard Brake Drum Diameter	Minimum Lining Thickness Front	Minimum Lining Thickness Rear
1988	900	65–80	NA	0.461①	—	—	—	—
	9000	76–90	NA	0.787②	0.768③	—	0.039	0.039
1989	900	65–80	NA	0.461①	—	—	—	—
	9000	76–90	NA	0.787②	0.768③	—	0.039	0.039
1990	900	65–80	NA	0.461①	—	—	—	—
	9000	76–90	NA	0.787②	0.768③	—	0.039	0.039
1991-92	900	65–80	NA	0.461①	—	—	—	—
	9000	76–90	NA	0.787②	0.768③	—	0.039	0.039

NA—Not available
① 0.374 Rear
② 0.295 Rear
③ 0.276 Rear

WHEEL ALIGNMENT

		Caster		Camber			Steering Axis
Year	Model	Range (deg.)	Preferred Setting (deg.)	Range (deg.)	Preferred Setting (deg.)	Toe-in (in.)	Inclination (deg.)
1988	900	1 1/2–2 1/2	2	0–1	1/2	5/64	NA
	9000	1 1/8–2 1/8	1 5/8	1 1/8N–1/8N	5/8N	1/16	NA
1989	900	1 1/2–2 1/2	2	0–1	1/2	5/64	NA
	9000	1 1/8–2 1/8	1 5/8	1 1/8N–1/8N	5/8N	1/16	NA
1990	900	1 1/2–2 1/2	2	0–1	1/2	5/64	NA
	9000	1 1/8–2 1/8	1 5/8	1 1/8N–1/8N	5/8N	1/16	NA
1991–92	900	1 1/2–2 1/2	2	0–1	1/2	5/64	NA
	9000	1 1/8–2 1/8	1 5/8	1 1/8N–1/8N	5/8N	1/16	NA

NA—Not available
N—Negative

ENGINE MECHANICAL

NOTE: Disconnecting the negative battery cable on some vehicles may interfere with the functions of the on board computer systems and may require the computer to undergo a relearning process, once the negative battery cable is reconnected.

Engine Assembly

REMOVAL & INSTALLATION

900 Series

8 VALVE ENGINE

NOTE: The engine and transmission should be removed as a unit.

1. Disconnect the negative battery cable. Drain the radiator.
2. Disconnect the windshield washer hose, unbolt the hood hinge links and remove the hood from the vehicle.
3. If equipped with power steering, disconnect the lines at the servo pump.

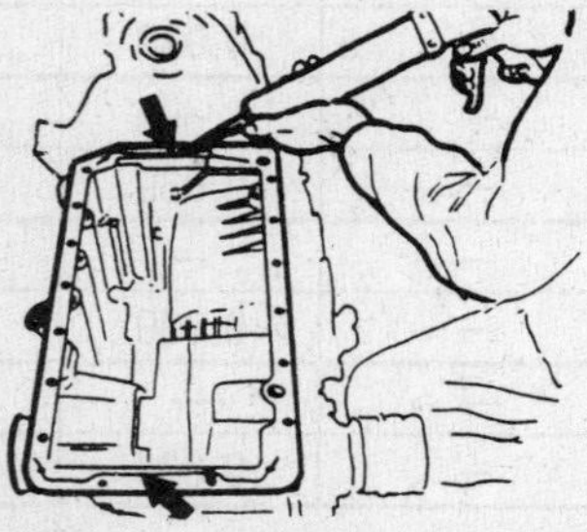

Sealer should only be applied to the grooves at each end of the steel engine-to-transaxle gasket

4. Disconnect the positive battery lead at the starter. Remove the radiator hoses. Remove the engine ground wire. Disconnect the temperature transmitter cable. Remove the coil.
5. Disconnect the cable harness from the clutch cover. If equipped with manual transmission, disconnect the hydraulic line from the clutch slave cylinder and plug the lines.
6. Disconnect the CI system electrical connections from the warm up regulator, thermo-time switch cold start valve and the auxiliary air valve. On catalytic converter equipped vehicles, also disconnect the oxygen sensor and the throttle switch cables.
7. Disconnect the oil pressure transmitter cable. Loosen the fuel line connections at the fuel distributor. Remove the air filter along with the mixture control unit.
8. Disconnect the throttle cable. Disconnect the hose at the expansion tank. Disconnect the heater hoses at the heater. Disconnect the brake vacuum hose.
9. Remove the clips and remove the bellows from the inner drivers.
10. Place the spacer (Saab tool 83–93–209) or equivalent between the upper control arm underside and the vehicle body.

NOTE: Insert the tool from the engine compartment side. The spacer makes the front suspension unloaded when the vehicle is raised.

11. Lift the vehicle and support it safely.
12. Remove the lower end piece from the control arm. Pull out the steering knuckle assembly and support the end piece against the control arm outer end.
13. If equipped with manual transmission, put the gear lever in neutral. Remove the nut and tap out the taper pin in the gear shift rod joint. Separate the joint from the gear shift rod.
14. If equipped with automatic transmission, remove the retaining screw from the gear selector cable at the transmission. Withdraw the cable with the gear selector rod in its extreme forward position **P**. Slide back the spring loaded sleeve on the gear shift rod and unhook the end of the cable.
15. Separate the exhaust pipe from the exhaust manifold. Disconnect the speedometer cable from the transmission.
16. Remove the rear engine mounting bolts. Slacken the front engine mounting nut so the mounting can be lifted out of the bracket.
17. Attach the hoist to the 2 lugs on the engine and raise the assembly slightly. Move the assembly to one side and free the 2 U-joints.
18. Carefully remove the unit from the vehicle.

To install:

19. Install the engine and transmission assembly. Ensure that the U-joints are connected.
20. Install the front and rear engine mounts, exhaust pipes, speedometer cable, gear selector cable (automatic transmission) or gear shift rod (manual transmission).
21. Assemble the steering knuckle and control arm. Connect the throttle cable, all heater hoses, oil pressure switch, fuel lines and all CI system electrical connections.
22. Connect the clutch master cylinder, install the battery and battery cables and connect all other components previously disconnected.
23. Fill the radiator with coolant, the engine with oil and the transaxle with fluid. Start the engine and allow it to reach operating temperature.

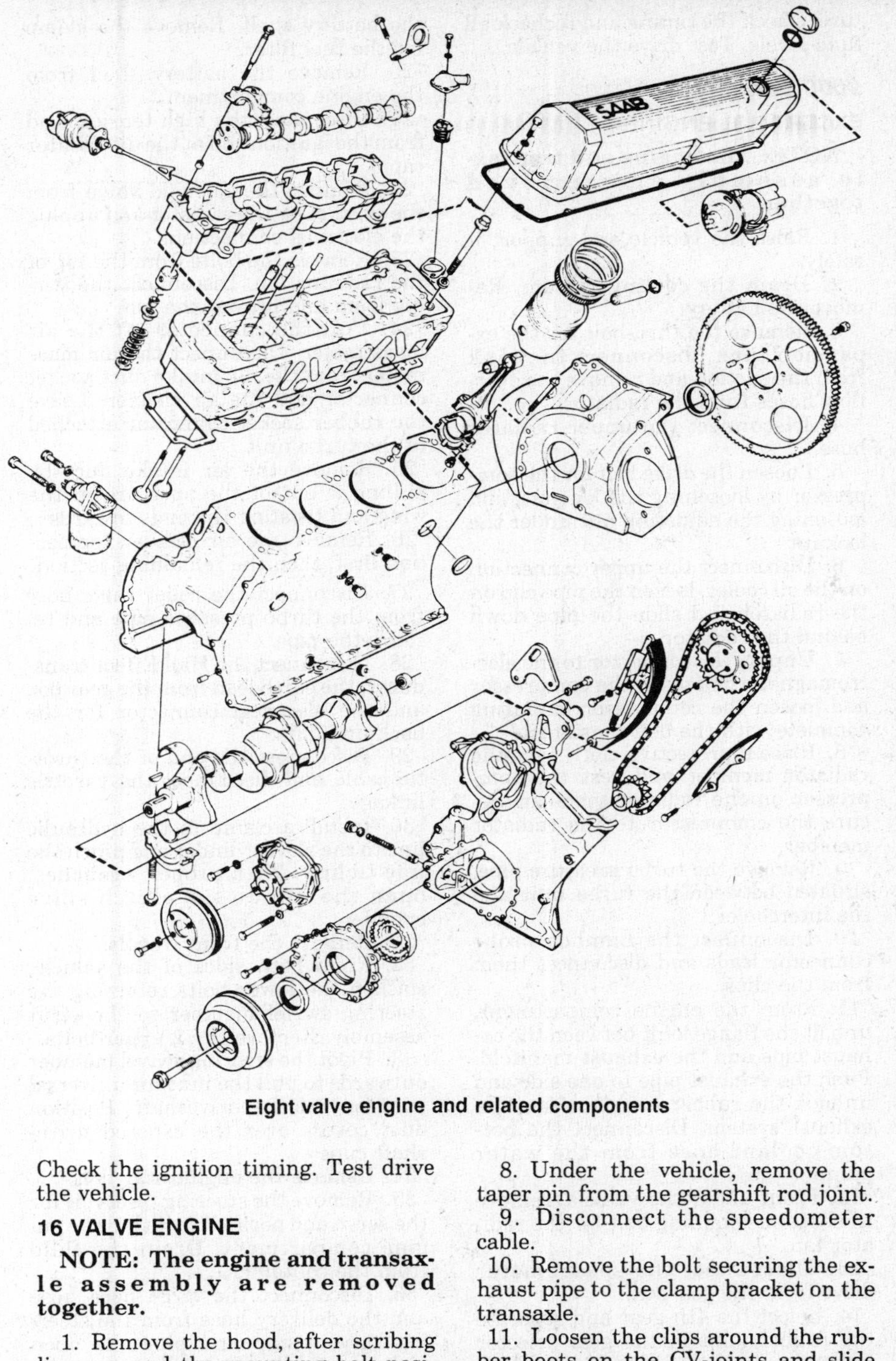

Eight valve engine and related components

Check the ignition timing. Test drive the vehicle.

16 VALVE ENGINE

NOTE: The engine and transaxle assembly are removed together.

1. Remove the hood, after scribing lines around the mounting bolt positions to aid later refitting.
2. Install Saab special tool 83-93-209, or equivalent, under the upper control arm on the right side.
3. Disconnect and remove the battery.
4. Drain the engine coolant.
5. Slacken the wheel nuts on the right front wheel.
6. Raise and safely support the vehicle.
7. Put the transmission selector into **R**.
8. Under the vehicle, remove the taper pin from the gearshift rod joint.
9. Disconnect the speedometer cable.
10. Remove the bolt securing the exhaust pipe to the clamp bracket on the transaxle.
11. Loosen the clips around the rubber boots on the CV-joints and slide the boots clear, this operation can also be done from above.
12. On the right side of the vehicle, remove the front wheel.
13. Separate the end piece from the lower control arm.
14. Separate the universal joint and position the knuckle in front of the driver. Support the end piece against the outer end of the control arm.
15. Disconnect the positive lead from the battery and free it from the clips holding it to the body. Disconnect the ground cable from the transaxle.
16. Disconnect the starter motor leads.
17. Unbolt the exhaust pipe from the exhaust manifold.
18. Disconnect the pressure pipe from the steering servo pump and have a plug handy to prevent oil escaping from the pipe. Take care not to drip oil onto the engine mounting and control arm rubbers.
19. From the left side of the vehicle, disconnect the cooling system hoses at the following connections, the heat exchanger valve, the expansion tank, the bottom of the radiator and the thermostat housing.
20. Disconnect the left fuel injection system cable harness as follows, at the air mass meter sensor, at the throttle switch, at the A.I.C. actuator, at the injectors, at the the NTC resistor (thermostatic switch) and at the ground points on the front lifting lug. Use the proper tool to release the tension in the springs on the terminal blocks.
21. Disconnect the block and plug connector (ground lead). Disconnect the lead at the alternator and the green/white cable to the positive terminal on the regulator. Disconnect the ground (black) cable. Disconnect the black cable from the oil pressure switch. Disconnect the cable for the A.I.C. actuator. Disconnect the yellow/white cable from the temperature transmitter. Disconnect the gray cable from the knock detector. Release the cable harness from the clip on the fuel injection manifold, from the rear of the engine and from the coolant hose between the engine and the expansion tank.
22. Withdraw the loose cables and guide the harness unit out of the engine compartment. Place it on top of the power distribution unit.
23. Remove the adjusting bolt in the alternator bracket, remove the drive belts and lift off the alternator.
24. Disconnect the brake servo hose from the intake manifold. Disconnect the throttle cable and sheath.
25. Remove the air conditioner compressor and bracket from the block. Place them on the filter housing for the heater system. Secure the alternator so it will not drop or become damaged.
26. Disconnect the fuel lines at their connections at the front of the fuel injection manifold and on the fuel pressure regulator.
27. Remove the coil.
28. Disconnect the turbo pressure line from the turbo compressor and the intercooler/throttle housing.
29. Remove the auxiliary fan.
30. Remove the air mass meter together with the suction pipe for the

turbo unit. Disconnect the hoses at the solenoid valve and the crankcase ventilation at the suction pipe.

31. Disconnect the cables from the Hall transmitter and coil in the distributor. Free the Hall transmitter cable from the clips on the clutch cover.

32. Disconnect the solenoid valve hoses from the connections on the turbo unit and charging pressure regulator.

33. Disconnect the hydraulic hose from the clutch slave cylinder. Plug the hose to stop fluid from escaping.

34. Remove the engine mounting bolts.

35. Attach suitable lifting equipment to the engine lifting hooks. Raise the engine until the left, inner CV-joint can be separated.

36. Raise the engine to enable the hoses on the oil cooler to be disconnected.

NOTE: When lifting the engine out of the vehicle, keep it close to the fire wall to prevent the radiator and solenoid valve from being damaged by the front engine mounting.

37. Disconnect the hose to the power steering pump and drain the oil in the system.

To install:

38. Before installation, check that the inner CV-joint boots are packed with the correct grease.

39. Suspend the engine from the lifting gear. Adjust the lifting gear so the front engine mounting is slightly lower than the rear mounting.

40. Lower the engine into the engine compartment until the hoses to the oil cooler and servo pump can be connected.

41. Guide the engine into position, attending to the following items in order, the front engine mounting, left inner CV-joint and right inner CV-joint. Lower the engine until it rests on the rear engine mountings and install the mounting bolts. Unhook the lifting gear and unbolt the lifting lug from the water pump.

42. Connect the clutch master cylinder and all turbo unit connections.

43. Install the distributor, coil and all wiring for the electronic ignition.

44. Install the air mass meter, auxiliary fan, fuel lines, air conditioner, throttle cable and brake booster.

45. Install the alternator, drive belts, all electrical connections, cooling system hoses, exhaust system, battery and cables.

46. Assemble the steering knuckle and halfshafts.

47. Fill the radiator with coolant, the engine with oil and the transmission with fluid. Start the engine and allow it to reach operating temperature. Check the timing and recheck all fluid levels. Test drive the vehcile.

9000 Series

EXCEPT 2.3L ENGINE

NOTE: The engine and transaxle assembly are removed together.

1. Raise the vehicle and support it safely.

2. Drain the cooling system. Remove the battery.

3. Remove the thru-bolt for the expansion tank, disconnect the tank from the suction and remove the overflow hoses from the radiator.

4. Disconnect the upper radiator hose.

5. Loosen the drive belt for the compressor by loosening the locknut, and loosening the adjusting nut under the locknut.

6. Disconnect the upper connection on the oil cooler, loosen the pipe clip on the radiator and slide the pipe down behind the radiator.

7. Unplug the connector to the electromagnetic clutch on the compressor and loosen the compressor mounting complete with the belt tensioner.

8. Place a protective cloth over the radiator member and rest the compressor on the radiator member. Secure the compressor to the radiator member.

9. Remove the turbo pressure pipe, situated between the turbo unit and the intercooler.

10. Disconnect the Lambda probe connector leads and disconnect them from the clips.

11. From the engine compartment, unbolt the flange joint between the exhaust pipe and the exhaust manifold. Push the exhaust pipe to one side and unhook the rubber hangers from the exhaust system. Disconnect the bottom coolant hose from the water pump.

12. From under the vehicle, remove the bottom retaining bolt for the radiator fan.

13. Disconnect the speedometer drive from the gearbox.

14. Select the 4th gear and separate the rubber joint in the gear selector linkage.

15. Remove the clips on the rubber gaiters over the inboard universal joints and slide the gaiters off the drive axles.

16. Disconnect the electrical leads from the alternator and the starter motor. Unplug the connector for the oil pressure switch.

17. Remove the clips and remove the top radiator hose.

18. Disconnect the top radiator hose at the cylinder head.

19. Unscrew the junction block from the battery shelf. Remove the clamp for the fuel filter.

20. Remove the battery shelf from the engine compartment.

21. Disconnect the high tension lead from the ignition coil at the distributor cap.

22. Remove the solenoid valve from the bracket on the radiator and unplug the electrical connections.

23. Remove the bolts from the top of the radiator fan. Disconnect the wiring loam and lift out the fan.

24. Pull the connector off the air mass meter. Disconnect the air mass meter from the air intake duct socket connector and the air cleaner. Leave the rubber socket connector attached to the turbo unit.

25. Remove the air intake duct by pulling it out of the aperture in the wing and twisting the ends inwards.

26. Remove the air cleaner top section first, then the remaining section.

27. Disconnect the relief valve hose from the turbo pressure pipe and remove the pipe.

28. Disconnect the Hall Effect transducer, the earth lead from the gear box and the electrical connector for the back-up lights.

29. Disconnect the end of the throttle cable and disconnect the throttle linkage.

30. Install a clamp to the hydraulic line to the slave cylinder and pinch the line tightly. With proper wrenches, open the line to the clutch slave cylinder.

31. Remove the front wheels.

32. From both sides of the vehicle, slacken the lower bolts retaining the steering swivel member to the strut assembly. Remove the 2 upper bolts.

33. Pivot the steering swivel member outwards to pull the inboard universal joint out of the driveshaft. Position dust covers over the exposed driveshaft cups.

34. Remove the engine stay bolt.

35. Remove the steering reservoir for the servo and position it within the engine compartment. Drain the fluid from the container.

36. Disconnect the large bore hose and the delivery hose from the steering servo pump and plug the open ends.

37. Disconnect the fuel return lie from the pressure regulator.

38. Remove the nut from the rear engine mounting and back off the front mount bolts a few turns.

39. Attach the lifting sling (Saab 83–92–409) to the rear lifting lug.

40. Lift the engine sufficiently to provide access for the removal of the components located between the engine and the fire wall.

41. Disconnect the vacuum hoses from the inlet manifold.

42. Remove the coolant hoses running between the heat exchanger and the water pump pipe.
43. Separate the coupling between the fuel pipe and the fuel injection manifold. Do not allow the fuel to spill or collect.
44. Cut the clips securing the wiring looms to the oil pipe, water pipe, inlet manifold steady bar and the oil supply pipe.
45. Unclip the wiring loom to the fuel injection manifold.
46. Disconnect the grounding connections and the electrical connectors from the wiring harness.
47. Unbolt the air cooled oil cooler and place it on top of the engine. The 2 lower bolts need only be loosened.
48. Carefully remove the engine from the vehicle, taking care not to damage the radiator.

To install:

49. Install the engine in the vehicle and connect all the engine mounts. Tighten the bolts to specification.
50. Connect the oil cooler lines, all electrical connections, fuel lines, vacuum and coolant hoses and power steering lines.
51. Assemble the steering knuckle and install the halfshafts. Connect the clutch master cylinder, throttle cable, ignition system, turbo system components, air mass meter, radiator fan, battery and exhaust system.
52. Install all accesory drive belts and tighten to specification.
53. Fill the radiator with coolant, the engine with oil and the transmission with fluid. Start the engine and allow it to reach operating temperature. Check the ignition timing and all fluid levels. Test drive the vehicle.

2.3L ENGINE

1. Raise and support the vehicle safely. Remove the inner fender wells and right side fender. Remove the middle filler panel from under the spoiler. Lower the vehicle.
2. Drain the coolant. Disconnect the battery cables and remove the battery. Remove the brace bar for the ABS unit. Unplug the 3 connectors for the DI system and release the clips for the leads from the bracket attached to the battery shelf.
3. Disconnect the positive battery cable from the terminal block and remove the terminal block. Disconnect the batter negative cable from the fender. Remove the battery shelf and tuck the leads out of the way.
4. Unplug the connector for the knock detector in the block. Disconnect the top radiator hose.
5. Remove the air mass meter and the brace bar for the air intake silencer. Release the toggle clips on the air cleaner and the throttle housing. Lift off the air intake complete with silencer and air mass meter.
6. Unplug the connectors for the washer fluid level sensor, remove the hose from the reservior, remove the retaining screws and remove the reservior.
7. Remove the clip, lift up the throttle lever and disconnect the cable from the linkage. Disconnect the fuel return hose from the fuel pressure regulator and tuck out of the way. Place a rag under the connector and disconnect the fuel supply hose from the injection rail. Pull out the end of the hose and tuck it out of the way.
8. Remove the cover over the space behind the false bulkhead panel. Remov ethe rubber moulding from the edge of the panel. Remove the clip securing the engine wiring loom to the bulkhead. Remove the securing boltss and lift out the bulkhead.
9. Unplug the connector in the engine wiring loom and tuck the loom away on top of the engine. Unplug the connector for the road speed sensor lead. Pull the cable out of the grommet in the bottom of the bulkhead and tuck the lead away at the top of the engine.
10. Disconnect the heater box hoses from the cylinder head and tuck them away behind the brake fluid reservior. Unscrew the clip securing the kickdown cable to the pupe from the steering servo pump.
11. Clean the surrounding areas and disconnect the hoses from the oil cooler for the automatic transmission. Plug the ports in the transmission and the fittings in the hoses. Secure the hoses to the radiator crossmember.
12. Unplug the connector from the brake fluid reservior. Remove the nut from the gear selector tod linkage. Slacken the cable and tuck it behind the brake fluid reservoir.
13. Disconnect the hoses from the expansion tank. Remove the securing bolt, snip through the tie, unplug the connector and lift the tank out. Use a spanner wrench to release the tension applied bu the belt tensioner and applyu a hard upward pull on the belt. As the belt loosens, install locking pin 83-94-488 in the tensioner. Ease the belt off the air conditioner compressor pulley.
14. Install a protective steel panel over the oil cooler and cover the right section of the radiator crossmember with rags. Unplug the connector from the air conditioner compressor, remove the securing bolts and lift the compressor onto the radiator crossmember. Secure the compressor with a piece of wire.
15. Disconnect the top radiator hose from the water pump. Remove the clip and disconnect the bottom radiator hose from the clamp. Disconnect the oxygen sensor by unpluging the connector under the inlet manifold.
16. Snip through the ties securing the servo hoses, charcoal canister hose and wiring loom to the top torque arm. Disconnect the radio suppressor lead from the torque arm and remove the arm. Disconnect the hose to the charcoal canister from the inlet manifold and tuck it out of the way.
17. Unbolt the exhaust pipe from the exhaust manifold. Unplug the connector from the oil level sensor and place the lead along the edge of the fender. Remove the top securing bolt and loosen the 2 bolts at the bottom of the oil cooler. Snip through any ties and tie the cooler to the engine.
18. Remove the securing bolt for the power steering fluid reservior, lower the reservoir and siphon off the fluid. Disconnect the servo hose from the reservior and tuck the end under the inlet manifold. Stand the reservoir on the bulkhead.
19. Raise and support the vehicle safely. Disconnect the steering servo delivery pipe from the pump. Use a second spanner across the flats to stop the fitting turning. Install a plug in the end of the pipe and wipe up any spills immediately. Remove the pipe clip from the support bearing rear engine mount.
20. Remove the nuts from the engine mountings on the right side. Remove the clips for the CV-boots. Loosen the bottom bolts securing the struts to the steering swivel member and remove the top bolts. Pull away the steering swivel members to separate the driveshaft joints. Lower the vehicle.
21. Unbolt the left engine mount. Disconnect the struts from the hood and install extensions (83-94-439) onto the ends. Attach an engine lift to the eyes on the engine. Pay particular attention to the following components when removing the engine: ABS unit, kickdown cable, cooling fan, steering servo pump, alternator pulley.

To install:

22. Install the engine assembly into the vehicle. Insert the bolt through the left engine mount prior to lowering the assembly. Pay attention that the following components are kept clear of the engine: ABS unit, kickdown cable, cooling fan, steering servo pump and the alternator pulley.
23. Tighten the engine mountings. Check that the CV-joints are greased and install the halfshafts on the transmission. Install the struts and tighten the bolts to 58–77 ft. lbs. (78–105 Nm). Install the clips on the CV-boots.
24. Install the filler panel under the spoiler. Connect the steering servo pipe to the pump and secure the pipe to the bearing bracket. Lower the vehicle.

25. Connect the charcoal canister, power steering fluid reservoir, oil level sensor and oil cooler, exhaust system and radiator hoses. Install the air conditioner compressor and tighten the bolts to 15 ft. lbs. (20 Nm).

26. Install the serpentine belt and release the locking pin in the tensioner. Connect the radiator hose to the expansion tank and install the tank in the vehicle.

27. Install the oxygen sensor, engine torque arm, gear selector linkage, road speed sensor, heater box hoses, throttle cable, knock detector, false bulkhead, washer hose, fuel supply and return hoses, air intake and brace bar and battery.

28. Connect the DI wiring. Install the bracket for the ABS and washer fluid reservoir. Remove the hood strut extensions.

29. Raise and support the vehicle safely. Install the right wheel well. Install the wheels and tighten the lugs to 96 ft. lbs. (130 Nm).

30. Refill the cooling system, power steering fluid, washer fluid, engine oil and transmission fluid.

31. Start the vehicle and allow it to reach operating temperature. Check that all systems are working properly.

Cylinder Head

REMOVAL & INSTALLATION

900 Series

8 VALVE ENGINE

1. Disconnect the negative battery cable. Drain the radiator.

2. Remove the rubber bellows from between the air flow sensor and the throttle valve housing and disconnect the throttle cable from the throttle valve housing.

3. Disconnect the cable from the temperature transmitter. Remove the vacuum hose of the power brake booster from the intake manifold.

4. Disconnect the fuel lines from the fuel distributor to the injection valves. Tape the ends of the lines to prevent dirt from entering the system. Remove the bracket from the throttle valve housing mounting.

5. Remove the hose clamps at the connections to the thermostat housing, water pump and intake manifold.

6. Unbolt the exhaust pipe from the exhaust manifold.

7. Remove the distributor cap and ignition wires. Rotate the engine until cylinders No. 1 and No. 4 are at TDC. This must be done due to the design of the distributor driving dog which only allows the valve cover/distributor assembly to be removed with the engine in this position. Remove the valve cover.

8. Remove the camshaft sprocket bolts. Keep the chain on the sprocket and place sprocket/chain assembly between chain guide and tensioner. A center bolt is not used on the sprocket.

9. Remove the 2 bolts from the timing cover under the front of the head.

10. Remove the cylinder head bolts.

11. Raise the vehicle and support it safely. Place a support under the rear end of the engine. Remove the engine mounting bolt in the cylinder head.

12. Remove the screws in the transmission cover. Remove the cylinder head from the vehicle.

To install:

13. Be sure to use a new cylinder head gasket. Torque the bolts first to 44 ft. lbs. and then to 70 ft. lbs.

14. Make sure the markings on the camshaft and the bearing cap are aligned with one another.

15. Check that the flywheel mark is in line with the mark on the cylinder block and that the engine is set on No. 1 cylinder.

16. Install the 2 screws in the timing cover on the front of the cylinder head.

17. Install the timing chain and sprocket as follows:

a. Remove the tension from the chain tensioner with special tool 83-93-357 or equivalent.

b. Hook the tool into the catch of the tensioner and pull upwards.

c. Place the timing sprocket on the camshaft so the mark on the sprocket and the screw holes coincide. If necessary, move the chain to position.

d. Install the 3 retaining bolts in the sprocket and camshaft.

e. If the distributor is mounted to the valve cover, the rotor should be facing the line on the edge of the distributor housing.

16 VALVE ENGINE

1. Remove the hood after scribing reference marks next to the mounting bolts, to aid later installation.

2. Remove the battery.

3. Drain the coolant from the radiator and cylinder block.

4. Remove the exhaust manifold and turbo unit.

5. Remove the tensioning pulley and drive belt for the air conditioner compressor.

6. Slacken the securing bolts for the steering pump bracket, remove the drive belt and push the pump aside.

7. Undo the wiring harness clips on the cylinder head.

8. Remove the 2 bolts in the timing cover, which are screwed into the cylinder head from underneath.

9. Remove the bolts in the right-hand engine mounting which are

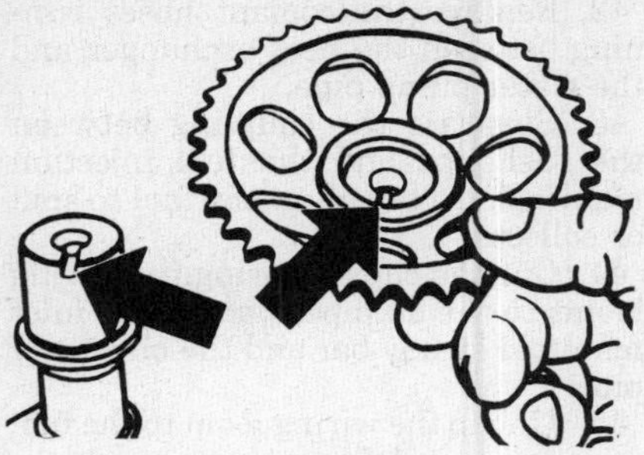

Alignment of cam gear to camshaft

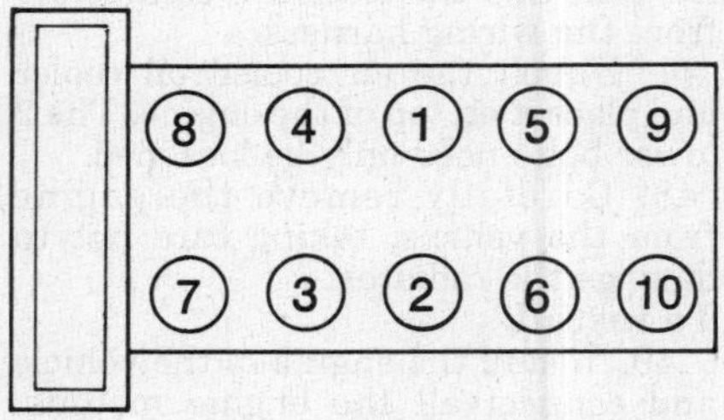

Cylinder head bolt torque sequence

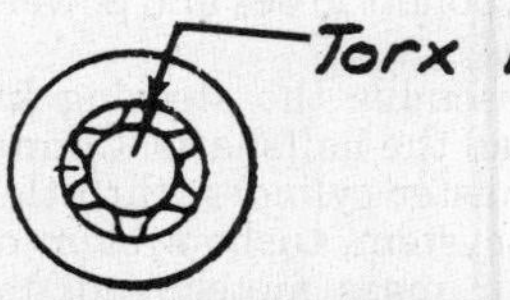

Top view of Torx head cylinder head bolts

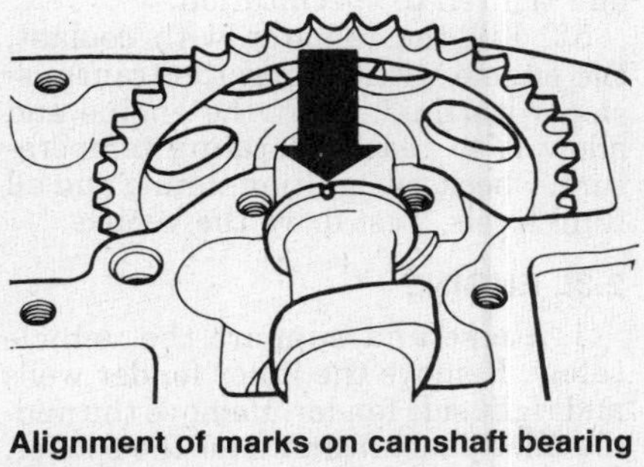

Alignment of marks on camshaft bearing caps

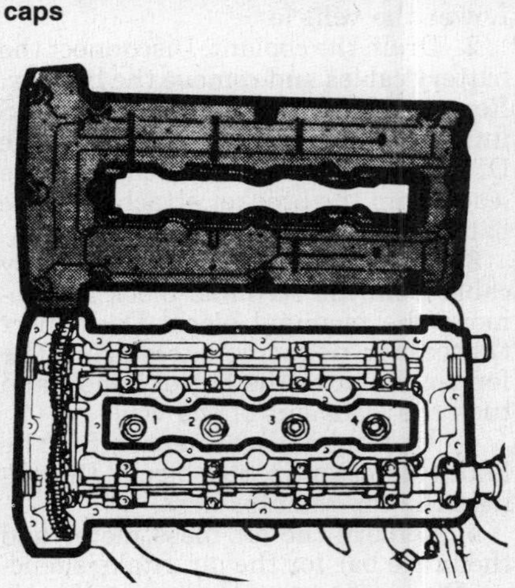

Top view of 16 valve cylinder head

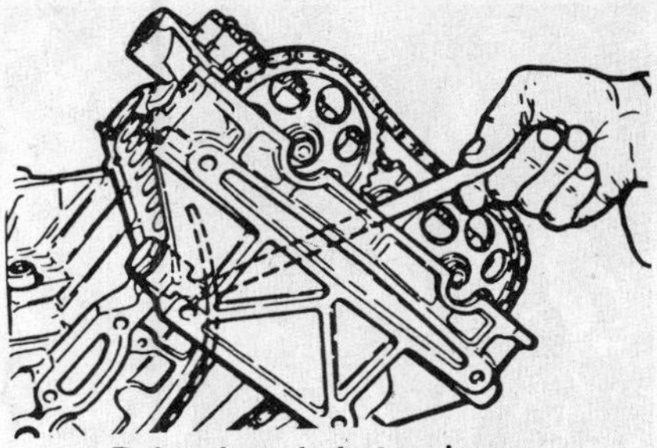

Releasing chain tensioner—16 valve engine

screwed into the cylinder head, together with the spacer sleeves.

10. Disconnect the hose between the thermostat housing and the radiator at the thermostat housing.

11. Remove the fuel pressure regulator and disconnect the ground leads for the fuel injection system.

12. Remove the AIC actuator. Remove the bracket for the air conditioning compressor from the cylinder head.

13. Remove the intake manifold complete with injectors and injection manifold.

14. Disconnect the lead from the temperature transmitter.

15. Remove the lid on the valve cover and the ignition cables together with the distributor cap.

16. Remove the valve cover. Disconnect the crankcase ventilation hose and remove the semi-circular rubber plug halves from the cylinder head.

17. Remove the air conditioning compressor and put it on the air intake for the heating system.

18. Align the timing marks on the crankshaft and camshafts. To do this, remove the cover on the transaxle bell housing which reveals the timing marks on the flywheel. Turn the engine so the **0** mark on the flywheel is aligned with the mark on the housing, or the endplate if the clutch cover has been removed. This makes certain that the pistons for No. 1 and 4 cylinders are at TDC.

19. Remove the cam chain tensioner.

20. Block up the engine to lift the cylinder head off the block. Remove the cylinder head bolts and siphon off the oil from the cylinder head.

21. Install a guide pin in one of the bolt holes and lift off the cylinder head, making sure the pivoting guide for the cam chain is not damaged.

To install:

22. Align the **0** mark on the flywheel with the timing mark on the housing. Align the marks on the camshafts with their respective timing marks.

23. Install the cylinder head gasket, making sure it is held in position by the guide sleeves in the cylinder head flange.

24. Install the guide pin (Saab special tool 83–92–128 or equivalent) and position the timing chain and pivoting guide.

25. Carefully install the cylinder head. Use the guide pin as a pivot for the head, which must be turned slightly to enable it to pass the pivoting guide. Thereafter, alignment will be determined by the guide sleeves.

26. Install the cylinder head bolts and tighten them in 3 stages. Stage 1, torque to 45 ft. lbs. evenly. Stage 2, torque to 63 ft. lbs. evenly. Stage 3, another 90 degrees (¼ turn). Retighten the bolts after the engine has reached normal operating temperature. Remember to install the 2 M8 sized bolts in the underside of the cylinder head.

27. Install the camshaft sprockets, fitting the sprocket for the exhaust cam first. Make sure the chain between the crankshaft sprocket and the camshaft sprocket is kept tight. Next install the intake cam sprocket. Keep the chain tight between the sprockets.

28. Lightly tighten the center bolts securing the camshaft sprockets. Adjust the chain tensioner and install it under tension. Tighten the bolt.

29. Release the tensioner by pressing the pivoting guide firmly against it. Thereafter, press the pivoting guide against the chain to put a basic tension on the chain.

30. Depress the pivoting guide to check that the tensioner is working. Rotate the crankshaft 2 complete turns clockwise, viewed from the transmission end. Check that the earlier settings of the crankshaft and camshaft timings have not changed. Tighten the cam sprocket bolts to 49 ft. lbs.

31. Continue the installation in the reverse order of the removal procedure.

9000 SERIES

1. Disconnect the negative battery cable. Raise and support the vehicle safely.

2. Remove the right front wheel assembly and the inner fender panel.

3. Drain the coolant. Remove the radiator expansion tank. Disconnect the steering servo reservoir and set aside. Leave the hoses attached.

4. Loosen the compressor drive belt and remove the belt.

5. Disconnect the electrical leads from the air compressor.

6. Unbolt the compressor from its mounting bracket. Disconnect the top pipe connecting on the air cooled oil cooler and push the pipe to one side. Rest the compressor on the radiator crossmember. Unbolt the compressor mounting bracket and remove it.

7. Unbolt the front exhaust pipe flange and unhook the rubber hangers.

8. Remove the steady bar for the turbo unit and the oil return pipe.

9. Disconnect the hose from the intercooler at the turbo unit. Disconnect the oil supply pipe from the turbo.

10. Disconnect the hose between the air mass meter and the turbo unit. Disconnect the coolant hose from the thermostat housing and the hose from the cylinder head.

11. Disconnect the oil supply hose or pipe so as not to obstruct the removal of the exhaust manifold. If necessary, remove the clip holding the pipe to the cylinder head and slave cylinder.

12. Unbolt and lift off the exhaust manifold complete with the turbo unit, pushing the oil supply pipe aside at the same time.

13. Disconnect the lead to the temperature transducer.

14. Remove the engine stay bracket from its attachment point on the wing.

15. Remove and remove the bolt securing the engine stay bracket to the cylinder head. Remove the intake manifold from the cylinder head.

16. Disconnect the breather hose for the crankcase ventilation from the camshaft cover.

17. Disconnect the vacuum hose and the Hall Effect transducer lead from the distributor and remove the distributor cap complete with the high tension leads.

18. Unscrew and remove the spark plug inspection plate and the clips for the high tension leads.

19. Remove the camshaft cover.

20. Align the crankshaft with the **0** timing mark and check that the camshaft timing marks also coincide. Remove the camshaft sprockets.

21. Remove the camshaft tensioner. Remove the 2 cylinder head bolts adjacent to the timing cover, which is accessible from below.

22. Disconnect the starter motor lead from the clip on the thermostat housing.

23. Remove the Torx® type cylinder head bolts.

24. Install a guide pin in the drilled hole in the right top corner of the cylinder head. Make sure the timing chain is positioned such that the pivoting chain guide will not obstruct the cylinder head and carefully lift the cylinder head from the engine block.

To install:

25. Before installation, clean both the cylinder head and the engine block surfaces. Install a new gasket. Be sure the crankshaft is aligned in the **0** position and that the camshafts are align with their respective timing marks.

NOTE: When the pistons of the No. 1 and No. 4 cylinders are at TDC, the crankshaft 0 mark on the flywheel must be align with the mark on the clutch cover or the end plate, if the clutch cover has been removed. The marks on the camshafts must be align with those on the cam bearing caps. This indicates the exhaust valves for No. 1 and No. 4 cylinders are closed.

26. Install a guide pin in the drilled hole in the top of the right corner of the cylinder head and lower the cylinder head carefully into position on the engine block. Locate the cylinder head on the guide sleeves.
27. Install the cylinder head bolts, tightening them in the correct sequence to the specified torque.
 a. Tightening sequence for the 2.0L engine:
 Stage 1–Torque to 44 ft. lbs.
 Stage 2–Torque to 67 ft. lbs.
 Stage 3–Run the engine to normal operating temperature and allow the engine to cool for 30 minutes.
 Stage 4–Slacken the bolts and retighten each bolt to 66 ft. lbs.
 Stage 5–Tighten by turning the bolts through a further 90 degrees (¼ turn).
 b. Tightening sequence for the 2.3L engine:
 Stage 1–Torque to 44 ft. lbs.
 Stage 2–Torque to 59 ft. lbs.
 Stage 3–Tighten by turning bolts through a further 90 degrees (¼ turn).
28. Position the inlet valve camshaft sprocket, followed by the exhaust valve camshaft sprocket. Be sure the chain is correctly positioned between the guides. Tighten the sprocket center bolts to 48 ft. lbs.
29. Install the timing chain tensioner. Advance the tensioner before installing it. Release the tensioner and rotate the crankshaft 2 revolutions. Make sure the camshaft and flywheel timing marks are correctly aligned.
30. Install the both halves of the split seal and the camshaft cover. Install the bolt at the distributor end and the middle bolt at the other end first. Tighten the bolts to 16 ft. lbs.
31. Check that the timing marks for the distributor rotor are aligned, Install the distributor cap and connect the lead for the Hall Effect transducer. Connect all vacuum hoses.
32. Connect the high tension leads to the spark plugs. Secure the leads in the clips. Install the inspection plate and tighten the retaining screws.
33. Install the clip securing the starter motor lead to the thermostat housing.
34. Install a new gasket on the inlet manifold and install the manifold in place. Install the top securing bolts first and then install the lower bolts, using an extension bar.
35. Install the bolt for the engine stay bracket to the cylinder head and position the stay bracket in place. Install a new gasket onto the exhaust manifold and position the exhaust manifold to the cylinder head.
36. Install the oil supply pipe. Install the clip and the slave cylinder bolt. Install the oil return line and the steady bar for the turbo unit.
37. Connect the hose between the turbo unit and the intercooler. Connect the cooler hose to the thermostat housing and the hose to the cylinder head.
38. Install the air mass meter socket connector into the turbo unit and tighten the clip. Connect the hose between the intercooler and the turbo unit.
39. Install and tighten the nuts securing the front section of the exhaust pipe to the turbo compressor. Bolt the air conditioing compressor mounting bracket onto the cylinder head and engine block.
40. Install the air conditioning compressor. Leave the coolant hose in the bracket when installing the compressor.
41. Connect the electrical leads and make sure the lead is clear of the compressor pulley. Install the steering servo reservoir. Install the coolant expansion tank and tighten the hose clip.
42. Connect the top pipe to the air cooled oil cooler and secure the cooler to the radiator. Install the overflow line between the expansion tank and the radiator.
43. Install the compressor belt, adjust the tension and tighten the belt tensioner bolt. Install the inner right wheel arch, and install the wheel.
44. Lower the vehicle and tighten the wheel. Connect the negative battery cable and fill the cooling system with coolant. Start the engine and test the engine operation.

Valve Lash

ADJUSTMENT

8 Valve Engine

1. Remove the valve cover. The pistons of cylinders No's 1 and 4 must be at TDC before distributor and valve cover can be removed.
2. Using an special Saab tool 8392185 or equivalent, rotate the crankshaft as necessary to position the high point of the camshaft lobe 180 degrees away from the valve depressor face, base circle of the cam lobe must contact the valve depressor, on the valve which the clearance is to be checked.

Checking valve clearance with feeler gauge

Valves	inches
Intake	0.008–0.010
Exhaust (except turbo)	0.0l6–0.018
Exhaust (turbo)	0.0l8–0.020

NOTE: The special crankshaft turning wrench fits the center screw of the crankshaft belt pulley at the dash panel.

3. Check the maximum and minimum clearances using a feeler gauge. The minimum feeler gauge should slip in, but the maximum feeler gauge should not.
4. Measure and record the clearance of all the valves in the same manner. Adjust the clearance of any valves that are not within specification.
5. To adjust the valves, remove the camshaft, tappets and adjusting pallets (shims) of any valves that need to be adjusted.
6. Using a micrometer, measure and record the thickness of the pallet (shim). This thickness plus the valve clearance adds up to the total distance between the valve and the cam.
7. The choice of the adjusting pallet (shim) is determined by the measured total distance between the valve depressor (tappet) and the cam, less the specified valve clearance for an intake or exhaust valve as the case may be.
8. Insert the new adjusting pallet (shim) an the valve depressor (tappet) and reinstall the camshaft.
9. Repeat the measurement procedure to insure that the clearances are correct.
10. Install the valve cover using a new valve cover gasket.

16 Valve Engine

Correct valve clearance is critical to the proper functioning of the engine. On the 2.3L engine, the valve clearance measurement is the distance between the tip of the valve stem and the camshaft bearing seat. Special tool 83-

93-753, or equivalent, must be used to ensure an accurate measurement.

1. Disconnect the negative battery cable.

2. Remove the camshafts and followers.

3. Place tool 83-93-753, or equivalent, across the camshaft bearing seats. Line up the instrument to read the depth to the tip of the valve stem.

4. Check that when the instrument is displaying the maximum depth reading of 0.807 in. (20.5 mm) it actually reaches the tip of the valve stem. This can be verified by noting a small clearance between the tool and the bottom of the bearing bore closest to the valve.

5. Check that instrument does not touch the tip of the valve stem when showing the minimum depth reading of 0.768 in. (19.5 mm). The valve clearance is correct when the reading obtained is between the minimum and maximum values.

6. If the valve clearance deviates from the specified checking values, adjustment must be made to the valve stem or the valve seat. This requires removal of the head.

7. After correct valve clearance is obtained, reinstall the camshafts and followers.

8. Connect the negative battery cable. Start the engine, check the timing and test drive the vehicle.

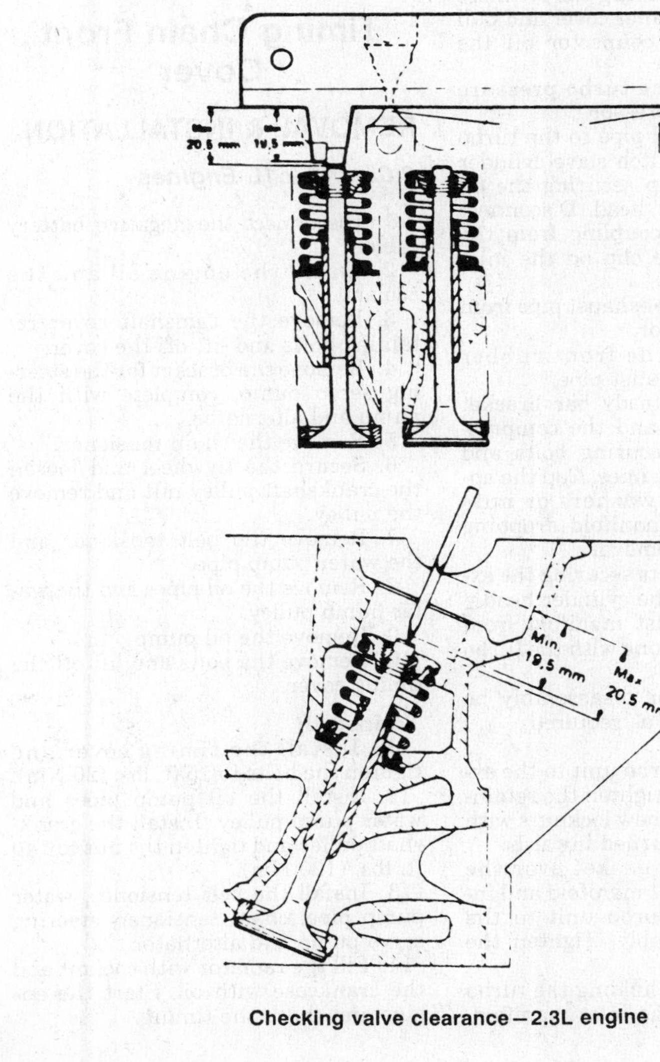

Checking valve clearance – 2.3L engine

Intake Manifold

REMOVAL & INSTALLATION

8 Valve Engine

1. Disconnect the negative battery cable. Disconnect all hoses, wires and connectors that would inhibit the intake manifold from being removed.

2. It may be necessary to remove the distributor cap and the ignition wires to gain clearance. Remove the throttle valve housing.

3. Remove the intake manifold retaining bolts. Remove the intake manifold from the engine.

To install:

4. Installation is the reverse of the removal procedure. Be sure the proper gasket is used. A coolant leakage could occur if the wrong one is used.

5. Install intake manifold and tighten bolts to 13 ft. lbs. (18 Nm).

16 Valve Engine

1. Disconnect the negative battery cable. Disconnect all hoses, wires and connectors that would inhibit the intake manifold being removed.

2. Remove the turbo pressure pipe, the lubricating oil pressure pipe and the return oil pipe.

3. Remove the intake manifold retaining bolts. Remove the intake manifold along with the injection manifold, injectors and the AIC regulator.

To install:

4. Installation is the reverse of the removal procedure. Be sure the proper gasket is used. A coolant leak could occur if the wrong one is used.

5. Install intake manifold and tighten bolts to 16 ft. lbs. (22 Nm).

Exhaust Manifold

REMOVAL & INSTALLATION

1. Disconnect the negative battery cable. Disconnect all necessary hoses, wires, and connectors that would inhibit the exhaust manifold from being removed.

2. Unbolt the exhaust pipe at the connecting flange.

3. If equipped with a heat shield, remove it.

4. Remove the exhaust manifold bolts. Remove the exhaust manifold from the vehicle.

To install:

5. Install the exhaust manifold. Apply an antiseize compound to the manifold bolts and tighten to 18 ft. lbs. (25 Nm) for turbo engines and 13 ft. lbs. (18 Nm) for all others.Installation is the reverse of removal.

6. Install the heat shield and connect all wires and hoses previously disconnected. Connect the negative battery cable.

Turbocharger

REMOVAL & INSTALLATION

900 Series

1. Disconnect the negative battery cable. Remove the charge pressure regulator and block off the exhaust pipe. Remove the battery as required. Remove the tension on compressor belt.

2. Disconnect the hose between the compressor and the throttle housing.

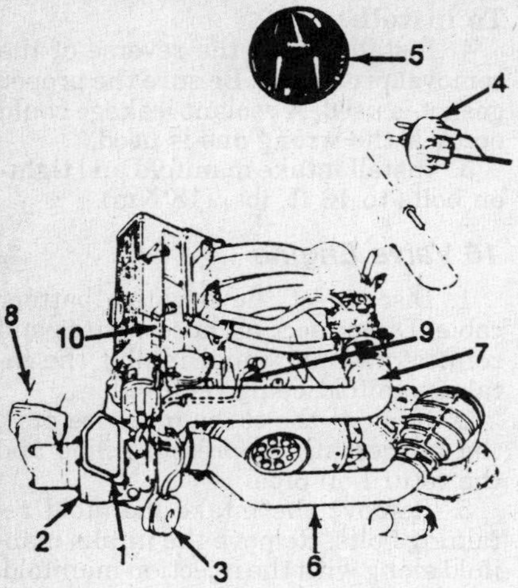

1. Turbocharger
2. Wastegate boost control
3. Diaphragm capsule
4. Over-pressure guard
5. Turbo gauge
6. Hose, air cleaner to turbocharger
7. Hose, turbocharger to inlet manifold
8. Exhaust outlet pipe
9. Oil supply line
10. Oil return line

Eight valve engine turbocharger assembly

3. Disconnect the oil supply line and the oil return line at the turbo unit.

4. Remove the retaining bolts securing the turbo to the exhaust manifold. Remove the turbo unit from the vehicle. Plug the holes in the turbo unit to prevent dirt from entering.

To install:

5. Installation is the reverse of the removal procedure.

6. Fill the lubricating inflow of the turbo unit with engine oil before connecting the oil return line at the turbo.

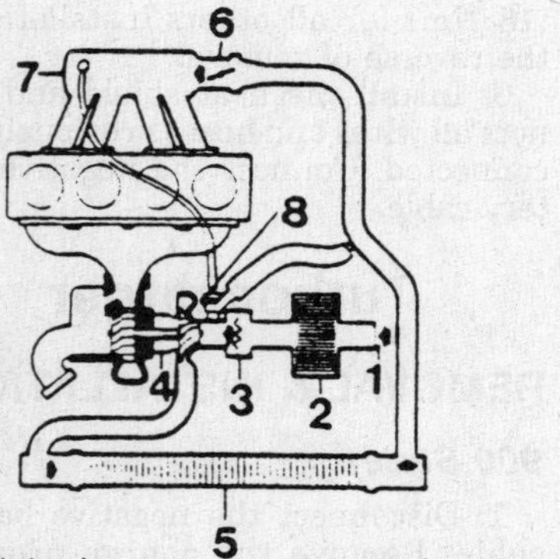

1. Air intake
2. Air cleaner
3. Air massmeter
4. Turbo unit
5. Intercooler
6. Throttle housing
7. Inlet manifold
8. Relief valve

Sixteen valve engine turbocharger assembly

7. Crank the engine for about 30 seconds with terminal 15 on the ignition coil disconnected. This will fill the lubricating system of the turbo before the engine is started.

9000 Series

1. Disconnect the negative battery cable. Release the tension on the compressor belt by slackening the belt tensioner.

2. Disconnect the top pipe coupling on the air cooled oil cooler and disconnect the clips securing the pipe to the radiator.

3. Remove the compressor mounting bolts. Insert a sheet of metal to protect the oil cooler and lift the compressor towards the expansion tank.

4. Remove the solenoid valve from its mounting on the radiator and disconnect the electrical leads.

5. Disconnect the electrical leads at the radiator fan. Unbolt and remove the fan.

6. Unplug the electrical connectors for the air mass meter. Disconnect the toggle fasteners securing the air mass meter to the air cleaner cover and pull the rubber socket connector off the turbo unit.

7. Disconnect the turbo pressure pipe from the compressor.

8. Remove the oil pipe to the turbo unit. Unbolt the clutch slave cylinder and remove the clip securing the oil pipe to the cylinder head. Disconnect the oil pipe banjo coupling from the block and undo the clip on the inlet manifold.

9. Disconnect the exhaust pipe from the turbo compressor.

10. Disconnect the front rubber hangers for the exhaust pipe.

11. Remove the steady bar bracket between the sump and the compressor. Remove the securing bolts and loosen the oil return lines. Cap the aperture to prevent washers or nuts from the exhaust manifold dropping inside during the removal.

12. Remove the nuts securing the exhaust manifold to the cylinder head.

13. Lift the exhaust manifold from the cylinder head, along with the turbo unit.

14. Should further disassembly be necessary, complete as required.

To install:

15. Position the turbo unit to the exhaust manifold and tighten the retaining nuts. Install the new locknuts with the locking flange turned inwards.

16. Install a new gasket over the studs for the exhaust manifold and install the manifold/turbo unit to the cylinder head assembly. Tighten the nuts to 30 ft. lbs.

17. Install the clip holding the turbo oil supply pipe to the inlet manifold. Connect and tighten the banjo coupling to the engine block. Make sure the copper washers are in good condition. Secure the pipe to the turbo unit.

18. Install the return oil pipe and the steady bar bracket between the turbo unit and the crankcase. Connect the rubber hangers for the front exhaust hanger.

19. Bolt the exhaust pipe to the turbo compressor. Use new locking nuts with the locking flanges turned outward. Tighten to 19 ft. lbs.

20. Install the turbo pressure pipe to the compressor and assemble the air mass meter and rubber socket connector between the air cleaner body and the inlet side of the turbo compressor.

21. Assemble the fan and solenoid valve, securing the electrical leads into their clips. Connect the return hose to the solenoid valve. Insert a piece of metal to protect the oil cooler and install the air conditioning compressor.

22. Reconnect the oil pipe to the oil cooler and secure the pipe clip to the radiator. Install the compressor belt and tighten it to specification.

Timing Chain Front Cover

REMOVAL & INSTALLATION

2.0L and 2.1L Engines

1. Disconnect the negative battery cable.

2. Drain the engine oil and the coolant.

3. Remove the camshaft cover retaining bolts and lift off the cover.

4. Remove the bracket for the steering servo pump, complete with the pump and alternator.

5. Remove the chain tensioner.

6. Secure the flywheel and loosen the crankshaft pulley nut and remove the pulley.

7. Remove the belt tensioner and the water pump pipe.

8. Remove the oil pipes and the water pump pulley.

9. Remove the oil pump.

10. Remove the bolts and lift off the timing cover.

To install:

11. Install the timing cover and tighten the bolts to 15 ft. lbs. (20 Nm).

12. Install the oil pump pipes and water pump pulley. Install the crankshaft pulley and tighten the nut to 140 ft. lbs. (190 Nm).

13. Install the belt tensioner, water pump pipe, chain tensioner, steering servo pump and alternator.

14. Fill the radiator with coolant and the crankcase with oil. Start the engine and check the timing.

2.3L Engine

1. Disconnect the negative battery cable.
2. Lock the flywheel using tool 83-93-993, or equavilent.
3. Raise and support the vehicle safely. Drain the coolant and the oil. Remove the right front wheel and wheel well.
4. Remove the serpentine belt and belt tensioner. Remove the tie bar between the wheel arch and the subframe.
5. Remove the steering servo pump, pump bracket and the alternator. Remove the top engine mounting bracket. Remove the torque arm.
6. Remove the engine mounting bracket, top belt tensioner bracket, air conditioner compressor and compressor bracket. Use a suitable rigid cover to prevent the oil cooler from being damaged.
7. Disconnect the coolant hoses and remove the water pump. Remove the crankshaft pulley and swivel the crankshaft sensor out of the way.
8. Move the coolant pipe aside and remove the oil pan. Remove the timing cover securing bolts. Note the locations of all bolts as they are of different lengths. Remove the bolts securing the timing cover to the cylinder head. Remove the timing cover.

To install:

9. Remove all traces of old sealant front the cover. Apply a 1 mm bead of anaerobic sealant to the flanges of the cover. Use sealant sparingly as excess sealant can get into the oilways and do serious damage to the engine.
10. Install the timing cover taking care not to damage the head gasket. Install the bolts in their correct positions and tighten to 15 ft. lbs. (20 Nm).
11. Install the oil pan and tighten bolts to 15 ft. lbs. (20 Nm). Install the coolant pipe and secure the crankshaft sensor.
12. Install the engine mountings, water pump and cooling hoses, air conditioner compressor, steering servo pump, pump bracket and the alternator. Install the top engine mounting bracket and torque arm.
13. Install the serpentine belt and belt tensioner. Remove the tie bar between the wheel arch and the subframe.
14. Install the wheel well and wheel. Fill the radiator with coolant and the engine with oil. Connect the battery. Start the engine and allow it to reach normal operating temperature. Check for leaks.

Front Cover Oil Seal

REPLACEMENT

8 Valve Engine

1. Disconnect the negative battery cable. Remove the alternator belt. If equipped with power steering or air conditioning, remove the required belts.
2. Remove the clutch cover (torque converter cover) and lock the crankshaft using Saab tool 83–92–987 or equivalent by locking the tool to the ring gear.
3. From under the vehicle, remove the pulley retaining bolt using Saab tool 83–92–961 or equivalent. Remove the pulley from the vehicle.
4. Pull off the old seal ring using a suitable tool.
5. Installation is the reverse of removal. Torque the retaining bolt to 137 ft. lbs.

16 Valve Engine

1. Disconnect the negative battery cable. Raise and support the vehicle safely.
2. Remove the right front wheel and tire assembly. remove the inner front fender panel.
3. Loosen and remove the drive belts.
4. Remove the retaining bolt for the crankshaft pulley.
5. Remove the crankshaft pulley.
6. Using a prybar, carefully remove the oil seal without marring the crankshaft stub end.
7. Install a new, oiled seal, using an appropriate seal installer.
8. Install the pulley and tighten the retaining bolt to 134 ft. lbs. (180 Nm).
9. Tighten the drive belts, using a belt tension gauge. (New belt – 180 lbs.; Used belt – 120 lbs.)
10. Install the forward section of the inner fender panel.
11. Replace the front wheel and lower the vehicle.

Timing Chain and Sprockets

REMOVAL & INSTALLATION

8 Valve Engine

1. Disconnect the negative battery cable. Remove the engine from the vehicle.
2. Remove the distributor cap and ignition wires. Rotate the engine until cylinders No. 1 and No. 4 are at TDC. This must be done due to the design of the distributor driving dog which only allows the valve cover/distributor assembly to be removed with the engine in this position.
3. Remove the valve cover assembly. Remove the sprocket from the camshaft and rest it on the chain tensioner and the chain guide.
4. Remove the cylinder head. Remove the crankshaft pulley and oil pump assembly. Remove the water pump and pulley assembly.
5. Remove the timing chain cover. Remove the timing chain and chain wheel from the engine. Remove the chain tensioner, if required.

NOTE: Engines beginning with engine No. E57340, are equipped with a new cam chain tensioner that requires a different release procedure. The new tensioner is also a direct replacement for the old style unit and may be used in all engines. A complete tensioner kit, consisting of the tensioner body and guide, must be used. To release the pressure on the cam chain, pivot the reverse latch on the tensioner body with a suitable prying tool. This will allow movement of the chain guide from point A to point B. When reinstalling the cam sprocket to the camshaft, the reverse latch must again be pivoted to release pressure on the chain guide.

To install:

6. To install the chain assembly, have the No. 1 piston at TDC and the camshaft in position for No. 1 cylinder firing position to be in its firing mode, before the cylinder head is installed.
7. Do not rotate either the camshaft or the crankshaft without the chain in place. Damage to the valves or pistons can occur, after the cylinder head is installed.
8. Replace the chain tensioner, if removed.
9. Place the camshaft sprocket to the chain and suspend it from the crankshaft sprocket. Position the chain between the chain guide and the tensioner.
10. Install the timing chain cover assembly while pulling up the chain to avoid being caught under the cover.
11. Install the water pump assembly and install the cylinder head. Torque the head bolts to specification.
12. Using the tensioner release tool, disengage the tensioner and install the cam sprocket and chain to the camshaft. Align the marks on the sprocket and the camshaft bearing.
13. Install the sprocket retaining bolts. Release the tensioner assembly. Install the oil pump assembly, seal and pulley.
14. Continue the installation as required. Do not use an early type inlet manifold gasket as coolant leakage could occur within the engine.

16 Valve Engine

1. Disconnect the negative battery cable. Remove the engine from the vehicle.
2. Remove the lid on the valve cover and remove the ignition wires. Remove the valve cover.

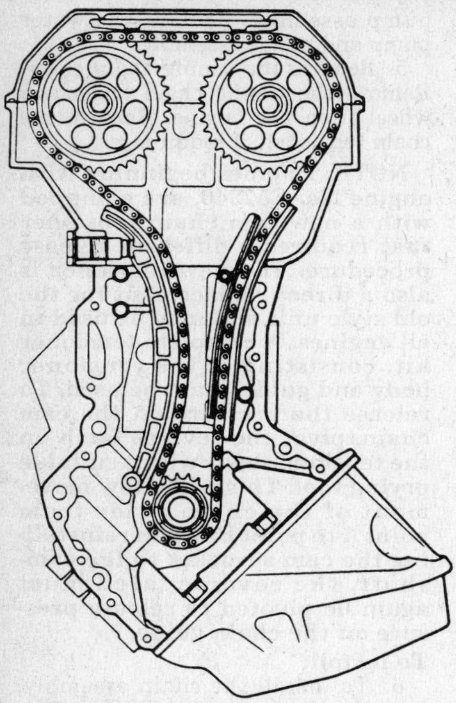

Timing mark location – 16 valve engine

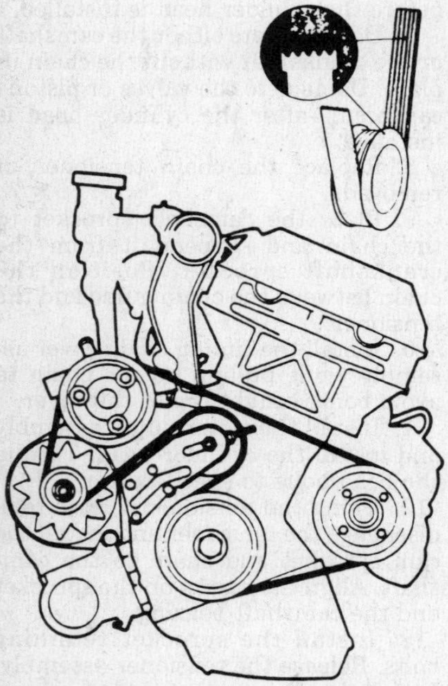

Serpentine drive belt adjustment

3. Position the crankshaft for TDC, with the 0 mark on the flywheel align with the timing mark on the transaxle end plate. These marks must be aligned before the timing chain is removed.

4. Remove the crankshaft pulley using a puller. Remove the water pump, located behind the crankshaft pulley. Remove the timing cover, 2 bolts of which are screwed into the underside of the cylinder head.

5. The cam chain and crankshaft timing sprocket should now both be visible. From above, release the timing chain tensioner by pressing the pivoting guide firmly against it. Remove the chain tensioner.

6. Using a special tool to hold the camshafts, remove the center bolts securing the camshaft sprockets. Throughout this procedure, keep the camshafts in their basic correct setting. If they are rotated out of position at any stage, especially without their sprockets and chain, the valves can be damaged.

7. Disconnect the timing chain from the sprockets and remove the chain, clearing it from the crankshaft sprockets.

To install:

8. To install the timing chain, place the chain around the crankshaft sprocket. Run the chain up through the opening in the cylinder head if not already done. Install the chain and sprocket on the exhaust cam first. Make sure the chain is taut between the crankshaft and camshaft sprockets. Install the bolts but do not tighten.

9. Install the chain and sprocket to the intake cam. Keep the chain taut between the cam sprockets while it is being installed. Install the bolts but do not tighten. Make sure the chain is seated in the guide tensioner grooves.

10. Tension the chain tensioner by fully depressing the piston and then rotating it to the locked position.

11. Install the chain tensioner with the piston under tension. Make sure the copper gasket is in good condition and that the sealing surface is clean and free from burrs.

12. Trigger the chain tensioner by pressing the pivoting chain guide against it, thereafter, press the pivoting guide against the chain to give the chain its basic tension. Check that the chain tensioner maintains tension on the chain when the pressure on the chain guide is released and that the basic setting stop for the tensioner holds the chain guide tight against the chain. A limited amount of play will be present until the hydraulic pressure takes over once the engine is running.

13. Check the setting by rotating the crankshaft 2 complete turns in its normal direction of rotation around to the timing mark. The basic setting of the cams should remain unaltered.

14. Lock the exhaust cam by using a wrench on the cast hex bolt and torque the sprocket bolt to 48 ft. lbs. Repeat this on the intake cam.

15. Complete the procedure on the intake cam sprocket. When loosening or torquing the sprocket center bolts, hold the cam still using a wrench installed over the flats on the camshaft. The accuracy ðf the timing chain adjustment will depend on the condition of the chain.

Camshaft

REMOVAL & INSTALLATION

8 Valve Engine

1. Disconnect the negative battery cable. Remove the distributor cap and ignition wires. Rotate the engine until cylinders No. 1 and No. 4 are at TDC. This must be done due to the design of the distributor driving dog which only allows the valve cover/distributor assembly to be removed with the engine in this position.

2. Remove the valve cover assembly.

3. Be sure both the crankshaft and camshaft are at the No. 1 cylinder firing mode and the indexing lines still are aligned.

4. Remove the camshaft sprocket, keeping the chain on the sprocket, Place the sprocket between the chain guide and tensioner.

5. Remove the camshaft bearing caps and lift the camshaft from the bearing assembly housing. The bearing assembly housing can then be removed, if necessary.

6. Installation is the reverse of the removal procedure. Be sure the timing marks are aligned.

16 Valve Engine

1. Disconnect the negative battery cable. Remove the engine from the vehicle.

2. Remove the lid on the valve cover. Disconnect the spark plug wires and vacuum hose from the distributor and remove the distributor cap.

3. Remove the valve cover and position the crankshaft for TDC. The 0 mark on the flywheel should align with the timing mark on the bell housing end plate.

4. Remove the distributor. Remove the oil pipe.

5. Remove the center bolts securing the camshaft sprockets. Use a proper holding tool to hold the camshafts from rotating. Always keep the camshafts in their correct basic setting. If the setting of the crankshaft or camshafts is altered at this stage the valves can be damaged.

6. Remove the camshaft timing chain tensioner. Remove the camshaft sprockets.

7. Remove the camshaft bearing caps. Keep them in correct order for later reassembly. Lift out the camshafts.

8. Installation is the reverse the removal procedure. When installing, the bearing caps marked 1–5 belong to the intake cam, while those marked 6–10 go with the exhaust cam. Torque the bearing cap bolts to 11 ft. lbs.

Balance Shafts

REMOVAL & INSTALLATION

1. Disconnect the negative battery cable.
2. Remove the timing cover noting the different bolt lengths.
3. Remove the top guide from the balance shaft chain.
4. Remove the chain tensioner and pivoting guide. Remove the idler wheel sprocket and the balance shaft chain.
5. Remove the oil pump drive dog and the sprocket from the crankshaft. Remove the pivoting chain guide for the timing chain.
6. Remove the fixed chain guide for both the timing and balance shaft chains. Remove the timing chain guard followed by the chain and sprocket.
7. Remove the balance shafts taking care not to damage the inner bearing shells.

To install:

8. Rotate the crankshaft to bring pistons No. 1 and 4 to TDC. Ensure that the timing cover flange is absolutely clean.
9. Lubricate the balance shaft journals and bearing housings. Insert the balance shafts into their respective tunnels, taking care not to damage the inner bearing shells. Tighten the bearing bolts to 9 ft. lbs. (12 Nm).

Note: The shaft with the smaller thrust ring, marked INL, is the one for the inlet side of the engine. The shaft with the larger thrust ring, marked EXH, is the one for the exhaust side of the engine.

10. Install the chain and sprocket on the crankshaft. Install the chain guard. Install the chain guide that serves both the timing and balance shaft chains.
11. Install the pivoting guide for the timing chain and the oil pump drive dog.
12. Install the balance shaft chain and idler wheel sprocket, ensuring that the timing marks on the bearing housing and sprocket ar in line. When installing, leave some slack in the chain in line with the tensioner, and keep the chain reasonably taut by means of the top chain guide.

NOTE: There is an alternate way of installing the balance shaft chain. Install the top chain guide first and then adjust the run of the chain around the sprockets. Adjusting the chain is easier this way, although it will be more awkward to install the idler wheel sprocket.

13. Cock the chain tensioner and insert a paper clip through the hole in the cylinder to prevent the tensioner being triggered. Before installing the tensioner, make sure that the plunger is turned to the position in which the spring acts fully on it.
14. Install the pivoting chain guide and the tensioner. Tighten the tensioner to 9 ft. lbs. (12 Nm). Install the top chain guide and trigger the tensioner by removing the paper clip.
15. Rotate the crankshaft a few times to ensure the chain is installed correctly.
16. Install the timing cover. Connect the negative battery cable. Start the engine, check the timing and test drive the vehicle.

Piston and Connecting Rod

POSITIONING

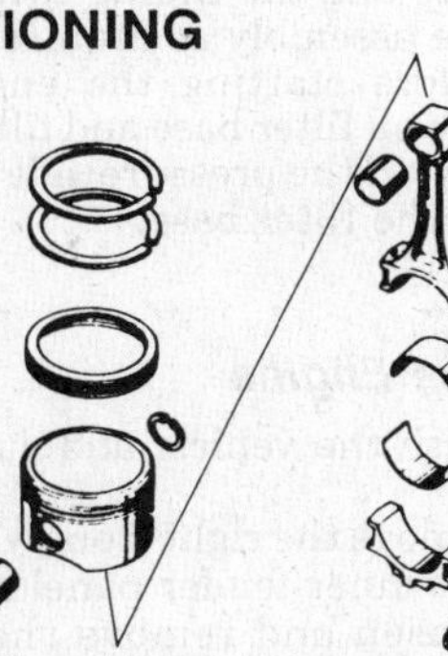

Piston and connecting rod assembly

ENGINE LUBRICATION

Oil Pan

REMOVAL & INSTALLATION

900 Series

On 900 Series vehicles the top of the transmission is used as the oil pan (sump). To remove the oil pan, remove the engine and transmission as an assembly. Then, separate the engine and transmission once out of the vehicle.

9000 Series

2.0L ENGINE

1. Disconnect the negative battery cable and drain the engine oil.
2. Remove the right front wheel and fender liner.
3. Remove the nut front the front and rear engine mount, and remove the bolt securing the torque arm to the top of the engine. Attach a engine lift and raise the engine off its mounts.
4. Remove the bottom bolt and the stud securing the sump to the transmission case. Remove the bracket for the rear engine mount and brace bar.
5. Remove the dipstick tube and unplug the oil level sensor on the pan. Remove the sensor.
6. Remove the bracket and oil return pipe for the turbo unit, if equipped. Fold down the edge of the splash plate and remove the 2 rubber plugs in the back of the transmission case. Remove the 2 bolts securing the sump to the block.
7. Remove the outer filler panel on the right side. Loosen the securing bolt for the right wheel well brace bar.
8. Remove the 2 bolts in the subframe mount. Pry the subframe away from its front mounting and insert a block of wood about 1.0 in. (3 cm) thick.
9. Remove the bolts securing the oil pan to the block and remove the pan.

To install:

10. Throughly clean the flanges on the pan and block. Apply Permatex® Ultra Blue Sealant, or equivalent, along the groove in the edge of the sump flange.
11. Install the oil pan and bolts. Tighten the bolts, starting in the middle, to 15 ft. lbs. (20 Nm). Install the 2 rubber plugs in the transmission case and reinstall the splash plate to its original position.
12. Remove the block of wood and install the bolts securing the subframe to the front mount. Tighten the attaching bolts for the wheel well liner brace bar.
13. Install the bracket and oil return pipe for the turbo unit. Make sure that the seal on the pipe is in place.
14. Install the dipstick tube complete with seal. Make sure that the rube is secured properly in the bracket at the top. Install the oil level sensor and plug on the connector.
15. Bolt the bracket and brace bar for the rear engine mounting into position. Install the bottom bolt and stud holding the transmission and oil pan together.
16. Lower the engine into position and install the torque arm. Install the front section of the exhaust pupe and the oxygen sensor. Install the front and rear engine mounts.
17. Install the filler panel, inner wheel well and wheel. Fill the engine with oil and run to normal operating temperature to check for leaks.

2.3L ENGINE

1. Disconnect the negative battery cable. Remove the oil dipstick, raise the vehicle and drain the oil.
2. Remove the right front wheel and inner wheel well. Remove the bolts in the front and rear engine mounts.
3. Remove the oxygen sensor and the front section of the exhaust pipe. Lower the vehicle.
4. Remove the tie rod between the wheel well and the subframe. Attach an engine crane and lift the engine slightly.
5. Remove the bottom bolt holding the transmission to the oil pan. Unplug the connector for the oil level sensor and remove the sensor.
6. Fold down the edge of the splash plate and remove the 2 rubber plugs in the back of the transmission case. Remove the 2 bolts securing the oil pan to the block under the plugs.
7. Remove the remaining oil pan bolts and using a drift, tap the guide sleeve into the block. Remove the oil pan from the back first.

To install:

8. Throughly clean the flanges on the sump and block using solvent. Applyu an even bead of Permatex® Ultra Blue sealant, or equivalent, along the oil pan flange.
9. Install the rubber seal for the oil strainer in the groove on the oil pan. Install the oil pan, front edge first and then the back. Install the bolts loosely. Tighten the bolts to 15 ft. lbs. (20 Nm) starting in the middle.

NOTE: The longer bolt with washer is installed on the right side.

10. Install the 2 rubber plugs in the back of the transmission and return the edge of the splash plate to its original position. Install the bolt securing the oil pan to the transmission case at the bottom and install the oil level sensor.
11. Align the engine over the mounts and lower it into position. Install the tie rod between the wheel well and the subframe. Install the dipstick.
12. Install the bolts in the front and rear engine mounts. Install the oxygen sensor and exhaust pipe.
13. Install the inner wheel well and wheel. Lower the vehicle, fill with oil and run the engine to normal operating temperature to check for leaks.

Oil Pump

REMOVAL & INSTALLATION

8 Valve Engine

The oil pump is a gear type pump and is driven by the crankshaft. It is positioned between the timing cover and crankshaft pulley. The pump assembly can be removed with the engine in the vehicle.

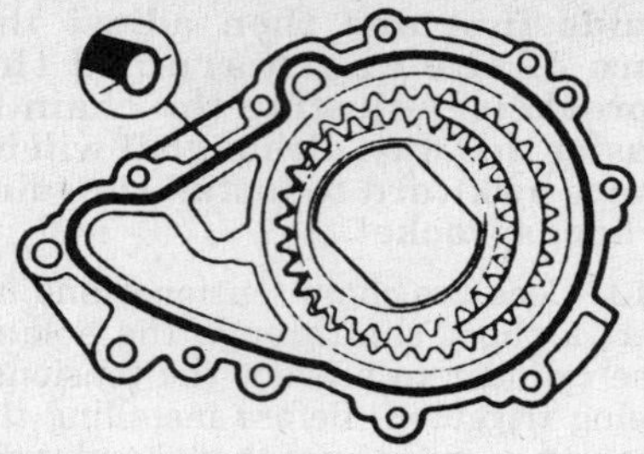

Oil pump cross-section—16 valve engine

1. Disconnect the negative battery cable. Remove the crankshaft pulley.
2. Lock the crankshaft in place by using a flywheel locking bracket.
3. Remove the oil pump retaining bolts. Remove the oil pump from the timing cover.

To install:

4. Before installation, prime the pump assembly and be sure the mark on the outer gear is visible.
5. Install a new gasket and install the pump and the timing cover. Complete the assembly as required.
6. Before starting the engine, remove the oil filter base and fill the passage way on the pressure side with oil. Replace the filter base.

16 Valve Engine

1. Raise the vehicle and support it safely.
2. Remove the right front wheel and the front inner fender panel section.
3. Loosen and remove the multi-groove belt. Loosen the compressor drive belt. Remove the idler pulley on 2.3L engines.
4. Remove the crankshaft pulley. On 2.3L engines, remove the crankshaft sensor bolts. It may be necessary to hold the crankshaft while removing the pulley bolt.
5. Remove the oil pump cover retaining bolts. Remove the oil pump taking care not to loose the spring.

To install:

6. Before installing the pump, install a new O-ring seal.
7. Install the pump to the engine and install the retaining bolts. Tighten the bolts to 6 ft. lbs. (8 Nm).
8. Install the pulley, crankshaft sensor, idler pulley and drive belts, as necessary. Tighten the pulley bolt to 140 ft. lbs. torque. Tighten the new drive belt to 180 lbs. strand tension or a used belt to 120 lbs. strand tension.
9. Install the inner fender panel and the right front wheel. Lower the vehicle and check oil pump operation.

Rear Main Bearing Oil Seal

REMOVAL & INSTALLATION

This seal is otherwise known as the crankshaft seal at the flywheel end. The seal can be changed with the engine in the vehicle, but the clutch and flywheel must first be removed.

1. Remove the transmission and the flywheel from the vehicle.
2. Pry the old seal ring from the crankshaft an appropriate tool. Take care not to damage the crankshaft or sealing flange.
3. Install the new seal with the spring ring turned inwards toward the crankshaft using tool 83-92-540, or equivalent.
4. Install the clutch and flywheel.

ENGINE COOLING

Radiator

REMOVAL & INSTALLATION

1. Disconnect the negative battery cable. Drain the radiator. As required, remove the radiator grille. As required disconnect and plug the transmission lines.
2. Disconnect the hoses from the radiator. Disconnect the electrical leads to the radiator fan and the auxiliary fan, if equipped.
3. Disconnect the electrical lead to the thermal switch and solenoid valve. Remove the ignition coil and solenoid valve from the bracket. Remove the oil cooler.
4. Remove the 2 bolts from the upper radiator support. Lift the radiator out of the vehicle by pulling the top of the radiator slightly backwards.
5. Installation is the reverse of the removal procedure.

Heater Core

REMOVAL & INSTALLATION

900 Series

1. Disconnect the negative battery cable. Remove the dash panel under the switches on the steering column and the lower section of the instrument panel.
2. Remove the air diffuser and retaining screws.

3. Remove the left defroster and speaker grille.
4. Remove the control rod from between the coolant shut off valve and the control rod by sliding the rod as far forward as it will go to free it from the knob, then pull it rearward to free it from the shut off valve.

NOTE: The plastic joint at the control knob is accessible from underneath once the switches below the heater controls have been moved backward.

5. Remove the lower section of the heater housing.
6. Drain the coolant and disconnect the hoses. Plug the ends of the hoses to prevent coolant from leaking into the compartment.
7. Separate the heater core from the housing and guide it backward and downward. It will be necessary to disconnect the brake pedal return spring and depress the brake pedal slightly.
8. The water valve and the heater core can be separated after their removal. Do not kink or break the capillary tube.
9. Installation is the reverse of the removal procedure.

9000 Series

1. Disconnect the negative battery cable. Remove the hood assembly.
2. Disconnect the wiper arms. Remove the covers on the evaporator and wiper motor. Unplug the connector for the fan control unit on vehicles with automatic climate control.
3. Remove the false fire wall panel. Drain the radiator.
4. Remove the plastic drainage tube moulding below the windshield moulding.
5. Remove the securing bolts the electronic ignition control unit and position it aside.
6. Remove the clip and unplug the connectors. Remove the complete wiper assembly.
7. Remove the rubber lead through panel for the coolant hoses. Drain cooling system. Disconnect the quick release couplings for the coolant hoses at the heat exchanger.
8. Remove the throttle dashpot assembly.
9. Remove the vacuum pump retaining screws. Position the pump aside.
10. Remove the evaporator body retaining screws and the clips for the refrigerant hoses.
11. Remove the lock washer and disconnect the cable for the temperature valve.
12. Carefully lift the evaporator and remove the clips on either side of the fan. Remove the complete fan assembly by twisting the fan diagonally upwards.
13. Remove the screw in the center of the casing. Release the clips and the grille at the discharge duct.
14. Separate the fan housing and undo the securing screw for the fan motor.
15. Lift the cover upward and withdraw the motor complete with the impeller.
16. Release the retaining clips and disconnect the hoses from the heater core.
17. Pull the heater core from the engine side of the fire wall.

To install:

18. Install the heater core, attach the retaining clips and connect the heater hoses.
19. Install the fan housing and motor. Install the evaporator and vacuum pump. Connect the temperature valve cable.
20. Install the throttle dashpot assembly. Install the wiper assembly and the drainage tube.
21. Install all covers, connect the wipers and connect all electrical connectors. Install the hood.
22. Fill the cooling system with coolant, connect the negative battery cable, start the engine and allow it to reach operating temperature. Check for leaks.

Water Pump

REMOVAL & INSTALLATION

8 Valve Engine

1. Disconnect the negative battery cable. Drain the cooling system.
2. Remove the necessary components in order to gain access to the water pump assembly.
3. Remove the drive belts. Remove the water pump pulley.
4. Remove the water pump retaining bolts. Remove the water pump from the engine.
5. Installation is the reverse of the removal procedure. Tighten the bolts to 15 ft. lbs. (20 Nm).

16 Valve Engine

1. Disconnect the negative battery cable. Raise and support the vehicle safely.
2. Remove the right front wheel assembly. Remove the front section of the inner fender panel.
3. Drain the engine coolant. Loosen the drive belts. Remove the water pump pulley and the belt tensioning pulley.
4. Remove the clips holding the oil lines at the oil cooler. Remove the clips securing the water pipe to the engine block. Disconnect the coolant hoses from the water pump.
5. Remove the bolt securing the water pump to the bracket. Remove the water pump.
6. Installation is the reverse of the removal procedure. Tighten the bolts to 15 ft. lbs. (20 Nm).

Thermostat

REMOVAL & INSTALLATION

Thermostats on both the 8 valve and 16 valve engines are located in housings on the fronts of the cylinder heads, facing their respective radiators. The housings are cast elbows, which are unbolted from the heads in order to gain access to the thermostats. When installing the new thermostat, always install with the spring facing down. Use sealing compound on the joining surfaces of the elbow and head. Tighten the bolts to 13 ft. lbs. (18 Nm).

Cooling System Bleeding

On 900 Series vehicles only, the bleeder nipple, located in the thermostat housing cover, should be opened when adding coolant to the system. The nipple should not be opened when the engine is running.

9000 Series vehicles do not require a special bleeding procedure. Fill the engine with coolant, start the engine and allow it to reach operating temperature (thermostat open) and check the coolant level. Add coolant as necessary.

ENGINE ELECTRICAL

NOTE: Disconnecting the negative battery cable on some vehicles may interfere with the functions of the on board computer systems and may require the computer to undergo a relearning process, once the negative battery cable is reconnected.

Distributor

REMOVAL

1. Rotate the engine until the No. 1 piston is at TDC of the compression stroke (crankshaft mark and timing mark aligned). Disconnect the negative battery cable. Remove all neces-

sary components in order to gain access to the distributor assembly.

2. Remove the distributor cap with the spark plug wires attached and position it out of the way. Disconnect the distributor wiring connector and vacuum hoses. Be sure to tag all the wires and vacuum lines for easy installation.

3. Matchmark the distributor housing, base and the engine block for installation reference. Remove the distributor base retaining nut. Remove the distributor.

INSTALLATION

Timing Not Disturbed

1. To install, double check that the crankshaft mark and timing mark are still aligned and that the engine has not been turned. Align the matchmark on the distributor housing and the engine block.
2. Install the distributor.
3. Connect all wiring and vacuum hoses. Reinstall distributor cap.
4. Connect the negative battery cable. Start engine and set timing. Then tighten distributor mounting nut.

Timing Disturbed

1. Remove the spark plug from No. 1 cylinder and position a compression gauge or a finger over the spark plug hole.
2. Slowly crank the engine until compression pressure starts to build.
3. Continue cranking the engine so the timing marks align with the TDC mark.
4. Install the distributor assembly while aligning the matchmark on the distributor and engine block.
5. Connect all wiring and vacuum hoses. Reinstall distributor cap.
6. Connect the negative battery cable. Start engine and set timing. Then tighten distributor mounting nut.

Distributorless Ignition

REMOVAL & INSTALLATION

Crankshaft Sensor

1. Disconnect the negative battery cable.
2. Install a flywheel lock (tool 83-93-993), or equivalent, to the flywheel.
3. Remove the front section of the right fender well. Remove the air conditioner belt.
4. Remove the center bolt and remove the idler wheel pulley. Remove the alternator belt.
5. Remove the crankshaft pulley. Apply gentile heat if necessary. Remove the crankshaft sensor. Disconnect the electrical connector and slip through the wire tie.

Distributor assembly location

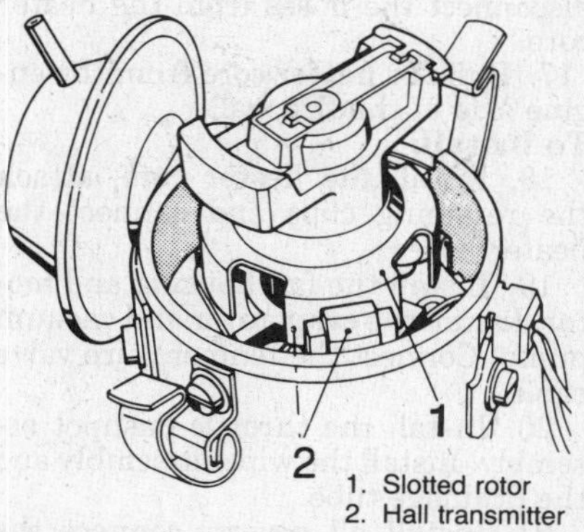

Distributor assembly with Hall effect pick up

To install:

6. Install the crankshaft pulley. Tighten the bolt to 140 ft. lbs. (190 Nm).
7. Install the crankshaft sensor and connect the electrical connector. The crankshaft sensor bolts must be installed using Loctite® 270, or equivalent.
8. Install the alternator belt, idler wheel pulley, air conditioner belt and right fender well.
9. Remove the flywheel lock and connect the negative battery cable.

Slotted Rotor

1. Disconnect the negative battery cable.
2. Install a flywheel lock (tool 83-93-993), or equivalent, to the flywheel.
3. Remove the front section of the right fender well. Remove the air conditioner belt.
4. Loosen the bolt and nut on the belt tensioner. Remove the alternator belt.
5. Remove the crankshaft pulley. Apply gentile heat if necessary. Remove the rotor.

To install:

6. Install the new rotor on the pulley. Install the crankshaft pulley and tighten the bolt to 140 ft. lbs. (190 Nm).
7. Install the alternator belt, idler wheel pulley, air conditioner belt and right fender well.
8. Remove the flywheel lock and connect the negative battery cable.

Ignition Coils

1. Disconnect the negative battery cable.
2. Remove the 4 bolts securing the ignition cartridge to the camshaft cover. Remove and invert the cartridge.
3. Remove the 6 screws from the coil shroud and remove the shroud.
4. Remove the ignition coil by lifting upward.
5. Installation is the reverse of removal. Tighten the ignition cartridge bolts to 9 ft. lbs. (12 Nm).

Ignition Timing

ADJUSTMENT

Conventional Method

1. Start the engine and allow it to reach normal operating temperature.
2. Apply the hand brake and position the gear selector in neutral, if equipped with a manual transmission, or **P**, if equipped with an automatic transmission. Turn off all accessories and stop the engine.
3. Install a tachometer (connect between the negative side of the coil and ground) and check the idle speed. Adjust as necessary.
4. Disconnect and plug the vacuum advance lines at the distributor, if equipped.
6. Stop the engine and connect the timing light. Connect the power leads to the battery and the inductive pickup to the No. 1 plug wire.
7. Start the engine and allow it to idle. Point the timing light at the mark on the front cover and read the basic timing by noting the position of the groove in the front pulley in relation to the timing mark or scale on the front cover. If the timing is incorrect, loosen the distributor mounting bolt. Turn the distributor slightly to adjust timing.
8. When the reading is correct, tighten the distributor mounting bolt and turn the engine **OFF**.
9. Turn the engine OFF. Reconnect the distributor advance line. Disconnect the timing light and tachometer.

Ignition Service Instrument Method

The equipment in the vehicle comprises a pin in the engine flywheel and a service socket in the clutch cover. The ignition service instrument is connected to the clutch cover by means of a special connector and the plug lead No. 1 cylinder by means of a terminal. The ignition service instrument is also connected to the ignition service socket at the fuse box and by means of an impulse transmitter at the plug lead for No. 1 cylinder.

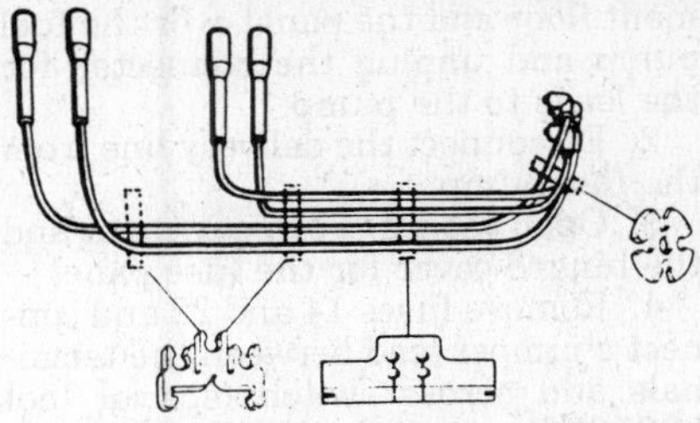
Secondary wiring routing – 8 valve engine

The Saab ignition service instrument consists of a tachometer, cam angle meter, stroboscope lamp and switch for operating the starter.

Alternator

PRECAUTIONS

Several precautions must be observed with alternator equipped vehicles to avoid damage to the unit.

- If the battery is removed for any reason, make sure it is reconnected with the correct polarity. Reversing the battery connections may result in damage to the one-way rectifiers.
- When utilizing a booster battery as a starting aid, always connect the positive to positive terminals and the negative terminal from the booster battery to a good engine ground on the vehicle being started.
- Never use a fast charger as a booster to start vehicles.
- Disconnect the battery cables when charging the battery with a fast charger.
- Never attempt to polarize the alternator.
- Do not use test lamps of more than 12 volts when checking diode continuity.
- Do not short across or ground any of the alternator terminals.
- The polarity of the battery, alternator and regulator must be matched and considered before making any electrical connections within the system.
- Never separate the alternator on an open circuit. Make sure all connections within the circuit are clean and tight.
- Disconnect the battery ground terminal when performing any service on electrical components.
- Disconnect the battery if arc welding is to be done on the vehicle.

BELT TENSION ADJUSTMENT

Adjust the alternator belt tension so the belt can be depressed about ½ in. at the midpoint of its longest straight run.

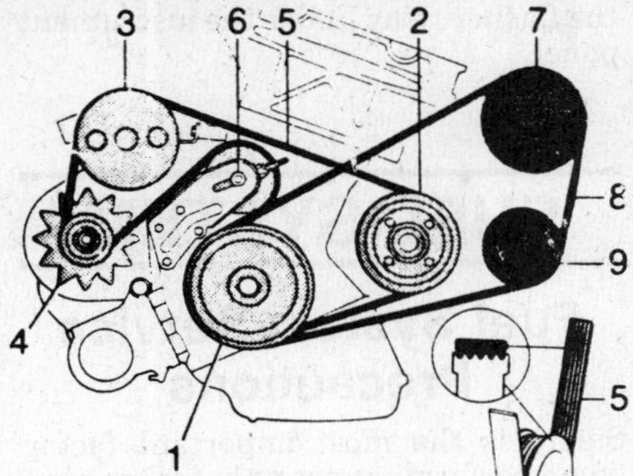

1. Crankshaft pulley
2. Water pump pulley
3. Steering servo pump pulley
4. Alternator pulley
5. Multi-groove belt
6. Belt tensioner
7. Compressor
8. Compressor belt
9. Adjusting device

Serpentine drive belt routing

REMOVAL & INSTALLATION

2.0L and 2.1L Engines

1. Disconnect the negative battery cable.
2. Raise and support the vehicle safely.
3. Remove the right front wheel assembly.
4. Remove the inner fender panel from the right fender.
5. Loosen the alternator belt and remove it from the alternator pulley.
6. Remove the alternator wire connections from the rear of the alternator.
7. Loosen the 2 securing bolts for the alternator.
8. Using a prybar, push the alternator to the left, pull the alternator forward and remove it from the vehicle.
9. Installation is the reverse of the removal procedure.

2.3L Engine

1. Disconnect the negative battery cable.
2. Remove the right front wheel and inner wheel well.
3. Remove the serpentine belt.
4. Remove the brace bar from between the wheel arch and the subframe.
5. Remove the 2 securing bolts and remove the poser steering pump by dropping it down and towards the rear.
6. Remove the 2 securing bolts and drop the alternator down and towards the rear. Remove the pump bracket.
7. Disconnect the electrical leads from the alternator and remove it by lifting it towards the front of the vehicle.

To install:

8. Connect the leads to the alternator and place it in position in the engine compartment.
9. Install the bracket for the power steering pump.
10. Install the top bolt in the alternator, leaving it loose. Align the alternator, secure it to the bracket and tighten the bolts.
11. Install the power steering pump, brace bar, serpentine belt, wheel well and wheel. Reconnect the negative battery cable.

Starter

REMOVAL & INSTALLATION

Except Turbocharged and 2.3L Engines

1. Disconnect the negative battery cable.
2. Remove the flywheel cover. Remove the gearbox dipstick, if equipped with manual transmission.
3. Remove the starter motor heat shield and the rear mounting bolts.
4. Disconnect the starter motor wires. Remove the front mounting bolts.
5. Remove the starter from the vehicle.
6. Installation is the reverse of removal.

Turbocharged Engine

1. Disconnect the negative battery cable. Remove the battery and the battery tray.
2. Remove the turbocharger suction pipe, preheater hose and the flywheel cover.
3. Remove the gearbox dipstick, if equipped with manual transmission. Remove the bracket and bolts between the turbocharger and the gearbox.
4. Disconnect the starter motor wires.
5. Loosen the oil return pipe on the turbocharger enough to allow it to be bent slightly.
6. Remove the starter motor heat shield and the rear mounting bolts.
7. Remove the front starter mounting bolts.
8. Remove the starter from the vehicle. The starter will have to be tilted downward and then lifted out forward.
9. Installation is the reverse the removal procedure. Be sure to use a new gasket on the oil return pipe connecting flange on the turbocharger.

2.3L Engine

1. Disconnect the negative battery cable.
2. Remove the rubber elbow from between the air mass meter and throttle housing. Remove the top bolt in the inlet manifold brace bar.

3. Loosen the top bolt for the starter motor using an appropriate wrench and then remove the bolt using a flexible socket extension.
4. Disconnect the electrical leads and remove the lower securing bolt from below. Move the brace bar out of the way.
5. Remove the starter motor by lifting it towards the rear and up between the inlet manifold and brake master cylinder.

To install:

6. Lift the starter motor into position. Install the top bolt first, leaving it loose, and then the bottom bolt using a flexible socket extension.
7. Ensure that the starter mtor is properly aligned and tighten the top bolt using an appropriate wrench.
8. Tighten the bottom bolt (with the brace bar installed) from underneath and reconnect the electrical connectors.
9. Tighten the top bolt in the inlet manifold brace bar. Install the rubber elbow. Reconnect the negative battery cable.

EMISSION CONTROLS

Please refer to "Emission Controls" in the Unit Repair section for system maintenance procedures. Due to the complex nature of modern electronic engine control systems, comprehensive diagnosis and testing procedures fall outside the confines of this repair manual. For complete information on diagnosis, testing and repair procedures concerning all modern engine and emission control systems, please refer to "Chilton's Guide to Fuel Injection and Electronic Engine Controls".

Emission Warning Lamps

RESETTING

On all vehicles except the Turbo, the **EXH** maintenance light on the dash comes on every 30,000 miles. The oxygen sensor should be changed at this time. On the Turbo vehicles, the **EXH** maintenance light is not used and the oxygen sensor is replaced every 60,000 miles.

To reset the mileage counter for the **EXH** light, press the reset button on the counter unit. The unit is located at the flasher relay under the instrument panel.

FUEL SYSTEM

Fuel System Service Precautions

Safety is the most important factor when performing not only fuel system maintenance but any type of maintenance. Failure to conduct maintenance and repairs in a safe manner may result in serious personal injury or death. Maintenance and testing of the vehicle's fuel system components can be accomplished safely and effectively by adhering to the following rules and guidelines.

- To avoid the possibility of fire and personal injury, always disconnect the negative battery cable unless the repair or test procedure requires that battery voltage be applied.
- Always relieve the fuel system pressure prior to disconnecting any fuel system component (injector, fuel rail, pressure regulator, etc.), fitting or fuel line connection. Exercise extreme caution whenever relieving fuel system pressure to avoid exposing skin, face and eyes to fuel spray. Please be advised that fuel under pressure may penetrate the skin or any part of the body that it contacts.
- Always place a shop towel or cloth around the fitting or connection prior to loosening to absorb any excess fuel due to spillage. Ensure that all fuel spillage (should it occur) is quickly removed from engine surfaces. Ensure that all fuel soaked cloths or towels are deposited into a suitable waste container.
- Always keep a dry chemical (Class B) fire extinguisher near the work area.
- Do not allow fuel spray or fuel vapors to come into contact with a spark or open flame.
- Always use a backup wrench when loosening and tightening fuel line connection fittings. This will prevent unnecessary stress and torsion to fuel line piping. Always follow the proper torque specifications.
- Always replace worn fuel fitting O-rings with new. Do not substitute fuel hose or equivalent where fuel pipe is installed.

RELIEVING FUEL SYSTEM PRESSURE

1988–89

1. Remove the luggage compartment floor and the panel over the fuel pump and unplug the connector for the leads to the pump.
2. Disconnect the delivery line from the fuel pump.
3. Open the glove compartment and the hinged cover for the fuse panel.
4. Remove fuses 14 and 22 and connect a jumper lead between the terminals and across switch (special tool 8393886), or equivalent, to provide power to the pump. Make sure the switch is **OFF**.
5. Start the pump by moving the switch to **ON**.
6. The pressure is released through the discharge outlet on the pump.

1990–92

1. Disconnect the negative battery cable.
2. Loosen the fitting on the fuel injection rail to release the pressure in the system. Soak up any escaping fuel with a rag. Tighten the fitting.

Fuel Tank

REMOVAL & INSTALLATION

NOTE: All vehicles are equipped with a plastic gas tank. Care should be exercised when removing the fuel pump from the plastic gas tank.

900 Series

1. Remove the luggage compartment floor, floor panel and cover over the fuel pump.
2. Using the electric fuel pump, drain the fuel tank. Remove the fuel pump relay and run the pump by connecting across fuse numbers 30 and 27 until the tank is empty.
3. Disconnect the negative battery cable. Remove the cover from the fuel gage sender unit and disconnect all electrical leads from the tank.
4. Raise and support the vehicle safely.
5. Remove the filler pipe and breather lines. Disconnect the pressure and return fuel lines.
6. Remove the securing straps and lower the tank.

To install:

7. Check that the rubber seal is undamaged and correctly fitted to the opening of the gauge sender unit. Check that the straps are correctly fitted and plug the ends of the filler pipe and breather with tape.
8. Lift the tank into position and support by the straps. Adjust the position of the tank and then tighten the straps. Remove the masking tape from the lines.
9. Reconnect all fuel lines. Ensure that the rubber grommet is in posi-

tion. Reconnect all electrical lines to the fuel tank and install the covers.

10. Lower the vehicle and connect the negative battery cable. Install the fuel pump relay, add fuel and start the vehicle. Check for leaks.

9000 Series

1. Disconnect the negative battery cable.
2. Remove the 2 screws and lift out the luggage compartment floor. Remove the fuel tank cover and disconnect all electrical leads to the fuel tank.
3. Disconnect the fuel return line and drain the fuel tank.
4. Raise and support the vehicle safely. Remove the left rear wheel. Disconnect the fuel filler pipe and breather hose. Cap the hoses with tape to prevent fuel leakage and the entrance of dirt.
5. Remove the handbrake cable and securing clip. Support the tank and remove the fuel tank strap attaching nuts. Lower and remove the tank.

To install:

6. Ensure that the anti-splash device (butterfly valve) is vertical when installing the tank.
7. Lift the fuel tank into place and secure with the straps. Position the tank and tighten the strap attaching bolts.
8. Connect all fuel and electrical lines previously disconnect.
9. Install the handbrake cable and rear wheel. Lower the vehicle. Install the fuel tank cover and luggage compartment floor.
10. Connect the negative battery cable. Add fuel to the tank and start the vehicle. Check for leaks.

Fuel Filter

REMOVAL & INSTALLATION

1. Disconnect the electrical connectors at the fuel pump. Remove the fuel pump fuse.
2. Crank the engine until fuel is exhausted from the system. Disconnect the negative battery cable.
3. Carefully remove the fuel filter fittings, using the proper wrench and covering it with a shop towel.
4. Remove the fuel filter assembly from the vehicle.
5. Installation is the reverse of the removal procedure. The filter is installed with arrows pointing in direction of flow.

Electric Fuel Pump

PRESSURE TESTING

Voltage Check

1. Remove the round cover plate from the top of the fuel pump.
2. Measure the voltage between the positive and negative terminals when the fuel pump is operating.
3. The lowest permissible voltage is 11.5 volts.

Capacity Check

NOTE: Be sure the fuel filter is not clogged and that the battery is fully charged.

1. Disconnect the return fuel pipe from the fuel distributor.
2. Connect the test pipe to the fuel distributor and place the other end in a suitable container.
3. On vehicles with the safety switch on the air flow sensor, remove the switch connector from the air flow sensor.
4. On vehicles with the fuel pump relay and the pulse sensor, remove the pump relay. Connect a jumper lead between terminals 30 and 87 on 900 Series vehicles.
5. Switch on the ignition and allow the pump to run for 30 seconds. Measure the quantity of fuel. The proper specification should be 900cc/30 seconds. This should be measured in the return line.

REMOVAL & INSTALLATION

1988–89

NOTE: All vehicles are equipped with a plastic gas tank. Care should be exercised when removing the fuel pump from the plastic gas tank.

1. Disconnect the electrical connectors at the fuel pump. Remove the fuel pump fuse.
2. Crank the engine until fuel is exhausted from the system. Disconnect the negative battery cable.
3. Remove the rear floor panel in the luggage compartment. Remove the valve cover from above the fuel pump.
4. Disconnect the electrical connections from the fuel pump.
5. Carefully disconnect the fuel lines from the fuel pump. Be sure to use the proper wrench and cover it with a shop towel while removing the connections.
6. Remove the fuel pump mounting clamp. Lift the fuel pump from the tank assembly.
7. Installation is the reverse of the removal procedure.

1990–92

1. Disconnect the negative battery cable.
2. Relieve the fuel system pressure.
3. Remove the luggage compartment floor and pump cover.
4. Release the clip and disconnect the electrical connector to the fuel pump.
5. Disconnect the fuel lines from the pump and tie out of the way.
6. Remove the fuel pump screw top using tool 83-94-462, or equivalent. Lift the pump and transfer to a container. Tilt the pump to pour out the remaining fuel.

To install:

7. Place a new O-ring in the tank and place the pump in position. Ensure that the alignment marks are in line. Tighten the fuel pump screw top to 40 ft. lbs. (50 Nm).
8. Place new O-rings inside the fuel line fittings and connect to the pump. Connect the electrical connector and install the saftey clip.
9. Connect the negative battery cable and check that the pump is working properly. Install the pump cover and floor panel.

Fuel Injection

IDLE SPEED AND MIXTURE ADJUSTMENT

8 Valve Engine

1. Run the engine until it reaches operating temperature.
2. Adjust the idle speed to 825–925 rpm.
3. If not equipped with a catalytic converter, remove the pulse/air hose and plug the air intake to the non-return valves. Connect the CO meter sensor to the exhaust pipe.
4. Remove the oxygen sensor wire.
5. If equipped with a catalytic converter, remove and plug the front exhaust pipe and connect the CO meter sensor to the pipe with the aid of a connecting piece. Remove the oxygen sensor wire.
6. Read and adjust the idle speed and CO valve as required. Before each reading, increase the engine speed and allow it to return to idle. Wait 30 seconds before taking the next CO reading.
7. Adjust the idle speed by turning the idle adjusting screw on the throttle valve housing.
8. Adjust the CO by turning the adjusting screw located on the fuel distributor clockwise for a richer mixture and counterclockwise for a leaner mixture. These adjustments affect each other, therefore these adjustments should be carried out in steps.
9. On catalyst equipped vehicles, connect the oxygen sensor wire and remove the CO meter probe from the front exhaust pipe connection. Install the plug in the front exhaust pipe. Insert the probe at the rear of the

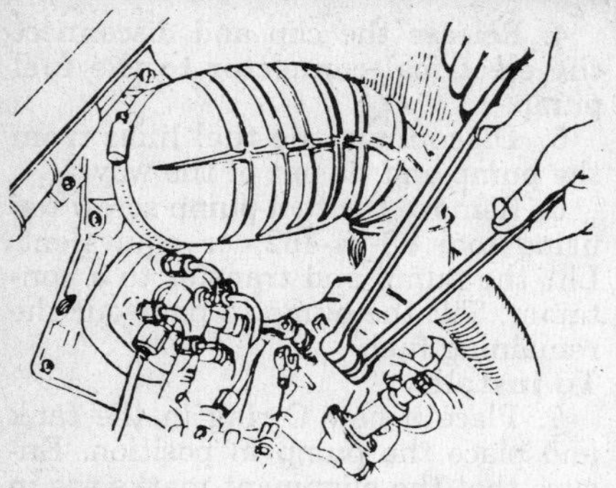

CIS injection idle speed adjustment

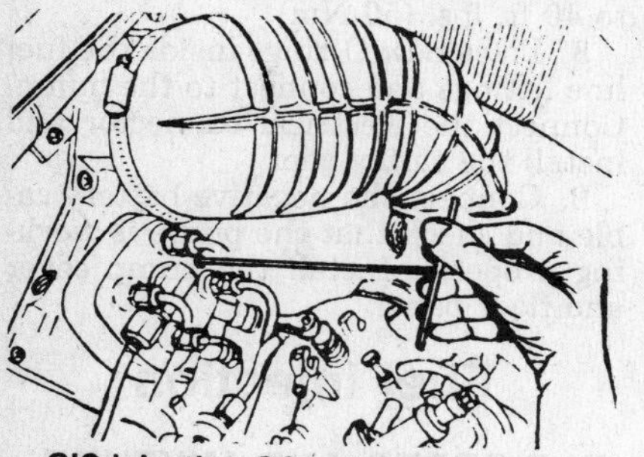

CIS injection CO value adjustment

tailpipe. The CO meter reading should be less than 0.3 percent with the engine at idle, and the engine and converter at normal operating temperature.

16 Valve Engine

1. Start the engine and allow it to run until it reaches normal operating temperature.
2. Connect a tachometer to the engine.
3. Ground the green/red lead of the single pole test socket on the right wheel housing to close the idling control valve. Use a jumper lead to do the grounding.
4. Set the idling speed to 775–825 rpm.
5. Disconnect the jumper lead from the test socket. Check that the engine speed changes and then settles down at 775–925 rpm.
6. CO value at simulated full load conditions should be 4.0–6.0 percent. Refer to underhood specifications label.

NOTE: Be careful not to confuse the connector for the throttle switch with that of the idling control valve, as this will destroy the electronic control unit.

Fuel Injector

REMOVAL & INSTALLTION

900 Series

1. Detach the crankcase ventilation hose from the camshaft cover.
2. Unplug the electrical connectors from the injectors. Free the wiring loom by by undoing the clip located at the fuel injection manifold.
3. Disconnect the banjo fittings at either end of the fuel injection manifold. Hold the injectors steady with a backup wrench.
4. Remove the bolts securing the fuel injection manifold to the inlet manifold. Lift off the fuel injection manifold complete with the injectors.
5. Slide off the clips located between the injectors and the manifold. Remove the injectors with a slight twist and a pull.

To install:

6. Prior to installation, check the O-rings on the injectors and replace any that are damaged.
7. Fit the injectors in the manifold. Check that the injectors are in the correct position and fully pushed into the inlet manifold.
8. The remainder of the installation procedure is the reverse of removal.

9000 Series

2.0L AND 2.1L ENGINE

1. Ensure that the area around the injectors is clean. Remove the false bulkhead panel.
2. Disconnect the fuel return line from the pressure regulator and the hose from the inlet manifold. Disconnect the line from the fuel filter.
3. Unplug the connectors on the injectors. Remove the fuel injection rail but leave the injectors in place.
4. Remove the O-ring seals for the injectors. Lift out the fuel rail complete with the injectors through the space between the inlet manifold and the bulkhead.

To install:

5. Release the clip to replace the injector.
6. Lubricate the injector O-ring with vaseline and then install the injector in the reverse order of removal.

2.3L ENGINE

1. Disconnect the fuel return line from the injection rail.
2. Disconnect the pulsator from the injection rail, snip through the tie and tuch the fuel line and pulsator out of the way.
3. Disconnect the vacuum hose from the pressure regulator. Unplug the connectors from the injectors.
4. Remove the bolts and lift off the injection rail complete with the injectors.
5. Remove the injector clips and then the injectors.

To install:

6. Apply a dab of vaseline to the O-rings for the injectors. Fit the injection rail complete with injectors on the inlet manifold and tighten the bolts.
7. Plug the connectors into the injectors. Connect the vacuum hose to the pressure regulator. Connect the fuel hose with pulsator to the injection rail and use a tie to secure the fuel hose.
8. Connect the fuel return line to the injection rail. Check that the system is working properly and inspect all connections for leaks.

DRIVE AXLE

Halfshaft

REMOVAL & INSTALLATION

900 Series

NOTE: The entire front axle assembly must be removed in order to remove the halfshaft from the vehicle.

1. Disconnect the negative battery cable. Remove the upper shock absorber bolt.
2. Raise the vehicle and support it safely. Remove the wheel and tire assembly.
3. Remove the brake housing and position it on the wheel housing to avoid damage to the brake hose. Remove the brake disc and parking brake assembly along with the cable.
4. Remove the large clamp from the rubber bellows on the inner universal joint. To separate the inner universal joint, install the cover in the rubber bellows to stop the needle bearings from falling out and to keep dirt from entering. Install the protective cap on the inner driver.
5. Disconnect the tie rod from the steering arm using the proper tool. Remove the nut on the upper ball joint. Remove the bolts from the lower control arm bracket.
6. Remove the halfshaft through the wheel housing and remove the entire front axle assembly.
7. If the differential bearing cap is to be removed, remove the retaining bolts and remove the cap and the inner drive using the proper removal tools.

To install:

8. Install the halfshaft through the wheel housing.
9. Install the lower control arm, upper ball joint and tie rod.
10. Install the rubber bellows on the halshafts after inserting them into the transaxle.
11. Install the brake system components previously removed. Install the

wheels and tires, and lower the vehicle.

12. Install the upper shock absorber bolt and connect the negative battery. Check the alignment and test drive the vehicle.

9000 Series

1. Disconnect the negative battery cable. Remove the hubcap and loosen the center axle nut. Raise and support the vehicle safely.
2. Remove the inner fender panel for working access.
3. Unbolt the MacPherson strut from the steering swivel member and detach the flexible brake hose from the clip on the strut.
4. Loosen the clip on the rubber boot on the inboard universal joint.
5. Separate the 2 halves of the joint. Install protective covers over the rubber boot and the drive axle.
6. Remove the hub center nut and withdraw the driveshaft from the steering swivel member.

To install:

7. Install the driveshaft and hub center nut. Tighten the hub center nut to 195–208 ft. lbs.
8. Join the 2 halves of the joint and install the rubber boot.
9. Install the MacPherson strut and tighten the strut-to-steering swivel bolts to 56–75 ft. lbs.
10. Install the brake hose and inner fender panel. Lower the vehicle and connect the negative battery cable. Check the alignment and test drive the vehicle.

CV-Boot

REMOVAL & INSTALLATION

1. Place spacer tool 83-93-209, or equivalent, between the underside of the top wishbone and the body.
2. Raise and support the vehicle safely. Remove the front wheel.
3. Remove the CV-clamps and loosen CV-boot.
4. Remove the halfshaft and steering knuckle assembly. Remove the CV-boot.

To install:

5. Install the new CV-boot on the intermediate driveshaft. Assemble the halfshaft and steering knuckle to the intermediate shaft.
6. Pack the CV-joint and boot with Esso Beacon EP2 grease, or equivalent.
7. Complete installation of the halfshaft and tighten the CV-boot clamps.
8. Install the front wheel, lower the vehicle and remove the spacer tool.

Front Wheel Hub, Knuckle and Bearings

REMOVAL & INSTALLATION

900 Series

NOTE: The entire front axle assembly must be removed from the vehicle when removing the wheel bearings.

1. Disconnect the negative battery cable. Remove the upper bolt of the shock absorber.
2. Raise the vehicle and support it safely. Remove the tire and wheel assembly.
3. Remove the brake housing and position it by the wheel housing to avoid damage to the brake hose. Remove the brake disc and parking brake assembly with the cable.
4. Remove the large clamp from the rubber bellows on the inner universal joint. To separate the inner universal joint, install the cover 7323736, or equivalent, in the rubber bellows to stop the needle bearing from falling out and to keep dirt from entering. Install the protective cap 7838469, or equivalent, on the inner drive.
5. Disconnect the tie rod from the steering arm using the proper tool. Remove the nut on the upper ball joint. Remove the bolts from the lower control arm bracket.
6. Remove the driveshaft through the wheel housing and remove the entire front axle assembly.
7. Place the steering knuckle housing in a press and press out the driveshaft.
8. Remove the lockring and press out the bearing using a suitable drift.

To install:

9. Install the bearing and secure the lockring. Using a press, install the driveshaft.
10. Install the driveshaft through the wheel housing. Install the lower control arm, upper ball joint and tie rod.
11. Install the rubber bellows on the driveshaft after installation.
12. Install all brake system components previously removed. Install the wheel and tire, and lower the vehicle.
13. Connect the negative battery cable, install the upper shock absorber bolt and check the alignment. Test drive the vehicle.

9000 Series

The front wheel bearing are double row angular contact bearings which are permanently lubricated and maintenance free. The bearings cannot be replaced individually. To remove the hub proceed as follows:

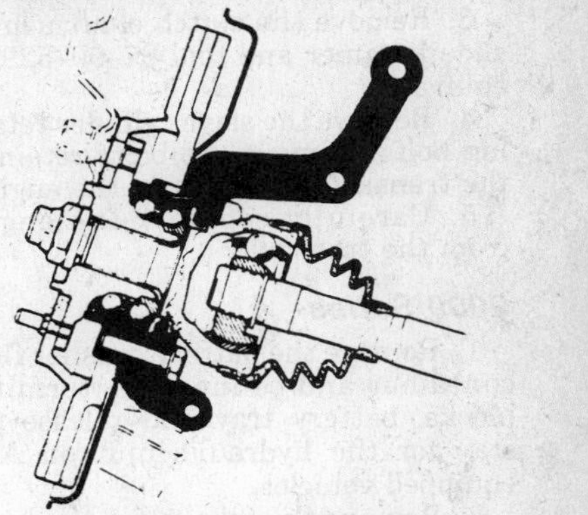

Front axle assembly—9000 Series

1. Loosen the hub center nut and the wheel bolts.
2. Raise the vehicle and support it safely.
3. Remove the tire and wheel assembly. Remove the hub center nut and thrust washer.
4. Remove the flexible brake hose from its support clip.
5. Unbolt the caliper and rest it upon the suspension arm.
6. Unscrew the locating stud for the disc and remove it from the hub.
7. Push in on the driveshaft. Remove the 4 bolts securing the hub to the steering swivel member.
8. Lift the hub and disc back plate from the suspension assembly. Renew the bearings or replace the hub.

To install:

9. The installation of the hub is in the reverse of the removal procedure.
10. Tighten the hub securing bolts to 40–43 ft. lbs., and the center hub nut to 195–208 ft. lbs.

MANUAL TRANSAXLE

For further information on transmissions/transaxles, please refer to "Chilton's Guide to Transmission Repair".

Transaxle Assembly

REMOVAL & INSTALLATION

On some engines, the top of the transmission case functions as the oil pan (sump) for the engine.

900 Series

1. Disconnect the negative battery cable. Remove the engine and transaxle from the vehicle as an assembly.
2. Position the engine and transaxle assembly in a suitable holding fixture. Drain the engine oil.

3. Remove the clutch shaft using a slide hammer and tool 87-90-529, or equivalent.

4. Remove the slave cylinder retaining bolts. Remove the bolts retaining the transaxle assembly to the engine.

5. Carefully separate the engine from the transaxle.

9000 Series

1. Remove the battery, washer fluid container and connectors, terminal blocks, battery tray and release the stay for the hydraulic unit on ABS equipped vehicles.

2. Remove the 8 bolts for the bulkhead cover. On 2.3L equipped vehicles, remove the rubber strip, lift the cover and disconnect the washer hoses from the nozzle and remove the cover.

3. On 2.3L eqquipped vehicles, separate the speedometer cable connector by first removing the washer hose and then the speedometer cable through the rubber grommet.

4. Disconnect the electrical connector from the air mass meter. On 2.3L eqquipped vehicles, disconnect the connector for the intake air temperature sensor and disconnect the hose on the delivery pipe from the by-pass valve.

5. On 2.3L equipped vehicles, Remove the delivery pipe between the throttle housing and the intercooler. Also, remove the nuts retaining the starter motor. Remove the starter and place it on the steering gear.

6. Separate the selector rod universal joint and selector rod. Pinch the slave cylinder pressure hose with clamping tongs and separate the pressure line.

7. Remove the upper bolts for the stay at the wheel housing. Release the left-hand engine mounting. Attach a sling to the engine lifting beam. Raise and support the vehicle safely.

8. Remove the left wheel and wheel housing liner. Disconnect the reverse light connector from the gearbox.

9. Separate the suspension arm from the ball joint. Remove the anti-roll bar. Remove the lower bolt for the stay at the wheel housing and the 3 bottom bolts from the joint between the engine and the gearbox.

10. Remove the center and left skirts under the spoiler. Separate the sub-frame at the front and rear. Lower the sub-frame. Remove the universal joint and lower the vehicle.

11. Sling the gearbox from the workshop hoist and remove the top nut and bolt from the joint face. Remove the gearbox and lower to the floor.

To install:

12. Prior to installation, ensure that the driveshaft is in position and the aluminum tube is pressed into the seal.

13. Slide the gearbox into position, guiding the driver and input into place. Install the top bolt and nut into the joint face. Release the gearbox from the hoist. Raise and support the vehicle safely.

14. Fit the 3 bottom bolts into the joint face and tighten to 40–74 ft. lbs. (54–100 Nm). Install the universal joint. Raise the sub-frame into position and secure.

15. Install the anti-roll bar, suspension to the ball joint and the bracket for the wheel housing stay. Do not tighten the wheel housing stay bolt. Lower the vehcile.

16. Remove the lifting beam. Install the starter and the top mounting of the wheel housing. Now tighten all wheel housing bolts.

17. Tighten the left engine mount. Raise and support the vehicle safely.

18. Connect the negative battery cable and reverse light switch. Install the wheel housing liners, left wheel, under car skirts, selector rod universal joint, slave cylinder pressure pipe (remove the clamping tongs), speedometer cable, washer hose and the remainder of the components removed.

19. Bleed the slave cylinder, check the oil level and road test the vehicle.

LINKAGE ADJUSTMENT

900 Series

1. Engage reverse gear and turn the ignition key to the **LOCKED** position.

2. Move the gear lever back and forth. The selector rod should then move 0.12–0.16 in. (3–4mm).

3. Adjust by moving the housing longitudinally. Use special tool 87-90-370, or equivalent, to loosen the gear lever housing bolts.

9000 Series

1. Lock the gear lever in reverse by inserting a 0.16 in. (4mm) drill bit through the locating holes in the gear lever and lever housing. Remove the rubber boot for access to the holes.

2. Connect the selector rod to the selector universal joint and tighten the pinch bolt to 22–25 ft. lbs. (30–33 Nm).

3. Remove the drill bit and install the rubber boot.

CLUTCH

Clutch Assembly

REMOVAL & INSTALLATION

900 Series

1. Disconnect the negative battery cable. Remove the clutch housing cover.

2. Install the spacer 83-90-023, or equivalent, between the clutch fork and the diaphragm spring. Keep the clutch pedal depressed when the ring is being installed.

3. Unhook the spring clip and remove the cover located in front of the clutch shaft. Remove the clutch shaft plastic propeller.

4. Remove the clutch shaft by means of an M8 bolt installed in the shaft end and tool 83-93-175, or equivalent. Withdraw the shaft as far as possible.

5. Remove the clutch slave cylinder retaining bolts.

6. Remove the clutch retaining bolts and remove the clutch, clutch disc and the slave cylinder complete with the clutch release bearing. Be sure the slave cylinder sleeve is not damaged by the clutch during the removal procedure.

7. Installation is the reverse of the removal procedure.

9000 Series

1. Disconnect the negative battery cable. Remove the transaxle assembly.

2. Install a flywheel locking tool, if available and remove the clutch assembly from the flywheel.

To install:

3. Use a centering arbor type tool or an appropriate input shaft to center the clutch plate to the flywheel.

4. Tighten the pressure plate bolts to 10.4–19.4 ft. lbs. Remove the flywheel lock, if used.

5. Slide the transaxle assembly over the locating dowels, engaging the transaxle input shaft into the clutch plate splines.

6. Secure the transaxle to the engine with the necessary attaching bolts. Remove the lifting sling from the transaxle.

7. Continue the installation in the reverse order of the removal procedure.

PEDAL HEIGHT/FREE-PLAY ADJUSTMENT

The clutch system in all vehicles is hydraulically operated. The slave cylinder acts directly on the release bearing, and adjustment of the clutch is automatic.

Clutch Master Cylinder

REMOVAL & INSTALLATION

1988–89

1. Disconnect the negative battery cable. Remove the clamp holding the

pipe from the cylinder at the body and remove the pipe at the cylinder.

2. Remove the left screen under the instrument panel.

3. Remove the pin holding the pushrod to the clutch pedal.

4. Remove the bolts inside the dash panel. Remove the clutch cylinder from inside the engine compartment.

5. Remove the hose from the fluid container and hang it aside so the fluid does not come out.

6. Installation is the reverse of the removal procedure. Tighten bolts to 6–7 ft. lbs. (8–10 Nm). Bleed the system.

1990–92

1. Remove the sound baffle. Remove the clip and withdraw the clevis pin from the master cylinder pushrod.

2. Place a suitable drip tray under the pressure pipe. Pinch the supply hose with clamping tongs. Disconnect the pressure pipe from the master cylinder.

3. Remove the master cylinder mounting bolts and remove the master cylinder.

4. Release the clamping tongs and disconnect the supply hose from the master cylinder.

To install:

5. Install the supply hose and pressure pipe nipple on the master cylinder.

6. Install the master cylinder on its mounting. Tighten the pressure pipe nipple and remove the clamping tongs.

7. Install the clevis pin and the sound baffle. Bleed the system and test the operation of the clutch.

Clutch Slave Cylinder

REMOVAL & INSTALLATION

900 Series

1. Disconnect the negative battery cable.

2. Remove the clutch assembly.

3. Remove the clutch release bearing together with the clutch slave cylinder.

4. Installation is the reverse of the removal procedure.

9000 Series

1. Disconnect the negative battery cable.

2. Remove the transaxle from the vehicle.

3. Remove the clutch release bearing. Disconnect the pressure pipe. Remove the bleed nipple.

4. Remove the retaining bolts that hold the slave cylinder in place.

5. Remove the clutch slave cylinder.

6. Installation is the reverse of the removal procedure.

1. Housing
2. Spring with seat
3. Sealing
4. Washer
5. Piston and rear seal
6. Pushrod assembly

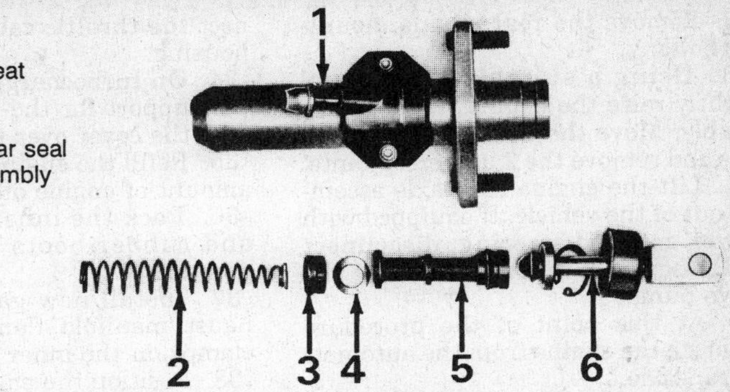

Exploded view of clutch master cylinder

Hydraulic Clutch System Bleeding

1. Connect a hose to the slave cylinder bleeder valve. Place the other end of the hose in a suitable jar partially filled with brake fluid.

2. Fill the master cylinder with brake fluid.

3. Open the bleeder valve on the slave cylinder a ½ turn.

4. Place a cooling system tester gauge over the opening of the master cylinder.

5. Pump the tester until all air has been expelled from the system.

6. Close the slave cylinder bleeder valve.

7. Check that all air has been removed from the system by depressing the clutch pedal.

AUTOMATIC TRANSAXLE

For further information on transmissions/transaxles, please refer to "Chilton's Guide to Transmission Repair".

Transaxle Assembly

REMOVAL & INSTALLATION

900 Series

NOTE: The engine and transaxle must be removed as an assembly. Removal of the engine by itself is not recommended.

1. Disconnect both battery cables. Drain the engine coolant.

2. Disconnect the windshield washer hose. Mark the engine hood hinges. Remove engine hood retaining bolts then hood assembly.

3. Disconnect and remove the following items:

a. Disconnect all electrical connections from the starter motor.

b. Disconnect the upper radiator hose.

c. Disconnect all ground leads.

d. Disconnect the temperature sending unit electrical connection.

e. Remove the ignition coil.

f. Disconnect the lower radiator hose.

g. Remove the air cleaner, air intake, preheater hose, crank case ventilation hose and intake hose.

h. Disconnect and plug the end of the fuel lines.

i. Disconnect the choke cable and the throttle cable.

j. Disconnect the hoses to the expansion tank.

k. Disconnect the oil pressure sending unit electrical connection.

l. Disconnect the alternator wiring harness.

m. Disconnect the heater hoses and the brake servo vacuum hoses.

4. On fuel injected vehicles, disconnect the electric wiring and fuel connections to the fuel injection system. Disconnect the flow meter and air cleaner with electrical connections.

5. If equipped with the APC system, disconnect the wiring to the solenoid valve. Remove the solenoid valve and the electrical connector to the knock sensor.

6. Remove the boot clips and rubber boots from the inner axle shafts.

7. Place special spacer tool 8393209 or equivalent between the underside of the upper frame and the vehicle body from the wheel housing side. The spacer tool relives the front suspension of load when the vehicle is raised.

8. Remove the lower end piece from the frame. Remove the steering knuckle package and support the end piece against the outer end of the frame.

9. Remove the gear selector cable retaining screw at the gearbox. Pull out the cable with the gear selector rod to its outer or **P** position. Move back the spring loaded sleeve on the gear selector rod and unhook the cable end piece.

10. Remove the exhaust pipe from the exhaust manifold.

11. Disconnect the speedometer cable from the transaxle.

12. Remove the rear engine mounting bolts.
13. Using a suitable lifting tool slightly raise the engine transaxle assembly. Move the unit slightly to the side and remove the 2 universal joints.
14. Lift the engine transaxle assembly out of the vehicle. If equipped with power assisted steering, disconnect and plug the 2 hydraulic lines at the servo pump.
15. At this point of the procedure separate the engine from the automatic transaxle.
16. Clean the outside of the engine and automatic transaxle and drain the oil out of the engine.
17. Remove the cover over the flywheel ring gear. On turbocharged vehicles remove the turbocharger support.
18. Remove the starter motor if necessary.
19. Disconnect the throttle cable at the throttle housing.
20. Remove all retaining bolts between the engine and transaxle and disconnect the hydraulic hoses from the oil cooler.
21. Remove the retaining bolts securing the ring gear to the torque converter.
22. Turn the flexplate, so the plate angles will be horizontal. Lift the engine carefully off the transaxle.
23. Install the torque converter support special tool 8790255 or equivalent.
24. Position the transaxle assembly on suitable workstand or holding fixture.

To install:

25. Before installing the transaxle to the engine make sure the mating surfaces are thoroughly clean.
26. Check that there are no cracks in the flexplate, of the engine particularly on a turbocharged engine.
27. Remove the torque converter support. Apply anti-corrosion grease to the center pin of the torque converter and the center of the flexplate. Make sure the 2 guide sleeves are installed into the gearcase.
28. Install a new sheet metal gasket to the joint face of the gearcase. Apply Bostik silicone compound part 2680 or equivalent, into the grooves in the gasket. Install the transaxle to the engine.
29. Position the flexplate so the sheetmetal angles are horizontal.
30. Take care not to damage the torque converter when lowering the engine onto the transaxle.
31. Apply thread sealing compound and tighten all retaining bolts.
32. Align the torque converter with the flexplate then gradually tighten the retaining bolts to 25–30 ft. lbs. (33–39 Nm).
33. Install the starter motor. Connect the throttle cable to the throttle housing.
34. On turbocharged vehicles install the support for the turbocharger. Install the cover over the ring gear.
35. Refill the engine with the correct amount of engine oil.
36. Pack the inner universal joins and rubber boots with a suitable grease.
37. Install new gaskets to the exhaust manifold flanges. Install new clamps on the inner axle shafts.
38. Position the engine transaxle assembly so the front mounting will engage in its bracket slightly before the rear mountings.
39. Lower the engine transaxle assembly into the vehicle. Guide the front mounting into the bracket and lower the rear of the engine transaxle assembly to approximately 2 inches (50–60mm) above the mountings.
40. Position the engine to the side, guide the left universal joint into position and then move the engine to the left.
41. Carefully lower the engine transaxle assembly and guide it onto the engine mountings, at the same time aligning the right universal joint with its axle shaft.
42. Align the exhaust system flanges. Check that the gaskets are correctly installed.
43. Install the right end piece to the frame. Check that the right universal joint is aligned with the axle shaft.
44. Install the bolts into the rear engine mountings and tighten all of the engine mountings and bolt together the exhaust flanges.
45. Reconnect the speedometer cable.
46. Reconnect the cable to the gear selector rod and bolt the cable end piece to the transaxle casing. Make sure gear selector operates properly.
47. Install the boot clamps to the inner universal joints. Remove the special spacer tool.
48. Connect and install the following items:
    a. Connect all electrical connections to the starter motor.
    b. Connect the upper radiator hose.
    c. Connect all ground leads.
    d. Connect the temperature sending unit electrical connection.
    e. Install the ignition coil.
    f. Connect the lower radiator hose.
    g. Install the air cleaner, air intake, preheater hose, crank case ventilation hose and intake hose.
    h. Connect the fuel lines.
    i. Connect the choke cable and the throttle cable.
    j. Connect the hoses to the expansion tank.
    k. Connect the oil pressure sending unit electrical connection.
    l. Connect the alternator wiring harness.
    m. Connect the heater hoses and the brake servo vacuum hoses.
49. On fuel injected vehicles, connect the electric wiring and fuel connections to the fuel injection system. Reconnect the flow meter and air cleaner electrical connections.

NOTE: On fuel injected engines make sure there is at least ½ in. (10mm) of clearance at the bottom of the throttle control assembly.

50. If equipped with the APC system, reconnect the wiring to the solenoid valve. Install the solenoid valve and the electrical connector to the knocking sensor.
51. Refill the coolant and bleed the cooling system through the bleeder nipple on the thermostat housing. If equipped with power assisted steering, connect the hydraulic lines to the servo pump.
52. Reconnect both battery cables. Refill the transaxle assembly with the correct amount of specified fluid.
53. Install the hood in the marked position and connect the windshield washer hose.
54. Road test the vehicle in all driving ranges for proper operation.

9000 Series

1. Disconnect the battery cables and remove the battery.
2. Remove the windshield washer fluid reservoir. Plug the outlet to keep the fluid from running out.
3. Open the terminal box and disconnect the cables.

NOTE: When removing the fuel filter, wrap a rag around the line to prevent fuel from spraying out. Use proper caution when working with the fuel system.

4. Release the fuel filter clamp and remove the filter. Remove the battery cable.
5. Disconnect the wiring from the air mass meter and remove the meter.
6. Disconnect the hose from the transaxle and the bypass hose from the turbocharger delivery pipe. Remove the delivery pipe.
7. Disconnect the throttle cable form the throttle housing.
8. Disconnect the gear selector lever cable from the selector lever. Do not separate the ball joint on the cable.
9. Disconnect the inlet hose from the oil cooler on top of the transaxle. Disconnect the selector lever from the transaxle.
10. Remove the return line from the

oil cooler. Place a drain pan under the transaxle to collect the oil.

11. Remove the clamp retaining the turbocharger oil supply line to the transaxle, if equipped.

12. Disconnect the speedometer cable from the transaxle. Remove the top retaining bolt for the starter motor.

13. Disconnect the starter motor stay from the intake manifold. Disconnect the top end of the starter stay from the wheel housing.

14. Place engine support yoke 83–93–977 or equivalent, in position on the wheel housing to support the engine when the transaxle has been removed.

15. Raise and safely support the vehicle. Remove the left front wheel and remove the inner fender liner.

16. Remove the starter motor and suspend it aside. Remove the bolts retaining the torque converter to the driveplate.

17. Remove the bolts retaining the ball joint to the suspension arm. Remove the anti-roll bar mounting nut from the suspension arm. Remove the 2 bolts retaining the anti-roll bar bearing.

18. Remove the front engine mount bolt.

19. Split the sub-frame at the front and slightly open the joint. Remove the 2 bolts at the rear of the sub-frame, 1 of the bolts retains the steering gear.

20. Remove the bolts for the rear sub-frame joint and remove the 2 bolts in the front corner. Lower the sub-frame from the vehicle.

21. Remove the clamps from both the left and right CV-joints, separate the joints. Allow the driveshafts to hang down.

22. Position a suitable lifting device under the transaxle and lower it from the vehicle.

To install:

23. Install a suitable converter holding device to prevent the converter from falling out during installation. With the transaxle on a suitable lifting device, raise it into position under the vehicle.

24. Guide the transaxle into position aligning the converter pin as a guide. Install 1 bolt to retain the assembly.

25. Reattach the driveshafts and install the clamps over the CV-boots.

26. Install the transaxle mounting bolt through the engine mount.

27. Raise the sub-frame assembly into position. Make sure the engine mount is in position.

28. Install all sub-frame bolts and the engine mount bolt.

29. Install the anti-roll bar and bearing into the suspension arm. Install all of the suspension arm bolts.

30. Install the torque converter-to-driveplate bolts. Use Loctite® 242, or equivalent, on the bolts.

31. Install the starter and stay. Lower the vehicle.

32. Remove the engine support tool. Install the upper starter mount bolts. Connect the speedometer cable.

33. Install the turbocharger oil pipe to the engine block.

34. Reconnect the selector lever cable. The selector should be in the **N** detent. Adjust the selector as needed.

35. Reconnect the oil cooler hose to the transaxle oil cooler.

36. Reconnect the turbocharger delivery pipe. Install the air mass meter.

37. Install the battery tray. Install the fuel filter.

38. Attach the battery cable to the battery tray. Reconnect the cables to the terminal box.

39. Install the windshield washer fluid bottle and connect the electrical leads.

40. Install the battery and reconnect the cables.

41. Raise and safely support the vehicle.

42. Install the inner fender cover. Install the wheel and tire assembly.

43. Lower the vehicle, refill the fluid in the transaxle.

44. Road test the vehicle and check the operation of the transaxle. Adjust the throttle linkage as needed. Check the fluid level.

SHIFT LINKAGE ADJUSTMENT

1. Remove the gear selector lever cover.

2. Slack off the gear selector lever housing nuts with tool 83–91–23 or equivalent.

3. Lift the gear selector lever housing and turn it so the adjustment nuts of the cable will be reachable.

4. Adjust the cable longer or shorter to bring the **N** and **D** clearance to specification.

5. Assemble the gear selector housing and check the clearance in **N** and **D**.

6. The proper setting of the selector cable can be accomplished by adding or removing shims at the transmission case end of the cable. A maximum of 3 shims may be used.

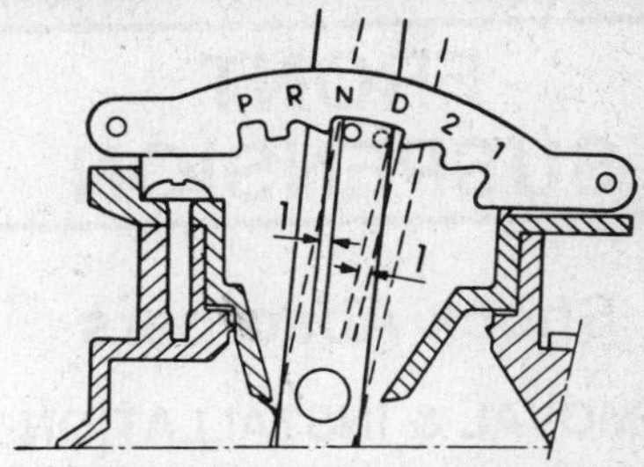

Manual linkage clearance, equal in "N" and "D"

THROTTLE CABLE ADJUSTMENT

900 Series

1. Remove the screw for the pressure tap on the transaxle and connect a pressure gauge. Block the drive wheels and apply the handbrake.

2. Start the engine and check that the idle speed is 850 rpm in **P**.

3. Disconnect the throttle cable from the spindle lever and ensure that it is not binding. If so, clean the throttle cable throughly and reconnect.

4. With the gear selector in **D**, check that the cable is released and adjust the throttle cable to obtain the lowest possible pressure.

5. Readjust the cable so the pressure increases to 1.4 psi.

6. With the gear selector in **P**, check that the pressure is now 59–69 psi. Pressure should not be allowed to exceed 69 psi.

7. Tighten the cable locknuts.

9000 Series

1. With the engine idling, check the clearance between the cable stop and the end of the throttle cable. The clearance should be 0.08–0.10 in. (2.0–2.5mm).

2. If clearance is not within specification, loosen the locknuts and adjust the cable.

3. Tighten the locknuts and recheck the clearance.

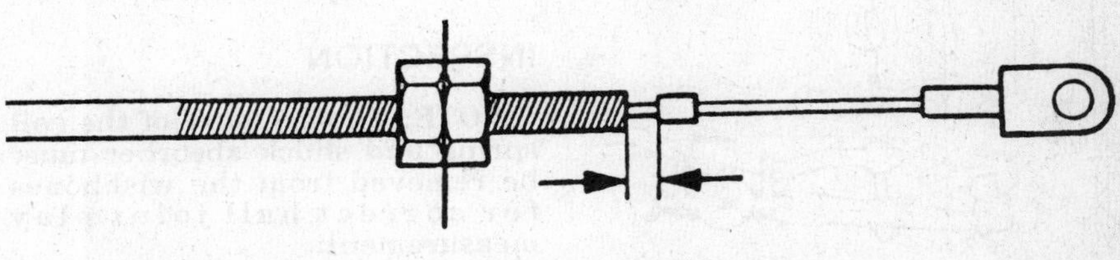
Throttle cable adjustment—9000 Series

FRONT SUSPENSION

Shock Absorbers

REMOVAL & INSTALLATION

900 Series

1. Disconnect the negative battery cable. Remove the upper shock absorber nut.
2. Raise the vehicle and support it safely. Remove the tire and wheel assembly.
3. Remove the shock absorber retaining bolts. Remove the shock from the vehicle. Save all washers and rubber parts.

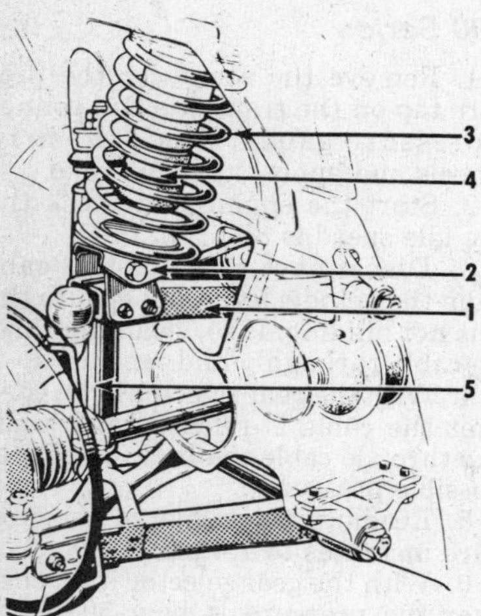

1. Upper control arm
2. Lower spring support
3. Coil spring
4. Rubber buffer
5. Shock absorber

Front suspension assembly

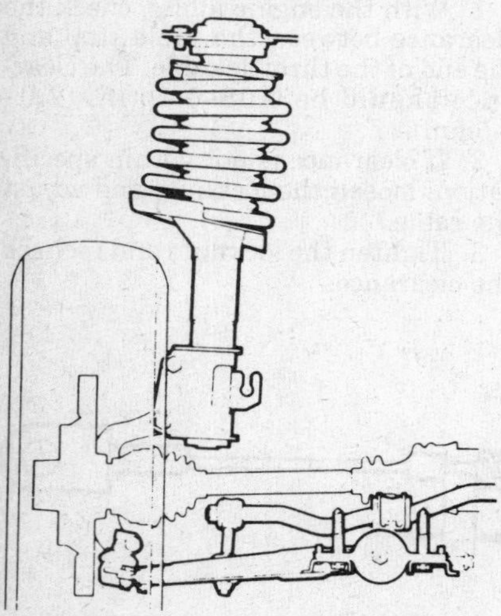

Front suspension assembly—9000 Series

4. Installation is the reverse of removal. Tighten the lower shock bolts to 21 ft. lbs. (95 Nm).

MacPherson Strut

REMOVAL & INSTALLATION

9000 Series

1. Disconnect the negative battery cable. Raise and support the vehicle safely. Remove the front tire and wheel assembly.
2. Remove the front brake hose from the retaining clip on the strut assembly.
3. Unbolt the strut from the steering swivel arm.
4. Remove the 3 retaining bolts from the top of the strut.
5. Remove the strut from the vehicle.
6. Installation is the reverse of the removal procedure. Tighten upper mounting bolts to 9–12 ft. lbs. (40–54 Nm); lower mounting bolts to 18–24 ft. lbs. (78–105 Nm).

Coil Springs

REMOVAL & INSTALLATION

900 Series

1. Disconnect the negative battery cable. Remove the upper shock absorber retaining nuts.
2. Raise and support the vehicle safely. Remove the tire and wheel assembly.
3. Install a spring compression tool or equivalent, engaging the upper shanks directly in the spring at the second free turn from the top of the lower shanks around the spring caps. These alignment shanks are located on the last turn of the spring with the color coded cup right beside the end of the coil.
4. Compress the spring at the top end, approximately 1½ in. If the upper spring attachment of the steel cone is left behind in the wheel housing, remove it.
5. Remove the spring and the steel cone from the vehicle.
6. Installation is the reverse of removal.

Ball Joints

INSPECTION

NOTE: The pressure of the coil spring and shock absorber must be removed from the wishbones for correct ball joint play measurement.

1. Position spacer 83-93-209, or equilavent, under the upper wishbone.
2. Check end play by compressing the ball joint with a pair of large pliers. Maximum end play is 0.08 in. (2.0mm).
3. Check radial play by applying pressure between the wishbone and the vertical link. Car should be taken so that the ball joint seal is not ripped. Maximum radial play is 0.04 in. (1.0mm).
4. Replace any ball joint that does not pass either test. Inspect the ball joint seals. Replace any that are damaged.

REMOVAL & INSTALLATION

1. Disconnect the negative battery cable. Raise and support the vehicle safely. Remove the tire and wheel assembly.
2. Remove the brake caliper and position it aside so the brake hose will not be damaged.
3. Remove the nut that holds the ball joint ball bolt to the steering knuckle housing.
4. Remove the ball joint from the control arm assembly.
5. Installation is the reverse of removal. Tighten the ball joint-to-lower control arm bolts to 6–8 ft. lbs. (25–34 Nm); the ball joint-to-steering knuckle bolts to 11–15 ft. lbs. (50–68 Nm).

Upper Control Arms

REMOVAL & INSTALLATION

900 Series

NOTE: To remove the left upper control arm, the engine must first be removed from the vehicle.

1. Raise the vehicle and support it safely.
2. Remove the tire and wheel assembly. Remove the shock absorber. Compress the coil spring, using a spring compression tool.
3. Remove the 2 bolts attaching the upper ball joint and lower spring seat to the upper control arm.
4. Remove the bolts from both upper control arm bearing brackets.
5. Remove the coil spring from the vehicle.
6. Remove the control arm and bearings from the vehicle. Save the spacers under the bearings and record the number of spacers used under each bearing.
7. Remove both of the bearing nuts. Now the bearings and bushings can be removed from the control arm.

To install:

8. Install the rubber bushings with tool 78-41-331, or equivalent. Spray the bushings with soapy water prior to

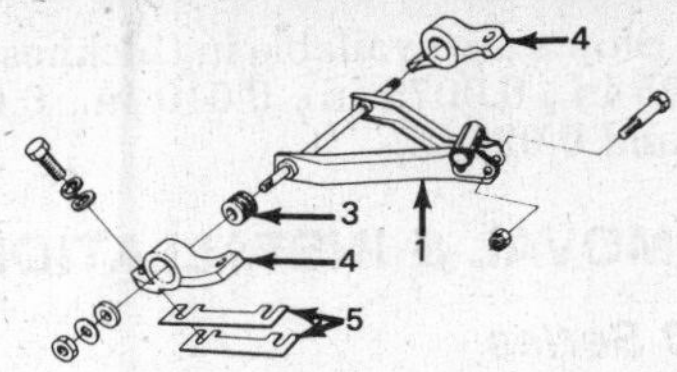

1. Upper control arm
3. Rubber bushing
4. Bearing
5. Spacers

Upper control arm assembly

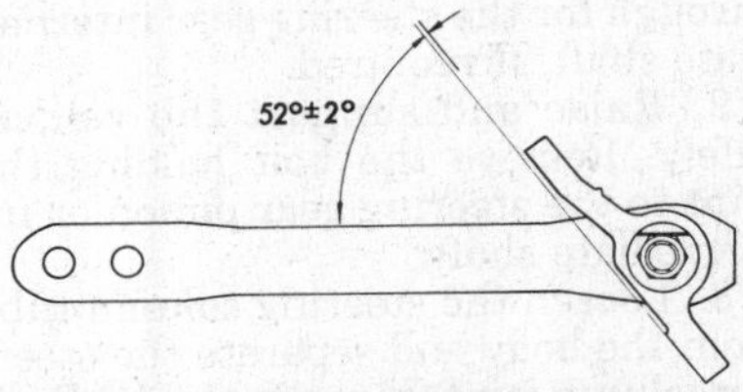

Checking the angle between the upper control arm and bearing

installation. Do not use oil or grease to aid installation.

9. Install the bearing to the control arm. Once the 2 nuts are tight, the angle between the control arm and the bearing should be 60–64 degrees. Tighten the bolts to 54–66 ft. lbs. (75–90 Nm).

10. Install the control arm and tighten into place. Install the upper ball joint to the control arm and install the coil spring. Check that the spring bushing is seated in the upper spring pocket.

11. Raise the outer end of the lower control arm slightly and install the shock absorber. Install the wheel and lower the vehicle.

12. Check the wheel alignment and test drive the vehicle.

Lower Control Arms

REMOVAL & INSTALLATION

900 Series

1. Disconnect the negative battery cable. Raise the vehicle and support it safely. Remove the tire and wheel assembly.

2. Disconnect the lower end of the shock absorber.

3. Remove the 2 bolts that attach the ball joint to the control arm.

4. Remove the lower control arm attaching bolts from under the engine compartment floor.

5. Remove the control arm and its attaching brackets from the vehicle.

6. Remove the control arm bearing nuts and remove the bearings from the control arm.

To install:

7. Install the rubber bushings with the aid of tool 78–41–349, or equivalent. Soap the bushings prior to installation. When both nuts are tightened, the angle between the control arm and bearing should be 16–20 degrees. Tighten the bolts to 70–77 ft. lbs. (100–120 Nm).

8. Install the control arm and secure the ball joint. Raise the control arms slightly and fit the shock absorber.

9. Install the wheel and lower the vehicle. Check the wheel alignment and test drive the vehcile.

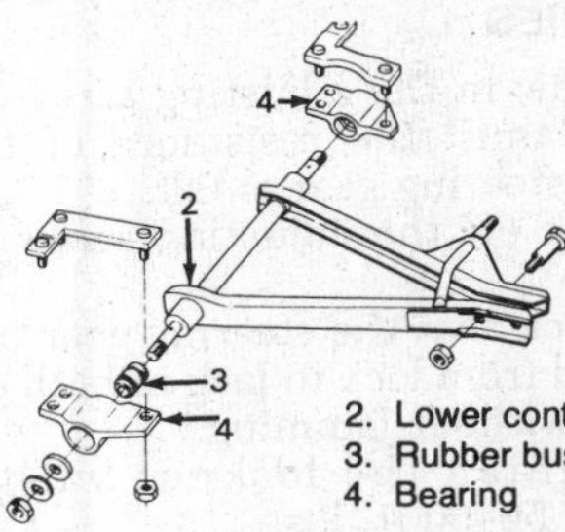

2. Lower control arm
3. Rubber bushing
4. Bearing

Lower control arm bushings—900 Series

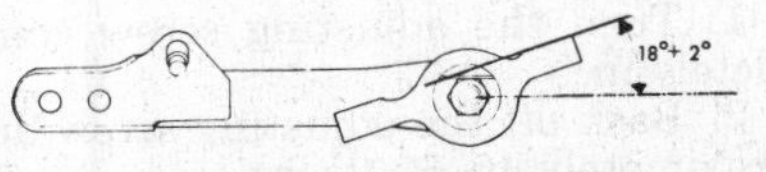

Checking the angle between the lower control arm and bearing

9000 Series

1. Disconnect the negative battery cable. Raise the vehicle and support it safely. Remove the tire and wheel assembly.

2. Remove the bolts securing the suspension arm to the ball joint.

3. Remove the nut from the bolt securing the suspension arm to the anti-roll bar link. Remove the upper securing bolt for the link.

4. Press down on the suspension arm and withdraw the anti-roll bar link.

5. Remove the nuts at the front of the suspension arm from the bolts securing the arm to the frame.

6. Remove the rear bolts securing the reinforcement member to the frame.

7. Remove the bolts securing the control arm rear pivot to the frame. Remove the control arm.

To install:

8. Installation is the reverse of removal. Leave the nuts for the bushings in the suspension arm rear pivot loose.

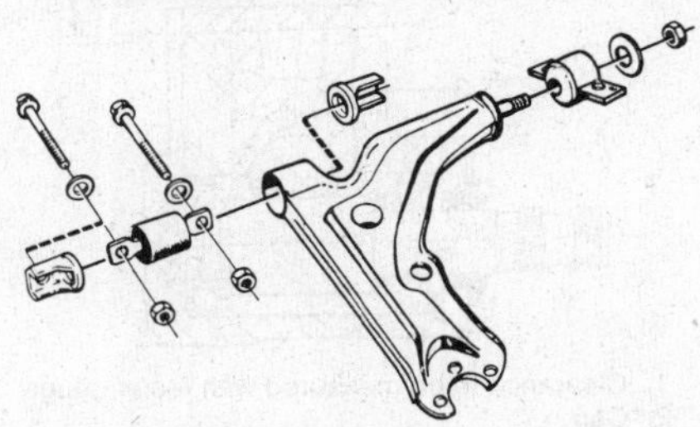

Lower control arm bushings—9000 Series

9. After the arm is installed and the remaining bolts in place, tighten the rear pivot bolts. Tighten the bolts to 16–22 ft. lbs. (71–96 Nm).

10. Check the wheel alignment and adjust as required, after the vehicle has been allowed to settle by bouncing or driving.

REAR SUSPENSION

Shock Absorbers

REMOVAL & INSTALLATION

1. Disconnect the negative battery cable. Raise the vehicle and support it safely.

2. Place a suitable jackstand under the rear axle to prevent it from dropping and stretching the brake lines.

3. Position a jack at the rear of the spring link. Remove the shock absorber retaining nuts.

4. Remove the bolts in the spring link mounting on the rear axle—anti-roll bar on the 9000 Series.

5. Lower the spring link so the shock absorber can be removed from the vehicle.

6. Installation is the reverse of removal.

Coil Springs

REMOVAL & INSTALLATION

1. Disconnect the negative battery cable. Raise and support the vehicle safely.

2. Remove the tire and wheel assembly. Position a jack under the

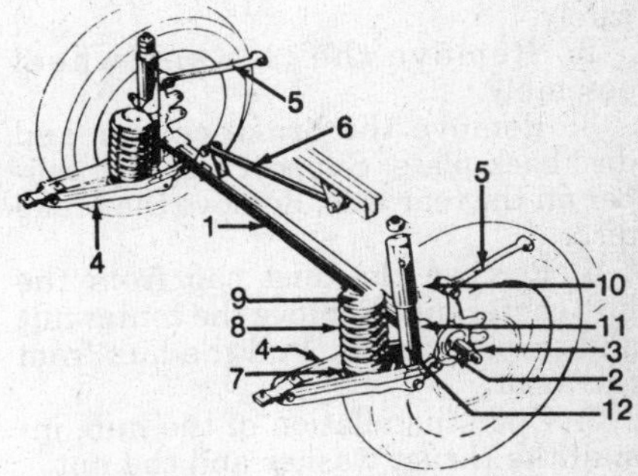

1. Rear axle
2. End piece
3. Stub axle
4. Spring links
5. Rear links
6. Cross bar
7. Spring seat
8. Coil spring
9. Spring insulator
10. Rubber buffer
11. Stop
12. Shock absorber

Rear suspension assembly

spring link and disconnect the lower end of the shock absorber.

3. From under the vehicle, remove the 2 locknuts that secure the front spring link bearing to the body of the vehicle.

4. Position a jackstand under the rear axle to prevent the brake lines from being damaged by the weight of the rear axle.

5. Lower the spring link so the spring can be removed from the vehicle together with the upper spring support and the rubber spacer at the lower spring seating which is retained by the spring tension.

6. Installation is the reverse of the removal procedure.

Rear Wheel Bearings

REMOVAL & INSTALLATION

900 Series

1. Slide the hub onto the stub axle and install the washer locknut.

2. Tighten the locknut to a torque of 210 ft. lbs. (300 Nm). If the part of the nut collar that had previously been staked aligns with the locking groove, install a new nut.

3. Stake the nut collar into the locking groove.

4. Complete the assembly by installing the dust cap and install the brake disc, brake housing, wheel and wheel nuts.

9000 Series

The wheel bearings are not press fitted on the outboard driveshaft or stub axle, but are incorporated in the hub. The wheel bearings are double row, angular contact bearings which are permanently lubricated and maintenance free. The bearings cannot be replaced individually.

1. Disconnect the negative battery cable. Raise the vehicle and support it safely.

2. Remove the tire and wheel assembly.

3. Remove the brake caliper and disc back plate. Support the disc caliper on the rear axle. Remove the brake disc.

4. Remove the dust cap from the hub center nut. Remove the center nut and thrust washer. Pull the hub from the axle.

5. Upon installation of the hub, install the thrust washer and the nut.

6. Tighten the center nut to a torque of 195–208 ft. lbs.

7. Lock the nut by using drift to punch the flange of the nut in the stub axle thread. Install the dust cap, the disc brake components, the wheel and lower the vehicle.

STEERING

Steering Wheel

CAUTION

On vehicles equipped with an air bag, the negative battery cable must be disconnected for 20 minuites before working on the system. Failure to do so may result in deployment of the air bag and possible personal injury.

REMOVAL & INSTALLATION

1. Disconnect the negative battery cable.

2. On some vehicles it will be necessary to remove the bottom cover of the steering wheel bearing.

3. Remove the steering wheel safety pad. Remove the steering wheel emblem. Remove the horn contact. Remove the steering wheel holding nut and washer.

4. Remove the steering wheel using the proper steering wheel removal tool.

5. Installation is the reverse of removal.

Manual Rack and Pinion

ADJUSTMENT

Radial Play

1. Install the plunger without the spring and screw on the cap without the gasket by hand until it butts against the plunger. Do not use a wrench, as the cap will become damaged.

2. Measure the clearance between the cap and the housing with a feeler gauge.

3. Add 0.002–0.006 in. to the measured clearance to allow for the play to be left between the plunger and cap after assembly. Measure the thickness of the gasket and shims with a micrometer. Shims are available in thickness of 0.005 in., 0.0075 in., 0.010 in., 0.015 in. and 0.020 in.

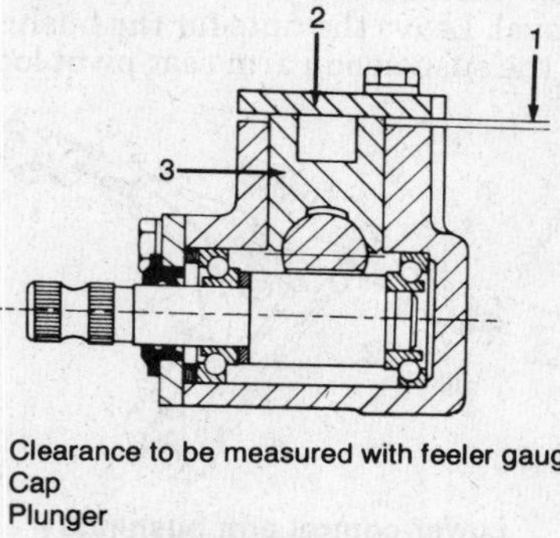

1. Clearance to be measured with feeler gauge
2. Cap
3. Plunger

Radial-play adjustment

REMOVAL & INSTALLATION

900 Series

1. Disconnect the negative battery cable. Remove the left screen under the instrument panel and loosen the rubber bellows at the body lead through for the steering gear intermediate shaft, if required.

2. Raise and support the vehicle safely. Remove the bolt holding the joint to the steering gear pinion or intermediate shaft.

3. Loosen the steering column tube from the body and separate the steering column joint from the pinion. Position the steering column so the wiring harness is not damaged.

4. Remove both tire and wheel assemblies. Remove the tie rod ends at the steering arms with the proper removal tool. Remove the 2 steering gear clamps.

5. Move the rack to the right as far as possible. Lift the steering gear to the right so the tie rod can be bent down in the opening of the engine compartment floor.

6. Pull the rack to the left and lift the steering gear down through the opening in the engine compartment floor.

7. Installation is the reverse of the removal procedure. Tighten the clamp bolt to 26–30 ft. lbs. (35–42 Nm); the body to steering gear bolts to 44–60 ft. lbs. (60–80 Nm); the tie rod end nuts to 37–44 ft. lbs. (50–60 Nm).

Power Rack and Pinion

ADJUSTMENT

Radial Play

900 SERIES

1. Screw in the adjusting screw all the way until the resistance of the twisting steering gear is felt.

2. Back off the adjusting screw ½ turn.

3. Check that the steering gear can be turned from lock to lock in both directions without jamming.

4. Tighten the locknut with a torque of 50–60 ft. lbs.

9000 SERIES

1. Turn the adjusting screw completely in.

2. Back off the adjusting screw approximately 40–60 ft. lbs.

3. Tighten the locknut to 47–54 ft. lbs. (65–75 Nm).

REMOVAL & INSTALLATION

900 Series

1. Disconnect the negative battery cable. Remove the left screen under the instrument panel and loosen the rubber bellows at the body lead through for the steering gear intermediate shaft, if required. Disconnect and plug the power steering fluid lines.
2. Raise and support the vehicle safely. Remove the bolt holding the joint to the steering gear pinion or intermediate shaft.
3. Loosen the steering column tube from the body and separate the steering column joint from the pinion. Position the steering column so the wiring harness is not damaged.
4. Remove both tire and wheel assemblies. Remove the tie rod ends at the steering arms with the proper removal tool. Remove the 2 steering gear clamps.
5. Move the rack to the right as far as possible. Lift the steering gear to the right so the tie rod can be bent down in the opening of the engine compartment floor.
6. Pull the rack to the left and lift the steering gear down through the opening in the engine compartment floor.
7. Installation is the reverse of the removal procedure. Tighten the clamp bolt to 26–30 ft. lbs. (35–42 Nm); the body to steering gear bolts to 44–60 ft. lbs. (60–80 Nm); the tie rod end nuts to 37–44 ft. lbs. (50–60 Nm); the hydraulic lines to 15–25 ft. lbs. (20–34 Nm).

9000 Series

1. Disconnect the negative battery cable. Remove the padding from under the instrument panel and the trim on the left side of the center tunnel, as required. Fold back the carpet where the steering column passes through the fire wall. Remove the rubber boot from the intermediate shaft.
2. Remove the pinch bolt in the lower clamp, loosen the bolt in the upper clamp and remove the intermediate shaft.

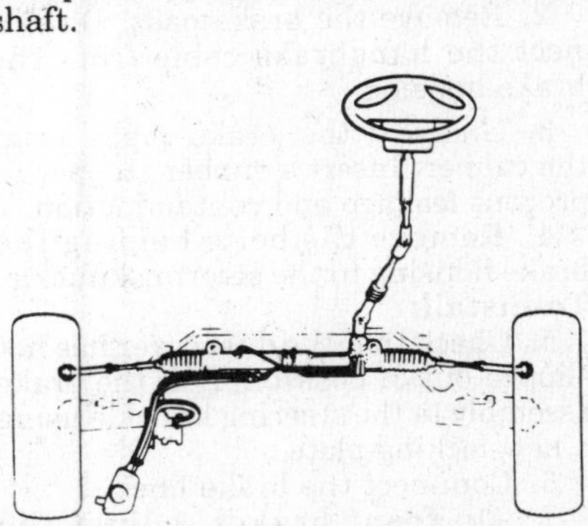

Power steering assembly – 9000 Series

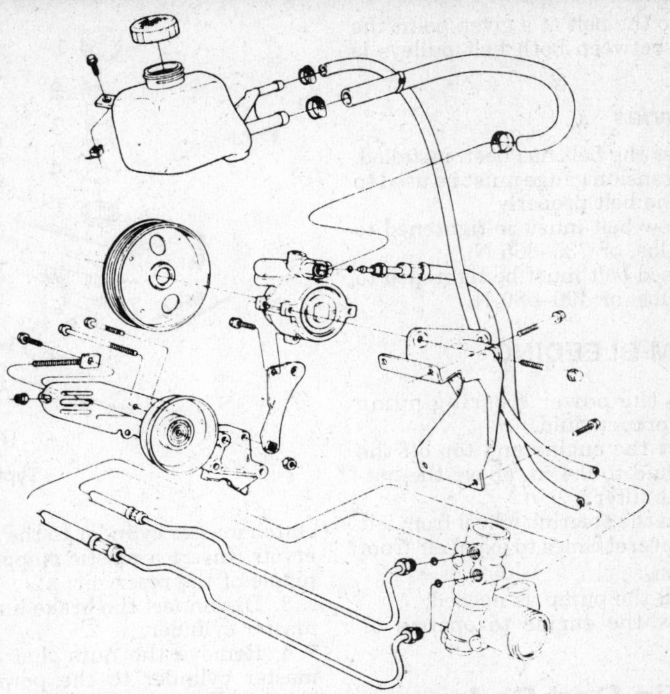

Power steering system and related components – 9000 Series

3. Remove the cover panel from the fire wall. Take care not to damage the gasket, seal and plastic bushing.
4. Raise and support the vehicle safely. Remove both tire and wheel assemblies.
5. Remove the rear section of the inner fender panel under the left fender.
6. Separate the left and right tie rod ends from the steering arms.
7. Drain the power steering fluid from the pump reservoir.
8. Disconnect the hoses from the pump and reservoir. Plug the openings to prevent fluid from leaking out and dirt from entering.
9. Remove the retaining bolts from the rack and pinion assembly.
10. Remove the vertical brace between the engine subframe and the body.
11. Lift out the rack and pinion unit through the left fender inner panel opening. Do not damage the rubber boots or brake hose.

To install:

12. Installation is the reverse of the removal procedure.
13. Tighten the rack and pinion gear securing bolts to 46–56 ft. lbs. (60–80 Nm); the steering column pinch bolt to 27–32 ft. lbs. (35–42 Nm).
14. Fill the reservoir and bleed the system by allowing the engine to run at idle.

Power Steering Pump

REMOVAL & INSTALLATION

900 Series

1. Disconnect the negative battery cable. Drain the fluid from the power steering pump.
2. Drain the coolant from the drain cock on the engine block and disconnect the hose from between the expansion tank and the water pump.
3. Disconnect the power steering pump hoses. Grip the hexagonal nipple on the pump when removing the delivery line.
4. Unbolt the pump unit from the bracket and the engine mounting. Remove the power steering belt. Remove the pump complete with its mounting.
5. Installation is the reverse of the removal procedure.

9000 Series

1. Disconnect the negative battery cable. Remove the fluid from the pump reservoir.
2. Raise and support the vehicle safely. Remove the right front wheel and the right inner fender panel.
3. Remove the drive belt. Remove the bracket for the engine oil filler pipe. Remove the engine stay bracket. Disconnect the hoses from the pump. Plug the openings.
4. Remove the pump retaining bolts. Remove the pump. Note that 1 bolt is located behind the pump pulley and is accessible only through the aperature in the pulley.
5. Installation is the reverse of the removal procedure.

BELT ADJUSTMENT

900 Series

Tighten the belt so when pressure is

applied to the belt at a given point the distance between both belt pulleys is 5–10mm.

9000 Series

1. After the belt has been installed, a strand tension gauge must be used to tighten the belt properly.
2. A new belt must be tightened to 170–200 lbs. or 735–865 N.
3. A used belt must be tightened to 110–130 lbs. or 490–580 N.

SYSTEM BLEEDING

1. Fill the power steering pump with the proper fluid.
2. Start the engine and top off the level of fluid to 0.4 in. above the bottom of the filter.
3. Turn the steering wheel from left to right several times to expel air from the system.
4. Refill the pump as needed.
5. Allow the engine to operate at idle.

Tie Rod Ends

REMOVAL & INSTALLATION

1. Disconnect the negative battery cable. Raise and support the vehicle safely.
2. Remove the tire and wheel assembly. Remove the nut.
3. Disconnect the ball joint bolt from the steering arm using the proper removal tool. Do not knock the ball joint bolt out, as this could cause damage to the ball joint and other related parts.
4. Back off the nut that locks the end assembly to the tie rod.
5. Unscrew the end assembly from the tie rod.
6. Installation is the reverse of the removal procedure. Check and adjust the toe-in as required.

BRAKES

For all brake system repair and service procedures not detailed below, please refer to "Brakes" in the Unit Repair section.

Master Cylinder

REMOVAL & INSTALLATION

1. Disconnect the negative battery cable. Disconnect the electrical connection to the brake warning switch.
2. Disconnect the hose from the clutch master cylinder to the fluid reservoir. Insert a plastic stopper in the nipple of the reservoir.
3. Disconnect the brake lines to the master cylinder.
4. Remove the nuts that hold the master cylinder to the power brake booster. Remove the master cylinder from the vehicle.
5. Installation is the reverse of removal. Bleed the system as required.

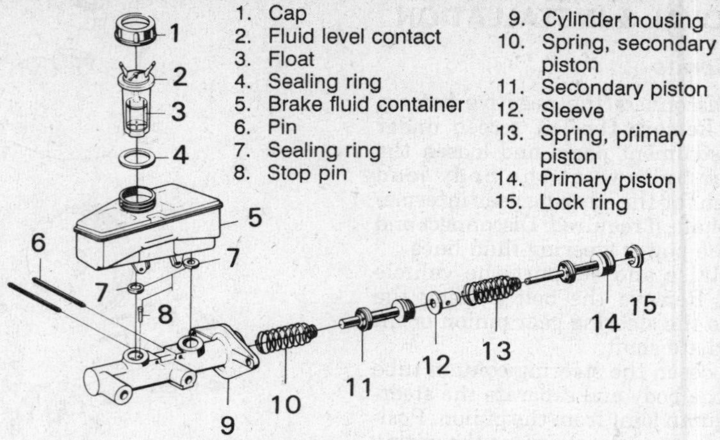

Typical brake master cylinder

Power Brake Booster

REMOVAL & INSTALLATION

1. Disconnect the negative battery cable. Remove the steering column bearing cover, ash tray and safety padding screw. Remove the upper circlip on the brake pedal pushrod, if equipped.
2. Remove the 2 electrical connections on the brake light switch. Remove the safety padding screws in the engine compartment.
3. Remove the vacuum hose from the non-return valve which is located on the vacuum booster.
4. Disconnect the brake lines and the electrical connections for the brake warning switch from the master cylinder. Disconnect the line to the clutch master cylinder from the fluid reservoir. Insert stoppers in the lines to prevent loss of the brake fluid.
5. Remove the cotter pin from the servo unit pushrod at the brake pedal.
6. Remove the vacuum booster together with the master cylinder and the bracket.

NOTE: The bracket is mounted on the dash panel with 4 bolts and nuts. Three of these bolts are accessible from under the passenger's compartment after removal of the screen section and parts of the dash panel insulation felt below the instrument panel. The 4th nut is accessible from the engine compartment by the bracket.

7. Separate the master cylinder and the bracket from the vacuum booster.
8. Installation is the reverse of removal. Tighten the mounting bolts to 16–22 ft. lbs. (22–30 Nm). Bleed the system as required.

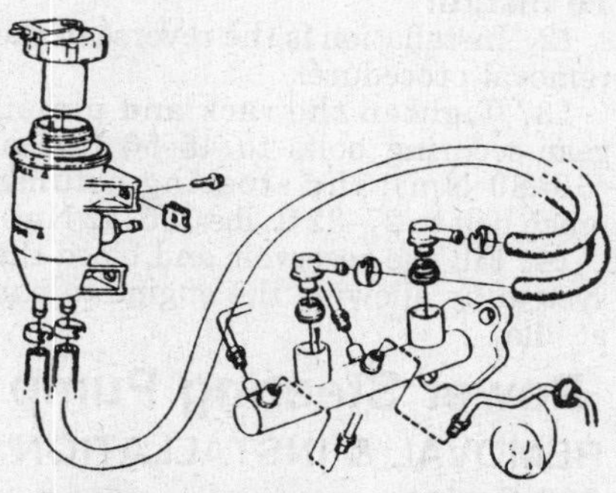

Brake master cylinder and related components – 9000 Series

Brake Caliper

REMOVAL & INSTALLATION

900 Series

1. Raise and support the vehicle safely. Remove the appropriate wheel.
2. Remove the brake pads. Disconnect the handbrake cable from the brake housing.
3. Unscrew the brake pipes from the caliper. Insert a rubber stopper to prevent leakage and contamination.
4. Remove the bolts holding the brake housing to the steering knuckle.

To install:

5. Check that the dust cover has not slipped out of position. Bolt the brake assembly to the steering knuckle using a new locking plate.
6. Connnect the brake lines.
7. On front brakes, adjust the handbrake cable so that the clearance

between the lever and the yoke is 0.019 in. (0.5 mm) with the handbrake in the off position.

8. Install the brake pads, top off the brake fluid and bleed the brake system.

9. With the engine off, pump the brake pedal to seat the pads. On front brakes, after pumping the pedal, pull the handbrake up 5 notches. Continue to pump the pedal until the handbrake operates after the lever has been pulled up 2–4 notches.

10. Install the wheel, lower the vehicle and test drive.

9000 Series

FRONT

1. Raise and support the vehicle safely. Remove the appropriate wheel.
2. Loosen the brake hose fitting on the caliper.
3. Remove the lower guide pin bolt and pivot the caliper up to remove the brake pads.
4. Remove the upper guide pin bolt. Disconnect the brake hose from the caliper and install a dust cap over the hose to prevent leakage and contamination. Remove the caliper assembly

To install:

5. Check that the guide pins slide freely in their bores and lubricate as necessary. Inspect the dust cover for damage and replace as necessary.
6. Reconnect the brake hose to the caliper but leave the fitting loose.
7. Install the caliper and tighten the upper guide pin bolt.
8. Install the brake pads. Tighten the lower guide pin bolt.
9. Tighten the brake hose at the caliper. Bleed the brake system.
10. Install the wheel, lower the vehicle and pump the brake pedal to seat the brake pads against the rotor.

REAR

1. Raise and support the vehicle safely. Remove the appropriate wheel.
2. Unhook the handbrake cable from the lever on the caliper. Fully depress the brake pedal and secure it in that position.
3. Loosen the brake hose fitting at the caliper. Remove the dust caps and unscrew the guide pins.
4. Remove the brake pad retaining clip and lift off the caliper body. Remove the brake pads.
5. Disconnect the brake hose from the caliper by rotating the caliper. Install a dust cap on the hose end to prevent leakage and contamination.

To install:

6. Check that the guide pins slide freely in their bores and lubricate as necessary. Inspect the dust cover for damage and replace as necessary.
7. Install the brake pads. Install the caliper on the rear axle and tighten the guide pins. Install the dust caps.
8. Install the retaining clip, tighten the brake hose fitting and reconnect the handbrake cable to the lever.
9. Adjust the handbrake, bleed the brake system and install the wheel. Lower the vehicle.
10. Pump the brake pedal to seat the brake pads on the rotor. Test drive the vehicle.

Disc Brake Pads

REMOVAL & INSTALLATION

900 Series

FRONT

1. Raise and support the vehicle and support it safely. Remove the wheel.
2. Clean the brake housing.
3. Rotate the brake disc so 1 of the recesses in the edge of the disc aligns with the brake pads.
4. Remove the damper spring, pin retaining clip and pad retaining pin. If the pad retaining pin is difficult to remove, use a tapping-out tool 83 90 270, or equivalent, and removal tool 89 96 175, or equivalent.
5. Withdraw the brake pads, if the pads are seating firmly, use pad extractor No. 89 95 771, or equivalent.

To install:

6. Siphon a sufficient quantity of brake fluid from the master cylinder reservoir to prevent the brake fluid from overflowing the master cylinder when installing new pads. This is necessary as the piston must be forced into the cylinder bore to provide sufficient clearance to install the pads.
7. Inspect the caliper and piston assembly for breaks, cracks or other damage. Overhaul or replace the caliper as necessary.
8. Push the piston next to the rotor back into the cylinder bore until the end of the piston is flush with the boot retaining ring. Rotate the piston using tool 89 96 043, or equivalent, while simultaneously pushing the piston back into the cylinder. The automatic handbrake is reset this way.

CAUTION

If the piston is pushed further than this, the seal will be damaged and the caliper assembly will have to be overhauled.

NOTE: Check that the position of the piston has not displaced the dust cover and the yoke moves easily in the groove on the brake housing.

9. Fit the new brake pads together with the pad retaining pin, pin retaining clip and damper spring.
10. Check the adjustment of the handbrake cable. Check the distance between the lever and the yoke. The clearance should be 0.019 in. (0.5mm) maximum, on both sides. If adjustment is required, make it on the adjustment nut on the handbrake lever.

NOTE: Note that the cable cross over, which implies that the right adjusting nut should be used to adjust the left brake mechanism and vice-versa.

11. Refill the master cylinder with fresh brake fluid.
12. With the engine switched off, pump the brake pedal repeatedly until the foot brake starts to operate.
13. Pull the handbrake lever up 5 notches. Continue to pump the brake pedal until the handbrake operates after having been pulled up a further 2–4 notches.
14. Install the tire and wheel assembly. Pump the brake pedal several times to bring the pads into adjustment. Road test the vehicle. If a firm pedal cannot be obtained, bleed the brakes.

REAR

1. Raise the vehicle and support it with safely. Remove the wheel.
2. Clean the brake housing.
3. Tap out the brake pad retaining pins using a 0.11 in. (2.5mm) drift. Save the retaining spring.
4. Withdraw the brake pads. If they are difficult to remove, use extractor tool 89 95 771, or equivalent.

To install:

5. Siphon a sufficient quantity of brake fluid from the master cylinder reservoir to prevent the brake fluid from overflowing the master cylinder when installing new pads. This is necessary as the piston must be forced into the cylinder bore to provide sufficient clearance to install the pads.
6. Inspect the caliper and piston assembly for breaks, cracks or other damage. Overhaul or replace the caliper as necessary. Check that the dust cover retainer is properly in position and the cover is in good condition.
7. Push the piston back into the cylinder bore just enough to allow clearance for installation of the new pads.
8. Fit the pad retaining pins and the pin retaining clip.
9. Refill the master cylinder to the correct level with the proper brake fluid.
10. Replace the wheel and lower the vehicle. Pump the brake pedal several times to bring the pads into correct adjustment. Road test the vehicle.

NOTE: If a firm pedal cannot be obtained, the system will require bleeding.

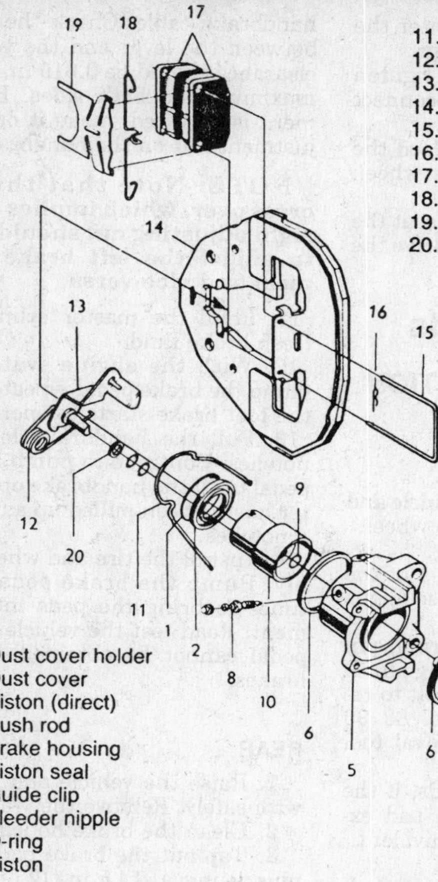

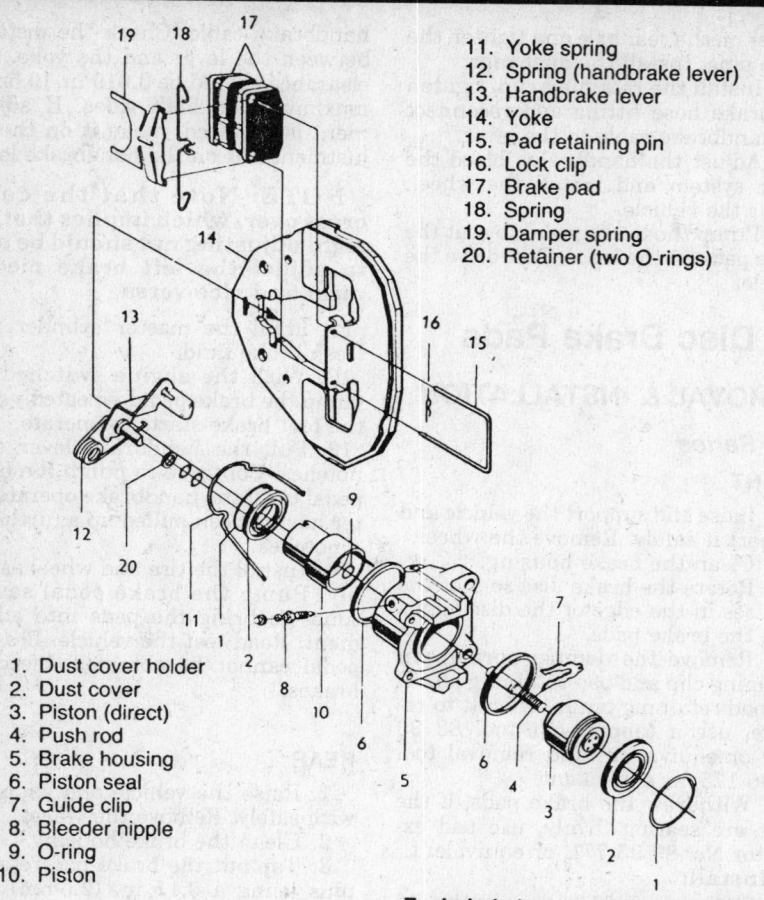

11. Yoke spring
12. Spring (handbrake lever)
13. Handbrake lever
14. Yoke
15. Pad retaining pin
16. Lock clip
17. Brake pad
18. Spring
19. Damper spring
20. Retainer (two O-rings)

1. Dust cover holder
2. Dust cover
3. Piston (direct)
4. Push rod
5. Brake housing
6. Piston seal
7. Guide clip
8. Bleeder nipple
9. O-ring
10. Piston

Exploded view of the front brake caliper

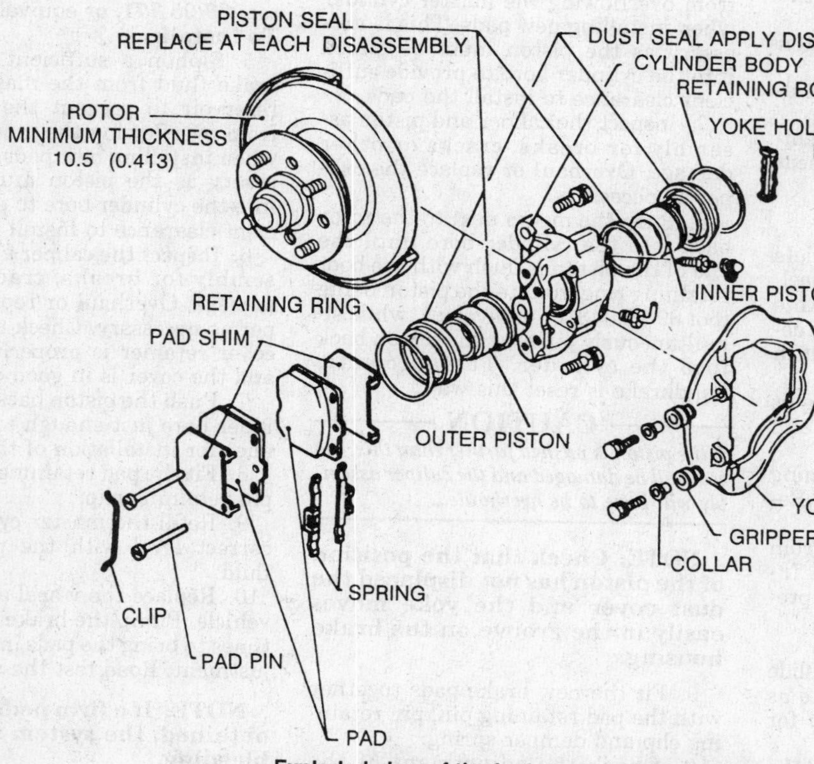

Exploded view of the brake caliper (sliding type)

9000 Series

FRONT

1. Raise and support the vehicle and support it safely. Remove the wheel.
2. Remove the lower guide pin bolt.
3. Pivot the caliper upwards and remove the pads.
4. Check that the guide pins slide freely and the dust covers are in good condition.
5. Clean the abutment surfaces between the pads and the carrier.

To install:

6. Fit the new pads and pivot the hydraulic body back to its normal position.
7. Refit and tighten the bolt in the lower guide pin.
8. Install the wheel and lower the vehicle.
9. Pump the brake pedal to move the pads to their operating positions.

REAR

1. Raise the vehicle and support it with safely. Remove the wheel.
2. Release the handbrake and remove the retaining spring.
3. Slide the handbrake cable out of the slot of the lever.
4. Remove the dust caps and then use a 7mm Allen key, hexagon bit adapter, to remove the guide pins.
5. Lift off the hydraulic body and remove the pads.

To install:

6. Remove the screw plug from the handbrake adjusting screw and screw the piston into the hydraulic body by means of the adjusting screw.
7. Place the new pads into position.
8. Replace the hydraulic body and install the guide pins complete with dust caps.
9. Install the retaining spring for the handbrake lever.
10. Screw the adjusting screw fully home and then back it off approximately ¼–½ turn. Check that the brake disc is running freely and refit the plug.
11. Install the handbrake cable to the lever and check with a feeler gauge, that the clearance between the lever and the stop is 0.02–0.06 in. (0.5–1.5mm) Adjust as necessary by means of the adjusting screw at the handbrake lever inside the vehicle.
12. Replace the wheel and lower the vehicle.

Brake Rotor

REMOVAL & INSTALLATION

1. Raise and support the vehicle safely. Remove the appropriate wheel.
2. On 900 Series rear rotors, apply the hand brake and loosen the locknut.
3. Remove the brake caliper and hand out of the way using a piece of wire. Do not allow the caliper to hang by the brake hose.
4. Remove the bolts holding the rotor to the hub. Remove the rotor.

To install:

5. Check the rotor runout. If not within specification have the rotor ground.
6. Installation is the reverse of removal. Tighten the rotor retaining bolts to 52–81 ft. lbs. (70–110 Nm).
7. On 900 Series rear rotors, tighten the locknut to 250–265 ft. lbs. (340–360 Nm). Lock the nut into position by peening it into the groove in the shaft with a punch.

Parking Brake Cable

ADJUSTMENT

900 Series

1988–90

Check the adjustment of the handbrake cable. Check the distance between the handbrake lever and the yoke: the clearance should be a maximum 0.019 in. and should be equal on both sides. Adjust as necessary using the adjustment nut on the handbrake lever. Note that the cables cross over therefore, the right adjustment nut should be used to adjust the left brake mechanism and vice-versa.

1991–92

Tilt up the rear seat (remove the seat on convertibles) and insert a 0.06 in. (1.5mm) feeler gauge between the handbrake lever and the stop. Screw the locknut down onto the adjusting sleeve until the feeler gauge slips out. Reinstall the rear seat. Apply and release the handbrake lever a few times to seat the cable.

9000 Series

The parking brake is self adjusting on the rear calipers, however if necessary, check with a feeler gauge, that the clearance between the lever and the stop is 0.02–0.06 in. (0.5–1.5mm) Adjust as necessary by means of the adjusting screw at the handbrake lever inside the vehicle.

Brake System Bleeding

The brake fluid reservoir is part of the normal brake system and is filled or checked in the usual manner. Always clean the reservoir cap and surrounding area thoroughly before removing the cap. Fill the reservoir only to the FULL or MAX mark; do not overfill. Use only fresh DOT 3 brake fluid from unopened containers. Do not use any fluid containing a petroleum base. Do not use any fluid which has been exposed to water or moisture. Failure to use the correct fluid will affect system function and component life.

Master Cylinder

If the master cylinder has been emptied of fluid, it must be bled separately from the rest of the system. Since the cylinder has no check valve, air can become trapped within it. To bleed the brake master cylinder after it has been drained:

1. Disconnect the 2 brake lines from the master cylinder. Plug the lines immediately. The brake fluid reservoir must be in place and connected to the master cylinder. Check the fluid level before beginning.
2. An assistant should slowly depress and hold the brake pedal.
3. With the pedal held down, use fingers to plug each outlet port on the master cylinder; release the brake pedal.
4. Repeat Steps 2 and 3 three or four times. The air will be bled from the cylinder.
5. Connect the brake lines to the master cylinder and tighten the fittings.
6. Start the engine, allowing the system to pressurize. Shut the ignition **OFF** and bleed the brake lines at the wheels.

Lines and Calipers

The brake system must be bled any time a line, hose or component is loosened or removed. Any air trapped within the lines can affect pedal feel and system function. Bleeding the system is performed in the usual manner with an assistant in the car to pump the brake pedal. Make certain the fluid level in the reservoir is maintained at or near correct levels during bleeding operations.

With the ignition **OFF**, depress the brake pedal several times until pedal feel changes to a noticeably stiffer resistance. Slowly pump the pedal a few more times; with the pedal depressed, loosen the bleeder screw ⅓–½ turn. Tighten the bleeder screw before the fluid pressure is gone. Release the pedal, pump again slowly, hold the pedal depressed and repeat the bleeding process until no air bubbles are seen in the fluid.

If all bleeding is necessary at all wheels, begin at the right rear, then the left front, left rear and right front wheels.

Anti-Lock Brake System Service

PRECAUTIONS

- If the vehicle is equipped with an air bag system, always properly disable the system before commencing work on the ABS system.
- Certain components within the ABS system are not intended to be serviced or repaired individually. Only those components with REMOVAL & installation procedures should be serviced.
- Do not use rubber hoses or other parts not specifically specified for the ABS system. When using repair kits, replace all parts included in the kit. Partial or incorrect repair may lead to functional problems and require the replacement of other components.
- Lubricate rubber parts with clean, fresh brake fluid to ease assembly. Do not use lubricated shop air to clean parts; damage to rubber components may result.
- Use only brake fluid from an unopened container. Use of suspect or contaminated brake fluid can reduce system performance and/or durability.
- A clean repair area is essential. Perform repairs after components have been thoroughly cleaned; use only denatured alcohol to clean components. Do not allow ABS components to come into contact with any substance containing mineral oil; this includes used shop rags.
- The control unit is a microproces-

sor similar to other computer units in the vehicle. Insure that the ignition switch is **OFF** before removing or installing controller harnesses. Avoid static electricity discharge at or near the controller.

- Never disconnect any electrical connection with the ignition switch **ON** unless instructed to do so in a test.
- Avoid touching connector pins with fingers.
- Leave new components and modules in the shipping package until ready to install them.
- To avoid static discharge, always touch a vehicle ground after sliding across a vehicle seat or walking across carpeted or vinyl floors.
- If any arc welding is to be done on the vehicle, the ABS control unit should be disconnected before welding operations begin.
- Never allow welding cables to lie on, near or across any vehicle electrical wiring.
- If the vehicle is to be baked after paint repairs, disconnect and remove the control unit from the vehicle.

RELIEVING ANTI-LOCK BRAKE SYSTEM PRESSURE

The anti-lock brake system contains brake fluid under extreme pressure. Parts of the hydraulic modulator may contain brake fluid at pressures up to 3045 psi (210 bar). Extreme care must be taken when working on lines and components.

Always discharge the system pressure completely before loosening any brake line or performing any component replacement. To discharge the system pressure:

1. Turn the ignition switch **OFF** during the procedure.
2. Step firmly on the brake pedal at least 25–30 times. The system is not completely discharged until a distinct change is felt in the brake pedal. The effort needed to press the pedal will clearly increase when the system pressure is released.
3. Once the pedal feel changes, pump the pedal a few more times.
4. Perform tests and/or repairs as necessary with the ignition **OFF** at all times. If the ignition is turned **ON**, the pump will run, repressurizing the system.

NOTE: A running production change was made on 9000 Series vehicles during model year 1989. Some procedures for vehicles before and after the change are different. The term early production refers to 9000 Series vehicles through VINs K1009731, K2004996 or K8000276. Late production begins with VINs K1009732, K2004997, K8000277 and covers all subsequent years and models.

Hydraulic Modulator

REMOVAL & INSTALLATION

900 Series

1. Disconnect the negative battery cable. Depressurize the system.
2. Thoroughly clean the unit, connections and surrounding surfaces to prevent dirt from entering the hydraulics.
3. Remove the center console.
4. Remove the padded trim panel under the dash.
5. Remove the heater duct.
6. Remove the insulation from behind the pedal assembly; disconnect the left defroster hose from the heater box.
7. Remove the retaining clip and remove the pin from the hydraulic unit pushrod.
8. Remove the air intake assembly.
9. Unbolt the coolant expansion tank; move it out of the way without disconnecting the hoses.
10. Disconnect the 4 electrical connectors from the hydraulic unit.
11. Unbolt the bracket between the hydraulic unit and the front assembly. Disconnect the wiring harness to the sensor and disconnect the ground strap at the hydraulic unit. Move the bracket and wiring out of the way.
12. Use a syringe or similar tool to remove as much fluid as possible from the hydraulic unit reservoir. It will not be possible to remove all the fluid.
13. For vehicles with manual transmissions, disconnect the hose running to the clutch system from the reservoir. Immediately plug the hose end; do not allow air to enter the clutch system.
14. Label or diagram the placement of the brake lines and large-diameter return line at the hydraulic unit. Disconnect the lines together to avoid undue strain on any one line. Immediately plug the lines and the ports on the valve block to prevent dirt from entering the system.
15. Inside the vehicle, remove the 4 nuts holding the hydraulic unit to the firewall. Remove the hydraulic unit by lifting it out of the engine compartment.

To install:

NOTE: During installation, it is possible for the brake switch and/or the cruise control switch to be pressed in inadvertently. If this occurs, the switch may be reset with a pair of pliers.

16. Place the hydraulic unit in position. Make certain the pushrod is correctly aligned with the pedal assembly.
17. Connect the pushrod with the pin and clip.
18. Install the 4 retaining nuts and tighten them to 19 ft. lbs. (26 Nm).
19. Connect the brake lines and the return line.
20. For manual transmission vehicles, connect the clutch hose to the reservoir, making sure no fluid is lost from the hose.
21. Connect the sensor connector and install the bracket at the front of the assembly. Connect the ground strap to the top bolt.
22. Connect the 4 electrical harnesses. Make sure the connectors are squarely placed and firmly fitted.
23. Install the coolant tank and reinstall the air intake.
24. Fill the hydraulic unit reservoir with clean, fresh DOT 4 fluid. Switch the ignition **ON** to confirm that the pump is working. The pump should shut off within 60–90 seconds when correct pressure is achieved. If the pump does not shut off within 120 seconds, shut the ignition **OFF** immediately and allow the pump at least 10 minutes to cool.
25. Inspect the system for leaks. Make certain the ABS and BRAKE warning lamps go off.
26. Connect the left defroster hose and reinstall the insulation at the pedals.
27. Install the heater duct; install the trim below the dash.
28. Install the center console.
29. Test drive the vehicle to confirm correct operation of the service brakes, the ABS and the clutch, if so equipped.

9000 Series

NOTE: A running production change was made during model year 1989. Some procedures for vehicles before and after the change are different. The term early production refers to 9000 Series vehicles through VINs K1009731, K2004996 or K8000276. Late production begins with VINs K1009732, K2004997, K8000277 and covers all subsequent years and models.

EARLY PRODUCTION

1. Disconnect the negative battery cable. Depressurize the system.
2. Thoroughly clean the unit, connections and surrounding surfaces to prevent dirt from entering the hydraulics.
3. Remove the 5 screws holding the lower dash panel below the steering column.
4. Remove the retaining clip hold-

ing the pin to the pushrod and remove the pin.

5. Disconnect and remove the battery.

6. Gain access to the hydraulic unit by removing the following components: the clip holding the wire harness to the battery shelf, the terminal block and harness from the battery shelf, the fuel filter, the battery shelf and the rubber elbow between the throttle housing and the inlet manifold.

7. Disconnect the 4 wiring connectors from the hydraulic unit. Release the wire harness retainers. Disconnect the ground strap from the hydraulic unit.

8. Remove the bracket from the hydraulic unit.

9. Use a syringe or similar tool to remove as much fluid as possible from the hydraulic unit reservoir. It will not be possible to remove all the fluid.

10. For vehicles with manual transmissions, remove the clutch fluid hose from the hydraulic unit reservoir. Immediately plug the end of the hose; do not allow air to enter the clutch system.

11. Raise and support the vehicle safely.

12. Remove the left front wheel and the left front wheel arch liner or inner fender.

13. Disconnect the electrical connector for the pump motor at the hydraulic unit.

14. Clean the brake fittings and valve block thoroughly. Disconnect the brake lines from the valve block and immediately plug the lines and ports to prevent the entry of dirt.

15. Remove the 4 nuts holding the hydraulic unit to the firewall. Lift the hydraulic unit out of the engine compartment.

To install:

NOTE: During reinstallation, it is possible for the brake switch and/or the cruise control switch to be pressed in inadvertently. If this occurs, the switch may be reset with a pair of pliers.

16. Place the hydraulic unit in position. Make certain the pushrod is correctly aligned with the pedal assembly.

17. Connect the pushrod with the pin and clip.

18. Install the 4 retaining nuts and tighten them to 19 ft. lbs. (26 Nm).

19. Reconnect the brake lines to the valve block.

20. Connect the pump motor harness at the hydraulic unit.

21. Install the bracket on the hydraulic unit; install the support and wire harness retainers.

22. Connect the 4 wiring harnesses to the hydraulic unit.

23. Install the rubber elbow at the throttle housing, the battery shelf, the fuel filter, the terminal block and the wire harness retainers.

24. Install the battery and connect the cables. Leave the ignition switched **OFF**.

25. For manual transmission vehicles, reconnect the fluid hose to the reservoir.

26. Fill the reservoir with DOT 4 fluid brake fluid.

27. Switch the ignition **ON** and confirm that the pump is working. The pump should shut off within 60–90 seconds when correct pressure is achieved. If the pump does not shut off within 120 seconds, shut the ignition **OFF** immediately and allow the pump at least 10 minutes to cool.

28. Check the system for leaks. Make certain the ABS and BRAKE warning lamps are off.

29. Bleed the brake system.

30. Reinstall the lower dash panel.

31. Install the fender liner and wheel. Lower the vehicle to the ground.

32. Test drive the vehicle to confirm correct operation of the service brakes, the ABS and the clutch, if so equipped.

LATE PRODUCTION

1. Disconnect the negative battery cable. Depressurize the system.

2. Thoroughly clean the unit, connections and surrounding surfaces to prevent dirt from entering the hydraulics.

3. Remove the 5 screws holding the lower dash panel below the steering column.

4. Remove the retaining clip holding the pin to the pushrod and remove the pin.

5. Disconnect and remove the battery.

6. Gain access to the hydraulic unit by removing the following components: the 2 clips holding the positive battery cable at the bottom of the battery shelf, the terminal block and wires at the front of the battery shelf, the bracket and connector(s) at the rear of the battery shelf and the fuel filter from the battery shelf.

7. Remove the battery shelf and move the fuel filter out of the way.

8. Use a syringe or similar tool to remove as much fluid as possible from the hydraulic unit reservoir. It will not be possible to remove all the fluid.

9. Raise and support the vehicle safely.

10. Remove the left front wheel and the left front wheel arch liner or inner fender. Remove the support between the fender and the subframe.

11. Disconnect the 5 wiring connectors from the hydraulic unit. Release the wire harness retainer and disconnect the ground strap from the hydraulic unit.

12. For vehicles with manual transmissions, remove the clutch fluid hose from the hydraulic unit reservoir. Immediately plug the end of the hose; do not allow air to enter the clutch system.

13. Clean the brake fittings and valve block thoroughly. Disconnect the brake lines from the valve block and immediately plug the lines and ports to prevent the entry of dirt.

14. Use an 8mm hex bit to remove the accumulator from the hydraulic unit.

15. Remove the 4 nuts holding the hydraulic unit to the firewall. Lift the hydraulic unit out of the engine compartment.

To install:

NOTE: During installation, it is possible for the brake switch and/or the cruise control switch to be pressed in inadvertently. If this occurs, the switch may be reset with a pair of pliers.

16. Place the hydraulic unit in position. Make certain the pushrod is correctly aligned with the pedal assembly.

17. Connect the pushrod with the pin and clip.

18. Install the 4 retaining nuts and tighten them to 19 ft. lbs. (26 Nm).

19. Install the accumulator. Tighten it to 28 ft. lbs. (38 Nm).

20. Reconnect the brake lines to the valve block.

21. For manual transmission vehicles, reconnect the fluid hose to the reservoir. Make certain there is fluid in the hose before connecting it.

22. Connect the ground strap.

23. Connect the wiring harnesses to the hydraulic unit.

24. Install the fender support.

25. Reinstall the battery shelf.

26. Install the 2 clips at the bottom of the battery shelf, the terminal block and harness, the bracket and connector(s) at the rear of the shelf and the fuel filter.

27. Install the battery and connect the cables. Leave the ignition switched **OFF**.

28. Fill the reservoir with DOT 4 fluid brake fluid.

29. Switch the ignition **ON** and confirm that the pump is working. The pump should shut off within 60–90 seconds when correct pressure is achieved. If the pump does not shut off within 120 seconds, shut the ignition **OFF** immediately and allow the pump at least 10 minutes to cool.

30. Check the system for leaks. Make certain the ABS and BRAKE warning lamps are off.

31. Bleed the brake system.

32. Reinstall the lower dash panel.

33. Install the fender liner and wheel. Lower the vehicle to the ground.
34. Test drive the vehicle to confirm correct operation of the service brakes, the ABS and the clutch, if so equipped.

Electronic Control Unit (ECU)

REMOVAL & INSTALLATION

900 Series

1. Disconnect the negative battery cable.
2. Unbolt the ECU from the left fender in the engine compartment.
3. Lift the unit and disconnect the wire harness.
4. When reinstalling, connect the wire harness first, fit the unit into position and secure the mounting bolts.
5. Connect the negative battery cable.

9000 Series

1. Disconnect the negative battery cable.
2. Remove the left cover on the firewall panel.
3. Loosen the clips holding the unit.
4. Lift the ECU and disconnect the wire harness.
5. Install in reverse order, making sure the clips are secure. Connect the negative battery cable.

CHASSIS ELECTRICAL

Air Bag

DISARMING

Always disconnect the negative battery cable and wait 20 minutes prior to servicing any component that may trigger the air bag. Do not use any diagnostic instruments that are battery powered to diagnose faults in the steering wheel or electronic control unit. Using such devices may trigger the air bag.

Heater Blower Motor

REMOVAL & INSTALLATION

Without Air Conditioning

900 SERIES

1. Disconnect the negative battery cable.
2. Remove the switch panel and the upper section of the instrument panel. The screws retaining the switch panel are of different lengths. The screws are marked with grooves. Note the position of the screws for reassembly.
3. Disconnect the electrical leads to the fan motor.
4. Remove the retaining screws for the right defroster valve housing.
5. Remove the fan retaining screws. Remove the fan from its housing.
6. The installation is in the reverse of the removal procedure.

9000 SERIES

1. Disconnect the negative battery cable.
2. Remove the cover from the windshield wipers.
3. Remove the fresh air filter assembly.
4. Unplug the connectors for the fan motor and fan resistors.
5. Disconnect the temperature control cable.
6. Release the clips on either side of the fan body and turn the body diagonally upwards.
7. Remove the screws in the center of the casing, release the clips and remove the grille from the discharge duct.
8. Separate the fan casing. Remove the screw securing the fan motor.
9. Lift the cover for the lead and withdraw the motor complete with impeller.
10. Installation is the reverse of the removal procedure.

With Air Conditioning

900 SERIES

1. Disconnect the negative battery cable.
2. Remove the switch panel and the upper section of the instrument panel. The screws are marked with grooves. Note the position of the screws for reassembly.
3. Disconnect the electrical leads to the fan motor.
4. Remove the retaining screws for the right defroster valve housing.
5. Remove the fan retaining screws. Remove the fan from its housing.
6. The installation is in the reverse of the removal procedure.

9000 SERIES

1. Disconnect the negative battery cable. Remove the hood assembly.
2. Disconnect the wiper arms. Remove the covers on the evaporator and wiper motor. Unplug the connector for the fan control unit on vehicles with automatic climate control.
3. Remove the false fire wall panel.
4. Remove the plastic drainage tube moulding below the windshield moulding.
5. Remove the securing bolts the electronic ignition control unit and position it aside.
6. Remove the clip and unplug the connectors. Remove the complete wiper assembly.
7. Remove the rubber lead through panel for the coolant hoses. Drain cooling system. Disconnect the quick release couplings for the coolant hoses at the heat exchanger.
8. Remove the throttle dashpot assembly.
9. Remove the vacuum pump retaining screws. Position the pump aside.
10. Remove the evaporator body retaining screws and the clips for the refrigerant hoses.
11. Remove the lock washer and disconnect the cable for the temperature valve.
12. Carefully lift the evaporator and remove the clips on either side of the fan. Remove the complete fan assembly by twisting the fan diagonally upwards.
13. Remove the screw in the center of the casing. Release the clips and the grille at the discharge duct.
14. Separate the fan housing and undo the securing screw for the fan motor.
15. Lift the cover upward and withdraw the motor complete with the impeller.
16. Installation is the reverse of removal. Be careful not to separate the connector for the radiator fan control when installing the false fire wall panel.

Windshield Wiper Motor

REMOVAL & INSTALLATION

900 Series

1. Disconnect the negative battery cable. Remove the wiper arms from the vehicle. Remove the rubber grommets.
2. Remove the 4 mounting screws. Disconnect the electrical lead. Remove the wiper unit from the vehicle.
3. Separate the wiper motor from the wiper assembly.
4. Installation is the reverse of the removal procedure.

9000 Series

1. Disconnect the negative battery cable.
2. Raise the covers on the wiper arms, remove the retaining nuts and lift the arms off.
3. Remove the rubber grommets

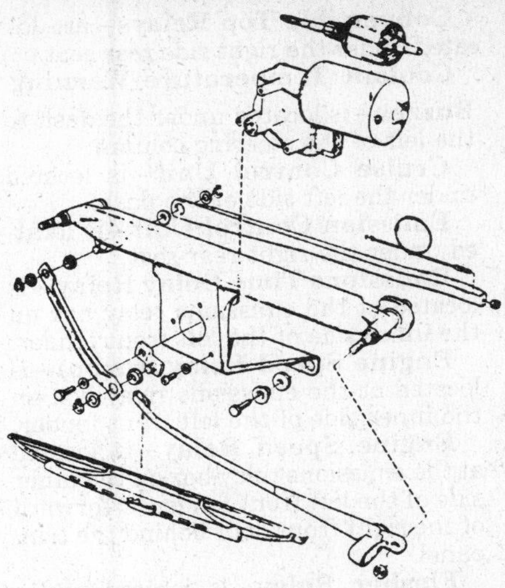

Windshield wiper and motor assembly

from the spindles and remove the 4 bulkhead panel bolts.
4. Lift the bulkhead panel from the vehicle.
5. Disconnect the electrical connector from the wiper motor.
6. Remove the spindle nuts and remove the 4 retaining bolts for the wiper motor bracket.
7. Push downward and pull forward on the pushrod for the left wiper.
8. Lift out the wiper motor assembly complete with the bracket and the pushrod linkage.
9. Installation is the reverse of the removal procedure.

Windshield Wiper Switch

REMOVAL & INSTALLATION

1. Disconnect the negative battery cable.
2. Pull the steering wheel as far forward as it will go. Remove the cover from under the steering column assembly.
3. Disconnect the electrical connector from the switch assembly.
4. Remove the switch retaining screws. Remove the switch from the vehicle.
5. Installation is the reverse of the removal procedure.

Instrument Cluster

REMOVAL & INSTALLATION

900 Series

1. Disconnect the negative battery cable. Remove the steering wheel.
2. Remove the 4 screws in the switch panel and tilt the panel back.
3. Remove the left speaker/defroster grille. Pull apart the instrument panel connectors. Disconnect the speedometer cable.
4. Remove the instrument panel retaining screws. Carefully remove the unit from the vehicle.
5. Installation is the reverse of the removal procedure.

9000 Series

1. Disconnect the negative battery cable.
2. Remove the speaker grilles on either side of the panel.
3. Unscrew the top section of the instrument panel, which is retained by 7 screws including 1 in the glove box.
4. Lift off the top instrument panel section.
5. Remove the air duct from the opening in the top.
6. Disconnect the speedometer cable, the vacuum hoses to the turbo pressure gauge and unplug all connectors to the display panel.
7. Remove the 2 screws of the instrument display panel.
8. Withdraw the instrument cluster through the top of the instrument panel.
9. Installation is the reverse of the removal procedure. Be sure the air duct fitting is tight when reassembling the duct tubing.

Radio

REMOVAL & INSTALLATION

1. Disconnect the negative battery cable.
2. Remove the instrument panel trim around the radio. It may be necessary to remove the radio tuning post nuts or attaching screws first.
3. Slowly remove the radio untill the wire connectors can be reached. Disconnect all wires from the radio. Remove the radio.

To install:

4. Connect all wires to the radio and place the radio in the dash.
5. Install the attaching screws or tuning post nuts. Install the instrument panel trim around the radio.
6. Connect the negative battery cable and test the radio.

Headlight Switch

REMOVAL & INSTALLATION

1. Disconnect the negative battery cable.
2. Pull the switch from its mounting on the instrument panel assembly.
3. Disconnect the electrical connectors from the switch.
4. Remove the switch from the vehicle.
5. Installation is the reverse of the removal procedure.

Combination Switch

REMOVAL & INSTALLATION

1. Disconnect the negative battery cable.
2. Remove the steering wheel.
3. Remove the cover under the bearing support.
4. Remove the combination switch retaining bolts and electrical connections.
5. Remove the switch from the vehicle.
6. Installation is the reverse of the removal procedure.

Ignition Switch

REMOVAL & INSTALLATION

900 Series

1. Disconnect the negative battery cable.
2. Remove the center console.
3. Disconnect the electrical connections from the switch.
4. Remove the assembly from the vehicle.
5. Installation is the reverse of removal.

9000 Series

1. Disconnect the negative battery cable.
2. Remove the steering wheel assembly.
3. Remove the cover panels from the wiper/washer and direction indicator switches.
4. Remove the upper section of the instrument panel. Remove the instrument cluster assembly.
5. Remove the clip securing the wiring loom and flexible ducts to the steering column.
6. Unplug the connector for the wipers, direction signals and leads for the horn switch and ignition switch.
7. Remove the pinch bolt in the upper joint, loosen the other bolts and withdraw the universal joint from the splines on the steering column shaft.
8. Remove the steering column wheel adjustment assembly by tapping out the roll pin and removing the nut and washer. Withdraw the shaft from the clamp and lift the upper section of the steering wheel adjustment assembly. Remove the 3 socket head bolts and lift off the holder for the directional indicator unit.
9. Remove the upper section of the steering column, removing the rubber bushing completely from the housing.

10. Remove the shake-proof washer. Remove the column bearing.
11. With the switch support out remove the socket head screws and remove the ignition switch.
12. To remove the cylinder, turn the ignition key to position **1**, press in on the locking tab and withdraw the cylinder.

To install:

13. Install the cylinder, socket head screws, shake-proof washer and column bearing. Install the rubber bushing and the upper section of the column.
14. Install the steering column wheel adjustment assembly using a new roll pin. Install the directional indicator unit.
15. Install the universal joint and pinch bolt. Connect all electrical connections. Install the cover panels.
16. Install the steering wheel and connect the negative battery cable.

Stoplight Switch

REMOVAL & INSTALLATION

1. Disconnect the negative battery cable.
2. Remove the necessary trim and padding to gain access to the switch assembly.
3. Disconnect the electrical connections from the switch assembly.
4. Remove the switch from its mounting.
5. Installation is the reverse of the removal procedure.

Neutral Safety Switch

ADJUSTMENT

1. Disconnect the wires from the switch. The wide terminals are for back-up lights and the narrow ones are for the starter motor.
2. Loosen the locknut and unscrew the switch 2 turns.
3. With the selector in **D**, connect a test light between the narrow terminals. The light should light up.
4. Screw in the switch until the light goes out. Mark that position on both the transmission and the switch.
5. Move the test light, the wide terminals and screw switch in until the light goes out again. Count the number between the lights going out.
6. Turn the switch to a point halfway between the 2 lights-out points.
7. Secure the locknut to 4–6 ft. lbs. torque. If the safety switch is locked too tight, it may be damaged.

Fuses, Circuit Breakers and Relays

LOCATION

900 Series

Fuses are located in the electrical distribution box on the left hand wheel housing under the hood. The oxygen sensor fuse is located on the right side of the engine compartment at the fresh air intake. The convertible top fuse is located in the engine compartment at the right side distribution block. Passive seat belt fuses are located under the left rear seat.

9000 Series

Fuses are located in a fuse box and can be reached through an access panel in the glove compartment. Fuses are also located in the electrical distribution box near the left headlight. ABS fuses are located in the engine compartment behind the firewall partition on the ABS relay and fuse board.

Fusible Links

LOCATION

Fusible links may be located at the battery, starter or alternator.

Relays and Computers

LOCATION

900 Series

Air conditioner Compressor Relay—is located at the electrical distribution box on the left hand wheel housing in the engine compartment.

APC Control Unit—is located in the engine compartment forward of the left wheel housing.

Central Door Lock Control Unit—is located under the right side of the dash.

Convertible Top Relays—are located under the right side rear seat.

Coolant Temperature Warning Buzzer—is located under the dash to the left of the steering column.

Cruise Control Unit—is located under the left side of the dash.

Emission Control Unit—is located under the right rear seat.

Emissions Time Delay Relay—is located at the emissions relay box on the inner side of the left front fender.

Engine Speed Relay (turbo)—is located at the emissions relay box on the inner side of the left front fender.

Engine Speed Relay—is located at the emissions relay box on the inner side of the left front fender or forward of the right front door behind the trim panel.

Flasher Relay—is located under the left side of the dash.

Fog Light Relay—is located at the emissions relay box or the electrical distribution box on the left side of the engine compartment.

Fuel Injection Control Unit—is located under the right side of the dash.

9000 Series

Air conditioner Compressor Relay—is located in the electrical distribution box behind the glove box.

ABS Control Unit—is located in the engine compartment behind the firewall partition above the ignition system control unit.

ABS Main Relay—is located in the engine compartment behind the firewall partition on the ABS relay and fuse board.

ABS Pump Relay—is located in the engine compartment behind the firewall partition on the ABS relay and fuse board.

Air Bag Control Unit—is located under the right side of the dash.

Burglar Alarm Electronic Control Unit—is located under the right side of the dash.

Cruise Control Unit—is located under the dash to the left of the steering column.

Daytime Driving Lights Relay (Canada)—is located in the electrical distribution box behind the glove box.

DI-APC Control Unit—is located under the left front seat.

Flasher Relay—is located in the electrical distribution box behind the glove box.

18

Subaru

Justy, Legacy, Loyale, StD, Xt—All Models

SERIAL NUMBER IDENTIFICATION

Vehicle Identification Plate

The vehicle identification plate is located on the bulkhead in the engine compartment.

Engine Number

The engine serial number is stamped on the front right side of the crankcase, on all engines except the 1.2L engine. On the 1.2L engine, the serial number is stamped at the right rear side of the engine below the cylinder head.

Vehicle Identification Number

The Vehicle Identification Number (VIN) is stamped on a plate located on the top of the dashboard on the drivers side and is visible through the windshield.

Transaxle Number

The transaxle serial number is located on a sticker fixed to the upper surface of the main case (manual transaxle) or to the converter housing (automatic transaxle).

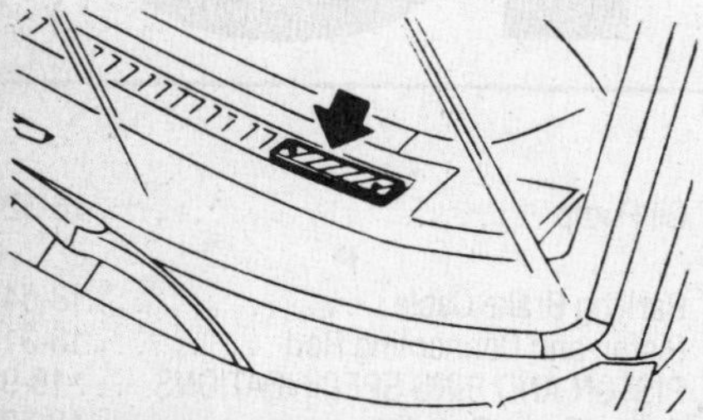

Vehicle Identification Plate is located on a plate attached to the bulkhead panel in the engine compartment

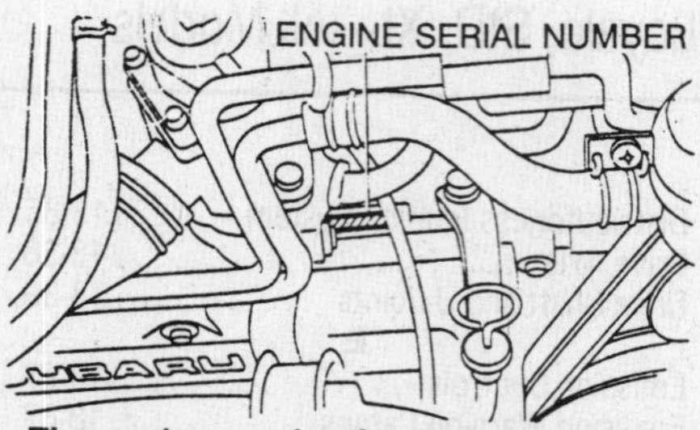

The engine number is stamped on the front right-side of the crankcase – except 1.2L engine

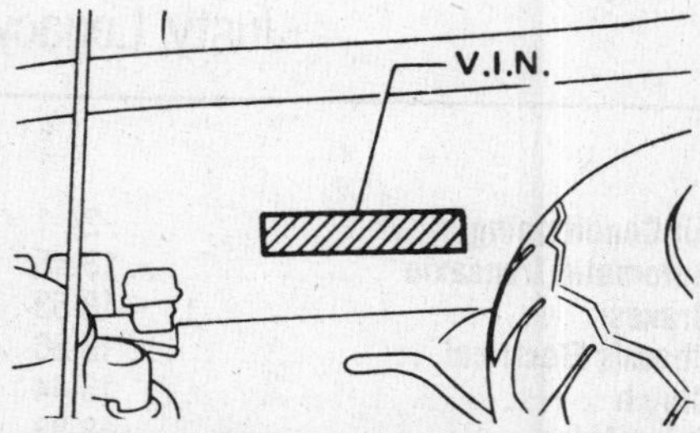

Vehicle Identification Number is located on the left side of the dash

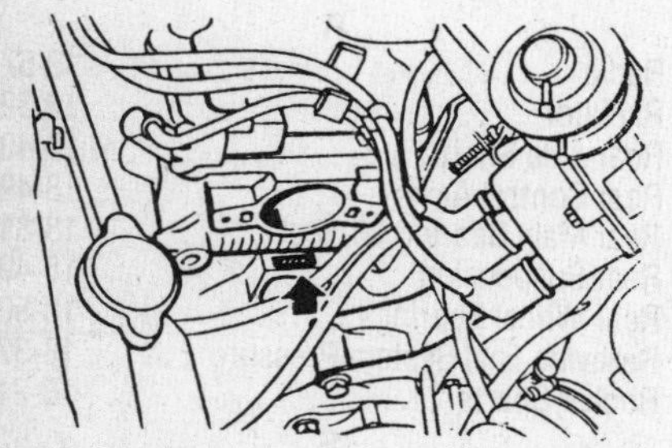

The engine number is stamped on the rear-side of the engine, below the cylinder head – 1.2L engine

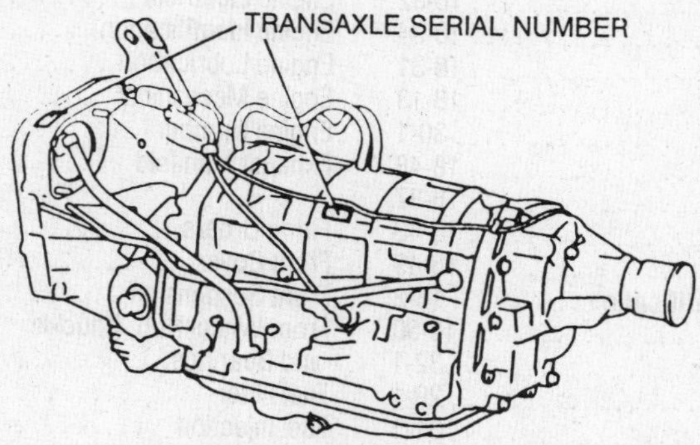

Transaxle identification plate location

ENGINE IDENTIFICATION

Year	Model	Engine Displacement cu. in. (cc/liter)	Engine Series Identification	No. of Cylinders	Engine Type
1988	Justy	73 (1200/1.2)	7	3	OHC
	Justy 4WD	73 (1200/1.2)	8	3	OHC
	STD	109 (1800/1.8)	4	4	OHC
	STD 4WD	109 (1800/1.8)	5 ①	4	OHC
	XT Coupe	109 (1800/1.8)	4	4	OHC
	XT Coupe 4WD	109 (1800/1.8)	7	4	OHC
	XT Coupe	163 (2700/2.7)	8	6	OHC
	XT6 Coupe 4WD	163 (2700/2.7)	9	6	OHC
1989	Justy	73 (1200/1.2)	7	3	OHC
	Justy 4WD	73 (1200/1.2)	8	3	OHC
	STD	109 (1800/1.8)	4	4	OHC

ENGINE IDENTIFICATION

Year	Model	Engine Displacement cu. in. (cc/liter)	Engine Series Identification	No. of Cylinders	Engine Type
1989	STD 4WD	109 (1800/1.8)	5②	4	OHC
	XT Coupe	109 (1800/1.8)	4	4	OHC
	XT Coupe 4WD	109 (1800/1.8)	7①	4	OHC
	XT Coupe	163 (2700/2.7)	8	6	OHC
	XT6 Coupe 4WD	163 (2700/2.7)	9①	6	OHC
	Legacy	135 (2200/2.2)	6	4	OHC
1990	Justy	73 (1200/1.2)	7	3	OHC
	Justy 4WD	73 (1200/1.2)	8	3	OHC
	Loyale	109 (1800/1.8)	4	4	OHC
	Loyale 4WD	109 (1800/1.8)	5	4	OHC
	XT Coupe	109 (1800/1.8)	4	4	OHC
	XT Coupe 4WD	109 (1800/1.8)	7	4	OHC
	XT Coupe	160 (2700/2.7)	8	6	OHC
	XT6 Coupe 4WD	163 (2700/2.7)	9	6	OHC
	Legacy	135 (2200/2.2)	6	4	OHC
1991-92	Justy	73 (1200/1.2)	7	3	OHC
	Justy 4WD	73 (1200/1.2)	8	3	OHC
	Loyale	109 (1800/1.8)	4	4	OHC
	Loyale 4WD	109 (1800/1.8)	5	4	OHC
	XT Coupe	109 (1800/1.8)	4	4	OHC
	XT Coupe 4WD	109 (1800/1.8)	7	4	OHC
	XT Coupe	160 (2700/2.7)	8	6	OHC
	XT6 Coupe 4WD	163 (2700/2.7)	9	6	OHC
	Legacy	135 (2200/2.2)	6	4	OHC

OHC—Overhead Camshaft
NOTE: STD designates—4 door sedan
Station wagon
Touring wagon
3 door wagon

① Air Suspension
② Without Air Suspension; Code 7 with Air Suspension

GENERAL ENGINE SPECIFICATIONS

Year	Model	Engine Displacement cu. in. (cc)	Fuel System Type	Net Horsepower @ rpm	Net Torque @ rpm (ft. lbs.)	Bore × Stroke (in.)	Compression Ratio	Oil Pressure @ 2000 rpm
1988	Justy	73 (1200)	EFC Carb.	66 @ 5200	70 @ 3200	3.07 × 3.27	9.0:1	35–40
	STD	109 (1800)	SPFI	84 @ 5200	137 @ 2800	3.62 × 2.64	9.5:1	57–64
	XT Coupe	109 (1800)	EEM-MPFI	97 @ 5200	103 @ 3200	3.62 × 2.64	7.7:1	57–64
	XT6 Coupe	163 (2700)	MPFI	145 @ 5200	156 @ 4000	3.62 × 2.64	9.5:1	57–64
1989	Justy③	73 (1200)	EFC Carb.	66 @ 5200	70 @ 3200	3.07 × 3.27	9.1:1	35–40
	STD	109 (1800)	EEM-SPFI	90 @ 5200	101 @ 2800	3.62 × 2.64	9.5:1	57–64
	STD	109 (1800)	EEM-MPFI Turbo	115 @ 5200	134 @ 2800	3.62 × 2.64	7.7:1	57–64
	XT Coupe	109 (1800)	EEM MPFI	97 @ 5200	103 @ 3200	3.62 × 2.64	9.5:1	57–64
	Legacy	135 (2200)	EEM-MPFI	130 @ 5600	137 @ 2400	3.82 × 2.95	9.5:1	②
	XT6 Coupe	163 (2700)	EEM-MPFI	145 @ 5200	156 @ 4000	3.62 × 2.64	9.5:1	57–64

GENERAL ENGINE SPECIFICATIONS

Year	Model	Engine Displacement cu. in. (cc)	Fuel System Type	Net Horsepower @ rpm	Net Torque @ rpm (ft. lbs.)	Bore × Stroke (in.)	Com-pression Ratio	Oil Pressure @ 2000 rpm
1990	Justy	73 (1200)	EFC Carb.	66 @ 5200	70 @ 3200	3.07 × 3.27	9.1:1	35–40
	Justy ①	73 (1200)	EEM-MPFI	66 @ 5200	70 @ 3200	3.07 × 3.27	9.1:1	35–40
	Loyale	109 (1800)	EEM-SPFI	90 @ 5200	101 @ 2800	3.62 × 2.64	9.5:1	57–64
	Loyale	109 (1800)	EEM-MPFI Turbo	115 @ 5200	134 @ 2800	3.62 × 2.64	7.7:1	57–64
	XT Coupe	109 (1800)	EEM MPFI	97 @ 5200	103 @ 3200	3.62 × 2.64	9.5:1	57–64
	Legacy	135 (2200)	EEM-MPFI	130 @ 5600	137 @ 2400	3.82 × 2.95	9.5:1	②
	XT6 Coupe	163 (2700)	EEM-MPFI	145 @ 5200	156 @ 4000	3.62 × 2.64	9.5:1	57–64
1991–92	Justy	73 (1200)	EFC Carb.	66 @ 5200	70 @ 3200	3.07 × 3.27	9.1:1	35–40
	Justy ①	73 (1200)	EEM-MPFI	73 @ 5600	71 @ 2800	3.07 × 3.27	9.1:1	35–40
	Loyale	109 (1800)	EEM-SPFI	90 @ 5200	101 @ 2800	3.62 × 2.64	9.5:1	57–64
	XT Coupe	109 (1800)	EEM MPFI	97 @ 5200	103 @ 3200	3.62 × 2.64	9.5:1	57–64
	XT6 Coupe	163 (2700)	EEM-MPFI	145 @ 5200	156 @ 4000	3.62 × 2.64	9.5:1	57–64
	Legacy	135 (2200)	EEM-MPFI	130 @ 5600	137 @ 2400	3.82 × 2.95	9.5:1	②
	Legacy	135 (2200)	EEM-MPFI Turbo	160 @ 5600	NA	3.82 × 2.95	8.0:1	②

NA—Not Available
EEM—Electronic Engine Managemnet
EFC—Electronic Fuel Control
MPFI—Multi-point Fuel Injection
SPFI—Single-point Fuel Injection
STD Designates—4 Door Sedan
Station Wagon
Touring Wagon
3 Door Wagon
① GL
② 14 psi @ 600 rpm
43 psi @ 5000 rpm
③ EEM-MPFI introduced mid year

ENGINE TUNE-UP SPECIFICATIONS

Year	Model	Engine Displacement cu. in. (cc)	Spark Plugs Type	Spark Plugs Gap (in.)	Ignition Timing (deg.) MT	Ignition Timing (deg.) AT	Com-pression Pressure (psi)	Fuel Pump (psi)	Idle Speed (rpm) MT	Idle Speed (rpm) AT	Valve Clearance In.	Valve Clearance Ex.
1988	Justy	73 (1200)	BPR6ES-11	0.040	5B	—	160	1.3–2.0	750–850	—	0.0051–0.0067	0.0091–0.0106
	STD	109 (1800)	BPR6ES-11	0.040	20B	20B	168	61–71 ④	600–800	700–900	Hyd.	Hyd.
	XT Coupe	109 (1800)	BPR6ES-11	0.040	20B	20B	168	61–71	600–800	700–900	Hyd.	Hyd.
	XT6 Coupe	163 (2700)	BPR6ES-11	0.040	20	20B	168	61–71	600–850	700–850	Hyd.	Hyd.
1989	Justy	73 (1200)	BPR6ES-11	0.040	5B	5B	160	1.3–2.0	750–850 ②	800–900 ③	0.0057–0.0067	0.0091–0.0106
	STD	109 (1800)	BPR6ES-11	0.040	20B	20B	168	61–71 ④	600–800	700–900	Hyd.	Hyd.
	XT Coupe	109 (1800)	BPR6ES-11	0.040	20B	20B	168	61–71	600–800	700–900	Hyd.	Hyd.
	XT6 Coupe	163 (2700)	BPR6ES-11	0.040	20B	20B	168	61–71	600–800	700–900	Hyd.	Hyd.
	Legacy	135 (2200)	BKR6E-11	0.040	20B	20B	168	36	600–800	700–800	Hyd.	Hyd.

ENGINE TUNE-UP SPECIFICATIONS

Year	Model	Engine Displacement cu. in. (cc)	Spark Plugs Type	Spark Plugs Gap (in.)	Ignition Timing (deg.) MT	Ignition Timing (deg.) AT	Com-pression Pressure (psi)	Fuel Pump (psi)	Idle Speed (rpm) MT	Idle Speed (rpm) AT	Valve Clearance In.	Valve Clearance Ex.
1990	Justy	73 (1200)	BPR6ES-11	0.040	5B	5B	160	1.3–2.0 ①	750–850 ②	800–900 ③	0.0057–0.0067	0.0091–0.0106
	Loyale	109 (1800)	BPR6ES-11	0.040	20B	20B	168 ⑤	61–71 ④	600–800	700–900	Hyd.	Hyd.
	XT Coupe	109 (1800)	BPR6ES-11	0.040	20B	20B	168	61–71	600–800	700–900	Hyd.	Hyd.
	XT6 Coupe	163 (2700)	BPR6ES-11	0.040	20B	20B	168	61–71	600–800	700–900	Hyd.	Hyd.
	Legacy	135 (2200)	BKR6E-11	0.040	20B	20B	168	36	600–800	700–800	Hyd.	Hyd.
1991	Justy	73 (1200)	BPR6ES-11	0.040	5B	5B	160	1.3–2.0 ①	750–850 ②	800–900 ③	0.0057–0.0067	0.0091–0.0106
	Loyale	109 (1800)	BPR6ES-11	0.040	20B	20B	168 ⑤	61–71 ④	600–800	700–900	Hyd.	Hyd.
	XT Coupe	109 (1800)	BPR6ES-11	0.040	20B	20B	168	61–71	600–800	700–900	Hyd.	Hyd.
	XT6 Coupe	163 (2700)	BPR6ES-11	0.040	20B	20B	168	61–71	600–800	700–900	Hyd.	Hyd.
	Legacy	135 (2200)	BKR6E-11	0.040	20B ⑥	20B ⑥	168	36	600–800	700–800	Hyd.	Hyd.
1992	SEE UNDERHOOD SPECIFICATIONS STICKER											

B—BTDC
① EEM-MPFI—61–71
② Idle-up system on—850–950
③ Idle-up system on—900–1000
④ SPFI—36–50
⑤ MPFI Turbo—145
⑥ 15B–Turbo

FIRING ORDERS

NOTE: To avoid confusion, always replace spark plug wires one at a time.

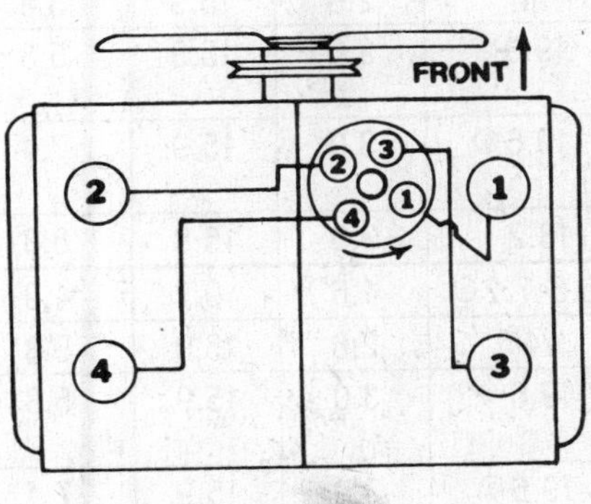

1.8L Engine
Engine Firing Order: 1–3–2–4
Distributor Rotation: Counterclockwise

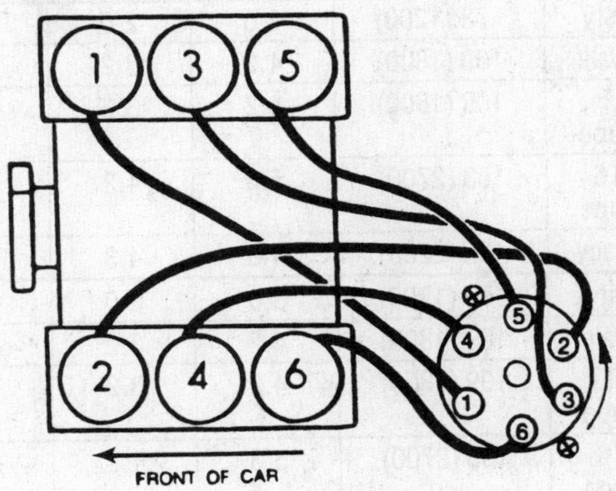

2.7L Engine
Engine Firing Order: 1–6–3–2–5–4
Distributor Rotation: Counterclockwise

FIRING ORDERS

NOTE: To avoid confusion, always replace spark plug wires one at a time.

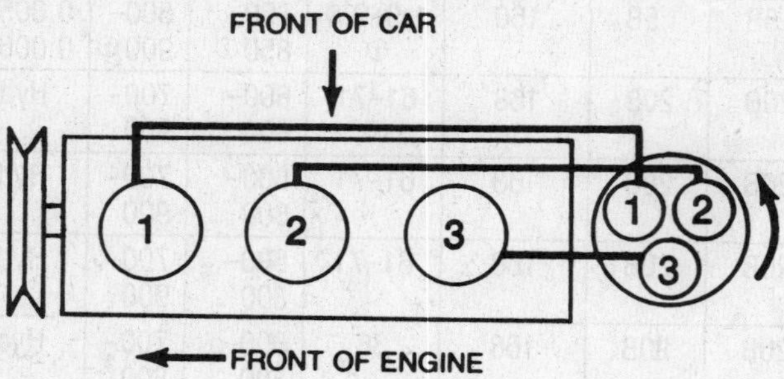

1.2L Engine
Engine Firing Order: 1–3–2
Distributor Rotation: Counterclockwise

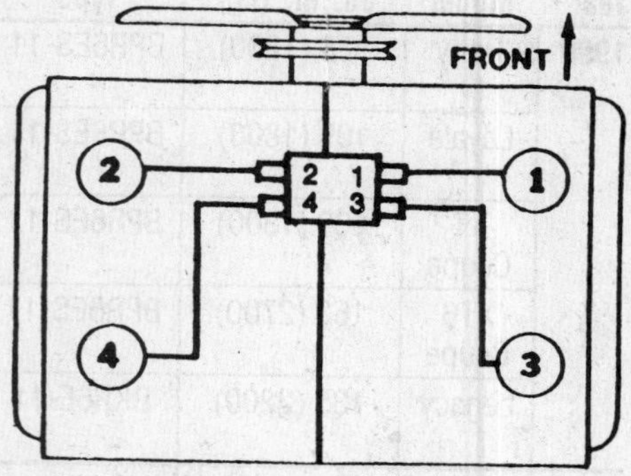

2.2L Engine
Engine Firing Order: 1–3–2–4
Distributorless Ignition System

CAPACITIES

Year	Model	Engine Displacement cu. in. (cc)	Engine Crankcase (qts.) with Filter	Engine Crankcase (qts.) without Filter	Transmission (pts.) 4-spd	Transmission (pts.) 5-Spd	Transmission (pts.) Auto.	Drive Axle (pts.)	Fuel Tank (gal.)	Cooling System (qts.)
1988	Justy	73 (1200)	3.0	2.0	—	4.8④	—	1.6	9.2	4.5
	STD	109 (1800)	4.2	3.2	—	6.0	10–13②	1.6	15.9	5.8
	XT Coupe	109 (1800)	4.2	3.2	—	5.4④	20	1.6	15.9	5.8
	XT6 Coupe	160 (2700)	4.2	3.2	—	5.4④	20	1.6	15.9	7.4
1989	Justy	73 (1200)	3.0	2.0	—	4.8④	6.6–7.2①	1.6	9.2	4.9
	STD	109 (1800)	4.2	3.2	—	④	14	2.6	15.9	5.8
	XT Coupe	109 (1800)	4.2	3.2	—	5.4④	19.6	3.0	15.9	5.8
	XT6 Coupe	163 (2700)	5.3	4.2	—	7.4	19.6③	3.0	15.9	7.4
	Legacy	135 (2200)	4.8	4.3	—	7.0⑤	18.2	3.0	15.9	6.3
1990	Justy	73 (1200)	3.0	2.0	—	4.8④	6.6–7.2①	1.6	9.2	4.9
	Loyale	109 (1800)	4.2	3.2	—	④	14	2.6	15.9	5.8
	XT Coupe	109 (1800)	4.2	3.2	—	5.4④	19.6	3.0	15.9	5.8
	XT6 Coupe	163 (2700)	5.3	4.2	—	7.4	19.6③	3.0	15.9	7.4
	Legacy	135 (2200)	4.8	4.3	—	7.0⑤	18.2	3.0	15.9	6.3
1991-92	Justy	73 (1200)	3.0	2.0	—	4.8④	6.6–7.2①	1.6	9.2	4.9
	Loyale	109 (1800)	4.2	3.2	—	④	14	2.6	15.9	5.8
	XT Coupe	109 (1800)	4.2	3.2	—	5.4④	19.6	3.0	15.9	5.8
	XT6 Coupe	163 (2700)	5.3	4.2	—	7.4	19.6③	3.0	15.9	7.4
	Legacy	135 (2200)	4.8	4.3	—	7.0⑤	18.2	3.0	15.9	6.3

① 4WD—11.9
② Differential—2.6
③ 4WD—29 pts.
④ 4WD—7 pts.
⑤ 4WD—7.4 pts.

CAMSHAFT SPECIFICATIONS

All measurements given in inches.

Year	Engine Displacement cu. in. (cc)	Journal Diameter 1	2	3	4	5	Lobe Lift In.	Ex.	Bearing Clearance	Camshaft End Play
1988	73 (1200)	—	—	—	—	—	1.4520–1.4528	1.4520–1.4528	—	0.0012–0.0150
	109 (1800)	1.4946–1.4953	1.9080–1.9087	1.8883–1.8890	①	—	1.5650–1.5689	1.5650–1.5689	0.0008–0.0021	0.0012–0.0102
	163 (2700)	1.4946–1.4953	1.9080–1.9087	1.8883–1.8890	1.8687–1.8693	①	1.5606–1.5646	1.5606–1.5646	0.0008–0.0021	0.0012–0.0102
1989	73 (1200)	—	—	—	—	—	1.4520–1.4528	1.4520–1.4528	—	0.0012–0.0150
	109 (1800)	1.4946–1.4953	1.9080–1.9087	1.8883–1.8890	①	—	1.5650–1.5689	1.5650–1.5689	0.0008–0.0021	0.0012–0.0102
	135 (2200)	②	③	④	—	—	1.2752–1.2791	1.2752–1.2791	0.0022–0.0035	0.0012–0.0102
	163 (2700)	1.4946–1.4953	1.9080–1.9087	1.8883–1.8890	1.8687–1.8693	①	1.5606–1.5646	1.5606–1.5646	0.0028	0.0012–0.0102
1990	73 (1200)	—	—	—	—	—	1.4520–1.4528	1.4520–1.4528	—	0.0012–0.0150
	109 (1800)	1.4946–1.4953	1.9080–1.9087	1.8883–1.8890	①	—	1.5650–1.5689	1.5650–1.5689	0.0008–0.0021	0.0012–0.0102
	135 (2200)	②	③	④	—	—	1.2752–1.2791	1.2752–1.2791	0.0022–0.0035	0.0012–0.0102
	163 (2700)	1.4946–1.4953	1.9080–1.9087	1.8883–1.8890	1.8687–1.8693	①	1.5606–1.5646	1.5606–1.5646	0.0028	0.0012–0.0102
1991-92	73 (1200)	—	—	—	—	—	1.4520–1.4528	1.4520–1.4528	—	0.0012–0.0150
	109 (1800)	1.4946–1.4953	1.9080–1.9087	1.8883–1.8890	①	—	1.5650–1.5689	1.5650–1.5689	0.0008–0.0021	0.0012–0.0102
	135 (2200)	②	③	④	—	—	1.2752–1.2791	1.2752–1.2791	0.0022–0.0035	0.0012–0.0102
	163 (2700)	1.4946–1.4953	1.9080–1.9087	1.8883–1.8890	1.8687–1.8693	①	1.5606–1.5646	1.5606–1.5646	0.0028	0.0012–0.0102

① Camshaft distributor LH journal: 1.5340–1.5346 in.
② RH front and LH rear—1.2573–1.2579 in.
③ RH and LH center—1.4738–1.4744 in.
④ RH rear and LH front—1.4935–1.4941 in.

CRANKSHAFT AND CONNECTING ROD SPECIFICATIONS

All measurements are given in inches.

Year	Engine Displacement cu. in. (cc)	Crankshaft Main Brg. Journal Dia.	Main Brg. Oil Clearance	Shaft End-play	Thrust on No.	Connecting Rod Journal Diameter	Oil Clearance	Side Clearance
1988	73 (1200)	1.6525–1.6529	0.0006–0.0018	0.0031–0.0070	4	1.6531–1.6535	0.0008–0.0021	0.0028–0.0118
	109 (1800)	①	②	0.0004–0.0037	2	1.7715–1.7720	0.0004–0.0021	0.0028–0.0130
	163 (2700)	③	④	0.0004–0.0037	3	1.7715–1.7720	0.0004–0.0028	0.0028–0.0130

CRANKSHAFT AND CONNECTING ROD SPECIFICATIONS

All measurements are given in inches.

Year	Engine Displacement cu. in. (cc)	Crankshaft: Main Brg. Journal Dia.	Crankshaft: Main Brg. Oil Clearance	Crankshaft: Shaft End-play	Crankshaft: Thrust on No.	Connecting Rod: Journal Diameter	Connecting Rod: Oil Clearance	Connecting Rod: Side Clearance
1989	73 (1200)	1.6525–1.6529	0.0006–0.0018	0.0031–0.0070	4	1.6531–1.6535	0.0008–0.0021	0.0028–0.0118
	109 (1800)	①	②	0.0004–0.0037	2	1.7715–1.7720	0.0004–0.0021	0.0028–0.0130
	135 (2200)	2.3616–2.3622	0.0004–0.0012	0.0012–0.0045	3	2.0466–2.0472	0.0005–0.0015	0.0028–0.0130
	163 (2700)	③	④	0.0004–0.0037	3	1.7715–1.7720	0.0004–0.0028	0.0028–0.0130
1990	73 (1200)	1.6525–1.6529	0.0006–0.0018	0.0031–0.0070	4	1.6531–1.6535	0.0008–0.0021	0.0028–0.0118
	109 (1800)	①	②	0.0004–0.0037	2	1.7715–1.7720	0.0004–0.0021	0.0028–0.0130
	135 (2200)	2.3616–2.3622	0.0004–0.0012	0.0012–0.0045	3	2.0466–2.0472	0.0005–0.0015	0.0028–0.0130
	163 (2700)	③	④	0.0004–0.0037	3	1.7715–1.7720	0.0004–0.0028	0.0028–0.0130
1991-92	73 (1200)	1.6525–1.6529	0.0006–0.0018	0.0031–0.0070	4	1.6531–1.6535	0.0008–0.0021	0.0028–0.0118
	109 (1800)	①	②	0.0004–0.0037	2	1.7715–1.7720	0.0004–0.0021	0.0028–0.0130
	135 (2200)	2.3616–2.3622	0.0004–0.0012	0.0012–0.0045	3	2.0466–2.0472	0.0005–0.0015	0.0028–0.0130
	163 (2700)	③	④	0.0004–0.0037	3	1.7715–1.7720	0.0004–0.0028	0.0028–0.0130

① Front—2.1637–2.1642
Center—2.1635–2.1642
Rear—2.1636–2.1642

② Front and Rear—0.0001–0.0014
Center—0.0003–0.0011

③ Front—2.1637–2.1642
Center Both—2.1635–2.1642
Rear—2.1636–2.1642

④ Front and Rear—0.0001–0.0014
Center Both—0.0003–0.0011

VALVE SPECIFICATIONS

Year	Engine Displacement cu. in. (cc)	Seat Angle (deg.)	Face Angle (deg.)	Spring Test Pressure (lbs.) ①	Spring Installed Height (in.) ①	Stem-to-Guide Clearance (in.) Intake	Stem-to-Guide Clearance (in.) Exhaust	Stem Diameter (in.) Intake	Stem Diameter (in.) Exhaust
1988	73 (1200)	45	45	④	125	0.0008–0.0020	0.0016–0.0028	0.2742–0.2748	0.2734–0.2740
	109 (1800)	45	45	45.2–51.8 @ 1.122 ③	—	0.0014–0.0026	0.0016–0.0028	0.2736–0.2742	0.2734–0.2740
	163 (2700)	45	45	45.2–51.8 @ 1.122 ②	—	0.0014–0.0026	0.0016–0.0028	0.2736–0.2742	0.2734–0.2740
1989	73 (1200)	45	45	④	125	0.0008–0.0020	0.0016–0.0028	0.2742–0.2748	0.2734–0.2740
	109 (1800)	45	45	45.2–51.8 @ 1.122 ③	—	0.0014–0.0026	0.0016–0.0028	0.2736–0.2742	0.2734–0.2740
	135 (2200)	45	45	⑤	—	0.0014–0.0024	0.0016–0.0026	0.2343–0.2348	0.2341–0.2346
	163 (2700)	45	45	45.2–51.8 @ 1.122 ②	—	0.0014–0.0026	0.0016–0.0028	0.2736–0.2742	0.2734–0.2740

VALVE SPECIFICATIONS

Year	Engine Displacement cu. in. (cc)	Seat Angle (deg.)	Face Angle (deg.)	Spring Test Pressure (lbs.) ①	Spring Installed Height (in.) ①	Stem-to-Guide Clearance (in.)		Stem Diameter (in.)	
						Intake	Exhaust	Intake	Exhaust
1990	73 (1200)	45	45	④	125	0.0008–0.0020	0.0016–0.0028	0.2742–0.2748	0.2734–0.2740
	109 (1800)	45	45	45.2–51.8 @ 1.122 ③	—	0.0014–0.0026	0.0016–0.0028	0.2736–0.2742	0.2734–0.2740
	135 (2200)	45	45	⑤	—	0.0014–0.0024	0.0016–0.0026	0.2343–0.2348	0.2341–0.2346
	163 (2700)	45	45	45.2–51.8 @ 1.122 ②	—	0.0014–0.0026	0.0016–0.0028	0.2736–0.2742	0.2734–0.2740
1991-92	73 (1200)	45	45	④	125	0.0008–0.0020	0.0016–0.0028	0.2742–0.2748	0.2734–0.2740
	109 (1800)	45	45	45.2–51.8 @ 1.122 ③	—	0.0014–0.0026	0.0016–0.0028	0.2736–0.2742	0.2734–0.2740
	135 (2200)	45	45	⑤	—	0.0014–0.0024	0.0016–0.0026	0.2343–0.2348	0.2341–0.2346
	163 (2700)	45	45	45.2–51.8 @ 1.122 ②	—	0.0014–0.0026	0.0016–0.0028	0.2736–0.2742	0.2734–0.2740

① All values are for inner spring
② Outer spring—100.5–115.5 @ 1.240
③ Outer spring—112.9–129.7 @ 1.240
④ Spring—112.8–129.8 @ 1.248
⑤ 1.457 in. @ 34.0–39.0 lbs.
1.154 in. @ 92.2–106.1 lbs.

PISTON AND RING SPECIFICATIONS

All measurements are given in inches.

Year	Engine Displacement cu. in. (cc)	Piston Clearance	Ring Gap			Ring Side Clearance		
			Top Compression	Bottom Compression	Oil Control	Top Compression	Bottom Compression	Oil Control
1988	73 (1200)	0.0015–0.0024	0.0079–0.0138	0.0079–0.0138	0.0120–0.0350 ①	0.0014–0.0030	0.0010–0.0026	Snug
	109 (1800)	0.0004–0.0016	0.0079–0.0138	0.0079–0.0138	0.0120–0.0350 ①	0.0016–0.0031	0.0012–0.0028	Snug
	163 (2700)	0.0006–0.0014	0.0079–0.0138	0.0079–0.0138	0.0120–0.0350 ①	0.0016–0.0031	0.0012–0.0028	Snug
1989	73 (1200)	0.0015–0.0028	0.0079–0.0138	0.0079–0.0138	0.0120–0.0350 ①	0.0014–0.0030	0.0010–0.0026	Snug
	109 (1800)	0.0006–0.0012 ②	0.0079–0.0138	0.0079–0.0138	0.0120–0.0350 ①	0.0016–0.0031	0.0012–0.0028	Snug
	132 (2200)	0.0004–0.0012	0.0079–0.0138	0.0079–0.0138	0.0076–0.0276 ①	0.0016–0.0031	0.0012–0.0028	Snug
	163 (2700)	0.0006–0.0014	0.0079–0.0138	0.0079–0.0138	0.0120–0.0350 ①	0.0016–0.0031	0.0012–0.0028	Snug

PISTON AND RING SPECIFICATIONS

All measurements are given in inches.

Year	Engine Displacement cu. in. (cc)	Piston Clearance	Ring Gap Top Compression	Ring Gap Bottom Compression	Ring Gap Oil Control	Ring Side Clearance Top Compression	Ring Side Clearance Bottom Compression	Ring Side Clearance Oil Control
1990	73 (1200)	0.0015–0.0028	0.0079–0.0138	0.0079–0.0138	0.0120–0.0350 ①	0.0014–0.0030	0.0010–0.0026	Snug
	109 (1800)	0.0006–0.0012 ②	0.0079–0.0138	0.0079–0.0138	0.0120–0.0350 ①	0.0016–0.0031	0.0012–0.0028	Snug
	132 (2200)	0.0004–0.0012	0.0079–0.0138	0.0079–0.0138	0.0076–0.0276 ①	0.0016–0.0031	0.0012–0.0028	Snug
	163 (2700)	0.0006–0.0014	0.0079–0.0138	0.0079–0.0138	0.0120–0.0350 ①	0.0016–0.0031	0.0012–0.0028	Snug
1991–92	73 (1200)	0.0015–0.0028	0.0079–0.0138	0.0079–0.0138	0.0120–0.0350 ①	0.0014–0.0030	0.0010–0.0026	Snug
	109 (1800)	0.0006–0.0012 ②	0.0079–0.0138	0.0079–0.0138	0.0120–0.0350 ①	0.0016–0.0031	0.0012–0.0028	Snug
	132 (2200)	0.0004–0.0012	0.0079–0.0138	0.0079–0.0138	0.0076–0.0276 ①	0.0016–0.0031	0.0012–0.0028	Snug
	163 (2700)	0.0006–0.0014	0.0079–0.0138	0.0079–0.0138	0.0120–0.0350 ①	0.0016–0.0031	0.0012–0.0028	Snug

① For rails only
② Non-turbo; 0.0004–0.0012 for turbo models

TORQUE SPECIFICATIONS

All readings in ft. lbs.

Year	Engine Displacement cu. in. (cc)	Cylinder Head Bolts	Main Bearing Bolts	Rod Bearing Bolts	Crankshaft Pulley Bolts	Flywheel Bolts	Manifold Intake	Manifold Exhaust	Spark Plugs
1988	73 (1200)	51–57 ②	30–35	29–33	47–54	65–71	14–22	14–22	13–17
	109 (1800)	44–50	30–35	29–31	66–79	51–55	13–16	19–22	13–17
	163 (2700)	44–50	30–35	29–31	66–79	51–55	13–16	19–22	13–17
1989	73 (1200)	51–57 ②	30–35	29–33	58–72	65–71 ③	14–22	14–22	13–17
	109 (1800)	47 ①	29–35	29–31	66–79	51–55	13–16	19–22	13–17
	132 (2200)	④	⑤	32–34	66–79	58–62	NA	19–26	14–22
	163 (2700)	47 ⑥	29–35	29–31	66–79	51–55	13–16	19–22	13–17
1990	73 (1200)	51–56 ②	30–35	29–33	58–72	65–71 ③	14–22	14–22	13–17
	109 (1800)	47 ①	29–35	29–31	66–79	51–55	13–16	19–22	13–17
	132 (2200)	④	⑤	32–34	66–79	58–62	NA	19–26	14–22
	163 (2700)	47 ⑥	29–35	29–31	66–79	51–55	13–16	19–22	13–17

TORQUE SPECIFICATIONS

All readings in ft. lbs.

Year	Engine Displacement cu. in. (cc)	Cylinder Head Bolts	Main Bearing Bolts	Rod Bearing Bolts	Crankshaft Pulley Bolts	Flywheel Bolts	Manifold		Spark Plugs
							Intake	Exhaust	
1991-92	73 (1200)	51–56 ②	30–35	29–33	58–72	65–71 ③	14–22	14–22	13–17
	109 (1800)	47 ①	29–35	29–31	66–79	51–55	13–16	19–22	13–17
	132 (2200)	④	⑤	32–34	66–79	58–62	NA	19–26	14–22
	163 (2700)	47 ⑥	29–35	29–31	66–79	51–55	13–16	19–22	13–17

NA Not available
① 1st step—22 ft. lbs.
2nd step—43 ft. lbs.
3rd step—47 ft. lbs.
② 1st step—29 ft. lbs.
2nd step—54 ft. lbs.
3rd step—back off 90 degrees or more in reverse of tightening sequence
4th step—54 ft. lbs.
③ ECVT—54–61 ft. lbs.
④ 1st step—51 ft. lbs.
2nd step—back off bolts by 180 degrees
3rd step—25 ft. lbs. (bolts 1 & 2)
4th step—14 ft. lbs. (bolts 3–6)
5th step—Tighten all bolts 80–90 degrees in sequence
6th step—Retighten all bolts 80–90 degrees in sequence; not to exceed 180 degrees in 5th and 6th step
⑤ See text
⑥ 1st step—29 ft. lbs.
2nd step—47 ft. lbs.
3rd step—back off 90 degrees or more in reverse of tightening sequence
4th step—44–50 ft. lbs.

BRAKE SPECIFICATIONS

All measurements in inches unless noted.

Year	Model	Lug Nut Torque (ft. lbs.)	Master Cylinder Bore	Brake Disc		Standard Brake Drum Diameter	Minimum Lining Thickness	
				Minimum Thickness	Maximum Runout		Front	Rear
1988	Justy	58–72	⑤	⑧	0.0060	7.09 ⑩	0.295 ①	0.067 ⑦
	STD	58–72	0.8125	0.630	0.0040	7.09 ⑩	0.295 ①	0.259B
	XT Coupe ⑭	58–72	0.8125	0.787 ③	0.0040	7.09 ⑩	0.295 ①	0.259 ④
1989	Justy	58–72	⑤	0.610 ⑧	0.0060	7.09 ⑩	0.295 ①	0.067 ⑦
	STD	58–72	⑨	0.630 ③	0.0040	7.09 ⑩	0.295	0.059 ②
	XT Coupe ⑭	58–72	⑨	0.630 ⑪	0.0040	7.09 ⑩	0.295	0.059 ② ④
	Legacy	58–72	⑬	0.870 ③ ⑫	0.0039	⑥	0.295	0.256
1990	Justy	58–72	⑤	0.610 ⑧	0.0060	7.09 ⑩	0.295 ①	0.067 ⑦
	Loyale	58–72	⑨	0.630 ③	0.0040	7.09 ⑩	0.295	0.059 ②
	XT Coupe ⑭	58–72	⑨	0.630 ⑪	0.0040	7.09 ⑩	0.295	0.059 ② ④
	Legacy	58–72	⑬	0.870 ③ ⑫	0.0039	⑥	0.295	0.256
1991-92	Justy	58–72	⑤	0.610 ⑧	0.0060	7.09 ⑩	0.295 ①	0.067 ⑦
	Loyale	58–72	⑨	0.630 ③	0.0040	7.09 ⑩	0.295	0.059 ②
	XT Coupe ⑭	58–72	⑨	0.630 ⑪	0.0040	7.09 ⑩	0.295	0.059 ② ④
	Legacy	58–72	⑬	0.870 ③ ⑫	0.0039	⑥	0.295	0.256

① Justy GL—0.315 in. (includes metal backing)
② Rear disc brake including metal backing
③ Rear disc brake—0.335 in. service limit
0.390 in standard
④ XT Coupe with 2700cc engine. Rear disc brake including metal backing—0.315
⑤ Small diuameter—$^{7}/_{8}$ in.
Large diameter—1 in.
⑥ Not applicable
⑦ Standard—0.173 in.
⑧ Standard—0.709 in.
⑨ Small diuameter—$^{13}/_{16}$ in.
Large diameter—1 in.
⑩ Service limit—7.17 in.
⑪ XT-6—0.787 in.
⑫ Standard thickness—0.940
⑬ L, LS models—1 in.
LX models—1$^{1}/_{16}$ in.
⑭ Includes XT6 Coupe

WHEEL ALIGNMENT

Year	Model	Caster Range (deg.)	Caster Preferred Setting (deg.)	Camber Range (deg.)	Camber Preferred Setting (deg.)	Toe-in (in.)	Steering Axis Inclination (deg.)
1988	2WD XT Coupe	3 15/16P–4 13/16P	4 1/16P	3/4N–3/4P	0	1/8–1/8	NA
	4WD XT Coupe (4 cylinder)	2 5/8P–4 1/8P	3 3/8P	1/16N–1 3/8P	5/8P	3/64–1/8	NA
	4WD XT Coupe (6 cylinder)	2 3/4P–4 1/4P	3 1/2P	1/16P–1 9/16P	13/16P	3/64–13/64	NA
	2WD Sedan	1 3/4P–3 1/4P	2 1/2P	0–1 1/2P	3/4P	13/64–3/64 (1)	NA
	4WD Sedan with Air Susp.	1 7/16P–2 15/16P	2 3/16P	7/16P–1 15/16P	1 13/16P	5/64–5/16 (1)	NA
	4WD Sedan without Air Susp.	1 1/16P–2 9/16P	1 13/16P	15/16P–2 7/16P	1 11/16P	5/64–5/16 (1)	NA
	2WD SW	1 5/16P–2 13/16P	2 1/16P	1/4P–1 3/4P	1P	13/64–3/64	NA
	4WD SW with Air Susp.	1 7/16P–2 15/16P	2 3/16P	7/16P–1 15/16P	1 13/16P	5/64–5/15 (1)	NA
	4WD SW	13/16P–2 5/16P	1 9/16P	15/16P–2 7/16P	1 3/4P	5/64–5/16	NA
	Justy	1 1/2P–3 1/2P	2 1/2P	5/16N–1 11/16P	11/16P	3/32–1/2 (1)	NA
1989	2WD XT Coupe	3 15/16P–4 13/16P	4 1/16P	3/4N–3/4P	0	1/8–1/8 (1)	NA
	4WD XT Coupe (4 cylinder)	2 5/8P–4 1/8P	3 3/8P	1/16N–1 3/8P	5/8P	3/8 (1) –1/8 (1)	NA
	4WD XT Coupe (6 cylinder)	2 3/4P–4 1/4P	3 1/2P	1/16P–1 9/16P	13/16P	3/8 (1) –1/8 (1)	NA
	2WD Sedan	1 3/4P–3 1/4P	2 1/2P	0–1 1/2P	3/4P	1/4–1/16 (1)	NA
	4WD Sedan with Air Susp.	1 7/16P–2 15/16P	2 3/16P	7/16P–1 15/16P	1 13/16P	1/16 (1) –3/16 (1)	NA
	4WD Sedan without Air Susp.	1 1/16P–2 9/16P	1 13/16P	15/16P–2 7/16P	1 11/16P	1/16 (1) –3/16 (1)	NA
	2WD SW	1 5/16P–2 13/16P	2 1/16P	1/4P–1 3/4P	1P	1/16 (1) –3/16 (1)	NA
	4WD SW with Air Susp.	1 7/16P–2 15/16P	2 3/16P	7/16P–1 15/16P	1 13/16P	1/16 (1) –3/16 (1)	NA
	4WD SW	13/16P–2 5/16P	1 9/16P	15/16P–2 7/16P	1 3/4P	1/16 (1) –3/16 (1)	NA
	Justy	1 1/2P–3 1/2P	2 1/2P	5/16N–1 11/16P	11/16P	5/16–1/16 (1)	NA
	Legacy FWD Sedan	2 1/16P–4 1/16P (3)	3 1/16P (4)	3/4N–1/4P	1/4N	1/16–1/16 (1)	NA
	Legacy 4WD Sedan (2)	2P–4P	3P	1/2N–1/2P	0	1/16–1/16 (1)	NA
1990	2WD XT Coupe	3 15/16P–4 13/16P	4 1/16P	3/4N–3/4P	0	1/8–1/8 (1)	NA
	4WD XT Coupe (4 cylinder)	2 5/8P–4 1/8P	3 3/8P	1/16N–1 3/8P	5/8P	3/8 (1) –1/8 (1)	NA
	4WD XT Coupe (6 cylinder)	2 3/4P–4 1/4P	3 1/2P	1/16P–1 9/16P	13/16P	3/8 (1) –1/8 (1)	NA
	2WD Loyale Sedan	1 3/4P–3 1/4P	2 1/2P	0–1 1/2P	3/4P	1/4–1/16 (1)	NA
	4WD Loyale Sedan without Air Susp.	1 1/16P–2 9/16P	1 13/16P	15/16P–2 7/16P	1 11/16P	1/16 (1) –3/16 (1)	NA
	2WD Loyale SW	1 5/16P–2 13/16P	2 1/16P	1/4P–1 3/4P	1P	1/16 (1) –3/16 (1)	NA
	4WD Loyale SW	13/16P–2 5/16P	1 9/16P	15/16P–2 7/16P	1 3/4P	1/16 (1) –3/16 (1)	NA
	Justy	1 1/2P–3 1/2P	2 1/2P	5/16N–1 11/16P	11/16P	5/16–1/16 (1)	NA
	Legacy FWD Sedan	2 1/16P–4 1/16P (3)	3 1/16P (4)	3/4N–1/4P	1/4N	1/16–1/16 (1)	NA

WHEEL ALIGNMENT

Year	Model	Caster Range (deg.)	Caster Preferred Setting (deg.)	Camber Range (deg.)	Camber Preferred Setting (deg.)	Toe-in (in.)	Steering Axis Inclination (deg.)
	Legacy 4WD Sedan ②	2P–4P	3P	1/2N–1/2P	0	1/16–1/16 ①	NA
	Legacy 4WD Wagon	1 3/4P–3 3/4P	2 3/4P	1/2N–1/2P	0	1/16–1/16 ①	NA
1991-92	2WD XT Coupe	3 15/16P–4 13/16P	4 1/16P	3/4N–3/4P	0	1/8–1/8 ①	NA
	4WD XT Coupe (4 cylinder)	2 5/8P–4 1/8P	3 3/8P	1/16N–1 3/8P	5/8P	3/8 ① –1/8 ①	NA
	4WD XT Coupe (6 cylinder)	2 3/4P–4 1/4P	3 1/2P	1/16P–1 9/16P	13/16P	3/8 ① –1/8 ①	NA
	2WD Loyale Sedan	1 3/4P–3 1/4P	2 1/2P	0–1 1/2P	3/4P	1/4–1/16 ①	NA
	4WD Loyale Sedan without Air Susp.	1 1/16P–2 9/16P	1 13/16P	15/16P–2 7/16P	1 11/16P	1/16 ① –3/16 ①	NA
	2WD Loyale SW	1 5/16P–2 13/16P	2 1/16P	1/4P–1 3/4P	1P	1/16 ① –3/16 ①	NA
	4WD Loyale SW	13/16P–2 5/16P	1 9/16P	15/16P–2 7/16P	1 3/4P	1/16 ① –3/16 ①	NA
	Justy	1 1/2P–3 1/2P	2 1/2P	5/16N–1 11/16P	11/16P	5/16–1/16 ①	NA
	Legacy FWD Sedan	2 1/16P–4 1/16P ③	3 1/16P ④	3/4N–1/4P	1/4N	1/16–1/16 ①	NA
	Legacy 4WD Sedan ②	2P–4P	3P	1/2N–1/2P	0	1/16–1/16 ①	NA
	Legacy 4WD Wagon	1 3/4P–3 3/4P	2 3/4P	1/2N–1/2P	0	1/16–1/16 ①	NA

Air Susp.—Air suspension
SW—Station Wagon
P—Positive
N—Negative
① Toe out
② Same with air suspension
③ Legacy FWD Wagon—1 13/16P–3 13/16P
④ Legacy FWD Wagon—2 13/16P

ENGINE MECHANICAL

NOTE: Disconnecting the negative battery cable on some vehicles may interfere with the functions of the on board computer systems and may require the computer to undergo a relearning process, once the negative battery cable is reconnected.

Engine Assembly

REMOVAL & INSTALLATION

Legacy

2.2L ENGINE

1. Open the hood and support with a suitable prop.
2. Relieve the fuel pressure.
3. Drain the coolant.
4. Disconnect the battery cables and remove the battery from the vehicle.
5. Remove the cooling system reservoir tank, radiator and cooling fan assembly.
6. Remove the air intake duct.
7. Disconnect the hoses, harness connectors and cables.
8. Remove the power steering pump.
9. If equipped with air conditioning, discharge refrigerant and remove pressure lines.
10. Remove the exhaust system.
11. Disconnect the engine mount from the front crossmember.
12. Remove the nuts which hold lower side of the engine to transaxle.
13. If equipped with automatic transaxle, disconnect the driveplate from the torque converter.
14. Remove the pitching stopper rod.
15. Support engine with a lifting device and wire ropes.
16. Support transaxle with a suitable jack.
17. Remove the bolts which hold upper side of engine to transaxle.
18. Remove the engine.

To install:

19. Install the engine to transmission and temporarily tighten the engine to transaxle nuts and bolts.
20. If equipped with automatic transaxle, install the torque converter bolts and tighten to 17–20 ft. lbs. (23–26 Nm).
21. Remove the jack and hoist.
22. Raise the vehicle and support safely. Tighten the nuts which hold the lower side of the engine to transmission and torque to 34–61 ft. lbs. (46–54 Nm).
23. Tighten the front engine mount nuts to 40–61 ft. lbs. (54–83 Nm).
24. Install the exhaust system. Tighten the front exhaust pipe to cylinder head and front exhaust pipe to hanger bracket to 18–25 ft. lbs. (25–34 Nm), front exhaust pipe to rear exhaust pipe to 9–17 ft. lbs. (13–23 Nm).
25. Tighten the nuts which hold the

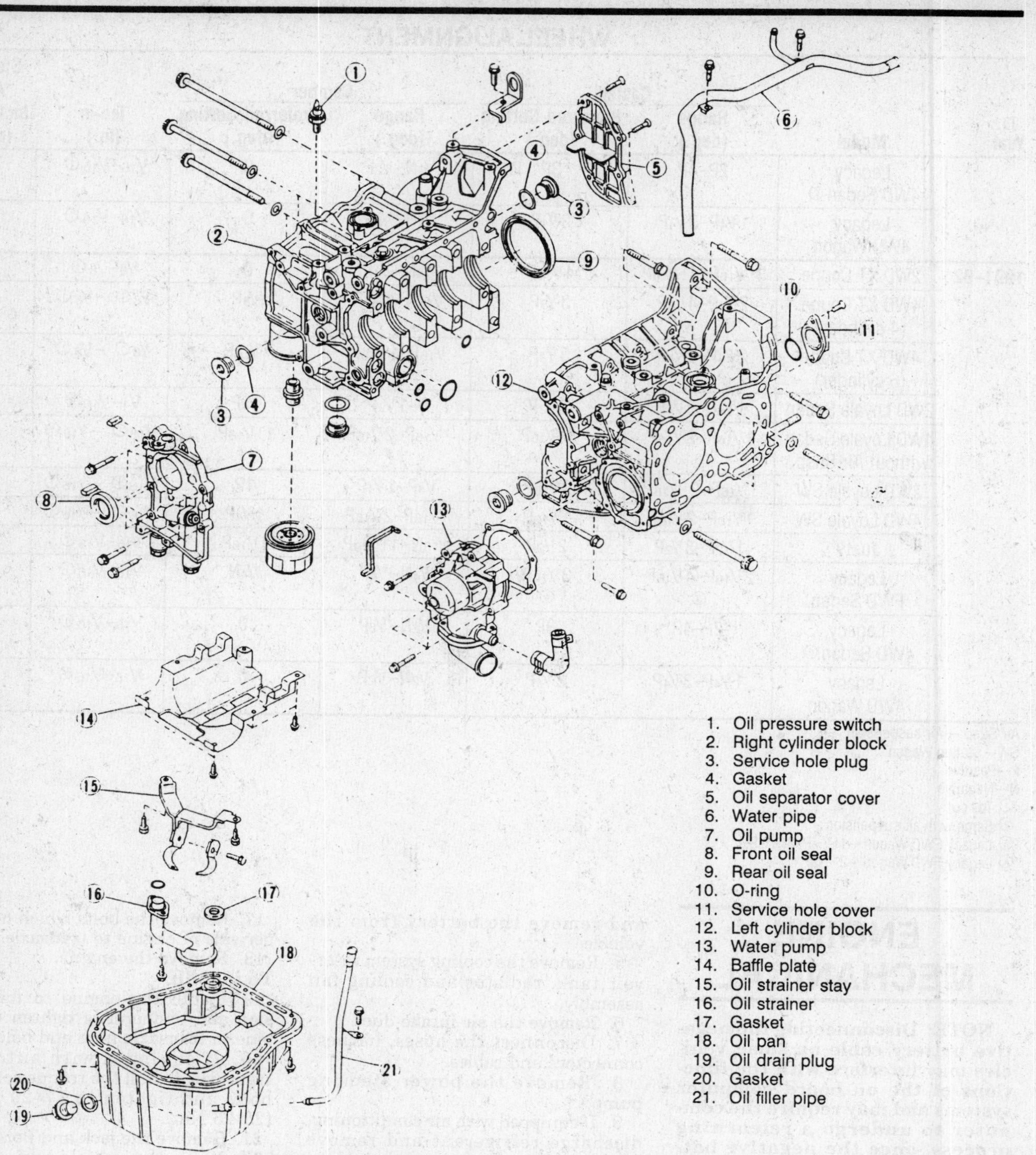

1. Oil pressure switch
2. Right cylinder block
3. Service hole plug
4. Gasket
5. Oil separator cover
6. Water pipe
7. Oil pump
8. Front oil seal
9. Rear oil seal
10. O-ring
11. Service hole cover
12. Left cylinder block
13. Water pump
14. Baffle plate
15. Oil strainer stay
16. Oil strainer
17. Gasket
18. Oil pan
19. Oil drain plug
20. Gasket
21. Oil filler pipe

Cylinder block assembly and components – 2.2L engine

upper side of the engine to transmission. Tighten the body side to 35–49 ft. lbs. (47–67 Nm) and engine or transmission side to 33–40 ft. lbs. (44–54 Nm).

26. Install the pitching stopper rod.
27. If equipped with air conditioning, install pressure hoses and belt.
28. Install the power steering pump. Torque to 22–36 ft. lbs. (29–49 Nm).
29. Connect all cables, hoses and harness connectors.
30. Install the air intake duct.
31. Install the cooling system.
32. Fill the cooling system with coolant.
33. Install the battery and connect the battery cables.
34. Check and correct the coolant and oil level.
35. If equipped with air conditioning, recharge the system with refrigerant.

Justy

NOTE: The engine and transaxle are removed as an assembly.

1. Disconnect the negative battery cable. Matchmark and remove the hood.
2. Drain the cooling system. Drain the engine oil. Drain the transaxle fluid.
3. Remove the front bumper and grille assembly.

4. Disconnect the radiator hoses. Remove the radiator retaining bolts. Remove the radiator from the vehicle.
5. Remove the air cleaner assembly. Disconnect the accelerator linkage. Disconnect the clutch cable, if equipped with manual transaxle.
6. Disconnect the heater hoses from the heater unit. Disconnect the brake booster hose. Disconnect the speedometer cable from the transaxle.
7. If equipped with 4WD, disconnect the hoses for the assembly. Disconnect any other required hoses and vacuum lines necessary to remove the engine.
8. Raise and support the vehicle safely. Disconnect the exhaust pipes from the their mounting. Disconnect the gearshift rod from the transaxle.
9. Matchmark and remove the driveshaft. Properly support the engine and transaxle assembly. Remove the center member and crossmember assembly.
10. Remove the engine mount retaining bolts. Lower the vehicle.
11. Using the proper lifting equipment, carefully remove the engine and transaxle assembly from the vehicle.

To install:
12. Lower the engine assembly into the vehicle. Install the engine mount retaining bolts.
13. Install the center member, crossmember and driveshaft. Install the exhaust pipes and gearshift rod.
14. Connect all vacuum lines and hoses to their proper locations. Connect the speedometer cable.
15. Install the accelerator linkage, clutch cable (manual transaxles) and the air cleaner.
16. Install the radiator. Fill the cooling system. Fill the engine with oil and the transaxle with fluid.
17. Install the negative battery cable. Install the hood.
18. Start the engine and allow it to reach normal operating temperature. Check for leaks. Test drive the vehicle.

XT Coupe

1.8L ENGINE

1. Properly relieve the fuel pump pressure. Disconnect the negative battery cable.
2. Matchmark and remove the hood. Drain the cooling system. Drain the engine oil. Properly discharge the air conditioning system, if equipped.
3. Remove the spare tire assembly. Remove the spare tire support. Remove the battery cable clamp assembly. Remove the air cleaner assembly.
4. Disconnect and plug the fuel line hoses. Disconnect the canister hoses and the brake booster hose.
5. If equipped with automatic transaxle, remove the diaphragm vacuum hose.
6. If equipped with a turbocharger, remove the vacuum switch hose at the heater vacuum tank. Remove the wastegate valve hoses.
7. If equipped with 4WD, remove the selective drive vacuum line and the differential lock vacuum hose.
8. Disconnect the electrical wiring harness connectors, the oxygen sensor electrical connector, the ignition coil wire and the distributor connector at the crank sensor.
9. Disconnect the alternator electrical connector, the air conditioning compressor connector, the pulse coil connector, the radiator fan motor connector and the thermo-switch connector.
10. Disconnect the accelerator cable. If equipped with manual transaxle, disconnect the hill holder cable connection on the clutch release fork.
11. If not equipped with a turbocharger, raise and support the vehicle safely. Remove the exhaust system. Lower the vehicle.
12. If equipped with a turbocharger, remove the accelerator cable cover and the top turbocharger cover assembly.
13. To remove the lower turbocharger cover assembly, raise and support the vehicle safely, loosen the front exhaust pipe connection. Lower the vehicle and remove the lower turbocharger cover from the center exhaust pipe.
14. Disconnect the center exhaust pipe to turbocharger connection. Raise and support the vehicle safely. Disconnect the center exhaust pipe to rear exhaust pipe connection. Remove the hanger bolt and disconnect the center exhaust pipe at the transaxle. Remove the center exhaust pipe.
15. Lower the vehicle. Remove the radiator. Disconnect and plug the heater hoses.
16. Disconnect and plug the air conditioning hoses at the compressor, if equipped. Remove the power steering pump assembly, if equipped.
17. Remove the engine pitching stopper rod. Remove the timing cover plate. Remove the retaining bolts that hold the torque converter to the driveplate, if equipped with automatic transaxle.
18. Properly support the engine and the transaxle assemblies, using the proper equipment.
19. Remove the bolts which hold the engine mount to the front crossmember. Remove the bolts which hold the lower side of the engine to the transaxle.
20. Install the proper engine lifting equipment to the engine. Remove the bolts that retain the upper side of the engine to the transaxle.
21. Carefully remove the engine from the vehicle. If equipped with manual transaxle, move the engine horizontally until the mainshaft is withdrawn from the clutch cover.

To install:
22. Install the engine paying close attention to the mainshaft position.
23. Install all engine attaching and mounting bolts. Reconnect the engine to the transaxle.
24. Reconnect the air conditioner lines and install the power steering pump. Install all hoses and vacuum lines previously disconnected.
25. Install the exhaust system and if equipped, the turbocharger. Install the accelerator cable and hill holder, if equipped.
26. Reconnect all electrical connections previously disconnected. Connect the fuel lines and install the spare tire assembly.
27. Fill the engine with oil, the radiator with coolant and the transaxle with fluid. Start the engine and allow it to reach operating temperature. Check for leaks. Test drive the vehicle.

2.7L ENGINE

1. Properly relieve the fuel system pressure. Disconnect the negative battery cable. Matchmark and remove the hood.
2. Properly discharge the air conditioning system, if equipped. Drain the engine oil. Drain the cooling system.
3. Disconnect the canister hose and the hose bracket. Disconnect and plug the fuel lines.
4. Disconnect the power brake vacuum line booster. If equipped with manual transaxle and 4WD, disconnect the differential lock vacuum hose.
5. Disconnect the engine wiring harness connectors, the oxygen sensor connector, the bypass air valve control connector, the ignition coil and the distributor connector to the crank sensor.
6. Disconnect the alternator connector, the air condition compressor connector, the engine ground connector, the radiator fan motor connector and the thermo-switch electrical connector.
7. Disconnect the accelerator cable. Disconnect the cruise control cable, if equipped. Disconnect and plug the heater hoses.
8. Disconnect the hill holder cable on the clutch release fork side of the assembly, if equipped with manual transaxle.
9. Raise and support the vehicle safely. Disconnect the front exhaust pipe from the engine.
10. Disconnect the front to rear exhaust pipe connection. Disconnect the front exhaust pipe at the transaxle and hanger locations.
11. Lower the vehicle. Disconnect

and plug the air conditioning compressor hoses.

12. Remove the radiator fan shroud assembly. If equipped with automatic transaxle, disconnect and plug the fluid lines. Remove the radiator.

13. Remove the timing hole plug. Remove the bolts that retain the torque converter to the driveplate, if equipped with automatic transaxle.

14. Remove the buffer rod mounting bolts. Remove the bolts that support the engine mount to the front crossmember. Remove the bolts that hold the lower side of the engine to the transaxle assembly.

15. Install the proper engine lifting equipment. Properly support the transaxle assembly.

16. Remove the bolts that retain the upper side of the engine to the transaxle

17. Carefully remove the engine from the vehicle. If equipped with manual transaxle, move the engine in the axial direction until the mainshaft is withdrawn from the clutch cover.

To install:

18. Install the engine taking care to align the mainshaft. Install all bolts and mountings that attach the engine to the transaxle and chassis. Reconnect the torque converter to the engine flywheel (automatic transaxle).

19. Install the radiator, radiator shroud and transaxle fluid cooler lines. Install the air conditioner compressor and lines. Install the exhaust system.

20. Install the hill holder, if equipped. Reconnect all vacuum lines and hoses previously disconnected. Reconnect all electrical connections taking care to clean all connectors and ground locations.

21. Install all fuel lines. Fill the engine with oil, the transmisson with fluid and the cooling system with coolant. Start the engine and allow it to reach operating temperature. Check for leaks. Test drive the vehicle.

Except Justy, XT Coupe and Legacy

1.8L ENGINE

1. If equipped with fuel injection, properly relieve the fuel pump pressure. Disconnect the negative battery cable.

2. Matchmark and remove the hood. Drain the cooling system. Drain the engine oil. Properly discharge the air conditioning system, if equipped.

3. Remove the spare tire assembly. Remove the spare tire support. Remove the battery cable clamp assembly. Remove the air cleaner assembly.

4. Disconnect and plug the fuel line hoses. Disconnect the canister hoses and the brake booster hose.

5. If equipped with automatic transaxle, remove the diaphragm vacuum hose.

6. If equipped with a turbocharger, remove the vacuum switch hose at the heater vacuum tank. Remove the wastegate valve hoses.

7. If equipped with 4WD, remove the selective drive vacuum line and the differential lock vacuum hose.

8. Disconnect the electrical wiring harness connectors, the oxygen sensor electrical connector, the ignition coil wire and the distributor connector at the crank sensor.

9. Disconnect the alternator electrical connector, the air conditioning compressor connector, the pulse coil connector, the radiator fan motor connector and the thermo-switch connector.

10. Disconnect the accelerator cable. If equipped with manual transaxle, disconnect the hill holder cable connection on the clutch release fork.

11. If not equipped with a turbocharger, raise and support the vehicle safely. Remove the exhaust system. Lower the vehicle.

12. If equipped with a turbocharger, remove the accelerator cable cover and the top turbocharger cover assembly.

13. To remove the lower turbocharger cover assembly, raise and support the vehicle safely, loosen the front exhaust pipe connection. Lower the vehicle and remove the lower turbocharger cover from the center exhaust pipe.

14. Disconnect the center exhaust pipe to turbocharger connection. Raise and support the vehicle safely. Disconnect the center exhaust pipe to rear exhaust pipe connection. Remove the hanger bolt and disconnect the center exhaust pipe at the transaxle. Remove the center exhaust pipe.

15. Lower the vehicle. Remove the radiator. Disconnect and plug the heater hoses.

16. Disconnect and plug the air conditioning hoses at the compressor, if equipped. Remove the power steering pump assembly, if equipped.

17. Remove the pitching stopper rod. Remove the timing cover plate. Remove the retaining bolts that hold the torque converter to the driveplate, if equipped with automatic transaxle.

18. Properly support the engine and the transaxle assemblies, using the proper equipment.

19. Remove the bolts which hold the engine mount to the front crossmember. Remove the bolts which hold the lower side of the engine to the transaxle.

20. Properly install the proper engine lifting equipment to the engine. Remove the bolts that retain the upper side of the engine to the transaxle.

21. Carefully remove the engine from the vehicle. If equipped with manual transaxle, move the engine horizontally until the mainshaft is withdrawn from the clutch cover.

To install:

22. Install the engine and attach all engine and transaxle mounts. Install all bolts attaching the engine to the transaxle. Install the torque converter bolts on automatic transaxle equipped vehicles.

23. Install the air conditioner compressor and reconnect the lines. Install the power steering pump. Install the radiator and reconnect the coolant hoses.

24. Install the exhaust system and turbocharger, if equipped. Install the accelerator cable and hill holder (manual transaxles).

25. Reconnect all vacuum lines, hoses and electrical connections previously disconnected. Reconnect the fuel lines.

26. Install the spare tire assembly and the hood. Fill the engine with oil, the transaxle with fluid and the cooling system with coolant. Start the engine and allow it to reach operating temperature. Check for leaks. Test drive the vehicle.

Cylinder Head

REMOVAL & INSTALLATION

Justy

1. Disconnect the negative battery cable. Drain the cooling system.

2. Remove the air cleaner assembly. Remove the drive belts. Remove the spark plug wires.

3. Position the engine at TDC with No. 1 cylinder on the compression stroke. Matchmark and remove the distributor assembly.

4. Remove the crankshaft pulley, using pulley removal tool 499205500 or equivalent. Remove the outer front timing belt cover.

5. Loosen the tensioner bolt and position it in the direction that loosens the belt. Tighten the tensioner bolt in that position.

6. Remove the camshaft driveplate. Mark the timing belt, in the direction of rotation, for reinstallation and than remove it from the engine.

7. Remove the tensioner and spring. Remove the camshaft pulley, using pulley removal tool 499205500 or equivalent. Remove the inner belt cover and cover mount.

8. Remove the PCV hose from the rocker arm cover. Remove the rocker arm cover retaining bolts. Remove the rocker arm cover from the engine. Remove the rocker arm assembly.

9. Remove the exhaust manifold retaining bolts. Remove the exhaust

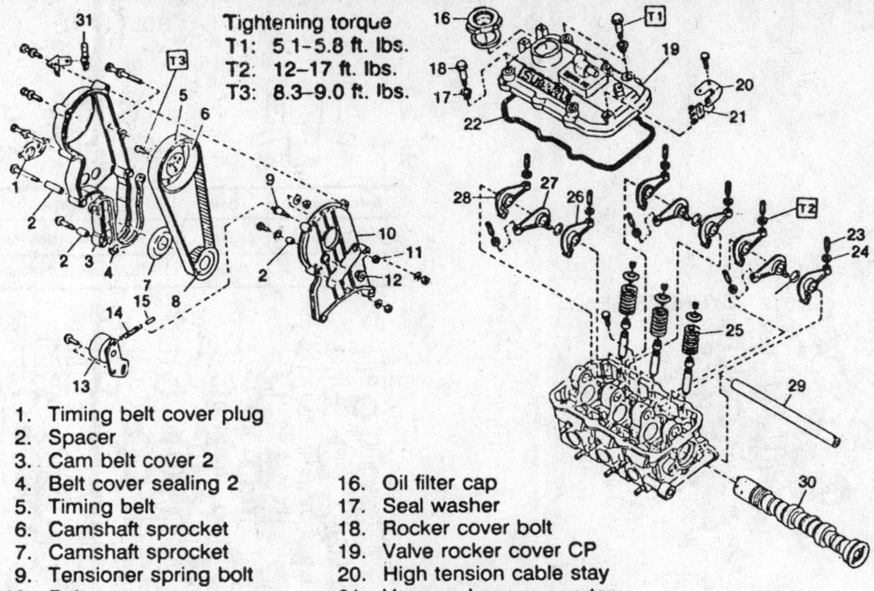

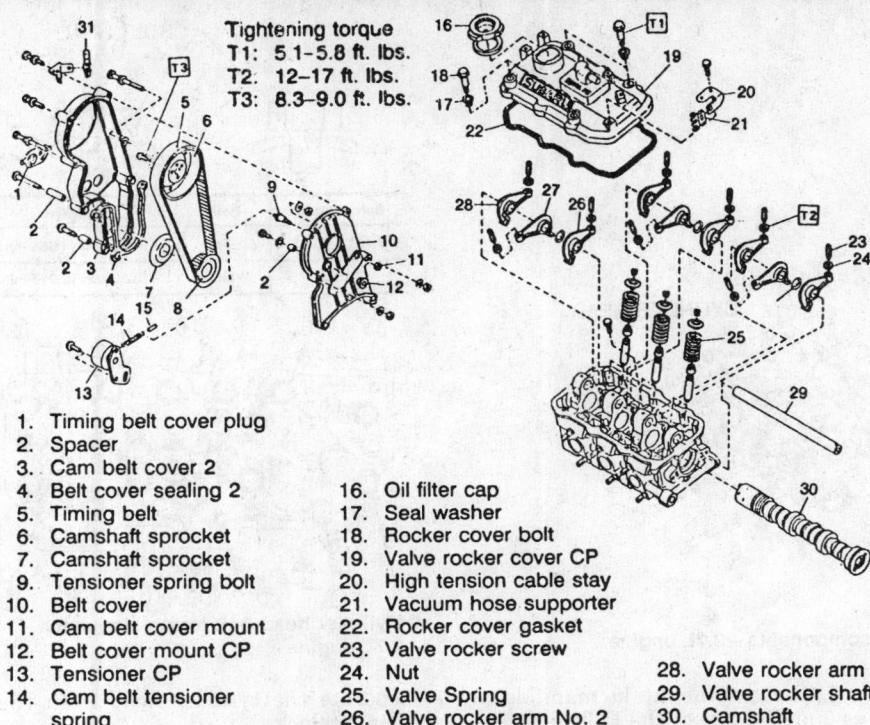

1. Timing belt cover plug
2. Spacer
3. Cam belt cover 2
4. Belt cover sealing 2
5. Timing belt
6. Camshaft sprocket
7. Camshaft sprocket
9. Tensioner spring bolt
10. Belt cover
11. Cam belt cover mount
12. Belt cover mount CP
13. Tensioner CP
14. Cam belt tensioner spring
15. Tensioner spring amper
16. Oil filter cap
17. Seal washer
18. Rocker cover bolt
19. Valve rocker cover CP
20. High tension cable stay
21. Vacuum hose supporter
22. Rocker cover gasket
23. Valve rocker screw
24. Nut
25. Valve Spring
26. Valve rocker arm No. 2
27. Valve rocker arm No. 3
28. Valve rocker arm
29. Valve rocker shaft
30. Camshaft
31. Stay

Cylinder head and related components – 1.2L engine

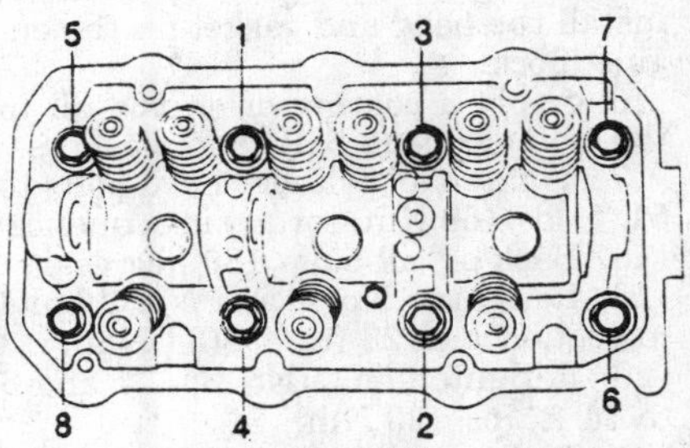

Cylinder head bolt torque sequence – 1.2L engine

manifold from the engine. Discard the gasket.

10. Disconnect all required electrical wiring and vacuum lines. Remove the air suction valve and pipe, if equipped.

11. Disconnect the accelerator linkage. Remove the intake manifold retaining bolts. Remove the intake manifold along with the carburetor. Discard the gasket.

12. Be sure the engine is cold before removing the cylinder head bolts. Loosen, than remove the cylinder head bolts. Carefully remove the cylinder head from the engine.

To install:

13. Installation is the reverse of the removal procedure. Be sure to use new gaskets or RTV sealant, as required.

14. Install the cylinder head and tighten the bolts as follows:

a. Step 1 – Torque all bolts in sequence to 29 ft. lbs. (39 Nm).

b. Step 2 – Torque all bolts in sequence to 54 ft. lbs. (73 Nm).

c. Step 3 – Loosen bolts 90 degrees or more in reverse order of tightening sequence.

d. Step 4 – Torque all bolts in sequence to 54 ft. lbs. (73 Nm).

15. Adjust the valves to specification, as required. Install the camshaft pulley, timing belt, tensioner and drive plate.

16. Install the distributor and all other components previously removed. Fill the cooling system with coolant and check the engine oil.

17. Start the engine, check the ignition timing, check for leaks and test drive the vehicle.

XT Coupe

1.8L AND 2.7L ENGINES

1. Disconnect the negative battery cable. Drain the cooling system. Properly discharge the air conditioning system, if equipped.

2. Remove the timing belt assemblies. Remove the camshaft assemblies.

3. On non air conditioning equipped vehicles, remove the bolt attaching the alternator bracket to the cylinder head and the bolt retaining the adjusting bar to the cylinder head.

4. Be sure the engine is cold before removing the intake manifold retaining bolts. Loosen, then remove the intake manifold bolts. Carefully remove the manifold from the engine.

5. Remove the water bypass line at the cylinder head. Remove the spark plugs.

6. Be sure the engine is cold before removing the cylinder head bolts. Loosen, then remove the cylinder head bolts. Carefully remove the cylinder head from the engine.

To install:

7. On 1.8L engines, install the cylinder head using a new gasket and tighten the bolts as follows:

a. Step 1 – Torque bolts in the proper sequence to 22 ft. lbs. (30 Nm).

b. Step 2 – Torque bolts in the proper sequence to 43 ft. lbs. (58 Nm).

c. Step 3 – Torque bolts in the proper sequence to 47 ft. lbs. (63 Nm).

8. On 2.7L engines, install the cylinder head using a new gasket and tighten the bolts as follows:

a. Step 1 – Torque all bolts in sequence to 29 ft. lbs. (39 Nm).

b. Step 2 – Torque all bolts in sequence to 47 ft. lbs. (63 Nm).

c. Step 3 – Loosen all bolts 90 degrees or more in the reverse order of tightening sequence.

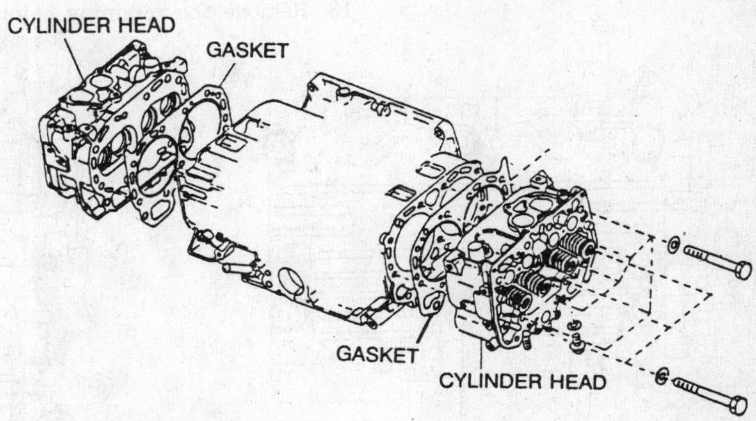

Cylinder head and related components – 1.8L engine

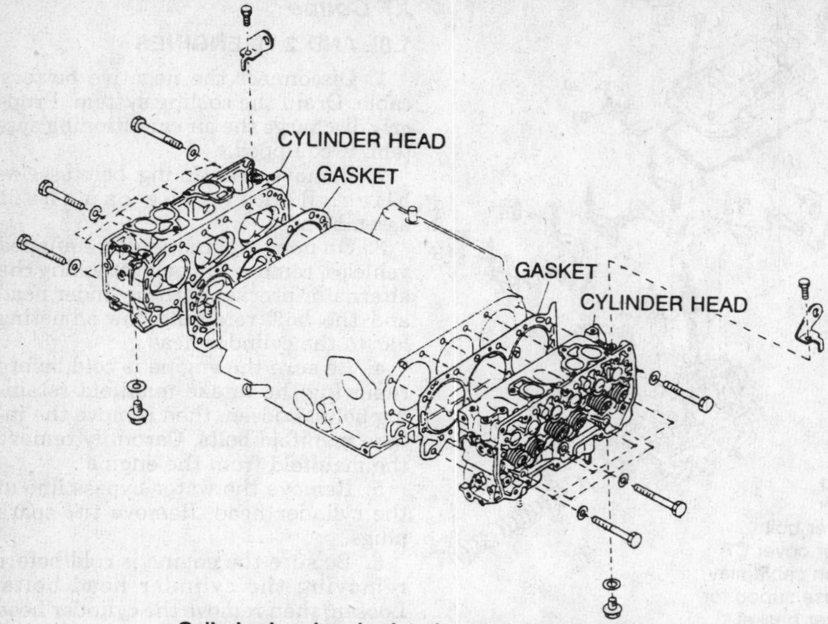

Cylinder head and related components – 2.7L engine

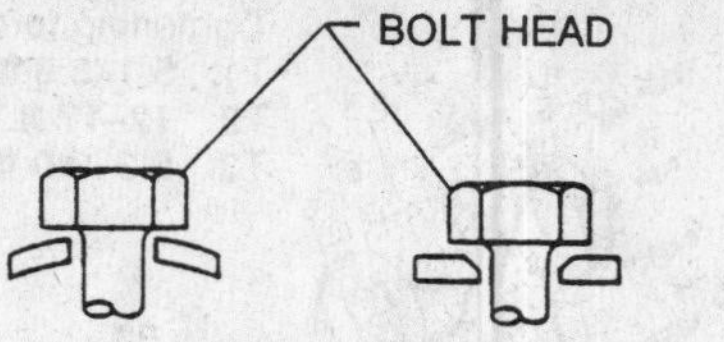

Bolt position	Color	Bolt length
①, ②, ⑨, ⑬	Silver	118.5 mm (4.665 in)
Others	Yellow	132.5 mm (5.217 in)

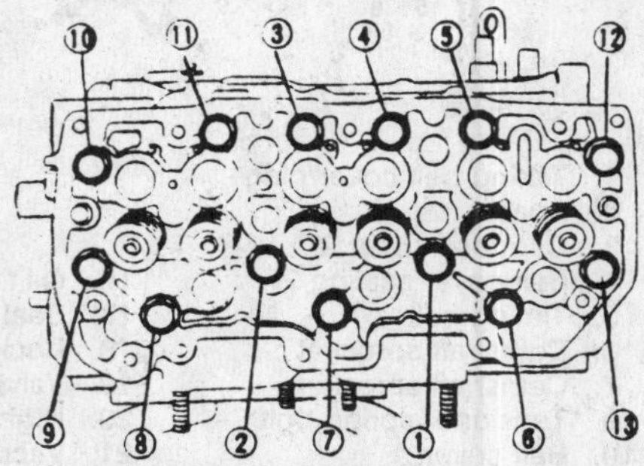

Cylinder head bolt torque sequence – 2.7L engine

d. Step 4 – Torque all bolt in sequence to 44–50 ft. lbs. (59–67 Nm).

9. Install the intake manifold and tighten bolts to 13–16 ft. lbs. (18–21 Nm).

10. Install the camshafts and timing belt. Install all other components previously disconnected. Fill the cooling system with fluid and check the engine oil.

11. Start the engine and check the ignition timing. Check for leaks and test drive the vehicle.

Legacy

2.2L ENGINE

1. Remove the negative battery cable. Remove the V-belt.
2. Remove the power steering pump assembly.
3. Remove the alternator and its bracket.
4. Remove the intake manifold cover and disconnect the PCV hose.
5. Disconnect the spark plug wire connectors and remove the connector from the bracket attaching bolt.
6. Remove the crankshaft and cam angle sensors. Remove the oil pressure switch connector.
7. Remove the knock sensor and the PCV blow-by hose.
8. Remove the intake manifold and gasket. Remove the water pipe.
9. Remove the timing belt, camshaft sprocket and related components.
10. Remove the left hand dipstick tube attaching bolt.
11. Remove the cylinder head bolts in sequence, leaving 2 top (1 and 3) loosely in block to keep cylinder head from falling.
12. If necessary, tap cylinder head with a plastic hammer to loosen from cylinder block.
13. Remove the retaining bolts and separate the cylinder head from the engine block.
14. Remove the cylinder head gasket.

To install:

15. Using new cylinder head gaskets, install the head and gasket on the engine block.
16. Apply a coating of engine oil to the cylinder head bolts and washers.
17. Tighten all cylinder head bolts to 51 ft. lbs. (69 Nm) torque in sequence.
18. Back off all bolts 180 degrees.
19. Retighten No. 1 and No. 2 bolts in sequence to 25 ft. lbs. (34 Nm).
20. Retighten bolts No. 3, 4, 5 and 6 to 14 ft. lbs. (20 Nm).
21. Tighten all bolts an additional 80–90 degrees in sequence. Do not tighten over 90 degrees.
22. Further tighten all bolts an additional 80–90 degrees in sequence.

NOTE: The total retightening range must not exceed 180 degrees maximum.

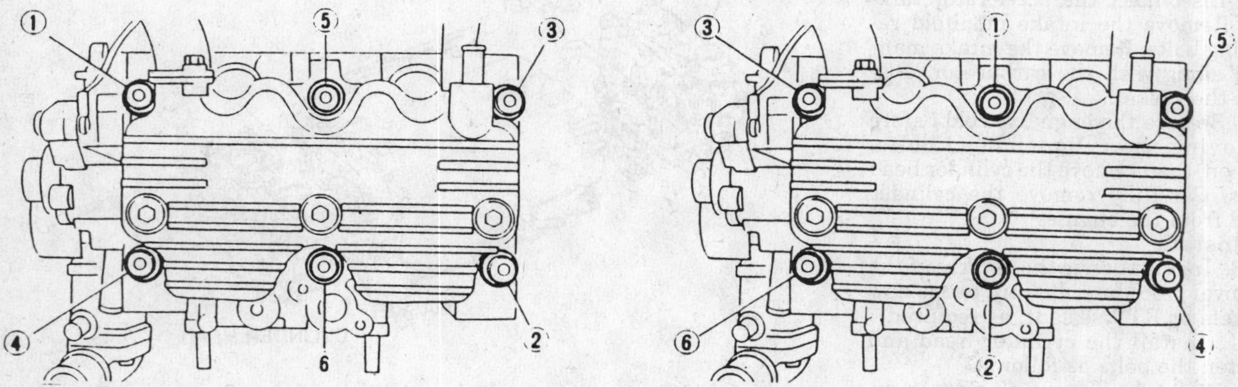

Cylinder head torque sequence – 2.2L engine

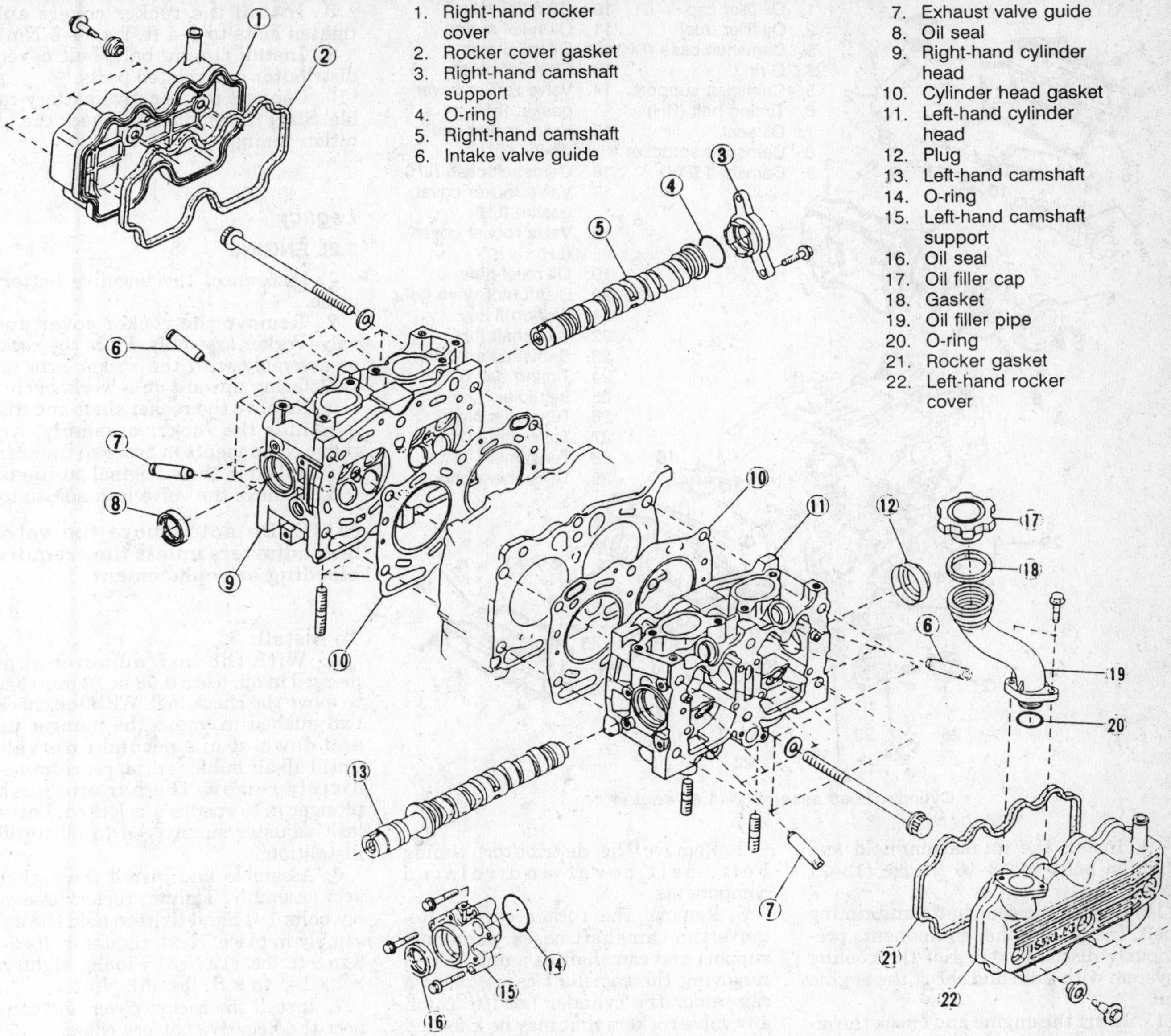

1. Right-hand rocker cover
2. Rocker cover gasket
3. Right-hand camshaft support
4. O-ring
5. Right-hand camshaft
6. Intake valve guide
7. Exhaust valve guide
8. Oil seal
9. Right-hand cylinder head
10. Cylinder head gasket
11. Left-hand cylinder head
12. Plug
13. Left-hand camshaft
14. O-ring
15. Left-hand camshaft support
16. Oil seal
17. Oil filler cap
18. Gasket
19. Oil filler pipe
20. O-ring
21. Rocker gasket
22. Left-hand rocker cover

2.2L engine cylinder head – exploded view

23. Complete the assembly in the reverse of the removal procedure.

Except Justy, XT Coupe and Legacy

1.8L ENGINE

1. Disconnect the negative battery cable. Drain the cooling system. Properly discharge the air conditioning system, if equipped.
2. Remove the timing belt assemblies. Remove the camshaft assemblies.
3. On non air conditioning equipped vehicles, remove the bolt attaching the alternator bracket to the cylinder head and the bolt retaining the adjusting bar to the cylinder head.
4. Be sure the engine is cold before removing the intake manifold retaining bolts. Loosen, than remove the intake manifold bolts. Carefully remove the manifold from the engine.
5. Remove the water bypass line at the cylinder head. Remove the spark plugs.
6. Be sure the engine is cold before removing the cylinder head bolts. Loosen, than remove the cylinder head bolts. Carefully remove the cylinder head from the engine.

To install:

7. Install the cylinder head using a new gasket and tighten the bolts as follows:
   a. Step 1 – Torque bolts in the proper sequence to 22 ft. lbs. (30 Nm).
   b. Step 2 – Torque bolts in the proper sequence to 43 ft. lbs. (58 Nm).
   c. Step 3 – Torque bolts in the proper sequence to 47 ft. lbs. (63 Nm).
8. Installation is the reverse of the removal procedure. Be sure to use new gaskets or RTV sealant, as required.

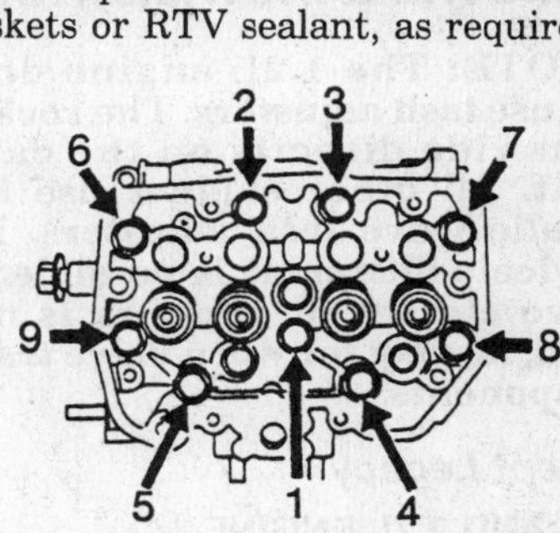

Cylinder head bolt torque sequence – 1.8L engine

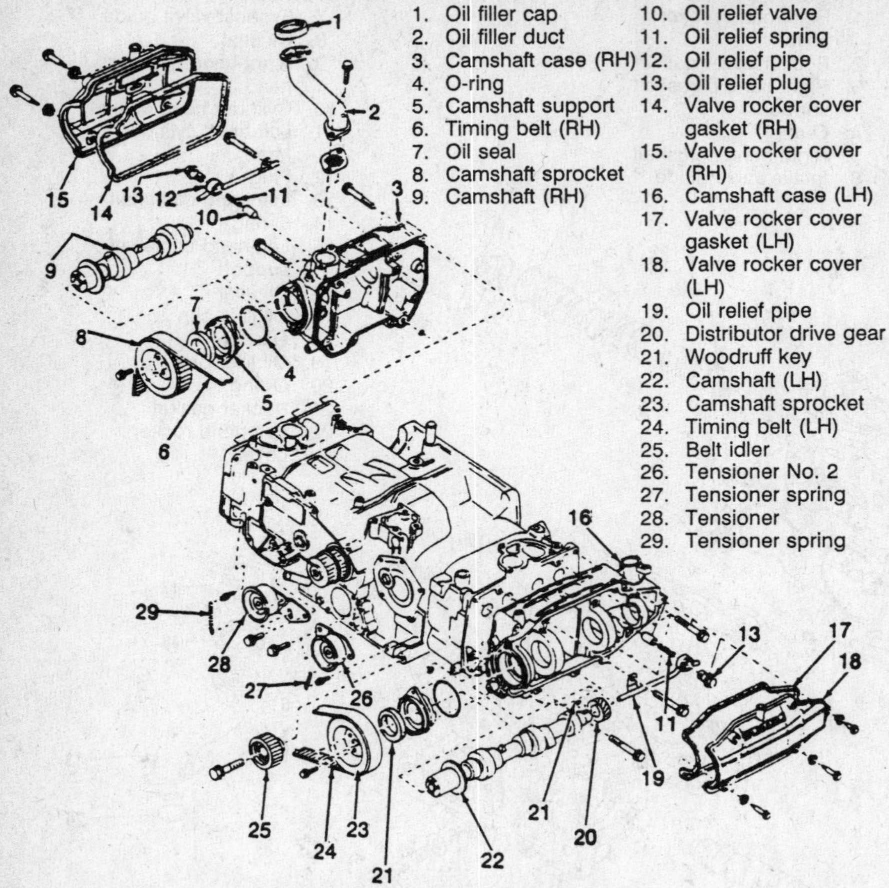

1. Oil filler cap
2. Oil filler duct
3. Camshaft case (RH)
4. O-ring
5. Camshaft support
6. Timing belt (RH)
7. Oil seal
8. Camshaft sprocket
9. Camshaft (RH)
10. Oil relief valve
11. Oil relief spring
12. Oil relief pipe
13. Oil relief plug
14. Valve rocker cover gasket (RH)
15. Valve rocker cover (RH)
16. Camshaft case (LH)
17. Valve rocker cover gasket (LH)
18. Valve rocker cover (LH)
19. Oil relief pipe
20. Distributor drive gear
21. Woodruff key
22. Camshaft (LH)
23. Camshaft sprocket
24. Timing belt (LH)
25. Belt idler
26. Tensioner No. 2
27. Tensioner spring
28. Tensioner
29. Tensioner spring

Cylinder head assembly – 1.8L engine

9. Install the intake manifold and tighten bolts to 13–16 ft. lbs. (18–21 Nm).
10. Install the camshafts and timing belt. Install all other components previously disconnected. Fill the cooling system with fluid and check the engine oil.
11. Start the engine and check the ignition timing. Check for leaks and test drive the vehicle.

Valve Lash Adjusters

REMOVAL & INSTALLATION

NOTE: The 1.2L engine does not use lash adjusters. The rocker arms ride directly on the camshaft. All other engines use hydraulic valve lash adjusters. No service adjustment is possible. If valve clatter or tapping is noticed, inspect for worn valve train components.

Except Legacy

1.8L AND 2.7L ENGINE

1. Disconnect the negative battery cable.
2. Remove the distributor, timing belt, belt cover and related components.
3. Remove the rocker covers. Remove the camshaft cases, camshaft support and camshaft as a unit. When removing the camshaft cases, place a rag under the cylinder head to catch any valve rockers that may be knocked loose.
4. Remove the rocker arms and valve lash adjusters from the cylinder head. Do not lay the adjusters down. Keep all components in the order of removal.

To install:

5. With the lash adjuster in a vertical position, push the adjuster pivot inward with a quick and hard motion. If the pivot is depressed more than 0.020 in. (0.5 mm), put the adjuster in a container of oil and pump the plunger until the depression is within specification. If the lash adjuster is still not within specification, replace it.
6. Insert the lash adjusters into the cylinder head. Apply grease to the valve rocker and install the rocker on the cylinder head.
7. Install the camshaft case assembly and tighten bolts to 12–14 ft. lbs. (17–20 Nm).
8. Install the rocker covers and tighten bolts to 3–4 ft. lbs. (4–5 Nm).
9. Install timing belt, belt cover, distributor and related parts.
10. Connect the negative battery cable. Start the engine and check the ignition timing.

Legacy

2.2L ENGINE

1. Disconnect the negative battery cable.
2. Remove the rocker cover and valve rocker assembly. Place the rocker assembly with the rocker arm air vent facing upward on a workbench.
3. Remove the rocker shaft and disassemble the rocker assembly. Arrange components in order so they can be installed in their original positions.
4. Remove the valve lash adjusters.

NOTE: Do not remove the valve lash adjusters unless they require bleeding or replacement.

To install:

5. With the lash adjuster submerged in oil, use a 0.08 in. (2 mm) bar to move the check ball. With the check ball pushed in, move the plunger up and down at one second intervals untill all air bubbles disappear. Immediately remove the bar and push plunger in to ensure it is locked. Leave lash adjuster submerged in oil untill installtion.
6. Assemble and install the rocker arm assembly. Tighten rocker assembly bolts 1–4 hand tight to hold the assembly in place. Next, tighten bolts 5–8 to 9 ft. lbs. (12 Nm). Finally, tighten bolts 1–4 to 9 ft. lbs. (12 Nm).
7. Install the rocker cover and connect the negative battery cable.
8. Start the engine, check the ignition timing and test drive the vehicle.

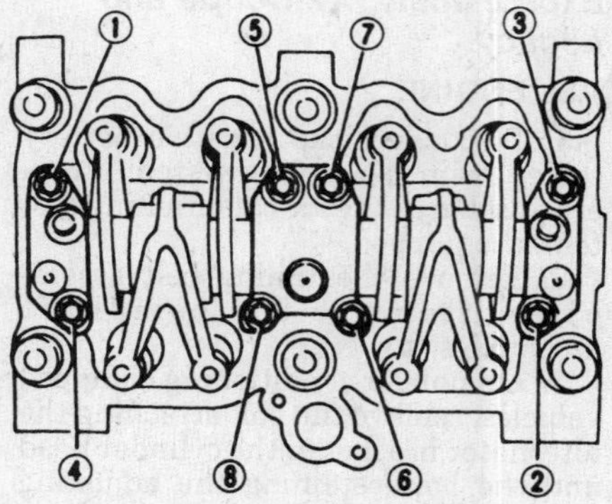

Rocker arm assembly torque sequence – 2.2L engine

Rocker Shafts

REMOVAL & INSTALLATION

Justy

1. Disconnect the negative battery cable. Remove the air cleaner assembly.
2. Remove the valve cover retaining bolts. Remove the valve cover from the engine.
3. Using the valve clearance adjuster tool 498767000 or equivalent, loosen the valve rocker nut and screw.
4. Remove the bolt from the valve rocker shaft journal. Pull the rocker shaft out from the cylinder head. Remove the spring washer and the valve rocker arms.

To install:

5. Install valve rocker arms. Slide the rocker shaft through the journals. Ensure that the alignment holes are in the correct position. If there is difficulty aligning the rocker shaft, use rocker shaft guide 498005400 or equivalent.
6. When the rocker shaft is aligned, install and tighten the rocker shaft bolts. Adjust the valve lash.
7. Install the valve cover and all other components previously remove. Start the engine, check the ignition timing and test drive the vehicle.

Intake Manifold

REMOVAL & INSTALLATION

Justy

1. Disconnect the negative battery cable. Drain the cooling system. Remove the air cleaner assembly. On fuel injected models, relieve the fuel system pressure.
2. Disconnect the fuel line and accelerator cable. Label and disconnect the required vacuum lines. Label and disconnect all required electrical connections. Remove the upper radiator hose.
3. Remove the necessary components in order to gain access to the intake manifold retaining bolts.
4. Remove the intake manifold retaining bolts. Remove the intake manifold and discard the gasket.

To install:

5. Clean all gasket material from the intake manifold and cylinder head.
6. Install the intake manifold using a new gasket. Tighten the intake manifold bolts to 14–22 ft. lbs. (19–30 Nm).
7. Reconnect all vacuum and electrical connections. Install all components previously removed.
8. Reconnect the radiator hose and fill the cooling system with coolant.
9. Reconnect the fuel lines and then the negative battery cable. Start the engine and check for leaks.

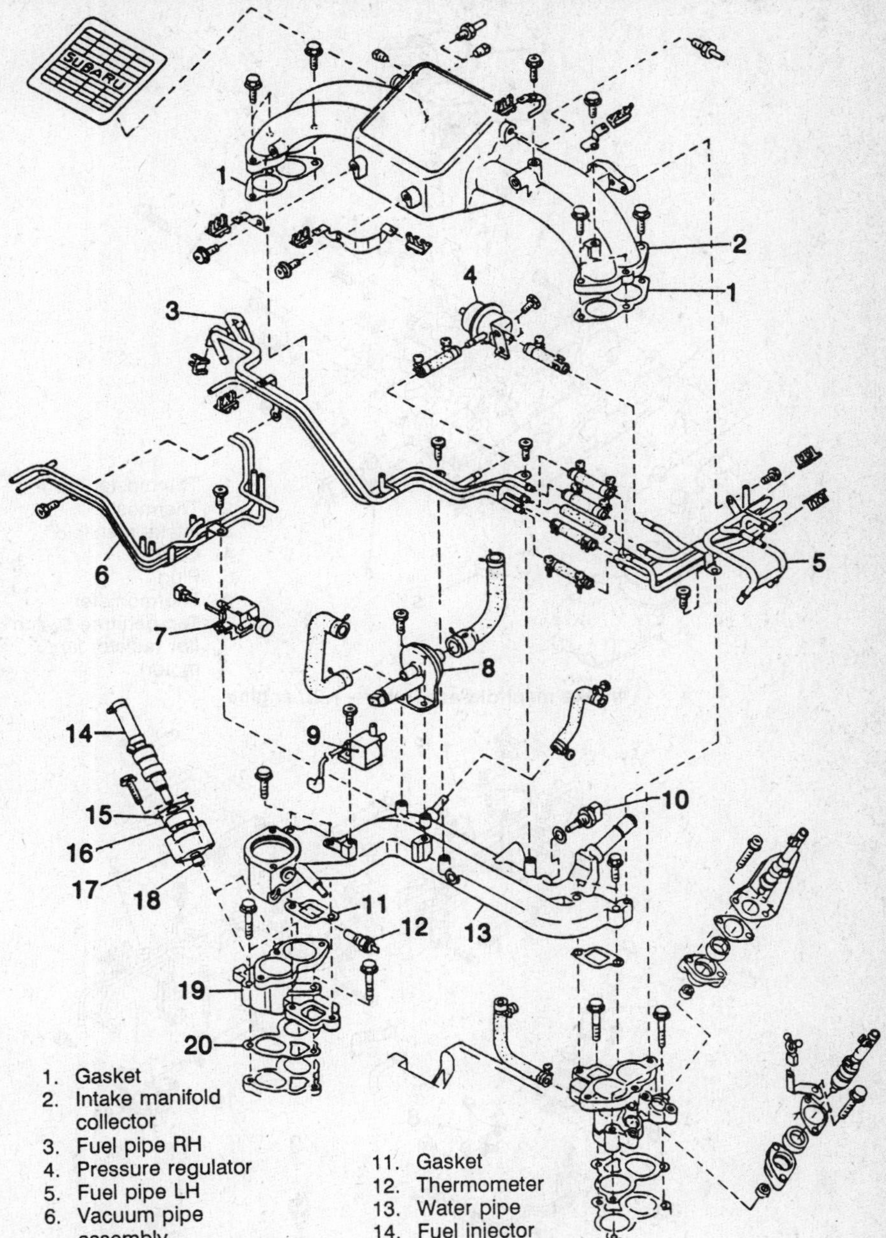

1. Gasket
2. Intake manifold collector
3. Fuel pipe RH
4. Pressure regulator
5. Fuel pipe LH
6. Vacuum pipe assembly
7. Purge control solenoid valve
8. Auxiliary air valve
9. EGR solenoid valve
10. Coolant thermosensor
11. Gasket
12. Thermometer
13. Water pipe
14. Fuel injector
15. Holder plate
16. Insulator
17. Holder
18. Seal
19. Intake manifold
20. Gasket

Intake manifold assembly – 1.8L engine with MPFI

XT Coupe

1.8L AND 2.7L ENGINE

1. Properly relieve the fuel system pressure. Disconnect the negative battery cable. Drain the cooling system. Remove the air cleaner assembly.
2. Remove the fuel pipe covers. Remove the fuel pipe assemblies.
3. Disconnect the required electrical connectors and vacuum hoses.
4. Remove the purge control solenoid valve. Remove the water pipe assembly.
5. Remove the necessary components in order to gain access to the intake manifold collector retaining bolts.
6. Remove the intake manifold collector retaining bolts. Remove the intake manifold collector from the engine. Discard the gaskets.

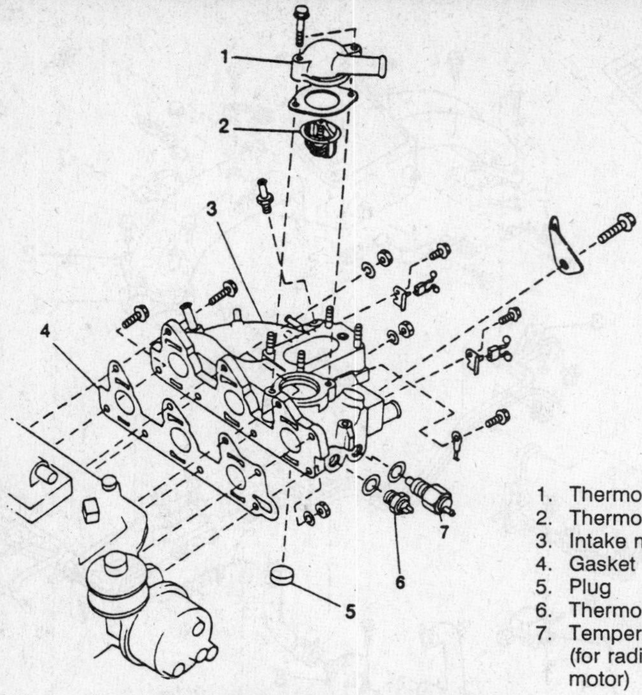

1. Thermostat cover
2. Thermostat
3. Intake manifold
4. Gasket
5. Plug
6. Thermometer
7. Temperature switch (for radiator fan motor)

Intake manifold assembly – 1.2L engine

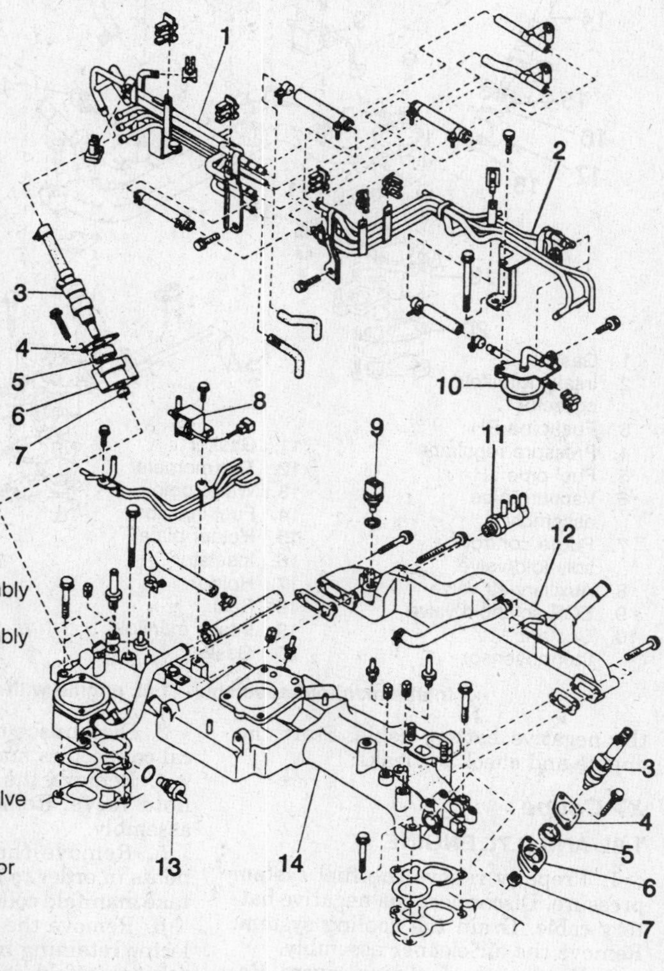

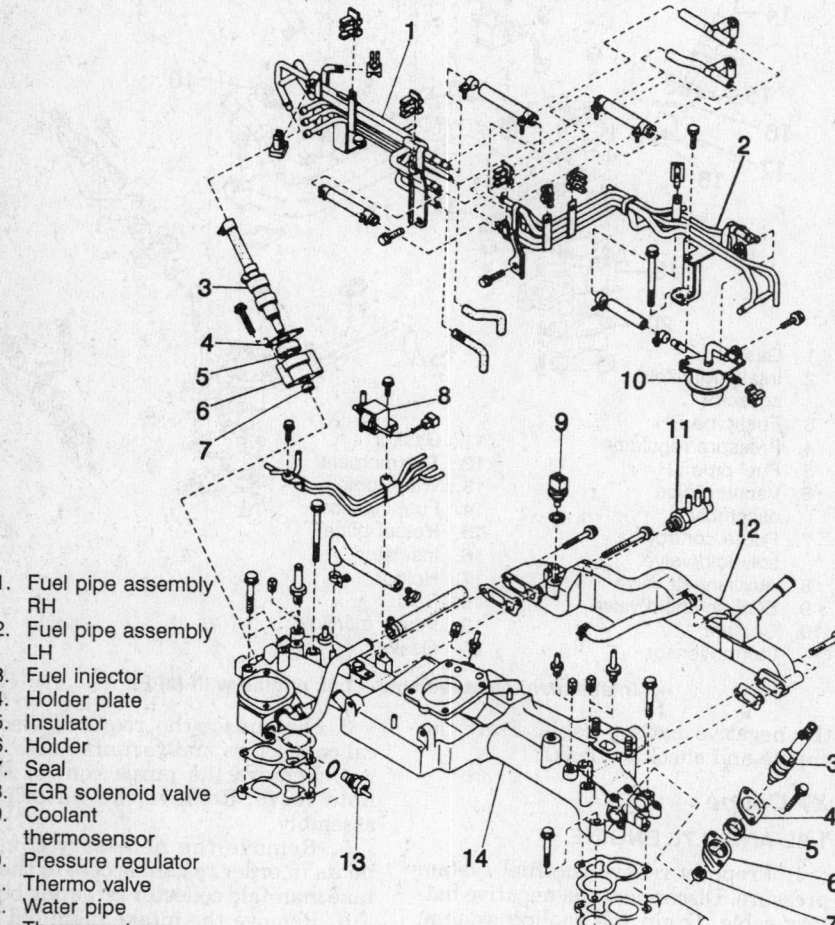

1. Fuel pipe assembly RH
2. Fuel pipe assembly LH
3. Fuel injector
4. Holder plate
5. Insulator
6. Holder
7. Seal
8. EGR solenoid valve
9. Coolant thermosensor
10. Pressure regulator
11. Thermo valve
12. Water pipe
13. Thermometer
14. Intake manifold

Intake manifold assembly – 1.8L engine with TBI

7. As required, remove the fuel injectors from their bores. Remove the intake manifold retaining bolts. Remove the intake manifold from the engine. Discard the gaskets.

To install:

8. Clean the mating surfaces of the intake manifold and cylinder head throughly.
9. Using a new gasket, install the intake manifold. Tighten the bolts to 13–16 ft. lbs. (18–22 Nm).
10. Install all components previously removed including the purge control solenoid valve and water pipe assembly.
11. Reconnect all vacuum and electrical connections. Fill the cooling system with coolant.
12. Reconnect the fuel lines and then the negative battery cable. Start the engine and check for leaks.

Legacy

2.2L ENGINE

1. Remove the negative battery cable. Remove the V-belt.
2. Remove the power steering pump assembly.
3. Remove the alternator and its bracket.
4. Remove the intake manifold cover and disconnect the PCV hose.
5. Disconnect the spark plug wire connectors and remove the connector from the bracket attaching bolt.
6. Remove the crankshaft and cam angle sensors. Remove the oil pressure switch connector.
7. Remove the knock sensor and the PCV blow-by hose.
8. Remove the intake manifold and gasket. Remove the water pipe.

To install:

9. Clean the mating surfaces of the intake manifold and the cylinder head throughly.
10. Install the intake manifold using a new gasket and tighten the bolts to specified torque. Connect the fuel lines.
11. Install the knock sensor, PVC hose, crankshaft and cam angle sensors, oil pressure switch connector, spark plug wire connectors, intake manifold cover, alternator, power steering pump and V-belt.
12. Check the coolant and oil level. Fill as necessary. Start the engine and check for leaks.

Except Justy, XT Coupe and Legacy

1.8L ENGINE

1. If equipped with fuel injection, properly relieve the fuel system pressure. Disconnect the negative battery cable. Remove the spare tire from the engine compartment.

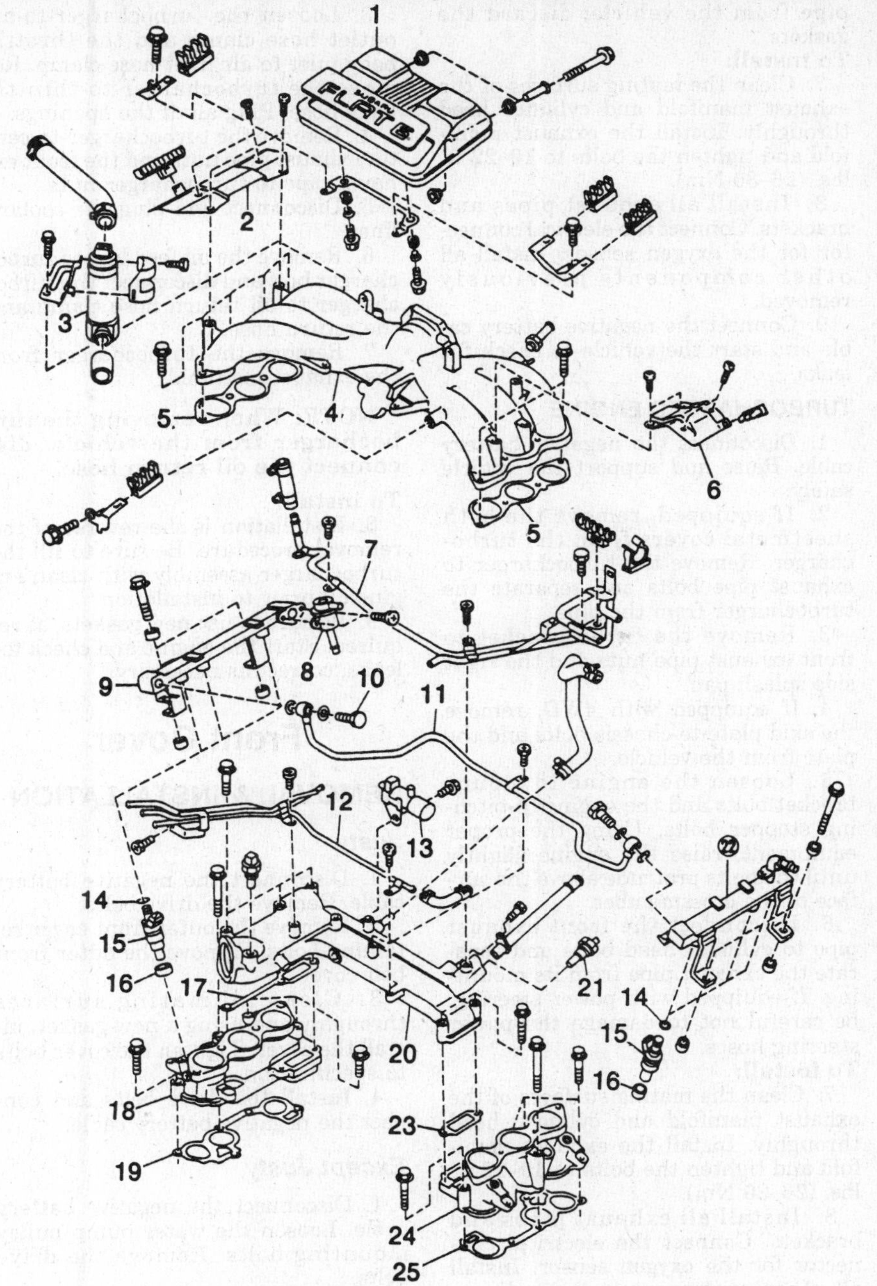

1. Intake manifold cover
2. Fuel pipe cover RH
3. Bypass air control valve
4. Intake manifold collector
5. Gasket
6. Fuel pipe cover LH
7. Fuel pipe assembly
8. Pressure regulator
9. Fuel pipe RH
10. Union bolt
11. Fuel Pipe
12. Fuel Pipe assembly
13. Purge control solenoid valve
14. O-ring
15. Fuel injector
16. Insulator
17. Gasket
18. Intake manifold RH
19. Gasket
20. Water pipe
21. Coolant thermostat
22. Fuel pipe LH
23. Gasket
24. Intake manifold LH
25. Gasket

Intake manifold assembly—2.7L engine

2. Disconnect the emission control system hoses, remove the mounting bracket screws and withdraw the air cleaner assembly.

3. If equipped with a turbocharger or a fuel injection system, loosen the hose clamps and remove the air intake duct.

4. Drain the cooling system. Remove the water hoses from the thermostat housing. Disconnect the thermo-switch connector.

5. If equipped with a distributor vacuum control valve, disconnect the hoses and electrical leads from it.

6. Disconnect the automatic choke to voltage regulator wire at the connector, the EGR solenoid wiring, if equipped and the EGR pipe.

7. Disconnect the throttle cable from it's bracket. If equipped with a carburetor, disconnect the fuel line.

8. If equipped with a turbocharger or fuel injection system, disconnect the hose clamps, pull off hoses and remove the fuel pressure regulator assembly.

9. If equipped with MPFI system, remove the fuel injectors from the intake manifold.

10. Remove the intake manifold to cylinder head bolts. Remove the intake manifold assembly from the engine. The air cleaner brackets will come off as the unit is unbolted. Make sure to note locations of these brackets and remove them.

To install:

11. Clean the mating surfaces of the intake manifold and cylinder head throughly. Install the intake manifold using a new gasket and tighten the bolts to 13–16 ft. lbs. (18–22 Nm).

12. Install the fuel injectors, regulator assembly and associated hoses, if equipped.

13. Install the throttle cable, all vacuum hoses and electrical connectors. Install all other components previously removed.

14. Connect the fuel lines and then reconnect the negative battery cable.

15. Start the engine, check for leaks and test drive the vehicle.

Exhaust Manifold

REMOVAL & INSTALLATION

Justy

1. Disconnect the negative battery cable. Remove the air cleaner assembly.

2. Raise and support the vehicle safely. Disconnect the exhaust manifold from the exhaust pipe. Lower the vehicle.

3. Disconnect the oxygen sensor electrical connector. Remove the exhaust manifold cover plate assembly.

4. Remove the exhaust manifold retaining bolts. Remove the exhaust manifold from the engine. Discard the gasket.

To install:

5. Clean the mating surfaces of the exhaust manifold and cylinder head throughly. Install the exhaust manifold and tighten the bolts to 14–22 ft. lbs. (19–30 Nm).

6. Install the oxygen sensor electrical connector and manifold cover plate. Raise the vehicle and install the exhaust pipe. Lower the vehicle.

7. Install the oxygen sensor electrical connector and air cleaner. Reconnect the negative battery cable.

XT Coupe and Legacy

1.8L AND 2.7L ENGINE

1. Disconnect the negative battery cable. Raise and support the vehicle safely.

2. Disconnect the oxygen sensor electrical connector.

3. Remove the exhaust manifold assembly to cylinder head retaining bolts.

4. Remove the exhaust manifold assembly to exhaust pipe retaining bolts.

5. If equipped with a turbocharger, remove the exhaust pipe to turbocharger assembly. Remove the turbocharger plate and gasket.

6. Remove the exhaust manifold assembly from the vehicle.

To install:

7. Clean the mating surfaces of the exhaust manifold and cylinder head throughly. Install the exhaust manifold and tighten the bolts to 19–22 ft. lbs. (26–30 Nm).

8. Install the turbocharger exhaust pipe assembly, oxygen sensor electrical connector and exhaust manifold to exhaust pipe bolts.

9. Lower the vehicle and reconnect the negative battery cable.

Except Justy, XT Coupe and Legacy

WITHOUT TURBOCHARGED ENGINE

1. Disconnect the negative battery cable. Raise and support the vehicle safely.

2. Disconnect the electrical connector from the oxygen sensor.

3. From the upper shell cover, remove the air duct.

4. Loosen the front exhaust pipe to cylinder head nuts.

5. Remove the front exhaust pipe to rear exhaust pipe nuts, then separate the pipes, discard the gasket.

6. Remove the front exhaust pipe to bracket bolt. While supporting the front exhaust pipe, remove the pipe to cylinder head nuts and front exhaust pipe from the vehicle; discard the gaskets.

To install:

7. Clean the mating surfaces of the exhaust manifold and cylinder head throughly. Install the exhaust manifold and tighten the bolts to 19–22 ft. lbs. (26–30 Nm).

8. Install all exhaust pipes and brackets. Connect the eletrical connector for the oxygen sensor. Install all other components previously removed.

9. Connect the negative battery cable and start the vehicle to check for leaks.

TURBOCHARGED ENGINE

1. Disconnect the negative battery cable. Raise and support the vehicle safely.

2. If equipped, remove the both sheetmetal covers from the turbocharger. Remove the turbocharger to exhaust pipe bolts and separate the turbocharger from the pipe.

3. Remove the turbo bracket to front exhaust pipe nuts and the right side splash pan.

4. If equipped with 4WD, remove the skid plate-to-chassis bolts and the plate from the vehicle.

5. Loosen the engine to mount bracket bolts and the engine-to-pitching stopper bolts. Using the proper equipment, raise the engine slightly until the bolts protrude above the surface of the crossmember.

6. Disconnect the front exhaust pipe to cylinder head bolts and separate the exhaust pipe from its mounting. If equipped with power steering, be careful not to damage the power steering hoses.

To install:

7. Clean the mating surfaces of the exhaust manifold and cylinder head throughly. Install the exhaust manifold and tighten the bolts to 19–22 ft. lbs. (26–30 Nm).

8. Install all exhaust pipes and brackets. Connect the electrical connector for the oxygen sensor. Install all other components previously removed in reverse order of removal.

9. Connect the negative battery cable and start the vehicle to check for leaks.

Turbocharger

REMOVAL & INSTALLATION

1. Disconnect the negative battery cable. Drain the cooling system. Remove the air cleaner.

2. Disconnect the airflow meter-to-turbocharger inlet clamp, then remove the air intake duct. Cover the airflow meter and turbocharger openings.

3. Loosen the turbocharger-to-air outlet hose clamp and the throttle body inlet to air inlet hose clamp. Remove the turbocharger-to-throttle body hose. Plug all of the openings.

4. Remove the turbocharger-to-center exhaust pipe nuts and the front exhaust pipe to turbocharger nuts.

5. Disconnect and plug the coolant lines.

6. Remove the oil feed line to turbocharger bolt and disconnect the turbocharger to oil return hose clamp and the return hose.

7. Remove the turbocharger from the exhaust manifold.

NOTE: When removing the turbocharger from the vehicle, disconnect the oil return hose.

To install:

8. Installation is the reverse of the removal procedure. Be sure to fill the turbocharger assembly with clean engine oil prior to installation.

9. Be sure to use new gaskets, as required. Start the engine and check for leaks, correct as necessary.

Front Cover

REMOVAL & INSTALLATION

Justy

1. Disconnect the negative battery cable. Remove the drive belts.

2. Remove the outer front cover retaining bolts. Remove the outer front belt cover.

3. Clean all mating surfaces throughly and using a new gasket, install the cover. Tighten the cover bolts to specification.

4. Install the drive belts and connect the negative battery cable.

Except Justy

1. Disconnect the negative battery cable. Loosen the water pump pulley mounting bolts. Remove the drive belts.

2. Remove the water pump pulley and pulley cover. Using the proper tools, remove the crankshaft pulley.

3. Disconnect the oil pressure switch electrical connector. Remove the oil level dipstick gauge and tube assembly.

4. Remove the right cover retaining bolts. Remove the right cover.

5. Remove the left cover retaining bolts. Remove the left cover.

6. Remove the center cover retaining bolts. Remove the center cover.

To install:

7. Clean all mating surfaces throughly. Using new gaskets, install the covers and tighten to bolts to 4 ft. lbs. (5 Nm).

8. Install the water pump, pump pulley and drive belts. Tighten the water pump bolts to 7–8 ft. lbs. (9–11 Nm). Connect the oil pressure switch and install the oil level dipstick.

FRONT OIL SEAL REPLACEMENT

1. Remove the timing belt cover assembly.
2. On the 1.8L engine, slide both the No. 1 and No. 2 crankshaft sprockets from the crankshaft. On the 1.2L engine, slide crankshaft sprocket from the crankshaft. When removing the crankshaft sprockets, be sure to remove the Woodruff® key from the crankshaft.
3. Using a small prybar, pry the front oil seal from the crankcase.

To install:

4. Using a new oil seal lubricated with engine oil, drive the new seal into the crankcase until it seats. When installing the new oil seal, be careful not to cut the sealing lips.
5. Reinstall the crankshaft sprockets and timing belt cover assembly.

Timing Belt and Tensioner

REMOVAL & INSTALLATION

Justy

1. Disconnect the negative battery cable. Remove the alternator drive belt. Loosen the crankshaft pulley bolts but do not remove.

NOTE: An access hole is provided in the wheelhouse panel to loosen and then remove the crankshaft pulley bolts.

2. Position the crankshaft with No. 3 cylinder at TDC.
3. Remove the crankshaft bolts and pulley, using pulley removal tool 499205500 or equivalent. Remove the outer front timing belt cover.
4. Loosen the tensioner bolt and position it in the direction that loosens the belt. Tighten the tensioner bolt in that position.
5. Remove the camshaft drive pulley plate. Mark the timing belt, if to be used again, in the direction of rotation for reinstallation. Remove the belt from the sprockets.
6. If necessary, remove the tensioner and spring. Remove the camshaft pulley, using pulley removal tool 499205500 or equivalent. Remove the inner belt cover and cover mount, only as required.

To install:

7. When installing the timing belt, rotate and align the matchmark of the camshaft driven pulley 0.120 in. diameter hole with the matchmark of the cam belt side cover.
8. Align the matchmark of the crankshaft drive pulley with the matchmark of the crankshaft cover. Install the camshaft drive belt.
9. Loosen the tensioner bolt ½ turn.
10. Tighten the tensioner bolt below the adjusting wheel first. Tighten the other bolt. Check to be sure all sprocket and housing matching marks are in agreement.
11. Install the camshaft drive pulley plate.
12. Install the cam belt cover.
13. Install the crankshaft pulley and bolts.
14. Tighten the crankshaft bolts to 58–72 ft. lbs. and complete any remaining assembly procedures, as required.

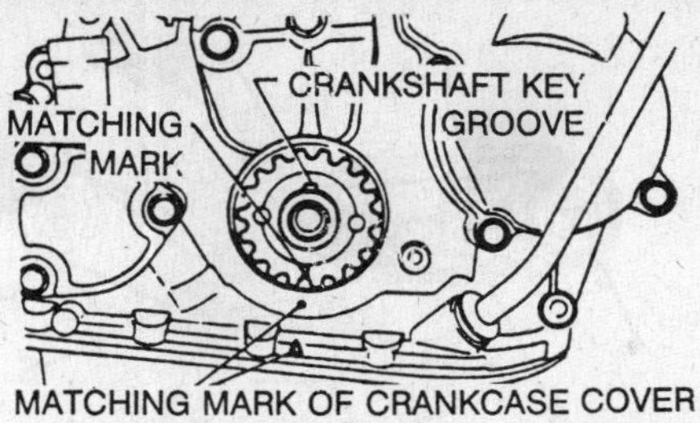

Crankshaft gear alignment—1.2L engine

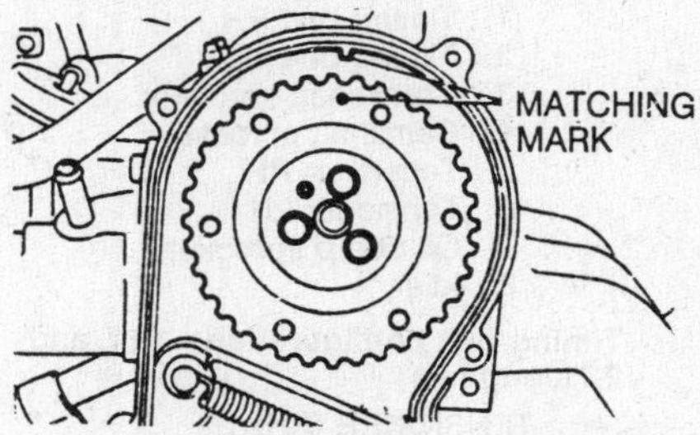

Camshaft gear alignment—1.2L engine

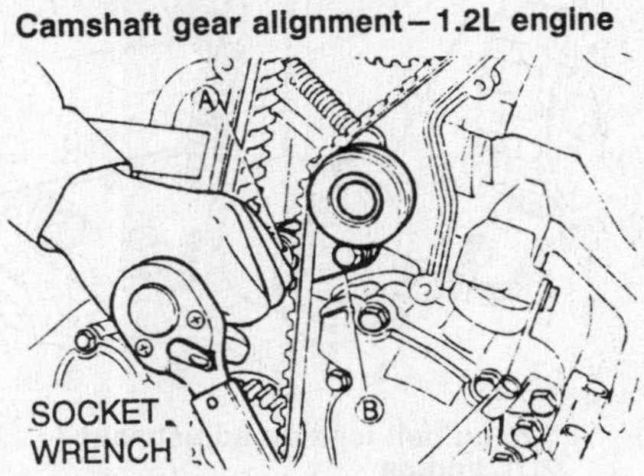

Timing belt tension adjustment—1.2L engine

XT Coupe

1.8L ENGINE

1. Disconnect the negative battery cable.
2. Raise the vehicle and support safely. Remove the underpans.
3. Remove the bolts from the underside of the radiator fan shroud.
4. Lower the vehicle and remove the fan shroud bolts and the shroud. Disconnect the fan motor connector.
5. Remove the alternator and the drive belt. Remove the dipstick tube and dipstick. Disconnect the oil pressure gauge harness.
6. Remove the water pump pulley and the air intake boot.
7. Attach a stopper tool to the flywheel or torque converter to prevent it from turning. With the crankshaft stationary, remove the crank pulley.
8. Remove the front left, right and center belt covers, in that order.
9. To remove the right belt, loosen tensioner No. 1 (right) mounting bolts on the No. 1 cylinder side. With the tensioner in the fully slackened position, tighten the mounting bolts.
10. Mark the belt in the direction of rotation and front and rear indicator, if to be reused. Remove the right timing belt.
11. Loosen tensioner No. 2 (left) mounting bolts on the No. 2 cylinder side. With the tensioner in the fully slackened position, tighten the mounting bolts.
12. Remove the right crankshaft sprocket (front). Mark the direction of rotation and front and rear indicator on the belt, if to be reused and remove the left timing belt.
13. If necessary, remove the left crankshaft sprocket (rear without dowel pin).
14. If necessary, remove the tensioner assemblies along with the tensioner spring. Remove the belt idler, as required.
15. If necessary and using a camshaft sprocket removal tool 499207000 or equivalent, remove the camshaft sprockets. As required, remove the inner belt covers.

To install:

16. Install the belt cover seals, as required. Install the camshaft sprockets, if removed.
17. If removed, install the right tensioner and attach the tensioner spring to the tensioner and position it to the right side cylinder block. Hand tighten the bolts. Attach the spring to the assembly and tighten the top bolt. Loosen it ½ turn. Push down on the tensioner until it stops, hand tighten the lower bolt.
18. To install the left tensioner, if removed, attach the tensioner spring to the tensioner and position it to the left side cylinder block. Hand tighten the bolts. Attach the spring to the assembly and tighten the top bolt. Loosen it ½ turn. Raise the tensioner until it stops, hand tighten the lower bolt.
19. Install the belt idler to the cylinder block. Install the crankshaft

sprockets. Temporarily install the crankshaft pulley.

20. To install the left timing belt, align the center of the 3 lines on the flywheel with the timing mark on the flywheel housing.

21. Align the timing mark on the camshaft sprocket with the notch in the belt cover.

22. Install the timing belt to the crankshaft sprocket (No. 2), the oil pump sprocket, belt idler pulley and the camshaft sprocket, in that order to prevent downward slackening of the belt.

NOTE: The No. 2 crankshaft sprocket is identified as the sprocket without a dowel pin.

23. Loosen the tensioner and ensure smooth movement for belt tension adjustment.

24. Using belt tension wrench tool 499437000 of equivalent, apply 33–55 lbs. of tension to the left camshaft sprocket in the counterclockwise direction.

25. While applying torque, tighten the lower tensioner retaining bolt 12–14 ft. lbs. Tighten the upper tensioner retaining bolt 12–14 ft. lbs. Be sure the timing marks remain in alignment.

26. To install the right belt, rotate the crankshaft one full turn clockwise from the position where the left timing belt was installed. Align the center of the 3 lines on the flywheel with the timing mark on the flywheel housing.

27. Align the timing mark on the camshaft sprocket with the notch in the belt cover. Install the right timing belt over the crankshaft sprocket (with dowel pin) and camshaft sprocket.

28. Using belt tension wrench tool 499437000 or equivalent, apply 33–55 lbs. of tension to the right camshaft sprocket in the counterclockwise direction. While applying torque, tighten the lower tensioner retaining bolt 12–14 ft. lbs. Tighten the upper tensioner retaining bolt 12–14 ft. lbs. Be certain the timing marks remain in alignment.

29. Remove the crankshaft pulley and install the necessary seals, if not in place and install the front center cover to the engine block.

30. Install the left and right belt covers.

31. Install the crankshaft pulley and while holding the crankshaft stationary, tighten the bolt to 66–79 ft. lbs.

32. Complete the assembly of the water pump pulley, the oil dipstick tube, the alternator drive belt and the radiator shroud. Complete any other Step that is needed to complete this procedure.

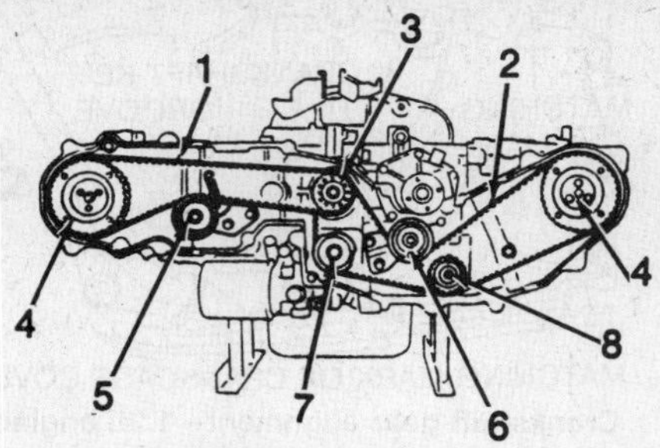

1. Timing belt RH
2. Timing belt LH
3. Crankshaft sprocket
4. Camshaft sprocket
5. Tensioner RH
6. Tensioner LH
7. Oil pump sprocket
8. Idler

Timing belt configuration—1.8L and 2.7L engines

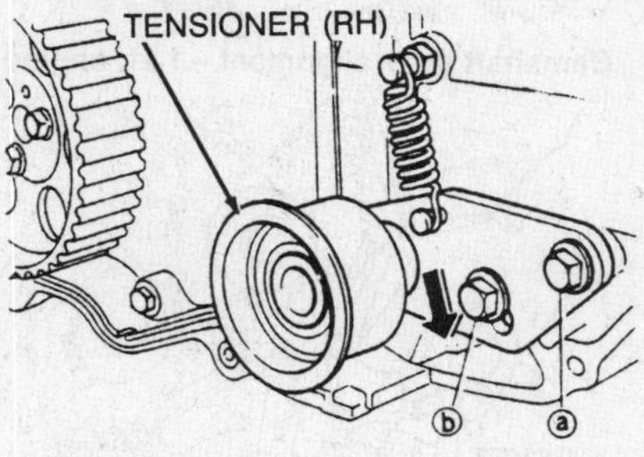

Timing belt tension adjustment—2.7L engine

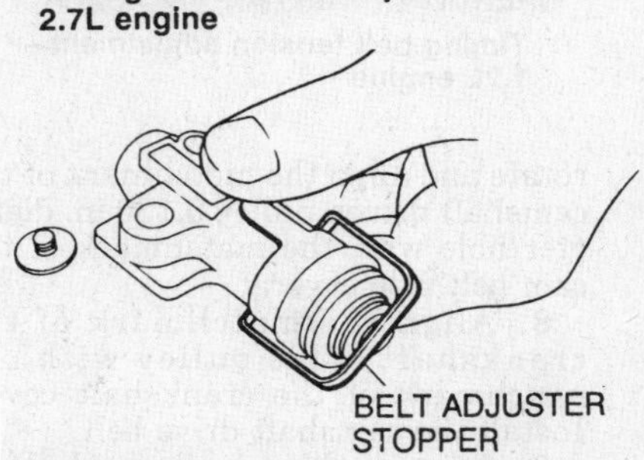

Installation of belt adjuster stopper clip—2.7L engine

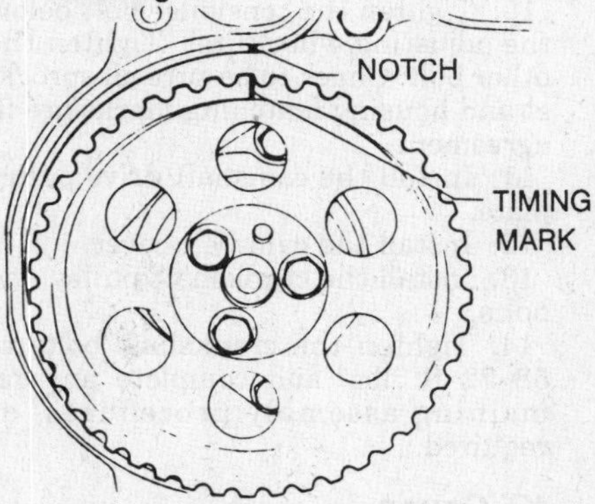

Camshaft gear alignment—2.7L engine

2.7L ENGINE

1. Disconnect the negative battery cable. Loosen the V-belt tensioner locknut fully counterclockwise.

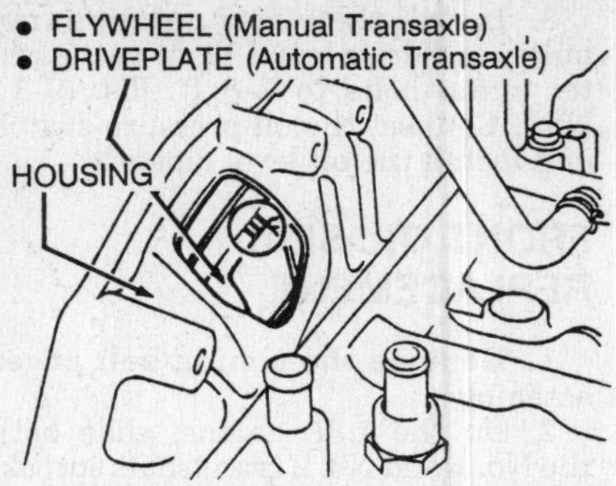

Flywheel timing mark alignment—2.7L engine

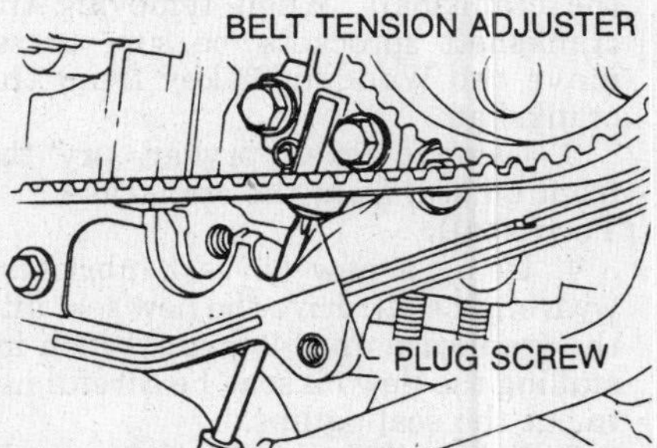

Removing the plug screw from the belt tension adjuster—2.7L engine

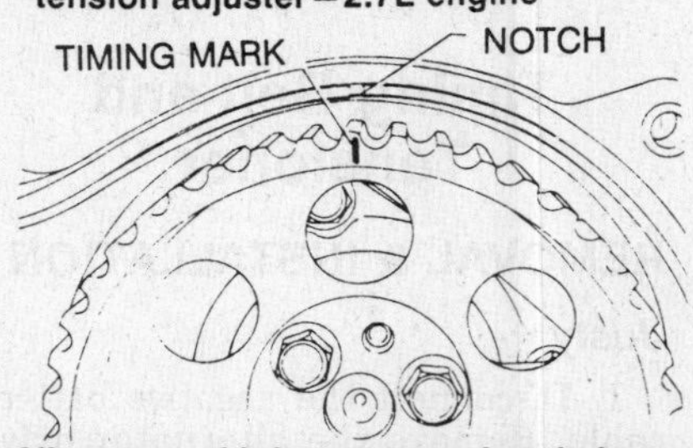

Alignment of left cam sprocket timing mark to notch in belt cover—2.7L engine

2. Loosen 2 alternator mounting bolts. Loosen water pump pulley mounting bolts. Remove the water pump pulley.

3. Lock the crankshaft and remove the crankshaft pulley.

4. Remove the right, left and center belt covers. To remove the right belt, loosen the tensioner mounting bolts on the No. 1 cylinder. With the tensioner in the fully slackened position, tighten the mounting bolts.

5. Mark the belt in the direction of rotation. Remove the right timing belt.

6. If necessary, remove the right tensioner assembly. Remove the crankshaft sprocket.

7. To remove the left belt, remove the idler pulley. Remove the rubber plug. Remove the plug screw from the belt tension adjuster lower side.

8. Position a tool into the hole in the bottom of the belt tension adjuster. Turn the screw clockwise to loosen the belt tension.

9. Install belt adjuster stopper tool 13082AA000. Remove the left belt tensioner assembly.

10. Mark the belt in the direction of rotation. Remove the left timing belt.

11. If necessary, remove the crankshaft sprocket and the idler pulley. Remove the belt tensioner assembly. Reinstall the plug screw in its mounting.

12. As required, remove the inner belt covers. As required, check the belt tensioners and replace, as necessary.

To install:

13. Align the center of the 3 lines on the flywheel with the timing mark on the flywheel housing. Install the camshaft sprockets, if removed and align the timing mark on the camshaft sprocket with the notch in the belt cover.

14. Install the hydraulic belt tensioner assembly. Remove the plug screw from the tensioner lower side. Install a prybar into the hole in the bottom of the tensioner and turn the screw clockwise to compress the rubber boot. Install the belt adjuster stopper tool.

NOTE: The clip furnished as a spare part can be used in place of the belt adjuster stopper tool.

15. Using a syringe type tool, add oil through the air vent hole on top of the rubber boot of the belt tensioner. Install the belt tensioner to 17–20 ft. lbs. and install the rubber plug. Install the idler pulley to 29–35 ft. lbs..

16. Install the crankshaft sprocket No. 2, which is identified by the absence of a dowel pin, onto the crankshaft. Be sure all marks are in alignment and install the timing belt from the crankshaft side. Be careful not to loosen the belt.

17. Install the left tensioner and be sure its movement is correct. Tighten to 29–35 ft. lbs.

18. Check for proper operation. Remove the belt adjuster stopper tool. Be sure the end of the left tensioner arm contacts the top of the belt tensioner adjuster.

19. To install the right belt, rotate the crankshaft 1 turn clockwise from the position where the left timing belt was installed. Align the center of the 3 lines on the flywheel with the timing mark on the flywheel housing.

20. Align the timing mark on the camshaft sprocket with the notch in the belt cover. Temporarily tighten the tensioner bolts while moving the right tensioner downward. Slightly loosen the higher bolt.

21. Install the crankshaft sprocket to the crankshaft. Install the timing belt from the crankshaft side.

22. Loosen the lower right tensioner assembly retaining bolt and apply tension to the belt.

23. Using belt tension wrench tool 499437100 or equivalent, apply 33–55 lbs. of tension to the right camshaft sprocket in the counterclockwise direction. While applying torque tighten the lower tensioner retaining bolt 17–20 ft. lbs. Tighten the upper tensioner retaining bolt 17–20 ft. lbs.

24. Using a belt tension wrench, apply 33–55 lbs. of torque to the belt and tighten the idler pulley.

25. Install the crankshaft pulley and tighten to 66–79 ft. lbs. Install the oil dipstick, if removed.

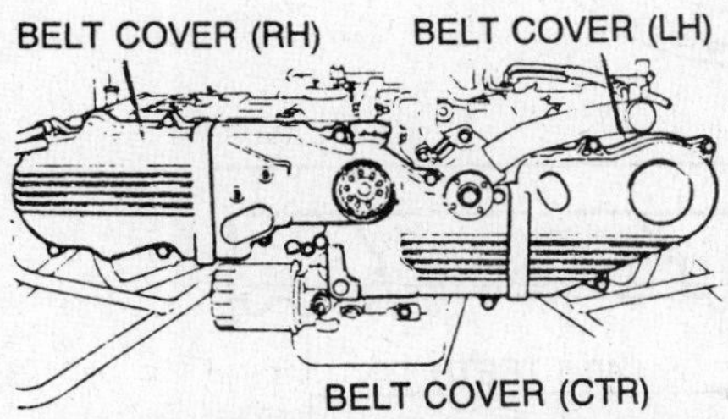

Left, center and right belt covers—typical of 1.8L, 2.2L and 2.7L engines

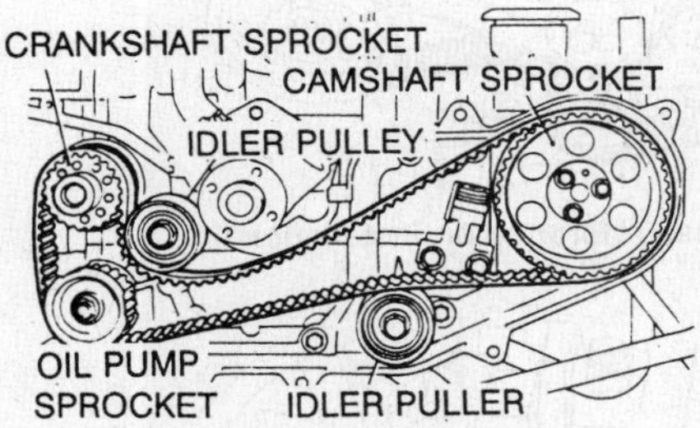

Installation of left timing belt—2.7L engine

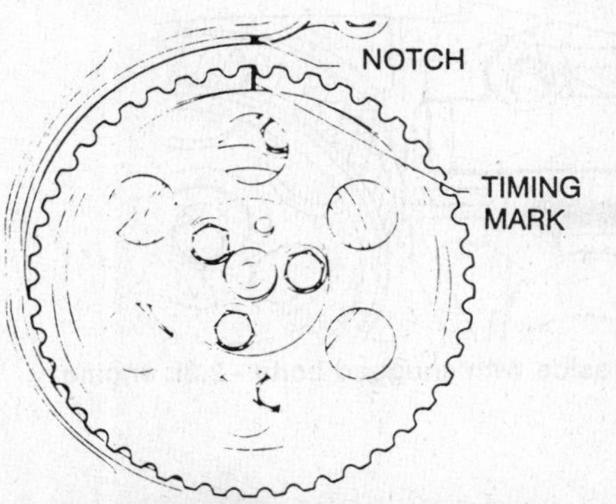

Alignment of right cam sprocket timing mark to notch in belt cover—2.7L engine

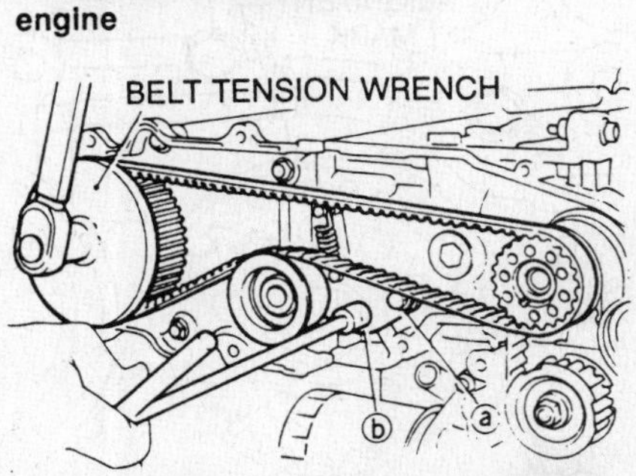

Adjustment of right timing belt tension with belt tension wrench—2.7L engine

26. Install the water pipe, the water pump pulley, the center, right and left covers, the air conditioning compressor and install the alternator.

27. Install the V-belt and adjust tension to 143–166 lbs. for a new belt or 99–143 lbs. for a used belt. Lock adjuster bolt by turning counterclockwise. Complete any further installation.

Legacy

2.2L ENGINE

The 2.2L OHC engine uses a single cam belt drive system with a serpentine type belt. The left side of the engine uses a hydraulic cam belt tensioner which is continuously self adjusting.

1. Disconnect the negative battery cable. Remove the accessories, alternator and brackets, the power steering pump and the air conditioning compressor.

2. Remove the crankshaft pulley bolt by using a special crankshaft holding tool.Remove the pulley.

3. Remove the cam belt covers. Align the crankshaft and camshaft sprocket so each cam sprocket notch aligns with the cam cover notches. Align the crankshaft sprocket top tooth notch, located at the rear of the tooth, with the notch on the crank angle sensor boss. Mark the 3 alignment points as well as the direction of the cam belt.

4. Loosen the tensioner adjusting bolts and remove the bottom 3 idlers, the cam belt and the cam belt tensioner. The cam sprockets can then be removed with a modified camshaft sprocket wrench tool.

5. If the sprockets are removed, note the reference sensor at the rear of the left cam sprocket.

To install:

6. Install the crankshaft sprocket and all of the idlers except for the lower right. Compress the hydraulic tensioner in a vise slowly and temporarily secure the plunger with a pin. Install the tensioner and the pulley.

7. After the cam belt components are installed, align the crankshaft sprocket notch on the rear sprocket tooth with the crank angle sensor boss. This places the sprocket notch in the 12 o'clock position.

8. Align the camshaft sprockets with the notches in the cam belt cover. As the directional marked belt is installed, align the marks on the belt with the crankshaft sprocket and the left camshaft sprocket. Install the lower right idler.

9. Load the tensioner by pushing it towards the crankshaft with a prybar and tighten the bolts. Remove the ten-

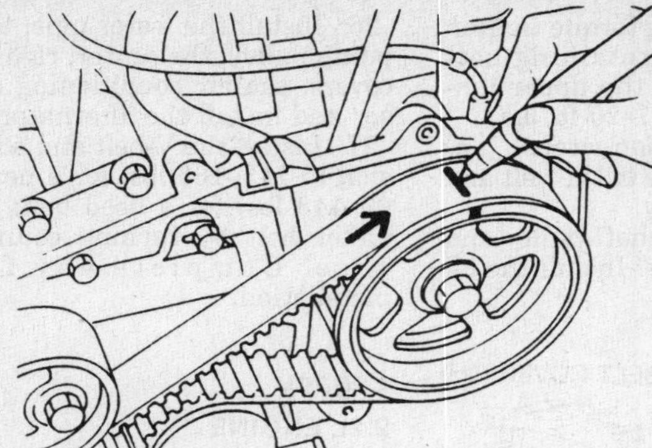

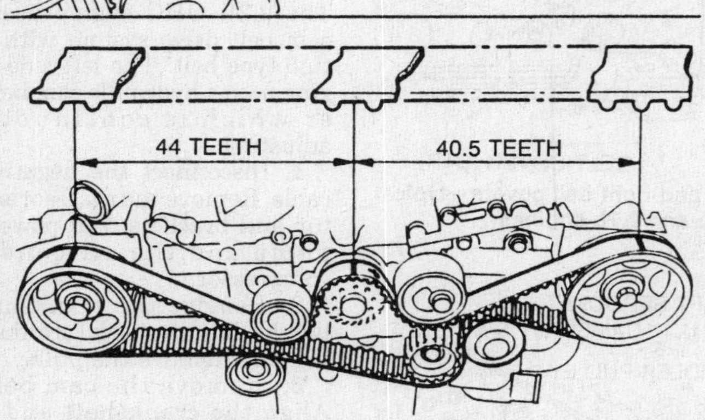

Mark the timing belt for easy reference on installation

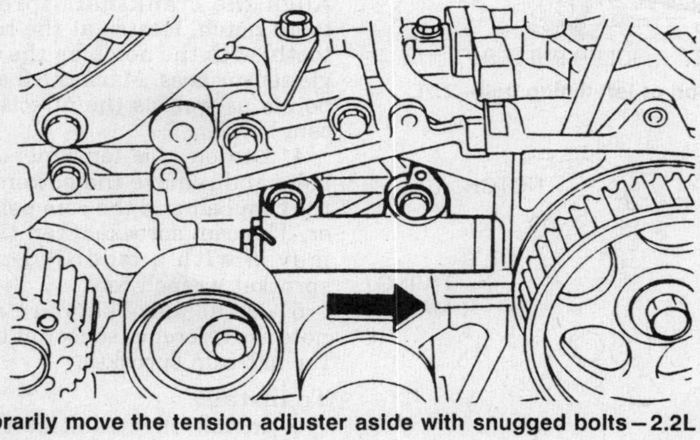

Temporarily move the tension adjuster aside with snugged bolts – 2.2L engine

sioner retention pin and the belt tension is automatically set. Rock the crankshaft back and forth 1 time to distribute the belt tension.

10. Verify the correctness of the timing by noting that the notches on the 2 cam pulleys and the notch on the crankshaft pulley all point to the 12 o'clock position when the belt is properly installed.

11. Complete the engine component assembly by installing the cam belt covers, the crankshaft pulley bolt and pulley and the remaining components.

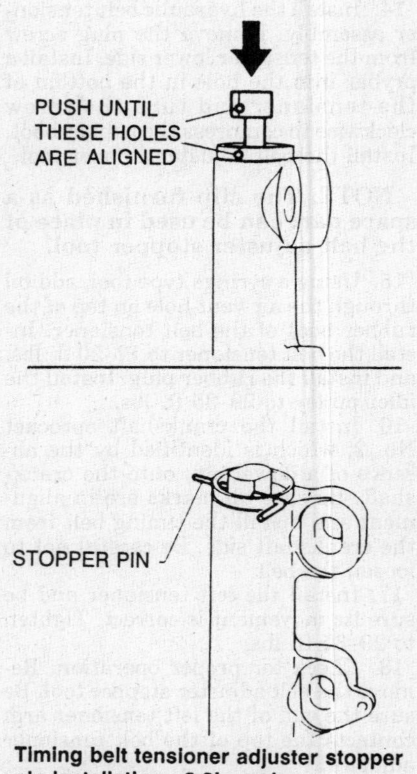

Timing belt tensioner adjuster stopper pin installation – 2.2L engine

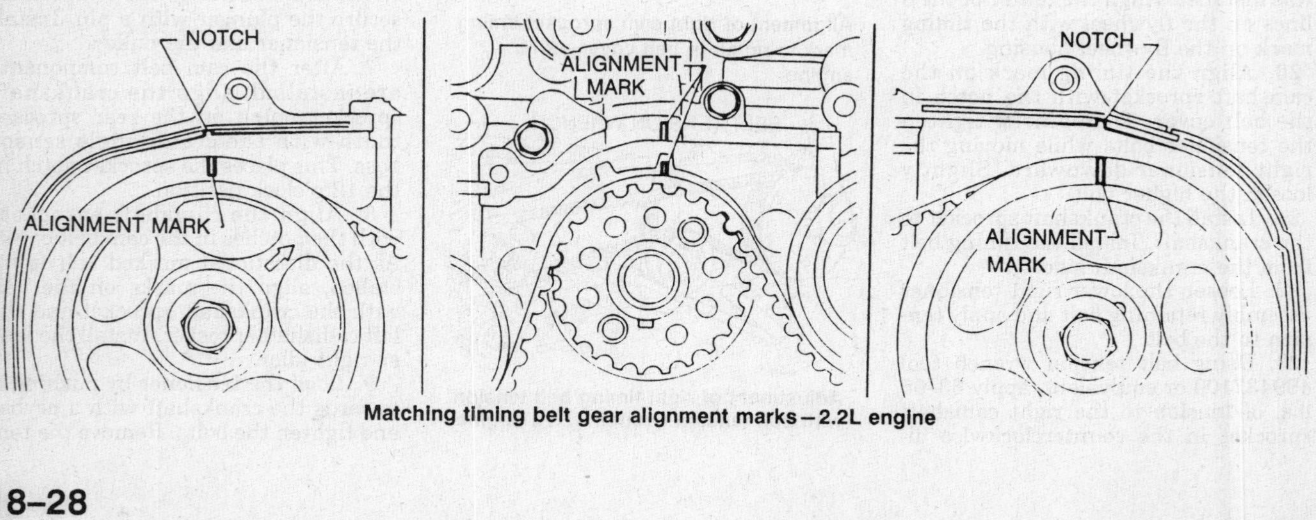

Matching timing belt gear alignment marks – 2.2L engine

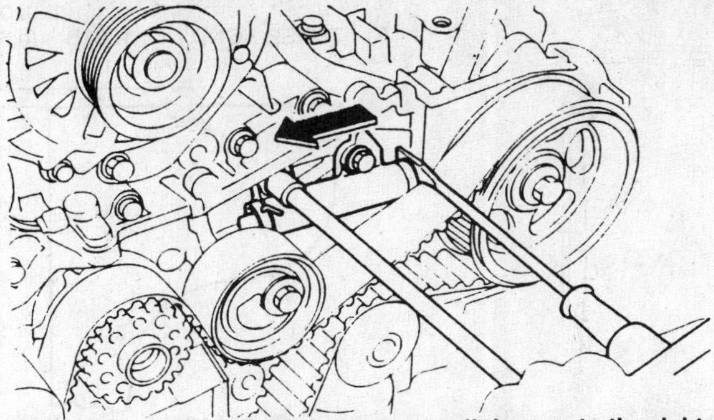

After belt installation, move the tension adjuster all the way to the right and tighten bolts – 2.2L engine

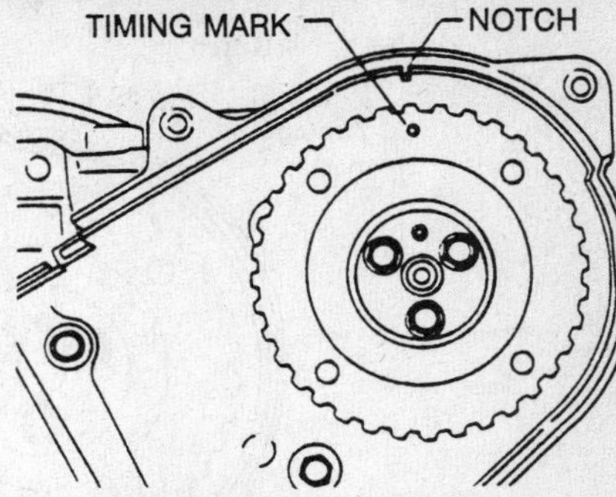

Left camshaft gear alignment – 1.8L engine

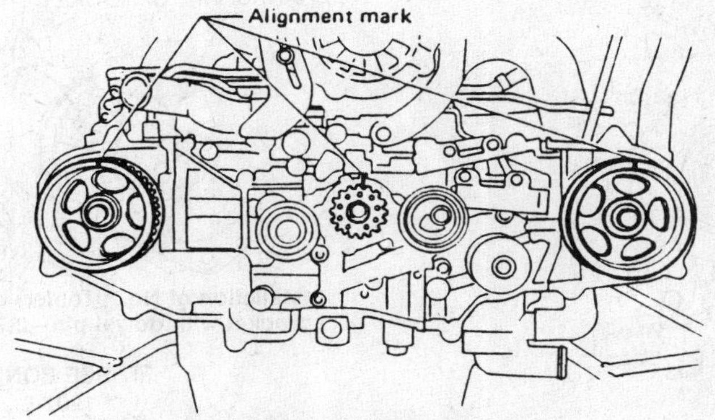

Alignment marks – 2.2L engine

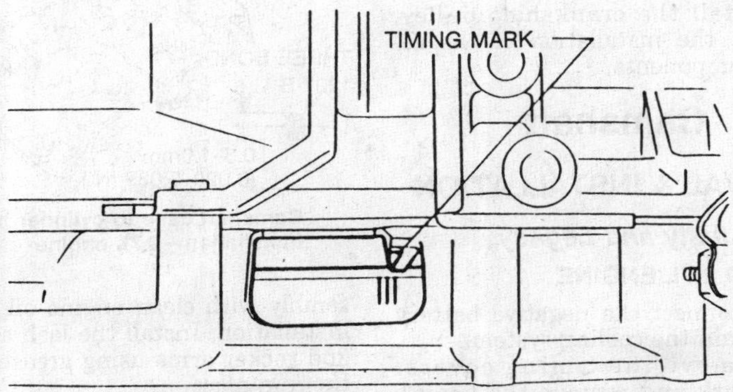

Flywheel alignment marks for timing belt servicing – 1.8L engine

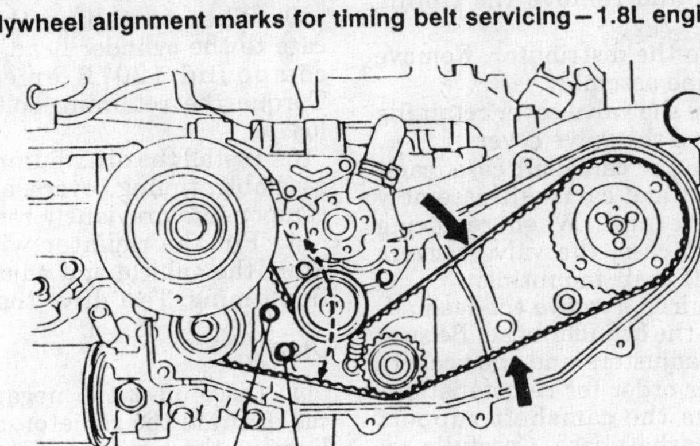

Left timing belt tensioner servicing – 1.8L engine

Except Justy, XT Coupe and Legacy

1.8L ENGINE

1. Disconnect the negative battery cable.
2. Remove the timing covers.
3. To remove the right belt, loosen tensioner No. 1 mounting bolts on the No. 1 cylinder. With the tensioner in the fully slackened position, tighten the mounting bolts.
4. Mark the belt in the direction of rotation. Remove the right timing belt.
5. Loosen tensioner No. 2 mounting bolts on the No. 2 cylinder. With the tensioner in the fully slackened position, tighten the mounting bolts.
6. Remove the right crankshaft sprocket. Matchmark and remove the left timing belt. Remove the left crankshaft sprocket.
7. Remove the tensioner assemblies along with the tensioner spring. Remove the belt idler.
8. Using camshaft sprocket removal tool 499207000 or equivalent and, if necessary, remove the camshaft sprockets. As required, remove the inner belt covers.

To install:

9. Install the camshaft sprockets, if removed.
10. To install the right tensioner, attach the tensioner spring to the tensioner and position it to the right side cylinder block. Hand tighten the bolts. Attach the spring to the assembly and tighten the top bolt. Loosen it ½ turn. Push down on the tensioner until it stops, hand tighten the lower bolt.
11. To install the left tensioner, attach the tensioner spring to the tensioner and position it to the left side cylinder block. Hand tighten the bolts. Attach the spring to the assembly and tighten the top bolt. Loosen it ½ turn. Raise the tensioner until it stops, hand tighten the lower bolt.
12. Install the belt idler to the cylinder block. Install the crankshaft sprockets. Temporarily install the crankshaft pulley.
13. To install the left timing belt,

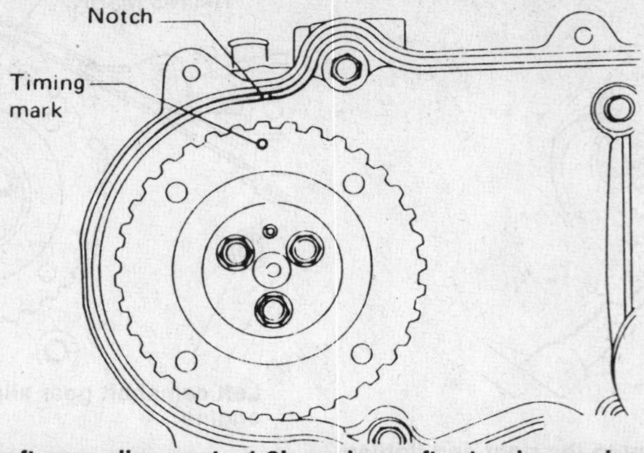

Right camshaft gear alignment – 1.8L engine – after turning engine 1 complete rotation with the left timing belt installed

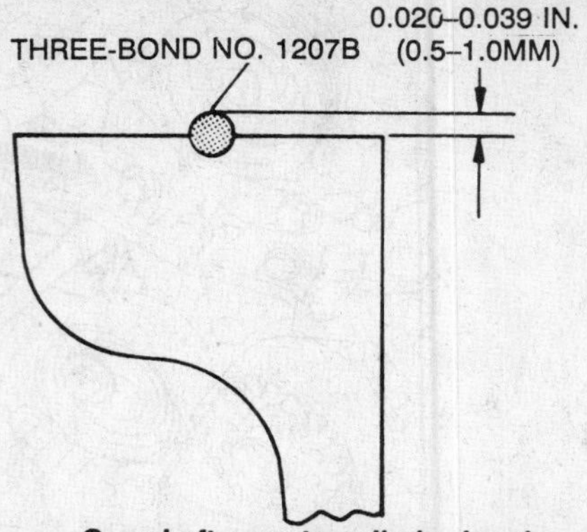

Camshaft case to cylinder head installation – 1.8L engine

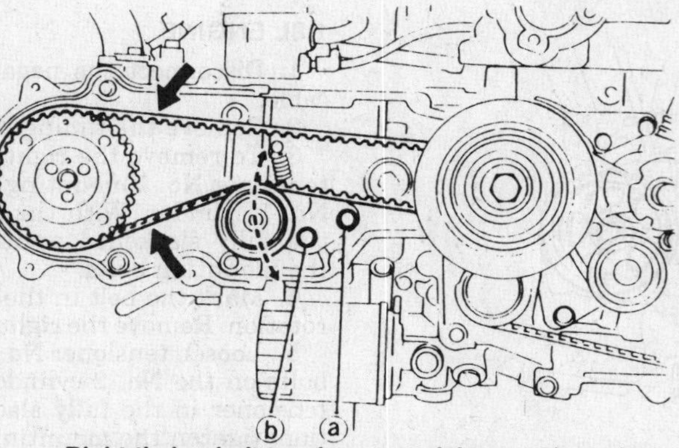

Right timing belt tensioner servicing – 1.8L engine

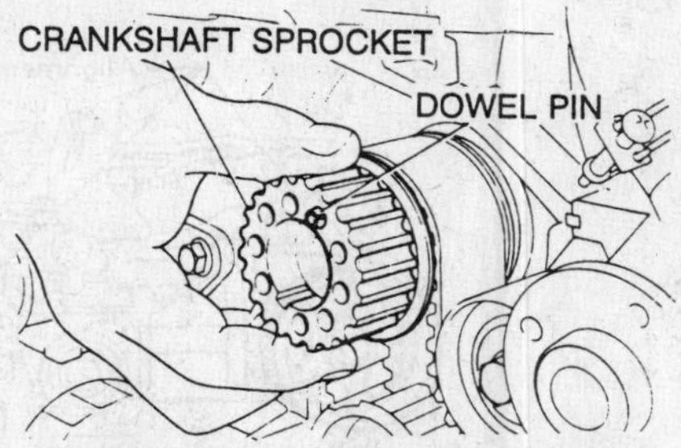

Installation of No. 1 (outer) crankshaft sprocket with dowel pin – 2.7L engine

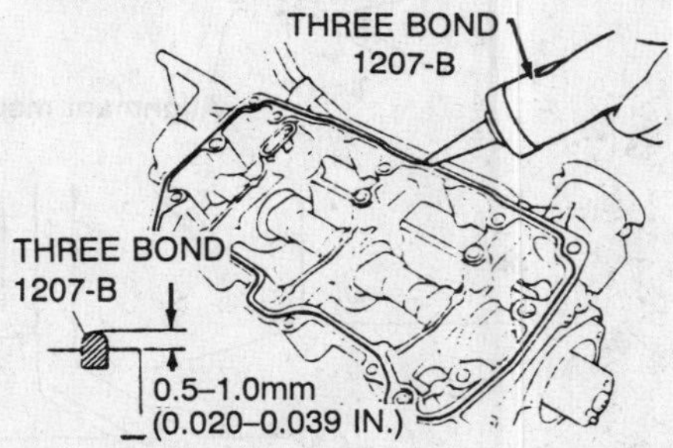

Camshaft case to cylinder head installation – 2.7L engine

align the center of the 3 lines on the flywheel with the timing mark on the flywheel housing. Align the timing mark on the camshaft sprocket with the notch in the belt cover. Install the timing belt to the No. 2 crankshaft gear (without dowel pin), oil pump gear and the belt idler, in that order.

14. Using belt tension wrench tool 499437000 or equivalent, apply 33–55 lbs. of tension to the left camshaft sprocket in the counterclockwise direction. While applying torque, tighten the lower tensioner retaining bolt 12–14 ft. lbs. Tighten the upper tensioner retaining bolt 12–14 ft. lbs.

15. To install the right belt, rotate the crankshaft 1 turn clockwise from the position where the left timing belt was installed. Align the center of the 3 lines on the flywheel with the timing mark on the flywheel housing.

16. Align the timing mark on the camshaft sprocket with the notch in the belt cover. Install the right timing belt.

17. Using belt tension wrench tool 499437000, apply 33–55 lbs. of tension to the right camshaft sprocket in the counterclockwise direction. While applying torque tighten the lower tensioner retaining bolt 12–14 ft. lbs. Tighten the upper tensioner retaining bolt 12–14 ft. lbs.

18. Install the crankshaft pulley. Continue the installation of the removed components.

Camshaft

REMOVAL & INSTALLATION

Except Justy and Legacy

1.8L AND 2.7L ENGINE

1. Disconnect the negative battery cable. Drain the cooling system.

2. Remove the timing covers. Matchmark and remove the timing belt.

3. Remove the distributor. Remove the water pipe assembly.

4. Remove the valve cover retaining bolts. Remove the valve covers.

5. Remove the camshaft case, camshaft support and camshaft assembly as a complete unit. When removing the camshaft case, the valve rockers may come off their mounting.

6. As required, remove the lash adjusters from the cylinder head. Be sure to keep the adjusters and the rockers in the proper order for reinstallation.

7. Remove the camshaft support from the camshaft case. Carefully remove the camshaft from its mounting.

To install:

8. Be sure to coat the camshaft assembly with clean engine oil prior to installation. Install the lash adjusters and rocker arms using grease to hold them in place.

9. When installing the camshaft case to the cylinder head, use sealing compound 1207B or equivalent. Torque the retaining bolts 17–20 ft. lbs.

10. Install the distributor, water pipe assembly, timing covers, and all other components previously removed.

11. Fill the radiator with coolant. Start the vehicle and adjust the ignition timing. Test drive the vehicle.

Justy

1. Disconnect the negative battery cable. Drain the radiator, as required. Remove the air cleaner assembly. Remove the drive belts. Properly discharge the air condition system, as required.

2. Remove the tensioner and spring. Remove the camshaft pulley, using pulley removal tool 499205500 or equivalent. Remove the inner belt cover and cover mount.

3. Remove the PCV hose from the rocker arm cover. Remove the rocker arm cover retaining bolts. Remove the rocker arm cover from the engine. Remove the rocker arm assembly.

4. As required, remove the radiator, air conditioning condenser and grille work in order to remove the camshaft from the cylinder head.

5. Carefully withdraw the camshaft from the cylinder head.

To install:

6. Before installing the camshaft be sure to coat it with clean engine oil. Install the camshaft, timing belt, rocker arms and rocker shaft.

7. Adjust the valves to specification and install the rocker cover using a new gasket.

8. The remainder of the installation procedure is the reverse of removal. Start the engine and adjust the ignition timing and idle speed. Test drive the vehicle.

Legacy

2.2L ENGINE

NOTE: It is assumed that engine has been removed from the vehicle.

1. Remove the timing belt covers, the timing belt, camshaft sprockets and other components necessary to expose the camshaft.

2. Remove the valve covers. Remove the rocker arm assembly bolts in sequence.

3. To remove the left camshaft, perform the following procedures:

   a. Remove the cam angle sensor.

   b. Remove the oil dipstick tube attaching bolt.

   c. Remove the camshaft support on the left side.

   d. Remove the O-ring.

   e. Remove the camshaft and seal(rear) from the left side.

4. To remove the right camshaft, perform the following procedures:

   a. Remove the camshaft support on the right side.

   b. Remove the O-ring.

   c. Remove the camshaft and seal (rear) from the right side. Remove the oil seal from the camshaft support.

To install:

5. To install the left camshaft, perform the following procedures:

   a. Lubricate the camshaft journals, install the oil seal (rear) and install the camshaft into the cylinder head.

   b. Install the O-ring into the camshaft support and install the support.

   c. Install oil seal into the camshaft support.

   d. Install the bolt into the dipstick tube and install the camshaft sensor.

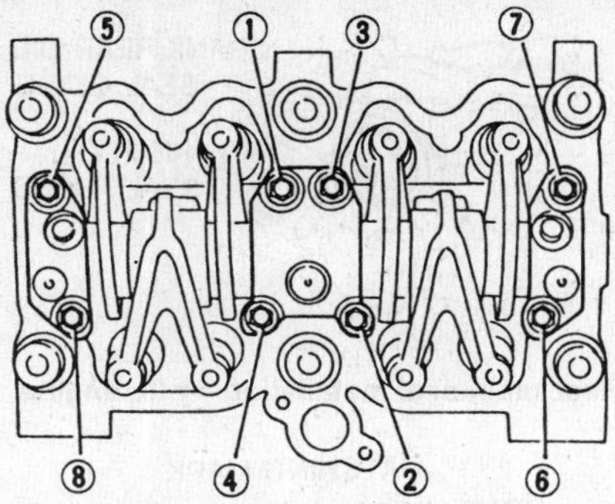

Rocker arm assembly bolt removal sequence – 2.2L engine

6. To install the right camshaft, perform the following procedures:

   a. Lubricate the camshaft journals and install the right camshaft.

   b. Install the O-ring into the camshaft support and install the support.

   c. Install a new oil seal in the rear of the cylinder head.

7. To complete assembly, perform the following procedures:

   a. Install the rocker arm assemblies and torque bolts to 9 ft. lbs., in proper sequence.

   b. Install the camshaft sprockets and related components. Install the timing belt and complete the assembly of covers.

Piston and Connecting Rod

POSITIONING

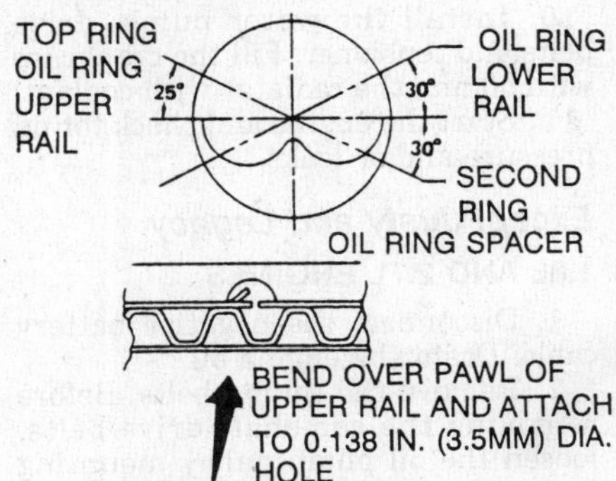

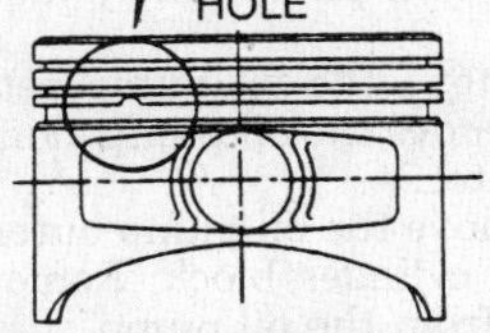
Piston identification – 1.8L and 2.7L engines – typical of 2.2L engine

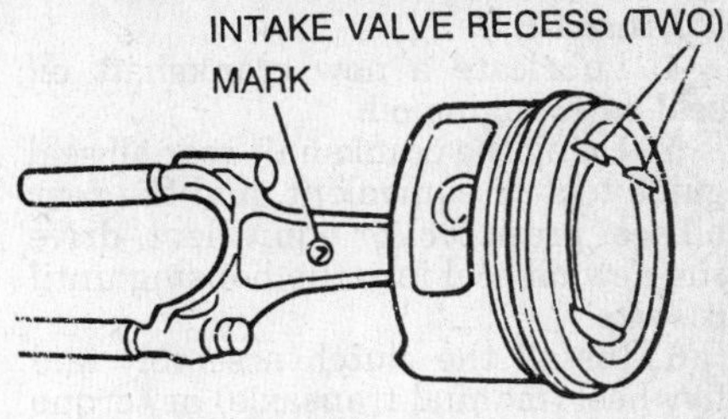

Piston identification – 1.2L engine

ENGINE LUBRICATION

Oil Pan

REMOVAL & INSTALLATION

NOTE: In most cases it is not necessary to remove the engine from the vehicle to remove the oil pan.

1. Disconnect the negative battery cable. Drain the engine oil.

2. Raise and support the vehicle safely. Remove the required components in order to gain access to the oil pan retaining bolts.

3. Loosen the engine mounts and raise the engine, as necessary, to gain access to the oil pan retaining bolts.

4. Remove the oil pan retaining bolts. Remove the oil pan from the engine.

To install:

5. Clean all mating surfaces throughly. Using a new gasket, install the oil pan and tighten the retaining bolts to 3–4 ft. lbs (4–5 Nm).

6. Lower the engine and tighten the mounting bolts. Reinstall all components previously removed. Lower the vehicle.

7. Fill the crankcase with oil. Connect the negative battery cable. Start the engine and check for leaks.

Rear Main Bearing Oil Seal

REMOVAL & INSTALLATION

1. Remove the engine from the vehicle and position it in a suitable holding fixture.

2. Remove the clutch assembly and the flywheel (manual transaxle) or the torque converter and flexplate (automatic transaxle) from the crankshaft.

3. Using a small prybar, pry the rear oil seal from the crankcase. Be careful not to damage the crankshaft or the crankcase housing.

To install:

4. Lubricate a new crankshaft oil seal with engine oil.
5. Using the crankshaft rear oil seal guide tool or equivalent, and the rear oil seal press tool or equivalent, drive the new oil seal into the housing until it seats.
6. Install the clutch assembly and flywheel (manual transaxle) or torque converter and flexplate (automatic transaxle). Install the engine in the vehicle.

Oil Pump

REMOVAL & INSTALLATION

Justy

1.2L ENGINE

1. Disconnect the negative battery cable. Drain the engine oil. Drain the cooling system.
2. Remove the oil dipstick, dipstick guide and guide sealing.
3. Remove the alternator. Remove the timing belt.
4. Raise and support the vehicle safely. Remove the oil pan. Lower the vehicle.
5. Remove the water pump cover. Remove the water pump impeller. When removing the impeller, lock the balance shaft using the proper tool.
6. Remove the crankcase cover retaining bolts. Remove the crankcase cover along with the oil pump assembly.

To install:

7. Disassemble the oil pump. Check the tip clearance of the rotors. Clearance should be 0.0008–0.0059 in. (0.02–0.15 mm). Check the side clearance between the oil pump inner rotor and the pump cover. Clearance should be 0.0020–0.0063 in. (0.05–0.16 mm). If not within specification, replace the oil pump.
8. Install the oil pump and crankcase. Install the water pump assembly.
9. Install the oil pan and lower the vehicle. Install the timing belt, alternator and oil dipstick.
10. Fill the engine with oil, start the engine and check for proper oil pressure.

Legacy

2.2L ENGINE

1. Drain the engine oil.
2. Drain coolant from engine.
3. Remove all belt covers, drive belts and other necessary components.
4. Remove the belt tensioner bracket.
5. Remove the water pump.
6. Remove the oil pump.

To install:

7. Disassemble the oil pump and scribe alignment marks on the inner and outer rotors for ease of reassembly.

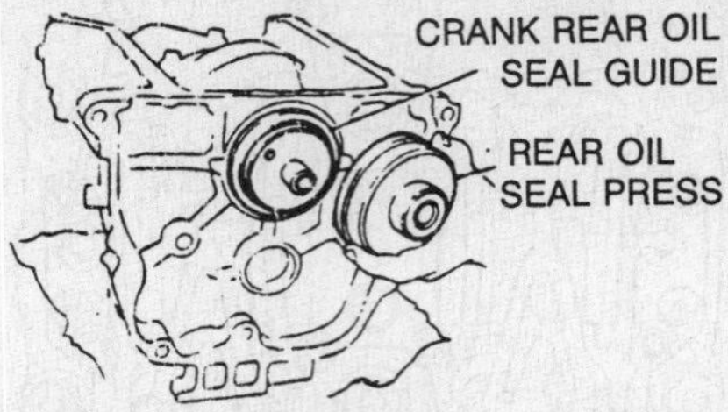

Rear main seal installation – 1.2L engine

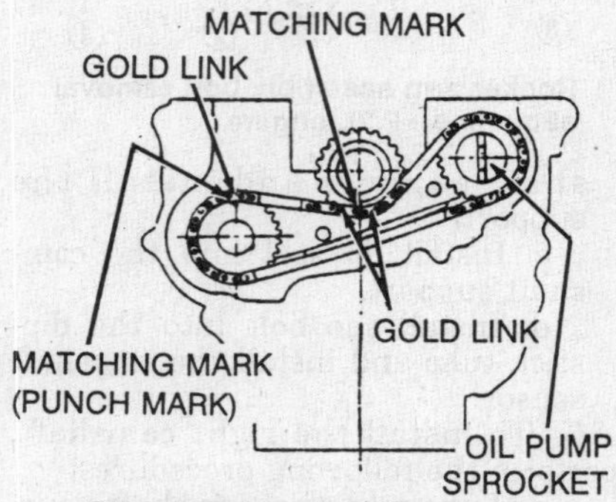

Oil pump sprocket alignment – 1.2L engine

8. Measure the top clearance of the rotors. Clearnace should be 0.0016–0.0055 in. (0.04–0.14 mm). Measure the clearance between the outer rotor and the cylinder block housing. Clearance should be 0.0039–0.0069 in. (0.10–0.175 mm). Measure the clearance between the oil pump inner rotor and the pump cover. Clearance should be 0.0008–0.0028 in. (0.02–0.07 mm). If not within specification, replace the pump body or rotors.
9. Assemble the oil pump and install on the cylinder block using three bond 1215, or equivalent. Tighten bolts to specification.
10. Install the water pump, drive belts and tensioner. Fill the crankcase with oil and the radiator with coolant.
11. Start the engine and check for oil pressure and/or leaks.

Except Justy and Legacy

1.8L AND 2.7L ENGINES

1. Disconnect the negative battery cable. Drain the engine oil.
2. Remove the timing belts. Before removing the camshaft drive belts, loosen the oil pump pulley mounting nut.
3. Remove the oil pump retaining bolts. Remove the oil pump along with the oil filter.
4. Remove the oil pump outer rotor from the cylinder block. Remove the oil filter from the oil pump.

To install:

5. Disassemble the oil pump. Check the case clearance by measuring the clearance between the outer rotor and the cylinder block rotor housing. The clearance should be 0.0039–0.0071 in. (0.10–0.18 mm). Check the side clearance as follows:
   a. Measure the depth of the rotor housing bore in the cylinder block (L).
   b. Measure the total height of the case projection (H1) plus oil pump inner and outer rotors (H2).
   c. Calculate the side clearance using the formula: $C = L - (H1 + H2)$.
6. The clearance should be 0.05–0.16 in. (0.0020–0.0063 mm). If not within specification, replace the oil pump rotors.
7. Install the oil pump assembly on the cylinder block. Install the timing belts and camshaft drive belts.
8. Fill the crankcase with oil. Start the engine and check for correct oil pressure.

ENGINE COOLING

Radiator

REMOVAL & INSTALLATION

Justy

1. Disconnect the negative battery cable. Drain the cooling system.
2. Disconnect the radiator fan electrical connection. Remove the upper and lower radiator hoses.
3. Remove the radiator mounting bolts. Remove the radiator from the vehicle.
4. Installation is the reverse of the removal procedure.
5. Fill the cooling system with coolant. Start the engine and allow it to reach operating temperature. Check for leaks.

Legacy

2.2L ENGINE

1. Disconnect the battery cables and remove the battery from the vehicle.
2. Drain the engine coolant and disconnect the radiator hoses at the engine.
3. Remove the V-belt cover.
4. Remove the reservoir overflow tank and hose.
5. Disconnect the radiator and air conditioning fan electrical wiring connectors.
6. Remove the radiator brackets at

the top radiator support panel. Move the radiator sightly to the left.

7. Disconnect the automatic transaxle cooler lines from the radiator. Allow the fluid to drain into a container.

8. Lift the radiator/fan assemblies up and from the vehicle.

To install:

9. Attach radiator mounting cushions to the pins on the lower side of the radiator.

10. Fit the cushions on the lower side of the radiator, into holes on the body side and install the radiator with inlet and outlet hoses attached.

11. Install the radiator brackets and tighten the attaching bolts.

12. Connect the electrical wiring, the inlet and outlet hoses, install the reservoir tank and overflow hose.

13. Install the V-belt and tighten. Install the battery and connect the cables.

14. Remove the air vent plug on the radiator. Install the coolant into the radiator.

15. Operate the engine and correct the coolant level as required. Open the air vent plug as required to remove air from system.

Except Justy and Legacy

1.8L AND 2.7L ENGINE

1. Disconnect the negative battery cable. Drain the cooling system.

2. Remove the radiator hoses. Disconnect and plug the automatic transaxle lines, if equipped.

3. Disconnect the main wire harness at the thermo-switch. Disconnect the fan electrical connection.

4. Disconnect the electrical wire from the secondary fan motor. Remove the upper and lower bolts retaining the shrouds to the radiator assembly.

5. Remove the shroud assembly along with the fan motor.

6. Remove the radiator mounting bolts. Remove the radiator from the vehicle.

7. Installation is the reverse of the removal procedure.

8. Fill the cooling system with coolant. Start the engine and allow it to reach operating temperature. Check for leaks.

Heater Core

REMOVAL & INSTALLATION

Justy

1. Disconnect the negative battery cable. Drain the cooling system.

2. Disconnect the heater hoses from the heater core assembly.

3. Pull off the right and left defroster ducts from the defroster nozzles. Pull the ducts from the heater unit.

4. Disconnect the electrical wires from the fan switch and the blower motor.

5. Disconnect the air mix cable from the heater unit. Disconnect the mode cable from the heater unit.

6. Remove the bolts that retain the heater unit to the instrument panel.

7. As required, for working clearance remove the glove box door assembly.

8. Disconnect the inside/outside air control cable from the blower assembly.

9. Remove the instrument panel assembly.

10. Remove the heater unit retaining bolts. Remove the heater unit from the vehicle.

11. Remove the heater core cushion. Loosen the heater core holder and than remove it. Pull the heater core from its mounting and remove it from the heater case.

To install:

12. Pressure test the new heater core prior to installation.

13. Install the heater core in the heater cushion, then install the heater core/unit assembly in the vehicle.

14. Install the instrument panel, control cables and glove box door. Connect the heater mode and control cables to the heater box. Connect all electrical connections previously disconnected.

15. Install the defroster nozzles and the heater hoses. Fill the cooling system with coolant.

16. Connect the negative battery cable, start the engine and check for leaks.

Except Justy

NOTE: Depending upon working clearance the air conditioning system may have to be discharged in order to service the blower motor. If this is the case, be sure to observe all the required safety precautions when discharging and recharging the air conditioning system.

1. Disconnect the negative battery cable. Drain the cooling system.

2. Disconnect the heater hoses from the heater core assembly.

3. Remove the instrument panel assembly.

4. Disconnect the electrical harness connector from the blower motor assembly. Disconnect the temperature control cable.

5. Remove the heater unit retaining bolts. Remove the heater unit from the vehicle.

6. Remove the heater core retaining connectors. Remove the heater core from its mounting.

To install:

7. Pressure test the heater core prior to installation. Install the heater core in the heater unit, and the heater unit in the vehicle.

8. Reconnect all electrical connections. Install the heater hoses.

9. Fill the cooling system with coolant. Connect the negative battery cable. Start the engine and bring to operating temperature. Check for leaks.

Water Pump

REMOVAL & INSTALLATION

Except Justy and Legacy

1.8L AND 2.7L ENGINES

1. Disconnect the negative battery cable. Drain the cooling system.

2. Disconnect the radiator hose and bypass hose from the water pump.

3. Loosen the pulley nuts. Loosen the alternator mounting bolts. Remove the drive belt.

4. Remove the timing belt cover. Remove the water pump retaining bolts. Remove the water pump from the engine.

5. Install the water pump. Be sure to use a new gasket or RTV sealant, as required.

6. Install the timing belt cover, alternator, drive belts and pulley.

7. Connect the radiator hoses and fill the cooling system with coolant. Start the engine and allow it to reach operating temperature. Check the coolant level.

Justy

1. Disconnect the negative battery cable. Drain the engine oil. Drain the cooling system.

2. Remove the oil dipstick, dipstick guide and guide sealing.

3. Remove the alternator. Remove the timing belt.

4. Raise and support the vehicle safely. Remove the oil pan. Lower the vehicle.

5. Remove the water pump cover. Remove the water pump impeller. When removing the impeller, lock the balance shaft using the proper tool.

6. Remove the crankcase cover retaining bolts. Remove the crankcase cover along with the remaining water pump assembly.

To install:

7. Install the crankcase cover and water pump assembly. Be sure to use new gaskets or RTV sealant, as required.

8. Install the oil pan, timing belt, alternator and dipstick.

9. Fill the crankcase with oil, the radiator with coolant and connect the negative battery cable. Start the en-

gine and allow it to reach normal operating temperature. Check the coolant level.

Legacy

2.2L ENGINE

1. Disconnect the negative cable from the battery.
2. Drain the cooling system.
3. Disconnect the lower radiator hose.
4. Disconnect the electrical connectors and remove the radiator fan motor assembly.
5. Remove the V-belts as necessary.
6. Remove the timing belt covers and the timing belt.
7. Remove the tensioner adjuster and the cam angle sensor.
8. Remove the left camshaft pulley.
9. Remove the left rear timing belt cover.
10. Remove the tensioner bracket.
11. Disconnect the radiator and heater hoses from the water pump assembly.
12. Remove the water pump assembly.

To install:

13. Inspect the water pump and replace, as required.
14. After the water pump is in place, complete the assembly of the components in the reverse of their removal.
15. Fill the cooling system and correct the level as required, during and after the warm-up.

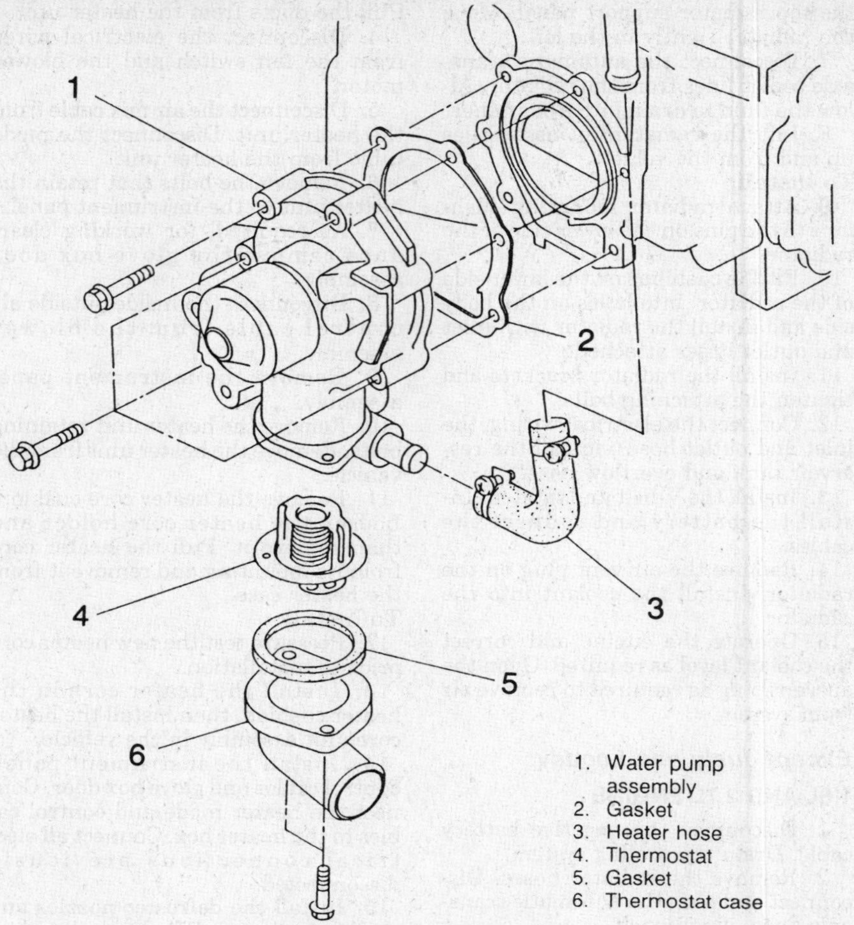

1. Water pump assembly
2. Gasket
3. Heater hose
4. Thermostat
5. Gasket
6. Thermostat case

Water pump servicing – 2.2L engine

Thermostat

REMOVAL & INSTALLATION

Except Justy and Legacy

1.8L AND 2.7L ENGINES

1. Disconnect the negative battery cable. Drain the cooling system.
2. Remove the thermostat housing retaining bolts. Remove the thermostat housing and cover assembly.
3. Remove the thermostat from the intake manifold.
4. Installation is the reverse of the removal procedure. Be sure to use a new gasket or RTV sealant, as required.

Justy

1. Disconnect the negative battery cable. Drain the cooling system.
2. Remove the thermostat housing retaining bolts. Remove the thermostat housing and cover assembly.
3. Remove the thermostat from the intake manifold.
4. Installation is the reverse of the removal procedure. Be sure to use a new gasket or RTV sealant, as required.

Legacy

2.2L ENGINE

1. Disconnect the negative battery cable. Drain the cooling system.
2. Remove the thermostat case cover and gasket. Pull out the thermostat.

To install:

3. Clean the mating surface of the thermostat case cover throughly.
4. Install the thermostat using a new gasket. The thermostat must be installed with the jiggle pin upward and to the front.
5. Fill the cooling system with coolant. Star the vehicle and allow it to reach operating temperature. Check the coolant level.

ENGINE ELECTRICAL

NOTE: Disconnecting the negative battery cable on some vehicles may interfere with the functions of the on board computer systems and may require the computer to undergo a relearning process, once the negative battery cable is reconnected.

Distributor

NOTE: The 1.8L and 2.7L engines with both the SPFI and MPFI systems use an LED and photodiode pulse pick-up in the distributor for cylinder and crankshaft position determination. The electronic ignition circuit operates in the same basic manner as the standard electronic distributors used on the remaining engines.

REMOVAL & INSTALLATION

Undisturbed Engine

1. Disconnect the negative battery cable. Remove the air cleaner assembly. If equipped, label and disconnect the hose from the distributor.
2. Disconnect the primary wire from the coil. If equipped with a breakerless ignition, disconnect the distrib-

utor electrical wiring connector from the vehicle wiring harness.

3. Disconnect the distributor cap retaining clamps or remove the screws and the cap from the distributor. Position the cap and ignition wires aside.

NOTE: If necessary to remove the ignition wires from the cap to provide room to remove the distributor, be sure to label the wires and the cap terminals for easy and accurate reinstallation.

4. Position the engine at TDC with No. 1 cylinder on the compression stroke or using chalk, mark the distributor rotor to distributor housing and the distributor housing to engine relationships.

5. Remove the distributor to engine hold-down bolt.

6. Remove the distributor from the engine, taking care not to damage or lose the O-ring.

NOTE: Do not disturb the engine while the distributor is removed. If the engine cranked or rotated while the distributor is removed, the engine will have to be retimed.

To install:

7. Position the distributor in the block, make sure the O-ring is in place, align the distributor rotor to housing marks and the distributor housing to engine marks.

NOTE: If equipped with an octane selector, install and tighten the hold-down bolt finger-tight.

8. To complete the installation, reverse the removal procedures. Recheck the ignition timing.

Disturbed Engine

1. Disconnect the negative battery cable. Remove the air cleaner assembly. If equipped, label and disconnect the hose from the distributor.

2. Disconnect the primary wire from the coil. If equipped with a breakerless ignition, disconnect the distributor electrical wiring connector from the vehicle wiring harness.

3. Disconnect the distributor cap retaining clamps or remove the screws and the cap from the distributor. Position the cap and ignition wires aside.

NOTE: If necessary to remove the ignition wires from the cap to provide room to remove the distributor, be sure to label the wires and the cap terminals for easy and accurate reinstallation.

4. Remove the distributor to engine hold-down bolt.

5. Remove the distributor from the engine, taking care not to damage or lose the O-ring.

To install:

6. If equipped, remove the plastic dust cover from the timing port on the flywheel housing.

7. Remove the No. 1 spark plug. Use a wrench on the crankshaft pulley bolt and place the transaxle in the **N** position and slowly rotate the engine until the TDC **0** degree mark on the flywheel aligns with the pointer.

8. If Step 7 is impractical for any reason, the following method can be used to get the No. 1 piston on TDC. Remove the 2 bolts that hold the right valve cover and remove the cover to expose the valves on No. 1 cylinder. Rotate the engine so the valves in No. 1 cylinder are closed and the TDC **0** degree mark on the flywheel aligns with the pointer.

9. Align the small depression on the distributor drive pinion with the mark on the distributor housing; this will align the rotor with the No. 1 spark plug terminal on the distributor cap.

NOTE: If equipped with an octane selector, set the pointer midway between the A and R. Make sure the O-ring is located in the proper position.

10. Align the distributor housing to engine matchmarks and install the distributor into the engine. Make sure the drive is engaged. Install the hold-down bolt finger-tight. Using a timing light, perform the ignition timing procedures.

11. To complete the installation, remove the timing light and reverse the removal procedures.

Distributorless Ignition

REMOVAL & INSTALLATION

Ignition Coil

2.2L ENGINE

1. Disconnect the battery negative terminal.

2. Remove the intake manifold cover.

3. Disconnect the wires from the ignition coil.

4. Remove the ignition coil.

5. To install, reverse the installation procedure.

Ignition Timing

ADJUSTMENT

NOTE: There are no timing procedures available for the 2.2L Legacy. The ignition system is distributorless and is operated via crankshaft and camshaft sensors, using the "waste type" spark system.

Justy

1. Connect test mode connectors (2 pin type, Green in color), located beneath the left side of the instrument panel.

2. Allow the engine to reach operating temperature. Adjust the idle speed to specification. Connect a timing light, according to manufacturer's instructions.

3. Start the engine and check the ignition timing. If timing is not within specification, loosen the distributor hold-down bolt.

4. Rotate the distributor until the correct timing specification is reached. Tighten the distributor hold-down bolt.

5. Disconnect the test mode connector.

XT Coupe

1. Allow the engine to reach operating temperature. Adjust the idle speed to specification. Connect a timing light, according to manufacturers instructions.

2. Be sure the idle contact of the throttle sensor is in the engaged position. Connect the test mode connectors together, located in the trunk area for both the 1.8L and 2.7L engines, using the MPFI system. The connectors will be found near of each other.

NOTE: The check engine warning light will come on; this does not indicate there is a problem. The ignition timing must not be adjusted and cannot be checked while the idle switch is disengaged or the test mode connectors disconnected.

3. If timing is not within specification, loosen the distributor hold-down bolt.

4. Rotate the distributor until the correct timing specification is reached. Tighten the distributor hold-down bolt.

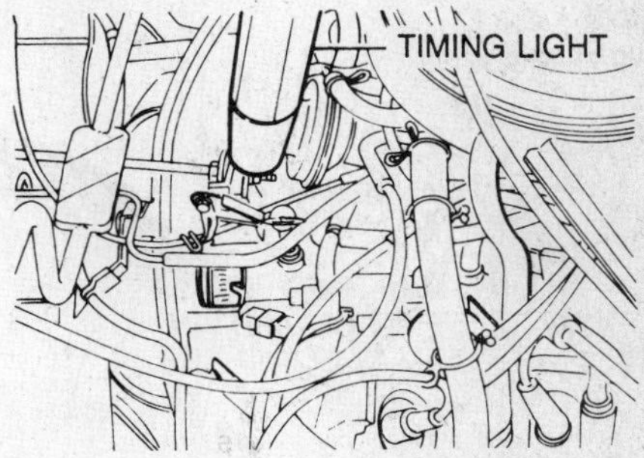

Timing mark location—except XT Coupe

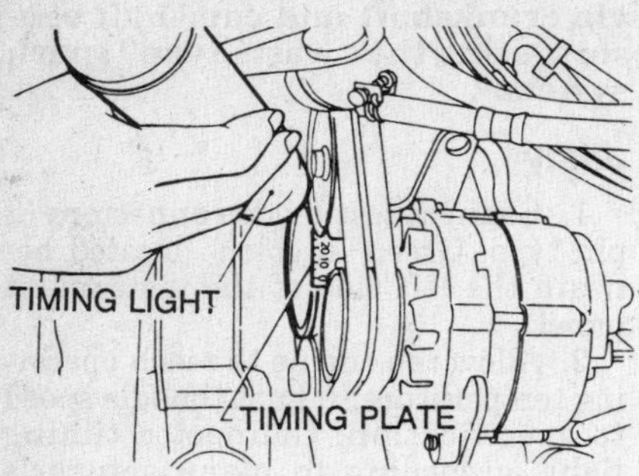

Timing mark location – 1.8L engine except XT Coupe

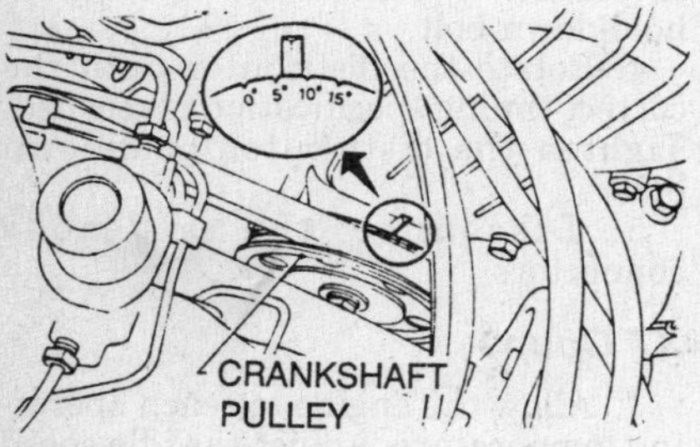

Timing mark location – 1.2L engine

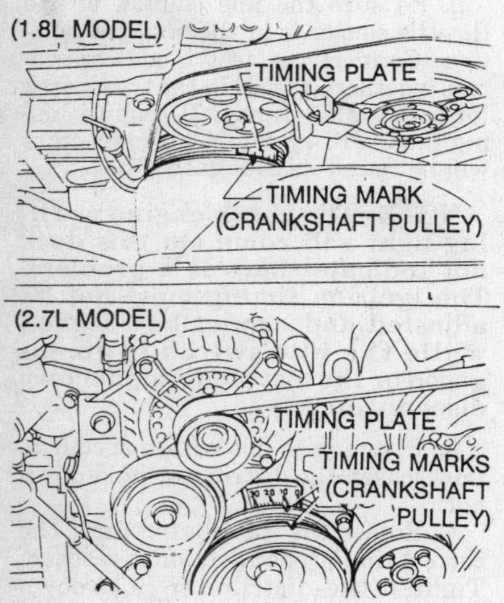

Timing mark locations – XT Coupe equipped with 1.8L and 2.7L engines

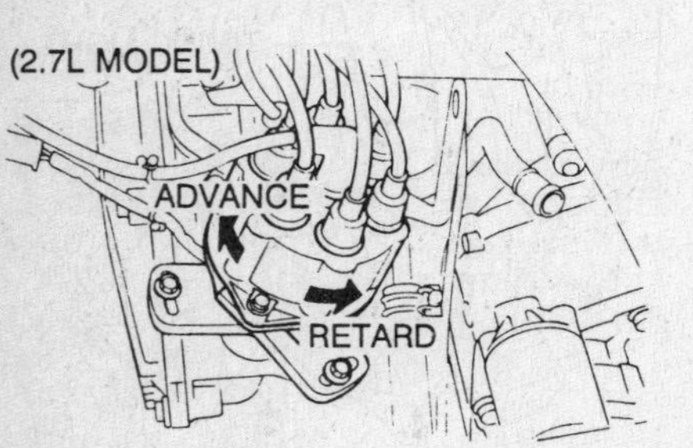

Distributor movement to change basic timing – 2.7L engine

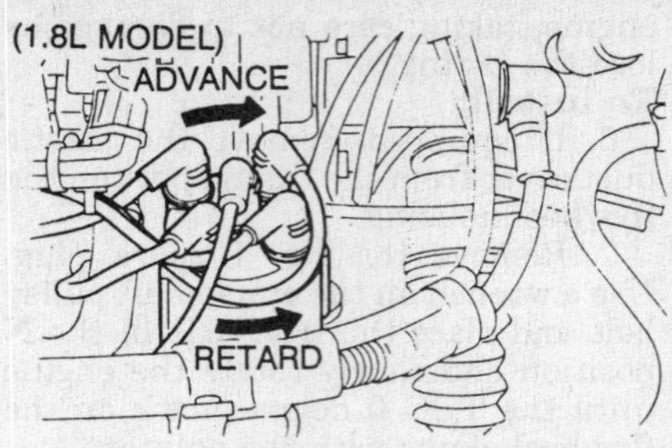

Distributor movement to change basic timing – 1.8L engine

Except Justy and XT Coupe

1. Allow the engine to reach operating temperature. Adjust the idle speed to specification. Connect a timing light, according to manufacturer's instructions.
2. If the engine is equipped with a carburetor, disconnect and plug the distributor vacuum line.
3. Be sure the idle switch is in the engaged position. Connect the test mode connectors, located under the left side of the dash for the MPFI system or on the left side of the engine compartment for the SPFI system. The connectors will be located side-by-side.

NOTE: The check engine warning light will come on; this does not indicate that there is a problem. The ignition timing must not be adjusted and cannot be checked while the idle switch is inoperative or the test mode connectors disconnected.

4. If timing is not within specification, loosen the distributor hold-down bolt.
5. Rotate the distributor until the correct timing specification is reached. Tighten the distributor hold-down bolt.

Alternator

PRECAUTIONS

Observing these precautions will ensure safe handling of the electrical system components and will avoid damage to the vehicle's electrical system.

- Be absolutely sure of the polarity of a booster battery before making connections. Connect the cables positive to positive and negative to negative. If jump starting, connect the positive cables first and the last connection to a ground on the body of the booster vehicle, so arcing cannot ignite the hydrogen gas that may have accumulated near the battery. Even a momentary connection of a booster battery with polarity reserved may damage the alternator diodes.
- Disconnect both vehicle battery cables before attempting to charge the battery.
- Never ground the alternator output or battery terminal. Be cautious when using metal tools around a battery to avoid creating a short circuit between the terminals.
- Never run an alternator without a load unless the field circuit is disconnected.
- Never attempt to polarize an alternator.
- Never disconnect any electrical components with the ignition switch turned ON.

BELT TENSION ADJUSTMENT

1. To adjust the belt tension, first loosen the alternator to bracket adjusting bolt.
2. Lift up on the alternator to increase the tension on the belt. When it takes moderate thumb pressure to move the longest span of belt ½ in., the tension adjustment is correct.
3. Tighten the adjusting bolt so the alternator will not move in the adjusting bracket.

REMOVAL & INSTALLATION

1. Disconnect the negative battery cable.
2. Label and disconnect the wiring from the alternator. Remove the necessary components in order to gain access to the alternator retaining bolts.
3. Remove the alternator retaining bolts.
4. Remove the drive belt. Remove the alternator from the vehicle.
5. Installation is the reverse of the removal procedure.

Starter

REMOVAL & INSTALLATION

1. Remove the spare tire from the engine compartment, as required.
2. Disconnect the negative battery cable. As required, raise and support the vehicle safely.
3. Disconnect the wiring harness from the starter.
4. Remove the starter retaining bolts. Remove the starter from its mounting.
5. Installation is the reverse of the removal procedure.

EMISSION CONTROLS

Please refer to "Emission Controls" in the Unit Repair section for system maintenance procedures. Due to the complex nature of modern electronic engine control systems, comprehensive diagnosis and testing procedures fall outside the confines of this repair manual. For complete information on diagnosis, testing and repair procedures concerning all modern engine and emission control systems, please refer to "Chilton's Guide to Fuel Injection and Electronic Engine Controls".

Emission Warning Lamps

RESETTING

Some vehicles are equipped with an EGR light that illuminates when the vehicle attains 60,000 miles (96,000 km). In order to reset the light for an other 60,000 mile increment, the following procedure must be done:

1. Remove the lower instrument panel cover, exposing the fuel panel.
2. Directly behind or along the side of the fuse panel, a blue 2 piece connector will be noted. Disconnect the 2 blue connectors.
3. Near the blue connectors will be a green connector that is not connected to any other wire.
4. Connect the green connector into the matching blue connector, thus resetting the emission light and recycling the system for another 60,000 mile increment.
5. Be sure the indicator light is out and reinstall the lower instrument panel cover.

FUEL SYSTEM

Relieving Fuel System Pressure

Fuel Injected Vehicles

1. Disconnect the electrical wiring connector from the fuel pump.
2. Start the engine. Once the engine has stopped, crank the engine for 5 seconds or more. If the engine starts, let the engine run until it stops.
3. Turn the ignition switch **OFF**.
4. Reconnect the electrical wiring connector of the fuel pump.

Fuel Filter

REMOVAL & INSTALLATION

Justy

1. Carefully relieve the fuel pump pressure. As required, raise and support the vehicle safely.
2. Remove the flange bolts and remove the lower fuel pump bracket assembly.
3. Disconnect and plug the fuel lines at the fuel filter.
4. Remove the fuel filter retaining bolts. Remove the fuel filter assembly from its mounting.
5. Installation is the reverse of the removal procedure.

Except Justy

FUEL INJECTED ENGINE

The fuel filter is located inside the engine compartment on the left fender assembly.

1. Properly relieve the fuel system pressure.
2. Disconnect the fuel lines from the fuel filter.
3. Pull the fuel filter from the bracket and remove it from the vehicle.
4. Installation is the reverse of the removal procedure. Start the engine and check for leaks.

CARBURETED ENGINE

1. Disconnect the negative battery cable.
2. Raise and support the vehicle safely.
3. Disconnect the fuel hoses from the fuel filter.
4. Pull the fuel filter from the bracket and remove it from the vehicle.
5. Installation is the reverse of the removal procedure.

Electric Fuel Pump

PRESSURE TESTING

1. Raise and support the vehicle safely.
2. Using a fuel pressure gauge, connect into the fuel line.
3. Turn the ignition switch to the **ON** position. Observe the fuel pressure, it should be:

   2.6–3.3 psi for carburetor equipped vehicle.

   61–71 psi for MPFI equipped vehicle.

   36–50 psi for SPFI equipped vehicle.
4. If the fuel pump does not meet specification, replace it.
5. After testing, disconnect the pressure gauge and reconnect the fuel line.

REMOVAL & INSTALLATION

Justy

1. Carefully relieve the fuel pump pressure. As required, raise and support the vehicle safely.
2. Remove the flange bolts and remove the lower fuel pump bracket assembly.
3. Disconnect and plug the fuel lines at the fuel pump assembly. Disconnect the fuel pump electrical connector.
4. Remove the fuel pump assembly retaining bolts. Remove the fuel pump assembly from its mounting.
5. Installation is the reverse of the removal procedure.

Except Justy

CARBURETED ENGINE

1. Carefully relieve the fuel pump pressure. As required, raise and support the vehicle safely.
2. Remove the flange bolts and remove the lower fuel pump bracket assembly.
3. Disconnect and plug the fuel lines at the fuel pump assembly. Disconnect the fuel pump electrical connector.
4. Remove the fuel pump assembly retaining bolts. Remove the fuel pump assembly from its mounting.
5. Installation is the reverse of the removal procedure.

FUEL INJECTED ENGINE

1. Properly relieve the fuel system pressure.
2. Raise and support the vehicle safely. Devise a clamp for the thicker hose leading to the pump and clamp it off a few inches from the nipple on the pump. This will prevent the fuel from running from the tank while the pump is disconnected.
3. Being careful not to bend the hose sharply, loosen the hose clamp and disconnect the large hose leading into the pump. Do the same with the outlet from the damper.
4. Remove the pump bracket to chassis retaining bolts. Remove the fuel pump and pump damper assembly.
5. Installation is the reverse of the removal procedure.

Carburetor

REMOVAL & INSTALLATION

Justy

1. Disconnect the negative battery cable. Remove the air cleaner assembly.
2. Disconnect the fuel line. Disconnect the return and vent line hoses.
3. Disconnect the main diaphragm, distributor vacuum line and canister vent hose.
4. Disconnect the idle solenoid valve wires and hoses. Disconnect the harness electrical connector.
5. Disconnect the primary and secondary air bleed hoses. Disconnect the accelerator cable from the throttle lever.
6. Remove the carburetor retaining bolts. remove the carburetor from its mounting. Discard the gasket.
7. Installation is the reverse of the removal procedure. Be sure to use a new base gasket.

Fuel Injection

IDLE MIXTURE ADJUSTMENT

SPFI or MPFI fuel injected vehicles are equipped with computer controlled idle mixture adjustment. No adjustments to the system are necessary.

Fuel Injector

REMOVAL & INSTALLATION

SPFI Engine

1. Disconnect the negative battery cable.
2. Remove the injector cap and gasket.
3. Hold the injector using pliers, then pull out the injector from the throttle body.
4. Remove the injector and O-ring from the throttle body.
5. To install, reverse the removal procedure.

MPFI Engine

1. Relieve the fuel pressure by disconnecting the fuel pump connector, starting the engine and allow the engine to run until it stalls. After stalling, crank the starter for approximately 5 seconds and turn ignition switch **OFF**.
2. Remove the spark plug caps.
3. Disconnect the connector from the fuel injector.

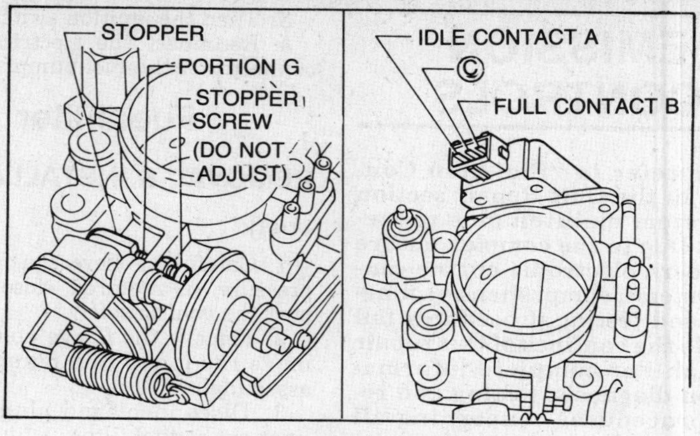

Fuel injection adjustment terminals

4. Remove the fuel injector cover.
5. Remove the injector while turning.

NOTE: Do not attempt to pry the injectors with a prybar or similar tool. Do not pinch the injector pin with pliers. Be careful not to damage the O-ring. If the injector is difficult to remove by hand, remove the injector and fuel pipe as a unit and push the injector out from the back side.

6. To install, reverse the removal procedure.

DRIVE AXLE

Halfshaft

REMOVAL & INSTALLATION

Justy

1. Raise and support the vehicle safely. Remove the tire and wheel assembly.
2. Remove the disc brake assembly. Remove the dust cover, cotter pin, castle nut, conical spring. Remove the center piece, using the proper tools.

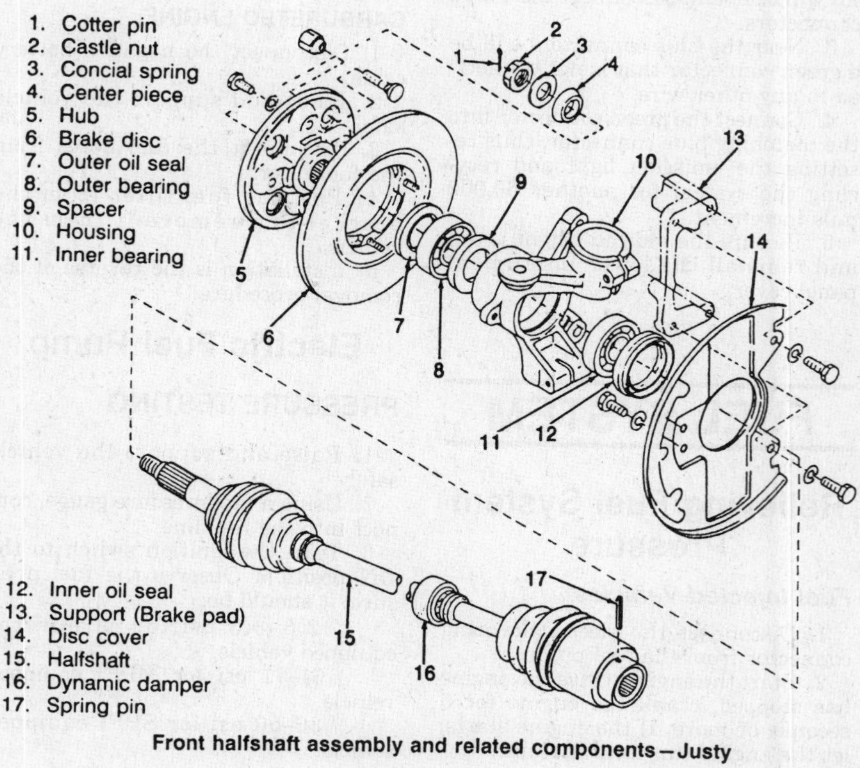

Front halfshaft assembly and related components – Justy

3. Pull the hub and disc assembly from the halfshaft. Remove the disc cover from the housing.

4. Drive out the spring pin connecting the halfshaft to the differential, using the proper tool.

5. Remove the cotter pin and the castle nut from the tie rod end ball joint.

6. Remove the tie rod end ball joint from the knuckle arm, using the proper puller.

7. Remove the bolt that retains the housing to the strut. Carefully push down the housing in order to remove it from the strut.

8. Remove the ball joint of the transverse link from the housing. Remove the housing and the halfshaft assembly as a complete unit.

9. Separate the housing from the halfshaft, using removal tools 922493000 and 921122000 or their equivalents.

To install:

10. Join the housing and halfshaft using installation tool 927210000.

11. Install housing and axle assembly to strut but do not tighten.

12. Install the dust seal on the spindle. Insert the halfshaft into the differential and install the spring pin. Lubricate the splines with grease.

13. Install the tie rod end ball joint and tighten the nut ot 18–22 ft. lbs. (25–29 Nm). Tighten the housing-to-strut bolt to 25–33 ft. lbs. (34–44 Nm).

14. Install the disc cover, hub, disc brake assembly and castle nut. Install the caliper assembly.

15. Install the wheel and tire, lower the vehicle and test drive.

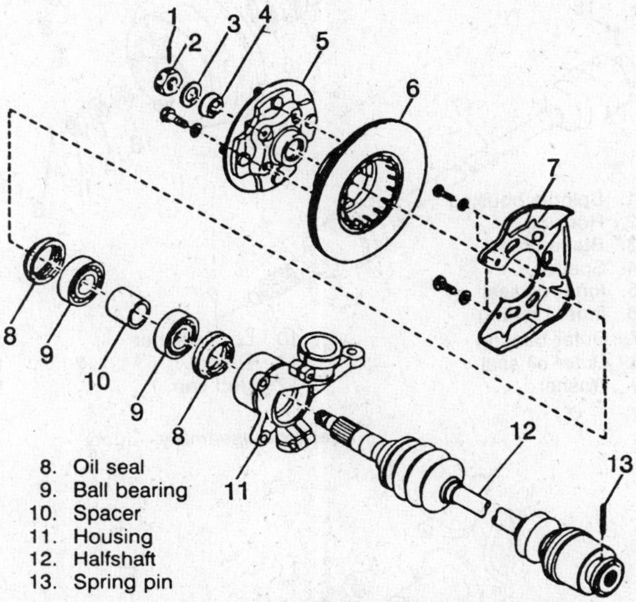

1. Cotter pin
2. Castle nut
3. Washer spring
4. Center piece
5. Hub
6. Brake disc
7. Disc cover
8. Oil seal
9. Ball bearing
10. Spacer
11. Housing
12. Halfshaft
13. Spring pin

Front halfshaft assembly and related components – except Justy and XT Coupe with 2.7L engine

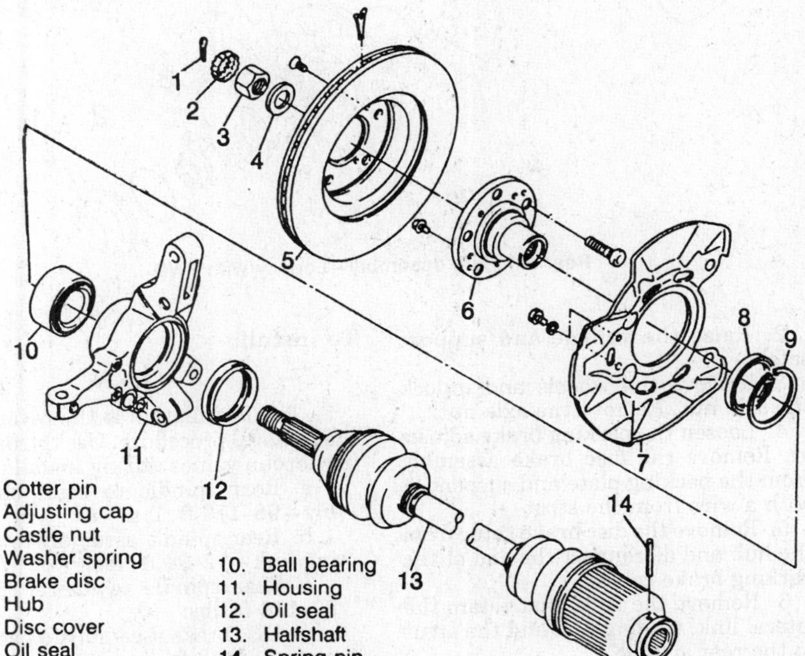

1. Cotter pin
2. Adjusting cap
3. Castle nut
4. Washer spring
5. Brake disc
6. Hub
7. Disc cover
8. Oil seal
9. Snap ring
10. Ball bearing
11. Housing
12. Oil seal
13. Halfshaft
14. Spring pin

Front halfshaft assembly and related components – XT Coupe with 2.7L engine

Except Justy

1. Release the parking brake. Raise and support the vehicle safely. Remove the tire and wheel assembly.

2. Pull out the parking brake cable outer clip from the caliper. Disconnect the parking brake cable end from the caliper lever.

3. Drive out the double offset joint spring pin, using the proper tools.

4. Loosen the 2 retaining bolts and remove the disc brake assembly from the housing. Remove the 2 bolts that connect the housing and the damper strut.

5. Remove the dust cover, cotter pin. Disconnect the tie rod end ball joint from the housing knuckle arm, using the proper puller tool.

6. Remove the halfshaft from the differential spindle along with the housing assembly.

7. Remove the housing from the halfshaft, using tool 926470000 or equivalent.

To install:

8. Install the halfshaft into the hub using installer 922431000, or equivalent. Take care not to damage the inner oil seal lip. Tighten the axle nut temporarily.

9. Install the double offset joint on the spindle and drive a new spring pin into place. Install the tie rod and tighten the nut to 61–83 ft. lbs. (83–113 Nm).

10. Install the axle nut and tighten to 137 ft. lbs. (186 Nm).

11. Install the disc brake assembly and parking brake. Install the wheel and tire.

12. Lower the vehicle and test drive.

Driveshaft

REMOVAL & INSTALLATION

4WD

1. Raise and support the vehicle safely.

2. Remove the driveshaft flange to rear differential flange bolts.

NOTE: If equipped with a center bearing, remove the center bearing to chassis bolts and lower the assembly from the vehicle.

3. Position a drain pan under the rear of the transaxle. Remove the driveshaft from the vehicle.

To install:

4. Install the driveshaft and tighten the flange bolts to 17–24 ft. lbs. (24–32 Nm).

5. If equipped with a center bearing, raise the assembly and install the center bearing bolts. Tighten to attaching bolts to 25–33 ft. lbs. (34–44 Nm).

6. Lower the vehicle, check the transaxle fluid level and test drive.

Rear Axle Shafts

REMOVAL & INSTALLATION

Justy

2WD

1. Raise and support the vehicle safely. Remove the tire and wheel assembly.

2. Remove the dust cap. Straighten the locking washer edge. Remove the nut, lock washer and washer.

3. Remove the brake drum. Be sure not to drop the outer bearing.

4. Remove the brake line bracket from the spindle housing.

5. Loosen the bolts and remove the brake assembly. Suspend the assembly aside with wire.

6. Using the proper tools, drive out the spring pin connecting the halfshaft assembly to the differential.

7. Remove the strut, lower link and trailing link. Pull the housing along with the halfshaft from its mounting.

8. Separate the housing from the halfshaft, using removal tools 922493000 and 921122000 or their equivalent.

To install:

9. Join the housing and halfshaft. Install the strut, lower link and trailing link. Install the spring pin connecting the halfshaft assembly to the differenital.

10. After tightening the rear axle halfshaft-to-axle housing nut, tighten the axle shaft nut 30 degrees further.

11. Install the brake assembly, brake line bracket and brake drum.

12. Install the wheel and tire, lower the vehicle and test drive.

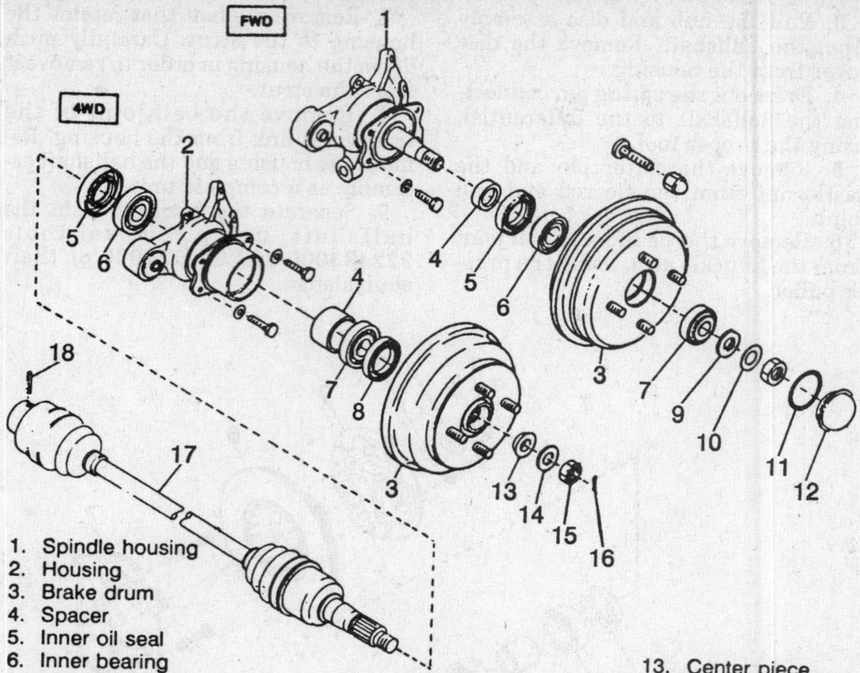

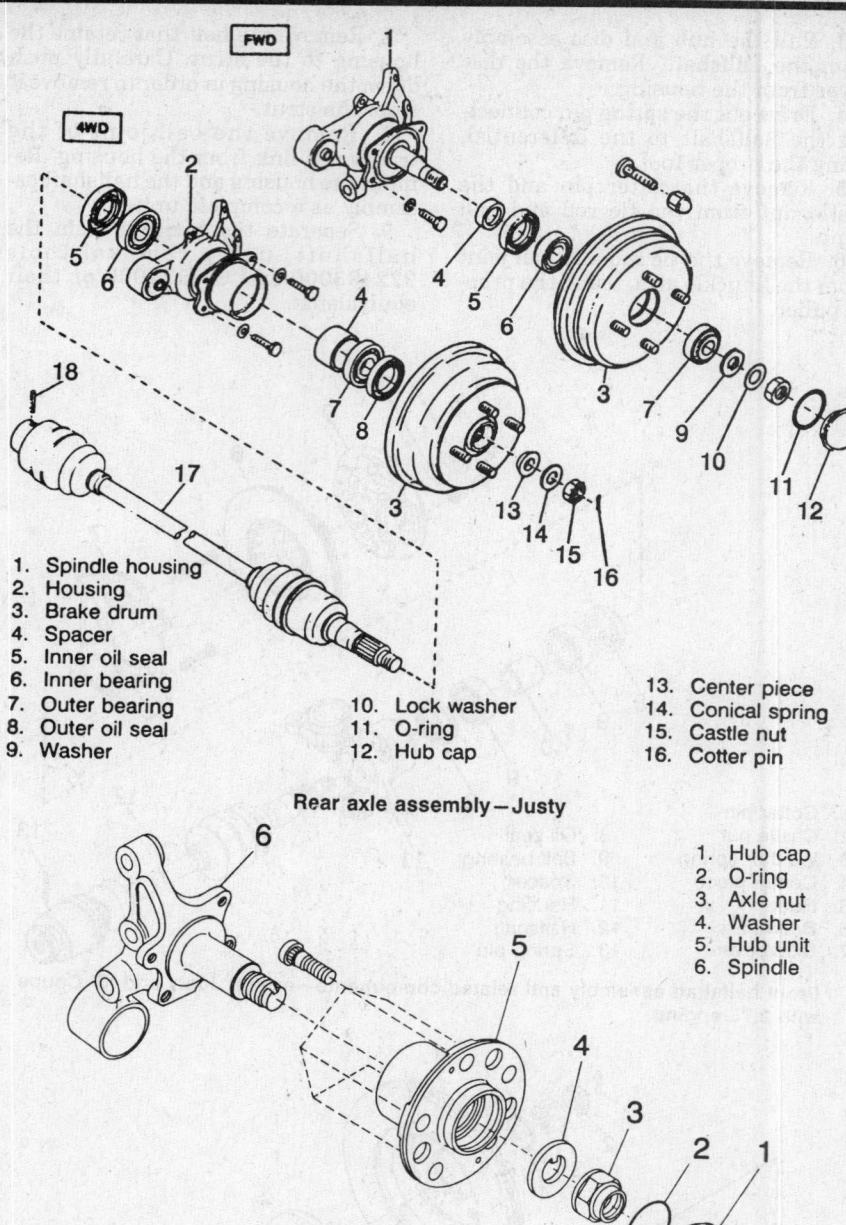

1. Spindle housing
2. Housing
3. Brake drum
4. Spacer
5. Inner oil seal
6. Inner bearing
7. Outer bearing
8. Outer oil seal
9. Washer
10. Lock washer
11. O-ring
12. Hub cap
13. Center piece
14. Conical spring
15. Castle nut
16. Cotter pin

Rear axle assembly – Justy

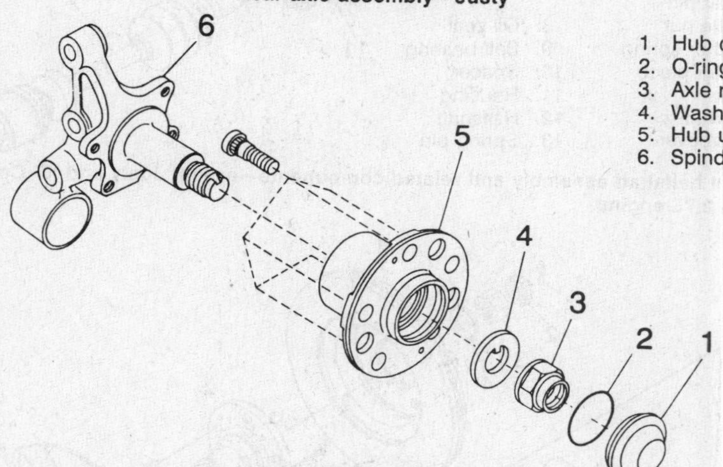

1. Hub cap
2. O-ring
3. Axle nut
4. Washer
5. Hub unit
6. Spindle

Rear axle/hub assembly – Legacy with FWD

Legacy

2WD

1. Disconnect the negative battery cable.

2. Raise the vehicle and support safely.

3. Remove the wheels and unlock the axle nut. Remove the axle nut.

4. Loosen the parking brake adjuster. Remove the disc brake assembly from the backing plate and suspend it with a wire from the strut.

5. Remove the disc brake rotor from the hub and disconnect the end of the parking brake cable.

6. Remove the bolts that retain the lateral link, trailing link and the strut to the rear spindle.

7. Remove the rear spindle, backing plate and hub as a unit.

To install:

8. The installation is the reverse of the removal procedure. Use the following torque values during installation.

a. Rear spindle to strut assembly – 98–119 ft. lbs.

b. Rear spindle assembly to trailing link – 72–94 ft. lbs.

c. Rear spindle to lateral link – 87–116 ft. lbs.

d. Disc brake assembly to backing plate – 34–43 ft. lbs.

e. Axle nut – 123–152 ft. lbs.

f. Wheel nuts – 58–72 ft. lbs.

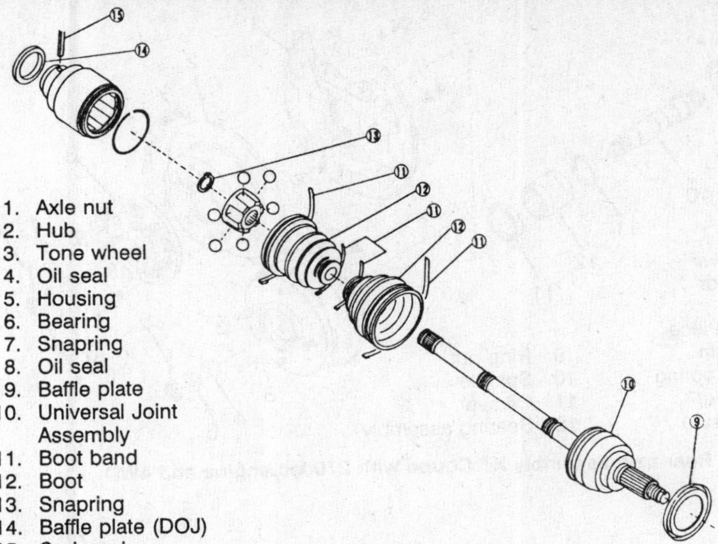

1. Axle nut
2. Hub
3. Tone wheel
4. Oil seal
5. Housing
6. Bearing
7. Snapring
8. Oil seal
9. Baffle plate
10. Universal Joint Assembly
11. Boot band
12. Boot
13. Snapring
14. Baffle plate (DOJ)
15. Spring pin

Front axle and hub assembly – Legacy

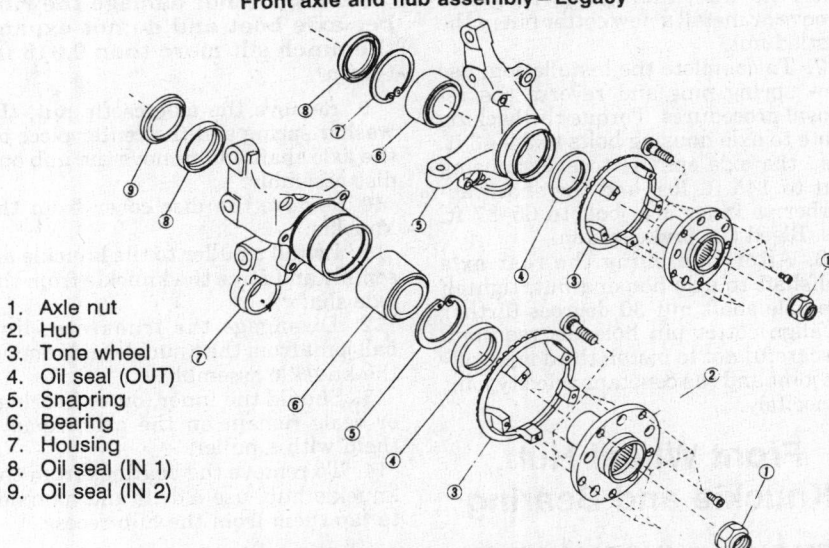

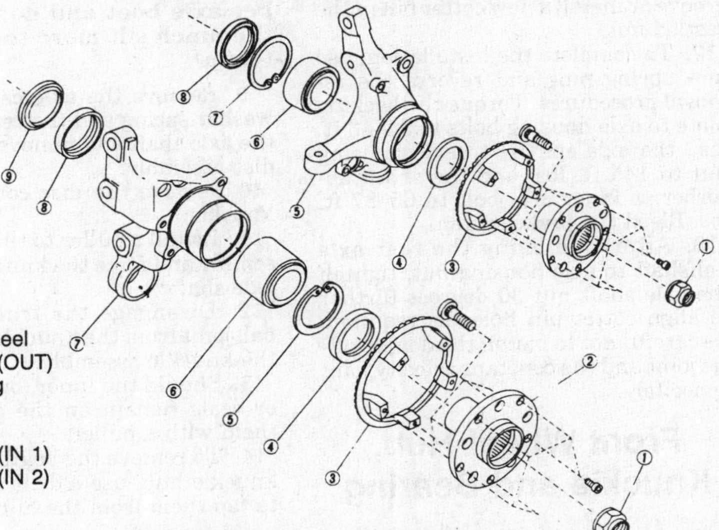

1. Axle nut
2. Hub
3. Tone wheel
4. Oil seal (OUT)
5. Snapring
6. Bearing
7. Housing
8. Oil seal (IN 1)
9. Oil seal (IN 2)

Rear hub assembly – Legacy with 4WD

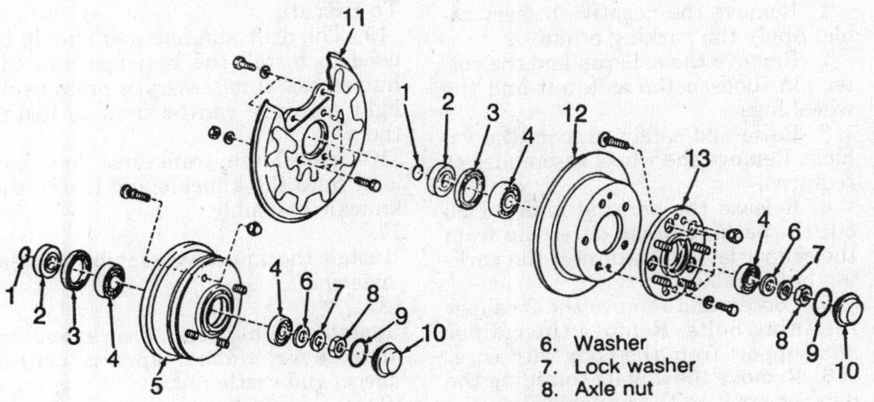

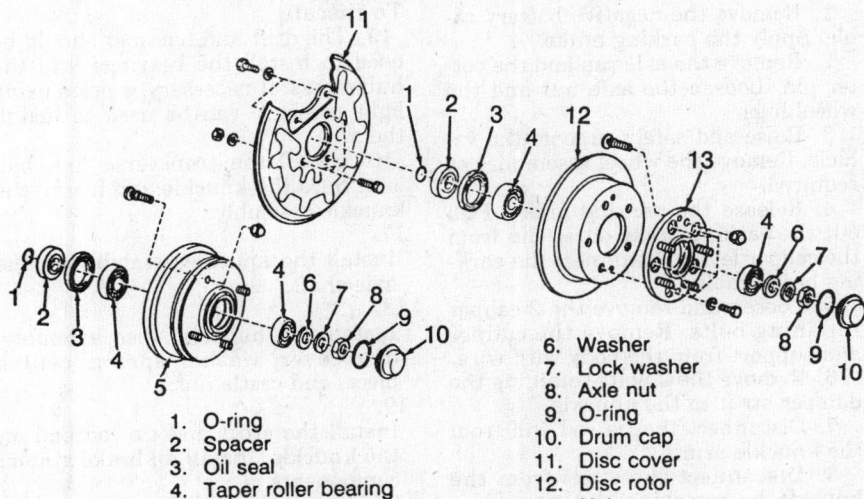

1. O-ring
2. Spacer
3. Oil seal
4. Taper roller bearing
5. Brake drum
6. Washer
7. Lock washer
8. Axle nut
9. O-ring
10. Drum cap
11. Disc cover
12. Disc rotor
13. Hub

Rear axle assembly XT Coupe with 2700cc engine and 2WD

4WD

1. Disconnect the negative battery cable.
2. Raise the vehicle and support safely. Remove the wheel assemblies.
3. Unlock axle nut and remove from axle.
4. Loosen the parking brake adjuster.
5. Remove the disc brake assembly and suspend it on a wire from the body or strut.
6. Remove the disc rotor from the hub and disconnect the end of the parking brake cable.
7. Remove the speed sensor from the backing plate, if equipped with Anti-lockBrake System (ABS).
8. Remove the bolts that secure the lateral link assembly and the trailing link assembly to the rear housing. Discard the self-locking nuts and replace with new nuts.
9. Remove the spring pin that secures the rear differential spindle to the inner CV-joint.
10. Remove the inner CV-joint and shaft from the differential spindle.
11. Disengage the rear drive shaft from the rear hub and remove the shaft.

To install:

12. When installing the shaft, reverse the removal procedures with the following additions:
    a. Use new seals.
    b. Using a new axle nut, pull the axle shaft through the hub splines.
    c. Install the axle shaft onto the differential spindle and install the spring pin into place.
    d. Using new nuts on the trailing link, tighten to 72–94 ft. lbs.
    e. Torque disc brake assembly to the rear housing assembly bolts/nuts to 34–43 ft. lbs.
    f. Torque the axle nut to 123–152 ft. lbs.
    g. Wheel nut torque to 58–72 ft. lbs.

Except Justy and Legacy

2WD

1. Raise and support the vehicle safely. Remove the tire and wheel assembly.
2. Remove the dust cap. Straighten the lock washer. Remove the nut, lock washer and washer.
3. Remove the brake drum. Be sure not to drop the outer bearing.
4. Remove the brake line bracket from the spindle housing.
5. Loosen the bolts and remove the brake assembly. Suspend the assembly aside with wire.
6. Remove the damper strut, lower link and trailing link.
7. Remove the spindle assembly re-

taining bolts. Remove the spindle from its mounting.

To install:

8. Install the spindle assembly, damper strut, lower link and trailing link. Tighten all bolts to specification.
9. Install the brake assembly and brake line bracket. Install the brake drum.
10. Install wheel and tire, lower the vehicle and test drive.

4WD

1. Firmly apply the parking brake.
2. Remove the rear wheel cap and the cotter pin, then loosen the castle nut.
3. Disconnect the shock absorber from the inner arm.
4. Loosen the crossmember outer bushing lock bolts. Remove the inner trailing arm to chassis bolt and the inner arm.
5. Raise and support the vehicle safely. Remove the rear wheel assemblies.
6. Using a 0.24 in. (6mm) diameter steel rod or a pin punch, drive the inner/outer spring pins from the double offset joints.
7. With the trailing arm fully lowered, remove the ball joint from the trailing arm spindle and the inner double offset joint and the differential spindle.
8. Remove the castle nut and the brake drum or rear wheel caliper If equipped, remove the brake caliper and properly position it aside. Do not disconnect the brake hose from the caliper.
9. Disconnect and plug the brake hose from the inner arm bracket.
10. If equipped with rear brake drums, remove the brake assembly from the trailing arm.
11. Disconnect the inner arm from the outer arm and remove the inner arm from the vehicle.
12. Secure the inner arm in a vise, then using a hammer and a punch, straighten the staked portion of the ring nut or remove the cotter pin from the castled nut. Using the wrench tool 925550000 or equivalent, remove the ring nut.
13. Using a plastic hammer on the outside of the spindle, drive it inward to remove it.
14. Clean, inspect and replace the necessary parts.

To install:

15. Using an arbor press and a piece of 1.38 in. (35mm) diameter pipe, insert the spindle from the inside and press the outer bearing's inner race from outside.
16. Using the wrench tool 925550000 or equivalent, torque the axle shaft ring nut to 127–163 ft. lbs. Using a punch and a hammer, stake the ring nut, facing the ring nut groove or install a new cotter pin in the castled nut.
17. To complete the installation, use new spring pins and reverse the removal procedures. Torque the backing plate to axle housing bolts to 34–43 ft. lbs., the axle spindle to axle housing nut to 145 ft. lbs. and the shock absorber to inner arm bolt to 65–87 ft. lbs. Bleed the brake system.
18. After tightening the rear axle halfshaft to axle housing nut, tighten the axle shaft nut 30 degrees further to align cotter pin holes as required. Be careful not to install the double offset joint and the constant velocity joint oppositely.

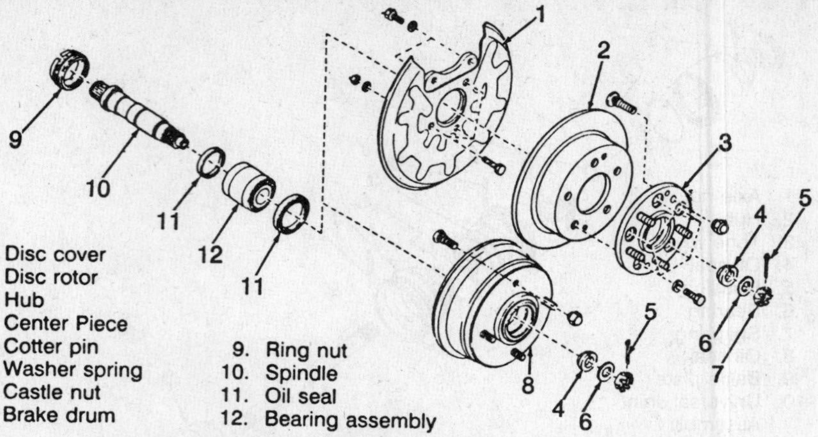

1. Disc cover
2. Disc rotor
3. Hub
4. Center Piece
5. Cotter pin
6. Washer spring
7. Castle nut
8. Brake drum
9. Ring nut
10. Spindle
11. Oil seal
12. Bearing assembly

Rear axle assembly XT Coupe with 2700cc engine and 4WD

Front Wheel Hub, Knuckle and Bearing

REMOVAL & INSTALLATION

1. Remove the negative battery cable. Apply the parking brake.
2. Remove the axle cap and the cotter pin. Loosen the axle nut and the wheel lugs.
3. Raise and safely support the vehicle. Remove the wheel assemblies as required.
4. Release the parking brake. Pull out the parking brake outer clip from the caliper lever. Disconnect the parking brake cable.
5. Loosen and remove the 2 caliper retaining bolts. Remove the caliper and support from the body with wire.
6. Remove the 2 bolts retaining the damper strut to the knuckle.
7. Disconnect the tie rod end from the knuckle arm.
8. Disconnect the strut from the knuckle by removing the pinch bolt and opening the pinch slit in the knuckle with an appropriate pry tool.

NOTE: Do not damage the rubber axle boot and do not expand the pinch slit more than 0.016 in. (4mm).

9. Remove the axle castle nut, the washer spring and the center piece on the axle shaft and remove the hub and disc assembly.
10 Remove the disc cover from the knuckle.
11. Install a puller to the knuckle assembly and force the knuckle from the axle shaft.
12. Disengage the transverse link ball joint from the knuckle and remove the knuckle assembly.
13. Should the inner, outer bearings or seals remain on the axle, remove them with a puller.
14. To remove the bearings from the knuckle hub, use a drift and hammer to tap them from the hub recess.

To install:

15. The drift and hammer should be used to install the bearings into the hub recess. If necessary, a press using light pressure can be used to install them.
16. Install the transverse link ball joint onto the knuckle and install the knuckle assembly.
17. Install the knuckle assembly on the axle shaft.
18. Install the hub and disc assembly, disc cover, washer spring, center piece, and castle nut.
19. Install the strut and tie rod end on the knuckle. Install all brake system components.
20. Install the wheels and lower the vehicle. Connect the negative battery cable. Adjust the parking brake.

MANUAL TRANSAXLE

REMOVAL & INSTALLATION

Justy

1. Disconnect the negative battery cable. Remove the air cleaner assembly. Raise and support the vehicle safely.
2. Disconnect the electrical wiring connectors from the starter. Remove the starter to transaxle bolts and the starter from the vehicle.
3. From the transaxle, disconnect the speedometer cable, the back-up light switch connector and the ground cable. If equipped with 4WD, remove the activation hoses from the actuator.
4. Disconnect the electrical connector between the ignition coil and the distributor.
5. Disconnect the clutch cable and the bracket from the transaxle. In place of the clutch cable bracket, install the lifting hook, or equivalent.
6. Removing the pitching stopper and brackets between the transaxle and chassis.
7. Install engine supporter tool 921540000 or equivalent.
8. Install the vertical hoist to T000100 transaxle lifting hook and raise the transaxle slightly.
9. From under the vehicle, remove the under covers.
10. Disconnect the rear exhaust pipe from the front exhaust pipe and the vehicle.
11. Remove the center crossmember to engine/transaxle assembly bolts.
12. Using a pin punch and a hammer, drive out the axle shaft to driveshaft spring pin. Discard the spring pin and separate the axle shaft.
13. Remove the transaxle mounting bracket.
14. Disconnect the gearshift rod and stay from the transaxle.
15. Properly support the engine assembly. Remove the transaxle to engine bolts.
16. Using the vertical hoist, lift the transaxle from the vehicle.

To install:

17. Install the transaxle assembly in the vehicle and install the transaxle-to-engine bolts. Install the gearshift rods on the transaxle.
18. Join the axle shaft and the differential. Install a new axle shaft spring pin. Install the center crossmember and tighten bolts to 27–49 ft. lbs. (37–67 Nm).
19. Install the rear exhaust pipe, engine under covers, pitching stopper and brackets, clutch cable, electrical connectors, speedometer cable, 4WD activation hoses, starter wires and starter-to-transaxle bolts.
20. Lower the vehicle, connect the negative battery cable, check the transaxle fluid and test drive the vehicle.

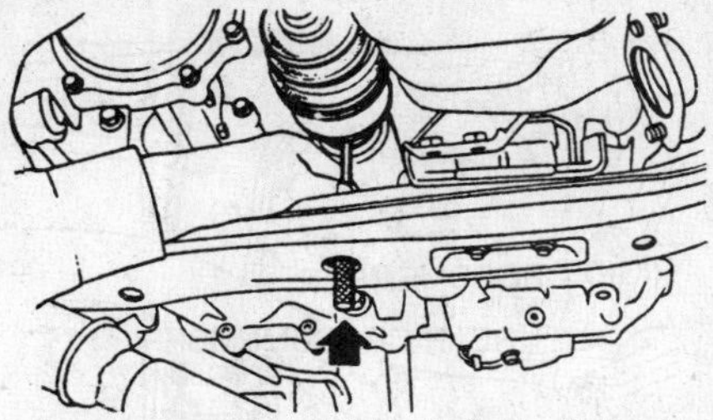

Removing the spring pin from the axle shaft – Justy

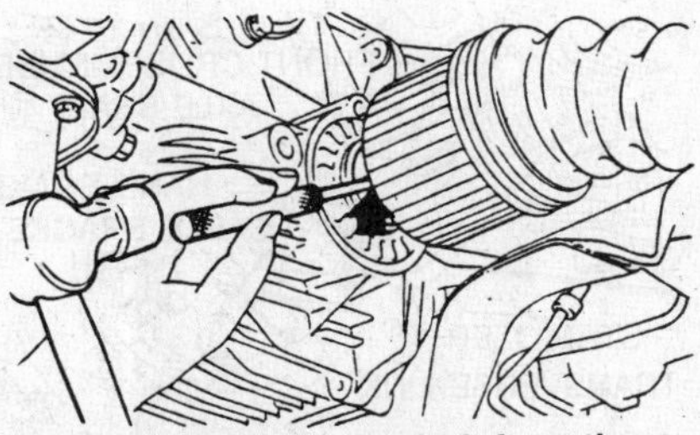

Separating the axle shaft from the driveshaft

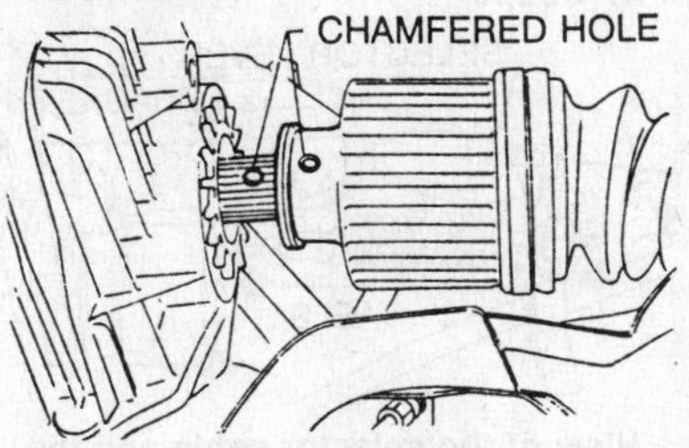

Aligning the chamfered holes of the axle shaft with the driveshaft

Legacy

1. Disconnect the negative battery cable. If equipped with a turbocharger, remove the manifold cover.
2. Remove the air intake duct. If equipped with a turbocharger, discharge the air conditioning and disconnect the air conditioner pressure hose. Remove the resonator chamber, air inlet and outlet ducts.
3. Disconnect all cables and harness connectors attached to the transaxle. Remove the starter.
4. Remove the pitching stopper rod and bracket. On turbocharged models, remove the clutch operating cylinder assembly and free the release fork. Remove the transaxle oil level gauge.
5. Remove the connector holder bracket. Remove the turbocharger cooling ducts and disconnect the center exhaust pipe from the turbocharger.
6. Remove the upper transaxle attaching bolts. Remove the driveshaft, gearshift system, front stabilizer and halfshafts.
7. Remove the nuts holding the lower side fo the engine to the transaxle. Install a transaxle jack. Remove the rear cushion rubber mounting nuts and rear crossmember. Remove the transaxle.
8. On turbocharged models, remove the release bearing from the clutch cover.

To install:

9. Install the transaxle and temporarily tighten bolts. Install the clutch release assembly on Turbo models.
10. Install the rear cushion and crossmember. Tighten cushion bolts to 20–35 ft. lbs. (27–47 Nm); crossmember front bolts to 87–116 ft. lbs. (118–157 Nm), rear 40–61 ft. lbs. (54–83 Nm). Tighten the transaxle to engine bolts to 34–40 ft. lbs. (46–54 Nm).
11. After tightening all bolts check that the release fork is in the proper position. Install the halfshafts, and temporarily install the transverse link and stabilizer.
12. Install the gear shift system and driveshaft. Lower the vehicle and tighten the transverse link to 43–51 ft. lbs. (59–69 Nm); stabilizer to 14–2 ft. lbs. (20–29 Nm).
13. Install the connector holder bracket, pitching stopper, tubrocharger cooling duct, clutch operating cylinder, air conditioner hoses, resonator, air inlet and outlet.
14. Install the starter assembly. Connect all previously disconnected harnesses and connectors.
15. Connect the negative battery cable, check the transaxle fluid level and test drive the vehicle.

Except Legacy and Justy

1.8L AND 2.7L ENGINES

1. Disconnect the negative battery cable. Remove the air cleaner assembly.
2. Remove the clutch cable and the hill holder cable. Remove the speedometer cable.
3. Remove the oxygen sensor electrical connector and the neutral switch connector.
4. If equipped with 4WD, remove the disconnect the electrical connections at the back up light and differential lock indicator switch assembly. Disconnect the differential lock vacuum hose.
5. Disconnect the starter electrical connections. Remove the starter retaining bolts. Remove the starter from the transaxle case.
6. Remove the air intake boot. Disconnect the pitching stopper rod from its mounting bracket. Remove the

right side engine to transaxle mounting bolt.

7. Install engine support bracket 927160000 and engine support tool 927150000 or their equivalents. Remove the buffer rod from the engine and body side bracket.

NOTE: Before attaching the special engine support tools, connect the adjuster to the buffer rod assembly on the right side of the engine.

8. Raise and support the vehicle safely.

9. Disconnect the exhaust pipes at the exhaust manifold flange. Remove the exhaust system up to the rear exhaust pipe assembly.

10. If equipped with 4WD, matchmark and remove the driveshaft. Remove the complete gear shift assembly.

11. Loosen the upper bolt and nut from the plate that secures the transverse link to the stabilizer. Remove the lower bolt and separate the link from the stabilizer.

12. Remove the right brake cable bracket from the transverse link. Remove the bolt retaining the link to the crossmember on each side.

13. Lower the transverse link. Using tool 398791700 or equivalent, remove the spring pin and separate the axle shaft from the driveshaft on each side of the assembly by pushing the rear of the tire outward.

14. Remove the engine to transaxle mounting bolts. Position the proper transaxle jack under the transaxle assembly.

15. Remove the rear cushion rubber mounting bolts. Remove the rear crossmember assembly.

16. Turn the engine support tool adjuster counterclockwise in order to slightly raise the engine.

17. Move the transaxle jack toward the rear of the vehicle until the mainshaft is withdrawn from the clutch cover.

18. Carefully remove the transaxle assembly from the vehicle.

To install:

19. Carefully raise the transaxle until the mainshaft is aligned with the clutch side. Install the engine to the transaxle and temporarily tight the mounting bolts.

20. Install the rear crossmember rubber cushion and tighten nuts to 20–35 ft. lbs. (27–47 Nm). Install the rear crossmember and tighten front bolts to 65–87 ft. lbs. (88–118 Nm); rear bolts to 27–49 ft. lbs. (37–67 Nm).

21. Tighten the engine to transaxle nuts to 34–40 ft. lbs. (46–54 Nm). Remove the transaxle jack.

22. Install the halfshaft into the differential and spring pin into place. Install the transverse link and stabilizer temporarily to the front crossmember. Install the brake cable bracket. Lower the vehicle and tighten transverse link bolt to 43–51 ft. lbs. (59–69 Nm); stabilizer bolts to 14–22 ft. lbs. (20–29 Nm).

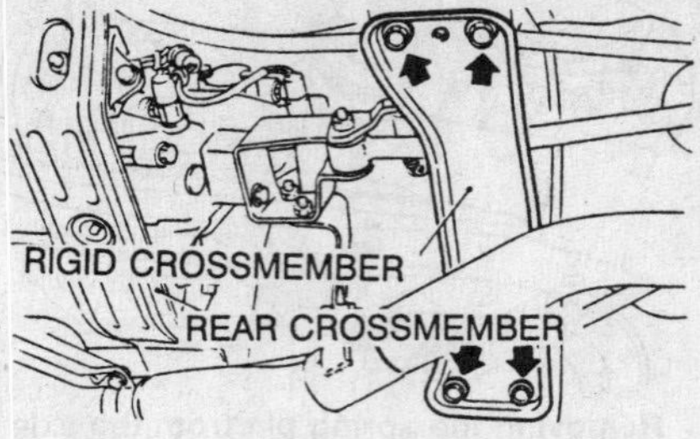

View of the rigid and rear crossmembers —XT Coupe

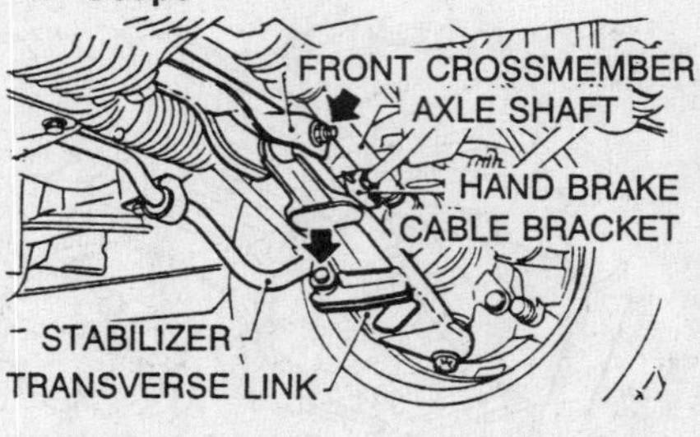

View of the front suspension assembly —XT Coupe

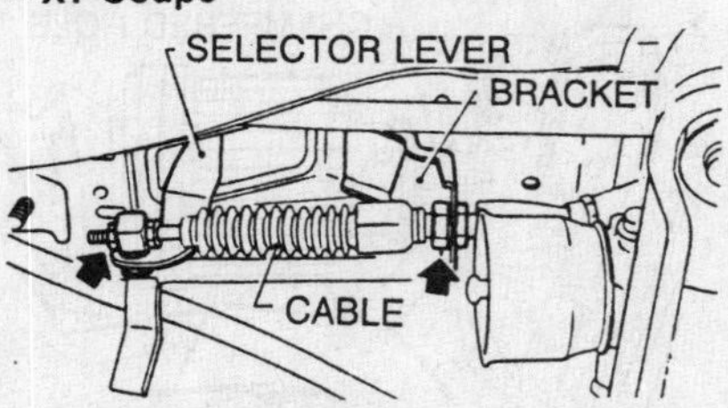

View of the selector cable and the selector cable bracket—XT Coupe

23. Install the gearshift system. Install the driveshaft (4WD vehicles). Install the starter, pitching stopper, timing hole plug, air intake boot and speedometer cable. Reconnect all electrical and vacuum connectors.

24. Connect the clutch cable and hill holder. Install the front exhaust pipe.

25. Connect the negative battery cable, check the transaxle fluid level and test drive the vehicle.

CLUTCH

REMOVAL & INSTALLATION

1. Remove the transaxle from the vehicle.

2. Gradually loosen the pressure plate to flywheel assembly bolts. Loosen the bolts 1 turn at a time, working around the pressure plate.

3. Remove the clutch plate and the disc from the vehicle.

4. Inspect the parts for wear or damage and replace any parts, as necessary.

5. Installation is the reverse of the removal procedure.

6. Use clutch disc guide tool 499747000 or equivalent, to align the clutch on the non-turbocharged 1.8L and 2.7L engines. Use tool 499747100 or equivalent, to align the clutch on the turbocharged 1.8L engine. Use tool 499745500 or equivalent, to align the clutch on the 1.2L engine.

7. When installing the clutch pressure plate assembly, make sure the marks on the flywheel and the clutch pressure plate assembly are at least 120 degrees apart. This is for purposes of balance. Also, make sure the clutch disc is installed properly, noting the **FRONT** and **REAR** markings.

FREE-PLAY ADJUSTMENT

1. Remove the clutch release fork return spring.

2. Loosen the cable locknut, then adjust the spherical nut so there is the following play between the spherical nut and the release fork seat.

1.8L and 2.7L engines, 2WD except turbocharger—0.08–0.12 in.

2WD/4WD turbocharged, 1.8L engine and the 4WD 2.7L engine—0.12–0.16 in.

1.2L engine—0.08–0.16 in.

3. Tighten the locknut and reconnect the release spring.

Clutch Cable

REMOVAL & INSTALLATION

The clutch cable is connected to the clutch pedal at 1 end and to the clutch release lever at the other end. The cable conduit is retained by a bolt and clamp on a bracket mounted on the flywheel housing.

1. If necessary, raise and support the vehicle safely.

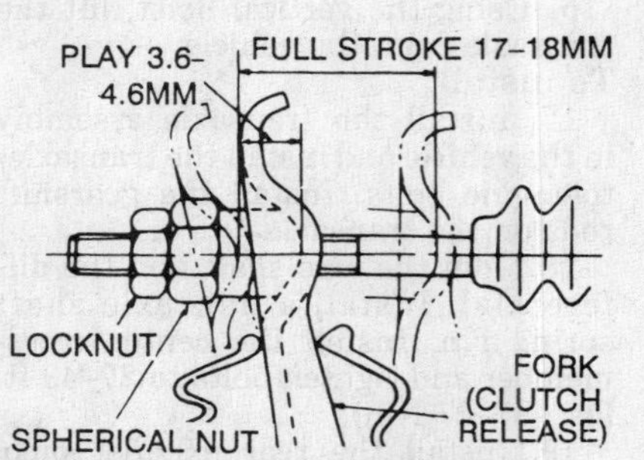

Clutch linkage free-play adjustment at the release fork

2. Disconnect both ends of the cable and the conduit, then remove the assembly from under the vehicle.
3. Using engine oil, lubricate the clutch cable. If the cable is defective, replace it.
4. Installation is the reverse the removal procedure.

CABLE ADJUSTMENT

The clutch cable can be adjusted at the cable bracket where the cable is attached to the side of the transaxle housing.

1. Remove the circlip and clamp.
2. Slide the cable end in the direction desired and then replace the circlip and clamp into the nearest gutters on the cable end.

NOTE: The cable should not be stretched out straight nor should it have right angle kinks in it. Any straightening should be gradual.

3. Check the clutch for proper operation.

AUTOMATIC TRANSAXLE

Transaxle

REMOVAL & INSTALLATION

Justy

WITH ECVT TRANSAXLE

NOTE: When removing and installing ECVT transaxle, always remove and install the engine and transaxle as an assembly.

1. Disconnect the negative battery cable. Drain the coolant by removing drain plug from radiator.
2. Remove the grille. Disconnect hoses and electric wiring from radiator and remove the radiator.
3. Remove front hood release cable and remove radiator upper support member. Disconnect horn and remove the air cleaner assembly.
4. Disconnect the following hoses and cables:
   a. Hoses from carburetor
   b. Hoses from the heater unit
   c. Hose for brake booster
   d. Clutch cable
   e. Accelerator cable
   f. Choke cable from carburetor, if equipped
   g. Speedometer cable
   h. Distributor wiring
5. Disconnect selector cable. Set selector lever at **N** position. Remove clip and detach selector cable from bracket. Remove snap pin, clevis pin and separate selector cable from transaxle.
6. Remove the pitching stopper from the bracket.
7. Disconnect the starter cable, engine wiring harness connectors, ground lead terminals and brush holder harness connector.
8. Remove the hanger from the rear of transaxle.
9. Remove under covers and remove the exhaust system.
10. Remove the driveshaft from transaxle.
11. Remove transverse link.
12. Remove the spring pin retaining the axle shaft by using a suitable tool and separate front axle shaft from the transaxle.
13. Remove engine and transaxle mounting brackets.
14. Raise the engine and remove center member and crossmember.
15. Lift up the engine/transaxle assembly carefully and remove it from the vehicle.

To install:

16. Position the engine/transaxle assembly in the vehicle. Install engine and transaxle mounting brackets.
17. Install center member and crossmember.
18. Install the axle shaft to transaxle with new spring pin.
19. Install gearshift rod and stay to transaxle.
20. Install the exhaust system. Connect driveshaft to transaxle.
21. Install transverse link and under covers to the vehicle.
22. Reconnect the pitching stopper to bracket.
23. Reconnect the following hoses and cables:
   a. Hoses to carburetor
   b. Hoses to the heater unit
   c. Hose to brake booster
   d. Clutch cable to transaxle
   e. Accelerator cable
   f. Choke cable from carburetor, if equipped
   g. Speedometer cable
   h. Distributor wiring
24. Reconnect the starter cable, engine wiring harness connectors, ground lead terminals and brush holder harness connector. Install the air cleaner assembly.
25. Install radiator upper member and connect hood release cable to lock assembly. Reconnect the horn.
26. Install the radiator and connect hoses and electric wiring. Attach grille to the vehicle.
27. Refill the coolant. Reconnect the battery cable.
28. Check all fluid levels. Road test vehicles for proper operation in all driving ranges.

XT Coupe

1.8L AND 2.7L ENGINES

1. Disconnect the negative battery cable. Remove the air cleaner assembly.
2. Remove the clutch cable and the hill holder cable. Remove the speedometer cable. Remove the torque converter from the drive plate.
3. Remove the oxygen sensor electrical connector and the neutral switch connector.
4. If equipped with 4WD, remove the disconnect the electrical connections at the back up light and differential lock indicator switch assembly. Disconnect the differential lock vacuum hose.
5. Disconnect the starter electrical connections. Remove the starter retaining bolts. Remove the starter from the transaxle case.
6. Remove the air intake boot. Disconnect the pitching stopper rod from its mounting bracket. Remove the right side engine to transaxle mounting bolt.
7. Install engine support bracket 927160000 and engine support tool 927150000 or their equivalents. Remove the buffer rod from the engine and body side bracket.

NOTE: Before attaching the special engine support tools, connect the adjuster to the buffer rod assembly on the right side of the engine.

8. Raise and support the vehicle safely.
9. Disconnect the exhaust pipes at the exhaust manifold flange. Remove the exhaust system up to the rear exhaust pipe assembly.
10. If equipped with 4WD, matchmark and remove the driveshaft. Remove the complete gear shift assembly.
11. Loosen the upper bolt and nut from the plate that secures the transverse link to the stabilizer. Remove the lower bolt and separate the link from the stabilizer.
12. Remove the right brake cable bracket from the transverse link. Remove the bolt retaining the link to the crossmember on each side.
13. Lower the transverse link. Using tool 398791700 or equivalent, remove the spring pin and separate the axle shaft from the driveshaft on each side of the assembly by pushing the rear of the tire outward.
14. Remove the engine to transaxle mounting bolts. Position the proper transaxle jack under the transaxle assembly. Disconnect the transaxle cooler lines.
15. Remove the rear cushion rubber

mounting bolts. Remove the rear crossmember assembly.

16. Turn the engine support tool adjuster counterclockwise in order to slightly raise the engine.

17. Move the transaxle jack toward the rear of the vehicle until the mainshaft is withdrawn from the clutch cover.

18. Carefully remove the transaxle assembly from the vehicle.

To install:

19. Carefully raise the transaxle until the mainshaft is aligned with the clutch side. Install the engine to the transaxle and temporarily tighten the mounting bolts.

20. Install the rear crossmember rubber cushion and tighten nuts to 20–35 ft. lbs. (27–47 Nm). Install the rear crossmember and tighten front bolts to 65–87 ft. lbs. (88–118 Nm); rear bolts to 27–49 ft. lbs. (37–67 Nm).

21. Tighten the engine-to-transaxle nuts to 34–40 ft. lbs. (46–54 Nm). Remove the transaxle jack.

22. Install the halfshaft into the differential and spring pin into place. Install the transverse link and stabilizer temporarily to the front crossmember. Install the brake cable bracket. Lower the vehicle and tighten transverse link bolt to 43–51 ft. lbs. (59–69 Nm); stabilizer bolts to 14–22 ft. lbs. (20–29 Nm).

23. Install the gearshift system. Install the driveshaft (4WD vehicles). Install the starter, pitching stopper, timing hole plug, air intake boot and speedometer cable. Reconnect all electrical and vacuum connectors.

24. Connect the clutch cable and hill holder. Install the front exhaust pipe. Connect the oil cooler lines. Install and tighten the torque converter mounting bolts to 17–20 ft. lbs. (23–26 Nm).

25. Connect the negative battery cable, check the transaxle fluid level and test drive the vehicle.

1800 Sedan/Station Wagon, Loyale and XT Coupe

2WD AND 4WD NON-ELECTRONIC 3 AND 4 SPEED TRANSAXLES

1. Disconnect the negative battery cable.

2. Remove clamp from spare tire supporter and remove the spare tire.

NOTE: Use care when removing spare tire assembly from the vehicle.

3. Remove spare tire supporter and battery clamp.

4. Remove speedometer cable and retaining clip. Before disconnecting speedometer cable, remove front exhaust pipe on 4 speed automatic transaxle.

5. Disconnect the following electrical harness connections on the 3 speed automatic transaxle:
   a. Oxygen sensor connector
   b. ATF temperature switch connector
   c. Kickdown solenoid valve connector
   d. 4WD solenoid valve connector on 4WD equipped vehicles

6. Disconnect the following electrical harness connections on the 4 speed automatic transaxle:
   a. Oxygen sensor connector
   b. Transaxle harness connector
   c. Inhibitor switch connector
   d. Revolution sensor connector on 4WD equipped vehicles

7. Disconnect the diaphragm vacuum hose on 3 speed automatic transaxle and 4WD vacuum hose on 4WD equipped vehicles.

8. Remove clip band which secures air breather hose to pitching stopper.

9. Remove the pitching stopper rod. Remove the starter.

10. Remove timing hole inspection plug and remove the 4 bolts which hold torque converter to driveplate.

11. Support the engine assembly with special engine support tool 926610000 or equivalent.

12. Remove engine-to-transaxle mounting nut and bolt on the right side.

13. Remove the exhaust system.

NOTE: Apply a penetrating oil or equivalent to all exhaust retaining nuts in advance to facilitate removal.

14. On turbocharged vehicles, remove accelerator cable cover and upper and lower turbocharger covers. Remove the center exhaust pipe at turbocharger location and at rear exhaust pipe. Remove any exhaust brackets or hangers that attach to the transaxle, as necessary.

15. On non-turbocharged vehicles, disconnect front exhaust pipe from the engine and from the rear exhaust pipe. Remove any exhaust brackets or hangers that attach to the transaxle as necessary.

16. Drain all transaxle fluid from the oil pan.

17. Remove the driveshaft on 4WD vehicles. Plug the opening at the rear of extension housing to prevent oil from flowing out.

18. Disconnect the linkage rod for a 3 speed or cable for a 4 speed. from the select lever.

19. Remove stabilizer from transverse link by loosening (not removing) nut and bolt on the lower side of plate.

20. Remove parking brake cable bracket from transverse link and bolt holding transverse link to crossmember on each side. Lower the transverse link.

21. Remove spring pin and separate axle shaft from transaxle on each side.

NOTE: Use a suitable tool to remove spring pin. Discard old spring pin and always install a new pin.

22. Disconnect the axle shaft from transaxle on each side. Be sure to remove axle shaft from transaxle by pushing the rear of tire outward.

23. Remove engine-to-transaxle mounting nuts.

24. Disconnect oil cooler hoses and oil supply pipe. Be careful not to damage the oil supply pipe O-ring.

25. Place transaxle jack or equivalent under transaxle. Always support transaxle case with a transaxle jack.

NOTE: Do not place jack under oil pan otherwise oil pan may be damaged.

26. Remove rear cushion rubber mounting nuts and rear crossmember. Move torque converter and transaxle as a unit away from the engine. Remove the transaxle.

To install:

27. Install transaxle to engine and temporarily tighten engine-to-transaxle mounting nuts.

28. Install rear crossmember to rear cushion rubber mounts. Align rear cushion guide with rear crossmember guide hole and tighten nuts.

29. Install rear crossmember to chassis. Be careful not to damage threads. Torque rear crossmember bolts to 39–49 ft. lbs.

30. Tighten engine to transaxle nuts on the lower side to 34–40 ft. lbs. Remove transaxle jack from the vehicle.

31. Install axle shaft to transaxle and install spring pin into place.

NOTE: Always use new spring pin. Be sure to align the halfshaft and shaft from the transaxle at chamfered holes and engage shaft splines correctly.

32. Install transverse link temporarily to front crossmember by using bolt and self-locking nut. Do not complete final torque at this point.

33. Install stabilizer temporarily to transverse link. Install parking brake cable bracket to transverse link.

34. Connect the linkage rod for a 3 speed or cable for a 4 speed to the select lever. Make sure the lever operates smoothly all across the operating range.

35. Install propeller shaft on 4WD vehicles. Torque propeller shaft to rear differential retaining bolts to 13–20 ft. lbs. and center bearing location retaining bolts to 25–33 ft. lbs.

36. Connect oil cooler hoses and oil supply pipe. Lower vehicle to floor.
37. Tighten transverse link to front crossmember mounting bolts and transverse link to stabilizer mounting bolts with the tires placed on the ground when the vehicle is not loaded. Tightening torque for transverse link to front crossmember (self-locking nuts) 43–51 ft. lbs. and transverse link to stabilizer 14–22 ft. lbs.
38. Tighten engine to transaxle nuts on the upper side to 34–40 ft. lbs.
39. Raise vehicle and safely support. Install exhaust system.

NOTE: Before installing exhaust system, connect speedometer cable on 4 speed vehicles.

40. On turbocharged vehicles, install the center exhaust pipe at turbocharger location and at rear exhaust pipe. Install any exhaust brackets or hangers that attach to the transaxle as necessary. Install upper and lower turbocharger covers and accelerator cable cover.
41. On non-turbocharged vehicles, connect front exhaust pipe to the engine and rear exhaust pipe. Install any exhaust brackets or hangers that attach to the transaxle as necessary.
42. Remove the special engine support tool. Install and tighten torque converter to driveplate mounting bolts to 17–20 ft. lbs.
43. Install timing hole inspection plug.
44. Install starter.
45. Install pitching stopper. Be sure to tighten the bolt for the body side first and then the 1 for engine or transaxle side. Tightening torque for chassis side is 27–49 ft. lbs. and for engine or transaxle side is 33–40 ft. lbs.
46. Reconnect the following electrical harness connections on the 3 speed automatic transaxle:
a. Oxygen sensor connector
b. ATF temperature switch connector
c. Kickdown solenoid valve connector
d. 4WD solenoid valve connector on 4WD equipped vehicles
47. Reconnect the following electrical harness connections on the 4 speed automatic transaxle:
a. Oxygen sensor connector
b. Transaxle harness connector
c. Inhibitor switch connector
d. Revolution sensor connector on 4WD equipped vehicles
48. Reconnect the diaphragm vacuum hose on 3 speed automatic transaxle and 4WD vacuum hose on 4WD equipped vehicles.
49. Secure air breather hose to pitching stopper with a clip band.
50. Reconnect the speedometer cable. Manually tighten cable nut all the way and then turn it approximately 30 degrees more with a tool.
51. Connect the battery ground cable. Refill and check transaxle oil level.
52. Install spare tire supporter and battery clamp. Install spare tire.
53. Road test vehicle for proper operation across all operating ranges.

XT Coupe and 1990 Legacy

4 SPEED ELECTRONIC TRANSAXLE

1. Disconnect the negative battery cable.
2. Remove speedometer cable or electronic wiring connector from speed sensor.
3. Disconnect the following electrical harness connections on the automatic transaxle:
a. Oxygen sensor connector
b. Transaxle harness connector
c. Inhibitor switch connector
d. Revolution sensor connector on 4WD equipped vehicles
e. Crankshaft and camshaft angle sensor connector on Legacy vehicles
f. Knock sensor connectors and transaxle ground terminal on Legacy vehicles
4. Remove clip band which secures air breather hose to pitching stopper.
5. Remove the starter and air intake boot.
6. Remove timing hole inspection plug and remove the 4 bolts which hold torque converter to driveplate.
7. Disconnect pitching stopper rod from bracket.
8. Remove engine to transaxle mounting nut and bolt on the right side.
9. Remove the buffer rod from the vehicle. Support the engine assembly with special engine support tool or equivalent.
10. Remove the exhaust system. Remove exhaust brackets or hangers that attach to the transaxle, as necessary.
11. Matchmark and remove the driveshaft on 4WD vehicles. Plug the opening at the rear of extension housing to prevent oil from flowing out.
12. Disconnect the gear shift cable from the transaxle select lever.
13. Remove stabilizer from transverse link.
14. Remove parking brake cable bracket from transverse link and bolt holding transverse link to crossmember on each side. Lower the transverse link.
15. Remove spring pin and separate halfshaft from transaxle on each side.

NOTE: Use a suitable tool to remove spring pin. Discard old spring pin and always install a new pin.

16. Disconnect the halfshaft from transaxle on each side. Be sure to remove axle shaft from transaxle by pushing the rear of tire outward.
17. Remove engine to transaxle mounting nuts.
18. Disconnect oil cooler hoses.
19. Place transaxle jack or equivalent, under transaxle. Always support transaxle case with a transmission jack.

NOTE: Do not place jack under oil pan otherwise oil pan may be damaged.

20. Remove rear cushion rubber mounting nuts and rear crossmember.
21. Move torque converter and transaxle as a unit away from the engine. Remove the transaxle.

To Install:

22. Install transaxle to engine and temporarily tighten engine to transaxle mounting nuts.
23. Install rear crossmember to rear cushion rubber mounts. Align rear cushion guide with rear crossmember guide hole and tighten nuts.
24. Install rear crossmember to chassis; be careful not to damage threads. Torque rear crossmember bolts to 39–49 ft. lbs.
25. Tighten engine to transaxle retaining nuts to 34–40 ft. lbs. Remove transaxle jack from the vehicle.
26. Remove the engine support tool and install buffer rod.
27. Install axle shaft to transaxle and install spring pin into place.

NOTE: Always use new spring pin. Be sure to align the axle shaft and shaft from the transaxle at chamfered holes and install shaft splines correctly.

28. Install transverse link temporarily to front crossmember by using bolt and self locking nut. Do not complete final torque at this point.
29. Install stabilizer temporarily to transverse link. Install parking brake cable bracket to transverse link.
30. Lower vehicle to floor. Tighten transverse link to front crossmember mounting bolts and transverse link to stabilizer mounting bolts with the tires placed on the ground when the vehicle is not loaded. Tightening torque for transverse link to front crossmember (self locking nuts) 43–51 ft. lbs. and transverse link to stabilizer 14–22 ft. lbs.
31. Raise and safely support the vehicle. Reconnect the gear shift cable to the select lever. Make sure the lever operates smoothly all across the operating range.
32. Install propeller shaft on 4WD vehicles. Torque propeller shaft-to-rear differential retaining bolts to

17–24 ft. lbs. and center bearing location retaining bolts to 25–33 ft. lbs.

33. Connect oil cooler hoses.
34. Tighten engine to transaxle bolts to 34–40 ft. lbs.
35. Install starter.
36. Install pitching stopper. Be sure to tighten the bolt for the body side first and then the 1 for engine or transaxle side. Tightening torque for chassis side is 27–49 ft. lbs. and for engine or transaxle side is 33–40 ft. lbs.
37. Install and tighten torque converter-to-driveplate mounting bolts to 17–20 ft. lbs.
38. Install timing hole inspection plug, air intake boot and air breather hose to pitching stopper.
39. Reconnect the following electrical harness connections on the automatic transaxle:
    a. Oxygen sensor connector
    b. Transaxle harness connector
    c. Inhibitor switch connector
    d. Revolution sensor connector on 4WD equipped vehicles
    e. Crankshaft and camshaft angle sensor connector on Legacy
    f. Knock sensor connectors and transaxle ground terminal on Legacy
40. Reconnect the speedometer cable. Manually tighten cable nut all the way and then turn it approximately 30 degrees more with a tool.
41. Install exhaust system and exhaust brackets or hangers that attach to the transaxle, as necessary.
42. Connect the battery ground cable. Refill and check transaxle oil level.
43. Road test vehicle for proper operation across all operating ranges.

FRONT SUSPENSION

MacPherson Strut

REMOVAL & INSTALLATION

Justy

1. Disconnect the negative battery cable. Remove the bolts that retain the strut assembly to the body.
2. Raise and support the vehicle safely. Remove the tire and wheel assembly.
3. Remove the brake hose from the brake hose bracket on the strut assembly. Remove the retaining bolt that retains the brake hose bracket to the strut.
4. Properly support the hub and disc assembly. Remove the retaining bolt from the strut to the housing.
5. Fit the proper tool into the housing slit and pull the strut assembly from the housing.
6. Remove the strut from the vehicle.

To install:

7. Install the strut and tighten the upper attaching nuts to 29–43 ft. lbs. (39–59 Nm); lower attaching bolts to 25–40 ft. lbs. (34–54 Nm).
8. Install the brake hose and bracket assembly. Install the wheel and lower the vehicle.

Except Justy

1. Disconnect the negative battery cable. If equipped with air suspension, remove the cover and the air line assembly.
2. Remove the bolts that retain the strut assembly to the body.
3. Raise and support the vehicle safely. Remove the tire and wheel assembly.
4. Disconnect the brake hose from the caliper body. Pull the brake hose retaining clip and remove the brake hose from the damper strut bracket.
5. Remove the bolt that retains the damper strut to the housing. Remove the bolt that retains the damper strut bracket to the housing.
6. Pull the strut assembly from the housing gradually and carefully, with the housing assembly in the downward position.
7. Remove the strut assembly from the vehicle.

To install:

8. Install the strut assembly on the vehicle. Tighten the strut attaching bolts to 28–37 ft. lbs. (38–50 Nm).
9. Install the brake hose on the caliper and bleed the brake system. If equipped with air suspension, install the air line assembly.
10. Install the wheel and tire. Lower the vehicle, connect the negative battery cable and test drive the vehicle.

Ball Joints

INSPECTION

1. Raise and support the vehicle safely.
2. Using a prybar, position it under the wheel, then pry upward on the wheel several times. If more than 0.012 in. (3mm) of movement is noticed at the ball joint it should be replaced.
3. Inspect the dust seal, if damaged it should be replaced.

REMOVAL & INSTALLATION

1. Raise and support the vehicle safely. Remove the tire and wheel assembly.
2. Properly support the lower control arm assembly. Remove the cotter pin and castle nut from the ball joint.
3. Disconnect the ball joint from the lower control arm assembly.
4. Remove the bolt retaining the ball joint to the housing. Remove the ball joint from the housing.

To install:

5. Install the ball joint into the housing and tighten the nut to 28–37 ft. lbs. (38–50 Nm).
6. Connect ball joint to the transverse link and tighten the castle nut to 29 ft. lbs. (39 Nm). Install the cotter pin in the castle nut.
7. Install the front wheels and lower the vehicle.

Lower Control Arm

REMOVAL & INSTALLATION

Justy

1. Raise and support the vehicle safely. Remove the tire and wheel assembly.
2. Properly support the lower control arm. Remove the brake hoses as necessary. Remove the bolt that retains the lower control arm to the crossmember.
3. Remove the ball joint retaining bolt and the stabilizer tension rods. Remove the ball joint, if required.
4. Remove the lower control arm from the vehicle.

To install:

5. Install the ball joint, if removed. Install the lower control arm. Tighten the crossmember-to-control arm bolt to 43–58 ft. lbs. (59–78 Nm) only after the vehicle is on the ground with the chassis loaded.
6. Install the castle nut on the ball joint and tighten to 29 ft. lbs. (39 Nm).
7. Temporarily install the tension rod-to-control arm bolt. Then, install the tension rod-to-bracket bolt and tighten to 40–54 ft. lbs. (54–74 Nm). Now, tighten the tension rod-to-control arm bolt to 54–69 ft. lbs. (74–93 Nm).

NOTE: It is very important that the tension rod bolts be tightened in the order given in the text. If the tightening sequence is reversed, the tension rod will interfere with the bracket causing unusual noise.

8. Install the wheels. If components were replaced, have the alignment checked.

Except Justy

1. Raise and support the vehicle safely. Remove the tire and wheel assembly.

2. As required, remove the parking brake cable from the lower control arm assembly.
3. Remove the bolt that retains the stabilizer assembly to the lower control arm.
4. Remove the front exhaust pipe, as necessary to gain working clearance.
5. Properly support the lower control arm assembly. Remove the ball joint from its mounting.
6. Remove the lower control arm to crossmember retaining bolt. Remove the lower control arm from the vehicle.

To install:

7. Install the lower control arm. Tighten the retaining bolt to 43–51 ft. lbs. (59–69 Nm) only after the vehicle is on the ground with the chassis loaded.
8. Install the ball joint and tighten the castle nut to 18–22 ft. lbs. (25–29 Nm). Install any exhaust system components previously removed.
9. Install the stabilizer assembly and tighten the bolts to 14–22 ft. lbs. (20–29 Nm). Install the parking brake.
10. Install the wheels, lower the vehicle and check the alignment.

REAR SUSPENSION

Shock Absorbers

REMOVAL & INSTALLATION

1. Raise and support the vehicle safely. Remove the tire and wheel assembly.
2. Properly support the rear axle assembly. Loosen the upper shock absorber to chassis nuts.
3. Remove the washer and the bushing, being sure to note their correct assembly sequence for installation.
4. Remove the shock absorber to trailing arm retaining bolt. Remove the shock absorber from its mounting.

To install:

5. Install the shock absorber and tighten the bolts to specification. Be sure to properly install the washers.
6. Do not fully tighten the upper mounting nuts until the lower shock nut has been installed with the washer and the pin shoulder contracting each other.
7. Install the wheels and lower the vehicle.

MacPherson Strut

REMOVAL & INSTALLATION

Justy

1. Raise and support the vehicle safely. Remove the tire and wheel assembly. Properly support the rear axle assembly.
2. From the upper portion of the strut mount, remove the trim cover.
3. Remove the strut to body retaining nut. Push the lower arm downward, and remove the coil spring.
4. Remove the strut to axle housing bolts. Remove the strut from the vehicle.

To install:

5. When installing coil spring, fit the lower rubber seat and coil spring end face in the coil spring seat mounting recess of the lower control arm.
6. Install the strut-to-housing bolt and tighten to 25–40 ft. lbs.
6. Install the wheels and remove the supports from under the rear axle. Lower the vehicle and test drive.

Except Justy

1. Raise and support the vehicle safely. Remove the tire and wheel assembly.
2. If equipped with air suspension, remove the cover and disconnect the air line.
3. Properly support the rear axle assembly. Remove the upper strut retaining bolts.
4. Remove the lower strut retaining bolts.
5. Remove the strut assembly from the vehicle.

To install:

6. Install the strut assembly and tighten the attaching bolts to 22–29 ft. lbs.
7. If equipped with air suspension, reconnect the air line. Install the wheel and tire assembly and remove the supports under the rear axle.
8. Lower the vehicle and test drive.

Springs

REMOVAL & INSTALLATION

Justy

1. Raise and support the vehicle safely. Remove the tire and wheel assembly. Properly support the rear axle assembly.
2. Remove the strut bolt trim cover and remove the strut upper bolts.

NOTE: The rear spring is under extreme tenison. Serious injury can result if the spring should fly out of the vehicle.

3. Place a floorjack under the control arm to prevent the spring from expanding. Remove the rear spindle to control arm bolt. Slowly lower the control arm until all spring pressure is released. Push the control arm downward, and remove the coil spring.

To install:

4. Place the spring in the holder cups. Ensure that the spring insulators are installed. Using a floor jack, lift up on the lower control arm to compress the spring. Install the control arm bolts and tighten to 54–69 ft. lbs. (74–93 Nm).
5. Install the rear strut and tighten the lower bolts to 72–87 ft. lbs. (98–118 Nm) and the upper nuts to 40–54 ft. lbs. (54–74 Nm).
6. Remove the rear axle supports and lower the vehicle.

Rear Control Arms

REMOVAL & INSTALLATION

Justy

1. Raise and support the vehicle safely. Remove the tire and wheel assembly.
2. Properly support the rear axle assembly. Remove the coil spring assembly.
3. Remove the control arm to crossmember bolt. Separate the control arm from the crossmember.
4. Remove the control arm to axle housing bolt Separate the control arm from the axle housing.
5. Remove the assembly from the vehicle.

To install:

6. Install the control arm on the axle housing and tighten the bolts to 54–69 ft. lbs. (74–93 Nm).
7. Install the control arm to crossmember bolt and tighten to 43–58 ft. lbs. (59–78 Nm).
8. Install the coil spring. Install the wheels, remove the rear axle supports and lower the vehicle.

Legacy

TRAILING LINK

1. Loosen the rear wheel lugs, raise and safely support the vehicle and remove the wheel assemblies.
2. Remove the rear parking brake clamps and the ABS sensors, as required.
3. Remove the bolts retaining the trailing link to the body.
4. Remove the bolts retaining the trailing link to the rear housing.
5. Remove the trailing link from the vehicle.
6. To install the trailing link, place in position and install the bolts at each end.

7. Torque the bolts to 72–94 ft. lbs.
8. Complete the assembly.

LATERAL LINK

1. Remove the stabilizer from the lateral link.
2. Remove the parking brake cable and the ABS sensor clamp from the trailing link, as required.
3. Loosen the bolts that secure the trailing link to the bracket and remove the bolts that retain the trailing link to the rear housing.
4. If equipped with 4WD, remove the Double Offset Joint (DOJ) pin and axle shaft to provide working space.
5. Remove the front lateral link from the rear crossmember.
6. Temporarily install front lateral link to the rear crossmember and remove the rear lateral link from the crossmember.
7. To install the link, reverse the removal procedure. Torque the bolts to the following specifications:
   a. 4WD – 61–83 ft. lbs.
   b. FWD – 87–116 ft. lbs.

Except Justy and Legacy

1. Raise and support the vehicle safely. Remove the tire and wheel assembly.
2. Properly support the rear axle assembly.
3. Remove the strut to lower control arm bolt and separate the strut from the lower control arm.
4. If equipped with 4WD, use a 0.24 in. (6mm) pin punch and drive the spring pins from the halfshaft-to-axle shaft and the halfshaft to differential assembly. While pushing downward on the inner arm, separate the halfshaft from the axle shaft. Pull the halfshaft from the differential and position it aside.
5. Disconnect and plug the brake hose from the brake line at the lower control arm.
6. Remove the outer arm-to-lower control arm bolts, then separate the lower control arm from the outer arm. Properly support the inner arm.
7. Remove the inner arm-to-crossmember bolt. Remove the lower control arm from the vehicle.

To install:

8. Install the inner arm and tighten the inner arm to crossmember bolt to 51–65 ft. lbs. (69–88 Nm). Install the outer arm and tighten the attachign bolts to 94–108 ft. lbs. (127–147 Nm).
9. Install the brake hose and line. If equipped with 4WD, reassemble the halfshaft to differential assembly using new spring pins.
10. Install the strut, remove the rear axle supports and lower the vehicle. As required, bleed the brake system.

Rear Wheel Bearings

ADJUSTMENT

2WD

1. Raise and support the vehicle safely. Remove the rear wheel assembly.
2. Temporarily tighten the axle nut to 36 ft. lbs. on all vehicles except Justy or to 29 ft. lbs. for Justy.
3. Turn the drum or disc back and forth several times to ensure that bearings are properly seated.
4. Turn the nut backwards $\frac{1}{8}$–$\frac{1}{10}$ turn in order to obtain the correct starting force.
5. Using a spring gauge at 90 degrees to the wheel lug, check the rotating force. Specifications should be 1.9–3.2 lbs. for all vehicles except Justy or for Justy are 3.1–4.4 lbs.
6. After the adjustment is completed, bend the lock washer. After installing a new O-ring to the grease cap, install the cap.

REMOVAL & INSTALLATION

1. Raise and support the vehicle safely. Remove the rear tire and wheel assembly.
2. If equipped with rear disc brakes, remove the caliper and properly support it.
3. Using a small prybar, remove the rear wheel grease cap.
4. Using a hammer and a punch, flatten the lock washer and loosen the axle nut. Remove the lock washer and the thrust plate. When removing the drum or disc, be careful not to drop the inner race from the outer bearing.

NOTE: If the brake drum on the Justy is difficult to remove, use wheel puller tool 9224930000 or equivalent, to remove the brake drum.

5. Using a gear puller, remove the spacer and the inner race of the inner bearing.
6. Using a brass drift and a hammer, drive the outer race of the inner bearing from the drum or disc.
7. Using a brass drift and a hammer, drive the outer race of the outer bearing from the drum or disc.

To install:

8. Clean and inspect the parts for damage, replace defective parts, if necessary.
9. Using bearing installation tool 925220000 or equivalent, for all vehicles except Justy or tool 922111000 or equivalent, for Justy, press the outer race of the inner bearing into the drum or disc until it seats against the shoulder.
10. When pressing the bearing, be sure not to exceed the load to the bearing, so as not to damage it.
11. Apply a small amount of grease to the oil seal lips, then install the oil seal until it is flush with the drum or disc.
12. Using bearing installation tool 921130000 or equivalent, for all vehicles except Justy or tool 922111000 or equivalent, for Justy, press the outer race of the outer bearing into the drum or disc until it seats against the shoulder.
13. Apply approximately $\frac{1}{8}$ oz. of wheel bearing grease to the inner and the outer bearings. Fill the disc or drum hub with 1 oz. of wheel bearing grease.
14. Install a new spacer O-ring, the spacer and the inner race of the inner bearing onto the trailing arm spindle.
15. When installing the spacer, be sure to face the stepped surface toward the bearing. Use a new thrust plate and lock washer.
16. To complete the installation, reverse the removal procedure. Adjust the wheel bearing.

STEERING

Steering Wheel

REMOVAL & INSTALLATION

1. Disconnect the negative battery cable.
2. Disconnect the horn lead from the wiring harness, located beneath the instrument panel. On the XT Coupe, remove the horn pad.

NOTE: If equipped with telescopic steering wheel, remove the telescopic lever assembly.

3. Working behind the steering wheel, remove the steering wheel cover to steering wheel screws. It may be necessary to lower the column from the dash by removing the screws.
4. Lift the crash pad assembly from the front of the wheel.
5. Matchmark the steering wheel and the column for installation.
6. Remove the steering wheel retaining nut. Using a steering wheel puller tool, remove the steering wheel from the column.

To install:

7. Install the steering wheel on the column in the same position as removed. Tighten the center nut to 22–29 ft. lbs. (29–39 Nm).

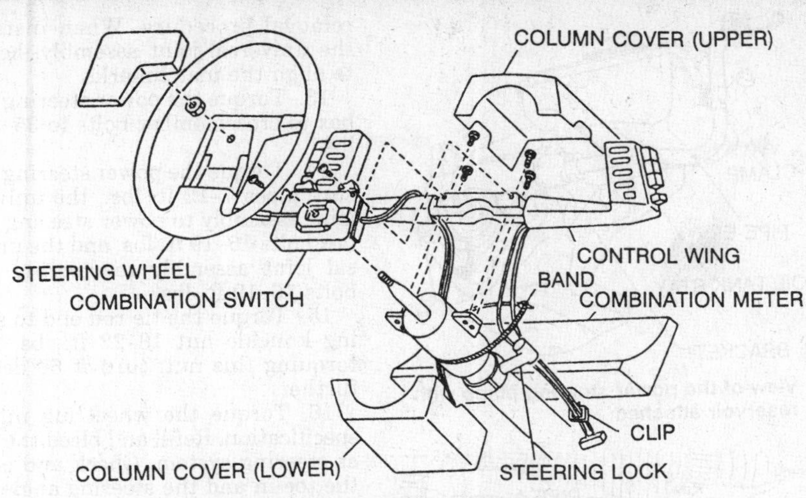

Typical steering wheel and related components

NOTE: Do not hammer on the steering wheel or the steering column, as damage to the collapsible column could result.

8. Install the crash pad assembly and wheel cover.
9. If the column was lowered, tighten the steering column-to-dash screws.
10. Install the telescopic lever, if removed. Connect the horn lead and install the horn pad, if removed.

Steering Column

REMOVAL & INSTALLATION

1. Disconnect the negative battery cable.
2. Remove the universal joint connecting bolts and then remove the universal joint.
3. Disconnect the ignition and combination switch electrical connectors under the dash. Disconnect any other electrical connector attached to the steering column.
4. Remove any other component which impedes the removal of the steering column. On XT Coupe, remove the speedometer cable.
5. Remove the steering shaft attaching bolts under the dash. Remove the steering column from inside the interior.
6. If equipped with telescopic column, set the steering column to its lowest position and install a tilt lock bolt from below the column bracket.

To install:

7. Install the steering column from the interior side and tighten attaching bolts to 14–22 ft. lbs. (20–29 Nm). Remove the telescopic lock bolt.
8. Install all under dash components previously removed and reconnect all electrical connectors.
9. Install the universal joint and tighten to 15–20 ft. lbs. (21–26 Nm). Connect the negative battery cable.
10. Check all operations of the steering column before driving vehicle.

Manual Steering Gear

REMOVAL & INSTALLATION

Justy

1. Disconnect the negative battery cable. Raise and support the vehicle safely. Remove the front tire and wheel assemblies.
2. Disconnect the universal joint coupling bolts. Remove the dust seal.
3. Using the proper tools, disconnect the tie rod ends from the knuckle arms.
4. Remove the steering gear retaining bolts. Lower the assembly and pull the pinion from the dust seal toward the engine compartment.
5. Remove the steering gear from the vehicle.

To install:

6. Installation is the reverse of the removal procedure.
7. Adjust the toe-in and the turning angles to specifications.
8. Tighten the tie rod end to 18–22 ft. lbs.

XT Coupe

1. Be sure the parking brake lever is in the released position. Disconnect the negative battery cable.
2. Raise and support the vehicle safely. Remove the front tire and wheel assemblies.
3. Remove the outer tie rod end cotter pin. Remove the castle nut. Using the proper tool, remove the tie rod end from the steering knuckle.
4. Remove the pinch bolt from the torque rod universal joint.

NOTE: Do not attempt to remove the steering gear assembly or crossmember with the pinch bolt installed to the torque rod universal joint.

5. Loosen the exhaust manifold retaining bolts. Lower the exhaust pipe.
6. Remove the steering gear retaining bolts.
7. Move the assembly toward the pinion. As the pinion shafts comes off the torque rod, rotate the steering gear rearward and remove it from the vehicle, toward the pinion.

To install:

8. Installation is the reverse of the removal procedure.
9. Adjust the toe-in and the turning angles to specifications.
10. Tighten the tie rod end to steering knuckle nuts to 18–22 ft. lbs.

Except Justy and XT Coupe

1. Disconnect the negative battery cable.
2. Raise and support the vehicle safely. Remove the front tire and wheel assemblies.
3. Remove the tie rod end cotter pin and loosen the castle nut. Using a ball joint puller, separate the tie rod ends from the housing knuckle arm.
4. If necessary, disconnect the handbrake cable hanger from the tie rod.
5. Remove the pinch bolt from the torque rod universal joint. Disconnect the pinion with the gearbox from the steering column.
6. If equipped with an hot air pipe, disconnect it.
7. Disconnect the exhaust manifold to engine bolts, pull downward on the exhaust manifold.
8. Remove the boot from the steering gear.
9. Remove the steering gear to crossmember bolts, pull downward on the steering gear to disconnect the pinion flange. Turn the gearbox rearward and remove it toward the left side.
10. When removing the gearbox, be careful not to damage the gearbox boot. Inspect the removed parts for wear or damage and if necessary, replace the parts.

To install:

11. To install, reverse the removal procedures. Torque the steering gearbox to crossmember bolts to 35–52 ft. lbs.
12. Torque the pinch bolt to universal joint to 15–20 ft. lbs.
13. Torque the exhaust manifold to engine bolts to 19–22 ft. lbs.
14. Torque the rubber coupling to steering gear bolts to 10–14.5 ft. lbs.
15. Torque the tie rod end to steering knuckle nut to 18–22 ft. lbs.

16. Adjust the toe-in and the turning angles to specifications.

17. When torquing the tie rod end to steering knuckle nuts, torque the nut 60 degrees turn further, after torquing to specification.

ADJUSTMENT

1. Tighten the backlash adjuster until it bottoms, back off the screw 15 degrees for Justy or 25 degrees for all vehicles except Justy.

2. Torque the locknut to 22–36 ft. lbs. for all XT Coupe and Justy or 36–47 ft. lbs. for all vehicles, except XT Coupe and Justy.

3. A clearance of 0.0025 in. is provided between the screw tip and the sleeve plate for Justy.

4. A clearance of 0.004 in. for all vehicles except Justy, is provided between the screw tip and the sleeve plate.

Power Steering Gear

REMOVAL & INSTALLATION

1. Disconnect the negative battery cable. Remove the spare tire. If equipped with a turbocharger, remove the spare tire support.

2. If necessary, disconnect the thermo-sensor connector.

3. Raise and support the vehicle safely. Remove the front tire and wheel assemblies.

4. Disconnect the electrical connector from the oxygen sensor. Remove the front exhaust pipe assembly. If equipped with an air stove, remove it.

5. Remove the tie rod end cotter pin and loosen the castle nut. Using a ball joint puller, separate the tie rod ends from the steering knuckle arm.

6. As required, remove the jack up plate and the clamp.

7. From the power steering gear, remove the center pressure pipe, connect a vinyl hose to the pipe and joint, then turn the steering wheel to discharge the fluid into a container.

NOTE: When discharging the power steering fluid, turn the steering wheel fully, left and right. Be sure to disconnect the other pipe and drain the fluid in the same manner.

8. Make alignment marks on the steering shaft universal joint assembly to power steering unit and the steering shaft to universal joint assembly. Remove the lower and upper universal joint to shaft bolts. Lift the universal joint assembly upward and secure it aside.

9. From the control valve of the gearbox assembly, remove the power steering C and D pressure pipes. Remove pipe D first and pipe C second.

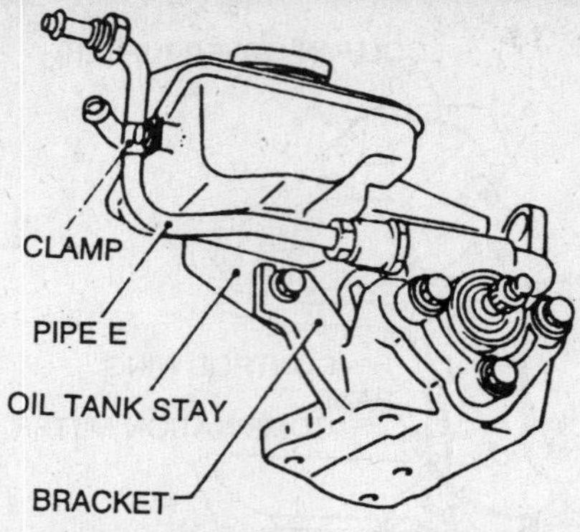

View of the power steering pump with reservoir attached

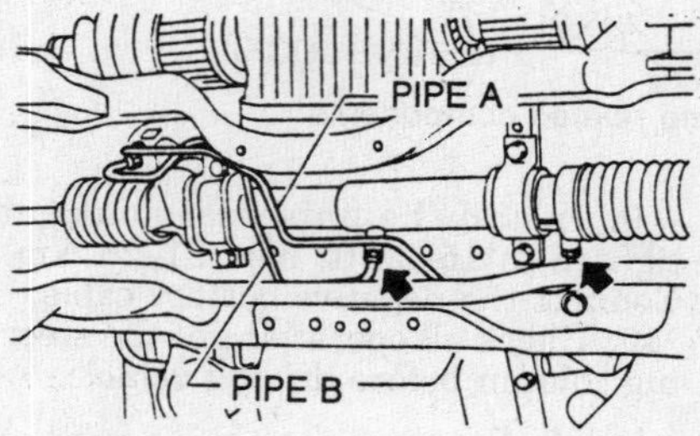

Remove the pressure lines to drain the power steering fluid

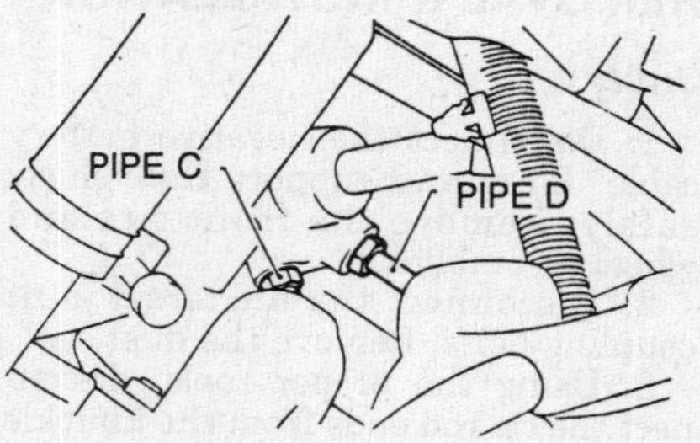

View of the power steering gear pressure lines

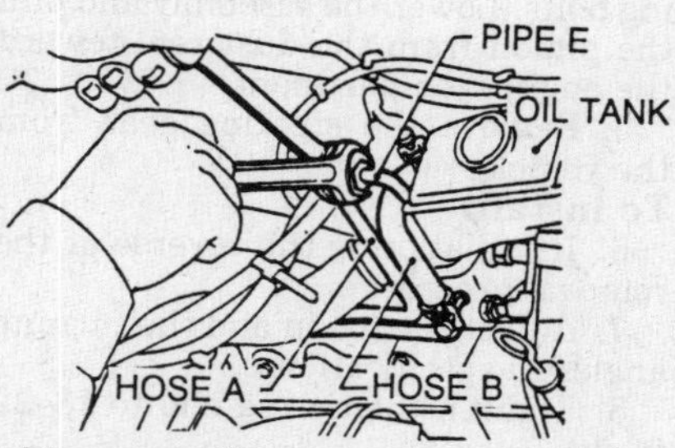

Disconnect the power steering pump hoses from the pressure lines

10. From the control valve of the gearbox assembly, remove the power steering **A** and **B** pressure pipes. Remove pipe **A** first and pipe **B** second.

11. Remove the power steering gearbox to crossmember assembly bolts. Remove the gearbox assembly from the vehicle.

To install:

12. Installation is the reverse of the removal procedure. When installing the universal joint assembly, be sure to align the matchmarks.

13. Torque the power steering gearbox to crossmember bolts to 35–52 ft. lbs.

14. Torque the power steering pressure pipes 7–12 ft. lbs., the universal joint assembly to power steering gearbox bolts 16–19 ft. lbs. and the universal joint assembly to steering shaft bolts 16–19 ft. lbs.

15. Torque the tie rod end to steering knuckle nut 18–22 ft. lbs. After torquing this nut, turn it 60 degrees further.

16. Torque the wheel lug nuts to specification. Refill and bleed the power steering system. Check and adjust the toe-in and the steering angle.

ADJUSTMENT

Tighten the backlash adjuster until it bottoms, back off the screw 30 degrees and torque the locknut to 22–36 ft. lbs., 0.0049 in. should be provided between the screw tip and the sleeve plate.

Power Steering Pump

REMOVAL & INSTALLATION

1. Disconnect the negative battery cable.

2. Using a siphon, drain the power steering fluid from the reservoir.

3. Loosen, but do not remove the power steering pump pulley nut. Loosen the pulley drive belts.

4. Remove the power steering pump pulley nut and the pulley.

5. Disconnect and plug the **A** pressure hose from the **E** pipe. Disconnect the **B** pressure hose from the oil tank.

6. When disconnecting the **A** hose, use wrenches to prevent the **E** pipe from twisting.

7. Remove the **E** hose to reservoir clamp. Loosen the reservoir to bracket bolt, then remove the **A** and **B** bolts on the upper part of the reservoir, this will allow the fluid to run out.

NOTE: To minimize the fluid loss from the reservoir, remove both bolts while the reservoir is pressed against the oil pump, then quickly remove the reservoir. It is a good idea to remove the pump and the reservoir as a unit, then separate the reservoir from the pump on a bench.

8. Remove the power steering pump to bracket bolts. Remove the pump from the vehicle.

To install:

9. Installation is the reverse of the

removal procedure; be sure to use new O-rings.

10. Torque the power steering pump to bracket bolts to 22–36 ft. lbs.

11. Torque the reservoir stay to bracket bolts to 14–17 ft. lbs.

12. Torque the reservoir to pump bolts to 14–22 ft. lbs.

13. Torque the pulley nut to pump nut to 31–46 ft. lbs.

14. Refill the power steering reservoir. Bleed the power steering system.

DRIVE BELT ADJUSTMENT

1. Using a pair of adjustable jawed pliers, with a piece of rag between the jaws, remove the idler cover cap by turning and pulling.

2. Turn the adjusting bolt until the correct belt tension is obtained. If removing the belt, loosen the adjusting bolt until the drive belt can be removed.

NOTE: The correct belt tension is obtained when the belt can be flexed 6–8 mm by applying finger pressure to the midpoint of the longest span.

3. After a new belt is installed and the correct tension obtained, replace the idler cap cover by pushing in and turning.

SYSTEM BLEEDING

1. Be sure the power steering reservoir is filled with fluid. Raise and support the vehicle safely.

2. With the engine running, turn the steering wheel back and forth, from lock to lock, until the air is removed from the fluid.

3. Lower the vehicle, recheck the reservoir fluid level and correct, as required.

Tie Rod Ends

REMOVAL & INSTALLATION

1. Raise and support the vehicle safely.

2. Remove the front tire and wheel assemblies.

3. Remove the cotter pin and castle nut from the tie rod end stud.

4. Using a ball joint puller, separate the tie rod end from the steering knuckle.

To install:

5. Install the tie rod and tighten the castle nut to 18–22 ft. lbs. Install a new cotter pin.

6. Install the front tire and lower the vehicle.

BRAKES

For all brake system repair and service procedures not detailed below, please refer to "Brakes" in the Unit Repair section.

Master Cylinder

REMOVAL & INSTALLATION

1. Disconnect the negative battery cable. Disconnect and plug the brake lines at the master cylinder.

2. It is advised to thoroughly drain the fluid from the master cylinder before performing any removal procedures.

3. If equipped with fluid level indicator, disconnect the electrical harness connector from the master cylinder.

4. Remove the master cylinder to power brake booster retaining nuts. Remove the master cylinder from its mounting.

To install:

5. Bench bleed the master cylinder prior to installtion.

6. Install the master cylinder on the power booster and tighten the nuts to 7–13 ft. lbs. (10–18 Nm).

7. Connect the fluid level indicator. Connect the brake lines and tighten the flarenut to 9–13 ft. lbs. (13–18 Nm).

8. Bleed the brake system as required.

Proportioning Valve

The proportioning valve is attached to a bracket and is located directly under the master cylinder. It's purpose is to provide even braking pressure to all of the wheels.

REMOVAL & INSTALLATION

1. Disconnect the negative battery cable. Disconnect and plug the brake tubes from the proportioning valve. If equipped with an electrical connector, disconnect it.

2. Remove the proportioning valve-to-bracket bolts. Remove the valve from the vehicle.

3. Installation is the reverse of the removal procedure. Tighten the flarenuts to 9–13 ft. lbs. (13–18 Nm); the proportioning valve attaching nuts to 15–21 ft. lbs. (20–29 Nm).

Power Brake Booster

REMOVAL & INSTALLATION

1. Disconnect the negative battery cable. Disconnect the vacuum hose from the power brake booster. If equipped, disconnect the connector for the brake fluid level indicator.

2. Remove the master cylinder from the brake booster. Depending upon the vehicle, it may not be necessary to completely remove the master cylinder. It may be possible to remove the retaining bolts and position the assembly aside.

3. Remove the brake pedal pushrod to power booster spring pin and clevis pin, then disconnect the pushrod from the brake pedal.

4. From under the dash, remove the power booster to firewall bolts.

5. Remove the brake booster assembly from the vehicle.

6. Installation is the reverse of the removal procedure. Tighten the attaching nuts to 9–17 ft. lbs. (13–23 Nm). Bleed the brake system, as required.

Brake Caliper

REMOVAL & INSTALLATION

Front

1. Raise and support the vehicle safely. Remove the front wheels.

2. Remove the brake hose from the caliper body and plug the hose to prevent the entrance of dirt or moisture.

3. Remove the hand brake cable and brake pads. Remove the caliper assembly by pulling it out of the support. Do not remove the guide pin unless it it damaged.

To install:

4. Rotate the piston until the notch at the head of the piston is vertical.

5. Install the hand brake cable and brake pads. Install the caliper assembly on the support and tighten the support bolt to 36–51 ft. lbs. (49–69 Nm).

6. Connect the brake hose and tighten the fitting to 11–15 ft. lbs. (15–21 Nm).

7. Bleed the brake system. Install the wheels and lower the vehicle. Check the fluid level in the master cylinder.

Rear

1. Raise and support the vehicle safely. Remove the rear wheels.

2. Disconnect and plug the brake hose from the caliper body.

3. Remove the bolts securing the caliper to the support and remove the caliper.

To install:

4. Install the caliper and tighten the attaching bolts to 34–43 ft. lbs. (46–58 Nm).

5. Connect the brake hose and

tighten the fitting to 12–14 ft. lbs. (16–20 Nm).

6. Bleed the brake system. Install the rear wheels and lower the vehicle. Check the fluid level in the master cylinder.

Disc Brake Pads

REMOVAL & INSTALLATION

Front

1. Raise and support the vehicle safely. Remove the wheel assemblies.
2. Release the parking brake and disconnect the cable from the caliper lever.
3. Remove the lock pin bolts from the lower front of the caliper.
4. Rotate the caliper on the support, swinging it upward and aside.
5. Remove the brake disc pads, noting the position of the shim pads and pad clips.

To install:

6. Inspect the brake rotor, calipers and retaining components. Correct as necessary.
7. Remove a small portion of brake fluid from the master cylinder reservoir. With an appropriate tool, turn the caliper piston clockwise into the cylinder bore and align the notches. Be sure the boot is not twisted or pinched.

NOTE: Do not force the piston straight into the caliper bore. The piston is mounted on a threaded spindle which will bend under pressure.

8. Install the new pads into the calipers, being sure all shims and clips are in their original positions.
9. Swing the calipers down into position and install the lock pin bolts. Tighten to 25–33 ft. lbs.
10. Reconnect the parking brake cable and fill the master cylinder reservoir.
11. Install the wheel assembly. Bleed the brakes as required and lower the vehicle. Road test the vehicle.

Rear

1. Raise and safely support the vehicle. Remove the wheel assemblies.
2. Disconnect the brake pad lining wear indicator, if equipped. Remove any anti-rattle springs or clips, if equipped.
3. Pull the caliper away from the center of the vehicle to push piston into caliper bore. Remove the caliper guide pins and remove the caliper from the rotor. Hang the caliper from the body with a support wire.
4. Slide the disc pads from the caliper, noting any shims or shields behind the pad.

NOTE: If equipped with parking brake, use a suitable tool to rotate the piston back into the caliper bore. If not equipped with parking brake, the piston can be pushed straight back into the bore.

5. Push the piston into the caliper bore. To install the pads, position any shims or shields in place and reverse the removal procedure.

Brake Shoes

REMOVAL & INSTALLATION

1. Raise and safely support the vehicle. Remove the rear wheels.
2. Remove the brake drums.
3. Remove the adjusting wedge spring and the upper and lower return springs.
4. Remove the hold-down springs.
5. Lift the brake shoes from the backing plate and disconnect the parking brake, if equipped.
6. Disconnect the rear shoe from the push bar.
7. Clamp the push bar in a vise and remove the tension spring and adjusting wedge.

To install:

8. The installation is the reverse of the removal procedure.
9. Center the brake shoes on the backing plate, making sure the adjusting wedge is fully released before installing the drum.
10. Install the drums and the wheel assemblies.
11. Apply the brakes several times to automatically adjust the shoes. If necessary, bleed the brake system to obtain proper brake operation.
12. Road test the vehicle as required.

Wheel Cylinder

REMOVAL & INSTALLATION

1. Raise and support the vehicle safely. Remove the wheel and tire assembly.
2. Remove the brake drum and the brake shoes from the backing plate.
3. Disconnect and plug the brake line at the back of the wheel cylinder.
4. Remove the wheel cylinder to backing plate bolts. Remove the wheel cylinder from the backing plate.
5. Installation is the reverse of the removal procedure. Tighten the wheel cylinder attaching nuts to 6–7 ft. lbs. (8–10 Nm). Be sure to bleed the brake system, as required.

Parking Brake Cable

ADJUSTMENT

1. Pull the parking brake lever up forcefully. Release it and repeat several times.
2. It should take the specified number of notches to apply the parking brake.

Except Justy – 3–5 notches
Justy – 6 notches

3. Loosen the locknut on the turnbuckle and adjust the length of the cable, so the parking brake is applied within specification.
4. Tighten the locknut and recheck operation of the parking brake lever.

Brake System Bleeding

There are some general rules for effectively bleeding the brakes. First, start with the brakes connected to the secondary (rear) chamber of the master cylinder. Second, the time interval between pumping the brake pedal should be approximately 3 seconds. Finally, the air bleeder on each brake should be opened only for 1–2 seconds.

Standard Brakes

1. Fit one end of a vinyl tube into the air bleeder and put the other end into a brake fluid container.
2. Starting with the wheel that is farthest from the master cylinder, slowly depress the brake pedal and keep it depressed. Then, open the air bleeder to discharge air together with the fluid. Keep the bleeder open only 1–2 seconds.
3. With the bleeder closed, slowly release the brake pedal and repeat the procedure in 3–4 seconds.
4. When all air has beed released from the system, check and fill the master cylinder with fluid.

Anti-Lock Brakes

1. Bleed the brakes at all 4 wheels as described above.
2. Attach the vinyl hose to the bleeders on top of the hydraulic unit. Bleed this port in the same fashion as the wheels. Move the hose to the other bleeder and repeat. These ports bleed the primary circuit.
3. Remove the cone screw from the secondary bleeder port and install a bleeder screw. Install the clear vinyl hose on the bleeder.
4. Open the bleeder and depress the brake pedal. Keep the bleeder open and intermittently apply an electrical

signal to the solenoid valve. To apply the signal:

a. Disconnect both battery terminals.

b. Disconnect the 2-pin and 12-pin connectors at the hydraulic assembly.

c. At the 12-pin connector, connect terminals 1 and 3 to battery ground and terminals 5 and 7 to the positive battery terminal. Take care not to short the terminals.

d. When the last connection is made, the electrical signal is transmitted to the solenoids. Do not send the signal for more than 5 seconds. Break the connections at the positive terminal after 2–3 seconds.

5. When the brake pedal moves to the end of its stroke, close the bleeder and allow the pedal to return slowly. If the electrical signal is not transmitted for any reason, the bleeder need not be closed before returning the pedal.

6. Repeat the above steps until the fluid tube contains no air.

7. With the electrical signal disconnected and the brake pedal released, remove the bleeder fitting and re-install the cone screw. Tighten to 6 ft. lbs. (8 Nm).

8. Repeat the procedure for the other secondary bleeder port. Both secondary ports must be bled.

9. Carefully remove the jumper wires and reconnect the connectors to the hydraulic unit. Connect the battery cables with the ignition **OFF**.

Anti-Lock System Brake Service

PRECAUTIONS

- Certain components within the ABS system are not intended to be serviced or repaired individually. Only those components with REMOVAL & installation procedures should be serviced.
- Do not use rubber hoses or other parts not specifically specified for the ABS system. When using repair kits, replace all parts included in the kit. Partial or incorrect repair may lead to functional problems and require the replacement of components.
- Lubricate rubber parts with clean, fresh brake fluid to ease assembly. Do not use lubricated shop air to clean parts; damage to rubber components may result.
- Use only DOT 3 brake fluid from an unopened container.
- If any hydraulic component or line is removed or replaced, it may be necessary to bleed the entire system.
- A clean repair area is essential. Always clean the reservoir and cap thoroughly before removing the cap. The slightest amount of dirt in the fluid may plug an orifice and impair the system function. Perform repairs after components have been thoroughly cleaned; use only denatured alcohol to clean components. Do not allow ABS components to come into contact with any substance containing mineral oil; this includes used shop rags.
- The Anti-Lock control unit is a microprocessor similar to other computer units in the vehicle. Ensure that the ignition switch is **OFF** before removing or installing controller harnesses. Avoid static electricity discharge at or near the controller.
- If any arc welding is to be done on the vehicle, the Anti-Lock Control Unit (ALCU) connectors should be disconnected before welding operations begin.

Hydraulic Unit

REMOVAL & INSTALLATION

1. Disconnect the negative battery cable. Disconnect the harness connectors at the hydraulic unit.
2. Remove the emission canister from the engine compartment.
3. Disconnect the inlet and outlet lines from the top of the actuator. Label the lines for installation. Immediately plug the lines and ports to prevent the entry of dirt.
4. Remove the screw holding the ABS relay cover and remove the cover. Remove the bolts holding the hydraulic unit bracket to the body. Not that one of these bolts has the pump motor ground attached.
5. Lift the actuator and bracket clear of the vehicle. Keep the unit upright at all times. The brackets and relays may be removed for transfer to a replacement unit.
6. Except for the 2 relays, the hydrauic unit contains no replaceable components. Never attempt to disassemble the unit.

To install:

7. Install the relays and brackets. The nuts on the bushing bolts holding the hydraulic unit to the brackets should be tightened to 6 ft. lbs. (8 Nm).
8. Install the hydraulic unit and brackets and tighten the nuts to 25 ft. lbs. (34 Nm). Make sure the ground is attached.
9. Check thar the relays are firmly seated and install the relay cover. Connect the brake lines and tighten to 11 ft. lbs. (15 Nm).
10. Install the canister in the engine compartment. Bleed all 4 wheels, then bleed the hydraulic actuator primary and secondary circuits.

Wheel Speed Sensor

REMOVAL & INSTALLATION

Front

1. Disconnect the negative battery cable. Disconnect the speed sensor harness in the engine compartment.
2. Remove the bolts holding the sensor harness brackets. Take careful note of placement and location of the harness retainers.
3. Remove the sensor retaining bolt at the front hub. Remove the front wheel speed sensor by lifting it straight out of the housing. Do not damage the tip of the sensor. Clean or replace as necessary.

To install:

4. Place the sensor into the mount and install the retaining bolts. Tighten to 10 ft. lbs. (14 Nm).
5. Remove the caliper and brake disc. Use a non-ferrous feller gauge to check the clearance between the top of the sensor and the tone wheel. Check the clearance at several locations on the hub. Clearance should be 0.039–0.059 in. (1.0–1.5 mm).
6. If the air gap is too small, the sensor may be raised using ABS sensor shims. Remove the sensor, install the shim and recheck. If the clearance is too large, check the tone wheel, hub and sensor for damage.
7. Once the air gap is correct, reset the retaining bolt to the correct torque.
8. Working from the sensor end, install each harness clip and retainer exactly as removed. Connect the sensor to the ABS harness in the engine compartment.

Rear

1. Remove the rear seat. Disconnect the ABS harness connector.
2. Remove the rear sensor harness retaining bracket from the trailing link. Remove any other retainers and take carefull note of the cable routing.
3. Remove the retaining bolt holding the sensor to the hub. Remove the rear wheel sensor by lifting straight out of the housing. Do not damage the tip of the sensor. Clean or replace as necessary.

To install:

4. Place the sensor into the mount and install the retaining bolts. Tighten to 10 ft. lbs. (14 Nm).
5. Remove the caliper and brake disc. Use a non-ferrous feller gauge to check the clearance between the top of the sensor and the tone wheel. Check the clearance at several locations on the hub. Clearance should be 0.031–0.051 in. (0.8.0–1.3 mm).
6. If the air gap is too small, the sen-

sor may be raised using ABS sensor shims. Remove the sensor, install the shim and recheck. If the clearance is too large, check the tone wheel, hub and sensor for damage.

7. Once the air gap is correct, reset the retaining bolt to the correct torque.

8. Working from the sensor end, install each harness clip and retainer exactly as removed. Connect the sensor to the ABS harness in the engine compartment.

CHASSIS ELECTRICAL

Blower Motor

REMOVAL & INSTALLATION

Justy

1. Disconnect the negative battery cable.
2. Remove the coupler that connects the instrument panel harness to the blower motor.
3. Remove the coupler that connects the resistor to the instrument panel harness.
4. Detach the blower assembly. Remove the screws retaining the blower motor to the blower assembly.
5. Remove the motor assembly. Remove the nut retaining the fan to the motor assembly.

To install:

6. Install the motor assembly and tighten the nuts to specificaiton.
7. Install the couplers that connect the resistor to the instrument panel and the instrument panel to the blower motor.
8. Connect the negative battery cable.

XT Coupe and Legacy

NOTE: Depending upon working clearance the air conditioning system may have to be discharged in order to service the blower motor. If this is the case, be sure to observe all the required safety precautions when discharging and recharging the air conditioning system.

1. Disconnect the negative battery cable.
2. Remove the lower instrument panel cover on the passenger side of the vehicle.
3. Remove the glove box assembly, as required for working clearance.
4. Remove the heater duct, if not equipped with air conditioning.
5. If equipped with air conditioning, separate the evaporator from the blower assembly.
6. Disconnect the blower motor harness and the resistor electrical harness connector.
7. Remove the blower motor retaining bolts. Remove the blower motor assembly from its mounting.

To install:

8. Install the blower motor retaining bolts and tighten to 4–7 ft. lbs. Connect the blower motor harness and the resistor electrical harness connector.
9. Install the evaporator to the blower assembly as required. Install the heater duct as required. Install the glove box.
10. Install the lower instrument panel cover and connect the negative battery cable.

Except Justy, XT Coupe and Legacy

NOTE: Depending upon working clearance the air conditioning system may have to be discharged in order to service the blower motor. If this is the case, be sure to observe all the required safety precautions when discharging and recharging the air conditioning system.

1. Disconnect the negative battery cable.
2. Remove the lower instrument panel cover on the passenger side of the vehicle. Remove the glove box assembly, as required for working clearance.
3. If equipped with a vacuum actuator, set the control lever to the **CIRC** position and disconnect the vacuum hose from the assembly. Remove the actuator from its mounting.
4. Remove the heater duct, if not equipped with air conditioning.
5. If equipped with air conditioning, separate the evaporator from the blower assembly.
6. Disconnect the blower motor harness and the resistor electrical harness connector.
7. Remove the blower motor retaining bolts. Remove the blower motor assembly from its mounting. As required, separate the fan from the blower motor.

To install:

8. Install the blower motor and tighten the retaining bolts to specificaiton. Connect the blower motor harness and the resistor electrical harness connector.
9. Install the evaporator to the blower assembly as required. Install the heater duct as required. Install the glove box.
10. Install the vacuum actuator and connect the vaccum line. Install the lower instrument panel cover and connect the negative battery cable.

Windshield Wiper Motor

REMOVAL & INSTALLATION

Justy and Legacy

1. Disconnect the negative battery cable.
2. At the wiper motor, disconnect the electrical connector.
3. Remove the wiper motor to cowl bolts.
4. Separate the wiper link from the motor.
5. If necessary, replace the wiper motor.

To install:

6. Install the wiper motor and tighten the cowl bolts Install the wiper link on the motor.
7. Connect the electrical connector and the negative battery cable. Check for proper operation.

Except Justy and Legacy

1. Disconnect the negative battery cable.
2. Remove the wiper blades from the wiper arms by pulling the retaining lever up and sliding the blade away from the arm.
3. Slide the covering boot up the wiper arm.
4. Remove the wiper arms to linkage nuts and the arms.
5. Disconnect the electrical wiring connectors from the wiper motor.
6. Remove the cowl to body screws and the cowl from the vehicle.
7. Find or fabricate a ring which has the same diameter as the outer diameter of the plastic joint that retains the linkage to the wiper motor. Force the ring down over the joint to force the 4 plastic retaining jaws inward, then disconnect and remove the linkage.
8. Remove the wiper motor to firewall bolts and the motor.

To install:

9. Install the wiper motor and tighten the attaching bolts.
10. Install the wiper linkage. Install the cowl. Connect the wiper electrical wiring harness.
11. Connect the negative battery cable and install the wiper arms after the ignition switch has been ON for a few seconds to put the linkage in Park position.

Instrument Cluster

REMOVAL & INSTALLATION

Justy

1. Disconnect the negative battery cable. Remove the steering wheel.
2. Remove the defroster duct assembly.
3. Disconnect the heater control cable from the inside/outside air selector rod at the heater unit.
4. Disconnect the speedometer cable. Disconnect the electrical harness connectors.
5. Remove the covers for the instrument cluster retaining bolts.
6. Remove the instrument cluster retaining bolts. Remove the instrument cluster from its mounting.

To install:

7. Install the instrument cluster and tighten the retaining bolts securely. Install the screw covers.
8. Connect the speedometer cable and electrical harness.
9. Install the heater control cable and the defroster duct assembly.
10. Install the steering wheel and connect the negative battery cable.

XT Coupe and Legacy

1. Disconnect the negative battery cable. Remove the lower cover on the driver's side. Remove the side ventilation duct.
2. Open the fuse box lid. Remove the fuse box to instrument panel screws and the fuse box.
3. Remove the lower cover on the passenger's side. Using a medium prybar, pry the upper cover, at 3 points, from the instrument panel.
4. Remove the console. Remove the steering column assembly, the combination meter and the control wing as a unit.
5. Disconnect the electrical harness connectors from the radio and other necessary components.
6. Remove the instrument panel to chassis bolts and the instrument panel from the vehicle.

To install:

7. Install the instrument panel and tighten the bolts securely. Connect the electrical harness connnectors.
8. Install the steering column and console. Install the lower and upper instrument panel covers.
9. Install the fuse box, the side ventilation duct and the lower instrument panel cover on the drivers side. Connect the negative battery cable.

Except Justy, XT Coupe and Legacy

1. Disconnect the negative battery cable.
2. Remove the bolts securing the steering column and pull it down.
3. Disconnect the electrical wiring connectors, then remove the cluster visor screws and the visor, except on GL and GLF.
4. On the GL and GLF, remove the center ventilator control lever by pulling it. Remove the 3 screws accessible through the ventilator grille to the right of the cluster and the 1 screw accessible through the grille on the left. Remove the visor.
5. On the station wagon 4WD GL, remove the turn signal light switch.
6. Remove the cluster retaining screws, then pull the cluster out far enough to disconnect the speedometer cable and electrical connectors from behind, then remove the cluster assembly from the vehicle.

To install:

7. Install the instrument cluster. Connect all electrical connectors and the speedometer cable. Tighten the attaching screws securely.
8. Install the vehtilator control lever on GL and GLF models. Install the turn signal light switch on 4WD GL models.
9. Install the cluster visor screws and the visor. Lift the steering column and tighten the bolts to specification. Connect the negative battery cable.

Radio

REMOVAL & INSTALLTION

1. Disconnect the negative battery cable.
2. Remove the radio panel and center dash panel cover. Remove the clock as necessary.
3. Remove the tapping screws and pull the radio out slightly. Disconnect the electrical connections and antenna feed. On XT Coupe, some connections are located at the floor under the rug. Remove the radio.

To install:

4. Install the radio. Connect all electrical connections and the antenna feeder wire. Tighten screws securely.
5. Install the radio and center dash panel covers. Install the clock as necessary. Connect the negative battery cable.

Combination Switch

REMOVAL & INSTALLATION

1. Disconnect the negative battery cable. Remove the lower cover to instrument panel screws and the lower cover.
2. Remove the covers to steering column screws and the upper and lower column covers.
3. Remove the steering wheel cover and the nut. Using a steering wheel puller tool, remove the steering wheel from the steering column.
4. Remove the electrical harness to steering column clip and band fitting, then disconnect the electrical connectors.
5. Remove the combination switch to control wing bracket screws. Remove the switch assembly from its mounting.

To install:

6. Install the combination switch and tighten the bracket screws securely. Connect the electrical harness.
7. Install the steering wheel assembly and tighten the center nut to specification. Install the steering column covers and the lower instrument panel cover.
8. Connect the negative battery cable. Check for proper operation.

Ignition Switch

REMOVAL & INSTALLATION

NOTE: The ignition switch is mounted to the steering column using shear bolts. These bolts are constructed so the heads shear off when the bolt is torqued.

1. Disconnect the negative battery cable. Remove the steering wheel.
2. Remove the upper and lower steering column covers from the steering column.
3. Remove the hazard knob.
4. Drill a pilot hole into the shear bolts, then using a screw extractor, remove the screws from the steering column.
5. Remove the ignition switch from the steering column.

To install:

6. Install the ignition switch using new shear bolts.
7. Install the hazard knob, steering bolumn covers and the steering wheel.
8. Connect the negative battery cable and check the switch for proper operation.

Stoplight Switch

REMOVAL & INSTALLATION

The stoplight switch is located on the brake pedal bracket, under the instrument panel. The switch is held in place by 2 locknuts, by which all adjustment is accomplished. To replace the switch, the wiring is disconnected, the locknuts are loosened and the switch removed.

After installation, the travel to operate the switch plunger should be adjusted to 0.071–0.130 in. (1.8–3.3 mm).

Neutral Safety Switch

ADJUSTMENT

This switch is mounted on the transaxle shift lever shaft, bolted to the transaxle. It also operates the back-up lights.

1. Remove the shift lever shaft nut.
2. Remove the shift lever from the shaft.
3. Make sure the slot in the shaft is vertical, **N** position.
4. Remove the switch mounting bolts but leave the switch in place.
5. Remove the setscrew from the lower face of the switch.
6. Insert a 0.059 in. drill bit through the set screw hole. Turn the switch slightly so the bit passes through into the back part of the switch.
7. Bolt the switch down.
8. Remove the drill bit and replace the set screw.
9. Install the lever and tighten the shaft nut.
10. Make sure the engine can start only in **P** or **N** and that the back-up lights turn functioning in **R**. Adjust the shift linkage, if necessary.

Fusible Links

Legacy

The main fusible link is located in the underhood fuse box.

Loyale

The fusible links are located in a box on the left fender apron, ahead of the shock tower.

Justy

The 30 and 60 amp main fuses are located in a holder on the left front shock tower, near the brake master cylinder.

XT

The fusible links are located in a box at the left side of the engine compartment, near the battery.

Circuit Breakers

The power window circuit breakers are located under the front seat.

Modules and Computers

Legacy

- **Air Suspension Control Module**—is located under the driver's seat.
- **Anti-lock Brake Control Module**—is located under front passenger seat.
- **Automatic Shoulder Belt Control Module (sedan)**—is located in left side of trunk, near left trunk lid hinge mount.
- **Automatic Shoulder Belt Control Module (wagon)**—is located on left rear shock tower, just below rear speaker.
- **Automatic Transaxle Control Module**—is located under left dash, immediately to left of steering column.
- **Check Connector (Black)**—is located behind left side of center console, under dashboard.
- **Cruise Control Module**—is located under right dash, to right of glove box.
- **Daytime Running Light Control Module (Canada)**—is located under right dash, above right kick panel.
- **Diagnosis Connector**—is located behind left side of center console, under dashboard.
- **ECU**—is located under left dash, to left of steering column.
- **Fuse Box**—is located under left side of instrument panel.
- **Illumination Control Unit**—is located behind center dashboard, to left of glovebox.
- **MPFI Control Module**—is located behind left dash, above fuse box.
- **Shift Lock Control Module**—is located under left center dash, behind left side of center console.
- **Sunroof Control Module**—is located in roof, at rear of sunroof.
- **Test Mode Connector (green)**—is located is located below right side of steering column.
- **Turn Signal/Flasher Module**—is located behind center dashboard, to left of glove box.
- **Wastegate Valve and Controller (1991)** —is located at exhaust side of turbocharger.

Loyale

- **Automatic Transaxle Control Module (4 speed)**—is located in left quarter panel.
- **MPFI Module**—is located under left dash, directly below the steering column.
- **Power Window Control Module**—is located under the front passenger seat.

Justy

- **Automatic Seat Belt Module**—is located behind right front kick panel.
- **Check Connector (under dash)**—is located under left side dashboard.
- **Check Connectors (ECVT)**—are located in ECVT module harness at the control module.
- **Check Connectors (underhood)**—are located in wire harness at left front shock tower.
- **Daytime Running Lamp Control Module (Canada)**—is located behind right front kick panel.
- **ECVT Module**—is located under left dash, above hood release.
- **EFC Control Module**—See Fuel Control Module
- **Front Wiper Intermittent Control Module**—is located behind the fusebox.
- **Fuel Control Module (EFC or MPFI)**—is located under left dash, near hood release.
- **Illumination Control Module**—is located on brake pedal bracket.
- **MPFI Control Module**—See Fuel Control Module
- **Read Memory Connector**—is located under left dashboard.
- **SELECT Diagnostic Connector**—is located in wire harness, just below ignition coil on firewall.
- **Shift Lock Module**—is located on brake pedal bracket.

XT

- **Automatic Seat Belt Control Module**—is located in right quarter panel, behind right door.
- **Automatic Transaxle Control Module**—is located in left quarter panel, behind left door.
- **Check Connectors (underhood, 17-pin and 11-pin)**—are located in wire harness at left side of firewall.
- **Cruise Control Module**—is located in right side dash, above glove box light assembly.
- **Illumination Control Module**—is located on top of steering column, under upper column cover.
- **Intermittent Wiper Module**—is located in center console, just ahead of shifter assembly.
- **Mode Control Module**—is located under left center dash, behind center console.
- **MPFI Control Module**—is located in front of trunk, mounted to underside of rear window shelf.
- **Pneumatic Suspension Control Module**—is located under the driver's seat.
- **Power Window Control Module**—is located under front passenger seat.
- **Read Memory Connector**—is located in wiring harness at front of trunk, under rear window shelf. The connectors are black.
- **Test Mode Connectors**—are located in front of trunk, under rear window shelf. The 2 connectors are green.

19

Toyota

Camry, Celica, Corolla, Cressida, MR2, Supra, Tercel—All Models

SERIAL NUMBER IDENTIFICATION

Vehicle Identification Plate

All vehicles have the Vehicle Identification Number (VIN) stamped on a plate which is attached to the left side of the instrument panel. This plate is visible through the windshield.

The serial number consists of a series identification number followed by a 6-digit production number.

Engine Number

Basically, Toyota uses 5 types of engines:

A-Series
3A-C
4A-LC
4A-F, 4A-FE
4A-GE
4A-GEC, 4A-GELC
E-Series
3E, 3E-E
M-Series
5M-GE, 7M-GE, 7M-GTE
S-Series
3S-FE, 3S-GE, 3S-GTE, 5S-FE
Z-Series
2VZ-FE

Engines within each series are similar, as the cylinder block designs are the same. Variances within each series may be due to ignition types, displacements and cylinder head design.

Serial numbers of the engines may be found on the following locations:

A-Series engines – stamped vertically on the left side rear of the engine block.

E-Series engines – stamped on the left side rear of the engine block.

M-Series engines – stamped horizontally on the passenger side of the engine block, behind the alternator.

S-Series engines – the serial number can be found on the rear left side of the block, under the thermostat housing.

Z-Series engines – stamped on the front, right (passenger) side of the cylinder block.

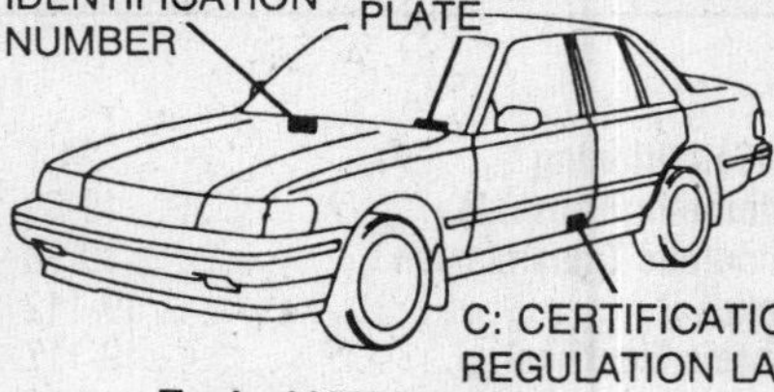

Typical VIN location

SPECIFICATIONS

ENGINE IDENTIFICATION

Year	Model	Engine Displacement cu. in. (cc/liter)	Engine Series Identification	No. of Cylinders	Engine Type
1988	Tercel	88.6 (1452/1.4)	3A-C	4	SOHC
		88.9 (1456/1.5)	3E	4	SOHC
	Corolla	97.0 (1587/1.6)	4A-LC	4	SOHC
		97.0 (1587/1.6)	4A-GEC, 4A-GELC	4	DOHC
		97.0 (1587/1.6)	4A-F	4	DOHC
	Camry	121.9 (1998/2.0)	3S-FE	4	DOHC
	Celica	121.9 (1998/2.0)	3S-FE, 3S-GE	4	DOHC
		121.9 (1998/2.0)	3S-GTE	4	DOHC, TURBO
	Supra	180.3 (2954/3.0)	7M-GE	6	DOHC
		180.3 (2954/3.0)	7M-GTE	6	DOHC, TURBO
	MR2	97.0 (1587/1.6)	4A-GELC	4	DOHC
		97.0 (1587/1.6)	4A-GZE	4	DOHC, SUPER
	Cressida	168.4 (2759/2.8)	5M-GE	6	DOHC
1989	Tercel	88.9 (1457/1.5)	3E	4	SOHC
	Corolla	97.0 (1587/1.6)	4A-GE	4	DOHC
		97.0 (1587/1.6)	4A-F, 4A-FE	4	DOHC
	Camry	121.9 (1998/2.0)	3S-FE	4	DOHC
		153.0 (2058/2.5)	2VZ-FE	6	DOHC
	Celica	121.9 (1998/2.0)	3S-FE, 3S-GE	4	DOHC
		121.9 (1998/2.0)	3S-GTE	4	DOHC, TURBO
	Supra	180.3 (2954/3.0)	7M-GE	6	DOHC
		180.3 (2954/3.0)	7M-GTE	6	DOHC, TURBO
	MR2	97.0 (1587/1.6)	4A-GELC	4	DOHC
		97.0 (1587/1.6)	4A-GZE	4	DOHC, SUPER

ENGINE IDENTIFICATION

Year	Model	Engine Displacement cu. in. (cc/liter)	Engine Series Identification	No. of Cylinders	Engine Type
1989	Cressida	180.3 (2954/3.0)	7M-GE	6	DOHC
1990	Tercel	88.9 (1457/1.5)	3E	4	SOHC
		88.9 (1457/1.5)	3E-E	4	SOHC
	Corolla	97.0 (1587/1.6)	4A-FE	4	DOHC
		97.0 (1587/1.6)	4A-GE	4	DOHC
	Camry	121.9 (1998/2.0)	3S-FE	4	DOHC
		153.0 (2058/2.5)	2VZ-FE	6	DOHC
	Celica	97.0 (1587/1.6)	4A-FE	4	DOHC
		121.9 (1998/2.0)	3S-GTE	4	DOHC, TURBO
		132.0 (2164/2.2)	5S-FE	4	DOHC
	Supra	180.3 (2954/3.0)	7M-GE	6	DOHC
		180.3 (2954/3.0)	7M-GTE	6	DOHC, TURBO
	MR2	121.9 (1998/2.0)	3S-GTE	4	DOHC, TURBO
		132.0 (2164/2.2)	5S-FE	4	DOHC
	Cressida	180.3 (2954/3.0)	7M-GE	6	DOHC
1991-92	Tercel	88.9 (1457/1.5)	3E-E	4	SOHC
	Corolla	97.0 (1587/1.6)	4A-FE	4	DOHC
		97.0 (1587/1.6)	4A-GE	4	DOHC
	Camry	121.9 (1998/2.0)	3S-FE	4	DOHC
		153.0 (2058/2.5)	2VZ-FE	6	DOHC
	Celica	97.0 (1587/1.6)	4A-FE	4	DOHC
		121.9 (1998/2.0)	3S-GTE	4	DOHC, TURBO
		132.0 (2164/2.2)	5S-FE	4	DOHC
	Supra	180.3 (2954/3.0)	7M-GE	6	DOHC
		180.3 (2954/3.0)	7M-GTE	6	DOHC, TURBO
	MR2	121.9 (1998/2.0)	3S-GTE	4	DOHC, TURBO
		132.0 (2164/2.2)	5S-FE	4	DOHC
	Cressida	180.3 (2954/3.0)	7M-GE	6	DOHC

OHV—Overhead Valves
SOHC—Single Overhead Camshaft
DOHC—Double Overhead Camshaft
TURBO—Turbocharged
SUPER—Supercharged

GENERAL ENGINE SPECIFICATIONS

Year	Model	Engine Displacement cu. in. (cc)	Fuel System Type	Net Horsepower @ rpm	Net Torque @ rpm (ft. lbs.)	Bore × Stroke (in.)	Com-pression Ratio	Oil Pressure ①
1988	Tercel	88.6 (1452)	2 bbl	62 @ 4800	76 @ 2800	3.05 × 3.03	9.0:1	4.3
		88.9 (1456)	2 bbl	78 @ 6000	87 @ 4000	2.87 × 3.43	9.3:1	4.3
	Corolla	4A-LC 97.0 (1587)	2 bbl	74 @ 5200	86 @ 2800	3.19 × 3.03	9.0:1	4.3
		4A-F 97.0 (1587)	2 bbl	90 @ 6000	95 @ 3600	3.19 × 3.03	9.5:1	4.3
		97.0 (1587)	EFI	116 @ 6600 ③	110 @ 4800 ④	3.19 × 3.03	9.4:1	4.3
	Camry	121.9 (1998)	EFI	115 @ 5200	124 @ 4400	3.39 × 3.39	9.3:1	4.3

GENERAL ENGINE SPECIFICATIONS

Year	Model	Engine Displacement cu. in. (cc)	Fuel System Type	Net Horsepower @ rpm	Net Torque @ rpm (ft. lbs.)	Bore × Stroke (in.)	Com-pression Ratio	Oil Pressure ①
1988	Celica	3S-FE 121.9 (1998)	EFI	115 @ 5200	124 @ 4400	3.39 × 3.39	9.3:1	4.3
		3S-GE 121.9 (1998)	EFI	135 @ 6000	125 @ 4800	3.39 × 3.39	9.2:1	4.3
		3S-GTE 121.9 (1998)	EFI	190 @ 6000	190 @ 3200	3.39 × 3.39	8.5:1	4.3
	Supra	7M-GE 180.3 (2954)	EFI	200 @ 6000	185 @ 4800	3.27 × 3.58	9.2:1	4.3
		7M-GTE 180.3 (2954)	EFI	230 @ 5600	246 @ 4000	3.27 × 3.58	8.4:1	4.3
	MR2	4A-GELC 97.0 (1587)	EFI	112 @ 6600	100 @ 4800	3.19 × 3.03	9.4:1	4.3
		4A-GZE 97.0 (1587)	EFI	145 @ 6400	140 @ 4000	3.19 × 3.03	8.0:1	4.3
	Cressida	168.4 (2759)	EFI	156 @ 5200	165 @ 4400	3.27 × 3.35	9.2:1	4.3
1989	Tercel	88.6 (1452)	2 bbl	78 @ 6000	87 @ 4000	2.87 × 3.43	9.3:1	4.3
	Corolla	4A-FE 97.0 (1587)	EFI	100 @ 5600	101 @ 4400	3.19 × 3.03	9.5:1	4.3
		4A-F 97.0 (1587)	2 bbl	90 @ 6000	95 @ 3600	3.20 × 3.00	9.5:1	4.3
		97.0 (1587)	EFI	116 @ 6600	100 @ 4800	3.19 × 3.03	9.4:1	4.3
	Camry	121.9 (1998)	EFI	115 @ 5200	124 @ 4400	3.39 × 3.39	9.3:1	4.3
		153.0 (2507)	EFI	153 @ 5600	155 @ 4400	3.44 × 2.74	9.0:1	4.3
	Celica	3S-FE 121.9 (1998)	EFI	115 @ 5200	124 @ 4400	3.39 × 3.39	9.3:1	4.3
		3S-GE 121.9 (1998)	EFI	135 @ 6000	125 @ 4800	3.39 × 3.39	9.2:1	4.3
		3S-GTE 121.9 (1998)	EFI	190 @ 6000	190 @ 3200	3.39 × 3.39	8.5:1	4.3
	Supra	7M-GE 180.3 (2954)	EFI	200 @ 6000	188 @ 3600	3.27 × 3.58	9.2:1	4.3
		7M-GTE 180.3 (2954)	EFI	232 @ 5600	254 @ 3200	3.27 × 3.58	8.4:1	4.3
	MR2	4A-GELC 97.0 (1587)	EFI	115 @ 6600	100 @ 4800	3.19 × 3.03	9.4:1	4.3
		4A-GZE 97.0 (1587)	EFI	145 @ 6400	140 @ 4000	3.19 × 3.03	8.0:1	4.3
	Cressida	180.3 (2954)	EFI	190 @ 5600	185 @ 4400	3.27 × 3.58	9.2:1	4.3
1990	Tercel	3E 88.9 (1457)	1 bbl	78 @ 6000	87 @ 4000	2.87 × 3.43	9.3:1	4.3
		3E-E 88.9 (1457)	EFI	82 @ 5200	89 @ 4400	2.87 × 3.43	9.3:1	4.3
	Corolla	4A-FE 97.0 (1587)	EFI	102 @ 5800	101 @ 4800	3.19 × 3.03	9.5:1	4.3
		4A-GE 97.0 (1587)	EFI	130 @ 6800	102 @ 5800	3.19 × 3.03	9.5:1	4.3

GENERAL ENGINE SPECIFICATIONS

Year	Model	Engine Displacement cu. in. (cc)	Fuel System Type	Net Horsepower @ rpm	Net Torque @ rpm (ft. lbs.)	Bore × Stroke (in.)	Com-pression Ratio	Oil Pressure ①
1990	Camry	3S-FE 121.9 (1998)	EFI	115 @ 5200	124 @ 4400	3.39 × 3.39	9.3:1	4.3
		2VZ-FE 153.0 (2508)	EFI	156 @ 5600	160 @ 4400	3.44 × 2.74	9.0:1	4.3
	Celica	4A-FE 97.0 (1587)	EFI	103 @ 6000 ⑤	102 @ 3200 ⑥	3.19 × 3.03	9.5:1	4.3
		5S-GTE 121.9 (1998)	EFI	200 @ 6000	200 @ 3200	3.39 × 3.39	9.5:1	4.3
		3S-FE 132.0 (2164)	EFI	130 @ 5400	140 @ 4400	3.43 × 3.58	9.5:1	4.3
	Supra	7M-GE 180.3 (2954)	EFI	200 @ 6000	188 @ 3600	3.27 × 3.58	9.2:1	4.3
		7M-GTE 180.3 (2954)	EFI	232 @ 5600	254 @ 3200	3.27 × 3.58	8.4:1	4.3
	MR2	3S-GTE 121.9 (1998)	EFI	200 @ 6000	200 @ 3200	3.39 × 3.39	9.5:1	4.3
		5S-FE 132.0 (2164)	EFI	130 @ 5400	140 @ 4400	3.43 × 3.58	9.5:1	4.3
	Cressida	180.3 (2954)	EFI	190 @ 5600	185 @ 4400	3.27 × 3.35	9.2:1	4.3
1991–92	Tercel	3E-E 88.9 (1457)	EFI	82 @ 5200	90 @ 4400	2.88 × 3.43	9.3:1	4.3
	Corolla	4A-FE 97.0 (1587)	EFI	102 @ 5800	101 @ 4800	3.19 × 3.03	9.5:1	4.3
		4A-GE 97.0 (1587)	EFI	130 @ 6800	105 @ 6000	3.19 × 3.03	10.3:1	4.3
	Camry	3S-FE 121.9 (1998)	EFI	115 @ 5200	124 @ 4400	3.39 × 3.39	9.3:1	4.3
		2VZ-FE 153.0 (2058)	EFI	156 @ 5600	160 @ 4400	3.44 × 2.74	9.0:1	4.3
	Celica	4A-FE 97.0 (1587)	EFI	103 @ 6000	102 @ 3200	3.19 × 3.10	9.5:1	4.3
		3S-GTE 121.9 (1998)	EFI	200 @ 6000	200 @ 3200	3.39 × 3.39	8.8:1	4.3
		5S-FE 132.0 (2164)	EFI	130 @ 5400	140 @ 4400	3.43 × 3.58	9.5:1	4.3
	Supra	7M-GE 180.3 (2954)	EFI	200 @ 6000	188 @ 3600	3.27 × 3.58	9.2:1	4.3
		7M-GTE 180.3 (2954)	EFI	232 @ 5600	254 @ 3200	3.27 × 3.58	8.4:1	4.3
	MR2	3S-GTE 121.9 (1998)	EFI	200 @ 6000	200 @ 3200	3.39 × 3.39	8.8:1	4.3
		5S-FE 132.0 (2164)	EFI	130 @ 5400	140 @ 4400	3.43 × 3.58	9.5:1	4.3
	Cressida	7M-GE 180.3 (2954)	EFI	190 @ 5600	185 @ 4400	3.27 × 3.58	9.2:1	4.3

EFI Electronic Fuel Injection
① At Idle
② FX-16: 108 @ 6600
③ FX-16: 110 @ 6600
④ FX-16: 98 @ 4800
⑤ California: 102 @ 5800
⑥ California: 101 @ 4800

GASOLINE ENGINE TUNE-UP SPECIFICATIONS

Year	Model	Engine Displacement cu. in. (cc)	Spark Plugs Type	Spark Plugs Gap (in.)	Ignition Timing (deg.) MT	Ignition Timing (deg.) AT	Compression Pressure (psi)	Fuel Pump (psi)	Idle Speed (rpm) MT	Idle Speed (rpm) AT	Valve Clearance In.	Valve Clearance Ex.
1988	Tercel	88.6 (1452)	BPR5EY-11 ①	0.043	5B	5B	178	2.6–3.5	650	900	0.008	0.012
		88.9 (1456)	BPR5EY-11	0.043	3B	3B	184	2.6–3.5	650	900	0.008	0.008
	Corolla	4A-LC 97.0 (1587)	BPR5EY-11	0.043	5B	5B	163	2.5–3.5	650	750	0.008	0.012
		4A-F 97.0 (1587)	BPR5EY-11	0.043	5B	5B	191	2.5–3.5	650	750	0.008	0.010
		4A-GE 97.0 (1587)	BCPR5EP-11	0.043	10B	10B	179	33–38	800	800	0.008	0.010
	Camry	121.9 (1998)	BCPR5EY-11	0.043	10B	10B	178	38–44	700	750	0.009	0.013
	Celica	3S-FE 121.9 (1998)	BCPR5EY-11	0.043	10B	10B	178	38–44	650	650	0.009	0.013
		3S-GE 121.9 (1998)	BCPR5EP-11	0.043	10B	10B	178	33–38	750	750	0.008	0.010
		3S-GTE 121.9 (1998)	BCPR5EP-8	0.031	10B	—	178	33–38	750	—	0.008	0.010
	Supra	7M-GE 180.3 (2954)	BCPR5EP-11	0.043	10B	10B	156	33–40	700	700	0.008	0.010
		7M-GTE 180.3 (2954)	BCPR6EP-N8	0.031	10B	10B	142	33–40	650	650	0.008	0.010
	MR2	4A-GE 97.0 (1587)	BCPR5EP-11	0.043	10B	10B	179	38–44	800	800	0.008	0.010
		4A-GZE 97.0 (1587)	BCPR6EP-11	0.043	10B	10B	156	33–38	800	800	0.008	0.010
	Cressida	168.4 (2759)	BPR5EP-11	0.043	—	10B	164	35–38	—	650	Hyd.	Hyd.
1989	Tercel	88.9 (1456)	BPR5EY-11	0.043	3B	3B	184	2.6–3.5	700	900	0.008	0.008
	Corolla	4A-FE 97.0 (1587)	BCPR5EY	0.031	10B	10B	191	38–44	800	800	0.008	0.010
		4A-F 97.0 (1587)	BCPR5EY-11	0.043	5B	5B	191	2.5–3.5	650	750	0.008	0.010
		4A-GE 97.0 (1587)	BCPR5EP-11	0.043	10B	10B	179	33–44	800	800	0.008	0.010
	Camry	121.9 (1998)	BCPR5EY-11	0.043	10B	10B	178	38–44	700	700	0.009	0.013
		153.0 (2507)	BCPR6E-11	0.043	10B	10B	142	38–44	700	700	0.007	0.013
	Celica	3S-FE 121.9 (1998)	BCPR5EY-11	0.043	10B	10B	178	38–44	700	700	0.009	0.013
		3S-GE 121.9 (1998)	BCPR5EP-11	0.043	10B	10B	178	33–38	750	750	0.008	0.010
		3S-GTE 121.9 (1998)	BCPR5EP-8	0.031	10B	—	178	33–38	750	—	0.008	0.010
	Supra	7M-GE 180.3 (2954)	BCPR5EP-11	0.043	10B	10B	156	38–44	700	700	0.008	0.010
		7M-GTE 180.3 (2954)	BCPR6EP-N8	0.031	10B	10B	142	33–40	650	650	0.008	0.010

GASOLINE ENGINE TUNE-UP SPECIFICATIONS

Year	Model	Engine Displacement cu. in. (cc)	Spark Plugs Type	Spark Plugs Gap (in.)	Ignition Timing (deg.) MT	Ignition Timing (deg.) AT	Compression Pressure (psi)	Fuel Pump (psi)	Idle Speed (rpm) MT	Idle Speed (rpm) AT	Valve Clearance In.	Valve Clearance Ex.
1989	MR2	4A-GE 97.0 (1587)	BCPR5EP-11	0.043	10B	10B	179	38–44	800	800	0.008	0.010
		4A-GZE 97.0 (1587)	BCPR6EP-11	0.043	10B	10B	156	33–38	800	800	0.008	0.010
	Cressida	180.3 (2954)	BCPR5EP-11	0.043	—	10B	156	38–44	—	700	0.008	0.010
1990	Tercel	3E 88.9 (1457)	5PR5EY-11	0.043	3B	3B	184	2.6–3.5	700	900	0.008	0.008
		3E-E 88.9 (1457)	5PR5EY-11	0.043	10B	10B	184	38–44	800	800	0.008	0.008
	Corolla	4A-FE 97.0 (1587)	BCPR5EY	0.031	10B	10B	190	38–44	②	②	0.006–0.010	0.008–0.012
		4A-CE 97.0 (1587)	BCPR5EY	0.031	10B	10B	190	38–44	800	800	0.006–0.010	0.008–0.012
	Camry	3S-FE 121.9 (1998)	BCPR5EY-11	0.043	10B	10B	178	38–44	650–750	650–750	0.007–0.011	0.011–0.015
		2VZ-FE 153.0 (2508)	BCPR6E-11	0.043	10B	10B	178	38–44	650–750	650–750	0.005–0.009	0.011–0.015
	Celica	4A-FE 97.0 (1587)	BCPR5EY	0.031	10B	10B	191	38–44	800	800	0.006–0.010	0.008–0.012
		3S-GTE 121.9 (1998)	BCPR5EP-8	0.031	10B	—	178	38–44	750	—	0.006–0.010	0.008–0.012
		5S-FE 132.0 (2164)	BPR5EYA-11	0.043	10B	10B	178	38–44	650–750 ③	650–750 ④	0.007–0.011	0.011–0.015
	Supra	7M-GE 180.3 (2954)	BCPR5EP-11	0.043	10B	10B	156	38–44	700	700	0.006–0.010	0.008–0.012
		7M-GTE 180.3 (2954)	BCPR6EP-N8	0.031	10B	10B	142	33–40	650	650	0.006–0.010	0.006–0.010
	MR2	3S-GTE 121.9 (1998)	BKR6EP-8	0.031	10B	10B	164	33–38	750–850	—	0.006–0.010	0.008–0.012
		5S-FE 132.0 (2164)	BKR5EYA-11	0.043	10B	10B	142	38–44	700–800 ⑤	650–750 ④	0.007–0.011	0.011–0.015
	Cressida	180.3 (2954)	BCPR5EP-11	0.043	—	10B	156	38–44	—	700	0.006–0.010	0.008–0.012
1991	Tercel	3E-E 88.9 (1457)	BPR5EY-11	0.043	10B	10B	184	33–37	750	800	0.008	0.008
	Corolla	4A-FE 97.0 (1587)	BCPR5EY	0.031	10B	10B	191	38–44	⑥	⑥	0.006–0.010	0.008–0.012
		4A-GE 97.0 (1587)	BKR6EP-8	0.031	10B	10B	190	38–44	800	800	0.006–0.010	0.008–0.012
	Camry	3S-FE 121.9 (1998)	BCPR5EY-11	0.043	10B	10B	178	38–44	650	650	0.007–0.011	0.011–0.015
		2VZ-FE 153.0 (2058)	BCPR6E-11	0.043	10B	10B	178	38–44	700	700	0.005–0.009	0.011–0.015

GASOLINE ENGINE TUNE-UP SPECIFICATIONS

Year	Model	Engine Displacement cu. in. (cc)	Spark Plugs Type	Spark Plugs Gap (in.)	Ignition Timing (deg.) MT	Ignition Timing (deg.) AT	Com-pression Pressure (psi)	Fuel Pump (psi)	Idle Speed (rpm) MT	Idle Speed (rpm) AT	Valve Clearance In.	Valve Clearance Ex.
1991	Celica	4A-FE 97.0 (1587)	BCPR5EY	0.031	10B	10B	191	38–44	800	800	0.006–0.010	0.008–0.012
		3S-GTE 121.9 (1998)	BKR6EP-8	0.031	10B	10B	164	33–38	800	800	0.006–0.010	0.008–0.012
		5S-FE 132.0 (2164)	BKR5EYA-11	0.043	10B	10B	178	38–44	⑦	750	0.007–0.011	0.011–0.015
	Supra	7M-GE 180.3 (2954)	BCPR5EP-11	0.043	10B	10B	171	38–44	700	700	0.006–0.010	0.008–0.012
		7M-GTE 180.3 (2954)	BCPR6EP-N8	0.031	10B	10B	142	33–40	650	650	0.006–0.010	0.008–0.012
	MR2	3S-GTE 121.9 (1998)	BKR6EP-8	0.031	10B	10B	164	33–38	800	800	0.006–0.010	0.008–0.012
		5S-FE 132.0 (2164)	BKR5EYA-11	0.043	10B	10B	178	38–44	⑧	⑧	0.007–0.011	0.011–0.015
	Cressida	7M-GE 180.3 (2954)	BCPR5EP-11	0.043	10B	10B	171	38–44	700	700	0.006–0.010	0.008–0.011
1992	SEE UNDERHOOD SPECIFICATIONS STICKER											

NOTE: The Underhood Specifications sticker often reflects tune-up specification changes made in production. Sticker figures must be used if they disagree with those in this chart.

MT Manual transmission
AT Automatic transmissino
NA Not adjustable
A After Top Dead Center
B Before Top Dead Center
Hyd. Hydraulic valve lash adjusters

① Can. wagon w/MT: BPR5EY; 0.031 in.
② 2WD: 700
4WD: 800
③ Canada: 750–850
④ Canada: 700–800
⑤ Canada: 800–900
⑥ 2WD Federal and Canada: 700
2WD California and 4WD: 800
⑦ USA: 700
⑧ USA M/T, Canada A/T: 750
USA A/T: 700
Canada M/T: 850

FIRING ORDERS

NOTE: To avoid confusion, always replace spark plug wires one at a time.

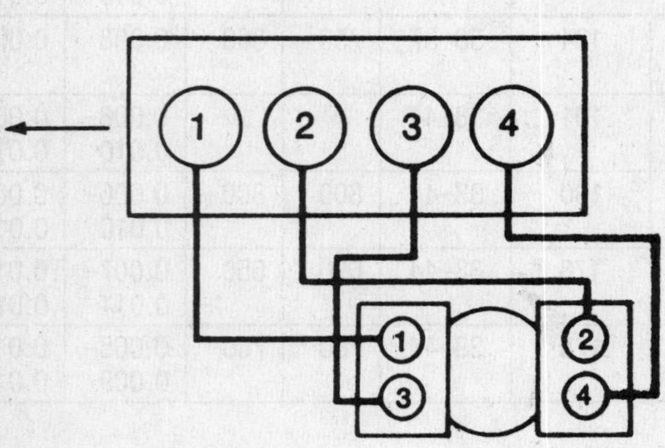

3A-C and 4A-F Engines
Engine Firing Order: 1-3-4-2
Distributor Rotation: Counterclockwise

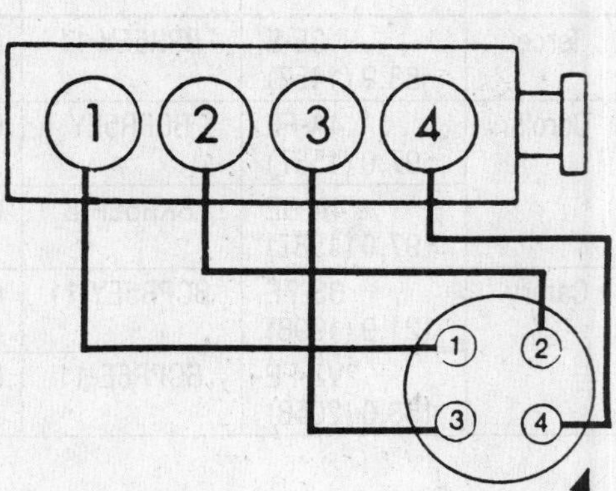

3E and 3E-E Engines
Engine Firing Order: 1-3-4-2
Distributor Rotation: Counterclockwise

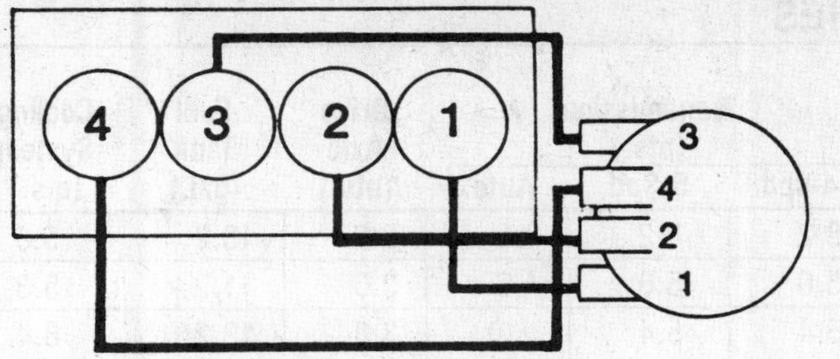

3S-FE, 3S-GE, 3S-GTE and 5S-FE Engines
Engine Firing Order: 1–3–4–2
Distributor Rotation: Counterclockwise

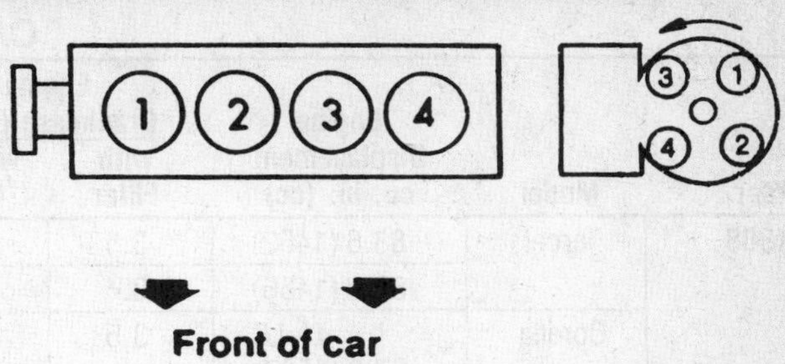

4A-F and 4A-FE Engines
Engine Firing Order: 1–3–4–2
Distributor Rotation: Counterclockwise

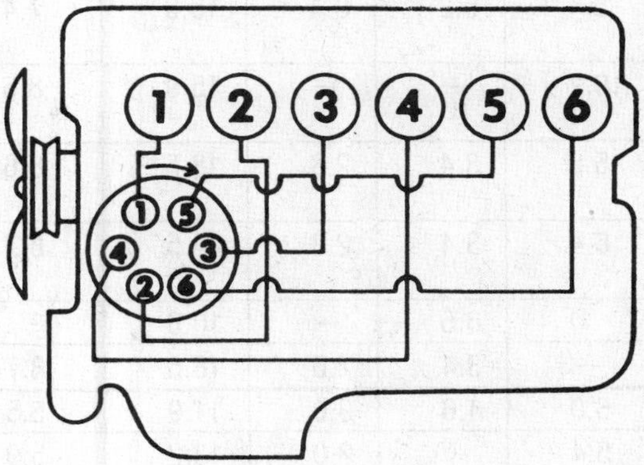

5M-GE Engine
Engine Firing Order: 1–5–3–6–2–4
Distributor Rotation: Clockwise

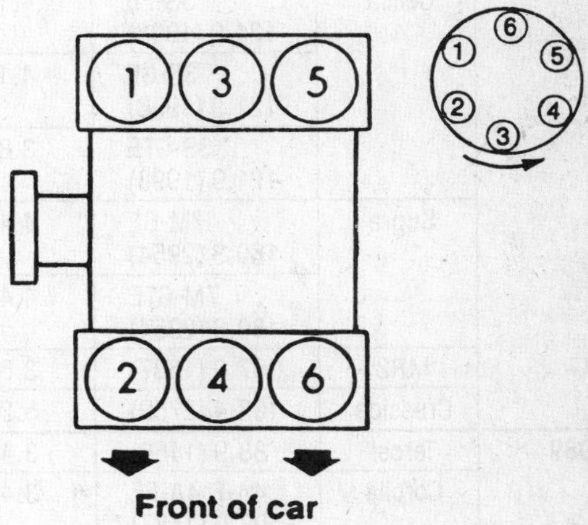

2VZ-FE Engine
Engine Firing Order: 1–2–3–4–5–6
Distributor Rotation: Counterclockwise

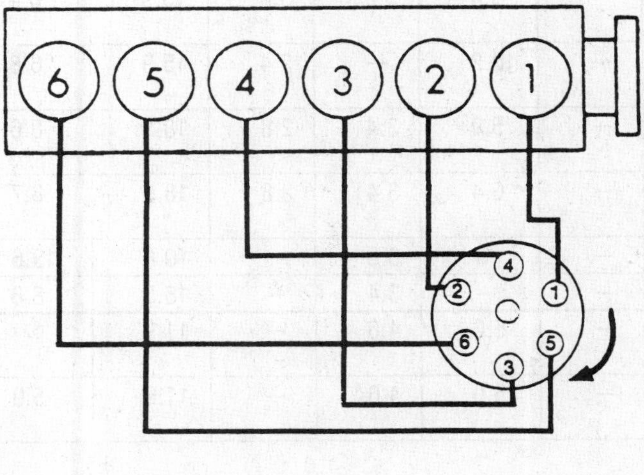

7M-GE Engine
Engine Firing Order: 1–5–3–6–2–4
Distributor Rotation: Clockwise

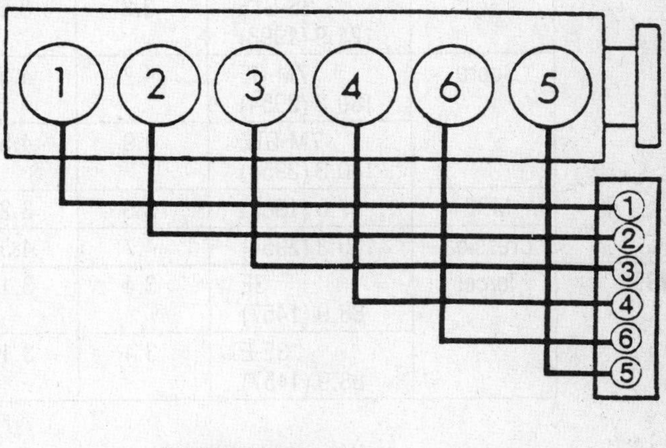

7M-GTE Engine
Engine Firing Order: 1–5–3–6–2–4
Distributorless Ignition System

CAPACITIES

Year	Model	Engine Displacement cu. in. (cc)	Engine Crankcase (qts.) with Filter	Engine Crankcase (qts.) without Filter	Transmission (pts.) 4-spd	Transmission (pts.) 5-Spd	Transmission (pts.) Auto.	Drive Axle (pts.)	Fuel Tank (gal.)	Cooling System (qts.)
1988	Tercel	88.6 (1452)	3.5	3.2	8.2	8.2	8.8	2.2	13.2	5.6
		88.9 (1456)	3.4	3.1	5.0	5.0	4.6	3.0	11.9	5.3
	Corolla	4A-LC 97.0 (1587)	3.5	3.2	5.4	5.4	①	3.0	13.2②	6.4
		4A-F 97.0 (1587)	3.3	3.2	5.4	5.4	①	3.0	13.2	6.3
		4A-GE 97.0 (1587)	3.9	3.5	5.4	5.4	①	3.0	13.2	6.3
	Camry	121.9 (1998)	4.1	3.9	5.4	5.4	5.2	3.4②	15.9	6.8
	Celica	3S-FE 121.9 (1998)	4.1	3.9	5.4	5.4	5.2	3.4	15.9	6.8
		3S-GE 121.9 (1998)	4.1	3.9	5.4	5.4	5.2	3.4	15.9	7.4
		3S-GTE 121.9 (1998)	3.8	3.6	—	10.2	—	—	15.9	8.5
	Supra	7M-GE 180.3 (2954)	4.4	4.1	—	5.0	3.4	2.8	18.5	8.6
		7M-GTE 180.3 (2954)	4.4	4.1	—	6.4	3.4	2.8	18.5	8.7
	MR2	97.0 (1587)	3.5	3.2	—	③	6.6	—	10.8	④
	Cressida	168.4 (2759)	5.2	4.9	—	—	3.4	2.6	18.5	8.7
1989	Tercel	88.9 (1456)	3.4	3.1	5.0	5.0	4.6	3.0	11.9	5.5
	Corolla	4A-F, 4A-FE 97.0 (1587)	3.4	3.2	5.4	5.4	①	3.0	13.2	5.9
		4A-GE 97.0 (1587)	3.9	3.6	5.4	5.4	①	3.0	13.2	6.3
	Camry	121.9 (1998)	4.1	3.9	—	5.4④	5.2	3.4②	15.9	6.8
		153.0 (2507)	4.1	3.9	—	—	5.2	2.2	15.9	9.0
	Celica	3S-FE 121.9 (1998)	4.1	3.9	—	5.4	5.2	3.4	15.9	6.6
		3S-GE 121.9 (1998)	4.1	3.8	—	5.4	5.2	3.4	15.9	6.4
		3S-GTE 121.9 (1998)	3.8	3.5	—	10.2	—	2.4	15.9	6.8
	Supra	7M-GE 180.3 (2954)	4.7	4.3	—	5.0	3.4	2.8	18.5	8.6
		7M-GTE 180.3 (2954)	4.9	4.5	—	6.4	3.4	2.8	18.5	8.7
	MR2	97.0 (1587)	3.5	3.2	—	③	6.6	—	10.8	13.6
	Cressida	180.3 (2954)	4.7	4.3	—	—	3.4	—	18.5	8.8
1990	Tercel	3E 88.9 (1457)	3.4	3.1	—	5.0	4.6	—	11.9	⑤
		3E-E 88.9 (1457)	3.4	3.1	—	5.0	4.6	—	11.9	5.9

CAPACITIES

Year	Model	Engine Displacement cu. in. (cc)	Engine Crankcase (qts.) with Filter	Engine Crankcase (qts.) without Filter	Transmission (pts.) 4-spd	Transmission (pts.) 5-Spd	Transmission (pts.) Auto.	Drive Axle (pts.)	Fuel Tank (gal.)	Cooling System (qts.)
1990	Corolla	4A-FE 97.0 (1587)	3.3	3.2	—	⑥	⑦	⑧	13.2	⑨
		4A-GE 97.0 (1587)	3.4	3.6	—	⑥	⑦	⑧	13.2	⑨
	Camry	3S-FE 121.9 (1998)	4.1	3.9	—	⑩	⑪	⑫	15.9	⑬
		2VZ-FE 132.0 (2508)	4.1	3.9	—	⑩	⑪	⑫	15.9	10.0
	Celica	4A-FE 97.0 (1587)	3.3	3.1	—	5.4	7.0	—	15.9	⑯
		3S-GTE 121.9 (1998)	3.8	3.5	—	10.2	—	2.4	15.9	6.8
		5S-FE 132.0 (2164)	⑭	⑮	—	5.4	7.0	—	15.9	⑰
	Supra	7M-GE 180.3 (2954)	4.7	4.3	—	5.0	3.4	1.8	18.5	8.6
		7M-GTE 180.3 (2954)	4.7	4.3	—	6.4	3.4	2.8	18.5	8.7
	MR2	3S-GTE 121.9 (1998)	4.1	3.8	—	5.4	7.0	—	10.8	14.4
		5S-FE 132.0 (2164)	4.4	4.0	—	8.8	7.0	—	10.8	13.7
	Cressida	180.3 (2954)	4.3	4.7	—	—	3.4	—	18.5	8.8
1991-92	Tercel	3E-E 88.9 (1457)	3.4	3.1	—	5.0	5.2	3.0	11.9	5.2⑱
	Corolla	4A-FE 97.0 (1587)	3.3	3.2	—	⑲	⑳	㉑	13.2	⑨
		4A-GE 97.0 (1587)	3.9	3.6	—	⑲	⑳	㉑	13.2	⑨
	Camry	3S-FE 121.9 (1998)	4.3	3.9	—	㉒	㉓	㉔	15.9	⑬
		2VZ-FE 153.0 (2058)	4.1	3.9	—	㉒	㉓	㉔	15.9	10.0
	Celica	4A-FE 97.0 (1587)	3.3	3.1	—	㉕	7.0	2.4	15.9	⑯
		3S-GTE 121.9 (1998)	4.1	3.8	—	㉕	7.0	2.4	15.9	6.9
		5S-FE 132.0 (2164)	⑭	⑮	—	㉕	7.0	2.4	15.9	⑰
	Supra	7M-GE 180.3 (2954)	4.7	4.3	—	5.0	3.4	2.8	18.5	8.6
		7M-GTE 180.3 (2954)	5.0	4.7	—	6.4	3.4	2.8	18.5	㉖
	MR2	3S-GTE 121.9 (1998)	4.1	3.8	—	5.4	7.0	—	10.8	14.4
		5S-FE 132.0 (2164)	4.4	4.0	—	8.8	7.0	—	10.8	13.7

CAPACITIES

Year	Model	Engine Displacement cu. in. (cc)	Engine Crankcase (qts.) with Filter	Engine Crankcase (qts.) without Filter	Transmission (pts.) 4-spd	Transmission (pts.) 5-Spd	Transmission (pts.) Auto.	Drive Axle (pts.)	Fuel Tank (gal.)	Cooling System (qts.)
1991-92	Cressida	7M-GE 180.3 (2954)	4.7	4.3	—	—	3.4	—	18.5	8.8

① A240E, A241H: 6.6
A131L: 5.2
② 4wd rear diff.: 2.4
③ C52: 5.4; E51: 8.8
④ 42d: 10.4
⑤ M/T: 5.5
M/T (EL31L-NGKB5A) 5.3
A/T: 5.4
⑥ C50, C52: 5.4
All-trac (E5SFS, ES7F5): 10.6
⑦ AL131L: 5.2
A240L: 6.6
A241H: 6.6
⑧ Transfer (A241H A/T only): 3.0
Rear differential: 2.4
⑨ M/T (except AE29L-ACMXRR): 5.9
M/T (AE29L-ACMXKK): 6.6
A/T: 6.4
⑩ S51: 5.4
ES6FS: 10.6
ES2: 8.8
⑪ A140L, A140E, A540E: 5.2
A540H: 7.0
⑫ Differential
SV21 A/T: 3.4
VZV21 A/T: 2.1
Transfer (A540H only): 1.48
⑬ M/T: 6.8
A/T (2WD): 6.7
A/T (4WD): 7.2
⑭ w/oil cooler: 4.4
w/out oil cooler: 4.3
⑮ w/oil cooler: 4.0
w/out oil cooler: 3.9
⑯ M/T: 5.5
A/T: 5.8
⑰ M/T: 6.6
A/T: 6.4
⑱ M/T: 5.2
A/T: 5.7
⑲ M/T (C50, C52): 5.4
M/T (E55F5, E57F5): 10.6
⑳ A131L: 5.2, A240L: 6.6, A241H:6.6
㉑ A131L only: 3.0
㉒ S51: 5.4, E52: 8.8
㉓ A140L, A140E: 0.8
A540E: 12.4, A540H: 14.8
㉔ A/T only SV21: 3.4, VZV21: 2.2
㉕ C52, S53: 5.4, E150F: 11.0
㉖ M/T: 8.7, A/T: 8.5

CAMSHAFT SPECIFICATIONS

All measurements given in inches.

Year	Engine Displacement cu. in. (cc)	Journal Diameter 1	2	3	4	5	6	7	Bearing Clearance	Camshaft End Play
1988	3A-C 88.6 (1452)	1.1015–1.1022	1.1015–1.1022	0.1015–1.1022	0.1015–1.1022	—	—	—	0.0015–0.0029	0.0031–0.0071
	3E 88.9 (1456)	1.0622–1.0628	1.0622–1.0628	1.0622–1.0628	1.0622–1.0628	—	—	—	0.0015–0.0029	0.0031–0.0071
	4A-F 97.0 (1587)	0.9035–0.9041 ③	0.9035–0.9041	0.9035–0.9041	0.9035–0.9041	—	—	—	0.0015–0.0028	②
	4A-LC 97.0 (1587)	1.1015–1.1022	1.1015–1.1022	0.1015–1.1022	0.1015–1.1022	—	—	—	0.0015–0.0029	0.0031–0.0071
	4A-GZE, 4A-GEC, 4A-GELC 97.0 (1587)	1.0610–1.0616	1.0610–1.0616	1.0610–1.0616	1.0610–1.0616	—	—	—	0.0014–0.0028	0.0031–0.0075
	3S-FE 121.9 (1998)	1.0614–1.0620	1.0614–1.0620	1.0614–1.0620	1.0614–1.0620	—	—	—	0.0010–0.0024	0.0018–0.0039
	3S-GE, 3S-GTE 121.9 (1998)	1.0614–1.0620	1.0614–1.0620	1.0614–1.0620	1.0614–1.0620	—	—	—	0.0010–0.0024	0.0047–0.0079 ④
	5M-GE 168.4 (2759)	1.4944–1.4951	1.6913–1.6919	1.7110–1.7116	1.7307–1.7313	1.7504–1.7510	1.7700–1.7707	1.7897–1.7907	0.0010–0.0026	0.0028–0.0098
	7M-GE, 7M-GTE 180.3 (2954)	1.0610–1.0616	1.0586–1.0620	1.0586–1.0620	1.0586–1.0620	1.0586–1.0620	1.0586–1.0620	1.0586–1.0620	0.0010–0.0037 ①	0.0031–0.0075

CAMSHAFT SPECIFICATIONS

All measurements given in inches.

Year	Engine Displacement cu. in. (cc)	Journal Diameter 1	2	3	4	5	6	7	Bearing Clearance	Camshaft End Play
1989	3E 88.9 (1456)	1.0622–1.0628	1.0622–1.0628	1.0622–1.0628	1.0622–1.0628	—	—	—	0.0015–0.0029	0.0031–0.0071
	4A-F, 4A-FE 97.0 (1587)	0.9035–0.9041 ③	0.9035–0.9041	0.9035–0.9041	0.9035–0.9041	—	—	—	0.0015–0.0028	②
	4A-GZE, 4A-GEC, 4A-GELC 97.0 (1587)	1.0610–1.0616	1.0610–1.0616	1.0610–1.0616	1.0610–1.0616	—	—	—	0.0014–0.0028	0.0031–0.0075
	3S-FE 121.9 (1998)	1.0614–1.0620	1.0614–1.0620	1.0614–1.0620	1.0614–1.0620	—	—	—	0.0010–0.0024	0.0018–0.0039
	3S-GE, 3S-GTE 121.9 (1998)	1.0614–1.0620	1.0614–1.0620	1.0614–1.0620	1.0614–1.0620	—	—	—	0.0010–0.0024	0.0047–0.0114
	2VZ-FE 153.0 (2507)	1.0610–1.0616	1.0610–1.0616	1.0610–1.0616	1.0610–1.0616	1.0610–1.0616	—	—	0.0014–0.0028	0.0012–0.0031
	7M-GE, 7M-GTE 180.3 (2954)	1.0610–1.0616	1.0586–1.0620	1.0586–1.0620	1.0586–1.0620	1.0586–1.0620	1.0586–1.0620	1.0586–1.0620	0.0010–0.0037 ①	0.0031–0.0075
1990	3E, 3E-E 88.9 (1457)	1.0622–1.0628	1.0622–1.0628	1.0622–1.0628	1.0622–1.0628	—	—	—	0.0015–0.0029	0.0031–0.0071
	4A-FE 97.0 (1587)	0.9035–0.9041 ③	0.9035–0.9041	0.9035–0.9041	0.9035–0.9041	—	—	—	0.0014–0.0028	②
	4A-GE 97.0 (1587)	1.0610–1.0616	1.0610–1.0616	1.0610–1.0616	1.0610–1.0616	—	—	—	0.0014–0.0028	0.0031–0.0075
	3S-FE 121.9 (1998)	1.0614–1.0620	1.0614–1.0620	1.0614–1.0620	1.0614–1.0620	—	—	—	0.0010–0.0024	⑤
	3S-GTE 121.9 (1998)	1.0614–1.0620	1.0614–1.0620	1.0614–1.0620	1.0614–1.0620	—	—	—	0.0010–0.0024	0.0047–0.0114
	5S-FE 132.0 (2164)	1.0614–1.0620	1.0614–1.0620	1.0614–1.0620	1.0614–1.0620	—	—	—	0.0010–0.0024	⑤
	2VZ-FE 153.0 (2507)	1.0610–1.0616	1.0610–1.0616	1.0610–1.0616	1.0610–1.0616	1.0610–1.0616	—	—	0.0014–0.0028	0.0012–0.0031
	7M-GE, 7M-GTE 180.3 (2954)	1.0610–1.0616	1.0586–1.0620	1.0586–1.0620	1.0586–1.0620	1.0586–1.0620	1.0586–1.0620	1.0586–1.0620	0.0010–0.0037 ①	0.0031–0.0075
1991-92	3E-E 88.9 (1457)	1.0622–1.0628	1.0622–1.0628	1.0622–1.0628	1.0622–1.0628	—	—	—	0.0015–0.0029	0.0031–0.0071
	4A-FE 97.0 (1587)	0.9035–0.9041 ③	0.9035–0.9041	0.9035–0.9041	0.9035–0.9041	—	—	—	0.0014–0.0028	②
	4A-GE 97.0 (1587)	1.0610–1.0616	1.0610–1.0616	1.0610–1.0616	1.0610–1.0616	—	—	—	0.0014–0.0028	0.0031–0.0071
	3S-FE 121.9 (1998)	1.0614–1.0620	1.0614–1.0620	1.0614–1.0620	1.0614–1.0620	—	—	—	0.0010–0.0024	⑤
	3S-GTE 121.9 (1998)	1.0614–1.0620	1.0614–1.0620	1.0614–1.0620	1.0614–1.0620	—	—	—	0.0010–0.0024	0.0047–0.0114
	5S-FE 132.0 (2164)	1.0614–1.0620	1.0614–1.0620	1.0614–1.0620	1.0614–1.0620	—	—	—	0.0010–0.0024	⑤
	2VZ-FE 153.0 (2507)	1.0610–1.0616	1.0610–1.0616	1.0610–1.0616	1.0610–1.0616	1.0610–1.0616	—	—	0.0014–0.0028	0.0012–0.0031
	7M-GE, 7M-GTE 180.3 (2954)	1.0610–1.0616	1.0586–1.0620	1.0586–1.0620	1.0586–1.0620	1.0586–1.0620	1.0586–1.0620	1.0586–1.0620	0.0010–0.0037 ①	0.0031–0.0075

① No. 1: 0.0014–0.0028
② Intake: 0.0012–0.0033; Exhaust: 0.0014–0.0035
③ Exhaust No. 1: 0.9822–0.9829
④ 35-GTE: 0.0039–0.0094
⑤ Intake: 0.0018–0.0039; Exhaust: 0.0012–0.0033

CRANKSHAFT AND CONNECTING ROD SPECIFICATIONS

All measurements are given in inches.

Year	Engine Displacement cu. in. (cc)	Crankshaft Main Brg. Journal Dia.	Crankshaft Main Brg. Oil Clearance	Crankshaft Shaft End-play	Crankshaft Thrust on No.	Connecting Rod Journal Diameter	Connecting Rod Oil Clearance	Connecting Rod Side Clearance
1988	3A-C 88.6 (1452)	1.8891–1.8898	0.0006–0.0013	0.0008–0.0087	3	1.5742–1.5748	0.0008–0.0020	0.0059–0.0098
	3E 88.9 (1456)	1.9683–1.9685	0.0006–0.0014	0.0008–0.0087	3	1.8110–1.8113	0.0006–0.0019	0.0059–0.0138
	3S-FE, 3S-GE, 3S-GTE 121.9 (1998)	2.1648–2.1653	0.0007–0.0015 ①	0.0008–0.0087	3	1.8892–1.8898	0.0009–1.0022	0.0063–0.0123
	4A-F, 4A-LC 97.0 (1587)	1.8891–1.8898	0.0006–0.0013	0.0008–0.0087	3	1.5742–1.5748	0.0008–0.0020	0.0059–0.0098
	4A-GEC, 4A-GELC 97.0 (1587)	1.8891–1.8898	0.0005–0.0015	0.0008–0.0087	3	1.5742–1.5748	0.0008–0.0020	0.0059–0.0098
	4A-GZE 97.0 (1587)	1.8891–1.8898	0.0006–0.0013	0.0008–0.0087	3	1.6529–1.6535	0.0008–0.0020	0.0059–0.0098
	5M-GE 168.4 (2759)	2.3625–2.3627	0.0012–0.0048	0.0020–0.0098	4	2.1659–2.1663	0.0008–0.0021	0.0063–0.0117
	7M-GE, 7M-GTE 180.3 (2954)	2.3625–2.3627	0.0012–0.0048	0.0020–0.0098	4	2.1659–2.1663	0.0008–0.0021	0.0063–0.0117
1989	3E 88.9 (1456)	1.9683–1.9685	0.0006–0.0014	0.0008–0.0087	3	1.8110–1.8113	0.0006–0.0019	0.0059–0.0138
	3S-FE, 3S-GE, 3S-GTE 121.9 (1998)	2.1649–2.1655	0.0006–0.0013 ①	0.0008–0.0087	3	1.8892–1.8898	0.0009–0.0022	0.0063–0.0123
	4A-F, 4A-FE 97.0 (1587)	1.8891–1.8898	0.0006–0.0013	0.0008–0.0087	3	1.5742–1.5748	0.0008–0.0020	0.0059–0.0098
	4A-GEC, 4A-GELC 97.0 (1587)	1.8891–1.8898	0.0006–0.0013	0.0008–0.0087	3	1.6529–1.6535	0.0008–0.0020	0.0059–0.0098
	4A-GZE 97.0 (1587)	1.8891–1.8898	0.0006–0.0013	0.0008–0.0087	3	1.6529–1.6535	0.0008–0.0020	0.0059–0.0098
	2VZ-FE 153.0 (2507)	2.5191–2.5197	0.0011–0.0022	0.0008–0.0087	3	1.8892–1.8898	0.0011–0.0026	0.0059–0.0130
	7M-GE, 7M-GTE 180.3 (2954)	2.3625–2.3627	0.0012–0.0019	0.0020–0.0098	4	2.1659–2.1663	0.0008–0.0021	0.0063–0.0117
1990	3E, 3E-E 88.9 (1457)	1.9683–1.9685	0.0006–0.0014	0.0008–0.0087	3	1.8110–1.8113	0.0006–0.0019	0.0059–0.0138
	4A-FE 97.0 (1587)	1.8891–1.8898	0.0006–0.0013	0.0008–0.0087	3	1.5742–1.5748	0.0008–0.0020	0.0059–0.0098
	4A-GE 97.0 (1587)	1.8891–1.8898	0.0006–0.0013	0.0008–0.0087	3	1.6529–1.6535	0.0008–0.0020	0.0059–0.0098
	3S-FE 121.9 (1998)	2.1649–2.1655	0.0010–0.0017	0.0008–0.0087	3	1.8892–1.8898	0.0009–0.0022	0.0063–0.0123
	3S-GTE 121.9 (1998)	2.1653–2.1655	0.0006–0.0013 ①	0.0008–0.0087	3	1.8892–1.8898	0.0009–0.0022	0.0063–0.0123
	5S-FE 132.0 (2164)	2.1653–2.1655	0.0006–0.0013 ①	0.0008–0.0087	3	1.8892–1.8898	0.0009–0.0022	0.0063–0.0123
	2VZ-FE 153.0 (2507)	2.5191–2.5197	0.0011–0.0022	0.0008–0.0087	3	1.8892–1.8898	0.0011–0.0026	0.0059–0.0123
	7M-GE, 7M-GTE 180.3 (2954)	2.3625–2.3627	0.0012–0.0019	0.0020–0.0098	4	2.0470–2.0474	0.0008–0.0021	0.0063–0.0117

CRANKSHAFT AND CONNECTING ROD SPECIFICATIONS

All measurements are given in inches.

Year	Engine Displacement cu. in. (cc)	Crankshaft Main Brg. Journal Dia.	Crankshaft Main Brg. Oil Clearance	Crankshaft Shaft End-play	Crankshaft Thrust on No.	Connecting Rod Journal Diameter	Connecting Rod Oil Clearance	Connecting Rod Side Clearance
1991-92	3E-E 88.9 (1457)	1.9683–1.9685	0.0006–0.0013	0.0008–0.0087	3	1.8110–1.8113	0.0006–0.0019	0.0059–0.0138
	4A-FE 97.0 (1587)	1.8891–1.8898	0.0006–0.0013	0.0008–0.0087	3	1.5742–1.5748	0.0008–0.0020	0.0059–0.0098
	4A-GE 97.0 (1587)	1.8891–1.8898	0.0006–0.0013	0.0008–0.0087	3	1.6529–1.6535	0.0008–0.0020	0.0059–0.0098
	3S-FE 121.9 (1998)	2.1649–2.1655	0.0010–0.0017	0.0008–0.0087	3	1.8892–1.8898	0.0009–0.0022	0.0063–0.0123
	3S-GTE 121.9 (1998)	2.1653–2.1655	0.0006–0.0013 ①	0.0008–0.0087	3	1.8892–1.8898	0.0009–0.0022	0.0063–0.0123
	5S-FE 132.0 (2164)	2.1653–2.1655	0.0006–0.0013 ①	0.0008–0.0087	3	1.8892–1.8898	0.0009–0.0022	0.0063–0.0123
	2VZ-FE 153.0 (2507)	2.5191–2.5197	0.0011–0.0022	0.0008–0.0087	3	1.8892–1.8898	0.0011–0.0026	0.0059–0.0123
	7M-GE, 7M-GTE 180.3 (2954)	2.3625–2.3627	0.0012–0.0019	0.0020–0.0098	4	2.0470–2.0474	0.0008–0.0021	0.0063–0.0117

① No. 3: 0.0011–0.0019 (1988)
No. 3: 0.0010–0.0017 (1989–91)

VALVE SPECIFICATIONS

Year	Engine Displacement cu. in. (cc)	Seat Angle (deg.)	Face Angle (deg.)	Spring Test Pressure (lbs.)	Spring Installed Height (in.)	Stem-to-Guide Clearance (in.) Intake	Stem-to-Guide Clearance (in.) Exhaust	Stem Diameter (in.) Intake	Stem Diameter (in.) Exhaust
1988	3A-C 88.6 (1452)	45	44.5	52.0	1.520	0.0010–0.0024	0.0012–0.0026	0.2744–0.2750	0.2742–0.2748
	3E 88.9 (1456)	45	44.5	35.1	1.384	0.0010–0.0024	0.0012–0.0026	0.2350–0.2356	0.2348–0.2354
	3S-FE 121.9 (1998)	45.5	44.5	39.6	1.366	0.0010–0.0024	0.0012–0.0026	0.2350–0.2356	0.2348–0.2354
	3S-GE, 3S-GTE 121.9 (1998)	45.5	44.5	38.6 ③	1.366	0.0010–0.0023	0.0012–0.0025	0.2346–0.2352	0.2344–0.2350
	4A-F 97.0 (1587)	45	44.5	34.8	1.366	0.0010–0.0024	0.0012–0.0026	0.2350–0.2356	0.2348–0.2354
	4A-LC 97.0 (1587)	45	44.5	52.0	1.520	0.0010–0.0024	0.0012–0.0026	0.2744–0.2750	0.2742–0.2748
	4A-GEC, 4A-GELC 4A-GZE 97.0 (1587)	45	44.5	35.9	1.366	0.0010–0.0024	0.0012–0.0026	0.2350–0.2356	0.2348–0.2354
	5M-GE 168.4 (2759)	45	44.5	①	②	0.0010–0.0024	0.0012–0.0026	0.3138–0.3144	0.3136–0.3142
	7M-GE, 7M-GTE 180.3 (2954)	45	44.5	35.0	1.378	0.0010–0.0024	0.0012–0.0026	0.2350–0.2356	0.2348–0.2354
1989	3E 88.9 (1456)	45	44.5	35.1	1.384	0.0010–0.0024	0.0012–0.0026	0.2350–0.2356	0.2348–0.2354
	3S-FE 121.9 (1998)	45.5	44.5	39.6	1.366	0.0010–0.0024	0.0012–0.0026	0.2350–0.2356	0.2348–0.2354

VALVE SPECIFICATIONS

Year	Engine Displacement cu. in. (cc)	Seat Angle (deg.)	Face Angle (deg.)	Spring Test Pressure (lbs.)	Spring Installed Height (in.)	Stem-to-Guide Clearance (in.) Intake	Stem-to-Guide Clearance (in.) Exhaust	Stem Diameter (in.) Intake	Stem Diameter (in.) Exhaust
1989	3S-GE, 3S-GTE 121.9 (1998)	45.5	44.5	38.6 ③	1.366	0.0010–0.0023	0.0012–0.0025	0.2346–0.2352	0.2344–0.2350
	4A-F, 4A-FE 97.0 (1587)	45	44.5	34.8	1.366	0.0010–0.0024	0.0012–0.0026	0.2350–0.2356	0.2348–0.2354
	4A-GEC, 4A-GZE 97.0 (1587)	45	44.5	34.7	1.366	0.0010–0.0024	0.0012–0.0026	0.2350–0.2356	0.2348–0.2354
	2VZ-FE 153.0 (2507)	45	44.5	41.0–47.2	1.331	0.0010–0.0024	0.0012–0.0026	0.2350–0.2356	0.2348–0.2354
	7M-GE, 7M-GTE 180.3 (2954)	45	44.5	35.0	1.378	0.0010–0.0024	0.0012–0.0026	0.2350–0.2356	0.2348–0.2354
1990	3E, 3E-E 88.9 (1457)	45	44.5	35.1	1.3842	0.0010–0.0024	0.0012–0.0026	0.2350–0.2356	0.2348–0.2354
	4A-FE 97.0 (1587)	45	45.5	34.8	1.366	0.0010–0.0024	0.0012–0.0026	0.2350–0.2356	0.2348–0.2354
	4A-GE 97.0 (1587)	45	44.5	35.9	1.366	0.0010–0.0024	0.0012–0.0026	0.2350–0.2356	0.2348–0.2354
	3S-FE 121.9 (1998)	45	45.5	42.5	1.366	0.0010–0.0024	0.0012–0.0026	0.2350–0.2356	0.2348–0.2354
	3S-GTE 121.9 (1998)	45	45.5	53.1	1.354	0.0010–0.0023	0.0012–0.0025	0.2346–0.2352	0.2344–0.2350
	5S-FE 132.0 (2164)	45	45.5	42.5	1.366	0.0010–0.0024	0.0012–0.0026	0.2350–0.2356	0.2348–0.2354
	2VZ-FE 153.0 (2507)	45	45.5	47.2	1.331	0.0010–0.0024	0.0012–0.0026	0.2350–0.2356	0.2348–0.2354
	7M-GE, 7M-GTE 180.3 (2954)	45	45.5	35	1.378	0.0010–0.0024	0.0012–0.0026	0.2350–0.2356	0.2348–0.2354
1991–92	3E-E 88.9 (1457)	45	44.5	35.1	1.3842	0.0010–0.0024	0.0012–0.0026	0.2350–0.2356	0.2348–0.2354
	4A-FE 97.0 (1587)	45	45.5	34.8	1.366	0.0010–0.0024	0.0012–0.0026	0.2350–0.2356	0.2348–0.2354
	4A-GE 97.0 (1587)	45	44.5	35.9	1.366	0.0010–0.0024	0.0012–0.0026	0.2350–0.2356	0.2348–0.2354
	3S-FE 121.9 (1998)	45	44.5	42.5	1.366	0.0010–0.0024	0.0012–0.0026	0.2350–0.2356	0.2348–0.2354
	3S-GTE 121.9 (1998)	45	44.5	53.1	1.354	0.0010–0.0023	0.0012–0.0025	0.2346–0.2352	0.2344–0.2350
	5S-FE 132.0 (2164)	45	44.5	42.5	1.366	0.0010–0.0023	0.0012–0.0026	0.2350–0.2356	0.2348–0.2354
	2VZ-FE 153.0 (2507)	45	44.5	47.2	1.331	0.0010–0.0024	0.0012–0.0026	0.2350–0.2356	0.2348–0.2354
	7M-GE, 7M-GTE 180.3 (2954)	45	44.5	35	1.378	0.0010–0.0024	0.0012–0.0026	0.2350–0.2356	0.2348–0.2354

① Intake: 76.5–84.4; Exhaust: 73.4–80.9
② Intake: 1.575; Exhaust: 1.693
③ 3S-GTE: 44.1

PISTON AND RING SPECIFICATIONS

All measurements are given in inches.

Year	Engine Displacement cu. in. (cc)	Piston Clearance	Ring Gap Top Compression	Ring Gap Bottom Compression	Ring Gap Oil Control	Ring Side Clearance Top Compression	Ring Side Clearance Bottom Compression	Ring Side Clearance Oil Control
1988	3A-C 88.6 (1452)	0.0039–0.0047	0.0079–0.0185	0.0079–0.0204	0.0118–0.0402	0.0016–0.0031	0.0012–0.0028	Snug
	3E 88.9 (1456)	0.0028–0.0035	0.0102–0.0142	0.0118–0.0177	0.0059–0.0157	0.0016–0.0031	0.0012–0.0028	Snug
	3S-FE 121.9 (1998)	0.0018–0.0026	0.0106–0.0205	0.0106–0.0209	0.0079–0.0323	0.0018–0.0028	0.0018–0.0028	Snug
	3S-GE 121.9 (1998)	0.0012–0.0020	0.0130–0.0264	0.0177–0.0323	0.0079–0.0283	0.0012–0.0028	0.0008–0.0024	Snug
	3S-GTE 121.9 (1998)	0.0012–0.0020	0.0130–0.0224	0.0177–0.0272	0.0079–0.0244	0.0015–0.0031	0.0012–0.0028	Snug
	4A-F 97.0 (1587)	0.0024–0.0031	0.0098–0.0138	0.0059–0.0118	0.0039–0.0236	0.0016–0.0031	0.0012–0.0028	Snug
	4A-LC 97.0 (1587)	0.0035–0.0043	0.0098–0.0138	0.0059–0.0165	0.0078–0.0276	0.0016–0.0031	0.0012–0.0028	Snug
	4A-GEC, 4A-GELC, 4A-GZE 97.0 (1587)	0.0039–0.0047 ①	0.0098–0.0185	0.0078–0.0118	②	0.0016–0.0031	0.0012–0.0028	Snug
	5M-GE 168.4 (2759)	0.0024–0.0031	0.0091–0.0150	0.0098–0.0209	0.0040–0.0201	0.0012–0.0028	0.0008–0.0024	Snug
	7M-GE 180.3 (2954)	0.0020–0.0028	0.0091–0.0150	0.0098–0.0209	0.0039–0.0157	0.0012–0.0028	0.0008–0.0024	Snug
	7M-GTE 180.3 (2954)	0.0028–0.0035	0.0114–0.0173	0.0098–0.0209	0.0039–0.0173	0.0012–0.0028	0.0008–0.0024	Snug
1989	3E 88.9 (1456)	0.0028–0.0035	0.0102–0.0142	0.0118–0.0177	0.0059–0.0157	0.0016–0.0031	0.0012–0.0028	Snug
	3S-FE 121.9 (1998)	0.0018–0.0026	0.0106–0.0197	0.0106–0.0201	0.0079–0.0217	0.0012–0.0028	0.0012–0.0028	Snug
	3S-GE 121.9 (1998)	0.0012–0.0020	0.0130–0.0217	0.0177–0.0276	0.0079–0.0236	0.0012–0.0028	0.0008–0.0024	Snug
	3S-GTE 121.9 (1998)	0.0012–0.0020	0.0130–0.0217	0.0177–0.0264	0.0079–0.0236	0.0015–0.0031	0.0012–0.0028	Snug
	4A-F 97.0 (1587)	0.0024–0.0031	0.0098–0.0138	0.0059–0.0118	0.0039–0.0236	0.0016–0.0031	0.0012–0.0028	Snug
	4A-FE 97.0 (1587)	0.0024–0.0031	0.0098–0.0138	0.0059–0.0118	0.0039–0.0236	0.0020–0.0031	0.0012–0.0028	Snug
	4A-GEC, 4A-GELC, 4A-GZE 97.0 (1587)	0.0039–0.0047 ①	0.0098–0.0185	0.0079–0.0165	②	0.0016–0.0031	0.0012–0.0028	Snug
	2VZ-FE 153.0 (2507)	0.0018–0.0026	0.0118–0.0205	0.0138–0.0236	0.0079–0.0217	0.0004–0.0031	0.0012–0.0028	Snug
	7M-GE 180.3 (2954)	0.0020–0.0028 ③	0.0091–0.0150	0.0098–0.0209	0.0039–0.0157	0.0012–0.0028	0.0008–0.0024	Snug
	7M-GTE 180.3 (2954)	0.0028–0.0035	0.0114–0.0173	0.0098–0.0209	0.0039–0.0173	0.0012–0.0028	0.0008–0.0024	Snug

PISTON AND RING SPECIFICATIONS

All measurements are given in inches.

Year	Engine Displacement cu. in. (cc)	Piston Clearance	Ring Gap Top Compression	Ring Gap Bottom Compression	Ring Gap Oil Control	Ring Side Clearance Top Compression	Ring Side Clearance Bottom Compression	Ring Side Clearance Oil Control
1990	3E, 3E-E 88.9 (1457)	0.0028–0.0035	0.0102–0.0189	0.0118–0.0224	0.0059–0.0205	0.0016–0.0031	0.0012–0.0028	Snug
	4A-FE 97.0 (1587)	0.0024–0.0031	0.0098–0.0177	0.0059–0.0157	0.0039–0.0276	0.0016–0.0031	0.0012–0.0028	Snug
	4A-GE 97.0 (1587)	0.0039–0.0047	0.0098–0.0185	0.0079–0.0165	0.0059–0.0205	0.0012–0.0031	0.0012–0.0028	Snug
	3S-FE 121.9 (1998)	0.0018–0.0026	0.0118–0.0205	0.0138–0.0236	0.0079–0.0217	0.0004–0.0031	0.0012–0.0028	Snug
	3S-GTE 121.9 (1998)	0.0028–0.0035	0.0130–0.0217	0.0177–0.0264	0.0079–0.0236	0.0016–0.0031	0.0012–0.0028	Snug
	5S-FE 132.0 (2164)	0.0031–0.0039	0.0106–0.0197	0.0138–0.0234	0.0079–0.0217	0.0012–0.0028	0.0012–0.0028	Snug
	2VZ-FE 153.0 (2507)	0.0018–0.0026	0.0118–0.0205	0.0138–0.0236	0.0079–0.0217	0.0004–0.0031	0.0012–0.0028	Snug
	7M-GE 180.3 (2954)	0.0031–0.0039	0.0091–0.0150	0.0098–0.0209	0.0039–0.0157	0.0012–0.0028	0.0008–0.0024	Snug
	7M-GTE 180.3 (2954)	0.0028–0.0035	0.0114–0.0173	0.0098–0.0209	0.0039–0.0173	0.0008–0.0024	0.0008–0.0024	Snug
1991-92	3E-E 88.9 (1457)	0.0028–0.0035	0.0102–0.0189	0.0118–0.0224	0.0059–0.0205	0.0016–0.0031	0.0012–0.0028	Snug
	4A-FE 97.0 (1587)	0.0024–0.0031	0.0098–0.0177	0.0059–0.0157	0.0039–0.0276	0.0016–0.0031	0.0012–0.0028	Snug
	4A-GE 97.0 (1587)	0.0039–0.0047	0.0098–0.0185	0.0079–0.0165	0.0059–0.0205	0.0012–0.0031	0.0012–0.0028	Snug
	3S-FE 121.9 (1998)	0.0018–0.0026	0.0118–0.0205	0.0138–0.0236	0.0079–0.0217	0.0004–0.0031	0.0012–0.0028	Snug
	3S-GTE 121.9 (1998)	0.0028–0.0035	0.0130–0.0217	0.0177–0.0264	0.0079–0.0236	0.0016–0.0031	0.0012–0.0028	Snug
	5S-FE 132.0 (2164)	0.0031–0.0039	0.0106–0.0197	0.0138–0.0234	0.0079–0.0217	0.0012–0.0028	0.0012–0.0028	Snug
	2VZ-FE 153.0 (2507)	0.0018–0.0026	0.0118–0.0205	0.0138–0.0236	0.0079–0.0217	0.0004–0.0031	0.0012–0.0028	Snug
	7M-GE 180.3 (2954)	0.0031–0.0039	0.0091–0.0150	0.0098–0.0209	0.0039–0.0157	0.0012–0.0028	0.0008–0.0024	Snug
	7M-GTE 180.3 (2954)	0.0028–0.0035	0.0114–0.0173	0.0098–0.0209	0.0039–0.0173	0.0012–0.0024	0.0008–0.0024	Snug

① 4A-GZE: 0.0047–0.0055

② Code T: 0.0059–0.0205
Code R: 0.0118–0.0402

③ 1990 Supra: 0.0031–0.0039

TORQUE SPECIFICATIONS

All readings in ft. lbs.

Year	Engine Displacement cu. in. (cc)	Cylinder Head Bolts	Main Bearing Bolts	Rod Bearing Bolts	Crankshaft Pulley Bolts	Flywheel Bolts	Manifold Intake	Manifold Exhaust	Spark Plugs
1988	3A-C 88.6 (1452)	40–47	40–47	34–39	80–94	55–61	15–21	15–21	16–20
	3E 88.9 (1456)	②	40–47	27–31	105–117	60–70	11–17	33–42	11–15
	3S-FE 121.9 (1998)	45–50	40–45	33–38	78–82	70–75	11–17	27–33	11–15
	3S-GE, 3S-GTE 121.9 (1998)	38–42	40–45	44–50	78–82	①	12–16	30–34 ③	11–15
	4A-F 97.0 (1587)	40–47	40–47	32–40	80–94	55–61	11–17	15–21	11–15
	4A-LC 97.0 (1587)	40–47	40–47	32–40	80–94	55–61	15–21	15–21	16–20
	4A-GEC, 4A-GELC 4A-GZE 97.0 (1587)	④	40–47	32–40	100–110	50–58	15–21	27–31	11–15
	5M-GE 168.4 (2759)	55–61	72–78	31–34	185–205	51–57	11–15	26–32	11–15
	7M-GE, 7M-GTE 180.2 (2954)	55–61	72–78	45–49	185–205	51–57	11–15	26–32	11–15
1989	3E 88.9 (1456)	②	40–47	27–31	105–117	60–70	11–17	33–42	11–15
	3S-FE 121.9 (1998)	45–50	40–45	33–38	78–82	70–75	11–17	27–33	11–15
	3S-GE, 3S-GTE 121.9 (1998)	38–42	40–45	44–50	78–82	①	12–16	30–34	11–15
	4A-F, 4A-FE 97.0 (1587)	40–47	40–47	32–40	80–94	55–61	11–17	15–21	11–15
	4A-GEC, 4A-GELC 4A-GZE 97.0 (1587)	②	40–47	④	95–105	50–58	18–22	27–31	11–15
	2VZ-FE 153.0 (2507)	②	43–47	16–20	176–186	58–64	11–15	26–32	11–15
	7M-GE, 7M-GTE 180.2 (2954)	55–61	72–78	45–49	185–205	51–57	11–15	26–32	11–15
1990	3E, 3E-E 88.9 (1457)	⑨	42	29	112	88	14	38	13
	4A-FE 97.0 (1587)	44 ⑩	44	36	87	58	14	18	13
	4A-GE 97.0 (1587)	④	44	29	101	54	9	29	13
	3S-FE 121.9 (1998)	⑥	43	36	80	⑤	14	36	13
	3S-GTE 121.9 (1998)	⑥	43	49	80	80	14	38	13
	5S-FE 132.0 (2164)	⑥	43	⑦	80	⑤	14	36	13
	2VZ-FE 153.0 (2507)	⑬	⑧	⑨	181	61	13	13	13

TORQUE SPECIFICATIONS

All readings in ft. lbs.

Year	Engine Displacement cu. in. (cc)	Cylinder Head Bolts	Main Bearing Bolts	Rod Bearing Bolts	Crankshaft Pulley Bolts	Flywheel Bolts	Manifold Intake	Manifold Exhaust	Spark Plugs
1990	7M-GE, 7M-GTE 180.3 (2954)	58	75	47	195	54	13	29	13
1991-92	3E-E 88.9 (1457)	⑨	42	29	112	65	14	35	13
	4A-FE 97.0 (1587)	44 ⑩	44	36	87	58	14	18	13
	4A-GE 97.0 (1587)	⑪	44	⑫	101	54	20	29	13
	3S-FE 121.9 (1998)	⑥	43	36	80	65	14	36	13
	3S-GTE 121.9 (1998)	⑥	43	49	80	80	14	38	13
	5S-FE 132.0 (2164)	⑥	43	⑦	80	65	14	36	13
	2VZ-FE 153.0 (2507)	⑬	⑧	⑦	181	61	13	29	13
	7M-GE, 7M-GTE 180.3 (2954)	58 ⑩	75	47	195	54	13	29	13

① New: 65; Used:63
② See text
③ 3S-GTE: 38
④ 29 ft.lbs. and an additional 90° turn
⑤ M/T: 65
A/T: 61
⑥ 36 ft. lbs. and an additional 90° turn
⑦ 18 ft. lbs. and an additional 90° turn
⑧ 45 ft. lbs. and an additional 90° turn
⑨ 22 ft. lbs., 36 ft. lbs. and an additional 90° turn
⑩ Torque in sequence, in 3 steps
⑪ 22 ft. lbs., an additional 90° turn plus an additional 90° turn
⑫ 29 ft. lbs. plus an additional 90° turn
⑬ 25 ft. lbs., an additional 90° turn plus an additional 90° turn

BRAKE SPECIFICATIONS

All measurements in inches unless noted.

Year	Model	Lug Nut Torque (ft. lbs.)	Master Cylinder Bore	Brake Disc Minimum Thickness	Brake Disc Maximum Runout	Standard Brake Drum Diameter	Minimum Lining Thickness Front	Minimum Lining Thickness Rear
1988	Tercel	65–86	①	0.394	0.006	7.126 ②	0.040	0.040
	Corolla	65–86	①	⑤	0.006	7.874	0.040	0.040
	Camry	65–86	①	0.945 ⑦	0.003	9.079	0.040	0.040
	Celica	65–86	①	0.827 ⑥	0.006	7.913	0.040	0.040
	Supra	65–86	①	0.827 ④	0.005	—	0.040	0.040
	MR2	65–86	①	0.827 ③	0.006	—	0.040	0.040
	Cressida	65–86	①	0.827 ④	0.006	—	0.040	0.040

BRAKE SPECIFICATIONS

All measurements in inches unless noted.

Year	Model	Lug Nut Torque (ft. lbs.)	Master Cylinder Bore	Brake Disc Minimum Thickness	Brake Disc Maximum Runout	Standard Brake Drum Diameter	Minimum Lining Thickness Front	Minimum Lining Thickness Rear
1989	Tercel	65–86	①	0.394	0.006	7.087 ②	0.040	0.040
	Corolla	65–86	①	⑤	0.004	7.874	0.040	0.040
	Camry	65–86	①	0.945 ⑦	0.003 ⑧	7.874	0.040	0.040
	Celica	65–86	①	0.827 ⑥	0.003	7.874	0.040	0.040
	Supra	65–86	①	0.827 ④	0.005	—	0.040	0.040
	MR2	65–86	①	0.827 ③	0.006	—	0.040	0.040
	Cressida	65–86	①	0.827 ④	0.003 ⑧	—	0.040	0.040
1990	Tercel	65–86	①	0.394	0.0059	7.087	0.394	0.039
	Corolla	65–86	①	⑨	⑩	7.874	0.039	0.039
	Camry	65–86	①	⑪	⑫	7.874	0.039	0.039
	Celica	65–86	①	⑬	⑫	7.874	0.039	0.039
	Supra	65–86	①	0.827 ⑥	0.005	—	0.039	0.039
	MR2	65–86	①	⑭	⑮	—	0.039	0.039
	Cressida	65–86	①	⑯	⑰	—	0.039	0.039
1991–92	Tercel	65–86	①	0.709	0.0035	7.087	0.039	0.039
	Corolla	65–86	①	⑨	⑩	7.874	0.039	0.039
	Camry	65–86	①	⑪	⑫	7.874	0.039	0.039
	Celica	65–86	①	⑲	⑫	7.874	0.039	0.039
	Supra	65–86	①	⑯	0.005	—	0.039	0.039
	MR2	65–86	①	⑭	⑮	—	0.039	0.039
	Cressida	65–86	①	⑯	⑰	—	0.039	0.039

① Not specified by the manufacturer
② Wagon & 4wd: 7.913
③ Rear disc: 0.354
④ Rear disc: 0.669
⑤ 1987 FWD (exc. FX16): 0.492
1988–89 FWD & FX16: 0.669
1987 RWD: 0.669
1987 Rear disc: 0.315
1988–89 Rear disc: 0.354
⑥ ABS or 4wd: 0.945
⑦ 4wd rear disc: 0.354
⑧ Rear disc: 0.006
⑨ Front:
4A-FE—0.669
4A-GE—0.827
Rear: 0.315
⑩ Front disc: 0.0035
Rear disc: 0.0039
⑪ Front: 0.945
Rear: 0.354
⑫ Front: 0.0028
Rear: 0.0059
⑬ Front: 0.787
Rear: 0.354
⑭ Front: 0.945
Rear: 0.591
⑮ Front: 0.0028
Rear: 0.0039
⑯ Front: 0.827
Rear: 0.669
⑰ Front: 0.0028
Rear: 0.0052
⑱ Front: 0.866
Rear: 0.709
⑲ Front
2WD: 0.787
4WD: 0.906
Rear: 0.354

WHEEL ALIGNMENT

Year	Model	Caster Range (deg.)	Caster Preferred Setting (deg.)	Camber Range (deg.)	Camber Preferred Setting (deg.)	Toe-in (in.)	Steering Axis Inclination (deg.)
1988	Tercel (Sedan)	③	①	3/4N–3/4P	0	0.08 out–0.08 in	11 1/2
	(Wagon)	1 11/16P–3 3/16P	2 1/4P	3/16N–1 5/16P	9/16P	0.12 out–0.04 in	12
	Corolla (FX/FX16)	1/8P–1 5/8P	7/8P	1N–1/2P	1/4N	0.04 out–0.12 in	12 1/2
	(4A-F)	⑤	⑥	15/16N–9/16P	3/16N	0–0.08	12 11/16
	(4A-GE)	9/16P–2 1/16P	1 5/16P	1N–1/2P	1/4N	0–0.08	12 13/16
	Camry (Sedan)	1 5/16P–2 7/16P	1 11/16P	3/16N–1 5/16P	9/16P	0.04 out–0.12 in	12 3/4
	(Wagon)	1/4P–1 3/4P	1P	1/4N–1 1/4P	1/2P	0.04 out–0.12 in	13
	Celica	7/16P–1 15/16P	1 3/16P	15/16N–9/16P	3/16N	0.08 out–0.08 in	13 1/2
	Supra	6 3/4P–8 1/4P	7 1/2P	13/16N–11/16P	1/16N	0.08 out–0.08 in	11
	MR2	4 15/16P–5 13/16P	5 1/16P	1/2N–1P	1/4P	0–0.08	12
	Cressida	4 1/16P–5 9/16P	4 13/16P	5/16N–1 3/16P	7/16P	0–0.16	10 1/2
1989	Tercel	③	④	3/4N–3/4P	0	0.08 out–0.08 in	11 1/2
	Corolla	③	⑥	15/16N–9/16P	3/16N	0–0.08	12 11/16
	(4A-GE)	9/16P–2 1/16P	1 5/16P	1N–1/2P	1/4N	0–0.08	12 13/16
	Camry (Sedan)	1 5/16P–2 7/16P	1 11/16P	3/16N–1 5/16P	9/16P	0.04 out–0.12 in	12 3/4
	(Wagon)	1/4P–1 3/4P	1P	1/4N–1 1/4P	1/2P	0.04 out–0.12 in	13
	Celica	7/16P–1 15/16P	1 3/16P	15/16N–9/16P	3/16N	0.08 out–0.08 in	13 1/2
	Supra	6 3/4P–8 1/4P	7 1/2P	13/16N–11/16P	1/16N	0.08 out–0.08 in	11
	MR2	4 15/16P–5 13/16P	5 1/16P	1/2N–1P	1/4P	0–0.08	12 1/16
	Cressida	4 1/16P–5 9/16P	4 13/16P	5/16N–1 3/16P	7/16P	0–0.16	10 1/2
1990	Tercel	③	④	3/4N–3/4P	0	0.08 out–0.08 in	11 1/2
	Corolla	9/16P–2 1/16P	1 5/16P	1N–1/2P	1/4N	0–5/32	12 13/16
	(4WD)	3/4P–1 3/4P	1 1/4P	5/16N–11/16P	3/16P	0–5/32	12
	Camry (Sedan)	1 3/16P–2 3/16P	1 11/16P	1/16N–1 1/16P	9/16P	0.04 out–0.04 in	12 3/4
	(Wagon)	1/2P–1 1/2P	1P	0–1P	1/2P	0.04 out–0.04 in.	12 13/16
	Celica	1/4P–1 3/4P	1P	15/16N–9/16P	3/16N	1/16–1/8	14 3/16
	Supra	7 3/16P–8 3/16P	7 11/16P	11/16N–5/16P	3/16N	3/64 out–3/64 in	10 15/16
	MR2	2P–13 1/4P	2 3/4P	1 13/32N–13/32N	1N	0.04 out–0.04 in	13
	Cressida	6 9/16P–8 1/16P	7 5/16P	0–1P	1/2P	3/32–1/4	13 3/16
1991–92	Tercel	1 3/4P–3 1/4P	2 1/2P	3/4N–3/4P	0	3/64 out–3/64 in	11 1/2
	Corolla (2WD)	9/16P–2 1/16P	1 5/16P	1N–1/2P	1/4N	0–5/64	12 5/8
	(4WD)	3/4P–1 3/4P	1 1/4P	5/16N–11/16P	3/16P	0–5/64	12
	Camry (Sedan)	1 3/16P–2 3/16P	1 11/16P	1/16P–1 1/16P	9/16P	0.04 out–0.04 in	12 3/4
	(Wagon)	1/2P–1 1/2P	1P	0–1P	1/2P	0.04 out–0.04 in.	12 13/16
	Celica	1/4P–1 3/4P	1P	15/16N–9/16P	3/16P	1/32–1/16	14 3/16
	Supra	7 3/16P–8 3/16P	7 11/16P	11/16N–5/16P	3/16N	3/64 out–3/64 in	10 15/16
	MR2	2P–13 1/4P	2 3/4P	1 13/32N–13/32N	1N	0.04 out–0.04 in	13

WHEEL ALIGNMENT

Year	Model	Caster Range (deg.)	Caster Preferred Setting (deg.)	Camber Range (deg.)	Camber Preferred Setting (deg.)	Toe-in (in.)	Steering Axis Inclination (deg.)
1991–92	Cressida	6 9/16P–8 1/16P	7 5/16P	0–1P	1/2P	3/32–1/4	13 3/16

N—Negative
P—Positive
① Man. Str.: 1/16N–1 1/3P
Pwr. Str.: 1 1/4P–3P
② Man. Str.: 2/3P
Pwr. Str.: 2 1/4P
③ Man. Str.: 1/4P–1 3/4P
Pwr. Str.: 1 3/4–3 1/4P
④ Man. Str.: 1P
Pwr. Str.: 2 1/2P
⑤ Exc. Coupe: 9/16P–2 1/16P
Coupe: 3/4P–2 1/4P
⑥ Exc. Coupe: 1 5/16P
Coupe: 1 1/2P

ENGINE MECHANICAL

NOTE: Disconnecting the negative battery cable on some vehicles may interfere with the functions of the on board computer systems and may require the computer to undergo a relearning process, once the negative battery cable is reconnected.

Engine Assembly

REMOVAL & INSTALLATION

2VZ-FE Engine

1. Disconnect the negative battery cable.
2. Remove the battery.
3. Drain the cooling system and engine oil.
4. Remove the hood.
5. Remove the ignition coil, igniter and bracket assembly.
6. Remove the radiator.
7. Remove the and coolant reservoir tank.
8. If equipped with automatic transaxle, disconnect the throttle cable from the throttle body.
9. If equipped with cruise control, remove the cruise control actuator and vacuum pump.
10. Remove the air cleaner assembly.
11. If equipped with manual transaxle, remove the clutch release cylinder. Position it aside with the hydraulic line still attached.
12. Disconnect the speedometer and transaxle control cables.
13. Remove the alternator and the belt adjusting bar.
14. Remove the air conditioning compressor and position it aside. Do not disconnect the refrigerant lines.
15. Disconnect the 2 water bypass hoses and fuel lines.
16. Tag and disconnect the brake booster, air conditioning control valve and charcoal canister vacuum hoses.
17. Tag and disconnect any additional wires and lines which may interfere with engine removal.
18. Raise and support the vehicle safely.
19. Remove the engine under covers.
20. Remove the lower suspension crossmember.
21. Remove the halfshafts.
22. Remove the power steering pump and position it aside without disconnecting the hydraulic lines.
23. Remove the front exhaust pipe.
24. Remove the engine mounting center member.
25. Remove the front, center and rear engine mount insulator and bracket assemblies.
26. Lower the vehicle. Remove the glove box and then tag and disconnect the 3 ECU connectors, the circuit opening, cowl wire and instrument wire connectors. Pull the main engine harness out through the firewall.
27. Remove the power steering reservoir tank and position it aside without disconnecting the hydraulic lines.
28. Remove the 2 right side engine mounting stays. Remove the left side engine mounting stay.

To install:

29. When installing the engine pay close attention to the following torque specifications.
    a. Transaxle-to-engine mounting bolts—56 ft. lbs. (77 Nm).
    b. Bracket-to-frame bolt—40 ft. lbs. (56 Nm).
    c. Engine mount bracket-to-block—28 ft. lbs. (39 Nm).
    d. Front and rear engine mounting through bolts and nuts—60 ft. lbs. (84 Nm).
    e. Right hand mounting rubber on body side—30 ft. lbs. (41 Nm). Right hand mounting rubber on engine side—45 ft. lbs. (62 Nm).
    f. Strut-to-body nuts—40 ft. lbs. (56 Nm).
    g. Tie rod end-to-steering knuckle ball joints—29 ft. lbs. (40 Nm).
30. Connect a suitable lifting device to the 2 engine hangers. If with ABS, remove the clamp bolts for the power steering oil cooler pipes. Remove the right and left engine mount insulators and their brackets. Slowly remove the engine/transaxle assembly as a unit.
31. Connect the battery cable, refill all fluids, start the engine and check for leaks.

3A-C Engine

1. Disconnect the negative battery cable.
2. Remove the battery and battery carrier.
3. Remove the hood.
4. Drain the cooling system.
5. Disconnect and plug the transaxle fluid lines, if equipped with automatic transaxle.
6. Remove the radiator.
7. Remove the windshield washer tank.
8. Disconnect the heater hoses.
9. If equipped with air conditioning, remove the condenser fan assembly.

10. If equipped with power steering, remove the power steering pump and position it aside. Plug the lines to prevent fluid leakage.
11. If equipped with air conditioning, remove the air conditioning compressor and position it aside. Leave the refrigerant lines connected.
12. Disconnect and tag the engine ground strap, the oxygen sensor wire, the distributor connector, the ground strap from the dash panel, the oil pressure switch wire, the coolant fan wire, the water temperature gauge wire, the back-up light switch and neutral safety switch wires.
13. Disconnect the accelerator cable. If equipped with automatic transaxle, disconnect the accelerator cable.
14. Disconnect and plug the fuel line hoses. Disconnect and tag all vacuum hoses.
15. Disconnect the air suction filter from the cylinder block.
16. Remove the transaxle upper mount bolts.
17. Raise and support the vehicle safely.
18. Remove the front exhaust pipe.
19. Remove the oil cooler lines, if equipped.
20. If equipped with manual transaxle, disconnect the clutch release cable.
21. Remove the stiffener plates.
22. Disconnect the engine mounting absorber.
23. Remove the engine mount bolts.
24. Remove the torque converter cover and torque converter bolts.
25. Position a lifting device under the transaxle assembly. Remove the lower transaxle retaining bolts. As required, remove the starter.
26. Properly support the engine/transaxle assembly. Connect a suitable lifting device to the engine lifting hooks. Carefully remove the engine from the vehicle.

To install:

27. Properly support the engine/transaxle assembly. Connect a suitable lifting device to the engine lifting hooks. Carefully install the engine into the vehicle.
28. Install the lower transaxle retaining bolts. As required, install the starter.
29. Install the torque converter bolts and cover.
30. Install the engine mount bolts.
31. Connect the engine mounting absorber.
32. Install the stiffener plates.
33. If equipped with manual transaxle, connect the clutch release cable.
34. Install the oil cooler lines, if equipped.
35. Install the front exhaust pipe.
36. Install the transaxle upper mount bolts.
37. Connect the air suction filter to the cylinder block.
38. Connect the fuel line hoses and all vacuum hoses.
39. Connect the accelerator cable. If equipped with automatic transaxle, connect the accelerator cable.
40. Connect the engine ground strap, the oxygen sensor wire, the distributor connector, the ground strap to the dash panel, the oil pressure switch wire, the coolant fan wire, the water temperature gauge wire, the back-up light switch and neutral safety switch wires.
41. If equipped with air conditioning, install the air conditioning compressor.
42. If equipped with power steering, install the power steering pump and position it aside.
43. If equipped with air conditioning, install the condenser fan assembly.
44. Connect the heater hoses.
45. Install the windshield washer tank.
46. Install the radiator.
47. Connect the transaxle fluid lines, if equipped with automatic transaxle.
48. Refill the cooling system.
49. With an assistant, install the hood.
50. Install the battery and battery carrier.
51. Connect the negative battery cable, start the engine and check for leaks.
  a. On manual transaxles, torque the flywheel bolts to 58 ft. lbs. (78 Nm).
  b. On automatic transaxles, torque the driveplate bolts to 47 ft. lbs. (64 Nm).
  c. Engine mount nuts to 29 ft. lbs. (39 Nm).
  d. Stiffener plate bolts to 29 ft. lbs. (34 Nm).
  e. 14mm upper and lower transaxle mount bolts to 29 ft. lbs. (39 Nm); 17mm upper and lower transaxle bolts to 43 ft. lbs. (39 Nm).

3E and 3E-E Engine

1. Disconnect the negative battery cable.
2. Remove the battery.
3. Remove the hood.
4. Remove the engine under covers.
5. Drain the cooling system.
6. If equipped with automatic transaxle, disconnect and plug the transaxle fluid lines.
7. Remove the radiator.
8. Remove the windshield washer tank.
9. Disconnect the heater hoses. On 3E-E engine, disconnect the PCV hoses and remove the air cleaner assembly with the air intake collector.
10. If equipped with cruise control, disconnect and remove the actuator assembly. Disconnect the accelerator cable.
11. If equipped with automatic transaxle, disconnect the accelerator cable.
12. Disconnect and plug the fuel line hoses.
13. Remove the charcoal canister assembly.
14. Disconnect the brake booster hose.
15. Disconnect the transaxle control cables.
16. Disconnect the speedometer cable.
17. If equipped with automatic transaxle, remove the clutch release cylinder and the selecting bell crank.
18. Disconnect the engine ground strap, the oxygen sensor wire, the oil pressure switch wire, the coolant fan wire, the water temperature gauge wire, the back-up light switch and neutral safety switch wires. Tag all wires.
19. Disconnect and tag all vacuum hoses.
20. Disconnect the wiring harness from the intake manifold. Remove the intake manifold ground strap. Disconnect the Cold Mixture Heater (CMH) connector, the alternator electrical connector, starter electrical wires and all other wires necessary to remove the engine.
21. Remove the Vacuum Switching Valve (VSV).
22. If equipped with power steering, remove the power steering pump and position it aside.
23. If equipped with air conditioning, remove the air conditioning compressor and position it aside. Leave the refrigerant lines connected.
24. Disconnect the exhaust pipe at the manifold.
25. Remove the halfshafts.
26. Support the engine/transaxle assembly properly.
27. Connect a suitable lifting device to the engine lifting hooks.
28. If equipped with manual transaxle, remove the rear mounting thru-bolt and the rear mounting assembly. If equipped with automatic transaxle, remove the front mounting thru-bolt and front mounting assembly.
29. Remove the right and left side mounting bolts and brackets.
30. Carefully lift the engine assembly out of the vehicle.

To install:

31. Lower the engine into the vehicle with the help of an assistant.
32. Connect all electrical and hoses fittings.
33. Install all necessary components.
34. During installation, observe the following torque specifications:
  a. On manual transaxles, torque the flywheel bolts to 65 ft. lbs. (88 Nm).

b. On automatic transaxles, torque the driveplate bolts to 13 ft. lbs. (18 Nm).

c. Left bracket and mounting insulator bolts to 35 ft. lbs. (48 Nm).

d. Right mounting insulator and front through bolts to 47 ft. lbs. (64 Nm); rear mounting insulator-to-body bolts to 54 ft. lbs. (73 Nm).

e. If equipped with automatic transaxle, torque the rear mounting insulator bolts to 43 ft. lbs. (58 Nm).

35. Connect the battery cable, refill all fluids, start the engine and check for leaks.

3S-FE Engine

CAMRY (2WD)

1. Disconnect the negative battery cable.
2. Remove the hood.
3. Drain the cooling system.
4. Remove the igniter and bracket assembly.
5. Tag and disconnect all vacuum hoses, electrical wires and cables that are necessary to remove the engine.
6. Remove the radiator and coolant reservoir tank.
7. If equipped with automatic transaxle, disconnect the throttle cable and bracket from the throttle body.
8. Disconnect the accelerator cable from the throttle body.
9. If equipped with cruise control, remove the cruise control actuator and bracket.
10. Disconnect the ground wire from the alternator upper bracket.
11. Remove the air cleaner assembly, air flow meter and air cleaner hose.
12. Remove the heater hoses.
13. Disconnect and plug the fuel lines.
14. Disconnect the speedometer cable.
15. If equipped with manual transaxle, remove the clutch release cylinder and tube bracket. Do not disconnect the tube from the bracket.
16. Disconnect the transaxle control cable.
17. If equipped with air conditioning, remove the air conditioning compressor and position it aside. Do not disconnect the refrigerant lines.
18. If equipped with power steering, remove the power steering pump and position it aside. Do not disconnect the lines.
19. Raise and support the vehicle safely.
20. Drain the engine oil.
21. Remove the engine under covers.
22. Remove the suspension lower crossmember.
23. Remove the halfshafts.
24. Disconnect the exhaust pipe from the catalytic converter.
25. Disconnect the engine mounting center crossmember member.
26. Lower the vehicle.
27. Tag and disconnect the ECU electrical connectors.
28. Connect a suitable lifting device to the engine. Raise the engine slightly and remove the engine retaining brackets and bolts.
29. Carefully remove the engine/transaxle assembly from the vehicle.

NOTE: Be careful not to hit the power steering gear housing or the neutral safety switch.

To install:

30. Lower the engine into the vehicle with the help of an assistant.
31. Connect all electrical and hoses fittings.
32. Install all necessary components.
33. During installation, observe the following torque specifications:

a. On manual transaxles, torque the flywheel bolts to 65 ft. lbs. (88 Nm).

b. On automatic transaxles, torque the driveplate bolts to 13 ft. lbs. (18 Nm).

c. Left bracket and mounting insulator bolts to 35 ft. lbs. (48 Nm).

d. Right mounting insulator and front through bolts to 47 ft. lbs. (64 Nm); rear mounting insulator-to-body bolts to 54 ft. lbs. (73 Nm).

e. If equipped with automatic transaxle, torque the rear mounting insulator bolts to 43 ft. lbs. (58 Nm).

34. Connect the battery cable, refill all fluids, start the engine and check for leaks.

CAMRY (AWD)

1. Disconnect the negative battery cable.
2. Drain the cooling system.
3. Remove the hood.
4. Disconnect the accelerator cable from the throttle body.
5. Remove the radiator and the coolant reservoir tank.
6. Disconnect the heater hoses.
7. Disconnect the inlet hose at the fuel filter. Disconnect the return hose at the fuel return pipe.
8. Disconnect and remove the cruise control actuator.
9. Remove the air cleaner assembly.
10. Remove the clutch slave cylinder and hose bracket without disconnecting the hydraulic line. Position the assembly aside.
11. Disconnect the speedometer cable and the transaxle control cables.
12. If equipped with air conditioning, disconnect and remove the compressor with the refrigerant lines still attached. Move the compressor aside.
13. Tag and disconnect all wires, connecters and vacuum lines necessary to remove engine.
14. Raise and support the vehicle safely.
15. Drain the engine oil and remove the engine under covers.
16. Remove the lower suspension crossmember and the halfshafts.
17. Disconnect and remove the driveshaft.
18. Remove the power steering pump with the hydraulic lines still attached and position it aside.
19. Remove the front exhaust pipe.
20. Remove the engine mounting center member and the stabilizer bar. Lower the vehicle.
21. Tag and disconnect the ECU connectors and pull them out through the firewall.
22. Remove the power steering pump reservoir tank.
23. Connect a suitable lifting device to the eyelets on the engine.
24. Remove the right side engine mount stay and then remove the insulator and bracket.
25. Remove the left side engine mount insulator and bracket.
26. Remove the engine and transaxle as an assembly.

NOTE: Be careful not to hit the power steering gear housing or the neutral safety switch.

To install:

27. Lower the engine into the vehicle with the help of an assistant.
28. Connect all electrical and hoses fittings.
29. Install all necessary components.
30. During installation, observe the following torque specifications:

a. Right and left engine mount bracket bolts and nuts to 38 ft. lbs. (52 Nm).

b. Right side engine mount stay bolt and nut to 54 ft. lbs. (73 Nm).

c. Engine mounting center member: member-to-body bolts—29 ft. lbs. (39 Nm); member-to-other bolts—38 ft. lbs. (52 Nm)

d. Lower crossmember bolts: outer—153 ft. lbs. (206 Nm); inner—29 ft. lbs. (39 Nm).

31. Connect the battery cable, refill all fluids, start the engine and check for leaks.

Celica

1. Disconnect the negative battery cable.
2. Remove the battery.
3. Remove the hood.
4. Drain the cooling system.
5. Tag and disconnect all vacuum hoses, electrical wires and cables that are necessary to remove the engine.
6. Disconnect the ignition coil connector and high tension wire from the coil.
7. Remove the suspension upper brace.
8. Remove the radiator.

9. Remove the coolant reservoir tank.
10. If equipped with automatic transaxle, disconnect the throttle cable and bracket from the throttle body.
11. Disconnect the accelerator cable from the throttle body.
12. If equipped with cruise control, remove the cruise control actuator and bracket.
13. Remove the oxygen sensor.
14. Remove the air cleaner assembly, air flow meter, air cleaner hose and air cleaner bracket.
15. Remove the igniter.
16. Remove the heater hoses.
17. Disconnect and plug the fuel lines.
18. Disconnect the speedometer cable.
19. If equipped with manual transaxle, remove the clutch release cylinder and tube bracket. Do not disconnect the tube from the bracket.
20. Disconnect the transaxle control cable.
21. Remove the air conditioning compressor and position it aside. Do not disconnect the refrigerant lines.
22. Raise and support the vehicle safely.
23. Drain the engine oil and transaxle fluid.
24. Remove the right under cover.
25. Remove the power steering pump and position it aside. Do not disconnect the lines.
26. Remove the suspension lower crossmember.
27. Remove the halfshafts.
28. Disconnect the exhaust pipe from the catalytic converter.
29. Remove the engine rear mounting bolt. Lower the vehicle.
30. Disconnect the ECU electrical connectors.
31. Remove the power steering pump reservoir mounting bolts.
32. Connect a suitable lifting device to the engine. Raise the engine slightly and remove the engine retaining brackets and bolts.
33. Carefully remove the engine/transaxle assembly from the vehicle. Be careful not to hit the power steering gear housing or the neutral safety switch.

To install:

34. Lower the engine into the vehicle with the help of an assistant.
35. Connect all electrical and hoses fittings.
36. Install all necessary components.
37. During installation, observe the following torque specifications:
    a. Right and left engine mount bracket bolts and nuts to 38 ft. lbs. (52 Nm).
    b. Right side engine mount stay bolt and nut to 54 ft. lbs. (73 Nm).
    c. Engine mounting center member: member-to-body bolts – 29 ft. lbs. (39 Nm); member-to-other bolts – 38 ft. lbs. (52 Nm)
    d. Lower crossmember bolts: outer – 153 ft. lbs. (206 Nm); inner – 29 ft. lbs. (39 Nm).
38. Connect the battery cable, refill all fluids, start the engine and check for leaks.

3S-GE Engine

1. Disconnect the negative battery cable.
2. Remove the battery.
3. Remove the hood.
4. Drain the cooling system.
5. Tag and disconnect the connector high tension lead at the ignition coil.
6. Remove the 4 bolts and 2 nuts securing the upper suspension brace. Remove the brace.
7. If equipped with automatic transaxle, disconnect the throttle cable and its bracket at the throttle body.
8. If equipped with manual transaxle, disconnect the throttle cable from the throttle body.
9. Remove the coolant overflow tank.
10. Remove the cruise control actuator and its bracket.
11. Remove the oxygen sensor.
12. Tag and disconnect the cooling fan leads at the radiator.
13. Disconnect the heater hoses.
14. If equipped with automatic transaxle, disconnect the fluid cooler lines.
15. Remove the radiator and the 2 supports.
16. Remove the air cleaner assembly and bracket.
17. Remove the igniter.
18. Tag, disconnect and plug the fuel hoses at the filter and fuel return pipe.
19. Disconnect the speedometer cable.
20. If equipped with manual transaxle, disconnect the transaxle control cable at the shift and selector levers. If equipped with automatic transaxle, disconnect the cable at the swivel and at the bracket and then remove it.
21. Unbolt the air conditioning compressor and position it aside with the refrigerant lines still attached.
22. Tag and disconnect any remaining wires or electrical leads. Tag and disconnect any remaining vacuum hoses.
23. Raise and support the vehicle safely.
24. Drain the engine oil.
25. Remove the right side engine under cover.
26. Remove the lower suspension crossmember.
27. Remove both halfshafts.
28. Unbolt the power steering pump. Disconnect the 2 vacuum hoses and remove the drive belt. Position the pump aside with the hydraulic lines still connected to it.
29. Disconnect the exhaust pipe at the manifold.
30. Remove the rear engine mount bolt. Lower the vehicle and then remove the front engine mount bolts.
31. Remove the power steering pump reservoir and position it aside.
32. Connect a suitable lifting device to the engine lifting hooks. Take up the engine's weight with the hoist and remove the right and left engine mounts.
33. Slowly and carefully, remove the engine and transaxle assembly. Be careful not to hit the power steering gear housing or the neutral safety switch.

To install:

34. Lower the engine into the vehicle with the help of an assistant.
35. Connect all electrical and hoses fittings.
36. Install all necessary components.
37. During installation, observe the following torque specifications:
    a. Right and left engine mount bracket bolts and nuts to 38 ft. lbs. (52 Nm).
    b. Right side engine mount stay bolt and nut to 54 ft. lbs. (73 Nm).
    c. Engine mounting center member: member-to-body bolts – 29 ft. lbs. (39 Nm); member-to-other bolts – 38 ft. lbs. (52 Nm)
    d. Lower crossmember bolts: outer – 153 ft. lbs. (206 Nm); inner – 29 ft. lbs. (39 Nm).
38. Connect the battery cable, refill all fluids, start the engine and check for leaks.

3S-GTE Engine

CELICA (2WD)

1. Disconnect the negative battery cable.
2. Remove the battery.
3. Drain the coolant from the engine and turbocharger intercooler.
4. Remove the hood.
5. Disconnect the accelerator cable at the throttle body.
6. Remove the radiator.
7. Disconnect the heater and intercooler hoses.
8. Disconnect the fuel inlet line at the fuel filter and the return line at the return pipe.
9. Remove the cruise control actuator and bracket.
10. Remove the air cleaner assembly.
11. Remove the clutch release cylinder and bracket without disconnecting the hydraulic line. Move it aside.
12. Disconnect the speedometer and transaxle control cables.
13. Remove the alternator.
14. Remove the air conditioning compressor and position it aside. Do

not disconnect the refrigerant lines.

15. Tag and disconnect any wires, cables, hoses, connectors and vacuum lines which might interfere with engine removal.

16. Raise and support the vehicle safely.

17. Drain the engine oil and remove the under covers.

18. Remove the lower suspension crossmember.

19. Remove the front halfshafts and the driveshaft.

20. Remove the power steering pump and bracket without disconnecting the hydraulic lines. Position the power steering pump aside.

21. Disconnect the front exhaust pipe at the manifold and tailpipe and remove it.

22. Remove the engine mounting center member and lower the vehicle.

23. Unplug the 3 ECU connectors, remove the 2 screws and pull the connectors out through the firewall.

24. Remove the power steering pump reservoir tank.

25. Connect a suitable lifting device to the lifting brackets on the engine. Remove the 2 bolts holding the right engine mount insulator to the mounting bracket. Remove the 4 bolts holding the left engine mount insulator to the mounting bracket and then lower the engine out of the vehicle.

To install:

26. Lower the engine into the vehicle with the help of an assistant.

27. Connect all electrical and hoses fittings.

28. Install all necessary components.

29. During installation, observe the following torque specifications:

a. Torque the right and left engine mount bracket bolts to 38 ft. lbs. (52 Nm).

b. When installing the engine mounting center member, tighten the outer bolts to 29 ft. lbs. (39 Nm), tighten the inner bolts to 38 ft. lbs. (52 Nm).

c. Tighten the front exhaust pipe bolts to 46 ft. lbs. (62 Nm).

d. When installing the lower suspension crossmember, tighten the outer bolts to 154 ft. lbs. (208 Nm), tighten the inner bolts to 29 ft. lbs. (39 Nm).

30. Connect the battery cable, refill all fluids, start the engine and check for leaks.

CELICA (AWD)

1. Disconnect the negative battery cable.
2. Remove the hood.
3. Raise and support the vehicle safely.
4. Remove the engine under covers.
5. Drain the cooling system, engine oil and transaxle fluid.
6. Remove the air cleaner assembly.
7. Disconnect the accelerator cable from the throttle body.
8. Remove the relay box from the battery. Disconnect the wires and connectors from the box.
9. Remove the air conditioning relay box from its mounting bracket.
10. Remove the injector solenoid resistor and fuel pump resistor from the engine compartment.
11. Remove the radiator and coolant overflow tank.
12. If equipped with cruise control, disconnect the wiring and remove the cruise control actuator.
13. Remove the wiper arms and outside windshield moulding. Then, remove the upper brace which is retained by 4 nuts and 2 bolts. The brace connects from the struts to the firewall.
14. Remove the ignition coil.
15. From inside the engine compartment, tag and disconnect all electrical wiring and vacuum hoses necessary to remove the engine.
16. Remove the engine wire bracket.
17. Remove the charcoal canister.
18. Disconnect the heater hoses.
19. Disconnect the speedometer cable from the transaxle.
20. Disconnect and plug the fuel hoses.
21. Remove the starter.
22. Remove the clutch release cylinder without disconnecting the hydraulic tube. Move the unit aside.
23. Disconnect the control cables from the transaxle.
24. Remove the turbocharger pressure sensor and air conditioning Air Switching Valve (ASV) from inside the engine compartment.
25. From the passenger compartment, unplug the connectors from the ECU, air conditioning amplifier and cowl wires. Pull the wiring harnesses through the firewall.
26. Remove the suspension lower crossmember.
27. Remove the front halfshafts and the driveshaft.
28. Remove the power steering pump and bracket without disconnecting the hydraulic lines. Position the pump aside.
29. Disconnect the front exhaust pipe at the manifold and tailpipe and remove it.
30. Remove the engine mounting center member and lower the vehicle. Unplug the 3 TCCS ECU connectors, remove the 2 screws and pull the connectors out through the firewall.
31. Remove the power steering pump reservoir tank.
32. Connect a suitable lifting device to the lifting brackets on the engine.
33. Remove the 2 bolts holding the right engine mount insulator to the mounting bracket. Remove the 4 bolts holding the left engine mount insulator to the mounting bracket.
34. Slowly and carefully, remove the engine and transaxle assembly for the top of the vehicle.

To install:

35. Lower the engine into the vehicle with the help of an assistant.

36. Connect all electrical and hoses fittings.

37. Install all necessary components.

38. During installation, observe the following torque specifications:

a. Torque the left mounting bracket-to-transaxle case bolts to 38 ft. lbs. (52 Nm).

b. Torque the left mounting insulator through bolt to 47 ft. lbs. (63 Nm) and 4 hex head bolts to 64 ft. lbs. (87 Nm).

c. Torque the right mounting insulator nuts to 38 ft. lbs. (52 Nm) and the thru bolt to 64 ft. lbs. (87 Nm).

d. Torque the front and rear bolts to 57 ft. lbs. (77 Nm).

e. Torque the front and rear engine mounting through bolts to 64 ft. lbs. (87 Nm).

f. Torque the lower crossmember nuts and bolts to 112 ft. lbs. (152 Nm).

g. Torque the suspension upper brace nuts to 47 ft. lbs. (64 Nm) and bolts to 15 ft. lbs. (21 Nm).

39. Connect the battery cable, refill all fluids, start the engine and check for leaks.

1990–92 MR2

1. Disconnect the negative battery cable.
2. Remove the hood and side panels.
3. Raise and support the vehicle safely.
4. Remove the engine under covers.
5. Drain the cooling system, engine oil and transaxle fluid.
6. Remove the suspension upper brace that criss-crosses from the struts to the firewall.
7. Remove the air cleaner assembly.
8. Remove both air connector tubes.
9. Disconnect the accelerator cable from the throttle body.
10. If equipped with cruise control, disconnect the wiring and remove the cruise control actuator and accelerator linkage assemblies.
11. Disconnect the brake booster vacuum hose.
12. Disconnect the ground strap connector.
13. Remove the check connector and turbocharger pressure sensor.
14. Remove the injector solenoid resistor, fuel pump relay, fuel pump resistor and the air conditioning vacuum switching valve.
15. Disconnect the filler and over-

flow hoses from the water filler connection. Remove the water filler from the engine.
16. Remove the engine relay box. Disconnect the wires and connectors from the box.
17. Remove the ignition coil and igniter.
18. From inside the luggage compartment, disconnect the wiring harnesses for the ECU, starter relay, cooling fan and engine wires.
19. Disconnect the starter wiring.
20. Disconnect the radiator hose from the water inlet.
21. Disconnect and plug the fuel inlet and return hoses.
22. Disconnect the radiator hoses from the water outlet housing.
23. Disconnect the heater hoses.
24. Disconnect the control cables from the transaxle.
25. Remove the tailpipe and front exhaust pipe.
26. Remove the engine compartment cooling fan.
27. Remove the idler pulley bracket and unbolt the air conditioning compressor. Move the compressor aside. Leave the refrigerant lines connected.
28. Remove the intercooler.
29. Remove the rear engine mounting insulator.
30. Disconnect the speedometer cable from the transaxle.
31. Disconnect the stabilizer link from the shock absorber.
32. Remove the wire clamp bolt and remove the ABS speed sensor.
33. Remove the lower suspension arms.
34. Remove the driveshafts.
35. Remove the 4 bolts and remove the lower crossmember.
36. Remove the front engine mounting insulator.
37. Remove the nut and bolt attaching the clutch release cylinder to the transaxle. Remove the mounting bracket bolts and remove the clutch release cylinder without disconnecting the hydraulic tube.
38. Remove the right and left engine mounting stays.
39. Remove the lateral control rod and air cleaner case bracket.
40. Connect a suitable lifting device to the engine hanger brackets. Tension the lifting device to support the weight of the engine, then remove the left and right mounting insulator fasteners; 2 bolts and 3 nuts for each insulator.
41. Carefully lower then raise the engine from the vehicle.

To install:

42. Lower the engine into the vehicle with the help of an assistant.
43. Connect all electrical and hoses fittings.
44. Install all necessary components.
45. During installation, observe the following torque specifications:
   a. Torque the rear engine mounting bracket-to-transaxle case bolts to 38 ft. lbs. (52 Nm) for the 14mm bolts and 57 ft. lbs. (77 Nm) for the 17mm bolts.
   b. Torque the left mounting insulator through bolt to 47 ft. lbs. (63 Nm) and 4 hex head bolts to 54 ft. lbs. (73 Nm).
   c. Torque the right mounting insulator nuts to 54 ft. lbs. (73 Nm) and the thru bolt to 64 ft. lbs. (87 Nm).
   d. Torque the front and rear bolts to 57 ft. lbs. (77 Nm).
   e. Torque the front and rear engine mounting through bolts to 64 ft. lbs. (87 Nm).
   f. Torque the lower crossmember nuts and bolts to 112 ft. lbs. (152 Nm).
   g. Torque the suspension upper brace nuts to 47 ft. lbs. (64 Nm) and bolts to 15 ft. lbs. (21 Nm).
46. Connect the battery cable, refill all fluids, start the engine and check for leaks.

1988 4A-LC Engine

1. Disconnect the negative battery cable.
2. Remove the battery.
3. Remove the hood.
4. Drain the cooling system and the oil.
5. Remove the air cleaner hose and air cleaner.
6. Remove the coolant reservoir tank, radiator and fan shroud.
7. Disconnect the actuator, accelerator and throttle cables at the carburetor.
8. Tag and disconnect all wires, hoses and lines which might interfere with engine removal.
9. Disconnect the fuel lines at the fuel pump.
10. Remove the heater hoses.
11. Remove the power steering pump and air conditioning compressor and position them aside. Do not disconnect the hydraulic or refrigerant lines.
12. Disconnect the speedometer cable at the transaxle.
13. Remove the clutch release cylinder without disconnecting the hydraulic line and position it aside.
14. Disconnect the shift control cable and then raise the vehicle and support it with safety stands.
15. Disconnect the exhaust pipe at the manifold.
16. Disconnect the front and rear engine mounts at the center member. Remove the center member.
17. Disconnect the halfshafts at the transaxle and lower the vehicle.
18. Connect a suitable lifting device to the lift brackets on the engine.
19. Remove the right engine mount thru-bolt. Remove the left engine mount and bracket.
20. Lift the engine out of the vehicle.

To install:

21. Lower the engine into the vehicle with the help of an assistant.
22. Connect all electrical and hoses fittings.
23. Install all necessary components.
24. During installation, observe the following torque specifications:
   a. Tighten the engine mount center member to 29 ft. lbs. (39 Nm).
   b. Tighten the front and rear engine mounts to 29 ft. lbs. (39 Nm).
   c. Tighten the exhaust pipe-to-manifold bolts to 46 ft. lbs. (62 Nm).
25. Connect the battery cable, refill all fluids, start the engine and check for leaks.

4A-F Engine

1. Disconnect the negative battery cable.
2. Remove the battery.
3. Remove the hood.
4. Remove the engine under covers.
5. Drain the cooling system, engine oil and transaxle oil.
6. Remove the air cleaner and air cleaner flexible hose.
7. Remove the coolant reservoir tank, radiator and cooling fan.
8. If equipped with automatic transaxle, disconnect the accelerator and throttle cables at the carburetor.
9. Disconnect all electrical wires and vacuum lines necessary to remove the engine.
10. Disconnect the fuel lines at the fuel pump. Plug the lines.
11. Disconnect the heater hoses at the water inlet housing.
12. Remove the power steering pump and set it aside with the hydraulic lines still attached.
13. Remove the air conditioning compressor and set it aside with the refrigerant lines still attached.
14. Disconnect the speedometer cable from the transaxle.
15. If equipped with manual transaxle, remove the clutch release cylinder and position it aside with the hydraulic lines still attached.
16. Disconnect the shift control cables.
17. Raise and support the vehicle safely.
18. Remove the 2 nuts from the flange and then disconnect the exhaust pipe at the manifold.
19. Disconnect the halfshafts at the transaxle.
20. Remove the 2 hole covers and then remove the front, center and rear engine mounts from the center member. Remove the 5 bolts and insulators and remove the center member.
21. Connect a suitable lifting device

to the lifting brackets on the engine.

22. Remove the 3 bolts and mounting stay. Remove the bolt, 2 nuts and the thru-bolt and pull out the right side engine mount. Remove the 2 bolts and the left mounting stay. Remove the 3 bolts and disconnect the left engine mount bracket from the transaxle.

23. Lift the engine/transaxle assembly out of the vehicle.

To install:

24. Lower the engine into the vehicle with the help of an assistant.

25. Connect all electrical and hoses fittings.

26. Install all necessary components.

27. During installation, observe the following torque specifications:

a. Torque the right engine mount insulator bolt to 47 ft. lbs. (64 Nm); tighten the nut to 38 ft. lbs. (52 Nm). Align the insulator with the bracket on the body and tighten the bolt to 64 ft. lbs. (87 Nm).

b. Align the left engine mount insulator bracket with the transaxle bracket and tighten the bolt to 35 ft. lbs. (48 Nm).

c. Install the right mounting stay and tighten the 3 bolts to 31 ft. lbs. (42 Nm). Install the left stay and tighten the 2 bolts to 15 ft. lbs. (21 Nm).

d. Install the engine center member and tighten the 5 bolts to 45 ft. lbs. (61 Nm).

e. Install the front and rear engine mounts and bolts. Align the bolts holes in the brackets with the center member and tighten the front mount bolts to 35 ft. lbs. (48 Nm); tighten the center and rear mounts to 38 ft. lbs. (52 Nm). Install the 2 hole covers and tighten the rear mounting bolt to 58 ft. lbs. (78 Nm).

28. Connect the battery cable, refill all fluids, start the engine and check for leaks.

4A-FE Engine

COROLLA

1. Disconnect the negative battery cable.

2. Remove the battery.

3. Remove the hood.

4. Remove the engine under covers.

5. Drain the cooling system, engine and transaxle oil.

6. Remove the air cleaner and air cleaner flexible hose.

7. Remove the coolant reservoir tank, radiator and cooling fan.

8. If equipped with automatic transaxle, disconnect the accelerator and throttle cables.

9. If equipped with cruise control, remove the cruise control actuator.

10. Disconnect the No. 2 junction block, the ground strap connector and the ground strap.

11. Disconnect the check, vacuum sensor and oxygen sensor connectors. Disconnect the air conditioning compressor wire.

12. Disconnect the vacuum hoses at the brake booster, power steering pump, vacuum sensor, charcoal canister and vacuum switch.

13. Disconnect the fuel lines at the fuel pump.

14. Disconnect the heater hoses at the water inlet housing.

15. Remove the power steering pump and set it aside with the hydraulic lines still attached.

16. Remove the air conditioning compressor and set it aside with the refrigerant lines still attached.

17. Disconnect the speedometer cable at the transaxle.

18. If equipped with manual transaxle, remove the clutch release cylinder and position it aside with the hydraulic lines still attached.

19. Disconnect the shift control cables.

20. Raise and support the vehicle safely.

21. Disconnect the oil cooler lines and the exhaust pipe (at the manifold).

22. Disconnect the halfshafts and the driveshaft at the transaxle.

23. Connect a suitable lifting device to the lifting brackets on the engine and raise it just enough to relieve pressure on the mounts.

24. Pull out the hole covers and remove the 5 bolts on the front and rear engine mounts. Remove the mounts from the center crossmember. Remove the 4 center crossmember bolts and the 8 bolts from the sub-frame. Remove the front and rear mounting bolts and then remove the member.

25. Remove the engine mount stay and the mount.

26. Remove the air cleaner bracket. Disconnect the left side mounting bracket from the transaxle bracket and then lift out the engine/transaxle assembly slowly and carefully.

To install:

27. During installation, observe the following torque specifications:

a. Torque the right engine mount insulator bolt and nuts to 38 ft. lbs. (52 Nm). Align the insulator with the bracket on the body and tighten the bolt to 64 ft. lbs. (87 Nm).

b. Align the left engine mount insulator bracket with the transaxle bracket and tighten the bolt to 35 ft. lbs. (48 Nm).

c. Install the left stay and tighten the 2 bolts to 15 ft. lbs. (21 Nm).

d. Install the engine center member and tighten the 5 bolts to 45 ft. lbs. (61 Nm).

e. Install the front and rear engine mounts and bolts. Align the bolts holes in the brackets with the center member and tighten the front mount bolts to 35 ft. lbs. (48 Nm); tighten the center and rear mounts to 42 ft. lbs. (57 Nm). Install the 8 sub-frame bolts and tighten the lower arm bolt to 152 ft. lbs. (206 Nm) and the rear bolt to 94 ft. lbs. (127 Nm).

1990–92 CELICA

1. Disconnect the negative battery cable.

2. Remove the battery.

3. Raise the vehicle and support safely.

4. Remove the engine under covers.

5. Drain the cooling system and engine oil.

6. Remove the air cleaner assembly along with its hose and any attachments.

7. Disconnect the accelerator and throttle cables at the bracket.

8. Remove the lower cover from the relay box. Disconnect the fusible link cassette and connectors. Remove the engine relay box.

9. Remove the air conditioning relay box from the bracket.

10. Remove the coolant reservoir tank, radiator and cooling fan.

11. Disconnect the check connector, vacuum sensor connector and ground strap from the left front fender apron. Remove the engine wiring bracket. Disconnect the noise filter assembly.

12. Remove the charcoal canister.

13. Disconnect the heater hose from the water inlet.

14. Disconnect the speedometer cable at the transaxle.

15. Disconnect the fuel hose.

16. If equipped with manual transaxle, remove the clutch release cylinder and position it aside with the hydraulic lines still attached.

17. Disconnect the shift control cables from the transaxle.

18. Tag and disconnect all vacuum hoses and electrical wires necessary to remove the engine.

19. Remove the suspension lower crossmember.

20. Disconnect the oxygen sensor connector.

21. Remove the front exhaust pipe assembly.

22. If equipped with automatic transaxle, disconnect control cable from engine mounting center member.

23. Remove the front halfshafts.

24. Unbolt the air conditioning compressor and wire it aside with the refrigerant lines still attached.

25. Remove the power steering pump assembly without disconnecting the hydraulic lines.

26. Connect a suitable lifting device to the engine lifting brackets. Tension the lifting device slightly to take the pressure off the mounts.

27. Remove the engine mounting center member.
28. Remove the front engine mounting insulator and bracket.
29. Remove the rear mounting insulator and bracket.
30. Disconnect the ground wire from the fender apron. Remove the ground strap from the transaxle.
31. Remove the right and left engine mounting stay.
32. Slowly and carefully, remove the engine and transaxle assembly from the top of the vehicle.

NOTE: Be careful not to hit the power steering gear housing or the neutral safety switch.

To install:

33. Lower the engine into the vehicle with the help of an assistant.
34. Connect all electrical and hoses fittings.
35. Install all necessary components.
36. During installation, observe the following torque specifications:
    a. Torque the left mounting bracket-to-transaxle case bolts to 38 ft. lbs. (52 Nm).
    b. Torque the left mounting insulator-to-bracket bolts to 35 ft. lbs. (48 Nm) and the thru bolt to 64 ft. lbs. (87 Nm).
    c. Torque the right engine stay bolts to 31 ft. lbs. and the left engine stay bolts to 15 ft. lbs. (21 Nm).
    d. Torque the front and rear engine mounting bracket and insulator fasteners to 57 ft. lbs. (77 Nm).
    e. Torque the engine center member bolts to 38 ft. lbs. (52 Nm) and the center member-to-insulator bolts to 47 ft. lbs. (64 Nm).
    f. Torque the front and rear engine mounting through bolts to 64 ft. lbs. (87 Nm).
37. Connect the battery cable, refill all fluids, start the engine and check for leaks.

4A-GEC and 4A-GELC Engine

1. Disconnect the negative battery cable.
2. Remove the battery.
3. Remove the air cleaner assembly.
4. Drain the cooling system and engine oil.
5. Remove the fuel tank protectors and the engine under cover.
6. Disconnect the accelerator cable.
7. If equipped with cruise control, disconnect the cruise control at the cable actuator. If equipped with automatic transaxle, disconnect the throttle cable.
8. Disconnect the heater hoses at the water inlet housing on the rear of the cylinder head cover. Disconnect the radiator hose and the air bleeder hose at the water inlet housing.
9. Disconnect and plug the fuel line at the fuel filter. Disconnect the fuel return hose. Tag and disconnect the vacuum hose at the charcoal canister.
10. Tag and disconnect the engine ground strap and the main wiring harness connector at the engine. Disconnect the back-up light switch connector as required.
11. Disconnect the speedometer cable.
12. Remove the transaxle gravel shield.
13. Remove the ground strap from the water inlet housing.
14. Remove the radiator overflow tank.
15. Remove the air conditioning and alternator drive belts. Remove the alternator.
16. Disconnect the radiator hose at the water outlet housing.
17. Tag and disconnect the 2 connectors at the igniter, the noise filter connector, the cooling fan electrical connector, the cylinder head ground strap, the air condition compressor connector and the high tension leads at the ignition coil.
18. Remove the rear luggage compartment trim.
19. Tag and disconnect the circuit opening relay connector, the ball connections at the electronic control unit and the electrical lead for the cooling fan computer.
20. Pull the main wiring harness out and through the engine compartment.
21. Remove the mounting bolts and remove the air conditioning compressor. Position it aside without disconnecting the refrigerant lines.
22. If equipped with a manual transaxle, disconnect the control cables from the outer shift lever and gear shift selector lever. If equipped with automatic transaxle, disconnect the control cable at the gear shift lever.
23. If equipped with a manual transaxle, remove the control cable bracket on the transaxle. Remove the clutch release cylinder.
24. Disconnect the engine oil cooler lines, if equipped. Disconnect the automatic transaxle fluid lines if equipped.
25. Remove the exhaust pipe assembly. Remove the oxygen sensor at the exhaust manifold.
26. If equipped with automatic transaxle, remove the mounting bolts and remove the stiffener plate at the transaxle. Remove the flywheel shield.
27. Remove the right halfshaft. Disconnect the left halfshaft from the side gear shaft and position it aside.
28. Remove the front and rear engine mount bolts. Place a block of wood on an hydraulic floor jack and carefully position the jack under the engine. Raise the jack just enough to ease the engine's weight on the mounts. Remove the right and left engine mounts.
29. Make sure there are no remaining wires or hoses connected to the engine and then slowly and carefully raise the vehicle while lowering the jack supporting the engine/transaxle assembly.

To install:

30. Lower the engine into the vehicle with the help of an assistant.
31. Connect all electrical and hoses fittings.
32. Install all necessary components.
33. During installation, observe the following torque specifications:
    a. Torque the right engine mount insulator bolt and nuts to 38 ft. lbs. (52 Nm). Align the insulator with the bracket on the body and tighten the bolt to 64 ft. lbs. (87 Nm).
    b. Align the left engine mount insulator bracket with the transaxle bracket and tighten the bolt to 35 ft. lbs. (48 Nm).
    c. Install the left stay and tighten the 2 bolts to 15 ft. lbs. (21 Nm).
    d. Install the engine center member and tighten the 5 bolts to 45 ft. lbs. (61 Nm).
    e. Install the front, center and rear engine mounts and bolts. Align the bolts holes in the brackets with the center member and tighten the front mount bolts to 35 ft. lbs. (48 Nm); tighten the center mounts to 38 ft. lbs. (52 Nm); and the rear mount bolts to 42 ft. lbs. (57 Nm).
    f. Bounce the engine several times to unload the front and rear mounts, for automatic transmission only, and then tighten the rear bolt to 64 ft. lbs. (87 Nm). Install the front bolt and tighten it to 64 ft. lbs. (87 Nm).
34. Connect the battery cable, refill all fluids, start the engine and check for leaks.

4A-GZE Engine

1. Disconnect the negative battery cable.
2. Raise and support the vehicle safely.
3. Remove the fuel tank protectors and the engine under cover.
4. Drain the cooling system and engine oil.
5. Remove the supercharger intercooler.
6. Remove the battery.
7. Disconnect the air conditioning idle-up, charcoal canister and cruise control vacuum hoses. Disconnect the air bleeder hose at the water inlet housing.
8. If equipped with automatic transmission, disconnect the throttle cable.
9. Disconnect the cruise control cable, the heater hoses and radiator hose.
10. Disconnect the fuel inlet and return lines.

11. Disconnect the speedometer cable and the brake booster vacuum line.
12. Remove the radiator reservoir tank. Disconnect the radiator hose at the water outlet housing.
13. Tag and disconnect any remaining hoses, wires or lines which may interfere with engine removal.
14. Remove the 5 clip fasteners and pull out the rear luggage compartment trim. Disconnect the connectors and pull the wiring harness through the engine compartment.
15. Remove the air conditioning compressor and position it aside with the refrigerant lines still connected.
16. Disconnect the shifter control cables.
17. Remove the clutch release cylinder and bracket and position it aside with the hydraulic lines still attached.
18. Disconnect the coolant lines at the engine and transmission oil coolers.
19. Remove the front exhaust pipe.
20. Remove the rear halfshafts.
21. Remove the front and rear engine mount insulators. Lower the engine slightly and then remove the right and left engine mounts. Carefully support the engine and raise the vehicle, over the engine.

To install:

22. Lower the engine into the vehicle with the help of an assistant.
23. Connect all electrical and hoses fittings.
24. Install all necessary components.
25. During installation, observe the following torque specifications:
    a. When installing the rear engine mount insulator, tighten the 10mm bolts to 38 ft. lbs. (52 Nm); tighten the 12mm bolts to 58 ft. lbs. (78 Nm).
    b. Install the front engine mount insulator to the body and tighten the inner bolt to 38 ft. lbs. (52 Nm), tighten the outer bolts to 54 ft. lbs. (73 Nm).
    c. Connect the mounting bracket to the insulator and install the thru-bolt. Bounce the engine several times and tighten the thru-bolt to 58 ft. lbs. (78 Nm).
26. Connect the battery cable, refill all fluids, start the engine and check for leaks.

5M-GE Engine

1. Disconnect the negative battery cable.
2. Remove the battery.
3. Remove the hood.
4. Remove the air cleaner assembly.
5. Remove the fan shroud.
6. Drain the cooling system.
7. Disconnect the upper and lower radiator hoses.
8. If equipped with automatic transmission, disconnect and plug the oil lines from the oil cooler.
9. Detach the hose which runs to the thermal expansion tank and remove the expansion tank from its mounting bracket.
10. Remove the radiator.
11. If equipped with automatic transmission, remove the throttle cable bracket from the cylinder head. Remove the accelerator and actuator cable bracket from the cylinder head.
12. Tag and disconnect the cylinder head ground cable, the oxygen sensor wire, the oil pressure sending unit, alternator wires, the high tension coil wire, the water temperature sending, the thermo-switch wires, the starter wires, the ECT connectors, the solenoid resistor wire connector and the knock sensor wire.
13. Tag and disconnect the brake booster vacuum hose from the air intake chamber, along with the EGR valve vacuum hose. Disconnect the actuator vacuum hose from the air intake chamber, if equipped with cruise control. Disconnect the heater bypass hoses from the engine.
14. Remove the glove box, and remove the ECU computer module. Disconnect the 3 connectors, and pull out the EFI wiring harness from the engine compartment side of the firewall.
15. Remove the 4 shroud and 4 fluid coupling screws, and the shroud and coupling as a unit.
16. Remove the engine undercover protector.
17. Disconnect the coolant reservoir hose. Remove the radiator and the coolant expansion tank.
18. Remove the air conditioning compressor drive belt, and remove the compressor mounting bolts. Without disconnecting the refrigerant hoses, lay the compressor aside and secure it.
19. Disconnect the power steering pump drive belt and remove the pump stay. Unbolt the pump and lay it aside without disconnecting the fluid hoses.
20. Remove the engine mounting bolts from each side of the engine. Remove the engine ground cable.
21. If equipped with manual transmission, remove the shift lever from the inside of the vehicle.
22. Raise and support the vehicle safely.
23. Drain the engine oil.
24. Disconnect the exhaust pipe from the exhaust manifold. Remove the exhaust pipe clamp from the transmission housing.
25. If equipped with manual transmission, remove the clutch slave cylinder.
26. Disconnect the speedometer cable at the transmission.
27. If equipped with automatic transmission, disconnect the shift linkage from the shift lever. If equipped with manual transmission, disconnect the wire from the back-up light switch.
28. Remove the stiffener plate from the ground cable.
29. Disconnect and plug the fuel line from the fuel filter and the return hose from the fuel hose support.
30. Remove the 2 bolts from the top and bottom of the steering universal, and remove the sliding yoke.
31. Disconnect the tie rod ends. Disconnect the pressure line mounting bolts from the front crossmember.
32. Remove the intermediate shaft from the driveshaft.
33. Position a jack under the transmission, with a wooden block between the 2 to prevent damage to the transmission case. Place a wooden block between the cowl panel and the cylinder head rear end to prevent damage to the heater hoses.
34. Unbolt the engine rear support member from the frame, along with the ground cable.
35. Make sure all wiring is disconnected, all hoses disconnected, and everything clear of the engine and transmission. Connect a suitable lifting device to the lift brackets on the engine, and carefully lift the engine and transmission up and out of the vehicle.

To install:

36. Lower the engine into the vehicle with the help of an assistant.
37. Connect all electrical and hoses fittings.
38. Install all necessary components.
39. During installation, observe the following torque specifications:
    a. When installing the rear engine mount insulator, tighten the 10mm bolts to 38 ft. lbs. (52 Nm); tighten the 12mm bolts to 58 ft. lbs. (78 Nm).
    b. Install the front engine mount insulator to the body and tighten the inner bolt to 38 ft. lbs. (52 Nm), tighten the outer bolts to 54 ft. lbs. (73 Nm).
    c. Connect the mounting bracket to the insulator and install the thru-bolt. Bounce the engine several times and tighten the thru-bolt to 58 ft. lbs. (78 Nm).
40. Connect the battery cable, refill all fluids, start the engine and check for leaks.

5S-FE Engine

1990–92 CELICA

1. Disconnect the negative battery cable.
2. Remove the battery.
3. Remove the hood.
4. Raise the vehicle and support safely.
5. Remove the engine under covers.

6. Drain the cooling system and engine oil.
7. Remove the air cleaner assembly along with hoses and any attachments.
8. Disconnect the accelerator and throttle cables at the bracket.
9. Remove the lower cover from the relay box. Disconnect the fusible link cassette and connectors. Remove the engine relay box.
10. Remove the air conditioning relay box from the bracket.
11. Remove the cruise control actuator assembly.
12. Remove the coolant reservoir tank, radiator and cooling fan.
13. Remove the 2 wiper arms and outside lower windshield moulding. Remove the suspension upper brace where it attaches to the struts and the firewall.
14. Remove the ignition coil assembly. Disconnect the check connector, igniter connector, vacuum sensor connector and ground strap from the left front fender apron. Remove the engine wiring bracket. Disconnect the noise filter assembly.
15. Remove the charcoal canister.
16. Disconnect the heater hose from the water inlet.
17. Disconnect the speedometer cable.
18. Disconnect the fuel hose.
19. If equipped with a manual transaxle, remove the clutch release cylinder and position it aside with the hydraulic lines still attached.
20. Disconnect the shift control cables from the transaxle.
21. Tag and disconnect the vacuum sensor hose from the gas filter on the air intake chamber, brake booster vacuum hose and air conditioning vacuum hoses on air intake chamber.
22. Disconnect 2 cowl wire connectors and engine wire clamp from engine fender apron.
23. Tag and disconnect: engine ECU connector, cowl wire connectors and air conditioning amplifier connector.
24. Remove the suspension lower crossmember.
25. Disconnect the oxygen sensor connector.
26. Remove all necessary brackets and retaining bolts.
27. Remove the front exhaust pipe assembly.
28. If equipped with automatic transaxle, disconnect control cable from engine mounting centermember. Remove the front halfshafts.
29. Unbolt the air conditioning compressor and then wire it aside with the refrigerant lines still attached.
30. Remove the power steering pump assembly without disconnecting the hydraulic lines.
31. Connect a suitable lifting device to the engine lifting brackets.
32. Remove the engine mounting center member.
33. Remove the front engine mounting insulator and bracket.
34. Remove the rear mounting insulator and bracket.
35. Disconnect the ground wire from the fender apron. Remove the ground strap from the transaxle.
36. Remove the right and left engine mounting stay.
37. Slowly and carefully, remove the engine and transaxle assembly from the top of the vehicle.

NOTE: Be careful not to hit the power steering gear housing or the neutral safety switch.

To install:

38. Lower the engine into the vehicle with the help of an assistant.
39. Connect all electrical and hoses fittings.
40. Install all necessary components.
41. During installation, observe the following torque specifications:
a. Torque the left mounting bracket-to-transaxle case bolts to 38 ft. lbs. (52 Nm).
b. Torque the left mounting insulator-to-bracket bolts to 35 ft. lbs. (48 Nm) and the thru bolt to 64 ft. lbs. (87 Nm).
c. Torque the right engine stay bolts to 31 ft. lbs. and the left engine stay bolts to 15 ft. lbs. (21 Nm).
d. Torque the front and rear engine mounting bracket and insulator fasteners to 57 ft. lbs. (77 Nm).
e. Torque the engine center member bolts to 38 ft. lbs. (52 Nm) and the centermember-to-insulator bolts to 47 ft. lbs. (64 Nm).
f. Torque the front and rear engine mounting through bolts to 64 ft. lbs. (87 Nm).
42. Connect the battery cable, refill all fluids, start the engine and check for leaks.

1990–92 MR2

1. Disconnect the negative battery cable.
2. Remove the hood and engine side panels.
3. Raise and support the vehicle safely.
4. Remove the engine under covers.
5. Drain the cooling system, engine oil and transaxle fluid.
6. Remove the suspension upper brace that criss-crosses from the struts to the firewall.
7. Remove the air cleaner assembly.
8. Disconnect the accelerator cable from the throttle body.
9. If equipped with cruise control, disconnect the wiring and remove the cruise control actuator and accelerator linkage assemblies.
10. Disconnect the brake booster vacuum hose.
11. Disconnect the ground strap connector.
12. Remove the check connector and vacuum sensor.
13. Remove the the air conditioning vacuum switching valve.
14. Disconnect the filler and overflow hoses from the water filler connection. Remove the water filler from the engine.
15. Remove the engine relay box. Disconnect the wires and connectors from the box.
16. Remove the ignition coil and igniter.
17. From inside the luggage compartment, disconnect the wiring harnesses for the ECU, starter relay, cooling fan and engine wires.
18. Disconnect the starter wiring.
19. Disconnect the radiator hose from the water inlet.
20. Disconnect and plug the fuel inlet and return hoses.
21. Disconnect the radiator hoses from the water outlet housing.
22. Disconnect the heater hoses.
23. Disconnect the control cables from the transaxle.
24. If equipped with automatic transaxle, disconnect and plug the oil cooler hoses.
25. Remove the front exhaust pipe.
26. Remove the driveshafts.
27. Remove the idler pulley bracket and unbolt the air conditioning compressor. Move the compressor aside. Leave the refrigerant lines connected.
28. Remove the front and rear engine mounting insulator.
29. Disconnect the speedometer cable from the transaxle.
30. If equipped with manual transaxle, remove the nut and bolt attaching the clutch release cylinder to the transaxle. Remove the mounting bracket bolts and remove the clutch release cylinder without disconnecting the hydraulic tube. If equipped with automatic transaxle, unbolt and remove the control cable bracket from the transaxle.
31. Remove the rear engine mounting bracket.
32. Remove the right and left engine mounting stays.
33. If equipped with manual transaxle, remove the lateral control rod and air cleaner case bracket. If equipped with automatic transaxle, unbolt the air cleaner case bracket, disconnect the charcoal canister tube and the ground strap from the transaxle.
34. Connect a suitable lifting device to the engine hanger brackets. Tension the lifting device to support the weight of the engine, then remove the left and right mounting insulator fasteners.

35. Carefully lower then raise the engine from the vehicle.

To install:

36. Lower the engine into the vehicle with the help of an assistant.

37. Connect all electrical and hoses fittings.

38. Install all necessary components.

39. During installation, observe the following torque specifications:

a. Torque the left mounting bracket-to-transaxle case bolts to 38 ft. lbs. (52 Nm).

b. Torque the left mounting insulator-to-bracket bolts to 35 ft. lbs. (48 Nm) and the thru bolt to 64 ft. lbs. (87 Nm).

c. Torque the right engine stay bolts to 31 ft. lbs. and the left engine stay bolts to 15 ft. lbs. (21 Nm).

d. Torque the front and rear engine mounting bracket and insulator fasteners to 57 ft. lbs. (77 Nm).

e. Torque the engine center member bolts to 38 ft. lbs. (52 Nm) and the center member-to-insulator bolts to 47 ft. lbs. (64 Nm).

f. Torque the front and rear engine mounting through bolts to 64 ft. lbs. (87 Nm).

40. Connect the battery cable, refill all fluids, start the engine and check for leaks.

7M-GE and 7M-GTE Engine

SUPRA

1. Disconnect the negative battery cable.
2. Remove the hood.
3. Raise and support the vehicle safely.
4. Remove the engine under cover.
5. Drain the cooling system and engine oil.
6. Remove the radiator.
7. On 7M-GE, remove the air cleaner assembly. On 7M-GTE, remove the No. 4 air cleaner pipe along with the No. 1 and 2 air cleaner hose.
8. Remove the No. 7 air cleaner hose with the air flow meter and air cleaner cap.
9. Remove the air conditioning belt. Remove the alternator drive belt, water pump pulley and fan assembly. Remove the power steering belt.
10. Disconnect the brake booster hose, the heater valve hose, the cruise control hose and the charcoal canister hose.
11. Remove the heater hoses.
12. Tag and disconnect all electrical wire and vacuum hoses necessary to remove the engine.
13. If equipped with cruise control, disconnect the cruise control cable.
14. Disconnect the accelerator cable.
15. If equipped with automatic transmission, disconnect the throttle cable.
16. Remove the air conditioning compressor. Position the unit aside. Do not disconnect the refrigerant lines.
17. On the 7M-GTE, remove the No. 6 air cleaner hose and the upper radiator outlet hose.
18. Remove the power steering pump. Position the unit aside; do not disconnect the hydraulic lines.
19. If equipped with manual transmission, remove the shift lever.
20. Disconnect the ground strap from the fuel hose clamp. On the 7M-GTE, remove the engine mounting absorber.
21. Disconnect and plug the fuel lines.
22. Raise the vehicle and support safely.
23. Remove the exhaust pipe.
24. Remove the driveshaft.
25. Disconnect the speedometer cable.
26. If equipped with automatic transmission, remove the shift linkage. If equipped with manual transmission, remove the clutch release cylinder.
27. Properly support the engine and transmission assembly. Remove the No. 1 front crossmember. Remove the engine retaining mounts.
28. Position a piece of wood between the engine firewall and the rear of the cylinder head to prevent damage to the heater hose.
29. Make sure there are no remaining wires or hoses connected to the engine and then slowly and carefully remove the engine and transmission from the vehicle.

To install:

30. Lower the engine into the vehicle with the help of an assistant.

31. Connect all electrical and hoses fittings.

32. Install all necessary components.

33. During installation, observe the following torque specifications:

a. Torque the left mounting bracket-to-transaxle case bolts to 38 ft. lbs. (52 Nm).

b. Torque the left mounting insulator-to-bracket bolts to 35 ft. lbs. (48 Nm) and the thru bolt to 64 ft. lbs. (87 Nm).

c. Torque the right engine stay bolts to 31 ft. lbs. and the left engine stay bolts to 15 ft. lbs. (21 Nm).

d. Torque the front and rear engine mounting bracket and insulator fasteners to 57 ft. lbs. (77 Nm).

e. Torque the engine center member bolts to 38 ft. lbs. (52 Nm) and the center member-to-insulator bolts to 47 ft. lbs. (64 Nm).

f. Torque the front and rear engine mounting through bolts to 64 ft. lbs. (87 Nm).

34. Connect the battery cable, refill all fluids, start the engine and check for leaks.

CRESSIDA

1. Disconnect the negative battery cable.
2. Drain the cooling system.
3. Remove the hood.
4. Remove the battery and tray.
5. Disconnect the accelerator, throttle and cruise control cables.
6. Remove the air cleaner assembly complete with the air flow meter, hoses and connector pipe.
7. Tag and disconnect all electrical wires and vacuum hoses necessary to remove the engine.
8. Remove the radiator.
9. Remove the drive belt and unbolt the air conditioning compressor. Position it aside and suspend it with wire. Do not disconnect the refrigerant lines.
10. Unbolt the power steering pump. Position it aside and suspend it with wire. Do not disconnect the hydraulic lines.
11. Remove the windshield washer fluid reservoir.
12. Remove the glove box and disconnect the 6 connectors from the main wiring harnerss and then pull the main wiring harness through the firewall and into the engine compartment.
13. Disconnect the heater hoses.
14. Raise the vehicle and support it safely.
15. Remove the engine under cover and drain the oil.
16. Disconnect the exhaust pipe at the manifold.
17. Disconnect the driveshaft at the transmission flange and position it aside.
18. Disconnect the speedometer cable and the transmission linkage.
19. Disconnect the starter lead and the ground lines at the stiffener plate and left side engine mount.
20. Disconnect and plug the fuel lines.
21. Remove the front wheels and then disconnect the power steering rack. Leave the hydraulic lines attached and lay the rack aside.
22. Loosen the 8 bolts and the ground strap and then remove the rear engine support.
23. Lower the vehicle and remove the 4 engine mount-to-suspension bolts. Attach an engine hoist to the 2 engine hangers and then slowly and carefully lift the engine out of the vehicle.

To install:

24. Lower the engine into the vehicle with the help of an assistant.

25. Connect all electrical and hoses fittings.

26. Install all necessary components.

27. During installation, observe the following torque specifications:

a. Torque the left mounting

bracket-to-transaxle case bolts to 38 ft. lbs. (52 Nm).

b. Torque the left mounting insulator-to-bracket bolts to 35 ft. lbs. (48 Nm) and the thru bolt to 64 ft. lbs. (87 Nm).

c. Torque the right engine stay bolts to 31 ft. lbs. and the left engine stay bolts to 15 ft. lbs. (21 Nm).

d. Torque the front and rear engine mounting bracket and insulator fasteners to 57 ft. lbs. (77 Nm).

e. Torque the engine center member bolts to 38 ft. lbs. (52 Nm) and the center member-to-insulator bolts to 47 ft. lbs. (64 Nm).

f. Torque the front and rear engine mounting through bolts to 64 ft. lbs. (87 Nm).

28. Connect the battery cable, refill all fluids, start the engine and check for leaks.

Cylinder Head

REMOVAL & INSTALLATION

2VZ-FE Engine

1. Disconnect the negative battery cable and drain the cooling system.
2. Disconnect the throttle cable at the throttle body. If equipped with cruise control, remove the cruise control actuator and vacuum pump.
3. Remove the air cleaner hose.
4. Raise the vehicle and support safely. Remove the engine undercovers.
5. Remove the lower suspension crossmember and the front exhaust pipe.
6. Remove the alternator. Remove the ISC valve.
7. Remove the throttle body, EGR pipe, EGR valve and vacuum modulator.
8. Remove the vacuum pipe and the distributor.
9. Remove the exhaust crossover pipe. Disconnect the cold start injector and then remove the injector tube.
10. Tag and disconnect all hoses leading to the air intake chamber and then remove the chamber.
11. Remove the fuel delivery pipes and the injectors.
12. Disconnect the water temperature sensor and remove the upper radiator hose. Remove the water outlet. Remove the water bypass outlet.
13. Loosen the 2 bolts and remove the cylinder head rear plate.
14. Remove the intake and exhaust manifolds.
15. Remove the timing belt, camshaft pulleys and the No. 2 idler pulley.
16. Remove the No. 3 timing belt cover.

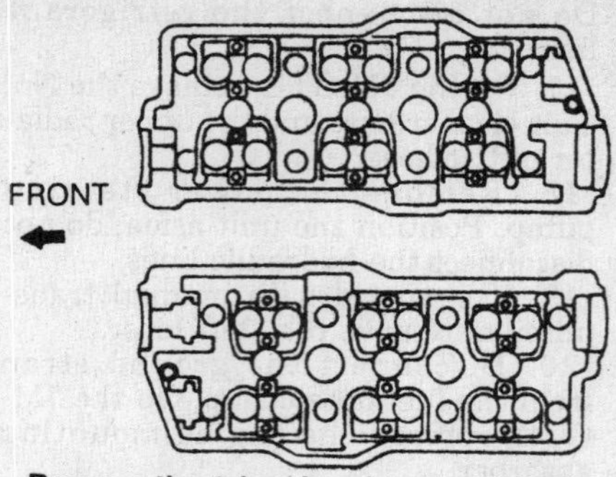

Remove the 2 hex head bolts—2VZ-FE engine

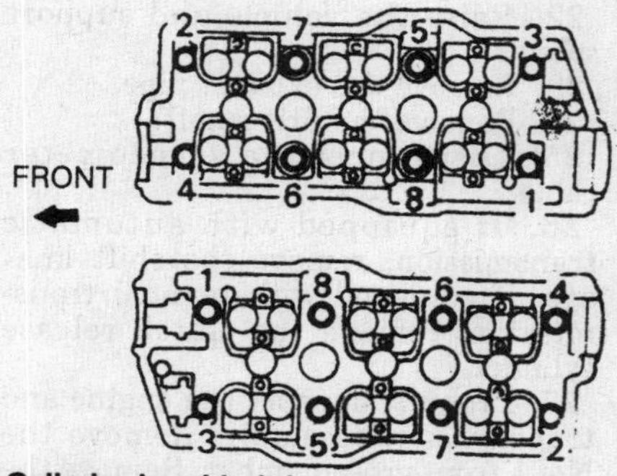

Cylinder head bolt loosening sequence—2VZ-FE engine

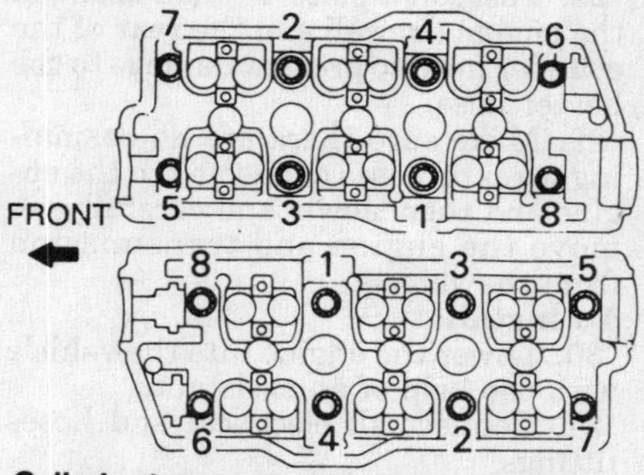

Cylinder head bolt tightening sequence—2VZ-FE engine

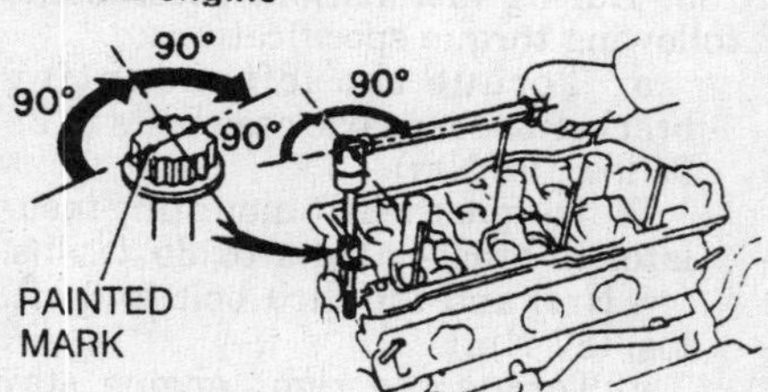

Angle torquing the cylinder head bolts—2VZ-FE engine

17. Remove the cylinder head covers and then remove the camshafts.
18. Remove the 2 hex head cylinder head bolts. Loosen and remove the remaining cylinder head bolts, in several stages, in the proper sequence. Remove the cylinder heads from the block.

To install:

19. Install new cylinder head gaskets on the block and then position the cylinder heads.
20. Torque the cylinder head bolts as follows:

a. Install the regular (12-sided) cylinder head bolts and tighten them to 25 ft. lbs. (34 Nm) in the the proper sequence.

b. Mark the front of each bolt with a dab of paint and then tighten each bolt, in order, an additional 90 degrees (the dab of paint will be at the 3 o'clock position.

c. Retighten all bolts, in sequence, an additional 90 degrees so the paint dab is now at 6 o'clock.

d. Coat the threads of the 2 hex head bolts with engine oil and install them. Tighten each bolt to 13 ft. lbs. (18 Nm).

21. Installation of the remaining components. Start the engine and check for leaks.

3A-C, 4A-C and 4A-LC Engines

EXCEPT 1988 COROLLA FX

1. Disconnect the negative battery cable. Remove the exhaust pipe from the manifold. Drain the cooling system.
2. Remove the air cleaner. Tag and disconnect all hoses necessary to remove the cylinder head.
3. Remove all linkages from the carburetor. Disconnect and plug the fuel lines at the cylinder head and manifold.
4. Remove the fuel pump, carburetor and intake manifold.
5. Remove the cylinder head cover. Note the position of the spark plug wires and remove them. Remove the spark plugs.
6. Set the No. 1 cylinder to TDC of its compression stroke. This is accomplished by removing the No. 1 spark plug, placing finger over the hole and then turning the crankshaft pulley until pressure is exerted against finger.
7. Remove the crankshaft pulley with the proper tool. Remove the water pump pulley. Remove the top and bottom timing cover.
8. Matchmark the camshaft pulley and timing belt for reassembly. Loosen the belt tensioner. Remove the water pump.
9. Remove the timing belt. Do not bend, twist, or turn the belt inside out.
10. Remove the rocker arm bolts and remove the rocker arms. Remove the camshaft pulley by holding the camshaft with a suitable tool and removing the belt in the pulley end of the shaft. Do not hold the cam on the lobes, as damage will result.
11. Remove the camshaft seal. Remove the camshaft bearing caps and set them down in the order they appear on the engine. Remove the camshaft.
12. Loosen the head bolts in the re-

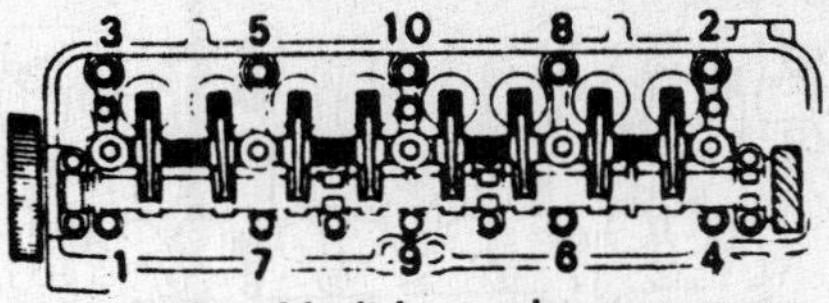

Cylinder head bolt loosening sequence—3A-C, 4A-C and 4A-LC engines

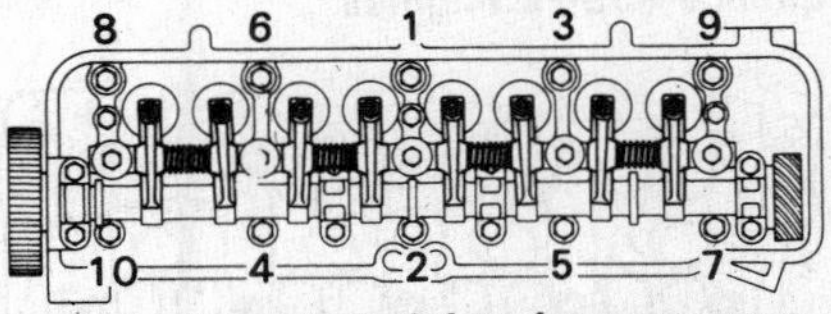

Cylinder head bolt tightening sequence—3A-C, 4A-C and 4A-LC engines

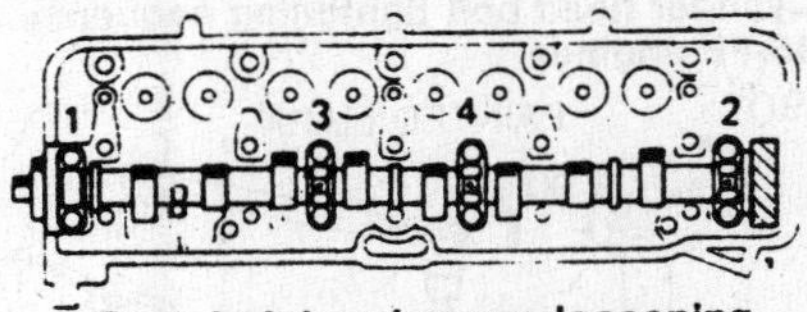

Camshaft bearing cap loosening sequence—3A-C, 4A-C and 4A-LC engines

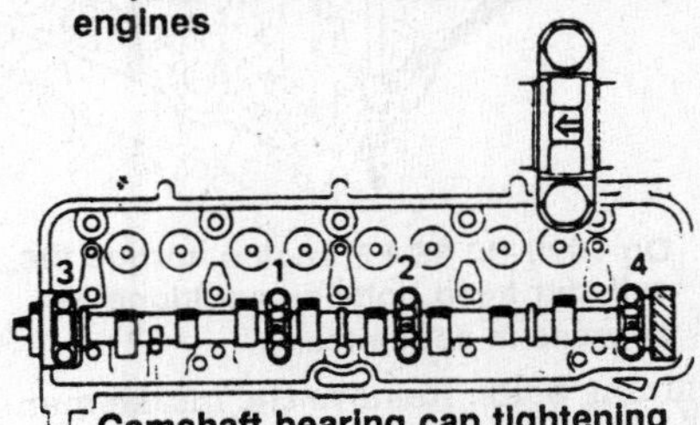

Camshaft bearing cap tightening sequence—3A-C, 4A-C and 4A-LC engines

verse order of the torque sequence. Lift the head directly up. Do not attempt to slide it off.

To install:

13. Clean the gasket mating surfaces. Use care not to damage the aluminum components. Lower the cylinder head onto the engine with the help of an assistant. Make sure the dowel pins are aligned and no hoses or wires are between the head and cylinder block.

14. During installation, observe the following torques:

Cam bearing caps 8–10 ft. lbs. (11–14 Nm).

Cam sprocket 29–39 ft. lbs. (39–53 Nm).

Crankshaft pulley 55–61 ft. lbs. (75–83 Nm)

Manifold bolts 15–21 ft. lbs. (20–29 Nm).

Rocker arm bolts 17–19 ft. lbs. (23–26 Nm).

Timing gear idler bolt 22–32 ft. lbs. (30–44 Nm).

Adjust belt tension 0.24–0.28 in. (6–7mm).

Adjust the valves.

15. Connect all electrical and hoses fittings.

16. Install all necessary components.

17. Connect the battery cable, refill all fluids, start the engine and check for leaks.

1988 COROLLA FX

1. Disconnect the negative battery cable. Raise the vehicle and support safely. Remove the engine under cover. Drain the cooling system and engine oil.

2. Disconnect the exhaust pipe at the manifold. Remove the air cleaner assembly.

3. If equipped with automatic transaxle, disconnect the accelerator and throttle cables. Disconnect and plug the fuel lines at the fuel pump.

4. Disconnect the water hose and remove the water outlet housing. Disconnect the water hoses at the cylinder head. Disconnect the water pump pulley.

5. If equipped with power steering, remove the pump stay. Remove the distributor and spark plugs.

6. Remove the cylinder head cover and gasket. Lift out the half circle plug.

7. Remove the No. 1 and No. 2 timing belt covers. With the engine at TDC of the compression stroke, matchmark the timing belt to the camshaft timing pulley and then slide it off the pulley. Be sure to secure the bottom of the belt so as not to lose valve timing.

8. Loosen each rocker arm support bolt a little at a time and in sequence. Remove the rocker arms.

9. Secure the camshaft, using the proper tool, and remove the camshaft timing pulley. Measure the camshaft thrust clearance. With the camshaft still secure, loosen the distributor drive gear bolt. Loosen the camshaft bearing cap bolts gradually and in sequence. Remove the camshaft.

10. Loosen and remove the cylinder head bolts gradually, in several stages. Remove the cylinder head.

To install:

11. Position the cylinder head on the block with a new gasket and tighten the head bolts in several stages, in the order shown, to a final torque of 43 ft. lbs. (59 Nm).

12. Install the distributor drive gear and plate washer to the camshaft, coat the bearing journals with clean engine oil and position the cam into the head. Position the bearing caps over each journal with the arrow pointing forward. Install a new oil seal and then tighten the cap bolts to 9 ft. lbs. (13 Nm) in the correct order.

13. Install the camshaft timing pulley and tighten the bolt to 34 ft. lbs. (47 Nm).

14. Install and tighten the rocker arm support bolts gradually in 3 passes and in the proper sequence. Tighten to 18 ft. lbs. (25 Nm).

NOTE: Loosen the rocker arm adjusting screw before installation.

15. Align the mark on the No. 1 camshaft bearing cap with the small hole in the timing pulley and install the timing belt so the marks made earlier are in alignment.

16. Connect all electrical and hoses fittings.

17. Install all necessary components.

18. Connect the battery cable, refill all fluids, start the engine and check for leaks.

3E Engine

1. Disconnect the negative battery cable.

2. Drain the cooling system.

3. On 1990–92 3E engine, remove the air cleaner assembly.

4. Remove the right engine under cover.

5. If equipped with power steering, remove the power steering pump and bracket. If equipped with air conditioning and without power steering, remove the idler pulley bracket.

6. Disconnect the radiator hoses. Disconnect the accelerator cable. If equipped with automatic transaxle, disconnect the throttle cable from the bracket mounted to the transaxle case.

7. Remove the timing belt and camshaft timing pulley. Disconnect the heater inlet hose. Disconnect and plug the fuel lines.

8. Remove the air suction hose and valve assembly. Disconnect the brake booster hose from the intake manifold. Disconnect the water inlet hose. Disconnect the intake manifold water hose from the intake manifold.

9. Tag and disconnect all electrical wires, vacuum lines and cables that will interfere with cylinder head removal.

10. Remove the EVAP, VSV and the No. 2 cold enrichment breaker valves. Disconnect the water bypass hoses from the carburetor. Remove the valve cover.

11. Disconnect the exhaust pipe. Remove the intake manifold stay and ground strap. Remove the wire harness clamp bolt from the intake manifold.

12. Measure the cylinder head camshaft thrust clearance using a dial indicator gauge. Standard clearance should be 0.0031–0.071 in. Maximum clearance should be 0.0098 in. If not within specification replace defective parts as required.

13. Loosen then remove the cylinder head bolts in 3 phases and in the proper sequence. Remove the cylinder head from the engine.

14. Installation is the reverse of the removal procedure. During installa-

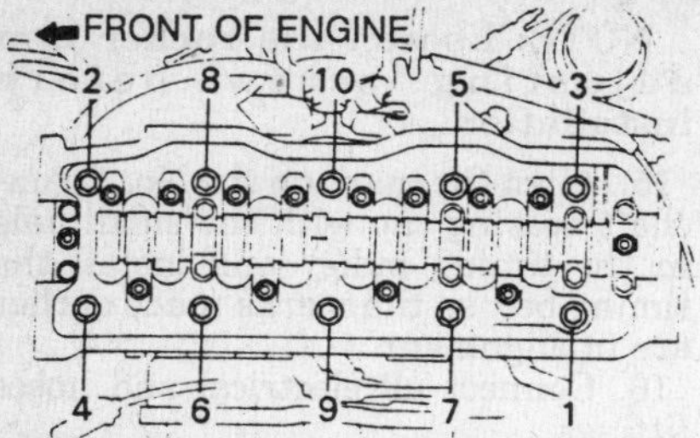

Cylinder head bolt loosening sequence – 3E and 3E-E engines

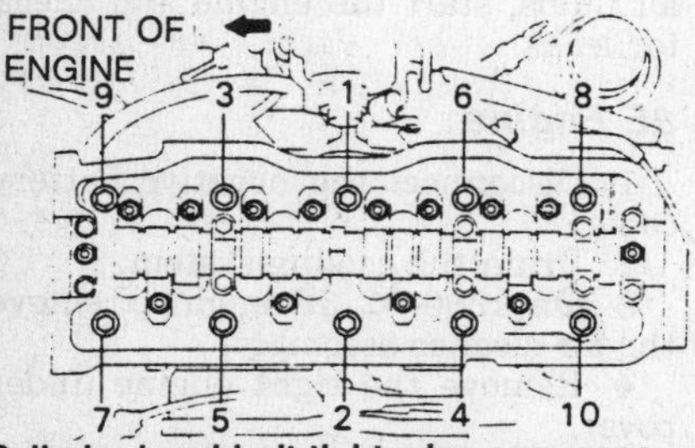

Cylinder head bolt tightening sequence – 3E and 3E-E engines

tion, use a new head gasket. Torque the cylinder head bolts as follows:

a. Tighten the cylinder head bolts in sequence to 22 ft. lbs. (29 Nm).

b. Tighten the bolts is sequence again to 36 ft. lbs. (49 Nm).

c. Retighten each bolt an additional 90 degree turn.

3E-E Engine

1. Disconnect the negative battery cable.
2. Remove the right engine under cover.
3. Drain the cooling system.
4. Disconnect the accelerator and throttle cables.
5. Remove the PCV hoses.
6. Remove the air cleaner and air intake collector assembly.
7. If equipped with power steering, remove the power steering pump and bracket. If equipped with air conditioning and without power steering remove the idler pulley bracket.
8. Remove the pulsation damper; disconnect the fuel inlet and return hoses from the delivery pipe. Plug the hoses to prevent fuel leakage.
9. Disconnect the radiator hoses.
10. Disconnect the heater and water inlet hoses.
11. From the water inlet pipe, disconnect the water bypass hose that connects to the auxiliary air valve.
12. Tag and disconnect all electrical wire and vacuum hoses that interfere with removal of the cylinder head.
13. Remove the exhaust pipe stay and disconnect the exhaust pipe from the manifold.
14. Remove the intake manifold stay.
15. Remove the timing belt and the camshaft timing pulley.
16. Remove the valve cover.
17. Loosen then remove the cylinder head bolts in 3 phases and in the proper sequence. Remove the cylinder head from the engine.

To install:

18. Clean the gasket mating surfaces. Use care not to damage the aluminum components. Lower the cylinder head onto the engine with the help of an assistant. Make sure the dowel pins are aligned and no hoses or wires are between the head and cylinder block.
19. During installation, observe the following torques:

a. Tighten the cylinder head bolts in sequence to 22 ft. lbs. (29 Nm).

b. Tighten the bolts is sequence again to 36 ft. lbs. (49 Nm).

c. Retighten each bolt an additional 90 degree turn each.

20. Connect all electrical and hoses fittings.
21. Install all necessary components.
22. Connect the battery cable, refill all fluids, start the engine and check for leaks.

3S-FE Engine

1. Disconnect the negative battery cable.
2. Drain the cooling system.
3. If equipped with automatic transaxle, disconnect the throttle cable and bracket from the throttle body.
4. Disconnect the accelerator cable and bracket from the throttle body and intake chamber. If equipped with cruise control, remove the actuator and bracket.
5. Remove the air cleaner hose and the alternator.
6. Remove the oil pressure gauge, engine hangers and alternator upper bracket. Raise the vehicle and support safely. Remove the right wheel and tire assembly.
7. Remove the right under cover. Remove the suspension lower crossmember. Disconnect the exhaust pipe from the catalytic converter. Separate the exhaust pipe from the catalytic converter.
8. Disconnect the water temperature sender gauge connector, water temperature sensor connector, cold start injector time switch connector, upper radiator hose, water hoses, and the emission control vacuum hoses.
9. Remove the water outlet and gaskets. Remove the distributor. Remove the water bypass pipe. Remove the EGR valve and modulator.
10. Remove the throttle body assembly. Remove the cold start injector pipe. Remove the air intake chamber air hose, the throttle body air hose, and the power steering pump hoses, if equipped. Remove the air tube.
11. Remove the intake manifold re-

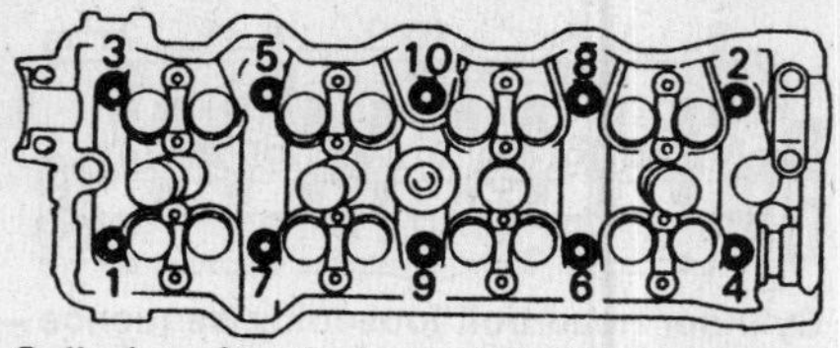

Cylinder head bolt LOOSENING sequence – 3S-FE engines

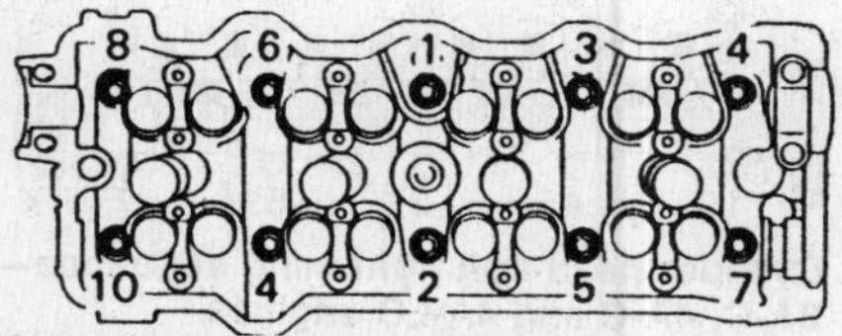

Cylinder head bolt tightening sequence – 3S-FE engines

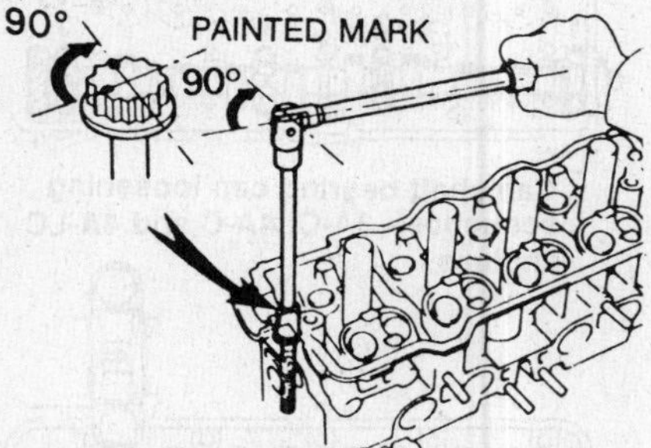

On 1990–92 3S-FE engine, torque the cylinder head bolts an additional 90 degrees in sequence.

taining bolts. Remove the intake manifold. Remove the fuel delivery pipe and the injectors. Remove the spark plugs.

12. Remove the camshaft timing pulley. Remove the No. 1 idler pulley and tension spring. Remove the No. 3 timing belt cover. Properly support the timing belt so meshing of the crankshaft timing pulley does not occur and the timing belt does not shift.
13. Remove the cylinder head cover. Arrange the grommets in order so they can be reinstalled in the correct order.
14. Remove the camshafts.
15. Loosen, then remove the cylinder head bolts in 3 phases and in the proper sequence. Remove the cylinder head from the engine.

To install:

16. Clean the gasket mating surfaces. Use care not to damage the aluminum components. Lower the cylinder head onto the engine with the help of an assistant. Make sure the dowel pins are aligned and no hoses or wires are between the head and cylinder block.
17. During installation, observe the following torques:

a. Apply a light coat of clean engine oil to the threads of the head bolts prior to installation.

b. On 1988–89 vehicles, torque the cylinder head to specification and in 3 phases to 47 ft. lbs. (64 Nm).

c. On 1990–92 vehicles, torque the cylinder head to specification and in 3 phases to 47 ft. lbs. (64 Nm). Then, mark the front of each cylinder head bolt with a dab of paint. Finally, re-torque the cylinder head bolts an additional 90 degrees in the proper sequence.

18. Connect all electrical and hoses fittings.

19. Install all necessary components.

20. Connect the battery cable, refill all fluids, start the engine and check for leaks.

3S-GE Engine

1. Disconnect the negative battery cable. Drain the cooling system.

2. Tag and disconnect the ignition coil connector and the spark plug wire at the ignition coil. Remove the 4 nuts and 2 bolts and lift out the upper suspension brace.

3. If equipped with automatic transaxle, disconnect the throttle cable with its bracket from the throttle body. Disconnect the accelerator cable from the throttle body. Remove the radiator overflow tank.

4. If equipped with cruise control, remove the cruise control actuator and its bracket.

5. Disconnect the air flow meter connector. Remove the air cleaner cap clips. Loosen the hose clamp and remove the air cleaner hose and the air flow meter along with the air cleaner top. Lift out the filter element and then remove the air cleaner case.

6. Tag and disconnect the oxygen sensor lead. Remove the 4 mounting bolts and remove the exhaust manifold heat insulator. Remove the alternator and bracket.

7. Raise the vehicle and support safely. Remove the right front wheel and tire assembly.

8. Remove the right side engine under cover and remove the lower suspension crossmember.

9. Disconnect the exhaust pipe at the manifold. Remove the exhaust manifold stay and the EGR pipe. Unbolt the manifold and remove it along with the lower heat insulator.

10. Remove the distributor. Disconnect the oil pressure switch connector.

11. Tag and disconnect all electrical leads and vacuum hoses at the water outlet. Remove the upper radiator hoses, the heater outlet hose and the water bypass hose. Remove the water outlet.

12. Disconnect the heater inlet hose and the water bypass hose and then remove the water bypass pipe.

13. Disconnect the throttle position sensor lead, the ventilation hose, the air valve hose and any emission control vacuum hoses at the throttle body. Remove the 4 bolts and lift out the throttle body.

14. Remove the forward engine hanger and the No. 2 intake manifold stay. Remove the EGR vacuum modulator.

15. Tag and disconnect any remaining vacuum hoses which may interfere with cylinder head removal. Tag and disconnect the fuel injector electrical leads at the injector.

16. Disconnect the fuel inlet hose at the fuel filter. Disconnect the fuel return hose at the return pipe.

17. Remove the No. 1 and No. 3 intake manifold stays. Tag and disconnect the 2 VSV connectors. Disconnect the 2 power steering vacuum hoses. Remove the intake manifold and the air control valve.

18. Remove the fuel delivery pipe with the injectors attached. Pull the 4 injector insulators out of the injector holes in the cylinder head.

19. Remove the cylinder head cover. Remove the spark plugs. Remove the No. 1 engine hanger.

20. Remove the power steering reservoir and position it aside with the hydraulic lines still attached.

21. Remove the camshaft timing pulleys. Remove the No. 1 idler pulley and tension spring.

22. Remove the bolt holding the No. 2 and No. 3 timing covers. Remove the 4 mounting bolts and remove the No. 3 timing cover.

23. Loosen and remove the camshaft bearing caps, in several stages, and in the proper sequence. Lift out the camshafts and the oil seal. When removing the camshaft bearing caps, keep them in the proper order.

24. Loosen and remove the cylinder head bolts, in several stages, and in the proper sequence. Remove the cylinder head.

To install:

25. Position the cylinder head onto the cylinder block with a new gasket. Lightly coat the cylinder head bolts with engine oil, install them into the head and tighten them in several passes, in the proper sequence, to 40 ft. lbs. (53 Nm).

26. Position the camshafts into the cylinder head so the No. 1 cam lobes are facing outward.

27. Apply silicone sealant to the outer edge of the mating surface on the No. 1 bearing cap only. Position the bearing caps over each journal with the arrows pointing forward and in numerical order from the front to the rear.

28. Lightly coat the cap bolt threads with engine oil. Tighten them in several stages, and in the proper sequence, to 14 ft. lbs. (19 Nm).

29. Check the camshaft thrust clearance. Coat the inside of a new oil seal with grease and carefully tap it onto the camshaft with a suitable drift. Install the No. 3 timing belt cover.

30. Connect the idler pulley tension spring to the pulley and the pin on the cylinder head. Install the idler pulley onto the pivot pin, force it to the left as far as it will go and tighten it. Make sure the tension spring is not out of the groove in the pin. Install the camshaft timing pulleys and the timing belt.

31. Installation of the remaining components is in the reverse order of removal. Tighten the lower suspension crossmember end bolts to 154 ft.

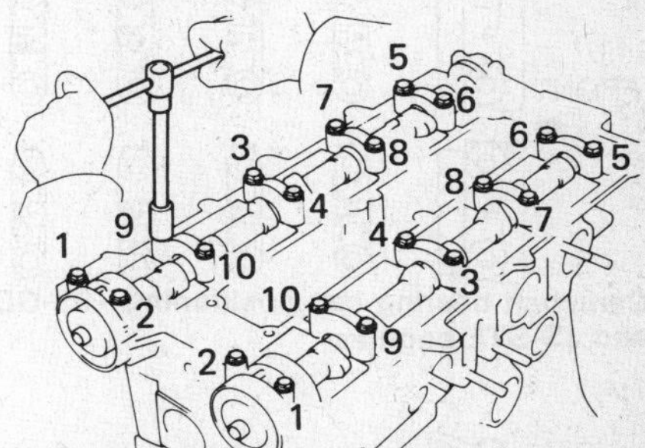

Remove the camshaft bearing caps in the order—3S-GE and 3S-GTE engines

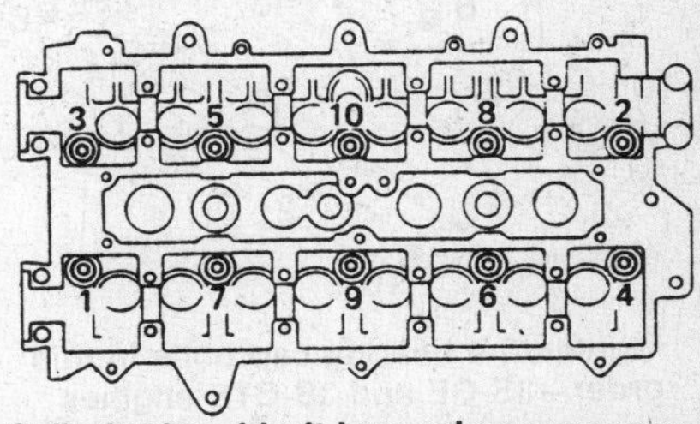

Cylinder head bolt loosening sequence—3S-GE and 3S-GTE engines

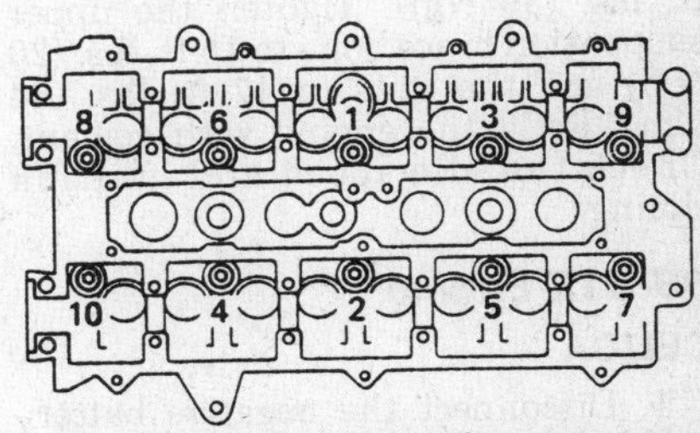

Cylinder head bolt tightening sequence—3S-GE and 3S-GTE engines

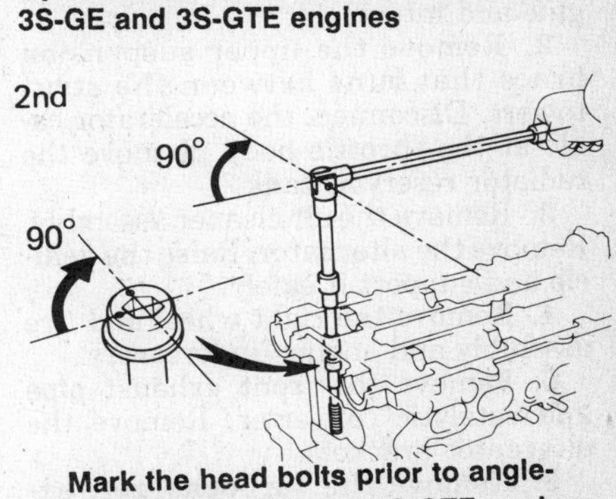

Mark the head bolts prior to angle-torquing—3S-GE and 3S-GTE engines

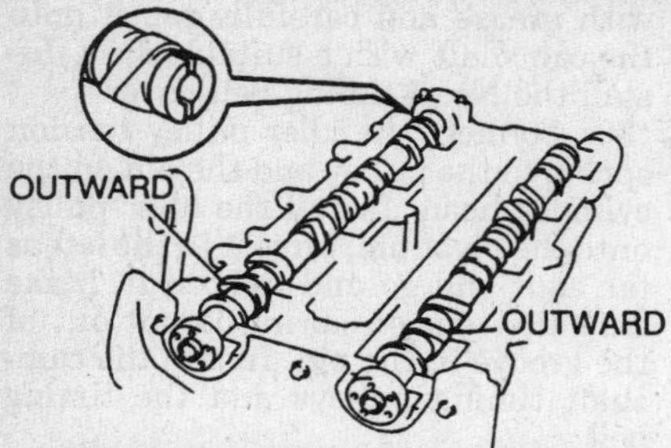

Camshaft positioning – 3S-GE and 3S-GTE engines

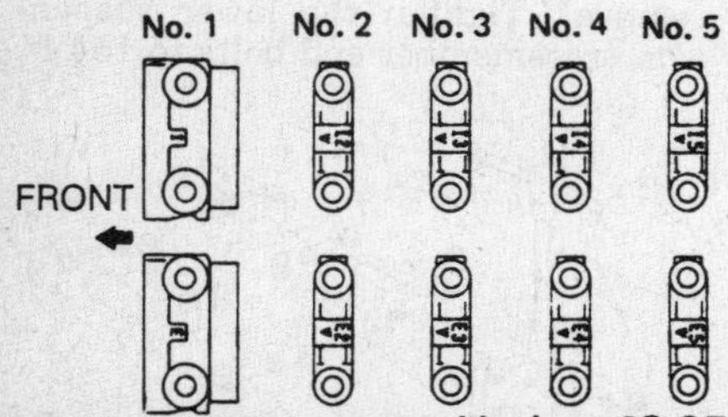

Camshaft bearing cap positioning – 3S-GE and 3S-GTE engines

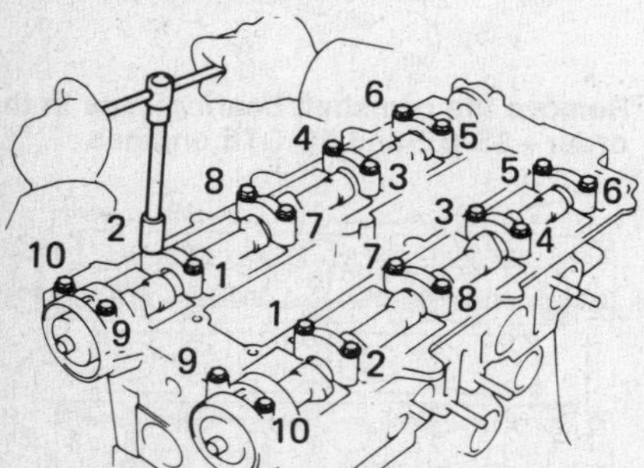
Tighten the bearing cap bolts in this order – 3S-GE and 3S-GTE engines

lbs. (209 Nm) and the center bolt to 29 ft. lbs. (39 Nm). Tighten the upper suspension brace bolts to 15 ft. lbs. (20 Nm) and the nuts to 47 ft. lbs. (64 Nm). Refill the engine with coolant. Check the idle speed and ignition timing.

3S-GTE Engine

CELICA

1. Disconnect the negative battery cable. Drain the coolant from the engine and intercooler.
2. Remove the upper suspension brace that runs between the strut towers. Disconnect the accelerator cable at the throttle body. Remove the radiator reservoir tank.
3. Remove the air cleaner assembly. Remove the alternator. Raise the vehicle and support it safely.
4. Remove the right wheel and tire assembly and engine undercovers.
5. Remove the front exhaust pipe and catalytic converter. Remove the alternator brackets.
6. Remove the turbocharger, exhaust manifold and distributor.
7. Disconnect the air hose and remove the No. 2 air pipe.
8. Remove the left engine hanger along with the reservoir tank. Remove the oil pressure switch.
9. Remove the water outlet housing and the water bypass pipe.
10. Remove the throttle body and disconnect the cold start injector lead.
11. Remove the EGR valve, vacuum modulator and EGR control Vacuum Switching Valve (VSV). Remove the delivery pipe and all injectors. Remove the vacuum pipe.
12. Remove the intake manifold stays and the No. 1 air pipe.
13. Disconnect the vacuum hose and remove the fuel pressure VSV. Remove the T-VIS Vacuum Switching (VSV), vacuum tank and the turbo pressure VSV.
14. Remove the intake manifold with the air control valve. Remove the power steering reservoir tank without disconnecting the hydraulic hoses. Position the pump aside.
15. Remove the spark plugs and the No. 2 front cover. Remove the timing belt and the PCV pipe.
16. Remove the cylinder head cover. Remove the camshaft timing pulleys and the No. 1 idler pulley. Remove the No. 3 timing belt cover.

NOTE: Secure the timing belt so the belt does not unmesh from the cranskahsft pulley.

17. Gradually loosen and remove the camshaft bearing cap bolts, in several passes, in the proper sequence. Remove the bearing caps, oil seals and lift out the camshafts.
18. Remove the right rear engine hanger. Remove the cylinder head bolts, in several stages, in the sequence and lift off the cylinder head.

To install:

19. Position the cylinder head and a new gasket on the block and torque the bolts as follows:
    a. Coat the head bolts with engine oil and tighten them in several passes, in sequence, to 40 ft. lbs. (54 Nm).
    b. Mark the front of each bolt with a dab of paint.
    c. Retighten the bolts an additional 90 degrees turn. The paint dabs should now all be at a 90 degree angle to the front of the head.
20. Install the right rear engine hanger and tighten it to 14 ft. lbs. (19 Nm).
21. Position the camshafts in the cylinder head with the No. 1 lobes facing outward. Coat the No. 1 bearing cap with seal packing and then install all the caps over the bearing journals. Coat the bearing cap bolts with engine oil and then tighten them to 14 ft. lbs. (19 Nm) in several stages, in the order shown. Grease 2 new oil seals and install them into the camshafts.
22. Install the No. 3 timing belt cover and the No. 1 idler pulley. Install the camshaft timing pulleys.
23. Install the cylinder head cover and the timing belt.
24. Install the remaining components, start the engine and check for leaks.

1990–92 MR2

1. Disconnect the negative battery cable.
2. Drain the coolant from the engine and intercooler.
3. Tag and disconnect all hoses, lines and wiring that interfere with removal of the turbocharger, exhaust manifold, intake manifold and cylinder head.
4. Remove the engine hood side panels.
4. Remove the upper suspension brace that runs between the strut towers.
5. Disconnect the accelerator cable at the throttle body.
6. If equipped with cruise control, remove the cruise control actuator and disconnect the accelerator linkage.
7. Remove the air cleaner cap.
8. Remove the right front engine hanger.
9. Remove the intercooler.
10. Remove the front exhaust pipe, catalytic converter and turbocharger.
11. Remove the throttle body and cold start injector.
12. Remove the exhaust manifold and distributor.
13. Remove the No. 2 air tube.
14. Remove the left engine hanger.
15. Remove the EGR vacuum modulator and Vacuum Switching Valve (VSV).
16. Remove the vacuum pipe, EGR valve and EGR pipe.
17. Remove the water outlet and housing.
18. Remove the oil pressure switch.
19. Remove the oil cooler.
20. Remove the water bypass pipe.
21. Remove the intake manifold stays and the No. 1 air pipe.
22. Remove the T-VIS Vacuum Switching (VSV), vacuum tank and the turbocharger pressure VSV.
23. Remove the intake manifold with the air control valve.
24. Remove the delivery pipe and fuel injectors.
25. Remove the cylinder head cover.
26. Remove the camshaft timing pulleys and the No. 1 idler pulley.
27. Remove the No. 3 timing belt cover.

NOTE: Secure the timing belt so the belt does not unmesh from the cranskahsft pulley.

28. Gradually loosen and remove the camshaft bearing cap bolts in several passes, in the proper sequence. Remove the bearing caps, oil seals and lift out the camshafts.

29. Remove the cylinder head bolts in several stages, in the the proper sequence and lift off the cylinder head from the alignment dowels on the block.

To install:

30. Position the cylinder head and a new gasket on the block.

31. Torque the cylinder head bolts as follows:

a. Coat the head bolts with engine oil and tighten them in several passes, in sequence to 36 ft. lbs. (49 Nm).

b. Mark the front of each bolt with a dab of paint.

c. Retighten the bolts an additional 90 degrees turn. The paint dabs should now all be at a 90 degree angle to the front of the head.

32. Position the camshafts in the cylinder head with the No. 1 lobes facing outward. Coat the No. 1 bearing cap with seal packing and then install all the caps over the bearing journals. Coat the bearing cap bolts with engine oil and then tighten them to 14 ft. lbs. (19 Nm) in several stages, in the order shown. Grease 2 new oil seals and install them into the camshafts.

33. Check and adjust the valve clearance, as necessary.

34. Install the No. 3 timing belt cover, No. 1 idler pulley and camshaft timing pulleys.

35. Install the cylinder head cover with 2 new gaskets and 12 new bolt seal washers. Torque the cover bolts to 21 inch lbs. (2.5 Nm).

36. Install the remaining components, start the engine and check for leaks.

4A-F Engine

1. Disconnect the negative battery cable. Drain the cooling system.

2. Remove the engine undercover and then disconnect the exhaust pipe at the manifold.

3. Remove the air cleaner and hoses. If equipped with automatic transaxle, disconnect the accelerator and throttle cables at the bracket.

4. Tag and disconnect all wires, lines and hoses that may interfere with removal of the exhaust manifold, intake manifold and cylinder head.

5. Disconnect the fuel lines at the fuel pump. Disconnect the heater hoses at the engine.

6. Disconnect the water hose and the bypass hose at the rear of the cylinder head. Remove the 2 bolts and pull off the water outlet pipe.

7. Remove the 2 mounting bolts and lift out the exhaust manifold stay. Remove the upper manifold insulator and then remove the exhaust manifold.

8. Remove the distributor.

9. Disconnect the 2 water hoses at the water inlet (front of head) and then remove the inlet housing.

10. Remove the fuel pump.

11. Disconnect the PCV and water hoses at the intake manifold. Remove the manifold stay and then remove the intake manifold and wire clamp.

12. Remove the drive belts and the power steering pump support.

13. Remove the spark plugs and the cylinder head cover.

14. Remove the No. 3 and No. 2 front covers. Turn the crankshaft pulley and align its groove with the **0** mark on the No. 1 front cover. Check that the camshaft pulley hole aligns with the mark on the No. 1 camshaft bearing cap (exhaust side).

15. Remove the plug from the No. 1 front cover and matchmark the timing belt to the camshaft pulley. Loosen the idler pulley mounting bolt and push the pulley to the left as far as it will go; tighten the bolt. Slide the timing belt off the camshaft pulley and support it so it won't fall into the case.

16. Remove the camshaft pulley and check the camshaft thrust clearance. Remove the camshafts.

17. Gradually loosen the cylinder head mounting bolts in several passes, in the sequence. Remove the cylinder head.

To install:

18. Clean the gasket mating surfaces. Use care not to damage the aluminum components. Lower the cylinder head onto the engine with the help of an assistant. Make sure the dowel pins are aligned and no hoses or wires are between the head and cylinder block.

19. Lightly coat the cylinder head bolts with engine oil and then install them. Tighten the bolts in 3 stages, in the proper sequence. On the final pass, torque the bolt to 44 ft. lbs. (60 Nm).

19. Position the camshafts into the cylinder head. Position the bearing caps over each journal with the arrows pointing forward.

20. Tighten each bearing cap a little at a time and in the reverse of the removal sequence. Tighten to 9 ft. lbs. (13 Nm) Recheck the camshaft endplay.

21. Install the camshaft timing pulleys making sure the camshaft knock pins and the matchmarks are in alignment. Lock each camshaft and tighten the pulley bolts to 43 ft. lbs. (59 Nm).

22. Align the matchmarks made during removal and then install the timing belt on the camshaft pulley. Loosen the idler pulley set bolt. Make sure the timing belt meshing at the crankshaft pulley does not shift.

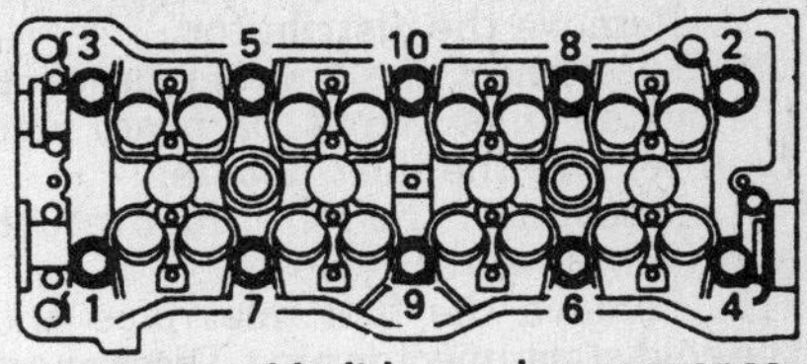

Cylinder head bolt loosening sequence—4A-F and 4A-FE engines (Corolla and Celica)

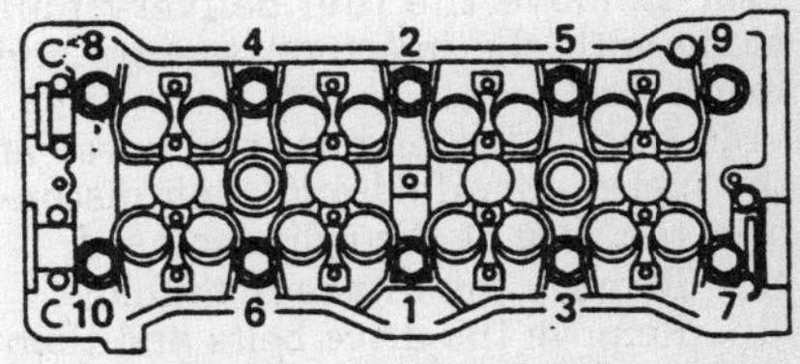

Cylinder head bolt tightening sequence—4A-F and 4A-FE engines (Corolla)

23. Rotate the crankshaft clockwise 2 revolutions from TDC to TDC. Make sure each pulley aligns with the marks made previously. If the marks are not in alignment, the valve timing is wrong. Shift the timing belt meshing slightly and then repeat Steps 21–23.

24. Tighten the set bolt on the timing belt idler pulley to 27 ft. lbs. (37 Nm). Measure the timing belt deflection at the top span between the 2 camshaft pulleys. It should deflect no more than 0.16 in. at 4.4 lbs. of pressure. If deflection is greater, readjust by using the idler pulley.

25. Install the remaining components, start the engine and check for leaks.

4A-FE Engine

COROLLA

1. Disconnect the negative battery cable at the battery. Drain the cooling system.

2. Remove the engine undercover and then disconnect the exhaust pipe at the manifold.

3. Remove the air cleaner and hoses; disconnect the intake air temperature sensor. Disconnect the accelerator and throttle cables at the bracket on vehicles with automatic transaxle.

4. Remove the cruise control actuator cable.

5. Tag and disconnect all wires, lines and hoses that may interfere with exhaust manifold, intake manifold and cylinder head removal.

6. Disconnect the fuel lines at the fuel pump. Disconnect the heater hoses at the engine.

7. Disconnect the water hose and the bypass hose at the rear of the cylinder head. Remove the 2 bolts and pull off the water outlet pipe.

8. Remove the 2 mounting bolts and lift out the exhaust manifold stay. Remove the upper manifold insulator and then remove the exhaust manifold.

9. Remove the distributor.
10. Disconnect the 2 water hoses at the water inlet (front of head) and then remove the inlet housing.
11. Disconnect the PCV, fuel return and vacuum sensing hoses.
12. Remove the fuel inlet pipe and the cold start injector pipe. Disconnect the 4 vacuum hoses and then remove the EGR vacuum modulator.
13. Remove the fuel delivery pipe along with the injectors, spacers and insulators.
14. Unbolt the engine wire cover at the intake manifold and then disconnect the wire at the cylinder head.
15. Remove the intake manifold.
16. Remove the drive belts and then remove the water pump.
17. Remove the spark plugs, cylinder head cover and semi-circular plug.
18. Remove the No. 3 and No. 2 front covers. Turn the crankshaft pulley and align its groove with the **0** mark on the No. 1 front cover. Check that the camshaft pulley hole aligns with the mark on the No. 1 camshaft bearing cap (exhaust side). If not, rotate the crankshaft 360 degrees until the marks are aligned.
19. Remove the plug from the No. 1 front cover and matchmark the timing belt to the camshaft pulley. Loosen the idler pulley mounting bolt and push the pulley to the left as far as it will go; tighten the bolt. Slide the timing belt off the camshaft pulley and support it so it won't fall into the case.
20. Remove the camshaft pulley and check the camshaft thrust clearance. Remove the camshafts.
21. Gradually loosen the cylinder head mounting bolts in several passes, in the the proper sequence. Remove the cylinder head.

NOTE: On 1990–92 vehicles, the cylinder head bolts on the right (intake) side of the cylinder head are 3.54 in. (90mm) and the bolts on the left (exhaust) side of the head are 4.25 in. (108mm). Label the bolts to ensure proper installation.

To install:

22. Position the cylinder head on the block with a new gasket. Lightly coat the cylinder head bolts with engine oil and then install them. Tighten the bolts in 3 stages, in the proper sequence. On the final pass, torque the bolt to 44 ft. lbs. (60 Nm).
23. Position the camshafts into the cylinder head. Position the bearing caps over each journal with the arrows pointing forward.
24. Tighten each bearing cap a little at a time and in the reverse of the removal sequence. Tighten to 9 ft. lbs. (13 Nm) and recheck the camshaft endplay.

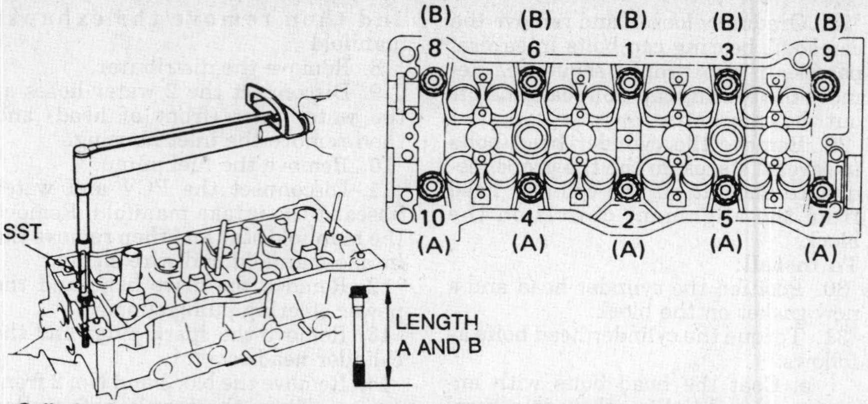

Cylinder head bolt tightening sequence and positioning on 1990–92 4A-FE engine (Corolla and Celica). The "A" bolts are 3.54 in. (90mm) and the "B" bolts are 4.25 in. (108mm)

25. Install the camshaft timing pulleys making sure the camshaft knock pins and the matchmarks are in alignment. Lock each camshaft and tighten the pulley bolts to 43 ft. lbs. (59 Nm).
26. Align the matchmarks made during removal and then install the timing belt on the camshaft pulley. Loosen the idler pulley set bolt. Make sure the timing belt meshing at the crankshaft pulley does not shift.
27. Rotate the crankshaft clockwise 2 revolutions from TDC to TDC. Make sure each pulley aligns with the marks made previously. If the marks are not in alignment, the valve timing is wrong. Shift the timing belt meshing slightly and then repeat Steps 24–26.
28. Tighten the set bolt on the timing belt idler pulley to 27 ft. lbs. (37 Nm). Measure the timing belt deflection at the top span between the 2 camshaft pulleys. It should deflect no more than 0.24 in. (6mm) at 4.4 lbs. of pressure. If deflection is greater, readjust by using the idler pulley.
29. Install the remaining components, start the engine and check for leaks.

1990–92 CELICA

1. Disconnect the negative battery cable. Drain the cooling system.
2. If equipped with automatic transaxle, disconnect the throttle cable and bracket from the throttle body.
3. Disconnect the accelerator cable and bracket from the throttle body.
4. Remove the air cleaner cap and hose.
5. Remove the engine under covers. Remove the suspension lower crossmember.
6. Disconnect all lines, hoses and electrical wires that interfere with exhaust manifold, intake manifold and cylinder head removal.
7. Disconnect the front exhaust pipe, distributor and exhaust manifold.
8. Remove the water outlet and gaskets. Remove the water inlet and inlet housing.
9. Unbolt and remove the power steering pump without disconnecting hoses.
10. Remove the throttle body, cold start injector pipe, cold start injector, delivery pipe and fuel injectors.
11. Remove the Air Control Valve (ACV) assembly. Disconnect engine wiring from the timing belt cover and from the intake manifold.
12. Remove the vacuum pipe, EGR vacuum modulator and EGR Vacuum Switching Valve (VSV) assembly.
13. Remove the EGR valve and gasket. Remove the water inlet pipe and fuel return hose from the fuel filter.
14. Remove the intake manifold with retaining manifold stay (bracket).
15. Remove the valve cover.
16. Remove the camshaft timing pulley, No. 1 idler pulley and tension spring and No. 3 timing belt cover. Properly support the timing belt so meshing of the crankshaft timing pulley does not occur and the timing belt does not shift.
17. Remove the fan belt adjusting bar, engine hangers and power steering drive belt adjusting strut or bracket.
18. Remove the camshafts. Make sure to uniformly loosen and remove bearing cap bolts in several phases and in the proper sequence when removing the camshafts.
19. Loosen then remove the cylinder head bolts in 3 phases and in the proper sequence. Remove the cylinder head from the engine.

NOTE: The cylinder head bolts on the right (intake) side of the cylinder head are 3.54 in. (90mm) and the bolts on the left (exhaust) side of the head are 4.25 in. (108mm). Label the bolts to ensure proper installation.

To install:

20. Install the cylinder head on the

cylinder head block. Place the cylinder head in position on the cylinder head gasket.

21. Apply a light coat of clean engine oil to the threads of the head bolts before installation. Tighten the bolts in 3 stages, in the proper sequence. On the final pass, torque the bolt to 44 ft. lbs. (60 Nm).

22. Installation of the camshafts and remaining components. Start the engine and check for leaks.

4A-GE, 4A-GEC and 4A-GELC Engines

1. Disconnect the negative battery cable. Remove the engine undercover. Drain the cooling system and engine oil.

2. Loosen the clamp and then disconnect the No. 1 air cleaner hose from the throttle body. Disconnect the actuator and accelerator cables from the bracket on the throttle body.

3. If equipped with power steering, Remove the power steering pump and its bracket. Position the pump aside with the hydraulic lines connected.

4. Loosen the water pump pulley set nuts. Remove the drive belt adjusting bolt and then remove the belt. Remove the set nuts and then remove the fluid coupling along with the fan and the water pump pulley.

5. Disconnect the upper radiator hose at the water outlet on the cylinder head. Disconnect the 2 heater hoses at the water bypass pipe and the cylinder head rear plate.

6. Remove the distributor. Remove the cold start injector pipe and the PCV hose from the cylinder head.

7. Remove the pulsation damper from the delivery pipe. Disconnect the fuel return hose from the pressure regulator.

8. Tag and disconnect all vacuum hoses which may interfere with cylinder head remvoval. Remove the wiring harness and the vacuum pipe from the No. 3 timing cover. Tag and disconnect all wires which might interfere with exhaust manifold, intake manifold and cylinder head removal.

9. Disconnect the exhaust bracket from the exhaust pipe. Disconnect the exhaust manifold from the exhaust pipe.

10. Remove the vacuum tank and the VCV valve. Remove the exhaust manifold.

11. Remove the 2 mounting bolts and remove the water outlet housing from the cylinder head with the No. 1 bypass pipe and gasket. Pull the No. 1 bypass pipe out of the housing.

12. Remove the fuel delivery pipe along with the fuel injectors.

13. Remove the intake manifold stay. Remove the intake manifold along with the air control valve.

14. Remove the cylinder head covers and their gaskets. Remove the spark plugs. Remove the No. 1 and No. 2 timing belt covers and their gaskets.

15. Rotate the crankshaft pulley until its groove is in alignment with the **0** mark on the No. 1 timing belt cover. Check that the valve lifters on the No. 1 cylinder are loose. If not, rotate the crankshaft 1 complete revolution (360 degrees).

16. Place matchmarks on the timing belt and 2 timing pulleys. Loosen the idler pulley bolts and move the pulley to the left as far as it will go and then retighten the bolt.

17. Remove the timing belt from the camshaft pulleys. When removing the timing belt, support the belt so the meshing of the crankshaft timing pulley and the timing belt does not shift. Never drop anything inside the timing case cover. Be sure the timing belt does not come in contact with dust or oil.

18. Lock the camshafts and remove the timing pulleys. Remove the No. 4 timing belt cover.

19. Using a dial indicator, measure the endplay of each camshaft. If not within specification, replace the thrust bearing.

20. Loosen each camshaft bearing cap bolt a little at a time and in the correct sequence. Remove the bearing caps, camshaft and oil seal.

21. Loosen the cylinder head bolts gradually in 3 stages, and in the proper order using the proper tool.

22. Remove the cylinder head.

NOTE: On 1990–92 engines, the cylinder head bolts on the right (intake) side of the cylinder head are 3.54 in. (90mm) and the bolts on the left (exhaust) side of the head are 4.25 in. (108mm). Label the bolts to ensure proper installation.

To install:

23. Position the cylinder head on the block with a new gasket. Lightly coat the cylinder head bolts with engine oil and then install the short head bolts on the intake side and the long ones on the exhaust side. Tighten the bolts in 3 stages, in the proper sequence. On the final pass, torque the bolt to 43 ft. lbs.

24. On 1988–92 engines, coat the head bolts with engine oil and tighten them in several passes, in the sequence shown to 22 ft. lbs. (29 Nm). Mark the front of each bolt with a dab of paint and then retighten the bolts a further 90 degree turn. The paint dabs should now all be at a 90 degree angle to the front of the head. Retighten the bolts one more time a further 90 degree turn. The paint dabs should now all be pointing toward the rear of the head.

25. Position the camshafts into the cylinder head. Position the bearing caps over each journal with the arrows pointing forward.

26. Tighten each bearing cap a little at a time and in the reverse of the removal sequence. Tighten to 9 ft. lbs. (13 Nm). Recheck the camshaft endplay.

27. Drive the camshaft oil seals onto the end of the camshafts using a suitable seal installer. Be careful not to install the oil seals crooked. Install the No. 4 timing belt cover.

28. Install the camshaft timing pulleys making sure the camshaft knock pins and the matchmarks are in alignment. Lock each camshaft and tighten the pulley bolts to 34 ft. lbs. (47 Nm).

29. Align the matchmarks made during removal and then install the timing belt on the camshaft pulley. Loosen the idler pulley set bolt. Make sure the timing belt meshing at the crankshaft pulley does not shift.

30. Rotate the crankshaft clockwise 2 revolutions from TDC to TDC. Make sure each pulley aligns with the marks made previously. If the marks are not in alignment, the valve timing is wrong. Shift the timing belt meshing slightly and then repeat Steps 28–30.

31. Tighten the set bolt on the timing belt idler pulley to 27 ft. lbs. (37 Nm). Measure the timing belt deflection at the top span between the 2 camshaft pulleys. It should deflect no more than 0.16 in. at 4.4 lbs. of pressure. If deflection is greater, readjust by using the idler pulley.

32. Install the remaining components, start the engine and check for leaks.

4A-GZE Engine

1. Disconnect the negative battery cable. Remove the hood and engine undercovers.

2. Drain the engine coolant and remove the intercooler. Remove the battery.

3. Disconnect the air bleeder hose at the water inlet housing. Disconnect the cruise control vacuum hose and the throttle cable.

4. Remove the air flow meter with the No. 3 air cleaner hose. Disconnect the accelerator cable.

5. Remove the accelerator link and disconnect the air conditioning idle-up vacuum hoses.

6. Disconnect the heater hose at the rear of the cylinder head. Disconnect the brake booster vacuum hose and remove the radiator reservoir tank. Disconnect the No. 1 radiator hose at the water outlet housing.

7. Tag and disconnect any remaining hoses, wires or connections which may interfere with exhaust manifold, intake manifold and cylinder head removal.

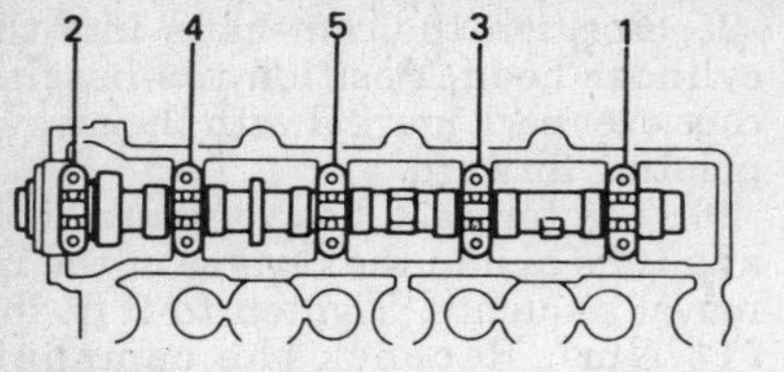

Loosen the camshaft bearing cap bolts in this order—4A-GE (all) and 4A-GZE engines

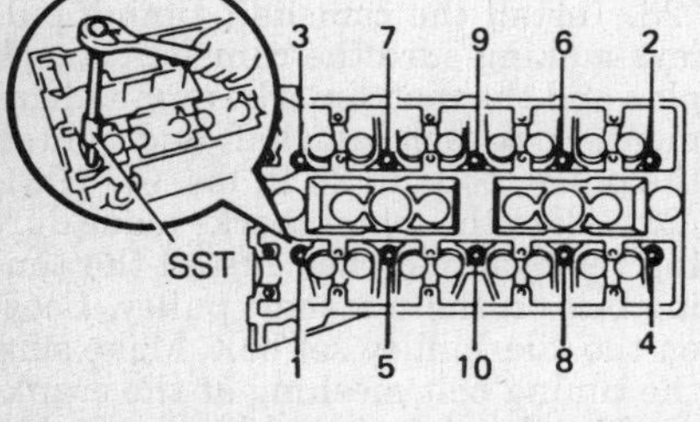

Cylinder head bolt loosening sequence—4A-GE and 4A-GZE engines

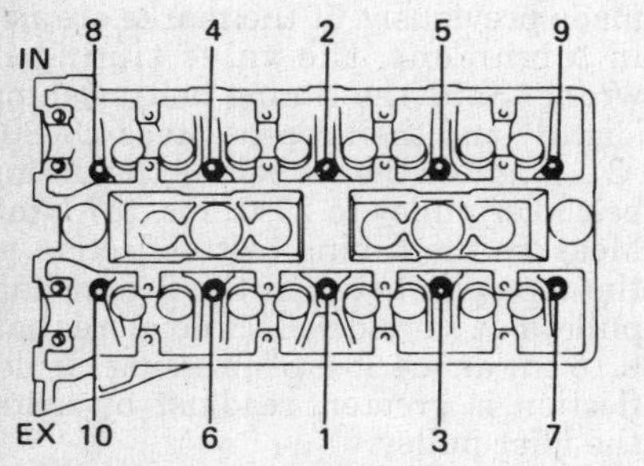

Cylinder head bolt tightening sequence—4A-GE and 4A-GZE engines

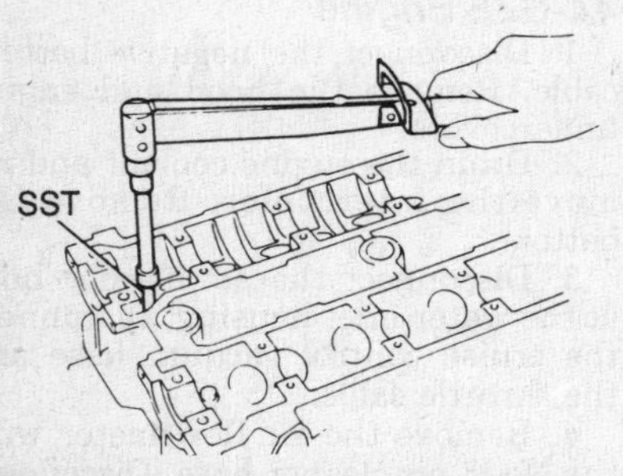

Cylinder head bolt tightening sequence—4A-GE (all) and 4A-GZE engines

Position the camshafts into the cylinder head as shown—4A-GE (all) and 4A-GZE engines

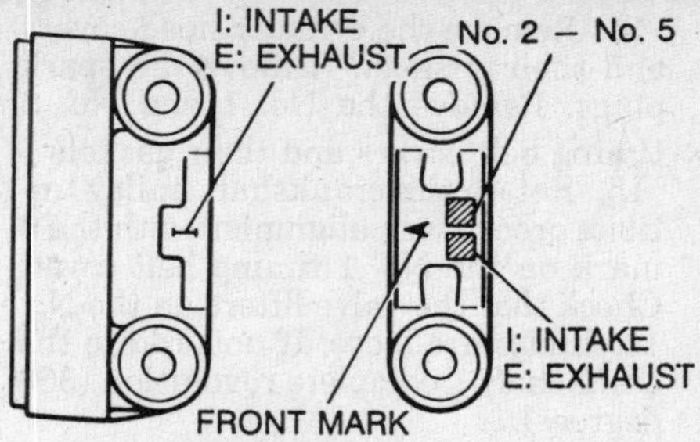

Camshaft bearing cap positioning (the arrows must always point forward)—4A-GE (all) and 4A-GZE engines

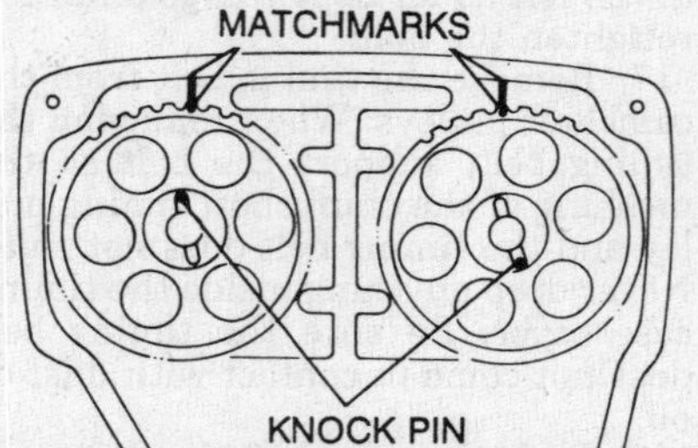

Align the camshaft knockpin with the camshaft timing pulley—4A-GE (all) and 4A-GZE engines

8. Remove all drive belts and then remove the water pump pulley. Remove the supercharger.

9. Remove the No. 2 air outlet duct and the No. 3 fuel pipe. Remove the No. 1 vacuum transmitting pipe.

10. Disconnect the No. 2 fuel hose and remove the cylinder head rear cover.

11. Remove the fuel delivery pipe and the fuel injectors.

12. Remove the water outlet with the bypass pipe. Remove the EGR valve with the pipe still attached.

13. Remove the intake manifold. Remove the front exhaust pipe.

14. Remove the air conditioning compressor and bracket. Remove the distributor and alternator with its bracket.

15. Remove the exhaust manifold.

16. Rotate the crankshaft pulley until its groove is in alignment with the **0** mark on the No. 1 timing belt cover. Check that the valve lifters on the No. 1 cylinder are loose. If not, rotate the crankshaft 1 complete revolution (360 degrees).

17. Place matchmarks on the timing belt and 2 timing pulleys. Loosen the idler pulley bolts and move the pulley to the left as far as it will go and then retighten the bolt.

18. Remove the timing belt from the camshaft pulleys. When removing the timing belt, support the belt so the meshing of the crankshaft timing pulley and the timing belt does not shift. Never drop anything inside the timing case cover. Be sure the timing belt does not come in contact with dust or oil.

19. Lock the camshafts and remove the timing pulleys. Remove the No. 4 timing belt cover.

20. Using a dial indicator, measure the endplay of each camshaft. If not within specification, replace the thrust bearing.

21. Loosen each camshaft bearing cap bolt a little at a time and in the sequence shown. Remove the bearing caps, camshaft and oil seal.

22. Loosen the cylinder head bolts gradually in 3 stages, and in the proper order, using the proper tool.

23. Remove the cylinder head from the vehicle.

To install:

24. Position the cylinder head on the block with a new gasket.

25. Torque the cylinder head bolts as follows:

a. Lightly coat the cylinder head bolts with engine oil and then install the short head bolts on the intake side and the long ones on the exhaust side.

b. Tighten them in several passes, in the proper sequence to 22 ft. lbs. (29 Nm).

c. Mark the front of each bolt with a dab of paint and then retighten the bolts a further 90 degree turn. The paint dabs should now all be at a 90 degree angle to the front of the head.

d. Retighten the bolts one more time a further 90 degree turn. The paint dabs should now all be pointing toward the rear of the head.

26. Position the camshafts into the cylinder head. Position the bearing caps over each journal with the arrows pointing forward.

27. Tighten each bearing cap a little at a time and in the reverse of the removal sequence. Tighten to 9 ft. lbs. (13 Nm). Recheck the camshaft endplay.

28. Drive the camshaft oil seals onto the end of the camshafts using a suitable seal installer. Be careful not to install the oil seals crooked. Install the No. 4 timing belt cover.

29. Install the camshaft timing pulleys making sure the camshaft knock pins and the matchmarks are in alignment. Lock each camshaft and tighten the pulley bolts to 34 ft. lbs. (47 Nm).

30. Align the matchmarks made during removal and then install the timing belt on the camshaft pulley. Loosen the idler pulley set bolt. Make sure the timing belt meshing at the crankshaft pulley does not shift.

31. Rotate the crankshaft clockwise 2 revolutions from TDC to TDC. Make sure each pulley aligns with the marks made previously. If the marks are not in alignment, the valve timing is wrong. Shift the timing belt meshing slightly and then repeat Steps 28–30.

32. Tighten the set bolt on the timing belt idler pulley to 27 ft. lbs. (37 Nm). Measure the timing belt deflection at the top span between the 2 camshaft pulleys. It should deflect no more than 0.16 in. (4mm). at 4.4 lbs. of pressure. If deflection is greater, readjust by using the idler pulley.

33. Install the remaining components, start the engine and check for leaks.

5M-GE Engine

1. Disconnect the negative battery cable. Drain the cooling system.

2. Disconnect the exhaust pipe from the exhaust manifold.

3. Remove the throttle cable bracket from the cylinder head if equipped with automatic transmission, and remove the accelerator and actuator cable bracket.

4. Tag and disconnect the ground cable, oxygen sensor wire, high tension coil wire, distributor connector, solenoid resistor wire connector and thermo- switch wire, if equipped with automatic transmission.

5. Tag and disconnect the brake booster vacuum hose, EGR valve vacuum hose, fuel hose from the intake manifold and actuator vacuum hose, if equipped with cruise control.

6. Disconnect the upper radiator hose from the thermostat housing, and disconnect the 2 heater hoses.

7. Disconnect the No. 1 air hose from the air intake connector. Remove the 2 clamp bolts, loosen the throttle body hose clamp and remove the air intake connector and the connector pipe.

8. Tag and disconnect all emission control hoses from the throttle body and air intake chamber, the 2 PCV hoses from the cam cover and the fuel hose from the fuel hose support.

9. Remove the air intake chamber stay and the vacuum pipe and ground cable. Remove the bolt that attaches the spark plug wire clip, leaving the wires attached to the clip. Remove the distributor from the cylinder head with the cap and wires attached, by removing the distributor holding bolt.

10. Tag and disconnect the cold start injector wire and disconnect the cold start injector fuel hose from the delivery pipe.

11. Loosen the nut of the EGR pipe, remove the 5 bolts and 2 nuts and remove the air intake chamber and gasket.

12. Remove the glove box and remove the ECU module. Disconnect the 3 connectors and pull the EFI (fuel injection) wire harness out through the engine side of the firewall.

13. Remove the pulsation damper and the No. 1 fuel pipe. Remove the water outlet housing by first loosening the clamp and disconnecting the water bypass hose.

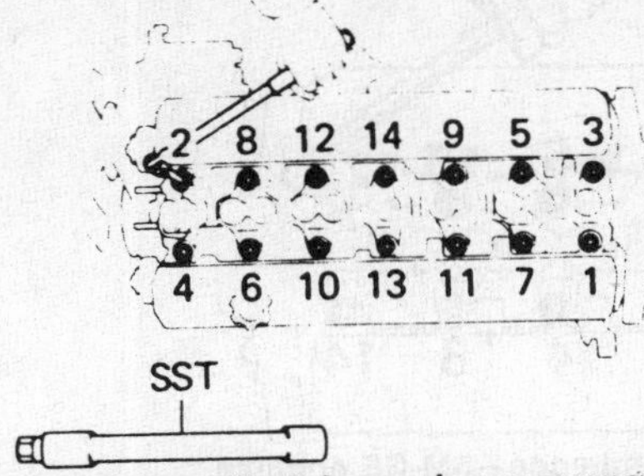

Cylinder head bolt loosening sequence—5M-GE engine

14. Remove the intake manifold.

15. Disconnect the power steering pump drive belt and remove the power steering pump without disconnecting the fluid hoses. Position the pump aside.

16. Disconnect the oxygen sensor connector and remove the exhaust manifold.

17. Remove the timing belt and camshaft timing gears. Remove the timing belt cover stay, and remove the oil pressure regulator and gasket. Remove the No. 2 timing belt cover and gasket.

18. Tag and disconnect any other wires, linkage and/or hoses still attached to the cylinder head or may interfere with its removal.

19. Carefully remove the 14 head bolts gradually in 2–3 passes and in the proper sequence.

20. Carefully lift the cylinder head from the dowels on the cylinder block and remove it.

21. Installation is the reverse of the removal procedure. Torque the cylinder head bolts, in several stages, to 58 ft. lbs. (78 Nm).

5S-FE Engine

1. Disconnect the negative battery cable. Drain the cooling system.

2. Tag and disconnect all lines, hoses and electrical wires that interfere with exhaust manifold, intake manifold and cylinder head removal.

3. MR2, remove the engine under covers, engine hood side panels and the brace that runs across the struts.

4. If equipped with automatic transaxle, disconnect the throttle cable and bracket from the throttle body.

5. Disconnect the accelerator cable and bracket from the throttle body and intake chamber.

6. If equipped with cruise control, remove the actuator and bracket.

7. Remove the air cleaner cap.

8. On Celica, remove the alternator and unbolt the air conditioning compressor from its mounting bracket. Leave the refrigerant lines connected and wire the compressor aside.

9.Remove the distributor.

10. On Celica, raise and support the vehicle safely. Remove the right tire and wheel assembly and engine under covers.

11. Remove the suspension lower crossmember.

12. Disconnect the exhaust pipe from the catalytic converter.

13. Remove the exhaust pipe and catalytic converter.

14. Remove the water outlet and the water bypass pipe.

15. Remove the throttle body and cold start injector. On Celica, remove the cold start injector pipe.

16. Remove the EGR valve and modulator.

17. On MR2, remove the fuel pressure Vacuum Switching Valve (VSV). On Celica and MR2, remove the EGR vacuum switching valve.

18. Remove the air intake chamber air hose, the throttle body air hose, and the power steering pump hoses, if equipped. On Celica, remove the air tube.

19. Remove the intake manifold.

20. Remove the fuel delivery pipe and the injectors.

21. Remove the camshaft timing pulley, No. 1 idler pulley and tension spring and No. 3 timing belt cover. Properly support the timing belt so meshing of the crankshaft timing pulley does not occur and the timing belt does not shift.

22. Remove the engine hangers and oil pressure switch. On Celica, remove the alternator bracket.

23. Remove the valve cover.

24. Remove the camshafts.

25. Loosen then remove the cylinder head bolts in 3 phases and in the proper sequence. Remove the cylinder head from the engine.

To install:

26. Install the cylinder head on the cylinder head block. Place the cylinder head in position on the cylinder head gasket.

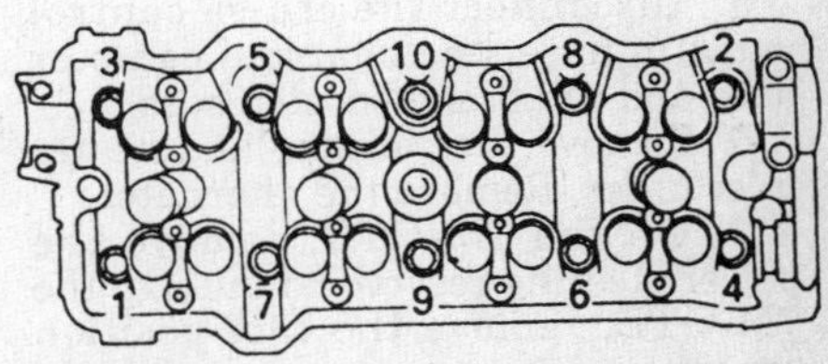

Cylinder head bolt loosening sequence—5S-FE engine

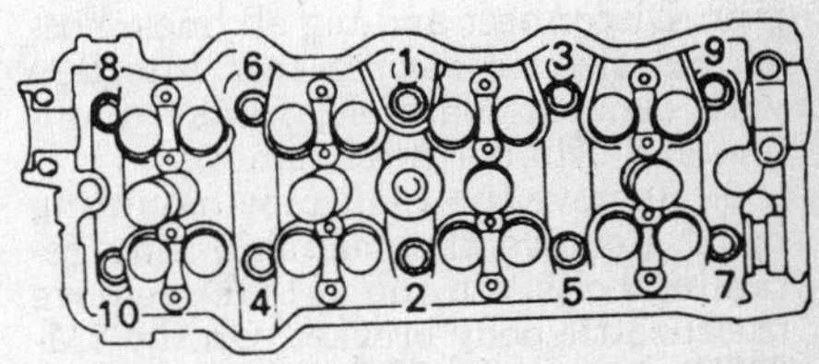

Cylinder head bolt tightening sequence—5S-FE engine

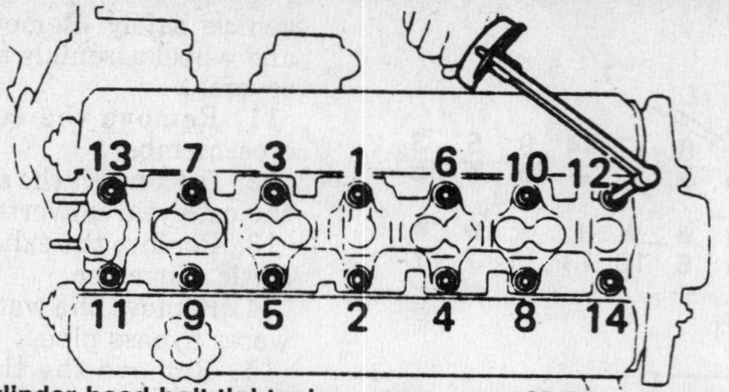

Cylinder head bolt tightening sequence—5M-GE engines

27. Torque the cylinder head bolts as follows:
  a. Apply a light coat of clean engine oil to the threads of the head bolts before installation.
  b. Tighten the bolts in several passes, in the proper sequence to 36 ft. lbs. (49 Nm).
  c. Mark the front of each bolt with a dab of paint.
  d. Retighten the bolts a further 90 degree turn. The paint dabs should now all be at a 90 degree angle to the front of the head.
28. Installation of the remaining components is the reverse of the removal precedure.

7M-GE and 7M-GTE Engine

1. Disconnect the negative battery cable. Drain the cooling system.
2. Disconnect the exhaust pipe from the exhaust manifold. Disconnect the cruise control cable, if equipped.
3. Disconnect the accelerator cable. Disconnect the throttle cable, if equipped with automatic transmission. Disconnect the engine ground strap.
4. On the 7M-GE, remove the No. 1 air cleaner hose along with the intake air pipe assembly. On the 7M-GTE, remove the No. 4 air cleaner pipe along with the No. 1 and No. 2 air cleaner hose.
5. Disconnect the cruise control vacuum hose, the charcoal canister hose and the brake booster hose.
6. Remove the radiator and heater inlet hoses. Remove the alternator.
7. On the 7M-GTE, remove the power steering reservoir tank. On the 7M-GTE, remove the cam position sensor.
8. Remove the air intake chamber with the connector. Remove the PCV pipe. Disconnect and tag all lines, hoses and electrical wires that interfere with exhaust manifold, intake manifold and cylinder head removal.
9. Remove the EGR pipe mounting bolts. Remove the manifold stay retaining bolts. On the 7M-GE, remove the throttle body bracket. On the 7M-GTE, remove the ISC pipe.
10. Remove the air intake connector mounting bolt (7M-GTE). On the 7M-GE, remove the cold start injector tube. On the 7M-GTE, disconnect the cold start injector. Disconnect the EGR vacuum modulator from the bracket.
11. Disconnect the engine wire from the clamps of the intake chamber. Remove the nuts and bolts, vacuum pipes and intake chamber with the connector and gasket.
12. On the 7M-GTE, remove the ignition coil and bracket.
13. Remove the pulsation damper, the VSV and the No. 1 fuel pipe. Remove the No. 2 and No. 3 fuel pipes. On the 7M-GTE, remove the auxiliary air pipe.
14. On the 7M-GE, remove the high tension wires and the distributor. Remove the oil dipstick. On the 7M-GTE, remove the turbocharger assembly.
15. Remove the exhaust manifold. Remove the water outlet housing. Remove the cylinder head covers. Remove the spark plugs.
16. Remove the timing belt and the camshaft timing pulleys. Remove the cylinder head retaining bolts gradually and in the proper sequence. carefully remove the cylinder head from the engine. As the cylinder head is lifted, separate the No. 5 water bypass line from its union.

To install:

17. Clean the gasket mating surfaces. Use care not to damage the aluminum components. Lower the cylinder head onto the engine with the help of an assistant. Make sure the dowel pins are aligned and no hoses or wires are between the head and cylinder block.
18. During installation, observe the following torques:
  Cylinder head bolts; 1st step to 20 ft. lbs. (27 Nm), 2nd step to 40 ft. lbs. (54 Nm) and 3rd step to 58 ft. lbs. (78 Nm).
  Cam bearing caps 8–10 ft. lbs. (11–14 Nm).
  Cam sprocket 29–39 ft. lbs. (39–53 Nm).
  Crankshaft pulley 55–61 ft. lbs. (75–83 Nm)
  Manifold bolts 15–21 ft. lbs. (20–29 Nm).

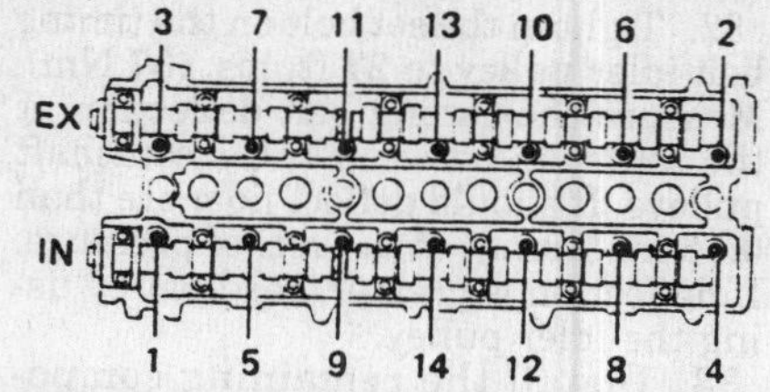

Cylinder head bolt loosening sequence—7M-GE and 7M-GTE engines

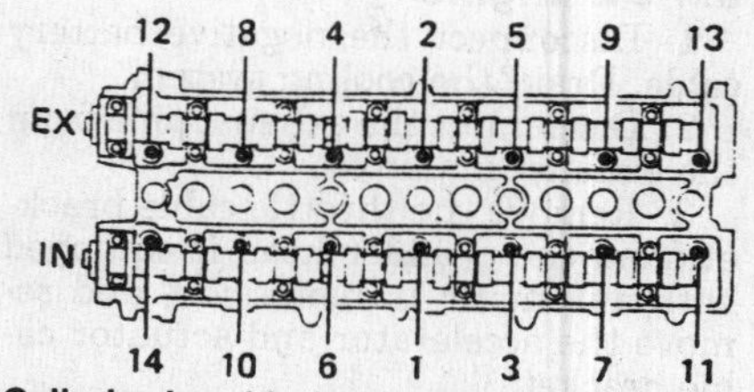

Cylinder head bolt tightening sequence—7M-GE and 7M-GTE engines

  Rocker arm bolts 17–19 ft. lbs. (23–26 Nm).
  Timing gear idler bolt 22–32 ft. lbs. (30–44 Nm).
  Adjust belt tension 0.24–0.28 in. (6–7mm).
  Adjust the valves.
19. Connect all electrical and hoses fittings.
20. Install all necessary components.
21. Connect the battery cable, refill all fluids, start the engine and check for leaks.

Valve Lifters

REMOVAL & INSTALLATION

1. Disconnect the negative battery cable.
2. Drain the cooling system.
3. Remove the valve cover(s).
4. Remove the camshafts.
5. Remove the valve lifters and shims from the cylinder head.
6. Label each lifter and shim with the respective cylinder head bore.
7. Inspect the lifters and shims for excessive wear. Replace worn lifters and shims as required.

To install:

8. Install the lifters and shims into the cylinder head. Make sure the lifter can be rotated freely by hand.
9. Install the camshafts.
10. Install the valve cover(s) using new gaskets and sealant, as required.
11. Fill the cooling system to the proper level.
12. Connect the negative battery cable.

Valve Lash

ADJUSTMENT

2VZ-FE Engine

1. Remove the air intake chamber and the cylinder head covers.

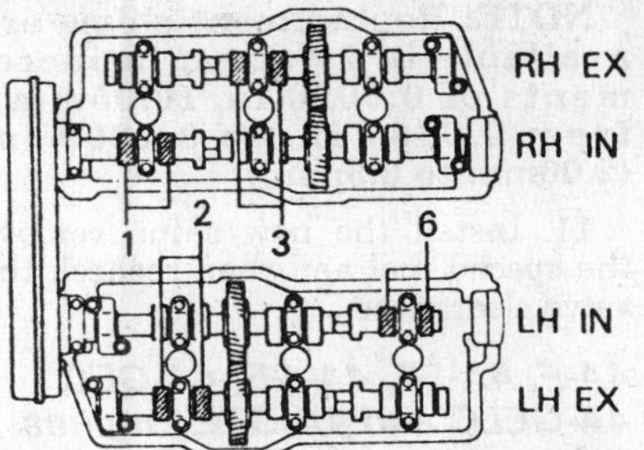

Adjust these valves first – 2VZ-FE engine

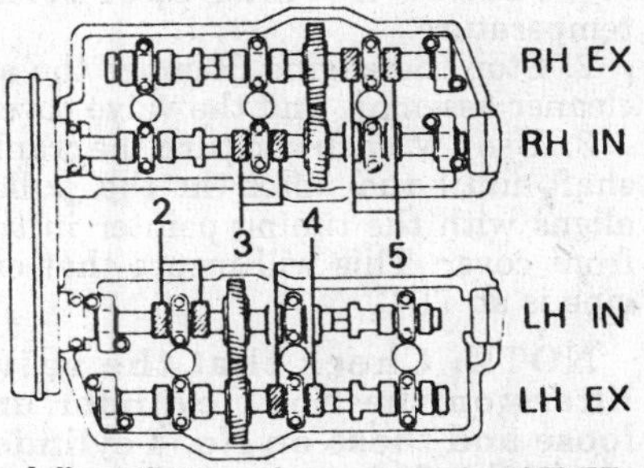

Adjust these valves second – 2VZ-FE engine

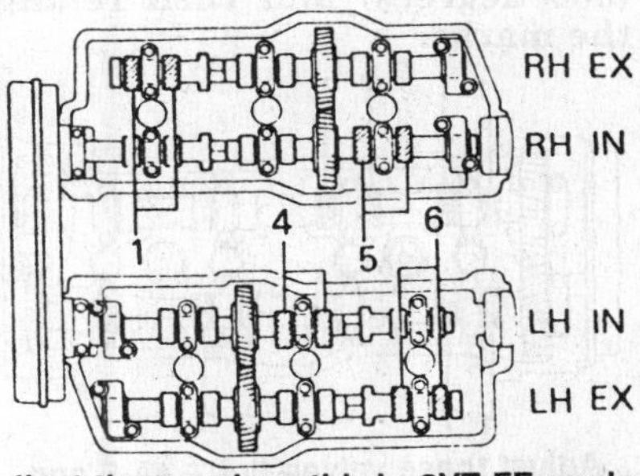

Adjust these valves third – 2VZ-FE engine

2. Use a wrench and turn the crankshaft until the notch in the pulley aligns with the timing mark **0** of the No. 1 timing belt cover. This will insure that No. 1 piston is at TDC of the compression stroke.

NOTE: Check that the valve lifters on the No. 1 (intake) cylinder are loose and those on No. 1 cylinder (exhaust) are tight. If not, turn the crankshaft 1 complete revolution (360 degrees) and then re-align the marks.

3. Using a flat feeler gauge measure the clearance between the camshaft lobe and the valve lifter on the first set of valves shown. This measurement should correspond to specification.

NOTE: If the measurement is within specifications, go on to the next step. If not, record the measurement taken for each individual valve.

4. Turn the crankshaft ⅔ revolution (240 degrees).

5. Measure the clearance of the second set of valves shown.

NOTE: If the measurement is within specifications, go on to the next step. If not, record the measurement taken for each individual valve.

6. Turn the crankshaft ⅔ revolution (240 degrees).

7. Measure the clearance of the third set of valves shown.

NOTE: If the measurement for this set of valves (and also the previous ones) is within specifications, go no further, the procedure is finished. If not, record the measurements and then proceed to Step 8.

8. Turn the crankshaft to position the intake camshaft lobe of the cylinder to be adjusted, upward.

9. Using a suitable tool, turn the valve lifter so the notch is easily accessible; it should be toward the spark plug.

10. Install tool 09248–55010 or equivalent, between the 2 camshafts lobes and then turn the handle so the tool presses down the valve lifter evenly.

11. Using a suitable tool and a magnet, remove the valve shims.

12. Measure the thickness of the old shim with a micrometer. Using this measurement and the clearance ones made earlier (from Step 3, 5 or 7), determine what size replacement shim will be required in order to bring the valve clearance into specification.

NOTE: Replacement shims are available in 17 sizes, in increments of 0.0020 in. (0.05mm), from 0.0984 in. to 0.1299 in. (2.50mm to 3.300mm)

13. Install the new shim, remove the special tool and then recheck the valve clearances.

3A-C, 3E, 3E-E, 4A-C and 4A-LC Engines

1. Start the engine and run it until it reaches normal operating temperature.

2. Stop the engine. Remove the air cleaner assembly and the cylinder head cover.

3. Turn the crankshaft until the point or notch on the pulley aligns with the **0** or **T** mark on the timing scale. This will set the engine to TDC of the compression stroke.

NOTE: Check that the rocker arms on the No. 1 cylinder are loose. If not, turn the crankshaft one complete revolution (360 degrees).

4. Retighten the cylinder head bolts to specifications. Also, retighten the valve rocker support bolts to the proper specifications.

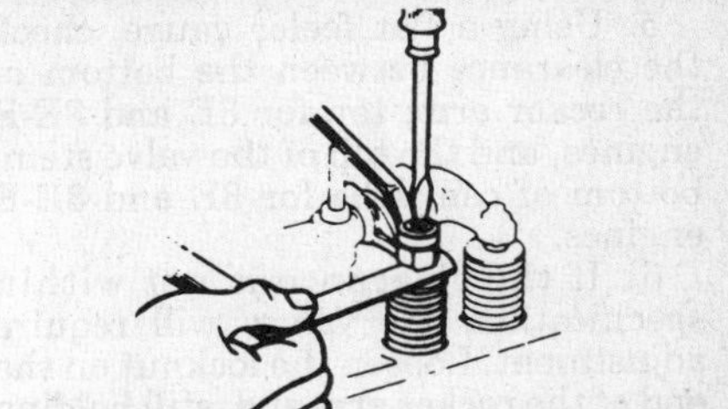
Valve lash adjustment – 3A-C, 4A-C and 4A-LC engines

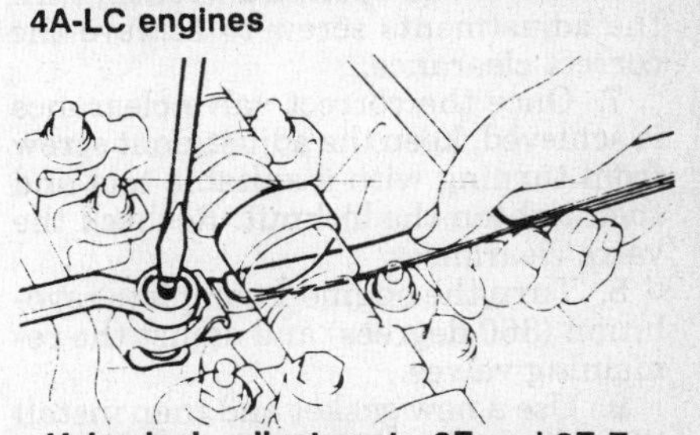
Valve lash adjustment – 3E and 3E-E engines

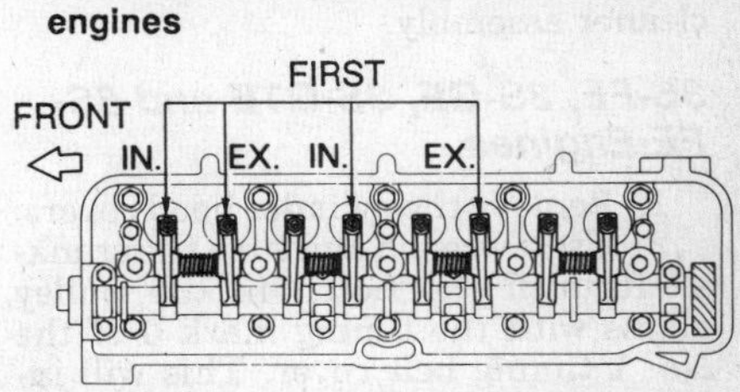

Adjust these valves first – 3A-C, 4A-C and 4A-LC engines

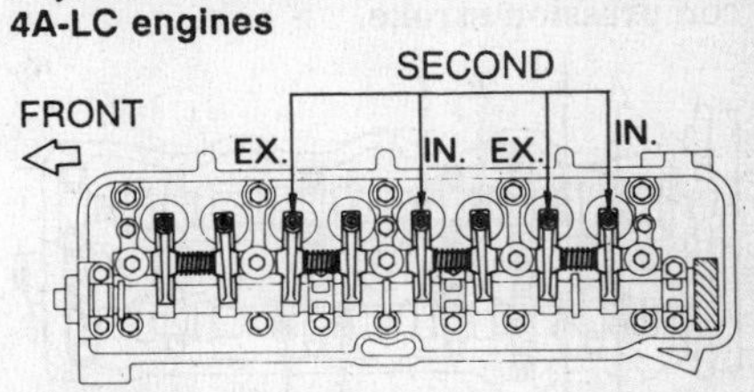

Adjust these valves second – 3A-C, 4A-C and 4A-LC engines

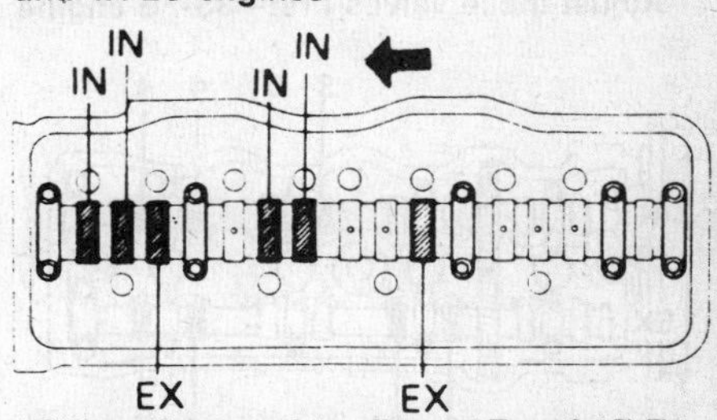

Adjust these valves first – 3E and 3E-E engines

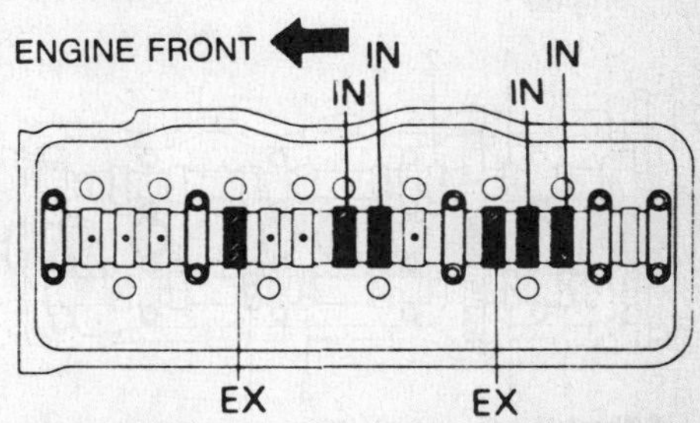

Adjust these valves second – 3E and 3E-E engines

5. Using a flat feeler gauge, check the clearance between the bottom of the rocker arm; top for 3E and 3E-E engines, and the top of the valve stem; bottom of cam lobe for 3E and 3E-E engines.

6. If the clearance is not within specification, the valves will require adjustment. Loosen the locknut on the end of the rocker arm and, still holding the nut with an open end wrench, turn the adjustments screw to achieve the correct clearance.

7. Once the correct valve clearance is achieved, keep the adjustment screw from turning with a suitable tool and then tighten the locknut. Recheck the valve clearances.

8. Turn the engine 1 complete revolution (360 degrees) and adjust the remaining valves.

9. Use a new gasket and then install the cylinder head cover. Install the air cleaner assembly.

3S-FE, 3S-GE, 3S-GTE and 5S-FE Engines

1. Remove the cylinder head covers.

2. Use a wrench and turn the crankshaft until the notch in the pulley aligns with the timing mark **0** of the No. 1 timing belt cover. This will insure that No. 1 piston is at TDC of the compression stroke.

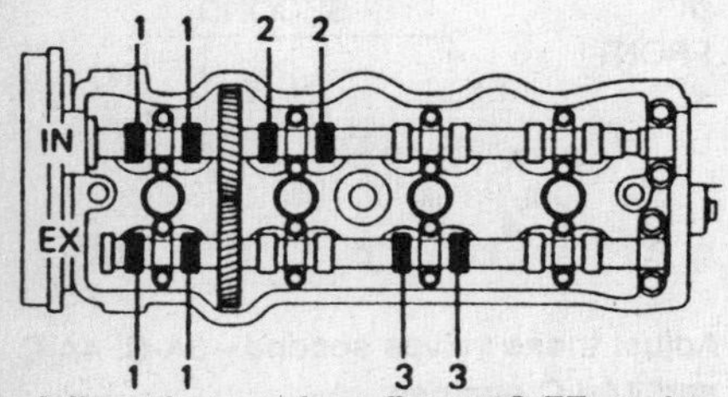

Adjust these valves first – 3S-FE engine

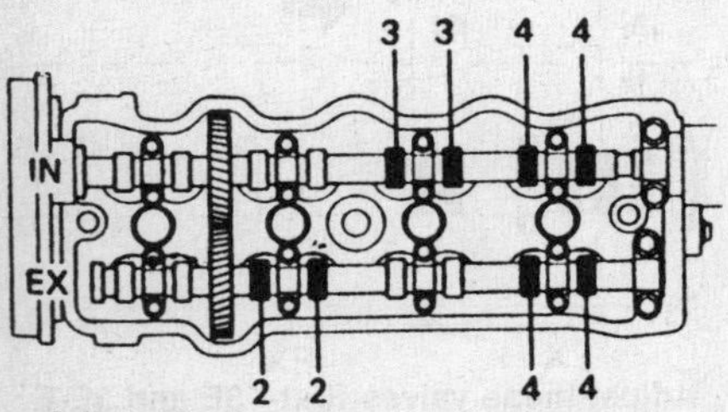

Adjust these valves second – 3S-FE engine

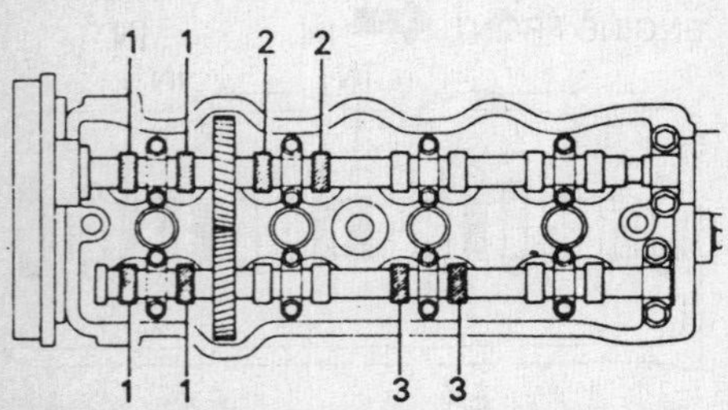

Adjust these valves first – 3S-GE, 3S-GTE and 5S-FE engines

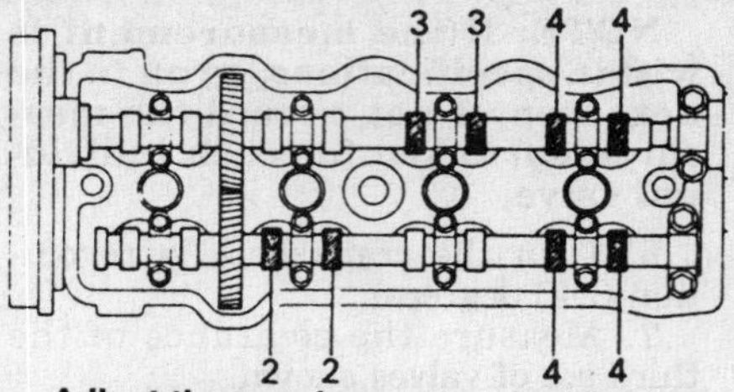

Adjust these valves second – 3S-GE, 3S-GTE and 5S-FE engines

NOTE: Check that the valve lifters on the No. 1 cylinder are loose and those on No. 4 cylinder are tight. If not, turn the crankshaft 1 complete revolution (360 degrees) and then realign the marks.

3. Using a flat feeler gauge measure the clearance between the camshaft lobe and the valve lifter on the first set of valves shown. This measurement should correspond to specification.

NOTE: If the measurement is within specifications, go on to the next step. If not, record the measurement taken for each individual valve.

4. Turn the crankshaft 1 complete revolution and realign the timing marks.

5. Measure the clearance of the second set of valves.

NOTE: If the measurement for this set of valves (and also the previous one) is within specifications, go no further, the procedure is finished. If not, record the measurements and then proceed to Step 6.

6. Turn the crankshaft to position the intake camshaft lobe of the cylinder to be adjusted, upward.

NOTE: Both intake and exhaust valve clearance may be adjusted at the same time, if required.

7. Using a suitable tool, turn the valve lifter so the notch is easily accessible.

8. Install tool 09248–70012 for 3S-GE or 09248–55010 for 3S-FE, 3S-GTE, 5S-FE, between the 2 camshaft lobes and then turn the handle so the tool presses down both (intake and exhaust) valve lifters evenly.

9. Using a suitable tool and a magnet, remove the valve shims.

10. Measure the thickness of the old shim with a micrometer. Using this measurement and the clearance ones made earlier (from Step 3 or 5), determine what size replacement shim will be required in order to bring the valve clearance into specification.

NOTE: Replacement shims are available in 27 sizes, in increments of 0.0020 in. (0.05mm), from 0.0787 in. to 0.1299 in. (2.00mm to 3.3mm).

11. Install the new shim, remove the special tool and then recheck the valve clearances.

4A-F, 4A-FE, 4A-GE, 4A-GEC, 4A-GELC and 4A-GZE Engines

1. Start the engine and run it until it reaches normal operating temperature.

2. Stop the engine. Remove the air cleaner assembly and the valve cover.

3. Use a wrench and turn the crankshaft until the notch in the pulley aligns with the timing pointer in the front cover. This will insure that engine is at TDC.

NOTE: Check that the valve lifters on the No. 1 cylinder are loose and those on No. 4 cylinder are tight. If not, turn the crankshaft one complete revolution (360 degrees) and then re-align the marks.

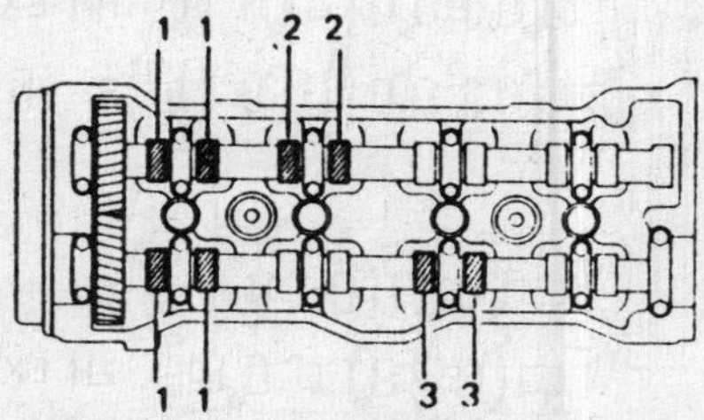

Adjust these valves first – 4A-F and 4A-FE engines

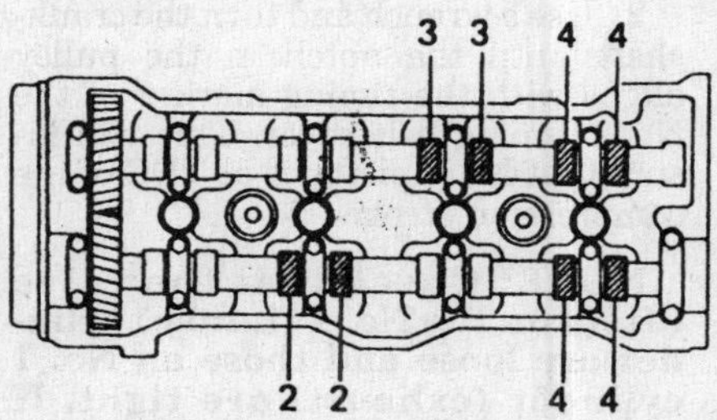

Adjust these valves second – 4A-F and 4A-FE engines

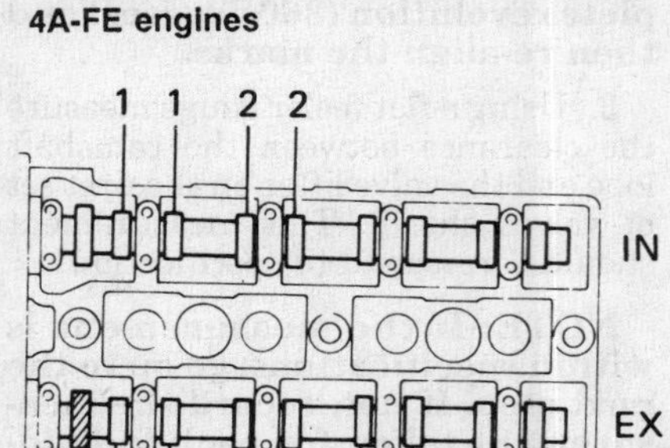

Adjust these valves first – 4A-GE, 4A-GEC and 4A-GELC (Corolla) engines

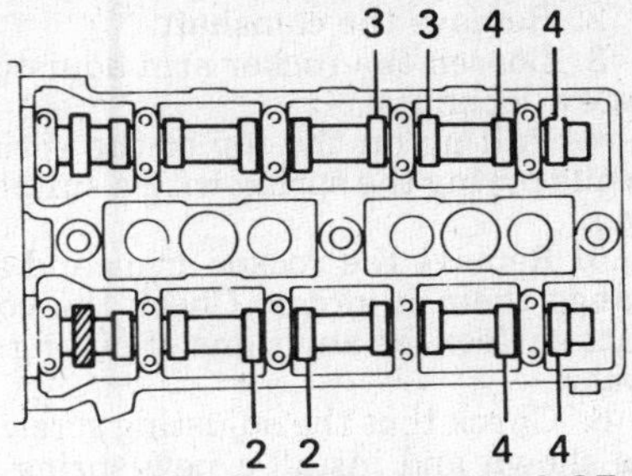

Adjust these valves second—4A-GE, 4A-GEC and 4A-GELC (Corolla) engines

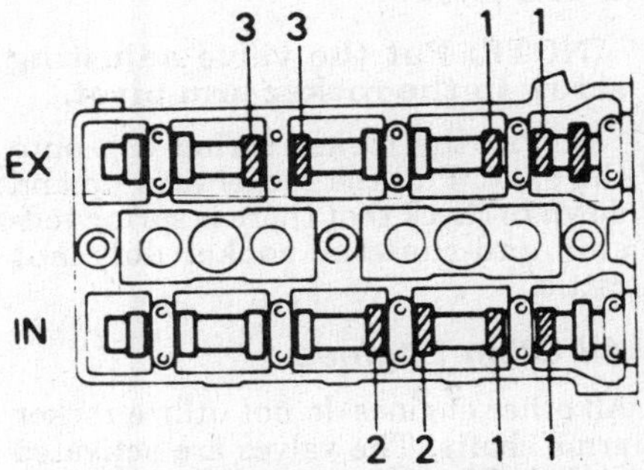

Adjust these valves first—4A-GE, 4A-GELC and 4A-GZE (MR2) engines

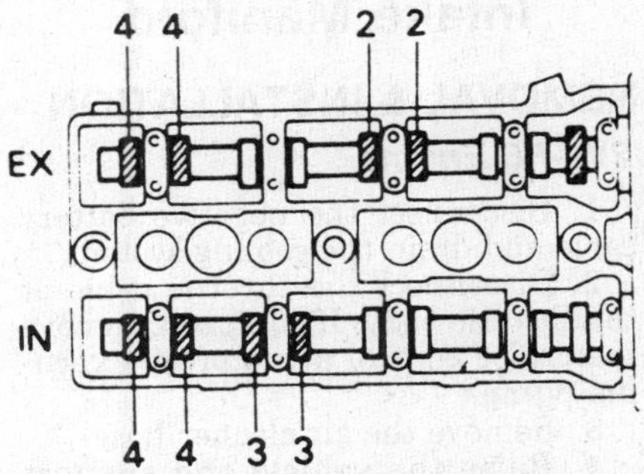

Adjust these valves second—4A-GE, 4A-GELC and 4A-GZE (MR2) engines

4. Using a flat feeler gauge measure the clearance between the camshaft lobe and the valve lifter. Check the first set of valves shown.

NOTE: If the measurement is within specifications, go on to the next step. If not, record the measurement taken for each individual valve.

5. Turn the crankshaft 1 complete revolution and realign the timing marks. Measure the clearance of the second set of valves shown.

NOTE: If the measurement for this set of valves (and also the previous one) is within specification, go no further, the procedure is finished. If not, record the measurements and then proceed to Step 6.

6. Turn the crankshaft to position the intake camshaft lobe of the cylinder to be adjusted, upward. Both intake and exhaust valve clearance may be adjusted at the same time, if required.

7. Using a suitable tool, turn the valve lifter so the notch is easily accessible.

8. Install tool 09248–70011 for 4A-GE or 09248–55010 4A-F, 4A-FE, 4A-GEC, 4A-GELC, 4A-GZE or equivalent, between the 2 camshafts lobes and then turn the handle so the tool presses down both (intake and exhaust) valve lifters evenly. On the 4A-GE, the tool will work on only one valve lifter at a time.

NOTE: On the 4A-FE, 4A-GEC, 4A-GELC and 4A-GZE, position the notch toward the spark plug before pressing down the valve lifter.

9. Using a suitable tool and a magnet, remove the valve shims.

10. Measure the thickness of the old shim with a micrometer. Using this measurement and the clearance of ones made earlier, determine what size replacement shim will be required in order to bring the valve clearance into specification.

11. Install the new shim, remove the special tool and then recheck the valve clearance.

5M-GE Engines

These engine is equipped with hydraulic lash adjusters in the valve train. The adjusters maintain a 0 clearance between the rocker arm and valve stem, no adjustment is possible or necessary.

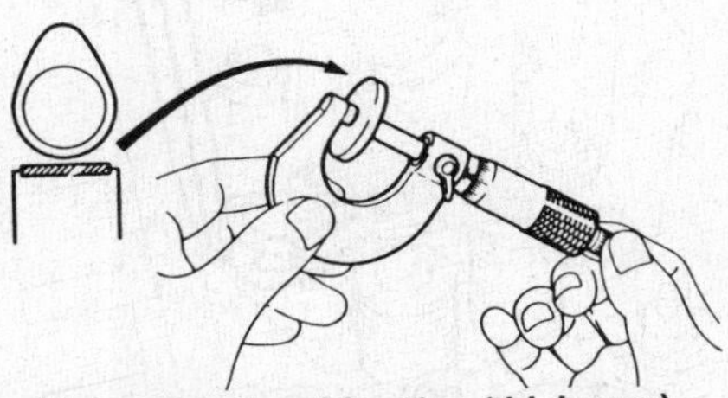
Measuring the shim size (thickness)

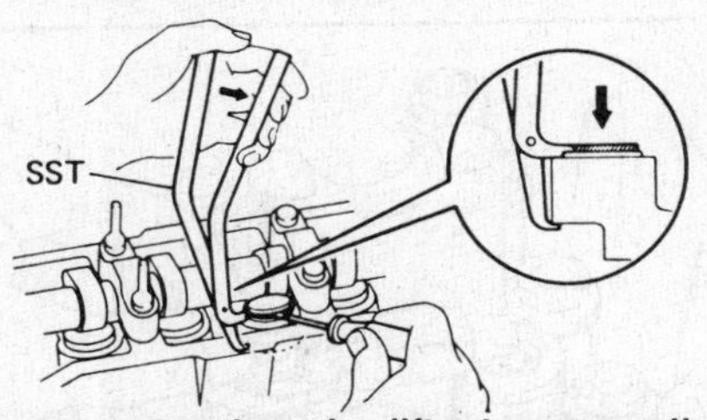

Depressing the valve lifter to remove the shim—3S-GE and 4A-GE engines

7M-GE and 7M-GTE Engines

1. Remove the cylinder head covers.
2. Use a wrench and turn the crankshaft until the notch in the pulley aligns with the timing mark **0** of the No. 1 timing belt cover. This will insure that engine is at TDC.

NOTE: Check that the valve lifters on the No. 1 cylinder are loose and those on No. 6 cylinder are tight. If not, turn the crankshaft 1 complete revolution (360 degrees) and then realign the marks.

3. Using a flat feeler gauge measure the clearance between the camshaft lobe and the valve lifter. This measurement should correspond to specification. Check the first set of valves shown.

NOTE: If the measurement is within specifications, go on to the next step. If not, record the measurement taken for each individual valve.

4. Turn the crankshaft ⅔ revolution (240 degrees).
5. Measure the clearance of the second set of valves shown.

NOTE: If the measurement is within specifications, go on to the next step. If not, record the measurement taken for each individual valve.

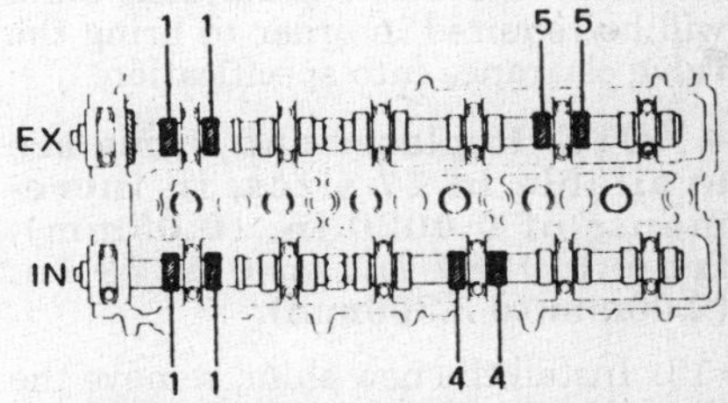

Adjust these valves first—7M-GE and 7M-GTE engines

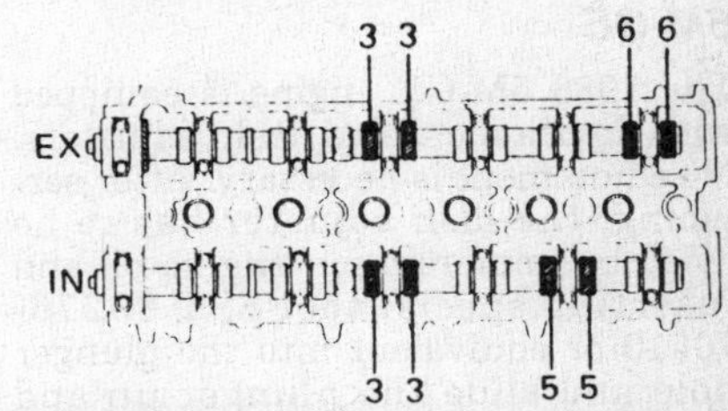

Adjust these valves second—7M-GE and 7M-GTE engines

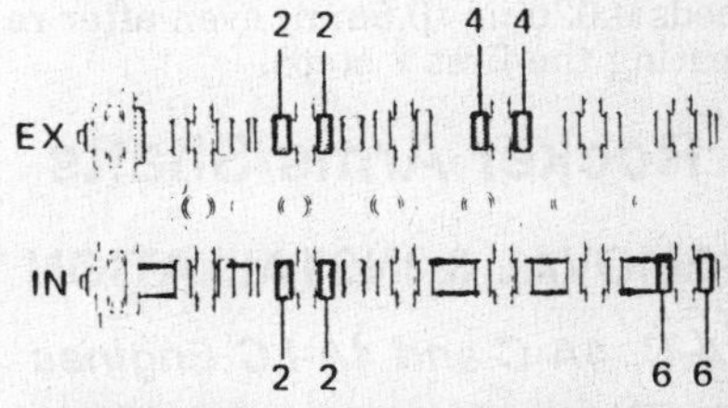

Adjust these valves third—7M-GE and 7M-GTE engines

6. Turn the crankshaft ⅔ revolution (240 degrees).
7. Measure the clearance of the third set of valves shown.

NOTE: If the measurement for this set of valves (and also the previous ones) is within specifications, go no further, the procedure is finished. If not, record the measurements and then proceed to Step 8.

8. Turn the crankshaft to position the intake camshaft lobe of the cylinder to be adjusted, upward.

NOTE: Both intake and exhaust valve clearance may be adjusted at the same time, if required.

9. Using a suitable tool, turn the valve lifter so the notch is easily accessible.
10. Install tool 09248–55010 or equivalent, between the 2 camshafts lobes and then turn the handle so the tool presses down both (intake and exhaust) valve lifters evenly.
11. Using a suitable tool and a magnet, remove the valve shims.
12. Measure the thickness of the old shim with a micrometer. Using this measurement and the clearance ones made earlier (from Step 3, 5 or 7), determine what size replacement shim will be required in order to bring the valve clearance into specification.

NOTE: Replacement shims are available in 17 sizes, in increments of 0.0020 in. (0.05mm), from 0.0787 in. to 0.1299 in. (2.00mm to 3.300mm).

13. Install the new shim, remove the special tool and then recheck the valve clearance.

5M-GE

The 1988 5M-GE engine is equipped with hydraulic valve lash adjusters. No adjustment is necessary. After servicing, the lash adjuster has to be bleed by immersing in engine oil and inserting special tool SST 09276-70010 or equivalent into the plunger hole and slide the plunger up and down several times while pushing down lightly on the check ball. Replace the adjuster if the plunger stroke exceeds 0.020 in. (0.5mm) even after repeating the first 2 steps.

Rocker Arms/Shafts

REMOVAL & INSTALLATION

3A-C, 4A-C and 4A-LC Engines

1. Disconnect the negative battery terminal.
2. Remove the air cleaner and all necessary hoses.
3. Remove all linkage from the carburetor.
4. Remove the valve cover and gasket.
5. Remove the rocker arm bolts.
6. Installation is the reverse of the removal procedure. Install a new valve cover gasket before replacing the valve cover. Tighten the rocker arm bolts to 17–19 ft. lbs. (23–26 Nm).

3E and 3E-E Engines

1. Disconnect the negative battery cable.

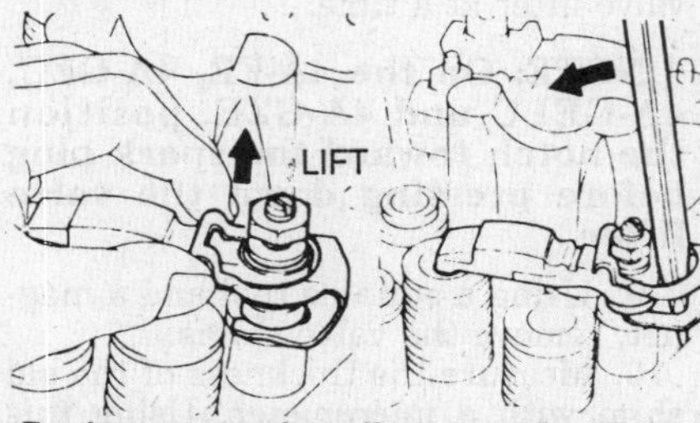

Rocker arm spring clip removal—3E and 3E-E engines

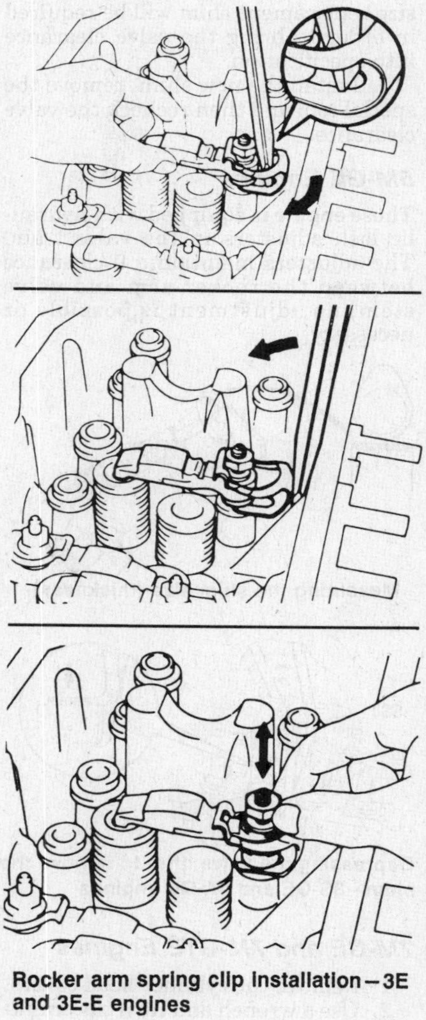

Rocker arm spring clip installation—3E and 3E-E engines

2. Remove the camshaft.
3. Loosen the rocker arm adjusting screw locknuts.
4. Pull up on the top of the spring while prying the spring with a suitable tool.
5. Remove the rocker arms and arrange them in order. Check the contact surface for any signs of pitting or wear.
6. Check that the adjusting screw is as shown and install a new spring to the rocker arm.
7. Press the bottom lip of the spring until it fits into the groove on the rocker arm pivot.

NOTE: Put the valve adjusting screw in the rocker arm pivot.

8. Pry the rocker spring clip onto the pivot. Pull the rocker arm up and down to check that there is spring tension and that the rocker does not rattle.

All Other Engines

All other engines do not utilize rocker arms shafts. The valves are activated directly by the camshaft via valve adjusting shims.

Intake Manifold

REMOVAL & INSTALLATION

2VZ-FE Engine

1. Disconnect the negative battery cable and drain the cooling system.
2. Disconnect the throttle cable at the throttle body. If equipped, remove the cruise control actuator and vacuum pump.
3. Remove the air cleaner hose.
4. Raise the vehicle and support safely. Remove the engine undercovers.
5. Remove the lower suspension crossmember and the front exhaust pipe.
6. Remove the alternator. Remove the ISC valve.
7. Remove the throttle body. Remove the EGR pipe, valve and vacuum modulator.
8. Remove the vacuum pipe and the distributor.
9. Remove the exhaust crossover pipe. Disconnect the cold start injector and then remove the injector tube.
10. Tag and disconnect all hoses leading to the air intake chamber and then remove the chamber.
11. Remove the fuel delivery pipes and the injectors.
12. Disconnect the water temperature sensor and remove the upper radiator hose. Remove the water outlet. Remove the water bypass outlet.
13. Loosen the 2 bolts and remove the cylinder head rear plate.

14. Remove the No. 2 idler pulley bracket stay.
15. Remove the 8 bolts and 4 nuts and lift out the intake manifold with its 2 gaskets.

To install:

16. Clean the gasket mating surfaces being careful not to damage the aluminum surfaces. Check the mating surfaces for warpage with a straight edge.
17. Match the old gasket with the new for an exact match. Use new gaskets when installing the manifold and tighten the bolts, from the center outward, to 13 ft. lbs. (18 Nm).
18. Install the remaining components, start the engine and check for leaks.

3E and 3E-E Engines

1. Disconnect the negative battery cable. Drain the coolant. Remove the air cleaner assembly.
2. Tag and disconnect all wires, hoses or cables that interfere with intake manifold removal.
3. Remove the necessary components in order to gain access to the intake manifold retaining bolts.
4. Remove the carburetor from the throttle body.
5. Disconnect the intake manifold water hoses.
6. Remove the intake manifold retaining bolts. Remove the intake manifold from the vehicle.

To install:

7. Clean the gasket mating surfaces being careful not to damage the aluminum surfaces. Check the mating surfaces for warpage with a straight edge.
8. Match the old gasket with the new for an exact match. Use new gaskets when installing the manifold and tighten the bolts, from the center outward, to 13 ft. lbs. (18 Nm).
9. Install the remaining components, start the engine and check for leaks.

3S-FE and 5S-FE Engines

1. Disconnect the negative battery cable. Drain the cooling system. Remove the air cleaner assembly.
2. Tag and disconnect all wires, hoses or cables that interfere with intake manifold removal.
3. Remove the necessary components in order to gain access to the intake manifold retaining bolts.
4. Remove the throttle body and cold start injector pipe.
5. Remove the air tube assembly. If equipped with power steering, remove the hoses before removing the air tube assembly.
6. Remove the intake manifold retaining bolts. Remove the intake manifold from the vehicle.

To install:

7. Clean the gasket mating surfaces being careful not to damage the aluminum surfaces. Check the mating surfaces for warpage with a straight edge.
8. Match the old gasket with the new for an exact match. Use new gaskets when installing the manifold and tighten the bolts, from the center outward. Tighten the intake manifold mounting bolts to 14 ft. lbs. (19 Nm). Tighten the 12mm manifold stay bolt to 14 ft. lbs. (19 Nm); tighten the 14mm bolts to 31 ft. lbs. (42 Nm).
9. Install the remaining components, start the engine and check for leaks.

3S-GE and 3S-GTE Engines

1. Disconnect the negative battery cable. Drain the cooling system. Remove the air cleaner assembly.
2. Tag and disconnect all wires, hoses or cables that interfere with intake manifold removal.
3. Remove the necessary components in order to gain access to the intake manifold retaining bolts.
4. Remove the intake manifold retaining bolts and nuts. Remove the intake manifold from the vehicle.

To install:

5. Clean the gasket mating surfaces being careful not to damage the aluminum surfaces. Check the mating surfaces for warpage with a straight edge.
6. Match the old gasket with the new for an exact match. Use new gaskets when installing the manifold and tighten the bolts, from the center outward, to 13 ft. lbs. (18 Nm).
7. Install the remaining components, start the engine and check for leaks.

4A-F, 4A-FE, 4A-GE, 4A-GEC, 4A-GELC and 4A-GZE Engines

1. Disconnect the negative battery cable. Drain the coolant. Remove the air cleaner assembly.
2. Tag wires, hoses or cables in the way of manifold removal.
3. Remove the necessary components in order to gain access to the intake manifold retaining bolts.
4. Remove the intake manifold retaining bolts. Remove the intake manifold from the vehicle.

To install:

5. Clean the gasket mating surfaces being careful not to damage the aluminum surfaces. Check the mating surfaces for warpage with a straight edge.
6. Match the old gasket with the new for an exact match. Use new gaskets when installing the manifold and tighten the bolts, from the center outward, to the proper specification.
7. Install the remaining components, start the engine and check for leaks.

5M-GE Engine

1. Disconnect the negative battery cable. Drain the engine coolant.
2. Tag and disconnect all wires, hoses or cables that interfere with intake manifold removal.
3. Remove the air intake chamber.
4. Disconnect and move the wiring away from the fuel delivery and injector pipe.
4. Remove the fuel injector and delivery pipe.
5. Remove the fuel pressure regulator, which is mounted on the center of the intake manifold.
6. Remove the EGR valve from the rear of the manifold.
7. Disconnect the radiator hoses, heater hoses, and vacuum lines from the intake manifold.
8. Remove the distributor cap and position it aside.
9. Remove the intake manifold retaining bolts. Remove the intake manifold and gasket from the engine.

To install:

10. Clean the gasket mating surfaces being careful not to damage the aluminum surfaces. Check the mating surfaces for warpage with a straight edge.
11. Match the old gasket with the new for an exact match. Use new gaskets when installing the manifold and tighten the bolts, from the center outward and torque the manifold fasteners to 10–15 ft. lbs. (13.6–20 Nm).
12. Install the remaining components, start the engine and check for leaks.

7M-GE and 7M-GTE Engines

1. Disconnect the negative battery cable. Drain the cooling system.
2. Remove the air cleaner assembly.
3. Tag and disconnect all wires, hoses or cables that interfere with intake manifold removal.
4. Remove the necessary components in order to gain access to the intake manifold retaining bolts.
5. Remove the air intake connector along with the air intake chamber assembly.
6. Remove the fuel delivery pipe with the injectors still attached.
7. Remove the intake manifold retaining bolts. Remove the intake manifold from the vehicle.

To install:

8. Clean the gasket mating surfaces being careful not to damage the aluminum surfaces. Check the mating surfaces for warpage with a straight edge.
9. Match the old gasket with the new for an exact match. Use new gaskets when installing the manifold and tighten the bolts, from the center outward, to 13 ft. lbs. (18 Nm).
10. Install the remaining components, start the engine and check for leaks.

Exhaust Manifold

REMOVAL & INSTALLATION

2VZ-FE Engine

1. Disconnect the negative battery cable and drain the cooling system.
2. Disconnect the throttle cable at the throttle body. If equipped with cruise control, remove the cruise control actuator and vacuum pump.
3. Remove the air cleaner hose.
4. Raise the vehicle and support safely. Remove the engine undercovers.
5. Remove the lower suspension crossmember and the front exhaust pipe.
6. Remove the alternator and the Idle Speed Control (ISC) valve.
7. Remove the throttle body, EGR pipe, EGR valve and vacuum modulator.
8. Remove the vacuum pipe and the distributor.
9. Remove the exhaust crossover pipe. Disconnect the cold start injector and then remove the injector tube.
10. Tag and disconnect all hoses leading to the air intake chamber and then remove the chamber.
11. Remove the fuel delivery pipes and the injectors.
12. Disconnect the water temperature sensor and remove the upper radiator hose. Remove the water outlet. Remove the water bypass outlet.
13. Loosen the 2 bolts and remove the cylinder head rear plate.
14. Remove the intake manifold.
15. Disconnect the oxygen sensor and then remove the outside heat insulator for the righ manifold. Remove the manifold and gasket and then remove the inner heat shield.
16. Remove the left side heat shield and then remove the manifold.

To install:

17. Clean the gasket mating surfaces being careful not to damage the aluminum surfaces. Check the mating surfaces for warpage with a straight edge.
18. Match the old gasket with the new for an exact match. Use new gaskets when installing the manifold and tighten the bolts, from the center outward, to 29 ft. lbs. (39 Nm).
19. Install the remaining components, start the engine and check for leaks.

3E and 3E-E Engines

1. Disconnect the negative battery cable. Remove the exhaust manifold heat insulator shield assembly.
2. Remove the necessary components in order to gain access to the exhaust manifold retaining bolts.
3. Disconnect the exhaust manifold bolts at the exhaust pipe. Disconnect the oxygen sensor electrical wire. It may be necessary to raise and support the vehicle safely before removing these bolts.

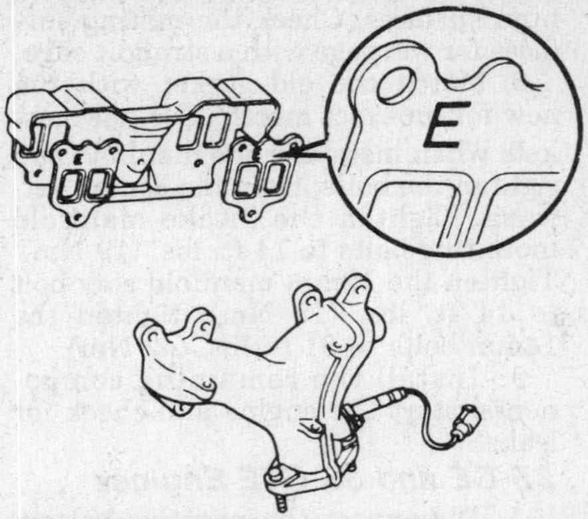

Exhaust manifold gasket installation—3E and 3E-E engines

4. Remove the exhaust manifold retaining bolts. Remove the exhaust manifold from the vehicle.

To install:

5. Clean the gasket mating surfaces being careful not to damage the aluminum surfaces. Check the mating surfaces for warpage with a straight edge.
6. Match the old gasket with the new for an exact match. During installation, the **E** mark on the gasket must face outward. Use new gaskets when installing the manifold and tighten the bolts, from the center outward, to 38 ft. lbs. (51 Nm).
7. Install the remaining components, start the engine and check for leaks.

3S-FE and 5S-FE Engines

1. Disconnect the negative battery cable.
2. Raise and support the vehicle safely.
3. Remove the exhaust manifold heat insulator shield assembly.
4. Remove the necessary components in order to gain access to the exhaust manifold retaining bolts.
5. Disconnect the exhaust manifold bolts at the exhaust pipe or catalytic converter. It may be necessary to raise and support the vehicle safely before removing these bolts.
6. Remove the exhaust manifold retaining bolts. Remove the exhaust manifold from the vehicle. On 5S-FE engine, the exhaust manifold and catalytic converter are removed as one unit.

To install:

7. Be sure to use new gaskets. On 5S-FE, the **R** mark should be toward the rear. On 3S-FE, tighten the exhaust manifold bolts to 31 ft. lbs. (41 Nm). On 5S-FE, torque the exhaust manifold retaining bolts to 36 ft. lbs. (49 Nm).

5M-GE, 3S-GE, 3S-GTE, 4A-F, 4A-FE, 4A-GE, 4A-GEC, 4A-GELC and 4A-GZE Engines

1. Disconnect the negative battery cable. Raise the vehicle and support safely. Remove the right gravel shield from under the vehicle.
2. Remove the throttle body and turbocharger.
3. Remove the exhaust pipe support stay. Unbolt the exhaust pipe from the exhaust manifold flange.
4. Disconnect the oxygen sensor connector. On the 3S-GE and 4A-F, remove the upper heat insulator.
5. Remove the manifold retaining nuts. Remove the exhaust manifold from the vehicle.

To install:

6. Clean the gasket mating surfaces being careful not to damage the aluminum surfaces. Check the mating surfaces for warpage with a straight edge.
7. Match the old gasket with the new for an exact match. Use new gaskets when installing the manifold and tighten the bolts, from the center outward, to the proper specification.
8. Install the remaining components, start the engine and check for leaks.

7M-GE and 7M-GTE Engines

1. Disconnect the negative battery cable. Remove the exhaust manifold heat insulator shield assembly, if equipped.
2. Remove the necessary components in order to gain access to the exhaust manifold retaining bolts.
3. Disconnect the exhaust manifold bolts at the exhaust pipe. It may be necessary to raise and support the vehicle safely before removing these bolts.
4. On 7M-GTE, remove the turbocharger.
5. Remove the exhaust manifold retaining bolts. Remove the exhaust manifold from the vehicle.

To install:

6. Clean the gasket mating surfaces being careful not to damage the aluminum surfaces. Check the mating surfaces for warpage with a straight edge.
7. Match the old gasket with the new for an exact match. Use new gaskets when installing the manifold and tighten the bolts, from the center outward, to 29 ft. lbs. (39 Nm).
8. Install the remaining components, start the engine and check for leaks.

Combination Manifold

REMOVAL & INSTALLATION

3A-C, 4A-C and 4A-LC Engines

1. Disconnect the negative battery cable.

2. Remove the air cleaner and all necessary hoses.
3. Remove all the carburetor linkages.
4. Remove the carburetor.
5. Remove the intake/exhaust manifold pipe.
6. Remove the intake/exhaust manifold.
7. Installation is the reverse of the removal procedure. Tighten the manifold bolts to 15–21 ft. lbs. (20–29 Nm).

Supercharger

REMOVAL & INSTALLATION

4A-GZE Engine

1. Disconnect the negative cable at the battery. Drain the cooling system and remove the radiator reservoir tank.
2. Remove the Vacuum Switching Valve (VSV) and the intercooler.
3. Remove the air flow meter with the No. 3 air cleaner hose. Disconnect the accelerator cable (rod) and throttle cable.
4. Disconnect the PCV, brake booster, ACV, air conditioning idle-up and emission control vacuum hoses.
5. Remove the No. 1 intake air connector pipe and its air hose.
6. Loosen the idler pulley locknut and adjusting bolt and remove the supercharger drive belt.
7. Disconnect the No. 2 and 3 water bypass hoses. Loosen the air hose clamp.
8. Remove the air inlet duct stay. Remove the throttle body.
9. Disconnect the ACV and supercharger connectors and the 2 ACV hoses. Remove the 2 nuts and the ACV. Remove the pivot bolt and nut, remove the 2 stud bolts and then rotate the assembly so the hub is facing upward; remove the supercharger.
10. Installation is the reverse of the removal procedure. Tighten the 2 stud bolts to 25 ft. lbs. (34 Nm).

Turbocharger

REMOVAL & INSTALLATION

3S-GTE Engine

CELICA

1. Disconnect the negative battery cable. Drain the coolant from the engine and intercooler.
2. Remove the air cleaner assembly.
3. Remove the catalytic converter and the oxygen sensor.
4. Disconnect the 2 intercooler water lines and the reservoir tank line. Loosen the clamps, disconnect the air hose and remove the intercooler.
5. Remove the alternator duct and the No. 2 alternator bracket.
6. Remove the turbocharger heat insulator and the turbocharger outlet elbow. Remove the turbocharger stay.
7. Remove the turbocharger.
8. Installation is in the reverse order of removal. Pour about 20cc of new oil into the turbocharger oil inlet and then spin the impeller to lubricate the bearing. Tighten the turbo-to-manifold bolts to 47 ft. lbs. (64 Nm).

1990–92 MR2

1. Disconnect the negative battery cable.
2. Drain the coolant from the engine and the intercooler.
3. Raise and support the vehicle safely.
4. Remove the engine under covers and engine hood side panels.
5. Tag and disconnect all water hoses, vacuum lines, air tubes, engine control cables, transaxle control cables and electrical wires that interfere with turbocharger removal.
6. Remove the brace that runs across the struts.
7. Remove the air cleaner assembly.
8. Remove the front exhaust pipe.
9. Discharge the air conditioning system. Disconnect the refrigerant hoses and electrical wiring from the compressor. Remove the compressor and idler pulley bracket from the engine.
10. Remove the front engine mounting insulator.
11. Remove the front mounting bracket and clutch release cylinder. Leave the hydraulic lines connected and position the release cylinder aside.
12. Remove the engine cooling fan.
13. Remove the catalytic converter.
14. Remove the Vacuum Transmitting Valve (VTV).
15. Remove the air bypass valve.
16. Remove the heat insulator from the turbocharger.
17. Remove the oxygen sensor.
18. Remove the heat insulators from the turbocharger outlet elbow.
19. Disconnect the oil hose from the turbocharger oil pipe.
20. Remove the turbocharger mounting stay.
21. Unbolt and remove the turbocharger oil pipe from the block.
22. Remove the 4 nuts and separate the turbocharger and gasket from the exhaust manifold.

To install:

23. Clean the turbocharger and exhaust manifold gasket surfaces.
24. Prior to installing the turbocharger, pour approximately 1.2 cu. in. (20cc) of new oil into the oil inlet and then turn the impeller wheel by hand a few times in order to lubricate the bearing.
25. Install a new gasket onto the exhaust manifold.
26. Mount the turbocharger onto the gasket and install the 4 nuts. Torque the nuts in a criss-cross pattern to 47 ft. lbs. (64 Nm).

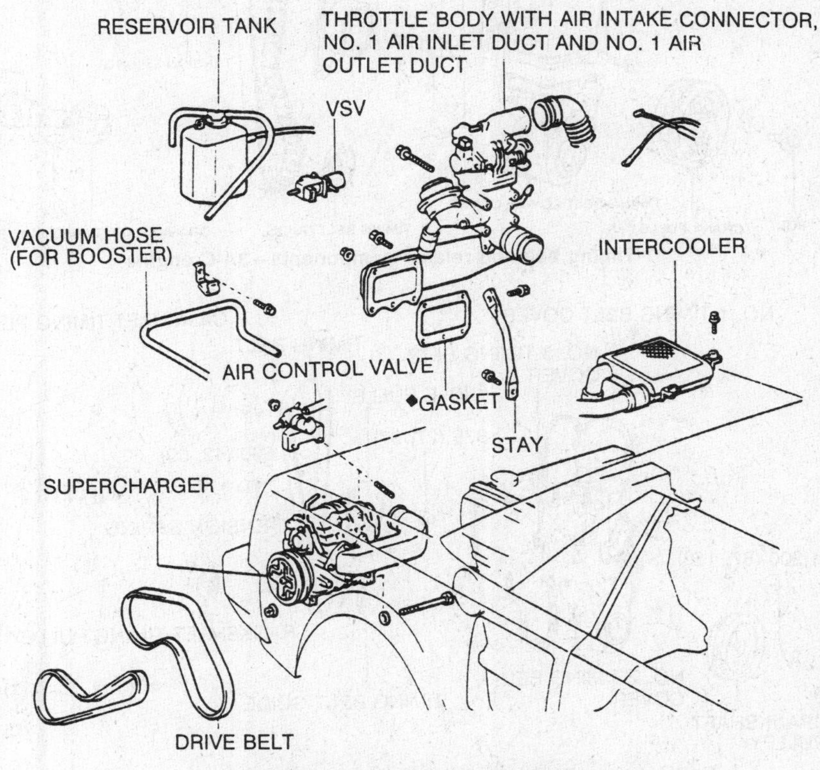

Supercharger and related components

27. Installation of the remaining components is the reverse of the removal procedure. Torque the oil pipe union bolt to the block to 38 ft. lbs. (51 Nm); stay-to-turbocharger bolts to 51 ft. lbs.; stay-to-block bolts to 43 ft. lbs. (59 Nm) and oxygen sensor to 33 ft. lbs. (44 Nm).

7M-GTE Engine

1. Disconnect the negative battery cable and drain the cooling system.
2. Remove the No. 4 air cleaner pipe with the No. 1 and No. 2 air cleaner hoses still attached.
3. Disconnect the 3 air hoses, the PCV hose and the electrical lead at the air flow meter. Disconnect the power steering idle-up air hose and then remove the No. 7 air cleaner hose with the air flow meter and cap still attached.
4. Disconnect the oxygen sensor and remove the turbocharger heat insulator.
5. Remove the oil dipstick guide.
6. Remove the No. 1 air cleaner pipe with the No. 6 air cleaner hose.
7. Disconnect the front exhaust pipe.
8. Remove the mounting nuts and union bolt for the turbocharger oil line. Remove the turbocharger stay.
9. Remove the No. 2 turbocharger stay. Disconnect the No. 1 turbocharger water hose at the water outlet housing. Disconnect the union pipe.
10. Remove the turbocharger and its gasket.

To install:

11. Prior to installing the turbocharger, pour approximately 1.2 cu. in. (20cc) of new oil into the oil inlet and then turn the impeller wheel by hand a few times in order to lubricate the bearing.
12. Position a new gasket with the protrusion pointing toward the rear and then install the turbocharger unit. Tighten the mounting bolts to 33 ft. lbs. (44 Nm). Tighten the union bolt to 25 ft. lbs. (34 Nm) and the nut to 9 ft. lbs. (13 Nm). The remainder of the installation is in the reverse order of removal.

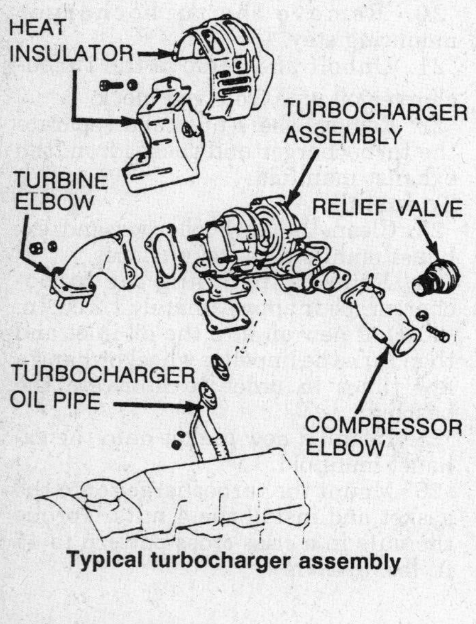

Typical turbocharger assembly

Timing Belt Front Cover

REMOVAL & INSTALLATION

3A-C, 4A-LC and 4A-C Engines

1. Disconnect the negative battery cable.
2. Remove all drive belts.
3. Bring the No. 1 cylinder to TDC on the compression stroke.
4. Remove the crankshaft pulley with a suitable puller.
5. Remove the water pump pulley.
6. Remove the upper and lower timing case covers.
7. Installation is the reverse of removal. Tighten the timing belt cover to 5–8 ft. lbs. (7–11 Nm).

3E, 3E-E and 3S-FE Engines

1. Disconnect the negative battery cable.
2. On 3E and 3E-E engines, remove the air cleaner assembly. On 3E-E engine, disconnect the accelerator and throttle cables.
3. Remove all drive belts.
4. On the 3S-FE engine remove the alternator, alternator bracket and right engine mounting stay (2WD). If equipped with cruise control remove the actuator and bracket assembly.
5. Raise and support the vehicle safely.
6. Remove the right tire and wheel assembly. Remove the right side engine under cover. Remove the right side engine mount insulator.
7. On the 3E and 3E-E engines, remove the cylinder head cover.
8. Set the No. 1 piston to TDC of the compression stroke and remove the crankshaft pulley using the proper tools.
9. Remove the engine front cover retaining bolts. Remove both front covers from the engine.

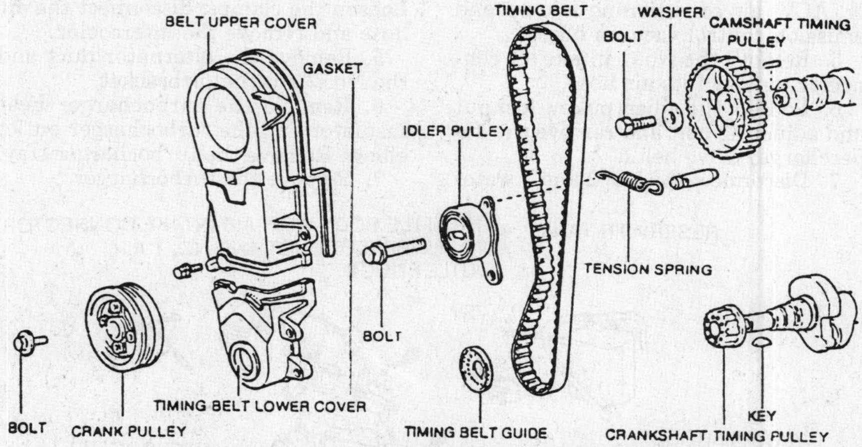

Timing belt and related components – 3A-C engine

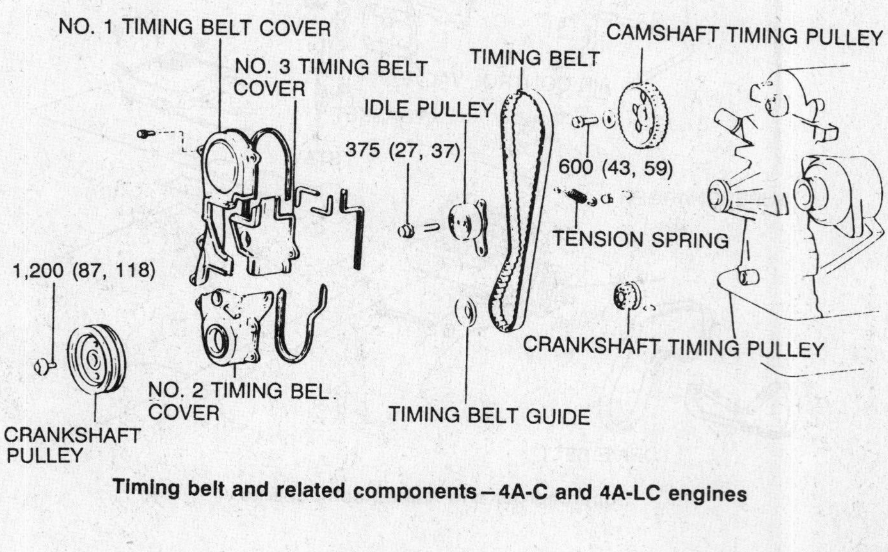

Timing belt and related components – 4A-C and 4A-LC engines

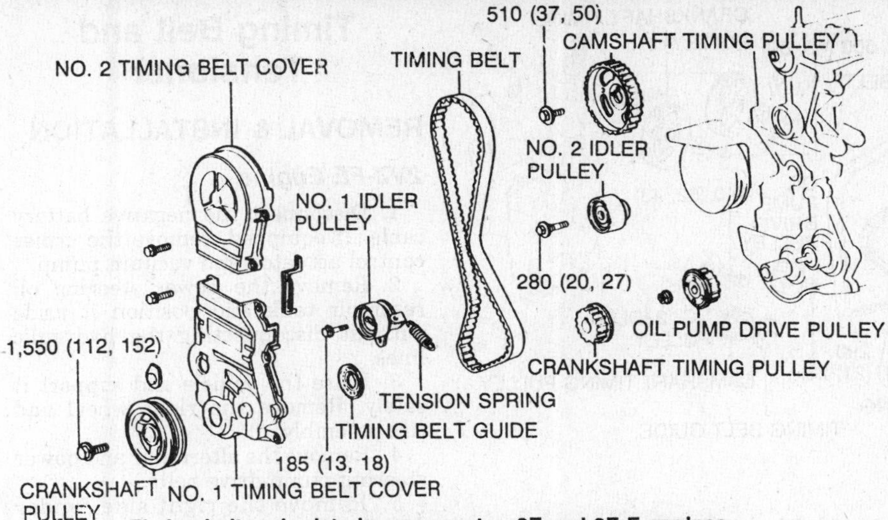

Timing belt and related components—3E and 3E-E engines

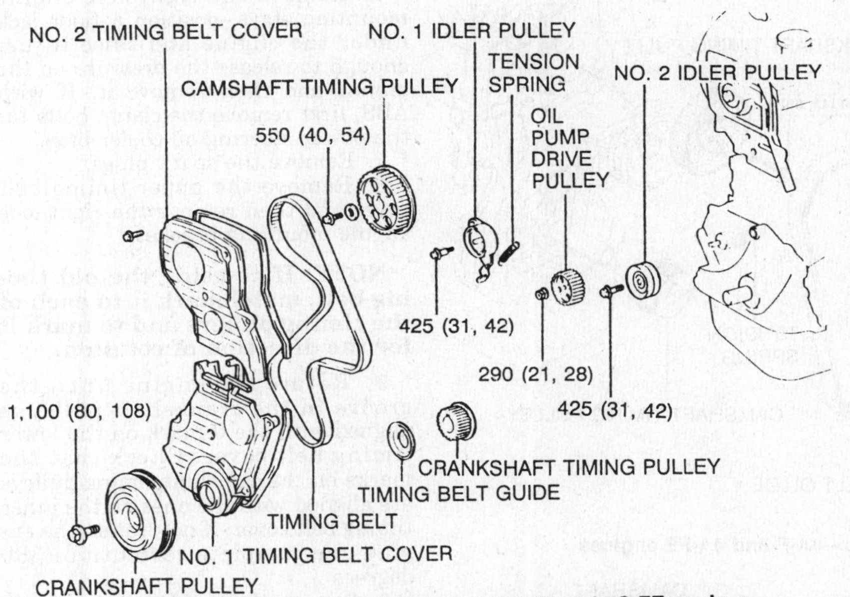

Timing belt and related components—3S-FE and 5S-FE engines

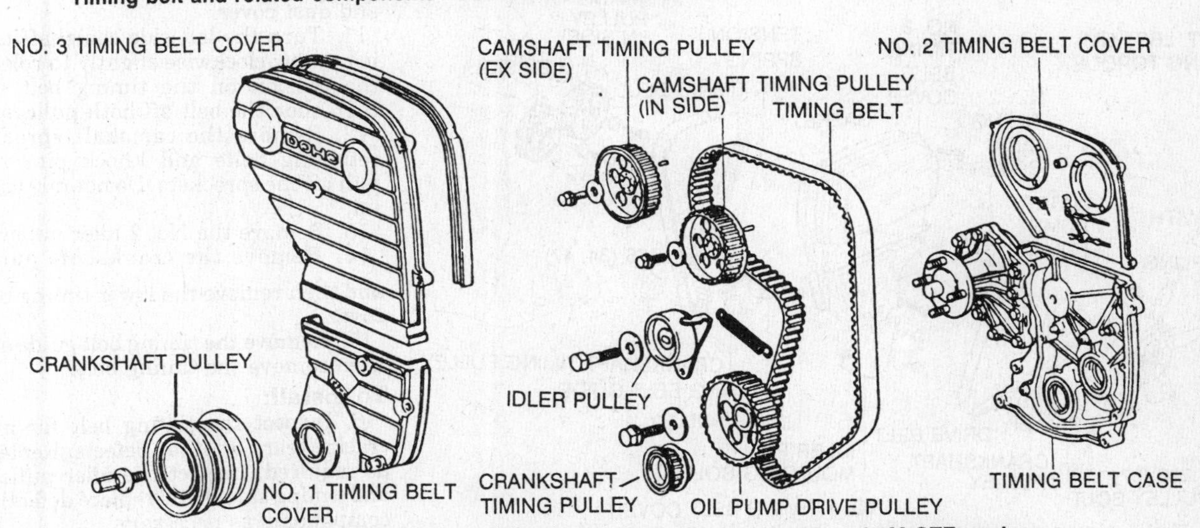

Timing belt and related components—5M-GE, 7M-GE and 7M-GTE engines

10. Installation is the reverse of the removal procedure. Use new cover seals. On 3E and 3E-E engines, torque the crankshaft pulley bolt to 112 ft. lbs. (154 Nm). On 3S-FE engine, torque the crankshaft pulley bolt to 80 ft. lbs. (108 Nm).

7M-GE and 7M-GTE Engines

1. Disconnect the negative battery cable.
2. Drain the cooling system.
3. Remove the radiator and water outlet.
4. Remove the spark plugs, drive belts and alternator.
5. Remove the upper timing belt cover and seal.
6. Set the No. 1 piston to TDC of the compression stroke.
7. Remove the timing belt from the camshaft sprockets.
8. Remove the crankshaft pulley using the proper tool.
9. Remove the lower timing belt cover and seal.
10. Installation is the reverse of the removal procedure. Use new cover seals. Torque the crankshaft pulley bolt to 195 ft. lbs. (265 Nm).

OIL SEAL REPLACEMENT

1. Remove the front cover.
2. Inspect the oil seal for signs of wear, leakage, or damage.
3. If worn, pry the old seal out. Remove it toward the front of the cover. Once the seal has been removed, it must be replaced.
4. Wipe the seal bore with a clean rag.
5. Drive the oil seal into place using a suitable seal installer. Work from the front of the cover. Be extremely careful not to damage the seal.
6. Install the front cover.

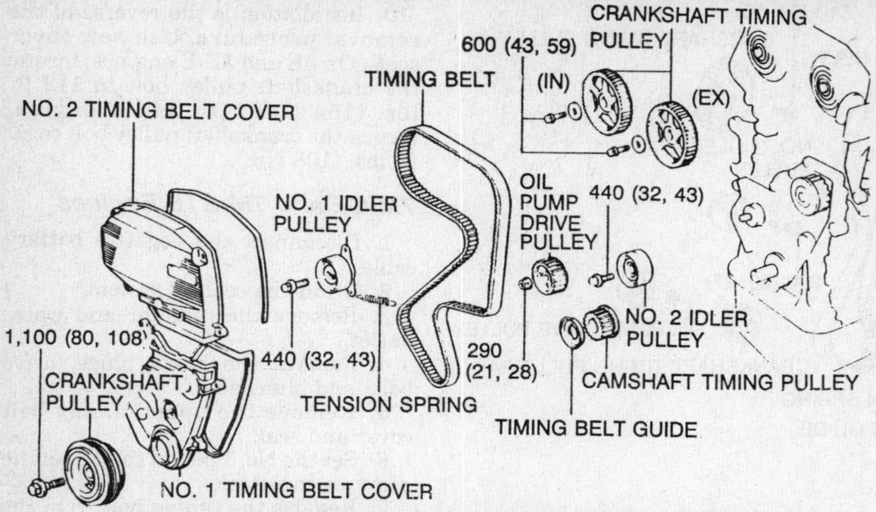

Timing belt and related components – 3S-GE and 3S-GTE engines

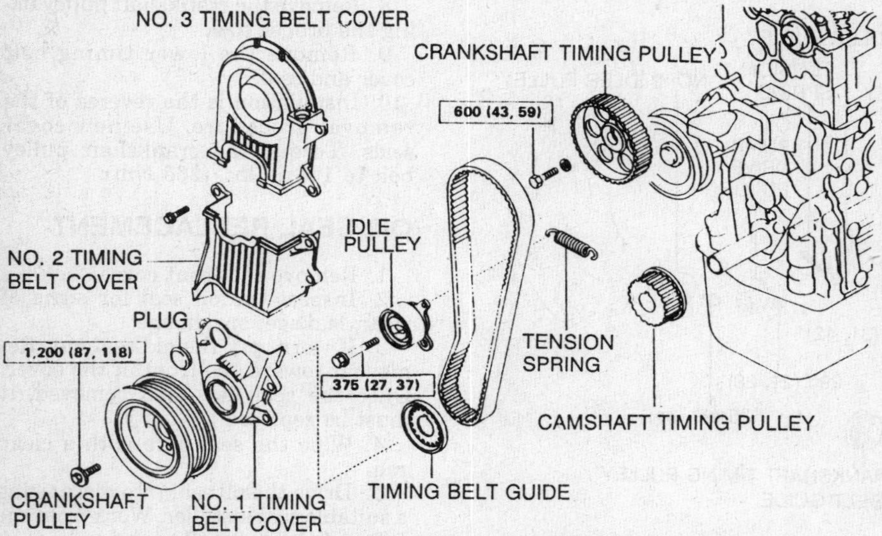

Timing belt and related components – 4A-F and 4A-FE engines

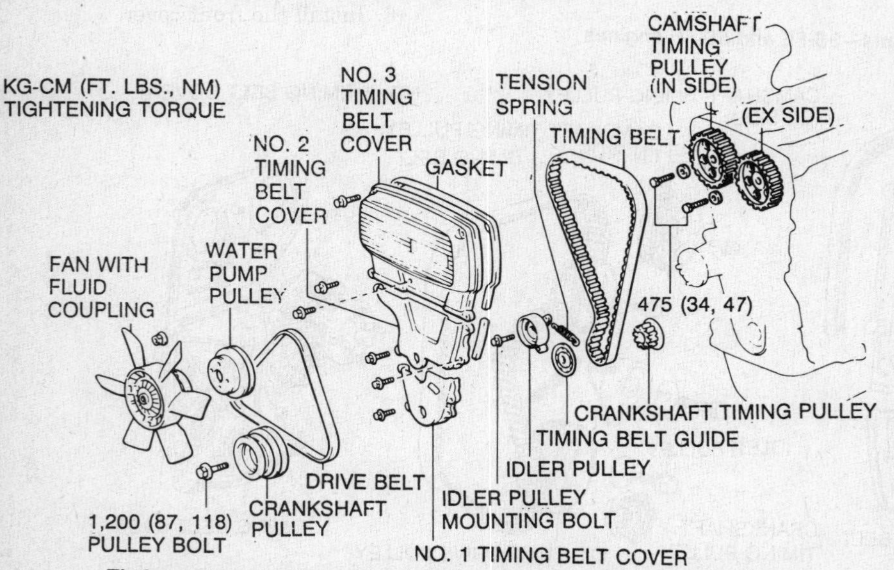

Timing belt and related components – 4A-GE and 4A-GZE engines

Timing Belt and Tensioner

REMOVAL & INSTALLATION

2VZ-FE Engine

1. Disconnect the negative battery cable. If equipped, remove the cruise control actuator and vacuum pump.
2. Remove the power steering oil reservoir tank and position it aside without disconnecting the hydraulic lines.
3. Raise the vehicle and support it safely. Remove the right wheel and tire assembly.
4. Remove the alternator and power steering pump drive belts.
5. Remove the right side fender apron seal.
6. Remove the right side engine mounting stays, position a floor jack under the engine and raise it just enough to release the pressure on the mount and then remove it. If with ABS, first remove the clamp bolts for the power steering oil cooler lines.
7. Remove the spark plugs.
8. Remove the upper timing belt cover and then remove the right side engine mounting bracket.

NOTE: If re-using the old timing belt, matchmark it to each of the timing pulleys and to mark it for the direction of rotation.

9. Rotate the engine until the groove in the crankshaft pulley is aligned with the **0** mark on the lower timing belt cover. Check that the marks on the camshaft timing pulleys are aligned with the ones on the inner timing belt cover; if not, rotate the engine one complete revolution (360 degrees).
10. Remove the timing belt tensioner and dust cover.
11. Turn the left side camshaft timing pulley clockwise slightly to release the tension on the timing belt and then slide the belt off both pulleys.
12. Remove the camshaft sprocket retaining bolts and knock pins and pull off the sprockets. Do not mix them up.
13. Remove the No. 2 idler pulley.
14. Remove the crankshaft pulley and then remove the lower timing belt cover.
15. Remove the timing belt guide and then remove the timing belt.

To install:

16. Inspect the timing belt for any cracks, tears or other defects. Replace as required. Inspect the idler pulleys and timing sprockets; replace defective components as necessary.
17. Install the timing belt over the

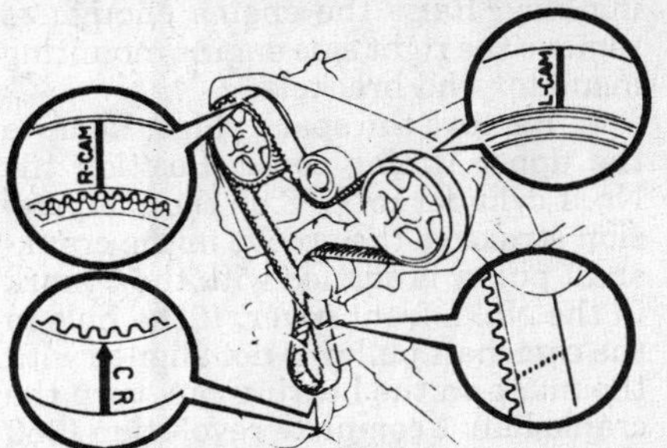

There should be installation marks on the timing belt – 2VZ-FE engine

Check that the marks on the sprockets and the No. 3 cover are aligned – 2VZ-FE engine

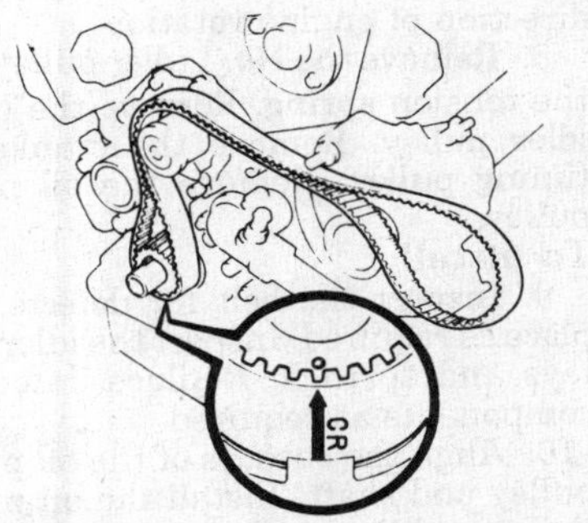

Installing the belt on the crankshaft – 2VZ-FE engine

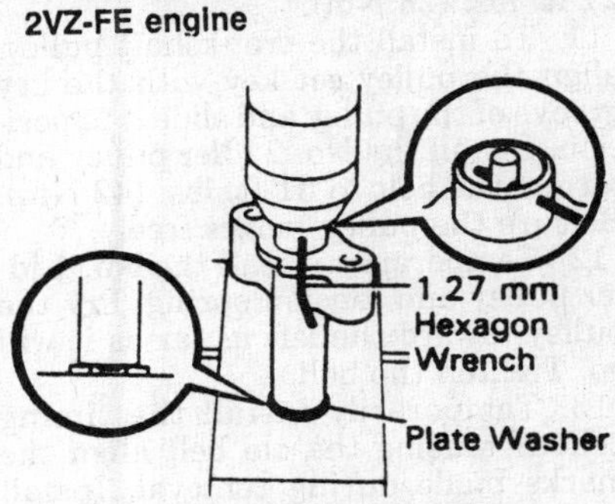

Set the timing belt tensioner – 2VZ-FE engine

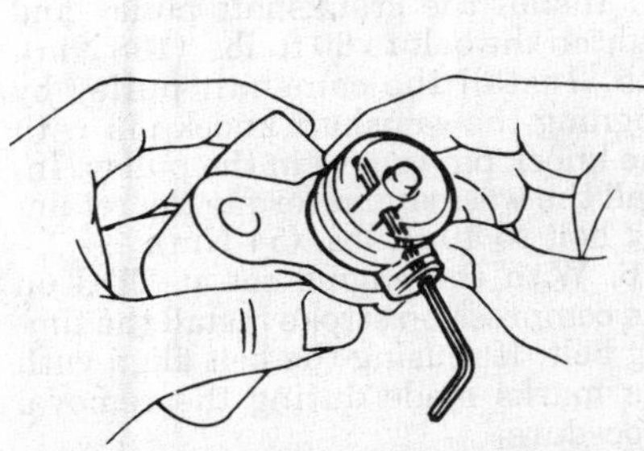
Install the hex wrench – 2VZ-FE engine

crankshaft sprocket so the mark on the belt aligns with the drilled mark on the sprocket. Install the belt over the No. 1 idler and water pump pulleys.

18. Install the timing belt guide so the cupped side faces outward. Install the No. 1 timing belt cover with a new gasket.

19. Install the crankshaft pulley so the set key is aligned with the groove in the shaft and then tighten the retaining bolt to 181 ft. lbs. (245 Nm).

20. Install the No. 2 idler pulley and tighten the bolt to 29 ft. lbs. (39 Nm). Check the pulley for smooth operation.

21. Install the left camshaft sprocket, flange side out, so the camshaft knock pin hole aligns with the groove in the sprocket. Install the knock pin and tighten the sprocket bolt to 80 ft. lbs. (108 Nm).

22. Set the No. 1 piston to TDC of the compression stroke again. Rotate the right camshaft until the knock pin hole is aligned with the timing mark on the No. 3 timing belt cover. Rotate the left camshaft sprocket until the timing mark on the sprocket is aligned with the one on the No. 3 cover.

23. Check that the mark on the timing belt aligns with the edge of the No. 1 timing belt cover. Rotate the left camshaft sprocket clockwise slightly so the mark on the timing belt will align with the timing mark on the sprocket, slide the belt over the sprocket and then align the mark on the sprocket with the one on the No. 3 belt cover. Check for tension between the crankshaft and camshaft sprockets.

24. Align the mark on the timing belt with the timing mark on the right camshaft sprocket.

25. Hang the belt over the sprocket, flange side inward, and then align the marks on the sprocket with the one on the No. 3 timing belt cover.

26. Slide the sprocket onto the camshaft so the knock pin hole and groove align. Install the knock pin and then tighten the retaining bolt to 80 ft. lbs. (108 Nm).

27. Install a plate washer between the timing belt tensioner and the cylinder block and then slowly press in the pushrod. When the holes in the pushrod and the housing align, insert a 1.27mm Allen wrench to retain the pushrod and then release the pressure. Install the tensioner and tighten the mounting bolts to 20 ft. lbs. (26 Nm); remove the Allen wrench.

28. Rotate the crankshaft pulley 2 complete revolutions clockwise and check that each pulley is still aligned with the timing marks. If not, remove the belt and start over again.

29. Installation of the remaining components is in the reverse order of removal. Make all necessary engine adjustments.

3A-C, 4A-C and 4A-LC Engines

1. Remove the timing belt upper and lower covers.

2. If the timing belt is to be reused, mark an arrow in the direction of engine revolution on its surface. Matchmark the belt to the pulleys at the proper locations.

3. Loosen the idler pulley bolt, push it to the left as far as it will go and then temporarily tighten it.

4. Remove the timing belt, idler pulley bolt, idler pulley and the return spring Do not bend, twist, or turn the belt inside out. Do not allow grease or water to come in contact with it.

5. Inspect the timing belt for cracks, missing teeth or overall wear. Replace as necessary. Install the return spring and idler pulley.

6. Install the timing belt. Align the marks made earlier, if reusing the old belt.

7. Adjust the idler pulley so the belt deflection is 0.24–0.28 in. (6–7mm) at 4.5 lbs. of pressure. Check the valve timing.

8. Installation of the remaining components is in the reverse order of removal.

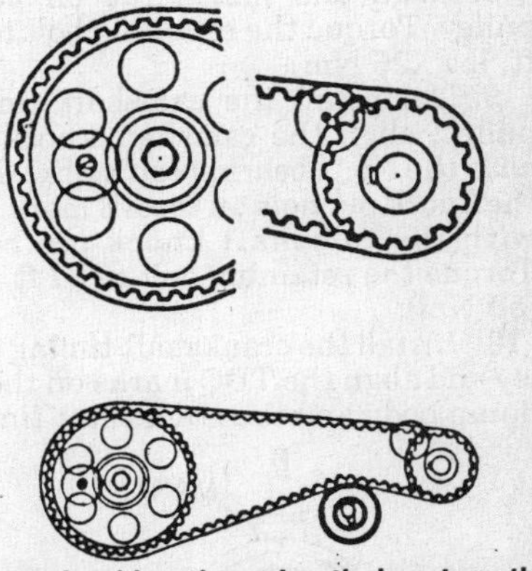
When checking the valve timing, turn the crankshaft 2 complete revolutions clockwise from TDC to TDC and make sure each pulley aligns with the marks shown

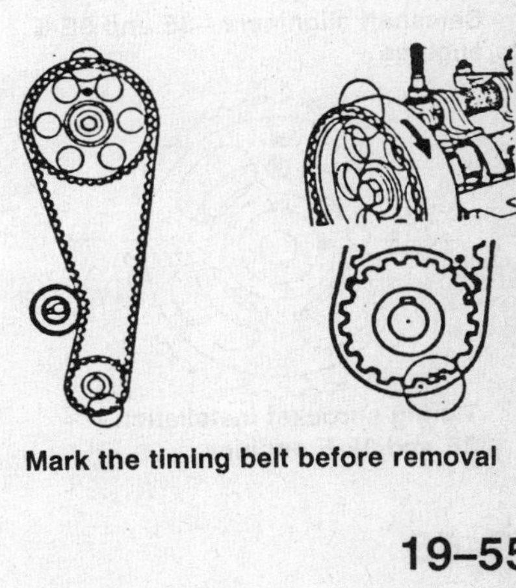
Mark the timing belt before removal

3E and 3E-E Engines

1. Disconnect the negative battery cable. Remove the right side engine under cover. On 3E-E engine, disconnect the accelerator and throttle cables.

2. Remove the drive belts, alternator and alternator bracket. Remove the air cleaner and air intake collector (3E-E) assemblies and spark plugs.

3. Raise the engine and remove the right side engine mounting insulator assembly.

4. Remove the cylinder head cover. Set the engine to TDC on the compression stroke. Remove the crankshaft pulley using the proper removal tool.

5. Remove both timing belt covers. Remove the timing belt guide. Remove the timing belt and the No. 1 idler pulley. If using the old belt matchmark it in the direction of engine rotation. Matchmark the pulleys.

6. Remove the tension spring. Remove the No. 2 idler pulley. Remove the crankshaft pulley, camshaft pulley and oil pump pulley using the proper tools.

To install:

7. Inspect the belt for defects. Replace as required. Inspect the idler pulleys and springs. Replace defective components as required.

8. Align and install the oil pump pulley. Torque the retaining bolt to 20 ft. lbs. (26 Nm).

9. To install the camshaft timing pulley, align the camshaft knock pin with the No. 1 bearing cap mark. Align the knock pin hole on the 3E mark side with the camshaft knock pin hole. Torque the retaining bolt to 37 ft. lbs. (50 Nm).

10. Install the crankshaft timing pulley and align the TDC marks on the oil pump body and the crankshaft timing pulley. Install the No. 1 idler pulley. Pry the idler pulley toward the left as far as it will go and temporarily tighten the retaining bolt.

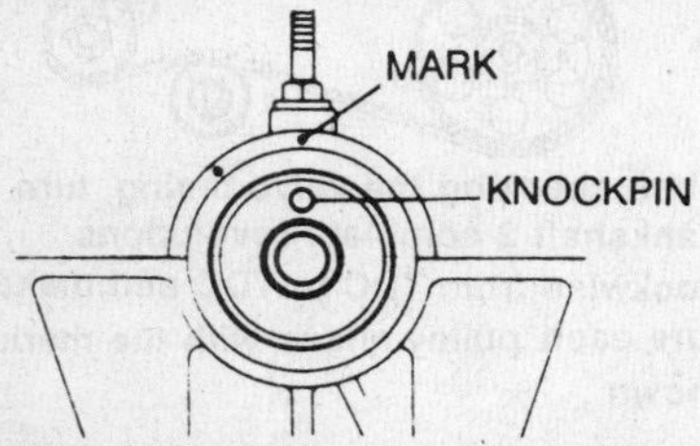

Camshaft alignment—3E and 3E-E engines

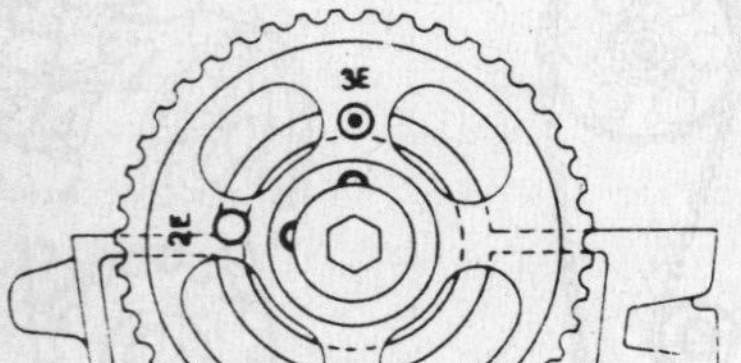

Timing sprocket installation—3E and 3E-E engines

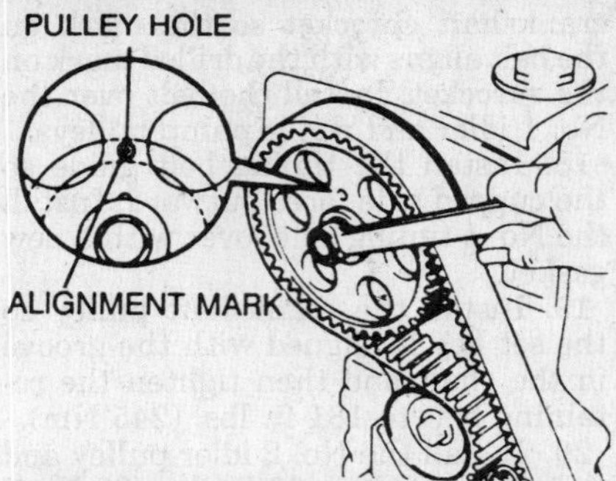

Timing sprocket alignment—3E and 3E-E engines

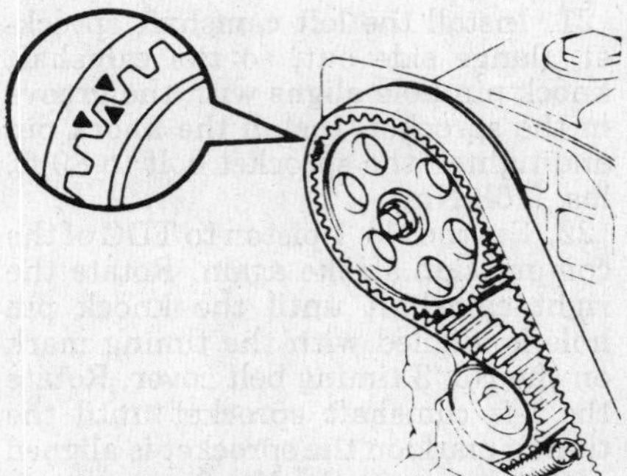
Timing belt alignment—3E and 3E-E engines

11. Install the No. 2 idler pulley and torque the retaining bolt to 20 ft. lbs. (27 Nm). Install the timing belt. If reusing the old belt align it with the marks made during the removal procedure.

12. Inspect the valve timing and the belt tension by loosening the No. 1 idler pulley set bolt. Temporarily install the crankshaft pulley bolt and turn the crankshaft 2 complete revolutions in the clockwise direction.

13. Check that each pulley aligns with the proper markings. Torque the No. 1 idler pulley bolt to 13 ft. lbs. (18 Nm). Check for proper belt tension. Install the belt guide.

14. Install the timing belt covers. Align and install the crankshaft pulley. Torque the retaining bolt to 112 ft. lbs. (152 Nm).

15. Installation of the remaining components is the reverse of the removal procedure.

3S-FE and 5S-FE Engines

1. Disconnect the negative battery cable. Raise and support the vehicle safely. Remove the right tire and wheel assembly.

2. If equipped, remove the cruise control actuator and bracket. Remove the drive belts.

3. Remove the alternator and alternator bracket and right engine mounting stay. Raise the engine enough to remove the right side engine mounting insulator and brackets.

4. Remove the spark plugs. Remove the upper timing cover. Position the No. 1 cylinder to TDC on the compression stroke so the groove in the crankshaft pulley is aligned with the **0** mark in the No. 1 front cover. If the hole in the camshaft pulley is not aligned with the mark on the bearing cap, turn the crankshaft 1 complete revolution (360 degrees).

5. If reusing the belt place matchmarks on the timing belt and the camshaft pulley. Loosen the mount bolt of the No. 1 idler pulley and position the pulley toward the left as far as it will go. Tighten the bolt. Remove the belt from the camshaft pulley.

6. Remove the camshaft pulley. Remove the crankshaft pulley using the proper removal tool. Remove the lower timing cover.

7. Remove the timing belt and the belt guide. If reusing the belt mark the belt and the crankshaft pulley in the direction of engine rotation.

8. Remove the No. 1 idler pulley and the tension spring. Remove the No. 2 idler pulley. Remove the crankshaft timing pulley. Remove the oil pump pulley.

To install:

9. Inspect the belt for defects. Replace as required. Inspect the idler pulleys and springs. Replace defective components as required.

10. Align the cutouts of the oil pump pulley and shaft. Install the oil pump pulley and torque the retaining nut to 21 ft. lbs. (28 Nm).

11. To install the crankshaft pulley, align the pulley set key with the key groove of the pulley and slide it in position. Install the No. 2 idler pulley and torque the bolt to 31 ft. lbs. (42 Nm). Be sure the pulley moves freely.

12. Temporarily install the No. 1 idler pulley and tension spring. Pry the pulley toward the left as far as it will go. Tighten the bolt.

13. Temporarily install the timing belt. If reusing the old belt align the marks made during removal. Install the timing belt guide.

14. Install the No. 1 timing belt cover. Install the crankshaft pulley and tighten the bolt to 80 ft. lbs. (108 Nm).

15. Install the camshaft pulley by aligning the camshaft knock pin with the knock pin groove in the pulley. Install the washer and torque the retaining bolt to 40 ft. lbs. (54 Nm).

16. With the engine set at TDC on the compression stroke install the timing belt. If reusing the belt align with the marks made during the removal procedure.

17. Once the belt is installed be sure

there is tension between the crankshaft pulley, water pump pulley and camshaft pulley. Loosen the No. 1 idler pulley mount bolt ½ turn. Turn the crankshaft pulley 2 revolutions from TDC to TDC, in the clockwise direction. Torque the No. 1 idler pulley mount bolt to 31 ft. lbs. (42 Nm).

18. Installation the remaining components is the reverse of the removal procedure.

3S-GE and 3S-GTE Engines

CELICA (2WD)

1. Disconnect the negative battery cable. Raise the vehicle and support safely. Remove the right front tire and wheel assembly. Remove the right side fender liner.
2. Remove the windshield washer and radiator reservoir tanks. Remove the cruise control actuator, if equipped.
3. Remove the power steering belt. Remove the power steering pump and position it aside with the hydraulic lines still attached.
4. Remove the alternator and support bracket. Remove the upper timing belt cover.
5. Set the No. 1 piston to TDC of the compression stroke by aligning the groove on the crankshaft pulley with the **0** mark on the lower timing belt cover. Check that the matchmarks on the 2 camshaft timing pulleys and the rear timing belt cover are aligned, if not, turn the crankshaft 1 complete revolution clockwise (360 degrees).
6. If the timing belt is to be reused, draw a directional arrow on it and matchmark the belt to the 2 camshaft pulleys. Loosen the No. 1 idler pulley bolt and shift the pulley as far left as possible; tighten the set bolt. Remove the timing belt from the 2 camshaft pulleys. Support the belt so the meshing of the belt with the remaining pulleys does not shift.
7. Carefully hold the camshafts with an adjustable wrench and remove the camshaft pulley set bolts. Remove the pulleys and their set pins.
8. Remove the crankshaft pulley. Remove the lower timing belt.
9. Remove the timing belt guide and then remove the timing belt from the remaining pulleys. Be sure to matchmark the belt to the pulleys if it is to be reused.
10. Remove the No. 1 idler pulley and the tension spring. Remove the No. 2 idler pulley, the crankshaft timing pulley and the oil pump pulley.

To install:

11. Install the oil pump pulley and tighten it to 21 ft. lbs. (28 Nm). Install the crankshaft timing pulley by sliding it onto the crankshaft over the Woodruff key. Install the No. 2 idler pulley and tighten it to 32 ft. lbs. (43 Nm).

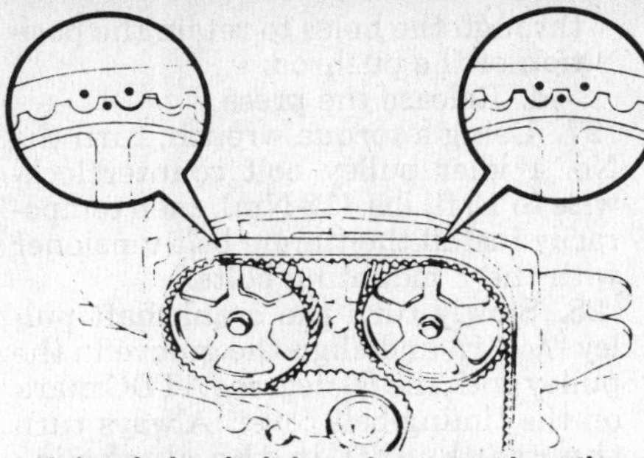

If the timing belt is to be reused on the 3S-GE engine, place matchmarks on the belt and pulleys

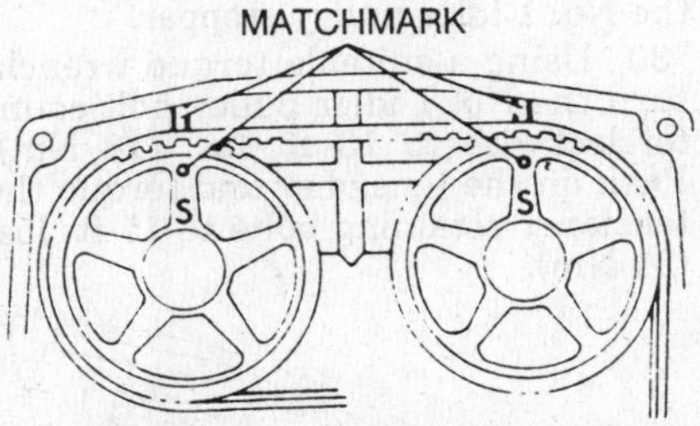

Align the matchmarks on the camshaft timing pulleys with those on the rear timing belt cover—3S-GE engine

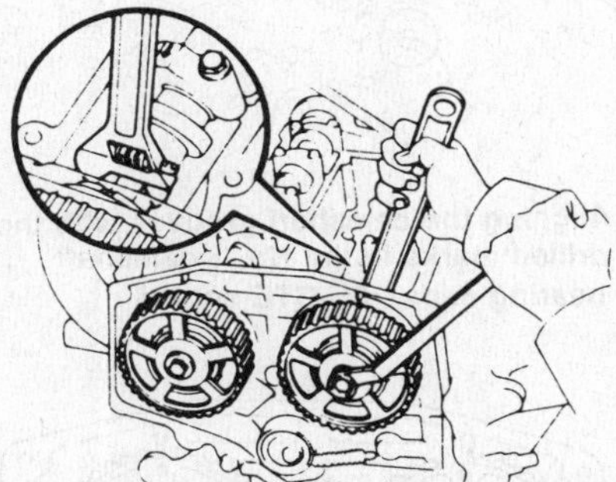

Hold the camshaft with an adjustable wrench when removing the camshaft timing pulley—3S-GE engine

12. Install the No. 1 idler pulley and the tension spring. Move the pulley as far to the left as it will go and then tighten it.
13. Install the timing belt on all pulleys except the 2 camshaft pulleys. Make sure the matchmarks made earlier are in alignment.
14. Install the timing belt guide with the cup side out. Install the lower timing belt cover and then install the crankshaft pulley. Tighten it to 80 ft. lbs. (108 Nm).
15. Check that the No. 1 cylinder is at TDC of the compression stroke for the crankshaft. The crankshaft pulley groove should be aligned with the **0** mark on the lower timing belt cover.
16. Check that the No. 1 cylinder is at TDC of the compression stroke for the camshaft.

NOTE: There are 2 types of camshafts, one with 2 holes on the timing pulley contact surface and one with 5 holes on the timing pulley contact surface. All replacement camshaft have 5 holes.

2 Hole: Using a wrench, turn the camshafts so the camshaft knock pin aligns with the matchmark on the rear

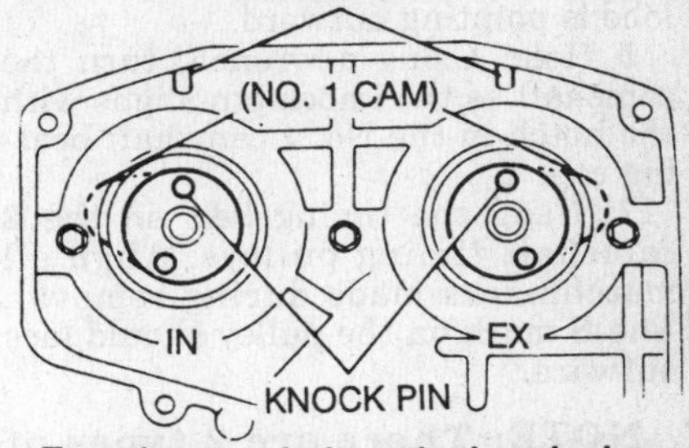

Turn the camshaft so the knock pins align with the matchmark on the rear timing cover and the No. 1 lobes are facing outward—3S-GA engine (2 hole type)

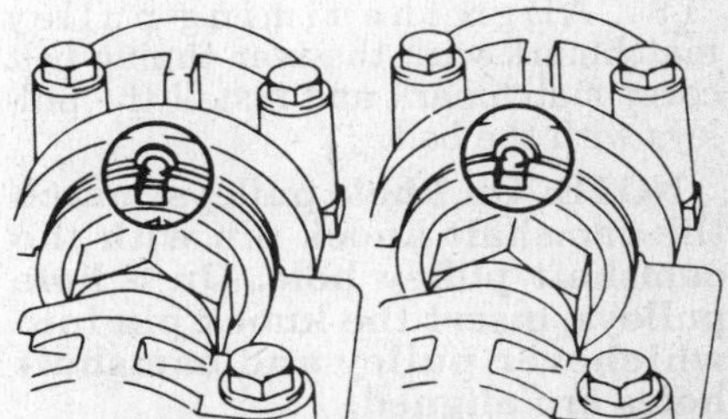

Align the knock pin and the No. 1 bearing cap mark—3S-GE engine (5 hole type)

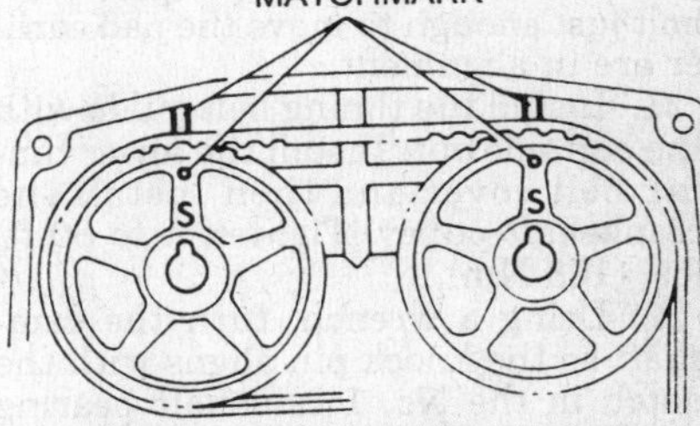

Align the camshaft knock pin with the hole in the camshaft timing pulley—3S-GE engine (1 hole type pulley only)

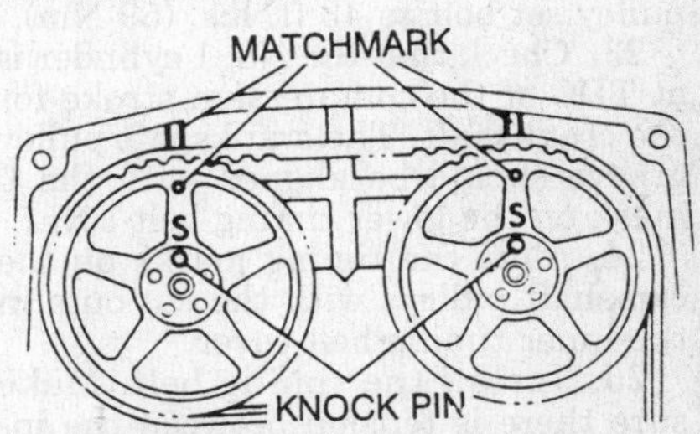

Insert the knock pin into whichever camshaft timing pulley and camshaft holes are aligned—3S-GE engine (5 hole type pulley only)

timing belt cover. And the No. 1 cam lobe is pointing outward.

5 Hole: Using a wrench, turn the camshaft so the knock pin aligns with the notch in the No. 1 camshaft bearing cap.

17. Hang the timing belt on the 2 camshaft timing pulleys. Align all matchmarks made during removal. The **S** mark on the pulley should face outward.

NOTE: There are 2 types of camshaft pulleys. One has 5 holes on the camshaft contact surface and one has 1 hole on the contact surface. All replacement pulleys have 5 holes.

18. Align the timing pulley matchmark with the rear timing belt cover matchmark and install the pulleys with the belt.

NOTE: On 1 hole pulleys, match the camshaft knock pin with the camshaft pulley hole. On 5 hole pulleys, insert the knock pin into whichever pulley and camshaft holes are aligned.

19. Hold the camshaft with an adjustable wrench and tighten the pulley set bolt to 43 ft. lbs. (59 Nm).

20. Loosen the No. 1 idler pulley set bolt just enough to move the pad earlier are in alignment.

21. Install the timing belt guide with the cup side out. Install the lower timing belt cover and then install the crankshaft pulley. Tighten it to 80 ft. lbs. (108 Nm).

22. Using a wrench, turn the camshaft so the knock pin aligns with the notch in the No. 1 camshaft bearing cap. Install the camshaft pulleys so the **S** mark faces up. Then, align the holes in the pulley and camshaft and insert the knockpin. Hold the camshaft with an adjustable wrench and tighten the pulley set bolt to 43 ft. lbs. (59 Nm).

23. Check that the No. 1 cylinder is at TDC of the compression stroke for the crankshaft. The crankshaft pulley groove should be aligned with the **0** mark on the lower timing belt cover.

24. Align the timing marks on the camshaft pulleys with the cut-outs in the inner timing belt cover.

25. Install the timing belt. Make sure there is tension between the intake camshaft pulley and the crankshaft pulley.

26. Set the timing belt tensioner as follows:

a. Mount the tensioner in a vise and align the holes in the tensioner pushrod with the holes in the housing.

b. Using a press, slowly press the pushrod into the housing until the holes are exactly aligned.

c. Insert a 1.27mm hex wrench through the holes to retain the position of the pushrod.

d. Release the press.

27. Using a torque wrench, turn the No. 1 idler pulley bolt counterclockwise to 13 ft. lbs. (18 Nm), then temporarily install the timing belt tensioner with the 2 mounting bolts.

28. Slowly turn the crankshaft pulley $^{5}/_{6}$ turn and align the groove in the pulley with the 60 degrees ATDC mark on the timing belt cover. Always turn the crankshaft in the clockwise direction.

29. Insert a 0.75 in. (1.90mm) feeler gauge between the tensioner body and the No. 1 idler pulley stopper.

30. Using a suitable torque wrench, turn the No. 1 idler pulley bolt counterclockwise to 13 ft. lbs. (18 Nm). Push on the tensioner and torque the tensioner retaining bolts to 15 ft. lbs. (21 Nm).

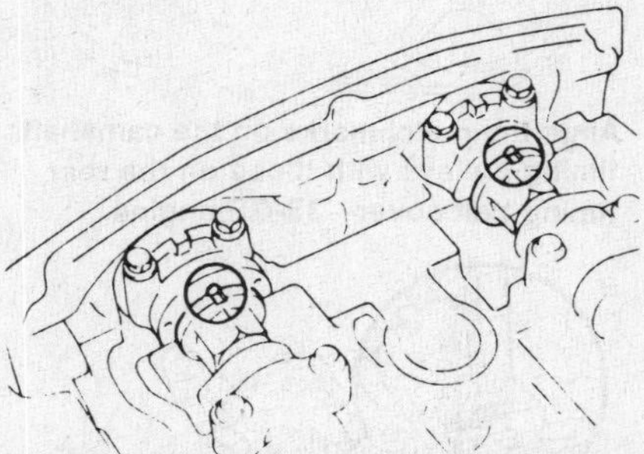

Aligning the camshaft grooves with the drilled marks in the No. 1 camshaft bearing caps – 3S-GTE engine

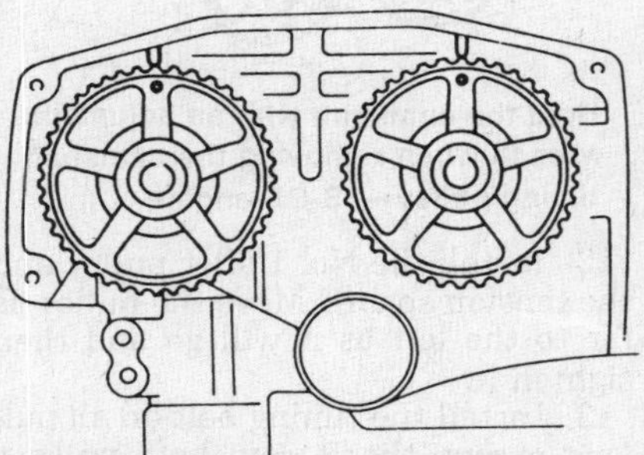

Aligning the camshaft and timing belt cover marks – 3S-GTE engine

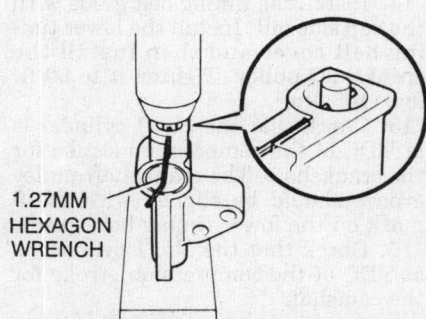

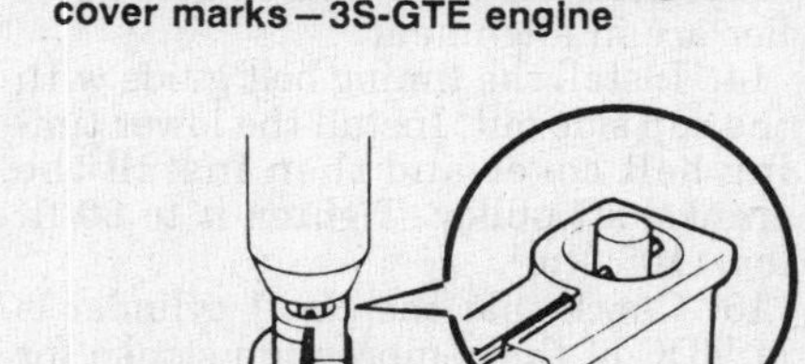

Setting the timing belt tensioner – 3S-GTE engine

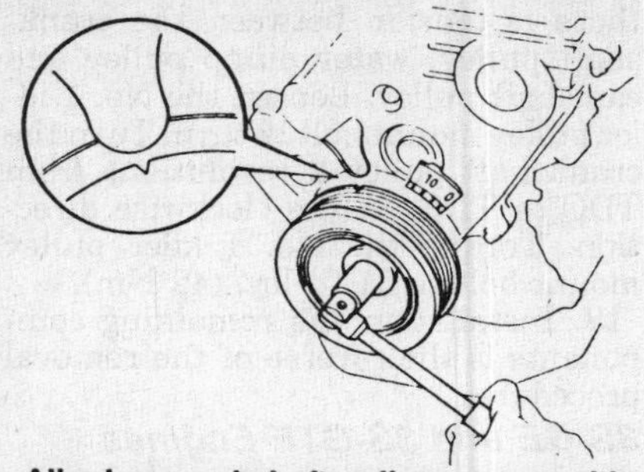

Aligning crankshaft pulley groove with 60 degree ATDC mark on the timing belt cover

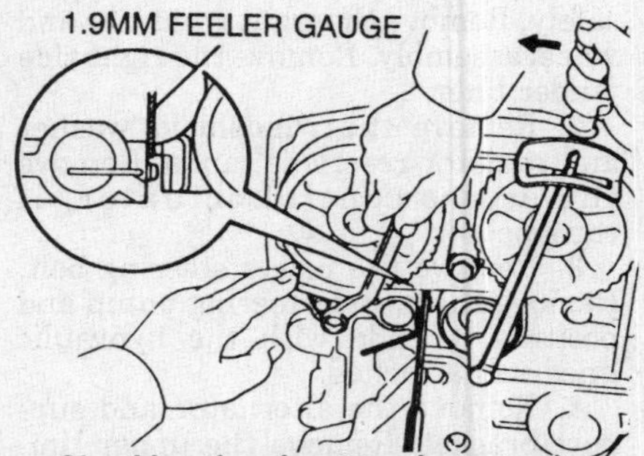

Checking the clearance between the tensioner body and the No. 1 idler pulley stopper

31. Remove the 1.27mm hex wrench from the tensioner.

32. Slowly turn the crankshaft pulley in the clockwise direction $^{5}/_{6}$ turn and align the groove in the pulley with the 60 degrees ATDC mark on the timing belt cover. Always turn the crankshaft in the clockwise direction.

33. Turn the No. 1 idler pulley bolt to 18 ft. lbs. (25 Nm).

34. Using a feeler gauge, check the clearance between the tensioner body and the No. 1 idler pulley stop. The clearance should be 0.071–0.087 in. (1.8–2.2mm). If the clearance is not as specified, remove the tensioner and reinstall it.

35. Installation of the remaining components is the reverse of the removal procedure.

4A-F and 4A-FE Engines

1. Raise and support the vehicle safely. Remove the right wheel and undercover. Remove the air cleaner.

2. Remove the drive belts. Remove the power steering pump and the air conditioning compressor, with brackets, and position them aside. Leave the hydraulic and refrigerant lines connected.

3. Remove the spark plugs and the cylinder head cover. Rotate the crankshaft pulley so the **0** mark is in alignment with the groove in the No. 1 front cover. Check that the lifters on the No. 1 cylinder are loose; if not, turn the crankshaft 1 complete revolution (360 degrees).

4. Position a floor jack under the engine and remove the right side engine mounting insulator.

5. Remove the water pump and crankshaft pulleys. The crankshaft pulley will require a 2-armed puller.

6. Loosen the 9 bolts and remove the No. 1, No. 2 and No. 3 front covers. Remove the timing belt guide.

7. Loosen the bolt on the idler pulley, push it to the left as far as it will go and then retighten it. If reusing the timing belt, draw an arrow on it in the direction of engine revolution (clockwise) and then matchmark the belt to the pulleys.

8. Remove the timing belt. Remove the idler pulley bolt, the pulley and the tension spring.

9. Remove the crankshaft timing pulley.

10. Lock the camshaft and remove the camshaft timing pulleys.

To install:

11. Install the camshaft timing pulley so it aligns with the knockpin on the exhaust camshaft. Tighten the pulley to 34 ft. lbs. (47 Nm)–1988; 43 ft. lbs. (59 Nm)–1989–92. Align the mark on the No. 1 camshaft bearing cap with the center of the small hole in the pulley.

12. Install the crankshaft timing pulley so the marks on the pulley and the oil pump body are in alignment.

13. Install the idler pulley and its tension spring, move it to the left as far as it will go and tighten it temporarily.

14. Align the matchmarks made during removal and then install the timing belt on the camshaft pulley. Loosen the idler pulley set bolt. Make sure the timing belt meshing at the crankshaft pulley does not shift.

15. Rotate the crankshaft clockwise 2 revolutions from TDC to TDC. Make sure each pulley aligns with the marks made previously. If the marks are not in alignment, the valve timing is wrong. Shift the timing belt meshing slightly and then repeat Steps 14–15.

16. Tighten the set bolt on the timing belt idler pulley to 27 ft. lbs. (37 Nm). Measure the timing belt deflection at the top span between the 2 camshaft pulleys. It should deflect no more than 0.16 in. (4mm) at 4.4 lbs. of pressure–1988; 0.24 in. (7mm) at 4.4 lbs. of pressure–1989–92. If deflection is greater, readjust by using the idler pulley.

17. Installation of the remaining components is the reverse of the removal procedure.

4A-GE, 4A-GEC, 4A-GELC and 4A-GZE Engines

1. Disconnect the negative battery cable. Disconnect the No. 2 air cleaner hose from the air cleaner.

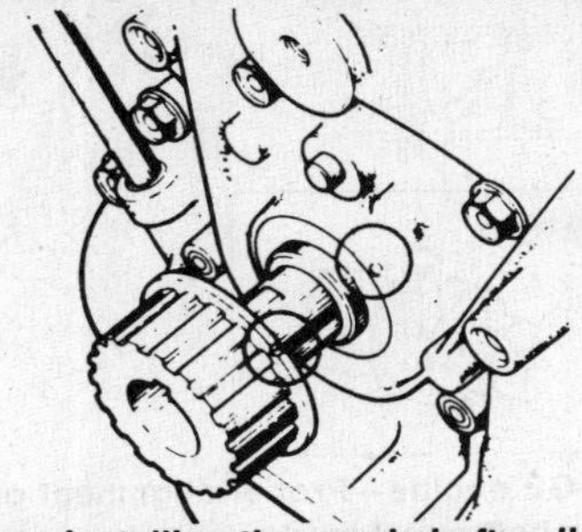

When installing the crankshaft pulley, make sure the TDC marks on the oil pump body and the pulley are in alignment–4A-GE engine

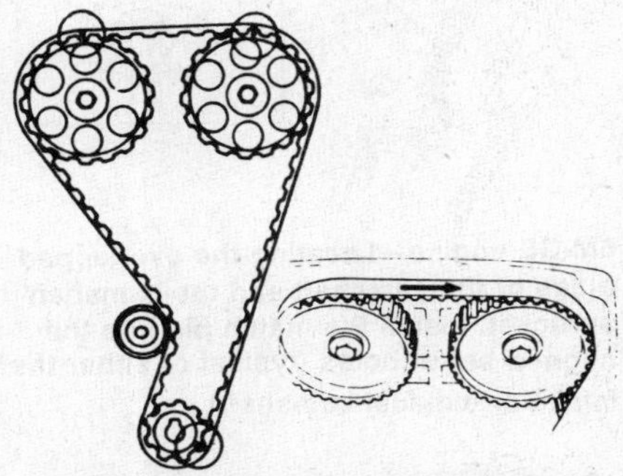

If the timing belt is to be reused, draw a directional arrow and matchmark the belt to the pulleys as shown–4A-GE engine

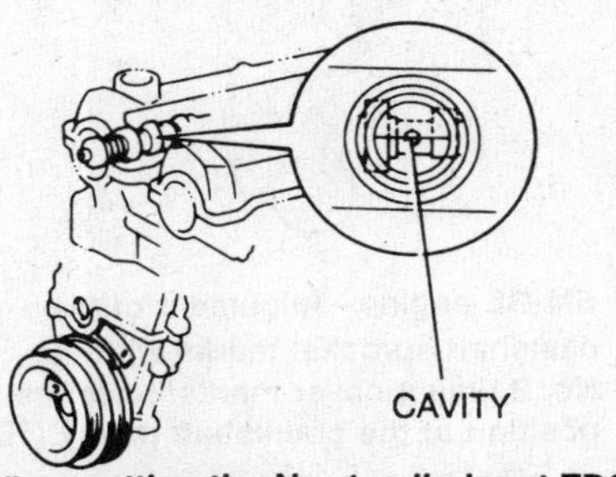

When setting the No. 1 cylinder at TDC on the 4A-GE engine, remove the oil filler cap and check that the cavity in the camshaft is visible

2. If equipped with power steering, remove the power steering pump and position it aside. Do not disconnect the pump hydraulic lines.

3. Loosen the water pump pulley set nuts, remove the drive belt adjusting bolt and then remove the drive belt. Remove the set nuts and then remove the fluid coupling along with the fan and the water pump pulley.

4. Remove the spark plugs. Rotate the crankshaft pulley so the groove on it is in alignment with the **0** mark on the No. 1 timing belt cover. Remove the oil filler cap and check that the cavity in the camshaft is visible. If not, turn the camshaft 1 complete revolution (360 degrees).

5. On the MR2 and 1988–92 Corolla, remove the right side engine mount insulator.

6. Lock the crankshaft pulley and remove the pulley bolt. Using a gear puller, remove the crankshaft pulley. Remove the three timing belt covers and their gaskets. Remove the timing belt guide.

7. Loosen the bolt on the idler pulley, push it to the left as far as it will go and then retighten it. If reusing the timing belt, draw an arrow on it in the direction of engine revolution (clockwise) and then matchmark the belt to the pulleys.

8. Remove the timing belt. Remove the idler pulley bolt, the pulley and the tension spring.

9. Remove the cylinder head covers, lock the camshaft and remove the camshaft timing pulleys.

To install:

10. Install the camshaft timing pulleys and cylinder head covers. Tighten the pulley to 34 ft. lbs. (47 Nm).

11. Install the crankshaft timing pulley so the marks on the pulley and the oil pump body are in alignment.

12. Install the idler pulley and its tension spring, move it to the left as far as it will go and tighten it temporarily.

13. Install the timing belt. If the old one is being used, align all the marks made during removal.

14. Installation of the remaining components is the reverse order of the removal procedure.

5M-GE Engine

1. Disconnect the negative battery cable.

2. Loosen the mounting bolts of each of the crankshaft-driven components at the front of the engine and remove the drive belts.

3. Rotate the crankshaft in order to set the No. 1 cylinder to TDC of its compression stroke (both valves of the No. 1 cylinder closed, and TDC marks aligned).

4. Remove the upper and front timing belt cover and gaskets.

5. Loosen the idler pulley bolt and lever the idler pulley toward the alternator side of the engine in order to relieve the tension on the timing belt. Hand tighten the idler pulley bolt.

6. Remove the timing belt from the camshaft pulleys.

7. Remove the camshaft timing pulleys as follows. Hold the pulleys stationary with a spanner wrench. Remove the center pulley bolt. Do not attempt to use timing belt tension as a tool to remove the center pulley bolts, as the belt could become damaged.

NOTE: Do not interchange the intake and exhaust timing pulleys, as they differ for use with each camshaft.

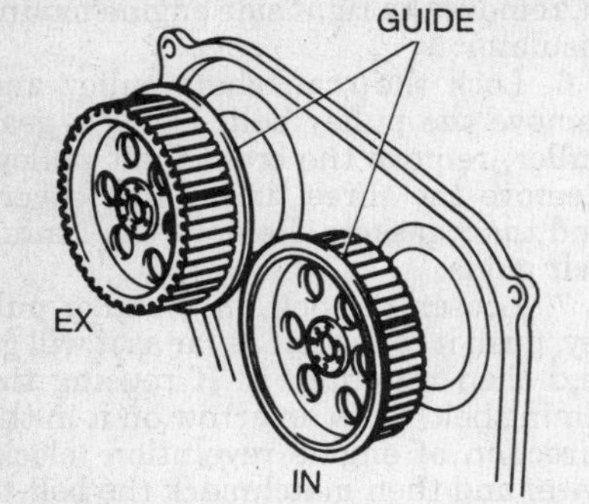

5M-GE engines—When installing the camshaft sprockets, be sure that the guides are positioned as shown. (IN—intake camshaft sprocket; EX—exhaust camshaft sprocket)

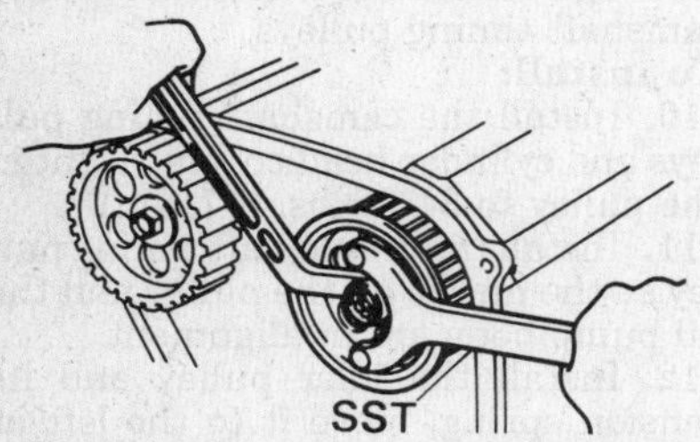

5M-GE engine—Use a spanner wrench 9SST, as shown, to hold the camshaft sprocket while loosening the camshaft sprocket bolt. Do not attempt to use belt tension to hold the sprocket in place while removing the camshaft sprocket bolt

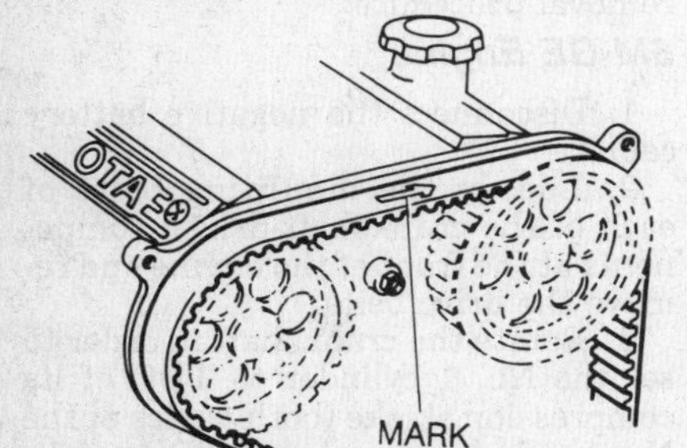

5M-GE engine—Paint a mark on the timing belt prior to belt removal to indicate the belts direction of normal rotation. Point the mark in the same direction if the belt is to be reinstalled

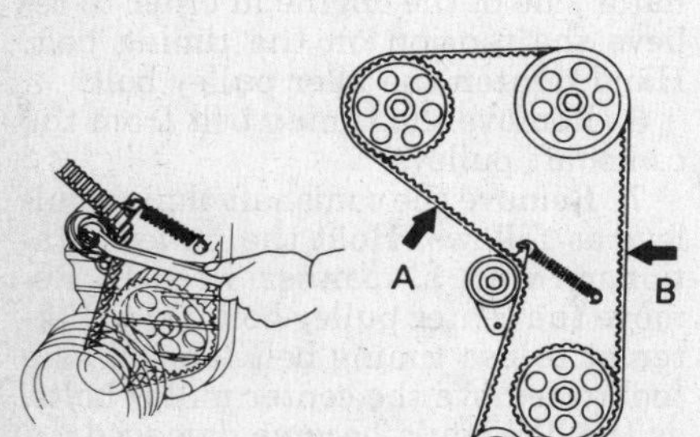

5M-GE engine—When adjusting the timing belt tension, be sure the tension at "A" is the same as that at "B"

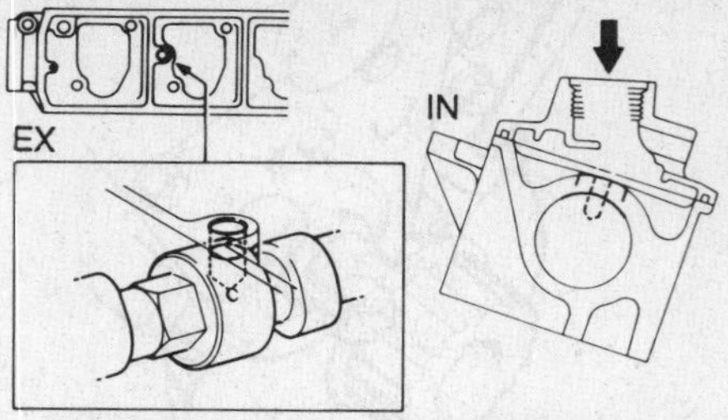

5M-GE engine—Proper alignment of the camshaft matchmarks with the match holes of the camshaft housings (IN—intake; EX—exhaust)

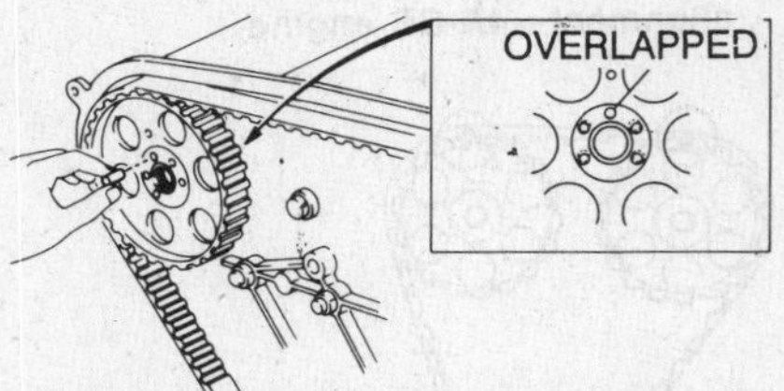

5M-GE engine—Locating the overlapped holes of the camshaft and the camshaft sprocket. Install the match pin into the aligned set of holes (typical of either the intake or exhaust camshaft)

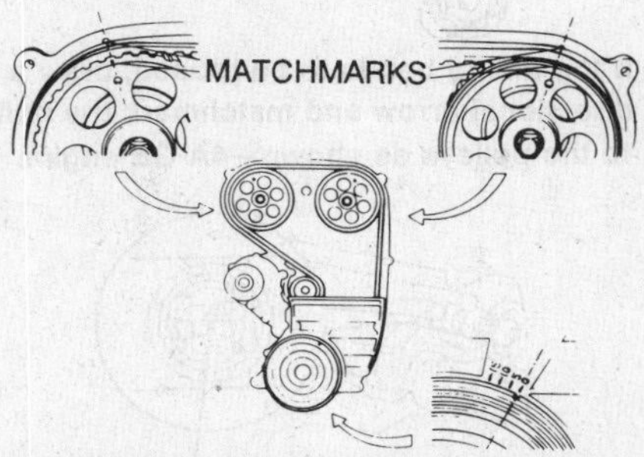

5M-GE engine—Alignment of the camshaft sprocket marks with the No. 2 timing cover marks. Note the position of the crankshaft pulley (TDC)

8. Remove the center crankshaft pulley bolt. Using a puller, remove the crankshaft pulley.

9. Using chalk or crayon, mark the timing belt to indicate its direction of rotation. This mark must face the same direction during installation of the belt.

10. Remove the lower timing belt cover, then the belt.

11. If damaged, the crankshaft pulley can be removed using a puller; the oil pump driveshaft pulley can be removed in the same manner as the camshaft pulleys.

12. Inspect the timing belt for damage, such as cuts, cracks, missing teeth, abrasions, nicks, etc. If the belt teeth are damaged, check that the camshafts rotate freely and correct as necessary.

13. Should damage be evident on the belt face, check the idler pulley belt surface for damage. If damage is present on one side of the belt only, check the belt guide and the alignment of each pulley. If the belt teeth are excessively worn, check the timing belt cover gasket for damage and/or proper installation.

14. Check the idler pulley for damage and smoothness of rotation. Also check the free length of the tension spring, which should be 2.8 in. (71mm), measured between the inside of each end "clip". Replace the spring if the length exceeds this limit.

To install:

15. Install the crankshaft and oil pump driveshaft if these items were removed previously. Torque the oil pump driveshaft center pulley bolt to 16 ft. lbs. (22 Nm). The crankshaft pulley must be evenly driven into place.

16. Install the idler pulley and the tension spring. Lever the pulley towards the alternator side of the engine and tighten the bolt.

17. Check the mark made during Step 9 of removal and temporarily install the timing belt on the crankshaft pulley. The mark must face in the same direction as it did originally.

18. Install the lower timing belt cover. Install the crankshaft pulley and torque the center pulley bolt to 195 ft. lbs. (265 Nm).

19. Remove the oil filter cap of the intake camshaft cover, and the complete camshaft cover on the exhaust side.

20. Check that the match holes of both No. 2 camshaft journals are visible through the camshaft housing match holes. If necessary, temporarily install the camshaft pulley and guide pin, and rotate the camshaft(s) until the holes are aligned.

21. Install the timing pulleys. Note that the belt guide of the exhaust camshaft pulley should be positioned towards the engine; the belt guide of the intake camshaft pulley should be positioned away from the engine. Do not yet install the pulley retaining bolts.

22. Align the following marks. Each camshaft pulley mark must be aligned with its respective mark on the rear, upper timing belt cover. Align the crankshaft pulley notch with the TDC (0) mark of the timing tab.

NOTE: The No. 1 cylinder must be positioned at TDC on its compression stroke.

23. Install the timing belt.

24. Loosen the idler pulley bolt and tension the timing belt. The timing belt tension must be the same between the exhaust camshaft pulley and the crankshaft pulley, as it is between the intake camshaft pulley and the oil pump driveshaft pulley.

25. There are 5 pin holes on each camshaft and each timing pulley. On

the exhaust side: Install the match pin into the one hole of the pulley which is aligned with one of the camshaft pin holes. Repeat this on the intake side. Only one of the holes of each side should be aligned to allow insertion of the match pins.

26. Using a spanner wrench to hold the camshaft pulleys, install and tighten the camshaft pulley bolts. These bolts should be torqued to 51 ft. lbs. (69 Nm).

27. Install the exhaust camshaft cover, using a new gasket. Install the oil filler cap. Install the timing belt cover and gasket.

28. Install and adjust the drive belts at the front of the engine. Reconnect the battery cable.

7M-GE and 7M-GTE Engines

1. Disconnect the negative battery cable. Drain the cooling system. Remove the radiator. Remove the water outlet.
2. Remove the spark plugs. Remove the drive belts. Remove the No. 3 timing belt cover.
3. Position the engine at TDC on the compression stroke. Remove the timing belt from the camshaft sprockets. If reusing the belt, matchmark the belt and the sprockets in the direction of engine rotation.
4. Remove the camshaft pulleys. Remove the crankshaft pulley using the proper removal tools. Remove the power steering air pipe, if equipped.
5. If equipped with air conditioning remove the compressor and position it aside. Do not disconnect the refrigerant lines.
6. Remove the No. 1 timing belt cover. Remove the timing belt. Remove the idler pulley and the tension spring. Remove the oil pump drive pulley.

To install:

7. Inspect the belt for defects. Replace as required. Inspect the idler pulleys and springs. Replace defective components as required.
8. Install the oil pump drive pulley and retaining bolt. Tighten the bolt to 16 ft. lbs. (22 Nm).
9. Install the crankshaft timing pulley. Temporarily install the idler pulley and tension spring. Tighten the assembly to 36 ft. lbs. (49 Nm). Pry the idler pulley toward the left as far as it will go and temporarily tighten the bolt.
10. Temporarily install the timing belt. If reusing the old belt install it using the marks made during the removal procedure. Install the No. 1 timing belt cover.
11. If equipped with air conditioning, install the compressor assembly. If equipped, install the power steering air pipe.
12. Align the set key with the key groove and install the crankshaft pulley and torque the retaining bolt to 195 ft. lbs. (265 Nm).
13. Install the camshaft timing pulleys. Torque the retaining bolts to 36 ft. lbs. (49 Nm).
14. Loosen the idler pulley bolt. Install the timing belt to the INTAKE side and the EXHAUST side. Tighten the idler pulley bolt to 36 ft. lbs. (49 Nm).
15. Make sure the timing belt tension **A** is equal to the timing belt tension **B**. If not adjust the idler pulley. Turn the engine 2 complete revolutions in the clockwise direction and check to see that everything is aligned properly.
16. Turn both the intake and exhaust camshaft pulleys inward at the same time to loosen the timing belt between the 2 sprockets. Belt deflection should be 4.4–6.6 lbs. If not adjust the idler pulley.
17. Install the remaining components, start the engine and check for leaks and proper operation.

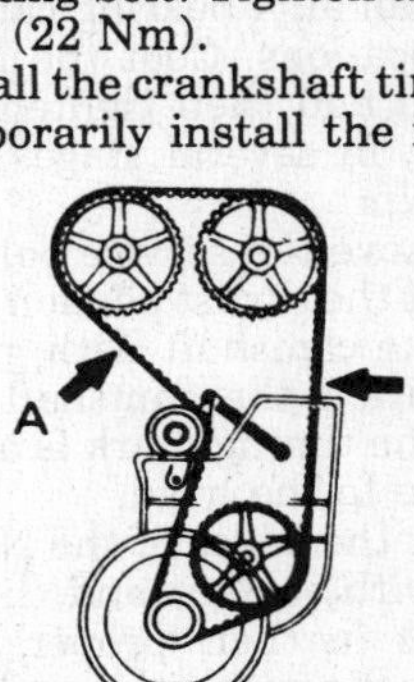

7M-GE and 7M-GTE engines timing belt tension check

Timing Sprockets

REMOVAL & INSTALLATION

Timing sprocket/pulley removal and installation procedures are detailed within the individual Timing Belt sections.

Camshaft

REMOVAL & INSTALLATION

2VZ-FE Engine

NOTE: Due to a nominal thrust clearance, the camshafts must be held absolutely level during removal. If not, the section of the cylinder head receiving the thrust may crack or be damaged, thus causing the camshaft to break.

1. Remove the cylinder head(s).
2. Rotate the exhaust camshaft in the right cylinder head until the 2 pointed marks on the camshaft drive and driven gears are aligned.
3. Secure the exhaust camshaft sub-gear to the driven gear with bolt. This is important as it will eliminate the torsional spring force of the sub-gear.
4. Loosen the bearing cap bolts in the proper sequence and then remove the 4 bearing caps and the right side exhaust camshaft.
5. Loosen the bearing cap bolts in the proper sequence and then remove the 5 bearing caps and the right side intake camshaft.

NOTE: Be sure to arrange all the bearing caps in their proper order.

6. Rotate the exhaust camshaft in the left cylinder head until the pointed mark on the camshaft drive and driven gears are aligned.
7. Secure the exhaust camshaft sub-gear to the driven gear with bolt. This is important as it will eliminate the torsional spring force of the sub-gear.
8. Loosen the bearing cap bolts in the proper sequence and then remove the 4 bearing caps and the left side exhaust camshaft.
9. Loosen the bearing cap bolts in the proper sequence and then remove the 5 bearing caps and the left side intake camshaft.

NOTE: Be sure to arrange all the bearing caps in their proper order.

To install:

10. Coat the thrust portion of the right side intake camshaft with suitable grease and then position the camshaft into the head so the 2 timing marks are at a 90 degree angle to the head.

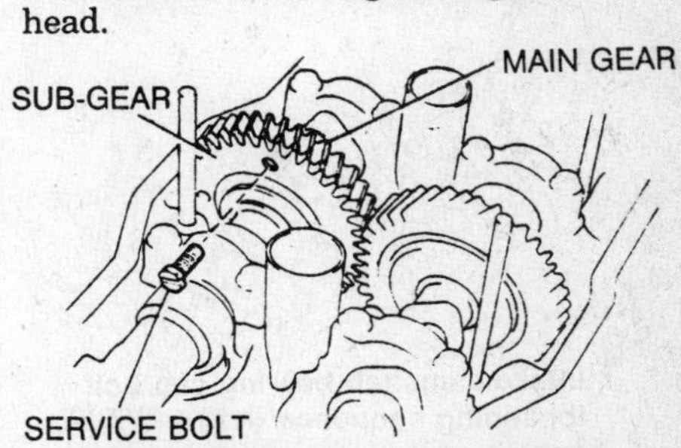

Install a service bolt in the exhaust camshaft sub-gear (right)—2VZ-FE engine

Exhaust camshaft bearing cap bolt loosening sequence (right)—2VZ-FE engine

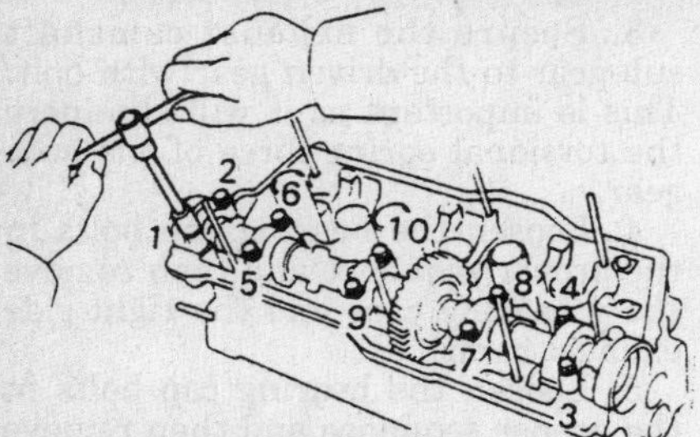
Intake camshaft bearing cap bolt loosening sequence (right) – 2VZ-Fe engine

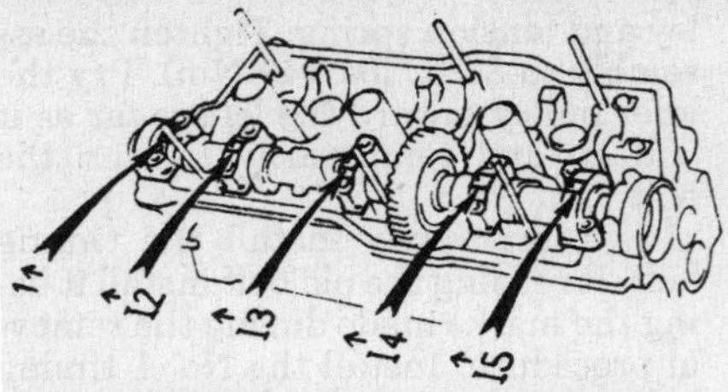
Intake camshaft bearing cap installation (right) – 2VZ-FE engine

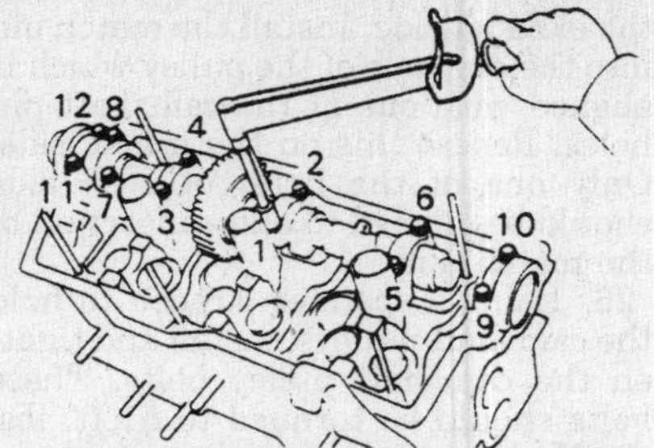
Intake camshaft bearing cap bolt tightening sequence (left) – 2VZ-FE engine

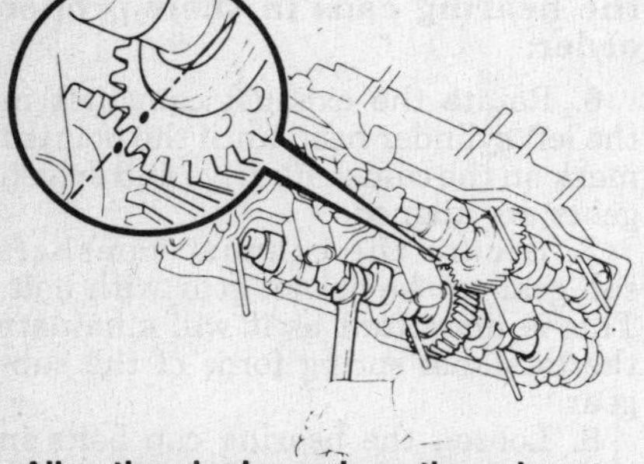
Align the single mark on the exhaust camshaft (left) – 2VZ-FE engine

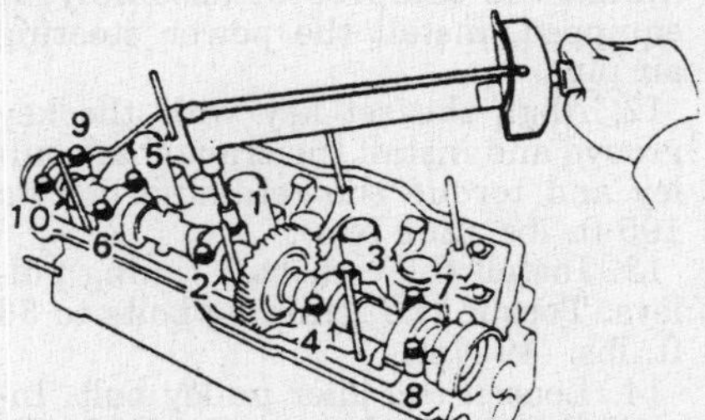
Intake camshaft bearing cap bolt tightening sequence (right) – 2VZ-FE engine

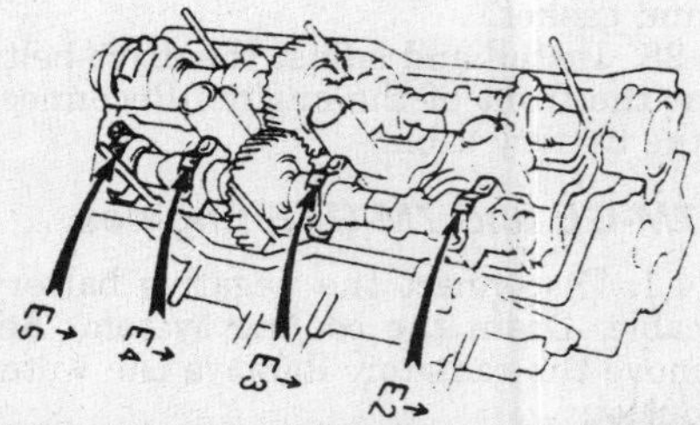
Exhaust camshaft bearing cap installation (left) – 2VZ-FE engine

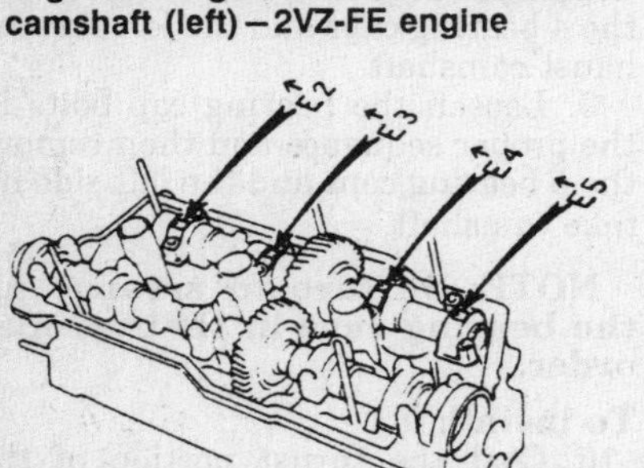
Exhaust camshaft bearing cap installation (right) – 2VZ-FE engine

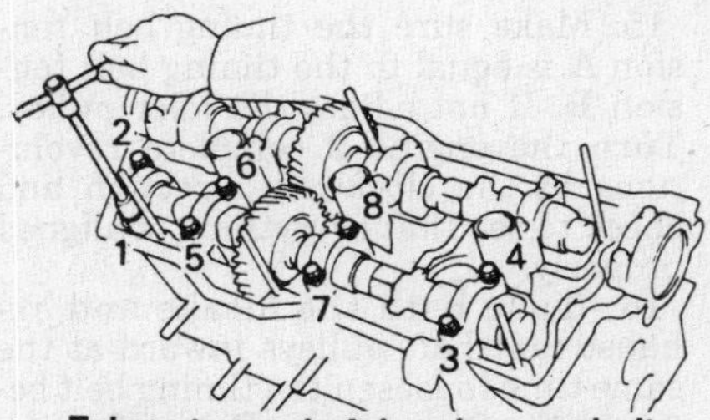
Exhaust camshaft bearing cap bolt loosening sequence (left) – 2VZ-FE engine

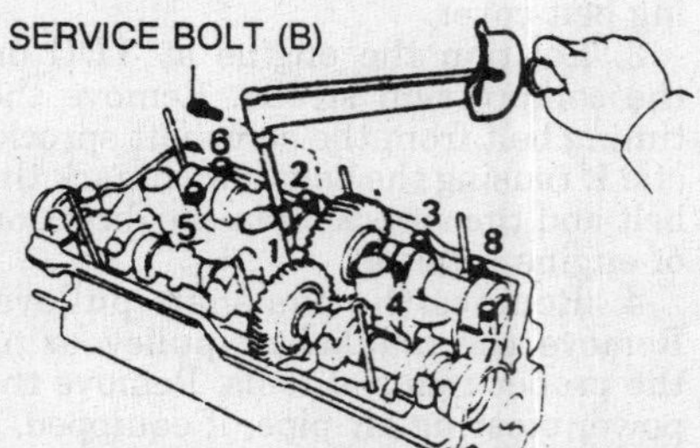

Exhaust camshaft bearing cap bolt tightening sequence (left) – 2VZ-FE engine

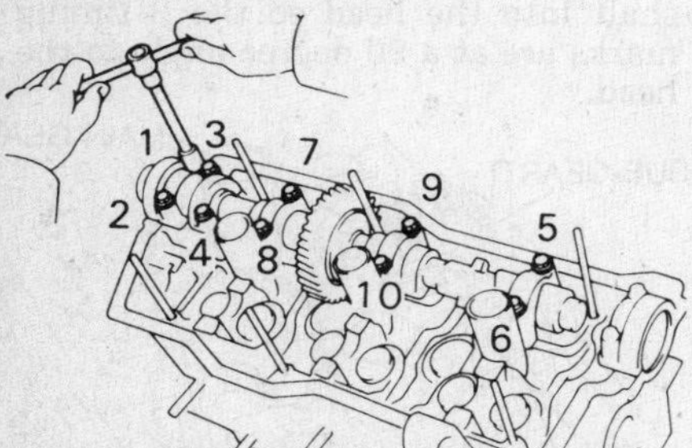
Intake camshaft bearing cap bolt loosening sequence (left) – 2VZ-FE engine

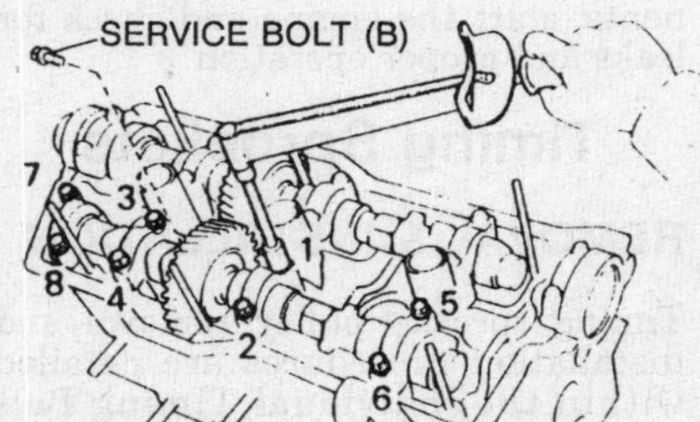

Exhaust camshaft bearing cap bolt tightening sequence (right) – 2VZ-FE engine

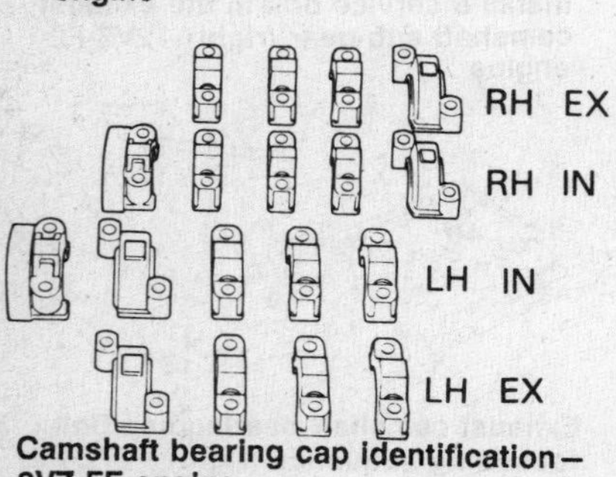

Camshaft bearing cap identification – 2VZ-FE engine

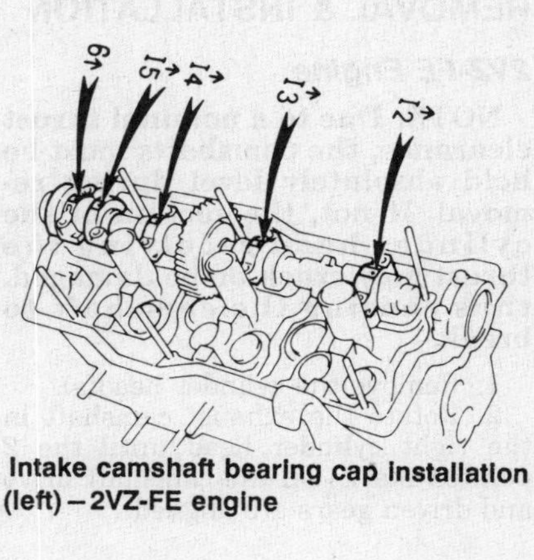
Intake camshaft bearing cap installation (left) – 2VZ-FE engine

11. Coat the edges of the No. 1 bearing cap with sealant and then install all 5 caps in their proper locations. Coat the bolts with engine oil and then tighten them, in sequence, in several stages, to 12 ft. lbs. (16 Nm).
12. Coat the thrust portion of the right side exhaust camshaft with grease and then position the camshaft into the head so the 2 timing marks align with those on the intake shaft.
13. Install all 4 bearing caps in their proper locations. Coat the bolts with engine oil and then tighten them, in sequence, in several stages, to 12 ft. lbs. (16 Nm).
14. Remove the service bolt.
15. Coat the thrust portion of the left side intake camshaft with grease and then position the camshaft into the head so the timing mark is at a 90 degree angle to the head.
16. Coat the edges of the No. 1 bearing cap with sealant and then install all 5 caps in their proper locations. Coat the bolts with engine oil and then tighten them, in sequence, in several stages, to 12 ft. lbs. (16 Nm).

17. Coat the thrust portion of the left side exhaust camshaft with grease and then position the camshaft into the head so the timing mark aligns with the one on the intake shaft.

18. Install all 4 bearing caps in their proper locations. Coat the bolts with engine oil and then tighten them, in sequence, in several stages, to 12 ft. lbs. (16 Nm).

19. Remove the service bolt.

20. Installation of the remaining components is in the reverse order of removal.

3S-FE and 5S-FE Engines

1. Remove the cylinder head.

2. To remove the exhaust camshaft, set the knock pin of the exhaust camshaft at 10–45 degree BTDC of camshaft angle. This angle will help to lift the exhaust camshaft level and evenly by pushing No. 2 and No. 4 cylinder camshaft lobes of the exhaust camshaft toward their valve lifters.

3. Secure the exhaust camshaft sub-gear to the main gear using a service bolt. When removing the exhaust camshaft be sure the torsional spring force of the sub-gear has been eliminated.

4. Remove the No. 1 and No. 2 rear bearing cap bolts and remove the cap. Uniformly loosen and remove bearing cap bolts No. 3 to No. 8 in several passes and in the proper sequence. Do not remove bearing cap bolts No. 9 and 10 at this time. Remove the No. 1, 2 and 4 bearing caps.

5. Alternately loosen and remove bearing cap bolts No. 9 and 10. As these bolts are loosened check to see that the camshaft is being lifted out straight and level.

NOTE: If the camshaft is not lifted out straight and level retighten No. 9 and 10 bearing cap bolts. Reverse Steps 4 through 1, than start over from Step 3. Do not attempt to pry the camshaft from its mounting.

6. Remove the exhaust camshaft from the engine.

7. To remove the intake camshaft, set the knock pin of the intake camshaft at 80–115 degrees BTDC of camshaft angle. This angle will help to lift the intake camshaft level and evenly by pushing No. 1 and No. 3 cylinder camshaft lobes of the intake camshaft toward their valve lifters.

8. Remove the No. 1 and No. 2 front bearing cap bolts and remove the front bearing cap and oil seal. If the cap will not come apart easily, leave it in place without the bolts.

9. Uniformly loosen and remove bearing cap bolts No. 3 to No. 8 in several phases and in the proper sequence. Do not remove bearing cap bolts No. 9 and 10 at this time. Remove No. 1, 3 and 4 bearing caps.

10. Alternately loosen and remove bearing cap bolts No. 9 and 10. As these bolts are loosened and after breaking the adhesion on the front bearing cap, check to see that the camshaft is being lifted out straight and level.

NOTE: If the camshaft is not lifted out straight and level retighten No. 9 and 10 bearing cap bolts. Reverse Steps 10 through 7, than start over from Step 8. Do not attempt to pry the camshaft from its mounting.

11. Remove the intake camshaft from the engine.

To install:

12. Before installing the intake camshaft, apply multi-purpose grease to the thrust portion of the camshaft. Position the camshaft at 80 degrees BTDC of camshaft angle on the cylinder head. Apply seal packing kit 08826–00080 or equivalent, and apply it to the front bearing cap. Coat the bearing cap bolts with clean engine oil. Uniformly and in several phases tighten the camshaft bearing caps to 14 ft. lbs. (19 Nm).

13. To install the exhaust camshaft, set the knock pin of the camshaft at 10 degrees BTDC of camshaft angle. Apply multi-purpose grease to the thrust portion of the camshaft. Position the exhaust camshaft gear with the intake camshaft gear so the timing marks are in alignment with one another. Be sure to use the proper alignment marks on the gears. Do not use the assembly reference marks.

14. Turn the intake camshaft clockwise or counterclockwise little by little until the exhaust camshaft sits in the bearing journals evenly without rocking the camshaft on the bearing journals.

15. Coat the bearing cap bolts with clean engine oil. Uniformly and in several phases tighten the camshaft bearing caps to 14 ft. lbs. (19 Nm). Remove the service bolt from the assembly.

16. Installation of the remaining components is the reverse of the removal procedure.

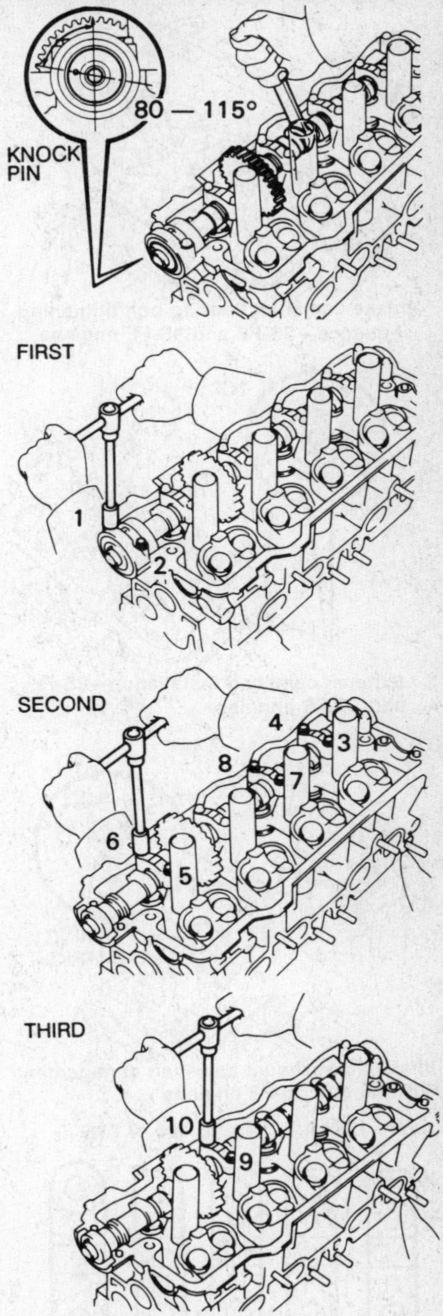

Intake camshaft removal procedure 3S-FE and 5S-FE engines

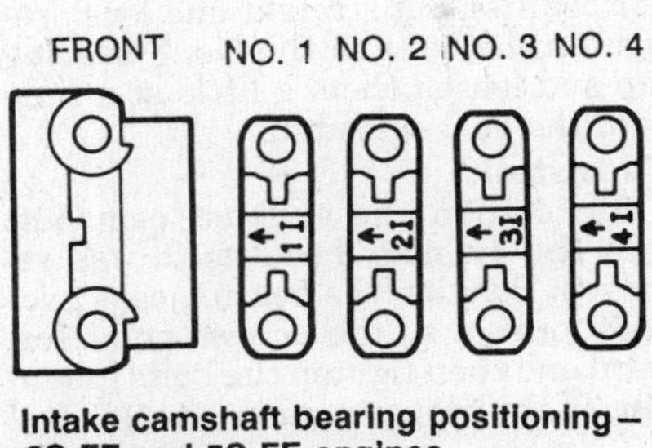

Intake camshaft bearing positioning–3S-FE and 5S-FE engines

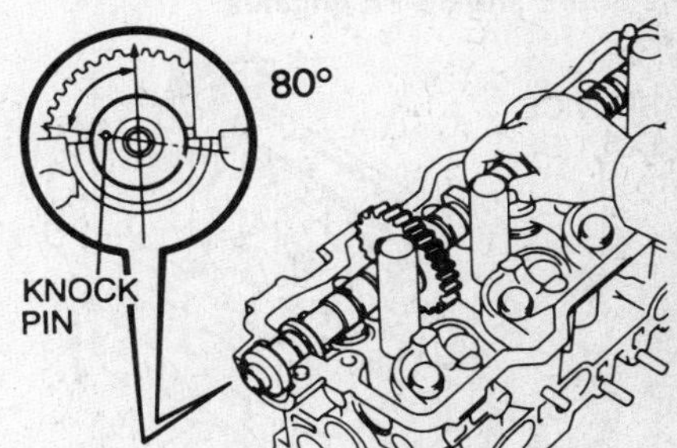

Installing the intake camshaft–3S-FE and 5S-FE engines

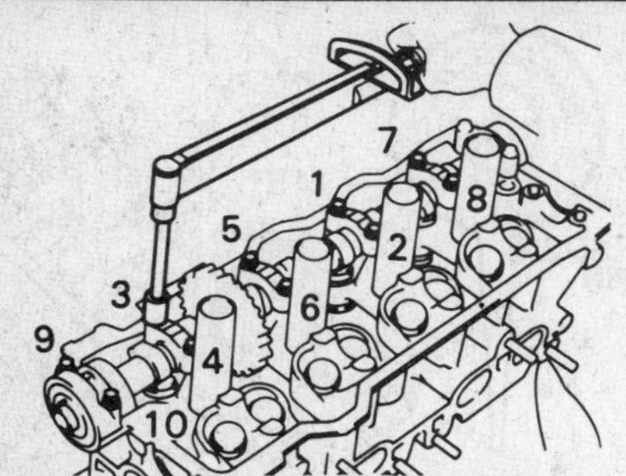

Intake camshaft bearing bolt tightening sequence – 3S-FE and 5S-FE engines

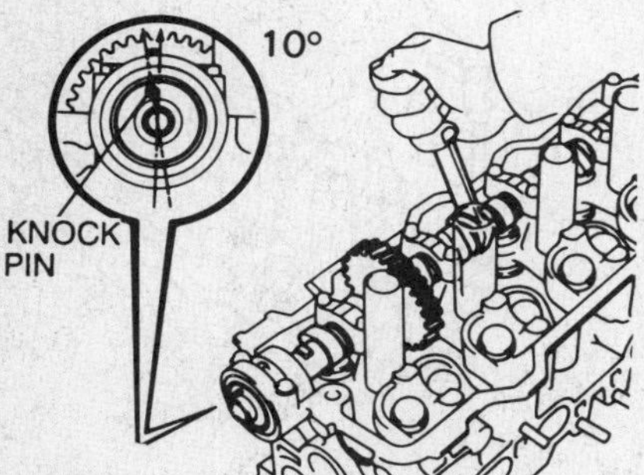

Exhaust camshaft installation – 3S-FE and 5S-FE engines

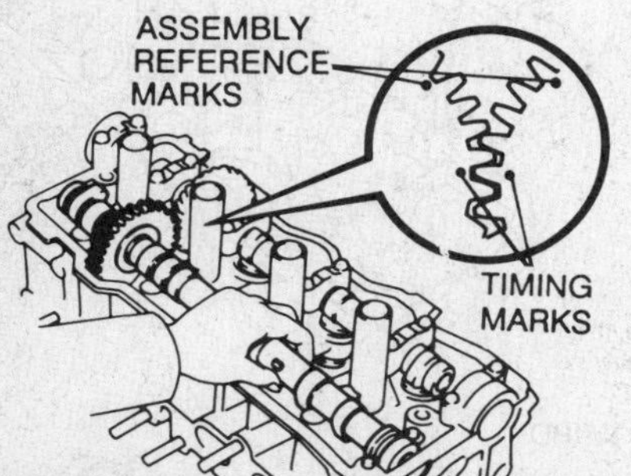

Intake and exhaust camshaft engagement – 3S-FE and 5S-FE engines

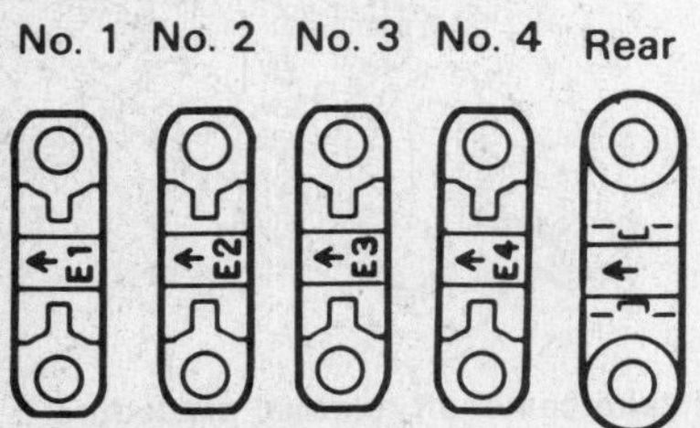

Exhaust camshaft bearing cap positioning – 3S-FE and 5S-FE engines

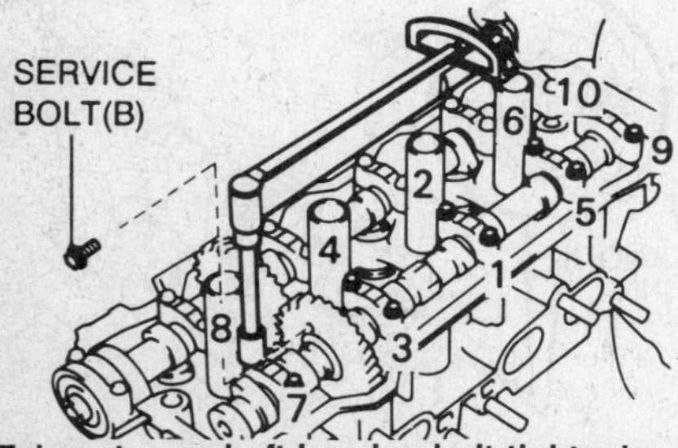

Exhaust camshaft bearing bolt tightening sequence – 3S-FE and 5S-FE engines

4A-F and 4A-FE Engines

1. Disconnect the negative battery cable. Drain the cooling system.
2. Remove the spark plugs and the cylinder head cover.
3. Remove the No. 3 and No. 2 front covers. Turn the crankshaft pulley and align its groove with the **0** mark on the No. 1 front cover. Check that the camshaft pulley hole aligns with the mark on the No. 1 camshaft bearing cap (exhaust side).
4. Remove the plug from the No. 1 front cover and matchmark the timing belt to the camshaft pulley. Loosen the idler pulley mounting bolt and push the pulley to the left as far as it will go; tighten the bolt. Slide the timing belt off the camshaft pulley and support it so it won't fall into the case.
5. Remove the camshaft pulley and check the camshaft thrust clearance. Remove the camshafts.

NOTE: Due to the relatively small amount of camshaft thrust clearance, the camshaft must be held level during removal. If the camshaft is not level on removal, the portion of the head receiving the thrust may crack or be damaged.

6. Set the service bolt hole on the intake camshaft gear (the one not attached to the timing pulley) at the 12 o'clock position so the Nos. 1 and 3 cylinder camshaft lobed can push their lifters evenly. Loosen the No. 1 bearing caps on each camshaft a little at a time and remove them.
7. Secure the intake camshaft subgear to the main gear with a service bolt to eliminate any torsional spring force. Loosen the remaining bearing caps a little at a time, in the proper sequence and remove the intake camshaft. If the camshaft cannot be lifted out straight and level, retighten the bolts in the No. 3 bearing cap and loosen them a little at a time with the gear pulled up.
8. Turn the exhaust camshaft approximately 105 degrees so the knock pin is about 5 minutes before the 6:30 o'clock position. Loosen the remaining bearing caps a little at a time, in the proper sequence and remove the exhaust camshaft. If the camshaft can not be lifted out straight and level, retighten the bolts in the No. 3 bearing cap and loosen them a little at a time with the gear pulled up.

To install:

9. Position the exhaust camshaft into the cylinder head as it was removed. Position the bearing caps over each journal so the arrows point forward and then tighten the bolts gradually, in the proper sequence to 9 ft. lbs. (13 Nm).
10. Coat the lip of a new oil seal with MP grease and drive it into the camshaft.
11. Set the knock pin on the exhaust camshaft so it is just above the edge of the cylinder head and engage the intake camshaft gear to the exhaust gear so the mark on each gear is in alignment. Roll the intake camshaft down onto the bearing journals while engaging the gears with each other.
12. Position the bearing caps over each journal on the intake camshaft so the arrows point forward and then tighten the bolts gradually, in the proper sequence to 9 ft. lbs. (13 Nm).
13. Remove the service bolt and install the No. 1 intake bearing cap. If it does not fit properly, pry the camshaft gear backwards until it does. Tighten the bolts to 9 ft. lbs. (13 Nm).
14. Rotate the camshafts 1 revolution (360 degrees) from TDC to TDC and check that the marks on the 2 gears are still aligned.

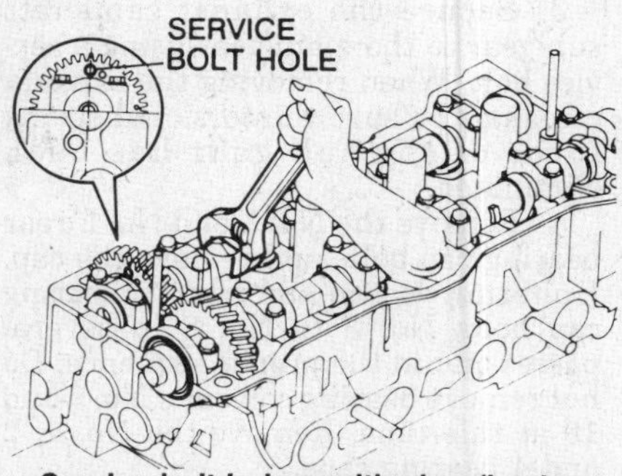

Service bolt hole positioning (intake camshaft) – 4A-F and 4A-FE engines

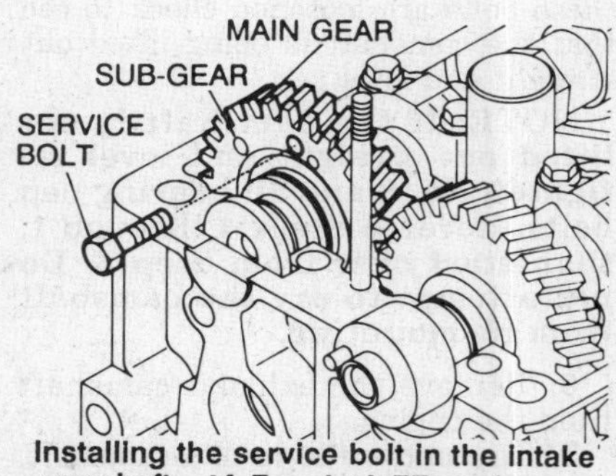

Installing the service bolt in the intake camshaft – 4A-F and 4A-FE engines

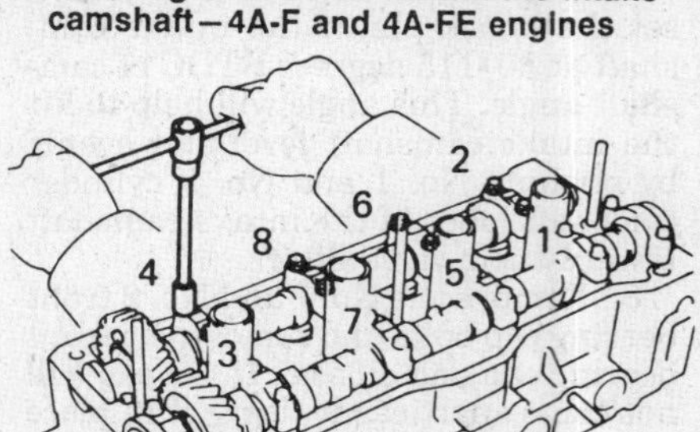

Intake camshaft bearing cap loosening sequence – 4A-FE engine (1990–92)

15. Install the camshaft timing pulley making sure the camshaft knock pins and the matchmarks are in alignment. Lock each camshaft and tighten the pulley bolts to 43 ft. lbs. (59 Nm).

16. Align the matchmarks made during removal and then install the timing belt on the camshaft pulley. Loosen the idler pulley set bolt. Make sure the timing belt meshing at the crankshaft pulley does not shift.

17. Rotate the crankshaft clockwise 2 revolutions from TDC to TDC. Make sure each pulley aligns with the marks made previously.

18. Tighten the set bolt on the timing belt idler pulley to 27 ft. lbs. (37 Nm). Measure the timing belt deflection at the top span between the 2 camshaft pulleys. It should deflect no more than 0.24 in. (7mm) at 4.4 lbs. of pressure. If deflection is greater, readjust by using the idler pulley.

19. Installation of the remaining components is the reverse of the removal procedure.

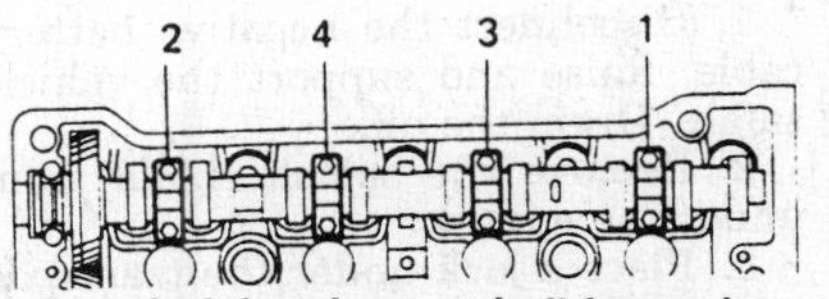

Camshaft bearing cap bolt loosening sequence – 4A-F and 4A-FE engines (1988–89)

Exhaust camshaft bearing cap loosening sequence – 4A-FE engine (1990–92)

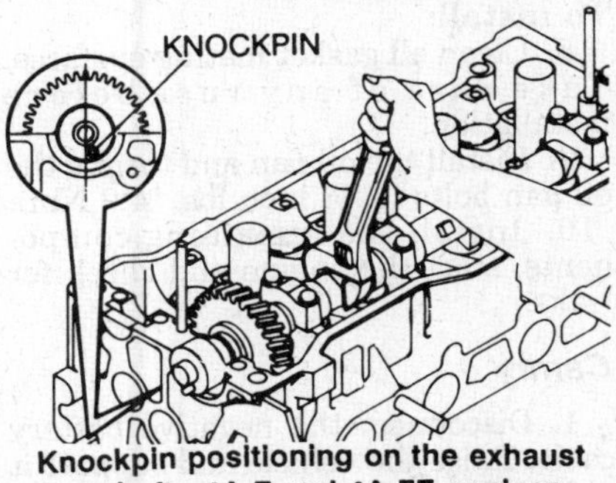

Knockpin positioning on the exhaust camshaft – 4A-F and 4A-FE engines

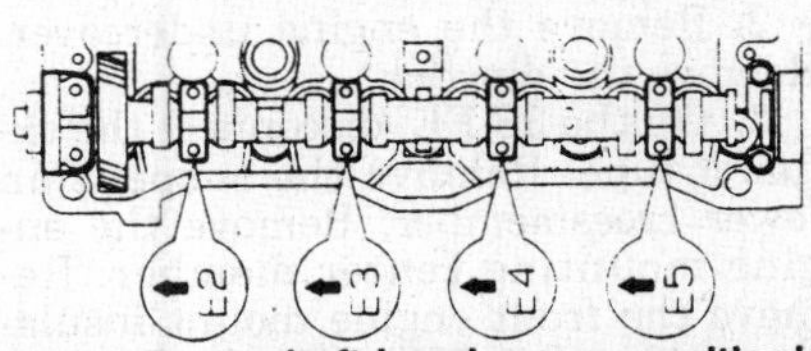

Exhaust camshaft bearing cap positioning – 4A-F and 4A-FE engines

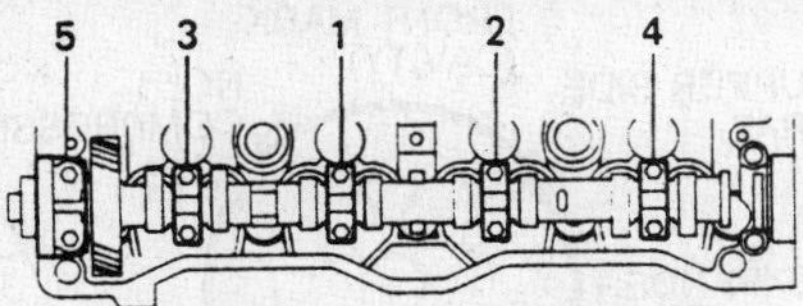

Camshaft bearing cap bolt tightening sequence – 4A-F and 4A-FE engines (1988–89)

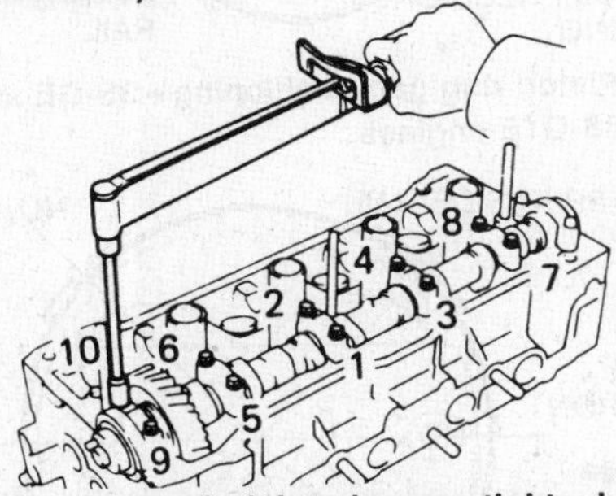

Exhaust camshaft bearing cap tightening sequence – 4A-FE engine (1990–92)

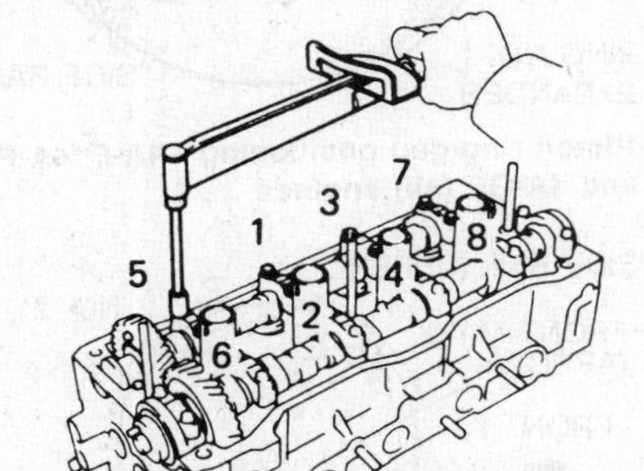

Intake camshaft bearing cap tightening sequence – 4A-FE engine (1990–92)

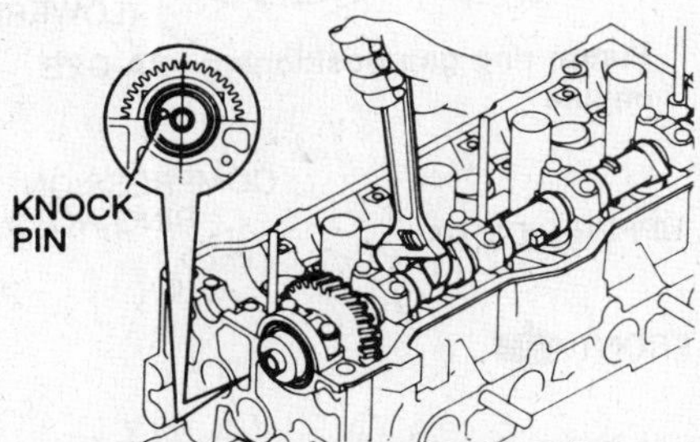

Turn the exhaust camshaft until the knockpin is here – 4A-F and 4A-FE engines

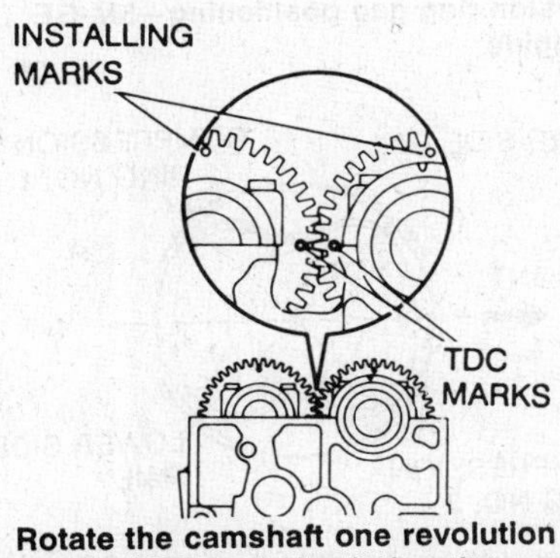

Rotate the camshaft one revolution from TDC to TDC and check that the marks are lined up – 4A-F and 4A-FE engines

Intake camshaft bearing cap positioning – 4A-F and 4A-FE engines

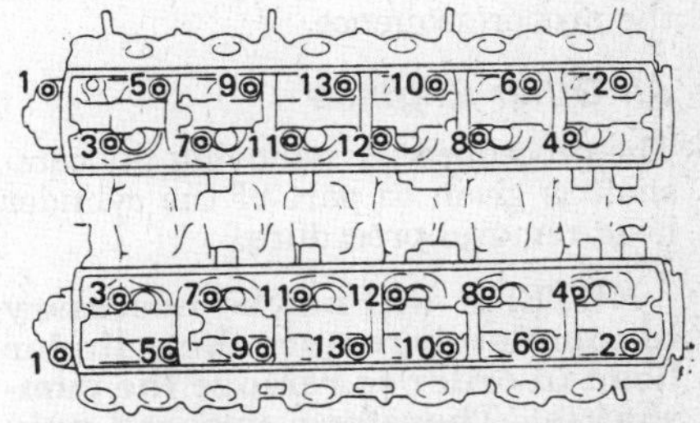

5M-FE engine camshaft housing bolt removal sequence. Loosen bolts gradually on three passes

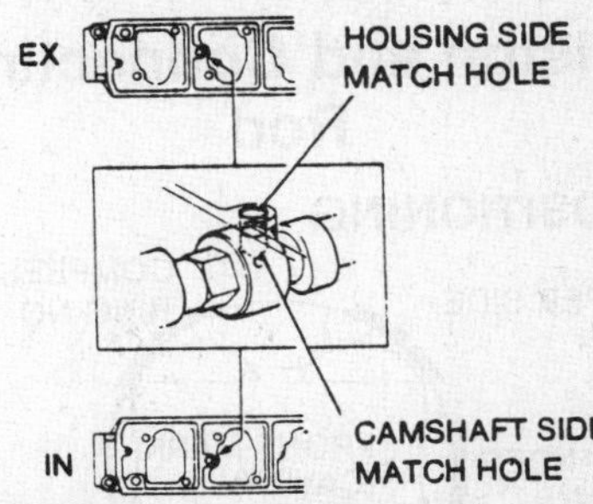

5M-GE engine camshaft housing torque sequence

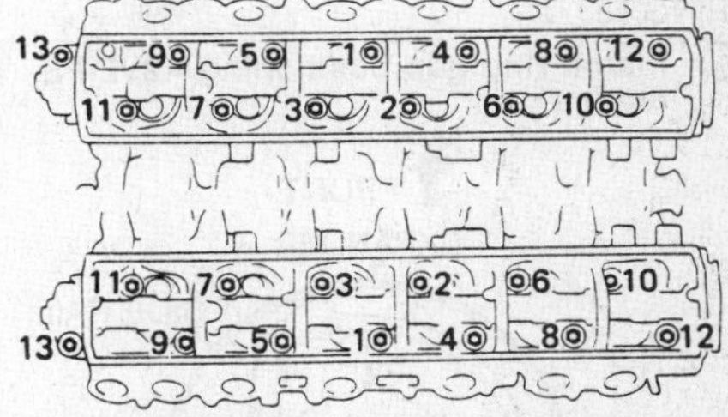

Before installing the camshaft housings, align the match hole on each No. 2 cam journal with the hole in the housing

5M-GE Engine

1. Remove the 2 camshaft covers.

2. Remove the timing belt assembly.

3. Following the sequence shown, loosen the camshaft housing nuts and bolts in 3 passes. Remove the housings (with camshafts) from the cylinder head.

4. Remove the camshaft housing rear covers. Squirt clean oil down around the cam journals in the housing, to lubricate the lobes, oil seals and bearings as the cam is removed. Begin to pull the camshaft out of the back of the housing slowly, turning and pulling. Remove the cam completely.

5. To install, lubricate the entire camshaft with clean oil. Insert the cam into the housing from the back, and slowly turn it and push it into the housing. Install new O-rings and the housing end covers.

6. Installation of the remaining components is in the reverse order of removal. Tighten camshaft housing bolts to 15–17 ft. lbs. (20–23 Nm) in the proper sequence.

All Other Engines

The procedure for removing the camshaft is given as part of the cylinder head removal procedure.

NOTE: It will not be necessary to completely remove the cylinder head in order to remove the camshaft(s). Therefore, proceed only as far as necessary, to remove the camshaft, with the cylinder head removal procedure.

Piston and Connecting Rod

POSITIONING

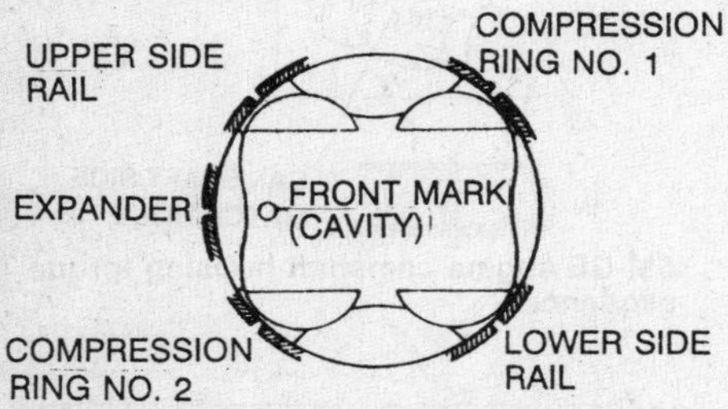

Piston ring gap positioning – 2VZ-FE engine

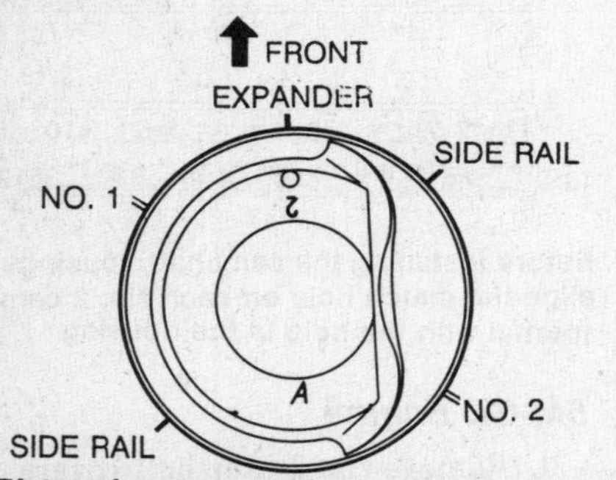

Piston ring gap positioning – 3A-C, 4A-C and 4A-LC engines

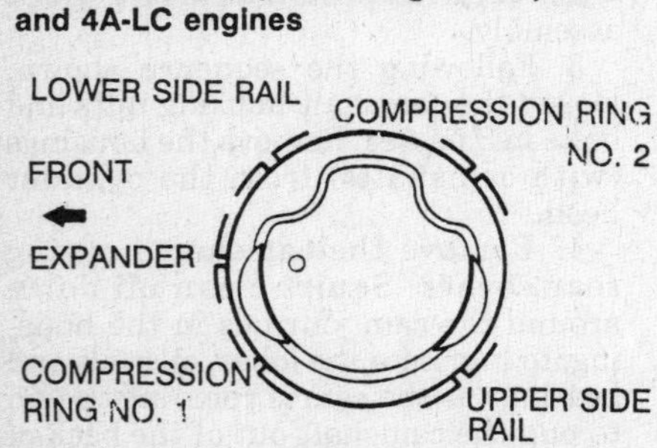

Piston ring gap positioning – 3E and 3E-E engines

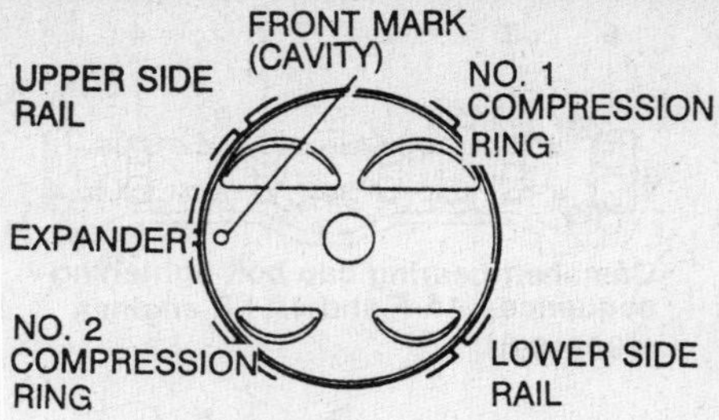

Piston ring gap positioning – 3S-GE and 3S-GTE engines

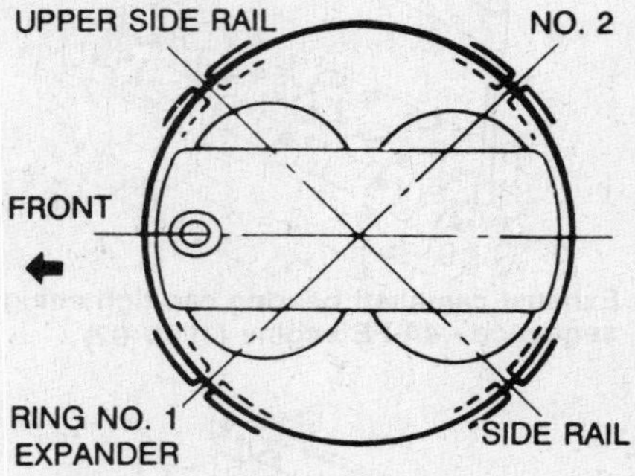

Piston ring gap positioning – 4A-F, 4A-FE and 4A-GE (all) engines

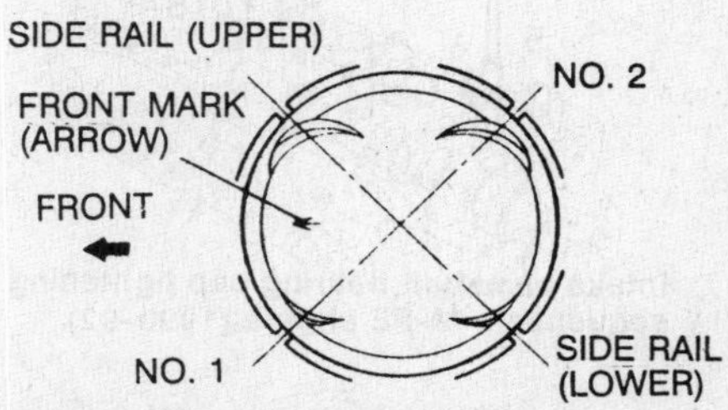

Piston ring gap positioning – 4A-GZE engine

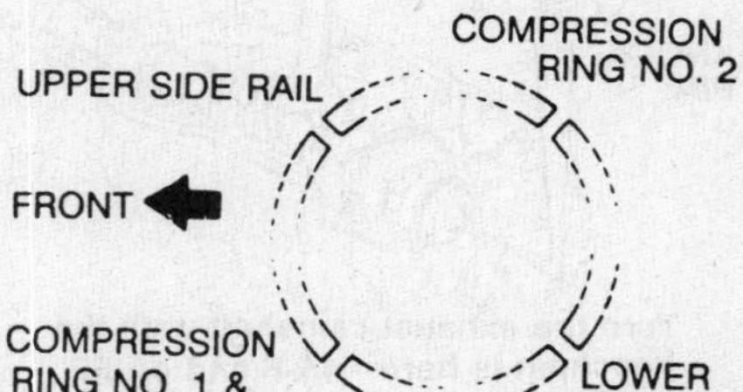

Piston ring gap positioning – 5M-GE engine

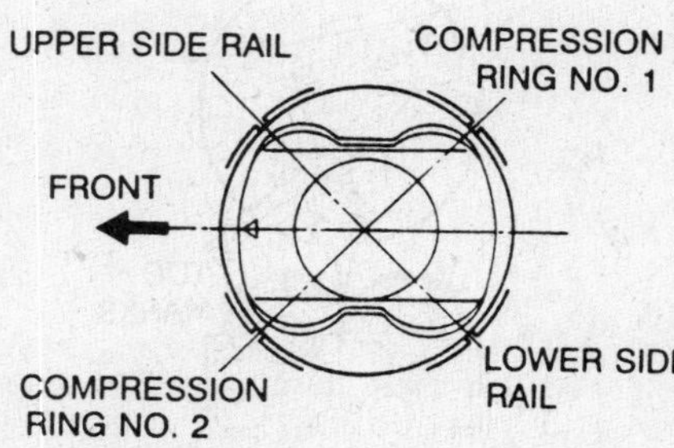

Piston ring gap positioning – 7M-GE and 7M-GTE engines

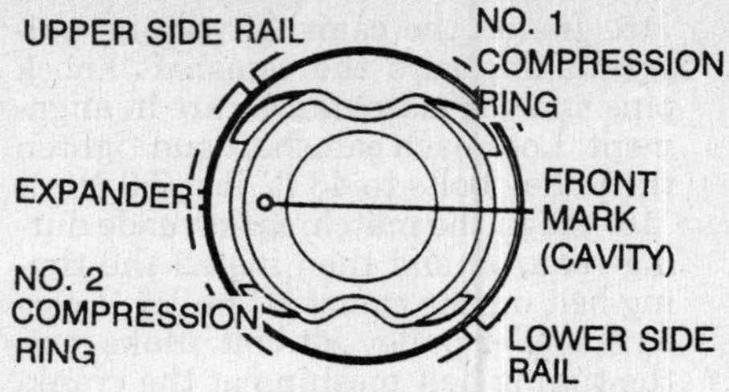

Piston ring gap positioning – 3S-FE and 5S-FE engines

ENGINE LUBRICATION

Oil Pan

REMOVAL & INSTALLATION

Corolla

1. Disconnect the negative battery cable. Raise and support the vehicle safely. Drain the oil.
2. Remove the splash shield from under the engine.
3. Place a jack under the transaxle to support it.
4. Remove the bolts which secure the engine rear supporting crossmember to the chassis. On the 4A-GE, remove the center mounting and stiffener plate.
5. Raise the jack under the transaxle, slightly.
6. Remove the front exhaust pipe.
7. Remove the oil pan retaining bolts. Remove the oil pan from the vehicle. If the oil pan does not come out easily, it may be necessary to unbolt the rear engine mounts from the crossmember.

To install:

8. Clean all gasket mating surfaces. Take care of any rust before installation.
9. Install the oil pan and torque the oil pan bolts to 43 inch lbs. (4.9 Nm).
10. Install the remaining components, start the engine and check for leaks.

Camry

1. Disconnect the negative battery cable. Raise the vehicle and support it safely. Drain the oil.
2. Remove the engine undercover. Remove the dipstick.
3. On the 3S-FE, disconnect the exhaust pipe. Remove the suspension lower crossmember. Remove the engine mounting center member. Remove the front engine mount insulator and bracket; 2VZ-FE only. Remove the stiffener plate.

4. Remove the oil pan retaining bolts. Remove the oil pan.

To install:

5. Clean the gasket mating surfaces. Always use a new pan gasket. Some engines were assembled using RTV gasket material in place of a conventional gasket. In that case, apply a thin (5mm) bead of RTV material to the groove around the pan mating surface.

6. Assemble the pan within 15 minutes. Torque pan bolts to 48 inch lbs. (5.4 Nm). On the 2VZ-FE, tighten the pan bolts to 52 inch lbs. (5.9 Nm).

Cressida, Celica and Supra

3S-FE, 5S-FE, 3S-GE AND 3S-GTE ENGINES

1. Disconnect the negative battery cable. Raise the vehicle and support it safely. Drain the engine oil.

2. Remove the engine undercovers.

3. On the 3S-GE and 5S-FE, disconnect the exhaust pipe from the exhaust manifold.

4. Remove the lower suspension crossmember. Remove the engine mounting center member.

5. Remove the engine stiffener plate and the oil level gauge.

6. Remove the oil pan retaining bolts. Remove the oil pan.

To install:

7. Clean all gasket mating surfaces. Take care of any rust before installation..

8. Apply a 5mm bead of RTV gasket material to the groove around the pan flange. Apply the oil within 3 minutes of application.

9. Install the oil pan and torque the oil pan bolts to 48 inch lbs. (5.4 Nm).

10. Install the remaining components, start the engine and check for leaks.

5M-GE ENGINE

1. Disconnect the negative battery cable. Raise the vehicle and support it safely. Drain the oil and cooling system.

2. Remove the air cleaner assembly. Mark any disconnected lines and/or hoses for easy reassembly. Remove the oil level gauge.

3. Disconnect the upper radiator hose at the radiator. Loosen the drive belts.

4. Remove the fan shroud bolts. Remove the 4 fluid coupling flange attaching nuts, then remove the fluid coupling along with the fan and the fan shroud.

5. Remove the engine undercover. Remove the exhaust pipe clamp bolt from the exhaust pipe stay.

6. Remove the 2 stiffener plates from the exhaust pipe. If equipped with manual transaxle, remove the clutch housing undercover.

7. Remove the 4 engine mount bolts from each side of the engine.

8. Place a jack under the transaxle and raise the engine about 1¾ in.

9. Remove the oil pan retaining bolts. Remove the oil pan from the engine.

To install:

10. Clean all gasket mating surfaces. Take care of any rust before installation.

11. Use a new oil pan gasket during installation. Apply a small amount of sealer to the oil pan gasket at each of the 4 corners of the oil pan.

12. Torque the oil pan fasteners to 57–82 inch lbs. (8–11 Nm).

13. Install the remaining components, start the engine and check for leaks.

Supra

7M-GE AND 7M-GTE ENGINES

1. Disconnect the negative battery cable. Remove the hood.

2. Raise the vehicle and support it safely. Remove the engine under cover. Drain the engine oil.

3. If equipped with automatic transmission, remove the fluid cooler hose clamp.

4. Remove the No. 1 front suspension crossmember. Remove the front exhaust pipe bracket and stiffener plates.

5. On the 7M-GTE disconnect the engine oil cooler hose from the engine oil pan.

6. Remove the brake hose brackets and clips. Disconnect the intermediate shaft. Disconnect the stabilizer bar links from the lower control arms.

7. Properly support the engine assembly. Remove the engine mounting bolts. Remove the TEMS actuator assembly.

8. Remove the shock absorbers from the body. Disconnect the front suspension member.

9. Remove the oil pan retaining bolts. Remove the oil pan from the engine.

To install:

10. Clean all gasket mating surfaces. Take care of any rust before installation.

11. Install the oil pan and torque the oil pan bolts to 9 ft. lbs. (13 Nm).

12. Install the remaining components, start the engine and check for leaks.

Cressida

7M-GE ENGINE

1. Disconnect the negative battery cable and drain the cooling system.

2. Raise and safely support the vehicle. Remove the engine under cover and drain the oil.

3. Disconnect the front exhaust pipe at the manifold and at the main tube and remove it.

4. Disconnect the automatic transmission oil cooler pipe.

5. Remove the 9 bolts, ground strap, exhaust pipe stay and the engine rear end plate and then remove the stiffener plates.

6. Loosen the bolt and disconnect the intermediate shaft.

7. Disconnect the front suspension crossmember at the front engine mounts. Position a floor jack under the crossmember, remove the remaining mounting bolts and then lower the crossmember.

8. Remove the pan retaining bolts and then carefully pry the pan from the cylinder block.

To install:

9. Clean all gasket mating surfaces. Take care of any rust before installation.

10. Install the oil pan and torque the oil pan bolts to 9 ft. lbs. (13 Nm).

11. Install the remaining components, start the engine and check for leaks.

Tercel

3A-C ENGINE

1. Disconnect the negative battery cable. Drain the cooling system. Remove the radiator.

2. Raise the vehicle and support it safely. Drain the engine oil.

3. Remove the engine under cover. Remove the stabilizer bracket bolts and lower the stabilizer assembly. Remove the right and left stiffener plates.

4. Remove the oil pan retaining bolts. Remove the oil pan from the vehicle.

To install:

5. Clean all gasket mating surfaces. Take care of any rust before installation.

6. Install the oil pan and torque the oil pan bolts to 43 inch lbs. (4.9 Nm).

7. Install the remaining components, start the engine and check for leaks.

3E AND 3E-E ENGINES

1. Disconnect the negative battery terminal. Raise the vehicle and support it safely. Drain the oil.

2. Remove the right engine under cover. Remove the sway bar and any other necessary steering linkage parts.

3. Disconnect the exhaust pipe from the manifold. Raise the engine enough to take the weight off it.

4. Remove the timing belt.

5. Continue to raise the engine enough to remove the oil pan. Remove the oil pan retaining bolts. Remove the oil pan.

To install:

6. Clean all gasket mating surfaces.

Take care of any rust before installation.

7. Install the oil pan and torque the oil pan bolts to 43 inch lbs. (4.9 Nm).

8. Install the remaining components, start the engine and check for leaks.

MR2

4A-GELC AND 4A-GZE ENGINES

1. Disconnect the negative battery cable. Raise and support the vehicle safely. Drain the engine oil.

2. Remove the exhaust manifold pipe. Remove the timing belt. Remove the crankshaft timing pulley.

3. Support the weight of the engine with a floor jack and then remove the right side engine mount.

4. Remove the oil pan retaining bolts. Remove the oil pan.

To install:

5. Clean all gasket mating surfaces. Take care of any rust before installation.

6. Apply a 5mm bead of RTV gasket material to the groove around the pan flange. Apply the oil pan within 3 minutes of application and tighten the mounting bolts and nuts to 43 inch lbs. (6 Nm).

7. Install the remaining components, start the engine and check for leaks.

3S-GTE AND 5S-FE ENGINES

1. Disconnect the negative battery cable.

2. Drain the engine oil and remove the engine under covers.

3. Remove the right engine hood side panel.

4. Remove the brace that runs across the struts.

5. If equipped with cruise control, remove the cruise control actuator assembly and disconnect the accelerator linkage.

6. Remove the front exhaust pipe.

7. On 3S-GTE with air conditioning, unbolt the compressor and move it aside. Leave the refrigerant lines connected. On 5S-FE, remove the air conditioner idler pulley.

8. On 3S-GTE, remove the catalytic converter and the intercooler.

9. Remove the stiffener plate.

10. On 3S-GTE, disconnect the turbocharger outlet hose where it connects to the oil pan.

11. Remove the dipstick.

12. Remove the oil pan 17 bolts and 2 nuts that attach the oil pan to the block.

13. Insert a suitable seal cutting tool between the oil pan and the block. Work the tool around the pan to break the sealant. Remove the oil pan.

To install:

14. Clean all gasket mating surfaces. Take care of any rust before installation.

15. Apply a 5mm bead of RTV gasket material to the groove around the pan flange. Apply the oil pan within 5 minutes of application and tighten the mounting bolts and nuts to 43 inch lbs. (6 Nm).

16. Install the remaining components, start the engine and check for leaks.

Oil Pump

REMOVAL & INSTALLATION

2VZ-FE, 3E, 3E-E and 3S-FE Engines

1. Remove the oil pan. Remove the oil strainer. On the 3E, remove the dipstick.

2. Raise the engine using a chain hoist. Remove the timing belt and pulleys.

3. On the 2VZ-FE, remove the alternator and the air conditioning compressor and bracket. Do not disconnect the refrigerant lines.

4. Remove the oil pump from the engine.

5. Installation is the reverse of the removal procedure. Clean the gasket mating surfaces. Pack the oil pump cavities with petroleum jelly. Start the engine and check for oil pressure.

3A-C, 3S-GE, 3S-GTE (Except MR2), 4A-F and 4A-GE Engines

1. Remove the fan shroud. Raise and support the vehicle safely.

2. Drain the oil. On the Tercel, drain the coolant and remove the radiator.

3. Remove the oil pan and the oil strainer. Remove the oil pan baffle plate on the 4A-GE. Remove the crankshaft pulley and the timing belt. Remove the oil dipstick guide and then the dipstick.

4. Remove the mounting bolts and then use a rubber mallet to carefully tap the oil pump body from the cylinder block.

To install:

5. To install, position a new gasket on the cylinder block.

6. Position the oil pump on the block so the teeth on the pump drive gear are engaged with the teeth of the crankshaft gear.

7. Clean the gasket mating surfaces. Pack the oil pump cavities with petroleum jelly. Start the engine and check for oil pressure.

8. Installation of the remaining components is the reverse of the removal procedure.

3S-GTE (1990–92 MR2) and 5S-FE (1990–92 Celica and MR2) Engines

1. Disconnect the negative battery cable.

2. Drain the engine oil.

3. Remove the oil pan.

4. Remove the oil pump strainer and baffle plate.

5. Connect a suitable lifting device to the engine and raise the engine a small amount.

6. Remove the timing belt.

7. Remove the No. 2 idler pulley, crankshaft timing pulley and oil pump pulley.

8. Remove the oil pump retaining bolts.

9. Remove the oil pump and gasket by carefully tapping on the outside of the oil pump body. Discard the gasket.

NOTE: One of the oil pump bolts is longer than the rest. Make sure this bolt is identified so it may be installed in the original location.

To install:

10. Clean the gasket mating surfaces. Pack the oil pump cavities with petroleum jelly.

11. Use a new oil pump gasket. On 3S-GTE, torque the oil pump bolts to 69 inch lbs. (8 Nm). On 5S-FE, torque the bolts to 82 inch lbs. (9.3 Nm).

12. Installation of the remaining components, start the engine and check for leaks.

All Other Engines

1. Remove the oil pan.

2. Unbolt the oil pump retaining bolts. Remove the oil pump from the engine.

To install:

3. Clean the gasket mating surfaces. Pack the oil pump cavities with petroleum jelly. Start the engine and check for oil pressure.

Rear Main Bearing Oil Seal

REMOVAL & INSTALLATION

NOTE: The 3A-C engine must be removed from the vehicle before this procedure can be attempted.

1. Remove the transmission or transaxle.

2. Remove the clutch cover assembly and flywheel.

3. Remove the oil seal retaining plate, complete with the oil seal.

4. Using a suitable tool pry the old seal from the retaining plate. Be careful not to damage the plate.

5. Install the new seal, carefully, by using a block of wood to drift it into place. Do not damage the seal as a leak will result.

6. Lubricate the lips of the seal with multipurpose grease. Installation is the reverse of removal.

ENGINE COOLING

Radiator

REMOVAL & INSTALLATION

1. Disconnect the negative battery cable.
2. Drain the cooling system.
3. On 1990–92 MR2, remove the front under covers.
4. Remove the radiator hoses.
5. If equipped with an automatic transmission or transaxle, disconnect and plug the oil cooler lines.
6. Remove the ignition coil, ignitor and bracket assembly on the 2VZ-FE.
7. Remove the hood lock from the radiator upper support, as required. It may be necessary to remove the grille in order to gain access to the hood lock/radiator support assembly.
8. Remove the fan shroud, as required. If equipped with an electric fan (2 on the MR2), disconnect the wiring harness and thermo-switch connectors.
9. Disconnect the overflow hose from the thermal expansion tank and remove the tank from its bracket.
10. Unbolt and remove the radiator upper support.
11. Remove the radiator retaining bolts. Raise the radiator and cooling fan(s) from the lower supports and remove from vehicle.

To install:

12. Lower the radiator and cooling fan(s) onto the lower supports and install the retaining bolts.
13. Install the radiator upper support.
14. Mount the thermal expansion tank and connect the overflow hose.
15. Connect the cooling fan wiring harnesses and thermo-switch connectors. Install the fan shroud, if removed.
16. Install the hood lock and grille, if removed.
17. On 2VZ-FE, install the ignition coil, igniter and bracket assembly.
18. If equipped automatic transmission or transaxle, connect the oil cooler lines.
19. Install the radiator hoses.
20. On 1990–92 MR2, install the front under covers.
21. Fill the cooling system to the proper level.
22. Connect the negative battery cable. Start the engine and check for leaks.

Heater Core

NOTE: On some vehicles, the air conditioning assembly is integral with the heater assembly (including the heater core) and therefore the heater core removal may differ from the procedures detailed below. In some case it may be necessary to remove the air conditioning/heater housing and assembly to remove the heater core. Due to the lack of information available at the time of this publication, a general heater core removal and installation procedure is outlined for each vehicle. The removal steps can be altered as required.

REMOVAL & INSTALLATION

Tercel

1988–90

1. Disconnect the negative battery terminal.
2. Drain the radiator.
3. Remove the ash tray and retainer.
4. Remove the rear heater duct (optional).
5. Remove the left and right side defroster ducts.
6. Remove the under tray (optional).
7. Remove the glove box.
8. Remove the main air duct.
9. Disconnect the radio and remove it.
10. Disconnect the heater control cables and remove them. Mark each cable with the control lever that it connects to.
11. Disconnect the heater hoses.
12. Remove the front and rear air ducts.
13. Disconnect the electrical connectors and vacuum hoses going to the heater unit.
14. Remove the heater bolts and remove the heater. Slide the heater to the right side of vehicle to remove it.
15. Remove the heater core.

To install:

16. Install the heater core into the housing.
17. Fill the cooling system to the proper level. Operate the heater and check for leaks.
18. Install the ramaining components, start the engine and check for proper operation.

1991–92

1. Disconnect the negative battery cable and drain the cooling system.
2. Remove the safety pad from the instrument panel.
3. If equipped with air conditioning, recover the refrigerant from the air conditioning system.
4. Disconnect and cap the evaporator hoses.
5. Remove the instrument lower finish panel and disconnect the harness from the heater and air conditioning assemblies.
6. Remove the air conditioning amplifier.
7. Remove the evaporator housing by removing the 3 screws.
8. Disconnect the heater hoses from the heater core.
9. Remove the heater/air conditioning control assembly.
10. Remove the heater register center duct, instrument panel reinforcements and remove the heater assembly.
11. Remove the screws and clips from the heater assembly case halves. Remove the core from the case.

To install:

12. Install the heater core into the case and install the retainers.
13. Install the heater case assembly and instrument panel components.
14. Install the air conditioning assembly.
15. Connect all hoses and harnesses.
16. Evacuate, recharge and leak test the air conditioning system.
17. Refill the cooling system, start the engine and check for leaks.

Corolla

1. Disconnect the negative battery cable.
2. Drain the cooling system.
3. Remove the gear shift knob and console as necessary.
4. Tag and disconnect the vacuum hoses from heater housing assembly.
5. Remove the under tray or package tray from the right side of the vehicle.
6. Release the 2 clamps and remove the blower duct from the right side of the heater housing.
7. Remove any interfering air ducts.
8. Disconnect the 2 water (heater) hoses from the rear of the heater housing.
9. Tag and disconnect all wires and cables leading from the heater housing and position them aside.
10. Remove all mounting bolts and then remove the heater housing carefully toward the rear of the vehicle.
11. Remove the heater housing assembly from the vehicle. Remove any retaining brackets or hardware that may retain the heater core to the heater housing. Grasp the heater core by the end plate and carefully pull it out of the heater housing.

To install:

12. Install the heater core into the heater housing, make sure to clean heater housing of all dirt, leaves, etc. before heater core installation.
13. Fill the cooling system to the proper level. Operate the heater and check for leaks.
14. Install the remaining compo-

nents, start the engine and check for leaks after the cooling system has pressurized.

Camry

1. Drain the cooling system.
2. Remove the console, if equipped, by removing the shift knob (manual), wiring connector, and console attaching screws.
3. Remove the carpeting from the tunnel.
4. If necessary, remove the cigarette lighter and ash tray.
5. Remove the package tray, if it makes access to the heater core difficult.
6. Remove the bottom cover/intake assembly screws and withdraw the assembly.
7. Remove the cover from the water valve.
8. Remove the water valve.
9. Remove the hose clamps and remove the hoses from the core.
10. Remove the heater core.

To install:

11. Install the heater core into the heater housing, make sure to clean heater housing of all dirt, leaves, etc. before heater core installation.
12. Fill the cooling system to the proper level. Operate the heater and check for leaks.
13. Install the remaining components, start the engine and check for leaks after the cooling system has pressurized.

Cressida

1. Disconnect the negative battery cable.
2. Drain the cooling system.
3. Remove the hood release and the fuel lid release levers.
4. Remove the left instrument panel undercover and lower center pad. Remove the finish plate, then remove the radio assembly.
5. Remove the heater control knobs, heater control panel and ashtray.
6. Remove the right side instrument panel undercover, glove box door and glove box.
7. Remove the front pillar garnish, cluster finish panel and instrument cluster gauge assembly.
8. Remove the safety pad and side defroster hose. Remove the heater assembly air ducts.
9. Remove the lower pad reinforcement and remove the front seats. Remove the center console assembly and the cowl side trim panel.
10. Remove the scuff plate, then position the floor carpeting aside. Remove the rear heater duct, if equipped, and heater control assembly.
11. Disconnect the heater hoses from the heater core assembly, remove the heater core grommet.
12. Remove the blower motor duct, center duct and instrument panel brace. Remove the heater core assembly from the vehicle.
13. Remove the nuts securing the heater core to the heater core assembly and remove the heater core.

To install:

14. Install the heater core into the heater housing, make sure to clean heater housing of all dirt, leaves, etc. before heater core installation.
15. Fill the cooling system to the proper level. Operate the heater and check for leaks.
16. Install the remaining components, start the engine and check for leaks after the cooling system has pressurized.

Corolla

1. Disconnect the negative battery cable.
2. Drain the cooling system.
3. Remove the center console, scuff plate and front seats.
4. Position the floor carpet aside and remove the heater duct, if equipped.
5. Remove the under tray, glove box and blower duct.
6. On the Corolla station wagon and sedan vehicles, remove the following components:
   a. Remove the heater control knobs and lens. Remove the cluster lower center panel finish, ashtray and heater control assembly.
   b. Remove the instrument cluster finish panel, radio and air ducts.
7. On the Corolla coupe and liftback vehicles, remove the following components:
   a. The instrument cluster finish panel, instrument cluster, radio trim panel and radio.
   b. Ashtray, heater control knobs, heater control panel, heater control assembly and air duct.
8. Disconnect the heater hoses from the heater core assembly and remove the heater hose grommet.
9. Remove the heater core assembly retaining screws and remove the heater core assembly from the vehicle.
10. Remove the heater core from the heater core assembly.

To install:

11. Install the heater core into the heater housing, make sure to clean heater housing of all dirt, leaves, etc. before heater core installation.
12. Fill the cooling system to the proper level. Operate the heater and check for leaks.
13. Install the remaining components, start the engine and check for leaks after the cooling system has pressurized.

MR2

1988–89

1. Disconnect the negative battery cable.
2. Drain the cooling system.
3. Disconnect the heater hose at the engine compartment.
4. Remove the clips retaining the lower part of the heater unit case, then remove the lower part of the case.
5. Using a suitable tool, carefully pry open the lower part of the heater unit case.
6. Remove the heater core assembly from the heater unit case.
7. Installation is the reverse of the removal procedure. Fill the cooling system to the proper level. Operate the heater and check for leaks.

1990–92

1. Disconnect the negative battery cable.
2. Drain the engine cooling system and discharge the air conditioning system. Plug the open refrigerant lines to prevent contamination.
3. Remove the door scuff plate and kick panels.

CAUTION

Use extreme caution when working around the SRS system. Accidental air bag deployment may occur and cause personal injury. Work must be started after approximately 20 seconds or longer from the time the ignition switch is turned to the Lock position and the negative battery terminal cable is disconnected from the battery. The air bag system is equipped with a backup power source so that if work is started within 20 seconds of disconnecting the battery cable , the air bag may be deployed.

5. Remove the steering wheels and column cover.
6. Remove the rear console box, console upper panel and console box.
7. Remove the instrument panel lower finish panel and backing plate.
8. Remove the heater duct.
9. Remove the combination switch from the steering column and remove the turn signal bracket.
10. Remove the glove box under cover and glove box assembly.
11. Remove the center cluster finish panel.
12. Remove the instrument cluster finish panel by inserting a taped suitable tool under the panel and pry outward.
13. Remove the instrument panel, pull out and disconnect any harnesses.
14. Remove the radio, heater control, ash tray retainer and clock.
15. Remove the side defroster nozzle and bracket.
16. Disconnect the steering column and remove.
17. Remove the instrument panel re-

tainers and carefully remove the panel from the vehicle. That was't so bad, was it?
18. Disconnect the heater hoses.
19. Remove the heater case retainers and remove the assembly from the vehicle.
20. Remove the case half screws and clips. Separate the 2 halves and remove the heater core.

To install:

21. Install the heater core. Make sure the gaskets are in place. Assemble the case.
22. Install the heater assembly and case retainers.
23. Connect the heater hoses. Fill the cooling system and check for leaks before installing the instrument panel.
24. Install the instrument panel and retainers. Be careful not to damage the panel.
25. Connect the steering column.
26. Install the side defroster nozzle and bracket.
27. Install the radio, heater control, ash tray retainer and clock.
28. Install the instrument cluster and connect harnesses.
29. Install the instrument cluster finish panel by snapping into place.
30. Install the center cluster finish panel.
31. Install the glove box under cover and glove box assembly.
32. Install the turn signal bracket and combination switch to the steering column.
33. Install the heater duct.
34. Install the instrument panel lower finish panel and backing plate.
35. Install the rear console box, console upper panel and console box.
36. Install the steering wheel and column cover.
37. Install the door scuff plate and kick panels.
38. Evacuate, recharge and leak test the air conditioning system.
39. Connect the negative battery cable.

Water Pump

REMOVAL & INSTALLATION

1. Disconnect the negative battery cable.

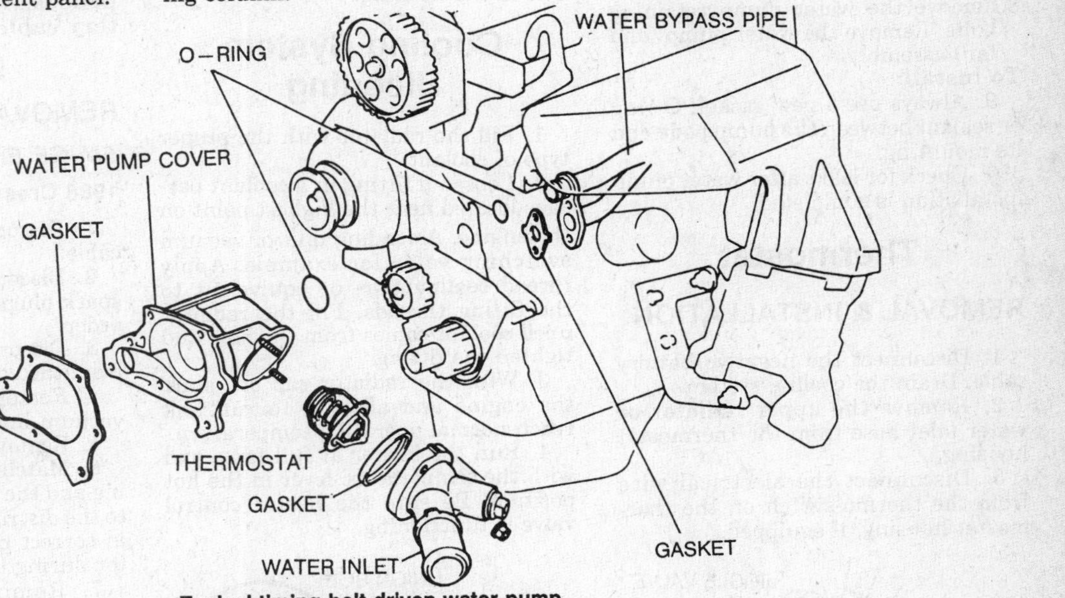

Typical timing belt driven water pump

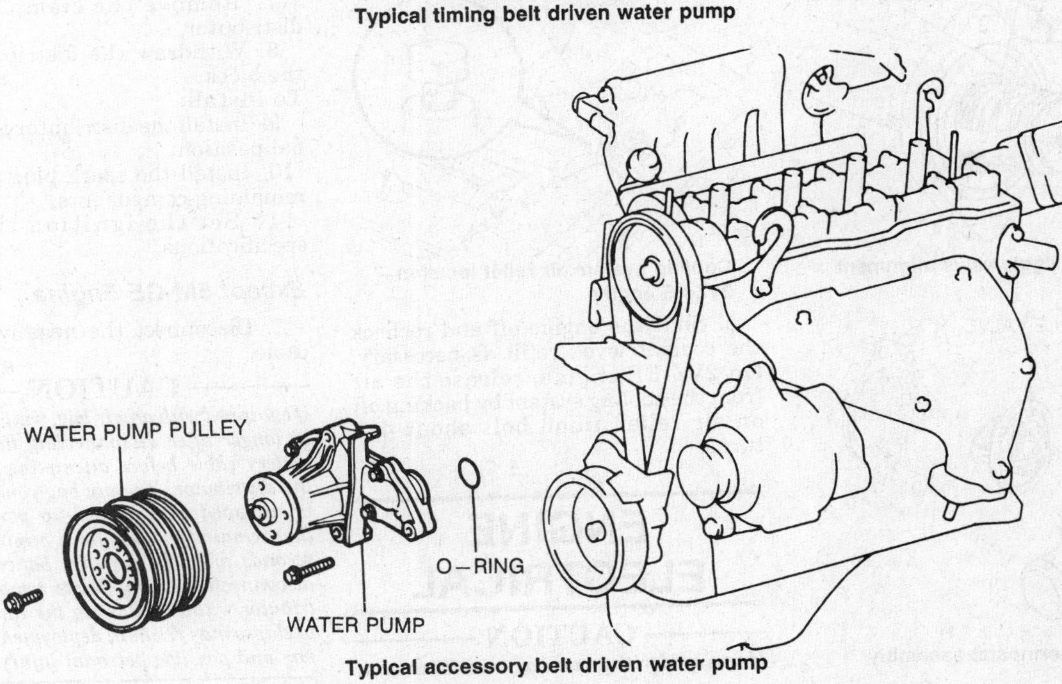

Typical accessory belt driven water pump

2. Drain the cooling system.

2. Remove the fan shroud retaining bolts and remove the fan shroud, if equipped (RWD vehicles). Loosen and remove all drive belts.

3. Remove all necessary components in order to gain access to the water pump retaining bolts.

4. On some vehicles, it will be necessary to remove the timing covers. On the 1988–92 Camry and 1990–92 MR2 and 1990–92 Celica (5S-FE), remove the timing belt and pulleys.

5. Remove the oil dipstick for 4A series engines.

6. As required, remove the complete air cleaner assembly.

7. Remove all hoses and inlet pipe from the water pump assembly.

8. Remove the water pump retaining bolts. Remove the water pump (and fan) assembly.

To install:

9. Always use a new gasket, O-ring or sealant between the pump body and its mounting.

10. Check for leaks after water pump installation is completed.

Thermostat

REMOVAL & INSTALLATION

1. Disconnect the negative battery cable. Drain the cooling system.

2. Remove the upper radiator or water inlet hose from the thermostat housing.

3. Disconnect the electrical wire from the thermo-switch on the thermostat housing, if equipped.

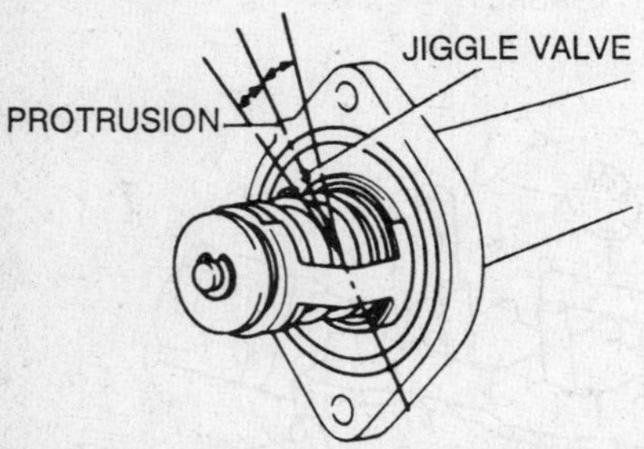

Thermostat jiggle valve alignment

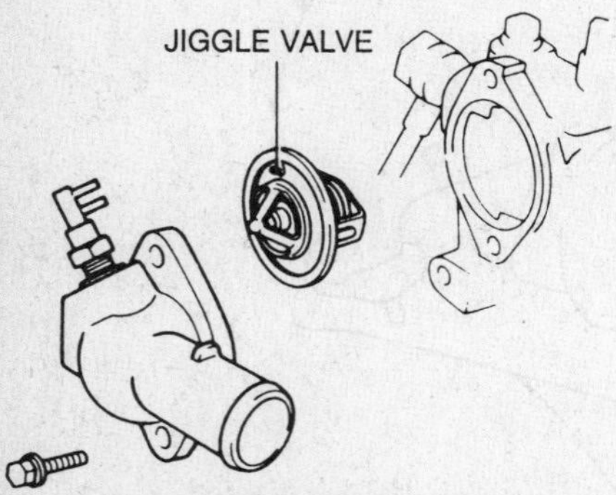

Typical thermostat assembly

4. Remove the thermostat housing retaining bolts. Remove the thermostat housing from the engine.

5. Remove the thermostat.

To install:

6. Clean the gasket mating surfaces. Be sure to use a new thermostat gasket. Be sure the thermostat is installed with the spring pointing down and the jiggle valve up. On the 3S-FE, 5S-FE, 4A-FE (type A thermostat) and 3S-GTE engines, align the jiggle valve with the protrusion on the thermostat housing. The jiggle valve may be aligned within 5–10 degrees on either side of the protrusion. On the 2VZ-FE and 4A-FE (type B thermostat), align the jiggle valve with the upper stud bolt in the housing.

Cooling System Bleeding

1. Fill the radiator with the proper type of coolant.

2. Loosen a fitting in a coolant passage located near the highest point on the engine. A sending unit of vacuum switching valve for example. Apply thread sealing tape or equivalent to the fitting threads. Fill the radiator until coolant comes from the hole and tighten the fitting.

3. With the radiator cap off, start the engine and allow it to run and reach normal operating temperature.

4. Run the heater at full force and with the temperature lever in the hot position. Be sure the heater control valve is functioning.

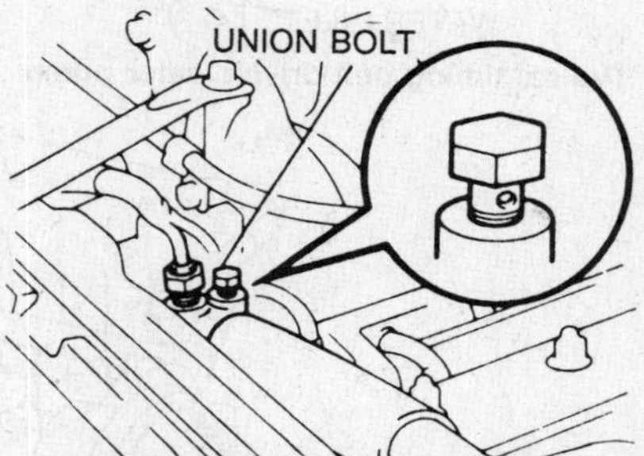

Cooling system air relief location—2VZ-FE engine

5. Shut the engine off and recheck the coolant level, refill as necessary. On 2VZ-FE engine, release the air from the cooling system by backing off on air relief union bolt about 4–5 turns.

ENGINE ELECTRICAL

CAUTION

If equipped with an air bag, wait 20 seconds or longer after disconnecting the negative battery cable before attempting to remove or service any electrical component. The air bag control system is equipped with a back-up power source that remains charged for a minimum of 20 seconds after the negative battery cable is disconnected. Attempting to remove or service an electrical component without allowing the time interval to elapse may result in deployment of the air bag and possible personal injury.

NOTE: Disconnecting the negative battery cable on some vehicles may interfere with the functions of the on board computer systems and may require the computer to undergo a relearning process, once the negative battery cable is reconnected.

Distributor

REMOVAL

5M-GE Engine

1988 Cressida

1. Disconnect the negative battery cable.

2. Disconnect the cables from the spark plugs, after marking the wiring order.

3. Disconnect the high tension cable from the coil.

4. Remove the primary wire and the vacuum line from the distributor.

5. Remove the distributor cap.

6. Matchmark the distributor housing and the engine block and the rotor to the distributor housing; this will aid in correct positioning of the distributor during installation.

7. Remove the clamp from the distributor.

8. Withdraw the distributor from the block.

To install:

9. Install the distributor to its original position.

10. Install the spark plug wires and remaining components.

11. Set the ignition timing to specifications.

Except 5M-GE Engine

1. Disconnect the negative battery cable.

CAUTION

If equipped with an air bag, wait 20 seconds or longer after disconnecting the negative battery cable before attempting to remove the distributor. The air bag control system is equipped with a back-up power source that remains charged for a minimum of 20 seconds after the negative battery cable is disconnected. Attempting to remove the distributor without allowing the time interval to elapse may result in deployment of the air bag and possible personal injury.

2. Disconnect the electrical leads and spark plug wires from the distributor.
3. Remove the water-proof cover, if installed.
4. On the Supra, with the 7M-GE engine, remove the oil filler cap and rotate the crankshaft clockwise until the nose of the camshaft is visible through the hole. Turn the crankshaft counterclockwise 120 degrees. Now, turn it clockwise 10–40 degrees until the TDC marks on the front cover and the crankshaft pulley are aligned.
5. Remove the intercooler on the 3S-GTE engine.
6. Matchmark the distributor housing and the engine block and the rotor to the distributor housing; this will aid in correct positioning of the distributor during installation.
7. Remove the hold-down bolts and pull the distributor from the engine.

To install:

8. Install the distributor to its original position.
9. Install the spark plug wires and remaining components.
10. Set the ignition timing to specifications.

INSTALLATION

Timing Not Disturbed

5M-GE ENGINE (1988 CRESSIDA)

1. Insert the distributor in the block and align the matchmarks.
2. Engage the distributor drive with the oil pump driveshaft; make sure the gear teeth mesh properly.
3. Install the distributor clamp, cap, high tension wire, primary wire and vacuum line.
4. Install the wires on the spark plugs.
5. Start the engine. Check the timing and adjust, as necessary.

EXCEPT 5M-GE ENGINE

1. Install a new distributor housing O-ring. Apply a thin coat of clean engine oil to the new O-ring before installation.
2. Insert the distributor in the block and align the matchmarks on the housing and the rotor made during removal.
3. Install the distributor hold-down bolts.
4. On 3S-GTE engine, install the intercooler.
4. Install the water-proof cover, if removed.
5. Connect the electrical leads and spark plug wires to the distributor.
6. Connect the negative battery cable.
7. Set the timing.

Timing Disturbed

5M-GE ENGINE (1988 CRESSIDA)

1. Determine Top Dead Center (TDC) of the No. 1 cylinder's compression stroke by removing the spark plug from the No. 1 cylinder and placing a finger or a compression gauge over the spark plug hole. Crank the engine slowly until compression pressure starts to build up. Continue cranking the engine until the timing marks indicate TDC or 0 degrees.
2. Remove the oil filler cap. Looking into the camshaft housing with the aid of a flashlight, check to make sure the match hole on the second (No. 2) journal of the camshaft housing is aligned with the hole in the No. 2 journal of the camshaft. If the holes are not aligned, rotate the camshaft 1 full turn.
3. Install a new O-ring on the distributor cap shaft; make sure the distributor cap is still removed at this time. Align the matchmark on the distributor drive gear with that of the distributor housing.
4. Insert the distributor into the camshaft housing, align the center of the mounting flange with the bolt hole in the side of the housing.
5. Align the rotor tooth in the distributor with the pickup coil. Temporarily install the distributor pinch bolt. Install the distributor cap and install the oil filler cap.
6. Connect the cables to the spark plugs in the proper order by using the marks made during removal. Install the high tension wire on the coil.
7. Start the engine. Adjust the ignition timing.

EXCEPT 5M-GE ENGINE

1. Set the engine at TDC of the No. 1 cylinder's firing stroke. This can be accomplished by removing the No. 1 spark plug and turning the engine by hand with a finger over the spark plug hole. As No. 1 is coming up on its firing stroke, pressure will be felt. Make sure the timing marks are set as follows:
   a. On 4A-GE, 4A-GEC and 4A-GELC, align the groove on the crankshaft pulley with the 0 mark on the No. 1 timing cover.
   b. For all, except the Supra (7M-GE), 1989–92 Cressida, MR2, Corolla (4A-GE) and Camry, coat the spi-

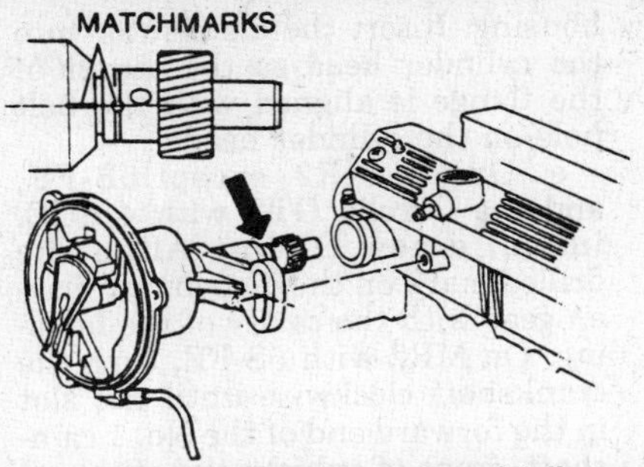

Align the matchmarks on the distributor gear and housing—5M-GE engine

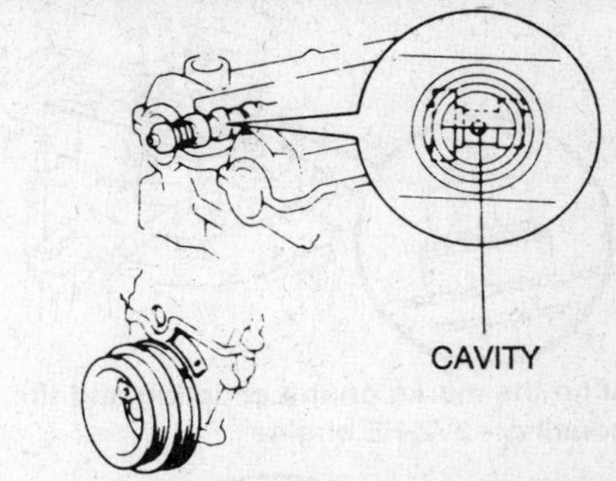

Setting the No. 1 cylinder to TDC of the compression stroke—4A-GE engine

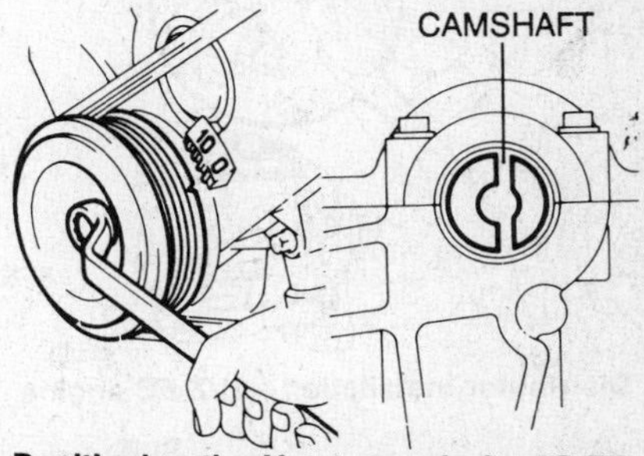

Positioning the No. 1 camshaft—3S-FE, 3S-GE and 3S-GTE engines

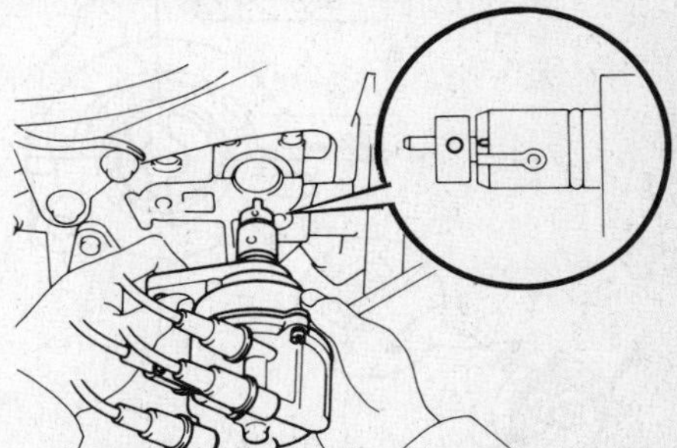

Distributor alignment—3S-FE and 3S-GE engines

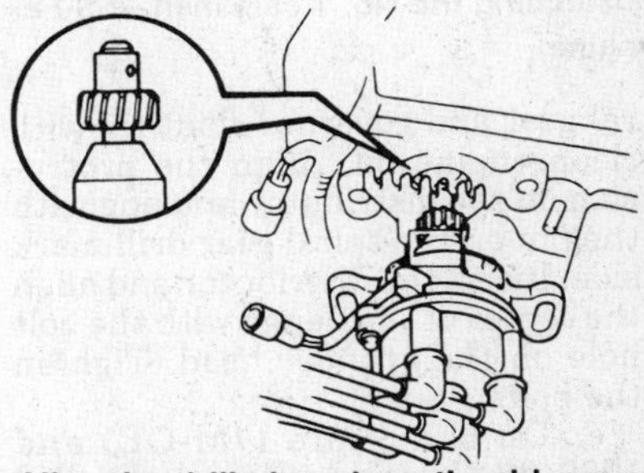

Align the drilled mark on the drive gear with the cavity of the housing—4A-GE engine

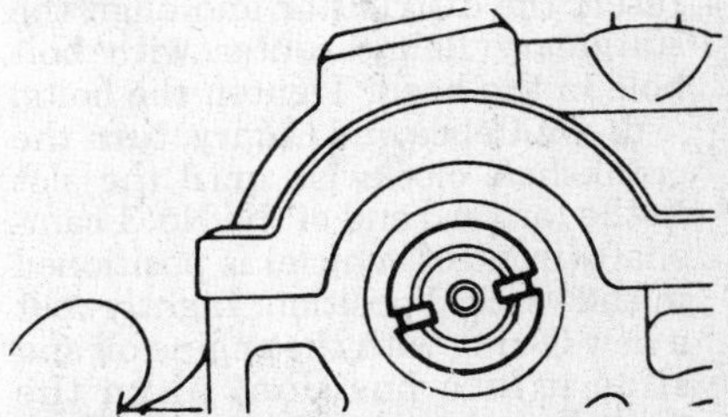

Position the camshaft slit as shown—4A-F engine

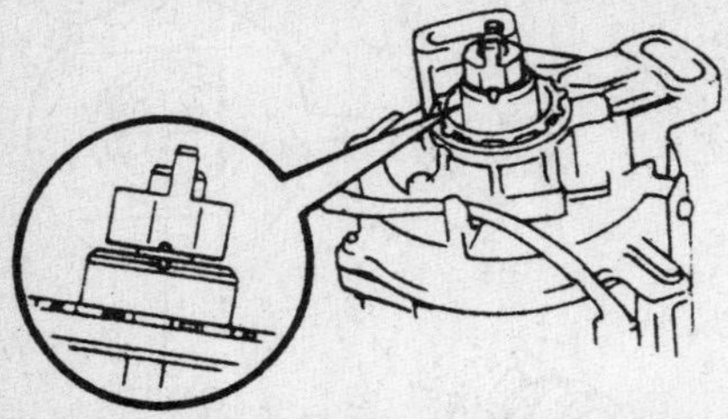

Align the marks on the coupling and the housing – 2VZ-FE engine

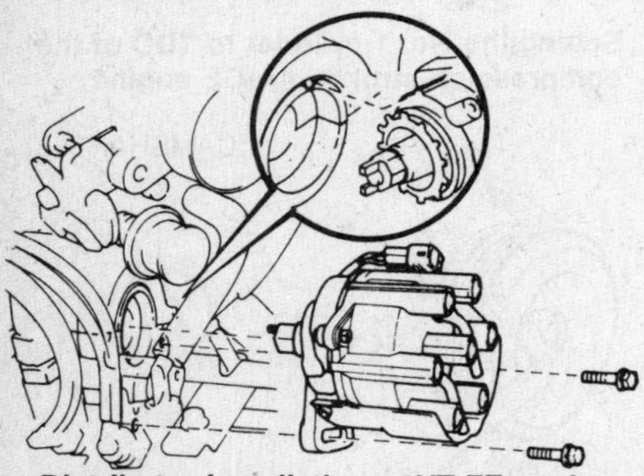

Distributor installation – 2VZ-FE engine

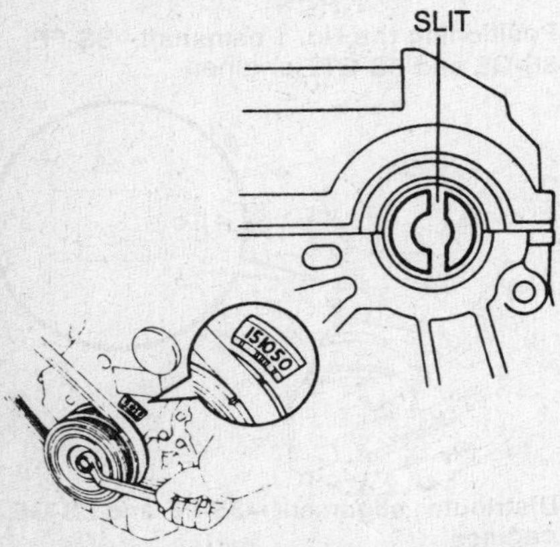

Positioning the No. 1 camshaft – 5S-FE engine

ral gear and governor shaft tip with clean engine oil. Align the protrusion on the distributor housing with the pin on the spiral gear drill mark side. Insert the distributor and align the center of the flange with the bolt hole on the cylinder head. Tighten the bolts.

c. On the Supra (7M-GE) and 1989–92 Cressida, align the drilled mark on the driven gear with the groove on the distributor housing. Insert the distributor and align the stationary flange center with bolt hole in the head. Tighten the bolts.

d. On Celica and Camry, turn the crankshaft clockwise until the slot in the forward end of the No. 1 camshaft (front of vehicle) is positioned in the vertical position. Lightly coat a new O-ring with the engine oil and slide it into position. Align the drilled mark or cutout, on the coupling with the notch of the shaft housing. Insert the distributor into the cylinder head so the center of the flange is aligned with the bolt hole on the cylinder head.

e. On the MR2, except 5S-FE, and the Corolla GTS with 4A-GE, install a new O-ring. Align the drilled mark on the distributor driven gear with the cavity of the housing. On MR2 with 5S-FE, turn the crankshaft clockwise until the slot in the forward end of the No. 1 camshaft, front of vehicle, is positioned in the vertical position. Then, align the cutout portion of the coupling with the groove in the housing. Insert the distributor and align the center of the flange with the bolt hole on the cylinder head. Tighten the hold-down bolts.

f. On the Corolla (4A-F, 4A-FE), install a new O-ring. Align the protrusion on the distributor housing with the groove of the coupling side. On the 4A-FE, align the center of the flange with the bolt hole on the cylinder head. Tighten the hold-down bolts.

g. On Camry with 2VZ-FE, align the cut-out marks of the coupling and the housing and then insert the distributor so the line on the housing and the cut-out on the distributor attachment cap are aligned. Tighten the hold-down bolts.

2. Connect the spark plug wires; check the idle speed and the ignition timing.

Distributorless Ignition

REMOVAL & INSTALLATION

Camshaft Position Sensor

SUPRA (7M-GTE)

1. Disconnect the negative battery cable.
2. Disconnect the cam position sensor connector.
3. Remove the oil filler cap.
4. Look into the oil filler opening with a flashlight and rotate the crankshaft clockwise until the the nose of the cam can be seen.
5. Once the nose of the cam comes into view, rotate the crankshaft approximately 120 degrees counterclockwise.
6. Turn the crankshaft approximately 10–40 degrees clockwise until the TDC mark on the timing belt cover is aligned with the TDC mark on the crankshaft pulley; the engine is now at TDC. Don't move it from this position.
7. Remove the No. 4 air cleaner pipe with the No. 1 and No. 2 air cleaner hoses.
8. Disconnect the 3 air hoses and the PCV hose.
9. Disconnect the air flow meter connector.
10. Disconnect the power steering idle up air hose.
11. Remove the air flow meter mounting bolt and attendant hose clamps. Remove the No. 7 air cleaner hose, air flow meter and air cleaner cap as a unit.
12. Unbolt and remove the power steering reservoir tank. Leave the hoses connected and move the tank aside.
13. Remove the cam position sensor hold-down bolt.
14. Withdraw the cam position sensor from the cylinder head.
15. Remove the cam position sensor O-ring. Discard the O-ring and replace with new.

To install:

16. Install a new O-ring.
17. Align the drilled mark on the driven gear with the groove of the housing.
18. Insert the cam position sensor into the cylinder head so the center of the sensor flange is aligned with the bolt hole in the head.
19. Lightly tighten the hold-down bolt.

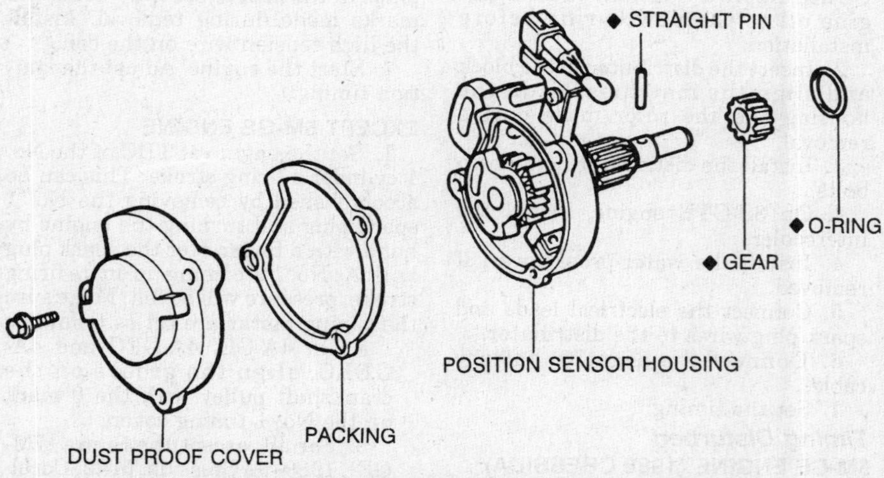

Cam position sensor exploded view – 7M-GTE engine

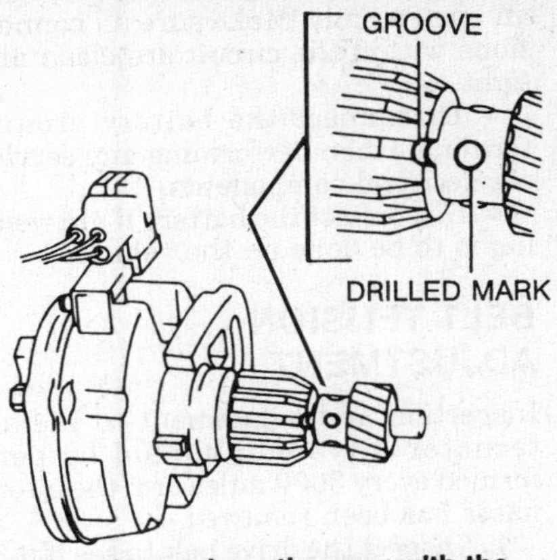

Align the mark on the gear with the groove in the housing—7M-GTE engine

20. Install the power steering reservoir tank.
21. Install the air cleaner cap, air flow meter and No. 7 air cleaner hose. Install the mounting bolt and tighten the clamps.
22. Connect the power steering idle up air hose.
23. Connect the air flow meter connector.
24. Connect the PCV hose and 3 air hoses.
25. Install the air cleaner pip and No. 1 and No. 2 air cleaner hoses.
26. Connect the cam position sensor connector.
27. Start and warm up the engine. Adjust the timing.

Ignition Timing

ADJUSTMENT

3A-C, 4A-LC and 4A-F Engine

1. Warm up the engine and set the parking brake. Connect a tachometer and check the engine speed to make sure it is within specifications. Adjust as required.
2. Connect the dwell meter or tachometer to the negative side of the coil, not to the distributor primary lead, damage to the ignition control will result.
3. All engines require a special type of tachometer which hooks up to the service connector wire coming out of the distributor.
4. Connect a timing light to the engine, as outlined in the instructions supplied by the manufacturer of the light.
5. Disconnect and plug the vacuum line from the distributor vacuum unit. If a vacuum advance/retard distributor is used, disconnect and plug both vacuum lines from the distributor.
6. Allow the engine to run at the specified idle speed with the gear shift in **N** for a manual transmission or **D** for an automatic transmission. Be sure the parking brake is firmly set and the wheels are chocked.
7. Point the timing light at the timing marks. With the engine at idle, timing should be at the specification. If not, loosen the pinch bolt at the base and rotate the distributor to advance or retard the timing, as required.
8. Stop the engine and tighten the pinch bolt. Start the engine and recheck the timing. Stop the engine and disconnect the timing light and the tachometer. Connect the vacuum line(s) to the vacuum and advance unit.

Except 3E and 3E-E Engine

1. Connect a timing light to the engine following the manufacturer's instructions. On the 7M-GTE, connect the timing light pick-up to the No. 6 spark plug wire.
2. The engines require a special type of tachometer which hooks up to the service connector wire coming from the distributor.
3. Start the engine and run it at idle. Remove the rubber cap from the check connector or open the lid; short the connector at terminals **T** and $\mathbf{E_1}$ on 1988 California vehicles. On all 1989–92 vehicles, with the 4A-GE, 4A-FE, 3S-FE, 3S-GTE, 5S-FE, 2VZ-FE and 7M-GE, short the $\mathbf{TE_1}$ and $\mathbf{E_1}$ terminals.
4. Loosen the distributor pinch bolt so the distributor can be turned. Aim the timing light at the marks on the crankshaft pulley and slowly turn the distributor until the timing mark is aligned. Tighten the distributor pinch bolt. Unshort the connector.

NOTE: The 7M-GTE utilizes a cam position sensor in place of a distributor. Turn this the same as a distributor.

3E Engine

1. Remove the cap. Using the proper tachometer, connect the test probe of the tachometer to the service probe connector at the Integrated Ignition Assembly (IIA).
2. Disconnect the vacuum hose from the IIA sub-diaphragm and plug it.
3. With the engine idling and the electric fan off, check the timing.
4. Loosen the hold-down bolt and adjust the timing, as required.
5. Retighten the hold-down bolt and recheck the ignition timing.

3E-E Engine

1. Warm up the engine to normal operating temperature.
2. Connect a tachometer and timing light to the engine. The tachometer may be connected either to the service

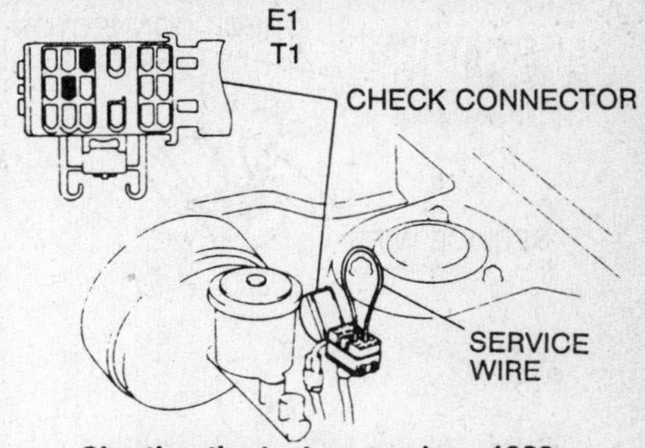

Shorting the test connector—1989 2VZ-FE engine

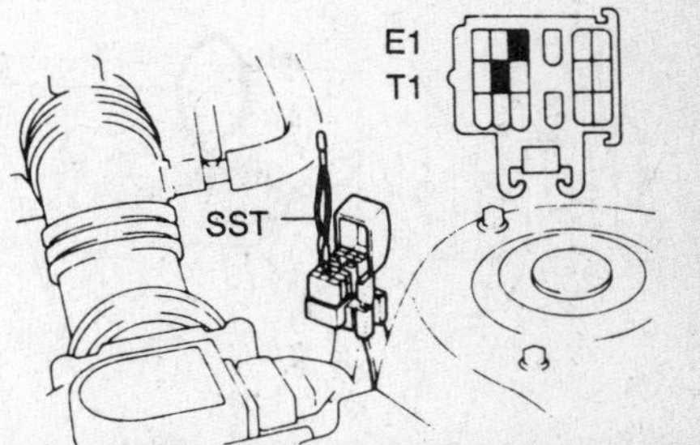

Shorting the test connector—1990–92 2VZ-FE engine

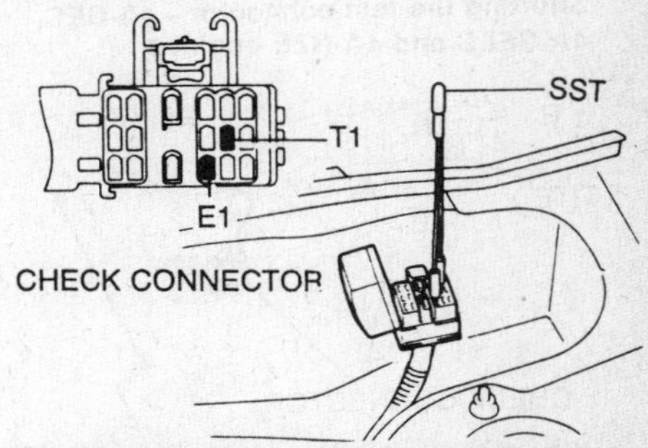

Shorting the test connector—3E-E engine

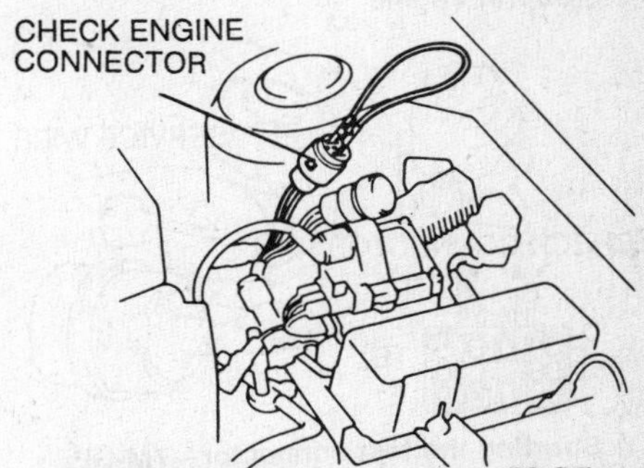

Shorting the test connector—5M-GE (Supra) engine

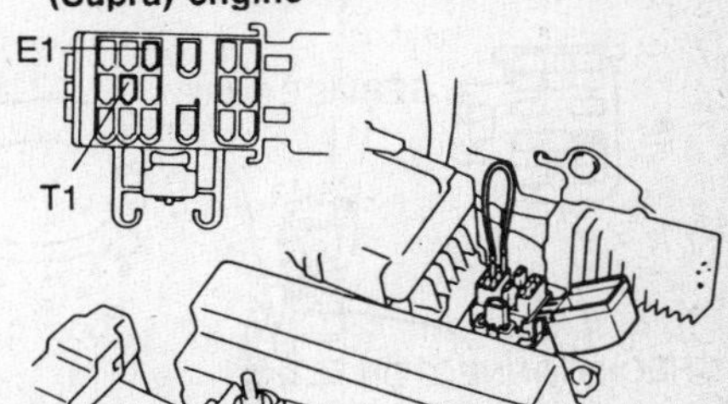

Shorting the test connector—5M-GE (Cressida) engine

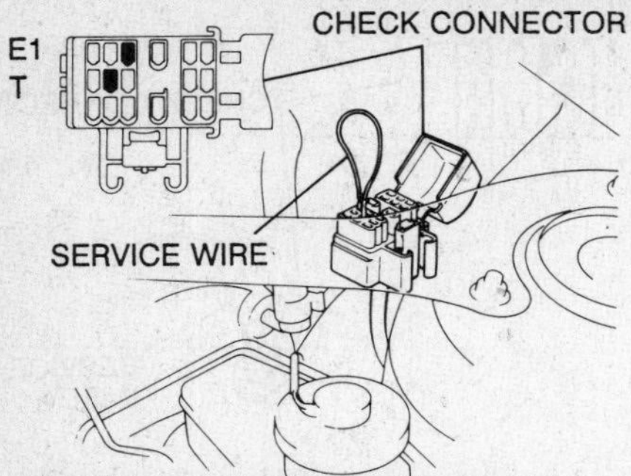

Shorting the test connector—3S-FE, 3S-GE and 3S-GTE engines

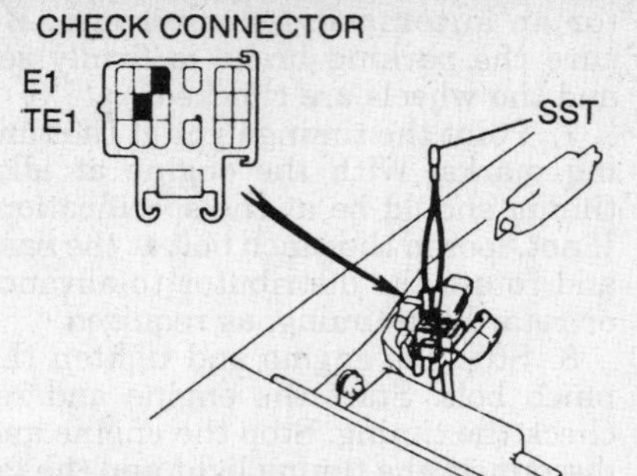

Shorting the test connector—5S-FE engine

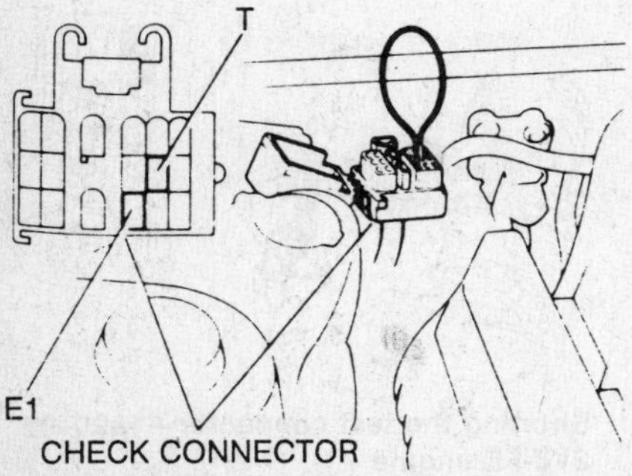

Shorting the test connector—4A-GEC, 4A-GELC and 4A-GZE engines

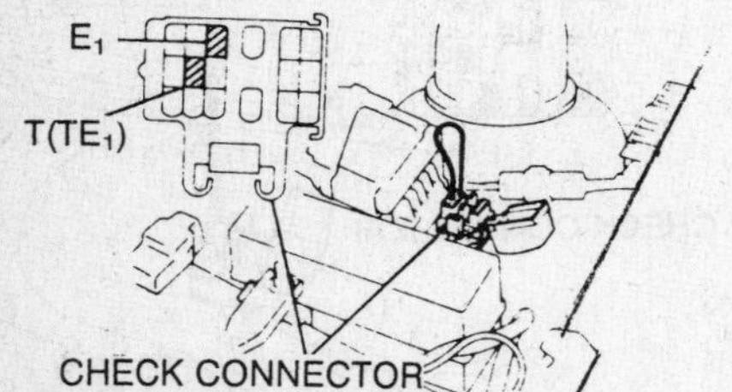

Shorting the test connector—7M-GE (Supra) engine

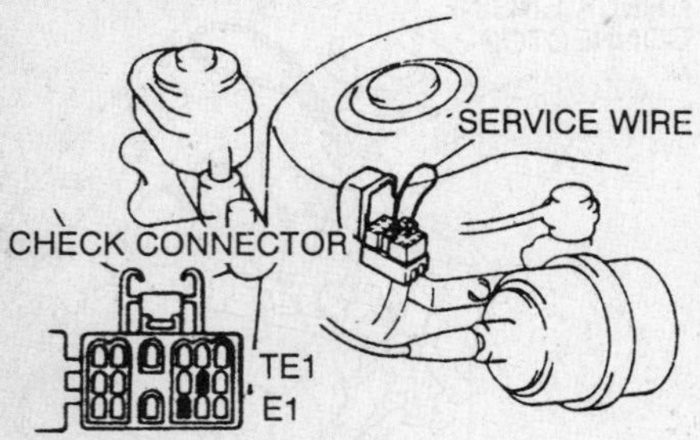

Shorting the test connector—7M-GE engine (Cressida)

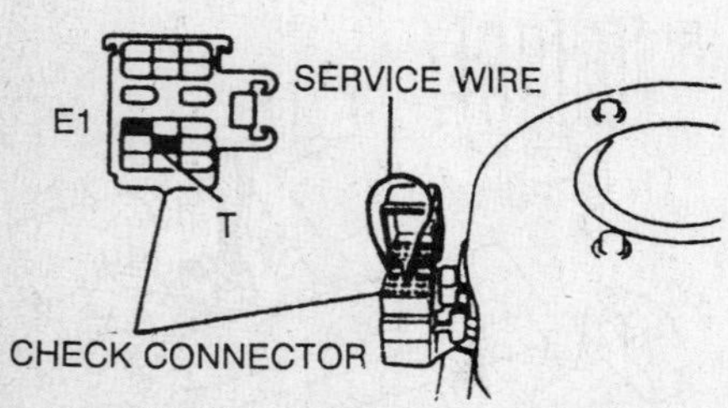

Shorting the test connector—4A-FE engine

connector of the distributor or to the **IG** terminal of the check connector.

NOTE: The engines require a special type of tachometer which hooks up to the service connector wire coming from the distributor.

3. Open the lid on the check connector and short the connector at terminals **T** and **E_1**.

4. Loosen the distributor pinch bolt so the distributor can be turned. Aim the timing light at the marks on the crankshaft pulley and slowly turn the distributor until the timing mark is aligned. Tighten the distributor pinch bolt. Unshort the connector.

Alternator

PRECAUTIONS

Several precautions must be observed with alternator equipped vehicles to avoid damage to the unit.

- If the battery is removed for any reason, make sure it is reconnected with the correct polarity. Reversing the battery connections may result in damage to the one-way rectifiers.
- When utilizing a booster battery as a starting aid, always connect the positive to positive terminals and the negative terminal from the booster battery to a good engine ground on the vehicle being started.
- Never use a fast charger as a booster to start vehicles.
- Disconnect the battery cables when charging the battery with a fast charger.
- Never attempt to polarize the alternator.
- Do not use test lamps of more than 12 volts when checking diode continuity.
- Do not short across or ground any of the alternator terminals.
- The polarity of the battery, alternator and regulator must be matched and considered before making any electrical connections within the system.
- Never separate the alternator on an open circuit. Make sure all connections within the circuit are clean and tight.
- Disconnect the battery ground terminal when performing any service on electrical components.
- Disconnect the battery if arc welding is to be done on the vehicle.

BELT TENSION ADJUSTMENT

Inspection and adjustment to the alternator drive belt should be performed every 3000 miles or if the alternator has been removed.

1. Inspect the drive belt to see if it is cracked or worn; be sure it's surfaces are free of grease or oil.
2. Push down on the belt halfway between the fan and the alternator pulleys or crankshaft pulley with thumb pressure; belt deflection should be ³⁄₈–½ in.
3. If the belt tension requires adjustment, loosen the adjusting link bolt and move the alternator until the proper belt tension is obtained.
4. Do not over-tighten the belt, as damage to the alternator bearings could result. Tighten the adjusting link bolt.
5. Drive the vehicle and re-check the belt tension; adjust, as necessary.

REMOVAL & INSTALLATION

Except Celica AWD

NOTE: On some vehicles, the alternator is mounted very low on the engine. On these vehicles it may be necessary to remove the gravel shield and work from under the vehicle in order to gain access to the alternator.

1. Disconnect the negative battery cable.
2. Remove the air cleaner, if necessary, to gain access to the alternator.
3. Remove the power steering or air conditioning drive belts, as required.
4. Unfasten the bolts which attach the adjusting link to the alternator. Remove the alternator drive belt.

NOTE: On the 2VZ-FE, remove the No. 2 right side mounting stay.

5. Unfasten and tag the alternator bolt and withdraw the alternator from it's bracket.
6. Installation is the reverse of the removal procedure. After installing the alternator, adjust the belt tension.

Celica AWD

1. Disconnect the negative battery cable.
2. Remove the lower alternator duct.

3. Loosen the idler pulley bolt.
4. Loosen the adjusting bolt and remove the drive belt.
5. Disconnect the alternator connectors, alternator lead wire, air conditioning compressor connector, water temperature switch connector and oxygen sensor connector.
6. Unbolt and disconnect the ground strap and engine wire from the brackets.
7. Unbolt and remove the alternator bracket.
8. Unbolt and remove the alternator.
9. Remove the upper alternator duct and disconnect the lead wire.

To install:

10. Connect the lead wire and attach the alternator duct.
11. Install the alternator. Torque the 12mm bolt to 14 ft. lbs. (19 Nm) and the 14mm bolt to 38 ft. lbs. (52 Nm).
12. Install the alternator bracket. Torque the turbine outlet elbow bolt to 32 ft. lbs. (43 Nm) and the bracket bolt to 39 ft. lbs. (29 Nm).
13. Install the engine wire and ground strap.
14. Connect the alternator and engine wiring.
15. Install the drive belt and adjust the drive belt tension.
16. Install the lower alternator duct.
17. Connect the negative battery cable.

Starter

REMOVAL & INSTALLATION

1. Disconnect the negative battery cable. Disconnect the cable which runs from the starter to the battery, at the battery end.
2. Remove the air cleaner assembly, if necessary, to gain access to the starter.
3. If equipped with an automatic transmission/transaxle, it may be necessary to disconnect the throttle linkage connecting rod or the transmission/tranasaxle oil filler tube.
4. On the 3S-GE, disconnect the exhaust pipe at the manifold. On the Camry with 2VZ-FE, remove the ignitor bracket. On 1990–92 Celica with 3S-GTE, remove the engine compartment relay box and the battery. On 1990–92 Celica with 4A-FE, remove the lower suspension crossmember and the air cleaner cap. On 1990–92 Celica with 5S-FE, cruise control and ABS, remove the engine compartment relay box and the cruise control actuator. On 1990–92 Corolla with 4A-GE, remove both engine undercovers, front exhaust pipe and electric cooling fan.
5. Disconnect all of the wiring at the starter. Remove the starter retaining bolts. Remove the starter from the vehicle.
6. Installation is the reverse of the removal procedure.

FUEL SYSTEM

Fuel System Service Precautions

Safety is the most important factor when performing not only fuel system maintenance but any type of maintenance. Failure to conduct maintenance and repairs in a safe manner may result in serious personal injury or death. Maintenance and testing of the vehicle's fuel system components can be accomplished safely and effectively by adhering to the following rules and guidelines.

- To avoid the possibility of fire and personal injury, always disconnect the negative battery cable unless the repair or test procedure requires that battery voltage be applied.
- Always relieve the fuel system pressure prior to disconnecting any fuel system component (injector, fuel rail, pressure regulator, etc.), fitting or fuel line connection. Exercise extreme caution whenever relieving fuel system pressure to avoid exposing skin, face and eyes to fuel spray. Please be advised that fuel under pressure may penetrate the skin or any part of the body that it contacts.
- Always place a shop towel or cloth around the fitting or connection prior to loosening to absorb any excess fuel due to spillage. Ensure that all fuel spillage (should it occur) is quickly removed from engine surfaces. Ensure that all fuel soaked cloths or towels are deposited into a suitable waste container.
- Always keep a dry chemical (Class B) fire extinguisher near the work area.
- Do not allow fuel spray or fuel vapors to come into contact with a spark or open flame.
- Always use a backup wrench when loosening and tightening fuel line connection fittings. This will prevent unnecessary stress and torsion to fuel line piping. Always follow the proper torque specifications.
- Always replace worn fuel fitting O-rings with new. Do not substitute fuel hose or equivalent where fuel pipe is installed.

RELIEVING FUEL SYSTEM PRESSURE

1. Remove the fuel pump fuse from the fuse block, fuel pump relay or disconnect the harness connector at the tank while engine is running.
2. It should run and then stall when the fuel in the lines is exhausted. When the engine stops, crank the starter for about 3 seconds to make sure all pressure in the fuel lines is released.
3. Install the fuel pump fuse, relay or harness connector after repair is made.

Fuel Tank

REMOVAL & INSTALLATION

Carbureted Engines

1. Disconnect the negative battery cable. Relieve the fuel pressure.
2. Drain the fuel using an approved pump and container.
3. Label and disconnect the fuel filler, breather, tank-to-evaporator, filler hose and tank-to-return pipe hoses.
4. Raise the vehicle and support safely. Disconnect the tank wiring and remove the under cover, if equipped.
5. Place a floor jack under the tank, remove the retaining bolts and lower the tank far enough to disconnect any hoses and electrical connectors still connected.

To install:

6. Raise the tank with the jack and connect the hoses and electrical connectors. Install the retaining bolts. Torque the retainer to 15 ft. lbs. (20 Nm).
7. Connect the wiring and fuel hoses.
8. Connect the breather hose and fuel filler hose.
9. Connect the battery cable, fill the tank with fuel and check for leaks.

Fuel Injected Engines

1. Disconnect the negative battery cable. Relieve the fuel system pressure.
2. Drain the tank with an approved pump and container.
3. Raise the vehicle and support safely. Remove the rear and center exhaust pipes.
4. Remove the driveshaft for AWD and RWD vehicles. Disconnect the fuel filler and air breather hose.
5. Disconnect the feed, return and evaporative hoses.
6. Disconnect the parking brake cable bracket and return spring.
7. Disconnect the tank harness connectors.
8. Place a floor jack under the tank and remove the tank retainers. Lower the tank and disconnect any wiring or hoses.

To install:

9. Raise the tank into position and

install the tank retainers. Torque the retainers to 15 ft. lbs. (20 Nm).

10. Connect the tank harnesses and hoses.

11. Connect the parking brake cable bracket and return spring.

12. Connect the feed, return and evaporative hoses.

13. Install the driveshaft, if removed. Connect the fuel filler and air breather hoses.

14. Install the rear and center exhaust pipes.

15. Refill the tank and check for leaks.

16. Connect the battery cable.

Fuel Filter

REMOVAL & INSTALLATION

1. Disconnect the negative battery cable. Unbolt the retaining screws and remove the protective shield for the fuel filter.

2. Place a pan under the delivery pipe to catch the dripping fuel and slowly loosen the union bolt or flare nut to bleed off the fuel pressure.

3. Drain the remaining fuel.

4. Disconnect and plug the inlet line.

5. Unbolt and remove the fuel filter.

NOTE: When tightening the fuel line bolts to the fuel filter, use a torque wrench. The tightening torque is very important, as under or over tightening may cause fuel leakage. Insure that there is no fuel line interference and that there is sufficient clearance between it and any other parts.

6. Coat the flare nut, union nut and bolt threads with engine oil.

7. Hand tighten the inlet line to the fuel filter.

8. Install the fuel filter and then tighten the inlet bolt to 22 ft. lbs. (30 Nm).

9. Reconnect the delivery pipe using new gaskets and then tighten the union bolt to 22 ft. lbs. (30 Nm).

10. Run the engine for a few minutes and check for any fuel leaks.

11. Install the protective shield.

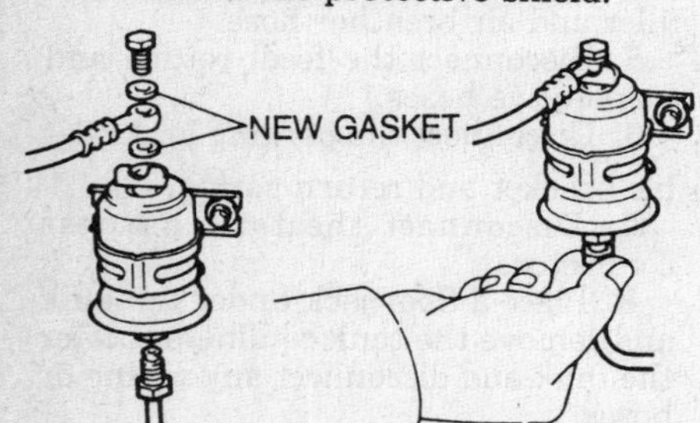

Fuel filter assembly

Mechanical Fuel Pump

The 3E, 3A-C, 4A-C and 4A-F engines use a mechanical type fuel pump. It is located on the right rear of the cylinder head.

PRESSURE TESTING

1. Remove the line which runs from the fuel pump to the carburetor.

2. Attach a pressure gauge to the outlet side of the pump.

3. Run the engine and check the pressure.

4. Check the pressure. It should be 2.6–3.5 psi.

5. If the pressure is below the specifications, check for restrictions or replace the pump.

6. Reconnect the carburetor line.

REMOVAL & INSTALLATION

1. Disconnect and plug the fuel lines to the pump.

2. Remove the nuts which hold the pump to the cylinder head.

3. Remove the pump assembly.

4. Installation is the reverse of removal. Always use a new gasket when installing a fuel pump.

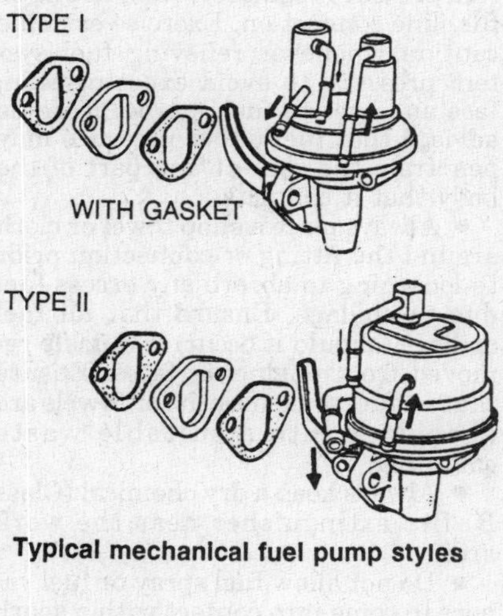

Typical mechanical fuel pump styles

Electric Fuel Pump

PRESSURE TESTING

CAUTION

Do not operate the fuel pump unless it is immersed in gasoline and connected to its resistor.

1. Turn the ignition switch to the **ON** position, but don't start the engine.

2. Remove the rubber cap from the fuel pump check connector and short terminals **Fp** and **+B** with a jumper wire.

NOTE: The check connector on all engines is in a small plastic box with a flip-up lid; it is found near the strut tower or battery. The box is roughly the same size and shape for every engine and terminals Fp and +B are always in the same location.

3. Check that there is pressure in the hose to the cold start injector. On 4A-FE and 4A-GE engines, check for pressure at the regulator fuel return hose. On 2VZ-FE, 3S-FE, 5S-FE, 3S-GE 4A-GE and 3S-GTE engines, check for pressure at the fuel filter hose.

NOTE: At this time, fuel return noise from the pressure regulator should be audible.

4. If no pressure can be felt in the line, check the fuses and all other related electrical connections. If everything is all right, the fuel pump will probably require replacement.

5. Remove the jumper wire, reinstall the rubber cap and turn off the ignition switch.

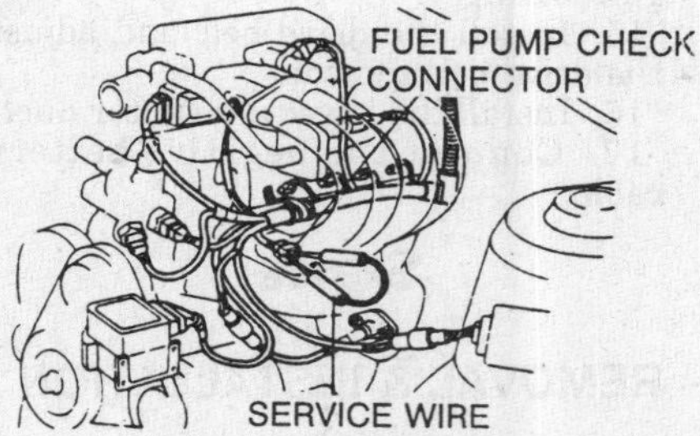

Shorting the fuel pump check connector – typical

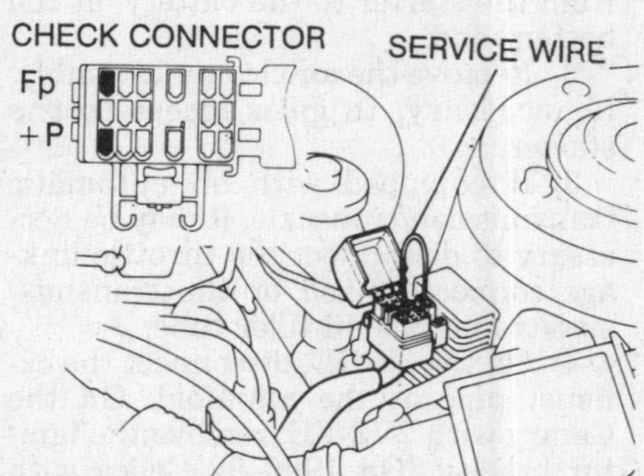

Shorting the fuel pump check connector (3S-GE shown, check connector terminals are in same location for all engines)

REMOVAL & INSTALLATION

The fuel pump is mounted inside the fuel tank on all vehicles. On all vehicles except 1990 Celica (non-turbocharged) and 1991–92 Tercel removal of the fuel tank is necessary to remove the fuel pump. On 1990–92 Celica non-turbocharged and 1991–92 Tercel, access to the pump is gained by removing the rear seat cushion.

Except 1990–92 Celica Non-Turbocharged and 1990–92 MR2

1. Disconnect the negative battery cable.
2. Drain the fuel from the tank and then remove the fuel tank.
3. Remove the bolts and then pull the fuel pump bracket up and out of the fuel tank.
4. Remove the mounting nuts then tag and disconnect the wires at the fuel pump.
5. Pull the fuel pump out of the lower side of the bracket. Disconnect the pump from the fuel hose.
6. Remove the rubber cushion and the clip. Disconnect the fuel pump filter from the pump.
7. Installation is in the reverse order of removal procedure. Use a new fuel bracket gasket.

1990–92 Celica Non-Turbocharged and Tercel

1. Disconnect the negative battery cable.
2. Remove the rear seat cushion.
3. Remove the 5 retaining screws and floor service hole cover.
4. Disconnect all the electrical fuel pump connections at the fuel pump assembly.
5. Disconnect the fuel pipe and hose from the fuel pump bracket. Remove the fuel pump bracket assembly from the fuel tank. Remove the fuel pump from the fuel bracket.
6. Installation is the the reverse of the removal procedure. Use a new fuel bracket gasket.

1990–92 MR2

1. Disconnect the negative battery cable.
2. Drain the fuel tank into a suitable container.
3. Remove the console boxes, left lower instrument panel finish panel, ash tray and retainer.
4. Remove the center instrument panel finish panel.
5. Disconnect the fuel pump connector and fuel sender.
6. Remove the 2 screws and floor service hole cover.
7. Remove the engine under covers, front luggage under cover and fuel tank protectors.
8. Remove the parking brake intermediate lever and center floor crossmember.
9. Remove the air conditioning and radiator hoses from the body and move out of the way. If necessary, disconnect the hoses after draining the radiator and discharging the air conditioning system.
10. Remove the fuel tank heat insulators and disconnect the fuel hoses and tubes.
11. Using a suitable jack, remove the tank retainers and lower the tank.
12. Remove the pump assembly from the tank. Remove the pump from the bracket.

To install:

13. Install a new gasket, pump and retaining screws. Torque the screws to 35 inch lbs. (3.4 Nm).
14. Install the fuel tank using a suitable jack.
15. Connect all wiring and fuel hoses to the tank.
16. Torque the tank retainers to 22 ft. lbs. (29 Nm).
17. Install the remaining components, refill the tank and check for leaks.
18. Connect the battery cable.

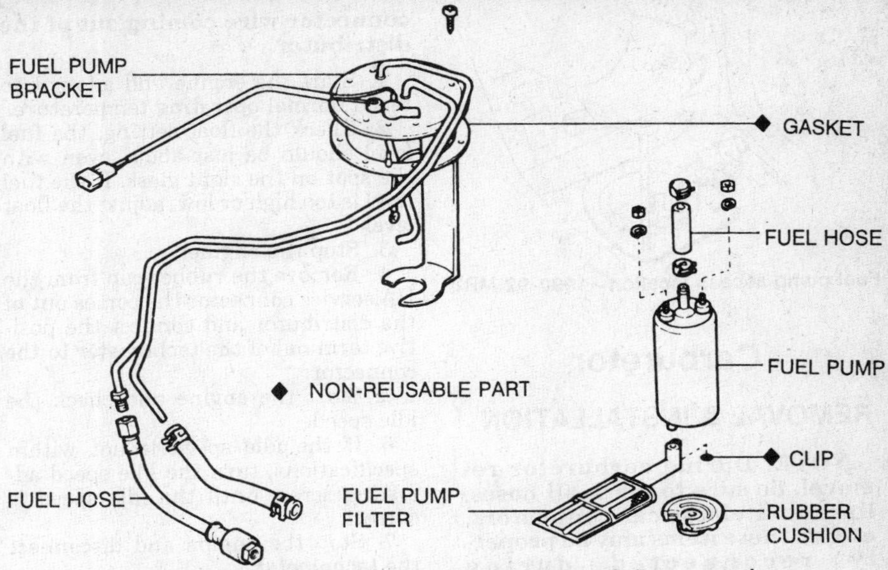

Typical fuel pump assembly – except 1990–92 Celica non-turbo

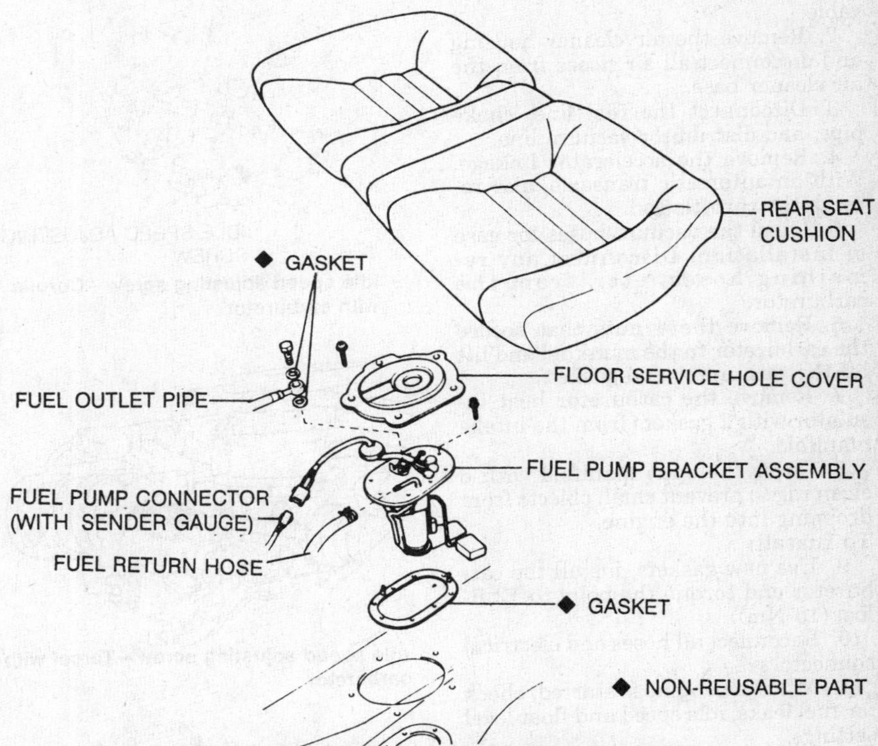

Fuel pump assembly – 1990–92 Celica non-turbo

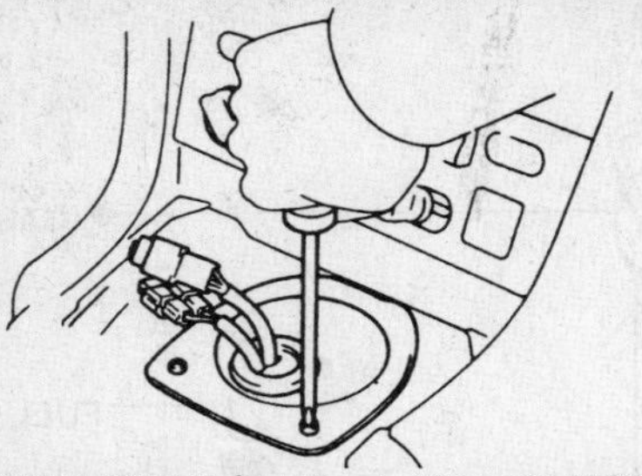

Fuel pump access location—1990–92 MR2

Carburetor

REMOVAL & INSTALLATION

NOTE: During carburetor removal, be sure to mark all hoses, lines and electrical connectors, etc., so these items may be properly reconnected during installation.

1. Disconnect the negative battery cable.
2. Remove the air cleaner housing and disconnect all air hoses from the air cleaner base.
3. Disconnect the fuel line, choke pipe, and distributor vacuum line.
4. Remove the accelerator linkage. With an automatic transaxle, also remove the throttle rod.
5. Label the vacuum hoses for ease of installation. Disconnect any remaining hoses, etc., from the carburetor.
6. Remove the 4 nuts that secure the carburetor to the manifold and lift off the carburetor and gasket.
7. Remove the carburetor heat insulator with 2 gaskets from the intake manifold.
8. Cover the open manifold with a clean rag to prevent small objects from dropping into the engine.

To install:

9. Use new gaskets, install the carburetor and torque the bolts to 12 ft. lbs. (15 Nm).
10. Reconnect all hoses and electrical connectors.
11. After the engine is started, check for fuel leaks, idle speed and float level settings.

IDLE SPEED ADJUSTMENT

The idle speed and mixture should be adjusted under the following conditions: the air cleaner must be installed, the choke fully opened, the transmission should be in **N** and all electrical accessories (including the electric engine cooling fan) should be turned off.

NOTE: All carbureted engines require a special type of tachometer which hooks up to the service connector wire coming out of the distributor.

1. Start the engine and allow it to reach normal operating temperature.
2. Check the float setting; the fuel level should be just about even with the spot on the sight glass. If the fuel level is too high or low, adjust the float level.
3. Stop the engine.
4. Remove the rubber cap from the IIA service connector the comes out of the distributor and connect the positive terminal of the tachometer to the connector.
5. Start the engine and check the idle speed.
6. If the idle speed is not within specifications, turn the idle speed adjusting screw until the idle speed is correct.
7. Stop the engine and disconnect the tachometer.

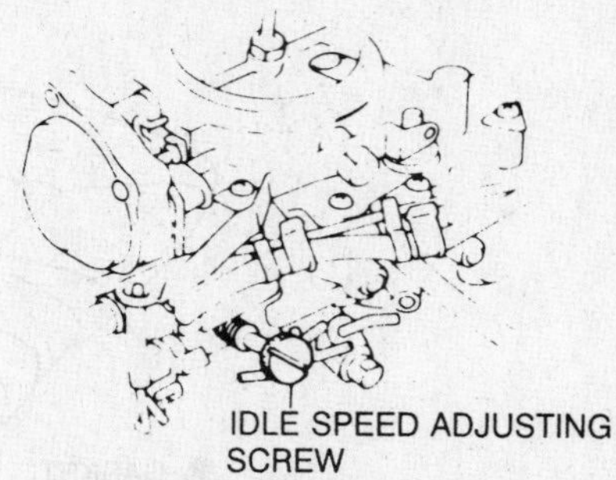

Idle speed adjusting screw—Corolla with carburetor

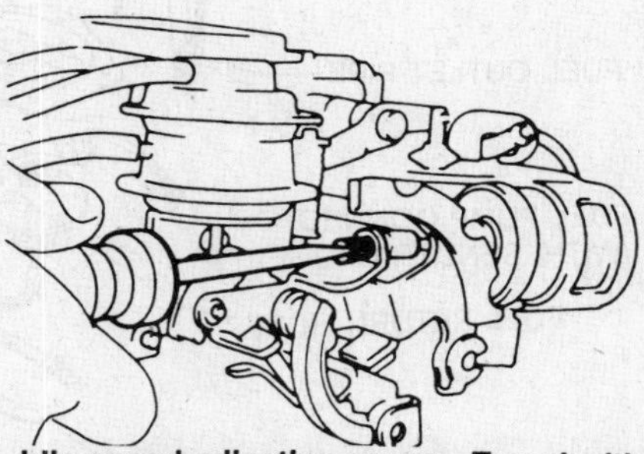

Idle speed adjusting screw—Tercel with carburetor

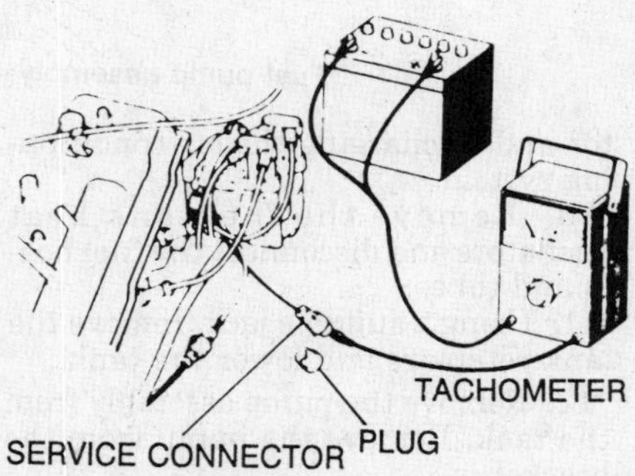

Tachometer hook-up—Corolla and Tercel with carburetor

SERVICE ADJUSTMENTS

For all carburetor service Adjustment procedures and Specifications, please refer to "Carburetor Service" in the Unit Repair Section.

Fuel Injection

IDLE SPEED ADJUSTMENT

Idle speed adjustment is performed under the following conditions:

Engine at normal operating temperature.

Air cleaner installed.

Air pipes and hoses of the air induction and EGR systems properly connected.

All vacuum lines and electrical wires connected and plugged in properly.

All electrical accessories in the **OFF** position.

Transaxle in the **N** position.

Supra, 1989–92 Cressida, Camry (2VZ-FE Engine) and 1990 Celica (3S-GTE Engine)

Idle speed is controlled by the Electronic Control Unit (ECU) and is not adjustable.

1988 Cressida

1. Connect a tachometer and timing light to the engine.
2. Start the engine and warm up to normal operating temperature.
3. Check the ignition timing. Adjust as necessary.
4. Check the idle speed.
5. If the idle speed is not within specifications, adjust by turning the idle speed adjusting screw on the throttle body.
6. Stop the engine and disconnect the tachometer.

Tercel

1. Run the engine until it reaches normal operating temperature. The cooling fan must not be running during the idle speed adjustment.
2. Connect a tachometer to the engine.

NOTE: Do not allow the tachometer or coil terminals to be grounded. This will damage the injection system.

3. On 3E-E engine, disconnect the idle up Vacuum Switching Valve (VSV) connector.
4. Run the engine at 2500 rpm for 2 minutes.
5. Adjust the idle speed by turning the idle speed adjusting screw.
6. On 3E-E engine, connect the idle

up Vacuum Switching Valve (VSV) connector.

7. Disconnect the tachometer.

Corolla

1. Run the engine until it reaches normal operating temperature. The cooling fan must not be running during the idle speed adjustment.

2. Connect a tachometer to the engine.

NOTE: Do not allow the tachometer or coil terminals to be grounded. This will damage the injection system.

3. Run the engine at 2500 rpm for 2 minutes.

4. On 1990 vehicles, short the check connector at terminals $\mathbf{TE_1}$ and $\mathbf{E_1}$ using a suitable jumper wire.

5. Adjust the idle speed by turning the idle speed adjusting screw.

6. On 1990–92 vehicles, remove the jumper wire from the connector terminals.

7. Disconnect the tachometer.

Camry and 1988 Celica (3S-FE Engine)

1. Run the engine until it reaches normal operating temperature. The cooling fan must not be running during the idle speed adjustment.

2. Connect a tachometer to the engine.

NOTE: Do not allow the tachometer or coil terminals to be grounded. This will damage the injection system.

3. On 1988 Camry, short the check connector at terminals **T** and $\mathbf{E_1}$ using a suitable jumper wire. On 1989–92 Camry, short the check connector at terminals $\mathbf{TE_1}$ and $\mathbf{E_1}$. On Celica, short the check connector at terminals $\mathbf{TE_1}$ (California) or **T** (except California) and $\mathbf{E_1}$.

4. Run the engine at 1000–3000 rpm for 5 seconds and return the engine to idle.

5. Adjust the idle speed by turning the idle speed adjusting screw.

6. Remove the jumper wire from the connector terminals.

7. Disconnect the tachometer.

1988–89 Celica (3S-GE Engine)

1. Run the engine until it reaches normal operating temperature. The cooling fan must not be running during the idle speed adjustment.

2. Connect a tachometer to the engine.

NOTE: Do not allow the tachometer or coil terminals to be grounded. This will damage the injection system.

3. Run the engine at 2500 rpm for 2 minutes.

4. Pinch the No. 1 air intake chamber vacuum hose.

5. Adjust the idle speed by turning the idle speed adjusting screw.

6. Release the No. 1 vacuum hose.

7. Disconnect the tachometer.

1990–92 MR2 (5S-FE Engine) and 1990–92 Celica (4A-FE and 5S-FE Engine)

1. Run the engine until it reaches normal operating temperature. The cooling fan must not be running during the idle speed adjustment.

2. Connect a tachometer to the engine.

NOTE: Do not allow the tachometer or coil terminals to be grounded. This will damage the injection system.

3. On 4A-FE engine, run the engine at 2500 rpm for 2 minutes. On 5S-FE engine, run the engine at 1000–3000 rpm for 5 seconds. Allow the engine to return to idle.

4. Short the check connector at terminals $\mathbf{TE_1}$ and $\mathbf{E_1}$ using a suitable jumper wire.

5. Adjust the idle speed by turning the idle speed adjusting screw.

6. Remove the jumper wire from the connector terminals.

7. Disconnect the tachometer.

MR2

4A-GE ENGINE

1. Run the engine until it reaches normal operating temperature. The cooling fan must not be running during the idle speed adjustment.

2. Connect a tachometer to the engine.

NOTE: Do not allow the tachometer or coil terminals to be grounded. This will damage the injection system.

3. Run the engine at 2500 rpm for 2 minutes.

4. Adjust the idle speed by turning the idle speed adjusting screw.

5. Disconnect the tachometer.

4A-GZE ENGINE

1. Run the engine until it reaches normal operating temperature. The cooling fan must not be running during the idle speed adjustment.

2. Connect a tachometer to the engine.

NOTE: Do not allow the tachometer or coil terminals to be grounded. This will damage the injection system.

3. Short the check connector at terminals **T** and $\mathbf{E_1}$ using a suitable jumper wire.

4. Adjust the idle speed by turning the idle speed adjusting screw.

5. Remove the jumper wire from the connector terminals.

6. Disconnect the tachometer.

3S-GTE

1. Adjust the idle speed with the engine at normal operating temperature, air cleaner installed, all accessories switched **OFF** and the transmission in **N**.

2. Connect a tachometer to the **IG** terminal of the check connector.

3. Using a jumper wire SST 09843-18020, jump the **TE1** and **E1** terminals of the check connector.

4. The idle should be 700 rpm (US with auto trans), 750 rpm (US with manual, Canada with automatic transaxle) and 850 rpm (Canada with manual transaxle).

5. If not within specifications, turn the idle adjusting screw at the throttle body to specifications.

6. Turn the ignition switch OFF and disconnect all test equipment.

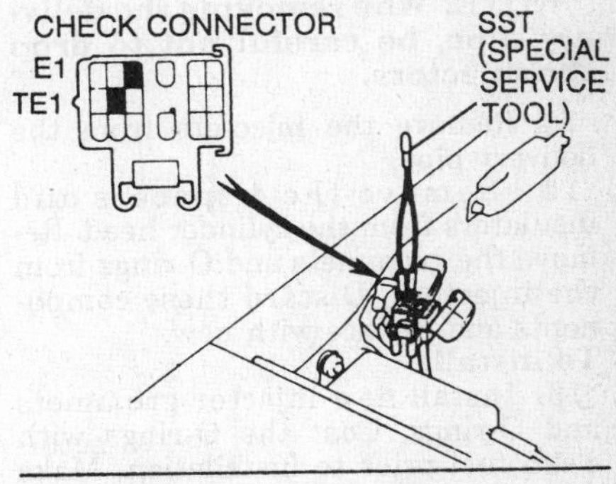

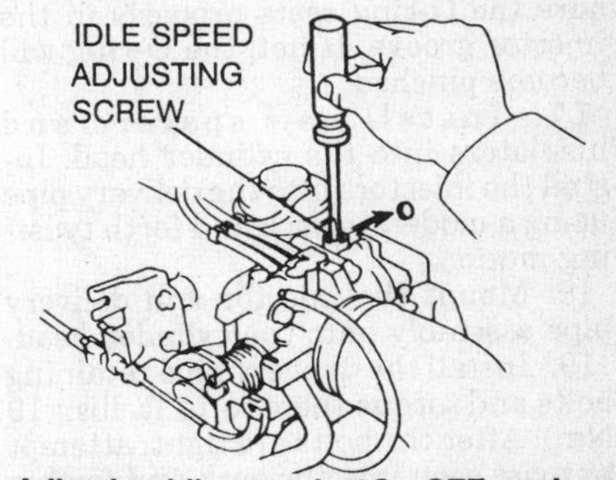

Adjusting idle speed – 3S – GTE engines

IDLE MIXTURE ADJUSTMENT

On all fuel injected engines, the idle mixture is controlled by the Electronic Control Unit (ECU) and is not adjustable.

Fuel Injector

REMOVAL & INSTALLATION

Tercel

1. Relieve the fuel system pressure.

2. Disconnect the negative battery cable.
3. Disconnect and tag the PCV and vacuum hoses.
4. Remove the air intake connector.
5. Disconnect the accelerator and throttle cables.
6. Disconnect the vacuum sensing hose.
7. Remove the 2 bolts remove the dashpot and link bracket. Disconnect the spring from the dashpot and throttle linkage.
8. Remove the pulsation damper and disconnect the fuel inlet hose from it.
9. Remove the clamp and disconnect the fuel return hose.
10. Remove the cold start injector pipe.
11. Disconnect the injector harness connectors.
12. Remove the 2 bolts attaching the fuel delivery pipe to the cylinder head.
13. Pull the delivery pipe and fuel injectors from the cylinder head.

NOTE: Whe removing the delivery pipe, be careful not to drop the injectors.

14. Remove the injectors from the delivery pipe.
15. Remove the 4 spacers and insulators from the cylinder head. Remove the grommets and O-rings from the injectors. Discard these components and replace with new.

To install:

16. Install new injector grommets and O-rings. Coat the O-rings with clean fuel prior to installation. Make sure the O-ring seats properly in the injector groove. If not, the O-ring will become pinched.
17. Install new spacers and insulators into the cylinder head. Install the injectors into the delivery pipe using a moderate back and forth twisting motion.
18. Mount the injector and delivery pipe assembly onto the cylinder head.
19. Install the delivery pipe retaining bolts and torque them to 14 ft. lbs. (19 Nm). After the bolts are tight, attempt to twist each injector back and forth a small amount by hand. The injectors should rotate smoothly. If not, the injector O-ring are probably not installed properly. Replace the O-rings as required.
20. Connect the injector harness connectors.
21. Install the cold start injector pipe.
22. Connect the fuel return hose to the delivery pipe.
23. Connect the fuel inlet hose to the delivery pipe using new gaskets. Install the pulsation damper and torque to 22 ft. lbs. (29 Nm).
24. Connect the spring to the throttle linkage and dashpot. Install the dashpot and link bracket.
25. Connect the vacuum sensing hose.
26. Connect the throttle and accelerator cables.
27. Install the air intake connector.
28. Connect the PCV hoses.
29. Connect the negative battery cable.
30. Start the engine and check for fuel leaks.

Corolla and 1988–89 MR2

1. Relieve the fuel system pressure.
2. Disconnect the negative battery cable. On MR2 with 4A-GZE engine, remove the throttle body.
3. Disconnect and tag the vacuum and PCV hoses. On MR2 with 4A-GZE engine, loosen the air outlet duct and remove the throttle cable bracket.
4. Disconnect the fuel return hose from the pressure regulator. On MR2 with 4A-GZE engine, remove the fuel pressure regulator.
5. Disconnect the injector harness connectors.
6. Remove the cold start injector pipe.
7. Disconnect the fuel inlet pipe.
8. Remove the delivery pipe bolts and remove the delivery pipe and fuel injectors from the cylinder head.

NOTE: When removing the delivery pipe, be careful not to drop the injectors.

9. On Corrola and MR2 with 4A-GE engine, remove the 2 spacers and 2 insulators from the cylinder head. On MR2 with 4A-GZE engine, there are 3 spacers and 4 insulators.
10. Remove the injectors from the delivery pipe. Remove the O-rings and grommets from the injectors and discard them.

To install:

11. Install new injector grommets and O-rings. Coat the O-rings with clean fuel prior to installation. Make sure the O-ring seats properly in the injector groove. If not, the O-ring will become pinched.
12. Install new spacers and insulators into the cylinder head.
13. Install the injectors into the delivery pipe using a moderate back and forth twisting motion.
14. Mount the injector and delivery pipe assembly onto the cylinder head.
15. Install the delivery pipe retaining bolts and torque them to 11–13 ft. lbs. (15–17 Nm). After the bolts are tight, attempt to twist each injector back and forth a small amount. The injectors should rotate smoothly. If not, the injector O-ring are probably not installed properly. Replace the O-ring(s) as required.
16. Connect the fuel inlet pipe using new gaskets. Torque the union bolt to 22 ft. lbs. (29 Nm).
17. Install the cold start injector pipe.
18. Connect the injector harness connectors.
19. Connect the fuel return hose to the fuel pressure regulator.
20. Connect the vacuum and PCV hoses.
21. Connect the negative battery cable.
22. Start the engine and check for leaks.

Camry and Celica (3S-FE Engine)

1. Relieve the fuel system pressure and disconnect the negative battery cable
2. Remove the cold start injector pipe.
3. Disconnect the vacuum sensing hose from the fuel pressure regulator.
4. Disconnect the injector harness connectors.
5. Disconnect the hose from fuel return pipe.
6. Remove the fuel pressure pulsation damper.
7. Remove the 2 bolts and the delivery pipe together with the 2 injectors attached.

NOTE: When removing the delivery pipe, be careful not to drop the injectors.

8. Remove the 4 insulators and the 2 spacers from the cylinder head. Pull out the four injectors from the delivery pipe.

To install:

9. Insert 4 new insulators and 2 spacers into the injector holes in the cylinder head.
10. Install the grommet and a new O-ring to the delivery pipe end of each injector.
11. Apply a thin coat of fuel to the O-ring on each injector and then press them into the delivery pipe.
12. Install the injectors together with the delivery pipe into the cylinder head. Tighten the 2 mounting bolts to 9 ft. lbs. (13 Nm). After the bolts are tight, attempt to twist each injector back and forth a small amount. The injectors should rotate smoothly. If not, the injector O-ring are probably not installed properly. Replace the O-ring(s) as required.
13. Install the fuel pressure pulsation damper with 2 new gaskets on the union bolt.
14. Connect the hose to the fuel return pipe.
15. Connect the injector harness connectors.
16. Connect the vacuum sensing hose to the fuel pressure regulator.

17. Install the cold start injector pipe.
18. Connect the negative battery cable.
19. Start the engine and check for leaks.

Camry (2VZ-FE Engine)

1. Relieve the fuel system pressure.
2. Disconnect the negative battery cable.
3. Drain the cooling system.
4. If equipped with automatic transaxle, disconnect the throttle cable from the bracket and throttle body.
5. Remove the air cleaner cap, air flow meter and air cleaner flexible hose.
6. Disconnect and tag all interfering vacuum hoses and electrical wiring harness connectors.
7. Remove the right engine mounting stay.
8. Disconnect the cold start injector connector.
9. Disconnect the cold start injector tube.
10. Disconnect the EGR pipe.
11. Remove the engine hanger and air intake chamber stay.
12. Remove the air intake chamber.
13. Disconnect harness connectors from the tops of the injectors.
14. Disconnect the wiring harness clamps from the left delivery pipe.
15. Disconnect the fuel hoses from the pressure regulator, fuel filter and delivery pipes.
16. Unbolt and pull the left and right delivery pipes with fuel injectors from the intake manifold.

NOTE: When removing the delivery pipes, be careful not to drop the injectors.

17. Remove the injectors from the delivery pipes.
18. Remove the 6 insulator and 4 spacers from the injector openings.

To install:

19. Insert 6 new insulators and 4 spacers into the injector openings in the intake manifold.
20. Install the grommet and a new O-ring to the delivery pipe end of each injector.
21. Apply a thin coat of fuel to the O-ring on each injector and then press them into the delivery pipe.
22. Mount the injectors together with the delivery pipes onto the intake manifold. Tighten the mounting bolts to 9 ft. lbs. (13 Nm). After the bolts are tight, attempt to twist each injector back and forth a small amount. The injectors should rotate smoothly. If not, the injector O-ring are probably not installed properly. Replace the O-ring(s) as required.
23. Connect the fuel hoses to the pressure regulator, fuel filter and delivery pipes. Use new gaskets on union bolt connections.
24. Connect the wiring harness clamps to the left delivery pipe.
25. Connect the harness connectors to the tops of the injectors.
26. Install the air intake chamber. Torque the mounting nuts to 32 ft. lbs. (43 Nm).
27. Install the air intake chamber stay and engine hanger.
28. Connect the EGR pipe and torque the pipe union nut to 58 ft. lbs. (78 Nm).
29. Connect the cold start injector tube and plug in the connector.
30. Install the right engine mounting stay.
31. Connect all vacuum hoses and electrical wiring harness connectors.
32. Install the air cleaner cap, air flow meter and air cleaner flexible hose.
33. If equipped with automatic transaxle, connect the throttle cable to the cable bracket and throttle body.
34. Fill the cooling system and connect the negative battery cable.
35. Start the engine and check for leaks.

Celica

3S-GE ENGINE

1. Disconnect the negative battery cable.
2. Relieve the fuel system pressure.
3. Drain the cooling system.
4. Disconnect the throttle cable and the accelerator cable from the throttle linkage.
5. Disconnect the ignition coil connector and the high tension cord, then remove the suspension upper brace.
6. Disconnect the air cleaner hose.
7. Remove the ignitor.
8. Remove the throttle body.
9. Remove the No. 2 engine hanger and the No. 2 intake manifold stay.
10. Loosen the union nut of the EGR pipe.
11. Remove the cold start injector pipe.
12. Remove the EGR modulator.
13. Tag and disconnect all hoses and wires which interfere with injector removal.
14. Raise and support the vehicle safely.
15. Remove the suspension lower crossmember.
16. Disconnect the exhaust pipe.
17. Remove the No. 1 and the No. 3 intake manifold stays.
18. Disconnect the ground strap.
19. Remove the intake manifold.
20. Remove the 3 bolts and the delivery pipe with the injectors.

NOTE: When removing the delivery pipes, be careful not to drop the injectors.

21. Remove the injectors from the delivery pipes.
22. Remove the insulators and spacers from the injector openings.

To install:

23. Insert new insulators and spacers into the injector openings in the intake manifold.
24. Install the grommet and a new O-ring to the delivery pipe end of each injector.
25. Apply a thin coat of fuel to the O-ring on each injector and then press them into the delivery pipe.
26. Mount the injectors together with the delivery pipes onto the intake manifold. Tighten the mounting bolts to 9 ft. lbs. (13 Nm). After the bolts are tight, attempt to twist each injector back and forth a small amount. The injectors should rotate smoothly. If not, the injector O-ring are probably not installed properly. Replace the O-ring(s) as required.
27. Installation of the remaining components is the reverse of the removal procedure. Fill the cooling system. Start the engine and check for fuel leaks.

3S-GTE ENGINE

1. Relieve the fuel system pressure and disconnect the negative battery cable.
2. Remove the throttle body.
3. Remove the fuel pressure regulator.
4. Remove the EGR vacuum modulator
5. Disconnect the electrical connections from fuel injectors.
6. Remove the pulsation damper. Disconnect fuel inlet hose from the delivery pipe.
7. Disconnect the fuel return hose from the return pipe.
8. Remove the delivery pipe and fuel injectors and related components (insulators, spacers, O-ring and grommet).

NOTE: When removing the delivery pipes, be careful not to drop the injectors.

9. Remove the injectors from the delivery pipes.
10. Remove the insulators and spacers from the injector openings.

To install:

11. Insert new insulators and spacers into the injector openings in the intake manifold.
12. Install the grommet and a new O-ring to the delivery pipe end of each injector.
13. Apply a thin coat of fuel to the O-ring on each injector and then press them into the delivery pipe.
14. Mount the injectors together with the delivery pipes onto the intake manifold. Tighten the mounting bolts

to 9 ft. lbs. (13 Nm). After the bolts are tight, attempt to twist each injector back and forth a small amount. The injectors should rotate smoothly. If not, the injector O-ring are probably not installed properly. Replace the O-ring(s) as required.

15. Installation of the remaining components is the reverse of the removal procedure. Start the engine and check for fuel leaks.

MR2

3S-GTE ENGINE

1. Relieve the fuel system pressure.
2. Disconnect the negative battery cable.
3. Remove the throttle body.
4. Remove the left engine hood side panel.
5. Remove the air cleaner.
6. Remove the charcoal canister.
7. Remove the EGR vacuum switching valve, vacuum modulator, EGR valve and pipe.
8. Remove the cold start injector pipe and cold start injector.
9. Remove the Idle Speed Control (ISC) water bypass hoses and air hoses.
10. Disconnect the vacuum sensing hose from the vacuum sensing pipe on the injector cover.
11. Disconnect the harness connectors from the tops of the injectors.
12. Disconnect the 2 wire clamps from the mounting bolts on the No. 2 timing cover. Disconnect the 2 wire clamps from the wire brackets on the intake manifold.
13. Disconnect the fuel inlet hose from the fuel filter.
14. Disconnect the fuel return hose from the fuel pressure regulator.
15. Remove the bolt that attaches the fuel inlet hose to the water outlet.
16. Remove the 3 bolts holding the delivery pipe to the cylinder head.
17. Disconnect the fuel inlet hose from the delivery pipe.
18. Remove the delivery pipe and fuel injectors and related components (4 insulators, 3 spacers and injector O-ring and grommets).

NOTE: When removing the delivery pipes, be careful not to drop the injectors.

19. Disconnect the vacuum sensing hose from the pressure regulator and remove the cover plate from the delivery pipe. Remove the injectors from the delivery pipe using the proper tool.

To install:

20. Insert 4 new insulators and 3 spacers into the injector openings.
21. Install the grommet and a new O-ring to the delivery pipe end of each injector.
22. Apply a thin coat of fuel to the O-ring on each injector and then press them into the delivery pipe. Make sure the injector connectors are positioned correctly.
23. Mount the injectors together with the delivery pipes. Tighten the mounting bolts to 14 ft. lbs. (19 Nm).
24. Installation of the remaining components is the reverse of the removal procedure. The injector harness connectors are color coded. The No. 1 and No. 3 injector connectors are brown and the No. 2 and No. 4 connectors are grey. Start the engine and check for fuel leaks.

Celica and MR2

5S-FE ENGINE

1. Disconnect the negative battery cable.
2. Remove the throttle body.
3. On MR2, remove the engine hood side panels, air cleaner and cruise control actuator.
4. Remove the cold start injector pipe.
5. On Celica, remove the fuel pressure regulator. On MR2, disconnect the brake booster vacuum hose from the intake manifold.
6. Disconnect necessary engine wiring and remove the left and right accelerator brackets.
7. Disconnect the electrical connectors from fuel injectors.
8. Disconnect wire retaining clamps from the No. 2 timing belt cover and and intake manifold as necessary to gain access for removal/installation of fuel injectors.
9. Disconnect the fuel return hose from the return pipe.
10. Remove the delivery pipe and fuel injectors and related components (insulators, spacers, O-ring and grommet).

NOTE: When removing the delivery pipes, be careful not to drop the injectors.

11. Remove injectors from the delivery pipe.

To install:

12. Insert new insulators and spacers into the injector openings in the intake manifold.
13. Install the grommet and a new O-ring to the delivery pipe end of each injector.
14. Apply a thin coat of fuel to the O-ring on each injector and then press them into the delivery pipe.
15. Mount the injectors together with the delivery pipes onto the intake manifold. Tighten the mounting bolts to 9 ft. lbs. (13 Nm). After the bolts are tight, attempt to twist each injector back and forth a small amount. The injectors should rotate smoothly. If not, the injector O-ring are probably not installed properly. Replace the O-ring(s) as required.
16. Installation of the remaining components is the reverse of the removal procedure. Start the engine and check for leaks.

Supra

1. Relieve the fuel system pressure.
2. Disconnect the negative battery cable.
3. Drain the cooling system.
4. Tag and disconnect all hoses and wires which interfere with injector removal.
5. Disconnect accelerator connecting rod.
6. On 7M-GE engine, remove the air intake connector. On 7M-GTE engine, remove the throttle body.
7. Remove the ISC valve and gasket.
8. Disconnect the injector connectors.
9. Disconnect the cold start injector tube from the delivery pipe.
10. Remove the pulsation damper and the 2 gaskets.
11. Remove the union bolts and 2 gaskets from the fuel return pipe support.
12. Remove the clamp bolts from the No. 1 fuel pipe and Vacuum Switching Valve (VSV).
13. Remove the union bolts and 2 gaskets from the pressure regulator.
14. Disconnect the fuel hose from the No. 2 fuel pipe.
15. Remove the clamp bolt and the return fuel pipe.
16. Loosen the locknut and remove the pressure regulator.
17. Remove the 3 bolts, and then remove the delivery pipe with the injectors.

NOTE: When removing the delivery pipe, be careful not to drop the injectors.

18. Remove the 6 insulators and the 3 spacers from the cylinder head, then pull out the injectors from the delivery pipe.

To install:

19. Before installing, apply a thin coat of gasoline to the O-ring on each injector and then press them into the delivery pipe.
20. Insert 6 new insulators into the injector hole of the cylinder head.
21. Install the black rings on the upper portion of each of the 3 spacers, then install the spacers on the delivery pipe mounting hole of the cylinder head.
22. Install the 3 spacers and bolts and torque to 13 ft. lbs. (After the bolts are tight, attempt to twist each injector back and forth a small amount. The injectors should rotate smoothly. If not, the injector O-ring are probably

not installed properly. Replace the O-ring(s) as required.

23. Fully loosen the locknut of the pressure regulator. Push the pressure regulator completely into the delivery pipe by hand, then turn the regulator counterclockwise until the outlet faces outward in the correct position. Torque the locknut to 18 ft. lbs. (24 Nm).

24. Install the No. 2 fuel pipe and clamp bolt.

25. Connect the fuel hose.

26. Install the union bolt and 2 new gaskets to the pressure regulator and torque the union bolt to 18 ft. lbs. (24 Nm).

27. Install the No. 1 fuel pipe, Vacuum Switching Valve (VSV) and clamp bolt.

28. Install the union bolt and 2 new gaskets to the support pipe and torque the union bolts to 22 ft. lbs. (30 Nm).

29. Install the pulsation damper and 2 new gaskets and torque to 29 ft. lbs. (39 Nm).

30. Connect the injector connectors.

31. Install the Idle Speed Control (ISC) valve with a new gasket and torque to 9 ft. lbs. (13 Nm).

32. Install the throttle body or the air intake connector.

33. Connect the accelerator connecting rod.

34. Connect all vacuum hoses and electrical wires.

35. Refill the cooling system and connect the negative battery cable.

36. Start the engine and check for leaks.

Cressida

5M-GE ENGINE

1. Disconnect the negative battery cable.
2. Relieve the fuel system pressure.
3. Remove the air intake chamber.
4. Remove the distributor.
5. Remove the fuel pipe.
6. Unplug the wiring connectors from the tops of the fuel injectors and remove the 2 plastic clamps that hold the wiring harness to the fuel delivery pipe.
7. Unscrew the 4 mounting bolts and remove the delivery pipe with the injectors attached. Do not remove the injector cover.

NOTE: When removing the delivery pipe, be careful not to drop the injectors.

8. Pull the injectors out of the delivery pipe.

To install:

9. Insert 6 new insulators into the injector holes on the intake manifold.
10. Install the grommet and a new O-ring to the delivery pipe end of each injector.
11. Apply a thin coat of gasoline to the O-ring on each injector and then press them into the delivery pipe.
12. Install the injectors together with the delivery pipe in the intake manifold. Tighten the mounting bolts to 13 ft. lbs. (17 Nm). After the bolts are tight, attempt to twist each injector back and forth a small amount. The injectors should rotate smoothly. If not, the injector O-ring are probably not installed properly. Replace the O-ring(s) as required.
13. Secure the injector wiring harness to the delivery pipe with the plastic clamps. Connect the harness connectors to the tops of the injectors.
14. Install the fuel pipe, distributor and air intake chamber.
15. Connect the negative battery cable.
16. Start the engine and check for leaks.

7M-GE ENGINE

1. Disconnect the negative battery cable and drain the cooling system.
2. Relieve the fuel system pressure.
3. Remove the throttle body.
4. Remove the Idle Speed Control (ISC) valve.
5. Disconnect the injector harness connectors.
6. Disconnect the cold start injector from the delivery pipe.
7. Disconnect the EGR Vacuum Switching Valve (VSV) connector.
8. Remove the union bolt and 2 gaskets from the delivery pipe and fuel filter.
9. Remove the clamp bolt and remove the No. 1 fuel pipe with the vacuum switching valve.
10. Disconnect the No. 3 PCV hose.
11. Disconnect the vacuum sensing hose.
12. Disconnect the fuel hose from the No. 2 fuel pipe.
13. Remove the union bolt and 2 gaskets from the pressure regulator.
14. Remove the clamp bolts and remove the No. 2 fuel pipe.
15. Loosen the locknut and remove the fuel pressure regulator.
16. Remove the delivery pipe attaching bolts. Remove the delivery pipe with the 6 fuel injectors.
17. Remove the 6 insulators and 3 spacers from the cylinder head.
18. Remove the injectors from the delivery pipe.

To install:

19. Install new injector grommets and O-rings. Coat the O-rings with clean fuel prior to installation. Make sure the O-ring seats properly in the injector groove. If not, the O-ring will become pinched.
20. Install new insulators into the cylinder head. Install the black rings on the upper portion of each spacer. Then, install the spacers into the mounting holes in the head.
21. Install the injectors into the delivery pipe using a moderate back and forth twisting motion.
22. Mount the injector and delivery pipe assembly onto the cylinder head. Make sure the injector connectors are facing up.
23. Install the delivery pipe retaining bolts and torque them to 13 ft. lbs. (18 Nm). After the bolts are tight, attempt to twist each injector back and forth a small amount by hand. The injectors should rotate smoothly. If not, the injector O-ring are probably not installed properly. Replace the O-rings as required.
24. Install the fuel pipes and pressure regulator.
25. Connect the vacuum hoses and injector harness connectors. Install the Idle Speed Control (ISC) valve and throttle body.
26. Fill the cooling system to the proper level and connect the negative battery cable.
27. Start the engine and check for leaks.

DRIVE AXLE

Halfshaft

REMOVAL & INSTALLATION

Tercel

1. Raise the vehicle and support it safely.
2. Remove the cotter pin and locknut cap.
3. Have an assistant step on the brake pedal and at the same time, loosen the bearing locknut.
4. Remove the brake caliper and position it aside. Remove the brake disc.
5. Remove the cotter pin and nut from the tie rod end. Using a suitable puller, disconnect the tie rod end from the steering knuckle.
6. Matchmark the lower strut mounting bracket where it attaches to the steering knuckle, remove the mounting bolts and then disconnect the steering knuckle from the strut bracket.
7. Using a suitable puller, pull the axle hub off the outer halfshaft end.
8. Remove the stiffener plate from the left side of the transaxle assembly.
9. Using the proper tool, tap the halfshaft out of the transaxle casing.

To install:

NOTE: Be sure to cover the halfshaft input opening.

10. During installation, observe the following:

a. Coat the oil seal in the transaxle input opening with MP grease before inserting the halfshaft.

b. On 1988–92 vehicles, torque the bolts to 166 ft. lbs.

c. Tighten the tie rod end nut to 36 ft. lbs. (49 Nm).

d. Tighten the bearing locknut to 137 ft. lbs. (186 Nm).

e. Tighten the stiffener plate bolts to 29 ft. lbs. (39 Nm).

f. Check the front wheel alignment.

Corolla FX/FX16
1988 Corolla (4A-GE)

1. Raise and support the vehicle safely.
2. Remove the cotter pin, locknut cap and locknut from the hub.
3. Remove the engine under cover. Remove the 6 nuts attaching the halfshaft (front driveshaft) to the transaxle (differential side gear).
4. Remove the brake caliper from the steering knuckle and support it aside with wire. Remove the rotor disc.
5. Disconnect the steering knuckle from the lower arm by removing the bolt and 2 nuts, then disconnect the lower arm from the steering knuckle.
6. Using a suitable puller, pull the axle hub from the halfshaft. Be sure to cover the dust boot with a shop rag to prevent damage to the the boot.

To install:

7. During installation, observe the following:

a. Torque the steering knuckle to 47–64 ft. lbs. (64 Nm) on 1988 FX and 105 ft. lbs. (142 Nm) on 1988 sedan and wagon.

b. Torque the caliper bolts to 65 ft. lbs. (88 Nm).

c. Torque the bearing nut to 137 ft. lbs. (186 Nm).

d. Torque the halfshaft nuts to 27 ft. lbs. (36 Nm).

1988–92 Corolla (Except 1988 4A-GE)

1. Raise and safely support the vehicle.
2. Remove the cotter pin and locknut cap.
3. Have an assistant step on the brake pedal and at the same time, loosen the bearing locknut.
4. Remove the engine undercovers and then drain the gear oil or fluid.
5. Remove the cotter pin and nut from the tie rod end. Using a suitable puller, disconnect the tie rod end from the steering knuckle.
6. Remove the mounting bolts and then disconnect the steering knuckle from the lower control arm.
7. Use a rubber mallet and drive the outer end of the shaft out of the axle hub.
8. Using the proper tools, tap or pry the halfshaft out of the transaxle casing.

To install:

NOTE: Be sure to cover the halfshaft input opening.

9. During installation, observe the following:

a. Coat the oil seal in the transaxle input hole with MP grease before inserting the halfshaft.

b. Tighten the steering knuckle-to-lower arm bolts to 105 ft. lbs. (142 Nm).

c. Tighten the tie rod end nut to 36 ft. lbs. (49 Nm).

d. Tighten the bearing locknut to 137 ft. lbs. (186 Nm).

e. Check that there is 0.08–0.12 in. (2–3mm) axial play on each shaft.

f. Check the front wheel alignment.

1988–89 Celica (2WD)

1. Raise and safely support the vehicle.
2. Remove the wheels.
3. Remove the cotter pin, cap and locknut from the hub.
4. Remove the engine under covers.
5. Drain the transmission fluid or the differential fluid on the GTS.
6. Remove the transaxle gravel shield on the GTS.
7. Loosen the 6 nuts attaching the inner end of the halfshaft to transaxle (all except Celica GTS).

NOTE: Wrap the exposed end of the halfshaft in an old shop cloth to prevent damage to it.

8. Remove the cotter pin from the tie end rod and then press the tie rod out of the steering knuckle. Remove the bolt and 2 nuts and disconnect the steering knuckle from the lower arm control.
9. On all but the GTS, use a 2-armed gear puller or equivalent, and press the halfshaft out of the steering knuckle.
10. On the GTS, mark a spot somewhere on the left halfshaft and measure the distance between the spot and the transaxle case. Using the proper tool, pull the halfshaft out of the transaxle.
11. On the GTS, use a 2-armed puller and press the outer end of the right halfshaft out of the steering knuckle. Use a pair of pliers to remove the snap-ring at the inner end and pull the halfshaft out of the center driveshaft.
12. On all but the GT-S, remove the snap-ring on the center shaft with a pair of pliers and then pull the center shaft out of the transaxle case.

To install:

13. When installing the center driveshaft on ST and GT vehicles, coat the transaxle oil seal with grease, insert the halfshaft through the bearing bracket and secure it with a new snap-ring.
14. Repeat Step 13 when installing the inner end of the right halfshaft on the GTS.
15. On the right halfshaft of the GTS, use a new snap-ring, coat the transaxle oil seal with grease and then press the inner end of the shaft into the differential housing. Check that the measurement made in Step 10 is the same. Check that there is 0.08–0.11 in. (2–3mm) of axial play. Check also that the halfshaft will not come out by trying to pull it back by hand.
16. Press the outer end of each halfshaft into the steering knuckle on the GTS.
17. On the ST and GT, press the outer end of the halfshafts into the steering knuckle and then finger-tighten the nuts on the inner end.
18. Connect the steering knuckle to the lower control arm and tighten the bolts to 94 ft. lbs. (127 Nm).
19. Connect the tie rod end to the steering knuckle and tighten the nut to 36 ft. lbs. (49 Nm). Install a new cotter pin.
20. Tighten the hub locknut to 137 ft. lbs. (186 Nm) while depressing the brake pedal. Install the cap and use a new cotter pin.
21. On the ST and GT, tighten the 6 nuts on the inner halfshaft ends to 27 ft. lbs. (36 Nm) while depressing the brake pedal.
22. Install the transaxle gravel shield on the GTS.
23. Fill the transaxle with gear oil or fluid.
24. Install the engine under cover.

1990 Celica (2WD)

NOTE: The hub bearing can be damaged if it is subjected to the vehicle weight such as moving the vehicle when the halfshaft is removed. If with ABS, after disconnecting halfshaft, work carefully so as not damage the sensor rotor serrations on the halfshaft.

1. Raise and safely support the vehicle. Remove the wheels.
2. Remove the cotter pin, cap and locknut (loosen locknut while depressing brake pedal) from the hub.
3. Remove the engine under covers.
4. Drain the transaxle fluid.
5. Remove the brake caliper and rotor disc.
6. Disconnect the tie rod end (remove cotter pin and nut) from the steering knuckle.
7. Disconnect steering knuckle from the lower arm.
8. Remove the halfshaft from the

steering knuckle using a suitable puller. Cover the halfshaft boot with shop cloth or equivalent to protect it from damage.

9. Remove the left side halfshaft using the proper tool.

10. Remove the the right side halfshaft. On the 5S-FE engine, remove the 2 bolts of the center bearing bracket and pull out the halfshaft with center bearing case and center halfshaft. On the 4A-FE engine, use a suitable brass punch tap out the right side halfshaft.

To install:

11. Install the left side halfshaft. Apply grease to the transaxle oil seal lip. Position the new snap-ring opening side facing downward using brass punch, tap halfshaft in until it makes contact with the pinion shaft. Install the outboard joint side of the halfshaft to the axle hub.

12. Install right side halfshaft on the 5S-FE engine using the following procedure:

a. Apply grease to the transaxle oil seal lip.

b. Insert the center halfshaft with the right side to the transaxle through the bearing bracket. When inserting the halfshaft, insert so the straight pin on the center bearing case aligns with the hole on the bearing bracket.

c. Install retaining bolts and torque to 47 ft. lbs. (64 Nm).

d. Install the outboard joint side of the halfshaft to the axle hub.

13. Install right side halfshaft on the 4A-FE engine using the following procedure:

a. Apply grease to the transaxle oil seal lip.

b. Position the new snap-ring opening side facing downward using brass punch, tap halfshaft in until it makes contact with the pinion shaft.

c. Install the outboard joint side of the halfshaft to the axle hub.

14. Check that the halfshaft will not come out by trying to pull it by hand.

15. Connect the steering knuckle to the lower control arm and tighten the bolts to 94 ft. lbs. (128 Nm).

16. Connect the tie rod end to the steering knuckle and tighten the nut to 36 ft. lbs. (49 Nm). Install a new cotter pin.

17. Install all necessary brake components. Tighten the hub locknut to 137 ft. lbs. (186 Nm) while depressing the brake pedal. Install the cap and use a new cotter pin.

18. Fill the transaxle to the proper level. Install the engine under cover. Check front wheel alignment.

Celica (AWD)

FRONT

NOTE: The hub bearing can be damaged if it is subjected to the vehicle weight such as moving the vehicle when the halfshaft is removed. On 1990–92 vehicles with ABS, after disconnecting halfshaft work carefully so as not damage the sensor rotor serrations on the halfshaft.

1. Raise and support the vehicle safely.

2. Remove the wheels.

3. Remove the cotter pin, cap and locknut from the hub.

4. Remove the transaxle gravel shield, if with manual transmission. Remove the engine under cover and front fender apron seal.

5. Remove the cotter pin and nut from the tie rod end and then disconnect it from the steering knuckle.

6. Remove the bolt and 2 nuts and disconnect the steering knuckle from the lower control arm.

7. Loosen the 6 nuts attaching the inner end of the halfshaft to the transaxle side gear shaft.

8. Grasp the halfshaft and push the axle carrier outward until the shaft can be removed from the side gear shaft.

NOTE: Wrap the exposed end of the halfshaft in an old shop cloth to prevent damage to it.

9. Use a rubber mallet and tap the outer end of the shaft from the axle hub.

To install:

10. Press the outer end of the halfshaft into the axle hub, position the inner end and install the 6 nuts finger-tight.

11. Connect the tie rod end to the steering knuckle and tighten the nut to 36 ft. lbs. (49 Nm). Install a new cotter pin. If the cotter pin holes do not align, tighten the nut until they align. Never loosen it.

12. Connect the steering knuckle to the lower control arm and tighten to 94 ft. lbs. (127 Nm).

13. Tighten the 6 inner shaft mounting nuts to 48 ft. lbs. (65 Nm). Measure the distance between the right and left side shafts; it must be less then 27.75 in. (704.7mm).

14. With the brake pedal depressed, install the bearing locknut and tighten it to 137 ft. lbs. (186 Nm). Install the cap and a new cotter pin.

15. Install the wheels and lower the vehicle.

REAR

1. Raise and safely support the vehicle. Remove the wheels.

2. Remove the cotter pin, locknut cap and bearing nut.

3. Scribe matchmarks on the inner joint tulip and the side gear shaft flange. Loosen and remove the 4 nuts.

4. Disconnect the inner end of the shaft by punching it upward and then pull the outer end from the axle carrier. Remove the halfshaft.

To install:

5. Position the halfshaft into the axle carrier and pull the inner end down until the matchmarks are aligned.

6. Connect the halfshaft to the side gear shaft and tighten the nuts to 51 ft. lbs. (69 Nm).

7. Install the bearing nut and tighten it to 137 ft. lbs. (186 Nm) with the brake pedal depressed. Install the cap and a new cotter pin.

8. Install the wheels and lower the vehicle.

Camry (2WD)

1988

1. Raise and support the vehicle safely.

2. Remove the wheels.

3. Remove the cotter pin, cap and locknut from the hub.

4. Remove the transaxle gravel shield, if with manual transaxle. Remove the engine under cover and front fender apron seal.

5. Loosen the 6 nuts attaching the inner end of the halfshaft to the transaxle or center shaft.

NOTE: Wrap the exposed end of the halfshaft in an old shop cloth to prevent damage to it.

6. Remove the brake caliper with the hydraulic line still attached, position it aside and suspend it with a wire. Remove the rotor.

7. Remove the 2 bolts attaching the ball joint to the steering knuckle. Pull the lower control arm down while pulling the strut outward; this will disconnect the inner end of the halfshaft from the transaxle.

8. Using a 2-armed puller, or the like, press the outer end of the halfshaft from the steering knuckle and then remove the halfshaft.

9. Drain the transaxle fluid, remove the snap-ring with pliers and pull the shaft out of the transaxle case.

To install:

10. When installing the center halfshaft, coat the transaxle oil seal with grease, insert the halfshaft through the bearing bracket and secure it with a new snap-ring.

11. Press the outer end of the halfshaft into the steering knuckle, position the inner end and install the 6 nuts finger-tight.

12. Reconnect the ball joint to the steering knuckle, if disconnected, and tighten the bolts to 94 ft. lbs. (127 Nm).

13. Install the rotor and brake cali-

per. Tighten the caliper-to-knuckle bolts to 65 ft. lbs. (88 Nm).

14. Tighten the wheel bearing locknut to 137 ft. lbs. (186 Nm) while depressing the brake pedal. Install the locknut cap and use a new cotter pin.

15. Tighten the 6 inner end nuts to 27 ft. lbs. (36 Nm) while depressing the brake pedal.

16. Install the transaxle gravel shield, if equipped.

17. Fill the transaxle to the proper level.

1989–92

1. Raise and support the vehicle safely.

2. Remove the front wheels.

3. Remove the cotter pin, cap and locknut from the hub.

4. Remove the engine under covers.

5. Drain the transmission fluid or the differential fluid on the wagon.

6. Remove the transaxle gravel shield on the wagon.

7. Loosen the 6 nuts attaching the inner end of the halfshaft to transaxle, all except wagon.

NOTE: Wrap the exposed end of the halfshaft in an old shop cloth to prevent damage to it.

8. Remove the cotter pin from the tie end rod and then press the tie rod out of the steering knuckle. Remove the bolt and 2 nuts and disconnect the steering knuckle from the lower arm control.

9. On all except 4 cylinder wagon, use a 2-armed gear puller or equivalent, and press the halfshaft out of the steering knuckle.

10. On the 4 cylinder wagon, mark a spot somewhere on the left halfshaft and measure the distance between the spot and the transaxle case. Using the proper tool, pull the halfshaft out of the transaxle.

11. On the 4 cylinder wagon, use a 2-armed puller and press the outer end of the right halfshaft out of the steering knuckle. Remove the snap-ring at the inner end and pull the halfshaft out of the center driveshaft.

12. On all except the 4 cylinder wagon, remove the snap-ring on the center shaft and pull the center shaft out of the transaxle case.

To install:

13. When installing the center driveshaft on sedan and V6 vehicles, coat the transaxle oil seal with grease, insert the halfshaft through the bearing bracket and secure it with a new snap-ring.

14. Repeat Step 13 when installing the inner end of the right halfshaft on the 4 cylinder wagon.

15. On the right halfshaft of the 4 cylinder wagon, use a new snap-ring, coat the transaxle oil seal with grease and then press the inner end of the shaft into the differential housing. Check that the measurement made in Step 10 is the same. Check that there is 0.08–0.12 in. (2–3mm) of axial play. Check also that the halfshaft will not come out by trying to pull it by hand.

16. Press the outer end of each halfshaft into the steering knuckle on the 4 cylinder wagon.

17. On all except the 4 cylinder wagon, press the outer end of the halfshafts into the steering knuckle and then finger-tighten the nuts on the inner end.

18. Connect the steering knuckle to the lower control arm and tighten the bolts to 83 ft. lbs. (113 Nm).

19. Connect the tie rod end to the steering knuckle and tighten the nut to 36 ft. lbs. (49 Nm). Install a new cotter pin.

20. Tighten the hub locknut to 137 ft. lbs. (186 Nm) while depressing the brake pedal. Install the cap and use a new cotter pin.

21. On all except the 4 cylinder wagon, tighten the 6 nuts on the inner halfshaft ends to 27 ft. lbs. (36 Nm) while depressing the brake pedal.

22. Install the transaxle gravel shield on the wagon.

23. Fill the transaxle with gear oil or fluid.

24. Install the engine under cover.

Camry (AWD)

FRONT – 1988

1. Raise and safely support the vehicle.

2. Remove the wheels.

3. Remove the cotter pin, cap and locknut from the hub.

4. Remove the transaxle gravel shield. Remove the engine under cover and front fender apron seal.

5. Remove the cotter pin and nut from the tie rod end and then disconnect it from the steering knuckle.

6. Remove the bolt and 2 nuts and disconnect the steering knuckle from the lower control arm.

7. Loosen the 6 nuts attaching the inner end of the halfshaft to the transaxle side gear shaft.

8. Grasp the halfshaft and push the axle carrier outward until the shaft can be removed from the side gear shaft.

NOTE: Wrap the exposed end of the halfshaft in an old shop cloth to prevent damage to it.

9. Use a rubber mallet and tap the outer end of the shaft from the axle hub.

To install:

10. Press the outer end of the halfshaft into the axle hub, position the inner end and install the 6 nuts finger-tight.

11. Connect the tie rod end to the steering knuckle and tighten the nut to 36 ft. lbs. (49 Nm). Install a new cotter pin. If the cotter pin holes do not align, tighten the nut until they align. Never loosen it.

12. Connect the steering knuckle to the lower control arm and tighten to 94 ft. lbs. (127 Nm).

13. Tighten the 6 inner shaft mounting nuts to 48 ft. lbs. (65 Nm). Measure the distance between the right and left side shafts; it must be less then 27.75 in. (704.7mm).

14. With the brake pedal depressed, install the bearing locknut and tighten it to 137 ft. lbs. (186 Nm). Install the cap and a new cotter pin.

15. Install the wheels and lower the vehicle.

FRONT – 1989–92

1. Raise and support the vehicle safely.

2. Remove the wheels.

3. Remove the cotter pin, cap and locknut from the hub.

4. Remove the engine undercovers.

5. Disconnect the tie rod end from the steering knuckle.

6. Disconnect the lower control arm at the steering knuckle and pull it down and aside.

7. Use a plastic hammer and carefully tap the outer end of the halfshaft until it frees itself from the axle hub.

8. Cover the outer boot with a rag and then remove the inner end of the halfshaft from the transaxle. Use the proper tools.

To install:

9. Coat the lip of the oil seal with grease and then carefully drive the inner end of the shaft into the transaxle until it makes contact with the pinion shaft.

NOTE: Be careful not to damage the boots when installing the halfshafts; also, position the boot snap-ring so the opening is facing downward.

10. Put the outer end of each shaft into the axle hub, being careful not to damage the boots.

11. Check that there is 0.08–0.12 in. (2–3mm) of axial play. Check also that the halfshaft will not come out by hand.

12. Connect the lower control arm to the steering arm and tighten the bolt to 83 ft. lbs. (113 Nm).

13. Connect the tie rod to the steering knuckle and tighten the nut to 36 ft. lbs. (49 Nm). Use a new cotter pin to secure it.

14. Install the axle bearing locknut and tighten it to 137 ft. lbs. (186 Nm) while stepping on the brake pedal. Install the locknut cap and then a new cotter pin.

15. Fill the transaxle with gear oil or fluid, install the undercovers and wheels. Lower the vehicle and check the front end alignment.

REAR—1988–92

1. Raise and support the vehicle safely. Remove the wheels.
2. Remove the cotter pin, locknut cap and bearing nut.
3. Scribe matchmarks on the inner joint tulip and the side gear shaft flange. Loosen and remove the 4 nuts.
4. Disconnect the inner end of the shaft by punching it upward and then pull the outer end from the axle carrier. Remove the halfshaft.
5. Position the halfshaft into the axle carrier and pull the inner end down until the matchmarks are aligned.
6. Connect the halfshaft to the side gear shaft and tighten the nuts to 51 ft. lbs. (69 Nm).
7. Install the bearing nut and tighten it to 137 ft. lbs. (186 Nm) with the brake pedal depressed. Install the cap and a new cotter pin.
8. Install the wheels and lower the vehicle.

MR2

NOTE: On 1990–92 vehicles with ABS, after disconnecting halfshaft, work carefully so as not damage the sensor rotor serrations on the halfshaft.

1. Raise and support the vehicle safely.
2. Remove the wheels.
3. Remove the cotter pin, cap and locknut from the hub.
4. Remove the transaxle gravel shield.
5. Loosen the 6 nuts attaching the inner end of the halfshaft to transaxle.

NOTE: Wrap the exposed end of the halfshaft in an old shop cloth to prevent damage to it.

6. If equipped with automatic transaxle, remove the 2 bolts holding the ball joint to the rear axle carrier and disconnect the lower arm from the rear axle carrier.
7. Also, if equipped with automatic transaxle, remove the cotter pin and nut using the proper tool. Disconnect the suspension arm from the rear axle carrier.
8. While holding the halfshaft, knock the outer end of the wheel hub assembly. Remove the halfshaft.

To install:

9. Press the outer end of the halfshaft into the wheel hub assembly.
10. Position the inner end of the halfshaft and install the 6 nuts finger-tight.
11. Install the transaxle gravel shield.
12. Tighten the wheel bearing locknut to 137 ft. lbs. (186 Nm) while depressing the brake pedal. Install the locknut cap and use a new cotter pin.
13. Tighten the 6 inner end nuts to 27 ft. lbs. (36 Nm) while depressing the brake pedal. Torque the suspension arm nut to 36 ft. lbs. (49 Nm) and the lower arm to rear axle carrier to 83 ft. lbs. (113 Nm).
14. Fill the transaxle to the proper level.

1988 Cressida

1. Raise and support the vehicle safely.
2. Place matchmarks on the halfshaft and flanges.
3. Remove the 4 nuts retaining the halfshaft to the differential and disconnect the halfshaft from the differential.
4. Remove the 4 nuts retaining the halfshaft to the axle shaft and disconnect the halfshaft from the axle shaft. Remove the halfshaft from the under the vehicle.
5. Installation is the reverse order of the removal procedure. Be sure to align the matchmarks on the halfshaft and torque the retaining nuts to 51 ft. lbs. (69 Nm).

Supra and 1989–92 Cressida

1. Raise and support the vehicle safely. Remove the rear wheels.
2. Using a suitable jack, raise the No. 2 suspension arm until it is horizontal. Matchmarks the rear halfshaft to the side gear shaft flange.
3. Remove the 6 retaining nuts (while an assistant is depressing the brake pedal) and disconnect the rear halfshaft from the differential.
4. Remove the cotter pin and locknut cap. Loosen and remove the bearing locknut.
5. Using a suitable hammer, tap out the rear halfshaft.
6. Installation is the reverse order of the removal procedure. Tighten the bearing locknut to 203 ft. lbs. (275 Nm) and the 6 halfshaft retaining bolts to 51 ft. lbs. (69 Nm).

CV-Boot

REMOVAL & INSTALLATION

1. Mount the halfshaft in a suitable holding fixture.
2. Remove the inboard joint boot clamps.
3. Place matchmarks on the inboard joint tulip and tripod.
4. Remove the inboard joint tulip from the halfshaft.
5. Remove the tripod joint snap-ring.
6. Place matchmarks on the shaft and tripod.
7. Using a brass punch or equivalent remove the tripod joint from the halfshaft.
8. Remove inboard joint boot.
9. Remove the outboard joint boot clamps and boot.
10. Installation is the reverse of the removal procedures. Pack all CV-joints with suitable grease. Use new boot retaining clamps and snap-rings as necessary.

Driveshaft and U-Joints

REMOVAL & INSTALLATION

1. Raise and support the rear axle housing safely.
2. Matchmark the driveshaft and companion flange. Unfasten the bolts which attach the driveshaft universal joint yoke flange to the mounting flange on the differential drive pinion.
3. If equipped with 3 universal joints, perform the following:
   a. Remove the driveshaft sub-assembly from the U-joint sleeve yoke.
   b. Remove the center support bearing from its bracket.
4. Remove the driveshaft end from the transmission.
5. Plug the transmission opening to keep the transmission oil from running out.
6. Remove the driveshaft.

To install:

7. Apply multipurpose grease on the section of the U-joint sleeve which is to be inserted into the transmission.
8. Insert the driveshaft sleeve into the transmission.

NOTE: Be careful not to damage any of the seals.

9. If equipped with 3 U-joints and center bearings, perform the following:
   a. Adjust the center bearing clearance with no load placed on the drive line components; the top of the rubber center cushion should be 0.04 in. (1mm) behind the center of the elongated bolt hole.
   b. Install the center bearing assembly. Use the same number of washers on the center bearing brackets as were removed.
   c. Matchmark the arrow marks on the driveshaft and grease fittings.
10. Align the matchmarks. Secure the U-joint flange to the differential pinion flange with the mounting bolts.

NOTE: Be sure the bolts are of the same type as those removed and that they are tightened securely.

11. Remove the axle housing supports and lower the vehicle.
12. Tighten the center bearing-to-bracket bolts to 30 ft. lbs. (40 Nm) on 1988 Cressida, 27 ft. lbs. (37 Nm) on 1989–92 Cressida or 36 ft. lbs. (49 Nm) on Supra. Tighten the flange bolts to 31 ft. lbs. (42 Nm) on 1988 Cressida or 54 ft. lbs. (74 Nm) on Supra and 1989–92 Cressida.

1988–92 Camry (AWD), 1988–92 Celica (AWD) 1989–92 Corolla (AWD), 1988 Tercel Wagon (AWD)

1. Matchmark the front driveshaft flange and the front center bearing flange. Remove the 4 bolts, washers and nuts and disconnect the rear end of the front driveshaft from the front center bearing flange. Pull the shaft out of the transfer case and remove it. Plug the transfer case to prevent leakage.
2. Depress the brake pedal and loosen the cross groove set bolts ½ turn. These bolts are at the front edge of the rear driveshaft; rear edge of the rear center bearing.
3. Matchmark the rear flange of the rear driveshaft to the differential pinion flange and then disconnect them.
4. Remove the 2 mounting bolts from the front and rear center bearings and then remove the 2 center bearings, intermediate shaft and rear driveshaft as an assembly.
5. Matchmark the universal joint and the rear center bearing flange, remove the bolts and separate the rear driveshaft from the rear center bearing.
6. Pull the front and rear center bearings from the intermediate shaft.

To install:

7. Install the 2 center bearings onto the intermediate shaft ends and then temporarily install the assembly.
8. Align the matchmarks and connect the rear driveshaft to the differential. Tighten the bolts to 54 ft. lbs. (74 Nm); 27 ft. lbs. (37 Nm) on the Corolla.
9. Press the front driveshaft yoke into the transfer case, align the matchmarks at the rear of the shaft with those on the front center bearing flange and tighten the bolts to 54 ft. lbs. (74 Nm); 27 ft. lbs. (37 Nm) on the Corolla.
10. With the front edge of the rear driveshaft in position, depress the brake pedal and tighten the cross groove joint set bolts to 20 ft. lbs. (27 Nm) on all except 1989–92 Celica. On 1989–92 Celica, torque the bolts to 48 ft. lbs. (65 Nm).
11. With the vehicle in an unladen condition, adjust the distance between the rear edge of the boot cover and the rear driveshaft to 2.58–2.78 in. (65.5–70.5mm) on all except 1989–92 Celica. On 1989–92 Celica adjust the distance to 2.85–3.05 in. (72.5–77.5mm).
12. With the vehicle in an unladen condition, adjust the distance between the rear side of the center bearing housing and the rear side of the cushion to 0.45–0.53 in. (11.5–13.5mm).
13. Tighten the center bearing mounting bolts to 27 ft. lbs. (37 Nm). Make sure the center line of the bracket is at right angles to the shaft axial direction.

Rear Axle Shaft, Bearing and Seal

NOTE: These service procedures apply to rear wheel drive, four wheel drive vehicles and MR2 only. For rear axle shaft service front wheel drive vehicles (except AWD), refer to "Rear Axle Hub, Carrier and Bearing" in the Rear Suspension section.

REMOVAL & INSTALLATION

Corolla (AWD) and Tercel (AWD)

1. Raise and support the vehicle safely.
2. Drain the oil from the axle housing.
3. Remove the rear wheels.
4. Punch matchmarks on the brake drum and the axle shaft to maintain rotational balance.
5. Remove the brake drum and related components.
6. Remove the rear bearing retaining nut.
7. Remove the backing plate attachment nuts through the access holes in the rear axle shaft flange.
8. Use a slide hammer with a suitable adapter to withdraw the axle shaft from its housing.

NOTE: Use care not to damage the oil seal when removing the axle shaft.

9. Repeat the procedure for the axle shaft on the opposite side.

NOTE: Be careful not to mix the components of the 2 sides.

10. Installation is performed in the reverse order of removal. Coat the lips of the rear housing oil seal with multipurpose grease prior to installation of the rear axle shaft. Always use new nuts, as they are the self-locking type.

1988 Cressida

1. Raise and safely support the vehicle.

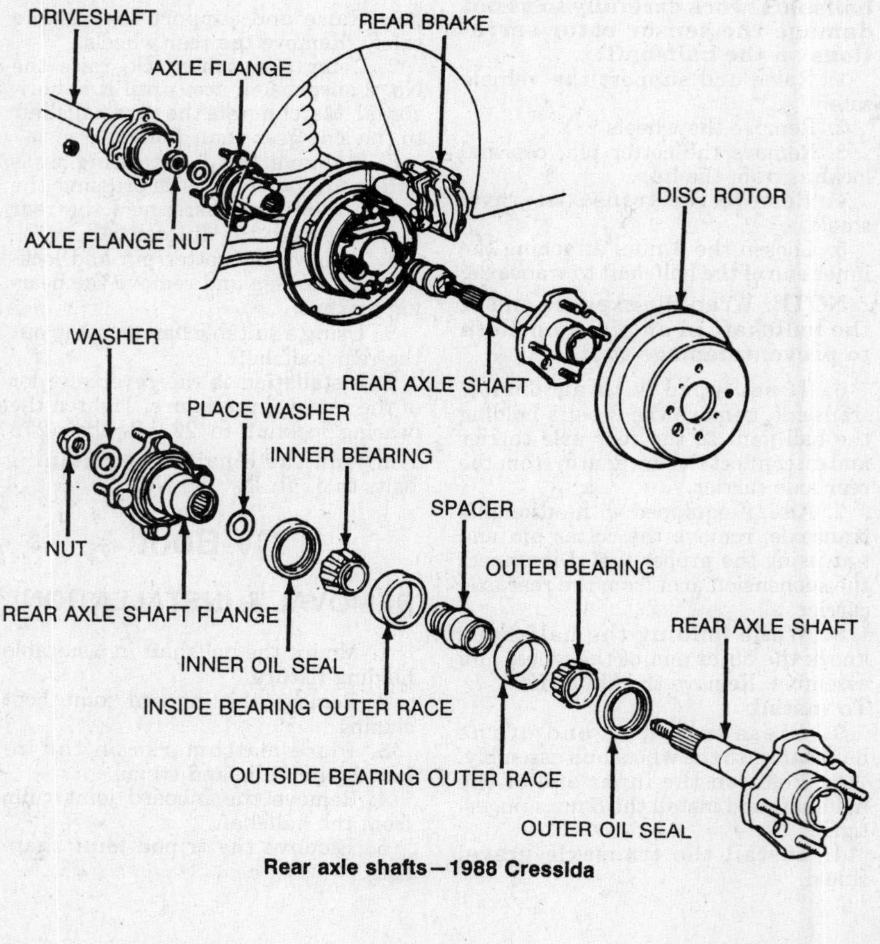

Rear axle shafts – 1988 Cressida

2. Disconnect the halfshaft from the axle flange and lower the halfshaft aside.
3. Apply the parking brake completely; pulled up as far as possible.
4. Remove the axle flange nut.

NOTE: The axle flange nut is staked in place. It will be necessary to loosen the staked part of the nut with a hammer and chisel, prior to loosening the nut.

5. Using the proper tools, disconnect the axle flange from the axle shaft. Be careful not to lose the plate washer from the bearing side of the flange.
6. Remove the parking brake shoes.
7. Using the proper tools, pull out the rear axle shaft, along with the oil seal and outer bearing.
8. Clean and inspect the bearings, races, and seal. If these parts are in good condition, repack the bearings with MP grease No. 2 and proceed to Step 15 to install the axle shaft.
9. Using a hammer and chisel, increase the clearance between the axle shaft hub and the outer bearing.
10. Using a puller installed with the jaws in the gap made in Step 9, pull the outer bearing from the axle shaft and remove the oil seal.
11. Drive the outer bearing race out of the hub with a brass drift and a hammer.

NOTE: Bearing and races must be replaced in matched sets. Do not use a new bearing with an old race or vice-versa.

12. Drive the new outer bearing race into the axle shaft hub until it is completely seated.

NOTE: The inner bearing race is replaced in the same manner as Steps 11 and 12.

13. Repack and install both bearings into the hub, being careful not to mis the bearings. The bearings should be packed with No. 2 multi-purpose grease.
14. Drive the seals into place. The inner seal should be driven to a depth of 1.22 in.; the outer to 0.217 in.

To install:

15. Apply a thin coat of grease to the axle shaft flange. Install the rear axle shaft into the housing and install the flange with the plate washer.
16. Using the proper tools, draw the axle shaft into the flange.
17. Install a new axle shaft flange nut. Torque the nut to 22–36 ft. lbs. (30–49 Nm). There should be no horizontal play evident at the axle shaft.
18. Turn the axle shaft back and forth and retorque the nut to 58 ft. lbs. (78 Nm).

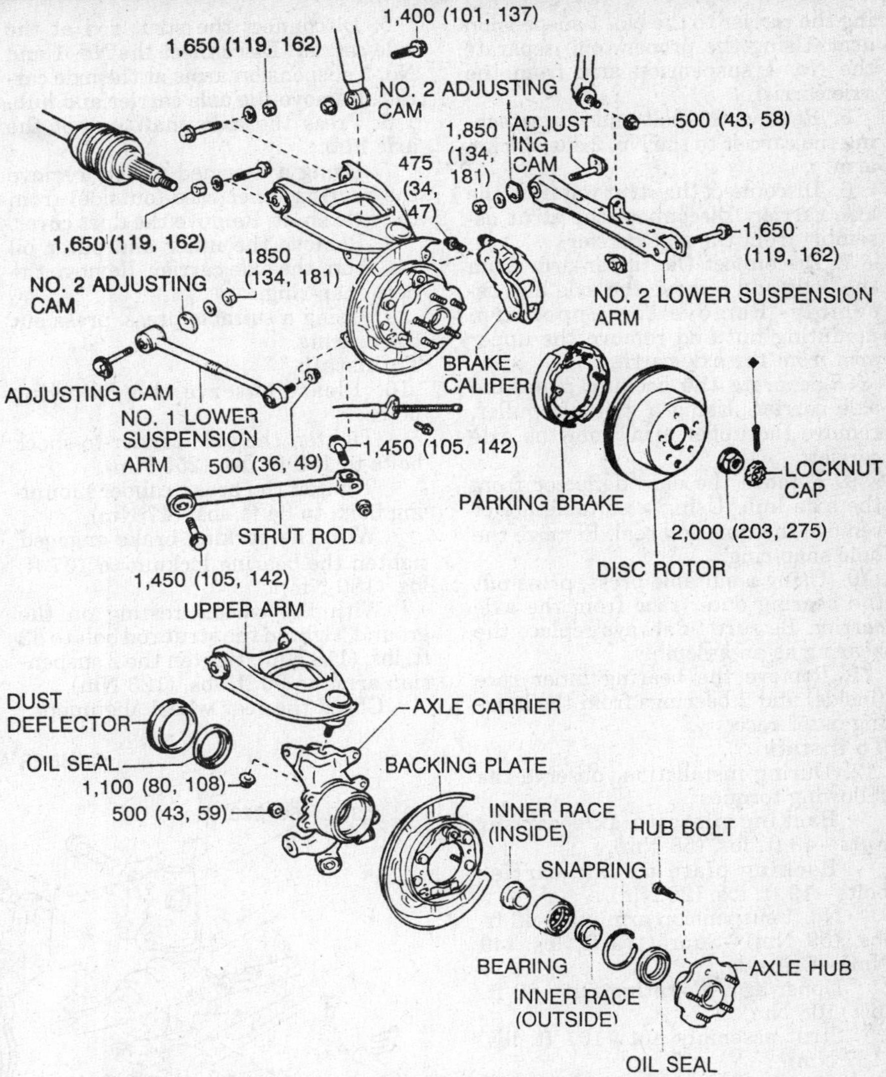

Rear axle, hub and bearing—1988–92 Cressida

19. Using a torque wrench, check the amount of torque required to turn the axle shaft. The correct rotational torque is 0.9–3.5 inch lbs.

NOTE: The shaft should be turned at a rate of 6 seconds per turn to attain a true rotational torque reading.

20. If the rotational torque is less than specified, tighten the nut 5–10 degrees at a time until the proper rotational torque is reached. Do not tighten the nut to more than 145 ft. lbs. (196 Nm).
21. If the rotational torque is greater than specified, replace the bearing spacer and repeat Steps 18–20, if necessary.
22. After the proper rotational torque is reached, restake the nut into position.
23. Install the parking brake shoes.
24. Connect the axle driveshaft to the flange and torque the nuts to 51 ft. lbs. (69 Nm).

NOTE: If the maximum torque is exceeded while retightening the nut, replace the bearing spacer and repeat Steps 18–20. Do not back off the axle shaft nut to reduce the rotational torque.

25. Install the rear wheel and lower the vehicle.

Supra and 1989–92 Cressida

1. Raise and support the vehicle safely.
2. Remove the rear wheel and tire assembly. Remove the disc brake caliper from the rear axle carrier and suspend it with wire. Remove the rotor disc.
3. Remove the rear driveshaft. Disconnect the parking brake cable assembly.
4. Remove the bolt and nut attach-

ing the carrier to the No. 1 suspension arm. Using the proper tool, separate the No. 1 suspension arm from the axle carrier.

5. Remove the bolt and nut attaching the carrier to the No. 2 suspension arm.

6. Disconnect the strut rod from the axle carrier. Disconnect the strut assembly from the axle carrier.

7. Disconnect the upper arm from the body and remove the axle hub assembly. Remove the upper arm mounting nut and remove the upper arm from the axle carrier.

8. Separate the backing plate and axle carrier. Using a suitable puller, remove the upper arm from the axle carrier.

9. Remove the dust deflector from the axle hub. Using a suitable puller remove the inner oil seal. Remove the hole snap-ring.

10. Using a suitable press, press out the bearing outer race from the axle carrier. Be sure to always replace the bearing as an assembly.

11. Remove the bearing inner race (inside) and 2 bearings from the bearing outer race.

To install:

12. During installation, observe the following torques:

Backing plate to axle carrier nuts – 43 ft. lbs. (58 Nm).

Backing plate to axle carrier bolts – 19 ft. lbs. (26 Nm).

No. 1 suspension arm nut – 43 ft. lbs. (59 Nm) – Supra. 36 ft. lbs. (49 Nm) – Cressida.

Upper arm mounting nut – 80 ft. lbs. (108 Nm).

Strut assembly nut – 101 ft. lbs. (137 Nm).

Upper arm to body bolt – 121 ft. lbs. (164 Nm) for Supra or 119 ft. lbs. (162 Nm) for Cressida.

No. 2 suspension arm to axle carrier – 121 ft. lbs. (164 Nm) for Supra or 119 ft. lbs. (162 Nm) for Cressida.

Strut rod to axle carrier – 121 ft. lbs. (164 Nm) for Supra. 105 ft. lbs. or (142 Nm) for Cressida.

Disc brake caliper bolts – 34 ft. lbs. (47 Nm).

1988–92 Camry (AWD) and Celica (AWD)

1. Raise the vehicle and support it safely.

2. Remove the rear wheel. Remove the disc brake caliper from the rear axle carrier and suspend it with wire. Remove the rotor disc.

3. Remove the rear halfshaft. Disconnect the parking brake cable assembly and remove the cable.

4. Remove the 2 axle carrier set nuts and the 2 bolts and then remove the camber adjusting cam.

5. Disconnect the strut rod at the axle carrier. Disconnect the No. 1 and No. 2 suspension arms at the axle carrier. Remove the axle carrier and hub.

6. Press the axle shaft out of the axle hub.

7. Using a 2-armed puller, remove the bearing inner race (outside) from the axle shaft. Remove the dust cover.

8. Remove the inner and outer oil seal from the axle carrier. Remove the hole snap-ring.

9. Using a suitable press, press out the bearing.

To install:

10. Please observe the following notes:

Tighten the axle carrier-to-shock bolts to 188 ft. lbs. (255 Nm).

Tighten the brake caliper mounting bolts to 34 ft. lbs. (47 Nm).

With the parking brake engaged, tighten the bearing locknut to 137 ft. lbs. (186 Nm).

With the wheels resting on the ground, tighten the strut rod bolt to 83 ft. lbs. (113 Nm); tighten the 2 suspension arms to 90 ft. lbs. (123 Nm).

Check the rear wheel alignment.

MR2

1. Raise and support the vehicle safely.

2. Remove the rear wheel and tire assembly. Remove the cotter pin, bearing locknut cap and bearing locknut.

3. Disconnect the parking brake cable. Remove the disc brake caliper from the rear axle carrier and suspend it with wire. Remove the rotor disc.

4. Disconnect the rear axle carrier from the lower arm. Remove the cotter pin and nut from the suspension arm. If equipped with ABS, remove the speed sensor from the axle carrier.

5. Using a suitable tool separate the suspension arm from the rear axle carrier.

6. Place matchmarks on the strut lower bracket and camber adjusting cam.

7. Remove the 2 axle carrier set nuts and 2 bolts with the camber adjusting cam. Remove the rear axle carrier and axle hub.

8. Remove the dust deflector from the axle hub. Using a suitable puller remove the inner oil seal. Remove the hole snap-ring.

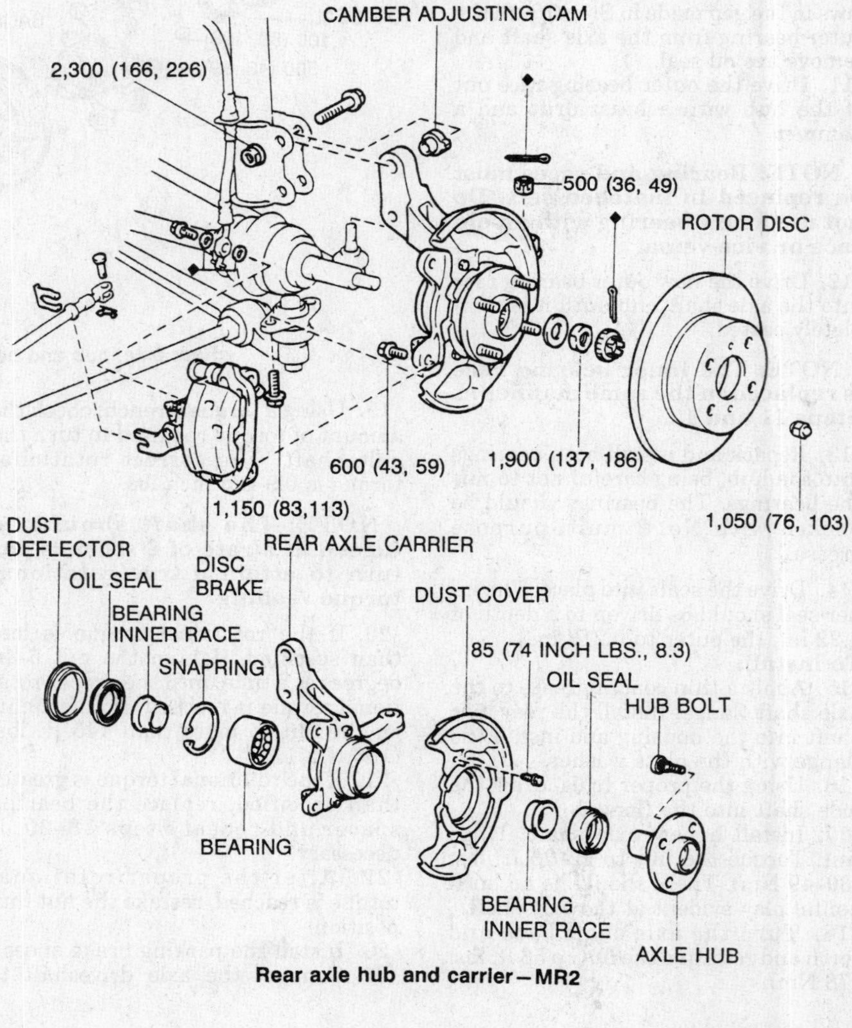

Rear axle hub and carrier – MR2

9. Remove the 3 bolts holding the disc brake dust cover to the rear axle carrier. Using a suitable puller remove the axle hub from the rear axle carrier.

10. Remove the bearing inner (inside) race. Using a suitable puller remove the bearing inner race (outside) from the rear axle hub.

11. Using a suitable puller remove the outer oil seal.

12. Remove the hub bearing by first placing the removed inner race (outside) in the bearing and using a suitable press, press out the bearing. Be sure to always replace the bearing as an assembly.

To install:

13. Installation is the reverse order of the removal procedure. During installation, observe the following torque specifications:

Two camber adjusting cam set bolts – 166 ft. lbs. (226 Nm).

Suspension arm nut – 36 ft. lbs. (49 Nm).

Rear axle carrier to the lower arm – 59 ft. lbs. (1988–92: 83 ft. lbs.).

Brake caliper – 43 ft. lbs. (59 Nm).

Bearing locknut – 137 ft. lbs. (186 Nm) all except 3S-GTE engine. On 3S-GTE engine torque the rear wheel bearing locknut to 217 ft. lbs. (294 Nm).

If equipped with ABS, torque the wheel sensor bolt to 74 inch lbs. (8.3 Nm).

Front Wheel Hub, Knuckle and Bearings

REMOVAL & INSTALLATION

Front Wheel Drive Vehicles Only

1. Raise and support the vehicle safely. Remove the front wheels.

2. Remove the cotter pin from the bearing locknut cap and then remove the cap.

3. Depress the brake pedal and loosen the bearing locknut.

4. Remove the brake caliper mounting nuts, position the caliper aside with the hydraulic line still attached and suspend it with a wire.

5. Remove the brake disc.

6. Remove the cotter pin and nut from the tie rod end and then, using a tie rod end removal tool, remove the tie rod.

7. Place matchmarks on the shock absorber lower mounting bracket and the camber adjustment cam, remove the bolts and separate the steering knuckle from the strut.

8. Remove the 2 ball joint attaching nuts and disconnect the lower control arm from the steering knuckle.

9. Carefully grasp the axle hub and knuckle assembly and pull it out from the halfshaft using the proper tool.

NOTE: Cover the halfshaft boot with a shop rag to protect it from any damage.

10. Clamp the steering knuckle in a vise and remove the dust deflector. Remove the nut holding the steering knuckle to the ball joint. Press the ball joint out of the steering knuckle.

11. Remove the dust deflector from the hub.

12. Pry out the bearing inner oil seal and then remove the hole snap-ring.

13. Remove the 3 bolts attaching the steering knuckle to the disc brake dust cover.

14. Remove the axle hub from the steering knuckle using the proper tool.

15. Remove the bearing inner race (inside).

16. Remove the bearing inner race (outside).

17. Remove the oil seal from the knuckle.

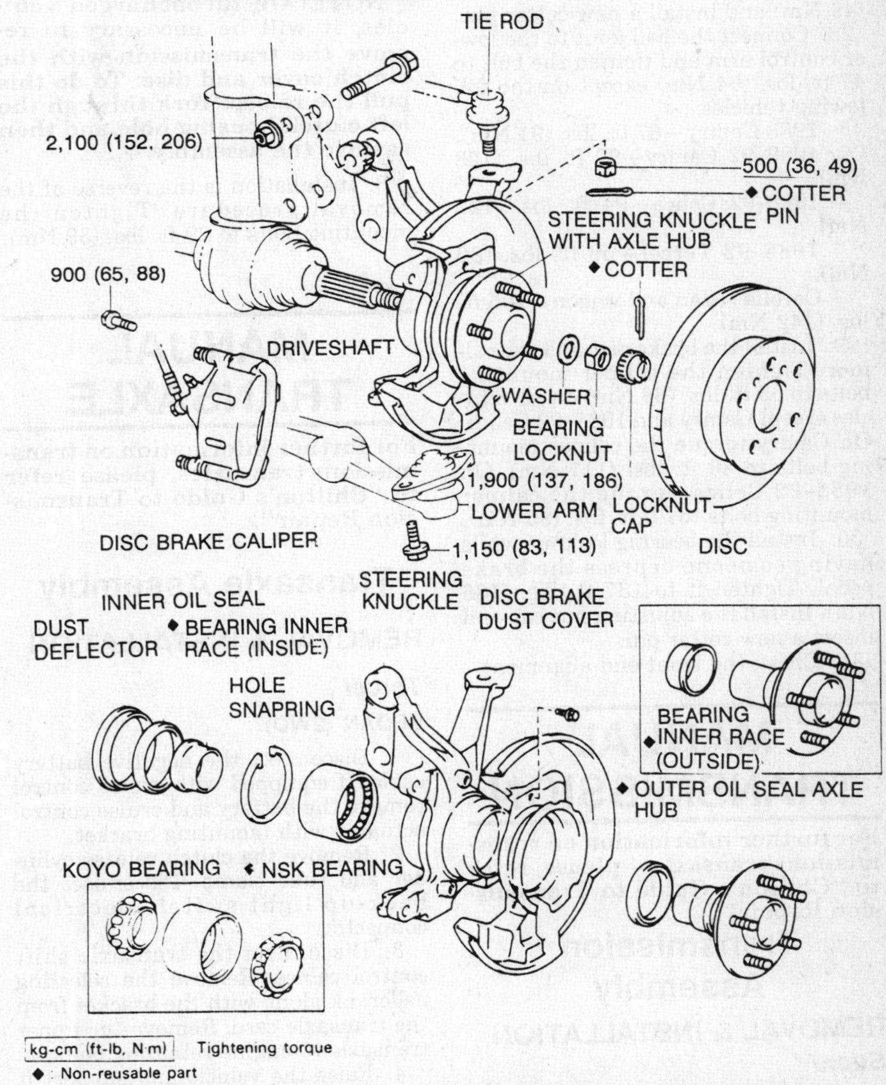

Front axle hub and steering knuckle assembly – front wheel drive models

18. Position an old bearing inner race (outside) on the bearing and then use a hammer and a drift to carefully knock the bearing out of the knuckle.

To install:

19. Press a new bearing into the steering knuckle.

20. Using a suitable oil seal installation tool, drive a new oil seal into the knuckle.

21. Install the disc brake dust cover onto the knuckle using liquid sealant.

22. Apply grease between the oil seal lip, oil seal and the bearing and then press the axle hub into the steering knuckle.

23. Install a new hole snap-ring into the knuckle.

24. Press a new oil seal onto the knuckle and coat the contact surface of the seal and the halfshaft with grease. Press a new dust deflector into the knuckle.

25. Position the ball joint on the steering knuckle and tighten the nut to 14 ft. lbs. (20 Nm). Remove the nut, install a new one and tighten it to 82 ft. lbs. (111 Nm). On 1988 Camry and Corolla torque the nut to 94 ft. lbs. (127 Nm).

26. Connect the knuckle assembly to the lower strut bracket. Insert the mounting bolts from the rear and make sure the matchmarks made earlier are in alignment. Tighten the nuts as follows:

105 ft. lbs. (142 Nm) on 1988 Celica.

188 ft. lbs. (255 Nm) on 1989–92 Celica.

166 ft. lbs. (226 Nm) on 1988 Camry and 1988–92 Tercel.

224 ft. lbs. (304 Nm) on 1989–92 Camry.

194 ft. lbs. (263 Nm) on Corolla sedan and wagon.

27. Connect the tie rod end to the knuckle, tighten the nut to 36 ft. lbs. (49 Nm) and install a new cotter pin.

28. Connect the ball joint to the lower control arm and tighten the bolt to 47 ft. lbs. (64 Nm) except on the following vehicles:

1988 Camry – 67 ft. lbs. (91 Nm).

1989–92 Camry – 90 ft. lbs. (123 Nm).

1988–92 Celica – 94 ft. lbs. (122 Nm).

1988–92 Tercel – 59 ft. lbs. (80 Nm).

Corolla sedan and wagon – 105 ft. lbs. (142 Nm).

29. Install the brake disc and the caliper. Tighten the caliper mounting bolts to 65 ft. lbs. (88 Nm) on all vehicles except Camry and 1988–92 Celica. On Camry torque the caliper mounting bolts to 86 ft. lbs. (117 Nm). On 1988–92 Celica, torque the caliper mounting bolts to 70 ft. lbs. (95 Nm).

30. Install the bearing locknut while having someone depress the brake pedal. Tighten it to 137 ft. lbs. (186 Nm). Install the adjusting nut cap and insert a new cotter pin.

31. Check the front end alignment.

MANUAL TRANSMISSION

For further information on transmissions/transaxles, please refer to "Chilton's Guide to Transmission Repair".

Transmission Assembly

REMOVAL & INSTALLATION

Supra

1. Disconnect the negative battery cable. Remove the center console trim panel. Remove the shift lever.

2. Raise and support the vehicle safely. Drain the transmission fluid. Remove the driveshaft.

3. Disconnect the front exhaust pipe from the tailpipe. Remove the front exhaust pipe.

4. Disconnect the speedometer cable. Disconnect the back-up light switch electrical connector. If equipped with ABS, disconnect the rear speed sensor electrical connector.

5. Remove the clutch release cylinder. Remove the starter assembly.

6. Support the engine and the transmission using the proper equipment. Remove the transmission support crossmember.

7. Remove the transmission mounting bolts. Remove the flywheel housing bolts. Carefully, move the transmission rearward, down, and out of the vehicle.

NOTE: On turbocharged vehicles, it will be necessary to remove the transmission with the clutch cover and disc. To do this pull the release fork through the left clutch housing hole and then remove the assembly.

8. Installation is the reverse of the removal procedure. Tighten the mounting bolts to 29 ft. lbs. (39 Nm).

MANUAL TRANSAXLE

For further information on transmissions/transaxles, please refer to "Chilton's Guide to Transmission Repair".

Transaxle Assembly

REMOVAL & INSTALLATION

Tercel

SEDAN (2WD)

1. Disconnect the negative battery cable. If equipped with cruise control remove the battery and cruise control actuator with mounting bracket.

2. Remove the clutch release cylinder and tube clamp. Disconnect the back-up light switch electrical connector.

3. Disconnect the transaxle shift control cables. Remove the selecting bellcrank along with the bracket from the transaxle case. Remove the upper transaxle-to-engine retaining bolts.

4. Raise the vehicle and support it safely. Remove the under covers. Drain the transaxle fluid. Disconnect the speedometer cable.

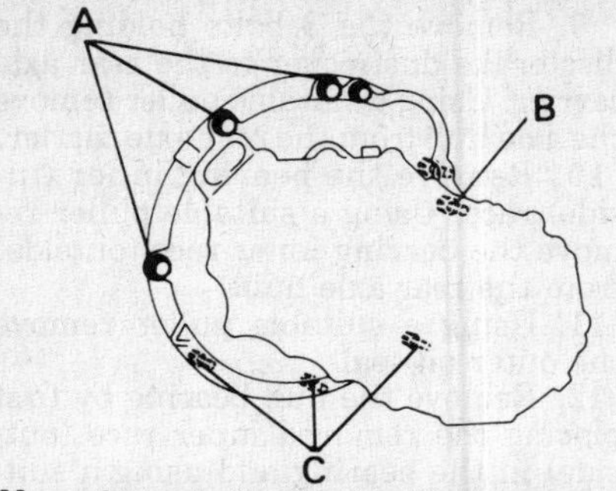

Manual transaxle mounting bolt locations on Tercel (2WD)

5. Disconnect both halfshafts. Remove the engine rear mounting brackets. Remove the starter assembly.

6. Support the engine and transaxle assembly using the proper equipment. Disconnect the left engine mounting.

7. Remove the remaining engine-to-transaxle retaining bolts. Carefully remove the transaxle assembly from the vehicle.

8. Installation is the reverse of the removal procedure. Tighten the transaxle-to-engine bolts to 47 ft. lbs. (64 Nm) – **A**; 34 ft. lbs. (46 Nm) – **B**; and 65 inch lbs. (7.4 Nm) – **C**. Tighten the front engine mount bracket bolts to 43 ft. lbs. (58 Nm). Tighten the rear engine mount bracket bolts to 21 ft. lbs. (28 Nm). Tighten the left engine mount bolts to 35 ft. lbs. (48 Nm) and the front and rear mount bolts to 47 ft. lbs. (64 Nm).

1988 WAGON (AWD)

1. Disconnect the negative battery cable. Remove the air cleaner assembly.

2. Remove the console. Remove the shift lever assembly. Remove the upper engine-to-transaxle retaining bolts.

3. Raise and support the vehicle safely. Drain the transaxle fluid. Remove both halfshafts.

4. On some vehicles, it will be necessary to remove the catalytic converter air inlet pipe. Remove the front exhaust pipe.

5. Disconnect the selector rod from the rear drive shift link lever. Disconnect the speedometer cable. Remove the right stiffener plate.

6. Disconnect the back-up light switch electrical connector. Disconnect the AWD and the low gear indicator switch electrical connectors.

7. Support the engine and the transaxle using the proper equipment. Remove the remaining transaxle-to-engine retaining bolts. Remove the rear crossmember assembly.

NOTE: Properly position a piece of wood between the engine

and the firewall so the assembly will not make contact with the power brake booster when it is removed.

8. Carefully remove the transaxle assembly from the vehicle.

9. Installation is the reverse of the removal procedure. Tighten the 14mm transaxle-to-engine bolts to 29 ft. lbs. (39 Nm) and the 17mm bolts to 43 ft. lbs. (59 Nm). Tighten the rear support member bolts to 70 ft. lbs. (95 Nm) and the right stiffener plate bolts to 29 ft. lbs. (39 Nm).

Corolla

2WD (EXCEPT C52 TRANSAXLE)

1. Disconnect the negative battery cable. Remove the air cleaner assembly.

2. Disconnect the back-up light switch electrical connector. Remove the speedometer cable. Disconnect the transmission control cables.

3. Raise the vehicle and support safely. Remove the water inlet from the transaxle. Remove the clutch release cylinder.

4. Remove the under cover. Remove the front and rear mounting. Remove the engine mounting center member.

5. Disconnect the halfshaft from the transaxle. Disconnect the steering knuckle from the lower control arm. Pull the steering knuckle outward and remove the left halfshaft.

6. Remove the starter. Disconnect the ground strap. Remove the No. 2 engine rear plate.

7. Support the engine and the transaxle using the proper equipment. Remove the left engine mounting.

8. Remove the engine-to-transaxle retaining bolts. Carefully remove the transaxle assembly from the vehicle.

9. Installation is the reverse of the removal procedure. Tighten the 12mm engine-to-transaxle bolts to 47 ft. lbs. (64 Nm) and the 10mm bolts to 34 ft. lbs. (46 Nm). Tighten the left engine mount bolts to 38 ft. lbs. (52 Nm). Tighten the front and rear engine mount bolts and the engine mounting center member bolts to 29 ft. lbs. (39 Nm).

2WD (C52 TRANSAXLE)

1. Disconnect the negative battery cable. Drain the radiator. Remove the air cleaner assembly. Remove the engine cooling fan assembly.

2. Disconnect the oxygen sensor electrical connector and the back-up light switch connector.

3. Remove the clutch release cylinder. It may be possible to leave the fluid lines attached to the cylinder.

4. Disconnect the water inlet from the transaxle. Disconnect the transaxle control cables. Disconnect the speedometer cable. Disconnect the ground cable.

5. Remove the starter. Remove the engine under covers. Remove the front exhaust pipe.

6. Disconnect the front and rear engine mountings. Remove the engine mounting center member.

7. Remove the left front wheel. Loosen the 6 nuts while depressing the brake pedal. Disconnect the halfshaft from the side gear shaft. Disconnect the lower ball joint from the lower control arm. Pull the shock absorber outward. Remove the halfshaft.

8. Support the engine and the transaxle assembly using the proper equipment. Remove the No. 2 engine rear plate. Remove the left engine mounting.

9. Remove the transaxle retaining bolts. Carefully remove the transaxle assembly from the vehicle.

10. Installation is the reverse of the removal procedure. Tighten the 12mm engine-to-transaxle bolts to 47 ft. lbs. (64 Nm) and the 10mm bolts to 34 ft. lbs. (46 Nm). Tighten the left engine mount bolts to 38 ft. lbs. (52 Nm). Tighten the front and rear engine mount bolts and the engine mounting center member bolts to 29 ft. lbs. (39 Nm).

AWD

1. Remove the engine and transaxle as an assembly.

2. Remove the rear end plate.

3. Disconnect the vacuum lines and then remove the transfer case vacuum actuator.

4. Remove the right and center transfer case stiffener plates.

5. Pull the transaxle out slowly until there is approximately 2.36–3.15 in. (60–80mm) clearance between the transaxle and the engine.

6. Turn the output shaft in a clockwise direction and then remove the transaxle.

To install:

7. Install the transaxle assembly to the engine and tighten the 10mm bolts to 34 ft. lbs. (46 Nm). Tighten the 12mm bolts to 47 ft. lbs. (64 Nm).

8. Tighten the 8mm stiffener plate bolts to 14 ft. lbs. (20 Nm) and the 10mm bolts to 27 ft. lbs. (37 Nm).

9. Tighten the rear end plate mounting bolts to 17 ft. lbs. (23 Nm).

10. Install the engine/transaxle assembly.

Camry

2WD

1. Disconnect the negative battery cable. Remove the clutch release cylinder and tube clamp. Remove the clutch tube bracket.

2. Disconnect the control cables. Disconnect the back-up light switch electrical connector. Remove the ground strap.

3. Remove the starter assembly. Remove the transaxle upper mounting bolts.

4. Raise and support the vehicle safely. Remove the under covers. Drain the transaxle fluid. Disconnect the speedometer cable.

5. Remove the suspension lower crossmember. Remove the engine mounting centermember.

6. Disconnect both driveshafts. Remove the center driveshaft. Disconnect the left steering knuckle from the lower control arm. Remove the stabilizer bar.

7. Properly support the engine and remove the left engine mount.

8. Properly support the transaxle assembly. Remove the engine-to-transaxle bolts, lower the left side of the engine and carefully ease the transaxle out of the engine compartment.

To install:

9. Install the transaxle and tighten the 12mm mounting bolts to 47 ft. lbs. (64 Nm) and the 10mm bolts to 34 ft. lbs. (46 Nm).

10. Tighten the left engine mount to 38 ft. lbs. (52 Nm). Tighten the 4 center engine mount bolts to 29 ft. lbs. (39 Nm). Tighten the front and rear engine mount bolts to 32 ft. lbs. (43 Nm).

11. Tighten the lower crossmember bolts to 153 ft. lbs. (207 Nm) – 4 outer bolts; and, 29 ft. lbs. (39 Nm) – 2 inner bolts.

AWD

1. Remove the engine/transaxle assembly.

2. Remove the transfer case stiffener plate and the exhaust pipe front brake.

3. Remove the left stiffener plate and the front engine mount.

4. Remove the left engine mount bracket and then separate the transaxle from the engine.

5. Installation is in the reverse order of removal. Tighten the 12mm transaxle-to-engine mounting bolts to 47 ft. lbs. (64 Nm) and the 10mm bolts to 34 ft. lbs. (46 Nm).

Celica

2WD

1. Disconnect the negative battery cable. On some vehicles, it may be necessary to remove the battery. Remove the air cleaner assembly.

2. Remove the clutch tube bracket. Disconnect the back-up light switch at the transaxle. Disconnect the speedometer and the engine ground strap.

3. Disconnect the transaxle control cable and position them aside.

4. Unbolt the clutch release cylin-

der. It may be possible to position it aside with the hydraulic line still attached.

5. Remove the upper transaxle retaining bolts. Raise the vehicle and support safely. Remove the engine undercover. Drain the transaxle fluid.

6. Disconnect the exhaust pipe from the manifold. Remove the lower suspension crossmember. Remove the starter assembly.

7. Properly support the engine and transaxle assembly. Remove the front and rear transaxle mounts. Remove the center engine mount.

8. Disconnect both halfshafts at the transaxle. Unbolt the steering knuckle from the suspension arm and pull it outward. Remove the left halfshaft.

9. On some vehicles, remove the No. 2 rear engine plate. With the engine properly supported remove the left engine mount.

10. Remove the engine-to-transaxle bolts, lower the left side of the engine and carefully ease the transaxle out of the engine compartment.

To install:

11. Install the transaxle and tighten the 12mm engine-to-transaxle bolts to 47 ft. lbs. (64 Nm) and the 10mm bolts to 34 ft. lbs. (46 Nm).

12. Tighten the left engine mount bolts to 38 ft. lbs. (52 Nm).

13. Tighten the center member bolts and the front and rear engine mount bolts to 29 ft. lbs. (39 Nm).

AWD

1. Remove the engine and transaxle assembly.

2. Separate the transaxle from the engine.

3. Installation is in the reverse order of removal. Tighten the 12mm engine-to-transaxle bolts to 47 ft. lbs. (64 Nm) and the 10mm bolts to 34 ft. lbs. (46 Nm). Tighten the left engine mount bolts to 38 ft. lbs. (52 Nm). Tighten the center member bolts and the front and rear engine mount bolts to 29 ft. lbs. (39 Nm).

MR2

1988–89

1. Disconnect the negative battery cable. Drain the radiator. Raise and support the vehicle safely. Drain the transaxle fluid.

2. Disconnect the back-up light switch and the speedometer cable at the transaxle. On 4A-GZE engine, remove the intercooler.

3. Loosen the mounting bolts and remove the water inlet from the transaxle.

4. Remove the engine undercover. Remove the fuel tank protector.

5. Disconnect the transaxle control cables at the transaxle and position them aside.

6. Remove the water hose clamp from the control cable bracket and then remove the No. 2 control cable bracket.

7. Remove the main control cable bracket and the clutch release cylinder. Position these components aside.

8. Disconnect the exhaust pipe from the manifold, remove the pipe bracket from the chassis and then remove the exhaust pipe assembly from the bracket.

9. Remove the transaxle protector. Disconnect the halfshaft from the side gear shaft. Remove the starter assembly.

10. Remove the No. 2 engine rear plate. Remove the front and rear engine mounts from the body.

11. Properly support the engine and remove the left engine mount.

12. Properly support the transaxle assembly. Remove the engine-to-transaxle bolts, lower the left side of the engine and carefully ease the transaxle out of the engine compartment.

13. Remove the side gear shaft from the transaxle.

14. Installation is the reverse of the removal procedure. Tighten the 12mm engine-to-transaxle bolts to 47 ft. lbs. (64 Nm) and the 10mm bolts to 34 ft. lbs. (46 Nm). On 1988–89 vehicles, tighten the left and rear engine mounts to 38 ft. lbs. (52 Nm). On 1990–92 vehicles, tighten the front engine mounting bracket thru-bolt to 71 ft. lbs. (96 Nm) and the rear engine mounting thru-bolt to 64 ft. lbs. (87 Nm).

1990–92

1. Disconnect the negative battery cable.

2. Raise the vehicle and support safely. Drain the transaxle fluid.

3. Remove the rear wheels and under covers.

4. Disconnect the halfshafts from the knuckle and stabilizer bar from the knuckle.

5. Disconnect the exhaust pipe from the manifold.

6. Remove the air cleaner.

7. Disconnect the shift cables, clutch release cylinder and speedometer cable.

8. Remove the starter motor, rear end plate and engine stiffener plate.

9. Install an engine holding fixture to support the engine and transaxle. Remove the engine and transaxle mounting brackets.

10. Place a suitable transaxle jack under the assembly, remove the bell housing bolts and remove the transaxle from the vehicle.

To install:

11. Place a suitable transaxle jack under the assembly, Install the transaxle bolts and torque to 47 ft. lbs. (64 Nm).

12. Install the engine and transaxle mounting brackets. Torque the engine mounts to 71 ft. lbs. (96 Nm), transaxle bolts to 47 ft. lbs. (64 Nm) and the stiffener plate bolts to 27 ft. lbs. (37 Nm).

13. Install the starter motor, rear end plate and engine stiffener plate.

14. Connect the shift cables, clutch release cylinder and speedometer cable.

15. Install the air cleaner.

16. Connect the exhaust pipe to the manifold and torque to 46 ft. lbs. (62 Nm).

17. Connect the halfshafts to the knuckle and stabilizer bar to the knuckle.

18. Install the rear wheels and under covers.

19. Lower the vehicle safely.

20. Connect the negative battery cable, refill the transaxle and check for leaks.

LINKAGE ADJUSTMENT

Manual transmission linkage adjustments are neither possible or necessary. Damaged bushings can cause shift cable problems.

CLUTCH

Clutch Assembly

REMOVAL & INSTALLATION

1. Disconnect the negative battery cable.

2. Remove the transmission or transaxle assembly from the vehicle.

NOTE: On some 1988–90 Supra's, the clutch assembly is removed along with the transmission. On the 1989–92 Corolla (AWD) and the 1988–92 Camry and Celica AWD, the engine and transaxle are removed from the vehicle as an assembly.

3. Matchmark the clutch cover to the flywheel.

4. Remove the clutch pressure plate retaining bolts small amounts in a criss-cross pattern to relieve the clutch disc spring tension.

5. Remove the clutch cover.

6. Remove the clutch disc.

7. Remove the retaining clip and withdraw the release bearing.

8. Remove the release fork and boot assembly.

To install:

9. Using a suitable clutch disc align-

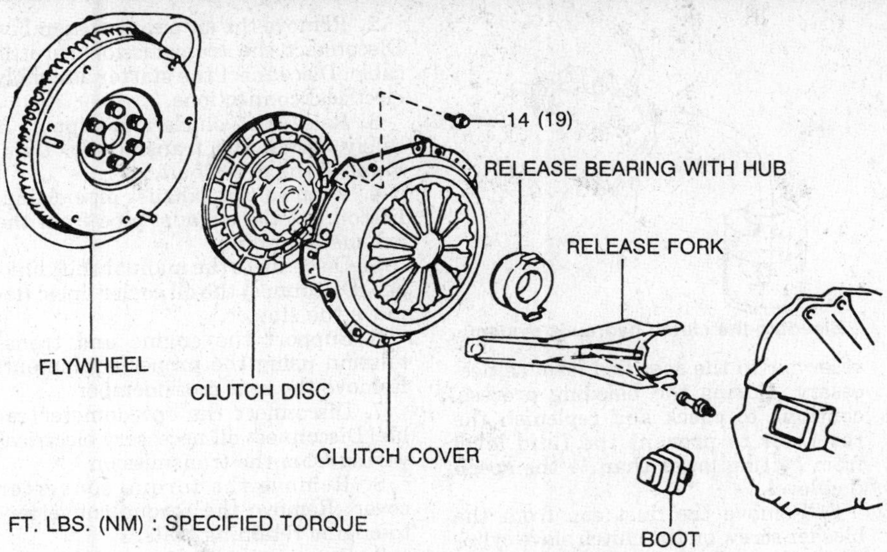

Tighten the clutch cover retaining bolts in a criss-cross pattern

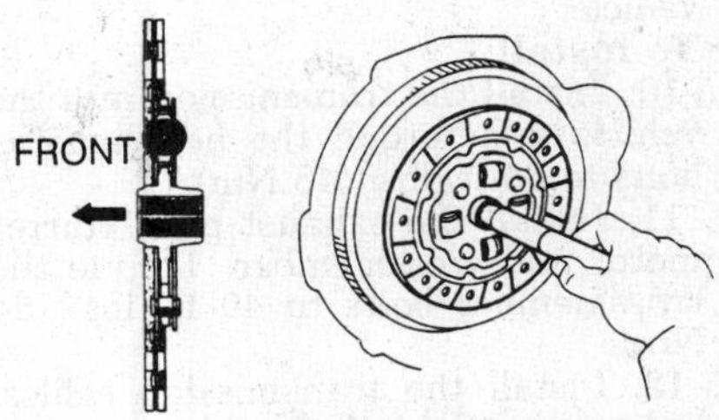

Use a clutch pilot tool to center the clutch disc on the flywheel

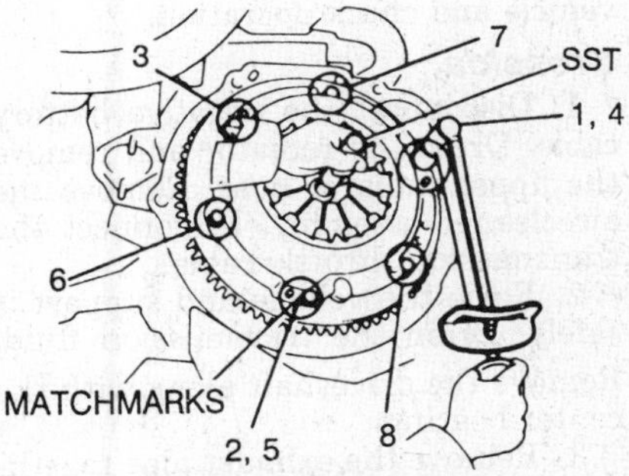

Typical clutch assembly

ment tool, install the clutch disc onto the flywheel.

10. Position the clutch cover onto the flywheel and align the matchmarks.
11. Install the clutch cover retaining bolts. Torque the bolts in a criss-cross pattern to 14 ft. lbs. (19 Nm).
12. Lubricate the release fork pivot and contact points, release bearing, bearing hub and input shaft spline surfaces with a suitable molybdenum disulphide lithium based or multi-purpose grease.
13. Install the boot, release fork, hub and bearing assemblies.
14. Install the transmission or transaxle.

PEDAL HEIGHT/FREE-PLAY ADJUSTMENT

Except 1988 Tercel Wagon

1. Adjust the clearance between the master cylinder piston and the pushrod to specification by loosening the pushrod locknut and rotating the pushrod while depressing the clutch pedal lightly.
2. Tighten the locknut when finished the adjustment.
3. Adjust the release cylinder free-play by loosening the release cylinder pushrod locknut and rotating the pushrod until proper specification is obtained.
4. Measure the clutch pedal free-play after performing the adjustments. If it fails to fall within specification, repeat the procedure. Pedal free-play specifications are as follows:

1988–92 Tercel – 0.2–0.59 in. (5–15mm)

1988 Corolla (RWD W/4A-LC) – 0.51–0.91 in. (13–23mm)

1988 Corolla (RWD W/4A-GE) – 0.2–0.59 in. (5–15mm)

1988 Corolla (FWD) – 0.28–0.67 in. (7–17mm)

1988–92 Corolla – 0.2–0.59 in. (5–15mm)

Celica – 0.2–0.59 in. (5–15mm)

Supra – 0.2–0.59 in. (5–15mm)

1988 Cressida – 0.2–0.59 in. (5–15mm)

MR2 – 0.2–0.59 in. (5–15mm).

1988 Tercel Wagon

1. Depress the clutch pedal several times.
2. Depress the clutch pedal until resistance is felt. Free-play should be within specification.

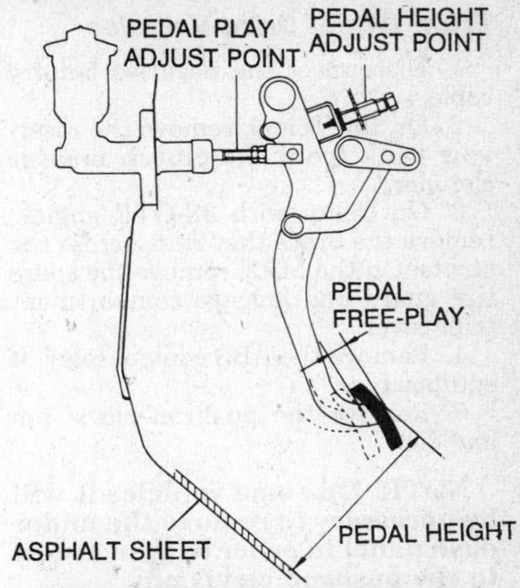

Typical clutch pedal adjustment points

3. Check the clutch release sector pawl. Six notches should remain between the pawl and the end of the sector. If less than 6 notches, replace the clutch disc. If the clutch disc has been replaced, the pawl should be 3–10 notches.
4. To obtain either the used or new position, change the position of the E-ring.

Clutch Cable

REMOVAL & INSTALLATION

1988 Tercel Wagon

1. Disconnect the negative battery cable.
2. Disconnect the sector tension spring from the clutch pedal.
3. Disconnect the clutch release cable from the release fork lever.
4. Turn the release sector toward the front side and disconnect the release cable from the release sector. Remove the release cable.
5. Installation is the reverse of the removal procedure.

Clutch Master Cylinder

REMOVAL & INSTALLATION

Rear Wheel Drive Vehicles

1. Disconnect the negative battery cable. Remove the pushrod clevis pin and clip.

NOTE: On some vehicles, it will be necessary to remove the under dash panel in order to gain access to the pushrod clevis pin.

2. Disconnect the fluid line.
3. Unbolts and remove the clutch master cylinder.
4. Installation is the reverse of the removal procedure. Bleed the system.

Front Wheel Drive Vehicles

1. Disconnect the negative battery cable.
2. On the Tercel, remove the reservoir tank from the clutch master clyinder.
3. On Celica with 3S-GTE engine, remove the brace that runs across the struts. On the MR2, remove the spare tire guard and luggage compartment trim cover.
4. Remove the ABS control relay, if equipped.
5. Remove the pushrod clevis pin and clip.

NOTE: On some vehicles it will be necessary to remove the under dash panel in order to gain access to the pushrod clevis pin.

6. On the 1988–92 Corolla (4A-GE) remove the brake booster.
7. Disconnect the fluid line and plug the end of the line to prevent leakage.
8. Unbolt and remove the clutch master cylinder.
9. Installation is the reverse of the removal procedure. Bleed the system.

Clutch Slave Cylinder

REMOVAL & INSTALLATION

Except 1990–92 MR2

1. Disconnect the negative battery cable. Raise and support the vehicle safely.
2. Remove the gravel shield, if equipped.
3. Disconnect the fluid line.
4. Remove the slave cylinder retaining bolts.
5. Remove the clutch slave cylinder from the vehicle.
6. Installation is the reverse of the removal procedure. Bleed the system.

1990–92 MR2

1. Disconnect the negative battery cable.
2. Raise and support the vehicle safely.
3. Remove the engine under cover.
4. Disconnect the control cables from the transaxle.
5. Disconnect the fluid line.
6. Support the engine and transaxle.
7. From the engine side, remove the engine front mounting bracket bolts.
8. Unbolt and remove the clutch slave cylinder.
9. Installation is the reverse of the removal procedure. Bleed the system.

Hydraulic Clutch System Bleeding

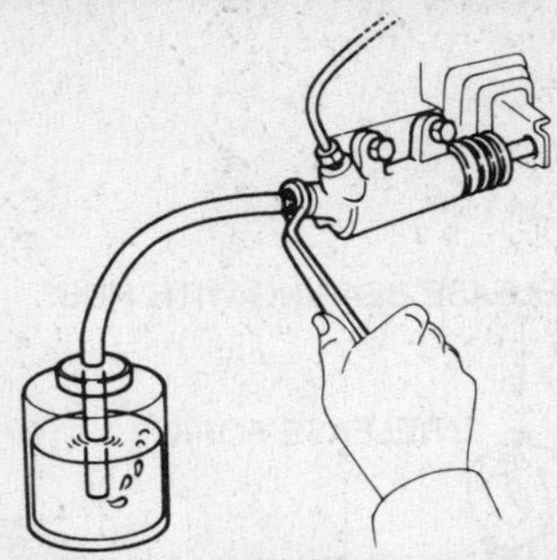

Bleeding the clutch hydraulic system

1. Check and fill the clutch fluid reservoir to the specified level as necessary. During the bleeding process, continue to check and replenish the reservoir to prevent the fluid level from getting lower than ½ the specified level.
2. Remove the dust cap from the bleeder screw on the clutch slave cylinder and connect a tube to the bleeder screw and insert the other end of the tube into a clean glass or metal container.

NOTE: Take precautionary measures to prevent the brake fluid from getting on any painted surfaces.

3. Pump the clutch pedal several times, hold it down and loosen the bleeder screw slowly.
4. Tighten the bleeder screw and release the clutch pedal gradually. Repeat this operation until air bubbles disappear from the brake fluid being expelled out through the bleeder screw.
5. Repeat until all evidence of air bubbles completely disappears from the fluid being pumped out of the tube.
6. When the air is completely removed tighten the bleeder screw and replace the dust cap.
7. Check and refill the master cylinder reservoir, as necessary.
8. Depress the clutch pedal several times to check the operation of the clutch and check for leaks.

AUTOMATIC TRANSMISSION

For further information on transmissions/transaxles, please refer to "Chilton's Guide to Transmission Repair".

Transmission Assembly

REMOVAL & INSTALLATION

Corolla

1. Disconnect the negative battery cable.
2. Remove the air cleaner assembly. Disconnect the transmission throttle cable. Disconnect the starter assembly electrical connections.
3. Raise the vehicle and support it safely. Drain the transmission fluid. Remove the driveshaft.
4. Remove the exhaust pipe clamp. Disconnect the exhaust pipe from the exhaust manifold.
5. Disconnect the manual shift linkage. Disconnect the oil cooler lines. Remove the starter.
6. Support the engine and transmission using the proper equipment. Remove the rear crossmember.
7. Disconnect the speedometer cable. Disconnect all necessary electrical wiring from the transmission.
8. Remove the torque converter cover. Remove the torque converter-to-engine retaining bolts.
9. Remove the bolts retaining the transmission to the engine. carefully remove the transmission from the vehicle.

To install:

10. Install the transmission into the vehicle and torque the bell housing bolts to 35 ft. lbs. (48 Nm).
11. Install the exhaust pipe, starter motor and crossmember. Torque the crossmember bolts to 40 ft. lbs. (54 Nm).
12. Install the transmission cables, linkage and driveshaft.
13. Refill the transmission with the approved fluid and check for leaks.
14. Lower the vehicle. Road test the vehicle and check operation.

Cressida

1. Disconnect the negative battery cable. Drain the radiator and remove the upper radiator hose. Remove the air cleaner assembly. Disconnect the transmission throttle cable.
2. Raise the vehicle and support it safely. Drain the transmission fluid. Remove the driveshaft along with the center bearing.
3. Remove the exhaust pipe together with the catalytic converter. Disconnect the manual shift linkage. Remove the speedometer cable.
4. Disconnect the oil cooler lines. As necessary, remove the transmission oil filler tube. As required, remove the starter assembly. Remove the speedometer cable.
5. Remove both stiffener plates and the catalytic converter cover from the transmission housing and cylinder block.
6. Support the engine and transmission properly. Remove the rear crossmember.
7. Remove the torque converter cover. Remove the torque converter-to-engine retaining bolts.
8. Remove the bolts retaining the

transmission to the engine. carefully remove the transmission from the vehicle.

To install:

9. Install the transmission into the vehicle and torque the bell housing bolts to 35 ft. lbs. (48 Nm).
10. Install the exhaust pipe, starter motor and crossmember. Torque the crossmember bolts to 47 ft. lbs. (64 Nm). Torque the torque converter bolts to 20 ft. lbs. (27 Nm).
11. Install the transmission cables, linkage and driveshaft.
12. Refill the transmission with the approved fluid and check for leaks.
13. Lower the vehicle. Road test the vehicle and check operation.

Supra

1. Disconnect the negative battery cable. Remove the air cleaner assembly. Disconnect the transmission throttle cable.
2. Raise the vehicle and support it safely. Drain the transmission fluid. Disconnect the electrical connectors for the neutral safety switch and back-up lights.
3. Remove the intermediate driveshaft along with the center bearing. Disconnect the exhaust pipe from the tail pipe.
4. Disconnect the transmission oil cooler lines. Plug the lines to prevent leakage. Disconnect the manual shift linkage and speedometer cable.
5. Remove the exhaust pipe bracket and torque converter cover.
6. Remove both stiffener brackets.
7. Support the engine and transmission using the proper equipment. Remove the rear crossmember.
8. Remove the engine under cover. Remove the torque converter-to-engine retaining bolts. Remove the starter.
9. Remove the bolts retaining the transmission to the engine. carefully remove the transmission from the vehicle.

To install:

10. Install the transmission into the vehicle and torque the bell housing bolts to 35 ft. lbs. (48 Nm).
11. Install the exhaust pipe, starter motor and crossmember. Torque the crossmember bolts to 40 ft. lbs. (54 Nm).
12. Install the transmission cables, linkage and driveshaft. Adjust the throttle cable.
13. Refill the transmission with the approved fluid and check for leaks.
14. Lower the vehicle. Road test the vehicle and check operation.

SHIFT LINKAGE ADJUSTMENT

Corolla

1. Loosen the nut on the shift linkage.
2. Push the selector lever all the way to the rear of the vehicle.
3. Return the lever 2 notches to the **N** shift position.
4. While holding the selector lever slightly toward the **R** shift position tighten the connecting rod nut.

Cressida and Supra

1. Loosen the nut on the shift linkage. Push the selector lever all the way to the rear of the vehicle.
2. Return the lever 2 notches to the **N** shift position.
3. While holding the selector lever slightly toward the **R** shift position, tighten the connecting rod nut.

THROTTLE LINKAGE ADJUSTMENT

1. Remove the air cleaner.
2. Confirm that the accelerator linkage opens the throttle fully. Adjust the linkage as necessary.
3. Peel the rubber dust boot back from the throttle cable.
4. Loosen the adjustment nuts on the throttle cable bracket (cylinder head cover) just enough to allow cable housing movement.
5. Have an assistant depress the accelerator pedal fully.
6. Adjust the cable housing so the gap between its end and the cable stop collar is 0.04 in. (1mm).
7. Tighten the adjustment nuts. Make sure the adjustment has not changed. Install the dust boot and the air cleaner.

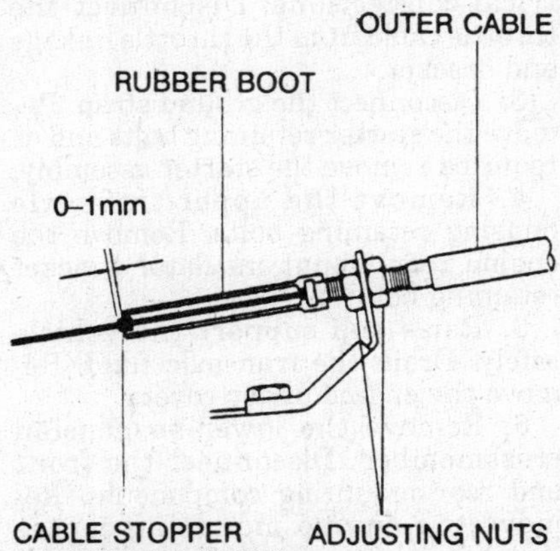

Throttle cable adjustment

AUTOMATIC TRANSAXLE

For further information on transmissions/transaxles, please refer to "Chilton's Guide to Transmission Repair".

Transaxle Assembly

REMOVAL & INSTALLATION

Tercel

1. Disconnect the negative battery cable. Drain the radiator and remove the upper radiator hose, as required. Remove the air cleaner assembly.
2. Raise the vehicle and support it safely. Remove both halfshafts. Drain the fluid from the transaxle and differential, if equipped.
3. Remove the torque converter cover. Remove the bolts that retain the torque converter to the crankshaft. Remove the exhaust pipe. Remove the shift lever rod.
4. Remove the speedometer cable and back-up light connector. If equipped with AWD, remove the electrical solenoid connector. Disconnect and remove all throttle linkage.
5. Remove the fluid lines from the transaxle. Remove the starter assembly, as required. On AWD vehicles, remove the rear driveshaft.
6. Support the engine and transaxle using a suitable jack. Remove the rear crossmember.
7. Remove the transaxle-to-engine retaining bolts. Separate the transaxle from the engine and carefully remove it from the vehicle.

To install:

8. Install the transaxle and tighten the transaxle-to-engine bolts to 47 ft. lbs. (64 Nm). Tighten the left engine mount bracket bolts to 32 ft. lbs. (43 Nm). Tighten the rear engine mount bracket bolts to 43 ft. lbs. (58 Nm). Tighten the torque converter mounting bolts to 13 ft. lbs. (18 Nm).
9. Install the transaxle cables, linkage and halfshafts.
10. Refill the transaxle with the approved fluid and check for leaks.
11. Lower the vehicle. Road test the vehicle and check operation.

Corolla

1. Disconnect the negative battery cable. Remove the air cleaner.
2. Disconnect the neutral start switch. Disconnect the speedometer cable.
3. Disconnect the shift control cable and throttle linkage.
4. Disconnect the oil cooler hose. Plug the end of the hose to prevent leakage.
5. Drain the radiator and remove the water inlet pipe.
6. Raise and support the vehicle safely. Drain the transaxle fluid. As required remove the exhaust front pipe.
7. Remove the engine undercover. Remove the front and rear transaxle mounts.
8. Support the engine and transaxle

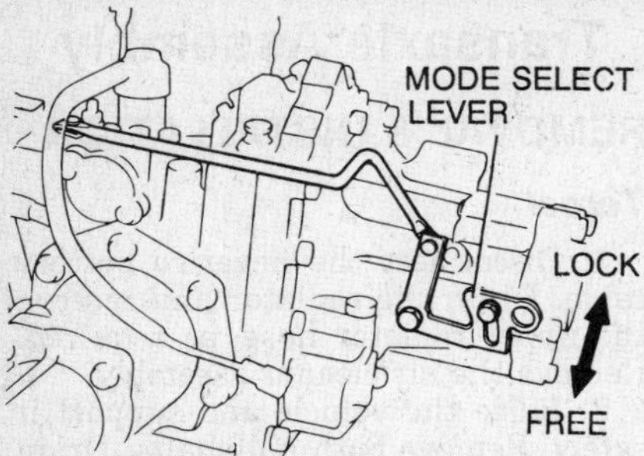

Setting the mode selector—Corolla w/A241H

using the proper equipment. Remove the engine center support member.

9. Remove the halfshafts. Remove the starter assembly. Remove the steering knuckles, as required.

10. Remove the flywheel cover plate. Remove the torque converter bolts.

11. Remove the left engine mount. Remove the transaxle-to-engine bolts. Slowly and carefully back the transaxle away from the engine. Lower the assembly to the floor.

To install:

12. Install the transaxle and tighten the transaxle-to-engine bolts to 47 ft. lbs. (64 Nm). Tighten the left engine mount bracket bolts to 32 ft. lbs. (43 Nm). Tighten the rear engine mount bracket bolts to 43 ft. lbs. (58 Nm). Tighten the torque converter mounting bolts to 13 ft. lbs. (18 Nm).

13. When installing the A241H vehicle transaxle on AWD vehicles, be sure the mode selector lever is positioned in the **FREE** mode and attach the lock bolt.

14. Install the transaxle cables, linkage and halfshafts.

15. Refill the transaxle with the approved fluid and check for leaks.

16. Lower the vehicle. Road test the vehicle and check operation.

Camry

1. Disconnect the negative battery cable. Remove the air flow meter and the air cleaner assembly.

2. Disconnect the transaxle wire connector. Disconnect the neutral safety switch electrical connector.

3. Disconnect the transaxle ground strap. Disconnect the throttle cable from the throttle linkage.

4. Remove the transaxle case protector. Disconnect the speedometer cable and control cable.

5. Disconnect the oil cooler hoses. Remove the upper starter retaining bolts, as required remove the starter assembly. Remove the upper transaxle housing bolts. Remove the engine rear mount insulator bracket set bolt.

6. Raise and support the vehicle safely. Drain the transaxle fluid.

7. Remove the left front fender apron seal. Disconnect both driveshafts.

8. Remove the suspension lower crossmember assembly. Remove the center driveshaft.

9. Remove the engine mounting center crossmember. Remove the stabilizer bar. Remove the left steering knuckle from the lower control arm.

10. Remove the torque converter cover. Remove the torque converter retaining bolts.

11. Properly support the engine and transaxle assembly. Remove the rear engine mounting bolts. Remove the remaining transaxle to engine retaining bolts.

12. Carefully remove the transaxle assembly from the vehicle.

To install:

13. Install the transaxle and tighten the 12mm transaxle housing bolts to 47 ft. lbs. (64 Nm); tighten the 10mm bolts to 34 ft. lbs. (46 Nm). Tighten the rear engine mount set bolts to 38 ft. lbs. (52 Nm). Tighten the torque converter mounting bolts to 20 ft. lbs. (27 Nm).

14. Install the transaxle cables, linkage and halfshafts.

15. Refill the transaxle with the approved fluid and check for leaks.

16. Lower the vehicle. Road test the vehicle and check operation.

Celica

1. Disconnect the negative battery cable. Remove the air flow meter and the air cleaner hose.

2. Disconnect the speedometer cable. Remove the starter assembly electrical connections. Disconnect the throttle cable from the throttle linkage and bracket.

3. Disconnect the ground strap. Remove the starter retaining bolts and as required remove the starter assembly.

4. Remove the upper transaxle housing retaining bolts. Remove the engine rear mount insulator bracket retaining bolt.

5. Raise and support the vehicle safely. Drain the transaxle fluid. Remove the engine under covers.

6. Remove the lower suspension crossmember. Disconnect the front and rear mounting components. Remove the engine mounting center member.

7. Remove the left halfshaft. Disconnect the right halfshaft.

8. Disconnect the exhaust pipe from the manifold. Remove the stiffener plate. Disconnect the control cable.

9. Disconnect the oil cooler hoses. Remove the torque converter cover. Remove the torque converter retaining bolts.

10. Support the engine and transaxle assembly, using the proper equipment. Remove the transaxle-to-engine retaining bolts. Disconnect the front and rear transmission mount bolts.

11. Carefully lower the transaxle assembly to the floor.

To install:

12. Install the transaxle and tighten the 12mm engine-to-transaxle bolts to 47 ft. lbs. (64 Nm) and the 10mm bolts to 34 ft. lbs. (46 Nm). Tighten the torque converter bolts to 20 ft. lbs. (27 Nm). Tighten the left engine mount bracket bolts to 32 ft. lbs. (43 Nm). Tighten the rear engine mount bracket bolts to 43 ft. lbs. (58 Nm).

13. Install the transaxle cables, linkage and halfshafts.

14. Refill the transaxle with the approved fluid and check for leaks.

15. Lower the vehicle. Road test the vehicle and check operation.

MR2

1. Disconnect the negative battery cable. Remove the air flow meter and the air cleaner hose.

2. Remove the intercooler on the 4A-GZE.

3. Remove the water inlet set bolts. Disconnect the ground strap. Remove the transaxle mounting set bolt.

4. Disconnect the speedometer cable at the transaxle. Disconnect the throttle cable from the throttle linkage and the bracket.

5. Raise and support the vehicle safely. Drain the transaxle fluid. Remove the left tire.

6. Remove the transaxle gravel shield. Disconnect the speedometer cable at the transaxle assembly.

7. Disconnect the oil cooler lines at the transaxle. Remove the transaxle control cable clip and retainer and then disconnect the cable from the bracket. Remove the bracket.

8. Remove the starter assembly. Disconnect the exhaust pipe at the manifold. Remove the pipe.

9. Remove the stiffener plate. Remove the rear engine end plate. Remove the torque converter cover. Remove the torque converter retaining bolts.

10. Disconnect both the right and left halfshafts from their side gear shafts. Depress and hold the brake pedal while removing the halfshaft retaining nuts. Properly position the halfshaft aside.

11. Disconnect the suspension arm from the rear axle carrier, using the proper tools. Disconnect the rear axle carrier from the lower control arm.

12. Disconnect the halfshaft from the side gear shaft. Properly position the driveshaft aside.

13. Support the engine and transaxle assembly, using the proper equipment. Remove the transaxle-to-engine retaining bolts. Disconnect the front and rear transmission mount bolts.

14. Carefully lower the transaxle assembly to the floor.

To install:

15. Install the transaxle and tighten the transaxle-to-engine bolts to 47 ft. lbs. (64 Nm). Tighten the left engine mount bracket bolts to 32 ft. lbs. (43 Nm). Tighten the rear engine mount bracket bolts to 43 ft. lbs. (58 Nm). Tighten the torque converter mounting bolts to 20 ft. lbs. (27 Nm).
16. Install the transaxle cables, linkage and halfshafts.
17. Refill the transaxle with the approved fluid and check for leaks.
18. Lower the vehicle. Road test the vehicle and check operation.

SHIFT LINKAGE ADJUSTMENT

Tercel

SEDAN

1. Loosen the swivel nut on the selector lever.
2. Push the lever fully toward the right side of the vehicle.
3. Return the lever 2 notches to the **N** position.
4. Set the shift lever in the **N** position.
5. While holding the selector lever slightly toward the **R** shift position tighten the swivel nut to 48 inch lbs. (5.4 Nm).

WAGON

1. Push the selector lever all the way to the rear of the vehicle.
2. Return the lever 2 notches to the **N** shift position.
3. Set the selector lever in the **N** position.
4. While holding the selector lever slightly toward the **R** shift position tighten the connecting rod nut.

Corolla, Camry, MR2 and Celica

1. Loosen the swivel nut on the selector lever.
2. Push the lever fully toward the right side of the vehicle.
3. Return the lever 2 notches to the **N** position.
4. Set the selector lever in the **N** position.
5. While holding the selector lever slightly toward the **R** shift position tighten the swivel nut to 48 inch lbs. (5.4 Nm).

THROTTLE LINKAGE ADJUSTMENT

1. Remove the air cleaner.
2. Confirm that the accelerator linkage opens the throttle fully. Adjust the linkage as necessary.
3. Peel the rubber dust boot back from the throttle cable.

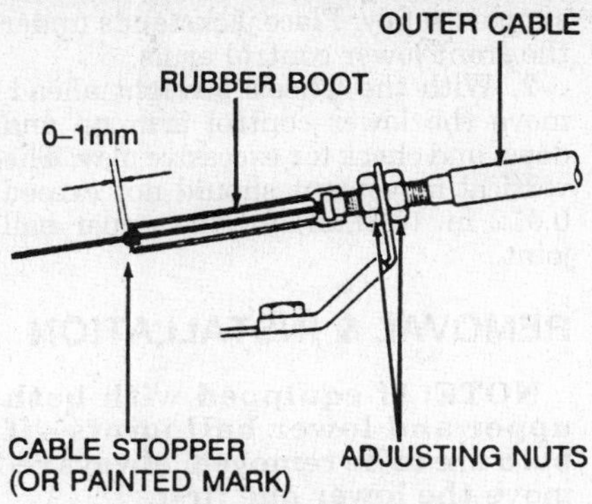

Throttle cable adjustment

4. Loosen the adjustment nuts on the throttle cable bracket (cylinder head cover) just enough to allow cable housing movement.
5. Have an assistant depress the accelerator pedal fully.
6. Adjust the cable housing so the distance between its end and the cable stop collar is 0.04 in.
7. Tighten the adjustment nuts. Make sure the adjustment has not changed. Install the dust boot and the air cleaner.

FRONT SUSPENSION

MacPherson Strut

REMOVAL & INSTALLATION

1. Remove the hubcap and loosen the lug nuts.
2. Raise and support the vehicle safely.

NOTE: Do not support the weight of the vehicle on the suspension arm; the arm will deform under its weight.

3. Unfasten the lug nuts and remove the wheel.
4. Remove the union bolt and 2 washers and disconnect the front brake line from the disc brake caliper. Remove the clip from the brake hose and pull off the brake hose from the brake hose bracket.
5. Remove the caliper and wire it aside. Matchmark on the strut lower bracket and camber adjust cam, if equipped. Remove the 2 bolts and nuts which attach the strut lower end to the steering knuckle lower arm.
6. Disconnect and remove the TEMS actuator from the top of the strut on late Cressida and Supra.
7. Unbolt the upper control arm where it attaches to the body on 1987–89 Supra.

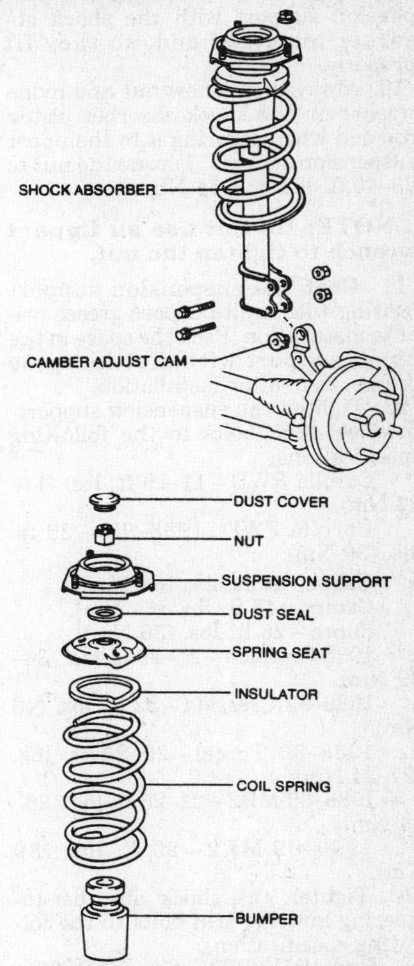

Strut used on all front wheel drive vehicles

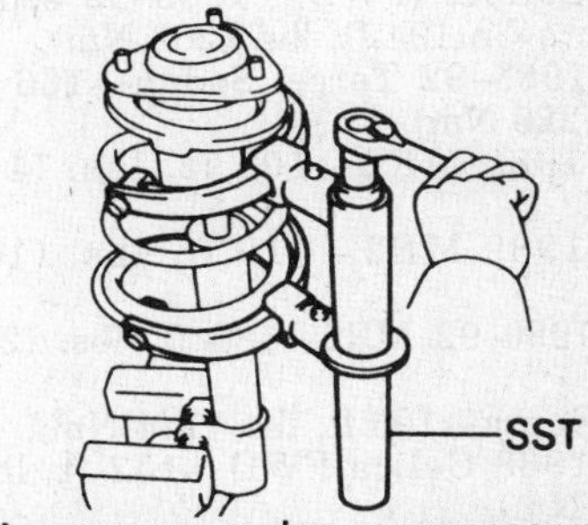

Always use a spring compressor when servicing the MacPherson strut

8. Remove the 3 nuts (4 nuts on the FX vehicles) which secure the upper strut mounting plate to the top of the wheel arch.

NOTE: Press down on the suspension lower arm, in order to remove the strut assembly. This must be done to clear the collars on the steering knuckle arm bolt holes when removing the shock/spring assembly.

To install:

9. Align the hole in the upper sus-

pension support with the shock absorber piston or end, so they fit properly.

10. Always use a new nut and nylon washer on the shock absorber piston rod end when securing it to the upper suspension support. Torque the nut to 29–40 ft. lbs. (39–54 Nm).

NOTE: Do not use an impact wrench to tighten the nut.

11. Coat the suspension support bearing with multipurpose grease prior to installation. Pack the space in the upper support with multipurpose grease, also, after installation.

12. Tighten the suspension support-to-wheel arch bolts to the following specifications:

Corolla RWD – 11–16 ft. lbs. (11–22 Nm).

Corolla FWD (1988–90) – 29 ft. lbs. (39 Nm).

Celica – 47 ft. lbs. (64 Nm).

Camry – 47 ft. lbs. (64 Nm).

Supra – 26 ft. lbs. (35 Nm).

1988 Cressida – 25–29 ft. lbs. (34–39 Nm).

1989–92 Cressida – 32 ft. lbs. (43 Nm).

1988–90 Tercel – 20–25 ft. lbs. (27–34 Nm).

1988–89 MR2 – 21–25 ft. lbs. (28–34 Nm).

1990–92 MR2 – 29 ft. lbs. (39 Nm).

13. Tighten the shock absorber-to-steering knuckle arm bolts to the following specifications:

Corolla (RWD) – 50–65 ft. lbs. (68–88 Nm).

Corolla (FWD) – 1988–92 sedan and wagon 194 ft. lbs. (263 Nm).

1988–92 Tercel Sedan – 166 ft. lbs. (226 Nm).

1988 MR2 – 105 ft. lbs. (143 Nm).

1989 MR2 – 119 ft. lbs. (162 Nm).

1990–92 MR2 – 188 ft. lbs. (255 Nm).

Supra – 106 ft. lbs. (144 Nm).

1988 Celica FWD – 152 ft. lbs. (204 Nm).

1989–92 Celica FWD – 188 ft. lbs. (255 Nm).

Cressida – 80 ft. lbs. (109 Nm).

1988 Camry – 166 ft. lbs. (226 Nm).

1989–92 Camry – 224 ft. lbs. (304 Nm).

All others – 65 ft. lbs. (88 Nm).

14. Adjust the front wheel bearing preload.

15. Bleed the brake system.

Upper Ball Joints

INSPECTION

1. Raise the front of the vehicle and support safely. Place jackstands under the front lower control arms.

2. With the wheels straight ahead, move the lower control arm up and down and check for excessive play. The vertical movement should not exceed 0.012 in. (0.3mm) for the upper ball joint.

REMOVAL & INSTALLATION

NOTE: If equipped with both upper and lower ball joints – if both are to be removed, always remove the lower one first.

Supra

The ball joint is an integral component of the upper control arm. Ball joint replacement requires that the entire arm assembly be replaced. Please refer to Upper Control Arm.

Lower Control Arm/ Ball Joints

INSPECTION

Corolla (RWD) and Cressida

Raise the front end and position a piece of wood under the wheel and lower the vehicle until there is an ½ load on the strut. Check the front wheel play. Replace the lower ball joint if the play at the wheel rim exceeds 0.1 in. (2.54mm) vertical motion or 0.25 in. (6mm) horizontal motion. Be sure the dust covers are not torn and that they are securely glued to the ball joints.

NOTE: Do not raise the control arm on Corolla or Cressida vehicles; damage to the arm will result.

Tercel, Camry, Corolla (FWD), Celica, Supra and MR2

1. Raise the vehicle and place wooden blocks under the front wheels. The block height should be 7.09–7.87 in. (180–200 Nm).

2. Use jack stands for additional safety.

3. Make sure the front wheels are in a straight forward position.

4. Check the wheels.

5. Lower the jack until there is approximately half a load on the front springs.

6. Move the lower control arm up and down to check that there is no ball joint vertical play. Replace if any play exists.

REMOVAL & INSTALLATION

NOTE: If equipped with both upper and lower ball joints, if both ball joints are to be removed, always remove the lower and then the upper ball joint.

Corolla (RWD)

The ball joint and control arm cannot be separated from each other. If one fails, then both must be replaced as an assembly, in the following manner:

1. Remove the stabilizer bar securing bolts.

2. Unfasten the torque strut mounting bolts.

3. Remove the control arm mounting bolt and detach the arm from the front suspension member.

4. Remove the steering knuckle arm from the control arm with a ball joint puller.

5. Inspect the suspension components, which were removed for wear or damage. Replace any parts, as required.

To install:

6. During installation, please observe the following:

a. When installing the control arm on the suspension member, tighten the bolts partially at first.

b. Complete the assembly procedure and lower the vehicle to the ground.

c. Bounce the front of the vehicle several times. Allow the suspension to settle, then tighten the lower control arm bolts to 51–65 ft. lbs. (69–88 Nm).

d. Lubricate the ball joint.

e. Check the front end alignment.

Celica

1. Raise and support the vehicle safely. Remove the wheels.

2. Disconnect the lower control arm from the steering knuckle.

3. Remove the nut and disconnect the stabilizer bar from the control arm.

4. On all but the left-side control arm, if with automatic transmissions, remove the control arm front set nut and washer. Remove the rear bracket bolts and then remove the arm.

5. On the left arm, if with automatic transmissions, remove the control arm front set nut and washer. Remove the 4 bolts and 2 nuts that attach the lower suspension crossmember to the frame and remove the crossmember. Remove the bolt and nut and lift out the lower arm with the lower arm shaft.

To install:

6. On all but the left-side control arm, if with automatic transmissions, install the lower control arm shaft washer with the tapered side toward the body. Install the lower arm with the bracket and then temporarily install the washer and nut to the lower arm shaft and bracket bolts.

7. On the left-side arm, if with automatic transmissions, position the washer on the lower arm shaft and then install them to the lower arm. Temporarily install the washer and nut to the shaft with the tapered side toward the body. Install the lower arm with the shaft to the body and temporarily install the rear brackets. Install the bolt and nut to the lower arm shaft and tighten them to 154 ft. lbs. (208 Nm). Install the crossmember to the body and tighten the 4 bolts to 154 ft. lbs. (208 Nm). Tighten the 2 nuts to 29 ft. lbs. (39 Nm).

8. Connect the lower arm to the steering knuckle and tighten the bolt and 2 nuts to 94 ft. lbs. (127 Nm).

9. Connect the stabilizer bar to the control arm and tighten the nut to 26 ft. lbs. (35 Nm).

10. Install the wheel, lower the vehicle and bounce it several times to set the suspension.

11. Tighten the front set nut to 156 ft. lbs. (212 Nm). Tighten the rear bracket bolts to 72 ft. lbs. (98 Nm).

MR2

1. Raise the vehicle and support it safely. Remove the wheel.

2. Remove the cotter pin and castle nut and then press the lower arm out of the ball joint.

3. Press the ball joint out of the steering knuckle.

4. Remove the 2 nuts and disconnect the strut bar from the control arm.

5. Remove the lower control arm-to-body bolt and remove the arm.

To install:

6. When installing the lower arm, position it in the strut bar and tighten the nuts finger-tight. Do the same thing with the arm-to-body bolt.

7. Connect the control arm to the ball joint and tighten the castle nut to 58 ft. lbs. (78 Nm). Install a new cotter pin.

8. Tighten the strut bar-to-arm bolts to 83 ft. lbs. (113 Nm).

9. Install the tires, lower the vehicle and bounce it several times to set the suspension.

10. Tighten the control arm-to-body bolt to 94 ft. lbs. (127 Nm) on 1988–89 vehicles and 87 ft. lbs. (118 Nm) on 1990–92 vehicles. Check the wheel alignment.

Cressida

1. Raise and support the vehicle safely. Remove the wheels.

2. Remove the 2 knuckle arm-to-strut bolts, pull down on the control arm and disconnect it and the knuckle arm from the strut.

3. Remove the cotter pin and nut and press the tie rod off the knuckle arm.

4. Remove the nut attaching the stabilizer bar to the control arm and disconnect the bar.

5. Remove the 2 nuts and then disconnect the strut bar from the control arm.

6. Disconnect the control arm from the crossmember and remove it and the rack boot protector as an assembly.

7. Remove the cotter pin and nut and then press the knuckle arm off the control arm.

To install:

8. Press the knuckle arm into the control arm and then install the assembly into the crossmember.

9. Connect the stabilizer bar to the control arm and tighten the nut to 13 ft. lbs. (18 Nm).

10. Connect the strut bar to the control arm and tighten the nuts to 54 ft. lbs. (73 Nm) for 1988 and 76 ft. lbs. (103 Nm) for 1989–92.

11. Connect the knuckle arm to the strut housing and tighten the bolts to 80 ft. lbs. (108 Nm).

12. Install the wheel and lower the vehicle. Bounce the vehicle several times to set the suspension and then tighten the control arm-to-body bolt to 80 ft. lbs. (108 Nm) for 1988 or 121 ft. lbs. (164 Nm) for 1989–92.

13. Check the front wheel alignment.

Supra

NOTE: This procedure is for ball joint removal only. To remove the lower control arm, please refer to the Lower Control Arm procedure.

1. Raise and support the vehicle safely.

2. Remove the wheels.

3. Remove the steering knuckle and then remove the upper control arm.

4. Remove the lower ball joint mounting nuts and the bolt. Remove the attachment plate.

5. Remove the lower ball joint.

6. Installation is in the reverse order of removal. Tighten the ball joint mounting bolt and nuts to 94 ft. lbs. (127 Nm).

Tercel

1. Raise and support the vehicle safely. Remove the wheels.

2. Remove the 2 bolts attaching the ball joint to the steering knuckle.

3. Remove the stabilizer bar nut, retainer and cushion.

4. Raise the opposite wheel until the body of the vehicle just lifts off the supports.

5. Loosen the lower control arm mounting bolt, wiggle the arm back and forth and then remove the bolt. Disconnect the lower control arm from the stabilizer bar.

NOTE: When removing the lower control arm, be careful not to lose the caster adjustment spacer.

6. Carefully mount the lower control arm in a vise and then, using a ball joint removal tool, disconnect the ball joint from the arm.

To install:

7. Tighten the ball joint-to-control arm nut to 51–65 ft. lbs. (69–88 Nm) and use a new cotter pin.

8. Tighten the steering knuckle-to-control arm bolts to 59 ft. lbs. (80 Nm).

9. Before tightening the stabilizer bar nuts, mount the wheels and lower the vehicle. Bounce the vehicle several times to settle the suspension and then tighten the stabilizer bolts to 66–90 ft. lbs. (90–122 Nm).

10. Tighten the arm-to-body bolts to 83 ft. lbs.

11. Check the front end alignment.

Corolla (FWD)

1. On all vehicles, except the left side on those with automatic transaxle, perform the following:

   a. Remove the bolt and 2 nuts attaching the ball joint to the lower arm and disconnect the lower arm from the steering knuckle.

   b. On 4A-F and 4A-FE, remove the nut holding the stabilizer bar to the lower arm and disconnect the bar from the arm.

   c. On 4A-GE, remove the lower nut on the stabilizer bar link and disconnect the link from the arm.

   d. Remove the rear bracket bolts and nut. Remove the lower arm front mounting bolt.

   e. Remove the rear bracket and the stabilizer bar bracket and lift out the lower control arm.

2. To remove the left control arm, if with automatic transaxle, perform the following:

   a. Disconnect the arm at the steering knuckle.

   b. Disconnect the stabilizer bar at the lower arm.

   c. Remove the lower arm rear brackets. Move the stabilizer bar toward the rear and remove the bracket.

   d. Remove the 6 bolts and 2 nuts and remove the suspension crossmember with the lower arm.

   e. Remove the lower arm from the crossmember.

To install:

3. To install the left control arm, if with automatic transaxle, install the lower arm on the crossmember and install the assembly to the body.

4. On all others, install the lower arm to the body, move the stabilizer bar into position and install the front mounting bolt. Install the stabilizer bar and rear brackets.

5. Connect the lower arm to the steering knuckle and tighten the bolts to 105 ft. lbs. (142 Nm).

6. On 4A-F, connect the stabilizer bar to the lower arm and tighten the nut to 13 ft. lbs. (18 Nm).

7. On 4A-GE, connect the stabilizer bar link to the lower arm and tighten the nut to 26 ft. lbs. (35 Nm).

8. Lower the vehicle and bounce it several times to stabilize the suspension. Tighten the lower arm front bolt to 174 ft. lbs. (235 Nm) for 1988 or 152 ft. lbs. (206 Nm) for 1989–92. Tighten the rear bracket bolts to 94 ft. lbs. (127 Nm) on the lower arm side, 37 ft. lbs. (50 Nm) on the stabilizer bar side and tighten the small bolt and nut to 14 ft. lbs. (19 Nm).

9. Check the front end alignment.

Camry

1. Raise and support the vehicle safely. Remove the wheels.

2. Remove the 2 bolts attaching the ball joint to the steering knuckle.

3. Remove the stabilizer bar nut, retainer and cushion.

4. Remove the nut attaching the lower arm shaft to the lower arm.

5. Remove the lower suspension crossmember (2 bolts and 4 nuts).

6. Remove the lower control arm and lower arm shaft as an assembly.

7. Grip the lower arm assembly in a vise and remove the ball joint cotter pin and retaining nut. With a ball joint removal tool, pull the ball joint out of the control arm.

To install:

8. Position the ball joint in the lower arm and tighten the nut to 67 ft. lbs. (91 Nm) for 1988 or 90 ft. lbs. (123 Nm) for 1989–92. Install a new cotter pin.

9. Install the lower arm to the stabilizer bar and then install the lower arm shaft to the body. Install the lower arm nut and retainer. Screw on a new stabilizer bar end nut and retainer.

10. Connect the ball joint to the steering knuckle and tighten the bolts to 94 ft. lbs. (127 Nm) for 1988 or 83 ft. lbs. (113 Nm) for 1989–92.

11. Install the suspension lower crossmember. Tighten the inner bolts to 32 ft. lbs. (43 Nm) and the outer ones to 153 ft. lbs. (207 Nm).

12. Install the wheels and lower the vehicle. Bounce it several times to set the suspension.

13. Tighten the stabilizer bar end nut and the lower arm shaft-to-lower arm bolt to 156 ft. lbs. (212 Nm).

Upper Control Arm

REMOVAL & INSTALLATION

Supra

1. Raise and support the vehicle safely. Remove the wheels.

2. Unclip the brake hose bracket at the steering knuckle, remove the retaining nut and press the upper arm out of the knuckle.

3. Remove the upper mounting bolt and nut and lift out the upper control arm.

4. Connect the upper arm to the body. Connect the arm to the steering knuckle.

5. Install the wheels and lower the vehicle. Bounce it several times to set the suspension and then tighten the arm-to-knuckle nut to 80 ft. lbs. (108 Nm). Tighten the arm-to-body bolt to 121 ft. lbs. (164 Nm).

Lower Control Arms

REMOVAL & INSTALLATION

Tercel and Celica

1. Raise the vehicle and support safely. Remove the front wheels.

2. Disconnect the lower arm from the steering knuckle, at the ball joint.

3. Disconnect the stabilizer bar from the lower arm.

4. Remove the lower arm by loosening the front bolt. Remove the bracket bolts, stabilizer bracket and front bolt.

5. Remove the lower arm.

To install:

6. Install the lower arm assemblies. Loosely install the lower arm bushing bolts and clamp.

7. Install the ball joint bolts and torque to 59 ft. lbs. (80 Nm) for the Tercel and 94 ft. lbs. (127 Nm) for the Celica.

8. Install the stabilizer and torque to 76 ft. lbs. (103 Nm) for the Tercel and 26 ft. lbs. (34 Nm) for the Celica.

9. Install the front wheels and lower the vehicle. Bounce up and down to stabilize the suspension. With the vehicle weight on the suspension, torque the front side to 105 ft. lbs. (142 Nm) for the Tercel and 156 ft. lbs. (212 Nm) for the Celica. Torque the rear bracket bolts to 94 ft. lbs. (126 Nm) for the Tercel and 72 ft. lbs. (96 Nm) for the Celica.

10. Align the front end.

Corolla (RWD)

1. Raise and support the vehicle safely.

2. Remove the wheel.

3. Disconnect the steering knuckle from the control arm.

4. Disconnect the tie rod, stabilizer bar and strut bar from the control arm.

5. Remove the control arm mounting bolts, and remove the arm.

6. Install in reverse of above. Tighten, but do not torque fasteners until vehicle is on the ground.

7. Lower vehicle to ground, rock it from side-to-side several times and torque control arm mounting bolts to 51–65 ft. lbs. (68–88 Nm), stabilizer bar to 16 ft. lbs. (22 Nm), strut bar to 40 ft. lbs. (54 Nm) and shock absorber to 65 ft. lbs. (88 Nm).

8. Align the front end.

Corolla (FWD)

1. Raise the vehicle and support safely. Remove the front wheels.

2. Disconnect the lower arm from the steering knuckle, at the ball joint.

3. Disconnect the stabilizer bar from the lower arm.

4. Remove the lower arm by loosening the front bolt. Remove the bracket bolts and nuts, rear and stabilizer bracket and front bolt for all FWD Corollas, except left side with an automatic transaxle.

5. Remove the lower arms with the suspension crossmember for FWD Corollas with automatic transaxle. Remove the 4 left and right lower arm rear brackets, stabilizer bar bracket and 6 crossmember bolts.

To install:

6. Install the crossmember and lower arm assemblies. Loosely install the lower arm bushing bolts.

7. Install the ball joint bolts and torque to 105 ft. lbs. (142 Nm).

8. Install the stabilizer and torque to 26 ft. lbs. (35 Nm).

9. Install the front wheels and lower the vehicle. Bounce up and down to stabilize the suspension. With the vehicle weight on the suspension, torque the lower control arm bushing bolts to 174 ft. lbs. (235 Nm) for the front bolt, 94 ft. lbs. (126 Nm) for the lower arm side, 37 ft. lbs. (50 Nm) for the stabilizer bar side and 14 ft. lbs. (19 Nm) for the small bolt and nut.

10. Align the front end.

Camry

1. Raise the vehicle and support safely. Remove the front wheels.

2. Disconnect the lower arm from the steering knuckle, at the ball joint.

3. Disconnect the stabilizer bar from the lower arm.

4. Remove the crossmember and lower arm as an assembly. Remove the lower suspension with the lower suspension arm shaft.

To install:

5. Install the crossmember and lower arm assemblies. Loosely install the lower arm bushing bolts. Torque the crossmember bolts to 112 ft. lbs. (152 Nm).

6. Install the ball joint bolts and torque to 90 ft. lbs. (123 Nm).

7. Install the stabilizer nut loosely.

8. Install the front wheels and lower the vehicle. Bounce up and down to stabilize the suspension. With the ve-

hicle weight on the suspension, torque the lower control arm bushing and stabilizer nuts to 156 ft. lbs. (212 Nm).

9. Align the front end.

Supra

1. Raise and support the vehicle safely. Remove the wheels.
2. Disconnect the stabilizer bar link from the lower control arm. Remove the locknut and press the ball joint out of the steering knuckle.
3. Disconnect the lower control arm at the strut. Matchmark the front and rear adjusting cams to the body. Remove the nuts and cams and then remove the lower arm.
4. Unbolt the ball joint from the control arm.

To install:

5. Install the ball joint to the arm and tighten the nuts to 94 ft. lbs. (127 Nm).
6. Position the lower control arm and install the adjusting cams and nuts finger-tight.
7. Connect the ball joint to the steering knuckle and tighten a conventional nut to 14 ft. lbs. (20 Nm). Install a locknut on top of the other and tighten it to 107 ft. lbs. (145 Nm).
8. Tighten the arm-to-strut bolt to 106 ft. lbs. (143 Nm). Tighten the stabilizer bar link nut to 47 ft. lbs. (64 Nm).
9. Install the wheels and lower the vehicle. Bounce the vehicle several times to set the suspension. Align the matchmarks on the adjusting cams and the body and tighten them to 177 ft. lbs. (240 Nm). Check the front alignment.

MR2

1. Raise the vehicle and support safely. Remove the front wheels.
2. Disconnect the lower arm from the steering knuckle, at the ball joint using a ball joint separator.
3. Disconnect the stabilizer and strut bar from the lower arm.
4. Remove the lower arm by loosening the front bolt.

To install:

5. Install the lower arm assembly. Loosely install the lower arm bushing bolt.
6. Install the ball joint bolts and torque to 58 ft. lbs. (78 Nm).
7. Install the stabilizer and torque to 47 ft. lbs. (64 Nm). Install the strut bar bolts and torque to 83 ft. lbs. (113 Nm).
8. Install the front wheels and lower the vehicle. Bounce up and down to stabilize the suspension. With the vehicle weight on the suspension, torque the lower control arm bushing bolts to 87 ft. lbs. (118 Nm).
9. Align the front end.

Front Wheel Bearings

NOTE: These procedures apply to rear wheel drive vehicles only. For front wheel bearing service on front wheel drive vehicles, refer to the "Drive Axle" section.

ADJUSTMENT

NOTE: The Supra uses preset hub bearings. The adjustment is obtained through the hub nut torque. The hub assembly has to be removed from the vehicle to adjust the hub nut. Make sure the bearing axial play does not exceed 0.0020 in. (0.05mm).

1. With the front hub/disc assembly installed, tighten the castellated nut to 19–23 ft. lbs. (26–31 Nm).
2. Rotate the disc back and forth, 2–3 times, to allow the bearing to seat properly.
3. Loosen the castellated nut until it is only finger-tight.
4. Tighten the nut firmly, using a box wrench.
5. Measure the bearing preload with a spring scale attached to a wheel mounting stud. Check it against the specifications.
6. Install the cotter pin.

NOTE: If the hole does not align with the nut (or cap) holes, tighten the nut slightly until it does.

7. Finish installing the brake components and the wheel.

REMOVAL & INSTALLATION

Except Supra

1. Raise the vehicle and support safely. Remove the disc/hub assembly.
2. If either the disc or the entire hub assembly is to be replaced, unbolt the hub from the disc.

NOTE: If only the bearings are to be replaced, do not separate the disc and hub.

3. Using a brass rod as a drift, tap the inner bearings cone out. Remove the oil seal and the inner bearings.

NOTE: Throw the old oil seal away.

4. Drive out the inner bearing cup.
5. Drive out the outer bearing cup. Inspect the bearings and the hub for signs of wear or damage. Replace components, as necessary.

To install:

6. Install the inner bearing cup an then the outer bearing cup, by driving them into place.

NOTE: Use care not to cock the bearing cups in the hub.

7. Pack the bearings, hub inner well and grease cap with multipurpose grease.
8. Install the inner bearing into the hub.
9. Carefully install a new oil seal with a soft drift.
10. Install the hub on the spindle. Be sure to install all of the washers and nuts which were removed.
11. Adjust the bearing preload.
12. Install the caliper assembly and the wheel.

Supra

1. Raise the vehicle and support safely.
2. Remove the front wheel, brake caliper and disc. Hange the caliper from the suspension with a piece of wire. Remove the hose bracket from the knuckle.
3. Disconnect the tie rod from the knuckle using a tie rod separator, or equivalent.
4. Disconnect the steering knuckle from the upper suspension arm using a ball joint separator, or equivalent.
5. Remove the steering knuckle from the lower ball joint using a ball joint separator.
6. Remove the bearing cap from the rear of the knuckle with a suitable prybar.
7. Remove the axle hub lock nut using a hammer and chisel to loosen the stacked part of the lock nut.
8. Press the axle hub out of the knuckle using a bearing puller or press.
9. Remove the hub bearing inner race by using a bearing puller SST 09950–20017 or equivalent.
10. Remove the 4 dust cover bolts and cover.
11. Remove the snapring and press out the hub bearing.

To install:

12. Press in the hub bearing and install the snapring.
13. Install the outer seal and disc dust cover. Torque the cover bolts to 14 ft. lbs. (19 Nm).
14. Install the axle hub using a press.
15. Install the axle lock nut and torque to 147 ft. lbs. (199 Nm). Stake the lock nut using a hammer and chisel.
16. Install the bearing cap.
17. Install the axle hub onto the vehicle. Torque the upper ball joint to 76 ft. lbs. (103 Nm), lower ball joint to 92 ft. lbs. (125 Nm) and tie rod end to 36 ft. lbs. (49 Nm).
18. Install the brake rotor, caliper and front wheel.
19. Lower the vehicle and have the front end aligned.

REAR SUSPENSION

Shock Absorbers

REMOVAL & INSTALLATION

Corolla (RWD) and 1989–92 Corolla (AWD)

1. Raise and support the vehicle safely. Support the rear axle.
2. Unfasten the upper shock absorber retaining nuts. It may be necessary to hold the shock absorber shaft with a suitable tool while removing the top retaining nut.

NOTE: Always remove and install the shock absorbers one at a time. Do not allow the rear axle to hang in place as this may cause undue damage.

3. Remove the lower shock retaining nut where it attaches to the rear axle housing.
4. Remove the shock absorber.
5. Inspect the shock for wear, leaks or other signs of damage.
6. Installation is in the reverse order of removal. During installation, observe the following:
   a. Tighten the upper retaining nuts to 18 ft. lbs. (25 Nm).
   b. Tighten the lower retaining nuts to 27 ft. lbs. (37 Nm).

1988 Cressida

1. Raise and support the differential housing and the suspension control arms safely.
2. Remove the brake hose clips. Disconnect the stabilizer bar end.
3. Disconnect the halfshaft at the CV-joint on the wheel side.
4. With a jackstand under the suspension control arm, unbolt the shock absorber at its lower end. Using a prybar to keep the shaft from turning, remove the nut holding the shock absorber to its upper mounting. If with TEMS, disconnect the actuator and remove it. Remove the shock.
5. Installation is in the reverse order of removal. Torque the halfshaft nuts to 44–57 ft. lbs. (60–78 Nm), the upper shock mounting nut to 14–22 ft. lbs. (19–30 Nm) and the lower shock mounting nut to 22–32 ft. lbs. (30–44 Nm).

MacPherson Strut

REMOVAL & INSTALLATION

Tercel

1. Working inside the vehicle, remove the shock absorber cover and package tray bracket.
2. Raise the rear of the vehicle and support safely. Remove the wheel.
3. Disconnect the brake line from the wheel cylinder, if necessary. Disconnect the brake line from the flexible hose at the mounting bracket on the strut tube. Disconnect the flexible hose from the strut.
4. Loosen the nut holding the suspension support to the shock absorber; do not remove the nut.
5. Remove the bolts and nuts mounting on the strut on the axle carrier and then disconnect the strut.
6. Remove the 3 upper strut mounting nuts and carefully remove the strut assembly.

To install:

7. Install the strut assembly into the vehicle. During installation, observe the following torque specification:
   a. Tighten the upper strut retaining nuts to 23 ft. lbs. (31 Nm).
   b. Tighten the lower strut-to-axle carrier bolts to 105 ft. lbs. (143 Nm).
   c. Tighten the nut holding the suspension support to the shock absorber to 36 ft. lbs. (49 Nm).
   d. Tighten the strut-to-axle beam nut to 47 ft. lbs. (64 Nm) on 1988–92 vehicles.
8. Bleed the brakes.

Camry and Corolla (2WD)

1. On the 4-door sedan, remove the package tray and vent duct.
2. On the hatchback, remove the speaker grilles.
3. Disconnect the brake line from the wheel cylinder.
4. Remove the brake line from the brake hose.
5. Disconnect the brake hose from its bracket on the strut.
6. Remove the strut suspension support cover. Loosen, but do not remove, the nut holding the suspension support to the strut.
7. Unbolt the strut from the rear arm and or axle carrier.
8. Unbolt the strut from the body.

To install:

9. Install the strut assembly. During installation, please observe the following torque specifications:
   a. Tighten the strut-to-body bolts to 23 ft. lbs. (31 Nm) on 1988 Camry or 29 ft. lbs. (39 Nm) on 1988–89 Corolla sedan/wagon and 1989–92 Camry).
   b. Tighten the strut-to-axle carrier bolts 166 ft. lbs. on 1988–92 Camry and 105 ft. lbs. (143 Nm) on Corolla.
   c. Tighten the the suspension support-to-strut nut to 36 ft. lbs. (49 Nm).
10. Refill and bleed the brake system.

Celica (FWD)

1. Raise and support the vehicle safely. Position an hydraulic jack under the rear hub assembly; raise it just enough to support the assembly.
2. On the liftback, remove the rear speaker grilles.
3. On the coupe, remove the suspension service hole cover.
4. On the ST and GT vehicles, disconnect and plug the brake line at the backing plate. Remove the clip and E-ring and then disconnect the brake hose and tube from the strut housing.
5. On the GTS, remove the union bolts and gaskets and disconnect the brake line from the brake cylinder. Remove the clip and E-ring from the strut and then disconnect the brake hose from the strut housing.
6. Loosen, but do not remove, the nut attaching the suspension support to the strut.

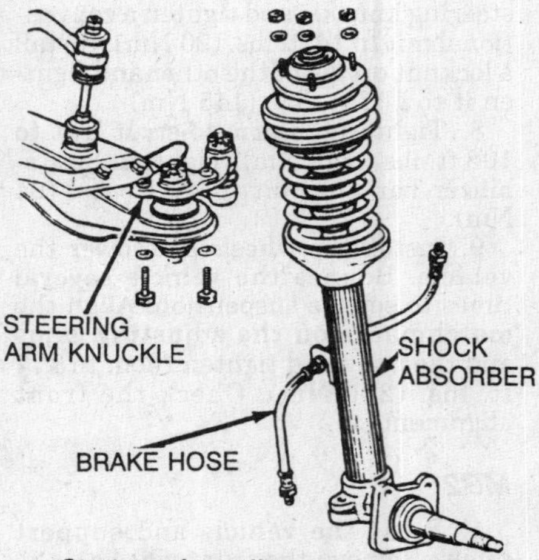

Strut used on all rear wheel drive vehicles

7. Disconnect the stabilizer bar at the lower end of the strut housing.
8. Disconnect the strut at the axle carrier.
9. Remove the 3 strut-to-body bolts and then remove the strut.

To install:

10. Tighten the upper strut-to-body nuts to 23 ft. lbs. (31 Nm).
11. Tighten the lower strut-to-carrier bolts to 119 ft. lbs. (162 Nm).
12. Connect the stabilizer bar to the strut and tighten the bolts to 26 ft. lbs. (35 Nm).
13. Tighten the strut holding nut to 36 ft. lbs. (49 Nm). Install the dust cover onto the suspension support.
14. Reconnect the brake line and hose. Bleed the system, lower the vehicle and check the rear wheel alignment.

MR2

1. Raise and support the vehicle safely. Position a hydraulic floor jack under the rear hub assembly; raise it just enough to support the assembly.
2. Remove the union bolts and gaskets and disconnect the brake line from the brake cylinder. Remove the clip and E-ring from the strut and then disconnect the brake hose from the strut housing.
3. Matchmark the lower strut bracket and the camber adjusting cam, remove the 2 axle carrier bolts and the adjusting cam and disconnect the strut from the carrier.
4. Remove the engine hood side panel.
5. Remove the 3 upper strut-to-body nuts and then remove the strut.

To install:

6. Position the strut and tighten the upper mounting nuts to 23 ft. lbs. (31 Nm).
7. Install the engine hood side panel.
8. Connect the axle carrier to the lower strut bracket. Insert the mounting bolts from the rear and align the matchmarks made in Step 3. Tighten the nuts to 105 ft. lbs. (142 Nm) on 1988 vehicles and 166 ft. lbs. (226 Nm) on 1989–92 vehicles.
9. Connect the brake line, bleed the system and check rear wheel alignment.

Supra and 1989–92 Cressida

1. Raise and support the vehicle safely. Remove the wheels.
2. Remove the speaker grille and interior quarter panel trim, if equipped with TEMS.
3. Disconnect the strut from the axle carrier.
4. Remove the strut cap. Remove the Toyota Electronic Modulated Suspension (TEMS) actuator.
5. Remove the 3 strut mounting nuts from the body and remove the strut assembly.
6. Mount the strut assembly in a vise. Using a spring compressor, compress the coil spring.
7. Remove the strut suspension support nut. Remove the strut suspension support, remove the coil spring and bumper.

To install:

8. Mount the strut in a vise. Using a spring compressor, compress the coil spring.
9. Install the bumper to the strut, align the coil spring end with the lower seat hollow and install the coil spring.
10. Align the strut suspension support hole and piston rod and install it. Align the suspension support with the strut lower bushing.
11. Install the strut suspension support nut and torque it to 20 ft. lbs. (27 Nm). Connect the strut assembly with the 3 retaining nuts and torque them to 10 ft. lbs. (14 Nm).
12. Connect the strut assembly to the axle carrier and torque it to 101 ft. lbs. (137 Nm).
13. Install the TEMS actuator and strut cap. Install the quarter panel trim panel and speaker grille.

Coil Springs

REMOVAL & INSTALLATION

1. Loosen the rear wheel lug nuts.
2. Raise and support the rear axle housing and frame safely.
3. Remove the lug nuts and wheel.
4. If equipped, disconnect the rear stabilizer bar from the axle housing or suspension arm on 1988–89 Supra and 1988 Cressida. Remove the bolt holding the stabilizer bar bushing to the rear axle housing.
5. Unfasten the lower shock absorber end. On the Corolla RWD, Corolla FWD (AWD) and Tercel AWD, disconnect the lateral control rod from the axle.

NOTE: On Supra and 1988 Cressida with IRS suspension, remove the rear halfshafts.

6. Slowly lower the jack under the rear axle housing until the axle is at the bottom of its travel.
7. Withdraw the coil spring, complete with its insulator.
8. Inspect the coil spring and insulator for wear, cracks or weakness; replace either or both, as necessary.
9. Installation is the reverse of the removal procedure.

Lower Control Arms

REMOVAL & INSTALLATION

MR2

1. Raise and support the vehicle safely. Remove the wheels.
2. Remove the cotter pin and retaining nut from the bottom of the ball joint stem. Using a ball joint removal tool, press the ball joint out of the control arm.
3. Remove the strut rod nut and retainer from the lower control arm.
4. Remove the bolt holding the lower control arm to the body. Remove the cushion and then disconnect the lower arm from the strut rod. Remove the lower control arm.

To install:

5. Connect the lower arm to the strut rod. Install the strut rod nut, cushion and retainer.
6. Connect the lower arm to the body and install the retaining nut fingertight.
7. Connect the lower control arm to the ball joint and tighten the retaining nut to 67 ft. lbs. (91 Nm). Install a new cotter pin.

NOTE: If the holes do not align when installing the new cotter pin, tighten the nut until they are aligned. Do not loosen the nut!

8. Install the wheel and lower the vehicle. Tighten the strut rod nut to 86 ft. lbs. (117 Nm) and the arm-to-body bolt to 94 ft. lbs. (127 Nm).

Celica, Camry and Corolla

1. Raise the vehicle and support safely. Remove the rear wheels.
2. Remove the nut from the axle carrier.
3. Place the matchmarks to the toe adjust cam and suspension member.
4. Remove the service hole cover and loosen the bolt and remove the toe adjust plate No. 2.
5. Remove the bolt with toe adjust cam and disconnect the suspension arm and remove.

To install:

6. Face the mark on the suspension arms to the rearward of the vehicle. Install the bushing with the slit side towards the rear and the small paint spot to the outside of the vehicle for the Camry.
7. Install the stamped suspension arm with the identification mark **L** for left and **R** for right. Temporarily install the suspension arms with the bolt, washer and nut. Do not tighten at this time.
8. Loosely install the bolt into the axle carrier.
9. Install the rear wheels and lower the vehicle. Bounce the suspension up and down a few times.
10. Torque suspension components with the vehicle weight loading the suspension.
11. Torque the suspension arm bolts to 83 ft. lbs. (113 Nm) and the nut to 166 ft. lbs. (226 Nm).
12. Have the wheel alignment checked.

Supra and 1989–92 Cressida

1. Raise and support the vehicle safely. Remove the wheels.
2. Remove the halfshaft.
3. Remove the nut and disconnect the No. 1 lower arm from the axle carrier. Matchmark the adjusting cam to the body, remove the cam and bolt and then lift out the No. 1 arm.
4. Remove the bolt and nut and disconnect the No. 2 lower arm from the axle carrier. Matchmark the adjusting cam to the body, remove the cam and bolt and then lift out the No. 2 arm.

To install:

5. Position the No. 2 arm and in-

stall the adjusting cam and bolt so the matchmarks are in alignment. Connect the arm to the axle carrier.

6. Position the No. 1 arm and install the adjusting cam and bolt so the matchmarks are in alignment. Connect the arm to the axle carrier. Use a new nut and tighten it to 43 ft. lbs. (59 Nm).

7. Install the halfshaft.

8. Install the wheels and lower the vehicle. Bounce it several times to set the suspension and then tighten the body-to-arm bolts and nuts to 136 ft. lbs. (184 Nm) on Supra or 134 ft. lbs. (181 Nm) on Cressida. Tighten the No. 2 arm-to-carrier bolt to 121 ft. lbs. (164 Nm) on Supra or 119 ft. lbs. (162 Nm) – Cressida. Tighten the No. 1 arm-to-carrier nut to 36 ft. lbs. (49 Nm).

9. Check the rear wheel alignment.

Upper Control Arm

REMOVAL & INSTALLATION

Supra and 1989–92 Cressida

1. Raise and support the rear of the vehicle safely. Remove the wheels.
2. Unbolt the brake caliper and suspend aside. Remove the halfshaft.
3. Disconnect the parking brake cable at the equalizer. Remove the 2 cable brackets from the body and then pull the cable through the suspension member.
4. Disconnect the 2 lower arms and the strut rod at the axle carrier. Disconnect the lower strut mount.
5. Disconnect the upper arm at the body and remove the axle hub assembly.
6. Remove the upper arm mounting nut. Remove the backing plate mounting nuts and separate the plate from the carrier. Press the upper arm out of the axle carrier.

To install:

7. Connect the upper arm to the body.
8. Connect the axle hub assembly to the arm with a new nut.
9. Connect the No. 1 lower control arm with a new nut and tighten it to 43 ft. lbs. (59 Nm) on Supra and 36 ft. lbs. (49 Nm) on Cressida. Connect the No. 2 lower arm and the strut rod.
10. Tighten the upper arm mounting nut to 80 ft. lbs. (108 Nm). Tighten the strut to 101 ft. lbs. (137 Nm).
11. Reconnect the parking brake cable and install the halfshaft. Install the brake caliper and tighten the bolts to 34 ft. lbs. (47 Nm).
12. Install the wheels and lower the vehicle. Bounce it several times to set the suspension and then tighten the upper arm-to-body bolt, the No. 2 lower arm-to-carrier and the strut rod to 121 ft. lbs. (164 Nm) on Supra and 119 ft. lbs. (162 Nm) on Cressida.

Upper and Lower Control Arms

REMOVAL & INSTALLATION

Tercel Wagon (AWD) and 1989–92 Corolla (AWD)

1. Raise and support the vehicle safely. Remove the wheels and support the rear axle safely.
2. Remove the upper control arm-to-body bolt. Remove the upper arm-to-axle bolt and lift out the upper control arm.
3. Remove the lower control arm-to-body bolt. Remove the lower arm-to-axle bolt and lift out the lower control arm.

To install:

4. Install the upper control arm with the nuts and bolts just snugged down.
5. Install the lower control arm with the nuts and bolts just snugged down.
6. Install the wheels, remove the safety stands and floor jack and then lower the vehicle.
7. Bounce the vehicle several times to stabilize the suspension and then raise the axle housing until the body is free.
8. Tighten all bolts to 83 ft. lbs. (113 Nm) on the Tercel and 72 ft. lbs. (98 Nm) on the Corolla.

Rear Wheel Bearings

NOTE: These procedures apply to front wheel drive vehicles only. For information on all rear wheel drive vehicles and all AWD vehicles, please refer to Rear Axle Shaft in the Drive Axle section.

REMOVAL & INSTALLATION

1988–89 Tercel Sedan

1. Raise and support the vehicle safely.
2. Remove the rear wheels.
3. Remove the brake drums.
4. Remove the locknut cap and cotter pin. Pry off the locknut and then remove the locknut itself.
5. Pull off the axle hub along with the outer wheel bearing and thrust washer.
6. Pry the inner bearing oil seal out of the brake drum and then remove the inner bearing.
7. Using a brass drift and hammer, drive out the bearing races.

To install:

8. Press new outer bearing races into the axle hub and fill it and the bearing cap with grease.
9. Pack the bearing with grease.
10. Position the inner bearing into the hub and then drive in a new oil seal. Coat the seal with grease.
11. Position the axle hub/brake drum onto the axle shaft. Install the outer bearing, fill the hole with grease and position the thrust washer. Install the bearing locknut and tighten it to 22 ft. lbs. (29 Nm).
12. Spin the axle hub several times to snug down the bearing and then loosen the bearing locknut until it can be turned by hand.

NOTE: There must be absolutely no brake drag at this time.

13. Retighten the bearing locknut until there is a bearing preload of 0.9–2.2 lbs. (3.2–9.8 N) while turning the wheel.
14. Install the locknut lock, a new cotter pin and the cap. If the cotter pin hole does not align properly, align the holes by tightening the nut.
15. Lower the vehicle.

Camry (2WD), Celica (2WD) Corolla (2WD) and 1990–92 Tercel

1. Raise and support the vehicle safely.
2. Remove the rear wheel and tire assembly.
3. Remove the brake drum. On the Corolla FX, remove the disc brake caliper from the axle carrier and suspend it with a wire.
4. Disconnect and plug the brake line at the backing plate.
5. Remove the 4 axle hub-to-carrier bolts and slide off the hub and brake assembly. Remove the O-ring from the backing plate.
6. Remove the bolt and nut attaching the carrier to the strut rod.
7. Remove the bolt and nut attaching the carrier to the No. 1 suspension arm.
8. Remove the bolt and nut attaching the carrier to the No. 2 suspension arm.
9. Unbolt the carrier from the rear strut tube and remove the carrier.
10. Using a hammer and cold chisel, loosen the staked part of the hub nut and remove the nut.
11. Using a 2-armed puller or the like, press the axle shaft from the hub.
12. Remove the bearing inner race (inside).
13. Using a 2-armed puller again, pull off the bearing inner race (outside) over the bearing and then press it out of the hub.

To install:

14. Position a new bearing inner race (outside) on the bearing and then

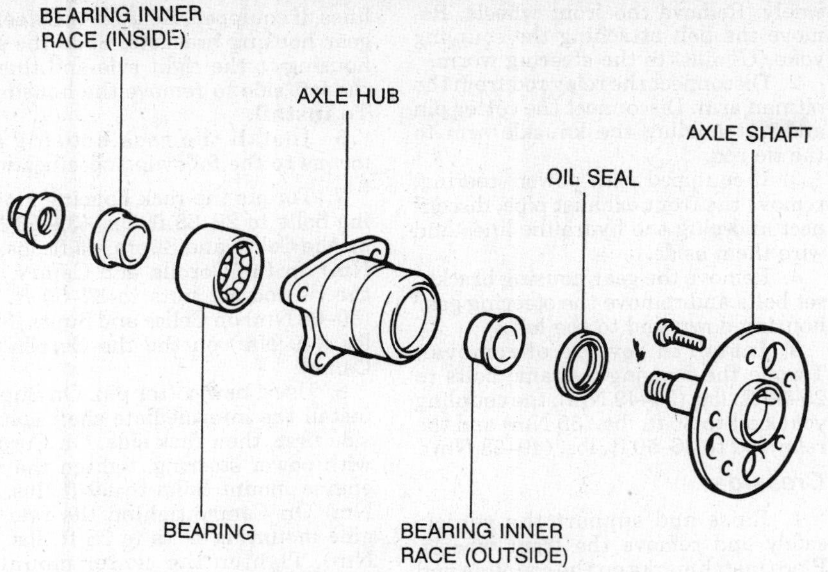

Pressed axle hub bearing assembly

press a new oil seal into the hub. Coat the lip of the seal with grease.

15. Position a new bearing inner race (inside) on the bearing and then press the inner race with the hub onto the axle shaft.

16. Install the nut and tighten it to 90 ft. lbs. (123 Nm). Stake the nut with a brass drift.

17. Position the axle carrier on the strut tube and tighten the nuts to 119 ft. lbs. (162 Nm), 105 ft. lbs. (142 Nm) on Corolla FX and Corolla sedan and wagon; 166 ft. lbs. (226 Nm) on Camry.

18. Install the bolt and nut attaching the carrier to the No. 2 suspension arm; finger-tighten it only.

NOTE: Make sure the lip of the nut is in the hole on the arm.

19. Repeat Step 5 for the No. 1 suspension arm.

NOTE: Make sure the lip of the nut is in the hole on the arm.

20. Install the strut rod-to-carrier bolt so the lip of the nut is in the groove on the bracket.

21. Install a new O-ring onto the axle carrier. Install the axle hub and brake backing plate. Tighten the 4 bolts to 59 ft. lbs. (80 Nm).

22. Reconnect the brake line, install the brake drum and then bleed the brakes.

23. Lower the vehicle and bounce it a few times to set the rear suspension.

24. Tighten the suspension arm bolts and the strut rod bolt to 64 ft. lbs. (87 Nm). On the Camry and Celica, tighten the strut rod mounting bolts to 83 ft. lbs. (113 Nm) and the suspension arm bolts to 134 ft. lbs. (181 Nm).

STEERING

Steering Wheel

CAUTION

On vehicles equipped with an air bag, the negative battery cable must be disconnected, before working on the system. Failure to do so may result in deployment of the air bag and possible personal injury.

REMOVAL & INSTALLATION

Without Air Bag

CAUTION

Do not attempt to remove or install the steering wheel by hammering on it. Damage to the energy-absorbing steering column could result.

1. Disconnect the negative battery cable. Position the front wheels straight ahead.

2. Unfasten the horn and turn signal multi-connector(s) at the base of the steering column shroud.

3. If equipped with a 3 spoked wheel, loosen the trim pad retaining screws from the back side of the steering wheel. The 2 spoke steering wheel is removed in the same manner as the three spoke, except that the trim pad should be pried off with a small prybar. Remove the pad by lifting it toward the top of the wheel.

4. Lift the trim pad and horn button assembly from the wheel.

5. Remove the steering wheel hub retaining nut.

6. Scribe matchmarks on the hub and shaft to aid in correct installation.

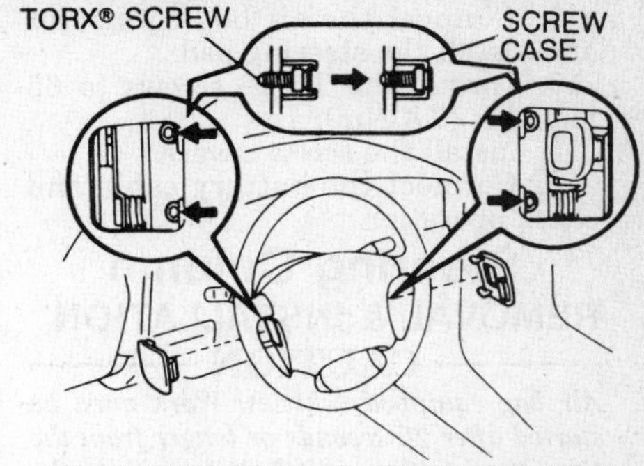

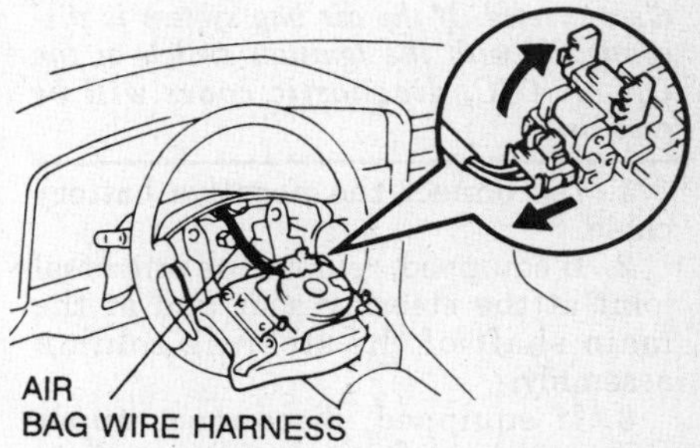

Removing the air bag

7. Use a suitable puller to remove the steering wheel.

8. Installation is the reverse of removal. Tighten the wheel retaining nut to 25 ft. lbs. (34 Nm).

With Air Bag

CAUTION

Air bag equipped vehicles: Work must be started after 20 seconds or longer from the time the ignition switch is turned to the LOCK and the negative battery terminal is disconnected. If the air bag system is disconnected with the ignition switch at the ON or ACC, diagnostic codes will be recorded.

1. Disconnect the negative battery cable.

2. Place the front wheels facing straight ahead.

3. Remove the steering wheel screw covers.

4. Using a Torx® wrench T30, loosen the screws until the groove trailing the screw circumference catches on the screw case.

5. Pull the wheel pad out from the steering wheel and disconnect the air bag connector.

6. Remove the steering wheel nut. Place matchmarks on the wheel and steering shaft.

7. Using a steering wheel puller SST 09213-31021 or equivalent, remove the steering wheel.

To install:

8. Install the steering wheel, align the matchmarks and torque the nut to 25 ft. lbs. (34 Nm).

9. Connect the air bag connector and install the steering pad.
10. Torque the Torx® screws to 65 inch lbs. (7.4 Nm).
11. Install the screw covers.
12. Connect the battery cable and check operation.

Steering Column

REMOVAL & INSTALLATION

CAUTION

Air bag equipped vehicles: Work must be started after 20 seconds or longer from the time the ignition switch is turned to the LOCK and the negative battery terminal is disconnected. If the air bag system is disconnected with the ignition switch at the ON or ACC, diagnostic codes will be recorded.

1. Disconnect the negative battery cable.
2. If equipped, remove the universal joint at the steering gear and at the main shaft of the steering column assembly.
3. If equipped, disconnect upper universal joint from the intermediate shaft.
4. Remove the steering wheel.
5. Remove the instrument lower finish panels, air ducts and column covers.
6. Disconnect all electrical connections for ignition switch and combination switch. Remove the combination switch, as necessary.
7. Loosen column hole cover clamp bolt. Remove the support mounting bolt. Remove the 4 column tube mounting bolts. Pull out steering column. On some vehicles, remove the bolts from the column hole cover plate and remove 2 column bracket mounting nuts. Turn the steering column assembly clockwise and remove it from the vehicle, as necessary.

To install:

8. Place the the steering column assembly in the installed position. Tighten all necessary mounting nuts (torque evenly). Install and tighten all column cover bolts. Tighten column hole cover clamp bolt, as necessary.
9. Install combination switch. Reconnect all electrical connectors.
10. Install instrument lower finish panels, air duct and column covers.
11. Install or connect the universal joint. Insure that the retaining bolts are installed through both shaft grooves.
12. Install steering wheel and connect the negative battery cable.

Manual and Power Steering Rack

REMOVAL & INSTALLATION

Corolla (RWD)

1. Raise and support the vehicle safely. Remove the front wheels. Remove the bolt attaching the coupling yoke (U-joint) to the steering worm.
2. Disconnect the relay rod from the pitman arm. Disconnect the cotter pin and nut holding the knuckle arm to the tie rod.
3. If equipped with power steering, remove the front exhaust pipe, disconnect and plug the hydraulic lines and wire them aside.
4. Remove the gear housing bracket set bolts and remove the steering gear housing down and to the left.
5. Install in reverse of removal. Torque the housing-to-frame bolts to 25–36 ft. lbs. (34–49 Nm); the coupling yoke bolt to 26 ft. lbs. (35 Nm) and the relay rod to 36–50 ft. lbs. (49–68 Nm).

Cressida

1. Raise and support the vehicle safely and remove the front wheels. Place matchmarks on the coupling and steering column shaft. Disconnect the solenoid connectors.
2. Disconnect the Pitman arm from the relay rod using a tie rod puller on the Pitman arm set nut. Disconnect the tie rod ends from the steering knuckles.
3. Remove the steering damper, if equipped.
4. Disconnect the steering gearbox at the coupling. Unbolt the gearbox from the chassis and remove. Remove the grommets from the gear housing.
5. Installation is in the reverse order of removal, with the exception of first aligning the matchmarks and connecting the steering shaft to the coupling before bolting the gearbox into the vehicle permanently. Tighten the steering damper bolts to 20 ft. lbs. (26 Nm). Tighten the tie rod ends to 43 ft. lbs. (59 Nm). Tighten the mounting bracket bolts to 56 ft. lbs. (76 Nm).

Corolla (FWD), Camry and Supra

1. Raise and support the vehicle safely. Remove the front wheels. Open the hood. Remove the 2 set bolts, and remove the sliding yoke from between the steering rack housing and the steering column shaft. On Supra, unbolt and remove the intermediate shaft (rack housing side first).
2. Remove the cotter pin and nut holding the knuckle arm to the tie rod end. Using a tie rod puller, disconnect the tie rod end from the knuckle arm.
3. On Corolla and Camry with power steering, remove the lower crossmember, remove the engine under cover, center engine mount member and the rear engine mount.
4. On Supra with turbocharger, remove the No. 1 air intake connector with No. 2 air hose.
5. Disconnect the power steering lines, if equipped. Remove the steering gear housing brackets. Slide the gear housing to the right side and then to the left side to remove the housing.

To install:

6. Install the rack housing and torque to the following specifications.
7. Torque the rack housing mounting bolts to 29–39 ft. lbs. (39–53 Nm) on the Celica and Supra; 43 ft. lbs. (58 Nm) on the Corolla and Camry, and the tie rod set nuts to 37–50 ft. lbs. (50–68 Nm) on Celica and Supra; 36 ft. lbs. (49 Nm) on the the Corolla and Camry.
8. Use a new cotter pin. On Supras, install the intermediate shaft column side first, then rack side. On Corollas with power steering, tighten the rear engine mount bolts to 29 ft. lbs. (39 Nm) On Camry tighten the rear engine mounting bolts to 38 ft. lbs. (52 Nm). Tighten the center mounting member to 29 ft. lbs. (39 Nm).
9. On power steering-equipped vehicles, bleed the power steering system and check for fluid leaks.
10. Adjust toe-in on all vehicles.

Celica

1. Raise and support the vehicle safely. Remove the front wheels.
2. Remove the both engine under covers.
3. Remove the 2 bolts that connect the steering column U-joint to the rack and then disconnect the column from the rack.
4. Remove the cotter pin and nut and then using a tie rod end removal tool, disconnect the tie rod end from the steering knuckle.
5. Remove the lower suspension crossmember.
6. Remove the mounting bolts and remove the center engine mount member.
7. Disconnect the exhaust pipe from the manifold. Position it aside.
8. Tag and disconnect the 2 hydraulic lines. Position them aside and suspend on a wire.
9. Remove the rear engine mount bracket.
10. Remove the mounting bolts and brackets and lower steering rack from the vehicle.

To install:

11. Position the rack assembly, install the grommets and brackets and then tighten the 2 bolts and 2 nuts to 43 ft. lbs. (59 Nm).
12. Install the rear engine mount bracket and tighten the 2 bolts to 38 ft. lbs. (52 Nm).
13. Connect the hydraulic lines and tighten the union nuts to 29 ft. lbs. (39 Nm).
14. Connect the exhaust pipe to the manifold.

15. Install the center engine mount member and tighten the bolts to 29 ft. lbs. (39 Nm).
16. Install the lower crossmember and tighten the 5 outer bolts to 154 ft. lbs. (208 Nm). Tighten the center bolts to 29 ft. lbs.
17. Installation of the remaining components is in the reverse order of removal. Tighten the tie rod end nuts to 36 ft. lbs. (49 Nm) and use a new cotter pin. Tighten the steering column U-joint bolts to 26 ft. lbs. (35 Nm). Fill the power steering pump to the proper level, bleed the system and check the wheel alignment.

1988–89 MR2 and Tercel

1. Raise and support the vehicle safely.
2. Remove the front wheels and the engine under cover.
3. Place matchmarks on the main shaft, joint yoke and pinion shaft. Remove the intermediate shaft from the worm gear shaft.
4. If equipped with power steering, disconnect the 2 hydraulic lines. Position them aside and suspend on a wire.
5. Remove both tie rod ends.
6. Remove the lower suspension crossmember. Remove the center floor crossmember.
7. Remove the rack housing bracket mounting bolts and brackets.

NOTE: Be careful not to damage the rubber boots.

8. Remove the steering linkage.
9. Installation is the reverse of the removal procedure. Fill the power steering pump to the proper level and bleed the system. Check the wheel alignment.

1990–92 MR2

1. Disconnect the negative battery cable.
2. Position the front wheels straight ahead.
3. Raise the vehicle and support safely. Remove the front wheels and front luggage under cover.
4. Remove the dust cover. Matchmark the universal joint and control valve shaft for installation.
5. Disconnect the tie rod ends using a tie rod separator or equivalent.
6. Disconnect the power steering hoses and drain the fluid into a container.
7. Remove the housing-to-frame retaining bolts and remove the assembly.

To install:

8. Install the assembly and torque the retaining bolts to 32 ft. lbs. (43 Nm).
9. Connect the tie rods and torque to 36 ft. lbs. (49 Nm).
10. Connect the universal joint and torque to 26 ft. lbs. (35 Nm).
11. Install the dust cover and connect the power steering hoses.
12. Connect the battery cable. Start the engine and bleed the system.
13. Install the front wheels and lower the vehicle.

Power Steering Pump

REMOVAL & INSTALLATION

Tercel, Corolla, Camry, Celica, 1988–89 MR2 and 1988 Cressida

1. Raise and support the vehicle safely. Remove the fan shroud.
2. On Camry and Celica, remove the right front wheel and the engine under cover. Remove the lower suspension crossmember.
3. Unfasten the nut from the center of the pump pulley. Disconnect the vacuum hose from the air control valve, if equipped.

NOTE: Use the drive belt as a brake to keep the pulley from rotating.

4. Withdraw the drive belt. On some vehicles, it may be necessary to remove the pulley in order to remove the drive belt.
5. If equipped with an idler pulley and on the Corolla FX, push on the drive belt to hold the pulley in place and remove the pulley set nut. Loosen the idler pulley set nut and adjusting bolt. Remove the drive belt and loosen the drive pulley to remove the Woodruff key.
6. Remove the pulley and the Woodruff key from the pump shaft.
7. Detach and plug the intake and outlet hoses from the pump reservoir.

NOTE: Tie the hose ends up high so the fluid cannot flow out of them. Drain or plug the pump to prevent fluid leakage.

8. Remove the bolt from the rear mounting brace.
9. Remove the front bracket bolts and withdraw the pump.

To install:

10. Tighten the pump pulley mounting bolt to 25–39 ft. lbs. (34–53 Nm).
11. Tighten the 5 outer mounting bolts on the lower crossmember to 154 ft. lbs. (209 Nm). On Celica, tighten the center bolt to 29 ft. lbs. (39 Nm).
12. Adjust the pump drive belt tension. The belt should deflect 0.13–0.93 in. under thumb pressure applied midway between the air pump and the power steering pump.
13. Fill the reservoir with Dexron® II automatic transmission fluid. Bleed the air from the system.

Supra and 1989–92 Cressida

7M-GE ENGINE

1. Raise and support the vehicle safely. Drain the fluid from the reservoir tank.
2. Disconnect the air hose from the air control tank. Disconnect the return hose from the reservoir tank.
3. Remove the engine under cover. Disconnect and plug the pressure hose from the power steering pump.
4. Holding the power steering pump pulley, remove the pulley set nut. Remove the drive belt adjusting nut.
5. Remove the power steering pump set bolt. Remove the drive belt, pulley and Woodruff key.
6. Disconnect the oil cooler hose bracket from the power steering pump. Remove the drive belt adjust bolt and remove the power steering set bolt and power steering pump.
7. Installation is the reverse order of the removal procedure. Be sure to bleed the system upon completion of the installation procedure.

7M-GTE ENGINE

1. Raise and support the vehicle safely. Drain the fluid from the reservoir tank.
2. Remove the No. 1 and No. 2 air hoses with the No. 4 air cleaner pipe.
3. Disconnect the connector from the air flow meter. Remove the air flow meter installation bolt. Loosen the 5 clamps and disconnect the air hoses, release the 3 clips on the air cleaner case. Loosen the No. 7 air hose clamp and remove the No. 7 air cleaner hose with the air flow meter.
4. Remove the oil reservoir tank with bracket. Disconnect the 2 air hoses from the air control valve on the power steering pump.
5. Remove the adjusting strut. Remove the engine under cover.
6. Holding the power steering pump pulley, remove the pulley set nut. Remove the drive belt adjusting nut.
7. Remove the power steering pump set bolt. Remove the drive belt, pulley and Woodruff key.
8. Disconnect and plug the pressure hose from the power steering pump.
9. Remove the power steering set bolt and power steering pump.
10. Installation is the reverse order of the removal procedure. Be sure to bleed the system upon completion of the installation procedure.

1990–92 MR2

The power steering pump is not driven by a conventional drive belt. Instead the pump is driven by an electric motor. The pump and motor are combined as one unit. Removal and installation is as follows:

1. Disconnect the negative battery cable.

2. Raise and support the vehicle safely.
3. Remove the front luggage under cover.
4. Remove the pump shield and rear stay.
5. Disconnect the hydraulic lines from the pump. Plug the lines to prevent the loss of power steering fluid.
6. Disconnect the electrical wires from the top of the motor.
7. Remove the pump mounting bolts, bushings and spacers. Check the bushings for cracks and deformation. Replace as necessary.
8. Remove the pump and motor assembly.

To install:

9. Position the pump and install the mounting bolts, bushings and spacers. Torque the bolts to 19 ft. lbs. (25 Nm).
10. Connect the electrical wires to the top of the motor.
11. Connect the hydraulic lines.
12. Install the rear stay and pump shield.
13. Install the front luggage under cover.
14. Lower the vehicle and connect the negative battery cable.
15. Fill the power steering reservoir to the proper level and bleed the system.

BELT ADJUSTMENT

1. Inspect the power steering drive belt to see that it is not cracked or worn. Be sure its surfaces are free of grease or oil.
2. Push down on the belt halfway between the fan and the alternator pulleys (or crankshaft pulley) with thumb pressure. Belt deflection should be 3/8–1/2 in. (10–13mm).
3. If the belt tension requires adjustment, loosen the adjusting link bolt and move the power steering pump until the proper belt tension is obtained.
4. Do not over-tighten the belt, as damage to the power steering pump bearings could result. Tighten the adjusting link bolt.
5. Drive the vehicle and re-check the belt tension. Adjust as necessary.

SYSTEM BLEEDING

1. Raise and support the vehicle safely.
2. Fill the pump reservoir with the proper fluid.
3. Rotate the steering wheel from lock-to-lock several times. Add fluid if necessary.
4. With the steering wheel turned fully to one lock, crank the starter while watching the fluid level in the reservoir.

NOTE: Do not start the engine. Operate the starter with a remote starter switch or have an assistant do it from inside the vehicle. Do not run the starter for prolonged periods.

5. Repeat Step 4 with the steering wheel turned to the opposite lock.
6. Start the engine. With the engine idling, turn the steering wheel from lock-to-lock several times.
7. Lower the front of the vehicle and repeat Step 6.
8. Center the wheel at the midpoint of its travel. Stop the engine.
9. The fluid level should not have risen more than 0.2 in. (5mm). If it does, repeat Step 7.
10. Check for fluid leakage.

Tie Rod Ends

REMOVAL & INSTALLATION

1. Scribe alignment marks on the tie rod and rack end.
2. Working at the steering knuckle arm, pull out the cotter pin and then remove the castellated nut.
3. Using a tie rod end puller, disconnect the tie rod from the steering knuckle arm.
4. Repeat the first 2 steps on the other end of the tie rod (where it attaches to the relay rod or steering rack).

To install:

5. Align the alignment marks on the tie rod and rack end.
6. Install the tie rod end.
7. Tighten the tie rod end nuts to 36 ft. lbs. (49 Nm).Install a new cotter pin.
8. Install the front wheels and lower the vehicle.

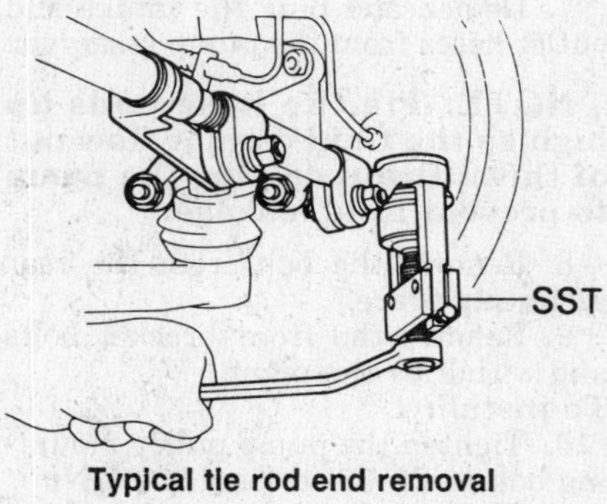

Typical tie rod end removal

BRAKES

For all brake system repair and service procedures not detailed below, please refer to "Brakes" in the Unit Repair section.

Master Cylinder and ABS Actuator

REMOVAL & INSTALLATION

1. Disconnect the negative battery cable. Label and disconnect the electrical connectors.
2. Remove the fluid in the master cylinder with a suitable syringe.
3. On MR2, remove the luggage compartment.
4. Disconnect the hydraulic lines from the master cylinder. Plug the ends of the lines to prevent loss of fluid.
5. Detach the hydraulic fluid pressure differential switch wiring connectors.
6. Loosen the master cylinder reservoir mounting nuts.
7. Unfasten the nuts and remove the master cylinder or ABS actuator assembly from the power brake unit.

To install:

8. Before tightening the master cylinder mounting nuts or bolts, screw the hydraulic line into the cylinder body a few turns.
9. Install the master cylinder or actuator. Torque the hydraulic lines to 11 ft. lbs. (15 Nm) and the mounting nuts to 25 ft. lbs. (34 Nm).
10. After installation is completed, bleed the master cylinder and the brake system.

Proportioning Valve

A proportioning valve is used to reduce the hydraulic pressure to the rear brakes because of weight transfer during high speed stops. This helps to keep the rear brakes from locking up by improving front to rear brake balance.

REMOVAL & INSTALLATION

1. Disconnect the brake lines from the valve unions.
2. Remove the valve mounting bolt, if used, and remove the valve.

NOTE: If the proportioning valve is defective, it must be replaced as an assembly; it cannot be rebuilt.

3. Installation is the reverse of removal. Bleed the brake system after it is completed.

Power Brake Booster

REMOVAL & INSTALLATION

1. Disconnect the negative battery cable. Remove the master cylinder and disconnect the vacuum hose from the brake booster.

2. Remove the instrument lower finish panel, as required.
3. On the MR2, remove the wheel guard, instrument lower finish panel and air duct.
4. Remove the brake pedal return spring.
5. Remove clip and clevis pin.
6. Remove the brake booster nuts and clevis pin.
7. Pull out the brake booster and gasket.
8. Installation is the reverse order of the removal procedure. Bleed the brake system. Torque the nuts to 25 ft. lbs. (34 Nm).

Brake Caliper

REMOVAL & INSTALLATION

1. Raise and support the vehicle safely.
2. Remove the front or rear wheels.
3. Disconnect the brake hose from the caliper. Plug the end of the hose to prevent loss of fluid.
4. Remove the bolts that attach the caliper to the torque plate.
5. Lift up and remove the caliper assembly.
6. Installation is the reverse of the removal procedure. Grease the caliper slides and bolts with Lithium grease or equivalent. Torque the caliper bolts to 20–27 ft. lbs. (27–41 Nm) on front disc brakes and 14 ft. lbs. (20 Nm) for rear disc brakes. Fill and bleed the system.

Disc Brake Pads

REMOVAL & INSTALLATION

MR2 – Rear, Cressida – Front and Rear and 1988–92 Celica (AWD) – Front

1. Raise and support the vehicle safely.
2. Remove the wheels.
3. Siphon a sufficient quantity of brake fluid from the master cylinder reservoir to prevent any brake fluid from overflowing the master cylinder when removing or installing new pads. This is necessary as the piston must be forced into the caliper bore to provide sufficient clearance when installing the pads.
4. Grasp the caliper from behind and carefully pull it to seat the piston in its bore.
5. Loosen and remove the lower caliper slide pin (mounting bolt).
6. Swivel the caliper upward and aside, exposing the brake pads. Do not disconnect the brake line.
7. Slide out the old brake pads along with any anti-squeal shims, anti-rattle springs, pad wear indicators, pad guide plates and pad support plates. Take great care to note the position of all assorted pad hardware.
8. Check the brake disc (rotor) for thickness and run-out. Inspect the caliper and piston assembly for breaks, cracks, fluid seepage or other damage. Overhaul or replace as necessary.

To install:

9. Install the pad support plates, anti-rattle springs or guide plates into the torque plate.
10. Install the pad wear indicators onto the pad.
11. Install the anti-squeal shims on the outside of each pad and then install the pad assemblies into the torque plate.
12. Swivel the caliper back down over the pads. If it won't fit, use a C-clamp or hammer handle and carefully force the piston into its bore.
13. On the MR2 rear disc, turn the piston clockwise while pushing it in until it locks in place. Fit the protrusion on the inner pad into the groove in the piston stopper and then install the caliper. Be careful not to pinch the boot.
14. Install and tighten the lower slide pin or mounting bolt.
15. Install the parking brake cable on the Corolla with rear discs and then adjust the automatic adjuster by pulling and releasing the parking brake lever several times.
16. Install the wheel and lower the vehicle. Check the brake fluid level. Adjust the parking brake on rear disc brake vehicles.

MR2 – Front, Tercel – Front Camry – Front and Rear Supra – Front and Rear 1988–92 Celica (AWD) – Rear Celica (2WD) – Front and Rear Corolla (FWD) – Front and Rear

1. Raise and support the vehicle safely.
2. Remove the wheels.
3. Siphon a sufficient quantity of brake fluid from the master cylinder reservoir to prevent any brake fluid from overflowing the master cylinder when removing or installing new pads. This is necessary as the piston must be forced into the caliper bore to provide sufficient clearance when installing the pads.
4. Grasp the caliper from behind and carefully pull it to seat the piston in its bore.
5. Loosen and remove the 2 caliper mounting pins (bolts) and then remove the caliper assembly. Position it aside. Do not disconnect the brake line.
6. Slide out the old brake pads along with any anti-squeal shims, springs, pad wear indicators and pad support plates. Make sure to note the position of all assorted pad hardware.

To install:

7. Check the brake disc (rotor) for thickness and run-out. Inspect the caliper and piston assembly for breaks, cracks, fluid seepage or other damage. Overhaul or replace as necessary.
8. Install the pad support plates into the torque plate.

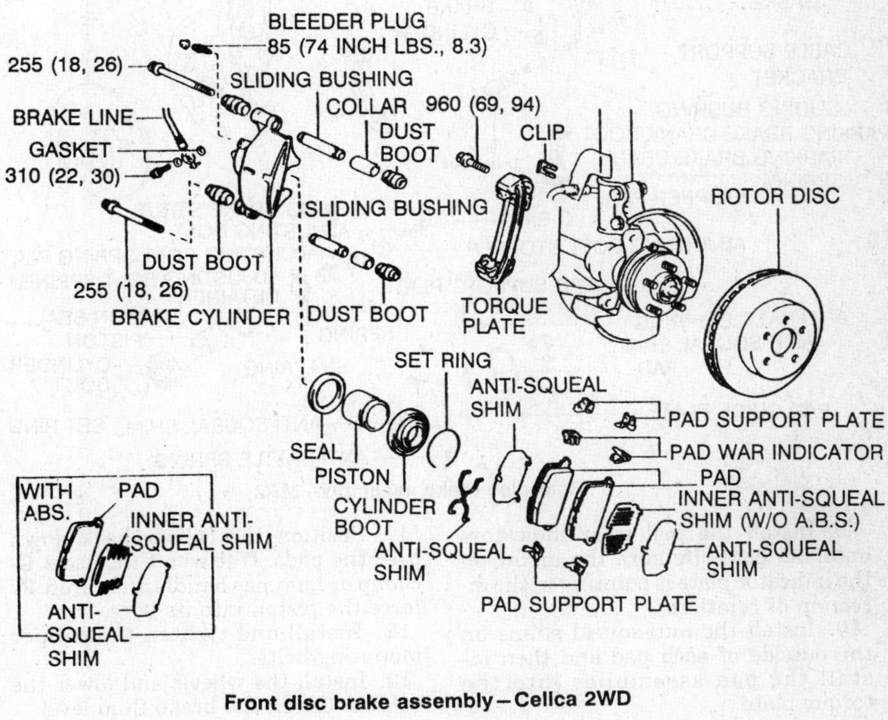

Front disc brake assembly – Celica 2WD

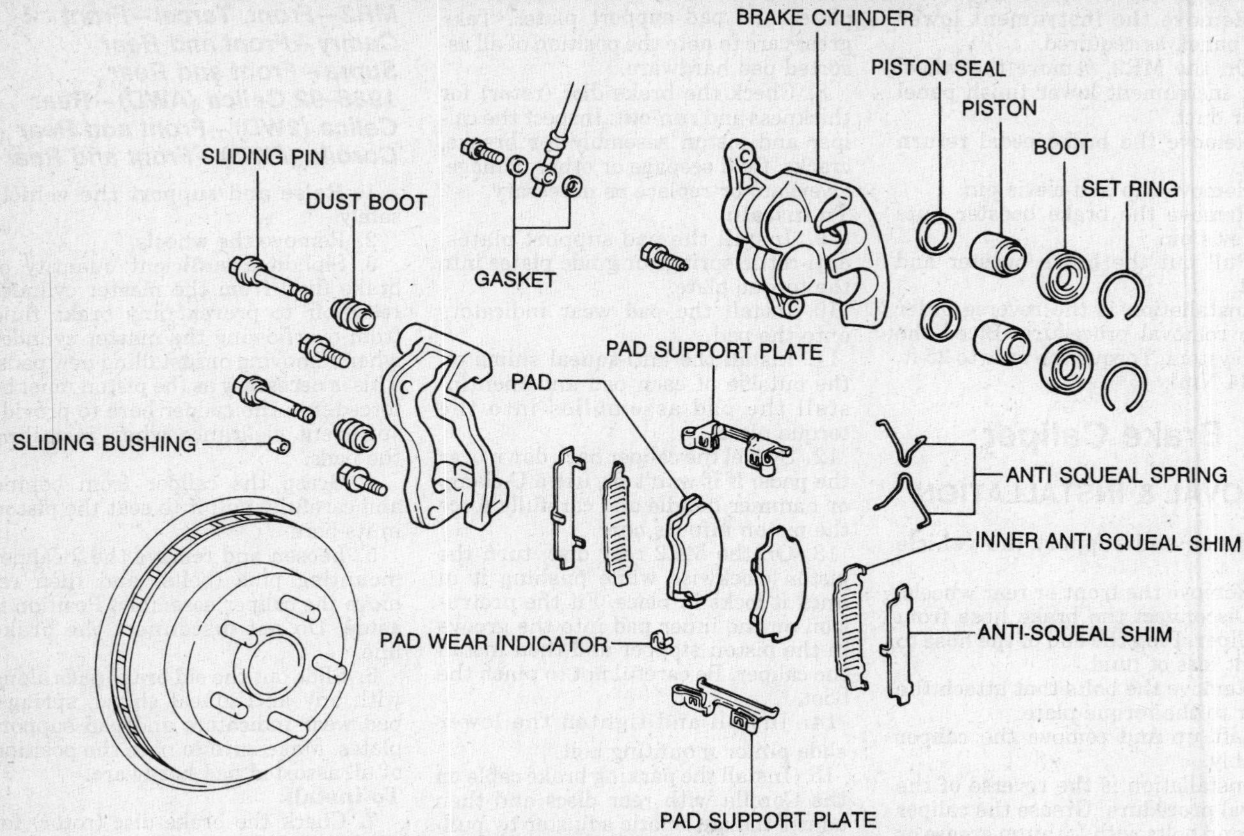

Front disc brake assembly—1990–92 MR2 with 3S-GTE engines

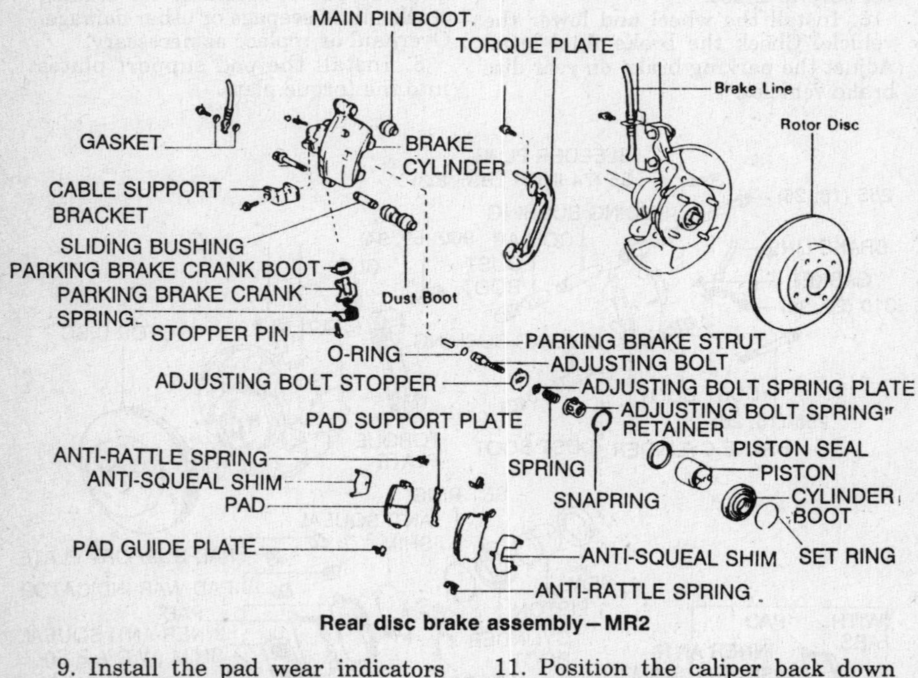

Rear disc brake assembly—MR2

9. Install the pad wear indicators onto the pads. Be sure the arrow on the indicator plate is pointing in the direction of rotation.

10. Install the anti-squeal shims on the outside of each pad and then install the pad assemblies into the torque plate.

11. Position the caliper back down over the pads. If it won't fit, use a C-clamp or hammer handle and carefully force the piston into its bore.

12. Install and tighten the caliper mounting bolts.

13. Install the wheels and lower the vehicle. Check the brake fluid level.

Brake Rotor

REMOVAL & INSTALLATION

1. Raise and support the vehicle safely.
2. Remove the wheels.
3. Temporarily attach 2 lug nuts onto the rotor disc.
4. Unbolt the torque plate from the steering knuckle.
5. Remove the lug nuts and pull the rotor from the wheel hub.
6. Installation is the reverse of the removal procedure.

Brake Drums

REMOVAL & INSTALLATION

1. Raise and support the vehicle safely.
2. Remove the wheels.
3. Remove the brake drum. Tap the drum lightly with a rubber mallet in order to free it. If the brake drum cannot be removed easily, insert a small prybar through the hole in the backing plate and hold the automatic adjuster lever away from the adjusting bolt. Using another prybar, relieve the brake shoe tension by rotating the adjusting bolt (star wheel) in a clockwise direction. If the drum still will not come off,

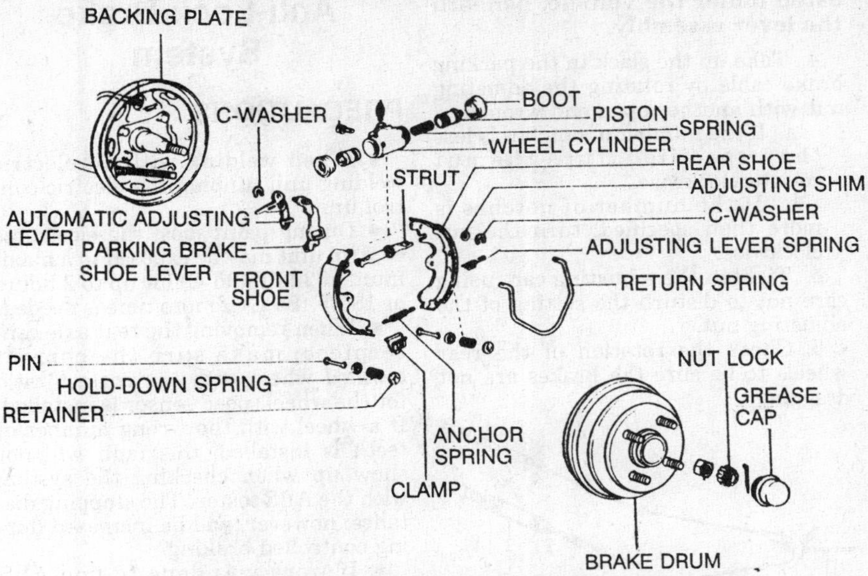

Exploded view of the rear drum brake—Tercel sedan

use a puller; but, first make sure the parking brake is released.

4. Installation is the reverse of the removal procedure.

Brake Shoes

REMOVAL & INSTALLATION

Tercel Sedan, Corolla (RWD) 1988 Cressida

1. Raise and support the vehicle safely. Remove the wheels.
2. Remove the brake drums.

NOTE: Do not depress the brake pedal once the brake drum has been removed.

3. Carefully unhook the tension spring from the leading (front) brake shoe. On the Tercel, its a return spring; also remove the clamp.
4. Press the hold-down spring retainer in and turn the pin.
5. Remove the hold-down spring, retainers and the pin. Pull out the brake shoe and unhook the anchor spring from the lower edge.
6. Remove the hold-down spring from the trailing (rear) shoe. Pull the shoe out with the adjuster strut, automatic adjuster assembly and springs attached and disconnect the parking brake cable. Remove the tension/return and anchor springs from the rear shoe.
7. Remove the adjusting strut. Unhook the adjusting lever spring from the rear shoe and then remove the automatic adjuster assembly by popping out the C-clip.

To install:

8. Inspect the shoes for signs of unusual wear or scoring.
9. Check the wheel cylinder for any sign of fluid seepage or frozen pistons.
10. Clean and inspect the brake backing plate and all other components. Check that the brake drum inner diameter is within specified limits. Lubricate the backing plate bosses and the anchor plate.
11. Mount the automatic adjuster assembly onto a new rear brake shoe. Make sure the C-clip fits properly. Connect the adjusting strut and install the spring.
12. Connect the parking brake cable to the rear shoe and then position the shoe so the lower end rides in the anchor plate and the upper end is against the boot in the wheel cylinder. Install the pin and the hold-down spring. Press the retainer down over the pin and rotate the pin so the crimped edge is held by the retainer. Install the anchor spring between the front and rear shoes and then stretch the spring enough so the front shoe will fit as the rear did in Step 10. Install the hold-down spring, pin and retainer. Stretch the tension/return spring between the 2 shoes and connect it so it rides freely. Don't forget the return spring clamp on the Tercel.
13. Check that the automatic adjuster is operating properly; the adjusting bolt should turn when the parking brake lever, in the brake assembly, not in the vehicle!, is moved. Adjust the strut as short as possible and then install the brake drum. Set and release the parking brake several times.
14. Install the wheel and lower the vehicle. Check the level of brake fluid in the master cylinder.

Camry, Celica, Tercel Wagon and Corolla (FWD)

1. Raise and support the vehicle safely. Remove the wheels.
2. Remove the brake drums.

NOTE: Do not depress the brake pedal once the brake drum has been removed.

3. Carefully unhook the return spring from the leading (front) brake shoe. Grasp the hold-down spring pin with pliers and turn it until its in line with the slot in the hold-down spring. Remove the hold-down spring and the pin. Pull out the brake shoe and unhook the anchor spring from the lower edge.
4. Remove the hold-down spring from the trailing (rear) shoe. Pull the shoe out with the adjuster strut, automatic adjuster assembly and springs attached and disconnect the parking brake cable. Unhook the return spring and then remove the adjusting strut. Remove the anchor spring.
5. Remove the adjusting strut. Unhook the adjusting lever spring from the rear shoe and then remove the automatic adjuster assembly by popping out the C-clip.

To install:

6. Inspect the shoes for signs of unusual wear or scoring.
7. Check the wheel cylinder for any sign of fluid seepage or frozen pistons.
8. Clean and inspect the brake backing plate and all other components. Check that the brake drum inner diameter is within specified limits. Lubricate the backing plate bosses and the anchor plate.
9. Mount the automatic adjuster assembly onto a new rear brake shoe. Make sure the C-clip fits properly. Connect the adjusting strut/return spring and then install the adjusting spring.
10. Connect the parking brake cable to the rear shoe and then position the shoe so the lower end rides in the anchor plate and the upper end is against the boot in the wheel cylinder. Install the pin and the hold-down spring. Rotate the pin so the crimped edge is held by the retainer.
11. Install the anchor spring between the front and rear shoes and then stretch the spring enough so the front shoe will fit as the rear did in Step 10. Install the hold-down spring and pin. Connect the return spring/adjusting strut between the 2 shoes and connect it so it rides freely.
12. Check that the automatic adjuster is operating properly; the adjusting bolt should turn when the parking brake lever (in the brake assembly, not in the vehicle!) is moved. Adjust the strut as short as possible and then in-

stall the brake drum. Set and release the parking brake several times.

13. Install the wheel and lower the vehicle. Check the level of brake fluid in the master cylinder.

Wheel Cylinder

REMOVAL & INSTALLATION

1. Plug the master cylinder inlet to prevent hydraulic fluid from leaking.
2. Remove the brake drums and shoes.
3. Working from behind the backing plate, disconnect the hydraulic line from the wheel cylinder.
4. Unfasten the screws retaining the wheel cylinder and withdraw the cylinder.

To install:

5. Installation is performed in the reverse order of removal. However, once the hydraulic line has been disconnected from the wheel cylinder, the union seat must be replaced. To replace the seat, proceed in the following manner:
   a. Use a screw extractor with a diameter of 0.1 in. or equivalent, having reverse threads, to remove the union seat from the wheel cylinder.
   b. Drive in the new union seat with a $^5/_{16}$ in. bar, used as a drift.
6. Bleed the brake system after completing wheel cylinder, brake shoe and drum installation.

Parking Brake Cable

ADJUSTMENT

1. Slowly pull the parking brake lever upward, without depressing the button on the end of it, while counting the number of notches required until the parking brake is applied.

NOTE: Two "clicks" are equal to 1 notch.

2. Check the number of notches against the following specifications.
   - Tercel (2WD) – 5–8 clicks.
   - 1988–92 Tercel (2WD) – 7–9 clicks.
   - 1988 Tercel (AWD) – 6–8 clicks.
   - 1988–90 Celica – 4–7 clicks.
   - 1988–90 Corolla FWD – 4–7 clicks (FX16 and vehicles w/rear disc – 5–8).
   - MR2 – 5–8 clicks.
   - Cressida – 5–8 clicks.
   - Camry – 5–8 clicks.
   - Supra – 5–8 clicks.
3. If the brake system requires adjustment, loosen the cable adjusting nut cap which is located at the rear of the parking brake lever.

NOTE: On some vehicles, the adjustment and lock nuts are located under the vehicle, beneath the lever assembly.

4. Take up the slack in the parking brake cable by rotating the adjusting nut with another open end wrench.
   a. If the number of notches is less than specified, turn the nut counterclockwise.
   b. If the number of notches is more than specified, turn the nut clockwise.
5. Tighten the adjusting cap, using care not to disturb the setting of the adjusting nut.
6. Check the rotation of the rear wheels to be sure the brakes are not dragging.

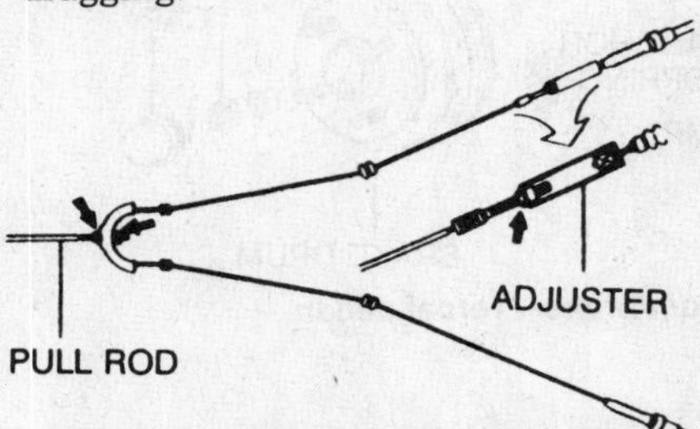

Parking brake adjustment – Cressida

Brake System Bleeding

Master Cylinder

1. Check the fluid level in the master cylinder. Add fluid as necessary. Never allow the master cylinder to run dry.
2. Disconnect the brake tubes from the master cylinder.
3. Slowly depress the brake pedal and hold it.
4. Close off the outlet opening on the master cylinder with finger pressure and release the brake pedal.
5. Repeat Steps 3 and 4 several times to bleed all the air from the master cylinder.

Brake Lines

1. Bleed the caliper or wheel cylinder with the longest hydraulic line. Connect a vinyl tube the to the bleeder screw on the brake cylinder and submerge the other end of the tube in a transparent container half filled with clean brake fluid.
2. Pump the brake pedal several times and loosen the bleeder screw with the pedal held down.
3. When brake fluid stops coming out of the tube, tighten the bleeder screw and release the brake pedal.
4. Repeat Steps 2 and 3 until no air bubbles can be seen in the container.
5. Repeat the procedure for each wheel.
6. Check the level in the master cylinder. Add fluid as necessary.

Anti-Lock Brake System

PRECAUTIONS

- When welding with an electric welding unit, unplug the electric control unit.
- During paint jobs, the electronic control unit may be exposed to a maximum of 203°F (95°C) for up to 2 hours or 185°F (85°C) if more time is needed.
- When removing the rear axle centerpiece, make sure the correct toothed wheel with the correct ratio for the wheel speed sensor is installed. If a wheel with the wrong number of teeth is installed, this fault will not show up when checking the system with the ABS tester. The stopping distance, however, will be increased during controlled braking.
- If work was done to non-ABS brake components, a simple operational test will be sufficient. This means that after driving about 5 mph, the warning light on the instrument panel should go out if the ABS system is intact.
- If ABS components have been replaced, the entire system should be checked using the appropriate tester in combination with brake test bench or an adaptor in combination with a multimeter.

RELIEVING ANTI-LOCK BRAKE SYSTEM PRESSURE

Pump the brake pedal at least 20 times with the ignition key in the OFF position. Place a shop rag around the hydraulic line fitting and wear safety glasses when disconnecting the hydraulic lines.

Anti-Lock Brake Actuator

REMOVAL & INSTALLATION

1. Disconnect the negative battery cable.
2. Remove the brake fluid from the master cylinder with a syringe.
3. Remove the plastic cover from the actuator.
4. Disconnect the hydraulic lines from the actuator. Plug the lines to prevent loss of fluid.
5. Disconnect the electrical connectors from the actuator.
6. Remove the actuator from the bracket.
7. Installation is the reverse of the removal procedure. Fill and bleed the system.
8. Connect the battery cable, start

the engine and check brake system operation before driving the vehicle.

CHASSIS ELECTRICAL

CAUTION

On vehicles equipped with an air bag, the negative battery cable must be disconnected, before working on the system. Failure to do so may result in deployment of the air bag and possible personal injury.

Air Bag

DISARMING

Work must be started after about 20 seconds or longer from the time the ignition switch is turned to the **LOCK** position and the battery cable is disconnected from the battery.

Heater Blower Motor

NOTE: On most vehicles, the air conditioner assembly is integral with the heater assembly (including the blower motor) and therefore the blower motor removal may differ from the procedures detailed below. In some case it may be necessary to remove the air conditioning-heater housing and assembly to remove the blower motor. Due to the lack of information available at the time of this publication, a general blower motor removal and installation procedure is outlined. The removal steps can be altered, as required.

REMOVAL & INSTALLATION

Celica and Supra

1. Disconnect the negative battery cable.
2. Working from under the instrument panel, unfasten the defroster hoses from the heater box.
3. Unplug the multi-connector.
4. Loosen the mounting screws and withdraw the blower assembly.
5. Installation is the reverse of the removal procedure.
6. Check the blower for proper operation at all speeds.

Cressida

1. Disconnect the negative battery cable.
2. Remove the instrument panel undercover and cowl side trim panel.
3. Remove the air duct and the glove box.
4. Disconnect the heater control cable from the blower motor and remove the blower duct.
5. Disconnect the heater relay from the heater relay electrical connector.
6. Remove the retaining screws from the blower motor assembly. Remove the assembly from the vehicle.
7. Remove the blower motor from the blower motor assembly.
8. Installation is the reverse order of the removal procedure.
9. Check the blower for proper operation at all speeds.

Corolla and Tercel

1. Disconnect the negative battery cable.
2. Remove the under tray, if equipped.
3. Remove the blower duct and air duct. Before removing the air duct, remember to remove the 2 attaching clamps.
4. Remove the glove box and the heater control cable.
5. Disconnect the electrical connector on the blower motor.
6. Remove the blower motor retaining bolts and remove the blower motor.
7. Installation is the reverse of the removal procedure.
8. Check the blower for proper operation at all speeds.

Camry and MR2

1. Disconnect the negative battery cable.
2. Remove the 3 screws attaching the retainer.
3. Remove the glove box. Remove the duct between the blower motor assembly and the heater assembly.
4. Disconnect the blower motor wire connector at the blower motor case.
5. Disconnect the air source selector control cable at the blower motor assembly.
6. Loosen the nuts and bolts attaching the blower motor to the blower case, remove the blower motor from the vehicle.
7. Installation is the reverse of the removal procedure.
8. Check the blower for proper operation at all speeds.

Front Windshield Wiper Motor

REMOVAL & INSTALLATION

Tercel and Corolla

1. Disconnect the negative battery terminal.
2. Remove the wiper arm.
3. Insert a small prybar between the linkage and the motor. Pry up to separate the linkage from the motor.
4. Disconnect the electrical connector from the motor.
5. Remove the mounting bolts and remove the motor.
6. Installation is the reverse of the removal procedure.

Celica, Supra, Camry and Cressida

1. Remove the access hole cover.
2. Separate the wiper and motor by prying gently with a small prybar.
3. Remove the left and right cowl ventilators.
4. Remove the wiper arms and the linkage mounting nuts. Push the linkage pivot ports into the ventilators.
5. Loosen the wiper link connectors at their ends and with the linkage from the cowl ventilator.
6. Start the wiper motor and turn the ignition key to the **OFF** when the crank is a position best suited for removal of the motor.

NOTE: The wiper motor is difficult to remove when it is in the parked position. If the motor is turned off at the wiper switch, it will automatically return to this position.

7. Unplug the wiper motor connector.
8. Loosen the motor mounting bolts and withdraw the motor.

To install:

9. Be sure to install the wiper motor with it in the park position by connecting the multi-connector and operating the wiper control switch.
10. Assemble the crank, connect the wiring and check operation.

MR2

1. With the wiper arms in the up position and the wiper switch on **LOW**, turn the ignition switch to the **OFF** position.
2. Disconnect the negative battery cable. Disconnect the wiper motor electrical connector, then remove the light retractor relay from the wiper bracket.
3. Remove the wiper motor set bolts. Manually lower the wiper arms, then connect the wiper link hook to the dash panel service hole.
4. Disconnect the wiper motor link. Remove the wiper motor attaching bolts then remove the wiper motor.
5. Installation is the reverse order of the removal procedure.

Rear Wiper Motor

REMOVAL & INSTALLATION

1. Disconnect the negative battery cable.

2. Remove the wiper arm and rear door trim cover.
3. Disconnect the wiper motor wire connector.
4. Remove the wiper motor bracket attaching bolts and the wiper motor along with the bracket.
5. Installation is the reverse of the removal procedure.

Instrument Cluster

REMOVAL & INSTALLATION

Cressida

1. Disconnect the negative battery cable.
2. Remove the cluster finish panel.
3. Loosen the instrument cluster retaining screws and tilt the panel forward.
4. Detach the speedometer cable and wiring connectors.
5. Remove the entire cluster assembly.
6. Remove the instruments from the panel as required.
7. Installation is the reverse of the removal proecedure.

Tercel and Corolla

1. Disconnect the negative battery cable.
2. Remove the steering column cover.

NOTE: Be careful not to damage the collapsible steering column mechanism.

3. On Tercel, remove the heater control knob and the center instrument cluster finish panel.
4. Remove the switches and hole cover from the cluster hood.
5. Remove the cluster hood.
6. Remove the cluster attaching screws and pull the unit forward.
7. Disconnect the speedometer and any other electrical connections that are necessary.
8. Remove the instruments from the panel as required.
9. Installation is the reverse of the removal procedure.

Camry, Celica and Supra

1. Disconnect the negative battery cable.
2. Remove the fuse box cover from under the left side of the instrument panel.
3. Remove the heater control knobs.
4. Carefully pry off the heater control panel.
5. Remove the cluster hood.
6. Unscrew the cluster finish panel retaining screws and pull out the bottom of the panel.
7. Unplug the electrical connectors and disconnect the speedometer cable.
8. Remove the instrument cluster.
9. Remove the instruments from the panel as required.
10. Installation is the reverse of the removal procedure.

MR2

1. Disconnect the negative battery cable.
2. Remove the steering column covers.
3. Pull the rheostat knob from the cluster finish panel and remove the nut from the rheostat. Remove the cluster finish panel and disconnect the rheostat multi-connector.
4. Remove the cluster hood.
5. Remove the cluster attaching screws and pull the unit forward.
6. Disconnect the speedometer and any other electrical connections that are necessary.
7. Remove the instruments from the panel as required.
8. Installation is the reverse of the removal procedure.

Radio

REMOVAL & INSTALLATION

Celica, Supra, Camry and Cressida

1. Disconnect the negative battery cable.
2. Remove the ash tray and cigarette lighter, if required.
3. Remove the radio finish panel.
4. Detach the antenna lead from the jack on the radio case.
5. Remove the cowl air intake duct.
6. Detach the power and speaker leads. Label the leads for assembly.
7. Remove the radio support nuts and bolts.
8. Remove the radio from beneath the dashboard.
9. Installation is the reverse of the removal procedure.

Corolla and Tercel

1. Disconnect the negative battery cable.
2. Remove the 2 screws from the top of the dashboard center trim panel.
3. Lift the center panel out far enough to gain access to the cigarette lighter wiring and disconnect the wiring. Remove the trim panel.
4. Unfasten the screws which secure the radio to the instrument panel braces.
5. Lift out the radio and disconnect the leads from it. Remove the radio.
6. Installation is the reverse of the removal procedure.

MR2

1. Disconnect the negative battery cable.
2. Remove the ash tray and cigarette lighter.
3. Remove the radio finish panel.
4. Remove the radio bolts and pull the radio out far enough to disconnect all necessary electrical wiring.
5. Remove the radio.
6. Installation is the reverse of the removal procedure.

Concealed Headlights

CAUTION

Before attempting to manually operate the concealed (retractable) headlights, first pull the fuse or disconnect the negative battery cable. Otherwise the headlights and motor shaft may suddenly move and catch hand and fingers. When opening and closing retractable headlights, make sure nobody is near them, otherwise personal injury may result.

MANUAL OPERATION

NOTE: If the headlights are frozen and inoperative, carefully melt the ice before attempting the manual operation procedure. Operation of a frozen headlight will drain the battery and may cause damage to the motor and operating linkages.

1. Switch the headlight and retractable headlight switches to the **OFF** position.
2. Pull the retractable headlight fuse or disconnect the negative battery cable.
3. Remove the rubber cap from the manual operation knob.
4. Manually turn the knob clockwise until the headlights are in the desired position (open or closed).
5. Install the rubber cap.
6. Install the fuse.
7. Make sure the lights work properly.

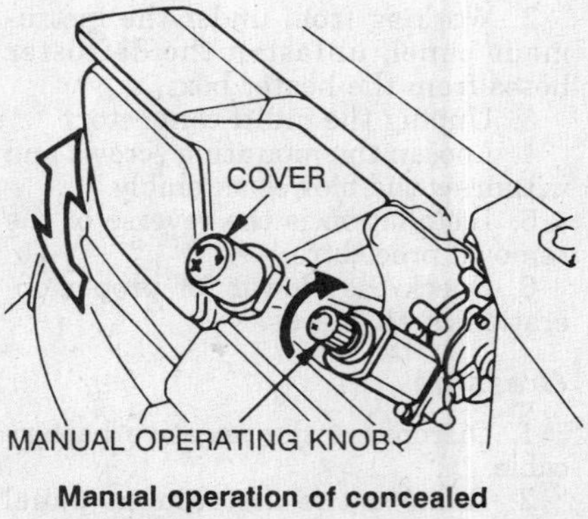

Manual operation of concealed headlights

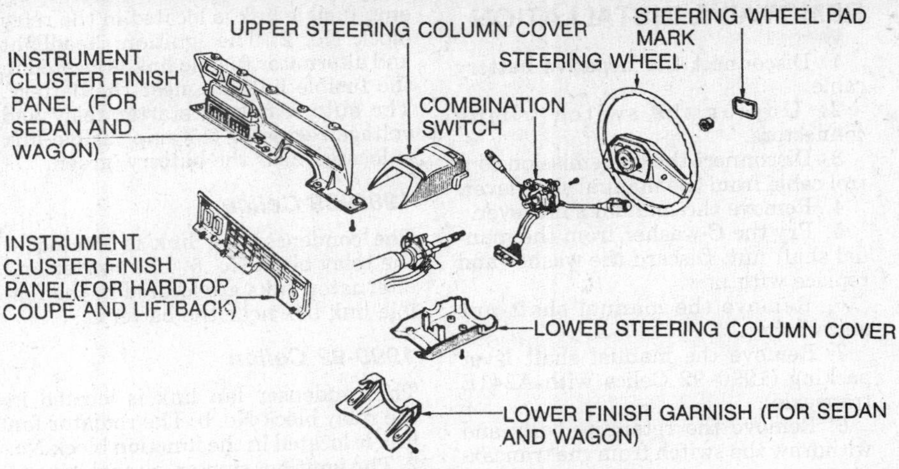

Typical combination switch mounting

Combination Switch

REMOVAL & INSTALLATION

1. Disconnect the negative battery cable.
2. Remove the steering column garnish.
3. Remove the upper and lower steering column covers.
4. Remove the steering wheel.
5. Trace the switch wiring harness to the multi-connector. Push in the lock levers and pull apart the connectors.
6. If equipped with electronic modulated suspension (TEMS), remove the steering sensor. On air bag equipped vehicles, disconnect the cable connectors, remove the spiral cable housing attaching screws and slide the cable assembly from the front of the combination switch.
7. Unscrew the mounting screws and slide the combination switch from the steering column.
8. Installation is the reverse of the removal procedure. Check all switch functions for proper operation.

Ignition Lock/Switch

REMOVAL & INSTALLATION

1. Disconnect the negative battery cable.
2. Unfasten the ignition switch connector under the instrument panel.
3. Remove the screws which secure the upper and lower halves of the steering column cover.
4. Turn the lock cylinder to the **ACC** position with the ignition key.
5. Push the lock cylinder stop in with a small, round object (cotter pin, punch, etc.).

NOTE: On some vehicles, it may be necessary to remove the steering wheel and combination switch first.

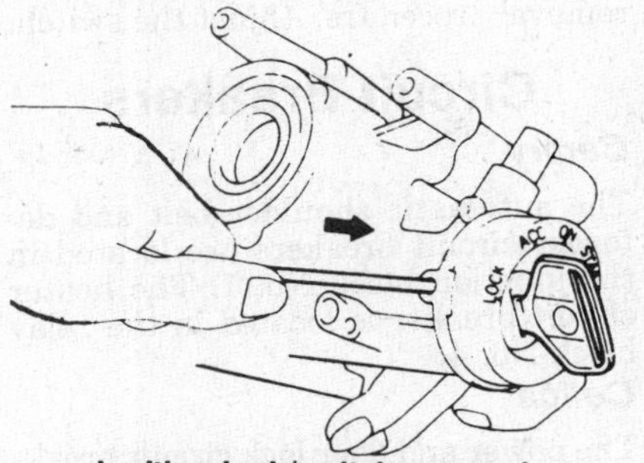

Ignition lock/switch removal

6. Withdraw the lock cylinder from the lock housing while depressing the stop tab.
7. To remove the ignition switch, unfasten its securing screws and withdraw the switch from the lock housing.

To install:

8. Align the locking cam with the hole in the ignition switch and insert the switch into the lock housing.
9. Secure the switch with its screw(s).
10. Make sure both the lock cylinder and column lock are in the **ACC** position. Slide the cylinder into the lock housing until the stop tab engages the hole in the lock.
11. Install the steering column covers.
12. Connect the ignition switch connector.
13. Connect the negative battery cable.

Stoplight Switch

ADJUSTMENT

1. Remove the instrument lower finish panel and the air duct if required to gain access to the stoplight switch.
2. Disconnect the stoplight switch connector.
3. Loosen the switch locknut.
4. Turn the stoplight switch until the end of the switch lightly contacts the pedal stopper.
5. Hold the switch and tighten the locknut.
6. Connect the switch connector.
7. Depress the brake pedal and verify that the brake lights illuminate.
8. Install the air duct and the lower finish panel, if removed.

REMOVAL & INSTALLATION

1. Disconnect the negative battery cable.
2. Remove the instrument lower finish panel and the air duct if required to gain access to the stoplight switch.
3. Disconnect the stoplight switch connector.
4. Remove the switch mounting nut, then slide the switch from the mounting bracket on the pedal.

To install:

5. Install the switch into the mounting bracket and adjust.
6. Connect the switch connector.
7. Depress the brake pedal and verify that the brake lights illuminate.
8. Install the air duct and the lower finish panel, if removed.

Clutch Switch

ADJUSTMENT

1. Attempt to start the engine when the clutch pedal is released. The engine should not start.
2. Depress the clutch pedal fully and attempt to start the engine. The engine should start.
3. If the engine does not start, depress the clutch pedal fully. With the clutch pedal depressed, loosen the switch locknut.
4. Use the adjusting nut to turn the switch until the tip of the switch contacts the clutch pedal stop.
5. Tighten the locknut and attempt to start the engine. Re-adjust as necessary.

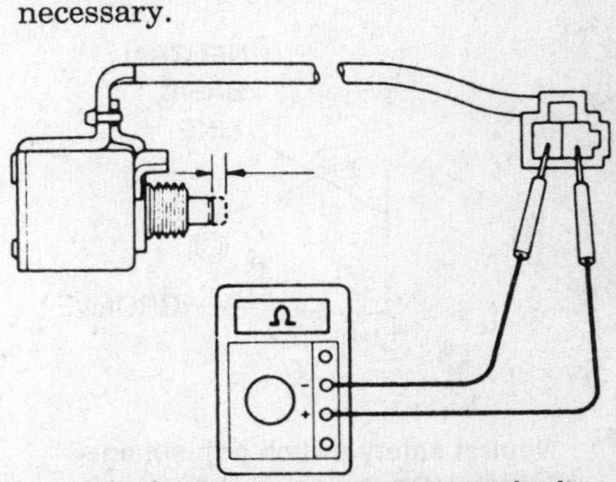

Checking clutch start switch continuity

6. If the switch cannot be adjusted, check the switch continuity with a suitable ohmmeter. There should be continuity between the switch terminals when the switch is on (tip pushed in) and no continuity when the switch is off (tip released). If the continuity is not as specified, replace the switch.

REMOVAL & INSTALLATION

1. Disconnect the negative battery cable.
2. Disconnect the switch connector.
3. Remove the switch adjusting nut.
4. Withdraw the switch from the mounting bracket.
5. Installation is the reverse of the removal procedure. Adjust the switch.

Neutral Safety Switch

The shift lever is adjusted properly if the engine will not start in any position other than **N** or **P**.

ADJUSTMENT

1. Loosen the neutral start switch bolt. Position the selector in the **N** position.
2. Align the switch shaft groove with the neutral base line which is located on the switch.
3. Tighten the bolt to 48 inch lbs. (5.4 Nm). on all vehicles except Tercel wagon. On the Tercel wagon, tighten to 9 ft. lbs. (13 Nm).

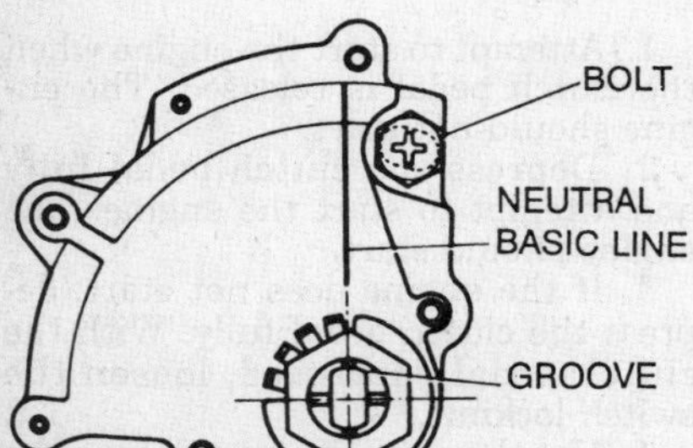

Neutral safely switch adjustment—Cressida and Supra

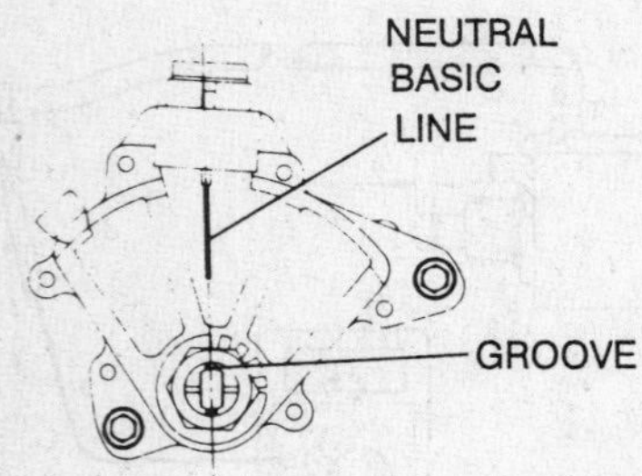

Neutral safety switch adjustment—Tercel, MR2, Camry and Celica

REMOVAL & INSTALLATION

1. Disconnect the negative battery cable.
2. Unplug the switch wiring connectors.
3. Disconnect the transmission control cable from the manual shift lever.
4. Remove the manual shift lever.
5. Pry the C-washer from the manual shaft nut. Discard the washer and replace with new.
6. Remove the manual shaft nut and washer.
7. Remove the manual shaft lever packing (1990–92 Celica with A241E transaxle).
8. Remove the retaining bolts and withdraw the switch from the transaxle case.
9. Installatión is the reverse of the removal procedure. Adjust the switch.

Circuit Breakers

Camry

The automatic shoulder belt and defogger circuit breakers are located in the junction block No. 1. The heater circuit breaker is located in the relay block No. 4.

Celica

The power and door lock circuit breakers are located in the relay block No. 1. The defogger circuit breaker is located in the junction block No. 1. The heater circuit breaker is located in the relay block No. 4.

Corolla

The heater circuit breaker is located in the relay block No. 4. The defogger and power circuit breakers are located in the junction block No. 1, left kick panel.

Cressida and Supra

The heater circuit breaker is located in the relay block No. 4. The defogger and power circuit breakers are located in the junction block No. 1, left kick panel. The automatic shoulder belt and door lock circuit breakers are located in the relay block No. 3, left kick panel.

MR2

The door lock circuit breaker is located in the relay block No. 1, left kick panel, upper section.

Tercel

The defogger circuit breaker is located in the junction block No. 1, left kick panel.

Fusible Links

Camry

The condenser fan and radiator fan 30 amp fusible link is located in the relay block No. 2. The ignition, headlight and alternator fusible link is located in the fusible link box near the battery. The auto seat belt, starter relay and voltage regulator 0.5 amp fusible link is located near the battery, green.

1988–89 Celica

The condenser fan link is located in the relay block No. 5. The starter and alternator links are located in the fusible link box near the battery.

1990-92 Celica

The condenser fan link is located in the relay block No. 5. The radiator fan link is located in the junction block No. 2. The ignition, starter, alternator and Anti-lock brake system links are located in the fusible link box near the battery.

Corolla

The condenser fan link is located in the relay block No. 5. The radiator fan link is located in the junction block No. 2. The A/C 7.5 amp fuse is located in the bottom of the relay block No. 4.

Cressida

The condenser fan, ignition switch, Anti-lock brake and alternator link is located in the junction block No. 2, near the battery.

1988–89 MR2

The heater, ignition switch and headlight links are located in the relay block No. 2, left engine compartment. The A/C and heater fuses are located in the relay block No. 4, right kick panel.

1990–92 MR2

The ignition switch, heater and alternator links are located in the relay block No. 5, right side luggage compartment. The Anti-lock Brake System (ABS) link is located in the relay block No. 5, right side luggage compartment.

Supra

The ECU-battery 15 amp Fuse is located in the fuse block. The ignition switch, condenser fan, Anti-lock Brake System (ABS) and alternator link is located in the junction block No. 2, left fender.

1988–90 Tercel

The condenser fan and radiator fan links are located in the relay block No. 2, left inner fender. The A/C 10 amp fuse is located in the relay block No. 4.

1991–92 Tercel

The defogger and heater links are located in the relay block No. 6, left kick panel. The ignition switch and alternator links are located in the relay block No. 2.

Relays, Sensors, Modules and Computer Location

Camry

- **A/C Fan Relays**—are located in the relay block No. 2.
- **Air Flow Meter**—is located on the air cleaner assembly.
- **Air Intake Temperature Sensor**—is located at the air cleaner assembly.
- **Anti-lock Brake System Control Unit**—is located on the right side rear compartment.
- **Blower Control Relay**—is located in the relay block No. 4, on the right side kick panel.
- **Coolant Temperature Sensor**—is located in the thermostat housing, to the right of the start injector time switch.
- **Diagnostic Check Connector**—is located near the master cylinder.
- **EFI Main Relay**—is located in the relay block No. 2, a round relay with 4 wires.
- **EFI Water Temperature Sensor**—is located at the intake manifold.
- **Electronic Controlled Transmission (ECT) Control Unit (4WD)**—is located behind the console.
- **Electronic Controlled Transmission (ECT) Control Unit (FWD)**—is located behind the instrument panel, to the right.
- **Engine Electronic Control Unit (ECU)**—is located behind the radio.
- **Fuel Pump Relay**—refer to the circuit opening relay.
- **Hazard Flasher**—is located within the turn signal flasher.
- **Knock Sensor**—is threaded into the cylinder block, under the intake manifold.
- **Start Injector Time Switch**—is located in the thermostat housing, to the left of the coolant temperature sensor.
- **Throttle Position Sensor**—is located on the throttle body.
- **Turn Signal Flasher**—is located in the relay block No. 1, left kick panel, upper section, 3 wires.

1988–89 Celica

- **A/C Dual and High Pressure Switches**—is located at the right inner fender.
- **A/C Clutch Relay**—is located in the relay block No. 5.
- **A/C Fan Relays**—are located in the relay block No. 5.
- **Air Flow Meter**—is located on the air cleaner assembly.
- **Air Intake Temperature Sensor**—is located at the air cleaner assembly.
- **Anti-lock Brake System Computer (FWD)**—is located on top of the ECU.
- **Blower Relay**—is located behind the relay block No. 4.
- **Cold Start Injector**—is located at the air intake plenum.
- **Diagnostic Check Connector**—is located rearward of the left strut tower.
- **EFI Resistor**—is located rearward of the EGR valve.
- **EFI Water Temperature Sensor**—is located at the intake manifold.
- **Electronic Controlled Transmission (ECT) Control Unit**—is located under console, in front of shifter.
- **Engine Electronic Control Unit (ECU)**—is located under the radio, behind the console.
- **Fuel Pump Control Relay**—is located in engine compartment, center of firewall.
- **Hazard Flasher**—is located within the turn signal flasher.
- **Heater and Horn Relays**—are located in the relay block No. 4.
- **Ignitor**—is located in the center of the firewall.
- **Knock Sensor**—is threaded into the cylinder block, under the intake manifold.
- **Oil Pressure Switch**—is located at the transmission side of the engine.
- **Start Injector Time Switch**—is located in the thermostat housing, to the left of the coolant temperature sensor.
- **Throttle Position Sensor**—is located on the throttle body.
- **Turn Signal Flasher**—is located in the relay block No. 1, upper position.
- **Vehicle Speed Sensor**—is located inside the speedometer head.
- **Water Temperature Sender**—is located on the left side of the engine cylinder head.

1990–92 Celica

- **Air Flow Meter and Air Intake Temperature Sensor**—is located on the air cleaner assembly.
- **Anti-lock Brake System Computer**—is located in the rear seat area, right side.
- **Blower Control Relay**—is located at the blower housing.
- **Cold Start Injector**—is located at the air intake plenum.
- **Coolant Temperature Sensor**—is located in the thermostat housing.
- **Diagnostic Check Connector**—is located rearward of the left strut tower.
- **EFI Water Temperature Sensor**—is located at the left side of the cylinder head, above the transmission.
- **Engine Control Unit (ECU)**—is located behind the console.
- **Fuel Pump Relay**—refer to the circuit opening relay.
- **Ignitor**—is located at the firewall.
- **Knock Sensor**—is threaded into the cylinder block.
- **Oil Pressure Switch**—is located at the driver's side of the engine.
- **Radiator Fan Relay**—is located in the junction block No. 2.
- **Start Injector Time Switch**—is located in the thermostat housing.
- **Throttle Position Sensor**—is located on the throttle body.
- **Water Temperature Sender**—is located at the driver's side of the engine.

Corolla

- **Air Flow Meter and Air Intake Temperature Sensor**—is located on the air cleaner assembly.
- **Cold Start Injector**—is located at the air intake plenum.
- **Coolant Temperature Sensor**—is located in the thermostat housing.
- **Diagnostic Check Connector**—is located rearward of the left strut tower.
- **EFI Water Temperature Sensor**—is located at the left side of the cylinder head.
- **Engine Electronic Control Unit (ECU)**—is located behind the radio.
- **Oil Pressure Switch**—is located at the front of the engine, near the distributor.
- **Start Injector Time Switch**—is located in the thermostat housing.
- **Throttle Position Sensor**—is located on the throttle body.
- **Turn Signal Flasher**—is located to the right of the junction block No. 1, under the instrument panel.
- **Hazard Flasher**—is located within the turn signal flasher.
- **Water Temperature Sender (4A-GE engine)**—is located at the right side of the intake manifold.

Cressida

- **Air Flow Meter and Air Intake Temperature Sensor**—is located on the air cleaner assembly.
- **Anti-lock Brake System (ABS) Control Unit**—is located behind the trunk trim panel, on the right side.
- **Diagnostic Check Connector**—is located rearward of the left strut tower.

• **EFI Water Temperature Sensor**—is located at the right side of the cylinder head.

• **Engine and Transmission Control Unit**—is located behind the glove compartment.

• **Fuel Pump Relay**—is located in the relay block No. 4, right kick panel.

• **Hazard Flasher**—is located within the turn signal flasher.

• **Oil Pressure Switch**—is located at the right side of the engine.

• **Starter Relay (US)**—is located in the relay block No. 3.

• **Throttle Position Sensor**—is located on the throttle body.

• **Toyota Diagnostic Communication Link (TDCL)**—is located the the left of the steering column.

• **Turn Signal Flasher**—is located in the relay block No. 3, left kick panel.

• **Water Temperature Sender**—is located at the right side of the engine.

1988–89 MR2

• **Air Flow Meter and Air Intake Temperature Sensor**—is located on the air cleaner assembly.

• **Blower Resistor**—is located behind the instrument panel, in the center.

• **Coolant Temperature Sensor**—is located in the thermostat housing.

• **Diagnostic Check Connector**—is located at the firewall.

• **EFI Water Temperature Sensor**—is located at the cylinder head, left side.

• **Engine Electronic Control Unit (ECU)**—is located on center of the firewall.

• **Hazard Flasher**—is located within the turn signal flasher.

• **Heater Relay**—is located in the relay block No. 4,

• **Oil Pressure Switch**—is located at the rear of the engine, on the right side.

• **Throttle Position Sensor**—is located on the throttle body.

• **Turn Signal Flasher**—is located behind the instrument panel, on the left side.

• **Water Temperature Sender**—is located at the right side of the intake manifold.

1990–92 MR2

• **Anti-lock Brake System (ABS) Control Unit**—is located under the instrument panel.

• **Blower Resistor**—is located behind the instrument panel.

• **Coolant Temperature Sensor**—is located in the thermostat housing.

• **Cooling Fan Relay**—is located in the relay block No. 2.

• **Diagnostic Check Connector**—is located at the rear firewall, on the left side.

• **EFI Water Temperature Sensor**—is located at the cylinder head.

• **Engine Electronic Control Unit (ECU)**—is located on the firewall.

• **Oil Pressure Switch**—is located at the rear of the engine, on the left side.

• **Turn Signal Flasher**—is located in the relay block No. 1, left kick panel, at the top.

Supra

• **Air Flow Meter**—is located on the air cleaner assembly.

• **Anti-lock Brake System (ABS) Control Unit**—is located behind the glove compartment.

• **Blower Resistor**—is located near the relay block.

• **Diagnostic Check Connector**—is located at the left strut tower.

• **EFI Water Temperature Sensor**—is located under the upper radiator hose.

• **Electronic Fuel Injection (EFI) Relay**—is located in the junction block No. 2.

• **Engine Electronic Control Unit (ECU)**—is located behind the glove compartment.

• **Fuel Pump Relay**—is located at the right strut tower.

• **Oil Pressure Switch**—is located at the right of the engine, near the throttle body.

• **Start Injector Time Switch**—is located in the thermostat housing.

• **Turn Signal and Hazard Flasher**—is located in the relay block No. 5, left kick panel.

• **Water Temperature Sender**—is located at the right side of the intake manifold.

Volkswagen 20

Cabriolet, Corrado, Fox, Golf, GTi, Jetta, Passat, Quantum, Scirocco—All Models

SERIAL NUMBER IDENTIFICATION

Vehicle Identification Plate

All vehicles have an identification plate bearing the chassis number on the top of the dash board at the driver's side. This plate is easily visible through the windshield and aids in rapid identification. The VIN indicates such information as model year, type, date of manufacture, etc. This information is more easily read from the Vehicle Identification Label in the luggage compartment. That label also provides engine and transaxle code numbers, paint, interior and option code numbers. Since the manufacturer sometimes makes production changes in mid-model year, this information is sometimes required when locating parts.

Engine Number

The diesel engine number with it's 2 letter code is stamped on the block between the injection pump and the vacuum pump. On gasoline engine vehicles, the engine code/number is stamped on the block just below the cylinder head, near the No. 2 or No. 3 spark plug.

SPECIFICATIONS

ENGINE IDENTIFICATION

Year	Model	Engine Displacement cu. in. (cc/liter)	Engine Series Identification	No. of Cylinders	Engine Type
1988	Jetta	109 (1780/1.8)	RV	4	SOHC
	Jetta GL	109 (1780/1.8)	PF	4	SOHC
	Jetta GLI 16V	109 (1780/1.8)	PL	4	DOHC
	Jetta Carat	109 (1780/1.8)	PF	4	DOHC
	Quantum GL5	136 (2226/2.2)	KX	5	SOHC
	Quantum Syncro	136 (2226/2.2)	JT	5	SOHC
	Scirocco-16V	109 (1780/1.8)	PL	4	DOHC
	Cabriolet	109 (1780/1.8)	JH	4	SOHC
	Golf/GL	109 (1780/1.8)	RV	4	SOHC
	Golf/GT	109 (1780/1.8)	PF	4	SOHC
	Golf GTI 16V	109 (1780/1.8)	PL	4	DOHC
	Fox/GL	109 (1780/1.8)	UM	4	SOHC
1989	Jetta	109 (1780/1.8)	RV	4	SOHC
	Jetta Diesel	97 (1588/1.6)	ME	4	Water cooled in-line Diesel
	Jetta GL	109 (1780/1.8)	PF	4	SOHC
	Jetta GLI 16V	109 (1780/1.8)	PL	4	DOHC
	Jetta Carat	109 (1780/1.8)	PF	4	DOHC
	Quantum GL5	136 (2226/2.2)	KX	5	SOHC
	Quantum Syncro	136 (2226/2.2)	JT	5	SOHC
	Scirocco-16V	109 (1780/1.8)	PL	4	DOHC
	Cabriolet	109 (1780/1.8)	JH	4	SOHC
	Golf/GL	109 (1780/1.8)	RV	4	SOHC
	Golf/GT	109 (1780/1.8)	PF	4	SOHC
	Golf GTI 16V	109 (1780/1.8)	PL	4	DOHC
	Fox/GL	109 (1780/1.8)	UM	4	SOHC
1990	Jetta GL	109 (1780/1.8)	RV	4	SOHC
	Jetta GLI 16V	121 (1984/2.0)	9A	4	DOHC
	Jetta Carat	109 (1780/1.8)	PF	4	DOHC
	Jetta Diesel	97 (1588/1.6)	ME	4	SOHC
	Golf/GL	109 (1780/1.8)	RV	4	SOHC

ENGINE IDENTIFICATION

Year	Model	Engine Displacement cu. in. (cc/liter)	Engine Series Identification	No. of Cylinders	Engine Type
1990	GTI	109 (1780/1.8)	PF	4	SOHC
	GTI 16V	121 (1984/2.0)	9A	4	DOHC
	Cabriolet	109 (1780/1.8)	JH	4	SOHC
	Fox	109 (1780/1.8)	UM, JN	4	SOHC
	Passat, GL	121 (1984/2.0)	9A	4	DOHC
	Corrado ①	109 (1780/1.8)	PG	4	SOHC G-Charger
1991-92	Jetta GL	109 (1780/1.8)	RV	4	SOHC
	Jetta GLI 16V	121 (1984/2.0)	9A	4	DOHC
	Jetta Carat	109 (1780/1.8)	PF	4	DOHC
	Jetta ECO Diesel	97 (1588/1.6)	IV	4	Turbo Diesel
	Golf/GL	109 (1780/1.8)	RV	4	SOHC
	GTI	109 (1780/1.8)	PF	4	SOHC
	GTI 16V	121 (1984/2.0)	9A	4	DOHC
	Cabriolet	109 (1780/1.8)	JH	4	SOHC
	Fox	109 (1780/1.8)	UM, JN	4	SOHC
	Passat, GL	121 (1984/2.0)	9A	4	DOHC
	Corrado ①	109 (1780/1.8)	PG	4	SOHC G-Charger

SOHC—Single Overhead Camshaft DOHC—Double Overhead Camshaft ① Supercharged

GENERAL ENGINE SPECIFICATIONS

Year	Model	Engine Displacement cu. in. (cc)	Fuel System Type	Net Horsepower @ rpm	Net Torque @ rpm (ft. lbs.)	Bore × Stroke (in.)	Com-pression Ratio	Oil Pressure @ rpm
1988	Jetta	109 (1780)	Digifant II	100 @ 5400	107 @ 3400	3.19 × 3.40	10.0:1	28 @ 2000
	Jetta GL	109 (1780)	Digifant II	105 @ 5400	110 @ 3400	3.19 × 3.40	10.0:1	28 @ 2000
	Jetta GLI 16V	109 (1780)	CIS-E Fuel Inj.	123 @ 5800	120 @ 4250	3.19 × 3.40	10.0:1	28 @ 2000
	Jetta Carat	109 (1780)	Digifant II	105 @ 5400	110 @ 3400	3.19 × 3.40	10.0:1	28 @ 2000
	Quantum GL5	136 (2226)	CIS-E Fuel Inj.	110 @ 5500	122 @ 2400	3.19 × 3.40	8.5:1	28 @ 2000
	Quantum Syncro	136 (2226)	CIS-E Fuel Inj.	115 @ 5500	126 @ 3000	3.19 × 3.40	8.5:1	28 @ 2000
	Scirocco 16V	109 (1780)	CIS-E Fuel Inj.	123 @ 5800	120 @ 4250	3.19 × 3.40	10.0:1	28 @ 2000
	Cabriolet	109 (1780)	CIS-E Fuel Inj.	90 @ 5500	100 @ 3000	3.19 × 3.40	9.0:1	28 @ 2000
	Golf GL	109 (1780)	Digifant II	100 @ 5400	107 @ 3400	3.19 × 3.40	10.0:1	28 @ 2000
	Golf GT	109 (1780)	Digifant II	105 @ 5400	110 @ 3400	3.19 × 3.40	10.0:1	28 @ 2000
	Golf GTI 16V	109 (1780)	CIS-E Fuel Inj.	123 @ 5800	120 @ 4250	3.19 × 3.40	10.0:1	28 @ 2000
	Fox/GL	109 (1780)	CIS-E Fuel Inj.	81 @ 5500	93 @ 3250	3.19 × 3.40	9.0:1	28 @ 2000
1989	Jetta	109 (1780)	Digifant II	100 @ 5400	107 @ 3400	3.19 × 3.40	10.0:1	28 @ 2000
	Jetta (Diesel)	97.0 (1588)	Fuel Inj.	52 @ 4800	72 @ 2000	3.01 × 3.40	23.0:1	28 @ 2000
	Jetta GL	109 (1780)	Digifant II	105 @ 5400	110 @ 3400	3.19 × 3.40	10.0:1	28 @ 2000
	Jetta GLI 16V	109 (1780)	CIS-E Fuel Inj.	123 @ 5800	120 @ 4250	3.19 × 3.40	10.0:1	28 @ 2000
	Jetta Carat	109 (1780)	Digifant II	105 @ 5400	110 @ 3400	3.19 × 3.40	10.0:1	28 @ 2000
	Quantum GL5	136 (2226)	CIS-E Fuel Inj.	110 @ 5500	122 @ 2400	3.19 × 3.40	8.5:1	28 @ 2000
	Quantum Syncro	136 (2226)	CIS-E Fuel Inj.	115 @ 5500	126 @ 3000	3.19 × 3.40	8.5:1	28 @ 2000
	Scirocco 16V	109 (1780)	CIS-E Fuel Inj.	123 @ 5800	120 @ 4250	3.19 × 3.40	10.0:1	28 @ 2000
	Cabriolet	109 (1780)	CIS-E Fuel Inj.	90 @ 5500	100 @ 3000	3.19 × 3.40	9.0:1	28 @ 2000
	Golf GL	109 (1780)	Digifant II	100 @ 5400	107 @ 3400	3.19 × 3.40	10.0:1	28 @ 2000

GENERAL ENGINE SPECIFICATIONS

Year	Model	Engine Displacement cu. in. (cc)	Fuel System Type	Net Horsepower @ rpm	Net Torque @ rpm (ft. lbs.)	Bore × Stroke (in.)	Com-pression Ratio	Oil Pressure @ rpm
1989	Golf GT	109 (1780)	Digifant II	105 @ 5400	110 @ 3400	3.19 × 3.40	10.0:1	28 @ 2000
	Golf GTI 16V	109 (1780)	CIS-E Fuel Inj.	123 @ 5800	120 @ 4250	3.19 × 3.40	10.0:1	28 @ 2000
	Fox/GL	109 (1780)	CIS-E Fuel Inj.	81 @ 5500	93 @ 3250	3.19 × 3.40	9.0:1	28 @ 2000
1990	Jetta GL	109 (1780)	Digifant II	105 @ 5400	110 @ 3400	3.19 × 3.40	10.0:1	28 @ 2000
	Jetta GLI 16V	121 (1984)	CIS-E Fuel Inj.	134 @ 5800	133 @ 4400	3.25 × 3.65	10.8:1	28 @ 2000
	Jetta Carat	109 (1780)	Digifant II	105 @ 5400	110 @ 3400	3.19 × 3.40	10.0:1	28 @ 2000
	Jetta Diesel	97.0 (1588)	Fuel Inj.	52 @ 4800	71 @ 2000	3.01 × 3.40	23.0:1	28 @ 2000
	Cabriolet	109 (1780)	CIS-E Fuel Inj.	94 @ 5400	100 @ 3000	3.19 × 3.40	10.0:1	28 @ 2000
	Golf GL	109 (1780)	Digifant II	100 @ 5400	107 @ 3400	3.19 × 3.40	10.0:1	28 @ 2000
	GTI	109 (1780)	Digifant II	105 @ 5400	110 @ 3400	3.19 × 3.40	10.0:1	28 @ 2000
	GTI 16V	121 (1984)	CIS-E Fuel Inj.	134 @ 5800	133 @ 4400	3.25 × 3.65	10.8:1	28 @ 2000
	Fox/GL	109 (1780)	CIS-E Fuel Inj.	81 @ 5500	93 @ 3250	3.19 × 3.40	9.0:1	28 @ 2000
	Passat GL	121 (1984)	CIS-E Fuel Inj.	134 @ 5800	133 @ 4400	3.25 × 3.65	10.8:1	28 @ 2000
	Corrado	109 (1780)	Digifant II	158 @ 5600	166 @ 4000	3.19 × 3.40	8.0:1	28 @ 2000
1991–92	Jetta GL	109 (1780)	Digifant II	105 @ 5400	110 @ 3400	3.19 × 3.40	10.0:1	28 @ 2000
	Jetta GLI 16V	121 (1984)	CIS-E Fuel Inj.	134 @ 5800	133 @ 4400	3.25 × 3.65	10.8:1	28 @ 2000
	Jetta Carat	109 (1780)	Digifant II	105 @ 5400	110 @ 3400	3.19 × 3.40	10.0:1	28 @ 2000
	Jetta ECO Diesel	97.0 (1588)	Fuel Inj.	52 @ 4800	71 @ 2000	3.01 × 3.40	23.0:1	28 @ 2000
	Cabriolet	109 (1780)	CIS-E Fuel Inj.	94 @ 5400	100 @ 3000	3.19 × 3.40	10.0:1	28 @ 2000
	Golf GL	109 (1780)	Digifant II	100 @ 5400	107 @ 3400	3.19 × 3.40	10.0:1	28 @ 2000
	GTI	109 (1780)	Digifant II	105 @ 5400	110 @ 3400	3.19 × 3.40	10.0:1	28 @ 2000
	GTI 16V	121 (1984)	CIS-E Fuel Inj.	134 @ 5800	133 @ 4400	3.25 × 3.65	10.8:1	28 @ 2000
	Fox/GL	109 (1780)	CIS-E Fuel Inj.	81 @ 5500	93 @ 3250	3.19 × 3.40	9.0:1	28 @ 2000
	Passat GL	121 (1984)	CIS-E Fuel Inj.	134 @ 5800	133 @ 4400	3.25 × 3.65	10.8:1	28 @ 2000
	Corrado	109 (1780)	Digifant II	158 @ 5600	166 @ 4000	3.19 × 3.40	8.0:1	28 @ 2000

GASOLINE ENGINE TUNE-UP SPECIFICATIONS

Year	Model	Engine Displacement cu. in. (cc)	Spark Plugs		Ignition Timing (deg.)		Com-pression Pressure (psi)	Fuel Pump (psi)	Idle Speed (rpm)		Valve Clearance	
			Type	Gap (in.)	MT	AT			MT	AT	In. ②	Ex. ②
1988	Jetta	109 (1780)	WR7DS	0.024–0.032	6 BTDC @ Idle	6 BTDC @ Idle	131–174	NA	800–900	800–900	Hyd.	Hyd.
	Jetta GL	109 (1780)	W7DTC	0.027–0.035	6 BTDC @ Idle	6 BTDC @ Idle	131–174	NA	800–900	800–900	Hyd.	Hyd.
	Jetta GLI 16V	109 (1780)	F6DTC	0.027–0.035	6 BTDC @ Idle	6 BTDC @ Idle	145–189	NA	800–900	800–900	Hyd.	Hyd.
	Jetta Carat	109 (1780)	W7DTC	0.027–0.035	6 BTDC @ Idle	6 BTDC @ Idle	131–174	NA	800–900	800–900	Hyd.	Hyd.
	Quantum GL5	136 (2226)	WR7DS	0.027–0.035	6 BTDC @ Idle	3 ATDC @ Idle	131–174	NA	750–850	750–850	Hyd.	Hyd.
	Quantum Syncro	136 (2226)	WR7DS	0.027–0.035	6 BTDC @ Idle	3 ATDC @ Idle	131–174	NA	750–850	750–850	Hyd.	Hyd.
	Scirocco 16V	109 (1780)	F6DTC	0.027–0.035	6 BTDC @ Idle	—	145–189	NA	800–900	800–900	Hyd.	Hyd.

GASOLINE ENGINE TUNE-UP SPECIFICATIONS

Year	Model	Engine Displacement cu. in. (cc)	Spark Plugs Type	Spark Plugs Gap (in.)	Ignition Timing (deg.) MT	Ignition Timing (deg.) AT	Compression Pressure (psi)	Fuel Pump (psi)	Idle Speed (rpm) MT	Idle Speed (rpm) AT	Valve Clearance In. ②	Valve Clearance Ex. ②
1988	Cabriolet	109 (1780)	W7DTC	0.027–0.035	6 BTDC @ Idle	6 BTDC @ Idle	131–174	NA	850–1000	850–1000	Hyd.	Hyd.
	Golf GL	109 (1780)	WR7DS	0.024–0.032	6 BTDC @ Idle	6 BTDC @ Idle	131–174	NA	800–900	800–900	Hyd.	Hyd.
	Golf GT	109 (1780)	W7DTC	0.027–0.035	6 BTDC @ Idle	6 BTDC @ Idle	131–174	NA	800–900	800–900	Hyd.	Hyd.
	Golf GTI 16V	109 (1780)	F6DTC	0.027–0.035	6 BTDC @ Idle	—	145–189	NA	800–900	800–900	Hyd.	Hyd.
	Fox/GL	109 (1780)	W7DTC	0.027–0.035	6 BTDC @ Idle	6 BTDC @ Idle	131–174	NA	800–1000	800–1000	Hyd.	Hyd.
1989	Jetta	109 (1780)	WR7DS	0.024–0.032	6 BTDC @ Idle	6 BTDC @ Idle	131–174	NA	800–900	800–900	Hyd.	Hyd.
	Jetta GL	109 (1780)	W7DTC	0.027–0.035	6 BTDC @ Idle	6 BTDC @ Idle	131–174	NA	800–900	800–900	Hyd.	Hyd.
	Jetta GLI 16V	109 (1780)	F6DTC	0.027–0.035	6 BTDC @ Idle	6 BTDC @ Idle	145–189	NA	800–900	800–900	Hyd.	Hyd.
	Jetta Carat	109 (1780)	W7DTC	0.027–0.035	6 BTDC @ Idle	6 BTDC @ Idle	131–174	NA	800–900	800–900	Hyd.	Hyd.
	Quantum GL5	136 (2226)	WR7DS	0.027–0.035	6 BTDC @ Idle	3 ATDC @ Idle	131–174	NA	750–850	750–850	Hyd.	Hyd.
	Quantum Syncro	136 (2226)	WR7DS	0.027–0.035	6 BTDC @ Idle	3 ATDC @ Idle	131–174	NA	750–850	750–850	Hyd.	Hyd.
	Scirocco 16V	109 (1780)	F6DTC	0.027–0.035	6 BTDC @ Idle	6 BTDC @ Idle	145–189	NA	800–900	800–900	Hyd.	Hyd.
	Cabriolet	109 (1780)	W7DTC	0.027–0.035	6 BTDC @ Idle	6 BTDC @ Idle	131–174	NA	850–1000	850–1000	Hyd.	Hyd.
	Golf GL	109 (1780)	WR7DS	0.024–0.032	6 BTDC @ Idle	6 BTDC @ Idle	131–174	NA	800–900	800–900	Hyd.	Hyd.
	Golf GT	109 (1780)	W7DTC	0.027–0.035	6 BTDC @ Idle	6 BTDC @ Idle	131–174	NA	800–900	800–900	Hyd.	Hyd.
	Golf GTI 16V	109 (1780)	F6DTC	0.027–0.035	6 BTDC @ Idle	—	145–189	NA	800–900	800–900	Hyd.	Hyd.
	Fox/GL	109 (1780)	W7DTC	0.027–0.035	6 BTDC @ Idle	6 BTDC @ Idle	131–174	NA	800–1000	800–1000	Hyd.	Hyd.
1990	Jetta GL	109 (1780)	W7DTC	0.024–0.032	6 BTDC @ Idle	6 BTDC @ Idle	145–189	NA	800–900	800–1000	Hyd.	Hyd.
	Jetta GLI 16V	121 (1984)	F6DSR ①	0.024–0.032	6 BTDC @ Idle	—	NA	NA	800–900	—	Hyd.	Hyd.
	Jetta Carat	109 (1780)	W7DCO	0.024–0.032	6 BTDC @ Idle	6 BTDC @ Idle	145–189	NA	800–900	800–1000	Hyd.	Hyd.
	Golf GL	109 (1780)	W7DCO	0.024–0.032	6 BTDC @ Idle	6 BTDC @ Idle	145–189	NA	800–900	800–1000	Hyd.	Hyd.
	GTI	109 (1780)	W7DCO	0.024–0.032	6 BTDC @ Idle	—	145–189	NA	800–900	—	Hyd.	Hyd.
	GTI 16V	121 (1984)	F6DSR ①	0.027–0.035	6 BTDC @ Idle	—	NA	NA	800–900	—	Hyd.	Hyd.
	Fox	109 (1780)	W7RDS	0.024–0.032	6 BTDC @ Idle	6 BTDC @ Idle	131–174	NA	800–1000	800–1000	Hyd.	Hyd.

GASOLINE ENGINE TUNE-UP SPECIFICATIONS

Year	Model	Engine Displacement cu. in. (cc)	Spark Plugs Type	Spark Plugs Gap (in.)	Ignition Timing (deg.) MT	Ignition Timing (deg.) AT	Compression Pressure (psi)	Fuel Pump (psi)	Idle Speed (rpm) MT	Idle Speed (rpm) AT	Valve Clearance In. ②	Valve Clearance Ex. ②
1990	Cabriolet	109 (1780)	W7DCO	0.024–0.032	6 BTDC @ Idle	6 BTDC @ Idle	145–189	NA	800–1000	800–1000	Hyd.	Hyd.
	Passat	121 (1984)	F6DSR ①	0.028–0.032	6 BTDC @ Idle	6 BTDC @ Idle	NA	NA	800–900	800–1000	Hyd.	Hyd.
	Corrado	109 (1780)	W6DPO	0.028–0.031	6 BTDC @ Idle	—	116–174	NA	800–900	—	Hyd.	Hyd.
1991	Jetta GL	109 (1780)	W7DTC	0.024–0.032	6 BTDC @ Idle	6 BTDC @ Idle	145–189	NA	800–900	800–1000	Hyd.	Hyd.
	Jetta GLI 16V	121 (1984)	F6DSR ①	0.024–0.032	6 BTDC @ Idle	—	NA	NA	800–900	—	Hyd.	Hyd.
	Jetta Carat	109 (1780)	W7DCO	0.024–0.032	6 BTDC @ Idle	6 BTDC @ Idle	145–189	NA	800–900	800–1000	Hyd.	Hyd.
	Golf GL	109 (1780)	W7DCO	0.024–0.032	6 BTDC @ Idle	6 BTDC @ Idle	145–189	NA	800–900	800–1000	Hyd.	Hyd.
	GTI	109 (1780)	W7DCO	0.024–0.032	6 BTDC @ Idle	—	145–189	NA	800–900	—	Hyd.	Hyd.
	GTI 16V	121 (1984)	F6DSR ①	0.027–0.035	6 BTDC @ Idle	—	NA	NA	800–900	—	Hyd.	Hyd.
	Fox	109 (1780)	W7RDS	0.024–0.032	6 BTDC @ Idle	6 BTDC @ Idle	131–174	NA	800–1000	800–1000	Hyd.	Hyd.
	Cabriolet	109 (1780)	W7DCO	0.024–0.032	6 BTDC @ Idle	6 BTDC @ Idle	145–189	NA	800–1000	800–1000	Hyd.	Hyd.
	Passat	121 (1984)	F6DSR ①	0.028–0.032	6 BTDC @ Idle	6 BTDC @ Idle	NA	NA	800–900	800–1000	Hyd.	Hyd.
	Corrado	109 (1780)	W6DPO	0.028–0.031	6 BTDC @ Idle	6 BTDC @ Idle	116–174	NA	800–900	800–900	Hyd.	Hyd.
1992	REFER TO UNDERHOOD STICKER											

DIESEL ENGINE TUNE-UP SPECIFICATIONS

Year	Engine Displacement cu. in. (cc)	Valve Clearance ① Intake (in.)	Valve Clearance ① Exhaust (in.)	Intake Valve Opens (deg.)	Injection Pump Setting (deg.)	Injection Nozzle Pressure (psi) New	Injection Nozzle Pressure (psi) Used	Idle Speed (rpm)	Cranking Compression Pressure (psi)
1989	97.0 (1588)	0.008–0.012 ①	0.016–0.020 ①	N.A.	Align Marks	1885 ③	1706 ③	800–850 ② ④	406 minimum
1990	97.0 (1588)	0.008–0.012 ①	0.016–0.020 ①	N.A.	Align Marks	1885 ③	1706 ③	800–850 ② ④	406 minimum
1991	97.0 (1588)	0.008–0.012 ①	0.016–0.020 ①	N.A.	Align Marks	NA	NA	800–850	406 minimum
1992	REFER TO UNDERHOOD STICKER								

N.A. Not Available

① Warm clearance given.
Cold clearance:
Intake—0.006–0.010
Exhaust—0.014–0.018

② Volkswagen has lowered the idle speed on early models to this specification.

③ Turbo diesel—
New—2306
Used—2139

④ Turbo diesel—900–1000

⑤ ECO diesel has a catalytic converter and is different from Turbo Diesel.

FIRING ORDERS

NOTE: To avoid confusion, always replace spark plug wires one at a time.

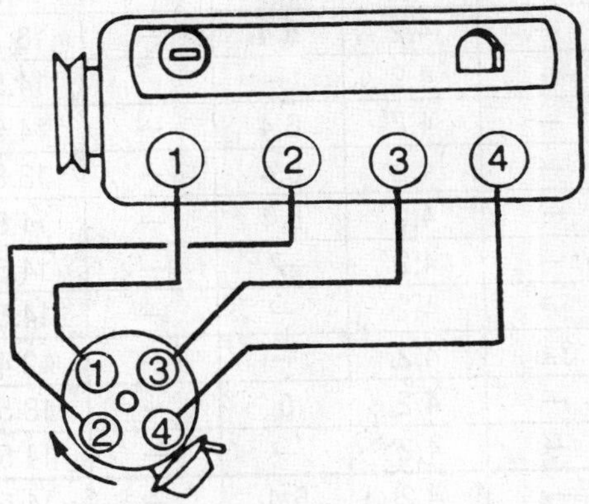

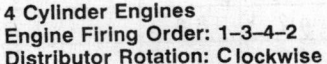

4 Cylinder Engines
Engine Firing Order: 1–3–4–2
Distributor Rotation: Clockwise

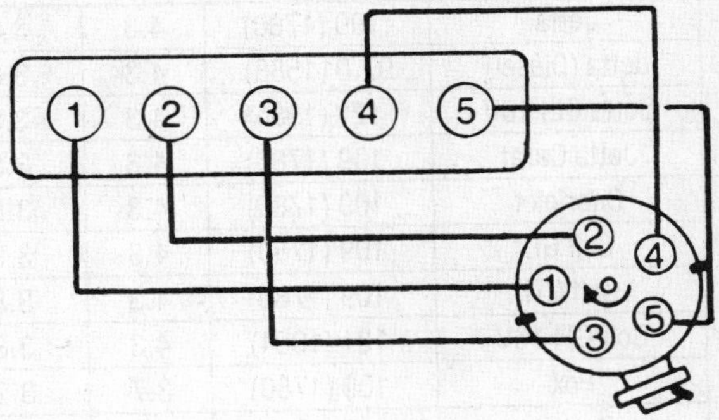

5 Cylinder Engine
Engine Firing Order: 1–2–4–5–3
Distributor Rotation: Clockwise

CAPACITIES

Year	Model	Engine Displacement cu. in. (cc)	Engine Crankcase (qts.) with Filter	Engine Crankcase (qts.) without Filter	Transmission (pts.) 4-Spd	Transmission (pts.) 5-Spd	Transmission (pts.) Auto.	Drive Axle (pts.)	Fuel Tank (gals.)	Cooling System (qts.)
1988	Jetta	109 (1780)	4.3	3.8	—	4.2	6.4	—	14.5	7.3
	Jetta GL	109 (1780)	4.3	3.8	—	4.2	6.4	—	14.5	7.3
	Jetta GLI 16V	109 (1780)	4.3	3.8	—	4.2	6.4	—	14.5	7.3
	Jetta Carat	109 (1780)	4.3	3.8	—	4.2	6.4	—	14.5	7.3
	Quantum GL5	136 (2226)	4.0	3.8	—	4.2	6.4	—	15.8	8.5
	Quantum Syncro	136 (2226)	4.0	3.5	—	5.0	6.4	1.2	18.5	8.5
	Scirocco 16V	109 (1780)	4.3	3.8	—	4.2	6.4	—	13.8	5.1
	Cabriolet	109 (1780)	4.3	3.8	—	4.2	6.4	—	13.8	5.1
	Golf GL	109 (1780)	4.3	3.8	—	4.2	6.4	—	14.5	7.3
	Golf GT	109 (1780)	4.3	3.8	—	4.2	6.4	—	14.5	7.3
	Golf GTI 16V	109 (1780)	4.3	3.8	—	4.2	6.4	—	14.5	7.3
	Fox/GL	109 (1780)	3.7	3.2	3.6	—	—	—	12.4	6.9
1989	Jetta	109 (1780)	4.3	3.8	—	4.2	6.4	—	14.5	7.3
	Jetta (Diesel)	97.0 (1588)	4.8	4.3	—	4.2	6.4	—	13.7	7.3
	Jetta GL	109 (1780)	4.3	3.8	—	4.2	6.4	—	14.5	7.3
	Jetta GLI 16V	109 (1780)	4.3	3.8	—	4.2	6.4	—	14.5	7.3
	Jetta Carat	109 (1780)	4.3	3.8	—	4.2	6.4	—	14.5	7.3
	Quantum GL5	136 (2226)	4.0	3.8	—	4.2	6.4	—	15.8	8.5
	Quantum Syncro	136 (2226)	4.0	3.5	—	5.0	6.4	1.2	18.5	8.5
	Scirocco 16V	109 (1780)	4.3	3.8	—	4.2	6.4	—	13.8	5.1
	Cabriolet	109 (1780)	4.3	3.8	—	4.2	6.4	—	13.8	5.1
	Golf GL	109 (1780)	4.3	3.8	—	4.2	6.4	—	14.5	7.3
	Golf GT	109 (1780)	4.3	3.8	—	4.2	6.4	—	14.5	7.3
	Golf GTI 16V	109 (1780)	4.3	3.8	—	4.2	6.4	—	14.5	7.3
	Fox/GL	109 (1780)	3.7	3.2	3.6	—	—	—	12.4	6.9

CAPACITIES

Year	Model	Engine Displacement cu. in. (cc)	Engine Crankcase (qts.) with Filter	Engine Crankcase (qts.) without Filter	Transmission (pts.) 4-Spd	Transmission (pts.) 5-Spd	Transmission (pts.) Auto.	Drive Axle (pts.)	Fuel Tank (gals.)	Cooling System (qts.)
1990	Jetta	109 (1780)	4.3	3.8	—	4.2	6.4	—	14.5	7.3
	Jetta (Diesel)	97.0 (1588)	4.3	3.8	—	4.2	6.4	—	13.7	7.3
	Jetta GLI 16V	121 (1984)	4.3	3.8	—	4.2	—		14.5	7.3
	Jetta Carat	109 (1780)	4.3	3.8	—	4.2	6.4	—	14.5	7.3
	Cabriolet	109 (1780)	4.3	3.8	—	4.2	6.4	—	13.8	5.1
	Golf GL	109 (1780)	4.3	3.8	—	4.2	6.4	—	14.5	7.3
	Golf GTI	109 (1780)	4.3	3.8	—	4.2	—	—	14.5	7.3
	Golf GTI 16V	121 (1984)	4.3	3.8	—	4.2	—	—	14.5	7.3
	Fox	109 (1780)	3.7	3.2	3.6	4.2	—	—	12.4	6.4 ①
	Passat	121 (1984)	4.3	3.8	—	4.2	6	—	18.5	7.3
	Corrado	109 (1780)	4.3	3.8	—	4.2	—	—	14.5	7.3
1991–92	Jetta	109 (1780)	4.3	3.8	—	4.2	6.4	—	14.5	7.3
	Jetta Diesel	97.0 (1588)	4.3	3.8	—	4.2	6.4	—	13.7	7.3
	Jetta GLI 16V	121 (1984)	4.3	3.8	—	4.2	—		14.5	7.3
	Jetta Carat	109 (1780)	4.3	3.8	—	4.2	6.4	—	14.5	7.3
	Cabriolet	109 (1780)	4.3	3.8	—	4.2	6.4	—	13.8	5.1
	Golf GL	109 (1780)	4.3	3.8	—	4.2	6.4	—	14.5	7.3
	Golf GTI	109 (1780)	4.3	3.8	—	4.2	—	—	14.5	7.3
	Golf GTI 16V	121 (1984)	4.3	3.8	—	4.2	—	—	14.5	7.3
	Fox	109 (1780)	3.7	3.2	3.6	4.2	—	—	12.4	6.4 ①
	Passat	121 (1984)	4.3	3.8	—	4.2	6	—	18.5	7.3
	Corrado	109 (1780)	4.3	3.8	—	4.2	NA	—	14.5	7.3

NA—Not Available
① With A/C, 6.9 qts.

CRANKSHAFT AND CONNECTING ROD SPECIFICATIONS

All measurements are given in inches.

Year	Engine Displacement cu. in. (cc)	Crankshaft Main Brg. Journal Dia.	Crankshaft Main Brg. Oil Clearance	Crankshaft Shaft End-play	Crankshaft Thrust on No.	Connecting Rod Journal Diameter	Connecting Rod Oil Clearance	Connecting Rod Side Clearance
1988	109.0 (1780)	2.126	0.001–0.003	0.003–0.007	3	1.881	0.0049 ①	0.015 ①
	136.0 (2226)	2.282	0.001–0.003	0.003–0.007	3	1.881	0.0006–0.002	0.015 ①
1989	97.0 (1588)	2.126	0.001–0.003	0.003–0.007	3	1.881	0.0049 ①	0.015 ①
	109.0 (1780)	2.126	0.001–0.003	0.003–0.007	3	1.881	0.0049 ①	0.015 ①
	136.0 (2226)	2.282	0.001–0.003	0.003–0.007	3	1.881	0.0006–0.002	0.015 ①
1990	97.0 (1588)	2.126	0.001–0.003	0.003–0.007	3	1.881	0.0049 ①	0.015 ①
	109.0 (1780)	2.126	0.001–0.003	0.003–0.007	3	1.881	0.0049 ①	0.015 ①
	121.0 (1984)	2.126	0.001–0.003	0.003–0.007	3	1.881	0.0049 ①	0.015 ①

CRANKSHAFT AND CONNECTING ROD SPECIFICATIONS

All measurements are given in inches.

Year	Engine Displacement cu. in. (cc)	Crankshaft Main Brg. Journal Dia.	Main Brg. Oil Clearance	Shaft End-play	Thrust on No.	Connecting Rod Journal Diameter	Oil Clearance	Side Clearance
1991-92	97.0 (1588)	2.126	0.001–0.003	0.003–0.007	3	1.881	0.0049 ①	0.015 ①
	109.0 (1780)	2.126	0.001–0.003	0.003–0.007	3	1.881	0.0049 ①	0.015 ①
	121.0 (1984)	2.126	0.001–0.003	0.003–0.007	3	1.881	0.0049 ①	0.015 ①

NOTE: Main and connecting rod bearings are available in 3 undersizes.
① Wear limit
② Bearings marked with blue dot. Bearing marked with red dot—2.3988-2.3991.

VALVE SPECIFICATIONS

Year	Engine Displacement cu. in. (cc)	Seat Angle (deg.)	Face Angle (deg.)	Spring Test Pressure (lbs.)	Spring Installed Height (in.)	Stem-to-Guide Clearance (in.) Intake	Stem-to-Guide Clearance (in.) Exhaust	Stem Diameter (in.) Intake	Stem Diameter (in.) Exhaust
1988	109.0 (1780)	45	45	NA	NA	0.039 ① max	0.051 ① max	0.3140	0.3130
	136.0 (2226)	45	45	NA	NA	0.039 ① max	0.051 ① max	0.3140	0.3130
1989	97.0 Diesel (1588)	45	45	96–106 @ 0.92	NA	0.051 ① max	0.051 ① max	0.3140	0.3130
	109.0 (1780)	45	45	NA	NA	0.039 ① max	0.051 ① max	0.3140	0.3130
	136.0 (2226)	45	45	NA	NA	0.039 ① max	0.051 ① max	0.3140	0.3130
1990	97.0 (1588)	45	45	96–106 @ 0.92	NA	0.051 ① max	0.051 ① max	0.3140	0.3130
	109.0 (1780)	45	45	NA	NA	0.039 ① max	0.051 ① max	0.3140	0.3130
	121.0 (1984)	45	45	NA	NA	0.039 ① max	0.051 ① max	0.3140	0.3130
1991-92	97.0 (1588)	45	45	96–106 @ 0.92	NA	0.051 ① max	0.051 ① max	0.3140	0.3130
	109.0 (1780)	45	45	NA	NA	0.039 ① max	0.051 ① max	0.3140	0.3130
	121.0 (1984)	45	45	NA	NA	0.039 ① max	0.051 ① max	0.3140	0.3130

NOTE: Exhaust valves must be ground by hand
NA Not available
① Measure at valve face end, stem flush with cam end of guide opening.

PISTON AND RING SPECIFICATIONS

All measurements are given in inches.

Year	Engine Displacement cu. in. (cc)	Piston Clearance	Ring Gap Top Compression	Ring Gap Bottom Compression	Ring Gap Oil Control	Ring Side Clearance Top Compression	Ring Side Clearance Bottom Compression	Ring Side Clearance Oil Control
1988	109.0 (1780)	0.0010–0.0030	0.0120–0.0180	0.0120–0.0180	0.0120–0.0180	0.0008–0.0020	0.0008–0.0020	0.0008–0.0020
	136.0 (2226)	0.0011	0.0100–0.0200	0.0100–0.0200	0.0100–0.0200	0.0008–0.0030	0.0008–0.0030	0.0008–0.0030

PISTON AND RING SPECIFICATIONS

All measurements are given in inches.

Year	Engine Displacement cu. in. (cc)	Piston Clearance	Ring Gap Top Compression	Ring Gap Bottom Compression	Ring Gap Oil Control	Ring Side Clearance Top Compression	Ring Side Clearance Bottom Compression	Ring Side Clearance Oil Control
1989	97.0 (1588)	0.001–0.003	0.0120–0.0200	0.0120–0.0200	0.0100–0.0180	0.0020–0.0040	0.0020–0.0030	0.0010–0.0020
	109.0 (1780)	0.0010–0.0030	0.0120–0.0180	0.0120–0.0180	0.0120–0.0180	0.0008–0.0020	0.0008–0.0020	0.0008–0.0020
	136.0 (2226)	0.0011	0.0100–0.0200	0.0100–0.0200	0.0100–0.0200	0.0008–0.0030	0.0008–0.0030	0.0008–0.0030
1990	97.0 (1588)	0.0010–0.0030	0.0120–0.0180	0.0120–0.0180	0.0100–0.0180	0.0020–0.0040	0.0020–0.0030	0.0010–0.0020
	109.0 (1780) (Exc. Corrado)	0.0010–0.0030	0.0120–0.0180	0.0120–0.0180	0.0120–0.0180	0.0008–0.0020	0.0008–0.0020	0.0008–0.0020
	109.0 (1780) (Corrado)	0.0010–0.0020	0.0060–0.0140	0.0600–0.0140	0.0100–0.0200	0.0010–0.0030	0.0010–0.0030	0.0010–0.0020
	121.0 (1984)	0.0010–0.0200	0.0070–0.0150	0.0070–0.0150	0.0090–0.0190	0.0008–0.0027	0.0008–0.0027	0.0008–0.0020
1991–92	97.0 (1588)	0.0010–0.0030	0.0120–0.0180	0.0120–0.0180	0.0100–0.0180	0.0020–0.0040	0.0020–0.0030	0.0010–0.0020
	109.0 (1780) (Exc. Corrado)	0.0010–0.0030	0.0120–0.0180	0.0120–0.0180	0.0120–0.0180	0.0008–0.0020	0.0008–0.0020	0.0008–0.0020
	109.0 (1780) (Corrado)	0.0010–0.0020	0.0060–0.0140	0.0600–0.0140	0.0100–0.0200	0.0010–0.0030	0.0010–0.0030	0.0010–0.0020
	121.0 (1984)	0.0010–0.0200	0.0070–0.0150	0.0070–0.0150	0.0090–0.0190	0.0008–0.0027	0.0008–0.0027	0.0008–0.0020

TORQUE SPECIFICATIONS

All readings in ft. lbs.

Year	Engine Displacement cu. in. (cc)	Cylinder Head Bolts	Main Bearing Bolts	Rod Bearing Bolts ①	Crankshaft Pulley Bolts	Flywheel Bolts	Manifold Intake	Manifold Exhaust	Spark Plugs
1988	109.0 (1780)	②	47	33 ③	145 ④	14 ⑥	18	18	14
	136.0 (2226)	②	47	33 ③	253	14	18	18	14
1989	97.0 Diesel (1588)	① ②	47	33 ③	130	14	18	18	14
	109.0 (1780)	②	47	33 ③	145 ④	14 ⑥	18	18	14
	136.0 (2226)	②	47	33 ③	253	14	18	18	14
1990	97.0 Diesel (1588)	① ②	47	33 ③	253	14	18	18	14
	109.0 (1780) (Exc. Corrado)	②	47	22 ③	④	14 ⑥	18	18	14
	109.0 (1780)	②	47	22 ① ⑤	④	72	11	18	14
	121.0 (1984)	②	47	22 ③	④	74	15	18	14

TORQUE SPECIFICATIONS

All readings in ft. lbs.

Year	Engine Displacement cu. in. (cc)	Cylinder Head Bolts	Main Bearing Bolts	Rod Bearing Bolts ①	Crankshaft Pulley Bolts	Flywheel Bolts	Manifold Intake	Manifold Exhaust	Spark Plugs
1991-92	97.0 Diesel (1588)	① ②	47	33 ③	253	14	18	18	14
	109.0 (1780) (Exc. Corrado)	②	47	22 ③	④	14 ⑥	18	18	14
	109.0 (1780)	②	47	22 ① ⑤	④	72	11	18	14
	121.0 (1984)	②	47	22 ③	④	74	15	18	14

① Always use new bolts.
② With 12 points (polygon) head bolts
Torque in 4 steps:
1st step—29 ft./lbs.
2nd step—43 ft./lbs.
3rd step—additional 1/4 turn (180 degrees) further in one movement (two 90 degree turns are permissible)
Note tightening sequence
Do not retorque at 1000 miles
With 6 point (hex) head bolts
Torque in steps to 54 ft. lbs. with engine cold, when engine is warmed up, torque to 61 ft. lbs. Head bolts must be retorqued after 1000 miles.
③ Stretch bolts: 22 ft. lbs. plus 1/4 (90 degree) turn.
④ Engine UM up to JN707651
133 ft. lbs.
From Engine JN707652
66 ft. lbs. plus 1/2 turn (180°)
⑤ Stretch Bolts: 22 ft. lbs. plus 1/2 (180°) turn.
⑥ Auto. trans., 72 ft. lbs.

BRAKE SPECIFICATIONS

All measurements in inches unless noted.

Year	Model	Lug Nut Torque (ft. lbs.)	Master Cylinder Bore	Brake Disc Minimum Thickness	Brake Disc Maximum Runout	Maximum Brake Drum Diameter	Minimum Lining Thickness Front	Minimum Lining Thickness Rear
1988	Jetta	81	0.820	0.393 ② ④	0.002	7.100	0.276	0.098 ③
	Quantum	81	0.820	0.410 ② ④	0.002	7.900	0.276	0.098 ③
	Scirocco	81	0.820	0.410 ② ④	0.002	—	0.276	③
	Cabriolet	81	0.820	0.393 ② ④	0.002	7.100	0.276	0.098 ③
	Golf	81	0.820	0.393 ② ④	0.002	7.100	0.276	0.098 ③
	Fox/GL	81	0.820	0.393	0.002	7.100	0.276	0.098 ③
1989	Jetta	81	0.820	0.393 ② ④	0.002	7.900	0.276	0.098 ③
	Quantum	81	0.820	0.410 ② ④	0.002	7.900	0.276	0.098 ③
	Scirocco	81	0.820	0.410 ② ④	0.002	—	0.276	③
	Cabriolet	81	0.820	0.393 ② ④	0.002	7.100	0.276	0.098
	Golf	81	0.820	0.393 ② ④	0.002	7.100	0.276	0.098 ③
	Fox/GL	81	0.820	0.393	0.002	7.100	0.276	0.098
1990	Jetta	81	0.809	0.393 ② ④	0.002	7.087	0.276	0.098
	Jetta ABS	81	0.874	0.787 ④	0.001	—	0.276	0.276
	Golf	81	0.809	0.393 ② ④	0.002	7.087	0.276	0.098
	Fox	81	0.809	0.393	0.002	7.087 ⑤	0.276	0.098
	Passat	81	0.874 ①	0.708 ④	0.001	—	0.276	0.276
	Corrado	81	0.874 ①	0.787 ④	0.001	—	0.276	0.276
1991-92	Jetta	81	0.809	0.393 ② ④	0.002	7.087	0.276	0.098
	Jetta ABS	81	0.874	0.787 ④	0.001	—	0.276	0.276
	Golf	81	0.809	0.393 ② ④	0.002	7.087	0.276	0.098

BRAKE SPECIFICATIONS

All measurements in inches unless noted.

Year	Model	Lug Nut Torque (ft. lbs.)	Master Cylinder Bore	Brake Disc Minimum Thickness	Brake Disc Maximum Runout	Maximum Brake Drum Diameter	Minimum Lining Thickness Front	Minimum Lining Thickness Rear
	Fox	81	0.809	0.393	0.002	7.087 ⑤	0.276	0.098
	Passat	81	0.874 ①	0.708 ④	0.001	—	0.276	0.276
	Corrado	81	0.874 ①	0.787 ④	0.001	—	0.276	0.276

NOTE: Minimum lining thickness is as recommended by manufacturer. Due to variations in state inspection regulations, the minimum thickness may be different than that recommended by the manufacturer.

① Non ABS; NA
② Vented discs; 0.708
③ Disc brake: 0.276
④ Rear disc brake: 0.315
⑤ Wagon; 7.913

WHEEL ALIGNMENT

Year	Model	Caster ① Range (deg.)	Caster ① Preferred Setting (deg.)	Camber Range (deg.)	Camber Preferred Setting (deg.)	Toe-in (in.)	Steering Axis Inclination (deg.)
1988	Jetta GL	±30′	1°30′	±20′	−30′	0°±10′	NA
	Jetta GLI	±30′	1°35′	±20′	−35′	0°±10′	NA
	Jetta GLI 16V	±30′	1°35′	±20′	−40′	0°±10′	NA
	Quantum	±30′	+30′ ②	±30′	−40′	+10′ ±10′	NA
	Quantum Syncro	±20′	1°	±30′	−20′	+12′ ±5′	NA
	Scirocco 16V	±30′	1°50′	±30′	+20′	−5′ to −30′	NA
	Cabriolet	±30′	1°50′	±30′	+20′	−5′ to −30′	NA
	Golf GL	±30′	1°30′	±20′	−30′	0° to ±10′	NA
	Golf GTI	±30′	1°35′	±20′	−35′	0° to ±10′	NA
	Golf GTI 16V	±30′	1°35′	±20′	−40′	0° to ±10′	NA
	Fox	±20′	2°	±20′	−30′	−20′ to 0°	NA
	Fox Wagon	±20′	1°45′	±20′	−30′	−20′ to 0°	NA
1989	Jetta GL	±30′	1°30′	±20′	−30′	0°±10′	NA
	Jetta GLI	±30′	1°35′	±20′	−35′	0°±10′	NA
	Jetta GLI 16V	±30′	1°35′	±20′	−40′	0°±10′	NA
	Quantum	±30′	+30′ ②	±30′	−40′	+10′ ±10′	NA
	Quantum Syncro	±20′	1°	±30′	−20′	+12′ ±5′	NA
	Scirocco 16V	±30′	1°50′	±30′	+20′	−5′ to −30′	NA
	Cabriolet	±30′	1°50′	±30′	+20′	−5′ to −30′	NA
	Golf GL	±30′	1°30′	±20′	−30′	0° to ±10′	NA
	Golf GTI	±30′	1°35′	±20′	−35′	0° to ±10′	NA
	Golf GTI 16V	±30′	1°35′	±20′	−40′	0° to ±10′	NA
	Fox	±20′	2°	±20′	−30′	−20′ to 0°	NA
	Fox Wagon	±20′	1°45′	±20′	−30′	−20′ to 0°	NA
1990	Jetta GL	±30′	1°30′	±20′	−30′	0°±10′	NA
	Jetta GLI 16V	±30′	1°35′	±20′	−40′	0°±10′	NA
	Scirocco 16V	±30′	1°50′	±30′	+20′	−5′ to −30′	NA
	Cabriolet	±30′	1°50′	±30′	+20′	−5′ to −30′	NA
	Golf GL	±30′	1°30′	±20′	−30′	0° to ±10′	NA
	Golf GTI	±30′	1°35′	±20′	−35′	0° to ±10′	NA

WHEEL ALIGNMENT

Year	Model	Caster ① Range (deg.)	Caster ① Preferred Setting (deg.)	Camber Range (deg.)	Camber Preferred Setting (deg.)	Toe-in (in.)	Steering Axis Inclination (deg.)
	Golf GTI 16V	±30′	1°35′	±20′	−40′	0° to ±10′	NA
	Fox	±20′	2°	±20′	−30′	−20′ to 0°	NA
	Fox Wagon	±20′	1°45′	±20′	−30′	−20′ to 0°	NA
	Passat	±30′	+1°40′	±20′	−1°20′	0°±10′	NA
	Corrado	±30′	+1°35′	±20′	−40′	0°±10′	NA
1991-92	Jetta GL	±30′	1°30′	±20′	−30′	0°±10′	NA
	Jetta GLI 16V	±30′	1°35′	±20′	−40′	0°±10′	NA
	Scirocco 16V	±30′	1°50′	±30′	+20′	−5′ to −30′	NA
	Cabriolet	±30′	1°50′	±30′	+20′	−5′ to −30′	NA
	Golf GL	±30′	1°30′	±20′	−30′	0° to ±10′	NA
	Golf GTI	±30′	1°35′	±20′	−35′	0° to ±10′	NA
	Golf GTI 16V	±30′	1°35′	±20′	−40′	0° to ±10′	NA
	Fox	±20′	2°	±20′	−30′	−20′ to 0°	NA
	Fox Wagon	±20′	1°45′	±20′	−30′	−20′ to 0°	NA
	Passat	±30′	+1°40′	±20′	−1°20′	0°±10′	NA
	Corrado	±30′	+1°35′	±20′	−40′	0°±10′	NA

① Not adjustable

② With link rod stabilizer bar: + 55′ to +1°55′

GASOLINE ENGINE MECHANICAL

Engine Assembly

REMOVAL & INSTALLATION

Except Quantum, Fox and Passat Synchro

NOTE: The engine and transaxle are lifted as an assembly from the vehicle.

CAUTION

Use care when disconnecting the fuel lines. Fuel under pressure may still be in the lines and, if sprayed, may cause fire or personal injury.

1. Disconnect the battery cables and remove the battery.
2. Open the fuel filler cap to relieve tank pressure and then relieve the fuel system pressure.
3. Remove the air intake duct between the fuel distributor and the throttle body. On Corrado, remove the air tubing from the G-charger and the intercooler. At the throttle body, pull back the accelerator cable clip and disconnect the cable from the ball. Loosen the accelerator cable locknut and remove the cable from the cylinder head cover.
4. Remove the radiator cap and set the heater temperature control to full hot. Place a pan under the thermostat housing and remove the thermostat flange to drain the coolant.
5. Remove the upper radiator hose and disconnect the wiring from the radiator fan motor and switches. Remove the mounting nuts or bolts and lift out the radiator and fan shroud as an assembly.
6. Except Corrado, at the front of the vehicle remove the apron, the trim and the grille. Disconnect the headlight electrical connectors and the hood release cable from the hood latch assembly.
7. Begin disconnecting electrical connections and vacuum lines, carefully labeling each one. Don't forget ground connections that are screwed to the body.

NOTE: If equipped with power steering, remove pump and reservoir and set them aside; do not disconnect the fluid lines. If equipped with air conditioning, remove the compressor and set it aside without disconnecting the lines.

8. On CIS fuel systems, much of the system can be removed as a unit without disconnecting fuel lines. Remove the injectors from their holes and protect them with caps. Remove the cold start injector and warm-up regulator, if equipped, and disconnect the fuel supply and return lines from the fuel distributor.
9. On vehicles with CIS fuel injection, unsnap the clips holding the air cleaner housing together and lift the fuel distributor/air sensor assembly from the vehicle with all the other fuel lines attached.

NOTE: If equipped with an automatic transaxle, place the selector lever in the P position.

10. On vehicles with cable shift linkage, disconnect the shift linkage cables and remove the clutch slave cylinder from the transaxle without disconnecting the line and set it aside.
11. On vehicles with rod shift linkage, remove the 2 rods with the plastic socket ends and unbolt the remaining linkage from the transaxle case as required. Disconnect the clutch cable, lift it from the case and set it aside.
12. Disconnect the electrical connectors from the starter, the back-up light switch and the ground cable from the transaxle. Remove the speedometer cable from the transaxle and plug the hole in the case.

13. On vehicles with automatic transaxle, disconnect the cable from the actuating lever and remove it from the bracket.

14. Attach an engine sling tool VW–2024A or equivalent, to the engine and attach the sling to a suitable lifting device. On 16V engine, remove the idle stabilizer valve and the upper intake manifold to attach the sling.

15. Unbolt the exhaust pipe from the manifold or remove the spring clamps holding the exhaust pipe to the manifold and lower the pipe.

NOTE: On some models, special tools are required for removing and installing the exhaust pipe-to-manifold spring clamps; VW3140/1 and /2 or equivalent. This is a set of different sized wedges for spreading the spring clamps in steps. The installed spring clamp has considerable tension and could cause damage or injury if not properly removed. Clamps with wedges installed are also under high tension and should be handled carefully.

16. Unbolt the halfshafts from the flanges and hang them from the body with wire.

17. Make sure everything is disconnected and unbolt the mounts. Remove the starter first and the front mount with it.

18. With all mounts unbolted, slightly lower the engine/transaxle assembly and tilt it towards the transaxle side. Then carefully lift the assembly from the vehicle.

To install:

19. Carefully install the engine/transaxle assembly and make sure all mounts are securely bolted to the engine/transaxle. Start all nuts and bolts that secure the mounts to the body but don't tighten them yet.

20. With all mounts installed and the engine safely in the vehicle, allow some slack in the lifting equipment. With the vehicle safely supported, shake the engine/transaxle as a unit to settle it in the mounts. Torque all mounting bolts, starting at the rear and working forward. Torque to 33 ft. lbs. (41 Nm) for 10mm bolts or 54 ft. lbs. (73 Nm) for 12mm bolts.

21. Install the starter and torque the bolts to 33 ft. lbs. (45 Nm).

22. Connect the halfshafts to the flanges and torque the bolts to 33 ft. lbs. (45 Nm).

23. Install the exhaust pipe and use new self locking nuts to secure the flange. Torque the nuts to 30 ft. lbs. (40 Nm). On vehicles equipped with spring clamps, the clamps can be used again.

24. Connect the shift linkage and install the clutch cable or slave cylinder. Adjust the clutch and shift linkage as required.

25. Install the fuel system components.

26. Install the air conditioning compressor and/or power steering pump, if equipped. Install and adjust the drive belts.

27. Connect the wiring and vacuum hoses.

28. Install the radiator, fan and heater hoses. Use a new O-ring on the thermostat and torque the thermostat housing bolts to 7 ft. lbs. (10 Nm).

29. Fill and bleed the cooling system. Check the adjustment of the acclerator cable.

30. Install any body parts that were removed.

Quantum

1. The engine is lifted out of the vehicle without the transaxle. Disconnect the negative battery cable.

2. Move the heater control valve to fully open and remove the radiator cap.

3. At the power steering pump, remove the drive belt cover, the drive belt, the mounting bolts and the pump. Move the pump aside with the hoses connected.

4. Remove the grille and the radiator cover.

5. Remove the lower radiator hose and drain the coolant.

6. Remove the front bumper with the energy absorber.

7. Remove the vacuum hoses from the intake manifold, the upper radiator hose, the radiator hose from the thermostat housing and the heater hose, drain the remaining coolant.

8. Disconnect the electrical connectors from the oil pressure switch, the control pressure regulator and the thermo-time switches.

9. Remove the cylinder head cover ground wire.

10. Remove the control pressure regulator, leaving the lines attached. On the linkage rod near it, remove the ball joint circlip and disconnect the pushrod.

11. Remove the alternator drive belt, the bracket bolts and the alternator assembly.

12. Remove the air duct and the front engine stop.

13. Disconnect the electrical connectors from the cold start valve, the frequency valve and the throttle switch. Disconnect the electrical leads at the idle stabilizer valve, the Hall sender at the distributor and the oxygen sensor.

14. Remove the accelerator cable circlip and disconnect the cable rod from the throttle body.

15. Remove the distributor cap, the cold start valve and the vacuum hose from the thermo valve.

16. Remove the fuel injection cooling hose.

17. Remove the fuel injectors from the intake manifold; leave the fuel lines connected.

NOTE: When removing the fuel injectors and the cold start valve, place caps on the ends to protect them from damage.

18. Remove the air filter housing bolts and the filter.

19. Remove the heater hoses and, if equipped with an automatic transaxle, disconnect the oil cooler hoses.

20. If equipped with air conditioning, remove the drive belt and disconnect the wiring to the compressor. Unbolt the compressor bracket from the engine and secure the compressor aside without disconnecting the freon lines.

21. Attach the supporting tool 2084 or equivalent, to the crankshaft pulley to keep the engine from turning and remove the center crankshaft pulley bolt.

22. Of the 4 crankshaft pulley bolts, remove 2 and loosen 2. To loosen the pulley, tap lightly on the remaining bolts. Remove the bolts and the pulley. Do not remove the camshaft belt drive sprocket.

23. Remove the front engine mount and the subframe-to-body bolts. Remove the exhaust pipe from the exhaust manifold and the support bracket.

24. Disconnect the starter cables and remove the starter.

25. Working through the starter hole, remove the torque converter to flywheel bolts. Unhook the shift rod clip and disconnect the rod.

26. Remove the rubber plugs from the left side frame member. Using the support tool VW 785/1 or equivalent, connect it to the transaxle and to the frame member, then adjust to make contact with the transaxle. This tool is to hold the transaxle in place while the engine is out of the vehicle.

27. Remove the upper engine-to-transaxle bolts; leave 1 bolt in place.

28. Attach the engine support tool US 1105 and the lift tool 9019 or equivalent, to the engine.

NOTE: If equipped with an automatic transaxle, secure the torque converter before removing the engine from the transaxle.

29. Remove the remaining engine-to-transmisssion bolts and engine mount bolts and lift the engine, while prying the engine apart from the transaxle. Remove the engine from the vehicle by twisting the front slightly to the left and lift straight out.

To install:

30. Carefully install the engine and bolt it to the transaxle. Torque the transaxle-to-engine bolts to 22 ft. lbs. (30 Nm) for 8mm bolts, 32 ft. lbs. (43 Nm) for 10mm bolts or 43 ft. lbs. (58 Nm) for 12mm bolts.
31. Make sure all mounts are securely bolted to the engine/transaxle. Start all nuts and bolts that secure the mounts to the body but don't tighten them yet.
32. With all mounts installed and the engine safely in the vehicle, allow some slack in the lifting equipment. With the vehicle safely supported, shake the engine/transaxle as a unit to settle it in the mounts. Torque all mounting bolts to 32 ft. lbs. (43 Nm), starting at the rear and working forward. Torque the subframe-to-body bolts – 28 ft. lbs. (35 Nm), plus ¼ turn.
33. On vehicles with automatic transaxle, install the torque converter-to-drive plate bolts and torque to 22 ft. lbs. (30 Nm).
34. Install the starter and torque the bolts to 33 ft. lbs. (45 Nm).
35. Install the front crankshaft pulley and torque the center bolt to 331 ft. lbs. (450 Nm). Torque the 4 smaller bolts to 14 ft. lbs. (19 Nm).

NOTE: When installing the crankshaft pulley, align the matchmark on the sprocket with the mark on the pulley. When installing the crankshaft bolt, coat the threads with Loctite® thread sealer or equivalent.

36. Install the air conditioner compressor and power steering pump. Install and adjust the tension on the drive belts.
37. Install the fuel system components.
38. Connect all wiring and vacuum hoses.
39. Install the cooling system hoses and refill the system.
40. Fill and bleed the cooling system. Check the adjustment of the accelerator cable.
41. Install all body parts that were removed.
42. After the engine is started and all adjustments have been made, if there is excessive vibration, loosen all the engine mount bolts and and retorque them with the engine running at idle.

Fox

1. The engine is lifted from the vehicle without the transaxle. Disconnect the battery ground cable and remove the battery.
2. Open the heater valve and the cap on the coolant expansion tank. Drain the coolant by removing the bottom hose. Disconnect the electrical connector from the radiator cooling fan.
3. Remove the radiator and fan as an assembly.
4. If equipped with air conditioning, remove the compressor and condenser and place them aside without disconnecting any refrigerant lines.
5. Detach and label all the electrical wires and vacuum lines connecting the engine to the body.
6. Much of the fuel system can be removed as a unit without disconnecting fuel lines, Remove the injectors from their holes and protect them with caps. Remove the cold start injector and warm-up regulator, if equipped, disconnect the throttle cable and remove the air duct. Lay these aside without disconnecting the fuel lines.
7. Disconnect the speedometer cable from the transaxle and plug the hole. Detach the clutch cable.
8. Loosen the charcoal filter clamp and move the filter to the rear of the engine compartment.
9. Remove the upper engine-to-transaxle bolts.
10. Remove the left and right engine mounting nuts.
11. Remove the front engine stop and the starter.
12. Remove the clutch cover and the 2 lower engine-to-transaxle bolts.
13. Disconnect the exhaust pipe from the manifold at the flange. Then remove the bolt from the exhaust pipe

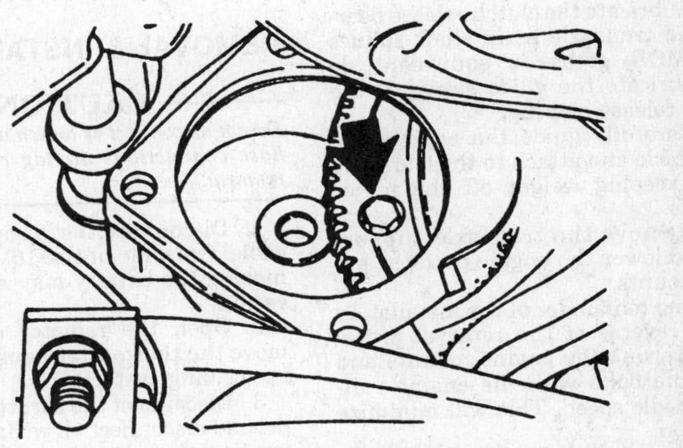

Remove the torque converter–to–flywheel bolts through the starter hole

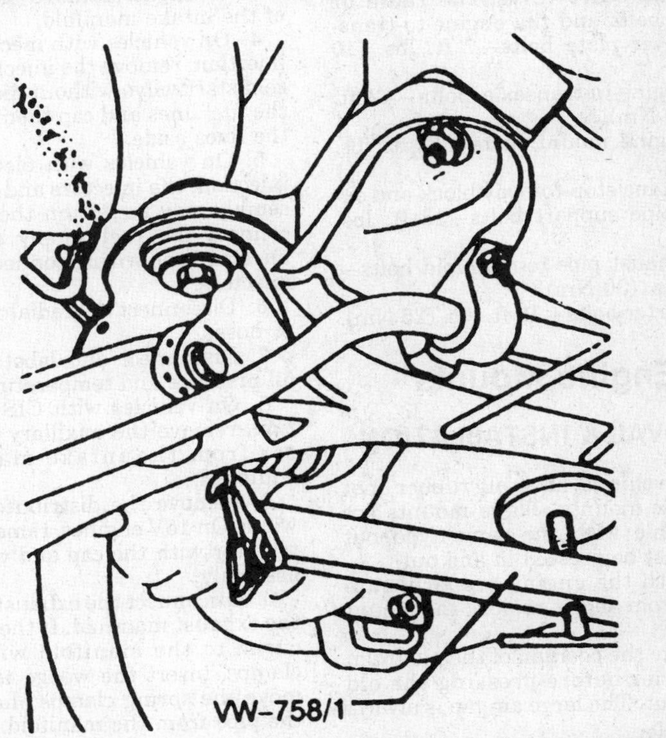

Support tool holds transaxle in place when removing Fox engine

support and remove the exhaust pipe from the manifold.

14. Install transaxle support bar VW–758/1 or equivalent, with slight preload. This is to hold the transaxle in place while the engine is out.

15. Install sling US–1105 or equivalent, on the engine lifting eyes, located on the left side of the cylinder head.

16. Lift the engine until its weight is taken off the engine mounts.

17. Adjust the support bar to contact the transaxle.

18. Separate the engine and transaxle.

19. Carefully lift the engine out of the engine compartment so as not to damage the transaxle main shaft, clutch and body.

To install:

20. Lubricate the clutch release bearing and transaxle main shaft splines with MOS_2 grease or equivalent; do not lubricate the guide sleeve or the clutch release bearing.

21. Carefully guide the engine into the vehicle and attach to the transaxle while keeping weight off the motor mounts.

22. Remove the transaxle support bar and lower the engine onto the engine mounts.

23. The remainder of the installation is the reverse of the removal procedure. Torque the engine mounts and subframe bolts with the engine running at idle speed. This will minimize vibration.

24. Torque the following:

Cold start valve, the radiator mount bolts and the engine-to-transaxle cover plate bolts – 7 ft. lbs. (10 Nm).

Engine-to-transaxle bolts – 42 ft. lbs. (55 Nm).

Engine mount bolts – 30 ft. lbs. (40 Nm).

Engine stop-to-body block and exhaust pipe support bolts – 18 ft. lbs. (25 Nm).

Exhaust pipe-to-manifold bolts – 22 ft. lbs. (30 Nm).

Starter bolts – 18 ft. lbs. (25 Nm).

Engine Mounts

REMOVAL & INSTALLATION

Earlier vehicles have all rubber, not hydraulic mounts. These mounts are replaceable with the same type but they must be pressed in and out.

1. With the engine properly supported from above, remove the mount carrier.

2. Note the position of the mount in the carrier before pressing the old mount out. The large air gap is always at the top.

3. Reinstall the carrier and mount and center the mount in the bracket on the frame while tightening the bolts.

ENGINE ALIGNMENT

If there is excessive engine vibration, before removing mounts, an engine alignment procedure may cure the problem. Loosen all the bolts that go into the rubber mounts themselves. With the vehicle safely supported, shake the engine/transaxle as a unit to settle it in the mounts. Retorque all mounting bolts, starting at the rear and working forward. This procedure helps to minimize vibration.

Cylinder Head

REMOVAL & INSTALLATION

CAUTION

Do not disconnect or loosen any refrigerant hose connections during cylinder head removal.

1. Disconnect the negative battery cable. On some of the 16V models, removing the battery may make the job easier.

2. Open the radiator cap and remove the thermostat housing to drain the cooling system.

3. Disconnect the throttle cable. Label and disconnect all wiring and vacuum lines from the intake manifold. On the 16V engine, remove the upper half of the intake manifold.

4. On vehicles with mechanical fuel injection, remove the injectors and the cold start valve without disconnecting the fuel lines and cap them. Secure all the lines aside.

5. On vehicles with electronic fuel injection, the injectors and fuel rail assembly may be left on the head. Disconnect the fuel supply and return lines and the wiring connector for the injectors.

6. Disconnect the radiator and heater hoses.

7. Disconnect and label wiring for oil pressure and temperature sensors.

8. On vehicles with CIS fuel injection, remove the auxiliary air regulator from the intake manifold, if equipped.

9. Remove the distributor cap and wires. On 16V engines, remove the distributor with the cap and wires as an assembly.

10. Disconnect the exhaust pipe from the exhaust manifold. If the pipe is secured to the manifold with spring clamps, insert the wedge tools to remove the spring clamps and separate the pipe from the manifold.

NOTE: Special tools are required for removing and installing the clamps; VW3140/1 and /2 or equivalent. This is a set of different sized wedges for spreading the spring clamps in steps. The installed spring clamp has considerable tension and could cause damage or injury if not properly removed. Clamps with wedges installed are also under high tension and should be handled carefully.

11. Remove the EGR pipe from the exhaust manifold, if equipped.

12. Remove the accessory drive belts and any accessory that is bolted to the head. On Corrado, a special clamping tool (3191) is required to remove the spring loaded belt tensioner.

13. Turn the engine to TDC of No. 1 cylinder, if possible, and remove the cylinder head cover, timing belt cover and belt.

14. Loosen the cylinder head bolts in the reverse of the tightening sequence.

15. Remove the bolts and lift the head straight off.

To install:

16. Before reinstalling the head, check the flatness of the head and block in both width and length, then diagonally from each corner.

17. Install the new cylinder head gasket with the word TOP or OBEN facing upward; do not use any sealing compound.

18. Carefully set the head on and install bolts No. 10 and 8 on 4 cylinder engines, or 9 and 11 on 5 cylinder engines. These holes are smaller and will properly locate the gasket and cylinder head.

19. Install the remaining bolts. Torque the bolts in sequence in 3 steps: 29 ft. lbs. (39 Nm), 44 ft. lbs. (60 Nm) and an additional ½ turn. Two quarter turns are allowed.

20. Install the camshaft drive belt and adjust the tension.

21. Connect the exhaust pipe to the manifold. Use new gaskets and self locking nuts and torque to 18 ft. lbs. (25 Nm). On vehicles that use spring clamps, install the clamps and carefully remove the wedge tools.

22. Connect the EGR pipe, if equipped.

23. Install the fuel injection equipment and connect the wiring.

24. On 16V engines, install the upper intake manifold.

25. Install the ignition system components.

26. Connect the throttle cable and all wiring and vacuum lines.

27. Connect the cooling system hoses and install the thermostat with a new O-ring. Torque the housing bolts to 7 ft. lbs. (10 Nm). Refill the cooling system.

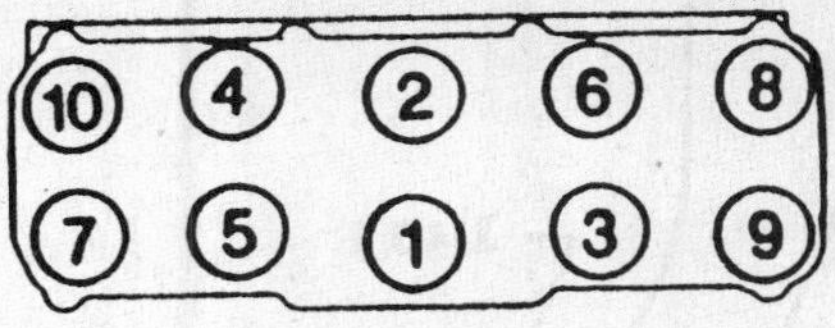

Head bolt torque sequence—4 cylinder engine

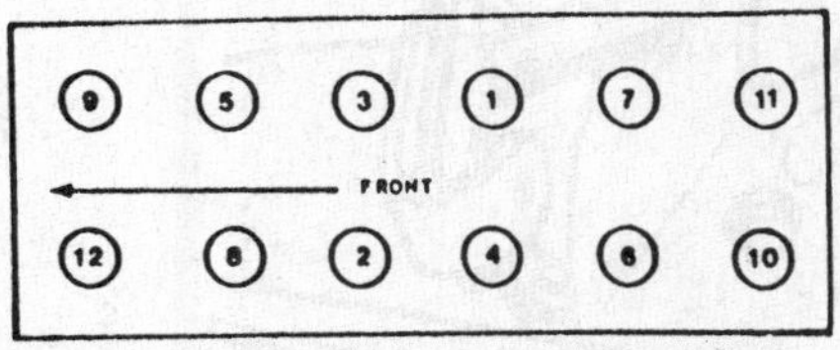

Head bolt torque sequence—5 cylinder engine

28. Install the accessory drive belts and adjust the tension.
29. When everything has been properly installed and connected, be sure to change the oil and filter before starting the engine.

Valve Lifters

REMOVAL & INSTALLATION

1. Remove the camshaft(s).
2. The valve lifters can be easily lifted out of the head by hand. Place hydraulic lifters camshaft side down on a clean surface. Keep all lifters in order so they can be installed in the same position.

To install:

3. Make sure the engine is not at TDC of any cylinder.
4. Set the lifters in place and carefully install the camshaft. Allow the lifters to bleed down for 30 minutes before turning the engine or the valves may hit the pistons.

Valve Lash

ADJUSTMENT

Most vehicles now have hydraulic valve lifters and require no adjustment. On these vehicles there will be a sticker under the hood indicating hydraulic lifters. The overhead camshaft acts directly on the valves through bucket-type camshaft followers which fit over the springs and valves. With solid lifters, adjustment is made with an adjusting disc (shim) which fits into the camshaft follower. Different thickness discs result in changes in valve clearance.

NOTE: VW recommends that 2 special tools (VW 546) and the special pliers (VW 208) or equivalent, be used to remove and install the adjustment discs. One is a prybar to compress the valve springs and the other a pair of special pliers to remove the disc. Care must be taken not to gouge the camshaft lobes. The camshaft follower has 2 slots which permit the disc to be lifted out.

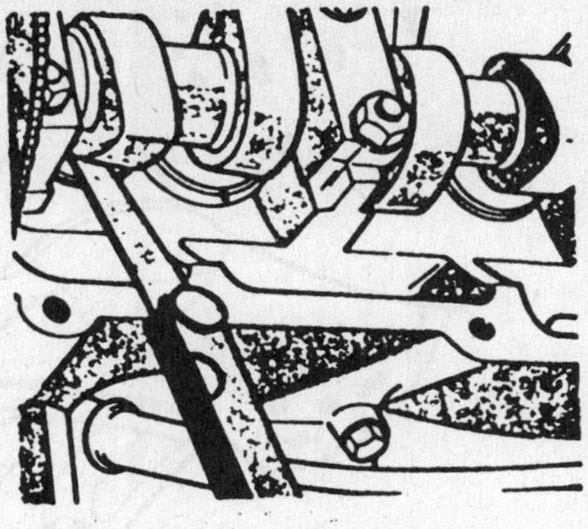

Checking valve lash adjustment

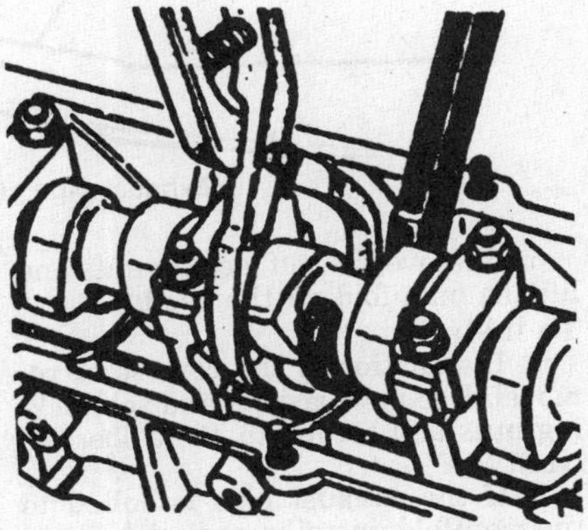

Using the special pliers to remove the adjustment shim

1. Valve clearance is checked with the engine moderately warm; coolant temperature should be about 95°F (35°C). Remove the accelerator linkage and the upper drive belt cover. On diesel engine, remove the air cleaner and any hoses or lines as needed.
2. Remove the cylinder head cover. Valve clearance is checked in the firing order 1–3–4–2 for the 4 cylinder and 1–2–4–5–3 for the 5 cylinder engine, with the piston of the cylinder being checked at TDC of the compression stroke. Both valves will be closed at this position and the camshaft lobes will be pointing straight up.
3. Turn the crankshaft pulley bolt with a socket wrench to position the camshaft for checking.

NOTE: Do not turn the camshaft by the camshaft sprocket mounting bolt, this will stretch the drive belt or cause the camshaft to jump timing. When turning the crankshaft pulley bolt, turn it clockwise only.

4. With the No. 1 piston at TDC of the compression stroke, determine the clearance with a feeler gauge. Intake clearance should be 0.008–0.012 in. (0.2–0.3mm); exhaust clearance should be 0.016–0.020 in. (0.4–0.5mm).
5. Check the other cylinders in the firing order, turning the crankshaft to bring each particular piston to the top of the compression stroke (both valves closed). Record the individual clearances.
6. If measured clearance is within tolerance levels of 0.002 in. (0.05mm), it is not necessary to replace the adjusting discs.
7. If adjustment is necessary, the discs will have to be removed and replaced with thicker or thinner ones which will yield the correct clearance. Discs are available in 0.002 in. (0.05mm) increments from 0.12–0.17 in. (3.05–4.32mm).

NOTE: The thickness of the adjusting discs is etched on 1 side. When installing, the marks must face the camshaft followers. Discs can be reused if they are not worn or damaged.

8. To remove the discs, turn the camshaft followers so the grooves are accessible when the prybar is depressed.
9. Press the camshaft follower down with the prybar and remove the adjusting discs with the proper tool.
10. Replace the adjustment discs as necessary. If the measured clearance is larger than the given tolerance, remove the existing disc and insert a thicker one to bring the clearance up to specification. If it is smaller, insert a thinner one.
11. Recheck all valve clearances after adjustment.
12. Install the valve cover. Connect the accelerator linkage and hoses.

Intake Manifold

REMOVAL & INSTALLATION

1. Disconnect the negative battery cable. Remove the air duct from the throttle valve body and disconnect the accelerator cable.
2. Remove the cylinder head cover.
3. On electronic fuel injection systems, remove the idle stabilizer valve, fuel pump pressure switch and the fuel injector wiring harness.
4. Disconnect and label the vacuum and the emission control hoses.
5. Disconnect and label all electrical lines.

6. Remove the injectors and disconnect the line from the cold start valve.
7. On mechanical fuel injection systems, remove the auxiliary air regulator.
8. If equipped, disconnect the EGR pipe from the exhaust manifold.
9. Loosen and remove the retaining bolts and lift off the manifold.

NOTE: The intake manifold on the 16V engine is removed in 2 halves (upper and lower). The upper half is removed first. The gasket between the halves should be replaced if the manifold is separated. Torque the upper-to-lower bolts to 15 ft. lbs. (20 Nm).

10. Installation is the reverse of removal. Use new gaskets and torque the manifold bolts to 18 ft. lbs. (25 Nm).

Exhaust Manifold

REMOVAL & INSTALLATION

NOTE: On some models, special tools are required for removing and installing the exhaust pipe–to–manifold spring clamps; VW3140/1 and /2 or equivalent. This is a set of different sized wedges for spreading the spring clamps in steps. The installed spring clamp has considerable tension and could cause damage or injury if not properly removed. Clamps with wedges installed are also under high tension and should be handled carefully.

1. Disconnect the negative battery cable and remove any heat shields that may be in the way.
2. Remove the emissions sample tap and, if equipped, disconnect the EGR pipe from the exhaust manifold.
3. On models with manifold studs, remove the self locking nuts and lower the exhaust pipe.
4. Expand the spring clamp by pushing the exhaust pipe to one side and insert the starter wedge into the clamp all the way up to the shoulder.
5. Push the pipe to the other side and install another wedge in the opposite clamp. Continue to work the pipe side to side while pushing the wedges into the clamps until the clamps are spread far enough to lift off easily.

CAUTION

The removed spring clamps with wedges in them are very highly loaded and, if misshandled, could fly apart with enough force to cause serious injury. Store the removed clamps in a safe area where they won't be disturbed.

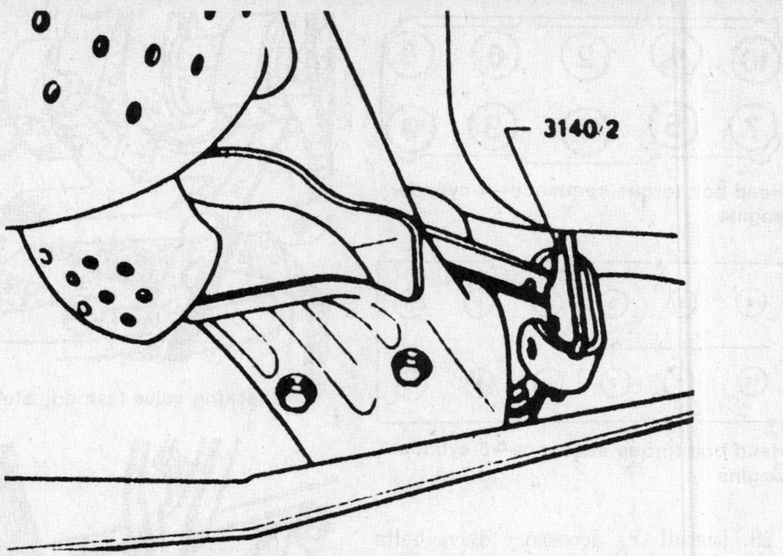

Exhaust pipe clamp removal tools

5. Remove the self locking nuts and lift the manifold off the head.

To install:

6. Installation is the reverse of removal. Use new gaskets and self locking nuts and torque to 18 ft. lbs. (25 Nm).
7. If the exhaust pipe is bolted to the manifold, install a new gasket and use new self locking nuts. Torque the nuts to 30 ft. lbs. (40 Nm).
8. If equipped with spring clamps, hold the pipe in position with a new gasket and install the clamps. Carefully remove the wedge tools.

Supercharger

REMOVAL & INSTALLATION

Corrado

The Corrado has a belt driven supercharger with an intercooler for supplying up to 11.6 psi (0.8 BAR) of boost. The belt is the same serpentine ribbed belt used to drive the other engine accessories. Belt tension is maintained with a spring loaded automatic belt tensioner. Releasing the tension to remove the belt requires a special clamping tool, VW3191 or equivalent. The supercharger cannot be repaired; leaking or otherwise faulty units must be replaced.

1. Install clamping tool and compress the belt tensioner. Remove the belt from the supercharger pulley.
2. Remove the connector hose and silencer from the outlet side of the supercharger and remove the 2 upper inlet hoses.
3. Remove the front and rear mounting bolts and carefully lift the supercharger onto the top of the engine.
4. Allow the oil to drain back into the engine for a few minutes, then remove the oil lines and take the supercharger out of the vehicle.
5. Installation is the reverse of removal. Start the fittings for the oil lines but don't tighten them until the unit is bolted in place. Be sure to use new sealing rings. Torque the following:

Supercharger mounting bolts – 18 ft. lbs. (25 Nm).

Mounting bracket-to-engine bolt – 33 ft. lbs. (45 Nm).

Oil line fittings – 11 ft. lbs. (15 Nm).

Timing Belt Front Cover

REMOVAL & INSTALLATION

1. Disconnect the negative battery cable.
2. The accessory drive belts must be removed. On Corrado, this requires special tool 3191 or equivalent, to compress the spring loaded belt tensioner.
3. To remove the crankshaft accessory drive pulley, hold the center crankshaft sprocket bolt with a socket and loosen the pulley bolts.
4. The cover is now accessible. It comes off in 2 pieces, remove the upper half first. Take note of any special spacers or other hardware.
5. Installation is the reverse of removal.

FRONT OIL SEAL REPLACEMENT

Except 5 Cylinder Engine

1. Remove the timing belt cover and the timing belt.

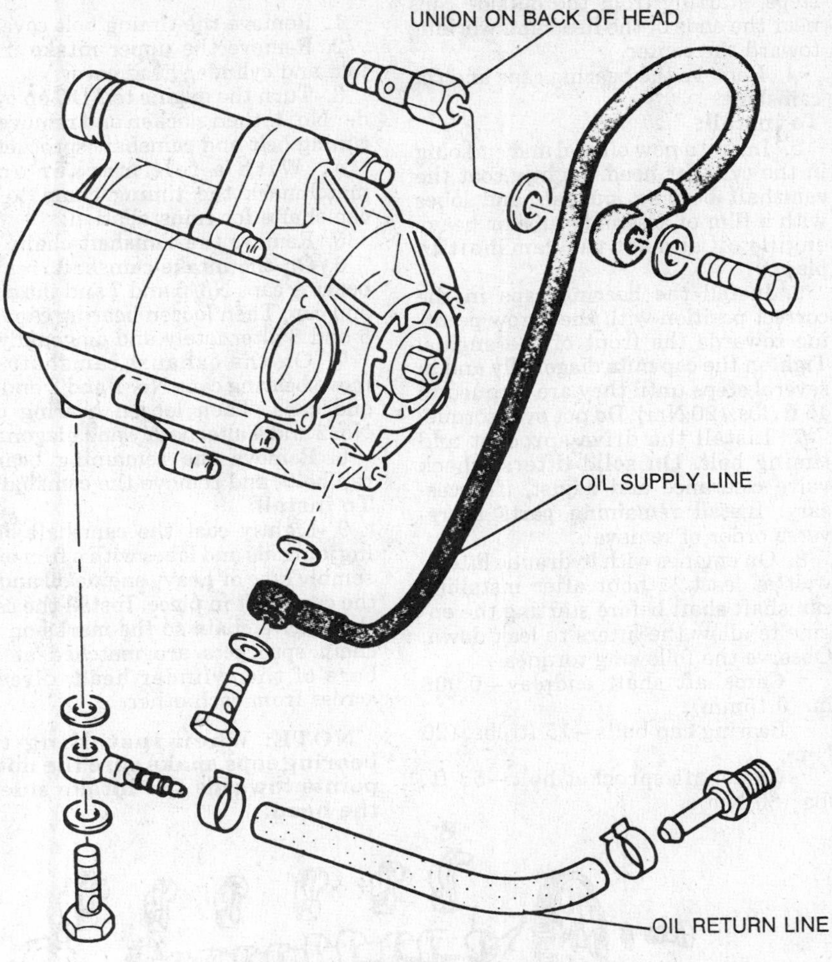

Supercharger oil supply and return lines on Corrado

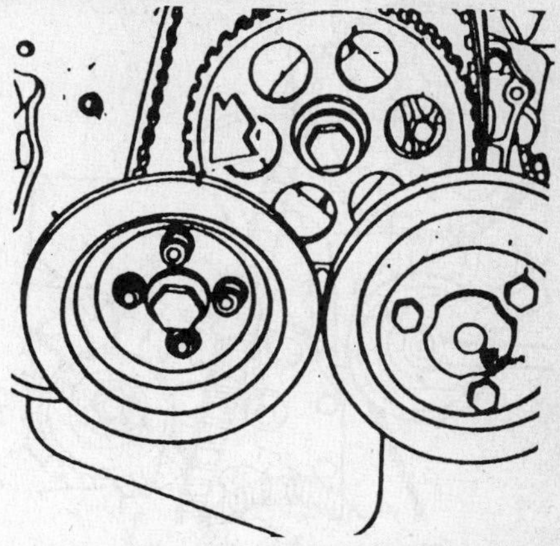

Timing marks on crankshaft pulley and intermediate shaft sprocket

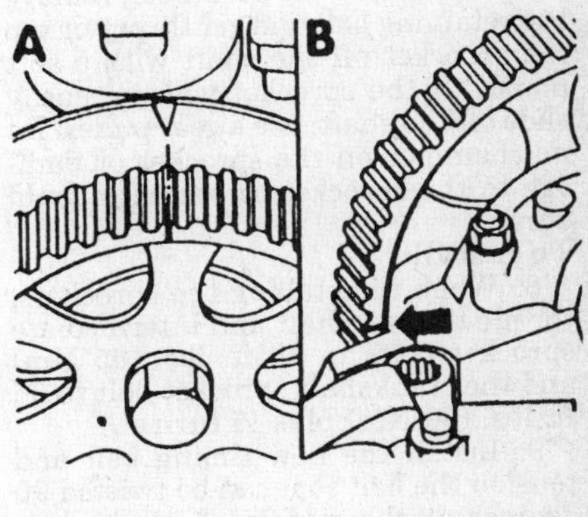

Timing marks on camshaft sprockets: A – 16V engine, B – 8 valve engine

2. Remove the crankshaft sprocket.
3. Using a small prybar, pry the seal from the carrier or use the seal extractor tool VW–10–219 or equivalent, to pull out the seal.

NOTE: When removing the seal, be careful not to damage the carrier.

To install:

4. Lubricate the new seal lips and use the seal installation tool VW–10–203 or equivalent, to press the new seal into the carrier.
5. Install the crankshaft sprocket and torque the bolt to 133 ft. lbs. (180 Nm) for 12mm hex head bolts, or 66 ft. lbs. (90 Nm) plus ½ turn for all 12 sided bolts.
6. Install the timing belt and check the ignition timing.

5 Cylinder Engine

1. Remove the timing belt cover and remove the timing belt.
2. Remove the crankshaft sprocket.
3. Install the hex head bolt of the seal removal and installation tool VW–3083 or equivalent, into the seal extractor guide VW–2085, or equavilant.
4. Attach the tools to the oil seal and pull the seal from the carrier.

To install:

5. Slide the sleeve of the installation tool onto the crankshaft journal, lubricate the seal and slide it over the sleeve. Install the thrust sleeve against the oil seal and press it in until seated.
6. Install the crankshaft sprocket and pulley and torque the bolt to 258 ft. lbs. (350 Nm).
7. Install the timing belt and check the timing.

Timing Belt and Tensioner

REMOVAL & INSTALLATION

NOTE: Do not turn the engine or camshaft with the camshaft drive belt removed. The pistons will contact the valves and cause internal engine damage.

1. Disconnect the negative battery cable and remove the accessory drive belts, crankshaft pulley and the timing belt cover(s).
2. Temporarily reinstall the crankshaft pulley bolt, if removed and turn the engine to TDC of No. 1 cylinder. The mark on the camshaft sprocket should be aligned with the mark on the rear drive belt cover, if equipped, or the edge of the cylinder head.
3. On 8 valve engines, the notch on the crankshaft pulley should align with the dot on the intermediate shaft sprocket. Remove distributor cap, the rotor should be pointing toward the No. 1 mark on the rim of the distributor housing.
4. On 4 cylinder engines, loosen the locknut on the tensioner pulley and turn the tensioner counterclockwise to relieve the tension on the timing belt.
5. On 5 cylinder engines, loosen the water pump bolts and turn the pump clockwise to relieve timing belt tension.

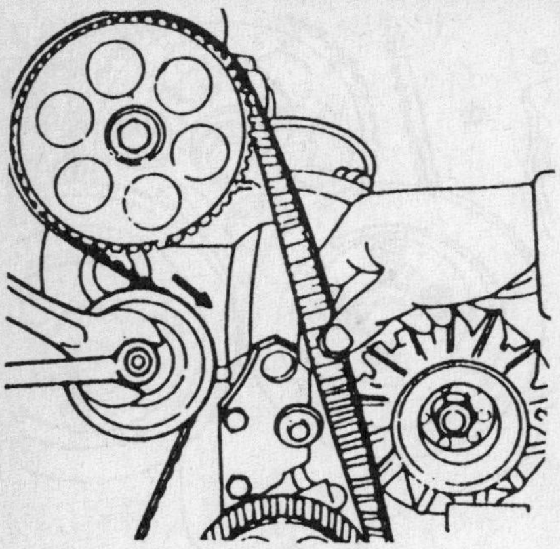

Checking belt tension

6. Slide the timing belt from the sprockets.

7. To remove the sprockets, remove the retaining bolt and gently pry or tap the sprocket off the shaft with a soft mallet. If the sprocket will not easily slide off the shaft, use a gear puller. Do not hammer on the sprocket or damage to the sprocket or bearings could occur.

To install:

8. When reinstalling the sprockets, torque the camshaft and intermediate sprocket bolts to 48 ft. lbs. (65 Nm) and the crankshaft sprocket bolt to 66 ft. lbs. (90 Nm) plus ½ turn.

9. Install the new timing belt and tension the belt so it can be twisted 90 degrees at the middle of it's longest section, between the camshaft and intermediate sprockets.

10. Recheck the alignment of the timing marks and, if correct, turn the engine 2 full revolutions to return to TDC of No. 1 cylinder. Recheck belt tension and timing marks. Readjust as required. Torque the tensioner nut to 33 ft. lbs. (45 Nm).

11. Reinstall the belt cover and accessory drive belts.

Camshaft

REMOVAL & INSTALLATION

SOHC Engine

1. Disconnect the negative battery cable. Remove the timing belt cover(s), the timing belt, camshaft sprocket and cylinder head cover.

2. Number the bearing caps from front to back. If the cap does not already have one, scribe an arrow pointing towards the front of the engine. The caps are offset and must be installed correctly. Factory numbers on the caps are not always on the same side.

3. Remove the front and rear bearing caps. Loosen the remaining bearing cap nuts diagonally, in several steps, starting from the outside caps near the ends of the head and working toward the center.

4. Remove the bearing caps and the camshaft.

To install:

5. Install a new oil seal and end plug in the cylinder head. Lightly coat the camshaft bearing journals and lobes with a film of assembly lube or heavy engine oil and set the camshaft in place.

6. Install the bearing caps in the correct position with the arrow pointing towards the front of the engine. Tighten the cap nuts diagonally and in several steps until they are torqued to 15 ft. lbs. (20 Nm). Do not over torque.

7. Install the drive sprocket and timing belt. On solid lifters, check valve clearance and adjust, if necessary. Install remaining parts in reverse order of removal.

8. On engines with hydraulic lifters, wait at least ½ hour after installing camshaft shaft before starting the engine to allow the lifters to leak down. Observe the following torques:

Camshaft shaft endplay – 0.006 in. (0.15mm).

Bearing cap bolts – 15 ft. lbs. (20 Nm).

Camshaft sprocket bolt – 58 ft. lbs. (80 Nm).

DOHC Engine

1. Remove the timing belt cover.

2. Remove the upper intake manifold and cylinder head cover.

3. Turn the engine to TDC on cylinder No. 1, then slacken and remove the timing belt and camshaft sprocket.

4. With a felt marker only, matchmark the timing chain to the camshafts for reinstallation.

5. Remove the camshaft chain.

6. On the intake camshaft, remove bearing caps No. 5 and 7 and the chain end cap. Then loosen bearing caps No. 6 and 8 alternately and diagonally.

7. On the exhaust camshaft, remove bearing caps No. 1 and 3 and the end caps. Then loosen bearing caps No. 2 and 4 alternately and diagonally.

8. Remove the remaining bearing cap bolts and remove the camshaft.

To install:

9. Lightly coat the camshaft bearing journals and lobes with a film of assembly lube or heavy engine oil and set the camshaft in place. Install the camshaft drive chain so the marks on the chain sprockets are matched at the base of the cylinder head, directly across from each other.

NOTE: When installing the bearing caps, make sure the notch points towards the intake side of the head.

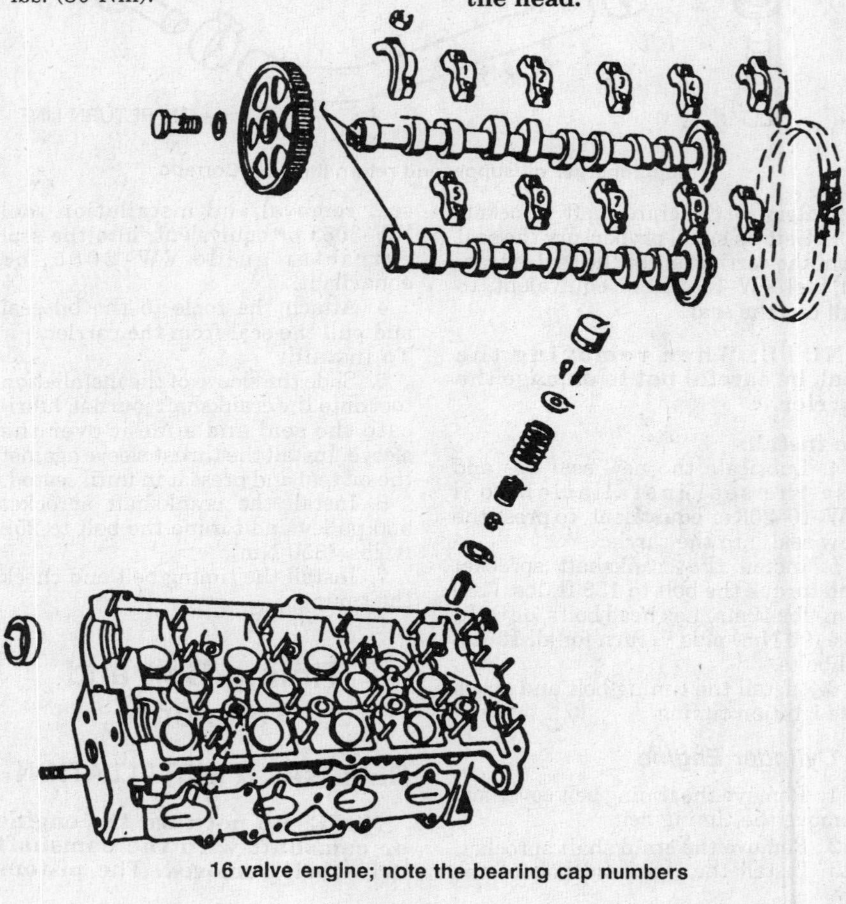

16 valve engine; note the bearing cap numbers

Sixteen valve engine camshaft shaft alignment

10. On the intake camshaft, install and torque bearing caps No. 6 and 8 alternately and diagonally.
11. Install and torque the remaining intake camshaft bearing caps.
12. On the exhaust camshaft, torque bearing caps No. 2 and 4 alternately and diagonally.
13. Install and torque the remaining exhaust camshaft bearing caps.
14. Install the drive sprocket and timing belt.
15. Install remaining parts in reverse order of removal. Wait at least ½ hour after installing camshaft shafts before starting the engine to allow the hydraulic lifters to leak down.

Camshaft shaft end play—0.006 in. (0.15mm).

Bearing cap bolts—11 ft. lbs. (15 Nm).

camshaft shaft sprocket bolt—48 ft. lbs. (65 Nm).

Piston and Connecting Rod

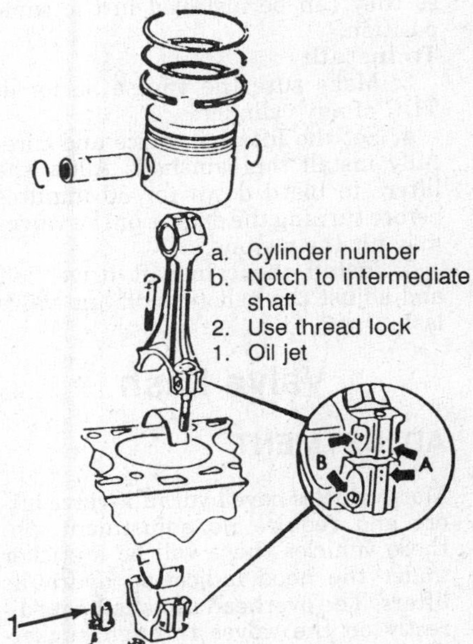

Arrow on piston points towards the camshaft drive belt

DIESEL ENGINE MECHANICAL

Engine Assembly

REMOVAL & INSTALLATION

1. The engine and transaxle are lifted from the vehicle as an assembly. Disconnect the battery cables and remove the battery.
2. Open the fuel filler cap to relieve tank pressure, then loosen the fuel filter fitting to relieve system pressure. Be sure to take the appropriate fire safety precautions.
3. Remove the air filter and disconnect the accelerator cable from the injection pump.
4. Remove the radiator cap. Turn the heater temperature control to full hot and remove the thermostat housing to drain the coolant.
5. Remove the upper radiator hose and disconnect the wiring from the radiator fan motor and switches. Remove the mounting nuts or bolts and lift out the radiator and fan shroud as an assembly.
6. Begin disconnecting electrical connections and vacuum lines, carefully labeling each one. Don't forget ground connections that are screwed to the body.

NOTE: If equipped with power steering, remove pump and reservoir and set them aside. Do not disconnect the fluid lines. If equipped with air conditioning, remove the compressor and set it aside without disconnecting the lines.

7. Carefully disconnect the fuel delivery lines from the pump and injectors and remove them as an assembly. Set the lines where they will stay clean and cap the injectors and pump fittings to keep them clean.
8. Disconnect the fuel inlet and outlet lines from the pump and plug the holes to keep the pump clean. Note the outlet fitting has a special orifice.
9. On turbocharged engines, disconnect the exhaust pipe and the oil lines from the turbocharger and cap the oil line fittings on the turbocharger. Unbolt the turbocharger and lift it out of the engine.
10. If equipped with an automatic transaxle, place the selector lever in **P** and disconnect the selector cable at the transaxle.
11. On manual transaxle shift linkage, remove the 2 rods with the plastic socket ends and unbolt the remaining linkage from the case as required. Disconnect the clutch cable, lift it out of the case and set it aside.
12. Disconnect the wiring from the starter, the backup light switch and the ground cable from the transaxle. Remove the speedometer cable from the transaxle and plug the hole in the case.
13. Attach an engine sling tool VW-2024A or equivalent, to the engine and attach the sling to a suitable lifting device.
14. Remove the nuts or spring clamps holding the exhaust pipe to the manifold or turbo charger.

NOTE: On some models, special tools are required for removing and installing the exhaust pipe-to-manifold spring clamps; VW3140/1 and /2 or equivalent. This is a set of different sized wedges for spreading the spring clamps in steps. The installed spring clamp has considerable tension and could cause damage or injury if not properly removed. Clamps with wedges installed are also under high tension and should be handled carefully.

15. Unbolt the halfshafts from the flanges and hang them from the body with wire.
16. Make sure everything is disconnected and unbolt the mounts. Remove the starter first and the front mount with it.
17. With all mounts unbolted, slightly lower the engine/transaxle assembly and tilt it towards the transaxle side. Then carefully lift the assembly out of the vehicle.

To install:

18. Carefully install the engine/transaxle assembly and make sure all mounts are securely bolted to the engine/transaxle. Start all nuts and bolts that secure the mounts to the body but don't tighten them yet.
19. With all mounts installed and the engine safely in the vehicle, allow some slack in the lifting equipment. With the vehicle safely supported, shake the engine/transaxle as a unit to settle it in the mounts. Torque all mounting bolts, starting at the rear and working forward. Torque to 33 ft. lbs. (41 Nm) for 10mm bolts or 54 ft. lbs. (73 Nm) for 12mm bolts.
20. Install the starter and torque the bolts to 33 ft. lbs. (45 Nm).
21. Connect the halfshafts to the flanges and torque the bolts to 33 ft. lbs. (45 Nm).
23. Install the exhaust pipe and use new self locking nuts to secure the flange. Torque the nuts to 30 ft. lbs. (40 Nm). On vehicles equipped with spring clamps, the clamps can be used again.
24. Connect the shift linkage and the

clutch cable, if equipped. Make any necessary adjustments.

25. Install the fuel injector lines and torque to 18 ft. lbs. (25 Nm). Be careful not to over torque the line nuts. If a line is damaged or clogged, replace all lines as a set.

26. Connect the inlet and outlet lines to the injector pump. Note the special outlet fitting has the word "OUT" printed on the top. Use new gaskets.

27. Install the air conditioning compressor and/or power steering pump, if equipped. Install and adjust the drive belts.

27. Connect the wiring and vacuum hoses.

28. Install the radiator, fan and heater hoses. Use a new O-ring on the thermostat and torque the thermostat housing bolts to 7 ft. lbs. (10 Nm).

29. Fill and bleed the cooling system. Check the adjustment of the accelerator cable.

ENGINE ALIGNMENT

After reinstalling all the mounts and mounting bolts, loosen all the bolts that go into the rubber mounts themselves. With the vehicle safely supported, shake the engine/transaxle as a unit to settle it in the mounts. Retorque all mounting bolts, starting at the rear and working forward. This procedure helps to minimize vibration.

Cylinder Head

REMOVAL & INSTALLATION

NOTE: The cylinder head bolts on all Diesel vehicles are stretch bolts and must be replaced when removed.

1. Disconnect the battery ground cable.

2. Remove the thermostat and drain the cooling system.

3. Remove the fuel lines from the injectors and the pump as an assembly. Put the lines where they'll stay clean and protect the injector and pump fittings with caps.

4. Disconnect the radiator and heater hoses.

5. Disconnect all vacuum and electrical connections and carefully label for installation.

6. On turbocharged vehicles, unbolt the exhaust pipe and oil lines from the turbocharger and remove the turbocharger.

7. On non-turbocharged vehicles, remove the air cleaner and disconnect the exhaust pipe from the manifold.

NOTE: On some models, special tools are required for removing and installing the exhaust pipe-to-manifold spring clamps; VW3140/1 and /2 or equivalent. This is a set of different sized wedges for spreading the spring clamps in steps. The installed spring clamp has considerable tension and could cause damage or injury if not properly removed. Clamps with wedges installed are also under high tension and should be handled carefully.

8. Remove the cylinder head cover and camshaft drive belt cover.

9. Turn the engine to TDC of No. 1 cylinder, if possible, and remove the camshaft drive belt.

10. Remove the head bolts in the reverse order of installation sequence and lift the head out of the vehicle. The torque sequence is the same as for gasoline engines.

To install:

11. On these engines, the pistons actually project above the deck of the block. If the crank shaft and pistons are not to be removed, examine the old head gasket to see how many notches are on the edge near the oil return hole, between No. 2 and 3 cylinders. Replace the gasket with the same thickness.

Measure the piston pop-up to determine head gasket thickness

12. If the pistons were removed or if the old gasket in not available, the piston height (pop up) must be measured to select the proper head gasket. Use a dial indicator or caliper to obtain the measurement.

A. Pop-up on engines with solid lifters:
- 0.026–0.031 in. (0.67–0.80mm)—1 notch
- 0.032–0.035 in. (0.81–0.90mm)—2 notches
- 0.036–0.040 in. (0.91–1.02mm)—3 notches

B. Pop-up on engines with hydraulic lifters:
- 0.026–0.034 in. (0.66–0.86mm)—1 notch
- 0.034–0.035 in. (0.87–0.90mm)—2 notches
- 0.036–0.040 in. (0.91–1.02mm)—3 notches

13. Install the new cylinder head gasket with the word TOP or OBEN facing upward. Do not use any sealing compound.

14. Turn the crankshaft to TDC of No. 1 cylinder, then back about ¼ turn to bring all pistons about even.

15. Carefully lower the head on and install new head bolts into No. 8 and 10 first. These holes are smaller and will properly locate the gasket and cylinder head.

16. Install the remaining bolts and torque in the proper sequence in 3 steps: 29 ft. lbs. (40 Nm), 44 ft. lbs. (60 Nm), then a full ½ turn more. Two quarter turns are allowed.

17. Installation of the remaining parts is the reverse of removal, be sure to change the oil and filter. Install the camshaft drive belt and set injection pump timing.

18. Install the fuel injector lines and torque to 18 ft. lbs. (25 Nm). Be careful not to over torque the line nuts. If a line is damaged or clogged, replace all lines as a set.

19. After the engine has be run about 1000 miles, the cylinder head bolts must be re-torqued. Remove the cylinder head cover and turn each head bolt, in sequence, an additional ¼ turn in 1 movement. This can be done on a cold or warm engine.

Valve Lifters

REMOVAL & INSTALLATION

1. Remove the camshaft.

2. The valve lifters can be easily lifted out of the head by hand. Place hydraulic lifters camshaft side down on a clean surface. Keep all lifters in order so they can be installed in the same position.

To install:

3. Make sure the engine is not at TDC of any cylinder.

4. Set the lifters in place and carefully install the camshaft. Allow the lifters to bleed down for 30 minutes before turning the engine or the valves may hit the pistons.

5. Install the camshaft drive belt and adjust the belt tension and valve lash.

Valve Lash

ADJUSTMENT

Most vehicles have hydraulic valve lifters and require no adjustment. On these vehicles there will be a sticker under the hood indicating hydraulic lifters. The overhead camshaft acts directly on the valves through bucket-type camshaft followers which fit over the springs and valves. On solid lifters, adjustment is made with an adjusting disc (shim) which fits into the camshaft follower. Different thickness discs result in changes in valve clearance.

NOTE: The manufacturer recommends that 2 special tools VW 546 and the special pliers VW 208 or equivalent, be used to remove and install the adjustment discs. One is a prybar to compress the valve springs and the other a pair of special pliers to remove the disc. Care must be taken not to gouge the camshaftshaft lobes. The camshaft follower has 2 slots which permit the disc to be lifted out.

1. Valve clearance is checked with the engine moderately warm; coolant temperature should be about 95°F (35°C). Remove the air cleaner and any hoses, lines or cables as needed.
2. Remove the cylinder head cover. Valve clearance is checked in the firing order 1–3–4–2, with the piston of the cylinder being checked at TDC of the compression stroke. Both valves will be closed at this position and the camshaft lobes will be pointing straight up.
3. Turn the crankshaft pulley bolt with a socket wrench to position the camshaft for checking.

NOTE: Do not turn the engine by the camshaft sprocket mounting bolt, this will stretch the drive belt. When turning the crankshaft pulley bolt, turn it clockwise only.

4. With the No. 1 piston at TDC of the compression stroke, determine the clearance with a feeler gauge. Intake clearance should be 0.008–0.012 in. (0.2–0.3mm); exhaust clearance should be 0.016–0.020 in. (0.4–0.5mm).
5. Check the other cylinders in the firing order, turning the crankshaft to bring each particular piston to the top of the compression stroke. Record the individual clearances.
6. If measured clearance is within tolerance levels of 0.002 in., it is not necessary to replace the adjusting discs.
7. If adjustment is necessary, the discs will have to be removed and replaced with thicker or thinner ones which will yield the correct clearance. Discs are available in 0.002 in. increments from 0.12–0.17 in.

NOTE: The thickness of the adjusting discs is etched on 1 side. When installing, the marks must face the camshaft followers. Discs can be reused if they are not worn or damaged.

8. To remove the discs, turn the camshaft followers so the grooves are accessible when the prybar is depressed.
9. Press the camshaft follower down with the prybar and remove the adjusting discs with the proper tool.

NOTE: Do not press the camshaft follower down with that piston at TDC. The valves will contact the piston and may damage the engine. Turn the engine ¼ turn past TDC and press the tappet down.

10. Replace the adjustment discs as necessary to bring the clearance within the 0.002 in. tolerance level. If the measured clearance is larger than the given tolerance, remove the existing disc and insert a thicker one to bring the clearance up to specification. If it is smaller, insert a thinner one.
11. Recheck all valve clearances after adjustment.
12. Install the valve cover. Connect the accelerator linkage and hoses.

Exhaust Manifold

REMOVAL & INSTALLATION

NOTE: On some models, special tools are required for removing and installing the exhaust pipe-to-manifold spring clamps; VW3140/1 and /2 or equivalent. This is a set of different sized wedges for spreading the spring clamps in steps. The installed spring clamp has considerable tension and could cause damage or injury if not properly removed. Clamps with wedges installed are

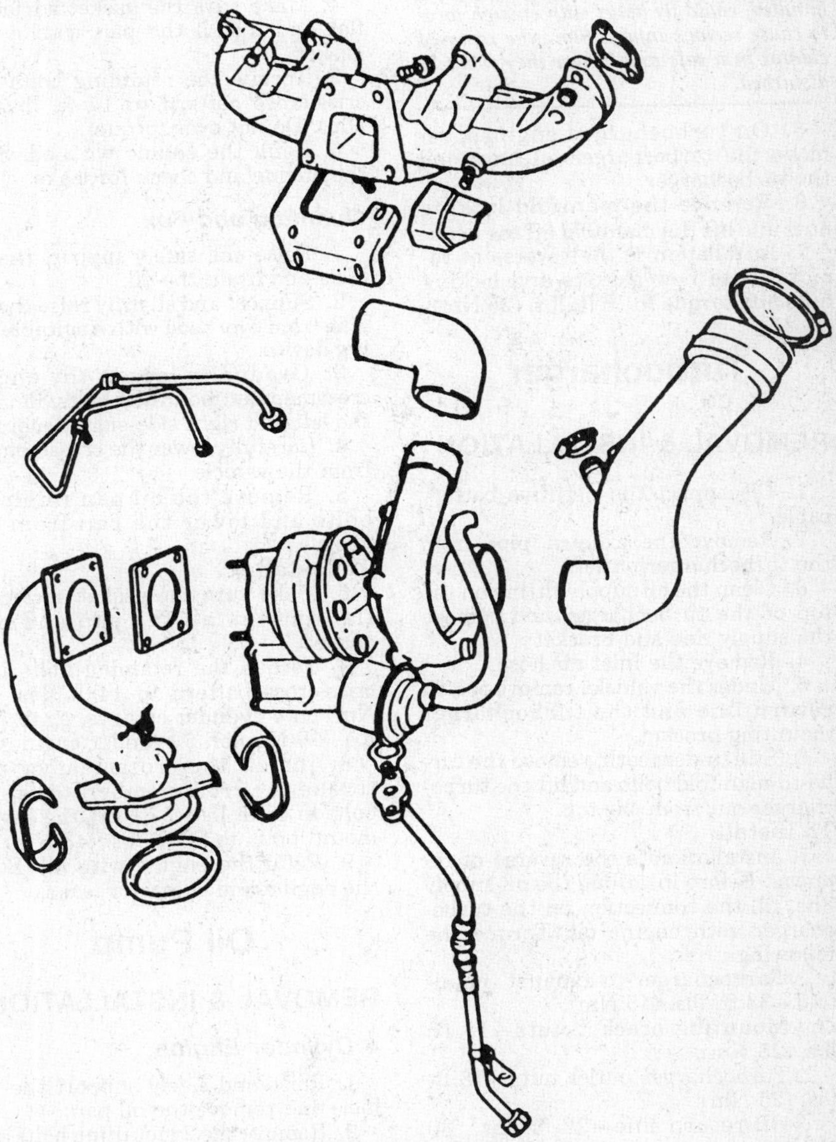

Diesel engine turbocharger

also under high tension and should be handled carefully.

1. Disconnect the negative battery cable and remove any heat shields that may be in the way.
2. On turbocharged engines, unbolt the exhaust pipe from the turbocharger outlet.
3. On non-turbocharged engines, expand the spring clamp by pushing the exhaust pipe to one side and insert the starter wedge into the clamp all the way up to the shoulder.
4. Push the pipe to the other side and install another wedge in the opposite clamp. Continue to work the pipe side to side while pushing the wedges into the clamps until the clamps are spread far enough to lift off easily.

CAUTION

The removed spring clamps with wedges in them are very highly loaded and, if miss handled, could fly apart with enough force to cause serious injury. Store the removed clamps in a safe area where they won't be disturbed.

5. On turbocharged engines, remove the turbocharger oil lines and the turbocharger.
6. Remove the manifold locking nuts and lift the manifold off the head.
7. Installation is the reverse of removal. Use new gaskets and locking nuts and torque to 18 ft. lbs. (25 Nm).

Turbocharger

REMOVAL & INSTALLATION

1. Disconnect the negative battery cable.
2. Remove the exhaust pipe from the turbocharger outlet.
3. Clean the oil supply fitting on the top of the turbocharger and remove the supply line and bracket.
4. Remove the inlet air hose.
5. Under the vehicle, remove the oil return line and the turbocharger mounting bracket.
6. Still underneath, remove the turbo-to-manifold bolts and lift the turbocharger out from the top.

To install:

7. Installation is the reverse of removal. Before installing the oil supply line, fill the connection on the turbocharger with engine oil. Torque the following:

Turbocharger-to-exhaust manifold—33 ft. lbs. (45 Nm).

Mounting bracket nuts—18 ft. lbs. (25 Nm).

Turbocharger outlet nuts—18 ft. lbs. (25 Nm).

Oil return line—22 ft. lbs. (30 Nm).

ENGINE LUBRICATION

Oil Pan

REMOVAL & INSTALLATION

Except Quantum and Fox

The oil pan can be removed with the engine in the vehicle, however on the Quantum and Fox, the engine must be raised off it's mounts.

1. Raise and safely support the vehicle and drain the oil.
2. Loosen and remove the bolts retaining the oil pan.
3. Lower the pan from the engine.

To install:

4. Make sure the gasket surface is flat and install the pan with a new gasket.
5. Torque the retaining bolts in a criss-cross pattern to 14 ft. lbs. (20 Nm). Do not over torque.
6. Refill the engine with oil. Start the engine and check for leaks.

Quantum and Fox

1. Raise and safely support the vehicle and drain the oil.
2. Support and slightly raise the engine from overhead with a suitable lifting device.
3. Gradually loosen the engine crossmember mounting bolts. Remove the left and right side engine mounts.
4. Carefully lower the crossmember from the vehicle.
5. Remove the oil pan retaining bolts and lower the pan from the vehicle.

To install:

6. Make sure the gasket surface is flat and install the pan and new gasket.
7. Torque the retaining bolts in a criss-cross pattern to 14 ft. lbs. (20 Nm) on 4 cylinder engines, or to 7 ft. lbs. (10 Nm) on 5 cylinder engines.
8. Install the crossmember and torque the crossmember-to-frame bolts to 42 ft. lbs. (57 Nm), the engine mount bolts to 32 ft. lbs. (43 Nm).
9. Refill the engine with oil. Start the engine and check for leaks.

Oil Pump

REMOVAL & INSTALLATION

4 Cylinder Engine

1. Raise and safely support the vehicle and remove the oil pan.
2. Remove the 2 mounting bolts and lower the pump from the engine.
3. Remove the bottom cover and disassemble the pump. The pressure relief valve is in the bottom cover.
4. Clean and inspect all parts for wear and replace as needed.
5. After reassembling the pump, prime it with oil and install in the reverse order of removal.
6. Observe the following torques:

Oil pump bottom cover bolts—7 ft. lbs. (10 Nm).

Oil pump suction foot bolts—7 ft. lbs. (10 Nm).

Oil pump retainting bolts—15 ft. lbs. (20 Nm).

5 Cylinder Engine

On these vehicles, the oil pump is on the front of the engine, behind the front crankshaft sprocket. However it is still necessary to remove the oil pan.

1. Remove the accessory drive belts and the steering pump. Lay the pump aside without disconnecting the lines.
2. Remove the timing belt cover and the upper radiator cover and turn the engine to TDC of No. 1 cylinder. Check to see that the mark on the camshaft drive sprocket aligns with the top edge of the head.
3. Loosen the water pump bolts and pry the pump to loosen the camshaft drive belt.
4. Remove the drive belt sprocket from the crankshaft; do not turn the engine with the camshaft drive belt removed. The pistons may contact the valves and cause engine damage.
5. Remove the oil dip stick and drain the crankcase.
6. Support and slightly raise the engine with a suitable lifting device.
7. Gradually loosen the engine crossmember mounting bolts. Remove the left and right side engine mounts.
8. Carefully lower the crossmember from the vehicle.
9. Remove the oil pan and the oil pickup tube.
10. Remove the oil pump from the front of the engine.
11. When disassembling the pump, note the mark on the gears faces the rear end cover of the pump.

To install:

12. Clean all parts and check the pump for wear or damage, replace parts as necessary. The gears can be replaced as a set and the rear cover should be replaced if it's scored.
13. Reassemble the pump and pack it with petroleum jelly.
14. Installation is the reverse of removal. Torque the following:

Oil pump cover and pickup tube—7 ft. lbs. (10 Nm).

Oil pump mounting bolts—15 ft. lbs. (20 Nm).

Oil pan—7 ft. lbs. (10 Nm).

Crossmember bolts – 42 ft. lbs. (57 Nm).

Motor mount bolts – 32 ft. lbs. (43 Nm).

Front crankshaft sprocket – 258 ft. lbs. (350 Nm).

OIL PUMP INSPECTION

4 Cylinder Engine

1. With the oil pump on the bench and the bottom cover removed, insert a feeler gauge between the gear teeth. A new pump will have 0.002 in. (0.05mm) clearance, the wear limit is 0.008 in. (0.20mm).
2. Place a straight edge across the gears and the pump housing. Use a feeler gauge to measure the distance between the gears and the straight edge; the axial play of the gears. It should be no more than 0.006 in. (0.15mm).
3. The pressure relief valve is in the lower housing.

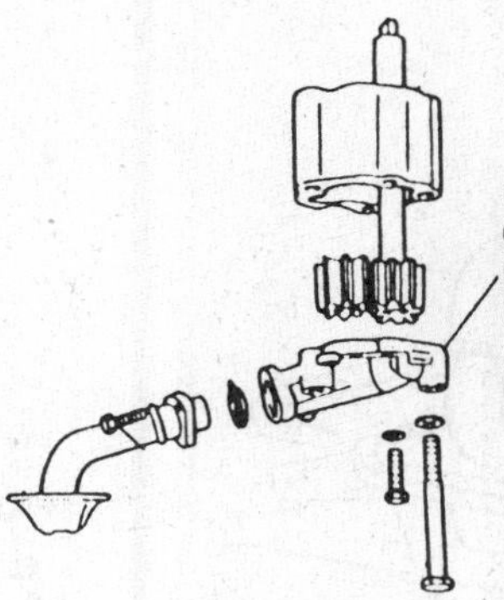

4 Cylinder oil pump

5 Cylinder Engine

The oil pump cannot be checked for clearance or axial play other than to inspect for score marks on the cover or outer housing. If any scoring is found, replace the pump an an assembly. The pressure relief valve is in the outer housing.

Rear Main Bearing Oil Seal

REMOVAL & INSTALLATION

The rear main oil seal is located in a housing on the rear of the cylinder block. To replace the seal on all vehicles it is necessary to remove the transaxle and flywheel.

1. Remove the transaxle and flywheel.
2. Using a small prybar tool VW-2086 or equivalent on 5 cylinder engine, VW-10-221 or equivalent on 4 cylinder engine, pry the old seal out of the support ring.

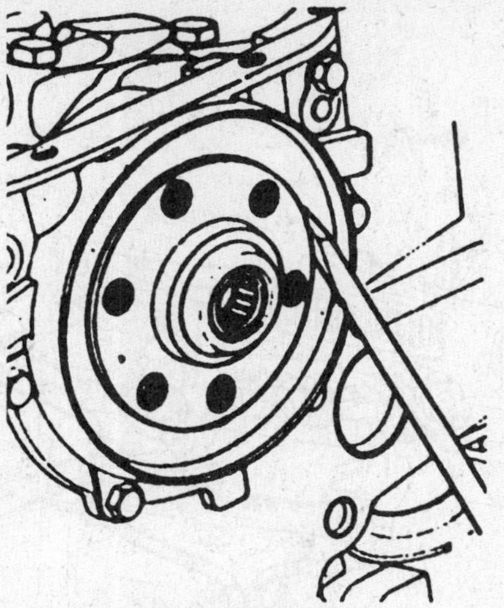

Rear main oil seal removal

3. To install, lightly oil the new seal and press it into place using tool VW-2003/2A or equivalent, to start the seal and tool VW-2003/1 or equivalent, to seat the seal. Be careful not to damage the seal or score the crankshaft.
4. Install the flywheel and transaxle.

ENGINE COOLING

Radiator

REMOVAL & INSTALLATION

NOTE: When replacing coolant/antifreeze, only a phosphate-free product must be used to help prevent damage to the water jacket sealing surfaces of the cylinder head. Other types of coolant may cause corrosion of the cooling system thus leading to engine overheating and damage.

1. To drain the cooling system, remove the thermostat housing from under the water pump housing.
2. Disconnect the wiring on the radiator for the thermostatic switch and electric fan(s).
3. Remove the upper and lower hoses and the overflow hose.
4. There will be 1 or 2 bolted clips holding the top of the radiator. Remove these clips and carefully lift the radiator and fan assembly up and out of the vehicle. Be careful not to lose the rubber washers on the bottom locating studs.
5. The fan and shroud can be unbolted and removed as an assembly.
6. Installation is the reverse of removal. Torque the clip bolts to 7 ft. lbs. (10 Nm).

Electric Cooling Fan

TESTING

On all vehicles, the fan is operated by a thermo switch screwed into the radiator side tank, which turns on to complete a ground circuit. The fan will operate with or without the ignition switch **ON**.

1. Disconnect the plug at the thermo switch.
2. Jumper the connections on the plug. The fan should run.
3. If the fan does not run, check for voltage at one side of the plug. If there is power there, the problem is either in the wiring between the plug and the fan or the fan itself.
4. If the fan runs during the test but not with a coolant temperature of 200°F (98°C), replace the switch.
5. The fan and shroud can be removed as an assembly by unbolting the top cover and the shroud and carefully lifting the shroud up and out.

Heater Core

REMOVAL & INSTALLATION

The heater core is contained in the fresh air/heater box located in the center of the dashboard. On air conditioned vehicles, the evaporator is also located in the heater box. On Quantum, the evaporator is located under the hood separate from the heater box.

Fox and Quantum

1. The entire dashboard and heater assembly must be removed. Disconnect the negative battery cable.
2. Drain the engine coolant or clamp the heater hoses.
3. Disconnect the heater hoses at the firewall and plug the core fittings.
4. Inside the vehicle, remove the knee bar, if equipped, the shifter handle and boot, the center console and the temperature controls from the dash.
5. Remove the left and right air distribution ducts.
6. In the engine compartment, remove the cowl cover and remove the air distribution housing cover.
7. Inside the vehicle, remove the lower housing retaining clips and remove the housing.

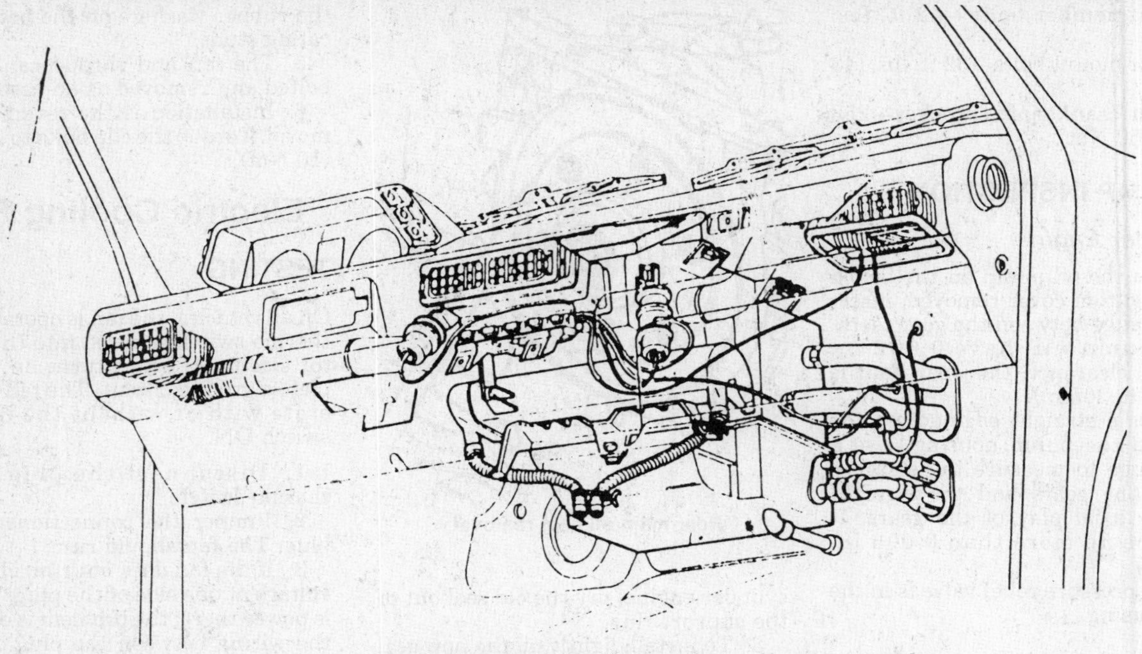

Fox heater and air conditioning assembly

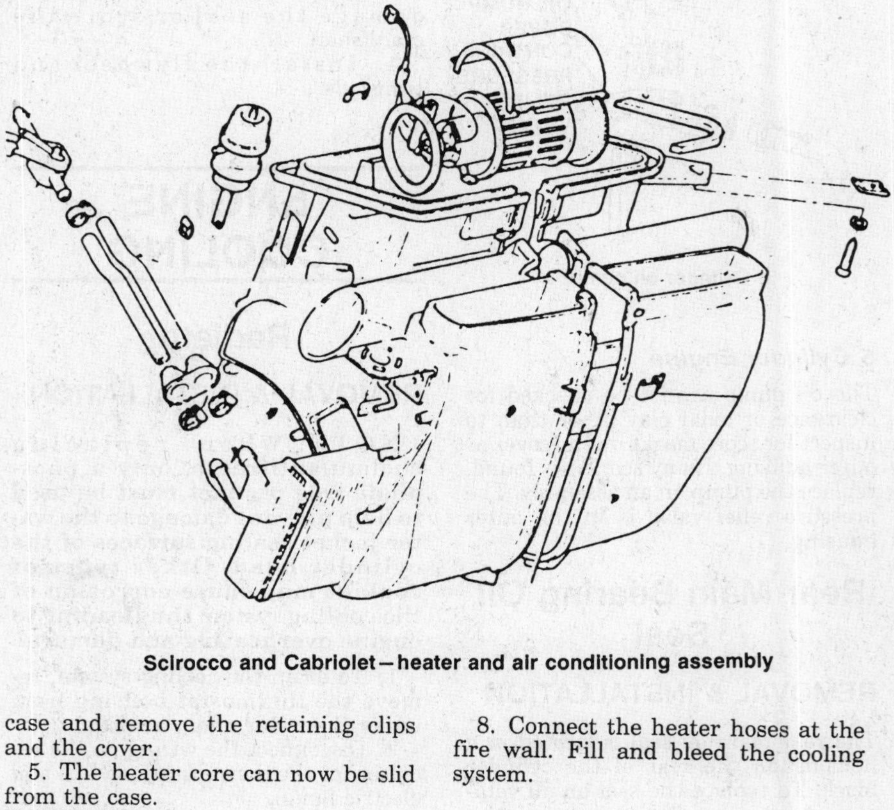

Scirocco and Cabriolet—heater and air conditioning assembly

NOTE: If equipped with air conditioning the heater box also contains the air conditioning system evaporator mounted in the lower housing cover. When removing the lower cover, lay the cover and evaporator aside without removing the refrigerant lines.

8. Remove the bolts retaining the heater case and remove the case.
9. Remove the clips holding the case together and split the case, the heater core can now be removed.

To install:

10. To install, insert the heater core into the case and reassemble the case.
11. Install the case into the vehicle. Attach the lower heater case cover to the heater case. Install the air distribution ducts and the control cables.
12. Install the center console, etc. and reconnect the heater inlet hoses.
13. Install the air distribution housing cover and cowl and refill the cooling system.

Scirocco and Cabriolet

1. Disconnect the negative battery cable.
2. Drain the cooling system or clamp the heater hoses.
3. Disconnect the heater hoses at the firewall and plug the core fittings to prevent coolant leakage inside the vehicle.
4. Inside the vehicle, remove the center console side panels and all ducting from the heater box. Locate the heater core cover on the side of the case and remove the retaining clips and the cover.
5. The heater core can now be slid from the case.

To install:

6. Insert the heater core into the case. Install the heater core cover, making sure the gasket on the cover is properly fitted.
7. Connect the ducting and assemble the console.
8. Connect the heater hoses at the fire wall. Fill and bleed the cooling system.

Passat and Corrado

1. The dashboard and air distribution assembly must be removed. Disconnect the negative battery cable.
2. Drain the cooling system and dis-

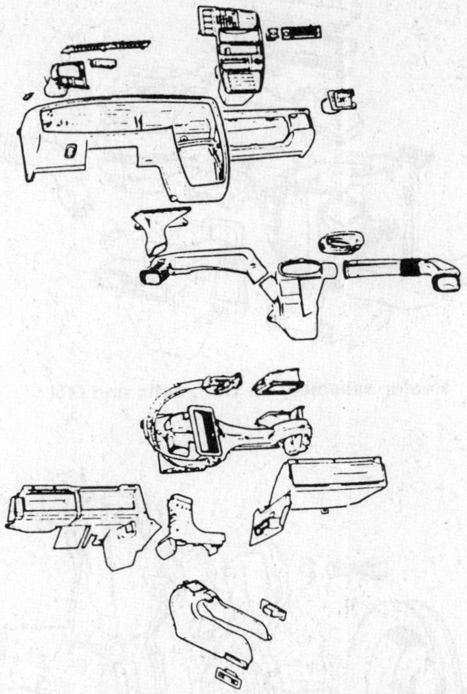

Dashboard must be removed to remove heater assembly; Corrado shown

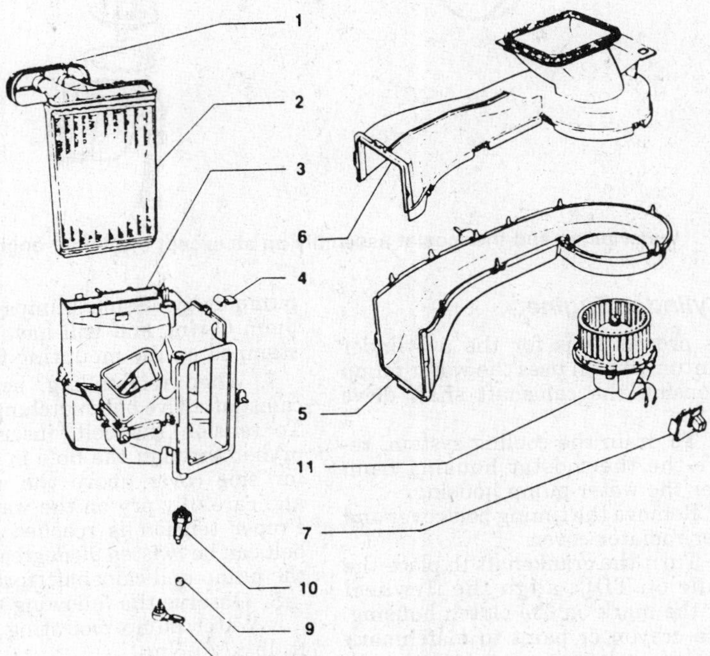

1. Seal
2. Heater core
3. Air distribution housing
4. Retaining clip
5. Lower air duct
6. Upper air duct
7. Blower motor series resistor
8. Temperature flap lever
9. Lever bushing
10. Mode lever flap
11. Blower motor

Remove but do not disassemble the air distribution housing

connect the heater hoses from the firewall.

3. Remove the steering wheel, instrument cluster and trim panel around the cluster.
4. Remove the knee bar from below the dashboard.
5. Remove the cassette storage drawers, if equipped, from the center console. Pull the knobs off the heater controls and remove the radio.
6. Carefully pry out the heater control trim plate and remove the control assembly. Remove the center console trim plate.
7. Carefully pry out the air vents at the ends of the dashboard and remove the glove compartment.
8. Pry off the caps from the screws at each end of the dashboard and remove the bolts.
9. At the upper rear portion of the dashboard, remove the 2 nuts. These can only be seen with the aid of a mirror.
10. Remove the nut at the back of the center console and pull the dashboard slightly out to disconnect the remaining wires. Remove the dashboard.
11. Remove the air distribution assembly and remove the heater core.

To install:

12. If the retaining lugs are broken off, the heater core can be secured in place with screws. Install the heater core and the air distribution assembly. Make sure the seal is in good condition, replace if necessary.
13. Install the dashboard and connect the wiring.
14. Install the instrument cluster and steering wheel.
15. Fill and bleed the cooling system.

Golf, Jetta and GTI

1. The dashboard and air distribution assembly must be removed. Disconnect the negative battery cable.
2. Drain the cooling system and disconnect the heater hoses from the firewall.
3. Properly discharge the air conditioning system.
4. Remove the gear shift knob and boot and remove the center console.
5. Remove the steering wheel.
6. Remove the knee bar from below the dashboard.
7. Remove the steering column support bracket and lower the column.
8. Pull the knobs off the heater controls and remove the control assembly and the radio.
9. Remove the headlight switch and switch blanks to gain access to the screws. Remove the instrument cluster and trim panel around the cluster.
10. Remove the glove compartment.
11. At the firewall, remove the plastic tray and remove the 2 nuts holding the top of the dashboard.

12. Remove the main fuse panel and disconnect the plugs at the back. Disconnect the ground wires.

13. Disconnect any remaining wiring from the dashboard and remove the 4 last screws; 1 at each end and 1 at each end of the instrument cluster area. Remove the dashboard.

14. Disconnect the ducts and remove the heater housing. Remove the screws and slide the heater core out of the housing.

To install:

15. Install the heater core and make sure the housing seals and gaskets are in good condition. Replace as necessary.

16. With the heater housing properly installed, install the dashboard and connect the wiring. Install the steering column bracket bolts.

17. Install the fuse panel and connect the wiring.

18. Install the glove compartment, shift boot and knob and steering wheel.

19. When the interior has been assembled, evacuate and recharge the air conditioner and fill and bleed the cooling system.

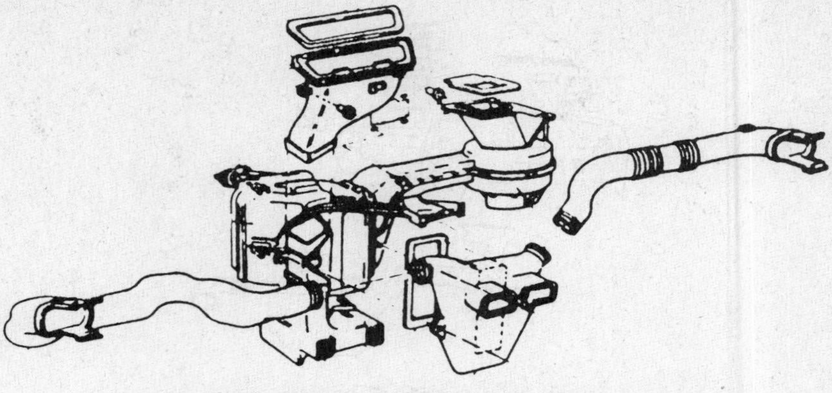

Heater assembly on Golf, Jetta and GTI

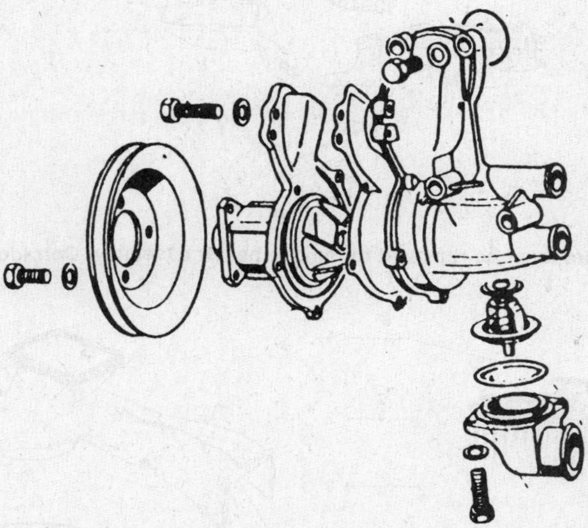

Water pump and thermostat assembly on all except 5 cylinder engine

Water Pump

REMOVAL & INSTALLATION

Except Corrado, 5 Cylinder and Diesel Engines

1. To drain the cooling system, remove the thermostat housing from under the water pump housing.

2. Raise and safely support the vehicle. Loosen but don't remove the bolts holding the pulley to the water pump.

3. Remove the timing belt cover.

4. Loosen the alternator and/or steering pump as required to remove the water pump drive belt.

5. Remove the water pump pulley. On some vehicles, the crankshaft pulley must also be removed.

6. All the bolts are now accessible and the water pump can be removed from it's housing.

To install:

7. Be sure to clean the housing before installing the new gasket. Install the pump into the housing and torque the pump-to-housing bolts to 7 ft. lbs. (10 Nm).

8. Install the water pump drive pulley and torque the bolts to 15 ft. lbs. (20 Nm). If the crankshaft drive pulley was removed, install it and torque the bolts to 15 ft. lbs. (20 Nm).

9. Adjust drive belt tension and install the thermostat and housing. Torque the bolts to 7 ft. lbs. (10 Nm).

5 Cylinder Engine

This procedure is for the 5 cylinder Quantum, which uses the water pump to tension the camshaft shaft drive belt.

1. To drain the cooling system, remove the thermostat housing from under the water pump housing.

2. Remove the timing belt cover and upper radiator cover.

3. Turn the crankshaft to place the engine on TDC; align the flywheel with the mark on the clutch housing. Use a crayon or paint to matchmark the timing belt to the camshaft sprocket and crankshaft sprocket.

4. Loosen the water pump to relieve the tension on the timing belt and remove the belt.

5. All the bolts are now accessible and the water pump can be removed from it's housing.

To install:

6. When installing the new pump, check the O-ring groove size on the pump body. Some pumps require a 5mm O-ring and will have the No. 5 stamped on the mounting flange.

7. When reinstalling, be sure the camshaft drive belt matchmarks align. To tension the belt, insert a small prybar through the hole in the radiator side cover above the alternator, and carefully pry on the water pump. Proper tension is reached when the belt can be twisted 90 degrees between the pump and camshaft sprockets.

8. Observe the following torques:

Water pump mounting bolts—15 ft. lbs. (20 Nm).

Thermostat housing bolts—7 ft. lbs. (10 Nm).

Water pump pulley bolts—18 ft. lbs. (25 Nm).

Corrado

WITHOUT AIR CONDITIONING

The water pump is driven by the same belt that drives the alternator and supercharger. On vehicles with air condi-

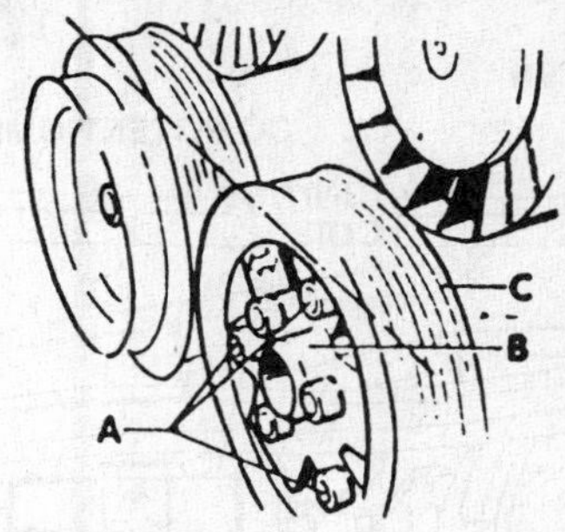

LOOSEN BOLTS A
TURN INNER PART B TO ALIGN PULLEYS C

Water pump pulley alignment

tioning, the water pump and steering pump use the same V-belt.

1. To drain the cooling system, remove the thermostat housing from under the water pump housing.
2. Loosen the water pump pulley bolts but don't remove the pulley yet.
3. Install the belt tensioner holding tool VW3191 or equivalent, to compress the tensioner and loosen the belt.
4. Loosen the power steering pump and remove it's drive belt.
5. With the water pump pulley removed, the pump bolts are now accessible. Remove the pump.

To install:

6. Clean the housing and use a new gasket when installing the pump. Torque the bolts to 7 ft. lbs. (10 Nm).
7. The water pump and crank shaft pulleys must be aligned. Loosen the outer section bolts and turn the outer section relative to the inner section until the 2 pulleys are aligned; water pump pulley moves in and out. Torque the bolts to 18 ft. lbs. (25 Nm).
8. Remove the belt tensioner tool and complete the reassembly. Torque the thermostat housing bolts to 7 ft. lbs. (10 Nm) and fill the cooling system.

WITH AIR CONDITIONING

1. To drain the cooling system, remove the thermostat housing from under the water pump housing.
2. Raise and safely support the vehicle.
3. Working under the vehicle, loosen but don't remove the bolts holding the pulley to the water pump.
4. Loosen the power steering pump and remove the drive belt.
5. Remove the water pump pulley and remove the pump.

To install:

6. Installation is the reverse of removal. Be sure to clean the pump housing before installing the pump with a new gasket. Torque the following:

Water pump-to-housing—7 ft. lbs. (10 Nm).
Water pump drive pulley—15 ft. lbs. (20 Nm).
Thermostat housing—7 ft. lbs. (10 Nm).
Steering pump bolts—18 ft. lbs. (25 Nm).

Diesel Engine

On some engines, the belt tension is adjusted with shims between the outer and inner halves of the pulley. On others, the alternator swivels to adjust belt tension.

1. To drain the cooling system, remove the thermostat housing from under the water pump housing.
2. Raise and safely support the vehicle.
3. Working under the vehicle, loosen but don't remove the bolts holding the pulley to the water pump.
4. On vehicles with a movable alternator, loosen the alternator and remove the drive belt.
5. Remove the water pump pulley and remove the pump.

To install:

6. Installation is the reverse of removal. Be sure to clean the pump housing before installing the new gasket. Torque the following:

Water pump-to-housing—7 ft. lbs. (10 Nm).
Water pump drive pulley—15 ft. lbs. (20 Nm).
Thermostat housing—7 ft. lbs. (10 Nm).
Alternator mounting bolts—18 ft. lbs. (25 Nm).

Thermostat

REMOVAL & INSTALLATION

The thermostat is the lowest point in the cooling system and is on the bottom of the water pump housing. Removing the thermostat is the only way to completely drain the coolant.

1. With a catch pan under the vehicle, loosen the bolts on the thermostat housing.
2. Remove the cap from the overflow bottle and allow the coolant to drain completely.
3. When the coolant is drained, remove the housing and clean both mating surfaces.
4. When installing, the thermostat spring goes up into the water pump housing and the new O-ring goes onto the thermostat. Torque the bolts to 7 ft. lbs. (10 Nm).
5. Refill the cooling system and check for leaks.

ENGINE ELECTRICAL

Distributor

REMOVAL & INSTALLATION

1. Disconnect the coil high tension wire and the connector plug at the distributor. Disconnect vacuum lines, if equipped.
2. Unsnap the cap retainer clips, and remove the cap and static shield as a unit.
3. At the front crankshaft pulley bolt, turn the engine to Top Dead Center (TDC) on No. 1 cylinder. Make a chalk or paint mark where the rotor points to the rim of the distributor; some vehicles already have a notch there. Also matchmark the distributor to the engine block or head.
4. Remove the bolt and distributor clamp and lift the distributor straight out.

NOTE: On some vehicles, the distributor engages it's drive with an offset slot and is easy to reinstall in the reverse order of removal, even if the drive has been disturbed.

To install:

5. If the engine has been disturbed with the distributor removed, turn the engine to TDC on cylinder No. 1 and align the timing marks.
6. With the rotor on the distributor, align the rotor with the matchmark for No. 1 spark plug wire and insert the distributor into it's mounting hole. Turn the rotor slightly back and forth while gently pushing in on the distributor to make sure the drive engages.
7. Except on 16V engines, if the drive does not engage, remove the distributor and, with a suitable tool, turn the oil pump drive so it is parallel with the crankshaft and try again.
8. With the distributor installed, align the matchmarks and reinstall the hold-down clamp and bolt, connector plug, cap and static shield, and high tension wire.
9. Check and adjust the ignition timing.

Ignition Timing

ADJUSTMENT

NOTE: The manufacturer specifies timing, idle speed and CO value all be adjusted together.

1. Run the engine to normal operating temperature, stop engine and con-

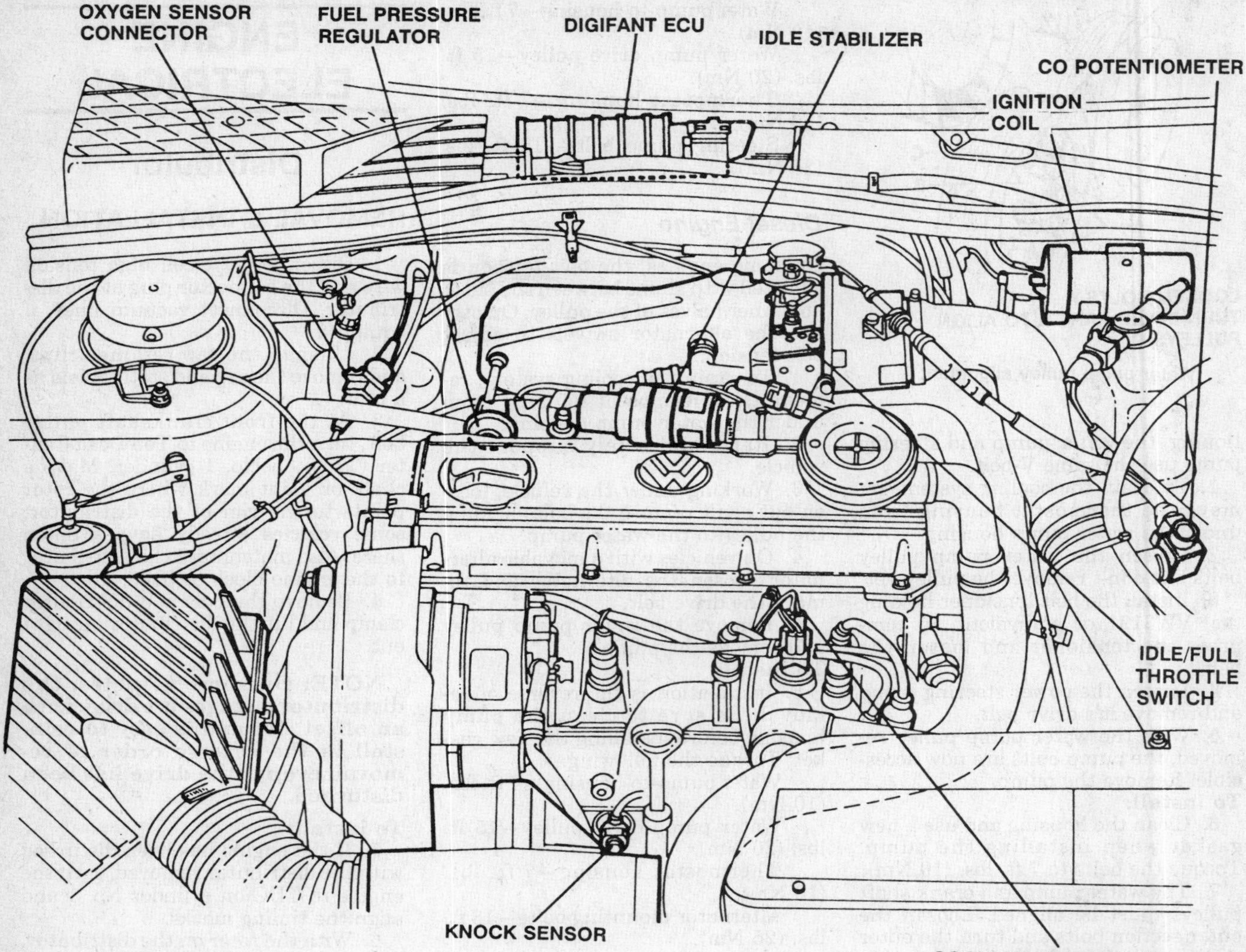

Digifant Engine Management System–Corrado shown

nect a tachometer and timing light according to manufacturer's instructions.

2. If equipped, disconnect both plugs from the idle stabilizer and plug them together. On engines with Digifant engine management systems, with the ignition **ON**, engine not running, verify that the idle stabilizer valve hums or buzzes. Do not disconnect any vacuum lines from the distributor.

3. Start the engine. On vehicles with Digifant engine management systems, disconnect the blue coolant temperature sensor plug.

4. Turn off all electrical equipment and set the idle speed.

5. Remove the timing mark cover from the top of the bell housing at the flywheel end of the engine and with the engine running, shine the timing light at the marks on the flywheel. If adjustment is required, loosen the distributor clamp bolt and rotate the distributor as needed to align the correct timing marks.

6. Stop the engine and reconnect plugs.

Alternator

PRECAUTIONS

- Before doing any work on any electrical system, always stop the engine and disconnect the battery cables. Also disconnect the battery and ABS control unit, if equipped, if any electric welding is to be done on the vehicle.
- Electronic parts and systems that can be easily and permanently damaged through careless use of electric welding, charging, soldering or test equipment. Carefully follow manufacturer's instructions when using such equipment.
- If equipped with electronically theft-protected radios, obtain the security code before disconnecting the battery.

BELT TENSION ADJUSTMENT

Except Corrado

1. Loosen both upper alternator bracket bolts.

2. Loosen the lower alternator pivot bolt. This is a long bolt with a 6mm socket head which should be accessible with the proper tool without removing the timing belt shield.

3. Do not use a prybar to tighten the belt. It is easy to gain enough tension pulling the alternator by hand against

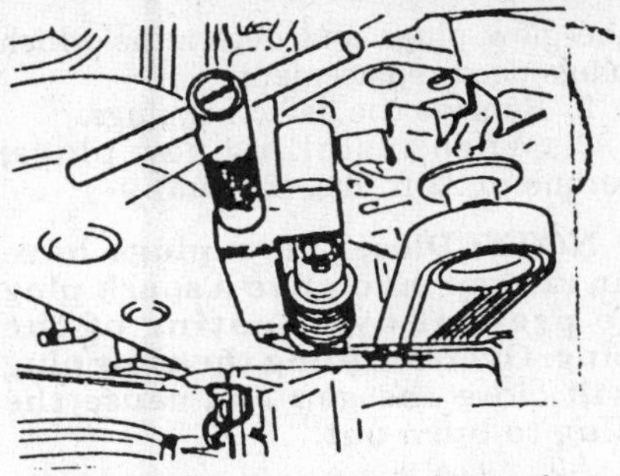

Belt tensioner compressing tool installed

the belt. Some vehicles have a toothed rack for setting belt tension.

4. Proper belt tension is attained when moderate finger pressure deflects the belt midway between the pulleys about 0.200 in. (5mm).

5. Securely tighten the mounting bolts.

Corrado

The vehicle is equipped with a serpentine ribbed belt which drives the alternator, supercharger and air conditioner. On vehicles without air conditioning, this belt drives the water pump instead. Belt tension is maintained with a spring loaded belt tension damper and idler pulley. Before installing the damper onto it's mounting bolts, it must be compressed 5 times to evacuate air, using VW tool 3191 or equivalent.

REMOVAL & INSTALLATION

Except Corrado

The alternator and voltage regulator are all in the same housing but the regulator can be removed separately. The brushes are on the regulator and should project no less than 0.200 in. (5mm) from the regulator. If the brushes are smaller, replace the regulator.

1. Disconnect the battery cables.
2. Remove the multi-connector plug and/or wires from the alternator and tag them for correct reinstallation.
3. Remove both upper alternator mounting bolts and bracket.
4. Remove the lower alternator pivot bolt. This is a long bolt with a 6mm socket head which should be accessible with the proper tool without removing the timing belt shield. Remove the alternator.
5. Installation is the reverse of removal. When reinstalling the belt, with moderate finger pressure the belt should deflect about 0.200 in. (5mm).

Corrado

The alternator and voltage regulator are all in the same housing but the reg-

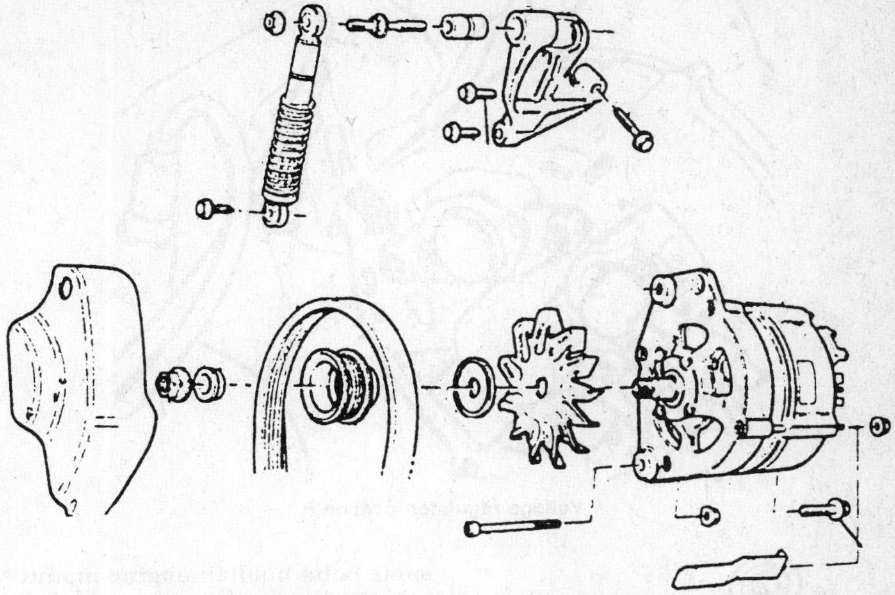

Corrado belt tensioner assembly

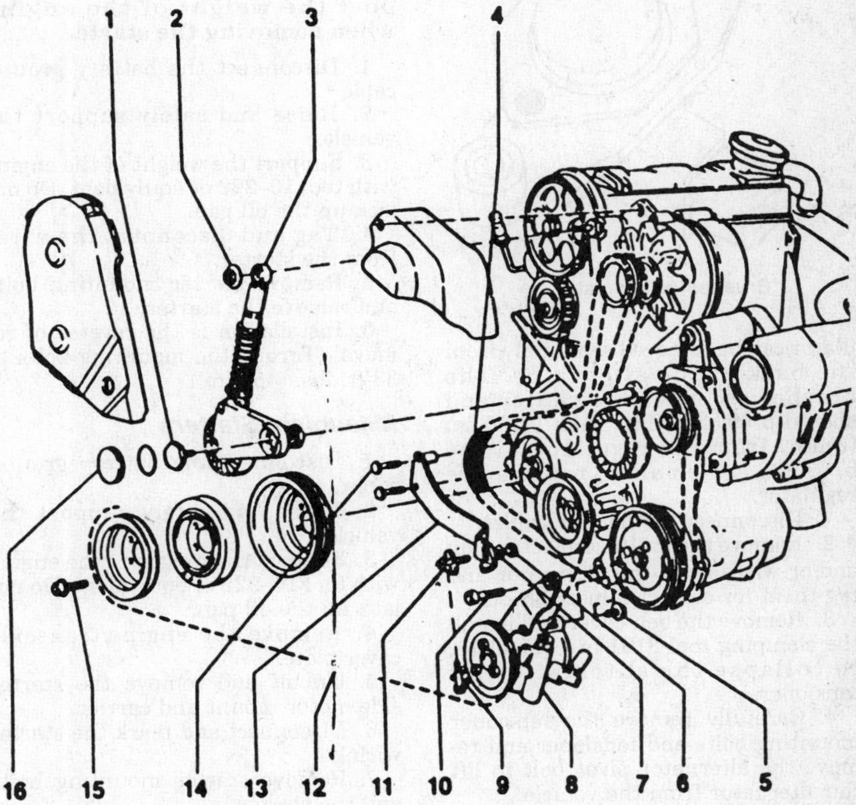

1. Shoulder bolt
2. Nut with washer
3. Belt tensioning damper
4. Ribbed belt
5. Shoulder bolt with point

6
7
8. Power steering pump
9. Adjusting bolt
11. Eye bolt
12. V–belt

Corrado belt lay-out–with air conditioning

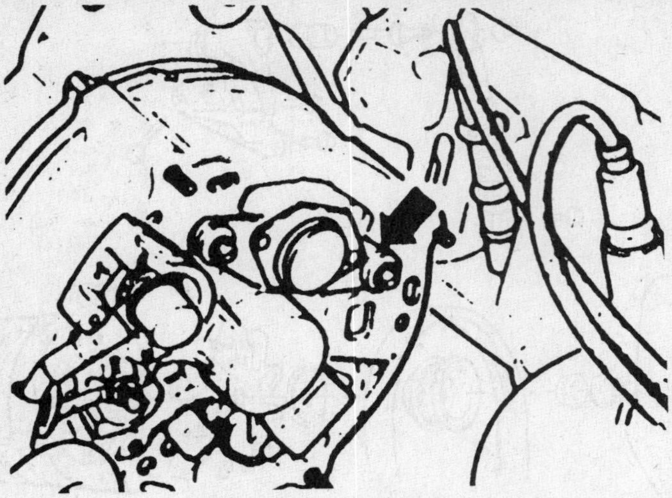

Voltage regulator location

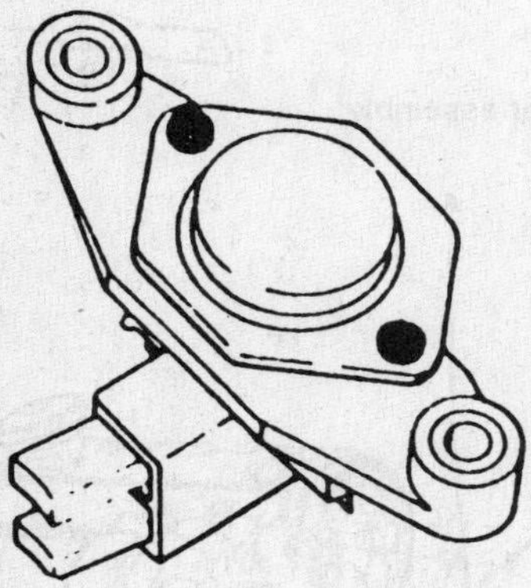

Brushes on regulator

ulator can be removed separately from the back of the alternator. The brushes are on the regulator and should project no less than 0.200 in. (5mm) from the regulator. If the brushes are smaller, replace the regulator.

1. Disconnect the battery cables.
2. Remove the multi-connector plug and/or wires from the alternator and tag them for correct reinstallation.
3. Remove the belt cover and install the clamping tool 3191 or equivalent, to collapse the automatic belt tensioner.
4. Carefully remove the tensioner mounting bolts and tensioner and remove the alternator pivot bolt to lift the alternator from the vehicle.
5. Installation is the reverse of removal.

Starter

REMOVAL & INSTALLATION

Bosch Starters

NOTE: On some vehicles, the same bolts hold an engine mount and the starter. On some vehicles, the starter bolts also hold the engine and transaxle together. Support the weight of the engine when removing the starter.

1. Disconnect the battery ground cable.
2. Raise and safely support the vehicle.
3. Support the weight of the engine with tool 10–222 or equivalent. Do not jack up the oil pan.
4. Tag and disconnect the wires from the starter.
5. Remove starter mounting bolts and remove the starter.
6. Installation is the reverse of removal. Torque the mounting bolts to 33 ft. lbs. (45 Nm).

Mitsubishi Starters

1. Disconnect the battery ground cable.
2. Raise and safely support the vehicle.
3. Support the weight of the engine with tool 10–222 or equivalent. Do not jack up the oil pan.
4. Remove the engine/transaxle cover plate.
5. Unbolt and remove the starter side motor mount and carrier.
6. Disconnect and mark the starter wiring.
7. Remove starter mounting bolts and the starter.
8. Installation is the reverse of removal. Torque the mounting bolts to 33 ft. lbs. (45 Nm).

Diesel Glow Plugs

REMOVAL & INSTALLATION

1. Remove the busbar connecting the glow plugs and determine which plugs need replacement.
2. Remove the defective plugs.
3. When installing new plugs, torque to 22 ft. lbs. (30 Nm).

NOTE: Diesel glow plugs have an air gap much like a spark plug to prevent overheating of the plug. Over-torquing the glow plug will close the gap and cause the plug to burn out.

TESTING

1. Disconnect the engine temperature sensor.
2. Connect a test light between No. 4 cylinder glow plug and ground. The glow plugs are connected by a flat, coated busbar, located near the bottom of the cylinder head.
3. Turn the ignition key **ON**; the test light should light, then go out after 10–30 seconds.
4. If there is no voltage, possible problems include a blown fuse, lack of power to or from the glow plug relay (check wiring) or the relay itself.
5. To test each plug individually, disconnect the wire and remove the busbar from the glow plugs.
6. Connect an ohmmeter to each glow plug connection or use a test light. Each plug must have continuity to ground. The engine will probably start with one defective glow plug, but it will make a lot of smoke.

EMISSION CONTROLS

Please refer to "Emission Controls" in the Unit Repair section for system maintenance procedures. Due to the complex nature of modern electronic engine control systems, comprehensive diagnosis and testing procedures fall outside the confines of this repair manual. For complete information on diagnosis, testing and repair procedures concerning all modern engine and emission control systems, please refer to "Chilton's Guide to Fuel Injection and Electronic Engine Controls".

Emission Warning Lamps

RESETTING

On models so equipped, the OXS

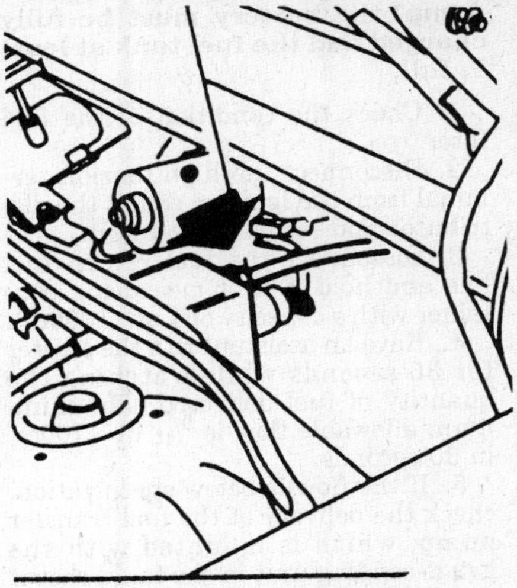

Oxygen sensor light reset button

warning light on the dash will turn on when it is time to replace the oxygen sensor. This is usually at 30,000 mile intervals. Under the hood near the wiper motor, is a black box with the speedometer cable connected to it. This is the mileage counter for the oxygen sensor that turns on the warning light. To reset the counter, find the white button on the box and push it in with a pen, listening (feeling) for the click.

GASOLINE FUEL SYSTEM

Fuel System Service Precautions

- Do not allow fuel spray or fuel vapors to come into contact with a heating element or open flame. Do not smoke while working on the fuel system.
- Always disconnect the negative battery cable unless the repair or test procedure requires that battery voltage be applied.
- Always relieve the fuel system pressure prior to disconnecting any fitting or fuel line connection.
- To control fuel spray when relieving system pressure, place a shop towel around the fitting prior to loosening to catch the spray. Ensure that all fuel spillage is quickly wiped up and that all fuel soaked rags are deposited into a proper fire safety container.
- Always keep a dry chemical (Class B) fire extinguisher near the work area.
- Always use a backup wrench when loosening and tightening fuel line fittings. Always follow the proper torque specifications.
- Do not re-use fuel system gaskets and O-rings, replace with new ones. Do not substitute fuel hose where fuel pipe is installed.

RELIEVING FUEL SYSTEM PRESSURE

On CIS systems, fuel pressure can be vented at the cold start injector line, either at the fuel distributor end or the injector end. Lay a rag over the fitting and use a socket or line wrench to crack the fitting.

On Digifant systems, pressure can be vented at the fuel pump run-on switch in front of the throttle body. Lay a rag over the switch and loosen the clamp.

Fuel Tank

REMOVAL & INSTALLATION

Except Fox and Quantum

1. Disconnect the negative battery cable. Remove the access panel under the rear seat or in the luggage compartment and disconnect the gauge sending unit wiring and hoses.
2. Raise and safely support the vehicle and drain the fuel tank.
3. On Scirocco and Cabriolet, remove the right rear inner fender and disconnect the breather hose from the filler. Remove but do not disconnect the gravity valve.
4. Detatch the fuel pump bracket from the body and lower the pump enough to disconnect the fuel hoses from the tank.

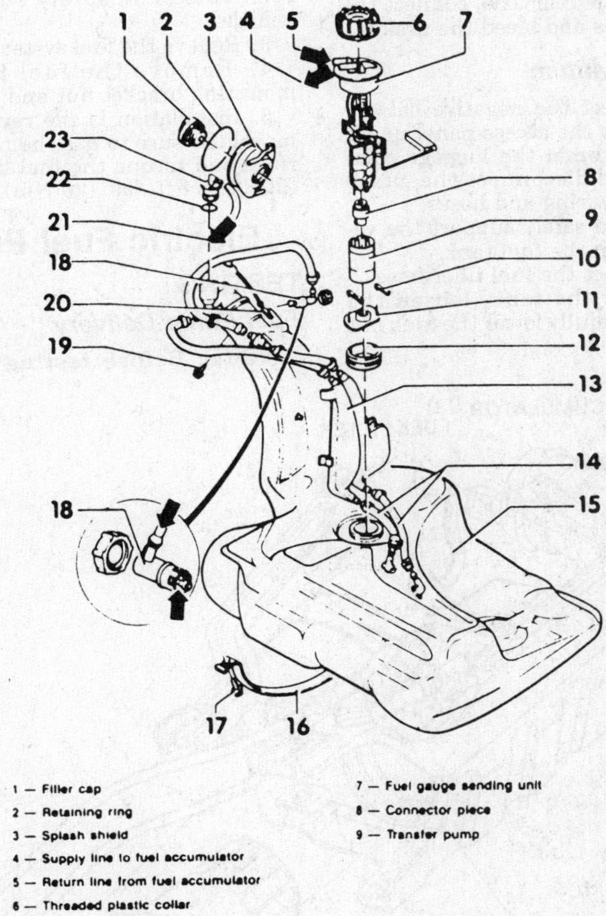

Fuel tank assembly on vehicles with plastic tank

5. On Scirocco and Cabriolet, the rear axle must be dropped out of the way. Disconnect the brake hydraulic hoses at both sides of the rear axle.
6. Detatch the rear axle from the body on both sides and let it hang on the parking brake cables.
7. Unhook the muffler supports and pull the large hose from the filler neck.
8. Support the tank, loosen the straps and carefully lower the tank out of the vehicle.

To install:

9. If a new tank is being installed, glue new foam strips to the tank in the same location as the old ones. Position the tank and connect the wiring and hoses to the sending unit.
10. Secure the tank in place with the straps and connect all hoses. Position the clamps so they do not contact the body.
11. Coat the tank with a rust protector or undercoating.
12. Install the rear axle, connect the hydraulic lines and bleed the brakes.

Fox and Quantum

1. Disconnect the negative battery cable. Remove the access panel under the rear seat or in the luggage compartment and disconnect the gauge sending unit wiring and hoses.
2. Raise and safely support the vehicle and drain the fuel tank.
3. Disconnect the fuel filler hose.
4. Support the tank, loosen the straps and carefully lower the tank out of the vehicle.
5. Installation is the reverse of removal.

Fuel Filter

REMOVAL & INSTALLATION

On the Digifant system, the fuel filter is a lifetime filter and only needs to be changed in the event of contamination. It is mounted under the vehicle, near the rear axle. It is in assembly with the pump, accumulator and reservoir, but can usually be removed separately.

On the CIS system, the fuel filter is mounted under the hood, sometimes on the fuel distributor. To make the job easier, open the clips holding the air filter housing and lift the whole assembly.

1. Disconnect the negative battery cable.
2. On vehicles with the Digifant system, raise and safely support the vehicle.
3. Relieve the fuel system pressure.
4. Remove the fuel lines, the mounting bracket nut and the filter.
5. Installation is the reverse of removal. Be sure to use the new sealing rings and torque the fuel lines to the filter to 14 ft. lbs. (20 Nm).

Electric Fuel Pump

TESTING

Fuel Pump Delivery

NOTE: Before testing the fuel pump, the battery must be fully charged and the fuel tank at least 1/4 full.

1. Check the condition of the fuel filter.
2. Disconnect the high tension terminal from the ignition coil at the distributor and securely ground it.
3. Disconnect the fuel return fuel line and hold it in a measuring container with a capacity of 1 qt. (1000cc).
4. Have an assistant run the starter for 30 seconds while watching the quantity of fuel delivered. The minimum allowable flow is 9/10 qt. (760cc) in 30 seconds.
5. If the flow is below specification, check the delivery of the fuel transfer pump, which is mounted with the gauge sending unit in the tank, except Quantum, which has only the main pump in the tank.
6. Under the rear seat or under the luggage compartment, remove the cover to expose the hoses and wires to the pump/sending unit.
7. Disconnect the output hose from the tank unit and plug it. Install a temporary fuel line and put the other end into the measuring container.
8. Have an assistant operate the starter for 10 seconds and measure the fuel delivered. The specification is about 10 oz. (300cc).
9. If the transfer pump is good, check for a dirty fuel filter, blocked lines or blocked fuel tank strainer, if equipped. If all of these are in good condition but the quantity measured is Step 4 is low, replace the main pump.

REMOVAL & INSTALLATION

NOTE: The main fuel pump is located under the vehicle in front of the rear axle or in front of the tank on the right side.

1. Disconnect the negative battery cable.
2. Raise and safely support the vehicle.
3. Disconnect the electrical connector.
4. Relieve the fuel system pressure.
5. Remove the mounting bolts and the fuel pump.
6. Installation is the reverse of removal. Be sure to use new sealing rings and/or gaskets.

Transfer Pump

REMOVAL & INSTALLATION

1. Disconnect the negative battery cable.
2. Under the rear seat or in the rear of the vehicle, pull back the carpet and

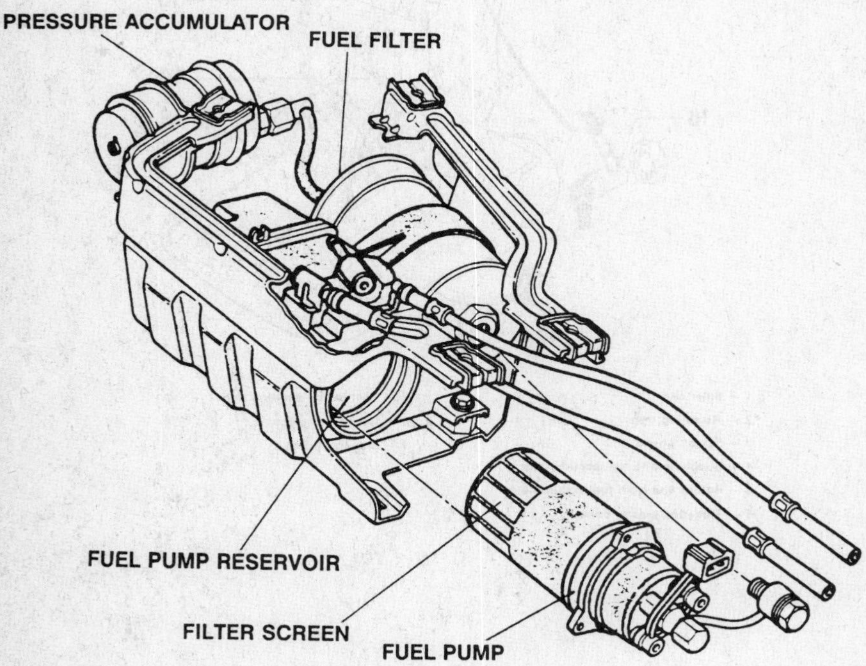

Fuel pump and reservoir assembly on Digifant system

remove the access plate from the floor (3 screws).

3. Disconnect the electrical connector and remove the fuel hoses from the sending unit.

4. Unscrew the cap and carefully lift the sending unit from the fuel tank. Note the orientation of the float in the tank.

5. Remove the transfer pump from the sending unit.

6. Installation is the reverse of removal. Be sure the float points the same way and use a new O-ring at the sending unit cap.

Fuel Injection

IDLE SPEED ADJUSTMENT

NOTE: The manufacturer specifies that timing, idle speed and CO value must all be adjusted together.

1. Run the engine to normal operating temperature, stop engine and connect a tachometer and timing light according to manufacturer's instructions. Remove the timing mark cover from the top of the bell housing.

Engine Code		% CO Value
PF		0.3–1.1
RV		0.3–1.1
9A		0.2–1.2
UM		0.3–1.2
JN		0.3–1.2
PG		0.3–1.1
JH	Checking	0.3–3.0
	Setting	0.8–1.2

2. On vehicles with the CIS system, if equipped, disconnect both plugs from the electronic idle stabilizer and plug them together.

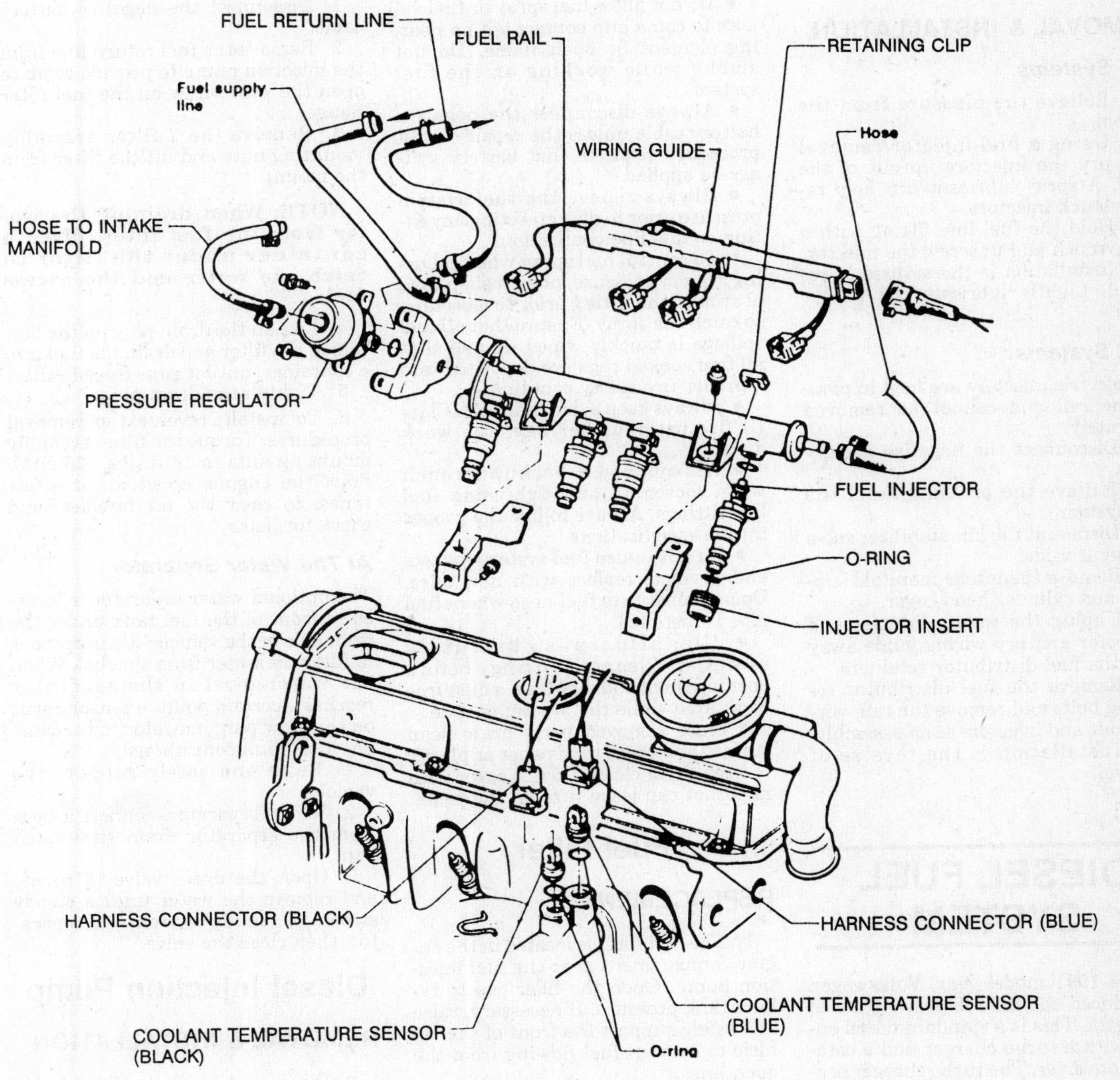

Fuel injector removal on Digifant system

3. On engines with Digifant engine management systems, with the ignition **ON**, engine not running, verify that the idle stabilizer valve hums or buzzes. Disconnect the blue temperature sensor plug.

4. Turn off all electrical equipment, start the engine and set the idle speed to 800–950 rpm. Do not disconnect any vacuum lines.

5. With the engine running, shine the timing light at the marks on the flywheel. If adjustment is required, loosen the distributor clamp bolt and rotate the distributor as needed to correctly align the timing marks.

Fuel Injector

REMOVAL & INSTALLATION

CIS Systems

1. Relieve the pressure from the system.
2. Using a fuel injector removal tool, pry the injectors up out of the head. A spray lubricant can help release stuck injectors.
3. Hold the fuel line fitting with a line wrench and unscrew the injector.
4. Installation is the reverse of removal. Lightly lubricate the rubber rings.

AFC Systems

The electric injectors are held in place by the rail and cannot be removed separately.

1. Disconnect the negative battery cable.
2. Relieve the pressure from the fuel system.
3. Dismount the idle stabilizer valve and lay it aside.
4. Remove the intake manifold supports and cylinder head cover.
5. Unplug the wiring harness end connector and pry wiring guide away from the fuel distributor retainers.
6. Remove the fuel distributor retaining bolts and remove the rail, wiring guide and injectors as an assembly.
7. Installation is the reverse of removal.

DIESEL FUEL SYSTEM

In the 1991 model year, Volkswagen introduced the ECO Diesel engine in the Jetta. This is a standard diesel engine with a turbo charger and a catalytic converter. The turbo charger provides only about 6 psi boost but about a 40 percent increase in air flowing through the engine. This provides a modest power increase, but the objective is to greatly improve the engine's emissions performance. Since the system is designed for improved emissions rather than power, there is no fuel enrichment device on the injection pump. This engine should be considered non-turbocharged and all specifications partaining to non-turbo charged engines also apply to the ECO Diesel.

Fuel System Service Precautions

- Do not allow fuel spray or fuel vapors to come into contact with a heating element or open flame. Do not smoke while working on the fuel system.
- Always disconnect the negative battery cable unless the repair or test procedure requires that battery voltage be applied.
- Always relieve the fuel system pressure prior to disconnecting any fitting or fuel line connection.
- To control fuel spray when relieving system pressure, place a shop towel around the fitting prior to loosening to catch the spray. Ensure that all fuel spillage is quickly wiped up and that all fuel soaked rags are deposited into a proper fire safety container.
- Always keep a dry chemical (Class B) fire extinguisher near the work area.
- Always use a backup wrench when loosening and tightening fuel line fittings. Always follow the proper torque specifications.
- Do not re-use fuel system gaskets and O-rings, replace with new ones. Do not substitute fuel hose where fuel pipe is installed.
- Cleanliness is absolutely escential. Clean all fittings before opening them and maintain a dust free work area while the system is open.
- Place removed parts on a clean surface and cover with paper or plastic to keep them clean. Do not cover with rags that can leave fuzz on the parts.

Fuel Filter

REPLACEMENT

1. The fuel filter is located in the engine compartment near the fuel injection pump. Open the filler cap to relieve tank pressure. If necessary, raise and safely support the front of the vehicle to prevent fuel flowing from the open lines.
2. Remove the fuel lines from the filter.
3. Loosen the mounting clamp or screws and lift the filter assembly straight up.

NOTE: When installing a new filter, partly fill it with diesel fuel.

4. Installation is the reverse of removal. Make sure the flow direction is correct. Torque the filter assembly to mount nuts to 18 ft. lbs. (25 Nm). Start the engine, accelerate it a few times, to clear the air bubbles, and check for fuel leaks.

DRAINING WATER

At The Filter

1. Disconnect the negative battery cable.
2. Remove the fuel return line from the injection pump to provide room to open the vent screw on the fuel filter flange.
3. Remove the 2 filter assembly mounting nuts and lift the filter from the mount.

NOTE: When draining the water from the fuel filter, place a container under the filter to catch the water and the excess fuel.

4. Loosen the drain plug on the bottom of the filter and drain the fuel into a container, until it runs free of water.
5. Tighten the drain plug.
6. To install, reverse the removal procedures. Torque the filter assembly mounting nuts to 18 ft. lbs. (25 Nm). Start the engine, accelerate it a few times, to clear the air bubbles, and check for leaks.

At The Water Separator

The optional water separator is located in front of the fuel tank under the right side of the vehicle; it's purpose is to filter the water from the fuel. When the water level in the separator reaches a certain point, a sensor turns on the glow plug indicator light, causing it to blink continuously.

1. Raise and safely support the vehicle.
2. At the separator, connect a hose from the separator drain to a catch pan.
3. Open the drain valve (3 turns) and remove the water until a steady stream of fuel flows from the separator, then close the valve.

Diesel Injection Pump

REMOVAL & INSTALLATION

1. Disconnect the negative battery cable and remove the air cleaner, cylinder head cover and timing belt cover.

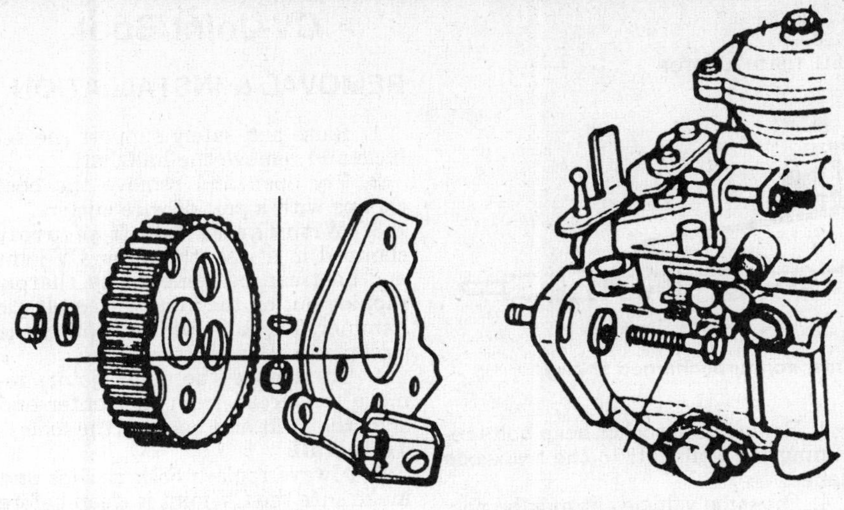

Align timing marks on the pump housing, mounting flange and drive sprocket

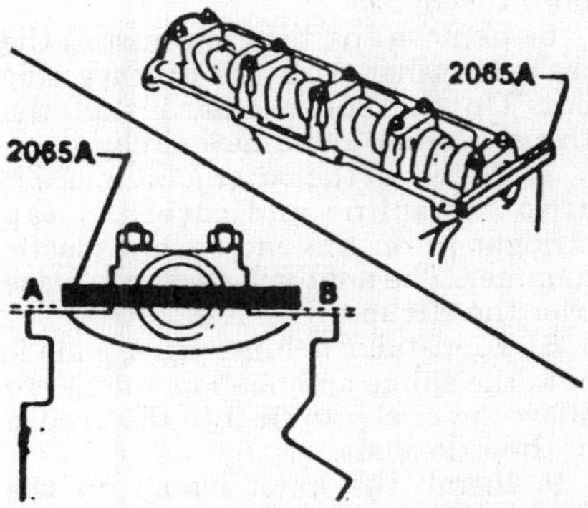

Install camshaft holding bar on Diesel engine when removing injection pump

2. Turn the engine to TDC of No. 1 cylinder and insert a setting bar into the slot on the front of the camshaft, VW tool 2065A or equivalent, to hold the camshaft in place. Remove the camshaft drive belt and sprocket. Be careful to not turn the engine while the belt is removed.

3. Loosen the pump drive sprocket nut but don't remove it yet. Install a puller on the sprocket and apply moderate tension.

4. Rap the puller bolt with light hammer taps until the sprocket jumps off the tapered shaft, then remove the puller and sprocket. Be careful not to lose the Woodruff key.

5. Hold the pump fittings with a wrench and using a line wrench, remove the injection lines from the pump. Do not bend or distort these lines. If necessary, remove the lines from the injectors also and set them aside as an assembly.

6. Disconnect the control cables, fuel solenoid wire and fuel supply and return lines.

7. Remove the pump mounting bolts and lift the pump from the vehicle.

To install:

8. When reinstalling, align the marks on the top of the mounting flange and the pump and torque the mounting bolts to 18 ft. lbs. (25 Nm).

9. Install the Woodruff key and sprocket and torque the nut to 33 ft. lbs. (45 Nm).

10. When reinstalling the supply and return lines, be sure the fitting marked OUT is used for the return line. This fitting has an orifice and must be in the correct place. Use new gaskets.

11. Turn the pump sprocket so the mark aligns with the mark on the side of the mounting flange and insert a pin through the hole in the sprocket to hold it in place.

12. Install the camshaft drive sprocket and belt and set the belt tension. Tension the drive belt by turning the tensioner pulley clockwise until belt can be flexed ½ in. (13mm) between the camshaft and the pump sprockets. Remove the pin.

13. Remove the camshaft holding bar. Turn the engine through 2 full turns, return to TDC of No. 1 cylinder and recheck the belt tension and camshaft timing. Set the injection timing and reinstall the lines and control cables.

Diesel Injection Timing

ADJUSTMENT

1. Turn the engine to TDC of No. 1 cylinder.

2. Be sure the manual cold start control is pushed in all the way and the pump lever is on the low idle stop.

3. Remove the center plug on the pump head and install the adapter tool VW-2066 or equivalent, and a dial indicator. Preload the dial indicator to 2.5mm.

4. Slowly turn the engine counterclockwise until the dial gauge stops moving, then zero the dial indicator. This is the bottom of the pump stroke.

5. Turn the engine clockwise until the TDC mark on the flywheel aligns with the pointer on the bell housing.

6. The dial indicator should read 0.83–0.97mm for non-turbocharged engines or 0.95–1.05mm for turbocharged engines.

7. If adjustment is required, remove the timing belt cover and loosen the pump mounting bolts without turning the engine.

8. Turn the pump body to attain the correct setting: non-turbocharged setting is 0.90mm or turbocharged setting is 1.00mm.

9. Torque the mounting bolts to 18 ft. lbs. (25 Nm) and turn the engine backwards about 1 turn. Turn the engine forwards to TDC of No. 1 cylinder and recheck the dial indicator.

10. When the correct setting is reached on the dial indicator, reinstall the belt cover and the center plug on the pump. Use a new copper gasket.

IDLE SPEED ADJUSTMENT

Diesel engines have both an idle speed and a maximum speed adjustment. The maximum speed adjustment is a high idle speed that prevents the engine from over-revving when the control lever is in the full speed position but there is no load on the engine. No increase in power is available through this adjustment. The adjusters are located side by side on top of the injection pump. The screw closest to the engine is the low-idle adjustment, the outer screw is the high-idle adjustment.

1. If the vehicle has no tachometer, connect a suitable diesel engine tachometer as per the manufacturer's instructions.

2. Run the engine to normal operating temperature.

3. If equipped, make sure the manual cold start knob is pushed in all the way.

4. Loosen the locknut and set the low idle to 900–1000 rpm, at a point where there is no vibration.

5. When tightening the locknut, apply a thread sealer to prevent the screw from vibrating loose.

6. Advance the control lever (throttle) to full speed. The high idle speed is 5300–5400 rpm for non-turbocharged engines or 5050–5100 rpm for turbocharged engines. Adjust as needed and secure the locknut with sealer.

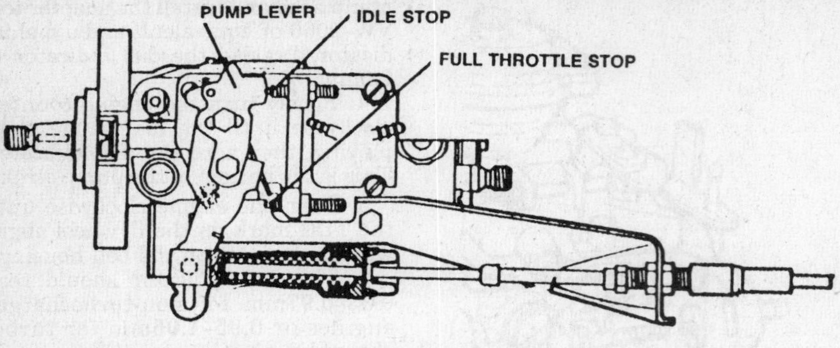

Diesel pump idle speed adjustments, non-turbocharged shown

Fuel Injector

REMOVAL & INSTALLATION

Faulty injectors can be located by loosening each pipe union one at a time with the engine at a slightly fast idle. If the engine speed remains constant after loosening a pipe union, that injector is faulty.

1. Remove the lines from all the injectors using a line wrench.
2. Carefully remove the injectors using a clean socket.
3. With a magnet or small pick, remove the heat shields from the injector holes and discard them. New heat shields must be installed.
4. Install the new heat shield and torque the injectors to 51 ft. lbs. (70 Nm). Torque the pipe unions to 18 ft. lbs. (25 Nm). Do not over torque the pipe unions or the flares will be flattened and a leak could result.

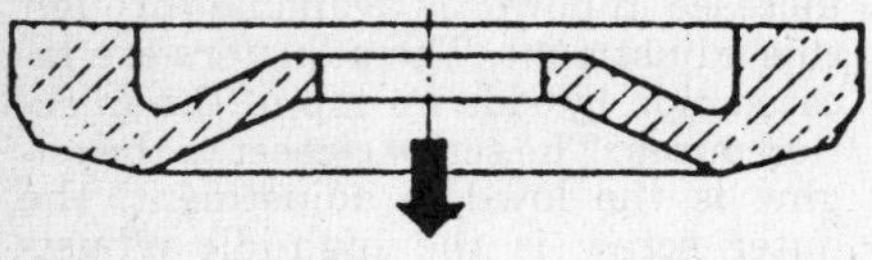

Diesel injector heat shield; concave side towards the engine

DRIVE AXLE

Halfshaft

REMOVAL & INSTALLATION

NOTE: When loosening or tightening axle nut, make sure the vehicle is on the ground. Axle nut torque is high enough that attempting to loosen it may cause the vehicle to fall off the support.

1. With the vehicle on the ground, remove the front axle nut.
2. Raise and safely support vehicle and remove the front wheels.
3. Remove the socket head bolts retaining the halfshaft to the transaxle flange.
4. On some vehicles, it may be necessary to separate the strut from the control arm.
   a. On Fox and Quamtum, matchmark the ball joint to the control arm and disconnect the strut there.
   b. On Passat and Corrado, remove the 2 bolts securing the strut to the steering knuckle to remove the left side halfshaft.
   c. On all other vehicles, remove the ball joint clamping bolt and push the control arm down, away from the ball joint.
5. Remove the transaxle side of the halfshaft from the drive flange and secure it out of the way. Do not let it hang unsupported.
6. Push the halfshaft out of the hub. A wheel puller may be required.

To install:

7. Fit the halfshaft to the drive flange and install the bolts. It is not necessary to torque them yet.
8. Apply a thread locking compound to the outer ¼ in. of the spline. Slip the spline through the hub and loosely install a new axle nut.
9. Assemble the front suspension, being careful to align the matchmarks.
   a. On Fox and Quantum, torque the ball joint–to–control arm nuts to 47 ft. lbs. (65 Nm).
   b. On Passat and Corrado, torque the strut–to–knuckle bolts to 70 ft. lbs. (95 Nm).
   c. On all other models, torque the ball joint clamping bolt to 37 ft. lbs. (50 Nm).
10. Install the wheel and hold it to keep the axle from turning. Torque the inner axle bolts to 33 ft. lbs. (45 Nm).
11. With the vehicle on the ground, torque the axle nut to 175 ft. lbs. (240 Nm) on Fox and Quantum and Scirocco/Cabriolet or 195 ft. lbs. (265 Nm) on Golf, Jetta, Passat and Corrado.
12. Check and adjust the front wheel alignment.

CV-Joint/Boot

REMOVAL & INSTALLATION

1. Raise and safely support the vehicle and remove the halfshaft.
2. Pry open and remove the boot clamps with a pair of wire cutters.
3. With the halfshaft securely clamped in a vise, the outer CV-joint and boot can be removed by sharply rapping out on the joint with a plastic hammer. The joint will snap off of the circlip and slide off the axle.
4. To remove the inner joint, remove the circlip from the center and slide the joint and boot off the axle.

To install:

5. Always replace both circlips and make sure the CV-joint is clean before installation. Wrap a piece of black electrical tape around the shaft splines and slip the inner clamp and the boot onto the shaft.
6. Remove the tape and install the dished washer with the concave side out. On the outer joint, install the thrust washer and a new circlip.
7. To install the outer joint, place it onto the spline and carefully tap straight in on the end with a plastic hammer. The joint will click into place over the circlip.
8. To install the inner joint, slide it onto the spline and push in enough to allow the circlip to fit into the groove in the axle shaft.
9. Install the inner clamp on the boot and fill the boot with special CV-joint grease. Do not use any other type of grease.
10. On the inner joint, stick the gasket to the joint before installing the halfshaft into the vehicle. Install the outer boot clamp and install the halfshaft.

Driveshaft and U-Joints

REMOVAL & INSTALLATION

Quantum Synchro

The Quantum Syncro has a driveshaft connecting the transaxle to the rear differential. A special tool, VW3139 or equivalent, is used as a retainer to keep the shaft from flexing and overloading the center U-joint and bearing when the shaft is removed. Use of this tool or a suitable substitute is required for proper alignment during installation. No parts are available for repairs, the shaft must be replaced as a unit.

When installing the driveshaft mounted in the retaining tool, the distance from the center bearing bracket to the body must be the same on both

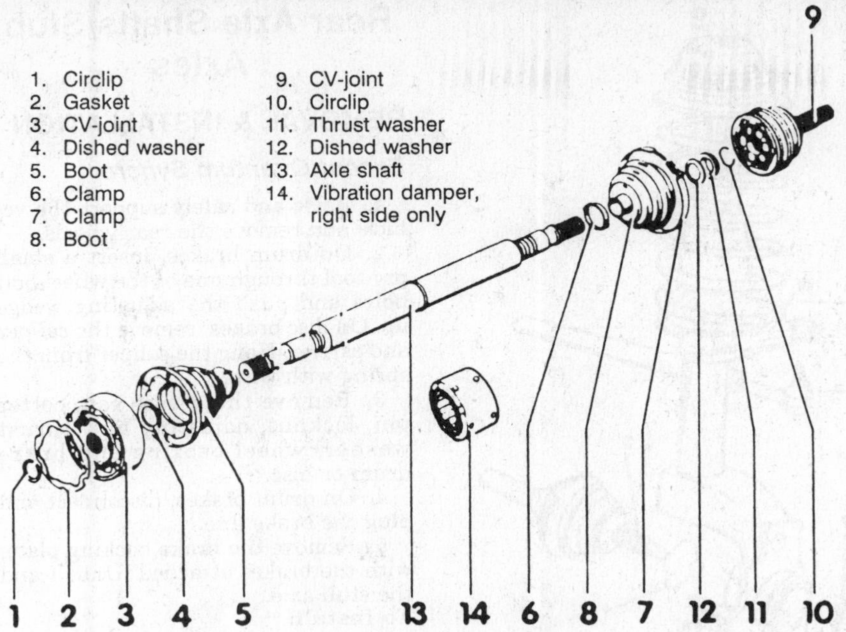

Remove the CV-joints to replace the boots

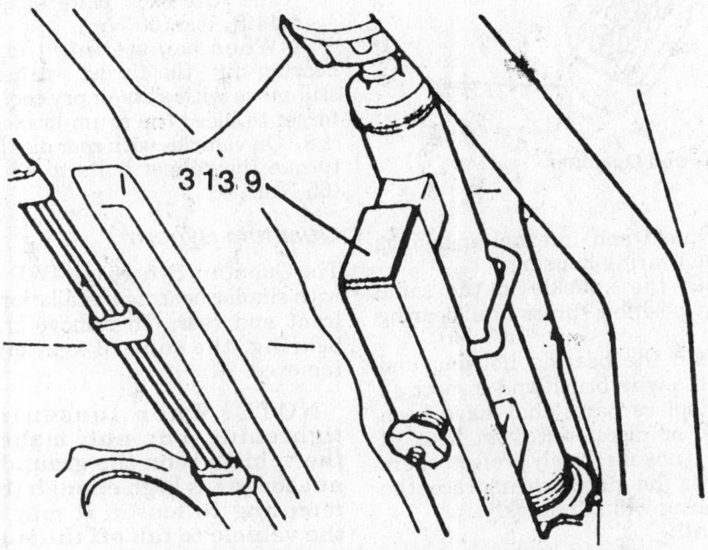

Drive shaft alignment tool in place—Quantum Synchro

sides. Shims are available for adjusting this distance.

1. Raise and safely support the vehicle.
2. Mark the position of the driveshaft to the transaxle flange and the rear differential flange.
3. Install the dirveshaft support tool.
4. Remove the center bearing bolts. Remove the driveshaft flange mounting bolts from both ends and remove the driveshaft from the vehicle.

To install:

5. Align scribe marks and install driveshaft. Torque the flange bolts to 40 ft. lbs. (55 Nm).
6. Measure the distance between the center bearing support bolt holes and the body. Use shims to make the bolt hole distance equal on both sides.
7. Install the center bearing support bolts and torque to 14 ft. lbs. (20 Nm). Remove the support tool.

Front Steering Knuckle

REMOVAL & INSTALLATION

On Fox and Quantum, the strut must be removed but a spring compressor is not needed. On all models, the hub and bearing are pressed into the knuckle and the bearing cannot be reused once the hub has been removed.

NOTE: When loosening or tightening axle nut, make sure the vehicle is on the ground. Axle nut torque is high enough that attempting to loosen it may cause the vehicle to fall off the support.

1. With the vehicle on the ground, remove the front axle nut.
2. Raise and safely support the vehicle and remove the front wheels. On Fox and Quantum, remove the strut.
3. Detatch the brake line from the strut and remove the caliper. Hang it from the body with wire.
4. Remove the caliper carrier and brake rotor.
5. Remove the cotter pin and nut and press out the tie rod end. A small puller is required.
6. Remove the ball joint clamp bolt and push the control arm down to disengage the ball joint.
7. Front wheel camber is set with eccentric washers on the bolts holding the bearing housing to the strut. Clean and mark the position of these washers so they can be reinstalled in the same position.
8. Remove the bolts and take the knuckle and bearing housing off the strut.

To install:

9. Fit the knuckle to the strut and install the bolts. Align the marks and torque the nuts to 70 ft. lbs . (95 Nm).
10. Make sure the axle splines are clean and apply a bead of thread locking compound to the outer portion. Slid the axle into the hub and install a new axle nut. Do not torque it yet.
11. Fit the lower ball joint in place and install the clamp bolt. Torque it to 37 ft. lbs. (50 Nm).
12. Connect the tie rod and torque the nut to 26 ft. lbs. (35 Nm), then tighten as required to install a new cotter pin.

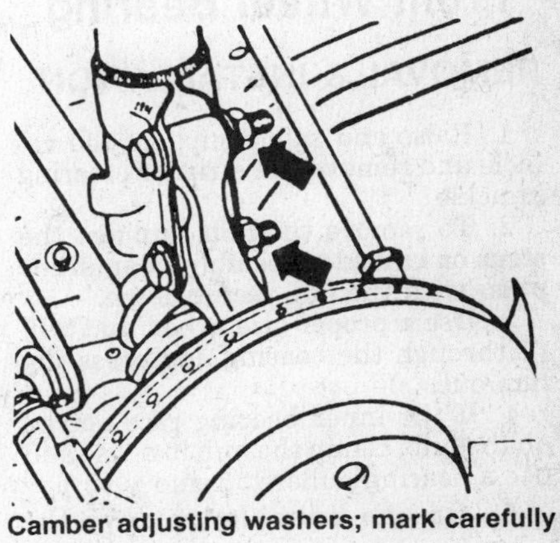

Camber adjusting washers; mark carefully

1. Cotter pin
2. Tie rod
3. Axle driveshaft
4. Circlip
5. Retainer nut
6. Brake caliper
7. Wheel bearing
8. Hub
9. Brake disc
10. Axle nut

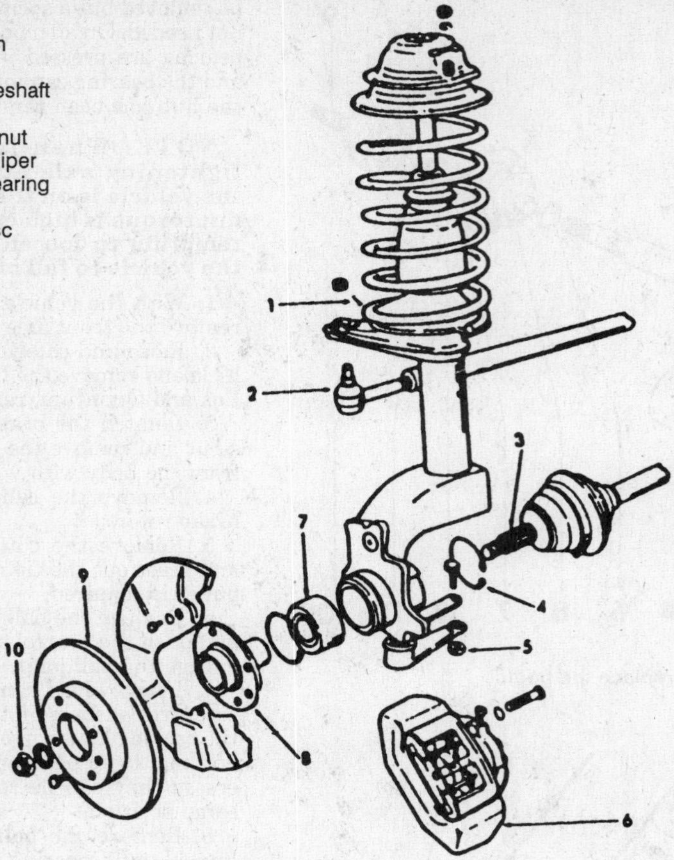

Front suspension components—Fox and Quantum

13. Install the brake disc and caliper. Torque the carrier bolts to 92 ft. lbs. (125 Nm) and the caliper guide bolts to 26 ft. lbs. (35 Nm). Secure the brake line in place.

14. With the wheel installed and the vehicle on the ground, torque the axle nut:

Scirocco, Cabriolet, Fox and Quantum—175 ft. lbs. (237 Nm)

Passat, Corrado, Golf and Jetta—195 ft. lbs. (265 Nm)

Front Wheel Bearing

REMOVAL & INSTALLATION

1. Raise and safely support the vehicle and remove the strut or steering knuckle.
2. To remove the hub, support the strut or knuckle assembly in an arbor press with the hub facing down.
3. Use a proper size arbor that will fit through the bearing and press the hub out.
4. If the inner bearing race stayed on the hub, clamp the hub in a vise and use a bearing puller to remove it.
5. On the knuckle, remove the splash shield and internal snaprings from the bearing housing.
6. With the knuckle in the same pressing position, press the bearing out.
7. Clean the bearing housing and hub with a wire brush and inspect all parts. Replace parts that have been distorted or discolored from heat. If the hub is not absolutely prefect where it contacts the inner bearing race, the new bearing will fail quickly.

To install:

8. The new bearing is pressed in from the hub side. Install the snapring and support the bearing housing on the press.
9. Using the old bearing as a press tool, press the new bearing into the housing up against the snapring. Make sure the press tool contacts only the outer race of the bearing.
10. Install the outer snapring and splash shield.
11. Support the inner race on the press and press the hub into the bearing. Make sure the inner race is supported or the bearing fail quickly.
12. Install the strut or knuckle and be sure to torque the axle nut before allowing the vehicle to roll.

Rear Axle Shafts/Stub Axles

REMOVAL & INSTALLATION

Except Quantum Syncro

1. Raise and safely support the vehicle and remove the rear wheels.
2. On drum brakes, insert a small pry tool through one of the wheel bolt holes and push the adjusting wedge up. On disc brakes, remove the caliper and carrier. Hang the caliper from the spring with wire.
3. Remove the grease cap, cotter pin, locknut, adjusting nut, thrust washer, wheel bearing and brake drum or disc.
4. On drum brakes, disconnect and plug the brake line.
5. Remove the brake backing plate, with the brakes attached. Unbolt and the stub axle.

To install:

6. Install the back plate and stub axle and torque the bolts:

Stub axle/Back plate on Golf and Jetta—52 ft. lbs. (70 Nm).

Stub axle/Back plate on all others—44 ft. lbs. (60 Nm).

7. When reinstalling the wheel bearing nut, the thrust washer must still move with a small pry tool. Don't forget to bleed the drum brakes.
8. On vehicles with rear disc brakes, torque the caliper bolts to 48 ft. lbs. (65 Nm).

Quantum Syncro

The Quantum Syncro is a 4WD vehicle with similar bearing installations both front and rear. To remove the rear bearing, the control arm must be removed.

NOTE: When loosening or tightening axle nut, make sure the vehicle is on the ground. Axle nut torque is high enough that attempting to loosen it may cause the vehicle to fall off the support.

1. With the vehicle on the ground, loosen the axle nut.
2. Raise and safely support the vehicle and remove the wheels.
3. Remove the bolts retaining the axle shaft to the differential and slide the axle shaft out of the hub. If necessary, use a bearing puller to push the axle out of the hub.
4. Remove the disc brake caliper and support aside, do not hang it by the brake line.
5. Unbolt the stabilizer bar from the control arm.
6. Remove the lower shock mount bolt and the control arm mounting bolts to remove the arm from the vehicle.

7. Remove the brake splash shield. To remove the hub, support the arm in an arbor press with the hub facing down.
8. Use a proper size arbor that will fit through the bearing and press the hub out.
9. If the inner bearing race stayed on the hub, clamp the hub in a vise and use a bearing puller to remove it.
10. Remove the snapring and press the bearing out.
11. Clean the bearing housing and hub with a wire brush and inspect all parts. Replace parts that have been distorted or discolored from heat. If the hub is not absolutely prefect where it contacts the inner bearing race, the new bearing will fail quickly.

To install:

12.. The new bearing is pressed in from the hub side. Install the snapring and support the bearing on the press. Lubricate the bearing housing and press it onto the outer bearing race. Install the outer snapring.
13. Support the hub on the press table and press the inner race of the bearing onto the hub. Make sure the press tool contacts only the inner race or the bearing will be destroyed.
14. Install the splash shield.
15. Install the control arm onto the vehicle and torque the following in order:

Control arm bolts—88 ft. lbs. (120 Nm).
Lower shock mount—48 ft. lbs. (65 Nm).
Stabilizer bar brackets—18 ft. lbs. (25 Nm).
Brake caliper—48 ft. lbs. (65 Nm).
Inner axle CV-joint bolts—33 ft. lbs. (45 Nm).

16. With the wheel installed and the vehicle on the ground, install a new axle nut and torque it to 170 ft. lbs. (230 Nm).

Pinion Seal

REMOVAL & INSTALLATION

Quantum Synchro

1. Raise and safely support the vehicle.
2. Mark the position of the driveshaft to the transaxle and differential flanges. Install the driveshaft alignment tool and remove the driveshaft.
3. Mark the position of the pinion nut to the pinion flange and remove the nut.
4. Use a puller tool to remove the pinion flange, then remove the pinion seal.

To install:

5. Pack the new seal with grease and install it into the differential with an appropriate seal installation tool.
6. Install the pinion flange. Clean the pinion threads and apply a thin coat of thread locking compound.
7. Install the pinion nut and tighten it to exactly the position marked. If a different nut is used or if the original position cannot be found, the pinion bearing pre-load will be incorrect and the differential must be removed for service.
8. Install and align the driveshaft.

Differential Carrier

REMOVAL & INSTALLATION

1. Raise and safely support the vehicle.
2. Mark the position of the driveshaft to the transaxle and differential flanges. Install the driveshaft alignment tool and remove the driveshaft.
3. Remove the bolts securing the halfshafts to the drive flanges. Hang the halfshafts from the body with wire.
4. Disconnect the vacuum hoses and wiring.
5. Support the differential housing and remove the axle beam bolts and install long M12 X 110mm bolts into the axle beam bolt locations to hold the axle beam in place. Do not let the axle beam down more than about 2 inches (50mm) or the brake lines will be damaged.
6. Remove the rear bracket from the differential carrier. Unbolt the differential carrier from the top or the axle beam and lower it out of the vehicle.

To install:

7. Fit the differential carrier to the axle beam and install the bolts. Torque to 59 ft. lbs. (80 Nm). Install the rear bracket and torque the bolts to 33 ft. lbs. (45 Nm).
8. Raise the assembly and install the normal axle beam–to–body bolts. Torque to 96 ft. lbs. (130 Nm).
9. Install the halfshafts and torque the bolts to 33 ft. lbs. (45 Nm).
10. Connect the vacuum hoses and wiring.
11. Install and properly align the driveshaft and remove the alignment

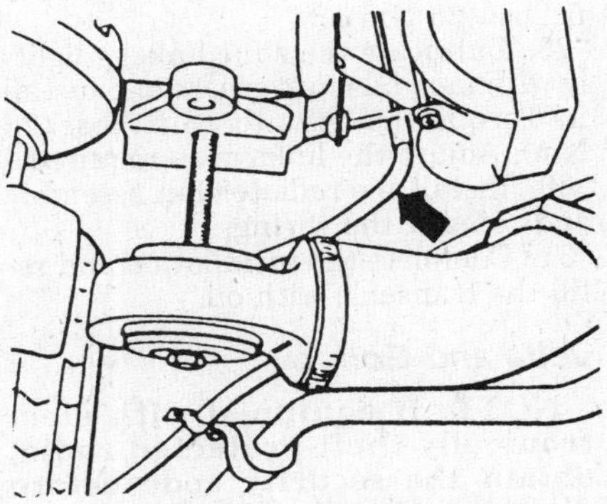

Install long bolts to hold axle mounting while removing differential carrier from Quantum Synchro

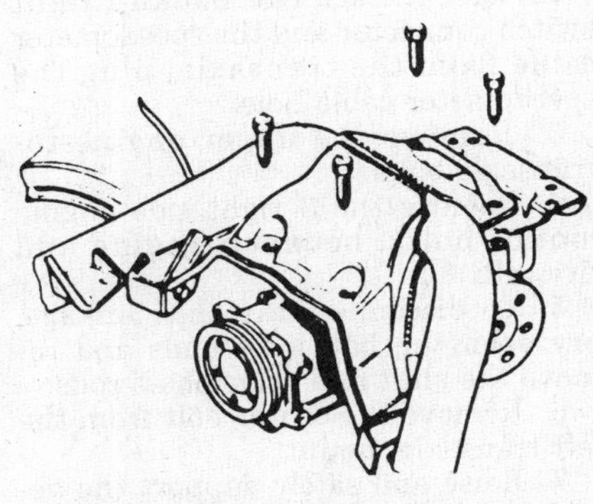

Removing differential carrier from Quantum Synchro

MANUAL TRANSAXLE

For further information on transmissions/transaxles, please refer to "Chilton's Guide to Transmission Repair".

Transaxle Assembly

REMOVAL & INSTALLATION

Passat and Corrado

NOTE: If equipped with electronically theft-protected radio, obtain the security code before disconnecting the battery.

1. Disconnect the negative battery cable.
2. On Corrado, remove the intercooler tubing.
3. Disconnect the backup light switch connector and the speedometer cable from the transaxle, plug the speedometer cable hole.
4. Remove the clutch slave cylinder without disconnecting the hydraulic line. Hang the cylinder from the body with wire.
5. On the cable shift linkage, remove the backup light switch bracket. Disconnect the cable from the relay lever but remove the gearshift lever with the cable still attached. Remove the cable support and set the cables aside.
6. If necessary, remove the intake hose from the air flow sensor.
7. Remove the upper transaxle-to-engine bolts.
8. Raise and safely support the ve-

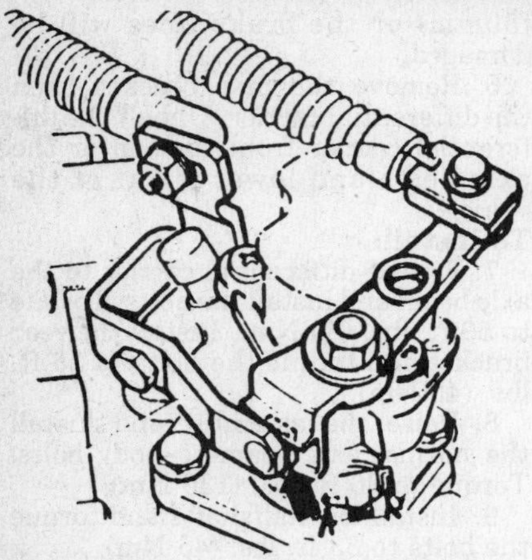

Cable shift linkage at transaxle, the relay lever is on the left

hicle and remove the front wheels. Connect the engine sling tool VW–10–222A or equivalent, to the loop in the cylinder head and just take the weight of the engine off the mounts. On 16V engine, the idle stabilizer valve must be removed to attach the tool. Do not try to support the engine from below.

9. Remove the drain plug and drain the oil from the transaxle. Dispose of the oil properly.
10. Remove the starter and front mount.
11. Remove the 3 bolts from the right side mount, between engine and firewall.
12. Remove the large center bolt from the left side transaxle mount.

NOTE: On vehicles with ABS, this bolt can be reached by removing the cooling system overflow bottle.

13. Remove the radiator fan shroud and fan as an assembly.
14. Remove the long transaxle support bracket which connects the front and rear mounts on the left side.
15. Remove the heat shield for the right side inner CV-joint.
16. Disconnect the halfshafts from the output flanges and hang them from the body.
17. Remove the left rear transaxle mount. It may be necessary to push the engine/transaxle rearward to get the lower bolt out.
18. Lower the transaxle slightly.
19. Remove the bell housing cover and position a jack under the transaxle.
20. Remove the last transaxle-to-engine bolts and gently pry the transaxle away from the engine. Lower it carefully from the vehicle.

To install:

21. Press the clutch release lever towards the transaxle housing and secure it with a pin or 8mm bolt.
22. Coat the input shaft lightly with molybdenum grease and carefully fit the transaxle in place. If necessary, put the transaxle in any gear and turn an output flange to align the input shaft spline with the clutch spline.
23. Install the engine-to-transaxle bolts and torque to 59 ft. lbs. (80 Nm).
24. When installing the mounts to the transaxle, torque the left and rear bracket–to–transaxle bolts to 18 ft. lbs. (25 Nm). Torque the remaining mount–to–transaxle bolts to 44 ft. lbs. (60 Nm). Don't forget the balance weight. Install but do not torque the bolts that go into the rubber mounts.
25. Install the starter and front mount.
26. With all mounts installed and the transaxle safely in the vehicle, allow some slack in the lifting equipment. With the vehicle safely supported, shake the engine/transaxle as a unit to settle it in the mounts. Torque all mounting bolts, starting at the rear and working forward. Torque the bolts that go into the rubber transaxle mounts to 44 ft. lbs. (60 Nm).
27. Install the halfshafts and torque the bolts to 33 ft. lbs. (45 Nm). Install the heat shield.
28. Remove the pin or bolt from the release lever and install the clutch slave cylinder. Torque the bolts to 18 ft. lbs. (25 Nm).
29. Lubricate the shift linkage lightly with molybdenum grease and install it. Torque the bolts to 18 ft. lbs. (25 Nm). Adjust the linkage as required.
30. Install the radiator fan assembly and connect the wiring.
31. Complete the installation and refill the transaxle with oil.

Jetta and Golf

NOTE: If equipped with electronically theft-protected radio, obtain the security code before disconnecting the battery.

1. Disconnect the negative battery cable.
2. Disconnect the backup light switch connector and the speedometer cable from the transaxle; plug the speedometer cable hole.
3. Remove the upper engine-to-transaxle bolts.
4. Remove the 3 right side engine mount bolts, between engine and firewall.
5. To disconnect the shift linkage, pry open the ball joint ends and remove the shift and relay shaft rods.
6. Remove the center bolt from the left transaxle mount.
7. Raise and safely support the vehicle and remove the front wheels. Connect the engine sling tool VW–10–222A or equivalent, to the loop in the cylinder head and just take the weight of the engine off the mounts. On 16V engine, the idle stabilizer valve must be removed to attach the tool. Do not try to support the engine from below.
8. Remove the drain plug and drain the oil from the transaxle. Dispose of the oil properly.
9. Remove the left inner fender liner.
10. Disconnect the halfshafts from the inner drive flanges and hang them from the body.
11. Remove the clutch cover plate and the small plate behind the right halfshaft flange.
12. Remove the starter and front engine mount.
13. Disconnect the clutch cable and remove it from the transaxle housing.
14. Remove the remaining transaxle mount bolts and mounts.
15. Place a jack under the transaxle and remove the last bolts holding it to the engine. Carefully pry the transaxle away from the engine and lower it from the vehicle.

To install:

16. Coat the input shaft lightly with molybdenum grease and carefully fit the transaxle in place. If necessary, put the transaxle in any gear and turn an output flange to align the input shaft spline with the clutch spline.
17. Install the engine-to-transaxle bolts and torque to 55 ft. lbs. (75 Nm).
18. When installing the mounts to the transaxle, torque the rear bracket–to–engine bolts and the transaxle support bolts to 18 ft. lbs. (25 Nm). Torque the left bracket–to–transaxle bolts to 25 ft. lbs. (35 Nm) and the remaining mounting bolts to 44 ft. lbs. (60 Nm). Install but do not torque the bolts that go into the rubber mounts.
19. Install the starter and front mount.
20. With all mounts installed and the transaxle safely in the vehicle, allow some slack in the lifting equipment. With the vehicle safely supported, shake the engine/transaxle as a unit to settle it in the mounts. Torque all mounting bolts, starting at the rear and working forward. Torque the bolts that go into the rubber mounts to 44 ft. lbs. (60 Nm).
21. Install the halfshafts and torque the bolts to 33 ft. lbs. (45 Nm). Install the clutch cover plates.
22. Connect the shift linkage and clutch cable and adjust as required.
23. Install the inner fender and complete the remaining installation. Refill the transaxle with oil.

Scirocco and Cabriolet

NOTE: If equipped with elec-

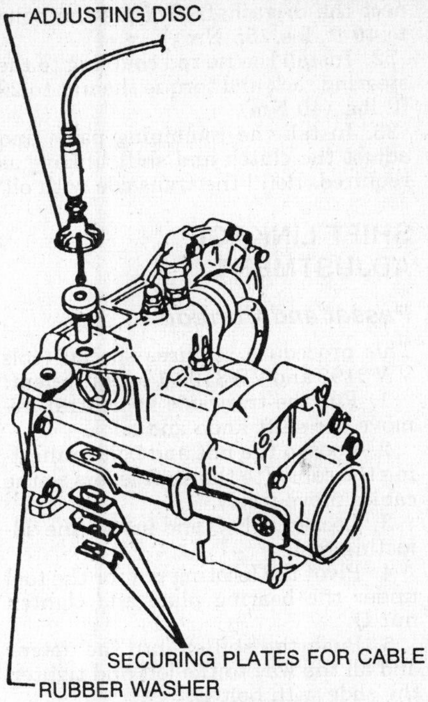

Clutch cable attachment

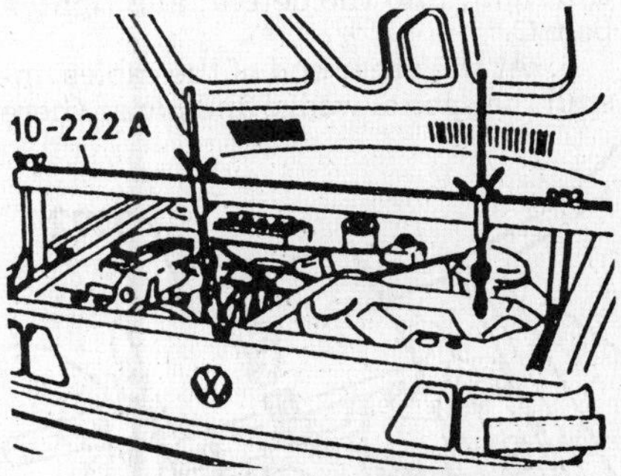

Supporting the engine to remove the transaxle

tronically theft-protected radio, obtain the security code before disconnecting the battery.

1. Disconnect the negative battery cable.
2. Disconnect the backup light switch connector and the speedometer cable from the transaxle, plug the speedometer cable hole.
3. Turn the engine to align the timing marks to TDC.
4. To disconnect the shift linkage, pry open the ball joint ends and remove both selector rods. Remove the pin, disconnect the relay rod and put the pin back in the hole on the rod for safe keeping.
5. Raise and safely support the vehicle and remove the front wheels. Connect the engine sling tool VW-10-222A or equivalent, to the loop in the cylinder head and just take the weight of the engine off the mounts. On 16V engine, the idle stabilizer valve must be removed to attach the tool. Do not try to support the engine from below.
6. Remove the drain plug and drain the oil from the transaxle. Dispose of the oil properly.
7. Detach the clutch cable from the linkage and remove it from the transaxle case.
8. Remove the starter and front engine mount.
9. Remove the small cover behind the right halfshaft flange and remove the clutch cover plate.
10. Disconnect the halfshafts from the drive flanges and hang them up with wire.
11. Remove the long center bolt from the left side transaxle mount.
12. Remove the entire rear mount assembly from the body and differential housing.
13. Lower the engine hoist enough to let the left mount free of the body and remove the mount from the transaxle.
14. Place a transaxle support jack under the transaxle, remove all the transaxle-to-engine bolts and carefully pry the transaxle away from the engine. Lower the transaxle from under the vehicle.

To install:

15. Coat the input shaft lightly with molybdenum grease and carefully fit the transaxle in place. If necessary, put the transaxle in any gear and turn an output flange to align the input shaft spline with the clutch spline.
16. Install the engine-to-transaxle bolts and torque to 55 ft. lbs. (75 Nm).
17. When installing the mounts to the transaxle, torque the bolts to 33 ft. lbs. (45 Nm). Install but do not torque the bolts that go into the rubber mounts.
18. Install the starter and front mount.
19. With all mounts installed and the transaxle safely in the vehicle, allow some slack in the lifting equipment. With the vehicle safely supported, shake the engine/transaxle as a unit to settle it in the mounts. Torque all mounting bolts, starting at the rear and working forward. Torque the bolts that go into the rubber mounts to 25 ft. lbs. (35 Nm). Torque the front mount bolts to 38 ft. lbs. (52 Nm).
20. Install the halfshafts and torque the bolts to 33 ft. lbs. (45 Nm). Install the clutch cover plates.
21. Connect the shift linkage and clutch cable and adjust as required.
22. Complete the remaining installation and refill the transaxle with oil.

Fox

NOTE: If equipped with electronically theft-protected radio, obtain the security code before disconnecting the battery.

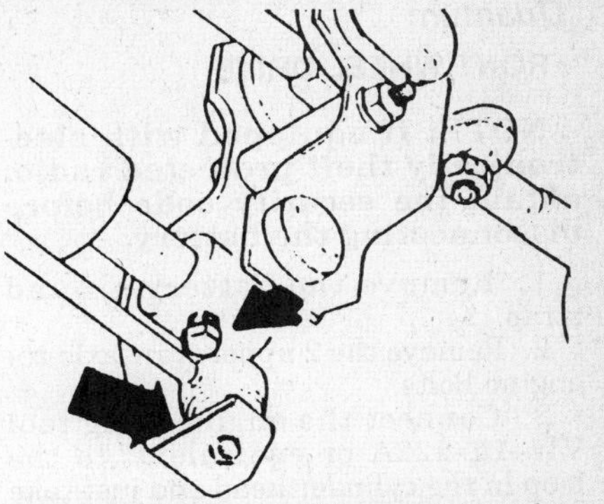
Fox and Quantum shifter control rod joint

1. Raise and safely support the vehicle and remove the front wheels.
2. Remove the drain plug and drain the oil from the transaxle. Dispose of the oil properly.
3. Disconnect the battery ground cable.
4. Disconnect the clutch cable.
5. Disconnect the exhaust pipe from the manifold.
6. Disconnect the speedometer cable and backup light switch.
7. Remove the bolt on the shift linkage, pry the control rod joint off and push the shift linkage coupling off the transaxle.
8. Detach the halfshafts from the transaxle.
9. Remove the starter and clutch cover plate.
10. Remove the exhaust pipe bracket from the transaxle and remove the pipe at the catalytic converter.
11. Support the transaxle with a jack.
12. Remove the transaxle crossmember and front mount bolts.
13. Remove the engine-to-transaxle bolts.
14. Carefully pry the transaxle away from the engine and lower it from the vehicle.

To install:

15. Coat the input shaft lightly with molybdenum grease and carefully fit the transaxle in place. If necessary, put the transaxle in any gear and turn an output flange to align the input shaft spline with the clutch spline.
16. Install the engine-to-transaxle bolts and torque to 40 ft. lbs. (55 Nm).
17. Torque the crossmember-to-body bolts to 47 ft. lbs. (65 Nm).
18. Install the rubber mount and torque the bracket bolts to 18 ft. lbs. (25 Nm) and the mount-to-body bolts to 80 ft. lbs. (110 Nm).
19. Connect the halfshafts to the drive flanges and torque the bolts to 33 ft. lbs. (45 Nm).
20. Install the remaining parts and adjust the clutch and shift linkage as required. Refill the transaxle with oil.

Quantum

FRONT WHEEL DRIVE

NOTE: If equipped with electronically theft-protected radio, obtain the security code before disconnecting the battery.

1. Remove the battery ground strap.
2. Remove the 2 upper transaxle-to-engine bolts.
3. Connect the engine sling tool VW–10–222A or equivalent, to the loop in the cylinder head and just take the weight of the engine off the mounts.
4. Raise and safely support the vehicle and remove the front wheels.
5. Remove the drain plug and drain the oil from the transaxle. Dispose of the oil properly.
6. On vehicles with hydraulic clutch, take the clip off the slave cylinder, drive out the roll pin and set the cylinder aside without opening the hydraulic line.
7. Remove the bolts to remove the halfshafts from the drive flanges.
8. Remove the front exhaust pipe, bracket and heat shield. Disconnect the backup light wires.
9. Remove the steering rack bracket.
10. Remove the bolt on the shift linkage, pry the control rod joint off and push the shift linkage coupling off the transaxle.
11. Remove the clutch cover, starter and lower transaxle-to-engine bolts.
12. Place a jack under the transaxle and remove the mounts.
13. Gently pry the transaxle away from the engine and lower it from the vehicle.

To install:

14. Coat the input shaft lightly with molybdenum grease and carefully fit the transaxle in place. If necessary, put the transaxle in any gear and turn an output flange to align the input shaft spline with the clutch spline.
15. Install the engine-to-transaxle bolts and torque to 40 ft. lbs. (55 Nm).
16. Install the transaxle mounts and torque the mount–to–transaxle bolts to 29 ft. lbs. (40 Nm) and the mount–to–body bolts to 80 ft. lbs. (110 Nm). Torque the subframe–to–body bolts to 52 ft. lbs. (70 Nm).
17. Connect the halfshafts to the drive flanges and torque the 10mm bolts to 59 ft. lbs. (80 Nm), the 8mm bolts to 33 ft. lbs. (45 Nm).
18. Install the steering rack bracket but do not tighten the bolts until the vehicle is resting on all 4 wheels. Torque the bolts to 33 ft. lbs. (45 Nm).
19. Install the remaining parts and adjust the clutch and shift linkage as required. Refill the transaxle with oil.

SYNCRO

NOTE: If equipped with electronically theft-protected radio, obtain the security code before disconnecting the battery.

1. Raise and safely support the vehicle and remove the front wheels.
2. Remove the drain plug and drain the oil from the transaxle. Dispose of the oil properly.
3. Disconnect the battery ground strap.
4. Remove the front muffler and exhaust header.
5. Disconnect the tie rod coupling from the steering rack.
6. Disconnect the speedometer cable.
7. Remove the locking pin and clutch slave cylinder without loosening the hydraulic line.
8. Remove the bolt on the shift linkage, pry the control rod joint off and push the shift linkage coupling off the transaxle.
9. Connect the engine sling tool VW–10–222A or equivalent, to the loop in the cylinder head and just take the weight of the engine off the mounts.
10. Mark the position of the driveshaft to the transaxle flange and disconnect the driveshaft. If the driveshaft is to be removed, install the driveshaft support tool.
11. Disconnect the wiring for the backup lights and differential lock and the vacuum lines from the servo.
12. Remove the starter and clutch cover plate.
13. Remove the right side transaxle mount and disconnect the halfshaft.
14. Place a jack under the transaxle and disconnect the left side mount and halfshaft.
15. Remove the front engine stop (rubber bumper) and the remaining engine-to-transaxle bolts.
16. Carefully pry the transaxle away from the engine and lower it from the vehicle.

To install:

17. Coat the input shaft lightly with molybdenum grease and carefully fit the transaxle in place. If necessary, put the transaxle in any gear and turn an output flange to align the input shaft spline with the clutch spline.
18. Install the engine-to-transaxle bolts and torque to 43 ft. lbs. (60 Nm).
19. Install the transaxle mounts and torque the mount–to–transaxle bolts to 29 ft. lbs. (40 Nm) and the mount–to–subframe bolts to 32 ft. lbs. (45 Nm).
20. Connect the halfshafts to the drive flanges and torque the bolts to 33 ft. lbs. (45 Nm).
21. Align the matchmarks and connect the driveshaft. Torque the bolts to 40 ft. lbs. (55 Nm).
22. Install the tie rod coupling to the steering rack and torque the nut to 29 ft. lbs. (40 Nm).
23. Install the remaining parts and adjust the clutch and shift linkage as required. Refill the transaxle with oil.

SHIFT LINKAGE ADJUSTMENT

Passat and Corrado

This procedure requires special tools VW 3193 and VW3192/1 or equivalent.

1. Put the transaxle in neutral, remove the shift knob and boot.
2. Loosen the nut and bolt connecting the cables to the shift levers so the cables move freely.
3. Loosen bolt **C** and install the adjusting tool.
4. Pivot the locating pin for the tool under the bearing plate and tighten nut **D**.
5. Push the shifter into the detent and all the way to the left and tighten the slide with bolt **E**.
5. Push the shifter all the way to the right, into the detent, and tighten bolt **C**.
6. At the other end of the cables, install the special wedge and pin so there

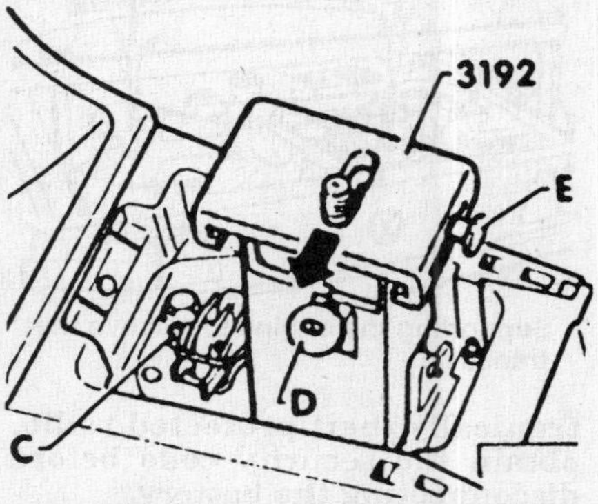

Passat and Corrado shifter adjustment

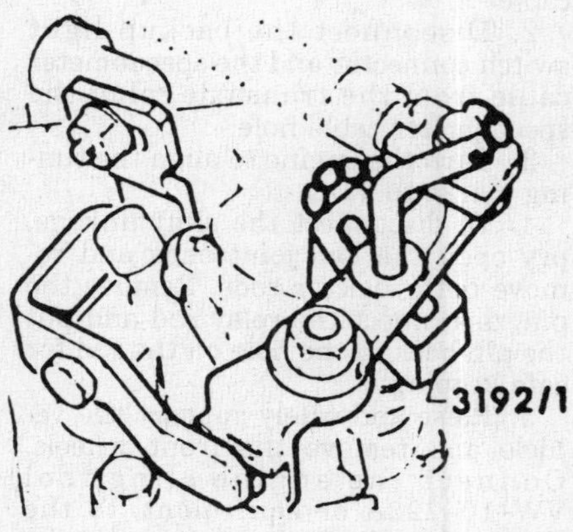

Passat and Corrado adjusting wedge in place

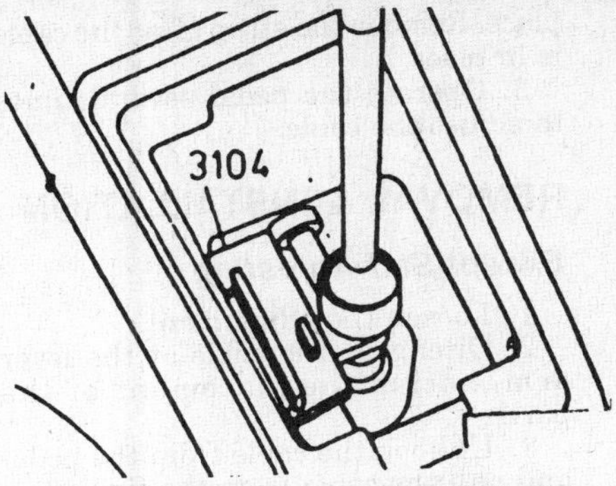

Golf and Jetta shifter adjusting tool

is no play in the lever but the lever is not raised.

7. The linkage is now set in place. Tighten the cables to the levers and remove the tools to check shifter operation.

Golf and Jetta

This procedure requires special tool VW 3104 or equivalent.

1. Put the transaxle in neutral.
2. Under the vehicle, loosen the clamp on the shifter rod so the shifter moves freely on the rod.
3. Remove the shifter knob and the boot.
4. Position the gauge alignment tool VW–3104 or equivalent, on the shifting mechanism and lock it in place.
5. Align the shift rod with the selector lever and torque the clamp to 19 ft. lbs. (26 Nm). The shifter linkage must not be under load during the adjustment.
6. Check shifter operation.

Scirocco and Cabriolet

1. Remove the shifter knob and boot.
2. Align the holes of the lever housing plate with the holes of the lever bearing plate. Check shifter operation.
3. If further adjustment is required, working under the vehicle remove the boot and loosen the shift rod clamp so the shifter moves easily on the rod.
4. Center the shift finger (fore and aft) in the lockout plate and move the shifter so the finger is disengaged from the lock out by $^{9}/_{16}$ in. (15mm).
5. Tighten the rod clamp to 14 ft. lbs. (20 Nm) and check shifter operation. If operation is spongy or binding, readjust the lock out finger to ½ in. (13mm).

Fox

1. Shift into neutral.
2. Remove the gear shift lever knob and shift boot.
3. Loosen the clamp nuts and check that shift finger slides freely on the shift rod.
4. Move the gear shift lever to the right side, between 3rd and 4th gear position. The gear shift lever should remain perpendicular to the ball housing.
5. With the inner shift lever in neutral and the gear shift lever between 3rd and 4th gear, tighten the clamp nut.
6. Check the engagement of all gears, including reverse and make sure the gear shift lever moves freely.

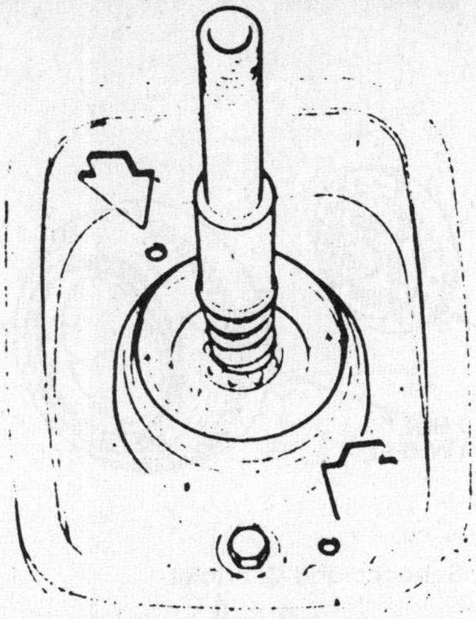
Align holes in shift plate; Quantum, Cabriolet and Scirocco

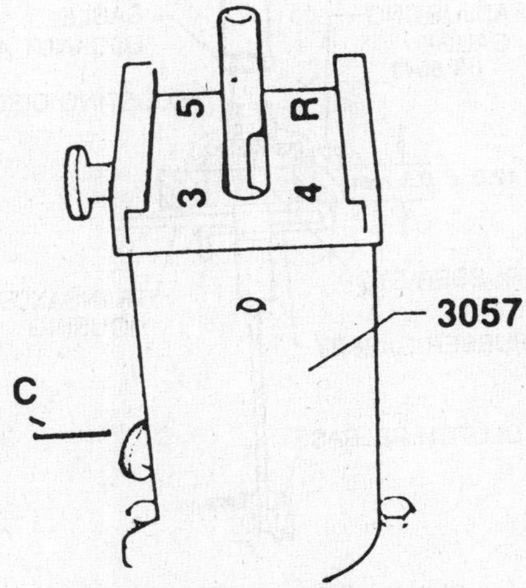

Shifter alignment tool in place on Quantum

Quantum

This procedure requires special tool VW–3057 or equivalent.

1. Place the lever in neutral.
2. Raise and safely support the vehicle and loosen the clamp nut.
3. Inside the vehicle, remove the gear lever knob and the shift boot. It is not necessary to remove the console. Align the centering holes of the lever housing and the lever bearing housing.
4. Install the tool with the locating pin toward the front. Push the lever to the left side of the tool cut-out. Tighten the lower knurled knob to secure the tool.
5. Move the top slide of the tool to the left stop and tighten the upper knurled knob.
6. Push the shift lever to the right side of the cutout. Align the shift rod and shift finger under the vehicle and tighten the clamp nut. Remove the tool.
7. Place the lever in first gear. Press the lever to the left side against the stop. Release the lever; it should spring back ¼–½ in. If not, move the lever housing slightly sideways to correct. Check that all gears can be engaged easily, particularly reverse.

CLUTCH

Clutch Assembly

REMOVAL & INSTALLATION

Jetta, Golf, Scirocco and Cabriolet

1. Raise and safely support the vehicle and remove the transaxle.
2. Attach a toothed flywheel holder tool VW–558 or equivalent, to the flywheel and gradually loosen the flywheel-to-pressure plate bolts a few turns at a time. Use a criss-cross pattern to prevent distortion.
3. Remove the flywheel and the clutch disc.
4. Use a small prybar to remove the release plate retaining ring. Remove the release plate.

To install:

5. Use new bolts to attach the pressure plate to the crankshaft. Use a thread locking compound and torque the bolts in a diagonal pattern to 72 ft. lbs. (100 Nm).
6. Lightly lubricate the clutch disc splines, release plate contact surface and pushrod socket with multi-purpose grease. Install the release plate, retaining ring and clutch disc.
7. Install a centering tool VW–547 or equivalent, to align the clutch disc.
8. Install the flywheel, tightening the bolts 1–2 turns at a time in a criss-cross pattern to prevent distortion. Torque the bolts to 14 ft. lbs. (20 Nm).
9. Remove the alignment tool, reinstall the transaxle and adjust the clutch cable.

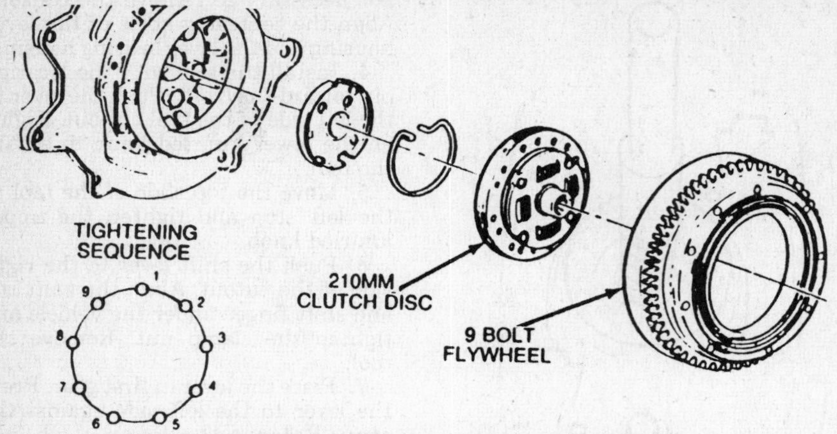

Clutch assembly on Golf, Jetta, Scirocco and Cabriolet

Fox, Quantum, Passat and Corrado

1. Raise and safely support the vehicle and remove the transaxle.
2. Matchmark the flywheel and pressure plate if the pressure plate is going to be reused.
3. Gradually loosen the pressure plate bolts 1–2 turns at a time in a criss-cross pattern to prevent distortion.
4. Remove the pressure plate and disc.
5. Check the clutch disc for uneven or excessive lining wear. Examine the pressure plate for cracking, scorching or scoring. Replace any questionable components.

To install:

6. Install the clutch disc and pressure plate with the springs on the disc towards the plate. Use an alignment tool to keep the clutch disc centered.
7. Gradually tighten the pressure plate-to-flywheel bolts in a criss-cross pattern. Tighten the bolts to 18 ft. lbs. (24 Nm).
8. Install the clutch release bearing.
9. Install the transaxle.

PEDAL HEIGHT/FREE-PLAY ADJUSTMENT

Hydraulic Clutch

If equipped with hydraulic clutch linkage, the slave cylinder has a bleeder screw to purge air from the system. The clutch pedal linkage rod is adjustable to maintain proper pedal height; 10mm above brake pedal.

Adjustable Cable

NOTE: On cable clutches, there is a special tool US5043 or equivalent, which makes it easier to determine proper adjustment, however proper adjustment can be accomplished without it.

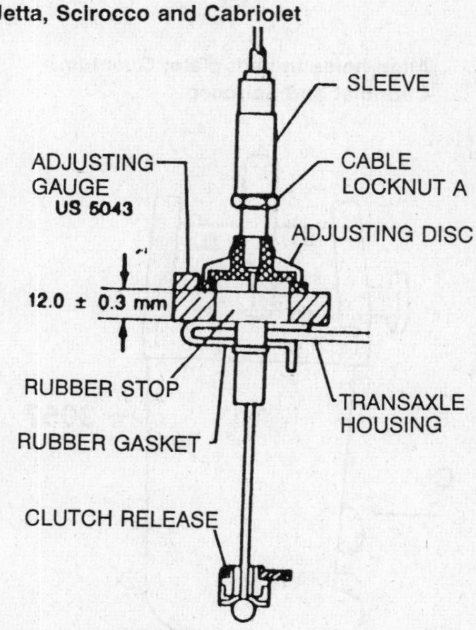

Clutch adjstment on cable clutch

1. Depress the clutch pedal several times.
2. Pull the cable adjusting sleeve up at the transaxle until resistance is felt and insert the gauge or measure the clearance.
3. Loosen the locknut and turn the adjusting sleeve until there is no free play at the gauge. Without the gauge, this distance should be 0.472 in. (12mm).
4. Tighten the locknut and operate the pedal several times. Recheck the adjustment.

Clutch Cable

ADJUSTMENT

Self-Adjusting Cables

1. If the original cable is being reinstalled, compress the spring and hold the cable in place on the transaxle. A 2nd person is required to attach the cable to the clutch lever.
2. If a new cable is being installed, there is a strap holding the spring in place. Remove the strap after the cable is in place.
3. Operate the pedal several times to adjust the cable.

REMOVAL & INSTALLATION

Except Self-Adjusting

1. Loosen the adjustment.
2. Disengage the cable at the lever arm, noting the placement of the parts.
3. Unhook the cable from the pedal and pull the cable from the firewall.

To install:

4. Grease the pedal end and install and connect the new cable. Adjust the pedal free-play.

Self-Adjusting

1. Depress the pedal several times.
2. Compress the spring located under the boot at the top of the adjuster mechanism and remove the cable at the release lever, noting the placement of the parts.
3. Unhook the cable from the pedal and pull the cable from the firewall.

To install:

4. Grease the pedal end and install the new cable onto the pedal. Compress the spring and have a helper pull the cable down and install to the release lever.
5. If the adjuster spring is retained by a strap, remove the strap after cable installation.
6. Depress the clutch pedal several times to adjust the cable.

Clutch Master Cylinder

REMOVAL & INSTALLATION

The clutch master cylinder is located on the fire wall below the brake master cylinder. The clutch slave cylinder is located on top of the transaxle. The clutch master cylinder is supplied fluid from the brake fluid reservoir. Whenever any part of the system is removed or replaced the system must be bled to remove any air that may be in the lines.

1. Remove the windshield washer bottle.
2. Remove the pressure line from the rear of the clutch master cylinder and plug the fitting.
3. Disconnect the fluid supply hose from the brake fluid reservoir.
4. Inside the vehicle, disconnect the pushrod from the clutch pedal by removing the clip on the retaining pin.
5. Remove the 2 mounting nuts and remove the clutch master cylinder from the vehicle.

To install:

6. Insert the pushrod through the

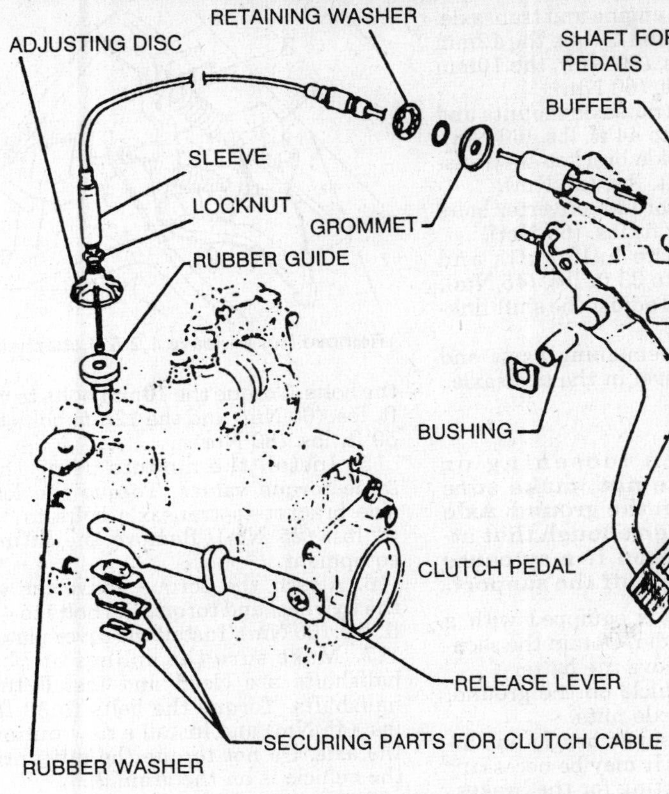

Clutch pedal and cable assembly

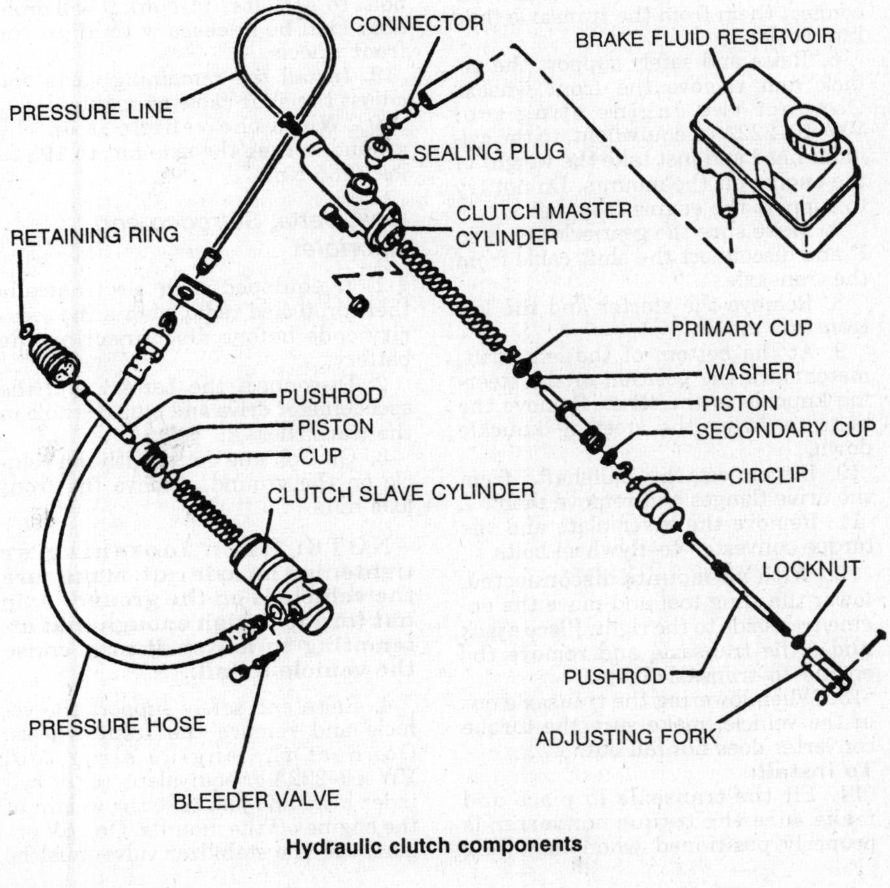

Hydraulic clutch components

firewall, install new nuts and torque to 5 ft. lbs. (7 Nm). Pin the rod to the pedal and install the clip.

7. Connect the supply line to the brake master cylinder and install the pressure line to the rear of the clutch master cylinder.

8. Fill the brake reservoir and bleed the clutch system.

Clutch Slave Cylinder

REMOVAL & INSTALLATION

1. Raise and safely support the vehicle.
2. Disconnect and plug the pressure line to the slave cylinder.
3. Remove the slave cylinder by removing the spring pin and clip from the transaxle.

To install:

4. Align the slave cylinder on the transaxle housing and insert the spring pin and clip. Bolt the cylinder in place.
5. Connect the pressure line and lower the vehicle.
6. Fill the brake reservoir and bleed the system.

Hydraulic Clutch System Bleeding

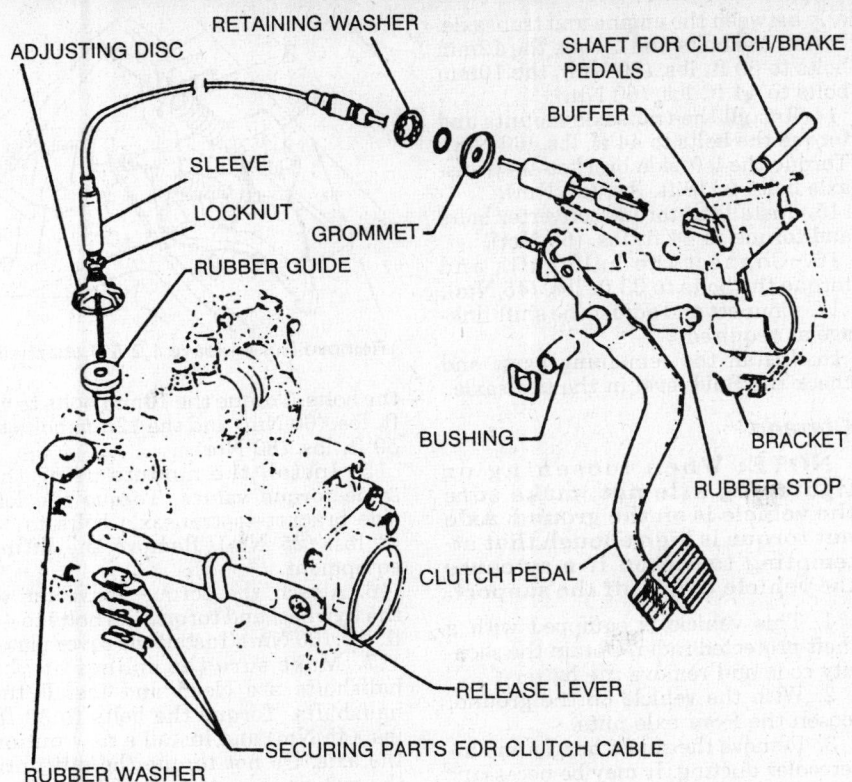

1. The clutch and brakes share the same reservoir. Clean all dirt and grease from the cap to make sure no foreign subtances enter the system.
2. Remove the cap and diaphragm and fill the reservoir to the top with the approved DOT brake fluid. Fully loosen the bleed screw which is in the slave cylinder body next to the inlet connection.
3. At this point bubbles of air will appear at the bleed screw outlet. When the slave cylinder is full and a steady stream of fluid comes out of the slave cylinder bleeder, tighten the bleed screw.
4. Refill the reservoir and cap it. Exert a light load of about 20 lbs. to the slave cylinder piston by pushing the release lever towards the cylinder and loosen the bleed screw. Maintain a constant light load, fluid and any air that is left will be expelled through the bleed port. Tighten the bleed screw when a steady flow of fluid with no air is being expelled.
5. Fill the reservoir fluid level back to normal capacity and if necessary repeat Step 4.
6. Exert a light load to the release lever but do not open the bleeder screw as the piston in the slave cylinder will move slowly down the bore. Repeat this operation 2–3 times; the fluid movement will force any air left in the

system into the reservoir. The hydraulic system should now be fully bled.

7. Check the the operation of the clutch hydraulic system and repeat this procedure, if necessary. Check the pushrod travel at the slave cylinder to insure the minimum travel 0.57 in. (15mm).

AUTOMATIC TRANSAXLE

For further information on transmissions/transaxles, please refer to "Chilton's Guide to Transmission Repair".

Transaxle Assembly

REMOVAL & INSTALLATION

Passat

1. If equipped with electronically theft-protected radio, obtain the security code before disconnecting the battery.
2. Remove the battery and disconnect the wiring from the transaxle.
3. Remove the upper engine-to-transaxle bolts.
4. Raise and safely support the vehicle and remove the front wheels. Connect the engine sling tool VW–10–222A or equivalent, to the cylinder head and just take the weight of the engine off the mounts. On Passat, the idle stabilizer valve must be removed to attach the tool. Do not try to support the engine from below.
5. Put the shifter in **P** and disconnect the shift cable.
6. Clamp and remove the hoses at the transaxle cooler.
7. Remove the starter and the engine's left and right mounts.
8. Remove the skid plate and disconnect the halfshafts from the drive flanges. Hang them from the body with wire.
9. Remove the torque converter plate and turn the engine as needed to remove the torque converter-to-flywheel bolts.
10. Remove the remaining transaxle mounts and lower the hoist slightly.
11. Support the transaxle with a jack and remove the remaining engine-to-transaxle bolts. Be careful to secure the torque converter so it does not fall out of the transaxle.
12. Carefully lower the transaxle out of the vehicle.

To install:

13. Fit the transaxle into the vehicle and make sure the guide pins fit properly between the engine and transaxle. Install the bolts and torque the 12mm bolts to 59 ft. lbs. (80 Nm), the 10mm bolts to 44 ft. lbs. (60 Nm).
14. Install the transaxle mounts and torque the bolts to 44 ft. lbs. (60 Nm). Torque the left side bracket–to–transaxle bolts to 18 ft. lbs. (25 Nm).
15. Install the torque converter bolts and torque to 44 ft. lbs. (60 Nm).
16. Connect the halfshafts and torque the bolts to 33 ft. lbs. (45 Nm).
17. Connect and adjust the shift linkage as required.
18. Install the remaining parts and check the fluid level in the transaxle.

Corrado

NOTE: When loosening or tightening axle nut, make sure the vehicle is on the ground. Axle nut torque is high enough that attempting to loosen it may cause the vehicle to fall off the support.

1. This vehicle is equipped with a theft protected radio. Obtain the security code and remove the battery.
2. With the vehicle on the ground, loosen the front axle nuts.
3. Remove the supercharger and intercooler ducting. It may be necessary to remove the ducting for the brakes.
4. Disconnect the wiring from the transaxle.
5. Clamp the coolant hoses and disconnect them from the transaxle fluid intercooler.
6. Raise and safely support the vehicle and remove the front wheels. Connect the engine sling tool VW–10–222A or equivalent, to the cylinder head and just take the weight of the engine off the mounts. Do not try to support the engine from below.
7. Make sure the gear selector is in **P** and disconnect the shift cable from the transaxle.
8. Remove the starter and the left transaxle mount.
9. At the bottom of the left strut, matchmark the position of the steering knuckle to the strut. Remove the bolts to swing the steering knuckle down.
10. Disconnect the halfshafts from the drive flanges and remove them.
11. Remove the cover plate and the torque converter–to–flywheel bolts.
12. With all mounts disconnected, lower the sling tool and move the engine/transaxle to the right. Place a jack under the transaxle and remove the engine–to–transaxle bolts.
13. When lowering the transaxle out of the vehicle, make sure the torque converter does not fall out.

To install:

14. Fit the transaxle in place and make sure the torque converter is properly positioned when installing

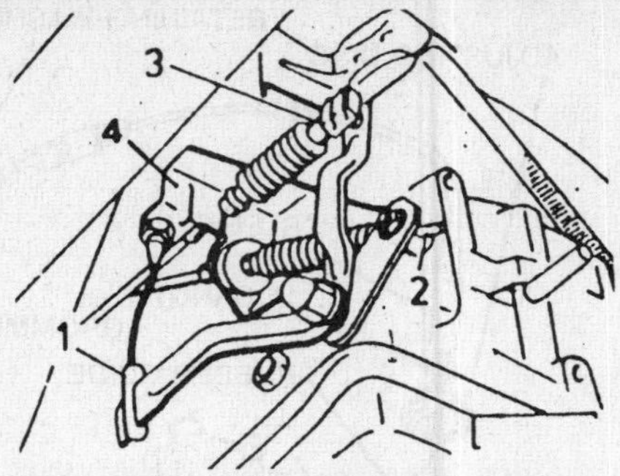

Remove No. 4—leave 1,2 & 3 attached

the bolts. Torque the 10mm bolts to 44 ft. lbs. (60 Nm) and the 12mm bolts to 59 ft. lbs. (80 Nm).

15. Install the mounts using the same torque values. Torque the left side bracket–to–transaxle bolts to 18 ft. lbs. (25 Nm). Remove the lifting equipment.
16. Attach the torque converter to the flywheel and torque the bolts to 44 ft. lbs. (60 Nm). Install the cover plate.
17. Make sure the splines on the halfshafts are clean and install the halfshafts. Torque the bolts to 33 ft. lbs. (45 Nm) and install a new nut on the axle. Do not torque the nut until the vehicle is on the ground.
18. Reassemble the left steering knuckle to the strut, making sure to align the matchmarks. Torque the bolts to 70 ft. lbs. (95 Nm). It will probably still be necessary to align the front wheels.
19. Install the remaining parts and adjust the shift cable as required.
20. When the vehicle is on the ground, torque the axle nut to 195 ft. lbs. (265 Nm).

Golf, Jetta, Scirocco and Cabriolet

1. If equipped with electronically theft-protected radio, obtain the security code before disconnecting the battery.
2. Disconnect the battery and the speedometer drive and plug the hole in the transaxle.
3. On Golf and Jetta, with the vehicle on the ground, remove the front axle nuts.

NOTE: When loosening or tightening an axle nut, make sure the vehicle is on the ground. Axle nut torque is high enough that attempting to loosen it may cause the vehicle to fall.

4. Raise and safely support the vehicle and remove the front wheels. Connect the engine sling tool VW–10–222A or equivalent, to the cylinder head and just take the weight of the engine off the mounts. On 16V engine, the idle stabilizer valve must be

removed to attach the tool. Do not try to support the engine from below.

5. Remove the driver's side rear transaxle mount and support bracket.

6. On Golf and Jetta, remove the front mount bolts from the transaxle and from the body and remove the mount as a complete assembly.

7. Remove the selector and accelerator cables from the transaxle lever but leave them attached to the bracket. Remove the bracket assembly to save the adjustment.

8. Unbolt the halfshafts from the drive flanges. On Golf and Jetta, the shafts must be removed, which may require separating the ball joints from the wheel bearing housing to gain the necessary clearance. Remove the ball joint clamping bolt.

9. Remove the heat shield and brackets and remove the starter. On Scirocco and Cabriolet, the front mount comes off with the starter.

10. Turn the engine as needed to remove the torque converter-to-flywheel bolts.

11. Remove the remaining transaxle mounts and, on Golf and Jetta, the subframe bolts and allow the subframe to hang free.

12. Support the transaxle with a jack and remove the remaining engine-to-transaxle bolts. Be careful to secure the torque converter so it does not fall out of the transaxle.

13. Carefully lower the transaxle from the vehicle.

To install:

14. When reinstalling, make sure the torque converter is fully seated on the pump shaft splines. The converter should be recessed into the bell housing and turn by hand. Keep checking that it still turns while drawing the engine and transaxle together with the bolts.

15. Install the engine–to–transaxle bolts and torque to 55 ft. lbs. (75 Nm).

16. Install all mount and subframe bolts before tightening any on them. Tighten the bolts starting at the rear and work forward. Torque the smaller bolts to 25 ft. lbs. (34 Nm) and the larger bolts to 58 ft. lbs. (80 Nm). Remove the lifting equipment when all mounts are installed.

17. Install the torque converter–to–flywheel bolts and torque them to 26 ft. lbs. (35 Nm).

18. Install the starter and torque the bolts to 14 ft. lbs. (20 Nm). Install the heat shields.

19. If the halfshafts were removed, make sure the splines are clean and apply a thread locking compound to the splines before sliding it into the hub. Connect the halfshafts to the drive flanges and torque the bolts to 37 ft. lbs. (50 Nm). Install new axle nuts but do not fully torque them until the vehicle is on the ground.

20. If removed, fit the ball joints to the control arm and torque the clamping bolt to 37 ft. lbs. (50 Nm).

21. Connect and adjust the shift linkage as required.

22. When assembly is complete and the vehicle is on its wheels, torque the axle nuts to 195 ft. lbs. (265 Nm).

SHIFT/THROTTLE LINKAGE ADJUSTMENT

1. With the engine warm and the gear selector in **P**, loosen the adjusting nut and disconnect the accelerator pedal cable from the transaxle.

2. On the intake plenum, loosen the nuts on the cable bracket and move the sleeve away from the throttle to take up any play. The throttle must remain closed.

3. Turn the nut on the throttle side of the bracket up to the bracket and tighten the other nut against the bracket. Be sure the throttle is still against it's stop.

4. Reconnect the cable to the transaxle and have an assistant push the gas pedal to the floor.

5. Push the transaxle lever against the stop and turn the adjusting nut to remove all slack from the cable. Tighten the locknut, release the pedal and push it again to check adjustment.

FRONT SUSPENSION

MacPherson Strut

REMOVAL & INSTALLATION

Except Fox and Quantum

NOTE: When loosening or tightening axle nut, make sure the vehicle is on the ground. Axle nut torque is high enough that attempting to loosen it may cause the vehicle to fall off the support.

1. With the vehicle on the ground, remove the front axle nut.

2. Raise and safely support the vehicle and remove the front wheels.

3. Detatch the brake line from the strut and remove the caliper. Hang it from the body with wire.

4. Clean and matchmark the position of the strut to the wheel bearing housing for reassembly.

5. Remove the bolts and push the wheel bearing housing down away from the strut.

CAUTION

On Scirocco and Cabriolet, do not remove the large nut in the center of the top bearing. The spring will be released while still compressed.

6. On Scirocco and Cabriolet, remove the nuts holding the rubber strut bearing and lower the strut from the vehicle. On other vehicles, remove the large center nut to lower the strut from the vehicle.

To install:

7. Place the strut into the fender and install the nuts. On Scirocco and Cabriolet, torque the 3 nuts to 14 ft. lbs. (20 Nm). On all other models, torque the center nut to 44 ft. lbs. (60 Nm).

8. Fit the wheel bearing housing into the strut and torque the bolts to 70 ft. lbs. (95 Nm).

9. Install the brake caliper and torque the bolts to 44 ft. lbs. (60 Nm).

10. When assembly is complete and the vehicle is on the ground, torque the axle nut to 145 ft. lbs. (196 Nm) for M18 nut or 175 ft. lbs. (237 Nm) for M20 nut.

Fox and Quantum

NOTE: When loosening or tightening axle nut, make sure the vehicle is on the ground. Axle nut torque is high enough that attempting to loosen it may cause the vehicle to fall off the support.

1. With the vehicle on the ground, remove the front axle nut.

2. Raise and safely support the vehicle and remove the wheels.

3. Remove the brake caliper from the strut and hang from the body it with wire. Detach the brake line from the strut and remove the rotor.

4. At the tie rod end, remove the cotter pin and the castellated nut and remove the end from the strut with a puller.

5. Loosen the stabilizer bar bushings and detach the end from the strut being removed.

6. Remove the ball joint clamp bolt and push the control arm down to disengage the ball joint from the strut.

7. On some vehicles, the halfshaft spline is secured in the hub with thread sealer. The best way to remove it is to push it out with a wheel puller. Do not use heat, this will ruin the bearing. Pull the strut away from the halfshaft.

8. Remove the upper strut-to-fender retaining nut and lower the strut assembly down and out of the vehicle.

To install:

9. Install the upper end of the strut and torque the nut to 44 ft. lbs. (60 Nm).

10. Make sure the axle splines are clean and apply fresh thread sealer to the outer end. Insert the axle through the hub and install a new axle nut. Do not torque the nut until the vehicle is on the ground.

11. Fit the ball joint into the strut and torque the campling bolt to 37 ft. lbs. (50 Nm) on Quantum, 44 ft. lbs. (60 Nm) on Fox.

12. Lightly lubricate the stabilizer arm bushings with silicone and install them. Torque the bolts to 15 ft. lbs. (20 Nm).

13. Install the brake calipers and torque the bolts to 44 ft. lbs. (60 Nm).

14. When the assembly is complete and the vehicle is on the ground, torque the axle nut to 170 ft. lbs. (230 Nm).

Lower Ball Joints

INSPECTION

1. To check the ball joint, raise and safely support the vehicle. Let the front wheels hang free.

2. Insert a prybar between the control arm and the ball joint clamping bolt. Be careful to not damage the ball joint boot.

3. Measure the play between the bottom of the ball joint and the clamping bolt with a caliper. Total must not exceed 0.100 in. (2.5mm).

REMOVAL & INSTALLATION

1. Raise and safely support the vehicle, allowing the front wheels to hang. Remove the front wheels.

2. On Fox and Quantum, matchmark the ball joint-to-control arm position.

3. Remove the ball joint clamping bolt.

4. Pry the lower control arm down to remove the ball joint from the strut.

5. Remove the ball joint-to-lower control arm retaining nuts and bolts or drill out the rivets with a ¼ in. (6mm) drill.

6. Remove the ball joint assembly.

To install:

7. On Fox and Quantum, install the ball joint in the reverse order of removal. If no parts were installed other than the ball joint, align the matchmarks. No camber adjustment is necessary if this is done. Pull the ball joint into alignment with pliers. Tighten the 2 control arm-to-ball joint bolts to 47 ft. lbs. (64 Nm) and the ball joint clamping bolt to 37 ft. lbs. (50 Nm) on Quantum, or 44 ft. lbs. (60 Nm) on Fox.

8. On all other vehicles, bolt the new ball joint in place. Torque the bolts to 18 ft. lbs. (25 Nm) and the ball joint clamping bolt to 37 ft. lbs. (50 Nm).

Lower Control Arm

REMOVAL & INSTALLATION

NOTE: When removing the driver's side control arm on Scirocco/Cabriolet vehicles equipped with an automatic transaxle, it may be necessary to lift the engine/transaxle. First support the engine from above or below. Remove the front left engine mounting nut and bolt, remove the rear mount and raise the engine to expose the front control arm bolt.

1. Raise and safely support the vehicle and remove the wheels.

2. Remove the ball joint clamping bolt and pry the control arm down.

3. Remove the rubber bushings to unfasten the stabilizer bar.

4. Remove the control arm mounting bolts and remove the control arm.

To install:

5. Installation is the reverse of removal. Torque the following components:

Fox and Quantum control arm bushing bolts—40 ft. lbs. (55 Nm).

Scirocco/Cabriolet control arm bushing bolts—50 ft. lbs. (68 Nm).

All others: front bushing bolts—96 ft. lbs. (130 Nm), rear bolts—59 ft. lbs. (80 Nm).

Stabilizer bar link rods—18 ft. lbs. (25 Nm).

Stabilizer bar bushing clamp bolts—32 ft. lbs. (43 Nm).

Ball joint clamping bolt—37 ft. lbs. (50 Nm) for 8mm bolt or 44 ft. lbs. (60 Nm) for 10mm bolt.

REAR SUSPENSION

Shock Absorbers

REMOVAL & INSTALLATION

NOTE: Do not remove both suspension struts at the same time as this will overload the axle beam bushings and brake lines.

1. Working inside the vehicle, remove the cap from the top shock mount and note the way the washers and bushings stack up.

2. Remove the upper strut bolts–to-body bolts.

3. Slowly lift the vehicle until the wheels are slightly off the ground.

4. Unbolt the strut from the axle and carefully remove the strut from the vehicle. It may be necessary to press the axle down slightly when removing the strut.

5. Installation is the reverse of removal. Torque the strut-to-body bolts to 26 ft. lbs. (35 Nm) and the strut-to-axle bolts to 77 ft. lbs. (105 Nm).

Rear Control Arms

REMOVAL & INSTALLATION

Quantum Syncro

The Quantum Syncro is a 4WD vehicle with similar bearing installations both front and rear. To remove the rear bearing, the control arm must be removed.

NOTE: When loosening or tightening axle nut, make sure the vehicle is on the ground. Axle nut torque is high enough that attempting to loosen it may cause the vehicle to fall off the support.

1. With the vehicle on the ground, loosen the axle nut.

2. Raise and safely support the vehicle and remove the wheels.

3. Remove the bolts retaining the axle shaft to the differential and slide the axle shaft from the vehicle.

4. Remove the disc brake caliper and hang it from the body with wire. Do not hang it by the brake lines.

5. Unbolt the stabilizer bar from the control arm.

6. Remove the lower shock mount bolt and the control arm mounting bolts and remove the arm from the vehicle.

To install:

7. Install the control arm onto the vehicle and torque the following in order:

Control arm bolts—88 ft. lbs. (120 Nm).

Lower shock mount—48 ft. lbs. (65 Nm).

Stabilizer bar brackets—18 ft. lbs. (25 Nm).

Brake caliper—48 ft. lbs. (65 Nm).

Inner axle CV-joint bolts—33 ft. lbs. (45 Nm).

8. With the wheel installed and the vehicle on the ground, install a new axle nut and torque it to 170 ft. lbs. (230 Nm).

Rear Wheel Bearings

REMOVAL & INSTALLATION

Except Quantum Synchro

1. Raise and safely support the vehicle and remove the rear wheels.

2. On drum brakes, insert a small pry tool through a wheel bolt hole and push up on the adjusting wedge to slacken the rear brake adjustment.
3. On disc brakes, remove the caliper.
4. Remove the grease cap, cotter pin, locking ring, axle nut and thrust washer. Carefully remove the bearing and put all these parts where they will stay clean.
5. Before installing, pack the bearing. If any brake dust has fallen onto the axle, wipe off all the axle grease and put on new high temperature bearing grease.
6. Installation is reverse of removal.

Quantum Synchro

1. Raise and safely support the vehicle and remove the rear control arm.
2. Remove the brake splash shield. To remove the hub, support the arm in an arbor press with the hub facing down.
3. Use a proper size arbor that will fit through the bearing and press the hub out.
4. If the inner bearing race stayed on the hub, clamp the hub in a vise and use a bearing puller to remove it.
5. Remove the snapring and press the bearing out.
6. Clean the bearing housing and hub with a wire brush and inspect all parts. Replace parts that have been distorted or discolored from heat. If the hub is not absolutely prefect where it contacts the inner bearing race, the new bearing will fail quickly.

To install:

7. The new bearing is pressed in from the hub side. Install the snapring and support the bearing on the press. Lubricate the bearing housing and press it onto the outer bearing race. Install the outer snapring.
8. Support the hub on the press table and press the inner race of the bearing onto the hub. Make sure the press tool contacts only the inner race or the bearing will be destroyed.
9. Install the splash shield and install the control arm onto the vehicle. No bearing adjustment is required.

ADJUSTMENT

1. When adjusting the bearing nut, the thrust washer must still be movable with light effort with a small pry tool.
2. When installing the locking ring, keep trying different positions of the ring on the nut until the cotter pin goes into the hole. Don't turn the nut to align the locking ring with the hole in the axle. Use a new cotter pin. Install the grease cap with a rubber hammer.

Rear Axle Assembly

REMOVAL & INSTALLATION

Except Syncro

1. Raise and safely support the vehicle and remove the rear wheels.
2. Remove the rear brake caliper or drum.
3. Disconnect the brake line and remove the caliper or back plate (with brakes attached) from the vehicle.
4. Disconnect the other end of the brake line and unclip the brake line and parking brake cable from the axle. Unhook the brake pressure regulator spring from the bracket.
5. Support one side of the axle beam so it does not fall and remove the lower shock mount bolts from both sides.
6. Unless that is the part being repaired, avoid removing the axle bushing brackets. Removing these will mean aligning the rear bushings upon reassembly.
7. Remove the bolt from the center of each bushing and lower the axle from the vehicle.

To install:

8. Install the axle but do not torque the bushing bolts yet. They should be torqued with the vehicle on the ground to properly align the bushings.
9. Install the brakes, connect the hydraulic line and bleed the brakes.
10. With the vehicle on the ground, torque the right side axle bushing bolt first, then pry the left side bushing slightly towards the center of the vehicle and torque the left side.
11. Torque the following:
    Passat and Quantum axle bushing bolts—52 ft. lbs. (70 Nm).
    All other axle bushing bolts—44 ft. lbs. (60 Nm).
    Passat lower shock bolt—77 ft. lbs. (105 Nm).
    Golf/Jetta and Corrado shock mount—52 ft. lbs. (70 Nm).
    Scirocco/Cabriolet shock mount—32 ft. lbs. (45 Nm).
    Fox/Quantum shock mount—52 ft. lbs. (70 Nm).

STEERING

Steering Wheel

CAUTION

Failure to properly disarm the air bag system may result in accidental deployment of the air bag and possible personal injury.

The air bag system is equipped with a back-up power supply. The battery must be disconnected for more than 20 minutes before the power supply is fully discharged and the system is considered disarmed. A memory saver device will keep the power supply charged.

REMOVAL & INSTALLATION

With Air Bag

1. Disconnect battery, wait more than 20 minutes.
2. Remove the Torx® head screws at the back of the steering wheel.
3. Carefully detach the air bag unit from the wheel and disconnect the wire at the center.
4. Place unit in a safe place, horn pad up.
5. Point the front wheels straight ahead, remove the ignition key to lock the steering column and remove the nut and spring washer.
6. Mark the position of the wheel to the spline and pull the wheel straight off.
7. Installation is the reverse of removal. Torque the steering wheel nut to 30 ft. lbs. (40 Nm). When installing the air bag, use new Torx® screws and tighten to 7.5 ft. lbs. (10 Nm.). Do not over torque or the air bag may not function properly.

Without Air Bag

1. Remove the horn pad.
2. Remove the ignition key and turn the wheel until it locks.
3. Remove the center nut and matchmark the wheel to the splines.
4. Pull the wheel straight off.
5. Installation is the reverse of removal. Torque the center nut to 30 ft. lbs. (40 Nm).

Steering Column

REMOVAL & INSTALLATION

CAUTION

If equipped with an air bag, the negative battery cable must be disconnected for more than 20 minutes to disarm the system. Failure to do so may result in deployment of the air bag and possible personal injury.

1. Disconnect the negative battery cable and remove the steering wheel, turn signal and wiper switches.
2. At the bottom end of the column, locate the universal joint and remove the clamp bolt.
3. If equipped, pry out the leaf spring that secures the lower end of the column near the floor.
4. Remove the bolts holding the column to the dash and remove the column as an assembly. On some vehicles, the column is attached to the dash with shear bolts with heads that

break off at a specific torque. Removing these bolts is similar to removing broken studs. If a drill is being used, be careful not to damage the threads in the dash. These bolts are torqued to only 18 ft. lbs. (25 Nm).

5. Installation is the reverse of removal.

NOTE: Adjustable height steering columns are not repairable and must be replaced as a unit.

Manual Steering Rack

ADJUSTMENT

On some vehicles, there is a rack and pinion free-play adjustment screw and locknut, however this is not always accessible with the rack installed in the vehicle. Loosen the locknut and adjust the screw to allow smooth, non-binding movement of the rack.

REMOVAL & INSTALLATION

1. Raise and safely support the vehicle.
2. Remove both front wheels and disengage both tie rod ends.
3. At the steering column, remove the boot clamp, push the boot towards the body and remove the clamp bolt from the universal joint.
4. Remove the rack mounting nuts and remove the rack from it's mounts.
5. At this point on some vehicles, the rack cannot be removed from the body. Support the engine/transaxle and remove the subframe bolts or the rear transaxle mount and bracket to allow the rack to move towards the rear.
6. Installation is the reverse of removal. Torque the subframe bolts to 96 ft. lbs. (130 Nm).

Power Steering Rack

ADJUSTMENT

On some vehicles, there is a rack and pinion free-play adjustment screw and locknut, however this is not always accessible with the rack installed in the vehicle. Loosen the locknut and adjust the screw to allow smooth, non-binding movement of the rack.

REMOVAL & INSTALLATION

1. Raise and safely support the vehicle.
2. Remove both front wheels and disengage both tie rod ends.
3. Remove the low pressure (sucton) hose from the pump and drain the system into a catch pan. Properly discard the fluid.
4. At the steering column, remove the boot clamp, push the boot towards the body and remove the clamp bolt from the universal joint.
5. On Scirocco and Cabriolet, remove the exhaust manifold and shift linkage bracket.
6. Remove the rack mounting nuts and remove the rack from it's mounts.
7. At this point on some vehicles, the rack cannot be removed from the body. Support the engine/transaxle and remove the subframe bolts to allow the rack to move towards the rear. On Scirocco and Cabriolet, remove the transaxle mount and bracket.
8. Disconnect the power steering hydraulic lines and remove the rack.
9. Installation is reverse of removal. Torque the subframe bolts to 96 ft. lbs. (130 Nm). Don't forget to refill and bleed the system using the correct steering fluid.

Power Steering Pump

REMOVAL & INSTALLATION

1. Remove the suction hose and the pressure line from the pump, drain the fluid into a catch pan. Properly discard the fluid.
2. Loosen the tensioning bolt at the front of the tensioning bracket and remove the drive belt from the pump drive pulley.
3. Remove the pump mounting bolts and lift the pump from the vehicle.
4. To install, reverse the removal procedures. Torque the mounting bolts to 15 ft. lbs. (20 Nm). Tension the drive belt. Fill the reservoir with approved power steering fluid and bleed the system.

BELT ADJUSTMENT

To tension the drive belt, adjust the tensioner bolt so the belt will flex ½ in. under light thumb pressure.

SYSTEM BLEEDING

1. With the wheels turned all the way to the left, add power steering flu-

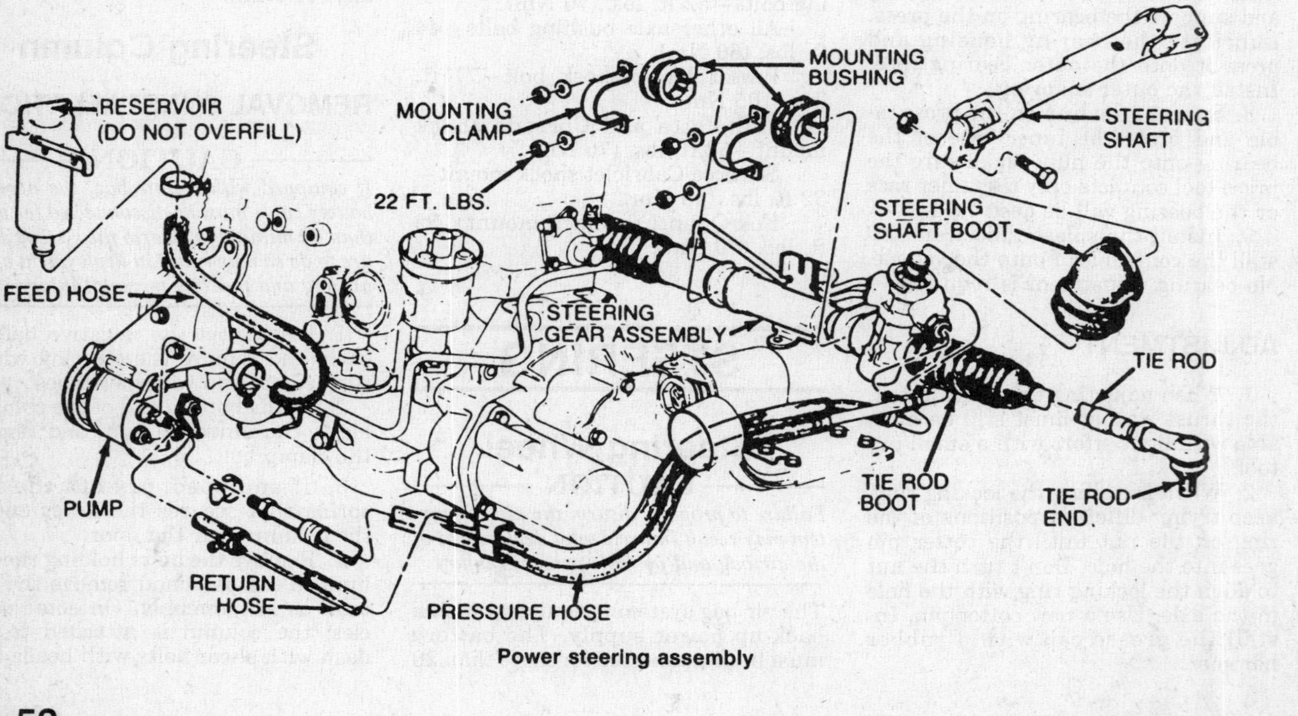

Power steering assembly

id to the **COLD** mark on the fluid level indicator.

2. Start the engine and run at fast idle momentarily, shut engine off and recheck fluid level. If necessary add fluid to to bring level to the **COLD** mark.
3. Start the engine and bleed the system by turning the wheels from side to side without hitting the stops.

NOTE: Fluid with air in it has a light tan or red appearance.

4. Return the wheels to the center position and keep the engine running for 2–3 minutes.
5. Road test the vehicle and recheck the fluid level making sure it is at the **HOT** mark.

Tie Rod Ends

REMOVAL & INSTALLATION

1. Raise and safely support the vehicle and remove the front wheels.
2. Remove the cotter pin and nut and press out the tie rod end. A small puller is required.
3. Hold the tie rod with a small pipe wrench or locking pliers and loosen the locking nut.
4. Back the nut away from the rod end far enough to mark the threads at the rod end with a crayon or chaulk, then unscrew the end from the tie rod.

To install:

5. When installing the new tie rod end, screw it onto the rod up to the mark on the threads. When tightening the locknut, hold the tie rod end and tighten the nut securely against it.
6. Reinstall the tie rod end into the steering knuckle and install the nut. Torque the nut to 22 ft. lbs. (30 Nm), then tighten as required to install a new cotter pin.
7. With the wheels on, lower the vehicle and roll it back and forth to settle the suspension. Check toe adjustment

NOTE: On some vehicles, only the right tie rod is adjustable.

BRAKES

CAUTION

Vehicles equipped with anti-lock brakes have extremely high fluid pressure in the system at all times. Do not disconnect any hydraulic fittings or open a bleeder without properly relieving the system pressure. Failure to properly discharge the system pressure may result in severe personal injury.

For all brake system repair and service procedures not detailed below, please refer to "Brakes" in the Unit Repair section.

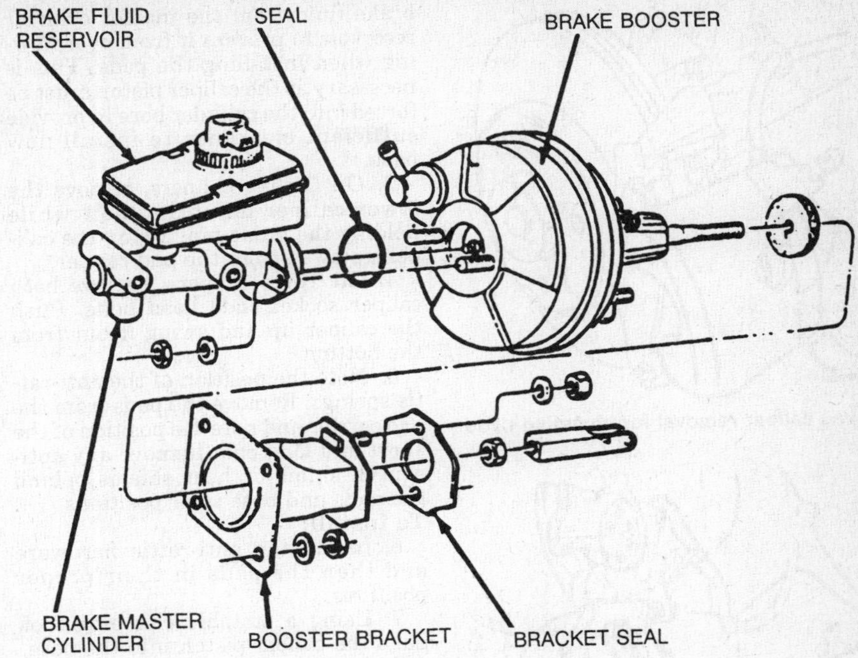

Master clyinder and brake booster

Master Cylinder

REMOVAL & INSTALLATION

Without ABS

1. Disconnect and plug the brake lines.
2. Disconnect the electrical plug from the sending unit for the low fluid switch.
3. Remove the 2 master cylinder mounting nuts and remove the master cylinder and reservoir.
4. The reservoir is held into the master cylinder by a press fit into rubber sealing plugs and can easily be pulled off. To reinstall, moisten the plugs with brake fluid and press it on.

To install:

5. Position the master cylinder and reservoir assembly onto the mounting studs on the booster and install the washers and nuts. Tighten the nuts to 15 ft. lbs. (20 Nm).
6. Remove the plugs and connect the brake lines.
7. Fill the reservoir and bleed the entire brake system.

Proportioning Valve

REMOVAL & INSTALLATION

1. Raise and safely support the vehicle.
2. Disconnect the spring and relieve the pressure by pushing the lever towards the axle.

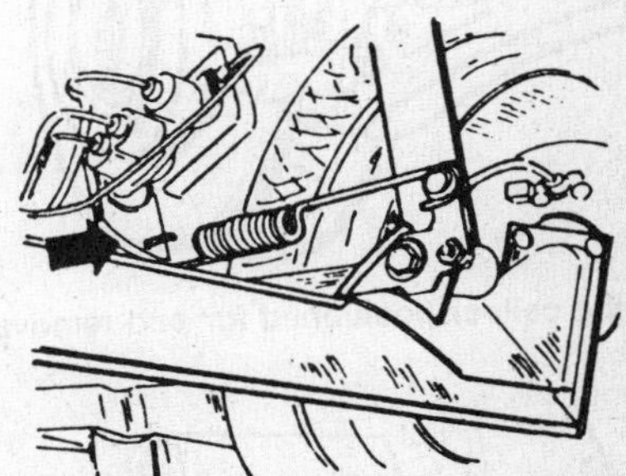

Relieving pressure at the proportioning valve—push lever toward axle

3. Using a line wrench, loosen the lines to the proportioning valve.
4. Remove the retaining nuts and remove the proportioning valve from the frame.
5. Installation is the reverse of removal. Bleed the brake system.

Power Brake Booster

REMOVAL & INSTALLATION

Without ABS

1. Remove the master cylinder from the booster.
2. At the pedals, remove the clevis pin on the end of the booster pushrod by unclipping it and pulling it from the clevis.
3. Disconnect the vacuum hose from the booster.
4. Unbolt the booster; remove the 2 nuts under the dashboard or the 4 nuts holding the booster to its bracket. Remove the booster.

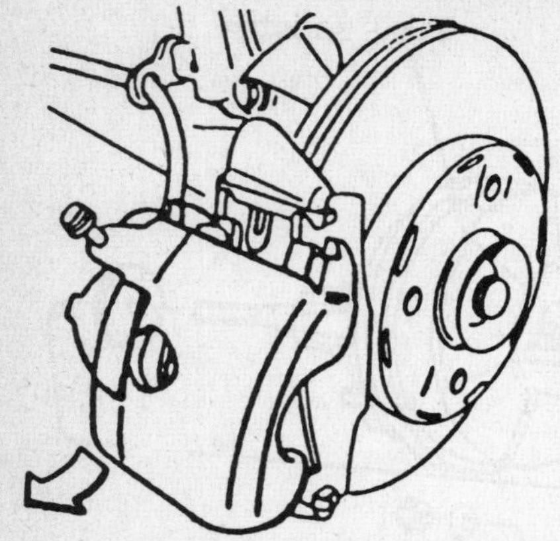

Teves caliper removal for changing pads

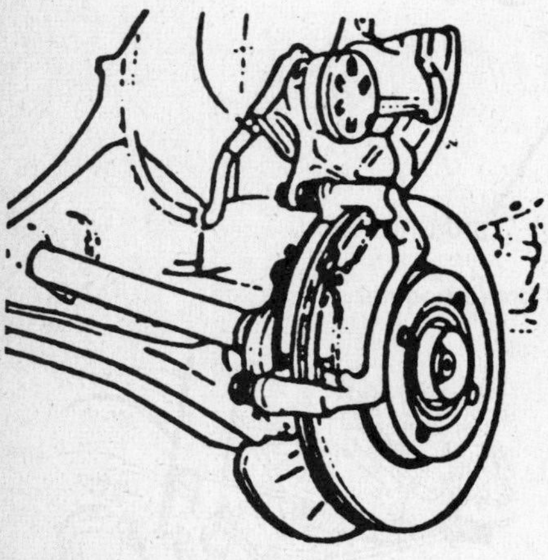

Girling caliper positioned for pad removal

Collapsing the caliper piston

5. Installation is the reverse of removal. Install the master cylinder and bleed the system.

Disc Brake Caliper and Pads

REMOVAL & INSTALLATION

1. Raise and safely support the vehicle and remove the front wheels.
2. Remove a sufficient quantity of brake fluid from the master cylinder reservoir to prevent it from over flowing when installing the pads. This is necessary as the caliper piston must be forced into the cylinder bore to provide sufficient clearance to install new pads.
3. On Girling calipers, remove the lower caliper mounting bolt while holding the guide pin. Swing the caliper up away from the pad carrier.
4. On Teves calipers, remove both caliper socket slide head bolts. Push the caliper up and swing it out from the bottom.
5. Note the position of the anti-rattle springs. Remove the pads from the pad carrier and note the position of the short and long pad. Remove any anti-squeek shims or heat shields behind the pads and note their positions.

To install:

6. Install the anti-rattle hardware and then the pads in their proper positions.
7. Using a suitable clamping tool, push the caliper piston into the bore.
8. Install any pad shims and install the caliper. Torque the caliper mounting bolts to 26 ft. lbs. (35 Nm).
9. Refill the master cylinder with fresh brake fluid.
10. Install the wheels and then pump the brake pedal several times to bring the pads into adjustment. Road test the vehicle.

Brake Rotor

REMOVAL & INSTALLATION

Front Disc Brakes

1. Raise and safely support vehicle and remove the wheel.
2. Remove the brake caliper, pads and the pad carrier.
3. With the wheel removed, the rotor is held in place only with a countersunk screw threaded into the hub. Remove the screw and slide the rotor off.

To install:

4. When reinstalling, torque the carrier bolts to 53 ft. lbs. (70 Nm). Before installing the rotor screw, lightly coat the threads with an anti-seize lubricant.
5. Reinstall the brake pads and caliper and pump the brake pedal several times to bring the pads into adjustment. Road test the vehicle.

Rear Disc Brakes

1. Raise and safely support the vehicle and remove the wheels.
2. Remove a sufficient quantity of brake fluid from the master cylinder reservoir to prevent it from over flowing when installing the pads. This is necessary as the caliper piston must be forced into the cylinder bore to provide sufficient clearance to install new pads.
3. Remove the parking brake cable clip from the caliper. Remove the parking brake cable.
4. Remove the upper mounting bolt from the brake caliper.
5. Swing the housing downward and remove the brake pads.
6. Check the rotor for scoring and resurface or replace as necessary. Check the caliper for fluid leaks or cracked boots. If any damage is found, the caliper will require overhauling or replacement.

To install:

NOTE: When replacing brake pads, always replace both pads on both sides of the vehicle. Mixed pads will cause uneven braking.

7. Retract the piston into the housing by rotating the piston clockwise using a 12mm Allen wrench.
8. Carefully clean the anchor plate and install the new brake pads onto the pad carrier.
9. Install the caliper to the pad carrier using a new self locking bolt and torque to 26 ft. lbs. (35 Nm).
10. Fasten the hand brake cable to the caliper. It may be necessary to back off the adjustment nuts at the hand brake handle.
11. Fill the reservoir with brake fluid and pump the brake pedal about 40 times with the engine off to set the piston. Setting the piston with the power assist could cause the piston to jam.
12. Check the parking brake operation, adjust the cable if necessary.
13. Road test the vehicle.

Brake Drums

REMOVAL & INSTALLATION

1. Raise and safely support vehicle and remove the rear wheels.
2. Insert a small pry tool through a wheel bolt hole and push up on the adjusting wedge to slacken the rear brake adjustment.
3. Remove the grease cap, cotter pin, locking ring, axle nut and thrust washer. Carefully remove the bearing and put all these parts where they will stay clean.
4. Carefully remove the drum.
5. Before installing, if any brake dust has fallen onto the axle, wipe off all the axle grease and apply a coat of new high temperature bearing. Install the parts in the reverse order of removal.

NOTE: When tightening the axle nut, the thrust washer must still be movable with a small pry

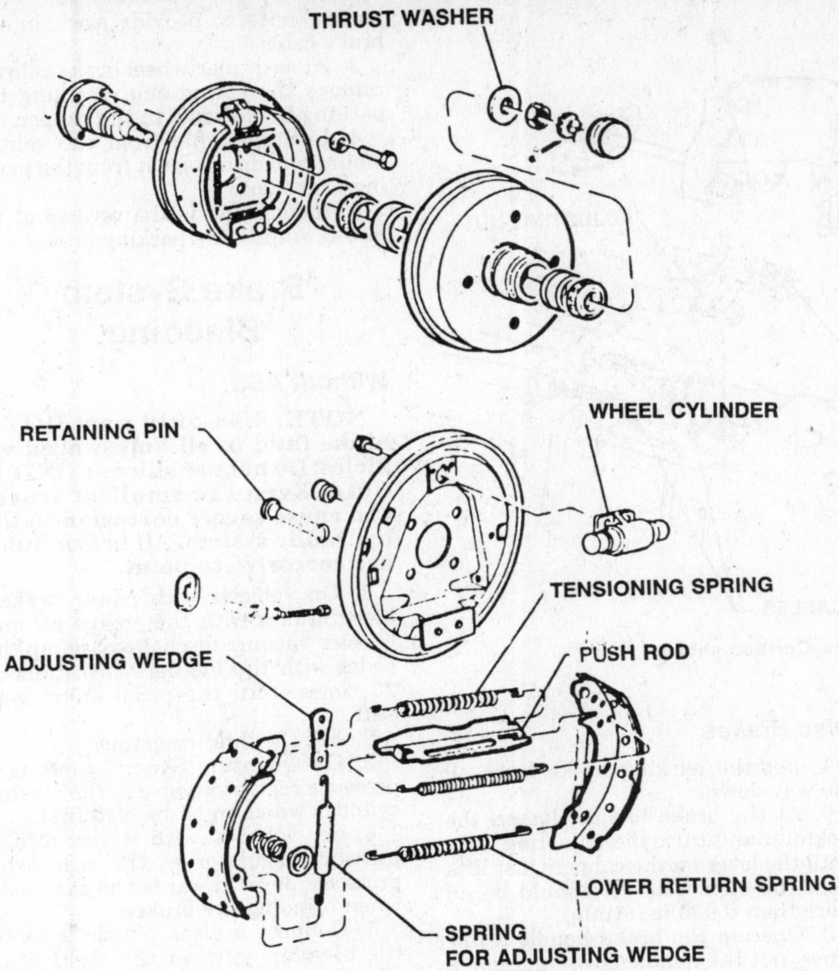

Rear brake drum removal and brake assembly

tool. Spin the drum and check that the thrust washer can still be moved.

6. When installing the locking ring, keep trying different positions of the ring on the nut until the cotter pin goes into the hole. Don't turn the nut to align the locking ring with the hole in the axle. Use a new cotter pin. Install the grease cap with a rubber hammer.

Brake Shoes

REMOVAL & INSTALLATION

1. Raise and safely support the vehicle and remove the rear wheels.
2. Remove the rear brake drum.
3. Remove the spring retainers by holding the pin behind the back plate, push in on the retainer and turn it 1/4 turn.
4. Remove the shoes from the back plate by pulling first 1 shoe, then the other against the upper spring and from it's wheel cylinder slot. Detatch the parking brake cable from the brake lever. The entire shoe assembly should now be free of the vehicle.
5. Carefully note the position of each spring, as spring shapes and positions have varied from vehicle to vehicle and year to year.
6. Clamp the pushrod in a vise and begin removing the springs, starting with the lower return spring, adjusting wedge spring, upper return spring and then the tensioning spring and adjusting wedge.
7. On most vehicles, the parking brake lever must be removed from the old shoes and reused. When new parts are purchased, don't forget the clip that holds the parking brake lever pin in place.

To install:

8. Check the wheel cylinder for frozen pistons or leaks. If any defects are found, replace the wheel cylinder.
9. Inspect the springs. If the springs are damaged or show signs of overheating they should be replaced. Indications of overheated springs are discoloration and distortion.
10. Inspect the brake drum and recondition or replace as necessary.
11. Clean the back plate and lubricate the shoe contact points with a suitable brake lubricant.
12. With the push rod clamped in a vise, attach the front brake shoe and tensioning spring.
13. Insert the adjusting wedge between the front shoe and pushrod so its lug is pointing toward the backing plate.
14. Remove the parking brake lever from the old shoe and attach it onto the new rear brake shoe.
15. Put the rear brake shoe and parking brake lever assembly onto the pushrod and hook up the spring.
16. Connect the parking brake cable to the lever and place the whole assembly onto the backing plate.
17. Install the hold-down springs.
18. Install the upper and lower return springs.
19. Install the adjusting wedge spring.
20. Center the brake shoes on the backing plate making sure the adjusting wedge is fully released (all the way up) before installing the drum.
21. Install the drum and wheel assembly.
22. Apply the brake pedal a few times to bring the brake shoe into adjustment.
23. If the wheel cylinder was replaced, bleed the system.
24. Road test the vehicle.

Wheel Cylinder

REMOVAL & INSTALLATION

1. Raise and safely support the vehicle and remove the wheel, drum and brake shoes.
2. Loosen the brake line on the rear of the cylinder but do not pull the line away from the cylinder or it may bend.
3. Remove the bolts and lockwashers that attach the wheel cylinder to the backing plate and remove the cylinder.
4. Position the new wheel cylinder on the backing plate and install the cylinder attaching bolts and lockwashers. Torque to 6 ft. lbs. (8 Nm).
5. Attach the brake line.
6. Install the brakes and bleed the system.
7. Road test the vehicle.

Parking Brake Cable

ADJUSTMENT

Fox and Quantum parking brake adjustment is made at the cable compensator, which is attached to the lever under the vehicle. On all other vehi-

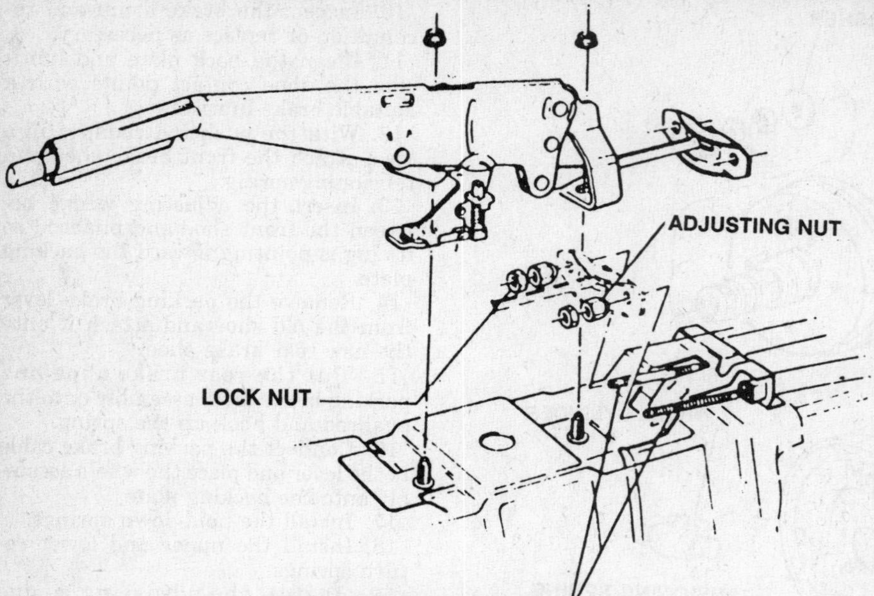

Parking brake adjusters–Corrado shown

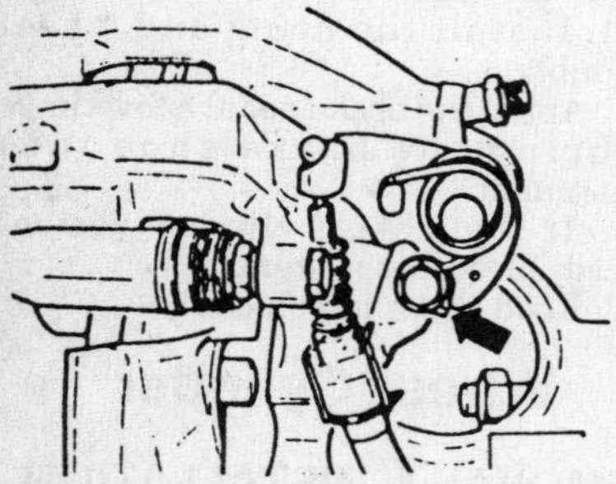
Lever moved off stop (arrow) no more than ¼ in. (1.0mm)

cles, the cable end nuts are in front of or behind the hand brake lever. Adjustment is performed at the cable end nuts.

Drum Brakes

1. Raise and safely support the vehicle.
2. Apply the parking brake so the lever is on the 2nd notch.
3. The Fox and Quantum adjustment is made directly under the passenger compartment, under a heat shield.
4. Tighten the compensator nut or adjusting nuts until both rear wheels can not be turned by hand.
5. Release the parking brake lever and check that both wheels can be easily turned.
6. Lubricate the Fox and Quantum compensator with chassis grease and reinstall the heat shield.

Disc Brakes

1. Let the parking brake lever all the way down.
2. At the brake handle, loosen the locknuts and turn the adjusting nuts until the lever on the calipers just lifts from the stop. The lift should be no more than 0.040 in. (1mm).
3. Operate the brake handle a few times, put the handle down and make sure both rear wheels still turn freely.
4. Retighten the locknuts at the handle.

REMOVAL & INSTALLATION

Rear Drum Brakes

1. Block the front wheels and release the hand brake.
2. Raise and safely support the rear of the vehicle.
3. Remove the rear brake shoes.
4. Remove the brake cable assembly from the back plates.
5. On Fox and Quantum, unhook the cable from the compensator.
6. On all others, remove the cable adjusting nut(s) and detach the cable guides from the floor pan.
7. Pull the cables out from under the vehicle.
8. Installation is the reverse of removal. Adjust the brakes and parking brake and road test the vehicle.

Disc Brakes

1. Raise and safely support the vehicle.
2. Release the parking brake. It may be necessary to unscrew the adjusting nuts to provide slack in the brake cable.
3. At each rear wheel brake caliper, remove the spring clip retaining the parking brake cable to the caliper.
4. Lift the cable from the caliper mount and disengage it from the parking brake lever.
5. Installation is the reverse of removal. Adjust the parking brake.

Brake System Bleeding

Without ABS

NOTE: Use only new DOT 4 brake fluid in all Volkswagen vehicles. Do not use silicone (DOT 5) fluid. Even the smallest traces can cause severe corrosion to the hydraulic system. All brake fluids are corrosive to paint.

1. On vehicles with power brakes, bleed brakes with the engine off and booster vacuum discharged; pump the pedal with the bleeders closed about 20 times until the pedal effort gets stiff.
2. Fill the fluid reservoir.
3. On Quantum Syncro, there is a pressure regulator valve at the master cylinder which must be bled first.
4. On vehicles with a rear brake pressure regulator at the rear axle, press the lever towards the rear axle when bleeding the brakes.
5. Connect a clear plastic tube to the bleeder valve at the right rear wheel, with the other end in a clean container.
6. Using either a power bleeder or an assistant pumping the pedal, open the bleeder valve until there are no air bubbles in the fluid stream. Be careful not to let the reservoir run out.
7. Repeat the procedure in sequence at the left rear, right front, left front: working farthest from the master cylinder to the nearest.

Anti-Lock Brake System Service

Vehicles with anti-lock brake systems (ABS) have an electronic fault memory and an indicator light on the instrument panel. When the engine is first started, the light will go on to indicate the system is pressurizing and performing a self diagnostic check. After the system is at full pressure, the light will go out. If it remains lit, there is a fault in the system. The fault memory can only be accessed with the VW tester VAG 1551 or VAG 1598, or equivalent. Be sure to unplug the ABS control unit connector and ground before doing any electric welding on the vehicle.

CAUTION

The ABS modulator assembly is capable of self pressurizing to more than 3000 psi. Serious injury may result if the brake service is attempted without disabling and depressurizing the system.

RELIEVING ANTI-LOCK BRAKE SYSTEM PRESSURE

With the ignition off, pump the brake pedal 25–35 times to depressurize the system. The system will recharge itself via the electric pump as soon as the ignition is turned on. Disconnect the pump or the battery to prevent unintended pressurization. The system can then be bled normally.

Modulator Assembly

REMOVAL & INSTALLATION

1. Turn the ignition **OFF** and depress the brake pedal 25–35 times to depressurize the modulator assembly. Disconnect the pump or battery to prevent unintended pressurization.
2. Inside the vehicle, under the left rear seat for Passat, behind the left kick panel for Corrado or near the right tail light for Golf and Jetta, locate and disconnect the ABS control unit and the ground connection.
3. Remove the brake fluid from the reservoir with a suction pump.
4. Disconnect the brake lines from the modulator assembly and protect it's connections from contamination with suitable plugs.
5. Working inside the vehicle, remove the left shelf under the dash to gain access to the brake pedal linkage. Remove the clevis bolt and disconnect the pedal.
6. Remove the locknuts and remove the pressure modulator.
7. Installation is the reverse of removal. Use new locknuts and torque to 18 ft. lbs. (25 Nm). Refill the reservoir with new brake fluid and bleed the system.

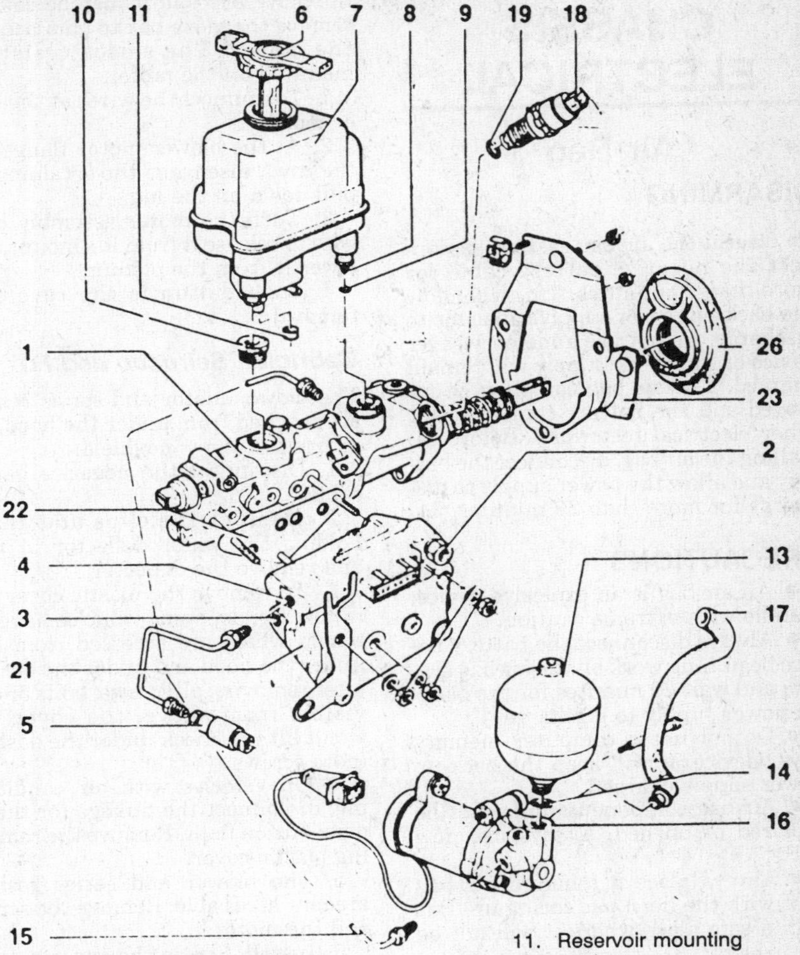

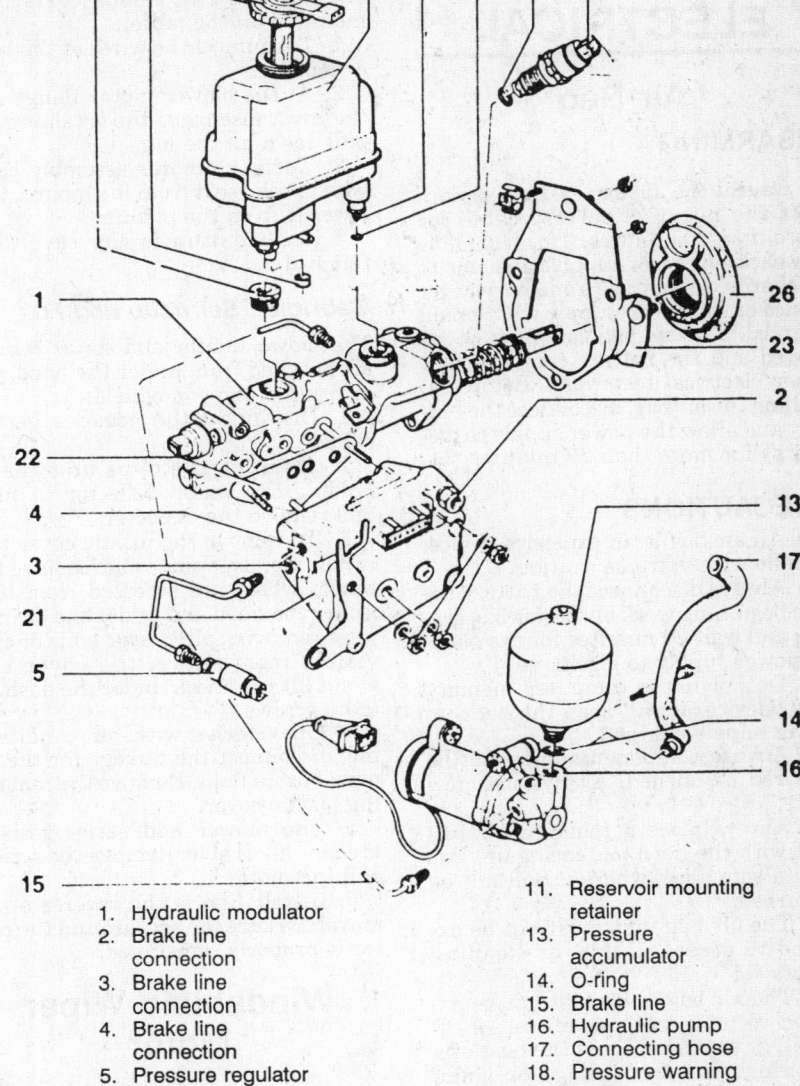

1. Hydraulic modulator
2. Brake line connection
3. Brake line connection
4. Brake line connection
5. Pressure regulator valve
6. Cap
7. Fluid reservoir
8. Sleeve
9. O-ring
10. Sealing plug
11. Reservoir mounting retainer
13. Pressure accumulator
14. O-ring
15. Brake line
16. Hydraulic pump
17. Connecting hose
18. Pressure warning switch
19. Seal
21. Hydraulic modulator
22. Gasket
23. Bracket
26. Gasket

ABS brake modulator assembly

Wheel Sensor

In addition to the pressure modulator and electronic control unit, the ABS system includes a wheel speed sensor. These sensors feed a speed signal to the control unit, which compares all the speed signals. Brake fluid pressure is modified as needed to prevent wheel lockup.

REMOVAL & INSTALLATION

1. Raise and safely support the vehicle.
2. Remove the wheel and unbolt and remove the sensor from the wheel bearing housing.
3. The rotor portion of the sensor assembly is secured to the inside of the wheel hub. To remove the front rotor, the hub must be pressed out of the front wheel bearing.
4. On the rear wheels, the sensor is bolted to the stub axle just above the axle beam mounting pad. The sensor rotor is pressed into the brake disc. To remove it:
   a. Remove the wheel bearing and the brake rotor.
   b. Insert a drift pin through the wheel bolt holes and gently tap the speed sensor rotor out a little bit at each hole, much like removing an inner wheel bearing race.

To install:

5. When reinstalling, use a suitable sleeve to drive the speed sensor rotor into the disc evenly. When the cover ring is installed, the distance from the ring to the splash shield should be 0.375 in. (9.5mm).
6. When reinstalling the sensor, use a dry lubricant on the sides of the sensor and torque the bolt to 7 ft. lbs. (10 Nm).

CHASSIS ELECTRICAL

Air Bag

DISARMING

To disarm the air bag system, disconnect the negative battery cable for more than 20 minutes. This will allow the back-up power supply capacitor to discharge. Do not use a memory saver device or the power supply will remain charged. The air bag can then be removed and the battery connected for other electrical test work. Before installing the air bag, disconnect the battery and allow the power supply to discharge for more than 20 minutes.

PRECAUTIONS

- An air bag is an explosive device. Handle with extreme caution.
- Always disconnect the battery before beginning work on the air bag system and wait 20 minutes for the back-up power supply to discharge.
- Do not use a computer memory saver device. It will keep the back-up power supply charged.
- Air bag components must not be repaired or opened. Always use new parts.
- Always place a removed air bag unit with the horn pad facing up. Put it in a safe place where it will not be disturbed.
- The air bag unit must not be exposed to grease, fluids, or cleaning agents.
- The air bag unit must not be exposed to temperatures above 194°F (90°C) at any time. Even the heat of a soldering iron can damage or ignite the charge.
- Any testing on the air bag system must be done with the air bag installed in the vehicle. Use only Volkswagen approved test equipment and procedures specified with that equipment's instruction manual.
- Storage and transport of air bags is subject to rules governing explosive devices and should be done only in the original package.
- Failure to follow proper safety precautions may result in personal injury through accidental firing of the air bag, or through failure of the air bag in an accident.

Heater Blower Motor

REMOVAL & INSTALLATION

Golf, Jetta, Corrado, Passat

The blower motor is located behind the glove box and it may be easier to remove the glove box to gain access to the motor. The series resistor is mounted on the motor.

1. Disconnect the wires at the blower motor.
2. At the blower motor flange near the cowl, disengage the retaining lug; pull down on the lug.
3. Turn the motor assembly clockwise to release it from it's mount, then lower it from the plenum.
4. Installation is the reverse of removal.

Cabriolet, Scirocco and Fox

The blower motor and series resistor are reached from under the hood, just in front of the windshield.

1. Disconnect the negative battery cable.
2. Remove the clips and gasket holding the water deflector in place and remove the deflector.
3. To remove the plastic cover that is now visible. Some vehicles have fasteners which are accessed from both under the hood and under the dash. If after removing all screws, bolts or clips visible from above, the cover still won't lift off, check under the dash for more screws.
4. On vehicles with air conditioning, disconnect the linkage for the air distribution flaps. Remove the remaining plastic cover.
5. The blower and series resistor are now accessible. Remove the screws and the motor.
6. Installation is the reverse of removal. Be sure the seal around the motor is properly reinstalled.

Windshield Wiper Motor

REMOVAL & INSTALLATION

Except Quantum

When removing the wiper motor, leave the mounting frame in place. If possible, do not remove the wiper drive crank from the motor shaft.

1. Disconnect the negative battery cable and unplug the multi-connector from the wiper motor.
2. Disconnect the crank arm from the wiper arm assembly.
3. Remove the retaining nut and the crank arm from the wiper motor shaft.
4. Remove the motor mounting bolts and the motor from the vehicle.

To install:

5. Temporarily connect the multi-connector and run the motor. Turn the switch **OFF** and disconnect the power after the motor stops; it will stop in the park position.
6. Connect the crank arm to the wiper assembly and reverse the removal procedures.

Quantum

1. Disconnect the negative battery cable and unplug the multi-connector from the wiper motor.
2. Remove the 3 motor-to-linkage bracket retaining screws.
3. Carefully pry the motor crank from the 2 linkage arms.
4. Remove the motor from the vehicle.
5. Installation is the reverse of removal. The crank arm should be at a right angle to the motor.

Windshield Wiper and Turn Signal Switch

REMOVAL & INSTALLATION

CAUTION

If equipped with an air bag, the negative battery cable must be disconnected for 20 minutes before working on the system. Failure to do so may result in deployment of the air bag and possible personal injury.

1. Disconnect the negative battery cable and remove the steering wheel.
2. Remove the screws securing the

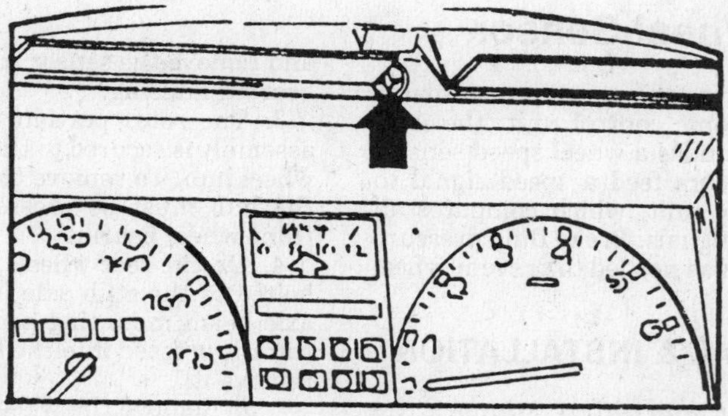

Tip down cluster panel, revealing Phillips screw

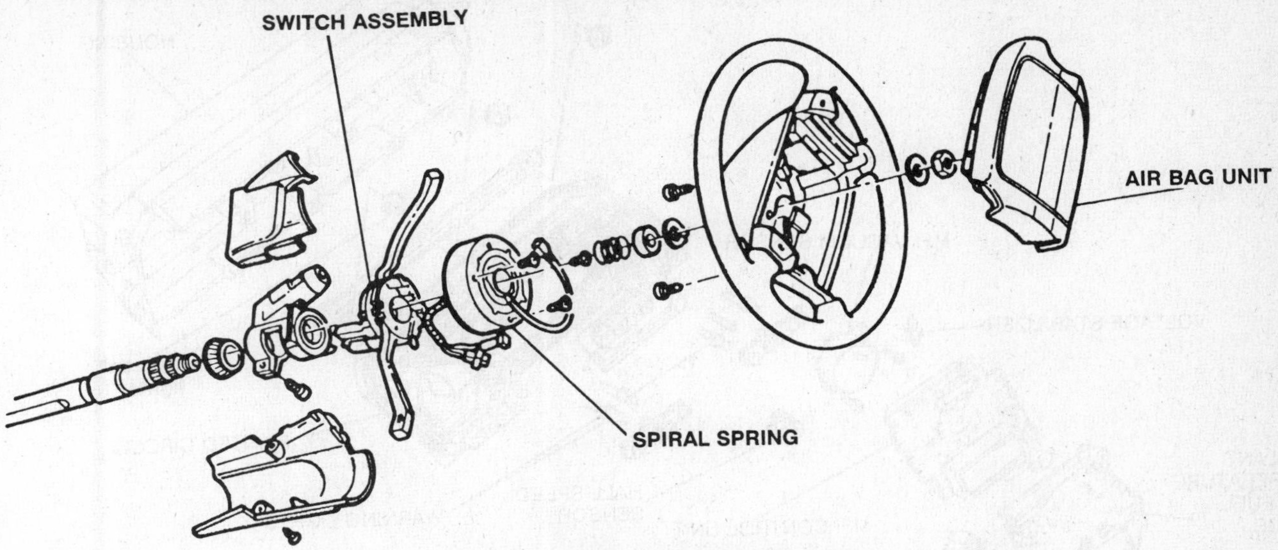

Steering column assembly with air bag

Instrument cluster removal–Scirocco (Quantum similar)

steering column covers and remove the covers.
3. Remove the 3 retaining screws and remove the turn signal switch.
4. Remove the screws to remove windshield wiper/washer switch and carefully disconnect the wires.
5. Installation is the reverse of the removal procedure.

Instrument Cluster

REMOVAL & INSTALLATION

Quantum

1. Disconnect the negative battery cable.
2. Carefully pry off the switch trim below the instrument cluster.
3. Pull heater the control knobs off and press out the heater control plate.
4. Remove 2 Phillips head screws holding the heater control trim to panel.
5. Remove the 7 Phillips screws around the perimeter of the instrument cluster.
6. Disconnect all wiring to switches and warning lights. Remove all trim panels.
7. Start to pull down on the instrument cluster and remove the screws at the top of the cluster.
8. Tip out the top of the instrument cluster.
9. Remove the speedometer cable by twisting the tabs of the plastic fixture around the end of the cable.
10. Disconnect the multi-point connector and remove instrument cluster.

To install:

11. Connect the speedometer cable and the multi-point connector.
12. Fit the instrument cluster into place and install the screws.
13. Connect the switches and warning lights and connect the battery to test the system before completing installation of the trim panels.

Golf, Fox, GTI, Jetta, Scirocco and Cabriolet

1. Disconnect the negative battery cable and pull off the temperature control knobs and levers, except Scirocco.
2. Unclip the heater control trim plate, separate the electrical connectors and remove the plate, except Scirocco.
3. Remove the retaining screws and the instrument panel trim plate.
4. Remove the retaining screws and pull out the instrument panel.
5. Squeeze the clips on the speedometer cable head and remove the cable from the instrument cluster.
6. Disconnect all of the vacuum hose and the electrical connections.
7. Installation is the reverse of removal.

Corrado and Passat

1. Disconnect negative battery cable.
2. Peel off the horn button cover, starting at the bottom and remove the steering wheel.
3. Remove the screw trim caps, trim screws and cluster trim.
4. Unscrew the trip odometer reset button.
5. Remove the cover screws (1) and cover, then remove the cluster screws (2) and cluster.
6. Carefully disconnect the vacuum line and multi-point connector.
7. Installation is the reverse of removal.

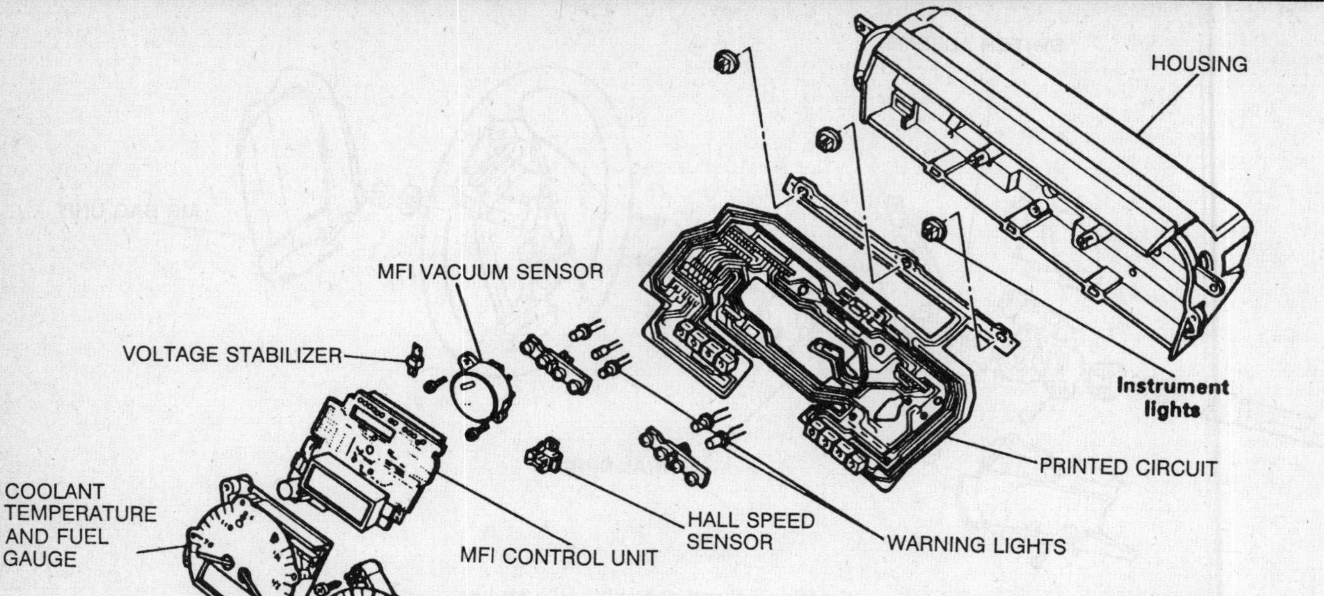

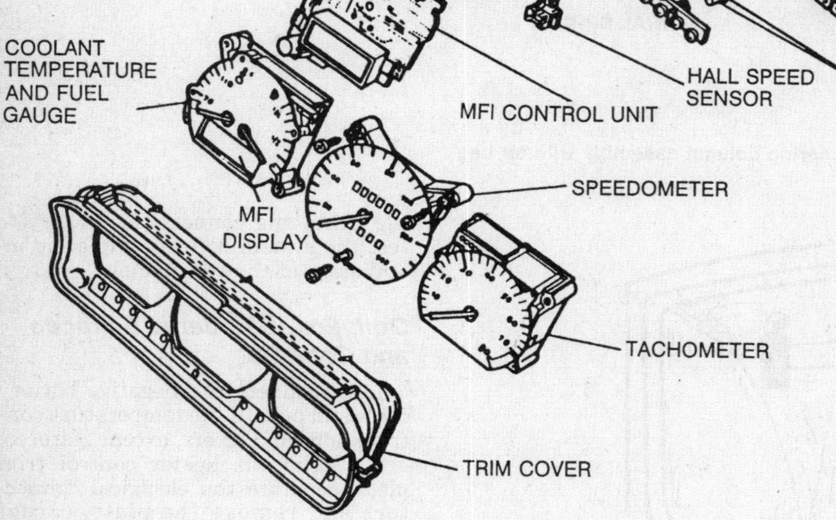

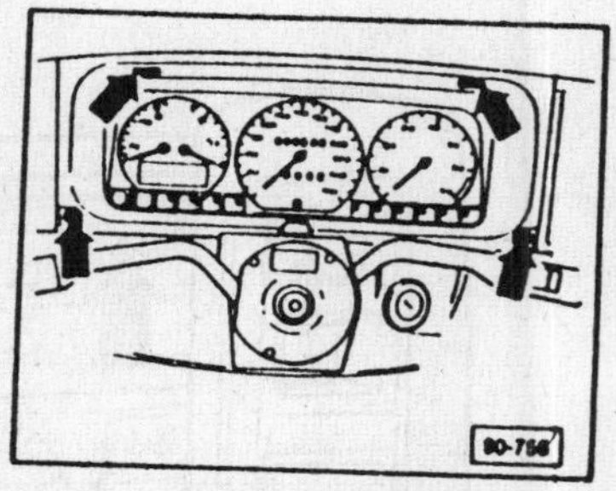

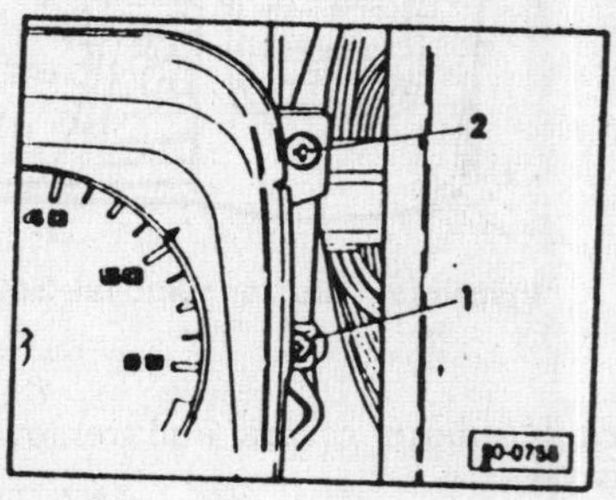

Instrument cluster assembly

Radio

REMOVAL & INSTALLATION

1. On vehicles with a theft protected radio, obtain the reset code.
2. Insert the removal tools into the sides of the face plate.
3. Push the tools away from each other to release the spring clips.
4. Pull the radio out far enough to unplug the wiring in the back.

To install:

5. Connect the wiring and push the radio part way into the slot.
6. Switch the radio **ON**. If it works properly, push it all the way into the slot to lock it in place.
7. To enter the reset code on theft protected radios, with the switch **ON**, push and hold the **AM/FM** and **SCAN** buttons at the same time. The display will change to **CODE**, then change again to **1000**. Release the buttons.
8. Use the first 4 station preset buttons to enter the security code, which will appear in the display.

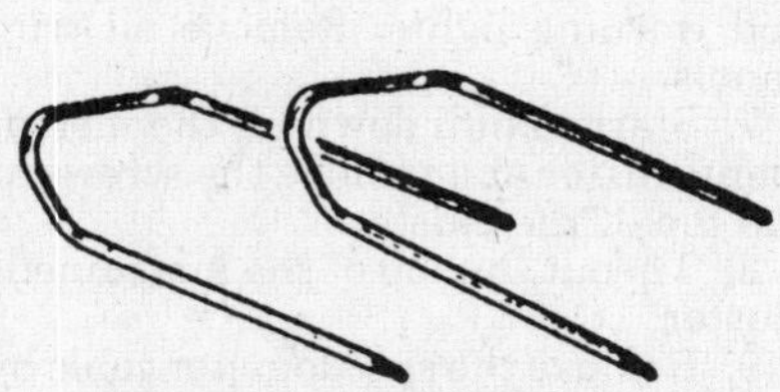

Special tool for radio removal

9. When the code is correctly entered, push and hold the **AM/FM** and **SCAN** buttons again. The word **SAF** will appear in the display to indicate the code is correct and accepted. If the word does not appear, the code is incorrect.
10. When the buttons are released, the station frequency will appear in the display. The radio is now ready for station programing.

Headlight Switch

REMOVAL & INSTALLATION

1. To remove, disconnect the battery.

2. Carefully pry on one side of the switch, then the other, to walk the switch from it's position; be careful not to damage the dash padding.

3. To reinstall, reconnect the wires and push the switch back into it's position.

Ignition Lock/Switch

REMOVAL & INSTALLATION

CAUTION

If equipped with an air bag, the negative battery cable must be disconnected for 20 minutes before working on the system. Failure to do so may result in deployment of the air bag and possible personal injury.

1. Disconnect the negative battery cable and remove steering wheel.
2. Remove bottom switch cover, turn signal and wiper switches.
3. Some vehicles have a locking ring and spring below the switches. To remove, place a thin tube or pipe over the splines and push lightly against the spring. Remove the locking ring with snapring pliers and allow the spring to push the locking ring and spacer up. Lift out the remaining parts.
4. Unplug the ignition switch, remove the socket-head screw and remove the entire lock and support ring housing.
5. To remove the lock cylinder, carefully drill a ⅛ in. hole into the housing at the spot indicated, insert a key into the lock and remove the key and cylinder.

To install:

6. The cylinder is held in place with a spring loaded detent that automatically clicks into the drilled hole when the cylinder is pushed into the housing.

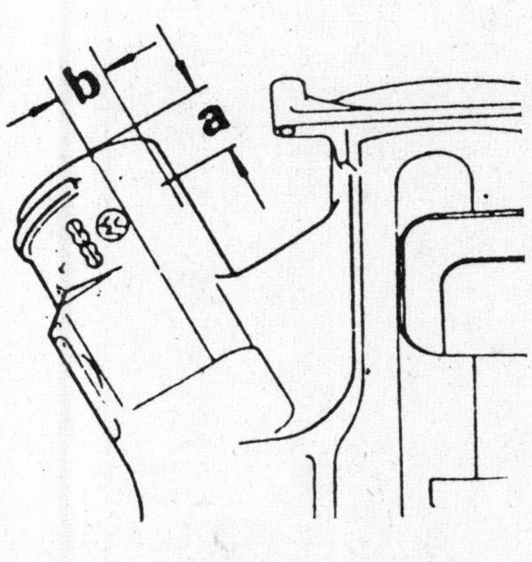

Dimension A 0.470 in. (12mm)— dimension B 0.390 in. (10mm)

Clutch Switch

ADJUSTMENT

On vehicles with cruise control, the system is deactivated by a vacuum switch connected to the clutch linkage when the clutch pedal is pushed. There is no adjustment or repair possible. If the switch malfunctions, it must be replaced.

Neutral Safety Switch

ADJUSTMENT

Neutral safety switches, in all vehicles, are at the shifter inside the vehicle. Adjustment is accomplished by moving the switch on the slots so the starter will operate only in **P** or **N**.

All newer vehicles have an automatic shift lock that prevents the shifter

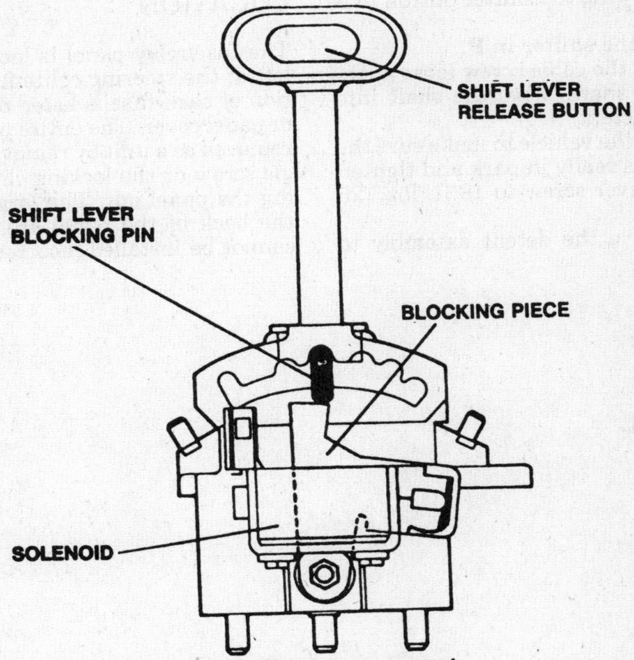

Automatic shift lock activated

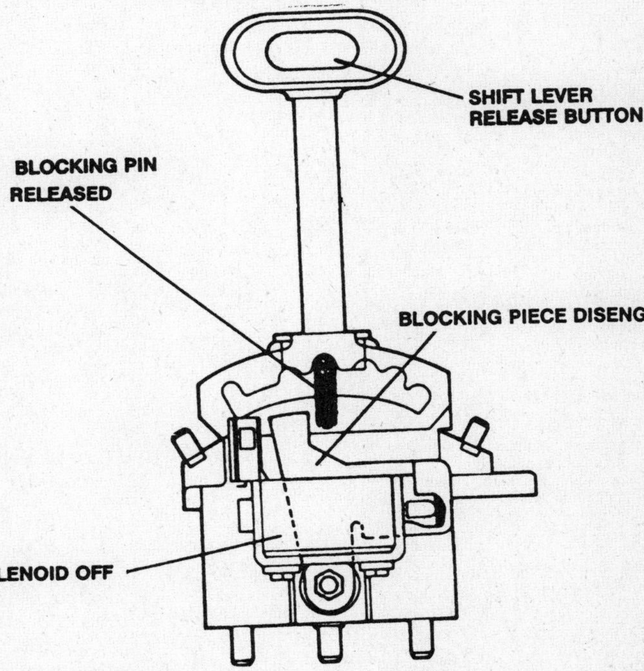

Automatic shift lock released

from moving out of **P** or **N** with the engine running unless the brake pedal is pushed. If the vehicle speed is over 3 mph, the locking system will not activate. There is also a 1 second delay when shifting into **N**.

With the ignition **ON** and the shifter in **P** or **N**, a solenoid is activated and a blocking piece prevents the locking pin from moving when the shifter button is pushed. When the brake pedal is pushed, the solenoid is deactivated and the spring loaded blocking piece moves away, allowing the shifter button to be pushed in.

1. Put the shifter in **P**.
2. With the cable screw loose at the gear lever shaft, move the shaft into the park detent.
3. Rock the vehicle to make sure the transaxle is really in park and tighten the shift lever screw to 18 ft. lbs. (25 Nm).
4. Remove the detent assembly to adjust the solenoid switch. With the solenoid off there should be a gap of 0.012 in. (0.3mm) between the pushrod and blocking piece.
5. With 12 volts supplied to the solenoid, the shift lever should be held in **P** or **N** until the brake pedal is pushed.

Fuses, Circuit Breakers and Relays

LOCATION

The fuse/relay panel is located to the left of the steering column. The function of each fuse is listed on the shelf or panel cover. The entire panel can be removed as a unit by removing the single screw or the locking clips and lifting the panel out. The large plugs on the back of the panel are keyed and cannot be installed incorrectly.

Computers

LOCATION

The engine management computers are located under the fresh air intake louvers in the hood. With the hood up, remove the plastic rain shield to expose the computer, wiper gear, idle stabilizer, if equipped, and on some vehicles, access to the blower motor. When removing the computer, first disconnect the battery. If equipped with an electronically theft-protected radio, obtain the security code before disconnecting the battery.

Flashers

LOCATION

The flasher is always on the relay assembly but its location varies from year-to-year, even on the same vehicle. It is always one of the corner locations.

Volvo 21

240, 740, 760, 780, 940, Coupe—All Models

SERIAL NUMBER IDENTIFICATION

Vehicle Identification Plate

The VIN plate on all vehicles is located on the top left surface of the dash and is also stamped on the right door pillar. Emission control information is on a label located on the left shock tower under the hood. There is also a vehicle plate on the right shock tower that includes the VIN number, engine type, emission equipment, vehicle weights and color codes.

Engine Number

Engine type designation, part number and serial number identification, for the B230F (Turbo and Non-Turbo) and B234F, is stamped into the left side of the block just below the head. An information tag is also affixed to the timing cover.

Engine type designation, part number and serial number identification for the B280F is mounted on a plate between the intake manifold and the water pump.

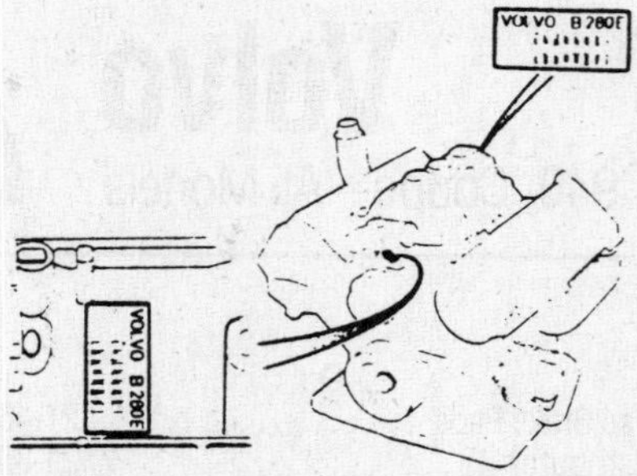

B280F engine identification location

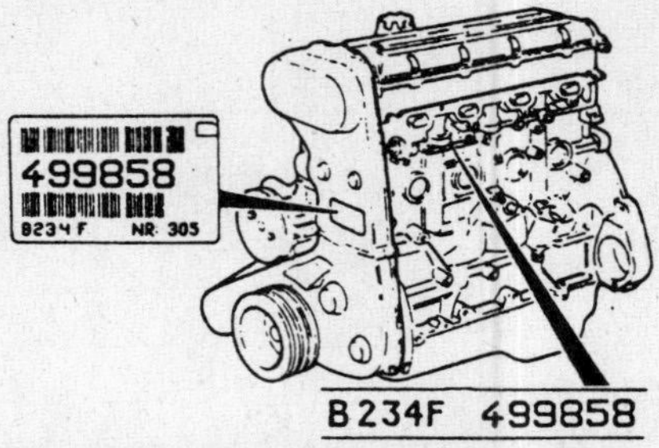

B234F, B230F and B230F-Turbo engine identification location

Transmission Number

The transmission type designation, serial number and part number appear on a metal plate riveted to the underside of the transmission. The final drive reduction ratio, part number and serial number are found on a metal plate riveted to the left side of the differential.

SPECIFICATIONS

ENGINE IDENTIFICATION

Year	Model	Engine Displacement cu. in. (cc/liter)	Engine Series Identification	No. of Cylinders	Engine Type
1988	240 DL	140 (2320/2.3)	B230F	4	OHC
	240 GL	140 (2320/2.3)	B230F	4	OHC
	740 GL	140 (2320/2.3)	B230F	4	OHC
	740 GLE	140 (2320/2.3)	B230F	4	OHC
	740 Turbo	140 (2320/2.3)	B230F-Turbo	4	OHC
	760 GLE	174 (2849/2.9)	B280F	6	OHC
	760 Turbo	140 (2320/2.3)	B230F-Turbo	4	OHC
	780	174 (2849/2.9)	B280F	6	OHC
1989	240 DL	140 (2320/2.3)	B230F	4	OHC
	240 GL	140 (2320/2.3)	B230F	4	OHC
	740 GL	140 (2320/2.3)	B230F	4	OHC
	740 GLE	140 (2320/2.3)	B234F	4	① DOHC
	740 Turbo	140 (2320/2.3)	B230F-Turbo	4	OHC
	760 GLE	174 (2849/2.9)	B280F	6	OHC
	760 Turbo	140 (2320/2.3)	B230F-Turbo	4	OHC
	780	174 (2849/2.9)	B280F	6	OHC

ENGINE IDENTIFICATION

Year	Model	Engine Displacement cu. in. (cc/liter)	Engine Series Identification	No. of Cylinders	Engine Type
1990	240 DL	140 (2320/2.3)	B230F	4	OHC
	240 GL	140 (2320/2.3)	B230F	4	OHC
	740 GL	140 (2320/2.3)	B230F	4	OHC
	740 GLE	140 (2320/2.3)	B234F	4	① DOHC
	740 Turbo	140 (2320/2.3)	B230F-Turbo	4	OHC
	760 GLE	174 (2849/2.9)	B280F	6	OHC
	760 Turbo	140 (2320/2.3)	B230F-Turbo	4	OHC
	780	174 (2849/2.9)	B280F	6	OHC
	780 Turbo	140 (2320/2.3)	B230F-Turbo	4	OHC
1991-92	240 DL	140 (2320/2.3)	B230F	4	OHC
	240 GL	140 (2320/2.3)	B230F	4	OHC
	740 GL	140 (2320/2.3)	B230F	4	OHC
	740 Turbo	140 (2320/2.3)	B-230F-Turbo	4	OHC
	940 SE	140 (2320/2.3)	B-230F-Turbo	4	OHC
	940 Turbo	140 (2320/2.3)	B-230F-Turbo	4	OHC
	940 GLE	140 (2320/2.3)	B-234F	4	① DOHC
	Coupe	140 (2320/2.3)	B230F-Turbo	4	OHC

OHC—Overhead Camshaft
DOHC—Double Overhead Camshafts
① 16 Valve Engine

GENERAL ENGINE SPECIFICATIONS

Year	Model	Engine Displacement cu. in. (cc)	Fuel System Type	Net Horsepower @ rpm	Net Torque @ rpm (ft. lbs.)	Bore × Stroke (in.)	Com-pression Ratio	Oil Pressure @ rpm
1988	240 DL	140 (2320) B230F	LH	114 @ 5400	136 @ 2750	3.78 × 3.15	9.8:1	35–85 @ 2000
	240 GL	140 (2320) B230F	LH	114 @ 5400	136 @ 2750	3.78 × 3.15	9.8:1	35–85 @ 2000
	740 GL	140 (2320) B230F	LH	114 @ 5400	136 @ 2750	3.78 × 3.15	9.8:1	35–85 @ 2000
	740 GLE	140 (2320) B230F	LH	114 @ 5400	136 @ 2750	3.78 × 3.15	9.8:1	35–85 @ 2000
	740 Turbo	140 (2320) B230F-Turbo	LH	160 @ 5300	187 @ 2900	3.78 × 3.15	8.7:1	35–85 @ 2000
	760 GLE	174 (2849) B280F	LH	146 @ 5100	173 @ 3750	3.58 × 2.86	9.5:1	57 @ 3000
	760 Turbo	140 (2320) B230F-Turbo	LH	160 @ 5300	187 @ 2900	3.78 × 3.15	8.7:1	35–85 @ 2000
	780	174 (2849) B280F	LH	146 @ 5100	173 @ 3750	3.58 × 2.86	9.5:1	57 @ 3000
1989	240 DL	140 (2320) B230F	LH	114 @ 5400	136 @ 2750	3.78 × 3.15	9.8:1	35–85 @ 2000
	240 GL	140 (2320) B230F	LH	114 @ 5400	136 @ 2750	3.78 × 3.15	9.8:1	35–85 @ 2000
	740 GL	140 (2320) B230F	LH	114 @ 5400	136 @ 2750	3.78 × 3.15	9.8:1	35–85 @ 2000

GENERAL ENGINE SPECIFICATIONS

Year	Model	Engine Displacement cu. in. (cc)	Fuel System Type	Net Horsepower @ rpm	Net Torque @ rpm (ft. lbs.)	Bore × Stroke (in.)	Compression Ratio	Oil Pressure @ rpm
1989	740 GLE	140 (2320) B234F	LH	153 @ 5700	150 @ 4450	3.78 × 3.15	10.0:1	73 @ 3000 ①
	740 Turbo	140 (2320) B230F-Turbo	LH	160 @ 5300	187 @ 2900	3.78 × 3.15	8.7:1	35–85 @ 2000
	760 GLE	174 (2849) B280F	LH	146 @ 5100	173 @ 3750	3.58 × 2.86	9.5:1	57 @ 3000
	760 Turbo	140 (2320) B230F-Turbo	LH	160 @ 5300	187 @ 2900	3.78 × 3.15	8.7:1	35–85 @ 2000
	780	174 (2849) B280F	LH	146 @ 5100	173 @ 3750	3.58 × 2.86	9.5:1	57 @ 3000
	780 Turbo	140 (2320) B230F-Turbo	LH	175 @ 5300	187 @ 2900	3.78 × 3.15	8.7:1	35–85 @ 2000
1990	240 DL	140 (2320) B230F	LH	114 @ 5400	136 @ 2750	3.78 × 3.15	9.8:1	35–85 @ 2000
	240 GL	140 (2320) B230F	LH	114 @ 5400	136 @ 2750	3.78 × 3.15	9.8:1	35–85 @ 2000
	740 GL	140 (2320) B230F	LH	114 @ 5400	136 @ 2750	3.78 × 3.15	9.8:1	35–85 @ 2000
	740 GLE	140 (2320) B234F	LH	153 @ 5700	150 @ 4450	3.78 × 3.15	10.0:1	73 @ 3000 ①
	740 Turbo	140 (2320) B230F-Turbo	LH	160 @ 5300	187 @ 2900	3.78 × 3.15	8.7:1	35–85 @ 2000
	760 GLE	174 (2849) B280F	LH	146 @ 5100	173 @ 3750	3.58 × 2.86	9.5:1	57 @ 3000
	760 Turbo	140 (2320) B230F-Turbo	LH	160 @ 5300	187 @ 2900	3.78 × 3.15	8.7:1	35–85 @ 2000
	780	174 (2849) B280F	LH	146 @ 5100	173 @ 3750	3.58 × 2.86	9.5:1	57 @ 3000
	780 Turbo	140 (2320) B230F-Turbo	LH	175 @ 5300	187 @ 2900	3.78 × 3.15	8.7:1	35–85 @ 2000
1991-92	240 DL	140 (2320) B230F	LH	114 @ 5400	136 @ 2750	3.78 × 3.15	9.8:1	35–85 @ 2000
	240 GL	140 (2320) B230F	LH	114 @ 5400	136 @ 2750	3.78 × 3.15	9.8:1	35–85 @ 2000
	740 GL	140 (2320) B230F	LH	114 @ 5400	136 @ 2750	3.78 × 3.15	9.8:1	35–85 @ 2000
	740 Turbo	140 (2320) B230F-Turbo	LH	162 @ 5300	195 @ 2900	3.78 × 3.15	8.7:1	35–85 @ 2000
	940 SE	140 (2320) B230F-Turbo	LH	162 @ 4000	195 @ 3450	3.78 × 3.15	8.7:1	35–85 @ 2000
	940 Turbo	140 (2320) B230F-Turbo	LH	162 @ 4800	195 @ 3450	3.78 × 3.15	8.7:1	35–85 @ 2000
	940 GLE	140 (2320) B234F	LH	153 @ 5700	150 @ 4450	3.78 × 3.15	10.0:1	73 @ 3000 ①
	Coupe	140 (2320) B230F-Turbo	LH	175 @ 5300	187 @ 2900	3.78 × 3.15	8.7:1	35–85 @ 2000

LH LH-Jetronic Injection
① 16 Valve Engine

ENGINE TUNE-UP SPECIFICATIONS

Year	Model	Engine Displacement cu. in. (cc)	Spark Plugs Type	Spark Plugs Gap (in.)	Ignition Timing ② (deg.) MT	Ignition Timing ② (deg.) AT	Compression Pressure (psi)	Fuel Pump (psi)	Idle Speed (rpm) MT	Idle Speed (rpm) AT	Valve Clearance In.	Valve Clearance Ex.
1988	240 DL	140 (2320) B230F	WR7DC	0.030	12B ①	12B ①	NA	36	750	750	0.014–0.016	0.014–0.016
	240 GL	140 (2320) B230F	WR7DC	0.030	12B ①	12B ①	NA	36	750	750	0.014–0.016	0.014–0.016
	740 GL	140 (2320) B230F	WR7DC	0.030	12B ①	12B ①	NA	36	750	750	0.014–0.016	0.014–0.016
	740 GLE	140 (2320) B230F	WR7DC	0.030	12B ①	12B ①	NA	36	750	750	0.014–0.016	0.014–0.016
	740 Turbo	140 (2320) B230F-Turbo	WR7DC	0.026	12B ①	12B ①	NA	43	750	750	0.014–0.016	0.014–0.016
	760 GLE	174 (2849) B280F	HR6DC	0.026	16B ①	16B ①	NA	35	750	750	0.004–0.006	0.010–0.012
	760 Turbo	140 (2320) B230F-Turbo	WR7DC	0.026	12B ①	12B ①	NA	43	750	750	0.014–0.016	0.014–0.016
	780	174 (2849) B280F	HR6DC	0.026	16B ①	16B ①	NA	35	750	750	0.004–0.006	0.010–0.012
1989	240 DL	140 (2320) B230F	WR7DC	0.030	12B ①	12B ①	NA	36	750	750	0.014–0.016	0.014–0.016
	240 GL	140 (2320) B230F	WR7DC	0.030	12B ①	12B ①	NA	36	750	750	0.014–0.016	0.014–0.016
	740 GL	140 (2320) B230F	WR7DC	0.030	12B ①	12B ①	NA	36	750	750	0.014–0.016	0.014–0.016
	740 GLE	140 (2320) B234F	WR7DC	0.030	15B ③	15B ③	NA	36	850	850	Hyd.	Hyd. ④
	740 Turbo	140 (2320) B230F-Turbo	WR7DC	0.026	12B ①	12B ①	NA	43	750	750	0.014–0.016	0.014–0.016
	760 GLE	174 (2849) B280F	HR6DC	0.026	16B ①	16B ①	NA	35	750	750	0.004–0.006	0.010–0.012
	760 Turbo	140 (2320) B230F-Turbo	WR7DC	0.026	12B ①	12B ①	NA	43	750	750	0.014–0.016	0.014–0.016
	780	174 (2849) B280F	HR6DC	0.026	16B ①	16B ①	NA	35	750	750	0.004–0.006	0.010–0.012
	780 Turbo	140 (2320) B230F-Turbo	WR7DC	0.026	12B ①	12B ①	NA	43	750	750	0.014–0.016	0.014–0.016
1990	240 DL	140 (2320) B230F	WR7DC	0.030	12B ①	12B ①	NA	36	750	750	0.014–0.016	0.014–0.016
	240 GL	140 (2320) B230F	WR7DC	0.030	12B ①	12B ①	NA	36	750	750	0.014–0.016	0.014–0.016
	740 GL	140 (2320) B230F	WR7DC	0.030	12B ①	12B ①	NA	36	750	750	0.014–0.016	0.014–0.016
	740 GLE	140 (2320) B234F	WR7DC	0.030	15B ③	15B ③	NA	36	850	850	Hyd.	Hyd. ④
	740 Turbo	140 (2320) B230F-Turbo	WR7DC	0.026	12B ①	12B ①	NA	43	750	750	0.014–0.016	0.014–0.016
	760 GLE	174 (2849) B280F	HR6DC	0.026	16B ①	16B ①	NA	35	750	750	0.004–0.006	0.010–0.012
	760 Turbo	140 (2320) B230F-Turbo	WR7DC	0.026	12B ①	12B ①	NA	43	750	750	0.014–0.016	0.014–0.016

ENGINE TUNE-UP SPECIFICATIONS

Year	Model	Engine Displacement cu. in. (cc)	Spark Plugs Type	Spark Plugs Gap (in.)	Ignition Timing ② (deg.) MT	Ignition Timing ② (deg.) AT	Compression Pressure (psi)	Fuel Pump (psi)	Idle Speed (rpm) MT	Idle Speed (rpm) AT	Valve Clearance In.	Valve Clearance Ex.
1990	780	174 (2849) B280F	HR6DC	0.026	16B ①	16B ①	NA	35	750	750	0.004–0.006	0.010–0.012
	780 Turbo	140 (2320) B230F-Turbo	WR7DC	0.026	12B ①	12B ①	NA	43	750	750	0.014–0.016	0.014–0.016
1991	240 DL	140 (2320) B230F	WR7DC	0.030	12B ①	12B ①	NA	36	750	750	0.014–0.016	0.014–0.016
	240 GL	140 (2320) B230F	WR7DC	0.030	12B ①	12B ①	NA	36	750	750	0.014–0.016	0.014–0.016
	740 GL	140 (2320) B230F	WR7DC	0.030	12B ①	12B ①	NA	36	750	750	0.014–0.016	0.014–0.016
	740 Turbo	140 (2320) B230F-Turbo	WR7DC	0.026	12B ①	12B ①	NA	43	750	750	0.014–0.016	0.014–0.016
	940 SE	140 (2320) B230F-Turbo	WR7DC	0.026	12B ①	12B ①	NA	43	750	750	0.014–0.016	0.014–0.016
	940 Turbo	140 (2320) B230F-Turbo	WR7DC	0.026	12B ①	12B ①	NA	43	750	750	0.014–0.016	0.014–0.016
	940 GLE	140 (2320) B234F	WR7DC	0.030	15B ③	15B ③	NA	36	850	850	Hyd.	Hyd. ④
	Coupe	140 (2320) B230F-Turbo	WR7DC	0.026	12B ①	12B ①	NA	43	750	750	0.014–0.016	0.014–0.016
1992	ALL	SEE UNDERHOOD SPECIFICATIONS STICKER										

NA—Not Available
① @ 750 rpm
② Vacuum advance disconnected, A/C turned off
③ @ 850 rpm
④ 16 Valve Engine

FIRING ORDERS

NOTE: To avoid confusion, always replace spark plug wires one at a time.

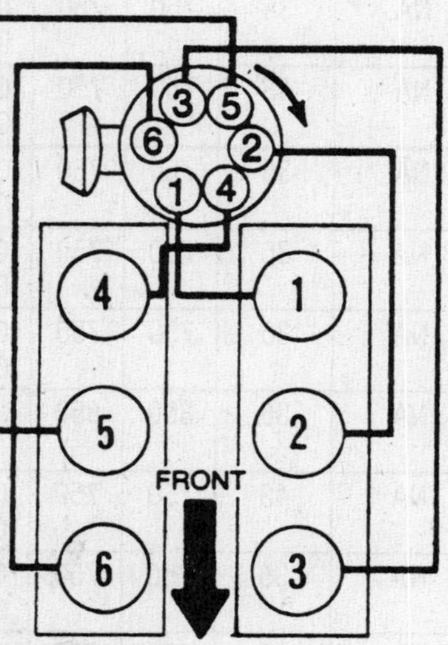

B280F – 6 Cylinder
Firing Order: 1-6-3-5-2-4
Distributor Rotation: Clockwise

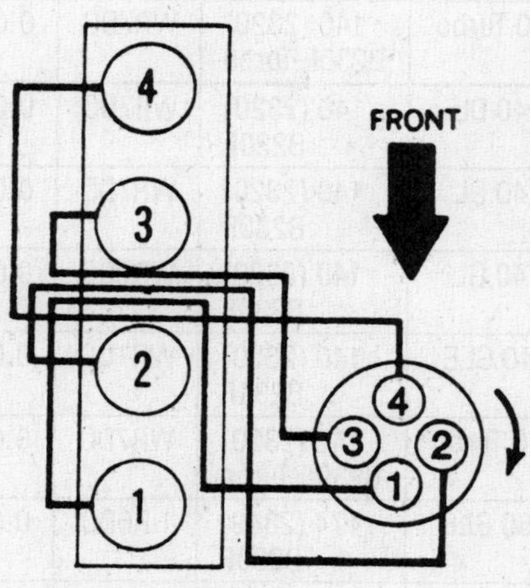

B230F and B234F – 4 Cylinder
Firing Order: 1-3-4-2
Distributor Rotation: Clockwise

CAPACITIES

Year	Model	Engine Displacement cu. in. (cc)	Engine Crankcase (qts.)① with Filter	Engine Crankcase (qts.)① without Filter	Transmission (pts) 4-Spd	Transmission (pts) 5-Spd	Transmission (pts) Auto.	Drive Axle (pts.)	Fuel Tank (gal.)	Cooling System (qts.)
1988	240 DL	140 (2320) B230F	4.0	3.5	—	3.2	15.8	②	15.8	10.0
	240 GL	140 (2320) B230F	4.0	3.5	—	3.2	15.8	②	15.8	10.0
	740 GL	140 (2320) B230F	4.1	3.6	—	3.2	15.8	②	15.8	10.0
	740 GLE	140 (2320) B230F	4.1	3.6	—	3.2	15.8	②	15.8	10.0
	740 Turbo	140 (2320) B230F-Turbo	4.1	3.6	—	4.8	15.8	②	15.8	10.0
	760 GLE	174 (2849) B280F	6.3	5.8	—	—	15.8	②	21.1	10.5
	760 Turbo	140 (2320) B230F-Turbo	4.1	3.6	—	—	15.8	②	21.1	10.0
	780	174 (2849) B280F	6.3	5.8	—	—	15.8	②	21.1	10.5
1989	240 DL	140 (2320) B230F	4.0	3.5	—	3.2	15.8	②	15.8	10.0
	240 GL	140 (2320) B230F	4.0	3.5	—	3.2	15.8	②	15.8	10.0
	740 GL	140 (2320) B230F	4.1	3.6	—	3.2	15.8	②	15.8	10.0
	740 GLE	140 (2320) B234F	4.2	3.7	—	4.8	15.8	②	15.8	10.0
	740 Turbo	140 (2320) B230F-Turbo	4.1	3.6	—	4.8	15.8	②	15.8	10.0
	760 GLE	174 (2849) B280F	6.3	5.8	—	—	15.8	②	21.1	10.5
	760 Turbo	140 (2320) B230F-Turbo	4.1	3.6	—	—	15.8	②	21.1	10.0
	780	174 (2849) B280F	6.3	5.8	—	—	15.8	②	21.1	10.5
	780 Turbo	140 (2320) B230F-Turbo	4.1	3.6	—	—	15.8	②	21.1	10.5
1990	240 DL	140 (2320) B230F	4.0	3.5	—	3.2	15.8	②	15.8	10.0
	240 GL	140 (2320) B230F	4.0	3.5	—	3.2	15.8	②	15.8	10.0
	740 GL	140 (2320) B230F	4.1	3.6	—	3.2	15.8	②	15.8	10.0
	740 GLE	140 (2320) B234F	4.2	3.7	—	4.8	15.8	②	15.8	10.0
	740 Turbo	140 (2320) B230F-Turbo	4.1	3.6	—	4.8	15.8	②	15.8	10.0
	760 GLE	174 (2849) B280F	6.3	5.8	—	—	15.8	②	21.1	10.5
	760 Turbo	140 (2320) B230F-Turbo	4.1	3.6	—	—	15.8	②	21.1	10.0

CAPACITIES

Year	Model	Engine Displacement cu. in. (cc)	Engine Crankcase (qts.) ① with Filter	without Filter	Transmission (pts) 4-Spd	5-Spd	Auto.	Drive Axle (pts.)	Fuel Tank (gal.)	Cooling System (qts.)
1990	780	174 (2849) B280F	6.3	5.8	—	—	15.8	②	21.1	10.5
	780 Turbo	140 (2320) B230F-Turbo	4.1	3.6	—	—	15.8	②	21.1	10.5
1991-92	240 DL	140 (2320) B230F	4.0	3.5	—	3.2	15.8	②	15.8	10.0
	240 GL	140 (2320) B230F	4.0	3.5	—	3.2	15.8	②	15.8	10.0
	740 GL	140 (2320) B230F	4.1	3.6	—	3.2	15.8	②	15.8	10.0
	740 Turbo	140 (2320) B230F-Turbo	4.1	3.6	—	4.8	15.8	②	15.8	10.0
	940 SE	140 (2320) B230F-Turbo	4.1	3.6	—	4.8	15.8	②	15.8	10.0
	940 Turbo	140 (2320) B230F-Turbo	4.1	3.6	—	4.8	15.8	②	15.8	10.0
	940 GLE	140 (2320) B234F	4.2	3.7	—	4.8	15.8	②	15.8	10.5
	Coupe	140 (2320) B230F-Turbo	4.1	3.6	—	—	15.8	②	21.1	10.5

① Models with turbo—add 0.6 qt. if oil cooler has been drained
② 1030 axle—2.8 pts.
1031 axle—3.4 pts.

CRANKSHAFT AND CONNECTING ROD SPECIFICATIONS

All measurements are given in inches.

Year	Engine Displacement cu. in. (cc)	Crankshaft Main Brg. Journal Dia.	Main Brg. Oil Clearance	Shaft End-play	Thrust on No.	Connecting Rod Journal Diameter	Oil Clearance	Side Clearance
1988	140 (2320) B230F	2.4981–2.4986	0.0011–0.0033	0.0015–0.0058	5	2.1255–2.1260	0.0009–0.0028	0.006–0.014
	140 (2320) B230F-Turbo	2.4981–2.4986	0.0011–0.0033	0.0015–0.0058	5	2.1255–2.1260	0.0009–0.0028	0.006–0.014
	174 (2849) B280F	2.7583	0.0035	0.0106	4	2.0585	0.0031	0.015
1989	140 (2320) B230F	2.4981–2.4986	0.0011–0.0033	0.0015–0.0058	5	2.1255–2.1260	0.0009–0.0028	0.006–0.014
	140 (2320) B230F-Turbo	2.4981–2.4986	0.0011–0.0033	0.0015–0.0058	5	2.1255–2.1260	0.0009–0.0028	0.006–0.014
	140 (2320) B234F	1.9640–1.9648	0.0011–0.0033	0.0015–0.0058	5	2.0472–2.0476	0.0009–0.0028	0.006–0.018
	174 (2849) B280F	2.7583	0.0035	0.0106	4	2.0585	0.0031	0.015

CRANKSHAFT AND CONNECTING ROD SPECIFICATIONS

All measurements are given in inches.

Year	Engine Displacement cu. in. (cc)	Crankshaft				Connecting Rod		
		Main Brg. Journal Dia.	Main Brg. Oil Clearance	Shaft End-play	Thrust on No.	Journal Diameter	Oil Clearance	Side Clearance
1990	140 (2320) B230F	2.4981–2.4986	0.0011–0.0033	0.0015–0.0058	5	2.1255–2.1260	0.0009–0.0028	0.006–0.014
	140 (2320) B230F-Turbo	2.4981–2.4986	0.0011–0.0033	0.0015–0.0058	5	2.1255–2.1260	0.0009–0.0028	0.006–0.014
	140 (2320) B234F	1.9640–1.9648	0.0011–0.0033	0.0015–0.0058	5	2.0472–2.0476	0.0009–0.0028	0.006–0.018
	174 (2849) B280F	2.7583	0.0035	0.0106	4	2.0585	0.0031	0.015
1991–92	140 (2320) B230F	2.4981–2.4986	0.0011–0.0033	0.0015–0.0058	5	2.1255–2.1260	0.0009–0.0028	0.006–0.014
	140 (2320) B230F-Turbo	2.4981–2.4986	0.0011–0.0033	0.0015–0.0058	5	2.1255–2.1260	0.0009–0.0028	0.006–0.014
	140 (2320) B234F	1.9640–1.9648	0.0011–0.0033	0.0015–0.0058	5	2.0472–2.0476	0.0009–0.0028	0.006–0.018

VALVE SPECIFICATIONS

Year	Engine Displacement cu. in. (cc)	Seat Angle (deg.)	Face Angle (deg.)	Spring Test Pressure (lbs. @ in.)	Spring Installed Height (in.)	Stem-to-Guide Clearance (in.)		Stem Diameter (in.)	
						Intake	Exhaust	Intake	Exhaust
1988	140 (2320) B230F	45	44.5	158 @ 1.08	1.79	0.0012–0.0024	0.0024–0.0036	0.3132–0.3138	0.3128–0.3134
	140 (2320) B230F-Turbo	45	44.5	158 @ 1.08	1.79	0.0012–0.0024	0.0024–0.0036	0.3132–0.3138	0.3128–0.3134
	174 (2849) B280F	45	44.5	143 @ 1.18	1.85	①	①	②	②
1989	140 (2320) B230F	45	44.5	158 @ 1.08	1.79	0.0012–0.0024	0.0024–0.0036	0.3132–0.3138	0.3128–0.3134
	140 (2320) B234F	45	44.5	144 @ 1.04	1.69	0.0012–0.0024	0.0016–0.0028	NA	NA
	140 (2320) B230F-Turbo	45	44.5	158 @ 1.08	1.79	0.0012–0.0024	0.0024–0.0036	0.3132–0.3138	0.3128–0.3134
	174 (2849) B280F	45	44.5	143 @ 1.18	1.85	①	①	②	②
1990	140 (2320) B230F	45	44.5	158 @ 1.08	1.79	0.0012–0.0024	0.0024–0.0036	0.3132–0.3138	0.3128–0.3134
	140 (2320) B234F	45	44.5	144 @ 1.04	1.69	0.0012–0.0024	0.0016–0.0028	NA	NA
	140 (2320) B230F-Turbo	45	44.5	158 @ 1.08	1.79	0.0012–0.0024	0.0024–0.0036	0.3132–0.3138	0.3128–0.3134
	174 (2849) B280F	45	44.5	143 @ 1.18	1.85	①	①	②	②

VALVE SPECIFICATIONS

Year	Engine Displacement cu. in. (cc)	Seat Angle (deg.)	Face Angle (deg.)	Spring Test Pressure (lbs. @ in.)	Spring Installed Height (in.)	Stem-to-Guide Clearance (in.)		Stem Diameter (in.)	
						Intake	Exhaust	Intake	Exhaust
1991-92	140 (2320) B230F	45	44.5	158 @ 1.08	1.79	0.0012–0.0024	0.0024–0.0036	0.3132–0.3138	0.3128–0.3134
	140 (2320) B234F	45	44.5	144 @ 1.04	1.69	0.0012–0.0024	0.0016–0.0028	NA	NA
	140 (2320) B230F-Turbo	45	44.5	158 @ 1.08	1.79	0.0012–0.0024	0.0024–0.0036	0.3132–0.3138	0.3128–0.3134

NOTE: Exhaust valves for turbo engines are stellite coated and must not be machined. They may be ground against the valve seat.

NA—Not available

① Tapered valve guide ID—0.3150–0.3158

② Tapered valve stem
Intake
Base—0.3135–0.3141
Top—3139–0.3145
Exhaust
Base—0.3127–0.3133
Top—3136–0.3141

PISTON AND RING SPECIFICATIONS

All measurements are given in inches.

Year	Engine Displacement cu. in. (cc)	Piston Clearance	Ring Gap			Ring Side Clearance		
			Top Compression	Bottom Compression	Oil Control	Top Compression	Bottom Compression	Oil Control
1988	140 (2320) B230F	0.0004–0.0012	0.0118–0.0217	0.0118–0.0217	0.0118–0.0236	0.0024–0.0036	0.0016–0.0028	0.0012–0.0026
	140 (2320) B230F-Turbo	0.0004–0.0012	0.0118–0.0217	0.0118–0.0217	0.0118–0.0236	0.0024–0.0036	0.0016–0.0028	0.0012–0.0026
	174 (2849) B280F	0.0007–0.0015	0.0157–0.0236	0.0157–0.0236	0.0157–0.0570	0.0017–0.0029	0.0009–0.0021	0.0003–0.0091
1989	140 (2320) B230F	0.0004–0.0012	0.0118–0.0217	0.0118–0.0217	0.0118–0.0236	0.0024–0.0036	0.0016–0.0028	0.0012–0.0026
	140 (2320) B234F	0.0004–0.0012	0.0120–0.0220	0.0120–0.0220	0.0120–0.0240	0.0024–0.0036	0.0016–0.0028	0.0012–0.0026
	140 (2320) B230F-Turbo	0.0004–0.0012	0.0118–0.0217	0.0118–0.0217	0.0118–0.0236	0.0024–0.0036	0.0016–0.0028	0.0012–0.0026
1990	140 (2320) B230F	0.0004–0.0012	0.0118–0.0217	0.0118–0.0217	0.0118–0.0236	0.0024–0.0036	0.0016–0.0028	0.0012–0.0026
	140 (2320) B234F	0.0004–0.0012	0.0120–0.0220	0.0120–0.0220	0.0120–0.0240	0.0024–0.0036	0.0016–0.0028	0.0012–0.0026
	140 (2320) B230F-Turbo	0.0004–0.0012	0.0118–0.0217	0.0118–0.0217	0.0118–0.0236	0.0024–0.0036	0.0016–0.0028	0.0012–0.0026
	174 (2849) B280F	0.0007–0.0015	0.0157–0.0236	0.0157–0.0236	0.0157–0.0570	0.0017–0.0029	0.0009–0.0021	0.0003–0.0091
1991-92	140 (2320) B230F	0.0004–0.0012	0.0118–0.0217	0.0118–0.0217	0.0118–0.0236	0.0024–0.0036	0.0016–0.0028	0.0012–0.0026
	140 (2320) B234F	0.0004–0.0012	0.0120–0.0220	0.0120–0.0220	0.0120–0.0240	0.0024–0.0036	0.0016–0.0028	0.0012–0.0026
	140 (2320) B230F-Turbo	0.0004–0.0012	0.0118–0.0217	0.0118–0.0217	0.0118–0.0236	0.0024–0.0036	0.0016–0.0028	0.0012–0.0026

TORQUE SPECIFICATIONS

All readings in ft. lbs.

Year	Engine Displacement cu. in. (cc)	Cylinder Head Bolts	Main Bearing Bolts	Rod Bearing Bolts	Crankshaft Pulley Bolts	Flywheel Bolts	Manifold Intake	Manifold Exhaust	Spark Plugs
1988	140 (2320) B230F	②	80	14 ③	43 ④	47–54	12	12	18
	140 (2320) B230F-Turbo	②	80	14 ③	43 ④	47–54	12	12	18
	174 (2849) B280F	⑤	①	33–37	177–206	33–37	7–11	7–11	8–11
1989	140 (2320) B230F	②	80	14 ③	43 ④	47–54	12	12	18
	140 (2320) B234F	⑥	80	15 ③	44 ④	47–54	12	12	14–22
	140 (2320) B230F-Turbo	②	80	14 ③	43 ④	47–54	12	12	18
	174 (2849) B280F	⑤	①	33–37	177–206	33–37	7–11	7–11	8–11
1990	140 (2320) B230F	②	80	14 ③	43 ④	47–54	12	12	18
	140 (2320) B234F	⑥	80	15 ③	44 ④	47–54	12	12	14–22
	140 (2320) B230F-Turbo	②	80	14 ③	43 ④	47–54	12	12	18
	174 (2849) B280F	⑦	①	33–37	177–206	33–37	7–11	7–11	8–11
1991-92	140 (2320) B230F	②	80	14 ③	43 ④	47–54	12	12	18
	140 (2320) B234F	⑥	80	15 ③	44 ④	47–54	12	12	14–22
	140 (2320) B230F-Turbo	②	80	14 ③	43 ④	47–54	12	12	18

① Torque main bearing nuts to 22 ft. lbs., in sequence. Then slacken 1st nut ½ turn, tighten to 22–26 ft. lbs., and protractor torque to 73–77°. Repeat for remaining nuts following the sequence.

② Torque head bolts in three stages; first, tighten in sequence to 15 ft. lbs., then to 44 ft. lbs. Protractor (angle) tighten 90°more in one movement.

③ Angle—tighten 90°

④ Angle—tighten 60°

⑤ Torque all head bolts in sequence to 44 ft. lbs. (60 Nm), then loosen bolt No. 1 and retorque it to 15 ft. lbs. (20Nm), then tighten it to 106°; repeat for all bolts following number sequence. Loosen and tighten one bolt at a time. Run engine to operating temperature. Then let cool for 2 hours. Finally, tighten each bolt in sequence an additional 45°.

⑥ Torque head bolts in 3 stages; first tighten in sequence to 15 ft. lbs., then to 30 ft. lbs. Protractor (angle) tighten 115° more in one movement.

⑦ Asbestos-free gaskets, fixed-washer bolts
Tighten all bolts in stages
1 Tighten bolts to 60 Nm (44 ft. lb.)
2 a Loosen bolts
b Tighten bolts to 40 Nm (30 ft. lb.)
c Angle-tighten bolts 160°–180°
3 Adjust valves

BRAKE SPECIFICATIONS

All measurements in inches unless noted.

Year	Model	Lug Nut Torque (ft. lbs.)	Master Cylinder Bore	Brake Disc Minimum Thickness	Brake Disc Maximum Runout	Standard Brake Drum Diameter	Minimum Lining Thickness Front	Minimum Lining Thickness Rear
1988	240 DL	88	0.878	① (F) 0.330 (R)	0.004 (F) 0.004 (R)	—	0.060	0.060
	240 GL	88	0.878	① (F) 0.330 (R)	0.003 (F) 0.004 (R)	—	0.060	0.060
	740 GL	63	②	③ (F) 0.330 (R)	0.003 (F) 0.004 (R)	—	0.118	0.078

BRAKE SPECIFICATIONS

All measurements in inches unless noted.

Year	Model	Lug Nut Torque (ft. lbs.)	Master Cylinder Bore	Brake Disc Minimum Thickness	Brake Disc Maximum Runout	Standard Brake Drum Diameter	Minimum Lining Thickness Front	Minimum Lining Thickness Rear
1988	740 GLE	63	②	③ (F) 0.330 (R)	0.003 (F) 0.004 (R)	—	0.118	0.078
	740 Turbo	63	②	③ (F) 0.330 (R)	0.003 (F) 0.004 (R)	—	0.118	0.078
	760 GLE	63	②	③ (F) 0.330 (R)	0.003 (F) 0.004 (R)	—	0.118	0.078
	760 Turbo	63	②	③ (F) 0.330 (R)	0.003 (F) 0.004 (R)	—	0.118	0.078
	780	63	②	③ (F) 0.330 (R)	0.003 (F) 0.004 (R)	—	0.118	0.078
1989	240 DL	88	0.878	① (F) 0.330 (R)	0.004 (F) 0.004 (R)	—	0.060	0.060
	240 GL	88	0.878	① (F) 0.330 (R)	0.003 (F) 0.004 (R)	—	0.060	0.060
	740 GL	63	②	③ (F) 0.330 (R)	0.003 (F) 0.004 (R)	—	0.118	0.078
	740 GLE	63	②	③ (F) 0.330 (R)	0.003 (F) 0.004 (R)	—	0.118	0.078
	740 Turbo	63	②	③ (F) 0.330 (R)	0.003 (F) 0.004 (R)	—	0.118	0.078
	760 GLE	63	②	③ (F) 0.330 (R)	0.003 (F) 0.004 (R)	—	0.118	0.078
	760 Turbo	63	②	③ (F) 0.330 (R)	0.003 (F) 0.004 (R)	—	0.118	0.078
	780	63	②	③ (F) 0.330 (R)	0.003 (F) 0.004 (R)	—	0.118	0.078
1990	240 DL	88	0.878	① (F) 0.330 (R)	0.004 (F) 0.004 (R)	—	0.060	0.060
	240 GL	88	0.878	① (F) 0.330 (R)	0.003 (F) 0.004 (R)	—	0.060	0.060
	740 GL	63	②	③ (F) 0.330 (R)	0.003 (F) 0.004 (R)	—	0.118	0.078
	740 GLE	63	②	③ (F) 0.330 (R)	0.003 (F) 0.004 (R)	—	0.118	0.078
	740 Turbo	63	②	③ (F) 0.330 (R)	0.003 (F) 0.004 (R)	—	0.118	0.078
	760 GLE	63	②	③ (F) ④ 0.330 (R)	0.003 (F) ⑤ 0.004 (R)	—	0.118	0.078
	760 Turbo	63	②	③ (F) ④ 0.330 (R)	0.003 (F) ⑤ 0.004 (R)	—	0.118	0.078
	780	63	②	③ (F) ④ 0.330 (R)	0.003 (F) ⑤ 0.004 (R)	—	0.118	0.078

BRAKE SPECIFICATIONS

All measurements in inches unless noted.

Year	Model	Lug Nut Torque (ft. lbs.)	Master Cylinder Bore	Brake Disc Minimum Thickness	Brake Disc Maximum Runout	Standard Brake Drum Diameter	Minimum Lining Thickness Front	Minimum Lining Thickness Rear
1991-92	240 DL	88	0.878	① (F) 0.330 (R)	0.004 (F) 0.004 (R)	—	0.060	0.060
	240 GL	88	0.878	① (F) 0.330 (R)	0.003 (F) 0.004 (R)	—	0.060	0.060
	740 GL	63	②	③ (F) 0.330 (R)	0.003 (F) 0.004 (R)	—	0.118	0.078
	740 Turbo	63	②	③ (F) 0.330 (R)	0.003 (F) 0.004 (R)	—	0.118	0.078
	940 SE	63	②	③ (F) 0.330 (R)	0.003 (F) 0.004 (R)	—	0.118	0.078
	940 Turbo	63	②	③ (F) 0.330 (R)	0.003 (F) 0.004 (R)	—	0.118	0.078
	940 GLE	63	②	③ (F) 0.330 (R)	0.003 (F) 0.004 (R)	—	0.118	0.078
	Coupe	63	②	③ (F) ④ 0.330 (R)	0.003 (F) ⑤ 0.004 (R)	—	0.118	0.078

(F)—Front
(R)—Rear
① Ventilated—0.820
Non-ventilated—0.536

② Early type—0.878
Late type—0.938
③ Ventilated—0.788
Non-ventilated—0.433

④ Multi link rear suspension
0.314
⑤ Multi link rear suspension
0.003

WHEEL ALIGNMENT

Year	Model	Caster Range (deg.)	Caster Preferred Setting (deg.)	Camber Range (deg.)	Camber Preferred Setting (deg.)	Toe-in (in.)	Steering Axis Inclination (deg.)
1988	240 DL	3P–4P	—	1/4P–3/4P	1/2P	1/8	12
	240 GL	3P–4P	—	1/4P–3/4P	1/2P	1/8	12
	740 GL	4 1/2P–5 1/2P	—	3/16N–13/16P	—	9/64	NA
	740 GLE	4 1/2P–5 1/2P	—	3/16N–13/16P	—	9/64	NA
	740 Turbo	4 1/2P–5 1/2P	—	3/16N–13/16P	—	9/64	NA
	760 GLE	4 1/2P–5 1/2P	—	3/16N–13/16P	—	9/64	NA
	760 Turbo	4 1/2P–5 1/2P	—	3/16N–13/16P	—	9/64	NA
	780	4 1/2P–5 1/2P	—	3/16N–13/16P	—	9/64	NA
1989	240 DL	3P–4P	—	1/4P–3/4P	1/2P	1/8	12
	240 GL	3P–4P	—	1/4P–3/4P	1/2P	1/8	12
	740 GL	4 1/2P–5 1/2P	—	3/16N–13/16P	—	9/64	NA
	740 GLE	4 1/2P–5 1/2P	—	3/16N–13/16P	—	9/64	NA
	740 Turbo	4 1/2P–5 1/2P	—	3/16N–13/16P	—	9/64	NA
	760 GLE	4 1/2P–5 1/2P	—	3/16N–13/16P	—	9/64	NA
	760 Turbo	4 1/2P–5 1/2P	—	3/16N–13/16P	—	9/64	NA
	780	4 1/2P–5 1/2P	—	3/16N–13/16P	—	9/64	NA

WHEEL ALIGNMENT

Year	Model	Caster Range (deg.)	Caster Preferred Setting (deg.)	Camber Range (deg.)	Camber Preferred Setting (deg.)	Toe-in (in.)	Steering Axis Inclination (deg.)
1990	240 DL	3P–4P	—	1/4P–3/4P	1/2P	1/8	12
	240 GL	3P–4P	—	1/4P–3/4P	1/2P	1/8	12
	740 GL	4 1/2P–5 1/2P	—	3/16N–13/16P	—	9/64	NA
	740 GLE	4 1/2P–5 1/2P	—	3/16N–13/16P	—	9/64	NA
	740 Turbo	4 1/2P–5 1/2P	—	3/16N–13/16P	—	9/64	NA
	760 GLE	4 1/2P–5 1/2P	—	3/16N–13/16P	—	9/64	NA
	760 Turbo	4 1/2P–5 1/2P	—	3/16N–13/16P	—	9/64	NA
	780	4 1/2P–5 1/2P	—	3/16N–13/16P	—	9/64	NA
1991-92	240 DL	3P–4P	—	1/4P–3/4P	1/2P	1/8	12
	240 GL	3P–4P	—	1/4P–3/4P	1/2P	1/8	12
	740 GL	4 1/2P–5 1/2P	—	3/16N–13/16P	—	9/64	NA
	740 Turbo	4 1/2P–5 1/2P	—	3/16N–13/16P	—	9/64	NA
	940 SE	4 1/2P–5 1/2P	—	3/16N–13/16P	—	9/64	NA
	940 Turbo	4 1/2P–5 1/2P	—	3/16N–13/16P	—	9/64	NA
	940 GLE	4 1/2P–5 1/2P	—	3/16N–13/16P	—	9/64	NA
	Coupe	4 1/2P–5 1/2P	—	3/16N–13/16P	—	9/64	NA

NA—Not Available
N—Negative
P—Positive

ENGINE MECHANICAL

NOTE: Disconnecting the negative battery cable on some vehicles may interfere with the functions of the on board computer systems and may require the computer to undergo a relearning process, once the negative battery cable is reconnected.

Engine Assembly

REMOVAL & INSTALLATION

B230F Engine

1. If equipped with manual transmission, remove the 4 retaining clips and lift up the shifter boot. Then, remove the snapring from the shifter.
2. Remove the battery.
3. Disconnect the windshield washer hose and engine compartment light wire. Scribe marks around the hood mount brackets on the under-side of the hood for later alignment. Remove the hood.
4. Remove the overflow tank cap. Drain the cooling system.
5. Remove the upper and lower radiator hoses. Disconnect the overflow hoses at the radiator. Disconnect the PCV hose at the cylinder head.
6. If equipped with automatic transmission disconnect the oil cooler lines at the radiator.
7. Remove the radiator and fan shroud.
8. Remove the air cleaner assembly and hoses.
9. Disconnect the hoses at the air pump. Remove the air pump and drive belt, if equipped.
10. Disconnect the vacuum pump hoses and remove the vacuum pump. disconnect the power brake booster vacuum hose.
11. Remove the power steering pump, drive belt and bracket. Position aside.
12. If equipped with air conditioning, remove the crankshaft pulley and compressor drive belt. Then, install the pulley again for reference. Remove the air conditioning wire connector and the compressor from its bracket and position aside. Remove the bracket.
13. Disconnect the vacuum hoses from the engine. Disconnect the carbon canister hoses.
14. Disconnect the distributor wire connector, high tension lead, starter cables and the clutch cable clamp.
15. Disconnect the wiring harness at the voltage regulator. Disconnect the throttle cable at the pulley and the wire for the air conditioning at the manifold solenoid.
16. Remove the gas cap. Disconnect the fuel lines at the filter and return pipe.
17. At the firewall, disconnect the electrical connectors for the ballast resistor and relays. Disconnect the heater hoses.
18. Disconnect the micro-switch connectors at the intake manifold and all remaining harness connectors to the engine.
19. Drain the crankcase.
20. Remove the exhaust manifold flange retaining nuts. Loosen the exhaust pipe clamp bolts and remove the bracket for the front exhaust pipe mount.
21. From underneath, remove the front motor mount bolts.
22. If equipped with automatic transmission, place the gear selector lever in **P** and disconnect the gear shift control rod from the transmission.
23. On manual transmission vehicles, disconnect the clutch cable. Then, loosen the set screw, drive out the pivot pin and remove the shifter from the control rod.
24. Disconnect the speedometer and the driveshaft from the transmission.
25. On overdrive equipped vehicles, disconnect the control wire from the shifter.
26. Raise and support the vehicle

safely. Then, using a floor jack and a wooden block, support the weight of the engine beneath the transmission.

27. Remove the bolts for the rear transmission mount. Remove the transmission support crossmember.

28. Lift out the engine using the proper lifting equipment.

To install:

29. If detached, join the engine and transmission. Install the engine assembly in the vehicle and tighten all engine mounting bolts. Install the transmission crossmember and remove the floorjack.

30. Install the driveshaft, speedometer cable, clutch cable (manual transmission) and gear selector mechanism.

31. Install the exhaust system. On turbocharger equipped vehicles, install the turbocharger and related exhaust pipes.

32. Install the air conditioner compressor and related accessory drive units. Install all accessory drive belts and tighten to the proper tension. Install the vacuum pump.

33. Install the radiator and shroud. Install all vacuum, coolant and fuel lines and hoses. Connect all electrical connectors previously disconnected.

34. Install the hood, windsheild wipers, battery and any other component previously removed.

35. Fill the engine with oil, the radiator with coolant and the transmission with fluid.

36. Adjust the reversing lock clamp and the gear selector. Adjust the throttle valve/pulley, automatic transmission kick-down cable and link rod.

37. Start the engine and allow it to reach operating temperature. Check the ignition timing and adjust the engine idle. Check for leaks.

B234F Engine

1. Disconnect the battery, negative cable first.

2. Disconnect the ground connection at the top of the side frame rail.

3. Release the bolted joint at the exhaust manifold front bracket.

4. Attach the sling or lifting equipment to the rear of the motor and support the motor from above. Release any wiring harnesses from their clips and place the wiring aside of the lifting gear.

5. Remove the splash guard under the engine, drain the engine oil and remove the air intake duct.

6. Undo the wiring clips on the front crossmember and right frame rail. Release the battery from the clips and work the wiring free of the roll bar.

7. If equipped with air conditioning, remove the compressor from its mount and position it aside. Do not disconnect any lines or hoses from the compressor.

8. Remove the bottom nut on the left engine mount.

9. On manual transmission vehicles, remove the clutch slave cylinder and position aside. Be careful of the rubber boot; it retains the piston within the cylinder.

10. Separate the front and rear universal joints. Unbolt the center support bearing and withdraw the driveshaft toward the rear of the vehicle.

11. Cut the rear cable tie holding the transmission wiring and separate the connectors.

12. For vehicles with manual transmission, the gear lever is removed by removing the locking bolt, removing the pivot pin between the lever and the selector rod and removing the circlip from the lever sleeve. Push the shift lever up and remove the bushings. For vehicles with automatic transmissions, the selector lever is disconnected by removing the clips from the joints between the lever and the selector rod. Withdraw the arm from the mounting.

13. Release the bolted joint at the front of the catalytic converter and release the oxygen sensor wire from the rear clip.

14. Remove the front exhaust pipe by removing the bolts at its joint to the exhaust manifold.

15. If equipped with automatic transmission, disconnect the oil lines at the transmission and plug the lines.

16. Remove the transmission crossmember. As soon as it is removed, position a floor jack below the transmission to support it.

NOTE: The following steps are in the upper engine area. It may be helpful to temporarily remove the hoist equipment for access. The hoist will need to be reinstalled later in the removal.

17. Remove the upper heat shield from the exhaust manifold. Remove the air hose from the lower heat shield.

18. Remove the top nut from the right motor mount.

19. Open the draincock on the right side of the engine block and drain the coolant into a container.

20. Label and remove the wiring from the distributor cap. Remove the cap and rotor and disconnect the braided engine ground wire.

21. Disconnect the wire to terminal 1 on the coil. Separate the wiring connectors on the right shock tower and release the cable clips on the firewall. Free the wiring from the clips.

22. Disconnect the heater hoses on the left firewall.

23. Release the fuel line connection at the left firewall and attend to any fuel spillage immediately. Plug the fuel lines.

24. Disconnect the wiring connector on the left side of the firewall and free the wires from the clips.

25. Disconnect the air mass meter, its wiring and the hoses connected to the air intake.

26. Release the throttle cable from the pulley.

27. Remove the vacuum hose to the brake booster from the intake manifold. Remove the evaporation hose from the intake manifold and the return line from the fuel distributor.

28. At the left shock tower, release the engine wiring harness from its clips and disconnect the wiring connectors. Remove the power steering reservoir from its clips.

29. Disconnect the coolant hoses at the thermostat housing and at the water pump.

30. Remove the drive belts.

31. Remove the radiator fan, the fan shroud and the drive pulley.

32. Remove the power steering pump from its mount. Place the pump on paper or rags atop the left shock tower. Do not disconnect any hoses from the pump.

33. If the lifting equipment was re-

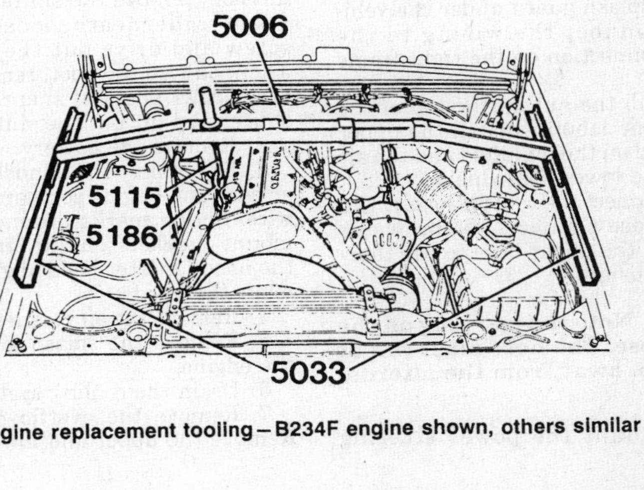

Engine replacement tooling—B234F engine shown, others similar

moved earlier, reconnect it.
34. Check the surroundings of the engine and transmission unit. With the exception of the jack and the motor mounts, there should be nothing connecting the engine/transmission assembly to the body of the vehicle. Take slight tension on the hoist and check that the engine is balanced. Reposition the lift points if the engine is not balanced.
35. Lift out the engine and the gearbox, being very careful of the radiator and surrounding components. Support the engine on appropriate stand.

To install:

36. When reinstalling, check the position and security of the hoist equipment. Lift the engine and gearbox into place in the vehicle.
37. Guide the engine mounts into place and support the transmission on the floor jack.
38. Replace the transmission crossmember and make sure the wiring for the oxygen sensor runs above the crossmember. Remove the floor jack when the crossmember is secure. The engine hoisting equipment may also be removed.
39. Use a new gasket and attach the exhaust pipe to the manifold. Attach the wire to the oxygen sensor.
40. Reconnect the shifting mechanism to the transmission.
41. Reconnect the transmission wiring and secure the harness with new wire ties.
42. Install the driveshaft. Tighten the front and rear universal joints and attach the center support bearing.
43. On manual transmission vehicle, connect the clutch slave cylinder. On automatic transmissions, connect the oil cooler lines.
44. Install the lower nut for the left motor mount. On vehicles with air conditioning, remount the compressor on its brackets.
45. Track the wiring between the anti-roll bar and the front crossmember. Install the cable clips on the crossmember and right side frame rail. Install the splash guard under the vehicle. Reconnect the wiring to the ground connection on the right frame rail.
46. Install the nut on the top of the right engine mount. Install the upper heat shield on the manifold and the air tube to the lower heat shield.
47. Reconnect the coolant hoses. The bottom hose connects to the water pump and the upper hose to the thermostat housing.

NOTE: Note the marking on the upper hose. The hose must run at least 1 in. away from the alternator belt.

48. Remount the power steering pump. Install its belt and the air conditioning belt, if equipped, and adjust to the correct tension.
49. Install the fan, pulley and shroud. Secure the wiring below the fan with new wire ties. Install the drive belt and adjust to the correct tension.
50. Reconnect the rear wiring harnesses on the firewall. Plug all connectors carefully and secure harnesses within the clips. Don't forget the wire to terminal 1 on the coil.
51. Reinstall the distributor rotor, cap and wires. Connect the braided engine ground cable.
52. Reconnect the wiring at the left shock tower. Make sure the wiring is secure in its clips. Install the power steering reservoir.
53. At the intake manifold, connect the vacuum line to the brake booster, the evaporation line and the return line for the fuel distributor.
54. At the left side of the firewall, attach the heater hoses and connect the fuel line.
55. Reattach the throttle cable to the pulley.
56. Install the air mass meter with its hoses and connections.
57. Fill the engine with proper coolant, set the heater to its hottest setting and check the system for leaks.
58. Install the engine oil.
59. Reconnect the battery leads (positive first) and the protective cap on the terminals.
60. Double check all installation items, paying particular attention to loose hoses or hanging wires, untightened nuts, poor routing of hoses and wires (too tight or rubbing) and tools left in the engine area.
61. Start the engine and check for leaks. This engine may be somewhat noisy when started; the noise will disappear as the tappets fill with oil.

B280F Engine

1. If equipped with manual transmission, remove the shifter assembly. From underneath, loosen the set screw and drive out the pivot pin. Then, pull up the boot, remove the reverse pawl bracket, snapring for the shifter and lift out the shifter.
2. Remove the battery.
3. Disconnect the windshield washer hose and engine compartment light wire. Scribe marks around the hood mount brackets on the underside of the hood for later hood alignment. Remove the hood.
4. Remove the air cleaner assembly.
5. Remove the splash guard under the engine.
6. Drain the cooling system.
7. Remove the overflow tank cap. Remove the upper and lower radiator hoses and disconnect the overflow hoses at the radiator.
8. If equipped with automatic transmission, disconnect the transmission cooler lines at the radiator.
9. Remove the radiator and fan shroud.
10. Disconnect the heater hoses, power brake hose at the intake manifold and the vacuum pump hose at the pump. Remove the vacuum pump and O-ring in the valve cover. Remove the gas cap.
11. At the firewall disconnect the fuel lines at the filter and return pipe, disconnect the relay connectors and all other wire connectors. Disconnect the distributor wires.

CAUTION

Use caution when disconnecting the fuel lines. The fuel lines may be under high pressure.

12. Disconnect the evaporative control carbon canister hoses and the vacuum hose at the EGR valve.
13. Disconnect the voltage regulator wire connector.
14. Disconnect the throttle cable and kickdown cable, on automatic transmission vehicles, the vacuum amplifier hose at the T-pipe and the hoses at the thermostat.
15. Disconnect the air pump hose at the backfire valve, the solenoid valve wire and the micro-switch wire.
16. Remove the exhaust manifold flange retaining nuts (both sides).
17. If equipped with air conditioning, remove the compressor and drive belt and place it aside. Do not disconnect the refrigerant hoses.
18. Drain the crankcase.
19. Remove the power steering pump, drive belt and bracket. Position aside.
20. From underneath, remove the retaining nuts for the front motor mounts.
21. Remove, as required, the front exhaust pipe.
22. On 49 states vehicles, remove the front exhaust pipe hangers and clamps and allow the system to hang.
23. If equipped with automatic transmission, place the shift lever in **P**. Disconnect the shift control lever at the transmission.
24. On manual transmission vehicles, disconnect the clutch cylinder from the bell housing. Leave the cylinder connected; secure it to the vehicle.
25. Disconnect the speedometer cable and driveshaft at the transmission.
26. Raise and safely support the vehicle. Place jackstands under the reinforced box member area to the rear of each front jacking attachment. Then, using a floor jack and a thick, wide wooden block, support the weight of the engine under the oil pan.

27. Remove the bolts for the rear transmission mount. Remove the transmission support crossmember.
28. Lift out the engine and transmission as a unit.

To install:

29. If separated, join the engine and transmission. Install the engine assembly in the vehicle and tighten all mounting bolts to specification. Install the transmission crossmember.
30. Install the driveshaft and speedometer cable. On manual transmission vehicles, install the clutch assembly and gear shift assembly. On automatic transmission vehicles install the gear selector assembly.
31. Install the exhaust system. Install all air conditioning compressor, power steering pump, air pump and alternator. Install and tighten all accessory drive belts to specification.
32. Connect the throttle and kickdown cables. Connect the charcoal canister and evaporative emissions control hoses.
33. Install the fuel lines and filter. Install the heater hoses and vacuum pump hoses.
34. Install the radiator, shroud and radiator hoses. Connect all other hoses or lines previously disconnected. Connect all electrical connections.
35. Install the hood, battery and gear selector.
36. Fill the cooling system with coolant, the engine with oil and the transmission with fluid. Start the engine and bring it to operating temperature. Adjust the timing and idle speed, as necessary, and check for leaks.

Cylinder Head

REMOVAL & INSTALLATION

B230F Engine

1. Disconnect the battery.
2. Remove the overflow tank cap and drain the coolant. Disconnect the upper radiator hose.
3. Remove the distributor cap and wires.
4. Remove the PCV hoses.
5. Remove the EGR valve and vacuum pump.
6. Remove the air pump, if equipped, and air injection manifold. Disconnect and remove all hoses to the turbocharger, if equipped. Plug all open hoses and holes immediately.
7. Remove the exhaust manifold and header pipe bracket.
8. Remove the intake manifold. Disconnect the manifold brace and the hose clamp to the bellows for the fuel injection air/flow unit. Disconnect the throttle cable and all vacuum hoses and electrical connectors to the fuel injection unit.
9. Remove the fuel injectors.
10. Remove the valve cover.
11. Loosen the fan shroud and remove the fan. Remove the shroud. Remove the upper belts and pulleys.
12. Remove the timing belt cover. Remove the timing belt.
13. Remove the camshaft, if necessary.
14. Loosen the cylinder head bolts in the reverse order of the torquing sequence. Remove the cylinder head.

To install:

15. Check the position of the crankshaft. No. 1 piston should be at TDC. Check the position of the camshaft for cylinder No. 1. Both lobes should be in such a position that if the head were install, the valves would be closed.
16. Install the cylinder head gasket and the cylinder head. Ensure that the O-ring for the water pump is in place. Apply a light coat of oil to the head bolts and install.
17. Tighten the head bolts in three stages using the proper sequence.
    a. Stage 1 – Tighten all bolts to 14 ft. lbs. (20 Nm).
    b. Stage 2 – Tighten all bolts to 43 ft. lbs. (60 Nm).
    c. Stage 3 – Angle tighten all bolts an additional 90 degrees.
18. Install the camshaft, camshaft gear and spacer, as required. Do not allow the cam to turn during installation. Set the timing belt tensioner and install the timing belt. Remove the tool from the belt tensioner to tension the belt.
19. Rotate the engine one full turn. Loosen the tensioner bolt one full turn and retighten to set the tensioner. Adjust the valve lash.
20. Install the fan shroud and fan. Install the accessory drive belts and pulleys.
21. Install the intake manifold, fuel injection system, throttle cable and valve covers.
22. Install the exhaust manifold and header pipe. Install the air pump assembly. If equipped with a turbocharger, install the turbocharger and related parts.
23. Install the EGR valve, vacuum pump, PCV hoses, distributor cap and wires, overflow tank and battery.
24. Fill the radiator with coolant, check the engine oil and transmission fluid. Start the engine and allow it to reach operating temperature. Check the timing.

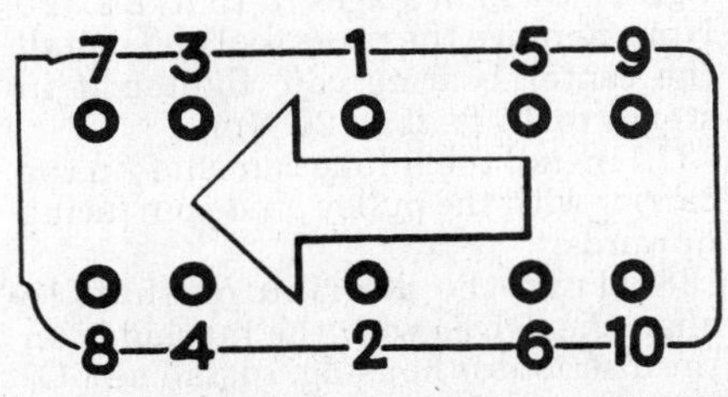

Head bolt tightening sequence for B230F and B280F. When removing, loosen the bolts in reverse order

B234F Engine

NOTE: The use of the correct special tools or their equivalent, is required for this procedure.

1. Disconnect the negative battery cable.
2. Remove the heat shield over the exhaust manifold.
3. Remove the cap from the expansion tank and open the draincock on the right side of the motor. Collect the drained coolant in a suitable container.
4. Unbolt the exhaust pipe from the bracket, remove the manifold nuts and remove the manifold from the head.
5. On the left side of the motor, remove the support under the intake manifold. and remove the bottom bolt in the cylinder block.
6. Remove the manifold intact and tie it or support it safely.
7. Disconnect the temperature sensor connectors, the heating hose under cylinders No. 3 and 4 and the upper radiator hose at the thermostat.
8. Remove the upper and lower timing belt covers.
9. Align the camshaft and crankshaft marks. Turn the engine to TDC, of the compression stroke, on cylinder No. 1 and make sure the pulley marks and the crank marks align.
10. Remove the protective cap over the timing belt tensioner locknut. Loosen the locknut, compress the tensioner, to release tension on the belts, and retighten the locknut, holding the tensioner in place.
11. Remove the timing belt from the camshafts. Do not crease or fold the belt.

NOTE: The camshafts and the crankshaft must not be moved when the belt is removed.

12. Remove the timing belt idler pulleys.
13. Remove the camshaft drive pulleys. Use a counterhold wrench to prevent the cam from turning.
14. Remove the plate or panel behind the pulleys. Remove the cover plate for the ignition wires. Label and disconnect the ignition wiring from the spark plugs and the distributor cap; remove the coil wire from the distributor cap.
15. Remove the valve cover and gasket. Clean the surfaces of any gasket remains.
16. Remove the distributor housing from the camshaft carrier. Remove the ignition wire clip next to the left bolt.
17. Plug the spark plug holes with crumpled paper. Remove the center

bearing cap for each camshaft. Remove the third nut in the center. Mark the cam bearing caps for proper reinstallation.

18. Install a camshaft press tool 5021 or similar, on the exhaust side cam in place of the removed bearing cap. When it is securely in place, remove the remaining bearing caps and nuts. Remove the tool and remove the exhaust camshaft.

19. Remove the intake camshaft in identical fashion.

NOTE: Label or identify each cam and its bearing caps. All removed components should be kept in order.

20. Using a magnet or a small suction cup, remove the tappets. Store them upside down, to prevent oil drainage, and keep them in order; they are not interchangeable.

21. Remove the remaining 4 nuts in the center of the cam carrier and detach the carrier from the head. If it is stuck, tap it very gently with a plastic mallet. Remove the O-rings around the spark plug holes.

22. Wipe the remaining oil off the cylinder head and remove the bolts in order. When all the bolts are removed, the cylinder head may be lifted free of the vehicle.

NOTE: The head is aluminum. Support it on clean wood blocks or similar to avoid scoring the face.

23. Clean the camshaft carrier and the head assembly of all gasket material and sealer. Carefully scrape the joint surfaces with a plastic scraper. Do not use metal tools to scrape or clean. Wash the surfaces with a degreasing compound and blow the surfaces completely dry. Inspect the head bolts for any sign of stretching or elongation in the midsection. If this is observed or suspected, discard the bolt. Bolts may not be used more than 5 times.

To install:

24. Install the new head gasket and a new O-ring for the water pump. Carefully place the cylinder head into position; do not damage the gasket.

25. Clean the head bolts and apply a light coat of oil. Install them and tighten, in sequence, in 3 steps: to 15 ft. lbs. (20 Nm), then all to 30 ft. lbs. (41 Nm). Third Step is to tighten each bolt through 115 degree of arc in 1 continuous motion. The use of protractor fitting tool 5098 is strongly recommended for this task.

26. Install the exhaust manifold with a new gasket. Attach the front exhaust pipe to its bracket and install the heat shields.

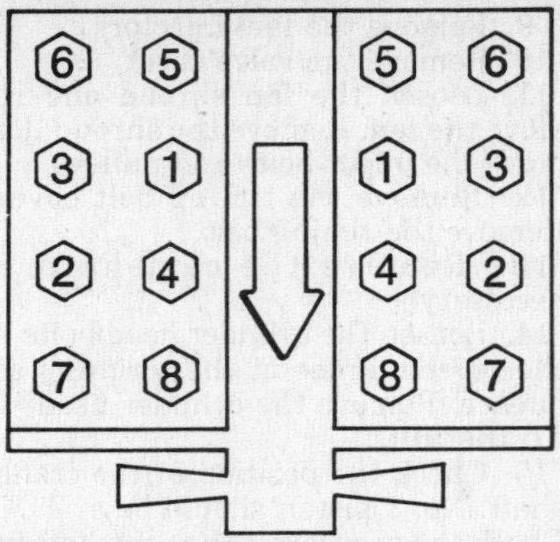

Head bolt tightening sequence for B234F. When removing, loosen the bolts in reverse order

27. On the left side of the motor, connect the temperature sensors, the heating hose under cylinders 3 and 4 and the upper coolant hose to the thermostat.

28. Fill the cooling system and check carefully for leaks, particularly around the head to block joint.

29. Install the intake manifold with a new gasket. Tighten the bottom bolts a few turns and place the manifold in position. Tighten all the bolts from the center outwards.

30. Reattach the support under the intake manifold and the cable clip. Double check all connections on and around the intake manifold.

31. Apply liquid sealing compound to the camshaft carrier. Use a small paint roller and coat the surfaces which match to the head and the bearing cap joint faces.

32. Install the cam carrier on the head and secure it with 4 of the 5 center nuts tightened to 15 ft. lbs. (20 Nm); do not install the middle nut.

33. Oil all matching surfaces on the cam carrier, bearing caps and tappets.

34. Insert the tappets; they must be inserted in their original order and place.

35. Install the exhaust side camshaft by placing it in the carrier with the pulley guide pin facing up. Using the rear bearing cap as a guide, press the cam into place with the press tool. Install the bearing caps in the original order.

36. Install the bearing cap nuts and tighten them in stages to 15 ft. lbs. (20 Nm). Remove the press tool and install the center bearing cap; tighten it in stages to 15 ft. lbs. (20 Nm).

37. Install the intake camshaft in the carrier with the pulley guide pin facing upwards.

38. Turn the distributor shaft to align the driver with the markings on the distributor housing. Install new O-rings on the housing and rotor shaft.

39. Using the rear bearing cap as a guide, press the cam into place with the press tool. Install the bearing caps in the original order.

40. Install the bearing cap nuts and tighten them in stages to 15 ft. lbs. (20 Nm). Remove the press tool and install the center bearing cap; tighten it in stages to 15 ft. lbs. (20 Nm).

41. Install the center nut in the cam carrier and tighten it to 15 ft. lbs. (20 Nm).

42. Double check the tightness of all the camshaft carrier nuts and the bearing cap nuts. All should be 15 ft. lbs.; do not overtighten.

43. Reinstall the distributor, connect the coil wire and install the ignition wire clip at the left bolt. Remove the paper plugs from the spark plug holes.

44. Use a silicone sealer and apply to the front and rear camshaft bearing caps. Install new gaskets for the valve cover and the spark plug wells. Install the spark plug gasket with the arrow pointing towards the front of the vehicle and the word "UP" facing up. Make sure the valve cover gasket is correctly positioned and install the valve cover.

45. Reconnect the ground wire at the distributor.

46. Install the ignition wires and the cover plate.

47. Using a compression seal driver tool 5025 or similar, install the oil seals for the front of each camshaft. Camshafts must not be allowed to turn during this operation.

48. Install the upper backing plate over the ends of the camshafts and adjust the plate so the cams are centered in the holes.

49. Replace the idler pulleys and tighten their mounts to 18.5 ft. lbs. (25 Nm).

50. Install the camshaft drive pulleys, using a counterhold to prevent the cams from turning.

51. Making sure the camshaft pulleys are properly aligned with the marks on the backing plate, position the timing belt so the double mark on the belt coincides exactly with the top mark on the belt guide plate, at the top of the crankshaft. Place the belt onto the cam pulleys and make sure the single marks on the belt line up exactly with the marks on the pulleys. Fit the belt over the idler pulleys; right side idler first, then the left.

52. Double check that the engine is on TDC, of the compression stroke, for cylinder No. 1 and that all the belt markings line up as they should.

53. Loosen the tensioner locknut. Rotate the crankshaft clockwise 1 full turn until the belt markings again coincide with the pulley markings.

NOTE: The engine must not be rotated counterclockwise while the tensioner is loose.

54. Turn the crankshaft smoothly clockwise until the pulley marks are 1½ teeth beyond the marks on the backing plate.
55. Tighten the tensioner locknut. Install the lower timing belt cover.
56. Install the radiator fan and pulley, the alternator drive belt and the negative battery cable.
57. Double check all installation items, paying particular attention to loose hoses or hanging wires, untightened nuts, poor routing of hoses and wires (too tight or rubbing) and tools left in the engine area.
58. Start the engine and allow it to run until the thermostat opens. Use extreme caution; the timing belt is exposed.

NOTE: This engine may be somewhat noisy when started. The noise will subside as oil reaches the tappets. Do not exceed 2500 rpm while the tappets are noisy.

59. Shut the engine off, rotate the crankshaft to bring the engine to TDC, of the compression stroke, of cylinder No. 1 and use tool 998 8500 or equivalent, to check the belt tension. Correct deflection is 5.5 ± 0.2 units when measured between the exhaust camshaft pulley and the idler. If the tension is not correct, repeat Steps 51–54.
60. Install the upper timing belt cover. Start the engine and final check all functions.

B280F Engine

1. Disconnect the battery. Drain the coolant.
2. Remove the air cleaner assembly and all attaching hoses.
3. Disconnect the throttle cable. On automatic transmission equipped vehicles, disconnect the kickdown cable.
4. Disconnect the EGR vacuum hose and remove the pipe between the EGR valve and manifold.
5. Remove the oil filler cap and cover the hole with a rag. Disconnect the PCV pipe(s) from the intake manifold.
6. Remove the front section of the intake manifold.
7. Disconnect the electrical connector and fuel line at the cold start injector. Disconnect the vacuum hose, both fuel lines. and the electrical connector from the control pressure regulator.
8. Disconnect the hose, pipe and electrical connector from the auxiliary air valve. Remove the auxiliary air valve.
9. Disconnect the electrical connector from the fuel distributor. Remove the wire loom from the intake manifolds. Disconnect the spark plug wires.
10. Disconnect the fuel injectors from their holders.
11. Disconnect the distributor vacuum hose, carbon filter hose and diverter valve hose from the intake manifold. Also, disconnect the power brake hose and heater hose at the intake manifold.
12. Disconnect the throttle control link from it pulley.
13. If equipped with an EGR vacuum amplifier, disconnect the wires from the throttle micro-switch and solenoid valve.
14. At the firewall, disconnect the fuel lines from the fuel filter and return line.
15. Remove the 2 attaching screws and lift out the fuel distributor and throttle housing assembly.
16. If not equipped with an EGR vacuum amplifier, disconnect the EGR valve hose from under the throttle housing.
17. Remove the cold start injector, rubber ring and pipe.
18. Remove the 4 retaining bolts and lift off the intake manifold. Remove the rubber rings.
19. Remove the splash guard under the engine.
20. If removing the left cylinder head, remove the air pump from its bracket.
21. Remove the vacuum pump and O-ring in the valve cover. Remove the vacuum hose from the wax thermostat.
22. If removing the right cylinder head, disconnect the upper radiator hose.
23. On air conditioned vehicles, remove the air conditioning compressor and secure it aside. Do not disconnect the refrigerant lines.
24. Disconnect the distributor leads and remove the distributor. Remove the EGR valve, bracket and pipe. At the firewall, disconnect the electrical connectors at the relays.
25. On air conditioned vehicles, remove the rear compressor bracket.
26. Disconnect the coolant hose(s) from the water pump to the cylinder head(s). If removing the left cylinder head disconnect the lower radiator hose at the water pump.
27. Disconnect the air injection system supply hose from the applicable cylinder head. Separate the air manifold at the rear of the engine. If removing the left cylinder head, remove the backfire valve and air hose.
28. Remove the valve cover(s).
29. On the left cylinder head, remove the Allen head screw and 4 upper bolts to the timing gear cover. On the right cylinder head, remove the 4 upper bolts to the timing gear cover and the front cover plate.
30. From under the vehicle, remove the exhaust pipe clamps for both header pipes.
31. If removing the right cylinder head, remove the retainer bracket bolts and pull the dipstick tube out of the crankcase.
32. Remove the applicable exhaust manifold(s).
33. Remove the cover plate at the rear of the cylinder head.
34. Rotate the camshaft sprocket, for the applicable cylinder head, into position so the large sprocket hole aligns with the rocker arm shaft. With the camshaft in this position, loosen the cylinder head bolts, in sequence, same sequence as tightening, and remove the rocker arm and shaft assembly.
35. Loosen the camshaft retaining fork bolt, directly in back of sprocket, and slide the fork away from the camshaft.
36. Next, it is necessary to hold the cam chain stretched during camshaft removal. Otherwise, the chain tensioner will automatically take up the slack, making it impossible to reinstall the sprocket on the cam without removing the timing chain cover to loosen the tensioner device. To accomplish this, a sprocket retainer tool 999 5104 is installed over the sprocket with 2 bolts in the top of the timing chain cover. A bolt is then screwed into the sprocket to hold it in place.
37. Remove the camshaft sprocket center bolt and push the camshaft to the rear, so it clears the sprocket.
38. Remove the cylinder head.

NOTE: Do not remove the cylinder head by pulling straight up. Instead, lever the head off by inserting 2 spare head bolts into the front and rear inboard cylinder head bolt holes and pulling toward the applicable wheel housing. Otherwise, the cylinder liners may be pulled up, breaking the lower liner seal and leaking coolant into the crankcase. If any do pull up, new liner seals must be used and the crankcase completely drained. If the head(s) seem

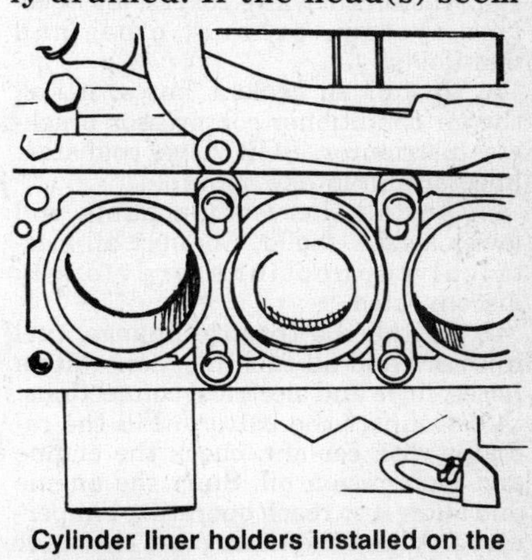

Cylinder liner holders installed on the B280F engine

stuck, gently tap around the edges of the head(s) with a rubber mallet, to break the joint.

39. Remove the head gasket. Clean the contact surfaces with a plastic scraper and lacquer thinner.
40. If the head is going to be off for any length of time, install liner holders tool 999 5093 or 2 strips of thick stock steel with holes for the head bolts, so the liners stay pressed down against their seals. Install the holders widthwise between the middle 4 head bolt holes.

To install:

41. If the dowels at the outboard corners of the block have slipped down, use a pair of needle-nose pliers to retrieve them. Prop them up with an 1/8 in. (3 mm). Remember to keep the timing chain taunt during cylinder head installation.
42. Remove the liner holders and install the head gaskets. The left and right head gaskets are different, ensure the correct one is installed. Install the cylinder head.
43. Install the camshaft and remove the timing chain retainer tool. Install the head bolts finger tight after lubricating with oil. Tighten the bolts as follows:
    a. Stage 1 – Tighten all bolts to 7 ft. lbs. (31 Nm).
    b. Stage 2 – Tighten all bolts to 22 ft. lbs. (97 Nm).
    c. Stage 3 – Tighten all bolts to 44 ft. lbs. (195 Nm).
    d. Wait 30 minutes, loosen all bolts in the same order of torquing and tighten to 13 ft. lbs. (58 Nm).
44. Install the camshaft center bolt and tighten to 52–66 ft. lbs. Install the timing gear case and rear cylinder head covers. Check and adjust the valve lash. After adjusting valve lash, turn the engine to TDC on No. 1 piston.
45. Install the valve covers, air injection system, exhaust pipes and manifolds.
46. Install all coolant hoses, install the air conditioner compressor brackets, distributor, EGR valve, cold start injector and intake manifold.
47. Install the vacuum pump and lower splash sheild. Connect all electrical connections previously disconnected.
48. Install the throttle linkage, fuel injectors and all fuel injection system hoses, lines and electrical connections.
49. Connect the battery. Fill the radiator with coolant, check the engine and trainmision oil. Start the engine and allow it to reach operating temperature. Adjust the timing and check for leaks.

Valve Lash

NOTE: The B234F engine uses hydraulic lash adjusters which do not require adjustment.

ADJUSTMENT

B230F Engines

1. Remove the valve cover. Scribe chalk marks on the distributor body indicating each of the 4 spark plug wire leads in the cap. Remove the distributor cap.
2. Crank over the engine with a remote starter switch or with a wrench on the crankshaft pulley center bolt (22mm hex) until the engine is in the firing position for No. 1 cylinder. At this point, the **0** degree or TDC mark on the crankshaft pulley is aligned with the timing pointer, the rotor is pointing at the No. 1 spark plug wire cap position and the camshaft lobes for No. 1 cylinder are pointing at the 10 o'clock and 2 o'clock positions. At this point, the clearance between the cam lobe and valve depressor (tappet) may be checked for the intake and exhaust valve of cylinder No. 1, using a feeler gauge. When checking clearance, the wear limit is 0.012–0.018 in. (0.3–0.4mm) for a cold engine and 0.012–0.020 in. (0.3–0.5mm) for a hot one 176°F (80°C).
3. Repeat Step 2 for cylinders No. 3, 4 and 2, in that order. Each time, rotate the crankshaft pulley 180 degrees so the rotor is pointing to the spark plug wire cap position for that cylinder and the cam lobes are pointing at the 10 and 2 o'clock positions for the valves of that cylinder.
4. If any of the valve clearance measurements are outside the wear limit, remove the old valve adjusting disc and install a new one to bring the clearance within specifications. First, rotate the valve depressors (tappets) until their notches are at a right angle to the engine center line. Attach valve depressor tool 999 5022 or equivalent, to the camshaft and screw down the tool spindle until the depressor (tappet) groove is just above the edge of its bore and still accessible with the special pliers tool 999 5026.
5. Remove the valve adjusting disc and measure with a micrometer. The valve clearance should be set to these tolerances: 0.014–0.016 in. (0.35–0.40mm) for a cold engine and 0.016–0.018 in. (0.40–0.45mm) for a hot one. So, if the measured clearance had been 0.019 in. (0.48mm) and the desired clearance 0.016 in. (0.40mm), for a net difference of 0.003 in. (0.076mm), then the new valve adjusting disc should be 0.003 in. (0.076mm) thicker than the old one to take up the clearance. Valve adjusting discs are available from in sizes 0.130–0.180 in. (3.3–4.6mm), in 0.002 in. (0.050mm) increments. Always oil the new disc and install it with the marks facing down.
6. Remove the valve tappet depressor tool. Rotate the engine a few times and recheck clearance. Install the valve cover with a new gasket.

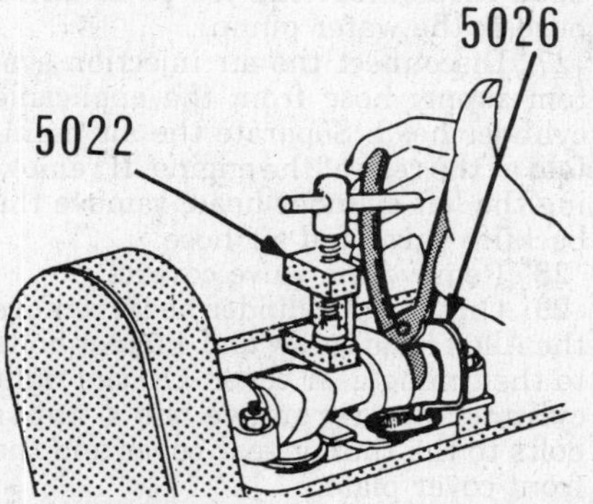

B230F valve adjustment tools – tappet depressor (5022) and shim pliers (5026)

B280F Engine

1. In order to gain access to the valve covers, disconnect or remove the following:
    a. Air conditioning compressor from bracket; do not disconnect refrigerant hoses
    b. EGR valve and hoses
    c. Air conditioning compressor bracket
    d. Fuel injection control pressure regulator
    e. Air pump
    f. Vacuum pump
    g. Hoses and wires from solenoid valve, California only
2. Using a 36mm hex socket on the crankshaft pulley bolt, rotate the crankshaft to the No. 1 cylinder TDC position, of the compression stroke. At this point the **0** mark on the timing plate aligns with the crankshaft pulley notch, the distributor rotor is pointing to the No. 1 cylinder spark plug wire cap position and both valves for No. 1 cylinder have clearance. At this position, adjust the intake valves of cylinders No. 1, 2 and 4; the exhaust valves of cylinders No. 1, 3 and 6. Insert a feeler gauge between the rocker arm and valve stem. Loosen the locknut and turn the adjusting screw in the required direction. Tighten the locknut and recheck clearance.

Clearance for B280F should be:

Cold engine

Intake – 0.004–0.006 in. (0.10–0.15mm)

Exhaust – 0.010–0.012 in. (0.25–0.30mm)

Hot engine

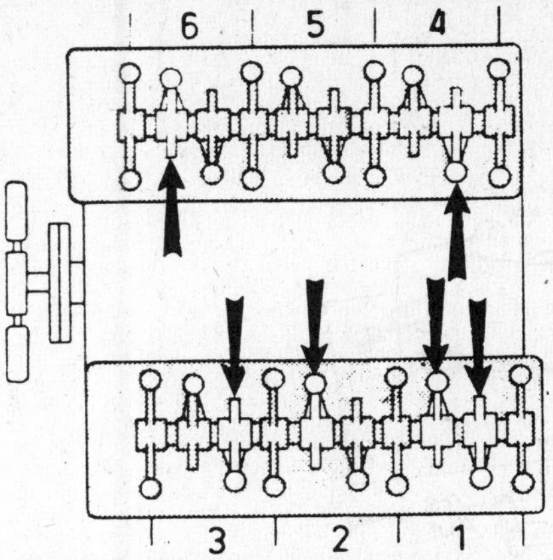

Adjust these valves (arrow) with the No. 1 cylinder at TDC—B280F engine

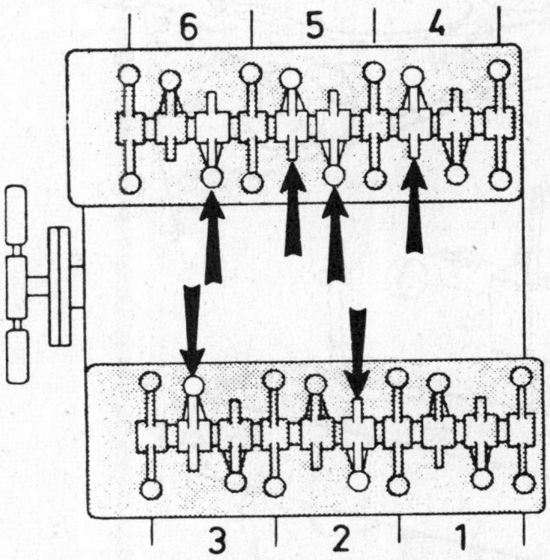

Adjust these valves (arrow) after rotating the engine 360 degrees

Intake—0.006–0.008 in. (0.15–0.20mm)

Exhaust—0.012–0.014 in. (0.30–0.35mm)

3. Rotate the crankshaft pulley 1 full 360 degrees turn to adjust the remaining valves. At this point, the **0** mark will again align with the pulley notch, the rotor is pointing 180 degrees opposite its former position and the No. 1 cylinder rockers contact the ramps of the camshaft. At this position, adjust the intake valves of cylinders No. 3, 5 and 6; the exhaust valves of cylinders No. 2, 4 and 5.

4. Install the valve covers with new gaskets. Connect all disconnected equipment.

Rocker Shafts

REMOVAL & INSTALLATION

B280F Engine

1. Disconnect the negative battery cable.
2. Remove the air cleaner assembly.
3. Disconnect the air pump bracket.
4. Remove the left valve cover, if necessary.
5. Tie the upper radiator hose aside and remove the oil filler cap and carbon canister hose.
6. On air conditioned vehicles, remove the air conditioning compressor from it bracket. Do not disconnect the hoses.
7. Remove the EGR valve.
8. Remove the air conditioning compressor rear bracket.
9. Remove the control pressure regulator.
10. Disconnect any hoses or wires in the way. Remove the right valve cover, if necessary.

NOTE: Do not jar the head while the rocker and bolts are loose, as the cylinder liner O-ring seals may break, necessitating the teardown of the engine.

11. The rocker arm bolts double as cylinder head bolts. Loosen the head bolts in the reverse order of the torquing sequence. If removing both rocker shafts, mark them left and right.

To install:

12. Install the rocker shafts. Follow cylinder head installation procedure for proper torque specification and sequence. Adjust the valve lash.
13. Install the valve covers, EGR valve, control pressure regulator, air conditioning compressor and bracket and air pump.
14. Connect all fuel, coolant and vacuum lines previously disconnected. Connect all electrical connections previously disconnected. Connect the battery.
15. Start the engine and allow it to reach operating temperature. Adjust the timing and check for leaks.

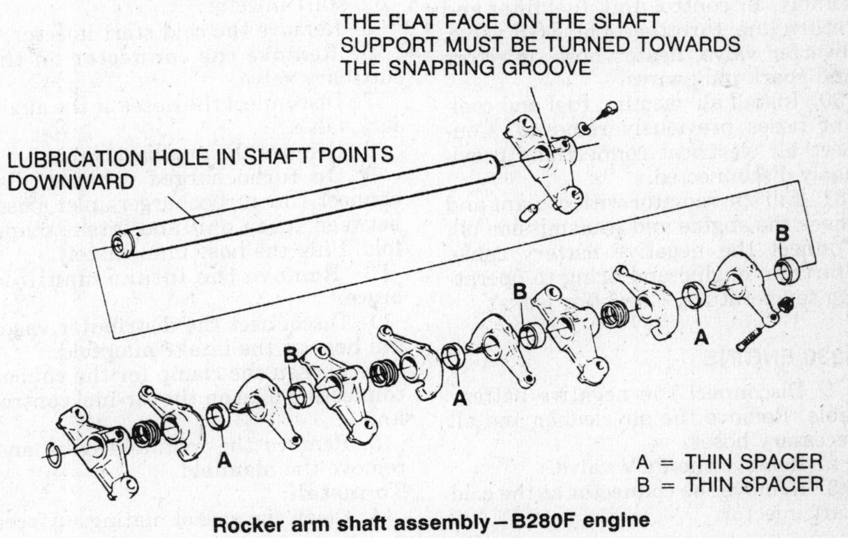

Rocker arm shaft assembly—B280F engine

Intake Manifold

REMOVAL & INSTALLATION

Inlet Duct

1. Disconnect the negative battery cable.
2. Disconnect the throttle and downshift linkage. Remove from the inlet duct, the positive crankcase ventilation, distributor advance, pressure sensor for electronic fuel injection Models only, and power brake hoses.
3. On electronic fuel injected Models, disconnect the contact for the throttle valve switch and remove the ground cable for the inlet duct.
4. Remove the bolts for the inlet duct stay. Remove the inlet duct-to-cylinder head retaining nuts and slide the inlet duct off the studs. Discard the old gasket.
5. To install, reverse the removal procedure. Use a new inlet duct gasket. Torque the nuts to 13–16 ft. lbs. (18–22 Nm).

Intake Manifold

B280F ENGINE

1. Disconnect the negative battery cable. Remove the air cleaner and all necessary hoses.
2. Drain the radiator coolant.
3. Remove the throttle cable from the pulley and bracket.
4. On automatic transmission vehicles, remove the throttle cable that is connected to the transmission.
5. Remove the EGR pipe from the EGR valve to the manifold.
6. Disconnect the EGR vacuum line.
7. Remove the oil filler cap and PCV valve.

NOTE: Cover the oil cap opening with a rag to keep dirt out.

8. Remove the front manifold bolts and remove the front section of the manifold.
9. Disconnect the cold start connector, fuel line and injector.
10. Disconnect the pressure control regulator vacuum lines, fuel lines and the connector.
11. Remove the auxiliary valve and its necessary piping.
12. Disconnect the electrical connections at the air fuel control unit.
13. Remove all 6 spark plug wires.
14. Remove all 6 injectors.
15. Move the wiring harness to the outside of the manifold.
16. Disconnect the vacuum hose at the distributor and the intake manifold.
17. Disconnect the heater hose at the intake manifold.
18. Disconnect the hose to the diverter valve.
19. Disconnect the vacuum hose to the power brake booster.
20. Disconnect the throttle cable link.
21. Disconnect the wires to the micro-switch.
22. Pull the wires away from the intake manifold.
23. Remove the fuel filter line and the return line.
24. Remove the air control unit.
25. Disconnect the vacuum hose from the throttle valve housing.
26. Remove the pipe and cold start injector assembly.
27. Remove the intake manifold from the vehicle.

To install:

28. Clean all gasket mating surfaces throughly. Install the intake manifold using new gaskets and tighten the bolts to 7–11 ft. lbs. (10–15 Nm).
29. Install the cold start injector assembly, air control unit, fuel filter and return line, throttle cable, EGR valve, diverter valve, heater hose, injectors and spark plug wires.
30. Install all vacuum, fuel and coolant hoses previously removed. Connect all electrical connections previously disconnected.
31. Fill the radiator with coolant and check the engine and transmission oil. Connect the negative battery cable. Start the engine and bring to operating temperature. Check for leaks.

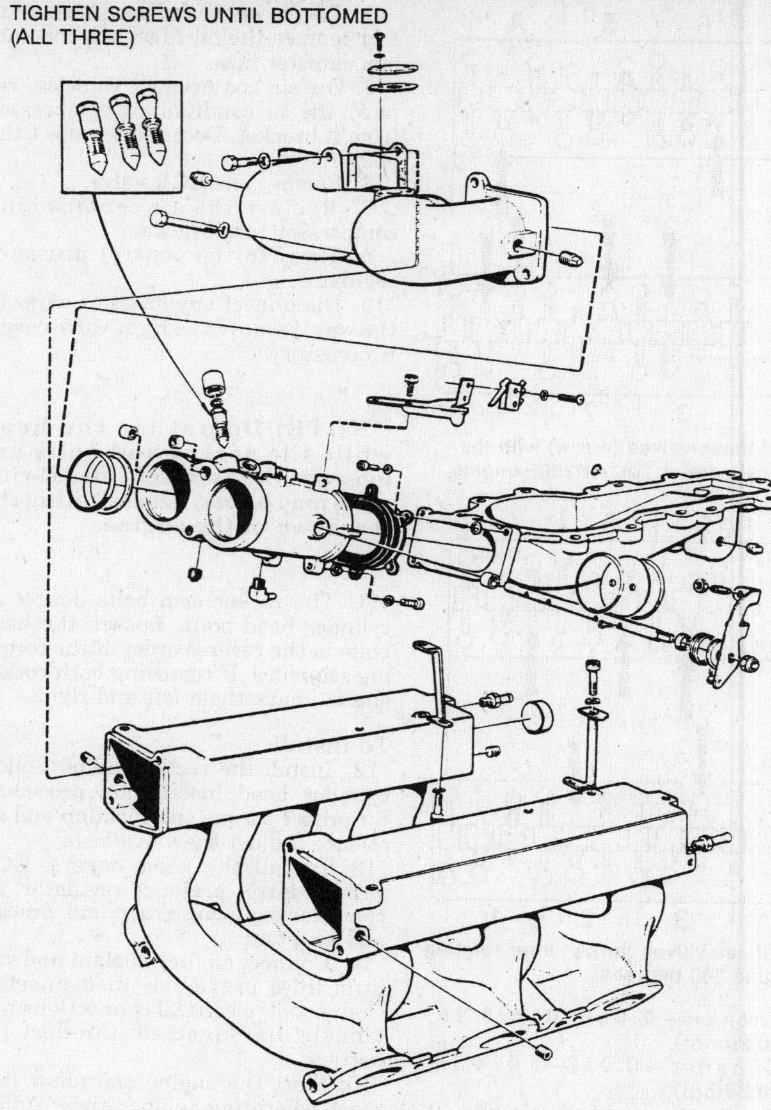
TIGHTEN SCREWS UNTIL BOTTOMED (ALL THREE)

Intake manifold assembly—B280F engine

B230 ENGINE

1. Disconnect the negative battery cable. Remove the air cleaner and all necessary hoses.
2. Remove the PCV valve.
3. Remove the connector at the cold start injector.
4. Remove the fuel hose from the cold start injector.
5. Remove the cold start injector.
6. Remove the connector on the auxiliary valve.
7. Disconnect the hoses at the auxiliary valve.
8. Remove the auxiliary valve.
9. On turbocharged vehicles, disconnect the turbocharger inlet hose, between turbo unit and intake manifold. Plug the hose immediately.
10. Remove the intake manifold brace.
11. Disconnect the distributor vacuum hose at the intake manifold.
12. Loosen the clamp for the rubber connecting pipe on the air-fuel control unit.
13. Remove the manifold bolts and remove the manifold.

To install:

14. Clean the gasket mating surfaces throughly. Install the intake manifold, using new gaskets, and tighten the bolts to 15 ft. lbs. (20 Nm).
15. Install the intake manifold brace, air-fuel control unit connecting pipe, turbocharger inlet hose (if equipped), auxiliary valve, cold start injector, fuel hose and PCV valve.
16. Connect all vacuum, fuel and coolant hoses previously disconnected. Connect all electrical connectors previously disconnected.
17. Connect the negative battery cable, start the engine and bring it to operating temperature. Adjust the timing and check for leaks.

B234F ENGINE

1. Remove the air mass meter and the air intake hose.
2. Detach the throttle pulley from the intake manifold and remove the link rod from the throttle lever.

3. Separate the throttle housing from the intake manifold and cut the cable tie holding the wiring to the vacuum hose connections.

4. Disconnect the lines and hoses from the manifold, including the brake booster vacuum hose, the evaporation line, the oil trap, the fuel pressure regulator line and the air control valve line. If equipped with a vacuum tank, disconnect its line at the manifold.

5. Disconnect the fuel return line at the distribution pipe. Disconnect the wiring to the injectors and remove the distribution pipe and injectors. Immediately protect these components from the entry of any dirt.

6. Unbolt and remove the intake manifold from the engine.

To install:

7. If installing a new manifold, it is necessary to transfer the various hose nipples and plugs to the new part. Install the manifold with a new gasket. Starting with the center bolts and working outward, tighten the bolts to 15 ft. lbs. (20 Nm).

8. Reconnect the hoses to their proper ports.

9. Position the injector wiring between cylinders 2 and 3 and reinstall the fuel distributor rail and the injectors. Tighten the pipe and the ground wires to the block. Connect the fuel pressure regulator line to the intake manifold.

10. Install the throttle pulley and connect the link rod.

11. Install the throttle housing with a new gasket. Check the operation of the throttle stops and switches.

12. Install the air mass meter and air inlet hose.

Exhaust Manifold

REMOVAL & INSTALLATION

B280F Engine

1. Raise and support the vehicle safely.

2. Unbolt the crossover pipe from the left and right side of the exhaust manifolds, if equipped.

NOTE: If the vehicle has the Y-type exhaust pipe disconnect this pipe at the left and right manifolds.

3. Remove any other necessary hardware.

4. Remove the left and right side manifolds.

5. Installation is the reverse of removal.

NOTE: Always use new gaskets when reinstalling the manifolds.

6. Torque the manifold bolts to 7–11 ft. lbs. (10–15 Nm).

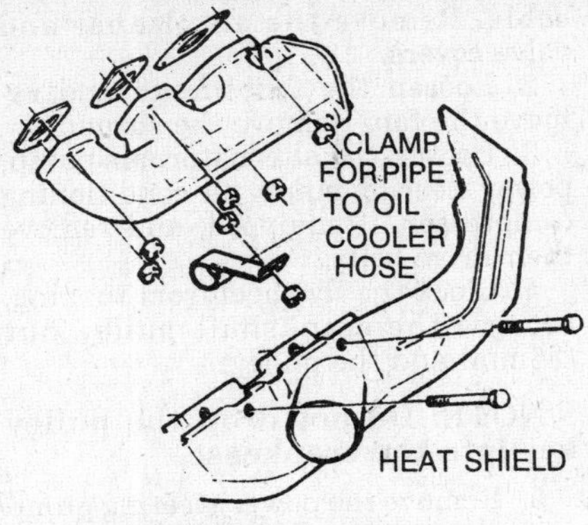

Exhaust manifold assembly—B280F engine

B230 Engine

1. Disconnect the negative battery cable. Remove the air cleaner and all necessary hoses.

2. Remove the EGR valve pipe from the manifold.

3. Remove the exhaust pipe from the exhaust manifold.

4. Remove the manifold bolts and remove the manifold.

NOTE: Remember to install new manifold gaskets before installing the manifold.

5. Installation is the reverse of removal.

6. Torque the manifold bolts to 10–20 ft. lbs. (14–27 Nm).

B234F Engine

1. Disconnect the front exhaust pipe from the manifold. Disconnect the catalytic converter from the front muffler.

2. Remove the heat shields (top and bottom) from the manifold and remove the air preheat hose.

3. Disconnect the front exhaust pipe from the bracket on the bell housing.

4. Unbolt the exhaust manifold and remove it from the vehicle.

5. Install the manifold with a new gasket and tighten the bolts to 15 ft. lbs. (20 Nm).

6. Install the front exhaust pipe with a new gasket; tighten the joint to the manifold to 20 ft. lbs. (27 Nm). Reattach the catalytic converter to the front muffler.

7. Install the heat shields and the preheat hose.

Turbocharger

REMOVAL & INSTALLATION

B230F–Turbocharged Engine

1. Disconnect the battery ground cable.

2. Disconnect expansion tank from retainer. Remove expansion tank retainer.

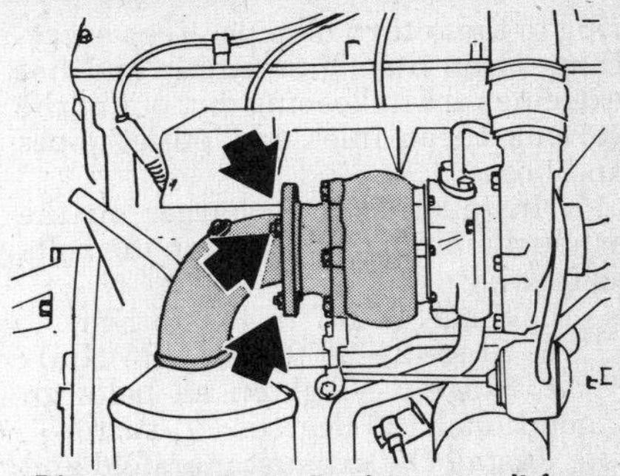

Disconnect the turbocharger unit from the exhaust manifold—B230F Turbo engine

3. Remove preheater hose to the air cleaner. Remove the pipe and rubber bellows between the air/fuel control unit and the turbocharger unit. Pull out the crankcase ventilation hose from the pipe.

4. Remove the pipe and pipe connector between the turbocharger unit and the intake manifold.

NOTE: Cover the turbocharger intake and outlet ports to keep dirt out of the system.

5. Disconnect the exhaust pipe and secure it aside.

6. Disconnect the spark plug wires at the plugs.

7. Remove the upper heat shield. Remove the brace between the turbocharger unit and the manifold.

8. Remove the lower heat shield by removing the 1 retaining screw under the manifold.

9. Remove the oil pipe clamp, retaining screws on the turbo unit and the pipe connection screw in the cylinder block under the manifold. Do not allow any dirt to enter the oilways.

10. Remove the manifold retaining screws and washers. Let 1 nut remain in position to keep the manifold in position.

11. Remove the oil delivery pipe. Cover the opening on the turbo unit.

12. Disconnect the air/fuel control unit by loosening the clamps. Move the unit with the lower section of the air cleaner up to the right side wheel housing. Place a cover over the wheel housing as protection.

13. Remove the air cleaner filter.

14. Remove the remaining nut and washer on the manifold. Lift the assembly forward and up. Remove the manifold gaskets. Disconnect the return oil pipe O-ring from the cylinder block.

15. Disconnect the turbocharger unit from the manifold.

To install:

16. Be sure to use a new gasket for the exhaust manifold and a new O-

ring to the return oil pipe. Keep everything clean during assembly and use extreme care in keeping dirt out of the various turbo inlet and outlet pipes and hoses.

17. Install the turbocharger on the exhaust manifold and tighten the bolts as follows:

a. Stage 1 – 0.7 ft. lbs. (3 Nm)
b. Stage 2 – 30 ft. lbs. (133 Nm)
c. Stage 3 – Tighten all bolts an additional 120 degrees ($^{3}/_{4}$ turn).

18. Install the exhaust manifold and turbocharger assembly on the engine. Connnect all oil pipes from and to the turbocharger using new O-rings.

19. Install the air/fuel control unit and air cleaner. Install the heat sheilds, spark plug wires, exhaust pipes, preheater assembly and expansion tank. Connect the negative battery cable.

20. Disconnect the wire at terminal 15 (brown) of the ignition coil. Use the ignition key to turn the engine over for about 30 seconds. This circulated the oil within the turbocharger, providing proper start-up lubrication.

21. Turn the ignition **OFF**, reconnect the coil wire, start the engine and allow it to idle for a few minutes prior to test driving.

Timing Belt Front Cover

REMOVAL & INSTALLATION

B230F Engine

1. Disconnect the negative battery cable. Loosen the fan shroud and remove the fan. Remove the shroud.
2. Loosen the alternator, power steering pump, if equipped, and air conditioning compressor, if equipped, and remove their drive belts.
3. Remove the water pump pulley.
4. Remove the 4 retaining bolts and lift off the timing belt cover.

To install:

5. Clean all gasket mating surfaces throughly. Install the timing belt cover using a new gasket. Tighten bolts to specification.
6. Install the water pump pulley, drive belts, air conditioning compressor, power steering pump and alternator.
7. Install the fan and shroud. Install the accessory drive belts. Connect the negative battery cable. Start the engine and check for leaks.

Timing Chain Front Cover

REMOVAL & INSTALLATION

B280F Engine

1. Disconnect the negative battery cable. Remove the air cleaner and valve covers.
2. Loosen the fan shroud and remove the fan. Remove the shroud.
3. Loosen the alternator, air pump, power steering pump, air conditioning compressor, if equipped, and remove their drive belts.
4. Block the flywheel from turning, remove the crankshaft pulley nut (36mm) and the pulley.

NOTE: Do not drop the pulley key into the crankcase.

5. Remove the power steering pump and place aside. Remove the pump bracket.
6. Remove the timing chain cover retaining bolts, 25–11mm hex bolts, tap and remove the cover.

To install:

7. Clean the gasket contact surfaces. Place the upper gasket on the cover and the lower gasket on the block. Install the cover and tighten to 7–11 ft. lbs. (10–15 Nm). Trim the gaskets flush with the valve cover.
8. Install a new crankshaft seal.
9. Block the flywheel, install the pulley, key and tighten the 36mm nut to 118–132 ft. lbs. (160–180 Nm).
10. Install the power steering pump, pump bracket, alternator, air pump, power steering pump and air conditioning compressor.
11. Install the fan and shroud. Install the accessory drive belts. Connect the negative battery cable. Start the engine and check for leaks.

Timing Chain and Sprockets

REMOVAL & INSTALLATION

B280F Engines

1. Remove the timing chain cover.
2. Remove the oil pump sprocket and drive chain.
3. Slacken the tension in both camshaft timing chains by rotating each tensioner lock ¼ turn counterclockwise and pushing the rubbing block piston.

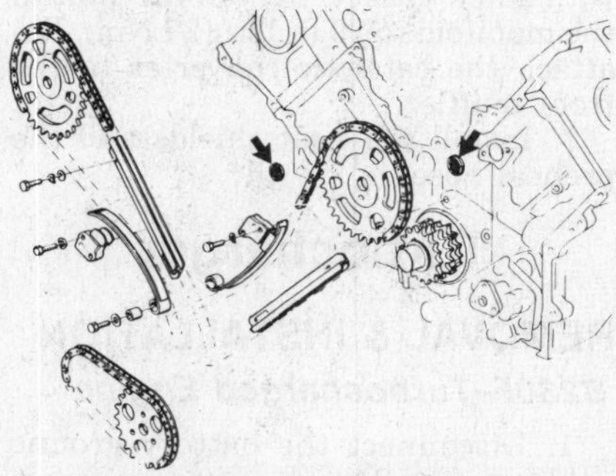

Timing chain tensioner and chain assembly – B280F engine

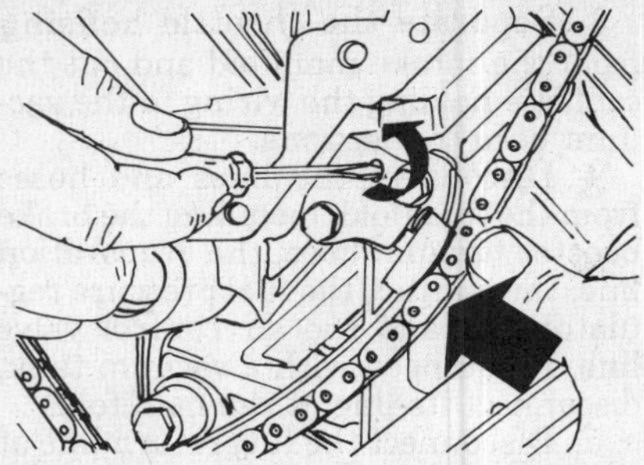

Relieving chain tension – B280F engine

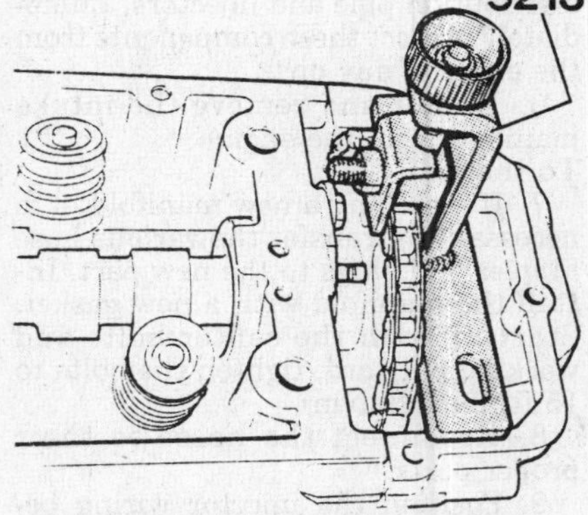

Timing chain gear holding tool 5213 – B280F engine

4. Remove both chain tensioners. Remove the 2 curved and the 2 straight chain damper/runners.
5. Remove the camshaft sprocket retaining bolt, 10mm Allen head, and the sprocket and chain assembly. Repeat for the other side.

To install:

6. Install the chain tensioners and tighten to 5 ft. lbs. (7 Nm). Install the curved chain damper/runners and tighten to 7–11 ft. lbs. (10–15 Nm). Install the straight chain damper/runners and torque to 5 ft. lbs. (7 Nm).
7. First install the left (driver) side camshaft sprocket and chain:

a. Rotate the crankshaft, using crankshaft nut, if necessary, until the crankshaft key is pointing directly to the left side camshaft and the left side camshaft key groove is pointing straight-up (12 o'clock).
b. Place the chain on the left side sprocket so the sprocket notchmark is centered precisely between the 2 white lines on the chain.
c. Position the chain on the crankshaft sprocket (inner), making sure the other white line on the chain aligns with the crankshaft sprocket notch.
d. While holding the left side chain and sprockets in this position, install the sprocket and chain on the left side camshaft, chain stretched on tension side, so the sprocket pin fits into the camshaft recess.
e. Tighten the sprocket center

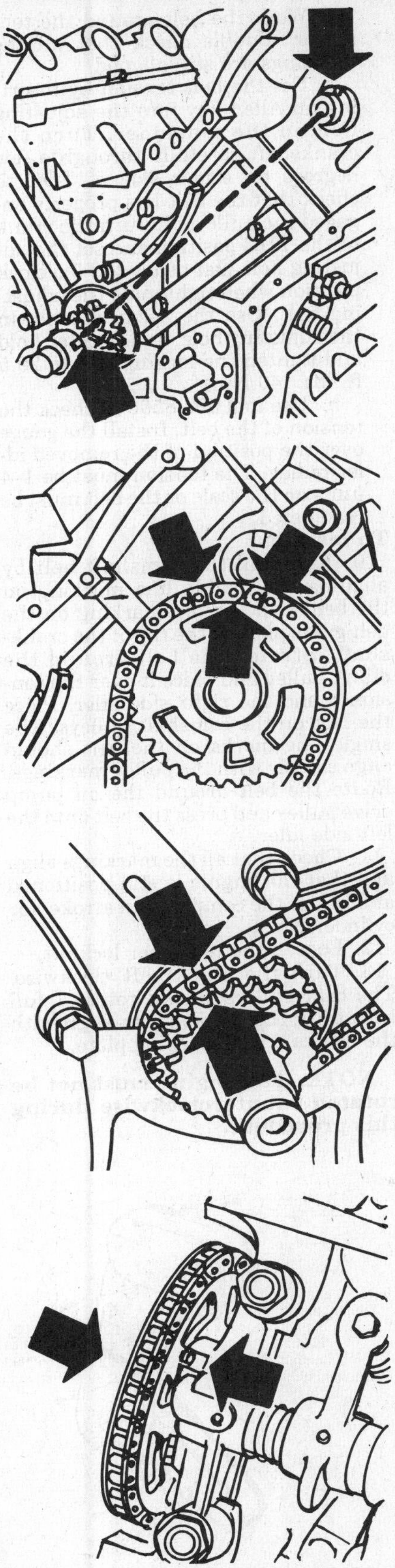

Left side camshaft timing chain installation sequence— B280F engine

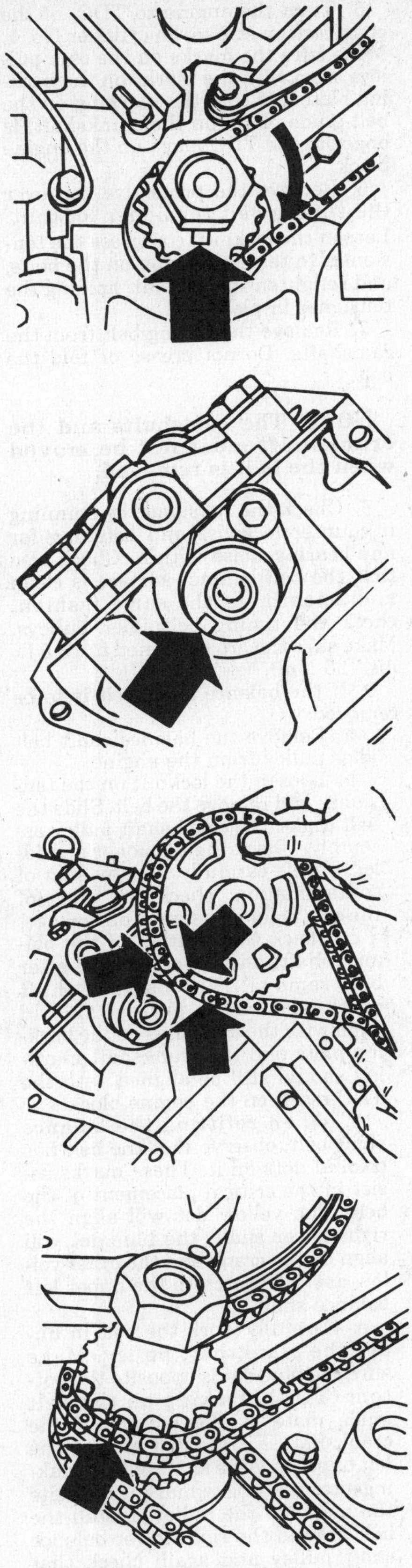

Right side camshaft timing chain installation sequence – B280F engine

bolt to 51–59 ft. lbs. (69–80 Nm); use a suitable tool to keep cam from turning.

8. To install the right side camshaft sprocket and chain:

a. Rotate the crankshaft clockwise until the crankshaft key points straight down (6 o'clock).

b. Align the camshaft key groove so it is pointing halfway between the 8 and 9 o'clock positions; at this position, the No. 6 cylinder rocker arms will rock.

c. Place the chain on the right side sprocket so the sprocket notchmark is centered precisely between the 2 white lines on the chain.

d. Then, position the chain on the middle crankshaft sprocket, making sure the other white line aligns with the crankshaft sprocket notch.

e. Install the sprocket and chain on the camshaft so the sprocket notch fits into the camshaft recess.

f. Tighten the sprocket nut to 51–59 ft. lbs. (69–80 Nm).

9. Rotate the chain tensioners ¼ turn clockwise each. The chains are tensioned by rotating the crankshaft 2 full turns clockwise. Recheck to make sure the alignment marks coincide.

10. Install the oil pump sprocket and chain.

11. Install the timing chain cover.

Timing Belt and Tensioner

REMOVAL & INSTALLATION

B230 Engine

1. Remove the timing belt cover.

2. To remove the tension from the belt, loosen the nut for the tensioner and press the idler roller back. The tension spring can be locked in this po-

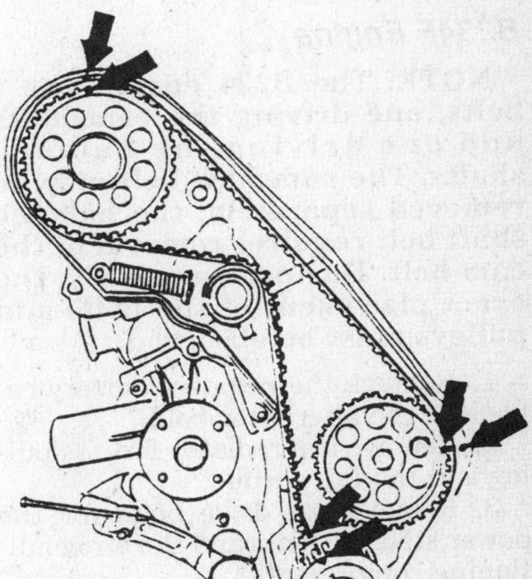

Timing belt alignment – B230F engine

sition by inserting the shank end of a 3mm drill through the pusher rod.

3. Remove the 6 retaining bolts and the crankshaft pulley.

4. Remove the belt, taking care not to bend it at any sharp angles. The belt should be replaced at 45,000 mile intervals, if it becomes oil soaked or frayed or if it is on a vehicle that has been sitting idle for any length of time.

To install:

5. If the crankshaft, idler shaft or camshaft were disturbed while the belt was out, align each shaft with is corresponding index mark to assure proper valve timing and ignition timing, as follows:

a. Rotate the crankshaft so the notch in the convex crankshaft gear belt guide aligns with the embossed mark on the front cover (12 o'clock position).

b. Rotate the idler shaft so the dot on the idler shaft drive sprocket aligns with the notch on the timing belt rear cover (4 o'clock position).

c. Rotate the camshaft so the notch in the camshaft sprocket inner belt guide aligns with the notch in the forward edge of the valve cover (12 o'clock position).

6. Install the timing belt (don't use any sharp tools) over the sprockets and then over the tensioner roller. New belts have yellow marks. The 2 lines on the drive belt should fit toward the crankshaft marks. The next mark should then fit toward the intermediate shaft marks, etc. Loosen the tensioner nut and let the spring tension automatically take up the slack. Tighten the tensioner nut to 37 ft. lbs. (51 Nm).

7. Rotate the crankshaft 1 full revolution clockwise and make sure the timing marks still align.

8. Install the drive belts, radiator fan and shroud. Connect the negative battery cable.

B234F Engine

NOTE: The B234 engine has 2 belts, one driving the camshafts and one driving the balance shafts. The camshaft belt may be removed separately; the balance shaft belt requires removal of the cam belt. During reassembly, the exact placement of the belts and pulleys must be observed.

1. Remove the negative battery cable and the alternator belt.

2. Remove the radiator fan, its pulley and the fan shroud.

3. Remove the drive belts for the power steering belts and the air conditioning compressor.

4. Beginning with the top cover, remove the retaining bolts and remove the timing belt covers.

5. Turn the engine to TDC, of the compression stroke, on cylinder No. 1. Make sure the marks on the cam pulleys align with the marks on the backing plate and that the marking on the belt guide plate (on the crankshaft) is opposite the TDC mark on the engine block.

6. Remove the protective cap over the timing belt tensioner locknut. Loosen the locknut, compress the tensioner, to release tension on the belts, and retighten the locknut, holding the tensioner in place.

7. Remove the timing belt from the camshafts. Do not crease or fold the belt.

NOTE: The camshafts and the crankshaft must not be moved when the belt is removed.

8. Check the tensioner by spinning it counterclockwise and listening for any bearing noise within. Check also that the belt contact surface is clean and smooth. In the same fashion, check the timing belt idler pulleys. Make sure the are tightened to 18.5 ft. lbs. (25 Nm).

9. If the balance shaft belt is to be removed:

a. Remove the balance shaft belt idler pulley from the engine.

b. Loosen the locknut on the tensioner and remove the belt. Slide the belt under the crankshaft pulley assembly. Check the tensioner and idler wheels carefully for any sign of contamination; check the ends of the shafts for any sign of oil leakage.

c. Check the position of the balance shafts and the crankshaft after belt removal. The balance shaft markings on the pulleys should align with the markings on the backing plate and the crankshaft marking should still be aligned with the TDC mark on the engine block.

d. When refitting the balance shaft belt, observe that the belt has colored dots on it. These marks assist in the critical placement of the belt. The yellow dot will align the right lower shaft, the blue dot will align on the crank and the other yellow dot will match to the upper left balance shaft.

e. Carefully work the belt in under the crankshaft pulley. Make sure the blue dot is opposite the bottom (TDC) marking on the belt guide plate at the bottom of the crankshaft. Fit the belt around the left upper balance shaft pulley, making sure the yellow mark is opposite the mark on the pulley. Install the belt around the right lower balance shaft pulley and again check that the mark on the belt aligns with the mark on the pulley.

f. Work the belt around the tensioner. Double check that all the markings are still aligned.

g. Set the belt tension by inserting an Allen key into the adjusting hole in the tensioner. Turn the crankshaft carefully through a few degrees on either side of TDC to check that the belt has properly engaged the pulleys. Return the crank to the TDC position and set the adjusting hole just below the 3 o'clock position when tightening the adjusting bolt. Use the Allen wrench, in the adjusting hole, as a counter hold and tighten the locking bolt to 29.5 ft. lbs. (40 Nm).

h. Use tool 998 8500 to check the tension of the belt. Install the gauge over the position of the removed idler pulley. The tension must be 1–4 units on the scale or the belt must be readjusted.

To install:

10. Reinstall the camshaft belt by aligning the double line marking on the belt with the top marking on the belt guide plate at the top of the crankshaft. Stretch the belt around the crank pulley and place it over the tensioner and the right side idler. Place the belt on the camshaft pulleys. The single line marks on the belt should align exactly with the pulley markings. Route the belt around the oil pump drive pulley and press the belt onto the left side idler.

11. Check that all the markings align and that the engine is still positioned at TDC, of the compression stroke, for cylinder No. 1.

12. Loosen the tensioner locknut.

13. Turn the crankshaft clockwise. The cam pulleys should rotate 1 full turn until the marks again align with the marks on the backing plate.

NOTE: The engine must not be rotated counterclockwise during this procedure.

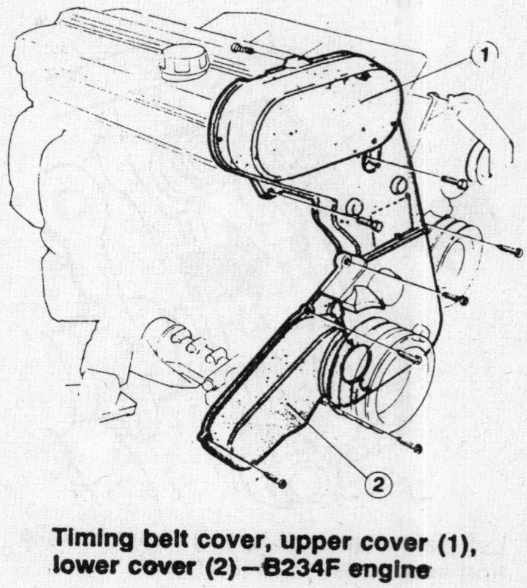

Timing belt cover, upper cover (1), lower cover (2)—B234F engine

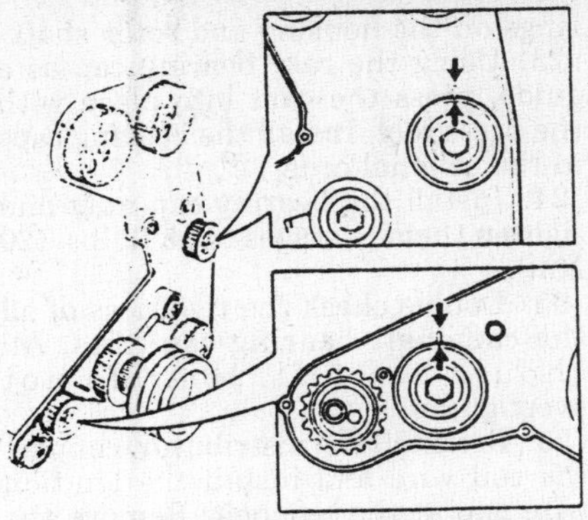

Balance shaft alignment—B234F engine

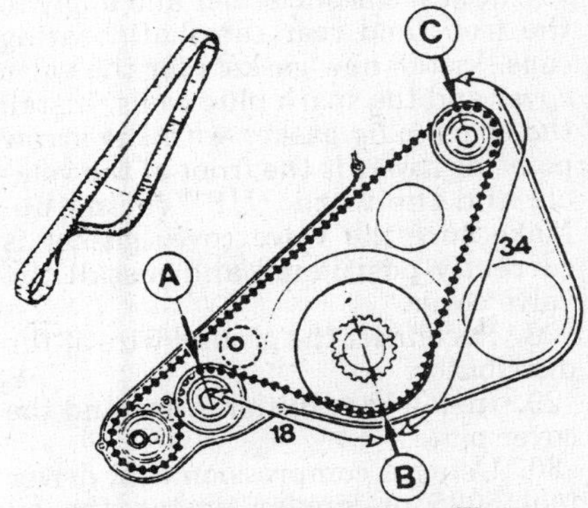

Balance shaft belt markings. There should be 18 teeth between A and B, 34 teeth between B and C—B234F engine

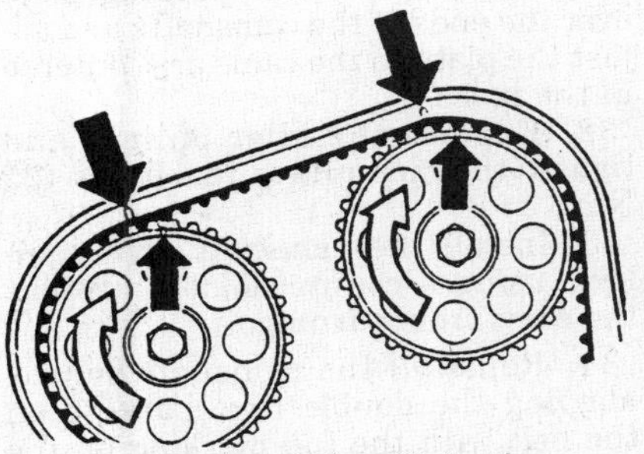

Rotate the engine 1½ teeth—B234F engine

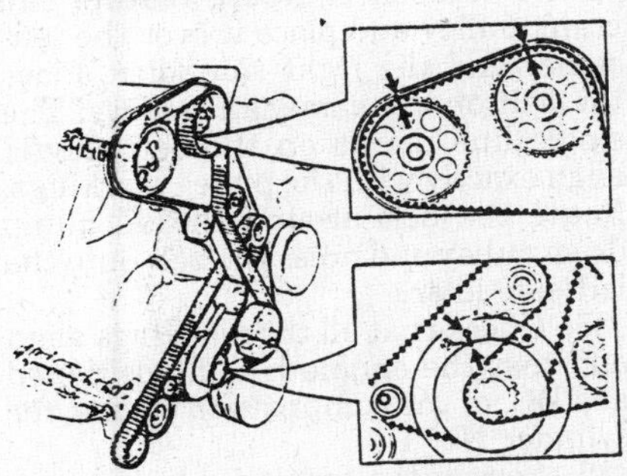

Timing mark alignment—B234F engine

14. Smoothly rotate the crankshaft further clockwise until the cam pulley markings are 1½ teeth beyond the marks on the backing plate. Tighten the tensioner locknut.
15. Check the tension on the balance

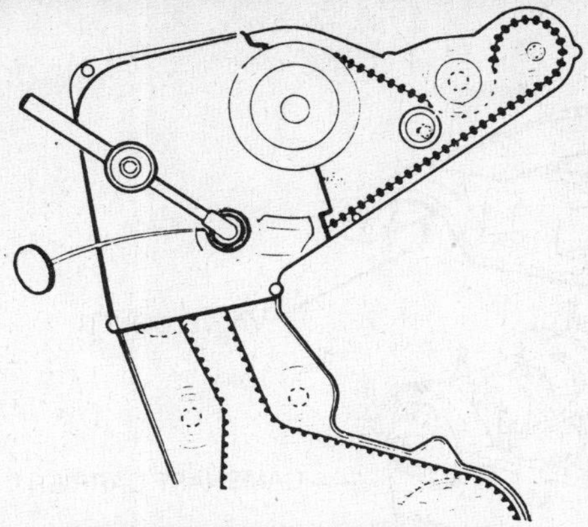

Timing belt tensioner adjustment—B234F engine

shaft belt; it should now be 3.8 units. If the tension is too low, adjust the tensioner clockwise. If the tension is too high, repeat Step 8g.
16. Check the belt guide for the balance shaft belt and make sure it is properly seated. Install the center timing belt cover, the one that covers the tensioner, the fan shroud, fan pulley and fan. Install all the drive belts and connect the battery cable.
17. Double check all installation items, paying particular attention to loose hoses or hanging wires, untightened nuts, poor routing of hoses and wires (too tight or rubbing) and tools left in the engine area.
18. Start the engine and allow it to run until the thermostat opens.

CAUTION

The upper and lower timing belt covers are still removed. The belt and pulleys are exposed and moving at high speed.

19. Shut the motor off and bring the motor to TDC, of the compression stroke, on cylinder No. 1.
20. Check the tension of the camshaft belt. Position the gauge between the right (exhaust) cam pulley and the idler. Belt tension must be 5.5 ± 0.2 units. If the belt needs adjustment, remove the rubber cap over the tensioner locknut, cap is located on the timing belt cover, and loosen the locknut.
21. Insert a suitable tool between the tensioner wheel and the spring carrier pin to hold the tensioner. If the belt needs to be tightened, move the roller to adjust the tension to 6.0 units. If the belt is too tight, adjust to obtain a reading of 5.0 units on the gauge. Tighten the tensioner locknut.
22. Rotate the crankshaft so the cam pulleys move through 1 full revolution and recheck the tension on the camshaft belt. It should now be 5.5 ± 0.2 units. Install the plastic plug over the tensioner bolt.
23. Final check the tension on the balance shaft belt by fitting the gauge and turning the tensioner clockwise. Only small movements are needed. After any needed readjustments, rotate the crankshaft clockwise through 1 full revolution and recheck the balance shaft belt. The tension should now be on the final specification of 4.9 ± 0.2 units.
24. Install the idler pulley for the balance shaft belt. Reinstall the upper and lower timing belt covers.
25. Start the engine and final check performance.

Camshaft

REMOVAL & INSTALLATION

B230F Engine

1. Remove the timing belt cover.
2. Remove the valve cover.
3. Remove the camshaft center bearing cap. Install camshaft press tool 5021 over the center bearing journal to hold the camshaft in place while removing the other bearing caps.
4. Remove the 4 remaining bearing caps.
5. Remove the seal from the forward edge of the camshaft.
6. Release camshaft press tool and lift out the camshaft.

To install:

7. Install the camshaft after lubricating with oil. Install the camshaft seal.
8. Install the camshaft press tool and install the camshaft bearing caps. Tighten bolts to 14 ft. lbs. (20 Nm).
9. Install the valve cover and timing belt cover.

B234F Engine

NOTE: The use of the correct special tools or their equivalent is required for this procedure.

1. Disconnect the negative battery cable.
2. Remove the alternator drive belt, the radiator fan and its pulley.
3. Remove the upper and lower timing belt covers.
4. Align the camshaft and crankshaft marks. Turn the engine to TDC, of the compression stroke, on cylinder No. 1 and make sure the pulley marks and the crank marks align with their matching marks on either the backing plate (cam pulleys) or the belt guide plate (crankshaft).
5. Remove the protective cap over the timing belt tensioner locknut. Loosen the locknut, compress the tensioner, to release tension on the belts, and retighten the locknut, holding the tensioner in place.
6. Remove the timing belt from the camshafts; do not crease or fold the belt.

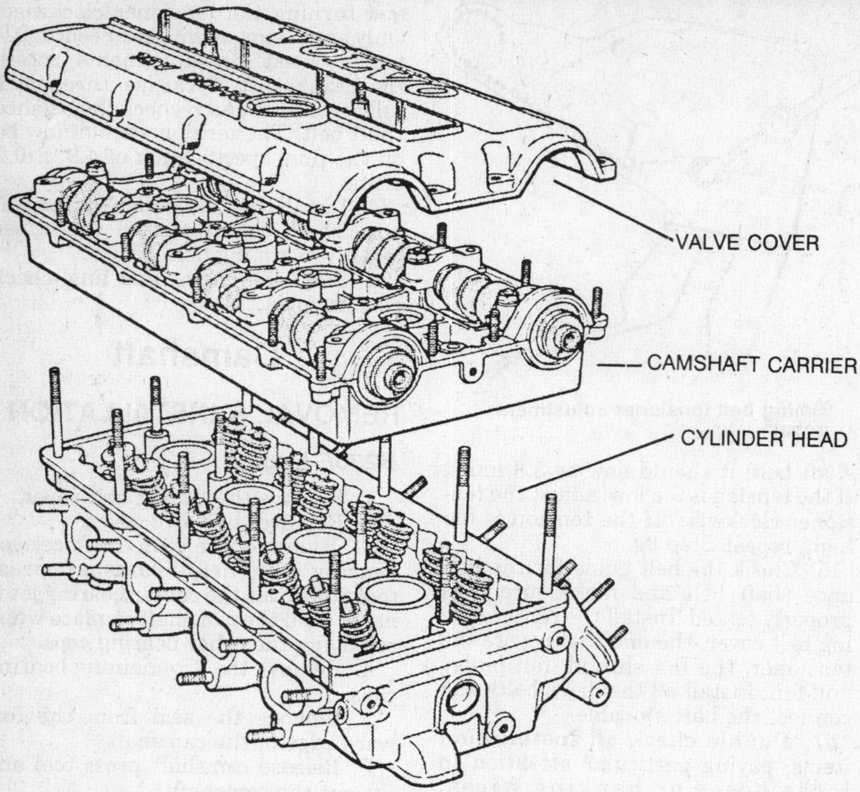

Camshaft carrier assembly – B234F engine

NOTE: The camshafts and the crankshaft must not be moved when the belt is removed.

7. Remove the timing belt idler pulleys.
8. Remove the camshaft drive pulleys. Use a counterhold wrench to prevent the cam from turning.
9. Remove the plate or panel behind the pulleys. Remove the cover plate for the ignition wires. Label and disconnect the ignition wiring from the spark plugs and the distributor cap; remove the coil wire from the distributor cap.
10. Remove the valve cover and gasket. Clean the surfaces of any gasket remains.
11. Remove the distributor housing from the camshaft carrier. Remove the ignition wire clip next to the left bolt.
12. Plug the spark plug holes with crumpled paper. Remove the center bearing cap for each camshaft. Mark the cam bearing caps for proper reinstallation.
13. Install a camshaft press tool 5021 or similar, on the exhaust side cam in place of the removed bearing cap. When it is securely in place, remove the remaining bearing caps and nuts. Remove the tool and remove the exhaust camshaft.
14. Remove the intake camshaft in identical fashion.

NOTE: Label or identify each cam and its bearing caps. All removed components should be kept in order.

15. Using a magnet or a small suction cup, remove the tappets. Store them upside down, to prevent oil drainage, and keep them in order; they are not interchangeable.

To install:

16. Clean and inspect the camshaft carrier and tappet bores for any sign of wear or scoring.
17. Oil all matching surfaces on the cam carrier, bearing caps and tappets.
18. Insert the tappets; they must be inserted in their original order and place.
19. Install the exhaust side camshaft by placing it in the carrier with the pulley guide pin facing up. Using the rear bearing cap as a guide, press the cam into place with the press tool. Install the bearing caps in the original order.
20. Install the bearing cap nuts and tighten them in stages to 15 ft. lbs. (20 Nm). Remove the press tool and install the center bearing cap; tighten it in stages to 15 ft. lbs. (20 Nm).
21. Install the intake camshaft in the carrier with the pulley guide pin facing upwards.
22. Turn the distributor shaft to align the driver with the markings on the distributor housing. Install new O-rings on the housing and rotor shaft.
23. Using the rear bearing cap as a guide, press the cam into place with the press tool. Install the bearing caps in the original order.
24. Install the bearing cap nuts and tighten them in stages to 15 ft. lbs. (20 Nm).
25. Double check the tightness of all the camshaft bearing cap nuts. All should be 15 ft. lbs.; do not overtighten.
26. Reinstall the distributor, connect the coil wire and install the ignition wire clip at the left bolt. Remove the paper plugs from the spark plug holes.
27. Use a silicone sealer and apply to the front and rear camshaft bearing caps. Install new gaskets for the valve cover and the spark plug wells. Install the spark plug gasket with the arrow pointing towards the front of the vehicle and the word "UP" facing up. Make sure the valve cover gasket is correctly positioned and install the valve cover.
28. Reconnect the ground wire at the distributor.
29. Install the ignition wires and the cover plate.
30. Using a compression seal driver tool 5025 or similar, install the oil seals for the front of each camshaft. Camshafts must not be allowed to turn during this operation.
31. Install the upper backing plate over the ends of the camshafts and adjust the plate so the cams are centered in the holes.
32. Replace the idler pulleys and tighten their mounts to 18.5 ft. lbs. (25 Nm).
33. Install the camshaft drive pulleys, using a counterhold to prevent the cams from turning.
34. Reinstall the camshaft belt by aligning the double line marking on the belt with the top marking on the belt guide plate at the top of the crankshaft. Stretch the belt around the crank pulley and place it over the tensioner and the right side idler. Place the belt on the camshaft pulleys. The single line marks on the belt should align exactly with the pulley markings. Route the belt around the oil pump drive pulley and press the belt onto the left side idler.
35. Check that all the markings align and that the engine is still positioned at TDC, of the compression stroke, for cylinder No. 1.
36. Loosen the tensioner locknut.
37. Turn the crankshaft clockwise. The cam pulleys should rotate 1 full turn until the marks again align with the marks on the backing plate.

NOTE: The engine must not be rotated counterclockwise during this procedure.

38. Smoothly rotate the crankshaft further clockwise until the cam pulley markings are 1½ teeth beyond the marks on the backing plate. Tighten the tensioner locknut.
39. Reinstall the fan pulley and fan. Install all the drive belts and connect the battery cable.
40. Double check all installation items, paying particular attention to loose hoses or hanging wires, untightened nuts, poor routing of hoses and wires (too tight or rubbing) and tools left in the engine area.
41. Start the engine and allow it to run until the thermostat opens.

CAUTION

The upper and lower timing belt covers are still removed. The belt and pulleys are exposed and moving at high speed.

NOTE: This engine may be somewhat noisy when started. The noise will subside as oil reaches the tappets. Do not exceed 2500 rpm while the tappets are noisy.

42. Shut the motor off and bring the motor to TDC, of the compression stroke, on cylinder No. 1.
43. Check the tension of the camshaft belt. Position the gauge between the right (exhaust) cam pulley and the idler. Belt tension must be 5.5 ± 0.2 units. If the belt needs adjustment, remove the rubber cap over the tensioner locknut and loosen the locknut.
44. Insert a suitable tool between the tensioner wheel and the spring carrier pin to hold the tensioner. If the belt needs to be tightened, move the roller to adjust the tension to 6.0 units. If the belt is too tight, adjust to obtain a reading of 5.0 units on the gauge. Tighten the tensioner locknut and remove the suitable tool.
45. Rotate the crankshaft so the cam pulleys move through 1 full revolution and recheck the tension on the camshaft belt. It should now be 5.5 ± 0.2 units. Install the plastic plug over the tensioner bolt.
46. Reinstall the remaining belt covers. Start the engine and final check performance.

B280F Engine

1. Remove the cylinder head.
2. Remove the camshaft rear cover plate.
3. Remove the camshaft retaining fork at the front of the cylinder head.
4. Pull the camshaft out the rear of the head.

To install:

5. Oil the camshaft and followers and install. Tighten the camshaft retaining bolt to 7–11 ft. lbs.
6. Install the camshaft retaining fork, install the rear cover plate and install the cylinder head.

Balance Shafts

REMOVAL & INSTALLATION

B234 Engine

NOTE: The use of the correct special tools or their equivalent is required for this procedure.

LEFT SHAFT AND HOUSING

1. Remove the timing and balance shaft belts.
2. Use a counterhold tool 5362 and remove the left side balance shaft pulley.
3. Remove the air mass meter and inlet hose.
4. Unfasten the bracket under the intake manifold and remove the bracket holding the alternator and power steering pump. These may be swung aside and tied with wire to the left shock tower.
5. Remove the bolts securing the balance shaft housing to the block. Using an extractor tool 5376 or similar, carefully separate the housing from the block. The housing must be removed evenly from both its front and rear mounts.

To install:

6. Clean the joint faces on the cylinder block. Place new O-rings in the grooves around the oil passages on the housing. The rings can be held in place with a light coating of grease.
7. Install the balance shaft housing. Make absolutely sure the housing is evenly mounted on the front and rear mountings. Tighten the bolts alternately in a diagonal pattern. Tighten each bolt ½ turn at a time; tighten them to 15 ft. lbs. (20 Nm). When all the bolts are at 15 ft. lbs. (20 Nm), loosen them individually and tighten each one to 7.5 ft. lbs. (10 Nm) plug 90 degrees of rotation.

NOTE: Make certain the shaft does not seize within the housing during installation.

8. If the halves of the housing were split apart during the repair, tighten the joint bolts to 6 ft. lbs. (8 Nm).
9. Install the drive pulley. Use a counterholding tool. Note that the pulley has a slot which will align with the guide on the shaft. The shallow side of the pulley faces inward, toward the engine. Tighten the center bolt for the pulley to 37 ft. lbs. (50 Nm).
10. Install the bracket for the alternator and power steering pump. Double check their connections and hoses. Attach the support under the intake manifold and don't forget the wire clamp on the bottom bolt.
11. Install the air mass meter and its intake hose.
12. Install the balance shaft belt and camshaft belt.

RIGHT SHAFT AND HOUSING

1. Remove the timing and balance shaft belts.
2. Use a counterhold tool 5362 and remove the left side balance shaft pulley.
3. Remove the balance shaft belt tensioner and remove the bolt running through the backing plate to the balance shaft housing.
4. Remove the air mass meter and its air inlet hose.
5. Remove the air preheat hose from the bottom heat shield at the exhaust manifold. Remove the nuts holding the right engine mount to the crossmember.
6. Connect a hoist or engine lift apparatus to the top of the engine. Lift the engine at the right side, being careful to maintain clearance between the brake master cylinder and the intake manifold.
7. Remove the complete motor mount from the block, including the pad and lower mounting plate.
8. Remove the bolts securing the balance shaft housing to the block. Using a extractor tool 5376 or similar, carefully separate the housing from the block. The housing must be removed evenly from both its front and rear mounts.

To install:

9. Clean the joint faces on the cylinder block. Place new O-rings in the grooves around the oil passages on the housing. The rings can be held in place with a light coating of grease.
10. Install the balance shaft housing. Make absolutely sure the housing is evenly mounted on the front and rear mountings. Tighten the bolts alternately in a diagonal pattern. Tighten each bolt ½ turn at a time; tighten them to 15 ft. lbs. (20 Nm). When all the bolts are at 15 ft. lbs. (20 Nm), loosen them individually and tighten each one to 7.5 ft. lbs. (10 Nm) plus 90 degrees of rotation.

NOTE: Make certain the shaft does not seize within the housing during installation.

11. If the halves of the housing were split apart during the repair, tighten the joint bolts to 6 ft. lbs. (8 Nm).
12. Install the drive pulley. Use a counterholding tool. Note that the pulley has a slot which will align with the guide on the shaft. The shallow side of the pulley faces inward, toward the engine. Tighten the center bolt for the pulley to 37 ft. lbs. (50 Nm).
13. Install the engine mount onto the block.

14. Using the studs on the crossmember as a guide, lower the engine into place on the front crossmember. When the engine is correctly seated, the lifting apparatus may be removed.
15. Reinstall the air mass meter and its air intake hose.
16. Reinstall the motor mount bolts and the air preheat tube at the lower part of the exhaust manifold.
17. Install the bolt through the backing plate and into the balance shaft housing. Reinstall the belt tensioner, tightening the bolt so the pulley is movable when the belt is in position.
18. Reinstall the balance shaft and camshaft belts.

Piston and Connecting Rod

POSITIONING

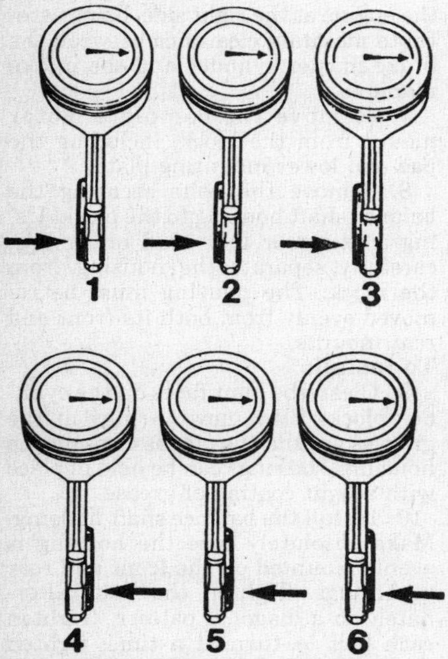

Piston and connecting rod position – B280F engine

ENGINE LUBRICATION

Oil Pan

REMOVAL & INSTALLATION

B230F Engine

1. Disconnect the negative battery cable. Raise and support the vehicle safely.
2. Drain the engine oil.

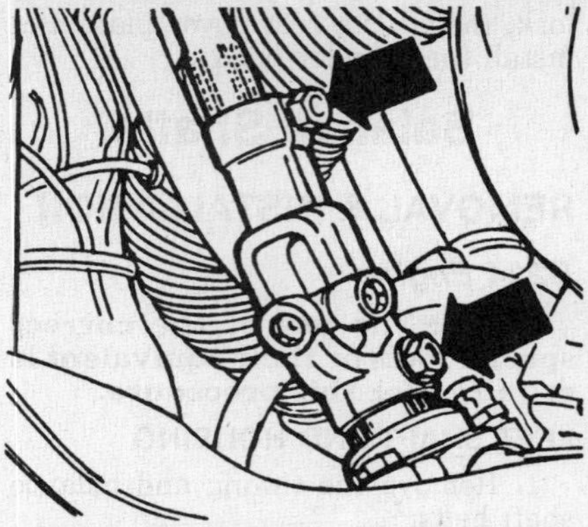
Steering yoke removal; arrows indicate the retaining nuts

3. Remove the splash guard.
4. Remove the engine mount retaining nuts.
5. Remove the lower bolt and loosen the top bolt on the steering column yoke.
6. Slide the yoke assembly up on the steering shaft.
7. Raise and safely support the front of the engine.
8. Remove the retaining bolts for the front axle crossmember.
9. Remove the crossmember.
10. Remove the left engine mount.
11. Remove the pan support bracket.
12. Remove the pan bolts and remove the pan.

To install:

13. Clean the gasket mating surfaces throughly. Install the oil pan and using new gaskets, tighten the bolts to 8 ft. lbs. (11 Nm).
14. Lower the engine and install all engine mounts. Install the front crossmember and install the bolts.
15. Install the yoke assembly on the steering shaft and tighten the bolts to 18 ft. lbs. (24 Nm).
16. Install the splash guard, lower the vehicle and connect the negative battery cable. Fill the engine with oil.
17. Start the engine and allow it to reach operating temperature. Check for leaks.

B234F Engine

1. Raise and safely support the vehicle. Disconnect the negative battery cable and remove the engine oil dipstick.
2. Remove the air mass meter and air inlet hose. Loosen the fan shroud.
3. Remove the bolts at both ends of the crossmember.
4. Fit a chain hoist or lifting apparatus to the top of the engine and relieve the weight of the engine by lifting the at the front.
5. At the right motor mount, unbolt the bottom mounting plate from the crossmember. At the left motor mount, unbolt the upper mounting plate from the cylinder block.
6. Drain the engine oil and replace the drain bolt when the pan is empty. Use a new washer and tighten the bolt to 44 ft. lbs. (60 Nm).
7. Remove the splash guard from under the engine, the bottom nut for the left motor mount and the wiring harness bracket from the transmission cover.
8. At the steering shaft, remove the lower clamping bolt and loosen the upper bolt. Matchmark the position of the splined joint and slide the fitting up the steering shaft.
9. Remove the rubber bump-stop on the front crossmember and remove the reinforcing bracket between the engine and transmission.
10. Disassemble the bolted joint at the front of the catalytic converter.
11. Carefully elevate the engine with the hoist. Make very certain that no hoses or wires are strained and that clearance is maintained at the firewall. Raise the motor only enough to perform the next Steps of the procedure.
12. Remove the left motor mount.
13. Unbolt and remove the oil pan. It will need to be lifted and turned during removal.

To install:

14. Clean the gasket surfaces and install the new gasket so the small tab on the gasket is on the same side as the starter. Lift the pan into place, install the retaining bolts and tighten them to 8 ft. lbs. (11 Nm).
15. Install the reinforcing bracket between the engine and transmission. Attach it first to the transmission and then to the engine block. Tighten the bracket in stages so all the bolts pull up evenly.
16. Install the bump-stop on the front crossmember. Lift the crossmember into position against the side rails, install the bolts and tighten only a few turns to hold it in place.
17. When all the bolts are installed, tighten the crossmember bolts to 70 ft. lbs. (95 Nm). Install the left motor mount and secure the plate to the cylinder block. Don't forget to attach the cable clip on the upper bolt.
18. Paying close attention to the placement of the motor mounts, lower the engine into position. When the engine is correctly seated, the lifting equipment may be removed from the vehicle.
19. At the right motor mount, tighten the plate onto the crossmember. Check the connection of the air preheat tube at the exhaust manifold.
20. Tighten the fan shroud. Adjust the position of the bottom bracket as needed.
21. Reconnect the wiring harness

bracket at the transmission, the bolted joint at the front of the catalytic converter and install the splashguard under the engine.

22. Tighten the left motor mount.
23. Observing the markings made earlier, reassemble the steering shafts. Insert and tighten the bottom bolt to 15 ft. lbs. (20 Nm). Tighen the upper bolt the same. Don't forget to install the small spring clips on the bolts.
24. Install the air mass meter and its hoses and connectors.
25. Fill the engine with the correct amount of oil and reinstall the dipstick.
26. Lower the vehicle, reconnect the battery cable and start the engine. Check for leaks.

B280F Engine

1. Disconnect the negative battery cable. Raise and support the vehicle safely. Remove the splash guard.
2. Drain the crankcase.
3. Remove the oil pan retaining bolts. Swivel the pan past the stabilizer bar and remove.

To install:

4. Clean the gasket mating surfaces throughly. Install the oil pan, using a new gasket, and tighten the bolts to 6–8 ft. lbs.
5. Install the splash guard, lower the vehicle and fill the crankcase with oil. Connect the negative battery cable. Start the engine and allow it to reach operating temperature. Check for leaks.

Oil Pump

REMOVAL & INSTALLATION

B230F Engine

1. Remove the oil pan.
2. Remove the 2 oil pump retaining bolts. Remove the oil pump and pull the delivery tube from the block.
3. When installing, use new sealing rings at either end of the delivery tube.4. Install the pump with the delivery tube attached. Align the pipe to the block so that the seal does not become damaged. Tighten the two oil pump retaining bolts.
5. Attach the clamp for the oil trap drain hose to the oil pump bolts. Make sure the hose is securely clamped behind the oil pump shoulder. Do not shorten the hose.

B234F Engine

1. Remove the timing belt.
2. Using a counterholding tool 5039 or similar, remove the oil pump drive pulley.
3. Thoroughly clean the area around the oil pump. Place sheets of newspaper or a container on the splash guard to contain any spillage and remove the oil pump mounting bolts. Remove the pump from the engine.
4. Remove the seal from the groove in the block. Clean the area with solvent, making certain there are no particles of dirt trapped in the pump area.

To install:

5. Install the new seal in the groove and install the new oil pump. Lubricate the pump with clean engine oil before installation. Tighten the mounting bolts to 7.5 ft. lbs. (10 Nm).
6. Using the counterhold, install the drive pulley and tighten the center bolt to 15 ft. lbs. (20 Nm) plug 60 degrees of rotation.
7. Clean the area of any oil spillage; remove the paper or container from the splash guard.
8. Install the timing belt.

B280F Engine

The oil pump body is cast integrally with the cylinder block. It is chain driven by a separate sprocket on the crankshaft and is located behind the timing chain cover. The pick-up screen and tube are serviced by removing the oil pan. To check the pump gears or remove the oil pump cover:

1. Disconnect the negative battery cable. Remove the air cleaner and valve covers.
2. Loosen the fan shroud and remove the fan. Remove the shroud.
3. Loosen the alternator, air pump, power steering pump, air conditioning compressor, if equipped, and remove their drive belts.
4. Block the flywheel from turning and remove the 36mm bolt and the crankshaft pulley.

NOTE: Do not drop key into crankcase.

5. Remove the timing gear cover (25 bolts).
6. Remove the oil pump drive sprocket and chain.
7. Remove the oil pump cover and gears.

To install:

8. Prime the pump, remove all air by filling it with clean engine oil and operating the pump by hands, before installation. Install the oil pump gears and cover. Install the oil pump drive sprocket and chain.
9. Install the timing gear cover, crankshaft pulley, alternator, air pump, power steering pump, air conditioning compressor and all accessory drive belts.
10. Remove the flywheel block and install the valve covers. Connect the negative battery cable.

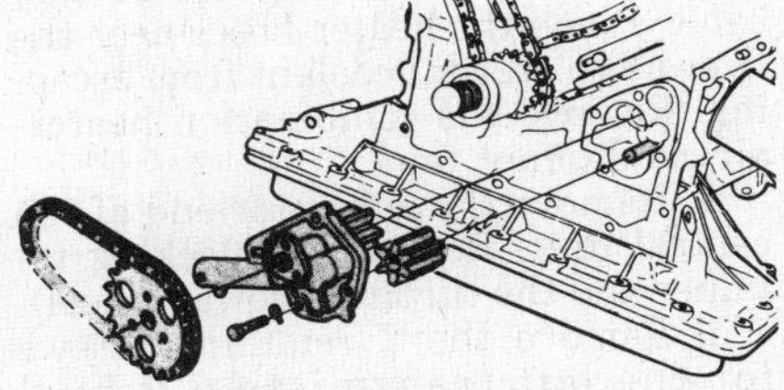

Oil pump installation – B280F engine

Rear Main Bearing Oil Seal

REMOVAL & INSTALLATION

B230F and B234F Engines

1. Disconnect the negative battery terminal.
2. Remove the transmission.
3. Remove the clutch and pressure plate, if equipped.
4. Remove the pilot bearing snapring and remove the bearing.
5. Remove the flywheel or driveplate which ever is applicable.

NOTE: Be careful not to press in the activator pins for the timing device.

6. Remove the rear oil pan brace.
7. Remove the 2 center bolts from the pan that bolt into the seal housing.
8. Loosen 2 bolts on either side of the 2 in the seal housing.
9. Remove the 6 seal housing bolts and remove the seal housing.

NOTE: Be careful not to damage the oil pan gasket when removing the seal housing.

10. Remove the seal using special tool 2817 or a suitable replacement.

To install:

11. Use a new gasket on the seal housing and coat the seal with oil prior to installation. Install the seal.
12. Install the seal housing and tighten the bolts to specification.
13. Install the rear oil pan brace and flywheel. Torque the flywheel to 47–54 ft. lbs. (64–73 Nm). When installing

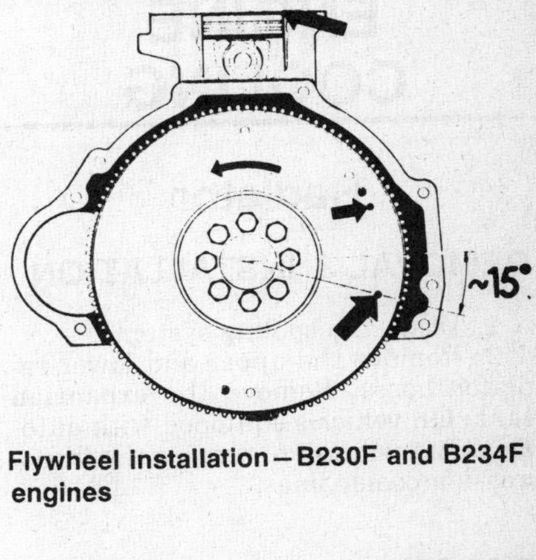

Flywheel installation – B230F and B234F engines

the flywheel turn the crankshaft to bring the No. 1 piston to TDC. The lower flywheel pin should be installed approximately 15 degrees from the horizontal and opposite the starter.

14. Install the pilot bearing. Install the clutch assembly and transmission, as required.

15. Connect the negative battery cable. Fill the transmission with fluid. Start the engine and allow it to reach operating temperature. Check for leaks.

B280F Engine

1. Disconnect the negative battery terminal.
2. Remove the transmission.
3. Remove the clutch and pressure plate, if equipped.
4. Remove the flywheel or driveplate, on automatic transmissions.

NOTE: On automatic transmissions remove the crankshaft spacer.

5. Remove the 2 rear pan bolts.
6. Remove the bolts in the seal housing and then the housing.

NOTE: Gently remove the housing so as not to damage the oil pan gasket.

7. Using tool 5107, remove the old seal.

To install:

8. Using the seal tool, install the new seal. Install the seal housing and tighten the seal housing bolts to 7–11 ft. lbs. (10–15 Nm).
9. Install the rear oil pan bolts. Install the flywheel and clutch assembly, as required. Tighten the flywheel bolts to 33–37 ft. lbs. (45–50 Nm).
10. Install the transmission and connect the negative battery cable. Fill the transmission with oil. Start the engine and allow it to reach operating temperature. Check for leaks.

ENGINE COOLING

Radiator

REMOVAL & INSTALLATION

1. Drain the cooling system.
2. Remove the upper and lower radiator hoses. Remove the expansion tank. On vehicles equipped with automatic transmission, remove the transmission cooler lines.
3. Disconnect the electrical connectors for the coolant sensor and the electric cooling fan.
4. Remove the retaining bolts for the radiator shroud and remove the radiator.

To install:

5. Install the radiator and tighten the bolts for the radiator shroud. Ensure that the radiator is positioned for maximum airflow. Connect the electrical connectors.
6. Install the radiator hoses, expansion tank and transmission cooler lines.
7. Fill the radiator with coolant and check the transmission fluid. Start the engine and allow it to reach operating temperature. Check the coolant level and transmission fluid.

Heater Core

REMOVAL & INSTALLATION

Without Air Conditioning

1. Disconnect the negative battery cable. Pinch the heater hoses near the firewall to prevent coolant from escaping. Remove the heater unit.
2. Place the unit on its side with the control valve facing upward. Remove the spring clips and separate the housing halves.
3. Disconnect the capillary tube from the heater core and then lift out the core.

To install:

4. Install the heater core. Take care to transfer the foam plastic packing to the new heater core and to install the fragile capillary tube carefully on the core.
5. Join the housing halves and install the spring clips. Install the heater unit.
6. Connect the negative battery cable. Start the engine and check the coolant level after it has reached operating temperature.

With Air Conditioning

240

CAUTION

Do not disconnect the refrigerant lines from the air conditioning system. These lines carry a dangerous refrigerant, the gas R-12.

1. Disconnect the negative battery cable. Pinch the heater hoses near the firewall to prevent coolant from escaping. Remove the combination heater-air conditioner unit.
2. Remove the left outer end of the central unit. Remove the locking retainer and the turbine (blower wheel).
3. Remove the 2 retaining screws for the left transmission tunnel bracket.
4. Remove the lockring for the left intake shutter shaft.
5. Remove the 3 retaining screws and lift off the inner end.
6. Remove the 3 retaining screws for the fan motor retainer.
7. Disconnect the heater hoses at the heater core.
8. Remove the clamps which retain the central unit halves together, lift off the left half and remove the heater core.

To install:

9. Install the heater core. Take care to transfer the foam plastic packing to the new heater core. Join the heater unit halves and install the spring clips.
10. Install the fan motor, inner end and intake shutter shaft. Install the outer end of the central unit, locking retainer and turbine.
11. Install the combination heater-air conditioner unit. Connect the heater hoses at the heater core.
12. Connect the negative battery cable. Start the engine and allow it to reach operating temperature. Check the coolant level.

EXCEPT 240

1. Disconnect the negative battery cable.
2. Pinch the hoses to the heater core near the firewall in the engine compartment. Use locking pliers. Make sure the hoses are pinched sufficiently so the hose is completely blocked off. Remove the hose clamps on the engine compartment side of the hoses, close to the firewall.
3. Press down the clip under the ashtray and pull the tray out. Remove the cigarette lighter and the storage compartment.
4. Remove the engine console around the shift lever and parking brake. Unplug the connector.
5. Remove the panel under the driver's side dashboard and remove the air duct to the steering column outlet.
6. Pull down the driver's side floor mat and remove the front and rear edge side panel screws. Remove the panels.
7. On the passenger's side, remove the 3 clips that fasten the panel under the glove compartment and remove the panel. Remove the glove compartment and its lighting.
8. Pull down the floor mat on the right side and remove the front and rear edge side panel screws.
9. Remove the radio compartment by pressing forward on the inner wall and removing the screw.
10. Remove the screws inside the center console and remove the side panel screws and the panels.
11. Remove the panel around the heater control. Remove the radio compartment console and remove the con-

trol panel. Free the central electrical unit and remove the mounting.

12. Remove the center panel vent and the screw holding the distribution unit. Mark all air ducts to the panel vents and to the distribution unit with tape for later installation and remove the ducts.

13. Remove the vacuum hoses from the vacuum motors.

14. Remove the distribution unit. Remove the heater core retaining clips and remove the heater core.

To install:

15. Install the heater core and fasten using the retaining clips. Install the distribution unit and connect the vacuum hoses to the motors. On climate unit-equipped vehicles, connect the red hose to the upper shutter for the panel vents and the light brown hose to the lower shutter. Connect the yellow and blue hoses to the floor/defrost shutter, the yellow to the lower one. On automatic climate control-equipped vehicles, connect the red hose to the upper shutter for the panel vent and the blue hose to the defrost vent. Connect the light brown hose to the lower shutter for the panel unit.

16. Install all ducts in their marked positions. Install the center panel vent, heater control panel and radio compartment console.

17. Install the center console and replace the floor mat. Install the glove compartment and clip the under dash panel into place. Install the front and rear edge side panels. Install the engine console around the shift lever and parking brake.

18. Install the ashtray. From the engine side connect the heater hoses to the heater core. Connect the negative battery cable.

19. Start the engine, allow it to reach operating temperature, check the coolant level and check for leaks.

Water Pump

REMOVAL & INSTALLATION

B230F and B234F Engines

1. Disconnect the negative battery cable. Remove the overflow tank cap. Drain the cooling system.
2. Remove the fan and fan shroud.
3. Remove the drive belts and the water pump pulley.
4. If necessary, remove the timing belt cover.
5. Remove the lower radiator hose.
6. Remove the retaining bolt for the coolant pipe, beneath exhaust manifold, and pull the pipe rearward.
7. Remove the 6 retaining bolts and lift off the water pump.

To install:

8. Clean the gasket contact surfaces

Ensure the O-ring at the lower lip of the water pump is in good condition. Replace if damaged

thoroughly and use a new gasket and O-rings, especially between the cylinder head and top of water pump.

9. Install the water pump and tighten the bolts to 11–15 ft. lbs. Install the coolant pipe, lower radiator hose, timing belt cover (as necessary), accessory drive belts and water pump pulley.

10. Install the fan and fan shroud. Fill the coolant system with coolant. Start the engine and allow it to reach operating temperature. Check for leaks. Add coolant as necessary.

B280F Engine

1. Disconnect the negative battery cable. On some varients of this engine it may be necessary to remove the front and main sections of the intake manifold.
2. Remove the overflow tank cap and drain the cooling system.
3. Disconnect both radiator hoses. On automatic transmission vehicles, disconnect the transmission cooler lines at the radiator. Disconnect the fan shroud. Remove the radiator and fan shroud.
4. Remove the fan.
5. Remove the hoses from the water pump to each cylinder head.
6. Remove the fan belts. Remove the water pump pulley.
7. Loosen the hose clamps at the rear of the water pump.
8. Remove the water pump from the block (3 bolts).

To install:

9. Transfer the thermal time lender and temperature sensor to the new water pump.
10. Transfer the thermostat cover, thermostat and rear pump cover to the new pump.
11. Install the new pump and tighten the bolts to 11–15 ft. lbs. Install the clamps, water pump pulley and fan belts.
12. Install the hoses that reach to each cylinder head. Install the fan, shroud and radiator. If equipped with an automatic transmission, connect the transmission cooler lines. Install the intake manifold, as necessary.
13. Connect the negative battery cable and fill the radiator with coolant. Start the engine and allow it to reach operating temperature. Check for leaks.

Thermostat

REMOVAL & INSTALLATION

1. Disconnect the lower radiator hose and drain the cooling system.
2. Remove the 2 bolts securing the thermostat housing to the cylinder head and carefully lift the housing free.
3. Remove all old gasket material from the mating surfaces and remove the thermostat.
4. Test the operation of the thermostat by immersing it in a container of heated water. Replace any thermostat that does not open at the correct temperature.
5. Place the thermostat, with a new gasket, in the cylinder head. Fit the thermostat housing to the head and hand-tighten the 2 bolts until snug. Do not tighten the bolts more than ¼ turn past snug.
6. Connect the lower radiator hose and replace the coolant.

Cooling System Bleeding

1. Fill the radiator with the proper type of coolant.
2. With the radiator cap off, start the engine and allow it to reach normal operating temperature.
3. Run the heater at full force and with the temperature lever in the **HOT** position. Be sure the heater control valve is functioning.
4. Shut the engine off and recheck the coolant level, refill as necessary.

ENGINE ELECTRICAL

NOTE: Disconnecting the negative battery cable on some vehicles may interfere with the functions of the on board computer systems and may require the computer to undergo a relearning process, once the negative battery cable is reconnected.

Distributor

REMOVAL

1. Unsnap the distributor cap clasps and remove the cap.
2. Crank the engine until No. 1 cyl-

inder is at Top Dead Center (TDC) of the compression stroke. At this point, the rotor should point to the spark plug wire socket for No. 1 cylinder and the 0 degree timing mark on the crankshaft damper should be aligned with the pointer. For ease of assembly, scribe a chalk mark on the distributor housing to note the position of the rotor.

3. Disconnect the negative battery terminal. Disconnect the primary lead from the coil at its terminal on the distributor housing. Disconnect the plug for the triggering contacts. On all models, remove the retaining screw for the primary voltage wire connector and pull it from the distributor housing.

4. Remove the vacuum hose(s) from the regulator.

5. Remove the distributor attaching screw and lift out the distributor.

Installation

1. When ready to install the distributor, if the engine has been disturbed, find TDC, of the compression stroke, for No. 1 cylinder. If the engine has not been disturbed, install the distributor with the rotor pointing to the No. 1 cylinder spark plug wire socket or the chalkmark made prior to removal. What is necessary is that the rotor aligns with the mark made prior to removal after the distributor is bolted down.

2. Connect the primary lead to its terminal on the distributor housing. On electronic fuel injected vehicles, connect the plug for the triggering contacts. Push the primary voltage wire connector into its slot in the distributor housing and tighten the retaining screw.

3. Connect the vacuum hose(s) to the vacuum regulator, if equipped.

4. Install the distributor cap and secure the clasps. Proceed to set the ignition timing. Tighten the distributor attaching screw.

Alternator

PRECAUTIONS

Several precautions must be observed with alternator equipped vehicles to avoid damage to the unit.

- If the battery is removed for any reason, make sure it is reconnected with the correct polarity. Reversing the battery connections may result in damage to the one-way rectifiers.
- When utilizing a booster battery as a starting aid, always connect the positive to positive terminals and the negative terminal from the booster battery to a good engine ground on the vehicle being started.
- Never use a fast charger as a booster to start vehicles.
- Disconnect the battery cables when charging the battery with a fast charger.
- Never attempt to polarize the alternator.
- Do not use test lamps of more than 12 volts when checking diode continuity.
- Do not short across or ground any of the alternator terminals.
- The polarity of the battery, alternator and regulator must be matched and considered before making any electrical connections within the system.
- Never separate the alternator on an open circuit. Make sure all connections within the circuit are clean and tight.
- Disconnect the battery ground terminal when performing any service on electrical components.
- Disconnect the battery if arc welding is to be done on the vehicle.

BELT TENSION ADJUSTMENT

Accessory drive belt tension is correct when the deflection made with light finger pressure on the belt at a midway point is about ½ in. Any belt that is glazed, frayed or stretched so it cannot be tightened sufficiently must be replaced.

Incorrect belt tension is corrected by moving the driven accessory (alternator, air pump, power steering pump or air conditioning compressor) away from or toward the driving pulley. Loosen the mounting and adjusting bolts on the respective accessory and tighten them, once the belt tension is correct. Never position a metal prybar on the rear end of the alternator, air pump or power steering pump housing, they can be deformed easily.

REMOVAL & INSTALLATION

1. Disconnect the negative battery cable.

2. Disconnect the electrical leads to the alternator. Remove all necessary components in order to gain access to the alternator retaining bolts.

3. Remove the adjusting arm-to-alternator bolt and the adjusting arm-to-engine bolt.

4. Remove the alternator mounting bolt.

5. Remove the fan belt and lift the alternator forward and out.

To install:

6. Install the alternator and fan belt. Tension the fan belt to specification.

7. Connect the electrical leads to the alternator and install all other previously removed components.

8. Connect the negative battery cable, start the vehicle and test the charging system.

Starter

REMOVAL & INSTALLATION

1. Disconnect the negative battery cable.

2. Disconnect the leads from the starter motor. Remove the necessary components in order to gain access to the starter retaining bolts. Raise and safely support the vehicle.

3. Remove the bolts retaining the starter motor to the flywheel housing and lift it off.

To install:

4. Position the starter motor to the flywheel housing and install the retaining bolts finger-tight. Apply locking compound to the threads and tighten the bolts to approximately 25 ft. lbs. (34 Nm).

5. Connect the starter motor leads and the negative battery cable.

EMISSION CONTROLS

Please refer to "Emission Controls" in the Unit Repair section for system maintenance procedures. Due to the complex nature of modern electronic engine control systems, comprehensive diagnosis and testing procedures fall outside the confines of this repair manual. For complete information on diagnosis, testing and repair procedures concerning all modern engine and emission control systems, please refer to "Chilton's Guide to Fuel Injection and Electronic Engine Controls".

Emission Warning Lamps

RESETTING

On most vehicles, there is an "Engine Service" or a "Lamda Sond" light on the instrument panel that lights up at specific intervals or if the on-board computer detects a fault. Re-setting this lamp is a simple procedure. Remove the sound proofing on the firewall above the pedals and remove the screws and catches at the base of the instrument panel. Tilt the panel out and locate the red re-set button.

FUEL SYSTEM

Fuel System Service Precautions

Safety is the most important factor when performing not only fuel system maintenance but any type of maintenance. Failure to conduct maintenance and repairs in a safe manner may result in serious personal injury or death. Maintenance and testing of the vehicle's fuel system components can be accomplished safely and effectively by adhering to the following rules and guidelines.

- To avoid the possibility of fire and personal injury, always disconnect the negative battery cable unless the repair or test procedure requires that battery voltage be applied.
- Always relieve the fuel system pressure prior to disconnecting any fuel system component (injector, fuel rail, pressure regulator, etc.), fitting or fuel line connection. Exercise extreme caution whenever relieving fuel system pressure to avoid exposing skin, face and eyes to fuel spray. Please be advised that fuel under pressure may penetrate the skin or any part of the body that it contacts.
- Always place a shop towel or cloth around the fitting or connection prior to loosening to absorb any excess fuel due to spillage. Ensure that all fuel spillage (should it occur) is quickly removed from engine surfaces. Ensure that all fuel soaked cloths or towels are deposited into a suitable waste container.
- Always keep a dry chemical (Class B) fire extinguisher near the work area.
- Do not allow fuel spray or fuel vapors to come into contact with a spark or open flame.
- Always use a backup wrench when loosening and tightening fuel line connection fittings. This will prevent unnecessary stress and torsion to fuel line piping. Always follow the proper torque specifications.
- Always replace worn fuel fitting O-rings with new. Do not substitute fuel hose or equivalent, where fuel pipe is installed.

RELIEVING FUEL SYSTEM PRESSURE

1. Place the proper size wrenches onto the fuel filter fittings.
2. Place a shop towel or rag around the fuel filter fittings and wrenches.
3. Slowly loosen the fuel line at the fuel filter until all pressure is relieved.
4. Tighten fuel filter fittings.

Fuel Tank

REMOVAL & INSTALLATION

1. Disconnect the negative battery cable. Raise and support the vehicle safely.
2. Release the fuel system pressure. Drain the fuel tank completely.

CAUTION

When performing this procedure, always have a dry-chemical fire extinguisher handy. Fuel vapors are extremely explosive.

3. In the trunk, remove the panels which cover the filler hose. It may be necessary to remove the spare tire on some vehicles. Roll back the carpet and remove the access panel cover.
4. Disconnect the fuel filler pipe connection. Label and disconnect all fuel lines leading to the fuel tank. Label and disconnect all electrical connectors at the fuel tank.
5. Position a floorjack under the tank, using a large piece of wood as a cushion between the fuel tank and the floorjack. Raise the jack so that it just contacts the tank.
6. On vehicles equipped with a saddle type fuel tank (driveshaft runs through a tunnel in the fuel tank), matchmark the driveshaft and remove.
7. Remove any shields or protective covers on the tank. Loosen and remove the tank retaining bolts. Lower the jack slowly and inspect for any obstructions.
8. Once on the ground, remove the fuel sender unit by unscrewing the lock-ring at the top of the tank. Drain the remaining fuel from the tank.

To install:

9. Install the fuel sender unit and tighten the lock-ring. Install the protective sheilds and raise the fuel tank into position. Install and tighten the attaching bolts.
10. Remove the floorjack. Install the driveshafts. Connect all fuel and electrical lines leading to the fuel tank.
11. Install the protective panel in the trunk and replace the spare tire (as required) and the carpet.
12. Lower the vehicle, connect the negative battery cable, turn the ignition key **ON** and check for leaks.

Fuel Filter

REMOVAL & INSTALLATION

On earlier vehicles, the fuel filter is mounted under the hood, on or near the air flow sensor. It is also near the battery, so disconnect the negative terminal before changing the filter. Be sure to use the new gaskets provided with the new filter.

On newer vehicles the fuel filter is a "lifetime" filter and only needs to be changed in the event of contamination. It is mounted under the vehicle, in assembly with the fuel pump, accumulator and reservoir but can usually be removed separately.

1. Disconnect the negative battery cable.
2. Raise and safely support the vehicle, if changing under vehicle filter.
3. Relieve the fuel system pressure.

CAUTION

When relieving the pressure in the fuel system, place a container under the filter to catch the excess fuel.

4. Remove the fuel lines, the mounting bracket nut and the filter.

To install:

NOTE: Fuel flow direction arrow is marked on the new (and old) filter. Arrow follows direction from fuel tank to engine.

5. Install the new filter in the proper direction. Be sure to use the new sealing rings and torque the fuel lines to the filter to 14 ft. lbs. (20 Nm).
6. Connect the negative battery cable, turn the ignition **ON** and check for leaks.

Electric Fuel Pump

REMOVAL & INSTALLATION

NOTE: The main fuel pump is located under the vehicle in front of the left rear wheel or at the crossmember. There is also a transfer pump located inside the fuel tank.

1. Disconnect the negative battery cable.
2. Raise and safely support the vehicle.
3. Disconnect the electrical connector.
4. Relieve the fuel system pressure.

CAUTION

When relieving the pressure in the fuel system, place a container under the fuel pump to catch the excess fuel.

5. Remove the mounting bolts and the fuel pump.
6. When installing, be sure to use new sealing rings and/or gaskets.

Transfer Pump

REMOVAL & INSTALLATION

1. Disconnect the negative battery cable. Raise and support the vehicle safely. Relieve the fuel system pressure.

2. Remove the fuel tank.
3. Loosen the lock-ring at the top of the fuel tank and remove the sending unit with the transfer pump attached. Note the direction of the float in the tank.
4. Remove the transfer pump from the sending unit.

To install:

5. Install the transfer pump on the sending unit. Install the sending unit in the fuel tank and tighten the lock-ring to specification. Do not overtighten the lock-ring as the plastic threads on some fuel tanks are easily stripped.
6. Install the fuel tank in the vehicle.
7. Lower the vehicle. Connect the negative battery cable, start the engine and check for leaks. Check the system fuel pressure.

Fuel Injection

IDLE SPEED AND MIXTURE ADJUSTMENT

Fuel injection is set at the factory and controlled by an Electronic Control Unit (ECU).

Fuel Injector

REMOVAL & INSTALLATION

1. Relieve the fuel system pressure. Disconnect all fuel and vacuum connections.
2. Loosen the attaching bolts and remove the fuel distribution rail and injectors as an assembly.
3. Remove the injector by pulling it from the injection rail.
4. When installing injectors, check the rubber O-rings for damage and replace as necessary. Lubricate the O-rings with petroleum jelly prior to installaiton.
5. Install the fuel rail assembly and tighten all attaching bolts. Connect all fuel and vacuum lines.

DRIVE AXLE

Halfshaft

REMOVAL & INSTALLATION

Multi-link Suspension

1. Raise and support the vehicle safely.
2. Remove the rear wheels and loosen the halfshaft retaining nut in the center of the wheel bearing housing.

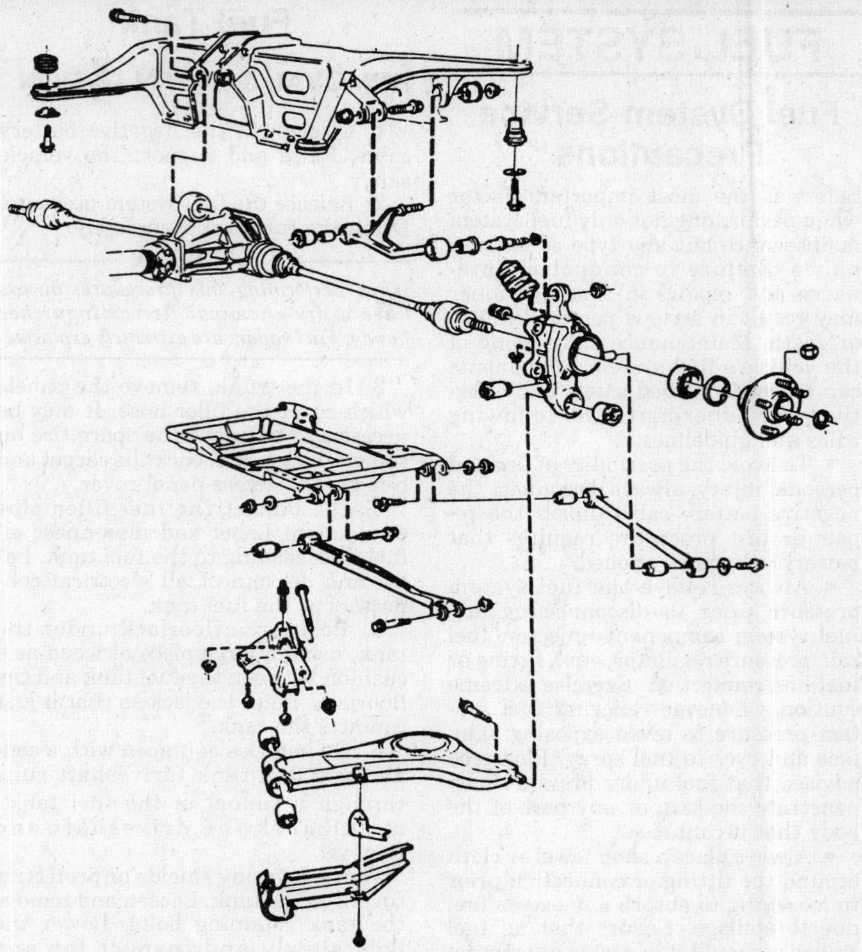

Volvo Multi-link differential—940 and Coupe vehicles

3. Remove the bolts holding the lower section of the differential. Remove the lower section with the trailing links attached.
4. Remove the bolts holding the halfshaft to the differential and remove the shaft from the wheelbearing housing. Inspect the CV-boots for damage and replace as necessary.

To install:

5. Install the wheel end of the halfshaft first, then install the differential end. Using new bolts, tighten to 70 ft. lbs. (94 Nm).
6. Temporarily install two long 12 mm bolts in the center holes of the lower differential section and install. The bolts will hold the lower section in place for alignment. Alignment of the panel is critical to proper rear wheel alignment.
7. Once aligned, install the 8 attaching bolts and tighten to 52 ft. lbs. (70 Nm) plus 30 degrees of rotation.
8. Install a new halfshaft retaining nut and tighten to 103 ft. lbs. (139 Nm) plus 60 degrees of rotation.
9. Install the wheel and lower the vehicle.

Driveshaft and U-Joints

REMOVAL & INSTALLATION

1. Raise and safely support the vehicle.
2. Mark the relative positions of the driveshaft yokes and transmission and differential housing flanges for purposes of assembly. Remove the nuts and bolts which retain the front and rear driveshaft sections to the transmission and differential housing flanges, respectively. Remove the support bearing housing from the driveshaft tunnel and lower the driveshaft and universal joint assembly as a unit.
3. Pry up the lock washer and remove the support bearing retaining nut. Pull off the rear section of the driveshaft with the intermediate universal joint and splined shaft of the front section. The support bearing may now be pressed off the driveshaft.
4. Remove the support from it housing.
5. Inspect the driveshaft sections

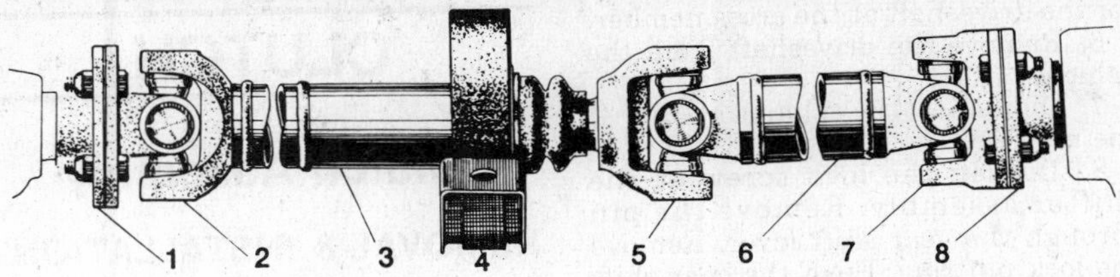

1. Flange on transmission
2. Front universal joint
3. Front section of driveshaft
4. Support bearing
5. Intermediate universal joint
6. Rear section of driveshaft
7. Rear universal joint
8. Flange on rear axle

Driveshaft with support bearing

for straightness. Using a dial indicator or rolling the shafts along a flat surface, make sure the driveshaft out-of-round does not exceed 0.010 in. (0.25mm). Do not attempt to straighten a damaged shaft. Any shaft exceeding 0.010 in. (0.25mm) out-of-round will cause substantial vibration and must be replaced. Also, inspect the support bearing by pressing the races against each other by hand and turning them in opposite directions. If the bearing binds at any point, it must be discarded and replaced.

To install:

6. Install the support bearing into its housing.
7. Press the support bearing and housing onto the front driveshaft section. Push the splined shaft of the front section, with the intermediate universal joint and rear driveshaft section, into the splined sleeve of the front section. Install the retaining nut and lock washer for the support bearing.
8. Taking note of the alignment marks made prior to removal, position the driveshaft and universal joint assembly to its flange connections and install but do not tighten its retaining nuts and bolts. Position the support bearing housing to the driveshaft tunnel and install the retaining nut. Tighten the nuts which retain the driveshaft sections to the transmission and differential housing flanges to a torque of 25–30 ft. lbs. (34–40 Nm).
9. Lower the vehicle. Road test the vehicle and check for driveline vibrations and noise.

Rear Axle Shaft, Bearing and Seal

REMOVAL & INSTALLATION

Except Multi-link Rear Suspension

1. Raise and safely support the vehicle.
2. Remove the applicable wheel and tire assembly.
3. Place a wooden block under the brake pedal, plug the master cylinder reservoir vent hole; remove and plug the brake line from the caliper. Be careful not to allow any brake fluid to spill onto the disc or pads. Remove the 2 bolts which retain brake caliper to the axle housing and lift off the caliper. Lift off the brake disc.
4. Remove the thrust washer bolts through the holes in the axle shaft flange. Using a slide hammer, remove the axle shaft, bearing and oil seal assembly. If possible, pull out the shaft by temporarily reinstalling the brake disc and using this to grab on to while pulling out the axle shaft.
5. Using an arbor press, remove the axle shaft bearing and its locking ring from the axle shaft. Remove and discard the old oil seal.

To install:

6. Fill the space between the lips of the new oil seal with wheel bearing grease. Position the new seal on the axle shaft. Using an arbor press, install the bearing with a new locking ring, onto the axle shaft.
7. Thoroughly pack the bearing with wheel bearing grease. Install the axle shaft into the housing, rotating it so it indexes with the differential. Install the bolts for the thrust washer and tighten to 36 ft. lbs. (50 Nm).
8. Install the brake disc. Position the brake caliper to its retainer on the axle housing and install the 2 retaining bolts. Torque the caliper retaining bolts to 45–50 ft. lbs. (61–68 Nm).
9. Unplug the brake line and connect it to the caliper. Bleed the caliper of all air trapped in the system.
10. Position the wheel and tire assembly on its lugs and hand-tighten the lug nuts. Remove the jack stands and lower the vehicle. Torque the lug nuts to 70–100 ft. lbs. (95–135 Nm).

Multi-Link Rear Suspension

1. Raise and support the vehicle safely.
2. Remove the wheel, brake caliper, rotor and handbrake cable. Hang the caliper out of the way on a piece of wire. Mark the position of the rotor prior to removal.
3. Remove the support arm, lower link arm and track rod. Use a puller to remove the track rod.
4. Remove the upper link. The wheel bearing housing can now be removed. There are shims between the bearing housing and the upper link arm. Collect them when the housing is removed.
5. Mount the housing in a vise and using tool 5340, or equivalent, apply a counter hold to the bearing housing. Press out the hub with a proper sized drift.
6. Remove the circlip holding the bearing and press the bearing out. Using tools 2722 and 5310, or equivalent, pull the inner ring off the hub.

To install:

7. Press in the bearing and install the circlip. Press on the inner ring. A counter hold must be used to avoid damaging the bearing.
8. Install the wheel bearing housing on the halfshaft. Install the shims between the upper link. Install the retaining nut.
9. Pull the wheel bearing housing outward at the top and tighten the upper link arm nut to 85 ft. lbs. (115 Nm). Pulling the housing outward insures proper alignment.
10. Tilt the bearing housing outward and install the lower link. Push the bottom of the bearing housing inwards and tighten the bolt to 36 ft. lbs. (48 Nm) plus 90 degrees of rotation.
11. Install the support arm and tighten to 44 ft. lbs. (59 Nm) plus 90 degrees of rotation. Instal the track rod and tighten to 63 ft. lbs. (85 Nm).
12. Install the handbrake cable. Install the rotor and caliper. Tighten the caliper bolts to 44 ft. lbs. (59 Nm).
13. Tighten the center halfshaft nut to 103 ft. lbs. (139 Nm) plus 60 degrees of rotation. Install the wheel and lower the vehicle.
14. Perform a rear wheel alignment.

MANUAL TRANSMISSION

For further information on transmissions/transaxles, please refer to "Chilton's Guide to Transmission Repair".

Transmission Assembly

REMOVAL & INSTALLATION

The transmission or the transmission

overdrive assembly may be removed with the engine installed in the vehicle.

240

1. Disconnect the negative battery cable. Disconnect the back-up light connector at the firewall.
2. Raise and safely support the vehicle. Loosen the set screw and drive out the pin for the shifter rod. Disconnect the shift lever from the rod.
3. Inside the vehicle, pull up the shift boot. Remove the fork for the reverse gear detent. Remove the snapring and lift up the shifter.
4. Disconnect the clutch cable and return spring at the throw-out fork and flywheel housing.
5. Disconnect the exhaust pipe bracket(s) from the flywheel cover. Remove the oil pan splash guard.
6. Using a floor jack and a block of wood, support the engine under the oil pan. Remove the transmission support crossmember.
7. Disconnect the driveshaft. Disconnect the speedometer cable. If equipped, disconnect the overdrive wire.
8. Remove the starter retaining bolts and pull free of the flywheel housing.
9. Support the transmission using another floor jack. Remove the flywheel bellhousing-to-engine bolts and remove the transmission.

To install:

10. Install the transmission in the vehicle and tighten the flywheel housing-to-engine bolts to 25–35 ft. lbs. (34–47 Nm).
11. Install the starter, driveshaft, speedometer cable and overdrive wire (if equipped).
12. Install the transmission crossmember and remove the floorjack. Install the oil pan splash guard and exhaust pipe brackets.
13. Connect the clutch cable and return spring. Assemble the gear shift lever.
14. Lower the vehicle. Connect the back-up light connector and negative battery cable.
15. Check the transmission oil level. Adjust the clutch as required

Except 240

1. Disconnect the negative battery cable.
2. Attach a lifting beam tool 5006, to the rear of the engine. This will support the engine once the transmission is removed.
3. Raise and support the vehicle safely.
4. disconnect the driveshaft at the transmission flange.
5. Disconnect the support bearing for the driveshaft at the crossmember.
6. Remove the driveshaft from the vehicle.
7. Disconnect the exhaust system at the muffler.
8. Loosen the lock screw at the shifter assembly. Remove the pin through the gear shift lever. Remove the lock pin ring. Push the gear shift lever up.
9. Remove the transmission crossmember and bracket. Cut the wire straps and disconnect the wires at the transmission.
10. Disconnect the clutch cable at the clutch slave cylinder.
11. Disconnect the exhaust system attachment at the transmission cover.
12. Position a floorjack jack under the transmission assembly. Remove the transmission to engine retaining bolts. Remove the transmission from the vehicle.

To install:

13. Install the transmission in the vehicle and tighten 30 ft. lbs. (41 Nm).
14. Install the exhaust system attachment, clutch cable and slave cylinder.
15. Install the crossmember and bracket. Connect all electrical connections.
16. Install the shifter assembly. Install the remainder of the exhaust system.
17. Install the driveshaft. Lower the vehicle and disconnect the lifting beam.
18. Connect the negative battery cable, check the transmission oil level and test drive the vehicle. Adjust the clutch as required.

LINKAGE ADJUSTMENT

Reverse gear detent clearance is the only adjustment that can be made to the shift linkage. Remove the shift lever cover, trim frame and ash tray assembly. Engage 1st gear and adjust the clearance between the detent plate and the gear shift lever. Also check clearance should be 0.004–0.06 in. (0.1–1.5mm).

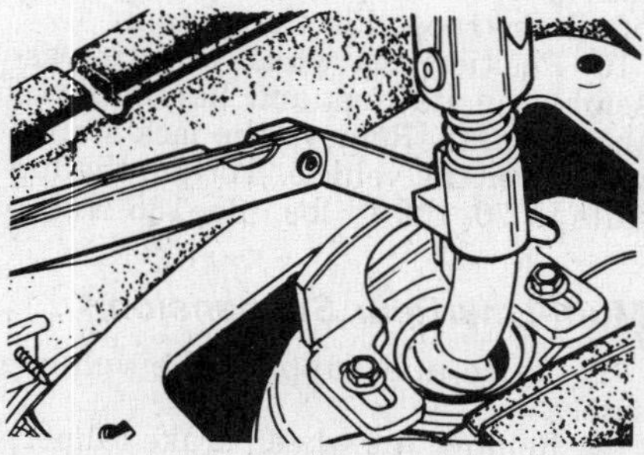

Reverse gear detent clearance adjustment—manual transmission

CLUTCH

Clutch Assembly

REMOVAL & INSTALLATION

1. Remove the transmission.
2. Scribe alignment marks on the clutch and flywheel. In order to prevent warpage, slowly loosen the bolts which retain the clutch to the flywheel diagonally in rotation. Remove the bolts and lift off the clutch and pressure plate.
3. Inspect the clutch assembly.
4. Clean the pressure plate and flywheel throughly with solvent prior to installation.
5. Position the clutch assembly, the longest side of the hub facing backwards, to the flywheel and align the bolt holes. Insert a pilot shaft (centering mandrel or drift) or an input shaft from an old transmission of the same type, through the clutch assembly and flywheel so the flywheel pilot bearing is centered.
6. Install the 6 bolts which retain the clutch assembly to the flywheel and tighten them diagonally in rotation, a few turns at a time. After all the bolts are tightened, remove the pilot shaft (centering mandrel).
7. Install the transmission.
8. If equipped with a hydraulic clutch, bleed the clutch.

Clutch Cable

REMOVAL & INSTALLATION

1. Raise and support the vehicle safely.
2. Disconnect the clutch cable from the clutch fork. Some vehicles are equipped with a release bearing that rotates. These vehicles will have a clutch return spring at the pedal assembly.
3. Disconnect the cable at the pedal assembly. Remove the cable. On some early vehicle vehicles, the clutch cable is fitted with a weight. Do not replace the weight when installing the new cable.

To install:

4. Install the cable in the vehicle.
5. Connect the clutch cable at the clutch fork and the pedal assembly.
6. Adjust the clutch cable.

CLUTCH ADJUSTMENT

1. Clutch play is adjusted under the vehicle at the clutch fork.
2. Loosen the locknut on the fork side of the cable bracket and turn the

adjust nut until the proper play is achieved. Tighten the locknut.

NOTE: Vehicles equipped with hydraulic clutch assemblies are not adjustable.

3. Clutch play for all engines except Turbo is 0.12–0.2 in. (3–5mm). Turbo clutch play (free movement rearward) is 0.04–0.12 in. (1–3mm).

Clutch Master Cylinder

REMOVAL & INSTALLATION

1. Remove the panel under the instrument panel. Remove the locking spring and pin from the clutch pedal assembly.
2. Disconnect the hose from the clutch fluid reservoir.
3. Unscrew the nipple from the cylinder housing. Place a container under the cylinder to catch the fluid that will spill out. Unbolt and remove the cylinder housing.
4. Installation is the reverse of removal. Make sure there is 0.04 in. (1mm) clearance between the pushrod and the pistons and adjust, if necessary. Fill the reservoir with DOT 4 brake fluid and bleed the system.

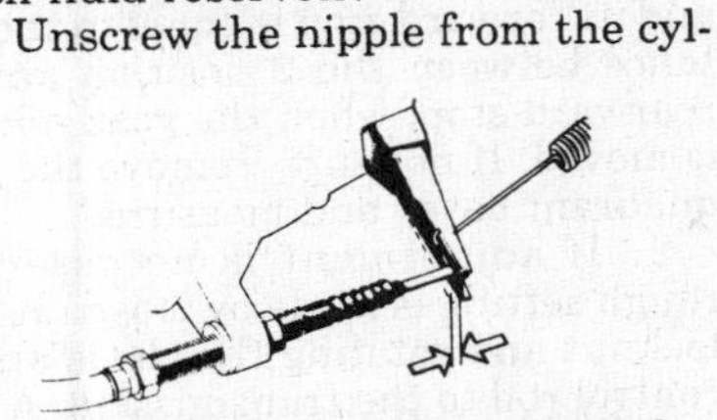

Adjusting the cable operated clutch

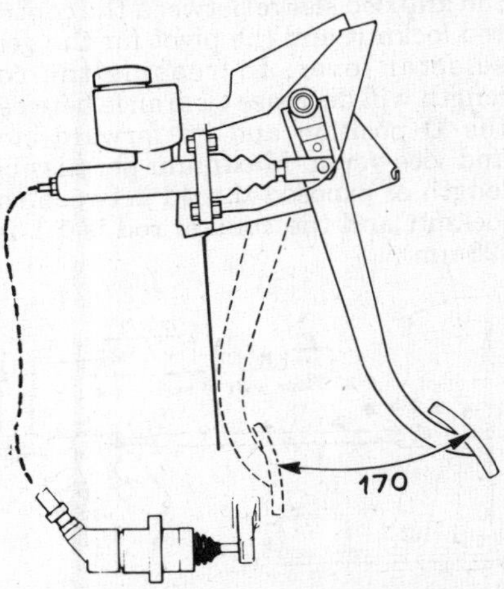

Checking the hydraulic clutch travel

Clutch Slave Cylinder

REMOVAL & INSTALLATION

1. Raise and support the vehicle safely.
2. Disconnect the fluid line a the cylinder.
3. Unbolt the cylinder from the flywheel housing.
4. Installation is the reverse of removal.
5. Bleed the hydraulic clutch system.

Hydraulic Clutch System Bleeding

The hydraulic clutch system should be bled any time the hoses have been loosened or any component replaced. The bleeding process is similar to bleeding the brake system. The purpose is to remove any air trapped within the lines.

Add brake fluid to the reservoir. Attach a length of hose to the bleeder nipple on the slave cylinder (at the transmission) and put the other end in a clear glass jar. Put enough brake fluid in the jar to cover the engd of the hose.

Have an assistant press the clutch pedal to the floor. Open the bleeder screw on the slave cylinder. Close the bleeder while the pedal is still depressed and repeat the process. As the bleeder is released each time, note the fluid in the jar. When no bubbles are seen, the system is bled. Tighten the bleeder fitting and remove the hose.

During bleeding, note the amount of fluid remaining in the reservoir. If the fluid level drops below the minimum level of the reservoir, refill with fluid before proceeding.

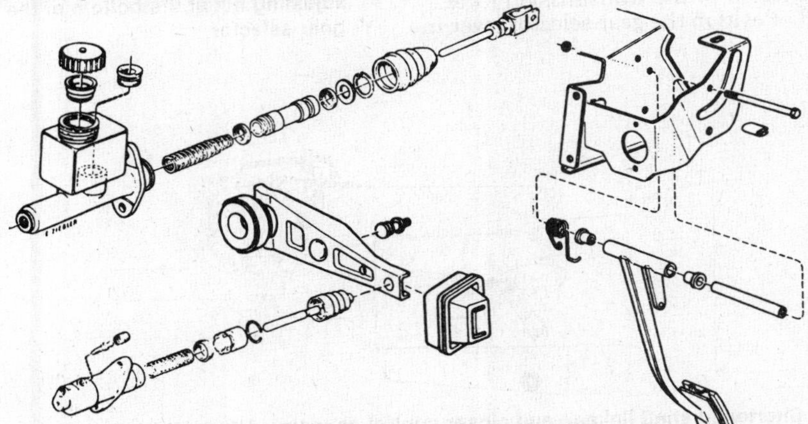

Hydraulic clutch assembly

AUTOMATIC TRANSMISSION

For further information on transmissions/transaxles, please refer to "Chilton's Guide to Transmission Repair".

Transmission Assembly

REMOVAL & INSTALLATION

240

1. Disconnect the negative battery cable. Remove the dipstick and filler pipe clamp.
2. Remove the bracket and throttle cable from the dashboard and throttle control, respectively.
3. Disconnect the exhaust pipe at the manifold.
4. Raise and safely support the vehicle.
5. Drain the fluid into a clean container.
6. Disconnect the driveshaft from the transmission flange.
7. Disconnect the selector lever controls and remove the reinforcing bracket from the pan.
8. Remove the torque converter attaching bolts.
9. Support the transmission with a jack equipped with a holding fixture.
10. Remove the crossmember.
11. Disconnect the exhaust pipe brackets and remove the speedometer cable form the case.
12. Remove the filler pipe.
13. Place a wooden block between the engine and firewall and lower the jack until the engine is against the block.

NOTE: If the battery cable appears to stretch to much, remove it.

14. Disconnect the starter wires, remove the converter housing bolts and pull the transmission backwards to clear the guide pins. Remove the transmission assembly.

To install:

15. Install the transmission assembly and tighten 14mm bolts to 35 ft. lbs. (47 Nm). Connect the starter wires.

16. Install the battery cable, if removed. Install the filler pipe, exhaust system, speedometer cable and transmission crossmember.
17. Install the torque converter attaching bolts. Install the selector lever controls and bracket.
18. Install the driveshaft and lower the vehicle. Install the throttle control and bracket. Connect the negative battery cable.
19. Fill the transmission with fluid, start the engine and check for leaks. Recheck the transmission fluid level after the vehicle reaches operating temperature.

Except 240

1. Disconnect the negative battery cable. Place the gear selector in the **P** position.
2. Disconnect the kickdown cable at the throttle pulley on the engine. Disconnect the battery ground cable.
3. Disconnect the oil filler tube at the oil pan and drain the transmission oil.
4. Disconnect the control rod at the transmission lever and disconnect the reaction rod at the transmission housing.
5. On AW 71 transmissions, disconnect the wire at the solenoid; slightly to the rear of the transmission-to-driveshaft flange.
6. Matchmark the transmission-to-driveshaft flange and unbolt the driveshaft.
7. Remove the transmission crossmember assembly.
8. Disconnect the exhaust pipe at the joint and remove the exhaust pipe bracket from the exhaust pipe. Remove the rear engine mount with the exhaust pipe bracket.
9. On B280F equipped vehicles, remove the bolts retaining the starter motor.
10. Remove the cover for the alternate starter motor location and the cover plate at the torque converter housing bottom on B280F equipped vehicles.
11. Disconnect the oil cooler lines at the transmission.
12. Remove the 2 upper screws at the torque converter cover. Remove the oil filler tube.
13. Place a transmission jack or a standard hydraulic floor jack under the transmission.
14. Remove the screws retaining the torque converter to the driveplate. Pry the torque converter back from the driveplate with a small prybar.
15. Slowly lower the transmission when pulling it back to clear the input shaft. Do not tilt the transmission forward or the torque converter may slide off.

To install:

16. Install the transmission and tighten the attaching bolts. Install the torque converter bolts. Install the torque converter cover and filler tube.
17. Connect the oil cooler lines. On B280F equipped vehicles, install the alternate starter and converter housing cover.
18. On B280F equipped vehicles, install the starter motor. Install the exhaust pipe along with the rear engine mount.
19. Install the exhaust pipe and bracket along with the rear engine mount.
20. Install the driveshaft, connect the solenoid wire (AW71 transmission), the control rod and reaction rod. Move the gear selector to the **P** position before attaching the control rod.
21. Install the oil filler tube, kickdown cable and connect the negative battery cable. Adjust the gear shift linkage and the kickdown cable.
22. Fill the transmission with fluid, start the engine and check for leaks. Recheck the transmission fluid level after the vehicle reaches operating temperature.

SHIFT LINKAGE ADJUSTMENT

240 Except BW55 Transmission

NOTE: On AW70/AW71 transmissions, the gear selector shift console has been moved forward and the shift linkage has been shortened.

1. Disconnect the shift rod from the transmission lever. Place both the transmission lever and the gear selector lever in the **2** position.
2. Adjust the length of the shift control rod so a small clearance of 0.04 in. (1mm) is obtained between the gear selector lever inhibitor and the inhibitor plate, when the shift control rod is connected to the transmission lever.
3. Position the gear selector lever in **D** and make sure a similar small clearance of 0.04 in. (1mm) exists between the lever inhibitor and the inhibitor plate. Disconnect the shift control rod from the transmission lever and adjust, if necessary.
4. Lock the control rod bolt with its safety clasp and tighten the locknut. Make sure the control rod lug follows with the transmission lever.
5. After moving the transmission lever to the **P** and **1** positions, make sure the clearances remain the same. In addition, make sure the output shaft is locked with the selector lever in the **P** position.

240 with BW55 Transmission

1. With the engine off, check that the distance between the **D** position and its forward stop is equal to the distance between the **2** position and its rearward stop, when the gear selector is moved. If not sure, remove the gear quadrant cover and measure.
2. If adjustment is necessary, a rough setting is made by loosening the locknut and rotating the clevis on the control rod to the transmission. A fine adjustment can be made by rotating the knurled sleeve between the control rod locknut and the pivot for the gear selector lever. Increasing the rod length will decrease clearance between the **D** position and its forward stop and vice-versa. Maximum permissible length of exposed thread between the locknut and the control rod is 1.1 in. (28mm).

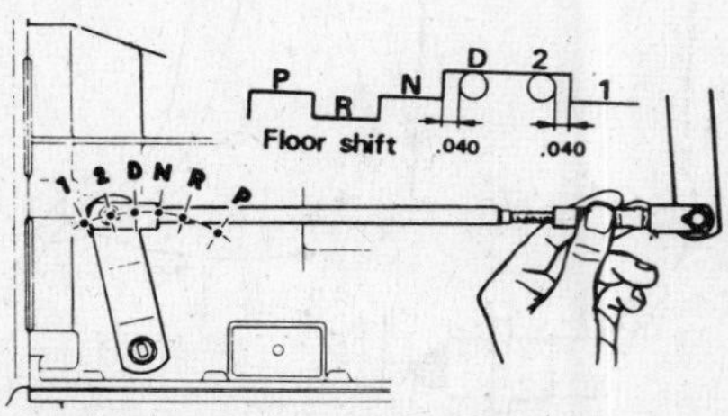

Adjust the shift linkage using the adjusting nut at the bottom of the gear selector

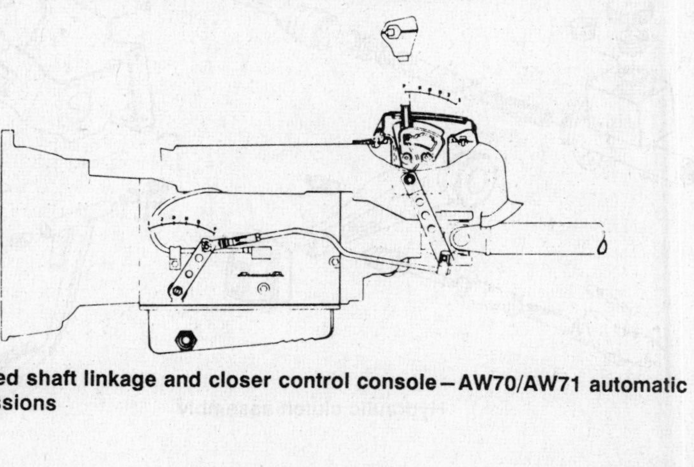
Shortened shaft linkage and closer control console—AW70/AW71 automatic transmissions

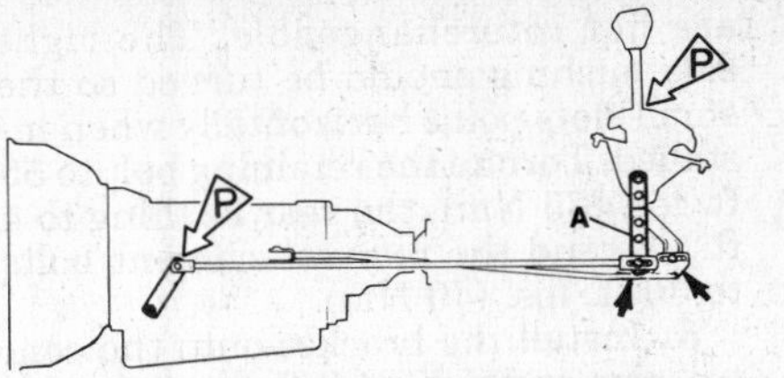

700/900 Series automatic transmission gear linkage. 'A' is the adjusting rod arm; arrows point to the locknuts on the adjustment (left) and reaction rod (right) arms

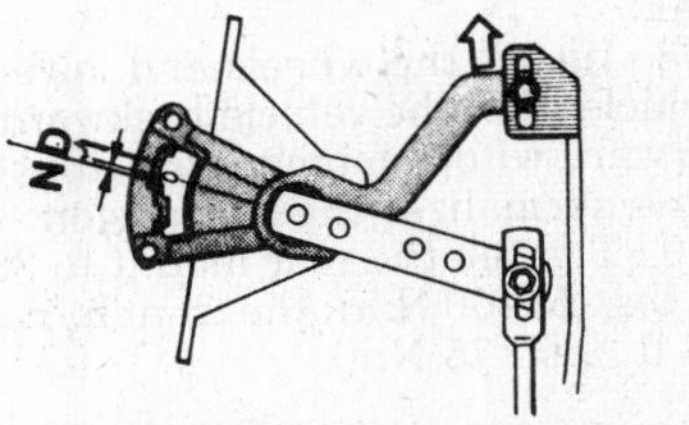

The clearance between the D and N position on the gear selector should be the same or less than the clearance between the 2 and 1 position

Except 240 vehicles

NOTE: Before adjusting the shift linkage, make sure the starter motor operates only in P or N positions; that the back-up lights light up only in R; that the shift lever is vertical in P with the vehicle level; that the clearance between D and N is the same or less than the clearance between 2 and 1.

ADJUSTING CLEARANCE

1. Check that the clearance between **D** and **N** is the same or less than the clearance between **2** and **1** on the shift lever. If clearance is correct, tighten the locknut to 12–17 ft. lbs. 16–23 Nm). If clearance is not correct, adjust as follows:

2. If no clearance is felt in **D**, move the reaction rod arm rearwards about 0.08 in. (2mm).

3. If no clearance is felt in position **2**, move the reaction rod arm forwards about 0.12 in. (3mm). Tighten the locknut.

4. After adjustment, check that the vehicle starts only in **P** or **N** and that the back-up light does not light in **R**, reduce clearance in **D** by moving the rod arm forward slightly.

FRONT SUSPENSION

MacPherson Strut

REMOVAL & INSTALLATION

1. Remove the hub cap and loosen the lug nuts a few turns.
2. Firmly apply the parking brake and place blocks in back of the rear wheels.
3. Raise and safely support the vehicle or using a floor jack at the center of the front crossmember. When the wheels are 2–3 in. (50–76mm) off the ground, the vehicle is high enough. Place jackstands under the front jacking points. Then, remove the floor jack from the crossmember, if used, and reposition it under the applicable lower control arm to provide support at the outer end. Remove the wheel and tire assembly.
4. Using a boll joint puller, disconnect the steering rod from the steering arm.
5. Disconnect the stabilizer bar at the link upper attachment.
6. Remove the bolt retaining the brake line bracket to the fender well.
7. Open the hood and remove the cover for the strut assembly upper attachment.
8. While keeping the strut from turning, loosen and remove the nut for the upper attachment.
9. Before lowering the strut assembly, wire or tie the strut to some stationary component or use a holding fixture such as SVO 5045, to prevent the strut from traveling down too far and damaging the hydraulic brake lines. Then lower the jack supporting the lower arm and allow the strut to tilt out to about a 60 degree angle. At this angle, the top of the strut assembly should just protrude past the wheel well, allowing removal of the strut from the top.

To install:

10. Carefully lift and guide the strut assembly into its upper attachment in the spring tower. Connect the stabilizer bar to the stabilizer link. Guide the shock absorber spindle into the upper attachment and raise the jack under the lower control are. Install the washer and nut on top of the shock absorber spindle. While holding the spindle from turning, tighten the nut to 15–25 ft. lbs. (20–34 Nm). Install the cover.
11. Attach the brake line bracket to its mount. Tighten the nut retaining the stabilizer bar to the link. Connect the steering rod at the steering arm.
12. Install the wheel and tire assembly. Lower the vehicle. Jounce the suspension a few times and then road test.

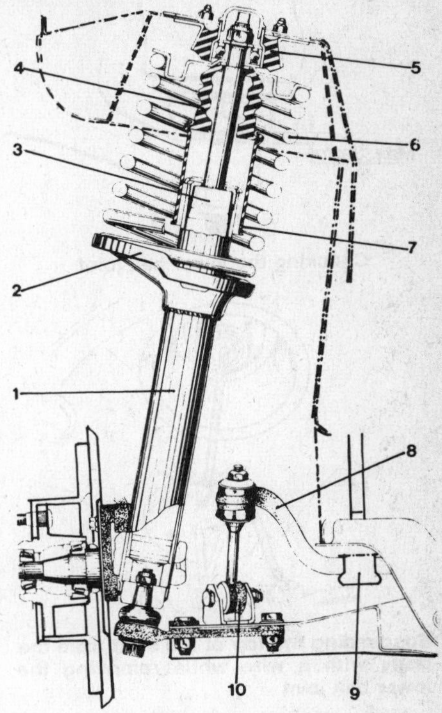

1. Strut assembly
2. Lower spring support
3. Shock absorber
4. Rubber bumper
5. Upper attachment
6. Coil spring
7. Ruber sleeve, protecting the shock absorber
8. Stabilizer bar
9. Stabilizer bar attachment
10. Stabilizer link

Front suspension – 240 shown others similar

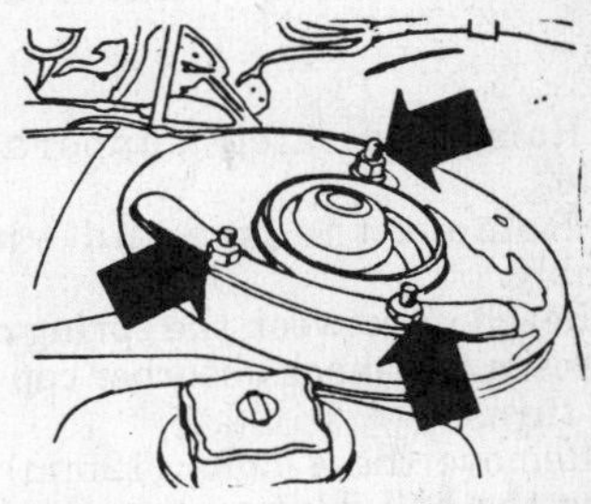

Loosen the upper strut nuts to adjust the camber

Lower Ball Joints

INSPECTION

Maximum axial play with normally loaded front end is 0.12 in. (3mm). Maximum radial play is 0.02 in. (0.5mm).

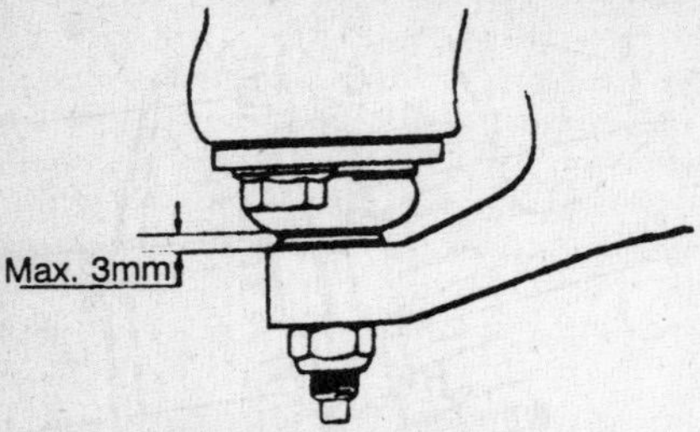

Checking the lower ball joint

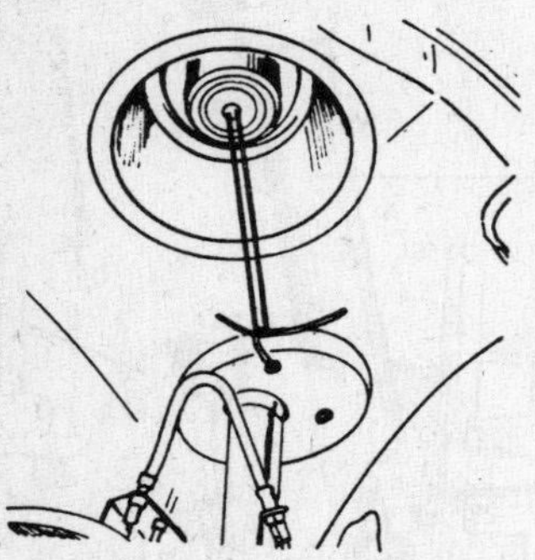
Suspending the top of the strut from the body with a wire while removing the lower ball joint

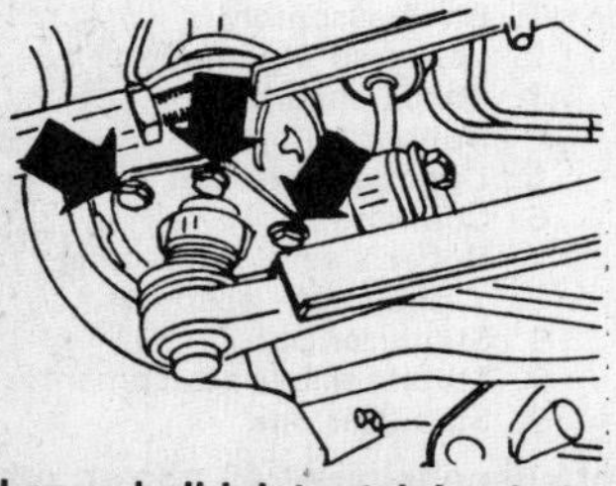
Lower ball joint retaining bolts

REMOVAL & INSTALLATION

240

1. Raise and safely support the vehicle.
2. Remove the tire and wheel assembly.
3. Reach in between the spring coils and loosen the shock absorber cap nut a few turns.
4. Remove the 4 bolts (12mm) retaining the ball joint seat to the bottom of the strut.
5. Remove the 3 nuts (19mm) retaining the ball joint to the lower control arm.
6. Place the ball joint and attachment assembly in a vise and remove the 19mm nut from the ball joint stud. Then, drive out the old ball joint.

To install:

7. Install the new ball joint in the attachment and tighten the stud nut to 35–50 ft. lbs. (47–68 Nm).

NOTE: On vehicles with power steering, the ball joint are different for the left and right side. Compared to previous years, the ball joint is 0.393 in. (1mm) forward in control rod attachment. It is therefore most important that these ball joints are installed on the correct side.

8. Attach the ball joint assembly to the strut. Tighten to 15–20 ft. lbs. (20–27 Nm).
9. Attach the ball joint assembly to the control arm. Tighten to 70–95 ft. lbs. (95–130 Nm).
10. Tighten the shock absorber cap nut. Install the wheel and tire. Lower the vehicle and road-test.

Except 240

1. Raise and safely support the vehicle. Remove the wheel.
2. Remove the bolt connecting the anti-roll bar link to the control arm.
3. Remove the cotter pin for the ball joint stud and remove the nut.
4. Using a ball joint press, remove the ball joint from the control arm. Make sure the press is located directly in line with the stud and that the rubber grease boot is not damaged by the puller.
5. Remove the bolts holding the ball joint to the spring strut. Press the control arm down and remove the ball joint.

To install:

6. When installing the new ball joint, always use new bolts and coat all threads with a liquid thread sealer. Torque bolts to 22 ft. lbs. (30 Nm), checking that the bolt heads sit flat on the ball joint, then angle-tighten (protractor-torque) 90 degrees torque the nut holding the control arm ball joint stud to 44 ft. lbs. (60 Nm). Use a new cotter pin on the ball joint stud and install the anti-roll bar link.
7. Install the nut holding the control arm ball joint stud and tighten to 44 ft. lbs. (59 Nm). Always use a new cotter pin on the ball joint stud.
8. Install the anti-roll bar and wheel. Lower the vehicle and check the alignment.

Lower Control Arms

REMOVAL & INSTALLATION

240

1. Raise and support the vehicle safely. Remove the front wheel.
2. Disconnect the stabilizer link at the control arm. Remove the ball joint.
3. Remove the control arm rear attachment plate, then remove the front retaining bolt.
4. Remove the control arm.

To install:

5. If the bushings are to be replaced, note that the right and left bushings are not interchangeable. The right side bushing should be turned so the small slots point horizontally when installed. Torque the retaining bolt to 55 ft. lbs. (75 Nm), the rear bushing to 4 ft. lbs. and the rear attachment bolts to 30 ft. lbs. (40 Nm).
6. Install the bracket onto the control arm and tighten it finger tight. Install the control arm and tighten the bolts a few turns. Install the stabilizer link and tighten the nuts loosely.
7. Install the ball joint and tighten the attaching bolts to 25–35 ft. lbs. (34–47 Nm). Tighten the stabilizer link.
8. Install the wheels and lower the vehicle. Roll the vehicle backward and forward while bouncing the front end. This normalizes the suspension.
9. Tighten the rear mount to 38–44 ft. lbs. (51–60 Nm); the front mount to 55 ft. lbs. (75 Nm).

Except 240

1. Raise and support the vehicle safely. Remove the front wheels.
2. Remove the ball joint nut. Disconnect the stabilizer link and strut bolt. Remove the front bushing.
3. Remove the ball joint. Unbolt the control arm at the crossmember and remove.

To install:

4. If the bushings are to be replaced, use an appropriate press and install the bushings from the front side of the arm.
5. Install the control arm over the end of the strut rod but do not tighten into place.
6. Install the ball joint and tighten to 44 ft. lbs. (60 Nm). Install the bushing, washer and bolt for the strut rod. Tighten to 70 ft. lbs. (94 Nm).
7. Install the stabilizer link and tighten to 63 ft. lbs. (85 Nm).
8. Install the wheels and lower the vehicle. Roll the vehicle backward and forward while bouncing the front end. This normalizes the suspension.
9. Tighten the control arm bolts to 63 ft. lbs. (85 Nm).

Stabilizer Bar

REMOVAL & INSTALLATION

1. Raise and support the vehicle safely. Remove the front wheels. Remove the splash guard, if equipped.
2. Remove the nut holding the upper sway bar link.
3. Remove the bolts for the retaining brackets and remove the sway bar.

To install:

4. Replace the link bushings if worn. Reconnect the lower link to the sway bar.
5. Install the sway bar and tighten

the retaining brackets. Install the links and tighten the upper link nut until 1.65 in. (42 mm) remains between the upper and lower surfaces of the washers.

6. Install the splash guard, if equipped. Install the wheels and lower the vehicle.

Front Wheel Bearings

REPLACEMENT AND ADJUSTMENT

1. Remove the hub cap and loosen the lug nuts a few turns.
2. Firmly apply the parking brake. Raise and safely support the vehicle. Support the lower control arms. Remove the wheel and tire assembly.
3. Remove the front caliper.
4. Pry off the grease cap from the hub. Remove the cotter pin and castle nut. Use a hub puller to pull off the hub. On the 760, remove the brake disc. If the inner bearing remains lodged on the stub axle, remove it with a puller.
5. Using a drift, remove the inner and outer bearing rings.

To install:

6. Thoroughly clean the hub, brake disc and grease cap.
7. Press in the new inner and outer bearing rings with a drift.
8. Press grease into both bearing with a bearing packer. If one is not available, pack the bearings with as much wheel bearing grease as possible by hand. Also coat the outsides of the bearings and the outer rings pressed into the hub. Fill the recess in the hub with grease up to the smallest diameter on the outer ring for the outer bearing. Place the inner bearing in position in the hub and press its seal in with a drift. The felt ring should be thoroughly coated with light engine oil.
9. Place the hub onto the stub axle. Install the outer bearing washer and castle nut.
10. Adjust the front wheel bearings by tightening the castle nut to 45 ft. lbs. (60 Nm) to seat the bearings. Then, back off the nut ⅓ of a turn counterclockwise. Torque the nut to 1 ft. lbs. (1.5 Nm). If the nut slot does not align with the hole in the stub axle, tighten the nut until the cotter pin may be installed. Make sure the wheel spins freely without any side play.
11. Fill the grease cap halfway with wheel bearing grease and install it on the hub.
12. Install the front caliper.
13. Install the wheel and tire assembly. Lower the vehicle. Tighten the lug nut to 70–100 ft. lbs. (95–135 Nm) and install the hub cap.

REAR SUSPENSION

Shock Absorbers

REMOVAL & INSTALLATION

1. Remove the hub cap and loosen the lug nuts a few turns. Raise and safely support the vehicle to unload the shock absorbers. Remove the wheel and tire assembly.

2. Remove the nuts and bolts which retain the shock absorber to its upper and lower attachments and remove the shock absorber. Make sure the spacing sleeve, inside the axle support arm for the lower attachment, is not misplaced.

3. The damping effect of the shock absorber may be tested by securing the lower attachment in a vise and extending and compressing it. A properly operating shock absorber should offer approximately 3 times as much resistance to extending the unit as compressing it. Replace the shock absorber if it does not function as above or if it fixed rubber bushings are damaged. Replace any leaking shock absorber.

To install:

4. Position the shock absorber to its upper and lower attachments. Make sure the spacing sleeve is installed inside the axle support (trailing) arm and is aligned with the lower attachment bolt hole.

5. Install the retaining nuts and bolts and torque to 63 ft. lbs. (85 Nm). On some vehicles, the shock fits inside the support arm.

6. Install the wheel and tire assembly. Lower the vehicle. Tighten the lug nuts to 70–100 ft. lbs. (95–135 Nm) and install the hub cap.

Coil Springs

REMOVAL & INSTALLATION

Except Multi-Link Suspension

1. Remove the hub cap and loosen the lug nuts a few turns. Raise and safely support the vehicle. Remove the wheel and tire assembly.
2. Place a hydraulic jack under the rear axle housing and raise the housing sufficiently to compress the spring. Loosen the nuts for the upper and lower spring attachments.

CAUTION

Due to the fact that the spring is compressed under several hundred pounds of pressure, when it is freed from its lower attachment, it will attempt to suddenly spring back to its extended position. It is therefore imperative that the axle housing be lowered with extreme care until the spring is fully extended. As an added safety measure, a chain may be attached to the lower spring coil and secured to the axle housing.

3. Disconnect the shock absorber at its upper attachment. Carefully lower the jack and axle housing until the spring is fully extended. Remove the spring.

To install:

4. Position the retaining bolt and inner washer, for the upper attachment, inside the spring and then, while holding the outer washer and rubber spacer to the upper body attachment, install the spring and inner washer to the upper attachment (sandwiching the rubber spacer) and tighten the retaining bolt.
5. Raise the jack and secure the bottom of the spring to its lower attachment with the washer and retaining bolt.
6. Connect the shock absorber to its upper attachment. Install the wheel and tire assembly.
7. Lower the vehicle. Tighten the lug nuts to 70–100 ft. lbs. (95–135 Nm) and install the hub cap.

Multi-Link Suspension

NOTE: To properly remove and install the rear coil springs the rear support arm assembly must be removed.

1. Raise and support the vehicle safely. Remove the rear wheels and support arm guards.
2. Remove the retaining bolts at the front and rear of the support arm. Separate the rear end of the support arm from the wheel bearing housing.
3. Place a jack with fixture 5972, or equivalent, under the support arm and clamp into place.
4. Remove the retaining bolt at the top of the damper and lower the support arm complete with the spring and damper.

To install:

5. Lift the assembly into place and tighten the upper damper bolt to 62 ft. lbs. (85 Nm).
6. Replace the mounting bolt and nut at the front of the support arm. Tighten the large nut to 51 ft. lbs. (70 Nm) plus 90 degrees of rotation. Tighten the other bolts to 35 ft. lbs. (48 Nm).
7. Tap the support arm in at the rear and tighten the bolt to 44 ft. lbs.

(60 Nm) plus 90 degrees rotation.
8. Replace the control arm guard and wheels. Lower the vehicle.

Rear Control Arm

REMOVAL & INSTALLATION

Except Multi-Link Suspension

1. Raise and support the vehicle safely. Remove the rear wheels.
2. Matchmark and disconnect the driveshaft.
3. Remove the rear axle housing assembly.
4. Unbolt the rear control arm from the axle housing.

To install:

5. Install the control arm on the axle housing and tighten the bolt finger tight.
6. Raise the assembly into postion and install the axle-to-frame bolts finger tight.
7. Install all other components and lower the vehicle. Tighten all control arm bolts to 85 ft. lbs. (115 Nm).

Multi-Link Suspension

1. Raise and support the vehicle safely. Remove the rear wheels.
2. Remove the brake caliper and tie it out of the way. Remove the bolt holding the rear support arm to the wheel housing and tap the support arm loose.
3. Remove the nut and bolt holding the lower control arm to the wheel bearing housing. Remove the track rod.
4. Remove the upper control arm from the wheel bearing housing. Note the number of shims between the upper control arm and the bearing housing.
5. Remove the nuts and bolts holding the upper control arm to the rear axle member. Remove the control arm from the vehicle.

To install:

6. Install the control arm and hand tighten the nuts and bolts attaching it to the rear axle. Install the spacers at the wheel bearing housing, position the arm and install the nut holding the arm to the housing.
7. Tighten the rear most nut at the axle support to 62 ft. lbs. (84 Nm). Tighten the front nut and bolt to 51 ft. lbs. (69 Nm) plus 60 degrees of rotation.
8. Pull the top of the wheel bearing housing outward and tighten upper control arm nut to 84 ft. lbs. (113 Nm). Pull the wheel bearing housing out and install the lower control arm. Do not tighten at this time.
9. Pull the wheel bearing housing inward and tighten the control arm nut to 37 ft. lbs. (50 Nm) plus 90 degrees of rotation.
10. Install the support arm and tighten to 44 ft. lbs. (59 Nm) plus 90 degrees of rotation. Install the track rod and tighten to 62 ft. lbs. (84 Nm).
11. Install the brake caliper and wheel. Lower the vehicle. perform a rear wheel alignment.

STEERING

Steering Wheel

CAUTION

On vehicles equipped with an air bag, the negative battery cable must be disconnected, before working on the system. Failure to do so may result in deployment of the air bag and possible personal injury.

REMOVAL & INSTALLATION

NOTE: The use of a knock-off type steering wheel puller or the use of a hammer may damage the collapsible column and is not recommended.

240

1. Disconnect the negative battery cable.
2. Remove the retaining screws for the upper half of the molded turn signal housing and lift off the housing.
3. Pry off the steering wheel impact pad.
4. Disconnect the horn plug contact.
5. Remove the steering wheel nut.
6. With the front wheels pointing straight-ahead and the steering wheel centered, install a steering wheel puller. Use a universal type puller, such as SVO 2263.

To install:

7. Make sure the front wheels are pointing straight ahead, then place the centered steering wheel on the column with the plug contact to the left. Install the nut and tighten to 20–30 ft. lbs. (27–40 Nm).
8. Connect the horn plug contact and install the impact pad.
9. Install the upper turn signal housing half.
10. Connect the negative battery cable and test the operation of the horn.

Except 240

1. Disconnect the negative battery cable.
2. Gently pry up the lower edge of the steering wheel center pad and remove it.
3. Unscrew the steering wheel center nut and remove the wheel using a suitable puller.
4. When installing, torque the center nut to 26 ft. lbs. (35 Nm).

Manual Steering Rack

REMOVAL & INSTALLATION

1. Disconnect the negative battery cable. Remove the lock bolt and nut from the column flange, at the steering gear. Bend apart the flange slightly with a suitable tool.
2. Raise and safely support the vehicle. Remove the front wheels.
3. Disconnect the steering rods from the steering arms, using a ball joint puller.
4. Remove the splash guard.
5. Disconnect the steering gear from the front axle member.
6. Disconnect the steering gear from the steering gear flange. Remove steering gear.

To install:

7. Install rubber spacers and plates for the steering gear attachment points.
8. Position the steering gear and guide the pinion shaft into the steering shaft flange. The recess on the pinion shaft should be aligned towards the lock bolt opening in the flange.
9. Attach the steering gear to the front axle member. Check that the U-bolts are aligned in the plate slots. Install flat washers and nuts.
10. Install the splash guard.
11. Connect the steering rods to the steering arms.
12. Install the front wheels and lower the vehicle.
13. Install the lock bolt for the steering shaft flange.

Power Steering Rack

REMOVAL & INSTALLATION

1. Disconnect the negative battery cable. Loosen the steering column shaft flange from the pinion shaft. Remove the lock bolt and bend apart the flange slightly.
2. Raise and safely support the vehicle. Remove the front wheels.
3. Disconnect the steering rods from the steering arms, with a ball joint puller.
4. Remove the splash guard.
5. Disconnect the hoses at the steering gear. Install protective plugs in the hose connections.
6. Remove the steering gear from the front axle member.
7. Remove the steering gear by pulling down until it is free from the steering shaft flange. On the 740 and 760

GLE, disconnect the lower steering shaft from the steering gear by removing the snaprings from the clamps. Loosen the upper clamp bolt, remove the lower clamp bolt and slide the joint up on the shaft. Then remove the unit on the left side of the vehicle.

To install:

8. Position the steering gear and attach the pinion shaft to the steering shaft flange.
9. Install right side U-bolt and bracket but do not tighten the nuts.
10. Install left side retaining bolts and tighten. Tighten the U-bolt nuts.
11. Connect the steering rods to the steering arms.
12. Install the lock bolt on the steering column flange.
13. Connect the return and pressure hoses to the steering gear.

Power Steering Pump

REMOVAL & INSTALLATION

1. Disconnect the negative battery cable. Remove all dirt and grease from around the suction line connections and from around the delivery line of the pump housing.
2. Using a container to catch any power steering fluid that might run out, disconnect the lines and plug them to prevent dirt from entering the system.
3. Remove the tensioning bolt and the attaching bolts.
4. Clear the pump free of the fan belt and lift it out.

To install:

5. If a new pump is to be used, the old brackets, fitting and pulley must be transferred from the old unit. The pulley may be removed with a puller and pressed on the pump shaft with a press tool. Under no circumstances should the pulley be hammered on, as this will damage the pump bearings.
6. To install, place the pump in position and loosely fit the attaching bolts. Connect the lines to the pump with new seals.
7. Place the fan belt onto the pulley and adjust the fan belt tension.
8. Tighten the tensioning bolt and the attaching bolts.
9. Fill the reservoir with Type A automatic transmission fluid and bleed the system.

SYSTEM BLEEDING

1. Fill the reservoir up to the edge with Automatic Transmission Fluid Type A. Raise and safely support the vehicle. Place the transmission in neutral and apply the parking brake.
2. Keeping a can of ATF Type A within easy reach, start the engine and fill the reservoir as the level drops.
3. When the reservoir level has stopped dropping, slowly turn the steering wheel from lock to lock several reservoir, if necessary.
4. Locate the bleeder screw on the power steering gear. Open the bleeder screw ½–1 turn and close it when oil starts flowing out.
5. Continue to turn the steering wheel slowly until the fluid in the reservoir is free of air bubbles.
6. Stop the engine and observe the oil level in the reservoir. If the oil level rises more than ¼ in. past the level mark, air still remains the system. Continue bleeding until the level rise is correct.
7. Lower the vehicle.

Tie Rod Ends

REMOVAL & INSTALLATION

The ball joints of the tie rod may be replaced individually. After the ball joint is disconnected, the locknut on the tie rod is loosened and the clamp bolt released. The ball joint is then screwed out of the tie rod, taking note of the number of turns. The new ball joint is screwed in the same number of turns, the clamp bolt and locknut tightened. The ball joint is locked to the rod with 55–65 ft. lbs. (75–88 Nm) of torque. The new ball joint is pressed into its connection and the ball stud not tightened to 23–27 ft. lbs. (31–37 Nm).

After reconditioning of the rods and joints, the wheel alignment must be adjusted.

BRAKES

For all brake system repair and service procedures not detailed below, please refer to "Brakes" in the Unit Repair section.

Master Cylinder

REMOVAL & INSTALLATION

1. Disconnect the negative battery cable. To prevent brake fluid form spilling onto and damaging the paint, place a protective cover over the fender apron and rags under the master cylinder.
2. Disconnect and plug the brake lines from the master cylinder.
3. Remove the nuts which retain the master cylinder and reservoir assembly to the vacuum booster and lift the assembly forward, being careful not to spill any fluid on the fender. Empty out and discard the brake fluid.

NOTE: Do not depress the brake pedal while the master cylinder is removed.

4. In order for the master cylinder to function properly when installed to the vacuum booster, the adjusting nut for the thrust rod of the booster must not prevent the primary piston of the master cylinder from returning to its resting position. A clearance (C) of 0.004–0.04 in. (0.1–1.0mm) is required between the thrust rod and primary piston with the master cylinder installed. The clearance may be adjusted by rotating the adjusting nut for the booster thrust rod in the required direction. To determine what the clearance (C) will be when the master cylinder and booster are connected, first measure the distance (A) between the face of the attaching flange and the center of the primary piston on the master cylinder, then measure the distance (B) that the thrust rod protrudes from the fixed surface of the booster, making sure the thrust rod is depressed fully with a partial vacuum existing in the booster. When measurement is subtracted from measurement (A), clearance (C) should be obtained. If not, adjust the length of the thrust rod by turning the adjusting screw to suit. After the final adjustment is obtained apply a few drops of locking compound, such as Loctite®, to the adjusting nut.
5. Position the master cylinder and reservoir assembly onto the studs for the booster and install the washers and nuts. Tighten the nuts to 17 ft. lbs. (23 Nm).
6. Remove the plugs and connect the brake lines.
7. Bleed the entire brake system.

Proportioning Valve

REMOVAL & INSTALLATION

1. Unscrew, disconnect and plug the brake pipe from the master cylinder, at the valve connection.
2. Slacken the connection for the flexible brake hose to the rear wheel a maximum of ¼ turn.
3. Remove the bolt(s) which retain the valve to the underbody and unscrew the valve from the rear brake hose.
4. To install the valve, place a new seal on it and screw the valve onto the rear brake hose and hand tighten. Secure the valve to the underbody with the retaining bolt(s).
5. Connect the brake pipe and tighten both connections, making sure there is no tension on the flexible rear hose.
6. Bleed the brake system.

Power Brake Booster

REMOVAL & INSTALLATION

1. Disconnect the negative battery cable.
2. Remove the master cylinder to power booster retaining bolts and position the master cylinder aside. Be careful not to damage the brake lines.
3. Disconnect the vacuum assist hose, from the booster.
4. From inside the vehicle, disconnect the brake pedal rod.
5. Remove the power booster retaining bolts. Remove the power booster from the vehicle.
6. Installation is the reverse of the removal procedure.

Brake Caliper

REMOVAL & INSTALLATION

1. Raise and support the vehicle safely. Remove the wheels.
2. Plug the reservoir cap vent hole. Label and disconnect the brake lines at the caliper. Plug the lines to prevent the entry of dirt.

NOTE: On rear brake calipers, disconnect the brake hose at the frame, remove the caliper and then disconnect the hose from the caliper.

3. Remove the 2 caliper attaching bolts and lif the unit off the retainer.

To install:

4. Check the mating surfaces of the caliper and retainer to ensure they are clean. Always use new retaining bolts and lightly coat them with locking compound.
5. Install the brake pads making sure that the caliper is parallel to the disc and that the disc can rotate freely. Position the caliper to its retainer over the disc and install the 2 retaining bolts. Tighten the bolts to 65–70 ft. lbs. (88–95 Nm) on the 240; 25 ft. lbs. (34 Nm) on all other vehicles.
6. Connect the brake lines to the caliper. Unplug the reservoir cap vent hole.
7. Install the wheels and lower the vehilce. Bleed the brake system.

Brake Rotor

REMOVAL & INSTALLATION

240

1. Raise and support the vehicle safely. Remove the wheels.
2. Remove the brake caliper but do not disconnect the the brake hose. Hang the caliper out of the way.
3. Loosen and remove the small retaining screws on the face of the disc. Remove the disc.

To install:

4. Ensure that the disc is sitting squarely on the mount and install the retaining screws.
5. Install the caliper and pads. Check that the disc can turn freely and that the caliper is seated.
6. Install the wheel and lower the vehicle.

Except 240

1. Raise and support the vehicle safely. Remove the wheels.
2. Remove the caliper and pads. Do not disconnect the brake hose, hang it out of the way.

NOTE: Late vehicle vehicles equipped with Multi-Link Suspension have a small stud threaded into the disc. While helping to locate the wheel, this stud also retains the disc to the hub. Do not loosen the large center hub to remove the disc.

3. Use a 10 mm allen wrench to disconnect the claiper bracket. Remove the center grease cap, the cotter pin and the castle nut. Remove the outer wheel bearing.
4. Remove the brake disc and the inner wheel bearing. It may be necessary to use a bearing puller, tool 2722, or equivalent.

NOTE: Vehicles equipped with ABS have the pulse wheel mounted within the disc. This toothed wheel must be removed and transferred to a new rotor if one is being installed. Use a universal gear puller and carefully lift off the pulse wheel. Use a bearing installation tool and a press to install the pulse wheel on the new disc.

To install:

5. Reassemble the wheel bearings and pack with grease. Install the inner bearing in the hub. Using a seal installation tool, install the new grease seal. Ensure that the face of the seal is even with the hub.
6. Install the brake disc, the outer wheel bearing and the castle nut. Rotate the disc while tightening the nut to 41 ft. lbs. (55 Nm). Loosen the nut ½ turn.
7. nstall the brake caliper. Use new attaching bolts and tighten them to 72 ft. lbs. (97 Nm). Install the brake pads.
8. Install the wheel and lower the vehicle.

Disc Brake Pads

REMOVAL & INSTALLATION

NOTE: The brake pads should be replaced when there is approximately 0.12 in. (3mm) of the lining left. The linings should under no circumstances be less than 0.06 in. (1.5mm).

Front Brakes

ATE TYPE

1. Raise the vehicle and support safely.
2. Mark the position of the wheels on the hubs and remove the front wheels.
3. Remove the retaining pins using a punch.
4. Remove the retaining spring.
5. Remove the brake pads and identify the pads, if they are to be reused.

To install:

6. To install, compress the pistons using a pair of pliers of special tool 2809 or equivalent.
7. Install the brake pads.
8. Install 1 retaining pin and a new retaining spring.
9. Install the other retaining pin.
10. Check the brake fluid level and pump the brake pedal.
11. Install the front wheel assemblies and lower the vehicle.

CAUTION

Check the brake pedal operation prior to driving the vehicle.

GIRLING TYPE

1. Raise the vehicle and support safely.
2. Mark the position of the wheels on the hubs and remove the front wheels.
3. Remove the spring clips.
4. Remove the retaining pins.
5. Remove the retaining springs.
6. Remove the brake pads and identify the pads if they are to be reused.

To install:

7. To install, compress the pistons using a pair of pliers of special tool 2809 or equivalent.
8. Install the brake pads.
9. Install the retaining springs.
10. Install the retaining pins.
11. Install the spring clips.
12. Check the brake fluid level and pump the brake pedal.
13. Install the front wheel assemblies and lower the vehicle.

CAUTION

Check the brake pedal operation prior to driving the vehicle.

Rear Brakes

ATE TYPE

1. Raise the vehicle and support safely.
2. Mark the position of the wheels on the hubs and remove the rear wheels.
3. Remove the retaining pins using a punch.
4. Remove the retaining spring.
5. Remove the brake pads and identify the pads if they are to be reused.

To install:

6. To install, compress the pistons using a pair of pliers of special tool 2809 or equivalent.
7. Install the brake pads.
8. Install 1 retaining pin and a new retaining spring.
9. Install the other retaining pin.
10. Check the brake fluid level and pump the brake pedal.
11. Install the rear wheel assemblies and lower the vehicle.

CAUTION

Check the brake pedal operation prior to driving the vehicle.

GIRLING TYPE

1. Raise the vehicle and support safely.
2. Mark the position of the wheels on the hubs and remove the rear wheels.
3. Remove the retaining spring.
4. Remove the spring clips.
5. Remove the retaining pins.
6. Remove the retaining springs.
7. Remove the brake pads and identify the pads if they are to be reused.
8. Remove any shims or damper washers fitted between the pads and the caliper pistons.

To install:

9. To install, compress the pistons using a pair of pliers of special tool 2809 or equivalent.
10. Install any shims or damper washers.
11. Install the brake pads.
12. Install the springs.
13. Install the retaining pins.
14. Install the spring clips.
15. Install the retaining spring.

NOTE: If damper washers have been fitted, make sure the large flat side faces the piston. A feeler gauge can be used to fit the washers.

16. Check the brake fluid level and pump the brake pedal.
17. Install the rear wheel assemblies and lower the vehicle.

CAUTION

Check the brake pedal operation prior to driving the vehicle.

Parking Brake Cable

ADJUSTMENT

1. Remove the rear ashtray, between the front seat backs, or the rear of the center console.
2. Tighten the parking brake cable adjusting screw so the brake is fully applied when pulled up 2–3 notches.
3. If one cable is stretched more than the other, they can be individually adjusted by removing the parking brake cover (2 screws) and turning the individual cable adjusting nut at the front of each yoke pivot.
4. Install the ashtray and parking brake cover, if equipped.

Brake System Bleeding

Whenever a spongy brake pedal indicates that there is air in the system, or when any part of the hydraulic system has been removed for service, the system must be bled. In addition, if the level in the master cylinder reservoir is allowed to drop below the minimum mark for too long a period of time, air may enter the system, necessitating bleeding.

If only one caliper is removed for servicing, it is usually necessary to bleed only that unit. If, however, the master cylinder, varning valve, or any of the main system lines are removed, the entire system must be bled.

Be careful not to spill any brake fluid onto the brake surfaces or the paint. When bleeding the entire system, the rear of the car should be raised higher than the front. Only use brake fluid bearing the designation DOT 4.

NOTE: The following procedure is acceptable for use on vehicles with and without ABS.

1. Check to make sure that floor mats are not obstructing pedal travel. Full pedal travel should be 6 in. (150 mm).
2. Clean the cap and top of the master cylinder reservoir, and make sure that the vent hole in the cap is open. Fill the reservoir to the maximum mark. Never allow the level to drop below the minimum mark during bleeding.
3. If only one brake caliper or line was removed, it will usually suffice to bleed only that wheel. Otherwise, prepare to bleed the entire system begining at the left front wheel.
4. Remove the protective cap for the cleeder and fit a $^5/_{16}$ line wrench on the nipple. Install a tight plastic hose onto the nipple and insert the other end of the hose into a glass bottle containing clean brake fluid. The hose must hang down below the surface of the fluid, or air will be sucked into the system when the brake pedal is released.
5. Open the bleeder screw a maximum of ½ turn. Have a helper slowly depress the pedal until it bottoms, pause a second, and then quickly release the pedal. This should be repeated until the fluid flowing into the bottle is completely free of air bubbles. During this procedure, check the master cylinder reservoir frequently. When completed, press the pedal to the bottom of its stroke and tighten the bleeder screw. Install the protective cap.

NOTE: On late vehicle vehicles, the front calipers are equipped with 2–3 bleeder screws. Attach one hose to each screw and submerge in brake fluid.

6. If the entire system is to be bled, use the same procedure for the remaining bleeder screws at the right front, left rear and right rear wheel. Follow this order specifically. If the pedal still feels spongy after bleeding the entire system, repeat the bleeding sequence.
7. Fill the reservoir to the maximum line. Turn the ignition **ON** but do not start the engine. Apply moderate force to the brake pedal. The pedal must travel no more than 2.4 in. (61 mm) without ABS; 2.17 in. (55 mm) with ABS. The brake warning light (and ABS warning light) must not be on.

Anti-Lock Brake System Service

SYSTEM PRECAUTIONS

- If the vehicle is equipped with an air bag system, always properly disable the system before commencing work on the ABS system.
- Certain components within the ABS system are not intended to be serviced or repaired individually. Only those components with removal and installation procedures should be serviced.
- Do not use rubber hoses or other parts not specifically specified for the ABS system. When using repair kits, replace all parts included in the kit. Partial or incorrect repair may lead to functional problems and require the replacement of other components.
- Lubricate rubber parts with clean, fresh brake fluid to ease assembly. Do not use lubricated shop air to clean parts; damage to rubber components may result.
- Use only brake fluid from an un-

opened container. Use of suspect or contaminated brake fluid can reduce system performance and/or durability.

• A clean repair area is essential. Perform repairs after components have been thoroughly cleaned. Do not allow ABS components to come into contact with any substance containing mineral oil; this includes used shop rags.

• The control unit is a microprocessor similar to other computer units in the vehicle. Insure that the ignition switch is **OFF** before removing or installing controller harnesses. Avoid static electricity discharge at or near the controller.

• Never disconnect any electrical connection with the ignition switch **ON** unless instructed to do so in a test.

• Avoid touching connector pins with fingers.

• Leave new components and modules in the shipping package until ready to install them.

• To avoid static discharge, always touch a vehicle ground after sliding across a vehicle seat or walking across carpeted or vinyl floors.

• If any arc welding is to be done on the vehicle, the ABS control unit should be disconnected before welding operations begin.

• Never allow welding cables to lie on, near or across any vehicle electrical wiring.

• f the vehicle is to be baked after paint repairs, disconnect and remove the control unit from the vehicle.

Hydraulic Modulator

REMOVAL & INSTALLATION

1. Disconnect the negative battery cable.
2. Remove the cover from the hydraulic modulator.
3. Remove both relays from the top of the unit; disconnect the wiring connector at the unit.
4. Disconnect the ground strap from the hydraulic modulator.
5. Place rags or towels around the unit to absorb brake fluid which will be spilled.
6. Clean the line connections thoroughly. Label each line using the letters marked on the hydraulic modulator (V, H, l, r, h).
7. Remove the brake lines from the modulator. Remove the bolt from the modulator support and push the support to the right. Remove the hydraulic modulator.

To install:

8. If a new modulator is being installed, remove the hexagonal plugs from the old unit and install on the new unit. Check that the rubber pads are not damaged; install the rubber pads onto the hexagonal plugs.
9. Install the modulator and tighten the support. If installing a new unit, remove the plugs from the brake line ports.
10. Reconnect the brake lines according to the labels made at removal. The lines must be in their exact original positions.
11. Remove the rags from the work area and dispose of them properly.
12. Install the relays on the hydraulic modulator.
13. Connect the wiring harness and the ground strap.
14. Install the cover on the unit.
15. Bleed the brake system. Vehicles with hydraulic clutches may require bleeding of the clutch system as well.
16. When bleeding is complete, test the brake system by having an assistant press hard on the brake pedal; keep it depressed for 30 seconds. During the 30 second period, check that no leakage occurs at the brake line connections on the hydraulic modulator.
17. Connect the negative battery cable. Test drive the vehicle, confirming system function.

Wheel Speed Sensors

REMOVAL & INSTALLATION

Front

1. Raise and safely support the front of the vehicle.
2. Remove the tire and wheel.
3. With the ignition switch **OFF**, disconnect the wheel speed sensor lead from the ABS harness. Remove any retaining bolts or clips holding the harness in place.

NOTE: Clips and retainers must be reinstalled in their exact original location. Take careful note of the position of each retainer and of the correct harness routing during removal.

4. Remove the single bolt holding the speed sensor.
5. Carefully remove the sensor straight out of its mount. Do not subject the sensor to shock or vibration; protect the tip of the sensor at all times.

To install:

6. Fit the sensor into position. Make certain the sensor sits flush against the mounting surface; it must not be crooked.
7. Install the retaining bolt.
8. Route the sensor cable correctly and install the harness clips and retainers. The cable must be in its original position and completely clear of moving components.
9. Connect the sensor cable to the ABS harness.
10. Install the wheel and tire.
11. Lower the vehicle to the ground.

Rear Without Multi-Link Suspension

1. Raise and safely support the vehicle.
2. Disconnect the sensor connector from the harness.
3. Remove the clips and retainers holding the sensor wire to the axle. Take note of the routing of the sensor wire; exact reinstallation is required.
4. Remove the retaining bolt holding the sensor to the differential housing.
5. Remove the sensor straight out of the housing; protect the tip from impact.
6. Reinstall in reverse order.

Rear With Multi-Link Suspension

1. Remove the spare tire and fold back the trunk carpet to expose the fuel filler pipe. Remove the cover(s) from the filler pipe.
2. Break the seal on the speed sensor harness connector and disconnect the sensor from the ABS harness.
3. Press the rubber grommet free of the bodywork and feed the sensor harness to the outside of the vehicle.
4. Raise and safely support the vehicle.
5. Install a jack with support fixture 5972 or its equivalent under the rear axle.
6. Remove the 2 bolts on each side of the rear axle assembly which hold the member to the body. Lower the rear axle slightly, but do not allow the drive shaft to press against the fuel tank.
7. Disconnect the right brake wire from its attachment.
8. Remove the sensor cable from the retaining clips and clamps. Take note of the routing of the cable; it must be reinstalled in its exact original position.
9. Clean the sensor area; remove the retaining bolts and remove the sensor. Protect the tip from damage or impact.

To install:

10. Apply a light coat of oil to the O-ring on the new sensor. Fit the sensor into place without damaging the tip. Tighten the retaining bolts to 7.5 ft. lbs. (10 Nm).
11. Install the sensor harness into the cable retainers, making certain it is routed correctly and out of the way of all moving parts.
12. Feed the cable through the body and secure the grommet.
13. Connect the right brake wire.

14. Raise the rear axle assembly and install the 4 bolts. Tighten each bolt to 52 ft. lbs. (70 Nm), then angle tighten each an additional 60 degrees.
15. Lower the vehicle to the ground.
16. Connect the sensor wiring harness to the ABS harness in the trunk and reseal the connector. Clamp the cable to the filler pipe.
17. Install the filler covers, reposition the carpet and install the spare tire.
18. Test drive the vehicle, confirming correct function of the ABS system and the dashboard warning lamp.

CHASSIS ELECTRICAL

Air Bag

DISARMING

Vehicles equipped with an air bag must be disarmed prior to performing service on the airbag or related systems. Disconnect the negative battery cable and keep the ignition in the **OFF** position before attempting to service these components. Failure to do so may result in deployment of the air bag and possible personal injury.

Heater Blower Motor

REMOVAL & INSTALLATION

240

WITHOUT AIR CONDITIONING

1. Disconnect the negative battery cable. Remove the heater unit.
2. Place the unit on its side with the control valve facing upward. Remove the spring clips and separate the housing halves.
3. Lift out the old fan motor and replace it with a new unit, making sure the support leg without the "foot" points to the output for the defroster channel.
4. Assemble the heater housing halves with new spring clips and seal the joint without clips with soft sealing compound.
5. Install the heater unit.

WITH AIR CONDITIONING

In order to remove the blower motor, both the right and left blower wheels must first be removed. The heater unit does not have to be removed.

1. Disconnect the negative battery cable.
2. Lift the carpet and remove the central unit side panels.
3. Remove the retaining screws for the control panel and move the panel as for back on the transmission tunnel as the electrical cables will permit.
4. Remove the attaching screws for the rear seat heater ducts and disconnect the ducts from the central unit.
5. Remove the instrument cluster.
6. Remove the glovebox by unscrewing the 4 attaching screws, removing the glovebox door stop and disconnecting the wires from the glovebox courtesy light. Remove the molded dashboard padding from under the glovebox.
7. Disconnect the vacuum hoses to the left and right defroster nozzle vacuum motors, then remove the nozzles and the left and right air ducts.
8. Remove the air hoses between the left and right inside air vents.
9. Remove the clamps on the central unit outer ends and remove the ends.
10. Pry off the locking retainer for the turbines (blower wheels) and remove both left and right blower wheels.
11. Position the heater control valve capillary tube aside.
12. Remove the left inner end (blower housing) from the central unit.
13. Unscrew the 3 retaining screws and remove the fan motor retainer.
14. Disconnect the plug contact from the fan motor control panel. Release the tabs of electric cables from the plug contact, remove the rubber grommet and pull the electrical cables down through the central unit right opening.
15. Remove the fan motor from the left opening.

To install:

16. Install the fan motor and connect all electrical connectors. Install the fan motor retainer. Install the left end (blower housing) on the central unit.
17. Install the heater control valve capillary tube. Install the left and right blower wheels and lock in place with the retainer. Install the central unit outer ends and clamps. Install the air hoses.
18. Install the vacuum hoses to the defroster nozzles and air ducts. Install the instrumen cluster. Install the rear seat heater ducts.
19. Install the control panel and install the retaining screws. Install the central unit side covers. Install the carpet.
20. Connect the negative battery cable. Test the operation of all functions.

Except 240

1. Disconnect the negative battery cable. Remove the panel under the glove compartment.
2. Unfasten the screws securing the fan motor and lower the motor. Disconnect the hose for air cooling on the motor and disconnect the wiring.
3. Remove the motor and fan.

To install:

4. Reconnect the wiring to the fan motor.
5. Spread a sealer around the mounting face of the fan mounting flange and install the fan motor.
6.Reconnect the hose for cooling and check fan operation. Reinstall the panel under the glove compartment.
7.Connect the negative batery cable.

Windshield Wiper Motor

REMOVAL & INSTALLATION

240

1. Disconnect the negative battery cable.
2. Disconnect the drive link from the wiper motor lever by unsnapping the locking tab under the dashboard.
3. Open the hood and disconnect the plug contact from the motor, located on the firewall.
4. Remove the 3 attaching screws and lift out the motor.
5. Reverse the above procedure to install, taking care to transfer the rubber seal, rubber damper and spacer sleeves to the new motor.

Except 240

1. Disconnect the negative battery cable. Remove the wiper arms.
2. Lift up the hood to its uppermost position by pushing the catch on the hood hinges.
3. Remove the plastic clips and screw securing the wiper mechanism cover plate. Remove the cover plate by lifting it upwards and forwards. Close the hood.
4. Remove the cover below the windshield.
5. Unbolt the motor from its mount. Disconnect the motor wires at the connectors.

To install:

6. Install the motor on its mount and connect the electrical wires. Install the cover below the windsheild.
7. With the hood closed, install the wiper mechanism cover plate. Open the hood and install the screws. Install the wipers and connect the negative battery cable.

Instrument Cluster

REMOVAL & INSTALLATION

1. Disconnect the negative battery

cable. Remove the soundproofing above the foot pedals.

2. Remove the 2 catches and screws holding the panel.

3. Press the instrument panel forwards. Remove the panel from the dash.

4. Disconnect the connectors and remove the instrument panel completely.

5. Installation is the reverse of removal.

cable. Remove the soundproofing above the foot pedals.

2. Remove the 2 catches and screws holding the panel.

3. Press the instrument panel forwards. Remove the panel from the dash.

4. Disconnect the connectors and remove the instrument panel completely.

5. Installation is the reverse of removal.

Radio

REMOVAL & INSTALLATION

1. Disconnect the negative battery cable.

2. Remove the radio control knobs by pulling them straight out.

3. Remove the control shaft retaining nuts.

4. Disconnect the speaker wires, power lean (either at the fuse box or the inline fuse connection), and the antenna cable from its jack on the radio.

5. Remove the hardware which attaches the radio to its mounting brackets, then slide the radio back and down from the dash.

To install:

6. Install the radio in the dash. Connect all the mounting brackets and hardware. Tighten bolts securely.

7. Connect all electrical connections. Ensure that the connections are free from corrosion.

8. Install the control shaft retaining nuts and tighten securely. Install the control knobs.

9. Connect the negative battery cable.

Combination Switch

REMOVAL & INSTALLATION

1. Disconnect the negative battery cable.

2. Remove the steering wheel.

3. Remove the upper and lower steering column casings.

4. Unscrew the switch/lever assembly.

5. Disconnect the wires from the switches.

6. Installation is the reverse of the removal procedure. Test the operation of the switch prior to installing the protective covers.

Ignition Lock/Switch

REMOVAL & INSTALLATION

240

1. Remove noise insulation panel and center side panel.

2. Disconnect the wires from the switch.

3. Pry out the switch with a suitable tool.

4. Install in reverse of removal.

Except 240

1. Remove the sound proofing under the instrument panel.

2. Disconnect the connector from the ignition switch.

3. Remove the upper steering column casing and the panel around the ignition switch.

4. Loosen the mounting screw for the switch.

5. Insert the key and turn it to the **START** position. Through the hole under the holder, press in the catch and remove the ignition switch.

To install:

6. Insert the key, turn and depress the locking tab. Remove the key.

7. Position the switch and release the locking tab by inserting the key. Tighten the mounting screw.

8. Install the steering column casing and the panel around the ignition switch. Connect the electrical connector. Install the sound proofing.

Stoplight Switch

ADJUSTMENT

The stoplight switch should be adjusted so that the taillights come on when the brake pedal is depressed by $^3/_8$–$^1/_2$ inch. Loosen the locknut and adjust the switch. Tighten the locknut and check the operation of the switch.

REMOVAL & INSTALLATION

1. Disconnect the negative battery cable.

2. Disconnect the electrical wiring on the switch.

3. Loosen the locknut and unscrew the switch.

4. Installation is the reverse of removal. Adjust the switch to specification.

Neutral Safety Switch

ADJUSTMENT

All vehicles have an adjustable switch, located under the shifter quadrant on the tunnel. To adjust:

1. Remove the shifter quadrant cover.

2. Place the shifter lever in **P**. Check that the round switch contact centers over the indicating line for **P**. If not, loosen the 2 switch mounting screws and align the switch.

3. Place the shifter lever in **N**. Repeat the check and adjust as necessary.

4. Finally check that the engine starts only in **P** or **N** and check that the back-up lights work only in **R**.

Fuses and Circuit Breakers

200 Series

Fuses are located at the left side kick panel. An in-line fuse serving the Jetronic fuel injection system is located on the left hand wheel housing by the ignition coil. An in-line fuse serving the EZ-116-K ignition system is located at the left hand wheel housing by the ignition coil.

940 AND 700 Series

Fuses are located under the center of the dash attached to the side of the relay box or at the left kick panel.

Fusible Links

Fusible links may be located at the battery, starter or alternator.

Unit Repair Sections

22 Tools and Equipment

In addition to the normal assortment of screwdrivers and pliers, automotive service work requires an investment in wrenches, sockets and the handles needed to drive them, and various measuring tools such as torque wrenches and feeler gauges.

The best approach to gathering the required equipment is to proceed slowly, buying high-quality tools as they are needed. An initial investment should be made in a set of quality wrenches, ranging in size from $\frac{1}{4}$ inch to one inch, if your car has standard bolts, or from 5mm to 19mm if your car has metric fasteners. High quality forged wrenches are available in three styles; open end, box end, and combination open/box end. The combination tools are generally the most desirable as a starter set; the wrenches shown in the illustration are of the combination type.

NOTE: Many later model American cars use both metric and standard nuts and bolts.

The other set of tools inevitably required is a ratchet handle and socket set. This set should have the same size range as your wrench set. The ratchet, extension, and flex drives fro the sockets are available in many sizes; it is advisable to choose a $\frac{3}{8}$ inch drive set initially. One break in the inch/metric sizing war is that metric-sized sockets sold in the U.S. have inch-sized drive ($\frac{1}{4}$, $\frac{3}{8}$, $\frac{1}{2}$, etc.). Sockets are available in six and twelve point versions; six point types are generally cheaper and are a good choice for a first set.

The choice of a drive handle for the sockets should be made with some care. If this is your first set, take the plunge and invest in a flexhead ratchet; it will get into many places otherwise accessible only through a long chain of universal joints, extensions and adapters. An alternative is a flex handle; such a tool is shown in the illustration, below the ratchet handle. In addition to the range of sockets mentioned, a rubber-lined spark plug socket should be purchased. Spark plugs have either a $\frac{13}{16}$ or a $\frac{5}{8}$ inch hex; get the correct socket for the plugs in your car.

The most important thing to consider when purchasing hand tools is quality. Don't be misled by the low cost of "bargain" tools. Forged wrenches, tempered screwdriver blades, and fine tooth ratchets are a much better investment than their less expensive counterparts. The skinned knuckles and frustration inflicted by poor quality tools make any job an unhappy core. Another consideration is that quality tools sold by reputable firms come with an on-the-spot replacement guarantee; if the tool breaks, you get a new one, no questions asked.

The tools needed for basic maintenance jobs, in addition to those just mentioned, include:

1. Jackstands, for support;
2. Oil filter wrench;
3. Oil filler spout or funnel;
4. Grease gun;
5. Battery hydrometer;
6. Battery post and clamp cleaner;
7. Container for draining oil;
8. Many rags for the inevitable spills.

In addition to these items there are several others which are not absolutely necessary, but handy to have around. These include a transmission funnel and filler tube, a drop (trouble) light on a long cord, an adjustable wrench (crescent wrench), and slip joint pliers.

A more extensive list of tools, suitable for tune-up work, can be drawn up easily. While the tools involved are slightly more sophisticated, they need not be outrageously expensive. For example, there are several inexpensive tach/dwell meters on the market that are every bit as good for the average mechanic as a $100.00 professional model. The key to these purchases is to make them with an eye towards adaptability and wide range. Using the tach/dwell meter example again, if the model you buy runs up to at least 1,500 rpm on the tachometer scale, the dwell meter works on 4, 6, or 8 cylinder engines, and the tachometer unit is adaptable to both conventional and electronic ignitions, it will serve for a long time on a variety of automobiles. A basic list of tune-up tools could include:

1. A tach/dwell meter;
2. Spark plug gauge and gapping tool;
3. Feeler blades;
4. Timing light.

In this list, the choice of a timing light should be made carefully. A light which works on the DC current supplied by the car battery is the best choice; it should have a xenon tube for brightness. If your car has electronic ignition, the light should have an inductive pick-up (the timing light illustrated has one of these), and since nearly all cars will have electronic ignition in the future, this feature is a reasonable one to look for.

In addition to these basic tools, there are several other tools and gauges you may find useful. These include:

1. A compression gauge. The screw-in type is slower to use, but eliminates the possibility of a faulty reading due to escaping pressure.
2. A manifold vacuum gauge.
3. A test light.
4. An induction meter. This is used

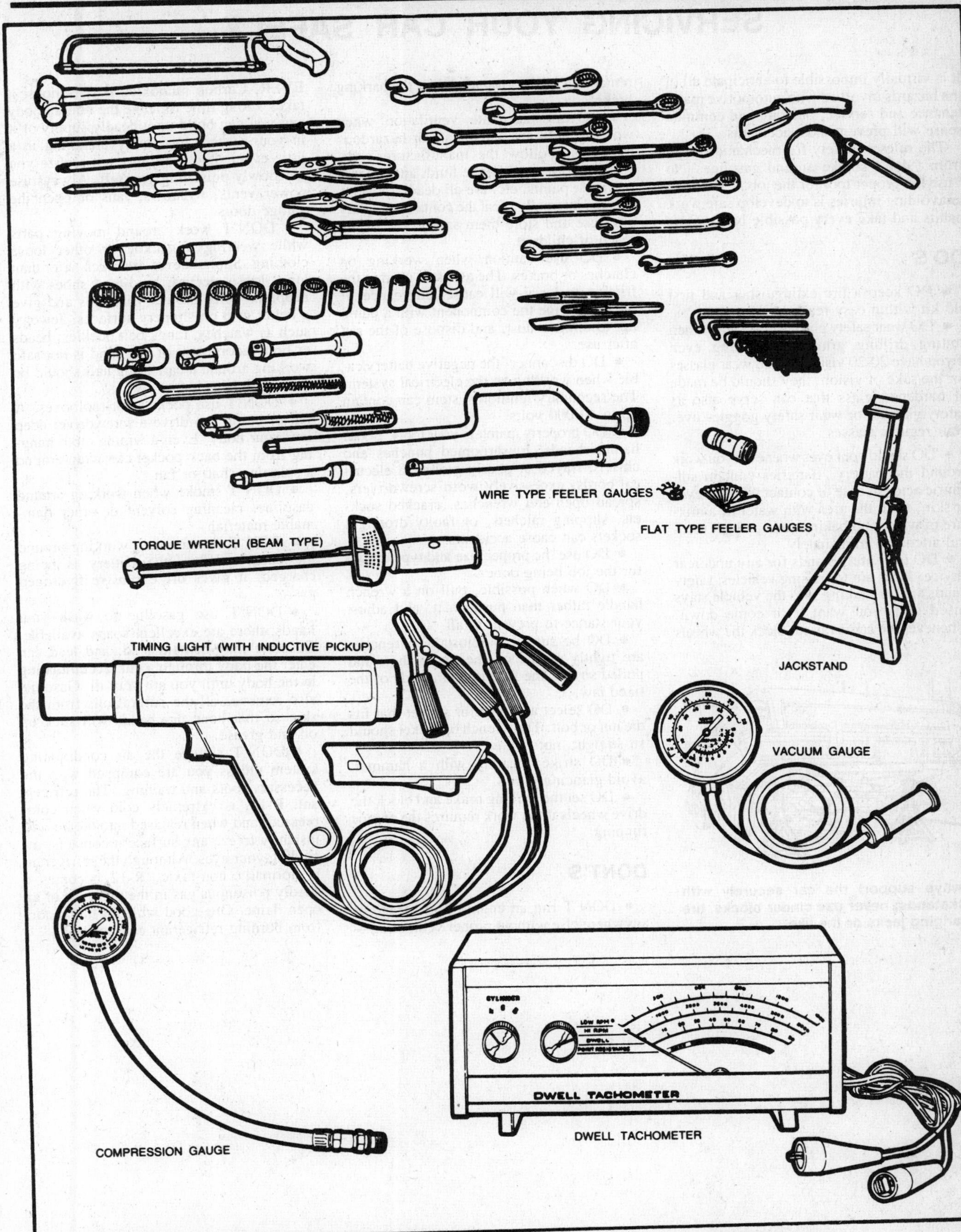

A basic tool collection will handle almost any automotive repair work

SERVICING YOUR CAR SAFELY

It is virtually impossible to anticipate all of the hazards involved with automotive maintenance and service, but care and common sense will prevent most accidents.

The rules of safety for mechanics range from "don't smoke around gasoline," to "use the proper tool for the job." The trick to avoiding injuries is to develop safe work habits and take every possible precaution.

DO'S

- DO keep a fire extinguisher and first aid kit within easy reach.
- DO wear safety glasses or goggles when cutting, drilling, grinding, or prying, even if you have 20-20 vision. If you wear glasses for the sake of vision, they should be made of hardened glass that can serve also as safety glasses, or wear safety goggles over your regular glasses.
- DO shield your eyes whenever you work around the battery. Batteries contain sulphuric acid. In case of contact with the eyes or skin, flush the area with water or a mixture of water and baking soda and get medical attention immediately.
- DO use safety stands for any undercar service. Jacks are for raising vehicles; safety stands are for making sure the vehicle stays raised until you want it to come down. Whenever the car is raised, block the wheels remaining on the ground and set the parking brake.

Always support the car securely with jackstands; never use cinder blocks, tire changing jacks or the like

- DO use adequate ventilation when working with any chemicals or hazardous materials. Follow the manufacturer's directions for usage. Brake fluid, anti-freeze, solvents, paints, etc. are all deadly poisons if taken internally. Seal the containers tightly after use and store them safely, out of the reach of children.
- DO use caution when working on clutches or brakes. The asbestos used in the friction material will cause lung cancer if inhaled. Wipe the component with a damp rag to remove dust, and dispose of the rag after use.
- DO disconnect the negative battery cable when working on the electrical system. The secondary ignition system can contain up to 40,000 volts.
- DO properly maintain your tools. Loose hammerheads, mushroomed punches and chisels, frayed or poorly grounded electrical cords, excessively worn screwdrivers, spread open-end wrenches, cracked sockets, slipping ratchets, or faulty droplight sockets can cause accidents.
- DO use the proper size and type of tool for the job being done.
- DO when possible, pull on a wrench handle rather than push on it, and adjust your stance to prevent a fall.
- DO be sure that adjustable wrenches are tightly closed on the nut or bolt and pulled so that the face is on the side of the fixed jaw.
- DO select a wrench or socket that fits the nut or bolt. The wrench or socket should sit straight, not cocked.
- DO strike squarely with a hammer; avoid glancing blows.
- DO set the parking brake and block the drive wheels if the work requires the engine running.

DONT'S

- DON'T run an engine in a garage or anywhere else without proper ventilation—EVER! Carbon monoxide is poisonous; it takes a long time to leave the human body and you can build up a deadly supply of it in your system by simply breathing in a little every day. You may not realize you are slowly poisoning yourself. Always use power vents, windows, fans or open the garage doors.
- DON'T work around moving parts while wearing a necktie or other loose clothing. Short sleeves are much safer than long, loose sleeves; hard-toed shoes with neoprene soles protect your toes and give a better grip on slippery surfaces. Jewelry such as watches, fancy belt buckles, beads or body adornment of any kind is not safe working around a car. Long hair should be hidden under a hat or cap.
- DON'T use pockets for toolboxes. A fall or bump can drive a screwdriver deep into your body. Even a wiping cloth hanging from the back pocket can wrap around a spinning shaft or fan.
- DON'T smoke when working around gasoline, cleaning solvent or other flammable material.
- DON'T smoke when working around the battery. When the battery is being charged, it gives off explosive hydrogen gas.
- DON'T use gasoline to wash your hands; there are excellent soaps available. Gasoline may contain lead, and lead can enter the body through a cut, accumulating in the body until you are very ill. Gasoline also removes all the natural oils from the skin so that bone dry hands will suck up oil and grease.
- DON'T service the air conditioning system unless you are equipped with the necessary tools and training. The refrigerant, R-12, is extremely cold when compressed, and when released into the air will instantly freeze any surface it contacts, including your eyes. Although the refrigerant is normally non-toxic, R-12 becomes a deadly poisonous gas in the presence of an open flame. One good whiff of the vapors from burning refrigerant can be fatal.

Basic Maintenance 23

INTRODUCTION

Routine maintenance is probably the most important part of automobile care and the easiest to neglect. A regular program aimed at monitoring essential systems ensures that all components are in good and safe working order, and can prevent small problems from developing into major headaches. Routine maintenance also pays big dividends in keeping major repair costs at a minimum and extending the life of the car.

The owner's manual that came with your car includes a maintenance schedule, indicating service intervals in numbers of months or thousand of miles. This schedule should always be followed. We have provided, in each section, a guide to service intervals based on an averaging of manufacturer's recommendations. In most cases, the suggested interval offered here will be close to that given by the manufacturer of your car, but the manufacturer's schedule should always take precedence.

We have divided the maintenance work to be done into three categories: Under Hood, Under Car, and Exterior. The checks in each section require only a few minutes of attention every few weeks; the services to be performed can be easily accomplished in a morning. The most important part of any maintenance program is regularity. The few minutes or occasional morning spent on these seemingly trivial tasks will forestall or eliminate major problems later.

UNDER HOOD

Automatic Transmission, Automatic Transaxle

The fluid level in the automatic transmission or transaxle should be checked every three months or 6000 miles. All automatic transmissions have a dipstick for fluid level checks.

1. Drive the car until it is at normal operating temperature. The level should not be checked immediately after the car has been driven for a long time at high speed, or in city traffic in hot weather; in those cases, the transmission should be given a half hour to cool down.
2. Stop the car, apply the parking brake, then shift slowly through all gear positions, ending in Park. Leave the engine running.
3. Remove the dipstick, wipe it clean, then reinsert it, pushing it fully home.
4. Pull the dipstick again and, holding it horizontally, read the fluid level.
5. Cautiously feel the end of the dipstick to determine the temperature. Most dipsticks are marked with both cool and hot levels. If the fluid is not up to the correct level, more will have to be added.
6. Fluid is added through the dipstick tube. You will probably need the aid of a spout or a long-necked funnel. Be sure that whatever you pour through is perfectly clean and dry. Fluid recommendations can be found in the owner's manual.

Add fluid slowly, and in small amounts, checking the level frequently between additions. Do not overfill, which will cause foaming, fluid loss, slippage, and possible transmission damage.

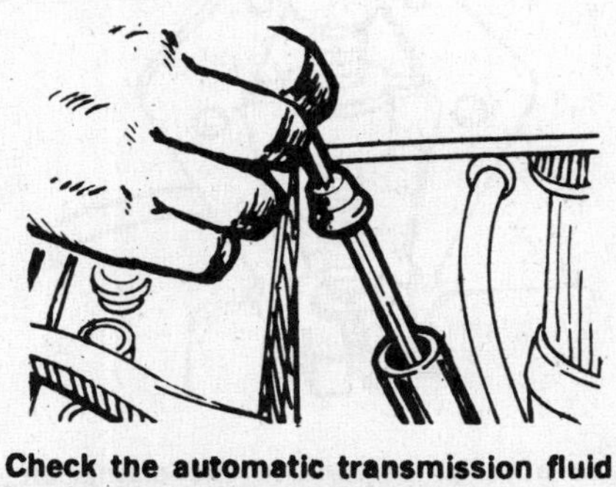

Check the automatic transmission fluid level with the dipstick provided

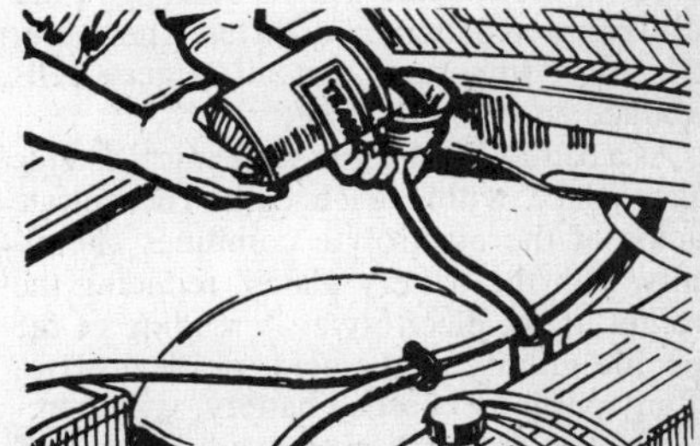

Fill the automatic transmission through the dipstick tube

Battery

FLUID LEVEL (EXCEPT "MAINTENANCE FREE" BATTERIES)

Check the battery electrolyte level at least once a month, or more often in hot weather or during periods of extended car operation. The level can be checked through the case on translucent polypropylene batteries; the cell caps must be removed on other models. The electrolyte level in each cell should be kept filled to the split ring inside, or the line marked on the outside of the case.

If the level is low, add only distilled water, or colorless, odorless drinking water, through

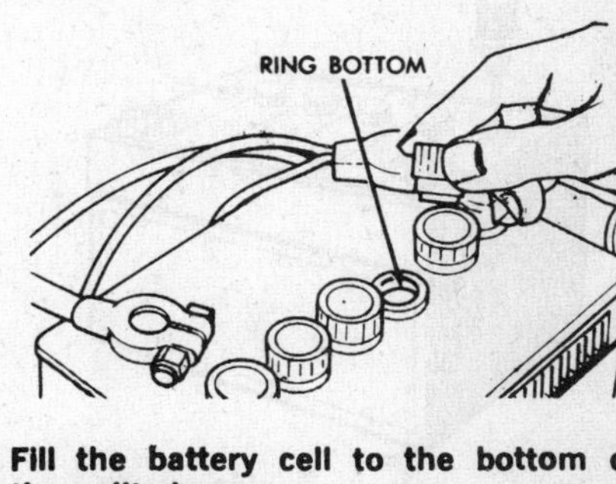

Fill the battery cell to the bottom of the split ring

the opening until the level is correct. Each cell is completely separate from the others, so each must be checked and filled individually.

If water is added in freezing weather, the car should be driven several miles to allow the water to mix with the electrolyte. Otherwise, the battery could freeze.

SPECIFIC GRAVITY (EXCEPT "MAINTENANCE FREE" BATTERIES)

While not technically exact, a practical measurement of the chemical condition of the battery is indicated by measuring the specific gravity of the acid (electrolyte) contained in each cell. The electrolyte in a fully charged battery is usually between 1.260 and 1.280 times as heavy as pure water at the same temperature (80°F). Variations in the specific gravity readings for a fully charged battery may differ. Therefore, it is most important that all battery cells produce an equal reading.

As a battery discharges, a chemical change takes place within each cell. The sulfate factor of the electrolyte combines chemically with the battery plates, reducing the weight of the electrolyte. A reading of the specific gravity of the acid, or electrolyte, of any partially charged battery, will therefore be less than that taken in a fully charged one.

The hydrometer is the instrument used for determining the specific gravity of liquids. The battery hydrometer is readily available from many sources, including local auto replacement parts stores. The following chart gives an indication of specific gravity value, related to battery charge condition. If, after charging, the specific gravity between any two cells varies more than 50 points (.050), the battery is probably bad.

Specific Gravity Reading	Charged Condition
1.260–1.280	Fully charged
1.230–1.250	Three-quarter charged
1.200–1.220	One-half charged
1.170–1.190	One-quarter charged
1.140–1.160	Just about flat
1.110–1.130	All the way down

CABLES AND CLAMPS

Once a year, the battery terminals and the cable clamps should be cleaned. Loosen the clamps and remove the cables, negative cable first. On batteries with posts on top, the use of a puller specially made for the purpose is recommended. These are inexpensive, and available in auto parts stores. Side terminal battery cables are secured with a bolt.

Clean the cable clamps and the battery terminal with a wire brush until all corrosion, grease, etc. is removed and the metal is shiny. It is especially important to clean the inside of the clamp thoroughly, since a small deposit of foreign material or oxidation there will prevent a sound electrical connection and inhibit either starting or charging. Special tools are available for cleaning these parts, one type for conventional batteries and another type for side terminal batteries.

Before installing the cables, loosen the battery hold-down clamp or strap, remove the battery and check the battery tray. Clear it of any debris, and check it for soundness. Rust should be wire brushed away, and the metal given a coat of anti-rust paint. Replace the battery and tighten the hold-down clamp or strap securely, but be careful not to overtighten, which will crack the battery case.

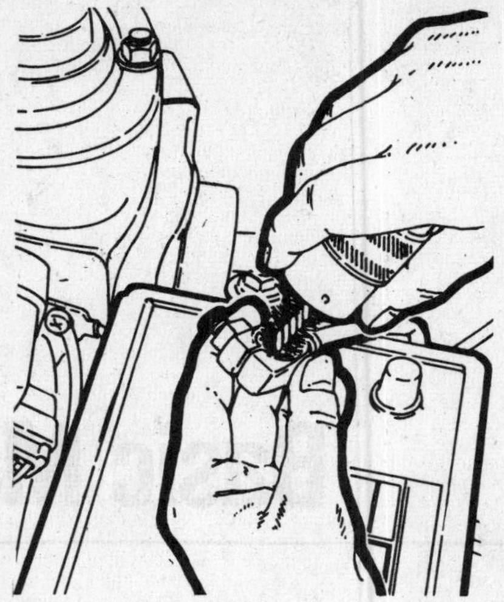

Clean the clamp with a wire brush

After the clamps and terminals are clean, reinstall the cables, negative cable last; do not hammer on the clamps to install. Tighten the clamps securely, but do not distort them. Give the clamps and terminals a thin external coat of grease after installation, to retard corrosion.

Check the cables at the same time that the terminals are cleaned. If the cable insulation is cracked or broken, or if the ends are frayed, the cable should be replaced with a new cable of the same length and gauge.

NOTE: Keep flame or sparks away from the battery; it gives off explosive hydrogen gas. Battery electrolyte contains sulphuric acid. If you should splash any on your skin or in your eyes, flush the affected area with plenty of clear water; if it lands in your eyes, get medical help immediately.

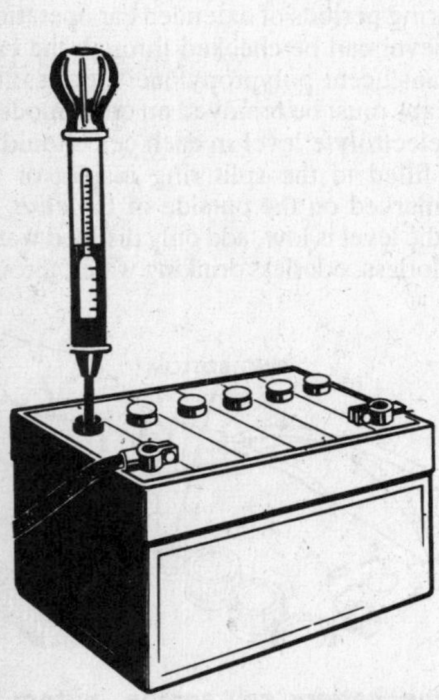

Testing battery specific gravity

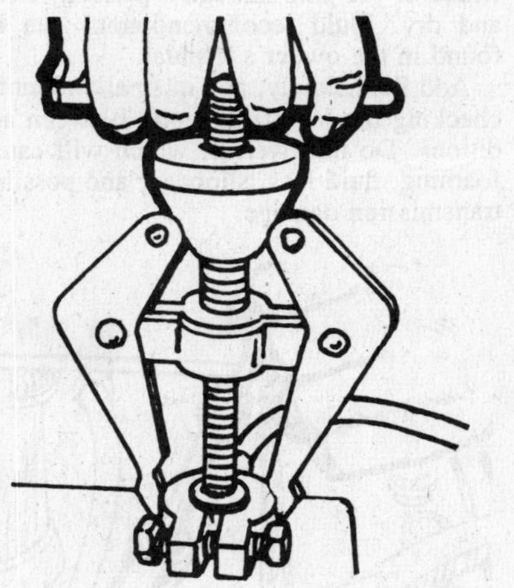

Use a puller to remove the clamp on post-type batteries

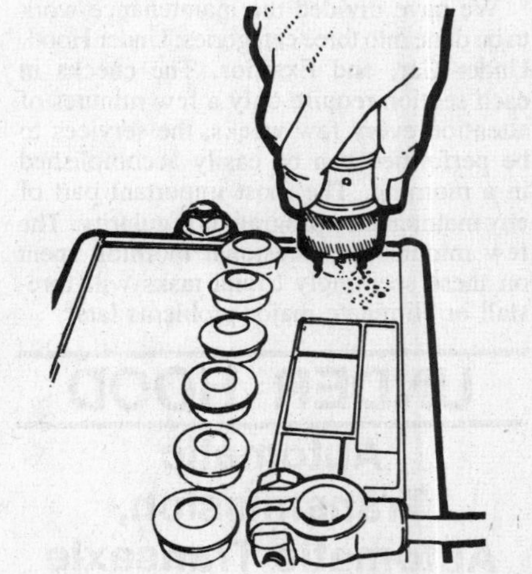

The posts are easily cleaned with a wire brush, or the battery post tool shown

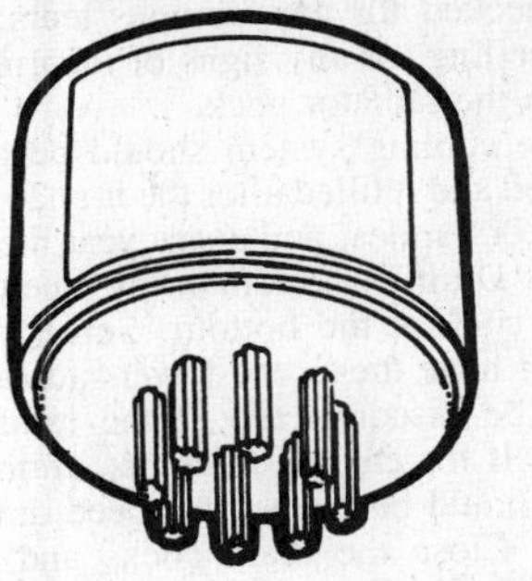

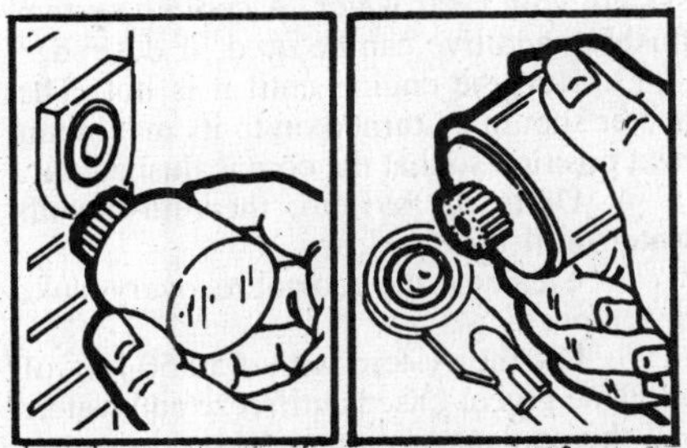

A special tool is required to clean the terminals and clamps on side terminal batteries

Brake Fluid

Once a month, the fluid level in the brake master cylinder should be checked.

1. Park the car on a level surface.
2. Clean off the master cylinder cover before removal. Some covers are retained by a bolt. Some of the newer master cylinders with plastic reservoirs have screw caps. Remove the cover, being careful not to drop or tear the rubber diaphragm which will probably be underneath. Be careful also not to drip any brake fluid on painted surfaces, as it eats paint.

NOTE: Brake fluid absorbs moisture from the air, which reduces effectiveness and will corrode brake parts once in the system. Never leave the master cylinder or the brake fluid container uncovered for any longer than necessary.

3. The fluid level should be about 1/4 inch below the lip of the master cylinder well.
4. If fluid addition is necessary, use only extra heavy duty disc brake fluid meeting DOT 3 or DOT 4 specifications. The fluid should be reasonably fresh, because brake fluid deteriorates with age.
5. Replace the cover, making sure that the diaphragm is correctly seated.

If the brake fluid is constantly low, the system should be checked for leaks. However, it is normal for the fluid level to fall gradually as the disc brake pads wear; expect the fluid level to drop about 1/8 inch for every 10,000 miles of wear.

Belt Tension

Every six months or 12,000 miles, check the water pump, alternator, power steering pump, air pump, and air conditioning compressor drive belts for proper tension. Also look for signs of wear, fraying, separation, glazing and so on, and replace the belts as required.

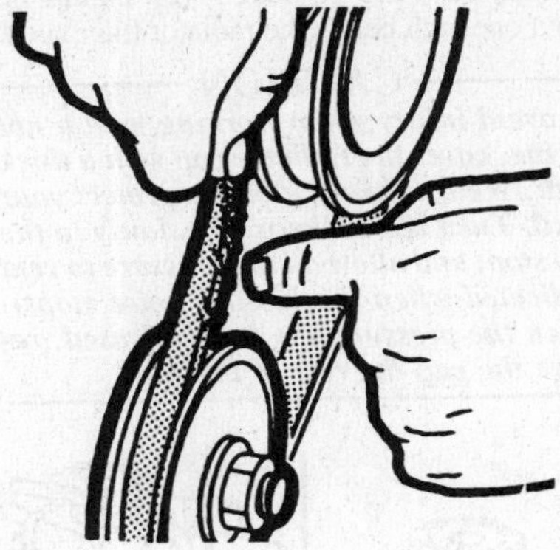

Check the belts for wear

Belt tension should be checked with a gauge made for the purpose. If a gauge is not available, tension can be checked with moderate thumb pressure applied to the belt at its longest span midway between pulleys. If the belt has a free span less than twelve inches, it should deflect approximately 1/8–1/4 inch. If the span is longer than twelve inches, deflection can range between 1/8 and 3/8 inches.

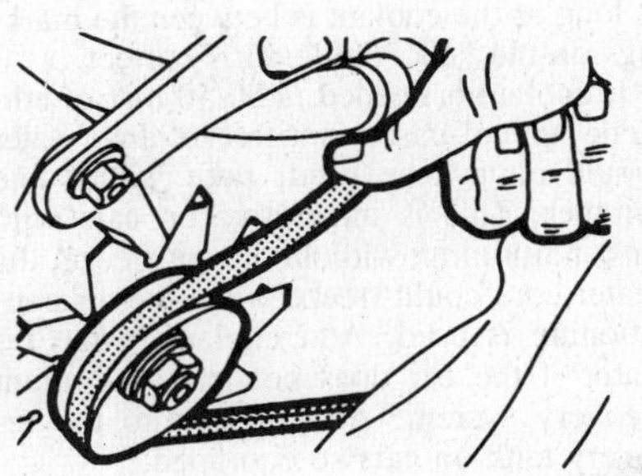

Check the belt tension at the middle of the longest span between pulleys

To adjust or replace belts:

1. Loosen the driven accessory's pivot and mounting bolts. Some air conditioning compressor belts are tensioned by an idler pulley; in this case, loosen the idler pulley and use a 1/2 in. drive ratchet in the square hole provided to lever the idler pulley up or down.
2. Move the accessory toward or away from the engine until the tension is correct. You can use a wooden hammer handle or broomstick as a lever, but do not use anything metallic.
3. Tighten the bolts and recheck the tension. If new bolts have been installed, run the engine for a few minutes, then recheck and readjust as necessary.

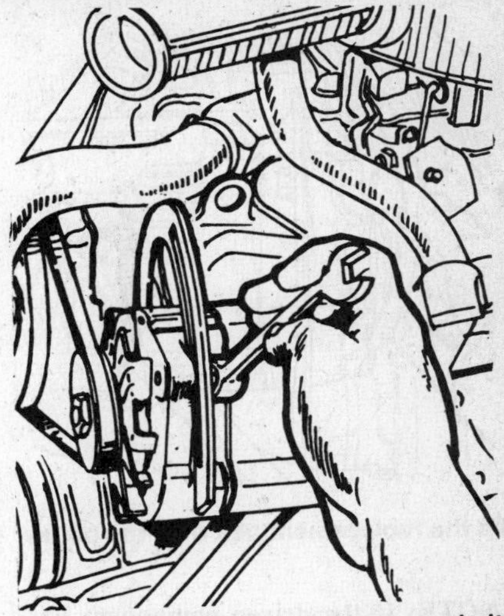

To either adjust or remove a belt, loosen the driven component's adjusting bolt

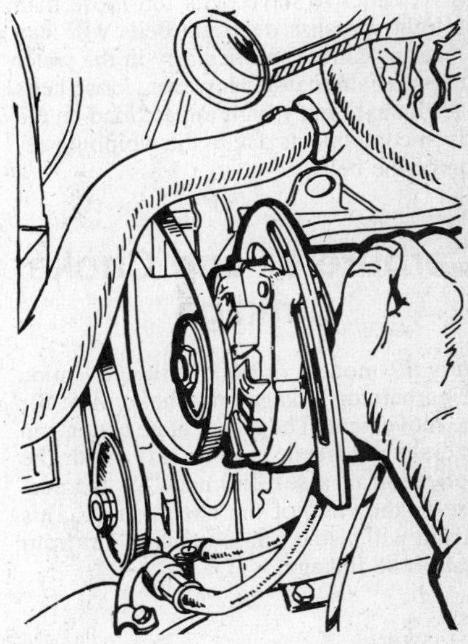

Push the component toward the engine to remove the belt

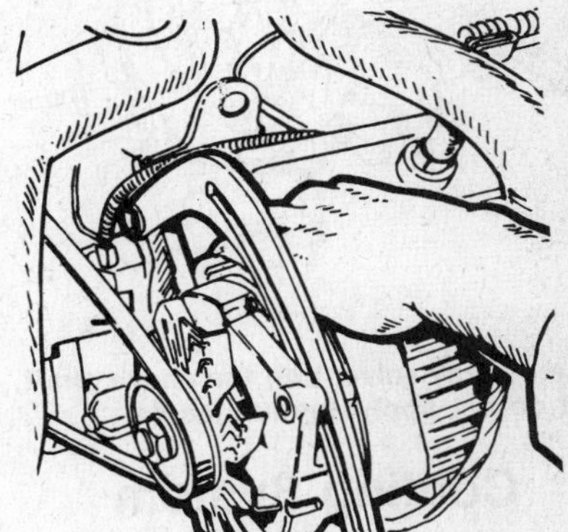

Pull outwards on the component to tension the belt, then tighten the bolts; recheck the belt tension after tightening

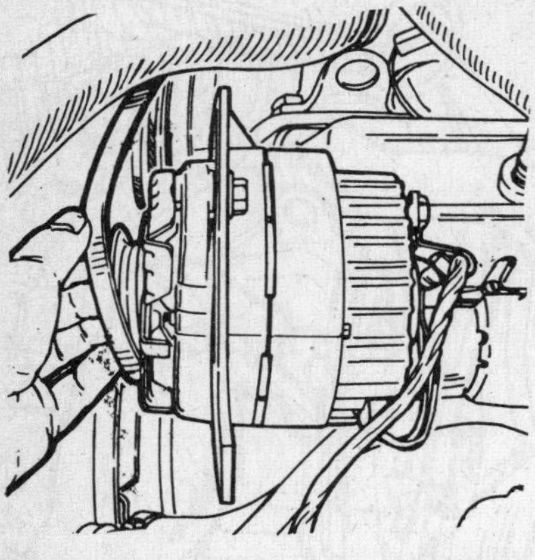

Slip the replacement belt over the pulley

NOTE: If the driven component has two drive belts, the belts should be replaced in pairs to maintain proper tension.

It is better to have belts too loose than too tight, because overtight belts will lead to bearing failure, particularly in the water pump and alternator. However, loose belts place an extremely high impact load on the driven components due to the whipping action of the belt.

Carburetor and Choke Linkage

Every 12 months or 6000 miles, examine the carburetor linkage and choke plate for free movement. The choke plate action can generally be freed, if necessary, with the application of a solvent made for the purpose to the ends of the choke shaft. This solvent will also clean grease and dirt from the throttle linkage.

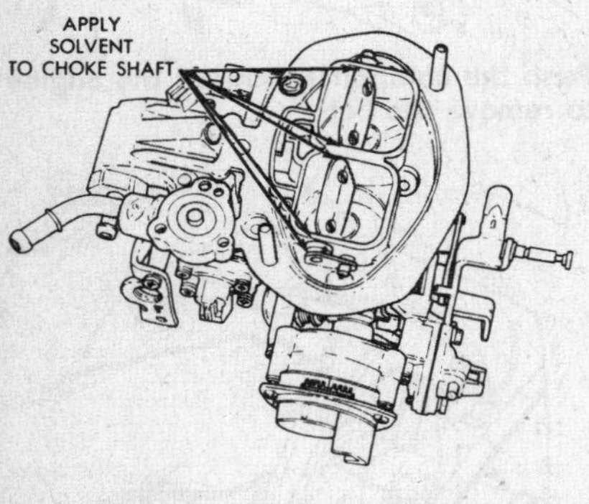

Use a spray solvent on the choke shaft, but do not apply any lubricants

Cooling System

Once a month, the engine coolant level should be checked. On cars without a coolant recovery system, this should only be done when the engine is cold. Remove the radiator cap; the coolant level should be about one inch below the radiator filler neck.

CAUTION

To avoid injury when working with a hot engine, cover the radiator cap with a thick cloth. Wear a heavy glove to protect your hand. Turn the radiator cap slowly to the first stop, and allow all the pressure to vent (indicated when the hissing noise stops). When the pressure has been released, remove the cap the rest of the way.

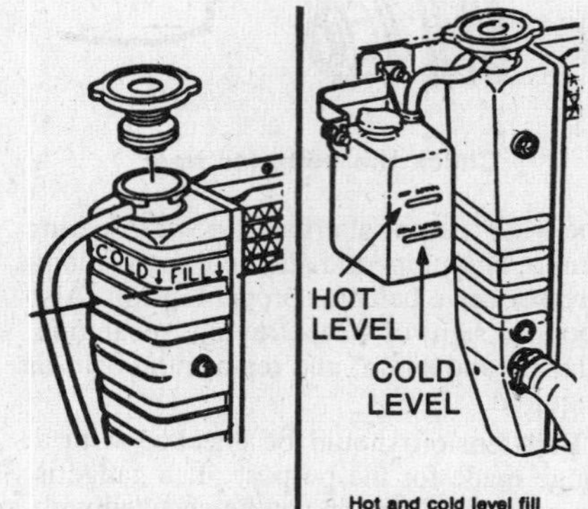

Fill level mark on crossflow radiator.

Hot and cold level fill marks, constant-full system.

Proper coolant level is about one inch below the radiator neck, or between the lines on the recovery tank

On cars with a coolant recovery tank, coolant should be visible within the tank; as long as the coolant is between the markings on the tank, the level is correct.

If coolant is needed, a 50/50 mix of ethylene glycol-based antifreeze and water should always be used, both winter and summer. This is imperative on cars with air conditioning; without the antifreeze, the heater core could freeze when the air conditioning is used. Add coolant to the radiator if the car does not have a coolant recovery system. Add coolant to the recovery tank on cars so equipped.

The radiator hoses and clamps and the radiator cap should be checked at the same time as the coolant level. Hoses which are brittle, cracked, or swollen should be replaced. Clamps should be checked for tightness (screwdriver tight only—do not allow the clamp to cut into the hose or crush the fitting). The radiator cap gasket should be checked for any obvious tears, cracks or swelling, or any signs of incorrect seating in the radiator neck.

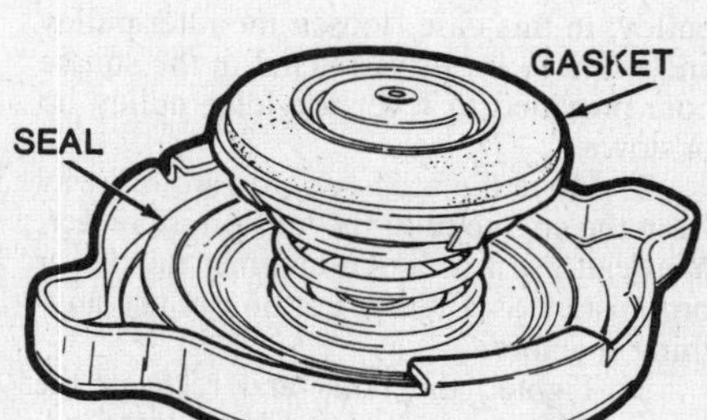

Check the radiator cap gasket and sealing surface

The cooling system should be drained, flushed and refilled after the first 24 months or 24,000 miles, and every year thereafter.

1. Drain the radiator by opening the drain cock at the bottom. Some radiators do not have these; the lower radiator hose must be disconnected at the radiator instead. If the engine block has drain plugs, they should be opened to speed draining.
2. Close the drain cocks and fill the system with clear water. A cooling system flushing additive can be used, if desired.
3. Run the engine until it is hot. The heater should be turned on to its maximum heat position so that the core is flushed out.
4. Drain the system, then flush with water until it runs clear.
5. Clean out the coolant recovery tank, if equipped.
6. Fill the system with a 50/50 mix of ethylene glycol-based antifreeze and water. Fill the coolant recovery tank midway between the marks with this mixture also.
7. Run the engine until it is hot, then let it cool and top up the radiator or coolant recovery tank as necessary with the antifreeze/water mixture.

Heat Riser

The heat riser is a thermostatically or vacuum operated valve in the exhaust manifold (not all cars have one). It closes when the engine is warming up, in order to preheat the incoming fuel/air mixture. If it sticks open, the result will be frequent stalling during warmup, especially in cold and damp weather. If it sticks shut, the result will be a rough idle after the engine is warm.

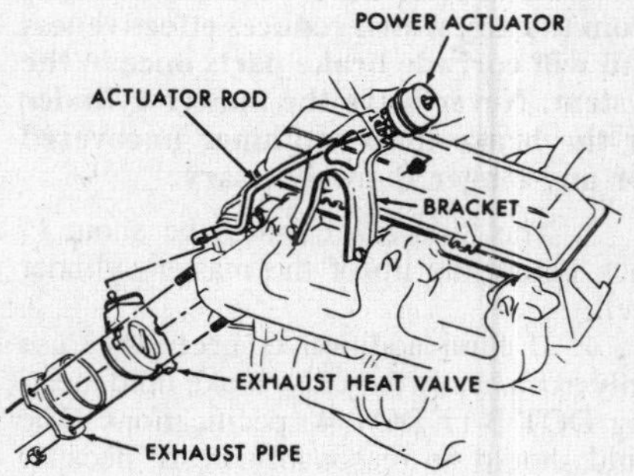

Exploded view of a vacuum-operated heat riser

The heat riser should move freely. It can be checked easily when the engine is cold by giving the counterweight on the valve shaft a twirl, or pulling the vacuum rod to open and shut the valve. If the valve is sticking or binding, a quick shot of solvent made for the purpose will free it up. This solvent should be applied every six months or 6000 miles to keep the valve free. If the valve is still stuck after application of the solvent, sometimes rapping the end of the

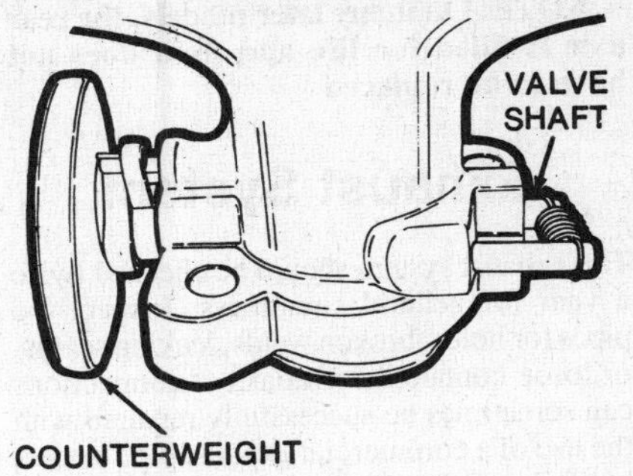

Thermostatically-operated heat control valve

shaft lightly with a hammer will break it loose. Otherwise, the components will have to be removed for further repairs.

Ignition Cables

The ignition system (points, condenser, rotor, spark plugs, etc.) receives regular attention in the form of a tune-up, and thus is not covered here. But one of the most commonly overlooked components is the ignition cable, or spark plug wire.

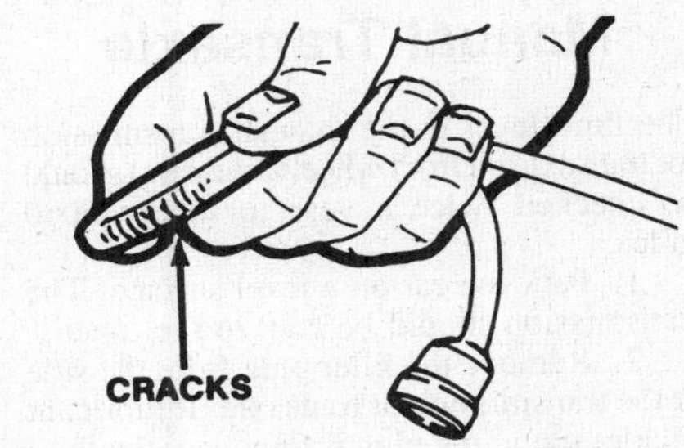

Inspect the ignition cables for cracks or breaks in the insulation

Although they rarely show any visible signs of deterioration, the ignition cables should be checked at every tune-up, and replaced at least every 50,000 miles. Cracking and embrittlement are of course obvious signs of wear, but most newer cables have silicone insulation and thus are not prone to display these conditions.

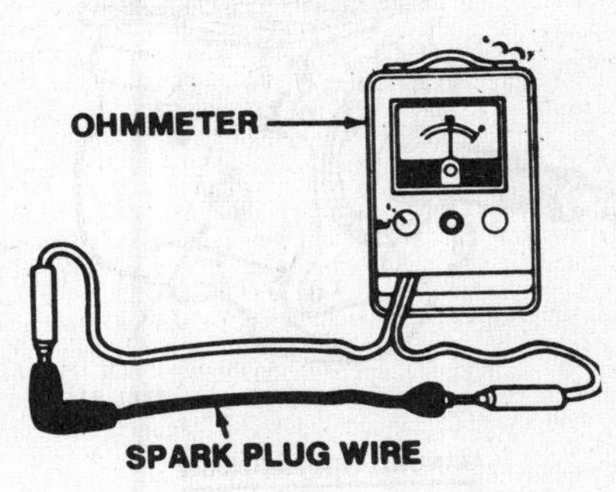

Test the ignition cables witl meter. Conventional ignitic should be removed from the cap, but electronic ignitl should first be tested throug

The most reliable way to check the cables is with an ohmmeter. On conventional ignitions, the resistance should be less than 7,000 ohms per foot (wire removed). On cars with electronic ignitions, it is generally recommended to leave the wire attached to the distributor cap; test with one lead from the ohmmeter connected to the corresponding terminal in the distributor cap, the other lead touched to the disconnected end of the cable at the spark plug. Then, if resistance seems close to the limit, remove the wire from the cap and retest. In general, the spark plug wires on electronic ignitions should be replaced if the total resistance is over 36,000 ohms.

Always replace the cables with new ones of the same type. Replace the wires one at a time, working from the longest to the shortest.

Oil Level

The engine oil should be checked on a regular basis, ideally at each fuel stop, or once a week. It is best to check when the engine is at operating temperature, but checking the level immediately after shutting off the engine will give a false reading, because all of the oil will not yet have drained back into the crankcase. The car should be parked on a level surface to obtain an accurate reading.

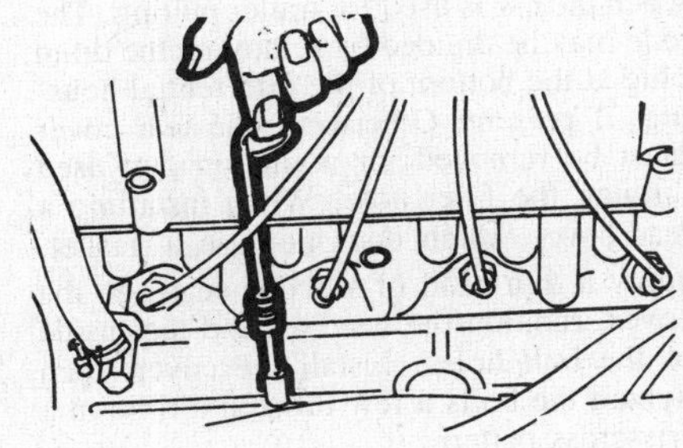

Check the engine oil level with the dipstick

1. Remove the oil dipstick. Wipe it clean, then replace it, seating it firmly.
2. Remove the dipstick again and hold it horizontally to prevent the oil from running. The level should be between the "Add" and "Full" marks on the dipstick. The dipstick may be marked "Add" and "Full," "Add" and "Safe," or may have lines scribed on it; in any case, the oil level should be above the lower marking.
3. If the oil is below the lower mark, enough oil should be added to the engine to raise the level to the upper mark. The markings are usually spaced so that one quart of oil will raise the level from the "Add" mark to the "Full" mark. Oil is added through the capped opening in the valve cover. Only oils labeled SF (gasoline engines) or CC (diesel engines) should be

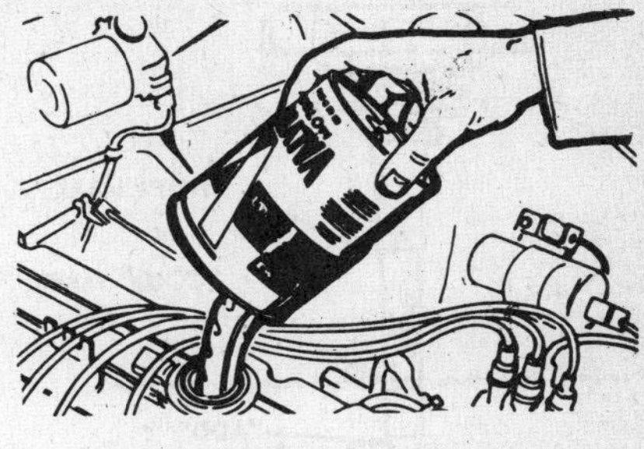

Add oil through the valve cover

used; select a viscosity that will be compatible with the temperatures expected until the next drain interval.

4. Replace the dipstick, then check the level again after any additions of oil. Be careful not to overfill, which will lead to leakage and seal damage.

Power Steering

The power steering fluid level is usually checked with a dipstick inserted into the pump reservoir. The dipstick may be attached to the reservoir cap, or inserted into a tube on the pump body. The level should be checked at every oil change. On some models, the power steering reservoir is translucent, allowing the level to be checked through the sides of the container without removing the cap. On others, the reservoir is a metal canister with a wingnut-attached

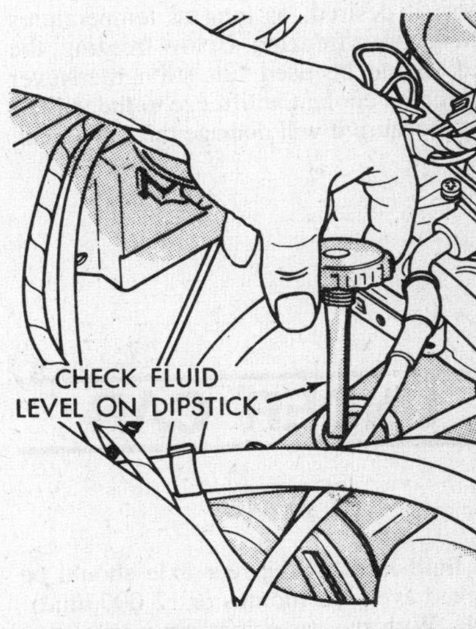

The power steering fluid level on many models is checked by means of a dipstick installed in the reservoir

cap. After the cap is removed, the level is checked with the scribed lines on the inside of the container.

On most models, the fluid level may be checked with the fluid either warm or cold. If checked with the fluid cold, the level will be slightly lower than with the fluid warm. If doubts arise about the specific procedures

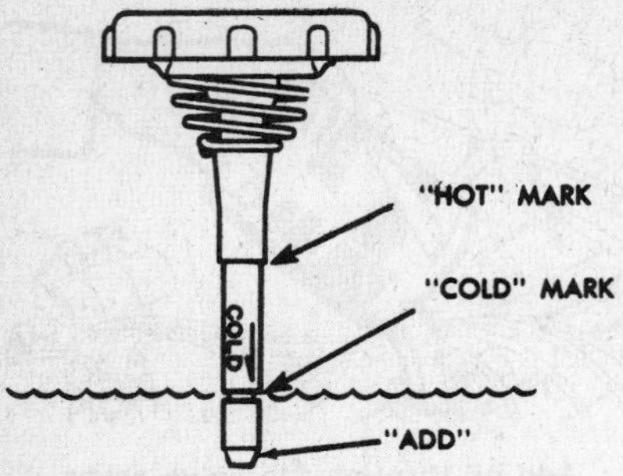

Typical power steering dipstick markings

for the car being checked, consult the owner's manual.

1. On all models, with the engine off, remove the dipstick, remove the cap or check the level through the side of the reservoir. If warm, the level should be between the "Hot" and "Cold" marks or even with the scribed line in the reservoir. If the fluid is cold, the level should be slightly lower.

2. If the level is low, add power steering fluid until the correct level is reached. Do not overfill the reservoir.

Windshield Washer Fluid

Check the fluid level in the windshield washer tank at every oil level check. The fluid can be mixed in a 50% solution with water, if desired, as long as temperatures remain above freezing. Below freezing, the fluid should be used full strength. Never add engine coolant antifreeze to the washer fluid, because it will damage the car's paint.

UNDER CAR

Axle

The fluid level in the rear axle should be checked every 12 months or 12,000 miles.

1. With the car parked on a level surface, remove the filler plug. The plug can be found either in the rear cover of the differential, or on the front of the pinion housing.

2. If lubricant trickles out when the plug is removed, the level is correct. If not, stick your finger in the hole (watch out for sharp threads); the fluid level should be even with edge of the filler hole.

3. If lubricant is needed, use SAE 80W-90 GL-5 gear oil (SAE 80W GL-5 in very cold climates) to fill standard axles. Limited

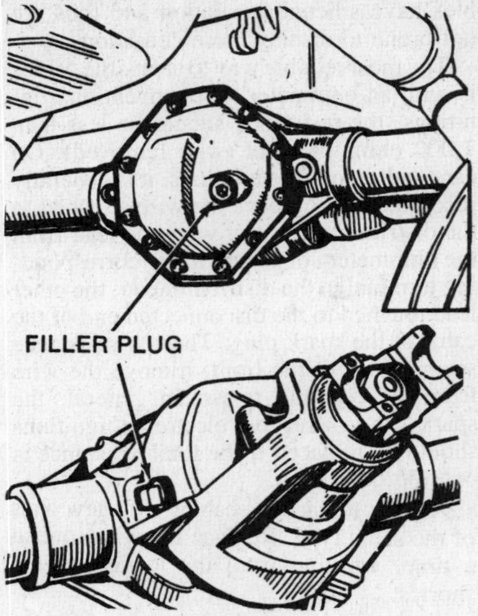

Rear axle filler plug locations

slip axles require a special lubricant, available in auto parts stores.

4. When the level is correct, install the plug and tighten until snug. Do not overtighten.

Standard axles should be drained and refilled with fresh lubricant every 15,000 miles when the car is used to pull a trailer. Limited slip axles should be drained and refilled at the first 7500 miles; the limited slip lubricant should be changed every 7500 miles when the car is used for trailer pulling. The axle may be drained by removing the drain plug at the bottom of the differential housing, if present. Otherwise, the rear cover must be removed, or a suction gun used through the filler hole. When installing a rear cover which does not use a gasket, apply a thin bead of silicone sealer to the cover, running the bead around the inside of the bolt holes. Install the cover, then tighten the bolts a few turns at a time in a crisscross pattern.

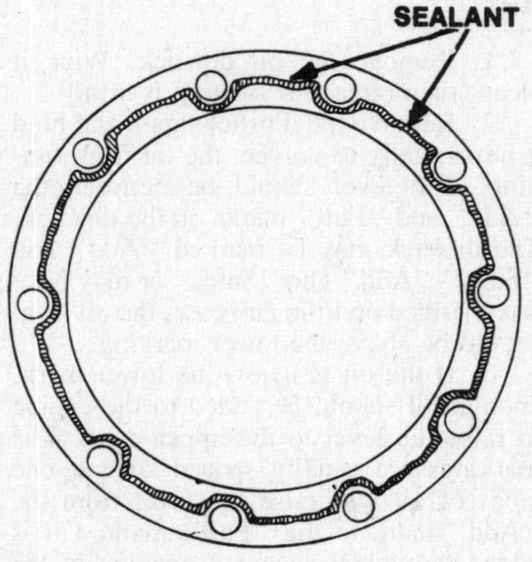

Apply a bead of silicone sealer to the rear cover if no gasket is used

NOTE: On many later models, the rear axle is filled for life and fluid does not have to be replaced.

Exhaust System

The exhaust system should be checked twice a year for general soundness. Inspect the pipes for holes, broken welds, leaking seams, or loose connections. Leaks at connections can sometimes be successfully repaired with the use of a commercial exhaust pipe sealer, but holes or breaks warrant replacement of the part. The exhaust pipe hangers and straps should be examined for any breaks or cracks; replace these as necessary. Some slight cracking of rubber hangers is normal, but deep cracks or cuts are cause for replacement.

CAUTION

Check the exhaust system only when it is cold. The temperature on an exhaust system using a catalytic converter can reach 1000°F after only a short period of engine operation.

Manual Transmission, Manual Transaxle

The fluid level in the manual transmission (or transaxle on front wheel drive cars) should be checked twice a year, or every 6000 miles.

1. Park the car on a level surface. The transmission should be cool to the touch.

2. Remove the filler plug from the side of the transmission or transaxle. If lubricant trickles out as the plug is removed, the fluid level is correct. If not, stick your finger into the hole (watch out for sharp threads); the lubricant should be right up to the edge of the filler hole.

3. If lubricant is needed, consult the owner's manual for the correct weight and type of fluid.

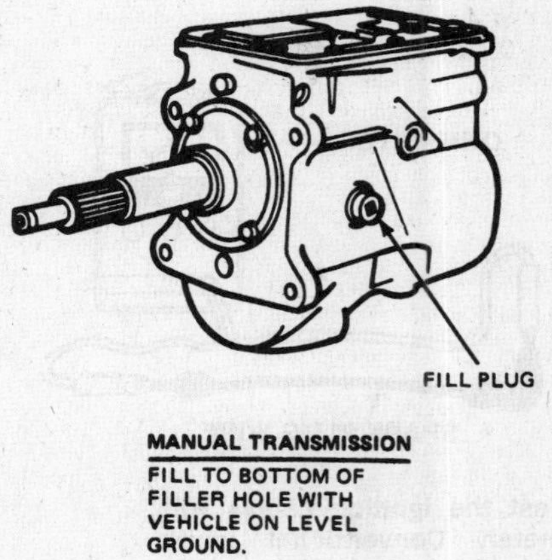

Typical manual transmission filler plug location

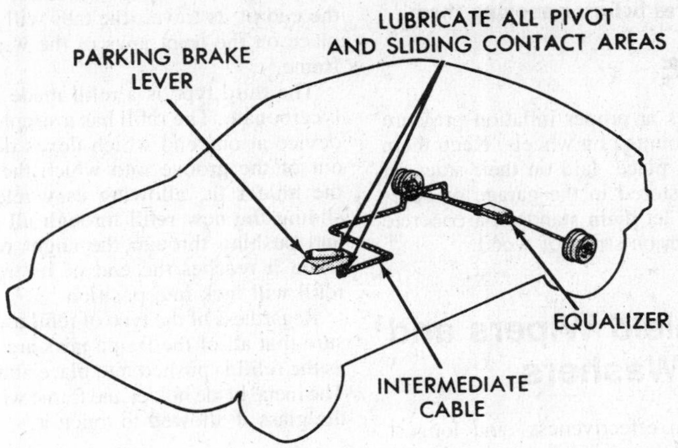

Lubricate the parking brake cable with white waterproof grease

NOTE: Some manual transmission/transaxle assemblies are filled with automatic transmission fluid rather than gear oil. Consult the owner's manual for lubricant information.

4. When the level is correct, install the filler plug and tighten until snug.

Parking Brake Linkage

The parking brake cable assembly should be inspected twice a year for fraying, kinks, and binding. A smooth white waterproof lubricant should be applied at the same time to all pivot points and areas in sliding contact.

Suspension Lubrication

Depending on the year of manufacture, there may be as many as twelve grease fittings on the suspension parts, or as few as two. Typical locations for grease nipples are on the ball joints, control arm pivot points, steering linkage, and the tie rod ends.

Lubricate these fittings with a small hand operated grease gun filled with EP chassis lubricant. Pump grease into the fitting slowly, until it begins to ooze out around the joint, or until the grease begins to expand the rubber boot around the fitting. Be extremely careful not to rupture any seals or boots, as this will lead to lubricant loss and contamination of the parts involved.

Occasionally, the grease nipples may become clogged with dirt or hardened grease. If so, unscrew them with a wrench of the proper size and clean them out with solvent. When reinstalled, they may be covered with plastic caps made for the purpose, or a piece of aluminum foil.

The chassis and suspension parts should be lubricated once a year, or every 7500 miles, whichever comes first.

Transfer Case

The transfer case on the four wheel drive Subaru shares a common lubricant supply with the transmission, therefore the transfer case lubricant supply does not have to be checked separately.

EXTERIOR

Drain Holes and Underbody

Most cars have drain holes spaced along the lower edge of the rocker panels and doors. These holes should be cleared of any debris or rust twice a year. A small screwdriver can be used to open plugged drain holes.

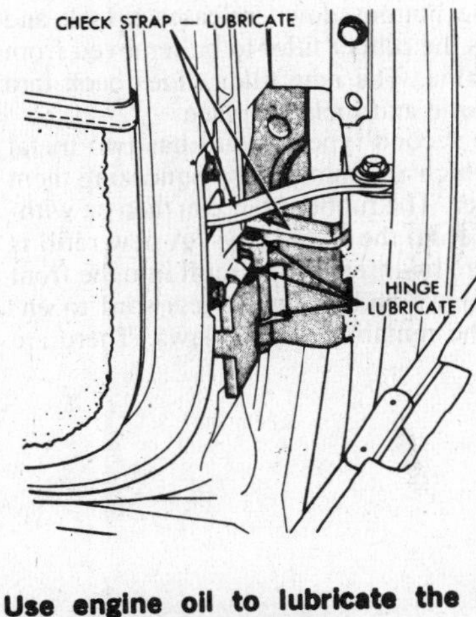

Use engine oil to lubricate the door, hood, and trunk hinges

Every spring, the underbody should be flushed with clear water to remove deposits of mud, road salt, and debris. It is advisable to loosen any packed-in sediment before flushing to assure a more thorough cleaning.

Hinges and Locks

Once a year, the door, hood, and trunk hinges, and all locks should be lubricated to ensure smooth operation. The hinge points should be lightly oiled. Lock cylinders may be easily lubricated with a shot of silicone spray directed into the keyhole. Silicone lubricant also works well on the door latch mechanisms, and keeps the door, trunk, and window weatherseals pliable when applied in a light film.

Tires

Tires should be checked weekly for proper air pressure. A chart, located either in the glove compartment or on the driver's or passenger's door, gives the recommended inflation pressures. Maximum fuel economy and tire life will result if the pressure is maintained at the highest figure given on the chart. Pressures should be checked before driving since pressure can increase as

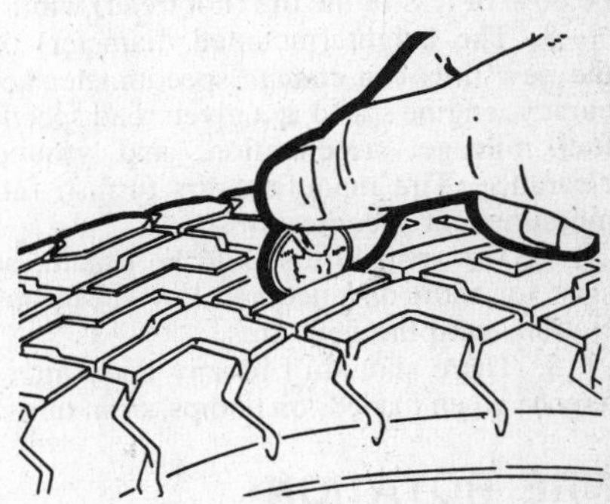

Tire tread depth can be checked with a penny. If the top of Lincoln's head is visible, the tires are due for replacement

much as six pounds per square inch (psi) due to heat buildup. It is a good idea to have your own accurate pressure gauge, because not all gauges on service station air pumps can be trusted. When checking pressures, do not neglect the spare tire. Note that some spare tires require pressures considerably higher than those used in the other tires.

While you are about the task of checking air pressure, inspect the tire treads for cuts, bruises and other damage. Check the air valves to be sure that they are tight. Replace any missing valve caps.

Check the tires for uneven wear that might indicate the need for front end alignment or tire rotation. Tires should be replaced when a tread wear indicator appears as a solid band across the tread.

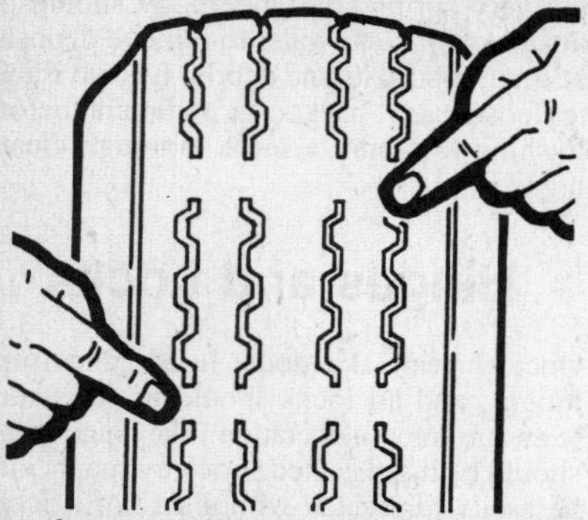

Tread wear indicators will appear as a band across the tire when the tread has worn out.

When buying new tires, give some thought to the following points, especially if you are considering a switch to larger tires or a different profile series:

1. All four tires must be of the same construction type. This rule cannot be violated. Radial, bias, and bias-belted tires must not be mixed.

2. The wheels should be the correct width for the tire. Tire dealers have charts of tire and rim compatibility. A mismatch will cause sloppy handling and rapid tire wear. The tread width should match the rim width (inside bead to inside bead) within an inch. For radial tires, the rim width should be 80% or less of the tire (not tread) width.

3. The height (mounted diameter) of the new tires can change speedometer accuracy, engine speed at a given road speed, fuel mileage, acceleration, and ground clearance. Tire manufacturers furnish full measurement specifications.

4. The spare tire should be usable, at least for short distance and low speed operation, with the new tires.

5. There shouldn't be any body interference when loaded, on bumps, or in turns.

TIRE ROTATION

Tire rotation is recommended every 6000 miles or so, to obtain maximum tire wear. The pattern you use depends on whether or not your car has a usable spare. Radial tires should not be cross-switched (from one side of the car to the other); they last longer if their direction of rotation is not changed. Snow tires sometimes have directional arrows molded onto the side of their carcass; the arrow shows the direction of rotation. They will wear very rapidly if the rotation is reversed. Studded tires will lose their studs if their rotational direction is reversed.

NOTE: Mark the wheel position or direction of rotation on radial tires or studded snow tires before removing them.

STORAGE

Store the tires at proper inflation pressure if they are mounted on wheels. Keep them in a cool dry place, laid on their sides. If the tires are stored in the garage or basement, do not let them stand on a concrete floor; set them on strips of wood.

Windshield Wipers and Washers

For maximum effectiveness, and longest element life, the windshield and wiper blades should be kept clean. Dirt, tree sap, road tar and so on will cause streaking, smearing and blade deterioration if left on the glass. It is advisable to wash the windshield carefully with a commercial glass cleaner at least once a month. Wipe off the rubber blades with the wet rag afterwards. For access to the blades on wiper systems which park below the hood line, turn the ignition key to "On" and run the wipers to the center of the windshield. Shut the wipers off with the ignition key, not the wiper switch. Do not attempt to move the wipers by hand; damage to the motor and drive mechanism will result.

If the blades are found to be cracked, broken or torn, they should be replaced immediately. Replacement intervals will vary with usage, although ozone deterioration usually limits blade life to about one year. If the wiper pattern is smeared or streaked, or if the blade chatters across the glass, the elements should be replaced. It is easiest and most sensible to replace the elements in pairs.

There are basically three different types of refills, which differ in their method of replacement. One type has two release buttons, approximately one-third of the way up from the ends of the blade frame. Pushing the buttons down releases a lock and allows the rubber filler to be removed from the frame. The new filler slides back into the frame and locks in place.

The second type of refill has two metal tabs which are unlocked by squeezing them together. The rubber filler can then be withdrawn from the frame jaws. A new refill is installed by inserting the refill into the front frame jaws and sliding it rearward to engage the remaining frame jaws. There are usually four jaws; be certain when installing that the refill is engaged in all of them. At the end of its travel, the tabs will lock into place on the front jaws of the wiper blade frame.

The third type is a refill made from polycarbonate. The refill has a simple locking device at one end which flexes downward out of the groove into which the jaws of the holder fit, allowing easy release. By sliding the new refill through all the jaws and pushing through the slight resistance when it reaches the end of its travel, the refill will lock into position.

Regardless of the type of refill used, make sure that all of the frame jaws are engaged as the refill is pushed into place and locked. The metal blade holder and frame will scratch the glass if allowed to touch it.

WASHER NOZZLE ADJUSTMENT

Centered Single Post–Non-Adjustable Nozzles

This type is usually located on the rear center of the hood panel, directly in front of the windshield. By loosening the body retaining nut from under the hood, the nozzle body can be turned to provide the best spray discharge to cover the windshield. Tighten the retaining nut while holding the nozzle in position.

Centered Single Post–Adjustable Nozzles

This nozzle is adjusted with a wrench, screwdriver, or pliers. If the nozzle has no gripping area, the adjustment is made by inserting a stiff wire into the nozzle opening and moving the nozzle in the direction desired. When using the wire as an adjuster tool, do not force the nozzle; the wire can be broken within the nozzle opening.

Individual Nozzles

A tab is usually fastened to the nozzle stem to assist in turning the nozzle in the desired direction. If a tab is not present, use a pair of pliers to gently move the nozzle.

Wiper Arm Nozzles

No adjustment is necessary on this type of nozzle, because the opening is centered on the wiper arm and moves along with the arm.

Air Conditioning Service 24

AIR CONDITIONING SYSTEMS

Automotive air conditioning systems are basic in design and operation, but many different components are used by the vehicle manufacturers to operate and control the systems to their specifications.

Basic System

The basic air conditioning system utilizes the compressor, condenser, evaporator, receiver-drier, expansion valve and a thermostatic or ambient type switch to control evaporator freeze-up. The controls are manually operated and the unit is basic in design. This system is usually installed as an add-on or after-market unit. A sight glass may be used in the system.

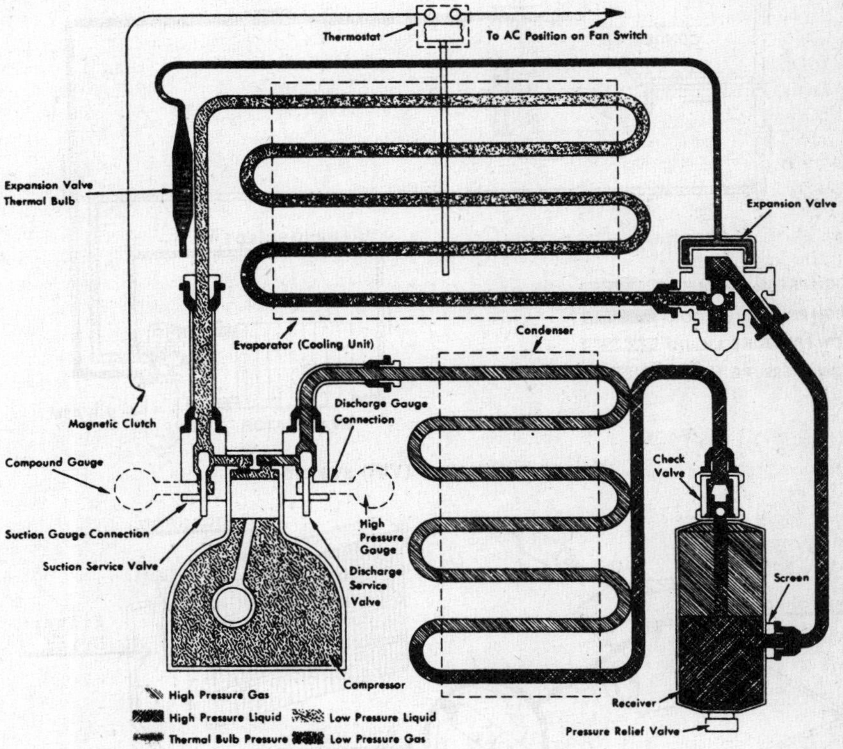

Basic air conditioning system

P.O.A. System

The P.O.A. (pilot operated absolute) suction throttling valve system contains the compressor condenser, evaporator, receiver-drier, expansion valve and a suction throttling valve. The suction throttling valve is used to keep the refrigerant gas in the evaporator at a pressure which will not allow the temperature of the evaporator core surface to go below 32 degrees F., thus preventing evaporator freeze-up. For the system to operate effectively, an equalizer line is connected between the suction side of the suction throttling valve and the expansion valve diaphragm. This modifies the operation of the expansion valve which now is controlled by the evaporator outlet temperature and compression suction pressure.

When a crank type compressor is used with the P.O.A. system, an accumulator is placed between the evaporator and the compressor. The accumulator operates as its name implies, accumulating any liquid refrigerant that may have passed from the evaporator and to prevent its moving to the compressor as a liquid, which may, in its form, cause internal compressor damage. A sight glass is normally used in this system.

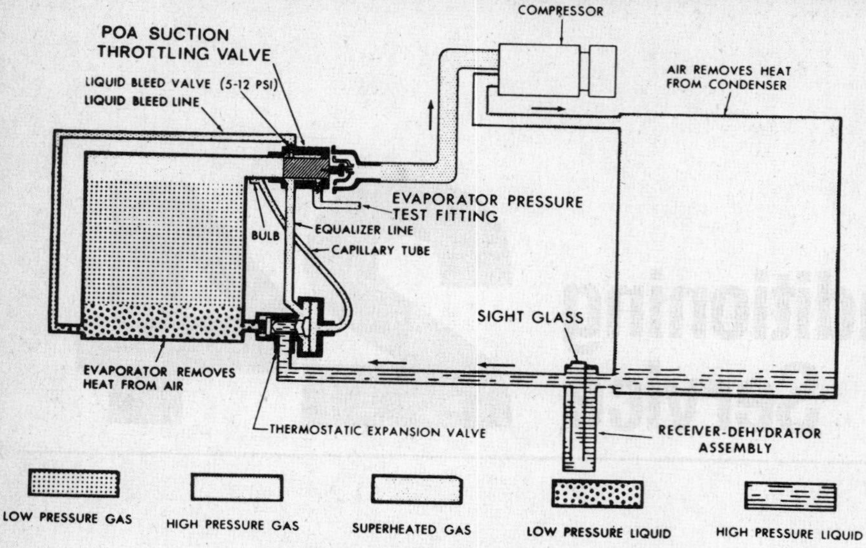

Pilot Operated Absolute (POA) system

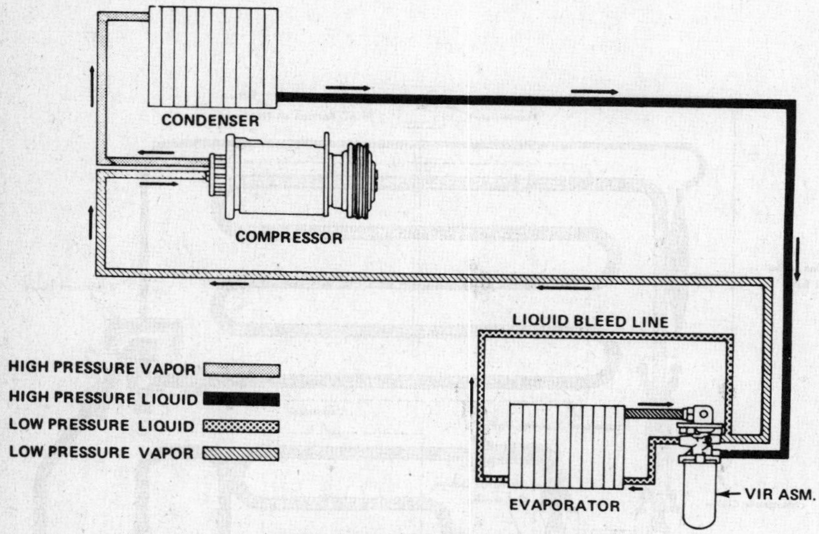

Valves In Receiver (VIR) system

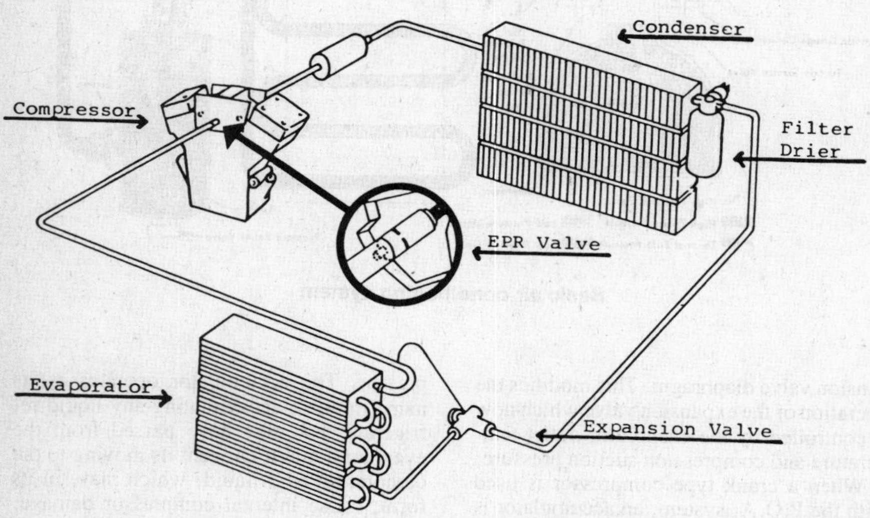

Evaporator Pressure Regulator (EPR) system

V.I.R. System

The V.I.R. system contains the compressor evaporator, condenser, muffler and a unit containing the P.O.A. valve, expansion valve and the receiver-drier. This unit is called the V.I.R. (valves in receiver) assembly. A muffler is normally used with this system and is located between the compressor and the condenser to absorb the compressor pulsations.

The V.I.R. assembly eliminates the outside equalizer line between the outlet of the P.O.A. valve and the expansion valve. The equalizer is now a drilled orfice in the wall between the P.O.A. valve and the expansion valve cavities of the V.I.R. housing. Should the valve prove defective during tests, the unit should be replaced, as it is not repairable or adjustable. A sight glass is normally used with this system.

E.P.R. System

The E.P.R. (evaporator pressure regulator) system includes the condenser, muffler, low pressure shut off valve receiver-drier, expansion valve, evaporator and a V-block, reciprocating crank type compressor. The E.P.R. valve is mounted on the suction side of the compressor and operates in conjunction with the expansion valve assembly, to regulate the flow of refrigerant from the evaporator to the compressor, under light air conditioning loads. By regulating the refrigerant flow, the evaporator temperature is controlled and freezing of the evaporator is prevented.

In contrast to other systems, the E.P.R. system uses the reheat procedure to control the temperature of the air, after it is cooled by passing through the evaporator fins. A manually controlled operating lever is connected to the heater water flow control valve and to a blend air door and the opening of the blend door proportions the amount of air around and through the heater core to control the mix of the cool and hot air for the desired inside temperature. A sight glass is used with this system.

Two types of expansion valves are used with this system. The first type has a capillary tube, mounted in a well on the suction line. The second type has no capillary tube, but senses the need to meter refrigerant into the evaporator by an internal sensing tube. This type of expansion valve is called the "H" type.

"H" Valve System

As was described in the E.P.R. system, the "H" expansion valve can be used with the E.P.R. valve, located in the V-block, reciprocating crank type compressor, to control the amount of refrigerant metered into the evaporator and to control the temperature of the evaporator coils to prevent freeze-up of the condensed moisture. However, when the "H" valve is used with the three piston, axial compressor, a cycling switch is used to control the temperature of the evaporator to prevent freeze-up, rather than the E.P.R. valve, as used with the reciprocating crank type compressor. This can be called the "H" valve system for explanation purposes only and should be recognized as such. The "H" system uses the same components as the other systems, basically the compressor (axial type), condenser, evaporator, expansion valve without a capillary tube ("H" type), receiver-drier, muffler and a low pressure shut off valve. The cycling clutch switch uses a capillary tube, attached to the surface of the suction line, to sense the need for refrigerant movement and compressor operation, therefore causing the electrical clutch pulley and coil to operate the compressor on demand from the cycling switch and to open the circuit to the coil when the demand is not needed. A sight glass is used with this system.

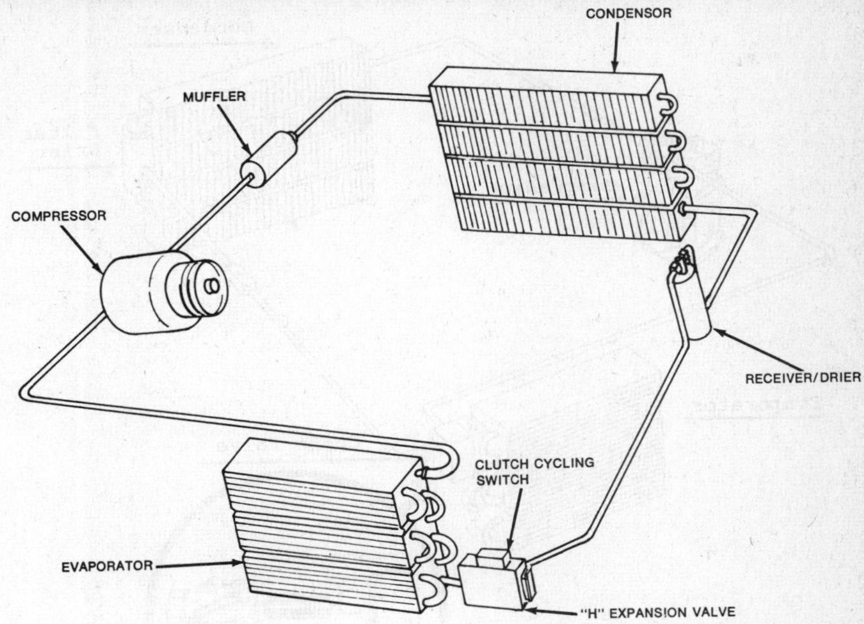

H type expansion valve system

CCOT System

The CCOT (cycling clutch orifice tube) system includes the compressor, condensor, evaporator, an accumulator-drier, a clutch cycling switch with a capillary tube, and a fixed orifce tube, mounted to the evaporator, replacing the expansion valve.

The clutch cycling switch with a temperature probing capillary tube, cycles the compressor clutch off and on as required to maintain a selected comfortable temperature within the vehicle, while preventing evaporator freeze-up. Full control of the system is maintained through the use of a selector control, mounted in the dash assembly. The selector control makes use of a vacuum supply and electrical switches to operate mode doors and the blower motor. A sight glass is not used in this system and one should not be installed. When charging the system, the correct quantity of refrigerant must be installed by measurement.

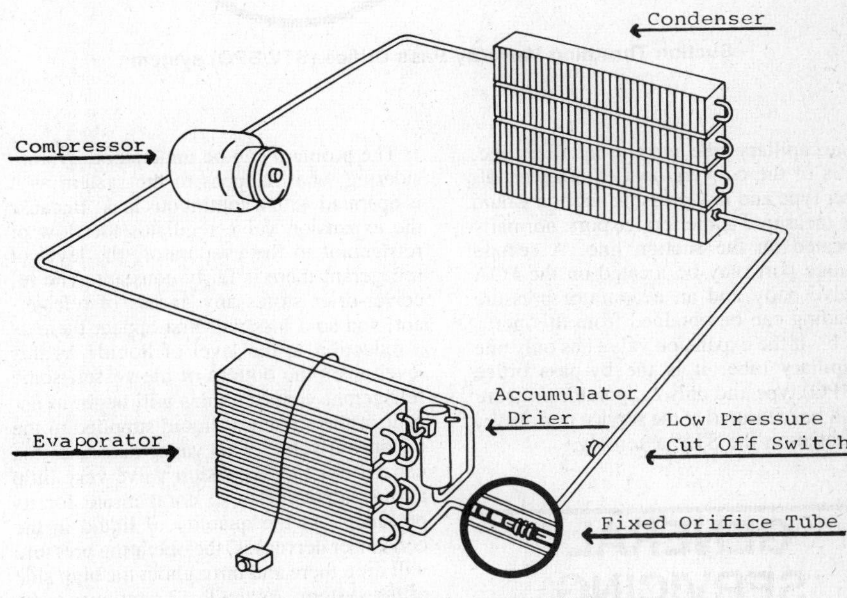

Cycling Clutch Orifice Tube (CCOT) system

STV/BPO System

The STV/BPO (suction throttling valve/bypass orifice) system uses either two types of external expansion valves or a mini-combination valve assembly contains an expansion valve, suction throttling valve and a service port. The expansion valve is of the "H" block design and is used to regulate the flow of refrigerant into the evaporator core. It is also the dividing point for the high and low pressure within the system. The suction throttling valve is used to control the evaporator pressure and to prevent coil freeze-up. The suction throttling starts when the compressor suction pressure decreases below the valve setting. The compressor suction pressure can continue to drop, but the evaporator pressure is held steady by the controlling or throttling action of the STV. A pressure differential valve is used within the combination valve assembly, to allow oil–laden refrigerant to by-pass the restriction formed when the STV assembly is closed, to assure oil return to the compressor during times of reduced heat loads on the system. The by-pass valve remains closed under high heat loads since ample oil is moving through the system and compressor.

Evaporator pressure can only be measured on this system and a special type connector must be used to attach the high pressure gauge line to the service gauge port.

When either of the external type expansion valves are used, separate suction throttling valves are used. The operation of each is basically the same as the components of the combination valve assembly.

The type of external expansion valve used with the system will dictate either low suction or evaporator pressure measurements from the gauge service ports. To determine the pressure measurement that may be obtained from the system, examine the external expansion valve for one of the following conditions:

a. Should the expansion valve have

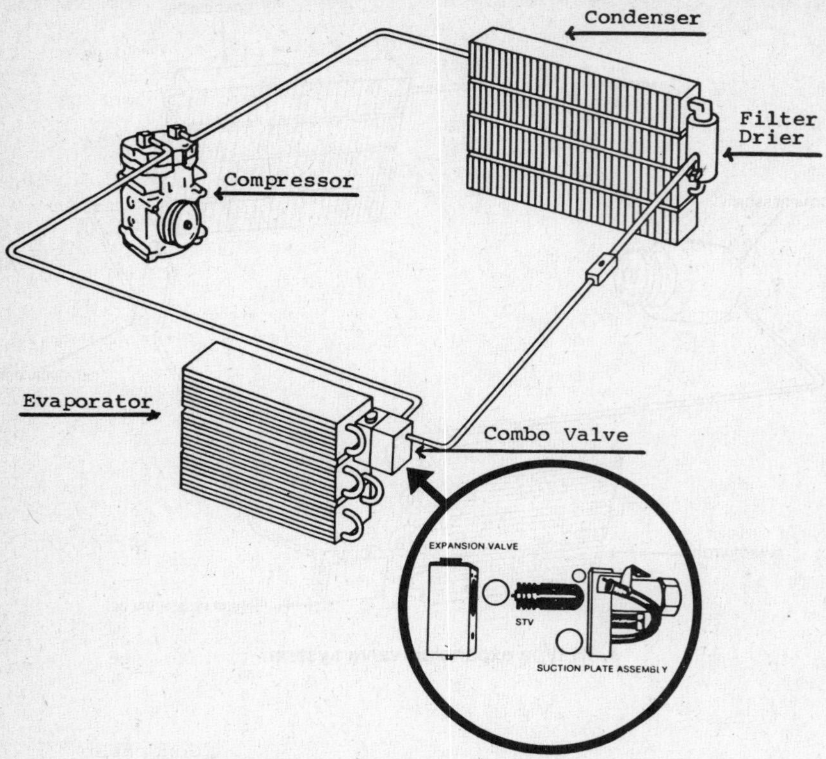

Suction Throttling Valve/By-Pass Orifice (STV/BPO) system

one capillary tube and one equalizer line, it is of the conventional external equalizer type and low pressure suction would be measured at the service port, normally located on the suction line. A second gauge port may be located on the POA valve body and an evaporator pressure reading can be obtained from this port.

b. If the expansion valve has only one capillary tube, it is the by-pass orfice (BPO) type and only evaporator pressure will be measured at the service port valve, located on the STV assembly.

GENERAL SERVICING PROCEDURES

The most important aspect of air conditioning service is the maintenance of a pure and adequate charge of refrigerant in the system. A refrigeration system cannot function properly if a significant percentage of the charge is lost. Leaks are common because the severe vibration encountered in an automobile can easily cause a sufficient cracking or loosening of the air conditioning fittings; as a result, the extreme operating pressures of the system force refrigerant out.

The problem can be understood by considering what happens to the system as it is operated with a continuous leak. Because the expansion valve regulates the flow of refrigerant to the evaporator, the level of refrigerant there is fairly constant. The receiver-drier stores any excess of refrigerant, and so a loss will first appear there as a reduction in the level of liquid. As this level nears the bottom of the vessel, some refrigerant vapor bubbles will begin to appear in the stream of liquid supplied to the expansion valve. This vapor decreases the capacity of the expansion valve very little as the valve opens to compensate for its presence. As the quantity of liquid in the condenser decreases, the operating pressure will drop there and throughout the high side of the system. As the R-12 continues to be expelled, the pressure available to force the liquid through the expansion valve will continue to decrease, and, eventually, the valve's orifice will prove to be too much of a restriction for adquate flow even with the needle fully withdrawn.

At this point, low side pressure will start to drop, and severe reduction in cooling capacity, marked by freeze-up of the evaporator coil, will result. Eventually, the operating pressure of the evaporator will be lower than the pressure of the atmosphere surrounding it, and air will be drawn into the system wherever there are leaks in the low side.

Because all atmospheric air contains at least some moisture, water will enter the system and mix with the R-12 and the oil. Trace amounts of moisture will cause sludging of the oil, and corrosion of the system. Saturation and clogging of the filter-drier, and freezing of the expansion valve orifice will eventually result. As air fills the system to a greater and greater extent, it will interfere more and more with the normal flows of refrigerant and heat.

From this description, it should be obvious that much of the repairman's time will be spent detecting leaks, repairing them, and then restoring the purity and quantity of the refrigerant charge. A list of general precautions that should be observed while doing this follows:

1. Keep all tools as clean and dry as possible.
2. Thoroughly purge the service gauges and hoses of air and moisture before connecting them to the system. Keep them capped when not in use.
3. Thoroughly clean any refrigerant fitting before disconnecting it in order to minimize the entrance of dirt into the system.
4. Plan any operation that requires opening the system beforehand, in order to minimize the length of time it will be exposed to open air. Cap or seal the open ends to minimize the entrance of foreign material.
5. When adding oil, pour it through an extremely clean and dry tube or funnel. Keep the oil capped whenever possible. Do not use oil that has not been kept tightly sealed.
6. Use only refrigerant 12. Purchase refrigerant intended for use in only automatic air conditioning systems. Avoid the use of refrigerant-12 that may be packaged for another use, such as cleaning, or powering a horn, as it is impure.
7. Completely evacuate any system that has been opened to replace a component, or that has leaked sufficiently to draw in moisture and air. This requires evacuating air and moisture with a good vacuum pump for at least one hour.

If a system has been open for a considerable length of time it may be advisable to evacuate the system for up to 12 hours (overnight).

8. Use a wrench on both halves of a fitting that is to be disconnected, so as to avoid placing torque on any of the refrigerant lines.
9. When overhauling a compressor, pour some of the oil into a clean glass and inspect it. If there is evidence of dirt or metal particles, or both, flush all refrigerant components with clean refrigerant before evacuating and recharging the system. In addition, if metal particles are present, the compressor should be replaced.
10. Schrader valves may leak only when under full operating pressure. Therefore, if leakage is suspected but cannot be located, operate the system with a full charge of refrigerant and look for leaks from all Schrader valves. Replace any faulty valves.

Additional Preventive Maintenance Checks

ANTIFREEZE

In order to prevent heater core freeze-up during A/C operation, it is necessary to maintain permanent type antifreeze protection of +15 degrees F. or lower. A reading of −15 degrees F. is ideal since this protection also supplies sufficient corrosion inhibitors for the protection of the engine cooling system.

NOTE: The same antifreeze should not be used longer than the manufacturer specifies.

RADIATOR CAP

For efficient operation of an air conditioned car's cooling system, the radiator cap should have a holding pressure which meets manufacturer's specifications. A cap which fails to hold these pressures should be replaced.

CONDENSER

Any obstruction of, or damage to, the condenser configuration will restrict the air flow which is essential to its efficient operation. It is therefore a good rule to keep this unit clean and in proper physical shape.

NOTE: Bug screens are regarded as obstructions.

CONDENSATION DRAIN TUBE

This single molded drain tube expels the condensation, which accumulates on the bottom of the evaporator housing, into the engine compartment.

If this tube is obstructed, the air conditioning performance can be restricted and condensation buildup can spill over onto the vehicle's floor.

Safety Precautions

Because of the importance of the necessary safety precautions that must be exercised when working with air conditioning systems and R-12 refrigerant, a recap of the safety precautions are outlined.

1. Avoid contact with a charged refrigeration system, even when working on another part of the air conditioning system or vehicle. If a heavy tool comes into contact with a section of copper tubing or a heat exchanger, it can easily cause the relatively soft material to rupture.
2. When it is necessary to apply force to a fitting which contains refrigerant, as when checking that all system couplings are securely tightened, use a wrench on both parts of the fitting involved, if possible. This will avoid putting torque on refrigerant tubing.
(It is advisable, when possible, to use tube or line wrenches when tightening these flare nut fittings.
3. Do not attempt to discharge the system by merely loosening a fitting, or removing the service valve caps and cracking these valves. Precise control is possible only when using the service gauges. Place a rag under the open end of the center charging hose while discharging the system to catch any drops of liquid that might escape. Wear protective gloves when connecting or disconnecting service gauge hoses.
4. Discharge the system only in a well ventilated area, as high concentrations of the gas can exclude oxygen and act as an anesthetic. When leak testing or soldering, this is particularly important, as toxic gas is formed when R-12 contacts any flame.
5. Never start a system without first verifying that both service valves are back-seated, if equipped, and that all fittings throughout the system are snugly connected.
6. Avoid applying heat to any refrigerant line or storage vessel. Charging may be aided by using water heated to less than 125° to warm the refrigerant container. Never allow a refrigerant storage container to sit out in the sun, or near any other source of heat, such as a radiator.
7. Always wear goggles when working on a system to protect the eyes. If refrigerant contacts the eyes, it is advisable in all cases to see a physician as soon as possible.
8. Frostbite from liquid refrigerant should be treated by first gradually warming the area with cool water, and then gently applying petroleum jelly. *A physician should be consulted.*
9. Always keep refrigerant drum fittings capped when not in use. Avoid sudden shock to the drum, which might occur from dropping it, or from banging a heavy tool against it. *Never carry a drum in the passenger compartment of a car.*
10. Always completely discharge the system before painting the vehicle (if the paint is to be baked on), or before welding anywhere near refrigerant lines.

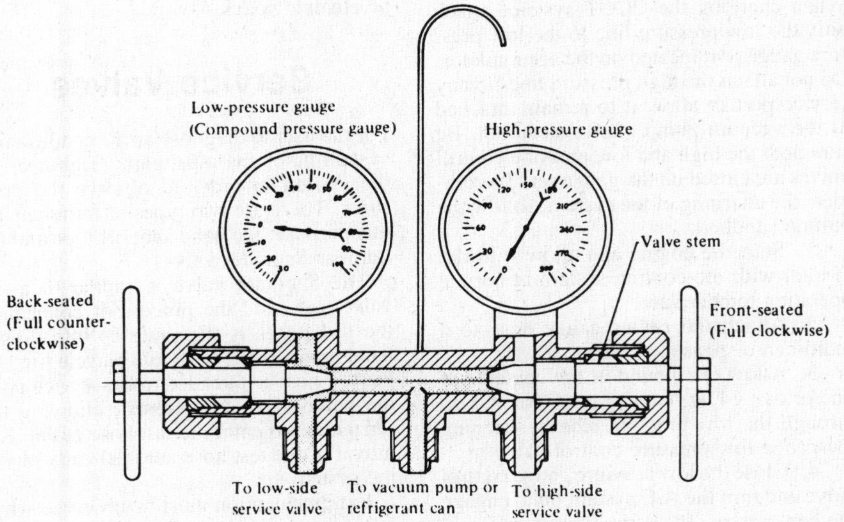

Typical manifold gauge set

AIR CONDITIONING TOOLS AND GAUGES

Test Gauges

Most of the service work performed on any air conditioning system requires the use of a set of two gauges, one for the high (head) pressure side of the system, the other for the low (suction) side.

The low side gauge records both pressure and vacuum. Vacuum readings are calibrated from 0–30 inches and the pressure graduations read from 0 to no less than 60 psi.

The high side guage measures pressure from 0 to at least 600 psi.

Both gauges are threaded into a manifold that contains two hand shut-off valves. Proper manipulation of these valves and the use of the attached-test hoses allow the user to perform the following services:

1. Test high and low side pressures.
2. Remove air, moisture, and contaminated refrigerant.
3. Purge the system of (refrigerant).
4. Charge the system (with refrigerant).

The manifold valves are designed so they have no direct effect on gauge readings, but serve only to provide for, or cut off, flow of refrigerant through the manifold. During all testing and hook-up operations, the valves are kept in a closed position to

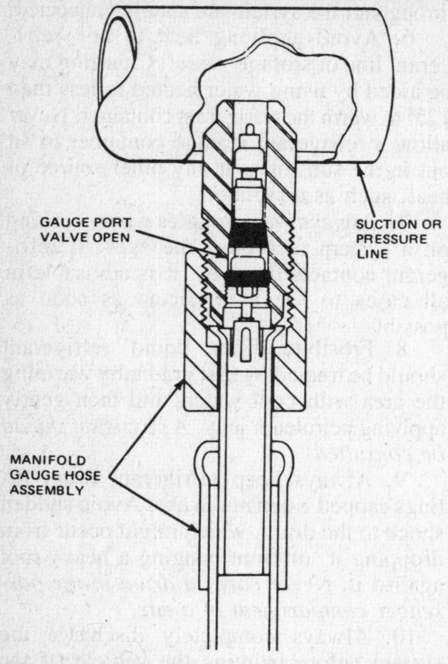

Manifold gauge hose connected to a Schraeder type service port

avoid disturbing the refrigeration system. The valves are opened only to purge the system of refrigerant or to charge it.

When purging the system, the center hose is uncapped at the lower end, and both valves are cracked open slightly. This allows refrigerant pressure to force the entire contents of the system out through the center hose. During charging, the valve on the high side of the manifold is closed, and the valve on the low side is cracked open. Under these conditions, the low pressure in the evaporator will draw refrigerant from the relatively warm refrigerant storage container into the system.

SYSTEMS WITH A SIGHT GLASS

Air conditioning systems that use a sight glass as a means to check the refrigerant level should be carefully checked to avoid under or over charging. The gauge set should be attached to the system for verification of pressures.

To check the system with the sight glass, clean the glass and start the vehicle engine. Operate the air conditioning controls on maximum for approximately five minutes to stabilize the system. The room temperature should be above 70 degrees. Check the sight glass for one of the following conditions:

1. If the sight glass is clear, the compressor clutch is engaged, the compressor discharge line is warm and the compressor inlet line is cool, the system has a full charge of refrigerant.
2. If the sight glass is clear, the compressor clutch is engaged and there is no significant temperature difference between the compressor inlet and discharge lines, the system is empty or nearly empty. By having the gauge set attached to the system, a measurement can be taken. If the gauge reads less than 25 psi, the low pressure cut-off protection switch has failed.
3. If the sight glass is clear and the compressor clutch is disengaged, the clutch is defective, or the clutch circuit is open, or the system is out of refrigerant. Bypass the low pressure cut-off switch momentarily to determine the cause.
4. If the sight glass shows foam or bubbles, the system can be low on refrigerant. Occasional foam or bubbles is normal when the room temperature is above 110 degrees or below 70 degrees. To verify, increase the engine speed to approximately 1500 rpm and block the airflow through the condensor in order to increase the compressor discharge pressure to 225–250 psi. If the sight glass still shows bubbles or foam, the refrigerant level is low.

CAUTION

Do not operate the vehicle engine any longer than necessary with the condensor airflow blocked. This blocking action also blocks the cooling system radiator and will cause the system to overheat rapidly.

When the system is low on refrigerant, a leak is present or the system was not properly charged. Use a leak detector and locate the problem area and repair. If no leakage is found, charge the system to its capacity.

CAUTION

It is not advisable to add refrigerant to a system utilizing the suction throttling valve and a sight glass, because the amount of refrigerant required to remove the foam or bubbles will result in an overcharge and potentially damage system components.

CCOT SYSTEM

When charging the CCOT system, attach only the low pressure line to the low pressure gauge port located on the accumulator. Do not attach the high pressure lines to any service port or allow it to remain attached to the vacuum pump after evacuation. Be sure both the high and low pressure control valves are closed on the gauge set. To complete the charging of the system, follow the outline supplied.

1. Start the engine and allow it to run at idle, with the cooling system at normal operating temperature.
2. Attach the center gauge hose to a multi-can dispenser.
3. Allow one pound or the contents of one or two 14 oz. cans to enter the system through the low pressure side by opening the gauge low pressure control valve.
4. Close the low pressure gauge control valve and turn the A/C system on to engage the compressor. Place the blower motor in its high mode.
5. Open the low pressure gauge control valve and draw the remaining charge into the system.
6. Close the low pressure gauge control valve and the refrigerant source valve on the multi-can dispenser. Remove the low pressure hose from the accumulator quickly to avoid loss of refrigerant through the Schrader valve.
7. Install the protective cap on the gauge port and check the system for leakage.
8. Test the system for proper operation.

Leak Testing the System

There are several methods of detecting leaks in an air conditioning system; among them, the two most popular are (1) halide leak-detection or the "open flame method," and (2) electronic leak-detection.

The halide leak detection is a torch like device which produces a yellow-green color when refrigerant is introduced into the flame at the burner. A brilliant blue or violet color indicates the presence of large amounts of refrigerant at the burner. A small leak will cause the flame to turn a yellow-green color.

An electronic leak detector is a small portable electronic device with an extended probe. With the unit activated, the probe is passed along those components of the system which contain refrigerant. If a leak is detected, the unit will sound an alarm signal or activate a display signal depending on the manufacturer's design. It is advisable to follow the manufacturer's instructions as the design and function of the detection may vary significantly.

NOTE: Caution should be taken to operate either type of detector in well ventilated areas, so as to reduce the chance of personal injury, which may result from coming in contact with poisonous gases produced when R-12 is exposed to flame or electric spark.

Service Valves

For the user to diagnose an air conditioning system he or she must gain "entrance" to the system in order to observe the pressures. There are two types of terminals for this purpose, the hand shut off type and the familiar Schrader valve.

The Schrader valve is similar to a tire valve stem and the process of connecting the test hoses is the same as threading a hand pump outlet hose to a bicycle tire. As the test hose is threaded to the service port, the valve core is depressed, allowing the refrigerant to enter the test hose outlet. Removal of the test hose automatically closes the system.

Extreme caution must be observed when removing test hoses from the Schrader valves as some refrigerant will normally escape,

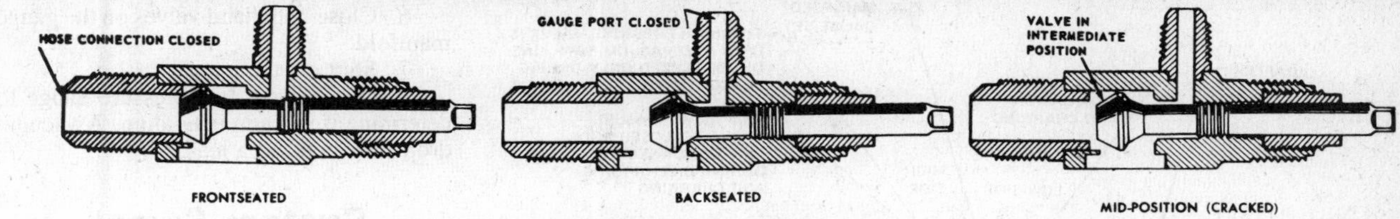

Manual service valve positions

usually under high pressure (observe safety precautions).

Some systems have hand shut-off valves (the stem can be rotated with a special racheting box wrench) that can be positioned in the following three ways:

1. FRONT SEATED—Rotated to full clockwise position.
   a. Refrigerant will not flow to the compressor, but will reach the test gauge port. COMPRESSOR WILL BE DAMAGED IF SYSTEM IS TURNED ON IN THIS POSITION.
   b. The compressor is now isolated and ready for service. However, care must be exercised when removing service valves from the compressor as a residue of refrigerant may still be present within the compressor. Therefore, remove service valves slowly, observing all safety precautions.
2. BACK SEATED—Rotated to full counterclockwise position. Normal position for system while in operation. Refrigerant flows to compressor but not to test gauge.
3. MID-POSITION (CRACKED)—Refrigerant flows to entire system. Gauge port (with hose connected) open for testing.

USING THE MANIFOLD GAUGES

The following are step-by-step procedures to guide the user to correct gauge usage.

1. WEAR GOGGLES OR FACE SHIELD DURING ALL TESTING OPERATIONS. BACKSEAT HAND SHUT-OFF TYPE SERVICE VALVES.
2. Remove caps from the high and low side of the service ports. Make sure both gauge valves are closed.
3. Connect the low side test hose to the service valve that leads to the evaporator (located between the evaporator outlet and the compressor).
4. Attach the high side test hose to the service valve that leads to the condenser.
5. Mid-position hand shutoff type service valves.
6. Start the engine and allow for warm-up. All testing and charging of the system should be done after the engine and system have reached normal operation temperatures (except when using certain charging stations).
7. Adjust the air conditioner controls to maximum cold.
8. Observe the gauge readings. When the gauges are not being used it is a good idea to:
   a. Keep both hand valves in the closed position.
   b. Attach both ends of the high and low service hoses to the manifold if extra outlets are present on the manifold, or plug them if not. Also, keep the center charging hose attached to an empty refrigerant can. This extra precaution will reduce the possibility of moisture entering the gauges. If air and moisture have gotten into the gauges, purge the hoses by supplying refrigerant under pressure to the center hose with both gauge valves open and all openings unplugged.

BAR GAUGE MANIFOLD AND COMPRESSOR SERVICE VALVE SETTINGS

Condition	Manifold Valves	Compressor Valves
Testing System	Both fully closed	Both cracked off backseat
Depressurizing System	Both cracked open	Both at mid position
Evacuating the system	Both wide open	Both at mid position
Charging in gas form with compressor running	High pressure valve closed Low pressure valve cracked	High pressure valve cracked off backseat Low pressure valve at mid position
Charging in liquid form with compressor off	Low pressure valve closed High pressure valve wide open	Both valves mid positioned

Note: A very small leak, causing system discharge about every two weeks, can be caused by a leaky Schrader type service valve. Check these valves with extra care when testing for a small leak.

DISCHARGING, EVACUATING AND CHARGING

Discharging the System

CAUTION

Perform this operation in a well-ventilated area.

When it is necessary to remove (purge) the refrigerant pressurized in the system, follow this procedure:

1. Operate the air conditioner for at least 10 minutes.
2. Attach the gauges, shut off the engine and air conditioner.
3. Place a container or rag at the outlet of the center charging hose on the gauge. The refrigerant will be discharged there and this precaution will avoid its uncontrolled exposure.
4. Open the low side hand valve on the gauge slightly.
5. Open the high side hand valve slightly.

NOTE: Too rapid a purging process will be identified by the appearance of an oil foam. If this occurs, close the hand valves a little more until this condition stops.

6. Close both hand valves on the gauge set when the pressures read 0 and all the refrigerant has left the system.

Evacuating the System

Before charging any system it is necessary to purge the refrigerant and draw out the trapped moisture with a suitable vacuum pump. Failure to do so will result in ineffective charging and possible damage to the system.

Use this hook-up for the proper evacuation procedure:

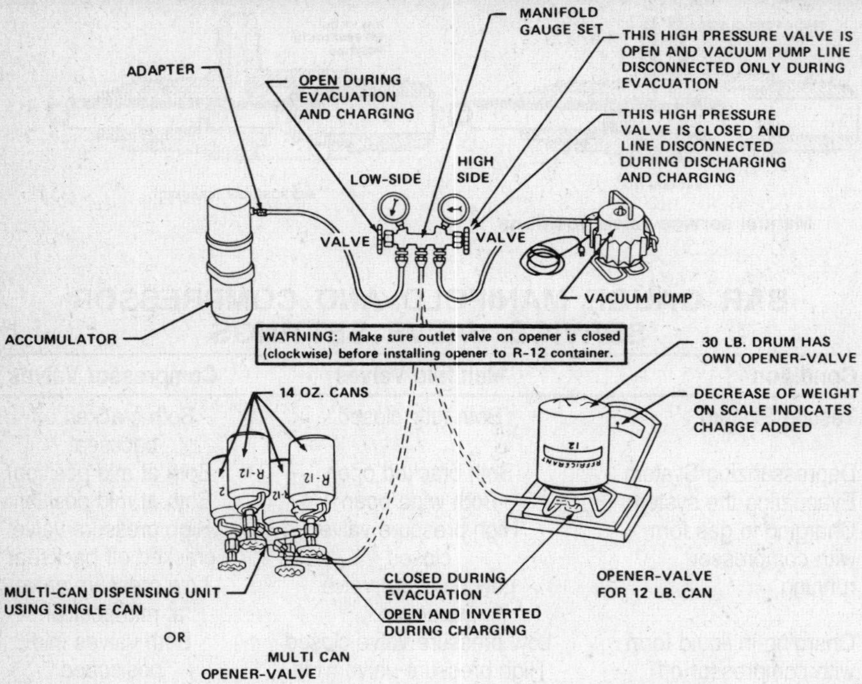

Typical gauge connections for discharge, evacuation and charging the system

1. Connect both service gauge hoses to the high and low service outlets.
2. Open both high and low side hand valves on the gauge manifold.
3. Open both service valves a slight amount (from back seated position), allow the refrigerant to discharge from the system.
4. Install the center charging hose of the gauge set to the vacuum pump.
5. Operate the vacuum pump for at least one hour (if the system has been subjected to open conditions for a prolonged period of time, it may be necessary to "pump the system down" overnight. Refer to the "System Sweep" procedure).

NOTE: If the low pressure gauge does not show at least 28" hg. within 5 minutes, check the system for a leak or loose gauge connectors.

6. Close both hand valves on the gauge manifold.
7. Shut off the pump.
8. Observe the low pressure gauge to determine if vacuum is holding. A vacuum drop may indicate a leak.

System Sweep

An efficient vacuum pump can remove all the air contained in a contaminated air conditioning system very quickly because of its vapor state. Moisture, however, is far more difficult to remove because the vacuum must force the liquid to evaporate before it will be able to remove it from the system. If a system has become severely contaminated, as, for example, it might become after all the charge was lost in conjunction with vehicle accident damage, moisture removal is extremely time consuming. A vacuum pump could remove all of the moisture only if it were operated for 12 hours or more.

Under these conditions, sweeping the system with refrigerant will speed the process of moisture removal considerably. To sweep, follow the following procedure:

1. Connect a vacuum pump to the gauges, operate it until vacuum ceases to increase, then continue operation for ten more minutes.
2. Charge the system with 50% of its rated refrigerant capacity.
3. Operate the system at fast idle for ten minutes.
4. Discharge the system.
5. Repeat twice the process of charging to 50% capacity, running the system for ten minutes, and discharging it, for a total of three sweeps.
6. Replace the drier.
7. Pump the system down as detailed in Step 1.
8. Charge the system.

Charging the System

CAUTION

Never attempt to charge the system by opening the high pressure gauge control while the compressor is operating. The compressor accumulating pressure can burst the refrigerant container, causing severe personal injuries.

BASIC SYSTEM

In this procedure the refrigerant enters the suction side of the system as a vapor while the compressor is running. Before proceeding, the system should be in a partial vacuum after adequate evacuation. Both hand valves on the gauge manifold should be closed.

1. Attach both test hoses to their respective service valve ports. Mid-position manually operated service valves, if present.

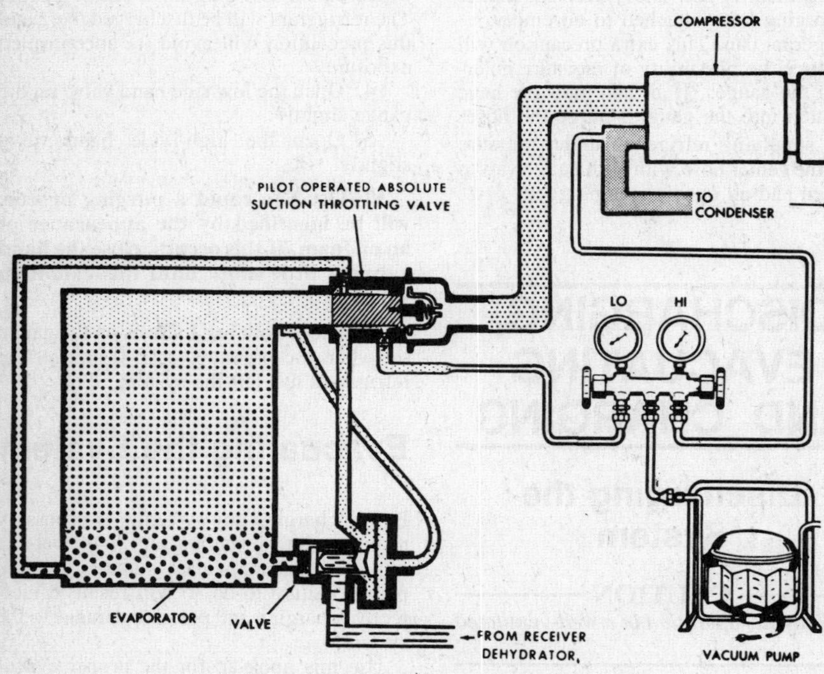

Schematic for evacuating the system

2. Install a dispensing valve (closed position) on the refrigerant container (single and multiple refrigerant manifolds are available to accommodate one to four 15 oz. cans).

3. Attach the center charging hose to the refrigerant container valve.

4. Open the dispensing valve on the refrigerant can.

5. Loosen the center charging hose coupler where it connects to the gauge manifold to allow the escaping refrigerant to purge the hose of contaminants.

6. Tighten the center charging hose connection.

7. Purge the low pressure test hose at the gauge manifold.

8. Start the engine, roll down the windows and adjust the air conditioner to maximum cooling. The engine should be at normal operating temperature before proceeding. The heated environment helps the liquid vaporize more efficiently.

9. Crack open the low side hand valve on the manifold. Manipulate the valve so that the refrigerant that enters the system does not cause the low side pressure to exceed 40 psi. Too sudden a surge may permit the entrance of unwanted liquid to the compressor. Since liquids cannot be compressed, the compressor will suffer damage if compelled to attempt it. If the suction side of the system remains in a vacuum, the system is blocked. Locate and correct the condition before proceeding any further.

NOTE: Placing the refrigerant can in a container of warm water (no hotter than 125° F) will speed the charging process. Slight agitation of the can is helpful too, but be careful not to turn the can upside down.

Some manufacturers allow for a partial charging of the A/C system in the form of a liquid (can inverted and compressor off) by opening the high side gauge valve only, and putting the high side compressor service valve in the middle position (if so equipped). The remainder of the refrigerant is then added in the form of a gas in the normal manner, through the suction side only.

SYSTEMS WITHOUT SIGHT GLASS, EXCEPT CCOT SYSTEM

The following procedure can be used to quickly determine whether or not an air conditioning system has the proper charge of refrigerant (providing ambient temperature is above 70° F. or 21° C.). This check can be made in a manner of minutes, thus facilitating system diagnosis by pinpointing the problem to the amount of charge in the system or by eliminating this possibility from the overall checkout.

1. Engine must be warm (thermostat open).
2. Hood and body doors open.
3. Selector lever set at NORM.
4. Temperature lever at COLD.
5. Blower on HI.
6. Normal engine idle.
7. Hand-feel the temperature of the evaporator inlet and outlet pipes with the compressor engaged.

a. Both same temperature or some degree cooler than ambient—proper condition: check for other problems.

b. Inlet pipe cooler than outlet pipe—low refrigerant charge.

- Add a slight amount of refrigerant until both pipes feel the same.
- Then add 15 oz. (1 can) additional refrigerant.

c. Inlet pipe has front accumulation—outlet pipe warmer: proceed as in Step b above.

If during the charging process the head pressure exceeds 200 psi, place an electric fan in front of the car and direct the turbulent air to the condenser. If no fan is available, repeatedly pour cool water over the top of the condenser. These cooling actions may be necessary on an extremely warm day to help dissipate the heat emitted by the engine during idle.

If this fails and pressure on the discharge side continues to rise, the system may be overcharged or the engine might be overheating. *Never* allow head pressure to go beyond 240 psi. during charging. If this condition occurs, stop the engine, find and correct the problem.

8. Continue dispensing refrigerant until the container is no longer cool to the touch. On a humid day, the outside of the container will frost. When the frost disappears the can is usually empty. To detach the dispensing can:

a. close the low pressure test gauge hand valve.

b. crack open the low pressure test hose at the manifold until the remaining pressure escapes.

c. tighten the hose coupler.

d. loosen the hose coupler connected to the refrigerant can.

e. discard the empty can and repeat Steps 2–8.

9. Continue to add refrigerant to the required capacity of the system. (Usually marked on the compressor).

CAUTION

DO NOT OVERCHARGE. This condition is usually indicated by an abnormally high side pressure reading and a noisy compressor resulting in ineffective cooling and damage to the system.

SYSTEMS WITH A SIGHT GLASS

Air conditioning systems that use a sight glass as a means to check the refrigerant level should be carefully checked to avoid under or over charging. The gauge set should be attached to the system for verification of pressures.

To check the system with the sight glass, clean the glass and start the vehicle engine. Operate the air conditioning controls on maximum for approximately five minutes

Check item / Amount of refrigerant	Almost no refrigerant	Insufficient	Suitable	Too much refrigerant
Temperature of high pressure and low pressure lines.	Almost no difference between high pressure and low pressure side temperature.	High pressure side is warm and low pressure side is fairly cold.	High pressure side is hot and low pressure side is cold.	High pressure side is abnormally hot.
State in sight glass.	Bubbles flow continuously. **Bubbles will disappear and something like mist will flow when refrigerant is nearly gone.**	The bubbles are seen at intervals of 1 - 2 seconds.	Almost transparent. Bubbles may appear when engine speed is raised and lowered. **No clear difference exists betwen these two conditions.**	No bubbles can be seen.
	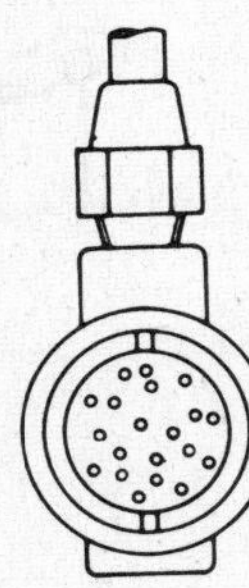	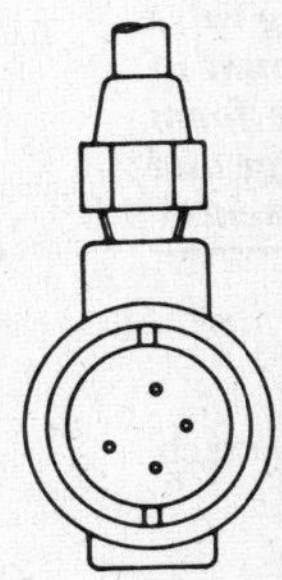	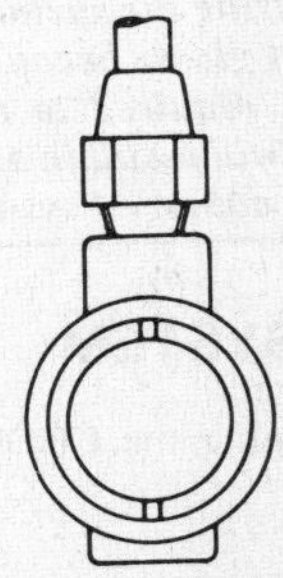	
Pressure of system.	High pressure side is abnormally low.	Both pressure on high and low pressure sides are slightly low.	Both pressures on high and low pressure sides are normal.	Both pressures on high and low pressure sides are abnormally high.
Repair.	**Stop compressor immediately** and conduct an overall check.	Check for gas leakage, repair as required, replenish and charge system.		Discharge refrigerant from service valve of low pressure side.

Using a sight glass to determine the relative refrigerant charge

to stabilize the system. The room temperature should be above 70 degrees. Check the sight glass for one of the following conditions:

1. If the sight glass is clear, the compressor clutch is engaged, the compressor discharge line is warm and the compressor inlet line is cool, the system has a full charge of refrigerant.

2. If the sight glass is clear, the compressor clutch is engaged and there is no significant temperature difference between the compressor inlet and discharge lines, the system is empty or nearly empty. By having the gauge set attached to the system, a measurement can be taken. If the gauge reads less than 25 psi, the low pressure cut-off protection switch has failed.

3. If the sight glass is clear and the compressor clutch is disengaged, the clutch is defective, or the clutch circuit is open, or the system is out of refrigerant. By-pass the low pressure cut-off switch momentarily to determine the cause.

4. If the sight glass shows foam or bubbles, the system can be low on refrigerant. Occasional foam or bubbles is normal when the room temperature is above 110 degrees or below 70 degrees. To verify, increase the engine speed to approximately 1500 rpm and block the airflow through the condenser in order to increase the compressor discharge pressure to 225–250 psi. If the sight glass still shows bubbles or foam, the refrigerant level is low.

CAUTION

Do not operate the vehicle engine any longer than necessary with the condenser airflow blocked. This blocking action also blocks the cooling system radiator and will cause the system to overheat rapidly.

When the system is low on refrigerant, a leak is present or the system was not properly charged. Use a leak detector and locate the problem area and repair. If no leakage is found, charge the system to its capacity.

CAUTION

It is not advisable to add refrigerant to a system utilizing the suction throttling valve and a sight glass, because the amount of refrigerant required to remove the foam or bubbles will result in an overcharge and potentially damaged system components.

CCOT SYSTEM

When charging the CCOT system, attach only the low pressure line to the low pressure gauge port located on the accumulator. Do not attach the high pressure line to any service port or allow it to remain attached to the vacuum pump after evacuation. Be sure both the high and the low pressure control valves are closed on the gauge set. To complete the charging of the system, follow the outline supplied.

1. Start the engine and allow it to run at idle, with the cooling system at normal operating temperature.

2. Attach the center gauge hose to a single or multi-can dispenser.

3. With the multi-can dispenser inverted, allow one pound or the contents of one or two 14 oz. cans to enter the system through the low pressure side by opening the gauge low pressure control valve.

4. Close the low pressure gauge control valve and turn the A/C system on to engage the compressor. Place the blower motor in its high mode.

5. Open the low pressure gauge control valve and draw the remaining charge into the system.

6. Close the low pressure gauge control valve and the refrigerant source valve, on the multi-can dispenser. Remove the low pressure hose from the accumulator quickly to avoid loss of refrigerant through the Schrader valve.

7. Install the protective cap on the gauge port and check the system for leakage.

8. Test the system for proper operation.

Leak Testing the System

There are several methods of detecting leaks in an air conditioning system; among them, the two most popular are (1) halide leak-detection or the "open flame method," and (2) electronic leak-detection.

The halide leak detection is a torch like device which produces a yellow-green color when refrigerant is introduced into the flame at the burner. A purple or violet color indicates the presence of large amounts of refrigerant at the burner.

An electronic leak detector is a small portable electronic device with an extended probe. With the unit activated, the probe is passed along those components of the system which contain refrigerant. If a leak is detected, the unit will sound an alarm signal or activate a display signal depending on the manufacturer's design. It is advisable to follow the manufacturer's instructions as the design and function of the detection may vary significantly.

CAUTION

Caution should be taken to operate either type of detector in well ventilated areas, so as to reduce the chance of personal injury, which may result from coming in contact with poisonous gases produced when R-12 is exposed to flame or electric spark.

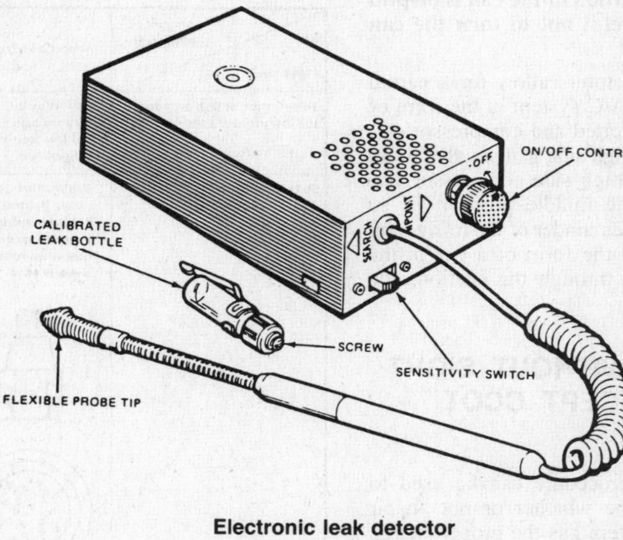

Electronic leak detector

Diesel Service 25

NOTE: Most procedures associated with diesel engined cars are similar to gas engined cars, although many parts of the diesel engine are unique compared to their gas engine counterparts. Standard maintenance and service procedures are given here while component removal, installation and adjustment procedures unique to diesel engines can be found in the appropriate section.

HOW THE DIESEL ENGINE WORKS

Four-stroke diesels require four piston strokes for the complete cycle of actions, exactly like a gasoline engine. The difference lies in how the fuel mixture is ignited. A diesel engine does not rely on a conventional spark ignition to ignite the fuel mixture for the power stroke. Instead, a diesel relies on the heat produced by compressing air in the combustion chamber to ignite the fuel and produce a power stroke. This is known as a compression-ignition engine. No fuel enters the cylinder on the intake stroke, only air. At the end of the compression stroke, fuel is sprayed into the precombustion chamber (prechamber). The mixture ignites and spreads out into the main combustion chamber, forcing the piston downward (power stroke). The fuel/air mixture ignites because of the very high combustion chamber temperatures generated by the extraordinarily high compression ratios used in diesel engines. Typically, the compression ratios used in automotive diesels run anywhere from 16:1 to 23:1. A typical spark-ignition engine has a ratio of about 8:1. This is why a spark-ignition engine which continues to run after you have shut off the engine is said to be "dieseling". It is running on combustion chamber heat alone.

Designing an engine to ignite on its own combustion chamber heat poses certain problems. For instance, although a diesel engine has no need for a coil, spark plugs, or a distributor, it does need what are known as "glow plugs". These superficially resemble spark plugs, but are only used to warm the combustion chambers when the engine is cold. Without these plugs, cold starting would be impossible, due to the enormously high compression ratios and the characteristics of the diesel fuel itself.

All diesel engines use fuel injection, be-

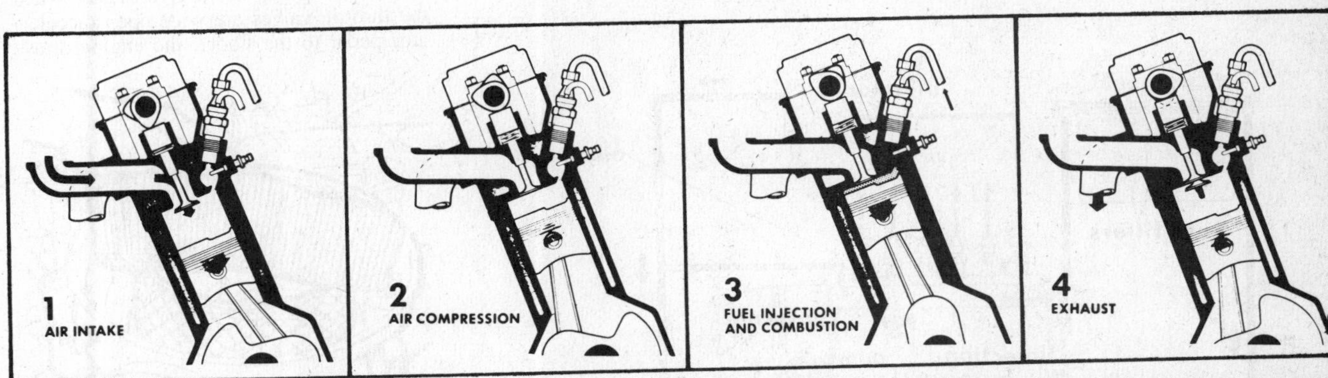

4-stroke diesel engine cycle. At *air intake* (1), rotation of the crankshaft drives a toothed belt that turns the camshaft, opening the intake valve. As the piston moves down, a vacuum is created, sucking fresh air into the cylinder, past the open intake valve. *Air compression* (2): As the piston moves up, both valves are closed, and the air is compressed about 23 times smaller than its original volume. The compressed air reaches a temperature of about 1,650°F., far above the temperature needed to ignite diesel fuel. *Fuel injection and compression* (3): As the piston reaches the top of the stroke, the air temperature is at its maximum. A fine mist of fuel is sprayed into the prechamber, where it ignites, and the flame front spreads rapidly into the combustion chamber. The piston is forced downward by the pressure (about 500 psi) of expanding gases. *Exhaust* (4): As the energy of combustion is spent and the piston begins to move upward again, the exhaust valve opens, and burnt gases are forced out past the open valve. As the piston starts down, the exhaust valve closes, the intake valve opens, and the air intake stroke begins again.

Increasingly, modern diesel engines are being equipped with turbochargers, exhaust gas–driven devices that force more air into the engine to increase power output

cause unlike spark-ignited engines, the fuel cannot be drawn through the intake tract

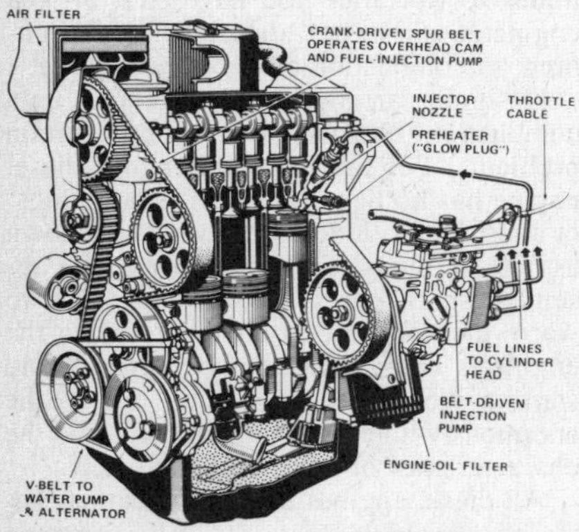

Cutaway view of typical 4-cylinder diesel engine.

and into the cylinders. The introduction of fuel into a diesel engine must be precisely timed so that each cylinder "fires" at the proper moment. Also, the fuel injection pressure (at the cylinder) must be great enough to overcome the high compression pressures, and properly atomize the fuel without the aid of a moving air mass (as in a carbureted gas engine). It is not uncommon for diesel engine fuel injection pressures to be set at 1500–1700 psi.

Diesel engines share many of their basic mechanical components with gasoline engines, though the cylinder block, head(s), crankshaft, connecting rods, pistons, etc., are manufactured to be much stronger for use in diesel engines. The additional strength of the components is necessary due to the very high cylinder pressure generated within the diesel engine.

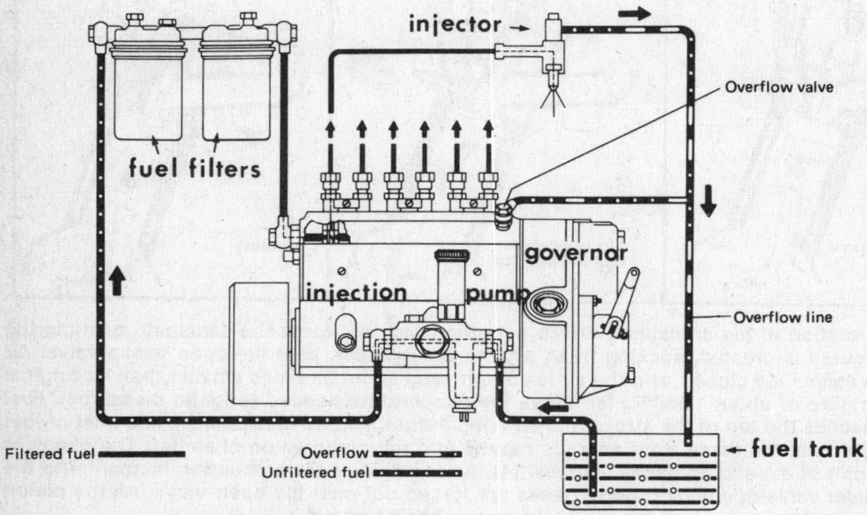

Typical diesel engine fuel system schematic

Maintenance and Service Procedures

Maintenance procedures for the diesel engine generally fall into three categories:

1. Fuel system
2. Starting system
3. Engine mechanical systems

Of these, the fuel system is usually the most likely source of engine troubles, and should be high on the list for regular maintenance attention.

FUEL SYSTEM

The typical diesel engine fuel system consists of fuel tank, fuel feed and return lines, mechanical fuel injection pump, fuel injectors and lines, and a large capacity fuel filter. On some models, the engine may also be equipped with a small, low pressure fuel pump which feeds the injection pump.

In addition to these, the air intake system (air cleaner, inlet manifold) should be checked over regularly to insure unrestricted air flow into the cylinders.

In operation, fuel is sucked out of the fuel tank by the injection pump (or its feed pump) and fed by the injection pump to the injectors in the cylinder head at a very high pressure. Before the fuel is allowed to enter the main injection pump, it passes through a specially built fuel filter which traps solid particles (and water on some models) in the fuel. Fuel that is not used is pumped back to the fuel tank through the fuel return lines. This recirculated fuel helps cool the injection pump.

Air Cleaner

On a gasoline engine, the volume of air taken in by the engine is controlled by throttle valves. When the throttle valves are closed (engine idling), air intake is restricted. When the throttle valves are wide open (accelerator pedal to the floor), the engine draws

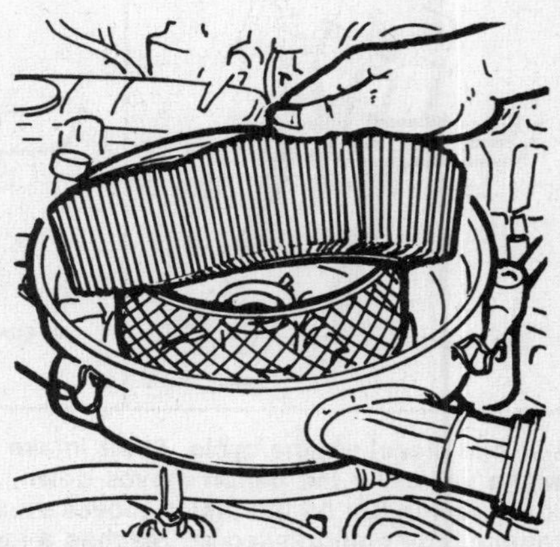

Because a greater quantity of air passes through the diesel engine, air filter maintenance is particularly important. Most diesel air filters on passenger cars are similar to their counterparts on gasoline engines.

in the maximum amount of air it possibly can. This applies to both carbureted and fuel injected gasoline engines.

The speed (rpm) of a diesel engine is controlled by the quantity of fuel which is injected into the engine; no air metering restrictions (throttle valves) are used. Because of this, diesel engines ingest as much air as they possibly can under all conditions. A much greater volume of air passes through the air cleaner of a diesel per mile, therefore, diesel air filters must either be larger or the filter replacement intervals more frequent than those of a similarly sized gasoline engine.

One word of caution: never remove the air cleaner on a diesel with the engine running, and never run the engine with the air cleaner removed. The volume of air drawn through the inlet manifold is very great, and, because the inlet manifold is unobstructed, anything drawn into the inlet manifold (air cleaner wing nut, etc.) goes straight to the combustion chambers, where it can cause major engine damage.

Fuel Filter

The diesel engine fuel filter is usually larger than the filter used on gasoline engines. The extra capacity is needed to trap the suspended particles in diesel fuel, which is generally "dirtier" than gasoline.

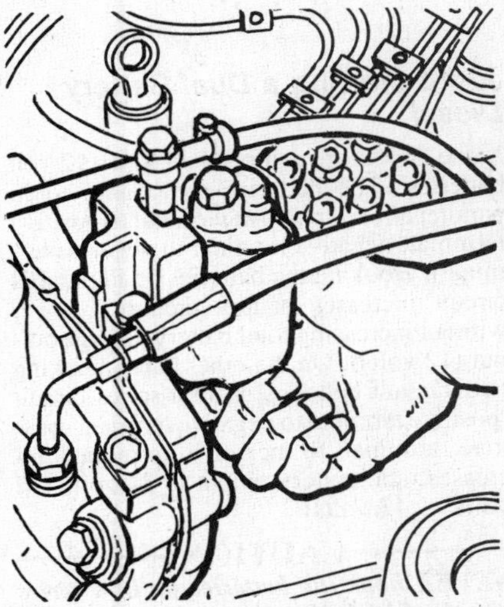

Many diesel engines use a spin-on type primary fuel filter.

On some engines, the fuel filter looks like a second engine oil filter, and is removed and installed in the same manner as the canister-type oil filter.

The fuel filter must be changed according to the manufacturer's suggested interval. See the owner's manual for information.

After installing the fuel filter start the engine and check for leaks. Run the engine for about two minutes, then stop the engine for the same amount of time to allow any air trapped in the injection system to bleed off.

Many diesels also have a small, in-tank filter which is usually maintenance-free.

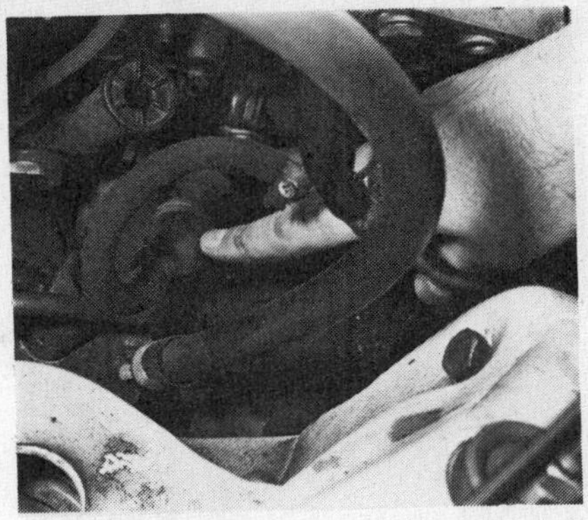

A smaller, in-line secondary filter is used on many engines.

Check the tightness of the clamps securing the injector lines. Note that the injector lines are all the same length.

Mercedes-Benz diesel engines use this stop switch, which shuts off fuel delivery

Water In Fuel

Diesel fuel is a hydrophilic fluid, that is, it naturally attracts water. Since diesel fuel and water do not mix, the water remains floating beneath the fuel at the bottom of the tank. This water must be removed every now and then, or it will be sucked into the fuel circuit and pass through the injection system, causing corrosion and possible component failure (injection pumps can cost up to $1,000). Water in the fuel system will also cause the engine to run poorly, if at all.

Most diesel fuel tanks are equipped with a separator which can isolate from 1 to 3 gallons of water from the fuel.

Many diesels are also equipped with "Water in Fuel" lights in the dashboard which warn of the presence of water in the fuel tank. These warning systems can be installed on models not so equipped.

On some diesels, there is a water catcher in the bottom of the fuel filter which can easily be bled off. In addition, there are several bolt-on water filters on the market which attach to the fuel line under the hood and separate water from the fuel. Depending on which kind you buy, draining water from the system is simply a matter of opening the petcock at the bottom of the filter and letting the water drain out, or, if money is no object, a separator is available on which water is drained from the filter simply by activating a switch on the dashboard.

Removing Water from the Fuel Tank

Treat diesel fuel with the same respect you would gasoline, and after the procedure, properly dispose of the fuel.

1. Remove the fuel tank cap.
2. Connect a pump or siphon hose to the ¼ in. fuel return hose (smaller of the two fuel hoses) above the rear axle, or under the hood near the fuel pump (on the passenger's side of the engine, near the front).
3. Siphon until all water is removed from the tank. Do not use your mouth to create siphon vacuum, EVER! The best method is to siphon the water into a large capacity see-through container. The water will collect at the bottom of the container.
4. When all water has been removed from the tank, be sure to reinstall the fuel return hose and fuel cap.

NOTE: If the entire fuel system (not just the tank) is contaminated by water, the vehicle must be stopped immediately and the fuel system must be purged. This includes draining and removing the fuel tank, blowing low pressure compressed air backwards through the fuel feed and return lines, and bleeding the water out of all injection components. This job should be referred to a qualified technician.

Cold Weather Fuel System Maintenance

CAUTION

NEVER use "starting aids" (e.g.—ether) to help start a diesel engine—serious engine damage will result.

As will be explained later under "Fuel Recommendations", diesel fuel tends to become "cloudy", or thicker, as the temperature drops. The thicker the diesel fuel becomes, the slower it flows through the fuel system, until finally it stops flowing altogether somewhere near the bottom of the thermometer.

One way to fight sluggish fuel flow is to use winterized blends of diesel fuel, straight No. 1 diesel fuel or add cold weather additives to the fuel to improve flow in cold weather.

NOTE: Consult your owners manual for recommendations and be sure to use a fuel conditioner compatible with water separators.

Another way is to install an aftermarket fuel system pre-heater. These are generally canisters which connect into the fuel line and use coolant from the engine cooling system to heat the fuel before it reaches the injection pump. The one drawback with this system is the engine must be started before the pre-heater begins to work. Also available are electric fuel warmers. These preheat the fuel going into the filter and can be used in conjunction with the coolant-type fuel heater.

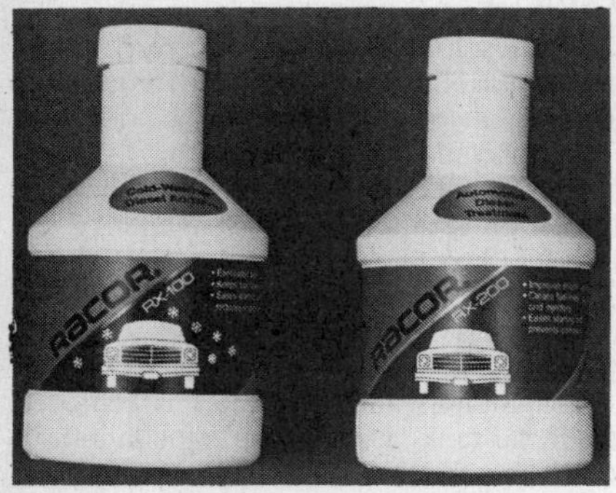

Cold weather additives and fuel conditioners can help improve cold weather flow of diesel fuel.

Some manufacturers offer an optional electric diesel fuel heater and engine block heaters. The fuel heater is thermostatically controlled to heat the fuel before it enters the fuel filter when fuel temperature is 20°F or lower. The fuel heater works only when the ignition key is in the RUN position. On these models, the fuel tank filter has a bypass valve which allows fuel to flow to the heater when the tank filter is covered with fuel wax. The engine block heater is equipped with an electrical cord wrapped up in the engine compartment. The cord

Some diesel engines come equipped with a built-in heating system to keep the engine warm in cold temperatures.

Most OEM heaters work from 110-volt house current.

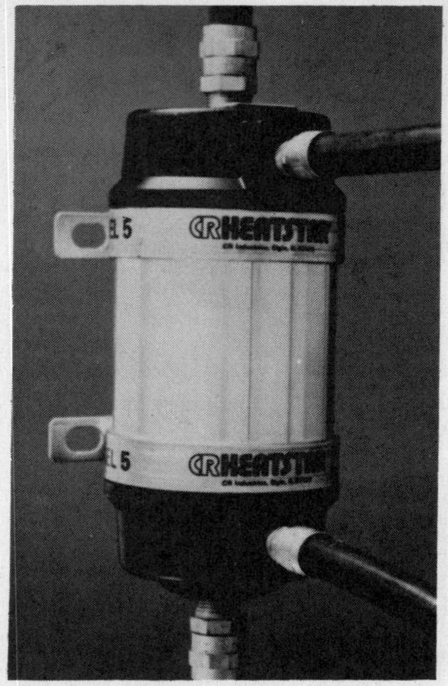

Some aftermarket diesel fuel warmers are thermostatically controlled heat exchangers that use engine coolant to keep diesel fuel above its "cloud point," the temperature at which it gels and forms wax that can clog a fuel system.

plugs into regular 110 volt household current. The block heater can be used, according to the type of oil in the crankcase, up to eight hours or overnight to warm up the block.

STARTING SYSTEM

The diesel starting system includes one (sometimes two) heavy duty batteries, the starter, and the glow plug circuit. In addition to the heavy duty battery(ies), the majority of diesel engines also have starters and battery cables designed specifically as heavy duty items for diesel usage only. Because of the high compression of any diesel, the torque required to turn the engine is much greater than a gasoline engine. The starter must be powerful enough to handle the increased load; the battery cables must be thick enough to withstand the heat generated by the starter load.

For battery maintenance, see the regular "Maintenance" section. Jump starting procedures for a dual battery car are given below. Starter maintenance is included in the appropriate car section.

The glow plug circuit is used on the diesel to initially start the engine. When the ignition switch is turned to the ON position, a light will come on in the instrument panel signalling that the glow plugs are preheating the combustion chambers. After a certain interval (depending on how cold the engine is), the light will go off. This signals that the starter may be engaged and the engine started. If the glow plug circuit malfunctions, especially in cold weather, the engine will be almost impossible to start.

CAUTION

NEVER use "starting aids" (e.g.—ether) to help start a diesel engine—serious engine damage will result.

Glow Plug Testing

To test each individual glow plug, disconnect the busbar and/or wire connector from the glow plug and connect a test light between the glow plug terminal and the positive battery terminal. If the test light lights, the glow plug is working. Replace individual glow plugs which do not work.

NOTE: Some diesel engines are equipped with either "slow glow" or "fast glow" glow plugs. Do not attempt to interchange any parts of these two glow plug systems.

To test the glow plug circuit, connect a test light to the terminal of one of the glow plugs (glow plug wiring still attached) and turn the ignition to the heating position. The test light should light for a short while. If not, the glow plug circuit is malfunctioning and must be diagnosed and repaired.

NOTE: Perform this operation on a cold engine only.

Jump-Starting a Dual Battery Diesel

Some diesels are equipped with two 12 volt batteries. The batteries are connected in parallel circuit (positive terminal to positive terminal, negative terminal to negative terminal). Hooking the batteries up in parallel circuit increases battery cranking power without increasing total battery voltage output (12 volts). On the other hand, hooking two 12 volt batteries up in a series circuit (positive terminal to negative terminal, positive terminal to negative terminal) increases total battery output to 24 volts (12 volts + 12 volts).

CAUTION

NEVER hook the batteries up in a series circuit; SEVERE electrical system damage will result.

In the event that a dual battery diesel must be jumped started, use the following procedure.

1. Open the hood and locate the batteries.
2. Position the donor car so that the jumper cables will reach from its battery (must be 12 volt, negative ground) to the appropriate battery in the diesel. Do not allow the cars to touch.
3. Shut off all electrical equipment on both vehicles. Turn off the engine of the donor car, set the parking brakes on both vehicles and block the wheels. Also, make sure both vehicles are in Neutral (manual

transmission models) or Park (automatic transmission models).

4. Using the jumper cables, connect the positive (+) terminal of the donor car battery to the positive terminal of one (not both) of the diesel batteries.

5. Using the second jumper cable, connect the negative (−) terminal of the donor battery to a solid, stationary, metallic point on the diesel (alternator bracket, engine block, etc.). Be very careful to keep the jumper cables away from moving parts (cooling fan, alternator belt, etc.) on both vehicles.

6. Start the engine of the donor car and run it at moderate speed.

7. Start the engine of the diesel.

8. When the diesel starts, disconnect the battery cables in the reverse order of attachment.

ENGINE MECHANICAL SYSTEMS

Included are engine lubrication and engine compression.

Although diesel engines are very low in carbon monoxide (CO) and hydrocarbon (HC) emissions, ''particulate'' emission output is very high from diesel engines. This is evident from the black smoke emitted by diesels, which is most noticeable during hard acceleration or high engine loads. The particulates are made up of mostly soot (carbon) and sulpher particles. The majority of these particulates are released into the atmosphere. However, some of the particulate matter, because it is produced within the engines cylinders, is left inside the engine and gradually contaminates the engine oil. This contamination makes the oil corrosive, due to the sulpher, and abrasive, due to the carbon. Serious engine damage will result if these contaminants continue to accumulate in the oil. Engine oil and filters of diesel engines must be changed more frequently than those of gasoline engines, due to the increased rate at which the contaminants form in the diesel. Consult the ''Maintenance'' section for oil and filter change procedures. The manufacturer's recommended oil change interval will be given in the owner's manual. An explanation of diesel engine oils is given at the end of this section.

As explained earlier, very high cylinder compression is the key to the operation of the diesel engine. The normal compression of most gasoline engines will rarely exceed 180 psi; whereas with diesel engines, compression pressures of 350–400 psi are commonplace.

CAUTION

DO NOT attempt to check the compression of a diesel engine with a standard compression gauge—personal injury could result. A special, high pressure compression gauge is needed to safely check the compression of any diesel.

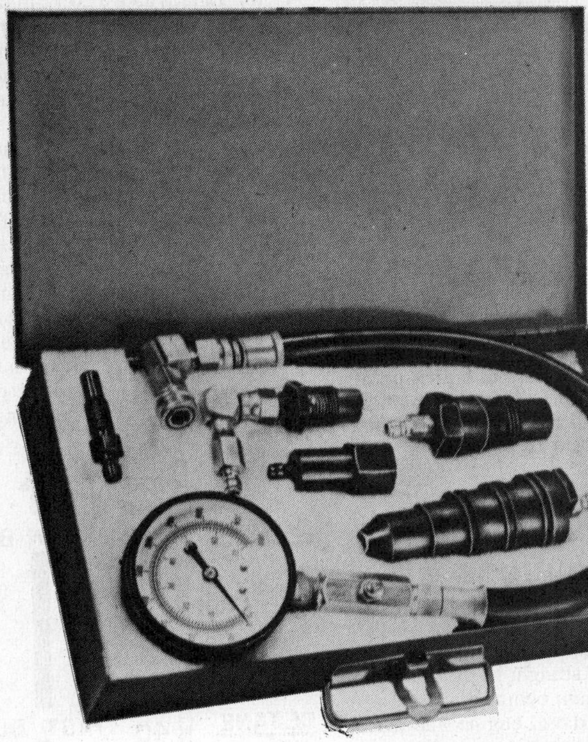

A diesel compression tester kit with adaptors (Courtesy S & G Tools).

Compression Test

1. Remove the air cleaner.

2. Disconnect the wire from the fuel shutoff solenoid terminal of the injection pump.

3. Disconnect the wires from the glow plugs and remove all glow plugs.

4. Screw compression gauge into the glow plug hole in the cylinder being checked.

5. Crank the engine, allowing six ''puffs'' for each cylinder.

The lowest reading cylinder should not be less than 70% of the highest, and no cylinder should be less than 275 pounds.

Idle Speed Adjustments

Idle speed adjustment procedures for individual diesel engines are given in the car section. Consult the following section for procedures to measure idle speed

Connecting a Tachometer to a Diesel Engine

As mentioned earlier, the diesel engine does not require an electrical ignition system. Because of this, problems arise when attempts are made to connect a tachometer to the engine for the purpose of idle adjustments, etc. The average gasoline engine tachometer senses the ignition spark pulses and converts them into a readable engine rpm signal. This type of tachometer is useless on the diesel engine, because of the diesel's compression ignition system.

There are several magnetic and photoelectric tachometers available from various tool manufacturers which were designed specifically for use with the diesel engine. These units can run into a little more money than the average do-it-yourselfer may be willing to spend, in which case any adjustments requiring the monitoring of engine rpm should be performed by a competent service technician.

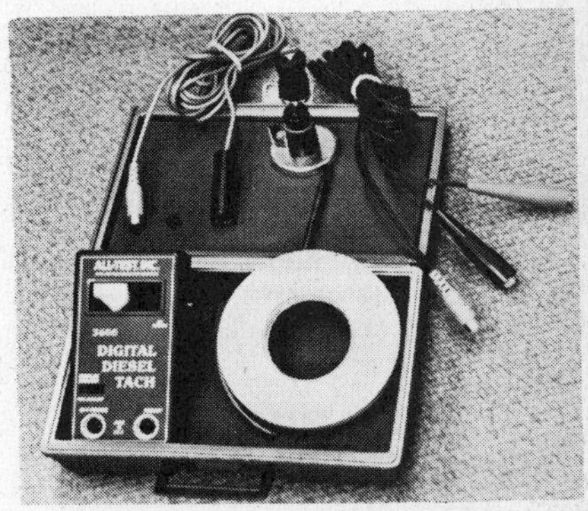

The newest equipment for measuring idle speed on a diesel engine includes (clockwise from lower left) a digital diesel tach display, photomagnetic pick-up with display input, magnetic swivel base (holder), DC power source for the display unit and a roll of magnetic tape.

The magnetic tape is attached to any moving part (such as the balancer). The pieces of tape must be at least 6 inches apart. Aim the photomagnetic pick-up at the moving object and adjust the position of the pick-up until the "on-target" light is lit. Flip the switch to TACH and read the rpm.

Diesel Engine Precautions

- Never run the engine with the air cleaner removed: if anything is sucked into the inlet manifold it will go straight to the combustion chambers, or jam behind a valve.
- Never wash a diesel engine: the reaction of a warm fuel injection pump to cold (or even warm) water can ruin the pump.
- Never operate a diesel engine with one or more fuel injectors removed unless fully familiar with injector testing procedures: some diesel injection pumps spray fuel at up to 1400 psi—enough pressure to allow the fuel to penetrate your skin.
- Do not skip engine oil and filter changes.
- Strictly follow the manufacturer's oil and fuel recommendations as given in the owner's manual.
- Do not use home heating oil as fuel for your diesel.
- Do not use "starting aids" (e.g.—ether) in the automotive diesel engine, as these "aids" can cause severe internal engine damage.
- Do not run a diesel engine with the "Water in Fuel" warning light on in the dashboard.
- If removing water from the fuel tank yourself, use the same caution you would use when working around gasoline engine fuel components.
- Do not allow diesel fuel to come in contact with rubber hoses or components on the engine, as it can damage them.

Fuel and Oil Recommendations

FUEL

Fuel makers produce two grades of diesel fuel, No. 1 and No. 2, for use in automotive diesel engines. Generally speaking, No. 2 fuel is recommended over No. 1 for driving in temperatures above 20°F. In fact, in many areas, No. 2 diesel is the only fuel available. By comparison, No. 2 diesel fuel is less volatile than No. 1 fuel, and gives better fuel economy. No. 2 fuel is also a better injection pump lubricant.

Two important characteristics of diesel fuel are its cetane number and its viscosity.

The cetane number of a diesel fuel refers to the ease with which a diesel fuel ignites. High cetane numbers mean that the fuel will ignite with relative ease or that it ignites well at low temperatures. Naturally, the lower the cetane number, the higher the temperature must be to ignite the fuel. Most commercial fuels have cetane numbers that range from 35 to 65. No. 1 diesel fuel generally has a higher cetane rating than No. 2 fuel.

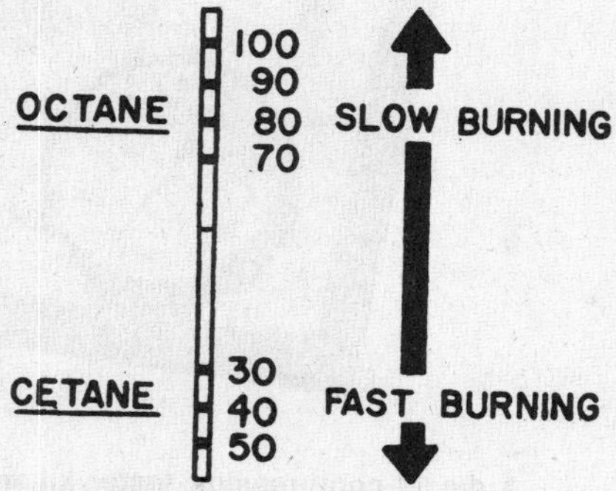

Cetane (diesel engine) versus octane (gasoline engine) ratings. The higher the cetane number, the faster the fuel burns

Viscosity is the ability of a liquid, in this case diesel fuel, to flow. Using straight No. 2 diesel fuel below 20°F can cause problems, because this fuel tends to become cloudy, meaning wax crystals begin forming in the fuel. In extreme cold weather, No. 2 fuel can stop flowing altogether. In either case, fuel flow is restricted, which can result in a "no start" condition or poor engine performance. Fuel manufacturers often "winterize" No. 2 diesel fuel by using various fuel additives and blends (No. 1 diesel fuel, kerosene, etc.) to lower its winter-time viscosity. Generally speaking, though, No. 1 diesel fuel is more satisfactory in extremely cold weather.

NOTE: No. 1 and No. 2 diesel fuels will mix and burn with no ill effects, although the engine manufacturer will undoubtedly recommend one or the other. Consult the owner's manual for information.

Depending on local climate, most fuel manufacturers make winterized No. 2 fuel available seasonally.

Many automobile manufacturers publish pamphlets giving the locations of diesel fuel stations nationwide. Contact the local dealer for information.

Do not substitute home heating oil for automotive diesel fuel. While in some cases, home heating oil refinement levels equal those of diesel fuel, many times they are far below diesel engine requirements. The result of using "dirty" home heating oil will be a clogged fuel system, in which case the entire system may have to be dismantled and cleaned.

One more word on diesel fuels. Don't thin diesel fuel with gasoline in cold weather. The lighter gasoline, which is more explosive, will cause rough running at the very least, and may cause extensive engine damage if enough is used.

OIL

Diesel engines require different engine oil from those used in gasoline engines. Besides doing the things gasoline engine oil does, diesel oil must also deal with increased engine heat and the diesel blow-by gases, which create sulphuric acid, a high corrosive.

Under the American Petroleum Institute (API) classifications, gasoline engine oil codes begin with an "S", and diesel engine oil codes begin with a "C". This first letter designation is followed by a second letter code which explains what type of service (heavy, moderate, light) the oil is meant for. For example, the top of a typical oil can will include: "API SERVICES SC, SD, SE, CA, CB, CC". This means the oil in the can is a good, moderate duty engine oil when used in a diesel engine.

It should be noted here that the further

COMPARISON OF #1 AND #2 DIESEL FUEL

Requirement	1-D	2-D
Flash Point, °F minimum	100	125
Cetane Number, minimum	40	40
Viscosity at 100°F, Centistokes		
Minimum	1.4	2.0
Maximum	2.5	4.3
Water and Sediment, % by volume maximum	Trace	0.05
Sulfur, % by weight maximum	0.5	0.5
Ash, % by weight maximum	0.01	0.01

Flash Point: The temperature at which diesel fuel ignites when exposed to a flame *in the open air.*
Cetane Number: See text

down the alphabet the second letter of the API classification is, the greater the oil's protective qualities are (CD is the severest duty diesel engine oil, CA is the lightest duty oil, etc.). The same is true for gasoline engine oil classifications (SF is the severest duty gasoline engine oil, SA is the lightest duty oil, etc.).

Many diesel manufacturers recommend an oil with both gasoline and diesel engine API classifications. Consult the owner's manual for specifications.

The top of the oil can will also contain an SAE (Society of Automotive Engineers) designation, which gives the oil's viscosity. A typical designation will be: SAE 10W-30, which means the oil is a "winter" viscosity oil, meaning it will flow and give protection at low temperatures.

On the diesel engine, oil viscosity is critical, because the diesel is much harder to start (due to its higher compression) than a gasoline engine. Obviously, if you fill the crankcase with a very heavy oil during winter (SAE 20W-50, for example), the starter is going to require a lot of current from the battery to turn the engine. And, since batteries don't function well in cold weather in the first place, you may find yourself stranded some morning. Consult the owner's manual for recommended oil specifications for the climate you live in.

LUBE OIL ANALYSIS

From an oil sample a laboratory can diagnose many potential engine problems—from piston wear to impending bearing failure. What's more, the laboratory can spot them quicker, and with greater accuracy. Just as easily, the lab can give the diesel a clean bill of health, saving the car owner unnecessary servicing and other routine preventive maintenance, costly in time and money.

There's nothing new about engine lube oil analysis. Thousands of the nation's trucks and buses regularly have their engine's lube oil analyzed by laboratories specializing in this type of work. What is new is the availability of lube oil analysis to individual vehicle owners rather than, as before, almost exclusively to companies operating fleets of diesel equipment.

Lube oil analysis can be a valuable indicator of internal engine condition.

Here's how lube oil analysis works. You write one of the several laboratories that offer individual diesel vehicle owners lube analysis service. By return mail you'll receive an oil sampling kit. It will probably contain a two-ounce plastic oil sampling container with a screw-on plastic top. Instructions tell you how to take the sample. Usually, a lab-bound sample of diesel lube oil may be taken in any of three ways, but always right after the engine has been shut off, so that the sampled oil is as close as possible to normal engine operating temperature. That's important to assure that the lab's test will be accurate. Oil samples can be taken during normal oil changes, when lube oil is drained anyway. Between oil changes, a sample can be drawn from the engine through the dipstick tube (where you normally check the oil's level). In drawing an oil sample from the dipstick tube, a small suction bulb fitted with a length of disposable tubing is used. The tubing is merely inserted into the dipstick tube, the suction bulb depressed, and the oil sample drawn. The third method of sampling is by loosening the drain plug on the engine's bypass oil filter (if your diesel has one). A little oil is caught in the lube sampling container. In all cases, extreme cleanliness is a must, so as not to contaminate the sample with dirt, grease, or other substances not actually found inside the engine. For example, using a rag that contains solvents, metal filings, or other impurities can contaminate the oil sample, leading to false and even alarming lab reports. A bit of technique is required: In taking a sample of lube oil during a routine oil drain, about half of the crankcase's lube oil should be allowed to drain out before the sample is taken. The sample taken, the date, make and model of the engine, its mileage, mileage since last oil change, and sometimes oil type are noted on the container's label, and the container is mailed to the laboratory.

Shortly, you'll receive the lab's report, which, based on a number of tests, including spectrochemical analysis (using a spectrometer, which can detect the presence of virtually all basic elements and contaminants), tells what's in the oil in what quantities and analyzes both the probable source of what was found and whether it indicates trouble. For one example, the finding of more than trace amounts of copper in an oil sample may strongly point to excessive bearing wear in a particular diesel whose bearings contain copper. Some analyses report on as many as eighteen basic elements that may be found in a diesel's lube oil sample, and in the report's "recommendation" may pinpoint their probable source—as, "indicates piston ring wear." Also indicated is the presence of such contaminants as water, solids (the products of oxidation and engine blow-by), and fuel dilution. Noted, too, is the lubricity of the sample—whether, or not, in the lab's opinion, it is still doing its internal engine lubricating job.

NOTE: Never use lube analysis and a lab's report of "good oil" to extend, beyond the manufacturer's recommendation, the mileage period between oil changes. Follow the manufacturers recommendations.

The more frequently an engine is lube-sampled, the more accurate and meaningful the lab's reports. Infrequent samplings, although they can spot sudden, unusual changes in internal engine condition, may fail to show the gradual deterioration of engine parts. Ideally, you should have the laboratory analyze a lube sample every other oil change. For most automobile diesels, that's every 6,000 miles. Analysis costs from $7 to $11 per sample. Drive an average 18,000 miles a year and you'd change your diesel's oil three times. In that time, you'd submit three samples to the lab at an annual lube analysis cost of $21 to $33.

Aftermarket Fuel System Accessories

Due to reasons described previously, most diesel engine problems can be attributed to either fuel contamination or cold weather fuel performance characteristics. Diesel-engined vehicle manufacturers have designed and installed various systems to combat these problems, but ultimately, their best efforts are limited by cost.

Inconvenience is a major concern to diesel owners. If water accumulates (in substantial quantities) in the diesel fuel system, the fuel and water must be siphoned from the fuel tank and purged from the remainder of the fuel system. It goes without saying that this operation is a messy, time-consuming process. Even if the vehicle is equipped with a water/fuel separator having a drain valve, the owner must manually open the valve from either under the hood or beneath the vehicle.

Although the fuel filter installed by the manufacturer offers adequate performance when maintained properly, the addition of another, separate diesel fuel filter is a wise improvement.

If you live in an extremely cold climate, you've probably experienced cold starting problems due to fuel "waxing", plugged filters, "gelled" fuel, etc. If your vehicle is not factory-equipped with the optional fuel line or cylinder block heaters, these

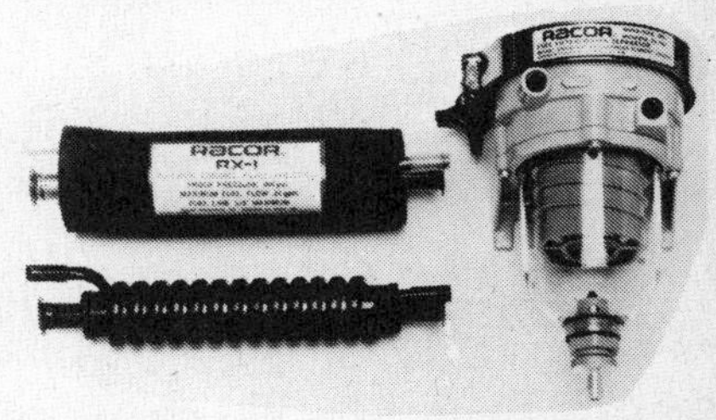

Aftermarket fuel filter/water separator and fuel line heater

heaters can be purchased from the aftermarket (retail auto parts manufacturers). The installation of either of these items can improve cold-starting dramatically.

WATER/FUEL SEPARATORS

Centrifugal Action

Sometimes referred to as a "cyclonic" water/fuel separator, this device uses baffles which spin the fuel as it comes through the separator inlet. Since water is heavier than diesel fuel, the water will spin away from the fuel, sink to the bottom of the separator, and collect in the sediment bowl.

This type of separator is most efficient in dealing with large water droplets. If the water is in emulsion with the fuel, that is, if the water is equally dispersed through the fuel in very small droplets, some of the water will remain with the fuel to travel through the fuel system.

Coalescing Action

In this type of separator, the fuel must pass through a coalescent filtering media before proceeding through the fuel system. The idea behind the coalescent media is to trap even the smallest droplets of water on the media. As the small droplets combine into larger, heavier droplets, gravity acts on the droplets to pull them downward, off of the media and into the sediment bowl.

FUEL FILTER/SEPARATOR COMBINATION UNITS

Most separators of either the centrifugal or coalescent types are available with disposeable fuel filtering elements which are built into the separator unit. If your car already has a large, disposeable filter, it would probably be more cost-effective to stay with a separator only, and to change the factory-equipped filter at the recommended intervals. Should your vehicle have a fairly small filter, and/or an inconveniently located water drain (or none at all), choose the filter/separator combination. The filter/separator offers both increased fuel filtering ability and efficient water separation.

Convenience Add-Ons

Available with many separators and filter/separators are items such as dash-mounted water-in-fuel indicator lamps, audible water-in-fuel alarms, and dash-controlled water ejection systems. A properly chosen system would warn you of water in the fuel, and allow you to eject the water by simply "flipping" a dash-mounted switch.

Installing a Separator

Clear installation instructions and the necessary installation parts will be provided with the separator kit. Follow those instructions exactly. A general list of suggestions follows:

1. Fuel additives should not be used unless approved by the separator manufacturer.
2. Do not install a separator within 4" of any exhaust system component.
3. If plastic fittings are supplied with the kit, do not replace them with metal fittings. Also, use extreme caution when tightening the fittings, especially those made of plastic.
4. Use a fuel-proof sealer on all fitting threads, only if the threads are not factory-coated with sealer.
5. Use only fuel-proof hoses for the installation.
6. Do not eliminate the original equipment fuel filter, even if a filter/separator is installed.
7. For new car warranty purposes, a filter/separator should be located BEFORE the original equipment filter. The fuel must pass through the original filter last, before entering the fuel injection pump.
8. If any type of fuel line heater is installed, it is best to position the heater between the fuel tank and the separator inlet.
9. To ease the job of the separator, the separator should be installed between the fuel transfer pump and the tank (unless the separator manufacturer specifies otherwise). Fuel and water which have been churned through the fuel transfer pump will be more difficult to separate.
10. Be sure that any wiring (for warning lamps, water ejection, etc.) is routed and connected properly. If the wiring must pass through a drilled hole, be sure to use a rubber grommet between the drilled component(s) and the wire to prevent damage to the wire.

FUEL LINE HEATERS

Two popular types of fuel line heaters are available for diesel passenger cars. Both types raise the temperature of the fuel to prevent "waxing" and "gelling" of the fuel in the lines during cold weather operation. One type uses engine coolant as a heating source. In order for this type to heat the fuel, the engine must first be started and allowed to run until the coolant temperature increases. Though this type of heater will usually increase fuel mileage, it offers no aid in starting ability.

The other type of heater uses a 12V DC electric heating element. This type is recommended, due to its ability to warm the fuel BEFORE the engine is started. This type of heater will also usually increase the overall fuel mileage.

Installation

Follow the manufacturer's instructions exactly. Also, see suggestions 5, 8, and 9 under "Separator Installation".

CYLINDER BLOCK HEATERS

A cylinder block heater electrically (usually 110V house current) heats the engine coolant, which in turn warms the cylinder block, heads, and engine oil. In this case, the warmth is not used to alter the characteristics of the fuel. Block heaters offer two main advantages when starting a diesel in cold weather:

1. The reduced viscosity (thinning) of the engine oil from the warmth allows the engine to be "turned over" easier (and faster) by the starter. Less strain is imposed on the starting system.
2. Because the diesel relies on the heat of compression to ignite the fuel, the increase in the base combustion chamber temperature results in a higher tempearture during compression. This allows the fuel to ignite easier than if just the glow plugs were used.

Installation

Most cylinder block heaters replace one of the existing freeze (or expansion) plugs of the cylinder block. Follow the manufacturers installation instructions exactly. Also, refer to the manufacturers recommendations for usage.

Carburetor Service 26

Functions

Gasoline is the source of fuel for power in the automobile engine and the carburetor is the mechanism which automatically mixes liquid fuel with air in the correct proportions to provide the desired power output from the engine. The carburetor performs this function by metering, atomizing, and mixing fuel with air flowing through the engine.

A carburetor also regulates the volume of air-to-fuel mixture which enters the engine. It is the carburetor's regulation of the mixture flow which gives the operator control of the engine speed.

METERING

The automotive internal conbustion engine operates efficiently within a relatively small range of air-to-fuel ratios. It is the function of the carburetor to meter the fuel in exact proportions to the air flowing into the engine, so that the optimum ratio of air-to-fuel is maintained under all operating conditions. Regulations governing exhaust gas emissions have made the proper metering of fuel by the carburetor an increasingly important factor. Too rich a mixture will result in poor fuel economy and increased emissions, while too lean a mixture will result in loss of power and generally poor performance.

Carburetors are matched to engines so that metering can be accomplished by using carefully calibrated metering jets which allow fuel to enter the engine at a rate proportional to the engine's ability to draw air.

ATOMIZATION

The liquid fuel must be broken up into small particles so that it will more readily mix with air and vaporize. The more contact the fuel has with the air, the better the vaporization. Atomization can be accomplished in two ways: air may be drawn into a stream of fuel which will cause a turbulence and break the solid stream of fuel into smaller particles; or a nozzle can be positioned at the point of highest air velocity in the carburetor and the fuel will be torn into a fine spray as it enters the air stream.

DISTRIBUTION

The carburetor is the primary device involved in the distribution of fuel to the engine. The more efficiently fuel and air are combined in the carburetor, the smoother the flow of vaporized mixture through the intake manifold to each combustion chamber. Hence, the importance of the carburetor in fuel distribution.

Principles

VACUUM

All carburetors operate on the basic principle of pressure difference. Any pressure less than atmospheric pressure is considered vacuum or a low pressure area. In the engine, as the piston moves down on the intake stroke with the intake valve open, a partial vacuum is created in the intake manifold. The farther the piston travels downward, the greater the vacuum created in the manifold. As vacuum increases in the manifold, a difference in pressure occurs between the carburetor and cylinder. The carburetor is positioned in such a way that the high pressure above it, and the vacuum or low pressure beneath it, causes air to be drawn through it. Fuel and air always move from high to low pressure areas.

VENTURI PRINCIPLE

To obtain greater pressure drop at the tip of the fuel nozzle so that fuel will flow, the principle of increasing the air velocity to create a low pressure area is used. The device used to increase the velocity of the air flowing through the carburetor is called a venturi. A venturi is a specially designed restriction placed on the air flow. In order for the air to pass through the restriction, it must accelerate, causing a pressure drop or vacuum as it passes.

Circuits

FLOAT CIRCUIT

The float circuit includes the float, float bowl, and a needle valve and seat. This circuit controls the amount of gas allowed to flow into the carburetor.

As the fuel level rises, it causes the float to rise which pushes the needle valve into its seat. As soon as the valve and seat make contact, the flow of gas is cut off from the fuel inlet. When the level of fuel drops, the float sinks and releases the needle valve from its seat which allows the gas to flow in. In actual operation, the fuel is maintained at practically a constant level. The float tends to hold the needle valve partly closed so that the incoming fuel just balances the fuel being withdrawn.

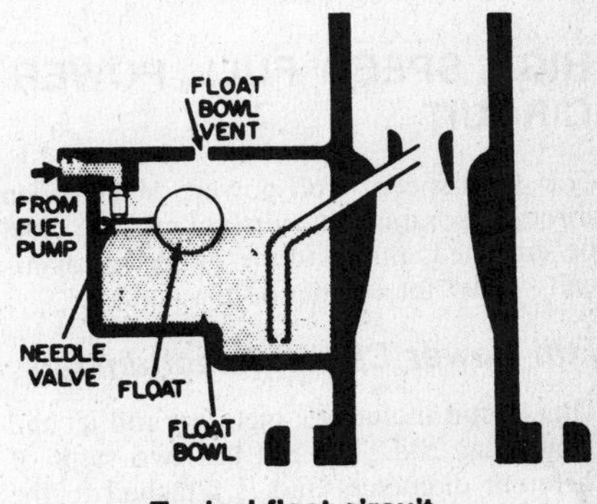

Typical float circuit

IDLE AND LOW SPEED CIRCUIT

When the throttle is closed or only slightly opened, the air speed is low and practically no vacuum develops in the venturi. This means that the fuel nozzle will not feed. Thus, the carburetor must have another circuit to supply fuel during operation with a closed or slightly opened throttle.

This circuit is called the idle and low speed circuit. It consists of passages in which air and gas can flow beneath the throttle plate. With the throttle plate closed, there is high vacuum from the intake manifold. Atmospheric pressure pushes the air/fuel mixture through the passages of the idle and low speed circuit and past the tapered point of the idle adjustment screw, which regulates engine idle mixture volume.

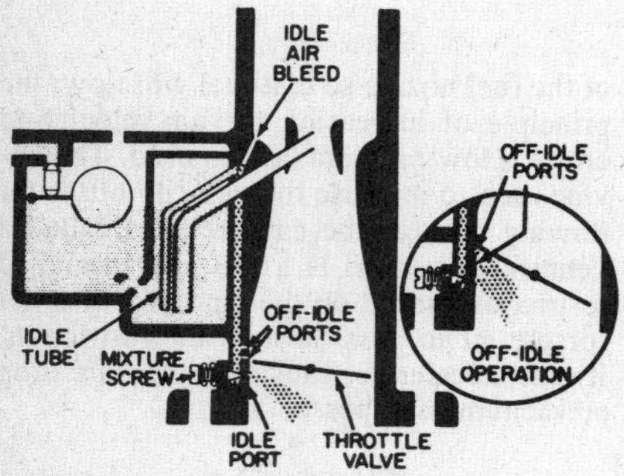

Typical idle and low speed circuit

HIGH SPEED PARTIAL LOAD CIRCUIT

When the throttle plate is opened sufficiently, there is little difference in vacuum between the upper and lower part of the air horn. Thus, little air/fuel mixture will discharge from the low speed and idle circuit. However, under this condition enough air is moving through the air horn to produce vacuum in the venturi to cause the main nozzle or high speed nozzle to discharge fuel. The circuit from the float bowl to the main nozzle is called the high speed partial load circuit. A nearly constant air/fuel ratio is maintained by this circuit from part to full-throttle.

HIGH SPEED FULL POWER CIRCUIT

For high-speed, full-power, wide open throttle operation, the air/fuel mixture must be enriched; this is done either mechanically or by intake manifold vacuum.

Full Power Circuit (Mechanical)

This circuit includes a metering rod jet and a metering rod. The rod has two steps of different diameters and is attached to the throttle linkage.

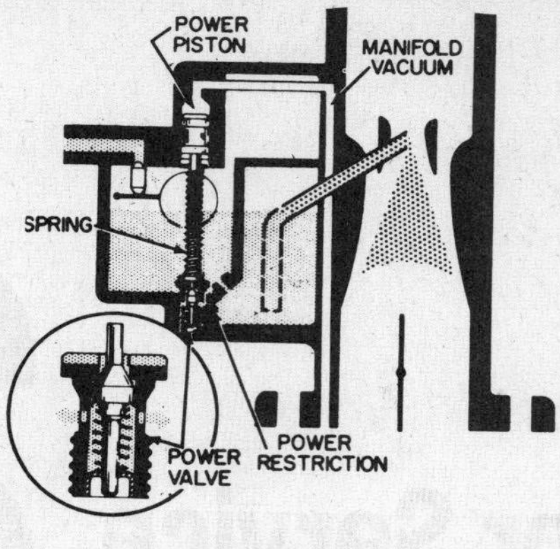

Typical power circuit

When the throttle is wide open, the metering rod is lifted, bringing the smaller diameter of the rod into the jet. When the throttle is partly closed, the larger diameter of the metering rod is in the jet. This restricts fuel flow to the main nozzle but adequate amounts of fuel do flow for part-throttle operation.

Full Power Circuit (Vacuum)

This circuit is operated by intake manifold vacuum. It includes a vacuum diaphragm or piston linked to a valve.

When the throttle is opened so that intake manifold vacuum is reduced, the spring raises the diaphragm or piston. This allows more fuel to flow in, either by lifting a metering rod or by opening a power valve.

ACCELERATOR PUMP CIRCUIT

For acceleration, the carburetor must deliver additional fuel. A sudden inrush of air is caused by rapid acceleration or applying full throttle.

When the throttle is opened, the pump lever pushes the plunger down and this forces fuel to flow through the accelerator pump circuit and out the pump jet. This fuel enters the air passage through the carburetor to supply additional fuel demands.

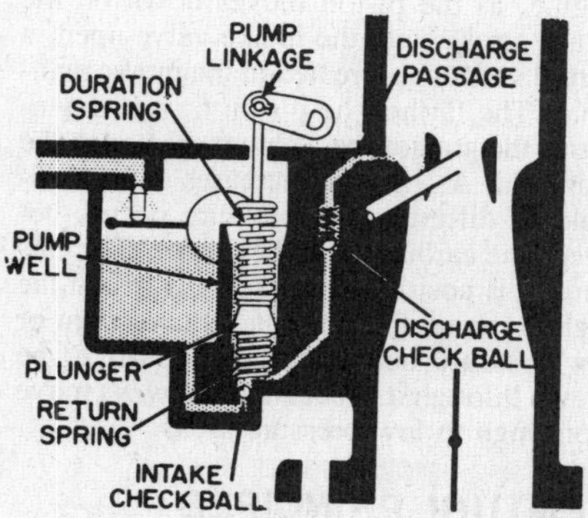

Typical accelerator pump circuit

CHOKE

When starting an engine, it is necessary to increase the amount of fuel delivered to the intake manifold. This increase is controlled by the choke.

The choke consists of a valve in the top of the air horn controlled mechanically by an automatic device. When the choke valve is closed, only a small amount of air can get past it. When the engine is cranked, a fairly high vacuum develops in the air horn. This vacuum causes the main nozzle to discharge a heavy stream of fuel. The quantity delivered is sufficient to produce the correct air/fuel mixture needed for starting the engine. The choke is released either manually or by heat from the engine.

OVERHAUL

Generally, when a carburetor requires major service, a rebuilt one is purchased on an exchange basis, or a kit may be bought for overhauling the carburetor.

The kit contains the necessary parts and some form of instructions for carburetor rebuilding. The instructions may vary between a simple exploded view and detailed step-by-step rebuilding instructions. Unless you are familiar with carburetor overhaul, the latter should be used.

There are some general overhaul procedures which should always be observed:

Efficient carburetion depends greatly on careful cleaning and inspection during overhaul since dirt, gum, water, or varnish in or on the carburetor parts are often responsible for poor performance.

Overhaul your carburetor in a clean, dust-free area. Carefully disassemble the carburetor, referring often to the exploded views. Keep all similar and lookalike parts segregated during disassembly and cleaning to avoid accidental interchange during assembly. Make a note of all jet sizes.

When the carburetor is disassembled, wash all parts except diaphragms, electric choke units, pump plunger, and any other plastic, leather, fiber, or rubber parts in clean carburetor solvent. Do not leave parts in the solvent any longer than is necessary to sufficiently loosen the deposits. Excessive cleaning may remove the special finish from the float bowl and choke valve bodies, leaving these parts unfit for service. Rinse all parts in clean solvent and blow them dry with compressed air or allow them to air dry. Wipe clean all cork, plastic, leather, and fiber parts with a clean, lint-free cloth.

Blow out all passages and jets with compressed air and be sure that there are no restrictions or blockages. Never use wire or similar tools to clean jets, fuel passages, or air bleeds. Clean all jets and valves separately to avoid accidental interchange.

Check all parts for wear or damage. If wear or damage is found, replace the defective parts. Especially check the following:

1. Check the float needle and seat for wear. If wear is found, replace the complete assembly.
2. Check the float hinge pin for wear and the float(s) for dents or distortion. Replace the float if fuel has leaked into it.
3. Check the throttle and choke shaft bores for wear or an out-of-round condition. Damage or wear to the throttle arm, shaft, or shaft bore will often require replacement of the throttle body. These parts require a close tolerance of fit; wear may allow air leakage, which could affect starting and idling.

NOTE: Throttle shafts and bushings are not included in overhaul kits. They can be purchased separately.

4. Inspect the idle mixture adjusting needles for burrs or grooves. Any such condition requires replacement of the needle, since you will not be able to obtain a satisfactory idle.
5. Test the accelerator pump check valves. They should pass air one way but not the other. Test for proper seating by blowing and sucking on the valve. Replace the valve if necessary. If the valve is satisfactory, wash the valve again to remove breath moisture.
6. Check the bowl cover for warped surfaces with a straightedge.
7. Closely inspect the valves and seats for wear and damage, replacing as necessary.
8. After the carburetor is assembled, check the choke valve for freedom of operation.

Carburetor overhaul kits are recommended for each overhaul. These kits contain all gaskets and new parts to replace those that deteriorate most rapidly. Failure to replace all parts supplied with the kit (especially gaskets) can result in poor performance later.

Some carburetor manufacturers supply overhaul kits of three basic types: minor repair; major repair; and gasket kits. Basically, they contain the following:

Minor Repair Kits:
- All gaskets
- Float needle valve
- Volume control screw
- All diaphragms
- Spring for the pump diaphragm

Major Repair Kits:
- All jets and gaskets
- All diaphragms
- Float needle valve
- Volume control screw
- Pump ball valve
- Main jet carrier
- Float
- Complete intermediate rod
- Intermediate pump lever
- Complete injector tube
- Some cover hold-down screws and washers

Gasket Kits:
- All gaskets

After cleaning and checking all components, reassemble the carburetor, using new parts and referring to the exploded view. When reassembling, make sure that all screws and jets are tight in their seats, but do not overtighten, as the tips will be distorted. Tighten all screws gradually, in rotation. Do not tighten needle valves into their seats; uneven jetting will result. Always use new gaskets. Be sure to adjust the float level when reassembling.

Stromberg Carburetors Only

The preceding information applies to Stromberg carburetors also, but the following, additional suggestions should be followed.

1. Soak the small cork gaskets (jet gland washers) in penetrating oil or hot water for at least a half hour prior to assembly, or they will invariably split.
2. When the jet is fully assembled, the jet tube should be a close fit without any lateral play, but it should be free to move smoothly. A few drops of oil, or polishing of the tube may be necessary to achieve this.
3. If the jet sealing ring washer is made of cork, soak it in hot water for a minute or two prior to installation.
4. Adjust the float height.
5. Center the jet so that the piston will fall freely (when raised) and seat with a distinct click. If the jet is not centered properly, it will hang up in the tube.

TROUBLESHOOTING

NOTE: Carburetor problems cannot be isolated effectively unless all other engine systems are functioning correctly and the engine is properly tuned.

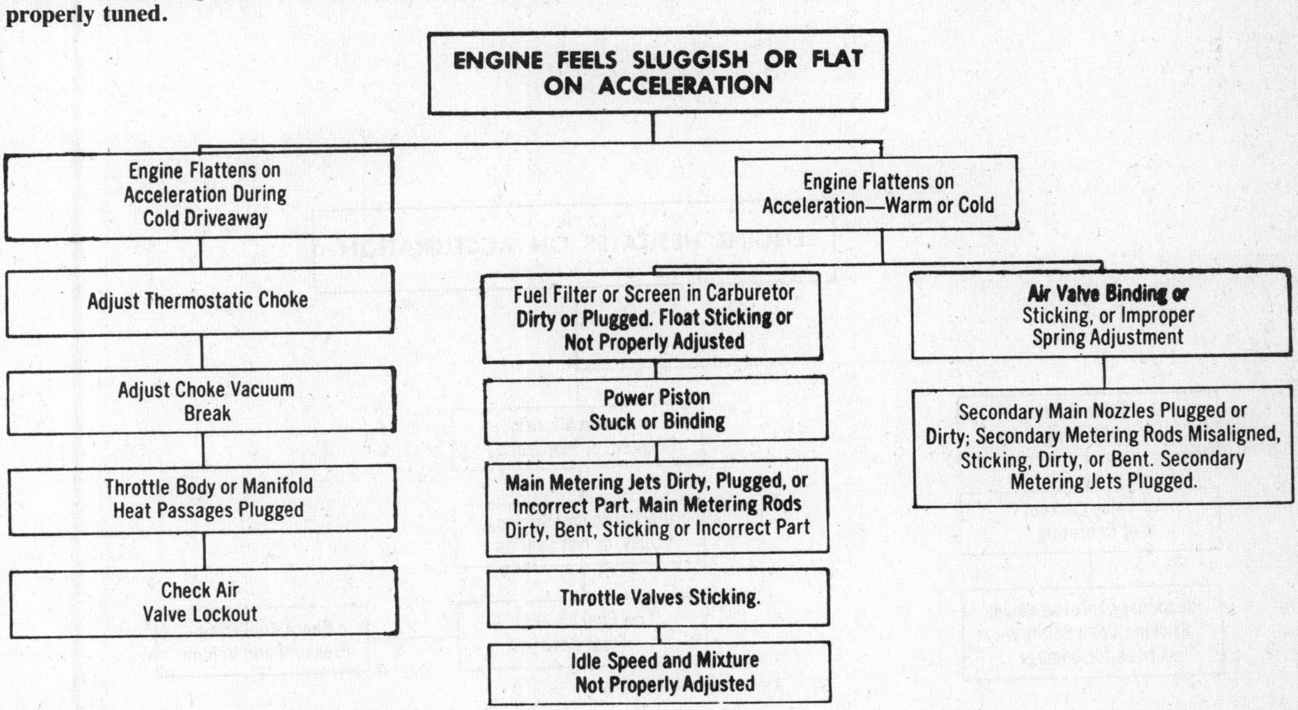

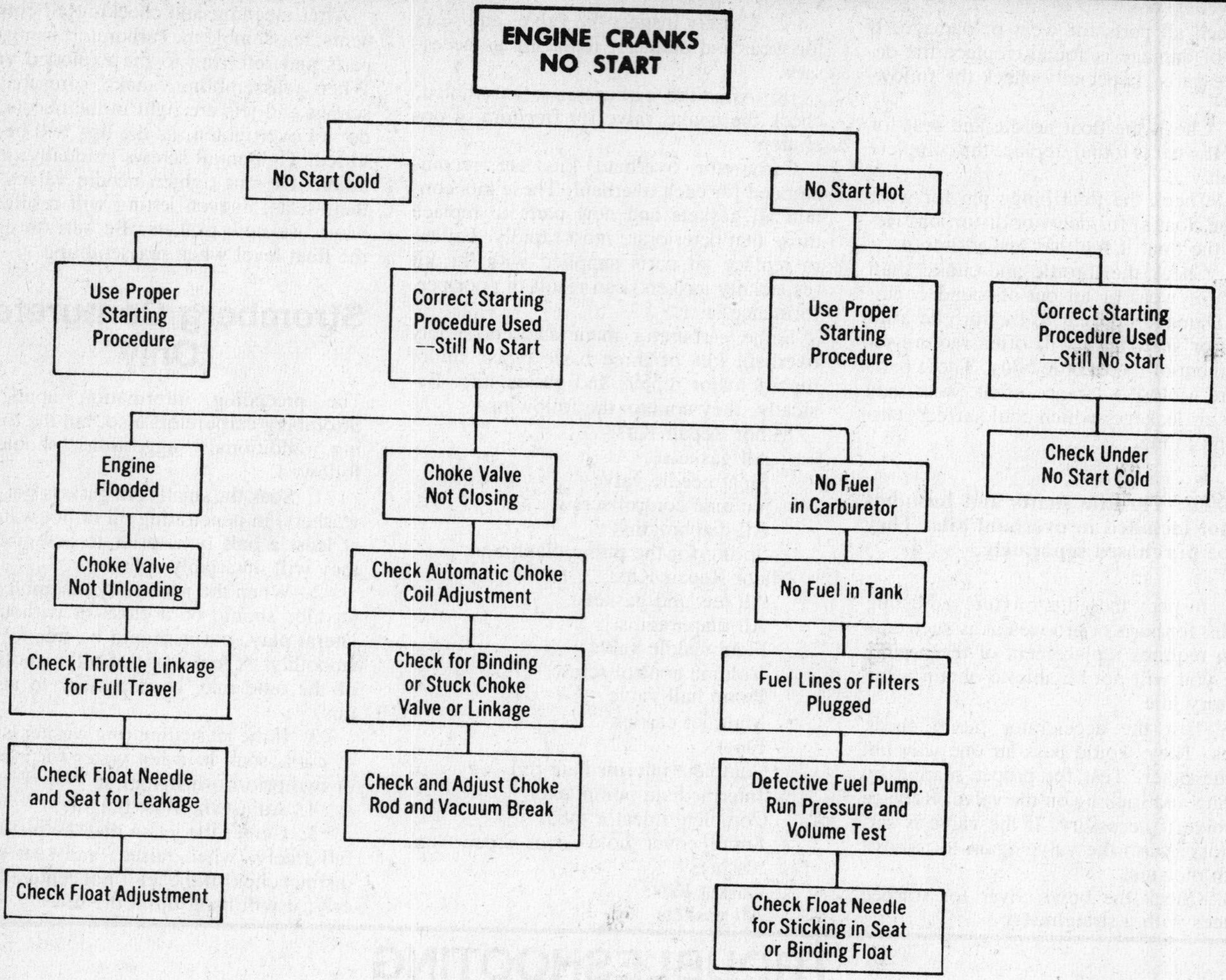
ENGINE CRANKS NO START
No Start Cold
No Start Hot
Use Proper Starting Procedure
Correct Starting Procedure Used —Still No Start
Use Proper Starting Procedure
Correct Starting Procedure Used —Still No Start
Engine Flooded
Choke Valve Not Unloading
Check Throttle Linkage for Full Travel
Check Float Needle and Seat for Leakage
Check Float Adjustment
Choke Valve Not Closing
Check Automatic Choke Coil Adjustment
Check for Binding or Stuck Choke Valve or Linkage
Check and Adjust Choke Rod and Vacuum Break
No Fuel in Carburetor
No Fuel in Tank
Fuel Lines or Filters Plugged
Defective Fuel Pump. Run Pressure and Volume Test
Check Float Needle for Sticking in Seat or Binding Float
Check Under No Start Cold

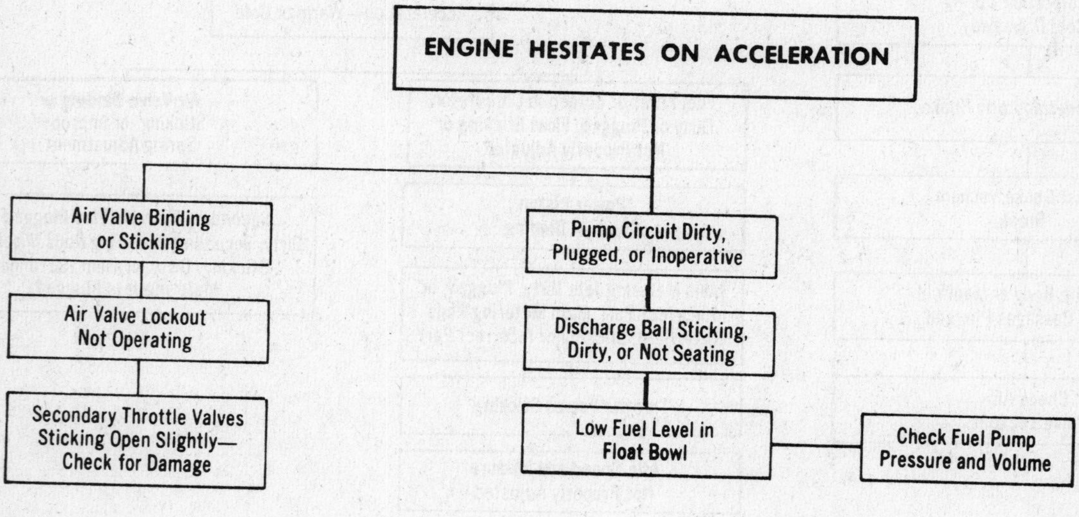
ENGINE HESITATES ON ACCELERATION
Air Valve Binding or Sticking
Air Valve Lockout Not Operating
Secondary Throttle Valves Sticking Open Slightly— Check for Damage
Pump Circuit Dirty, Plugged, or Inoperative
Discharge Ball Sticking, Dirty, or Not Seating
Low Fuel Level in Float Bowl
Check Fuel Pump Pressure and Volume

Emission Controls 27

GENERAL INFORMATION

The earth's atmosphere, at or near sea level, consists of approximately 78% nitrogen, 21% oxygen and 1% other gases. If it were possible to remain in this state, 100% clean air would result. However, many varied causes allow other gases and particulates to mix with clean air, causing the air to become unclean or polluted. Some of these pollutants are visible while others are invisible, with each having the capability of causing distress to the eyes, ears, throat, skin and respiratory system. These pollutants can also cause damage to the environment and to the many man made objects that are exposed to the elements. To better understand the causes of air pollution, pollutants can be categorized into 3 separate types, natural, industrial and automotive.

Natural Polution

This type of pollution has been present on earth before man appeared and is still a factor to be considered when discussing air pollution, although it causes only a small percentage of the present overall pollution problem existing in our country. It is the direct result of decaying organic matter, windborn smoke and particulates from such natural events as forest fires, volcanic ash, sand and dust which can spread over a large area of the countryside.

Industrial Polution

This type of pollution is caused primarily by industrial processes which are the burning of coal, oil and natural gas. The by-product of which in turn produces smoke and fumes. This type of polution occurs most severely during still, damp and cool weather. Working with Federal, State and Local mandated rules, regulations and by carefully monitoring the emissions, industries have greatly reduced the amount of pollutant emitted from their industrial sources.

Automotive Polution

This type of air pollution is the automotive emissions. The emissions from the internal combustion engine were not an appreciable problem years ago because of the small number of registered vehicles and the nation's small highway system. However, during the early 1950's, the trend of the American people was to move from the cities to the surrounding suburbs. This caused an immediate problem in the transportation area because the majority of the suburbs were not afforded mass transit conveniences. This lack of transportation created an attractive market for the automobile, which resulted in a dramatic increase in the number of vehicles produced and sold, along with a marked increase in highway construction between the cities and the suburbs. Multi-car families emerged with much emphasis placed on the individual vehicle per family member. As the increase in vehicle ownership and usage occurred, so did the pollutant levels in and around the cities. It was noted that a fog and smoke type haze was being formed and at times, remained in suspension over the cities and did not quickly dissipate. At first this smog, was thought to result from industrial pollution, but it was determined that the automobile emission was largely to blame.

CATEGORIZING VEHICLE EMISSIONS AND CONTROLS

To recognize the sources and methods used to control vehicle emission, three major categories have been established. They are crankcase emissions and controls, fuel evaporative

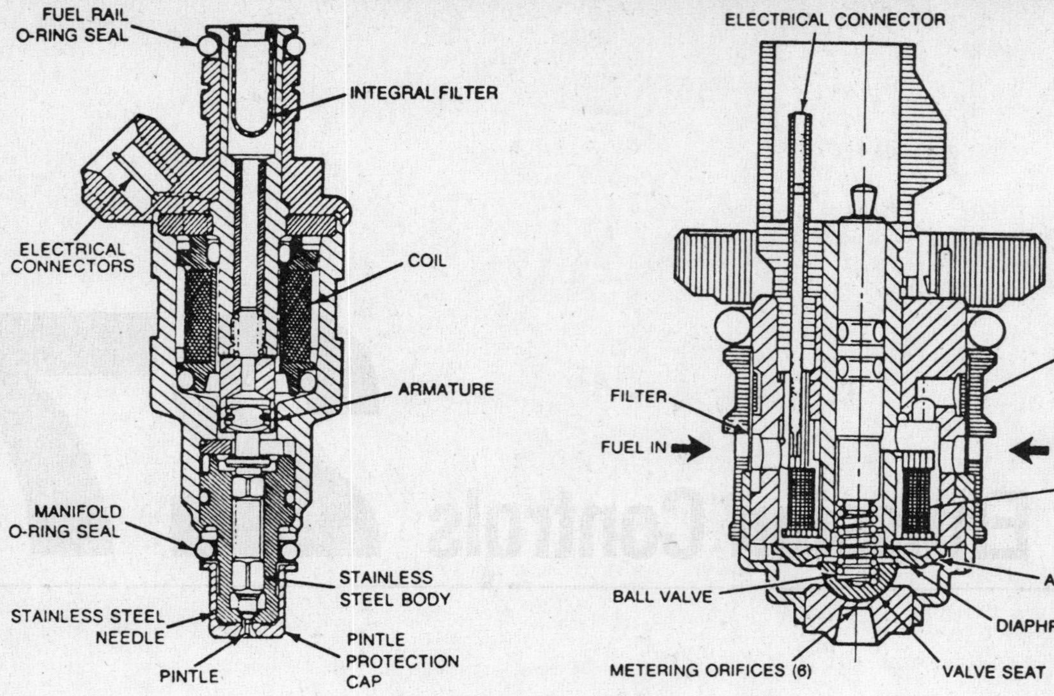

Sectional view of typical fuel injectors

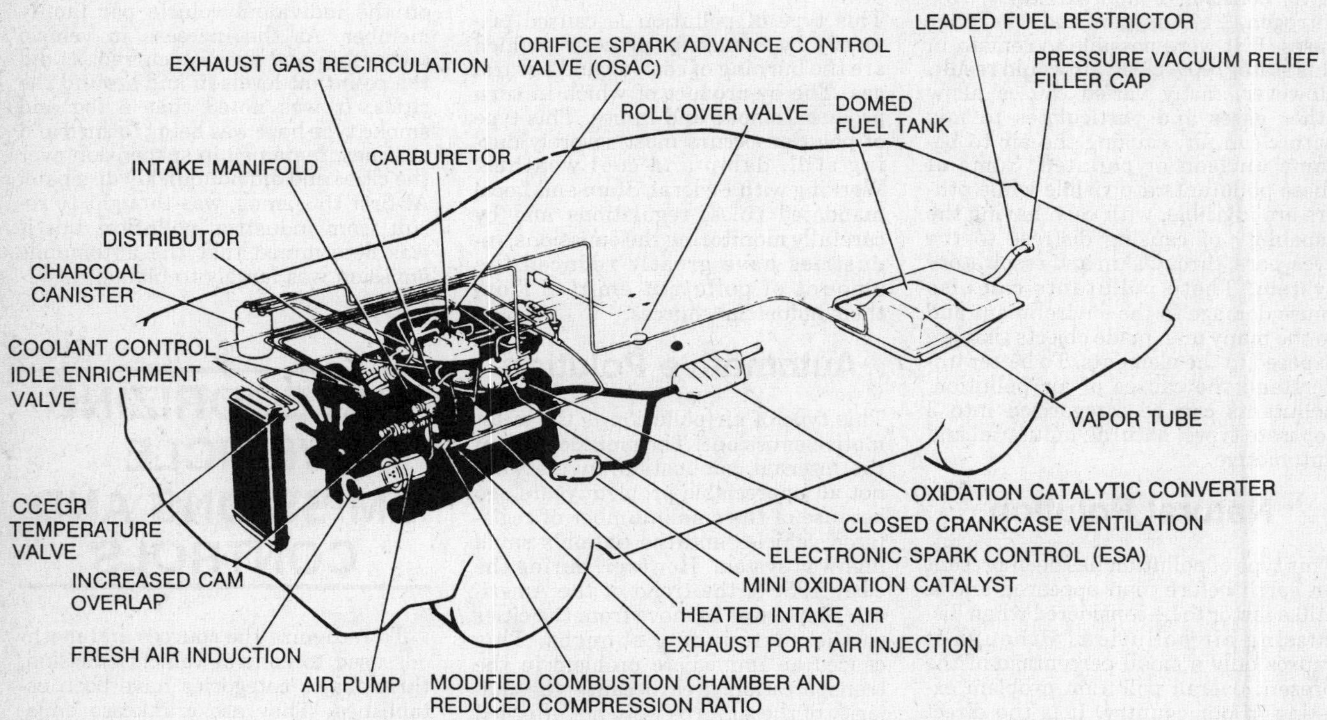

Typical emission control system component schematic—carbureted vehicles

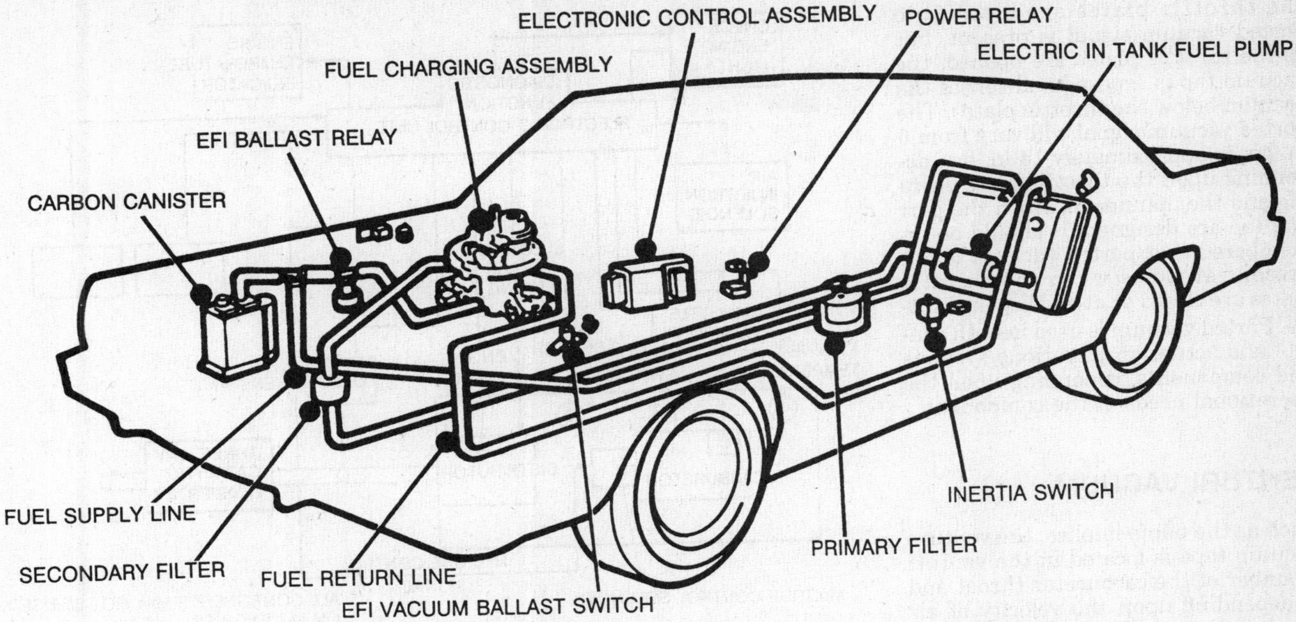

Typical emission control system component schematic—fuel injected vehicle

emissions and controls and exhaust emissions and controls.

Regardless of the manufacturer, or type of emission control device that is used, a means of actuation must be applied to the device in order for it to operate at a specific time or temperature during either the vehicle operation modes or during the combustion process. The actuating methods commonly used are vacuum, electrical, temperature and mechanical.

Vacuum Sources

The most common method of component actuation is by engine vacuum and has been used since the conception of emission controls. Three major sources of vacuum are obtained from the engine are manifold vacuum, ported vacuum and venturi vacuum. However, with the increased use of smaller engines, the demand for engine vacuum could not be totally supplied by the engine. Vacuum pumps were added to the engine and assisted in supplying the necessary required vacuum needs.

MANIFOLD VACUUM

The engine could be considered a large vacuum pump by having a constant negative pressure developed within the intake manifold as the engine is operated. The tap for this type vacuum is taken from the below the throttle plates or directly from the intake manifold. This source of vacuum will vary in strength between 17–22 in. hg. at idle, to approximately 0 in. hg. at wide open throttle. With the engine in a deceleration mode, the manifold vacuum will be at its highest, which is above idle specifications. Manifold vacuum, normally in conjunction with a vacuum reservoir or amplifier to insure an adequate vacuum volume, is used to actuate the emission control components rapidly.

PORTED VACUUM

The ported vacuum tap is located di-

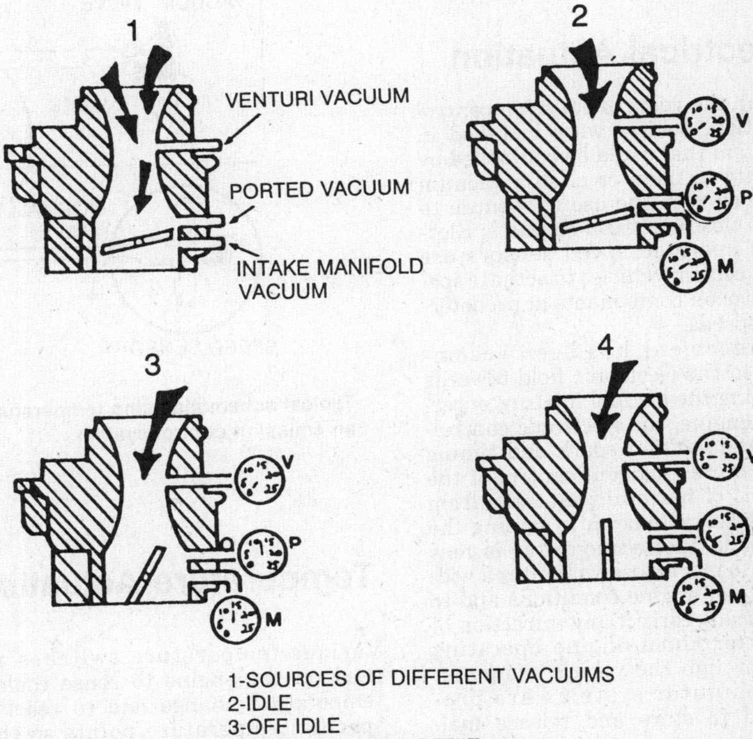

Engine vacuum sources

rectly above the throttle plates. When the throttle plates are closed, no ported vacuum signal is present, but as the throttle plates are opened, the vacuum tap is exposed and senses the vacuum below the throttle plates. The ported vacuum signal will vary from 0 in. hg. to approximately 14 in. hg., depending upon the throttle plate opening and the manner in which the port and tap are designed. It should be remembered that ported vacuum is not present at times when the throttle plates are closed or at wide open throttle. Ported vacuum is used in both control and actuation of various systems and components, depending upon the operational needs of the component.

VENTURI VACUUM

Such as the name implies, the venture vacuum tape is located in the venturi chamber of the carburetor throat and is depending upon the velocity of air flowing through the venturi chamber. An example would be as the throttle plates are opened and the velocity of the air flow increases, so would the venturi vacuum signal. The venturi vacuum varies from 0 in. hg. to approximately 4 in. hg. The venturi vacuum is normally used as a control or triggering signal to a component, so that ported or manifold vacuum can be applied to an emission control system.

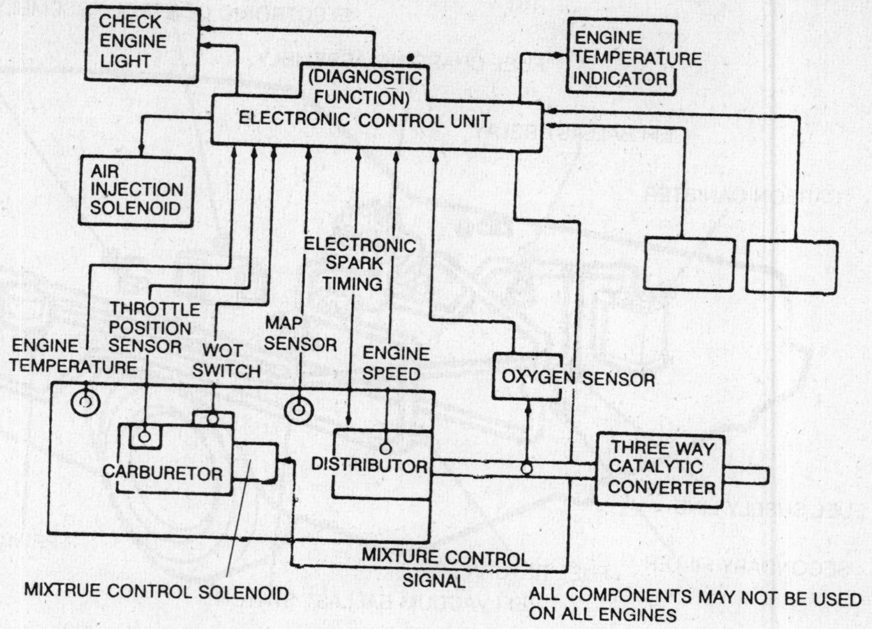

Typical schematic using electrical actuation to operate an emission control system

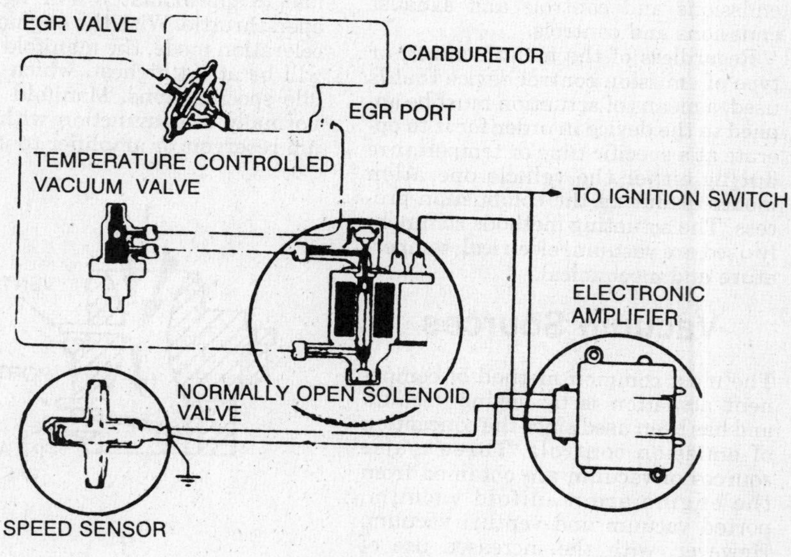

Typical schematic using temperature, electronics and manual actuation to operate an emission control system

Electrical Actuation

Through the years of emission control application, devices were installed in the system that could be controlled by electricity to open or close a vacuum passage or with the use of vacuum to open or close electrical contacts. Electrically operated speed sensors are used on certain vehicles to actuate specific emission components at predetermined speeds.

Advancement has been accomplished in the electronic field towards controlling the air/fuel mixture entering the engine, the electronic controlling of the ignition spark and timing and the more stringent control of the emissions of harmful pollutants from the exhaust system while gaining the optimum in fuel economy. It is now possible to more closely monitor a wider variety of engine conditions and to electronically correct any infraction of a pre-determined engine operating mode, through the vehicle computer. Most computer systems are programmed to store and release malfunctioning or system defects information by electronically probing its many circuits.

Temperature Actuation

Various temperature switches are used on the engine to sense coolant temperature change and to react at specific temperature points so that vacuum passages or electrical circuits can be opened or closed. This type of switch can control numerous vacuum or electrical circuits from a single supply source.

Mechanical Actuation

Mechanical switches are used to control the opening or closing of vacuum passages or the operation of electrical switches through the use of linkages, transmission shift rails or through manual operation.

CRANKCASE EMISSION CONTROL SYSTEM

System Description

Crankcase emissions are responsible for approximately 20% of all harmful automotive pollutants before any emission controls were installed on the vehicles. Crankcase emissions are the result of compression gasses being forced past the piston rings on both the compression and power strokes, resulting in an accumulation of gases, known as blowby gases, in the engine crankcase. These blowby gases become mixed with vapors from the agitated lubricating oil and must be relieved from the crankcase area to prevent internal engine pressures from building up.

Prior to the early 60's a road draft tube was used to ventilate the crankcase, which allowed the pollutants to be emitted into the atmosphere. With the installation of a regulating valve and necessary plumbing, the road draft tube was eliminated and the gases routed to the air intake area, to be drawn into the engine to be reburned with the air/fuel mixture. At first, engine vacuum was used as the controlling factor to draw the crankcase gases into the engine, but was found that the vacuum source varied at the wrong times. Different systems were experimented with, some with flow control valves while others merely direct the gases to the air cleaner assembly. Other systems had open breather caps to allow fresh air to enter the system while others had sealed breather caps with the fresh air supply being tapped from the air cleaner snorkel.

By 1968, all vehicles manufactured in the United States were equipped with a closed crankcase ventilation system, which did not allow any of the blowby gases and oil vapors to escape into the atmosphere. It is a closed system which utilizes a flow regulating valve called the positive crankcase ventilation valve, PCV valve, or may use a restrictor orifice in place of the PCV valve.

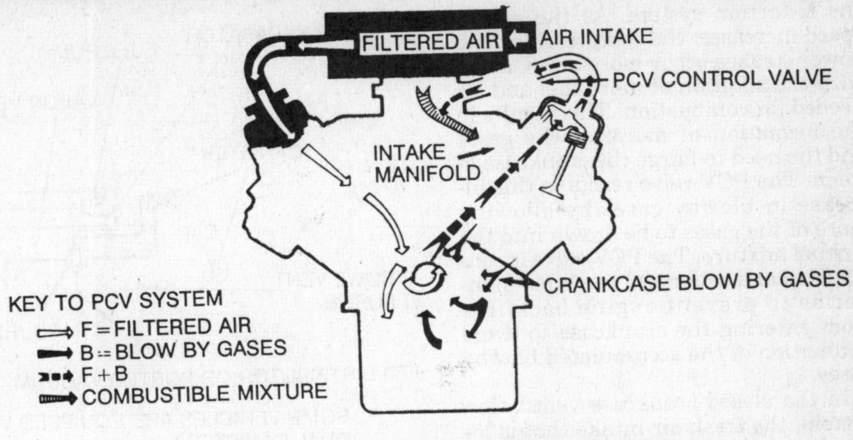

Typical closed crankcase ventilation control system

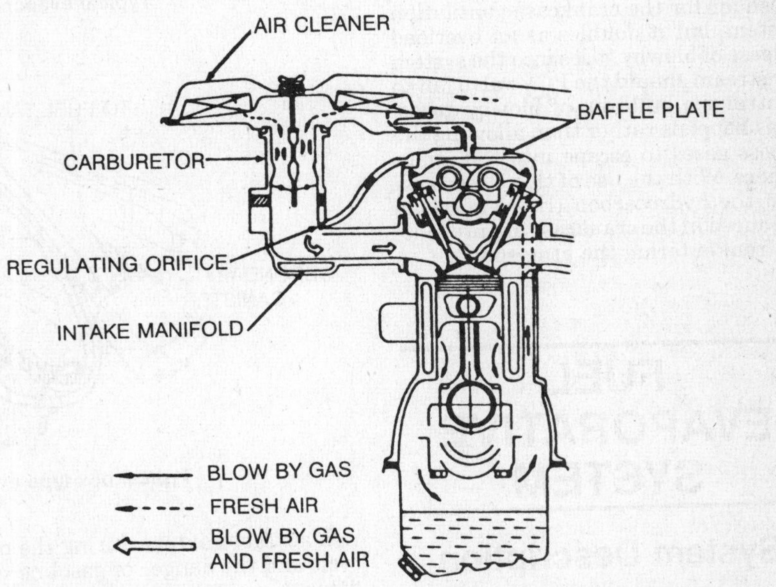

Positive crankcase ventilation system using a regulating orifice rather that a PCV valve

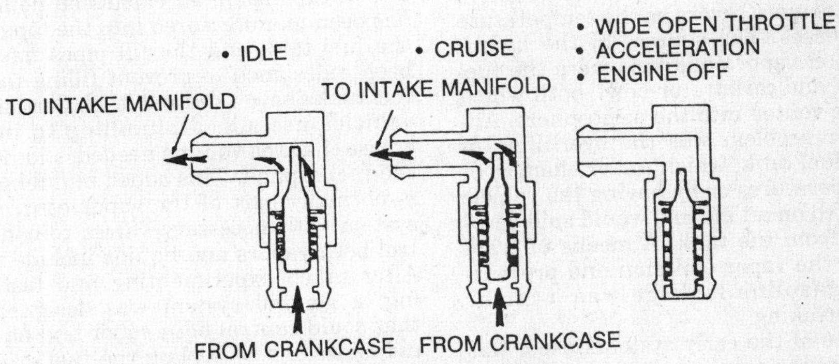

PCV valve operation

System Operation

The PCV valve is constructed and calibrated to perform the task of metering the gases from the crankcase as required and is matched to engine operation in the following manner. When the engine is idling, only a small amount of air and fuel is needed for combustion, resulting in a small amount of blowby gases being produced because the compression and power strokes are not occuring as frequently as at higher speeds. The PCV valve reacts to this lack of blowby gases and tends to restrict the flow into the induction system. As the engine speed increases, the compressions and power strokes occur more often, along with the addition of more fuel and air needed for combustion. This results in the formation of more blowby gases and the need to purge the crankcase of them. The PCV valve reacts to this increase in blowby gases by allowing more of the gases to be drawn into the air/fuel mixture. The PCV valve is constructed and calibrated in such a manner as to prevent engine backfires from entering the crankcase to avoid detonation of the accumulated blowby gases.

In the closed crankcase ventilation system, the fresh air intake that is located in the air cleaner or snorkel, has a dual role. Not only is it a source of fresh air for the crankcase ventilation system, but it doubles as an overload release of blowby gases into the system air stream should the PCV valve fail to control the build up of blowby gases. This happens rather than allowing the excess gases to escape into the atmosphere. With the use of this closed system the hydrocarbon (HC) emissions produced in the crankcase are prevented from entering the atmosphere.

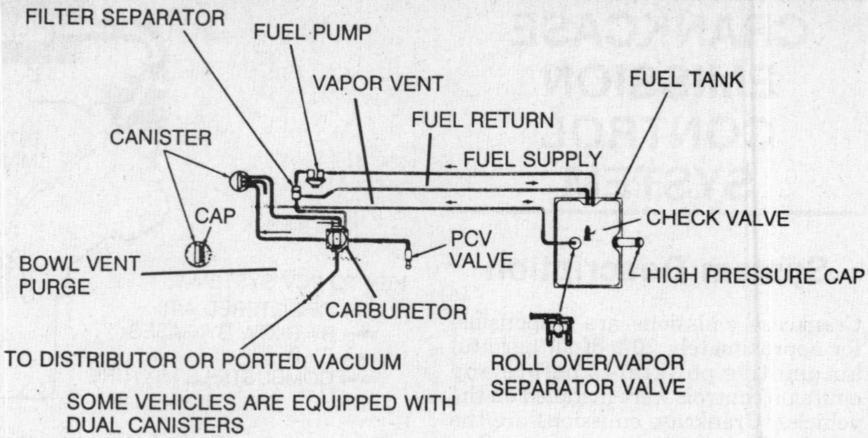

Typical evaporative control system schematic

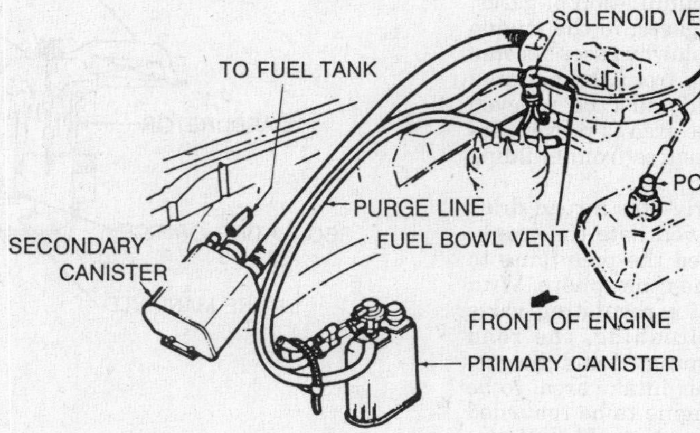

Typical box type evaporative control system schematic

FUEL EVAPORATIVE SYSTEM

System Description

Fuel evaporation vapors are found to account for approximately 20% of the total automotive emission problem and is more severe as the temperature increases. The sources of the hydrocarbon vapor emissions were the fuel tank and carburetor bowl, both which were vented into the atmosphere. Another problem was the overfilling of the fuel tank, which under changes of temperatures or by having the vehicle parked on an incline, would spill gasoline from the tank. A means of trapping the vapor emission and preventing gasoline leakage was a major undertaking.

One of the early systems used, was the engine crankcase to store the fuel vapors when the engine was not running. When the engine was started, the vapors were purged from the crankcase by the positive crankcase ventilation system. Certain drawbacks were noted in this system, some of which were the dilution of engine lubricating oils with gasoline, an over rich air/fuel mixture during the purge cycle and the danger of gasoline vapor detonation within the crankcase during engine start up.

To prevent fuel loss from the tank due to expansion, an expansion dome has been manufactured into the top of the fuel tank and the fill pipes have been redesigned to prevent filling the fuel tank above a desired level. Certain vehicles use added plumbing to increase the area volume needed, should the fuel expand. This added plumbing is normally part of the vapor control system with necessary valves to control both vapors and liquids included. After much experimenting and testing, a general system was designed that could control both vapor and liquid emissions by sealing the fuel system from the atmosphere. Although each vehicle manufacturer has designed their own vapor control system, similar components are used, resulting in systems that are basically the same in the manner or vapor collection and storage. However, the manner in which the vapors are purged may vary greatly.

System Components

SEALED FUEL TANK CAP

The first step in sealing the fuel system was to replace the vented fuel tank filler cap with a sealed cap. The venting of the fuel tank is accomplished through another component of the system which controls the vapor emission by storage. To prevent damage to the tank should excessive internal or external pressure exist due to this closed system, a pressure/vacuum relief valve is incorporated in the sealing cap. A tank pressure of ½–1 psi. can exist in the tank and is controlled by the relief valve.

FUEL TANK

The fuel tank has been redesigned to provide approximately 10% of the total tank volume for expansion space, should the fuel expand due to temperature changes. An overfill protector is provided by the filler neck to assure the expansion space is maintained

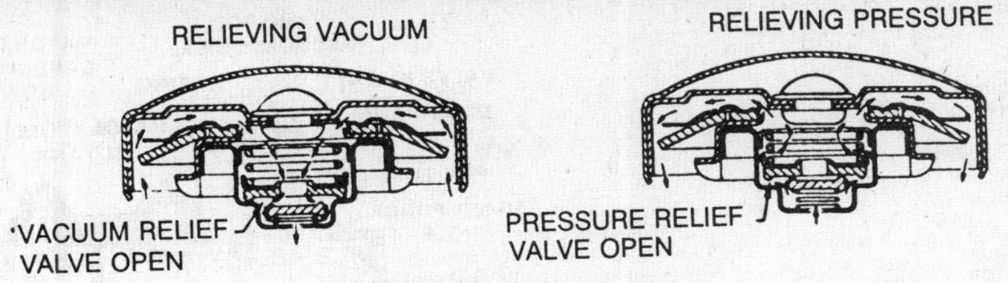

Sealed fuel tank cap operation

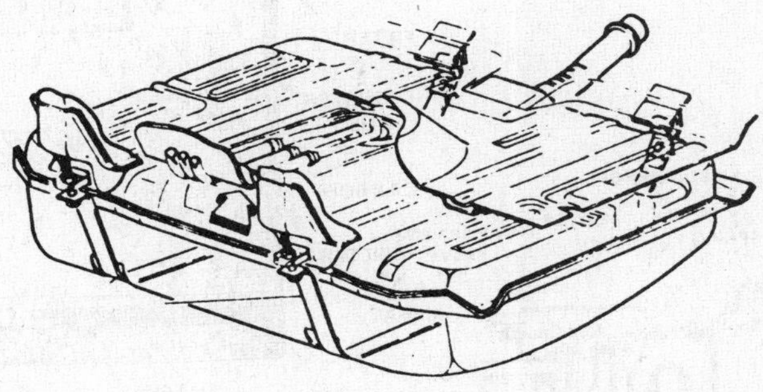

Typical emission related fuel tank

during the tank filling. The tank vapor venting is controlled by having a vent tap at or near the expansion chamber dome. A foam type filter or a vapor separator is used to allow the vapors to pass, but prevents the passage of liquid, which then returns to the fuel tank.

LIQUID/VAPOR SEPARATOR

Most all vehicles will have the liquid/vapor separator assembly within the vent lines. The purpose of the liquid/vapor separator is to assist in controlling fuel expansion and to allow the vapors to pass to a storage point while returning the liquid fuel back to the tank. Other vehicles may be equipped with separate expansion chambers, separate evaporation chambers and with one or more check valves in the lines. These components are usually located near the fuel tank. Added controls, either separate or as a part of another fuel evaporative control component, are installed to prevent the loss of fuel from the carburetor or throttle body and the fuel tank during a vehicle rollover situation.

CHARCOAL CANISTER

One of the most important components of the fuel evaporation system is the canister of activated charcoal. Activated charcoal has the capabilities to absorb and store fuel vapors. The charcoal can be purged with fresh air and cleaned of its vapors, resulting in its reuse many times. When the vehicle is not running, the fuel vapors are routed to the canister where they are absorbed and stored by the activated charcoal. When the engine is started, either engine vacuum or the air flow through the air cleaner, draws the vapors from the charcoal by allowing a metered amount of air to pass over and through the charcoal and then routed into the engine's induction system.

Three different types of purge methods are used to cleanse the vapors from the canister. The first method is the variable purge which draws the vapor laden air into the air cleaner from the canister. The variable air flow through the canister is dependent upon the air flow through the air cleaner and into the engine. The second method is called the demand purge system which connects the canister to the fuel metering system. As the throttle is opened, the engine vacuum draws the cleansing air through the canister and into the engine air flow. This type of system only operates when the throttle plates are open, therefore preventing the vapors from entering the fuel sytem metering device when the engine is idling. The third system is a combination of a constant and demand purge. One purge line is routed to the PCV line which provides a continual purge, but by having a restriction in the passage, the flow rate is controlled. Another passage is routed to the fuel metering device. However, as the throttle plates are opened, the engine vacuum acts upon the restriction in the PCV line, opening the restriction and allowing more vapors and air to flow through the PCV system and into the intake air/fuel flow.

With the increased use of fuel injection and electronics, the vapors are purged from the canister at specific engine rpm and temperatures, so as not to upset the controlled air/fuel mixture and cause increased emissions from the combustion chambers.

CARBURETOR BOWL VENTS

The carburetor bowl, regardless of its size, will allow hydrocarbons (HC) emissions to occur. To prevent this, a

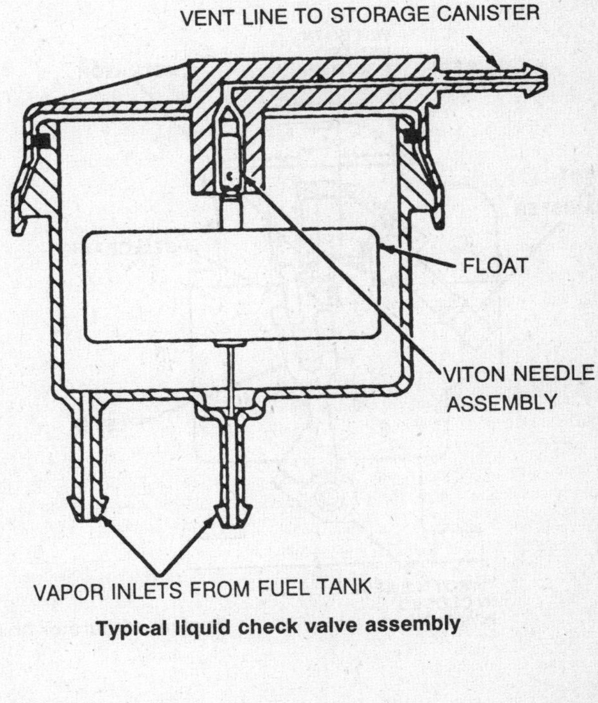

Typical liquid check valve assembly

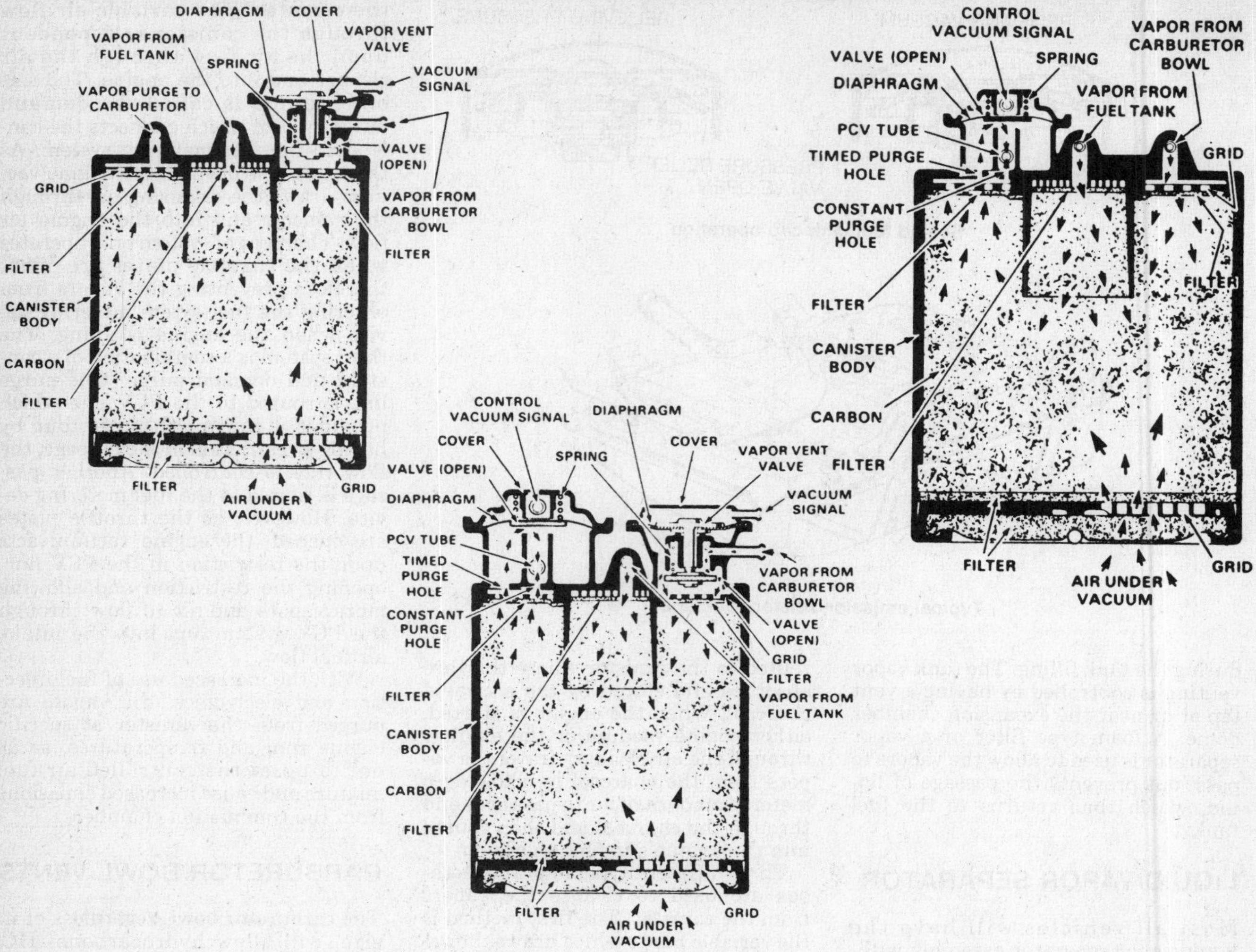

Typical vapor canisters used in the evaporative emission control system

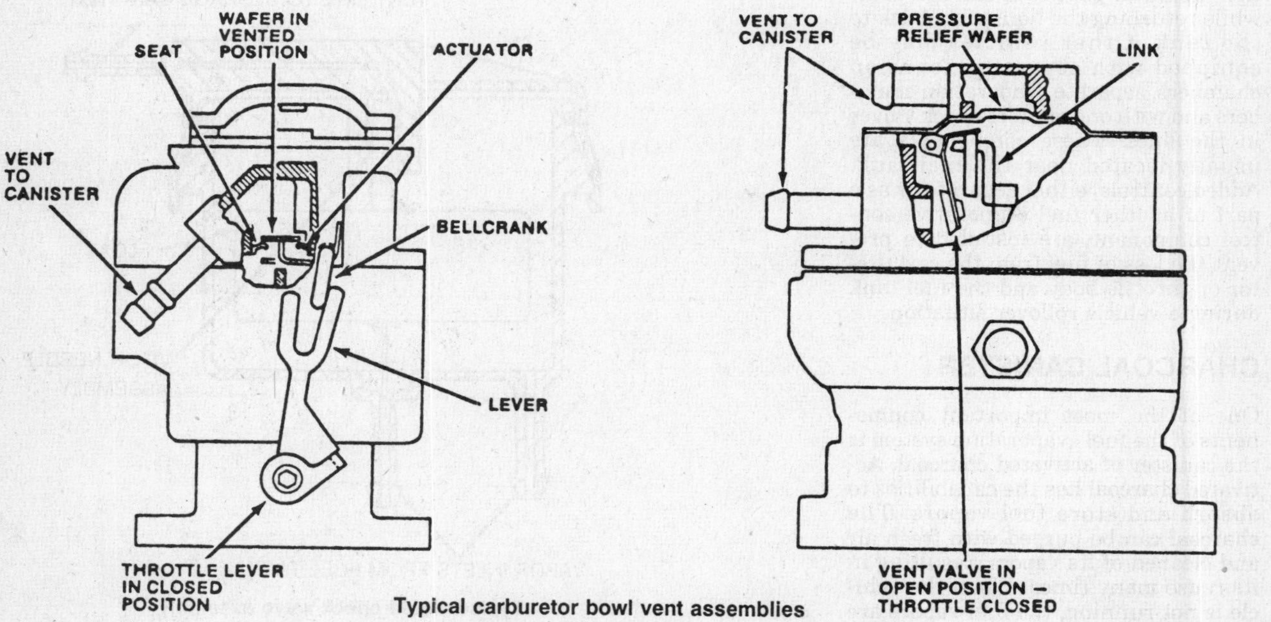

Typical carburetor bowl vent assemblies

valve normally called the anti-percolation valve, has been located at the top of the fuel bowl. During periods when the engine is idling or stopped, the valve opens the line to the charcoal canister, which receives the fuel vapors from the carburetor bowl. The vapors are then treated the same as the vapors from the fuel tank. The fuel bowl is vented internally during higher engine speeds, allowing no passage of hydrocarbon (HC) emission into the atmosphere. The carburetor bowl vent is operated either manually or electronically.

EXHAUST EMISSION CONTROL SYSTEM

System Description

The exhaust emission control system encompasses the automotive engine from the entrance of air into the engine's induction system until the exhaust byproduct of the combustion process emerges from the tail pipe. The engine exhaust was found to be responsible for approximately 60% of all automobile emissions before any pollution controls were installed on the engines. Through the trial and error period of the late 60's and early 70's many different systems were used, some separately and others in conjunction with other systems. While a number of the controls were dropped, others were refined and improved, resulting in greater emission control and driveability.

The exhaust emission system will include the following subsystems.

1. Thermostatically controlled air cleaner.
2. Air injection systems.
3. Ignition timing controls.
4. Increased temperature control.
5. Transmission or speed controlled spark system.
6. Exhaust gas recirculation system.
7. Catalytic converter and exhaust system.
8. Automotive engine feed back carburetor systems.
9. Carburetor and choke modifications.
10. Fuel injection systems.
11. Computer controlled engine controls.

System Components

THERMOSTATICALLY CONTROLLED AIR CLEANER

One of the first exhaust emission controls to be installed on the automobile engine was the thermostatically controlled air cleaner assembly. It has been modified through the years to accomodate many applications, but its main function remains, to maintain a minimum temperature of 100° F for air entering the induction system, to provide good driveability with a leaner air/fuel mixture which will help to reduce the exhaust emission of hydrocarbons (HC) and carbon monoxide (CO).

A damper valve, located in the snorkel or nozzle of the air cleaner housing, is operated by a thermostat or a vacuum motor, to provide either preheated air from a shroud around the exhaust manifold or unheated underhood air to the induction system. When the inducted air temperature is 85° F or below, the damper valve is fully closed to outside air. With the damper closed, the cool underhood air flows between the exhaust manifold and the shroud surrounding the manifold, where the air is heated. This preheated air then flows up through the hot air duct to the air cleaner and into the induction system. As the temperature of the inside air rises between 85–100° F, the thermostat or vacuum motor which is controlled by engine vacuum through a temperature sensor moves the damper valve partially open, blending the heated air with the outside air thus preventing the air temperature from becoming too hot. When the temperature in the air cleaner rises above 130° F, the thermostat or temperature sensor controlled vacuum motor opens the damper valve, assisted by a damper door spring, to outside air.

Regardless of the name given to this

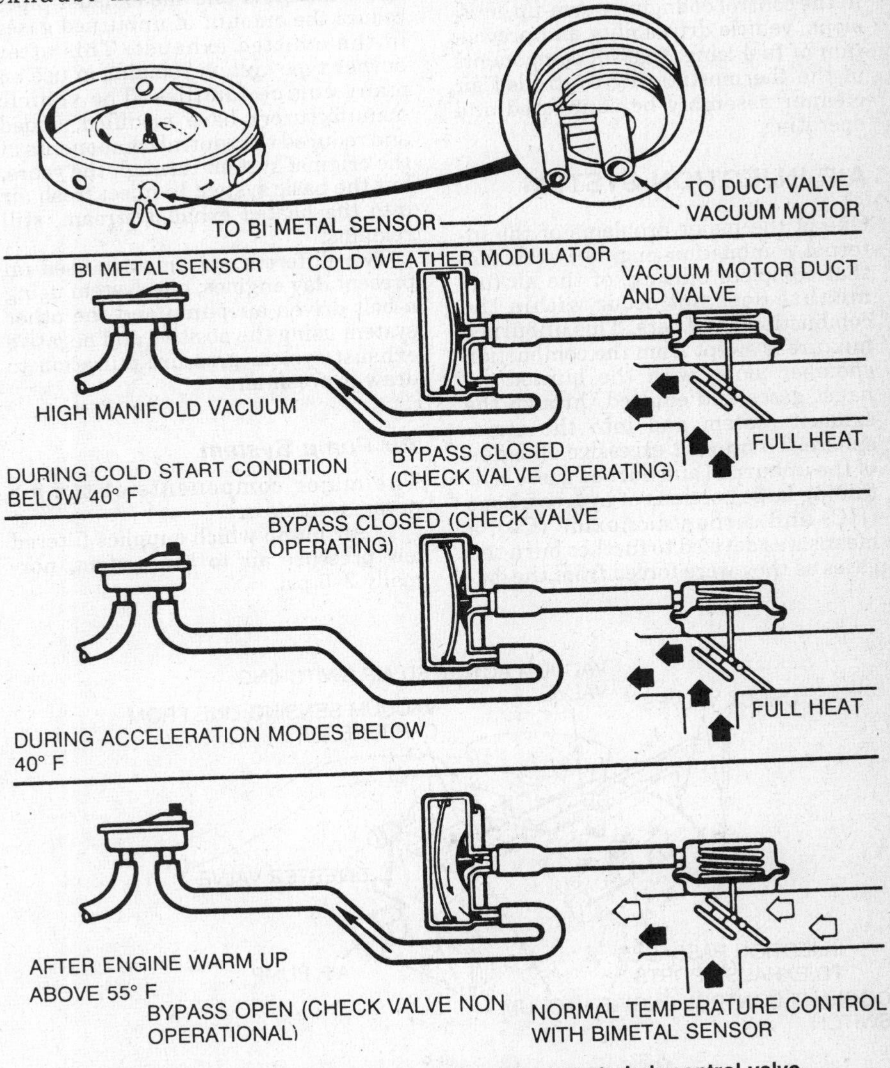

Operating sequences of the vacuum sensor operated air control valve

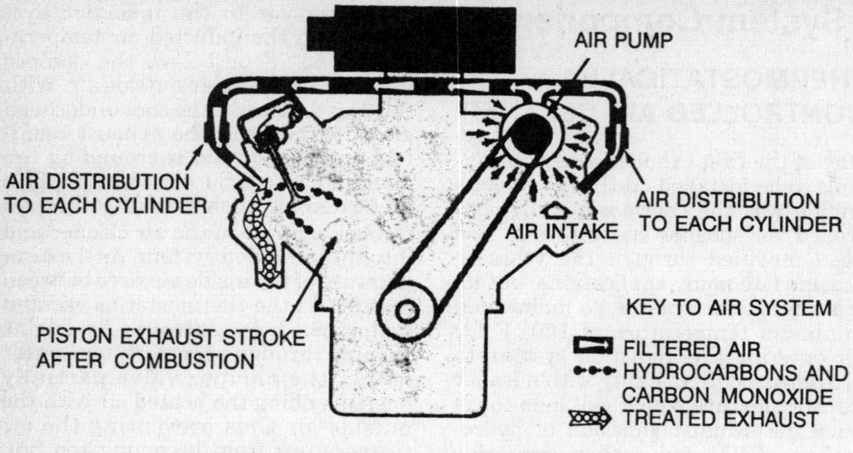

Typical air injection system flow schematic

system by the vehicle manufacturer, its basic function and operation has remained the same throughout its application years. It is extremely important in the control of engine warm-up emissions, vehicle driveability and prevention of fuel icing, that all components of the thermostatically controlled air cleaner assembly be connected and operating.

AIR INJECTION SYSTEM

One of the major problems of the internal combustion engine is the fact that complete burning of the air/fuel mixture does not occur within the combustion chambers. This unburned mixture is swept from the combustion chamber along with the burned exhaust gases and emitted through the exhaust system and into the atmosphere. To prevent excessive emission of the unburned and burned gases containing large portions of hydrocarbons (HC) and carbon monoxide (C0), a means was devised to further burn the gases as they were forced from the cylinders by injecting fresh, oxygen laden air into the heated exhaust gas stream.

The addition of oxygen to the exhaust manifold and the exhaust pipes reducs the amount of unburned gases in the emitted exhaust. This after burner type system remains in use on many engine families. The vehicle manufacturers have modified, added and reduced the control components of the original system through the years, but the basic system to inject fresh air into the heated exhaust stream, still remains.

Two different systems are used on present day engines, one system using a belt driven air pump and the other system using the positive and negative exhaust system pressure pulsation to draw in fresh air.

Air Pump System

The major components of the air pump systems are.

1. Air pump which supplies filtered low pressure air to the system, normally 2–5 psi.
2. Diverter valve which diverts air pump output air to the atmosphere during deceleration to prevent backfire. A pressure relief valve is incorporated to protect the system.
3. Check valve which prevents hot exhaust gases from entering the system.
4. Air manifold which distributes the fresh air to each exhaust port of other area of the exhaust system.
5. Air nozzle which injects the air into the exhaust system.
6. Manifold vacuum signal line which senses manifold vacuum to activate the diverter valve.

During the engine's normal operating condition, the air pump is supplying fresh air to the diverter valve which passes the air on to the air nozzles for injection into the exhaust stream. As the air is mixed with the hot exhaust gases, further combustion takes place in the exhaust system. During periods of deceleration, high manifold vacuum and a rich mixture is present in the combustion chambers. If fresh air was to be injected into the exhaust system during this condition, backfiring would occur. The diverter valve senses the high intake manifold vacuum and vents the fresh air from the air pump to the atmosphere. A switch valve is used by some manufacturers to redirect the fresh air from the exhaust valve port area to another location down stream in the exhaust system when the engine has warmed up to normal operating temperature, in order to avoid increasing the oxides of nitrogen (NOx) emissions while relying upon the heat of the exhaust to further the burning of the hydrocar-

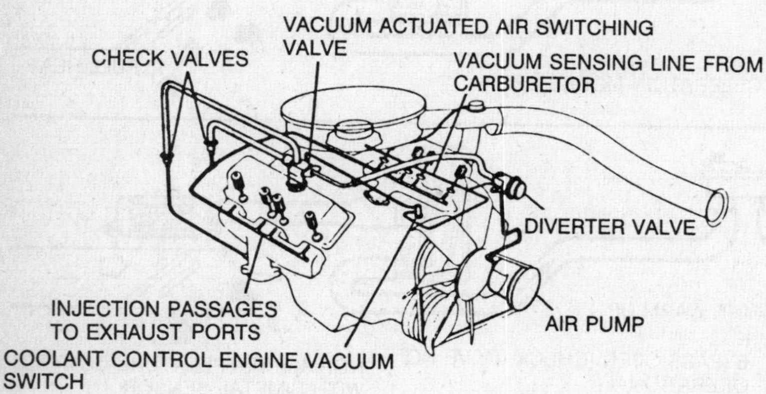

Major components of a typical air injection system

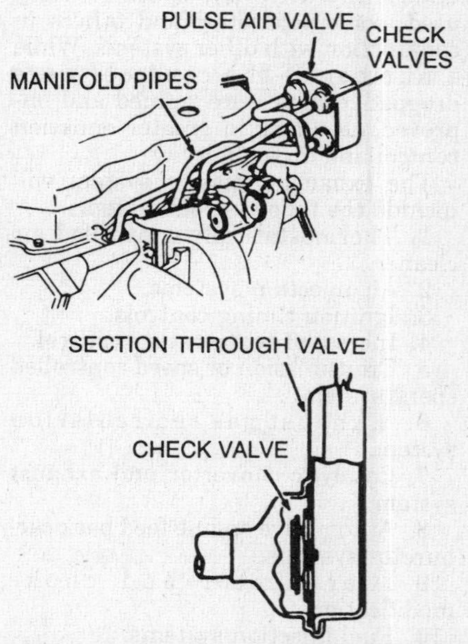

Major components of a typical pulse air system

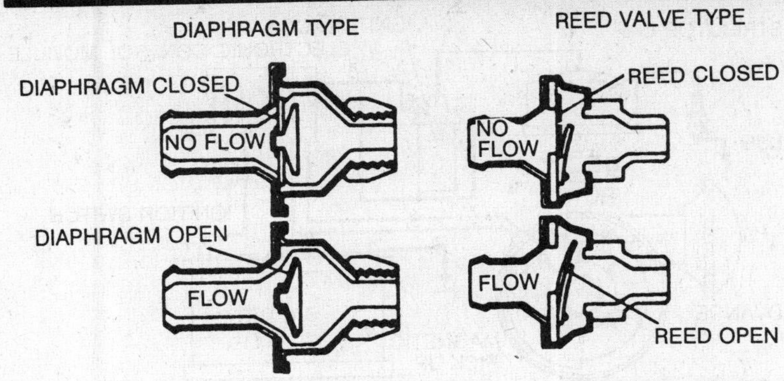

Crossection of operating diaphragm and reed type check valves

bons (HC) at the exhaust valve ports. Since the addition of the three-way catalytic converters and oxygen sensors to the exhaust system, the air injection system has been modified and electronically controlled to critically monitor the entrance of fresh air into the exhaust stream during different engine operational modes and at different locations.

Pulse Air Injection System

The pulse air injection system does not use an air pump, but relies on the positive high pressure and negative low pressure pulses of the exhaust flow from the engine to operate one or more one-way check valves, which allows filtered fresh air to enter the exhaust stream at periods of negative low pressure and to prevent the leakage of exhaust gases back through the inlet air tubing. The fresh air induction through the one-way valve is normally accomplished at idle or slightly above.

IGNITION TIMING CONTROLS

The purpose of the timing or spark

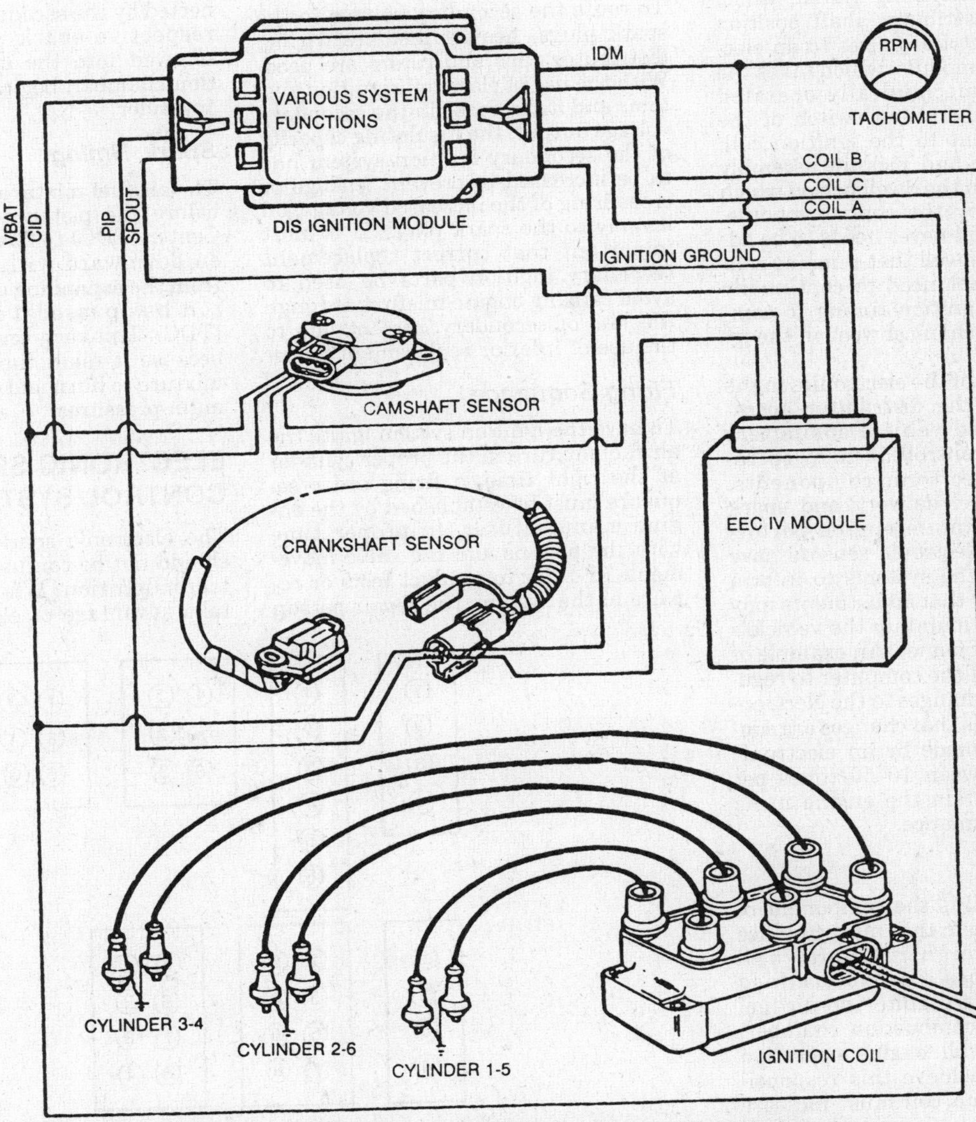

Typical Ford Motor Company EEC-IV system—schematic layout

controls is to fire the air/fuel mixture in the combustion chamber at a specific time in order to derive the most power from the mixture, while burning it as completely as possible to rid it of excess hydrocarbons (HC) and still maintain a combustion chamber temperature that prevents excess formations of nitrogen oxides (NOx).

Electronic Ignition

To increase the ignition system's durability to over 50,000 miles of engine operation and to eliminate adjustment or replacement of the contact point sets, and electronically controlled ignition system was introduced as standard equipment on American made automobiles, beginning with the 1975 model year.

The distributor primary circuit was changed from a breaker plate and cam assembly in the distributor, to a magnetic signal generating system which detects the distributor shaft position and sends electrical pulses to an electronic control module, which takes the place of the mechanically operated point set, in the off-on switch of the primary current to the ignition coil. The armature and pick-up assembly has no effect on the dwell period which is controlled by the control module. Therefore, dwell never needs to be adjusted. With a dwell that remains constant, we do not need to continually read just the ignition timing to compensate for mechanical wear in the ignition system.

With the use of the electronics in the controlling of the distributor operations, on-board vehicle computers were added to control the many operations of the electronic components, such as the fuel delivery and spark timing for optimum engine performance. Many electronic sensors have been added to the systems to inform the computer(s) that adjustments may be necessary to maintain the vehicle's electronic performance. An example of the quickness of the computer to regulate and direct changes to the electronic components, is that changes are continually being made by an electronic component between 10–50 times per second to maintain the engine in its optimum performance.

Ignition Coil

The ignition coil is the component of the ignition system that must produce voltage of enough intensity and strength to cross a predetermined spark plug gap to ignite the air/fuel mixture in the combustion chamber, under any and all engine operation conditions. To achieve this responsibility, the ignition coil must increase the primary voltage form an average of 12 volts to a secondary voltage, through induction, as high as 30,000 volts.

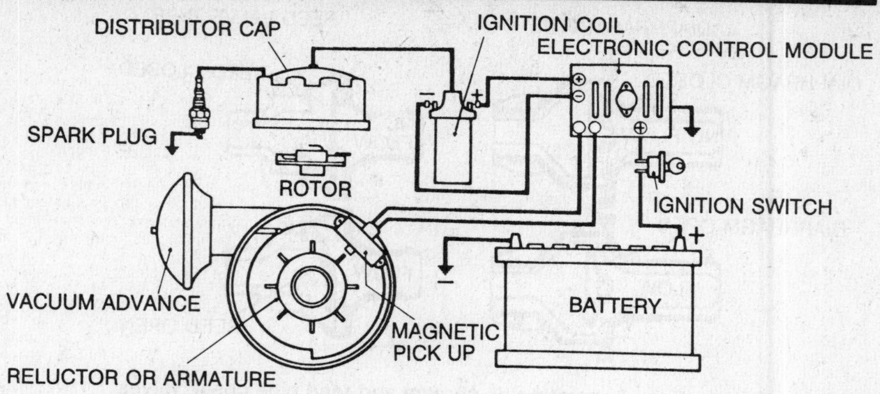

Typical electronic ignition system electrical schematic

Distribution of Spark

To route the secondary voltage to the spark plugs, heavily insulate wiring, distributor caps and rotors are used. With the use of electronic ignition systems and its increase in the secondary voltage output, the insulating capacity of the secondary ignition system had to be increased to prevent leakage or crossfiring of the increased voltage, on its way to the spark plugs. It is most important that correct replacement secondary ingition parts be used to avoid causing engine misfires through the loss of secondary current due to the use of inferior replacement parts.

Firing Sequence

To have the ignition system ignite the air/fuel mixture in the proper cylinder at the right time, a firing order sequence must be established by the engine manufacturer, in proper time with the pistons and the valve movement. In order to conduct tests or repairs to the system, the repair person must know the firing order sequence, the cylinder numbering order, the rotation of the distributor rotor and the location of the corresponding electrical terminals on the distributor cap, connected by the secondary cables to their respective spark plugs, which are screwed into the cylinder's combustion chamber, beginning with the No. 1cylinder.

Spark Timing

The air/fuel mixture must be ingnited before the piston reaches Top Dead Center (TDC) to have the piston driven downward with maximum force from the expanding gases after the piston has passed Top Dead Center (TDC). This early ignition is necessary because it takes time for the air/fuel mixture to burn and develop the maximum pressures.

ELECTRONIC SPARK CONTROL SYSTEM

The electronic spark control system should not be confused with the electronic ignition. While both systems take advantage of electronics to per-

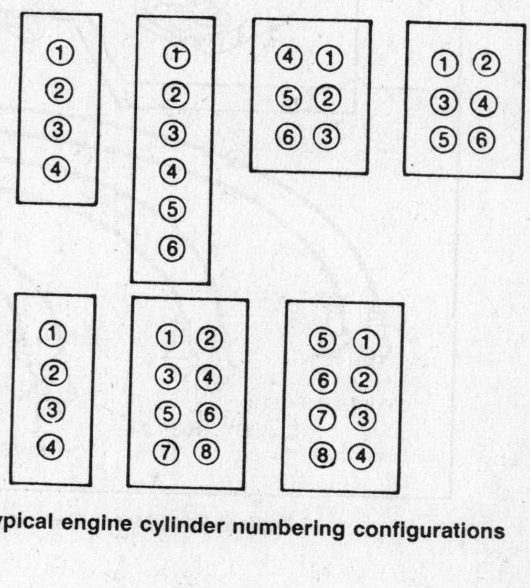

Typical engine cylinder numbering configurations

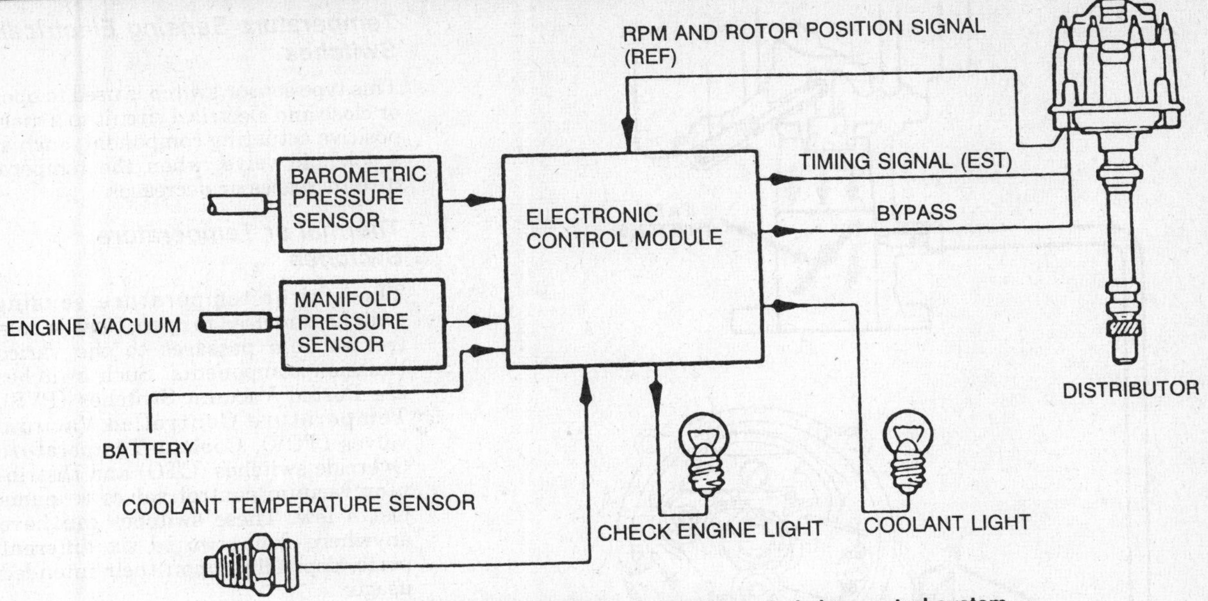

Typical first generation electronic spark timing control system

form specific duties in the ignition systems, their mode of operation differs greatly. The electronic spark control system does not use the conventional vacuum or centrifugal advance mechanism, but relies on sensors to monitor the critical and fast changing variables that affect engine performance, such as engine speed, engine spark timing, intake manifold vacuum, throttle plate postitioning and the rate of plate change, inducted air temperature and coolant temperature, to name a few. The computer receives signals from the sensors and with in mill-seconds, computes the signals to determine how the engine is operating and either advances or retards the spark to meet the engine's operating conditioning.

It should be noted that the spark timing is not based on a constant curve, but is an infinitely variable ignition system that relates to the engine speed and load requirements. The spark control system vary from manufacturer to manufacturer. One system may not signal the computer during the cranking mode, but rely on a predetermined initial timing position to fire the spark plugs, while another system relies on a second pick-up sensor in the distributor or on the crankshaft, to sense the piston position and signal the computer, which in turn signals the coil to fire the spark plugs. The computer spark control systems are also used in conjunction with carburetor or fuel injection electronic metering systems. Certain systems also have the capabilities of self-diagnosis to aid in the repair of malfunctioning internal circuits.

INCREASED TEMPERATURE CONTROLS

As we know, the cooling system's main function is to remove excess heat from the combustion area of the cylinder head and engine block. Increased temperature rated thermostats were installed as more emission controls were placed on the engines, with an average norm of approximately 195° F. This higher regulated temperature aids in the reduction of the combustion chamber's quench area, resulting in a cleaner burning air/fuel mixture. With the use of many temperature sensing components that control the operation of emission control system, it is important that the cooling system of the vehicle be properly maintained.

ELECTRICAL SENSORS AND VACUUM SWITCHING UNITS

With the use of both vacuum and electronics to operate the various emission control components under different temperature and operating modes, numerous sensors and switches are controlled by the engine coolant and ambient air temperatures. A sensor can be used separately to open or close electrical circuits, to send electrical impulses to a computer, to be used in conjunction with a switching unit to open or close vacuum passages, to control the switching form one vacuum source to another, to modulate a vacuum circuit by bleeding a metered amount of air into a system or opening a vacuum passage to the atmosphere.

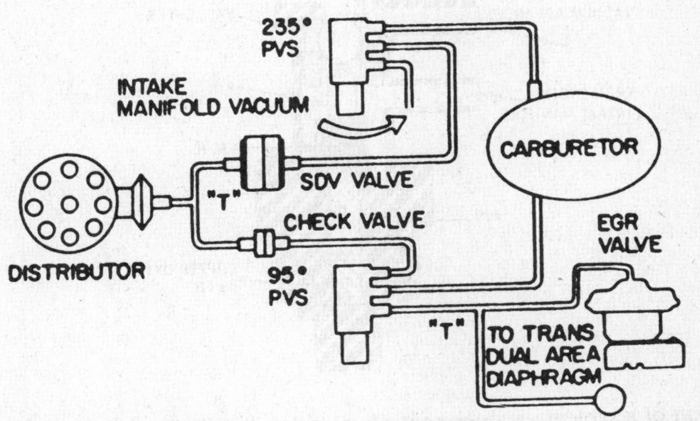

Typical temperature sensing valves used to control vacuum actuated components

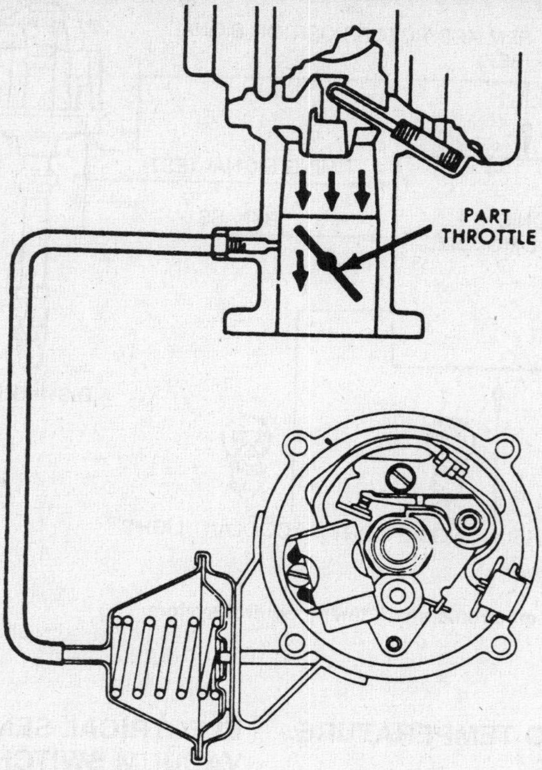

Crossection of a typical vacuum advance mechanism

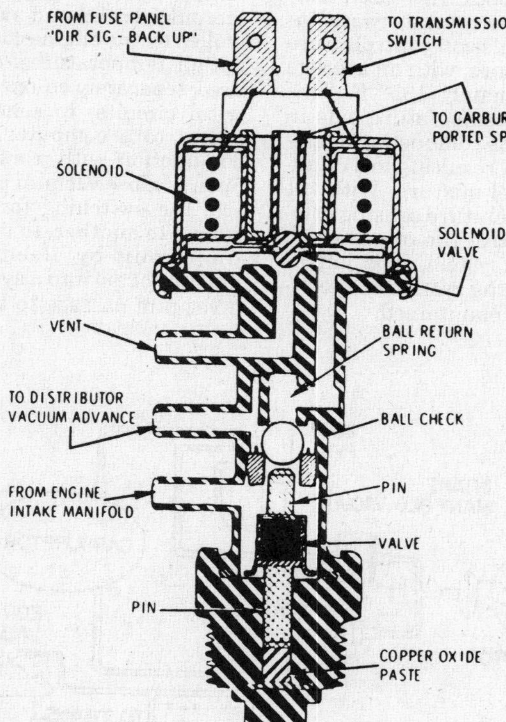

Crossection of a typical vacuum control valve for distributor vacuum advance operation

Temperature Sensing Electrical Switches

This type sensor/switch is used to open or close and electrical circuit to a more positive actuating component, such as a solenoid valve, when the temperature increases or decreases.

Thermal or Temperature Switches

Thermal or temperature sensing switches are used to open, close or control vacuum passages to the varied emission components. Such switches are Ported Vacuum Switches (PVS), Temperature Controlled Vacuum valves (TCV), Coolant Temperature Override switches (CTO) and Distributor vacuum control valves to name just a few. These switches can have anywhere from two to six different ports, depending upon their intended usage.

Presure or Vacuum Switches

These switches are normally used where pressure or vacuum sensing is needed. The switches are normally one-way valves, allowing pressure or vacuum to move in one direction only. Vacuum delay valves are considered to be classified as this type valve.

Mechanical or Motion switches

This type of switch is normally operated by the oil pressure of an automatic transmission or by a shifting rail of a standard transmission. It can also be located on the speedometer cable, reacting as a small generator to produce an electrical current signal, in direct proportion to the speed of the vehicle. This type of switch normally opens or closes a specific electrical circuit.

EXHAUST GAS RECIRCULATION SYSTEM (EGR)

In their attempt to reduce hydrocarbons (HC) emission, the manufacturers increased the combustion chamber temperatures to more thoroughly burn the air/fuel mixture. With the increase in temperature and pressure, another pollutant was created, oxides of nitrogen (N0x).

At temperatures below 2500° F, nitrogen remains an inert gas, but with combustion temperatures reaching as high as 4500° F, the nitrogen combines with the oxygen in the air/fuel mixture, resulting in the formation of oxides of nitrogen (NOx) a harmful pollutant. Through experimentation, it was found that a portion of the exhaust gases could be redirected into the combustion chamber, along with

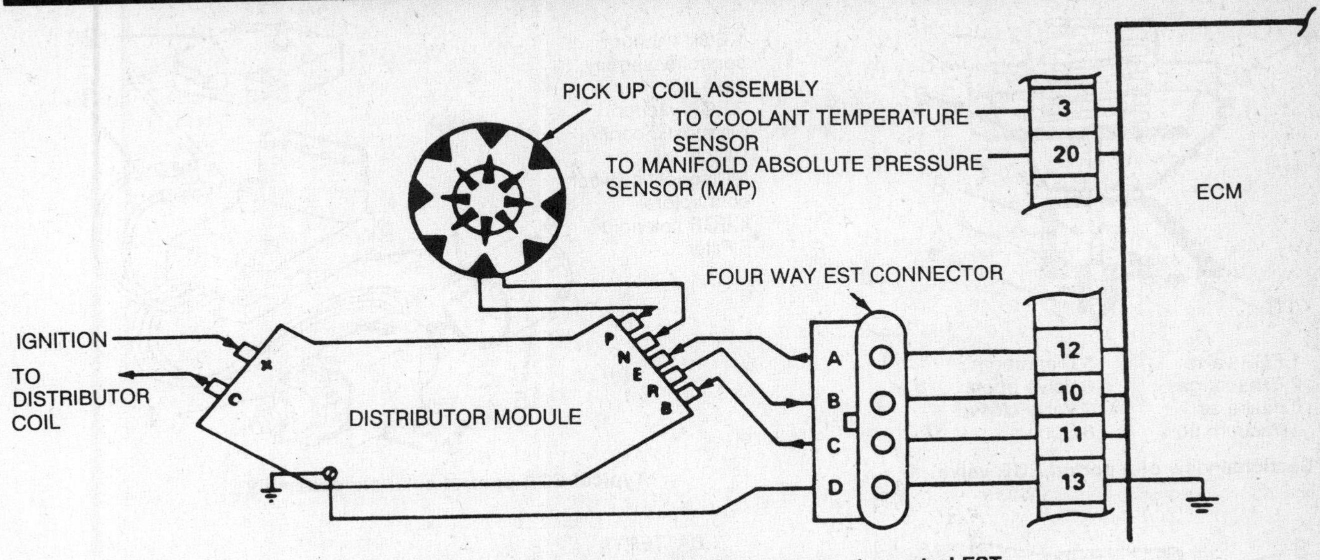

Typical usage of vacuum and temperature sensors to control EST

the inducted air/fuel mixture, resulting in the temperature of the burning process being lowered, causing a reduction in the emission of the oxides of nitrogen (NOx). The amount of the recirculated exhaust gases has to be carefully controlled. If too much exhaust gas is supplied at the wrong time, the engine may stall and be very rough at idle. If not enough exhaust gas is recirculated, the oxides of nitrogen (NOx) will not be reduced.

EGR Valves

Varied types of EGR valves are used with different control components, so that the proper amount of recirculated gas is directed into the air/fuel mixture at a specific time. The EGR valves are vacuum operated, either by intake manifold of by ported vacuum. With the increased use of electronics, sensors and controlling solenoids are used to regulate the operation of the EGR valve by the on-board computer. When the electronics are not used, the EGR controls are normally operated by venturi vacuum, ported vacuum or by an exhaust backpressure sensor.

To properly control the EGR system and to recirculate the exhaust gas only at a specific time, many different metering EGR valves are used, along with the necessary components such as vacuum amplifiers, temperature override switches, backpressure sensors, vacuum bias valves and timers. When vacuum amplifiers are used, the control or signalling vacuum is venturi vacuum, which is zero at idle and at its maximum (approximately 4 in. hg.) during heavy loads, paralleling the need for exhaust gas recirculation. When ported vacuum is used, the position of the throttle plate regulates the amount of vacuum available to the EGR valve. The vacuum is zero when the throttle plates are closed or in the wide open position, again paralleling the need for exhaust gas recirculation.

Because of the different types of EGR valves used, such as single diaphragm, double diaphragm, negative backpressure or positive backpressure units, the correct valve must be replaced on a specific engine in order for the EGR system to function properly.

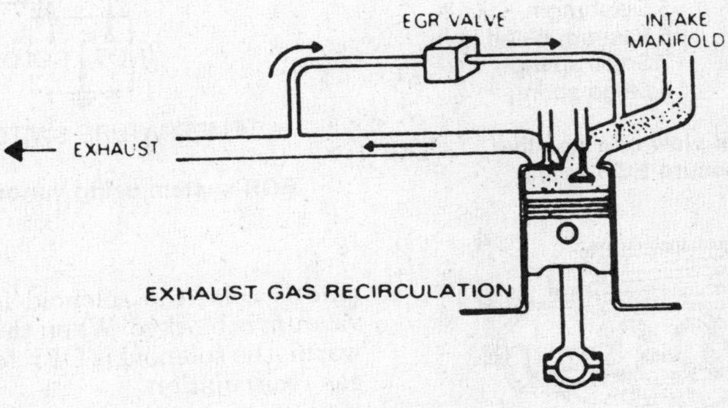

Typical EGR system schematic

1. EGR valve
2. Exhaust gas
3. Intake air
4. Diaphragm
5. Electronic control module
6. Manifold vacuum
7. Throttle position sensor
8. Manifold pressure sensor
9. coolant temperature sensor
10. EGR control solenoid

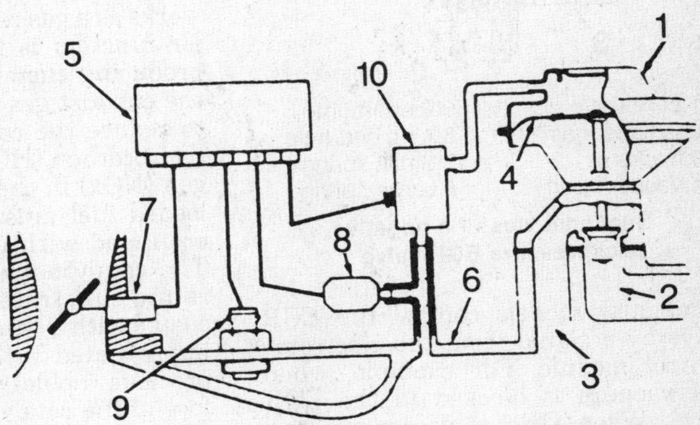

Typical operation of a vacuum solenoid controlled EGR valve

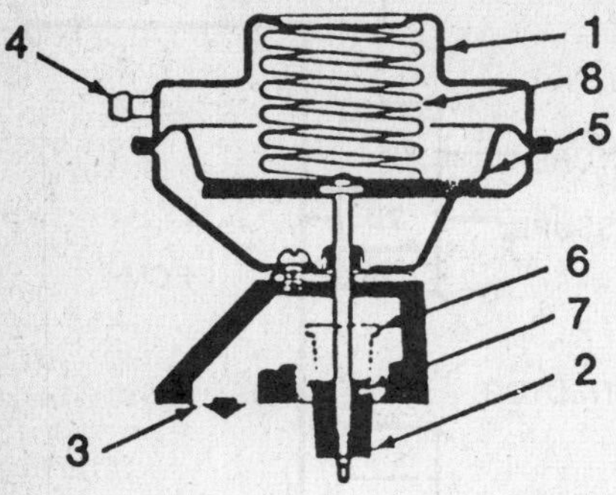

1.EGR valve
2.Exhaust gas
3.Intake air
4.Vacuum port
5.Diaphragm
6.Valve open
7.Valve closed
8.Spring

Sectional view of a ported EGR valve

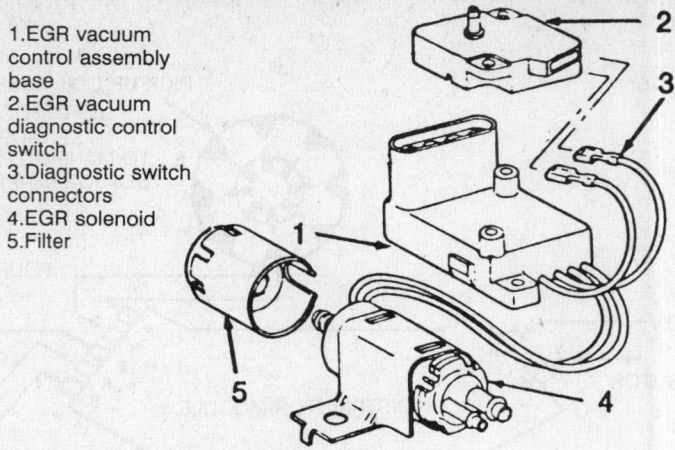

1.EGR vacuum control assembly base
2.EGR vacuum diagnostic control switch
3.Diagnostic switch connectors
4.EGR solenoid
5.Filter

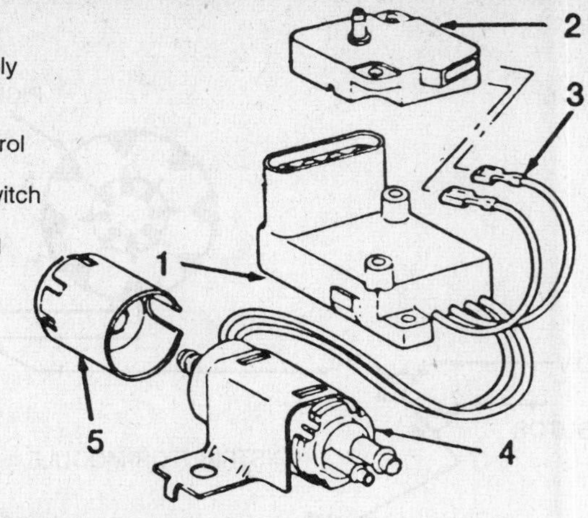

Typical EGR control solenoid assembly

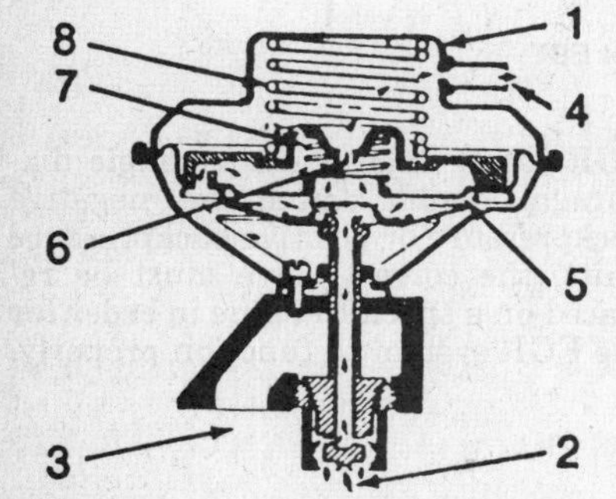

1.EGR valve
2.Exhaust gas
3.Intake air
4.Vacuum port
5.Diaphragm
6.Vacuum bleed hole
7.Small spring
8.Large spring

Sectional view of a positive backpressure EGR valve

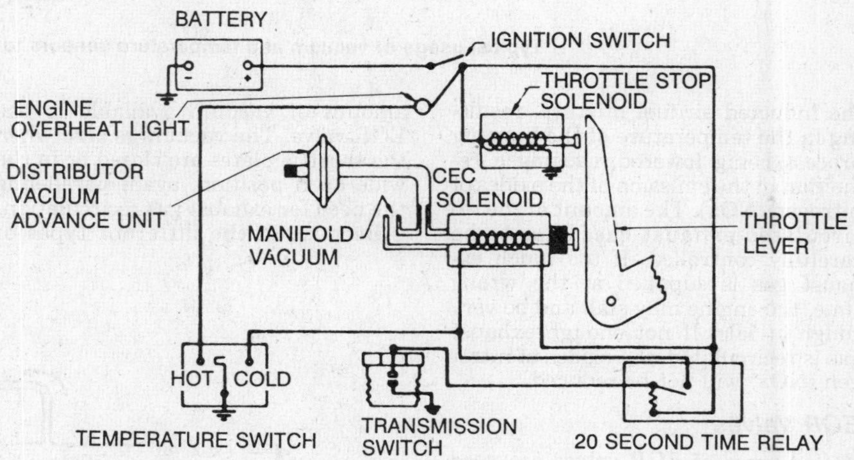

EGR system using various types of emission sensing and actuating controls

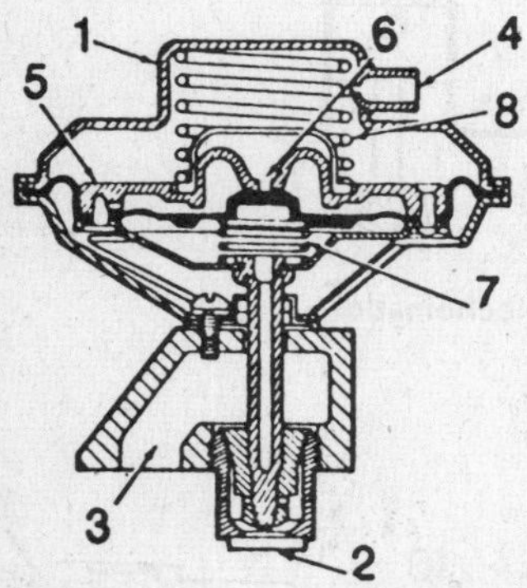

1.EGR valve
2.Exhaust gas
3.Intake air
4.Vacuum port
5.Diaphragm
6.Air bleed hole
7.Small spring
8.Large spring

Sectional vies of a negative backpressure EGR valve

Some new models control the EGR valve by a signal from the electronic control module. For example, when ON vacuum is blocked to the EGR valve. When OFF it passes vacuum and vacuum is allowed. When the engine is cold, the solenoid is ON and vacuum is blocked. When the engine is warm, the solenoid is OFF for exhaust gas recirculation.

CATALYTIC CONVERTERS

The catalytic converters are mounted in the engine exhaust stream and works as a gas reactor in which its major function is to speed up the heat producing chemical reaction between the exhaust gas components, in order to reduce the carbon monoxide (C0), hydrocarbon (HC) and oxides of nitrogen (NOx) in the engine exhaust. Unleaded fuel must be used in vehicles equipped with catalytic converters. The catalyst material is either a ceramic substrate or pellets that are coated with a base of alumina and then impregnated with catalyticaly active, precious (noble) metals. It is the surface of the catalyst material that controls the heat producing chemical reaction.

Two main types of converters are used. The first type contains, platinum and palladium to effectively catalyze the oxidation of the hydrocarbons (HC) and carbon monoxide (C0). The second type converter used is considered a three-way catalyst, containing a small percentage of platinum and a greater percentage of rhodium in the front part of the converters to reduce the oxides of nitrogen (N0x), while platinum and palladium are used in the rear section to oxidize the hydrocarbons (HC) and carbon monoxide (C0), as was done in the two way converters.

Oxidizing Catalytic Converters

The converters do not operate unless there is sufficient oxygen in the exhaust stream. It is extremely important that the proper amount of oxygen is supplied at all times. This is accomplished by a secondary air source, provided by either an air pump system or a pulse air type system. The catalytic

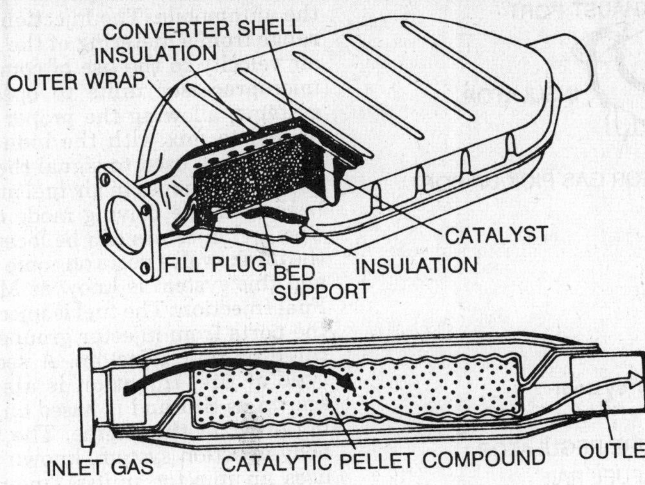

Pellet type catalytic converter

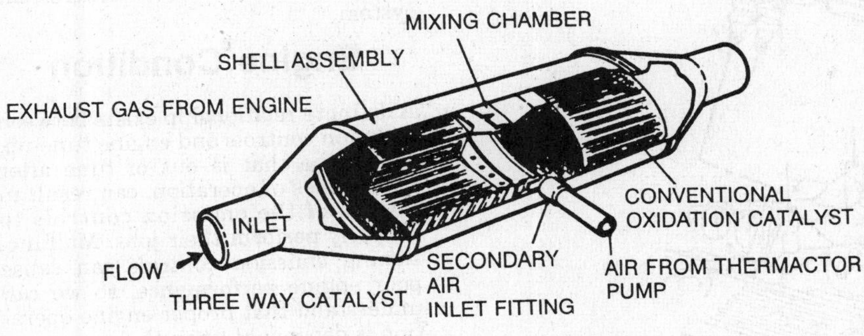

Sectional view of a three way catalytic converter

converter system is protected by several devices that block out the secondary air supply when the engine is laboring under any abnormal hot or cold operating situations, preventing converter overheating and burnout. Converter temperatures are normally between 900–1500° F with peak temperatures around 1800° F. Should the converter be supplied too rich a mixture of hydrocarbons (HC), such as would result from a misfiring spark plug or stuck choke valve, along with an oversupply of fresh air, the converter temperature would increase sharply, causing a burnout of the catalyst material. Because of the need to quickly heat the converters units, smaller or mini converters are placed in the exhaust stream before the main converter to preheat the exhaust gas.

Three-way Catalytic Converters

The three-way catalytic converters use a combination of catalyst which produce two different chemical reactions, oxidation and reduction. By adding fresh air to the unburned hydrocarbons (HC) and carbon monoxide (C0) within the converter, the oxidizing of combustion process takes place. Just the reverse process is required to lower the oxides of nitrogen (N0x) emissions. The oxides of nitrogen (N0x) already contains excessive oxygen and the process of separating the excess oxygen from the nitrogen is called a re-

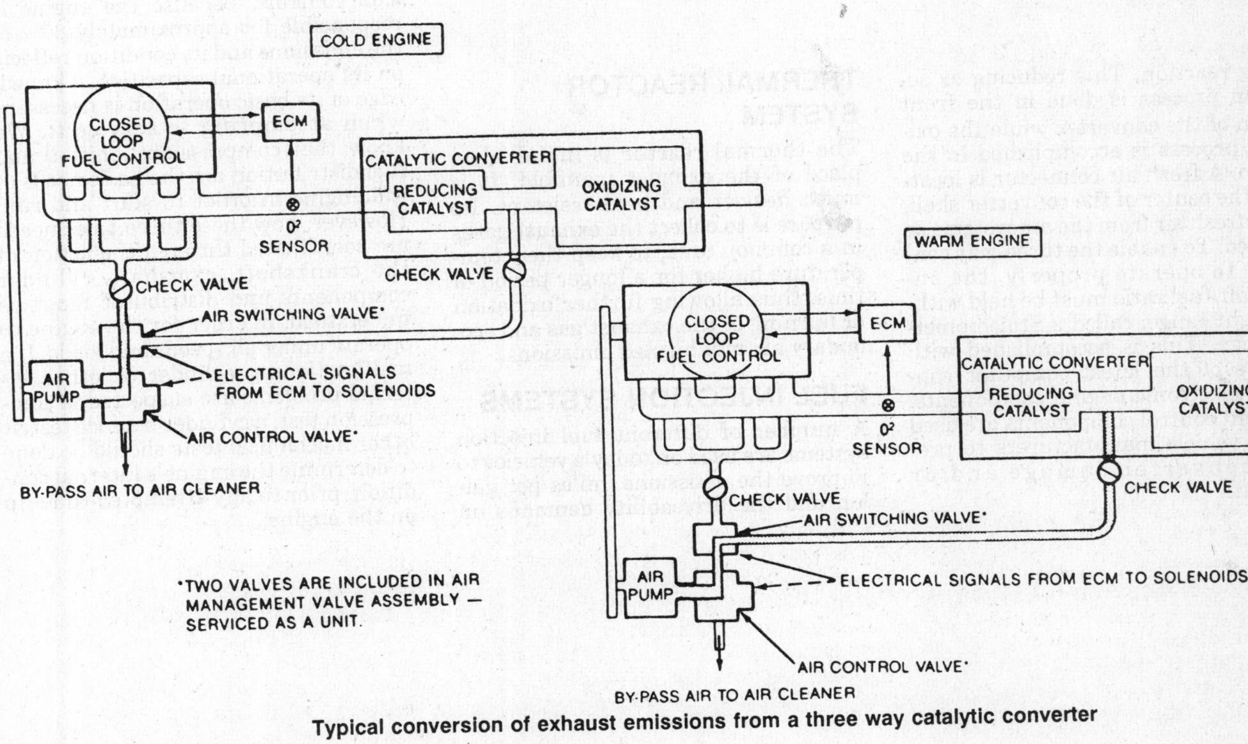

Typical conversion of exhaust emissions from a three way catalytic converter

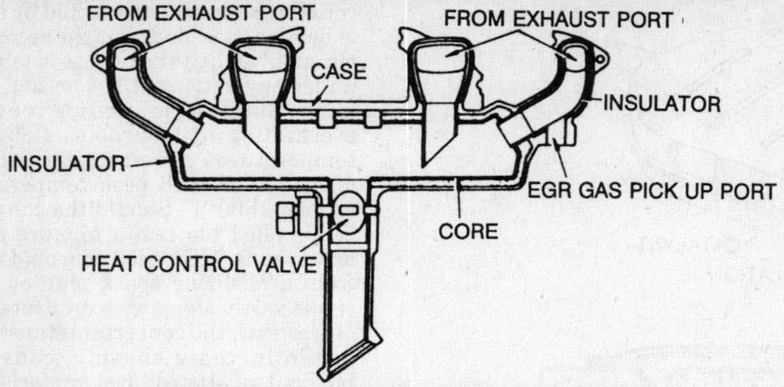

Sectional view of a typical thermal reactor system

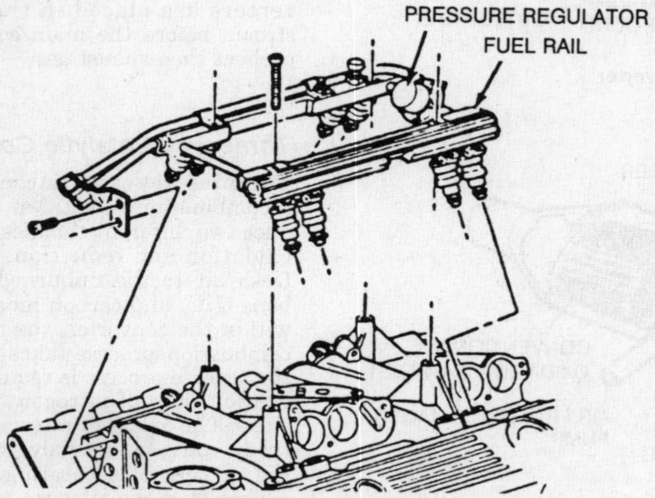

Fuel rail type fuel injection system

ducing reaction. This reducing or reduction process is done in the front section of the converter while the oxidizing process is accomplished in the section. A fresh air connector is located on the center of the converter shell, to add fresh air from the air systems as required. To enable the three-way converter to operate properly, the engine's air/fuel ratio must be held within a tight range, called a Stoichiometric range. This is accomplished with the use of the latest computer controlled electronic engine components. Different control components are used by the vehicle manufacturers to prevent converter damage and/or burnout.

THERMAL REACTOR SYSTEM

The thermal reactor is installed in place of the exhaust manifold. It is much heavier and heat resistant. Its purpose is to collect the exhaust gases in a common area, to keep their temperature higher for a longer period of time, thus allowing further oxidation or burning of the exhaust gas and secondary air mix burned emissions.

FUEL INJECTION SYSTEMS

A number of different fuel injection systems are used on today's vehicles to improve the emissions, miles per gallon and the driveability demands on the automobile. The injection systems range from a metering of the inducted air velocity to the use of computer or microprocessor units to operate the systems, allowing the proper amount of fuel to mix with the inducted air. Sensors are used to signal the control unit that changes in air/fuel mixture is needed as the driving mode changes. The fuel injectors can be located near the intake valve ports on some systems and this system is know as Multiport Fuel Injection. The fuel is injected into the ports from injector groups of two, three or four injectors. A sequential type of fuel injection is also used, known as SFI and is based on the firing order of the engine. The throttle body injection system, known as TBI, uses an injector unit(s), mounted in the throttle body to inject fuel as needed into the intake manifold. An electronic control module is used, in conjunction with sensors to control the system.

Engine Condition

A definate relationship exists between emission controls and engine tune-up. An engine that is out of tune after many miles of operation, can result in failure of the emission controls to properly perform their jobs. Malfunctioning emission controls can cause poor engine performance, so we can understand that proper engine operation is dependent upon thorough testing and servicing of all components related to both performance and emission controls. Because the engine is responsible for approximately 80% of the emissions and its condition reflects on its operational capacities, a knowledge of its basic operation is necessary when attempting to service it. We know that compression, ignition and fuel distribution are the basic needs of the engine in order to start and run. However, how these three basic needs are coordinated through the action of the crankshaft, camshaft, cylinder components and distributor must be understood, in order for the engine to operate under all speeds and loads. If a malfunctioning cylinder or other internal problems are suspected, a compression test, a cylinder balance test or other mechanical tests should be done to determine the engine's internal condition, prior to any attempted tune-up on the engine.

Engine Controls 28

ENGINE ELECTRONICS

History

In the latter part of the 1960's, Robert Bosch introduced the first true electronically controlled engine with an on-board computer. Today, almost every car produced has some kind electronic engine control. The once mechanically controlled engine functions of early model cars are all but extinct.

The first system, Bosch D-Jetronic, is comprised of electrically energized fuel injectors in which the injection time is controlled by an electronic control unit (ECU). The early system delivered a basic quantity of fuel and varied from this point depending upon engine load, engine speed and engine temperature.

Since the early days of the ECU, the controls have become more complex, with a much greater amount of computer memory and even the ability to learn.

In this section, the topics will include different types of electronically controlled fuel induction, spark control, the sensors and switches that provide the ECU with information, other non-engine related controls that the ECU might supply and some ECU self-diagnostics.

The most common fuel induction system with an ECU is electronic fuel injection. In this system, fuel can be delivered many different ways. One of which is the single point injection (SPI) were one or two injectors are mounted on a throttle body assembly. Fuel is delivered constantly through the injector(s), but in varying quantities. The SPI system very much resembles a carbureted system. SPI is more commonly known as throttle body injection (TBI). Another fuel injection system is multi-point injection (MPI). This system supplies one injector for each cylinder, usually positioned in the intake manifold, just above the intake valve. In MPI, fuel can be injected in two ways. One is to energize a group of injectors, thus atomizing fuel in the intake manifold and storing it for a short time until the intake valve opens. The second way is to sequentially energize each cylinder's injector as the intake valve is opened. This injection is the more efficient, effective and more complex system.

Another fuel induction system utilizing an ECU is the feedback carburetor (FBC). A conventional carburetor is still used, but it has a more precise air/fuel mixture control which is achieved through an integral mixture control solenoid. The solenoid is energized on and off by the ECU to maintain mixture demand. The ECU calculates air/fuel mixture demand changes by the data it receives through remote sensors. The most important sensor (and makes the system possible) is an oxygen (O_2) sensor (which will be discussed later in this section). The ECU monitors the exhaust gases for rich/lean conditions by way of the O_2 sensor and, in turn, controls the air/fuel mixture by increasing or decreasing the duty cycles (on and off) to the mixture control solenoid for an optimum 14.7:1 air/fuel ratio.

ECU Self-Diagnostics

The ECU can detect a malfunction or abnormality in the sensors or in the ECU itself and display a warning light on the instrument panel when it does. When this occurs, the ECU stores a trouble code for future system diagnosis. If the problem is sever enough to where it inhibits closed loop operation, the ECU will assume a backup system. This fail-safe circuit is pre-programmed into the ECU for minimal driveability operation so the vehicle can be driven to a nearby service facility. The trouble codes are usually a two digit numbers identified by the number of diagnostic LED or check engine light flashes. The trouble codes assist the service technician in isolating a faulty circuit or component within the system.

Electronic Data Sensors

The engine control system consists of various data sensors. Although data sensor names and applications vary from system to system, the most common input sensors/switches are:

- oxygen (O_2) sensor
- coolant temperature sensor
- manifold air pressure (MAP) sensor
- vehicle speed sensor (VSS)
- throttle position sensor (TPS)
- engine speed reference or distributor reference (rpm)
- air flow sensor
- air intake temperature sensor
- crankshaft sensor
- detonation (knock) sensor
- throttle body temperature sensor

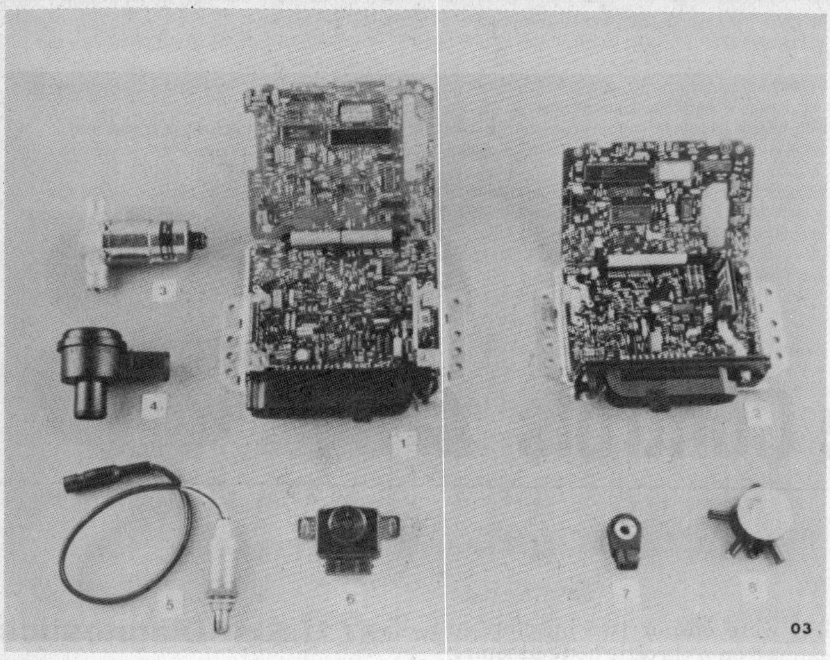

Engine electronic components

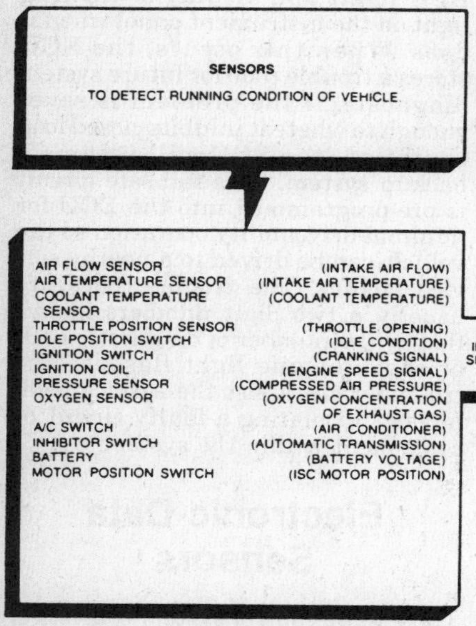

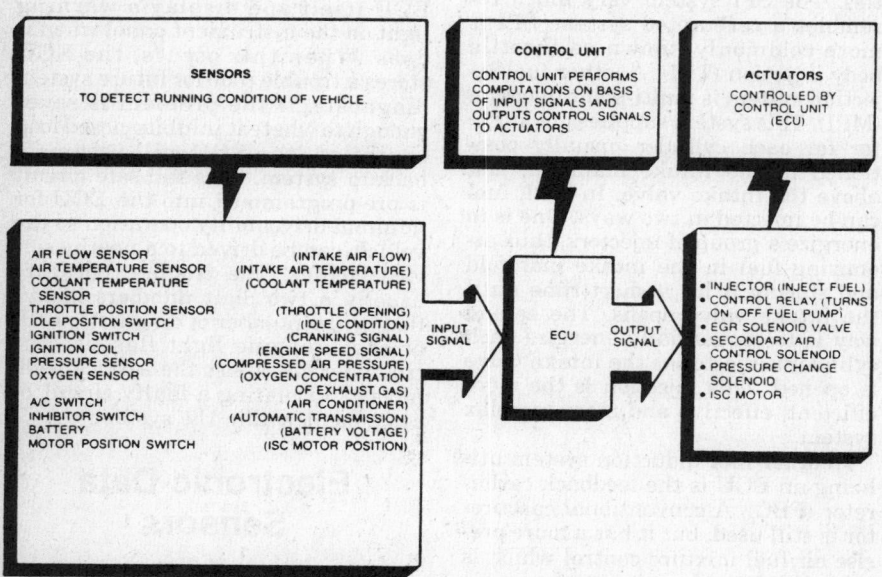

Typical fuel control system with ECU

- throttle idle switch
- transmission or drive switch
- a/c compressor clutch switch
- power steering pump switch
- altitude or barometric pressure sensor
- wide open throttle switch

Electronically Controlled Devices

Some of the output devices that the ECU may control vary from system to system, but the most common output or ECU controlled devices are:

- fuel injector(s)
- air/fuel mixture solenoid
- fuel pump relay
- a/c compressor clutch relay
- idle air control (IAC) valve
- idle speed control (ISC) motor
- ignition spark/timing
- canister purge solenoid
- torque converter clutch solenoid (automatic transmission)
- air management system (air induction)
- idle-up or throttle kicker solenoid
- alternator field control (charging system)
- turbocharger boost wastegate
- cooling fan relay

Component Description

THROTTLE BODY

The throttle body, in most fuel injected systems, is usually an alumunum housing that consists of one or two throttle blades which are attached to a throttle shaft. The housing has a throttle position sensor (TPS) sensor, idle air control motor and, in some cases, throttle body temperature sensor. On SPI systems, the housing also has an injector(s) and (in some cases) a fuel pressure regulator. The throttle body throttle blade controls the amount of air that enters the engine as well as the amount of vacuum.

ELECTRONIC CONTROL UNIT (ECU)

The ECU monitors and controls all engine control functions. The ECU consists of input and output devices, a central processing unit, a power supply and various memory banks. The input and output devices of the ECU convert electrical signals received by the data sensors and switches to the digital signal that are used by the central processing unit. The central processing unit receives digital signals that are used to perform all mathematical computations and logic functions necessary to deliver proper air/fuel mixture. The central processing unit is also responsible for calculating spark timing information. The main source of power that allows the ECU to function is generated from the battery of the vehicle and transported through the ignition system. The memory bank of the ECU is programmed with exact information that is used by the ECU during the open loop mode. This data is also used when a sensor of other component fails, allowing the vehicle to be driven to a repair facility.

CALIBRATION ASSSEMBLY OR PROM (PROGRAMMABLE READ ONLY MEMORY)

Some vehicle manufactures use one

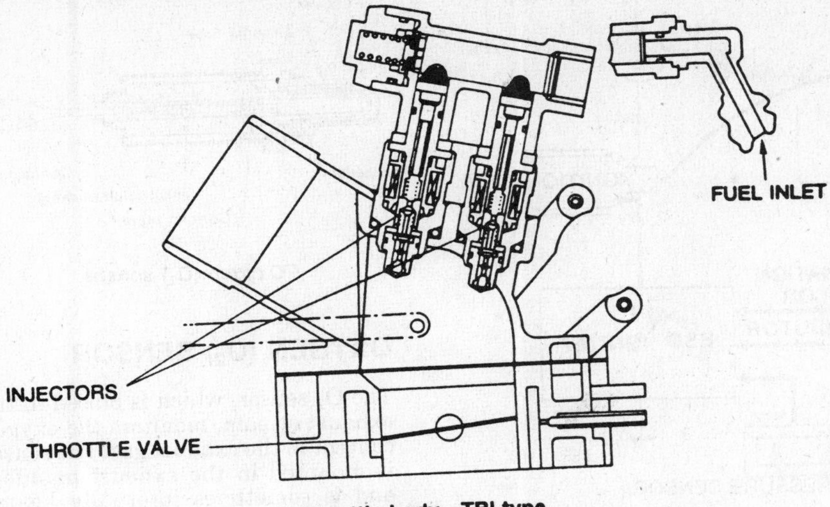

Throttle body—TBI type

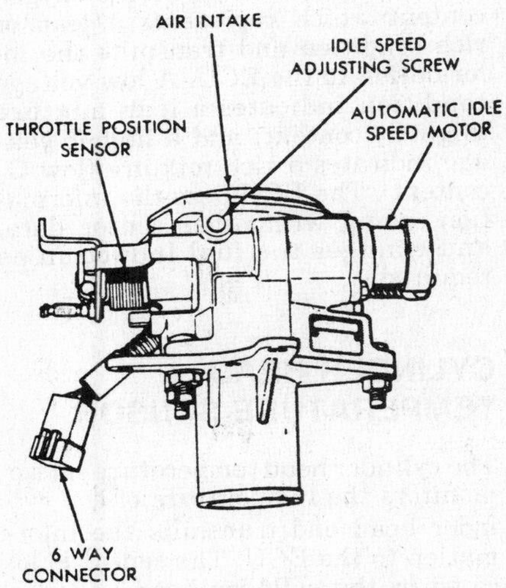

Throttle body—MFI type

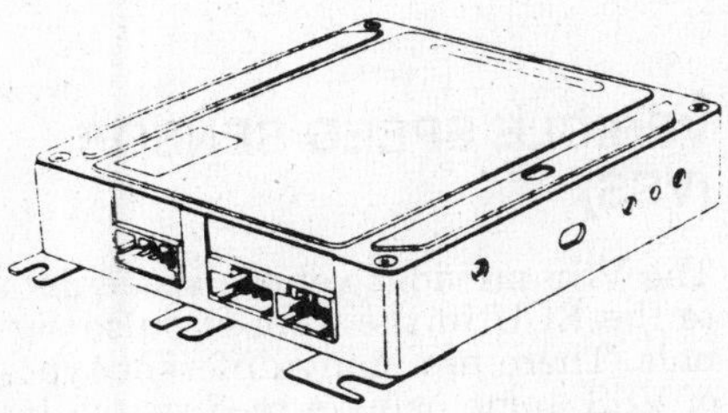
Electronic control unit (ECU)

ECU for several different model vehicles. This interchangeable ECU is possible through the use of a calibration assembly or prom. Information about the vehicle's engine, transmission, body and drive axle ratio are programmed and permanently stored into the assembly. If the battery supply should become disconnected from the ECU, the data stored into the assembly is not lost.

ELECTRONIC SPARK CONTROL (ESC)

The vehicles equipped with an ESC have the ability to change the ignition timing under any and all operating conditions. Data from various remote sensors (coolant temperature, throttle position, rpm, etc.) is transmitted to the ESC. The ESC computes the information and triggers the ignition spark at precisely the right instant. Some ESC systems (ie.,turbocharged engines) use a detonation (knock) sensor which senses pre-ignition and transmits the information to the ESC. The ESC modifies spark advance and boost pressure in order to eliminate knock.

MASS AIR FLOW SENSOR

The mass air flow (MAF) sensor is only incorporated in some Multi-point fuel injection systems. The MAF sensor is a very complex device which measures the air mass of the engine intake. Because the air mass is always changing with temperature, humidity and altitude, the fuel delivery rate must be adjusted to compensate for these changes so that a precise fuel mixture can be maintained.

AIR TEMPERATURE SENSOR

The air temperature sensor is located in the air stream of the air flow meter. The sensor supplies incoming air temperature information to the ECU. The ECU uses this data, along with other data, to regulate fuel injection rate.

THROTTLE POSITION SENSOR (TPS)

The TPS can be either a switch (or a combination of switches) or a variable resistor which is much more accurate in throttle position. The switch type TPS consists of switches that open and close at different throttle positions (usually at idle and wide open throttle) and sends the information to the ECU. The variable resistor type receives a reference voltage from the ECU and responds back to the ECU with a proportional voltage directly related to the position of the throttle plate.

ENGINE COOLANT TEMPERATURE SENSOR

The coolant temperature sensor is located in the engine coolant passage, usually located in the intake manifold. The sensor is resistor based and changes resistance as coolant temper-

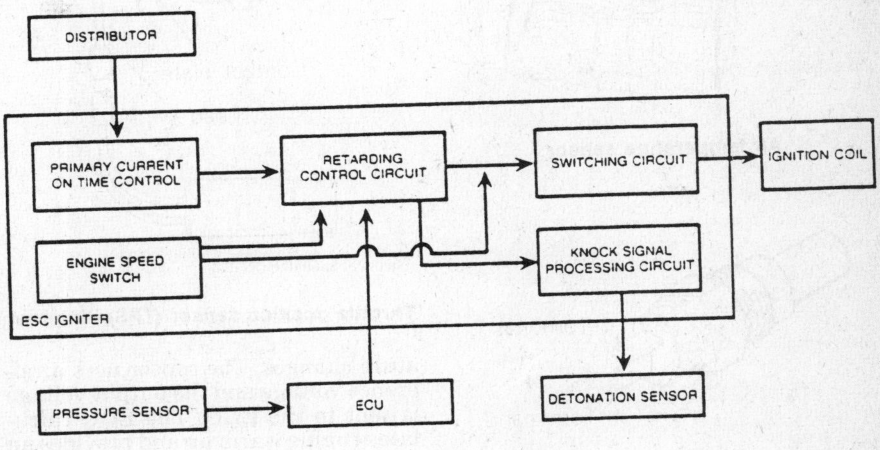

Electronic spark control (ESC) ignition with detonation sensor

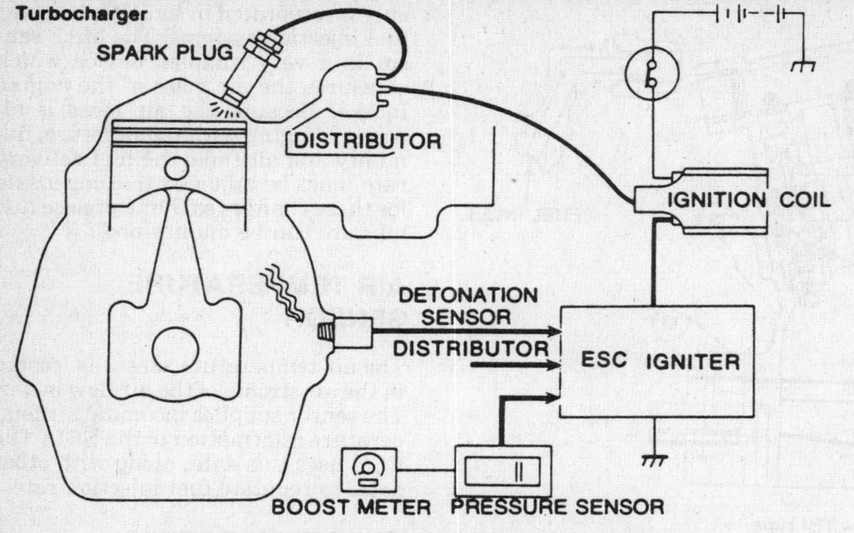

Electronic spark control (ESC) ignition system

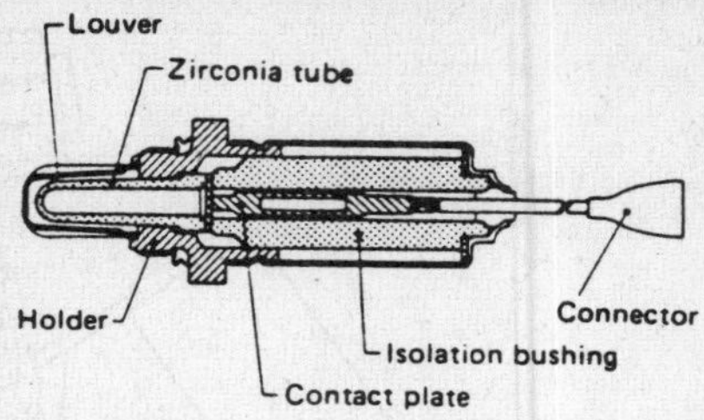

Oxygen (O_2) sensor

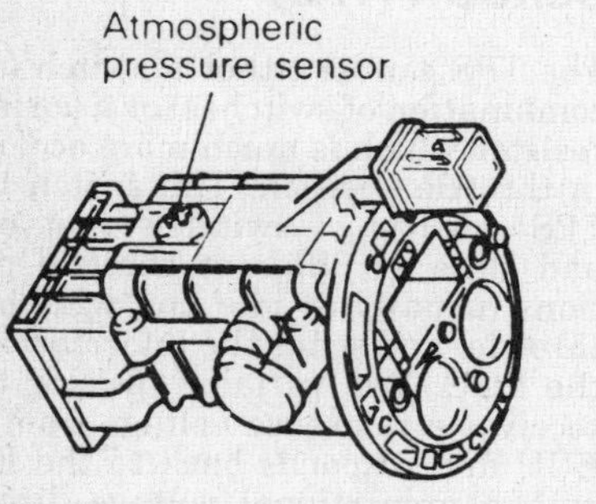

Mass air flow (MAF) sensor

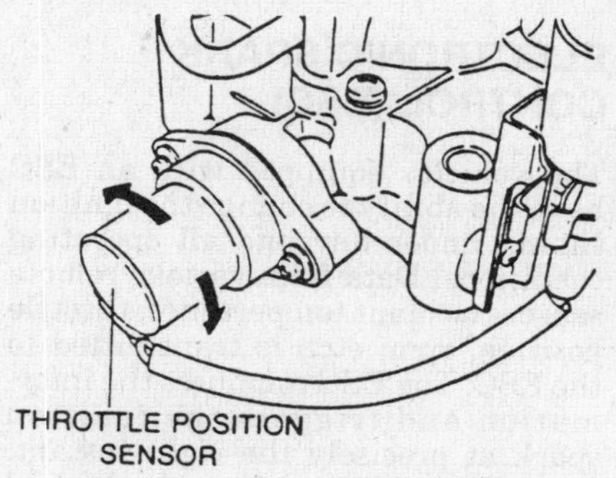

Throttle position sensor (TPS)

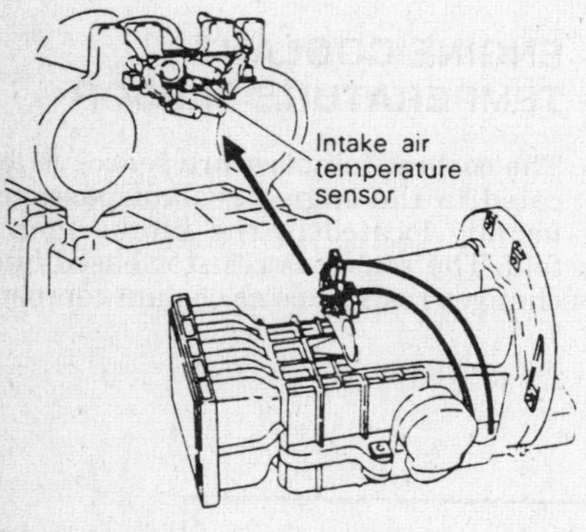

Air temperature sensor

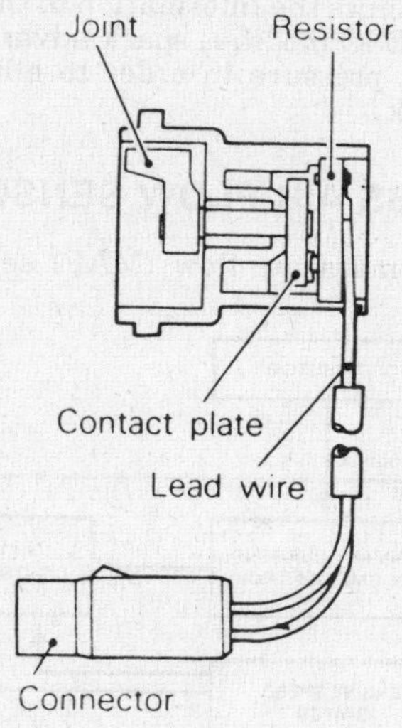

Throttle position sensor (TPS) – Internal

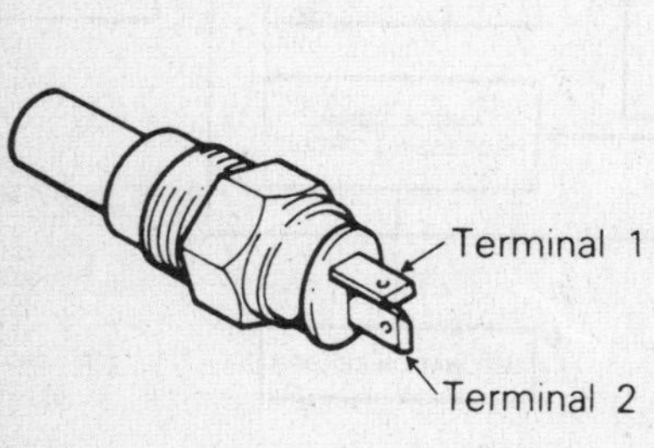

Coolant temperature sensor

ature changes. The sensor uses a reference voltage and the output voltage is sent to the ECU. The ECU calculates engine warm up and provides an optimum fuel enrichment when the engine is cold.

OXYGEN (O_2) SENSOR

The O_2 sensor, which is placed in the exhaust stream, monitors the oxygen content in the exhaust gas. The sensor is mounted in the exhaust manifold and is sometimes internally heated electrically for faster switching to the closed loop mode. The sensor produces a voltage proportional to the oxygen content which represents a lean or rich condition and transmits the information to the ECU. A low voltage condition indicates a lean mixture (high O_2 content) and a higher voltage indicates a rich mixture (low O_2 content). The ECU uses the information, along with other sensor data, and changes the fuel induction as required.

CYLINDER HEAD TEMPERATURE SENSOR

The cylinder head temperature sensor monitors the temperature of the cylinder head and transmits the information to the ECU. The sensor is located in the cylinder head and is a temperature sensitive resistive unit known as a thermistor.

VEHICLE SPEED SENSOR (VSS)

The VSS provides vehicle speed data to the ECU in the form of pulse signals. There are many different types of VSS, some using a reed switch installed in the speedometer unit and others using a optical type. In the optical type a light emitting diode (LED) is used to transmit light and photo diode receives the light. A shutter device, which is usually in-line with the speedometer cable, allows the LED light to reach the photo diode in vehicle speed related pulses. The reed switch type relies on a reed switch that opens and closes by way of a rotating magnet. The magnet rotates proportionally with the vehicle speed.

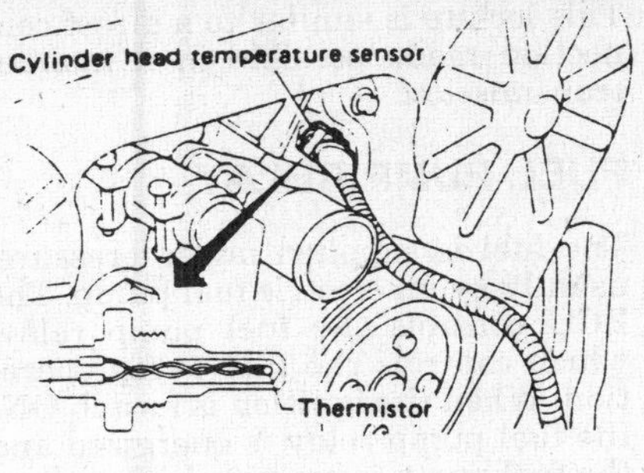

Cylinder head temperature sensor

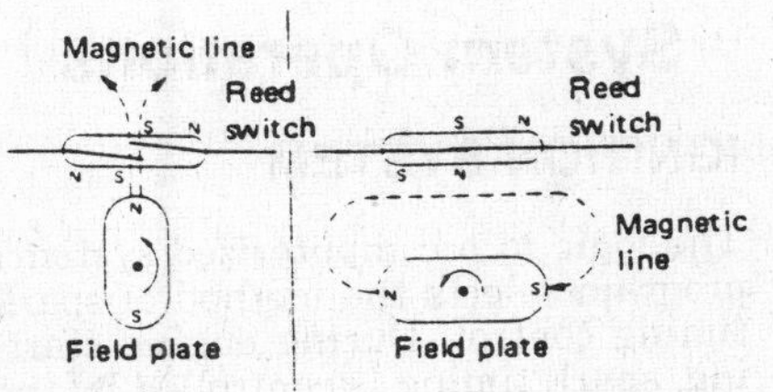

Vehicle speed sensor (VSS) – reed switch type

MANIFOLD AIR PRESSURE (MAP) SENSOR

The MAP sensor is a device that monitors manifold absolute pressure. The sensor is mounted remotely and senses vacuum through a connecting hose. The MAP sensor has a reference voltage from the ECU and transmits remaining voltage to the ECU to calculate engine load. The ECU uses this data along with other data to determine fuel demands.

DETONATION (KNOCK) SENSOR

The detonation sensor generates a signal when pre-ignition (knock) occurs in one or more combustion chambers. The sensor is made of a material that is sensitive to oscillation that the engine knock produces and sends signals to the ECU. The ECU, in turn, delays the ignition signal which retards the ignition timing and continues to do this until the engine knock ceases.

CRANKSHAFT (REFERENCE MARK) SENSOR

The crankshaft sensor may be located at either the rear of the engine, at the flywheel, at the front of the engine, near the crankshaft pulley or mounted in the distributor. The sensor detects crankshaft position in relation to top dead center and transmits the signals to ECU.

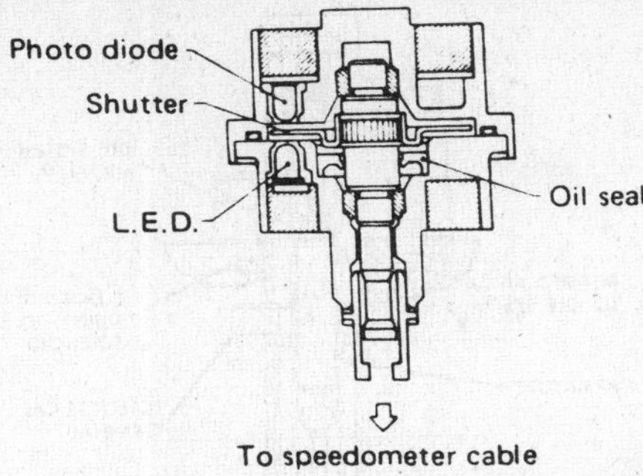

Vehicle speed sensor (VSS) – photo type

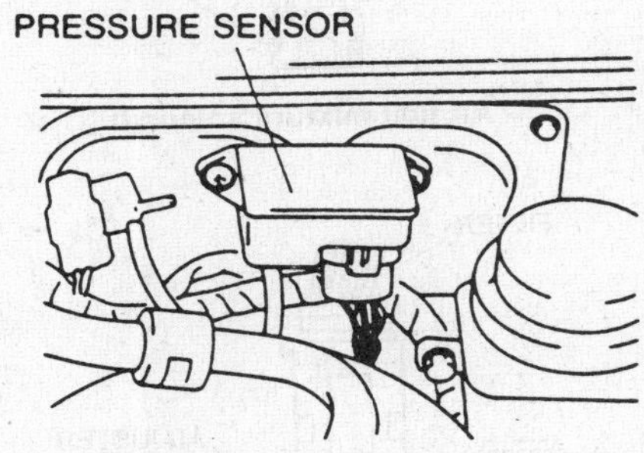

Manifold air pressure (MAP) sensor

Detonation (knock) sensor

IDLE SPEED CONTROL (ISC) MOTOR

The ISC is sometimes included on a feedback carburetor system and mounted to the side of the carburetor. The motor driven ISC maintains a steady idle by way of the ECU. When an added load is put on the engine (air conditioning or when the vehicle is in drive) the ECU can increase the idle via the ISC by extending a plunger which opens the throttle valve.

AIR/FUEL MIXTURE SOLENOID

The air/fuel mixture solenoid on a feedback carburetor operates in conjuntion with the fixed metering jets and/or the manually adjustable

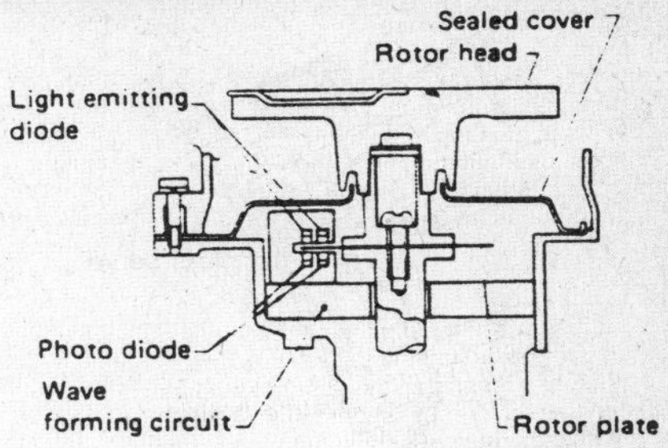

Crankshaft position sensor – distributor mounted

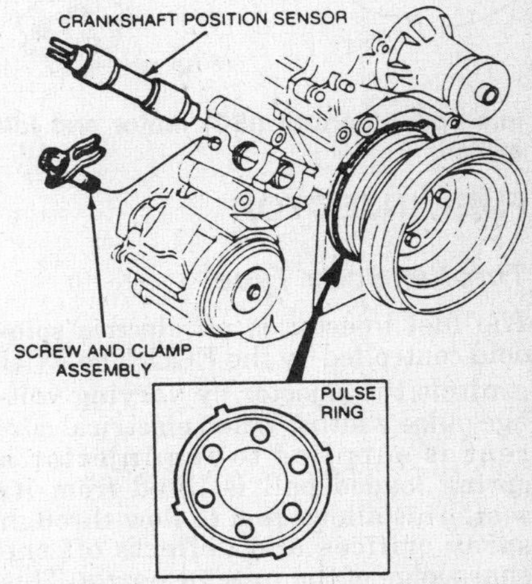

Crankshaft position sensor – near crankshaft pulley

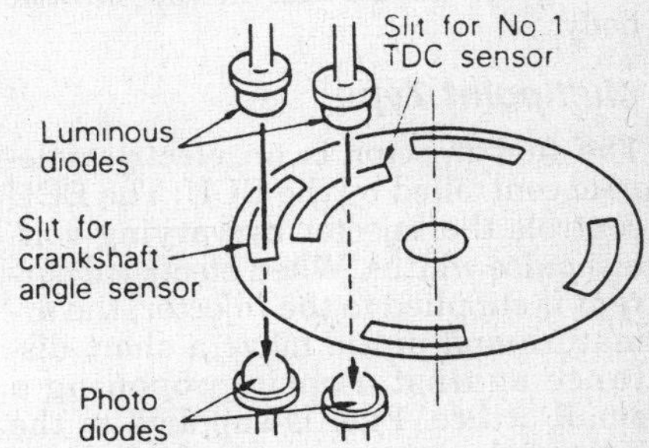

Crankshaft position sensor – optical type

idle speed mixture screw. The ECU energizes and de-energizes the solenoid in the closed loop mode. The solenoid usually controls a fixed air bleed and/or fuel discharge port.

IDLE AIR CONTROL (IAC)

The IAC in a fuel injection system controls the air flow around the throttle plate by extending and retracting a bypass valve in the bypass port. The ECU controls the valve by sending voltage pulses called counts or steps to increase or decrease the bypass air flow, thus increasing and decreasing the idle speed.

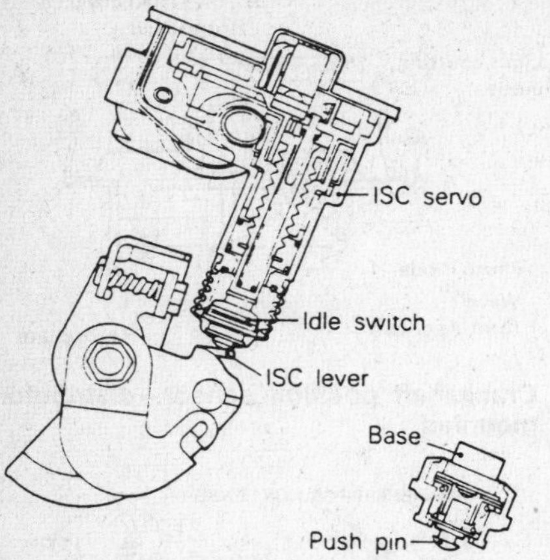

Idle speed control (ISC) motor and idle switch

FUEL INJECTOR

Throttle Body Type

The fuel injector is an electric solenoid controlled by the ECU. The ECU controls the injector by varying voltage pulse widths. When electrical current is supplied to the injector a spring loaded ball is lifted from its seat. This allows fuel to flow through spray orifices and deflects off the sharp edge of the injector nozzle. This action causes the fuel to form a 45° cone shaped spray pattern before entering the air stream in the throttle body.

Multipoint Type

The fuel injector is an electric solenoid controlled by the ECU. The ECU controls the injector by varying voltage pulse widths. When electrical current is supplied to the injector, the armature and pintle move a short distance against a spring, opening a small orifice. Fuel is supplied to the inlet of the injector by the fuel pump, then passes through the injector, around the pintle and out the orifice. Since the fuel is under high pressure, a fine spray is developed in the shape of a hollow cone. The injector, through this spraying action, atomizes the fuel and distributes it into the air entering the combustion chamber.

TORQUE CONVERTER CLUTCH (TCC) SOLENOID

The TCC solenoid is used on some automatic transmission, which allows for better fuel economy. When certain engine and vehicle speeds have been met, the ECU energizes the solenoid. This allows transmission fluid to flow into passages in the torque converter, which causes the converter to lock up.

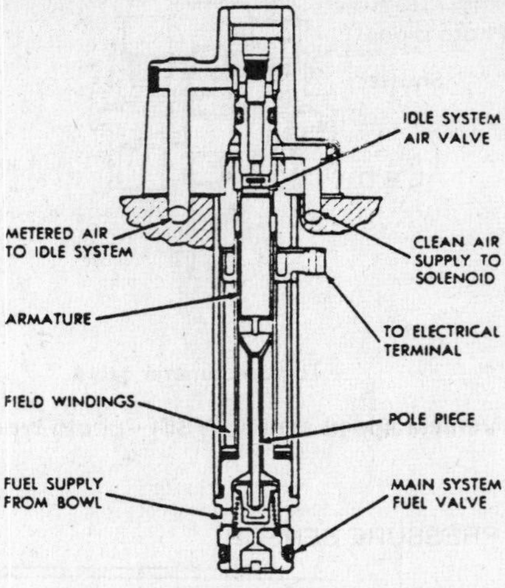

Air/fuel mixture solenoid

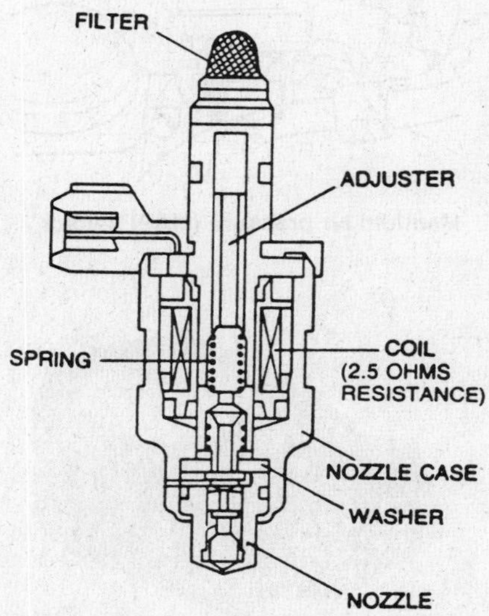

Fuel injector – TBI type

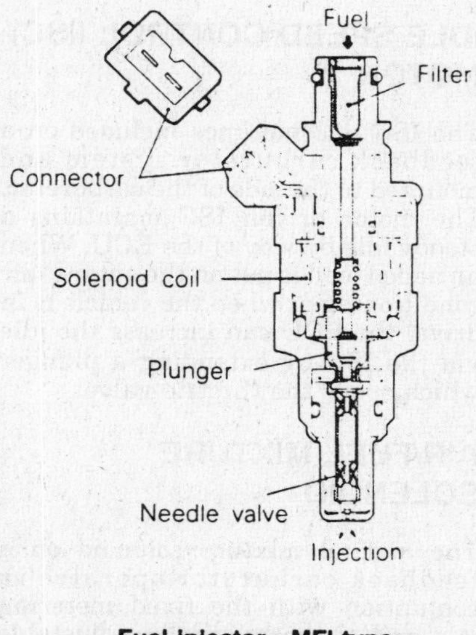

Fuel injector – MFI type

This lockup is similar to a direct connection made possible in a manual transmission.

FUEL PUMP RELAY

The fuel is supplied under pressure, usually by an electric fuel pump. The ECU controls the fuel pump relay, which controls the fuel pump operation. When the ignition is switch ON, the fuel pump relay is energized and the fuel pump is activated. The pump primes the fuel system with fuel to a pre-determined pressure.

System Operations

IGNITION SYSTEM

The logic in a computerized system's program selects the method of spark timing control. During engine starting, spark timing is controlled by the mechanical setting of the distributor. Once the engine is running, spark timing is turned over to the ECU. This scheme ensures that the car will start regardless of whether the electronic control system is working or not.

The goal of electronic spark timing is to produce maximum engine power by adjustment the advance of the ignition firing in relationship to top dead center (TDC). The spark timing can be chosen to produce the best engine power with input variables of engine rpm, engine coolant temperature, initial and operating manifold or barometric pressure.

The total spark advance is determined by computing the information received from the various engine sensors which affect spark timing. The processor will then adjust the timing according to information that has been calibrated in it. The processor has programmed into it specific information on:

Warm-Up Spark Advance – this is used when the engine is cold, since a greater amount of advance is required while the engine warms up.

Special Spark Advance – to improve fuel economy during steady driving conditions.

Spark Advance Due to Barometric Pressure – this is used when barometric pressure exceeds a preset calibrated amount.

All of this information is then added together and the initial mechanical advance is subtracted to determine the final spark advance.

The processor receives a timing pulse from a sensor which indicates crankshaft position for top dead center and engines rpm. The processor makes a decision based upon this information and the information that was calibrated into it. At that time, the computer sends a pulse to the ignition actuator circuit, which opens the ignition coil primary circuit to generate a secondary voltage pulse to fire the spark plugs. In some cases, the circuitry to open the primary of the coil may be in the computerized controller. The spark selection is performed mechanically by the distributor and rotor contacts as it is done in a non-electronic controlled system.

The ignition timing works along with electronic fuel control to control emissions and provide for optimum fuel economy and driveability because engine power, fuel economy and emissions are dependent on spark advance of the engine timing.

The system just described is considered to operate in open-loop. There are some electronically controlled ignition systems which receive an input from a knock sensor. These systems operate in a closed-loop mode which allows the ignition system to monitor the engine for mechanical changes, such as engine knock.

Engine knock is a condition where the air/fuel mixture in the cylinder does not burn normally. The pressure rise during this burning is so rapid compared to normal combustion that it is accompanied by an audible "knock".

Through some low level knock is acceptable, it is important to avoid excessive knock. To control engine knock, a knock sensor is installed in the engine or intake manifold. This helps to detect excessive engine knock.

The knock sensor is a tuned accelerometer and produces an output voltage depending on the amount of engine vibration occurring in a certain frequency band. When the processor receives a signal from the knock sensor, it retards the spark advance until the knocking stops and then starts increasing it again. This cycle is repeated as long as engine knock occurs.

FUEL CONTROL

In order for the processor to control fuel, it requires a sensor or sensors to monitor the state of the engine, and one or more actuators to do the actual controlling. The sensors measure: exhaust gas oxygen, manifold or barometric absolute pressure, engine rpm and speed, inlet air and coolant temperatures. Actuators are energized to control the air/fuel ratio.

The primary purpose of this control system is to maintain air/fuel ratio at or near 14.7:1 ratio. This is accomplished in two modes (during normal

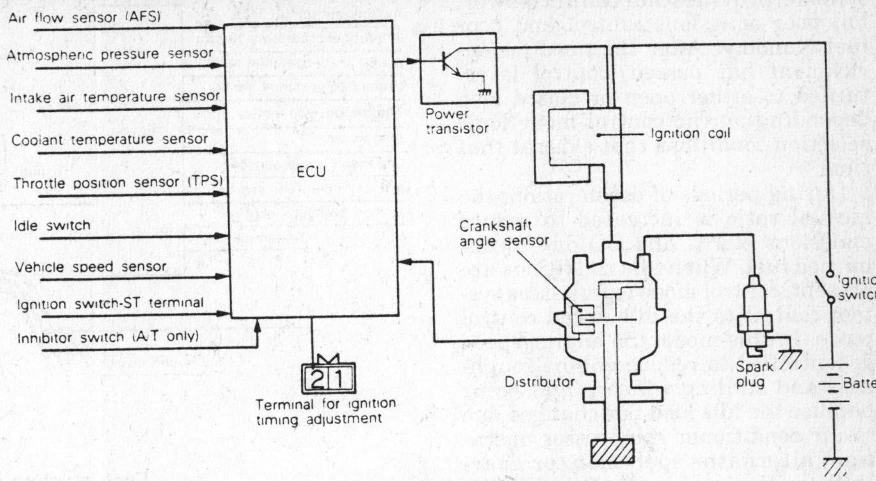

Ignition system with ECU

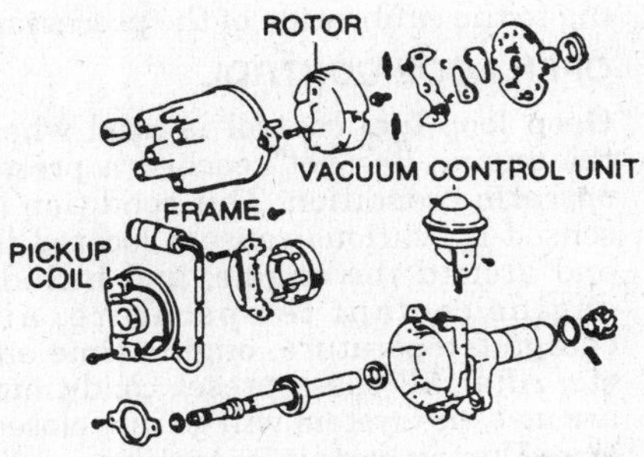

Electronic distributor assembly

engine operation) open and closed loop. The electronic fuel control system can operate in closed loop only when certain conditions are satisfied. Open loop mode is employed whenever these conditions are not satisfied. However, for either mode, the exhaust emissions will satisfy federal requirements if the average air/fuel ratio is held within the tolerance limits.

In addition to open and closed loop control modes, a practical fuel control system has other operating modes depending on engine conditions. These handle such conditions as starting, rapid acceleration or heavy load, sudden deceleration, idling, etc.

An automotive engine has various operating modes as the operating conditions change. Preprogrammed into the processor, control logic determines the operating mode from the engine conditions that exist. From these engine conditions, the system determines which operating modes are to be performed.

There are seven different engine operating modes which affect fuel control: engine crank, engine warmup, open loop, closed loop control, hard acceleration, deceleration and idle. The program for mode control logic determines the engine operating mode by reading various sensors.

When the ignition switch is initially switched on, the mode control logic automatically selects an engine-start control scheme which provides the low air/fuel ratio required for starting the engine. Once the engine rpm rises above the cranking value, the controller identifies the engine-started mode and passes control to the program for the engine warm-up mode. This operating mode keeps the air/fuel ratio low to prevent engine stall during cool weather until engine coolant temperature rises above a preset value.

When the coolant temperature rises, the mode control logic directs the system to operate in the open loop control mode until a certain time has elapsed and the exhaust gas sensor warms up enough to provide accurate readings. This condition is detected by monitoring the exhaust gas sensor's output for voltage readings above a certain minimum air/fuel mixture voltage set point. When the sensor has indicated a rich mixture a certain number of times (depending on calibration) and after the engine has been in open loop for a specific time, the control mode logic selects the closed loop mode for the system. The engine remains in the closed loop mode until either the exhaust gas sensor cools and fails to switch (from rich to lean) for a certain length of time, or a hard acceleration or deceleration occurs. If the sensor cools, the control mode logic selects the open loop mode again.

During hard acceleration of heavy engine loads, the control mode logic chooses a scheme which provides a rich air/fuel mixture for the duration of the acceleration or heavy load. This scheme provides maximum power, but poor emissions control and poor fuel economy. After the need for enrichment has passed, control is returned to either open or closed loop depending on the control mode logic selection conditions that exist at that time.

During periods of deceleration, the air/fuel ratio is increased to reduce emissions of HC and CO due to unburned fuel. When idle conditions are present, control mode logic passes system control to the idle speed control mode. In this mode, the engine speed is controlled to reduce engine roughness and stalling which might occur because the idle load has changed due to air conditioner compressor operation, alternator operation, or gearshift positioning from PARK or NEUTRAL to DRIVE.

Engine Crank

While the engine is being cranked, the fuel control system must provide an intake air/fuel ratio anywhere from 2:1 to 12:1, depending on engine temperature. Low temperatures affect the carburetor's ability to atomize or mix the incoming air and fuel. At low temperature, the fuel tends to form into large droplets. The larger fuel droplets tend to increase the apparent air/fuel ratio because the amount of usable fuel in the air is reduced, therefore, the system must provide a decreased air/fuel ratio to provide the engine with a more combustible air/fuel mixture. The engine temperature is read by the processor through an analog to digital converter from a temperature sensor in the engine water coolant passage. The processor's calibration determines what the proper air/fuel ratio must be at that temperature. The air/fuel is determined and controlled as in the open loop mode.

Engine Warm-up

While the engine is warming up, an enriched air/fuel ratio is still needed to keep it running smoothly, but the required air/fuel ratio changes as the temperature increases. Therefore, the fuel control system will stay in the open loop mode, but the air/fuel ratio commands continue to be altered due to the temperature changes. The emphasis in this control mode is on rapid and smooth engine warm-up. Fuel economy and emission control are still a secondary concern. The controller determines the warm-up time period based on the coolant temperature when the warm-up mode was selected. Naturally, an initially cold engine requires a longer warm-up time than a warm engine. The time allowed by

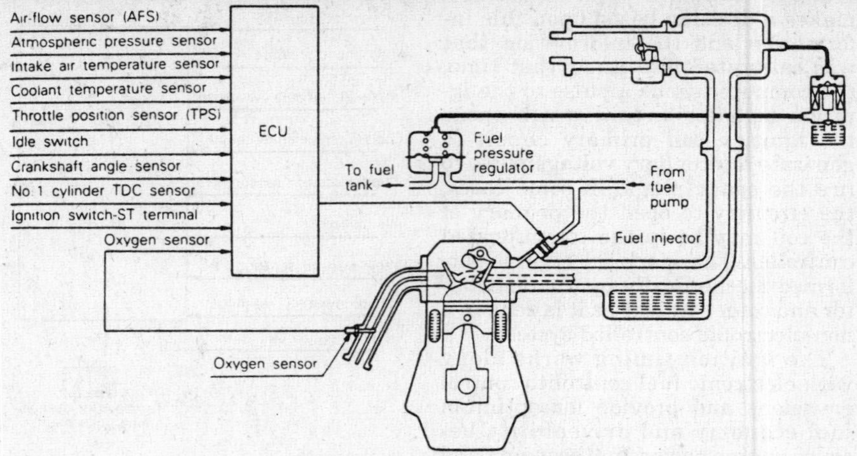

Fuel injection system with ECU

the controller timer is chosen according to the calibration of the processor.

OPEN LOOP CONTROL

Open loop fuel control is used when the engine has not reached a preset operating condition. This condition is sensed by various sensors located in and around the engine, and include engine coolant temperature, air charge temperature, engine time on, etc. After all these preset conditions are met, the system will go into closed loop. During certain operating conditions, such as a wide open throttle condition the system will go back into open loop.

CLOSED LOOP CONTROL

Closed loop fuel control is selected when the engine is warm and the exhaust gas oxygen sensor exceeds its minimum operating temperature. The intake air/fuel ratio is controlled in a closed loop by measuring the exhaust gas at the exhaust manifold and altering the input fuel flow rate or the air entering the main metering systems (depending on the type of fuel system used).

ACCELERATION ENRICHMENT (OPEN LOOP)

During periods of heavy engine load, such as wide open acceleration, fuel control is adjusted to provide an enriched ratio to maximize engine power while neglecting fuel economy and emission.

The computer detects this condition by reading the throttle position sensor voltage or the MAP sensor. Low intake manifold vacuum or throttle position corresponds to heavy engine loads. The fuel control system controller responds by increasing the amount of fuel to enter the intake manifold or to decrease the amount of air in the main metering system. This

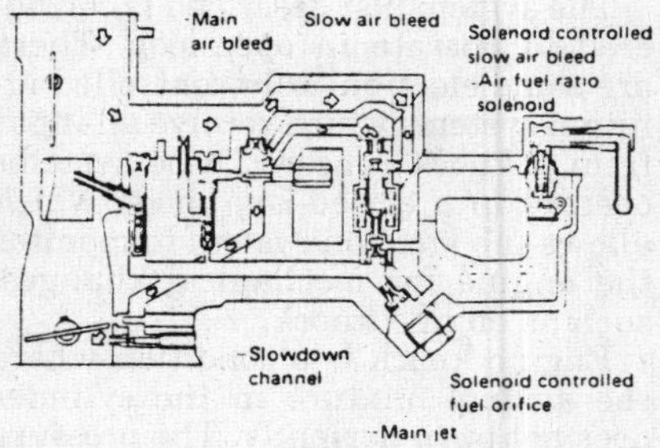

Feedback carburetor air flow

enrichment allows the engine to operate with a power greater than that allowed when emissions and fuel economy are controlled within specifications.

DECELERATION AND IDLE SPEED CONTROL (OPEN LOOP)

During periods of light engine load and high rpm, such as during closed throttle deceleration, coasting or engine idle, the engine requires a very lean air/fuel ratio to reduce excess emissions of HC and CO. Deceleration is indicated by a sudden increase in manifold vacuum and throttle position, indicating a closed throttle. When these conditions are detected by the processor, it computes a change in the amount of fuel required or amount of air entering the main or idle speed passages (depending on type of fuel system used). On certain engine applications with electronic fuel injection, the fuel may even be turned completely off during closed throttle deceleration.

Idle speed control is used to prevent engine stall during idle. The goal is to allow the engine to idle at as low an rpm as possible, yet keeping the engine from running rough and stalling when power takeoff accessories such as air conditioning compressor is turned on.

Turbocharging 29

DESCRIPTION

A turbocharger is an exhaust-driven turbine which drives a centrifugal compressor wheel. The compressor is usually located between the air cleaner and the engine's intake manifold, while the turbine is located between the exhaust manifold and the muffler. Primarily, the turbocharger compresses the air entering the engine, forcing more air into the cylinders. This allows the engine to efficiently burn more fuel, thereby producing more horsepower.

All of the exhaust gases pass through the turbine housing. The expansion of these gases, acting on the turbine wheel, causes it to turn. After passing through the turbine the exhaust gases are routed to the atmosphere through the exhaust system. On some non-automotive applications, the turbocharger provides sufficient muffling of the exhaust noises to eliminate the need for a muffler.

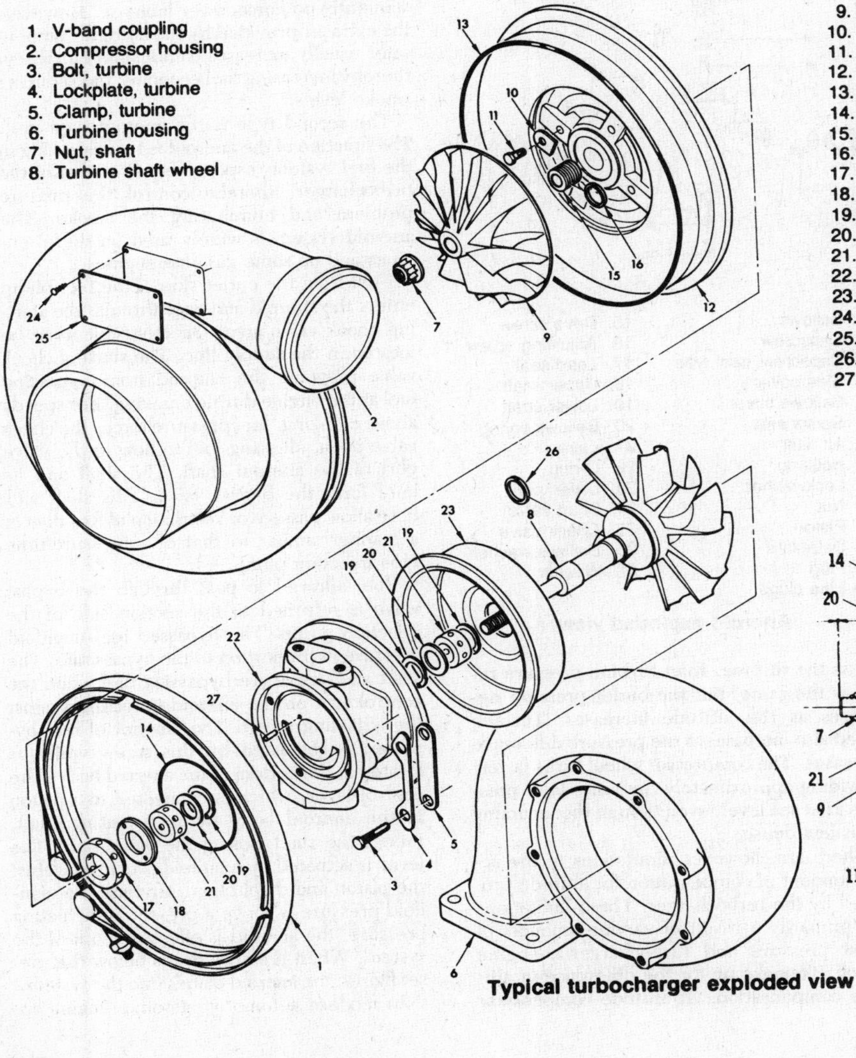

1. V-band coupling
2. Compressor housing
3. Bolt, turbine
4. Lockplate, turbine
5. Clamp, turbine
6. Turbine housing
7. Nut, shaft
8. Turbine shaft wheel
9. Compressor wheel
10. Lockplate, backplate
11. Bolt, backplate
12. Backplate (vaneless on some models)
13. O-ring (used on vaned backplate)
14. Seal ring
15. Seal spacer
16. Piston ring
17. Thrust collar
18. Inboard thrust washer
19. Bearing retainer
20. Bearing washer
21. Bearing
22. Center housing
23. Shroud, turbine
24. Drive screw
25. Nameplate
26. Piston ring, turbine
27. Pin

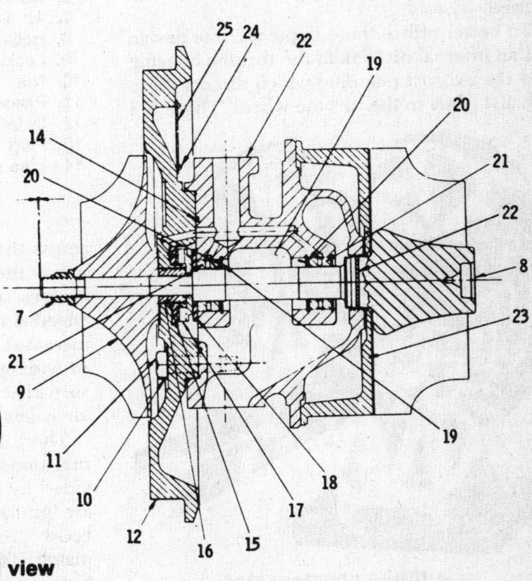

Typical turbocharger exploded view

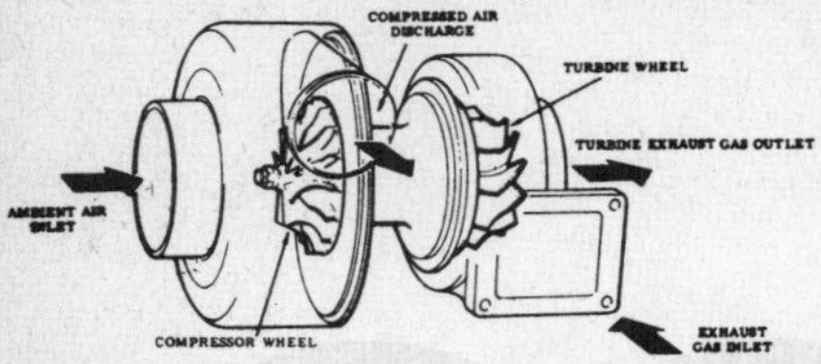

Typical turbocharger air flow schematic

The turbine also functions as a spark arrester. For example, the US Department of Agriculture recognizes the turbocharger as an adequate spark arrester for forestry operations.

OPERATION

The compressor and turbine are each enclosed in their own housings and are directly connected by a shaft. The housings are constructed of light alloy and are designed for maximum heat dissipation. The only power loss from the turbine to the compressor is the slight friction of the shaft journal bearings. Air is drawn in through the filtered intake system, compressed by the compressor wheel and discharged into the intake manifold. The extra charge of air provided by the turbocharger allows more fuel to be burned, providing more power.

As engine speed increases, the length of time the intake valves are open decreases, giving the air less time to fill the cylinders. On an engine running at 2500 rpm, the intake valves are open less than 0.017 second. The air drawn into a naturally aspirated engine's cylinder is less than atmospheric pressure. Turbochargers pack air into the cylinder at greater than atmospheric pressure at all speeds. The flow of exhaust gas from each cylinder occurs intermittantly as the exhaust valve opens. This results in fluctuating gas pressures, also known as pulse energy, at the turbine inlet. With a conventional turbine housing, only a small amount of pulse energy is used.

To better utilize these impulses, one design has an internal division in the turbine housing and the exhaust manifold which directs these exhaust gases to the turbine wheel. There is a separate passage for each half of the engine cylinder exhaust.

On some four and six cylinder engines built to accommodate turbochargers, there is a separate passage for the front two or three cylinders and another for the rear half.

By using a fully divided exhaust system combined with a dual scroll turbine housing, the result is a highly effective nozzle velocity. This produces higher turbine speeds and manifold pressures than can be obtained with an undivided system.

At high altitudes, a naturally aspirated engine drops 3% in horsepower per 1000 feet elevation due to a 3% decrease in air density per 1000 feet.

With a turbocharged engine, an increase in altitude also increases the pressure drop across the turbine. Inlet turbine pressure remains the same, but the outlet pressure decreases as the altitude increases. Turbine speed also increases as the pressure difference increases. The compressor wheel turns faster, providing approximately the same inlet pressure as at sea level, even though the incoming air is less dense.

There are, however, limitations to the actual amount of compensation for altitude provided by the turbocharger. These limitations are primarily a result of varying amounts of boost pressure and turbocharger-to-engine match. To make up for the difference in altitude compensation, an altitude compensator is added to the system. During rapid acceleration or rapid engine load changes, the turbocharger speed, reflected in manifold pressure, inherently lags behind the power or fuel demand exercised by the opening of the throttle. This lag does not exist in the fuel system, so an overly rich mixture accompanied by heavy smoke occurs until the turbocharger catches up.

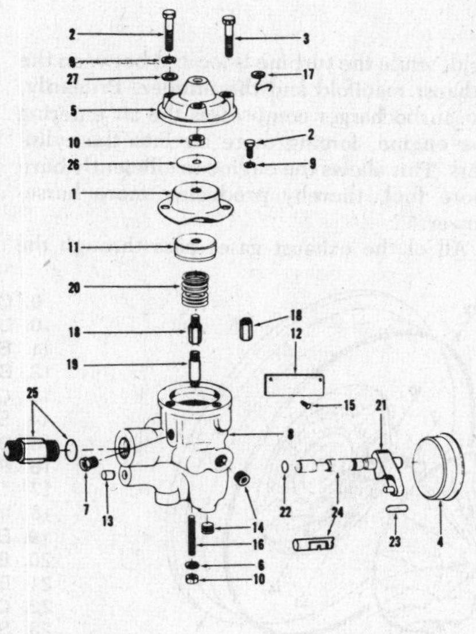

1. Bellows
2. Capscrew
3. Capscrew, seal type
4. Side cover
5. Bellows cover
6. Screw seal
7. Air filter
8. Housing
9. Lockwasher
10. Nut
11. Piston
12. Dataplate
13. Plug
14. Pipe plugs
15. Drive screw
16. Adjusting screw
17. Lead seal
18. Upper shaft
19. Lower shaft
20. Bellows spring
21. Lever
22. O-ring
23. Lever, pin
24. Shaft valve
25. Check valve
26. Bellows washer
27. Washer

Aneroid exploded view

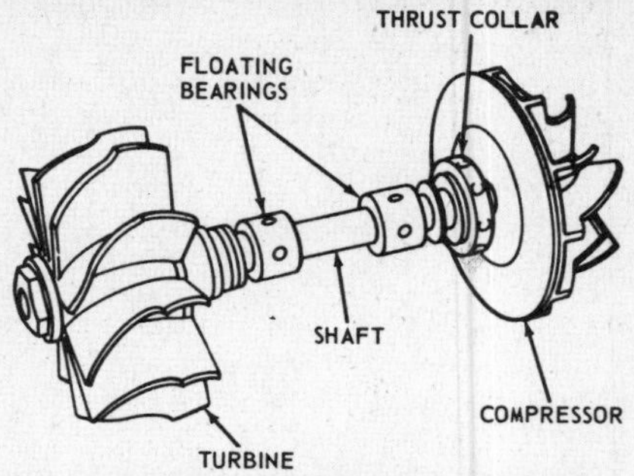

Basic parts of the turbocharger

On diesel engines, two types of altitude compensators are used. One is a compressed air type which is very similar in appearance to the turbocharger. This type supplies compressed air to the intake manifold at a pressure about equal to sea level pressure. There is no increase of fuel for combustion and consequently no horsepower increase. However, the extra air provided by the altitude compensator usually increases combustion efficiency, thereby increasing fuel economy and reducing smoke levels.

The second type is the aneroid type unit. The function of the aneroid is to create a lag in the fuel system response equal to that of the turbocharger, thereby control the mixture problem and eliminating the smoke. The aneroid system is widely used on diesel engines and on some gasoline engines.

Fuel from the outlet side of the fuel pump enters the aneroid and goes through the starting check valve area. On others, it must be located in the supply line. The starting check valve prevents the aneroid from bypassing fuel at the engine during cranking. For speeds above cranking, fuel pressure forces the check valve open, allowing fuel to flow to the valve port of the aneroid shaft. The shaft and its bore form the bypass valve. This shaft and bore allow passage or restriction of fuel flow in a manner similar to that of a pressure/time type injection pump.

Fuel allowed to pass through the bypass valve is returned to the suction side of the injection pump. The bypassed fuel manifold pressure in proportion to the bypass rate. The shaft and sleeve are bypassing fuel when the control arm on the aneroid is resting against the adjusting screw. The amount of fuel bypassed is regulated by this screw which is located at the bottom of the aneroid body. The control lever, which is connected to a piston in the aneroid body by an actuating shaft, rotates the shaft closing the valve port. The lever is actuated by manifold pressure against the piston and diaphragm. Anytime the manifold pressure is above a present air actuating pressure, the aneroid is effectively out of the system. When pressure drops below the preset figure, the aneroid comes into the system.

In modern automotive gasoline engine ap-

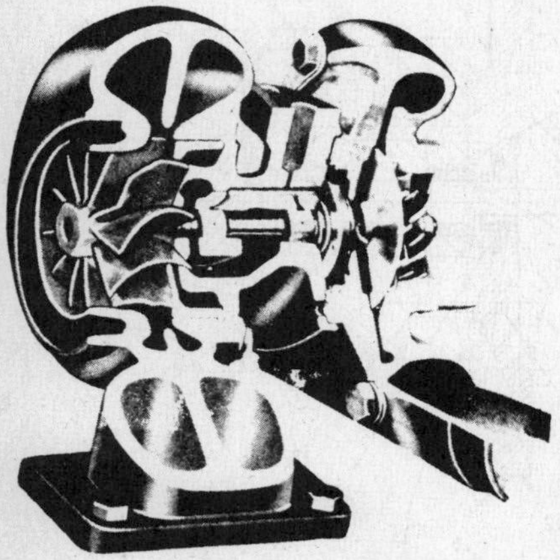

Altitude compensator

plications with their stricter emission control standards, turbocharger lag is compensated for by means of modified spark control and/or an enrichment vacuum regulator system. The spark control system changes the ignition timing on demand and the vacuum regulator system regulates vacuum flow at the carburetor through a remote power enrichment port.

Some engines, particularly passenger car applications, in which boost pressure must be held at low levels, utilize a wastegate unit. Since turbocharger operation is self-perpetuating, unchecked operation will increase boost pressure beyond the operating capabilities of these engines. Some method of limiting this boost increases must be used. The principle means is by the inclusion of a wastegate in the system. The wastegate, usually located in the outlet elbow assembly, is activated when boost pressure reaches a predetermined level (usually 3-7 psi depending on application). The wastegate opens and bypasses exhaust flow around the turbine.

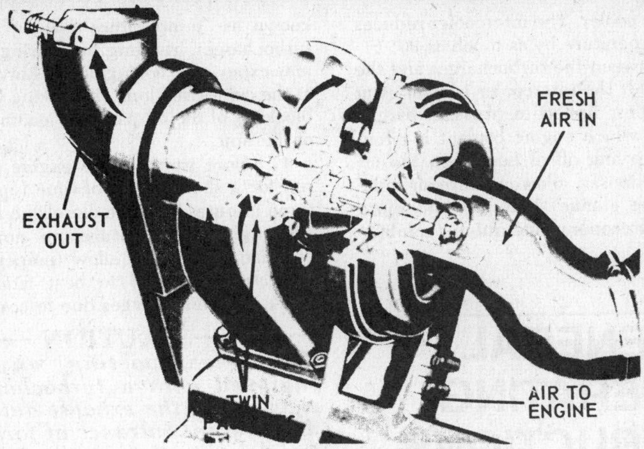

Twin passage turbine

LUBRICATION

Since turbine speeds routinely reach 140,000 rpm, adequate lubrication is vitally important. Turbochargers are lubricated by engine oil. Depending on the application, the lubrication may be either pressure-fed or gravity-fed. In areas of very heavy load or when shut-down after peak operation is routine, pressure feeding, sometimes with a separate oil pump, is used. In cases where a separate oil pump is used, the pump continues operating during spin-down. Since all parts of the rotating assemblies are protected by a film of oil, no metal-to-metal contact occurs. Consequently, no appreciable wear should occur. If a constant supply of clean engine oil is maintained, bearing life should be indefinite. If the unit has floating sleeve type bearings, they provide oil clearance between the bearing and housing as well as between the bearing and shaft. When the turbocharger is operating, this allows the bearing to turn as the shaft turns. All clearances in the turbocharger are closely controlled and carefully machined. Any dirt in the oil will adversely affect service life of the working parts. Oil and filter changes should occur regularly. Some manufacturers recommend more frequent oil changes for turbocharged engines. In any case, on turbocharged engines, the oil filter(s) should ALWAYS be changed with the engine oil. ALWAYS use oil of the recommended viscosity for that particular engine application. Check the owner's manual for your engine or vehicle for recommended intervals and proper viscosity.

TWO-CYCLE APPLICATIONS

Turbochargers may be used in addition to the regular scavenging process. In these cases, the air is drawn into the blower or scavenging pump and then transferred to the turbocharger where it is compressed and forced into the engine.

At light loads, there is little energy available to drive the turbocharger. The mechanically driven blower alone supplies scavenging air to cylinders. At increased loads, the turbocharger speeds up and takes in a sufficient amount of additional air to allow the inlet pressure to drop to atmospheric levels, causing the blower check valve to open. At this engine speed, the blower becomes unloaded, saving engine power, and the turbocharger enters the load range where it alone can provide scavenging and turbocharging. Under ideal conditions, the engine starting air contains enough energy to start the turbocharger and also supply enough air for combustion. In some applications, however, turbochargers can be equipped with additional methods for supplying necessary scavenging air while the engine is being started. This can be accomplished either mechanically by coupling the turbocharger to the crankshaft in such a manner that it is mechanically driven during starting and automatically disconnects when exhaust pressure is high enough or by jet air starting, where air is blown through jets into the turbocharger turbine or compressor. The air passing through the compressor also aids in scavenging during starting.

INTERCOOLERS

When the air passing through the compressor is compressed it becomes heated and expands. Expanding air is less dense, therefore less air is forced into the engine. This helps defeat the turbocharging process. To overcome this condition, some engine applications use a heat exchanger, also known as an inter-

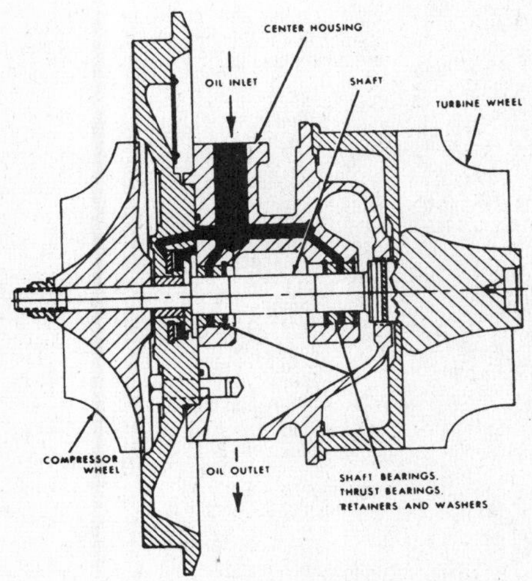

Turbocharger oil flow

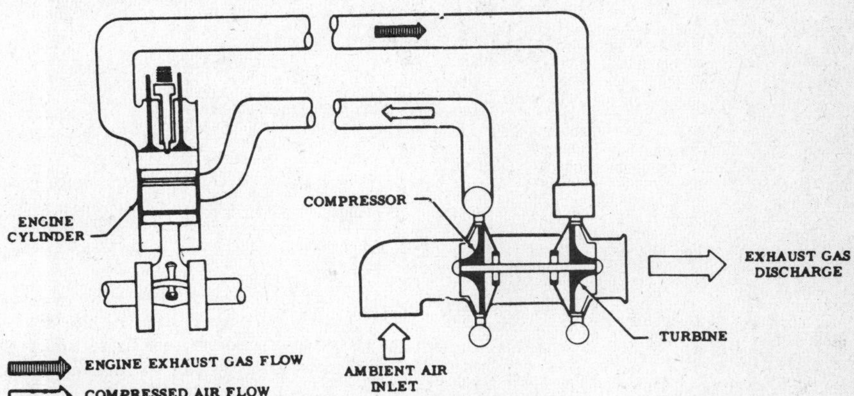

Two cycle air flow

cooler or after cooler. The intercooler reduces intake air temperature by as much as 90° F.

Located between the turbocharger and the intake manifold, the intercooler is a series of connected tubes, finned to provide dissipation, through which engine coolant is circulated. The carrying off of heat from the air makes the air denser, allowing more air to be forced into the engine. This provides more power, greater economy and quieter combustion.

GENERAL OPERATING INSTRUCTIONS FOR TURBOCHARGED ENGINES

1. After starting the engine, make sure there is sufficient oil pressure before accelerating or applying load.
2. When starting in cold weather, allow the engine to run a sufficient length of time (up to five minutes for diesel engines in extreme cold) before applying load or accelerating. This will insure adequate lubrication.
3. Should the engine stall at normal operating temperature, restart it immediately. This will prevent a rapid rise in the turbocharger known as "temperature soaking". Also, the turbocharger, running hot during operation, may experience coking due to hot oil build-up in the center section. This coking will cause a blockage of the oil passages leading to failure of the unit.
4. Before stopping the engine, allow it to run for a short length of time (up to two or three minutes for some diesels) to allow internal engine temperatures to normalize or equalize. Failure to allow temperature normalization can lead to heat fatique and/or blockage of oil passages due to coking.

CAUTION

When transporting an engine equipped with a turbocharger, always cover the exhaust outlet. This will prevent entrance of foreign material and/or the rotation of the turbine. Turbine rotation on a stopped engine could lead to bearing failure since no lubricating oil will be provided.

PREVENTIVE MAINTENANCE

1. Inspect all mountings and connections regularly to make sure they are secure and no leakage is present.
2. Make certain that there is no restriction in air flow at the crankcase ventilation system.
3. Run the engine at various, normal operating speeds and listen for unusual noises at the turbocharger.

Turbochargers normally emit a shrill whistle or whine. Bearings about to fail also emit a shrill whine, somewhat different from normal turbocharger noise. Try to distinguish between the two.

NOTE: After engine shut-off, the turbocharger will whine during rundown. Don't confuse this with bearing failure noise. Grating or scraping noises could indicate improper turbine or compressor wheel-to-housing clearances. If any such noises are heard, the unit should be removed for inspection.

4. Check the unit for unusual vibrations during operation.
5. Check for unusual smoking under load conditions. Excessive smoke means an incorrect air/fuel ratio.
6. Inspect and replace the air filter according to your owner's manual recommendations.

TROUBLE-SHOOTING

The turbocharger is a relatively simple unit. Most problems occur in other parts of the engine such as the lubrication system or the fuel system. With proper routine maintenance, the unit should give troublefree operation.

Engine Rebuilding 30

This section describes, in detail, the procedures involved in rebuilding a typical engine. The procedures are basically identical to those used in rebuilding engines of nearly all design and configurations.

The section is divided into two parts. The first, Cylinder Head Reconditioning, assumes that the cylinder head is removed from the engine, all manifolds are removed, and the cylinder head is on a workbench. The camshaft should be removed from overhead cam cylinder heads. The second section, Cylinder Block Reconditioning, covers the block, pistons, connecting rods and crankshaft. It is assumed that the engine is mounted on a work stand, and the cylinder head and all accessories are removed.

Procedures are identified as follows:

Unmarked—Basic procedures that must be performed in order to successfully complete the rebuilding process.

Starred (*)—Procedures that should be performed to ensure maximum performance and engine life.

Double starred (**)—Procedures that may be performed to increase engine performance and reliability.

In many cases, a choice of methods is also provided. Methods are identified in the same manner as procedures. The choice of method for a procedure is at the discretion of the user.

The tools required for the basic rebuilding procedure should, with minor exceptions, be those included in a mechanic's tool kit. An accurate torque wrench, and a dial indicator (reading in thousandths) mounted on a universal base should be available. Special tools, where required, all are readily available from the major tool suppliers. The services of a competent automotive machine shop must also be readily available.

When assembling the engine, any parts that will be in frictional contact must be prelubricated, to provide protection on initial start-up. Any product specifically formulated for this purpose may be used. NOTE: *Do not use engine oil*. Where semi-permanent (locked but removable) installation of bolts or nuts is desired, threads should be cleaned and coated with Loctite® or a similar product (non-hardening).

Aluminum has become increasingly popular for use in engines, due to its low weight and excellent heat transfer characteristics. The following precautions must be observed when handling aluminum engine parts:

—Never hot-tank aluminum parts.

—Remove all aluminum parts (identification tags, etc.) from engine parts before hot-tanking (otherwise they will be removed during the process).

—Always coat threads lightly with engine oil or anti-seize compounds before installation, to prevent seizure.

—Never over-torque bolts or spark plugs in aluminum threads. Should stripping occur, threads can be restored using any of a number of thread repair kits available (see next section).

Magnaflux and Zyglo are inspection techniques used to locate material flaws, such as stress cracks. Magnafluxing coats the part with fine magnetic particles, and subjects the part to a magnetic field. Cracks cause breaks in the magnetic field, which are outlined by the particles. Since Magnaflux is a magnetic process, it is applicable only to ferrous materials. The Zyglo process coats the material with a fluorescent dye penetrant, and then subjects it to blacklight inspection, under which cracks glow brightly. Parts made of any material may be tested using Zyglo. While Magnaflux and Zyglo are excellent for general inspection, and locating hidden defects, specific checks of suspected cracks may be made at lower cost and more readily using spot check dye. The dye is sprayed onto the suspected area, wiped off, and the area is then sprayed with a developer. Cracks then will show up brightly. Spot check dyes will only indicate surface cracks; therefore, structural cracks below the surface may escape detection. When questionable, the part should be tested using Magnaflux or Zyglo.

REPAIRING DAMAGED THREADS

Several methods of repairing damaged threads are available. Heli-Coil® (shown here), Keenserts® and Microdot® are among the most widely used. All involve basically the same principle—drilling out stripped threads, tapping the hole and installing a prewound insert— making welding, plugging and oversize fasteners unnecessary.

Two types of thread repair inserts are usually supplied—a standard type for most Inch Coarse, Inch Fine, Metric Coarse and Metric Fine thread sizes and a spark plug type to fit most spark plug port sizes. Consult the individual manufacturer's catalog to determine exact applications. Typical thread repair kits will contain a selection of prewound threaded inserts, a tap (corresponding to the outside diameter threads of the insert) and an installation tool. Most manufacturers also supply blister-packed thread repair inserts separately and a master kit with a variety of taps and inserts plus installation tools.

Before effecting a repair to a threaded hole, remove any snapped, broken or damaged bolts or studs. Penetrating oil can be used to free frozen threads; the offending item can be removed with locking pliers or with a screw or stud extractor. After the hole is clear, the thread can be repaired as follows.

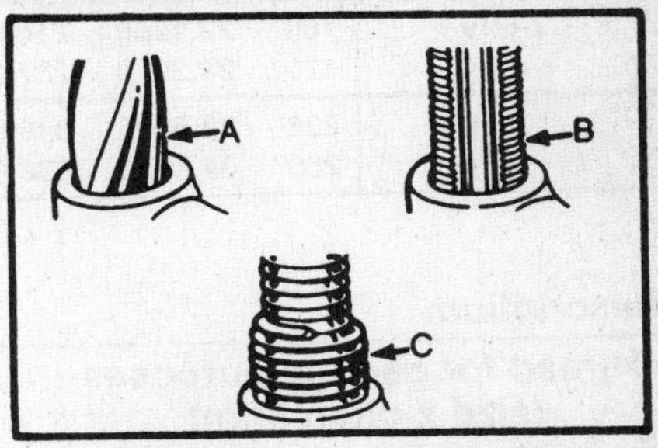

A. Drill out the damaged threads with the specified drill. Drill completely through the hole or to the bottom of a blind hole.

B. With the tap supplied tap the hole to receive the threaded insert. Keep the tap well oiled and back it out frequently to avoid clogging the threads.

C. Screw the threaded insert onto the installation tool until the tang engages the slot. Screw the insert into the tapped hole until it is ¼–½ turn below the top surface. After installation, break the tang off with a hammer and punch.

STANDARD TORQUE SPECIFICATIONS AND CAPSCREW MARKINGS

Newton-Meter has been designated as the world standard for measuring torque and will gradually replace the foot-pound and kilogram-meter torque measuring standard. Torquing tools are still being manufactured with foot-pounds and kilogram-meter scales, along with the new Newton-Meter standard. To assist the repairman, foot-pounds, kilogram-meter and Newton-Meter are listed in the following charts, and should be followed as applicable.

U.S. BOLTS

SAE Grade Number	1 or 2			5			6 or 7			8		
Capscrew Head Markings Manufacturer's marks may vary. Three-line markings on heads below indicate SAE Grade 5. 	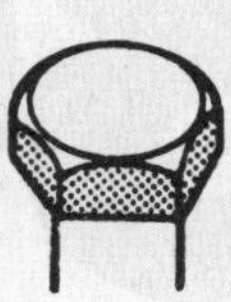											
Usage	Used Frequently			Used Frequently			Used at Times			Used at Times		
Quality of Material	Indeterminate			Minimum Commercial			Medium Commercial			Best Commercial		
Capacity Body Size	Torque			Torque			Torque			Torque		
(inches)-(thread)	Ft-Lb	kgm	Nm	Ft-Lb	kgm	Nm	Ft-Lb	kgm	Nm	Ft-Lb	kgm	Nm
1/4-20	5	0.6915	6.7791	8	1.1064	10.8465	10	1.3630	13.5582	12	1.6596	16.2698
-28	6	0.8298	8.1349	10	1.3830	13.5582				14	1.9362	18.9815
5/16-18	11	1.5213	14.9140	17	2.3511	23.0489	19	2.6277	25.7605	24	3.3192	32.5396
-24	13	1.7979	17.6256	19	2.6277	25.7605				27	3.7341	36.6071
3/8-16	18	2.4894	24.4047	31	4.2873	42.0304	34	4.7022	46.0978	44	6.0852	59.6560
-24	20	2.7660	27.1164	35	4.8405	47.4536				49	6.7767	66.4351
7/16-14	28	3.8132	37.9629	49	6.7767	66.4351	55	7.6065	74.5700	70	9.6810	94.9073
-20	30	4.1490	40.6745	55	7.6065	74.5700				78	10.7874	105.7538
1/2-13	39	5.3937	52.8769	75	10.3725	101.6863	85	11.7555	115.2445	105	14.5215	142.3609
-20	41	5.6703	55.5885	85	11.7555	115.2445				120	16.5860	162.6960
9/16-12	51	7.0533	69.1467	110	15.2130	149.1380	120	16.5960	162.6960	155	21.4365	210.1490
-18	55	7.6065	74.5700	120	16.5960	162.6960				170	23.5110	230.4860
5/8-11	83	11.4789	112.5329	150	20.7450	203.3700	167	23.0961	226.4186	210	29.0430	284.7180
-18	95	13.1385	128.8027	170	23.5110	230.4860				240	33.1920	325.3920
3/4-10	105	14.5215	142.3609	270	37.3410	366.0660	280	38.7240	379.6240	375	51.8625	508.4250
-16	115	15.9045	155.9170	295	40.7985	399.9610				420	58.0860	568.4360
7/8-9	160	22.1280	216.9280	395	54.6285	535.5410	440	60.8520	596.5520	605	83.6715	820.2590
-14	175	24.2025	237.2650	435	60.1605	589.7730				675	93.3525	915.1650
1-8	236	32.5005	318.6130	590	81.5970	799.9220	660	91.2780	894.8280	910	125.8530	1233.7780
-14	250	34.5750	338.9500	660	91.2780	849.8280				990	136.9170	1342.2420

METRIC BOLTS

Description	Torque ft-lbs. (Nm)			
Thread for general purposes (size x pitch (mm)	Head Mark 4		Head Mark 7	
6 x 1.0	2.2 to 2.9	(3.0 to 3.9)	3.6 to 5.8	(4.9 to 7.8)
8 x 1.25	5.8 to 8.7	(7.9 to 12)	9.4 to 14	(13 to 19)
10 x 1.25	12 to 17	(16 to 23)	20 to 29	(27 to 39)
12 x 1.25	21 to 32	(29 to 43)	35 to 53	(47 to 72)
14 x 1.5	35 to 52	(48 to 70)	57 to 85	(77 to 110)
16 x 1.5	51 to 77	(67 to 100)	90 to 120	(130 to 160)
18 x 1.5	74 to 110	(100 to 150)	130 to 170	(180 to 230)
20 x 1.5	110 to 140	(150 to 190)	190 to 240	(160 to 320)
22 x 1.5	150 to 190	(200 to 260)	250 to 320	(340 to 430)
24 x 1.5	190 to 240	(260 to 320)	310 to 410	(420 to 550)

CAUTION: Bolts threaded into aluminum require much less torque

NOTE: This engine rebuilding section is a guide to accepted rebuilding procedures. Typical examples of standard rebuilding procedures are illustrated.

CYLINDER HEAD RECONDITIONING

Procedure	Method
Identify the valves:	Invert the cylinder head, and number the valve faces front to rear, using a permanent felt-tip marker.
Remove the rocker arms (OHV engines only):	Remove the rocker arms with shaft(s) or balls and nuts. Wire the sets of rockers, balls and nuts together, and identify according to the corresponding valve.
Remove the camshaft (OHC engines only):	See the engine service procedures earlier in this book for details concerning specific engines.
Remove the valves and springs:	Using an appropriate valve spring compressor (depending on the configuration of the cylinder head), compress the valve springs. Lift out the keepers with needlenose pliers, release the compressor, and remove the valve, spring, and spring retainer.
Remove glow plugs and fuel injectors (Diesel engines only):	Label and remove all fuel injectors and glow plugs from the head. Glow plugs unscrew. See the appropriate car section for injector removal. Inspect glow plugs for bulges, cracks or signs of melting. Clean injector tips with a steel brush, then inspect for evidence of melting.
**Remove pre-combustion chamber inserts (Diesel engines only):	**Remove the pre-combustion chambers using a hammer and a thin, blunt brass drift, inserted through the injector hole (or glow plug hole, whichever is more convenient). If chamber is to be reused, carefully remove all carbon from it. NOTE: *Remove chamber only if being replaced, if a glow plug tip has broken off and must be removed, or if chamber is obviously damaged or loose.*

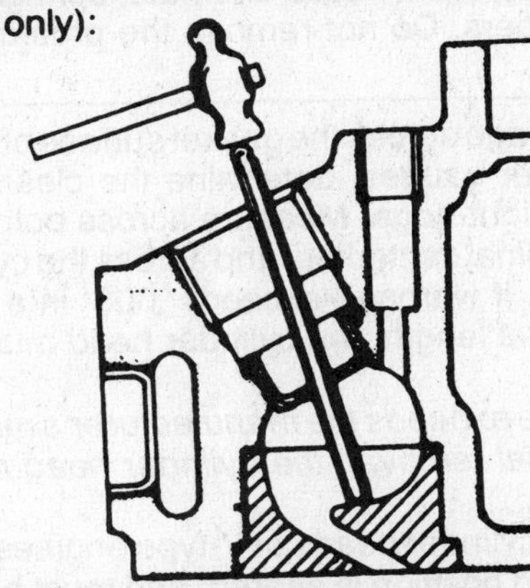

Removing pre-combustion chamber with a drift (© G.M. Corp.)

Procedure	Method
Check the valve stem-to-guide clearance:	Clean the valve stem with lacquer thinner or a similar solvent to remove all gum and varnish. Clean the valve guides using solvent and an expanding wire-type valve guide cleaner. Mount a dial indicator so that the stem is at 90° to the valve stem, as close to the valve guide as possible. Move the valve off its seat, and measure the valve guide-to-stem clearance by rocking the stem back and forth to actuate the dial indicator. Measure the valve stems using a micrometer, and compare to specifications, to determine whether stem or guide wear is responsible for excessive clearance.

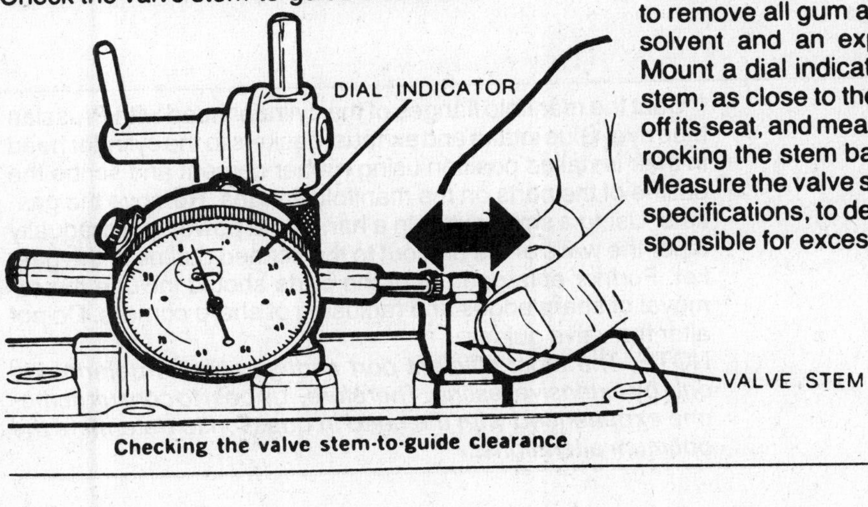

Checking the valve stem-to-guide clearance

CYLINDER HEAD RECONDITIONING

Procedure	Method
De-carbon the cylinder head and valves:	Chip carbon away from the valve heads, combustion chambers, and ports, using a chisel made of hardwood. Remove the remaining deposits with a stiff wire brush. NOTE: *Ensure that the deposits are actually removed, rather than burnished.*

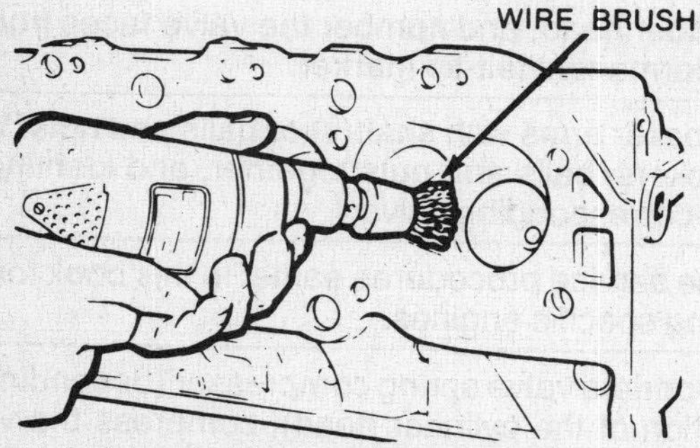

Removing carbon from the cylinder head

Procedure	Method
Hot-tank the cylinder head (cast iron heads only): CAUTION: *Do not hot-tank aluminum parts.*	Have the cylinder head hot-tanked to remove grease, corrosion, and scale from the water passages. NOTE: *In the case of overhead cam cylinder heads, consult the operator to determine whether the camshaft bearings will be damaged by the caustic solution.*
Degrease the remaining cylinder head parts:	Using solvent (i.e., Gunk), clean the rockers, rocker shaft(s) (where applicable), rocker balls and nuts, springs, spring retainers, and keepers. Do not remove the protective coating from the springs.
Check the cylinder head for warpage:	Place a straight-edge across the gasket surface of the cylinder head. Using feeler gauges, determine the clearance at the center of the straight-edge. Measure across both diagonals, along the longitudinal centerline, and across the cylinder head at several points. If warpage exceeds .003′ in a 6′ span, or .006′ over the total length, the cylinder head must be resurfaced. NOTE: *If warpage exceeds the manufacturer's maximum tolerance for material removal, the cylinder head must be replaced.* When milling the cylinder heads of V-type engines, the intake manifold mounting position is altered, and must be corrected by milling the manifold flange a proportionate amount.

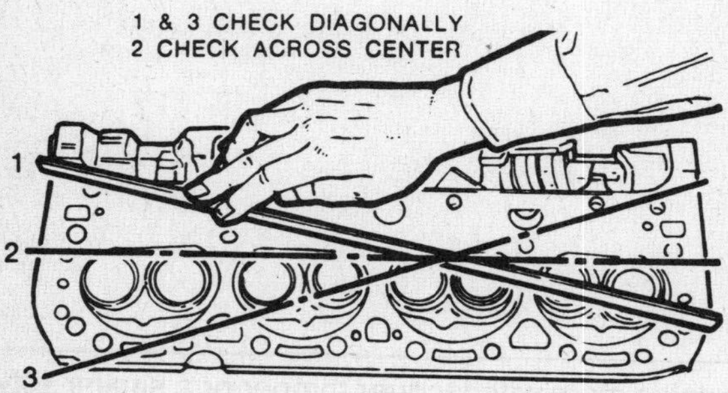

Checking cylinder head for warpage

Procedure	Method
**Porting and gasket matching:	**Coat the manifold flanges of the cylinder head with Prussian blue dye. Glue intake and exhaust gaskets to the cylinder head in their installed position using rubber cement and scribe the outline of the ports on the manifold flanges. Remove the gaskets. Using a small cutter in a hand-held power tool gradually taper the walls of the port out to the scribed outline of the gasket. Further enlargement of the ports should include the removal of sharp edges and radiusing of sharp corners. Do not alter the valve guides. NOTE: *The most efficient port configuration is determined only by extensive testing. Therefore, it is best to consult someone experienced with the head in question to determine the optimum alterations.*

CYLINDER HEAD RECONDITIONING

Procedure | **Method**

*Knurling the valve guides:

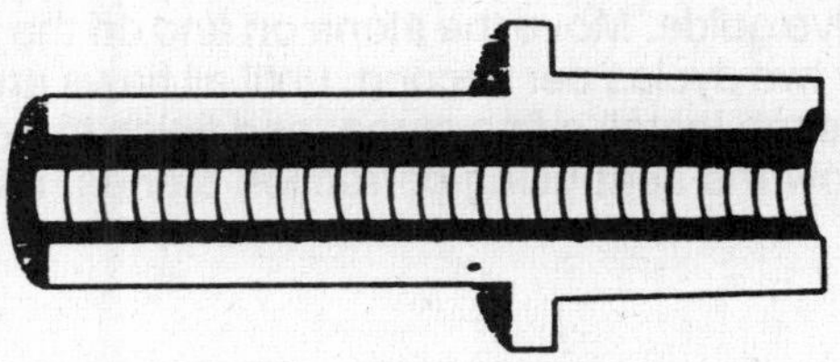

Cut-away view of a knurled valve guide

*Valve guides which are not excessively worn or distorted may, in some cases, be knurled rather than replaced. Knurling is a process in which metal is displaced and raised, thereby reducing clearance. Knurling also provides excellent oil control. The possibility of knurling rather than replacing valve guides should be discussed with a machinist.

Replacing the valve guides:
NOTE: *Valve guides should only be replaced if damaged or if an oversize valve stem is not available.*

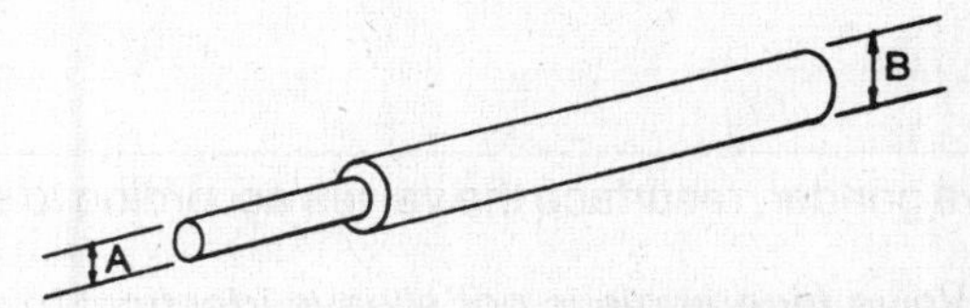

Valve guide removal tool

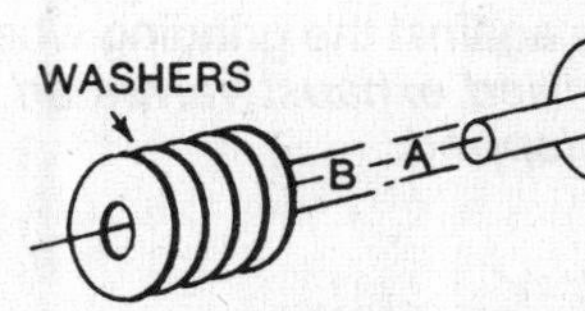

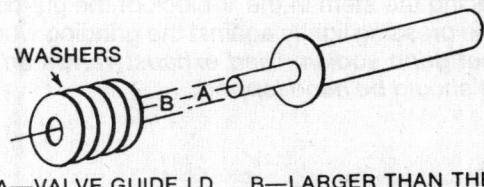

Valve guide installation tool (with washers used for installation)

Depending on the type of cylinder head, valve guides may be pressed, hammered, or shrunk in. In cases where the guides are shrunk into the head, replacement should be left to an equipped machine shop. In other cases, the guides are replaced as follows: Press or tap the valve guides out of the head using a stepped drift (see illustration). Determine the height above the boss that the guide must extend, and obtain a stack of washers, their I.D. similar to the guide's O.D., of that height. Place the stack of washers on the guide, and insert the guide into the boss.

NOTE: *Valve guides are often tapered or beveled for installation.*

Using the stepped installation tool (see illustration), press or tap the guides into position. Ream the guides according to the size of the valve stem.

Replacing valve seat inserts:

Replacement of valve seat inserts which are worn beyond resurfacing or broken, if feasible, must be done by a machine shop.

Resurfacing the valve seats using reamers:

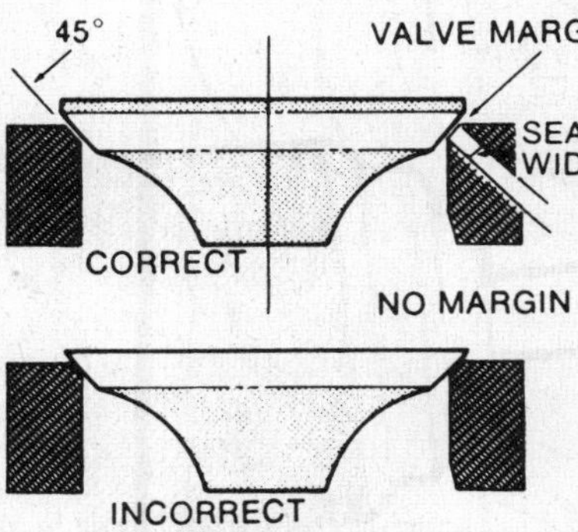

Valve seat width and centering

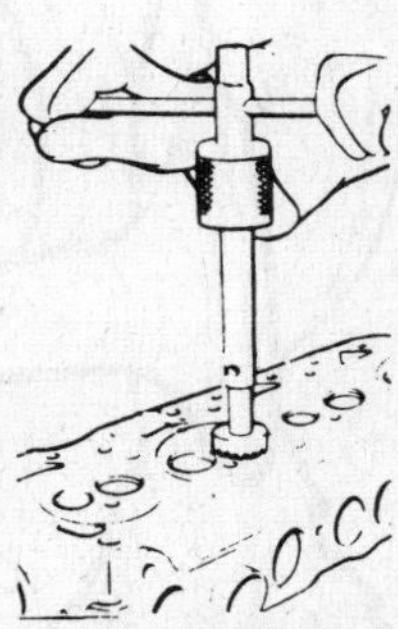

Reaming the valve seat

Select a reamer of the correct seat angle, slightly larger than the diameter of the valve seat, and assemble it with a pilot of the correct size. Install the pilot into the valve guide, and using steady pressure, turn the reamer clockwise.

CAUTION: *Do not turn the reamer counterclockwise.*

Remove only as much material as necessary to clean the seat. Check the concentricity of the seat (see below). If the dye method is not used, coat the valve face with Prussian blue dye, install and rotate it on the valve seat. Using the dye marked area as a centering guide, center and narrow the valve seat to specifications with correction cutters.

NOTE: *When no specifications are available, minimum seat width for exhaust valves should be 5/64", intake valves 1/16".*

After making correction cuts, check the position of the valve seat on the valve face using Prussian blue dye.

NOTE: *Do not cut induction hardened seats; they must be ground.*

CYLINDER HEAD RECONDITIONING

Procedure	Method
*Resurfacing the valve seats using a grinder:	*Select a pilot of the correct size, and a coarse stone of the correct seat angle. Lubricate the pilot if necessary, and install the tool in the valve guide. Move the stone on and off the seat at approximately two cycles per second, until all flaws are removed from the seat. Install a fine stone, and finish the seat. Center and narrow the seat using correction stones, as described above.
Resurfacing (grinding) the valve face:	Using a valve grinder, resurface the valves according to specifications. CAUTION: *Valve face angle is not always identical to valve seat angle.* A minimum margin of 1/32" should remain after grinding the valve. The valve stem top should also be squared and resurfaced, by placing the stem in the V-block of the grinder, and turning it while pressing lightly against the grinding wheel. NOTE: *Do not grind sodium filled exhaust valves on a machine. These should be hand lapped.*

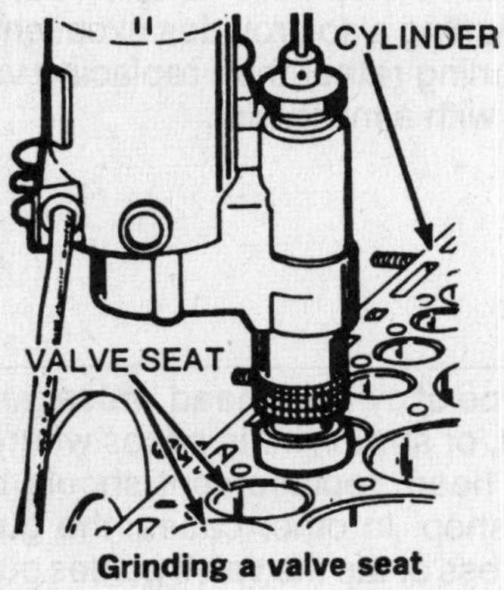

Grinding a valve seat

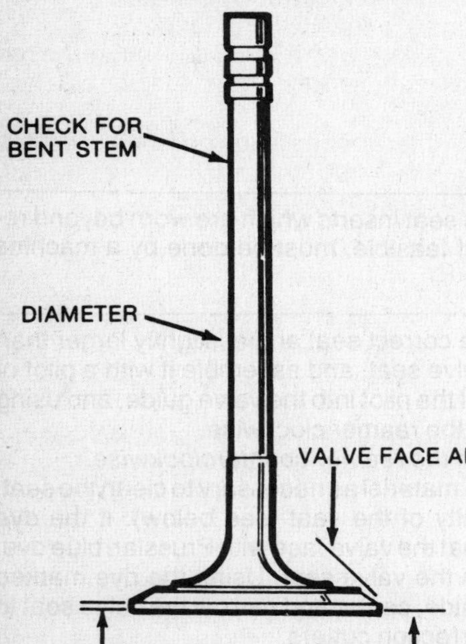

Critical valve dimensions

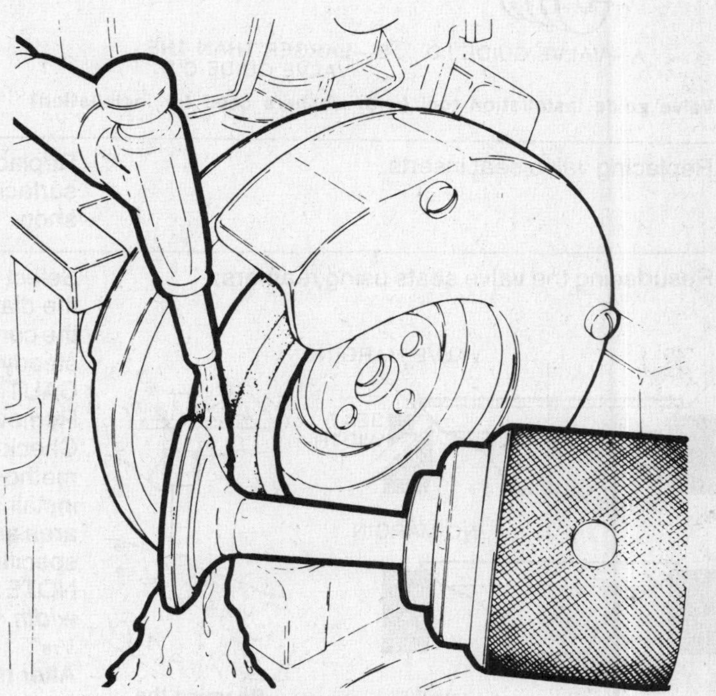
Valve grinding by machine

CYLINDER HEAD RECONDITIONING

Procedure	Method
Checking the valve seat concentricity:	Coat the valve face with Prussian blue dye, install the valve, and rotate it on the valve seat. If the entire seat becomes coated, and the valve is known to be concentric, the seat is concentric. *Install the dial gauge pilot into the guide, and rest the arm on the valve seat. Zero the gauge, and rotate the arm around the seat. Run-out should not exceed .002".

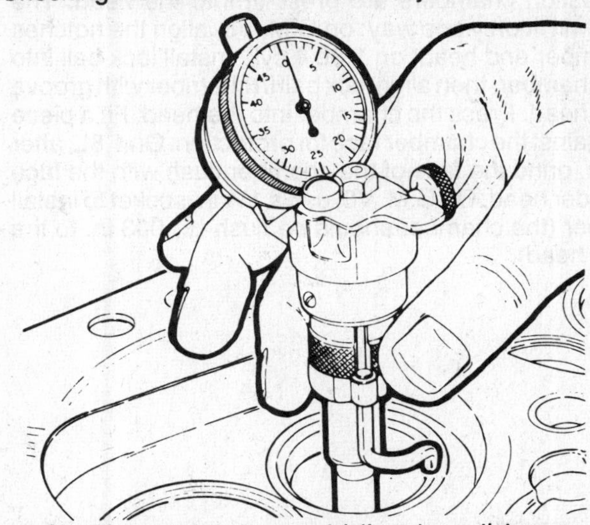

Checking valve seat concentricity using a dial gauge

Procedure	Method
*Lapping the valves: NOTE: *Valve lapping is done to ensure efficient sealing of resurfaced valves and seats.*	*Invert the cylinder head, lightly lubricate the valve stems, and install the valves in the head as numbered. Coat valve seats with fine grinding compound, and attach the lapping tool suction cup to a valve head. NOTE: *Moisten the suction cup.* Rotate the tool between the palms, changing position and lifting the tool often to prevent grooving. Lap the valve until a smooth, polished seat is evident. Remove the valve and tool, and rinse away all traces of grinding compound. **Fasten a suction cup to a piece of drill rod, and mount the rod in a hand drill. Proceed as above, using the hand drill as a lapping tool. CAUTION: *Due to the higher speeds involved when using the hand drill, care must be exercised to avoid grooving the seat.* Lift the tool and change direction of rotation often.

Hand lapping the valves

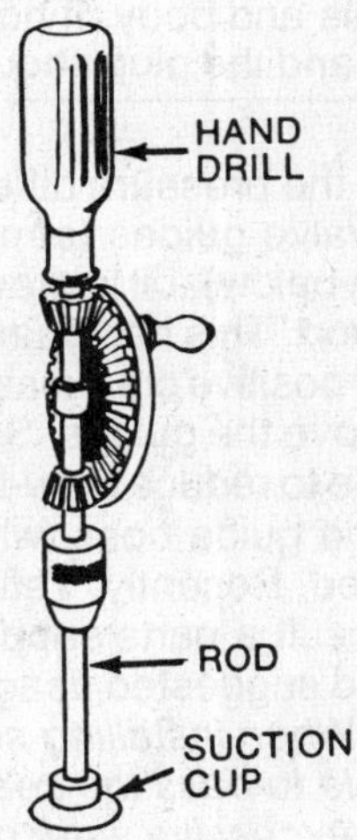

Home made mechanical valve lapping tool

Procedure	Method
Check the valve springs:	Place the spring on a flat surface next to a square. Measure the height of the spring, and rotate it against the edge of the square to measure distortion. If spring height varies (by comparison) by more than 1/16" or if distortion exceeds 1/16", replace the spring. **In addition to evaluating the spring as above, test the spring pressure at the installed and compressed (installed height minus valve lift) height using a valve spring tester. Springs used on small displacement engines (up to 3 liters) should be ∓ 1 lb. of all other springs in either position. A tolerance of ∓ 5 lbs. is permissible on larger engines.

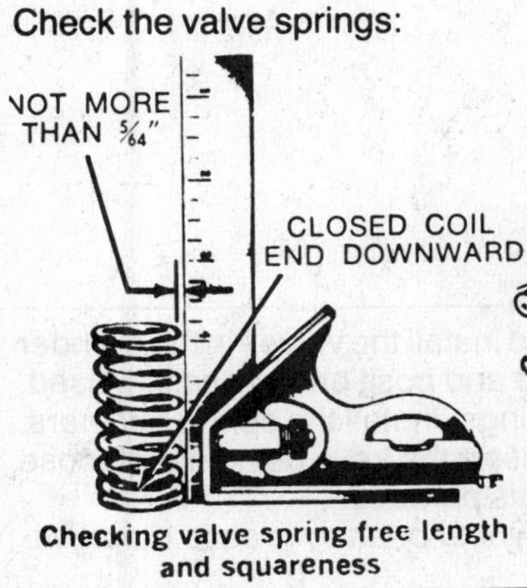

Checking valve spring free length and squareness

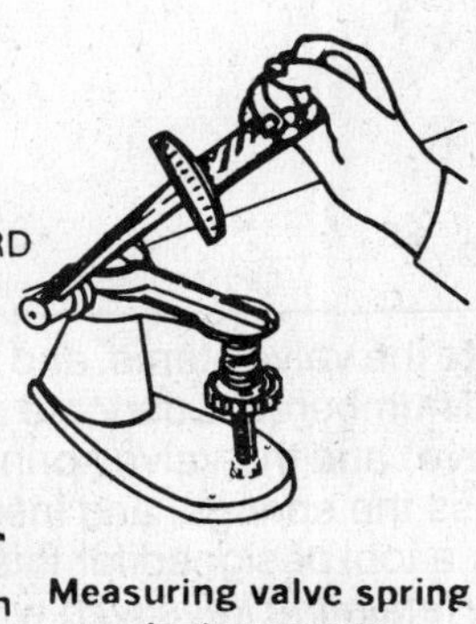

Measuring valve spring test pressure

CYLINDER HEAD RECONDITIONING

Procedure	Method
Install pre-combustion chambers (Diesel engines only)	Pre-combustion chambers are press-fit into the head. The chambers will fit only one way: on G.M. V8, align the notches in the chamber and head; on 1.8L 4 cyl., install lock ball into groove in chamber, then align lock ball in chamber with groove in cylinder head. Press the chamber into the head. Fit a piece of metal against the chamber face for protection. On 1.8L, after installation, grind the face of the chamber flush with the face of the cylinder head. On G.M. V8, use a 1¼ in. socket to install the chamber (the chamber should be flush ± .003 in. to the face of the head).
Install fuel injectors and glow plugs (Diesel engines)	Before installing glow plugs, check for continuity across plug terminals and body. If no continuity exists, the heater wire is broken and the plug should be replaced.
*Install valve stem seals:	*Due to the pressure differential that exists at the ends of the intake valve guides (atmospheric pressure above, manifold vacuum below), oil is drawn through the valve guides into the intake port. This has been alleviated somewhat since the addition of positive crankcase ventilation, which lowers the pressure above the guides. Several types of valve stem seals are available to reduce blow-by. Certain seals simply slip over the stem and guide boss, while others require that the boss be machined. Recently, Teflon guide seals have become popular. Consult a parts supplier or machinist concerning availability and suggested usages. NOTE: *When installing seals, ensure that a small amount of oil is able to pass the seal to lubricate the valve guides; otherwise, excessive wear may result.*
Install the valves:	Lubricate the valve stems, and install the valves in the cylinder head as numbered. Lubricate and position the seals (if used, see above) and the valve springs. Install the spring retainers, compress the springs, and insert the keys using needlenose pliers or a tool designed for this purpose. NOTE: *Retain the keys with wheel bearing grease during installation.*

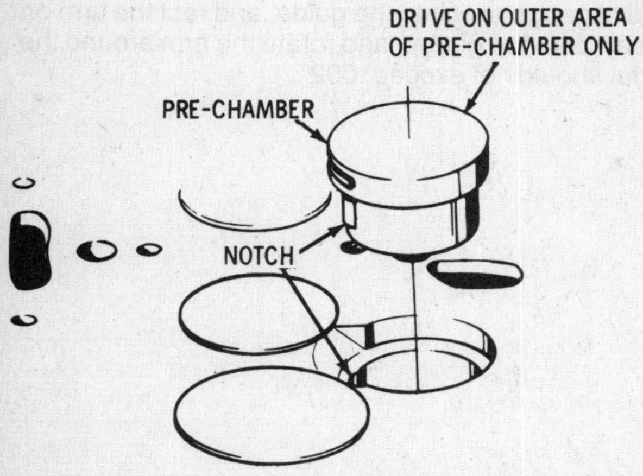

Align the notches to install the pre-combustion chamber (© G.M. Corp.)

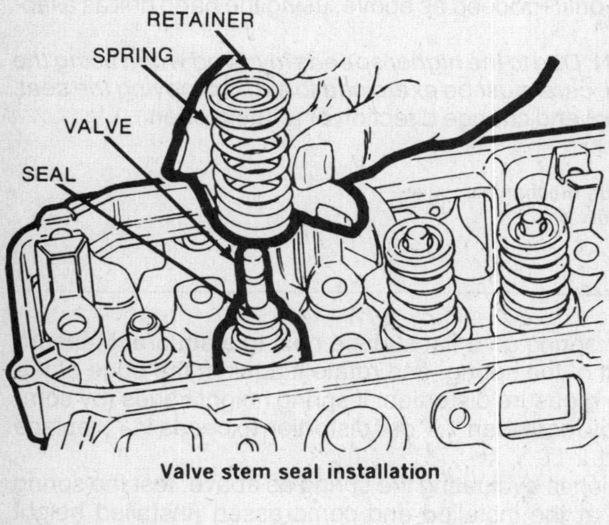

Valve stem seal installation

CYLINDER HEAD RECONDITIONING

Procedure	Method
Check valve spring installed height:	Measure the distance between the spring pad and the lower edge of the spring retainer, and compare to specifications. If the installed height is incorrect, add shim washers between the spring pad and the spring. CAUTION: *Use only washers designed for this purpose.*

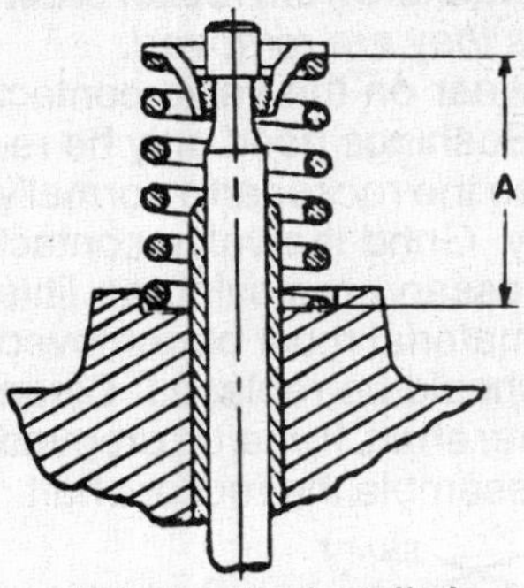

Valve spring installed height dimension

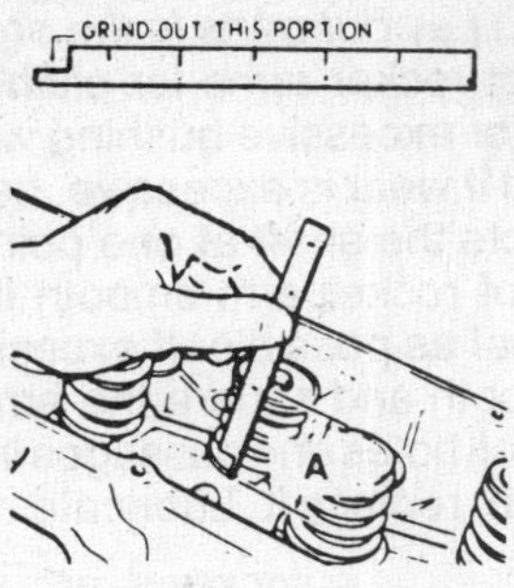

Measuring valve spring installed height

Procedure	Method
Install the camshaft (OHC engines only) and check end play:	See the engine service procedures earlier in this book for details concerning specific engines.
Inspect the rocker arms, balls, studs, and nuts (OHV engines only):	Visually inspect the rocker arms, balls, studs, and nuts for cracks, galling, burning, scoring or wear. If all parts are intact, liberally lubricate the rocker arms and balls, and install them on the cylinder head. If wear is noted on a rocker arm at the point of valve contact, grind it smooth and square, removing as little material as possible. Replace the rocker arm if excessively worn. If a rocker stud shows signs of wear, it must be replaced (see below). If a rocker nut shows stress cracks, replace it. If an exhaust ball is galled or burned, substitute the intake ball from the same cylinder (if it is intact), and install a new intake ball. NOTE: *Avoid using new rocker balls on exhaust valves.*

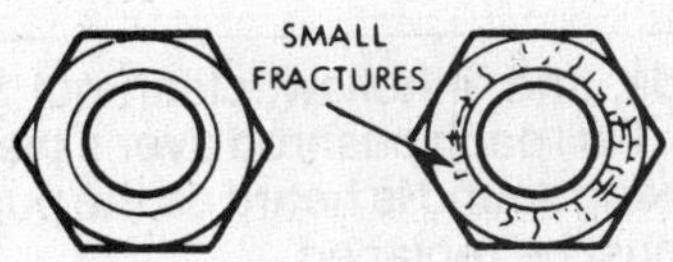

Stress cracks in the rocker nuts

Procedure	Method
Replacing rocker studs (OHV engines only):	In order to remove a threaded stud, lock two nuts on the stud, and unscrew the stud using the lower nut. Coat the lower threads of the new stud with Loctite®, and install. Two alternative methods are available for replacing pressed in studs. Remove the damaged stud using a stack of washers and a nut (see illustration). In the first, the boss is reamed .005–.006″ oversize, and an oversize stud pressed in. Control the stud extension over the boss using washers, in the same manner as valve guides. Before installing the stud, coat it with white lead and grease. To retain the stud more positively drill a hole through the stud and boss, and install a roll pin. In the second method, the boss is tapped, and a threaded stud installed. Retain the stud using Loctite® Stud and Bearing Mount.

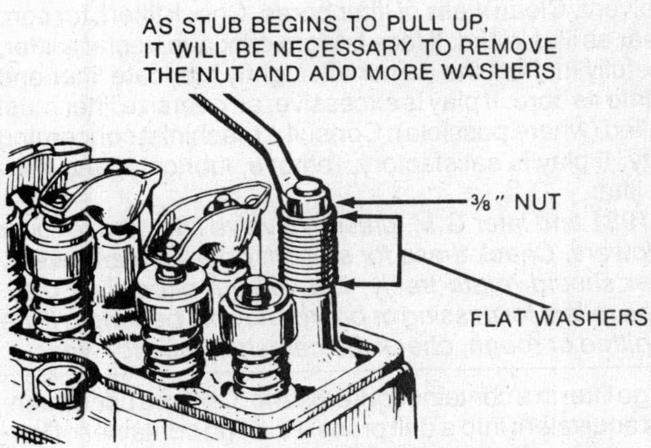

Extracting a pressed-in rocker stud

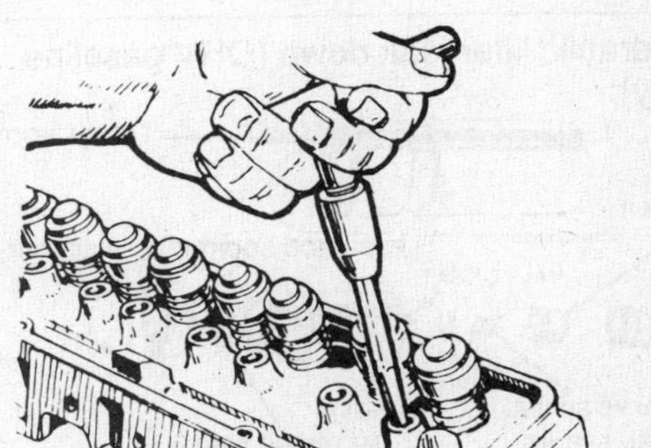
Reaming the stud bore for oversize rocker studs

CYLINDER HEAD RECONDITIONING

Procedure | **Method**

Inspect the rocker shaft(s) and rocker arms (OHV engines only):

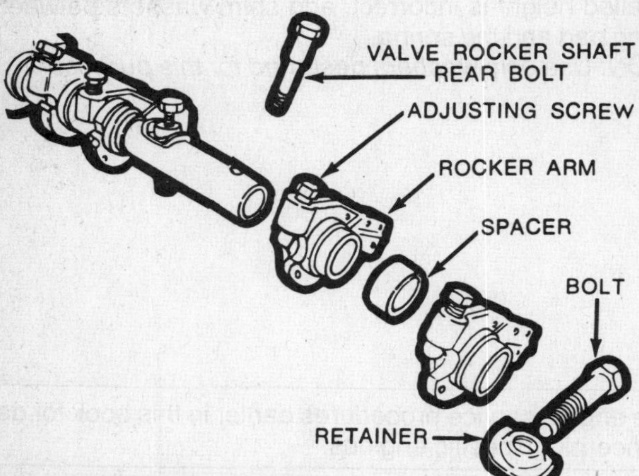

Disassemble the rocker shaft for inspection

Remove rocker arms, springs and washers from rocker shaft. NOTE: *Lay out parts in the order as they are removed.* Inspect rocker arms for pitting or wear on the valve contact point, or excessive bushing wear. Bushings need only be replaced if wear is excessive, because the rocker arm normally contacts the shaft at one point only. Grind the valve contact point of rocker arm smooth if necessary, removing as little material as possible. If excessive material must be removed to smooth and square the arm, it should be replaced. Clean out all oil holes and passages in rocker shaft. If shaft is grooved or worn, replace it. Lubricate and assemble the rocker shaft.

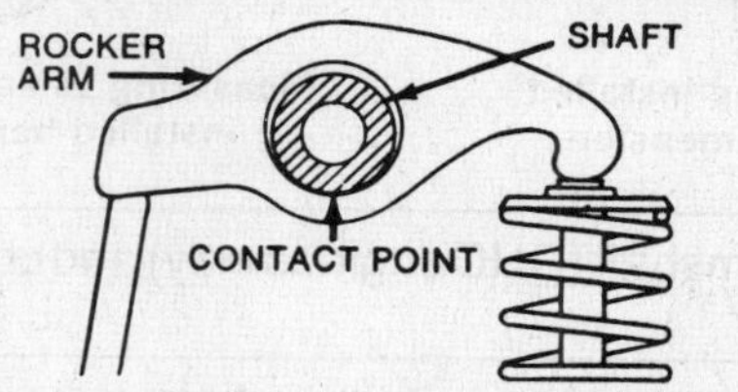

Rocker arm-to-rocker shaft contact area

Inspect the camshaft bushings and the camshaft (OHC engines):

See next section.

Inspect the pushrods (OHV engines only):

Remove the pushrods, and, if hollow, clean out the oil passages using fine wire. Roll each pushrod over a piece of clean glass. If a distinct clicking sound is heard as the pushrod rolls, the rod is bent, and must be replaced.

*The length of all pushrods must be equal. Measure the length of the pushrods, compare to specifications, and replace as necessary.

Inspect the valve lifters (OHV engines only):

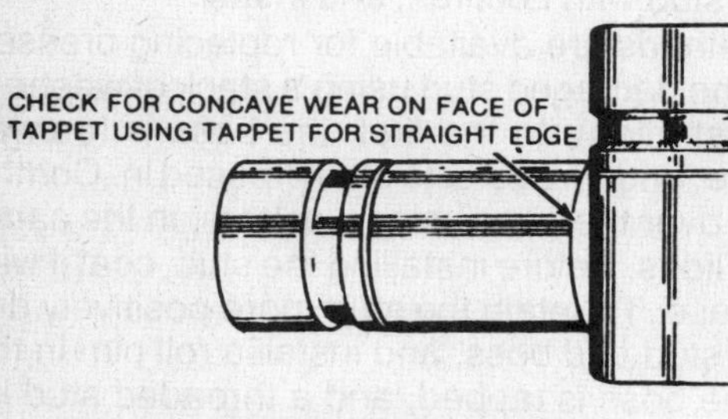

Checking the lifter face

Remove lifters from their bores, and remove gum and varnish, using solvent. Clean walls of lifter bores. Check lifters for concave wear as illustrated. If face is worn concave, replace lifter, and carefully inspect the camshaft. Lightly lubricate lifter and insert it into its bore. If play is excessive, an oversize lifter must be installed (where possible). Consult a machinist concerning feasibility. If play is satisfactory, remove, lubricate, and reinstall the lifter.

NOTE: *1981 and later G.M. diesel V8 valve lifters have roller cam followers. Check these for smooth operation and wear. The roller should rotate freely, but without excessive play. Check the rollers for missing or broken needle bearings. If the roller is pitted or rough, check the camshaft lobe for wear.*

***Testing hydraulic lifter leak down (OHV gasoline engines only):**

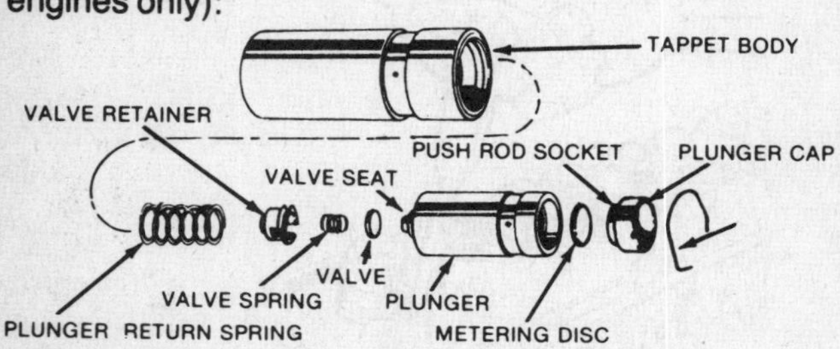

Typical exploded view of hydraulic valve lifter

Submerge lifter in a container of kerosene. Chuck a used pushrod or its equivalent into a drill press. Position container of kerosene so pushrod acts on the lifter plunger. Pump lifter with the drill press, until resistance increases. Pump several more times to bleed any air out of lifter. Apply very firm, constant pressure to the lifter, and observe rate at which fluid bleeds out of lifter. If the fluid bleeds very quickly (less than 15 seconds), lifter is defective. If the time exceeds 60 seconds, lifter is sticking. In either case, recondition or replace lifter. If lifter is operating properly (leak down time 15–60 seconds), lubricate and install it.

CYLINDER HEAD RECONDITIONING

Procedure	Method
Bleed the hydraulic lifters (diesel engines only):	After the cylinder heads are installed on G.M. V8 diesels, the valve lifters must be bled down before the crankshaft is turned. Failure to bleed down the lifters will cause damage to the valve train. See diesel engine rocker arm replacement procedure in Oldsmobile 88, 98, etc. car section for procedures. NOTE: *When installing new lifters, prime by working the lifter plunger while submerged in clean kerosene or diesel fuel.*

CYLINDER BLOCK RECONDITIONING

Procedure — **Method**

Checking the main bearing clearance:

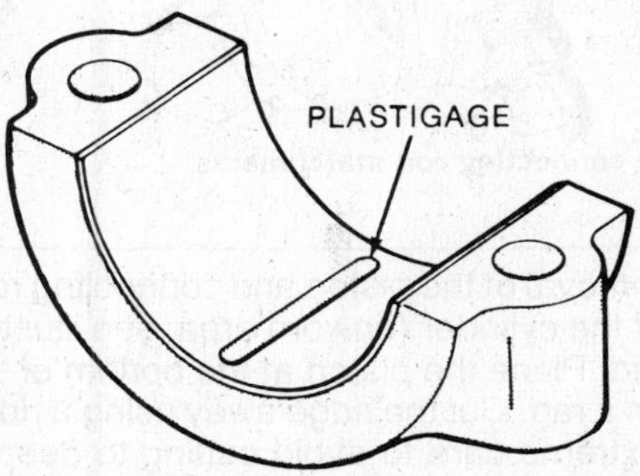

Plastigage® installed on the lower bearing shell

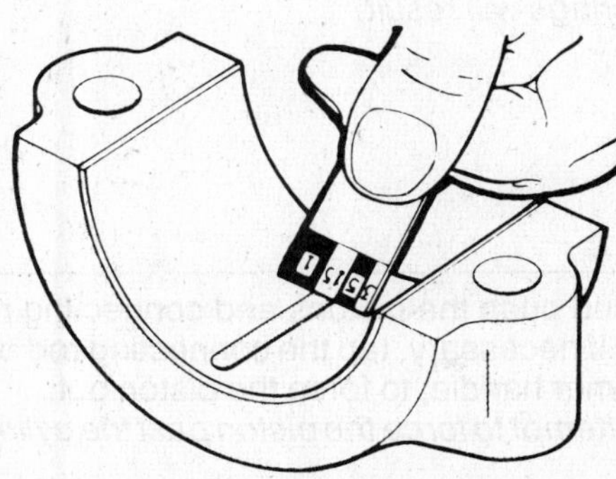

Measuring Plastigage® to determine bearing clearancc

Invert engine, and remove cap from the bearing to be checked. Using a clean, dry rag, thoroughly clean all oil from crankshaft journal and bearing insert.

NOTE: *Plastigage is soluble in oil; therefore, oil on the journal or bearing could result in erroneous readings.*

Place a piece of Plastigage along the full length of journal, reinstall cap, and torque to specifications. Remove bearing cap, and determine bearing clearance by comparing width of Plastigage to the scale on Plastigage envelope. Journal taper is determined by comparing width of the Plastigage strip near its ends. Rotate crankshaft 90° and retest, to determine journal eccentricity.

NOTE: *Do not rotate crankshaft with Plastigage installed.*

If bearing insert and journal appear intact, and are within tolerances, no further main bearing service is required. If bearing or journal appear defective, cause of failure should be determined before replacement.

*Remove crankshaft from block (see below). Measure the main bearing journals at each end twice (90° apart) using a micrometer, to determine diameter, journal taper and eccentricity. If journals are within tolerances, reinstall bearing caps at their specified torque. Using a telescope gauge and micrometer, measure bearing I.D. parallel to piston axis and at 30° on each side of piston axis. Subtract journal O.D. from bearing I.D. to determine oil clearance. If crankshaft journals appear defective, or do no meet tolerances, there is no need to measure bearings; for the crankshaft will require grinding and/or undersize bearings will be required. If bearing appears defective, cause for failure should be determined prior to replacement.

Checking the connecting rod bearing clearance:

Connecting rod bearing clearance is checked in the same manner as main bearing clearance, using Plastigage. Before removing the crankshaft, connecting rod side clearance also should be measured and recorded.

*Checking connecting rod bearing clearance, using a micrometer, is identical to checking main bearing clearance. If no other service is required, the piston and rod assemblies need not be removed.

CYLINDER BLOCK RECONDITIONING

Procedure	Method
Removing the crankshaft:	Using a punch, mark the corresponding main bearing caps and saddles according to position (i.e., one punch on the front main cap and saddle, two on the second, three on the third, etc.). Using number stamps, identify the corresponding connecting rods and caps, according to cylinder (if no numbers are present). Remove the main and connecting rod caps, and place sleeves of plastic tubing over the connecting rod bolts, to protect the journals as the crankshaft is removed. Lift the crankshaft out of the block.

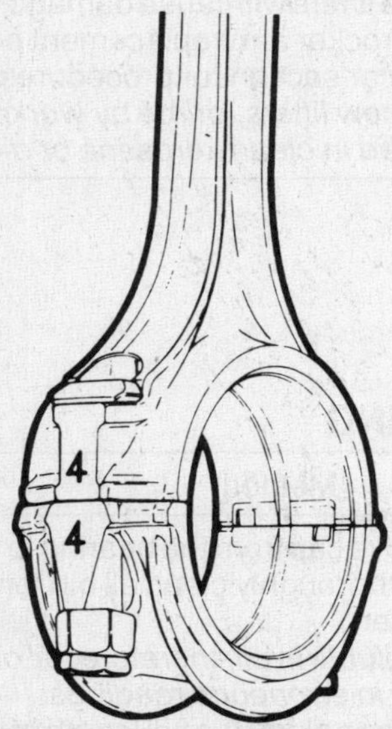

Connecting rod matched to cylinder with a number stamp

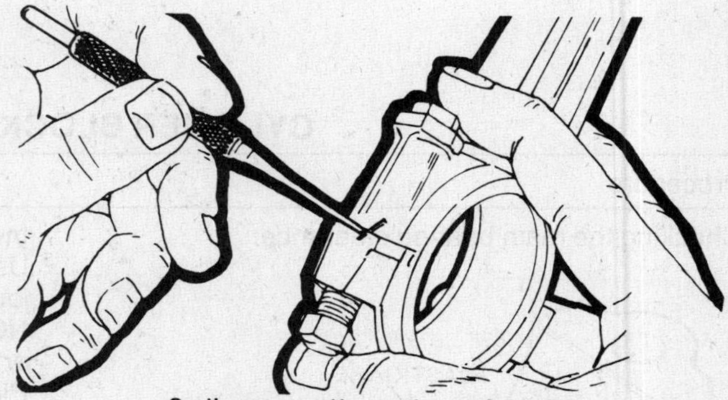
Scribe connecting rod matchmarks

Procedure	Method
Remove the ridge from the top of the cylinder:	In order to facilitate removal of the piston and connecting rod, the ridge at the top of the cylinder (unworn area; see illustration) must be removed. Place the piston at the bottom of the bore, and cover it with a rag. Cut the ridge away using a ridge reamer, exercising extreme care to avoid cutting to deeply. Remove the rag, and remove cuttings that remain on the piston. CAUTION: *If the ridge is not removed, and new rings are installed, damage to rings will result.*

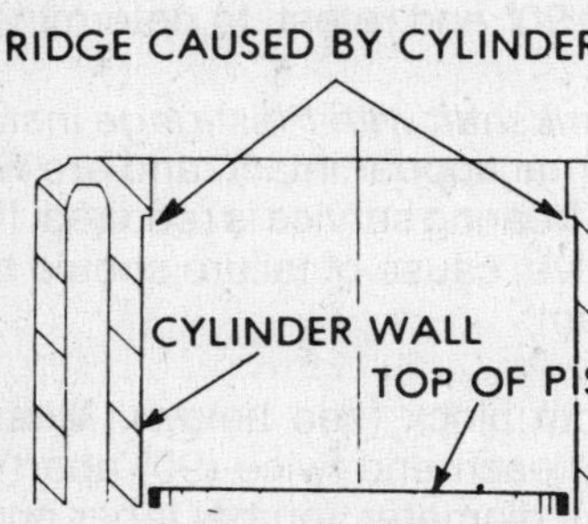

Cylinder bore ridge

Procedure	Method
Removing the piston and connecting rod:	Invert the engine, and push the pistons and connecting rods out of the cylinders. If necessary, tap the connecting rod boss with a wooden hammer handle, to force the piston out. CAUTION: *Do not attempt to force the piston past the cylinder ridge* (see above).

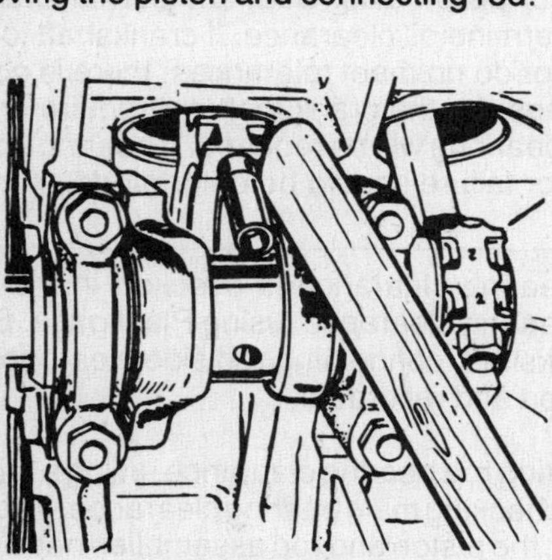
Removing the piston

CYLINDER BLOCK RECONDITIONING

Procedure	Method
Service the crankshaft:	Ensure that all oil holes and passages in the crankshaft are open and free of sludge. If necessary, have the crankshaft ground to the largest possible undersize. **Have the crankshaft Magnafluxed, to locate stress cracks. Consult a machinist concerning additional service procedures, such as surface hardening (e.g., nitriding, Tuftriding) to improve wear characteristics, cross drilling and chamfering the oil holes to improve lubrication, and balancing.
Removing freeze plugs:	Drill a small hole in the middle of the freeze plugs. Thread a large sheet metal screw into the hole and remove the plug with a slide hammer.
Remove the oil gallery plugs:	Threaded plugs should be removed using an appropriate (usually square) wrench. To remove soft, pressed in plugs, drill a hole in the plug, and thread in a sheet metal screw. Pull the plug out by the screw using pliers.
Hot-tank the block: NOTE: *Do not hot-tank aluminum parts.*	Have the block hot-tanked to remove grease, corrosion, and scale from the water jackets. NOTE: *Consult the operator to determine whether the camshaft bearings will be damaged during the hot-tank process.*
Check the block for cracks:	Visually inspect the block for cracks or chips. The most common locations are as follows: Adjacent to freeze plugs. Between the cylinders and water jackets. Adjacent to the main bearing saddles. At the extreme bottom of the cylinders. Check only suspected cracks using spot check dye (see introduction). If a crack is located, consult a machinist concerning possible repairs. **Magnaflux the block to locate hidden cracks. If cracks are located, consult a machinist about feasibility of repair.
Install the oil gallery plugs and freeze plugs:	Coat freeze plugs with sealer and tap into position using a piece of pipe, slightly smaller than the plug, as a driver. To ensure retention, stake the edges of the plugs. Coat threaded oil gallery plugs with sealer and install. Drive replacement soft plugs into block using a large drift as a driver. *Rather than reinstalling lead plugs, drill and tap the holes, and install threaded plugs.
*Check the deck height:	*The deck height is the distance from the crankshaft centerline to the block deck. To measure, invert the engine, and install the crankshaft, retaining it with the center main cap. Measure the distance from the crankshaft journal to the block deck, parallel to the cylinder centerline. Measure the diameter of the end (front and rear) main journals, parallel to the centerline of the cylinders, divide the diameter in half, and subtract it from the previous measurement. The results of the front and rear measurements should be identical. If the difference exceeds .005", the deck height should be corrected. NOTE: *Block deck height and warpage should be corrected at the same time.*

CYLINDER BLOCK RECONDITIONING

Procedure	Method
Check the block deck for warpage:	Using a straightedge and feeler gauges, check the block deck for warpage in the same manner that the cylinder head is checked (see Cylinder Head Reconditioning). If warpage exceeds specifications, have the deck resurfaced. NOTE: *In certain cases a specification for total material removal (Cylinder head and block deck) is provided. This specification must not be exceeded.*
Check the bore diameter and surface:	Visually inspect the cylinder bores for roughness, scoring, or scuffing. If evident, the cylinder bore must be bored or honed oversize to eliminate imperfections, and the smallest possible oversize piston used. The new pistons should be given to the machinist with the block, so that the cylinders can be bored or honed exactly to the piston size (plus clearance). If no flaws are evident, measure the bore diameter using a telescope gauge and micrometer, or dial guage, parallel and perpendicular to the engine centerline, at the top (below the ridge) and bottom of the bore. Subtract the bottom measurements from the top to determine taper, and the parallel to the centerline measurements from the perpendicular measurements to determine eccentricity. If the measurements are not within specifications, the cylinder must be bored or honed, and an oversize piston installed. If the measurements are within specifications the cylinder may be used as is, with only finish honing (see below). NOTE: *Prior to boring, check the block deck warpage, height and bearing alignment.* CAUTION: *The 4 cyl. 140 G.M. engine cylinder walls are impregnated with silicone. Boring or honing can be done only by a shop with the proper equipment.*

Measuring the cylinder bore with a dial gauge

TELESCOPE GAUGE 90° FROM PISTON PIN

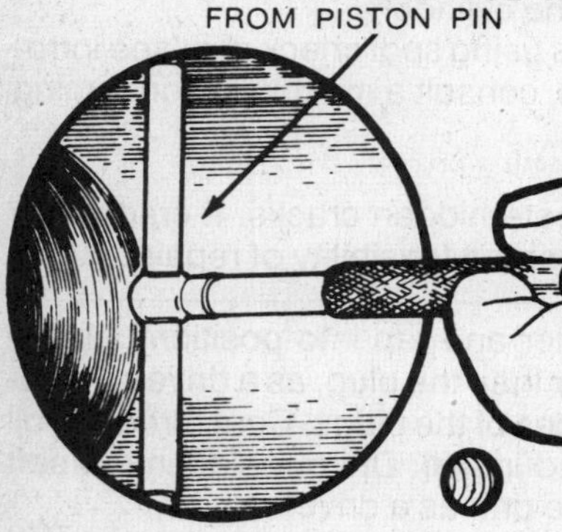

Measuring cylinder bore with a telescope gauge

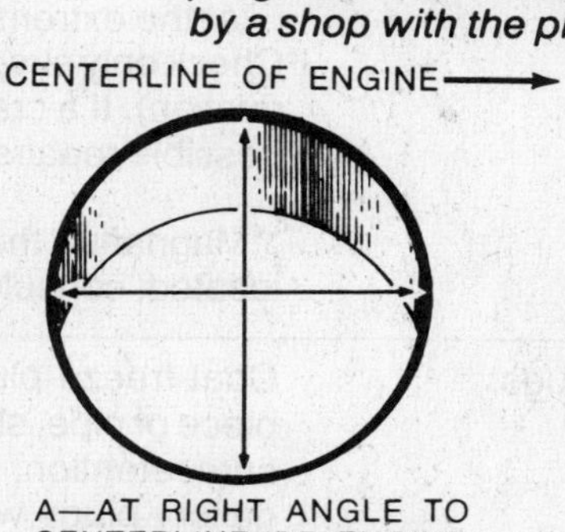

Cylinder bore measuring points

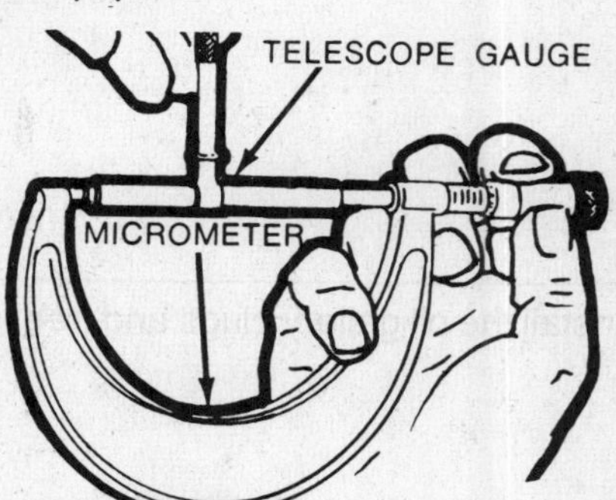

Determining cylinder bore by measuring telescope gauge with a micrometer

Procedure	Method
Check the cylinder block bearing alignment:	Remove the upper bearing inserts. Place a straightedge in the bearing saddles along the centerline of the crankshaft. If clearance exists between the straightedge and the center saddle, the block must be alignbored.

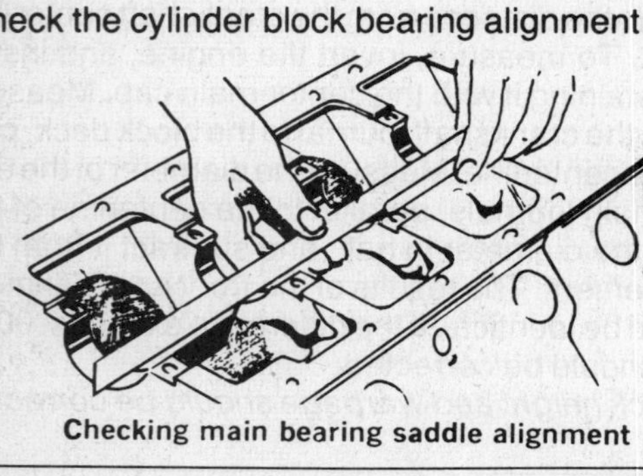

Checking main bearing saddle alignment

CYLINDER BLOCK RECONDITIONING

Procedure	Method
Clean and inspect the pistons and connecting rods:	Using a ring expander, remove the rings from the piston. Remove the retaining rings (if so equipped) and remove piston pin. NOTE: *If the piston pin must be pressed out, determine the proper method and use the proper tools; otherwise the piston will distort.* Clean the ring grooves using an appropriate tool, exercising care to avoid cutting too deeply. Thoroughly clean all carbon and varnish from the piston with solvent. CAUTION: *Do not use a wire brush or caustic solvent on pistons.* Inspect the pistons for scuffing, scoring, cracks, pitting, or excessive ring groove wear. If wear is evident, the piston must be replaced. Check the connecting rod length by measuring the rod from the inside of the large end to the inside of the small end using calipers (see illustration). All connecting rods should be equal length. Replace any rod that differs from the others in the engine. *Have the connecting rod alignment checked in an alignment fixture by a machinist. Replace any twisted or bent rods. *Magnaflux the connecting rods to locate stress cracks. If cracks are found, replace the connecting rod.

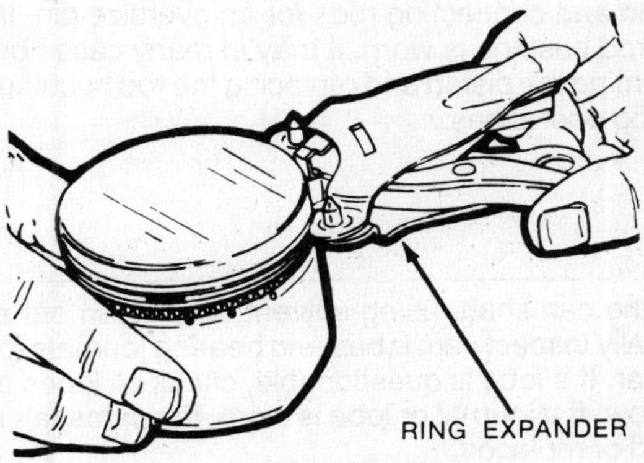

Removing the piston rings

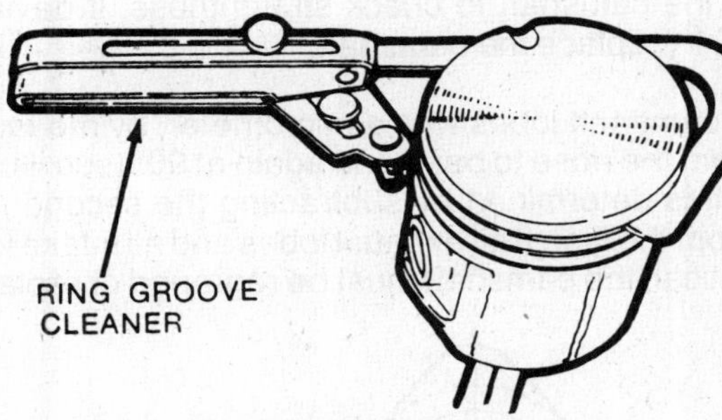

Cleaning the piston ring grooves

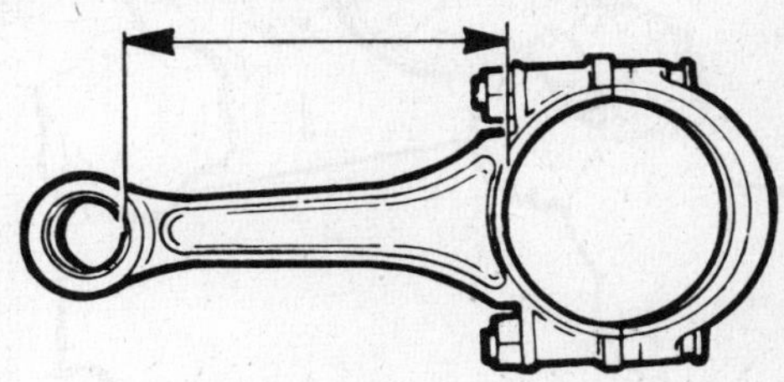

Check the connecting rod length (arrow)

Procedure	Method
Fit the pistons to the cylinders:	Using a telescope gauge and micrometer, or a dial gauge, measure the cylinder bore diameter perpendicular to the piston pin, $2\frac{1}{2}$° below the deck. Measure the piston perpendicular to its pin on the skirt. The difference between the two measurements is the piston clearance. If the clearance is within specifications or slightly below (after boring or honing), finish honing is all that is required. If the clearance is excessive, try to obtain a slightly larger piston to bring clearance within specifications. Where this is not possible, obtain the first oversize piston, and hone (or if necessary, bore) the cylinder to size.

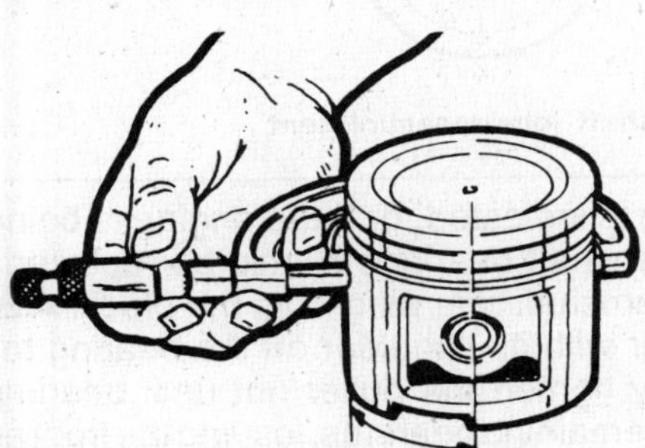

Measuring the piston prior to fitting

Procedure	Method
Assemble the pistons and connecting rods:	Inspect piston pin, connecting rod small end bushing, and piston bore for galling, scoring, or excessive wear. If evident, replace defective part(s). Measure the I.D. of the piston boss and connecting rod small end, and the O.D. of the piston pin. If within specifications, assemble piston pin and rod. CAUTION: *If piston pin must be pressed in, determine the proper method and use the proper tools; otherwise the piston will distort.*

CYLINDER BLOCK RECONDITIONING

Procedure | **Method**

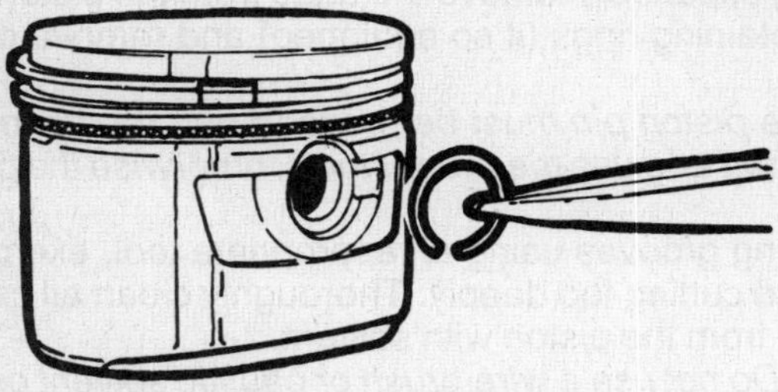

Installing piston pin lock rings

Install the lock rings; ensure that they seat properly. If the parts are not within specifications, determine the service method for the type of engine. In some cases, piston and pin are serviced as an assembly when either is defective. Others specify reaming the piston and connecting rods for an oversize pin. If the connecting rod bushing is worn, it may in many cases be replaced. Reaming the piston and replacing the rod bushing are machine shop operations.

Clean and inspect the camshaft:

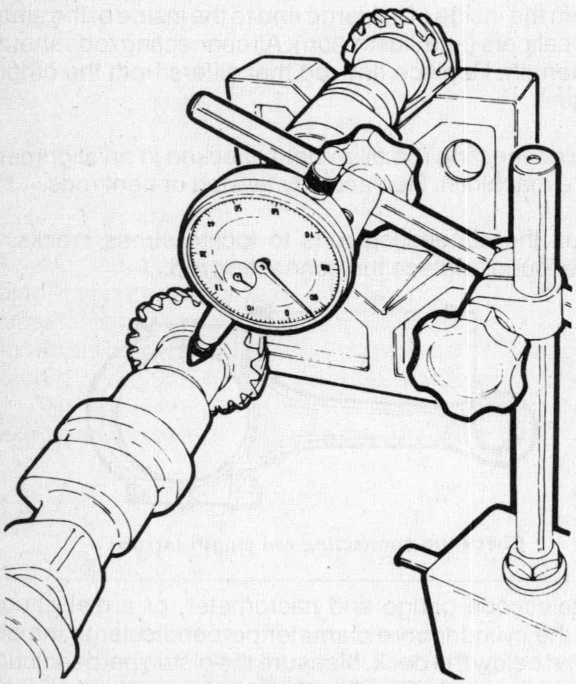

Checking the camshaft for straightness

Degrease the camshaft, using solvent, and clean out all oil holes. Visually inspect cam lobes and bearing journals for excessive wear. If a lobe is questionable, check all lobes as indicated below. If a journal or lobe is worn, the camshaft must be reground or replaced.

NOTE: *If a journal is worn, there is a good chance that the bushings are worn.*

If lobes and journals appear intact, place the front and rear journals in V-blocks, and rest a dial indicator on the center journal. Rotate the camshaft to check straightness. If deviation exceeds .001°, replace the camshaft.

*Check the camshaft lobes with a micrometer, by measuring the lobes from the nose to base and again at 90° (see illustration). The lift is determined by subtracting the second measurement from the first. If all exhaust lobes and all intake lobes are not identical, the camshaft must be reground or replaced.

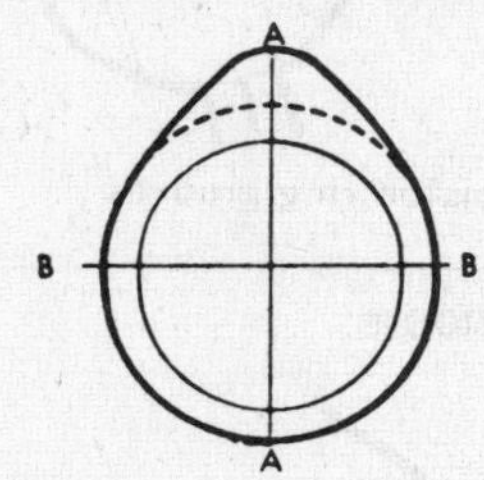

Camshaft lobe measurement

Replace the camshaft bearings (OHV engines only):

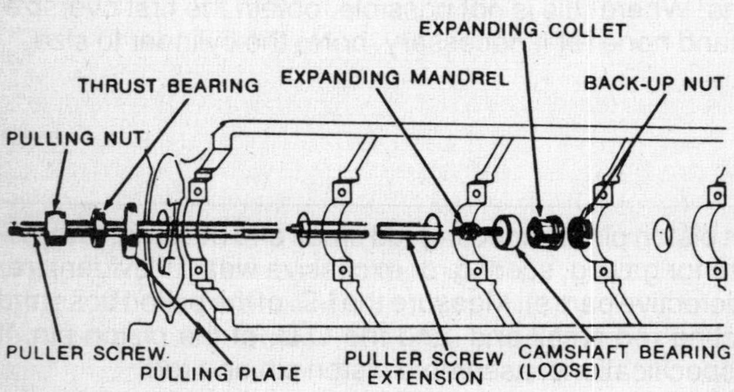

Camshaft removal and installation tool (typical)

If excessive wear is indicated, or if the engine is being completely rebuilt, camshaft bearings should be replaced as follows: Drive the camshaft rear plug from the block. Assemble the removal puller with its shoulder on the bearing to be removed. Gradually tighten the puller nut until bearing is removed. Remove remaining bearings, leaving the front and rear for last. To remove front and rear bearings, reverse position of the tool, so as to pull the bearings in toward the center of the block. Leave the tool in this position, pilot the new front and rear bearings on the installer, and pull them into position: Return the tool to its original position and pull remaining bearings into postion.

NOTE: *Ensure that oil holes align when installing bearings.*

Replace camshaft rear plug, and stake it into position to aid retention.

CYLINDER BLOCK RECONDITIONING

Procedure	Method
Finish hone the cylinders:	Chuck a flexible drive hone into a power drill, and insert it into the cylinder. Start the hone, and move it up and down the cylinder at a rate which will produce approximately a 60° cross-hatch pattern (see illustration). NOTE: *Do not extend the hone below the cylinder bore.* After developing the pattern, remove the hone and recheck piston fit. Wash the cylinders with a detergent and water solution to remove abrasive dust, dry, and wipe several times with a rag soaked in engine oil.
Check piston ring end-gap:	Compress the piston rings to be used in a cylinder, one at a time, into that cylinder, and press them approximately 1″ below the deck with an inverted piston. Using feeler gauges, measure the ring end-gap, and compare to specifications. Pull the ring out of the cylinder and file the ends with a fine file to obtain proper clearance. CAUTION: *If inadequate ring end-gap is utilized, ring breakage will result.*
Install the piston rings:	Inspect the ring grooves in the piston for excessive wear or taper. If necessary, recut the groove(s) for use with an overwidth ring or a standard ring and spacer. If the groove is worn uniformly, overwidth rings, or standard rings and spacers may be installed without recutting. Roll the outside of the ring around the groove to check for burrs or deposits. If any are found, remove with a fine file. Hold the ring in the groove, and measure side clearance. If necessary, correct as indicated above. NOTE: *Always install any additional spacers above the piston ring.* The ring groove must be deep enough to allow the ring to seat below the lands (see illustration). In many cases, a "go-no-go" depth gauge will be provided with the piston rings. Shallow grooves may be corrected by recutting, while deep grooves require some type of filler or expander behind the piston. Consult the piston ring supplier concerning the suggested method. Install the rings on the piston, lowest ring first, using a ring expander. NOTE: *Position the ring markings as specified by the manufacturer (see car section).*

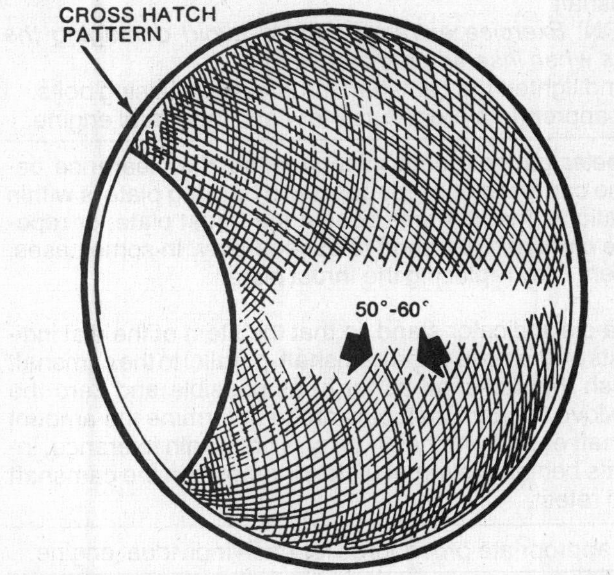

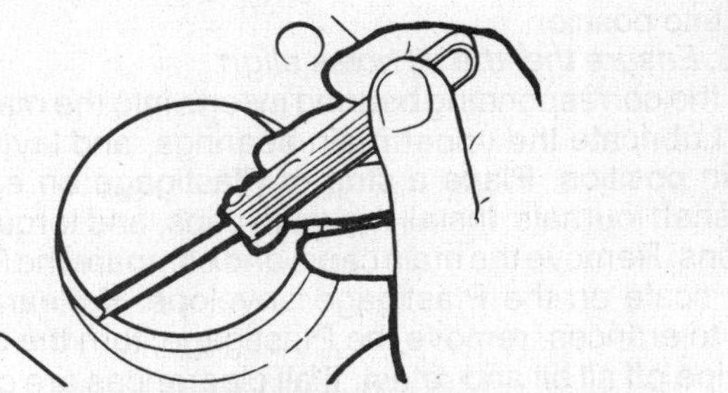

Checking ring end-gap

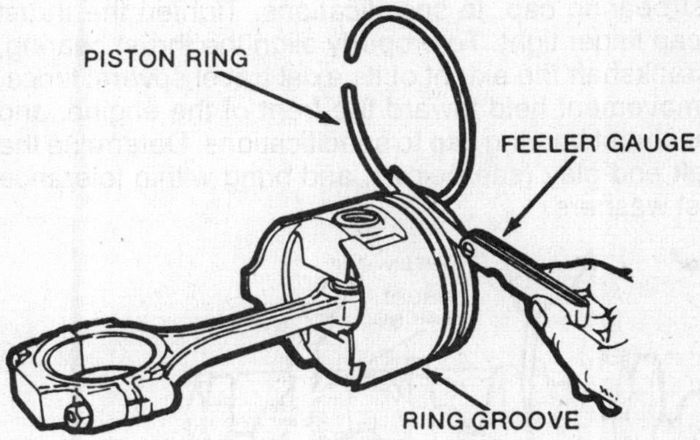

Checking ring side clearance

CYLINDER BLOCK RECONDITIONING

Procedure	Method
Install the camshaft (OHV engines only):	Liberally lubricate the camshaft lobes and journals, and install the camshaft. CAUTION: *Exercise extreme care to avoid damaging the bearings when inserting the camshaft.* Install and tighten the camshaft thrust plate retaining bolts. See the appropriate procedures for each individual engine.
Check camshaft end-play (OHV engines only):	Using feeler gauges, determine whether the clearance between the camshaft boss (or gear) and backing plate is within specifications. Install shims behind the thrust plate, or reposition the camshaft gear and retest end-play. In some cases, adjustment is by replacing the thrust plate. *Mount a dial indicator stand so that the stem of the dial indicator rests on the nose of the camshaft, parallel to the camshaft axis. Push the camshaft as far in as possible and zero the gauge. Move the camshaft outward to determine the amount of camshaft endplay. If the endplay is not within tolerance, install shims behind the thrust plate, or reposition the camshaft gear and retest.

Checking camshaft end-play with a feeler gauge

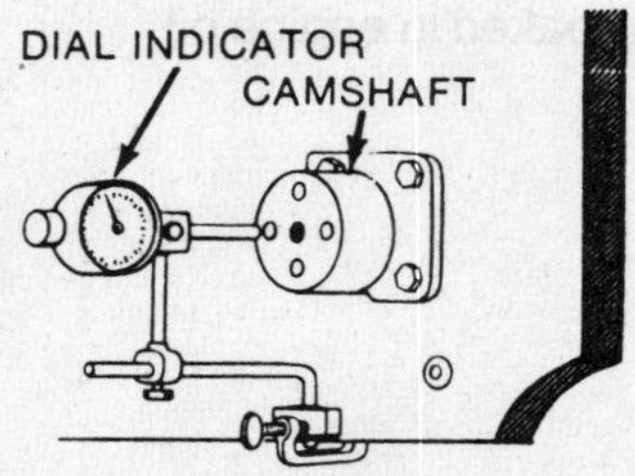

Checking camshaft end-play with a dial indicator

Procedure	Method
Install the rear main seal (where applicable):	See the appropriate procedures for each individual engine.
Install the crankshaft:	Thoroughly clean the main bearing saddles and caps. Place the upper halves of the bearing inserts on the saddles and press into position. NOTE: *Ensure that the oil holes align.* Press the corresponding bearing inserts into the main bearing caps. Lubricate the upper main bearings, and lay the crankshaft in position. Place a strip of Plastigage on each of the crankshaft journals, install the main caps, and torque to specifications. Remove the main caps, and compare the Plastigage to the scale on the Plastigage envelope. If clearances are within tolerances, remove the Plastigage, turn the crankshaft 90°, wipe off all oil and retest. If all clearances are correct, remove all Plastigage, thoroughly lubricate the main caps and bearing journals, and install the main caps. If clearances are not within tolerance, the upper bearing inserts may be removed, without removing the crankshaft, using a bearing roll out pin (see illustration). Roll in a bearing that will provide proper clearance, and retest. Torque all main caps, excluding the thrust bearing cap, to specifications. Tighten the thrust bearing cap finger tight. To properly align the thrust bearing, pry the crankshaft the extent of its axial travel several times, the last movement held toward the front of the engine, and torque the thrust bearing cap to specifications. Determine the crankshaft end-play (see below), and bring within tolerance with thrust washers.

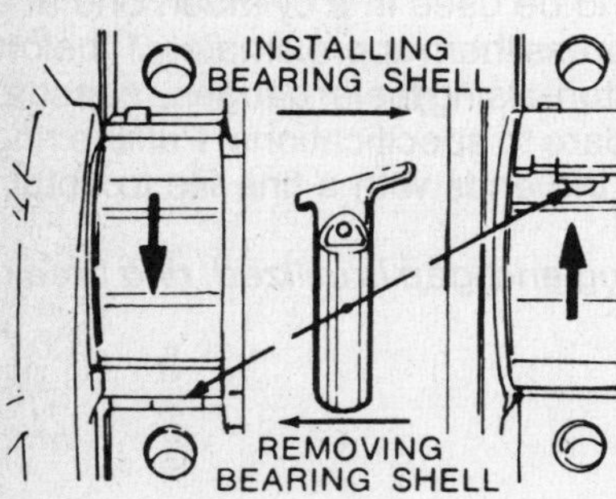

Removal and installation of upper bearing insert using a roll-out pin

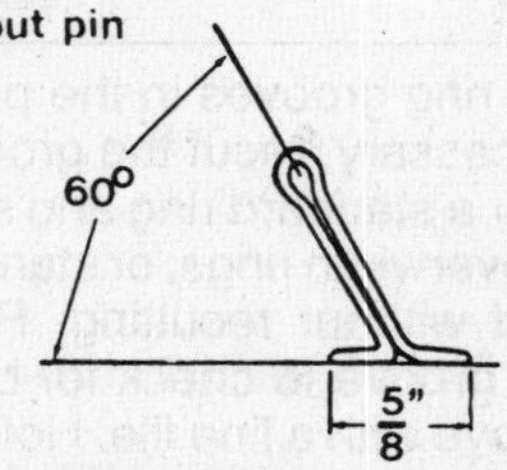

Home-made bearing roll-out pin

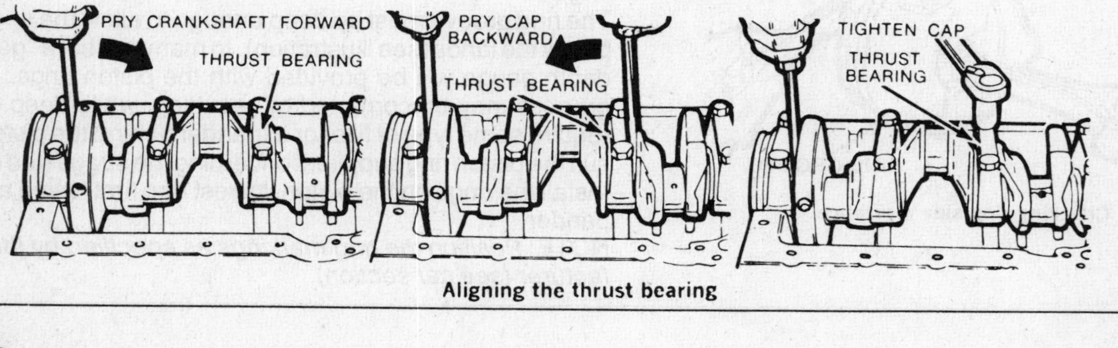

Aligning the thrust bearing

CYLINDER BLOCK RECONDITIONING

Procedure	Method
Measure crankshaft end-play:	Mount a dial indicator stand on the front of the block, with the dial indicator stem resting on the nose of the crankshaft, parallel to the crankshaft axis. Pry the crankshaft the extent of its travel rearward, and zero the indicator. Pry the crankshaft forward and record crankshaft end-play. NOTE: *Crankshaft end-play also may be measured at the thrust bearing, using feeler gauges* (see illustration).

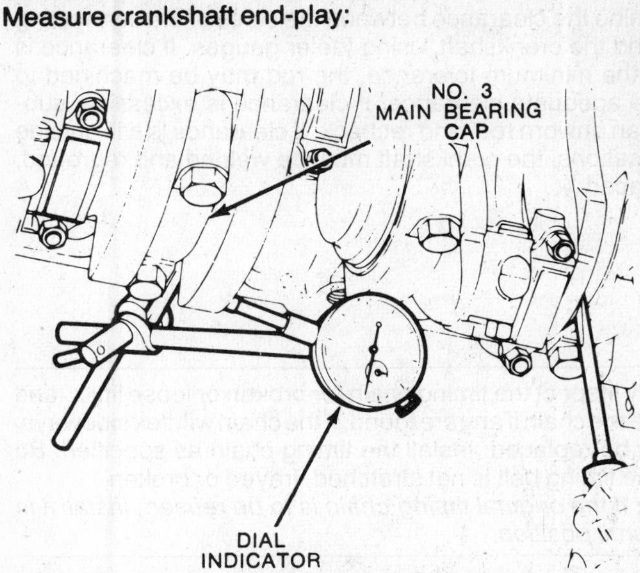

Checking crankshaft end-play with a dial indicator

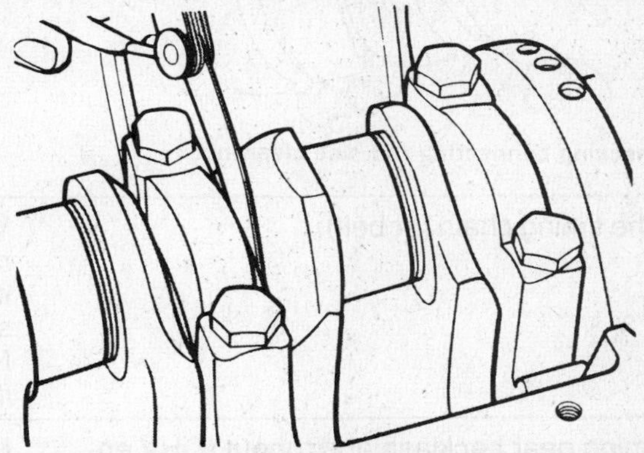

Checking crankshaft end-play with a feeler gauge

Procedure	Method
Install the pistons:	Press the upper connecting rod bearing halves into the connecting rods, and the lower halves into the connecting rod caps. Position the piston ring gaps according to specifications (see car section), and lubricate the pistons. Install a ring compressor on a piston, and press two long (8") pieces of plastic tubing over the rod bolts. Using the tubes as a guide, press the pistons into the bores and onto the crankshaft with a wooden hammer handle. After seating the rod on the crankshaft journal, remove the tubes and install the cap finger tight. Install the remaining pistons in the same manner. Invert the engine and check the bearing clearance at two points (90° apart) on each journal with Plastigage. NOTE: *Do not turn the crankshaft with Plastigage installed.* If clearance is within tolerances, remove *all* Plastigage, thoroughly lubricate the journals, and torque the rod caps to specifications. If clearance is not within specifications, install different thickness bearing inserts and recheck. CAUTION: *Never shim or file the connecting rods or caps.* Always install plastic tube sleeves over the rod bolts when the caps are not installed, to protect the crankshaft journals.

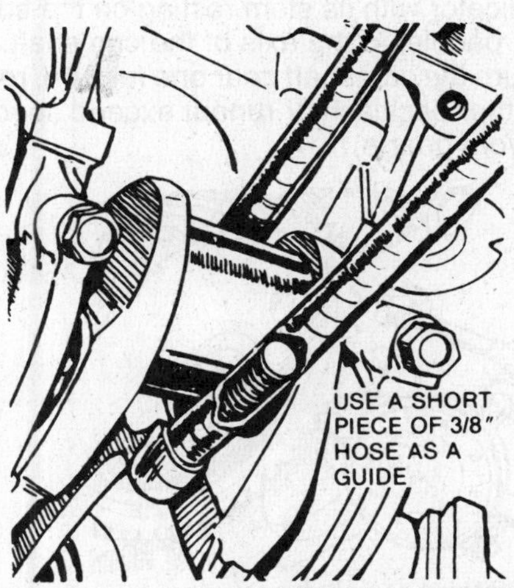

Tubing used to protect crankshaft journals and cylinder walls during piston installation

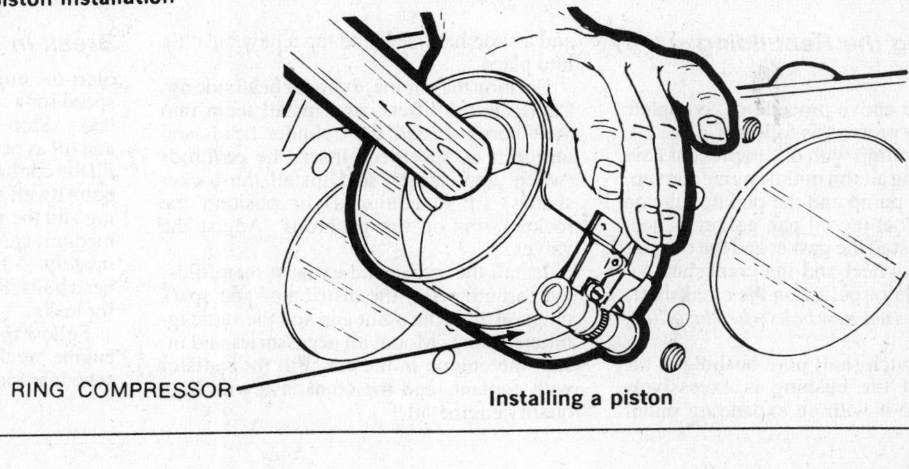

Installing a piston

CYLINDER BLOCK RECONDITIONING

Procedure	Method
Check connecting rod side clearance:	Determine the clearance between the sides of the connecting rods and the crankshaft, using feeler gauges. If clearance is below the minimum tolerance, the rod may be machined to provide adequate clearance. If clearance is excessive, substitute an unworn rod, and recheck. If clearance is still outside specifications, the crankshaft must be welded and reground, or replaced.
Inspect the timing chain (or belt):	Visually inspect the timing chain for broken or loose links, and replace the chain if any are found. If the chain will flex sideways, it must be replaced. Install the timing chain as specified. Be sure the timing belt is not stretched, frayed or broken. NOTE: *If the original timing chain is to be reused, install it in its original position.*
Check timing gear backlash and runout (OHV engines):	Mount a dial indicator with its stem resting on a tooth of the camshaft gear (as illustrated). Rotate the gear until all slack is removed, and zero the indicator. Rotate the gear in the opposite direction until slack is removed, and record gear backlash. Mount the indicator with its stem resting on the edge of the camshaft gear, parallel to the axis of the camshaft. Zero the indicator, and turn the camshaft gear one full turn, recording the runout. If either backlash or runout exceed specifications, replace the worn gear(s).

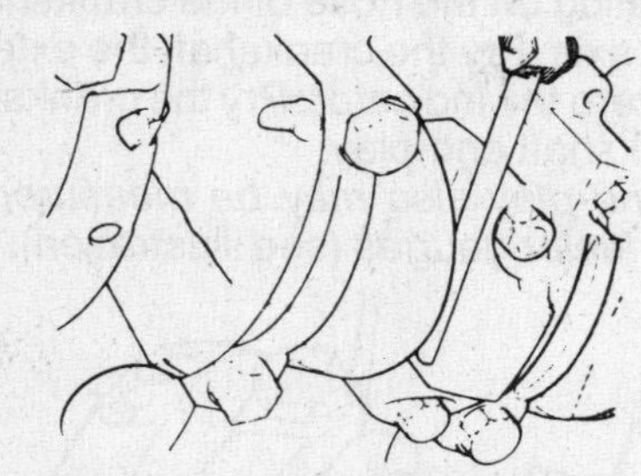

Checking connecting rod side clearance

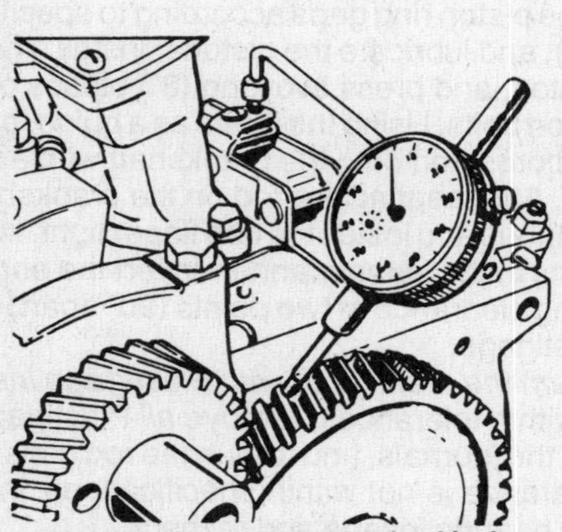

Checking camshaft gear backlash

Checking camshaft gear runout

Completing the Rebuilding Process

Following the above procedures, complete the rebuilding process as follows:

Fill the oil pump with oil, to prevent cavitating (sucking air) on initial engine start up. Install the oil pump and the pickup tube on the engine. Coat the oil pan gasket as necessary, and install the gasket and the oil pan. Mount the flywheel and the crankshaft vibration damper or pulley on the crankshaft. NOTE: *Always use new bolts when installing the flywheel.*
Inspect the clutch shaft pilot bushing in the crankshaft. If the bushing is excessively worn, remove it with an expanding puller and a slide hammer, and tap a new bushing into place.

Position the engine, cylinder head side up. Lubricate the lifters, and install them into their bores. Install the cylinder head, and torque it as specified. Insert the pushrods (where applicable), and install the rocker shaft(s) (if so equipped) or position the rocker arms on the pushrods. Adjust the valves.

Install the intake and exhaust manifolds, the carburetor(s), the distributor and spark plugs. Adjust the point gap and the static ignition timing. Mount all accessories and install the engine in the car. Fill the radiator with coolant, and the crankcase with high quality engine oil.

Break-in Procedure

Start the engine, and allow it to run at low speed for a few minutes, while checking for leaks. Stop the engine, check the oil level, and fill as necessary. Restart the engine, and fill the cooling system to capacity. Check the point dwell angle and adjust the ignition timing and the valves. Run the engine at low to medium speed (800–2500 rpm) for approximately ½ hour, and retorque the cylinder head bolts. Road test the car, and check again for leaks.

Follow the manufacturer's recommended engine break-in procedure and maintenance schedule for new engines.

CV-Joint/U-Joint Overhaul 31

CONSTANT VELOCITY JOINTS

Front wheel drive vehicles present several unique problems to engineers because the driveshaft must do three things, simultaneously. It must allow the wheels to turn for steering, telescope to compensate for road surface vibrations, and it must transmit torque continuously without vibration.

To compensate for these three factors a two-joint driveshaft allows the front wheels to perform these functions. This driveshaft mates disc type straight groove ball joint design with the bell type Rzeppa CV universal joint.

The Rzeppa joint on the outboard end of each driveshaft provides steering ability by allowing drive wheels to steer up to 43° while transmitting all available torque to the wheels. The inboard joint allows telescoping (up to 1½″) through the rolling action of balls in straight grooves and operates at angles up to 20°. The combined action of these two ball type u-joints eliminates vibration.

The typical front wheel drive vehicle uses two driveshaft assemblies—one to each driving wheel. Each assembly has a CV–joint at the wheel end called the outboard joint. A second joint on each shaft located at the transaxle end is called the inboard joint. This joint may be either the ball or tripode type. It allows the slip motion required when the driveshaft must shorten or lengthen in response to suspension action when traveling over an irregular surface.

Constant velocity joints are precision machined parts that have difficult jobs to perform in a hostile environment. They are exposed to heat, shock, torque, and many thousands of miles of service. For this reason, the lubricants used are specially formulated to be compatible with the rubber boot and give proper lubrication. Most CV-joint repair kits have this special lubricant included.

NOTE: Wear pattern in a used ball or tripode CV-joint are impossible to match during reassembly. If there are any signs of wear, abnormal operating noise, corrosion, heat discoloration, the joint must be replaced.

TROUBLESHOOTING

Noises from the engine, drive axles, suspension and steering in the front drive cars can be misleading to the untrained ear. Ideally a smooth road serves best for detecting operating condition(s) that cause noise.

- A humming noise could indicate that early stage of insufficient or incorrect lubricant.
- Worn driveshaft joints will cause a continuous knock at low speeds.

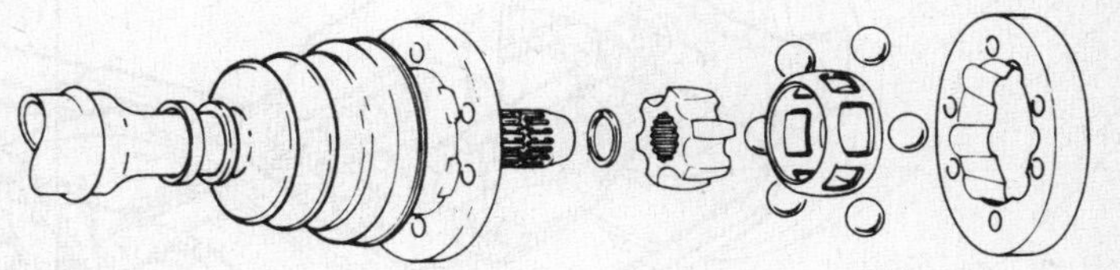

Ball style (Rzeppa) plunging CV joint

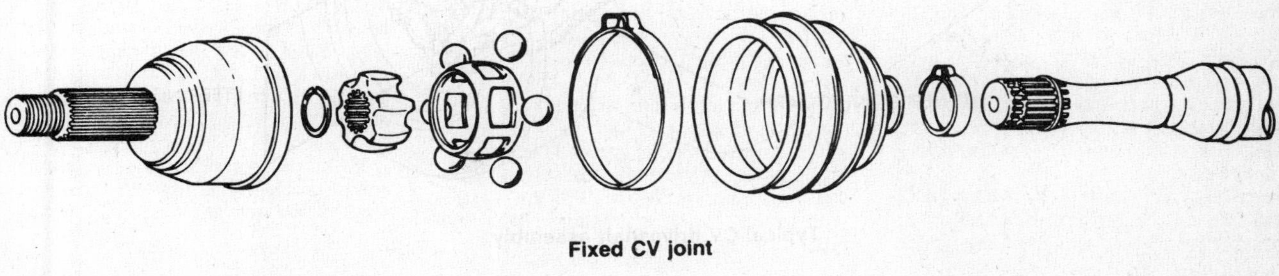

Fixed CV joint

• A popping or clicking sound on sharp turns indicates trouble in the outer or wheel end joint.

• The clunk noise at acceleration from coasting or deceleration from a load pull indicates two possibilities—damaged inner or transaxle joint or differential problem(s).

• An inner joint will create a vibration during acceleration due to plunging action hanging up and releasing repeatedly. Probable cause would be foreign particles or lack of lubrication, or improper assembly.

• Remember that tires, suspension, engine, and exhaust system are all up front to add their noises.

• Make a check with front wheels elevated off ground. Spin the wheels by hand to determine if wheel bearing could be noisy or if out of round tires are causing vibration. Many wheel bearings are prelubed and sealed at the factory.

CAUTION

Personal injury can occur from spinning wheels by engine power. Spinning a wheel at excess speed may cause damage to CV-joints that could be operating at angles too steep when wheels are allowed to hang. Over speeding might also cause damage to tires and the differential.

SHAFT REMOVAL

1. Remove the hub nut and discard it.
2. Drain the lubricant from the transaxle.

CAUTION

The lubricant may be hot.

3. The speedometer pinion gear assembly must be removed before the right drive shaft can be removed. (Automatic transaxles only).
4. Rotate the driveshaft to view the circlip.
5. Compress the circlip tangs with needle nose pliers as you pry into the side gear. This compresses the circlip in position for shaft removal later. Keep an awl between the differential pinion shaft and the end face of the shaft to prevent circlip reentry to the groove.
6. Remove the ball joint clamp bolt. Drop the lower arm too allow clearance. This will permit the front wheel to swing free.
7. Pull the outer splined shaft from the wheel hub, when swinging wheel hub away. Do not pull on the shaft. Grasp the joint housing.
8. Remove the inner joint by pulling outward on the inner joint housing. Do not pull the shaft.

NOTE: Do not allow the assembly to hang at either end. This can jam the CV-joint and cause vibration during operation. If necessary, support the shaft at either end by rope or wire.

INNER JOINT/BOOT

9. Place the assembly in a vise. Care must be taken not the crush the tubular shafts. Some shafts are solid steel.
10. If the inner joint needs replacement, cut the small rubber clamp, large metal clamp, and remove the rubber boot. These items must be discarded.
11. Inspect for internal wear and/or damage.
12. Clean the grease by hand from inside the joint housing and around the 3 ball trunnion assembly to inspect. Mark the tripod and housing for proper reassembly, if it is to be reinstalled.
13. To replace the boot, CV-joint, or both, remove the snap ring from the groove and tap the trunnion lightly with a brass drift pin. Leave the tripode bearings on the

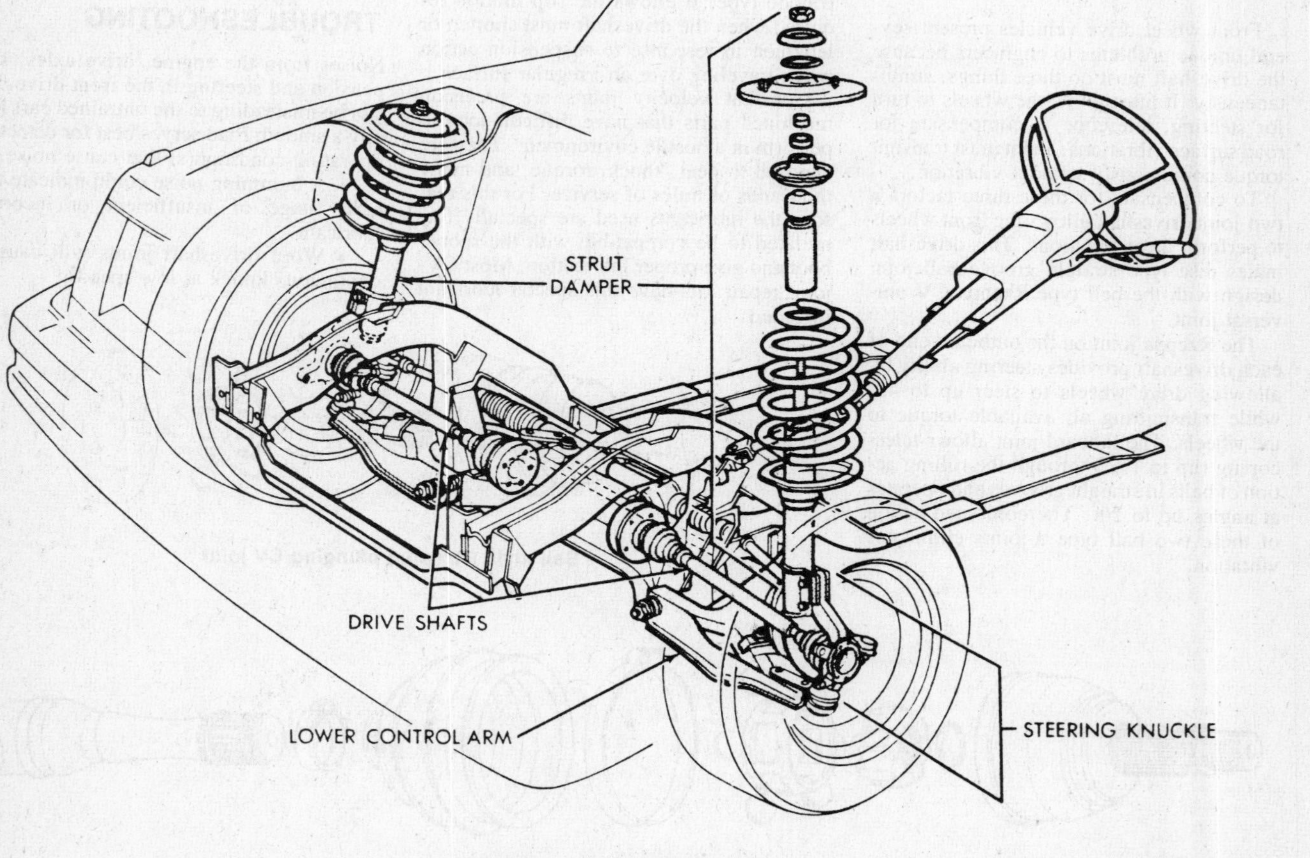

Typical CV driveshaft assembly

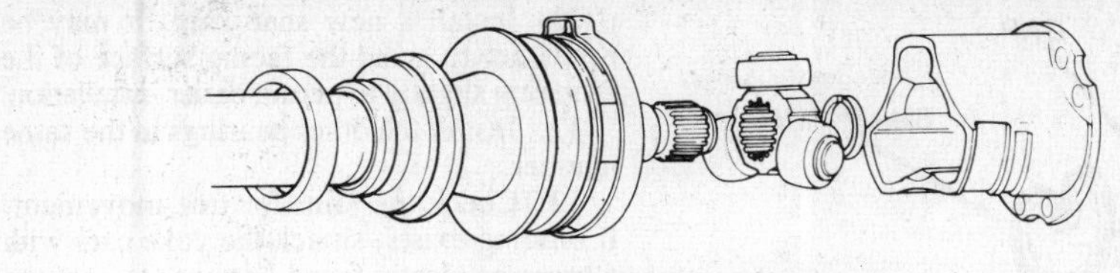

Closed tulip plunging CV joint

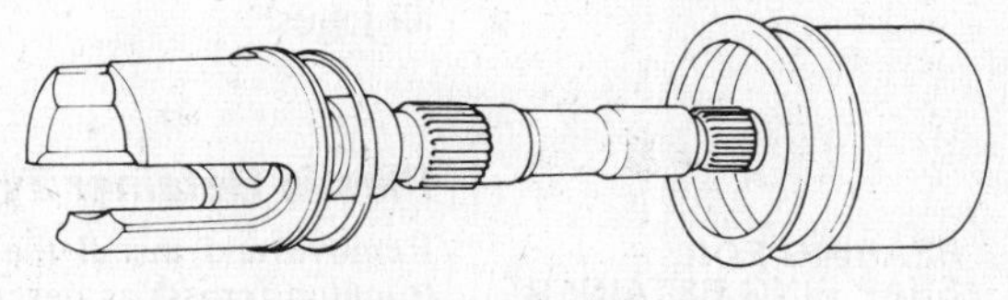

Open tulip plunging CV joint

trunnion. Care must be taken to support the bearings as they may fall off.

14. Installation is the reverse of removal with the following recommendations: When reinstalling the tripode on the shaft place the chamfer face toward the retainer groove. The grease provided with the repair kit must be used. It can not be substituted with any other type grease.

OUTER JOINT/BOOT

1. Place the shaft in a vise. Be careful not to over tighten the vise thereby damaging the shaft.
2. Remove the boot and clamps. Discard these parts.
3. Using a soft hammer rap sharply on the housing. This forces the inner race over the internal circlip. Never remove the slinger from the housing.
4. Remove and discard the circlip. A new one is included with the boot kit. Leave the lock ring in place.

NOTE: Never disassemble the cage and balls from the housing. Reuse the joint assembly with a new boot kit, unless the grease is contaminated and prior diagnosis indicated trouble. In that case replace the joint and boot.

5. Installation is the reverse of removal.

UNIVERSAL JOINTS

U-joint is mechanic's jargon for universal joint. U-joints should not be confused with U-bolts, which are U-shaped bolts used to connect U-joints to the differential pinion flange.

Universal joints provide flexibility between the driveshaft and axle housing to accommodate changes in the angle between them (changes of length are accommodated by the sliding splined yoke between the driveshaft and transmission). The engine and transmission are mounted rigidly on the car frame, while the driving wheels are free to move up and down in relation to the frame. The angles between the transmission, driveshaft and axle change constantly as the car responds to various road conditions.

To give flexibility and still transmit power as smoothly as possible, several types of universal joints are used.

The most common type of universal joint is the cross and yoke type. Yokes are used on the ends of the driveshaft with the yoke arms opposite each other. Another yoke is used opposite the driveshaft and when placed together, both yokes engage a center member, or cross, with four arms spaced 90° apart (the U-joint cross is alternately referred to as a spider, and the arms are called trunnions). A bearing cup (or cap) is used on each arm of the cross to accommodate movement as the driveshaft rotates. The bearings used are needle bearings.

A conventional universal joint will cause the driveshaft to speed up and slow down through each revolution and cause a corresponding change in the velocity of the driven shaft. This change in speed causes natural vibrations to occur through the driveline, necessitating a third type of universal joint: the constant velocity joint. A rolling ball moves in a curved groove, located between two yoke-and-cross universal joints, connected to each other by a coupling yoke. The result is a uniform motion as the driveshaft rotates, avoiding the fluctuations in driveshaft speed. This type of joint is found in cars with sharp driveline angles, or where the extra measure of isolation is desirable.

CROSS AND YOKE U-JOINT OVERHAUL

There are two types of cross and yoke U-joints. One type retains the cross within the yoke with C-shaped snap rings. The second type of joint is held together by injection molded plastic retainer rings. The second type cannot be reassembled with the same parts, once disassembled. However, repair kits are available.

Snap-Ring Type

1. Remove the driveshaft. For the correct procedure, see the car section for the model you are working on.
2. If the front yoke is to be disassembled, matchmark the driveshaft and sliding splined yoke (transmission yoke) so that driveline balance is preserved upon reassembly. Remove the snap rings which retain the bearing caps.
3. Select two sockets, one small enough

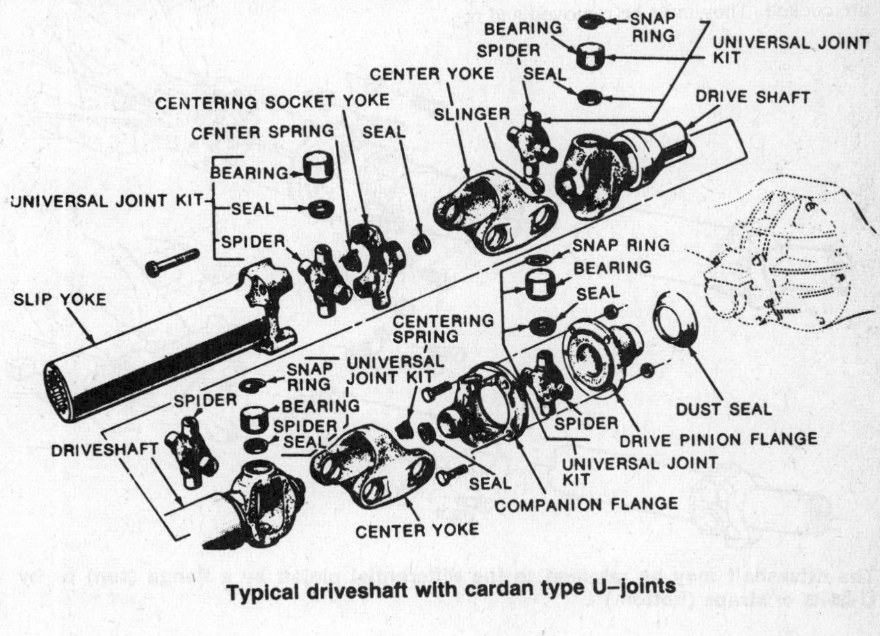

Typical driveshaft with cardan type U-joints

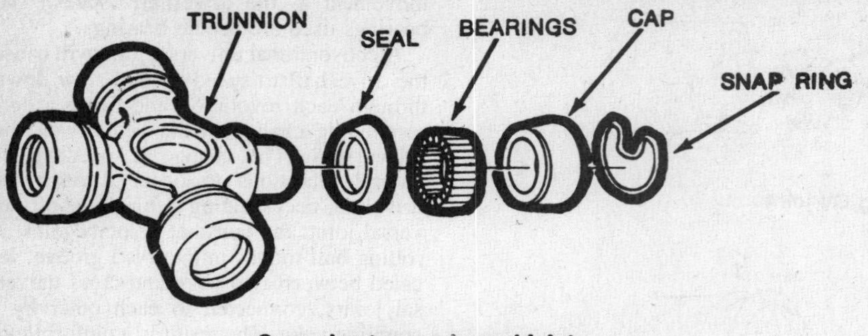

Snap ring type universal joint

to pass through the yoke holes for the bearing caps, the other large enough to receive the bearing cap.

4. Using a vise or a press, position the small and large sockets on either side of the U-joint. Press in on the smaller socket so that it presses the opposite bearing cap out of the yoke and into the larger socket. If the cap does not come all the way out, grasp it with a pair of pliers and work it out.

5. Reverse the position of the sockets so that the smaller socket presses on the cross. Press the other bearing cap out of the yoke.

6. Repeat the procedure on the other bearings.

7. To install, grease the bearing caps and needles thoroughly if they are not pregreased. Start a new bearing cap into one side of the yoke. Position the cross in the yoke.

8. Select two sockets small enough to pass through the yoke holes. Put the sockets against the cross and the cap, and press the bearing cap ¼ inch below the surface of the yoke. If there is a sudden increase in the force needed to press the cap into place, or if the cross starts to bind, the bearings are cocked. They must be removed and re-

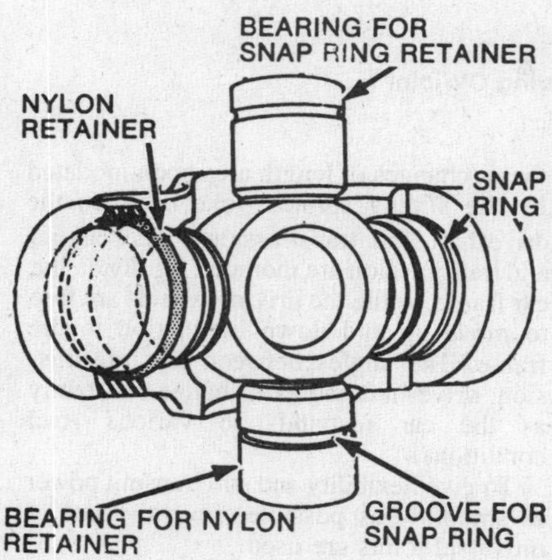

U-joint locking methods

started in the yoke. Failure to do so will greatly reduce the life of the bearing.

9. Install a new snap ring.

10. Start a new bearing into the opposite side. Place a socket on it and press in until the opposite bearing contacts the snap ring.

11. Install a new snap ring. It may be necessary to grind the facing surface of the snap ring slightly to permit easier installation.

12. Install the other bearings in the same manner.

13. Check the joint for free movement. If binding exists, smack the yoke ears with a brass or plastic faced hammer to seat the bearing needles. Do not strike the bearings, and support the shaft firmly. Do not install the driveshaft until free movement exists at all joints.

Plastic Retainer Type

Remove and install the bearing caps and trunnion (cross) as described for the snap-ring type universal joints. On an original universal joint, however, the bearing caps will be secured in the yokes with injected plastic. The plastic will shear when the bearing caps are pressed. Service snap-rings are installed in the groove on the inside (of yoke) of the installed caps.

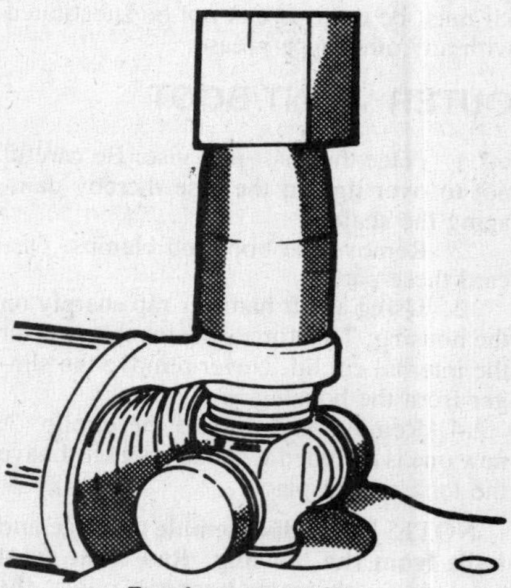

Bearing removal

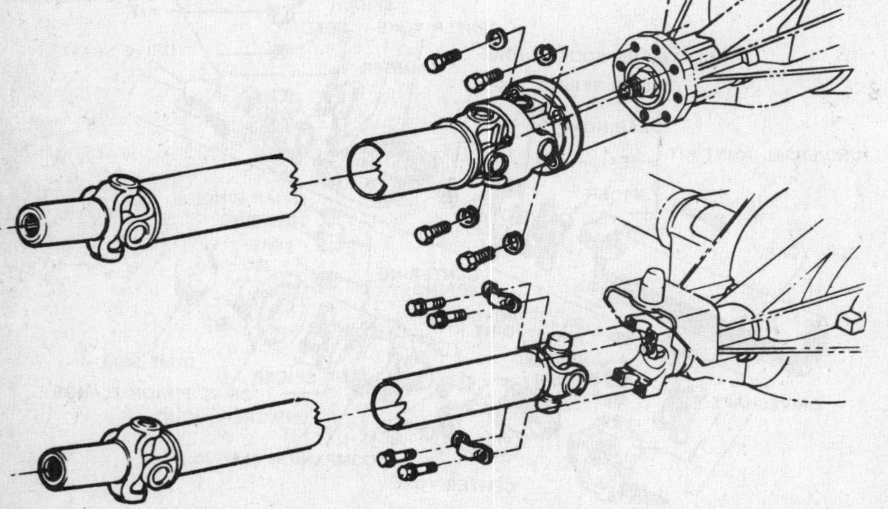

The driveshaft may be retained to the differential pinion by a flange (top) or by U-bolts or straps (bottom)

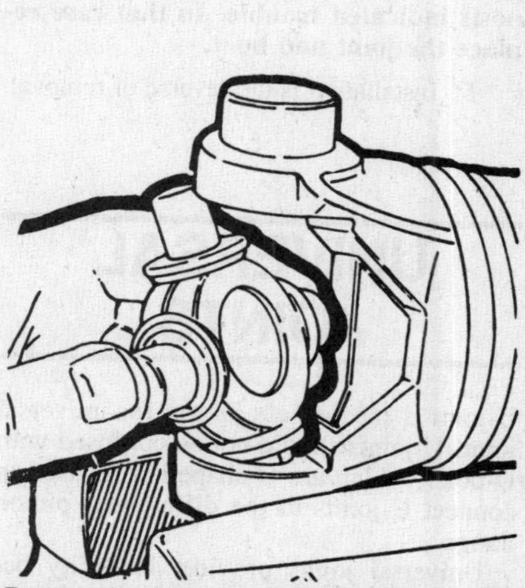

Press a bearing cap into the yoke, then install the cross

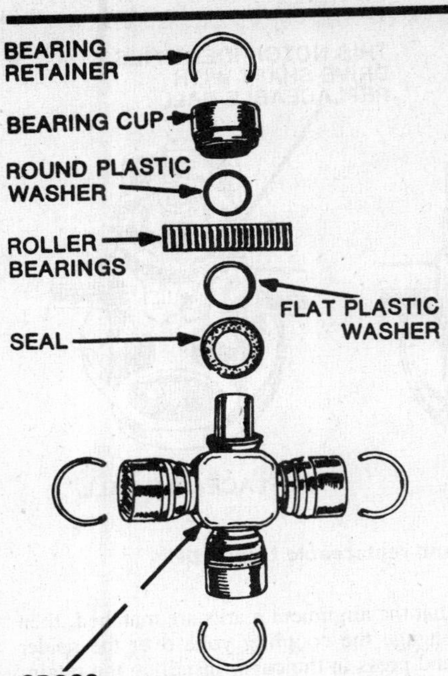

Plastic retainer U-joint repair kit components

NOTE: The plastic which retains the bearing will be sheared when the bearing cup is pressed out. Be sure to remove the remains of the plastic retainer from the ears of the yoke. It is easier to remove the remains if a small pin or punch is first driven through the injection holes in the yoke. Failure to remove all of the plastic remains may prevent the bearing cups from being pressed into place and the bearing retainers from being properly seated.

CARDAN TYPE U-JOINT OVERHAUL

Some with Cardan type U-joints use snap rings to retain the bearing cups in the yokes. Other cars have plastic retainers. Be sure to obtain the correct rebuilding kit.

1. Use a punch to mark the coupling yoke and the adjoining yokes before disassembly, to ensure proper reassembly and driveline balance.

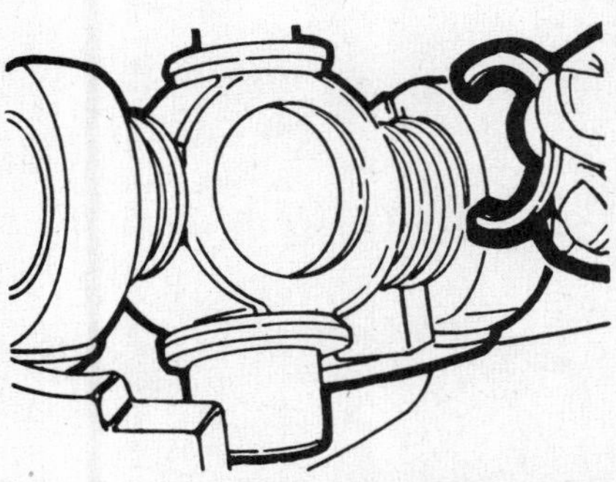

Service snap rings are installed inside the yoke

2. It is easiest to remove the bearings from the coupling yoke first. Follow the order indicated in the illustration.
3. Support the driveshaft horizontally on a press stand, or on the workbench if a vise is being used.
4. If snap rings are used to retain the bearing cups, remove them. Place the rear ear of the coupling yoke over a socket large enough to receive the cup. Place a smaller socket, or a cross press made for the purpose, over the opposite cup. Press the bearing cup out of the coupling yoke ear. If the cup is not completely removed, insert a spacer and complete the operation, or grasp the cup with a pair of slip joint pliers and work it out. If the cups are retained by plastic, this will shear the retainers. Remove any bits of plastic.
5. Rotate the driveshaft and repeat the operation on the opposite cup.
6. Disengage the trunnions of the spider, still attached to the flanged yoke, from the coupling yoke, and pull the flanged yoke and spider from the center ball on the ball support tube yoke.

NOTE: The joint between the shaft and coupling yoke can be serviced without disassembly of the joint between the coupling yoke and flanged yoke.

7. Pry the seal from the ball cavity, remove the washers, spring and three seats. Examine the ball stud seat and the ball stud for scores or wear. Worn parts can be replaced with a kit. Clean the ball seat cavity and fill it with grease. Install the spring, washer, ball seats, and spacer (washer) over the ball.
8. To assemble, insert one bearing cup

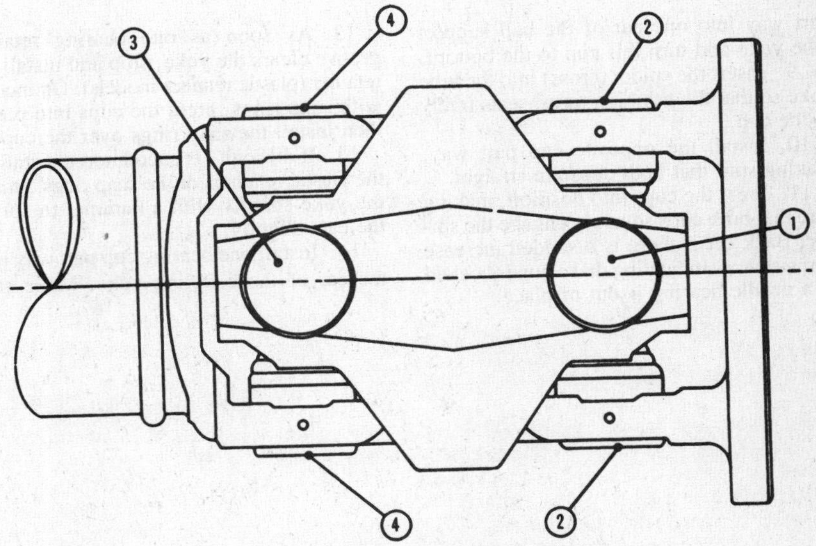

Cardan joint disassembly sequence

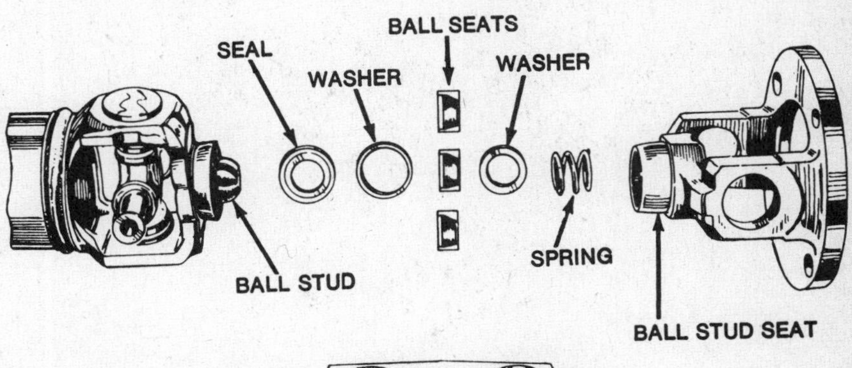

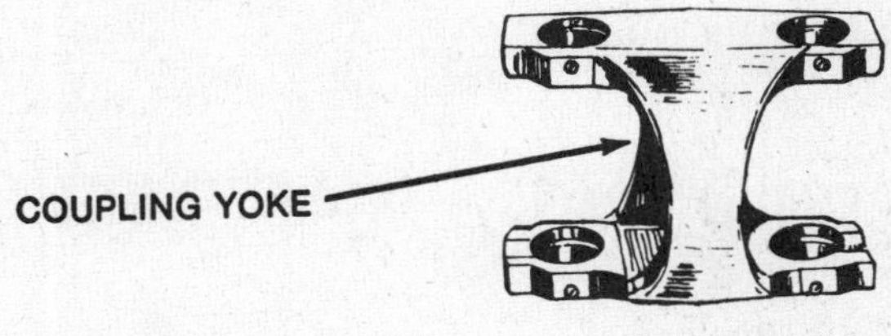

Cardan type joint

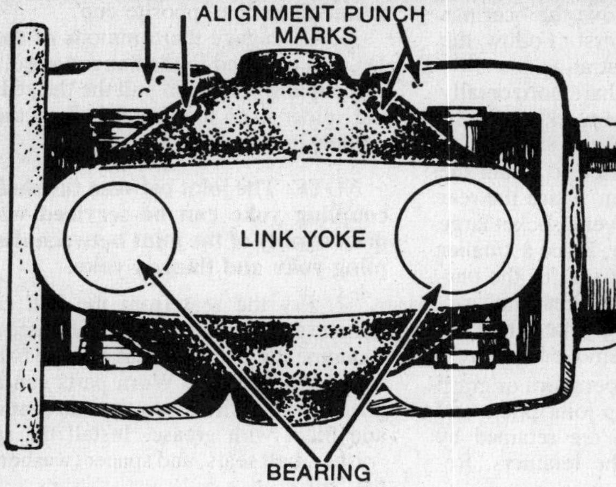

Match marks for double cardan joint

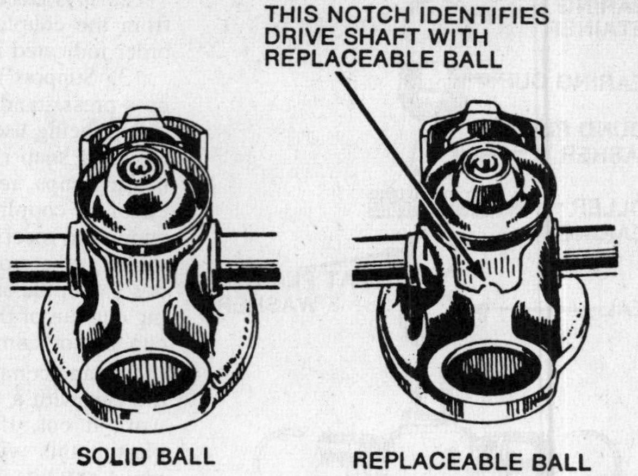

Solid and replaceable U-joint balls

part way into one ear of the ball support tube yoke and turn this cup to the bottom.

9. Insert the spider (cross) into the tube yoke so that the trunnion (arm) seats freely in the cup.

10. Install the opposite cup part way, making sure that both cups are straight.

11. Press the cups into position, making sure that both cups squarely engage the spider. Back off if there is a sudden increase in resistance, indicating that a cup is cocked or a needle bearing is out of place.

12. As soon as one bearing retainer groove clears the yoke, stop and install the retainer (plastic retainer models). On models with snap rings, press the cups into place, then install the snap rings over the cups.

13. If difficulty is encountered installing the plastic retainers or the snap rings, smack the yoke sharply with a hammer to spring the ears slightly.

14. Install one bearing cup part way into the ear of the coupling yoke. Make sure that the alignment marks are matched, then engage the coupling yoke over the spider and press in the cups, installing the retainers or snap rings as before.

15. Install the cups and spider into the flanged yoke as with the previous yoke.

NOTE: The flange yoke should snap over center to the right or left and up or down by the pressure of the ball seat spring.

Strut Overhaul 32

STRUT SERVICE AND REPAIR

MacPherson struts are appearing on the front (and rear) wheels of more and more cars. The strut design takes up less room in the engine compartment, compared to a conventional upper and lower arm with shock absorber arrangement. The trend toward smaller, lighter and more efficient vehicles mandates the use of a strut suspension to permit more room for engine accessories and front wheel drive components.

Strut Suspension Design

In a conventional front suspension, the wheel is attached to a spindle, which is in turn connected to upper and lower control arms through upper and lower ball joints. A coil spring between the control arms (sometimes on top of the upper arm) supports the weight of the vehicle and a shock absorber controls rebound and dampens oscillations.

In a MacPherson strut type suspension, the strut performs a shock dampening function like a shock absorber, but unlike a conventional shock absorber the strut is a structural part of the vehicle's suspension.

The strut assembly usually contains a spring seat to retain the coil spring that supports the vehicle's weight. The shock absorber is built into the body of the strut housing. The strut is normally attached at the bottom to the lower control arm and at

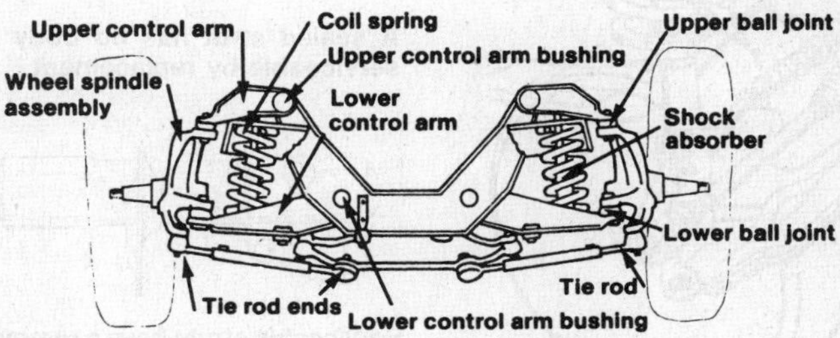

Conventional upper and lower arm suspension

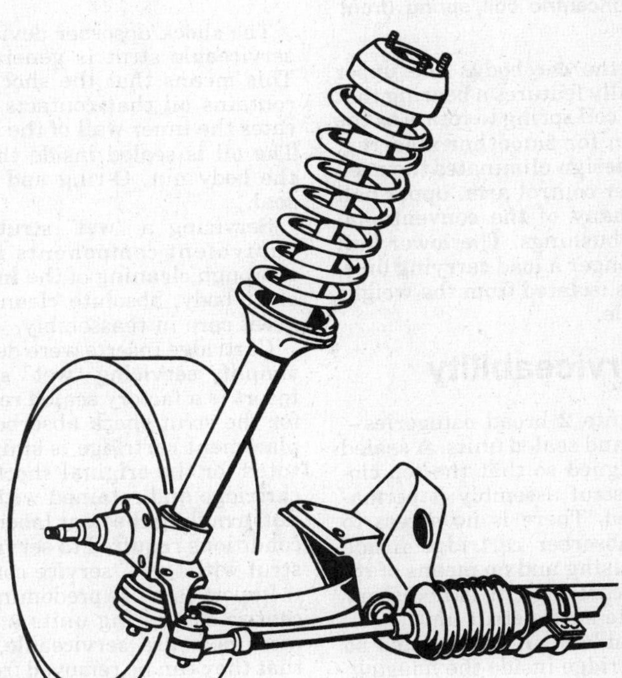

Strut with concentric coil spring (rear wheel drive)

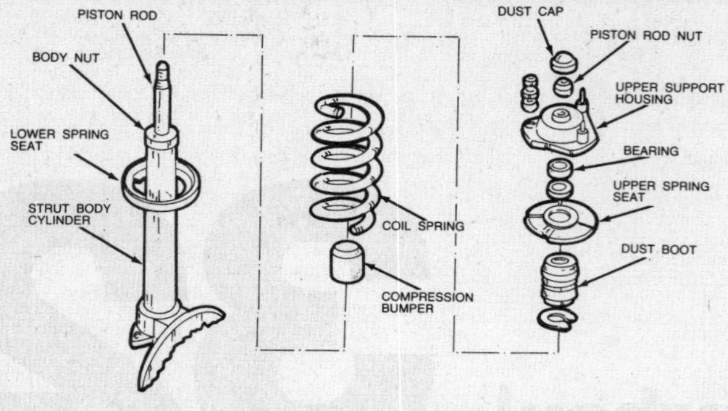

Exploded view of a typical strut

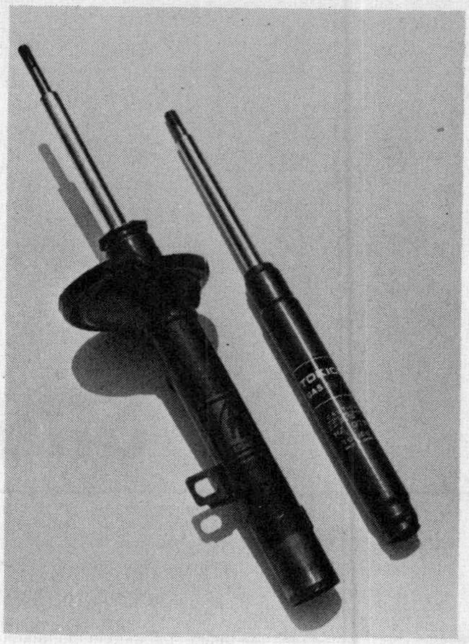

A replacement sealed strut on the left, compared to a replacement strut cartridge used on serviceable type struts on the right

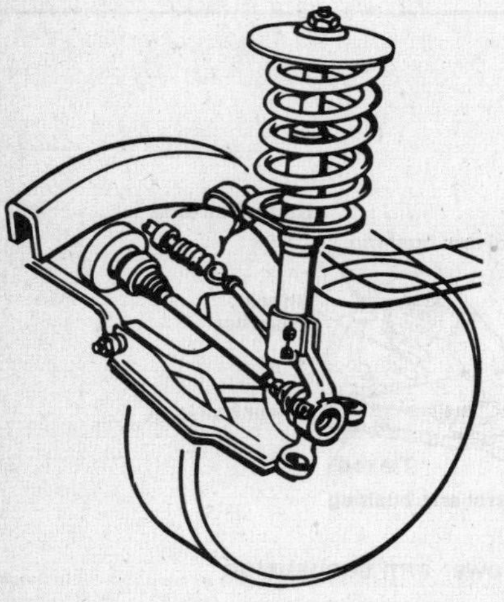

Strut with concentric coil spring (front wheel drive)

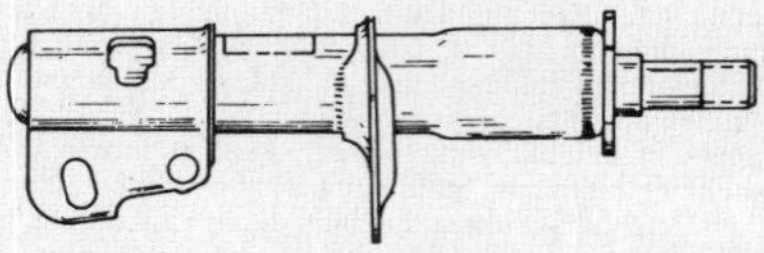

A sealed strut has no body nut and is serviceable by replacement

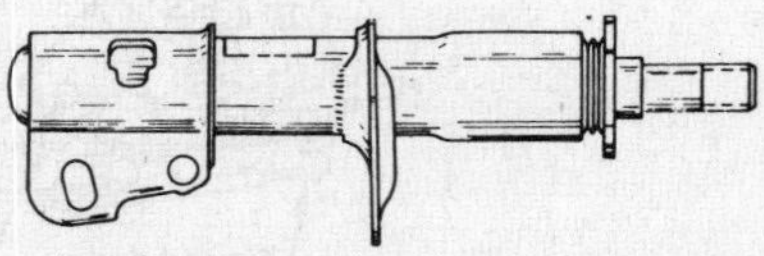

Serviceable struts have a removeable body nut to allow replacement of the strut cartridge

the top to the car body. The upper mount usually features a bearing that permits the coil spring to rotate as the wheels turn for smoother steering. The entire design eliminates the need for the upper control arm, upper ball joint and many of the conventional suspension bushings. The lower ball joint is no longer a load carrying unit, because it is isolated from the weight of the vehicle.

Serviceability

Struts fall into 2 broad categories—serviceable and sealed units. A sealed strut is designed so that the top closure of the strut assembly is permanently sealed. There is no access to the shock absorber cartridge inside the strut housing and no means of replacing the cartridge. It is necessary to replace the entire strut unit.

A serviceable strut is designed so that the cartridge inside the housing, that provides the shock absorbing function, can be replaced with a new cartridge. Serviceable struts use a threaded body nut in place of a sealed cap to retain the cartridge.

The shock absorber device inside a serviceable strut is generally "wet." This means that the shock absorber contains oil that contacts and lubricates the inner wall of the strut body. The oil is sealed inside the strut by the body nut, O-ring and piston rod seal.

Servicing a "wet" strut with the equivalent components involves a thorough cleaning of the inside of the strut body, absolute cleanliness and great care in reassembly.

Cartridge inserts were developed to simplify servicing "wet" struts. The insert is a factory sealed replacement for the strut shock absorber. The replacement cartridge is simply substituted for the original shock absorber cartridge and retained with the body nut, avoiding the near laboratory-like conditions required to service a "wet" strut with "wet" service components.

Import cars use predominantly concentric coil spring units and, for the most part are serviceable, meaning that they can be removed from the vehicle, disassembled and the shock absorber cartridge replaced in the old housing. Both OEM and aftermarket replacement cartriges can be used in these struts if they are serviceable.

Exceptions to the serviceable struts include some of the later model import cars, but even on these cars OEM struts can be replaced with aftermarket sealed strut assemblies.

WHEEL ALIGNMENT

It is not always necessary to re-align the wheels after struts are serviced. If care is taken matchmarking affected components and in reassembling, alignment may be unaffected. However, if wheels were not in proper alignment prior to service, or if the entire strut assembly was replaced, a wheel alignment check should be made. Generally, only camber is adjustable, and then only within a narrow range.

Do not attempt to bend components to correct wheel alignment.

Since the majority of OEM struts are serviced by replacement, most manufacturers recommend wheel alignment following strut replacement.

On most serviceable import struts, the position of the upper bearing plate or lower mount can be matchmarked and wheel alignment will be maintained during reassembly.

Tools

Without the right tools, a strut job will take longer than necessary and can be dangerous.

A normal selection of hand tools such as open end and box wrenches, sockets, pliers, screwdrivers and hammers are necessary to work on struts.

Extensions and universal joints will help reach tight spots. Be sure to have both metric and inch-sized wrenches on hand.

In addition to the normal handtools, some sort of spanner is necessary to remove the body nut on serviceable struts. Sometimes a pipe wrench can be used successfully. Also a strut vise should be used to avoid damage to the strut housing during the overhaul procedure.

Strut and cartridge replacement requires a spring compressor.

Makeshift tools for compressing coil springs—threaded rod, chains, wire or other methods—should never be used. The coil spring is under tremendous compression and can fly off causing personal injury and damage to equipment. Use only a good quality spring compressor such as described below.

Economy, or manual, spring compressors are the least expensive but more time consuming to use. Angle hooks grasp the spring coils and must be compressed with a wrench. For those who service struts infrequently, this is probably the wisest investment for purchase.

CAUTION

When using an "economy type" spring compressor be certain to install J-Bolts or U-Bolts around the coil spring-to-the-tool. This is to prevent the tool from slipping off the spring and causing personal injury.

Other manual spring compressors (jaws type) are faster to operate, have a more positive gripping action and can be used on or off the car. These types are probably not cost effective for the do-it-yourselfer, but can be rented from auto supply stores for single-time use. These are also safer to use than the "economy type" spring compressors.

MAINTAINING WHEEL ALIGNMENT

The location and method of adjusting wheel alignment determines the components that must be match-marked to maintain wheel alignment. There are 4 basic methods of adjusting wheel alignment. Almost all cars use one of these or a slight variation.

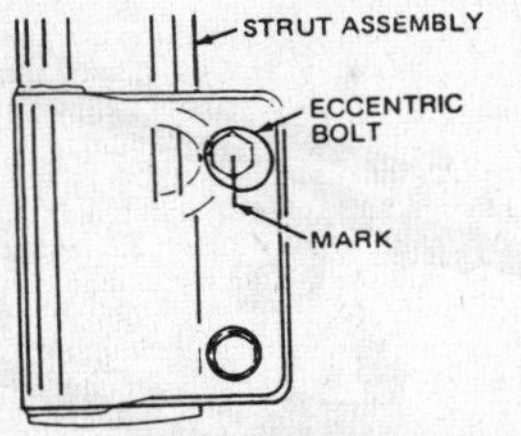

Mark the eccentric (camber adjusting bolt) relative to the clevis mounting bracket.

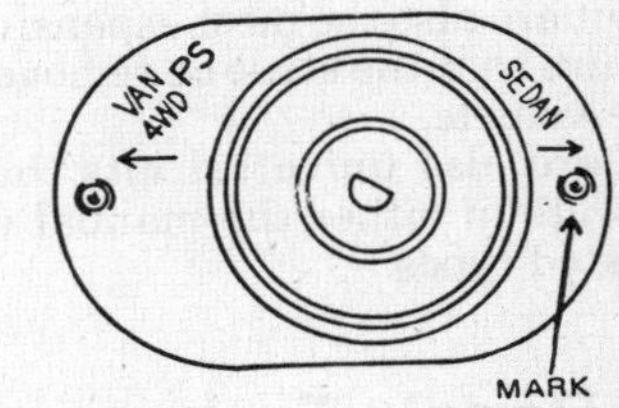

Mark the mounting stud that faces the front of the vehicle. This type of bracket is reversible for varying applications.

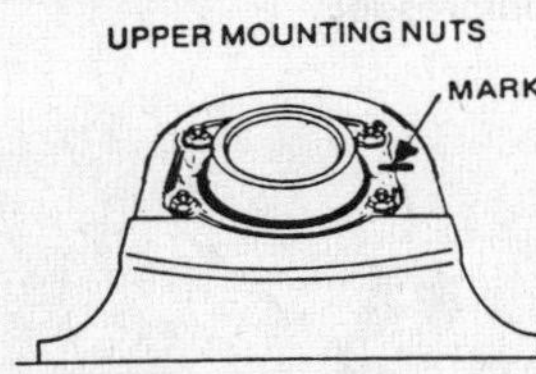

Mark the upper support housing relative to the inner fender before removing the strut from the upper mount.

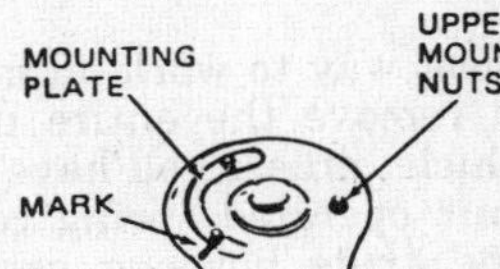

Mark the location of the mounting plate relative to the location on the inner fender.

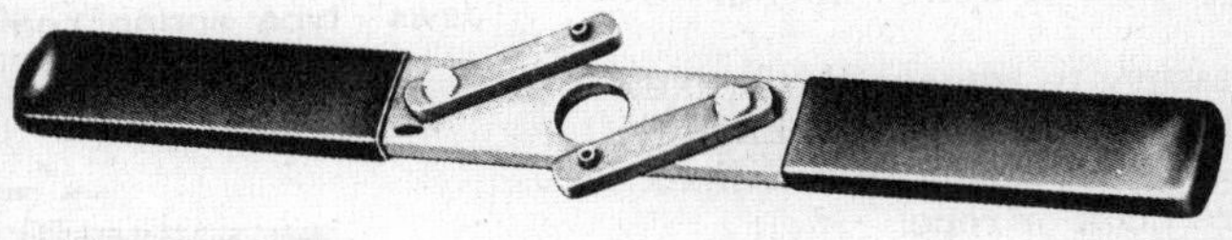

A simple spanner wrench designed for use with body nuts equipped with recessed lugs. A pipe wrench is a frequent substitute

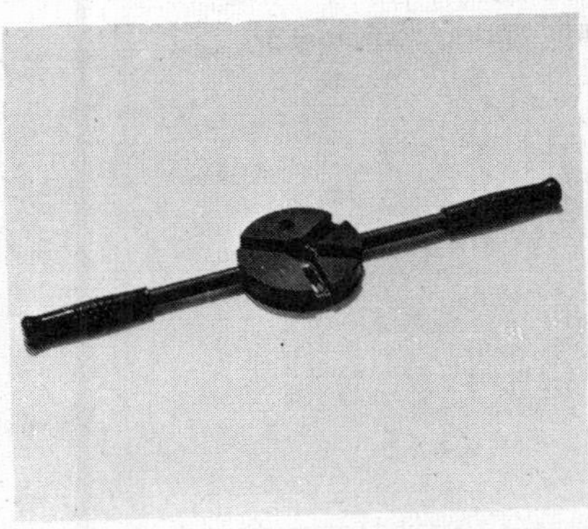

This type of spanner wrench comes with adapter inserts for various applications of body nuts. A torque wrench can be used with this spanner for tightening the strut body nut

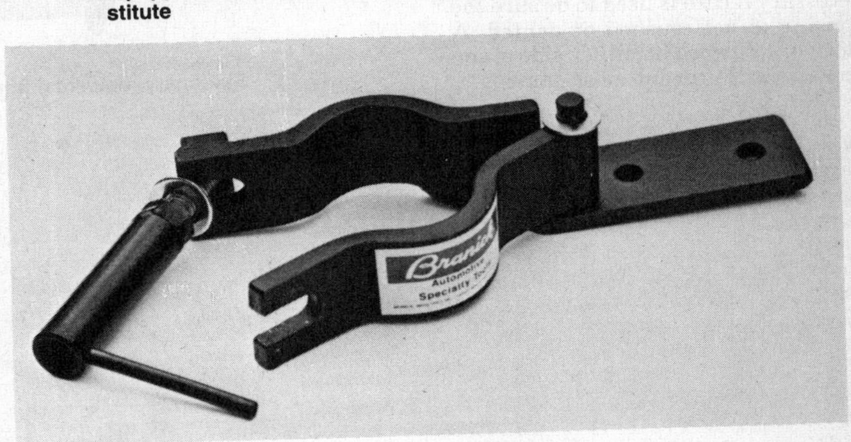

A strut vise should be used to prevent damage to the strut housing during overhaul. It may be placed in a bench vise or mounted to a workbench

For high volume work, compressors that are pneumatically or hydraulically operated are best. Air operated compressors are suitable for all types of struts (through use of adaptors), are lightweight and can be used on or off the vehicle. Bench mounted hydraulically operated units are probably the safest, but are also the most expensive and require that the strut be removed from the vehicle.

There are also universal kits that fit all struts in either the manual or air operated types.

Repair Tips

• Make sure you have all the tools you'll need. NEVER IMPROVISE A SPRING COMPRESSOR.

• Normally both front struts should be repaired or replaced at the same time.

• The easiest way to work on most struts is to remove the entire unit from the vehicle, unless you have access to an air operated spring compressor. Some struts, however, can be repaired while installed on the vehicle.

• Always read the instructions packaged with any replacement parts. In particular, note whether the body nut is supplied new or re-used.

• Mark the position(s) of any bearing plate nuts or cam bolts to assure proper alignment after installation.

• Be sure to protect the rubber boot on the drive axle of front wheel drive cars.

• If necessary to remove the brake caliper, do not let the caliper hang by the brake hose. Suspend the caliper from a wire hook or rope.

• Be careful in clamping a strut in a vise. Special strut vises are available to hold struts, but are not absolutely necessary if care is used to be sure the housing is not crushed or dented. A block of soft wood on either side of the housing will prevent most damage.

• Use a spring compressor to relieve tension from the spring. Be sure to clean and lubricate the screw threads, particularly on hand operated (manual) spring compressors.

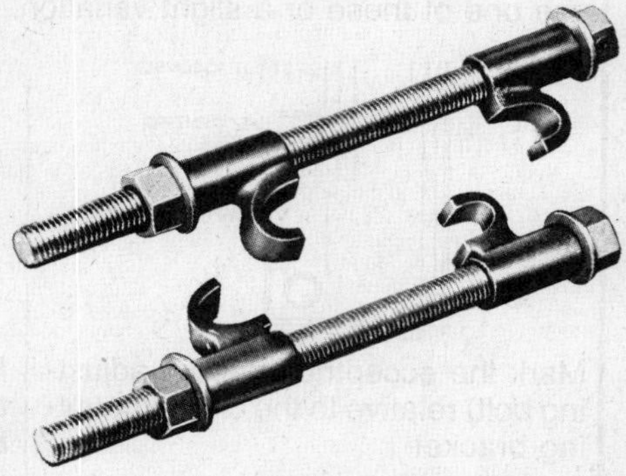

An economical manual spring compressor

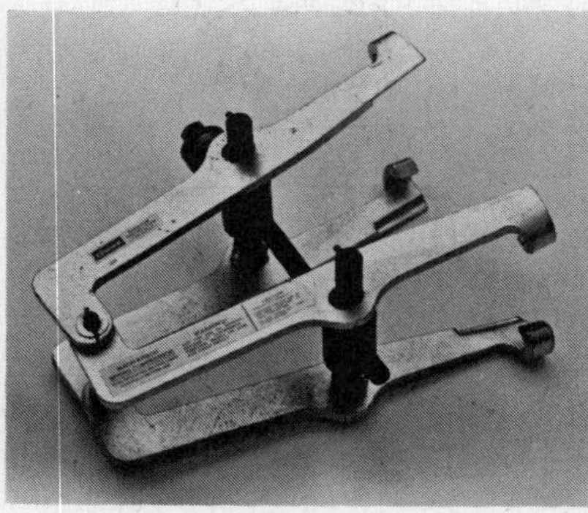

"Jaws" type spring compressor

Some springs have a special coating that should not be scuffed.

• If you are replacing the strut cartridge, clean the inside of the strut housing and the body nut threads before replacing the oil and installing a new cartridge.

• Be sure to use OEM quality fasteners any time a fastener is replaced.

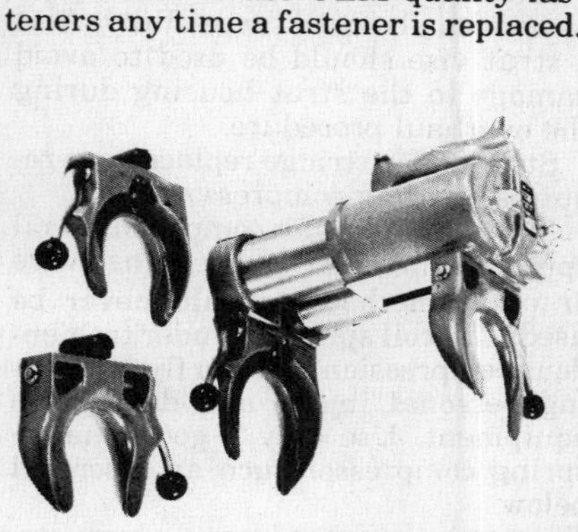

Lightweight, air operated, portable spring compressor can be used on or off the vehicle. Extra shoes are available to handle all strut applications

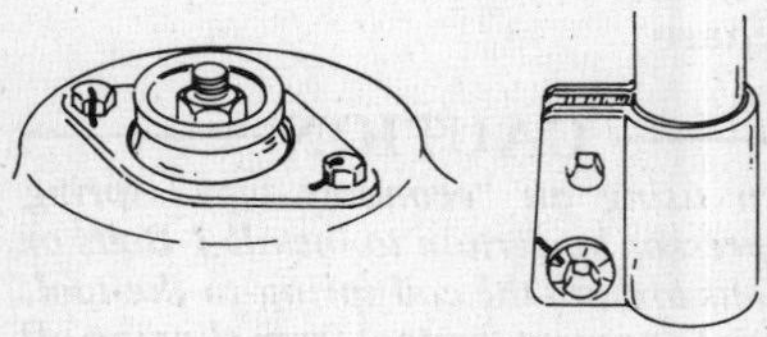

Mark the position of the attachments that control wheel alignment. See Maintaining Wheel Alignment earlier in this section

Stationary, universal pneumatic spring compressor

STRUT OVERHAUL

Following is a typical overhaul procedure of a serviceable MacPherson strut, after having removed the strut from the vehicle. The vehicle should be firmly and safely supported on jackstands. If it is necessary, to separate the brake line from the strut for strut removal, the brakes will have to be bled after reinstallation. Examine the strut assembly for damage, dented strut body, spring seat, broken or missing strut mounting parts. Any of these will require replacement of the complete assembly. Also inspect other suspension components for wear or damage. See the manufacturer's car section for specific MacPherson strut removal and installation procedures.

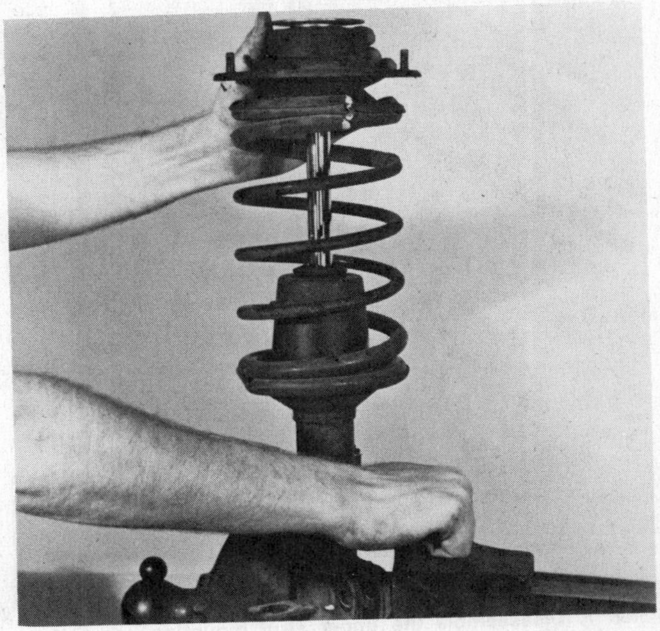

Step 1. **To make service easier, clamp the strut in a strut vise. The strut vise is designed to clamp the strut tight without damage to the strut cylinder. It is very handy for strut work and can be used in a bench vise or mounted to a workbench**

Step 3. **Position the spring compressor on the spring. Turn the load screw to open or close the compressor until the maximum number of spring coils can be engaged**

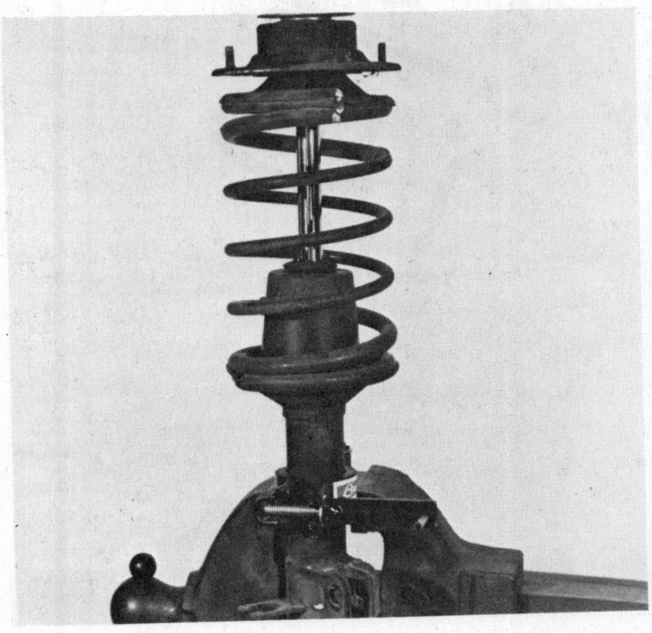

Step 2. **Matchmark the upper end of the coil spring and bearing plate to avoid confusion during reassembly**

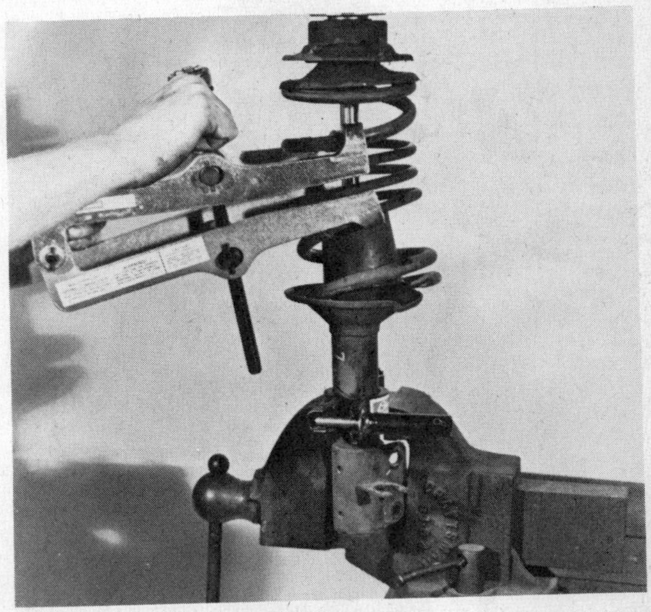

Step 4. **Tighten the load screw until the coil spring is loose from the spring seat. There is no need to compress the spring further than this point**

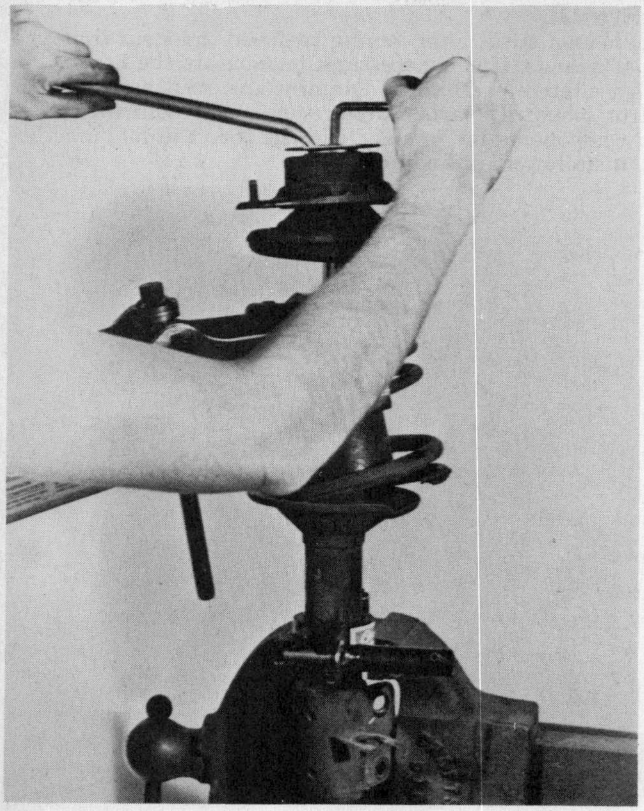

Step 5. **Using an offset wrench and an Allen wrench, loosen the piston rod nut**

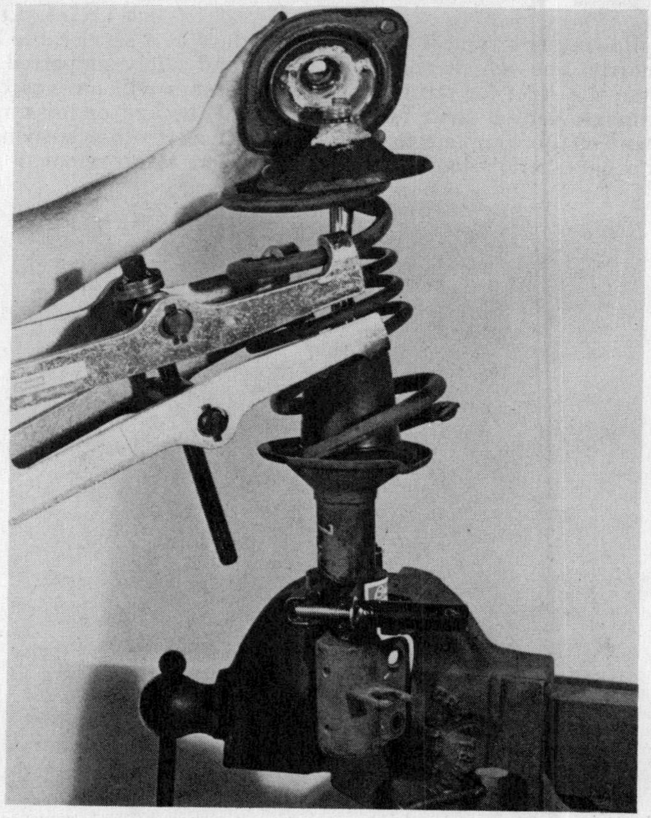

Step 7. **Disassemble the upper strut mounting parts. Keep the mounting parts in order of their removal. They'll be reassembled in reverse order**

Step 6. **Remove the piston rod nut from the strut piston rod**

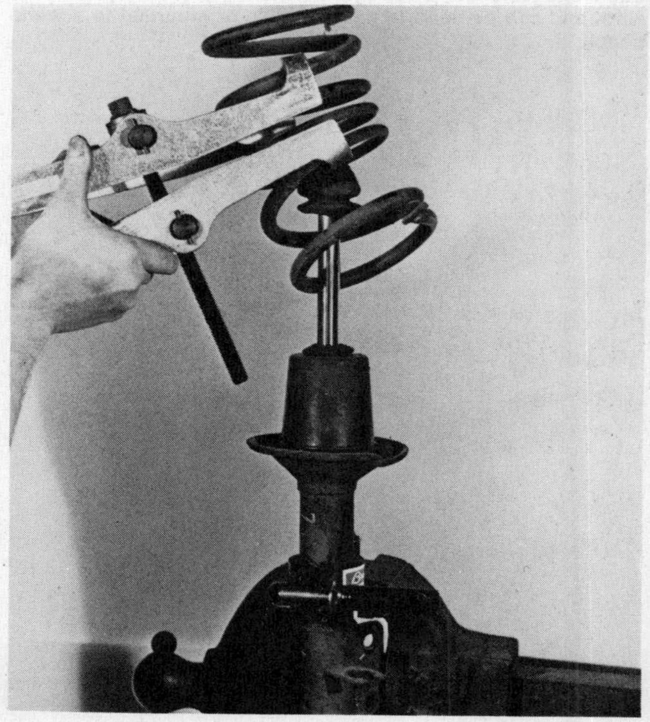

Step 8. **Remove the coil spring and compressor from the strut. There is no need to remove the compressor from the coil spring**

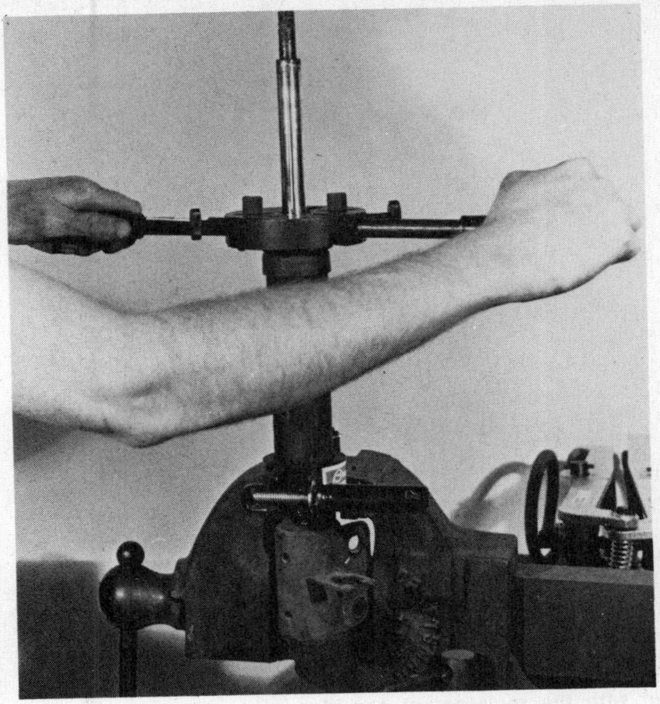

Step 9. **Use a spanner wrench or a pipe wrench to loosen the body nut**

Step 10. **Remove the body nut and discard if a new body nut came with the replacement cartridge. If not, save the body nut**

Step 11. **Grasp the piston rod and pull the cartridge out of the housing. Remove it slowly to avoid splashing oil. Be sure all pieces are removed from the housing**

Step 12. **Pour all of the strut fluid into a suitable container, clean the inside of the strut cylinder, and inspect the cylinder for dents and to insure that all loose parts have been remove from inside the strut body**

Step 13. **Refill the strut housing with approximately one once of the original oil or fresh oil. The oil helps dissipate internal cartridge heat during operation and results in a much cooler running, longer lasting unit. Do not overfill with oil—otherwise the oil may leak at the body nut after it expands when heated**

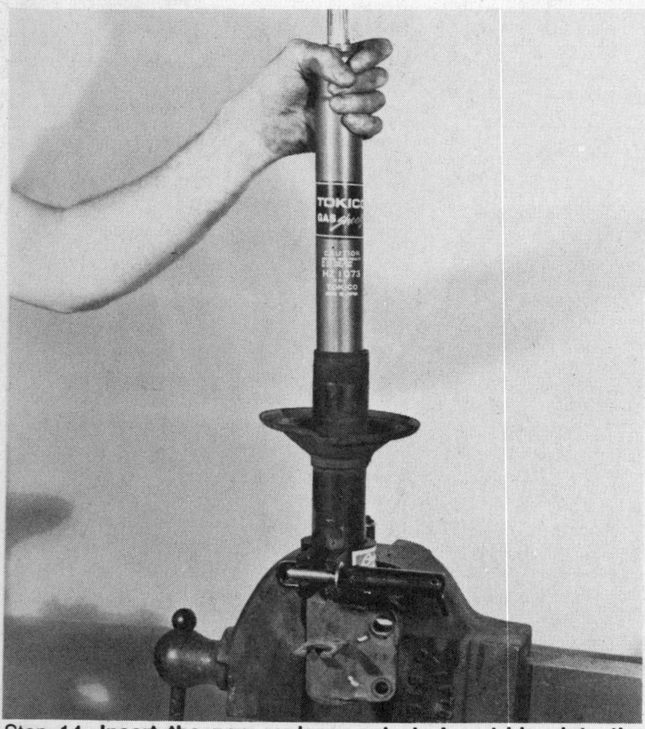

Step 14. **Insert the new replacement strut cartridge into the strut body**

Step 15. **Insert any special bushings which should be included with the replacement cartridge**

Step 16. **Place the body nut on the strut housing and start it by hand. Be sure not to cross-thread it**

Step 17. **Tighten the body nut cap securely**

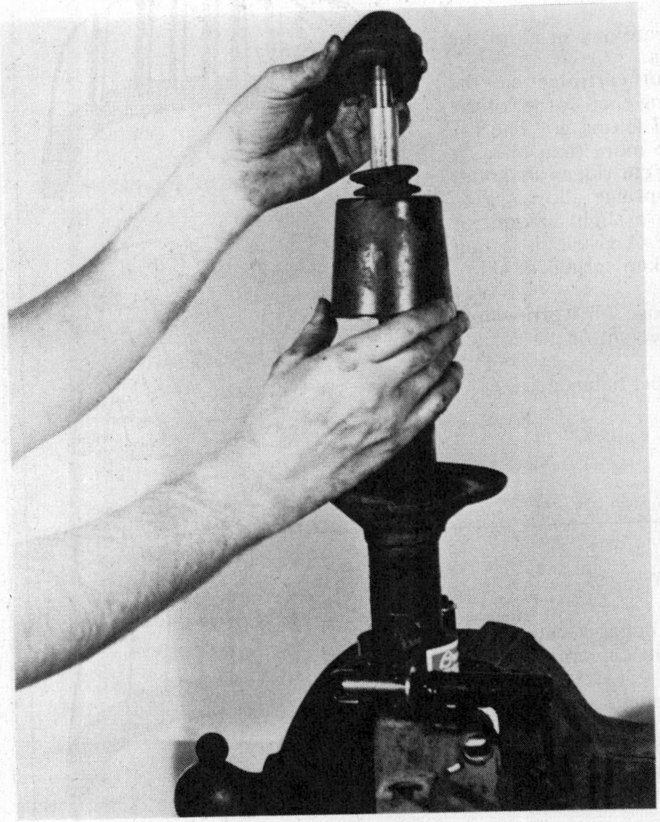

Step 18. **Inspect the upper strut mounting parts prior to reassembly. Replace any damaged components**

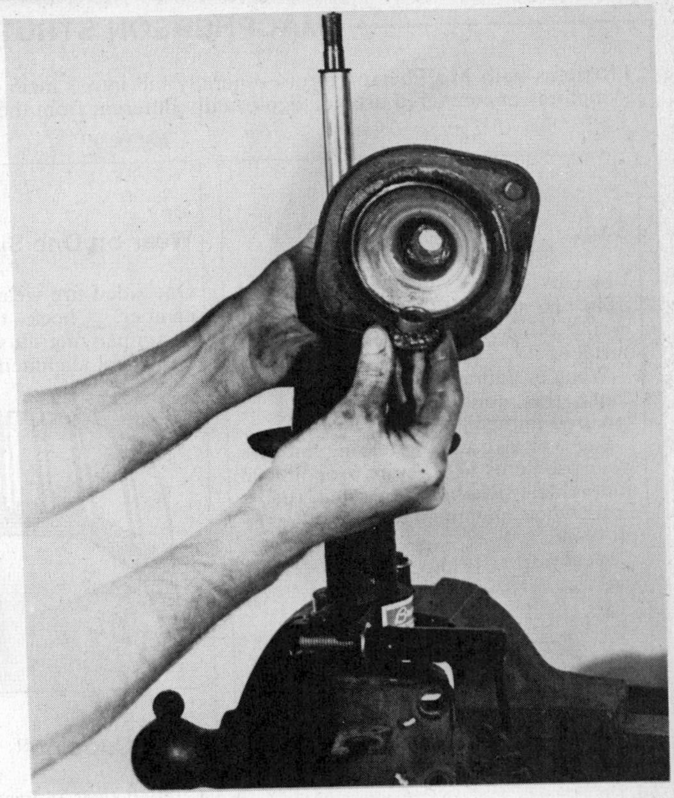

Step 19. **Repack the upper strut bearing with grease. Replace if excessive play is apparent**

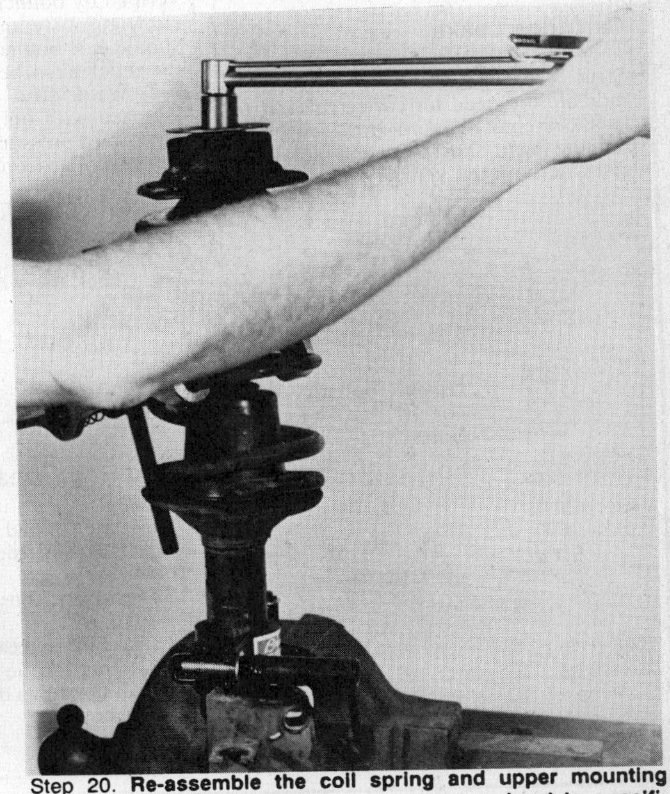

Step 20. **Re-assemble the coil spring and upper mounting parts in reverse order. Tighten the piston rod nut to specification and remove the spring compressor. Install the strut in the vehicle. See the car section for details**

MACPHERSON STRUT PROBLEM DIAGNOSIS

Problems with MacPherson struts generally fall into 3 main categories: suspension, tire wear and steering. In general, the symptoms encountered are not significantly different from those encountered on conventional suspensions.

Suspension

Sag

Vehicle "sag" is a visible tilt of the car from one side to the other or one end to the other while parked on a level surface.

Weak or damaged strut springs could cause this condition and should be repaired immediately.

Sag will also cause steering and tire wear problems to be more pronounced and vehicle instability on rough roads. Front wheel alignment will not solve the problem.

Weak strut springs increase vehicle sag. See "Tire Cupping".

Cartridge Leaks

Strut cartridge leaks (not seepage) indicate the need for cartridge or strut replacement. Be sure the leakage is coming from the strut, and not from elsewhere on the vehicle.

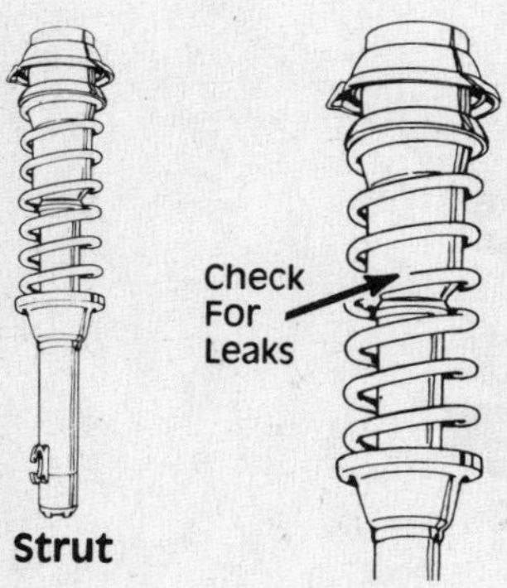

Abnormal Tire Wear

Wear on One Side

One sided tire wear indicates incorrect camber. Check the causes in the accompanying illustration and be sure the wheel alignment is correct.

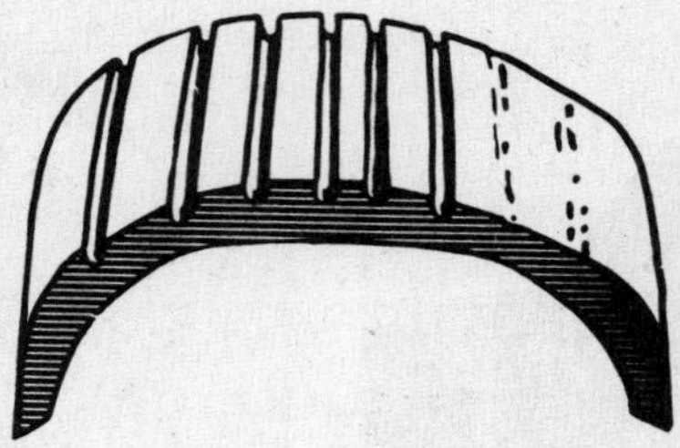

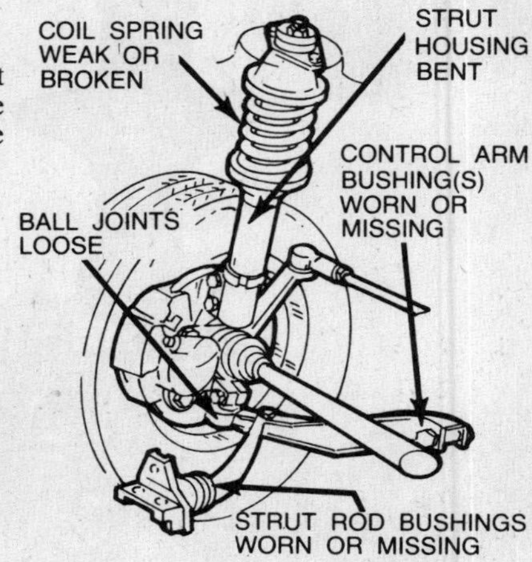

Tire "Cupping"

Cupped tires indicate any or all of the following problems.

1. A weak strut cartridge can be verified by bouncing each corner of the car vigorously and letting go. The car should not bounce more than once, if the shock absorber cartridges are good.
2. Weak strut springs allow sag to increase with only a slight amount of downward pressure. A visual inspection will reveal any broken springs or shiny spots.
3. Check for loose or worn wheel bearings with the weight of the car off of the wheel.
4. Check the wheel balance.

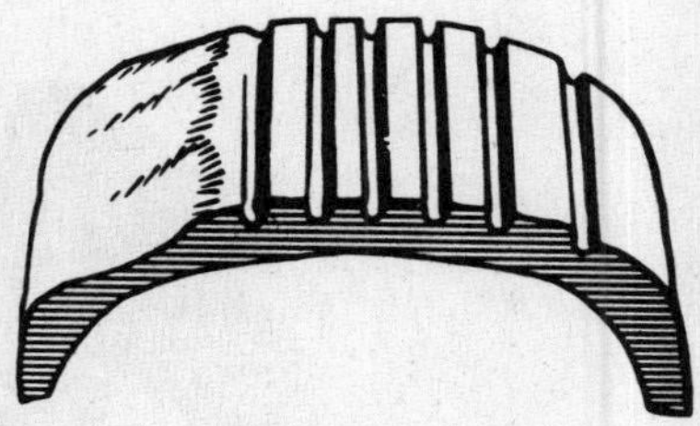

Tread Edge Wear

Wear along tread edges (feathering) indicates a suspension or steering system problem.

1. Strut rod bushings are worn or missing.
2. Tie rod end wear can be determined by grabbing the tie rod end firmly and forcing it up, down or sideways to check for lost motion.

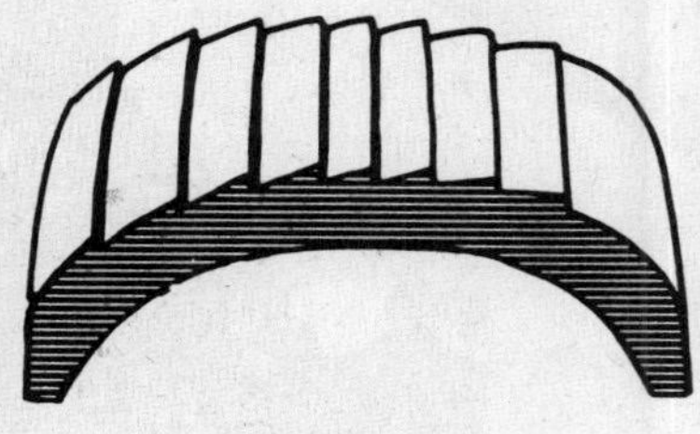

MACPHERSON STRUT PROBLEM DIAGNOSIS

Problems with MacPherson struts generally fall into 3 main categories: suspension, tire wear and steering. In general, the symptoms encountered are not significantly different from those encountered on conventional suspensions.

Steering

Tires

Both front tires should match and both rear tires should match. Be sure air pressure is correct.

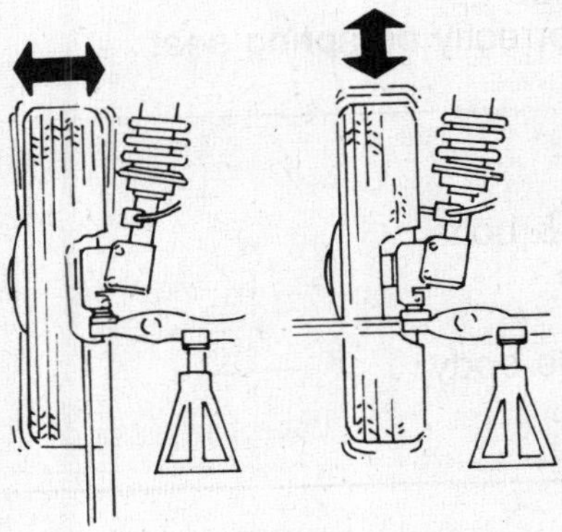

Strut Rod Bushings

Grasp the strut rod and shake it. Any noticeable play indicates excessive wear and need for parts replacement.

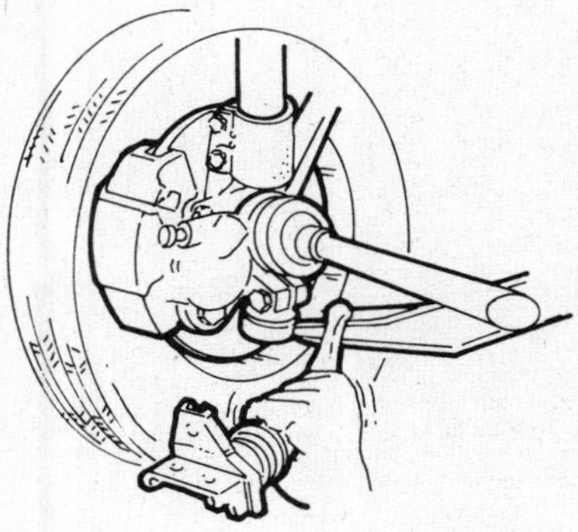

Ball Joints

Support the car under the frame or crossmember so that the jack does not interfere with the control arm. Rock the tire in and out and up and down. Excessive movement means that both ball joints should be replaced.

Struts with lower weight-carrying ball joints should be supported at the outer edge of the lower control arm. These vehicles usually have wear indicating ball joints that can be checked visually.

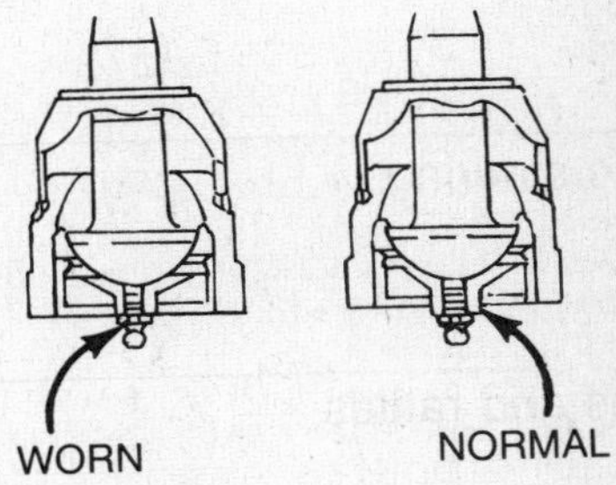

Stabilizer Bar Bushings

Check for worn bushings or lost motion with the vehicle level and the weight evenly distributed on all wheels.

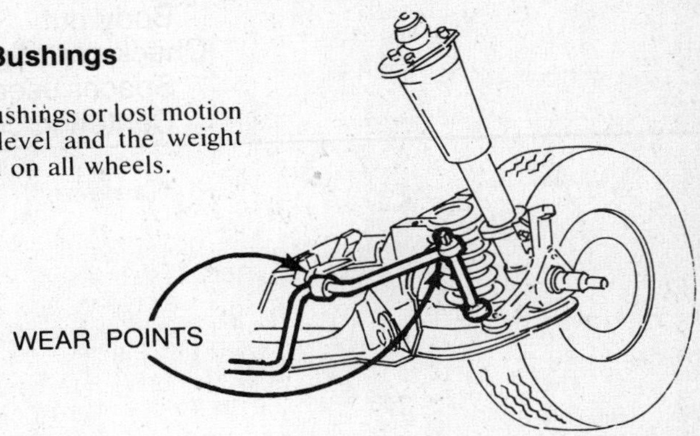

Control Arm Bushings

Support the car under the frame or body and remove the weight from the wheel and control arm. Check for free-play in the bushings at the pivot point, using a pry bar.

NOTE: Some control arm bushings are serviceable only by replacing the entire arm.

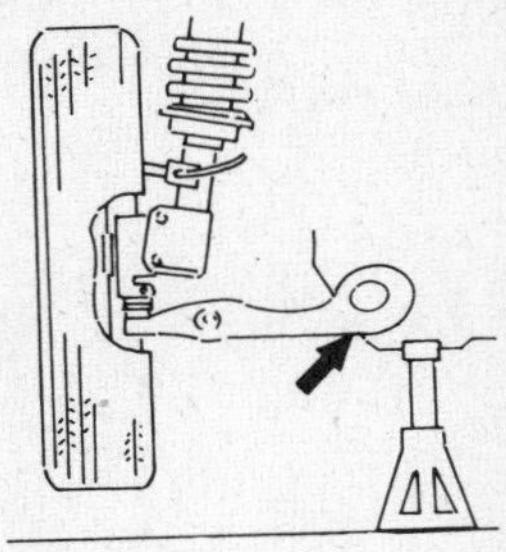

Strut Assembly

Check the strut assembly for cracks or dents in the housing. Look for worn, bent or loose piston rods or dents that will inhibit piston rod movement.

Steering Gear

Check for worn steering gear or loose or worn mounting bolts and bushings.

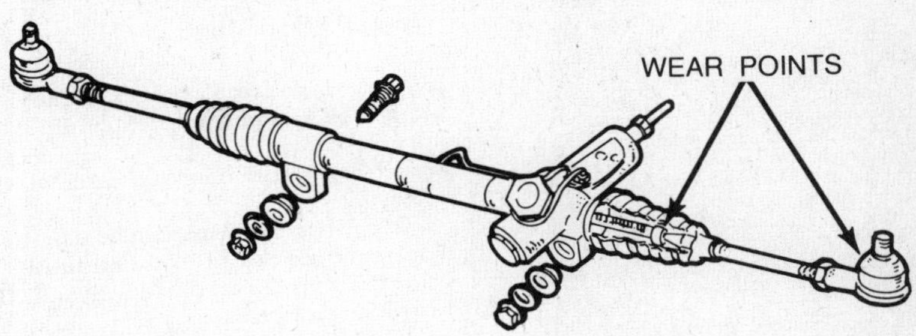

ROAD TEST TROUBLESHOOTING

Following are possible solutions to common potential problems which might be noticed during the road test after strut service is completed. Many are not exclusively strut service related.

Problem	Correction

Brake pedal low or soft	Bleed brakes
	Check for leaks
	  Brake lines
	  Wheel cylinder
	  Caliper piston seal

Erratic steering	Check upper support housing components for proper assembly Check spring assembly right side up Check for spring helix riding correctly on spring seat Check wheel alignment

Noises and rattles	Check torques
	  Piston rod nut
	  Upper support housing nuts & bolts
	  Lower mounting nuts & bolts
	  Body nut
	Check cartridge assembly in the body
	  Spacer used
	  Centering collar used

Brakes 33

BRAKE SYSTEM

Understanding the Brakes

HYDRAULIC SYSTEM

Basic Operating Principles

Hydraulic systems are used to actuate the brakes of all modern automobiles. The system transports the power required to force the frictional surfaces of the braking system together from the pedal to the individual brake units at each wheel. A hydraulic system is used for 2 reasons. First, fluid under pressure can be carried to all parts of an automobile by small hoses—some of which are flexible—without taking up significant amount of room or posing routing problems. Second, a great mechanical advantage can be given to the brake pedal end of the system, and the foot pressure required to actuate the brakes can be reduced by making the surface area of the master cylinder pistons smaller than that of any of the pistons in the wheel cylinders or calipers.

The master cylinder consists of a fluid reservoir and a double cylinder and piston assembly. Double type master cylinders are designed to separate 2 two-wheel braking systems hydraulically in case of a leak. The standard approach has been to utilize 2 separate two-wheel circuits; 1 for the front wheels and 1 for the rear wheels.

Most newer models now use a diagonally split system; i.e. 1 front wheel and the opposite rear wheel make up 1 braking circuit, while the remaining circuit consists of the other front wheel and its opposite side rear wheel.

Steel lines carry the brake fluid to a point on the vehicle's frame near each of the vehicle's wheels. The fluid is then carried to the wheel cylinders and/or calipers by flexible tubes in order to allow for suspension and steering movements.

The hydraulic system operates as follows: When at rest, the entire system, from the piston(s) in the master cylinder to those in the wheel cylinders or calipers, is full of brake fluid. Upon application of the brake pedal, fluid trapped in front of the master cylinder piston(s) is forced through the lines to the slave cylinders (wheel cylinders or calipers). Here, it forces the pistons outward, in the case of drum brakes, and inward toward the disc, in the case of disc brakes. The motion of the pistons is opposed by return springs mounted outside the cylinders in the drum brakes, and by internal springs or seals, in disc brakes.

Upon release of the brake pedal, a spring located inside the master cylinder immediately returns the master cylinder pistons to the normal position. The pistons contain check valves and the master cylinder has compensating ports drilled into it. These are uncovered as the pistons reach their normal position. The piston check valves allow fluid to flow toward the wheel cylinders or calipers as the pistons withdraw. Then, as the return springs force the shoes into the released position, the excess fluid flows back to the reservoir through the compensating ports. It is during the time the pedal is in the released position that any fluid that has leaked out of the system will be replaced through the compensating ports.

Dual circuit master cylinders employ 2 pistons, located 1 behind the other, in the same cylinder. The primary piston is actuated directly by mechanical linkage from the brake pedal. The secondary piston is actuated by fluid trapped between the 2 pistons. If a leak develops in the front of the secondary piston, it moves forward until it bottoms against the front of the master cylinder, and the fluid trapped between the pistons will operate 1 side of the split system. If the other side of the system develops a leak, the primary piston will move forward until direct contact with the secondary piston takes place, and it will force the secondary piston to actuate the other side of the split system. In either case, the brake pedal moves farther when the brakes are applied, and less braking power is available.

All dual circuit systems use a distributor switch to warn the driver when only half of the braking system is operational. This switch is located in

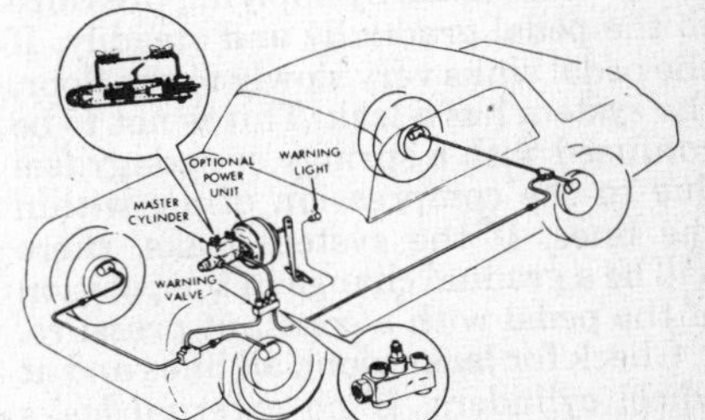

Dual braking system—front-to-rear split

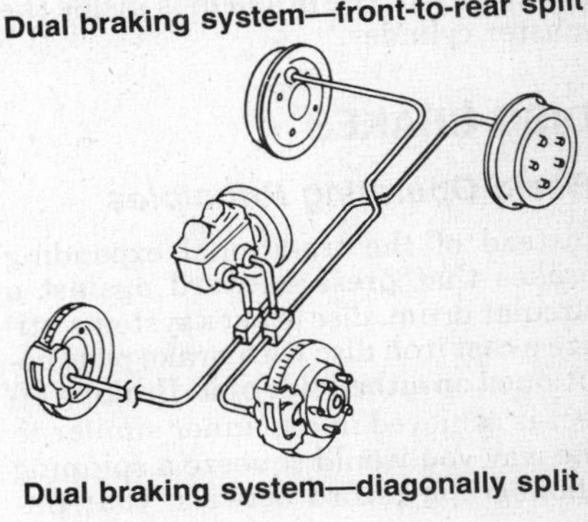
Dual braking system—diagonally split

a valve body which is mounted on the firewall, or the frame below the master cylinder. A hydraulic piston receives pressure from both circuits, each circuit's pressure being applied to 1 end of the piston. When the pressures are in balance, the piston remains stationary. When 1 circuit has a leak, however, the greater pressure in that circuit during application of the brakes will push the piston to 1 side, closing the distributor switch and activating the brake warning light.

In disc brake systems, this valve body also contains a metering valve and, in some cases, a proportioning valve (or valves). The metering valve keeps pressure from traveling to the disc brakes on the front wheels until the brake shoes or pads on the rear wheels have contacted the drums or rotors, ensuring that the front brakes will never be used alone. The proportioning valve throttles the pressure to the rear brakes so as to avoid rear wheel lockup during very hard braking.

These valves may be tested by removing the lines to the front and rear brake systems and installing special brake pressure testing gauges. Front and rear system pressures are then compared as the pedal is gradually depressed. Specifications vary with the manufacturer and design of the brake system.

Brake system warning lights may be tested by depressing the brake pedal and holding it while opening 1 of the wheel cylinder bleeder screws. If this does not cause the light to go on, substitute a new lamp, make continuity checks, and, finally, replace the switch as necessary.

The hydraulic system may be checked for leaks by applying pressure to the pedal gradually and steadily. If the pedal sinks very slowly to the floor, the system has a leak. This is not to be confused with a springy or spongy feel due to the compression of air within the lines. If the system leaks, there will be a gradual change in the position of the pedal with a constant pressure.

Check for leaks along all lines and at wheel cylinders. If no external leaks are apparent, the problem is inside the master cylinder.

DISC BRAKES

Basic Operating Principles

Instead of the traditional expanding brakes that press outward against a circular drum, disc brakes systems utilize a cast iron disc with brake pads positioned on either side of it. Braking effect is achieved in a manner similar to the way you would squeeze a spinning phonograph record between your fingers. The disc (rotor) is a one-piece casting which may be equipped with cooling fins between the 2 braking surfaces. The fins (if equipped) enable air to circulate between the braking surfaces making them less sensitive to heat buildup and more resistant to fade. Dirt and water do not affect braking action since contaminants are thrown off by the centrifugal action of the rotor or scraped off by the pads. Also, the equal clamping action of the 2 brake pads tends to ensure uniform, straight-line stops. All disc brakes are inherently self-adjusting.

There are 3 general types of disc brake:

1. A fixed caliper, 2 or 4-piston type
2. A floating caliper, single piston or double piston back-to-back type
3. A sliding caliper, single piston or double piston back-to-back type

The fixed caliper design uses 1 or 2 pistons mounted on either side of the rotor (in each side of the caliper). The caliper is mounted rigidly and does not move.

The sliding and floating designs are quite similar. In fact, these 2 types are often lumped together. In both designs, the pad on the inside of the rotor is moved into contact with the rotor by hydraulic force. The caliper, which is not held in a fixed position, moves slightly, bringing the outside pad into contact with the rotor. There are various methods of attaching floating calipers. Some pivot at the bottom or top, and some slide on mounting bolts. In any event, the end result is the same.

DRUM BRAKES

Basic Operating Principles

Drum brakes employ 2 brakes shoes mounted on a stationary backing plate. These shoes are positioned inside a circular cast iron (or aluminum) drum which rotates with the wheel assembly. The shoes are held in place by springs; this allows them to slide toward the drums (when they are applied) while keeping the lining and drums in alignment. The shoes are actuated by a wheel cylinder which is mounted at the top of the backing plate. When the brakes are applied, hydraulic pressure forces the wheel cylinder's 2 actuating links outward. Since these links bear directly against the top of the brake shoes, the tops of the shoes are then forced outward against the inner side of the drum. This action forces the bottom of the 2 shoes to contact the brake drum by rotating the entire assembly slightly (known as servo action). When pressure within the wheel cylinder is relaxed, return springs pull the shoes back away from the drum.

Most modern drum brakes are designed to self-adjust themselves during application when the vehicle is moving in reverse. This motion causes both shoes to rotate very slightly with the drum, rocking an adjusting lever, thereby causing rotation of the adjusting screw by means of a star wheel.

POWER BRAKE BOOSTERS

Power brakes operate just as standard brake systems except in the actuation of the master cylinder pistons. A vacuum diaphragm is located on the front of the master cylinder and assists the drive in applying the brakes, reducing both the effort and travel he must put into moving the brake pedal.

The vacuum diaphragm housing is connected to the intake manifold by a vacuum hose. A check valve is placed at the point where the hose enters the diaphragm housing, so that during pe-

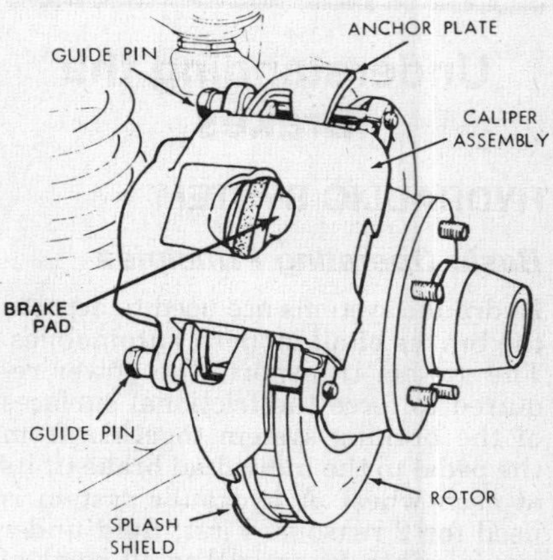

Typical disc brake assembly

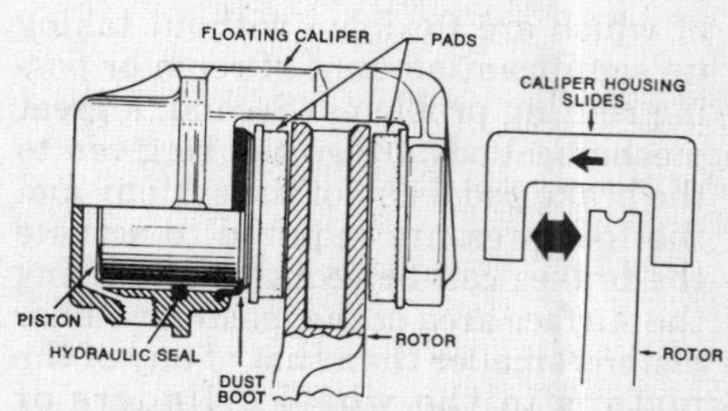

Typical floating caliper disc brake (sliding caliper similar)

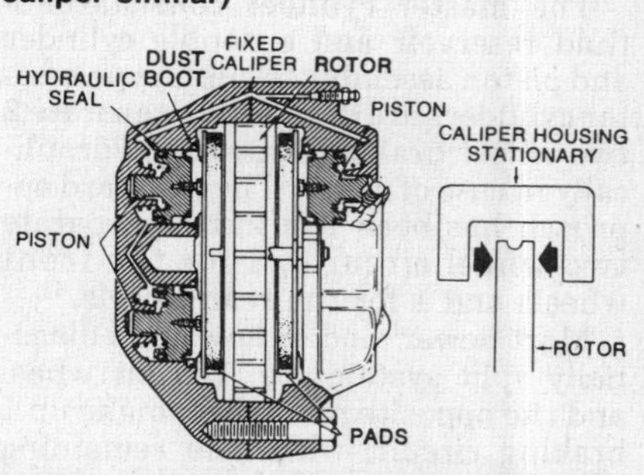

Typical fixed caliper disc brake (four piston shown)

riods of low manifold vacuum, brake assist vacuum will not be lost.

Depressing the brake pedal closes off the vacuum source and allows atmospheric pressure to enter on 1 side of the diaphragm. This causes the master cylinder pistons to move and apply the brakes. When the brake pedal is released, vacuum is applied to both sides of the diaphragm, and return springs return the diaphragm and master cylinder pistons to the released position. If the vacuum fails, the brake pedal rod will butt against the end of the master cylinder actuating rod, and direct mechanical application will occur as the pedal is depressed.

HYDRAULIC CYLINDERS AND VALVES

Master Cylinders

The master cylinder is a type of hydraulic pump that is operated by a push rod attached to the brake pedal or by a push rod that is part of the power brake booster. The cylinder provides a means of converting mechanical force into hydraulic pressure.

DUAL MASTER CYLINDER

In this type there are 2 separate hydraulic pressure systems. 1 of the hydraulic systems may be connected to the front brakes, and the other to the rear brakes, or the system may connect diagonal wheels. If 1 system fails, the other system remains operational, thus providing an additional safety measure. There are 2 distinct fluid reservoirs and each has a vent and replenishing port that leads into the cylinder bore. These ports have been called compensating and inlet ports or bypass ports, and the terms have been used inconsistently causing confusion. The terms "vents" and "replenishing ports" are now standardized S.A.E. terms. An airtight seal for the reservoir is provided in the form of a rubber diaphragm, which is held in place by a metal cover. A bail type retainer or a bolt usually holds the cover on the reservoirs. The cover is vented to permit atmospheric pressure to enter above the diaphragm. The diaphragm prevents moisture and debris from contaminating the fluid. The cylinder bore contains the return springs, 2 pistons, and the seals. The piston stop bolt (if present) may be assembled in a thread hole in the bottom of the cylinder.

Some master cylinders have the piston stop bolt assembled in a threaded hole in the side of the bore or in the bottom of the front reservoir, and others do not have stop bolts at all. Do not install a stop bolt in the reservoir of a master cylinder if 1 was not originally there. Some cylinders have a tapped hole, but no bolt was ever installed in production. *This was done on purpose, and is not an error.*

A retaining ring fits into a groove near the end of the bore and holds the piston assemblies in the cylinder bore.

Dual System—Applied

When the brake pedal is depressed, the push rod moves the primary piston forward in the cylinder bore. The primary vent port is sealed off by the lip of the primary cup. As a result, a solid column of fluid is created between the primary and secondary pistons.

With the help of the primary piston return spring, this column moves the secondary piston forward in the cylinder bore. This closes the secondary vent port. When both ports are closed, any further movement of the pushrod and pistons serves to increase the hydraulic pressure in the area ahead of each piston. This pressure is then transmitted through the 2 hydraulic brake systems to the brakes at each wheel.

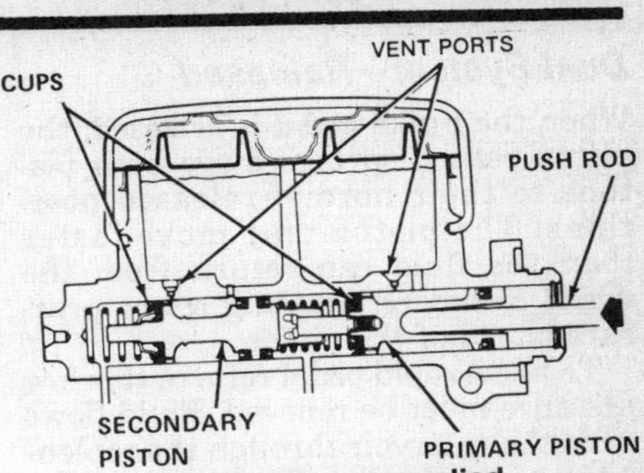

Dual system—applied

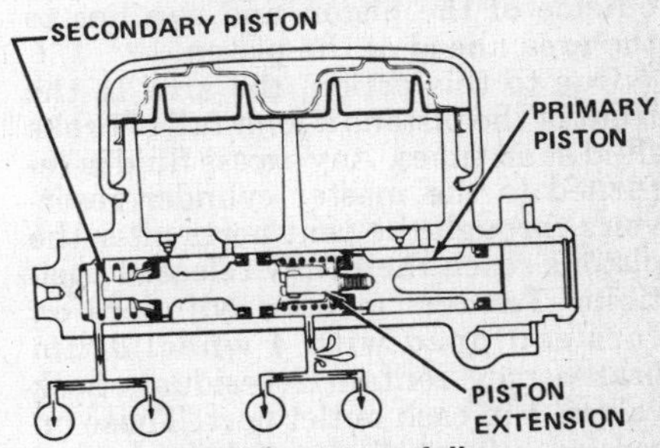

Primary system failure

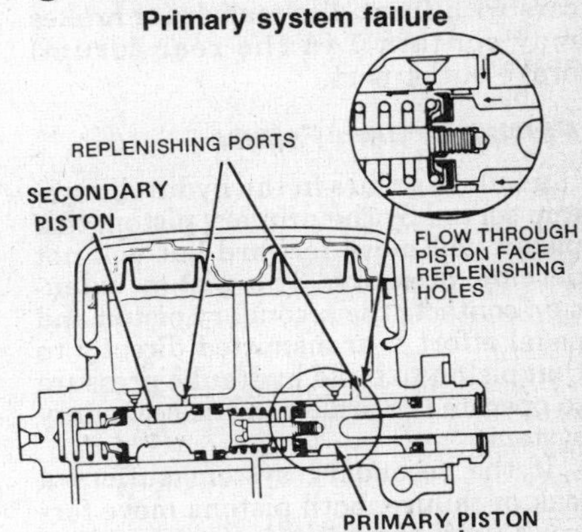

Dual system—released

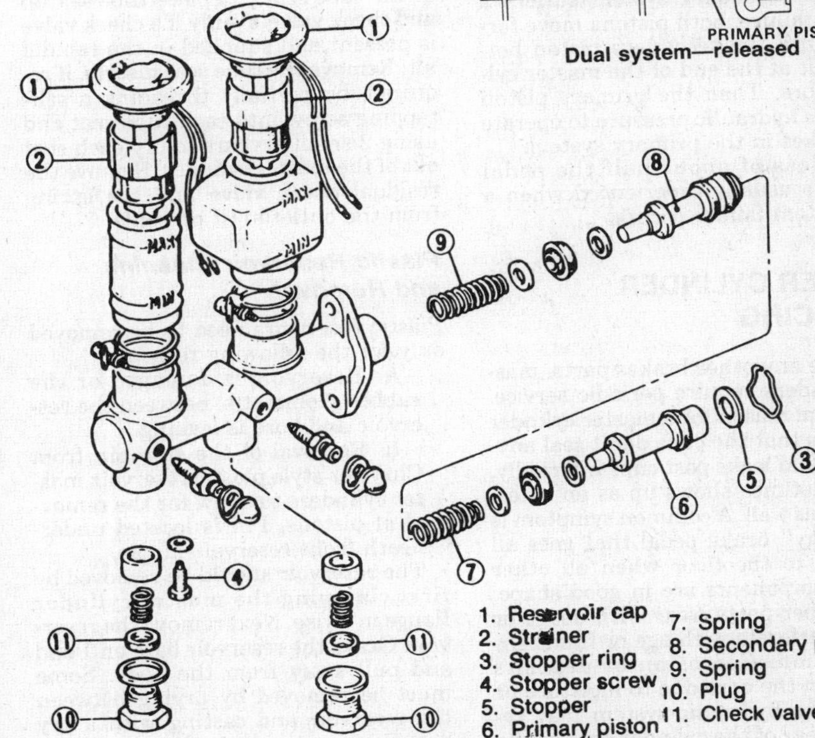

1. Reservoir cap
2. Strainer
3. Stopper ring
4. Stopper screw
5. Stopper
6. Primary piston
7. Spring
8. Secondary piston
9. Spring
10. Plug
11. Check valve

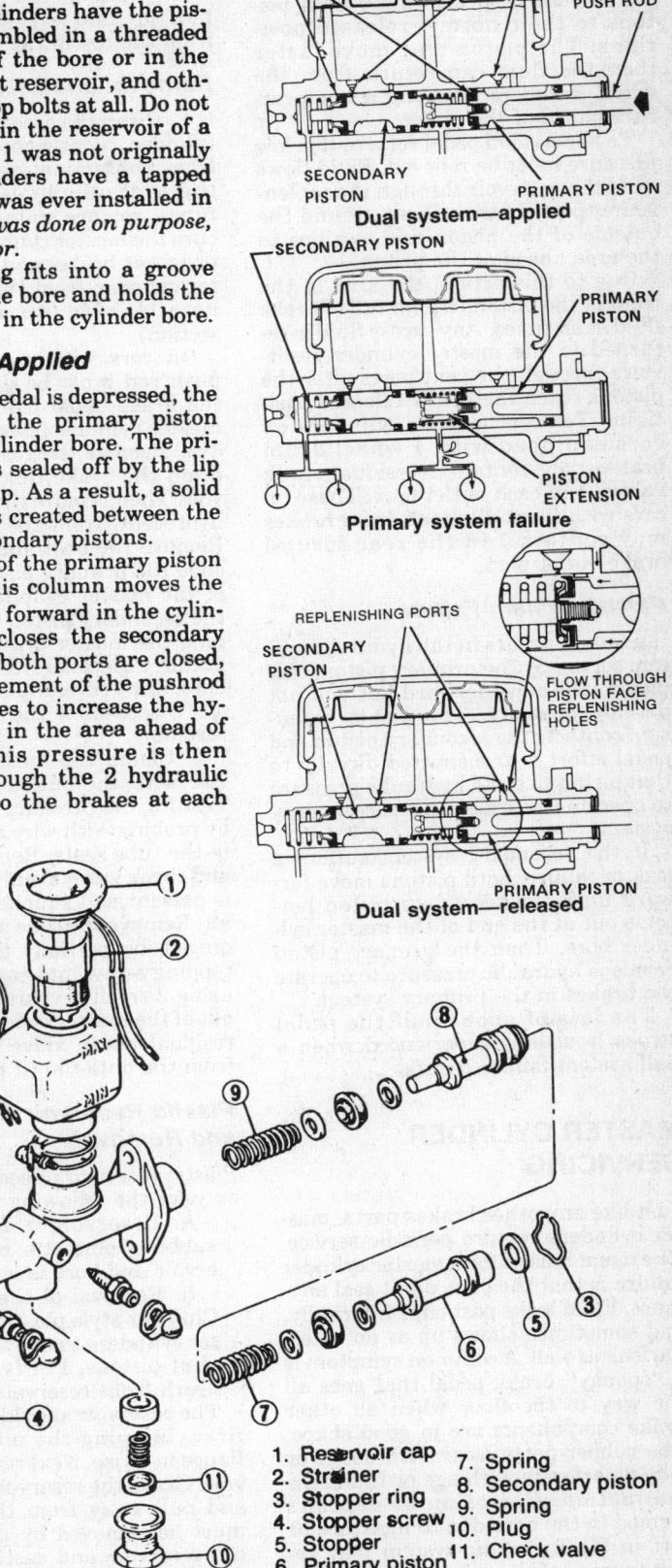

Exploded view of a dual system master cylinder

Dual System—Released

When the brake pedal is released, the piston return springs move both pistons to their normal released positions. The piston may move faster than the fluid can return from the wheel cylinders, creating a low pressure ahead of the piston.

To allow rapid pedal return, this low pressure must be relieved. Fluid flows from the reservoir through the replenishing port. It then flows around the outside of the piston and cup lips to the area ahead of the piston.

Due to this action, the area in the front of the pistons if kept full of brake fluid at all times. Any excess fluid is returned to the master cylinder reservoirs through the vent ports after the pistons reach their fully released positions. Tandem master cylinders on cars equipped with 4 wheel drum brakes may contain 2 residual check valves, 1 in each outlet port. Those on cars with front disc/rear drum brakes may contain 1 in the rear (drum) brake outlet port.

Partial System Failure

If a failure occurs in the hydraulic system served by the primary piston, this piston will move forward but will not develop pressure. The piston extension contacts the secondary piston and pedal effort is transmitted directly to that piston to build hydraulic pressure to operate the brakes in the secondary system.

If the secondary system suffers a leak or failure, both pistons move forward until the secondary piston bottoms out at the end of the master cylinder bore. Then the primary piston develops hydraulic pressure to operate the brakes in the primary system.

The loss of about half the pedal stroke is usually experienced when a half system failure occurs.

MASTER CYLINDER SERVICING

Just like any other brakes parts, master cylinders require periodic service. The usual reason for a master cylinder failure is that the cups don't seal anymore. Fluid leaks past cups internally, and sometimes shows up as an external leak as well. A common symptom is a "spongy" brake pedal that goes all the way to the floor when all other brake components are in good shape. The rubber parts wear with usage or may deteriorate with age or fluid contamination. Corrosion or deposits formed in the bore due to moisture or dirt in the hydraulic system may result in wear of the cylinder bore or the parts therein. Also, the fluid levels in the reservoirs should be checked periodically. Whenever needed, clean brake fluid should be added to maintain the fluid level ¼–½ in. (6–13mm) from the top of the reservoir.

Removal and Disassembly

1. Clean the area around the master cylinder to prevent dirt and grease from contaminating the cylinder or the hydraulic lines. Disconnect the tubes, remove nuts or bolts that secure the master cylinder to the firewall or power brake, and remove the master cylinder from the car (for further details, refer to appropriate car section).

On cars with manual brakes, the push rod must be disconnected from the brake pedal before removing the master cylinder from the car.

2. Remove the reservoir cover, and drain the brake fluid from the reservoir. Then remove the piston stop bolt, if present, from the master cylinder. Remove the boot and snap ring, then slide the primary piston assembly out of the master cylinder. Next, remove the secondary piston assembly by tapping the master cylinder, or by using needle nose pliers to pull it from its bore, or by carefully using compressed air. Disassemble the secondary piston assembly.

3. Clamp the master cylinder in a vise with the outlet ports facing up. Test for the presence of a check valve by probing with wire through the hole in the tube seats. Replace tube seat(s) and check valve(s) only if a check valve is present and supplied in the rebuild kit. Remove the tube seat inserts, if required, by partially threading a self-tapping screw into each tube seat and using 2 small prybars to pry each seat out of the master cylinder. Remove the residual check valve and the spring from the outlet(s) (if present).

Plastic Reservoir Cleaning and Removal

Plastic reservoirs need to be removed only for the following reasons:

a. Reservoir is damaged or the rubber grommet(s) between the reservoir and bore is leaking.

b. Removal of the stop pin from Chrysler style plastic reservoir master cylinders to allow for the removal of pistons. Pin is located underneath front reservoir nipple.

The reservoir should be removed by first clamping the master cylinder flange in a vise. Next remove the reservoir. Grasp the reservoir base on 1 end and pull away from the body. Some must be removed by prying between the reservoir and casting with a pry bar. Grommets can be reused if they are in good condition. Whether or not the reservoir is removed, it and the covers or caps should be thoroughly cleaned.

Cleaning and Inspection

Thoroughly clean the master cylinder and any other parts to be reused in clean alcohol. DO NOT USE PETROLEUM PRODUCTS FOR CLEANING. If the bore is not badly scored, rusted or corroded, it is possible to rebuild the master cylinder in some cases. A slight bit of honing is permissible to clean cups are facing.

CAUTION

Aluminum cylinder bores cannot be honed. The cylinder MUST be replaced if the bore is scored.

Lubricate all new rubber parts with brake fluid or brake system assembly lubricant.

CAST IRON BORE CLEAN-UP

Crocus cloth or an approved cylinder hone should be used to remove lightly pitted, scored, or corroded areas from the bore.

CAUTION

If an aluminum master cylinder has pits or scratches in the bore, it must be replaced.

Brake fluid can be used as a lubricant while honing lightly. The master cylinder should be replaced if it cannot be cleaned up readily. After using the crocus cloth or a hone, the master cylinder should be thoroughly washed in clean alcohol or brake fluid to remove all dust and grit. If alcohol is used, dry parts thoroughly before reinstalling.

CAUTION

Other solvents should not be used.

Then the clearance between the bore wall and the piston (primary piston of a dual system master cylinder) should be checked. If a narrow (⅛–¼ in. wide) 0.006 in. (0.15mm) feeler gauge can be inserted between the wall and a new piston, the clearance is excessive, and the master cylinder should be replaced. The maximum clearance allowed for units containing pistons without replenishing holes is 0.009 in. (0.23mm).

ALUMINUM BORE CLEAN-UP

Inspect the bore for scoring, corrosion and pitting. If the bore is scored or badly pitted and corroded the assembly should be replaced. *Under no conditions should the bore be cleaned with an abrasive material.* This will remove the wear and corrosion resistant anodized surface. Clean the bore with a clean piece of cloth around a wooden dowel and wash thoroughly with alcohol. Do not confuse bore discoloration or staining with corrosion.

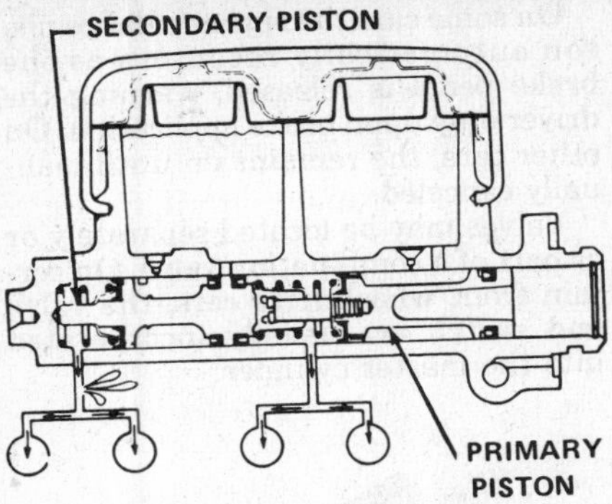

Secondary system failure

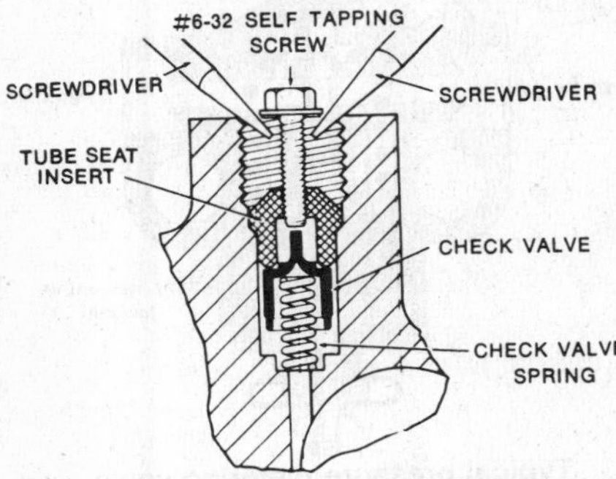

Removing the inserts from the master cylinder ports

Reassembly and Installation

1. Carefully install the new cups or seals in the same positions and in reverse order of removal.
2. Use brake fluid or assembly fluid very generously to keep from damaging the seals.
3. Placing the small end of the pressure spring into the secondary piston retainer, slide the assembly into the cylinder bore, taking care not to nick or gouge any rubber part.
4. Place the spring retainer of the primary piston assembly over the secondary piston shoulder and push both assemblies into the bore.
5. Install and tighten the piston retaining screw and gasket, while holding the pistons in their seated positions. At the same time, reinstall any piston snap rings.
6. Install the residual check valve and spring in the proper master cylinder outlet (or both outlets, if originally present). If the tube seat inserts were removed, install new seats in both fluid outlets making sure that they are securely seated.

Bleeding and Checking

1. Bleed the hydraulic system as described later in this section.

NOTE: Be sure to bench bleed a rebuilt or new master cylinder before installation.

2. Check master cylinder vent port clearance by watching for a spurt of brake fluid in both reservoir vent holes when the brake pedal is slightly depressed, indicating proper port clearance.

Master Cylinder Push Rod Adjustment

After assembly of the master cylinder to the power section, the piston cup in the hydraulic cylinder should just clear the compensating port hole when the brake pedal is fully released. If the push rod is too long, it will hold the piston over the port.

A push rod that is too short, will give too much loose travel (excessive pedal play).

Apply the brakes and release the pedal all the way observing brake fluid flow back into the master cylinder.

A full flow indicates the piston is coming back far enough to release the fluid.

A slow return of fluid indicates the piston is not coming back far enough to clear the ports. The push rod adjustment is too tight, and should be shortened.

Wheel Cylinders

DRUM BRAKE WHEEL CYLINDER

The wheel cylinder performs in response to the master cylinder. It receives fluid from the hydraulic hose through its inlet port. As the pressure increases, the wheel cylinder cups and pistons are forced apart. As a result, the hydraulic pressure is converted into mechanical force acting on the brake shoes. The wheel cylinder size may vary from front to rear. The variation in wheel cylinder size (diameter) is 1 of the factors controlling the distribution of braking force in a vehicle.

WHEEL CYLINDER OPERATION

The space between the caps in the cylinder bore must remain filled with fluid at all times. After depressing the brake pedal, additional brake fluid is forced into the cylinder bore. As a result of this, cups and pistons move outward in the cylinder bore pushing the shoe links an the brake shoes outward to contact the drum and apply the brakes.

On some designs, the end of the shoe web bears directly against the pistons and therefore, shoe links are not used.

SERVICE PROCEDURES

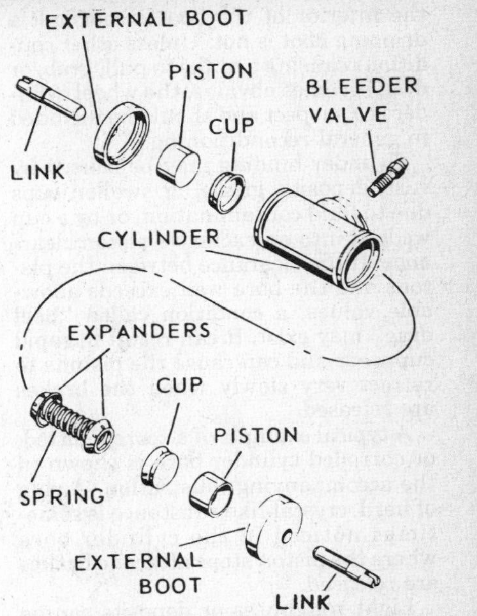

Typical wheel cylinder components

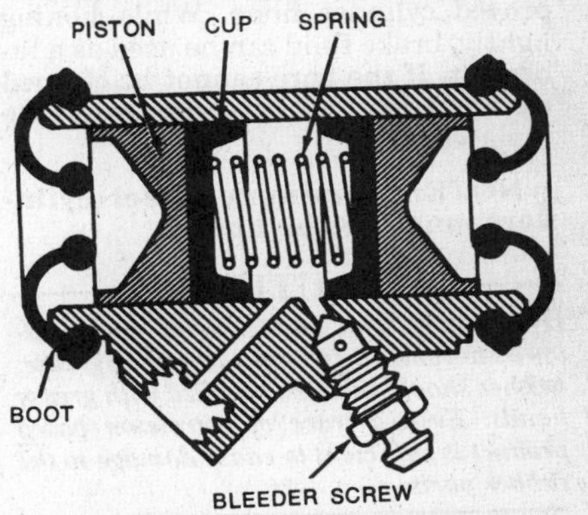

Double piston wheel cylinder

Wheel cylinders may need reconditioning or replacement whenever the brake shoes are replaced or when required to correct a leak condition. On many designs, the wheel cylinders can be disassembled without removing them from the backing plate. On some designs, however, the cylinder is mounted in an indention in the backing plate or a cylinder piston stop is welded to the backing plate. When servicing brakes of this type, the cylinder must be removed from the backing plate before being disassembled.

Diagnostic Inspection and Cleaning

Leaks which coat the boot and the cylinder with fluid, or result in a dropped reservoir fluid level, or dampen or stain the brake linings are dangerous. Such leaks can cause the brakes to "grab" or fail and should be immediately corrected. A leakage, not immediately apparent, can be detected by pulling back the cylinder boot. A small amount of fluid seepage dampening the interior of the boot is normal; a dripping boot is not. Unless other conditions causing a brake to pull, grab, or drag becomes obvious, the wheel cylinder is a suspect and should be included in general reconditioning.

Cylinder binding may be caused by rust, deposits, grime, or swollen cups due to fluid contamination, or by a cup wedged into an excessive piston clearance. If the clearance between the pistons and the bore wall exceeds allowable values, a condition called "heel drag" may exist. It can result in rapid cup wear and can cause the pistons to retract very slowly when the brakes are released.

A typical example of a scored, pitted, or corroded cylinder bore is shown in the accompanying illustration. A ring of hard, crystal-like substance is sometimes noticed in the cylinder bore where the piston stops after the brakes are released.

Light roughness or deposits can be removed with crocus cloth or an approved cylinder hone. While honing lightly, brake fluid can be used as a lubricant. If the bore cannot be cleaned up readily, the cylinder must be replaced.

NOTE: Aluminum wheel cylinders must not be honed.

CAUTION

Hydraulic system parts should not be allowed to come in contact with oil or grease, neither should those be handled with greasy hands. Even a trace of petroleum based product is sufficient to cause damage to the rubber parts.

Reconditioning Wheel Cylinders

It is common practice to recondition a wheel cylinder without dismounting it, however some brakes are equipped with external piston stops which prevent disassembly unless the cylinder is removed. In order to dismount, remove the shoe springs and spread the shoes apart, disconnect the brake line, remove the mounting bolts or retaining clips, and pull the cylinder free.

Pull the protective dust boots off the cylinder. Internal parts should slide out, or be picked out easily. Parts can be driven out with a wooden dowel, or blown out at low pressure by applying compressed air to the fluid inlet port. Parts which cannot be removed easily indicate they are damaged beyond repair and the cylinder should be replaced.

Clean the cylinder and the parts in alcohol and/or brake fluid (do NOT use gasoline or other petroleum based products). Use only lint-free wiping cloths. Crocus cloth can be used to clean minute scratches, signs of rust, corrosion or discoloration from the cylinder bore and pistons. Slide the cloth in a circular rather than a lengthwise motion. A clean-up hone may be used. After a cylinder has been honed, inspect it for excessive piston clearance and remove any burrs formed on the edge of fluid intake or bleeder screw ports.

CAUTION

Do not rebuild aluminum cylinders.

To check the maximum piston clearance, place a 1/4 in. (6mm) wide strip of feeler shim lengthwise in the cylinder bore.

If the piston an be inserted with the shim in place, the cylinder is oversize, and should be discarded. Depending upon the cylinder bore diameter, the shim (or the feeler gauge) thickness can vary as follows:

Assemble the cylinder with the internal parts, making sure that the cylinder wall is wet with brake fluid. Insert the cups and pistons from each end of a double-end cylinder; do not slide them through the cylinder. Cup lips should always face inward.

Hydraulic Control Valves

PRESSURE DIFFERENTIAL VALVE

The pressure differential valve activates a dash panel warning light if pressure loss in the brake system occurs. If pressure loss occurs in 1/2 of the split system, the other system's normal pressure causes the piston in the switch to compress a spring until it touches an electrical contact. This causes the warning lamp on the dash panel to light, thus warning the driver of possible brake failure.

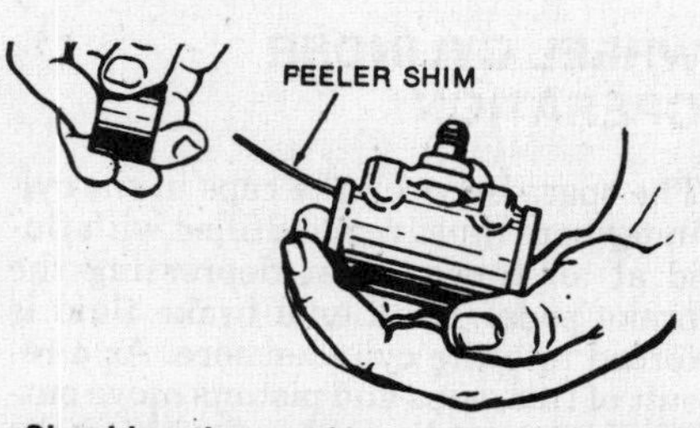

Checking the maximum piston clearance

Cylinder Bore	Shim
3/4 in.–1 3/16 in. (19–30mm)	.006" (.15mm)
1 1/4 in.–1 7/16 in. (32–37mm)	.007 in. (.18mm)
1 1/2 in. up (38mm)	.008 in. (.2mm)

On some cars the spring balance piston automatically recenters as the brake pedal is released, warning the driver only upon brake application. On other cars, the remains on until manually canceled.

Valves may be located separately or as part of a combination valve. On certain front wheel drive cars, the valve and switch are usually incorporated into the master cylinder.

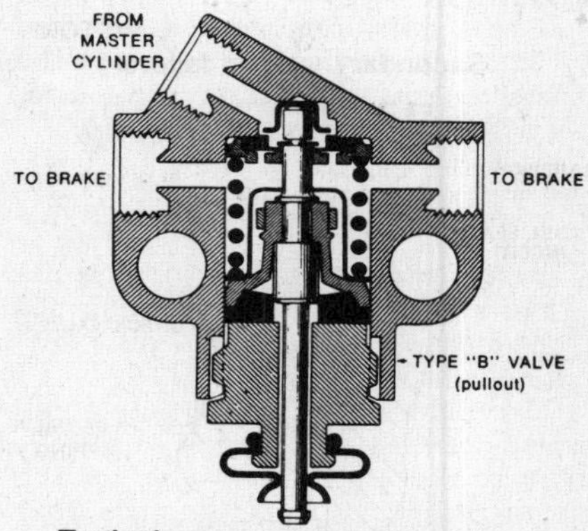

Typical pressure metering valve

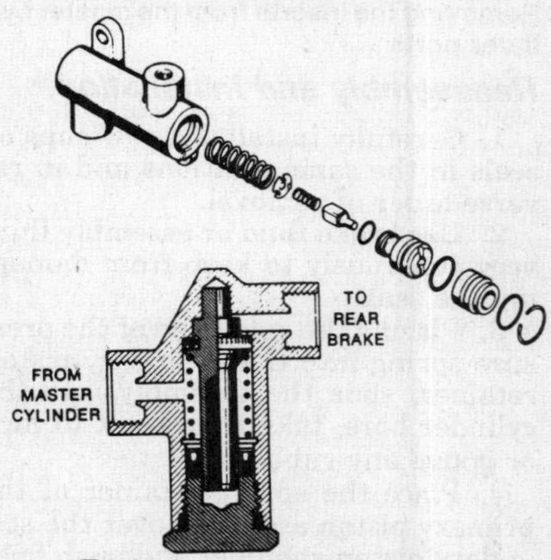

Typical proportioning valve

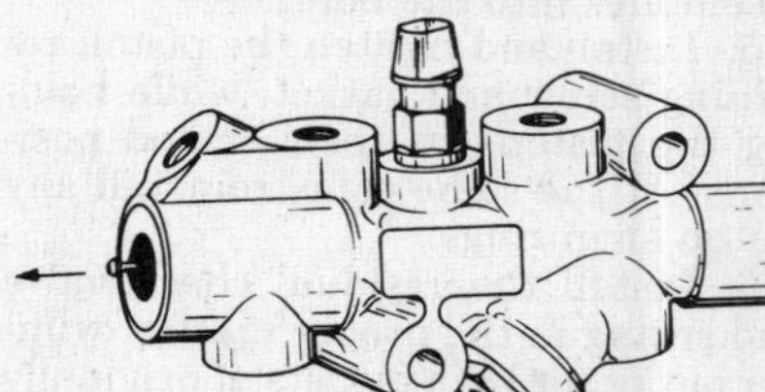

3–way combination valve

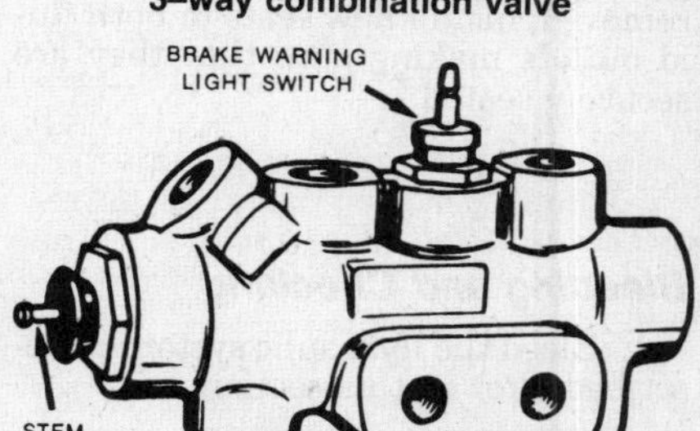

Two–way combination valve (metering and brake warning light switch)